Soliton Phenomenology

Mathematics and Its Applications (*Soviet Series*)

Volume 33

Soliton Phenomenology

by

Vladimir G. Makhankov

Laboratory of Computing Techniques and Automation,
Joint Institute for Nuclear Research, Moscow, U.S.S.R.

KLUWER ACADEMIC PUBLISHERS

DORDRECHT / BOSTON / LONDON

Library of Congress Cataloging-in-Publication Data

Makhan'kov, V. G.
Soliton phenomenology / Vladimir G. Makhankov.
p. cm. -- (Mathematics and its applications (Soviet series))
Includes bibliographies and index.
ISBN 9027728305
1. Solitons. 2. Phenomenological theory (Physics) 3. Nonlinear theories. I. Title. II. Series: Mathematics and its applications (Kluwer Academic Publishers). Soviet series.
QC174.26.W28M35 1989
530.1'4--dc19 88-23162
CIP

ISBN 90-277-2830-5

Published by Kluwer Academic Publishers,
P.O. Box 17, 3300 AA Dordrecht, The Netherlands.

Kluwer Academic Publishers incorporates
the publishing programmes of
D. Reidel, Martinus Nijhoff, Dr W. Junk and MTP Press.

Sold and distributed in the U.S.A. and Canada
by Kluwer Academic Publishers,
101 Philip Drive, Norwell, MA 02061, U.S.A.

In all other countries, sold and distributed
by Kluwer Academic Publishers Group,
P.O. Box 322, 3300 AH Dordrecht, The Netherlands.

Printed on acid-free paper

Printed in the Netherlands

SERIES EDITOR'S PREFACE

'Et moi, ..., si j'avait su comment en revenir, je n'y serais point allé.'

Jules Verne

The series is divergent; therefore we may be able to do something with it.

O. Heaviside

One service mathematics has rendered the human race. It has put common sense back where it belongs, on the topmost shelf next to the dusty canister labelled 'discarded non-sense'.

Eric T. Bell

Mathematics is a tool for thought. A highly necessary tool in a world where both feedback and non-linearities abound. Similarly, all kinds of parts of mathematics serve as tools for other parts and for other sciences.

Applying a simple rewriting rule to the quote on the right above one finds such statements as: 'One service topology has rendered mathematical physics ...'; 'One service logic has rendered computer science ...'; 'One service category theory has rendered mathematics ...'. All arguably true. And all statements obtainable this way form part of the raison d'être of this series.

This series, *Mathematics and Its Applications*, started in 1977. Now that over one hundred volumes have appeared it seems opportune to reexamine its scope. At the time I wrote

> "Growing specialization and diversification have brought a host of monographs and textbooks on increasingly specialized topics. However, the 'tree' of knowledge of mathematics and related fields does not grow only by putting forth new branches. It also happens, quite often in fact, that branches which were thought to be completely disparate are suddenly seen to be related. Further, the kind and level of sophistication of mathematics applied in various sciences has changed drastically in recent years: measure theory is used (non-trivially) in regional and theoretical economics; algebraic geometry interacts with physics; the Minkowsky lemma, coding theory and the structure of water meet one another in packing and covering theory; quantum fields, crystal defects and mathematical programming profit from homotopy theory; Lie algebras are relevant to filtering; and prediction and electrical engineering can use Stein spaces. And in addition to this there are such new emerging subdisciplines as 'experimental mathematics', 'CFD', 'completely integrable systems', 'chaos, synergetics and large-scale order', which are almost impossible to fit into the existing classification schemes. They draw upon widely different sections of mathematics."

By and large, all this still applies today. It is still true that at first sight mathematics seems rather fragmented and that to find, see, and exploit the deeper underlying interrelations more effort is needed and so are books that can help mathematicians and scientists do so. Accordingly MIA will continue to try to make such books available.

If anything, the description I gave in 1977 is now an understatement. To the examples of interaction areas one should add string theory where Riemann surfaces, algebraic geometry, modular functions, knots, quantum field theory, Kac-Moody algebras, monstrous moonshine (and more) all come together. And to the examples of things which can be usefully applied let me add the topic 'finite geometry'; a combination of words which sounds like it might not even exist, let alone be applicable. And yet it is being applied: to statistics via designs, to radar/sonar detection arrays (via finite projective planes), and to bus connections of VLSI chips (via difference sets). There seems to be no part of (so-called pure) mathematics that is not in immediate danger of being applied. And, accordingly, the applied mathematician needs to be aware of much more. Besides analysis and numerics, the traditional workhorses, he may need all kinds of combinatorics, algebra, probability, and so on.

In addition, the applied scientist needs to cope increasingly with the nonlinear world and the extra mathematical sophistication that this requires. For that is where the rewards are. Linear models are honest and a bit sad and depressing: proportional efforts and results. It is in the non-linear world that infinitesimal inputs may result in macroscopic outputs (or vice versa). To appreciate what I am hinting at: if electronics were linear we would have no fun with transistors and computers; we would have no TV; in fact you would not be reading these lines.

There is also no safety in ignoring such outlandish things as nonstandard analysis, superspace and anticommuting integration, *p*-adic and ultrametric space. All three have applications in both electrical engineering and physics. Once, complex numbers were equally outlandish, but they frequently proved the shortest path between 'real' results. Similarly, the first two topics named have already provided a number of 'wormhole' paths. There is no telling where all this is leading - fortunately.

Thus the original scope of the series, which for various (sound) reasons now comprises five subseries: white (Japan), yellow (China), red (USSR), blue (Eastern Europe), and green (everything else), still applies. It has been enlarged a bit to include books treating of the tools from one subdiscipline which are used in others. Thus the series still aims at books dealing with:

- a central concept which plays an important role in several different mathematical and/or scientific specialization areas;
- new applications of the results and ideas from one area of scientific endeavour into another;
- influences which the results, problems and concepts of one field of enquiry have, and have had, on the development of another.

Solitons, i.e. solitary localized waves with particle-like behaviour, and multi-solitons occur virtually everywhere. There is a good reason for that in that there is a solid, albeit somewhat heuristic argument which says that for wave-like phenomena the 'soliton approximation' is the next one after the linear one. It is also not too difficult via some searching in the voluminous literature - many hundreds of papers on solitons each year - to write down a long list of equations which admit soliton solutions and which model phenomena ranging over all the physical, biological, chemical and geological sciences as well as engineering. Yet, when lecturing on (mathematical) aspects of solitons I have found it not so easy to go beyond listing these equations. Largely because of lack of a book like this, which discusses where and how solitons arise, how they behave, whether or not they are stable and in what sense, which discusses approximate solitons and solitons in multidimensional spaces (which in a first simple natural formulation cannot exist), which discusses soliton (computer) experiments; all this for a wide range of phenomena especially in connection with solid state physics. There is a great deal of analytic material as well as there is especially a considerable collection of challenges for theoretical understanding. A great deal of the material covered in this book has not appeared in the monographic literature before. The phenomenology of solitons is very rich indeed.

The shortest path between two truths in the real domain passes through the complex domain.

J. Hadamard

La physique ne nous donne pas seulement l'occasion de résoudre des problèmes ... elle nous fait pressentir la solution.

H. Poincaré

Never lend books, for no one ever returns them; the only books I have in my library are books that other folk have lent me.

Anatole France

The function of an expert is not to be more right than other people, but to be wrong for more sophisticated reasons.

David Butler

Bussum, July 1990

Michiel Hazewinkel

Table of Contents

Preface

This book was conceived as a result of discussions with Professor Michiel Hazewinkel. It was also designed under a quite noticeable influence of him. Therefore, the responsibility for this book should be shared as follows: all its possible virtues must be equally allocated, and all its shortcomings, errors and omissions are due to me.

I would especially like to thank Drs. Aleksey Makhankov and Yurij Katyshev for their appreciable assistance in preparing and partially translating the manuscript. I would also like to thank Dr. David J. Larner and Kluwer Academic Publishers company for their unlimited patience while several deadlines came and went.

Thanks are also due to Mrs. Valeriya Sarantsev, since her assistance allowed me to considerably shorten the manuscript preparation time, and my wife Galia Makhankov for her patience during the time of my work on the book (more than three years).

In conclusion, I would like to thank Academician N.N. Bogolubov for his support, especially in the bad time after my illness which prolonged the work.

Introduction

The rapid development of concepts and methods associated with the investigation of nonlinear phenomena and their applications in all fields of modern physics have led to universality of the nonlinear approach. In addition to nonlinear differential equations describing a broad class of stochastic systems, the theory of integrable systems has evolved into a specific branch of mathematical physics.

The successes of physical theory are often decisively determined by structural morphology, i.e. a happy choice of model, with the constituent elements and interactions between them making a leading contribution to the phenomenon being described.

A remarkable achievement of the 20th century is the development of quantum field theory, the ideas and techniques of which are to be found throughout the entire spectrum of modern physics from condensed matter theory and plasma theory to particle physics and gravitation. The overwhelming majority of results (such as quantum electrodynamics with its impressively precise results) has been obtained here within the framework of perturbation theory where free field quanta corresponding to free particles are the morphologic units. The Fourier (linear) transformation representing any solution of the free equation in the form of a set of noninteracting harmonics (particles) is a mathematical tool relating to such an approach. The interaction is taken into account by perturbation theory which assumes that the coupling constant is small. However, such an approach has been found to be unsatisfactory in the construction of a theory of a number of both applied and fundamental physical phenomena. These include superconductivity and the phase transition problem, non-Abelian gauge theories in the infrared region and confinement, not to mention nonlinear problems of fluid and gas dynamics. Thus, for example, attempts to construct a theory of strong (and even weak) interactions in the manner of quantum electrodynamics has led in practice to insurmountable difficulties. This is not simply a matter of the interaction constant value, but that the approaches required are essentially different. A different morphological structure of the theory is necessary and the electroweak theory which resulted, based on the concept of non-Abelian gauge fields with broken symmetry, turns out to be essentially nonlinear.

Wherever possible, the mathematical basis of the new approach is a 'nonlinear Fourier transformation' (the inverse scattering method, etc.) which gives the solution to a nonlinear equation in the form of a set of both linear and nonlinear eigenmodes. The latter are particle-like entities and sometimes are linear mode bound states of a special type, viz. solitons. These essentially nonlinear objects which differ in principle from the linear modes may not be inserted into conventional perturbation theory and have as fundamental a role as the linear modes. In integrable systems, this follows immediately from the fact that contributions of various modes (linear and nonlinear) to the Hamiltonian (and all the other integrals of motion) are completely separated. Many such systems simulate fairly well the actual situation with due regard to reasonable assumptions. As to their statistical properties, the nonlinear components appear like noninteracting ideal gases of particles. This enables their contributions to correlation functions and Green functions to be calculated and a soliton thermodynamics to be constructed.

The essentially important characteristic of solitons is their stability region which determines the boundaries of the validity of such a description and the phase transition to a new state. Also this problem is connected with the extent to which the nonlinear systems is close to some completely integrable system. Therefore, in studies of nonlinear equations which describe real systems it is necessary to find every possible integrable reduction. Then, using perturbation theory of integrable systems one can take into account corrections related to deviations from the integrability. These deviations play the role of the interaction parameter of modern theories.

Such a study is most 'easily' carried out when a given deviation can be separated in the form of a right-hand side having a small parameter. However, we emphasize that such an investigation must be undertaken carefully since it is possible to obtain a 'solution' which does not satisfy the initial equation. In addition, in near integrable systems 'soliton' solutions are possible which essentially differ from well-known solutions to integrable analogs. This new approach to the study of physical systems finds application in practically all fields of physics. In condensed matter theory, for example enormous progress has been made in the description of structure phase transitions [1], a number of quasi-one-dimensional biological and organic systems [2] and in particular, in magnetic structures [3-9]. In field theory, soliton models (and their 'bag' limits) of particles [10], monopoles [11,12], classical solutions of Euclidean equations of motion (gravitational and Yang-Mills instantons, merons, etc.) [11-13] have been exhaustively studied.

In the first case, integrable reductions are the sine-Gordon, Landau-Lifshitz and nonlinear Schrödinger equations. In the second case, these are the self-duality equation, the Ernst equation or the axially symmetric stationary Einstein equations [14,15], and the equation for monopoles in the Prasad-Sommerfield limit [16].

The integrability property enables the construction of N-soliton solutions [16-18], the Bäcklund transformations, etc.

Essentially new types of integrable systems which have been studied in recent years are systems with internal degrees of freedom. The phenomena of fusion and decay of solitons [19], the exchange of internal ('color') degrees of freedom [20], etc., are possible in these systems. Such equations arise naturally in many physical problems.

Experimental studies of magnetic crystals show that most have a layered or multichain structure [21]. For the majority, the interaction between layers or chains has a considerable effect on the crystal dynamical behavior as a whole. Typical specimens are crystals of the salts $CrCl_2$, $CuCl_2$, $PbNiCl_3$, $CsNiCl_3$ [21]. Similar structures may be also found in organic compounds in the form of molecular chains [22]. The theoretical description of such structures is based upon a multicomponent generalization of the Heisenberg spin model [23]. The introduction of 'color' degrees of freedom for interacting spins in one-dimensional chains enables multilayer quasi-two-dimensional systems with weak coupling to be described. It is also established that the one-dimensional Hubbard model with a half-filled band corresponds to the two-component Heisenberg spin chain [24] with an intercomponent interaction. The multicomponent spin chain which corresponds to some generalized Hubbard model may be used to describe collective excitations and also their statistical properties in systems having several kinds of spins [25].

In this case, it is most difficult to model the low temperature region, where interaction between layers is essential [21]. In this region, the crystal dynamical behavior is defined by different nonlinear structures. The description, and the taking into account, of these structures are significant problems of condensed matter theory.

A similar situation also exists in quantum field theory. Modern unified field theories are essentially nonlinear (non-Abelian) and have no sufficiently rich set of fundamental fields to describe the particle spectrum in the linear approximation [26,27]. Therefore progress in modern theories such as extended supergravity [28] is determined, to a considerable extent, by the supply of inherent nonlinear objects, viz. monopoles, solitons, domain walls, and so forth.

A new and unexpected property of some quasi-one-dimensional magnetic systems and multidimensional field theoretic models is the global noncompact internal symmetry. Thus for the one-dimensional Hubbard model in the long-wave approximation [29] this group is $G = U(1,1)$ [30], for the extended $N = 4$ supergravity, $G = U(1,1)$ and at $N = 8$ we have $G = E_7$ [31]. The main feature of such models is the sign uncertainty of energy and a number of the 'no-go' theorems (concerning the existence and stability of solitons, e.g. the Derrick theorem) does not hold and the solution spectrum broadens significantly. The peculiarity of noncompact models became mostly evident in the study of excitations (soliton, i.e., hole-like and Bogolubov ones) of the stable condensate.

As a result of a great deal of work all over the world, the properties of solitons in plane (x,t) geometry are understood quite well. The time has come to proceed to more realistic and intricate multi-dimensional worlds. This transition - as might be expected - is nontrivial. Here, the soliton stability problem went ahead. In the early sixties, Derrick and Hobart proved the following theorem: In a space of more than one dimension, there do not exist stable stationary soliton-like solutions in the framework of conventional relativistically invariant nonlinear theories (without internal symmetries and differential interactions).

It is now necessary to determine what we shall understand by the expression *soliton in a non-one-dimensional space*. Our definition of the soliton is similar to the well-known field-theoretic concept of a quasi-particle solution: The soliton is a field configuration distinct from a vacuum one, described by a wave equation solution, localized in space and possessing finite energy.

In fact, the Derrick-Hobart theorem states that in the stationary case the Hamiltonian of a given system, $H[\varphi] = \text{const.}$, considered as a functional of the field determines a surface which cannot be a valley; it is either a hill or, at best, a saddle in functional space. Therefore, to stabilize the system additional constraints (integral of motion) are required. Nonlinear theory has been developed most consistently and completely in two-dimensional space-time, where the majority of results has been obtained analytically, especially for integrable systems. In higher space-time dimensions, with rare exceptions, only existence and stability problems can be studied analytically. Soliton dynamics, i.e. the processes of their formation and interaction still remains the domain of numerical experiments.

As a result we can now formulate the research program:

(1) Determination and investigation of linear oscillation modes (if any).

(2) Analysis of the existence conditions of particle-like solutions (PLS).

(3) Studies of their stability.

(4) Studies of the dynamics of solitons (PLS).

(5) Calculation of the PLS form factors.

(6) Constructing the soliton statistical model (kinetic equation) and finding their dynamical structure factors (DSF).

Item (1) is standard and reduces to investigating the dispersion properties of the system. If linearization is carried about a nontrivial state of the system (1) tends to (3). In general, the remaining items listed above are essentially different for various models and theories.

At the same time, a variety of models arising in theoretical physics have a lot in common. We can demonstrate this by an example of the classical field theory with the Lagrangian [32]

$$\mathfrak{L} = \mathfrak{L}_{\mathrm{d}} + U[(\bar{\phi}\phi)] \tag{0.1}$$

where $\mathfrak{L}_{\mathrm{d}}$ (the differential part) is $(\bar{\mu}\phi^{\mu})$ in relativistic theory and $\frac{i}{2}(\bar{\phi}\phi_t - \bar{\phi}_t\phi) + \bar{\phi}_k\phi^k$ in the nonrelativistic one, $\mu = 0, 1, \cdots, D$, $k = 1, \cdots, D$, and the potential U has the expansion

$$U \to \epsilon[m^2(\bar{\phi}\phi) - g^2(\bar{\phi}\phi)^2 + \cdots], \quad \phi \to 0. \tag{0.2}$$

Here ϕ is a scalar or vector field,

$$\phi = \begin{bmatrix} \phi_1 \\ \cdot \\ \cdot \\ \cdot \\ \phi_n \end{bmatrix}, \quad \phi^+ = (\phi_1^*, \cdots, \phi_n^*), \quad \bar{\phi} = \phi^+ \Gamma_0$$

$\Gamma_0 = \mathrm{diag}(1, \cdots, 1, -1, \cdots, -1)$ is a metric of the isotopic space $C_{p,q}$, and $R_{1,D}$ is a configurational space.

From eqns. (0.1) and (0.2) it follows that the theory is invariant under the global group $U(p,q)$.

In expansion (0.2) we only retain the first nonlinear term since, in this case, via the scale transformation $x \to \alpha x'$, $t \to \beta t'$, $\phi \to \gamma \phi'$ one can remove from the equation all the parameters including the small one:

$$\Box\phi + \epsilon(1 - |\phi|^2)\phi = 0, \quad \Box = \partial_t^2 - \Delta. \tag{0.3}$$

The x- and t-independent solution to eqn. (0.3) is called a stationary (vacuum) state of the system.* It is clear that two such solutions $\phi_1 = 0$, $|\phi_2| = 1$ satisfy eqn. (0.3).

Theory (0.1) includes two classes of models:

$$\epsilon = \pm 1.$$

The first class $(\epsilon = 1)$.

By linearizing eqn. (0.3) about ϕ_1 and ϕ_2 it is straightforward to find the dispersion relations

$$\omega_1^2 = \kappa^2 + 1 \tag{0.4}$$

$$\omega_2^2 = \kappa^2 - 2$$

whereby it follows that in this case the stable vacuum is trivial, $\phi_1 = 0$ (the symmetry is not broken). This class describes a number of physical models considered below and contains as mathematical models the ϕ_+^4 and SG theories. Soliton-like solutions (SLS) are elementary excitations over the trivial vacuum.

---

*In some cases such a state is described by the plane wave: $\phi_v = \phi_0 \exp\{i(kx - \omega t + \delta)\}$.

The secondary class ($\epsilon = 1$).
Now eqns. (0.4) become $\omega_1^2 = \kappa^2 - 1$, $\omega_2^2 = \kappa^2 + 2$, i.e. the stable vacuum is nontrivial, $\phi_v \equiv \phi_2 = \text{const.}$ (the symmetry is spontaneously broken at temperatures less than some critical one T_{cr}). In a series of nonrelativistic models such a vacuum describes the Bogolubov condensate at $T = 0$. The ϕ^4_- theory and models with non-compact internal symmetry are members of this class. The $\phi^4 - \phi^6$ model and other $P(\phi^2)$ models keep somewhat aloof, for two or more stable vacuum states can exist simultaneously in their framework [33].

Below we shall consider predominantly the results obtained in this field in JINR (Dubna) by the author and his colleagues in recent years. Part of the results on analytic and numerical analysis of nonintegrable nonlinear differential equations obtained in the 1970ties and the early 1980ties has been presented in the reviews [34], and nonlinear effects in quasi-one-dimensional models of condensed matter theory were given in the review [35].

Various aspects of the theory of solitons, integrable systems in general, the increasingly growing area of their applications have been reported in many monographs, reviews and articles (a great number of them have been discussed in detail or cited in [34,35]. Therefore, below, we shall follow the above-formulated program where possible. This determines the scenario. On the basis of a number of examples from condensed matter physics (plasmas and nuclei) we show how field-theoretical models arise which belong to a certain class.

In subsequent chapters, we consider various models and how far one can proceed in the way outlined above. We always endeavor to expound topics by way of problems studied in recent literature. This is also related to such traditional techniques as the inverse scattering transform. One succeeds to study analytically only a very small number of nonlinear differential equations. The overwhelming majority of these that describe physically interesting models (especially in the dimensions $D \geqslant 2$) may be analyzed numerically, as a rule. In the following, we therefore discuss computer experiments quite frequently.

The statistical subject distribution of the papers which had appeared by the mid-eighties and containing the term soliton (Dr. Yu. Katyshev and I have examined about twenty leading periodicals and many hundreds of preprints) is tabulated below:

Total	1360 papers
Condensed Matter	40%
Pure Mathematics	24%
Plasmas	7%
Others	29%

This distribution (certainly time dependent) has also prompted, in a sense, the organization of this book, i.e. as far as arranging the contents of each chapter and the degree of detail given for various topics was concerned. Thus, in Chapter 1 we consider at sufficient length some models of condensed matter and nuclear theories which can be described by nonlinear differential equations. A number of these equations have been obtained earlier, for instance, in plasma theory. We give less attention to their derivation in terms of plasma physics because this has been discussed in a series of reviews and books. We therefore dwell rather on the common properties of the corresponding solutions, as well as on their splitting (distinctive) features, i.e. on a comparative analysis.

Finally, a few words about the formula numbering used throughout the book. When we refer to a formula within the chapter in which that formula appears we do not include the chapter number: when we quote a formula which appears in a different chapter we include the chapter number.

References

1. J. Krumhansl and J. Schrieffer, *Dynamics and statistical mechanics of a one-dimensional model Hamiltonian for structure phase transition.* Phys. Rev. 1975, 11B, pp. 3535.
2. A.S. Davydov , *Solitons in quasi-one-dimensional structures.* Uspekhi Fiz. Nauk, 1982, 138, pp. 603 (in Russian).
3. A.R. Bishop and T. Schneider (eds.), *Solitons and Condensed Matter Physics.* Springer Solid State Sciences, V. 8 (Springer-Verlag, Berlin, 1978).
4. H. Mikeska, *Solitons in ferromagnet with easy plane anisotropy,* J. Phys., 1978, 11C, pp. L129.
5. V.K. Fedyanin, *Dynamical formfactor of neutron scattering on solitons in quasi-one-dimensional magnets,* JMMM, 1983, 31-34, pp. 1237.
6. J. Bernasconi and T. Schneider (eds.), *Physics in One Dimension,* Springer

Solid State Sciences, V. 23 (Springer-Verlag, Berlin, 1981).

7. K. Maki, *Solitons in low temperature physics. In: Progress in Low Temperature Physics.* Ed. D.F. Brewer, V. 8. North-Holland, Amsterdam, 1982.

8. A.M. Kosevich, B.A. Ivanov and A.S. Kovalev. *Nonlinear Waves of Magnetization. Dynamical and Topological Solitons.* Naukova Dumka, Kiev, 1983 (in Russian).

9. M. Steiner. *Solitons in 1-D magnets.* JMMM, 1983, 31-34, pp. 1277.

10. T.D. Lee. *Nontopological solitons and applications to hadrons.* Physica Scripta, 1979, 20, p. 440.

11. R. Jackiw. *Quantum meaning of classical field theory.* Rev. Mod. Phys., 1977, 49, p. 681.

12. A. Actor. *Classical solitons of $SU(2)$ Yang-Mills theories.* Rev. Mod. Phys., 1979, 51, p. 461.

13. S. Hawking. *Euclidean quantum gravity.* In: Recent Developments in Gravitation, 1978 Cargèse Lectures on Gravitation. Ed. M. Lévy and S. Desr, Plenum Press, N.Y. 1979, p. 145.

14. F.J. Ernst. *New formulation of the axially symmetric gravitational field problem.* Phys. Rev., 1968, 167, p. 1175.

15. D. Maison. *On the complete integrability of the stationary, axially symmetric Einstein equations.* J. Math. Phys., 1979, 20, p. 871.

16. M.K. Prasad. *Instantons and monopoles in Yang-Mills gauge field theories.* Physica, 1980, 1D, p. 167.

17. V.A. Belinskii and V.E. Zakharov. *Integration of Einstein equations by the inverse scattering problem method and calculation of exact soliton solutions.* Sov. Phys. JETP, 1978, 48, p. 985; Stationary gravitational solitons with axial symmetry. Sov. Phys. JETP, 1979, 50, p. 1.

18. P. Forgàcs and Z. Horváth and L. Palla. *Soliton theoretic framework for generating multimonopoles.* Ann. Phys., 1981, 136, p. 371.

19. V.E. Zakharov and A.V. Mikhailov. *Relativistically invariant two-dimensional models of field theory integrable by the inverse problem method.* Sov. Phys. JETP, 1978, 47, p. 1017.

20. S.V. Manakov. *Towards a theory of two-dimensional stationary self-focusing*

electromagnetic waves. Zh. Eksp. Teor. Fiz., 1973, 65, p. 505 (in Russian).

21. L.J. de Jongh and A.R. Miedema. *Experiments on simple magnetic model systems.* Adv. in Phys., 1974, 23, p. 1.

22. A.A. Ovchinnikov et al. *Theory of one-dimensional Mott semiconductors and the electronic structure of long molecules with conjugated bonds.* Uspekhi Fiz. Nauk, 1972, 108, p. 81.

23. M. Ito. *Transition temperature of layered system of isotopic spin with* $n \geqslant 3$ *components.* Prog. Theor. Phys., 1981, 65, p. 1773.

    A.B. Zolotovitskii and V.P. Kalashnikov. *Dispersion and damping of the Goldstone mode in a multicomponent system.* Teor. Mat. Fiz., 1981, 49, p. 273 (in Russian).

24. H. Shiba. *Thermodynamics properties of the one-dimensional half-filled-band Hubbard model II. Application of the grand canonical method.* Prog. Theor. Phys., 1972, 48, p. 2171.

    R.A. Bari. *Classical linear-chain Hubbard model: metal-insulator transition.* Phys. Rev., 1973, 7B, p. 4318.

    I. Egri. *Band structure of a one-dimensional Peierls-Hubbard model.* Solid State Comm., 1975, 17, p. 441.

25. V.P. Kalashnikov and N.V. Kozhevnikov. *The dynamical behavior of two-component spin system in a varying magnetic field.* Teor. Mat. Fiz., 1978, 37, p.402 (in Russian).

    A.B. Zolotovitskii and V.P. Kalashnikov. *Relaxation of Goldstone magnons in a ferrimagnetic semiconductor.* Phys. Lett., 1982, 88A, p. 135.

    M.I. Auslender et al. *Dispersion and damping of acoustic magnons in a multicomponent collinear magnetic at low temperatures.* Teor. Mat. Fiz., 1982, 51, p. 111 (in Russian).

    Y. Lepine. *Spin-Peierls transition of the anisotropic XY model in a magnetic field.* Phys. Rev., 1981, 24B, p. 5242.

26. M. Gell-Mann et al. *Complex spinors and unified theory of supergravity.* In: Supergravity, North-Holland, Amsterdam, 1979.

27. M. Gûnaydin. *Unitary realizations of the noncompact symmetry groups of supergravity.* CERN, TH. 3222, Geneva, 1981.

28. P. van Nieuwenhuizen. *Supergravity*. Phys. Rep., 1981, 68, p. 189.

29. U. Lindner and V.K. Fedyanin. *Solitons in a one-dimensional modified Hubbard model.* Phys. Stat. Sol., 1978, 89(b), p. 123.

30. V.G. Makhankov, N.V. Makhaldiani and O. Pashaev. *On the integrability and isotopic structure of the one-dimensional Hubbard model in the long wave approximation.* Phys. Lett., 1981, 81A, p. 161.

31. E. Cremmer, B. Julia and J. Scherk. *Supergravity theory in 11 dimensions.* Phys. Lett., 1978, 76B, p. 409.

    E. Cremmer and B. Julia. *The SO(8) supergravity.* Nucl. Phys., 1979, 159B, p. 141.

32. V.G. Makhankov. *Dynamical structure factors and clusterization in a class of relativistic field theory models.* JINR, p2-82-248, Dubna, 1982 (in Russian);

    Clusterization in a class of models of classical field theory. In: *Nonlinear and Turbulent Processes in Physics,* Ed. R.Z. Sagdeev, Gordon & Breach, Harwood Academic Publishers, N.Y. 1984, V. 3, p. 1471-1479.

33. I.V. Barashenkov and V.G. Makhankov. *Soliton-like excitations in a one-dimensional nuclear matter.* JINR, E2-84-173, Dubna, 1984.

34. V.G. Makhankov. *Dynamics of classical solitons.* Phys. Rep., 1978, 35, p. 1-128;

    Computer experiments in soliton theory. CPC, 1980, 21, p. 1-49.

    Solitons and numerical experiments. Sov. J. Part. Nucl., 1983, 14, p. 50-75.

35. V.G. Makhankov and V.K. Fedyanin. *Nonlinear effects in quasi-one-dimensional models of condensed matter theory.* Phys. Rep., 1984, 104, p. 1-86.

Part I. Quantum Systems and Classical Behaviour

Chapter 1.

SOME PHYSICAL MODELS AND NONLINEAR DIFFERENTIAL EQUATIONS

During recent years, concept has arisen of a new type of collective excitation in ordered media: the so-called particle-like or soliton-like excitation. There exist many interesting phenomena associated with the behaviour of ferromagnets at low temperatures which are related to such excitations. It should be emphasized that these phenomena have a macroscopical quasiclassical character, which means that an adequate classical (or quasiclassical) description of the behavior of such quantum systems must exist. The Heisenberg model provides the basis for the theoretical study of a wide class of ferromagnets at the quantum level. The question arises of formulating a consistent 'reduction procedure' connecting quantum statistical models and classical field models.

Below, in this chapter, by the use of simple examples from condensed matter physics, plasma and nuclear physics, we show how the reduction procedure works.

§1. Magnetic chain (the Heisenberg model)

Quantum statistical models used to be described by Hamiltonians written in terms of Bose, Fermi or spin vectors. In the first and second cases, the following Hamiltonians are usually discussed:

$$\hat{H} = E_0+\epsilon\sum_j \hat{n}_j + J\sum_j (a_j^+ a_{j+1} + a_{j+1}^+ a_j) + g\sum_j \hat{n}_j \hat{n}_{j+1}, \tag{1.1}$$

$$\hat{n}_j = a_j^+ a_j,$$

where ϵ, J, g are, respectively, the energy level, exchanging integrals, and nonlinearity parameter; a_j^+ and a_j are the creation and annihilation operators in the Heisenberg representation, related to the ith lattice cell (in configuration or momentum space), and acting in the Fock space. They obey the following commutation relations

$$[a_i,a_j^+] = \delta_{ij}, \ \ [a_i,a_j] = [a_i^+,a_j^+] = 0 \tag{1.2}$$

in the case of Bose systems (the Heisenberg-Weil algebra) and

$$\{a_i,a_j^+\} \equiv a_ia_j^+ + a_j^+ a_i = \delta_{ij}, \ \ \{a_i,a_j\} = \{a_i^+,a_j^+\} = 0 \tag{1.3}$$

for Fermi systems (in what follows we denote them by c_j^+, c_j). The vacuum state in both cases is the vector $|0>$ such that

$$a_j|0> = 0. \tag{1.4}$$

In terms of the spin operators $\hat{S}_i^a(a=x,y,z;i=\overline{1,N})$ with the algebra

$$[\hat{S}_i^a,\hat{S}_j^b] = \delta_{ij}\epsilon_{abc}\hat{S}_j^c \tag{1.5}$$

is given the Hamiltonian of the spin or quasi-spin systems.

The most popular such system is the Heisenberg model of the ferromagnet

$$\hat{H}_G = -\frac{1}{2}\sum_{ij}\{J_{ij}^{(1)}\hat{S}_i^x\hat{S}_j^x + J_{ij}^{(2)}\hat{S}_i^y\hat{S}_j^y + J_{ij}^{(3)}\hat{S}_i^z\hat{S}_j^z\}. \tag{1.6}$$

We consider its simplified version, the single-axis anisotropy model, $J_{ij}^{(4)} = J_{ij}^{(2)}$, with the nearest-neighbour interaction, $J_{ij}^{(a)} = J_j^{(a)} \neq 0$ if $j = i \pm 1$ and $J_{ij}^{(\alpha)} = 0$ otherwise. Then

$$\hat{H}_s = \sum_j\{J_j^{(1)}(\hat{S}_j^x\hat{S}_{j+1}^x + \hat{S}_j^y\hat{S}_{j+1}^y) + J_j^{(3)}\hat{S}_j^z\hat{S}_{j+1}^z\}. \tag{1.7}$$

This Hamiltonian can be represented in a more convenient form by introducing the operators

$$\hat{S}_j^{\mp} = \hat{S}_j^x \pm i\hat{S}_j^y, \ \ \hat{S}_j^{(0)} = \hat{S}_j^z \tag{1.8}$$

with the algebra

$$[\hat{S}_j^+,\hat{S}_{ik}^-] = 2\delta_{jk}\hat{S}_j^{(0)}, \ \ [\hat{S}_j^{(0)},\hat{S}_k^{\pm}] = \pm\delta_{jk}\hat{S}_j^{\mp} \tag{1.9}$$

and denoting $J_j^{(1)} = J_j$, $J_j^{(3)} = \tilde{J}_j$. Then we have

$$\hat{H}_s = -\sum_j\{\frac{1}{2}J_j(\hat{S}_j^+\hat{S}_{j+1}^- + \text{h.c.}) + \tilde{J}_j\hat{S}_j^z\hat{S}_{j+1}^z\}. \tag{1.10}$$

We have now two variants of the Hamiltonian which describe the whole class of physical quantum statistical models. Our concern will be with the classical (quasiclassical) behaviour of such systems. To describe it, one should formulate an adequate 'reduction procedure' that connects quantum q-number models (1) or (10) with classical c-number models. There are several variants of such a procedure:

(1) A crude but simple variant is a transition from operators a_j^+, a_j or $\hat{S}_j^a$ to c-numbers $\bar{\alpha}_j, \alpha_j$ or S_j^a. This is met quite frequently, even in the case of $s = 1/2$ and in the Fermi operators (see e.g. [1,2] and references cited therein, also [3] and Chapter 2);

(2) A finer approach consists of constructing a set of trial (single-particle) states, and the original Hamiltonian is averaged over these states (see [16] and references therein);

(3) This variant implies usage as the trial states the Heisenberg-Weyl or spin (generalized) coherent states (CS).

In what follows, we shall consider, as a rule, one-dimensional models. In this case, there is an exact transformation (Jordan-Wigner) connecting the Fermi-operators c_j, c_j^+ with the Spin operators $\hat{S}_j^-$, $\hat{S}_j^+$ at $s = 1/2$ (in this case they become the Pauli operators):

$$\hat{S}_j^+ = \exp\{-i\pi \sum_{k=1}^{j-1} c_k^+ c_k\} c_j, \quad \hat{S}_j^- = c_j^+ \exp\{i\pi \sum_{k=1}^{j-1} c_k^+ c_k\} \tag{1.11}$$

$$\hat{S}_j^z = \frac{1}{2} - c_j^+ c_j.$$

It means that for $D = 1$ models we can confine ourselves to investigations of spin and Bose Hamiltonians.

Consider model (10). The most well-known method of treating such a Hamiltonian is to 'bosonize' it via the Holstein-Primakoff transformations (or Dyson-Maleev and others). Recall that the basis element of these is the choice of the quantization axis, OZ:

$$\hat{S}_j^+ = \sqrt{2s - \hat{n}_j}\, a_j, \quad \hat{S}_j^- = a_j^+ \sqrt{2s - \hat{n}_j} \tag{1.12}$$

$$\hat{S}_j^z = s - \hat{n}_j, \quad \hat{n}_j = a_j^+ a_j.$$

As usual, we call (10) the isotropic Heisenberg Model (IHM) if $J = \tilde{J}$; the easy-axis Heisenberg Model (EAHM) if $\tilde{J} > J$, and the easy-plane Heisenberg Model (EPHM) if $J > \tilde{J}$. We give now the results of the work [4].

The IHM Hamiltonian possesses $SU(2)$ symmetry, or it is invariant under the transformation, $R \in SU(2)$

$$R^{-1}\hat{H}R = \hat{H}, \quad R = \prod_j \exp(i\vec{\theta}\hat{\mathbf{S}}).$$

The symmetry of the EPHM and EAHM Hamiltonian is $\mathbb{Z}_2 \otimes U(1)$.

Ground states of these models have essentially different symmetries. The ferromagnetic state in the first case (IHM) distinguishes the axis of spontaneous magnetization (in any case for large S) and reduces the initial symmetry to $U(1)$. In the second case, the axis of spontaneous magnetization coincides with the easy axis. In the third case, the axis of spontaneous mangetization lays in the easy-plane, the initial symmetry reduces to $\mathbb{Z}_2$. Thus the ferromagnetic ground state distinguishes a physical axis, the spontaneous ferromagnetization axis. On the other hand, the Holstein-Primakoff transformations are based on the choice of a mathematical axis: the quantization axis.

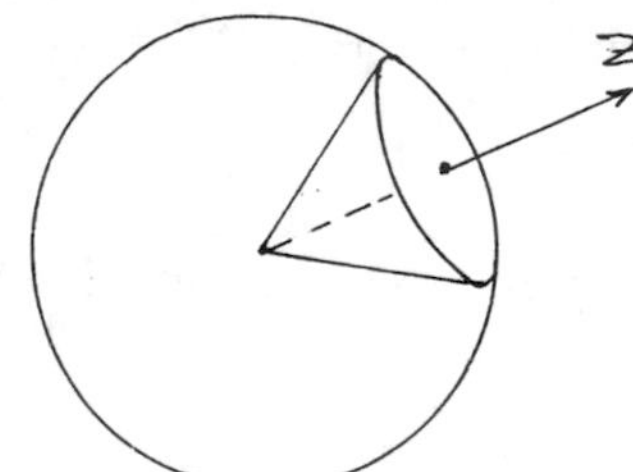

Figure 1.1. The direction of the physical axis is arbitrary. Let this be along the Z-axis. The Holstein-Primakoff transformation describes oscillations of the spin inside the cone.

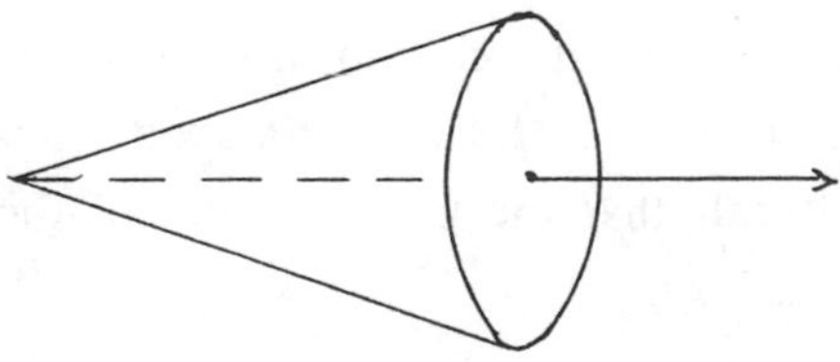

Figure 1.2. The Z-axis is along the easy axis.

Let us look at the relation in which they are connected in all three variants:

1. Isotropic Heisenberg Model. The direction of the physical axis is arbitrary. Let this be along the OX-axis. The Holstein-Primakoff transformations describe oscillations of the spin inside the cone (see Fig. 1.1).

2. Easy-axis Heisenberg Model. Select OZ-axis along the easy (Fig. 1.2).

3. Easy-plane Heisenberg Model. Here we have two directions:

(a) The axis perpendicular to the easy-plane.

(b) The spontaneous magnetization axis laying in the easy-plane.

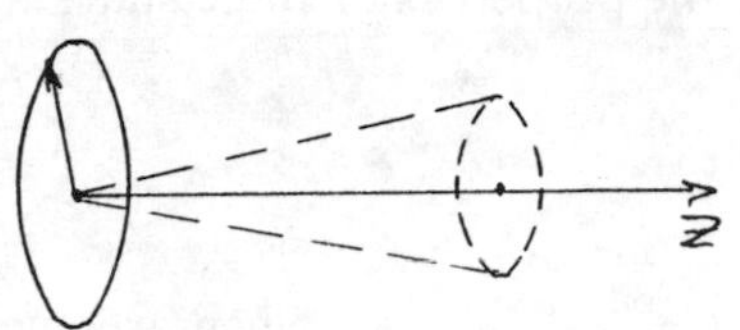

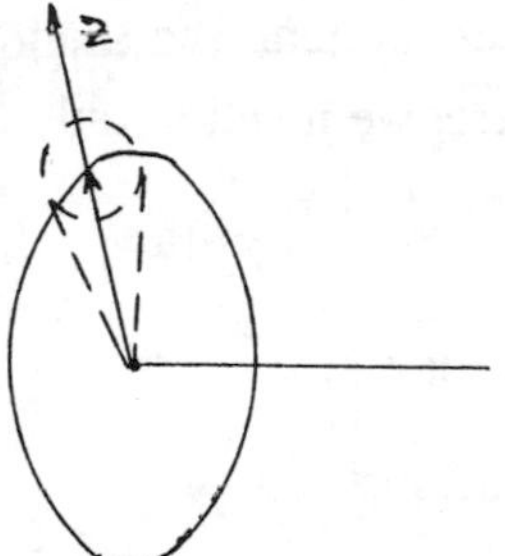

Figure 1.3. (a) The Z-axis is perpendicular to the easy plane.

(b) The spontaneous magnetization axis lies in the easy plane.

Generally speaking we can use both (a) and (b) as the quantization axis. (Figs. 1.3(a) and 1.3(b).) With the help of the Holstein-Primakoff transformation for the IHM, EAHM, EPHM (a) models we get the Hamiltonian $\hat{H}$ in terms of Bose operators:

$$\hat{H} = \hat{H}_2 + \hat{H}_4 + \hat{H}_6 \tag{1.13}$$

$$\hat{H}_2 = -\frac{s}{2}\sum_j J(a^+_{j+1}a_j + a^+_j a_{j+1}) - \frac{s^2}{2}\sum_j \tilde{J} + \frac{s}{2}\sum \tilde{J}(a^+_j a_i + a^+_{j+1}a_{j+1}),$$

$$\hat{H}_4 = \frac{1}{8}\sum_j J\left[(a^+_j)^2 a_j a_{j+1} + a^+_j a^+_{j+1}(a_{j+1})^2 + (a_{j+1}a^+_j)(a_j)^2 + (a^+_{j+1})^2 a_{j+1}a_j\right] -$$

$$-\frac{1}{2}\sum_j \tilde{J}(a^+_j a^+_{j+1} a_j a_{j+1}),$$

$$\hat{H}_6 = -\frac{1}{128s}\sum_j J\left[2a^+_{j+1}a^+_j a^+_{j+1} a_j a_{j+1} a_j + 2(a^+_j)^2 a^+_{j+1} a_j (a_{j+1})^2 -\right.$$

$$- a^+_{j+1}(a^+_j)^2(a_j)^3 - (a^+_{j+1})^3(a_{j+1})^2 a_j -$$

$$-(a_j^+)^3(a_j)^2a_{j+1}-a_j^+(a_{j+1}^+)^2(a_{j+1})^3\Big].$$

The Hamiltonian (13) is given in the normal (Wick) form and we can average it over the Heisenberg-Weyl (Glauber) coherent states

$$|\alpha> = e^{\alpha a^+ - \bar{\alpha}a}|0> = e^{-\frac{1}{2}|\alpha|^2+\alpha a^+}|0> = e^{-\frac{1}{2}|\alpha|^2}\sum_{n=0}^{\infty}\frac{\alpha^n}{\sqrt{n!}}|n> \tag{1.14}$$

where $\bar{\alpha}$ is the complex-conjugate, and $|n> = \sqrt{n!}(a^+)^n|0>$ are the n-particle orthogonal normalized states, generating the Fock space. The coherent states (14) determined by the points of the complex plane α (phase plane) form an overcomplete system. We shall discuss in detail the properties of these states in Chapter 5; here, we mention only the following:

$$<\alpha|\alpha> = 1, \tag{1.15a}$$

an operator $\tilde{A}$, written in normal form $\hat{A} = \sum_{m,n} c_{mn}(a^+)^m a^n$, when averaged over these states is

$$<\alpha|\hat{A}|\alpha> \equiv A = \sum_{m,n} c_{mn}\bar{\alpha}^m\alpha^n, \tag{1.15b}$$

since the operators a_i^+, a_i commute in neighbouring lattice sites the Fock space of the total lattice splits into the direct sum of the single-site spaces, and the lattice coherent states into the direct product of the single-site coherent states:

$$|\alpha> = \prod_j|\alpha>_j = \prod_j e^{-\frac{1}{2}|\alpha_j|^2}e^{\alpha_j a_j^+}|0>_j. \tag{1.15c}$$

Making use of (15) one easily gets the classical lattice Hamiltonian $H = <\alpha|\hat{H}|\alpha>$. To obtain its continuous version, the standard transformation must be carried out:

$$\alpha_j = \alpha(ja_0) \equiv \alpha(\chi) \tag{1.16a}$$

$$\alpha_{j+1} = \alpha(\chi)+a_0\alpha'(\chi)+\frac{1}{2}a_0^2\alpha''(\chi)+\ldots, \quad a_0\to 0. \tag{1.16b}$$

We can neglect terms with derivatives higher than the second derivative if

$$a_0k_0 << 1 \tag{1.17}$$

where k_0 characterizes gradients of the functions under consideration $\alpha(\chi)$ (viz., maximal wave numbers or reverse packet widths). Besides, the following natural condition should hold

$$(a_0/\lambda) \ll 1 \qquad (\lambda \text{ is a wavelength}) \tag{1.18}$$

necessary to represent the lattice discrete function through continuous functions. Condition (17) is stronger than (18):

$$a_0 k_0 \simeq a_0 \frac{2\pi}{\lambda} = 2\pi\left(\frac{a_0}{\lambda}\right) \ll 1.$$

Thus the small dimensionless parameter $\kappa = a_0 k_0 \ll 1$ is one of the main parameters of the reduction procedure.

Up to $0(a_0^2)$ we obtain from (13)

$$H_{cl} = \frac{J}{a_0}\int d\chi\left\{-2s(1-\rho)|\alpha|^2 + A|\alpha|^4 + \frac{|\alpha|^2}{16s^2} + a_0^2\left[s|\alpha_\chi|^2 - \frac{A}{2}(|\alpha|_\chi^2)^2 + \right.\right.$$

$$\left.\left. + \frac{1}{4}(\bar{\alpha}^2\alpha_\chi^2 + \bar{\alpha}_\chi^2\alpha^2) + \frac{|\alpha|^2}{16s}(|\alpha|_\chi^2)^2\right]\right\}$$

where $\alpha = \alpha(\chi,t)$, $J - \tilde{J} = J(1-\rho)$, $A = 1 - \rho + \frac{1}{8s} + \frac{1}{32s^2}$.

Using the conventional assumption of small anisotropy (more for convenience than of necessity) [1a]

$$1 - \rho = -a_0^2\frac{\Delta}{2} \tag{1.19}$$

and omitting terms $\frac{a_0^2}{2}A(|\alpha|_\chi^2)^2$ and $\frac{a_0^2}{16s}|\alpha|^2(|\alpha|_\chi^2)^2$ small compared with $A|\alpha|^4$, we obtain

$$H_{cl} = a_0 sJ\int d\chi\left\{\Delta|\alpha|^2 + |\alpha_\chi|^2 + B|\alpha|^4 + \frac{1}{4s}(\bar{\alpha}^2\alpha_\chi^2 + \bar{\alpha}_\chi^2\alpha^2) - \right. \tag{1.20}$$

$$\left. - \frac{|\alpha|^6}{16s^2a_0^2}\right\}$$

where

$$B = \frac{A}{2sa_0^2} = -\frac{1}{2s}\left(\Delta - \frac{1}{4sa_0^2} - \frac{1}{16s^2a_0^2}\right).$$

By varying the Hamiltonian (20) one gets the equation*:

$$i\dot{\alpha} = \alpha_{xx} - \Delta\alpha - |\alpha|^2\alpha(2B + \frac{3|\alpha|^2}{16s^3a_0^2}) + \frac{1}{2s}(\alpha^2\bar{\alpha}_{xx} - \bar{\alpha}\alpha_x^2 + 2\alpha|\alpha_x|^2), \tag{1.21}$$

$$Jsa_0t \to t, \quad \dot{} = \partial_t$$

Consider the variant EPHM (b) when the quantization axis is along that of spontaneous magnetization laying in the easy-plane. In this case the Hamiltonian reads:

$$\hat{H}_s = \sum_n \hat{H}_n \equiv -J\sum_n[\hat{S}_n^z\hat{S}_{n+1}^z + \hat{S}_n^z\hat{S}_{n+1}^x + \rho\hat{S}_n^y\hat{S}_{n+1}^y].$$

Rewrite this in the form $\hat{H} = \hat{H}_+ + \hat{H}_-$, where

$$\hat{H}_+ = -\frac{1+\rho}{2}\sum_n[\frac{1}{2}(\hat{S}_n^+\hat{S}_{n+1}^- + \hat{S}_n^-\hat{S}_{n+1}^+) + \hat{S}_n^z\hat{S}_{n+1}^z],$$

$$\hat{H}_- = -\frac{1-\rho}{2}\sum_n[\frac{1}{2}(\hat{S}_n^+\hat{S}_{n+1}^+ + \hat{S}_n^-\hat{S}_{n+1}^-) + \hat{S}_n^z\hat{S}_{n+1}^z],$$

using the above procedure we have the following classical model:

$$H = H_+ + H_-$$

$$H_+ = \frac{1+\rho}{2}sa_0^2\int|\alpha_x|^2dx \tag{1.22a}$$

$$H_- = -\frac{1-\rho}{2}\cdot 2s\int[|\alpha|^2 - \frac{\alpha^2+\bar{\alpha}^2}{2}](1 - \frac{|\alpha|^2}{2s})dx. \tag{1.22b}$$

In eqn. (22b) terms of the order of $O(a_0^2)$ are dropped; which is correct at small anisotropy, see (19). Now we have the Hamiltonian

$$\tilde{H}_{cl} = sa_0J\int[|\alpha_x|^2 + \frac{\Delta}{2}(|\alpha|^2 - \frac{\alpha^2+\bar{\alpha}^2}{2})(1 - \frac{|\alpha|^2}{2s})]dx$$

and the equation of motion

$$i\dot{\alpha} = \alpha_{xx} + \frac{\Delta}{2}(\alpha - \bar{\alpha}) + \frac{\Delta}{2s}|\alpha|^2\alpha + \frac{\Delta}{8s}\alpha^3 + \frac{3\Delta}{8s}|\alpha|^2\bar{\alpha}. \tag{1.23}$$

*There is a technique for obtaining the dynamical equations of the kind (21) based on the averaged Hamiltonians. This is discussed in Chapter 5.

As might be expected, $U(1)$ symmetry is broken in eqn. (14) (see Fig. 3b). We have obtained for the case of EPHM ($\Delta<0$) two representations: (21) and (23).

Representation (21)-EPHM (a). Let us consider the ground state for model (21). This is defined by the minimum of the functional (20). Up to the terms, leading with respect to $(a_0^2k_0^2)$, the extremum of (20) is given by the equations

$$\alpha_x = 0,$$

$$\alpha\left\{-|\Delta|\,(1-\frac{|\alpha|^2}{s})+\frac{1}{4s}\frac{|\alpha|^2}{sa_0^2}(1+\frac{3}{4}\frac{|\alpha|^2}{s})\right\} = 0$$

which gives:

$$(1)\quad \alpha = 0$$

or

$$(2)\quad |\alpha|^2 \simeq s \;\text{ if }\; 2sa_0^2|\Delta| \equiv s\delta >> 1. \tag{1.24}$$

The point $\alpha = 0$, as can be easily checked, corresponds to the H_{cl} minimum at $\Delta>0$ and the circle $|\alpha|^2 = s$ at $\Delta<0$.

Upon substituting $\alpha(\chi) = 0$ into eqn. (12) we find $<|\hat{S}_j^z|> = s-|\alpha|^2 = s$ to be the easy-axis ferromagnetic state, in which all the spins are arranged along the OZ axis.

In the second case, we have $<|\hat{S}_j^z|> = 0$, i.e. the classical ground state (in fact, taking place in the approximation $\delta s >> 1$) would seem as if it corresponds to the easy-plane quantum ground state.

Linear oscillations about these states are defined by the dispersion formulae:

$$(1)\quad \omega = k^2+\Delta,$$

$$(2)\quad \omega = |k|\sqrt{k^2-2\Delta}. \tag{1.25a}$$

When $\Delta>0$, the second state is unstable against perturbations of $k^2>2\Delta$. When $\Delta<0$, formula (25a) becomes

$$\omega = |k|\sqrt{k^2+2|\Delta|}, \tag{1.25b}$$

i.e. the Bogolubov dispersion formula (for details, see Chapter 5). Formula (25) can also be obtained via substituting the expression $\alpha = \sqrt{s}(1+\eta(\chi,t))$ into (21) where $\eta(\chi,t) = \eta_0 \exp i(k\chi - \omega t)$ and linearizing the equation so derived in η_0.

Representation (23)-EPHM (b). To the easy-plane vacuum (ground state) in the frame of this approach corresponds the solution $\alpha = 0$ of eqn. (23). The dispersion formula for corresponding small oscillations is given by the equation

$$i\dot{\alpha} = \alpha_{xx} + \frac{\Delta}{2}(\alpha - \bar{\alpha})$$

which implies

$$\omega = |k|\sqrt{k^2 + |\Delta|}\,. \tag{1.25c}$$

It means that two ways of using the Holstein-Primakoff transformations in the case of the EPH-model leads to different results, even in the linear approximation. To check which of these results is correct we consider the equation

$$i\dot{\psi} = \psi_{xx} - 2\bar{\psi}\frac{\psi_x^2}{1+|\psi|^2} + \Delta\frac{|\psi|^2 - 1}{|\psi|^2 + 1}\psi \tag{1.26}$$

that is a stereographic projection of the Landau-Lifshitz equation. This equation will be derived via the so-called spin coherent states in Chapter V, see also [5].

In terms of the function ψ the averaged z-projection S^z of the spin is given by

$$<|\hat{S}^z|> = -s\frac{|\psi|^2 - a}{|\psi|^2 + 1}$$

and the easy-plane vacuum ($\Delta<0$) is now defined by $|\psi|^2 = 1$. Substituting $\psi = 1+\eta$ into (26) we have

$$i\dot{\eta} = \eta_{xx} - \frac{|\Delta|}{2}(\eta + \bar{\eta})$$

whereby

$$\omega = |k|\sqrt{k^2 + |\Delta|}$$

which coincides with the formula (25c) and is in support of version (23), at any rate for describing small oscillations about the vacuum. Thus we can conclude that to use the Holstein-Primakoff transformations one must investigate symmetry properties of the Hamiltonian and the ground state of the original quantum system. Otherwise classical models obtained via the Holstein-Primakoff transformation procedure should be thought of as phenomenological ones only.

To derive quantum corrections for finite $s \neq \infty$, we should truncate the series at the term naturally defined on the value of $1/s$ since, due to operator ordering, the series in $1/s$ so obtained are asymptotic.

2. Magnetic chain with the magnon-phonon interaction

In this case the Hamiltonian has the form

$$\hat{H} = \hat{H}_s + H_L \tag{1.27}$$

where

$$\hat{H}_s = -\frac{1}{2}\sum_{ij}\left[\frac{1}{2}J_{ij}(\hat{S}_i^+\hat{S}_j^- + \text{h.c.}) + \tilde{J}_{ij}\hat{S}_i^z\hat{S}_j^z\right]$$

describes the spin-spin interaction, as in the previous section, and $H_L = T+V$ describes the lattice oscillations. In the harmonic approximation

$$T = \frac{m}{2}\sum_j \dot{x}_j^2, \quad V = \frac{mu_0^2}{2a_0^2}\sum_j (x_{j+1}-x_j-a_0)^2$$

with m being a spin 'mass', and u_0 the sound velocity in the crystal. In the nearest-neighbour approximation via the Holstein-Primakoff representation (12), the Hamiltonian $\hat{H}_s$ can be expressed through the bosonic variables to give (13). Lattice oscillations are governed by the Hamiltonian equations

$$\dot{p}_j = -\partial H/\partial x_j, \quad \dot{x}_j = \frac{p_j}{m}, \quad H = \langle\varphi|\hat{H}|\varphi\rangle$$

taking into account dependence of the integrals on x_j in (13).

Let the exchange integral be of the form

$$J_{ii+1} = J(|x_{i+1}-x_i|) = J_0 - J_1(x_{i+1}-x_i-a_0)$$

and (the same for $\tilde{J}$), and the second term be small compared with the first one. Go to the continuum limit by (16) with second derivative accuracy. Averaging $\hat{H}$ over the coherent states (14) up to $O(s^{-1})$ we get:

$$H = \int dx \left[\alpha|\varphi_x|^2 + \beta|\varphi|^2 + gx_x|\varphi|^2 - \frac{\lambda}{2}|\varphi|^4 + \frac{\mu_0^2}{2}(x_x)^2\right]$$

where

$$\alpha = \frac{1}{2}J_0 s a_0^2, \quad \beta = -J_0(1-\rho)s, \quad g = J_1 s a_0(1-\rho), \quad \lambda = 2J_0(\rho-1).$$

When obtaining this expression, we take into account a term proportional to $a^+a\cdot s^{-1}$ not only in the operator $\hat{S}^z$ but also in the expansion of the operator $\hat{S}^+$ and $\hat{S}^-$, which leads to the renormalization of λ by -1 in the brackets. Usually one neglects these terms.

$$i\dot{\varphi} = -\alpha\varphi_{xx}+\beta\varphi+gx_x\varphi-\lambda|\varphi|^2\varphi \tag{1.28a}$$

$$m\ddot{x} = \mu_0^2 x_{xx}+g(|\varphi|^2)_x. \tag{1.28b}$$

Equations (28) are a generalization of Zakharov's equations [6] for Langmuir plasma waves interacting with ion-sound ones (in the following we give a simple derivation of these equations proposed earlier by the author in [7]). This system is nonintegrable, but possesses a number of quasi-soliton solutions. Consider two of its limiting cases.

In the quasi-static limit (or for waves propagating with velocities much smaller than the sound velocities, $u \ll u_0$)

$$x_x = -g(|\varphi|^2)_x\frac{1}{mu_0^2}$$

and

$$i\dot{\varphi} = -\alpha\varphi_{xx}+\beta\varphi-\frac{g^2}{mu_0^2}|\varphi|^2\varphi-\lambda|\varphi|^2\varphi,$$

i.e. we get the integrable $S3$ reduction.

In the ultrasound limit when the excitation moves with near-sound velocity, for example, in the positive direction of the x axis, the operator $\partial_t^2-u_0\partial_x^2$ can be approximately replaced by $-2u_0(\partial_t+u_0\partial_x)\partial_x$ [18], and we obtain the system

$$i\dot{\varphi} = -\alpha\varphi_{xx}+\beta\varphi+gn\varphi-\lambda|\varphi|^2\varphi \tag{1.29}$$

$$m(\partial_t+u_0\partial_3)n+\frac{g}{2u_0}(|\varphi|^2)_x = 0$$

which is a generalization (note that a term arises proportional to λ) of the Yajima-Oikawa system [9] (see also [10]). The latter (with $\lambda = 0$) is integrable by the inverse scattering method.

From the form of the Hamiltonian H and eqns. (28) the following facts arise:

(1) In the case of the isotropic XXX model ($\rho=1$) the spin-lattice interaction is 'switched off' in the order considered in S^{-1}, i.e. the isotropic Heisenberg model remains integrable also after the 'reduction'.* In this sense it is unique.

(2) In the case of 'easy axis'-type magnets ($\rho>1$) the reduction procedure gives $S3$ with attraction ($S3_+$).

(3) For 'easy plane'-type magnets we have the $S3_-$ model in the above sense (under the limitations mentioned in the previous section).

3. Nonlinearity of exchange integrals and phonon anharmonism in the Heisenberg model

As before, we consider the Hamiltonian of interacting spin and phonon subsystems following the 1982 paper [12]:

$$\hat{H} = \hat{H}_s + H_l$$

where

$$H_l = T+V$$

$$\hat{H}_s = -\frac{1}{4}\sum_{i\sigma}\left[J_{ii+\sigma}(\hat{S}_i^+\hat{S}_{i+\sigma}^- + \text{h.c.}) - 2\tilde{J}_{ii+\sigma}\hat{S}_i^z\hat{S}_{i+\sigma}^z\right] - \mu h\sum_i \hat{S}_i^z. \tag{1.30}$$

Here, as above, T and V are kinetic and potential energies of the lattice oscillations; $J_{ii+\sigma} \equiv J(|x_{i+\sigma}-x_i|)$ the exchange integrals with the symmetry property $J_{ij} = J_{ji}$; $\vec{B} = (0,0,h)$ is an external magnetic field and μ is the magnetic susceptibility.

Discussing an analogous system in the previous section we make the following assumptions:

(1) Phonons are considered in the harmonic approximation

$$V = \frac{mu_0^2}{2a_0^2}\sum_i (x_{i+1}-x_i-a_0)^2. \tag{1.31}$$

---

*Integrability of the XXZ model on the quantum lattice level has been established for $S = 1/2$ in [11] by means of the Bethe Ansatz.

(2) The exchange integrals are approximated by the linear form

$$J(|x_{i+1}-x_i|) = J_0 - J_1(x_{i+1}-x_i-a_0) \tag{1.32}$$

with

$$J_1 = -\partial J/\partial x_i|_{x_i=x_{i+1}-a_0}.$$

Consider the situation when these two assumptions are not true [12]. Thus, for example, putting in eqn. (28b) , $u^2 \to u_0^2$, we get either $\phi \to 0$ or $n \to \infty$ which implies that system (28) is incorrect in this limit.

Using again the Holstein-Primakoff representation and coherent states (14) instead of eqn. (30) we obtain a classical Hamiltonian

$$H = H_0 - \sum_i J_{ii+1}\Big[s(\overline{\varphi}_i\varphi_{i+1}+\overline{\varphi}_{i+1}\varphi_i) - \rho s(|\varphi_i|^2+|\varphi_{i+1}|^2) -$$

$$-\rho\Big\{|\varphi_i|^2|\varphi_{i+1}|^2 - \frac{1}{2\rho}\overline{\varphi}_{i+1}\varphi_i(|\varphi_i|^2+|\varphi_{i+1}|^2)\Big\}\Big] - \mu h\sum_i|\varphi_i|^2,$$

$$\rho = \tilde{J}/J > 0.$$

The lattice oscillation kinetic energy is taken as $T = \frac{m}{2}\Sigma_i\dot{x}_i^2$ and in potential energy we add the next anharmonic term:

$$V = \frac{mu_0^2}{2a_0^2}\sum_i(x_{i+1}-x_i-a_0)^2 + \frac{V_{III}}{3!}\sum_i(x_{i+1}-x_i-a_0)^3. \tag{1.33}$$

To separate the anharmonic effects and those of the exchange integral non-linearity, consider these integrals in the linear approximation (32). In the long-wave limit we have the expansion:

$$x_{i\pm1} = x \pm x_X a_0 + \frac{1}{2}x_{XX}a_0^2 \pm \frac{1}{6}x_{XXX}a_0^3 + \frac{1}{4!}x_{XXXX}a_0^4 + \ldots$$

In the expansion of $\phi_{i\pm1}$ we retain only terms not higher than ϕ_{XX}, supposing ϕ be of the same order as x_{XX}. Then the field equations generated by the Hamiltonian (30) under the above assumptions have the form

$$i\dot{\varphi} = -\alpha\varphi_{XX} - \tilde{\mu}\varphi + gx_X\varphi - \lambda|\varphi|^2\varphi \tag{1.34a}$$

$$m\ddot{x} = cx_{\chi\chi} + D(x_\chi^2)_\chi + Ex_{\chi\chi\chi\chi} + g(|\varphi|^2)_\chi \tag{1.34b}$$

where

$$c = mu_0^2 - V_{III}a_0^3, \quad D = \frac{a_0^3}{2}V_{III}, \quad E = \frac{1}{12}mu_0^2a_0^2, \quad \lambda = 2J_0(\rho-1) > 0$$

$$g = -J_1 s(\rho-1)a_0, \quad \alpha = \frac{1}{2}J_0 s a_0^2, \quad \tilde{\mu} = s(J_0 - J_1)(1-\rho) - h\mu$$

and we neglect the higher-order term $\frac{2}{3}a_0^3 sJ_1[(x\varphi)_{\chi\chi\chi} - \rho\varphi x_{\chi\chi\chi}]$ with respect to nonlinearity and dispersion.

Differentiating eqn. (34b) over χ and using the notation $n = x_\chi$ we have

$$i\dot{\varphi} = -\alpha\varphi_{\chi\chi} - \tilde{\mu}\varphi + gn\varphi - \lambda|\varphi|^2\varphi \tag{1.35a}$$

$$\ddot{n} = \frac{c}{m}n_{\chi\chi} + \frac{D}{m}(n^2)_{\chi\chi} + \frac{E}{m}\partial_\chi^4 n + \frac{g}{m}(|\varphi|^2)_{\chi\chi}. \tag{1.35b}$$

This system with $\lambda = 0$ has been obtained by Davydov and co-workers (see [5]), and in [13] to describe coupled ion-acoustic and Langmuir plasma waves. In the review [7] the case of $\lambda = 0$ is analyzed in much detail; however, since eqns. (35) differ from plasma equations by the last λ-term in eqn. (35a), we discuss below eqn. (35) at length (see also [12]).

Consider now nonlinear effects due to the exchange integral expansion up to the second order

$$J_{ii+\sigma} \simeq J_0 - J_1(x_{i+\sigma} - x_i - a_0) + J_2(x_{i+\sigma} - x_i - a_0)^2 \tag{1.36}$$

supposing the potential V to be harmonic. Using the procedure discussed above we find the field equation system [12]

$$i\dot{\varphi} = -\alpha\varphi_{\chi\chi} - \tilde{\mu}\varphi + gx_\chi\varphi + c_1(x_\chi)^2\varphi - \lambda|\varphi|^2\varphi \tag{1.37}$$

$$m\ddot{x} = cx_{\chi\chi} + g(|\varphi|^2)_\chi + c_1(x_\chi|\varphi|^2)_\chi$$

where $c = mu_0^2$, $c_1 = 4sJ_2(\rho-1)$ and the remaining parameters are the same as in eqn. (34).

This system has the simplest form in the quasi-stationary limit $|m\ddot{x}| \ll |cx_{\chi\chi}|$:

$$i\dot{\varphi} = -\alpha\varphi_{\chi\chi} - \tilde{\mu}\varphi - d\frac{|\varphi|^2\varphi}{1 + b|\varphi|^2} \tag{1.38}$$

where $b = c_1/\sqrt{c}$, $d = 4[J_1 s^2(1-\rho)^2/c + \frac{1}{2}sJ_0]$ and higher-order terms are dropped. Equation (38) is the well-known NSE with saturable nonlinearity. This arose earlier in various branches of physics, particularly in nonlinear optics, and simulated saturation (decrease) effects of the nonlinear response of a medium in large electromagnetic fields (see also [7]).

If we take into account in the last term of eqn. (38) nonlinearity not higher than $O(b|\varphi|^2)$, then we get $(\varphi^4-\varphi^{6)})$ NSE which can be written in the conventional form [14]:

$$i\dot{\varphi}+\mu\varphi+\varphi_{xx}+(|\varphi|^2-\beta|\varphi|^4)\varphi = 0. \tag{1.39}$$

This equation, just as eqn. (38), possesses a number of remarkable properties which are interesting from the soliton phenomenology point of view and distinguish it from the usual $S3$ (see Chapters VI and X). It is now useful to emphasize, before we proceed, that the exchange integrals nonlinearity for a rigidly fixed lattice does not essentially change the system properties and leads only to a renormalization of the anisotropy coefficient ρ in the continuum representation.

4. Anisotropic magnetic chain in an external field breaking $U(1)$ (XY) symmetry

We use the Hamiltonian (Mikeska's model [17])

$$\hat{H} = -J\sum_i(\hat{\vec{S}}_i\hat{\vec{S}}_{i+1}+\delta\hat{S}_i^z\hat{S}_i^z)-g\mu h\sum_i\hat{S}_i^x \tag{1.40}$$

which corresponds to the Heisenberg ferromagnet with the anisotropy axis oriented along OZ in the external magnetic field $\vec{B} = \{h,0,0\}$. At sufficiently large spin we consider $\hat{S}_i^k$ to be classical fields depending on z and t, then the Hamiltonian (40) leads to the Landau-Lifshitz equation for the vector $\vec{S}$:

$$\dot{\vec{S}} = J\left\{a_0^2[\vec{S},\vec{S}_{zz}]+2\delta[\vec{S},\vec{e}_z](\vec{S},\vec{e}^z)\right\}+g\mu h[\vec{S},\vec{e}^x] \tag{1.41}$$

where μ is the Bohr magneton, g the gyromagnetic ratio and e^x, e^y, e^z is the coordinate system basis.

Represent the vector $\vec{S}$ in polar form

$$\vec{S} = s(\sin\theta\cos\varphi,\sin\theta\sin\varphi,\cos\theta). \tag{1.42}$$

When the external field vanishes, both Hamiltonian (40) and eqn. (41) are invariant under the vector $\vec{S}$ rotations in the XY plane (the $U(1)$ invariance) since only the combination $(S^x)^2+(S^y)^2$ appears in the Hamiltonian. Switching on the field $\vec{B}$ breaks this symmetry up, a preferred direction arises in the XY plane and the corresponding field acquires a 'mass'. Equation (41) in the component form reads

$$\begin{aligned}\dot{S}^x &= -Ja_0^2(S^zS^y_{zz}-S^yS^z_{zz})+J2\delta S^yS^z\\ \dot{S}^y &= -Ja_0^2(S^xS^z_{zz}-S^zS^x_{zz})-J2\delta S^xS^z+g\mu hS^z\\ \dot{S}^z &= Ja_0^2(S^xS^y_{zz}-S^yS^x_{zz})-g\mu hS^y.\end{aligned}$$

In the long-wave limit $(qa_0)^2 \ll |\Delta|$ in the first two equations one can neglect the terms proportional to a_0^2, to get

$$\begin{aligned}\dot{S}^x &= 2J\delta\cdot S^yS^z\\ \dot{S}^y &= -2J\delta\cdot S^xS^z+g\mu hS^z\\ \dot{S}^z &= Ja_0^2(S^xS^y_{zz}-S^yS^x_{zz})-g\mu hS^y.\end{aligned}$$

Using formula (42) in the approximation $\theta = \frac{\pi}{2}-\tilde{\theta}$, $\tilde{\theta} \ll 1$ we have

$$\begin{aligned}\dot{\varphi} &= -2sJ\delta\cdot\tilde{\theta}\\ \dot{\varphi} &= -2sJ\delta\cdot\tilde{\theta}+g\mu h\tilde{\theta}\\ \dot{\tilde{\theta}} &= a_0^2sJ\varphi_{zz}-g\mu h\sin\varphi \equiv a_0^2sJ(\varphi_{zz}-m^2\sin\varphi).\end{aligned}$$

The first and second equations of the system can be consistent in the small field limit $h \ll sJ\delta/g\mu$, which yields $\dot{\varphi} = -2sJ\tilde{\theta}\delta$ or $\ddot{\phi} = -2sJ\dot{\tilde{\theta}}\delta$, whereby

$$\frac{1}{c_0^2}\ddot{\varphi} = \varphi_{zz}-m^2\sin\varphi \tag{1.43}$$

where

$$m^2 = g\mu h/sJa_0^2, \quad c_0^2 = -2J^2s^2\delta a_0^2. \tag{1.44}$$

It follows that for easy-plane ferromagnets, $\delta<0$, eqn. (43) is the usual sine-Gordon equation. This result has been obtained in a somewhat different way by Mikeska in 1978 [17]. We give the conditions when the Sine-Gordon (43) is adequate to the original one (40)

$$q^2 a_0^2 << |\delta|, \tag{1.45}$$

$$h << s\frac{J|\delta|}{g\mu},$$

$$\theta^2 << 1, \quad s >> 1.$$

Later (see Chapter 13) we discuss some concrete results which can be obtained on the basis of model (43) and compare these with experimental data for some physical system ($CsNiF_3$).

5. Generalized Hubbard model

We give now the result by Fedyanin and Lindner [18,19] on reducing the so-called Hubbard model [20] to a system of classical differential equations.

It has been shown in [21-23] that the one-dimensional Hubbard model with one band is equivalent to the two-component Heisenberg chain with the spin Hamiltonian:

$$\hat{H}_s = \frac{1}{4}\sum_{\substack{j,\delta \\ \sigma=\pm 1}} J_{jj+\delta}(\hat{S}_{j,\sigma}^{-A}\hat{S}_{j+\delta,\sigma}^{+B} + \hat{S}_{j,\sigma}^{-B}\hat{S}_{j+\delta,\sigma}^{+A} + \text{h.c.}) \tag{1.46}$$

$$+\frac{\tilde{J}_0}{2}\sum_{\substack{j\in A \\ j+\delta\in B \\ \sigma \\ \text{and reverse}}}(\hat{S}_{j,\sigma}^{zA}\hat{S}_{j+\delta,\sigma}^{zB} + \hat{S}_{j,-\sigma}^{zA}\hat{S}_{j+\delta,\sigma}^{zB}) + \frac{U_0}{2}\sum_{\substack{j\in A\cup B \\ \sigma}}\hat{S}_{i,\sigma}^{z}\hat{S}_{j,-\sigma}^{z}.$$

Using the Jordan-Wigner transformation (11) we reduce the Hamiltonian to the standard form of the Hubbard model with the Coulomb repulsion of electrons in the neighbour sites [24]:

$$\hat{H} = H_0 + T + \hat{H}_s + V, \tag{1.47}$$

$$\hat{H}_s = \frac{1}{4}\sum_{\substack{\sigma=\pm 1 \\ j\in A, j+1\in B \\ \text{and reverse}}} J(|x_{j+1} - x_j|)(c_{j\sigma}^{+}c_{j+1,\sigma} + \text{h.c.}) + \frac{J_0}{2}\sum_{\substack{\sigma=\pm 1 \\ j\in A, j+1\in B \\ \text{and reverse}}} n_{j\sigma}n_{j+1,\sigma} +$$

$$+\frac{U_0}{2}\sum_{j\in A\cup B} n_{j\uparrow}n_{j\downarrow} - \mu\sum_{\substack{\sigma\pm 1 \\ j\in A\cup B}} n_{j\sigma}$$

$$H_0 = \frac{1}{8}(\tilde{J}_0 + U_0)N, \quad \mu = \frac{1}{8}(3\tilde{J}_0 + U_0), \quad T = \frac{m}{2}\sum_j \dot{x}_j^2, \tag{1.48}$$

$$V = \frac{m\omega_0^2}{2}\sum_j (x_{j+1} - x_j - a_0)^2.$$

The lattice field equations take the form

$$m\ddot{x}_j = m\omega_0^2\Delta^2 x_j - J_1\sum_\sigma(\bar{\varphi}_j^{A\sigma}\Delta\varphi_j^{B\sigma} + \varphi_j^{A\sigma}\Delta\bar{\varphi}_j^{B\sigma} + \bar{\varphi}_j^{B\sigma}\Delta\varphi_j^{A\sigma} + \varphi_j^{B\sigma}\Delta\bar{\varphi}_j^{A\sigma}) \tag{1.49a}$$

$$i\dot{\varphi}_j^{A\sigma} = J_0(\varphi_{j+i}^{B\sigma} + \varphi_{j-1}^{B\sigma}) + 2J_1[x_j\Delta\varphi_j^{B\sigma} - \Delta(x_j\varphi_j^{B\sigma})] - \mu\varphi_j^{A\sigma} + \tag{1.49b}$$

$$+\frac{1}{2}\tilde{J}_0\sum_{\sigma=\pm 1}(|\varphi_{j+1}^{B\sigma}|^2 + |\varphi_{j-1}^{B\sigma}|^2)\varphi_j^{A\sigma} + \frac{U_0}{2}|\varphi_j^{A,-\sigma}|^2\varphi^{A\sigma}$$

where

$$J(|x_{j+1} - x_j|) = J_0 - J_1(x_{j+1} - x_j - a_0), \quad 2\Delta\varphi_j \equiv \varphi_{j+1} - \varphi_{j-1},$$

$$\Delta^2 x_j \equiv x_{j+1} - 2x_j + x_{j-1}.$$

Here, the method of single-particle trial functions was applied to average $\hat{H}_s$

$$|\psi(t)> = \lambda^{-t}\left[1 + \sum_n(\alpha_n(t)a_n^+ - \bar{\alpha}_n(t)a_n)\right]|0>,$$

$$<\psi|a_n|\psi> = \frac{\alpha_n(t)}{|\lambda|^2} = \varphi_n(t)$$

(with $|\lambda|^2 = 1 + \Sigma_n|\alpha_n(t)|^2$) and the decoupling procedure

$$<\psi|a_n^+ a_n a_k|\psi> = <\psi|a_n^+|\psi><\psi|a_n|\psi><\psi|a_k|\psi>$$

was postulated with $\phi_n(t)$ being the Schrödinger probability amplitude [16].

Since the ground state in the Hubbard model can be antiferromagnetic we can take $<n_{j\sigma}^A> = <n_{j\pm 1,\sigma}^B>$ which causes the interaction term proportional to J_1 in eqn. (49a) to vanish. Therefore, in the long-wave approximation we have

$$\varphi_{j\pm 1}^{B\sigma} = \varphi_j^{A,-\sigma} \pm \Delta\varphi_j^{A,-\sigma} + \frac{1}{2}\Delta^2\varphi_j^{A,-\sigma} \tag{1.50}$$

since $\phi_j^{B\sigma} = 0$ if $j \in A$ and, vice versa, $\phi_j^{A\sigma} = 0$ if $j \in B$. As a result of these manipulations in the quasi-stationary limit we get

$$x_X = \frac{J_1}{m\omega_0^2}\sum_{\sigma=\pm 1}(\bar{\varphi}^\sigma\varphi^{-\sigma} + \bar{\varphi}^{-\sigma}\varphi^\sigma) + c_0$$

or for ϕ

$$i\dot{\varphi}^{\sigma} = J_0\varphi_{xx}^{-\sigma} + 2J_1(1-c_0)\varphi^{-\sigma} - \frac{(2J_1)^2}{m\omega_0^2}(\bar{\varphi}^{\sigma}\varphi^{-\sigma} + \text{c.c.})\varphi^{-\sigma} + \tilde{U}|\varphi^{-\sigma}|^2\varphi^{\sigma} - \mu\varphi^{\sigma}$$

with $\tilde{U} = \frac{1}{2}U_0 + \tilde{J}_0$. Having derived the latter equation, we neglect higher-order terms of the type $|\varphi^{-\sigma}|_{xx}^2\varphi^{\sigma}$ as being too small as far as nonlinearity and dispersion are concerned.

The equation for $\varphi^{-\sigma}$ is obtained via the change $\sigma \to -\sigma$. It is convenient in what follows to use the even and odd combinations of functions $\phi^{\sigma} = \binom{\varphi_\uparrow}{\varphi_\downarrow}$, i.e.

$$\psi^{(1)} = \varphi_\uparrow + \varphi_\downarrow, \; \psi^{(2)} = \varphi_\uparrow - \varphi_\downarrow, \; \varphi_\uparrow = \frac{1}{2}(\psi^{(1)} + \psi^{(2)}), \tag{1.51}$$

$$\varphi_\downarrow = \frac{1}{2}(\psi^{(1)} - \psi^{(2)})$$

which gives $(i = 1,2)$

$$i\dot{\psi}^{(i)} - J_0(\psi_{xx}^{(i)}\delta_{i1} - \psi_{xx}^{(i)}\delta_{i2}) + \mu_i\psi^{(i)} + \frac{g}{2}(|\psi^{(i)}|^2 - |\psi^{(i+1)}|^2)\psi^{(i)} + \tag{1.52}$$

$$+\frac{\tilde{U}}{4}(|\psi^{(i)}|^2\psi^{(i)} - \bar{\psi}^{(i)}(\psi^{(i+1)})^2) = 0$$

where $\mu_{1,2} = -\mu \pm 2J_1(1-c_0)$, $g = (2J_1)^2/m\omega_0^2$ and δ_{ij} is the Kronecker delta. If we neglect the last term on the right-hand side of eqn. (52) (which can be justified, for example, for high-frequency oscillating solutions $\psi^{(i)}$ with $\omega_1 \neq \omega_2$), the system under consideration is reduced to

$$i\dot{\psi}^{(i)} - \delta_{i1}J_0\psi_{xx}^{(i)} + \delta_{i2}J_0\psi_{xx}^{(i)} + \frac{g}{2}[\eta|\psi^{(i)}|^2 - |\psi^{(i+1)}|^2]\psi^{(i)} = 0 \tag{1.53}$$

at $\eta = 1 - \tilde{U}/2g$, $\psi^{(3)} \equiv \psi^{(1)}$. In the following chapters, we study soliton-like solutions to this system. The richest spectrum of soliton solutions is obtained for $\eta = 1$ (i.e. in the case of no electron-electron interaction, $U = 0$), when it becomes integrable. The integrability of the system

$$i\dot{\psi}^{(1)} + \psi_{xx}^{(1)} + (|\psi^{(1)}|^2 - |\psi^{(2)}|^2)\psi^{(1)} = 0 \tag{1.54}$$

$$i\dot{\psi}^{(2)} + \psi_{xx}^{(2)} + (|\psi^{(1)}|^2 - |\psi^{(2)}|^2)\psi^{(2)} = 0,$$

to which eqn. (53) is reduced via elementary transformations, was established in the paper by Makhankov, et al. [25]. In this paper, the Lax pair was obtained and some internal (isotopic) symmetry properties of the system were studied. This enabled quite a number of one-soliton solutions to be found [26]. In the subsequent work by Pashaev and the author (we consider these later) the study is extended to the nonlinear Schrödinger equation with the noncompact internal symmetry group $U(p,q)$:

$$i\dot{\psi}+\psi_{xx}+2g(\psi,\psi)\psi = 0 \tag{1.55}$$

where $\psi(x,t)$ is the column vector $(\psi)_a = \psi^{(a)}(x,t)$, ψ^{*} is the Dirac-conjugate row vector $\psi^{*} = \psi^{+}\gamma_0$, $\gamma_0 = \begin{bmatrix} ep & 0 \\ 0 & -e_q \end{bmatrix}$ and

$$(\psi,\psi) = \psi^{+}\gamma_0\psi = \sum_{a=1}^{p} |\psi^{(a)}|^2 - \sum_{a=p+1}^{n} |\psi^{(a)}|^2, \ p+q = n,$$

e_p is the p-dimensional unit matrix. It is necessary here to make some remarks about the previous five sections. In all the cases considered we have proceeded from a lattice quantum model expressed via spin operators. Then (apart from Section 4) a reduction procedure was used consisting of (a) the transition from spin variables to Bose or Fermi operator variables acting in Fock space, (b) introducing test Schrödinger functions (or coherent states) and, finally, (c) the passage to the continuum limit. All this introduces a fair number of uncertainties: generally speaking, using the Holstein-Primakoff transformations is appropriate at low temperatures and for large values of spin s; postulating (if needed) the decoupling procedure in the nonlinear term; the choice of the test functions is, in principle, an ambiguous procedure (probably, the best is the set of coherent states, at any rate bearing in mind the decoupling procedure), the passage to the continuum representation implies rejecting terms with higher derivatives. All this means that the original quantum lattice model and the final field (classical, quasi-classical or quantized) model are not, strictly speaking, consistent with each other.

The classical models obtained have, rather, to be considered as original phenomenological models, somehow connected with the initial quantum objects. If on both levels, such a remarkable property of the system as integrability becomes apparent (in the quantum lattice version this means, for example, the possibility to solve the problem via the Bethe Ansatz), then this can be taken to be a happy fact indicating a more 'intimate' closeness of the corresponding systems.

Therefore, in what follows even in the cases when we will study classical objects with a large degree of rigour it is necessary to realize that they are phenomenological models and can hardly pretend to be the object of a rigorous physical theory. Among such models there are equations such as eqn. (28) and (34) without the λ term obtained by Davydov et al. [27] and Fedyanin et al. [28] to describe soliton-like excitations in one-dimensional molecular crystals. One of the first papers in this field was published by Davydov and Kislukha [29], and in Ref. [30] the reduction procedure was formulated for the first time. Magnetic systems with phonon-magnon interactions have been also considered by Pushkarov and Pushkarov [31] and Yatsishin [32]. Finally in Ref. [33] an interesting result (from the point of view of the adequacy of the reduction procedure) has been obtained. The *XYZ* anisotropic Heisenberg ferromagnet

$$\hat{H} = J\sum_j(\chi\hat{S}_j^x\hat{S}_{j+1}^x + \eta\hat{S}_j^y\hat{S}_{j+1}^y + \hat{S}_j^z\hat{S}_{j+1}^z - \frac{1}{4})$$

was considered at $\chi \neq \eta \neq 1$.

After applying the Jordan-Wigner transformation to reduce the spin operators $\hat{S}^k$ to the Fermi operators c^+, c acting in Fock space, one obtains a Hamiltonian containing 'anomalous' terms of the form

$$J\frac{\chi-n}{2}\sum_j(c_{j+1}^+c_j^+ - c_{j+1}c_j)$$

and, as a result, the $S3$ nonintegrable variant:

$$i\dot{\psi} + \psi_{xx} + \alpha\bar{\psi}_x + |\psi|^2\psi = 0.$$

Thus, starting with the integrable quantum *XYZ* model we come to a nonintegrable classical object, i.e. in this instance, the reduction procedure 'loses' integrability. This is because this procedure uses essentially the information of the vacuum, over which excitations are considered, an incorrect choice of this gives rise to the wrong result (see, for example, Section 1).

6. The low-frequency wave interaction with a packet of high-frequency waves in plasmas

Here, largely following Ref. [7], we get a system of equations describing the interaction of the Langmuir and ion-sound waves in plasmas. The derivation has a quantitative character. Detailed and strict results are presented in the original papers [6] and [13].

In uniform plasmas without a magnetic field there are three oscillation modes, viz. two high-frequency and a low-frequency modes:

(1) Transverse oscillations (t) with the dispersion relation

$$\omega^t = (k^2c^2+\omega_p^2)^{1/2} \tag{1.56}$$

where $\omega_p = (4\pi ne^2/m_e)^{1/2}$, $\vec{k}$ and c are, respectively, the Langmuir frequency, the wavevector and the velocity of light.

(2) Longitudinal Langmuir oscillations (1) with the spectrum

$$\omega^l = \omega_p(1+\frac{3}{2}k^2d_e^2) \tag{1.57}$$

with $d_e = v_e/\omega_p$ being the Debye length, $v_e = (T_e/m_e)^{1/2}$ and the electron temperature T_e is measured in terms of energy;

(3) Longitudinal low-frequency ion-sound waves with

$$\omega^s = kv_s(1+k^2d_e^2)^{-1/2} \tag{1.58}$$

where $v_s = (Te/m_i)^{1/2}$ is the ion-sound velocity. The latter mode exists only in non-isothermal plasmas ($T_e >> T_i$).

The number of oscillation modes increases significantly in magnetized (and slightly non-uniform) plasmas. However, we will not discuss these here since there already exists an extensive literature devoted to this problem.

There exist a number of methods to obtain (with different degrees of rigour) nonlinear evolution equations describing wave phenomena in plasmas (see, for example, Ref. [7]). In recent years, the so-called reductive perturbation method [33,34] has become wide spread.

The Boussinesq equation can be obtained from the exact hydrodynamical equations

$$\vec{V}_t+(\vec{V}\vec{\nabla})\vec{V} = -\nabla\phi \tag{1.59}$$

$$\partial_t n_i+(\nabla n_i\vec{V}) = 0, \quad \nabla^2\phi = e^{\phi}-n_i.$$

We derive it here in a less rigorous but simpler and more descriptive way by using the well-known dispersion formula for s-waves (58)

$$\omega^2 = k^2(1+k^2)^{-1}$$

or

$$\omega^2(1+k^2) = k^2$$

which gives in a coordinate representation for the linear part of the operator

$$L = \partial_t^2(1-\nabla^2)-\nabla^2$$

or, in one-dimensional space,

$$L = \partial_t^2-\partial_x^2-\partial_x^2\partial_t^2.$$

The nonlinear term arises from $(\vec{V}\nabla)\vec{V}$ or $\frac{1}{2}\partial_x V^2$ in the one-dimensional case after differentiation with respect to x. As a result we get the so-called 'improved' version of the boussinesq equation (IB_q):

$$L\phi = (\phi^2)_{xx}. \tag{1.60}$$

This equation, just as its dispersion formula, approaches the usual Boussinesq equation when $k \ll 1$.

The modified 'improved' boussinesq equation has the form

$$L\phi = (\phi^3)_{xx} \tag{1.61}$$

and can be used along with the modified KdV equation to investigate the anharmonic lattice and nonlinear Alfvén waves. Similarly, the improved (regularized) variant of the KdV equation can be obtained. Using the formula $\omega = k(1+k^2)^{-1/2}$ we have in the small k limit

$$L = \partial_t+\partial_x-\partial_x^2\partial_t \tag{1.62}$$

and

$$L\phi = (\phi^2)_x.$$

This equation was derived in papers [35] and later studied by a number of authors.

The S3 equation. We make use of eqn. (57) and the fact that a high-frequency wave, propagating in a plasma, induces the Miller force [36] which expels the plasma from the region of field maximum

$$F_M \propto -\partial_x |E|^2$$

and leads to the plasma density variation. So the wave propagates in perturbed plasma with a density $n = n_0 + \delta n$, therefore, instead of eqn. (57) we have

$$\omega = \omega_p^0 + \frac{3}{2} k^2 d_e^2 \omega_p^0 + \frac{1}{2} \omega_p^0 \frac{\delta n}{n_0} \tag{1.63}$$

or

$$\delta\omega = \omega - \omega_p^0 = (\frac{3}{2} k^2 d_e^2 + \frac{1}{2} \frac{\delta n}{n_0}) \omega_p^0$$

which in coordinate representation gives for the high-frequency field envelope of l-waves

$$(i\partial_t + \partial_x^2 - \phi)\varphi = 0 \tag{1.64}$$

where we used the following dimensionless variables

$$t \leftarrow \frac{2}{3} \mu \omega_p t, \quad x \leftarrow \frac{2}{3} \sqrt{\mu}\, x / d_e, \quad \phi = \frac{3}{4\mu} \frac{\delta n}{n_0}, \quad |\varphi|^2 = \frac{3}{64} \frac{|E|^2}{\mu \pi n_0 T},$$

$$\mu = m_e / m_i.$$

To complete eqn. (64) it is necessary to describe the density perturbation ϕ. In a state close to equilibrium or, in the inertialess approximation, when the local equilibrium time is negligible, the Miller force is counterbalanced by the plasma pressure gradient, therefore

$$\delta n \propto -|E|^2 \tag{1.65}$$

and we obtain the $S3$ model (quasi-stationary limit).

The system of coupling Langmuir and ion-sound waves. If one takes into account the inertial delay time and the fact that in a quiescent plasma the density waves are the acoustic waves $(\partial_t^2 - \partial_x^2)\phi = 0$, then under the action of the Miller force in the right-hand side of the wave equation a source proportional to $\partial_x^2|\varphi|^2$ arises, viz.

$$\Box\phi \equiv (\partial_t^2-\partial_x^2)\phi = \partial_x^2|\varphi|^2. \tag{1.66}$$

(Equation (66) tends to eqn. (65) in the inertialess case.)

Remember that the procedure considered for obtaining system (64) and (66) gives the correct result only in the one-dimensional case. A more rigorous approach [6] (under the same assumptions) yields the system

$$\vec{\nabla}(-2i\vec{\nabla}\psi_t-\vec{\nabla}\Delta\psi+\phi\vec{\nabla}\psi) = 0 \tag{1.67}$$

$$(\partial_t^2-\Delta)\phi = \Delta(|\vec{\nabla}\psi|^2)$$

where $\psi(\vec{r},t)$ is now the high-frequency potential amplitude (not the field amplitude).

A system mathematically identical to (64) and (66) is also obtained in the study of whistler solitons propagating along the magnetic field. Their linear dispersion has the form

$$\omega^w = \frac{\omega_H}{\omega_p^2}k^2c^2 \tag{1.68}$$

(with $\omega_H = eB/m_ec$, and c being the speed of light) from which follows that ω^w depends strongly on the wave number k. This fact distinguishes helicons from Langmuir waves: they do not possess the rest mass which is the main fraction of the Langmuir plasmon energy $\omega^l \simeq \omega_p$. Making use of the qualitative method stated above, one easily finds the system of equations describing a narrow packet of whistlers $\Delta k/k << 1$:

$$i\dot{\varphi}+\varphi_{xx}+\phi\varphi = 0, \tag{1.69}$$

$$\Box\phi = \partial_x^2(|\varphi|^2),$$

here φ is the complex envelope of the helicon field

$$\varphi_x-i\varphi_y = \varphi\exp\{i\omega_0^w(k_0)t\}$$

$\omega_0^w = \omega^w|_{n=n_0}$ with the notation $x \leftarrow (\omega_p^2 v_s/\omega_H c^2)x$

$$t \leftarrow (\omega_p^2 v_s^2/\omega_H c^2)t, \quad \phi = (\omega_H^2 k_0^2 c^4/\omega_p^4 v_s^2)\frac{\delta n}{n_0}$$

$$|\varphi|^2 = \frac{k_0^2\omega_H^2 c^4}{4\pi n_0 m_i v_s^2 \omega_p^4}|\varphi|^2.$$

System (69) differs from the Langmuir system by the sign of the last term in the first equation which is associated with the dispersion formula (68). As a result, the helicon solitons move with the supersonic velocity $v > v_s$, and along with the soliton moves the density lump, not the well as in the case of the subsonic l-soliton. A rigorous derivation of eqn. (69) and a more detailed discussion of the physical model are given in Ref. [38]. Finally, bearing in mind that the dispersion of transverse electromagnetic waves in plasmas has the form of eqn. (56) for sufficiently long waves $kc \ll \omega_p$ we get again either the $S3$ equation or a system of the form of eqns. (64) and (66).

A model of coupled Langmuir and nonlinear ion-acoustic waves with dispersion; anharmonism and dispersion of low-frequency waves instead of the linear wave equation (66) give rise to

$$\phi_t + \phi_x + \alpha\phi_{xxx} + \beta(\phi^2)_x = -(|\varphi|^2)_x \tag{1.70}$$

(the Nishikawa-Hojo-Mima-Ikezi approximation [39]) or

$$(\Box - \alpha\partial_x^4)\phi - \beta\partial_x^2(\phi^2) = \partial_x^2(\varphi|^2 \tag{1.71}$$

(the author's approximation [13]). For ion-acoustic waves in plasmas the constants α and β are small parameters proportional to the ratio m_e/m_i:$\alpha = 4/3 m_e/m_i$, $\beta = 3\alpha$.

Both systems of equations (64) and (71) and (64) and (70) have a very universal nature (they arise in the study of different physical models) and possess interesting mathematical properties.

We mention finally still more nonlinear Schrödinger equations, which appeared for the first time perhaps in plasma physics. We consider the so-called *Derivative nonlinear Schrödinger equation*

$$i\dot{\psi} \pm \alpha\partial_\chi^2\psi + \frac{i}{4}\partial_\chi(|\psi|^2\psi) = 0 \tag{1.72}$$

describing the circularly polarized Alfvén wave propagating along the magnetic field. In eqn. (72), ψ is the complex magnetic field:

$$\psi = B_x \mp iB_y$$

and $\chi = z - v_a t$ is the coordinate in a system moving along the z axis with the

Alfvén velocity [40].

Equation (72) may be reduced to a more general form

$$i\dot{\varphi}+\alpha\partial_\eta^2\varphi-\frac{k}{4}|\varphi|^2\varphi+\frac{i}{4}\partial_\eta(|\varphi|^2\varphi) = 0, \tag{1.73}$$

via the substitution

$$\psi(\chi,t) = \varphi(\eta,t)\exp\{i(k\eta-\alpha k^2t)\}, \quad \eta = \chi-2\alpha kt.$$

There have been a number of papers [41,42] devoted to the investigation of eqns. (72) and (73). In the first work, the AKNS scheme has been adapted to solve eqn. (72), in the second the inverse scattering method (ISM) has been used to investigated eqn. (73), and it has been shown (apparently first) that a superposition of two Lax schemes solves the equation with the superposition of nonlinear terms such that each of them corresponds to one of the Lax schemes.

7. The ϕ^5 Schrödinger equation as a model to describe collective motions in nuclei

A system having a very general nature and occurring in different areas of physics is a non-relativistic version of the ϕ^6 theory. Unlike the ϕ^4 theory, this model is non-integrable but the polynomial order increase in the Lagrangian leads to the appearance of many new very interesting properties of the system such as, in particular, stable particle-like solutions in $D>1$ space dimensions.

Considerable interest in the study of nonlinear dynamical phenomena in many-nucleon systems has arisen from progress in the physics (and engineering) of heavy ions and is, therefore, further stimulated by the search for superheavy elements (the stability islands). Usually, collision processes of heavy ions are described on classical (hydrodynamical approach [40] and references therein) or quasi-classical (TDHF [41] or DTFT [42]) levels.

By using the results of Ref. [40] we show that in the quasi-classical limit the nuclear hydrodynamics equations with the Skyrme forces are reduced to the non-relativistic $\phi^4-\phi^6$ model for the corresponding choice of variables.

In the quasi-classical limit, the equations of nuclear hydrodynamics are (in conventional notation)

$$\rho_t+\vec{\nabla}(\rho\vec{V}) = 0, \tag{1.74}$$

$$\vec{V}_t + (\vec{V}\vec{\nabla})\vec{V} = \frac{\hbar^2}{4m^2}\vec{\nabla}\left[\frac{\Delta\rho}{\rho} - \frac{(\vec{\nabla}\rho)^2}{2\rho^2}\right] - \frac{1}{m}\vec{\nabla}\frac{\delta\epsilon}{\delta\rho},$$

and the interaction functional with the Skyrme forces is of the form

$$\epsilon[\rho] = -\frac{3}{8}|t_0|\int\rho^2 d^D x + \frac{t_3}{16}\int\rho^3 d^D x. \tag{1.75}$$

A discussion of the details of the eqn. (74) derivation with potential (75) have been given in Ref. [43] mentioned above, as well as in [46].

Introducing the function $\psi(\vec{x},t)$ by formulae

$$\rho(x,t) = |\psi(\vec{x},t)|^2, \quad \vec{V}(\vec{x},t) = \vec{\nabla}(\arg\psi(\vec{x},t))$$

from the second equation of (74) it is easy to show that ($\theta = \arg\psi$):

$$\theta_t + \frac{1}{2}(\vec{\nabla}\theta)^2 - \frac{\hbar^2}{4m^2}\left[\frac{\Delta\rho}{\rho} - \frac{(\vec{\nabla}\rho)^2}{2\rho^2}\right] + \frac{1}{m}\left[\frac{\delta\epsilon}{\delta\rho} - \lambda\right] = 0. \tag{1.76}$$

Equation (76) together with the first equation of (74) gives rise to

$$i\hbar\psi_t + \frac{\hbar^2}{2m}\Delta\psi - \left[\frac{\delta\epsilon[|\psi|^2]}{\delta|\psi|^2} - \lambda\right]\psi = 0$$

wherefrom using eqn. (75) we obtain

$$i\hbar\psi_t + \frac{\hbar^2}{2m}\Delta\psi + \left[\frac{3}{4}|t_0|\,|\psi|^2 - \frac{3}{16}t_3|\psi|^4\right)\psi = 0. \tag{1.77}$$

Thus - it should be stated once again - the equations of vortex-free (because $\vec{V} = \vec{\nabla}\theta$) nuclear hydrodynamics can be reduced to the NSF.

Let us now show how from the ϕ^6 nonlinear Schrödinger equation one may obtain the equations of hydrodynamics. We proceed from the dimensionless form of the 1*D* version of the nonlinear Schrödinger equation

$$i\psi_t + \psi_{xx} + \psi + (|\psi|^2 - \alpha|\psi|^4)\psi = 0. \tag{1.78}$$

Taking the ψ function in the form $\psi = \mathrm{R}e^{i\theta}$ from eqn. (78) we get

$$\dot{R} + 2\theta_x R_x + R\theta_{xx} = 0,$$

$$-R\dot{\theta} + R_{xx} - R\theta_x^2 - \frac{3}{16\alpha}R + R^3 - \alpha R^5 = 0.$$

Via the renormalization of R by $(3/4\alpha)^{1/2}$ and dividing the second equation by R we obtain

$$-\dot{\theta}-\theta_x^2+\frac{R_{xx}}{R}-\frac{1}{4}+R^2-\frac{3}{4}R^4 = 0.$$

Differentiating this with respect to x and going to the variables $\rho = R^2$ and $V = 2\theta_x$ we obtain the hydrodynamical-type equations

$$\dot{\rho}+(\rho V)_x = 0, \tag{1.79}$$

$$\dot{V}+VV_x-\partial_x\left[\frac{\rho_{xx}}{\rho}-\frac{1}{2}\frac{\rho_x^2}{\rho^2}+\rho-\frac{3}{4}\rho^2\right] = 0, \tag{1.80}$$

which coincide (allowing for notation) with eqns. (74) and (75). Notice that a one-dimensional version of eqn. (80) can also be written in the form of the conservation law

$$\dot{V}+\partial_x(\frac{1}{2}V^2-\frac{\rho_{xx}}{\rho}+\frac{1}{2}\frac{\rho_x^2}{\rho^2}-\rho+\frac{3}{4}\rho^2) = 0. \tag{1.81}$$

In what follows, we make use of these results to investigate elementary excitations in such a model. Note also that the nonlinear Schrödinger equation having a sufficiently arbitary nonlinearity can be written in the form of eqn. (74), viz. the equations of hydrodynamics (albeit vortex-free, since $\mathrm{rot}\,\vec{V} = 0$ and $\mathrm{rot}\,\vec{\nabla}\theta \equiv 0$).

8. 'Colour' generalization of a magnetic chain with magnon-phonon interaction

The starting point is now the Hamiltonian

$$\hat{H} = \hat{H}_s+H_L,$$

where

$$\hat{H}_s = -\frac{1}{2}\sum_{ij\alpha\beta}\left[\frac{1}{2}J_{ij}^{\alpha\beta}(\hat{S}_i^{+\alpha}\hat{S}_j^{-\beta}+\text{h.c.})+R_{ij}^{\alpha\beta}\hat{S}_j^{z\alpha}\hat{S}_j^{z\beta}\right], \tag{1.82}$$

$$H_L = T+U,\quad T = \frac{m}{2}\sum_j \dot{x}_j^2,\quad U = \frac{mu_0^2}{2a_0^2}\sum_j(x_{j+1}-x_j-a_0)^2,$$

which describes the interaction of spins of α different 'colours' $(\alpha=1,\ldots,n)$. One can obtain a colour generalization of the equations of Section 2. For that, just as in the previous section, one should neglect interaction between the 'colour' and the spatial degrees of freedom in the exchange integrals and consider only nearest-neighbour interactions

$$J_{ij}^{\alpha\beta} = T_{jj+\sigma}K^{\alpha\beta}, \quad R_{ij}^{\alpha\beta} = \tilde{J}_{jj+\sigma}L_1^{\alpha}L_2^{\beta}, \tag{1.83}$$

where $J_{jj+\sigma} \equiv J(|x_j - x_{j+\sigma}|)$ are the exchange integrals and $\sigma = 0, \pm 1$[*]. Then in terms of the Bose operators $a_i^{+\alpha}$ and a_i^{α} the Hamiltonian is

$$\hat{H} = H_0 - \frac{1}{2}\sum_{j\sigma}\Big\{sJ_{jj+\sigma}\sum_{\alpha\beta}K^{\alpha\beta}(a_j^{+\alpha}a_{j+\sigma}^{\beta} + \text{h.c.}) - \tag{1.84}$$

$$-\tilde{J}_{jj+\sigma}\Big[s\sum_{\alpha}(l_2L_1^{\alpha}a_j^{+\alpha}a_j^{\alpha} + l_1L_2^{\alpha}a_{j+\sigma}^{+\alpha}a_{j+\sigma}^{\alpha}) +$$

$$+\sum_{\alpha\beta}L_1^{\alpha}L_2^{\beta}\, a_j^{+\alpha}a_j^{\alpha}a_{j+\sigma}^{+\beta}a_{j+\sigma}^{\beta}\Big]\Big\} + H_L,$$

where $H_0 = -s/2J_0kN + s^2l_1l_2\sum_{j\sigma}\tilde{J}_{jj+\sigma}$, $k = TrK$, $l_i = TrL_i$, and where N is the total number of the lattice sites.

The evolution of the operator $a_n(t)$ is determined by the Heisenberg equation $i\hbar\dot{a}_n^{\alpha}(t) = [a_n^{\alpha}\hat{H}_s]$. Following the reduction procedure described in detail in Section 2 and taking into account that now

$$|\Psi(t)> = \prod_{\alpha=1}^{n}|\psi^{(\alpha)}(t)> \propto \prod_{\alpha=1}^{n}\prod_{j}e^{\varphi_j^{\alpha}(t)a_j^{+\alpha}}|0>_j$$

is the trial function suitable for describers low-laying excitations with due regard for the 'colour' degree of freedom, we obtain (with the same assumptions as in Section 2) the following system

$$\ddot{x} = u_0^2x_{XX} + \frac{s}{m}\sum_{\alpha\beta}\tilde{T}^{\alpha\beta}(\bar{\varphi}^{\alpha}\varphi^{\beta})_X \tag{1.84a}$$

$$i\dot{\varphi}^{\alpha} = -b\sum_{\beta}\Big[K_{(\alpha\beta)}\varphi_{XX}^{\beta} - sT_{\alpha\beta}\varphi^{\beta} + s\tilde{T}_{\alpha\beta}\varphi^{\beta}x_X\Big] - \tag{1.84b}$$

$$-\tilde{J}_0\sum_{\gamma}\Big[L_1^{\gamma}\bar{\varphi}^{\gamma}\varphi^{\gamma}\sum_{\beta}\delta^{\alpha\beta}L_2^{\beta}\varphi^{\beta} + (1\rightleftarrows 2)\Big],$$

---

[*]It should be also noted that the Hamiltonian H_L in eqn. (82) holds when only the acoustic phonon branch is in a crystal.

with

$$T_{\alpha\beta} = J_0 K_{(\alpha\beta)} - \tilde{J}_0(l_1 L_2^\alpha + l_2 L_1^\alpha)\delta_{\alpha\beta}, \quad b = J_0 \frac{s}{2},$$

$$\tilde{T}_{\alpha\beta} = J_1 K_{(\alpha\beta)} - \tilde{J}_1(l_1 L_2^\alpha + l_2 L_1^\alpha)\delta_{\alpha\beta},$$

$$J_{jj+\sigma} \simeq J_0 - J_1(x_{j+\sigma} - x_j - a_0), \quad \text{(the same for } \tilde{J}_{jj+\sigma}\text{)}.$$

$$K_{(\alpha\beta)} = K_{\alpha\beta} + K_{\beta\alpha}.$$

The reduction $L_1^\alpha = L_2^\alpha = 0$ ($\alpha = 1,...,n$) or $T_{\alpha\beta} = \mu K_{(\alpha\beta)}$

$$\tilde{T}_{\alpha\beta} = \nu K_{(\alpha\beta)}, \quad \nu \equiv J_1, \quad \mu = J_0, \quad K_{(\alpha\beta)} = \epsilon_\alpha \delta_{\alpha\beta}, \quad \epsilon_\alpha = \begin{cases} 1, & \alpha = 1,...,p \\ -1, & \alpha = p+1,...,n \end{cases}$$

instead of eqn. (84) gives

$$\ddot{n} = u_0^2 n_{XX} + \frac{s\nu}{m}(\psi,\psi)_{XX},$$

$$i\dot{\psi} = -b\psi_{XX} - s\mu\psi + s\nu n\psi,$$

where $n(\chi,t) = X_\chi$ and

$$\psi_\alpha(\chi,t) = \begin{cases} \varphi_\alpha(\chi,t), & \alpha = 1,...,p, \\ \varphi_\alpha^*(\chi,t), & \alpha = p+1,...,n \end{cases} \tag{1.85}$$

In the dimensionless variables

$$t \leftarrow (u_0^2/2b)t, \quad x = \chi \leftarrow (u_0/2b)\chi, \quad n \rightarrow (u_0^2/2bs\nu)n, \tag{1.85a}$$

$$\psi \rightarrow \left[\frac{u_0^4 m}{2bs^2\nu^2}\right]^{1/2} e^{is\mu t}\psi,$$

we obtain a $U(p,q)$ colour generalization of the Zakharov system

$$\Box n \equiv (\partial_t^2 - \partial_x^2)n = (\psi,\psi)_{xx}, \tag{1.86}$$

$$i\dot{\psi} + \frac{1}{2}\psi_{xx} - n\psi = 0.$$

The system is non-integrable, but is reduced to the integrable one in the 'relativistic' limit $v \rightarrow 1$. Consider this limit. Replacing the operator $\partial_t^2 - \partial_\chi^2$ by the operator $-2(\partial_t + \partial_\chi)\partial_\chi$, and integrating once, we have a 'colour' generalization of the

Yajima-Oikawa system:

$$n_t + u_0 n_\chi + \frac{s}{2mu_0}\sum_{\alpha\beta}\tilde{T}_{\alpha\beta}(\bar{\varphi}^\alpha\varphi^\beta)_\chi = 0, \tag{1.87a}$$

$$i\varphi_t^\alpha = -\sum_\beta \left[bK_{(\alpha\beta)}\varphi^\beta_{\chi\chi} + sT_{\alpha\beta}\varphi^\beta - s\tilde{T}_{\alpha\beta}\varphi^\beta n\right), \tag{1.87b}$$

if again

$$L_1^\alpha = L_2^\alpha = 0,\ \tilde{T}_{\alpha\beta} = J_1 K_{(\alpha\beta)},\ T_{\alpha\beta} = J_0 K_{(\alpha\beta)},\ (\alpha\beta = 1,\dots,n).$$

In the 'quasi-stationary' (inertialess) limit, eqn. (87a) is reduced to

$$n(\chi,t) = x_\chi = -\frac{s}{mu_0^2}\sum_{\alpha\beta}\tilde{T}_{\alpha\beta}(\bar{\varphi}^\alpha\varphi^\beta) + c, \tag{1.88}$$

which gives, instead of (87b), the system of equations only for functions $\varphi^\alpha(\chi,t)$:

$$i\varphi_t^\alpha = -\sum_\beta(bK_{(\alpha\beta)}\varphi^\beta_{\chi\chi} - M_{\alpha\beta}\varphi^\beta) - \kappa(\sum_{\gamma\delta}\tilde{T}_{\gamma\delta}\bar{\varphi}^\gamma\varphi^\delta)(\sum_\beta\tilde{T}_{\alpha\beta}\varphi^\beta) \tag{1.89}$$

$$-\tilde{J}_0\sum_\gamma\varphi^\alpha|\varphi^\gamma|^2(L_1^\gamma L_2^\kappa + L_2^\gamma L_1^\alpha),$$

where $M_{\alpha\beta} = s(T_{\alpha\beta} - c\tilde{T}_{\alpha\beta})$, $\kappa = s^2/mu_0^2$, and c is the constant of integration. This system is, in general, non-integrable. Therefore, we consider some of its reductions.

(1) Let the exchange integrals of colour degrees of freedom be proportional to each other

$$K_{(\alpha\beta)} = 2b_1 L_1^\beta\delta_{\alpha\beta} = 2b_2 L_2^\beta\delta_{\alpha\beta} \equiv \lambda_\beta\delta_{\alpha\beta},$$

then the system of equations of motion is generated by the Hamiltonian

$$H = \int d\chi\left[b(\varphi_\chi^+ K\varphi_\chi) - d(\varphi^+ K\varphi)^2 - \tilde{\mu}(\varphi^+ K\varphi)\right] \tag{1.90}$$

with

$$(\varphi^+ K\varphi) \equiv \sum_{\alpha\beta}\bar{\varphi}^\alpha K_{(\alpha\beta)}\varphi^\beta = \sum_\alpha \lambda^\alpha|\varphi^\alpha|^2,$$

$$d = x\nu^2 + \tilde{J}_0/2b_1b_2,\ \tilde{\mu} = s(\mu - c\nu),\ \ T_{\alpha\beta} = \mu\lambda_\beta\delta_{\alpha\beta},$$

$$\tilde{T}_{\alpha\beta} = \nu\lambda_\beta\delta_{\alpha\beta},\ \ M_{\alpha\beta}\tilde{\mu}\lambda_\beta\delta_{\alpha\beta},$$

$$\nu = J_1 - \frac{\tilde{J}_1}{2b_1b_2}(b_1l_1 + b_2l_2), \quad \mu = J_0 - \frac{\tilde{J}_0}{2b_1b_2}(b_1l_1 + b_2l_2),$$

and where $\varphi^\alpha(\chi,t)$ and, $\bar{\varphi}^\alpha(\chi,t)$ are the canonically conjugated variables

$$\{\varphi^\alpha(x),\bar{\varphi}^\beta(y)\} = i\delta^{\alpha\beta}\delta(x-y) \tag{1.91}$$

with the usual Poisson bracket

$$\{A,B\} = i\sum_{\alpha=1}^{n} \int_{-\infty}^{\infty} d\chi \left[\frac{\delta A}{\delta\varphi^\alpha(\chi)}\frac{\delta B}{\delta\bar{\varphi}^\alpha_{(\chi)}} - \frac{\delta B}{\delta\varphi^\alpha_{(\chi)}}\frac{\delta A}{\delta\bar{\varphi}^\alpha_{(\chi)}}\right]. \tag{1.92}$$

Introducing the variable

$$\varphi^{*\alpha}(\chi) = (\varphi^+(\chi)K)^\alpha = \bar{\varphi}^\beta(\chi)\lambda_\beta\delta_{\alpha\beta}$$

one can reduce Hamiltonian (90) and bracket (91) to the forms

$$H = \int d\chi[b(\varphi_\chi^+\varphi_\chi) - d(\varphi^*\varphi)^2 - \tilde{\mu}(\varphi^*\varphi)],$$

$$\{\varphi^\alpha(x),\varphi^{*\beta}(y)\} = i\delta^{\alpha\beta}\lambda_\beta\delta(x-y).$$

Then if we assume magnitudes of different colour interactions to be equal

$$\lambda_\alpha = \epsilon_\alpha = \begin{cases} 1, & \alpha = 1,\dots,p, \quad (p+q=n), \\ -1, & \alpha = p+1,\dots,p+q, \end{cases}$$

and consider the functions ψ_α via eqn. (85) and use the notation

$$(\Gamma_0)_{\alpha\beta} \equiv K_{(\alpha\beta)} = \epsilon_\alpha\delta_{\alpha\beta}, \quad \frac{d}{b} \equiv \kappa, \quad \frac{\mu'}{b} \equiv p, \quad H \leftarrow H/b,$$

we get

$$H = \int d\chi[(\psi_\chi,\psi_\chi) - \kappa(\psi,\psi)^2 - p(\psi,\psi)], \tag{1.93}$$

$$\{\psi^\alpha(x),\ \psi^{*\beta}(y)\} = i\delta^{\alpha\beta}\delta(x-y),$$

where

$$\psi^* = \psi^+\Gamma_0, \quad \Gamma_0 = \begin{bmatrix} e_p & 0 \\ 0 & e_q \end{bmatrix},$$

and

$$(\psi,\psi) = (\psi^+\Gamma_0\psi) = \sum_{\alpha=1}^{p} |\psi^\alpha|^2 - \sum_{\alpha=p+1}^{n} |\psi^\alpha|^2$$

is the $U(p,q)$ internal product. The system of equations

$$i\dot{\psi}+\psi_{xx}+2\kappa(\psi,\psi)\psi+\rho\psi = 0$$

corresponding to the Hamiltonian (93) is the $U(p,q)$ nonlinear Schrödinger equation describing an integrable hamiltonian system.

A similar reduction applied to system (87) gives

$$n_t+u_0 n_x+\frac{s\nu}{2mu_0}(\psi,\psi)_x = 0,$$

$$i\dot{\psi} = -b\psi_{xx}-s\mu\psi+n\psi s\nu,$$

or, in the variables (85a) with $m\rightarrow 2m$, the system

$$n_t+n_x+(\psi,\psi)_x = 0, \tag{1.94}$$

$$i\dot{\psi}+\frac{1}{2}\psi_{xx}-n\psi = 0.$$

Contrary to eqn. (86) this system is integrable by the inverse scattering method.

The above analysis implies that, even in the case of vanishing magnon-phonon interactions, an integrable version of the nonlinear Schrödinger equation arises, when intensities of the colour interaction are equal in terms of their absolute values. The same is valid for the generalized Yajima-Oikawa system as well.

We now turn to the question: which initial physical model of the magnet (82) corresponds to the classical $U(p,q)$ nonlinear Schrödinger equation obtained by the reduction procedure. In order to answer this, we consider the following subreduction

$$\tilde{J}_{jj+\sigma} = \rho J_{ii+\sigma}\frac{l}{b}, \quad K^{\alpha\beta} = L_1^\alpha\delta_{\alpha\beta} = L_2^\alpha\delta_{\alpha\beta} = (\Gamma_0)_{\alpha\beta},$$

in the Hamiltonian (82). Then we have

$$\hat{H}_s = -\frac{1}{2}\sum_{j,\sigma}U_{jj+\sigma}\sum_{\alpha=1}^{p}\left[\hat{S}_j^{+\alpha}\hat{S}_{j+\sigma}^{-\alpha}+\rho\hat{S}_j^{z\alpha}\sum_{\beta=1}^{p}\hat{S}_{j+\sigma}^{z\beta}\right]+ \tag{1.82a}$$

$$+\frac{1}{2}\sum_{j,\sigma}U_{jj+\sigma}\sum_{\alpha=p+1}^{n}\left[\hat{S}_j^{+\alpha}\hat{S}_{j+\sigma}^{-\alpha}-\rho\hat{S}_j^{z\alpha}\sum_{\beta=p+1}^{n}\hat{S}_{j+\sigma}^{z\beta}\right]+$$

$$+\frac{\rho}{2}\sum_{j,\sigma}U_{jj+\sigma}\left[\sum_{\alpha=1}^{p}\hat{S}_j^{z\alpha}\sum_{\beta=p+1}^{n}\hat{S}_{j+\sigma}^{z\beta}+\sum_{\alpha=p+1}^{n}\hat{S}_j^{z\alpha}\sum_{\beta=1}^{p}\hat{S}_{j+\sigma}^{z\beta}\right],$$

i.e. the Hamiltonian which describes many component mixture of ferromagnetic chains. The reduction procedure, when applied to (82a), gives the equation

$$i\dot{\varphi}^{\alpha} = -b\epsilon_{\alpha}\varphi^{\alpha}_{XX} - d\left[\sum_{\beta}\epsilon_{\beta}|\varphi^{\beta}|^2\right]\epsilon_{\alpha}\varphi^{\alpha} - \mu\epsilon_{\alpha}\varphi^{\alpha},$$

where

$$d = 2\tilde{\kappa}J_1[1-\rho\frac{l}{b}(p-q)]^2 + \rho\frac{l}{2b}J_0,\quad \mu = 2s(J_0 - cJ_1)[1-\rho\frac{l}{b}(p-q)],$$

which is none other than the $U(p,q)$ nonlinear Schrödinger equation (to see this, substitution (85) should be made). Let us consider some limiting cases:

(i) $p = n$, $q = 0$. This is the pure ferromagnetic system governed by a $U(n,0)$ nonlinear Schrödinger equation of attractive type,

$$i\dot{\varphi}^{\alpha} = -2b\varphi^{\alpha}_{XX} - \kappa\left[\sum_{\beta=1}^{n}|\varphi^{\beta}|^2\right]\varphi^{\alpha} - \mu\varphi^{\alpha},$$

in which $\kappa = 4\tilde{\kappa}J_1(1-\rho n\,l/b)^2 + \rho J_0\,l/2b$ and $\mu = 2s(J_0 - cJ_1)(1-\rho n\,l/b)$. If $n = 1$, this equation becomes the $U(1,0)$ nonlinear Schrodinger equation derived in [49] to describe magnetic crystals.

(ii) $p = 0$, $q = n$. For a purely 'antiferromagnetic' system we get the vector $U(0,n)$ nonlinear Schrödinger equation of repulsive type:

$$-i\dot{\varphi}^{\alpha} = -2b\varphi^{\alpha}_{XX} + \kappa\left[\sum_{\beta=1}^{n}|\varphi^{\beta}|^2\right]\varphi^{\alpha} - \mu\varphi^{\alpha},$$

with

$$\kappa = 4\tilde{\kappa}J_1(1+\rho\frac{l}{b}n)^2 + \rho\frac{l}{2b}J_0 \quad \text{and} \quad \mu = 2s(J_0 - cJ_1)(1+\rho\frac{l}{b}n).$$

(2) For a large number of real crystals, the interaction between colour components is altogether much weaker as compared with the interlattice interaction [50], but it is not negligible. To include the former, it is sensible in the colour space also to consider interactions only among nearest neighbours [51]. The corresponding reduction is

$$R^{\alpha\beta}_{ij} = \rho J^{\alpha\beta}_{ij},\quad J^{\alpha\beta}_{ij} = J_{jj+\sigma}M^{\alpha\beta} + J^1 U^{\alpha\beta}_{ij},$$

$$M^{\alpha\beta} = \delta^{\alpha\beta} + \epsilon\delta^{\beta,\alpha+\delta},\quad U^{\alpha\beta}_{ij} = \delta_{ij}\delta^{\beta,\alpha+\delta},$$

with $J^1/J \ll 1$, J and J^1 being, respectively, the intersite and interchain exchange integrals. In this way, Hamiltonian (82) is reduced to

$$\hat{H}_s = -\frac{1}{2}\sum_{j\sigma}\left[J_{jj+\sigma}\sum_{\alpha=1}^{n}\left\{\frac{1}{2}\left[\hat{S}_j^{+\alpha}\hat{S}_{j+\sigma}^{-\alpha}+\text{h.c.}\right]+\rho\hat{S}_j^{z\alpha}\hat{S}_{j+\sigma}^{z\alpha}\right\}\right.$$

$$-\frac{1}{2}J^1\sum_{\substack{\alpha=1\\ \delta=\pm 1}}^{n}\left\{\frac{1}{2}\left[\hat{S}_j^{+\alpha}\hat{S}_j^{-(\alpha+\delta)}+\text{h.c.}\right]+\rho\hat{S}_j^{z\alpha}\hat{S}_j^{z(\alpha+\delta)}\right\}$$

$$\left.-\frac{1}{2}\epsilon J_{jj+\sigma}\sum_{\alpha,\beta}\left\{\frac{1}{2}\left[\hat{S}_j^{+\alpha}\hat{S}_{j+\sigma}^{-(\alpha+\delta)}+\text{h.c.}\right]+\rho\hat{S}_j^{z\alpha}\hat{S}_{j+\sigma}^{z(\alpha+\delta)}\right\}\right].$$

In the XX model, when $\rho\to 0$, we obtain for the amplitude φ^α the equation

$$-i\dot{\varphi}^\alpha = \frac{sJ_0}{2}\sum_\beta\left[M_{(\alpha\beta)}\varphi_{XX}^\beta+M'_{(\alpha\beta)}\varphi^\beta\right]-\kappa\left(\sum_{\gamma\delta}M_{\gamma\delta}\bar{\varphi}^\gamma\varphi^\delta\right)\sum_\beta M_{\alpha\beta}\varphi^\beta,$$

where

$$M'_{(\alpha\beta)} = s(J_0-\epsilon J_1)M_{(\alpha\beta)}+J^1 s\delta^{\beta,\alpha+\delta}.$$

With this system, one can calculate corrections to the integrable system caused by the additional interchain interactions.

9. Multicolour Hubbard model

As was pointed out above (see Section 5), the one-dimensional one-band Hubbard model is equivalent to the two-component Heisenberg chain, therefore a reduction of the multicolour Heisenberg model (82) may be considered in the form of eqn. (83) with the additional assumptions

$$J_{jj+\sigma} = -J^0_{jj+\sigma},\quad J^0_0 \equiv 2t_0<0,\quad J^0_1 = 2I,\quad s = \frac{1}{2}, \tag{1.95}$$

$$K_{\alpha\beta} = (I\otimes\sigma_1)_{\alpha\beta},\quad \sigma_1 = \left(\begin{smallmatrix}0&1\\1&0\end{smallmatrix}\right),\quad I_{ab} = \delta_{ab}(a,b=1,\ldots,n).$$

Use the notation

$$\sigma = \begin{cases} k, & \text{for } k\in D^+,\ D^+ = [1,2,\ldots,\frac{n}{2}],\\ -(n+1-k), & \text{for } k\in D^-,\ D^- = [1+\frac{n}{2},\ldots,n],\end{cases}$$

with

$$\alpha,\beta = \begin{cases} 2k-1, & \text{for } \alpha,\beta \text{ odd,} \\ 2k, & \text{for } \alpha,\beta \text{ even,} \end{cases}$$

and put

$$\hat{S}_{2k-1} = \begin{cases} \hat{S}^A_\sigma, & \text{for } k \in D^+, \\ \hat{S}^A_{-\sigma}, & \text{for } k \in D^-, \end{cases}$$

$$\hat{S}_{2k} = \begin{cases} \hat{S}^B_\sigma, & \text{for } k \in D^+, \\ \hat{S}^B_{-\sigma}, & \text{for } k \in D^-, \end{cases}$$

For the matrix $R^{\alpha\beta}_{ij}$ let us take the reduction

$$R^{\alpha\beta}_{ij} \to -(U_0\delta_{ij}P^{\alpha\alpha}_{1\beta\beta} + \tilde{J}_{ij}P^{\alpha\alpha}_{2\beta\beta}), \tag{1.96}$$

where δ_{ij} is the Kronecker symbol,

$$\tilde{J}_{ij} = \begin{cases} \tilde{J}_0, & i,j \text{ for nearest neighbours,} \\ 0, & \text{otherwise,} \end{cases}$$

and

$$P^{\alpha\alpha}_{1\beta\beta} = L^{+\alpha\alpha}_A L^{-\beta\beta}_A + L^{+\alpha\alpha}_B L^{-\beta\beta}_B,$$

$$P^{\alpha\alpha}_{2\beta\beta} = (L^{+\alpha\alpha}_A L^{+\beta\beta}_B + L^{-\alpha\alpha}_A L^{-\beta\beta}_B) + (L^{+\alpha\alpha}_B L^{+\beta\beta}_A + L^{-\alpha\alpha}_B L^{-\beta\beta}_A),$$

with

$$L^{+\alpha\alpha}_{A(B)} = \begin{cases} (I \otimes e_{A(B)})_{\alpha\alpha}, & \text{for } k \in D^+, \\ 0, & \text{for } k \in D^-, \end{cases} \tag{1.97}$$

$$L^{-\beta\beta}_{A(B)} = \begin{cases} 0, & \text{for } k \in D^+, \\ (I \otimes e_{A(B)})_{\beta\beta}, & \text{for } k \in D^-, \end{cases}$$

$$e_A = \left(\begin{smallmatrix} 1 & 0 \\ 0 & 0 \end{smallmatrix}\right), \quad e_B = \left(\begin{smallmatrix} 0 & 0 \\ 0 & 1 \end{smallmatrix}\right).$$

The spin Hamiltonian (82) has the form

$$\hat{H}_s = \frac{1}{4}\sum_{j,\sigma} U_0 \sum_{\delta=1}^{n} J^0_{jj+\sigma} \left[\hat{S}^{-A}_{j,\delta}\hat{S}^{+B}_{j+\sigma,\delta} + \hat{S}^{-B}_{j,\delta}\hat{S}^{+A}_{j+\sigma,\delta} + \text{h.c.}\right] + \tag{1.98}$$

$$+\frac{1}{2}U_0 \sum_{j\in A\cup B} \sum_{\delta=1}^{n/2} \hat{S}^z_{j,\delta} \sum_{\delta'=1}^{n/2} \hat{S}^z_{j,-\delta'} +$$

$$+\frac{1}{2}\tilde{J}_0 \sum_{\substack{j\in A\\ j+\sigma\in B\\ \&\text{ v.v.}}} \left[\sum_{\delta=1}^{n/2} \hat{S}^{zA}_{j,\delta} \sum_{\delta'=1}^{n/2} \hat{S}^{zB}_{j+\sigma,\delta'} + \sum_{\delta=1}^{n/2} \hat{S}^{zA}_{j,-\delta} \sum_{\delta'=1}^{n/2} \hat{S}^{zB}_{j+\sigma,-\delta'}\right],$$

with the above assumptions and notations.

The generalized Jordan-Wigner transformation

$$\hat{S}^{+\alpha}_j = \exp\left\{-i\pi \sum_{\beta=1}^{\alpha-1} \sum_{k=1}^{j-1} c^{+\beta}_k c^{\beta}_k\right\} c^{\alpha}_j,$$

$$\hat{S}^{-\alpha}_j = (\hat{S}^{+\alpha}_j)^+, \quad \hat{S}^{z\alpha}_j = \frac{1}{2} - c^{+\alpha}_j c^{\alpha}_j,$$

via the colour Fermi operators

$$[c^{\alpha}_i, c^{+\beta}_j]_+ = \delta_{ij}\delta^{\alpha\beta}, \quad [c^{\alpha}_i, c^{\beta}_j]_+ = [c^{+\alpha}_i, c^{+\beta}_j]_+ = 0,$$

enables us to reduce the Hamiltonian (98) to a generalized multicomponent Hubbard model in the nearest-neighbour limit where lattice excitations are considered as classical fields:

$$\hat{H}_s = \frac{1}{4}\sum_{\delta=1}^{n/2} J^0_{jj+1}(c^{+A}_{j\delta} c^{B}_{j+1,\delta} + c^{+B}_{j\delta} c^{A}_{j+1,\delta} + \text{h.c.}) \tag{1.99}$$

$$+\frac{1}{2}U_0 \sum_{j\in A\cup B} \sum_{\delta=1}^{n/2} \hat{n}_{j,\delta} \sum_{\delta'=1}^{n/2} \hat{n}_{j,-\delta'} - \mu \sum_{j\in A\cup B} \sum_{\delta=1}^{n} \hat{n}_{j,\delta} +$$

$$+\frac{1}{2}\tilde{J}_0 \sum_{\substack{j\in A\\ j+\sigma\in B}} \sum_{\delta=1}^{n/2} \left[\hat{n}_{j,\delta} \sum_{\delta'=1}^{n/2} \hat{n}_{j+\sigma,\delta'} + \hat{n}_{j,-\delta} \sum_{\delta'=1}^{n/2} \hat{n}_{j+\sigma,-\delta'}\right] + H_0,$$

$$H_0 = \frac{1}{8}(\tilde{J}_0 + U_0)n^2 N, \quad \mu = \frac{1}{8}(3\tilde{J}_0 + U_0)n.$$

It is easy to check that for $\delta = \pm 1$ this formula becomes the Hamiltonian $\hat{H}_s$ of model (48).

Unlike the simple Hubbard model the situation for the ground state of its colour generalization is vague and it is of use to consider both ferromagnetic and antiferromagnetic states.

For the antiferromagnetic ground state, suppose, just as before,

$$<\hat{n}^{A}_{j,\delta}> = <\hat{n}^{B}_{j\pm1,\delta}>.$$

Also assume that condition (50) is valid for long-wave excitations over the anitferromagnetic vacuum and

$$\varphi^{B\delta}_{j\pm1} = \varphi^{A\delta}_{j} \pm \Delta\varphi^{A\delta}_{j} + \frac{1}{2}\Delta^2\varphi^{A\delta}_{j} \tag{1.100}$$

holds for the ferromagnetic one.

As a result, in the long-wave and quasi-stationary approximation we obtain

$$x_X = \frac{J_1}{m\omega_0^2}\sum_{\delta=1}^{n}(\bar{\varphi}^{\delta}\varphi^{-\delta} + \text{c.c.}) + c \tag{1.101}$$

for the antiferromagnetic vacuum, and

$$x_X = \frac{2J_1}{m\omega_0^2}\sum_{\delta=1}^{n}(\bar{\varphi}^{\delta}\varphi^{\delta}) + c \tag{1.101b}$$

for the ferromagnetic one.

Then the field equations read:

Antiferromagnetic:

$$i\dot{\varphi}^{\delta} = t_0\varphi^{-\delta}_{XX} + 2(t_0 + Ic)\varphi^{-\delta} - \frac{(2I)^2}{m\omega_0^2}\left[\sum_{\delta'=1}^{n/2}(\bar{\varphi}^{\delta'}\varphi^{-\delta'} + \text{c.c.})\right]\varphi^{-\delta}$$

$$-\mu\varphi^{\delta} + \tilde{U}\left\{\sum_{\delta'=1}^{n/2}|\varphi^{-\delta'}|^2\right\}\varphi^{\delta}, \tag{1.102}$$

with $\tilde{U} = \tilde{J}_0 + U_0/2$ in the same approximation as in Section 5,

Ferromagnetic:

$$i\dot{\varphi}^{\delta} = t_0\varphi^{\delta}_{XX} + 2(t_0 + Ic - \frac{\mu}{2})\varphi^{\delta} - \frac{(2I)^2}{m\omega_0^2}\left[\sum_{\delta'=1}^{n/2}(|\varphi^{\delta'}|^2 + |\varphi^{-\delta'}|^2)\right]\varphi^{\delta}$$

$$+\tilde{U}\left\{\sum_{\delta'=1}^{n}|\varphi^{-\delta'}|^2\right\}\varphi^{\delta}. \tag{1.103}$$

Different limiting versions of the Hubbard model and physical systems which correspond to them have been discussed by several authors (see, e.g., Refs. [22,23].

In particular, it has been shown that in the limit $U_0 \to \infty$ when the Coulomb repulsion plays a leading role, the model may be used to describe organic charge transfer salts of *TCNQ* (tetracyanoquinodimethan). In the opposite limit, $U_0 \to 0$, when the 'hopping' integral plays the central role, it describes mixed valency planar compounds of transition metals (*MVPC*). The first case corresponds to the Ising model, the second to the *XY* model*.

Equations (102) and (103) are non-integrable in general, however in the limit $\mu, \tilde{U} \to 0$ they become integrable. In fact, introducing linear combinations of the functions sought [18]:

$$u^\delta = \varphi^\delta + \varphi^{-\delta}, \quad v^\delta = \varphi^\delta - \varphi^{-\delta},$$

gives in this limit for an antiferromagnetic

$$i\dot{u}^\delta = t_0 u^\delta_{xx} + 2(t_0 + cI)u^\delta - \gamma \sum_{\delta'}^{n/2} (|u^{\delta'}|^2 - |v^{\delta'}|^2)u^\delta, \tag{1.104}$$

$$-i\dot{v}^\delta = t_0 v^\delta_{xx} + 2(t_0 + cI)v^\delta - \gamma \sum_{\delta'}^{n/2} (|u^{\delta'}|^2 - |v^{\delta'}|^2)v^\delta,$$

and for a ferromagnetic

$$i\dot{u}^\delta = t_0 u^\delta_{xx} + 2(t_0 + cI)u^\delta - \gamma \sum_{\delta'}^{n/2} (|u^{\delta'}|^2 + |v^{\delta'}|^2)u^\delta \tag{1.105}$$

$$i\dot{v}^\delta = t_0 v^\delta_{xx} + 2(t_0 + cI)v^\delta - \gamma \sum_{\delta'}^{n/2} (|u^{\delta'}|^2 + |v^{\delta'}|^2)v^\delta$$

where $\gamma = 4I^2/m\omega_0^2$, $\omega_0 = u_0/a_0$.

Introducing the vector functions $\psi^\alpha = (u^\delta, \bar{v}^\delta)^T$ in eqn. (104) and $\psi^\alpha = (u^\delta, v^\delta)^T$ in eqn. (105) we observe that under the above assumptions the colour Hubbard model gives the $U(n/2, n/2)$ nonlinear Schrödinger equation in the antiferromagnetic case, and the $U(n,0)$ nonlinear Schrödinger equation in the ferromagnetic case.

We have studied the many component spin system and found that under certain assumptions (the long-wavelength limit and so on) it may be associated with various field models with internal (colour) symmetries. Some of these models are integrable. These involve the nonlinear Schrödinger equation with $U(p,q)$ symmetry (obtained in the quasi-static limit) and the colour generalization of the Yajima-

*After discovery of High Temperature Superconductivity the Hubbard model gets popular and regarded as a model which will probably describe this phenomenon.

Oikawa system (derived in the near-sonic limit). Other non-integrable reductions may be considered in some sense as systems close to integrable ones.

All the equations obtained, besides the linear (phonon and magnon) solutions, admit essentially nonlinear (soliton) solutions as well. The properties of such solutions will be presented in detail in the following chapters.

Chapter 2.

PHYSICALLY INTERESTING NONLINEAR DIFFERENTIAL EQUATIONS

In this short chapter we give a list of nonlinear differential equations and their localized solutions (quasi-solitons) which, along with the KdV equation, often occur in various physical applications. This list is not complete of course. It is still being expanded, however sufficiently simple equations presented in the following can be used as mathematical models of quite different, in nature, physical systems. In this sense, and due to the mathematical properties of some of them, they fulfil the same role in the theory of nonlinear differential equations as do elliptic, parabolic and wave equations in the theory of linear partial differential equations and are therefore universal.

There is, however, an important distinction. There does not exist a university course, such as an 'Equations of Mathematical Physics' Course, in the theory of nonlinear partial differential equations, probably since even integrable equations possess their remarkable properties only when they are studied either on the whole axis or on the ring (i.e. under periodic boundary conditions). Setting arbitrary conditions on a finite interval is still not open to current exact methods of research. One particular exception are integrable relativistically invariant equations conserving integrability on the semi-axis, yet for a very specific class of physical problems only. The study of systems on finite intervals or with inhomogeneities (e.g. sourses), started some time ago, and is progressing rather slowly. Nevertheless, very interesting results of a computational and analytical nature have been obtained in this way. Certain aspects of this work are described below.

1. Equations with quadratic dispersion

This class of equations involves all the equations of the nonlinear Schrödinger equation type with the dispersion relation

$$i\psi_t + \psi_{xx} + \beta|\psi|^\nu \psi = 0 \tag{2.1}$$

or in a more general form*

$$i\psi_t + \psi_{xx} + \Phi(|\psi|)\psi = 0. \tag{2.2}$$

Soliton-like solutions to eqn. (1)

$$\psi_s = A\exp\{i[(\frac{v^2}{4} + \frac{2\beta}{2+\nu}A^\nu)t + \frac{v}{2}(x - vt) + \theta_0]\} \times \tag{2.3}$$

$$\times \operatorname{sech}^{2/\nu}[\frac{\nu}{2}A^{\nu/2}(\frac{2\beta}{2+\nu})^{1/2}(x - vt - x_0)]$$

exist at $\beta > 0$. The case $\beta < 0$ is more complicated and needs special consideration.

At $\nu = 2$, eqn. (1), or S3, is completely integrable and its envelope solitary waves are true solitons. For other values of ν, solutions of eqn (1) are only quasi-solitons.

Among the second-type equations is the ϕ^6 model mentioned earlier:

$$i\psi_t + \psi_{xx} + \alpha\psi + \psi(|\psi|^2 - |\psi|^4) = 0, \tag{2.4}$$

as well as equations with saturable nonlinearity

$$i\psi_t + \psi_{xx} + \frac{1}{\alpha}\psi[1 - \exp(-\alpha|\psi|^2)] = 0, \tag{2.5}$$

and

$$i\psi_t + \psi_{xx} + \alpha\psi + \frac{|\psi|^2}{1+|\psi|^2}\psi = 0, \tag{2.6}$$

frequently arising in physical applications. The first of these is used, for example, to study the behaviour of the sufficiently large amplitude packets of Langmuir waves in plasmas, $|\psi|^2 \simeq \alpha^{-1}$ (where $\alpha = \frac{4}{3}me/m_i$) near the stationary state [52].

*Here we give only one-dimensional versions of the corresponding equations. Their non-one-dimensional analogues will be presented later.

The second equation (and the first one) is applicable to simulate the highly nonlinear medium response to a powerful laser beam propagating in it [53].

This type of equation includes also the above-mentioned systems with $\Phi \propto -\delta n$ subject to one of the following equations

$$\Box\Phi = -\partial_x^2(|\psi|^2) \tag{2.7}$$

(Zakharov, 1972)

$$\partial_t\Phi + \partial_x(\Phi - |\psi|^2) = 0 \tag{2.8}$$

(Yajima-Oikawa, 1976)

$$(\partial_t + \alpha\partial_x^3)\Phi + \partial_x(\beta\Phi^2 - |\psi|^2 + \Phi) = 0 \tag{2.9}$$

(Nishikawa, et al., 1974)

$$(\Box - \alpha\partial_x^4)\Phi + \partial_x^2(\beta\Phi^2 + |\psi|^2) = 0 \tag{2.10}$$

(Makhankov, 1974).

In the first two models, low-frequency excitations are described by linear equations, but only the second one is integrable. In the others, low-frequency waves are nonlinear and, again, the second one is integrable with an appropriate choice of α and β [89]. The fact that model (9) is not integrable has been stated in 1983 by Benilov and Burtsev [54].

Soliton-like solutions of system (7) have the form

$$\psi_s = A\exp\{i(\frac{vx}{2} - \Omega t + \theta_0)\}\operatorname{sech}[\frac{A\gamma}{\sqrt{2}}(x - vt - x_0)] \tag{2.11}$$

$$\Phi = -\gamma^2|\psi_s|^2,\ \gamma^{-2} = 1 - v^2,\ \Omega = \frac{v^2}{4} - \frac{1}{2}\gamma^2 A^2.$$

Similar formulae correspond to one-soliton solutions of the Yajima-Oikama model (8). We shall yet turn to a discussion of solutions (11) and the like. Here we notice that systems (7) and (8) properly describe sufficiently slow small-amplitude solitons, $v < 1$. In the region $|v-1| \ll 1$, or for large soliton amplitudes A, it is necessary to use the nonlinear model (10).

The equations of the type considered include also

$$\psi_t + i\psi_{xx} + \frac{1}{4}\partial_x(|\psi|^2\psi) = 0 \tag{2.12}$$

and

$$\psi_t + i\psi_{xx} + \beta|\psi|^2\psi_x = 0. \tag{2.13}$$

The first equation, known as the derivative nonlinear Schrödinger equation (see [55] and (1.78)), was studied by a Japanese group [56]. Its soliton solutions describe an abrupt change of the plane wave phase accompanied by an amplitude variation localized at the same point in space. Kaup and Newell [57] modified the $ZS-AKNS$ scheme to include eqn. (12), too. Further developments in this direction have been described by Wadati, Konno, and Ichikawa (WKI) [58] where the superposition of the $ZS-AKNS$ and KN schemes was shown to enable the solution of a whole new class of equations of the nonlinear Schrödinger equation type with various nonlinearities

$$i\psi_t + \psi_{xx} \pm \alpha|\psi|^2\psi \mp i\beta(|\psi|^2\psi)_x = 0 \tag{2.14}$$

$$i\psi_t + \left[\frac{\psi}{(1\mp|\psi|^2)^{1/2}}\right]_{xx} = 0 \tag{2.15}$$

and systems

$$iq_t + q_{xx} + \alpha r q^2 - i\beta(rq^2)_x = 0 \tag{2.16}$$

$$ir_t - r_{xx} - \alpha r^2 q - i\beta(r^2 q)_x = 0$$

$$iq_t + \left[\frac{q}{(1-rq)^{1/2}}\right]_{xx} = 0 \tag{2.17}$$

$$ir_t - \left[\frac{r}{(1-rq)^{1/2}}\right]_{xx} = 0.$$

Equations (14) and (15) are obtained from eqns. (16) and (17) via the reduction $r = \pm\bar{q}$. Although the systems (16) and (17) as well as eqns. (14) and (15), derived from them, are very interesting mathematically (due to their integrability and exotic soliton solutions), they still do not possess the $S3$ generality because of the restrictions of their applications.

Equation (13) arises in the study of Bose gas models and is integrable, too [59]; moreover, the superposition of the $S3$ and eqn. (13)

$$i\psi_t + \psi_{xx} + \alpha|\psi|^2\psi + i\beta|\psi|^2\psi_x = 0 \tag{2.14a}$$

is integrable as well.

(b) Vector equations of the Landau-Lifshitz type. The latter, describing the dynamics of the magnetization vector, has been obtained by Landau and Lifshitz in 1935, and in a dissipation-free medium assumes the form

$$\partial_t \mathbf{M} = -g[\mathbf{M},\mathbf{H}_{\text{eff}}], \quad \mathbf{H}_{\text{eff}} = -\frac{\partial E}{\delta \mathbf{M}} \tag{2.18}$$

where g is a factor equal to $2\mu_0/\hbar$ for electron spin systems, H_{eff} is the effective magnetic field, $\mathbf{M}$ is the magnetic moment of the medium (crystal) unit volume, with $\mathbf{M}^2 = \text{const} = M_0^2$. In the crystal, $M_0 = 2\mu_0 s/a_0^2$ with μ_0 being the Bohr magneton, s the atomic spin, and a_0^3 the lattice cell volume.

The ferromagnetic energy can be expressed as a functional of $\mathbf{M}$ and its derivatives

$$E = \int W(M,\partial_{x_k} M_i) d^3 x. \tag{2.19}$$

The explicit form of the magnetic energy density is different for different models and agrees, in principle, with the microscopic Hamiltonian of the system. In the latter (quantum) case, a description of the crystal magnetic properties is usually developed using the Heisenberg model we have sketched above.

Equation (18) is the phenomenological in nature. As we saw previously, going from a microscopic description based on the Heisenberg model to a macroscopic one (eqn. (18)) is not trivial, and involves a reduction procedure the basis of which is a fairly complicated problem. We will not dwell on this problem but, rather, cite the book of Akhiezer et al. [60]. We only mention a formal procedure restoring such a passage, but not being its basis. A sufficiently detailed study of this procedure and a comprehensive review of nonlinear effects in magnetic media are presented in the nice book of Kosevich et al. [1] (for another approach, see Chapter V).

We deal here with the anisotropic Heisenberg ferromagnet (1.6) in the nearest-neighbour approximation written in the form

$$H = -\sum_{j\sigma}(J_1 \hat{S}_j^x \hat{S}_{j+\sigma}^x + J_2 \hat{S}_j^y \hat{S}_{j+\sigma}^y + J_3 \hat{S}_j^z \hat{S}_{j+\sigma}^z). \tag{2.20}$$

Let us perform the reduction procedure via the following scheme:

(1) Carry out a formal transition from the spin operator $\hat{\mathbf{S}}$ to the classical

magnetic moment

$$\hat{\mathbf{S}}_n \to \frac{a_0^3}{2\mu_0}\mathbf{M}(\mathbf{x}_n)$$

where $\mathbf{x}_n$ is the radius vector of the nth lattice site.

(2) Use the usual expansion of the function $\mathbf{M}(\mathbf{x})$

$$\mathbf{M}(\mathbf{x}+\vec{\sigma}) = \mathbf{M}(\mathbf{x})+\sigma_i\frac{\partial \mathbf{M}}{\partial x_i}+\frac{1}{2}\sigma_i\sigma_k\frac{\partial^2\mathbf{M}}{\partial x_i \partial x_k}+...$$

(3) Assuming the difference in the exchange integrals J_i to be small enough, represent them in the form

$$\mathbf{J} = J_0+\frac{a_0^2}{2}\mathbf{j}.$$

If we restrict ourselves to quadratic terms in $a_0^2(\sigma_i \sim a_0)$, and omit constant terms proportional to $\mathbf{M}^2$, then we get

$$W = \frac{\alpha}{2}(\partial x_i\mathbf{M})^2-\frac{\beta_1}{2}M_x^2-\frac{\beta_3}{2}M_z^2 \tag{2.20a}$$

where

$$\alpha = \frac{1}{a_0}J_0\left[\frac{a_0^3}{2\mu_0}\right]^2,\quad \beta_1 = \frac{j_1-j_2}{a_0}\left[\frac{a_0^3}{2\mu}\right]^2,\quad \beta_3 = \frac{j_3-j_2}{a_0}\left[\frac{a_0^3}{2\mu}\right]^2.$$

At $\beta_1 = 0$ a ferromagnet is called *uniaxial.*

(a) $\beta_3>0$. We have the *easy-axis* anisotropy. In the ground state the magnetization vector $\mathbf{M}$ is directed along this OZ-axis.

(b) $\beta_3<0$. This is the *easy-plane* anisotropy. In the ground state the vector $\mathbf{M}$ lies in a plane perpendicular to a preferred axis.

(4) By replacing the commutator in the Heisenberg equation

$$i\hbar\partial_t\hat{S}_i = [\hat{H},\hat{S}_i]$$

with the Poisson brackets and the Hamiltonian $\hat{H}$ by the quantity E, eqn. (19), with W defined via expression (20a) we get the Landau-Lifshitz equations in the conventional form (the one-dimensional version)

$$\frac{\hbar}{2\mu_0}\dot{\mathbf{M}}+\alpha[\mathbf{M}\partial_x^2\mathbf{M}]+\beta[\mathbf{M}\mathbf{e}_z](\mathbf{M}\mathbf{e}_z)+[\mathbf{M}\mathbf{e}_z]h = 0 \tag{2.21}$$

with the non-vanishing external magnetic field $\mathbf{h} = (0,0,h)$ and $\beta = \beta_3$.

Let us now describe a two-sublattice phenomenological model of an antiferromagnet. Such a system is specified by the two magnetization vectors $\mathbf{M}_1(\chi,t)$ and $\mathbf{M}_2(\chi,t)$, provided that $\mathbf{M}_1^2 = \mathbf{M}_2^2 = M_0^2$, which compensate each other $\mathbf{M}_1 = -\mathbf{M}_2$ at zero temperature in the antiferromagnetic state (in the absence of an external magnetic field, $h = 0$). Instead of $\mathbf{M}_1$ and $\mathbf{M}_2$ we can introduce the magnetization vector ($\mathbf{M} = \mathbf{M}_1 + \mathbf{M}_2$) and the antiferromagnetic vector ($\mathbf{L} = \mathbf{M}_1 - \mathbf{M}_2$) such that $(\mathbf{ML}) = 0$, $\mathbf{M}^2 + \mathbf{L}^2 = 4M_0^2$. In the ground state $\mathbf{M} = 0$, $\mathbf{L} = 2\mathbf{M}_1$.

The equations of motion have the form

$$\dot{\mathbf{M}} = -g([\mathbf{MH}_m] + [\mathbf{LH}_l]), \quad \mathbf{H}_m = -\frac{\delta E}{\delta \mathbf{M}} \tag{2.22a}$$

$$\dot{\mathbf{L}} = -g([\mathbf{LH}_m] + [\mathbf{MH}_l]), \quad \mathbf{H}_l = -\frac{\delta E}{\delta \mathbf{L}}. \tag{2.22b}$$

In the uniaxial antiferromagnet, the magnetic energy density is

$$2W = A\mathbf{M}^2 + \alpha_1(\partial_{x_i}\mathbf{L})^2 + \alpha_2(\partial_{x_i}\mathbf{M})^2 - \beta_1 L_z^2 - \beta_2 M_z^2 - (\mathbf{Mh}) \tag{2.23}$$

where A is the constant determined by the exchange integrals between the sublattices J.

Just as for ferromagnets, the uniaxial antiferromagnets fall into two groups: $\beta_1 > 0$, in the ground state the sublattice magnetic moments are oriented along the anisotropy axis Z and $\mathbf{M} = 0$ (the easy-axis antiferromagnet); $\beta_1 < 0$, in the ground state the vector $\mathbf{L}$ is perpendicular to the Z axis and $\mathbf{M} = 0$ (the easy-plane antiferromagnet).

2. Equations with 'linear' dispersion

The equations of this class include all the KdV-like equations, among which are also exotic ones. One can tentatively attribute to this class also the systems of equations describing n-wave interaction with linear dispersion $\omega \approx k + 0(k^3)$. We begin with the equation

$$(\partial_t + \partial_x + \partial_x^3)\varphi + \varphi^\nu \varphi_0 = 0. \tag{2.24}$$

As is well-known now, at $\nu = 1,2$ this equation is complete integrable and its solitons interact elastically. For $\nu \geqslant 3$ it loses this property, though it remains very 'near' to an integrable one. Soliton-like solutions to eqn. (24) form a two-parameter family and are

$$\varphi_s = (A \operatorname{sech}[\frac{A\nu}{\sqrt{2(\nu+1)(\nu+2)}}](x - \mathrm{v}t - x_0)])^{2/\nu} \tag{2.25}$$

$$\mathrm{v} = 1 + 2A^2[(\nu+1)(\nu+2)]^{-1}.$$

A symmetrical variant of (24), the so-called Boussinesq equation

$$\Box\varphi - \partial_x^4\varphi - \partial_x^2\varphi^2 = 0 \tag{2.24a}$$

possesses soliton-like solutions

$$\varphi_s = A \operatorname{sech}^2\left[\sqrt{\frac{A^2}{6}}(x - \mathrm{v}t - x_0)\right], \quad \mathrm{v} = \pm(1 + \frac{2}{3}A)^{1/2}. \tag{2.25a}$$

In addition, we cite some equations frequently appearing in the literature

$$(\partial_t + \partial_x)\varphi - \varphi_{xxt} + \varphi^\nu\varphi_x = 0. \tag{2.26}$$

For $\nu = 1$, this equation has been proposed by Peregrine to describe tidal waves [61] and then by Benjamin et al. [62] where it was called the regularized-long-wave equation. Its soliton-like solutions are also a two-parameter family, viz.,

$$\varphi_s = \left\{A \operatorname{sech}\left[\frac{A\nu}{\sqrt{2}}((\nu+1)(\nu+2) + 2A^2)^{-1/2}(x - \mathrm{v}t - x_0)\right]\right\}^{2/\nu} \tag{2.27}$$

$$\nu = 1 + 2A^2[(\nu+1)(\nu+2)]^{-1}.$$

One can consider other modifications of eqn. (26) such as

$$(\partial_t + \partial_x)\varphi + \varphi_{xtt} + \varphi^\nu\varphi_x = 0$$

$$(\partial_t + \partial x - \partial_t^3)\varphi + \varphi^\nu\varphi_x = 0.$$

These have a highly exotic dispersion and will not be discussed below.

The Benjamin-Ono equation [63,64]

$$\varphi_t + 6\varphi\varphi_x + \frac{P}{\pi}\int\frac{\varphi_{\chi\chi}}{\chi - x}d\chi \tag{2.28}$$

governs internal waves in stratified fluids of great depth. Its localized travelling waves (Lorentzian pulses) of the form

$$\varphi_s = \frac{2}{3}\frac{\mathrm{v}}{1+\mathrm{v}^2(x-\mathrm{v}t)^2}$$

are also true solitons. This fact was first established in numerical experiments by Meiss and Pereira [65], and multi-soliton solutions have been found by Joseph [66]. The problem has been completely solved by Ablovitz et al. [67] who studied the nonlinear intermediate longwave equation

$$\varphi_t + \frac{1}{\delta}\varphi_x + 6\varphi\varphi_x + T(\varphi_{xx}) = 0, \quad T(y) = \frac{1}{2\delta}\int\limits_{-\infty}^{\infty} \mathrm{cth}\left\{\frac{\pi(\chi - x)}{2\delta}\right\} y(\chi)d\chi \qquad (2.29)$$

and showed that it is an integrable system. The KdV and Benjamin-Ono equations are the opposite limiting cases of the general equation (29), respectively, at $\delta \to 0$ (shallow-water limit) and $\delta \to \infty$ (deep water). This class incorporates also a more exotic system embedded into the Wadati et al. [58] scheme

$$q_t = -\partial_x^2\left[\frac{q_x}{(1-rq)^{3/2}}\right], \quad r_t = -\partial_x^2\left[\frac{r_x}{(1-rq)^{3/2}}\right] \qquad (2.30)$$

and its reduction:

(1) $r = -q$

$$q_t + \partial_x^2\left[\frac{q_x}{(1+q^2)^{3/2}}\right] = 0 \qquad (2.31)$$

or

$$q_t + \partial_x^3\left[\frac{q}{(1+q^2)^{1/2}}\right]$$

(2) $r = -1$, $q = \varphi - 1$

$$\varphi_t = 2\partial_x^3(1/\sqrt{\varphi}) \qquad (2.32)$$

(the equation obtained by Dym). Let $\sqrt{\varphi} = \phi^{-1}$ then

$$\phi_t + \phi^3\partial_x^3\phi = 0 \qquad (2.33)$$

and finally

$$\phi_t + \phi_x + \phi^3 \phi_{xxx} = 0.$$

Physical applications of models (30)-(33) are limited in contrast to simple modifications of the KdV equation, which is similar to the case of eqn. (14)-(17) compared with the $S3$. Apparently, this is associated with the following fact. Usually such equations are obtained via an expansion of the initial equations in terms of nonlinearity and dispersion. The KdV- or $S3$-type equations are derived allowing for the first terms in such an expansion which, as a rule, is the case in uniform and isotropic systems without any intricate internal structure of anisotropy. In order to obtain 'exotic' equations, it is necessary that some of the first terms in the expansion vanish for some reason as is obvious from the example of eqn. (33) as compared with the KdV equation. Wadati et al. [58] found a system whose behaviour may be simulated by one of the exotic equations, viz. eqn. (31), and its modification can describe, with certain assumptions, the nonlinear transverse vibrations of an elastic rod or tense cord [68]. A similar exotic situation also occurs when an Alfvén wave propagates in anisotropic plasmas strictly along the magnetic field [55].

Sometimes complex modifications of the KdV equation are considered (cf. for example, [69])

$$(\partial_t + \partial_x^3)\psi + |\psi|^2 \psi_x = 0, \tag{2.34}$$

$$(\partial_t + \partial_x^3)\psi + (|\psi|^2 \psi)_x = 0. \tag{2.35}$$

The first of these is integrable as demonstrated by Hirota [70] via the inverse transform method[*)]. The second is non-integrable as has been illustrated in [69], so the initial packet decay and the interaction of quasi-solutions have a complicated inelastic character. In particular, equations of the type (34) and (35) arise in plasma theory [71].

In nonlinear optics, equations have long been known which describe the interactions of several waves. Especially intersting is the study of popular systems modelling three-wave interactions for a medium with quadratic nonlinearity. As a result of such an interaction the decay takes place of some 'pumpling' wave ψ_3 into two others, ψ_1, and ψ_2, and the inverse process of their fusion

---

[*)]Note that eqn. (34) is embedded into the standard $ZS-AKNS$ scheme.

$$\psi_3 \rightleftarrows \psi_1 + \psi_2 \tag{2.36}$$

(with ψ_i denoting complex envelopes).

The equations have the form

$$\begin{aligned}(\partial_t + v_1\partial_x)\psi_1 &= i\bar{\psi}_2\psi_3,\\ (\partial_t + v_2\partial_x)\psi_2 &= i\bar{\psi}_1\psi_3,\\ (\partial_t + v_3\partial_x)\psi_2 &= i\psi_1\psi_2.\end{aligned} \tag{2.37}$$

If the group velocities v_i of the waves satisfy the relation

$$\frac{v_2 - v_3}{v_1 - v_3} < 0, \tag{2.38}$$

then system (31) can be investigated via the Riemann problem [72].

Let ψ_1 be the pumping wave, then instead of eqn. (31) we get

$$\begin{aligned}(\partial_t + v_1\partial_x)\psi_1 &= i\psi_2\psi_3,\\ (\partial_t + v_2\partial_x)\psi_2 &= i\psi_1\bar{\psi}_3,\\ (\partial_t + v_3\partial_x)\psi_3 &= i\psi_1\bar{\psi}_2.\end{aligned} \tag{2.39}$$

When condition (38) holds, we again have an integrable system which, differs, however, from eqn. (37) since, in the first case, the pumping wave (ψ_3) decays into two countermoving waves and, in the second case, (ψ_1) into unidirection-moving waves. A system analogous to eqn. (39) arises also under the substitution $\psi_1 \rightleftarrows \psi_2$.

Finally, the system

$$\begin{aligned}(\partial_t + v_1\partial_x)\psi_1 &= i\bar{\psi}_2\bar{\psi}_3,\\ (\partial_t + v_2\partial_x)\psi_2 &= i\bar{\psi}_1\bar{\psi}_3,\\ (\partial_t + v_3\partial_x)\psi_3 &= i\bar{\psi}_1\bar{\psi}_2,\end{aligned} \tag{2.40}$$

describes an 'explosive' instability of three waves mutually amplifying each other. As is obvious from eqn. (40) in this case all three waves seem to be identical.

It is interesting to note that within the framework of the system (37), (39) and (40) there exist solitons of three types corresponding to three kinds of wave interactions. However, only in the first of these all three types of solitons are regular solutions with respect to x and t. Furthermore, their non-trivial interaction is possible; namely, the pumping wave (ψ_3) 'composite' soliton decays into solitons of the simple waves ψ_1 and ψ_2.

3. Relativistically-invariant equations

We proceed now to discuss the list of relativistically-invariant equations which occur quite often in the literature, especially in applications.

Usually the equation studied has the form:

$$\Box\psi = F(\psi) \tag{2.41}$$

where $F(\psi)$ is a polynomial, sine series or some other function of ψ. It is of great interest to note that eqn. (41) describes an integrable system if

$$\frac{d^2}{dy^2}F(y) = F(y),$$

see, for example [73] or

$$F(y) = \exp(y)+\exp(-2y),$$

see [74].

Unlike the majority of non-relativistic systems, in addition to bell solitons, equations of the kind (41) have kink-solitons, i.e. solutions representing a transition between two different asymptotic states. The most popular models of type (41) are the following:

(1) The Sine-Gordon equation

$$\Box y = -\sin y \tag{2.42}$$

is completely integrable [75] with well-known broad physical applications. A large number of original papers and several reviews have been devoted to them. Here we refer only to two [76]. In the framework of the Sine-Gordon equation the existence of quasi-solitons in real physical systems was brilliantly confirmed a century after Scott-Russell's observations, by the self-induced transparency of the propagation of ultra-short laser pulses in a two-level system*), the 'quantization' of

*)Note the self-induced transparency (as well as the KdV solitons and the *FPU* problem) was discovered from an analysis of numerical solutions to the equations which describe

the magnetic flux in the Josephson junction, and so on.

(2) The so-called ϕ^4 field theory with

$$F(y) = \begin{cases} -y+y^3 & (\phi^4_-) \qquad (2.43) \\ y-y^3 & (\phi^4_+) \qquad (2.44) \end{cases}$$

Equations (41) with the right-hand side (43) or (44), obtained via a Lagrangian in which the potential $U(y)$ is a quartic polynomial, has been carefully investigated for both real and complex fields (see Chapter XI and XII).

A more general approach is possible when $U(y)$ is a polynomial of some degree with respect to y^2, i.e. $V(y) = P_n(y^2)$. In particular:

(3) The ϕ^6 theory

$$F(y) = \alpha y - |y|^2 y + |y|^4 y \tag{2.45}$$

(see Chapter VII)

(4) A rational function of y^2, for example,

$$F(y) = \alpha y - \frac{|y|^2}{1+|y|^2} y$$

(see Chapters IX and XV)

(5) Logarithmic function

$$F(y) = \alpha y \ln(a^D |y|^2)$$

(see Chapter XV).

A great deal of computational work on the study of solutions to the multiple (non-integrable) Sine-Gordon equations has bee performed in Manchester and Los Alamos. In particular, the *DSG* equation

$$\Box y = \mp[\sin y + \frac{\lambda}{2} \sin \frac{y}{2}] \tag{2.46}$$

arising in studies of self-induced transparency in double-degenerate systems, has been investigated in [78] (see Chapter XI).

The next series of relativistically-invariant equations emerges as models of

---

optical pulse propagation' (The review by Lamb [67]).

classical and quantum field theories. Among these is the classical c-number massive Thirring model [79]:

$$(-i\gamma^{\mu}\partial_{\mu}+m)\psi(x,t)+g^{2}\gamma^{\mu}\psi(\psi^{*}\gamma_{\mu}\psi) = 0 \tag{2.47a}$$

where the spinor field ψ is the two component column with the components ψ_1 and ψ_2 and

$$\gamma^0 = \begin{bmatrix} 0 & 1 \\ 1 & 0 \end{bmatrix}, \quad \gamma^1 = \begin{bmatrix} 0 & -1 \\ 1 & 0 \end{bmatrix}, \quad \psi^+ = (\bar{\psi}_1, \bar{\psi}_2), \quad \psi^* = \psi^+\gamma^0.$$

Equations (47a) in explicit form is

$$-i(\partial_t+\partial_x)\psi_1+m\psi_2+2g^2|\psi_2|^2\psi_1 = 0, \tag{2.47b}$$

$$-i(\partial_t+\partial_x)\psi_2+m\psi_1+2g^2|\psi_1|^2\psi_2 = 0.$$

This system, describing a self-interacting Dirac Field in two-dimensional space-time, is integrable [79] and can be solved by the following linear problem:

$$A_1 = -\frac{i}{2}(\psi^*\gamma_1\psi)\gamma^5 - \begin{bmatrix} 0 & \lambda\bar{\psi}_2-\frac{1}{\lambda}\bar{\psi}_1 \\ \lambda\psi_2-\frac{1}{\lambda}\psi_1 & 0 \end{bmatrix} + \frac{i}{2}(\lambda^2-\frac{1}{\lambda^2})\gamma^5, \tag{2.48}$$

$$A_0 = \frac{i}{2}[(\psi^*\gamma_0\psi)+\lambda^2+\frac{1}{\lambda^2}]\gamma^5 - \begin{bmatrix} 0 & \lambda\bar{\psi}_2+\frac{1}{\lambda}\bar{\psi}_1 \\ \bar{\lambda\psi_2}+\frac{1}{\lambda}\psi_1 & 0 \end{bmatrix},$$

$$\gamma^5 = \gamma^0\gamma^1.$$

Later, the equivalence of the massive Thirring model and the Sine-Gordon equation on the quantum level has been stated [80]*), thus perhaps for the first time on the formal mathematical level the connection between Bose field solitons and fermions appears, which indicates likely (as yet at the model level) the secondary nature of fermions. Later on, these ideas have been widely used to study different models of the Kaluza-Klein type and compactification of 'superfluous' spatial variables.

We also consider the $SU(2)$ sigma model defined by the Lagrangian

*)Notice that for classical models on the whole axis such an equivalence is absent [81].

$$L = \frac{1}{2}[(\partial_\mu \mathbf{q})^2 + \lambda(\mathbf{q}^2 - 1)]$$

with λ being the Lagrange factor which implements the condition $\mathbf{q}^2 = 1$. The equations of motion have the form

$$\partial_\mu^2 \mathbf{q}(x,t) + (\partial_\mu \mathbf{q})^2 \mathbf{q} = 0 \tag{2.49}$$

or in the cone variables $\tau = (t+x)/2$, $\eta = (t-x)/2$

$$\partial_{\tau\eta}^2 \mathbf{q} + (\partial_\tau \mathbf{q} \cdot \partial_\eta \mathbf{q})\mathbf{q} = 0. \tag{2.50}$$

From eqn. (44) and $\mathbf{q}^2 = 1$, it follows immediately that $(\mathbf{q}_\tau^2)_\eta = (\mathbf{q}_\eta^2)_\tau = 0$ or $|\mathbf{q}|_\tau = f_1(\tau)$, $|\mathbf{q}_\eta| = f_2(\eta)$. By using conformal symmetry $\tau \to f(\tau)$, $\eta \to g(\eta)$ of eqn. (50) one may let $f_1 = f_2 = 1$, then the vectors $\mathbf{q}$, $\mathbf{q}_\tau$, $\mathbf{q}_\eta$, lie on the S^2 sphere and form the coordinate system basis. Furthermore, since $(\mathbf{q}\mathbf{q}_\eta) = (\mathbf{q}\mathbf{q}_\tau) = 0$ the only dynamical variable is the angle between q_τ and q_η, i.e. $\cos\alpha = (\mathbf{q}_\tau \mathbf{q}_\eta)$.

As demonstrated in [82] this angle satisfies the Sine-Gordon equation

$$\alpha_{\tau\eta} + \sin\alpha = 0. \tag{2.51}$$

In order to conclude the discussion of integrable relativistically-invariant systems consider the linear problem $i\psi_s = U\psi$, $i\psi_\eta = V\psi$, with

$$U = U_0 + \frac{U_1}{\lambda+1}, \quad V = V_0 + \frac{V_1}{\lambda-1}.$$

In the gauge $U_0 = V_0 = 0$, we have the Zakharov-Mikhailov system:

$$i\psi_\zeta = \frac{A}{\lambda+1}\psi, \quad i\psi_\eta = -\frac{B}{\lambda-1}\psi,$$

$(A = U_1,\ B = -V_1)$ and the equations of motion

$$A_\eta = \frac{i}{2}[A,B], \quad B_\zeta = -\frac{i}{2}[A,B]. \tag{2.52}$$

These equations are connected with the principal chiral field. Let the function $g(\zeta,\eta)$ take values in some Lie group G (e.g., of complex matrices). Consider the Lagrangian density

$$\mathfrak{L} = -\frac{1}{2}\mathrm{Tr}(g_\zeta g^{-1} \cdot g_\eta g^{-1}), \tag{2.53}$$

which is a double-side invariant form on G, i.e. that conserving its value at both left $(g \to hg)$ and right $(g \to gh)$ shifts, $h \in G$. The Lagrangian (53) generates the equations of motion

$$g_{\zeta\eta} = \frac{1}{2}(g_\zeta g^{-1} g_\eta + g_\eta g^{-1} g_\zeta) \tag{2.54}$$

which are usually called the equations of the principal chiral field on the group G. If the fields

$$A = ig_\zeta g^{-1}, \quad B = ig_\eta g^{-1} \tag{2.55}$$

belonging to the Lie algebra of the group G are introduced, we get the equations

$$A_\eta - B_\zeta = i[A,B]. \tag{2.56}$$

From eqn. (48), it also follows that

$$A_\eta + B_\zeta = 0. \tag{2.57}$$

The system of eqns. (56) and (57) is obviously equivalent to (46). Note that the fields (55) in gauge field theories are known as purely gauge fields. In field variables, the Lagrangian (53) is written as

$$\mathfrak{L} = \frac{1}{2}\mathrm{Tr}\,AB.$$

From eqns. (46) it follows that $\partial_\eta \mathrm{Tr}\,A = \partial_\zeta \mathrm{Tr}\,B = 0$ and generally $\partial_\eta \mathrm{Tr}\,A^n = \partial_\zeta \mathrm{Tr}\,B^n = 0$ at $n \in N$. Moreover, if $G = SU(n)$, then the matrices A and B are Hermitian and may be reduced to the diagonal form A_0 and B_0 (normal Jordan Form), and therefore eqns. (46) give $\partial_\eta A_0 = \partial_\zeta B_0 = 0$ or $A_0 = A_0(\zeta)$, $B_0 = B_0(\eta)$. The general form of A and B is

$$A = uA_0u^{-1}, \quad B = \mathrm{v}B_0\mathrm{v}^{-1}$$

with the matrices u and v defined up to multiplying from the right with an arbitrary diagonal unitary matrix h. Denoting the set of such matrices as $H \ni h$, we observe u and v together with A and B defined on a homogeneous space (coset space) $SU(N)/H$, - the so-called flag space. In the variables A_0 and B_0, the Hamiltonian of the system has the form

$$H = \frac{1}{2}\int \mathrm{Tr}(A_0^2(\zeta) + B_0^2(\eta))dx$$

i.e. vanishes with an accuracy to a constant (a consequence of a 'pure gauge').

A particular example is the $SU(2)$ group. In this case, H is a one-parameter subgroup (isomorphic with $U(1)$) and the flag space ($\dim SU(2)/U(1) = \dim SU(2) - \dim U(1) = 2$) is two-dimensional. Expanding A and B in terms of the $SU(2)$ algebra bases $A = \mathbf{A}\vec{\tau}$, $B = \mathbf{B}\vec{\tau}$ and finding that $\mathrm{Tr}A^2 = \mathrm{Tr}B^2 = \mathrm{const}$ (let this be equal to 2) we have $\mathbf{A}^2 = \mathbf{B}^2 = 1$ and $(\mathbf{A}\mathbf{A}_\xi) = (\mathbf{B}\mathbf{B}_\eta) = 0$. It is also easy to show that $(\mathbf{A}\mathbf{A}_\eta) = (\mathbf{B}\mathbf{B}_\xi) = (\mathbf{A}\mathbf{B}_\xi) = (\mathbf{B}\mathbf{A}_\eta) = 0$. Therefore, the vectors $\mathbf{A}_\eta$, $\mathbf{A}_\xi$, $\mathbf{B}_\xi$ lie in some plane perpendicular to $\mathbf{A}$, and the vectors $\mathbf{A}_\eta$, $\mathbf{B}_\xi$, $\mathbf{B}_\eta$ lie in a plane $\perp\mathbf{B}$. And so $\mathbf{A}_\eta$ and $\mathbf{B}_\xi$ are located at the intersection of these planes $\mathbf{A}_\eta = -\mathbf{B}_\xi$. The remaining parameters are the angle between $\mathbf{A}$ and $\mathbf{B}$, i.e. $\cos\alpha = (\mathbf{A}\mathbf{B})$ and the scalar products $(\mathbf{A}_\xi\mathbf{B}_\xi)$ and $(\mathbf{A}_\eta\mathbf{B}_\eta)$. By introducing the second angle β via the relations [72]

$$\beta_\xi = 2\frac{(\mathbf{A}_\xi\mathbf{B}_\xi)}{\sin^2\frac{\alpha}{2}}, \quad \beta_\eta = -2\frac{(\mathbf{A}_\eta\mathbf{B}_\eta)}{\sin^2\frac{\alpha}{2}}$$

we obtain the equations for α and β

$$\alpha_{\xi\eta} + \sin\alpha = \beta_\xi\beta_\eta\cdot(\sin\frac{\alpha}{2}/2\cos^3\frac{\alpha}{2}) \tag{2.58}$$

$$\beta_{\xi\eta} + \frac{1}{\sin\alpha}(\alpha_\eta\beta_\xi + \alpha_\xi\beta_\eta) = 0.$$

Equations (58) admit a further reduction $\beta = 0$, i.e. $(\mathbf{A}_\xi\mathbf{B}_\xi) = (\mathbf{A}_\eta\mathbf{B}_\eta) = 0$ then the angle α satisfies the Sine-Gordon equation.

4. Dynamical systems given by differential-difference equations

Usually classical lattice models arise in condensed matter physics due to the 'bosonization' procedure based upon the HP transformations and coherent (or test functions) representations. This procedure was considered to some extent in the previous chapter and some further aspects will be discussed in a special chapter devoted to exact results in soliton physics.

Sometimes the lattice models manifest themselves in classical physics and other fields. Thus, for example, in some approximations, when the Langmuir wave spectral structure is studied in weakly turbulent plasmas, the system

$$\dot{y}_n = y_n(y_{n+1} - y_{n-1}), \quad -\infty < n \leqslant \infty, \quad y_n(t) > 0. \tag{2.59}$$

However, systems of this type (in particular, the Volterra system) have long been known and used to simulate dynamics of populations in biology (the feeder-eater approximation), the stimulated Compton scattering and many other induced effects.

Equation (59) is derived usually from an integrodifferential equation of the type

$$\dot{y}(x,t) = y(x,t)\int_0^\infty dx'W(x,x')y(x',t) \tag{2.60}$$

when the kernel $W(x,x') = -W(x',x)$, proportional to the transition probability density, has a resonance denominator corresponding to some decay processes. Thus, in the case of the Langmuir spectra these are $l \to l+s$, in the case of photons $t \to t+s$ or $t \to t+l$ and so on. Then the kernel $W(x,x')$ degenerates into $W(x,x') = W_0(\delta(x-x'+\kappa)-\delta(x-x'-\kappa))$ and eqn. (60) becomes eqn. (59). If decay processes are forbidden by the conservation laws $\omega_1+\omega_2 \neq \omega_3$, $k_1+k_2 = k_3$, then in the case of induced scattering of Langmuir and electromagnetic waves by plasma particles the kernel $W(x,x')$ has the form:

$$W(x,x') = \partial_x G(x-x') \equiv \frac{W_0}{\sqrt{\pi}\Delta}\partial_x \exp[-\frac{(x-x')^2}{\Delta^2}] \tag{2.61}$$

(see, e.g., Refs. [83] or [84]: in the same papers one can also find expressions for W_0, Δ and x in terms of the plasma parameters).

As shown in Ref. [85], eqn. (58) with the boundary conditions $y_n \to \text{const}$, $n \to \pm\infty$ is a continuum analog to the KdV equation. Therefore, the initial condition $y(0)$ decay proceeds via the known KdV scheme. Through numerical experiments, Montes et al. [86], succeeded to find that eqn. (60) with the kernel (61) describes a system close to the integrable one (KdV?).

One of the first 'lattice' systems was a finite-difference analog of the Boussinesq (nonlinear string) equation which, according to numerical studies, indicated the unusual features (the Fermi-Pasta-Ulam problem, 1955):

$$\ddot{y}_n = f(y_{n+1}-y_n)-f(y_n-y_{n-1}) \tag{2.62}$$

where

$$\text{(a) } f(y) = y+\alpha y^2 \quad \text{or} \quad \text{(b) } f(y) = y+\beta y^3.$$

We note that in the continuum limit Eqn. (62) for $f(y) = y + \alpha y^2 + \beta y^3$ can be reduced to KdV or Bq (see, e.g., the monograph [34] mentioned earlier).

Substituting

$$f(y_n) = \exp(y_n) \tag{2.63}$$

into eqn. (62), we have a well-known integrable model, the Toda chain. In recent years, much work has been devoted to the study of various versions of this model and the results have been described in reviews [87].

Finally, in Ref. [88] a difference variant of the KdV-Burgers equation

$$\dot{y}_n = y_n(y_{n+1} - y_{n-1}) + \mu(y_{n+1} + y_{n-1} - 2y_n)$$

has been investigated for which deviations from integrability are manifested in oscillatory tails which appear in the quasi-soliton evolution process and their interaction (Chapters X and XII on soliton stability).

References

1. a) A.M. Kosevich, B.A. Ivanov, A.S. Kovalev. *Nonlinear waves of magnetization. Dynamical and topological solitons.* Naukova Dumka, Kiev, 1983.

   b) A.S. Davydov. *Solitons in molecular systems.* Naukova Dumka, Kiev, 1984.

2. K. Pushkarov, M. Primatarova. *Solitary clusters of spin deviations and lattice deformation in an anharmonic ferromagnet chain.* Phys. Stat. Sol. 1984, 123b, p. 573-584 (and references cited therein).

3. E.K. Sklyanin. *On complete integrability of Landau-Lifshitz eq. in modern problems of the magnetization theory.* Naukova Dumka, Kiev, 1986, p. 24.

4. A.V. Makhankov, V.G. Makhankov. *Spin coherent states, Holstein-Primakoff transformations for Heisenberg spin chain and Landau-Lifshitz equation status.* JINR, P17-87-295, Dubna 1987 and Phys. Stat. Sol. 1988, 145(b), p. 699-678.

5. V. Makhankov, R. Myrzakulov, A. Makhankov. *Generalized coherent states and the continuous Heisenberg XYZ model with one-ion anisotropy.* Physica Scripta, 1987, 35, p. 233-237.

6. V. Zakharov. *Collapse of Langmuir waves.* JETP, 1972, 62, p. 1745-1759.

7. V.G. Makhankov. *Dynamics of classical solitons.* Phys. Reports, 1978, 35, p. 1-128.

8. V.G. Makhankov. *Computer experiments in soliton theory.* Comp. Phys. Comm. 1980, 21, p. 1-49.

9. N. Yadjima, M. Oikawa. *Formation and interaction of sonic Langmuir solitons.* Prog. Theor. Phys. 1976, 56, p. 1719-1739. Inverse Scat. Method.

10. D. Benney. *Significant interactions between long and short waves.* Stud. Appl. Math. 1976, 55, p. 93; *A general theory for interactions between long and short waves.* 1977, 56, p. 81.

11. C.N. Yang, C.P. Yang. *One-dimensional chain of anisotropic spin-spin interactions.* Phys. Rev. 1966, 150, p. 321.

12. A. Kundu, V.G. Makhankov, O.K. Pashaev. *On nonlinear effects in magnetic chains.* Preprint JINR E17-82-602, Dubna 1982; Physica 1984, 11D, p. 375.

13. V.G. Makhankov. *On stationary solutions of Schrödinger equation with a selfconsistent potential satisfying Boussinesq equation.* Phys. Lett. 1974, 50A, p. 42.

14. I.V. Barashenkov, V.G. Makhankov. *Soliton-like excitations in one-dimensional nuclear matter.* Preprint JINR E2-84-173, Dubna 1984.

15. V.K. Fedyanin, V. Yushankhai. *One-dimensional anisotropic magnet in long-wave-length approximation.* Phys. Lett. 1979, 70A, p. 459.

16. V.G. Makhankov, V.K. Fedyanin. *Non-linear effects in quasi-one-dimensional models of condensed matter theory.* Phys. Rep. 1984, 104, p. 1-86.

17. H.J. Mikeska. *Solitons in ferromagnet with easy plane.* J. Phys. C. 1978, 11, p. L29.

18. V.K. Fedyanin, U. Lindner. *Solitons in a one-dimensional modified Hubbard model.* Phys. Stat. Sol. 1978, 89 (b), p. 123.

19. V.K. Fedyanin, U. Lindner. *Solitary solutions in a modified Hubbard chain.* Phys. Stat. Sol. 1979, 95 (b), p. K83; V.K. Fedyanin. Teor. Mat. Fiz. 1981, 46, p. 86.

20. J. Hubbard. *Electron correlations in narrow energy bounds.* Proc. Roy. Soc. London. 1963, A276, p. 238.

21. H. Shiba. *Thermodynamics properties of the one-dimensional half-filled-band*

Hubbard model. Progr. Theor. Phys. 1972, 48, p. 2171.

22. R.A. Bari. *Classical linear-chain Hubbard model: metal-insulator transition.* Phys. Rev. 1973, 7B, p. 4318.

23 I. Egri. *Band structure of a one-dimensional Peierls-Hubbard model.* Solid State Comm. 1975, 11, p. 441-444.

24. A.A. Ovchinnikov, et al. *Theory of one-dimensional Mott semiconductors and electron structure of long moleculae with conjugated bonds.* Sov. Phys. Uspekhi, 1972, 108, p. 81-112.

25. V.G. Makhankov, N.V. Makhaldiani, O.K. Pashaev. *On the integrability and isotopic structure of the one-dimensional Hubbard model in the long wave approximation.* Preprint JINR P2-80-233, Dubna 1980; Phys. Lett. 1981, 81A, p. 161.

26. V.G. Makhankov. *Quasi-classical solutions in the Lindner-Fedyanin model: 'Hole-like' excitations.* Phys. Lett. 1981, 81A, p. 156.

27. See the book by A.S. Davydov [1b] and his review Uspekhi Fiz. Nauk 1982, 138, p. 603.

28. V.K. Fedyanin, L. Yakushevich. *On exciton-phonon interaction in one-dimensional molecular crystals. I Linear approximation.* Teor. Mat. Fiz. 1977, 30, p. 133; ibid 1978, 37, p. 371.

29. A.S. Davydov, N.I. Kislukha. Phys. Stat. Sol. 1973, 59 (b), p. 456; *Solitons in one-dimensional molecular chains.* ZETF 1976, 71, p. 1090-1098.

30. V.K. Fedyanin, V.G. Makhankov, L. Yakushevich. *Exciton-phonon interaction in long-wave approximation.* Phys. Lett. 1977, 61A, p. 256.

31. D. Pushkarov, Kh. Pushkarov. *Solitons in one-dimensional ferromagnetic systems.* Phys. Stat. Sol. 1977, 81(b), p. 703; *Solitary defections in 1-D quantum crystal.* J. Phys. C. 1977, 10, p. 3711.

32. V. Yatzishin. *Solitary magnon in linear ferromagnetic chain.* Teor. Mat. Fiz. 1976, 32, p. 127.

33. B.D. Fried, Y.H. Ichikawa. *On the nonlinear Schrödinger equation for Langmuir waves.* J. Phys. Soc. Japan 1973, 34, p. 1073;

    K. Kodama, T.J. Taniuti. *Higher order approximation in the reductive perturbation method I and II.* J. Phys. Soc. Japan 1978, 45, p. 298.

34. for discussion on the Red. Part. Theory see an excellent book by R.K. Dodd et al. *Solitons and nonlinear wave eqns.* Academic Press, London, 1982.

35a. D.H. Peregrine. *Calculation of the development of an undulare bore.* J. Fluid Mech. 1966, 25, p. 321-330, see also

b. T.B. Benjamin, J.L. Bona, J.J. Mahony. *Model equation for long waves in nonlinear dispersive systems.* Phil. Trans. Roy. Soc. Lond. 1972, 272A, p. 47-78.

36. A.N. Gaponov, M.A. Miller. JETP 1958, 7, p. 168.

37. A. Kundu, V.G. Makhankov, O.K. Pashaev. *Integrable reductions of many component magnetic systems in (1,1) dimensions.* Physica Scripta 1983, 28, p. 229-234.

38. I.L. Bogolubsky, V.G. Makhankov. *Mechanism of energy transformation at helicon soliton formation and interaction.* Z. Tek. Fiz. 1976, 46, p. 447-451.

39. K. Nishikawa, et al. *Coupled nonlinear electron-plasma and ion-acoustic waves.* Phys. Rev. Lett. 1974, 33, p. 148.

40. A. Rogister. *Parallel propagation of nonlinear l.f. waves in high-β plasma.* Phys. Fluids 1971, 14, p. 2733-2739.

41. D. Kaup, A. Newell. *An exact solution for a derivative NLS eq.* J. Math. Phys. 1978, 19, p. 798.

42. Y. Ichikawa, K. Konno, M. Wadati, H. Sanuki. *Spiky soliton in circular polarized Alfven wave.* J. Phys. Soc. Japan 1980, 48, p. 279-286; see also Y. Ichikawa, K. Konno, M. Wadati. *New integrable nonlinear evolution eqns. leading to exotic solitons. In 'long-time prediction in dynamics'.* Wiley, N.Y. 1983, p. 345.

43. V.G. Kartavenko. *Soliton-like solutions in nuclear hydrodynamics.* Yad. Fiz. 1984, 40, p. 377-388.

44. J.W. Negele. Rev. Mod. Phys. 1982, 54, p. 913; see also S. Devi, M. Strayer, J. Irvine. *TDHF calculations for nuclear collisions: model studied of $\alpha-\alpha$ scattering.* J. Phys. 1979, G5, p. 281.

45. G. Holzwarth. *Static and dynamic Thomas-Fermi theory for nuclei.* Phys. Lett. 1977, 66B, p. 29.

46. R.V. Djolos, V.G. Kartavenko, V.P. Permyakov. *Nuclear hydrodynamics and collective density oscillations.* Yader. Fiz. 1981, 34, p. 144.

47. M. Girardeau. *Relationship between systems of impenetrable Bosons and Fermions in one-dimension.* J. Math. Phys. 1960, 1, p. 516.

E.H. Lieb, W. Liniger. *Exact analysis of an interacting Bose gas. I. The general solution and the ground state.* Phys. Rev. 1963, 130, p. 1605-1616.

48. P.P. Kulish, S.V. Manakov, L.D. Faddeev. *Comparison of the exact quantum and quasi-classical results for NLS.* Teor. Math. Fiz. 1976, 28, p. 38-45.

M. Ishikawa, H. Takayama. *Solitons in one-dimensional Bose system with repulsive delta-function interaction.* J. Phys. Soc. Japan 1980, 49, 1242.

49. V.K. Fedyanin, V.Yu. Yushankhai. *Possibility of observing the soliton like mode in* $[(CH_3)_4N][NiCl_3]$ *via neutron scattering experiments.* Phys. Lett. 1981, 85A, p. 100.

50. L.J. de Jongh, A.R. Miedema. *Experiments on simple magnetic model systems.* Adv. in Physics 1974, 23, p. 1-260.

51. M. Ito. *Transition temperature of layered system of isotopic spin with* $n \geqslant 3$ *components.* Progr. Theor. Phys. 1981, 65, p. 1773-1786.

52. E.L. Daves, J.H. Marburger. *Computer studies in self-focusing.* Phys. Rev. 1969, 179, p. 862-868.

53a. Ya.B. Zeldovich, Yu.P. Raizer. *Self-focusing of light. The Kerr-effect role and strictions.* JETP Lett. 1966, 3, p. 137-141.

b. N.G. Vakhitov, A.A. Kolokolov. *Stationary solutions to wave eq. with saturable nonlinearity.* Izvestiya Vuzov. Radiofizika. 1973, 16, p. 1020-1028.

54. E.S. Benilov, S.P. Burtzev. *To the integrability of the eqs. describing the Langmuir-wave - ion-acoustic-wave interaction.* Phys. Lett. 1983, 98A, p. 256-258.

55. A. Rogister. See [40].

56. Y.H. Ichikawa, K. Konno, M. Wadati, H. Sanuki. *Spiky soliton in circular polarized Alfven wave.* J. Phys. Soc. Japan. 1980, 48, p. 279-286.

57. D.J. Kaup, A.C. Newel. *An exact solution for a derivative nonlinear equation.* J. Math. Phys. 1987, 19, p. 798-801.

58. M. Wadati, K. Konno, Y.H. Ichikawa. *A generalization of inverse scattering method.* J. Phys. Soc. Japan. 1979, 46, p. 1965-1966; *New integrable nonlinear evolution eqs.* ibid, 1979, 47, p. 1698-1700.

59. A.V. Mikhailov, A.B. Shabat. *Integrability conditions for the system of two eqs. of the form* $u_t = A(u)u_{xx} + F(u, u_x)$. Teor. Mat. Fiz. 1985, 62, p. 163-185.

60. A.I. Akhiezer, V.G. Baryakhtar, S.V. Peletminskii. *Spin waves.* Nauka, Moscow, 1967 (translated by North-Holland, 1968).

61. D.H. Peregrin. See [35a].

62. T.B. Benjamin, J.L. Bona, J.J. Mahony. See [35b].

63. T.B. Benjamin. *Internal waves of finite amplitude and permanent form.* J. Fluid Mech. 1966, 25, p. 241-270. ibid, *Internal waves of permanent form in fluids of great depth.* 29, p. 559.

64. H. Ono. *Algebraic solitary waves in stratified fluids.* J. Phys. Soc. Japan. 1975, 39, p. 1082-1091.

65. J. Meiss, N. Pereira. *Internal wave solitons.* Phys. Fluids, 1978, 21, p. 700-702.

66. R.I. Joseph. *Multi-soliton-like solutions to the Benjamin-Ono eq.* J. Math. Phys. 1977, 18, p. 2251-2258.

67. M.J. Ablowitz, Y. Kodama, J. Satsuma. *On an internal wave eq. describing a stratified fluid with finite depth.* Phys. Lett. 1979, 73A, p. 283-286.

Y. Kodama, J. Satsuma, M.J. Ablowitz. *Nonlinear intermediate long-wave equation: Analysis and method of solution.* Phys. Rev. Lett. 1981, 46, p. 687-690.

68. Y.H. Ichikawa, K. Konno, M. Wadati. *Nonlinear transverse oscillations of elastic beams under tension.* J. Phys. Soc. Japan. 1981, 50, p. 1799-1802.

69a. B. Fornberg, G.B. Whitham. *A numerical and theoretical study of certain nonlinear wave phenomena.* Philos. Trans. Roy. Soc. L. 1978, 289, p. 373-404.

b. C. Karney, A. Sen, F. Chu. Preprint Plasma Phys. Lab. PPL-1455, N.Y. Princeton Univ. 1978. *The complex MkdVeq. a non-integrable evolution eq. in 'Solitons and cond. math. phys.'.* eds. A. Bishop and T. Schneider, Springer, N.Y. 1978, p. 71-75.

70. R. Hirota. *Exact solution of the modified Korteweg-de Vries eq. for multiple collisions of solitons.* J. Phys. Soc. Japan. 1972, 33, p. 1456-1458.

71. C. Karney, A. Sen, F. Chu. Preprint Plasma Phys. Lab. PPL-1452, N.Y. Princeton Univ. 1978.

72 *Theory of solitons. Inverse scattering method.* S.P. Novkov ed. Nauka, Moscow, 1980. a) p. 209-222, b) p. 234.

73. D.W. McLaughlin, A.C. Scott. *A restricted Bäcklund transformation.* J. Math. Phys. 1973, 14, p. 1877.

74. A.V. Mikhailov. *On integrability of* $D = 2$ *generalized Toda lattice.* JETP Lett. 1979, 30, p. 443-448.

75. M.J. Ablowitz, D.J. Kaup, A.C. Newel, H. Segur. *Method for solving the sine-Gordon eq.* Phys. Rev. Lett. 1973, 30, p. 1262-1264.

A.A. Takhtajan, L.D. Faddeev. *An essentially nonlinear one-dimensional model of classical field thoery.* Teor. Mat. Fiz. 1974, 21, p. 160-174.

76. A. Barone, F. Esposito, C.J. Magee, A.C. Scott. *Theory and applications of the SG eq.* Riv. Nuovo Cimento, 1971, 1, p. 227-267.

P.J. Caudrey, J.C. Eilbeck, J.D. Gibbon. *The Sine-Gordon eq. as a model classical field theory.* Nuovo Cimento. 1975, 25B, p. 497-512.

77. G.L. Lamb, Jr. *Analytical descriptions of ultrashort optical pulse propagation in a resonant medium.* Rev. Mod. Phys. 1971, 43, p. 99-124.

78. P. Dodd, R. Bullough, S. Duckworth. *Multisoliton solutions of nonlinear dispersive wave eqs. not solvable by the inverse method.* J. Phys. Math. Gen. 1975, 8A, p. L64.

P. Dodd, R. Bullough. Proc. Roy. Soc. L. 1976, 351A, p. 499; ibid. *Polynomial conserved densities for the SG eq.* 352, p. 481-503.

79. A.V. Mikhailov. *Complete integrability of the two-dimensional Thirring model.* JETP Lett. 1976, 23, p. 356.

80. S. Coleman. *Quantum SG eq. as the massive Thirring model.* Phys. Rev. 1975, 11D, p. 2088-2097.

S. Mandelstam. *Soliton operators for the quantized SG eq.* ibid, p. 3026-3030.

81. D.J. Kaup, A.C. Newel. *On the Coleman correspondence and the soliton of the MTM.* Lett. Nuovo Cimento. 1977, 20, p. 325-332.

82. K. Pohlmeyer. *Integrable Hamiltonian systems and interactions through quadratic constraints.* Comm. Math. Phys. 1976, 46, p. 707-718.

83. R. Sagdeev, A. Galeev. *Nonlinear Plasma Theory.* N.Y. Benjamin, 1969.

84. C. Montes. *Plasma Physics Nonlinear Theory and Experiments.* N.Y. Plenum,

1977, also Astroph. Jour. 1977, 216, p. 329.

85. S.V. Manakov. *On complete integrability and stochastization of discrete dynamical systems.* JETP, 1979, 67, p. 543-555.

86. C. Montes, J. Peyraud, M. Henon. *One-dimensional boson soliton collisions.* Phys. Fluids, 1979, 22, p. 176.

87. R.K. Dodd, see [34].

88. J. Fernandes, G. Reinisch. *Collapse of a Volterra soliton into a weak monotone shock wave.* Physica, 1978, 91A, p. 393.

89. I.M. Krichever. *Spectral theory of 'finite-zone' time-dependent Schrödinger operators.* Functional analysis and applications, 1986, 20, p. 42-54.

Part II. Some Exact Results in One-Dimensional Space

The theory of integrable systems has been considered in a large number of reviews and monographs. Therefore, we only consider below certain aspects of this theory. Since there is some controversy in defining integrable systems (as well as the term soliton) we apply the following:

(1) An integrable system is a system admitting the Lax representation (or zero curvature) for which one can obtain a countable set of integrals of motion and apply the inverse scattering method, the Riemann problem, the $\bar{\partial}$ problem, or the finite-zone integration method to study its dynamics, (c-integrability). As integrable systems are naturally regarded the systems which can be linearized through some substitution, (s-integrability).

(2) Following Faddeev and Zakharov, an integrable Hamiltonian system for which one can find the action-angle variables with which the Hamiltonian can be expressed, we call the complete (or completely) integrable system.

As has already been mentioned, an extensive review literature is now available (albeit, of a mathematical nature) [1] discussing, at sufficient length, studies of a broad class of integrable systems. Review papers are also devoted to some physical applications of integrable systems [2]. In this chapter, therefore, when we consider systems which are already fully discussed in the review literature, we do not go into details but rather restruct ourselves to giving only information necessary for further discussion. Results which are not sufficiently described in the literature are given later in much more detail. This is especially true for integrable systems with noncompact internal symmetry groups and their applications to the theory of a weakly non-ideal Bose gas.

Concluding this short introduction, let us illustrate using a nonlinear Schrödinger equation some properties of integrable systems which are important for what follows:

$$i\psi_t + \psi_{xx} + 2g|\psi|^2\psi = 0. \tag{3.1}$$

Equations integrable via the inverse scattering method possess the following general properties:

(1) As a rule, they describe Hamiltonian systems, i.e., systems whose behaviour is determined by a set of canonically conjugate variables. For these variables one can determine the Poisson bracket and express the equation in Hamiltonian form. In case (1), we have the variables ψ and $\bar{\psi}$ with the Poisson bracket

$$\{\psi(x),\bar{\psi}(y)\} = i\delta(x-y) \tag{3.2}$$

where

$$\{A,B\} = i\int_{-\infty}^{\infty} dx\left[\frac{\delta A}{\delta\psi}\frac{\delta B}{\delta\bar{\psi}}-\frac{\delta B}{\delta\psi}\frac{\delta A}{\delta\bar{\psi}}\right] \tag{3.2a}$$

the Hamiltonian

$$H = \int_{-\infty}^{\infty} dx[(\bar{\psi}_x\psi_x)-g(\bar{\psi}\psi)^2] \tag{3.3}$$

and the Hamiltonian equations of motion

$$i\psi_t = i\{H,\psi\} = \frac{\delta H}{\delta\bar{\psi}},\quad i\bar{\psi}_t = i\{H,\bar{\psi}\} = -\frac{\delta H}{\delta\psi} \tag{3.4}$$

(2) The equation of motion can be written either in the Lax form or in the form of a compatibility condition for certain overdetermined linear spectral problem. In the first case, a pair of operators (the Lax L,A pair) exists so that the operator equation

$$iL_t = [L,A] \tag{3.5a}$$

is equivalent to the original nonlinear equation. The operator L spectrum is time independent, and its eigenfunction satisfies the equation

$$iy_t = Ay+f(L)y. \tag{3.5b}$$

The functional $f(L)$ may be chosen as a matter of convenience (it also means that the (L,A) representation is not unique). For instance, for (1) we have

$$L = i\begin{bmatrix} 1+s & 0 \\ 0 & 1-s \end{bmatrix}\partial_x+\begin{bmatrix} 0 & \bar{\psi} \\ \psi & 0 \end{bmatrix}$$

$$A = -s\begin{bmatrix}1 & 0\\ 0 & 1\end{bmatrix}\partial_x^2 + \begin{bmatrix}|\psi|^2(1+s)^{-1} & +i\psi_x\\ -i\bar{\psi}_x & -|\psi|^2(1-s)^{-1}\end{bmatrix}$$

i.e., L and A are matrices of the second order, $g = (1-s^2)^{-1}$.

In the second case, let us write the following overdetermined system of linear matrix equations

$$y_x = A_1(x,t,\lambda)y, \quad y_t = A_0(x,t,\lambda)y. \tag{3.6}$$

Again, A_0 and A_1 are (2×2) matrices, and y are two-component columns. The compatibility condition of this system is obtained by differentiating the first equation of (6) with respect to t and the second one with respect to x and subtracting one from the other*)

$$A_{1t} - A_{0x} + [A_1, A_0] = 0. \tag{3.7}$$

It should be emphasized that the operators A_0 and A_1 depend not only on x and t, but also on some parameter, say λ, which is called the spectral parameter. Condition (7) must be satisfied indentically in λ. In the case of the nonlinear Schrödinger equation (1), the operators A_0 and A_1 have the form

$$A_1 = -i\lambda\sigma_3, \quad p = i\begin{bmatrix}0 & \bar{\psi}\\ \psi & 0\end{bmatrix}, \quad \begin{bmatrix}1 & 0\\ 0 & -1\end{bmatrix} \tag{3.8}$$

$$A_0 = B - 2\lambda P + 2i\lambda^2\sigma_3, \quad B = \begin{bmatrix}-i|\psi|^2 & \bar{\psi}_x\\ -i\psi_x & i|\psi|^2\end{bmatrix}.$$

The presence of a certain continuous time-independent parameter is a reflection of the fact that eqn. (1) describes a Hamiltonian system with an infinite number of degrees of freedom having an infinite set of conservation laws. In the case of systems with a finite number (N) of degrees of freedom one sometimes succeeds in finding N first integrals of motion†) with zero Poisson brackets (i.e. they are involutive). Such systems are called completely integrable. In fact, in this case a solution to the original equations can be in principle expressed by quadratures since, via

*) This equation is also known in literature as the zero curvature condition (detailed information is presented in the following).

†) It should be noted that due to an incorrectness in the review [2(d)] instead of N integrals $2N$ integrals appeared.

an appropriate canonical transformation, one may pass to action-angle variables. The angular variables are cyclic and the action variables then correspond to the integrals.

This concept has been generalized by Faddeev and Zakharov [3] for the field case. This means that:

(3) The systems described by the above equations possess an infinite countable set (or in the presence of the internal isotopic symmetry, a series of sets) of local (or nonlocal) conservation laws and integrals of motion. Usually the conservation laws can be written in the form

$$\dot{\rho}_n + \partial_x j_n = n, \quad n = 1,2,... \tag{3.9}$$

where the functionals ρ_n and j_n are polynomials in the field function and its spatial derivatives.

Integrating eqn. (9) over x we find the integrals of motion $I_n = \int_{-\infty}^{\infty} \rho_n(x,t)dx$ with $j_n \to 0$ when $|x| \to \infty$. If they are involutive and one can determine angular variables canonically conjugated to them, the corresponding system is called completely integrable (and the $I_n s$ play a role of the action variables).

(4) In some cases, one can solve the inverse scattering problem for the operator $L = i\sigma_3(\partial_x + i\lambda\sigma_3 - A_1)$, i.e. via scattering data find a potential whose role is played by the required function ψ. This means that for the equations discussed one can (in any case in principle) solve the Cauchy problem and the behaviour of the integrable systems is strictly determinate.

(5) The localized solutions (if they exist) to the integrable equations which correspond to the discrete spectrum of the operator L are usually called solitons. In the case of the simplest integrable systems, without internal symmetries, the soliton dynamics is trivial - they interact elastically (the first test of the integrability).

(6) For integrable systems, one can develop an exact quantum approach (usually connected with the Bethe Ansatz in some way or other) which enables one to find the ground state and excitation spectra of the system (see Faddeev et al. [14] and Thacker [5]), thereby connecting, with sufficient accuracy, classical description with quantum objects.

(7) Integrable systems may also be studied by the Riemann problem [1a], the finite-zone integration [6], the Darboux transformations [7], as well as by various

group-theoretical and algebraic-geometrical techniques [6,8] (see Chapter VIII and, for a more recent review, [17]).

In the following two chapters we consider the three equations most frequently occurring in applications: the nonlinear Schrödinger equation, the Landau-Lifshitz equation and the Sine-Gordon equation.

Chapter 3.

THE NONLINEAR SCHRÖDINGER EQUATION AND THE LANDAU-LIFSHITZ EQUATION

The following sections are devoted to the discussion of one of the most important properties of integrable systems, their gauge equivalence. As an example, we discuss two systems most frequently arising in condensed matter physics: the nonlinear Schrödinger equation and the Landau-Lifshitz equation. Let us establish the connection between them and show, via gauge symmetry (equivalence), one can generate solutions in the frame of a given system or find solutions to a given equation by the use of solutions of a different equation. We will show that a broad range of known symmetry transformations such as the Galilei, Bäcklund, etc., transformations are embedded into the gauge ones [12]. Writing the linear spectral problem in the form of the (A_0,A_1) pair allows a formulation in the language of gauge field theory and to apply the powerful apparatus of this theory. In particular, using algebraic-geometric concepts such as fibre spaces, and homogeneous and symmetric spaces, it is possible to give a consistent algebraic-geometric interpretation of the relations under consideration and even in 'geometrizing' the interaction.

At this point, it is useful to cite works which have made a dominant contribution to the topics being considered. Apparently, the first step in this direction was taken in the 1976 paper of Lakshmanan et al. [18], although at attempt was made there to determine the connection between the isotropic Landau-Lifshitz equation and the Korteweg-de Vries equation. Further discussion of the Heisenberg $SU(2)$ continuum XXX model and its connection to the nonlinear Schrödinger equation has been given by Lakshmanan [19], Corones [20], Takhtajan [21], and Borovik [24]. An equivalence of the $SU(2)$ isotropic Landau-Lifshitz equation and the nonlinear Schrödinger equation in terms of a linear spectral problem was established in the 1979 paper of Zakharov and Takhtajan [22]. A generalization of these results for higher compact symmetry groups was obtained by Honerkmap [100] and Orfanidis [12,13], and for noncompact groups by Pashaev et al. [9,23]. Finally, in the most general formulation in terms of the algebraic theory of gauge fields the problem has been considered by Fordy and Kulish [11] (see also [9] and [17]).

It should also be noted that for an understanding of this section one requires a knowledge of the elements of group theory, Lie algebras, and fibre space theory. We also assume a superficial knowledge of the inverse scattering method (the spectral transformation method). For subsequent chapters, except perhaps Chapter 5, this material is not necessary and may be omitted at a first reading. Nevertheless, some elementary concepts of the topics mentioned above are presented in the text as far as this is needed.

1. Nonlinear Schrödinger equation associated with a symmetric space

We formulate a nonlinear Schrödinger equation for a variable being an element of a tangent space of the manifold M which is considered to be a symmetric space [9,10].

More general manifolds are also feasible, i.e. reductive homogeneous spaces [11], and corresponding versions of the nonlinear Schrödinger equation. However, their physical meaning is not yet clear, and we confine ourselves to the equations obtained in Chapter 1.

Let us recall briefly the notion of a symmetric space. Consider the connected Lie group G with the algebra L. Let the algebra L be Z_2 graded (mod 2). This means that

$$L = L^0 \dot{+} L^1 \tag{3.10}$$

(the sign $\dot{+}$ denotes the direct sum of vector spaces) or $l = l^0 + l^1$ where $l \in L$, $l^0 \in L^0$, $l^1 \in L^1$

$$[L^0,L^0] \subset L^0, \ [L^0,L^1] \subset L^1, \ [L^1,L^1] \subset L^0$$

i.e. $[L^i,L^j] \subset L^{i+j (\mathrm{mod}\, 2)}$.

Here L^0 is the Lie algebra of the subgroup $H \subset G$. Then the coset space G/H is called a symmetric homogeneous space.

For simplicitly, we use a matrix representation corresponding to the algebra L of the group G. As G we choose the group $GL(N,C)$ with the algebra $gl(N,C)$.

Suppose that elements $l \in gl(N,C)$ are matrix functions of x and t, they define the symmetrical space M. The local geometry at a point $x(M \ni g(x))$ by virtue of the homogeneity is determined by the metric of the tangent space $\mathbb{R}^k_x$ at the point x $(k = \dim M)$. The latter space is identified with the space $L^1 \subset L$. The decomposition (10) $l = l^0 + 1^1$ has the form

$$l^0 = \begin{bmatrix} M_m & 0 \\ 0 & M_n \end{bmatrix} \in gl(m,c) \oplus gl(n,c), \quad l^1 = i \begin{bmatrix} 0 & R \\ Q & 0 \end{bmatrix} \equiv i\Psi \in L^1 \tag{3.11}$$

$(m+n = N)$. The internal degrees of freedom corresponding to the subgroups $GL(m,C)$ and $GL(n,C)$ are called 'flavour' and 'colour', respectively.

The generalized nonlinear Schrödinger equation for the element $\Psi(x) \in L^1$ has the form

$$-i\Gamma_{mn}\Psi_t + \Psi_{xx} + 2\Psi^3 = 0$$

where

$$\Gamma_{mn} = \begin{bmatrix} e_m & 0 \\ 0 & -e_n \end{bmatrix}, \quad e_m = \operatorname{diag} \underbrace{(1,\ldots,1)}_{m} . \tag{3.13}$$

System (12) is integrable. The corresponding spectral problem is given by the purely gauge potentials

$$A_1(x,t,\lambda) = \phi_x\phi^{-1}, \tag{3.14a}$$

$$A_0(x,t,\lambda) = \phi_t\phi^{-1}, \tag{3.14b}$$

valued in the Lie algebra of the gauge group G and having the form

$$A_1 \equiv U = -i\lambda\Sigma + i\Psi, \tag{3.15a}$$

$$A_0 \equiv V = \frac{m+n}{mn} i\lambda^2\Sigma - \frac{m+n}{mn} i\lambda\Psi - \Gamma_{mn}(i\Psi^2 - \Psi_x), \tag{3.15b}$$

where

$$\Sigma = \begin{bmatrix} 1/m\, e_m & 0 \\ 0 & -1/n\, e_n \end{bmatrix}. \tag{3.15c}$$

The fact that the connection curvature tensor

$$F_{\mu\nu} = \partial_\mu A_\nu - \partial_\nu A_\mu + [A_\nu, A_\mu]$$

vanishes, leads to the equation

$$U_t - V_x + [U,V] = 0 \tag{3.16}$$

which is equivalent to (12).

System (12) is Hamiltonian. Canonical variables may be obtained from the reducibility of the graded algebra (10). Its 'bosonic' sector, determined by L^0, is reduced by eqn. (11) and the 'fermionic' one splits into two components

$$L^1 = L^1_{(+)} \oplus L^1_{(-)}, \tag{3.17}$$

$$l^1 = i\Psi = i\Psi_{(+)} + i\Psi_{(-)} = i\left(\begin{smallmatrix}0 & 0\\ Q & 0\end{smallmatrix}\right) + i\left(\begin{smallmatrix}0 & R\\ 0 & 0\end{smallmatrix}\right),$$

which are nilpotent (as Grassmannian variables):

$$\Psi_{(+)}\Psi_{(+)} = \Psi_{(-)}\Psi_{(-)} = 0, \quad \text{but } \Psi_{(+)}\Psi_{(-)} \in L^0. \tag{3.18}$$

In these variables, eqn. (12) becomes

$$i\dot{\Psi}_{(+)} + \Psi_{(+)_{xx}} + 2\{\Psi_{(+)}, \Psi_{(-)}\}_+ \Psi_{(+)} = 0, \tag{3.19}$$

$$-i\dot{\Psi}_{(-)} + \Psi_{(-)_{xx}} + 2\{\Psi_{(+)}, \Psi_{(-)}\}_+ \Psi_{(-)} = 0,$$

where $\{\Psi_{(+)}, \Psi_{(-)}\}_+$ is the anticommutator, and $\Psi_{(+)}$ and $\Psi_{(-)}$ are like the Fermi operators.

The more conventional form of eqn. (12) in terms of the variables Q and R is

$$iQ_t + Q_{xx} + 2QRQ = 0, \tag{3.20}$$

$$-iR_t + R_{xx} + 2RQR = 0.$$

These variables as well as $\Psi_{(+)}$ and $\Psi_{(-)}$, are canonically conjugate

$$\{Q_{\alpha\beta}(x), R_{\gamma\delta}(y)\} = i\delta(x-y)\delta_{\alpha\delta}\delta_{\beta\gamma} \tag{3.21}$$

with respect to the Poisson bracket

$$\{A,B\} = iTr\int dx \left[\frac{\delta A}{\delta Q}\frac{\delta B}{\delta R} - \frac{\delta B}{\delta Q}\frac{\delta A}{\delta R}\right]$$

$$= i\int dx \left[\frac{\delta A}{\delta Q_{\alpha\beta}(x)}\frac{\delta B}{\delta R_{\beta\alpha}(x)} - \frac{\delta B}{\delta Q_{\alpha\beta}(x)}\frac{\delta A}{\delta R_{\beta\alpha}(x)}\right].$$

The Hamiltonian equations

$$i\dot{Q}_{\alpha\beta}(x) = i\{H,Q_{\alpha\beta}(x)\} = \frac{\delta H}{\delta R_{\alpha\beta}}, \quad i\dot{R}_{\beta\alpha}(x) = i\{H,R_{\beta\alpha}(x)\} = -\frac{\delta H}{\delta Q_{\alpha\beta}}$$

and the Hamiltonian

$$H = Tr\int dx(R_x Q_x - (RQ)^2) \tag{3.22}$$

give the system (20). Generally speaking, the Hamiltonian (22) is not Hermitian. Its Hermitian reduction is obtained under the standard condition

$$R = uQ^+ v \equiv \overset{\star}{Q} \tag{3.23}$$

with u and v being arbitrary Hermitian matrices of dimensions $m \times m$ and $n \times n$, respectively.

The linear problem (15) has m-fold degenerate eigenvalues (λ) and n-fold degenerate eigenvalues $(-\lambda)$ which is connected with the internal symmetry of system (12) and (20). In fact, $\Psi(x) \in L^1$ is an element of the tangent space $\mathbb{R}^k_x$ at the point x. The metric of $\mathbb{R}^k_x = L^1$ is invariant under the inner automorphisms $\Psi \to \Psi' = h\Psi h^{-1}$, where $h \in H = GL(m,C)\otimes GL(n,C)$. For $h_T = \exp(TB)$, $B \in L^0$, we have a transformation which maps the tangent space into itself $L^1 \to L^1$:

$$\Psi \to \Psi_T = h_T \Psi h_T^{-1},$$

$$ad\,B(\Psi) = [B,\Psi] = \left.\frac{d\Psi_T}{dT}\right|_{T=0}.$$

The scalar product $<\Psi,\Psi>$ on L^1 satisfies the relations

$$<\Psi_T,\Psi_T'> = <\Psi,\Psi'>$$

$$<[B,\Psi],\Psi'> + <\Psi,[B,\Psi']> = 0$$

where

$$<A,B> = Tr(ad\,A, ad\,B)$$

is the Killing form.

The metric on the tangent space in the case of eqn. (15) is given by the diagonal matrix $\Sigma\,(Tr\,\Sigma = 0)$.

The internal symmetry group (isogroup) H acts on the elements of L^1 and correspondingly on Q and R in a linear way. The transformation matrix

$$h = \begin{bmatrix} h_1 & 0 \\ 0 & h_2 \end{bmatrix} \in H = GL(m,c) \otimes GL(n,c), \tag{3.24}$$

where $h_1 \in GL(m,C)$ and $h_2 \in GL(n,C)$, commutes with $\Sigma: [h,\Sigma] = 0$ and the field variables are transformed as

$$R' = h_1 R h_2^{-1}, \quad Q' = h_2 Q h_1^{-1}.$$

In order to obtain the Noether conservation laws let us multiply the first equation of (20) by R from the left or the right, and the second equation of (20) by Q from the right or the left, respectively. Subtracting one from the other, we obtain the conserved Noether currents, $\partial_\mu J_\mu = 0$:

$$\begin{cases} J_0^{(n)} = RQ \\ J_1^{(n)} = i(R_x Q - RQ_x) \end{cases} \qquad \begin{cases} J_0^{(m)} = QR \\ J_1^{(m)} = i(Q_x R - QR_x) \end{cases}$$

and the integrals of motion

$$S_{\alpha\beta}^{(n)} = \int dx R_{\alpha\gamma}(x) Q_{\gamma\beta}(x) \tag{3.25a}$$

$$S_{\alpha\beta}^{(m)} = \int dx Q_{\alpha\gamma}(x) R_{\gamma\beta}(x) \tag{3.25b}$$

or, in the terms of the variable $\Psi \in L^1$,

$$S = \frac{1}{2} \int dx \{\Psi, \Psi\}_+ .$$

The integrals S commute with the Hamiltonian (in the sense of the Poisson bracket)

$$\{H, S_{\alpha\beta}^{(n)}\} = \{H, S_{\alpha\beta}^{(m)}\} = 0$$

and satisfy the commutation relations of the Lie algebra $gl(m,C) \oplus gl(n,C)$:

$$\{S_{\alpha\beta}, S_{\gamma\delta}\} = i(\delta_{\beta\gamma}\delta_{\alpha\beta} - \delta_{\alpha\gamma}\delta_{\beta\delta}). \tag{3.26}$$

The integrals (25a) correspond to the colourless states of a system with a definite flavour (the number of flavours is m) and (25b) corresponds to flavourless states with a definite colour. Notice that the states having a definite flavour and colour are not conserved and in the system there must be a nontrivial 'flavour' - and 'chromo' - dynamics of solitons.

Consider the Hermitian reduction (23). In this case, the integrals (25) satisfy the conjugation conditions

$$S^{(n)} = uS^{(n)+}u^{-1}, \quad S^{(m)} = \mathrm{v}^{-1}S^{(m)+}\mathrm{v}.$$

By diagonalization of the matrices u and v (this is done because of their Hermiticity) and the redefinition of $S^{(n)}$ and $S^{(m)}$ one obtains

$$\Gamma_{pq}S^{(n)+}\Gamma_{p,q} = S^{(n)}, \quad (p+q=n), \tag{3.27}$$

$$\Gamma_{ls}S^{(m)+}\Gamma_{ls} = S^{(m)}, \quad (l+s=m).$$

The matrix Γ_{mn} is given by eqn. (13).

The conditions of commutation (26) and symmetry (27) imply that the symmetry group generated by the charges $S^{(n)}$ and $S^{(m)}$ is $U(l,s)\otimes U(p,q)$. The Hermitian reduction (23) and the Hamiltonian are now

$$R = \Gamma_{ls}Q^{+}\Gamma_{pq} = \overset{*}{Q},$$

$$H = Tr\int dx[\overset{*}{Q}_x Q_x - (\overset{*}{Q}Q)^2].$$

We will now examine the geometrical meaning of the construction obtained. This is seen by making use of the gauge freedom concept for the spectral problem and correspondingly the gauge equivalence of integrable systems.

2. The sigma model representation of the nonlinear Schrödinger equation and the isotropic Landau-Lifshitz equation

The gauge equivalence concept of nonlinear evolution equations is based on their representation in the form of the zero curvature condition (the $U-V$ representation). Such a representation enables the study of nonlinear evolution equations and the theory of gauge fields and fibre spaces to be directly linked together.

Consider two linear problems (let $\Phi^{(i)} \in GL(N,C)$; $A_\mu \in gl(N,C)$ and $i = 1,2$; $\mu = 0,1$)

$$\partial_\mu \Phi^{(1)} = A_\mu^{(1)}(x,t,\lambda)\Phi^{(1)},$$

$$\partial_\mu \Phi^{(2)} = A_\mu^{(2)}(x,t,\lambda)\Phi^{(2)}.$$

These correspond to two nonlinear evolution equations

$$F^{(i)}_{\mu,\nu} = \partial_\nu A^{(i)}_\mu - \partial_\mu A^{(i)}_\nu + [A^{(i)}_\mu, A^{(i)}_\nu] = 0, \quad (i=1,2).$$

DEFINITION. Two nonlinear evolution equations are called gauge equivalent if the corresponding flat connections (potentials) $A^{(i)}_\mu$ are defined on a one-bundle and obtained one from the other via the gauge transformation, i.e.

$$A^{(1)}_\mu = gA^{(2)}_\mu g^{-1} + g_\mu g^{-1}, \quad g(x,t) \in GL(N,C), \tag{3.28}$$

with $\Phi^{(1)} = g\Phi^{(2)}$.

A whole class of symmetry transformations of integrable nonlinear evolution equations falls within the gauge transformation framework. The Bäcklund transformations, isotopic rotations, scale space-time (Galilean or Lorentz) transformations are realized as gauge transformations which correspond to the left group action [9,12]. Pohlmeyer's R transformation, related to the sigma model integrability, corresponds to the right group action [12].

Consider first the simplest gauge transformations: the isogroup transformations for a nonlinear Schrödinger equation. The global transformations (not affecting the bundle base) $h \in H$: $H = GL(m,C) \otimes GL(n,C)$ commute with Σ, $[h,\Sigma] = 0$. Due to this fact, the potentials and the equations of motion are transformed in a covariant way

$$A^{(1)}_\mu[\Psi] \to A^{(2)}_\mu[\Psi] = A^{(1)}_\mu[\Psi'] = A^{(1)}_\mu[h\Psi h^{-1}],$$

$$F^{(1)}_{\mu\nu}[\Psi] = 0 \to F^{(2)}_{\mu\nu}[\Psi] = F^{(1)}_{\mu\nu}[\Psi'] = F^{(1)}_{\mu\nu}[h\Psi h^{-1}] = 0,$$

(here $\Psi \in L^1$ is the sought function in the nonlinear Schrödinger equation), and

$$\Phi^{(1)} = h\Phi^{(2)}(x,t).$$

The Hamiltonian remains unchanged under such transformations so that $H = H'$, and therefore they allow (similar to the auto-Bäcklund transformations) a number of new solutions to be generated in the framework of the same nonlinear evolution equations.

More general gauge transformations correspond to the space-time ones

$$\Phi^{(1)}(x',t') = g(x,t)\Phi^{(2)}(x,t). \tag{3.29}$$

Consider these first via an example of the Galilei transformations, $x \to x' + vt$, $t \to t' = t$, in the framework of the nonlinear Schrödinger equation (12). In this case, the relations (28) and (29) are a local gauge transformation of the maximal Abelian subgroup, $U(1)^m \otimes U(1)^n \subset GL(m+n,C)$:

$$g(x,t) = \exp\left\{\frac{i}{2}\frac{mn}{m+n}\Sigma\theta(x,t)\right\}, \tag{3.30}$$

$$g(x,t) = \frac{v^2}{2}t - vx, \quad \Sigma = \begin{bmatrix} \frac{1}{m}e_m & 0 \\ 0 & -\frac{1}{n}e_n \end{bmatrix}.$$

The potentials $A_\mu^{(2)}$, eqns. (14) and (15) are then transformed into

$$A_1^{(1)} = i\Sigma(-\lambda + \frac{mn}{m+n}\frac{\theta x}{2}) + i\Psi' \tag{3.31}$$

$$A_0^{(1)} = i\Sigma\left[\frac{m+n}{mn}\lambda^2 + \frac{mn}{m+n}\frac{\theta_t}{2}\right] - i\Psi'\left[\frac{m+n}{mn}\lambda + \frac{1}{2}\theta_x\Gamma_{mn}\right] -$$

$$-\Gamma_{mn}(i\Psi'^2 - \Psi'_x),$$

$$\Psi' = \exp\{i\Gamma_{mn}\theta(x,t)/2\}\Psi.$$

Using the easily deduced relations

$$(\Psi')_x = (\Psi_x)' + [g_x g^{-1}, \Psi'], \ [g, \Sigma] = 0,$$

$$[g_x g^{-1}, \Psi'] = i\frac{mn}{m+n}\frac{\theta_x}{2}[\Sigma, \Psi'],$$

$$[\Sigma, \Psi'] = \frac{m+n}{mn}\Gamma_{mn}\Psi', \ [\Sigma, \Psi'^2] = 0,$$

$$[\Psi', \Gamma_{mn}] = -2\Gamma_{mn}\Psi',$$

$$[\Psi', \Gamma_{mn}\Psi'^2] = -2\Gamma_{mn}\Psi'^3,$$

for $\Psi'(x,t)$, we obtain from the condition $F_\mu^{(1)} = 0$

$$-i\Gamma_{mn}(\Psi'_t - v\Psi'_x) + \Psi'_{xx} + 2\Psi'^3 = 0,$$

and for the function $\Psi'(x',t')$ we have the nonlinear Schrödinger equation

$$-i\Gamma_{mn}\Psi'_{t'} + \Psi'_{x'x'} + 2\Psi'^3 = 0$$

which is identical to eqn. (12).

This means that the gauge transformations (28) - (30) correspond to the Galilei transformation, and in the new variables

$$R'(x',t) = R(x,t)\exp(i\theta(x,t)/2), \quad x' = x+vt,$$

$$Q'(x',t) = Q(x,t)\exp(-i\theta(x,t)/2), \quad t' = t,$$

eqn. (12) conserves its form.

Similarly, in the form of a local gauge transformation one can write the Bäcklund transformations for the nonlinear Schrödinger equation (see details in [12]).

Now we consider the transformations (28) and (29) of the group G realized on the Jost solutions of the linear problem (14) at the point $\lambda = \lambda_0$, i.e.

$$g(x,t,\lambda_0) = \Phi(x,t,\lambda=\lambda_0) \tag{3.32}$$

and

$$\partial_\mu g = A_\mu(x,t,\lambda=\lambda_0)g.$$

We regard the columns of the matrix $g(x,t;\lambda_0)$ as a local basis at the point (x,t) and the diagonal matrix Σ as a metric tensor of the flat homogeneous space G/H (H is the global group) which is invariant under the action of the local isogroup H. Let us introduce a tensor field, the metric tensor of the curved homogeneous space

$$S(x,t,\lambda_0) = g^{-1}(x,t,\lambda_0)\Sigma g(x,t,\lambda_0) \tag{3.33}$$

thereby localizing the isogroup H, because now $S \in G/H(x,t)$. As a result of such a transformation we have

$$\Phi_x^{(2)} - A_1^{(2)}(\lambda,\lambda_0)\Phi^{(2)}, \quad \Phi_t^{(2)} = A_0^{(2)}(\lambda,\lambda_0)\Phi^{(2)}, \tag{3.34}$$

$$\Phi^{(1)}(x,t,\lambda,\lambda_0) = g(x,t,\lambda_0)\Phi^{(2)}(x,t,\lambda,\lambda_0),$$

$$A_\mu^{(2)}(\lambda,\lambda_0) = g^{-1}A_\mu^{(1)}g - g^{-1}g_\mu,$$

where

$$A_1^{(2)}(x,t,\lambda,\lambda_0) = -i(\lambda-\lambda_0)S(x,t,\lambda_0), \tag{3.35a}$$

$$A_0^{(2)} = i\frac{m+n}{mn}(\lambda^2-\lambda_0^2)S - \frac{2mn}{m+n}(\lambda-\lambda_0)SS_x + \frac{n-m}{m+n}(\lambda-\lambda_0)S_x, \tag{3.35b}$$

when S satisfies the sigma model condition

$$S^2 = \frac{1}{mn}e_n + \frac{n-m}{mn}S, \quad N = m+n. \tag{3.36}$$

Using now that the connection curvature vanishes, $F^{(2)}_{\mu\nu} = 0$, we get a sigma model representation of the nonlinear Schrödinger equation which corresponds to a generalized isotropic Landau-Lifshitz equation (sometimes called a continuum Heisenberg model)*

$$S_t = \frac{1}{i}\frac{mn}{m+n}[S,S_{xx}] - 2\lambda_0\frac{m+n}{mn}S_x. \tag{3.37}$$

The ordinary isotropic Landau-Lifshitz equation corresponds to $G = SU(2)$ ($\lambda_0 = 0$, $m = n = 1$) and $S^2 = e_2$.

Thus we see that a gauge transformation of the form (32) gives a coupling between systems of different types described by different equations, thereby reminding one of the general Bäcklund transformations at the solution level.

Is it possible to cancel the term proportional to λ_0 in eqn. (37)? For vanishing boundary conditions of the Schrödinger function $\Psi_{|x|\to\infty} \to 0$ the system (12) is invariant under the Galilei transformations. Therefore, as we shall see below, the continuum spectrum of the linear problem (14b) lies on the whole real axis of the complex plane λ. The Jost solutions of the continuum spectrum only form the group G. In fact the functions $\Phi(x,t,\lambda)$ of the discrete spectrum have the asymptotics $\Phi_{|x|\to\infty} \to 0$ and hence have no inverse element Φ^{-1}.

Assuming $\lambda_0 = 0$ we obtain from eqn. (37)

$$S_t = \frac{mn}{m+n}\frac{1}{i}[S,S_{xx}]. \tag{3.38}$$

The same result can be obtained via the Galilei transformations $x' = x - vt$, $t' = t$. Thus eqn. (37) becomes the equation ($S'(x',t') = S(x',t')$):

$$S_{t'} - vS'_{x'} = \frac{mn}{m+n}\frac{1}{i}[S,S_{xx}] - 2\lambda_0\frac{m+n}{mn}S_{x'}. \tag{3.39}$$

*However it is necessary to notice the following difference. The Heisenberg model is a quantum system written in terms of spin operators, while both eqn. (37) and the Landau-Lifshitz equation describe classical systems. See Chapters 1, 2 and 5 on their possible relationship.

Putting $v = 2\lambda_0 m + n / mn$, we obtain eqn. (38).

The boundary conditions on Ψ and, correspondingly, on S are unchanged

$$\Psi'_{|x|\to\infty} \to 0, \quad S(x,t) \to \Sigma, \quad S'(x',t') \to \Sigma.$$

However, if vanishing boundary conditions are natural for compact isogroups H, the situation is quite different in the case of noncompact isogroups. The fact that their irreducible unitary representations are infinite dimensional leads to a problem with an infinite number of interacting particles or with nonvanishing boundary conditions.

To illustrate this, let us consider the $U(p,q)$ nonlinear Schrödinger equation

$$i\Psi_t + \Psi_{xx} + 2((\overset{*}{\Psi}\Psi) - \rho)\Psi = 0 \tag{3.40}$$

with the boundary conditions

$$\Psi_{x\to\pm\infty} \to \Psi_{\pm}, \quad \rho = (\overset{*}{\Psi}\Psi_{\pm}\Psi_{\pm}) = \text{const.} \tag{3.41a}$$

The Galilei transformations (30) do not change the form of eqn. (40), however the boundary conditions are altered

$$\Psi'_{x\to\pm\infty} \to \Psi_{\pm}\exp(i\theta(x,t)/2). \tag{3.41b}$$

The continuous spectrum of the linear problem (14b) lies now on the cuts along the real axis in the intervals $(-\infty, -\sqrt{\rho})$ and $(\sqrt{\rho}, +\infty)$. As a result we cannot choose the normalization point on the spectral paramtere λ_0 in the lacuna $(-\sqrt{\rho}, \sqrt{\rho})$, because $|\lambda_0| \geqslant \sqrt{\rho}$. Even though we can cancel in eqn. (39) the term proportional to λ_0 setting $v = 2\lambda_0(m+n)/mn$, the boundary conditions $S(x,t) \to S_0(x,t)$ become now $S'(x',t') \to S_0(x',t')$. This means that either the choice of λ_0 determines the boundary conditions for the corresponding isotropic Landau-Lifshitz equation or is dictated by them.

To observe more details of the gauge connection of the Schrödinger and Landau-Lifshitz equations and its geometrical meaning, let us consider now the inverse transformation from the isotropic Landau-Lifshitz equation to $S3$. In this case, we confine ourselves to the Hermitian reduction (23) and correspondingly to the pseudo-unitary group $G = SU(m,n)$. For that let us start with eqn. (37) at $\lambda_0 \neq 0$.

$$S_t = \frac{mn}{m+n}(-i)[S,S_{xx}] - 2\lambda_0 \frac{m+n}{mn} S_x$$

where

$$S = \sum_\alpha S^\alpha(x,t)\tau_\alpha,\quad \tau_\alpha \in su(m,n)$$

$$\tau_\alpha = (\lambda_1,\ldots,\lambda_l, -i\lambda_{l+1},\ldots,-i\lambda_{l+k}),\quad i\lambda_\alpha \in su(N),\quad N = m+n.$$

Then

$$su(m,n) = \tilde{k} \oplus \tilde{l},$$

$$\dim \tilde{k} = m^2+n^2-1 \equiv k,\quad \dim \tilde{l} = 2mn \equiv l.$$

The generators τ_α satisfy the relations*

$$\tau_\alpha \tau_\beta = g_{\alpha\beta} + d_{\alpha\beta\gamma}\tau_\gamma + i f_{\alpha\beta\gamma}\tau_\gamma,\quad Tr\tau_\alpha = 0 \tag{3.42}$$

therefore

$$\Gamma_0 S^+ \Gamma_0 = S,\quad TrS = 0,\quad \Gamma_0 = \begin{bmatrix} \Gamma_{ls} & 0 \\ 0 & \Gamma_{pq} \end{bmatrix}. \tag{3.43}$$

The sigma model condition (36), in component form implies that S satisfies the following relation:

$$g_{\alpha\beta}S^\alpha S^\beta = \frac{1}{mn},\quad d_{\alpha\beta\gamma}S^\alpha S^\beta = \frac{n-m}{mn}S^\gamma \tag{3.44}$$

with

$$g_{\alpha\beta} = (\underbrace{1,\ldots,1}_{l}, \underbrace{-1,\ldots,1}_{k}) = \Gamma_{lk}.$$

Let us consider at a fixed point (x,t) of the bundle base a set of basis vectors $(g_i)_j(x,t)$ that form the matrix $g \in SU(m,n)$ and diagonalize the metric

$$S(x,t) = g^{-1}(x,t)\Sigma g(x,t),\quad \Gamma_0 g^+ \Gamma_0 g = I. \tag{3.45}$$

This procedure is always possible if G is a connected Lie group (or S belongs to

*Our choice of the potentials A_μ in eqns. (14) and (15) corresponds to the tensor $g_{\alpha\beta}$ which differs in the case of $n = 1$ and $m = 2$ from the conventional one [13] by a factor of 2/3.

a connected component of the group unity). It follows from (45) that the diagonal metric Σ satisfies conditions (43) and (44), is real, and coincides with Σ defined earlier in (15d).

The subgroup $H \subset G$ of elements $h \in H$ retaining the matrix S uncharged is defined by the condition

$$\tilde{S} = \tilde{g}^{-1}\Sigma\tilde{g} = g^{-1}\Sigma g = S \text{ with } \tilde{g} = hg \tag{3.46}$$

or $[\Sigma,h] = 0$, i.e. $H = S(U(l,s) \otimes U(p,q))$, $(l+s=m, p+q=n)$ and coincides with the isotropy group of the linear problem in the space M (24). Thus M is isomorphic to G/H and in our case to

$$SU(m,n)/S(U(l,s) \otimes U(p,q)). \tag{3.47}$$

We then decompose the Lie algebra L of the group $G = SU(m,n)$ into two parts orthogonal with respect to the Killing metric (see eqn. (10))*)

$$L = L^0 \dot{+} L^1 \tag{3.10}$$

(L^0 is the Lie algebra of the subgroup H) so that

$$[L^i, L_j] \subset L^{i+j\,(\mathrm{mod}\,2)} \tag{3.48}$$

i.e. we introduce the $\mathbb{Z}_2$ grading. Such a decomposition for real semisimple Lie algebras is called the Cartan decomposition. The form of the commutation relations implies that there exists the involutive automorphism σ of the algebra L such that $\sigma^2 = 1$ and

$$\sigma(l^0) = l^0, \quad l^0 \in L^0, \tag{3.49}$$

$$\sigma(l^1) = -l^1, \; l^1 \in L^1.$$

For the real semisimple Lie algebra the opposite statement is also valid: for any involutive automorphism σ of the real semisimple Lie algebra there exists such a basis in L that eqn. (49) and the commutation relations (48) hold. The latter assertion follows from the fact that any involutive automorphism σ of the real semisimple Lie algebra can be reduced to a diagonal form with the elements $+1$

*)Such a decomposition is always possible for semisimple real Lie algebras through the Cartan theorem (see, for example [25]).

and -1.

The orthogonality of the vector spaces L^0 and L^1 follows from the invariance property of the Killing metric

$$(X,Y) = \mathrm{Tr}(adX, adY)$$

under the automorphism of the Lie algebra L

$$(\phi(X),\phi(Y)) = (X,Y).$$

In fact for σ, eqn. (49), we have

$$(l^0,l^1) = (\sigma(l^0),\ \sigma(l^1)) = -(l^0,l^1) = 0.$$

Note that the Killing form for the algebra $sl(n,C)$ is simply

$$(X,Y) = 2n\,\mathrm{Tr}(X\cdot Y). \tag{3.50}$$

We notice further that L^0 of the Cartan decomposition is a maximal compact subalgebra in L.

Finally, concluding this algebraic-geometric digression let us notice that in the case of maximally compact isotropy groups H the space G/H is a globally symmetrical Riemann space of constant curvature. The latter is of two types:

(1) G is compact (the curvature is nonnegative).

(2) G is noncompact (the curvature is nonpositive).

The first type includes, in particular, Grassmannian manifolds, projective spaces and spheres $SO(p+q)/SO(p)\otimes SO(q)$, $SU(p+q)/SU(p)\otimes SU(q)$, the second one involves their noncompact 'analogues' $SO(p,q)/SO(p)\otimes SO(q)$ (at $q=1$ we have the Lobachevsky space L^p), $SU(p,q)/SU(p)\otimes SU(q)$ (at $q=1$ we have the unit ball in C^p as a complex manifold). In the simplest low dimensional cases the first type includes the $SU(2)/U(1) = S^2$ sphere, the second one the $SU(1,1)/U(1) = L^2$ Lobachevsky plane. This digression is necessary for understanding the 'geometrization' of interactions in what follows.

We return to the decomposition (10) and represent the elements of L in the form $l = l_0 + l_1$ where

$$l_0 = \begin{bmatrix} M_m & 0 \\ 0 & M_n \end{bmatrix}, \quad l_1 = \begin{bmatrix} 0 & \overset{*}{l} \\ -l & 0 \end{bmatrix} \tag{3.51}$$

and M_m and M_n are $(m\times m)$ and $(n\times n)$ matrices correspondingly, l is an $(n\times m)$ matrix, and $\overset{*}{l} = \Gamma_{l,s} l^{+}\Gamma_{pq}$ is an $(m\times n)$ matrix. Introducing a chiral current $J_\mu = \partial_\mu g\cdot g^{-1}$ we may express $\partial_\mu S$ through g, Σ, and J_μ, viz.

$$\partial_\mu S = g^{-1}[\Sigma, J_\mu]g. \tag{3.52}$$

Let J_1 take values only in L^1

$$J_1 = \begin{pmatrix} 0 & i\overset{*}{\varphi} \\ i\varphi & 0 \end{pmatrix}. \tag{3.53}$$

One can find J_0 from the equation of motion (37) and the condition $g_{xt} = g_{tx}$, or in the variables J_μ

$$\partial_\mu J_\nu - \partial_\nu J_\mu + [J_\nu, J_\mu] = 0. \tag{3.54}$$

If one decomposes $J_0 = J_0^0 + J_0^1$ according to (51), then eqn. (54) and (37) are reduced to

$$J_{0x}^0 + [J_0^1, J_1] = 0, \tag{3.55a}$$

$$J_0^1 = -i\Gamma_{mn} J_{1x} - 2\lambda_0 \frac{m+n}{mn} J_1, \tag{3.55b}$$

respectively.

Substituting J_0^1 from (55b) into (55a) and integrating, we obtain

$$J_0 = \begin{pmatrix} -i(\overset{*}{\varphi}\varphi - \hat{\rho}_1) & \overset{*}{\varphi}_x - 2\lambda_0 \frac{m+n}{mn} i\overset{*}{\varphi} \\ -\varphi_x - 2\lambda_0 \frac{m+n}{mn} i\varphi & i(\varphi\overset{*}{\varphi} - \hat{\rho}_2) \end{pmatrix} \tag{3.56}$$

where $\hat{\rho}_1$ and $\hat{\rho}_2$ are the constant matrices of dimension $(m\times m)$ and $(n\times n)$, respectively, (the integration constants).

The linear (U,V) problem potentials for the generalized isotropic Landau-Lifshitz equation (see eqn. (35), $U = A_1$, $V = A_0$) are reduced under the gauge transformation $\Phi = g^{-1}\tilde{\Phi}$ to the form

$$\tilde{U} = gVg^{-1} + g_x g^{-1} = -i(\lambda-\lambda_0)\Sigma + J_1, \tag{3.57}$$

$$\tilde{V} = gVg^{-1} + g_t g^{-1} = i\frac{m+n}{mn}(\lambda^2 - \lambda_0^2)\Sigma - (\lambda-\lambda_0)\frac{m+n}{mn} J_1 + J_0.$$

The zero connection curvature condition $\tilde{F}_{\mu\nu} = 0$ gives the equation

$$i\varphi_t + \varphi_{xx} + 2\lambda_0 \frac{m+n}{mn} i\varphi_x + 2(\varphi\overset{*}{\varphi}\varphi - \rho\varphi) = 0. \tag{3.58}$$

Then the requirement $\hat{\rho}_1\varphi = \varphi\hat{\rho}_2$ leads to $\hat{\rho}_1 = \rho e_m$ and $\hat{\rho}_2 = \rho e_n$.

Applying the Galilei transformations to eqn. (58) and choosing $v = -2\lambda_0(m+n)/mn$ we get the conventional matrix nonlinear Schrödinger equation

$$i\varphi_t + \varphi_{xx} + 2(\varphi\overset{*}{\varphi}\varphi - \rho\varphi) = 0. \tag{3.59}$$

It is worth pointing out that the term with $\rho \neq 0$ necessary for nonvanishing boundary conditions (ρ is the condensate density) is usually introduced artificially. Now it arises automatically.

Using the relations (45) and (52) and after simple calculations we obtain the densities of the three first integrals for the nonlinear Schrödinger equation (20) in terms of $S \in G/H$:

$$N \equiv \mathrm{Tr}(RQ) = \left(\frac{mn}{m+n}\right)^2 \frac{1}{2} \mathrm{Tr}\, S_x^2. \tag{3.60}$$

$$\mathcal{P} \equiv \frac{1}{2i} \mathrm{Tr}(RQ_x - R_xQ) = \left(\frac{mn}{m+n}\right)^2 \frac{1}{2i} \mathrm{Tr}\left(\frac{m-n}{m+n} S_xS_{xx} + 2\frac{mn}{m+n} SS_xS_{xx}\right),$$

$$\mathcal{H} \equiv \mathrm{Tr}(R_xQ_x - (RQ)^2) = \left(\frac{mn}{m+n}\right)^2 \frac{1}{2} \mathrm{Tr}\left((S_{xx})^2 - 5\left(\frac{mn}{m+n}\right)^2 (S_x)^4\right).$$

As appears from the above, the nonlinear Schrödinger equation internal symmetry group coincides with H. The sigma model relations (36) defines unambiguously the maximal compact isotropy subgroup through the projectors on the space where the nonlinear Schrödinger equation field variables take the values

$$\hat{P}_{mn} = \frac{m}{m+n}(I - nS), \quad \hat{P}^2 = \hat{P}.$$

The constraint (36) modification leads to new systems which are gauge equivalent to the isotropic Landau-Lifshitz equation (see, e.g. [11,14]).

Notice that scalar fields which are valued in the space G/H where G is a noncompact group ($SU(1,1)$ or E_7) appear also in supergravity under dimensional reduction [15]. To avoid the possible problem of 'ghosts' connected with the noncompactness of G, the subgroup H is required to be localized and choosen as a maximal compact subgroup [16], thereby arriving, as we have already mentioned, at the globally symmetrical Riemann space (of constant curvature). Let us show how this fact follows from our example. The energy density for the isotropic Landau-Lifshitz equation (a continuum classical analogue of the Heisenberg chain) has the form (see eqn. (2.20) at $\beta_i = 0$)

$$W = \operatorname{Tr} S_x^2 = 2(\frac{m+n}{mn})^2 \operatorname{Tr} PQ = 2(\frac{m+n}{mn})^2 Tr\, \overset{*}{\varphi}\varphi. \tag{3.61}$$

In the vector version $G = SU(p+1,q)$, $H = S(U(1)\otimes U(p,q))$ we have

$$Tr\, \overset{*}{\varphi}\varphi = \sum_{a=1}^{p} |\varphi^a|^2 - \sum_{a=p+1}^{p+q} |\varphi^a|^2, \tag{3.62}$$

i.e. the energy density W, just as the metric, is not positive definite. However, the energy density together with the metric becomes negative (positive) definite, if $p = 0$ $(q=0)$, i.e. H is the maximal compact subgroup. In the case of $G = SU(1,q)$

$$W = \operatorname{Tr} S_x^2 = -2(\frac{m+n}{mn})^2 \sum_{a=1}^{a} |\varphi^a|^2$$

and the corresponding nonlinear Schrödinger equation describes a Bose gas with q colour degrees of freedom of a repulsive type (see the next chapter). This assumes a meaningful physical sense (when stationary states are possible in the system) only if the density of this gas ρ is nonzero, i.e. for an infinite number of particles. Then the energy may be made positive definite assuming $W' = \rho - \Sigma_{a=1}^{q} |\varphi^a|^2$. In such a Bose gas, bound states of 'hole' type can arise which can be interpreted as bound states of an infinite number of particles (see [17] and the next chapter).

In the case of the compact group $G = SU(p+1)$, $q = 0$ we arrive at a Bose gas of an attractive type with an arbitrary finite number of particles.

For low dimensional groups we have the consequences:

(a) $G = SU(2) \to S \in SU(2)/U(1) = S^2$ sphere, the space of constant positive curvature $\to$ a nonlinear Schrödinger equation of an attractive type.

(b) $G = SU(1,1) \to S \in SU(1,1)/U(1) = L^2$ pseudosphere, the space of constant negative curvature $\to$ a repulsive type nonlinear Schrödinger equation.

One may assume that a similar situation results in the more general case as well. In fact, Fordy and Kulish [11] have shown that, in the case of the Hermitian homogeneous symmetric space G/H with compact H (i.e. the space with constant curvature K) and the Killing form $g_{\alpha\beta}$ as a G-invariant Riemann metric, the corresponding vector nonlinear Schrödinger equation has the form ($G=SU(1+q)$ or $SU(1,q)$)

$$iq_t^\alpha+q_{xx}^\alpha+K\sum_{\beta,\gamma\in\theta^+}g_{\beta,-\gamma}q^\beta\bar{q}^\gamma q^\alpha=0 \tag{3.63}$$

where K is the Gaussian curvature and θ^+ is a subset of the algebra positive roots. It means that the coupling constant of the nonlinear Schrödinger equation is proportional to the symmetric space curvature. Thereby the sigma model representation of the nonlinear Schrödinger equation (the Landau-Lifshitz equation) enables the interaction in the nonideal Bose gas to be geometrized. We note also that the choice of the compact group $H\subset G$ leads to a Bose gas of one sort of particles (the symmetric space curvature determines the interaction type), but the choice of the noncompact group H already gives a mixture of two sorts of Bose gases. For example:

$$G=SU(2,1)\begin{cases} SU(2,1)/S(U(2)\otimes U(1))\rightarrow \begin{cases} U(0,2)\ \text{NSE} \\ \text{repulsive } \textit{Bose gas}\end{cases} \\ SU(2,1)/S(U(1,1)\otimes U(1))\rightarrow \begin{cases} U(1,1)\ \text{NSE} \\ \text{attractive } \textit{Bose gas} + \\ \text{repulsive } \textit{Bose gas}\end{cases}\end{cases}$$

Thus we can conclude that the noncompact group G realized in both linear and nonlinear fashion is related to a problem with an infinite number of particles which realizes its infinite-dimensional unitary representations. Factorization by the noncompact group $H=S(U(p,q)\otimes U(1))$ leads to a Bose gas of a mixed type.

Finally, we give examples of some reductions:

$SU(2)/S(U(1)\otimes U(1))\rightarrow U(1,0)$ $\quad SU(1,1)/S(U(1)\otimes U(1))\rightarrow U(0,1)$

$SU(3)/S(U(2)\otimes U(1))\rightarrow U(2,0)$ $\quad SU(2,1)/S(U(2)\otimes U(1))\rightarrow U(0,2)$

$SU(p+1,0)/S(U(p)\otimes U(1))\rightarrow U(p,0)$ $\quad SU(2,1)/S(U(1,1)\otimes U(1))\rightarrow U(1,1).$

3. Gauge connections of the Landau-Lifshitz equation with uniaxial anisotropy and the nonlinear schrödinger equation

In this section, we consider an example showing that the noncompact group to appear in the isotropic Landau-Lifshitz equation is natural and the non-Hermitian form of the nonlinear Schrödinger equation of Section 3.1 has a certain physical meaning.

Let us consider the Landau-Lifshitz equation with uniaxial anisotropy of the easy axis or easy plane types (see Chapter 2, eqn. (2.21) at $h_0=0$). In terms of the matrix variables $S = \Sigma_{\alpha=1}^{3} S^{\alpha}\sigma^{\alpha}$ this can be written in the form

$$2i\dot{S} = [S,S_{xx}]+\Delta[S,\sigma_3]\{S,\sigma_3\}_+ \tag{3.64}$$

with $\sigma_\alpha \in sl(2,C)$, $\Delta \in \mathbb{C}$.

The linear problem for eqn. (64) is defined by the gauge potentials [26,27]

$$U_1 = i\lambda S+\mu[\sigma_3,S], \tag{3.65}$$

$$V_1 = 2i\lambda^2 S+\lambda SS_x+2\lambda\mu[\sigma_3,S]-i\mu[\sigma_3,SS_x]+4i\mu^2\{\sigma_3,S\}_+\sigma_3,$$

where $\mu = \sqrt{\Delta} \in \mathbb{C}$ and $S^2 = I$. For real Δ, we have two cases, $\Delta<0$ and $\Delta>0$, which correspond to the easy plane and easy axis models. Then

$$\mu = \begin{cases} i\sqrt{|\Delta|}, & \Delta<0 \\ \sqrt{\Delta}, & \Delta>0. \end{cases} \tag{3.66}$$

It is easy to show that if $S^+ = S$ then $U_1 = -U_1^+$ (similarly for V_1) belongs to the algebra $su(2)$ only for real μ or positive Δ.

In the case of imaginary μ or $\Delta<0$, $U_1^+ \neq -U_1$ and is decomposed into two parts $U_1 = U_+ + U_-$ where $U_-^+ = -U_-$ and $\overset{*}{U}_+ \equiv \sigma_3 U_+^+ \sigma_3 = -U_+$. It means that $U_- \in su(2)$ and $U_+ \in su(1,1)$, i.e. $U_1 \in su(2)\dot{+}su(1,1)$ and in the case of arbitrary complex μ $U_1 \in sl(2,C)$. Let us show that the anisotropic Landau-Lifshitz equation is gauge equivalent to the isotropic case and so to the nonlinear Schrödinger equation [27]. However, the symmetries of the anisotropic and isotropic cases are, generally speaking, different. The scheme of the proof is the following

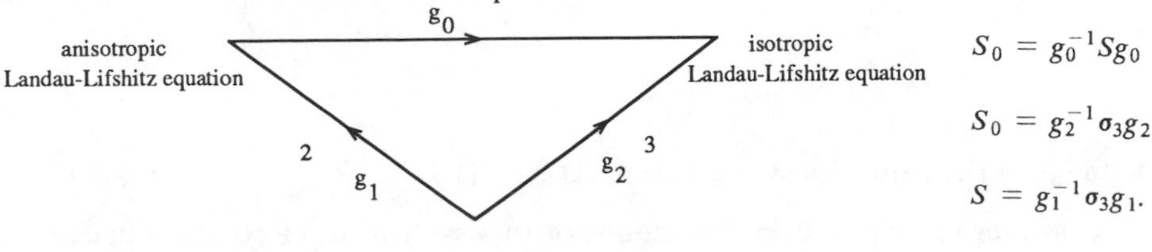

$$S_0 = g_0^{-1} S g_0$$

$$S_0 = g_2^{-1} \sigma_3 g_2$$

$$S = g_1^{-1} \sigma_3 g_1.$$

Consider the Jost solutions to the linear problem

$$\Phi_{1x}(x,t;\mu,\lambda) = U_1(x,t;\mu,\lambda)\Phi_1(x,t;\mu,\lambda), \tag{3.67a}$$

$$\Phi_{1t}(x,t;\mu,\lambda) = V_1(x,t;\mu,\lambda)\Phi_1(x,t;\mu,\lambda),$$

and

$$g_{0x} = U_1(\lambda_0) g_0, \tag{3.67b}$$

i.e. $g_0 \equiv \Phi(x,t;\mu,\lambda=\lambda_0)$. For g_0 to be a gauge group element λ_0 should belong to continuous spectrum of the operator L and cannot be chosen to always equal zero [9,17]. Apply the gauge transformation $\Phi_0 = g_0^{-1}\Phi_1$, then

$$U_0 = g_0^{-1} U_1 g_0 - g_0^{-1} g_{0x} = g_0^{-1}(U_1(\mu,\lambda) - U_1(\mu,\lambda_0)) g_0, \tag{3.68}$$

$$V_0 = g_0^{-1} V_1 g_0 - g_0^{-1} g_{0t} = g_0^{-1}(V_1(\mu,\lambda) - V_1(\mu,\lambda_0)) g_0.$$

By introducing the new variable

$$S_0(x,t;\lambda_0) = g_0^{-1} S g_0 \tag{3.69}$$

we obtain from eqn. (68)

$$U_0 = i(\lambda-\lambda_0) S_0, \quad V_0 = 2i(\lambda^2-\lambda_0^2) S_0 + (\lambda-\lambda_0) S_0 S_{0x}. \tag{3.70}$$

To obtain the latter formula we use the relations

$$S_0^2 = I, \quad g_0^{-1} S_x g_0 = S_{0x} - [S_0, g_0^{-1} g_{0x}],$$

$$g_0^{-1} S S_x g_0 = S_0 S_{0x} - S_0 [S_0, g_0^{-1} g_{0x}] = S_0 S_{0x} - 2\mu g_0^{-1} [\sigma_3, S] g_0,$$

$$g_0^{-1} g_{0x} = i\lambda_0 S + g_0^{-1} \mu [\sigma_3, S] g_0,$$

which follow from eqns. (67b), (68) and (69). Potentials (70) define (see eqn. (35)) the linear problem for the isotropic Landau-Lifshitz equation at $m = n = 1$, $\lambda \to -\lambda$ and $S \to S_0$. From the condition $F_{\mu\nu} = 0$ we have

$$S_{0t} = \frac{1}{2i}[S_0, S_{0xx}] + 4\lambda_0 S_{0x}$$

with $s_0 \in sl(2,C)$ since $U_1 = U_+ + U_- \in su(2)+su(1,1)$.

Consider some particular reductions of this system. If we confine ourselves to a Hermitian Hamiltonian of system (64), then $\mathrm{Im}\,\Delta = 0$ (but not μ). We choose the variants

$$\lambda_0 = 0, \quad \mu \neq 0, \quad \text{then } g_0 \in \begin{cases} SU(2) \text{ at } \Delta > 0 \\ SU(1,1) \text{ at } \Delta < 0 \end{cases} \tag{I}$$

$$\lambda_0 \neq 0, \quad \mu \neq 0, \quad \text{then } g_0 \in \begin{cases} SU(2) \text{ at } \Delta > 0 \\ SU(2,C) \text{ at } \Delta < 0 \end{cases} \tag{II}$$

(1) It follows thence that only at $\Delta > 0$ (easy-axis ferromagnet) we have the $SU(2)$ isotropic Landau-Lifshitz equation which is gauge equivalent to the $U(1,0)$ nonlinear Schrödinger equation with attraction [22].

(2) In the case $\Delta < 0$, (easy-plane ferromagnet) the isotropic magnet belongs to the noncompact group $SL(2,C)$ and the nonlinear Schrödinger equation corresponding to it is defined by a stationary (isotropy) subgroup over which factorization is performed. This also implies that such strange models, as the $SL(2,C)$ and $SU(1,1)$ isotropic Landau-Lifshitz and nonlinear Schrödinger equations associated with them become physically meaningful.

Consider the second case in more detail. Let us proceed to the direct correspondence between the anisotropic Landau-Lifshitz and nonlinear Schrödinger equations.

Let the element $g_2 \in G$ make the Gauge transformation between the isotropic Landau-Lifshitz and nonlinear Schrödinger equations, then

$$S_0(x,t,;\lambda) = g_2^{-1}(x,t;\lambda_0)\sigma_3 g_2(x,t;\lambda_0). \tag{3.71}$$

$$\partial_x g_2 \cdot g_2^{-1} = \begin{bmatrix} 0 & ir \\ ig & 0 \end{bmatrix}.$$

From eqns. (69) and (71) it follows that

$$S = g_0 S_0 g_0^{-1} = (g_2 g_0^{-1})^{-1} \sigma_3 g_2 g_0^{-1} = g_1^{-1} \sigma_3 g_1 \tag{3.72}$$

with the notation

$$g_1 = g_2 g_0^{-1}. \tag{3.73}$$

So g_1 diagonalizes S of the anisotropic Landau-Lifshitz equation and is the product of g_2 (which diagonalizes S_0 of the isotropic case) by the inverse of the Jost solution g_0 to the anisotropic Landau-Lifshitz equation linear problem. In other words, $g_2 = g_1 g_0$ and the Jost solution of the anisotropic linear problem couples g_1 and g_2. This connection is very essential and sets strong constraints on the symmetry group of gauge equivalent models.

In fact, the element g_0, as mentioned above, is defined by the symmetry group of the initial anisotropic models and by the sign of Δ. That the element g_2 belongs to some symmetry groups of the isotropic Landau-Lifshitz equation determines the form of an equivalent nonlinear Schrödinger equation, so, in particular:

$g_2 \in SU(2) \to U(1,0)$ nonlinear Schrödinger equation,

$g_2 \in SU(1,1) \to U(0,1)$ nonlinear Schrödinger equation.

There are also other possible (non-Hermitian) reductions. It follows from our considerations (see diagram above) that the paths $(1+2)$ and 3 must be consistent. Thus, if we have the conventional $SU(2)$ easy plane ferromagnet and $g_1 \in SU(2)$, then connection (73) is possible with additional constraints on:

(1) S and g_2 ($\partial_x S^z = 0$), or (2) g_0 and g_2 (an exotic model of nonlinear Schrödinger equation).

In the first version the relation of energy densities for the anisotropic and isotropic Landau-Lifshitz equations and the nonlinear Schrödinger equation particle density is of the form

$$\frac{1}{2} Tr(S_{0x})^2 = \frac{1}{2} Tr[S_x^2 - 8\mu\sigma_3 S_x + 8\mu^2(I - (S\sigma_3)^2)] = 4rq. \tag{3.74}$$

The term proportional to μ can lead to imaginary part of the Hamiltonian H_0 (isotropic Landau-Lifshitz equation) appearing and the same in the particle number of the nonlinear Schrödinger equation. So at $S \in SU(2)$, $\Delta < 0$ (μ is imaginary) $\partial_x S^z \sim \mathrm{Im}(rq) = \mathrm{Im}\, Tr(S_{0x})^2$. To the $U(1,0)$ and $U(0,1)$ nonlinear Schrödinger equation Hermitian variants correspond the reductions $rq = \pm |q|^2$ which, in the language of the gauge equivalent isotropic Landau-Lifshitz equation imply $S \in SU(2)$ and $S \in SU(1,1)$, respectively. To obtain a Hermitian version of the nonlinear Schrödinger equation (or the isotropic Landau-Lifshitz equation) from the anisotropic Landau-Lifshitz equation under consideration, $\partial_x S^z = 0$ is needed. Otherwise $(\partial_x S^z \neq 0)$ we obtain the isotropic model $S \in SL(2,C)/U(1)$ and the corresponding nonlinear Schrödinger equation is

$$iq_t + q_{xx} + 2(rq)q = 0, \quad -ir_t + r_{xx} + 2(rq)r = 0,$$

which is reduced to the (integrable) system

$$i\psi_t + \psi_{xx} + 2(\overset{*}{\psi}\psi)\psi + 2(\overset{*}{\psi}\sigma_1\psi)\sigma_1\psi = 0, \tag{3.75}$$

$$i\overset{*}{\psi}_t + \overset{*}{\psi}_{xx} + 2\overset{*}{\psi}(\overset{*}{\psi}\psi) + 2\overset{*}{\psi}\sigma_1(\overset{*}{\psi}\sigma_1\psi) = 0,$$

$$\overset{*}{\psi} = \psi^+\sigma_3, \quad \psi = (\phi_1, \phi_2)^{tr}, \quad \sigma_1 = \left(\begin{smallmatrix} 0 & 1 \\ 1 & 0 \end{smallmatrix}\right),$$

via the separation of the Hermitian and anti-Hermitian parts $q = \phi_1 + \phi_2$, $r = \bar{\phi}_1 - \bar{\phi}_2$. Its Hermitian subreduction is the $U(1,1)$ nonlinear Schrödinger equation

$$i\psi_t + \psi_{xx} + 2(\overset{*}{\psi}\psi)\psi = 0 \tag{3.76}$$

which is formulated on the algebra $SU(2,1)$. However, in this case, $\partial_x S^z = 0$ as well, which is most likely uninteresting from the physical point of view.

In [28] one more 'exotic' reduction of the easy plane anisotropic Landau-Lifshitz equation has been found:

$$i\phi_t + \phi_{xx} - 2(|\phi|^2 - \frac{\Delta}{2})\phi + \frac{|\phi_x|^2\phi}{\Delta - 2|\phi|^2} - \frac{\phi x}{2}\partial_x \ln(\frac{\Delta}{2} - |\phi|^2), \tag{3.77}$$

$$\phi(x,t)_{x\to\pm\infty} \to \sqrt{\Delta/2}\,\exp(i\alpha_\pm),$$

which follows from eqn. (75) through setting the constraint on the fields ϕ_1 and ϕ_2: $\phi_{2x} = \sqrt{\Delta/2 - |\phi_2|^2}\,\phi_1$, $(\phi \equiv \phi_1)$ and is related to the fact that S belongs to some unitary-pseudo-unitary mixture of $SL(2,C)$.

Concluding this chapter, let us mention that that algebraic-geometric (group) ideas developed in it, which enable us to formulate the nonlinear Schrödinger equation sigma model representation, may be naturally applied to still other integrable systems for which the (U,V) representation is known. This has been done [12] for a number of equations, namely the Sine-Gordon, nonlinear Schrödinger, Korteweg-de Vries and modified Korteweg de Vries equations.

Essentially this work is based on and sumarizes the results of previous work viz., the Pohlmeyer ($0(n)$ models and Sine-Gordon equation) [29] (see also [30]), Lakshman [19], Zakharov and Takhtajan [22] ($SU(2)$ isotropic Landau-Lifshitz equation and $U(1)$ nonlinear Schrödinger equation). In [31] a sigma model representation has been found for the Yajima-Oikawa system.

Chapter 4.

THE NONLINEAR SCHRÖDINGER EQUATION WITH $U(p,q)$ INTERNAL SYMMETRY AND THE SINE-GORDON EQUATION

In this chapter, we consider the mathematical aspects of a number of nonlinear Schrödinger equation models as complete integrable systems and their connection with the theory of a non-ideal Bose gas. We focus on the Hermitian models because these occur frequently in applications; also models having a non-compact internal symmetry and the peculiarities they possess are discussed in details. In conclusion, we present an outline of the Sine-Gordon equation.

1. Equations of motion and the internal symmetry group

Our concern here will be with the properties of the vector version of the nonlinear Schrödinger equation with $U(p,q)$ internal symmetry. We consider a system of n coupled equations for a set of complex valued functions defined on the whole axis $-\infty < x < +\infty$

$$i\psi_t^{(a)} + \psi_{xx}^{(a)} + 2g\left[\sum_{b=1}^{p} |\psi^{(b)}|^2 - \sum_{b=p+1}^{n} |\psi^{(b)}|^2\right]\psi^{(a)} = 0, \tag{4.1}$$

with $a = \overline{1,n}$; $p+q = n$.

With $n = 2$ we have the system we obtained in the first chapter for the Hubbard model (see (1.55)). Let us introduce the column vector ψ

$$\Psi = \begin{bmatrix} \psi^{(1)} \\ \cdot \\ \cdot \\ \cdot \\ \psi^{(n)} \end{bmatrix} \tag{4.2}$$

consisting of n complex functions $(\Psi)_a = \psi^{(a)}(x,t)$ and the Dirac conjugate line vector

$$\overset{*}{\Psi} = \Psi^+ \Gamma_{pq}, \tag{4.3}$$

with

$$\Gamma_{pq} = \begin{bmatrix} e_p & 0 \\ 0 & -e_q \end{bmatrix} \quad \text{(see (3.13)).}$$

We denote the internal product in the isotopic space as

$$(\Psi,\Psi) \equiv (\overset{*}{\Psi}\Psi) = \sum_{b=1}^{p} |\psi^{(b)}|^2 - \sum_{b=p+1}^{n} |\psi^{(b)}|^2 \tag{4.4}$$

then (1) reads

$$i\Psi_t + \Psi_{xx} + 2g(\Psi,\Psi)\Psi = 0 \tag{4.5}$$

and that conjugate to (1)

$$-i\overset{*}{\Psi}_t + \overset{*}{\Psi}_{xx} + 2g(\Psi,\Psi)\overset{*}{\Psi} = 0. \tag{4.6}$$

These equations can be derived from the Lagrangian density

$$\mathfrak{L} = \frac{i}{2}(\overset{*}{\Psi}\Psi_t - \overset{*}{\Psi}_t\Psi) - (\overset{*}{\Psi}_x\Psi_x) + g(\Psi,\Psi)^2 \tag{4.7}$$

or the Hamiltonian density

$$\mathcal{H} = (\overset{*}{\Psi}_x\Psi_x) - g(\overset{*}{\Psi}\Psi)^2). \tag{4.8}$$

Variables $\psi^{(a)}(x,t)$ and $\overset{*}{\psi}{}^{(a)}(x,t)$ are canonically conjugate in the sense of

$$\{\psi^{(a)}(x), \overset{*}{\psi}{}^{(b)}(y)\} = i\delta^{ab}\delta(x-y), \tag{4.9}$$

where the Poisson bracket has been given by relation (3.2a).

It follows from (7) and (8) that s,t-independent linear transformations $\Psi' = h\Psi$ conserving the bilinear form (Ψ,Ψ) are the symmetry transformations of the system. The pseudo-Hermitian form (4) is conserved under transformations of the pseudo-unitary group $U(p,q)$. The matrices $h \in U(p,q)$ then satisfy the conditions $(\Psi',\Psi') = (\overset{*}{\Psi}\overset{*}{h}{\cdot}h\Psi) = (\Psi,\Psi)$ or

$$\overset{*}{h}h = e, \tag{4.10}$$

where $\overset{*}{h} = \Gamma_{pq} h^{+} \Gamma_{pq}$ are the linear transformations of the n-dimensional complex space $\mathbb{C}^{p+q}$. The number of independent parameters of the $U(p,q)$ group is n^2 *) and there are n^2 conserved local currents (Noether's currents) J_μ^{ik} ($\mu=0,1; i,k=\overline{1,n}$) which are

$$J_0^{ik} = \overset{*}{\psi}{}^{(i)}\psi^{(k)}, \tag{4.11}$$

$$J_1^{ik} = i(\overset{*}{\psi}{}_x^{(i)}\psi^{(k)} - \overset{*}{\psi}{}^{(i)}\psi_x^{(k)}),$$

or in matrix form

$$J_0 = \overset{*}{\Psi}\otimes\Psi, \quad J_1 = i(\overset{*}{\Psi}_x\otimes\Psi - \overset{*}{\Psi}\otimes\Psi_x).$$

The corresponding charges

$$Q^{ik} = \int dx J_0^{ik} = \int dx \overset{*}{\psi}{}^{(i)}\psi^{(k)} \tag{4.12}$$

commute with the Hamiltonian, $\{Q^{ik},H\} = 0$, and satisfy the commutation relations of the Lie algebra $gl(p+q,C)$.

$$\{Q^{ik},Q^{jl}\} = \delta_{kj}Q^{il} - \delta_{il}Q^{jk} \tag{4.13}$$

and the conjugated condition $\overline{Q}^{ik} = \epsilon_{ik}Q^{ik}$ (no summation here), where

$$\epsilon_{ik} = \begin{cases} 1, & \text{for } 1 \leqslant i,\ k \leqslant p \ \text{ or } \ p+1 \leqslant j,\ k \leqslant n \\ -1, & \text{for } 1 \leqslant i \leqslant p < k \leqslant n \ \text{ or } \ 1 \leqslant k \leqslant p < i \leqslant n \end{cases}$$

The commutation relations (13) and the conjugation conditions imply the Q^{ik} form the algebra of the $U(p,q)$ group. By using these one can construct n^2 conserved Hermitian charges (see (1))

$$N_i \equiv N_{ii} = \int dx J_0^{ii} = Q^{ii}, \quad i = \overline{1,n}, \tag{4.14}$$

$$N_{ij} = \frac{1}{2}(Q^{ij}+Q^{ji}), \quad 1 \leqslant i,j \leqslant p; \quad C_{ij} = \frac{i}{2}(Q^{ij}-Q^{ji}), \quad p+1 \leqslant i,j \leqslant n,$$

$$T_{ij} = \frac{i}{2}(Q^{ij}+Q^{ji}), \quad 1 \leqslant i \leqslant p < j \leqslant n; \quad K_{ij} = \frac{1}{2}(Q^{ij}-Q^{ji}), \quad 1 \leqslant j \leqslant p < i \leqslant n.$$

---

*) Note that from system (1) we come to (10), so that the Lagrangian symmetry is $U(p,q)$.. If the linear problem (3.14) is connected to the group $SU(p+1,q)$, eqn. (5) becomes a vector with isotropy group nonlinear Schrödinger equation $S(U(1)\otimes U(p,q))$.

The diagonal 'charges' Q^{ii} are the numbers of i-particles and are positive for $1 \leqslant i \leqslant p$:

$$N_i \equiv Q^{ii} = \int_{-\infty}^{\infty} dx \, |\psi^{(i)}|^2 \tag{4.15a}$$

and negative for $p+1 \leqslant i \leqslant p+q = n$:

$$N_i = -\int_{-\infty}^{\infty} dx \, |\psi^{(i)}|^2. \tag{4.15b}$$

The physical meaning of these quantities will be made clear later. The off-diagonal charges generate a transformation h_1 mixing 'pure' states. They belong to the subalgebra $su(p,q) \subset u(p,q)$.

If we consider particular solutions of system (5)

$$\Psi_i = \begin{cases} \tilde{\Psi}_i, & 1 \leqslant i \leqslant p \\ \\ 0, & i \neq j \end{cases} \qquad \text{or} \qquad \begin{matrix} \Psi_i = 0, \quad i \neq j, \\ \\ \Psi_j = \tilde{\Psi}_j, \quad p+1 \leqslant j \leqslant n, \end{matrix}$$

then, making use of the transformations $h_1 \in SU(p,q)$ generated by charges N, C, T and K, we are able to construct a whole class of solutions of the same system. We utilize this possibility below to obtain several classes of soliton solutions of system (5).

The isotopic group $U(p,q)$ action is transferred into the linear problem via isotropy transformations $h \in S(U(1) \otimes U(p,q))$

$$h = \begin{bmatrix} 1 & 0 \\ 0 & h_1 \end{bmatrix}. \tag{4.16}$$

The connection coefficients A_0 and A_1 are transformed as

$$A_\mu \to A'_\mu = hA_\mu(\psi)\overset{*}{h} \tag{4.17}$$

with

$$\overset{*}{h} = \begin{bmatrix} 1 & 0 \\ 0 & \overset{*}{h}_1 \end{bmatrix} = \Gamma_0 h^{-1} \Gamma_0, \quad \Gamma_0 = \Gamma_{p+1q} = \begin{bmatrix} 1 & 0 \\ 0 & \Gamma_{pq} \end{bmatrix}.$$

It follows from the form of the operator

$$A_1(x,t;\lambda) = -i\lambda\Sigma + \Psi,$$

where

$$\Sigma = \begin{bmatrix} 1 & 0 \\ 0 & -1/n\, e_n \end{bmatrix}, \quad \Psi = i\begin{bmatrix} 0 & \overset{*}{\psi} \\ \psi & 0 \end{bmatrix},$$

that the fact that the matrix Σ is n-fold degenerated is connected directly with the isogroup properties (its algebra rank is two). The condition $A_1(h_1\Psi(x,t)) = hA_1(\psi)\overset{*}{h}$ holds only if $h\Sigma\overset{*}{h} = \Sigma$ or $\overset{*}{h}_1 h_1 = e$, implying $h_1 \in U(p,q)$[*)]. In other words, an internal symmetry causes the linear problem to be degenerated, this fact is evident if one recalls that Σ is a metric tensor of coset space, invariant under the isotropy group (see the preceding chapter).

2. $U(p,q)$ **nonlinear Schrödinger equation under trivial boundary conditions**

In fact, up to now we did not consider the boundary conditions which must be satisfied by the functions $\psi^{(\alpha)}$. Here we show that this point is very important since the boundary conditions will be also define the state and dynamics of the system. It is necessary right away to point out that in contrast to systems with a compact isospace (e.g., $U(n)$), the system (1) is more complicated and interesting since it allows a large set of boundary conditions for the field functions $\psi^{(a)}$. The simplest situation takes place under trivial boundary conditions, $\psi^{(a)}(\pm\infty) = 0$. The compact $U(2,0)$ version of the nonlinear Schrödinger equation has been studied in [33] and the noncompact $U(p,q)$ case in [32]. The structure of the transition matrix $S(\lambda)$ turns out to be nearly identical in both cases. To show this, let us rewrite eqn. (3.14b) of the linear spectral problem as follows

$$L\hat{\Phi} = \lambda\hat{\Phi}, \tag{4.18}$$

introducing the operator

$$L = i\Sigma\partial_x - i\Sigma\Psi \equiv i\Sigma(\partial_x + i\lambda\Sigma - A_1) \tag{4.18a}$$

and the solution matrix

[*)]In fact, $h_1 \in SU(p,q)$ in representation (16).

$$\hat{\Phi} = (\varphi_1,\dots,\varphi_{n+1}), \quad \varphi_i = (\varphi_i^{(1)},\dots,\varphi_i^{(n+1)})^{tr}$$

with φ_i being the $(n+1)$-component column corresponds to the i-th solution.

If we assume that the fundamental matrix solutions $\hat{\Phi}$ and $\hat{F}$ have the following asymptotic behaviours

$$f_i^{(k)} = \delta_{ik}\exp(-i\lambda\hat{\Sigma}_{ik}x), \quad x \to -\infty,$$

$$\varphi_i^{(k)} = \delta_{ik}\exp(-i\lambda\hat{\Sigma}_{ik}x), \quad x \to +\infty,$$

then we have the Jost matrix solutions. The completeness of each of these solutions implies their linear dependence

$$\hat{F}(x,\lambda) = \hat{\Phi}(x,\lambda)S(\lambda); \; S = \begin{bmatrix} S_{11} & S_{1\beta} \\ S_{\alpha 1} & S_{\alpha\beta} \end{bmatrix}. \tag{4.19}$$

The matrix $S(\lambda)$ as a function of the spectral parameter λ is called a transition matrix (sometimes the scattering matrix). In our case if $q \to 0$ at $x \to \pm\infty$ the matrix S has the following properties:

(1) unimodularity: $\det S(\xi) = 1$,

(2) pseudo-unitarity: $\overset{*}{S}S = e$,

(3) the element S_{11} and the block $S_{\alpha\beta}(\alpha,\beta{=}2,3,\dots,n+1)$ are conserved in time,

(4) the line $S_{1\beta}$ and the column $S_{\alpha 1}$ depend on time exponentially

$$S_{1\beta}(\lambda,t) = \exp\{i(\frac{n+1}{n})^2\lambda^2 t\}S_{1\beta}(\lambda,0),$$

$$S_{\alpha 1}(\lambda,t) = \exp\{-i(\frac{n+1}{n})^2\lambda^2 t\}S_{\alpha 1}(\lambda,0).$$

In principle, this enables the potential matrix $\hat{\Psi}$ to be reconstructed at any instant of time. The conserved elements of the matrix S, i.e. S_{11} and the block $S_{\alpha\beta}$, are analytical in the upper and lower half-plane of λ respectively. They generate an infinite series of integrals of motion. Once gauge transformations (16) from the isotropy group acting on the transition matrix

$$S(\lambda) \to S'(\lambda) = hS(\lambda)\overset{*}{h}$$

conserve the element $S_{11}(\lambda)$, the series of local polynomial integrals of motion so generated,

$$\ln S_{11}(\lambda) \simeq \sum_{k=1}^{\infty} (i\frac{n+1}{n}\lambda)^{-k} I_{11}^{(k)},$$

is invariant under the $U(p,q)$ group action (the isotopic group commutes with the infinite-dimensional Abelian group, generating the conservation laws). These integrals are a simple generalization of the known $U(1)$ nonlinear Schrödinger equation integrals of motion [33] and can be obtained by replacing the $U(1)$ internal product $|\psi|^2$ with the $U(p,q)$ one, (Ψ,Ψ). The latter is not valid for the conserved block $S_{\alpha\beta}(\lambda)$ and the series of non-local integrals of motion generated by it

$$\frac{S_{\alpha\beta}(\lambda)}{S_{\alpha\alpha}(\lambda)} \simeq \sum_{k=1}^{\infty} (i\frac{n+1}{n}\lambda)^{-k} I_{\alpha\beta}^{(k)}.$$

In fact, the local integrals $I_{11}^{(k)}$ are in involution with each other

$$\{I_{11}^{(k)}, I_{11}^{(l)}\} = 0 \quad k,l = 1,..., \tag{4.20}$$

and with the non-local integrals

$$\{I_{11}^{(k)}, I_{\alpha\beta}^{(l)}\} = 0 \tag{4.21}$$

the latter are not in involution and form a complicated algebraic structure (see [34]). Only with $k = 1$ are the integrals $I_{\alpha\beta}^{(1)}$ local and correspond to the conserving Noether currents. These currents are connected with the Lagrangian symmetry under the isotropy group, $H_1 = U(p,q)$ in our case. The integrals $I_{\alpha\beta}^{(1)}$ are subject to the commutation relations of the $su(p,q)$ algebra.

The Poisson brackets among the elements of the matrix $S(\lambda)$ have been calculated [32,35] to be of the form

$$\{S_{kl}(\lambda), S_{ps}(\mu)\} = \frac{n}{(n+1)(\lambda-\mu)} \lim_{x\to\infty} \Big[S_{pl}(\lambda) S_{ks}(\mu) e^{-i(\lambda-\mu)(\Sigma_{pp}-\Sigma_{kk})x} - S_{pl}(\mu) S_{ks}(\lambda) e^{-i(\lambda-\mu)(\Sigma_{ss}-\Sigma_{ll})x} \Big]. \tag{4.22}$$

The system of equations of the inverse problem for the vector (compact $U(n.0)$ version of the nonlinear Schrödinger equation [33] is naturally extended onto the case of the $U(p,q)$ nonlinear Schrödinger equation [32]. An essential difference arises when non-trivial boundary conditions are considered. To explain this we reduce the problem to a simpler one, but conserving in many respects the most essential features. Consider the plane wave solutions to the nonlinear Schrödinger equation

(we shall call them 'condensate' solutions)

$$\psi^{(a)} = c^{(a)}\exp\{i(\omega_a t - k_a x)\} \tag{4.23}$$

where $\omega_a = k_a^2 - (\overset{*}{c}c)$.

Following [36], one can show that the stability of the solutions under infinitesimal perturbations with frequency Ω and wavevector χ is determined by the dispersion formula

$$\Omega^2 = \chi^2[\chi^2 - 2(\overset{*}{c}c)]. \tag{4.24}$$

From (2.23), it follows that for the condensate stability the relation $(\overset{*}{c}c) < 0$ should hold: the isotopic space metric must be negatively definite

$$\sum_{a=1}^{p} |c^{(a)}|^2 < \sum_{a=p+1}^{n} |c^{(a)}|^2.$$

Hence stable solutions with nontrivial boundary conditions can exist only for systems with the noncompact $U(p,q)$ group or 'compact' $U(0,q)$ group. For example, in Manakov's $U(2,0)$ model $(g>0)$ and in the $U(1,1)$ model the solutions

$$\Psi = \begin{bmatrix} \psi^{(1)} \\ \psi^{(2)} \end{bmatrix} = \begin{matrix} c^{(1)}\operatorname{sech} ax \\ c^{(2)}\tanh ax \end{matrix} \quad a^2 = (c^{(1)})^2 + (c^{(2)})^2$$

exist, however, in the first case, this solution is unstable, since the condensate (or 'vacuum') over which it is constructed ($\psi^{(2)} = c^{(2)}$), is unstable. In the $U(1,1)$ and $U(0,2)$ models this solution is stable.

Generally, for nontrivial boundary conditions in fact we have a problem of infinite numbers of particles the interaction of which leads to the necessity of a renormalization of physical quantities. In this case, the language of noncompact groups is the most convenient. Therefore already for an example of the simplest pseudo-unitary group $U(1,1)$ we have a wealthy spectrum of stable particle-like excitations. Before we proceed to investigate them, we briefly recall the properties of the solutions to the simplest exact integrable unitary models $U(1,0)$ and $U(0,1)$ corresponding to the equations

$$i\psi_t + \psi_{xx} \pm 2g|\psi|^2\psi = 0 \tag{4.25}$$

with the signs plus and minus, respectively.

3. The $U(1,0)$ model

For functions decaying sufficiently rapidly at infinity the Cauchy problem is exact solvable - an initial packet breaks up into a set of solitons and a wave background. The former corresponds to the discrete part of the Lax operator (L) spectrum, the latter is relevant to the continuum part.

The integrals of particle number, momentum and energy written in field and action-angle variables are as follows:

$$N = \int |\psi|^2 dx = \int n(p)dp + \sum_{s=1} N_s, \tag{4.26}$$

$$P = -\frac{i}{2}\int (\psi^* \psi_x - \psi_x^* \psi)dx = \int pn(p)dp + \sum_{s=1} \frac{1}{2} v_s N_s,$$

$$E \equiv H = \int (|\psi_x|^2 - g|\psi|^4)dx = \int p^2 n(p)dp + \sum_{s=1} \frac{N_s}{12}(3v_s^2 - g^2 N_s^2).$$

The continuous action $n(\zeta)$ and angular $\varphi(\zeta)$ variables are related to the $\hat{S}$ scattering matrix elements by [38]

$$n(\zeta) = -\frac{2}{g\pi}\ln|S_{11}(-\frac{\zeta}{2})|, \quad \varphi(\zeta) = \arg S_{12}(-\frac{\zeta}{2}),$$

where

$$|S_{11}(\zeta)|^2 + |S_{12}(\zeta)|^2 = 1.$$

The discrete variables N_s, φ_s, v_s and ν_s are related to the discrete spectrum characteristics a_n and b_n of the Lax operator (which is non-self-adjoined in the $U(1,0)$ case). The quantities a_n, i.e. the Lax operator eigenvalues in the upper half-plane, are zeros of the S_{11} element of the scattering matrix S, the b_n are normalizing factors corresponding to them. The relationships are

$$N_s = \frac{4}{g}\operatorname{Im} a_n, \quad \varphi_s = \arg b_n,$$

$$v_s = \frac{4}{g}\operatorname{Re} a_n, \quad \nu_s = -\ln|b_n|.$$

From eqn. (2.26) it can be seen that the angular variables φ, φ_s, and ν_s are cyclic and the corresponding action functionals are integrals of motion. In quantum or quasi-classical language the continuous variables (wave background) correspond to 'microparticles' ('phonons', 'magnons' and so on) with the dispersion $E = p^2$ and mass 1/2. Macroparticle-solitons (i.e. the discrete variables) can be interpreted as a *special* form of bound states of N_s constituent 'particles' with mass 1/2. The soliton energy (spectrum) is

$$E_s = \frac{1}{N_s}p_s^2 - \frac{1}{12}g^2N_s^3, \quad (N_s \gg 1), \tag{4.27a)-(4.27b}$$

$$P_s = \frac{1}{2}v_sN_s, \quad M_s = \frac{1}{2}N_s. \tag{4.27c}$$

The second term of E_s is the binding energy and the first term represents the kinetic energy of the macroparticle-soliton with mass $N_s/2$.

Formulae (27) may be easily obtained via the Bohr-Sommerfeld quantization applied to the classical expression for the functionals E, P and M calculated for soliton solutions

$$\psi_s = \frac{c_s}{\sqrt{g}} \operatorname{sech}[c_s(x - v_st - x_0)]\exp\{i(\frac{1}{2}v_sx - \omega_st - \theta_0)\}, \tag{4.28}$$

with $c_s = g/2N_s$.

Now we mention another interesting fact. The energy functional can be also written in the differential form

$$dE_s = \mu dN_s \tag{4.29}$$

where $\mu = (1/4)(v_s^2 - g^2N_s^2)$ exactly coincides with the soliton solution frequency and corresponds to the energy per constituent in the soliton, i.e. is adequate for the chemical potential. Adding the term $g^2N_s/12$ to the right-hand side of eqn. (2.27), we get the binding energy

$$E_b = -\frac{g^2}{12}(N_s^2 - N_s) \tag{4.30}$$

reported in [39] for the case of an exact quantum problem of N_s Bose particles attracting each other with the Hamiltonian in configuration space

$$H = -\sum_i^{N_s}\left[\frac{\partial^2}{\partial x_i^2} + 2g\sum_j \delta(x_i - x_j)\right].$$

Note that this addition appears due to the ordering of the operators ψ and ψ^+ when one performs the exact quantization. From eqn. (30) it follows that for the exact quantization the lowest soliton energy at $N_s = 1$ coincides with the one-particle state energy of the continuous spectrum, i.e. the quantum soliton at $N_s = 1$ is reduced to a microparticle and, in the general case, it is their bound state. Thus one can relate the continuous spectrum quanta to soliton constituent bosons.

Concluding the $U(1,0)$ model section we consider a classical problem of plane wave dynamics

$$\psi = c_0 \exp\{i(\omega_0 t - k_0 x)\}$$

which corresponds to nontrivial boundary conditions in this system. From eqn. (24) we see that this is unstable and will be broken up into localized lumps with characteristic scale $l \simeq 1/c_0$ after a time period of order $\tau \simeq 1/c_0^2$ (the so-called modulational instability). A rather similar problem arises for periodic boundary conditions too. Assuming that the number of the lumps (solitons) produced is large, we can determine their width distribution function. To do this consider the spectral (2×2) problem (18) which in our case is

$$\phi_{1x} - i\chi\phi_1 = -\bar{\psi}\phi_2,$$

$$\phi_{2x} + i\chi\phi_2 = \psi\phi_1.$$

Differentiating the second of these equations with respect to x and eliminating the function ϕ_1 from the first equation, we get

$$\partial_x^2\phi_2 + (\chi^2 + |\psi|^2)\phi_2 = \frac{\psi_x}{\psi}(\phi_{2x} + i\chi\phi_2).$$

We then use the 'semiclassical' approximation. Letting $(\psi_x/\psi) \ll 1$ and omitting the right-hand side of this equation, we have

$$\partial_x^2\phi_2 + (\chi^2 + |\psi|^2)\phi_2 = 0$$

i.e. the Schrödinger equation, for which the quasi-classical quantization rule is

$$\int(\chi^2 + |\psi|^2)^{1/2} dx = 2\pi(n + \frac{1}{2}). \tag{4.31}$$

When a large number $n \gg 1$ of solitons are produced, their relative velocities are small, i.e. $\mathrm{Re}\chi \ll \mathrm{Im}\chi = c$ [37]. This also follows from eqn. (31) directly, and taking into account that

$$\rho(c) = -\frac{\partial n}{\partial c}$$

we get*) the soliton amplitude distribution function

$$\rho(c) = \frac{1}{2\pi L}\int_{-L}^{L}\frac{c\,dx}{(|\psi|^2 - c^2 + o(c^2))^{1/2}}$$

or, normalized to unity,

$$\rho(c) = \frac{1}{a_0}\frac{c}{\sqrt{a_0^2 - c^2}}. \tag{4.32}$$

Correspondingly, the soliton width distribution function is

$$\rho(l) = \frac{1}{l_0}\frac{(l_0/l)^2}{\sqrt{(l/l_0)^2 - 1}}, \quad l_0 = \frac{1}{a_0}. \tag{4.33}$$

Figure 4.1. shows these distributions.

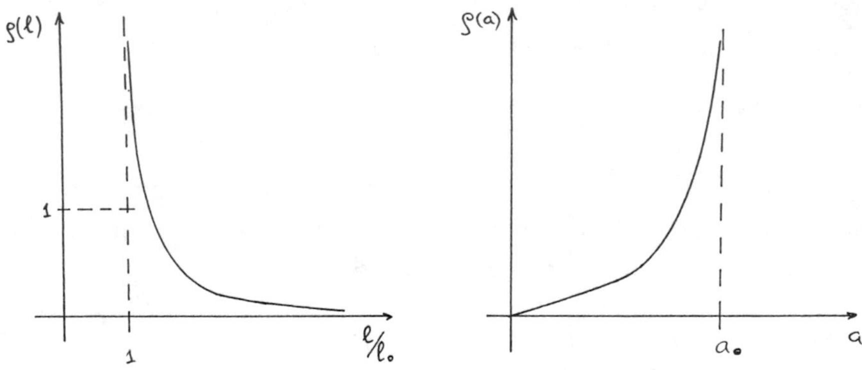

Figure 4.1. Distribution function of solitons ρ as a function of (a) their amplitude a, (b) their width l.

The average $<c>$ and root-mean-square $(<c^2>)^{1/2}$ soliton amplitudes and the dispersion $\Delta^2 = <c^2 - <c>^2>$ are easily calculated

$$<c> = \frac{\pi}{3}a_0, \quad \sqrt{<c^2>} = \sqrt{\frac{2}{3}}\,a_0, \tag{4.34}$$

*)Here we place the system into a symmetric box of length $2L$, as is usual in the study of plane waves.

$$\Delta^2 = (\frac{2}{3} - \frac{\pi^2}{16})a_0^2 \simeq 0.0498a_0^2,$$

i.e. $\Delta \simeq 0.223a_0$.

Thus we see that due to the modulational instability the plane wave decays into a set of solitons the amplitude of which are distributed according to eqn. (32). This means that:

(1) solitons having an amplitude greater than the decayed wave amplitude are absent, i.e. $c < a_0$;

(2) the average soliton amplitude is $<c> = 0.8a_0$;

(3) the distribution function $\rho(c)$ width is relatively small and equal to $\Delta \simeq <c>/4$.

Thus more than 60% of the solitons are located in the amplitude region $a_0 \leqslant c \simeq <c>$. This means that for semiquantitative estimates one may take the distribution function in the form $\rho(c) \simeq \delta(c - <c>)$. Finally, we emphasize that although the solitons considered are (on quantum and quasi-classical levels) bound states of a number of boson constituents (see eqns. (27) and (30)), when produced they have new properties radically differing from conventional bound states (such as, for example, nuclear nucleons). The integrability of the $S3$ equation provides the system with the conservation of additional quantities (quantum numbers: such as, for example, the total soliton number) like baryonic or leptonic charges leading to the soliton interaction elasticity. As a result, fusion processes of the latter producing 'heavier' ones are forbidden in spite of the energetic preference. This means that even at the classical level, the soliton possess an analogue of quantum properties which prevents the growth of a collapse-like instability. An arbitrary initial state[*)] decays into a set of solitons and a background. If the number of solitons is sufficiently large ($n \gg 1$), this set is described by the function $\rho(c)$ (32) or since $N_s = 1/2gc$

$$\rho(N_s) = \frac{1}{4}\frac{g^2}{c_0}\frac{N_s}{\sqrt{c_0^2 - \frac{1}{4}g^2N_s^2}}. \tag{4.35}$$

From this formula it follows that the maximum number of constituent bosons

*) Strictly speaking, this must be well localized.

which may be bound in the $S3$ system considered does not exceed the quantity $N_{\max} = (2/g)c_0$ defined by the initial conditions. As a result, in a rough approximation we have in the system the set of a large number of slowly moving macroparticles (solitons) with equal (in practice) masses $M \simeq 0.4N_{\max} \simeq 1/2<N>$.

Thus the simplest physical system the behaviour of which is described by the $U(1,0)$ model is a one-dimensional Bose gas with the δ-like pair interaction (attraction). In this case, for a given boson number N we obtain in the second quantization the $S3$ equation with the commutation relation $[\hat{\psi},\hat{\psi}^+] = \delta(x-y)$ in the quantum case and the Poisson bracket $\{\psi,\psi^*\} = i\delta(x-y)$ in the classical (and quasi-classical) case.

Some other systems are considered in Chapter 1. Manifestly, the dynamical and statistical features of all these systems are defined by the above-described behavior of solutions of the $U(0,1)$ model.

Concluding this subsection we note that, in addition to the soliton solutions considered above, in the $U(1,0)$ system there also exist quasi-periodic (bion) solutions. These may arise in the case when two or more zeros of $S_{11}(\chi)$ lie on a straight line which is parallel to the imaginary axis, i.e $\mathrm{Re}\chi_i = \mathrm{Re}\chi_k$ $(i{\neq}k)$, and are particular bound states of conventional solitons. However, since the mass defect of these objects is zero, $E_b = E_i+E_k$, they are unstable and decay into constituent solitons [40]. Then the picture considered naturally includes the 'bion' objects too.

4. The $U(0,1)$ model

In this case, the study is substantially more complicated. Now the L operator is self-adjoint and for zero asymptotics of the ψ at infinite it does not have a discrete spectrum. The behaviour of the solution as $t\to\infty$ has the form

$$\psi(x,t) \sim t^{-1/2}.$$

At nonzero ψ asymptotics the problem corresponds to a Bose gas with a finite density ρ_0 defined by the field function value at $|x|\to\infty$: $|\psi|^2\to\rho_0$. Sometimes, such a state is called the condensate, or because it has the lowest energy, the vacuum. In order to describe the condensate by a solution to the field equation, it is convenient to rewrite the latter in the form

$$i\psi_t+\psi_{xx}-2g(|\psi|^2-\rho_0)\psi = 0, \tag{4.36}$$

i.e. introducing into the equation the term $2g\rho\psi$ and, correspondingly, $-2g\rho|\psi|^2 \equiv -\mu|\psi|^2$ into the Hamiltonian. Here the quantity μ plays the role of a chemical potential. The condensate is described by the solution

$$\psi = \sqrt{\rho_0}\, e^{i\theta}$$

where $\theta = \text{const.}$

The integrals of particle number, momentum and energy are renormalized in this case as follows

$$\tilde{N} = \int_{\infty}^{\infty} (|\psi|^2 - \rho_0) dx, \tag{4.37}$$

$$P = \operatorname{Im} \int_{-\infty}^{\infty} (\psi^* \psi_x) dx + \rho_0 \alpha, \quad [38]$$

$$E = \int_{-\infty}^{\infty} [|\psi_x|^2 + g(|\psi|^2 - \rho_0)^2] dx$$

where α is the solution phase shift arising when the coordinate varies from $-\infty$ to $+\infty$

$$\psi(+\infty) = \sqrt{\rho_c}, \quad \psi(-\infty) = \sqrt{\rho_0}\, e^{i\alpha}.$$

In this case, again the L operator has a spectrum of continuous and discrete parts. Its continuous part corresponds to the Bogolubov excitation spectrum

$$\omega(p) = |p| \sqrt{p^2 + 4g\rho_0}, \tag{4.38}$$

$$p(\chi) = 2\lambda(\chi), \quad \lambda(\chi) = \sqrt{\chi^2 - g\rho_0}$$

and can be investigated by perturbation theory.

The discrete part corresponds to the so-called hole excitation mode obtained and studied for the first time in the quantum case by Lieb [41]. In the classical case, this mode is described by the complex kink solution

$$\psi_k = a_k th[a_k(x - v_k t - x_0)] + \frac{i}{2} v_k, \quad a_k = \sqrt{g\rho_0 - \frac{1}{4} v_k^2}. \tag{4.39}$$

The functionals $\tilde{N}, P$ and E are easily calculated for this solution to give

$$\tilde{N}_k = -\frac{2}{g} a_k, \tag{4.40}$$

$$P_k = \frac{2}{g}(g\rho_0 \arccos\frac{v_k}{2\sqrt{g\rho_0}} + \frac{1}{4}gv_k\bar{N}_k),$$

$$E_k = -\frac{1}{3}g^2N_k^3,$$

and $\alpha = 2\arccos(v_k/2\sqrt{g\rho_0})$.

The negative sign of $\tilde{N}$ should be regarded as a particle 'deficit' in the condensate, i.e. indicating the presence of holes. It is therefore natural to define the hole number as $(-\tilde{N}_k) = N_h > 0$.

The integrals of motion calculated via the action-angle variables have now the form [38]:

$$N = \int \frac{\chi}{\lambda(\chi)} n(\chi)d\chi + \sum_{k=1}^{n} N_k, \tag{4.41}$$

$$P = \int 2\lambda(\chi)n(\chi)d\chi + \sum_{k=1} P_k,$$

$$E = \int 4\lambda(\chi)\chi n(\chi)d\chi + \sum_{k=1} E_k.$$

A comparison of these with those obtained previously shows that their discrete soliton parts coincide. This means that from all possible soliton solutions of eqn. (36) for calculating the physical characteristics of the system $\tilde{N}, P$ and E one may use the complex kink having the form (39). Also note that the classical spectra (38) and (40) are reduced to the corresponding quantum case ones at $g \to 0$ (see [38]).

A characteristic feature of the hole solution (39) is the relation between its amplitude and phase. As a result, it is one-parameteric, contrary to the conventional solitons of the $U(1,0)$ equation. It leads to the velocity v_k dependence of the number N_n of holes bound in the double soliton: the greater the velocity, the smaller the number of holes bound in the buble. At $v_k = 2\sqrt{g\rho_0}$ the buble soliton disappears. The situation concerning the buble energy is analogous to this.

From eqn. (40) it also follows that the bubble energy is positive and that the ground state of the system (the condensate) has the lowest energy (for variables (37) it is zero). This means that, generally speaking, when some perturbation with energy E_0 is put in the condensate, it decays into two components. Part of the energy goes to the Bogolubov spectrum, the rest goes to the hole spectrum. The production of bubbles at rest demands the threshold energy

$$E_{\text{thr}} = \frac{8}{3g}(g\rho_0)^{3/2}n \tag{4.42}$$

where n is the number of bubbles.

A pair of oppositely moving bubbles is easier to produce since their energy is $E_{\text{pair}} = 16/3g\,(g\rho_0 - 1/4\,\mathrm{v}_k^2)^{3/2}$ and therefore the threshold is reduced by a factor $(1-\mathrm{v}_k^2/4g\rho_0)^{3/2}$. This means that for the production of a pair with $\mathrm{v}_k^2 \to 4\rho_0 g$ the energy threshold disappears as in the case of the Bogolubov spectrum at $p \to 0$. As a result, even for a low energy initial perturbation decay, both the hole mode and the Bogolubov mode are excited in the condensate, and therefore they are equivalent energetically. The basic difference between these modes consists in their topological structure. The Bogolubov plane wave excitation does not have such a structure*). The kink (39) has a nontrivial topology in the sense that the phase difference of this solution on the left and the right of the transition region is not zero:

$$\Delta\theta = \alpha \underset{|x|\to\infty}{\to} 2\arccos(\mathrm{v}_k/2\sqrt{g\rho}) \to \varphi, \quad \text{when } \mathrm{v}_k \to 0.$$

The results presented above (Sections 3 and 4) are immediately generalized for systems with higher, but compact symmetry groups of isotopic space, i.g. $U(2,0)$ and $U(0,2)$. As has already been mentioned, the first model has been investigated previously [42]. A study of the second model will be given later. Both may be considered as quasi-spin generalizations of the corresponding $U(1)$ models. They may describe, for example, two sublattice chains. We call them quasi-spin models since despite the presence of, say, two states in isotopic space (e.g. such as colours) the Poisson bracket of the field functions obeys a conventional Hiesenberg-Weil, algebra and the quantum operators corresponding to these functions satisfy the Bose commutation relations.

As a result of interactions between solitons in the framework of these models, the state vector may rotate in isospace (i.e. the solitons can change their colours) [42,43].

In those systems having a higher symmetry than $U(2)$ one may expect the

---

*)Or it possesses a trivial topology.

appearance of new qualitative effects such as decay and fusion of solitons (see, for example, studies of such phenomena in [33]).

5. The $U(1,1)$ model

We will now give the results reported in refs. [32,36,44] for a quasi-spin system with the noncompact group $U(1,1)$. It is obvious that like a centaur this exhibits a behaviour inherent to both $U(1,0)$ and $U(0,1)$ models, However, it also has new very specific properties. The energy of the system in this case is not positively definite, so the simplest solution to the $U(p,q)$ nonlinear Schrödinger equation with Hamiltonian (7b) is $\psi^{(i)}(x,t) = c^{(i)}(i=\overline{1,n})$ and the constants $c^{(i)}$ are subject to the condition: $\Sigma_{i=1}^{p}|c^{(i)}|^2 = \Sigma_{i=p+1}^{n}|c^{(i)}|^2$. The energy of the solution vanishes and we have an infinitely degenerate non-trivial 'vacuum' state. If, in the Hamiltonian (7b), the first p components are constants, $|\psi^{(i)}|^2 = \mu_i(i=\overline{1,p})$ we obtain a Hamiltonian of a repulsive-type nonlinear Schrödinger equation with the vacuum being renormalized (a compact subreduction $U(0,1)$)

$$H = \int dx \sum_{i=x}^{p} \{|\psi_x^{(i)}|^2 + (|\psi^{(i)}|^2 - \mu.$$

Both the above-mentioned properties do not belong to the compact $U(n)$ case.

In the case of the $U(1,1)$ model system (1) is reduced to two coupled Schrödinger equations

$$i\psi_t^{(1)} + \psi_{xx}^{(1)} + 2(|\psi^{(1)}|^2 - |\psi^{(2)}|^2)\psi^{(1)} = 0, \qquad (4.43)$$

$$i\psi_t^{(2)} + \psi_{xx}^{(2)} + 2(|\psi^{(1)}|^2 - |\psi^{(2)}|^2)\psi^{(2)} = 0.$$

Here, for simplicity, without loss of generality the coupling constant g is taken to be unity. The first three integrals of motion for the system (43) are

$$N_i = \int dx\,|\psi^{(i)}|^2, \quad P = \mathrm{Im}\int(\bar{\psi}^{(1)}\psi_x^{(1)} - \bar{\psi}^{(2)}\psi_x^{(2)}dx, \qquad (4.44)$$

$$E \equiv H = \int \mathcal{H} dx = \int \{|\psi_x^{(1)}|^2 - \psi_x^{(2)}|^2 - (|\psi^{(1)}|^2 - |\psi^{(2)}|^2)^2\} dx. \qquad (4.45)$$

Symmetry transformations $\psi' = h\psi$, which conserve the internal product $(\overset{*}{\psi}\psi)$, form the four-parameter pseudo-unitary group $U(1,1)$*). The matrix of transformation

*)The name of the nonlinear Schrödinger equation model studied is due to this fact. The notations $U(1,0)$ and $U(0,1)$ become clear for the corresponding $g>0$ and $g<0$ scalar

$h_1 \in SU(1,1)$ is of the form

$$h_{\cdot} = \begin{bmatrix} \alpha & \beta \\ \bar{\beta} & \bar{\alpha} \end{bmatrix}, \quad |\alpha|^2 - |\beta|^2 = 1 \tag{4.46}$$

and can be parametrized as

$$\alpha = \text{ch}\theta e^{i\varphi_\alpha}, \quad \beta = \text{sh}\theta e^{i\varphi_\beta}.$$

Using particular solutions of the $U(1,0)$ and $U(0,1)$ models, one may obtain particular solutions to the $U(1,1)$ system via the isotopic rotation R_1. Applying the R_1 operation (2.21) to the $U(1,0)$ solution vector

$$\psi = \begin{bmatrix} \psi_s \\ 0 \end{bmatrix}, \quad \psi_s = ae^{i\theta_1} \operatorname{sech} a\zeta, \quad \zeta = x - vt - x_0$$

we get

$$\psi = \begin{bmatrix} \alpha \\ \bar{\beta} \end{bmatrix} ae^{i\theta_1} \operatorname{sech} a\zeta \equiv \begin{bmatrix} c^{(1)} \\ c^{(2)} \end{bmatrix} e^{i\theta_1} \operatorname{sech} a\zeta \tag{4.47}$$

where $(\overset{*}{c}c) = a^2 > 0$, and $(\overset{*}{c}c) \equiv (c,c) = |c^{(1)}|^2 - |c^{(2)}|^2$.

Analogously, from the solution vector

$$\psi = \begin{bmatrix} 0 \\ \psi_k \end{bmatrix}, \quad \psi_k = be^{i\theta_2}(\text{th}\, b\zeta + i\frac{v}{2b})$$

we have

$$\psi = \begin{bmatrix} \beta \\ \alpha \end{bmatrix} b \cdot e^{i\theta_2}(\text{th}\, b\zeta + \frac{iv}{2b}) \equiv \begin{bmatrix} c^{(1)} \\ c^{(2)} \end{bmatrix} e^{i\theta_2}(\text{th}\, b\zeta + \frac{iv}{2b}) \tag{4.48}$$

with $(c,c) = -b^2 < 0$.

Note that isorotations may be used only for equations without the terms having chemical potentials, or in the case of $\mu_1 = \mu_2$, (in this case only the isogroup is not broken). To reduce the general system $\mu_1 \neq \mu_2$:

$$i\psi_t^{(j)} + \psi_{xx}^{(j)} + 2((\bar{\psi}\psi) + \mu_j)\psi^{(j)} = 0 \quad (j = 1,2) \tag{4.49}$$

---

variants of the nonlinear Schrödinger equation.

to the $U(1,1)$ symmetrical one, it is necessary to use the transformation

$$\psi^{(j)} \to \exp(2i\mu_j t)\varphi^{(j)}(x,t),$$

and then for the functions $\varphi^{(j)}(x,t)$ we have eqns. (43). The inverse transformation gives solutions to (49).

In the addition to these two types of solitons there exist others which cannot be obtained from the known 'compact' models by isorotation.

Let us consider the four types of naturally arising boundary conditions:

(1) Zero asymptotics at infinity for the ψ vector

$$\psi \to 0 \quad \text{as} \quad |x| \to \infty.$$

(2) The constant type

$$\psi^{(j)} \exp^{(i\omega_j t)} \to \begin{cases} a^{(j)}, & x \to +\infty \\ a^{(j)} e^{i\delta_j}, & x \to -\infty \end{cases} \quad (j=1,2).$$

(3) The quasi-constant type

$$\psi^{(1)} \to \begin{cases} a^{(1)}, & x \to +\infty \\ a^{(1)} e^{i\delta_1}, & x \to -\infty \end{cases} \quad \psi^{(2)} \exp\{i(\omega_2 t - \frac{v_0}{2}x)\} \to \begin{cases} a^{(2)}, & x \to +\infty \\ a^{(2)} e^{i\delta_2}, & x \to -\infty \end{cases}.$$

(4) The mixed type

$$\psi^{(i)} \to 0, \quad |x| \to \infty \quad \psi^{(2)} \to \begin{cases} a^{(2)}, & x \to +\infty \\ a^{(2)} e^{i\delta_2}, & x \to -\infty \end{cases}.$$

In the case of the first and second types of boundary conditions we get one-soliton solutions of the form (47) and (48), respectively (this problem has been discussed in detail in [32]).

In the first case, the solutions are constructed over the trivial (zero) vacuum $\psi_v = 0$ and are

$$\psi_l(x,t) = c\exp\{i(\frac{v}{2}x - \omega t + \theta_0)\}\operatorname{sech} a\zeta \tag{4.50}$$

where $\omega = v^2/4 - (c,c)$, $(c,c) = a^2 > 0$. This soliton is called a double drop ($2D$) according to [36].

In the second case, the solutions for systems (2.46) are constructed over the condensate

$$\psi_v^{(1)} = a^{(1)}, \ \psi_v^{(2)} = a^{(2)}\exp\{2i(\mu_2-\mu_1)t\}, \ -(a,a) = \mu_1$$

and are

$$\psi_{II}^{(1)} = c^{(1)}(\text{th}\, b\zeta + i\frac{\text{v}}{2b}), \tag{4.51}$$

$$\psi_{II}^{(1)} = c^{(2)}(\text{th}\, b\zeta + i\frac{\text{v}}{2b})\exp\{2i(\mu_2-\mu_1)t\}, \tag{4.51}$$

$$b^2 = -(c,c) = \mu_1 - \text{v}^2/4 > 0.$$

We call this kink soliton a double bubble (2B).

As has already been mentioned, both of these solutions are a trivial generalization of the corresponding solitons of the $U(1,0)$ and $U(0,1)$ models.

The third type of boundary conditions corresponds to the condensate (accurate to a constant phase):

$$\psi_v^{(1)} = a^{(1)}, \ \psi_v^{(2)} = a^{(2)}\exp\{i(\frac{\text{v}_0}{2}x - \omega_2 t)\}, \tag{4.52}$$

$$\omega_2 = 2(\mu_1-\mu_2) + \frac{\text{v}_0^2}{4}, \ (a,a) = -\mu_1 < 0.$$

This condensate may be considered as a mixture of two interacting Bose gases with attraction and repulsion between particles inside the components. The first component is at rest, the second component moves with velocity v_0. Over this condensate one may construct the following one-soliton solution:

$$\psi_{III}^{(1)} = c^{(1)}(\text{th}\, b\zeta + i\frac{\text{v}}{2b}), \tag{4.53}$$

$$\psi_{III}^{(2)} = c^{(2)}(\text{th}\, b\zeta + i\frac{\text{v}-\text{v}_0}{2b})\exp\{i(\frac{\text{v}_0}{2}x - \omega_2 t)\}$$

where again $b^2 = -(c,c) > 0$ and

$$(c^{(1)})^2(1+\text{v}^2/4b^2) = (a^{(1)})^2,$$

$$(c^{(2)})^2(1+(\text{v}-\text{v}_0)^2/4b^2) = (a^{(2)})^2.$$

The last two equations can be solved for the c vector 'length', i.e. b, defining the soliton width $l \sim b^{-1}$. Using the dimensionless variables

$$\alpha = a/\sqrt{\mu_1},\ \beta = v/2\sqrt{\mu_1},\ \beta_0 = v_0/2\sqrt{\mu_1},\ x = b/\sqrt{\mu_1},$$

we get the dispersion formula

$$2x_\pm^2 = 1-\beta_0^2-2\beta^2+2\beta\beta_0 \pm \{(\beta_0^2-\epsilon^2)(\beta_0^2-\epsilon^{-2})+4\beta\beta_0[\beta\beta_0-(\beta_0^2-\epsilon^{-2})]\}^{1/2}$$

where $(\alpha,\alpha) = -1$ and $\epsilon^2 = (a^{(2)}-a^{(1)})/(a^{(2)}+a^{(1)})$ defines the condensate density 'asymmetry'. In the papers [32,36] solutions of this type are studied in much detail. We note here only the following features:

(1) For the same values of the condensate parameters a and v_0 there are two solitons moving with velocity v: 'narrow' one having a width $l_+ \sim 1/x_+$ and a 'broad' one for which $l_- \sim 1/x_-$.

(2) In addition to the soliton upper bound velocity inherent to the hole mode (see eqns. (40) and (51))in the case considered at $\beta_0 > \epsilon$ the lower bound velocity arises and the allowed velocity range decreases with β_0. We call this soliton a double bubble in a beam system ($2B_0$).

The fourth type of boundary conditions defines the condensate

$$\psi_v^{(1)} = 0,\ \ |\psi_v^{(2)}| = a^{(2)} = \sqrt{\mu_2},$$

representing a Bose gas with repulsion and density μ_2. If some amount of the first (gravitating) gas $\psi^{(1)}$ is injected into it (as a perturbation) then in the system states will arise described by the solutions

$$\psi_{IV}^{(1)} = c^{(1)} \operatorname{sech} \kappa\zeta \cdot \exp\{i(\frac{v}{2}x-\omega t)\}, \tag{4.54}$$

$$\psi_{IV}^{(2)} = c^{(2)}(\operatorname{th} \kappa\zeta + iv/2x),$$

where

$$\omega = v^2/4-\kappa^2-2(\mu_2-\mu_1),\ \kappa^2 = (c^{(1)})^2+(c^{(2)})^2$$

and

$$(c^{(2)})^2(1+v^2/4\kappa) = (a^{(2)})^2 = \mu_2.$$

Now the dispersion formula has the form

$$2(c^{(2)})^2 = \mu_2-(c^{(1)})^2+\frac{v^2}{4}+[(\mu_2-(c^{(1)})^2+\frac{v^2}{4})^2+4(c^{(1)})^2\mu_2]^{1/2}$$

and describes the velocity dependence of the inverse soliton size $l^{-1} \sim \kappa$. Solution (54) is the first component drop in the second component bubble. This solution, as the preceding one, cannot be obtained by isorotation of some solution to the system with a compact isogroup. At the same time, applying the isorotation to soliton (54) yields a new type of solution whose behaviour is analogous to the bion type solution [32,36]. In fact,

$$\psi' = h_1 \psi = \begin{bmatrix} \alpha & \beta \\ \bar{\beta} & \bar{\alpha} \end{bmatrix} \psi(54) = \tag{4.55}$$

$$= \begin{bmatrix} \alpha c^{(1)} e^{i\theta} \operatorname{sech} \kappa\zeta + \beta c^{(2)} (\operatorname{th} \kappa\zeta + iv/2\kappa) \\ \bar{\beta} c^{(1)} e^{i\theta} \operatorname{sech} \kappa\zeta + \bar{\alpha} c^{(2)} (\operatorname{th} \kappa\zeta + iv/2\kappa) \end{bmatrix} =$$

$$= c^{(1)} e^{i\theta} \operatorname{sech} x\zeta \begin{bmatrix} \alpha \\ \bar{\beta} \end{bmatrix} + c^{(2)} (\operatorname{th} \kappa\zeta + iv/2\kappa) \begin{bmatrix} \beta \\ \bar{\alpha} \end{bmatrix}.$$

Further transformation $h_1(\alpha', \beta')$ applied to (55) leads naturally to renormalization of the constants α and β only (angles θ and φ_i superposition).

From (55) a peculiar superposition of two particular solutions (2B) and (2D) of the system (49) turns out to be its solution as well.

6. Quasi-classical quantization of the $U(1,1)$ nonlinear Schrödinger equation

The four types of solitons derived above can be quantized through the Bohr-Sommerfeld procedure. Applying this to the integral [36]

$$\frac{1}{2} \int_0^T dt \int dx \left[\frac{\delta \mathcal{L}}{\partial \dot{\psi}^{(i)}} \dot{\psi}^{(i)} + \text{c.c.} \right] = 2\pi n_i$$

one obtains the number-of-particles functional

$$N_i = -\int (|\psi^{(i)}(x,t)|^2 - |\psi^{(i)}(\infty)|^2) dx = n_i \gg 1, \tag{4.56}$$

as well as the spectra which correspond to the energy

$$E = \int (\mathcal{H}(x,t) - \mathcal{H}(\infty)) dx$$

to be, for the solutions obtained,

$$E^I = \frac{1}{4} N^I [v^2 - \frac{1}{3} (N^I)^2], \quad \omega = \frac{1}{4} [v^2 - (N^I)^2], \tag{4.57a}$$

$$N^{I} = N_1 - N_2 = 2(c,c)^{1/2};$$

$$E^{II} = \frac{1}{3}(N^{II})^3 + 2\frac{\mu_2-\mu_1}{\mu_1}(a^{(2)})^2 N^{II}, \tag{4.57b}$$

$$N^{II} = N_2 - N_1 = 2\sqrt{-(c,c)} = 2\sqrt{\mu_1 - v^2/4};$$

$$E^{III} = \frac{1}{3}(2b)^3 + 2[\mu_2 - \mu_1 + \frac{v_0}{4}(\frac{v_0}{2} - v)]N_2, \tag{4.57c}$$

$$N_1 = \frac{2(a^{(1)})^2}{b + v^2/4b}, \quad N_2 = \frac{2(a^{(2)})^2}{b + (v - v_0)^2/4b},$$

$$N_2 - N_1 = 2\sqrt{-(\overline{c}c)} = 2b;$$

$$E^{IV} = \frac{1}{3}(2\kappa)^3 - (3\mu_2 - 2\mu_1)N_1 - \mu_2(N_1)^2/N_1, \tag{4.57d}$$

$$N_1 = 2(c^{(1)})^2/\kappa, \quad N_2 = \frac{2(a^{(2)})^2}{\kappa + v^2/4\kappa}, \quad N_1 + N_2 = 2\kappa, \ \mu_2 \neq 0.$$

These formulae may be rewritten in the differential form

$$dE = \sum_i \tilde{\mu}_i dN_i \quad (i = 1,2), \tag{4.58}$$

i.e. in the form of a thermodynamic relation for a mixture of gases, and then the coefficients $\tilde{\mu}_i$ play a role of the corresponding chemical potentials. Conditions (56) and (58) mean that the soliton solutions obtained represent bound states of boson quanta of the fields $\psi^{(1)}$ and $\psi^{(2)}$ (holes or particles and holes) taking into account the reservation of their special behaviour stated before eqn. (35).

In the case under investigation, a method of quasi-classical quantization of the equations of the inverse problem is also applicable which leads to [*)]

$$\rho(\kappa) = |a_0|^{-1}\frac{\kappa}{\sqrt{(\overline{a}_0 a_0) - \kappa^2}}, \quad |a_0| = \sqrt{(\overline{a}_0 a_0)},$$

when $(\overline{a}_0 a_0) > 0$ and to

$$\rho(\kappa) = |a_0|^{-1}\frac{\kappa}{\sqrt{\kappa^2 + (\overline{a}_0 a_0)}}, \quad |a_0| = \sqrt{-(\overline{a}_0 a_0)},$$

[*)] The initial perturbation is assumed to be a very broad (in space) packet ($l_0 \gg x^{-1}$) with an almost constant amplitude and $\psi_x^{(0)}/\psi^{(0)} \ll 1$.

when $(\overline{a}_0 a_0) < 0$.

In the first case $\kappa = \mathrm{Im}\xi$ corresponds to the soliton amplitude and in the second case $\kappa = v/2$, i.e. we have the velocity distribution function of bubbles.

Concluding this section we note that the infinitely extended one-dimensional antigravitating Bose gas described by the field $\psi^{(2)}$ possesses an analogue of fermionic properties - a nontrivial physical vacuum - condensate and hole spectrum [41, 45,46]. In the framework of the continuum Hubbard model (see Chapter 1) it corresponds to the antisymmetric (in spin) wave function $\psi^{(2)} = \psi_{\uparrow} - \psi_{\downarrow}$ which describes the magnetic properties of crystals.

A successive and detailed account of the topics related to the quasi-classical quantization of solitons in magnetic systems can be found in [47] (see also [48,49], and the special review [50] devoted to these problems). Results obtained via the exact quantization technique (the quantum inverse scattering method) are given in the review [4b]. A more detailed discussion of these is beyond the scope of this book. We stress, however, that in the framework of complete integrable systems, quasi-classical spectra often coincide with exact ones, which is one of the remarkable features of such systems.

Finally, the $U(1,1)$ model considered above has not been studied completely on a quantum level, despite the fact that one of the basic constructions of the quantum inverse method, the R-matrix, was found in [51]. This is due to difficulties in constructing the vacuum state and the fact that the metric in the state space is indefinite.

7. The Sine-Gordon equation

It is convenient to write this equation in a form slightly different from eqn. (2.42) of chapter 2:

$$\Box\chi + \frac{m^2}{g}\sin g\chi = 0, \tag{4.59}$$

where m and g are the mass and coupling constants, respectively. There exist a large number of papers devoted to this equation, e.g. the well-known reviews [33,52] for the classical version and [4b] for the quantum one.

This equation is completely integrable and generated by the Hamiltonian

$$H = \frac{1}{g}\int(\frac{1}{2}\pi^2 + \frac{1}{2}(\varphi_x)^2 + 1 - \cos\varphi)dx \tag{4.60}$$

and the Poisson brackets

$$\{\pi(x), \varphi(y)\} = \delta(x-y)$$

with

$$\pi(x,t) = \varphi_t(x,t), \quad \varphi = g\chi, \quad m = 1.$$

The spectral problem for this equation in terms of the matrices U and V may be formulated as:

$$U = \frac{i\xi}{2}\sigma_3 + \frac{i}{4}(\varphi_x - \varphi_t)\sigma_1 - \frac{i}{8\xi}(\sigma_3\cos\varphi + \sigma_2\sin\varphi), \tag{4.61}$$

$$V = U + \frac{i}{4\xi}(\sigma_3\cos\varphi + \sigma_2\sin\varphi), \quad \sigma_1 = \begin{pmatrix} 0 & 1 \\ 1 & 0 \end{pmatrix}, \quad \sigma_2 = \begin{pmatrix} 0 & -i \\ i & 0 \end{pmatrix},$$

$$\sigma_3 = \begin{pmatrix} 1 & 0 \\ 0 & -1 \end{pmatrix}.$$

Sometimes eqn. (59) is written in the so-called cone variables

$$\varphi_{\xi t} = \sin\varphi. \tag{4.62}$$

In this case, the linear problem is defined by the matrices

$$U = i\xi\sigma_3 + \frac{i}{2}\varphi_x\sigma_1, \quad V = \frac{1}{4i\xi}(\sigma_3\cos\varphi + \sigma_2\sin\varphi). \tag{4.63}$$

As has been shown in [33], there exists a one-to-one correspondence of solutions to problems (61) and (63), therefore it is sufficient to consider only one of them. In contrast to the case of $S3$, the function φ sought for in the Sine-Gordon equation is real, which puts additional conditions on the scattering matrix elements

$$S_{11}(\xi) = \bar{S}_{11}(-\bar{\xi}), \quad S_{12}(\xi) = -\bar{S}_{12}(-\bar{\xi}), \quad S = \begin{bmatrix} S_{11} & S_{12} \\ -\bar{S}_{12} & \bar{S}_{11} \end{bmatrix},$$

and so the S_{11} zeros are symmetrically localized with respect to the imaginary axis. If $\mathrm{Re}\,\xi = 0$, then S_{11} is real. Here, as for the $S3$ case, we have the following evolution of S

$$\dot{S}_{11}(\xi) = 0, \quad S_{12}(\xi,t) = S_{12}(x,0)\exp(-t/2\xi),$$

for a continuous spectrum and

$$\dot{\xi}_n = 0, \quad S_{12}^{(n)}(t) = S_{12}^{(n)}(0)\exp(-t/2\xi_n),$$

where

$$S_{12}^{(n)} = S_{12}(\xi_n)$$

for a discrete case.

In the action-angle variables, momentum and energy are (see [33]) as follows:

$$P = \int_0^{\infty}(\frac{1}{4\xi}-\xi)\rho(\xi)d\xi - \sum_{k=1}\frac{2}{g}(e^{-p_k}-4e^{p_k}) - \sum_{k=1}\frac{4}{g}(e^{-n_k/4}-4e^{n_k/4})\sin\frac{g\eta_k}{16}, \tag{4.64}$$

$$H = \int_0^{\infty}(\frac{1}{4\xi}+\xi)\rho(\xi)d\xi + \sum_{k=1}\frac{2}{g}(e^{-p_k}+4e^{p_k}) + \sum_{k=1}\frac{4}{g}(e^{-n_k/4}+4e^{n_k/4})\sin\frac{g\eta_k}{16},$$

where

$$\rho(\xi) = -\frac{4}{\pi g\xi}\ln|S_{11}(\xi)|^2, \quad \kappa(\xi) = \arg S_{12}(\xi), \quad \xi > 0$$

for a continuous spectrum,

$$p_n = \ln|\xi_n|, \quad q_n = -\frac{8}{g}\ln|S_{12}^{(n)}|, \quad (\mathrm{Re}\,\xi_n = 0)$$

for the soliton part (kinks and antikinks) and

$$n_k = 4\ln|\xi_k|, \quad \nu_k = -\frac{4}{g}\ln|S_{12}^{(k)}|,$$

$$\eta_k = \frac{16}{g}\arg\xi_k, \quad \psi_k = \arg S_{12}^{(k)}, \quad (0 < \arg\xi_k < \frac{\pi}{2}),$$

for bions.

From these formulae it follows that the Sine-Gordon equation allows two types of solutions in the discrete spectrum depending on the position of the $S_{11}(\xi)$ function zeros[*)]:

[*)]In fact there are three of them, since solitons consist of kinks and antikinks depending on their topological charge. However, as the Hamiltonian is symmetric with respect to a discrete group related to this charge, the energy does not depend on Q_{top} and the number of kink and antikinks in the sums (64) depends on the initial and boundary conditions.

(1) Solitons (kinks (+) and antikinks (-))

$$\varphi_k(x,t) = 4\operatorname{arctg}\{\exp\{\pm\gamma(x - vt - x_0)\}\}, \quad v = \frac{4\lambda^2 - 1}{4\lambda^2 + 1}, \tag{4.65}$$

$$\zeta = i\lambda, \quad \operatorname{Re}\zeta = 0,$$

possessing the conserved topological charge

$$Q_{\text{top}} = \frac{1}{2\pi}\int_{-\infty}^{\infty} \varphi_x dx = \pm 1$$

respectively. The $s_{11}(\xi)$ zeros lie in the imaginary axis.

(2) bions (bound states of kink and antikink)

$$\varphi_b = 4\operatorname{arctg}\left\{\frac{\operatorname{Im}\xi}{|\operatorname{Re}\xi|}\frac{\sin(\frac{\operatorname{Re}\xi}{|\xi|}\gamma(t - vx - t_0))}{\operatorname{ch}(\frac{\operatorname{Im}\xi}{|\xi|}\gamma(x - vt - x_0))}\right\}, \quad v = \frac{4|\xi|^2 - 1}{4|\xi|^2 + 1}, \tag{4.66}$$

defined by two zeros of $s_{11}(\xi)$, ξ_1 and ξ_2 placed symmetrically with respect to the imaginary axis, i.e. $\bar{\xi}_1 = -\xi_2$.

It is advisable to use the notations

$$\omega = \frac{\operatorname{Re}\xi}{|\xi|},$$

then

$$\frac{\operatorname{Im}\xi}{|\xi|} = \sqrt{1-\omega^2}, \quad \frac{\operatorname{Im}\xi}{|\operatorname{Re}\xi|} = \sqrt{\frac{1}{\omega^2} - 1}$$

therefore

$$\sin\frac{g\eta}{16} = \frac{\operatorname{Im}\xi}{|\xi|} = \sqrt{1-\omega^2}, \quad \omega = \cos\frac{g\eta}{16}, \quad \gamma = (1-v^2)^{-1/2} = \frac{1+4|\xi|^2}{4|\xi|}.$$

In these variables, the energy and momentum of solitons and bions are expressed quite simply

$$E_k = M_k \cdot \gamma, \quad P_k = E_k v$$

where $M_k = 8/g$ is the rest mass of the solitons (kink or antikink) and

$$E_b = M_b \cdot \gamma, \quad P_p = E_b \cdot v \ \text{and} \ E_b = \frac{16}{g}\sqrt{1-\omega^2} = \frac{16}{g}\sin\frac{g\eta}{16}$$

is the bion rest mass. Notice that in the classical approach $E_b \to 0$ when $\omega^2 - 1$.

Also note that by virtue of the integrability of eqn. (59), both solitons and bions with zero topological charge are stable and interact elastically with linear waves and solitons and with each other. They may arise only through the decay of the initial perturbation and are not the result of a pair interaction of kink and antikink. This is the essential difference between the Sine-Gordon model and the relativistic φ^4 model, since in the letter the production of quasi-bions (long-lived bound states) in the kink-antikink interaction is possible, if their relative velocity is smaller than a critical one (here we omit delicate details of the interaction picture in the framework of the φ^4 theory; see the review [40] and chapter 11).

Here we briefly discuss the Sine-Gordon model quantization. Approximate quasiclassical spectra wave obtained in [53,54] (see also the review [50]).

The continuous spectrum quanta, 'phonons', are relativistic particles with mass $m = 1$, energy $m\gamma$ and momentum $m\gamma v$.

The kinks are massive particles with mass $M_k = 8m^3/g = -m/\pi = 8m/g$ where $\tilde{g}$ is the coupling constant renormalized by quantum effects,

$$\tilde{g} = \frac{g/m^2}{1 - g/8\pi m^2}.$$

The bions are peculiar bound state of 'phonons' (nonlinear phonons) and their spectrum is determined by the formula

$$M_b(n) = \frac{16m}{\tilde{g}} \sin\left[\frac{\tilde{g}\cdot n}{16}\right], \quad n = 1,2,\ldots < \frac{8\pi}{\tilde{g}}.$$

This means that:

(1) the maximum number of phonons which may be bound in the quantum bion is determined by the quantity $n_{\max} = 8\pi/\tilde{g}$;

(2) when $8\pi/\tilde{g} < 1$, there are no quantum bions.

An interesting situation occurs at $8\pi/\tilde{g} \gg 1$. In this case, in the lowest state $n = 1$ the bion corresponds to a free 'phonon', since $\sin(\tilde{g}/16 \simeq \tilde{g}/16$ and $M_b(1) \simeq m$ and to the bound state when $n \neq 1$.

Finally, notice that in the classical limit of weak coupling at $\tilde{g}\eta/16 \ll 1$ we have $\sin x \simeq x - x^3/6$ or $M_b \simeq \eta(1 - \tilde{g}^2/6(\eta/16)^2)$. The second term of this formula corresponds to the 'phonon' binding energy. Compare this energy with the constituent binding energy in the $S3$ soliton. In this case, assuming $\operatorname{arctg} x \simeq x$, $\gamma = 1+v^2/2$ we have, accurate to $0(x)$,

$$\varphi_b \simeq 4a\sin[\omega(vx - t(1+\frac{v^2}{2}))]\operatorname{sech}(a(x-vt)), \quad \omega \simeq 1-a^2/2$$

i.e.

$$\varphi_b \simeq 4a\frac{\sin(vx-\Omega t)}{\operatorname{ch}a(x-vt)} = 4a\operatorname{Im}\frac{e^{i(vx-\Omega t)}}{\operatorname{ch}a(x-vt)}$$

where $\Omega = 1+v^2/2-a^2/2$.

Comparing this expression with eqn. (28) ($g=1$) for the $S3$ soliton at $m = 1$ we see that

$$\varphi_b = 2\sqrt{2}\operatorname{Im}(e^{-it}\psi_{S3}),$$

and since $a = \eta/16$, $\eta = N_b$, $N_{S3} = 4a$, we find easily that binding energies of constituents in the bion and the soliton are the same (and $N_b = 4N_s$). Thus in the weak coupling (or small amplitude) and nonrelativistic velocity limit the Sine-Gordon bions are closely connected with the envelope solitons of the $S3$ equation*).

The observed feature of bions leads to the situation that as in the systems described by the Schrödinger equations, in relativistic theory when an external field acts on the systems, a cumulative effect of energy accumulation in the bionic excitations is possible even when the external field amplitude is less than the threshold of the kink-antikink pair production $E < 2M_k$. On account of this fact the discovery of long-lived bound states of solitons and the investigation of their properties are one of the most important parts of soliton theory (especially as regards non-one-dimensional models). For non-integrable models with polynomial or logarithmic Lagrangians (in particular the φ^4 theory) and also for the multiple Sine-Gordon equation exact analytical solutions describing the bound states have not been obtained. In this case, the main method of investigation is by numerical experiment (see the review [40] where properties of quasi-bions in some models as well as a number of approximate methods are described in detail).

*)Notice their difference. In the lowest quantum states the $S3$ soliton binding energy has the form $E_s \sim -N(N^2-1)$ and for the Sine-Gordon equation $E_b \sim -N^3$.

References

1. a) S.P. Novikov, S.V. Manakov, L.P. Pitaevskii and V.E. Zakharov. *Theory of solitons: The inverse scattering method.* N.Y. Consultants Bureau, 1984. 276 p. (Contemporary Soviet Mathematics.)

   b) R.K. Dodd, et al. see [34] part I.

   c) F. Calogero and A. Degasperis. *Spectral transform and solitons: Tools to solve and investigate nonlinear evolution eqs.* North-Holland, Amsterdam, 1982, 516 p. (Studies in Mathematics and its Applications v. 13.)

   d) *Solitons*. R.K. Bullough and P.J. Caudrey eds. Springer, N.Y. 1980, (Topics in Current Physics, Vol. 17).

2. a) *Solitons in action.* K. Lonngren and A. Scott eds. Academic, N.Y. 1978.

   b) *Solitons and condensed matter physics*. A.R. Bishop and T. Schneider eds. Springer, Berlin, N.Y. 1978.

   c) A.R. Bishop, J.A. Krumhansl and S.E. Trullinger. *Solitons in condensed matter: a paradigm.* Physica, 1980, 1D, p. 1-44.

   d) V.G. Makhankov, V.K. Fedyanin. *Non-linear effects in quasi-one-dimensional models of condensed matter theory.* Phys. Rep., 1984, 104, p. 1-86.

   e) Solitons in Physics. H. Wilhelmsson ed. Topical issue of Physica Scripta, 1979. 20, No. 3-4, 250 p.

3. V.E. Zakharov, L.D. Faddeev. *KdV equation: A complete integrable Hamiltonian system.* Funct. Anal. Appl. 1971, 5, p. 280-287.

4. A.G. Izergin, V.E. Korepin. *Quantum method of inverse problem.* Particles and Nuclei (Moscow), 1982, 13, p. 501-541.

5. H.B. Thacker. *Exact integrability in quantum field theory and statistical systems.* Rev. Mod. Phys. 1981, 53, p. 253-285.

6. I.M. Krichever. *Methods of algebraic geometry in the theory of non-linear equations.* Uspekhi Mat. Nauk, 1977, 32, p. 184-208.

7. M.A. Salle. *Darboux transformation for nonabelian and nonlocal equations of Toda chain type.* Teor. Mat. Fiz. 1982, 53, p. 227-237.

8. See, e.g. M.A. Olshanetsky, A.M. Perelomov. *Classical integrable finite-dimensional systems related to Lie algebras.* Phys. Rep. 1981, 71, p. 313-400 and also refs. [9] and [11].

9. V.G. Makhankov, O.K. Pashaev. *Nonlinear Schrödinger equation with noncompact isogroup.* Teor. Mat. Fiz. 1982, 53, p. 55-67; *On the gauge equivalence of the Landau-Lifshitz and NLS on symmetric spaces.* Phys. Lett. 1983, 95A, p. 95-100.

10. J. Honerkamp. *Gauge equivalence of exactly integrable field theoretic models.* THEP 79/5 Freiburg, 1979.

11. A.P. Fordy, P.P. Kulish. *Nonlinear Schrödinger equations and simple Lie algebras.* Comm. Math. Phys. 1983, 89, p. 427-443.

12. S.J. Orfanidis. *σ-models of nonlinear evolution equations.* Phys. Rev. 1980, 21D, p. 1513-1522.

13. S.J. Orfanidis. *$SU(n)$ Heisenberg spin chain.* Phys. Lett. 1980, 75A, p. 304-306.

14. R. Sasaki, Th.W. Ruijgrok. *An integrable $SU(3)$ spin chain.* Physica, 1982, 113A, p. 388-400.

15. E. Cremmer, B. Julia and J. Scherk. *Supergravity theory in 11 dimensions.* Phys. Lett. 1978, 76B, p. 409-412.

E. Cremmer and B. Julia. *The $SO(8)$ supergravity.* Nucl. Phys. 179, 159B, p. 141-212.

16. B. Julia and J.F. Luciani. *Nonlinear realizations of compact and noncompact gauge groups.* Phys. Lett., 1980, 90B, p. 270-274.

17. V.G. Makhankov, O.K. Pashaev and S.A. Sergeenkov. *Hole-like excitations in many component systems.* Physica Scripta, 1984, 29, p. 521-525 and Bose-gas models with noncompact symmetry groups. In: Proc. of the III International Symposium on Selected Topics in Statistical Mechanics, Dubna, August 1984. JINR, D17-84-850, Dubna, 1984, v. 2, p. 45-52.

18. M. Lakshmanan, T. Ruijgrok and C. Thomson. *On the dynamics of a continuum spin system.* Physica, 1976, 84A, p. 577-690.

19. M. Lakshmanan. *Continuum spin system as an exactly solvable dynamic system.* Phys. Lett. 1977, 61A, p. 53-54.

20. J. Corones. *Solitons as nonlinear magnons.* Phys. Rev. 1977, 16B, p. 1763-1765.

21. L. Takhtajan. *Integration of the continuum Heisenberg spin chain through the*

inverse scattering method. Phys. Lett. 1977, 64A, p.235-237.

22. V.E. Zakharov and L. Takhtajan. *Equivalence of NLS and Heisenberg ferromagnet.* Teor. Mat. Fiz. 1979, 38, p. 27-35.
23. A. Kundu, V. Makhankov and O. Pashaev. *On the gauge equivalence of the Landau-Lifshitz and NLS equations.* JINR E17-82-601, Dubna, 1982.
24. A. Borovik. *N-soliton solutions to nonlinear Landau-Lifshitz equation.* Pisma ZhETF, 1978, 28, p. 629-632.
25. A. Barut and R. Raczka. *Theory of group representations and applications.* PWN - Polish Scientific Publishers, Warszava 1977, v. 1, 455p.
26. A. Borovik. *Exact integration of the Landau-Lifshitz equation.* Solid State Comm. 1980, 34, p. 721.

    E. Sklyanin. *On complete integrability of the LL equation.* Preprint LOMI E-3-1979, Leningrad 1979, 32p. See also [3] part I.
27. A. Kundu and O. Pashaev. *Comments on the gauge equivalence between Heisenberg spin chains with single-site anisotropy and NLS.* J. Phys. 1983, 16C, p. L585-590.
28. V.P. Kotlyarov. *On the gauge equivalence of the equation of dynamics for uniaxial ferromagnets.* J. Phys. 1984, 17C, p. L139-L143.
29. K. Pohlmeyer. *Integrable Hamiltonian systems and interactions through quadratic constraints.* Comm. Math. Phys. 1976, 46, p. 707-718.
30. A. Neveu and N. Papanicolaou. *Integrability of the classical* $[\bar{\varphi}_i\psi_i]]_2^2$ *and* $[\bar{\psi}_i\psi_i]_2^2-[\bar{\psi}_i\gamma_5\psi_i]_2^2$ *interactions.* Comm. Math. Phys. 1978, 58, p. 31-64.
31. V. Makhankov and R. Myrzakulov. *σ-model representation of the Yajima-Oikawa equations.* JINR Comm. P5-84-719, Dubna, 1984.
32. V. Makhankov and O.K. Pashaev. *NLS equation with* $U(p,q)$ *isogroup.* Part I. General Analysis JINR E2-81-264, Dubna 1981; Part II. JINR E2-81-540, Dubna 1981, see also [9].
33. S.P. Novikov, et al. See [1a] part II.
34. V. Makhankov and O.K. Pashaev. *On properties of the NLS equation with* $U(p,q)$ *internal symmetry.* JINR Comm. E2-81-70, Dubna, 1981.
35. V. Makhankov and O.K. Pashaev. *A new integrable model of QFT in the state space with indefinite metric.* Phys. Lett. 1982, 89A, p. 218.

36. V. Makhankov. *Quasi-classical solutions in the Lindner-Fedyanin model - 'Hole' like excitations.* Phys. Lett. 1981, 81A, p. 156-160.

37. V. Zakharov and A. Shabat. *Exact theory of self-focusing and one-dimensional automodulation of waves in nonlinear media.* ZhETF 1971, 61, p. 118-134.

38. P. Kulish, S. Manakov and L. Faddeev. *Comparison of exact quantum and quasi-classical results for NLS.* Teor. Mat. Fiz. 1976, 28, p. 38. See [48] part I.

39. F.A. Berezin, G.P. Pokhil and V.M. Finkelberg. Vestnik Moscow University, 1964, ser. I, No. 1, p. 21.

40. V. Makhankov. See [8] part I.

41. E. Lieb. *Exact analysis of an interacting Bose gas II. The excitation spectrum.* Phys. Rev. 1963, p. 1616-1624.

42. P. Kulish. *Scattering of solitons with internal degrees of freedom.* In Problems of High Energy Physics and Quantum Field Theory. Proc. of II International Seminar, Protvino, 1979, p. 463-470.

Y. Ichikawa. *Topics on solitons in plasmas.* Physica Scripta 1979, 20, p. 296-305.

44. U. Lindner and V. Fedyanin. *Solitary solutions in a modified Hubbard chain.* Phys. Stat. Sol. 1979, 95 (b), p. K83-K85.

45. L. Faddeev. *Quantum complete integrable models of field theory.* JINR Comm. D2-12462, Dubna, 1979, p. 249-299.

46. E. Lieb and W. Liniger. *Exact analysis of an interacting Bose gas I.* The general solution and ground state. Phys. Rev. 1963, 130, p. 1605-1615.

47. A.M. Kosevich, et al. See [1a] part I.

48. A. Kosevich, B. Ivanov and A. Kovalev. *Nonlinear localized wave of ferromagnet magnetization as a bound state of a large number of magnons.* Pisma ZhETF 1977, 25, p. 516-520;

Magnon drops: a new type of collective excitation of ferromagnets. J. de Physique 1978, 39, C6-826; also Fiz. Niz. Temp. 1977, 3, p. 906-921.

49. V. Makhankov. See [7] part I.

50. L. Faddeev and V. Korepin. *Quantum theory of solitons.* Phys. Rep. 1978, 42, p. 1-87. See [4] part II.

51. V. Makhankov and O.K. Pashaev. *A new integrable model of QFT in the state space with indefinite metric,* see [35] part II.

52. A. Bishop. *Solitons in condensed matter physics.* Physica Scripta 1979, 20, p. 409-423; see also Soliton in action, eds. K. Lonngren and A. Scott (Academic Press N.Y. 1978) papers by A. Bishop (p. 72-101) D. McLaughlin and A. Scott (p. 201-255) and paper by P. Parmentier.

53. R. Dashen, B. Hasslacher and A. Neveu. *Particle spectrum in model field theories from semiclassical functional integral techniques.* Phys. Rev. 1975, 11D, p. 3424-3449.

54. V. Korepin and L. Faddeev. Teor. Mat. Fiz. 1975, 12, p. 143.

Part III. Noncompact Symmetries and Bose-Gas

Chapter 5.

DYNAMICAL SYMMETRY AND GENERALIZED COHERENT STATES

Bose gas and dynamical symmetry group

A nonideal Bose gas as a system which models various physical phenomena has been the subject of attention for a long time. Typical examples are superfluidity theory, theories of magnetism, nuclear matter and solid state. A great deal of work in both physics and mathematics has been done in studying the properties of such models. Taking into account only pair interactions of bosons, which is quite interesting from a physical point of view, in the case of one space dimension sometimes leads to exact mathematical solutions both on quantum and classical levels.

The simplest version of this model is a Bose gas with point-like pair interactions of particles. On the quasi-classical (and classical) level this is described by the cubic nonlinear Schrödinger equation $S3$:

$$i\psi_t+\psi_{xx}+\epsilon|\psi|^2\psi = 0, \tag{5.1}$$

where $|\psi|^2 = \rho$ is the boson density and $\epsilon = \pm 1$ for the attractive (repulsive) case, respectively.

As we have already seen in the preceding chapter, the linear problem for eqn. (1) is formulated on the algebra $su(2)$ in the first case $(\epsilon=+1)$ and $su(1,1)$ in the second case $(\epsilon=-1)$. Although in both cases eqs. (1) is $U(1)$ invariant, physical and especially mathematical formulations are essentially different. This is also true of the procedure for solving the Cauchy problem by means of the inverse spectral method.

The natural physical situation in the first case corresponds to the interaction of a finite number of bosons and, as a result, their peculiar bound states - solitons - can be produced. Such a physical picture is described by a boundary value problem with zero (trivial) boundary conditions on the whole axis or periodic conditions on the ring. The plane wave solutions (condensate) within the framework of the $\epsilon = 1$ model are found to be unstable.

In the second case, the situation is the opposite. A system of finite number of repelling bosons (trivial boundary conditions on the axis) is unstable: an arbitrary initial packet spreads out at the rate $\psi \propto t^{-1/2}$. Physically interesting is to consider stable condensate excitations (a system of an infinite number of particles), i.e. nontrivial (constant) boundary conditions at infinity or periodic conditions on the ring. The second situation usually occurs in the study of quantum models via the Bethe Ansatz or the quantum inverse method. In the latter case, the term condensate may be used only conventionally, since in such a formulation the Bose gas behaviour is fermionic-like (see [22]).

In the on-the-axis case, the Bose gas has a number of remarkable properties which definitely indicate the existence of a noncompact symmetry in the system. Among these properties, are a condensate state of an infinite number of particles and Bogolubov excitations connected with the hyperbolic canonical transformation. Finally, in the one-dimensional system*) there exists a 'hole' (Lieb's) excitation mode which corresponds to the hole-type soliton, the kink on the classical level (in the self-consistent field method). Note that the kink amplitude depends on the kink velocity and vanishes as this tends to the sound velocity. This property resembles the Lorentz contraction related to Lorentz group noncompactness.

The kink of the nonlinear Schrödinger equation describing a non-ideal repulsive Bose gas, as we noted above, may be thought of on the quasi-classical level as a bound state of either a finite number of holes [1] or an infinite number of particles [2]. The bound states of a finite number of particles in such a system are unstable and decay rapidly. These properties of the nonideal Bose gas indicate the presence of a noncompact symmetry in the system. All the unitary irreducible representations of noncompact groups are infinite-dimensional. These representations can be realized by introducing the finite-density condensate

$$\rho_0 = \lim_{\substack{N\to\infty \\ V\to\infty}} N/V \tag{5.2}$$

*) Discussions of hole excitations in more-dimensional systems can be found below.

in the quantum system and by imposing nontrivial boundary conditions on the field in the classical case.

In the following, using the results obtained in [1,2] we try to extract the group-theory structure of some models of a nonideal Bose gas connected with a non-Abelian dynamical group of linearized many-dimensional models and with the integrability of one-dimensional nonlinear models. Such consideration draws substantially on the results of Solomon [5,6], Kuratsuji and Suzuki [9], Gerry and Silverman [8], and Perelomov [7].

The dynamical symmetry group first arose in particle physics when various multiplets of particles were brought together in one irreducible representation of some noncompact group [3]. After that applications were found in a number of one-particle problems (such as the hydrogan atom, etc.) [4-6]. In the case of the ordinary symmetry groups, the Hamiltonian of the system commutes with all their generators. This means that energy degenerated states are transformed via the symmetry group representations. This is not the case for the dynamical symmetry group when the Hamiltonian of the system belongs simply to the group algebra, which is therefore denoted the spectrum generating algebra.

In some many-particle problems, the Hamiltonian of the system when linearized may be expressed through the generators of some dynamical symmetry group. Then the solution of the quantum mechanical eigenvalue problem reduces to a solution of the proper theoretic-group problem, i.e. to finding the irreducible representations of the dynamical group. In addition to yielding complete information about the spectra of the problem, the dynamical group makes it possible (1) to construct the natural coherent states for the initial quantum system [7], (2) to represent the Green functions via the Feynman integral over these states, and (3) to describe the classical behaviour of the system (see [9] for $SU(2)$ and [8] for $SU(1,1)$ models).

The theory of magnetism is especially interesting in this respect since there is a quantum microscopic theory based on the Heisenberg model, on the one hand, and a macroscopic theory of magnetism governed by the Landau-Lifshitz equations on the other hand (see chapters I and II). Sometimes one can establish such a correspondence, for example, for isotropic and easy-axis ferromagnets [10] but not yet for the antiferromagnet and some models of the anisotropic ferromagnet. As a first step in this direction one can construct the dynamical group and the coherent states of the proper spin models.

2. Quantum version (Generalized Coherent State)

In order to discuss quantum models it is necessary to introduce the concept of a generalized coherent state as formulated in a complete form by Perelomov [7b]. The reader interested in more details is referred to the review [7a] where one can find applications, as well as a historical and literature survey. In the following, we will use, to a considerable extent, the results of papers [7] illustrating general statements via examples of the $SU(2)$ and $SU(1,1)$ symmetries.

1. Let G be an arbitrary Lie group and let T be its irreducible unitary representation acting in a Hilbert space $\mathcal{H}$. Denote a vector of this space by $|\psi>$, the scalar product of two vectors by $<\varphi|\psi>$ and the projection operator on $|\psi>$ by $|\psi><\psi|$.

Fix a vector $|\psi_0> \in \mathcal{H}$. Let us consider the set of vectors $\{|\psi(g)>\}$ where $|\psi(g)> = T(g)|\psi_0>$ and g spans all the group G. The vectors which determine the same state, i.e. differing from one another only by a phase factor, are attributed to the same equivalence class $(|\psi(g_1)>\sim|\psi(g_2)>)$ if $|\psi(g_1)> = e^{i\theta}|\psi(g_2)>$, which is possible if $T(g_2^{-1}\cdot g_1)|\psi_0> = e^{i\theta}|\psi_0>$.

Let $H = \{h\}$ be the set of elements of the group G such that $T(h)|\psi_0> = e^{i\theta(h)}|\psi_0>$. This set can easily be shown to be a subgroup of the group G. It is natural to denote this as the stationary group of the vector $|\psi_0>$. From this construction we see that the vectors $|\psi(g)>$, corresponding to the elements of one left coset $g_{1h} \in g_1H$, only differ from one another by a phase factor and so determine the same state. Then distinct vectors (states) will correspond to the elements g_m of the coset space $M = G/H$ (a homogeneous space occurring in the preceding chapter). To describe a set of different states it is therefore sufficient to select one representative in each class. Geometrically, this construction can be interpreted in the following way: we can consider the group G as a fiber bundle with the base $M = G/H$ and the fiber H, then the choice of g_m corresponds to a certain cross-section of this fiber bundle.

This leads to a set of vectors (states) $\{|\psi_m>\}:|\psi_m> = T(g_m)|\psi_0>$, $g_m \in G/H$ which is called a system of coherent states on the group G with the reference vector $|\psi_0>$, or, alternatively, the generalized coherent states.

2. We show that ordinary coherent states introduced by Glauber are a

particular case of our generalized coherent states on the Heisenberg-Weil group with a three-parameter algebra defined, for example, by the Bose creation and annihilation operators

$$[a^+,a^+] = [a,a] = [a,I] = [a^+,I] = 0, \tag{5.3}$$

$$[a,a^+] = I.$$

The Heisenberg-Weil group W_1 is a group of transformations $T(g) = e^t e^{\alpha a^+ - \bar{\alpha} a}$, $g = g(t,\alpha)$, and the coherent states $\{|\alpha>\}$ are obtained by the action of the operator $T(g)$ on $|\psi_0> = |0>$, where $|0>$ is a vector satisfying the condition $a|a> = 0$ (the vacuum):

$$T(g)|0> = e^{it}|\alpha>, \quad |\alpha> = e^{\alpha a^+ - \bar{\alpha}a}|0> \equiv D(\alpha)|0>. \tag{5.4}$$

The coherent state $|\alpha>$ is specified by a complex number α, while the group element $g(t,\alpha) \in W_1$ is determined by three parameters. The fact that the number of parameters is reduced is connected with the factorization of W_1 by the stationary subgroup H which is now $U(1)$ and $h = e^{it} = g(t,0)$. The coherent state $|\alpha>$ is determined by a point in the homogeneous space W_1/H, i.e. the complex plane $\mathbb{C}^1$ in the case under consideration.

The Heisenberg-Weil coherent state $|\alpha>$ is often defined as an eigenvalue of the operator a

$$a|\alpha> = \alpha|\alpha>. \tag{5.5}$$

This can be expanded in terms of the usual n-quantum states $|n> = (n!)^{-1/2}(a^+)^n|0>$ forming the orthonormal basis:

$$|\alpha> = e^{-1/2|\alpha|^2}\sum_{n=0}^{\infty}\frac{\alpha^n}{\sqrt{n!}}|n>. \tag{5.6}$$

It is not difficult to show that definitions (4) and (5) are the same. To proceed, we use the known identity

$$e^A e^B = e^{1/2[A,B]}e^{A+B} \tag{5.7}$$

which holds (in particular, for the Heisenberg-Weil algebra (3)) when

$$[A,[A,B]] = [B,[A,B]] = 0$$

and we write $D(\alpha)$ in the normal form:

$$D(\alpha) = e^{\alpha a^+ - \bar{\alpha} a} = e^{-1/2|\alpha|^2} e^{\alpha a^+} e^{-\bar{\alpha} a}.$$

Then for the usual coherent states we have

$$|\alpha\rangle = D(\alpha)|0\rangle = e^{-1/2|\alpha|^2} e^{\alpha a^+} |0\rangle \tag{5.8}$$

from which expansion (6) follows immediately.

Further, notice that the commutators

$$[a^+, a^n] = -na^{n-1},$$

$$[a, (a^+)^n] = n(a^+)^{n-1},$$

are analogous to derivatives; then find

$$D(\alpha) a D^+(\alpha) = a - \alpha I, \tag{5.9}$$

$$D(\alpha) a^+ D^+(\alpha) = a^+ - \bar{\alpha} I,$$

whence

$$a|\alpha\rangle = aD(\alpha)|0\rangle = e^{-1/2|\alpha|^2} a e^{\alpha a^+} |0\rangle = e^{-1/2|\alpha|^2} \alpha e^{\alpha a^+} |0\rangle = \alpha|\alpha\rangle.$$

QED.

3. If we now take $SU(2)$ or $SU(1,1)$ as the group G then we get the corresponding generalized coherent states. Algebras of these groups are three-parameter and defined by the generators

$$J_\pm = J_1 \pm iJ_2, \quad J_0 = J_3$$

and

$$K_\pm = K_1 \pm iK_2, \quad K_0 = K_3$$

respectively. Their commutation relations are

$$[J_0, J_\pm] = \pm J_\pm, \quad [J_-, J_+] = -2J_0 \tag{5.10a}$$

and

$$[K_0, K_\pm] = \pm K_\pm, \quad [K_-, K_+] = 2K_0 \tag{5.10b}$$

while the operators

$$\hat{C}_1 = J_0^2 + \frac{1}{2}(J_+J_- + J_-J_+) \equiv J_0^2 + J_1^2 + J_2^2 \tag{5.11a}$$

and

$$\hat{C}_2 = K_0^2 - \frac{1}{2}(K_+K_- + K_-K_+) \equiv K_0^2 - K_1^2 - K_2^2 \tag{5.11b}$$

commute with all the operators J_i in the first case and K_i in the second case, i.e. these are the Casimir operators. According to Schur's lemma, for irreducible representations they are a multiple of the identity operator

$$\hat{C}_1 = j(j+1)I, \tag{5.12a}$$

$$\hat{C}_2 = k(k-1)I. \tag{5.12b}$$

Thus the $SU(2)$ and $SU(1,1)$ group representations are specified by a number, j in the first case and k in the second.

Let us paramterize $T(g)$ as follows

$$T_1(g) = \exp(\alpha J_+ - \bar{\alpha} J_- + i\lambda J_0), \quad |\alpha| \leqslant \frac{\pi}{2},$$

and analogously

$$T_2(g) = \exp(\alpha K_+ - \bar{\alpha} K_- + i\lambda K_0),$$

where λ is a real parameter.

If we now choose $|\psi_0>$ as an eigenvector of the operator J_0, $(K_0)^{*)}$ then the operator $e^{i\lambda J_0}$, $(e^{i\lambda K_0})^{*)}$ action on the vector $|\psi_0>$ leads simply to its phase shift, i.e. the elements $h = e^{i\lambda J_0}$, $(e^{i\lambda K_0})$ form the stationary group of the vector $|\psi_0>$.

This group coincides again with $U(1)$, so in the cases under consideration the generalized coherent states are defined on the homogeneous spaces $SU(2)/U(1)$ and $SU(1,1)/U(1)$, i.e. on the sphere S^2 in the first case and the pseudosphere $S^{1,1}$ in the second case. One can readily find that the geometrical constructions for the integrable $SU(2)$ (and $SU(1,1)$) Heisenberg model and the corresponding generalized coherent states are the same$^{\dagger)}$. The system of generalized coherent states can

---

*) This notation means either J_0 or K_0, $e^{i\lambda J_0}$ or $e^{i\lambda K_0}$ should be and so on.

†) There also exists a deeper (physical) relationship. We shall return to this in what follows.

be written in a form which is analogous to (4)

$$|\alpha\rangle_1 = D_1(\alpha)|\psi_0\rangle_1 \equiv e^{\alpha J_+ - \bar{\alpha} J_-}|\psi_0\rangle_1 \tag{5.13a}$$

and

$$|\alpha\rangle_2 = D_2(\alpha)|\psi_0\rangle_2 \equiv e^{\alpha K_+ - \bar{\alpha} K_-}|\psi_0\rangle_2. \tag{5.13b}$$

Notice that the operators $D_i(\alpha)$ do not form a group and their multiplication law is represented in the form

$$D_i(\alpha_1)D_i(\alpha_2) = D_i(\alpha_3)e^{i\phi(\alpha_1,\alpha_2)J_0,(K_0)}$$

(detailed results are presented in the review [7a]). The operators $D_i(\alpha)$ can be also written in the normal form (which is the Gauss decomposition)

$$D_1(\zeta) = e^{\zeta J_+}e^{\beta J_0}e^{-\bar{\zeta} J_-}, \tag{5.14a}$$

$$D_2(\zeta) = e^{\zeta K_+}e^{\beta J_0}e^{-\bar{\zeta} K_-}, \tag{5.14b}$$

where*)

$$\zeta = \frac{\alpha}{|\alpha|}tg|\alpha|,\ \beta = -2\ln\cos|\alpha| = \ln(1+|\zeta|^2) \tag{5.15a}$$

for $SU(2)$, and

$$\zeta = \frac{\alpha}{|\alpha|}th|\alpha|,\ \beta = -2\ln \mathrm{ch}|\alpha| = \ln(1-|\zeta|^2) \tag{5.15b}$$

for $SU(1,1)$.

There also exist other parametrizations of coherent state systems for the group $SU(2)$ and $SU(1,1)$. In particular, one of these is associated with the fact that the coherent states are determined by a point in the homogeneous space $SU(2)/U(1) = S^2$ in the first case and $S^{1,1}$ in the second case. This point is given by the unit vector $\vec{n} = (\sin\theta\cos\varphi, \sin\theta\sin\varphi, \cos\theta)$ on S^2: $\vec{n}^2 = n_1^2+n_2^2+n_3^2 = 1$ and by the pseudo-Euclidean unit vector $\vec{n} = (\mathrm{sh}\tau\cos\varphi, \mathrm{sh}\tau\sin\varphi, \mathrm{ch}\tau)$ on the hyperboloid $S^{1,1}$:

*) Equations (15) according to (14) do not depend on the representation (ζ,β do not depend on j or k) and are easily verified at $j = 1/2$ when $J_i = \sigma_i/2$, where σ_i are the Pauli matrices. For that it is necessary to use the 'matrix' exponential definition $e^A = I+A+(1/2!)A^2+...$

$n_3^2 - n_2^2 - n_1^2 = n^2 = 1$. Therefore

$$|\vec{n}\rangle_1 = e^{i\phi_1(n)} e^{-i\varphi J_0} e^{-i\theta J_2} |\psi_0\rangle_1, \tag{5.16a}$$

for $SU(2)$, and

$$|\vec{n}\rangle_2 = e^{i\phi_2(n)} e^{-i\varphi K_0} e^{-i\tau K_2} |\psi_0\rangle_2, \tag{5.16b}$$

for $SU(1,1)$.

Using the fact that $|\psi_0\rangle_i$ is an *E.V.* of the operator $J_0,(K_0)$ one can choose the phase $\phi_i(n)$ so that

$$|\vec{n}\rangle_1 = D_1(n)|\psi_0\rangle_1 = e^{i\theta(\vec{m}\vec{J})}|\psi_0\rangle_1,$$

$$|\vec{n}\rangle_2 = D_2(n)|\psi_0\rangle_2 = e^{i\tau(\vec{m}\vec{K})}|\psi_0\rangle_2.$$

$$\vec{m} = (\sin\varphi, -\cos\varphi, 0)$$

Formulae (5.13) can easily be obtained from these relations if one supposes

$$\alpha = -\frac{\theta}{2} e^{-i\varphi} \quad (\text{for } SU(2)), \tag{5.17a}$$

or

$$\alpha = -\frac{\tau}{2} e^{-i\varphi} \quad (\text{for } SU(1,1)). \tag{5.17b}$$

As a result we have three different, but equivalent forms of generalized coherent states for the groups $SU(2)$ and $SU(1,1)$: (13), (14) and (16). The transition from the angular variables on the homogeneous space G/H namely (θ,φ) or (τ,φ) to the variable ζ has a simple geometrical interpretation: it corresponds to a stereographic projection of the sphere S^2 or the hyperboloid $S^{1,1}$ onto the complex plane $\zeta = \xi + i\eta$ and a subsequent reflection with respect to the η-axis.

Further discussions are distinct for $SU(2)$ and $SU(1,1)$ coherent states.

In the first case, the states $|j,\mu\rangle$, with a certain spin projection on the X_3 axis, provide a basis in the unitary irreducible representation space $T^j(g)$ of the $SU(2)$ group. Especially simple is the system of coherent states for the reference vector[*)]

[*)] Later on we shall call them the spin coherent states.

$$|\psi_0\rangle_1 = |j, -j\rangle, \quad \text{i.e. } \mu - \quad j \tag{5.18}$$

so that $J_- |\psi_0\rangle = 0$.

In the second case, the group $SU(1,1)$ has the main series, two discrete series $T^{(+)}$, $T^{(-)}$ and the additional one of unitary irreducible representations. Respectively, one can construct a number of coherent state systems related to these series. In the following, we consider only those connected with the discrete series which can be realized via bosonic creation and annihilation operators. It is sufficient to consider only one of the two discrete series, say $T^{(+)}$, since all the results are easily carried over to $T^{(-)}$. The discrete series representations are infinite-dimensional though, in a sense, analogous to the finite-dimensional unitary irreducible representations of the group $SU(2)$. For example, the basis vector $|k,m\rangle$ in a space of the $T^{(+)}$ representation is labelled by an integer m going from zero to infinity. For the discrete series, k takes on values 1, 3/2, 2,... Taking into account that the group $SU(1,1)$ is infinite-connected, it is necessary to consider its universal covering group $\widetilde{SU(1,1)}$ (an infinite sheet surface). The representations of this already simply-connected group are determined also by the number k, however it is now an arbitrary non-negative number. In what follows, the basis vectors of the space $T^{(+)}$ are conveniently denoted by

$$|k,m\rangle \to |k,\mu\rangle, \quad \text{where } \mu = k + m, \tag{5.19}$$

so

$$K_0 |k,\mu\rangle = \mu |k,\mu\rangle.$$

We choose the vector $|\psi_0\rangle_2$ in the form[*)]

$$|\psi_0\rangle_2 = |k,k\rangle \tag{5.18b}$$

and have $K_- |\psi_0\rangle_2 = 0$

The coherent state systems constructed on vectors (18) via operators (14) have the form

$$|\zeta\rangle_1 = (1 + |\zeta|^2)^{-j} e^{\zeta J_+} |\psi_0\rangle_1 \quad \text{(spin coherent states)}, \tag{5.20a}$$

[*)]The system of $SU(1,1)$ generalized coherent systems constructed on the basis of the vector (18b) will be called later on the pseudospin coherent state.

$$|\zeta\rangle_2 = (1-|\zeta|^2)^k e^{\zeta K_+}|\psi_0\rangle_2 \text{ (pseudospin coherent states)}, \tag{5.20b}$$

The generalized coherent state systems (20) have very remarkable properties analogous to the properties of the usual Heisenberg-Weil coherent states.**) We review the most important of these without giving proofs.

(1) The operators $T(g)$ transfer one coherent state into another.

(2) The generalized coherent state systems are complete (more precisely, overcomplete).

(3) The coherent states are nonorthogonal to each other:

$$\langle\zeta'|\zeta\rangle_1 = [(1+|\zeta'|^2)(1+|\zeta|^2)]^{-j}(1+\bar{\zeta}'\zeta)^{2j}, \tag{5.21a}$$

$$\langle\zeta'|\zeta\rangle_2 = [(1-|\zeta'|^2)(1-|\zeta|^2)]^{k}(1-\bar{\zeta}'\zeta)^{-2k}. \tag{5.21b}$$

(4) The following property is one of the most important in what follows later on: the expansion of the unity (identity operator) takes place

$$\int d\mu_j(\zeta)|\zeta\rangle\langle\zeta| = I, \tag{5.22}$$

where

$$d\mu_j(\zeta) = \frac{2j+1}{\pi}\,\frac{d^2\zeta}{(1+|\zeta|^2)^2}, \tag{5.23}$$

and

$$\int d\mu_k(\zeta)|\zeta\rangle\langle\zeta| = I,$$

where

$$d\mu_k = \begin{cases} \dfrac{2k-1}{\pi}\,\dfrac{d^2\zeta}{(1-|\zeta|^2)^2}, & k \neq \dfrac{1}{2}, \\[2ex] \dfrac{1}{\pi}\,\dfrac{d^2\zeta}{(1-|\zeta|^2)^2}, & k = \dfrac{1}{2}, \text{ see [8]}. \end{cases} \tag{5.24}$$

(5) We give expressions for infinitesimal $J_i,(K_i)$ operators and their powers in the generalized coherent states representation.

**)We will denote the latter as either bosonic coherent states or simply coherent states.

We first recall results for the ordinary coherent states determined in the complex plane α:

$$|\alpha> = e^{-1/2|\alpha|^2} e^{\alpha a^+} |0>, \quad \text{(see (18))} \tag{5.25}$$

$$<\alpha|\beta> = \exp\{-\frac{1}{2}(|\alpha|^2+|\beta|^2-2\bar{\alpha}\beta)\},$$

$$|<\alpha|\beta>| = \exp\{-\frac{1}{2}|\alpha-\beta|^2\}.$$

Let

$$\hat{A} = \sum_{m,n} A_{mn}(a^+)^m a^n \tag{5.26}$$

be an operator written in the normal (Wick) form, then in the boson coherent states representation we have

$$Q_A(\alpha) \equiv <\alpha|\hat{A}|\alpha> = \sum_{m,n} A_{mn}(\bar{\alpha})^m \alpha^n \tag{5.27}$$

and for the $\hat{A}$ operator trace

$$\mathrm{Tr}\hat{A} = \int Q_A(\alpha)d\mu(\alpha), \quad d\mu(\alpha) = \frac{1}{\pi}d^2\alpha. \tag{5.28}$$

Equation (27) shows that replacing the Wick operators by the c-numbers in the boson coherent states representation does not require the invention of a decoupling procedure (see Chapter 1).

We shall see now that the situation changes in the case of generalized coherent states. For the operators J_i,(K_i) and their powers we have in the following:

$$<\zeta|J_0|\zeta> = -j\frac{1-|\zeta|^2}{1+|\zeta|^2}, \quad <\zeta'|J_+|\zeta> = 2j\frac{\bar{\zeta}'}{1+\bar{\zeta}'\zeta}, \tag{29a}$$

$$<\zeta'|J_-|\zeta> = 2j\frac{\zeta}{1+\bar{\zeta}'\zeta},$$

or $<\vec{n}|\vec{J}|\vec{n}> = -j\vec{n}$,

$$<\zeta|K_0|\zeta> = k\frac{1+|\zeta|^2}{1-|\zeta|^2}, \quad <\zeta|K_+|\zeta> = +2k\frac{\bar{\zeta}}{1-|\zeta|^2}, \tag{5.29b}$$

$$<\zeta|K_-|\zeta> = \overline{<\zeta|K_+|\zeta>},$$

or $<\vec{n}|\vec{K}|\vec{n}> = k\vec{n}$. Then for powers

$$<\zeta|J_0^2|\zeta> = \frac{j^2(1-|\zeta|^2)^2+2j|\zeta|^2}{(1+|\zeta|^2)^2} = (<|J_0|>)^2\left[1+\frac{2}{j}\,\frac{|\zeta|^2}{(1-|\zeta|^2)^2}\right], \tag{5.30}$$

$$<\zeta|J_+^n|\zeta> = 2j(2j-1)...(2j-n+1)\cdot(2j)^{-n}(<|J_+|>)^n,$$

$$<\zeta|J_-^n|\zeta> = \overline{<\zeta|J_+^n|\zeta>},$$

$$<\zeta|J_+J_-|\zeta> = <|J_+|><|J_-|>(1+\frac{|\zeta|^2}{2j}),$$

$$<\zeta|J_-J_+|\zeta> = <|J_+|><|J_-|>(1+\frac{|\zeta|^{-2}}{2j}),$$

$$<\zeta|J_\pm J_0|\zeta> = <|J_0|><|J_\pm|>(1+\frac{1}{j}\,\frac{|\zeta|^2}{1-|\zeta|^2}),$$

$$<\zeta|J_0J_\pm|\zeta> = <|J_0|><|J_\pm|>(1-\frac{1}{j}\,\frac{1}{1-|\zeta|^2}),$$

$$<\zeta|K_0^2|\zeta> = (<|K_0|>)^2(1+\frac{2}{k}\,\frac{|\zeta|^2}{(1+|\zeta|^2)^2}), \tag{5.30b}$$

$$<\zeta|K_+^n|\zeta> = 2k(2k+1)...(2k+n-1)\cdot(2k)^{-n}(<|K_+|>)^n,$$

$$<\zeta|K_+K_-|\zeta> = <|K_+|><|K_-|>(1+\frac{|\zeta|^2}{2k}),$$

$$<\zeta|K_-K_+|\zeta> = <|K_+|><|K_-|>(1+\frac{|\zeta|^{-2}}{2k}),$$

$$<\zeta|K_0K_\pm|\zeta> = <|K_0|><|K_\pm|>(1+\frac{k^{-1}}{1+|\zeta|^2}),$$

$$<\zeta|K_\pm K_0|\zeta> = <|K_0|><|K_\pm|>(1+\frac{1}{k}\,\frac{|\zeta|^2}{1+|\zeta|^2}),$$

All the formulae given above can be obtained from eqn. (21) and two simple relations:

(1) $$\frac{d}{d\zeta}\{(1+|\zeta|^2)^j<\zeta_1|\zeta>\} = <\zeta_1|\frac{d}{d\zeta}e^{\zeta J_+}|\psi_0> = (1+|\zeta|^2)^j<\zeta_1|J_+|\zeta>$$

or

$$<\zeta_1|J_+|\zeta> = (1+|\zeta|^2)^{-j}\frac{d}{d\zeta}\{(1+|\zeta|^2)^j<\zeta_1|\zeta>\}.$$

From the easily obtained relationship $e^{-\zeta J_+}J_0 e^{\zeta J_+} = J_0 + \zeta J_+$ one can obtain the following expression:

(2) $<\zeta_1|J_0|\zeta> = (1+|\zeta|^2)^{-j}<\zeta_1|e^{\zeta J_+}(J_0+\zeta J_+)|\psi_0> = -j<\zeta_1|\zeta>+\zeta<\zeta_1|J_+|\zeta>.$

So, for example, it is easy to verify that

$$\frac{d}{d\zeta}\{(1+|\zeta|^2)^j<\zeta_1|J_+|\zeta>\} = (1+|\zeta|^2)^j<\zeta_1|J_+^2|\zeta>,$$

$$\frac{d}{d\zeta}\{(1+|\zeta|^2)^j<\zeta_1|J_-|\zeta>\} = (1+|\zeta|^2)^j<\zeta_1|J_-J_+|\zeta>,$$

$$<\zeta_1|J_-J_+|\zeta> = <\zeta_1|J_+J_-|\zeta>-2<\zeta_1|J_0|\zeta>,$$

$$<\zeta_1|J_0^2|\zeta> = (1+|\zeta|^2)^{-j}<\zeta_1|J_0 e^{\zeta J_+}e^{-\zeta J_+}J_0 e^{\zeta J_+}|\psi_0>$$

$$= (1+|\zeta|^2)^{-j}<\zeta_1|e^{\zeta J_+}(J_0+\zeta J_+)^2|\psi_0>,$$

and so on.

From the above expressions for the various 'correlators' we see that the decoupling procedure in the generalized coherent states representation on the curved manifolds $SU(2)/U(1)$ and $SU(1,1)/U(1)$ already depends on the representation (the numbers j,k) and works properly only in the so-called semiclassical spin limit $j,k, \gg 1$.

A more detailed study shows that in the spin coherent states case, in correlators involving J_0 the circle $|\zeta|^2 = 1$ turns out to be an exception, in the vicinity of which the decoupling procedure works badly; for the correlators $<|J_+J_-|>$ and $<|J_-J_+|>$ we have $j \gg |\zeta|^2$ and $j \gg |\zeta|^{-2}$, respectively. This could be important in the study of Hamiltonians with terms biquadratic in a and a^+, since these can lead to terms quadratic in J_i or K_i.

Finally, we mention two more properties of the generalized coherent states useful for an understanding of further discussion.

(6) In the limit of large values of j (or k) the generalized coherent states tend to boson ones. In order o prove this we make the substitution

$$J_+,(K_+) \to \sqrt{2j}\,a^+, \quad \zeta \to \frac{\alpha}{\sqrt{2j}}, \quad (j \to k), \tag{5.31}$$

and assume $j, (k) \to \infty$. Then from eqn. (20a) we get

$$|\zeta>_1 \to \lim_{j\to\infty}(1+\frac{|\alpha|^2}{2j})^{-j}e^{\alpha a^+}|\psi_0>_1 = e^{-1/2|\alpha|^2}e^{\alpha a^+}|0>,$$

$$|\zeta>_2 \to \lim_{k\to\infty}(1-\frac{|\alpha|^2}{2k})^{k}e^{\alpha a^+}|\psi_0>_2 = e^{-1/2|\alpha|^2}e^{\alpha a^+}|0>,$$

i.e. eqn. (18).

(7) As can be seen, the generalized coherent states system depends essentially on the choice of the reference vector $|\psi_0>$. Usually this choice is determined for reasons of simplicity and also by the fact that the obtained states should be as close as possible to the classical ones. The latter means that these states have the minimal uncertainty (see [7a] and also [7c]). So, for the usual coherent states constructed on the vector $|\psi_0> = |0>$ (boson coherent states) we have

$$\Delta a\Delta a^+ \equiv <0|aa^+|0>-<0|a|0><0|a^+|0> = 1,$$

whereby, in particular, it follows that

$$\Delta p\Delta q = \frac{\hbar}{2} \tag{5.32}$$

with

$$\hat{p} = \sqrt{\frac{\hbar}{2}}\frac{a-a^+}{i}, \quad \hat{q} = \sqrt{\frac{\hbar}{2}}(a+a^+),$$

being the momentum and coordinate operators, respectively. In the generalized coherent states case on the groups $SU(2)$ and $SU(1,1)$ we consider as exceptional the states minimizing the uncertainty defined via the quadratic Casimir operator $\hat{C}^2$:

$$\hat{C}_2 - g^{lm}\hat{A}_l\hat{A}_m, \tag{5.33}$$

$$\Delta\hat{C}_2 = <\psi_0|\hat{C}_2|\psi_0>-g^{lm}<\psi_0|\hat{A}_l|\psi_0><\psi_0|\hat{A}_m|\psi_0>,$$

where $\hat{A}_l$ are the representation generators of the corresponding algebra, g^{lm} is the Killing metric. From eqn. (33), it follows that the vector $|\psi_0>$ corresponds to the minor weight of representation both in the $SU(2)$ group case ($|\psi_0>_1 = |j,-j>$, see eqn. (18a)) and in the $SU(1,1)$ group case ($|\psi_0>_2 = |k,k>$, eqn. (18b)) and satisfies the condition $\Delta\hat{C}_2 = \inf\{\Delta\hat{C}_2\}$. In fact, calculating $\hat{C}_2$ on the vectors $|\psi_0>$ we get

$$<\psi_0|J_0^2+\frac{1}{2}(J_+J_-+J_-J_+)|\psi_0> = j(j+1),$$

$$<\psi_0|J_\pm|\psi_0> = 0, \quad <\psi_0|J_0|\psi_0> = j,$$

and thence

$$\Delta\hat{C}_2 = j, \qquad (SU(2)), \tag{5.34}$$

$$|\Delta\hat{C}_2| = k, \qquad (SU(1,1)).$$

This result is intuitively connected with the fact that the minor weight vectors are 'vacuum' ones for the lowering operators $J_-(K_-)$

$$J_-|\psi_0>_1 = K_-|\psi_0>_2 = 0,$$

which makes the construction analogous to the Heisenberg-Weil one: $a|0> = 0$.

(5) In the above, we consider generalized coherent states for a system with one degree of freedom. Therefore the complex plane α, or two-dimensional manifolds, are the phrase space of such a system which follows, in particular, from the formula $\alpha = (q+ip)/\sqrt{2\hbar}$. The case of several (and in the limit, infinite) degrees of freedom (particles) is essentially analogous to the preceding one and many formulae are straightforward if the operators corresponding to different particles commute. The latter is valid for the Heisenberg-Weyl Bose operators a_j^+, a_j and for the Pauli (spin) operators S_j^l (or J_j^l):

$$[a_j,a_k^+] = \delta_{jk}I, \quad [a_j,a_k] = [a_j^+,a_k^+] = 0, \tag{5.35}$$

$$[S_l^i,S_m^j] = \delta_{\mathrm{lm}}C_k^{ij}S_m^k. \tag{5.36}$$

In what follows we briefly consider an extension of boson coherent states to the N-particle case since spin and pseudospin coherent states are extended in the same way.

The operators a_j, a_k^+ act in the Hilbert space $\mathcal{H}$ in which there exists the normalized vector $|0>$ cancelled by the operators a_j:

$$a_j|0> = 0, \quad <0|0> = 1.$$

Acting on $|0>$ by the operators a_j^+ we get a set of normalized vectors

$$|n_1,...,n_N> = \frac{(a_1^+)^{n_1}...(a_N^+)^{n_N}}{(n_1!...n_N!)^{1/2}}|0>. \tag{5.37}$$

One may introduce abbreviated notations for the sets of integers

$$[n] = (n_1,...,n_N), \tag{5.38}$$

$$[n]! = n_1!...n_N!,$$

$$(a^+)^{[n]} = (a_1^+)^{n_1}...(a_N^+)^{n_N}.$$

Then instead of eqn. (37) we have

$$|[n]> = \frac{(a^+)^{[n]}}{([n]!)^{1/2}}|0>. \tag{5.39}$$

The set of the vectors $\{|[n]>\}$ forms a basis in the space $\mathcal{H}$, which is now the Fock space.

Algebra (35a) is the $3N$-dimensional Heisenberg-Weil algebra. Its arbitrary element is

$$\hat{l} = \alpha a^+ - \bar{\alpha} a + isI, \tag{5.40}$$

where $\alpha = (\alpha_1,...,\alpha_N)$, $a = (a_1,...,a_N)$ are N-dimensional vectors and $\alpha a^+ = \Sigma_1^N \alpha_j a_j^+$ is their scalar product. An arbitrary group element is now

$$g = e^{\hat{l}} = \exp(isI)D(\alpha), \tag{5.41}$$

$$D(\alpha) = \exp(\alpha a^+ - \bar{\alpha} a).$$

Since the operators a_j^+, a_j with different indices commute, the algebra W_N splits into the direct sum of W_1 algebras and the W_N group into the direct product of W_1 groups:

$$w_N = \sum_{j=1}^{N} \oplus w_j, \quad W_N = \prod_{j=1}^{N} \otimes W_j.$$

In this context, we get a trivial extension of the formulae of the preceding section

$$|\alpha> = \exp(-\frac{|\alpha|^2}{2})\sum_{[n]}\frac{\alpha^{[n]}}{([n]!)^{1/2}}|[n]>, \tag{5.42}$$

$$d\mu(\alpha) = \pi^{-N}\prod_{j=1}^{N}\prod_{k=1}^{N}(d\mathrm{Re}\,\alpha_j)(d\mathrm{Im}\,\alpha_k).$$

This is analogous for the spin and pseudospin coherent states if we choose reference vectors so that

$$J_k^- |\psi_0>_1 = 0, \quad K_j^- |\psi_0>_2 = 0,$$

i.e. the vectors minimizing $\Sigma_j \Delta\hat{C}_{2j}$. Then, instead of the complex value ζ, we have a complex vector $\vec{\zeta} = (\zeta_1,...,\zeta_N)$, and so on. As a result eqn. (20) now reads as

$$|\vec{\zeta}>_1 = \prod_{k=1}^{N}(1+|\zeta_k|^2)^{-j_k} e^{\zeta_k J_k^{\prime}} |\psi_0>_1, \tag{5.44}$$

$$|\vec{\zeta}>_2 = \prod_{j=1}^{N}(1-|\zeta_j|^2)^{k_j} e^{\zeta_j K_j^{\prime}} |\psi_0>_2. \tag{5.45}$$

Calculating the averages of operators of the form $\hat{A}_j\hat{A}_k\hat{A}_m$ for different indices $j \neq k \neq m$ it is necessary to take them in the brackets

$$<\zeta_j\zeta_k\zeta_m|...|\zeta_m\zeta_k\zeta_j>. \tag{5.46}$$

Thus, one complex variable ζ_j corresponds to each degree of freedom and the phase space splits into a set of phase planes or curved phase manifolds (S_j^2 and $S_j^{1,1}$).

3. Quantum version. The representation in the form of a path integral over generalized coherent states

Consider the Hamiltonian $\hat{H}$ acting in the Hilbert space $\mathcal{H}$ defined above. Let $\hat{H}$ be represented in the form of a finite polynomial of the generators of the corresponding Lie group. Since, in this case, $\hat{H}$ conserves the quantum number j or k (i.e. does not change the representation), we consider the states with fixed j or k[*]. The transition amplitude from the coherent state $|\zeta>$ at time t to the state $|\zeta_1>$ at instant t_1 is given by

$$P(\zeta_1,t_1;\zeta,t) = <\zeta_1|\exp\{-\frac{i}{\hbar}\hat{H}(t_1-t)\}|\zeta>. \tag{5.47}$$

Divide up the interval t_1-t into n equal subintervals $\epsilon = (t_1-t)/n$ and pass to the limit of $n \to \infty$, then

$$P(\zeta_1,t_1;\zeta,t) = \lim_{n\to\infty} <\zeta_1|(1-\frac{i}{\hbar}\hat{H}\epsilon)^n|\zeta>$$

or

[*] For groups with a number of Casimir operators exceeding unity there are several such numbers.

$$P = \lim_{n\to\infty} <\zeta_1|(1-\frac{i}{\hbar}\hat{H}\epsilon_1)(1-\frac{i}{\hbar}\hat{H}\epsilon_2)...(1-\frac{i}{\hbar}\hat{H}\epsilon_n)|\zeta>.$$

Using the unity expansion (22) one may rewrite the preceding expression in the form

$$P = \lim_{n\to\infty}\sum_j\int...\int\left[\prod_{k=1}^{n-1}d\mu_j(\zeta_k)\right]\prod_{k=1}^{n}<\zeta_k|(1-\frac{i}{\hbar}\hat{H}\epsilon)|\zeta_{k-1}> \tag{5.48}$$

$$= \lim_{n\to\infty}\prod_j\int...\int\left[\prod_{k=1}^{n-1}d\mu_j(\zeta_k)\right]\left[\prod_{k=1}^{n}<\zeta_k|\zeta_{k-1}>\right]\prod_{k=1}^{n}\left\{1-\frac{i\epsilon}{\hbar}\frac{<\zeta_k|\hat{H}|\zeta_{k-1}>}{<\zeta_k|\zeta_{k-1}>}\right\}$$

where $\zeta_0 = \zeta$ and $\zeta_n = \zeta_1$. The expression in curly brackets in the limit of $\epsilon\to 0$ can be represented in the form

$$\{...\} = \exp\left\{-\frac{i\epsilon}{\hbar}\frac{<\zeta_k|\hat{H}|\zeta_{k-1}>}{<\zeta_k|\zeta_{k-1}>}\right\} \tag{5.49}$$

and

$$<\zeta_k|\zeta_{k-1}> = \exp\{\ln<\zeta_k|\zeta_{k-1}>\}.$$

By the use of the explicit expression (21a) for the correlator $<\zeta_k|\zeta_{k-1}>$, and setting $\zeta_{k-1} = \zeta_k-\Delta\zeta_k$, we have

$$<\zeta_k|\zeta_{k-1}> = 1-\partial_{\zeta'}<\zeta|\zeta'>|_{\zeta'=\zeta}\Delta\zeta-\partial_{\bar{\zeta}'}<\zeta|\zeta'>|_{\zeta'=\zeta}\Delta\bar{\zeta}$$

$$= 1+j\frac{\zeta_k\Delta\bar{\zeta}_k-\bar{\zeta}_k\Delta\zeta_k}{1+|\zeta_k|^2}+0((\Delta\zeta_k)^2)$$

or

$$<\zeta_k|\zeta_{k-1}> = \exp\left\{j\frac{\zeta_k\Delta\bar{\zeta}_k-\bar{\zeta}_k\Delta\zeta_k}{1+|\zeta_k|^2}\right\},$$

thence

$$\prod_{k=1}^{n}<\zeta_k|\zeta_{k-1}> = \exp\sum_1^n\epsilon\frac{1}{\epsilon}\ln<\zeta_k|\zeta_{k-1}> \tag{5.50}$$

$$\to\exp\left\{j\int_t^{t_1}(1+|\zeta(\tau)|^2)^{-1}\left[\zeta(\tau)\dot{\bar{\zeta}}(\tau)-\dot{\zeta}(\tau)\bar{\zeta}(\tau)\right]d\tau\right\}.$$

Substitution of eqns. (49) and (50) into eqn. (48) yields

$$P(\zeta_1,t_1;\zeta,t) = \lim_{n\to\infty}\sum_j \int\ldots\int\prod_{k=1}^{n-1} d\mu_j(\zeta_k)\times$$

$$\times\exp\left\{\frac{i\epsilon}{\hbar}\sum_{k=1}^{n}\frac{i\hbar j}{1+|\zeta_k|^2}\left[\bar{\zeta}_k\frac{\Delta\zeta_k}{\epsilon}-\zeta_k\frac{\Delta\bar{\zeta}_k}{\epsilon}\right]-<\zeta_k|\hat{H}|\zeta_k>\right\}.$$

In the continuum limit this formula can be written in the form of a functional integral

$$P(\zeta_1,\zeta;t_1-t) = \sum_j\int D\mu_j(\zeta)\exp\left\{\frac{i}{\hbar}\int_t^{t'}\mathcal{L}_j(\zeta,\bar{\zeta})d\tau\right\} \tag{5.51}$$

where

$$\mathcal{L}_j = j\frac{i\hbar}{1+|\zeta|^2}(\zeta\dot{\bar{\zeta}}-\bar{\zeta}\dot{\zeta})-\mathcal{H}_j(\zeta,\bar{\zeta}), \tag{5.52}$$

$$\mathcal{H}_j(\zeta,\bar{\zeta}) = <\zeta|\hat{H}|\zeta>,$$

$$|\zeta> = (1+|\zeta|^2)^{-j}e^{\zeta J_+}|\psi_0>,$$

or

$$P = \sum_j\int D\mu_j(\zeta)\exp(\frac{i}{\hbar}S_j) \tag{5.53}$$

and

$$S_j = \int_t^{t_1}\mathcal{L}_j d\tau.$$

In the case $S_j \gg \hbar$ the main contribution to the functional integral (53) comes from the stationary phase trajectory when $\delta S_j = 0$, i.e. to the quantum system under consideration may correspond a classical dynamical system with the Lagrangian $\mathcal{L}_j$. Analogous calculations in the $SU(1,1)$ case give

$$\mathcal{L}_k = \frac{i\hbar k}{1-|\zeta|^2}(\bar{\zeta}\dot{\zeta}-\dot{\bar{\zeta}}\zeta)-\mathcal{H}_k(\zeta,\bar{\zeta}). \tag{5.54}$$

The Hamiltonian equations of motion of the corresponding classical systems are

$$\dot{\zeta} = -i\frac{(1+|\zeta|^2)^2}{2\hbar j}\frac{\partial\mathcal{H}_j}{\partial\bar{\zeta}}, \tag{5.55}$$

$$\dot{\zeta} = i\frac{(1-|\zeta|^2)^2}{2\hbar k}\frac{\partial \mathcal{H}_k}{\partial \bar{\zeta}}.$$

The Poisson brackets for arbitrary functions A and B depending on ζ and $\bar{\zeta}$

$$\{A,B\}_j = i\frac{(1+|\zeta|^2)^2}{2\hbar j}\left[\frac{\partial A}{\partial \bar{\zeta}}\frac{\partial B}{\partial \zeta} - \frac{\partial A}{\partial \zeta}\frac{\partial B}{\partial \bar{\zeta}}\right], \tag{5.56a}$$

$$\{A,B\}_k = -i\frac{(1-|\zeta|^2)^2}{2\hbar k}\left[\frac{\partial A}{\partial \bar{\zeta}}\frac{\partial B}{\partial \zeta} - \frac{\partial A}{\partial \zeta}\frac{\partial B}{\partial \bar{\zeta}}\right], \tag{5.56b}$$

enable eqns. (55) to be rewritten in the standard Hamiltonian form

$$\dot{\zeta} = \{\zeta, \mathcal{H}_j\}, \quad \dot{\bar{\zeta}} = \{\bar{\zeta}, \mathcal{H}_j\}. \tag{5.57}$$

In the case of the boson coherent states in eqns. (55) and (56) one should replace the 'curved' metric by the flat one $(1 \pm |\zeta|^2)^2(2j)^{-1} \to 1$ and $\zeta \to \alpha$.

4. Quantum version. Some concrete models with dynamical symmetry

We now have everything required to formulate a quite general problem. Suppose we have some functional $\hat{H}$ expressed through operator variables $\hat{A}_i$ which belong to a representation of a Lie algebra L. Two variants are possible:

(1) $\hat{H}$ may be expressed in the form of a linear combination of $\hat{A}_i$, i.e. $\hat{H} \in L$.

(2) $\hat{H}$ is not linearized in the variables $\hat{A}_i$.

In the first case, L is called the spectrum generating algebra, and the corresponding Lie group the dynamical symmetry group. The generalized coherent states on G enable the quantum eigenvalue problem

$$\hat{H}_{\text{lin}}|\Psi_n\rangle = E_n|\Psi_n\rangle \tag{5.58}$$

to be reduced to a group theory problem, viz., to find unitary irreducible representations (usually a discrete series), so the problem proves to be exactly solvable.

In the second case, the situation is significantly complicated. However, when $\hat{H}$ is a polynomial in $\hat{A}_i$, by means of generalized coherent states one can obtain certain information about the system: at any rate via the methods developed above, one may construct its quasi-classical analogues described by nonlinear differential equations. In what follows, we discuss a number of problems of both the first and second type.

Dynamical symmetry

(1) A non-ideal Bose gas is described on the quantum level by the Hamiltonian in terms of the Heisenberg-Weil group generators (the Fourier representation)

$$\hat{H} = \sum_{\mathbf{k}} \epsilon_{\mathbf{k}} a_{\mathbf{k}}^{+} a_{\mathbf{k}} + \frac{1}{2V} \sum_{\mathbf{k},\mathbf{p},\mathbf{q}} U_{\mathbf{k}} a_{\mathbf{k}+\mathbf{p}}^{+} a_{\mathbf{k}-\mathbf{k}}^{+} a_{\mathbf{p}} a_{\mathbf{q}}, \quad \epsilon_{\mathbf{k}} = \frac{k^2}{2m}, \tag{5.59}$$

where V is the system volume.

Bogolubov's main assumption consists of the following: when the interboson interaction $U_{\mathbf{k}}$ vanishes, the ground state of the system is simply a set of particles with zero energy, $\epsilon_{\mathbf{k}} \to 0$ $(k \to 0)$, and at small $U_{\mathbf{k}}$ this state should be macroscopically occupied (a condensate in zeroth-order approximation - the Bose condensate), so that the operators a_0^{+} and a_0 can be regarded as the C-numbers

$$a_0 = a_0^{+} = \sqrt{N_0}, \quad N_0 = <a_0^{+} a_0>.$$

Then the reduced Hamiltonian has the form

$$\hat{H} = \frac{1}{2} N_0^2 \frac{U_0}{V} + \sum_{\mathbf{k}} \left[\epsilon_{\mathbf{k}} + N_0 \frac{V_0 + V_{\mathbf{k}}}{V} \right] a_{\mathbf{k}}^{+} a_{\mathbf{k}} + \frac{N_0}{2V} (a_{\mathbf{k}}^{+} a_{-\mathbf{k}}^{+} + a_{\mathbf{k}} a_{-\mathbf{k}}) \tag{5.60}$$

(the summation is taken over all $\mathbf{k}$ except $k = 0$). In passing from eqn. (59) to eqn. (60) we retain terms not higher than second order with respect to the operators $a_{\mathbf{k}}^{+}$ and $a_{\mathbf{k}}$. In this approximation

$$N = N_0 + \sum_{\mathbf{k}} a_{\mathbf{k}}^{+} a_{\mathbf{k}}$$

and eqn. (60) assumes the form

$$\hat{H} = \frac{N^2 U_0}{2V} + \sum_{\mathbf{k}} \left\{ (\epsilon_{\mathbf{k}} + \frac{N}{V} U_{\mathbf{k}}) a_{\mathbf{k}}^{+} a_{\mathbf{k}} + \frac{N}{2V} U_{\mathbf{k}} (a_{\mathbf{k}}^{+} a_{-\mathbf{k}}^{+} + a_{\mathbf{k}} a_{-\mathbf{k}}) \right\}. \tag{5.61}$$

Note the 'anomalous' Bogolubov terms of the form $a_{\mathbf{k}}^{+} a_{-\mathbf{k}}^{+}$, $a_{\mathbf{k}} a_{-\mathbf{k}}$ to be in (61) which we encountered in the first chapter (in the coordinate representation).

Let us introduce the operators

$$K_1^{(\mathbf{k})} = \frac{1}{2} (a_{\mathbf{k}}^{+} a_{-\mathbf{k}}^{+} + a_{\mathbf{k}} a_{-\mathbf{k}}), \tag{5.62}$$

$$K_2^{(\mathbf{k})} = -\frac{1}{2}(a_{\mathbf{k}}^+ a_{-\mathbf{k}}^+ - a_{\mathbf{k}} a_{\mathbf{k}}),$$

$$K_0^{(\mathbf{k})} = \frac{1}{2}(a_{\mathbf{k}}^+ a_{\mathbf{k}}^+ + a_{-\mathbf{k}} a_{-\mathbf{k}} + 1).$$

It is easy to show that these satisfy the commutation relations of the Lie algebra of the $SU(1,1)_k$ group (see eqn. (36b))*) In terms of these operators the Hamiltonian (61) is a linear function

$$\hat{H} = \sum_{\mathbf{k}} \oplus \frac{NU_{\mathbf{k}}}{V}\left[K_1^{(\mathbf{k})} + \mu_{\mathbf{k}}(K_0^{(\mathbf{k})} - \frac{1}{2})\right] + \frac{1}{2}\frac{N^2 U_2}{V}, \quad \mu_{\mathbf{k}} = 1 + \frac{\epsilon_{\mathbf{k}} V}{NU_{\mathbf{k}}}. \tag{5.63}$$

Thus the problem (58) is reduced to the eigenvalue problem of the operator

$$\hat{H} = \sum_{\mathbf{k}} \oplus(\alpha \,|\, K_0^{(\mathbf{k})} + \beta K_1^{(\mathbf{k})}) + \text{const.} \tag{5.64}$$

As is obvious from the above for the group $SU(1,1)$ there are two representations of the discrete series $T^{(+)}$ and $T^{(-)}$. For one of these, $T^{(+)}$, the spectrum is bounded below, and therefore, as we shall see, it is of physical interest. This representation can be realized via the pseudospin coherent states constructed above. Remember that for them there is the relation (19), i.e. $|\psi_0>$ is an e.v. of the operator K_0. The operator (64) may be simplified if one uses the unitary transformation

$$R_{\mathbf{k}}(\theta) = \exp\{-iK_2^{(\mathbf{k})}\theta_{\mathbf{k}}\}, \tag{5.65}$$

$$\tilde{K}_1^{(\mathbf{k})} = R_{\mathbf{k}} K_1^{(\mathbf{k})} R_{\mathbf{k}}^{-1} = \text{ch}\theta_{\mathbf{k}} K_1^{(\mathbf{k})} + \text{sh}\theta_{\mathbf{k}} K_0^{(\mathbf{k})},$$

$$\tilde{K}_0^{(\mathbf{k})} = R_{\mathbf{k}} K_0^{(\mathbf{k})} R_{\mathbf{k}}^{-1} = \text{sh}\theta_{\mathbf{k}} K_1^{(\mathbf{k})} + \text{ch}\theta_{\mathbf{k}} K_0^{(\mathbf{k})},$$

which corresponds to a rotation in the $SU(1,1)_k$ algebra space by the angle θ_k about the X_2 axis. The total rotation is described by the operator

$$R = \prod_{\mathbf{k}} \otimes R_{\mathbf{k}}(\theta) \tag{5.66}$$

and

$$\hat{\hat{H}} = R\hat{H}R^{-1} = \frac{N^2 U_0}{2V} + \sum_{\mathbf{k}} \oplus \frac{N}{V} U_{\mathbf{k}} \left[K_0^{(\mathbf{k})}(\mu_{\mathbf{k}}\text{ch}\theta_{\mathbf{k}} + \text{sh}\theta_{\mathbf{k}}) + K_1^{(\mathbf{k})}(\text{ch}\theta_{\mathbf{k}} + \mu_{\mathbf{k}}\text{sh}\theta_{\mathbf{k}}) - \frac{\mu_{\mathbf{k}}}{2}\right].$$

---

*)Notice that the analogous operators constructed via the Fermi operators C_k^+ and C_k generate the $SU(2)$ compact algebra.

Since $|\mathrm{th}\theta_{\mathbf{k}}| < 1$, depending on the sign of the $U_{\mathbf{k}}$ potential one may use the rotation to proceed either to $K_1^{(\mathbf{k})}$ or to $K_0^{(\mathbf{k})}$:

(1) $U < 0$, the attraction potential, $\mu_{\mathbf{k}} < 1$. Choose $\tanh\theta_{\mathbf{k}} = -\mu_{\mathbf{k}}$ then

$$\hat{\tilde{H}} = \sum_{\mathbf{k}} \oplus \frac{N}{V} U_{\mathbf{k}} (K_1^{(\mathbf{k})} \operatorname{sech}\theta_{\mathbf{k}} - \frac{\mu_{\mathbf{k}}}{2}) + \frac{N^2 U_0}{2V}, \quad \epsilon_{\mathbf{k}} < 2\frac{N}{V} |U_{\mathbf{k}}|.$$

In this case, the energy spectrum is continuous ($K_1^{(k)}$ generates a continuous series) and this means that the condensate obtained is unstable (at the classical level this corresponds to a plane wave instability within the $U(1)$ nonlinear Schrödinger equation framework, see Section 3).

(2) $U > 0$, the repulsive potential, $\mu_{\mathbf{k}} > 1$. Then let us take

$$\mathrm{ct}\theta_{\mathbf{k}} = -\mu_{\mathbf{k}} \tag{5.67}$$

then

$$\hat{\tilde{H}} = \sum_{\mathbf{k}} \oplus (E_{\mathbf{k}} K_0^{(\mathbf{k})} - \mu_{\mathbf{k}} \frac{N}{2V}) + \frac{N^2 U_0}{2V} \tag{5.68a}$$

where

$$E_{\mathbf{k}} = \frac{N}{V} U_{\mathbf{k}} \operatorname{csech}\theta_{\mathbf{k}} = (\epsilon_{\mathbf{k}}^2 + 2\epsilon_{\mathbf{k}} U_{\mathbf{k}} \frac{N}{V})^{1/2}. \tag{5.68b}$$

The Casimir operators $\hat{C}_{2\mathbf{k}}$ are given by

$$\hat{C}_{2\mathbf{k}} = \frac{1}{4}(\Delta_{\mathbf{k}}^2 - 1) = q_{\mathbf{k}}(q_{\mathbf{k}} - 1). \tag{5.69}$$

So the integrals of motion are differences of particle numbers in the states with opposite momenta

$$\Delta_{\mathbf{k}} = a_{\mathbf{k}}^{+} a_{\mathbf{k}} - a_{-\mathbf{k}}^{+} a_{-\mathbf{k}}.$$

Since the energy spectrum has to be bounded below, the only allowed representation is the $T^{(+)q_{\mathbf{k}}}$ series

$$\prod_{\mathbf{k}} \otimes T^{(+)q_{\mathbf{k}}} \quad \text{for } q_{\mathbf{k}} = \frac{1}{2}(1 + |\Delta_{\mathbf{k}}|). \tag{5.70}$$

The pseudospin coherent states are determined by the operators

$$D = \prod_{\mathbf{k}} \otimes D^{+}(q_{\mathbf{k}})$$

so that

$$K_0^{(\mathbf{k})}\,|[n]> = (n_{\mathbf{k}}+q_{\mathbf{k}})\,|[n]> \tag{5.71}$$

where again $|[n]> = \prod_{\mathbf{k}} |n_{\mathbf{k}}>$.

As a result from eqn. (61), the energy spectrum is given by

$$E([n]) = \sum_{\mathbf{k}} \left\{ E_{\mathbf{k}}(n_{\mathbf{k}} + \frac{1}{2}|\Delta_{\mathbf{k}}| + \frac{1}{2}) - \frac{\mu_{\mathbf{k}}}{2}\frac{N}{V}U_{\mathbf{k}} \right\} + \frac{N^2 U_0}{2V} \tag{5.72}$$

and e.f. of the Hamiltonian are pseudospin coherent states. It should be noticed that, in the case under consideration, the physical vacuum of the system (excitations over it are stable and correspond to a discrete energy spectrum) does not coincide with the reference vacuum $|\psi_0>$ by which we construct pseudospin coherent states. In fact, return to the problem (58) with the Hamiltonian (61) or (63), i.e.

$$\hat{H}\,|\Psi> = E\,|\Psi>. \tag{58a}$$

This problem has no discrete energy spectrum, and we perform the canonical transformation (66) which is a group theory analogue of the celebrated Bogolubov $u-v$ transformation. As a result

$$\tilde{H}\,|\tilde{\Psi}> = E\,|\tilde{\Psi}>, \quad |\tilde{\Psi}> = \prod_{\mathbf{k}} R_{\mathbf{k}}(\theta)\,|\Psi>,$$

with the Hamiltonian (68). For this problem, the representation of the series $T^{(+)q_{\mathbf{k}}}$ give the e.f. $|\tilde{\Psi}>$ of the Hamiltonian $\hat{H}$ with the discrete energy spectrum (72), then the lowest (vacuum) state appears at $n_{\mathbf{k}} = \Delta_{\mathbf{k}} = 0$, $q_{\mathbf{k}} = 1/2$ with energy

$$E([0]) = \frac{1}{2}\sum_{\mathbf{k}} \left\{ E_{\mathbf{k}} - (\epsilon_{\mathbf{k}} + \frac{N}{V}U_{\mathbf{k}}) \right\} + \frac{N^2 U_0}{2V}. \tag{5.73a}$$

This state - the Bogolubov condensate - corresponds to the e.f.

$$|\tilde{\Psi}_0> = |\frac{1}{2}, \frac{1}{2}> \equiv |\psi_0>.$$

Proceeding to the e.f. written in the initial variables, i.e. the variables of problem (58a), we obtain the Bogolubov condensate as

$$|\tilde{\Psi}_0> = \prod_{\mathbf{k}} R_{\mathbf{k}}(\theta)\,|\Psi_0> = |\psi_0>$$

or

$$|\Psi_0> = \prod_{\mathbf{k}} R_{\mathbf{k}}^{-1}(\theta)|\tilde{\Psi}_0> = \prod_{\mathbf{k}} R_{\mathbf{k}}^{-1}(\theta)|\psi_0> \tag{73b}$$

$$= \prod_{\mathbf{k}} \operatorname{sech}\frac{\theta_{\mathbf{k}}}{2} \exp\left\{\sum_{\mathbf{k}} \operatorname{th}\frac{\theta_{\mathbf{k}}}{2} a_{\mathbf{k}}^{+} a_{-\mathbf{k}}^{+}\right\}|\psi_0>.$$

Equation (73) is derived in the following way: in view of eqns. (72) and (70) the vacuum energy corresponds to the representation with $q_{\mathbf{k}} = 1/2$, therefore in our case the reference vector $|\psi_0>$ is $|1/2,1/2>$. Applying the Bogolubov rotation $R_{\mathbf{k}}^{-1}$ we get

$$|\Psi_0> = \prod_{\mathbf{k}} e^{i\theta_{\mathbf{k}} K_2^{(\mathbf{k})}} |\frac{1}{2}, \frac{1}{2}> \equiv \prod_{\mathbf{k}} e^{i\theta_{\mathbf{k}}} K_2^{(\mathbf{k})} |\psi_0>; \tag{5.74a}$$

an expression given by the parameterization (16b) at $\varphi = 0$, $\tau = -\theta_{\mathbf{k}}$. Passing to the paramterization (20b) via eqns. (15b) we have the formula

$$|\Psi_0> = \left\{\prod_{\mathbf{k}} (1-\zeta_{0\mathbf{k}}^2)^{1/2} \exp\left[\sum_{\mathbf{k}'} \zeta_{0\mathbf{k}'} a_{\mathbf{k}'}^{+} a_{-\mathbf{k}'}^{+}\right]\right\} |\frac{1}{2}, \frac{1}{2}> \tag{5.74b}$$

$$\zeta_{0\mathbf{k}} = -\operatorname{th}\frac{\tau_{\mathbf{k}}}{2} = \operatorname{th}(\theta_{\mathbf{k}}/2)$$

coinciding with eqn. (73b).

The Bogolubov condensate (73) is a rather complex state having some specific features. First, in the limit of vanishing interaction, we have: $-\operatorname{cth}\theta_{\mathbf{k}} = \mu_{\mathbf{k}} = 1+\epsilon_{\mathbf{k}} V/NU_R \sim U_k^{-1}$; $-\theta_{\mathbf{k}} \simeq U_{\mathbf{k}} \rightarrow 0$ and from eqn. (73) we get $|\Psi_0> \rightarrow |\psi_0>$, i.e. the ground state of noninteracting particles has energy

$$E_0 = \frac{1}{2}\sum_{\mathbf{k}}\left\{E_{\mathbf{k}} - (\epsilon_{\mathbf{k}} + \frac{N}{V} U_{\mathbf{k}})\right\} + \frac{1}{2}\frac{N^2 U_0}{V} \rightarrow 0 \text{ when } k \rightarrow 0$$

since, because of eqn. (68b)

$$E_{\mathbf{k}} = \sqrt{\epsilon_{\mathbf{k}}}(\epsilon_{\mathbf{k}} + 2\frac{N}{V} U_{\mathbf{k}})^{1/2} = |k| \left[\frac{k^2}{4m^2} + \frac{NU_{\mathbf{k}}}{mV}\right]^{1/2} \rightarrow \frac{k^2}{2m} = \epsilon_{\mathbf{k}}.$$

Notice that up to notations the spectrum (58b) is the Bogolubov spectrum obtained earlier in the quasi-classical (and classical) theory of a repulsive Bose gas when

$$U_{\mathbf{k}} = \int U(|\vec{x}|)e^{-i\mathbf{k}\vec{x}}d\vec{x} = 2g\int \delta(|\vec{x}|)e^{-i\mathbf{k}\vec{x}}d\vec{x} = 2g$$

i.e. within the $S3_{(-)}$. So the initial Bogolubov assumption is justified.

Further, in terms of pseudospin coherent states, the Bogolubov condensate is constructed within the framework of the $T^{(+)1/2}$ representation of the group $SU(1,1)$. Secondly, switching-on any weak interaction $U_k \neq 0$ leads to an essential reconstruction of the initial Bose condensate. In fact, expanding eqn. (73) in terms of orthonormal states $|[n]>$ we get

$$|\Psi_0> = \prod_{\mathbf{k}}(1-\zeta_{0\mathbf{k}}^2)^{1/2}\sum_n(\zeta_{0\mathbf{k}})^n\,|[n_{\mathbf{k}}]>, \tag{5.75}$$

where $|[n_{\mathbf{k}}]> = |n_{\mathbf{k}}>|n_{-\mathbf{k}}>$ at $n_{\mathbf{k}} = n_{-\mathbf{k}}$. From this expression we see that in the condensate, in the state with a given **k**, there are present all many-particle states $|[n_{\mathbf{k}}]>$ with probability $\omega_{n_k} = \zeta_{0\mathbf{k}}^{2n}(1-\zeta_{0\mathbf{k}}^2)$ (or for small $U_k \ll \epsilon_k V/N$, $\omega_{n_k} = (1-\theta_k^2)\theta_k^{2m}$. The Bogolubov condensate is given by the summation over all possible **k**. Thirdly, for every boson with momentum **k** in the condensate there exists a a boson with momentum $-\mathbf{k}$, i.e. the boson pairing effect arises.

Fourthly, the condensate reconstruction after the interaction switching-on is of a collective nature since, in both the spectrum (68b) and the quantity $\mu_{\mathbf{k}}$ determining $|\Psi_0>$, the interaction enters via the combination NU_k/V.

(2) The second example is a two-sublattice model of the Anderson XXZ antiferromagnet [12]:

$$\hat{H} = \sum_{\substack{i\in A\\ j\in B}}\left\{\frac{1}{2}(S_i^+S_j^+ + \text{h.c.}) - \rho S_i^z S_j^z\right\}. \tag{5.76}$$

'Bosonizing' this Hamiltonian via Holstein-Primakoff transformations we get in the approximation quadratic in the operators a_j^+ and a_j

$$\hat{H}_{\text{lin}} = \frac{1}{2}\sum_{ij}\{s(a_i^+a_j^+ + \text{h.c.}) + \rho s(a_i^+a_i + a_j^+a_j) - \rho s^2\},$$

or in the momentum representation

$$\hat{H}_{\text{lin}} = -\frac{1}{2}Nzs^2\rho + zs\rho\sum_{\mathbf{k}}N_{\mathbf{k}} + \frac{s}{2}\sum_{\mathbf{k},\vec{\delta}}\cos(\mathbf{k}\vec{\delta})(a_{\mathbf{k}}^+a_{-\mathbf{k}}^+ + a_{\mathbf{k}}a_{-\mathbf{k}}), \tag{5.77}$$

where z is the number of nearest neighbours, $\vec{\delta}$ is the lattice vector, s is the spin magnitude and ρ is the anisotropy. Let us introduce via eqn. (62) the operators $K_1^{(\mathbf{k})}$ and $K_2^{(\mathbf{k})}$ and $K_0^{(\mathbf{k})}$, and express through them the Hamiltonian

$$\hat{H}_{\text{lin}} = zs\rho\sum_{\mathbf{k}}(K_0^{(\mathbf{k})} - \frac{1}{z\rho}\sum_{\vec{\delta}}\cos(\mathbf{k}\vec{\delta})K_1^{(\mathbf{k})}) - \frac{1}{2}Nz\rho s(s+1),$$

i.e. we obtain again the Hamiltonian of the form $\hat{H} = H_0 + \Sigma_{\mathbf{k}} \oplus \hat{H}_{\mathbf{k}}$, $H_{\mathbf{k}} \in su(1,1)_{\mathbf{k}}$ (see eqn. (64)). This problem has already been solved in the preceding subsection. The characteristic parameters are

$$\text{th}\theta_{\mathbf{k}} = \frac{1}{z\rho}\sum_{\vec{\delta}}\cos(\mathbf{k}\vec{\delta}), \tag{5.78a}$$

$$E_{\mathbf{k}} = s(z^2\rho^2 - (\sum_{\vec{\delta}}\cos(\mathbf{k}\vec{\delta}))^2)^{1/2}. \tag{5.78b}$$

For a linear chain, $z = 2$, $\delta = \pm a_0$ and the spectrum is

$$E_k = 2s\sqrt{\rho^2 - \cos^2 ka_0} \tag{5.79a}$$

with the gap $E_0 = 2s\sqrt{\rho^2-1}$. Writing, as usual $\rho = 1+\Delta$, $|\Delta| \ll 1$ we get $E_0 = 2s\sqrt{2\Delta}$, i.e. the ground state is stable only for the easy-axis and isotropic variants $1 \gg \Delta \geqslant 0$. At $\rho = 1$ $(\Delta=0)$ the antiferromagnet spectrum is

$$\epsilon(k) = 2s\,|\sin ka_0|, \tag{5.79b}$$

a well-known spectrum of the Anderson model [12].

The construction of pseudospin coherent states and the condensate is completely analogous to that which has been described in the preceding subsection: they coincide up to replacing eqns. (67) and (68b) by eqns. (78) or (79). For example, for a chain

$$\zeta_{0k} = -\frac{1}{2}\text{th}\theta_k = \frac{|\sin ka_0| - 1}{\cos ka_0}.$$

(3) The last example deals with a phonon model in the harmonic approximation

$$\hat{H} = \sum_n \left[\frac{p_n^2}{2m} + \frac{m\omega_0^2 u_n^2}{2}\right] - \frac{m\omega_0^2}{2}\sum_n u_n u_{n+1},$$

where

$$u_n = (2m\omega_0)^{-1/2}(a_n^+ + a_n),\ p_n = i(2m\omega_0)^{-1/2}(a_n^+ - a_n),\ [a_n, a_m^+] = \delta_{nm}.$$

In the momentum representation, we have

$$\hat{H} = \frac{\omega_0}{4}\sum_k (2 - \cos ka_0)(a_k^+ a_k + a_{-k}^+ a_{-k} + 1) - \frac{\omega_0}{4}\sum_k \cos ka_0 (a_k^+ a_{-k}^+ + a_k a_{-k}),$$

which leads to the spectrum

$$E_k = \hbar\omega_k,\ \omega_k = \omega_0\sqrt{1 - \cos ka_0} = \sqrt{2}\,\omega_0 \left|\sin\frac{ka_0}{2}\right|. \tag{5.80}$$

Concluding this section on the basis of the concepts developed in Section 3 we discuss the nature of motion in the models studied with the noncompact dynamical symmetry $SU(1,1)_k$. From eqn. (54) we have the classical Lagrangian

$$\mathfrak{L}_{q_k} = \frac{i\hbar q_k}{1 - |\zeta_k|^2}(\dot{\bar{\zeta}}_k\zeta_k - \bar{\zeta}_k\dot{\zeta}_k) - \mathcal{H}_{q_k}(\zeta_k, \bar{\zeta}_k), \tag{5.81}$$

where $\mathcal{H}_{q_k} = <\zeta_k|\hat{H}|\zeta_k>$ is not altered under the Bogolubov canonical transformation (a rotation in the $SU(1,1)_k$ algebra space). Therefore we can use the 'rotated' Hamiltonian $\tilde{H}_k \propto K_0^{(k)}$ and 'rotated' pseudospin coherent states corresponding to it. Notice that we construct them, as usual on the reference vacuum $|\psi_0>$.

As a result, making use of eqn. (25b) we get the classical Lagrangian

$$\mathfrak{L}_k = \frac{i\hbar q_k}{1 - |\zeta_k|^2}\left\{\dot{\bar{\zeta}}_k\zeta_k - \bar{\zeta}_k\dot{\zeta}_k - \omega_k\frac{1 + |\zeta_k|^2}{1 - |\zeta_k|^2}\right\}, \tag{5.82}$$

where $\omega_k = E_k/\hbar$ is the frequency we dealt with in the examples (1)-(3). Varying eqn. (82) we have the classical equations of motion

$$\dot{\zeta}_k = i\omega_k\zeta_k. \tag{5.83}$$

From eqn. (83) we see that for each of the cases under consideration the motion of the k-th degree of freedom is a harmonic motion with frequency ω_k in the Lobachevsky plane (or in the hyperboloid) and does not depend on the representation. Thus, the phase space of the models under consideration in the momentum representation is a set of the Lobachevsky planes (proper for each k)

$$L = \dot{+}\sum_k L_k^1$$

and the dynamical group is $\prod_k \otimes SU(1,1)_k$. The question arises: which is the dynamical group that corresponds to these models in coordinate space? The spectrum generating algebra in k space is the current algebra

$$[J_\alpha^{(k)}, J_\beta^{(k')}] = iC_{\alpha\beta}^{\gamma} J_\gamma^{(k)} \delta_{-k'k}$$

with $C_{\alpha\beta}^{\gamma}$ being the $SU(1,1)$ algebra structure constants. Making the inverse Fourier transformation on a lattice we get as a dynamical symmetry algebra the Kac-Moody algebra (more precisely, one of its variants, the loop algebra)

$$[J_\alpha^n, J_\beta^m] = iC_{\alpha\beta}^{\gamma} J_\gamma^{n+m},$$

without the central extension since ' everything' is linearized (a continuous spectrum in the soliton sense).

5. Weakly nonideal Bose gas. A classical approach

We discuss here which construction corresponds on the classical level to the dynamical symmetry of a weakly non-ideal Bose gas. For that, rewrite eqn. (1) in the more convenient form

$$i\psi_t + \Delta\psi - 2(|\psi|^2 - \rho_0)\psi = 0, \tag{5.84}$$

where ρ_0 is the equilibrium density of the boson condensate (the three-dimensional case). The Hamiltonian is

$$H = \int d^3x \{(\nabla\bar{\psi}\nabla\psi) + (|\psi|^2 - \rho_0)^2\}.$$

Let us linearize eqn. (84) about the 'condensate'

$$\psi = \sqrt{\rho_0} + \epsilon a(\mathbf{k}, t), \quad (\epsilon \ll 1).$$

Then we get the Hamiltonian function in the momentum representation

$$H = \int \{(k^2 + 2\rho_0)\bar{a}(\mathbf{k})a(\mathbf{k}) + \rho_0[\bar{a}(\mathbf{k})\bar{a}(-\mathbf{k}) + a(\mathbf{k})a(-\mathbf{k})]\} d\mathbf{k} \tag{5.85}$$

with

$$\{a(\mathbf{k}), \bar{a}(\mathbf{k}')\} = i\delta(\mathbf{k} - \mathbf{k}').$$

Introduce new functions

$$J_1(\mathbf{k}) = \frac{1}{2}[\bar{a}(\mathbf{k})\bar{a}(-\mathbf{k})+a(\mathbf{k})a(-\mathbf{k})], \tag{5.86}$$

$$J_2(\mathbf{k}) = \frac{1}{2i}[\bar{a}(\mathbf{k})\bar{a}(-\mathbf{k})-a(\mathbf{k})a(-\mathbf{k})],$$

$$J_3(\mathbf{k}) = \frac{1}{2}[\bar{a}(\mathbf{k})\bar{a}(\mathbf{k})+a(-\mathbf{k})a(-\mathbf{k})],$$

constituting on the Poisson brackets $\{,\}$ the algebra $SU(1,1)_k$. These satisfy the relation

$$(J_3(\mathbf{k}))^2-(J_1(\mathbf{k}))^2-(J_2(\mathbf{k}))^2 = \Delta^2(\mathbf{k}), \tag{5.87}$$

where

$$\Delta(\mathbf{k}) = \frac{1}{2}[\bar{a}(\mathbf{k})a(\mathbf{k})-\bar{a}(-\mathbf{k})a(-\mathbf{k})],$$

which describes in the coordinates (J_1,J_2,J_3) a two-sheet hyperboloid, and their Poisson brackets with $\Delta^2(\mathbf{k})$ vanish (i.e. $\Delta^2(\mathbf{k})$ is the Casimir operator).

The Hamiltonian function (85) in these variables has the form:

$$H = \int\{(k^2+2\rho_0)J_3(\mathbf{k})+2\rho_0 J_1(\mathbf{k})\}d\mathbf{k}, \tag{5.88}$$

and the equations of motion

$$\dot{J}_1 = \{J_1,H\} = -(k^2+2\rho_0)J_2,$$

$$\dot{J}_2 = \{J_2,H\} = (k^2+2\rho_0)J_1-2\rho_0 J_3,$$

$$\dot{J}_3 = \{J_3,H\} = -2\rho_0 J_2,$$

lead to the harmonic oscillator equation for $J_2(\mathbf{k},t)$

$$\ddot{J}_2 = -\omega^2(\mathbf{k})J_2$$

where

$$\omega(\mathbf{k}) = |\mathbf{k}|\sqrt{k^2+4\rho_0} \tag{5.89}$$

is the Bogolubov frequency.

Choose a new coordinate system in the $SU(1,1)_k$ algebra space (J'_1,J'_2,J'_3) rotated about the axis J_2 by an imaginary angle θ with respect to the initial one

$$J_1(\mathbf{k}) = \text{ch}\theta(\mathbf{k})J'_1(\mathbf{k}) + \text{sh}\theta(\mathbf{k})J'_3(\mathbf{k}), \tag{5.90}$$

$$J_3 = \text{sh}\theta J'_1 + \text{ch}\theta J'_3,$$

with

$$\text{th}\theta = 2\rho_0(k^2+2\rho_0)^{-1}.$$

This transformation conserves the shape of the surface (87). The equations of motion

$$\dot{J}'_1 = -\omega J'_2, \quad \dot{J}'_2 = \omega J'_1, \quad \dot{J}'_3 = 0$$

give rise to the system of two uncoupled oscillators

$$\ddot{J}'_1 + \omega^2(k)J' = 0, \tag{5.91}$$

$$\ddot{J}'_2 + \omega^2(k)J'_2 = 0, \quad J'_3(k,t) = J'_3(\mathbf{k},0).$$

This means that our system traces on the hyperboloid (87) surface a circle with Bogolubov frequency $\omega(k)$ and the fixed value of $J'_3(\mathbf{k}) \equiv n(\mathbf{k})$. From (88) we find the energy

$$E = H = \int \omega(k)J'_3(\mathbf{k})d\mathbf{k} \equiv \int \omega(k)n(\mathbf{k})d\mathbf{k}, \tag{5.92}$$

which at $D = 1$ is exactly the nonlinear Schrödinger equation Hamiltonian in 'action-angle' variables in the solitonless sector that corresponds to the linear problem continuous spectrum (see eqn. (4.41)). The projection of the vector $J' = (J'_1, J'_2, J'_3)$ on the z-axis is the 'action' variable $n(\mathbf{k})$ which is the particle number density with the momentum $\mathbf{k}$ and the 'angular' variable describes its position in the circle $(J'_1)^2 + (J'_2)^2 = \text{const.}$ at a given instant of t:

$$\varphi(\mathbf{k},t) = \varphi(\mathbf{k},0) + \omega(k)t. \tag{5.93}$$

We see that the classical motion determined by the Hamiltonian (85) is a harmonic one on the hyperboloid surface and corresponds to a motion of a point in the phase space of the integrable system. Then the canonical transformation (90) appears to be connected with a direct scattering problem for nontrivial boundary conditions in the solitonless sector. However, one should note that this analogy is limited in the sense that the soliton asymptotic behaviour in the linear and nonlinear cases are distinct (this is especially clear for the phase [11]). Here we see that an analogue of the dynamical group present in the classical problem enables the geometrical structure of the model phase space (via Casimir operators) to be determined directly and to express the Hamiltonian through the action variable.

Chapter 6.

BOSE GAS, INTEGRABLE NONLINEAR SCHRÖDINGER EQUATIONS AND LANDAU-LIFSHITZ MODELS

1. Quantum models and nonlinear classical models corresponding to them. A new formulation of the reduction procedure.

We now turn to the second case of the general problem ($\hat{H}$ in the variables $\hat{A}_i$ is not linearized) with examples of models of the Heisenberg chain with the Hamiltonian:

$$\hat{H} = -\sum_n (I_x \hat{S}_n^x \hat{S}_{n+1}^x + I_y \hat{S}_n^y \hat{S}_{n+1}^y + I_z \hat{S}_n^z \hat{S}_{n+1}^z), \tag{6.1}$$

where $\hat{\mathbf{S}}_n$ are the spin (Pauli) operators commuting in different sites. This fact is very important, because the whole chain Gilbert space $\mathcal{H}$ is realized as the direct sum of Gilbert space of each site. In particular, this fact enables the Hamiltonian (1) to be considered using simple technique based on the representations of group direct product. Note, however, that unlike the first case where the representation was determined in a natural way (or even in a unique way) by the dynamical symmetry group of the system, in the general case such a choice is rather determined for the sake of convenience, simplicity (or even luck). In principle, the Hamiltonian $\hat{H}$ can be linearized but now in terms of a more general algebra (mixing sites) with the number of generators tending to infinity with the site number N. In this case, one loses the simplicity and insight connected with the factorization of representations of the general group into direct products.

The form of the Hamiltonian (1) suggests the system of generalized coherent states to be considered: these are representations of the rotation group $O(3)$ (or $SU(2)$). Let us consider a number of models.

(1) XXX versions (1) when $I_x = I_y = I_z = I_0$. We start with one term of the

sum in eqn. (1) and rewrite it in the form

$$\hat{\mathcal{H}}_m = -\frac{1}{2}(\hat{S}_m^+\hat{S}_{m+1}^- + \hat{S}_m^-\hat{S}_{m+1}^+) - \hat{S}_m^z\hat{S}_{m+1}^z. \tag{6.2}$$

Then $\hat{\mathcal{H}}_m$ is easily averaged over the spin coherent states of the $SU(2)$ group and we obtain (see the table of averages)

$$<m|<m+1|(-\hat{\mathcal{H}}_m)|m+1>|m> \tag{6.3}$$

$$= 2j^2\frac{\bar{\zeta}_m\zeta_{m+1}+\bar{\zeta}_{m+1}\zeta_m}{(1+|\zeta_m|^2)(1+|\zeta_{m+1}|^2)} + j^2\frac{(1-|\zeta_m|^2)(1-|\zeta_{m+1}|^2)}{(1+|\zeta_m|^2)(1+|\zeta_{m+1}|^2)}.$$

The Hamiltonian $<\hat{H}> = \Sigma_m<\hat{\mathcal{H}}_m>$ is a classical lattice analogue (in the sense defined above) of the initial quantum case (see also [27]). The next step in the reduction procedure formulated in Chapter 1 is to pass to the continuum limit by the relations

$$\zeta_{m+1} \simeq \zeta(x)+a_0\zeta'(x)+\frac{1}{2}a_0^2\zeta''(x)+\ldots,$$

$$|\zeta_{m+1}|^2 = |\zeta|^2+a_0(|\zeta|^2)'+\frac{1}{2}a_0^2(|\zeta|^2)''+\ldots, \qquad \sum_m \to \int\frac{dx}{a_0}.$$

This can now be carried out for eqn. (5.76). After simple manipulation we get

$$<\hat{H}> = 2j^2a_0I_0\int\frac{|\zeta'|^2dx}{(1+|\zeta|^2)^2}+\text{const.} \tag{6.4}$$

The Hamiltonian describes the motion of a free massless particle on the sphere S^2.

(2) The analogous procedure can be applied to the Hamiltonian (1) in the case of uniaxial anisotropy. Taking into account its small size we write (see Section 1.1)

$$I_3 = I_0+\frac{a_0^2}{2}\Delta, \tag{6.5}$$

then

$$<\hat{H}> = 2j^2a_0I_0\int\frac{|\zeta'|^2+\Delta|\zeta|^2}{(1+|\zeta|^2)^2}dx+\text{const.} \tag{6.6}$$

This means that the uniaxial anisotropy leads to the appearance of the mass term $m^2 = \Delta$, and in the easy axis case $(\Delta>0)$ the 'mass' is usual, but in the easy plane case $(\Delta<0)$ it is imaginary (symmetry breaking). In this case, the Lagrangian density reads

$$\mathcal{L} = -i\frac{j\hbar/a_0}{1+|\zeta|^2}(\bar{\zeta}\dot{\zeta}-\dot{\bar{\zeta}}\zeta)-2j^2a_0\frac{|\zeta'|^2+\Delta|\zeta|^2}{(1+|\zeta|^2)^2}. \tag{6.7}$$

Using this expression we obtain the equations of motion

$$\partial_t\frac{\partial\mathcal{L}}{\partial\dot{\zeta}}+\partial_x\frac{\partial\mathcal{L}}{\partial\zeta'}-\frac{\partial\mathcal{L}}{\partial\zeta} = 0 \text{ and } c.c.,$$

or, in dimensionless variables t,x,

$$i\dot{\zeta} = \zeta_{xx}-\frac{2\bar{\zeta}(\zeta_x)^2}{1+|\zeta|^2}+\Delta\frac{|\zeta|^2-1}{|\zeta|^2+1}. \tag{6.8}$$

The phenomenological uniaxial Landau-Lifshitz equation discussed in Chapters 1-3 to be a sigma model representation of the nonlinear Schrödinger equation has the form (see eqn. (3.64)):

$$\mathbf{S}_t = [\mathbf{S}\mathbf{S}_{xx}]+\Delta[\mathbf{S}\mathbf{e}_z](\mathbf{S}\mathbf{e}_z), \quad \mathbf{S}^2 = 1. \tag{6.9}$$

With the substitution

$$S_x+iS_y = \frac{2\psi}{1+|\psi|^2}, \quad S_z = \frac{1-|\psi|^2}{1+|\psi|^2}, \tag{6.10}$$

corresponding to the stereographic projection of the sphere (where the vector $\mathbf{S}$ takes its values) onto the plane $\psi = \psi_1+i\psi_2$ we find none but eqn. (5.90) up to the substitution $\psi\to\bar{\zeta}$.

As we have seen, eqn. (9) is integrable so is eqn. (8). This means that, starting with the integrable quantum Heisenberg XXZ model, we arrive at the classical anisotropic Landau-Lifshitz equation model integrable too. In contrast to Chapter 1 we use now quite a consistent reduction procedure based on the $SU(2)$ spin coherent states. Note that here a small value of anisotropy (5) is needed, however the procedure of deriving (6) is indifferent to the sign of Δ and can be regarded as the bases of a connection between the microscopic (1) and macroscopic (9) models. This fact is very remarkable and indicates that the identity of geometrical constructions of integrable Landau-Lifshitz models and spin coherent states is not apparently accidental.

(3) Consider the Heisenberg antiferromagnet model

$$\hat{H} = \sum_m \{\frac{1}{2}(\hat{S}_m^+ \hat{S}_{m+1}^- + h.c.) + \hat{S}_m^z \hat{S}_{m+1}^z\}. \tag{6.11}$$

Rewrite this via operators of the algebra $SU(1,1)$

$$K^{\mp} = i\hat{S}^{\mp}, \quad \hat{S}^z = K^0, \tag{6.12}$$

and obtain

$$\hat{H} = -\sum_m \{\frac{1}{2}(K_m^+ K_{m+1}^- + h.c.) - K_m^0 K_{m+1}^0\}. \tag{6.13}$$

Treating this model via the above scheme on the basis of pseudospin coherent states of the $SU(1,1)$ group, we arrive at the following classical version*)

$$\mathfrak{L} = i\frac{k\hbar/a_0}{1-|\zeta|^2}(\bar{\zeta}\dot{\zeta} - \dot{\bar{\zeta}}\zeta) - 2k^2 a_0 \frac{|\zeta_x|^2}{(1-|\zeta|^2)^2}. \tag{6.14}$$

Thus we have derived a Hamiltonian of a massless field taking values on the hyperboloid surface. Notice that the substitution (12) is needed to obtain the proper sign of the 'kinetic' term $|\zeta_x|^2$. The equations of motion generated by the Lagrangian (12) are

$$i\dot{\zeta} + \zeta_{xx} + 2\frac{\bar{\zeta}(\zeta_x)^2}{1-|\zeta|^2} = 0 \tag{6.15}$$

and correspond to the pseudostereographic projection

$$S_x + iS_y = \frac{2\zeta}{1-|\zeta|^2}, \quad S_z = \frac{1+|\zeta|^2}{1-|\zeta|^2} \tag{6.16}$$

of the isotropic Landau-Lifshitz equation on the group $SU(1,1)$ when the vector $\mathbf{S}$ lies on the hyperboloid

$$|\mathbf{S}|^2 = S_z^2 - S_x^2 - S_y^2 = 1.$$

To proceed further it should be emphasized that a connection between the quantum antiferromagnet model (13) and the isotropic Landau-Lifshitz noncompact model is still poorly explored. From the solution of the exact quantum problem

*)The vacuum (ground) state of this quantum model is very complicated which in all likelihood prompted Anderson to derive the two-sublattice model discussed in the preceding section.

(Des Cloizeaux-Pearson [29]) it is known that the vacuum structure of this model is complicated and the conventional classical phenomenology is rather constructed on the basis of Anderson's two-sublattice concept [12]. Now the noncompact isotropic Landau-Lifshitz model can be hardly considered as an adequate consistent alternative: it is however, a number of very interesting peculiar properties which make it worthwhile discussing here. Notice that eqns. (14) and (15) can be easily extended to the case of uniaxial anistropy with the same assumption as for a ferromagnet (5)

$$\hat{H} = -\sum_m \{ \frac{1}{2}(K_m^+ K_{m+1}^- + h.c.) - \rho K_m^0 K_{m+1}^0 \} \tag{6.17}$$

or

$$\mathfrak{L} = i\frac{k\hbar/a_0}{1-|\zeta|^2}(\bar{\zeta}\dot{\zeta} - \dot{\bar{\zeta}}\zeta) - 2k^2 a_0 \frac{|\zeta_x|^2 + \Delta|\zeta|^2}{(1-|\zeta|^2)^2}. \tag{6.18}$$

In terms of the functions S_i, eqn. (18) becomes the conventional uniaxial Landau-Lifshitz equation (see (3.64)):

$$2i\dot{S} = [S, S_{xx}] + \Delta[S,\sigma_3]\{S,\sigma_3\}_+, \tag{6.19}$$

$$S = \sum_{\alpha=1}^{3} S_\alpha \tau^\alpha, \quad \tau^\alpha \in su(1,1), \quad \Delta \in \mathbb{R}.$$

2. Nonlinear one-dimensional integrable models

We discuss here nonlinear one-dimensional models preserving integrability of linear systems as described above. Such a narrowing of model class is connected with the fact that a consistent theory exists only for integrable systems in one dimension. Non-integrable models are considered in subsequent chapters.

Recall some main results of the inverse scattering method for these models. For the problems on the whole axis the inverse scattering method formulation is essentially different for two types of boundary conditions: (1) zero ones, and (2) constant ones. In the first case, the inverse scattering problem is stated in the complex plane of the spectral parameter $\lambda = k^2$ [13a]. In the second case, i.e. for the condensate, on the two-sheet Riemannian surface $\xi = \sqrt{\rho_0 - \lambda^2}$ [14b] (the spectral problem algebra is 2×2). Further investigations [14] show that an even-dimensional block generalization of the second-type spectral problem is straightforward by passing to a many-sheet Riemannian surface. The main difficulties arise in the first odd versions of the $SU(2,1)$ spectral problem, where there are two branches of the continuous spectrum of the Lax operator L laying on the corresponding sheets of Riemannian surface, and the third branch in the complex plane. The inverse scattering problem methods developed up to now do not permit the overall study of such systems on the axis (even the simplest ones) due to the absence of analyticity of the monodromy matrix elements connected with the third continuous branch out the real axis. It is possible that the solution of this problem can be performed by finite-zone integration, when an analyticity band (Bargmann's band) arises. At the moment, some partial solutions to odd-dimensional spectral problems have been obtained [15,16].[*)]

We note the following characteristic features inherent to integrable Bose gas models.

(1) The existence of sigma-model representation enables interactions to be geometrized also in nonlinear integrable models (see Chapter 3), i.e. to connect the interaction type with the curvature sign of the surface of the corresponding projective space.

(2) Through the inverse scattering method one can investigate the stability of one-soliton solutions to the nonlinear Schrödinger equation with a noncompact group of internal symmetry. Other methods, including variational ones, seem to be powerless here because of the sign uncertainty of the isotopic space metric and hence the energy (including its kinetic part).

(3) The gauge equivalence of various integrable systems is their very powerful feature and enables solutions to be constructed of one system via solutions of another (in particular, see Chapter 3 on the connection between models of the nonlinear Schrödinger and Landau-Lifshitz equations).

We proceed to a more detailed study of the two simplest models.

[*)]After completing this manuscript, Dubrovin, Krichever, Malanyuk and the author developed a method which allows exact solutions (including multi-soliton ones) to be obtained for the class of Bose gas models described by vector versions of nonlinear Schrödinger equation with selfconsistent potentials [17].

3. The isotropic Landau-Lifshitz $SU(1,1)$ model

As has been pointed out noncompact sigma-models arise in the study of physical phenomena in various branches of science; in particular, in field theories of gravity [18] and extended supergravity [19], in string models [20] and Anderson's localization problem. The list is not yet complete. The classical continuum Heisenberg ferromagnet (Landau-Lifshitz) model can be formulated as a nonlinear $O(3)$ sigma-model and is equivalent to the $U(1,0)$ nonlinear Schrödinger equation. The vector of 'spin' or magnetic moment **S** moves on the sphere S^2, describing a loop when passing through the soliton solution. The equivalence of the $SU(2)$ isotropic Landau-Lifshitz model to the $U(1,0)$ nonlinear Schrödinger equation sets definite restrictions on the dynamics of the former. In particular, the boundary conditions on the matrix elements of S at infinity have to be diagonal:

$$S \to \sigma_3, \quad x \to \pm\infty, \tag{6.20}$$

and the plane spin wave, corresponding to the nonlinear Schrödinger equation condensate, turns out to be unstable and decays into a set of solitons. For readers wishing to know more about this topic, the book [21] is recommended.

The question arises: which physical sense may be related to the $SU(1,1)$ version of the isotropic Heisenberg model, formulated above and in [23], which is gauge equivalent to the $U(0,1)$ repulsive nonlinear Schrödinger equation? We discuss here the results of [23-25].

We start with eqn. (19) where

$$S(x,t) = \begin{bmatrix} S^z & iS^- \\ iS^+ & -S^z \end{bmatrix} \in su(1,1), \tag{6.21}$$

$$S^\pm = S^x \pm S^y,$$

satisfies the relations

$$\det S = -1, \quad S^2 = I. \tag{6.22}$$

This implies that the vector $\mathbf{S} = (S^x, S^y, S^z)$ lies on a two-sheet hyprboloid (the pseudosphere $S^{1,1}$)

$$(S^z)^2 - (S^x)^2 - (S^y)^2 = 1. \tag{6.23}$$

The equations of motion for the components of the vector **S** are given in the conventional way via the basis τ_α of the $su(1,1)$ algebra:

$$S(x,t) = \sum_1^3 S_\alpha(x,t)\tau_\alpha$$

where

$$\tau_\alpha\tau_\beta = g_{\alpha\beta} + iC_{\alpha\beta\gamma}\tau_\gamma,$$

with $g_{\alpha\beta} = \mathrm{diag}(-1,-1,1)$ being the Killing metric.

$$\mathrm{Tr}(\tau_\alpha\tau_\beta) = 2g_{\alpha\beta}, \quad [\tau_\alpha,\tau_\beta] = 2iC_{\alpha\beta\gamma}\tau_\gamma.$$

As a result, instead of eqn. (19) we get

$$S_t^\alpha = C_{\alpha\beta\gamma}S^\beta S_{xx}^\gamma. \tag{6.24}$$

The Poisson brackets associated with $L = su(1,1)$ are

$$\{S^\alpha(x),S^\beta(y)\} = -C_{\alpha\beta\gamma}S^\gamma(x)\delta(x-y), \tag{6.25}$$

i.e.

$$\{S^z(x),S^\pm(y)\} = \pm iS^\pm(x)\delta(x-y),$$

$$\{S^+(x),S^-(y)\} = 2iS^z(x)\delta(x-y).$$

Then the Hamiltonian equations

$$S_t^\alpha(x,t) = \{H,S^\alpha(x,t)\}$$

are just eqns. (24) if H is of the form

$$H = \frac{1}{2}\int dx S_x^\alpha S_x^\beta g_{\alpha\beta} = \frac{1}{4}\mathrm{Tr}\int dx(S_x)^2. \tag{6.26}$$

Introducing the hyperbolic angles τ and $\varphi(x,t)$

$$S^x = \mathrm{sh}\tau\cdot\cos\varphi, \tag{6.27}$$

$$S^y = \mathrm{sh}\tau\cdot\sin\varphi,$$

$$S^z = \mathrm{ch}\tau,$$

the Hamiltonian (26) may be rewritten as the energy of a scalar field

$$H = \frac{1}{2}\int dx(l_x)^2 \tag{6.28}$$

with $-dl^2 = d\tau^2 + \mathrm{sh}^2\tau d\varphi^2$ being the length element on the hyperboloid (23) in the Lobachevsky metric.

To make the picture more visual we pass to a pseudospherical projection of hyperboloid (23) onto the XY plane. Let us introduce a complex field by means of the relations

$$S^+ = \frac{\sqrt{\rho_0}\,\zeta}{\rho_0 - |\zeta|^2}, \quad S^z = \frac{\rho_0 + |\zeta|^2}{\rho_0 - |\zeta|^2}, \quad \zeta = \frac{S^+/\sqrt{\rho_0}}{\rho_0 + S^z}, \tag{6.29}$$

where ρ_0 is a constant ($\rho_0 > 0$). Now the hyperboloid upper sheet $S^z = (1 + S^+S^-)^{1/2} > 0$ is projected onto the opened disk, D_{in} of the radius $\sqrt{\rho_0}$ and the lowest sheet $S^z = -(1 + S^+S^-)^{1/2} < 0$ onto the outer part of the disk, D_{out} in the complex plane ζ. The upper pole $S^z = 1$ is projected onto the point $\zeta = 0$, and the lower one $S^z = -1$ onto infinity ($|\zeta| \to \infty$). When $S^z \to \infty$, we have $|\zeta| \to \sqrt{\rho_0} - 0$, and for $S^z \to -\infty$ we have $|\zeta| \to \sqrt{\rho_0} + 0$. Thus at the circle $|\zeta| = \sqrt{\rho_0}$ which the 'light cone' is projected onto, we can define the magnetization so that the *average* magnetization of the upper and lower sheets of the hyperboloid compensate each other

$$<M> = <S^z_{\text{up}} + S^z_d> = 0.$$

The Poisson brackets for the field $\zeta(x)$ and $\bar{\zeta}(x)$ we have the form (cf. eqn. (5.56b))

$$\{\zeta(x), \zeta(y)\} = (\rho_0 - |\zeta|^2)^2 \delta(x - y), \tag{6.30}$$

i.e. at $|\zeta|^2 \ll \rho_0$ they acquire the usual canonical form, just as in eqn. (5.56a). However, there is a distinction between these two models. In the noncompact case $SU(1,1)$ at $|\zeta|^2 \simeq \rho_0$ the Poisson bracket (30) vanishes, which is an extra confirmation of a macroscopic classical object such as the Bogolubov condensate appearing on the quantum level.

The equations of motion (24) in terms of ζ and $\bar{\zeta}$ are the same as eqn. (13), obtained earler via the pseudospin coherent states method from the Heisenberg antiferromagnet. The Hamiltonian (26) in these variables is equivalent to the one obtained from eqn. (14)

$$H = 2\rho_0 \int_{-\infty}^{\infty} dx \frac{|\zeta_x|^2}{(\rho_0 - |\zeta|^2)^2} = 2\rho_0 \int_{-\infty}^{\infty} dx g(\zeta, \bar{\zeta}) |\zeta_x|^2, \tag{6.31}$$

where $g(\zeta,\bar{\zeta}) = (\rho_0 - |\zeta|^2)^{-2}$ is the Poincaré metric in D_{in}.

As we have seen, there is a gauge correspondence between the $SU(1,1)$ isotropic Landau-Lifshitz and $U(0,1)$ nonlinear Schrödinger equations. Considering this in more detail, we recall briefly how this correspondence is established (see Chapter 3).

The linear problem has the form

$$\phi_{1x} = U_1(x,t;\lambda)\phi_1, \qquad (6.32)$$

$$\phi_{1t} = V_1(x,t;\lambda)\phi_1,$$

where ϕ_1 is a matrix solution and

$$U_1 = -i\lambda\sigma_3 + i\begin{pmatrix} 0 & \bar{\psi} \\ -\psi & 0 \end{pmatrix},$$

$$V_1 = \begin{pmatrix} 2i\lambda^2 + i(|\psi|^2 - \rho_0) & -2i\lambda\bar{\psi} + \bar{\psi}_x \\ 2i\lambda\psi + \psi_x & -2i\lambda^2 - i(|\psi|^2 - \rho_0) \end{pmatrix}.$$

The compatibility conditions

$$U_{1t} - V_{1x} + [U_1, V_1] = 0$$

give the equation

$$i\psi_t + \psi_{xx} - 2(|\psi|^2 - \rho_0)\psi = 0. \qquad (6.33)$$

Via the gauge transformation

$$U_2 = g^{-1}U_1 g - g^{-1}g_x, \quad \phi_2 = g^{-1}\phi_1, \qquad (6.34)$$

$$V_2 = g^{-1}V_1 g - g^{-1}g_t,$$

we obtain the system

$$\phi_{2x} = U_2\phi_2, \qquad (6.35)$$

$$\phi_{2t} = V_2\phi_2.$$

Let g be a normed Jost solution to system (32) at $\lambda = \lambda_0 > \sqrt{\rho_0}$,

$$g(x,t;\lambda_0) = \phi_1(x,t;\lambda=\lambda_0),$$

then operators (34) read

$$U_2 = -i(\lambda-\lambda_0)S, \tag{6.36}$$

$$V_2 = 2i(\lambda^2-\lambda_0^2)S-(\lambda-\lambda_0)S_x,$$

with

$$S(x,t) = g^{-1}(x,t)\sigma^3 g(x,t). \tag{6.37}$$

Then eqns. (35) give

$$S_t = \frac{1}{2i}[S,S_{xx}]-4\lambda_0 S_x.$$

In this equation one can remove the second term on the right-hand side by the Galillei transformation $t' = t,\ x' = x-4\lambda_0 t$ *with a corresponding change of boundary conditions.*

We shall deal further with the equation

$$S_t = \frac{1}{2i}[S,S_{xx}]. \tag{6.38}$$

Let us consider the Bose gas of constant density ρ_0. In other words, in the framework of eqn. (33) we have the boundary conditions

$$\psi(x,t)\to\psi_{\pm},\ \ \psi_x(x,t)\to 0 \ \text{ when } \ x\to\pm\infty$$

and $|\psi_+|^2 = |\psi_-|^2 = \rho_0$.

To obtain boundary conditions for the matrix S, soliton asymptotics of the linear spectral problem (32) are necessary. They have been given in [24,25] (see also Chapter 4). In contrast to the $SU(2)$ version of the isotropic Landau-Lifshitz equation, we get nondiagonal boundary conditions on $S(x,t)$

$$\underset{x\to\pm\infty}{S(x,t)\to S_{(\pm)}(x,t)} = \frac{1}{q}\begin{bmatrix} \sqrt{q^2+4\rho_0} & 2\psi_{\pm}e^{i\tilde{\theta}(x,t)} \\ -2\bar{\psi}_{\pm}e^{-i\tilde{\theta}(x,t)} & -\sqrt{q^2+4\rho_0}\end{bmatrix}, \tag{6.39}$$

where $\tilde{\theta}(x,t) = q\cdot(x+vt)$, $v = \sqrt{q^2+4\rho_0}$, and $q = 2\sqrt{\lambda_0^2-\rho_0}$ is the wave number, λ_0 the normalization point of the gauge transformation from the continuous spectrum ($|\lambda_0|>\sqrt{\rho_0}$). At $\rho_0\to 0$ we have well-known boundary conditions for the $SU(2)$ models, namely $S\to\sigma_3$.

Boundary conditions (38) imply a vector **S** rotation about the z-axis with frequency $\omega = qv$; that is the Bogolubov frequency in the theory of a weakly non-ideal Bose gas (see the preceding section). The **S** projection into the z-axis does not vary in time, and passing from $-\infty$ to $+\infty$ the S nondiagonal elements acquire a phase shift α so that

$$\psi_- = \psi_+ e^{i\alpha}. \tag{6.40}$$

Notice that solutions (38) satisfying the boundary conditions (39) have infinite energy (just as in the plane wave case). Only in the limit $q \gg \sqrt{\rho_0}$, $S_{(\pm)} \to \sigma_3$ (no condensate), does the energy become infinite. We have already encountered a similar situation when studying the $U(0,1)$ nonlinear Schrödinger equation in Subsection 4.4, where the particle number and the kink energy were infinite. To obtain finite results, the appropriate subtraction of condensate values has been used.

The gauge equivalence (Chapter 3) enables the correspondence between solutions to the nonlinear Schrödinger (33) and the isotropic Landau-Lifshitz equation (38) to be established via the formulae (see [25]):

$$\psi(x,t) = \frac{1}{2}(\tau_x + i\operatorname{sh}\tau\cdot\varphi_x)e^{i\alpha(x,t)}, \tag{6.41}$$

$$\alpha_x = \varphi_x \operatorname{ch}\tau, \quad \alpha_t = \varphi_t \operatorname{ch}\tau - \frac{1}{2}(\tau_x^2 + \varphi_x^2 \operatorname{sh}^2\tau - 4\rho_0),$$

or

$$\alpha = \varphi - \int_{-\infty}^{x} (1-\operatorname{ch}\tau)\varphi_\zeta d\zeta, \quad \text{(see, e.g., [21])},$$

where τ and φ present in the angular parametrization (27) the vector **S**. The first of these formulae is a natural generalization of the gauge coupling, $|\varphi|^2 = 1/4(\tau_x^2 + \varphi_x^2 \operatorname{sh}\tau)$. Substituting $\psi = |\psi| e^{i\alpha(x,t)}$ into eqn. (33), we obtain eqn. (41).

From eqn. (41) it is clear that to find solutions to the nonlinear Schrödinger equation via isotropic Landau-Lifshitz equation solutions one should merely differentiate and integrate. The inverse procedure is more complicated since it is connected with solving a system of nonlinear partial differential equations. However, as we have seen in Chapter 3, this difficulty can be avoided. In fact, knowing the Jost solutions to the nonlinear Schrödinger equation one can construct solutions to the isotropic Landau-Lifshitz equation. We shall focus our interest on soliton solutions to the $SU(1,1)$ isotropic Landau-Lifshitz model. For this purpose, we use the one-soliton Jost solution of the form ([24], see also Chapter 4)

$$\phi(x,t;\lambda,\lambda_0) = \begin{bmatrix} a & b \\ \bar{b} & \bar{a} \end{bmatrix}, \tag{6.42}$$

where

$$a(x,t;\lambda;\lambda_0) = e^{-i\zeta_0(x-2\lambda_0 t)}\left[1-\nu\frac{\rho_0+(\lambda-i\nu)(\lambda_0-\zeta_0)}{\rho_0(\nu+i\zeta_0)(1+e^{2z})}\right]\bar{\psi}_+, \tag{6.43}$$

$$b(x,t;\lambda,\lambda_0) = e^{-i\zeta_0(x-2\lambda_0 t)}\left[\lambda_0-\zeta_0-\frac{\nu(\lambda_0+\lambda-i\nu)}{(\nu-i\zeta_0)(1+e^{2z})}\right],$$

$$z = \nu[x-2(\lambda-2\lambda_0)t+x_0].$$

The amplitude and velocity of a kink in the nonlinear Schrödinger equation model (33) are determined by the spectral parameter λ, with $\nu = \sqrt{\rho_0-\lambda^2}$ being the kink amplitude and $v = 2\lambda$ its velocity. Formula (43) contains also a free parameter λ_0 which is the normalization point of transformation and belonging as before to the continuous spectrum $|\lambda_0| > \sqrt{\rho_0}$, and $\zeta_0 = \sqrt{\lambda_0^2-\rho_0}$. Thus the Jost solution (42) and (43) describes the scattering of the plane wave $e^{i\lambda_0 x}$ of the continuous spectrum by the soliton (kink) determined with $|\lambda| < \sqrt{\lambda_0}$. Since, in the noncompact model $|\lambda_0| > \sqrt{\rho_0}$, the gauge transformation *always* contains a free parameter λ_0*).

Substituting (42) into (37) we obtain the solution to (38)

$$S^z = \frac{\lambda_0}{\zeta_0} - \frac{\nu^2 \operatorname{sech}^2 z}{2\zeta_0(\lambda_0-\lambda)}, \tag{6.44}$$

$$S^+ = i\bar{\psi}_+ e^{-2i\zeta_0(x+2\lambda_0 t)}[(\lambda-\lambda_0)^2+(\nu\,\mathrm{th}z+i\zeta_0)^2]\frac{\rho_0+\lambda\lambda_0+i\nu\zeta_0}{2\rho_0\zeta_0(\nu+i\zeta_0)^2}. \tag{6.45}$$

This solution satisfies the boundary condition (39) if, in (40),

$$e^{i\alpha} = \frac{\psi_-}{\psi_+} = \frac{\rho_0-\lambda\lambda_0+i\nu\zeta_0}{\rho_0-\lambda\lambda_0-i\nu\zeta_0}.$$

---

*)As we shall see below, this fact is closely related to the complicated structure of the condensate in the framework of the noncompact model.

The latter is easily verified by a simple substitution. The Landau-Lifshitz equation, as an integrable system, has a countable set of integrals of motion. Let us consider the first three of these, which have obvious physical meaning, namely the magnetization projection on the ζ-axis, momentum and energy

$$M_z = \int dx(S^z(x,t)-S_0^z), \tag{6.46}$$

$$P = \int dx(\mathcal{P}(x,t)-\mathcal{P}_0),$$

$$H \equiv E = \int dx(\mathcal{H}(x,t)-\mathcal{H}_0),$$

where

$$\mathcal{P}(x,t) = \frac{i}{2}(1+S^z)^{-1}(S_x^+S^- - S_x^-S^+), \tag{6.47}$$

$$\mathcal{H}(x,t) = \frac{1}{2}[(S_x^z)^2 - S_x^+S_x^-],$$

$$J_0 = \lim_{|x|\to\infty} J(x,t), \quad J(x,t) = \{S^z, \mathcal{P}, \mathcal{H}\}.$$

Here, as in the case of the $U(0,1)$ nonlinear Schrödinger equation, to get finite values of M_z, P and H, we need to subtract infinite quantities $M_z^0 = \int S_0^z dx$ and $P^0 = \int \mathcal{P}_0 dx$, $E^0 = \int \mathcal{H}_0 dx$ (which are naturally attributed to the 'condensate'). The condensate describes the classical vacuum state and is specified by

$$S_0^z = \frac{\lambda_0}{\zeta_0}, \quad \mathcal{P}_0 = -2(\lambda_0-\zeta_0), \quad \mathcal{H}_0 = -2\rho_0, \tag{6.48a}$$

or, in the variables $k_0 = 2\sqrt{\lambda_0^2-\rho_0}$,

$$S_0^z = \frac{1}{k_0}\sqrt{k_0^2+4\rho_0}, \quad \mathcal{P}_0 = k_0 - \sqrt{k_0^2+4\rho_0}, \quad \mathcal{H}_0 = -2\rho_0. \tag{6.48b}$$

The latter relations imply that the condensate, i.e. the solution with the lowest energy in the system under consideration, to be infinitely degenerate over the wavbold etor k_0, and this is connected with the fact that the symmetry group is noncompact. In fact, let us consider a solution in the form of a plane wave for system (38) written in the variables S^z, S^+ and S^-

$$S_t^z = \frac{1}{2i}(S^+S_{xx}^- - S_{xx}^+S^-), \tag{6.49a}$$

$$S_t^+ = \frac{1}{i}(S_{xx}^zS^+ - S^zS_{xx}^+), \tag{6.49b}$$

$$S_t^- = -\frac{1}{i}(S_{xx}^z S^- - S^z S_{xx}^-). \tag{6.49c}$$

Substituting $S^\pm = S_0^\pm e^{\pm i(qx-\omega t)}$ into the first equation of (49) we get $S_t^z = 0$ i.e. $S^z = S_0^z = \text{const}$. Equations (49b) and (49c) lead both to the same result

$$\omega = q^2 S_0^z. \tag{6.50}$$

The condition $(S^z)^2-(S^x)^2-(S^y)^2 = 1$ gives $(S_0^z)^2 - S_0^+ S_0^- = 1$ or $S_0^z = (1+S_0^+ S_0^-)^{1/2}$, thence $\omega = q^2\sqrt{1+S_0^+ S_0^-}$. This means that without additional conditions the plane wave solution has the conventional dispersion $\omega \propto q^2$ and S_0^z is an arbitrary constant. This solution has, as its analogue, some 'exotic' solution to the $U(0,1)$ nonlinear Schrödinger equation. In order to obtain a solution satisfying the *normal* boundary conditions in the $U(0,1)$ nonlinear Schrödinger equation version, the gauge coupling of the $SU(1,1)$ isotropic Landau-Lifshitz equation and the $U(0,1)$ nonlinear Schrödinger equation should be allowed for; then formula (50) corresponds to the Bogolubov frequency

$$\omega = \omega_B = q\sqrt{q^2+4\rho_0} = q^2\sqrt{1+S_0^+ S_0^-},$$

whereby $q^2 S_0^+ S_0^- = 4\rho_0$ and we obtain the boundary conditions (39) since $|\psi_\pm|^2 = \rho_0$.

The quantities M_z, P and E for the solitons (44) and (45) are

$$M_z = \frac{\nu}{\zeta_0(\lambda_0-\lambda)}, \tag{6.51a}$$

$$P = 4\arcsin[\nu\{2(\lambda_0-\lambda)(\lambda_0+\zeta_0)\}^{-1/2}], \tag{6.51b}$$

$$E = 4\nu = 4\sqrt{\rho_0-\lambda^2}. \tag{6.51c}$$

They can also be expressed via the condensate wave number k_0

$$M_z(k_0,\lambda) = \frac{4\nu}{k_0}[(k_0^2+4\rho_0)^{1/2}-2\lambda]^{-1}, \tag{6.52a}$$

$$P(k_0,\lambda) = 4\arcsin\{2(\rho_0-\lambda^2)(\sqrt{k_0^2+4\rho_0}-2\lambda)^{-1}(\sqrt{k_0^2+4\rho_0}+k_0)^{-1}\}, \tag{6.52b}$$

$$E(k_0,\lambda) = 4\sqrt{\rho_0-\lambda^2}. \tag{6.52c}$$

The formulae obtained require comment. The plane wave solution (39) describes a 'spin' wave with the Bogolubov spectrum

$$\omega_B = q\sqrt{q^2+4\rho_0}. \tag{6.53}$$

Let us consider the way this solution is related to the quantum Bogolubov condensate (5.73). The structure of this state is complicated, but its 'observables' can be easily calculated, such as the energy density and average number of particles with the wave number k

$$<\Psi_0|N_k|\Psi_0> = <\Psi_0|a_k^+ a_k|\Psi_0>.$$

To obtain this expression we use the formula $e^{\beta_k a_k^+ a_{-k}^+} a_k e^{-\beta_k a_k^+ a_{-k}^+} = a_k - \beta_k a_{-k}^+$ and the reference vector $|\psi_0>$ such that $a_k|\psi_0> = 0$. As a result we have

$$<\Psi_0|N_k|\Psi_0> = \zeta_{0k}<\Psi_0|K_+|\Psi_0>$$

or, since all the correlators except the k-th one equal unity,

$$\zeta_{0k}<\zeta_{0k}|K_+^{(k)}|\zeta_{0k}> = 2j|\zeta_{0k}|^2(1-|\zeta_{0k}|^2)^{-1}\Big|_{j=1/2}$$

$$= |\zeta_{0k}|^2(1-|\zeta_{0k}|^2)^{-1}.$$

Here $\zeta_{0k} = \mathrm{th}\theta_k/2$, on the other hand, from (5.67)

$$\mathrm{cth}\theta_k = \mu_k = 1+(\epsilon_k V/NU_k),$$

hence in view of $\mathrm{th}\theta_k = 2\mathrm{th}\theta/2\cdot(1+\mathrm{th}^2\theta/2)^{-1}$ we have $\mathrm{th}\theta/2 = \mathrm{cth}\theta-(\mathrm{cth}^2\theta-1)^{1/2}$. Then

$$<\Psi_0|N_k|\Psi_0> = \frac{\mathrm{th}^2\theta_k/2}{1-\mathrm{th}^2\theta_k/2} = \frac{1}{2}\frac{1+\mathrm{th}^2\theta_k/2}{1-\mathrm{th}^2\theta_k/2}-\frac{1}{2}$$

$$\equiv <\Psi_0|K_z^{(k)}|\Psi_0>-\frac{1}{2}$$

i.e.

$$<\Psi_0|K_z^{(k)}|\Psi_0> = \frac{1}{2}+<\Psi_0|N_k|\Psi_0>. \tag{6.54}$$

Now $<K_z>$ is easily calculated giving

$$<\Psi_0|K_z^{(k)}|\Psi_0> = \frac{1}{2}\frac{NU_k+\epsilon_k V}{E_k V} \tag{6.55}$$

where $E_k = (\epsilon_k^2+2\epsilon_k NU_k/V)^{1/2}$ (see (5.68b)). As a result we get

$$<\Psi_0|N_k|\Psi_0> = \frac{1}{2}\frac{\epsilon_k - E_k + NU_k/V}{E_k}. \tag{6.56}$$

Let now N and V tend to infinity so that $\rho_0 = N/V \to \text{const.}$ (the thermodynamical limit). Consider a Bose gas model with the point-like pair interaction between particles of mass $m = 1/2$: $U = 2\Sigma_{i>j}\delta(x_i - x_j)$. Then $U_k = 2$, $NU_k/V = 2\rho_0$ and in liue of (56) we have

$$<\Psi_0|N_k|\Psi_0> = \frac{1}{2\omega_B}(2\rho_0 + k^2 - \omega_B), \tag{6.57}$$

$$<\Psi_0|N_z^{(k)}|\Psi_0> = \frac{1}{2\omega_B}(2\rho_0 + k^2), \quad \omega_B = k\sqrt{k^2 + 4\rho_0}. \tag{6.58}$$

The correspondence between formula (58) describing the average value of the operator $K_z^{(k)}$ in the Bogolubov condensate and the plane wave classical solution (39) can be obtained via the relation between the momentum of the latter and the condensate momentum k*):

$$(2\rho_0 + k^2)\omega_B^{-1} = \sqrt{q^2 + 4\rho_0}\, q^{-1},$$

or

$$q = \omega_B \rho_0^{-1/2} = k\rho_0^{-1/2}\sqrt{k^2 + 4\rho_0}. \tag{6.59a}$$

The inverse relation is

$$k^2 = \sqrt{\rho_0}\sqrt{q^2 + 4\rho_0} - 2\rho_0. \tag{6.59b}$$

Calculating $<K_{\pm}^{(k)}>$ and expressing these in terms of q one easily reconstructs the whole $S(x,t)$ matrix up to the phase $\theta(x,t)$ (1/2 is a relict of the quantum treatment and connected with thc referencbold etor taken in the form $|\psi_0> = |\frac{1}{2}, \frac{1}{2}>$)

$$S = \frac{1}{q}\begin{bmatrix} \sqrt{q^2 + 4\rho_0} & 2\sqrt{\rho_0}\, e^{i\theta} \\ 2\sqrt{\rho_0}\, e^{-i\theta} & -\sqrt{q^2 + 4\rho_0} \end{bmatrix}. \tag{6.60}$$

The phase $\theta(x,t)$ can be obtained through the conventional transformation of the averages $<K_i^{(q)}>$ (see [2b]),

*) In the region of small momenta, $q^2 \ll 4\rho_0$, both expressions are the same.

$$<K_i^{(q)}> = <\Psi_0| e^{-i\hat{P}x - i\hat{H}t} K_i^{(q)} e^{i\hat{H}t + i\hat{P}x} |\Psi_0>, \quad P = \sum_q q a_q^+ a_q,$$

whence

$$2<K_{\pm}^{(q)}(x,t)> = \frac{N}{V}\frac{\nu(q)}{E_q} e^{\mp i(E_q t + qx + \pi)},$$

$$2<K_0^{(q)}(x,t)> = E_q^{-1}\left[\frac{N}{V}\nu(q) + \frac{q^2}{2m}\right].$$

Thus there is correspondence between observables of Bogolubov condensate $<\Psi_0|K_i^{(q)}(x,t)|\Psi_0>$ and the quantitities S_i of the $SU(1,1)$ isotropic Landau-Lifshitz equation plane wave solution. The latter is attributed to the continuous spectrum of the inverse problem and can be naturally regarded as a classical analogue of a partial component of the Bogolubov condensate with momentum q. It is interesting to notice that solution (60) is given also by the gauge transformation starting with the simplest solution to the $U(0,1)$ nonlinear Schrödinger equation

$$\psi = \sqrt{\rho_0}\, e^{i\alpha}, \quad (\alpha = \text{const}), \tag{6.61}$$

and 'dressing' it by the continuous spectrum $e^{i\lambda_0 x}$ such that $q = 2\sqrt{\lambda_0^2 - \rho_0} \neq 0$ or $|\lambda_0| > \sqrt{\rho_0}$.

We have established the connection (at the level of observables) of a quantum problem on the Bogolubov condensate with its two nonlinear gauge-equivalent classical analogues in the solitonless sector. Note that both (60) and (61) are exact solutions of the corresponding nonlinear models.

Consider now the Bogolubov condensate energy

$$E_0 = \frac{1}{2}\sum_k \{E_k - (\epsilon_k + NU_k/V)\} + \frac{1}{2}N^2 U_0/V \rightarrow \tag{6.62}$$

$$\frac{1}{2}\sum_k \{k\sqrt{k^2 + 4\rho_0} - (k^2 + 2\rho_0)\} + \frac{1}{2}N\rho.$$

For small $k \ll 2\sqrt{\rho_0}$ the expression in curly brackets is reduced to $\{...\} \rightarrow -2\rho_0$, and at large k we get $\{\} \rightarrow -2\rho_0^2 k^{-2}$.

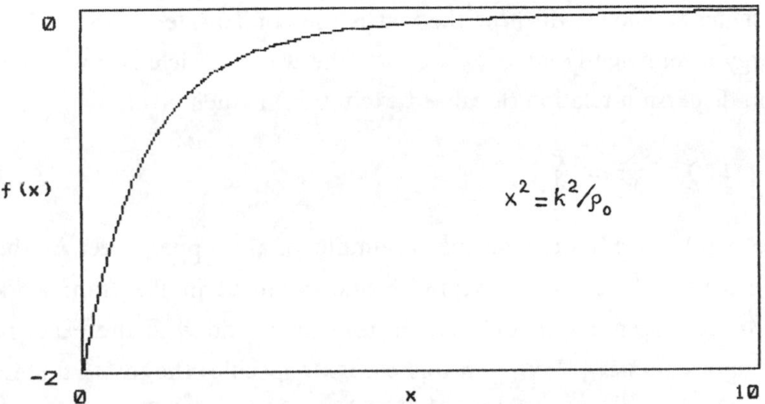

Figure 6.1. The Bogolubov condensate energy as a function of the wave number k.

Here again we have to make a reservation: as is evident from the Bogolubov dispersion relation $\omega_B = k\sqrt{k^2+4\rho_0}$ only particles having a sufficiently low energy $k^2 \leqslant 4\rho_0$ condense, i.e. are involved in the collective motion. The particles with energies $k^2 \gg 4\rho_0$ are in fact free with the conventional momentum dependence of energy $\epsilon_k \simeq k^2$ and do not fall in the condensate.

We now return to the nonlinear systems, viz. the $SU(1,1)$ isotropic Landau-Lifshitz equation and the $U(0,1)$ nonlinear Schrödinger equation which models nonperfect Bose gas behaviour on the classical level. We have already seen that, with respect to the spectrum, all these models give the same result in the solitonless sector. This is never valid when describing the Bogolubov condensate itself. The initial quantum model and the nonlinear $SU(1,1)$ classical model identically describe the Bogolubov condensate properties in terms of the operators $\hat{K}_i$ and $\hat{S}_i$ [*)] observables. However, making the transition to the $SU(1,1)$ classical model we lose information on the condensate fine structure (see (74b) and (75)). A still larger loss of information takes place when passing to a quasi-classical description via the $U(0,1)$ nonlinear Schrödinger equation. Here, only the Bogolubov spectrum remains for perturbations over the 'undressed' condensate $|\psi_0|^2 = \rho_0$.

We consider the dispersion relations for a soliton moving through the condensate (39), (49) or (50). Eliminating the parameter λ from (50) we get

$$E\cdot M_z = 8(1+\sqrt{1+4\rho_0 k_0^{-2}})\sin^2\frac{P}{4}. \tag{6.63}$$

[*)]The connection already mentioned in Section 5.2.

The parameter k_0 shows which of the waves the condensate consists of is excited. If its energy is sufficiently large, $k_0 \gg 2\sqrt{\rho_0}$ (the wave-particle is not collectivized), the soliton dispersion relation (localized excitation) is then given by

$$E(p) = \frac{16}{M_z}\sin^2\frac{P}{4}, \tag{6.64}$$

and corresponds to a state near the minimum of the upper sheet of the hyperboloid. This formula coincides with the one obtained in the framework of the compact $SU(2)$ Heisenberg model [35] (ferromagnet) and with the exact result for the Bethe spin complex [21,30]. In this case, M_z specifies the internal state (quantum number) of the soliton (the number of bound 'magnons') and is proportional to the 'macroparticle' (soliton) effective mass at small P:

$$E(p) \simeq P^2/M_z.$$

In the case $k_0 \ll 2\sqrt{\rho_0}$ (the wave-microparticle is collectivized), we have that $M_z \to \infty$ and is no longer a dynamical variable (recall that solution (39) describes the movement of the vector **S** with the Bogolubov frequency ω_B along a circle lying in a cross-section plane of the upper sheet of the hyperboloid $S_z^2 - S_x^2 - S_y^2 = 1$.) (See Fig. 6.2).

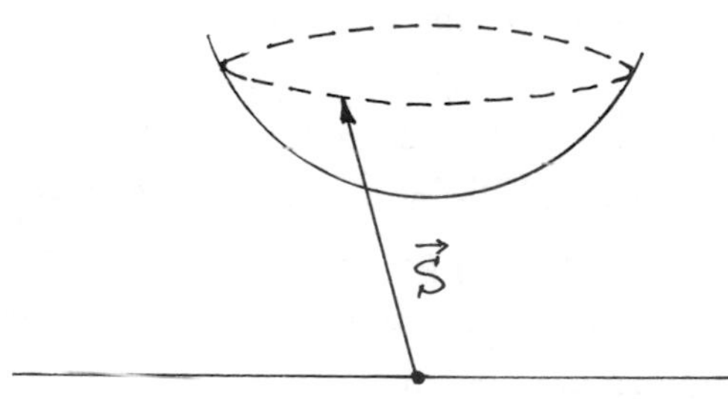

Figure 6.2.

Hence, excluding the parameter k_0 from the second and third equations of eqns. (52) we have

$$E(P) = 4\sqrt{\rho_0}\sin\frac{P}{2}, \tag{6.65}$$

with $P \in [0,2\pi]$.

Formula (65) up to a coefficient is the exact (nonlinear) dispersion law for the antiferromagnon - localized excitation of the Heisenberg antiferromagnet - obtained by Des Cloizeaux and Pearson [29] and also with the disperion, found above, of weak (linearized) excitations of the Anderson antiferromagnet (5.79b). For small P, we get $E(P) = 2\sqrt{\rho_0}\, P$, i.e. a one-magnon spectrum.

Let us summarize the results:

(1) The Bogolubov excitation mode in a repulsive Bose gas corresponds to a precession of the pseudospin vector on a hyperboloid surface (magnon).

(2) The hole (Lieb) mode described by a kink in the $U(0,1)$ nonlinear Schrödinger equation version corresponds to the pseudospin vector deviation form the precession plane (solitons are bound holes).

(3) The soliton spectrum has two remarkable limits: at low condensate density, $\rho_0 \to 0$, it tends to the soliton spectrum of the $O(3)$ Heisenberg ferromagnet (the magnon drop spectrum) (64); for excitations with energy less than the 'chemical potential', the pseudospin vector infinitely grows, the soliton spectrum (65) coincides with one for a hole in an antiferromagnet, then 'magnetizations' of the upper and lower hyperboloid sheets compensate each other when $q \to 0$ (the ground state).

Thus, just as the Bogolubov dispersion in two limiting cases describes particles with quadratic or linear dispersion, the soliton solution to the $SU(1,1)$ isotropic Landau-Lifshitz equation describes, respectively, the magnon spectrum of a ferromagnet or an antiferromagnet (let us note that such a behaviour in the linearized theory is attributed to ferrimagnets).

4. Bose gas models and nonlinear sigma-models. Summary

(1) The quasi-classical description of a non-ideal Bose gas is given by the nonlinear Schrödinger equation:

$$i\psi_t + \psi_{xx} + 2\epsilon|\psi|^2\psi = 0, \quad \epsilon = \pm 1. \tag{6.66}$$

The σ-model representations corresponding to eqn. (60) are associated with the compact $SU(2)/U(1)$ and the noncompact $SU(1,1)/U(1)$ manifolds, respectively.

Is there a connection between the condensate state and the non-compact symmetry of the system?

$<|\ddot{\Psi}|> \neq 0$ CONDENSATE <--------------> NON−COMPACT SYMMETRY GROUP G

Bogolubov (1947) showed that the canonical transformations nearby the condensate state are of hyperbolic type.

Solomon (1971) established that the Bogolubov transformations are the hyperbolic rotations in the space of dynamical symmetry group algebra of the linearized model and this group is noncompact $G = \prod_k \otimes SU(1,1)_k$.

Lieb (1963) discovered the existence of hole excitations in a one-dimensional repulsive Bose gas ($\epsilon = -1$). These correspond to vortices in a two-dimensional case.

Zakharov and Shabat (1973) have shown that the linear problem for eqn. (66) with $\epsilon = -1$ can be formulated on the $SU(1,1)$ algebra.

The discrete spectrum of this problem (dark solitons) corresponds to a hole mode.

In the quasi-classical approximation

DARK SOLITON ---→'BUBBLE', a bound state of a finite number of holes
or
a bound state of an infinite number of particles.

The bound state of a finite number of particles in the framework of eqn. (56) is stable in the first ($\epsilon = 1$) case and unstable in the second ($\epsilon = -1$) one.

(2) The group sense

COMPACT GROUPS	NONCOMPACT GROUPS
UNITARY REPRESENTATIONS	
FINITE DIMENSIONAL	INFINITE DIMENSIONAL

(3) Stable bound states

ATTRACTIVE NONLINEAR SCHRÖDINGER EQUATION FINITE NUMBER OF PARTICLES

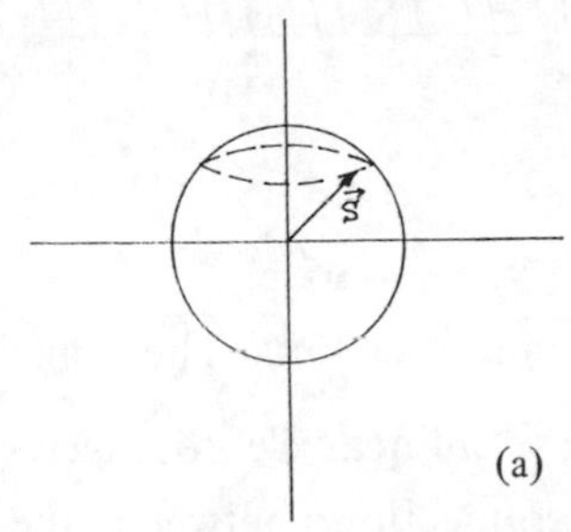

REPULSIVE NONLINEAR SCHRÖDINGER EQUATION INFINITE NUMBER OF PARTICLES

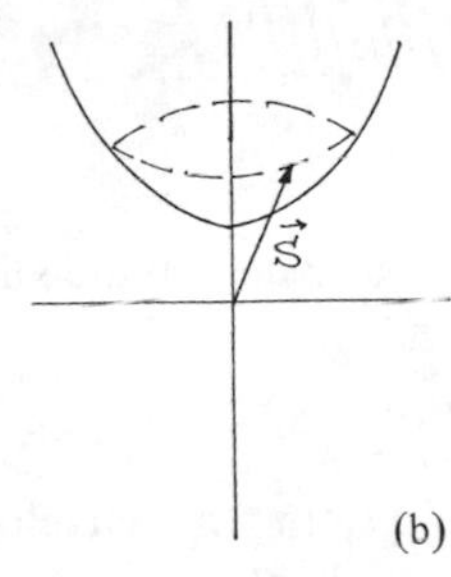

Figure 6.3. The movement of the vector S on (a) a compact manifold, the S^2 sphere, (b) a non-compact manifold, the upper sheet of a hyperboloid, $S^{1,1}$.

SYSTEMS WITH NONCOMPACT SYMMETRY

QUANTUM REALIZATION	CLASSICAL REALIZATION
BOSE GAS OF FINITE DENSITY $= \lim N/L$	NONTRIVIAL BOUNDARY CONDITIONS

5. The third version - The sigma-model representation connected with the nonlinear Schrödinger equation

$SU(2)$ isotropic Landau – Lifshitz equation $\qquad$ $SU(1,1)$ isotropic Landau – Lifshitz equation

$$\mathbf{S} = (S_x, S_y, S_z) \rightarrow S = \begin{pmatrix} S_z & S^- \\ S^+ & -S_z \end{pmatrix} \in su(2), \quad \mathbf{S} \rightarrow S = \begin{pmatrix} S_z & iS^+ \\ iS^+ & -S_z \end{pmatrix} \in su(1,1),$$

$$\mathbf{S}^2 = 1 \rightarrow \det S = -1, \quad \mathbf{S}^2 = 2S_z^2 - 1 \rightarrow \det S = -1.$$

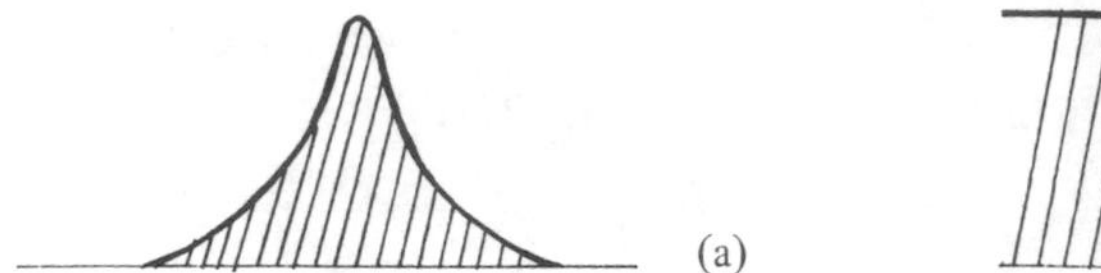
(a)

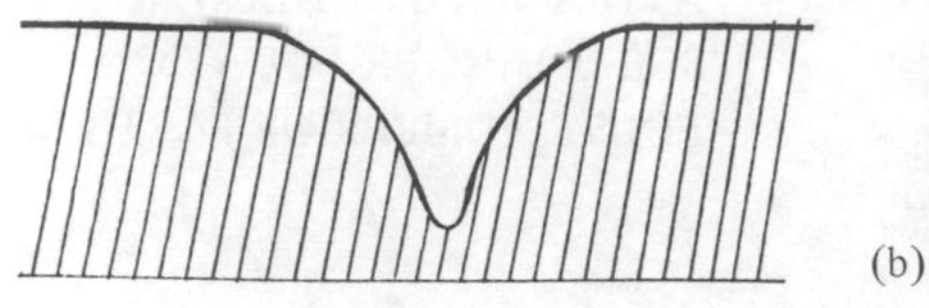
(b)

Figure 6.4. Bound states: (a) drop like, (b) bubble like.

The equation: $S_t = -\frac{i}{2}[S, S_{xx}]$ is the isotropic version. The small-amplitude approximation for the envelope leads to the nonlinear Schrödinger equation in more space dimensions. At $D = 1$, the connection between the nonlinear Schrödinger and the isotropic Landau-Lifshitz equations is exact. The spin wave in both cases is a precession of the vector **S** with a given S^z = const, radius $R^2 = S^+S^-$ and frequency $\omega = q^2S^z$; moreover, S^z can be a function of the wave number q. The spin wave in the nonlinear Schrödinger equation version answer the plane wave solutions

$$\psi = a \exp i(k_0 x - \omega_0 t), \tag{6.67}$$

with $\omega_0 = k_0^2 \mp 2a^2$. The gauge coupling (see (3.60))

$$\mathfrak{N} \equiv |\psi|^2 = \frac{1}{8}\mathrm{Tr}(S_x^2) = \frac{1}{4}[(S_x^z)^2 + S_x^+ S_x^-)], \tag{6.68}$$

$$\mathfrak{P} = -\frac{i}{2}(\bar{\psi}\psi_x - c.c.) = -\frac{i}{8}\mathrm{Tr}(S \cdot S_x \cdot S_{xx}),$$

$$\mathfrak{K} = |\psi_x|^2 - |\psi|^4 = \frac{1}{8}\mathrm{Tr}[(S_{xx})^2 - \frac{5}{4}(S_x)^4],$$

with, allowing for the physical meaning of the problem given in the nonlinear Schrödinger equation version, puts constraints on the isotropic Landau-Lifshitz equation solution choice. For example, for $|\psi|^2 = \rho_0$ from the first equation of (68) one easily obtains a connection between the spin wave and the Schrödinger wave $\rho_0 = \frac{1}{4}S_x^- S_x^+$.

Further, since

$$S^{\pm} = (1-(S^z)^2)^{1/2} e^{\mp i(qx-\omega t)}, \qquad SU(2) \text{ variant} \tag{6.69}$$

$$S^{\pm} = ((S^z)^2-1)^{1/2} e^{\mp i(qx-\omega t)}, \qquad SU(1,1) \text{ variant}$$

we have $4\rho_0 = \pm q^2(1-(S^z)^2)$, or

$$S^z = \frac{1}{q}\sqrt{q^2-4\rho_0}, \quad SU(2) \tag{6.70}$$

$$S^z = \frac{1}{q}\sqrt{q^2+4\rho_0}, \quad SU(1,1)$$

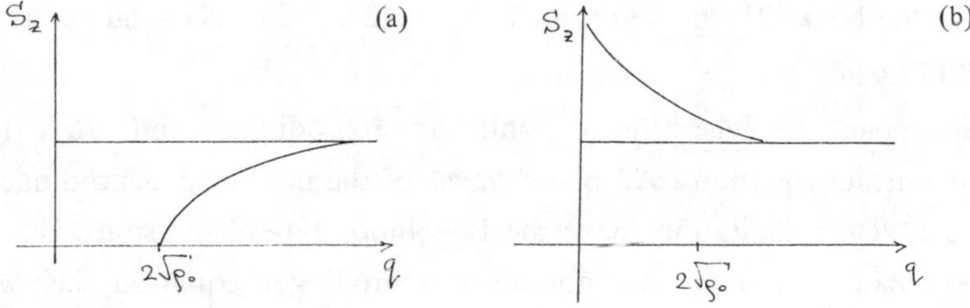

Figure 6.5. S^z as a function of q in (a) the compact case, (b) noncompact one.

By the second formula of (68) we also readily find a coupling between k_0 (the nonlinear Schrödinger equation wavbold etor) and q (the spin wavbold etor)

$$k_0 = q\text{ch}\theta = S^z q = \sqrt{q^2 \pm 4\rho_0}. \tag{6.71a}$$

Taking into account that $\omega_0 = k_0^2 \mp 2a^2$ and $\omega = q^2 S^z$ we have

$$\omega_0 = \omega S^z \mp 2(a^2-\rho_0) \ \text{(see (41))}. \tag{6.71b}$$

From (71a) it follows that only Schrödinger waves with $k_0 > 2\sqrt{\rho_0}$ have, in the $SU(1,1)$ case, spin-wave analogues of the form (39)*).

Also notice that in the nonlinear Schrödinger equation case the dispersion of a weak excitation of the wave (57), i.e. the condensate, is the Bogolubov type dispersion $\omega = 2k_0k + k\sqrt{k^2 \pm 4\rho_0}$ (ω and k are the frequency and the wavbold etor of the excitation). In the nonlinear spin case (Landau-Lifshitz equation) the waves themselves possess the Bogolubov dispersion due to the gauge coupling with the

*) Which completely corresponds to the above-obtained result on a possible 'dressing' of the Schrödinger solution via Jost functions only from the continuous spectrum.

nonlinear Schrödinger equation.

Summary

Compact models: $U(1,0)$ nonlinear Schrödinger and $SU(2)$ isotropic Landau-Lifshitz equations. All plane waves in the nonlinear Schrödinger equation version, i.e. with any k_0, are unstable with respect to excitations with $k<2\sqrt{\rho_0}$. The plane spin waves are unstable only for $k<2\sqrt{\rho_0}$. More details on the compact models $U(1,0)$ nonlinear Schrödinger and $SU(2)$ isotropic Landau-Lifshitz equations can be found in the book [21] or the original papers [28,31-35] (classical variant) and [27] (quantum one).

Noncompact models: $U(0,1)$ nonlinear Schrödinger and $SU(1,1)$ isotropic Landau-Lifshitz equations. All plane waves of the nonlinear Schrödinger equation are stable. Their excitations have the Bogolubov type dispersion. However, only 'dressed' plane waves of the nonlinear Schrödinger equation, i.e. waves with $k_0>2\sqrt{\rho_0}$, have analogues in the form of spin waves.

6. On the reduction procedure

In Chapter 1, we formulated a 'simplified' reduction procedure (i.e. a transition from a discrete quantum description to a classical continual one) employing, in the first stage, test Schrödinger functions. In the case of Bose operators, this scheme implies, in fact, using Heisenberg-Weil coherent states constructed on the conventional vacuum $|0>$ as the ground state. On the other hand, in Section 5.4 we have seen that even in a Bose gas model, governed by a 'linearized' Hamiltonian quadratic with respect to operators, in the presence of anomalous Bogolubov terms the system ground state is very complicated. It may be constructed by means of the spin coherent states and Bogolubov's rotation; to achieve this, the Hamiltonian is written in terms of Pauli operators. In nonlinear models, the latter can be made via Jordan-Wigner or Holstein-Primakoff transformations*). The spin coherent state minimizing the uncertainty relation are maximally classical

*) Let us note that in the one-dimensional case, the transformation between Fermi and Pauli operators is exact.

and enable a bridge to be constructed between quantum models (quantum statistical mechanics models) and classical models. In this sense, the reduction procedure based upon the generalized coherent states concept is most consistent. This fact has been noted in Gochev's papers [27a] where an attempt has been made to establish the connection between exact solutions to the quantum Heisenberg model (in the form of spin complexes and their associations) and quasi-classical solitons of the Landau-Lifshitz model.

A detailed discussion of this point can be found in the recent works by A. Makhankov and the author [45].

Chapter 7.

ϕ^6 THEORY AND BOSE DROPS

In this short chapter, we discuss one more Bose gas model which is very important from the applications point of view: the so-called phi-six (ϕ^6) theory. It is governed by an equation which can be written in two equivalent forms up to a scale transformation*)

$$i\psi_t + \Delta\psi \pm \psi + (|\psi|^2 - \alpha|\psi|^4)\psi = 0, \tag{7.1a}$$

$$i\psi_t + \Delta\psi + \alpha\psi + (|\psi|^2 - |\psi|^4)\psi = 0. \tag{7.1b}$$

We derived this in Chapter 1 after applying the reduction procedure to the Heisenberg ferromagnet and allowing for nonlinearity in the exchange integrals. The last term in (1) arises due to either spin-phonon interaction or the fact that the original operators were Pauli operators. In terms of a Bose gas eqn. (1) describes a quasi-classical gas of bosons with point-like attractive pair interaction and a three-particle repulsive interaction (e.g. the nucleons behaviour in the nuclei is similar for short-range interaction due to their Fermi statistics). Such an equation occurs also in nuclear hydrodynamics with Skyrme's forces [36], and its vector case - the multi-component generalization - arises in the time-dependent Hartree-Fock theory [37]. It may be mentioned that (1) is used to describe phenomenologically first-order phase transitions [38] in the study of soliton models of hadrons [39], etc. Apart from these purely physical aspects, eqn. (1) possesses a number of remarkable mathematical features:

(1) It describes possibly one of the simplest nonlinear models bearing stable multi-dimensional particle-like solutions.

(2) Within its framework, the particle-like and soliton-like solutions existence theorem were proved.

---

*)It is easily to check that there is a transition from one form to another if $\alpha > 0$ in eqn. (1a).

(3) Unlike the $S3$ model discussed above, both trivial and condensate boundary conditions are realizable here.

All this prompts us to give a more detailed discussion of eqn. (1) here.

1. General relations and solitons, drops (particle-like solutions)

To obtain a solution of eqn. (1) in explicit form, let us consider its one-dimensional version

$$i\psi_t + \psi_{xx} + \alpha\psi + \psi(|\psi|^2 - |\psi|^4) = 0. \tag{7.2}$$

The only parameter α in (2) is defined by the physical meaning of the model and can be here considered as positive as well as negative. In what follows, one more form of (2) will be of use, viz.,

$$i\varphi_t + \varphi_{xx} - \rho_0(\rho_0 + 2A)\varphi + 2(2\rho_0 + A)|\varphi|^2\varphi - 3|\varphi|^4\varphi = 0 \tag{7.3}$$

with $2\rho_0 + A > 0$, and A is given by

$$\rho_0^{-1} A_{1,2} = -2 - \frac{3}{4\alpha} \pm \frac{3}{4|\alpha|}\sqrt{1+4\alpha} \tag{7.4}$$

(here we follow works [40] and [41]).

Equation (3) is derived from (2) via the simple scale transformation

$$\varphi = [\frac{2}{3}(2\rho_0 + A)]^{1/2}\psi, \quad t' = \frac{3}{4}(A + 2\rho_0)^{-2}t, \quad x' = \frac{\sqrt{3}}{2}(A + 2\rho_0)^{-1}x.$$

Equation (3) is more convenient in studying condensate excitations whereas (2) enables solutions to be classified. Note also that eqn. (2) contains one parameter only, whereby solutions to (3) with the same A/ρ_0 are similar. Furthermore, in (3), the condensate state $|\varphi|^2 = \rho_0$ is written in an explicit form, which is readily checked. The condensate in terms of (2) can be found from the equation $\alpha + \rho_0 - \rho_0^2 = 0$ or $\rho_0 = \frac{1}{2}(1 + \sqrt{1+4\alpha})$. By introducing in (3) two parameters, A and ρ_0, we can break down this coupling; ρ_0 is now arbitrary but A/ρ_0 is given by relation (4). Thus, form (2) is convenient in describing systems under trivial boundary conditions, and can be called the 'drop' form, and (3) the 'condensate' form. We emphasize that the drop form is reduced to be condensate form if

$$\alpha \geqslant -\frac{1}{4}. \tag{7.5}$$

The term $\alpha\psi$ in (2) can be eliminated via the substitution

$$\psi(x,t) = e^{i\alpha t}\tilde{\psi}(x,t), \tag{7.6}$$

only under trivial (drop) boundary conditions. This fact was discussed above in detail.

Consider first the drop boundary conditions

$$\psi(x,t) \to 0 \quad \text{at} \quad |x| \to \infty. \tag{7.7}$$

The energy and particle number integrals are, in this case,

$$E = \int_{-\infty}^{\infty} dx \{ |\psi_x|^2 - \frac{1}{2}|\psi|^4 + \frac{1}{3}|\psi|^6 \}, \tag{7.8}$$

$$N = \int_{-\infty}^{\infty} dx\, |\psi|^2. \tag{7.9}$$

Equation (2) with (7) can be twice integrated (putting $\psi_t = 0$) to give

$$\psi_{dr} = e^{i\theta_0} \left[\frac{-4\alpha}{1 + \sqrt{1 + \frac{16}{3}\alpha\, \mathrm{ch}(2\sqrt{-\alpha}(x - x_0))}} \right]^{1/2}, \tag{7.10a}$$

so that a moving soliton is obtained via the Gallilei boost

$$\theta_0 \to \frac{\mathrm{v}}{2}x - \frac{\mathrm{v}^2}{4}t + \theta_0, \quad \mathrm{ch}(2\sqrt{-\alpha}(x - x_0)) \to \mathrm{ch}(2\sqrt{-\alpha}(x - \mathrm{v}t - x_0)). \tag{7.10b}$$

The soliton-like solution is seen from (10b) to depend on three parameters x_0, θ_0, v and exists when*)

$$-3/16 < \alpha < 0, \quad \text{(Region III see Section3).} \tag{7.11}$$

In this case, N and E are

$$N(\alpha) = \sqrt{3}\,\mathrm{arch} \left[(1 + \frac{16}{3}\alpha)^{-1/2} \right] \to 4\sqrt{-\alpha}\, \Big|_{\alpha \to 0}, \tag{7.12}$$

---

*)By means of transformation (6) one can introduce an extra parameter β, then the substitution $(-\alpha) \to \beta - \alpha$ should be carried out for the amplitude of (10). Such a substitution, however, renormalizes only the magnitude of α and we can always redenote $\alpha - \beta = \alpha'$.

$$E(\alpha) = (v^2 \quad \frac{3}{4} - \alpha)N(\alpha) + \frac{3}{4}\sqrt{-\alpha}. \tag{7.13}$$

When $\alpha \to -3/16, N$ grows infinitely with soliton width, and its amplitude tends to the constant value, $\psi_0 = \sqrt{3}/2$. Such behaviour of the solution to (10) is extremely proper, and implies that, as a result of an attractive pair force effect (the term proportional to $|\psi|^2\psi$), the particle density in the soliton increases to its peak value, then 3-body repulsion begins to act (at small distances). This compensates for the attractive forces and the increase in particle density so that further increase of the number of particles N gives rise to a growth in soliton size with the particle density and particle bound energy being constant. This bound state of a large number of bosons is naturally regarded as a drop of their condensed state ('fluid'). See Fig. 7.1.

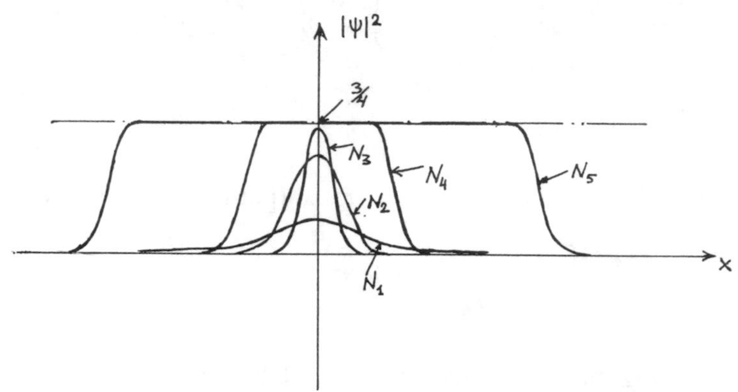

Figure 7.1. Bound states in the $\phi^3 - \phi^5$ nonlinear Schrödinger equation for various $\alpha \in (-3/16,0)$ (hence various particle numbers) $N_i = N(\alpha_i)$, $N_5 > N_4 > N_3 > N_2 > N_1$, $\alpha_5 \simeq -3/16$.

Note that such a situation is a characteristic feature of multi-nucleon systems [37b] and of the magnon gas in spin systems [42].

If $\alpha = -3/16$, the solution (10a) goes into the condensate

$$\psi_c = e^{i\theta_0}\sqrt{3}/2, \tag{7.14}$$

i.e. when $\alpha = -3/16$, $N = \infty$, we have a transition from the drop state to the condensate state. Such a behaviour of the model under consideration, along with the saturation effect, is one of the most remarkable of its features.

A numerical study of drop dynamics (in the 1-D case - slab dynamics), in particular for the vector version of eqn. (2), has been carried out by several authors (for details, see the review [37b] and the references therein, and also [43]).

2. Condensate states and their weak excitations

Let us consider eqn. (3) under the following nontrivial boundary conditions

$$|\varphi|^2 \to \rho_0. \tag{7.15}$$

Integrals of motion (8) and (9) are divergent in this case, so we rewrite them as follows

$$E = \int_{-\infty}^{\infty} dx\{|\varphi_x|^2 + (|\varphi|^2 - \rho_0)^2(|\varphi|^2 - A)\}, \tag{7.16a}$$

$$N = \int_{-\infty}^{\infty} dx\{|\varphi|^2 - \rho_0\}, \tag{7.16b}$$

with a real constant A.

(1) We investigate a dispersion of small oscillations about the condensate (15) via

$$\varphi(x,t) = \sqrt{\rho_0} + \zeta(x,t),$$

$$\zeta(x,t) = \eta_1 e^{i(kx-\omega t)} + \eta_2 e^{-i(kx-\omega t)},$$

then η_i satisfy the equations

$$\omega\eta_1 - k^2\eta_1 - 2\rho_0(\rho_0 - A)\eta_1 - 2\varphi_0^2(\rho_0 - A)\bar{\eta}_2 = 0,$$

$$-\omega\eta_2 - k^2\eta_2 - 2\rho_0(\rho_0 - A)\eta_2 - 2\varphi_0^2(\rho_0 - A)\bar{\eta}_1 = 0.$$

Wherefrom equating the determinant of the system to zero one obtains

$$\omega^2 = k^2[k^2 + 4\rho_0(\rho_0 - A)], \tag{7.17}$$

i.e. the Bogolubov dispersion law ignoring the fact that change of the interaction type causes the sound velocity to alter in such a condensate

$$v_s = \lim_{k\to\infty} \omega/k = 2\sqrt{\rho_0(\rho_0 - A)}. \tag{7.18}$$

From (17) and (18) we see that the condensate is considered to be stable when

$$\rho_0 \geqslant A. \tag{7.19}$$

(2) Small amplitude non-linear waves. Let us take φ in conventional form

$$\varphi(x,t) = \sqrt{\rho(x,t)}\exp(i\varphi(x,t)), \tag{7.20}$$

and introduce $V = \theta_x$. Then, from (3), the nuclear hydrodynamics equations follow:

$$\rho_t = -2(V\rho)_x, \tag{7.21a}$$

$$V_t = \frac{1}{2}\left[\frac{\rho_{xx}}{\rho} - \frac{1}{2}\frac{\rho_x^2}{\rho^2}\right]_x - 2VV_x + [(\rho_0-\rho)(3\rho-\rho_0-2A)]_x. \tag{7.21b}$$

To proceed further we simplify these equations by means of 'reductive perturbation theory' (see [44]). We choose new coordinates

$$\tau = \epsilon^{1/2}t,\ \ \xi = \sqrt{\epsilon}(x - v_s t),$$

and expand ρ and V in powers of ϵ about the condensate, $\rho = \rho_0\ V = 0$:

$$\rho = \rho_0 + \epsilon\rho_1 + \epsilon^2\rho_2 + \ldots,$$

$$V = \epsilon V_1 + \epsilon^2 V_2 + \ldots.$$

Substituting these in (21) and causing the coefficients at ϵ and ϵ^2 to vanish, we get

$$v_s\rho_1' = 2\rho_0 V_1', \qquad ' = \partial_\xi$$

$$v_s\rho_2' = \dot\rho_1 + 2(v_1\rho_1)' + 2\rho_0 V_2',\ \dot{} = \partial_\tau$$

$$v_s V_1' = 2(\rho_0 - A)\rho_1',$$

$$v_s V_2' = \dot V_1 + 2V_1V_1' - \rho_1'''/2\rho_0 + 2(\rho_0 - A)\rho_2' + 3(\rho_1^2)'.$$

The first (or third) of these equations gives the coupling $V_1 = v_s\rho_1/2\rho_0$ whereby the second and the fourth equations of the system are reduced to the Korteweg-de Vries equation

$$2\sqrt{\rho_0(\rho_0-A)}\,\dot\rho_1 - \frac{1}{2}\rho_1''' + 3(2\rho_0 - A)(\rho_1^2)' = 0. \tag{7.22a}$$

Via the scale transformation

$$\xi = \tilde{\xi}\,[2(2\rho_0 - A)]^{-1/2}, \quad \tau = \tilde{\tau}\frac{v_s}{2}(2\rho_0 - A)^{-3/2},$$

(22a) assumes the standard from

$$\partial_\tau \rho_1 - \partial_\xi^3 \rho_1 + 3(\rho_1^2)_{\tilde{\xi}} = 0, \tag{7.22b}$$

(we emphasize the relative sign of the dispersion term and the nonlinear term). One-soliton solutions of the latter are well-known:

$$\rho_1(\tilde{\xi},\tilde{\tau}) = -\frac{\tilde{b}}{2}\mathrm{sech}^2\left[\frac{\sqrt{\tilde{b}}}{2}(\tilde{\xi}+\tilde{b}\tilde{\tau} - \tilde{\xi}_0)\right], \tag{7.23a}$$

or in the original variables

$$\rho(x,t) = \left\{\rho_0 - \frac{b}{2\rho_0 - A}\mathrm{sech}^2\left[\sqrt{b}\,(x - v_s t + \frac{2b}{v_s}t - x_0)\right]\right\}^{1/2}, \tag{7.23b}$$

where $b = \tilde{b}/2(2\rho_0 - A) \geqslant 0$.

The solution describes a localized rarefraction wave moving in the condensate at a velocity close, but less than, the sound velocity. Such quasisolitons and the background will arise in an initial perturbation decay. Whereas the decay of a compression perturbation occurs in the solitonless part to give a dispersive packet of Bogolubov waves (which are sound waves at small wave numbers). Also note that the birth of the rarefraction soliton (23) is accompanied by the emission of 'linear' (Bogolubov type) waves moving in the same direction as the soliton but faster with the group velocity

$$v_{gr} = 2\frac{k^2 + \frac{1}{2}v_s^2}{\sqrt{k^2 + v_s^2}}.$$

These waves are usually called 'foregoers'*) This is simply a consequence of the Bogolubov type of dispersion law of the system, namely, at small k we have $\omega = k\sqrt{k^2 + v_s^2} \simeq k(v_s + \frac{1}{2}\frac{k^2}{v_s})$ that is the Korteweg-de Vries dispersion with an opposite sign in its cubic term.

---

*)The same picture takes place in the case of an arbitrary nonlinearity, when, however, there are one or several stable vacuum states (condensates).

3. Localized soliton like excitations of the condensate

We have written the system energy in the form of (16a) in order to make the energy of condensate excitations to be finite as well as the equations of motion to be satisfied. For that the condensate state should be the second-order zero of the polynomial part of the Hamiltonian density. Varying (16a) one can readily see eqn. (3) be the equation of motion. This fact provides the basis to write the equation of motion in such a form.

To find moving localized solutions of (3) we make the Ansatz (20) and the fact that $\varphi(x,t) = \varphi(\zeta)$, $\zeta = x - vt$; then we have $('\equiv d/d\zeta)$

$$\rho\rho'' - \frac{1}{2}\rho'^2 + v\theta'\rho^2 - (\theta')^2\rho^2 - \rho^2(\rho-\rho_0)(3\rho-\rho_0-2A) = 0, \tag{7.24}$$

$$\rho\theta' + \frac{v}{2}(\rho_0^2 - \rho) = 0.$$

Substituting the second equation in the first and integrating we get

$$\pm 2(\zeta-\zeta_0) = \int d(\rho-\rho_0)^{-1}(\rho^2 - A\rho - \frac{v^2}{4})^{-1/2}.$$

Denoting

$$r = \rho-\rho_0, \quad a = \rho_0^2 - A\rho_0 - \frac{v^2}{4} \equiv \frac{1}{4}(v_s^2 - v^2) \tag{7.25}$$

the integral assumes the conventional form

$$\pm 2(\zeta-\zeta_0) = \int \frac{dr}{r}(r^2 + (2\rho_0 - A)r + a)^{-1/2}.$$

Localized regular soliton-like solutions exist at $a > 0$ or $v^2 < v_s^2$. Thus, as in the case of the Gross-Pitaevskii model ($U(0,1)$ nonlinear Schrödinger equation) we obtain the condition for a hole-like mode to exist, yet a set of one-soliton solutions is here more interesting. Inversing integral (25), one has

$$r_\pm(\zeta) = \frac{\pm 2a}{\sqrt{A^2+v^2}\,\mathrm{ch}(2\sqrt{a}(\zeta-\zeta_0)) \pm (2\rho_0 - A)}. \tag{7.26}$$

The second of these solutions is singular, since $2\rho_0 > A$. The first one is regular and different in four parameter A regions:

$$(I)\ A > \rho_0, \qquad \alpha < -\frac{1}{4}, \tag{7.27a}$$

$$(II)\ \rho_0 > A > 0 \qquad \text{or in terms of } \alpha \qquad -\frac{3}{16} > \alpha > -\frac{1}{4}, \tag{7.27b}$$

$$A = 0 \qquad \alpha = -\frac{3}{16}, \tag{7.27c}$$

$$(III)\ -\frac{\rho_0}{2} < A < 0, \qquad 0 > \alpha > -\frac{3}{16}, \tag{7.27d}$$

$$(IV)\ A < -\frac{\rho_0}{2}, \qquad \alpha > 0. \tag{7.27e}$$

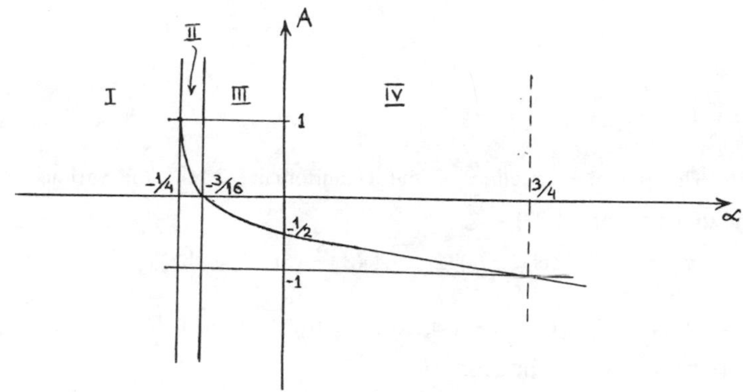

Figure 7.2. The dependence $A(\alpha)$, equation (74) and the regions (7.27).

In Fig. 7.2, the dependence $A(\alpha)$ and corresponding regions (27) are shown. The plots of the potential part of the Hamiltonian are depicted for various regions in Fig. 7.3.

In the second region, the soliton at rest is

$$\varphi_b = \sqrt{\rho_0}\, e^{i\theta_0} \frac{\text{ch}\{(\text{v}_s/2)(x-x_0)\}}{[\rho_0/A + \text{sh}^2\{\ \}]^{1/2}} \equiv e^{i\theta} \sqrt{A} \frac{\text{ch}\{\ \}}{(1+(A/\rho_0)\text{sh}^2\{\ \})^{1/2}}. \tag{7.28}$$

Since φ_b (28) is an even function of $(x-x_0)$ it describes a bubble in the condensate, whose depth depends on A: the smaller A the greater the rarefaction in the bubble.

In the third and fourth regions, $A<0$ we have

$$\varphi_k = \sqrt{\rho_0}\, e^{i\theta_0} \frac{\text{th}\{(\text{v}_s/2)(x-x_0)\}}{[1+(\rho_0/|A|)\text{sech}^2\{\ \}]^{1/2}} \equiv e^{i\theta} \sqrt{|A|} \frac{\text{sh}\{\ \}}{(1+(|A|/\rho_0)\text{ch}^2\{\ \})^{1/2}}. \tag{7.29}$$

wherefrom φ_k is an odd function of $(x-x_0)$ and has a kink form.

Phases of both solutions are found via eqn. (24) to give

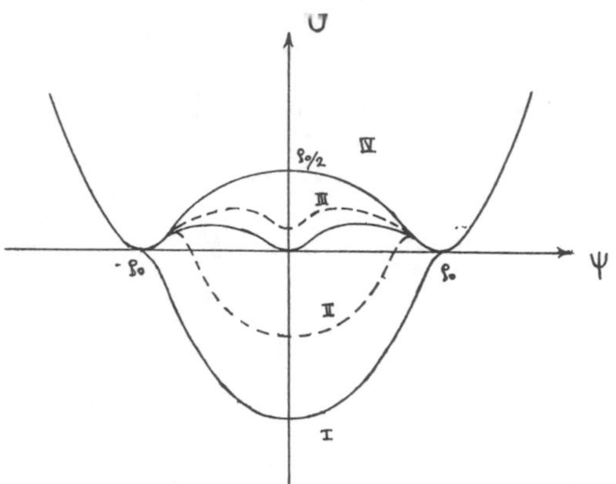

Figure 7.3. The potential fraction of the Hamiltonian (7.16a) for various α. I, II, III, IV correspond to the regions (7.27).

$$\varphi_{k,b} = e^{i\theta}\sqrt{2}\,\mathrm{ch}(\xi - i\mu)[(2\rho_0 - A)(A^2+v^2)^{-1/2} + \mathrm{ch}2\xi]^{-1/2}, \tag{7.30}$$

where $\xi = 1/2(v_s^2 - v^2)^{1/2}(x - vt - x_0)$, $\cos 2\mu = (\rho_0\sqrt{A^2+v^2})^{-1}(A\rho_0 + v^2/2)$, and kinks and bubble are specified by the sign of A.

When $A = 0$, or $\alpha = -3/16$, we get a solution under mixed boundary conditions

$$\varphi(x,t) \to 0, \quad \text{for} \quad x \to -\infty,$$

$$\varphi(x,t) \to \sqrt{\rho_0}\, e^{i\theta}, \quad \text{for} \quad x \to +\infty.$$

The energy of the system is readily verified to be finite (see (16a)) and the solution

$$\varphi_w(x) = e^{i\theta}\sqrt{\rho_0}\,(1 + 2\rho_0 e^{\pm 2\rho_0(x - x_0)})^{-1/2} \tag{7.31}$$

connects two stable vacuum states: the condensate $|\varphi|^2 = \rho_0$ and the trivial one $|\varphi|^2 = 0$. When θ = const., formula (31) describes a wave at rest, if $\theta = (v/2)x - \omega t$ the wave (31) moves together with the condensate (notice that, in the latter case, the linear term in the equation of motion changes along with A).

Integrals of the hole number and energy for solutions (30) are

$$N_{k,b} = -\mathrm{arch}[(A^2+v^2)^{-1/2}(2\rho_0 - A)],$$

$$E_{k,b} = \sqrt{a}\left(\frac{A}{2} + \rho_0\right) + \left[A\rho_0 + \frac{1}{4}(v^2 - A^2)\right]N_{k,b}.$$

Ultimately, we emphasize that for small amplitudes a (and hence velocities v close to the sound velocity v_s) the modules of (30) (see also formula (26))

$$\rho_{k,b}(x,t) = [\rho_0 - 2a\{\sqrt{a^2+v^2}\,\mathrm{ch}(2\sqrt{a}(\zeta-\zeta_0)) + 2\rho_0 - A\}^{-1}]^{1/2}$$

coincides with the approximate solution (23b) accurate to $0(a)$. To show this, note that $(2\rho_0 - A)^2 - (A^2+v^2) = 4a$, or $\sqrt{A^2+v^2} \simeq 2\rho_0 - A - 2a(2\rho_0 - A)^{-1}$ at $a \ll |2\rho_0 - A|$, therefore in the second term one can replace $\sqrt{a^2+v^2}$ by $2\rho_0 - A$ to get

$$\rho_{k,b} = \left\{\rho_0 - \frac{2a(2\rho_0 - A)^{-1}}{1+\mathrm{ch}(2\sqrt{a}(\zeta-\zeta_0))}\right\}^{1/2} \simeq \left\{\rho_0 - \frac{a(2\rho_0 - A)^{-1}}{\mathrm{ch}^2(\sqrt{a}(\zeta-\zeta_0))}\right\}^{1/2}$$

with $v = v_k - 2a/v_s$, $\tilde{b} = 2a(2\rho_0 - A)^{-1}$.

Besides the solutions given above, the ϕ^6 model possesses other localized solutions: in particular, the rational solution [40]. However, in the first region (see Fig. 7.3) there are no localized solutions.

References Part III

1. V. Makhankov, O.K. Pashaev and S. Sergeenkov. *Dynamical symmetry and spin waves of isotropic antiferromagnet.* JINR Rapid Comm. No 10-85, Dubna 1985, p. 45-52.

2. V. Makhankov and O.K. Pashaev. *Nonlinear integrable models with noncompact group in the Bose-gas theory.* In Proceedings of the III International Seminar on Group-Theoretical Methods in Physics, Yurmala (USSR) 1985, Nauka Moscow 1986, V.I., p. 346-355, also JINR Comm. P17-86-565, Dubna 1985.

   Noncompact magnets, Bogolubov condensate and Bose-gas models on the lattice. JINR P2-86-754, Dubna 1986; also see Sov. Doklady, 1988, 301, p. 1356-1361.

3. See the book *'High Energy Physics and the Theory of Elementary Particles'*, Naukova Dumka, Kiev, 1967.

4. I.A. Malkin and V.I. Manko. *Dynamical symmetries and the coherent states of the quantum systems.* Nauka, Moscow, 1979.

5. A.I. Solomon. *Group theory of superfluidity.* J. Math. Phys. 1971, 12, p. 390-394.

6. A.I. Solomon. *Dynamical group model of superfluid helium three.* J. Phys. 1981, 14A, p. 2177-2188.
7. a) A.M. Perelomov. *Generalized coherent coherent states.* Uspekhi Phys. Nauk, 1977, 123, p. 23-55.

   b) *Generalized coherent states and applications.* Nauka, Moscow 1987, p. 263.

   c) *Coherent states for arbitrary Lie group.* Comm. Math. Phys. 1972, 26, p. 222-236.
8. C. Gerry and S. Silverman. *Path integral for coherent states of the dynamical group $SU(1,1)$.* J. Math. Phys. 1982, 23, p. 1995-2003.
9. H. Kuratsuji and T. Suzuki. *Path integral in the representation of $SU(2)$ coherent state and classical dynamics in a generalized phase space.* J. Math. Phys. 1980, 21, p. 472-880.
10. C. Herring and C. Kittel. *On the theory of spin waves in ferromagnetic media.* Phys. Rev. 1951, 81, p. 869-880.

    I.M. Lifshitz and L.P. Pitaevskii. *Statistical physics.* Nauka, M, 1978, part II.
11. See [1a] part II, Chapter IV, formula (19). S.P. Novkov, et al.
12. P.W. Anderson. *An approximate quantum theory of the antiferromagnetic ground state.* Phys. Rev. 1952, 86, p. 694-701.
13. a) V. Zakharov and A. Shabat. See [37] part II.

    b) *On the interaction between solitons in a stable medium.* ZhETF 1973, 64, p. 1627-1639.
14. V. Gerdjikov and P.P. Kulish. *On multicomponent NLS under nonvanishing boundary conditions.* Zap. Nauch. Semin. LOMI 1983, 131, p. 34-36.
15. V. Makhankov, O. Pashaev and S. Sergeenkov. *Coloured solitons in a stable medium.* Phys. Lett. 1983, 98A, p. 227-232;

    Hole-like excitations in many-component systems. Physica Scripta, 1984, 29, p. 521-525, see [17] part II.
16. V. Makhankov, O. Pashaev, S. Sergeenkov and V. Fedyanin. *On interaction between solitons in the model of organic charge transfer salts.* JINR E17-83-755, Dubna 1983.
17. B.A. Dubrovin, I.M. Krichever, T. Malanyuk and V. Makhankov. *Exact solutions to Schrödinger equations with selfconsistent potentials.* JINR E5-87-

710, Dubna 1987; Sov. J. Part. and Nuclei., 1988.

18. F. Ernst. *New formulation of the axially symmetric gravitational field problem.* Phys. Rev. 1968, 167, p. 1175.

    D. Maison. *Stationary, axially symmetric Einstein spaces, a completely integrable Hamiltonian system?* MPI-PAE/PPh 14/78, München, 1978.

19. E. Cremmer and B. Julia. *The* $SO(8)$ *supergravity.* Nucl. Phys. 1979, 159B, p. 141-; Phys. Lett. 1978, 80, p. 48-.

    J. Ellis, M. Gaillard, M. Günaydin and B. Zumino. *Supersymmetry and noncompact groups in supergravity.* Nucl. Phys. 1983, 224B, p. 427-.

20. A. Zheltukhin. *Classical relativistic string as a two-dimensional* $SO(1,1)\times SO(2)$ *gauge model.* Teor. Mat. Fiz. 1982, 52, p. 73-88.

    V. Nesterenko. *A nonlinear two-dimensional σ-model in relativistic string theory.* Lett. Math. Phys. 1983, 7, p. 287-293.

21. A.M. Kosevich, et al. See [1a] part I.

22. E. Lieb and W. Liniger. *Exact analysis of an interacting Bose gas. I The general solution and the ground state.* Phys. Rev. 1963, 130, p. 1605-1616. See [46] part II.

23. V. Makhankov, O. Pashaev and S. Segeenkov. *Bose gas models with noncompact symmetry groups.* Proceedings III Intern. Symp. on Selected Topics in Stat. Mech. JINR D17-84-850, V. 2, p. 45-52, Dubna 1984. See [17] part II. Preprint JINR P2-84-513, Dubna 1984.

24. V. Makhankov, O. Pashaev and S. Segeenkov. $U(p,q)$ *NLS under nonvanishing boundary conditions.* JINR P2-83-186, Dubna 1983.

25. O.K. Pashaev and S. Sergeenkov. *Nonlinear σ-models with non-compact symmetry group and the theory of a nearly ideal Bose gas.* Phys. 1986, 137A, p. 282-294.

26. N.N. Bogolubov and N.N. Bogolubov Jr. *Introduction into quantum statistical mechanics.* Nauka, Moscow, 1984, 384 p.

27. I. Gochev. *Solitons and spin complexes in the anisotropic Heisenberg chain.* Phys. Lett. 1982, 89A, p. 31-32;

    Quantum domain wall and coherent states for the Heisenberg-Ising spin 1/2 chain. Phys. Lett. 1984, 104A, p. 36-37.

28. M. Lakshmanan, Th.W. Ruijgrok and C.J. Thompson. *On the dynamics of a continuum spin system.* Physica 1976, 84A, p. 577-590.

29. J. Des Cloizeaux and J.J. Pearson. *Spin-wave spectrum of the antiferromagnetic linear chain.* Phys. Rev. 1962, 128, p. 2131-2135.

30. H. Bethe. *Eigenwerte und Eigenfunction der linearen Atomkette.* Z. Phys. 1931, 71, p. 205-271, No. 2.

31. L.A. Takhtajan. *Integration of the continuum Heisenberg spin chain through the inverse scattering method.* Phys. Lett. 1977, 64A, p. 235-237.

32. M. Lakshmanan. *Continuum spin system as an exactly solvable dynamic system.* Phys. Lett. 1977, 61A, p. 53-54.

33. S.J. Orfanidis. *σ-models of nonlinear evolution equations.* Phys. Rev. 1980, 21D, p. 1513-1522; see [12] part II.

    $SU(n)$ Heisenberg spin chain. Phys. Lett. 1980, 75A, p. 304-306.

34. V. Zakharov and L. Takhtajan. *Equivalence of NLS and the Heisenberg ferromagnet equation.* Teor. Mat. Fiz. 1979, 38, p. 26-35, see [22] part II.

35. H.C. Fogedby. *Solitons and magnons in the classical Heisenberg chain.* J. Phys. 1980, 13A, p. 1467-1499.

36. V.G. Kartavenko. *Soliton like solutions in the nuclear hydrodynamics.* Yader. Fiz. 1984, 40, p. 377-388, see [43] part I.

37. S.K.R. Devi, M.R. Strayer and J.M. Irvine. *TDHF calculations for nuclear collisions. Model studies of alpha-alpha scattering.* J. Phys. 1979, 5G, p. 281- .

    b) see [44] part I.

38. L.D. Landau and E.M. Lifshitz. *Statistical Physics.* Ch. XIV, Gostekhizd, Moscow 1951;

    A.D. Bruce and R.A. Cowley. *Structural phase transitions.* Taylor and Francis Ltd., Londen 1981.

39. R. Friedberg, T.D. Lee and A. Sirlin. *Gauge-field non-topological solitons in three space-dimensions (II).* Nucl. Phys. 1976, 115B, p. 32-47.

40. I.V. Barashenkov and V. Makhankov. *Soliton like excitations in a one-dimensional nuclear matter.* See [14] part I.

41 I. Barashenkov, A.D. Gocheva, V. Makhankov and I.V. Puzynin. *Stability of solitary waves of Phi-6 NLS under non-vanishing boundary conditions.* JINR

E17-85-967, Dubna 1985, also Physica, 1989, 34D, p. 240-254.

42. A.M. Kosevich, et al. See [1a] part I.

43. P. Bonche, S. Koonin and J. Negele. *One-dimensional nuclear dynamics in the time-dependent Hartree-Fock approximation.* Phys. Rev. 1976, 13, p. 1226-1258.

44. R.K. Dodd, et al. See [34] part I.

45. A.V. Makhankov and V.G. Makhankov. *Spin coherent states, Holstein-Primakoff transformations for the Heisenberg ferromagnet and the status of Landau-Lifshitz equation.* Preprint JINR P17-87-295, Dubna 1987; Phys. Stat. Sol. (b), 1988, 145, p. 669-678.

    V.G. Makhankov, R. Myrzakulov and A.V. Makhankov. *Generalized coherent states and the continuous Heisenberg XYZ model with one-ion anisotropy.* Physica Scripta, 1987, 35, p. 233-237.

Part IV. Soliton-Like Solutions in One-Dimension

Chapter 8.

THE CLASS OF SOLITON SOLUTIONS TO THE VECTOR VERSION OF THE NONLINEAR SCHRÖDINGER EQUATION WITH SELF-CONSISTENT POTENTIALS

In this chapter we investigate in more detail one-soliton solutions of some vector nonlinear Schrödinger equations with self-consistent potentials. Several such studies have been described in the literature but are not yet well-known. This chapter is closely related to the preceding chapters which dealt with Bose gas models. Although the physical systems governed by these equations are different in nature, they have the same mathematical basis and are generated via the solution of the linear Schrödinger equation

$$(i\partial_t + \partial_x^2 - u)\psi = 0 \tag{8.1}$$

with the potential $u(x,t)$ satisfying some consistency conditions (or equations). We have already considered such equations in the preceding chapters: for the scalar variant, see the formulae (2.4), (2.1)-(2.10), of chapter 2, and for the vector variant see Chapter 3 and 4.

Below, for example, we shall discuss two models of the $U(n)$ nonlinear Schrödinger equation and a vector variant of the system (2.10).

1. Soliton solutions to the $U(n)$ nonlinear Schrödinger equation. Linearization method

Only considering one-soliton solutions, we make in (1) the conventional Ansatz:

$$\psi(x,t) = \varphi(x - vt)\cdot\exp\{i[\frac{v}{2}x - (\frac{v^2}{4} + \lambda)t]\} \tag{8.2}$$

obtaining

$$\varphi_{\zeta\zeta}-u\varphi = -\lambda\varphi, \quad (\zeta=x-vt), \tag{8.3}$$

i.e. the conventional Schrödinger equation. Consider solution (3) for the reflectionless (Bargmann) potential

$$u = -\frac{s(s+1)b^2}{\mathrm{ch}^2 b\zeta}. \tag{8.4}$$

It is known (e.g. from [1]) that the potential (4) for $s \in N$ has s discrete eigenvalues

$$\lambda_j = -j^2b^2, \quad (j=\overline{1,s}), \tag{8.5}$$

and j counts these levels.

To find the corresponding eigenfunctions we use the factorization method (see [1], p. 71-78 or [2]). Let us introduce the 'creation' and 'annihilation' operators with the relations

$$a_s^+ \varphi_{s,j} = \varphi_{s+1,j}, \tag{8.6}$$

$$a_s\varphi_{s,j} = \varphi_{s-1,j}, \tag{8.7}$$

where

$$a_s^+ = bs\,\mathrm{th}b\zeta-\frac{d}{d\zeta}, \quad a_s = bs\,\mathrm{th}b\zeta+\frac{d}{d\zeta}, \tag{8.8}$$

and $\varphi_{s,j}$ is the wave function with given s and j.

Fix also the condition

$$\varphi_{s,j} \equiv 0, \quad \text{for } s>n, \tag{8.9}$$

which means, as we shall see below, that the vector ψ is n-dimensional. The system (6), (7) and condition (9) determine all the solutions (3) with the potential (4) for $\lambda = \lambda_j$.

Let $s = j = n$, then from (6) we obtain

$$a_n^+ \varphi_{n,n} = 0. \tag{8.10}$$

The solution of the equation is

$$\varphi_{n,n} = \mathrm{sech}^n b\zeta. \tag{8.11}$$

For $s = n$, $j = n-1$ we have $\varphi_{n,n-1} = a_n^+ \varphi_{n-1,n-1}$ and so on. Finally, we obtain

$$\varphi_{s,j} = a_s^+ a_{s-1}^+ \ldots a_{j+1}^+ \varphi_{j,j} \equiv a_s^+ \ldots a_{j+1}^+ \operatorname{sech}^j b\zeta. \tag{8.12}$$

Let us find an explicit form $\varphi_{s,j}$ for $n = 1,2,3,4$.

$\underline{n = 1}$

$$\varphi_{1,1}(\zeta) = c_1 \operatorname{sech} b\zeta. \tag{8.13}$$

$\underline{n = 2}$

$$\varphi_{2,2}(\zeta) = c_2 \operatorname{sech}^2 b\zeta. \tag{8.14a}$$

From (6) and (13) we get

$$\varphi_{2,1} = a_2^+ \varphi_{1,1} = c_1 \operatorname{th} b\zeta \cdot \operatorname{sech} b\zeta. \tag{8.14b}$$

$\underline{n = 3}$

$$\varphi_{3,3} = c_3 \operatorname{sech}^3 b\zeta, \tag{8.15a}$$

$$\varphi_{3,2} = a_3^+ \varphi_{2,2} = c_2 \operatorname{th} b\zeta \cdot \operatorname{sech}^2 b\zeta, \tag{8.15b}$$

$$\varphi_{3,1} = a_3^+ \varphi_{2,1} = a_3^+ a_2^+ \varphi_{1,1} = c_1(4 - 5\operatorname{sech}^2 b\zeta)\operatorname{sech} b\zeta. \tag{8.15c}$$

$\underline{n = 4}$

$$\varphi_{4,4} = c_4 \operatorname{sech}^4 b\zeta, \tag{8.16a}$$

$$\varphi_{4,3} = a_4^+ \varphi_{3,3} = c_3 \operatorname{th} b\zeta \cdot \operatorname{sech}^3 b\zeta, \tag{8.16b}$$

$$\varphi_{4,2} = c_2(7\operatorname{sech}^2 b\zeta - 6)\operatorname{sech}^2 b\zeta, \tag{8.16c}$$

$$\varphi_{4,1} = c_1(7\operatorname{sech}^2 b\zeta - 4)\operatorname{th} b\zeta \cdot \operatorname{sech} b\zeta. \tag{8.16d}$$

Let us construct a solution to the $U(n)$ nonlinear Schrödinger equation making use of the consistency condition

$$u = -\sum_{j=1}^{n} |c_j|^2 \varphi_{n,j}^2. \tag{8.17}$$

This condition implies the reflectionless potential (4) be expanded over eigenfunction squares of discrete spectrum with positive coefficients.

For the case $n = 1$, we get the well-known

$$\psi_1 = ce^{i\theta_1} \operatorname{sech} b\zeta, \quad |c|^2 = 2b^2.$$

Here, and on $\theta_j = (x - \mathrm{v}/2t)\mathrm{v}/2 + j^2b^2t$.

For $n = 2$ there are two possible solutions

$$\psi_1 = e^{i\theta_1}\mathrm{sech}b\zeta\begin{pmatrix}\tilde{c}_1\\ \tilde{c}_2\end{pmatrix}, \quad |\tilde{c}_1|^2 + |\tilde{c}_2|^2 = 2b^2 \quad (\text{degenerate } in\ \lambda), \tag{8.18}$$

$$\psi = \begin{pmatrix}c_1e^{i\theta_1}\mathrm{sh}b\zeta\\ c_2e^{i\theta_2}\end{pmatrix}\mathrm{sech}^2b\zeta, \quad |c_1|^2 = |c_2|^2 = 6b^2. \tag{8.19}$$

For $n = 3$ four solutions exist. Three of these are degenerate in λ either completely (extension of the (18) variant)

$$\psi_1 = e^{i\theta_1}\begin{pmatrix}c_1\\ c_2\\ c_3\end{pmatrix}\mathrm{sech}b\zeta \quad |c_1|^2 + |c_2|^2 + |c_3|^2 = 2b^2, \tag{8.20}$$

or partially

$$\psi_2 = \begin{pmatrix}c_1e^{i\theta_1}\mathrm{sh}b\zeta\\ c_2e^{i\theta_1}\mathrm{sh}b\zeta\\ c_3e^{i\theta_2}\end{pmatrix}\mathrm{sech}^2b\zeta \quad |c_1|^2 + |c_2|^2 = |c_3|^2 = 6b^2,$$

$$\psi_3 = \begin{pmatrix}c_1e^{i\theta_1}\mathrm{sh}b\zeta\\ c_2e^{i\theta_2}\\ c_3e^{i\theta_2}\end{pmatrix}\mathrm{sech}^2b\zeta \quad |c_1|^2 = |c_2|^2 + |c_3|^2 = 6b^2, \tag{8.21}$$

and non-degenerate

$$\psi_4 = \begin{pmatrix}c_1e^{i\theta_1}(1-4\mathrm{sh}b\zeta)\\ c_2e^{i\theta_2}\mathrm{sh}b\zeta\\ c_3e^{i\theta_3}\end{pmatrix}\mathrm{sech}^3b\zeta \quad |c_1|^2 = \frac{|c_2|^2}{40} = \frac{|c_3|^2}{15} = \frac{3}{4}b^2$$

It is easy to obtain partially degenerate solutions of type (20) from the preceding solutions.

For $n = 4$ eight cases are possible. We only give the non-degenerate one:

$$\psi = \begin{pmatrix}c_1e^{i\theta_1}(1-\frac{4}{3}\mathrm{sh}^2b\zeta)\mathrm{sh}b\zeta\\ c_2e^{i\theta_2}(1-6\mathrm{sh}^2b\zeta)\\ c_3e^{i\theta_3}\mathrm{sh}b\zeta\\ c_4e^{i\theta_4}\end{pmatrix}\mathrm{sech}^4b\zeta. \tag{8.22}$$

Thus one can see that for $n = \overline{1,4}$ the consistency condition (17) is valid. Using the 'expansion of unity'

$$1 \equiv \frac{\mathrm{sh}^2 b\zeta + 1}{\mathrm{ch}^2 b\zeta},$$

it is straightforward to show that, reducing to a system of n linear equations with n unknown variables, we obtain the consistency condition equating the coefficients at equal powers of sh^{2k} from the identity

$$\frac{n(n+1)b^2}{\mathrm{ch}^2 b\zeta} \cdot \frac{(\mathrm{sh}^2 b\zeta + 1)^{n-1}}{\mathrm{ch}^{2n-2} b\zeta} = \sum_{j=1}^{n} |c_j|^2 \varphi_j^2. \tag{8.23}$$

This means that all the $j = \overline{1,s}$ eigenfunctions must be present in the expansion of the potential with given s, i.e. in the 's-well' all s levels should be occupied (although not necessarily once)*).

2. $U(2)$ Nonlinear Schrödinger equation. Dubrovin-Krichever technique

Let us discuss the $U(2)$ nonlinear Schrödinger equation version in more detail. First, note that just as in Chapter 4 (see Section 4.5, eqn. (4.55)) solutions (19) can be 'bred' by means of the isorotation

$$\psi = h_1 \psi$$

with

$$h_1 = \begin{bmatrix} \alpha & \beta \\ -\bar{\beta} & \bar{\alpha} \end{bmatrix} \in SU(2) \tag{8.24}$$

to have as a result the vector bions

$$\psi_b(x,t) = c \begin{bmatrix} c_1 \alpha e^{i\theta_1} - c_2 \bar{\beta} e^{i\theta_2} \mathrm{sh} b\zeta \\ c_1 \beta e^{i\theta_1} + c_2 \bar{\alpha} e^{i\theta_2} \mathrm{sh} b\zeta \end{bmatrix} \mathrm{sech}^2 b\zeta, \quad |c| = 1. \tag{8.25}$$

Now we can see how the $U(2)$ vector ψ (19) is obtained via the Dubrovin and

*) In fact, it is a consequence of the completeness of the eigenfunction squares system of the Sturm-Liouville operator, since for Bargmann's potentials the integral term in the expansion (23) proportional to reflection coefficient vanishes [18].

Krichever technique. Krichever has proposed [3,4] a method enabling soliton solutions (in particular, rational solutions) to be constructed of nonlinear equations admitted commutation (u,v) representations.*) Dubrovin and Krichever have developed this method and it is given in a complete form in the review paper [58] which also contains a number of original results. The approach does not use commutation relations and arises, in fact, from the algebraic-geometrical theory of integrable systems. The application of the latter is known to be useful for constructing periodical and quasi-periodical solutions to such systems yet, less known, is that the algebraic-geometrical technique enables as well to obtain quite effectively all the solutions ever known (multi-soliton and rational formulae and their combinations).

In its final form, if one ignores its origin, the technique is a generalization of the above linearization technique. In what follows, we only present an outline so that various theorems and propositions will be formulated but not proved.

Note also that the technique discussed is the most effective in studying models with noncompact symmetries, where condensate boundary conditions have physical sense and the conventional inverse scattering method is non-constructive. Moreover, it is, in fact, the technique to construct simultaneously integrable models associated with the time-dependent Schrödinger equation and its exact solutions. Further, we follow the paper [58].

The potential $u(x,t)$ of the non-stationary Schrödinger equation would be called the 'integrable' potential, associated with the rational algebraic curve in the equation

$$[i\partial_t - \partial_x^2 + u(x,t)]\psi(x,t,k) = 0 \tag{8.1}$$

has a solution of the form

$$\psi(x,t,k) = Q_N(x,t,\kappa)e^{ikx+ik^2t}, \tag{8.26}$$

where

$$Q_N(x,t,k) = k^N + a_1(x,t)k^{N-1} + \ldots + a_N(x,t)$$

*)Analogous approach was used by Cherednik in [5] to construct finite zone solutions for the nonlinear Schrödinger equation.

is a polynomial of some degree N.

In our considerations, formula (26) should, in fact, be regarded as an Ansatz. We focus here on the simplest variants of the consistency conditions and solutions.

To begin with we shall construct the complex integrable potentials. Let us set complex numbers $\kappa_1,\ldots,\kappa_M,(\alpha_{ij})$ where $i = \overline{1,N}$, $j = \overline{1,M}$. These are free parameters of our construction. For any set of these parameters we can uniquely determine the function $\psi(x,t,k)$ having the form (26) with the help of the following system of linear conditions

$$\sum_{j=1}^{M} \alpha_{ij}\psi(x,t,k)\Big|_{k=\kappa_j} = 0, \quad i = \overline{1,N}. \tag{8.27}$$

These conditions are N linear inhomogeneous equations resolvable if the corresponding matrix of coefficients $A(x,t)$ is non-degenerate. For that, the rank of the matrix (α_{ij}) must be equal to N.

THEOREM 1. *If the matrix $A(x,t)$ of the system is not identically singular (in x,t) then the function $\psi(x,t,k)$ of the form (26) under conditions (27) satisfies eqn. (1) when*

$$u(x,t) = 2i\partial_x a_1(x,t) = 2\partial_x^2 \ln\det A(x,t). \tag{8.28}$$

As was mentioned before, the potentials $u(x,t)$ corresponding to the arbitrary parameters $\kappa_j,(\alpha_{ij})$ are the complex and meromorphic functions of x,t. Now we shall describe the restrictions which are sufficient for reality and regularity (for real x,t) of the corresponding potentials $u(x,t)$. We shall assume that $M = 2N$ and that the values $\kappa_1,\ldots,\kappa_{2N}$ have non-zero imaginary parts and are subdivided into complex conjugate pairs

$$\kappa_{N+i} = \bar{\kappa}_i, \quad i = \overline{1,N}. \tag{8.29}$$

We can assume without loss of generality that the minor of the matrix $(\alpha_{ij}) \equiv (\alpha_{ij}^c)$ consisting of the columns with the numbers $j = N+1,\ldots,2N$ is non-singular. As follows from the previous remark the general case can be reduced to the case where this minor equals the unit matrix. In this case the conditions (27) have the form

$$\psi(\bar{\kappa}_i) = -\sum_{j=1}^{N} \alpha_{ij}\psi(\kappa_j), \quad i = \overline{1,N}, \tag{8.30}$$

where (α_{ij}) is a constant matrix of dimension $N \times N$. It is convenient to renormalize the function $\psi(x,t,k)$ in the following way

$$\Psi(x,t,k) = \frac{\psi(x,t,k)}{(k-\kappa_1)...(k-\kappa_N)} \equiv \left[1+\sum_{j=1}^{N}\frac{r_j(x,t)}{k-\kappa_j}\right]e^{ikx+ik^2t}. \tag{8.31}$$

The conditions (30) for this renormalized function Ψ take the form

$$\Psi(x,t,\bar{\kappa}_i) = -\sum_{j=1}^{N} C_{ij} \operatorname*{res}_{k=\kappa_j} \Psi(x,t,k), \tag{8.32}$$

where the constant matrix (C_{ij}) equals

$$C_{ij} = [R(\bar{\kappa}_i)]^{-1}\alpha_{ij}R'(\kappa_j), \quad i,j = \overline{1,Nm}, \tag{8.33}$$

$R(k) = (k-\kappa_1)...(k-\kappa_N)$, and the prime denotes the derivative with respect to k.

THEOREM 2. *Let the parameters* $\kappa_1,...,\kappa_N,(C_{ij})$ *which determine the function* $\Psi(x,t,k)$ *having the form (31) with the help of the conditions (32) satisfy the following requirements:*

(a) the matrix C_{ij} *is skew-Hermitian* $C_{ij} = -\bar{C}_{ij}$;

(b) if the points $\kappa_1,...,\kappa_N$ *are enumerated in such a way that*

$$\operatorname{Im}\kappa_i > 0, \quad i = 1,...,p; \quad \operatorname{Im}\kappa_i < 0, \quad i = p+1,...,N,$$

then the Hermitian matrix

$$(\frac{1}{i}C_{kl}), \quad 1 \leqslant k, \quad l \leqslant p,$$

must be positively defined and the Hermitian matrix

$$(\frac{1}{i}C_{kl}), \quad p+1 \leqslant k,l \leqslant N,$$

must be negatively defined (these matrices can be non-negative as well).

If these conditions regarding the parameters are fulfilled, the function $\psi(x,t,k)$ is the smooth function of real x,t for all $k \neq \kappa_j$ and satisfies eq. (1) with a real smooth potential $u(x,t)$. For these functions we have

$$\psi(x,t,k) = \frac{\det\hat{M}(x,t,k)}{\det M(x,t)}e^{ikx+ik^2t}, \tag{8.34}$$

$$u(x,t) = 2\partial_x^2 \ln \det M(x,t), \tag{8.35}$$

where

$$M_{ij}(x,t) = C_{ij} + \frac{e^{i(\overline{\omega}_i - \omega_j)}}{\overline{x}_i - \kappa_j}, \quad \omega_i = \kappa_i x + \kappa_i^2 t, i,j = \overline{1,N}, \tag{8.36}$$

$$\hat{M}_{ij} = M_{ij}:\ i,j = \overline{1,N};\ \hat{M}_{00} = 1,\ \ \hat{M}_{i0} = e^{i\overline{\omega}_i} \tag{8.37}$$

$$\hat{M}_{0i} = (k - \kappa_i)^{-1} e^{-i\omega_i},\ \ i = \overline{1,N}.$$

DEFINITION 1. We shall call the integrable potential $u(x,t)$, which is given by our construction with N parameters $\kappa_1,\ldots,\kappa_N$ and the $N \times N$ matrix C_{ij}, the N-soliton potential.

This definition coincides with the ordinary one for the scalar non-linear Schrödinger equation. In the vector case, our definition of the number of solitons does not always agree with the intuitive definition [26].

Let us determine in which case the two sets of the 'spectral data' κ_i, C_{ij} and κ_i', C_{ij}' determine the same Schrödinger operator and the same function $\Psi(x,t,k)$. Consider the relation (30). Represent the matrix (α_{ij}), which is related to C_{ij}, with the help of (33), in the block form

$$(\alpha_{ij}) = \begin{bmatrix} \alpha_+ & \beta \\ \gamma & \alpha_- \end{bmatrix},$$

where the matrix α_+ and α_- have the dimensions $p \times p$ and $(N-p) \times (N-p)$, respectively. Assume that the matrix α_- is invertible. Then the transformation $(\kappa_i,(\alpha_{ij})) \Rightarrow (\kappa_i',(\alpha_{ij}'))$, where

$$\kappa_i' = \begin{cases} \kappa_i, & i = \overline{1,p} \\ \overline{\kappa}_i, & i = \overline{p+1,N} \end{cases} \tag{8.38}$$

$$(\alpha_{ij}') = \begin{bmatrix} \alpha_+ - \beta\alpha_-^{-1}\gamma & -\beta\alpha_-^{-1} \\ \alpha_-^{-1}\gamma & \alpha_-^{-1} \end{bmatrix},$$

does not change the relations (2.11), which determine the function $\Psi(x,t,k)$. Hence, for the invertible minor α_- the points $\kappa_{p+1},\ldots,\kappa_N$ may be transformed from the lower to the upper half-plane without changing the Schrödinger operator and its eigenfunctions. For further discussion, it will be useful to introduce the functions:

$$\Psi_i = \Psi_i(x,t) = \underset{\kappa_i}{\mathrm{res}}\Psi(x,t,k) = r_i(x,t)e^{i\omega_i}, \quad \iota = \overline{1,N}. \tag{8.32a}$$

We shall present the transformations of the spectral data which correspond to the Galillei, scale and other simplest transformations of the Schrödinger operator:

(a) The Galillei transformation

$$x' = x + \mathrm{v}t, \quad t' = t. \tag{8.39}$$

In this case

$$\kappa_i' = \kappa_i - \frac{\mathrm{v}}{2}, \quad i = \overline{1,N}, \quad (C_{ij}') = (C_{ij}).$$

The corresponding potential and eigenfunctions are equal to

$$u(x,t) = u'(x',t'), \tag{8.40}$$

$$\Psi'(x',t',k') = \Psi(x,t,k)e^{-i\mathrm{v}/2(x+\mathrm{v}/2t)}, \quad k' = k - \frac{\mathrm{v}}{2}.$$

(b) Translations

$$x' = x + x_0, \quad t' = t + t_0. \tag{8.41}$$

In this case

$$\kappa_i' = \kappa_i, \quad i = \overline{1,N}, \tag{8.42a}$$

$$C_{ij}' = C_{ij}\exp\{i[(\overline{\kappa}_i - \kappa_j)x_0 + (\overline{\kappa}_i^2 - \kappa_j^2)t_0]\}, \quad i,j = \overline{1,N},$$

and

$$u'(x',t') = u(x,t), \tag{8.42b}$$

$$\Psi'(x',t',k') = \Psi(x,t,k)e^{ik(x_0+kt_0)}.$$

(c) The scaling transformations

$$x' = \lambda x, \quad t' = \lambda^2 t. \tag{8.43}$$

The corresponding transformation of the spectral data has the form

$$\kappa_i' = \lambda^{-1}\kappa_i, \quad i = \overline{1,N},$$

$$(C_{ij}') = (C_{ij}).$$

For the potential and eigenfunction we have

$$u'(x',t') = \lambda^{-2}u(x,t), \tag{8.44}$$

$$\Psi'(x',t',k') = \Psi(x,t,k), \;\; k' = \lambda^{-1}k.$$

(d) The space and time reflection

$$x' = -x, \quad t' = -t. \tag{8.45}$$

Then

$$\kappa_i' = \bar{\kappa}_i, \quad i = \overline{1,N}, \tag{2.46a}$$

$$C_{ij}' = \bar{C}_{ij}, \quad i,j = \overline{1,N}.$$

As this takes place

$$u'(x',t') = u(x,t), \;\; \Psi'(x',t',k') = \sqrt{\Psi(x,t,k)}, \;\; k' = \bar{k}. \tag{2.46b}$$

Let us discuss now the asymptotic behaviour of the solution under considera-tion: the potentials and the eigenfunctions.

To begin with, we shall consider the case $N = 1$. The system (8.32) is reduced to the equation

$$\left[C + \frac{e^{i(\bar{\omega}-\omega)}}{\bar{\kappa}-\kappa}\right]\Psi_1(x,t) = -e^{i\bar{\omega}}.$$

Here $\kappa = \kappa_1$ (let $\mathrm{Im}\,\kappa > 0$), $C = C_{11}$, $\mathrm{Re}\,C_{11} = 0$, $\mathrm{Im}\,C > 0$, (the case $C = 0$ is trivial), $\omega = \kappa x + \kappa^2 t$

$$\Psi_1(x,t) = -e^{i\bar{\omega}}\left[C + \frac{e^{i(\bar{\omega}-\omega)}}{\bar{\kappa}-\kappa}\right]^{-1}.$$

Denote $\kappa = \alpha + i\beta$. Then

$$\Psi_1(x,t) = \frac{i\beta}{\sqrt{(\bar{\kappa}-\kappa)c}} \cdot \frac{e^{i\alpha x + i(\alpha^2-\beta^2)t}}{\mathrm{ch}[\beta(x-x_0)+2\alpha\beta t]},$$

where

$$x_0 = \frac{1}{\beta}\ln\sqrt{(\bar{\kappa}-\kappa)c}.$$

For $r = r_1(x,t)$ we have the formula

$$r(x,t) = i\beta\{1+\mathrm{th}[\beta(x-x_0)+2\alpha\beta t]\}.$$

For the case $N = 1$, this corresponds to the well-known one-soliton potential of the Schrödinger equation which is decreasing in all directions except $x = -2\alpha t+\mathrm{const}$:

$$u(x,t) = 2i\partial_x r(x,t) = -2\beta^2\mathrm{ch}^{-2}[\beta(x-x_0)+2\alpha\beta t]. \tag{8.47a}$$

The eigenfunction of the corresponding Schrödinger operator has the form

$$\Psi(x,t,k) = \left[1+i\beta\frac{1+\mathrm{th}[\beta(x-x_0)+2\alpha\beta t]}{k-\kappa}\right]e^{ik(x+kt)}. \tag{8.47b}$$

In the case $N>1$, the asymptotic behaviour of $\Psi(x,t,k)$ for the general $(\kappa_i),(C_{ij})$ is too complicated to be analysed here. In the review by Dubrovin et al. mentioned above, some particular cases were studied and a peculiar case was found with the multi-solution asymptotics not reducing to pair-wise interactions.

3. The self-consistent conditions

The function $\Psi(x,t,k)$, which has been defined earlier in (31), can be represented in the neighbourhood of $k = \infty$ in the form

$$\Psi(x,t,\kappa) = (1+\sum_{s=1}^{\infty}\xi_s(x,t)k^{-s})e^{i\kappa(x+\kappa t)}. \tag{8.48}$$

(The first factor in (48) is the expansion in k^{-1} of the pre-exponential factor in (31)). From (31) it follows that:

$$\xi_1 = a_1 = \sum_{j=1}^{N} r_j, \quad \xi_1+\bar{\xi}_1 = 0.$$

Substitution of (48) into (1) gives us the equalities

$$i\dot{\xi}_s-2i\xi'_{s+1}-\xi''_s+u\xi_s = 0, \quad s = 0,1,\ldots;\ \xi_0 = 1. \tag{8.49}$$

(The dot denotes the time derivative and the prime denotes the x-derivative.)

Consider the meromorphic function

$$\Omega(x,t,\kappa) = \Psi(x,t,\kappa)\overline{\Psi(x,t,\bar{\kappa})}.$$

Expansion in the infinity has the form

$$\Omega(\kappa,t,\kappa) = 1+\sum_{s=2}^{\infty} J_s(x,t)k^{-s}.$$

The first few coefficients have the form

$$J_2 = \xi_2+\bar{\xi}_2-\xi_1^2,\quad J_3 = \xi_3+\bar{\xi}_3+\bar{\xi}_1(\bar{\xi}_2-\xi_2), \tag{8.50}$$

$$J_4 = \xi_4+\bar{\xi}_4+\xi_1(\bar{\xi}_3-\xi_3)+|\xi_2|^2.$$

Using eqns. (49) we find the representation of J_s in terms of the potential $u(x,t)$.

LEMMA 1. The following relations hold for any formal solution $\Psi(x,t,k)$ of eqn. (1) which has the form (48)

$$J_2(x,t) = \frac{1}{2}u(x,t)+C_2,\quad C_2 = \text{const}, \tag{8.51a}$$

$$\partial_x J_3(x,t) = \frac{1}{2}\dot{u}(x,t), \tag{8.51b}$$

$$\partial_x^2 J_4(x,t) = \frac{3}{8}\ddot{u}-\frac{1}{8}(u_{xxx}-6uu_x)_x. \tag{8.51c}$$

The relation (51a) was found in [5] and the relations (51b) and (51c) were found in [7]. The constant C_2 in (2.93) can be determined from the asymptotics of $\Omega(x,t,k)$ for $|x|\to\infty$. For example, in the case considered in the previous section where $\operatorname{Im}\kappa_i>0$ and the matrix C_{ij} is invertible we have $\Omega(x,t,k)\to 1$, $u(x,t)\to 0$, when $x\to\pm\infty$. Therefore, $C_2 = 0$.

The relations (51a-c) are the basis of the constructions of the solutions with all self-consistent conditions (see below (54a-c)). Let $E(k)$ be the rational function of the forms

$$E(k) = k+\sum_{i=1}^{n}\epsilon_i\frac{b_i^2}{k-k_i}, \tag{8.52a}$$

$$E(k) = k^2+\alpha k+\sum_{j=1}^{n}\epsilon_i\frac{b_i^2}{k-k_i}, \tag{8.52b}$$

$$E(k) = k^3+\beta k^2+\gamma k+\sum_{i=1}^{n}\epsilon_i\frac{b_i^2}{k-k_i}. \tag{8.52c}$$

Here $\alpha,\beta,\gamma,k_i,b_i$ are arbitrary real constants. The constants $\epsilon_i = \pm 1$. We shall denote

$$\Phi_i(x,t) = b_i \Psi(x,t,k_i), \quad i = \overline{1,n}.$$

The functions Φ_i satisfy the equation

$$i\dot{\Phi}_j - \Phi_j'' + u(x,t)\Phi_j = 0, \quad j = \overline{1,n}$$

by Definition 1. The functions $\Psi_j(x,t)$, which are determined with the help of (32a), satisfy the same equation

$$i\dot{\Psi}_j - \Psi_j'' + u(x,t)\Psi_j = 0, \quad j = \overline{1,N}$$

too.

THEOREM 3. *Let the functions $\Phi_i(x,t), \Psi_j(x,t)$ correspond to the set of data $\kappa_1,\ldots,\kappa_N$, C_{ij} and to the rational function which has one of the form (52a-c). Then they satisfy one of the self-consistent conditions:*

(1) *If $E(k)$ has the form (52a), then*

$$\frac{u}{2} + \sum_{i=1}^{n} \epsilon_i b_i^2 + c_2 = \sum_{i=1}^{n} \epsilon_i |\Phi_i(x,t)|^2 - \sum_{i,j=1}^{N} \overline{\Psi_i(x,t)} E_{ij} \Psi_j(x,t), \tag{8.54a}$$

where

$$E_{ij} = C_{ij}(\overline{E(\kappa_i)} - E(\kappa_j)), \quad i,j = \overline{1,N}. \tag{8.53}$$

(2) *If $E(k)$ has the form (52b) then*

$$\frac{\dot{u}}{2} + \alpha \frac{u'}{2} = \sum_{i=1}^{n} \epsilon_i |\Phi_i(x,t)|_x^2 - \left[\sum_{ij=1}^{N} \overline{\Psi_i(x,t)} E_{ij} \Psi_j(x,t) \right]_x. \tag{8.54b}$$

Here the matrix E_{ij} is the same as in (53).

(3) *If $E(k)$ has the form (52c), then*

$$\frac{3}{8}\ddot{u} - \frac{1}{8}(u_{xxx} - 6uu_x)_x + \beta \frac{\dot{u}_x}{2} + \gamma \frac{u_{xx}}{2} \tag{8.54c}$$

$$= \sum_{i=1}^{n} \epsilon_i |\Phi_i(x,t)|_{xx}^2 - \left[\sum_{ij=1}^{N} \overline{\Psi_i(x,t)} E_{ij} \Psi_j(x,t) \right]_{xx}$$

(the matrix E_{ij} is given with the help of (53)).

It must be mentioned that the matrix E_{ij} is Hermitian. Therefore, this matrix can be reduced to a diagonal or with the help of the linear transformation of $\Psi_1,...,\Psi_N$. By doing so we obtain the diagonal element equal to ± 1, or 0.

For the general data κ_i, C_{ij} and the rational functions of thc form (52a), (52c) the Hermitian forms of the right-hand sides of self-consistent conditions have large ranks which are equal to $N+n$. In the case of the self-consistent condition (54a), this means that the functions $\Phi_1,...,\Phi_n$, $\Psi_1,...,\Psi_N$ are the solutions to the $(N+n)$-component vector non-linear Schrödinger equation, the symmetry of which is determined by the signature of the Hermitian matrix

$$\begin{bmatrix} -\epsilon_1 & & & & 0 \\ & \cdot & & & \\ & & \cdot & & \\ & & & \cdot & \\ & & & -\epsilon_n & \\ 0 & & & & E_{ij} \end{bmatrix}. \tag{8.55}$$

The rank of this matrix decreases under some special conditions for the parameters of the construction. Consequently, the number of the component of the vector non-linear Schrödinger equations (and for other self-consistent conditions) will decrease.

The difference between the solutions (Ψ,Φ) of the self-consistent equation corresponding to the matrix (55) with the same rank and signature but with a different number n of poles of the functions $E(k)$ is as follows. As was shown in the review [58] the functions $\Psi_1,...,\Psi_N,\Phi_1,...,\Phi_n$ have different asymptotics for large x. The functions Φ_i have, as a rule, oscillating asymptotics but the functions $\Psi_j(x,t)$ are exponentially decreasing. This difference must be taken into account in the construction of the multi-soliton solutions, and the choice of the function $E(k)$ must depend on the required boundary conditions.

The formulae obtained are quite sufficient to construct as on exercise, the known one-soliton solutions to the scalar and the simplest vector version of the nonlinear Schrödinger equation. Some such examples are given in [58].

4. $U(2)$ **nonlinear Schrödinger equation. A modification of the Dubrovin-Krichever technique**

Here, to obtain the $U(2)$ nonlinear Schrödinger equation solutions, we employ the above ideas on equation linearization. Making use of the representation (48) and eqns. (49) one gets, bearing in mind $\zeta_0 \equiv 1$,

$$u = 2i\zeta_1', \tag{8.56}$$

$$-i\dot{\zeta}_1 + \zeta_1'' + 2i\zeta_2' - u\zeta_1 = 0. \tag{8.57}$$

Our concern is with the real potentials $u = \bar{u}$, so (56) leads to

$$\zeta_1 = -\bar{\zeta}_1, \tag{8.58a}$$

and from (57) one has

$$\zeta_2 = \bar{\zeta}_2. \tag{8.58b}$$

In the one-soliton sector it is possible to imply the Ansatz

$$\Psi = e^{i(kz + k^2t + \alpha(z,t))}(1 + \frac{\zeta(z)}{k} + \ldots + \frac{\zeta_N(z)}{k^N}) \tag{8.59}$$

with $z = x + vt$, $\alpha = (v/2)(z - v/2t)$, passing to the soliton rest frame. Equations (56), and (57) become ordinary differential equations. Once ζ_1 is purely imaginary we denote $\zeta_1 \to i/2\zeta_1$ and $\zeta_2 \to 1/4\zeta_2$, then

$$u = -\zeta_1', \ \ \zeta_1'' + \zeta_2' - u\zeta_1 = 0, \ \ \zeta_s \equiv 0, \ \ s \geqslant 3. \tag{8.60}$$

In our case $m = 2$, and we need the extra equation

$$\zeta_2'' - u\zeta_2 = 0. \tag{8.60a}$$

In the method described above, this system need not to be solved. However, in our case for a visual picture, we solve this system, finding, the values k_j via the boundary conditions. The system (60,60a), by integration, is reduced to the following form

$$\zeta_2 + \zeta_1' + \frac{1}{2}\zeta_1^2 = 2c_1,$$

$$\frac{1}{2}\zeta_2^2 + \zeta_1'\zeta_2 - \zeta_1\zeta_2' = 2c_2.$$

Eliminating ζ_2 and letting

$$\zeta_1 = 2(\ln y)', \tag{8.61}$$

one gets an equation linear in y:

$$y^{IV} - 2c_1 y'' + (c_1^2 - c_2)y = 0. \tag{8.62}$$

Its general solution contains four constants α_i.

To relate the results with those obtained in the preceding section we choose the following particular solution of type (61):

$$u = 2\kappa \mathrm{ch}\nu(z - z_0) + 2\nu \mathrm{ch}\kappa(z - z_0), \tag{8.63}$$

with $\kappa^2 = c_1 + \sqrt{c_2}$, $\nu^2 = c_1 - \sqrt{c_2}$. From (61), ζ_1 and ζ_2 are easily derived to give

$$\zeta_1 = 2\kappa\nu \frac{\mathrm{sh}\kappa z + \mathrm{sh}\nu z}{\kappa \mathrm{ch}\nu z + \nu \mathrm{ch}\kappa z}, \tag{8.64a}$$

$$\zeta_2 = (\kappa^2 - \nu^2) \frac{\kappa \mathrm{ch}\nu z - \nu \mathrm{ch}\kappa z}{\kappa \mathrm{ch}\nu z + \nu \mathrm{ch}\kappa z},$$

and then

$$\Phi(z,k) = 1 + \frac{i}{2k}\zeta_1 + \frac{1}{4k^2}\zeta_2, \tag{8.64b}$$

$$u = -\zeta_1' = -2\kappa\nu \frac{2\kappa\nu + (\kappa^2 + \nu^2)\mathrm{ch}\kappa z \cdot \mathrm{ch}\nu z - 2\kappa\nu \mathrm{sh}\kappa z \cdot \mathrm{sh}\nu z}{(\kappa \mathrm{ch}\nu z + \nu \mathrm{ch}\kappa z)^2}.$$

The boundary conditions $\Phi(z,k)|_{z \to \pm\infty} = 0$ give rise to the equation for finding k:

$$4k^2 + 2ik\zeta_1^{\mp} + \zeta_2^{\mp} = 0,$$

with $\zeta_{1,2}^{\mp} = \zeta_{1,2}|_{x \to \pm\infty}$. The solutions are

$$k = \pm\frac{i}{2}(\kappa \pm \nu). \tag{8.65}$$

Since the second two equations are given by the first two with a simultaneous sign change of κ and ν, we consider

$$k_1 = \frac{i}{2}(\kappa - \nu), \tag{8.66}$$

$$k_2 = \frac{i}{2}(\kappa + \nu).$$

The functions Ψ_1 and Ψ_2, defined by k_1 and k_2, are

$$\Psi_1(x,t) = e^{i\theta_1}\frac{\text{ch}\kappa z - \text{ch}\nu z + \text{sh}\kappa z + \text{sh}\nu z}{\kappa\text{ch}\nu z + \nu\text{ch}\kappa z}a_1, \tag{8.67}$$

$$\Psi_2(x,t) = e^{i\theta_2}\frac{\text{ch}\kappa z + \text{ch}\nu z + \text{sh}\kappa z + \text{sh}\nu z}{\kappa\text{ch}\nu z + \nu\text{ch}\kappa z}a_2,$$

$\theta_i = k_i z + k_i^2 + \alpha(z,t)$, a_i are constants. By substitution, $\kappa = 3\eta$, $\nu = \eta$, i.e.

$$k_1 = i\eta, \quad k_2 = 2i\eta, \tag{8.68}$$

the solutions (67) are reduced to those known from the preceding section:

$$\psi_1 = a_1 e^{-i\eta^2 t + i\alpha}\text{th}\eta(z - z_0)\cdot\text{sech}\eta(z - z_0), \tag{8.69}$$

$$\psi_2 = a_2 e^{-i4\eta^2 t + i\alpha}\text{sech}^2\eta(z - z_0),$$

and

$$u(x,t)\Big|_{\substack{\kappa=3\eta\\ \nu=\eta}} = -6\eta^2\text{sech}^2\eta(z - z_0), \tag{8.70}$$

which become solutions (19) if $|a_1|^2 = |a_2|^2 = 6\eta^2$. Let us emphasize that in getting (69) we obtain incidentally a new class of Bargmann's potentials (64b) parameterized by two constants κ and ν, as well as their wave functions when $\lambda_1 = 1/4(\kappa-\nu)^2$, $\lambda_2 = 1/4(\kappa+\nu)^2$. It may also be noticed that, to the above Bargmann's potential $u = n(n+1)b^2\text{sech}^2 bz$, corresponds the one with an equidistant spectrum k_i. In this case, in fact, once

$$\lambda_j = (jb)^2 = k_j^2,$$

then $k_j = jb$.

Let us check the consistency condition

$$|\psi_1|^2 + |\psi_2|^2 = -u. \tag{8.71}$$

We have from (67)

$$|\psi_1|^2 = 2a_1^2\frac{\text{ch}(\kappa+\nu)z - 1}{(\kappa\text{ch}\nu z + \nu\text{ch}\kappa z)^2}, \quad |\psi_2|^2 = 2a_2\frac{\text{ch}(\kappa-\nu)z + 1}{(\kappa\text{ch}\nu z + \nu\text{ch}\kappa z)^2},$$

and from (71)

$$a_1^2 = \frac{\kappa\nu}{2}(\kappa-\nu)^2, \quad a_2^2 = \frac{\kappa\nu}{2}(\kappa+\nu)^2, \tag{8.72}$$

and finally

$$\psi_1 = e^{-i(\kappa-\nu/2z)}\sqrt{\frac{\kappa\nu}{2}}(\kappa-\nu)\frac{\text{ch}\kappa z-\text{ch}\nu z+\text{sh}\kappa z+\text{sh}\nu z}{\kappa\text{ch}\nu z+\nu\text{ch}\kappa z}, \tag{8.73}$$

$$\psi_2 = e^{-i(\kappa+\nu/2z)}\sqrt{\frac{\kappa\nu}{2}}(\kappa+\nu)\frac{\text{ch}\kappa z+\text{ch}\nu z+\text{sh}\kappa z+\text{sh}\nu z}{\kappa\text{ch}\nu z+\nu\text{ch}\kappa z}.$$

Concluding this section let us consider the 'physical' characteristics of the $U(2)$ nonlinear Schrödinger equation solutions obtained, viz. the soliton energy and particle number:

$$E = \int_{-\infty}^{\infty}[(\psi_x,\psi_x)-\frac{1}{2}(\psi,\psi)^2]dx,$$

$$N = N_1+N_2, \quad N_i = \int_{-\infty}^{\infty}|\psi_i|^2dx.$$

There are two types of solitons (18) and (19). Corresponding integrals are readily calculated. For solution (18), it is the total particle number which only makes sense (by virtue of degeneration)

$$\tilde{N} = (|c_1|^2+|c_2|^2)\int_{-\infty}^{\infty}\text{ch}^{-2}bxdx = 4b.$$

For soliton (19) we have

$$N = N_1+N_2 \equiv 4b+8b = 12b.$$

For the energies, we have, respectively,

$$\tilde{E} = -\frac{4}{3}b^3 = -\frac{1}{48}\tilde{N}^3,$$

$$E = -12b^3 = -\frac{1}{144}N^3.$$

Comparing the energies of these two types of solutions (one-soliton and two-soliton ones according to our definition) for the same $N = \tilde{N}$ one sees that $E > \tilde{E}$. This fact implies that the stability problem becomes very essential: 'soliton' (19) may be unstable (with respect to initial or structural perturbations). Computer experiments carried out in JINR (Dubna) by Kholmurodov and the author show that soliton (18) is stable and that (19) decays into a sum of two, moving apart, scalar solitons of type:

$$\Psi \to (0, \mathrm{ch}^{-1})^{tr} + (\mathrm{ch}^{-1}, 0)^{tr}.$$

Since the time of the state (19) decay depends on the grid step (the discretization of the equation studied breaks down its integrability in our case), growing with its decrease, one can only state that solitons of type (19) are *structurally* unstable [59].

The following relation can also be derived easily (see Chapter 4, eqn. (4.58))

$$dE = \lambda_1 dN_1 + \lambda_2 dN_2,$$

with

$$\lambda_1 = -b^2 = -\frac{1}{144}N^2, \ \lambda_2 = -4b^2 = -\frac{1}{36}N^2,$$

resembling the thermodynamical relation for a mixture of gases.

5. $U(n)$ system with the Boussinesq potential

In this section, our concern will be with soliton solutions to the equations describing two wave (high- and low-frequency wave) interactions. These we have encountered in Chapters 1 and 2 when considering spin waves (or exciton waves) interacting with lattice oscillations (phonons) as well as the interaction of Langmuir and ion-sound waves in plasmas. All three systems are governed by the same mathematical model; that is, the coupled Schrödinger and nonlinear wave equations:

$$(i\partial_t - \partial_x^2 + u - \gamma|\psi|^2)\psi = 0, \tag{8.74a}$$

$$\Box u + (\alpha u^2 + \beta u_{xx})_{xx} = (|\psi|^2)_{xx}. \tag{8.74b}$$

In the general case, ψ is a vector function

$$\psi = (\psi^{(1)}, ..., \psi^{(n)})^{tr}$$

with $|\psi|^2$ replaced by the inner product $(\psi,\psi) \equiv (\psi^+ \gamma_0 \psi)$ and $\gamma_0 = \mathrm{diag}(1, ..., 1, -1, ..., -1)$ being the isospace metric.

Recall that constants α and β cannot be eliminated via scale transformations and the case $\alpha = 3$, $\beta = -1$, is integrable (Krichever's variant).

To find soliton-like solutions of system (74) one can employ the linearization method discussed in the preceding section. Moreover, we have solutions (19)-(22) to the linear Schrödinger equation necessary to do that. We simply need to satisfy the following consistency conditions

$$(v^2-1+\alpha u)u+\beta u_{xx} = \sum_{j=1}^{n} |c_j|^2\varphi_j^2, \tag{8.75}$$

$$-u+\gamma\sum_{j=1}^{n} |c_j|^2\varphi_j^2 = n(n+1)b^2\mathrm{ch}^{-2}bz. \tag{8.76}$$

Consider first the case $\gamma = 0$ (plasma problem); that is, we neglect a nonlinear wave frequency shift. Then, for a moving wave

$$\psi = e^{i\theta}\varphi(z),$$

$$u = u(z),$$

the system (54) assumes the form

$$\varphi_{zz}-u\varphi = -\lambda\varphi, \tag{8.77a}$$

$$\beta u_{zz}+\alpha u^2+(v^2-1)u = |\varphi|^2, \tag{8.77b}$$

of a nonlinear eigenvalue problem. The simplest class of solutions is obtained by assuming

$$u = -s(s+1)b^2\mathrm{sech}^2 bz. \tag{8.78}$$

As a result, we arrive at solutions (19)-(20) with the constants c_j given now by (77b) instead of (17). For the vector $U(2)$ nonlinear Schrödinger equation there are again two types of solutions:

1. **A shallow well, $s = 1$:**

$$\psi = \begin{pmatrix} c_1 \\ c_2 \end{pmatrix} e^{i\theta_1}\mathrm{sech}bz, \quad u = -2b^2\mathrm{sech}^2 bz, \tag{8.79}$$

$$|c_1|^2+|c_2|^2 = 2b^2(1-v^2-4\beta b^2).$$

The solution exists, if

$$\alpha+3\beta = 0, \quad \text{(strong parameter coupling)}$$

$$\beta < \frac{1-v^2}{4b^2}.$$

2. **A deep well,** $s = 2$:[*)]

$$\psi \begin{pmatrix} c_1 e^{i\theta_1} \mathrm{sh} bz \\ c_2 e^{i\theta_2} \end{pmatrix} \mathrm{sech}^2 bz, \quad u = -6b^2 \mathrm{sech}^2 bz, \tag{8.80}$$

$$|c_1|^2 = 6b^2(1-\mathrm{v}^2-4\beta b^2), \quad |c_2|^2 = 6b^2(1-\mathrm{v}^2+2b^2(\beta+3\alpha)).$$

Conditions for this solution to exist are

$$\beta < \frac{1-\mathrm{v}^2}{4b^2}, \quad \frac{1-\mathrm{v}^2}{2b^2} > -(3\alpha+\beta),$$

with

$$\theta_j = \frac{\mathrm{v}}{2}x + (\frac{\mathrm{v}^2}{4} + j^2 b^2)t + \theta_{0j}, \quad z = x + \mathrm{v}t + x_0,$$

in both cases (79) and (80).

If the consistency conditions are altered then new scalar solutions appear in the framework of (74). For example, one can easily verify that (79) is a solution to the scalar version of (74) if $c_1 = 0$ (or $c_2 = 0$). Similarly, (80) becomes a scalar solution to (54) when

$$(1) \quad c_2 = 0, \quad |c_1|^2 = -36b^4(\alpha+\beta), \quad b^2 = -\frac{1-\mathrm{v}^2}{2(3\alpha+\beta)}, \quad \alpha+\beta<0, \tag{8.81a}$$

$$(2) \quad c_1 = 0, \quad |c_2|^2 = 36b^4(\alpha+\beta), \quad b^2 = \frac{1-\mathrm{v}^2}{4\beta}, \quad \alpha+\beta>0. \tag{8.81b}$$

The existence conditions for both versions $U(1)$ and $U(2)$ of (74) show there are subsonic as well as supersonic solitons depending on the sign of the system parameters ratio α/β. Then, the $s = 1$ solitons (79) and the $s = 2$ solitons (80) exist simultaneously when $\alpha+3\beta = 0$ only. The regions of the existence of solutions (79) and (80) in the (α,β) plane are shown in Fig. 8.1. Note that the origin of coordinates corresponds to the Zakharov system and that the magnitude $a = 1-\mathrm{v}^2/2b^2$ is arbitrary.

[*)] It should be noted that solutions of such type for the plasma model ($\alpha=3\beta\leqslant 0$) have been shown to exist apparently for the first time in Refs. [15] and [16].

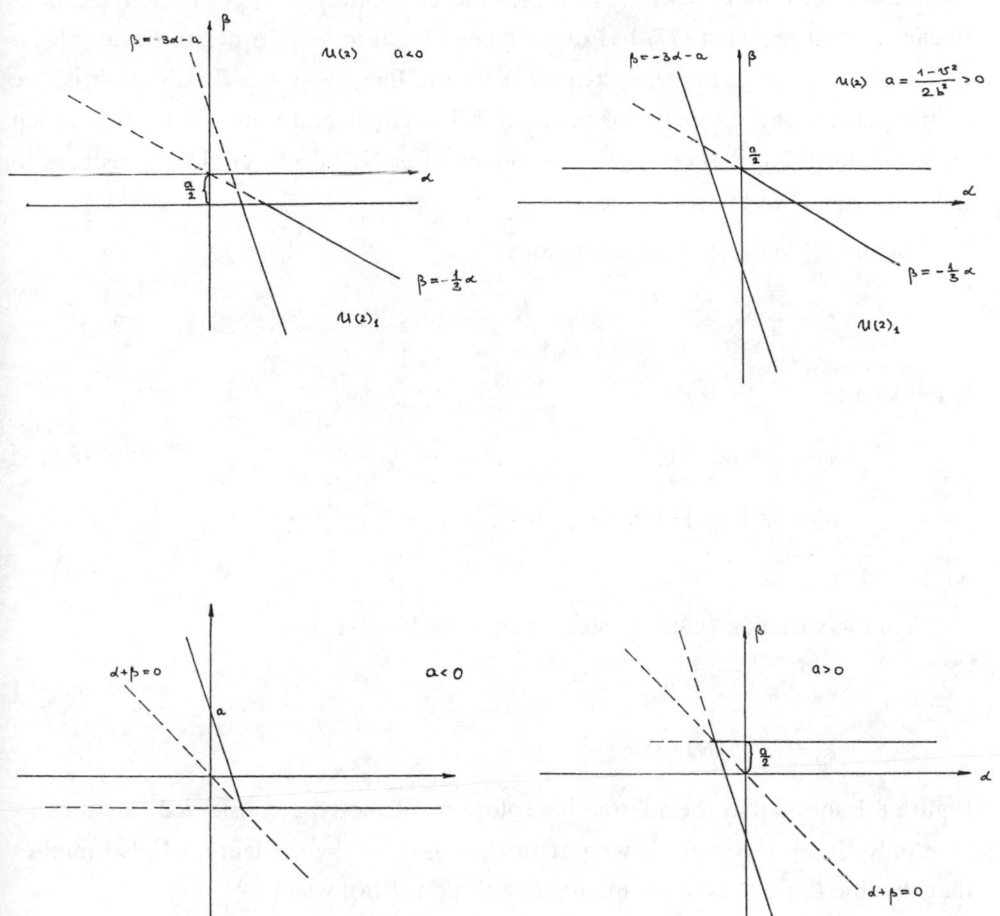

Figure 8.1. The regions of the existence of solutions (8.79) and (8.80) in the plane α, β, for various magnitudes of $a = (1 - v^2)(2b^2)^{-1}$.

In the same way, consistency conditions are resolved for $n = 3,4$, and so on, with an arbitrary metric signature. For details, see Ref. [8]. It should also be stressed that the general $s = 2$ solution is given by isorotation of (80) via a matrix

$h_1 \in SU(2)$ (see eqns. (24) and (25)).

We have obtained a particular class of soliton-like solutions to (57) (with arbitrary α and β) defined by the potential in the form (58). Eigenvalues are given by the known formula $\lambda_j = j^2 b^2$, $j = 1,\dots,s$. One can naturally suppose that the nonlinear spectral problem (77) has other types of solutions. To demonstrate this let us consider, as an example, system (77) on the line $\alpha - 3\beta = 0$. This option is due to the plasma physics model of interacting Langmuir and ion-sound waves which appeared for the first time in [9]. The model, after Nishikawa et al. [11], reduces to the same model in the soliton sector.

System (77) via scale transformations

$$\zeta = \frac{x}{\sqrt{|\beta\gamma|}}, \quad \eta = u|\beta|\gamma^2, \quad \psi^2 = \varphi^2|\beta|\gamma^4, \quad E = -\lambda|\beta|\gamma^2, \quad \gamma^{-2} = 1 - v^2$$

is reduced to

$$\psi_{\zeta\zeta} - \eta\psi + E\psi = 0, \tag{8.82a}$$

$$\eta_{\zeta\zeta} - \kappa\eta^2 + \mathrm{sgn}(1-v^2)\eta + \psi^2 = 0,$$

with $\kappa = \alpha/\beta$, $\beta < 0$.

Then the variant $\alpha = 3\beta < 0$, studied in Refs. [9-12] is

$$\psi_{\zeta\zeta} - \eta\psi + E\psi = 0, \tag{8.82b}$$

$$\eta_{\zeta\zeta} + 3\eta^2 + \mathrm{sgn}(1-v^2)\eta + \psi^2 = 0.$$

Figure 8.1 shows that the soliton-like solutions of the type considered exist in this case only if $a \simeq 1 - v^2 > 0$, so we put further $\mathrm{sgn}(1-v^2) = 1$. Figure 8.1 also implies that the line $\beta = \alpha/3$ (at $\alpha, \beta < 0$) crosses solution (81a), when

$$|c_1|^2 = 36b^4(\alpha+\beta) = 144b^4|\beta| = 9\frac{(1-v^2)^2}{25}, \tag{8.83a}$$

$$b^2 = \frac{1-v^2}{20|\beta|}, \quad \lambda = -b^2,$$

or in dimensionless variables

$$E = -\lambda|\beta|\gamma^2 = \frac{1}{20}. \tag{8.83b}$$

Solution (83) has been obtained in the author's paper [10] and independently by

Nishikawa et al. in [11]. In the latter, one should make the substitution $1-v^2 \simeq 2(1-v)$.

In Ref. [12], via a numerical study of the spectral problem (82b), the value $E = 1/20$ was shown to be an upper point of the distinct spectrum, which consists of a sequence of allowed bands with discrete eigenvalues, $E < 1/20$. The point $E = 0$ is a concentration point of this sequence (see Fig. 8.2).

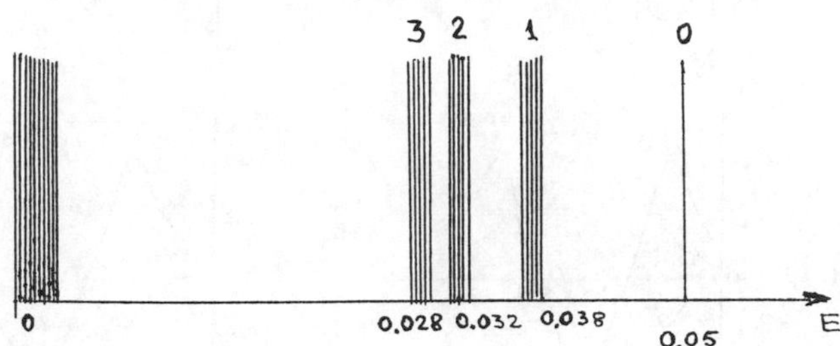

Figure 8.2. Eigenspectrum of the nonlinear spectral problem (82b), 1, 2, 3 denote the allowed bands.

Each of these bands is defined with the number j of potential $u(z)$ nodes related to one eigenfunction $\psi(z)$ node. So, in the upper point (83b) we have $j = 0$, in the first band $E_1 \simeq 0.038$, $j = 2$; then $E_2 \simeq 0.032$, $j = 4$; $E_3 \simeq 0.028$, $j = 6$. Thus the number of $u(z)$ nodes in the band equals twice the band number counting from the zero band $E_{\max} = 1/20$. The picture is shown qualitatively in Fig. 8.3 taken from [12].

Analogous results were recently derived by Kramer et al. in the work [13] for the spectral problem

$$\frac{1}{3}u_{xx} = \exp(u-\psi^2) - \frac{v}{\sqrt{v^2-2u}}, \tag{8.84}$$

$$\psi_{xx} + (1-\exp[u-\psi^2])\psi + \lambda\psi = 0,$$

which is reduced to (82a) via expanding exp with respect to ψ^2 and u, for $M \simeq 1$. Besides the zero band, the third band was found in their paper and it was shown that the distance between positive and negative peaks of the $\psi(z)$ function increases with the band number (see Fig. 8.4). The authors proclaimed that the distance enlarges infinitely when $E \to 0$, so that the positive and negative peaks behave

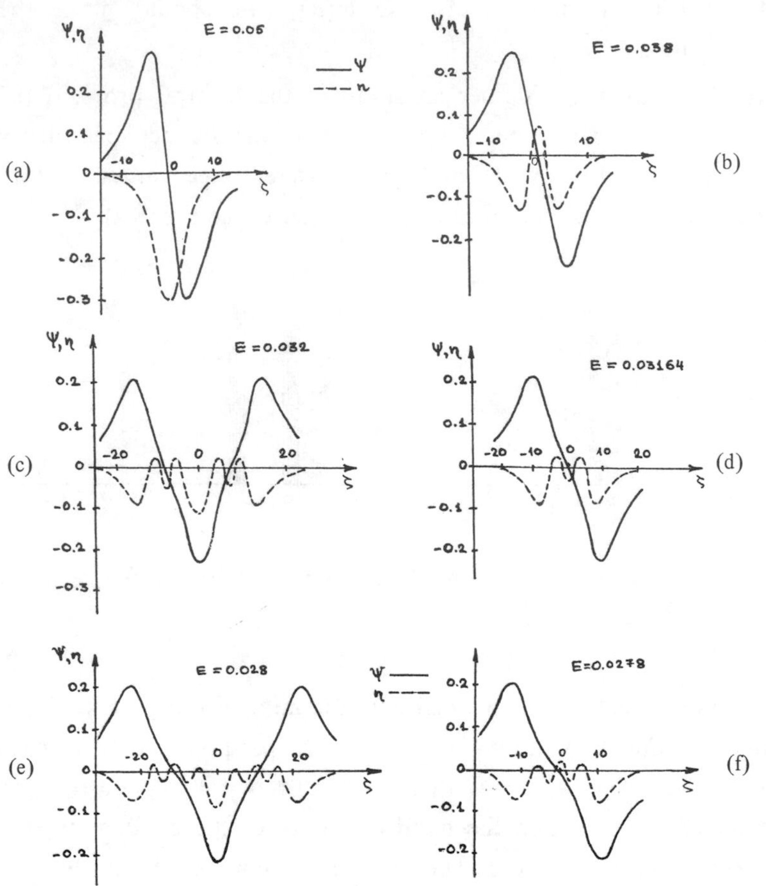

Figure 8.3. Solutions to (82b) in the allowed bands (a)→ the zero band, (b)→ the first one, (c), (d)→ the second one, (e), (f)→ the third one.

nearly as independent solitons (with phase shift $\theta_0 = \pi$). The potential amplitude $u(z)$ between these peaks tends to zero.

We note that the relation (83a) is replaced in their model by

$$b^2 = \frac{1-v^2}{20|\beta|v^2}, \quad (\text{for } \beta = 1/3), \tag{8.85}$$

(see (14) of paper [14]). Such a relation is also obtained in the framework of (74), if the second term in the left-hand side of eqn. (74b) is replaced by

$$(\alpha u^2 + \beta u_{xx})_{tt}. \tag{8.86}$$

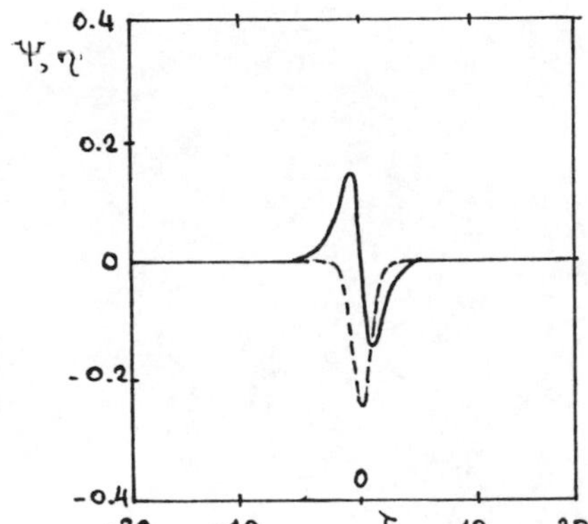

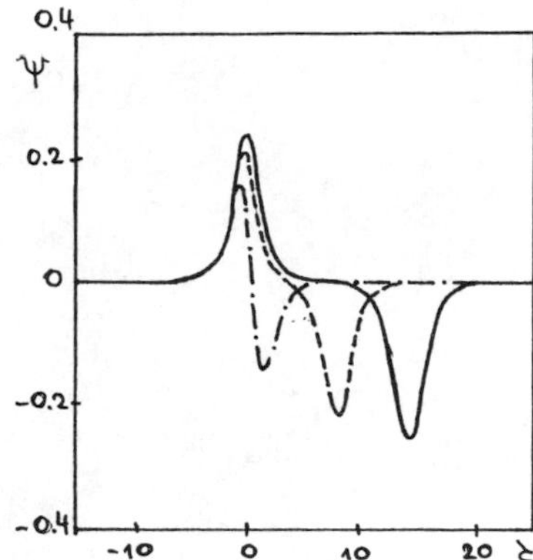

Figure 8.4. The distance between the positive and negative peaks as a function of the band number (a)→ the zero band.

This was pointed out in the author's review [41]. Such nonlinearity and dispersion of the plasma model is likely to be more adequate at small soliton velocities. Relation (85) may then be obtained easily from (83a), if one bears in mind that, in the soliton region, passing from (74b) to (86) only involves renormalization by v^2 of the constants α and β. The remaining results of paper [13] are in qualitative agreement with those derived in [12]. They imply that the system (74) (as well as (64)) possesses a countable set of soliton-like solutions consisting of coupled potential and wave function oscillations of the type shown in Fig. 8.3.

The stability problem of the soliton and soliton-like solutions described in this chapter needs to be thoroughly studied. We shall consider this in a later chapter.

Chapter 9.

THE EXISTENCE OF SOLITON-LIKE SOLUTIONS

In this chapter by way of an example of the ϕ^6-model we shall consider some problems concerning the existence of localized soliton-like solutions. The choice of the model is determined by its importance in applications (see Chapter 1 and 7) and by the rather rich set of solution-like solutions in vacuum and condensate. Firstly, we seek virial relations which exist for Lagrangian systems. From these relations the necessary conditions for the existence of soliton-like and particle-like solutions will follow*). We then formulate the concept of a mechanical analogy, whereby the conditions for the existence of soliton-like and particle-like solutions may be visualized and, in principle, can be developed for non-Lagrangian systems. Here we consider D-dimensional models.

1. Virial relations

Consider the equation

$$r^{1-D}\frac{d}{dr}(r^{D-1}\frac{d}{dr}\phi)-\kappa^2\phi+b\phi^{p-1}-c\phi^{q-1} = 0. \tag{9.1}$$

$\phi(r)$ is a real function, $\kappa^2 > 0$, p,q,b,c are real constants, and we consider D space dimensions. The search for either static or stationary, i.e. $\psi(x,t) = \phi(x)e^{-i\omega t}$, time-dependent solutions localized in space reduces to this equation (or to an equation of the same type). It is also obvious that up to a redenotation of κ^2 eqn. (1) appears in relativistic models of the nonlinear Klein-Gordon equation

$$\kappa^2 = m^2-\omega^2 \tag{9.2a}$$

*) Here and in what follows, the notion of soliton-like solutions is attributed chiefly to that describing localized excitations of the condensate (kinks, bubbles), and for particle-like solutions to localized excitation over a trivial vacuum.

and in nonrelativistic models of the nonlinear Schrödinger equation where

$$\kappa^2 = -\omega. \tag{9.2b}$$

Let us consider the conditions under which the soliton-like solutions of (1) exist. Note there is a great body of papers devoted to the problems of the existence of soliton-like solutions in the framework of various variants of the (1) model, which result perhaps from the pioneer work of Finkelstein and his co-authors [19]. Here, we only note works rather close to what follows [20-23].

The investigation of the conditions for soliton-like solutions of (1) to exist depends essentially on the asymptotics of $\phi(r)$. As in the above, we shall consider two cases which are interesting from a physical point of view:

$$\phi(r) \to 0_{r\to\infty} \qquad \text{(particle-like solutions)}, \tag{9.3}$$

$$\phi(r) \to \text{const}_{r\to\infty} \quad \text{(soliton-like solutions)}. \tag{9.4}$$

To begin with let us consider first (3). Zero asymptotics means that, for sufficiently large r, the behaviour of the function $\phi(r)$ is determined in fact by the linear part of eqn. (1). This means that

$$\phi(r) \sim r^{(1-D)/2}\exp(-\kappa r)$$

with D being the space dimension.

For $\kappa^2 \leqslant 0$, the modulus of ϕ decreasing at infinity is determined by the pre-exponential factor $\phi \sim r^{1-D/2}$; this is not enough to make the particle number and energy integrals finite. So for physically interesting solutions to exist it is necessary that

$$\kappa^2 > 0. \tag{9.2c}$$

Note, that eqn. (1) supplies an extremum to the Lagrangian

$$\mathfrak{L} = -\frac{1}{2}\int\left\{\phi_r^2 + \kappa^2\phi^2 - \frac{2b}{p}\phi^p + \frac{2c}{q}\phi^q\right\}d^Dr. \tag{9.5}$$

Introducing the functional

$$T = \frac{1}{2}\int\phi_r^2 r^{D-1}dr, \quad U_2 = \frac{\kappa^2}{2}\int\phi^2 r^{D-1}dr, \tag{9.6}$$

$$U_p = \frac{b}{p}\int \phi^p r^{D-1} dr, \quad U_q = \frac{c}{q}\int \phi^q r^{D-1} dr,$$

then

$$\mathcal{L} = -(T + U_2 - U_p + U_q). \tag{9.7}$$

After scale transformation

$$\tilde{\phi} = \frac{1}{\alpha}\phi(\lambda r) \tag{9.8}$$

one gets

$$\tilde{T} = \frac{1}{2}\int \tilde{\phi}_r^2 d^D r = \alpha^{-2}\lambda^{2-D} T.$$

Analogously

$$\tilde{U}_2 = \alpha^{-2}\lambda^{-D} U_2, \quad \tilde{U}_p = \alpha^{-p}\lambda^{-D} U_p, \quad \tilde{U}_q = \alpha^{-q}\lambda^{-D} U_q. \tag{9.9}$$

In new variables, $\mathcal{L}$ assumes the form

$$-\tilde{\mathcal{L}} = +\lambda^{-2}\lambda^{-D}(\lambda^2 T + U_2 - \alpha^{2-p} U_p + \alpha^{2-q} U_q). \tag{9.10}$$

Being an extremum of $\mathcal{L}$ at the point $\alpha = \lambda = 1$ (1) implies

$$\left.\frac{\partial \tilde{\mathcal{L}}}{\partial \alpha}\right|_{\alpha=1} = \left.\frac{\partial \tilde{\mathcal{L}}}{\partial \lambda}\right|_{\lambda=1} = 0. \tag{9.11}$$

From (11), two virial relations follow

$$2T + 2U_2 - pU_p + qU_q = 0, \tag{9.12}$$

$$(2-D)T + D(U_p - U_2 - U_q) = 0. \tag{9.13}$$

Solving these with respect to T and U_2 we obtain

$$4T = D\{(p-2)U_p - (q-2)U_q\}, \tag{9.14}$$

$$2U_2 = p\{(1 - p\frac{D-2}{2D})U_p - (1 - q\frac{D-2}{2D})U_q\}. \tag{9.15}$$

The relations (12) and (13) put two constraints on, in the general case, four functionals T, U_2, U_p, and U_q. The values T and U_2 are positive by definition (6) whereas relations (14) and (15) give

$$(p-2)U_p > (q-2)U_q, \tag{9.16}$$

$$(1-p\frac{D-2}{2D})U_p > (1-q\frac{D-2}{2D})U_q. \tag{9.17}$$

One can consider which limitations these conditions, necessary for the existence of particle-like solutions, give for different particular cases: $D = 1,2,3$ and U_2, U_p or U_q amount to zero. The results are given in Table 9.1.

	$b,c,m\neq 0$ $u_2\neq 0$	$m=U_2=0$	$b=U_p=0$	$c=U_p=0$	$m=b=0$ $u_2=0$ $U_p=0$	$m=c=0$ $U_2=U_1=$
$D=1$	$U_p > q\pm 2/p\pm 2\, U_q$	$U_q > U_p$, $p>q$	$q<-2$	$p>2$	$T=U_q$ $q=-2$	no
$D=2$	$U_p > U_q$, $U_p > q-2/p-2\, U_q$	$U_q = U_p$ $p>q$	no	$p>2$ $U_2=U_p$	no	no
$D=3$	$q>p\,(p,q>6)$, $2<p,q<6-$ no connection	$U_p > U_q$, no connection p and q	no	$6>p>2$	no	$p=6-$ instanton

Table 9.1

Model (1) is representative of a class of models, a potential part of the Lagrangian density of which is subject to the condition

$$U(\phi) = 0, \quad \text{for } \phi\to 0.$$

For this class of models one can obtain also virial relations of the type (12)-(13) and the conditions (2c) for localized solutions to exist. In the particular case $p = 4$, $q = 6$ we obtain

$$D = 1, \quad U_4 > 2U_6, \tag{9.18}$$

$$D = 2, \quad U_4 > 2U_6,$$

$$D = 3, \quad U_4 > 2U_6,$$

$$U_2 = \frac{1}{2}U_4, \quad T = \frac{3}{2}(U_4 - 2U_6), \tag{9.19}$$

$$U_2 = -\frac{\alpha}{2}\int\phi^2 r^2 dr, \quad U_4 = \frac{1}{4}\int\phi^4 r^2 dr, \quad U_6 = \frac{1}{6}\int\phi^6 r^2 dr. \tag{9.20}$$

Inequality (18) gives

$$\int \phi^4 r^2 dr > \frac{4}{3} \int \phi^6 r^2 dr. \tag{9.21}$$

As we have seen in the Chapter 7 solution (7.10a), for $\alpha \to -3/16$ is a drop of large size (radius) with constant density $\phi^2 = \phi_s^2$ and a thin transition layer $\Delta x \simeq (3)^{-1/2}$, so for the purpose of estimation this can be approximated by the step $\theta(x - x_0)$. Assume that for $D = 3$ the behaviour of the solution has the same character. In terms of α the density of the condensate is $\phi_c^2 = (3/4)$ for $D = 1$. Now estimation (21) is

$$\int \phi^4 r^2 dr > \frac{4}{3} \phi_s^2 \int \phi^4 r^2 dr,$$

or

$$\phi_s^2 < \frac{3}{4}, \tag{9.22}$$

i.e. we obtain for $D = 3$ the same magnitude of the maximal density (naturally there is a similar estimation for $D = 2$ too). Relation (19) enables an extra estimation to be made:

$$-\alpha \int \phi^2 r^2 dr = \frac{1}{4} \int \phi^4 r^2 dr, \text{ whence } -\alpha \int \phi^2 r^2 dr \simeq \frac{1}{4} \phi_s^2 \int \phi^2 r^2 dr$$

or

$$-\alpha \simeq \frac{1}{4} \phi_s^2 < \frac{3}{16}. \tag{9.23}$$

The last relation means that, in the three-dimensional case, bubble-like solutions exist for $\alpha > -3/16$. (It should be noted that in formulae (20) and (23) we have set $\alpha = -\kappa^2$). The second boundary for α follows strictly from the form of eqn. (1). Indeed, one can easily see that the drop-like solutions, i.e. those with $\phi(r) \to 0$ for $r \to \infty$ only exist for $\kappa^2 > 0$ (the $\kappa = 0$ point requires special consideration), because, in the small ϕ domain, terms proportional to ϕ^3 and ϕ^5 in the equation may be disregarded. Then one can readily find a solution of the linear equation. As a result, we have the existence region of drop-like solutions ($D = 3$)

$$0 > \alpha > -\frac{3}{16} \tag{9.24a}$$

or in terms of A

$$0 > A > -\frac{1}{2}. \tag{9.24b}$$

Now we proceed to study the existence conditions of bubble-like solutions (i.e. hole-like excitations of the condensate). There are analytical formulae for these solutions in one-dimension, see (7.28) and (7.30) at $A > 0$. The study of one-dimensional models shows also that eqn. (1) and the functional $\mathcal{L}$ have to be rewritten in terms of A (setting $\phi_c^2 = 1$)

$$\Delta\phi - (2A+1)\phi + 2(A+2)\phi^3 - 3\phi^5 = 0, \tag{9.25}$$

$$H = -L = \frac{1}{2}\int\{\phi_x^2 + (\phi^2-1)^2(\phi^2 - A)\}d^Dx. \tag{9.26}$$

Let us introduce the functionals

$$\tilde{T} = T, \ \tilde{U}_2 = \frac{2A+1}{2}\int(1-\phi^2)d^Dx, \tag{9.27}$$

$$\tilde{U}_4 = \frac{1}{2}(2+A)\int(1-\phi^4)d^Dx, \ \tilde{U}_6 = \frac{1}{2}\int(1-\phi^6)d^Dx,$$

and multiplying eqn. (25) by ϕ rewrite it in the form

$$\phi\Delta\phi - (2A+1)(\phi^2-1) + 2(2+A)(\phi^4-1) - 3(\phi^6-1) = 0.$$

Integrating it over d^Dx we obtain the first virial relation

$$\tilde{T} - \tilde{U}_2 + 2\tilde{U}_4 - 3\tilde{U}_6 = 0. \tag{9.28}$$

The second virial relation is obtained via the scale transformation $\phi \to \phi(\lambda r)$ described above and varying the 'energy' functional H. As a result we obtain

$$(2-D)\tilde{T} = DU \tag{9.29}$$

with

$$U = \frac{1}{2}\int(\phi^2-1)^2(\phi^2 - A)d^Dx,$$

or in terms of $\tilde{U}_2$, $\tilde{U}_4$ and $\tilde{U}_6$

$$(2-D)\tilde{T} - D(\tilde{U}_4 - \tilde{U}_2 - \tilde{U}_6) = 0. \tag{9.30}$$

(This means that $U = \tilde{U}_4 - \tilde{U}_2 - \tilde{U}_6$).

Relations (28) and (30) differ from (12) and (13) by redefinition of the functional U_i and changing their signs (note that $p = 4$, $q = 6$). As shown in Chapter 7 there are two types of localized excitations of the condensate at $D = 1$: bubbles and kinks which lie in the different domains of the parameter A (respectively α) divided by the point $A_0 = 0$. Let us see what can be concluded from the properties of solutions in the 3-dimensional case. Solving (28) and (30) we obtain as before $\tilde{U}_4 = 2\tilde{U}_2$, i.e. $(2+A)\int(1-\phi^2)(1+\phi^2)d^3x = (4A+2)\int(1-\phi^2)d^3x$. There is an estimation $(2+A)(1+\phi_b^2)\int(1-\phi^2)d^3x \simeq (4A+2)\int(1-\phi^2)d^3x$ for bubbles of large radius $r_0 \gg \Delta r$, $\phi^2 \simeq \phi_b^2 \ll 1$. Hence we get $(2+A)(1+\phi_b^2) \simeq 2+4A$ or $\phi_b^2 \simeq 3/2A$. The upper estimation for A is given by the second inequality $\tilde{U}_6 > \tilde{U}_2$, which gives $\int(1-\phi^2)(1+\phi^2+\phi^4)d^3x > (2A+1)\int(1-\phi^2)d^3x$. Whence we have $3 > 2A+1$ or $A < 1$ for small amplitude bubbles, $\phi_b^2 \to 1$. Thus, the existence of localized solutions with the condensate asymptotics $\phi_{r\to\infty} \to \mathrm{const}$ is given by

$$0 < A < 1. \tag{9.31}$$

Summarizing the results obtained in this section, we can say that in the 3-dimensional models of the ϕ^6 theory, stationary localized solutions to eqn. (25) exist for $A \in (-1/2, 1)$. The point $A_0 = 0$ divides the domains where the solutions are quite different: on the left from A_0, i.e. at $A \in (-1/2, 0)$, particle-like solutions are drops with zero asymptotics ($r \to \infty$), and on the right, at $A \in (0,1)$ soliton-like solutions are bubbles with the condensate asymptotics.

As an ancillary result, we have also obtained the existence conditions of particle-like solutions known from the early 1960s (Nehari, Zhidkov and others [24,25]) in the framework of the ϕ^n theory: $2 < n < 6$, for $D = 3$.

We shall now discuss a construction of localized solutions with the aid of the so-called mechanical analogy method.

2. Mechanical analogy method

Further we shall consider eqn. (1) for $p = 4$, and $q = 6$:

$$\phi'' + \frac{D-1}{r}\phi' - \kappa^2\phi + \phi^3 - \phi^4 = 0, \quad ({}' \equiv \frac{d}{dr}). \tag{9.32}$$

Transferring the second term into the right-hand side and multiplying the equation by ϕ' we obtain

$$\{(\phi')^2 - \kappa^2\phi^2 + \frac{1}{2}\phi^4 - \frac{1}{3}\phi^6\}' = -2\frac{D-1}{r}(\phi')^2. \tag{9.33}$$

Denote

$$U_p = \frac{1}{2}\{-\kappa^2\phi^2 + \frac{1}{2}\phi^4 - \frac{1}{3}\phi^6\}, \tag{9.34a}$$

$$E_b = \frac{1}{2}(\phi')^2 + U_p. \tag{9.34b}$$

Let us regard the field value ϕ at the point r as the position of a point particle of unit mass at time instant τ, moving in an external potential $U(\zeta)$ under the action of a frictional force, $F_{fr} = -(D-1)/\tau\zeta_\tau$. Then the solution $\phi(r)$ of eqn. (32) describes the particle trajectory $\zeta(\tau)$.

The relation (33) in one dimension is just the energy conservation law of the particle and in two and three dimensions it is the energy dissipation law. It is convenient to rewrite the potential $U(\zeta)$ in terms of the parameter A

$$U_p = -\frac{1}{2}(\zeta^2 - 1)^2(\zeta^2 - A). \tag{9.35}$$

It should be stressed that $U(\zeta)$ and the potential part of the energy density of the field model have opposite signs (cf. with (5), bearing in mind that, for stationary states, $\mathcal{L} = -H$). The plots $U(\zeta)$ for some values of A are given in the Fig. 9.1.

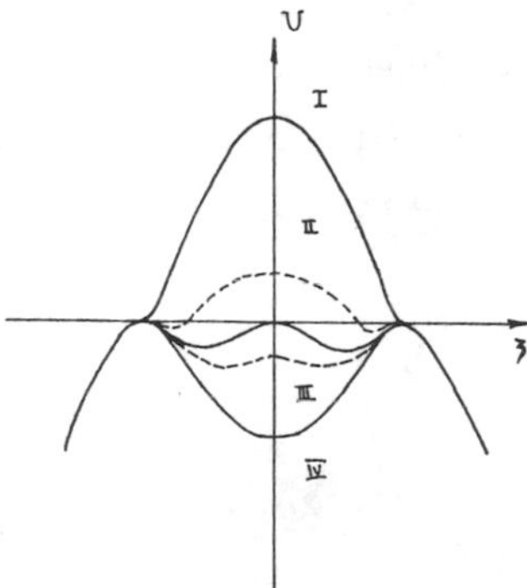

Figure 9.1. The plots $U(\zeta)$ for various A that correspond to characteristic regions of Chapter 8.

Fig. 9.2 shows the path curves of constant energy E in the phase $\zeta O \zeta_\tau$ plane in one dimension. One can see that finite particle motion in valleys and infinite ones are

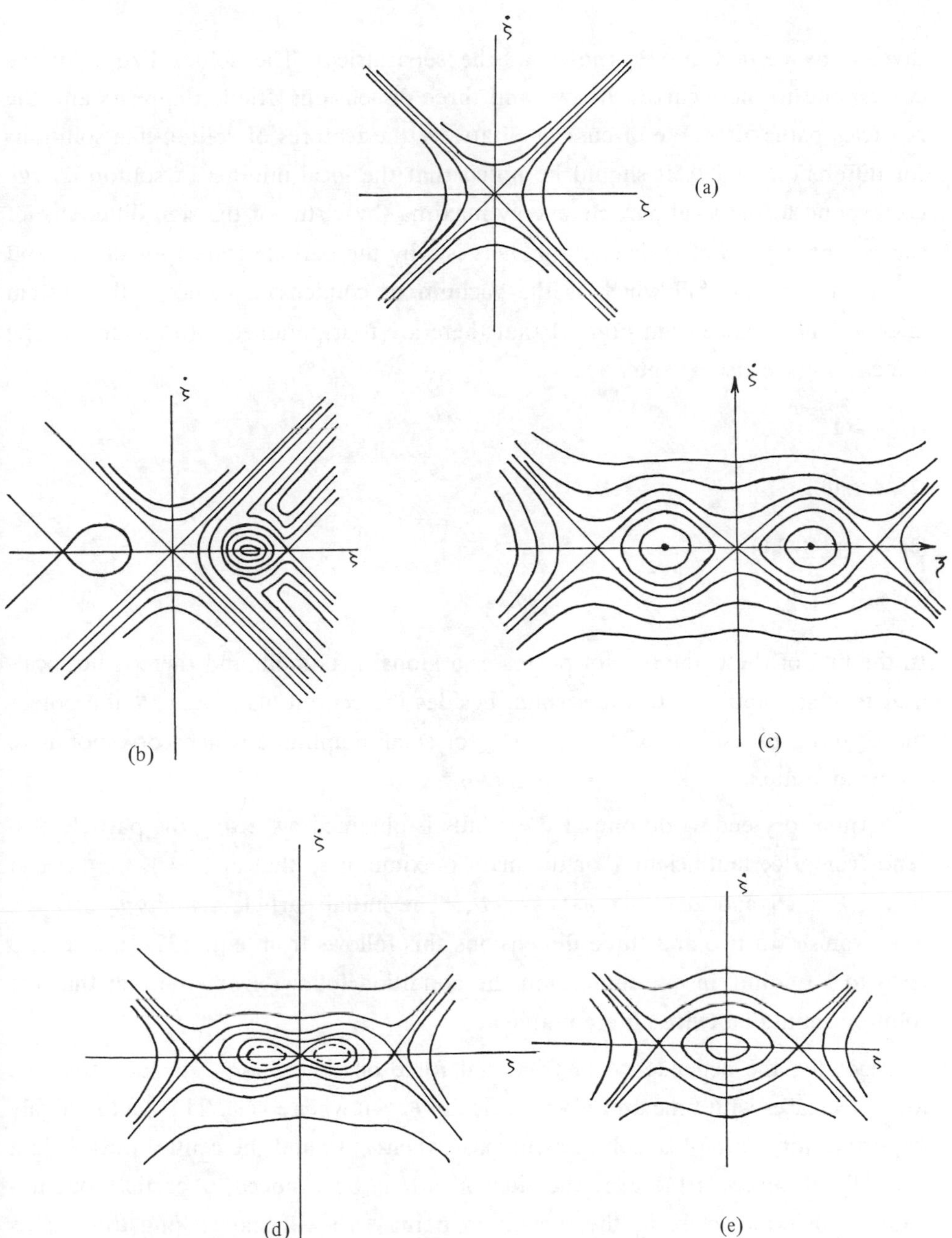

Figure 9.2. Potential reliefs for the above regions I - IV (Figure 9.1): (a) → I, (b) → II, (d) → III, (e) → IV, (c) corresponds to a special case $\alpha = -3/16$ ($A = 0$).

divided by certain curves known as the separatrices. The soliton-like solutions correspond to these curves. In two and three dimensions, friction appears and the particles paths alter. We discuss qualitatively the features of soliton-like solutions for different A and D. It should be noted that the local minima of soliton energy correspond to the local particle energy maxima (by virtue of the sign difference of the potential parts of their energy). This is why the particle trajectory should end on the top of the hill which is the vacuum or condensate value of the system energy. One can see from Fig. 9.1 that there are four parameter-A domains in one dimension (see also Chapter 7).

(1) $A \geqslant 1$

(2) $A \in (0,1)$

(3) $A \in (-1/2,0)$

(4) $A < -1/2$.

In the first of these, the U_p plot possesses a global maximum and there is no localized regular solutions. In the second, besides the central maximum U_0 the potential U_p has a two side-maxima $U_1 = U_{-1}$ of small amplitude which corresponds to the condensate.

A trajectory ending on one of these hills is obtained by letting the particle descend from a certain height h_i of the main maximum: so that at $D = 1$, $h_1 = U_1$; at $D = 2$, $h_2 > U_1$ and at $D = 3$, $h_3 > h_2 > U_1$. The initial particle velocity $\dot{\zeta}_0$ at $\tau = 0$ must vanish. In two and three dimensions this follows from eqn. (32) (the second term to be finite), in one dimension the condition follows from the fact that the solution should be twice differentiable at $x = 0$.

Secondly, the dependence of frictional force on time $F_{fr} \sim \tau^{-1}$ is such that the force decreases with time and asymptotically $F_{fr} \rightarrow 0$ when $\tau \rightarrow \infty$. This fact is highly important for soliton-like solutions to exist when $A \rightarrow 0$ and the central peak height only slightly exceeds those of the side ones $U_0 \gtrsim U_1$. Indeed, a particle put initially $(\tau = 0)$ rather close to the maximum point $\zeta \rightarrow 0$ will stay a long time in its vicinity since the acting force is proportional to the particle displacement $F \sim (\zeta - 0) \equiv \epsilon$ and $\tau \sim |\ln \epsilon|$. The last estimations follow directly from (35) for $\zeta \rightarrow 0$. This means, that the smaller the difference between the central and side peaks the closer should be the starting point of a particle to the maximum, for being a long time in the vicinity of the top the particle causes the frictional force to decrease by a sufficient amount. It also follows that the bubble amplitudes measured from the condensate value for all $D = 1,2,3$ are limited from above by unity in the selected scale, moreover $a(D=3) > a(D=2) > a(D=1)$ for the same A (see Fig. 9.3(a)) with $\zeta \to -\zeta$ changing.

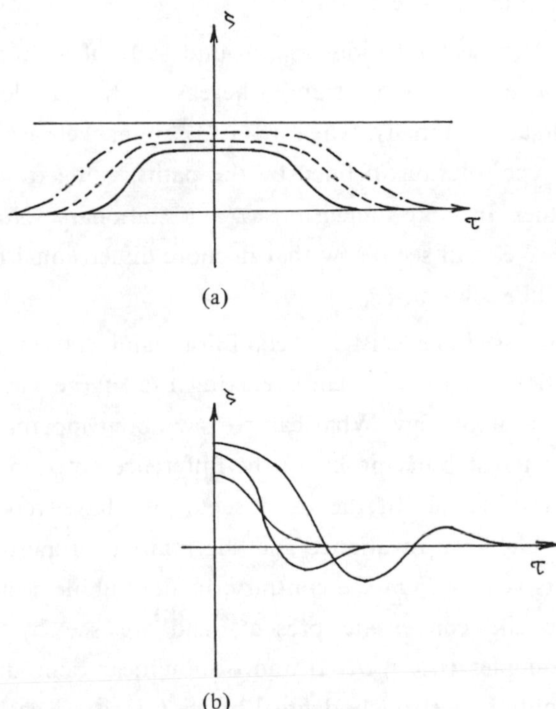

Figure 9.3. (a) Nodeless solutions for $D = 1,2,3$. Solid line $D = 1$, dash line $D = 2$, dash-dot line $D = 3$. (b) Solutions ($D = 2,3$) with nodes.

In the case $D = 1$ there is a step-like solution related to a path starting from the top of one hill and ending on the top of another (see Fig. 9.2(b)).

In the region $A \in (-1/2, 0)$ (Fig. 9.2(d)) the picture alters: soliton-like solutions are defined here with the paths starting on the slope of one of the side hills and ending on the top of the central one. In two- and three-dimensional models, in addition to a nodeless solution there exists a countable set of solutions labelled by the number of the function ϕ nodes (i.e. the number of times the particle path goes through the central hill top). Due to energy dissipation sooner or later the particle either goes down to a valley or stops on the central hill top. Such solutions for a given A are presented schematically in Fig. 9.3 (namely Fig. 9.3(a) shows nodeless solutions for different D; Fig. 9.3(b) solutions with the node number $n = 0, 1, 2$ for $D > 1$ and fixed A).

There is, however, an essential difference: node solutions exist in the drop sector and are absent in the bubble sector.

Thus, in the top sector besides the ground state of a drop there is its excited states (nonlinear excitation spectrum) whereas in the bubble sector there are no bubble excited states. Finally, when $A \in (-1/2, 0)$ as well as $A < -1/2$, there exist at $D = 1$ kink type solutions defined by the paths, going from the top of one hill to that of another. In more dimensions $(D > 1)$ stationary solutions of such a type are impossible. We shall see below that in more-dimensional models there appear, instead, vortex-like solutions.

We have discussed the existence conditions and certain features of localized *stationary* solutions of the ϕ^6 theory, making use of the virial relations together with the mechanical analogy. What can we say about moving drops and bubbles? Here (just as in the stability problem) the difference between them becomes more essential if not principal. In the drop sector we have repeatedly stressed this above: there is Gallilean invariance and the internal properties of a drop do not depend upon its velocity. On the contrary, in the bubble sector this invariance is broken due to the condensate present, and we see by a number of one-dimensional examples ($U(0,1), U(1,1)$ and ϕ^6-nonlinear Schrödinger equation) that the bubble amplitude is strongly defined by its velocity. Recall KdV bubbles with their amplitude decreasing with increasing velocity. The velocity of bubbles is bounded from above by the linear wave velocity. It is true for bubbles properly and for kinks too, and in the ϕ^6 model both bubble and kink become rarefaction solitons of KdV when $v \to v_{cr}$. We turn to the existence problem of more-dimensional soliton-like solutions in the chapter devoted to multi-dimensional solitons, where a more detailed reference list is also given.

Chapter 10.

SOLITON STABILITY

We first define the stability of soliton-like solutions. As has been mentioned above, these describe extremum states of some nonlinear system. There are two types of system stability: (1) with respect to a perturbation of the initial data, and (2) with respect to a perturbation of the evolution equation which describes the system behaviour (the structural stability).

In the first case, the problem has been investigated thoroughly. There are different methods and approaches, both linear and nonlinear. In the linear approximation, the stability problem is usually reduced to the study of the eigenvalue spectrum of the linearized evolution equation; in the nonlinear approximation, to the study of Lyapunov's inequality (see, for example, the reviews [14,26,27]). Sometimes the stability of soliton-like solutions follows from topological reasons [28].

In recent years, many papers have appeared in which structural stability in the framework of various evolution equations for different types of perturbations is investigated in some way or another. Attempts have been made to construct a general theory of soliton perturbations via the Green function method and the spectral transform (i.e. the transition from the (x,t) configurational space to the scattering-data space) on the basis of the well-known two-time formalism [29]. However, the methods derived are useful only for a very restricted class of perturbation functionals [30]. Therefore, in spite of impressive results obtained in this direction, the investigations are not complete, especially on the structural stability of systems *near* to integrable ones.

Note that there are rigorous results for the structural stability of Hamiltonian systems with a finite number of degrees of freedom (the so-called KAM theory named after Kolmogorov, Arnold and Moser). The main assertion of the KAM theory is: In the presence of a perturbation to an integrable system, the greater part (finite in measure) of the structure of its phase space is conserved if the conditions of linear independence (irrationality) of the nonperturbed system frequencies are fulfilled. This means that for small perturbations the greater part of the solutions remains quasi-periodic functions of the normal coordinates (action, angle). However, the rest of the phase space structure (where the KAM theorem conditions are not fulfilled) is destroyed and the system phase trajectories may wander throughout this region filling it densely (the so-called Arnold diffusion). Thus, a part of the orbits (quasi-periodic solutions) of the nonperturbed system becomes unstable and meander away to all the accessible area of the phase space. The stability (of instability) of the perturbed system depends on the region of the phase space where the unperturbed system was (see, for details, Ref. [98]). The question of the applicability of this result to infinite systems is open.

The problem of the stability of motion of integrable systems under non-Hamiltonian perturbations is particularly complicated.

From the point of view of computational science, both problems may be investigated by a unified approach: the initial-value problem. In the first case, one studies the evolution of a perturbed or unperturbed (a computer introduces perturbations by itself) *initial state* described by the unperturbed evolution equation and this state is a soliton solution whose stability is investigated.

In the second case, the evolution of an analogous initial state obeys the perturbed *equation*. For both cases, the solutions under investigation may depend on some slow time as a parameter. In the first case, we define as being stable the solutions for which initial perturbations do not grow with time during the evolution of the initial state. Note that in conformity with this definition it is also necessary to consider as stable weakly-radiating soliton-like solutions which are not destroyed by the initial perturbations.

It is natural to consider as *structurally stable* those solutions which conserve their shape for a sufficiently long time. The concept of 'a sufficiently long time' is determined by the scale of physical processes taking place in the system. Note that in this case part of the unperturbed solutions will be destroyed very quickly, then, instead, absolutely new types of solutions may appear, especially for perturbations of non-Hamiltonian type.

Below we give some examples of investigations in these two directions. The choice of these examples is very arbitrary and is of an illustrative character.

In the planar (x,t) case the majority of the systems studied has stable soliton or soliton-like solutions at any rate with respect to perturbations which do not change the symmetry of the system, (i.e. depending on x and t[*)]. The (longitudinal) stability of the true solitons follows, as was thought for a long time, from the integrability of the corresponding equations, e.g. $U(1)$ $S3$, KdV, MKdV and SG.

For example, via studying the asymptotic behaviour in time of a weak non-soliton perturbation to the kernel to the Marchenko equation, the $U(1,0)$ $S3$ soliton was shown to be stable and get released from the perturbation at the rate $\psi(t) \propto t^{-1/2}$ [32], and the bubble-like $U(0,1)$ soliton at the rate $\propto t^{-1}$ [33]. These results can be easily explained by considering the particle number integrals: $\int_{-\infty}^{\infty} |\psi|^2 dx$ in the first case, and $\int_{-\infty}^{\infty} (\rho_0 - |\psi|^2) dx$ in the second one. This can be done as a simple exercise.

Berryman [34] gives the only but very instructive, counter-example. He considers the longitudinal stability of solitons in the Boussinesq model and shows that these solitons are unstable with respect to infinitesimal short-wave perturbations of plane-wave type. Although this instability is, apparently, the result of 'bad' dispersion of the Boussinesq equation, $\omega^2 = k^2(1-k^2)$, this example shows that it is necessary to treat with definite care a deep-rooted opinion that the system integrability necessarily involves the stability of soliton, and especially multisoliton, solutions.

In fact, the second counterexample is related to the instability of the bound state of two solitons (bion) in the framework of the quite respectable (in the dispersion sense) $S3$ equation. As has been observed by Satsuma and Yajima in the computer experiment [35], the bion

$$\phi(x,t) = 4A \exp(i\frac{A^2 t}{2}) \frac{\mathrm{ch}3Ax + 3\mathrm{ch}Ax e^{4iA^2 t}}{\mathrm{ch}4Ax + 4\mathrm{ch}2Ax + 3\cos 4A^2 t}$$

decays into constituent solitons under the action of an initial perturbation with asymmetric imaginary part[†)].

Unstable soliton-like solutions arise for the nonlinear Klein-Gordon equation.

[*)] We call such perturbations longitudinal in contrast to transverse ones which violate the initial symmetry and depend for example, on x,y and t.
[†)] We underline, that the $S3$ bion does not have a mass defect, i.e. $E_b = E_{S1} + E_{S2}$ and is therefore unstable in perturbed systems. The fact that it arises in an initial packet decaying and lives quite a long time (!?) in the $S3$ system, is wonderful rather than lawful and is apparently connected with the higher integrals.

For the ϕ_+^4 field theory equation, when $F(\phi) = (1-|\phi|^2)\phi$, a possible instability of soliton-like solutions has been mentioned in the early paper of Zastavenko [36], where for complex solutions of the form $\phi = \chi(x)e^{-i\omega t}$ the instability region $0 \leqslant \omega^2 \leqslant 1/2$ has been found. In the study of such solutions by the computer the instability is displayed by strikingly, especially at $\omega \to 0$ (i.e. in the case of a real field function): even for very small perturbations introduced by the computer the soliton-like solution breaks up after several time steps, usuallly giving the χ singularity.

1. Stability of hole-like excitations in the ϕ^6 model of nonlinear Schrödinger equation. The spectral analysis.

In this section we demonstrate the technique to investigate stability of soliton-like objects via the spectral analysis. As an example, we consider the stability of ϕ^6 nonlinear Schrödinger equation bubbles. In Chapter 7 we have seen that the system described by eqn. (7.1) can be in two phase states depending on the parameter α value, viz. a gaseous state and condensed one. In the first case (particle-like solutions), the drops are stable. That is established via the Q-theorem (see the next section). Localized excitations of the condensate are of two types; the bubbles (7.28) with $\alpha \in (-1/4, -3/16)$ and the kinks (7.29) with $\alpha > -3/16$.

Let us consider solutions of the first type, since they have no topological structure and can be readily generalized into more dimensions. Moreover, we shall confine ourselves to the study of a bubble in rest (see (7.28))

$$\varphi_b(x) = \left\{ \frac{\rho_0 \mathrm{ch}^2(v_s x/2)}{\mathrm{sh}^2(v_s x/2) + \rho_0/A} \right\}^{1/2}. \tag{10.1}$$

The system with condensate present loses its Galillean invariance, and one cannot obtain a moving bubble from a stationary one via change of coordinate (such a change causes the boundary conditions to alter).

Let us linearize eqn. (7.3) with respect to a small (at least initially) perturbation $\delta\varphi(x,t)$ in the vicinity of the soliton (1) and assume that localized $\delta\varphi$ exist which grow exponentially with time:

$$\delta\varphi(x,t) = (f(x) + ig(x))e^{\nu t}, \tag{10.2}$$

where $f(x)$, $g(x)$ and ν are real, and

$$f(x) \to 0, \quad g(x) \to 0 \quad \text{for} \quad |x| \to \infty. \tag{10.3}$$

Putting

$$\varphi(x,t) = \varphi_b(x) + \overline{\delta}\varphi(x,t)$$

in eqn. (7.3) and considering only first order in $\delta\varphi$ gives the eigenvaluc problem

$$L_1 f(x) = -\nu g(x), \tag{10.4}$$

$$L_2 g(x) = \nu f(x),$$

where L_1 (upper) and L_2 (lower) operators are defined by

$$L_1 = -\frac{d^2}{dx^2} + \rho_0(2A + \rho_0) - 6(2\rho_0 + A)\varphi_b^2(x) + 15\varphi_b^4(x), \tag{10.5}$$

$$L_2 = -\frac{d^2}{dx^2} + \rho_0(2A + \rho_0) - 2(2\rho_0 + A)\varphi_b^2(x) + 3\varphi_b^4(x).$$

If the problem (3) and (4) has a solution $z \equiv (\nu, f, g)$ then $z' \equiv (-\nu, f, g)$ is a solution too. Therefore the existence of a real eigenvalue $\nu \neq 0$ will, irrespective of its sign, indicate exponential instability of the bubble (1).

The following fact is worth noting. One easily verifies that $L_2\varphi_b(x) = 0$ (this equality coincides with eqn. (73) for real static solutions). It follows that for a discrete eigenvalue $\nu \neq 0$

$$\nu \int_{-\infty}^{\infty} \varphi_b(x) f(x) dx = \int_{-\infty}^{\infty} \varphi_b(x) L_2 g(x) dx = \int_{-\infty}^{\infty} g L_2 \varphi_b dx = 0$$

by virtue of the self-conjugation of the operator L_2. This means that any solution $f(x)$, $g(x)$ to eqns. (4) with a rapidly decaying asymptotic satisfies the condition for normalization of the wave function to keep constant for $N[\varphi_b] = N[\varphi_b + \delta\varphi] + 0[(\delta\varphi)^2]$.

The system (3)-(4) was solved numerically in [37]. Leaving out details we give briefly the results. At $\rho_0 = 1$, the solution was obtained for several values of α for the interval $(-1/4, -3/16)$ or, as is just the same, $A \in (0,1)$: $A = 0.1, 0.2, ..., 0.9$. Standing bubbles (1) were shown to be 'Q-unstable' for all A's studied. The growth rate ν has a function of A is given in Fig. 10.1.

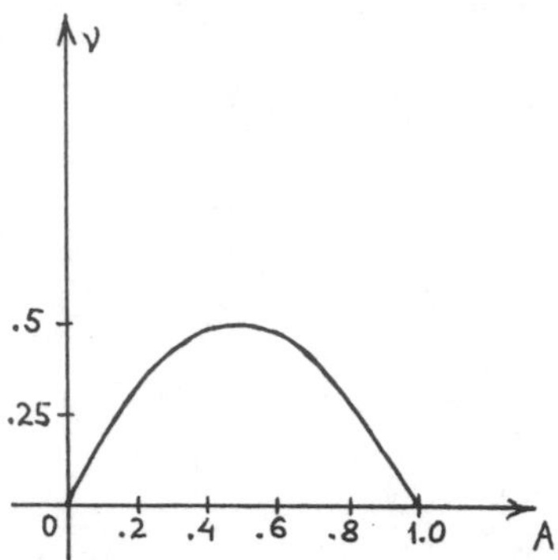

Figure 10.1. The growth rate of the bubble instability as a function of A.

The conclusion of instability which followed from the results of Ref. [37] only relates to bubbles at rest. The analysis of Chapter 7 implies small amplitude bubbles to be true solitons moving with near-sonic velocity and satisfying the KdV. These are stable according to [38]. So there is a critical velocity v_{cr} such that the bubble is stable at $v > v_{cr}$ and unstable at $v < v_{cr}$. Computer experiments carried out in Dubna JINR [60] gave the values for v_{cr}, viz.

$$v_{cr}(A = 0.2) \simeq 0.2, \quad v_{cr}(A = 0.5) \simeq 0.35, \quad v_{cr}(A = 0.8) \simeq 0.45.$$

According to their results, the main mode of bubble decay is its depth growing up to the vacuum state, $\varphi = 0$ with further size of the cavern increasing (infinitely?). This means that in the second region, $\alpha \in (-1/4, -3/16)$, see Fig. 7.3, the metastable condensate $|\psi|^2 = \rho_0$ is in fact, unstable with respect to the certain amplitude bubbles $a > a_{cr}(A)$ to evolve so that the system should proceed into the ground state $\psi = 0$ (the process is analogous to boiling, i.e. the first-kind phase transition).

A similar method was earlier applied by Anderson [39] to investigate the stability of soliton-drops in the ϕ^6 variant of the nonlinear Klein-Gordon equation. This method can also be applied in studying the infinitesimal stability of stationary states of various systems.

2. Stability of drop-like solitons. Variational methods

For Lagrangian systems, their static or stationary (for a definition see later) states are given by the extrema of the action

$$S = \int \mathcal{L} d^D x dt.$$

To study the stability of such states one can apply variational methods. Here we must make a reservation, these methods work for those systems whose differential part of the Hamiltonian density is sign definite. Then the stability problem is solved via a definition of the sign of the second variation (conventional or absolute) of the system energy

$$E = \int \mathcal{H} d^D x,$$

$$\mathcal{H} = \frac{\partial \mathcal{L}}{\partial \dot{\varphi}} \dot{\varphi} - \mathcal{L}.$$

With this method, Hobbart in 1963 and Derrick in 1964 proved the theorems which bear their names. This can be formulated as follows: in the framework of Lagrangian systems of more than one space dimension, there are no *stable* static states if the potential does not depend on the derivatives of the field function. The proof is very simple and we give it here making use of the results of the preceding chapter.

Consider the Hamiltonian

$$H \equiv E = \int \{ \frac{1}{2} (\chi')^2 + u(\chi) \} d^D x, \qquad (10.6)$$

with χ being a real scalar (or vector) field. The *static* states of the system are extremals of H, i.e. $\delta H = 0$, which leads to

$$\Delta^D \chi - \frac{\partial u}{\partial \chi} = 0. \qquad (10.7)$$

Denote these solutions by χ_0 and suppose their energy to be finite, which implies that χ_0 tends to its asymptotic value quite rapidly. Making use of (9.8) at $\alpha = 1$ and (9.11) in which $\mathcal{L} = -H$ we find the connection between the functionals

$$T = \frac{1}{2} \int (\chi')^2 d^D x \quad \text{and} \quad U = \int u(\chi) d^D x, \qquad (10.8)$$

$$(2-D)T = DU.$$

Calculating the second derivative of H with respect to λ one has (for $\chi = \chi_0$, or $\lambda = 1$)

$$\delta^2 H \sim \frac{d^2 H}{d\lambda^2}\Big|_{\lambda=1} = (2-D)(1-D)T + D(D+1)U. \tag{10.9}$$

Substituting (8) in eqn. (9) we get

$$\delta^2 H \sim 2(2-D)T, \tag{10.10}$$

i.e. the state χ_0 is a minimum of the energy (6), $\delta^2 H > 0$ only when $D < 2$. This means that for $D \geqslant 2$, the particle-like solution χ_0 is, generally speaking, unstable. In fact, this problem needs additional study since besides energy, the system can possess other integrals which should be taken into account in calculating $\delta^2 H$. So for the systems with an isotopic symmetry group the theorem can be transformed into the Q-theorem [40]. We give this here for the case of $U(1)$ symmetry when such an additional integral is the charge

$$Q = \operatorname{Im} \int \dot{\bar{\psi}} \psi d^D x. \tag{10.11}$$

We call complex field particle-like solutions of the form

$$\psi_s(x,t) = \varphi_s(x) e^{-i\omega t} \tag{10.12}$$

stationary (real field particle-like solutions periodic in time are naturally attributed to the stationary solutions). These tends to static ones for $\omega = 0$. It is possible to formulate uniformly the Q-theorem for the Schrödinger and Klein-Gordon models

$$i\dot{\psi} + \Delta\psi + F(|\psi|^2)\psi = 0, \tag{10.13}$$

$$(m^2 + \Box)\psi - F(|\psi|^2)\psi = 0. \tag{10.14}$$

The fact that the function F only depends on $|\psi|^2$ means that $U(1)$ symmetry is present in the system.

The particle number

$$N = \int |\psi|^2 d^D x \tag{10.15}$$

in the nonlinear Schrödinger equation case and the charge

$$Q = \omega\int|\varphi|^2 d^D x \tag{10.16}$$

in the nonlinear Klein-Gordon equation case are conserved.

THEOREM. *For stationary particle-like solutions* ψ_s *of eqns. (13) and (14) to be stable with fixed* N *or* Q *the following conditions are necessary and sufficient:*

(1) Test operators, i.e. the operators of eqns. (13) and (14) linearized about the particle-like solutions ψ_s *under consideration*

$$\hat{L}_s = -\Delta+\omega+\Phi(\varphi_s),$$

$$\Phi(\varphi_s) \equiv \frac{\partial F}{\partial \varphi}\bigg|_{\varphi_s},$$

$$\hat{L}_k = -\Delta+(m^2-\omega^2)+\Phi(\varphi_s),$$

possess not more than one negative eigenvalue.

(2) The eigenfunction related to it satisfies the condition

$$\int \delta\varphi \ \varphi_s d^D x \neq 0. \tag{10.17}$$

(3)

$$dN/d\omega < 0, \tag{10.18}$$

$$dQ/d\omega < 0. \tag{10.19}$$

In the first case for the particle-like solutions to exist ω *should be negative, in the second case* $m > \omega > 0$, *without loss of generality.*

The stability of stationary particle-like solutions with the charge or particle number fixed we call the *Q-stability*.

The proof of the Q-theorem for the $U(1)$ symmetry can be found in [40a] and [14]. The Q-stability of particle-like solutions for nonlinear Klein-Gordon equation is considered in [36], [40b] and $S3$ nonlinear Schrödinger equation in [40c]. The conditions of the Q-theorem are shown to be necessary in Ref. [41].

The Q-theorem is a particular case of the more general stability theorem by Liapunov. Liapunov's method to investigate the stability of motion of dynamical systems is based on searching some special functionals named the Liapunov functionals whose properties determine perturbed motion. Liapunov's method was developed for studying dynamical systems having a finite number of degrees of freedom. Later it was extended to continual (field) systems [42]. Considering solitons, one investigates the stability of their form; that is we speak of the stability of a set of soliton solutions $M(\psi_s)$ rather than of a solitary soliton. Elements of the set constitute a homogeneous space in which symmetry groups act transitively, and the solitary soliton breaks down these symmetries. Among such symmetry transformations are translations and rotations in space, rotations in isospace, and gauge transformations (if any) admitted by the equations of motion. In this way particle-like solutions $\tilde{\psi}_s$ is naturally called perturbed if $\tilde{\psi}_s \notin M(\psi_s)$. Fixing additional integrals defines possible trajectories T in M. These conditions used to be reflected in the option of a metric ρ which defines to what degree the perturbed solution deflects from the original one: it is determined so that $\rho(\psi'_s, \psi_s) = 0$ when $\psi'_s \in T$.

We define now that Liapunov stability means (i.e. stability with respect to the perturbation of initial data). Let us introduce the metrics ρ_0 and ρ, measuring the initial and current deviations of the perturbed solution from the original one such that $\rho \to 0$ when $\rho_0 \to 0$. Then the soliton ψ_s is *stable after Liapunov* with metrics ρ_0 and ρ; if $\forall \epsilon > 0$ there is $\delta(\epsilon) > 0$ such that from $\rho_0(\tilde{\psi}, \psi_s)\Big|_{t=0} < \delta$ follows $\rho(\tilde{\psi}, \psi_s) < \epsilon$ for any instant of time. Such stability is called often the orbital stability [43] (T is the orbit of a group) or stability with respect to the form [38]. *The Liapunov theorem* we give here without proof, too. For soliton-like solutions to be stable with metrics ρ_0 and ρ it is necessary and sufficient that Liapunov's functional $U[\psi]$ should exist, positive defined with respect to the metric ρ_0 in a vicinity of ψ_s (continuous with respect to the metric ρ_0, that is $U \to 0$ when $\rho_0(\psi, \psi_s) \to 0$) and *not growing with time along the trajectory of the system:* $U_t \leqslant 0$. (See [44] and [41].)

For relativistically invariant equations and other reversible theories integrals of motion can serve as Liapunov's functional.

The instability theorem (after Chetaev-Movchan) is as follows. In the above stability theorem we have to replace Liapunov's functional $U[\psi]$ by the Chetaev functional with the same properties as above except that there is at least one initial perturbation, small with respect to ρ_0, but $U_t > 0$ along the trajectory starting from this perturbed state.

These two theorems allow an analogue to be formulated of the Derrick-Hobart theorem applied to stationary solutions [41]:

Stationary particle-like solutions ψ_s are not energetically stable for they do not minimize the energy functional. However, these, can minimize the free energy functional

$$F = E - \omega Q \tag{10.20}$$

which appears for charge Q fixed, or

$$F = E - \omega N \tag{10.21}$$

for the normalization N fixed. In this case, ψ_s are conventionally stable. Taking as the Liapunov functionals the integrals F and defining the metrics ρ and ρ_0 by

$$\rho_0 = \|\psi(x,0) - \psi_s(x,0)\|_1, \tag{10.22}$$

$$\rho = \inf_{(\eta,\zeta)\in\mathbf{R}^2} \|\psi(x,t) - e^{i\eta}\psi_s(x-\zeta,t)\|_1,$$

we render the Liapunov theorem into the Q-theorem. Recall the norm $\|...\|$ to be the conventional norm in Sobolev space

$$\|f\|_s^2 = \sum_{j=0}^{2} \int_{-\infty}^{\infty} |f^{(j)}|^2 d^D x, \tag{10.23}$$

where $f \in L_2$ and $f^{(j)} = d^{(j)}f/dx^j$.

Taking

$$\rho = \inf_{\psi_s \in T} \|\psi(x,t) - \psi_s(x,t)\|_1, \tag{10.24}$$

with T being the orbit of the symmetry group one can extend the Q-theorem to vector fields ψ_a, $a = \overline{1,n}$ when the Lagrangian of the system is invariant under the action of some compact symmetry group (see, for example, [45]). In this case, we have instead of (19)

$$\omega_\alpha \omega_\beta \partial Q_\beta / \partial \omega_\alpha < 0$$

(condition (18) naturally holds). Let us list the conditions of Q-theorem applicability.

(1) It does not hold when regular solutions exist at discrete values of ω_α.

(2) If there are gauge fields the system should be studied from the beginning.

(3) In considering the stability of soliton-like solutions in the condensate, the

charge functional Q is not continuous over $\delta\varphi$ and Q-stability coincides with the absolute one. To stabilize soliton-like solutions at least one more *good* (i.e. continuous) integral must exist.

Just such a situation takes place in the preceding section in the case of bubble stability.

(4) Q-theorem (and other methods) using the energy integral to construct Liapunov's functional, is not valid for a system having noncompact internal symmetry, in particular for $U(p,q)$ nonlinear Schrödinger equation. As an example, one can consider the Hamiltonian

$$H = \frac{1}{2m}(p_1^2 - p_2^2) + (q_1^2 - q_2^2).$$

(5) When $D \geqslant 2$, Q-theorem is not applicable to nodal solutions. The following comments can be made:

(1) That regular solutions exist for discrete ω_s' means that the functional Q is not a continuous function of ω and (20) does not make sense.

(2) When proving the Q-theorem [40,41], a specific dependence F of ω is employed which can alter for nonzero gauge fields.

(3) The linear, with respect to $\delta\varphi$, functional

$$\delta Q = Q[\delta\varphi] = \omega \int \psi_s \delta\varphi d^D x$$

is no longer continuous at $\psi_s \to \psi_{\text{cond.}}$ since $\psi_s \notin L_2$. As a result (see [46]) ker $Q[\delta\varphi]$ is not closed and everywhere dense in L_2. The latter implies that for any $\tilde{\delta\varphi} \in L_2$ there exists a function $\delta\varphi$, satisfying the condition $Q[\delta\varphi] = 0$ (charge fixing), and quite close to $\tilde{\delta\varphi}$ with respect to the metric ρ.

Thus the Q-stability and the conditionless stability are here the same.

(4) An arbitrary solution to the equations considered (included stationary) does not provide the energy of the system $E[\psi]$ even with a local minimum [47]. It is impossible to minimize $E[\psi]$ via the fixing of additional constraints such as charge conservation. Therefore for those systems with a sign indefinite kinetic term the stability criterion, obtained via the spectral analysis of the linearized equation, differs essentially from the condition of energy minimizing. They are usually the same, however, for the models with a sign definite metric [48] (if gyroscopic forces are absent). A more detailed discussion of this topic can be found in [47].

(5) This concerns the stability of nodal solutions in two and three dimensions. Apart from nodeless particle-like solutions there exist a family of oscillating solutions labelled with the number of nodes $n = 0, 1, \ldots$. Even in the case of the nodeless $(n = 0)$ solution being Q-stable, solutions with $n \geqslant 1$ are unstable.

Summarizing the above results on stability one can conclude that in the general case the fact that the energy functional $E[\psi]$ is bound from below is neither sufficient nor necessary for particle-like solutions to be stable. Only for the 'compact' systems does such a condition become necessary (upon choosing an appropriate sign of energy), and for solutions, realizing this minimum, also sufficient (up to transformations of symmetry $\psi \in T$).

We apply the Q-theorem, for example, to the complex Schrödinger and Klein-Gordon equations with a nonlinear term of the form $\phi|\phi|^{\nu}$ (see, for example, Ref. [50])

$$i\phi_t + \phi_{xx} + |\phi|^{\nu}\phi = 0, \tag{10.25a}$$

$$(\Box + 1)\phi - |\phi|^{\nu}\phi = 0. \tag{10.25b}$$

In the first case, instead of the charge Q normalization of the wave function ϕ is conserved

$$\frac{d}{dt}\int |\phi|^2 dx \equiv \frac{dN}{dt} = 0,$$

and (see [19])

$$\frac{dN}{d\omega} < 0,$$

where $\omega < 0$. Substitution of $\phi(x,t) = \chi(x)e^{-i\omega t}$ into eqns. (25a,b) yields

$$-\chi_{xx} + \kappa^2\chi - \chi^{\nu+1} = 0, \quad \kappa^2 = \begin{cases} -\omega & (a), \\ 1 - \omega^2 & (b). \end{cases} \tag{10.26}$$

It can be shown that the conditions for applicability of the Q-theorem hold for eqn. (26). The ω dependence of N and Q remains to be found, and we find this by the scale transformation $x \to \kappa^{-1}\kappa$, $\chi \to \kappa^{2/\nu}\eta$ for which eqn. (26) has a form without $\varkappa$

$$-\eta_{\xi\xi} + \eta - \eta^{\nu+1} = 0.$$

Therefore

$$N = \int \chi^2 dx = \kappa^{(4-\nu)/\nu} \int \eta^2 d\xi \equiv \kappa^{(4-\nu)/\nu} C(\nu),$$

$$Q = \omega N = \omega \kappa^{(4-\nu)/\nu} C(\nu).$$

Calculating $dN/d\omega = -dN/d\kappa^2$ and $dQ/d\omega$, we get the corresponding regions of stability:

$$\nu < 4, \quad \omega = -\kappa^2 \text{ is arbitrary,} \tag{10.27a}$$

$$1 > \omega^2 > \frac{\nu}{4}. \tag{10.27b}$$

Thus, in both cases, stable soliton-like solutions exist at $\nu < 4$.

Earlier, in Chapter 8, we found a class of one-soliton solutions to the vector $S3$ version. Similar solutions obviously exist in the frame of the nonlinear Klein-Gordon equation. This can be established by using the technique developed in Chapter 8. Among these solutions there is the nodeless one ψ_0 possessing the lowest energy (completely degenerate). It is easy to verify that the Q-theorem conditions are satisfied [51] leading to its stability.

The opposite picture arises for non-degenerate vector solitons ψ_n with an energy higher then that of ψ_0, and a wave function having one or more nodes. The stability problem in such a case is not strictly solved, but one can expect instability since the solution ψ_n is a particular case of the n-soliton solution with zeros of the $S_{11}(\lambda)$ matrix element (the scattering matrix) specially located: these lie on a line parallel to the imaginary axis with constant spacing between them (i.e. with the same $\mathrm{Re}\lambda_i$). An initial condition, having moved one of those zeros from this line, gives rise to the decay of the soliton studied. Such an instability occurs during the evolution of two-soliton equation (bion-like) in the frame of the $S3$ nonlinear Schrödinger equation whose zeros lie on the line parallel to the $\mathrm{Im}\lambda$ axis. In [35] the bion

$$\psi(x,t) = 4A \frac{\mathrm{ch}3Ax + 3\mathrm{ch}Ax \cdot \exp(4iA^2 t)}{\mathrm{ch}4Ax + 4\mathrm{ch}2Ax + 3\cos 4A^2 t} \exp(i\frac{A^2}{2}t)$$

was investigated numerically, and, as we have already seen, decays into two constituent solitons under the action of an initial perturbation with an asymmetric imaginary part.

3. Structural stability

The notion, and features of the structural stability of soliton-like solutions and solitons, will be discussed in Chapter 11. In this section, we give only a very brief outline of some numerical investigations of this topic, and will raise the problem of soliton shape conservation under external perturbations.

From the early studies in this direction, there were the KdV and $S3$ soliton investigations with allowance for weak dissipation effects [52,53], linear and non-linear [54]. In Refs. [52] equations have been obtained via the Bogolubov two-time formalism which describe the time evolution of the KdV soliton parameters on the assumption that their form (functional dependence) does not change for various models of the damping*⁾. In Refs. [53] and [54] the influence of dissipation on the $S3$ solitons is studied numerically.

The already-mentioned initial investigations have shown that the soliton behaviour, (namely the conservation of its shape as a balance of dispersion and nonlinearity effects), depends very essentially on the type of perturbing term. This has been especially clearly observed in Pereira's paper [55], in which the behaviour of the $S3$ soliton has been studied for damping of power type $\gamma \propto \epsilon k^b$ (where k is the wave number). The author has shown that only for $b = 2$ (this is related to the $S3$ scale properties) does the soliton practically not change its shape with time. In other cases $(b \neq 2)$ soliton evolution takes place with a change of shape and this change increases with the coefficient ϵ in the damping term.

So, for $b = 4,3$ and 2 the inequalities $\epsilon \leqslant 0.01$, $\epsilon \leqslant 0.03$, $\epsilon \simeq 0.2$ have to hold in order to be able to neglect the change of soliton shape. In the KdV case the situation is more complicated. Even for weak perturbations of the KdV or $S3$ equations, solitons usually acquire tails (see, for example, Refs. [54] and [56]) or can change their shape [54,57]. We consider in detail the very interesting results obtained in this direction by the Japanese group [54].

In these papers the influence of Landau nonlinear damping on the dynamics of the $S3$ solitons of bell and kink types is studied, i.e. the initial value problem

$$F(\phi,\epsilon) \equiv i\phi_t + \frac{1}{2}\phi_{xx} + \alpha|\phi|^2\phi + \epsilon\phi P\int_{-\infty}^{\infty} \frac{|\phi(x',t)|^2}{x-x'}dx' = 0, \tag{10.28}$$

*⁾Later on, this method has been widely used in other papers (see, for example, Ref. [29b]).

$$\phi(x,0) = f(x),$$

(where ϵ is a constant and $\alpha = \pm 1$) is investigated. The function $f(x)$ is chosen to represent a soliton, a bound state of two solitons or a periodic wave. At $\alpha = 1$, as a result of modulational instability, bell solitons arise, at $\alpha = -1$ soliton solutions to $F(\phi,0) = 0$ are kinks of tanh type.

As computer experiments have shown, the presence of the integral term (perturbation) in eqn. (28) leads to very important qualitative effects: in the case of one soliton, $f(x) = A$ sech Ax, to its motion in the positive direction of the x-axis and the appearance of a tail behind (see Fig. 10.2), however, these effects are smaller, the smaller ϵ;

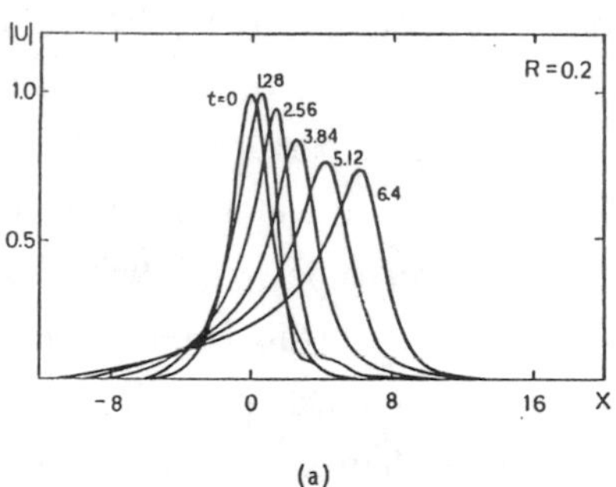

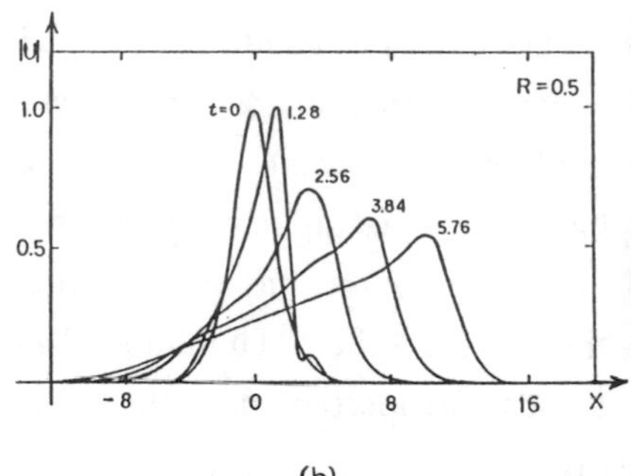

Figure 10.2. Instability of bell solutions under the action of the perturbation (10.28).

for the soliton bound state, $f(x) = 2A$ sech Ax, the perturbation leads to decay into constituent solitons. Perhaps this effect is related to the x-asymmetry of the perturbation in eqn. (28). In fact an analogous decay of the bound state (bion) had been observed earlier [55], even in the framework of the unperturbed integrable $S3$ equation in the presence of an initial perturbation with an asymmetric imaginary part, (unfortunately the perturbation amplitude was large).

Nonlinear Landau damping has a still greater effect upon kinks solitons ($\alpha = -1$), $\phi_k = A\,\mathrm{th}(Ax)\exp(i(vx - \omega t))$ ('dark' solitons in the witty terminology of the authors). Depending on the direction of kink motion its amplitude either decreases ($v<0$) or grows ($v>0$). In this connection the periodic waves $\phi(x,0) = 1 + A\cos\pi x$ break up asymmetrically into a series of (1) damping bell solitons (at $\alpha = 1$) or (2) growing 'dark' solitons (at $\alpha = 1$).

Ending this short section, we should say that the analytical study of the structural stability of solitons is mostly advanced for constant and localized perturbations, viz. $I[\varphi]\sim\delta(x-x_0)$. These results will be given in more detail in the next chapter.

References

1. R.K. Dodd, et al. see [34] part I.
2. Ph.M. Morse and H. Feshbach. *Methods of theoretical physics.* McGraw-Hill N.Y., London 1953, V, I and II.
3. I.M. Krichever. *On rational solutions of KP equation and integrable N-particle systems on line.* Funk. Anal. Priloz. 1978, 12, p. 76-78.
4. I.M. Krichever. *On rational solutions to Zakharov-Shabat equations for complete integrable N-particle systems on line.* Zap. Nauch. Semin. LOMI, 1979, 84, p. 117-130.
5. I.V. Cherednik. Funk. Anal. Priloz. 1978, 12, p. 45-52.
6. V.M. Eleonskii, I.M. Krichever and I.M. Kulagin. *Rational multi-soliton solutions of NLS.* Doklady AN 1986, 287, p. 606-610.
7. I.M. Krichever. *Spectral theory of 'finite-band' time-dependent Schrödinger operators. Non-stationary Peierls model.* Funk. Anal. Priloz. 1986, 20, p. 42-54. See [89] part I.
8. V. Makhankov, R. Myrzakulov and Yu.V. Katyshev. *Vector generalization of the system of equations for interacting h.f. and l.f. waves.* Preprint JINR P17-86-94, Dubna 1986. Teor. Mat. Fiz. 1987, 72, p. 22-34.
9. V.G. Makhankov. *On stationary solutions of Schrödinger equation with the selfconsistent potential satisfying the Boussinesq equation.* Phys. Lett. 1974, 50A, p. 42-44.
10. V. Makhankov. *Stationary solutions of coupled Schrödinger and Boussinesq equations and dynamics of Langmuir waves.* Preprint JINR E5-8390, Dubna 1974.
11. K. Nishikawa, et al. see [39] part I.
12. Ya. Bogomolov, I.A. Kolchugina, A.G. Litvak and A.M. Sergeev. *Near-sonic Langmuir solitons.* Phys. Lett. 1982, 91A, p. 447-450.
13. H. Kramer, E.W. Laedke and K.H. Spatschek. *Transition from standing to*

sonic Langmuir solitons. Phys. Rev. Lett. 1984, 52, p. 1226-1229.

14. V. Makhankov. See [7] part I.
15. K. Mima, K. Kato and K. Nishikawa. *Self-consistent stationary density cavities with many bounded Langmuir waves.* J. Phys. Soc. Jpn. 1977, 42, p. 290-296.
16. H. Hojo and K. Nishikawa. *On bounded Langmuir waves in self-consistent density cavity.* J. Phys. Soc. Jpn. 1977, 42, p. 1437-1438.
17. I.M. Krichever, et al. see [17] part III.
18. See, e.g. A.C. Newell. *The inverse scattering transform,* on R. Bullough and P. Caudrey eds. Solitons. Springer, Heidelberg 1980, p. 177-242, formula (6.170) p. 215.
19. R. Finkelstein, R. Levelier and M. Ruderman. Phys. Rev. 1951, 83, p. 326- .

    R. Finkelstein, C. Fronsdal and P. Kaus, ibid, 1956, 103, p. 1571- .
20. V.G. Makhankov. *On the existence of non-one-dimensional soliton-like solutions for some field theories.* Phys. Lett. 1977, 61A, p. 431-433.
21. I.V. Amirkhanov and E.P. Zhidkov. *Existence of positively-defined particle like solution of nonlinear equation of $\phi^4-\phi^6$ theory.* JINR P5-82-246, Dubna 1982.
22. V.G. Makhankov and P.E. Zhidkov. *On the existence of static hole-like solutions in a nonlinear Bose gas model.* Comm. of JINR P5-86-341, Dubna 1986.
23. R. Friedberg, T.D. Lee and A. Sirlin. *Class of scalar-field soliton solution in three space-dimensions.* Phys. Rev. 1976, 12D, p. 2739-2761.
24. L. Nehari. Proc. Roy. Irish Acad. 1963, 62A, p. 118- .
25. E.P. Shidkov and V.P. Shirikov. *On a boundary problem for the second order ordinary differential equation.* Z. Vych. Mat. Mat. Fiz. 1964, 4, p. 804- .
26. A. Scott, F. Chu and D. McLaughlin. *The soliton: A new concept in applied science.* Proc. IEEE 1973, 61, p. 1443-1483.
27. E. Laedke and K.H. Spatschek. *Nonlinear stability of envelope solitons.* Phys. Rev. Lett. 1978, 41, p. 1798-1801.

    Exact stability criteria for finite-amplitude solitons. Phys. Rev. Lett. 1972, 42 p. 1534-1537.
28. L.D. Faddeev. *In search for many-dimensional solitons,* in Proc. Non-loca

Non-linear and Non-renormalized Field Theories, Alushta 1976. JINR publishing D2-9788, Dubna 1976. p. 207-223.

A.S. Schvartz. *Topologically non-trivial solutions of classical equations and their role in QFT.* ibid p. 224-240.

29. a) E. Ott and R. Sudan. *Nonlinear theory of ion-acoustic waves with Landau damping.* Phys. Fluids 1969, 12, p. 3288-3294.

    b) J. Keener and D. McLaughlin. *Solitons under perturbations.* Phys. Rev. 1977, 16A, p. 777-790.

    c) M. Fogel, S. Trullinger, A. Bishop and J. Krumhansl. *Classical particle-like behavior of SG solitons in scattering potentials and applied fields.* Phys. Rev. Lett. 1976, 36, p. 1411-1414.

    d) Y. Ichikawa. *Topics on solitons in plasmas.* Phys. Scripta 1979, 20, p. 296-305.

    e) V.I. Karpman. *Soliton evolution in the presence of perturbation.* ibid. p. 462-478 and references therein.

    f) A. Bondeson, M. Lisak and D. Anderson. *Soliton perturbations. A variational principle for the soliton parameters.* ibid, p. 479-485 and Stability analysis of lower-hybrid cones. p. 343-345.

30. V.K. Fedyanin and V. Makhankov. *Soliton-like solutions in one-dimensional systems with resonance interaction.* Phys. Scripta 1979, 20, p. 552-557.

31. V Arnold. Sov. Math. Surveys 1963, 18, p. 9-36.

    J. Moser. *Nearly integrable and integrable systems.* Preprint N.Y. 1976 and Lectures on celestial mechanics. Springer, N.Y. 1971 (with C. Siegel).

32. V. Zakharov and A. Shabat. *Exact theory of two-dimensional self-focusing and one-dimensional self-modulation of waves in nonlinear media.* See [37] part II.

33. V. Makhankov, O.K. Pashaev and S. Sergeenkov. *Colour solutions in a stable medium. See [15] part III.*

34. J. Berryman. *Stability of solitary waves in shallow water.* Phys. Fluids 1976, 19, p. 771-777.

35. J. Satsuma and N. Yajima. *Initial value problems of one-dimensional self-modulation of nonlinear waves in dispersive media.* Suppl. Progr. Theor. Phys. 1974, 55, p. 284-306.

36. L.G Zastavenko. *Particle-like solutions of nonlinear wave equation.* Prik Mat. Mekh. 1965, 29 p. 430-439.

37. I.V. Barashenkov, A.D. Gocheva, V. Makhankov and I.V. Pusynin. *On stability of solitary waves of the* $\phi^4-\phi^6$ *NLS under non-vanishing boundary conditions.* JINR E17-85-967, Dubna 1985. See [41] part III.

38. T.B. Benjamin. *Stability of solitary waves.* Proc. Roy. Soc. Lond. 1972, 238, p. 153-183.

39. D. Anderson. *Stability of time-dependent particle-like solutions in nonlinear field theories. II.* J. Math. Phys. 1971, 12, p. 945-952.

    D. Anderson and G. Derrick. *Stability of time-dependent ...* part I. ibid, 1970, 11, p. 1336-1346.

40 a) V.G. Makhankov. *On stability of 'charged' solitons in the framework of Klein-Gordon equation with a saturable nonlinearity.* JINR Commun. P2-10362, Dubna, 1977.

    b) R. Friedberg, T.D. Lee and A. Sirlin. *Class of field soliton solutions in 3-D space.* Phys. Rev. 1976, 13D, p. 2739-2761. See [23] this part.

    c) Yu.P. Rybakov. *Conditional stability of regular solutions of nonlinear field theory.* In Problems of Gravity and Particles Theory. Atomizdat, Moscow 1979, p. 194-202.

    d) N.G. Vakhitov and A.A. Kolokolov. See [53b] part I.

41. Yu.P. Rybakov. *Stability of solitons.* In Problems of Gravity and Particles Theory, issue 16. Energoatomizdat, Moscow 1985.

42. T. Yoshizawa. *On the stability of solutions of a system of the differential equations.* Mem. Coll. Sci. Univ. Kyoto Math. ser. 1955, 24A, No. 1.

    A.M. SLobodkin. *On stability of conservative system equilibrium with infinite number of degrees of freedom.* Prikl. Mat. Mekh. 1962, 26, p. 356-358.

43. T. Cazenave and P.L. Lions. *Orbital stability of standing waves for some NLS.* Comm. Math. Phys. 1982, 85, p. 549-561.

44. A.A. Movchan. *Stability of processes with respect two metrics.* Pricl. Mat. Mekh. 1960, 24, p. 988-1001.

45. Yu.P. Rybakov. *On stability of multi-charged solitons.* In Problems of Gravity and Particles Theory, issue 14. Energoizdat Moscow, 1983.

46. L. Shvartz. *Analyse Mathématique.* Hermann, Paris 1967. Vol. I, ch. III.

47. I.V. Barashenkov. *Stability of solutions to nonlinear models possessing a sign-undefined metric.* Acta Phys. Austriaca 1983, 55, p. 155-165.

48. R. Jackiv and P. Rossi. *Stability and bifurcation in Yang-Mills theory.* Phys. Rev. 1980, 21D, p. 426-445.

49. A. Kummar, V. Nisichenko and Yu. Rybakov. *Stability of charged solitons.* Int. J. of Ther. Phys. 1979, 18, p. 425-432.

50. V. Katyshev, N. Makhaldiani and V. Makhankov. *On stability of soliton solutions to the NLS with nonlinearity* $|\psi|^{\nu}\psi$. Phys. Lett. 1978, 66A, p. 456-458.

51. I.V. Barashenkov. *Stability of 'coloured' solitons.* JINR Comm. P2-82-376, Dubna 1982.

52. E. Ott and R. Sudan. See [29a] this part.

53. A. Hasegawa and F. Tappert. *Transition of stationary nonlinear optical pulses in dispersive dielectric fibres. I and II.* Appl. Phys. Lett. 1973, 23, p. 142-144 and p. 171-172.

54. N. Yajima, M. Oikawa, J. Satsuma and C. Namba. Res. Inst. Appl. Phys. Repts. XXII, 1975, 70, p. 89.

55. N. Pereira. *Soliton in the dampled NLS.* Phys. Fluids 1977, 20, p. 1735-1743.

56. a) J. Fernandes and G. Reinish. See [88] part I.

b) J. Fernandes, G. Reinish, A. Bondeson and J. Weiland. *Collapse of a KdV soliton into a weak noise shelf.* Phys. Lett. 1978, 66A, p. 175-178.

c) S. Watanabe. *Soliton and generation of tail in nonlinear dispersive media with weak dissipation.* J. Phys. Soc. Jpn. 1978, 45, p. 276-282.

57. B. Cohen, K. Watson and B.J. West. *Some properties of deep water solitons.* Phys. Fluids 1976, 19, p. 345-354.

58. B.A. Dubrovin et al. See [17] part III, Sov. J. Particles & Nuclei. 1988, 19, p. 252-269.

59. V.G. Makhankov and Kh.T. Kholmurodov. *Numerical study of stability of vector* $U(2)$ *solitons.* JINR Rapid Communications. No. 5 [25]-87. Dubna, 1987.

60. I.V. Barashenkov and Kh. Kholmurodov. *Bose-gas with two- and three-body interactions: evolution of the unstable 'bubbles'.* JINR Communication, P17-86-698, Dubna, 1986 (in Russian).

Part V. Phenomenology of $D = 1$ Solitons

Chapter 11.

DYNAMICS OF THE FORMATION AND INTERACTION OF PLANE SOLITONS

This problem has been studied most extensively and consistently in the framework of completely-integrable equations where one often succeeds with the analytical solution of the Cauchy problem by means of the inverse method*). Note, once more, that these investigations were initiated by computer experiments on the Fermi-Pasta-Ulam problem and on the KdV, Sine-Gordon and $S3$ solitons. Recall that by analogy with mechanics one calls an equation integrable when it has an infinite set of (additive) integrals of motion. These integrals are constructed from powers of the field function and its derivatives and must be in involution. The first results obtained from investigations of the completely-integrable systems led to the conclusion that in their framework the solitary-wave interaction is reduced to only a shift in position and phase, leaving the wave shape and velocity unchanged. Thus (and hence the name) the soliton appeared to be a stable object like Kipling's 'cat that walked by himself'. This was indeed the case as long as one considered fields with a symmetry group not higher than $U(1)$.

In the beginning of 1978 a paper was published which showed that in the framework of an integrable system (the model of the main chiral field on the group $SU(N)$) the decay and mutual transformation of solitons are possible for $N \geqslant 3$ [11]. Thus it was shown that the original definition of the soliton concept was very narrow even on the set of integrable systems.

Consider, now, numerical experiments on soliton dynamics. With the exception of multi-soliton formulae the dynamics of soliton formation from arbitrary initial packets is known only asymptotically at $t \to \infty$ even for integrable systems, and,

*) In recent years an enormous number of papers (more than 1000) have appeared in this area. Also highly detailed reviews have been published [1-5] and books [6-10].

more particularly, in the framework of perturbed equations.

1. Computational procedures

First, a few words about numerical investigation techniques. As has been excellently formulated in the review report by Eilbeck [12] 'in the field of partial differential equations, numerical analysis is still almost as much an art as a science. For any special equation, the question of the 'best' numerical method is an extremely complicated one'. It is connected with many factors, in particular, the calculation accuracy required, the limitations of computer time and storage space, machine word length and so on. Moreover, very seldom may one find in the literature a comparison of numerical algorithms of different authors*) as well as detailed investigations of each given algorithm. Especially this is true for nonlinear equations. Therefore much further work is required both in theoretical studies of applied methods and in the analysis of their practical use. Henceforth, where possible, we shall dwell briefly on computational techniques.

Generally one uses two basic approaches to reduce the problem to one involving only a finite number of parameters: the function-approximation approach and the finite-difference approach.

In the first case, the exact solution $y(x,t)$ is approximated by an expression defined on a finite-dimensional subspace, e.g.

$$y(x,t) \simeq \tilde{y}(x,t) = \sum_{i=1}^{n} c_i(t) y_i(x),$$

where the $y_i(x)$ are appropriately-chosen basis functions (for more details see Ref. [12] and the references cited therein).

More frequently, the second approach is used to study partial differential equations where the exact solution $y(x,t)$ is approximated by the set of y_m^n defined at rectangular grid points in the x,t plane (or in the x,y,t cube) with steps h and τ, respectively, i.e. $x_m = hm$, $t_n = \tau n$. Various types of calculation scheme are used to find the time derivative, and these may be explicit or implicit. For this method of calculation whole science exists (we refer here only to the books [14]).

*) Exception is apparently a series of works by Ablowitz and Taha [13].

Nevertheless, in the study of *nonlinear* partial differential equations final recipes do not exist. Note only that the accuracy of the calculation sometimes depends essentially on the type of approximation of the nonlinear term: whether one takes the values at the point x_m, i.e., $F(y^n_m)$, or the half-sum at symmetric points $F(1/2\{y^n_{m+1}+y^n_{m-1}\})$ in x or in t, namely $F(1/2\{y^{n+1}_m+y^{n-1}_m\})$ and so on (see again Ref. [12]). We emphasize also that all methods used must obey the stability condition, i.e. $|y(x_m,t_n)-y^n_m| \leqslant N$ at $n \to \infty$. This condition together with the approximation-accuracy condition usually imposes limits both on the relation of the steps h and τ and on their values. There is a highly-promising approach in which one uses the coordinate Fourier transform of the solution sought for, and then solves the resulting ordinary differential equations by known difference schemes. By an inverse Fourier Transform (usually the so-called Fast Fourier Transform algorithm [15] is used) one gets the result sought for. We note here only three soliton papers in which this method has been successfully applied [16-18].

Sometimes it is possible to use a certain modified combination of these approaches.

2. KdV-like equations

(1) Some of the first nonlinear equations which displayed unusual properties in computer experiments were the Korteweg-de Vries and Sine-Gordon equations and a finite-difference analogue of the Boussinesq equation (or the nonlinear string equation, Fermi-Pasta-Vlam problem, 1955)*)

$$\frac{d^2}{dt^2}y_m = f(y_{m+1}-y_m)-f(y_m-y_{m-1}) \tag{11.1}$$

where

$$a) \;\; f(y) = y+\alpha y^2 \;\text{ or }\; b) \;\; f(y) = y+\beta y^2.$$

The FPU problem is closely related to how close system considered is to an integrable one, and hence to solitons (see, e.g., Ref [1]). Therefore, in what follows

*)Equation (1) may be reduced to the KdV or Boussinesq equations when $f(y) = y+\alpha y^2+\beta y^2$ at the transition to the continuum analogue, $h \to 0$ (see Ref. [10]).

we restrict ourselves to a discussion of soliton properties.

From the numerical experiment point of view, soliton and soliton-like solution dynamics are practically the same, both in the framework of the KdV equation and the similar eqns. (2.24) and (2.26) for identical initial conditions, φ_0. Depending on the shape of the function φ_0*⁾, zero, one, two, or more, solitons and some oscillatory tail are produced. This property of the KdV type equations together with an identical dispersion behaviour at $k \to 0$ led to the conclusion that they are all integrable. As a result, analytical and computational effort was wasted in vain to prove this fact. Note that the decay of the initial state into solitons is a characteristic feature of any quasi-integrable system.

In the case of soliton interaction, the picture changes qualitatively. The KdV and MKdV solitons,

$$\varphi_{(KdV)} = A^2 \operatorname{sech}^2 \left[\frac{A}{\sqrt{12}} (x - vt - x_0) \right], \quad v = 1 + \frac{1}{3} A^2, \tag{11.2a}$$

$$\varphi_{(MKdV)} = A \operatorname{sech} \left[\frac{A}{\sqrt{6}} (x - vt - x_0) \right], \quad v = 1 + \frac{1}{6} A, \tag{11.2b}$$

being true, suffer only a shift in their position [18], meanwhile solitons generated by the rest of the KdV-like equations (e.g., (2.24) for $\nu \neq 1,2$ and (2.26)), interact inelastically (see Refs. [17] and [20]). Although, the inelasticity of such an interaction is usually small it grows with ν and becomes evident at $\nu = 4$. It is very clearly observed for an example of eqn. (2.26) at $\nu = 1,2$. Apparently, for the first time, the equation

$$\varphi_t + \varphi_x + \varphi\varphi_x - \varphi_{xxt} = 0, \tag{11.3}$$

which is sometimes called the PBBM (after Peregrine [22] and Benjamin, Bona and Mahony [23]) or Regularized Long-Wave (RLW) equation, frustrated the integrability illusion. The difference scheme for this equation has been studied by Eilbeck and McGuire [21] and Abdulloev, Bogolubsky and the author [20]. As is known, in the numerical integration of the KdV equation there is a rather stiff relationship between the grid steps in the t and x directions: $\tau \ll ch^3$, or more

*⁾Wide-spread opinion that the number of solitons generated depends strongly on the integral $\int \varphi_0 dx$ is erroneous. From the initial state with $\int \varphi_0 dx = 0$ can be generated as many solitons as from that with $\int \varphi_0 dx \neq 0$ (see [155]).

exactly $\tau/h^3 \ll (4+h^2|\varphi_0|)^{-1}$ or $\tau/h^3 \leqslant |2-h^2\varphi_0|^{-1}$ (see in detail the review by Eilbeck [12]). In any case, experience shows that to reveal subtle effects in the numerical integration of the KdV equation is not an easy matter. In the study of the regularized long-wave equation (2) the situation changes for the better, and the stability condition of the difference scheme becomes $\tau \ll h$ [20][*)]. Nevertheless, for this case, some awkward aspects arise. Thus, for example, after the computer start with the exact (analytical) one-soliton solution,

$$\varphi_s = A\,\mathrm{ch}^{-2}\left\{\sqrt{\frac{A}{2(2A+3)}}\,(x-(1+\frac{2}{3}A)t-x_0)\right\} \tag{11.4}$$

this solution changes slightly (parts of a percent): the soliton 'breathes' and even alledgedly radiates faintly for a short time. Then the picture stabilizes. By virtue of this property of the RLW equation it was necessary to devise a rather complicated subtraction procedure [20] to reveal such a delicate effect as soliton interaction inelasticity. This attempt was encouraged by the results obtained for the analogous modification of the Boussinesq equation. As has already been mentioned, soliton interaction in the framework of the Bq equation was studied numerically only by the two-soliton formula [24]. Direct calculations on the basis of the Bq equation turned out to be practically impossible due to the instability of the linear spectrum $\omega^2 = k^2(1-k^2)$. This compelled us to modify the equation so that the linear spectrum coincides with the Bq spectrum in the limit $k \to 0$ and nevertheless is stable. Ion-acoustic waves in plasmas have just such a spectrum $\omega^2 = k^2/(1+k^2)$ which corresponds to the equation

$$(\Box - \partial_t^2\partial_x^2)\varphi - \partial_x^2\varphi^2 = 0.$$

This equation we have proposed to call the improved Boussinesq equation (IBq)[**)]. The numerical experiment [25] has shown, that the IBq solitons interact inelastically (see Fig. 11.1). In this paper, the computational algorithm is discussed. In spite of the obvious resemblance of the IBq and the RLW equations the inelasticities of the interaction of their solutions are substantially different. This is explained by the fact that the RLW solitons only overtake each other, but the IBq solitons may experience head-on collisions. One prefers to study the latter

---

*)It is possible that the name of this equation is associated with just this property. This equation was also used by Hammack [26] to simulate tsunami.

**)Note that it was known to Boussinesq too [27].

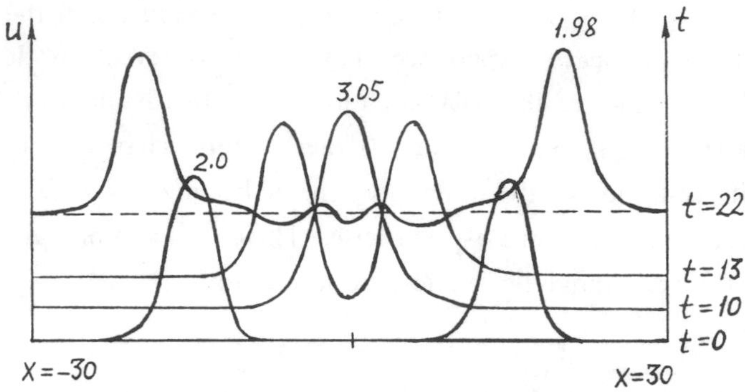

Figure 11.1. Dynamics of the nonelastic interaction of the improved version of the Boussinesq equation solitons ($A_1 = A_2 = 2$).

numerically due to their symmetry (leading to a saving of computer memory facility). Really, as was noted as early as 1975 in Refs. [28] and [29]. Soliton head-on collisions in the framework of various equations lead to a larger interaction inelasticity at the same soliton amplitudes than for unidirectional collisions. Therefore, in a nonintegrable system, the head-on collisions are mainly responsible for ergodization (for ion-acoustic waves in plasmas, see [28], for Langmuir waves see [29]).

Numerical experiments on KdV-like equations have revealed that quasi-soliton interaction inelasticity increases both with their amplitude and with the degree of nonlinearity in the $\nu > 1$ equation (e.g. for eqn. (1) at $\nu = 2.4$, see Ref. [17]).

Bona et al. [30] later derived a very perfect algorithm which permitted them to study the RLW soliton interaction dynamics in detail. Their results are in agreement with previous data [20,31]). In Fig. 11.2, taken from Ref. [30] a typical interaction of large amplitude solitons is shown.

Sometimes complex modifications of the KdV equation are considered (see, e.g. Refs. [17])*)

$$\psi_t + |\psi|^2\psi_x + \psi_{xxx} = 0, \tag{11.5a}$$

$$\psi_t + (|\psi|^2\psi)_x + \psi_{xxx} = 0. \tag{11.5b}$$

Equation (5a) is integrable as was shown by Hirota in 1972 via the inverse

*) Similar equations arise, for example, in plasma theory [32].

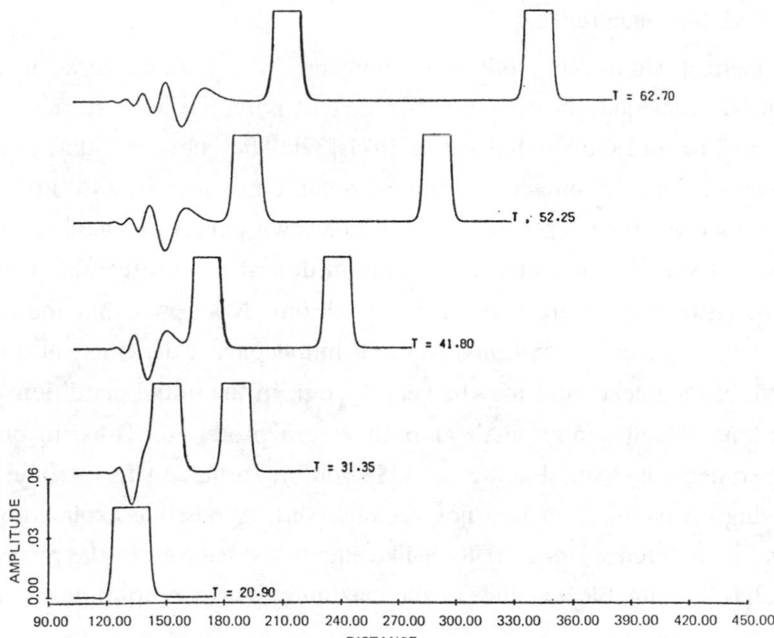

Figure 11.2. Interaction of solitons of the PBBM equation.

method [33]. Equation (5b), as was shown in [17,32] is nonintegrable, and the initial condition decay and quasi-soliton interaction have a rather complicated nature.

Also we, refer once more to a KdV-like equation which is sometimes called the Benjamin-Ono equation [34]

$$\varphi_t + 6\varphi\varphi_x + \frac{P}{\pi}\int_{-\infty}^{\infty} \frac{\varphi_{\xi\xi}}{\xi - x} d\xi = 0. \tag{11.6}$$

Numerical experiments by Meiss ans Pereira [35] indicated a soliton behaviour of travelling waves (Lorentzians),

$$\varphi_s = \frac{2}{3}\frac{v}{1+v^2(x-vt)^2},$$

both for collisions and an initial packet decay (multisoliton solutions have been found by Joseph [36])*).

*)The problem was completely solved in [40] where integrability of eqn. (6) was shown to

3. NSE-like equations

Numerical studies of soliton phenomena in the framework of the nonlinear Schrödinger equation with a self-consistent potential were started with the paper by Yajima and Outi published in 1971 [37]. They observed that envelope solitary waves of the Schrödinger equation with cubic nonlinearity ($S3$) interact elastically. This fact was then explained in the well-known papers of Zakharov and Shabat [6] where it was shown that the $S3$ equation describes an integrable system and solutions (2.3) at $\nu = 2$ are true envelope solitons. Note only that the formation time of the $S3$ soliton (or solitons) from an initial packet depends substantially on the form of the packet and may be very large if, in the initial condition, decay process solitons of sufficiently small amplitudes are produced. Thus, in one of the first experiments by Abdulloev et al. [38], soliton formation from some initial packet having Gaussian form was not yet observed. A possible explanation of this fact (and its difference from the KdV-like equations) is given in the review [39]. In Fig. 11.3(a), the time dependence of the maximum packet amplitude φ_{max} is given.

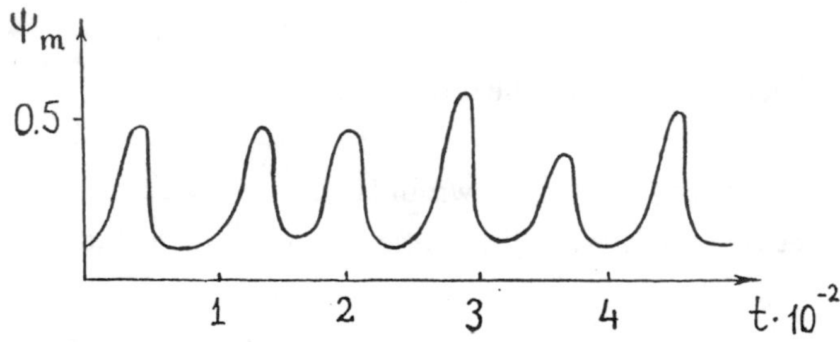

Figure 11.3a. Amplitude of a NSE wave packet as a function of time (at the point $x = 0$).

On the other hand, the fact that the initial packet does not decay into solitons for a long time may be related to the formation of a bound state of solitons, the bion (see in more detail Section 2.3).

The initial problem solution, i.e. the study of various initial packet decays in the framework of the $S3$ equation by the inverse method and the numerical experiment, is described in detail in the paper by Satsuma and Yajima [41].

The most comprehensive study of the Schrödinger equation (including numerical experiments) has been performed for the description of Langmuir waves and

be a particular case of the so-called 'nonlinear intermediate long-wave equation'.

turbulence in plasmas. Therefore we briefly note only the Langmuir soliton investigations (in other field of physics where this type of equation arises it is sufficiently only to change the terminology and notations).

The $S3$ equation

$$i\psi_t + \psi_{xx} + \alpha|\psi|^2\psi = 0 \tag{11.7}$$

has an enumerable set of integrals of motion and is integrable. In the theory of Langmuir turbulence it appears as a 'quasistatic limit' in the framework of Zakharov's system ($\alpha>0$)

$$i\psi_t + \psi_{xx} - \Phi\psi = 0, \tag{11.8a}$$

$$\Box\Phi = (|\psi|^2)_{xx}, \tag{11.8b}$$

at $\Phi_{tt} \to 0$ (here ψ is the complex amplitude of the Langmuir high-frequency field, Φ is the low-frequency variation of plasma density).

The initial value problem for this system as well as the interaction of two colliding solitons have been first investigated in [38]. Numerical experiments have shown that identical initial packets have absolutely different behaviours for the $S3$ equation and for the system (8).

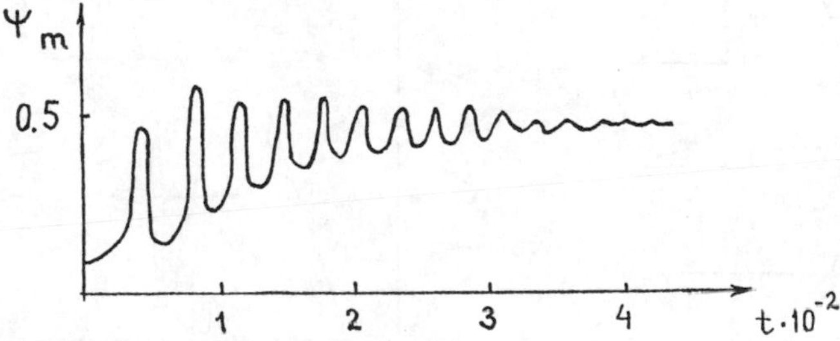

Figure 11.3b. Dependence of packet amplitude on time (in the origin) in the framework of Zakharov's system (11.8).

In Fig. 11.3(a) and (b) the results for such a calculation are shown for comparison: (a) corresponds to the $S3$ packet dynamics, (b) to the Zakharov system. In the second case, (see Fig. 11.3b taken from [42]), the quasi-soliton formation process proceeds sufficiently fast, an energy and momentum excess being taken away by the density waves Φ. A similar situation occurs in the quasi-soliton interaction too. The quasi-solitons of the system (8) display a rich picture both for interactions between them and with acoustic pulses of various shapes. These effects have been studied in much detail in [16,29,42]. In [29,42], elementary acts of interaction of solitons and acoustic pulses for the system (8) are simulated numerically, see Fig. 11.4.

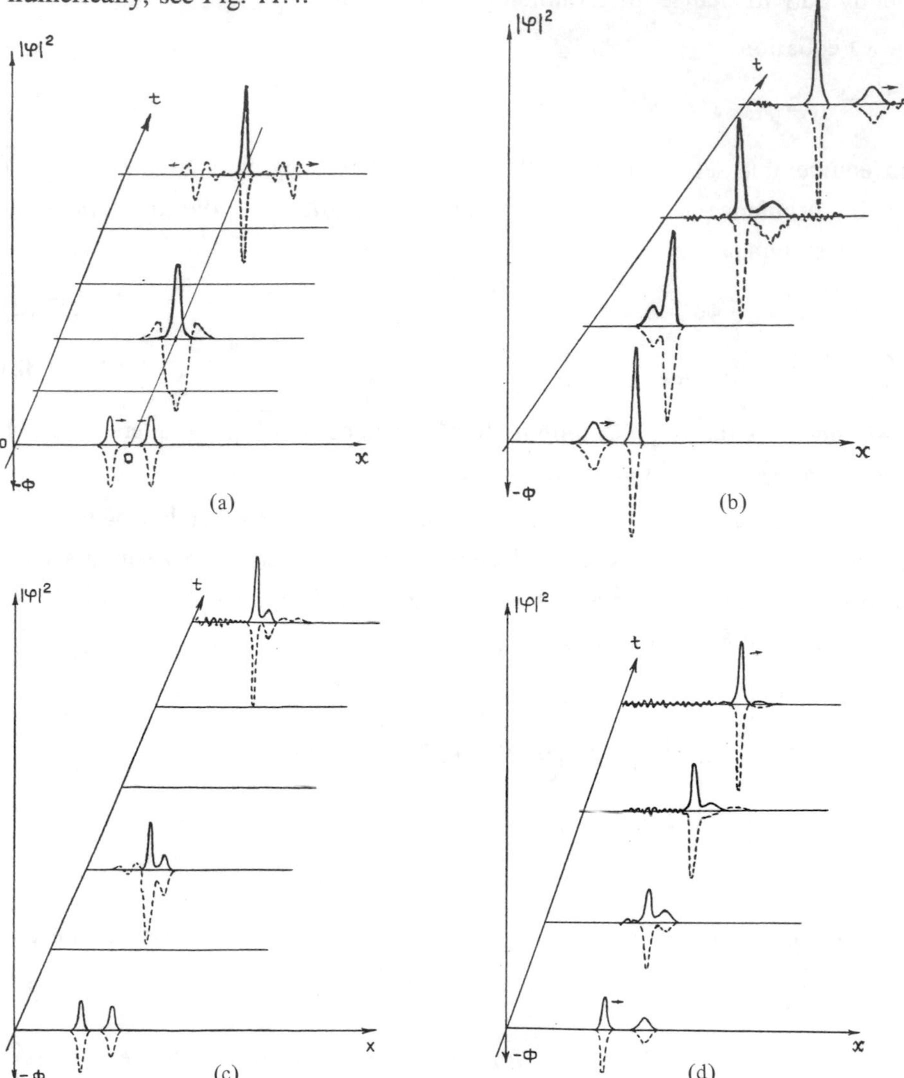

Figure 11.4a. Coalescence of two identical φ-solitons followed by Φ-wave radiation.

Figure 11.4b. Interaction of φ-solitons with different masses (a light soliton hits a heavy one).

Figure 11.4c. Interaction of solitons with approximately equal masses.

Figure 11.4d. Interaction of solitons of different masses.

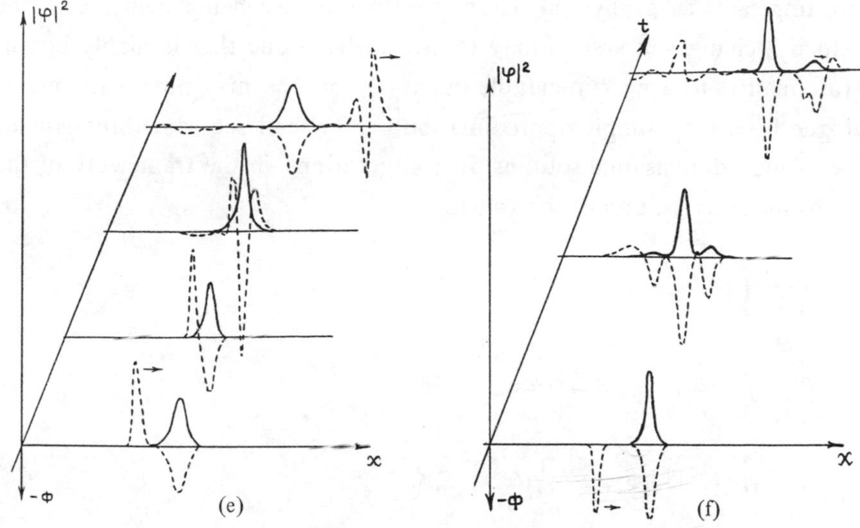

Figure 11.4e. Interaction of φ-soliton and Φ-pulse of compression.

Figure 11.4f. Interaction between φ-soliton and Φ-pulse of rarefraction.

On the basis of this an approximate kinetic equation for solitons is deduced and a solution is found describing the distribution of soliton widths [29] (see also Chapter 13).*) Moreover, it follows from the results of [42] that for a given amplitude of colliding solitons their interaction inelasticity decreases as their relative velocity Δv increases. As $\Delta v \to 2$ ($\Delta v = v_1 - v_2$, $v_1 = -v_2 \to 1$) the inelasticity becomes negligibly small. This result suggested to Yajima and Oikawa that the near-sonic system (8) may be completely integrable (history repeats itself). This has been splendidly proved by them in [43]. For this, as above, it is necessary to turn from eqn. (8b) to its right or left unidirectional version, i.e. to make the change $\square \to -2\partial_x(\partial_t + \partial_x)$ and to integrate once over x. Putting the integration constant equal to zero by virtue of the soliton conditions at infinity one obtains

$$(\partial_t + \partial_x)\Phi = --\partial_x |\psi|^2. \tag{11.8c}$$

As shown in [43], the system turns out to be integrable. Thus we come across a

*)Note that in these references the influence of phase relations on the soliton interaction is not studied.

very interesting fact: as a physical parameter (in the case being considered, the soliton velocity) changes, a system may transform from one that is highly nonintegrable (at small v) to a near-integrable one (v→1), so one may obtain its integrable analogue by a very simple approximation*). We shall see something similar in the case of multidimensional solitons. In computations, in the framework of the system (8a,b) the conservation of the values

$$N = \int_{-\infty}^{\infty} |\psi|^2 dx,$$

$$P = \int_{-\infty}^{\infty} \{i(\psi\bar{\psi}_x - \bar{\psi}\psi_x) + 2u\Phi\} dx,$$

$$H = \int_{-\infty}^{\infty} \{|\psi_x|^2 + \Phi|\psi|^2 + \frac{1}{2}(\Phi^2 + u^2)\} dx,$$

and $\Phi_t + u_x = 0$, which are, respectively, the particle number, momentum and energy of the system, is checked.

To complete the Schrödinger quasi-soliton topic, note a very interesting and, in a certain sense, paradoxical result. As we know, the near-sound solitons must be described by the system of the NSE with (1.70) or (1.74) at $(1-v^2) \to m_e/m_i$. In [45] an attempt is made to study an interaction of two colliding 'camel' solitons in the framework of the system of the Schrödinger and IBq equations (SIBq system)

$$i\psi_t + \psi_{xx} - \psi\Phi = 0,$$

$$(\Box - \frac{\epsilon}{3}\partial_x^2\partial_t^2)\Phi - \epsilon\partial_x^2(\Phi^2) = \partial_x^2(|\psi|^2).$$

From the results presented in Fig. 11.5, it seen that, unlike the Zakharov-system solitons, the two-humped near-sound solitons interact strongly both with each other and with acoustic pulses.

The IBq equation is not integrable, therefore it is very interesting to study a unidirectional interaction of the 'camel' solitons in the framework of the Nishikawa-Hojo-Mima-Ikezi (NHMI) system and (1.71). Without the terms $\beta(\Phi^2)_x$ and $\alpha\Phi_{xxx}$, the NHMI system turns into the integrable Yajima-Oikawa system. Is

*)We emphasize that the initial system (8a,b) also remains nonintegrable in the region v→1.

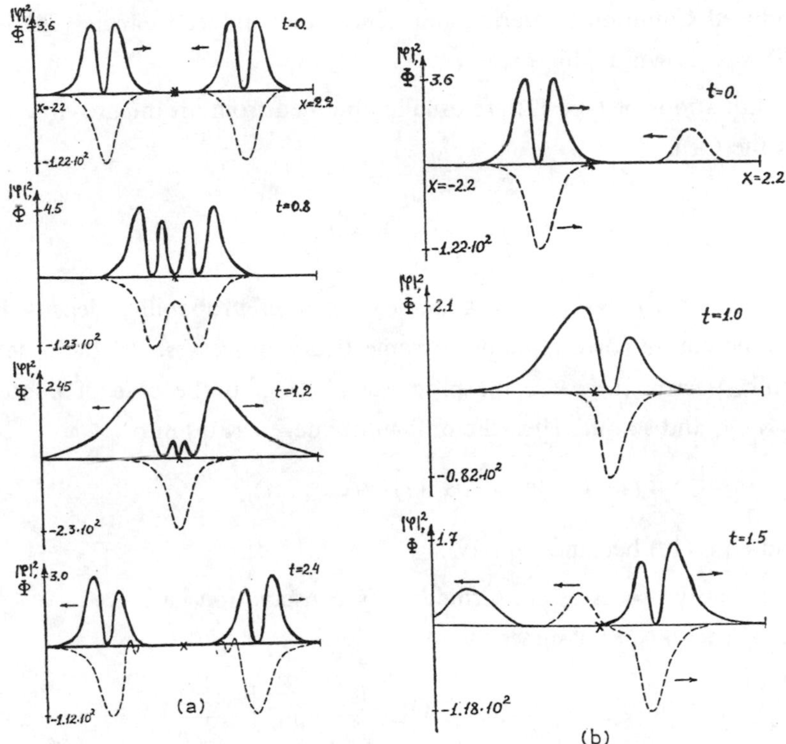

Figure 11.5a. Head-on collision of *SIBq* solitons;

Figure 11.5b. Interaction between an *SIBq* soliton and a compressional Φ-soliton.

the NHMI system consisting of two coupled integrable equations integrable for certain values of α and β? The results of [42] yield a disappointing answer, as does that of the analytical study by Benilov and Burtsev [45].

4. Equation for induced processes

We consider the equation

$$\frac{d}{dt}y_n = y_n(y_{n+1} - y_{n-1}). \tag{11.9}$$

After some approximations, systems of the form (9) arise in investigations of (a) the structure of Langmuir oscillation spectra in weakly turbulent plasmas, (b) interactions of populations in biology (the feeder-eater approximation), and (c) induced Compton scattering and many other induced effects. The integrability of (9) was shown in [46].

Equations of type (9) are usually obtained from an integro-differential equation of the form

$$\dot{y}(x,t) = y(x,t)\int_0^\infty dx' W(x,x')y(x',t), \tag{11.10}$$

where $W(x,x') = -W(x',x)$, and the transition probability density has resonance denominators corresponding to some decay processes. In the case of the Langmuir spectra, these are the processes $l \to l+s$, in the case of photons $t \to t+l$ or $t \to t+s$, and so on. Then the probability degenerates into

$$W(x,x') = W_0[\delta(x-x'+\kappa)-\delta(x-x'-\kappa)]$$

and eqn. (10) becomes eqn. (9).

If decay processes are forbidden by conservation laws $\omega_1+\omega_2 \neq \omega_3$, $k_1+k_2 = k_3$, the kernel $W(x,x')$ assumes the form

$$W(x,x') = \frac{d}{dx}G(x-x') \equiv \frac{W_0}{\sqrt{\pi}\Delta}\partial_x \exp\left[-\frac{(x-x')^2}{\Delta^2}\right] \tag{11.11}$$

in the case of the induced scattering of Langmuir and electromagnetic waves by plasma particles (see, e.g. [47,48], where one can find the expressions W_0, Δ and κ from the plasma parameters).

Manakov [46] showed that eqn. (19) with the boundary conditions $y_n \to \text{const}$, $n \to \pm\infty$ has the KdV equation as its continuum analogue. Therefore the decay of an initial condition $y_n(0)$ proceeds through the known KdV scheme.

The computer experiment by Montes et al. shows that eqn. (10) with the kernel (11) describes a system which is near integrable (KdV?). In [48], the problem of the initial condition (spectrum) $y(x,0)$ decay is studied, and it is shown that a quasi-soliton arises from a 'narrow' packet $\delta x \leqslant \Delta$ passing the stage of satellite transformation*).

A 'broad' initial packet ($\delta x > \Delta$) decays into a set of quasi-solitons distributed in

*)Note that for spherically-symmetric packets, this result was obtained earlier in [52], however, the authors did not attach a proper importance to it.

correspondence with their amplitude (or with velocity). In the subsequent paper by Montes et al. [49] it was shown that the solutions obtained earlier are quasi-soliton just as in the case of the RLW equation. Note that the interaction of solitons drawing together for the case of slightly different amplitudes resembles the satellite pumping-over, i.e. they remain well separated like the KdV solitons. In the case of triple quasi-soliton interaction the deviation from integrability becomes very essential: a fourth soliton of small amplitude is produced. This process is shown in Fig. 11.6 taken from the paper [49].

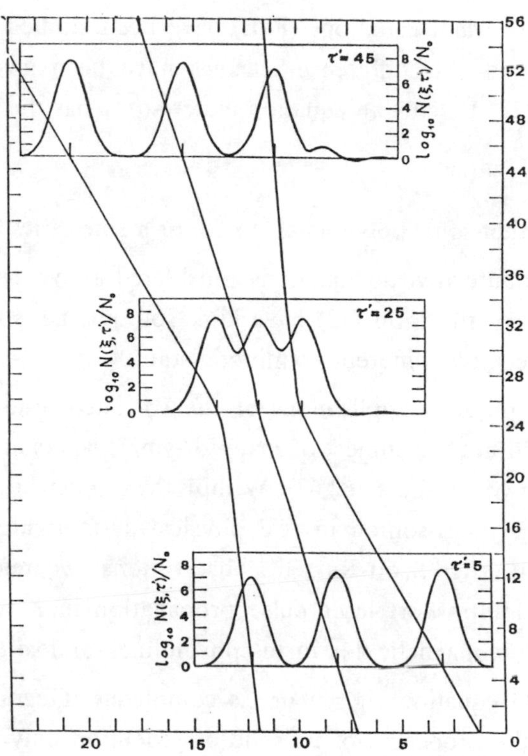

Figure 11.6. Production of a small amplitude soliton in the triple soliton collision.

Similar experiments were performed by another French group which studied quasi-soliton interactions in the framework of perturbed versions of eqns. (9)

$$\frac{d}{dt}y_n = y_n(y_{n+1} - y_{n-1}) + \mu(y_{n+1} + y_{n-1} - 2y_n),$$

and the KdV equation (the KdV-Burgers equation)

$$y_t + 6(y^2)_x + y_{xxx} = \mu y_{xx},$$

(see [50,51], respectively). In these cases as should be expected, the deviation from integrability is seen in the appearance of oscillatory tails during the quasi-soliton evolution process and their interaction.

5. Relativistically invariant equations

The formulation and interaction of planar solitons in the framework of nonlinear relativistically-invariant equations (RIE) have been studied in a series of papers. Discussions of possible difference schemes have been given by Eilbeck in the review paper [12]. Usually an equation under study has the form

$$\Box y = F(y), \tag{11.12}$$

where $F(y)$ is either some polynomial [53-56] or a sine series [53,57-61].

Unlike the nonrelativistic equations considered above, in addition to bell solitons, equations of the form (12) have kink solitons, i.e. solutions representing a transition between two different asymptotic states*).

Very broad physical applications of the so-called Sine-Gordon equation are well known, and are the subject of many original papers and several reviews. We mention only two of these [62,63]. Within the framework of this equation the existence of the quasi-soliton in real physical systems are brilliantly confirmed about 100 years after Scott-Russell's observations: we refer to the self-induced transparency in ultrashort laser pulse propagation in a two-level system**), the 'quantizing' of the magnetic flux in Josephson lines, and so on.

Since the SG equation $\Box y = -\sin y$ is completely integrable. We shall not discuss in detail the properties of its solution. Mention only that as well as in the case of KdV and $S3$, in the SG framework an arbitrary initial pulse breaks up into a set of solitons (it is often convenient to follow not the function $y(x,t)$ itself, but its derivatives y_x or y_t). Some very fine films are available which show various

*) Kink-like solutions may appear in the framework of the $S3$ equation at $\alpha < 0$ too.

**) Note that the self-induced transparency (as well as the KdV solitons and the FPU problem) 'was discovered from an analysis of numerical solutions of equations which describe optical pulse propogation' as Lamb said in his review [65] which incidentally, may be recommended as a good introduction to the field.

types of soliton, quasi-soliton and their bound state (bion) interaction [64]. In Fig. 11.7 taken from [12] the interaction of two bions is shown.

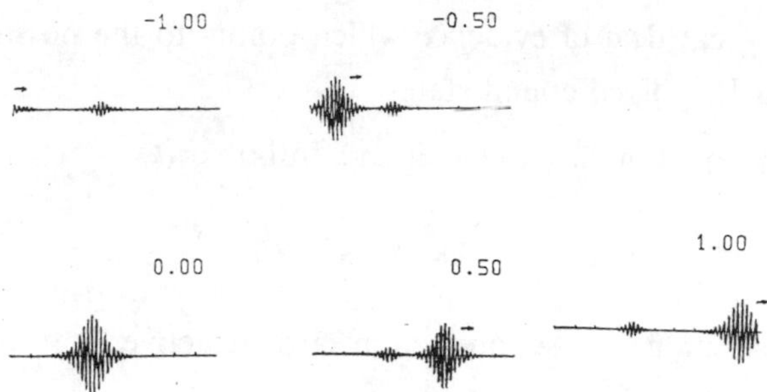

Figure 11.7. Bion-bion interaction in the *SG* model.

A great deal of computational work relating to the study of solutions of multiple SG equations has been carried out by different groups. In particular, the DSG equation

$$\Box y = \pm(\sin y \pm \frac{\lambda}{2}\sin\frac{y}{2}) \qquad (11.13)$$

arising in the study of self-induced transparency in double-degeneracy systems has been investigated in [53,58,61,66]*).

Equation (12), with

$$F(y) = \begin{array}{ll} -y+y^3 & (\varphi^4_-) \\ y-y^3 & (\varphi^4_+) \end{array} \qquad (11.14a\text{-}b)$$

(the so-called φ^4 theory) has been investigated in much detail both for real and complex functions in the papers [39], [53-56]. Equations (12) with the right-hand side (14) are obtained via a Lagrangian in which the potential $U(y)$ is a fourth-degree polynomial.

Quasi-soliton solutions have also been investigated in a more general approach when

$$U(y) \propto \sum_{n=1}^{m} c_n y^n$$

*)In [58], an initial pulse breakup into quasi-solitons in the framework of the TSG equation is investigated too.

(see [39, p. 50]).

In these papers, the inelasticity of the quasi-soliton interaction has been observed with a great deal of evidence which points to the possibility of the perturbation of their long-lived bound states.

The φ^4_- theory equation has kinks (k) and antikinks $(\bar{k})$

$$y_{k,\bar{k}}(x,t) = \pm \mathrm{th}\frac{\gamma}{2}(x-\mathrm{v}t-x_0),\ \ \gamma = (1-\mathrm{v}^2)^{-1/2}, \tag{11.15}$$

as quasi-soliton solutions whose interactions are attractive.

The φ^4_+ theory equation as well as the $S3$ equation may have sech and tanh quasi-soliton solutions. The sech solutions are stable only for the complex function $y = a(x)\exp(-i\omega t)$ when $\omega > 2^{-1/2}$. The interaction inelasticity of these quasi-solitons is the lower, the greater their relative velocity.

Finally, note that in computer experiments on φ^4_+ quasi-solitons, one has apparently encountered their instability for the first time.

6. Bound states of solitons (bions)

Bions are very interesting objects. If a soliton or a quasi-soliton may be treated as a bound state of, generally speaking, and infinite (in classical theory) number of constituents, it is natural to expect that solitons themselves may produce bound states.

Apparently, the first such states were described in the well-known paper [67], where a bisoliton (kink-antikink) for the SG equation was found (as in the review [39], we follow [63] by calling such solutions bions*)). This solution has the form

$$\varphi_b = 4\mathrm{arctg}[(\omega^{-2}-1)^{1/2}\mathrm{sech}z'\sin(z''+\delta)], \tag{11.16}$$

where

$$z' = \gamma\sqrt{1-\omega^2}\,(x-\mathrm{v}t-x_0),$$

$$z'' = \gamma\omega(\mathrm{v}x-t),$$

*)Such solutions are also sometimes called *breathers*.

$$\gamma^{-2} = 1 - v^2.$$

In the rest frame ($v=0$), this represents a solitary oscillating hump. In the laboratory system, where $v \neq 0$ a high-frequency filling of the hump appears which is connected with the emergence of the x-dependence under the sine sign. The approximate shape of such solutions is represented in Fig. 11.8.

Figure 11.8. A *SG* bion (a) - at rest, (b) moving one.

Before studying in detail the structure and dynamics of soliton bound states for the nonlinear Klein-Gordon equation (12) we mention briefly the possibility of such solutions in KdV-like models. As far as the author knows, bound states in KdV-like models have not been reported until now. Their absence in the case of integrable KdV and MKdV equations results directly from the solution of the inverse scattering problem, since only one soliton corresponds to each of the discrete levels. As regards the nonintegrable KdV-like equations, numerical experiments on soliton interactions carried out up to now have not given any positive result. Nevertheless, the question of the existence of bound states in the framework of nonintegrable KdV-like equations may be considered open, especially for complex functions and for large degrees of nonlinearity ν.

The bion solution (16) has two characteristic features which distinguish it from similar solutions in nonintegrable systems and in more than one space dimension. Firstly, a state so described cannot be obtained as a result of a kink-antikink interaction since, as a result of the SG integrability, the radiation does not appear in soliton interactions. Secondly, due to the absence of radiation the lifetime of the bion (16) is infinite. Finally, the SG bion interaction is elastic as is very clearly demonstrated in Fig. 11.7 taken from the report by Eilbeck [12].

In recent years, there have appeared a series of papers in which attention has been paid to the essential significance of bion type solutions in various one-dimensional physical models of condensed matter. We mention here the papers

[69-73] and the review [68]. Note the essential difference between kink soliton solutions and their bound states: the bions (and as we shall see below the quasi-bions: pulsons) as well as the envelope solitons have an internal oscillatory ('rotational') degree of freedom and, as a consequence, in the classical limit their rest mass

$$M_b = 2M\sqrt{1-\omega^2} \tag{11.17}$$

changes from zero to $2M$ due to the dependence on ω (here $M = 8$ is the rest mass of the kink or antikink in dimensionless variables). The analogy between the SG bions and the $S3$ envelope solitons turns out to be far deeper than may appear at first sight, as has been noted already in the 1976 paper by Kaup and the paper by Kaup and Newell (1978) [74]. In fact, for small amplitudes $\sqrt{\omega^{-2}-1} \ll 1$, the value of $\arctan(x)$ is approximately equal to x and eqn. (16) transforms with accuracy $0(x)$ into corresponding expressions for the $KG3$ envelope soliton $\varphi_b \simeq 2\sqrt{2}\,\mathrm{Im}\psi_{KG3}$ or $\mathrm{Re}\psi_{KG3}$ depending on the phase δ and at $v \ll 1$ we similarly get the $S3$ soliton transformed, i.e., $\varphi_b \simeq 2\sqrt{2}\,\mathrm{Re}\psi_{S3}e^{-it}$.

This feature of bions, as in systems described by Schrödinger and Klein-Gordon type equations, leads to the possibility of a cumulative effect of energy accumulation in bion excitations under the action of an external field, even when the external field amplitude is less than the kink-antikink pair production threshold, $E \ll 2M$ (for envelope solitons, see [29]).

The discovery of long-lived bound states of solitons, together with the investigation of their properties, constitute one of the most important problems of soliton theory (this is especially related to multidimensional models; see below). For nonintegrable models with polynomial or logarithmic Lagrangians (in particular φ^4 theory) as well as for the multiple SG equation, exact analytical solutions describing bound states have not been obtained. Here, as well as in the study of soliton interaction dynamics, computer experiments are of great importance.

Some indications for the possible existence of bound states are already contained in early numerical studies of initial-packet dynamics in the framework of the $S3$ equation. In [38,41], as a result of the evolution of an initial packet, a solution has been obtained which closely resembles a bound state from the point of view of modern concepts. It is possible that the long duration of nonlinear amplitude oscillations in the packet centre (see Fig. 11.3a) [38] is determined by this. In the earlier paper by Yajima and Outi [37] the decay of an initial Gaussian packet has also been studied. The production of three humps was observed: however, as these authors have pointed out, the calculation time was insufficient to yield an unambiguous answer to the question: are solitons produced? - or whether for a certain time the system will be in a quasi-periodic stage? Satsuma and Yajima [41], in a closely related paper, study the production of the $S3$ bion

$$\psi_b = 4\exp\left(-\frac{it}{2}\right)\frac{\mathrm{ch}3x + e^{-4it}3\mathrm{ch}x}{\mathrm{ch}4x + 4\mathrm{ch}2x + 3\cos 4t}$$

from the initial packet $\psi_0 = 2\mathrm{sech}(x)$. Unfortunately, the authors present only an initial stage of the packet evolution.

Full evidence supporting the existence of the bion comes from Kudryavtsev's computer experiments on kink interactions in the Higgs model (φ^4_- theory) [54]. These results have subsequently been confirmed and new interesting results have been obtained.

7. Kink-antikink interactions in the ϕ^4_- model

Since computer experiments on kink-antikink collisions in nonintegrable models have revealed a complicated and rich picture, we will consider these at some length. Kink-antikink collisions were studied in the framework of the Higgs model [53-56] (the ϕ^4_- model), modified SG [57] and DSG [53,58,61]. The phenomenology of such interactions has many common features (though there are peculiarities) that motivates us to discuss them with the example of the ϕ^4_- theory. This enables us to qualitatively identify the mechanism underlying the observed data.

In the numerical experiments by Kudryavtsev and others, $k\bar{k}$ collisions have been studied for the equation

$$(\Box - 1 + \phi^2)\phi = 0, \tag{11.18a}$$

with the Hamiltonian

$$H \equiv E = \int \mathcal{H} dx, \quad \mathcal{H} = \frac{1}{2}(\dot{\phi}^2 + \phi'^2 + \frac{1}{2}(\phi^2 - 1)^2). \tag{11.18b}$$

The initial state

$$\phi(x,0) = \text{th}\frac{\gamma}{\sqrt{2}}(x-x_0)-\text{th}\frac{\gamma}{\sqrt{2}}(x+x_0)-1, \tag{11.19}$$

$$\phi_t(x,0) = -\frac{\gamma}{\sqrt{2}}\text{v}\left[\text{sech}^2\frac{\gamma}{\sqrt{2}}(x-x_0)+\text{sech}^2\frac{\gamma}{\sqrt{2}}(x+x_0)\right],$$

describing a kink and an antikink moving from both infinities, towards each other, leads, with exponential accuracy ($\propto e^{-4x_0}$), to the solution of eqn. (18) in the form

$$\phi = \text{th}\frac{\gamma}{\sqrt{2}}(x-\text{v}t-x_0)-\text{th}\frac{\gamma}{\sqrt{2}}(x+\text{v}t+x_0)-1. \tag{11.20}$$

Inserting this into the computer we obtain the following picture. The final state turns out to depend on velocity of the colliding quasi-solitons. If the velocity v exceeds a certain critical value v_{cr}, the kink and antikink repel each other losing a certain fraction of their kinetic energy through radiation (see Fig. 11.9).

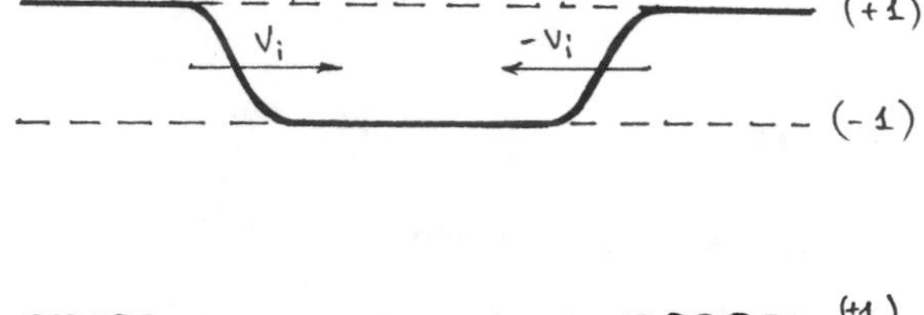

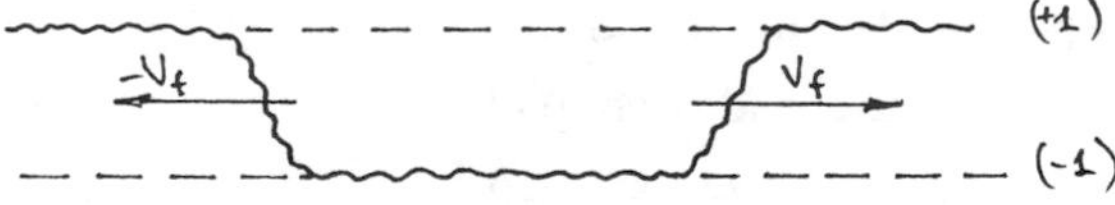

Figure 11.9. Kink-antikink interaction followed by radiation of linear waves in the framework of the ϕ^4_- model.

Moreover, this fraction is lower, the greater v exceeds v_{cr}, so that in the ultrarelativistic region, $\text{v}\to 1$, the collision is quasi-elastic.

At first an alternative interpretation seemed to be very simple: in the range $\text{v}<\text{v}_{cr}$, the $k\bar{k}$ pair loses sufficient energy to create their bound state, the ϕ^4_- bion (Fig. 11.10(a)). This was stated in early works [39,54]. The only shortcoming though would be that the value of v_{cr} differed in different numerical experiments and lay in the interval 0.2-0.25. More precise experiments [53,55,56,75] showed that such a discrepancy was not accidental but hid a very curious picture. It turned out that besides the upper critical velocity, $\text{v}_{up}\simeq 0.26$, there is a lower critical velocity, $\text{v}_l\simeq 0.19$, such that collisions of the $k\bar{k}$ pair with $\text{v}_{in}\geqslant \text{v}_{up}$ always led to kink and antikink reflecting from one another, and, in the case of $\text{v}_{in}\leqslant \text{v}_l$, with

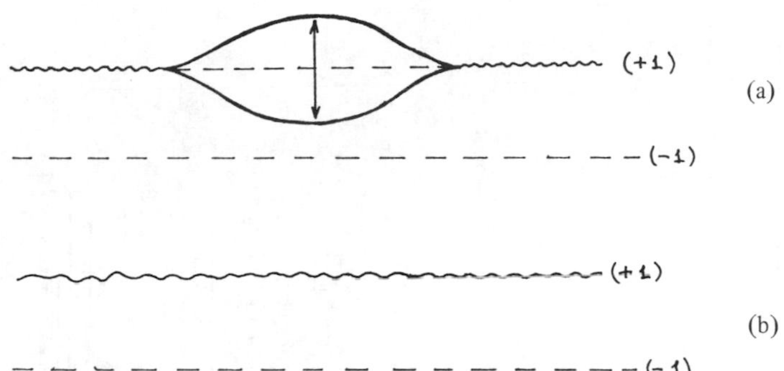

Figure 11.10a. ϕ^4_- bion production in the $k\bar{k}$ collision.

Figure 11.10b. Bionless final state.

their capture into a bound state. In the interval $v_{in} \in (v_l, v_{up})$ the ranges of initial velocities for which the collisions end in reflection (the reflection or resonance regions) alternate with ranges of incoming velocities for which the collisions end in kink-antikink capture (the capture regions). The first are also termed the reflection windows. Various authors give a different number of these (one in [53], three in [56], nine in [55]). Up to now, in [75] 35 such windows have been found and have corroborated the earlier claims. The broadest window is located near the v_l, with their width decreasing as v_{in} increases. This led to the conclusion [75] that the point v_{up} is a condensing point of the reflection windows (see Fig. 11.11). Campbell et al. have given a qualitative theory of the phenomenon [55], based in fact on the very simple observation that the ϕ^4_- kink-antikink interaction is a resonance one. We give here an outline of their basic ideas.

During the initial stage of collision part of the $k\bar{k}$ kinetc energy goes into the energy of localized excitations of k and $\bar{k}$. After kink-antikink bounce, the latter can be restored, at least in part, allowing the kinks to escape. For this to be true, there should be an extra degree of freedom, viz. a mode related to localized changing of the kink shape in addition to the Goldstone mode (translation mode) and continuous spectrum (radiation).

Let us consider the equation governing the excitation spectrum of the kink,

$$\phi = \phi_k + \delta\phi, \tag{11.21}$$

where ϕ_k is given by (15) and

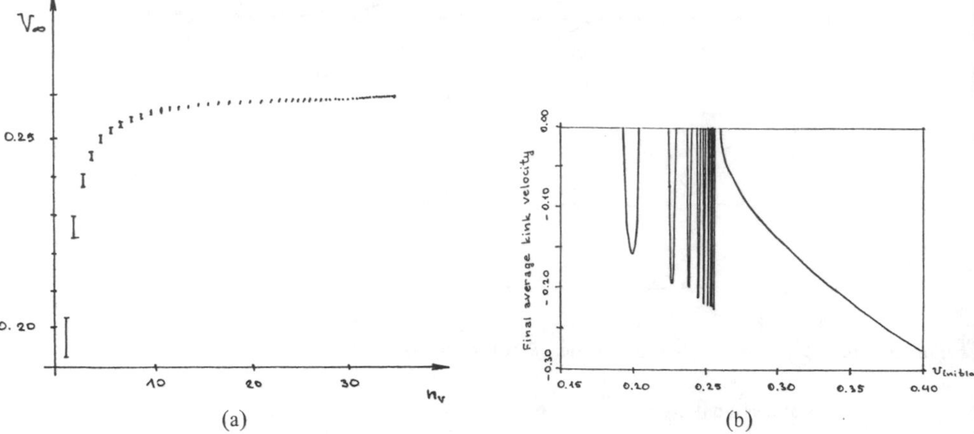

Figure 11.11a. Reflection (resonance) windows Δv_i as a function of the number of field small oscillations between bounces of kinks.

Figure 11.11b. Reflection windows in the v_{in}, v_{fin} plane ($v_{i,f}$ are kink velocities).

$$(\Box - 1 + 3\phi_k^2)\delta\phi = 0 \tag{11.22}$$

or for the Fourier transform

$$(-\partial_x^2 - 1 + 3\phi_k^2)\delta\phi = \omega^2\delta\phi.$$

This equation can be transformed to give the standard Schrödinger equation

$$(-\partial_x^2 + U)\delta\phi = (\omega^2 - 2)\delta\phi \tag{11.23}$$

with

$$U = \frac{3}{\text{ch}^2(x/\sqrt{2})}. \tag{11.24}$$

In Chapter 8 we have seen that potentials of the shape $m(m+1)b^2\text{sech}^2 bx$ are reflectionless (Bargmann's) and have just m discrete eigenvalues $\lambda_j = -b^2(m-j+1)^2$. In our case, $b = 2^{-1/2}$, $m = 2$ and $\lambda_0 = -2$, $\lambda_1 = -1/2$. Whence

$$\omega_0 = 0, \qquad \text{Goldstone mode,} \tag{11.25}$$

$$\omega_1 = \sqrt{\frac{3}{2}}, \text{localized shape vibration mode,}$$

$$\omega > \omega_c = \sqrt{2}, \text{continuous spectrum.}$$

One can readily check that Sine-Gordon kink excitations are governed by the equation

$$(\Box + \cos\phi_s)\delta\varphi = 0$$

or

$$-(\partial_x^2 + 2\mathrm{sech}^2 x)\delta\phi = (\omega^2 - 1)\delta\phi$$

and hence $m = 1$. The only discrete eigenvalue $\lambda_0 = -1$ $(\omega^2 = 0)$ corresponds to the Goldstone mode.

Considering ϕ^4_- kinks, we have already seen that there is a degree of freedom needed with frequency ω_1. Then energy transformation from the Goldstone mode into the ω_1 mode and vice versa will have a resonance character if

$$\omega_1 T = \delta + 2\pi n, \tag{11.26}$$

where T is the time between two successive impacts of k and $\bar{k}$, δ is some offset phase, n is an integer.

Condition (26) is the basic assumption of Campbell et al. [55] in constructing their phenomenological two bounce theory. From computer experiments, they found the time T between the two bounces in the centre of a two bounce windows as a function of the ordinal window number (or, to be more accurate, of the value $n = N + 2$) which is a straight line (Fig. 11.12).

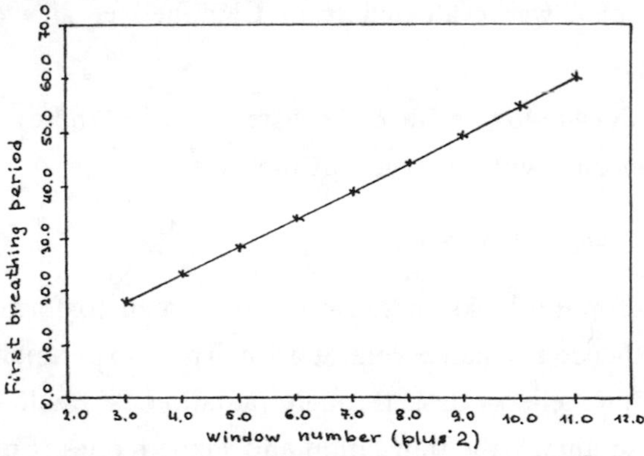

Figure 11.12. The time between the first and the second impacts of the $k\bar{k}$ pair at the centres of the windows.

Its slope $\simeq 5.2$ compares well with the value

$$(2\pi\omega_1^{-1}) = (2\pi\sqrt{2/3}) \simeq 5.13$$

expected from (26). From the intercept of the line we have

$$\delta \simeq 3.3 \ (\simeq\pi).$$

The second assumption verified in the numerical experiments is the relation between T and v

$$T(\text{vis}) = \alpha(\text{v}_{up}^2 - \text{v}^2)^{-1/2}. \tag{11.27}$$

Their results give $\alpha \simeq 3.0$ and the centres of two-bounce windows defined with

$$\text{v}_n^2 = \text{v}_{up}^2 - \frac{1.37}{(2n+\delta/\pi)^2} \simeq \text{v}_{up}^2 - \frac{1.37}{(2n+1)^2}. \tag{11.28}$$

It is interesting that the number of internal (small) bumps grows by one in going from one window to the next according to formula (26) (see Fig. 11.13). Making use of the formulae obtained and the natural relationship

$$\omega_1 \Delta T = \theta,$$

with θ being a certain fixed phase, we obtain the following estimation of the window widths (setting $\Delta T = (\partial T/\partial \text{v})\Delta \text{v}$)

$$\Delta \text{v}_n \sim 1.7\theta(2n+1)^{-3}. \tag{11.29}$$

The magnitude of θ was evaluated from Campbell et al.'s experiments*) to be $\theta \simeq 2$.

Note that only windows with $n \geqslant 3$ were noted in these experiments. This is presumably connected with the fact that the time

$$T_0 = 5\pi\omega_1^{-1} = (\delta + 4\pi)\omega_1^{-1}$$

is not sufficient for the kinks to separate far enough to form well-defined shape oscillations and hence resonance cannot arise. These experiments also showed that inside the windows energy loss through radiation is small compared with the energy distributed among the translation and shape modes (Fig. 11.14).

*)Here and throughout the section we mean *computer* experiments.

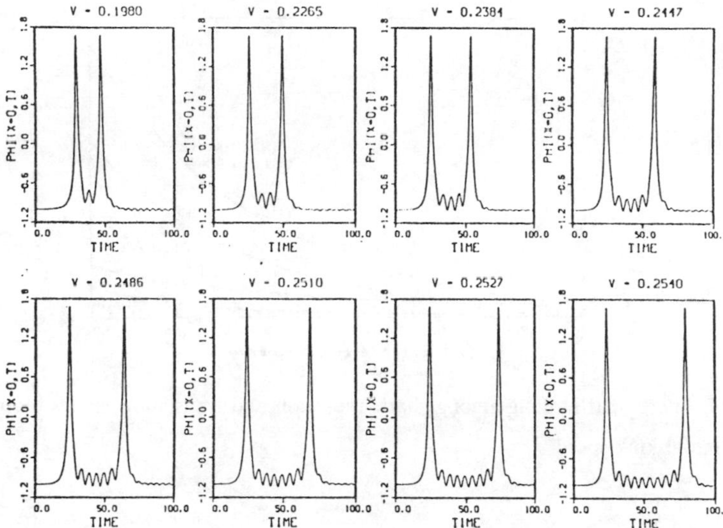

Figure 11.13a. The dependence of kink velocity v on time in the windows.

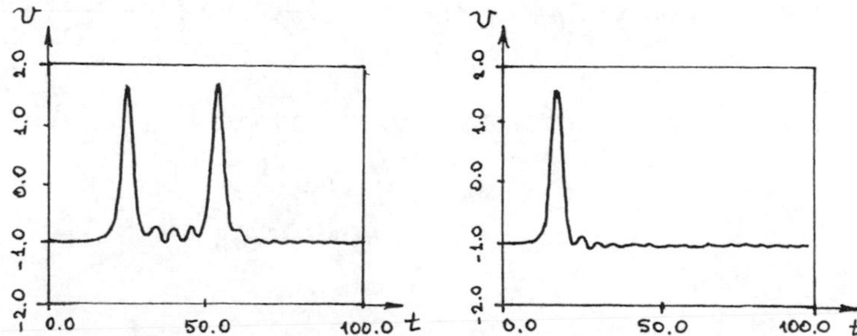

Figure 11.13b. The window and reflection plots.

Belova and Kudryavtsev using the idea of two interacting modes [75] constructed the model Hamiltonian

$$H = (M_k + I(R))\dot{R}^2 + U(R) + \dot{A}^2 + \omega_1^2 A - 2F(R)A. \tag{11.30}$$

In going from (18b) to (30) they make use of the Sugiyama Ansatz [76]:

$$\phi(x,t) = \phi_k(\frac{x+R}{\sqrt{2}}) - \phi_k(\frac{x-R}{\sqrt{2}}) - 1 + A(t)\left\{\delta\phi(\frac{x+R}{\sqrt{2}}) - \delta\phi(\frac{x-R}{\sqrt{2}})\right\}, \tag{11.31}$$

$$\delta\phi = 2^{-3/4}\sqrt{3}\,\mathrm{th}(\frac{x}{\sqrt{2}})\mathrm{sech}(\frac{x}{\sqrt{2}}),$$

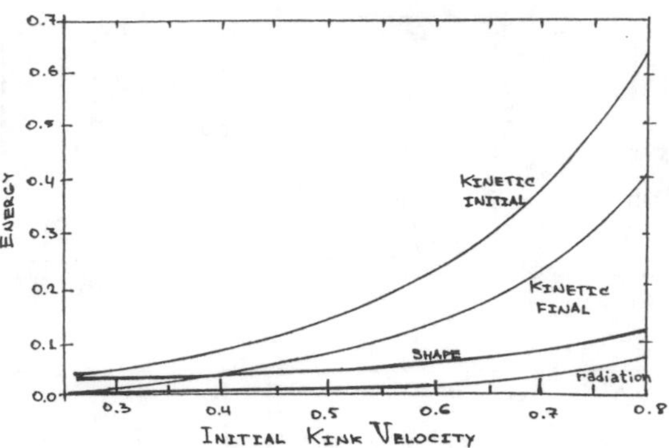

Figure 11.14. Energy partitioning among shape radiation and translation modes plotted as a function of the initial kink speed.

$$M_k = E_k(v=0) = 2\frac{\sqrt{2}}{3},$$

$$I(R) = \int_{-\infty}^{\infty} dx\phi_k'(x+R)\phi_k'(x-R) = 2\sqrt{2}\left\{1-\text{th}^2(\sqrt{2}R)\frac{\sqrt{2}R-\text{th}(\sqrt{2}R)}{\text{th}^3(\sqrt{2}R)}\right\},$$

$$U(R) = 4\sqrt{2}\left[-\frac{2}{3}+\sqrt{2}R+3\text{cth}(\sqrt{2}R)-(2+3\sqrt{2}R)\text{cth}^2(\sqrt{2}R)\right.$$
$$\left.+\,2\sqrt{2}R\text{cth}^3(\sqrt{2}R)\right],$$

$$F(R) = \pi\frac{\sqrt{3}}{2\sqrt{2}}\cdot\frac{3}{2}[1-\text{th}^2(\sqrt{2}R)]\text{th}^2(\sqrt{2}R).$$

Model (30) and (31) describes a dynamical system of two degrees of freedom (collective variables). Computer experiments [75] for this model display a picture very similar to that of the field model (16), and this confirms once again the phenomenological theory of Campbell et al. [56].

8. Kink-antikink collisions in the modified Sine-Gordon model

Another verification of the Campbell et al. phenomenology comes from studying the model in which kinks do not have a shape-vibration mode. Such a model was proposed in [77] and investigated in [57]. It is the modification of the SG with the potential

$$V(\phi) = \frac{(1-\alpha^2)(1-\cos\phi)}{1+\alpha^2+2\alpha\cos\phi} \tag{11.32}$$

in which a parameter $\alpha \in (-1,1)$ (Fig. 11.15).

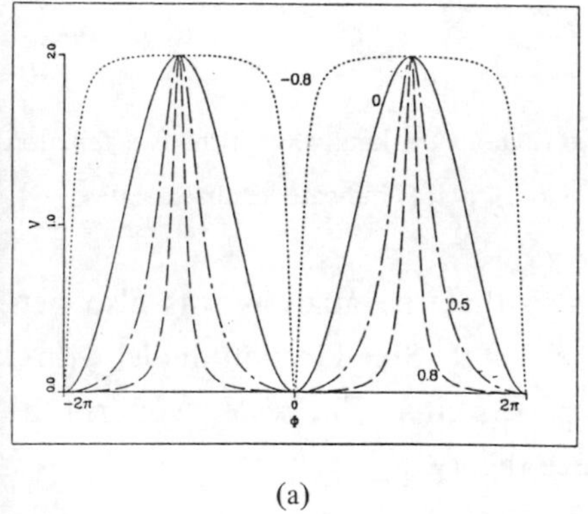

(a)

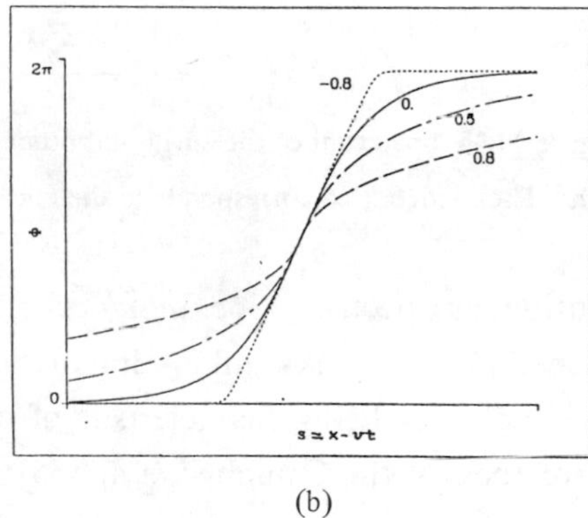

(b)

Figure 11.15a. The potential (11.32) of the modified *SG* for various α ($\alpha = 0, 0.5, 0.8$ and $\alpha = -0.8$).

Figure 11.15b. Corresponding kink profiles.

When $\alpha \to 0$ (32) tends to the SG potential (see Fig. 11.15).

One may expect that for $\alpha \ll 1$ certain features of the SG model remain in (32). Indeed, from Fig. 11.16 taken from [57] it follows that the shape vibration mode disappears for $\alpha > 0$ and also for small negative α. When $|\alpha|$ grows, in the system there appears one shape mode ($|\alpha| \simeq 0.06$), then two such modes ($|\alpha| \simeq 0.3$), three ($|\alpha| \simeq 0.5$), and so on. Results obtained by Campbell and Peyrard [57] enable the assertion with evidence that the above two-bounce theory embraces the essential nature of the phenomena ocurring in $k\bar{k}$ collisions. At any rate, in the region $\alpha \in (-0.05, 1)$ there are *no* shape modes *together with* resonance windows. This implies that possible resonance between the Goldstone mode and one from the

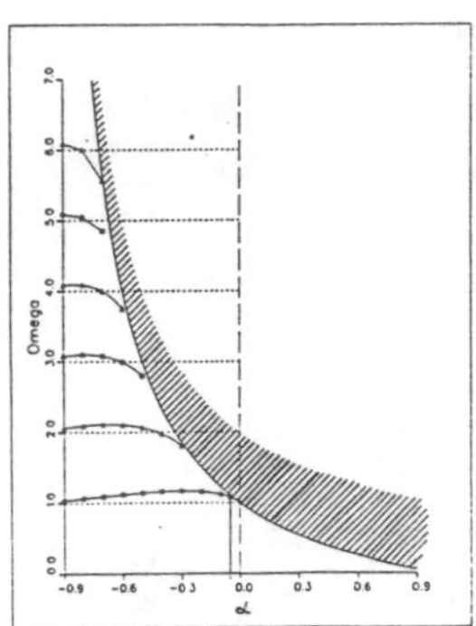

Figure 11.16. Spectrum of the small amplitude oscillations about the kink wave form as a function of α. The shaded area corresponds to the continuum ($\omega \geqslant \omega_c$), points indicate localized states.

continuous spectrum does not occur. Note also that the analysis was also performed [61] for kink-antikink interaction in the double Sine-Gordon model with a great accuracy being characteristic of the two papers cited. The results confirm the correctness of the Campbell et al. [55] resonance theory.

Finally, we should emphasize that the resonance of quasi-soliton interactions has been revealed numerically in the late 1970's in a number of non-one-dimensional classical field theories (see, for example, the review [39]). These results will be given in detail later.

9. Bions in ϕ^4_- theory

In the region $v_{in} < v_l$ as well as outside the resonance windows in the interval $v_{in} \subset (v_l, v_{up}), k\bar{k}$ collisions lead to bion creation. In this case, the mass defect, which is equal to $\Delta M = 2E_k - M_b$, is emitted through waves of the continuous spectrum (linear waves on the vacuum determined by the initial conditions). The kink or antikink energy E_k can be estimated from the conventional formula

$$E_k = \frac{1}{2}\int_{-\infty}^{\infty} [\phi_t^2 + \phi_k^2 + \frac{1}{2}(\phi^2 - 1)^2]dx = \frac{2\sqrt{2}}{3}\gamma \equiv M_k\gamma$$

and the mass M_b of the bion produced is less than the mass of two kinks $2M_k$. After a short relaxation the $\varphi(0,t)$ field oscillations in the bion become very regular (see Fig. 11.17) closely resembling the $\psi(0,t)$ amplitude plot for the $S3$ experiment

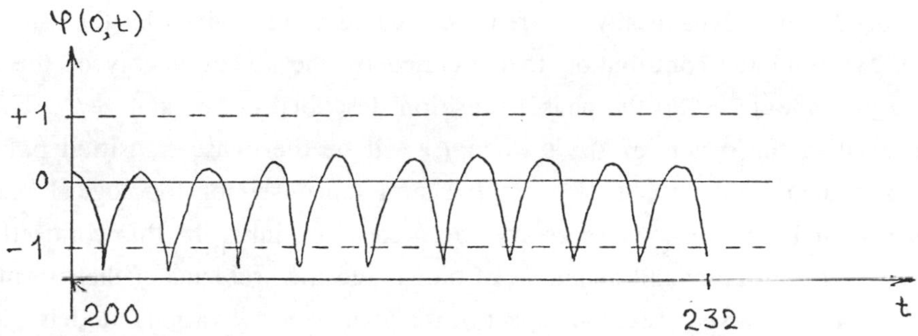

Figure 11.17. Oscillations of the field function in the bion regular phase.

(Fig. 11.3(a)). Unlike the $S3$ equation, the Higgs model is nonintegrable, and therefore the bion produced in the kink collision continues to emit small amplitude waves and to lose its energy. The large amplitude bion lifetime [39] turns out to be very long, $T \simeq 2.10^3\omega^{-1}$, and increases with decreasing amplitude.

These results imply that the bound states are not a privilege of the integrable $S3$ and SG equations, but may be produced also in quasi-soliton interactions in the framework of other equations.

In any case, one may assume that systems which are close to the integrable ones must possess solutions describing long-lived bound states.

Bion-type excitations can contribute greatly to the thermodynamics of the kink gas at very small temperatures. This is, firstly, due to the fact that they do not have rest mass and their frequency will be tuned on temperature T. For ϕ^4 bions, ω can be evaluated using the following formula

$$T \simeq 2M_k(1-\frac{1}{2}\omega^2)^{1/2}, \tag{11.33a}$$

whence

$$\omega \simeq \sqrt{2[1-(T/2M_k)^2]}, \quad M_k = 2\frac{\sqrt{2}}{3}. \tag{11.33b}$$

Secondly, the ideal gas of kinks at $T < T_{pt} \simeq M_k u_l^2$ $(u_l \simeq 0.2)$ will be unstable with respect to bion creation. Then the temperature T_{pt} can be regarded as the Curie point for the second kind phase transition associated with 'topology changing'. This transition may be considered as the transition from a gas of 'charged' particles k and $\bar{k}$ to a gas of neutral dipoles, bions. Note, however, when the kink gas is produced thermodynamically, the required temperature is given by the estimation $T_0 \simeq 2M_k$, and the contribution to it defined by the kinetic energy of the kinks $T \geqslant T_{pt}$, is about 2%. So the phase transition described occurs at $T \simeq T_0$. If kinks can exist in the system at $T \simeq 0$, then T_{pt} will be the phase transition point (ϕ^4 kinks and antikinks are always created simultaneously for topological reasons, since an infinite energy is necessary to create one kink). In this connection, a study of the structure and dynamics of bions, and the problems of their formation and stability, is a very essential aspect in the framework of various models.

10. Small-amplitude expansions

Consider approximate solutions for small-amplitude bions. Note the following however: the method of asymptotic expansion proposed here works in searching for *stationary* and *quasi-stationary* solutions of small amplitude. Here (as in the foregoing) stationary solutions are solutions of the type

$$\psi(x,t) = \phi(x)e^{-i\omega t},$$

or those periodic in time, $\psi(x,t) = \psi(x,t+T)$. Quasi-stationary solutions will be termed solutions having parameters which are depending on time (a slow time τ).

The technique, proposed by several authors [78-80] in various versions, is illustrated here with an example of the nonlinear Klein-Gordon equation for a real scalar field χ:

$$(\Box + 1 - \chi^2)\chi = 0. \tag{11.34}$$

In the small amplitude approximation, any relativistically invariant theory is reduced to (34) if the potential of the Hamiltonian

$$H = \frac{1}{2}\int(\dot{\chi}^2 + \chi'^2 + \chi^2 + 2U(\chi)]dx$$

has the expansion

$$U(\chi) \to -\chi^4 + 0(\chi^6).$$

Represent the function $\chi(x,t)$ as follows:

$$\chi(x,t) = a(x)\cos\omega t + b(x)\cos^3\omega t + ..., \quad |b| \ll |a|. \tag{11.35}$$

Inserting this into (34) we get

$$a_{xx} - \kappa^2 a + \frac{3}{4} a^3 = 0,$$

$$\kappa^2 = 1 - \omega^2.$$

Introducing $A^2 = 3/4a^2\kappa^{-2}$ gives the standard equation of the ϕ_+^4 theory,

$$A_{xx} - A + A^3 = 0, \tag{11.36}$$

with $A = \sqrt{2}\,\operatorname{sech}(x)$ as a particular solution. For small amplitudes $\kappa \ll 1$ we have

$$\chi(x,t) = \sqrt{\frac{4}{3}}\,\kappa A(\kappa x)\cos(\sqrt{1-\kappa^2}\,t) + \ldots, \tag{11.37}$$

$$b(x) = -\frac{1}{12\sqrt{3}}\kappa^3 A^3(\kappa x).$$

In the same approximation, the field energy

$$E \simeq \frac{1}{2}\int_{-\infty}^{\infty} (\dot{\chi}^2 + \chi^2)dx = \frac{1}{2}\int_{-\infty}^{\infty} [\chi(x,0)]^2 dx + 0(\chi^4) \tag{11.38}$$

and its density, $\mathcal{H}$, is time-independent.

We proceed to describe a small amplitude bion constructed on the vacuum $\chi_v = -1$. Let us write for this the field function $\chi(x,t) = \phi(x,t) - 1$, $|\phi(x,t)| \ll 1$, as a formal series in ϵ [79]

$$\phi(x,t) = \epsilon^2 g_0(x) + \sum_{n=0} + \tag{11.39}$$

$$+ [\epsilon^{2n+1} f_{2n+1}(x)\sin(2n+1)\omega t + \epsilon^{2n+2} g_{2n+2}(x)\cos(2n+2)\omega t]$$

and make the substitution

$$\tau = \sqrt{2}\,t(1+\epsilon^2)^{-1/2}, \quad \xi = \sqrt{2}\,\epsilon x(1+\epsilon^2)^{-1/2}.$$

For the function $A(\xi) = \sqrt{3/2}\,f_1(\xi)$, we have again eqn. (34) and the functions g_i are defined with the relations

$$g_1 = -\frac{3}{4} f_1^2, \quad g_2 = -\frac{1}{4} f_1^2, \ldots \tag{11.40}$$

Now we obtain the solution accurate to $0(\epsilon^2)$,

$$\phi(x,t) = \frac{2}{\sqrt{3}}\epsilon A(\xi)\sin\omega t - \epsilon^2 A(\xi)(1+\frac{1}{3}\cos 2\omega t), \tag{11.41}$$

$$\omega^2 = 2(1+\epsilon^2)^{-1} \simeq 2(1-\epsilon^2).$$

The bion energy is given by (38) again

$$E = \frac{4}{3}\epsilon^2 \int_{-\infty}^{\infty} \operatorname{sech}^2(\sqrt{2}\epsilon x)dx = \frac{4\sqrt{2}}{3}\epsilon = 2M_k(1-\frac{1}{2}\omega^2)^{1/2},$$

i.e. by the formula (33a).

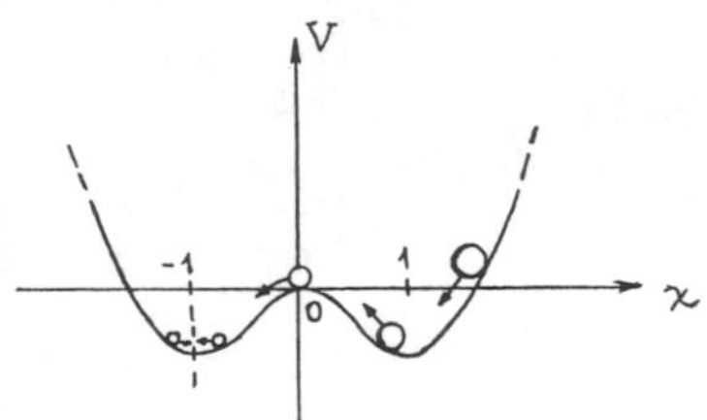

Figure 11.18. Picture qualitatively the evolution of a pulson in the ϕ^4_- model. The ball size is associated with its mass. V is the potential energy of the pulson as a function of the field amplitude (and hence mass). The vacuum $\varphi_v = -1$ is fixed by the boundary conditions.

Evolution in time of a Higgs (ϕ^4_-) bion is schematically shown in Fig. 11.18 (the dimension of the ball is related to the oscillation amplitude and thus to the bion mass). It is seen that, because of radiation, the heavy bion, losing its mass, overcomes a potential barrier and goes down into the left well. Formula (41) describes the bion already in this well.

Expansions (37) and (41) will be of use in our study of the properties of pulsons in $D > 1$ dimensions.

Chapter 12.

STRUCTURAL STABILITY AND PINNING OF SOLITONS

As was pointed out in the foregoing, in spite of considerable recent effort the problem of structural stability is far from complete both in the actual and conceptual state of affairs for the field system.

The first studies in this direction were the KdV and $S3$ soliton studies with allowance for weak dissipation effects. In particular, it was shown [81] that soliton behaviour, viz. the conservation of shape as a balance of dispersion and nonlinearity effects, depends very essentially on the type of perturbing term (see Section 10.4). Interesting results on the structural stability of solitons have also been obtained by the Japanese group [82].

Note that one of the specific features of perturbed solitons is that they acquire oscillatory tails.

We mention a series of papers devoted to the study of the SG soliton structural stability performed by Fogel et al. [85] (theory) and Currie et al. [86] (numerical experiments). These investigate the initial value problem

$$\Box\varphi + \omega_0^2(x)\sin\varphi + AF(x) = 0, \qquad (12.1)$$

$$\varphi(x, 0) = f(x).$$

The x-dependence of the system eigenfrequency ω_0 may simulate both the presence of impurities (nonuniformities) and its macrostructure (lamination, and so on). The last term of (1) models external influences (fields, currents, etc.). In [85], it is analytically predicted, as well as numerically confirmed in [86], that for sufficiently small perturbations the SG kink solitons behave in external fields as particles having an internal structure: they may be accelerated (decelerated), radiate, change their structure slightly with the accompaniment of transition radiation, and may be confined in some spatial region and then pushed out by switching off one of the plugs. The presence of radiation in the model (1) (as well as in its discrete analogue even for $d\omega_0/dx=0$) has a strong effect on the structure and especially on the dynamics of bound states (bions); their production, lifetime, break-up. Further, more comprehensive investigations [87,90] give a result somewhat analogous to that which we have already described concerning $k\bar{k}$ collisions in the ϕ^4 theory. It turns out that resonance phenomena manifest themselves in the interaction of a kink with impurities as a result of interaction of various excitation modes of the static kink. It is this fact which leads to a difference in behaviour of an *extended* kink and a *point-like* Newtonian particle.

Integrable models usually lose this property in the presence of boundaries and impurities (barring relativistically-invariant models defined on a half-axis, and a class of models of a very special kind obtained from homogeneous ones via various substitutions [88] and so on). Since impurities and boundaries are attributed to more or less realistic physical situations, the study of even the simplest soliton-bearing models in this case is not only of academic interest.

Usually, under the action of a weak, but long-in-time, perturbation, a soliton changes slowly (adiabatically) its characteristic parameters (the amplitude, velocity and so on) as well as its shape. The latter is associated with a breakdown of the balance of dispersion and nonlinear effects. Nevertheless, even in systems with dissipation and pumping can appear new stationary solutions of soliton-type (see, for example, [101,102]).

Models can be investigated in a most simple and complete way in which solitons interact with micro-impurities, i.e. impurities well localized in space. We will consider such systems later on. As in the case of ϕ^4 kink interaction, we can separate here two essentially different regions of parameters: In the first region, the interaction time of a soliton with a micro-impurity is small so that its energy, being positive, changes slightly upon interaction. We call this region, the '*passing*' one. In the second region, which we call the '*bound*' region, the behaviour of the soliton and, even more, its structure, are determined by the impurity (or by system boundaries). There is a very rich phenomonology of solitons in both regions which admit, in a number of cases, a substantial analytical description.

We begin with the bound region of parameters and consider some simplest examples which nevertheless give a clear understanding of the problems involved, the methods for solution, and some results.

1. Static bound states

Later we will use, to a great extent, the results of papers by Filippov, Galpern and coauthors [89-92].

Consider the model of a real scalar field

$$H = \int dx \{ \frac{1}{2}(\dot{\varphi}^2 + \varphi_x^2) + U(\varphi) + \mu\delta(x)F(\varphi)\}, \tag{12.2}$$

where the potential $U(\varphi)$ can contain a mass term $\propto \varphi^2$ and $F(\varphi)$ is a nonlinear function of $\varphi(x,t)$. The Euler-Lagrange equation is

$$\Box\varphi + \partial_\varphi U(\varphi) + \mu\delta(x)\partial_\varphi F(\varphi) = 0. \tag{12.3}$$

Static solutions follow from the condition $\partial_t\varphi = 0$.

We note that static localized solutions to (3) can exist even in the case where they disappear for $\mu = 0$. For example, let us consider the linear (at $\mu = 0$) Klein-Gordon equation ($U = \varphi^2$)

$$\Box\varphi + \varphi = -\mu\delta(x)F'(\varphi). \tag{12.4}$$

Its static solutions localized on the impurity are easily verified to be $\varphi(x) = \varphi_0 e^{-|x|}$. Combining solutions at $x = 0-\epsilon$ and $x = 0+\epsilon$, we get

$$2\varphi_0 = -\mu F'(\varphi_0), \tag{12.5}$$

whence we find $\varphi_0(\mu)$ with the energy

$$E(\varphi_0) \equiv H = \varphi_0^2 + \mu F(\varphi_0), \tag{12.6}$$

the extremum of which, $E'(\varphi_0) = 0$, gives eqn. (5). The latter can have more than one solution. To each of these corresponds its own state. States stable with respect to small field fluctuations obey the standard condition $E''(\varphi_0) > 0$, or

$$E''(\varphi_0) = 2 + \mu F''(\varphi_0(\mu)) > 0. \tag{12.7}$$

Values of μ in which $E''(\mu_i) = 0$ for $E''(\mu_i) \neq 0$ determine inflection points or bifurcations since, in these points, two solutions are created or annihilated with μ changing (see Fig. 12.1). One of these solutions, $E''(\mu_2) > 0$, is stable but the other, $E''(\mu_1) < 0$ is not.

One can easily check that, for example, when $F(\varphi) = \varphi^n$ stable solutions exist only for $n = 1$ and any μ. More sapid although simple example we have for

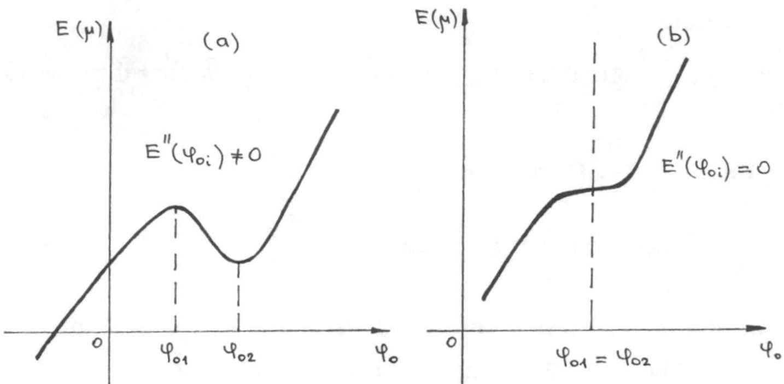

Figure 12.1a. Approximate plot of $E(\varphi)$ when $E^{\parallel}(\varphi_{0i}) \neq 0$.

Figure 12.1b. Approximate plot of $E(\varphi)$ at the bifurcation point.

$F(\varphi) = \dfrac{1}{3}\varphi^3 + a\varphi^2 + b\varphi$. In this case, from (5) we have

$$\varphi_0^{1,2} = -(1+\frac{1}{\mu})\pm[(a+\frac{1}{\mu})^2 - b]^{1/2}, \tag{12.8}$$

whence

$$E''(\varphi_0) = \mp 2\mu[(a+\frac{1}{\mu})^2 - b]^{1/2}. \tag{12.9}$$

From (9) it follows that the state $\varphi_0^{(2)}$ is stable, $\varphi_0^{(1)}$ is not, and that bifurcation points can only occur for $b > 0$ and are

$$\mu_1 = (b-a)^{-1}, \quad \mu_2 = -(b+a)^{-1}.$$

Four regions of the existence of solutions $\varphi_0^{1,2}$ defined by inequalities

$$\mu < \mu_1, \quad \mu > \mu_2,$$

are given in Table 12.1 and Fig. 12.2 (1-4) for various a and b. Other nonlinear functions $F(\varphi)$ may also be treated in a similar fashion. So for polynomials, beginning from the fifth order, the energy levels of locally stable states can intersect with μ changing (i.e. they can coincide for a specific μ). This picture is qualitatively given in Fig. 12.3. Bifurcation points are here μ_2 and μ_5. When $\mu = \mu_1 < \mu_2$, we have two solutions $\varphi_0^{(1)}$ and $\varphi_0^{(2)}$, with the second being unstable. At the point $\mu = \mu_2$, two new solutions, $\varphi_0^{(3)}$ and $\varphi_0^{(4)}$, appear and we have already two locally stable solutions $\varphi_0^{(2)}$ and $\varphi_0^{(4)}$, so that the energy of the latter state $E(\varphi_0^{(4)})$ at $\mu \geqslant \mu_2$ is

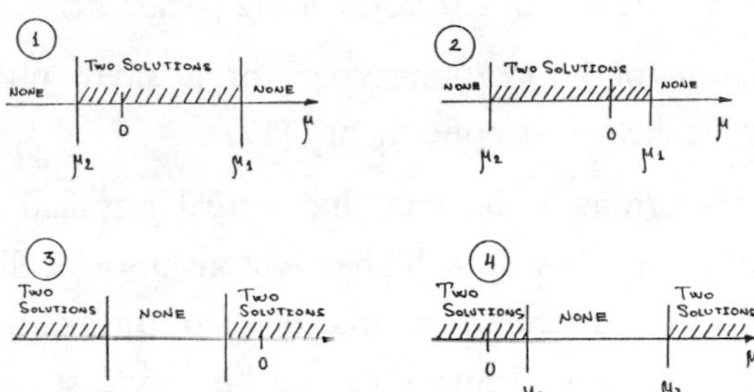

Figure 12.2. The regions of solution existence that correspond to those of Table 12.1.

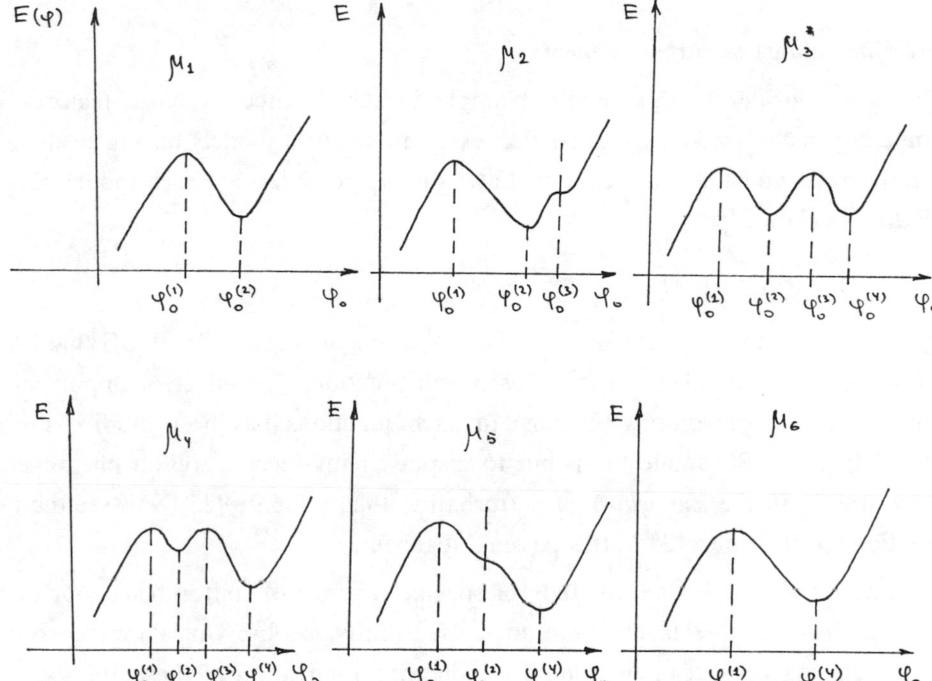

Figure 12.3. Approximate plots of $E(\varphi)$ for various μ.

higher than that of the former state, $E(\varphi_0^{(2)})$. When $\mu = \mu_3^*$, the energies of locally stable states become equal, and then, for $\mu = \mu_4 > \mu_3^*$, we have $E(\varphi_0^{(2)} > E(\varphi_0^{(4)})$, i.e. these states change places. At the point μ_5, solutions φ_0 (locally stable) and $\varphi_0^{(3)}$ (unstable) disappear (coalesce into one) so that at $\mu > \mu_5$ we have again two solutions $\varphi_0^{(1)}$ and $\varphi_0^{(4)}$. The latter appears instead of $\varphi_0^{(2)}$.

We describe here one of the possible ways for the creation (and annihilation) of static states in the system with *one* controlling parameter μ.

A full classification of bifurcation types for a small number of controlling parameters is given in the catastrophe theory [122].

What is really important is the fact that a real physical system can proceed from a locally stable state having a higher energy into locally stable states with lower energies due to, in particular, thermal or quantum fluctuations. In the vicinity of a bifurcation point, such transitions can be very fast and can be employed, for example, in high-speed memories, switches, and so on.

2. Bifurcational perturbation theory

We are considering this simple example in detail since its basic features are inherent to the general situation that exists in realistic models having to do with static states and their bifurcations. Later on, we consider one such model, viz. by letting in (2)

$$F(\varphi) = U(\varphi) = 1-\cos\varphi, \tag{12.10}$$

we arrive at the well-studied (in the homogeneous case, $\mu = 0$) SG model. It describes, in particular, the long Josephson junction. The effect of impurities on the behaviour of solitons (fluxons) in such junctions has been studied in [85-87,91-95]. The SG model turns out to display many-faceted soliton phenomenology already in the framework of perturbation theory [86,91,92,94,95] (see the next section which is devoted to the 'passing' region).

Interest in the theoretical study of specific features of soliton behaviour in the SG model continues to grow due to the feasibility of observing these experimentally, e.g. via laser scanning [96]. The potential application of these are also very tempting in computer technology [92,97].

We consider here the simplest configuration of a semi-bounded junction with a single microshort (microresistance) inhomogeneity. Microshort differs from microresistance with the sign of μ. We emphasize once again that the conventional perturbation theory of solitons [86,91,94,98] does not work in the *bound* region, when exploring static states and their bound states. The authors of [87,89,99] have developed a method which can be called a special bifurcational perturbational theory.

This is briefly as follows. Consider the system of solutions $\varphi_0^{(1)}$ and $\varphi_0^{(2)}$ which are created or disappear at the bifurcation point $\mu = \mu^*$. These are defined with the extrema of $E(\varphi_0)$ (see Fig. 12.3). Now return to the initial, i.e. time-dependent equation (3), and investigate the stability of solutions $\varphi_0^{(1)}$ and $\varphi_0^{(2)}$ via the standard linearization technique. Representing $\varphi^{(j)}$ as

$$\varphi^{(j)} = \varphi_0^{(j)} + \delta\varphi^{(j)},$$

$$\delta\varphi^{(j)} = e^{i\omega_j t}\delta\varphi^{(j)}(x),$$

we get the eigenvalue problem

$$\left[-\frac{d^2}{dx^2} + \partial^2_{\varphi_0^{(j)}} U(\varphi_0^{(j)}) + \mu\delta(x)\partial^2_{\varphi_0^{(j)}} F(\varphi_0^{(j)})\right]\delta\varphi_k^{(j)} = \omega_j^2(k)\delta\varphi_k^{(j)}. \tag{12.11}$$

The state $\varphi_0^{(2)}$ is stable when $\omega_2^2(k)$, i.e. *all* the eigenvalues $\lambda_2(k) = \omega_2^2(k)$ are positive. On the contrary, for unstable $\varphi_0^{(1)}$ we have $\omega_1^2(k) < 0$ and *all* the eigenvalues $\lambda_1(k)$ are negative*). The minor eigenvalue of $\lambda_2(k)$ and the major one of $\lambda_1(k)$ tend to zero, when $\mu \to \mu^*$ so that $(\lambda_1(\bar{k}) = \lambda_2(\tilde{k}))|_{\mu=\mu^*} = 0$. This means that the bifurcation point can be identified with a vanishing eigenvalue. We should note a reservation here; the system with a symmetry group has a set of zero (Goldstone) modes which have nothing to do with bifurcations. These modes should be eliminated. After that the problem to find bifurcation points (lines, surfaces) is reduced to searching for the function $\omega(\mu)$ and its zeros, since the states $\varphi(x)$ and their characteristic frequencies $\omega(\mu)$ alter continuously with parameter μ (in the general case, a vector parameter). The number of states, on the contrary, jumps when $\omega(\mu)$ goes through zero (it is sufficient to cause one frequency of the spectrum $\omega(\mu)$ to vanish).

Thus, those manifolds $\mu = \mu^*$ (i.e. points, lines, surfaces and so on) in the parameter space $\mu = (\mu_1,...,\mu_l)$, where one of the eigenvalues $\omega(\mu)$ vanishes, are the points, lines, surfaces, and so on, of bifurcations of the family of static states. Knowing these manifolds enables a qualitative description of the system considered to be obtained. In the vicinity of the manifolds one can construct expansions of the solution $\varphi(x,\mu)$, its energy $E(\mu)$ and the eigenvalues $\omega(\mu)$ as a formal series over powers of $\Delta\mu = \mu - \mu^*$,

---

*)Such a situation takes place only for the problem with a *single* controlling parameter. In many-parameter problems, a fraction of the eigenvalues can be positive in the vicinity of unstable equilibrium.

$$\varphi(x,\mu) = \varphi(x,\mu^*) + \sum_{n=1}^{\infty} \epsilon^n \phi_n(x), \quad \epsilon = \epsilon(\mu-\mu^*). \tag{12.12}$$

Inserting (11) in the equation of the boundary problem, one can find the functions $\phi_n(x)$. In example, (4), this reads

$$(-\frac{d^2}{dx^2}+1)\phi = -\mu_1\delta(x)F'(\phi),$$

with

$$\phi(x) \to_{|x|\to\infty} 0.$$

This one-parameter problem can be easily investigated. We therefore consider a somewhat more complicated example, but one intrinsic to the long Josephson junction with boundaries. Consider the SG model (10) under the boundary conditions*)

$$\phi'(0,t) = h, \tag{12.13a}$$

$$\phi'(l,t) = 0.$$

This means that on the left end of the junction we have a given constant magnetic field h, and on the right end, this is zero. Let the microshort be at the point x_1, then instead of $U(\phi)+\mu_1\delta(x)F(\phi)$ we have in (2) $(\mu_1 \to -\mu_1)$

$$[1-\mu_1\delta(x-x_1)](1-\cos\phi) \stackrel{\text{def}}{=} V(x,\phi).$$

Denoting $V^{(n)} = \partial_\phi^n V$ we have from (3) the equation governed static solutions

$$\phi'' = V^{(1)} \equiv [1-\mu_1\delta(x-x_1)]\sin\phi. \tag{12.14}$$

Equation (11) assumes the form $(\delta\varphi = \psi)$

$$\hat{L}\psi \equiv (-\partial_x^2 + V^{(2)})\psi = \omega^2\psi \tag{12.15}$$

with boundary conditions

$$\psi'(0) = \psi'(l) = 0. \tag{12.15a}$$

---

*) We take the magnitude $\phi'(l,t) = h_l$ equal to zero for the sake of keeping the analysis simple.

In this case, the controlling parameter space is two-dimensional $\mu = \{\mu_1, h\}$ so that fixing one of the parameters, say μ_1, we find the bifurcation point $h = h^*$ from the equation $\omega^2(h^*) = 0$. We denote the function related to h^* as $\phi_c(x) = \phi(x, h^*, \mu_1)$.

Inserting (12) into eqn. (14) one easily gets

$$\hat{L}\phi_1 = 0, \tag{12.16a}$$

$$\hat{L}\phi_2 = f_2 \equiv -\frac{1}{2} V_c^{(3)} \phi_1^2, \tag{12.16b}$$

$$\hat{L}\phi_3 = f_3 \equiv -\phi_1 (V_c^{(3)} \phi_2 + \frac{1}{6} V_c^{(4)} \phi_1^2), \tag{12.16c}$$

and so on, where $V_c^{(n)} = V^{(n)}(\phi_c, x)$.

Relation (13b) defines in a unique fashion the boundary condition on the right end, viz.

$$\phi_i'(l) = 0. \tag{12.17}$$

The condition on the left end, generally speaking, depends on the connection of the expansion parameter ϵ with the derivation $h - h^*$ where

$$h^* = \phi_c'(x)\Big|_{x-0}.$$

From (13a) and (12), we have in fact

$$\phi'(0) - h^* \equiv h - h^* = \sum_{n=1} \epsilon^n \phi_n'(0). \tag{12.18}$$

To satisfy this condition it is necessary that $\epsilon^n = h - h^* (n = 1,2,...)$.

For $n - 1$, we have

$$\phi_1'(0) = 1, \quad \phi_i'(0) = 0 \quad (i \geqslant 2); \tag{12.19}$$

for $n = 2$

$$\phi_2'(0) = 1, \quad \phi_i'(0) = 0 \quad (i \neq 2), \tag{12.20}$$

and so on.

Equation (15) at the bifurcation point ($\omega = 0$), is none other but eqn. (16a) for the function $\phi_i(x)$. Moreover, at this point, there is, by assumption, a solution of the boundary problem (15a). This implics that the boundary problem (16a) and

(19) has no solutions. The latter fact is readily checked using the following reasoning. Represent the general solution to eqn. (16a) in the form

$$\Psi = c_1\phi_1 + c_2\tilde{\phi}_1,$$

where ϕ_1 and $\tilde{\phi}_1$ are the fundamental solutions. The boundary problem (15a) implies

$$\left.\begin{aligned} \Psi'(l) &= c_1\phi_1'(l) + c_2\tilde{\phi}_1'(l) = 0 \\ \Psi'(0) &= c_1\phi_1'(0) + c_2\tilde{\phi}_1'(0) = 0 \end{aligned}\right\}. \tag{12.21}$$

Problem (17), (19) yields

$$\left.\begin{aligned} \Psi'(l) &= c_1\phi_1'(l) + c_2\tilde{\phi}_1'(l) = 0 \\ \Psi'(0) &= c_1\phi_1'(0) + c_2\tilde{\phi}_1'(0) = 0 \end{aligned}\right\}. \tag{12.22}$$

Systems (21) and (22) possess no common solutions, because the former is solvable when

$$\Delta = \begin{vmatrix} \phi_1'(l) & \tilde{\phi}_1'(l) \\ \phi_1'(0) & \tilde{\phi}_1'(0) \end{vmatrix} = 0$$

and the latter at $\Delta \neq 0$. Thus, expansion begins with

$$\epsilon = (h - h^*)^{1/2} \tag{12.23}$$

and the boundary condition at zero assumes the form (20). Consider this case. Take as the function $\phi_1(x)$ a solution of the boundary problem (16a), (20); the second solution is normalized so that $\tilde{\phi}_1'(0) = 1$, then a solution to (16b), and so on, is given by

$$\phi_n(x) = c_n\phi_1(x) - \phi_1^{-1}\int_x^l dy f_n\{\phi_n(x)\tilde{\phi}_1(y) - \tilde{\phi}_1(x)\phi_1(y)\}. \tag{12.24}$$

The condition $\phi_n'(l) = 0$ in this case is fulfilled automatically. The boundary condition on the left end, (20), enables the constants c_n to be found successively, step by step. From the equation $\phi_2'(0) = 1$ and (24) we have

$$\phi_1(0) = \int_0^l dx f_2 f_1 \overset{\text{def}}{\equiv} <f_2 f_1> = -\frac{1}{2}<V_c^3\phi_1^3> \tag{12.25}$$

accurate to the sign. Two solutions, $\phi_{\pm} = \phi_c \pm \epsilon\phi_1$, corresponding to different signs of $\phi_1(0)$, appear at the bifurcation point and coalesce for $\epsilon \to 0$. From the condition $\phi_3'(0) = 0$, one can obtain c_2, etc.*). Proceeding in the same manner, one can obtain expansions in powers of ϵ for ω^2 and the generalized energy

$$\mathfrak{F} = H + h\phi(0) \tag{12.26}$$

allowing for the energy of external fields which are necessary for the field h to be constant at the left junction end. Inserting the expansion

$$\psi = \sum_{n=0} \epsilon^n \psi_{(n)}, \quad \omega^2 = \sum_{n=1} \epsilon^n \omega^2_{(n)}, \quad V^{(2)} = V_c^{(2)} + \epsilon V_c^{(3)} \phi_1 + \ldots$$

in (15), we get for $\psi_{(n)}$ an integral representation of type (24); namely, from equation $\hat{L}\psi = \omega^2\psi$ follows

$$\hat{L}_c \psi_0 = 0,$$

$$\hat{L}_c \psi_1 = \omega_1^2 \psi_0 - V_c^{(3)} \phi_1 \psi_0,$$

which coincide with (16a) and (16b) if $\psi_0 \equiv \phi_1$ and $f_2 \equiv \psi_0(\omega^2_{(1)} - \phi_1 V_c^{(3)})$ therefore

$$\psi_1 = c\phi_1 - \phi_1^{-1}(0) \int_x^l (\omega^2_{(1)} - V_c^{(3)}\phi_1)\phi_1 \{\phi_1(x)\tilde{\phi}_1(y) - \tilde{\phi}_1(x)\phi_1(y)\} dy.$$

Making use of (15a) we get

$$\omega^2_{(1)} = -2\phi_1(0) < \phi_1^2 >. \tag{12.27}$$

To a first approximation, $\omega^2 = \omega^2_{(1)}$ and the solution with $\phi_1(0) > 0$ is unstable.

Expansion for $\mathfrak{F}$ is easily obtained, anticipating that $\partial\mathfrak{F}/\partial h = \phi(0)$, to give

$$\mathfrak{F} = \mathfrak{F}_c + \int_h^{h^*} dh \cdot \phi(0) = \mathfrak{F}_c + \epsilon^2 \phi_c(0) + \frac{2}{3}\epsilon^3 \phi_1(0) + \ldots$$

$$= \mathfrak{F}_c + (h - h^*)\phi_c(0) + \frac{2}{3}(h - h^*)^{3/2}\phi_1(0) + \ldots.$$

This implies that $\partial_h \mathfrak{F}$ is a continuous function, and $\partial_h^2 \mathfrak{F}$ has a singularity at $h \to h^*$, i.e. the bifurcation point under consideration is analogous to the point of a

*)One can verify, that if the expansion begins from $\epsilon = (h - h^*)^{1/3}$, then $\phi_3'(0) = 1$ and we have the unrealizable condition $< V_c^{(3)} \phi_1^{(3)} > = 0$.

second-kind phase transition*).

In this sense, bifurcational perturbation theory is a generalization of expansions in terms of the order parameter in a vicinity of phase transition points.

One can use bifurcational perturbation theory to perform quantitative calculations upon bifurcation manifolds and to find related ϕ_c.

3. Static states of the long Josephson junction with a single inhomogeneity

In this model we have eqn. (14) and the boundary conditions

$$\phi'(0,t) = h, \quad \phi(+\infty,t) = 2\pi n \quad (n \in \mathbb{Z}). \tag{12.28}$$

Upon integrating (14) once, we get in the regions of homogeneous

$$\frac{1}{4}\phi'^2 - \sin^2(\phi/2) = k^2 - 1 = \epsilon. \tag{12.29}$$

Following Fillipov and Galpern [89], we consider qualitatively the behaviour of solutions to problem (14), (28) via the phase diagram (in the ϕ', ϕ plane). The quantity ϵ is constant in the regions of homogeneity and, as we saw in Chapter 8, Section 2, this fulfils the role of energy in the mechanical analogy method, so we call it the 'mechenergy'. In Fig. 12.4, possible trajectories of the mechparticle are given in the plane ϕ',ϕ.

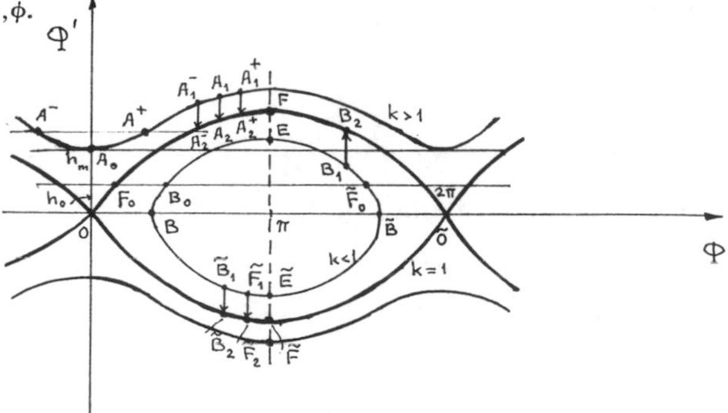

Figure 12.4. Possible trajectories of the mechparticle in the phase plane ϕ',ϕ.

When the inhomogeneity is absent ($\mu_1 = 0$), the solution (14), along with the junction state, are determined by ϵ and correspond to a trajectory in the plane

*)This analogy was pointed out in the paper by Fillipov and Galpern [89b].

(ϕ',ϕ). In particular, the state with $\epsilon = 0$ $(k^2=1)$ corresponds to a kink or an antikink (the separatrix OEO'). States with $\epsilon<0$ are described by cnoidal waves. When $\epsilon>0$, we have infinitely growing solutions. In the case of an inhomogeneity being present, $\mu_1 \neq 0$, a state is described with pieces of phase curves in the regions of homogeneity of the system according to eqn. (29). In going through inhomogeneity the field $\phi(x)$ remains continuous and the derivative $\phi'(x)$ has a jump. The jump is easy to find from the equation of motion and is

$$\phi'_+ - \phi'_- = -\mu_1 \sin\phi(x_1), \quad \phi_\pm = \phi(x_1 \pm 0). \tag{12.30}$$

From (29) it follows that at the left of x_1 we have

$$\frac{1}{4}\phi'^2 - \sin^2(\phi/2) = k^2 - 1, \tag{12.31a}$$

and at the right

$$\frac{1}{4}\phi'^2 - \sin^2(\phi/2) = k_1^2 - 1. \tag{12.31b}$$

By virtue of (30) and (31), the jump of k^2 is positive or negative depending on the sign of $\sin\phi(x_1)$, i.e.

$$\left.\begin{array}{l}\phi(x_1) \in (0,\pi) \Rightarrow \Delta\phi' < 0, \ \ \Delta k < 0 \\ \phi(x_1) \in (\pi,2\pi) \Rightarrow \Delta\phi' > 0, \ \ \Delta k < 0\end{array}\right. \text{ when } \phi' > 0$$

and vice versa for $\phi' < 0$.

Let us consider the upper half-plane of (ϕ',ϕ). Since the junction studied has not a right and any phase trajectory must end in zero or 2π (or possibly in $2\pi n$) as a part of the separatrix. This implies that for phase paths starting inside the separatrix (e.g. from the points B, B_0, and so on) $\Delta\phi' > 0$ and the point x_1 should lie to the right of a point $x_{\min}(\mu_1) > x_\pi$, where $\phi(x_\pi) = \pi$. On the contrary, for paths starting outside the separatrix, $\Delta\phi' < 0$ and x_1 lies to the left of $x_{\max}(\mu_1) < x_\pi$. Thus, for example, for $h = 0$ and given μ_1 there exists $x_{\min}(\mu_1)$ in which two solutions k and $\bar{k}$ appear in addition to the trivial one, $\phi = 0$. For x_1, growing solutions appear with several oscillations before approaching the asymptotics. Note that when $h = 0$ a symmetry exists between kinks and antikinks. When $h \neq 0$ the symmetry is violated together with energy degeneracy so that the kink is related to the trajectory $B_0 B_1 B_2 \tilde{O}$, and the antikink to the $\tilde{B}_0 \tilde{B}_1 \tilde{B}_2 O$ one. The point $x_{\min}(\mu_1)$ is a bifurcation point. If $h = 2\mathrm{sech} x_1$, then the path is related to the separatrix and the kink is localized at the point $x_1 = x_0$. When h increases (or decreases) this state is smoothly transformed into states of the type $A^+A_1^+A_2^+\tilde{O}$ (or $B_0B_1B_2\tilde{O}$). Let us see how, in this case, one can obtain the bifurcation point with respect to h at given μ_1 and x_1. Solutions to the SG equation in a region of homogeneity reads

$$\cos(\phi/2) = -k\,\mathrm{sn}(x+x_0,k) \tag{12.32}$$

where sn is the Jacobi elliptic sine.

Left of x_1 we have from (29)

$$\phi' = \pm[k_0^2-\cos^2(\phi/2)]^{1/2} = \pm 2k_0\sqrt{1-\mathrm{sn}^2} = \pm 2k_0\mathrm{cn}(x+x_0,k_0)$$

and at the left end $(x=0)$

$$\phi'(0) = h = 2k_0\mathrm{cn}(x_0,k_0) \tag{12.33}$$

for $h>0$ (due to the boundary condition).

Then from

$$\phi'(x_1+0)-\phi'(x_1-0) = -\mu_1\sin\phi(x_1)$$

one has

$$\sqrt{k_1^2-c_1^2}-\sqrt{k_0^2-c_1^2} = -\mu_1 s_1 c_1 \tag{12.34}$$

with $s_1 = \sin[\phi(x_1)/2]$, $c_1 = \cos[\phi(x_1)/2]$. As far as we consider the upper half-plane of (ϕ',ϕ), $\phi'>0$ and so the positive roots may be considered. Then $k_1^2 = 1$ and $k_0^2 = \epsilon+1$, and therefore from (34)

$$s_1-\sqrt{k_0^2-c_1^2} = -\mu_1 s_1 c_1$$

or

$$\mu_1 c_1(1-c_1^2)(2+\mu_1 c_1) = \epsilon = k_0^2-1 \tag{12.35}$$

i.e. we have obtained the equation which connects c_1 and ϵ. The magnitude of c_1 is given by (32) to be

$$c_1 = k_0\mathrm{sn}(x_0+x_1,k_0) \tag{12.36}$$

and x_0 is eliminated via (33).

By solving (35) taking into account (33) and (36) we obtain a function

$k_0(h,x_1,\mu_1)$ describing possible junction states for given h, x_1 and μ_1. It is easy to check that the function $k_0(h,x_1,\mu_1)$ is, in the general case, a many-valued function, i.e. there exists more than one value of k_0 for given h,x_1,μ_1 in certain areas of the parameter space. Equation (35) has, in fact, two roots, $c_1^{(+)}$ and $c_1^{(-)}$, if $|\epsilon| < \epsilon_m$ and $\epsilon_m = \epsilon(c_m)$ with c_m being a root of the equation $d\epsilon/dc_1|_{c_m} = 0$ (see Fig. 12.5).

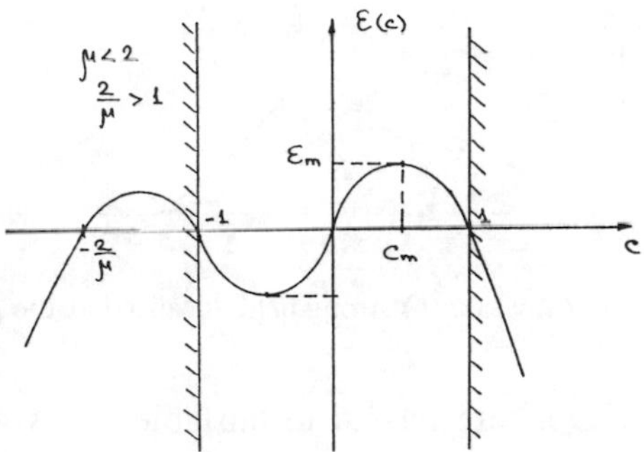

Figure 12.5. The energy of the mechparticle as a function of C.

When $|\epsilon| \to \epsilon_m$, these merge: $c_1^{(+)} = c_1^{(-)} = c_m$. Finally, if $|\epsilon| > \epsilon_m$, eqn. (35) has no solutions.

To find bifurcation surfaces in three-dimensional parameter space still remains not a simple problem; however, the spectrum of system states can be realized by the study of some cross-sections which are natural, from an experimental point of view. In [89b], static states were investigated in the plane (h,x_1), i.e. in the section $\mu_1 = \text{const}$. As a result, three bifurcation lines were obtained. To picture this, we simplify the problem and consider the dependence of the free energy $\mathcal{F}$ on h for a fixed value of x_1. This is shown in Fig. 12.6 taken from [89b].

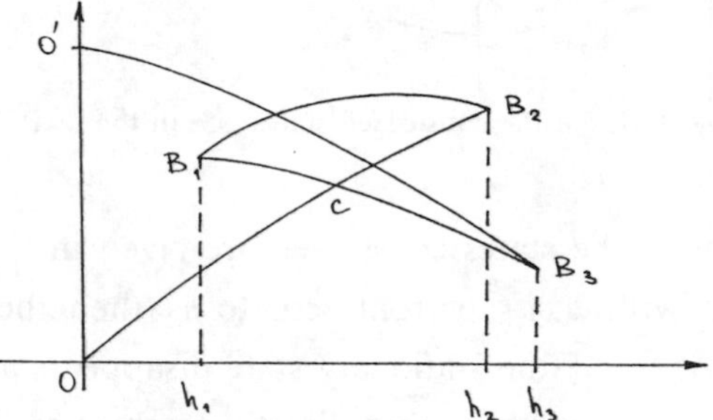

Figure 12.6. The dependence of free energy F on the field h.

The branch OB_2 corresponds to stable states with a weak field ϕ (Fig. 12.7(a)), and the branch B_1B_3 to stable states with a strong field ϕ localized near x_1 (bound state of a kink distorted with the left boundary) (Fig. 12.7(b)).

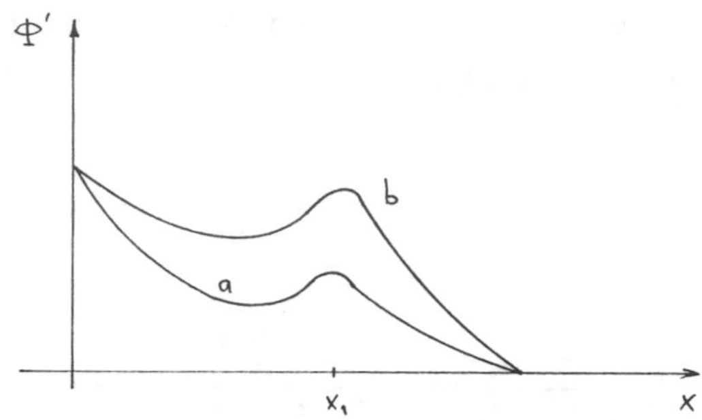

Figure 12.7. Stable states with (a) weak, (b) strong field, localized at the point x_1.

The branches B_1B_2 and $O'B_3$ are related to unstable states with an imaginary frequency ω_0. The values h_1, h_2 and h_3 are the bifurcation points in which the number of bound states changes by two as seen from Fig. 12.6. States relating to the broken line OCB_3 are the states with the lowest energy and hence globally stable. States related to the line B_1CB_2 correspond to a local energy minimum and hence are, generally speaking, metastable. Transitions from metastable states into stable ones will take place due to sufficiently large fluctuations.

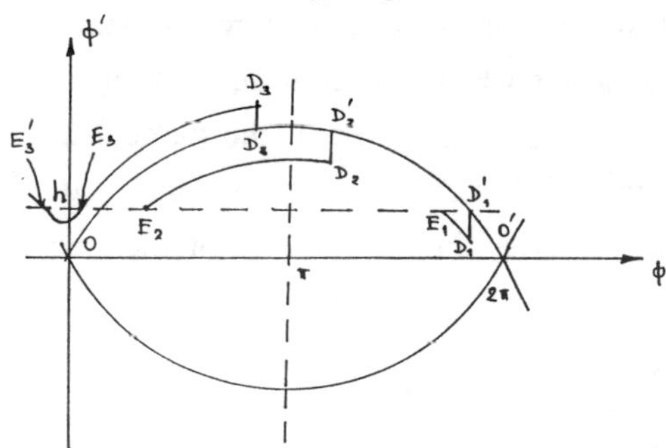

Figure 12.8. The phase trajectories of the considered states (see in the text).

The phase diagrams of the states considered are given in Fig. 12.8. Let h grow in the initial junction with $\phi(x) = 0$ from zero to h_2 then the system will move along the curve OB_2. At $h > h_2$ the stationary state disappears and the system goes into a state with the lowest energy either directly to the line B_1B_3 or through an intermediate (metastable) state on the line $O'B_3$. We also note that by means of an instant action the junction can be rendered from a state with lower energy into the locally stable state with higher energy.

In conclusion, it is of use to notice that the above results imply even quite small ($\mu \ll 1$) local perturbations to be able radically to alter the spectrum of localized soliton excitations of the system; in particular, new solutions appear which are both stable and unstable.

4. Passing region

The perturbation theory of solitons in the 'passing' region has been developed for several years, and many original papers and reviews have been devoted to it. Some early papers in this direction have already been mentioned in Chapter 10, Section 4 and at the beginning of this chapter. We cannot even briefly mention all the results recently obtained in this field (this would require a separate book) so much the more so because these are sometimes contradictary. Moreover these studies continue to develop explosively and there is no unified point of view either in a conceptual or a methodological sense.

We give here an outline of the most advanced and the most often applied methods, as well as recently obtained results.

(1) In studying the behaviour of solitons and their bound states under the action of a weak external force, either analytical or numerical methods are usually employed. As a rule, these are complementary. Analytical methods, being approximate, may be classified into two groups (approaches). Methods based on the technique of the inverse method (spectral transform) are attributed to the first group, the so-called direct methods to the second. Each possesses its own merits.

The approach based on the inverse method (we will call this the perturbation inverse method) enables the carrying out of detailed investigations of a great number of problems of perturbation theory, including those occurring in the solitonless sector. This means that one can treat the dynamics of a packet of continuous spectrum waves under perturbations.

Here we have to record a reservation; the perturbation inverse method is only applicable to a system integrable without perturbations. This fact may be regarded as a shortcoming of the method, as well as an advantage. A shortcoming, since its applicability is limited to the set of integrable systems; an advantage, since we learn much about these systems due to the inverse method.

In particular, the normal modes of integrable systems - their elementary excitations viz. solitons, their bound states, phonons, and so on, play the same role as normal modes in linear systems. Thus, arbitrary initial localized excitation splits up *asymptotically* into a superposition of elementary excitations. Polynomial integrals of motion (such as energy, momentum, particle numbers, etc.), expressed in terms of the action-angle variables, are the sums of normal mode contributions. Proceeding from configurational space to that of scattering data can be regarded as a generalized Fourier transformation for integrable systems [109].

The analogy between linear equations and nonlinear integrable equations persists yet further. As far as weak nonlinearity gives rise to the interaction of linear modes (quasi-particles), the perturbation which denies the integrability of nonlinear systems leads to the asymptotic interaction of nonlinear normal modes*). It is this fact which presents a bridge between weakly nonlinear equations and nearly integrable systems.

Thus, in the framework of nearly integrable systems (i.e. integrable in the absence of perturbations) can appear inelastic effects (e.g. radiation due to soliton collisions) making soliton phenomenology that much richer.

We saw in the first part of the book that a variety of physical models are governed by nearly integrable equations, therefore thorough studies of perturbed versions of the basic integrable equations such as NSE, KdV (and KdV-like equations) and SG are of great interest.

A second advantage of the perturbation inverse method is that it helps to visualize the interpretation of results in terms of the characteristics of the discrete and continuous spectra of the operator of the associated linear problem (scattering data). For example, movement in the complex plane of the pole of the reflection coefficient of the scattering matrix implies that the related soliton parameters alter. The appearance or disappearance of poles means the creation or annihilation of solitons under the influence of perturbations; the same as for bound states of solitons in NSE and SG systems.

This advantage can, however, become a shortcoming, especially in the bound parameter region described above. Here, even weak perturbations present can lead

*)When overlapping, localized modes interact in integrable systems too.

to qualitatively new solutions (including soliton-like ones) having their own excitation spectra. These solutions cannot be described in the framework of the original integrable system, and hence there is no place for them in the related concept of scattering data. We mention, as an example, the appearance of a localized bound state in the linear system with inhomogeneity (see model (4)). Therefore the results derived by means of the perturbation inverse method must be checked against, for example, a simple substitution, or direct methods.

The direct methods of perturbation theory also possess certain merits. These methods are rather straightforward for obtaining results in the adiabatic approximation, and they can be applied to investigate the structural stability of non-integrable systems (in particular, NKG and ϕ^5 NSE). However, they call for considerable *a priori* information and the study of effects which are beyond the scope of the adiabatic approximation (such as radiation, and the appearance of soliton oscillatory tails) also call for much effort. The one-soliton problems, as well as those concerning slightly overlapping solitons, are the simplest problems which can be solved via the direct perturbation methods. Even the problem of the structural stability of a bion requires averaging over its internal oscillations which cannot always be justified.

Last, but not least, come the numerical experiments. They were initiated and developed simultaneously with the direct analytical studies. Many qualitative results were obtained in this way. We present here an example showing the joint application of all three methods.

It is expedient to divide perturbations into two classes; regular and stochastic. The first, in turn, can also be subdivided into two groups: Hamiltonian (conservative) and dissipative perturbations. We give here a short list of recent results relating to soliton phenomenology in nearly integrable systems. Naturally, our emphasis will be on those equations found mostly in physical applications.

1. Nonlinear Schrödinger equations (S3)

We have already drawn attention to many of the seminal works done on this subject by various authors. Here we add that the behaviour of one-soliton, and slightly overlapping two soliton, solutions was studied under the action of weak regular perturbations in [103,104] via the perturbation inverse methods, and in

[105-107] via the direct methods (see, also, the surveys [108,109]). It was established that the action of the weak perturbation

$$F[\varphi] = \epsilon f(x,t) Q[\varphi], \quad \epsilon \ll 1, \tag{12.37}$$

with $f(x,t) \simeq$ const., on a soliton leads to its parameters as well as its shape altering adiabatically. Changes of shape are defined by first order corrections in ϵ and depend on the soliton variables, i.e. $\delta\varphi \propto \epsilon[\varphi_s]$. Allowance for second order corrections in ϵ as performed by Maslov in [110], does not qualitatively alter this conclusion.

The most interesting results are related to the interaction of solitons with each other and with bions (and also between bions) in the perturbed NSE and have been obtained in [113]. Since, in this case, an elastic channel is open there are possible interactions leading to NSE bion creation and decay, the pinning of solitons on an inhomogeneity, and so on. These processes have been found to be of a threshold character and the perturbation inverse method allows, in principle, an estimation of the threshold magnitude; secondly, the interaction picture depends substantially on the nature of the perturbation. This differs in the case of Hamiltonian, dissipative and stochastic, perturbations [14].

2. The Sine-Gordon equation

The perturbed SG equation is likely the most studied form of the nearly integrable equation. The reference list devoted to this numbers more than several hundred papers. We only mention here the reviews [115] where one can find specific references to the original papers, and several papers which have appeared more recently.

In the framework of the perturbed Sine-Gordon equation is also the inelastic channel along with those processes such as the creation and decay of a bion in kink-kink collisions and in kink-bion collisions. Breaking the weakly bound bion into a kink-antikink pair under the action of radiation (a nonsoliton wavetrain), a kink or another bion [113b], is also possible as is capture of a kink by a microinhomogeneity and the decay of a weakly bound bion due to the collision with such an inhomogeneity, and so on [113,116]. All these processes have a threshold character and the threshold characteristics can sometimes be evaluated via the perturbation inverse method.

Some such inelastic processes have been investigated in computer experiments (see, for example, [117,118]) and with the help of the perturbation inverse method. This presents the opportunity to compare results and draw certain conclusions. In particular, bion creation as a result of kink-antikink collisions and from continuous wavepackets has been most carefully studied [116].

It is useful to point out that caution is needed in considering the results obtained via the perturbation inverse method. Some misunderstandings have appeared in the study of one-soliton solutions to the perturbed NSE of the type

$$i\dot{\varphi}+\varphi_{xx}+|\varphi|^2\varphi+\epsilon\bar{\varphi}_x = 0. \tag{12.38}$$

which has been discussed in detail in the paper by Fedyanin and Makhankov [119] (see also the review [120]).

Here, we discuss the kink-antikink collision in the framework of the double Sine-Gordon equation:

$$\Box\varphi+\sin\varphi = \epsilon\sin(2\varphi). \tag{12.39}$$

In Ref. [113] an estimation was made applying the perturbation inverse method of the threshold velocity for colliding kinks to create a bion. This was (see (41) in [113b])

$$v_{thr} = \alpha\epsilon, \tag{12.40}$$

where $\alpha = 2.224$ is the constant estimated by the author of [113b] from a comparison of his result with that of the computer experiment*) of Ref. [121] for two values of ϵ, viz. $\epsilon = 0.1$ and $\epsilon = 0.2$. On the other hand, the results of [121] and especially [118] show that there is a critical value of ϵ, viz., $\epsilon_{cr} > 0$, above which ($\epsilon > \epsilon_{cr}$) the kink-antikink bound states do not exist, or are unstable and decay nearly aperiodically (during several periods).

The authors of [118] gave a simple qualitative explanation of this fact. Consider the small amplitude bion of eqn. (39). With the necessary accuracy, the sines in eqn. (39) can be expanded in series to give, for the coordinate fraction of the function $\varphi(x,t)$, the equation (see Chapter 11, Section 10 where we consider the

*)The constant may, in principle, be calculated by applying the perturbed inverse method.

asymptotic expansions)

$$f'' - (1-2\epsilon)f + \frac{1}{6}(1-8\epsilon)f^3 = 0.$$

This equation has localized solutions only when the coefficient of the cubic term is positive,

$$1-8\epsilon > 0, \tag{12.41}$$

and when the quasi-particles attract each other. From (41) it follows for

$$\epsilon > \epsilon_{cr} = 0.125$$

that the sign changes, and attractive forces are replaced by repulsive ones resulting in a break up of the soliton. The conclusion becomes quantitative only for small amplitude bions. However, computer experiments by Campbell, et al. [118] show that this conclusion also remains qualitatively effective for bions of quite large amplitudes. This means that the perturbation inverse method results can hardly be extrapolated in the region $\epsilon > 0.125$ and that the perturbation inverse method itself is insensitive to this qualitatively new effect.

We also note, in conclusion, that such qualitatively new effects can occur in the transition region of parameters and also in some other systems; in particular, when slow solitons and bions interact with micro-inhomogeneities.

Chapter 13.

DYNAMICAL STRUCTURE FACTORS OF SOLITON GAS

Taking an example from condensed matter physics, we discuss in this chapter how specific features of solitons occur in natural experiments.

Solitons have been encountered in one form or another in nature for time immemorial. Some civilizations have even been destroyed by the type of soliton known as tsunami (a Japanese name meaning waves in the harbour). We will not attempt to list here all possible soliton phenomena in nature. Rather, we shall focus on Mikeska's model for the $CsNiF_3$ crystal in an external magnetic field. This model was discussed at length in Chapter 1 where it was shown that it reduces to well-studied SG model (see Chapter 4) under certain assumptions. Making use of the ideal (dilute) soliton gas approximation - the phenomenological approach - we calculate the dynamical structure factors of the SG model and compare them with experimental data on $CsNiF_3$. These investigations have been carried out at the Joint Institute for Nuclear Research, Dubna, by Lisý, Fedyanin and the author in the early 1980s [123-125,148] and were, to a great extent, stimulated by the pioneering work of Krumhansl and Schrieffer [126], in which, apparently for the first time, the idea was formulated of the ideal soliton gas in the ϕ^4 model. The phenomenological approach emerged from this adea.

The growing interest in studying the SG Dynamical Structure Factors (DSF) is due to the experiments on neutron scattering on $CsNiF_3$ patterns in the parameter region for which Mikeska's model should be valid. For the investigation of various models, powerful experimental methods exist and foremost is the method of magnetic scattering of neutrons. Note that there exists a sufficient variety of quasi-one-dimensional magnets in nature [156]. One of the best studied of these is the ferromagnet $CsNiF_3$. A large number of papers has been devoted to this (see the references cited in [127]). These works were initiated in particular by the work of Mikeska [128] who suggested describing the dynamics of the 'classical' spins in $CsNiF_3$ in the presence of an external magnetic field by the Sine-Gordon (SG) equation. As is well known [129], this equation has three kinds of solutions (elementary excitations): magnons, kinks (and antikinks), and bions (breathers). The last two types of solutions are solitons in the strict sense [129]. In the 'action-angle' variables, the SG-Hamiltonian has been shown to be the sum of three Hamiltonians which describe a mixture of ideal gases of 'phonons', kinks and bions. By 'phonons' we mean the excitations described by solutions of the continuous spectrum (in the particular case we are considering we should refer to 'magnons'). Such solutions lie in an energy range higher than the soliton excitations; the 'optic magnons' with the spectrum $\omega_k = \sqrt{\omega_0^2 + c_0^2 k^2}$ are the solutions of the linearized SG-equation and do not satisfy the SG-equation.

Mikeska in [128] calculated the 'parallel' dynamic structure factor (DSF) $S_{11}(q,\omega)$ of the inelastic scattering of neutrons on kinks and antikinks in an approximation of the ideal nonrelativistic gas of these particle-like objects. His predictions have in general been confirmed by the experimental investigations of Steiner and Kjems [130]. With the purpose of improving agreement between theory and experiment, the theoretical concept has been developed in a number of papers (see, for example, [131]). In addition to the contribution of the 'parallel' dynamic structure factor (DSF) to the central peak the 'perpendicular' DSF ($S_\perp(q,\omega)$) has been calculated. Furthermore, agreement with experiment was slightly improved by the relativistic consideration [131] (a general formula for various structure factors of the scattering on the ideal gas of kinks is given in [124,136]). However, the dependence of the DSF on temperature (T) and the magnetic field (H) has not been satisfactorily explained and significant discrepancies remain between theory and experiment. Reiter [132] suggested describing the neutron experiments on $CsNiF_3$ not in the framework of the nonlinear SG-model but, rather, using the spin-wave theory. With the purpose of obtaining the spin-wave and kink contributions to the DSF, new experimental work have been carried out (for relevant references see [127]). In [127], the experimentally observed behaviour of the DSF is treated in such a way that the spin dynamics of $CsNiF_3$ in the symmetry-breaking magnetic field is a combination of spin-wave processes and kink-type excitations.

However, bions were not taken into account in [127] in the explanation of the experimental data although along with kinks they are stable solutions (solitons) of the SG-equation. Probably for the first time, the significance of bions for an explanation of the above mentioned experiments has been discussed in [133]. An attempt to calculate the bion contribution to the central peak has been made in [134]. The contribution of bions improved the agreement with experiment but only under additional condition that the largest size of the bion is finite. The entire structure and analytical formulae for the DSF were not obtained in [134]. The $S_{11}(q,\omega)$ was also considered in [135], however, the DSF was not obtained. It was concluded in [135] that bions do not contribute to $S_{\perp}(q,\omega)$. The bion contribution to S_{11} and $S_{\perp}$ was calculated to a first approximation (see below) in [123] (see also [136]). As was stated in [123,133,135,136], the activation energy of bions is zero so that their contribution to the dynamic and static characteristics of SG-systems may be significant. For some values of experimental parameters this contribution may dominate the kink contribution. Analytical calculations [123] have shown that:

(a) bions at low temperatures give a considerable contribution to the central peak;

(b) to a first approximation, S_{11} consists (in addition to the central peak) of two narrow side satellite peaks;

(c) $S_{\perp}$ makes a negligible contribution to the central peak but it forms two satellites.

It was supposed in [123] (as in [133,135]) that at low T the main contribution to the DSF is due to large-size bions. The results of [123,136] have been later generalized and specified in [137]. Exact analytical expressions were found for S_{11} and $S_{\perp}$ (in the ideal-gas phenomenology) in terms of functions of the internal frequency of the bion, an infinite series of satellites, 'even' (for S_{11}) and 'odd' (for $S_{\perp}$), with rapidly decreasing heights as the order number of the satellite grows. For the central peak and the first (most intense) satellites obtained earlier [123,136], corrections have been calculated. As a result, all characteristics of the satellites can be found, such as their positions, heights and widths. The dependence of these characteristics on the parameter of the system and on the experimental conditions can be explicitly determined. The qualitative behaviour of the satellites is in good agreement with that of the peaks obtained by numerical methods in [133].

Below we summarize the results of [123,136,137] and apply the formulae obtained to analyse the specific experiments [127] carried out on $CsNiF_3$.

The model

The quasi-one-dimensional ferromagnet $CsNiF_3$, with the easy-plane anisotropy in an external magnetic field H is usually described by the Heisenberg Hamiltonian

$$\mathcal{H} = -J\sum_i \vec{S}_i\vec{S}_{i+1} + D\sum_i (S_i^z)^2 - g\mu_B H \sum_i S_i^x, \tag{13.1}$$

with the parameters [127]

$$J = 23.6K, \quad D = 4.5K, \quad g = 2.25. \tag{13.1a}$$

For the temperature $T > 2.66K$ (the temperature of the transition from the three-dimensional antiferromagnet state) and $T < 24K$ (when the anisotropic model (1) is applicable) [127] $CsNiF_3$ is a good realization of a one-dimensional ferromagnet with plannar anisotropy. The spin dynamics of $CsNiF_3$ obeys apparently the classical description. This fact has been well established at least for $H = 0$ [126] (a detailed comparison between classical and quantum theories and experimental data has been carried out in [138]).

Assuming that the spin $\vec{S}_i$ is a classical vector we have, in spherical coordinates,

$$\vec{S}_i = S\{\cos\theta_i\cos\phi_i, \cos\theta_i\sin\phi_i, \sin\theta_i\}. \tag{13.2}$$

For small angles θ_i and if θ_i and ϕ_i slowly change across the chain, the following Hamiltonian can be obtained:

$$\mathcal{H} \approx Aa_0\sum_i\left[\frac{1}{2}\dot{\phi}_i^2 + \frac{c_0^2}{2a_0^2}(\phi_i - \phi_{i+1})^2 + \omega_0^2(1-\cos\phi_i)\right]. \tag{13.3}$$

Here, a_0 is the distance between the spins ($a_0 = 2.6\text{Å}$) and the relation between the parameters of the Hamiltonians (1) and (3) is as follows:

$$A = \hbar^2/2Da_0, \quad c_0^2 = 2DJS^2a_0^2\hbar^{-2}, \quad \omega_0^2 = 2DSg\mu_B H\hbar^{-2}. \tag{13.4}$$

Using the numerical values (1a) we find (in SI units)

$$A \doteq 3.44\times 10^{-37}, \quad c_0^2 \doteq 2.46\times 10^5, \quad \omega_0^2 \doteq 2.33\times 10^{22}H, \tag{13.4a}$$

where H is measured in kG. The constant A, having the dimension of mass x length, determines an energy scale; c_0 and ω_0 are the characteristic velocity and frequency of the system, respectively. Below, we shall often use the following important parameters:

$$\alpha = 16\beta A\omega_0 c_0 \approx 30.2\sqrt{H}/T, \quad d/a_0 = c_0/\omega_0 a_0 \approx 12.4/\sqrt{H}, \quad d \equiv c_0/\omega_0. \tag{13.5}$$

Here $\beta = (k_B T)^{-1}$. The parameter α defines the notion of 'low' temperatures, and the second parameter implies when the continuous approximation is valid. For typical experimental values of the magnetic field [130,131] ($H=5kG$) and for $T \simeq 10K$ the low-temperature condition ($\alpha \gg 1$) is satisfied as $\alpha \simeq 7$. The continuous approximation is valid if $d/a_0 \gg 1$. For the same H as above we have $d/a_0 \simeq 5.5$ and for $H \simeq 10kG$ (the typical H used in [127]) only $\simeq 3.9$.

The Hamiltonian (3) leads in the continuous approximation to the SG-equation for the angle $\phi(z,t)$:

$$\phi_{tt} - c_0^2\phi_{zz} + \omega_0^2 \sin\phi = 0, \tag{13.6}$$

under the additional assumption $2DS \gg g\mu_B H$. This inequality becomes worse if H increases and for $H \sim 10kG$ we have $2DS/g\mu_B H \sim 6$. The estimations given above show that in the case of the usual experimental temperatures and fields, eqn. (6) is not sufficiently adequate for the model (1). However, here we consider the SG-equation as a good one to describe the spin dynamics of $CsNiF_3$ over the range of the experimental conditions [127] and to discuss the dynamic structure factor of neutron scattering for such a system. Note that the above inequalities place strong limitations on the fields and temperatures for which the theory may be used to explain the experimental results. That is why the dependence of the DSF on q and ω is very important for comparison with experiment. In neutron experiments, these quantities can change over a wide range.

1. General technique to calculate the dynamical formfactor of solitons

The dynamical structure factor of a gas of soliton-like excitations is expressed as

$$S(q,\omega) = \overline{N} S_1(q,\omega), \tag{13.7}$$

where $\overline{N}$ is the average number of soliton in the system in equilibrium, and $S_1(q,\omega)$ is the dynamical formfactor (DFF) of a single soliton. The latter is the Fourier transform of the correlation function

$$S_1(x,t) = \langle \phi(x,t \,|\, M)\phi(0,0 \,|\, M) \rangle. \tag{13.8}$$

In order to construct this, functionals of soliton solutions are employed. In (8), M

stands for a set of free parameters of the solutions over which one should average (phase volume). For example, for a kink (or antikink) these are its initial position (x_0) and velocity (v). For solitons with an internal structure such as $S3$ drops or SG bions averaging goes also over their initial phase (θ_0) and internal frequency (Ω), i.e. in this case $M = \{x_0, v, \theta_0, \Omega\}$.

We define the soliton correlator

$$<F(x,t)> = Z_{1s}^{-1} \int dM_S F(x,t \mid M) e^{-\beta E_s(M)}, \tag{13.9}$$

where

$$Z_{1,s} = \int dM_s e^{-\beta E_s(M)} \tag{13.10}$$

is the ‘partition function’ of a single soliton, and E_s its energy given in terms of M. To obtain the dynamic structure factor we ought to have the Fourier transform of (9)

$$F(q,\omega) = \int <F> e^{i(qx-\omega t)} dxdt. \tag{13.11}$$

Using (9) and (11) we get

$$F(q,\omega) = Z_{1S}^{-1} \int dM_s \int dxdt e^{i(qx-\omega t)} F(x,t \mid M) e^{-\beta E_s}. \tag{13.12}$$

Anticipating the following dependence of a variety of the soliton solutions on x and t

$$\phi(x,t) = \varphi(x - vt - x_0) \exp\{i(kx - \Omega t)\},$$

we first take integrals over x and t going to

$$F(q,\omega \mid M) = (2\pi)^{-2} \int dxdt F(x,t \mid M) e^{i(qx-\omega t)}, \tag{13.13}$$

and then over dM_s

$$F(q,\omega) = Z_{1S}^{-1} \int dM_s F(q,\omega \mid M) e^{-\beta E_s}. \tag{13.14}$$

We see how the procedure works using, as the example, SG kinks (this was proposed in such a form by Fedyanin [124,125]). Thus,

$$F(x,t \mid M) = \phi(x,t \mid M) \cdot \phi(0,0 \mid M) \tag{13.15a}$$

with

$$\phi(x,t \mid M) = \phi\left[\frac{x - \mathrm{v}t - x_0)}{\Delta(\mathrm{v})}\right]. \tag{13.15b}$$

Introducing

$$\xi = \frac{x - x_0}{\Delta}, \quad \tau = \frac{\mathrm{v}t}{\Delta},$$

we have from (13)

$$F(q,\omega \mid M) = \frac{\Delta^2 e^{iqx_0}}{(2\pi \mathrm{v})^2}\phi(-\frac{x_0}{\Delta})\int d\xi d\tau \phi(\xi - \tau) e^{i\Delta(q\xi - \omega\tau/\mathrm{v})}.$$

This integral is readily carried out in the variables $r = 1/2(\xi + \tau)$, $\rho = \xi - \tau$, to give

$$F(q,\omega \mid M) = \Delta(2\pi \mid q \mid)^{-1}\phi(-\frac{x_0}{\Delta})f(q\Delta)e^{iqx_0}\delta(\mathrm{v} - \mathrm{v}_0), \tag{13.16}$$

$$\mathrm{v}_0 = \omega q^{-1},$$

where

$$f(\mathfrak{r}) = \int f(\rho) e^{i\mathfrak{r}\rho} d\rho. \tag{13.17}$$

In obtaining (16), the well-known formula

$$\delta(\phi(x)) = \sum_m \delta(x - x_m) \mid \phi'(x_m) \mid^{-1}$$

is employed with x_m being roots of the equation $\phi(x) = 0$.

Further integration over $dM_k \equiv dx_0 dp$ in (8) is now trivial

$$\int \phi(-\frac{x_0}{\Delta}) e^{iqx_0} dx_0 = \Delta f(q\Delta_0), \;\; \Delta_0 = \Delta(\mathrm{v}_0), \;\; p_0 = p(\mathrm{v}_0),$$

$$\int e^{-\beta E_k(p)}(\ldots)dp = p_0' e^{-\beta E_k(\mathrm{v}_0)}(..\mathrm{v}_0..),$$

to give as the result

$$S_{1k}(q,\omega) \equiv F(q,\omega) = \frac{p_0' \Delta_0^2}{2\pi \mid q \mid} Z_{1k}^{-1} f(\Delta_0 q) f(-\Delta_0 q) e^{-\beta E_k(\mathrm{v}_0)}. \tag{13.18}$$

In the same manner one can obtain the dynamical form factor for a *SG* bion (see [123,125]) however, the calculations become more tedious because of the averaging over θ_0 and Ω.

2. Dynamic structure factor scattering on a soliton gas. The SG model: phenomenological approach

In the gaseous approximation, the cross-section $\sigma_s(q,\omega)$ is expressed via the DSF, $S(q,\omega)$ as follows:

$$\sigma_s(q,\omega) = b^2\frac{k'}{k}S(q,\omega) \equiv b^2\frac{k'}{k}\overline{N}_s S_1(q,\omega), \tag{13.19}$$

where

$$q = k' - k, \quad \omega = E' - E,$$

are, respectively, the momentum and energy lost by the scattered particle, and b is the scattering length.

Fourier transform of spin pair correlation functions, i.e. the dynamical structure factor, $S(q,\omega)$ are found in experiments on inelastic neutron scattering. The DSF of various spin components can be independently measured

$$S^i(q,\omega) = \overline{N}_s S_1^i(q,\omega)$$

with

$$S_1^i(q,\omega) = (2\pi)^{-2}\int <S^i(z,t\,|\,M)S^i(0,0\,|\,M)> e^{i(qz-\omega t)}dz dt.$$

In Mikeska's model of $CsNiF_3$, the main parameters obtained from the experimental data are, as a result of the small value of the z spin component, 'parallel' (with respect to $\mathbf{H}\|0X$) and the 'perpendicular' DSFs, $S_\|$ and $S_\perp$. These are measured independently. Once

$$S = s\{\sin\theta\cos\varphi, \sin\theta\sin\varphi, \cos\theta\}$$

we can average the products

$$\cos\phi(z,t)\cdot\cos\phi(0,0) \quad \text{for} \quad S_\|,$$

and

$$\sin\phi(z,t)\cdot\sin\phi(0,0) \quad \text{for} \quad S_\perp.$$

In the SG model (6) there are two soliton modes:

(1) kinks

$$\varphi_k = 4\mathrm{arctg}\zeta_k, \quad \zeta_k = e^{\gamma d^{-1}(z - vt - z_0)} \tag{13.20}$$

(2) bions

$$\varphi_b = 4\mathrm{arctg}\zeta_b, \quad \zeta_b = \sqrt{1-\Omega^2}\,\sin\theta/\Omega\mathrm{ch}\mathfrak{z}, \tag{13.21}$$

$$\theta = \omega_0\Omega\gamma(t - vzc_0^{-2}) + \theta_0, \quad \mathfrak{z} = d^{-1}\sqrt{1-\Omega^2}\,\gamma(z - vt - z_0),$$

$$a_0 = \Omega^{-1}\sqrt{1-\Omega^2}, \quad \gamma^{-2} = 1 - v^2 c_0^{-2},$$

$$-\infty < z, \ z_0, \ t < +\infty, \quad -c_0 < v < c_0, \quad 0 \leqslant \theta_0 \leqslant 2\pi, \quad 0 < \Omega < 1.$$

1. Consider first the kink contribution to the DSF. Making use of the formulae

$$\cos\varphi_k = 1 - 2\mathrm{sech}^2[d^{-1}\gamma(z - vt - z_0)], \tag{13.22a}$$

$$\sin\varphi_k = 2\frac{\mathrm{sh}[d^{-1}\gamma(z - vt - z_0)]}{\mathrm{ch}^2[d^{-1}\gamma(z - vt - z_0)]}, \tag{13.22b}$$

and (18) we obtain the soliton part of the parallel DSF

$$S_{\parallel}(q,\omega) = \langle(1-\cos\varphi)(1-\cos\varphi_0)\rangle, \quad \varphi_0 = \varphi(0,0\,|\,M).$$

There are several terms in the parallel DSF. Some of these will renormalize the elastic component (Bragg's peak), $S_b \sim \delta(\omega)\delta(q)$ and the remaining terms (due to solitons) describe the quasi-elastic component of the central peak. The latter results in the central peak to be Gaussian in shape and is

$$S_{\parallel}(q,\omega) = \frac{16\bar{N}\gamma_0 M_k}{\mu^2\pi q}\left[\frac{x}{\mathrm{sh}x}\right]^2 Z_{ik}^{-1} e^{-\beta M_k c_0^2 \gamma_0}, \tag{13.23a}$$

where

$$Z_{ik} = 2L\int dp\, e^{-\beta E_k(v)} = \frac{4LE_k^0}{c_0} K_1(\beta E_k^0), \tag{13.23b}$$

$$E_k^0 = 8\mu c_0^2 \equiv M_k c_0^2, \quad x = (\pi q/2\mu\gamma_0),$$

and where L is the length of the system.

For an ideal gas in thermodynamical equilibrium $Z_1(T) = \bar{N}\cdot h$. Nevertheless, we introduce a phenomenological constant $k = h\bar{N}Z_1^{-1}$ to compensate for our lack of knowledge (an ideal gas $\Leftrightarrow$ a soliton gas?). In what follows, k should be best-fitted to experimental data.

The final result for $S_{\|}^{k}$, in dimensional variables, is

$$S_{\|}^{k}(q,\omega) = k16Ad^{3}\frac{|q|}{\hbar\gamma(v_{0})}e^{-\alpha\gamma_{0}/2}\mathrm{sh}^{-2}(x), \quad \mu = d^{-1}. \tag{13.24}$$

Here both kinks and antikinks are taken into account and $S_{k\perp}$ is obtained from (24) by substituting cosh instead of sinh which follows from (22) after integrating. Note that the magnon-kink interaction [139] somewhat alters the pre-exponential factor in (24)*⁾. However, this difference is almost unessential for real experimental parameters. Therefore, in further estimations of the kink contribution to the DSF, formula (24) will be used. As seen from (24), kinks contribute to the central peak only $\omega \in (-|q|c_0, |q|c_0)$. In the experiments [127], the observed side peaks were attributed to the spin-wave processes. However, such an interpretation does not lead to any satisfactory agreement with experiment for all the characteristics compared.

2. The DSF of bions will be calculated in the framework of the ideas gas model too.

The energy and the momentum of the bion are given by

$$E = \alpha\beta^{-1}\gamma\sqrt{1-\Omega^{2}}, \quad p = Evc_{0}^{-2}, \quad \alpha = 16\beta A\omega_{0}c_{0} = 16\beta\mu.$$

The form of the solution (21) results in difficulties in the calculation of the static and dynamic characteristics of bions compared with those of kinks: the kink solution depends only on a simple argument $\sim(z - vt - z_0)$. Now, the average over θ_0 becomes nontrivial and a new internal variable Ω appears. Although the averaging will be done for $\Omega \simeq 1$, we emphasize that the general formulae presented later for $S_{\|}$ and $S_{\perp}$ for bions, (27)-(29) and (40)-(41), are valid for all $\Omega \in (0,1)$.

The thermodynamics of bions is not so well developed as that for kinks. The calculation of the partition function of bions interacting with magnons led at first to a divergence [140]. In [141], this divergence has been eliminated and the partition function has been obtained in the lowest approximation (with an accuracy $O(\beta^{-2})$ at low temperatures). This result, however, does not correspond to exact

*⁾The validity of the inclusion of kink-magnon interactions in the fully integrable *SG*-model is discussed in [120].

calculations based on the transfer-integral method. There is no agreement for bion density in the literature. In the simplest model of an ideal gas, the basic thermodynamic functions of bions have been given in a number of papers (e.g. [123,137,142]). In these papers, two variants for the normalization of the partition function (and thus for the DSF) were used.

(a) In [123,136,137], the bion was considered as a purely classical object, and the variables θ_0 and $\theta = \arccos\Omega$ as parameters over which the classical average was carried out. The statistical integral of one bion from the subsystem of bions with fixed θ_0 and θ is given by $j(T;\theta,\theta_0) = h^{-1}\int dp dx_0 \exp(-\beta E) \equiv h^{-1}Z_1$. The number of bions in the subsystem is $N(T;\theta,\theta_j) = j(T;\theta,\theta_0)$ and their mean total number is obtained by averaging over θ_0 and θ.

(b) In [37], following [143], the canonical conjugate variables were considered, namely, (x_0,p) (describing the bion as a whole) and $(\theta_0, \varphi = 16Ac_0\theta, 0<\varphi<8\pi Ac_0)$ (describing the internal state of the bion). These variables determine the phase space of the bion. In such an approach, we have for the partition function of a single bion

$$\tilde{j}(T) = h^{-2}\int_0^L dx_0 \int_0^{2\pi} d\theta_0 \int_0^{8\pi Ac_0} d\varphi \int e^{-\beta E} dp = M\cdot 16LA\omega_0 h^{-1}[K_1 I_0 - I_1 K_0],$$

where L is the length of the system and the Bessel functions K and I depend on the argument $\alpha/2$. $\tilde{j}(T)$ differs from that of the variant (a) by the factor $M = 16\pi^2 Ac_0 h^{-1}$ (the total number of states which, in the quasiclassical quantization, corresponds to the internal state of the bion). The average value of $F(z_0,p,\theta,\theta_0)$ is now M times larger, too ($M \approx 40$ for $CsNiF_3$!). This contradiction makes it natural to introduce a phenomenological constant k especially for the bion gas. Note that the bion with limited phase space corresponding to its internal state has been considered in [140,141] (in [135], the DSF for bions with different Ω was expressed simply as an integral over all $(\omega_0\Omega)$).

We have for the bion density $\bar{n} = kj(T)/L$. At low temperatures this means that $\bar{n} \simeq k(8hA\omega_0 c_0^2\beta^2)^{-1}$, $\alpha \gg 1$ [123]. With the help of $j(T)$ it is easy to calculate other thermodynamic characteristics [123,136,137].

Let us now calculate the DSF of neutron scattering on the bion gas. $S_{\parallel}(q,\omega)$ is determined by the average of $\cos\phi(z,t)\cos\phi(0,0)$, where, according to (21),

$$1-\cos\phi_b = 8\zeta_b^2(1+\zeta_b^2)^{-2}. \tag{13.25}$$

Omitting terms which contribute to Bragg scattering, the parallel DSF is given by

$$S_{\|}(q,\omega) = k(2\pi)^{-2}h^{-1}\int dzdz_0dtdp\frac{d\theta_0}{2\pi}\frac{2}{\pi}\frac{d\Omega}{\sqrt{1-\Omega^2}}\exp[i(qz-\omega t) \tag{13.26}$$

$$-\beta E(p)](1-\cos\phi(z,t))(1-\cos\phi(0,0)).$$

In closed analytical form, the DSF of the bion gas is obtained by expanding the denominator in (25) in powers of the bion amplitude. At low temperatures (small amplitudes) only a few first terms play a role. Using the procedure described for the kink gas and performing an extra average over bion phases we have, in the approximation, $\cos\phi \simeq 1-8\epsilon^2$, $\sin\phi \simeq 4\epsilon$:

$$S_1(q,\omega) = Z_{1b}^{-1}\sum_{j=1,2}\frac{p'(\mathrm{v}_j)\Delta^2(\mathrm{v}_j)}{2\pi|q_j|}f(q\Delta(\mathrm{v}_j))f(-q\Delta(\mathrm{v}_j))e^{-\beta E_b(\mathrm{v}_j)},$$

where

$$q_j = q+\delta_{j1}\Omega\gamma_j\mathrm{v}_j-\delta_{j2}\Omega\gamma_j\mathrm{v}_j,\quad \mathrm{v}_j = q_j^{-1}(\omega+\delta_{j1}\Omega\gamma_j-\delta_{j2}\Omega\gamma_j),$$

and v_j are the solutions of the equations determined by δ-functions

$$\delta\left(\mathrm{v}-\frac{\omega+\gamma\Omega}{q+\mathrm{v}\gamma\Omega}\right) \text{ and } \delta\left(\mathrm{v}-\frac{\omega-\gamma\Omega}{q-\mathrm{v}\gamma\Omega}\right).$$

The parallel factor $S_{\|}$ contains several terms: those of the order $O(a_0^2)$ will renormalize the coefficient in the Bragg term $\sim\delta(\omega)\delta(q)$ (the same for the term of order $O(a_0^4)$ in the $\cos\phi_b$ expansion). The proper bion part of the correlator is given by both terms (after averaging over θ_0) since $\sin^2\theta = 1/2(1-\cos 2\theta)$:

$$<...>_M = 16a_0^4<\mathrm{sech}^2\mathfrak{x}\cdot\mathrm{sech}^2\mathfrak{x}_0>_{\mathfrak{x}_0}+16a_0^2<\cos 2\theta\cos 2\theta_0>_{\theta_0}\cdot<\mathrm{sech}^2\mathfrak{x}\cdot\mathrm{sech}^2\mathfrak{x}_0>_{\mathfrak{x}_0}.$$

The first term accurate to the coefficient $4a_0^4$ is $S_{\|}^k$ (23a) if we substitute in it $m = a_0\Omega$ instead of μ and $E_b^0 = M_bc_0^2$ instead of $E_k^0 = 8\mu c_0^2$. Denoting this by S_{cp} we have

$$S_{cp}(q,\omega,\Omega) = \frac{A_{cp}}{b_1}\left(\frac{x_1}{\mathrm{sh}x_1}\right)^2\exp(-\beta M_bc_0^2qb_1^{-1}), \tag{13.28a}$$

$$A_{cp} = \frac{32a_0^2M_b}{\pi\mu^2\Omega^2}\left(\frac{\bar{n}_b}{Z_{1b}}\right),\quad b_1 = \sqrt{q^2-\omega^2c_0^{-2}},\quad x_1 = \frac{\pi b_1}{2a_0\Omega}.$$

This bion contribution into the central peak is added to the contribution of the kinks $S_{\parallel}^{k}$.

The second term, the satellite term

$$S_{\parallel\text{sat}}(q,\omega,2\Omega) = \frac{A_{\parallel}}{b_2}\left[\frac{x_2}{\text{sh}x_2}\right]^2\left[\exp(-\beta M_b c_0^2\gamma_{\parallel}^{+})+\exp(-\beta M_b c_0^2\gamma_{\parallel}^{-})\right], \tag{13.29a}$$

$$A_{\parallel} = \frac{8a_0^2 M_b}{\pi\mu^2\Omega^2}\left[\frac{\bar{n}_b}{Z_{1b}}\right],\quad b_2 = \sqrt{q^2 - c_0^{-2}(\omega^2-4\Omega^2)},\quad x_2 = \frac{\pi b_2}{2a_0\Omega},$$

$$\gamma_{\parallel}^{\mp} = (q^2+4\Omega^2c_0^{-2})\,|\,qb_2 \pm 2\omega\Omega c_0^{-2}\,|^{-1},$$

describes the combinatorial scattering.

Similarly we get, for the perpendicular factor,

$$S_{\perp\text{sat}}(q,\omega,\Omega) = \frac{A_{\perp}}{b_{\perp}}(\text{ch}x_{\perp})^{-2}[\exp(-\beta E_b^0\gamma_{+})+\exp(-\beta E_b^0\gamma_{-})], \tag{13.41a}$$

$$A_{\perp} - \frac{2\pi M_b}{\mu^2\Omega^2}\left[\frac{\bar{n}_b}{Z_{1b}}\right],\quad b_{\perp} = \sqrt{q^2 - c_0^{-2}(\omega^2-\Omega^2)},\quad x_{\perp} = \frac{\pi b_{\perp}}{2a_0\Omega},$$

$$\gamma_{\pm} = (q^2+\Omega^2c_0^{-2})\,|\,qb_{\perp} \pm \omega\Omega c_0^{-2}\,|^{-1},$$

taking into account the next terms of the expansion in a_0^2 after some transformations integration can be carried out in (26) and $S_{\parallel}$ assumes the form (all these calculations are given in detail in [37]):

$$S_{\parallel}(q,\omega) = S_c(q,\omega) + \sum_{N=1}^{\infty} S_{\parallel\text{sat}}^{N}(q,\omega). \tag{13.27}$$

The first term, S_c, determines the direct contribution to the central peak, the rest (with $N = 1,2,\ldots$) as in the above will be described as satellites. The exact expressions for the central peak and the satellites have the form

$$S_c = k\frac{2^{10}Ac_0}{\pi^2 h\omega_0^2}\int dv\,d\Omega\frac{\gamma^2\exp[-\beta E(v)]}{\Omega(1-\Omega^2)}\left[\sum_{n=1}^{\infty}\frac{(-1)^n}{[(n-1)!]^2}\left[\frac{1-\Omega^2}{\Omega^2}\right]^n \times \right. \tag{13.28b}$$

$$\left.|\,\Gamma(n+ib/2)^2\right]^2\delta(a),$$

$$S_{\parallel\text{sat}}^{N} = k\frac{2^{10}Ac_0}{\pi^2 h\omega_0^2}\int\frac{dv\,d\Omega\gamma^2}{\Omega(1-\Omega^2)}e^{-\beta E(v)}\left[\sum_{n=N}^{\infty}\frac{(-1)^n n^2}{(n+N)!(n-N)!}\left[\frac{1-\Omega^2}{\Omega^2}\right]^n \times \right. \tag{13.29b}$$

$$|\Gamma(n+ib/2)|^2]^2[\delta(a-2N)+\delta(a+2N)].$$

In (28b) and (29b) the notation is as follows

$$a = \gamma(q\mathrm{v}-\omega)/\omega_0\Omega, \;\; b = \gamma d(q-\omega\mathrm{v}c_0^{-2})/(1-\Omega^2)^{1/2},$$

where $\Gamma(x)$ is the gamma-function and $\delta(x)$ is the Dirac delta-function. It must be noted that for small Ω, the series in (28b) and (29b) become inconvenient for the calculations: the terms of the series increase with n. However, as has been mentioned above, the main contribution to the scattering at low T is given by small amplitude bions ($\mathfrak{x} \ll 1, \Omega \sim 1$) [123,136] connected with the cut-ff factor $\exp[-\alpha\gamma(1-\Omega^2)^{1/2}]$ at $\alpha \gg 1$. In this sense, the solutions (28b) and (29b) determine corrections (formally all corrections) to the solution [123,136]. It is easy to recognize that already for $\Omega^2 > 1/2$ the series become nonformal. They converge well and their partial sums can be used. (Note that at small Ω a bion is not distinguished from a kink-antikink pair. We have already taken into account such states in the model of the ideal gas of kinks, antikinks and bions. Hence, the natural limitation arises for the smallest value of Ω: $\Omega > a_0/d$ [120]).

The essential feature of the scattering on the bions is the fact that the first satellites appear. The heights of the other satellites rapidly decrease as N increases [123] and we do not consider these here.

Consider first the central park in more detail. One integration in (28b) is easily performed due to the δ-functions. Integrating, for example, over v we obtain

$$S_c = k\frac{2^{10}Ad}{\pi^2 h\,|q|}\int d\Omega\frac{1-\Omega^2}{\Omega^4}\gamma(\mathrm{v}_0)e^{-\beta E(\mathrm{v}_0)}\left[\sum_{n=0}^{\infty}\frac{(-1)^n}{(n!)^2}\left(\frac{1-\Omega^2}{\Omega^2}\right)^n\times\right. \tag{13.30}$$

$$\left.|\Gamma(n+1+ib(\mathrm{v}_0)/2)|^2\right]^2, \quad \mathrm{v}_0 = \omega/q.$$

In a more compact form, this expression can be written using the hypergeometric function $F \equiv {}_2F_1$:

$$S_c = k\cdot 2^8 d^3\frac{A\,|q|}{h\gamma(\mathrm{v}_0)}\int d\Omega\frac{\exp[-\beta E(\mathrm{v}_0)]}{\sinh^2[\pi b(\mathrm{v}_0)/2]}\left|F\left(\frac{ib(\mathrm{v}_0)}{2},1+\frac{ib(\mathrm{v}_0)}{2};1;1-\Omega^2\right)\right|^2, \tag{13.30a}$$

or in a form suitable for the following estimations

$$S_c = k\frac{2^{10}Ad\delta^2}{\pi^2 h\,|q|\,\gamma(\mathrm{v}_0)}\int d\Omega g(\Omega)e^{f(\Omega)}, \tag{13.30b}$$

where

$$f(\Omega) = -\alpha\gamma(v_0)(1-\Omega^2)^{1/2} - \delta/\gamma(v_0)(1-\Omega^2)^{1/2},$$

$$g(\Omega) = \Omega^{-4}[1-\exp(-\delta/\gamma(v_0)(1-\Omega^2)^{1/2})]^{-2} \times$$

$$\left[1+\sum_{n=1}^{\infty}\frac{(-1)^n}{(n!)^2}\left[\frac{1-\Omega^2}{\Omega^2}\right]^n\prod_{l=1}^{n}(l^2+b^2/4)\right]^2.$$

Here, $\delta = \pi q d$ as in (24) for kinks. In [123,136] account was taken of the fact that for $\alpha \gg 1$ the function $f(\Omega)$ has a sharp maximum at $\Omega = \bar{\Omega} \sim 1$: $1-\bar{\Omega}^2 = \delta/\alpha\gamma^2(v_0)$. To estimate the integral in (30b) we thus use as in [123,136], the saddle-point method

$$S_c \approx k\frac{2^{10}A d\delta^{11/4}\exp[-2(\alpha\delta)^{1/2}]}{\pi^{3/2}h\,|q\,|\gamma^3(v_0)\alpha^{5/4}}\left[1+\sum_{n=1}^{\infty}\frac{(-1)^n}{\gamma^{2n}(v_0)}\prod_{l=1}^{n}\left[\frac{\delta}{\alpha}+\frac{\delta^2}{4\pi^2 l^2}\right]\right]^2, \tag{13.31}$$

$$2(\alpha\delta)^{1/2} \gg 1, \quad \delta/\alpha \ll 1.$$

The leading term of S_c (the factor before the square brackets) has been obtained in [123,136] (in these works the units $A = \omega_0 = c_0 = 1$ were used).

Let us now estimate $S^N_{\|sat}$ from (29b). We use the expression $\delta(\phi(x)) = \Sigma\delta(x-x_i)/|\phi'(x_i)|$, where x_i are the solutions of the equation $\phi(x) = 0$, and we specify $q, \omega \geqslant 0$. From the equations $a \pm 2N = 0$, we obtain the quadratic equations $qv-\omega\pm 2N\omega_0\Omega\gamma^{-1} = 0$. Using the solutions $v_{1,2}$ of these equations we find after intergration over v the parallel DSF, which, by analogy with S_c, can be written in terms of the hypergeometric functions. The solution is given by

$$S^N_{\|sat} = k\cdot Ah^{-1}(2^7 d)^2\int d\Omega G(\Omega)[e^{F_+(\Omega)}+e^{F_-(\Omega)}], \tag{13.32}$$

$$F_{\pm}(\Omega) = -\frac{B_N(\Omega)}{\sqrt{1-\Omega^2}} - \alpha\sqrt{1-\Omega^2}\frac{(qd)^2+(2N\Omega)^2}{|qdB_N/\pi\pm 2N\Omega\omega/\omega_0|},$$

$$G(\Omega) = \pi^3 B_N^{-3}(1-e^{-B_N/\sqrt{1-\Omega^2}})^{-2}\left[\sum_{n\geqslant N-1}\frac{(-1)^n(n+1)^2}{(n+1-N)!(n+1+N)!}\left[\frac{1-\Omega^2}{\Omega^2}\right]^{n+1}\times\right.$$

$$\left.\prod_{l=0}^{n}(l^2+\frac{B_N^2}{4\pi^2(1-\Omega^2)})\right]^2,$$

$$B_N(\Omega) = \pi[(qd)^2+(2N\Omega)^2-(\omega/\omega_0)^2]^{1/2}.$$

In [123,136], Ω-integration was carried out as in the case of S_c by the saddle-point method at $\bar{\Omega} \sim 1$. This is because, for large α and for ω not being too close to the frequency $[(qc_0)^2 + (2N\omega_0)^2]^{1/2}$, the functions $F_{\pm}(\Omega)$ have sharp maxima at $\Omega = \bar{\Omega}$.

Now we are able to estimate the 'effective length' of the bion. It is given by the formula $\Delta_B \approx d/\gamma\sqrt{1-\bar{\Omega}^2} = d(\alpha/\delta)^{1/2}$ (see (24)) and is of the order of the kink size ($\Delta_B \approx 2.2d$, $d = c_0/\omega_0$) for the typical experimental values [127] ($H = 10kG$, $T \sim 10K$ and $\delta = \pi qd = 1.95$). The bion density, determined from $n_B = 2/\alpha\beta c_0 h$, is only three times larger than the density of the kinks $n_k = (2/d)(\alpha/\pi)^{1/2}\exp(-\alpha/2)$. Thus the value of $n_B\Delta_B$ is of the order of 10^{-1} ($n_B\Delta_B$ is the relation between the whole length of the bions and the length of the system). In the first approximation we may then consider bions not to be overlapping and thus we may use the gaseous approach. Note that this indicates that the choice of the variant (a) of the normalization of the bion statistical sum must be preferred; which leads, incidentally, to better agreement with experiment.

In such a way the following features have been established in [136-137]:

(a) $S_{\parallel}$ are narrow peaks in frequency;

(b) at low temperatures the positions of the satellites are nearly independent of T but with T increasing the satellites move toward the central peak and their heights increase;

(c) the satellites are localized near the frequencies $\omega_N^{\parallel}$ given by

$$\omega_N^{\parallel} = [(qc_0)^2 + (2N\omega_0)^2]^{1/2}; \tag{13.33}$$

(d) the heights of the peaks strongly decrease with increasing N. The independence of the position of a satellite on q and H is immediately seen from (3) and (4):

(e) the satellites move from the central peak if q or H increases.

Although the method of estimations used in (32) allows us to draw some correct qualitative conclusions concerning the behaviour of the satellites, it only gives the rough behaviour of $S_{\parallel sat}^{N}$ at $\omega \to \omega_N^{\parallel}$ (this is especially true, as we shall see later, in the case of $S_{\perp\mathrm{sat}}$). Therefore, here we shall use another, more exact, method of estimation. Consider the first (most powerful) satellite ($N = 1$). The satellite contribution to the central peak is not zero but small, and the satellite does not vanish only in the vicinity of its maximum, i.e. near the frequency $\omega_1^{\parallel}$. Substituting in (2) at $N = 1$, $\omega > qc_0$,

$$y = B_1(\Omega)/B_1(1), \quad \Delta_\| = [(\omega_1^\|)^2 - \omega^2]^{1/2}/2\omega_0, \tag{13.34}$$

and we obtain, for $\Delta_\| \ll 1$, the following estimation:

$$S^1_{\|\mathrm{sat}} \approx k{\cdot}Ah^{-1}(2^5 d)^2 \Delta_\|^3 \int_0^1 dy{\cdot}y^2 \sinh^{-2}\left[\frac{\pi y}{(1-y^2)^{1/2}}\right] \exp\{-\alpha\Delta_\|\omega(1-y_2)^{1/2}/2\omega_0\}. \tag{13.35}$$

If $\alpha\omega\Delta_\|/2\omega_0 \gg 2\pi$ (which is possible for $\alpha \gg 1$) the main region of integration is near $y{\sim}1$ and the saddle-point method works well. We obtain, then, the previous formula [123,136]:

$$S^1_{\|\mathrm{sat}} \approx kAh^{-1}(2^6 d)^2(\pi\omega_0/\alpha\omega)^{5/4}\Delta_\|^{7/4}[1-\exp(\pi d\omega\Delta_\|/\omega_0)^{1/2}]^{-2}. \tag{13.36}$$

In the case for which $\alpha\Delta_\|\omega/2\omega_0 \ll 2\pi$, we have

$$S^1_{\|\mathrm{sat}} \approx k{\cdot}Ah^{-1}(2^6 d/10)^2\Delta_\|^3, \tag{13.37}$$

i.e. $S^1_{\|\mathrm{sat}} = 0$ if $\omega = \omega_1^\|$. In the region of intermediate $\Delta_\|$, where the maximum of the peak lies, the estimation can be written in the following term:

$$S^1_{\|\mathrm{sat}} \approx k{\cdot}Ah^{-1}(2^6 d/10)^2\Delta_\|^3\exp[-\alpha\omega\Delta_\|/2\omega_0]. \tag{13.38}$$

The maximum height of the first satellite

$$S^1_{\|\mathrm{sat}|\max} \approx 0.06k{\cdot}Ah^{-1}(2^5 d)^2(2\omega_0/\alpha\omega)^3 \tag{13.38a}$$

is reached at $(\alpha\omega\Delta_\|/2\omega_0)^2 \approx 10$. Thus, the distance from $\omega^\|_{\max}$ to $\omega_1^\|$ where $S^1_{\|\mathrm{sat}} = 0$ is approximately

$$\omega_1^\| - \omega^\|_{\max} \approx 2^3{\cdot}10\omega_0^4\alpha^{-2}(\omega_1^\|)^{-3} = 5\omega_1\alpha^{-2}[1-(qc_0/\omega_1^\|)^2]^2, \tag{13.39}$$

and is very small (numerical estimations will be given below).

Let us now calculate the perpendicular DSF, $S_\perp(q,\omega)$. In [123,137], it has been established that $S_\perp$ makes a very small contribution to the central peak consists of an infinite series of 'odd' satellites $S^N_{\perp\mathrm{sat}}$ located near the frequencies $\omega_N^\perp = [(qc_0)^2+(2N+1)^2\omega_0^2]^{1/2}$, $N = 0,1,\ldots(S^N_{\perp\mathrm{sat}} = 0$ if $\omega = \omega_N^\perp)$. $S_\perp$ is determined by averaging the product $\sin\phi(z,t)\sin\phi(0,0)$ where, according to (21), $\sin\phi = 4\mathfrak{x}(1-\mathfrak{x}^2)(1+\mathfrak{x}^2)^{-2}$. Substituting $\sin\phi$ into (26) we obtain [137]

$$S_\perp(q,\omega) = \sum_{N=0}^{\infty} S^N_{\perp\mathrm{sat}}(q,\omega), \tag{13.40}$$

where

$$S^{N}_{\perp\,\mathrm{sat}} = k\frac{2^6 d^2 A}{\pi h}\int\frac{d\Omega}{\Omega^2 B'_N}\left[\sum_{n\geqslant N}\frac{(-1)^n(2n+1)^2}{(n-N)!(n+1+N)!}\left(\frac{1-\Omega^2}{\Omega^2}\right)^n\times\right.$$

$$\left.|\,\Gamma(n+\frac{1}{2}+\frac{iB'_N}{2\pi(1-\Omega^2)^{1/2}})|^2\right]^2(e^{-\alpha E_+(1-\Omega^2)^{1/2}}+e^{-\alpha E_-(1-\Omega^2)^{1/2}})$$

$$B'_N(\Omega) = \pi[(qd)^2-(\omega/\omega_0)^2+(2N+1)^2\Omega^2]^{1/2}, \tag{13.41}$$

$$E_{\pm}(\Omega) = \frac{(qd)^2+(2N+1)^2\Omega^2}{|(2N+1)^2\Omega\omega/\omega_0\pm qdB'_N/\pi}.$$

The most noticeable contribution to $S_\perp$ is given by the first satellite ($N=0$). We consider this in more detail in the lowest approximation. The saddle-point method used in [123,136] for the analytical approximation of $S^1_{\perp\mathrm{sat}}$ must be applied with care because, near $\omega\to(q^2c_0^2+\omega_0^2)^{1/2}$, divergent expressions may appear [125].

In this case, two regions of ω must be considered:

(a) The region of the central peak $\omega<qc_0$. Here, the saddle-point method is well applicable and the contribution of $S^1_{\perp\mathrm{sat}}$ to the central peak is as follows:

$$S'_{\perp\mathrm{sat}} \approx k\cdot 2^3 h^{-1}\left[\frac{\pi^6 c_0^3}{A\omega_0^{13}\beta^5 B'_1(1)}\right]^{1/4}[E_+(1)]^{-5/4}(1+e^{\sqrt{\alpha\beta'_1(1)E_+(1)}})^{-2}+ \tag{13.42}$$

$$+[E_-(1)]^{-5/4}(1+e^{\sqrt{\alpha B'_1(1)E_-(1)}})^{-2}\}.$$

(b) The region $\omega\gtrsim qc_0$. In this region, near $\omega_1^\perp$, there is a satellite peak. Substituting

$$y = B'_1(\Omega)/B'_1(1),\ \text{and}\ \Delta_\perp = \omega_0^{-1}[(\omega_1^\perp)^2-\omega^2]^{1/2}, \tag{13.43}$$

this peak at $\Delta_\perp\ll 1$ is approximated by the formula

$$S^1_{\perp\mathrm{sat}} \approx k\cdot 2^9 A h^{-1} d^2\Delta_\perp\int_0^1 dy\cosh^{-2}\left[\frac{\pi y}{2(1-y^2)^{1/2}}\right]\times \tag{13.44}$$

$$\exp(-\alpha\omega\Delta_\perp(1-y^2)^{1/2}/\omega_0).$$

If $\pi\omega_0/\alpha\Delta_\perp\ll 1$ (the essential region of integration is near $y\sim 1$) by using the saddle-point method we obtain a result consistent with (2):

$$S^1_{\perp\mathrm{sat}} \approx k\cdot 2^9 h^{-1} A d^2\Delta_\perp^{-1/4}(\alpha\omega/\pi\omega_0)^{-5/4}[1+\exp(\pi\alpha\Delta_\perp\omega/\omega_0)^{1/2}]^{-2}. \tag{13.45}$$

In the very vicinity of $\omega_1^{\perp}$ $((\alpha\omega_{\perp}/\pi\omega_0) \ll 1)$ we have

$$S^1_{\perp \mathrm{sat}} \approx kAh^{-1}(2^3 d)^2 \Delta_{\perp}. \tag{13.46}$$

In the region of intermediate values of ω, $\alpha\omega\Delta_{\perp}/\pi\omega_0 \sim 1$, the satellite is approximated by the formula

$$S^1_{\perp \mathrm{sat}} \approx kAh^{-1}(2^3 d)^2 \Delta_{\perp} \exp(-\alpha\Delta_{\perp}\omega/\omega_0), \tag{13.47}$$

with the maximum at $\alpha\omega\Delta_{\perp}/\omega_0 \sim 1$ being

$$S^1_{\perp \mathrm{sat}|\max} \approx kA\,\omega_0(2^3 d)^2/eh\alpha\omega. \tag{13.48}$$

The difference $\omega_1^{\perp} - \omega_{\max}^{\perp}$ is now approximately

$$\omega_1^{\perp} - \omega_{\max}^{\perp} \approx \omega_0^4 \alpha^{-2}(\omega_1^{\perp})^{-3} = \omega_1^{\perp}\alpha^{-2}[1-(qc_0/\omega_1^{\perp})^2]^2. \tag{13.49}$$

3. $CsNiF_3$ and the SG model

In this section, we use the above expressions to estimate numerically the contribution of bions to the DSF in the case of $CsNiF_3$ using specific experimental conditions [127]. First, we compare the bion and kink contributions to the central peak. Using (24) and (31), the ratio of the maximum heights of $S_{k\|}(q,\omega)$ and $S_c(q,\omega)$ is written as

$$\frac{S_c(q,0)}{S_{k\|}} \approx k\frac{2^5\delta^{3/4}}{\pi^{1/2}\alpha^{5/4}}\sinh^2\frac{\delta}{2}\exp[\alpha/2-2(\alpha\delta)^{1/2}]\cdot[1-...]^2. \tag{13.50}$$

Here, α and δ are defined by (5) and (24), and the factor $[1-...]^2$ by (31) $(\omega=0)$. The applicability of (50) (in the idea-gas model) is restricted because the inequalities $2(\alpha\delta)^{1/2} \gg 1$ and $\delta/\alpha \ll 1$ must hold. In experiments [127], values of H, T and q are such that our approximations are quite justified. As an illustration, we give the table of the ratio $S_c(q,0)/S_{k\|}(1,0)$ for the experimental situation described in [127].

If the correction factor $[1-...]^2$ is supposed to be close to unity we see (for $k=1$) that $S_c/S_{k\|}$ as a function of α has a minimum at $\alpha=\alpha_{\min}$ $(\sqrt{\alpha_{\min}} = \sqrt{\delta}+\sqrt{\delta+5/2})$. Two threshold temperatures exist for which the bion contribution equals the kink contribution. Above one of these (T_1), and below the other (T_2), the contribution of bions will dominate. In the first experiments (see, e.g. [130,131]) typical values of H and q were $5Kg$ and $0.1\pi/a_0$, respectively. For

$T[K]$	α	$S_c(q,0)$ /	$S_{K\parallel}(q,0)$	$[1-...]^2$
		$k=1$	$k=40.7$	
9.0	10.6	0.027	1.10	0.59
6.0	15.9	0.075	3.05	0.66
4.2	22.7	0.177	7.20	0.70
3.2	29.8	0.657	26.70	0.73

Table 1. Example of the ratio of the bion and kink central peak heights. Parameters: $H = 10kG$, $q = 0.05\pi/a_0$, $\delta = 1.95$ [127].

these parameters and the experimental temperatures ($T = 6.3 - 14K$) the contribution of bions to the central peak, if estimated using (50), is found to be negligible (note that our approximations are rough in these cases). For the experimental conditions [127] and for $k = M \doteq 40.7$ the situation essentially alters. The heights of the bion central peak is always larger than that of the kinks: if T decreases, the bion contribution strongly predominates. Accounting for kinks only is absolutely insufficient in the ideal-gas phenomenology.

An important characteristic is the halfwidth of the central peak as well. For $\alpha \gg 1$ in the nonrelativistic approximation, the halfwidth of $S_{k\parallel}$ becomes $(\Delta\omega)_k \approx 2qc_0(\ln 2/\alpha)^{1/2}$. For $S_c(q,\omega)$, if $[1-...]^2 \approx 1$, the halfwidth $(\Delta\omega)_B$ is essentially determined by the factor $(1-\omega^2/q^2c_0^2)^{3/2}$, so $(\Delta\omega)_b \approx qc_0(1-1/2^{2/3})^{1/2}$. The correction $[1-...]^2$ is considerable at $\omega \to 0$, decreasing the height of the central peak; at $\omega \to qc_0$, $[1-...]^2 \to 1$. Thus, the width of the bion central peak is larger than that of the kinks. From the estimation $(\Delta\omega)_B/(\Delta\omega)_k \geqslant 0.37\alpha^{1/2}$, one can see that this ratio increases if T decreases.

Now consider the satellites. The first 'even' satellite is approximately described by (35)-(39). The value of $S^1_{\parallel\text{sat}}$ is negligible at $\omega \to 0$ and vanishes if $\omega = \omega_1^{\parallel}$. Qualitatively, the dependence of $\omega_1^{\parallel}$ on T, H and q has been discussed above. For the situation [127] (Table 1) the calculated value of $\omega_1^{\parallel} \approx 1.01 \times 10^{12} s^{-1} \approx 0.67meV$ is close to the experimental value. However, the results for the position and height of the first satellite differ from those determined experimentally. The position of the maximum of $S^1_{\parallel\text{sat}}$ given by (39) can be seen such that $\omega_1^{\parallel} - \omega_{\max}^{\parallel}$ is a maximum for $q \to 0$. In this case, $\omega_{\max}^{\parallel} = 2\omega_0(1-\delta/\alpha^2)$ and differs from $\omega_1^{\parallel}$ by only a few per cent. For the experimental situation discussed we obtain $\omega_1^{\|} - \omega_{max}^{\|\|} \approx 2.5 \times 10^{-2} meV (T=9K)$ and less (for smaller T's) as $\sim T^2$. The experimentally observed curve has a peak at $\omega \approx 0.4 meV$ and the decrease of T does noticeably alter the position of this maximum. Another discrepancy appears if we consider the height of the satellite. Thus, for the ratio $S^1_{\|sat|max} / S_{k\|}(q, 0)$ we have numerical values of 0.19 $(T=9K)$, 0.8 $(T=6K)$ and 8.2 $(T=4.2K)$ for $k = 1$ or 7.73, 32.5, and 333 for $k = M$, respectively (other parameters remain the same). Although this ratio increases if T decreases, such a strong increase has not been observed experimentally.

The first 'odd' satellite is described by (45)-(49) with a cut-off frequency $\omega_1^{\perp}$. The maximum value of $\omega_1^{\perp} - \omega_{max}^{\perp}$ equals ω_0 / α^2 and is even smaller (for the same T, H and q) than that for $S^1_{\|sat}$. The position of the $S^1_{\perp sat}$ peak, its dependence on T and its height differ from experimentally observed characteristics.

Thus, the phenomenology of the ideal gas of bions allows us to predict the satellite-like structure of the DSF of neutron scattering on $CsNiF_3$. In experimental work, satellites have been observed with properties analogous to those described above (for the satellites nearest to the central peak). But such a model does not give a quantitative agreement with experiment. Since a good quantitative agreement with experiment gives no theory employed for the explanation of the experimental data [127] (the kink theory of Mikeska [128], the spin-wave theory [132] and that which is discussed above), one can state that the mathematical models proposed to describe $CsNiF_3$ do not entirely satisfy the physical situation. The out-of-plane effects for spin-dynamics have been already discussed in [144], which lead to corrections of the light-axis type (Schrödinger type)*). Moreover, at low temperatures, quantum effects, which have not been considered here, may play a role. They place a restriction on the smallest mass of the bion. Note that the conditions which determine the low temperatures and the continuous approximation are only realized with a small margin. Finally, bions are the true solitons only in the framework of the pure SG-equation. In fact, in experiments as well as in

*) For $s = 1$, the spin goes out of the easy plane with a given probability, and a classical description based on the Landau-Lifshitz model is hardly adequate here. Recently in [157], [158] new four parameter models were proposed which correspond to the $s = 1$ Heisenberg ferromagnet averaged over $SU(3)/SU(2) \times U(1)$ generalized coherent states. There are governed by systems of four nonlinear equations which are reduced to the Landau-Lifshitz equation in a definite limit, only.

Mikeska's model, there exists a great number of perturbations. These presumably effect the behaviour of bions making their life-time finite which gives rise to the broadening of the resonance line to the Lorentz one and a shifting of its peak. Among such factors in particular is the spin-phonon interaction.

The intensity of the central peak (i.e. the static structure factor $S(q) = \int S(q,\omega)d\omega$) is in better agreement with experiment than the form of the line of $S(q,\omega)$. For the kink central peak we have, from (24) in the nonrelativistic approximation,

$$S_{k\parallel}(q) \approx 2^6\pi^{3/2}Ac_0^4q^2h^{-1}\omega_0^{-3}\alpha^{-1/2}\exp(-\alpha/2)\sinh^{-2}(\delta/2), \tag{13.51}$$

and $S_{k\perp}$ is obtained by replacing sinh → cosh. The bion contribution to $S(q)$ given by (31) is approximately

$$S_c(q) \approx k\cdot\frac{2^{11}Ac_0^2\delta^{11/4}}{\pi h\omega_0\alpha^{5/4}}\exp(-2(\alpha\delta)^{1/2}),\ \ \delta/\alpha \ll 1,\ \ 2(\alpha\delta)^{1/2} \gg 1. \tag{13.52}$$

This expression is not valid for very small q when the inequality $2(\alpha\delta)^{1/2} \gg 1$ does not hold. It is easy, however, to obtain the following more exact estimation if $q \to 0$:

$$S_c(0) = k\cdot 2^{11}\pi^{-2}Ah^{-1}dc_0\int_0^1\int_0^1 dx d\Omega(1-\Omega^2)(1-x^2)^{-1/2} \times \tag{13.53}$$

$$\exp[-\alpha(1-\Omega^2)^{1/2}/(1-x^2)^{1/2}]$$

$$= k\sqrt{2}\cdot 2^7\pi^{-1}Ah^{-1}dc_0\{5!!+9!!/2^3\alpha^2+\ldots\}/\alpha^2.$$

Figure 13.1 illustrates the $S(q)$ as a function of T. Bions and kinks are both taken into account. The result for kinks is practically the same in the ideal-gas approach as in the model of the gas of kinks interacting with magnons [139]. The account for bions leads to a slightly better (but not sufficient) agreement with experiment (for $k = 1$ and for $k = M$). In Figure 13.2, $S(q)$ is shown as a function of the wave vector. In this case (with the normalization point at $q = 0$) the agreement with experiment is even worse for bions. Figures 13.3 and 13.4 illustrate $S_\parallel(q,\omega)$ and $S_\perp(q,\omega)$ defined by kinks, antikinks and bions. In all these figures, the same normalization has been used as in [5]. We note that the use of $k = 16\pi^2Ac_0h^{-1}$ leads to a much poorer agreement with experiment that if $k = 1$.

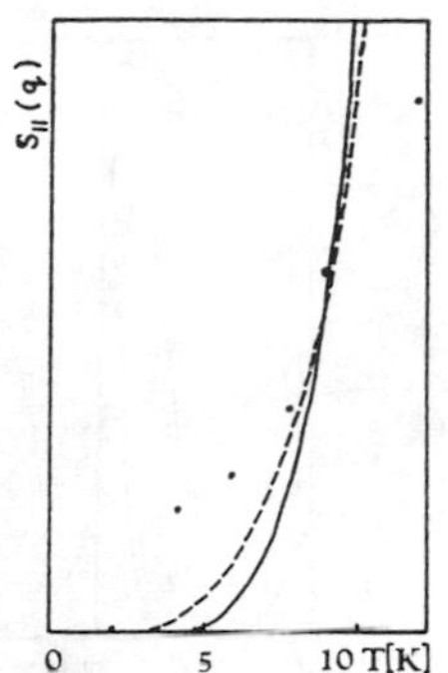

Figure 13.1. Dependence of $S_{\|}(q)$ on $T(S_{\|} = S_{k\|} + S_c)$ at $q = 0$, $H = 10kG$ and $k = 1$ (solid line) and $k = M$ (dashed line). Points are experimental data [127].

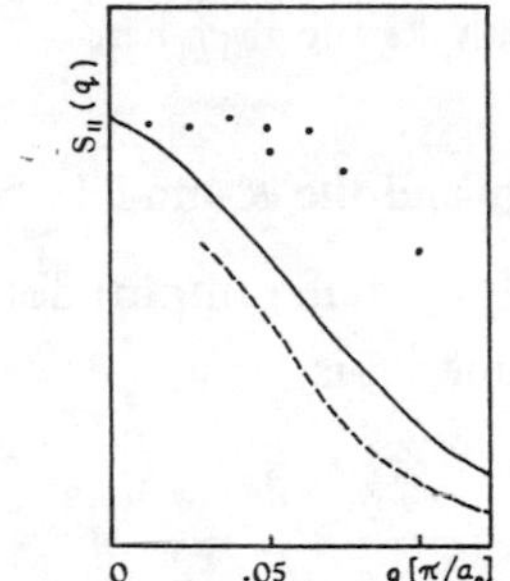

Figure 13.2. Dependence of $S_{\|}$ on q at $H = 10kG$, $T = 12K$ and $k = 1$ (solid line) and $k = M$ (dashed line). Points are experiment [127]. At $k = M$ due to the large value of $S_c(0)$ the discrepancy between the theory and experiment is significant. At $k = 1$ the bion influence is small: the resulting curve differs very slightly from $S_{k\|}(q)$.

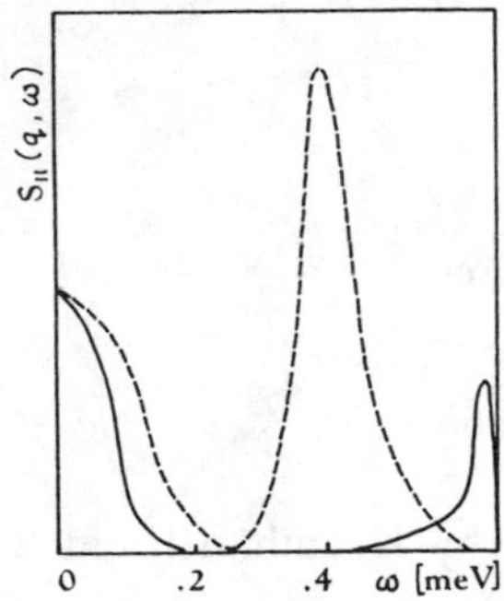

Figure 13.3. Dependence of $S_{\|}(q,\omega)$ on ω at $T = 6K$, $H = 10kG$ and $q = 0.05\pi/a_0$ taking into account kinks and bions ($k=1$). Solid line is theory, dashed line is experiment [127]. If $k = M$ the values of $S_{\|}$ in the central peak region slightly decrease and the satellite increases by about 14.8 times in comparison with the one given here.

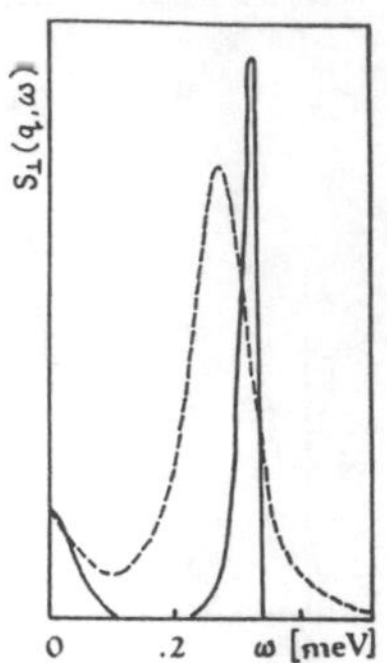

Figure 13.4. Dependence of $S_\perp(q,\omega)$ on ω at $T = 12K$, $H = 10kG$ and $q = 0.03\pi/a_0$. Solid line is theory, dashed line is experiment [127]. *CP* is due only to kinks. The satellite at $k = M$ increases by M times in comparison with the one given here.

4. The ideal gas phenomenology and the ϕ^4-model

At low temperature (hence for small amplitude) the (approximate) analytical description for bions in the ϕ^4 model is

$$\Box\chi \pm U(\chi) = 0,$$

where $U(\chi) = \mu^2\chi - gg\chi^3$ the upper sign is related to the ϕ^4_+ model, the lower to the ϕ^4_- one. Bion solutions of the ϕ^4_+ model accurate to the amplitude renormalization coincide with the small amplitude *SG* bions. Small amplitude bions of the ϕ^4_- model are given by the formula (11.41). Using this formula, one can easily obtain the DSF of kink and bions, $S = S_{cp} + S_1 + S_2$, as

$$S_{cp} = \frac{1}{4} S_{cp}(q,\omega)_{SG},$$

$$S_1 = \frac{1}{18} S_\perp(q,\omega,\bar{\Omega})_{SG}, \qquad \bar{\Omega} = \sqrt{2}\,\Omega_{SG}$$

$$S_2 = \frac{1}{36} S_\parallel(q,\omega,2\bar{\Omega})_{SG},$$

in the ϕ^4_- model, (employed by Krumhansl and Schrieffer to describe the structural phase transition). The phenomenology of the kink gas in this model has been discussed in Chapter 11. The existence of the resonance states (along with the kink shape mode of frequency ω_1) gives rise to the appearance of new satellite lines located near ω_1 and harmonics. So the DSF of the ϕ^4_- model, in addition to the central peak and the bion satellites, will contain the extra satellite the location of which is practically independent of temperature.

We conclude this section with the following remark. In studying soliton-bearing systems, one must take into account all the soliton-like excitations (at any rate, all the stable and long-lived modes).

Modes related to bound states of solitons and their localized excitations are much more difficult to consider due to their internal structure, therefore the temptation often arises to make recourse to the well-known picture of interacting linear waves (the picture of weak turbulence). This temptation is especially strong due to the fact that the bound states of solitons (bions, breathers) are frequently also specific bound states of a great number (in the quasi-classical limit) of linear waves (magnons, phonons, and so on), i.e. quasi-particles. By virtue of this fact, certain properties (including static ones) of the bion excitations masquerade as those of quasi-particle turbulence. Satellites in the SG DSF are sometimes attributed to such effects. At any rate, the position of the satellite centres at low temperatures (when $\bar{\Omega} \simeq \omega_0$) is well explained in terms of interacting spin waves. The movement of satellites with temperature, however is, different for the two mechanisms. The average bion frequency $\bar{\Omega}$ is defined by temperature (bions of such a frequency are most likely in the equilibrium bion gas) since, the virtue of the cut-off factor $\exp(-\beta E_b^0 \gamma_0)$ in formulae (35), (38) and (42)-(47), the estimation

$$2M_k^0 \gamma (1-\bar{\Omega}^2)^{1/2} \simeq T$$

may be applied whence $d\bar{\Omega}/dT < 0$. This implies the bion satellite position in DSF to move towards the central peak with temperature T. The processes of fusion and the decay of spin waves, governed by the conservation laws $\omega_1 = \omega_2 \pm \omega_3$, $k_1 = k_2 + k_3$, produce the frequencies $\omega \simeq \omega_0 n$ $(n \in \mathbb{Z})$ independent of T. Besides this, the allowed bion spectrum is $\omega = \omega_0 \Omega < \omega_0$ whereas for spin waves this is $\omega > \omega_0$.

5. Soliton gas kinetics

One of the important questions we have been only slightly concerned with above is the problem of the relaxation of the soliton gas to the equilibrium state. For example, in Chapter 11 we suppressed the lack of information in the constant k. It is obvious that this question is the simplest in the case of the kink gas. In recent years, a number of papers have been published by Baryakhtar and his colleagues which have been devoted to the non-equilibrium properties of the kink gas in magnetic systems [145].

In fact, they have considered the problem of how equilibrium is attained in a system of two gases, kinks and quasi-particles (magnons, phonons, and so on), viz. they have calculated the kinetic characteristics of the kink gas. Note that the interaction of kinks with linear waves in so far as the effects on the equilibrium properties are concerned has also been considered in the papers [139,146] and [126].

The problem of how equilibrium is reached in the kink-magnon system is similar to that considered by Bogolubov and Krylov in their 1939 paper [148]. They investigated the relaxation of a heavy gas impurity in a light gas when collisions between heavy particles can be neglected. The relaxation of a heavy particle is sufficient to be considered in a gas of light particles, i.e. the processes of the scattering of light particles by the heavy one should be studied. In integrable one-dimensional systems (such as the SG model) corresponding collision integrals in the kinetic equation vanish (reversibility) and stochastization appears together with interactions which destroy integrability. Even though such interactions are weak, they govern the process of attaining equilibrium; in particular, decceleration of domain walls [149] and collision integrals in the kinetic equation for kinks [145,150]. The consistent derivation of related formulae is beyond the scope of this book. By various examples, however, we survey recent tendencies in this direction.

Let us proceed to discuss briefly the results of paper [145]. Assuming the integral of kink-phonon collisions to be of the Focker-Planck type, the authors derived kinetic coefficients for the $D = 1$ SG and ϕ^4_- models. Both pure and perturbed, their versions were considered to be described by the Lagrangian

$$L = \frac{E}{c_0\omega_0}\int dx\{\frac{1}{2}(\dot{\phi}^2 - c_0^2\phi_x^2) - \omega_0^2 U(\phi)\}. \tag{13.54}$$

Representing the perturbed kink as $\phi(x,t) = \phi_0(\xi)+\varphi(x,t)$ one obtains the expansion of L in powers of φ:

$$L = L_0+L_1+L_2+....$$

The kink velocity is supposed to change *slightly* upon a single kink-phonon collision ($M_k \gg m$), and adiabatic perturbation theory is applied. Then the state of the system is given as an expansion over the eigenfunctions of the unperturbed

Hamiltonian, and the perturbation gives rise to a transition between these states.

The non-equilibrium thermodynamics of the kink gas is thereby described by the Focker-Planck equation

$$\dot{f}+\mathrm{v}\frac{\partial f}{\partial x} = -\frac{\partial}{\partial p}J(p), \tag{13.55}$$

for the distribution function $f(x,t,p)$, with

$$J(p) = A(p)f(p)+\frac{1}{2}\partial_p[B(p)f(p)] \tag{13.56}$$

being the dissipative current, and p and x the momentum and coordinate of the kink, respectively.

The coefficients A and B are such that $J(p)$ vanishes in the equilibrium state. The diffusion coefficient (D) and the viscosity coefficient (η) are expressed in terms of B:

$$\eta = (2nT)^{-1}\int_{-\infty}^{\infty} B(p)f_0(p)dp, \tag{13.57a}$$

$$D = 2T(nM_k)^{-1}\int_{-\infty}^{\infty} pf_0(p)dp\int^{p} dp' B^{-1}(p'), \tag{13.57b}$$

with $n(x) = \int f_0 dp$ being the probability for the kink to be at the point x, and f_0 is the equilibrium distribution function. The quantity B is expressed through the dissipative function

$$Q = \frac{1}{2}\frac{dE_k}{dt} = -\frac{1}{2}\frac{d}{dt}(Tr(\rho E)),$$

where E_k is the kink energy, E the operator of the magnon gas energy, and ρ the density matrix. The function Q can be constructed with pieces corresponding to the interactions of the kink with two magnons, three magnons and so on. In the particular case of $n = 2$

$$Q^{(2)} = \frac{\pi\mathrm{v}}{2\hbar}\sum_{k_1k_2} q_{12}\,|\psi(1,2)|^2(n_{k_1}-n_{k_2})\delta(\omega_2-\omega_1-\hbar q_{12}),$$

where $\psi(1,2)$ is an amplitude in the Hamiltonian of two-particle interaction, $\hbar q_{12}$ the momentum transferred, and n_{k_i} the magnon occupation numbers.

Two-magnon processes

1 A kink at rest ($v/c \to 0$)

The amplitude $\psi(1,2)$ for the ϕ^4 model is calculated, accurate to (v^2/c_0^2), to be [145]

$$\psi(1,2) = 0 + O(v^4/c_0^4) \text{ at } \omega_{k_1} = \omega_{k_2}.$$

This result holds in one dimension as well as in three dimensions. In the *SG* model, also $\psi(1,2) = 0$, which follows from the integrability of the $D = 1$ case. The fact that $\psi(1,2)$ vanishes in non-integrable models is extremely non-trivial. It means that the amplitude of magnon scattering on a kink (at rest) amounts to zero in the framework of both integrable and a number of non-integrable models, and this property remains valid in three dimensions.

2. Moving kink

The scattering amplitude $\psi(1,2)$ is naturally equal to zero in integrable models for any v/c. It also vanishes in the ϕ^4_- model ($D=1,3$) and the *SG* model ($D=3$), and the two-magnon contribution to relaxation processes is absent[*)] due to the Lorentz invariance of the models.

Three-magnon processes

This amplitude, $\psi(1,2,3)$, also vanishes in the $D = 1$ *SG* model. In the $D = 3$ *SG* model and the $D = 1,3$ ϕ^4_- models, $\psi(1,2,3) \neq 0$, and three-magnon processes lead to the kink gas relaxation. In the ϕ^4_- model, the following processes play a crucial role:

- the interaction of a kink with three magnons from a continuous spectrum (non-localized magnons)

$$H_1(3) = \sum_{k_1 k_2 k_3} [\psi(1,2,3) c^+_{k_1} c_{k_2} c_{k_3} + \text{h.c.}]$$

with the summation spanning all magnon wave vectors;

---

*)Note that the $\psi(1,2) \propto v^2 k_\perp^2$ in three-dimensional ferromagnet [149] (where $k_\perp$ is the magnon momentum in the kink plane).

- the interaction of a kink with a localized magnon and two non-localized ones

$$H_2(3) = \sum_{k_2 k_3} [\psi(1,2) c^+_{k_1} c_{k_2} c_0 + \text{h.c.}]$$

- the interaction of a kink with a non-localized magnon and two localized magnons

$$H_3(3) = \sum_k [\psi(k) c^+_0 c^+_0 c_k + \text{h.c.}].$$

The operators c_i and c_i^+ are time-independent and defined using the Bose operators of creation a_k^+ and annihilation a_k,

$$q_k = \left[\frac{\hbar}{2m\omega_k}\right]^{1/2} (a_k + a^+_{-k}), \quad \omega_k = \sqrt{4\omega_0^2 + k^2 c_0^2},$$

$$p_k = i\left[\frac{\hbar m \omega_k}{2}\right]^{1/2} (a_k - a^+_{-k}), \quad m = E_0 / c_0 \omega_0,$$

via a unitary operator $U(x_k)$,

$$c_k = U^+(x_k) a_k(x_k) U(x_k):$$

the evolution operator along the slow variable x_k (the so-called rotating axes representation [151]).

The authors of [145] give the following formulae for the quantity $B = B_1 + B_2 + B_3$ (see (57)) defining the kinetic coefficients

$$B_1 = 8\pi \sum_{k_1 k_2 k_3} |\psi(1,2,3)|^2 n_{k_1}(n_{k_2}+1)(n_{k_3}+1)(k_1 - k_2 - k_3)^2 \delta(\omega_1 - \omega_2 - \omega_3) \simeq$$

$$\simeq \frac{\hbar^2 \omega_0^3}{c_0^2} \left[\frac{\hbar\omega_0}{E_0}\right] \begin{cases} 0.15 (T/\hbar\omega_0) \exp(-4\beta\hbar\omega_0), & 4\beta\hbar\omega_0 \gg 1, \\ 2.5 \cdot 10^{-3} (T/\hbar\omega_0)^3, & 4\beta\hbar\omega_0 \ll 1, \end{cases}$$

$$B_2 = 4\pi \sum_{k_1 k_2} |\psi(1,2)|^2 n_{k_1}(n_{k_2}+1)(n_0+1)(k_1 - k_2)^2 \delta(\omega_1 - \omega_2 - \sqrt{3}\,\omega_0) \simeq$$

$$\simeq \frac{\hbar^2 \omega_0^3}{c_0^2} \left[\frac{\hbar\omega_0}{E_0}\right] \begin{cases} 0.2 (T/\hbar\omega_0) \exp[-\beta\hbar\omega_0 (2+\sqrt{3})], & 4\beta\hbar\omega_0 \gg 1, \\ 4.10^{-3} (T/\hbar\omega_0)^3, & 4\beta\hbar\omega_0 \ll 1, \end{cases}$$

$$B_3 = 8\pi \sum_k |\psi(k)|^2 n_k (n_0+1)^2 k^2 \delta(\omega_k - 2\sqrt{3}\,\omega_0) \simeq$$

$$\simeq \frac{\hbar^2\omega_0^3}{c_0^2}\left(\frac{\hbar\omega_0}{E_0}\right)7.7\cdot 10^{-3}\text{sh}^{-1}(\omega_0\hbar\beta)\text{sh}^{-2}(\omega_0\hbar\beta\sqrt{3}/2),$$

$$n_0 = n\sqrt{3}\,\omega_0.$$

From these formulae, at high temperatures, comparing the contributions of all the above processes, we obtain

$$B \simeq 6.2\times 10^{-3}(T^3/E_0c_0^2\omega_0^{-1}),\ \ 2\beta\hbar\omega_0 \ll 1. \tag{13.58a}$$

At low temperatures, B_j contains a cut-off factor, $\exp(-\alpha_j\beta\hbar\omega_0)$ (α_j is a numerical factor depending on j). Comparing the α_j, we find that $B_3 \gg B_2 \gg B_1$, so that

$$B \simeq 6.2\times 10^{-2}\left(\frac{\hbar^3\omega_0^4}{E_0c_0^2}\right)\exp(-2\sqrt{3}\,\hbar\omega_0\beta),\ \ \beta\hbar\omega_0 \gg 1. \tag{13.58b}$$

In the cited work, a contribution to B due to additional terms in the Lagrangian (54) is estimated such as

$$\Delta L_1 \simeq \int dx(\phi^2-1)^3$$

or

$$\Delta L_2 \simeq \int dx(\dot{\phi}^2-\alpha c_0^2\phi_x^2)^2,$$

and it is shown that in the ϕ^4 model such a contribution can be compared with, or even exceeds, that of three-magnon processes. In the $D = 1$ SG model, these terms are responsible for the relaxation of the system to equilibrium, and in $D = 3$ SG models these can compete with the three-magnon processes. Note also that the three-magnon processes vanishing in the $D = 1$ models show their integrability as well as does the existence of an additional (fourth) integral of motion. Formulae (58) give an estimation of the relaxation time of the kink gas after either its formation or after a sufficiently strong perturbation. It is seen that, at low temperatures, $\beta M_k \geqslant 1$, this time grows exponentially with T decreasing.

6. Turbulence of a soliton gas

We proceed now to a somewhat contrary example; the study of the kinetics of solitons in the NSE model with a selfconsistent potential [152]. The distribution function

$$\rho(a) = \frac{a}{a_0}(a_0^2 - a^2)^{-1/2} \tag{13.59}$$

of the soliton gas appearing due to the decay of a monochromatic wave in the $S3$ model was obtained in Chapter 4 (see (4.32), (4.33)). The method of quasi-classical quantization was applied to the linear problem. Some features of this function were discussed there in quite detail. It appears to be like the δ-function but with some distinctions. Due to the fact that inelastic interactions are forbidden in the integrable $S3$ system, the amplitude of the decaying wave sets an upper limit on the soliton amplitudes (see, for details, Section 4.3). How will the situation change in a non-integrable system, e.g. the system of the NSE plus wave equation (Chapter 1, Section 6)

$$i\dot{\psi}+\psi_{xx} - U\psi = 0, \tag{13.60}$$

$$\Box U = \partial_x^2 |\psi|^2.$$

Numerical studies [152,153] show that, in this system, possible are inelastic processes of interaction of solitons with each other and with sound pulses (Fig. 11.4). Three, most characteristic, groups of inelastic processes are distinguished:

(1) soliton pair interactions;

(2) interaction of a soliton with a sound pulse;

(3) three-soliton interactions.

In the first group, the phenomenology of the interactions is as follows: two solitons with similar amplitudes*) can coalesce; solitons with amplitudes differing by a factor more than 2 interact in a quasi-elastic fashion.

In the second group, inelastic processes are those of the interaction of a soliton with a well-localized rarefraction pulse.

In the third group, inelastic processes involving sound packets created in the

---

*) With due regard for the relativistic factor $\gamma^{-2} = 1 - v^2$.

preceding two-soliton fusion are possible. Such processes can result in a break-up of a third soliton if its amplitude does not exceed half the amplitude of the colliding solitons.

Employing this phenomenology, and following [152], we can infer qualitatively the behaviour of the soliton gas in the framework of system (60). To do this, write an equation expressing the balance of the number of solitons and find the distribution function of solitons over their amplitudes. Simplifying the problem, we proceed from a continuous amplitude spectrum to a discrete one (the amplitude lattice with spacing a_0).

Since inelastic processes of both fushion and fission of solitons take place if the ratio of soliton amplitudes is, to put it crudely, $2^n (n=0,1,2,...)$ we may regard the lattice as $2^n a_0$. Thus, calculations show that solitons of nearly equal amplitudes $a_0 (n=0)$ coalesce to produce sound which can break down solitons of half that amplitude $(a_0/2)$. In each fushion, sound waves are emitted with an energy capable of breaking down 64 solitons of half-amplitude. This estimation can be readily performed using the energy integral. In fact, (60) possesses three additive integrals of motion: the particle number integral

$$N = \int |\psi|^2 dx, \tag{13.61a}$$

the momentum integral

$$p = \int \{\mathrm{Im}(\bar{\psi}\psi_x) + Uu\} dx, \tag{13.61b}$$

and that of energy

$$E = \int \{ |\psi_x|^2 + U|\psi|^2 + \frac{1}{2}(U^2 + u^2)\} dx, \tag{13.61c}$$

with

$$U_t + u_x = 0.$$

Calculating the integral E at a given N for two solitons enables evaluation of the energy emitted in a single occurrence of fushion. Comparing this with the energy needed for the breaking down of the half-amplitude soliton $(N_1 = (1/2)N)$, gives the factor 64. Since only negative pulses can break down a soliton, the number of such occurrences does not exceed 32.

Let us consider the balance of these processes in a 'nearest-neighbour' approximation: we count solitons entering and leaving a given amplitude cell a. Let $\rho(a)$ be the soliton density of a given amplitude. Suppose also that these move with a near-sound velocity, $v \simeq 1$. Then $2\rho(a)$ solitons leave this ("a") state per unit time due to their coalescence*) and $16\times 2\ \rho(2a)$ due to fission by sound created in the coalescence of solitons of amplitude $2a$. In the state under consideration, $(1/2)2\rho(a/2)$ solitons arise due to the fusion of solitons of amplitude $(a/2)$, and $32\times 2\rho(4a)$ due to the fission of solitons of amplitude $2a$ by sound emitted in coalescing solitons of amplitude $4a$. As result, we have the equilibrium equation

$$\dot{\rho}(a) = 2\left[\frac{1}{2}\rho(\frac{a}{2})+32\rho(4a)-\rho(a)-16\rho(2a)\right], \tag{13.62}$$

with a stationary solution

$$\rho \propto a^{-5/2}. \tag{13.63}$$

This *crude* qualitative analysis predicts the appearance of solitons of amplitudes greater than the initial ones (due to processes of coalescence) in model (60) in contrast to the distribution (59) of the $S3$ model. The probability of their appearance decreases with amplitude (due to decay processes caused by the sound pulses). This means that, in the non-integrable model (60), there is an energy flow along amplitudes, enriching the spectrum in the large amplitude region. A stationary spectrum (63) occurs when solitons are produced in the given amplitude region (a_0) and disappear in the region of large amplitudes, $a \gg a_0$. The mechanism of the processes of soliton creation and annihilation is not essential in the inertial region considered above, and defined by the equilibrium equation (62).

Ultimately, it is to be noted that the picture drawn has a crude qualitative character and is given here to cast light on the extent to which the processes are complicated even for the 'simplest' non-integrable Bose-gas model (60).

As regard to the distribution (59) of the $S3$ model, we should note that even weak micro-impurities of type

$$i\dot{\psi}+\psi_{xx}+2|\psi|^2\psi = \epsilon\delta(x)\psi \tag{13.64}$$

lead to the dissipation of small amplitude solitons [154], resulting in the

*) On the assumption that every pair collision ends with coalescence.

distribution (59) growing slowly at small a_0.

References

1. A. Scott, F. Chu and D. McLaughlin. *The soliton: A new concept in applied science.* Proceed. IEEE 1973, 61, p. 1443-1483. see [26] part IV.

2. R. Miura. *The Korteveg- de Vries equation: A survey of results.* SIAM Review 1976, 18, p. 412-459.

3. Papers in:

   a) R. Miura ed. *Bäcklund Transformation, The Inverse Scattering Method, Solitons and Their Applications.* Lectures Notes in Mathematics Springer, Berlin, N.Y. 1976.

   b) F. Calogero ed. *Nonlinear Evolution Eqs. Solvable by the Spectral Transform.* Pitman, London 1978.

   c) R. Bullough and P. Caudrey eds. *Solitons.* Springer, Berling, N.Y. 1980.

4. V. Drinfeld, I. Krichever, Yu. Manin and S. Novikov. *Methods of algebraic geometry in contemporary mathematical physics.* in mathematical Physics Reviews. S. P. Novikov ed. Vol.I, Sect.C, 1980.

5. B.A. Dubrovin. et al. See [17] part III.

6. V. Zakharov, S. Manakov, S. P. Novikov, L. P. Pitaevskii. See [1a] part II.

7. C. L. Lamb Jr. *Elements of Solition Theory.* Wilcy- Interscience, N. Y. 1980.

8. M.J. Ablowitz, H. Segur. *Solitons and Inverse Scattering Transform.* SIAM, Philladel. 1981.

9. F. Calogero. A. Degasperis. *Spectral Transform and Solitons: Tool to Investigate and Solve NEE.* North- Holland, Amsterdam 1981. See [1c].

10. R.K. Dodd, et al. See [34] part I.

11. V. Zakharov, A. Mikhailov. *Relativistically- invariant 2-dim. field theory models integrating by the inverse transform.* ZhETF 1978, 74, p. 1953- 1973.

12. J. C. Eilbeck. *Numerical studies of solitons. in A.R. Bishop and T. Schneider eds. Solitons and Condensed Matter Physics.* Springer, Berlin, N.Y. 1978.

13. M. Ablowitz, T.R. Taha. *On analytical and numerical aspects of certa in non-linear evolution equations.* Preprints I.F.N.S. Clarkson College, Potsdam, N.Y.

1982:

I. Analytical No 14.

II. Numerical NLS eq. No 15,

III. Numerical KdV eq. No 16.

14. R. Richtmeyer, K. Morton. *Difference Method for Initial Value Problems 2nd Ed.* Interscience, London 1967.

    A. Mitchell. *Computational Methods in Partial Differential Equations.* J. Wiley, London 1969.

    W. Ames. *Numerical Methods for Partial Differential Equations, 2nd Ed.* Nelson, London 1977.

    D. Jacobs ed. *The state of the Art in Numerical Analysis.* Academic Press, London 1977.

15. J. Cooley, P.Lewis and P. Welch. IEEE Trans. Education E-12, 1969, No 1, p. 27-34.

16. N. Pereira, R. Sudan and J. Denavit. *Numerical simulations of 1-dim. solitons.* Phys. Fluids 1977,20,p. 271-281.

    Numerical study of 2-dim. generation and collapse of Langmuir solitons. ibid, p. 936-945.

17. B. Fornberg and G. Whitham. see [69a] part I.

18. J. Oficialski and I. Bialynicki-Birula. *Collisions of Gaussons.* Acta Phys. Pol. 1978, 9B, p. 759-775.

19. N. Zabusky and M. Kruskal. *Interaction of solitons in a colisionless plasma and recurrense of initial states.* Phys. Rev. Lett. 1965, 15, p. 240-243.

20. Kh. Abdulloev, I.L. Bogolubsky and V. Makhankov. *One more example of inelastic soliton interaction.* Phys. Lett., 1976, 56A, p. 427-428.

21. J. C. Eilbeck and G. Mcguire. *Numerical study of the RLWE.* I and II. J. Comp. Phys. 1975, 19, p. 43-57, and 1977, 23, p. 63-73.

22. D. Peregrin. see [35a] part I.

23. T. Benjamin, et al. see [35b] part I.

24. R. Hirota. *Exact N-soliton solution of the wave equation of long waves in shallow-water and in nonlinear lattices.* J. Math. Phys. 1973, 14, p. 810-815.

25. I. Bogolubsky. *Some examples of inelastic soliton interaction.* Comp. Phys.

Comm. 1977, 13, p. 149-155.

26. J. Hammack. *A note on tsunamis: Their generation and propagation in an ocean of uniform depth.* J. Fluid Mech. 1973, 60, p. 769-.
27. J. Boussinesq. Mem. Sci. (Paris) 1877, 23,p.1-.
28. T. Ogino and S. Takeda. *Computer simulation for the ion-acoustic solitons propagating in both directions.* J. Phys. Soc. Japan 1975, 39, p. 1365-1372.
29. L. Degtyarev, V. Makhankov and L. Rudakov. *Dynamics of formation and interaction of Langmuir solitons and strong turbulence.* JETP 1975 40, p. 264-275.
30. J. L. Bona, W. G. Pritchard and L. R. Scott. *Solitary wave interaction.* Phys. Fluids 1980, 23, p. 438- 441.
31. A. R. Santarelli. *Numerical analysis of the RLWE: Inelastic collision of solitary waves.* Nuovo. Cim. 1978, 46B, p. 179-188.
32. C. Karney, et al. see [69b] part I.
33. R. Hirota. see [70] part I.
34. H. Ono. see [64] part I.
35. J. Meiss and N. Pereira. see [65] part I.
36. R. Joseph. see [66] part I.
37. N. Yajima and A. Outi. *A new example of stable solitary waves.* Progr. Theor. Phys. 1971, 45, p. 1997-1998.
38. Kh. Abdulloev, L. Bogolubsky and V. Makhankov. *Dynamics of Langmuir turbulence. Formation and interaction of solitons.* Phys. lett. 1974, 48A, p. 161-163.
39. V. Makhankov. see [7] part I.
40. J. Satsuma, M. J. Ablowitz and Y. Kodama. *On an internal wave eq. describing a stratified fluid with finite depth.* Phys. Lett. 1979, 73A, p. 283-286. J. Math. Phys. 1982, 23, p. 564-.
41. J. Satsuma and N. Yajima. see [35] part IV.
42. Kh. Abdulloev, I. Bogolubsky and V. Makhankov. *Interaction of plane Langmuir solitons.* Nuclear Fusion 1975, 15, p. 21-26.
43. N. Yajima and M. Oikawa. see [9] part I.

44. E. Valeo and W. Kruer. *Solitons and resonant absorption.* Phys. Rev. Lett. 1974, 33, p. 750-753.

Y. Lee and G. Morales. Invited paper at Solition Conference in Tucson, Arizona, 1976 and refences therein.

45. E. Benilov and S. Burtsev. see [54] part I.

46. S. V. Manakov. see [85] part I.

47. R. Sagdeev and A. A. Galeev. *Nonlinear Plasma Theory.* Benjamin, N.Y. 1969, 92p.

48. C. Montes. *Plasma Physics. Nonlinear Theory and Experiments.* Plenum, N. Y. 1977, 222p. Astroph. Journal 1977, 216, p. 329.

49. C. Montes, et al. see [86] part I.

50. J. Fernandes and G. Reinish. see [88] part I.

51. J. Fernandes, et al. see [56b] part IV.

52. V. Makhankov and B. Shchinov. *Computer investigation of nonlinear dynamical problems of plasma theory.* Comp. Phys, Comm. 1972, 4, p. 327-332.

53. M. Ablowitz, M. Kruskal and J. Ladik. *Solitary wave collisions.* SIAM J. Appl. Math. 1979, 36, p. 428-437.

54. A. Kudryavtsev. *Soliton-like solutions for a Higgs scalar field.* JETP Lett. 1975, 22, p. 82-83.

55. D. K. Campbell, J. F. Schonfeld and C.A. Wingate. *Resonance Structure in kink-antikink interaction in ϕ^4-theory.* Physica 1983, 9D, p. 1-32.

56. M. Moshir. *Soliton-antisoliton scattering and capture in ϕ^4-theory.* Nucl. Phys. 1981, 185B, p. 318-332.

57. M. Peyrard and D.K. Campbell. *Kink-antikink interactions in a modified SG model.* Physica 1983, 9D, p. 33-51.

58. P. Dodd, R. Bullough and S. Duckworth. *Multisoliton solutions of nonlinear dispersive wave equations not solvable by the inverse method.* J. Phys. 1975, 8A, p. L64-L68. see [78] Part I.

59. S. Duckworth, R. Bullough, P. Caudrey, J. Gibbon. *Unusual soliton behaviour in the self-induced transparency of $Q(2)$ vibration-rotation transitions.* Phys. Lett. 1976, 57A, p. 19-22.

60. J. Perring and T. Skyrme. *A model unified field theory.* Nucl. Phys. 1962, 31,

p. 550-555.

61. D. Campbell, M. Peyrard and P. Sodano. *Kink-antikink interactions in the double SG equation.* Physica 1986, 19D, p. 165-205.

62. A. Barone, F. Esposito, C. Magee and A. Scott. *Theory and Applications of the SG equation.* Riv. Nuovo Cim. 1-71,1, p. 227-267.

63. P. Caudrey, J. Eilbeck, J. Gibbon. *The SG eq. as a model classical field theory.* Nuovo Cim. 1975, 25B, p. 497-512.

64. J.C. Eilbeck. *Kink collisions in the ϕ^4 model.* 1981, 16mm cine film approx. 4 mins.

    J. C. Eilbeck and P. Lomdahl. *SG-solitons,* 1981, 16mm cine film, sound track, approx, 12 mins.

65. G. Lamp Jr. *Analytic descriptions of ultrashort optical pulse propagation in a resonant medium.* J. Rev. Mod. Phys. 1971, 43, p. 99-124. see also [7] part V.

66. P. Dodd and R. Bullough. *Polynomial conserved densities for the SG equations.* Proc. Roy. Soc. London 1976, 351A, p. 499- and 1977, 352A, p. 481-503.

67. A. Seeger, H. Donth and A. Kochendorfer. *Theorie der Versetzungen in eindim. Atomreihen.* III Versetzungen, Eigenbewegungen und ihre Wechselwirkung. Z. für Physik 1953, 134, p. 173-193.

68. V. Makhankov and V. Fedyanin. see [16] part I.

69. A. Bishop. *Solitons in condensed matter physics.* Phys. Scripta 1979, 20, p. 409-423.

70. D. Kaup and A. Newell. *Solitons as particles and oscillations.* Proc. Roy. Soc. London 1978, 316A, p. 413-446.

71. T. Schneider and E. Stoll. *Molecular-dynamics investigation of structural phase transitions.* Phys. Rev. Lett. 1973, 31, p. 1254-1258.

    Molecular-dynamic study of a three-dim. nonlinear lattice model. Phys. Rev. 1978, 17B, p. 1302-1322.

72. E.Stoll, T. Schneider and A. Bishop. *Evidence for breather excitations in the SG chain.* Phys. Rev. Lett. 1979. 42, p. 937-939.

73. M. Fogel, S. Trullinger, A. Bishop and J. Krumhansl. *Dynamics of SG solitons in the presence of perturbations.* Phys. Rev. 1977, 15B, p. 1578-1592.

M. Rice, A. Bishop, J. Krumhansl and S. Trullinger. *Weakly pinned Fröhlich charge-density-wave condensates: A new nonlinear, current-carrying elementary excitation.* Phys. Rev. Lett. 1976, 36, p. 432-435.

74. a) D. Kaup. SIAM J. Appl. Math. 1976, 31, p. 121-133.

 b) D. Kaup and A. Newell. *Prediction of a nonlinear oscillating dipolar excitation in one-dim. condensates.* Preprint Clarkson Coll. of Technol. 1978.

75. T. I. Belova and A.E. Kudryavtsev. *Quasi-periodical orbits in the scalar classical ϕ^4-field theory.* Preprint ITEP- 94, Moscow 1985.

76. T. Sugiyama. *Kink-antikink collisions in the two-dimensional ϕ^4-model.* Progr. Theor. Phys. 1979, 61, p. 1550-1563.

77. M. Remoissenet and M. Peyrard. *A new simple model of a kink bearing Hamiltonian.* J. Phys. 1981, 14C, p. L481-L485.

78. V.M. Eleonskii and V.P. Silin. *Theory of waves near exact solutions of nonlinear electrodynamics and optics I and II.* JETP 1969, 29, p. 317-343 and 1970, 30, p. 262-272.

 A.M. Kosevich and A.S. Kovalev. *Self-localization of oscillations in a one-dimensional anharmonic chain.* JETP 1975, 40. p. 891-901.

79. R. Dashen, et al. see [53] part II.

80. I.L. Bogolubsky and V. Makhankov. *Dynamics of heavy spherically-symmetric pulsons.* JETP Lett. 1977, 25, p. 107-110.

81. a) E. Ott and R. Sudan. see [52] part IV.

 b) A. Hasegawa and F. Tappert. see [53] part IV.

82. N. Yajima and M. Oikawa, et al. see [54] part IV.

83. N. Pereira. see [55] part IV.

84. a) J. Fernandes and G. Reinish. see [88] part I.

 b) J. Fernandes, et al. see [56b] part IV.

 c) S. Watanabe. see [56c] part IV.

85. M. Fogel, et al. see [29c] part IV.

86. J. Currie, S. Trullinger, A. Bishop and J. Krumhansl. Phys. Rev. 1977, 15B, p. 5567- . See [139] this part.

87. A.T. Filippov. *Dynamics of formation of bound soliton states on the*

inhomogeneity of medium. Proc. III International Symposium on Celected Topics of Stat. Mechanics, JINR Publ. D17-84-850, V. 2, p. 281-287, Dubna 1985.

88. F. Calogero and A. Degasperis. *Exact solution via the spectral transform of a generalization with linearity x-dependent coefficient of the NLS.* Preprint 80, INFN, Roma, 1978; see also

    H.H. Chen and C.S. Liu. *The effects of gradual field gradients on a Langmuir soliton.* Phys. Rev. Lett. 1976, 37, 693-697.

89. Yu.S. Galpern and A.T. Filippov.

    a) *Bound states of solitons in long Josephson junction.* Pisma ZhETF 1982, 35, p. 470-472.

    b) *Bound states of solitons in non-uniform Josephson junction.* ZhETF 1984, 86, p. 1527- .

90. G.S. Kazacha, S. Serdyukova and A.T. Filippov. *Numerical simulations of fluxon movement in the system with micro-impurity.* Preprint JINR P11-84-76, Dubna, 1984, see also JINR P11-85-60, Dubna 1985.

91. *Solitons in action,* eds. K. Lonngren and A. Scott, Academic Press N.Y., London 1978.

92. T.A. Fulton, R.C. Dynes and P.W. Anderson. *The flux Shuttle- a Josephson junction shift register employing single flux quanta.* Proc. IEEE 1973, 61, p. 28-35.

93. C.S. Owen and D.J. Scalapino. *Vortex structure and critical currents in Josephson junctions.* Phys. Rev. 1967, 164, p. 538-544.

94. V.I. Karpman, N.A. Ryabova and V.V. Solovev. *Interaction of fluxons in long Josephson junctions.* ZhETF 1981, 81, p. 1327-1336.

96. C. Chi, M.M.T. Loy and D.C. Cronemeyer. *Optical probing technique for inhomogeneous superconditing films.* Appl. Phys. Lett. 1982, 40, p. 437-439.

    M. Scheuermann, J.R. Lhota, P.K. Kuo and J.T. Chen. *Direct probing by laser scanning of the current distribution and inhomogeneity of Josephson junctions.* Phys. Rev. Lett. 1983, 50, p. 74-77.

97. R.F. Broom, W. Kotyczka and A. Moser. IBM Journal Res. Dev. 1980, 24, p. 178.

98. V.I. Karpman. *Soliton evolution in the presence of perturbation.* Physica Scripta 1979, 20, p. 462-478.

99. T.L. Boyadjiev et al. *Bifurcations of bound states of fluxons in inhomogeneous Josephson junction of finite size.* JINR Comm. P11-85-807, Dubna 1985.

100. R. Jackiv and P. Rossi. see [48] part IV.

O. Rasizade. *Interaction of point-like charges with Higgs field in one-dimensional space.* Teor. Mat. Fiz. 1981, 48, p. 197-209; also

O. Rasizade. *Bifurcations and 'catastrophes' in the Higgs field with external charges.* Teor. Mat. Fiz. 1981, 49, p. 36-47.

101. O.A. Levring, M. Samuelsen and O. Olsen. *Exact and numerical solution to the perturbed SG equation.* Physica 1984, 11D, p. 349-358.

102. V.N. Kascheev. *Self-similar solutions to nonlinear equations with dissipation and constant external force.* Preprint LAFI-085, Riga, 1985; INIS Atomindex, 1987, v. 18, No. 8, p. 3523.

103. D.J. Kaup. *Perturbation expansion for the ZS inverse scattering transform.* SIAM J. Appl. Math. 1976, 31, p. 121-133.

104. V.I. Karpman and E. Maslov. Soliton perturbation theory. ZhETF 1977, 73, p. 538-559.

105. K.A. Gorshkov and L.A. Ostrovskii. *Interaction of solitons in nonintegrable systems: Direct perturbation method and applications.* Physica 1983, 3D, p. 428-438.

106. A. Bishop. *Solitons and physical perturbations.* In: Soliton in Action, eds. K. Longren and A.C. Scott. Academic Press, N.Y. 1978, p. 61-68.

107. A. Bondeson, et al. see [29f] part IV.

108. V.I. Karpman. see [29e] part IV.

109. A. Newell. see [18] part IV.

110. E.M. Maslov. *On the soliton perturbation theory in the second order approximation.* Teor. Mat. Fiz. 1980, p. 362-373.

111. D. Kaup and A.C. Newell. *Solitons as particles and oscillations.* Proc. Roy. Soc. London 1978, 361A, p. 413-446.

A.C. Newell. J. Math. Phys. 1977, 18, p. 922.

112. D.W. McLaughlin and A. Scott. *Perturbation analysis of fluxon dynamics.*

Phys. Rev. 1978, 18A, p. 1652-1680, see also in Solitons in Action, eds. K. Longren and A.C. Scott. Academic Press, N.Y. 1978, p. 201-256. A multi-soliton perturbation theory.

113. a) B. Malomed. *Inelastic interactions of solitons in nearly integrable systems. I.* Physica 1985, 15D, p. 374-384.

114. F. Abdullaev, A. Abdumalikov and A. Rakhmatov. *The dynamics of solitons in 1-D molecular crystals.* Phys. Stat. Sol. 1982, 112(b), p. K5-K10.

b) the same, part II, ibid, p. 385-401.

115. Yu. Kivshar. *On the soliton perturbation theory: SG equation.* Preprint 21-84, FTINT Kharkov, 1984.

D. McLaughlin and A. Scott. see [112] this part.

116. A. Kosevich and Yu. Kivshar. *Explanation of specific SG soliton dynamics in the presence of external perturbations.* Phys. Lett. 1983, 98A, p. 237-239; see also Fiz. Niz. Temp. 1982, 8, p. 1270-1284.

117. J.C. Eilbeck, P.S. Lomdahl and A. Newell. *Chaos in the inhomogeneously driven SG equation.* Phys. Lett. 1981, 87A, p. 1-4.

118. D.K. Campbell, et al. see [61] this part.

119. V.K. Fedyanin and V.G. Makhankov. *Soliton-like solutions in one-dimensional systems with resonance interaction.* Physica Scripta 1979, 20 p. 552-557.

120. V.G. Makhankov and V.K. Fedyanin. see [16] part I.

121. M. Peyrard and D.K. Campbell. see [57] this part.

122. V.I. Arnold. *Catastrophe Theory.* Springer, Heidelberg 1986.

123. V.G. Makhankov. *Dynamical structure factors and clasterization in a class of models of field theory.* JINR P2-82-248, Dubna, 1982, see also

V. Makhankov and V. Fedyanin. *Ideal gas of particle-like excitations at low temperatures.* Physica Scripta 1983, 28, p. 221-228.

124. V.K. Fedyanin. *Static and dynamical properties of ideal gas of particle-like excitations at low temperatures.* JINR Comm. P17-82-268 Dubna, 1982.

125. V. Lisy, V. Makhankov and V.K. Fedyanin. *Bion contribution to DSF of scattering of neutrons by $CsNiF_3$ patterns.* in Problems of Statistical Mechanics, JINR publ. D17-84-850, Dubna 1984, V.I., p. 443-449. also JINR P17-85-410,

Dubna 1985.

126. J.A. Krumhansl and T.R. Schrieffer. *Dynamics and statistical mechanics of a one-dimensional model Hamiltonian for structure phase transition.* Phys. Rev. 1975, 11B, p. 3535-3545.

127. M. Steiner. *Soliton in 1-D magnets.* J. Magn. Magn. Mat. 1983, 31-34, p. 1277.

M. Steiner, K. Kakurai and J.K. Kjems. *Experimental study of the spin dynamics in the 1-D ferromagnet with planar anisotropy,* $CsNiF_3$ *in an external magnetic field.* Z. Phys. 1983, 53B, p. 117-142.

128. H.J. Mikeska. see [17] part I.

129. M. Ablowitz, et al. see [75a] part I, and [75b].

130. J.K. Kjems and M. Steiner. *Evidence for soliton modes in the one-dimensional ferromagnet* $CsNiF_3$. Phys. Rev. Lett. 1978, 41, p. 1137-1140.

131. M. Steiner. *Neutron scattering observation of solitons.* J. Appl. Phys. 1979, 50, p. 7395-7400.

132. J. Reiter. *Have solitons been observed in* $CsNiF_3$*?* Phys. Rev. Lett. 1981, 46, p. 202-205, 518 (Erratum).

133. E. Stoll, T. Schneider and A.R. Bishop. *Evidence for breather excitations in the SG chain.* Phys. Rev. Lett. 1979, 42, p. 937-940.

134. J. Timonen and R.K. Bullough. *Breather contributions to the dynamical form factors of the SG systems* $CsNiF_3$ *and* $(CH_3)_4NMnCl_3$*(TMMC).* Phys. Lett. 1981, 82A, p. 183-187.

135. A.R. Bishop. *Nonlinear mode phenomenology for SG breather excitations.* J. Phys. 1981, 14A, p. 1417-1430.

136. V. Fedyanin and V. Makhankov. see [123b] this part.

137. V. Fedyanin and V. Lisy. *Bion contribution to the equilibrium and dynamical characteristics of quasi-one-dimensional systems.* Fiz. Niz. Temp. 1985, 11, p. 306-314.

138. J.K. Kjems, K. Kakurai and M. Steiner. *Neutron scattering study of spin fluctuations in* $CsNiF_3$ *without applied field.* J. Magn. Magn. Mat. 1983, 31-34, p. 1133-1134.

139. J.F. Currie, J. Crumhansl, A. Bishop and S. Trullinger. *Statistical mechanics*

of 1-D solitary-wave-bearing fields: Exact results and ideal-gas phenomenology. Phys. Rev. 1980, 22B, p. 477-496.

140. A.R. Bishop. *Statistical mechanics of solitons.* in Physics in one-dimension. Eds. J. Bernasconi and T. Schneider. Springer, Berlin 1981, p. 27-46.

141. N. Theodorakopoulos. *Thermodynamics of a SG breather gas.* Z. Phys. 1982, 46B, p. 367-370.

142. V. Lisy, V. Makhankov and V. Fedyanin. *Dynamical structure factors of the magnet $CsNiF_3$ and the continuous SG model.* JINR E17-85-410, Dubna 1985.

143. V. Korepin and L. Faddeev. see [54] part II.

144. P. Kumar. *Soliton instability in a one-dimensional magnet.* Phys. Rev. 1982, 25B, p. 483-486;

E. Magyari and H. Thomas. *Kink instability in planar ferromagnets.* Phys. Rev. 1982, 25B, p. 531-533.

145. V.G. Baryakhtar, B.A. Ivanov, A.L. Sukstanskii and E.V. Tartakovskaya. *Non-equilibrium thermodynamics of kink type soliton gas in quasi-one-dimensional systems.* Teor. Mat. Fiz. 1988, 74, p. 46-60.

146. T. Schneider and E. Stoll. *Classical statistical mechanics of the SG and ϕ^4 chain. II.* Dynamic properties. Phys. Rev. 1981, 23B, p. 4631-4660.

147. V. Makhankov and V. Fedyanin. see [16] part I.

148. N.N. Bogolubov and N.M. Krylov. *On the Fokker-Plank equation.* Notes of Math. Physics Chair. 1939, V. 4, p. 5-80.

149. B.A. Ivanov, Yu.N. Mitzai and N.V. Shakhova. ZhETF 1984, 87, p. 289-298.

150. V.G. Baryakhtar, I.V. Baryakhtar, B.A. Ivanov and A.L. Sukstanskii. *Kinetic properties of solitons interacting with thermostat.* in Problems of Statistical Mechanics, JINR publ. D17-81-758, Dubna 1981, p. 417-435.

151. A. Messia. *Quantum Mechanics.* Nauka, Moscow 1979, V. II, p. 238.

152. L.M. Degtyarev, V.G. Makhankov and L.I. Rudakov. *Dynamics of the formation and interaction of Langmuir solitons and strong turbulence.* Sov. Phys. JETP, 1975, 40, p. 264-275. see [29] part V.

153. Kh. Abdulloev, I. Bogolubskii and V. Makhankov. *Interaction of plane Langmuir solitons,* Nucl. Fusion 1975, 15, p. 21-26.

154. Yu. Kivshar, A.M. Kosevich and O.A. Chubykalo. *Scattering of bound*

quasiparticles (the dynamic solitons) by a point-like deffect in a one-dimensional system. Fiz. Niz. Temp. 1987, 13, No. 3.

155. V.G. Makhankov, O.K. Pashaev and Kh. Kholmurodov. *Threshold of KdV soliton production.* JINR Preprint E5-87-784, Physica Scripta, 1989, 39, p. 9-12.

156. M. Steiner, J. Villain and C.G. Windsor. Adv. Phys. 1976, 25, p. 87.

157. V.S. Ostrovskii. *Nonlinear dynamics of strongly anisotropic magnetics with spins* $S = 1$. Zh. Eks. Teor. Fiz. 1986, 91, p. 1690-1701.

158. Kh. Abdulloev, M. Agüero, A. Makhankov, V. Makhankov and Kh. Muminov. *Generalized spin coherent states as a tool to study quasiclassical behaviour of the Heisenberg Ferromagnets.* in V. Makhankov, V. Fedyanin and O. Pashaev edited 'Solitons and Applications'. Proceedings of the 4th international workshop, Dubna USSR (August 1989). World Scientific, Singapore, 1990. Preprint JINR E17-89-800, Dubna, 1989.

Part VI. Many - dimensional solitons

Chapter 14.

EXISTENCE AND STABILITY

In this part we consider which problems of soliton science discussed in previous chapters can be solved in more than one spatial dimension*). Here the situation differs radically from the one-dimensional case. As a rule, such systems can be obtained by applying the two-dimensional extension to the corresponding one-dimensional ones [1], and therefore are, generally speaking, weakly non-one-dimensional. In particular, the Kadomtsev-Petviashvili (KP) equation is the KdV two-dimensionalization. To date, this model has been sufficiently well explored via IST-type techniques and algebraic geometry [2].

However, the overwhelming majority of systems to be investigated are nonintegrable in the above sense. Therefore the programme formulated in the Introduction can only be accomplished for particular models or, at best, for classes of models.

Here, let us briefly recap the essential aspects of this programme and summarize the results already obtained.

The programme:

(1) The definition and study of linear oscillation modes.

(2) The formulation of the existence conditions of soliton-like solutions (SLS) or particle-like solutions (PLS).

(3) The examination of their stability.

(4) The dynamics of the SLS interacting with each other and with inhomogeneities (the bound states).

(5) Form-factors.

---

*)We employ conventional terminology: the space-time manifold is called the world.

(6) The statistical mechanics of SLS (PLS).

Let us begin with (2). Since, in $D \geqslant 2$ dimensions, with rare exceptions, explicit analytical solutions are not possible, the constructive way of proving the existence of soliton-like solutions is invalid. A numerical method to obtain such solutions remains, which needs independent evidence of their existence, since via this method one can obtain (naturally approximate) non-existence 'solutions'.

The stability (instability) of particle-like solutions in a number of models can be established via the Q-theorem or its generalisations (the variational approach) or by a linearization method (spectral analysis). In $D \geqslant 2$ dimensions, the latter method is hindered because of the absence of analytical formulae for soliton-like solutions. Finally, in the study of the behaviour of soliton-like solutions under various perturbations, the numerical method enables their stability (instability) in certain limits to be determined.

The dynamics of soliton-like solution interactions with each other and with inhomogeneities (and also their structural stability) in $D \geqslant 2$ dimensions still remains wholly in the domain of computer calculations. This has yielded a number of interesting results. These three aspects are of paramount importance for the whole programme in $D \geqslant 2$ dimensions. However, they may only be studied via computational experiments.

By means of several typical examples we show, in what follows, how various results in this field have been obtained, provide a discussion of these and formulate some unsolved problems.

Anticipating possible applications in condensed matter theory, plasma physics and other areas of theoretical physics, we consider those models which naturally (and most frequently) arise in such a framework: viz. the NSE, KG and SG models. In the discussion of these problems, we shall employ the results of Chapters 9-13.

1. Existence

In Chapter 9 we described two techniques which enable necessary and sufficient conditions for the existence of stationary soliton-like solutions in the framework of NSE and KGE to be established. In the simplest case of spherical or cylindrical symmetry these reduce to the existence conditions for solutions to the equation

$$\frac{d^2}{dx^2}y+\frac{D-1}{x}\frac{d}{dx}y+\frac{dU}{dy}y-\kappa^2 y = 0 \tag{14.1}$$

(where D is the space dimensionality) under the following boundary conditions:

$$(I) \quad y(x)|_{x\to\infty} = 0, \ y'(0) = y'(\infty) = 0, \ y(0) = \text{const.} \tag{14.2}$$

$$(II) \quad y(x)|_{x\to\infty} = y_0, \ y'(0) = y'(\infty) = 0, \ y(0) = \text{const.} \tag{14.3}$$

Equation (1) differs from that considered in Chapter 9 by the additional term $((D-1)/x)y'$. This leads to the appearance of the effective friction in the conservation law (9.33)

$$\frac{d}{d\tau}\mathcal{H} = -\frac{D-1}{\tau}y_\tau^2 \tag{14.4}$$

in the mechanical analogy method. We illustrate this here by an example of the model

$$U = \epsilon\frac{1}{p}y^p+\alpha\frac{1}{q}y^q \tag{14.5}$$

which we consider in the $D = 1$ variant.

From Table 9.1, via virial relations, one can establish the necessary conditions for the existence of particle-like solutions (boundary problem (2)) at $D>1$. So, for example, at $\alpha = 0$, $\epsilon = -1$ and $D = 3$ we have $1<p<6$: a result known from previous papers [3]. In the case of $D = 2$ we get $0<p<\infty$.

The mechanical analogy method or its various modifications, enable sufficient conditions for the existence of particle-like solutions to be determined. The nodeless solution to problem (2) exists only for a given value of $y(0) = a$ though dependent on the model (i.e. α and ϵ). Moreover, the dissipative term in eqn. (4) yields the appearance of a countable set of solutions with a definite number of nodes of the function $y(z)$. This statement, specifying the heart of the existence theorem, can be easily explained qualitatively by considering, for example, the model

$$y''+\frac{2}{x}y'-y+y^3 = 0, \tag{14.6a}$$

$$y'(0) = y(\infty) = 0, \ y(0) = a. \tag{14.6b}$$

The potential function through which a mechanical particle moves is given in Fig. 14.1.

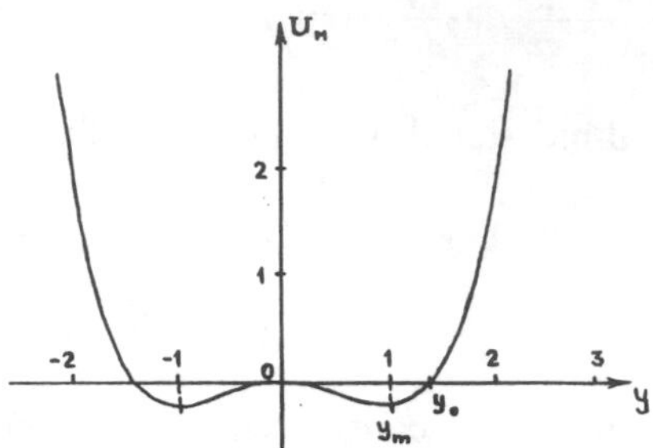

Figure 14.1. Potential in which the mechparticle moves.

In the absence of friction, as we have seen in Chapter 9, the particle with the initial amplitude a (less than y_0) and zero velocity $y'(0) = 0$ oscillates in the right-hand or left-hand wall, U_M. These oscillations are described by elliptic functions (the curve C_2 in Fig. 14.2). At $a = y_0$ the particle stops asymptotically at the hill top $y = 0$,

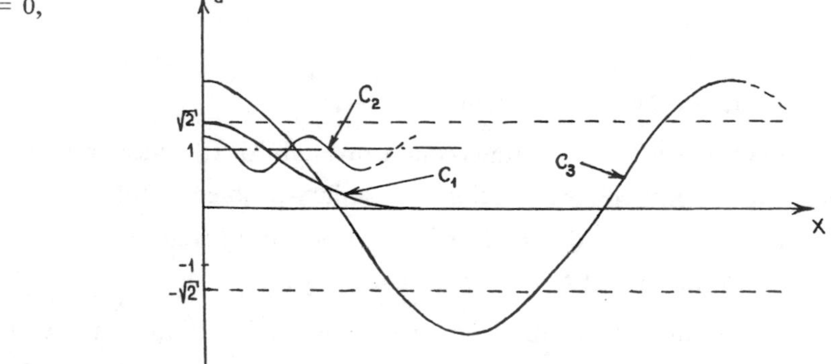

Figure 14.2. Solution of equation (14.6a) for $D = 1$.

which corresponds to the nodeless solution (the curve C_1 in Fig. 14.2). Finally for $a > y_0$ the particle oscillates from the right-hand wall to the left one, and vice versa, going over the middle barrier (the curve C_3). Friction (4) changes this picture in the following obvious way, see Fig. 14.3. Particles with $a \leqslant y_0$ stop with 'time' in the well (the right-hand or left-hand one) (the curve C_2^a). However, there exists such a value $y(0) = a_0 > y_0$ for which the particle (during its motion) loses exactly that energy so that it stops (asymptotically) at the barrier top $y = 0$ (the curve C_1^a). During the further increase of $y(0)$, the particle (sliding down on the

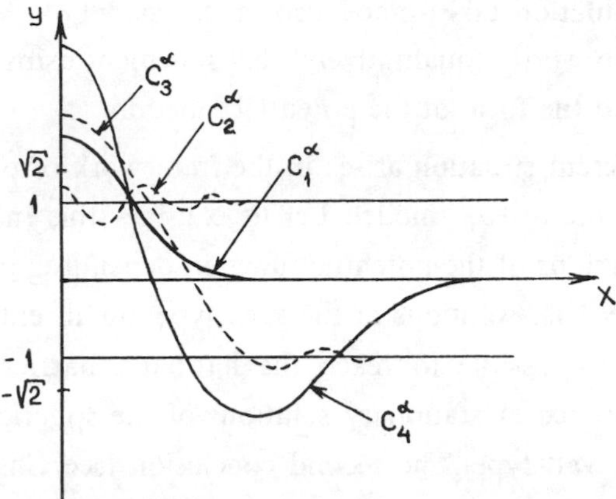

Figure 14.3. y_x, y phase plane for equation (14.6a) and related trajectories.

right-hand side) passes over the barrier at $y = 0$ and stops in the left-hand well (the curve C_3^α). For some value of $y(0) = a_1$, the particle (reflecting once from the left-hand wall) stops again at the top of the middle barrier which corresponds to the one-node solution (the curve C_4^α) and so on. This means that there exists a countable set of values a_i determining a countable set of solutions to problem (2) with a given number of nodes i (see Fig. 14.4 where the corresponding trajectories are presented).

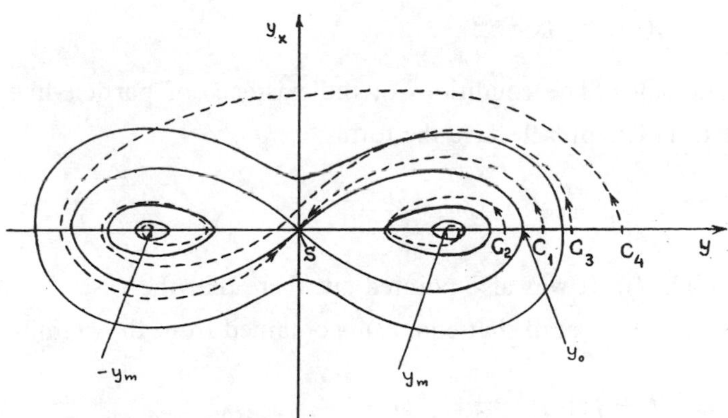

Figure 14.4

A similar situation takes place also in the model $\phi^4-\alpha|\phi|^6$ at $|\alpha|<3/16$. The reader can easily verify (qualitatively) this statement using the method of mechanical analogy and the form of the potential function.

A quite different situation arises in the framework of problem (3), i.e., for condensate BC in the $\phi^4-\alpha\phi^6$ model. Let us consider this in more detail. Figure 9.1 shows three versions of the potential function depending on α. The first conclusion following Fig. 9.1 is: solutions of the kink type are absent in the $D\geqslant 2$ cases (friction makes it impossible to reach the left-hand barrier). This means, in such models, the absence of stationary solutions of the spherically (cylindrically) symmetric domain wall type. The second conclusion (see Chapter 9) is: since in this formulation the particle stops (asymptotically) at the top of one of the side hills there exists for a given $1/4>|\alpha|>3/16$ only one value of $y(0)=a_0(\alpha)$ defining the bubble-type nodeless solution. The nodal solutions in the framework of the condensate formulation (3) do not exist (a rigorous proof of these statements is presented in a recent paper by P.E. Zhidkov and the author [4]).

We return to eqn. (1) and consider this at $\alpha=0$. We re-express this as:

$$x^{1-D}(x^{D-1}y')'-\kappa^2y+gy^{p-1}=0. \tag{14.7}$$

This equation describes stationary 'spherically' symmetric solutions to the NSE and NKG of the form

$$\psi(x,t)=y(x)e^{-i\omega t} \tag{14.8}$$

in D-dimensions. The conditions of the existence of particle-like solutions as one can see from Chapter 9 have the form

$$\kappa^2>0 \quad\text{and}\quad 2<p<\frac{2D}{D-2},$$

(see eqn. (9.17)). It was also pointed out there that the case $\kappa^2=0$ requires further investigation. We recall that eqn. (7) is obtained from the Hamiltonian

$$H=\int\{|\psi_x|^2-\frac{g}{2}|\psi|^4\}d^Dx$$

in the nonrelativistic NSE case and from

$$H=\int\{|\pi|^2+|\psi_x|^2+m^2|\psi|^2-\frac{1}{2}g|\psi|^4\}d^Dx$$

in the relativistic case with $\pi = \psi_t$.

Indeed for stationary solutions (8) both equations

$$i\psi_t + \Delta\psi + g|\psi|^2\psi = 0$$

and

$$\Box\psi + m^2\psi - g|\psi|^2\psi = 0$$

reduce to the same, eqn. (7), in which

$$\kappa^2 = \begin{cases} -\omega, & \text{for } NSE \\ m^2-\omega^2, & \text{for } KG \end{cases}.$$

Equation (7) is obtained by varying the action

$$S = \int (y_x^2 + \kappa^2 y - \frac{1}{p} g y^p) d^D x. \tag{14.9}$$

This action, at $\kappa^2 = 0$, possesses an additional invariance for certain values of p. To establish this, one may, for example, use the following reasoning. Let S be a dimensionless quantity; then, from the first term, it follows that the field y has the dimension

$$\{y\} \sim x^{2-D/2},$$

and the second term dimension is

$$\{gy^p\} \sim gx^{(D-p(D-2/2))};$$

hence, the coupling constant dimension is

$$\{g\} \sim x^{(p(D-2/2)-D)}.$$

It is easy to see that at

$$p = \frac{2D}{D-2}, \tag{14.10}$$

g becomes dimensionless. Therefore, in this case, the action (9) has no dimensional constants and is invariant under the transformation of the form

$$y \to \beta^{(2-D/2)} y(\beta x),$$

being a global change of the coordinate scale with a simultaneous consistent (global) change of the field gauge. Note that the condition (10) supporting the invariance of S exactly coincides with the particle-like solution existence condition (9.17) in the case $\kappa^2 = 0$. One can easily check that, in fact, particle-like solutions exist and are

$$y = \frac{c}{(1+x^2)^{(D-2/2)}}, \quad c = \left[\frac{D(D-2)}{g}\right]^{(D-2/4)}. \tag{14.11}$$

The constant c depends on the space dimensionality (e.g. $c = (3/g)^{1/4}$ at $D = 3$ and $c = (8/g)^{1/2}$ for $D = 4$). This may be easily shown. Equation (11) is a solution to eqn. (7) at $\kappa^2 = 0$ via the substitution. The solution of the (11)-type in a D-dimensional Euclidean space is sometimes called the instanton.

One more class of NSE and NKG models often arises in applications in physics. These are models with saturable or non-polynomial nonlinearity [7,8,21]

$$U = -ln(1+|\psi|^2)+(1-m^2)|\psi|^2. \tag{14.12}$$

Rewrite the equation with nonlinearity (12) in terms of the mechanical analogy ($\kappa^2=m^2-\omega^2$)

$$\frac{d}{dx}\{(y')^2-(\kappa^2-1)y^2-\ln(1+y^2)\} = -\frac{2}{x}(y')^2$$

and consider its left-hand side. The potential U_M has the from

$$U_M = -(\kappa^2-1)y^2-\ln(1+y^2). \tag{14.13}$$

The U_M extrema are at the points

$$y = 0, \; y_\pm = \pm\frac{\kappa}{\sqrt{1-\kappa^2}}. \tag{14.14}$$

It is easy to show that $y = 0$ is the maximum point of $U_M(0) = 0$ at $\kappa^2 > 0$, and therefore the particle may stop there if $\bar{y}$: $U_M(\bar{y}) > 0$. The latter is possible when

$$0<\kappa^2<1 \Rightarrow \omega^2<m^2<1+\omega^2. \tag{14.15}$$

These conditions are necessary for the existence of PLS (problem 2).

With the same reasoning as above, we make sure that under conditions (15) the equation has also a countable spectrum of nodal solutions $y_n(x)$ in addition to the nodeless solution.

The second class of models with non-polynomial nonlinearity (where a and b are constants)

$$U = b|\psi|^2 \ln(a^D |\psi|^2 e^{-1}) \tag{14.16}$$

is sometimes called the class of models with confinement since, within their framework, linear waves are absent. As can be established in a constructive way, these models have the solutions

$$y = (A\pi^{-1/2})^{D/2} \exp\{-b(\vec{x} - \vec{v}t - \vec{x}_0)^2\} \exp\{-i(\omega t - \vec{v}\vec{x} + \theta_0)\} \tag{14.17}$$

$$\omega = \frac{1}{2} v^2 + 2Db[1 - \ln(Aa\pi^{-1/2})]$$

for arbitrary D (for a physical meaning and for detailed results, see Refs. [9-11]).

Concluding this section let us note that $D = 1$ solutions of the kink type correspond to solutions of the vortex type in $D \geqslant 2$ space. Indeed, consider the $D = 2$ model of $S3_-$

$$i\psi_t + \Delta\psi - |\psi|^2 \psi = 0 \tag{14.18}$$

and write this equation in polar coordinates $(x,y) \to (r,\varphi)$. Assuming cylindrical symmetry let us look for a stationary solution in the form

$$\psi(x,t) = y(r) e^{i(\nu\varphi - \rho_0 t)}. \tag{14.19}$$

This satisfies the equation

$$y'' + \frac{1}{r} y' + (\rho_0 - \frac{\nu^2}{r^2}) y - y^3 = 0 \tag{14.20}$$

with the boundary condition

$$y^2|_{r \to +\infty} = \rho_0, \quad y'(\infty) = 0.$$

This means that at $r \to +\infty$ we have a rotating gas with density and angular velocity ρ_0. In the vortex centre, we have the expansion

$$y \sim r^{|\nu|}, \quad r \to 0,$$

i.e. at $\nu \neq 0$ in the vortex centre the density vanishes, satisfying

$$\rho(r) \sim r^{2|\nu|}, \quad r \to 0. \tag{14.21}$$

A solution to eqn. (20) at $\nu = 1$ has been numerically obtained in [12] and is presented in Fig. 14.5. It is clear that for large r one may neglect the terms $(1/r)y'$ and $(\nu^2/r^2)y$ in eqn. (20) and y tends to $\sqrt{\rho_0}$ asymptotically as the kink of the $S3$-model, and the plot $y(r)$ corresponds to half of the kink (in any case, at $\nu = 1$).

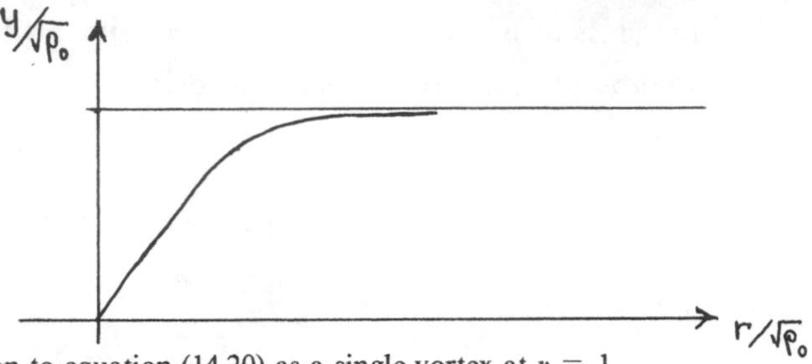

Figure 14.5. Solution to equation (14.20) as a single vortex at $\nu = 1$.

We have discussed issues associated with the existence of $D \geqslant 2$ dimensional static and stationary solutions to both NSE and NKG. A similar consideration can also be made for the SG model and it follows thence that there are no static localized $D \geqslant 2$ solutions (with corresponding symmetry) in the framework of the SG model. Stationary solutions (together with vortices) are possible in a rather exotic SG model of the form

$$H = \frac{1}{2}\int\{|\psi_t|^2 + |\psi_x|^2 + (1-\cos|\psi|)\}dx$$

(one of the feasible complex generalisations of SG).

The existence and properties of non-stationary solutions are discussed below. Next, we investigate the properties of quasi-stationary solutions (small-amplitude expansions).

2. Quasi-stationary solutions

In Section 10 of Chapter 11 we considered quasi-stationary solutions of small amplitudes in the framework of the NKG and SG equations and determined that models of such a type can be divided into two classes:

(1) with the trivial vacuum $\psi_v = 0$;

(2) with the condensate $\psi_v = \text{const}$.

The first case.

One ought to use an expansion of the type (10.35) and then in the D-dimensional space, eqn. (10.36) simply becomes

$$A_{xx}+\frac{D-1}{x}A_x-A+A^3=0. \tag{14.22}$$

We obtain again a countable spectrum of small amplitude solutions. In the case $D=3$, we obtain the amplitude relationships

$$A_0\simeq 4.34<A_1\simeq 14.10<A_2\simeq 29.13<\ldots.$$

Solutions of the form

$$\psi_n(r,t)=\sqrt{\frac{4}{3}}\kappa A_n(\kappa r)\cos(\sqrt{1-\kappa^2}\,t)\equiv \tag{14.23a}$$

$$\equiv \kappa_m\frac{A_n(\kappa r)}{A_n(0)}\cos(\sqrt{1-\kappa^2}\,t)$$

for the first three modes ($n=0,1,2$) have been explored in Dubna with the use of computers for the amplitudes $\kappa_m=0.2, 0.4$ and 0.7. At $\kappa_m\leqslant 0.4$ the 'pulson' (23) is the solution to the NKG equation

$$\Box\psi+\psi-\psi^3=0$$

to a high accuracy (errors are less than 1%). The radiation by small-amplitude ($\kappa_m^2\ll 1$) pulsons is very weak and its life-time τ is large: $\tau\gg 1$ as $\kappa_m\ll 1$ [20]. If $\kappa_m=0.7$, the pulson amplitude $\psi_m(t)$ slowly decreases, e.g. $\psi_m(t)=0.63$ at $t=80$ and the radius and the frequency grow. For the SG model the small-amplitude bion can be obtained making use of the expansion $\sin x = x-(1/6)x^3$ and eqn. (23c)

$$\chi_n(r,t)=\sqrt{6}\psi_n=\sqrt{8}\kappa A_n(\kappa r)\cos(\sqrt{1-\kappa^2}\,t). \tag{14.23b}$$

The second case.

There is an expansion (10.39) near the condensate (we set $\psi_v=-1$) and for $A(\xi)=\sqrt{(3/2)}f_1(\xi)$ we again get eqn. (22). Note that the masses $M_n=E_n$ of the three-dimensional pulsons

$$\psi_n(r,t)=\frac{2}{\sqrt{3}}\epsilon A_n(\epsilon r)\sin\omega t-\epsilon^2 A_n(\epsilon r)(1+\frac{1}{3}\cos 2\omega t) \tag{14.23c}$$

$$\epsilon=\omega^{-1}\sqrt{2-\omega^2}$$

for a few first values of $n = 0,1,2,3,4$ at small equal magnitudes of $\psi_m(0) = (2/\sqrt{3})\epsilon A_m(0)$ are related as $\int \psi_m^2 d^3x / A_m(0) = (1:2:3:4:9...)$. These pulsons also weakly radiate, at small amplitudes*).

3. Stability of many dimensional stationary solutions

In Chapter 10 (Sections 1 and 2) we have described the current situation in stability problems of SLS and PLS (bubbles and drops). Here, we note specific features of $D \geqslant 2$ dimensions.

Since most integrable models lose this property in proceeding to $D \geqslant 2$ dimensions the stability problem now becomes most important. The absence of analytical formulae for SLS (PLS) in the overwhelming majority of cases is the reason why variational and numerical methods are so effective.

Solitons may exist which have various types of symmetry such as planar, cylindrical, spherical, and so on, in systems with more than one space dimension. Planar solitons whose field function ϕ depends only on the coordinate along their motion, (e.g. the z axis, where the soliton wave-front is parallel to the XOY plane), are the most simple and most natural generalizations of one-dimensional solitons. Such solitons are indeed easily seen to be particular solutions of corresponding multidimensional equations by virtue of $\partial_x \phi = \partial_y \phi = 0$.

Naturally, there arises the question of stability of such solutions with respect to transverse perturbations, i.e. those depending on x and y. According to our point of view this means that we should solve the initial problem

$$M[\phi] = 0 \tag{14.24}$$

and

$$\phi = \psi_s(z, 0) + \delta\phi(r_\perp, z, 0)$$

where M is a nonlinear differential operator corresponding to some nonlinear partial equation, $\phi_s(z, 0)$ is a planar soliton solution of this equation, and $\delta\phi(r_\perp, z, 0)$ is a perturbation. Such stability problems concerning the planar Lagmuir solitons

*)Solutions of the pulson type with small amplitudes near $r = 0$ have been found accurate to $0(e^{-c/\epsilon})$.

and the collapse of Langmuir waves have been studied in a series of analytical and numerical works [13,14-16]. In these cases the operator M is defined by the system

$$\nabla(i\partial_t\vec{\varphi}+\nabla(\nabla\vec{\varphi})-\Phi\vec{\varphi}) = 0; \quad \Box\Phi = \Delta|\vec{\varphi}|^2, \tag{14.25}$$

$$\nabla\times\vec{\varphi} = 0; \quad \vec{\varphi} = \{\varphi_x,\varphi_y\}.$$

As an example of these investigations we consider the results of Ref. [13]. Thc authors used in their computations a two-space dimensional algorithm based on the Fourier transformation of system (25) under periodic boundary conditions along the y and z axes, and a 32×32 space grid.

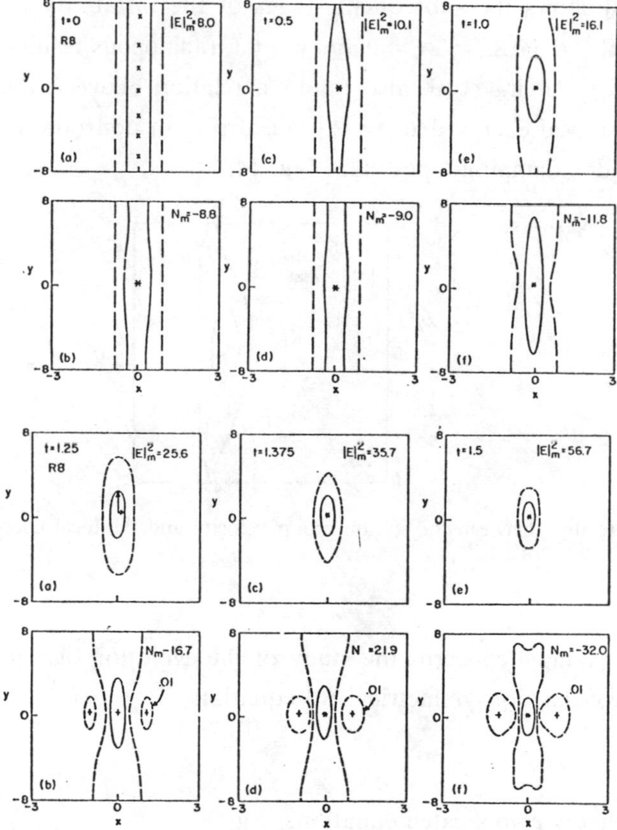

Figure 14.6. Instability (of the sausage type) of the planar Langmuir soliton (collapse).

The initial function $\vec{\varphi}$, the electric field, was chosen to be the planar soliton solution, and the perturbation to the potential Φ, i.e. the particle density, was

$$\Phi(y,z,0) = -(2A^2\mathrm{sech}^2 Ax)(1+2\epsilon\cos k_y y);$$

therefore neither the electric field nor the time derivative Φ_t were perturbed at the initial moment. The planar soliton, as might be expected, is unstable at sufficiently large amplitudes and wavelengths $l_y = 2\pi k_y^{-1}$. The stable region of the planar solitons is $k_y > \alpha A$ with α being a numerical factor of order unity in the computations [13]. After a short stage of selfconsistent reconstruction of the electric field, unstable solitons begin to collapse so that the amplitude φ grows infinitely (at any rate in the framework of the model considered). Figure 14.6 taken from [13] shows the subsequent stages of the instability of the stationary planar soliton at $l_y = 16$, $A = 2$, $v = 0$ and $\epsilon = 0.1$, (analogous results have also been obtained in [16]). Pereira et al. also found a relation between the soliton group velocity and the local energy density $|\varphi|^2$ occurring in nonstationary processes of collapse or Landau damping types*) (see Fig. 14.7).

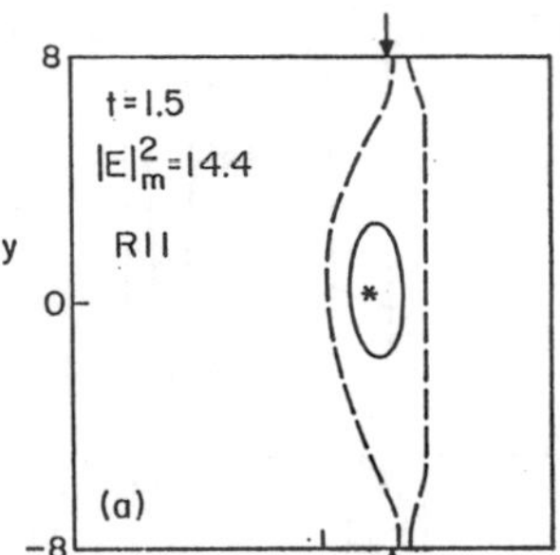

Figure 14.7. The relation between the soliton group velocity and the local energy density (asymmetric collapse).

The second example concerns the study of the evolution of a planar soliton in the framework of the 'nonsymmetrical' $S3$ equation

$$i\phi_t + \partial_{-}^2\phi + \phi|\phi|^2 = 0.$$

This and other more complicated equations, e.g.,

*) These authors also investigated the effect of Landau damping and pump-field on the collapse and showed the process of soliton birth and death to be periodic in time.

$$2i\phi_t + \partial_-^2 \phi = \frac{9}{2}\phi|\phi|^2 + 3\phi\Phi_y$$

$$\partial_+^2 \Phi = -3\partial_y |\phi|^2$$

occur when studying nonlinear two-dimensional plasma waves [14,19] and surface waves on water of finite depth [17,18]. The second system was shown by Anker and Freeman [18] to be integrable and allows the processes of soliton decay and fusion if the usual decay conditions $\omega_1+\omega_2+\omega_3 = 0$, $\vec{k}_1+\vec{k}_2+\vec{k}_3 = 0$ are satisfied. It should be noted that the stability problem is unfortunately left open (let us recall the very instructive example by Berryman).

The initial problem (24) for planar solitons of the first equation was numerically investigated by Pereira et al. [16] using the code of Ref. [13] described earlier. They revealed the stability of planar solitons of the form $\phi_s = \sqrt{2}\,\mathrm{sech} z$ with respect to perturbations even in z (of a sausage type) in agreement with results of [14,16]. However, these solitons turned out to be unstable to perturbations which were odd in z (snake type)

$$\delta\phi(y,z,t) = a\cos(k_y y)\cdot\partial_z\phi_s$$

with $a = 0.1$.

An initial linear stage of this instability corresponds closely to the theoretical predictions of Yajima [14] and has a characteristic exponential growth. When the soliton amplitude has reached a value of about half the unperturbed one the instability goes into a nonlinear regime as shown in Fig. 14.8. One can see that the instability leads to the soliton breaking up into pieces and to the soliton energy dispersing. Hence, we can call such an instability 'anticollapse'. The experimental conditions fail to give a definite unambiguous answer to the question, whether a definite nonlinear quasi-stationary state of soliton type may be achieved or whether the energy disperses quite uniformly over all space.

Planar bell-form solitons of relativistic Klein-Gordon equations are also subject to analogous types of transverse instability. Real Higgs and SG kinks as well as bell-solitons of the Kadomtsev-Petviashvili equation are stable.

It should be noted that planar solitons possess infinite energy charge, etc., like charged and current-carrying planes in electrodynamics. In this connection, it is natural in more than one space dimension to consider as soliton-like solutions only those field configurations which have infinite values of these quantities (i.e.,

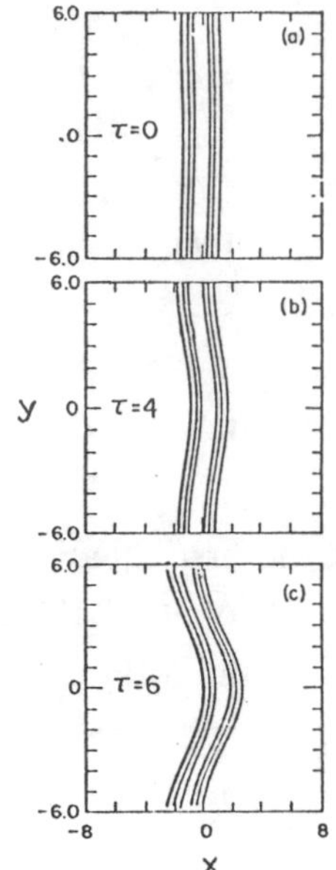

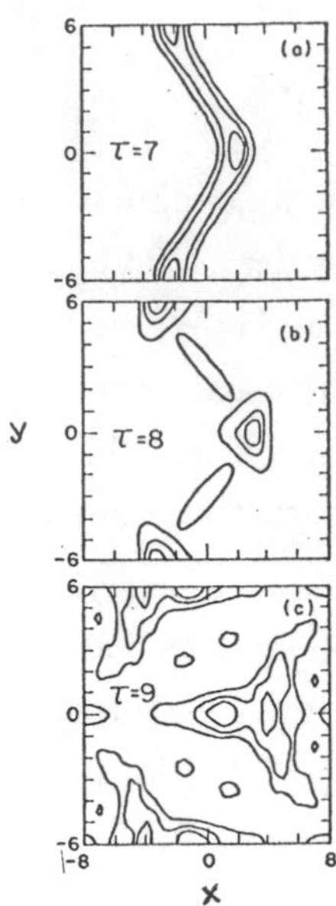

Figure 14.8. Nonlinear regime of the plane soliton instability in the asymmetric *NSE* (anticollapse).

particle-like solutions). For such solutions one can formulate a theorem of quasi-soliton stability (the Derrick-Hobart theorem). For systems with an isotopic symmetry group this theorem may be transformed to the Q-theorem [20], for which the quasi-soliton solution

$$\psi_s = R(r)e^{-i\omega t}$$

is stable when

$$dQ/d\omega < 0 \qquad (\text{for} RIE), \tag{14.26a}$$

$$dN/d\omega < 0 \qquad (\omega < 0 \text{ for } NSE), \tag{14.26b}$$

(and is unstable in the opposite case), where Q is the conserved quantity defined by the internal symmetry group. Thus, in the simplest case of $U(1)$ symmetry,

$$Q = \operatorname{Im}\int(\bar{\psi}_t\psi)d^Dx = \omega N,$$

where D is the space dimension. Using the scale transformations $x \to \kappa^{-1}x$, $R \to \kappa^{(2/n-2)}\chi$, it is easy to show that the stationary equation of the D-dimensional

ϕ_D^n-theory (in particular the ϕ_D^4 model) may be reduced (see Chapter 10) to

$$-\Delta_D \chi + \chi - \chi^{n-1} = 0,$$

with

$$N = \int R^2 d^D x = \kappa^B C(n), \ \beta \equiv \frac{4 - D(n-2)}{n-2}, \tag{14.27}$$

$$Q = \omega N.$$

Calculating $dN/d\omega = -dN/d\kappa^2$ and $dQ/d\omega$, we find the corresponding SLS stability regions

$$D < \frac{4}{n-2} \qquad \text{(NSE)},$$

$$1 > \omega^2 > \frac{n-2}{4-(n-2)(D-1)} \text{(NKG)}.$$

In both cases, we have

$$n < 2 + \frac{4}{D} \tag{14.28}$$

which leads to $n < 4$ at $D \geqslant 2$.

This means that PLS are unstable in the framework of the ϕ_D^4 theory at $D \geqslant 2$ in both the nonrelativistic (NSE) and the relativistic (NKG) versions.

Stable PLS have been found in the framework of a number of RI and non-relativistic models. In particular NSE and NKG with nonlinearity (12) have been investigated in the papers [7,8,21], $(\phi^4 - \alpha\phi^6)$ model in [20,22-24], the Gaussian model (16), in works [9-11] and the LLE model in [5,6]. In the Gaussian model, the stability of stationary solutions (17) ca be studied analytically. The inequality (26) of the Q-theorem along with eqn. (17) give $\omega^2 > 1/2$ [11]. For all the remaining models the stability regions are found via the numerical method: plots of the functions $Q(\omega)$ or $N(\omega)$ are to be found. As an example let us consider a RI model with saturable nonlinearity (12). In the papers [7,8], the plots $Q(\omega)$ were found numerically for several values of m. These graphs are presented in Fig. 14.9 together with a region in the plane m, ω (see eqn. (15)) where PLS exist: $\omega^2 < m^2 < 1 + \omega^2$ (Fig. 14.9a). The functions $Q_i(\omega, m = m_i)$ in Fig. 14.9b for $m_i = 0.4$; 0.6 and 1 correspond to three cross-sections along the lines III, II and I in Fig. 14.9a. The cross-hatched region in this figure corresponds to unstable PLS:

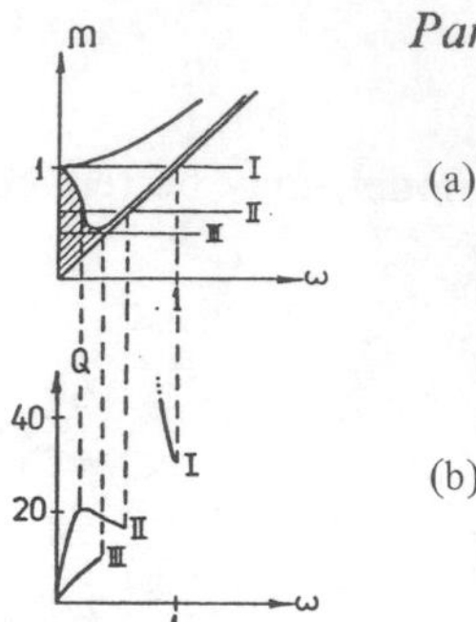

Figure 14.9a. Soliton existence and stability areas in the m,ω plane. The shaded area corresponds to unstable solitons as follow from Fig. 14.9b.

Figure 14.9b. The dependence Q on Ω correspond to the cross-sections along the lines I, II and III in Fig. 14.9a.

$dQ/d\omega > 0$. It is clear that the stability region $\omega_l < \omega < \omega_{up}$ contracts when m decreases and disappears at $m < m_{st} \simeq 0.48$. In the vicinity of the line $\omega = m(\kappa^2 \to 0)$ there is also a narrow instability band: $\omega_{up} < \omega < m$. This fact is easily established using the asymptotic expansions of the preceding section. Indeed, at small κ^2, after performing the corresponding scale transformations, only the first nonlinear term of the expansion of potential (12) does not contain the small parameter κ^2, i.e. effectively we have the ϕ^4 theory. In the framework of the latter, as we know (see eqn. (28) at $n = 4$, $\omega \simeq 1$),

$$dQ / d\omega \simeq \kappa^{2-D}$$

and small-amplitude PLS are stable only in the $D - 1$ variant accurate to the exponentially small emission (barring the SG model where the radiation is absent). In the above-mentioned papers the stability problem of PLS and bubbles has been also studied in the $\phi^4 - \phi^6$ model. In particular, for the nonrelativistic version of this model in [23] the dependence $Q(\omega)$ has been found for $D = 3$ PLS and in [24] results have been obtained indicating the absence of stable $D = 3$ bubbles, so that the problem of the existence of stable $D \geqslant 2$ bubbles is still open.

4. Static ring-shaped fluxons (the structure stability)

Here, as an example, we describe a system in which the existence of a stable static $(D=2)$-dimensional PLS is possible. This example can be considered as a two-dimensional generalization of the situation studied in Chapter 12. We mean the structural stability of solitons in a bound parameter region. We saw above that investigation of this region must be performed with special care since it is just here that such effects arise as bifurcations of solutions. Below we shall see that even for very small amplitudes of a localized perturbation, analogous effects are also possible in two-dimensions.

Consider a circular Josephson junction of radius R with a ring-shaped microinhomogeneity of radius r_0, then in the radially symmetric case for the phase difference of wave functions of superconducting electrons in the upper and lower superconductors we have the equation

$$\varphi'' + \frac{1}{r}\varphi' - (1-\mu\delta(r-r_0))\sin\varphi = 0, \tag{14.29a}$$

with the boundary conditions

$$\varphi'(r)|_{r=0} = \varphi'(r)|_{r=R} = 0, \tag{14.29b}$$

corresponding to the requirement of regularity of the magnetic field φ' at the centre and its vanishing at the junction boundary.

The condition of stability of the solution $\varphi_s(r)$ of eqn. (29) is reduced to the condition that E.V. spectrum should be positive with the boundary value problem

$$-\left[\frac{d^2}{dr^2} + \frac{1}{r}\frac{d}{dr} - (1-\mu\delta(r-r_0))\cos\varphi_s(r)\right]\psi(r) = \lambda\psi(r) \tag{14.30a}$$

$$\psi'(r)|_{\substack{r=0\\ r=R}} = 0. \tag{14.30b}$$

As we have seen, in a $(D=1)$-dimensional junction with boundaries and an inhomogeneity, the breeding (bifurcation) of solutions is possible with a smooth change of controlling parameters. In our case (problem (29)) those parameters are μ, r_0, R. Just as in the one-dimensional system, our chief aim is to determine bifurcation points in the space (μ, r_0, R). Usually, values of μ and R are limited by physical reasons: $\mu<2$, $R\leqslant 10$. On the other hand, the bifurcation only occurs at sufficiently large μ. It is easy to check [20,26] that when $\mu = 0$ the problem (29) has no stable static solutions. Qualitatively, this follows from the following simple reasoning. Consider the function

$$\varphi_s = 4\mathrm{arctg}[\exp(r-r_s)], \qquad (14.31)$$

$$\dot{\varphi}_s = 0,$$

to be an approximate solution of eqn. (29) at $R\to\infty$ and $r_s \gg 1$. It is easily seen that the energy of state (31) at $\mu = 0$,

$$E_s(\varphi) = 2\pi\int_0^R rdr[\frac{1}{2}(\varphi_r)^2+1-\cos\varphi], \qquad (14.32)$$

is $E_s = 16\pi r_s$. This energy is concentrated in the ring of radius r_s and width $\Delta r \simeq 1$. As a result, the ring (31) is subjected to the linear tension

$$F_s = -\frac{d}{dr_s}E_s = -16\pi \qquad (14.33)$$

squeezing the ring to the center (and thereby decreasing its potential energy). An approximate equation describing the ring compression is easily obtained via the energy conservation law $dE/dt = 0$ and the fact that at large $r_s \gg 1$ one may write E_s in the form

$$E_s \simeq (2\pi r_s)^{D-1}E_1 \equiv (2\pi r_s)^{D-1}8\gamma \qquad (14.34)$$

where $\gamma^{-2} = 1-v^2$, and D is the number of space dimensions. Differentiating eqn. (34) with respect to t and substituting $v = \dot{r}_s$ we get

$$\ddot{r}_s = -\frac{D-1}{r_s}(1-\dot{r}_s^2). \qquad (14.35a)$$

This equation has been obtained in a number of papers and, at $D = 3$, has the solution

$$r_s = r_{\mathrm{in}}cn\left[\frac{\sqrt{2}}{r_{\mathrm{in}}}t,\frac{1}{2}\right]. \qquad (14.35b)$$

Under the same assumptions, eqn. (35) also describes the Higgs ring, or bubble compression,

$$\chi = \mathrm{th}\left[\frac{1}{\sqrt{2}}(r-r_s)\right], \qquad (14.36)$$

$$(\Box - 1 + \chi^2)\chi = 0.$$

Let us find the force that acts on the ring due to the inhomogeneity assuming that the former is not deformed by the inhomogeneity. This assumption is confirmed by numerical studies of problem (29) performed in paper [27]. The interaction energy is

$$E_{\text{int}} = -2\pi \int \delta(r - r_0)(1 - \cos\varphi_s) r dr = \frac{4\pi\mu r_0}{\text{ch}^2(r_0 - r_s)} \equiv U_{\text{int}}.$$

Hence, it follows that the force acting on the ring due to the inhomogeneity is

$$F_{\text{int}} = -\frac{\partial}{\partial r_s} U_{\text{int}} = 8\pi\mu r_0 \frac{\text{th}(r_0 - r_s)}{\text{ch}^2(r_0 - r_s)}.$$

Equating F_{int} to F_s we get the ring equilibrium condition

$$\text{th}(r_0 - r_s) = \frac{2}{\mu r_0} \text{ch}^2(r_0 - r_s).$$

Hence, stationary states can exist at both $\mu > 0$ (the 'kink' is inside the inhomogeneity) and $\mu < 0$ (the 'kink' is outside the inhomogeneity and pushes off it). The maximum value of the attractive force is obtained from the condition ($\xi = r_0 - r_s$)

$$\frac{d}{d\xi} \frac{\text{th}\xi}{\text{ch}^2\xi} = 0 \Rightarrow \text{th}^2\xi|_{\max} = \frac{1}{3},$$

or

$$F_{\text{int}}^{\max} = \frac{2}{3\sqrt{3}}.$$

The condition of the existence of the stationary state is

$$|F_{\text{int}}^{\max}| > |F_s|, \tag{14.37}$$

$$\text{i.e.} \quad \mu > \frac{3\sqrt{3}}{r_0} \equiv \mu^*.$$

In Fig. 14.10 the behaviour of the total potential $U = E_s + U_{\text{int}}$ of the interaction of the ring with the inhomogeneity at fixed r_0 and various μ is qualitatively depicted. One can see that there exist three possibilities. At $\mu > \mu^*$, we have two static solutions related to two points r_1 and r_2 where $dU/dr_s = 0$. One of these solutions ($r_s = r_2$) is stable, the other not. At $\mu = \mu^*$, both stationary points merge into one

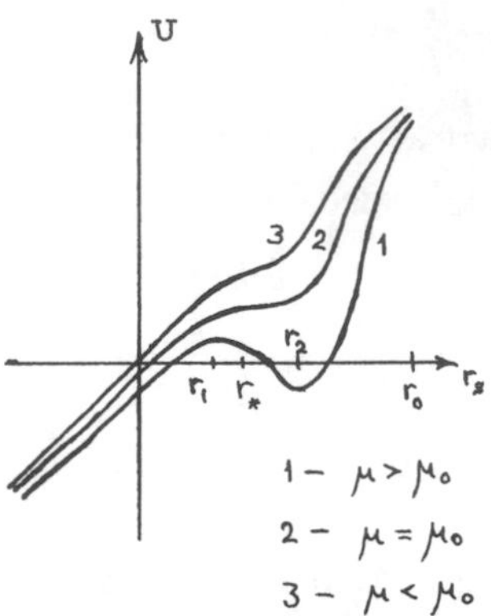

Figure 14.10. Approximate plots of the total potential $V(r_s)$ for various μ.

point $r_1 = r_2 = r^*$ where the function $U(r)$ has an inflection. An unstable solution remains: at $\mu < \mu^*$, static solutions disappear, i.e. the point μ^* is the bifurcation point.

In the argument put forward for $r_0 \gg 1$ the stable $\varphi_2(r)$ and unstable $\varphi_1(r)$ solutions coincide with each other in a form (the ring φ_s) differing only by the position (the unstable one is nearer to the ring centre at $\mu < 0$) and energy $E(r_s = r_1) > E(r_s = r_2)$. Numerical investigations of problems (29) and (30) performed in Dubna [27] have shown that in the finite interval, $R \leqslant 10$, and apart from the mentioned two solutions, a third unstable solution $\varphi_s(r)$ appears which exists at $\mu < \mu^*$, too. The dependence $\varphi_i(r)$ for all three obtained states is very similar qualitatively, and their energies are also close. The functions $\psi(r)$ are somewhat more different, however; the states differ most clearly in the values of λ. These features especially underline the difficulties of numerical and approximate (analytical) studies of static states and demonstrate the necessity of a joint solution of both boundary value problems (29) and (30). This manifests itself clearly near a bifurcation point where the states $\varphi_1(r)$ and $\varphi_2(r)$ become almost degenerate; whereas the corresponding values of λ differ quite strongly and depend on μ in different ways. Details of the simultaneous soliton of problems (29) and (30) can be found in [27]. Finally, note that the difference between $\mu^* = 0.838$ calculated via eqn. (37) and the critical value $\mu = 0.785$ obtained in numerical investigations is smaller than 7%.

Chapter 15

PULSONS AND Q-SOLITONS

1. Collapse of circular and spherical bubbles

Computer solitons of the problem (14.29) and (14.36) at $\mu = 0$ show that at the initial stage the bubble behaviour is very well described by eqn. (14.35b). Significant deviations begin in the region $r_s \cong 1$. In the bubble collapse process, the potential transforms into kinetic energy so that near the centre collapse is replaced by expansion, (it is if the bubble is 'reflected' from the centre), and a part of the energy is converted into radiation. On reaching the dimension $r_s = r_0$ which is somewhat smaller than the initial one r_{in} (due to radiation losses) the bubble collapses again, and so on. Its behaviour is very similar to that of a pendulum in the presence of friction. The number of oscillations depends on radiation losses. Bogolubsky and Makhankov [26] have performed numerical experiments for various values of r_{in}. The results show that a certain number of oscillations (usually 2-5) are sufficient for the emission to infinity of the main part of the bubble energy. In this series of experiments it has been noted that those bubbles with a sufficiently large initial dimension finish their life by forming some new long-lived objects. Bubbles with small dimensions disappear altogether. A second series of calculations has been performed to study the behaviour of the bubble (14.31) for the Sine-Gordon equation. Because of complete integrability of the Sine-Gordon equation in two-dimensional space-time it was assumed that the lifetime of the bubbles (14.31) would be very long. The results of the calculation refute these expectations: the radiation during the bubble collapse is very large and the picture is on the whole qualitatively analogous to that described above. Thus in these experiments the integrability of the Sine-Gordon equation in the (x,t) world does not manifest itself at all in real four-dimensional space-time.

More detailed studies display a number of specific features of the collapse and expansion Sine-Gordon and ϕ^4_- bubbles, as well as pulsons formed from them. Works which should be mentioned are those by Christiansen's group at Lyngby (see, for example, the review [29] where one can find a detailed reference list) and especially those by Geicke [30]. The results of these investigations lead to the inference that bubbles of large radii behave as quasi-solitons in the sense that when shrinking or expanding they interact practically the same as the plane kinks [29]. In [30], a more interesting and, at first glance, a somewhat paradoxical result was obtained in the numerical modelling of the collapse of ring-shape (i.e. $D = 2$ dimensional) bubbles. The following picture turns out to occur depending on the initial bubble radius and hence its energy (the behaviour of ϕ^4_- and Sine-Gordon bubbles is qualitatively the same):

(1) When the initial radii of bubbles are small enough, $r_{in} < 2$, stable pulsons are not formed (as was the case in three dimensions), and the energy of a shrunk bubble is radiated to infinity.

(2) For $2 \leqslant r_{in} \leqslant 3.5$, the shrunk bubble bears at the origin a pulson (we shall discuss its properties later).

(3) $4 \leqslant r_{in} \leqslant 33$, the shrunk bubble is not in practice reflected from the origin (the emitted energy is of the order of the bubble kinetic energy), and all its energy goes to infinity through the radiation of linear waves.

(4) In the region $r_{in} \cong 40$ (its width is unfortunately unknown) pulsons are again formed.

(5) The next region found by the author of [30] is $r_{in} \cong 93$ where pulsons are also produced.

In the last region, the bubble reflected from the origin expands, and then the returning effect occurs near $r_r \cong 39$, i.e. in the fourth region. In the region (intermediate) between the fourth and the fifth ones bubbles reflected from the origin come into the third region and pulsons are not formed. So, for example, the pulson with $r_{in} = 54$ returns to the region $r_r = 8$.

One can naturally expect that there should exist a countable set of those regions Δr_i starting from which the bubble will bear a pulson after a definite number of oscillations. In this connection the fourth region is related to the one reflection and the returning radius $r_i = 3$ [*)].

*) This value is estimated, for example, by extrapolating results $r_{in} = 54 \rightarrow r_r \cong 8$ and $r_{in} = 93 \rightarrow r_r \cong 39$ to find r_r as a function r_{in}.

All these results may be explained in terms of pulson stability.

2. Properties of pulsons

As was shown above, in the planar (x,t) world there are long-lived bound states of solitons (bions) in the framework of the Klein-Gordon equation. Naturally the question arises of the existence of analogues of the bion in two- and three-dimensional space. In accordance with the Derrick-Hobart theorem there are no stable stationary solitons for the Sine-Gordon and $\phi^4_\pm$ theory equations. In computer experiments carried out at Dubna in the framework of the ϕ^4_+ theory stable oscillatory solutions have not been observed [20] (see Fig. 15.1).

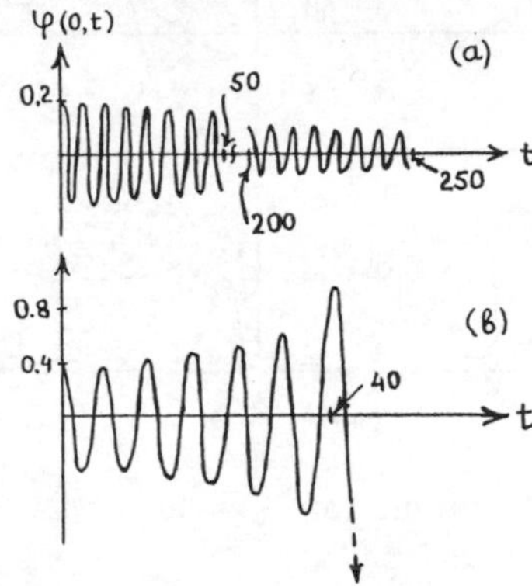

Figure 15.1. Time-dependence of the envelope of the oscillation amplitude of the field function in the origin (ϕ^4_+ model) (a) - dissipative mode, (b) - singular mode.

The situation changes in the case of the Sine-Gordon equation and ϕ^4_- theory. As we have already mentioned in the preceding section, in consequence of the evolution of a Higgs or Sine-Gordon bubble of a sufficiently large dimension (or mass, which is the same), in its centre a lump of field energy appears having a bell-shaped radial distribution. Moreover, the field function oscillates about the vacuum value fixed by boundary conditions (see Fig. 15.2). In the experiments described [28], this value was $\chi_\nu(t,\infty) = -1$ for the ϕ^4_- theory, and $\varphi^{SG}_\nu(t,\infty) = 0$ for the Sine-Gordon equation.

These energy lumps, names puslons, arise for rather arbitrary initial conditions proved that

$$\chi(0,0) > c_1 \geqslant 1,$$

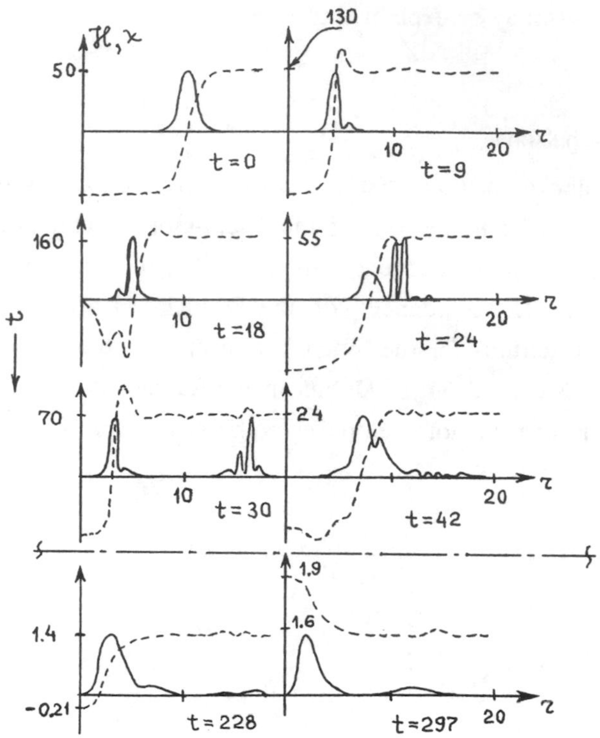

Figure 15.2. Evolution of the ϕ^4_- bubble for $R_0 = 10$.

$$\chi_{SG}(0,0) > c_2 \geqslant 2\pi.$$

The radiation of the pulsons is small, therefore their lifetime is sufficiently large, some thousands of oscillation periods (recall the Higgs bions in the planar world). These results have been confirmed later by the ITEP group (Moscow) [31] (for some detail see the review [20] for ϕ^4_- pulsons and by the Danish group [29] for Sine-Gordon pulsons.

For the multivacuum Sine-Gordon theory, Bogolubsky [32] observed heavy pulsons which have no analogues in the planar world. Their amplitudes are in the region $\chi_{SG}(0,t) \in (3\pi, 4\pi)$. In the region $(2\pi, 3\pi)$ these pulsons are unstable and quickly transform into the 2π pulsons considered above. We note that the pulsons described above are stable with respect to azimuthal perturbations [33].

In two dimensions, the pulson life-time grows by order as was shown by Geicke [30]. Hereafter, while describing $D = 2$ pulsons we will use the results of

[30].

The pulson behaviour is qualitatively described as follows. In the *first stage* ($t \leqslant 200$), the formation of the pulson occurs. After that, the *second stage* starts ($200 \leqslant t \leqslant 6 \times 10^3$) in which there are regular oscillations of the field function about the fixed vacuum with a period $T_p = 2\pi/\omega$. The period of the pulson depends on the model and slowly increases (up to that of linear oscillations, $T_l = 2\pi/\omega_l$) with a simultaneous decrease in pulsation amplitude. The latter causes the pulson dimension to grow. It is interesting to note that, although the asymptotic expansions (14.23) are, strictly speaking, valid for $\epsilon = \omega_l^{-1}(\omega_l^2 - \omega^2)^{1/2} \ll 1$, we can use them to estimate the adiabatic variation in time of the pulson amplitude via the formulae

$$\left[\frac{A(t_1)}{A(t_2)}\right]^2 \simeq \frac{1-(2\pi/T_1)^2}{1-(2\pi/T_2)^2} \qquad (SG) \tag{15.1a}$$

$$\left[\frac{A(t_1)}{A(t_2)}\right]^2 \simeq \left[\frac{B(t_1)}{B(t_2)}\right]^2 \simeq \frac{1-2(\pi/T_1)^2}{1-2(\pi/T_2)^2} \qquad (\phi^4_-). \tag{15.1b}$$

It is remarkable that the accuracy of this estimation is only about 1-2%. This fact is established from the results of [30] (see Tables II and III) and is attributed to the adiabatic nature of energy losses. It also means that $D \geqslant 2$ dimensional pulsons are the analogues of the one-dimensional bions.

In the regular stage, the amplitude of the weakly radiating pulson is periodically modulated. For the $D = 2$ case, the amplitude of modulation is about 1.5-2%. The period of modulation equals at the beginning $T_M \simeq (-7)T_p$ for Sine-Gordon pulsons and $T_M \simeq (5-6)T_p$ for ϕ^4_- ones and increases considerably ($\geqslant 1.5$) during the regular phase. We note the difference in behaviour of the Sine-Gordon and the ϕ^4_- pulsons. These are connected with different symmetries of the potential parts of the Hamiltonian. Oscillations in the Sine-Gordon pulson are symmetric with respect to the basic vacuum ($\varphi_v = 0$) (on which the pulson is formed). Large amplitude oscillations of the field function in the ϕ^4_- pulson are essentially asymmetric with respect to the basis vacuum ($\chi_v = -1$) so that the right amplitude is approximately twice that of the left one.

The regular phase of the pulson ends when its maximal amplitude is bigger than half a spacing between neighbouring wells of the potential: $B_\phi > 1$, $B_{SG} > \pi$. Upon reaching this limiting value, $B_\phi \simeq 1$, $B_{SG} \simeq \pi$, the third phase of the pulson starts. This phase was studied for $D = 3$ pulsons in [20,28]. Nothing similar is known to the author in two dimensions.

Let us point out some specific features of $D = 3$ pulsons:

(1) The phase of formation as well as the regular phase become approximately 5-7 times shorter.

(2) The modulation amplitude different for the Sine-Gordon and ϕ^4_- models is now larger in the beginning of the regular phase but rapidly diminishes, approaching zero at its end. The modulation period increases in time respectively from $T_M \simeq (8-10)T_p$ to $T_M \simeq (17-20)T_p$.

In the *third phase*, the amplitude $B(t)$ of the pulson decreases relatively rapidly ($\Delta t \simeq 10T_p$) up to a value $B \ll 1$ with its simultaneous 'swelling'.

Finally, the *fourth phase* starts in which oscillations in both the models become practically symmetric with respect to the basis vacuum: $\phi_v^{SG} = 0$, $\chi_v^{\phi} = -1$. In this phase, pulsons are well described by the small amplitude expansions (14.23).

3. Pulson stability

This problem is much complicated and yet unsolved. Up to now, qualitative reasoning has enabled a great number of numerical results to be explained from a unified point of view.

We discuss this problem, using as an example the model of a scalar relativistic field with saturable nonlinearity

$$\Box\psi + m^2\psi - \psi\frac{|\psi|^2}{1+|\psi|^2} = 0. \tag{15.2}$$

We saw above that *stable* 'charged' Q-solitons, $\psi(r,t\,|\,m) = \varphi(r,m)e^{-i\omega t}$ exist in this model for $m > m_{st} \simeq 0.48$ (see Chapter 14, Fig. 14.13a). The stable ω region has an upper as well as a lower limit: $\omega \in (\omega_d, \omega_u)$. Q-stability is connected with the additional integral of motion in the case of complex fields. By the Derick-Hobart theorem, model (2) does not possess *stable static* particle-like solutions, however by analogy with Q-stability one can suppose that slightly radiating real-field pulsons ($\bar{\psi}=\psi=\varphi$) will be stable by virtue of conserving an adiabatic invariant. It means that for such solutions an approximate integral of motion can exist

$$\frac{d}{dt}\int <\varphi^2> d^D x = 0, \quad <\varphi^2> = \frac{1}{T}\int_0^T \varphi^2 dt, \quad T > \frac{1}{\omega}, \tag{15.3}$$

with exponential accuracy under definite conditions [34]. Invariant (3) is evident to exist in the small amplitude case. For heavy pulsons with large amplitudes ($A \simeq 1$) this conjecture needs a verification.

In [35], real solutions ($\omega=0$) to (2) were investigated in two dimensions. The evolution of initial lumps $\varphi(r, m, \omega=0)$ governed by the Cauchy problem was studied:

$$\Box\varphi+\varphi[m^2-\varphi^2(1+\varphi^2)^{-1}] = 0,$$

$$\varphi(r,0)_{\mathrm{in}} = \psi(r,m,\omega,0), \quad Q = 0.$$

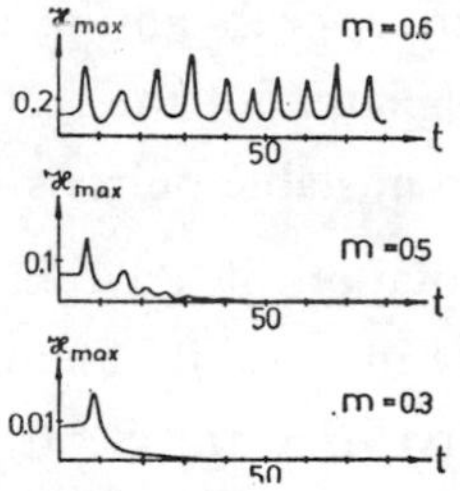

Figure 15.3. The energy density in the origin as a function of time for $m = 0.3, 05$ and 0.6.

The results are shown in Fig. 15.3 for the energy density at the origin $\mathcal{H}_{\max}(r=0,t)$ as a function of time at $m = 0.3$, 0.5 and 0.6. One can see that at $m > m_{\mathrm{cr}} \simeq 0.5$ the evolution of the initial lump $\varphi_{\mathrm{in}}(r)$ ends in the formation of a stable weakly radiating pulson. In the vicinity of $m_{\mathrm{cr}} \sim m$, pulsons begin to lose their stability so that their evolution splits into two phases: a short one ending in the pulson formation and a longer one related to its decay. Ultimately, when $m < m_{\mathrm{cr}}$ an initial lump decays aperiodically excluding the pulson phase.

It is very important to emphasize that the magnitudes of m_{st} and m_{cr} are the same with good accuracy [35]. Thus, stable chargeless pulsons occur in that region of ω where exist Q-stable solitons. In the region where complex solitons are unstable (Q-unstable) pulsons do not occur[*)].

In the evolution of Sine-Gordon and ϕ^4_- pulsons there is a third stage in which the pulson amplitude rapidly diminishes. This behaviour may be thought of as an instability of the pulson in the above sense. Then the stability interval is

---

[*)]For this reason apparently, numerical experiments on searching pulsons in $\varphi^4_+(D=3)$ theory gave no results.

$\omega \in (\omega_d, \omega_u)$. During the evolution of the pulson its frequency grows (its amplitude falls due to radiation) until it reaches the upper stability bound ω_u whereupon the pulson decays.

It must be emphasized once more that all the conclusions are of a qualitative (phenomenological) character.

From this point of view, Geicke's results may be qualitatively put into the following scheme. Depending on its initial radius, the pulson has a definite amplitude and, hence, frequency. Numerical experiments show that the stability region of ring-shape pulsons is the interval just as for pulsons of model (2) at $m \neq 1$, i.e. $\omega \in (\omega_d, \omega_u)$. When $\omega \notin (\omega_d, \omega_u)$ there appears an unstable pulson sometimes decaying aperiodically. This gives an explanation of the fact that there exists an energy gap (hence a gap of r_{in}) giving rise to unstable pulsons.

One may assume that the rotation of a weakly-nonsymmetric ϕ^4_- bubble can delay its collapse and essentially increase its lifetime (as far as we are aware, this idea has been surmised by Kudryavtsev and Scott independently). Meanwhile, our preliminary investigations have not led to positive results in the x,y,t case.

4. Pulson interaction

We have considered above those dynamic properties of the pulson which relate to their formation and stability. However, this is only the first step. The real dynamics in which specific soliton properties are completely displayed is their interaction.

Therefore, consider the results from two-dimensional cylindrically-symmetric (in the rest frame) $KG3$ (ϕ^4_+) pulson collisions obtained in Dubna [122]. In a series of calculations we study head-on collisions of two *unstable* $KG3$ pulsons of the form

$$\psi(x,y,t) = Af\left[u_0\sqrt{\gamma_i^2(x-\mathrm{v}_i t)^2+y^2}\right]\cos\left[\sqrt{1-u_0^2}\,\gamma_i(t-\mathrm{v}_i x)\right]$$

where v_i is the velocity of the ith pulson in units of the velocity of light, $\mathrm{v}_1 = -\mathrm{v}_2 = \mathrm{v} = 0.2, 0.3, 0.4, 0.6$; $\gamma_i^2 = (1-\mathrm{v}_i^2)^{-1}$ and the function $f(r)$ in the pulson rest frame is a solution of the boundary problem

$$f_{rr}+\frac{1}{r}f_r-f+f^3 = 0, \quad f_r(0) = f(\infty) = 0.$$

The collision time is chosen to be smaller than the development time of the instability. The pulson behaviour resembles the behaviour of one-dimensional SIBq-solitons or Higgs kinks. If the pulson velocity is greater than some critical value $v_{cr} \simeq 0.3$, they emerge from the interaction region and only then 'decay' via dissipative (spreading) (Fig. 15.4) or singular (collapse) (Fig. 15.5) modes.

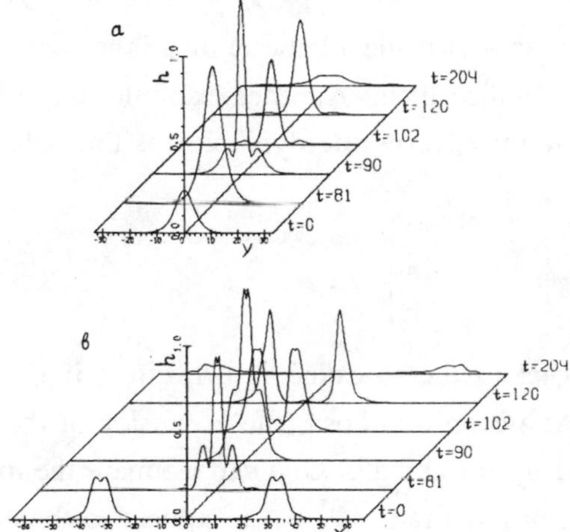

Figure 15.4. Interaction of two pulsons with $v > v_{cr}$ decaying afterwards.

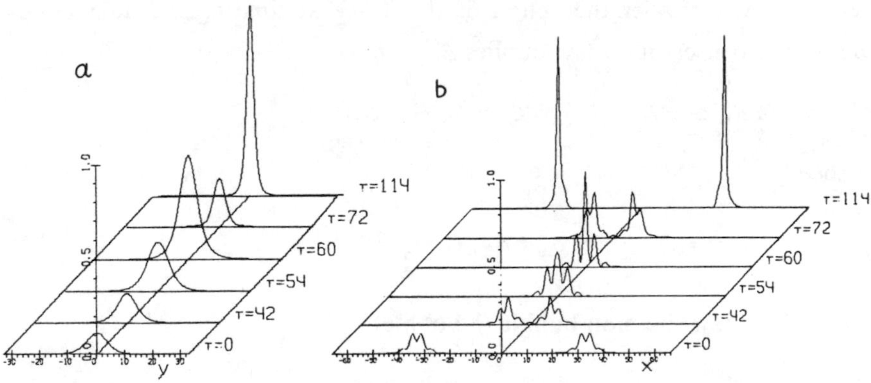

Figure 15.5. Interaction of two pulsons with $v > v_{cr}$ collapsing afterwards.

At $v \leqslant 0.3$ the pulsons merge into one pulson which afterwards collapse. A remarkable fact is that for $v > v_{cr}$ the number of unstable quasi-particles is conserved in the interaction; moreover, their lifetime is approximately the same as in the free state. This takes place despite the fact that the pulson collision is a large, almost self-consistent, perturbation of order unity for each of them. Here, already in the three-dimensional world we are apparently withness to a manifestation of 'soliton' properties (in the sense of Zabusky and Kruskal) by the unstable pulsons.

We have already seen that if the adiabatic invariant arises during the formation of pulsons this can stabilize them. Another example may be given in which the presence of this invariant governs the dynamics of the pulson. The kink-bubble (k-bubble)

$$\chi = 4\mathrm{arc}tg\left\{e^{-\gamma(r - r_{in} - v_0 t)}\right\}, \quad r_{in} \gg 1, \tag{15.4}$$

is known to be subject to the so-called return effect if $v_0 > 0$. Qualitatively this appears as follows. At a certain radius r_r the expansion of the bubble will stop and then will be followed by shrinking*). One can estimate the magnitude of r_r as follows. The bubble (4) of large radius $r_{in} \gg 1$ is subject to the action of the linear or surface tension force with the potential

$$U \simeq \pi(2r)^{D-1}$$

(where D stands for the number of dimensions). Initial inflation of the bubble therefore ceases under the action of this force at time t_r, and at a radius $r = r_r$. The energy conservation law implies $E_{in} = E_r$ or

$$E_{in} \simeq \pi(2r_{in})^{D-1}\gamma_{in}, \quad E_r = M_r \simeq \pi(2r_r)^{D-1},$$

whence

$$r_r = r_{in}\gamma^{1/D-1} \simeq (v_0 \ll 1)\; r_{in}(1+\frac{1}{2}v_0^2)^{1/D-1}. \tag{15.5}$$

Consider now the bion-bubble (b-bubble)

$$\chi = 4\mathrm{arc}tg\left\{\frac{(1-\Omega^2)^{1/2}}{\Omega}\;\frac{\cos\gamma(\Omega t - v_0 r)}{\mathrm{ch}\kappa(r - r_{in} - v_0 t}\right\}, \quad \kappa = \gamma(1-\Omega^2)^{1/2}. \tag{15.6}$$

*)Here, we omit the details of this process which one can find, for example, in [44].

It conserves like a bion the adiabatic invariant E/Ω for $\Omega \simeq 1$. This means that for small gradients $\partial_r \ln \chi = \kappa$ there is an asymptotic transformation from canonical variables $\dot{\chi}$, χ to variables ψ, $\bar{\psi}$ [34], if time dependence of the latter pair is

$$\psi = \varphi(r)e^{-i\Omega t}.$$

Moreover, for small κ, solution (6) is given of the $S3$ model

$$i\dot{\psi} + \psi_{rr} + \frac{D-1}{r}\psi_r + |\psi|^2\psi = 0, \tag{15.7}$$

with the integrals

$$N = \pi(2)^{D-1}\int |\psi|^2 r^{D-1} dr$$

and

$$E = \pi(2)^{D-1}\int \{|\psi_r|^2 - \frac{1}{2}|\psi|^4\} r^{D-1} dr.$$

Consider the behaviour with time of the quantity

$$B = \pi 2^{D-1}\int |\psi|^2 r^2 r^{D-1} dr = \langle r^2 \rangle N \geqslant 0 \tag{15.8}$$

proportional to the mean square radius of the lump. One can readily check that

$$\ddot{B}(t) = \pi 2^D (2-D)\int |\psi|^4 r^{D-1} dr + 8E. \tag{15.9}$$

It then follows from (9) that:

$\underline{D = 2}$. The $\ddot{B}$ sign is determined by the sign of E and $B(t) = 4Et^2 + c_1 t + c_2$. For $E < 0$, $B(t)$ vanishes on a finite timescale,

$$t_c = (8E)^{-1}(\sqrt{c_1^2 + 16c_2|E|} - c_1)$$

i.e. there is the collapse of the lump. At $t > t_c$, $B(t)$ is found to be negative, which is impossible by definition (8). The field becomes singular and the integral in (8) loses its sense. For $E > 0$, $B(t)$ grow monotonically and due to the conservation of N the packet infinitely expands.

$\underline{D = 3}$. $\ddot{B} < 8E$. The behaviour of the packet is similar to that above except that for its collapse $E < E_{cr} (E_{cr} > 0)$ is necessary, and the condition $E < 0$ is now sufficient rather than necessary. For $E > E_{cr}$, the packet will expand (see Fig. 15.6).

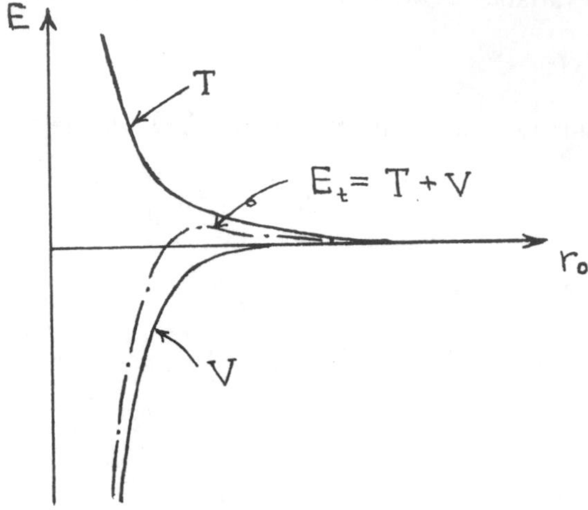

Figure 15.6. The total soliton energy E_s as a function of its radius for $S3$, $D = 3$.

The value of E_{cr} depends, generally speaking, on the initial packet state.

$D = 1$. $\ddot{B} > 8E$. Formula (9) allows only to realize the initial stage of the packet expansion or shrinking and now $E_{cr} < 0$. It is easy to verify that for the one-soliton solution (stationary)

$$\psi_s = ae^{-i\omega t}\operatorname{sech}\frac{ax}{\sqrt{2}}$$

we have $\ddot{B}\{\psi_s\} = 0$ (as might be expected).

By analyzing the function $B\{\psi\}$ one can determine the behaviour of the packets (bubbles)

$$\psi(r,t) = ae^{i(\mathrm{v}/2r-\omega t)}\operatorname{sech}\frac{a}{\sqrt{2}}(r-r_0-\mathrm{v}t), \quad r_0 \gg \frac{\sqrt{2}}{a}. \tag{15.10}$$

The real part of (10) for small amplitudes and velocity ($\mathrm{v} \ll 1$) coincides with (6) accurate to the renormalization of the amplitude and frequency[*)]. Up to the

[*)]This fact can be easily established by the reader (see, for example, [45]).

$O(\sqrt{2}/ar_0)$ terms we find the functionals

$$N = \pi(2r_0)^{D-1}N_1, \quad N_1 = 2\sqrt{2}\,a, \tag{15.11}$$

$$E = \frac{N}{4}(\mathrm{v}^2 - \frac{1}{12}N_1^2), \tag{5.12}$$

$$\int |\psi|^4 r^{D-1} dr = r_0^{D-1} N_1^3/12, \tag{15.13}$$

$$\ddot{B} = 2N(\mathrm{v}^2 - \frac{1}{6}N_1^2), \tag{15.14}$$

$$\omega = \frac{1}{4}\mathrm{v}^2 - \frac{1}{2}a^2 = \frac{1}{4}\mathrm{v}^2 - \frac{1}{16}N_1^2.$$

From the conservation of N follows

$$a = a_{\text{in}}(r_{\text{in}}/r)^{D-1}. \tag{15.15}$$

Formula (14) gives the critical value of the velocity separating shrinking bubbles from expanding ones

$$\mathrm{v}_{\text{cr}}^2 = \frac{4}{3}a^2. \tag{15.16}$$

The critical velocity (16) is also derived by energy considerations. A bubble at rest is under the action of the surface tension with force f_{st} defined by the free energy

$$F_0 = (E - \omega N)_{\mathrm{v}=0} = \pi\frac{2^{D-1}}{3}N_1 a^2 r_0^{D-1}$$

whereby

$$f_{st} = \frac{dF_0}{dr_0} = -\pi\frac{2^{D-1}}{3}(D-1)r_0^{D-2}N_1 a^2.$$

The bubble inflates when $T > F_0$($T = \pi 2^{D-3} r_0^{D-1} N_1 \mathrm{v}^2$ is the kinetic energy of expansion) which gives back (16). Since T and F depends on r in the same fashion this relation holds during the inflation (or collapse).

When applying these results to b-bubbles of the Sine-Gordon model one should bear in mind that the frequency - and hence the bound energy of kinks in the bion - will decrease (due to N=const) in the course of shrinking to cause the bion to split into a $k\bar{k}$ pair. The inverse process has to be during the expansion of a weakly bound pulson. These effects have been observed in numerical experiments, see, for example, [46].

In concluding this section we note one more non-trivial fact. For the generalized version of (7)

$$i\dot{\psi}+\psi_{rr}+\frac{D-1}{r}\psi_r-\frac{dV(\Phi)}{d\Phi}\psi = 0, \tag{15.17}$$

$$\Phi = |\psi|^2;\ \psi(r,t)\to 0,\ r\to\infty$$

formula (9) becomes*)

$$\ddot{B}(t) = 8E-4(D+2)\int U(|\psi|^2)dV+4D\int\frac{dU}{d\Phi}|\psi|^2dV. \tag{15.18}$$

Making use of (18) one can establish the process of self-localization of packets and draw quite general conclusions relating to their behaviour in time even of their stability.

(1) If $U(\Phi)\geqslant 0$ delocalization always takes place and there are no stable states.

(2) $U(\Phi)<0$ localization of wave packets is possible together with stationary states.

We illustrate the virtue of (18) by a very simple example of ϕ^n_D theory. Let $U(\Phi)$ be

$$U(|\psi|^2) = -\frac{2}{n}|\psi|^n,$$

then (18) yields

$$\ddot{B}(t) = 8E-\frac{8}{n}[\frac{D}{2}(n-2)-2]\int|\psi|^n dV. \tag{15.19}$$

We call the packets for which $\ddot{B}(t) = 0$ the stationary packets, and then the following proposition can be proved. Stationary packets are stable when $E<0$ and the second term in (19) is positive*).

We now check this proposition. In the case of ϕ^n_D theory this implies $2-D(n/2-1)>0$ or

*)For the sake of simplicity we assume packets to be centrally symmetric (for the more general case see [47]).

*)A somewhat similar proposition naturally takes place in the more general case [48].

$$1 < \frac{n}{2} < \frac{2}{D} + 1, \tag{15.20}$$

which is none other than the conditions for Q-stability of solitons of Chapter 14, Section 3 (see formula (14.23)).

Thus the dynamical equation (18) provides the opportunity to learn about the stability of packets.

Chapter 16.

INTERACTION OF Q-SOLITONS

1. Nonrelativistic models

Let us turn to the study of the dynamical properties of stable Q solitons in models with the simplest $U(1)$ internal symmetry. As has already been mentioned, stable nontopological solitons exist only in theories with internal symmetry. It is natural that investigations have been performed beginning with the simplest models (the $U(1)$ symmetry), and then higher symmetries have been considered (some of which can be spontaneously broken).

Note that by gradually making the models more complicated and properly choosing the parameters, Lee and Friedberg [37] were able to describe with satisfactory accuracy with respect to experimental data such properties of nucleons as the ratio of the magnetic moments of the proton and neutron, the ratio of the beta decay constants (axial to vector) and the root-mean-square proton radius.

If static properties of solitons may still be studied analytically in the framework of even complicated models, the interaction of multidimensional solitons can so far be investigated only by computers. Here the progress from simple models to more complicated ones is natural.

The simplest models studied in several papers in 1977-1979 were nonrelativistic models described by the Schrödinger equation with a nonlinear term of various forms. Among these investigations there are Tappert's very impressive computer-produced films showing interactions of cylindrical solitons in the framework of the Schrödinger equation with the exponential nonlinear term

$$i\phi_t+\Delta_{rr}\phi+\frac{\phi}{\alpha}(1-\exp\{-\alpha|\phi|^2\}) = 0$$

simulating the behaviour of Langmuir wave packets in the vicinity of stationary states.

In the paper [38] collisions of cylindrically-symmetric (cs) gaussons (terminology of the authors) have been studied for the equation

$$i\phi_t + (\frac{1}{2}\Delta_{rr} + b\ln[a^D|\phi|^2])\phi = 0 \tag{16.1}$$

where a and b are constants and D is the number of space dimensions. Equation (1) has the exact one-soliton solution of Gaussian shape

$$\phi(\vec{x},t) = (\frac{A}{\sqrt{\pi}})^{D/2}\exp\{i(-\omega t + \vec{v}\vec{x} + \theta_0)\}\exp\{-b(\vec{x} - \vec{v}t - \vec{x}_0)^2\}, \tag{16.2}$$

$$\omega = \frac{v^2}{2} + 2Db(1 - \ln\frac{aA}{\sqrt{\pi}}),$$

with velocity v and initial phase θ_0. In the cs geometry we get

$$\phi(\vec{r},t) = \frac{A}{\sqrt{\pi}}\exp\{i(-\omega t - \vec{v}\vec{r} + \theta_0)\}\exp\{-b(\vec{r} - \vec{v}t - \vec{r}_0)^2\}, \tag{16.3}$$

$$\omega = \frac{v^2}{2} + 4b(1 - \ln\frac{aA}{\sqrt{\pi}}).$$

For $A = e\sqrt{\pi}/a$ we have a particular solution $\omega = v^2/2$ which has been used in [38]. The numerical procedure consists of a formal solution of eqn. (1), neglecting the time dependence of $\phi(r,t)$ under the logarithm and keeping the time step δt small

$$\phi(x,y,t+\delta t) = \exp\{-ib\delta t\ln|\phi(x,y,t)|^2\}e^{-(i/2)\delta t\Delta_{rr}}\phi(x,y,t).$$

Then, in each of the steps, the Fourier transform is used so that

$$e^{-(i/2)\delta t\Delta_{rr}} \rightarrow \exp\{\frac{i}{2}(k_x^2 + k_y^2)\delta t\}.$$

Finally by the inverse transform one gets the function

$$F(x,y,t+\delta t) = e^{-(i/2)\delta t\Delta_{rr}}\phi(x,y,t)$$

which multiplied by $\exp\{-i\delta tb\ln|\phi|(x,y,t)|^2\}$ gives the solution sought for at the next time step.

The authors use a lattice containing 128×128 points setting $a = e$, $b = 400$, so that $|\phi|^2 \simeq 1$ and $\Delta r = 0.035$. The time step is $\delta t = 0.002$. At the initial time two identical gaussons are given moving towards each other with the impact parameter

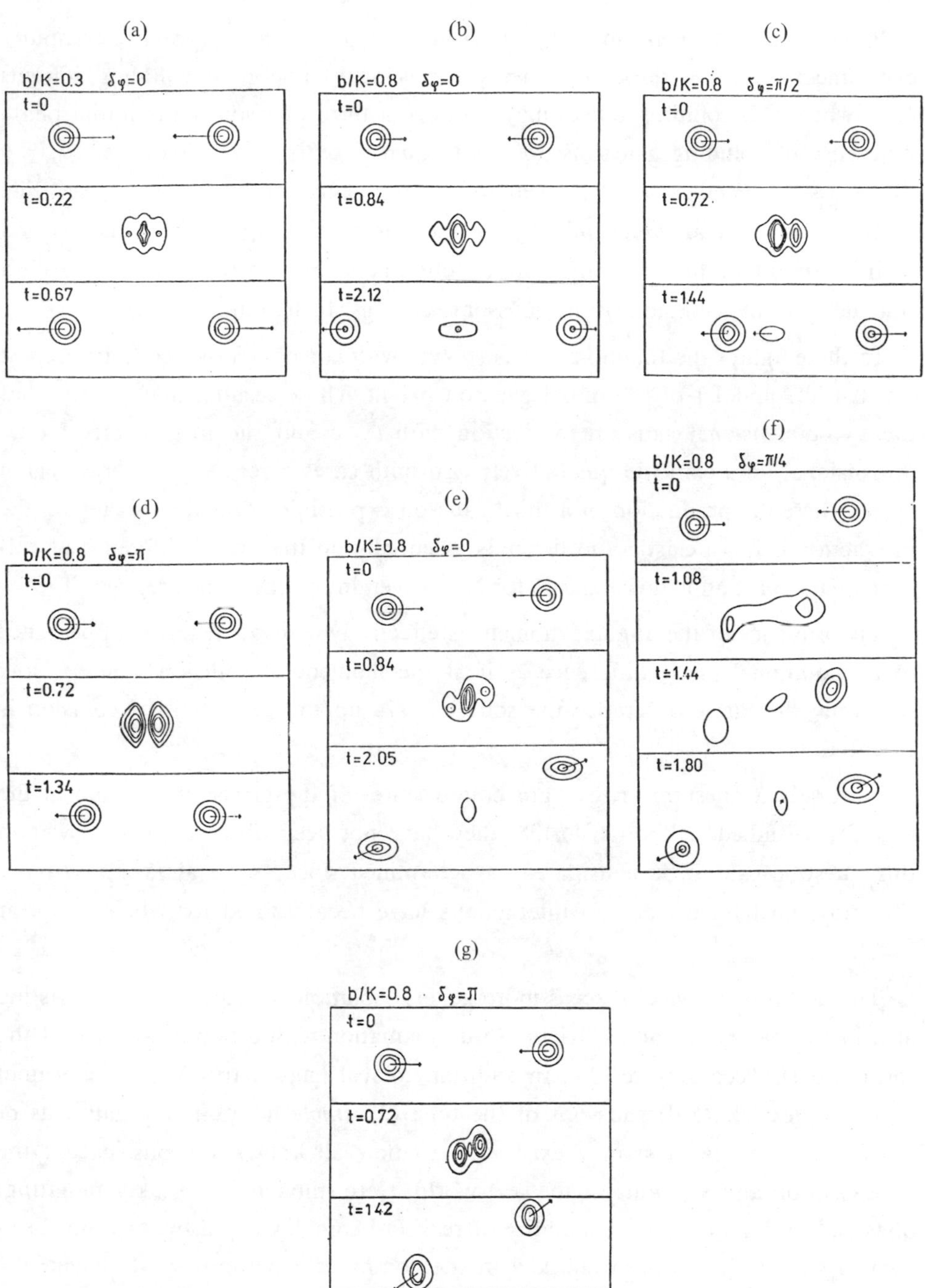

Figure 16.1. Two-gausson collisions taken from Ref. [38]. For details see the text.

(relative angular momentum $l=2vp$ which may or may not be zero. The computer experiments reveal a resonance energy (or velocity) region of colliding solitons ΔE_{res} where two colliding pulses may produce a third gausson if the initial phase difference of colliding gaussons does not equal π, see Fig. 16.1(b,c). At $\delta\theta_{in} = \pi$ the gaussons repel each other even in the resonance interval if $l = 0$, see Fig. 16.1(d). However at $l \neq 0$ and $\delta\theta_{in} = \pi$ there is an overlap of the gaussons and their scattering occurs as though an attraction acts between them (there is even a time delay in the emission of the gaussons; cf. Figs. 16.1(d) and 16.1(e)).

In these figures the function $|\phi|^2$ is shown, with constant value contour lines at 0.1, 0.4, 0.7 and 1.1 of the initial gausson height. These results show, firstly, that the two-dimensional gausson interaction with $l = 0$ and the interaction of one-dimensional ones coincide qualitatively. In both cases a resonance energy region exists where the production of a third gausson is possible. Outside this region the interaction is quasi-elastic and depends essentially on the phase difference of colliding gaussons and is quasi-elastic for $\delta\theta \to \pi$ even in the resonance region.

The presence of the angular moment is effectively equivalent to the appearance of an additional phase difference as if at the moment of collision $\delta\theta \neq \pi$. Note that only the phase difference the solitons gain up to the instant of collision is essential.

Naturally a question arises: are bound states of the gaussons possible in the model (1) studied? Although in [38] they have not been observed, the answer to this question cannot be considered as determined since as we already saw (eqn. (2)) only 'uncharged' gausson interactions have been studied for which $\omega = 0$ at $v = 0$.

The authors of [39,40] discuss more general solutions of gausson type (existing also in the framework of the Klein-Gordon equation with a nonlinear term of the form (3.5.1). According to [39], in addition to real (stationary in the rest frame) and complex (charged) gaussons of the type (2), *stable* nonradiating gaussons of pulsing type (G pulsons) may exist. Up to now all known pulsons, except the Sine-Gordon and $S3$ bions in the (x,t) world, were (though but weakly radiating) quasi-solitons. Although the stability of real and complex G pulsons has not been proved strictly, it has been checked in the computer experiments. Moreover, the radiation from the excited G pulson turns out to be very small [40]. In this context, the model with a nonlinear term of the form $\phi\ln|\phi|^2$ is until now unique and has features similar to the integrable (Sine-Gordon in the (x,t) world) ones.

In the third paper alpha-alpha scattering is studied using time-dependent Hartree-Fock theory. Rather convincing arguments are given in favour of reducing the three-dimensional problem to a two-dimensional one described by the nonrelativistic vector (Schrödinger) $\phi^4 - \alpha\phi^6$ model

$$i\phi_t + \Delta_{rr}\phi + a\phi + b\phi|\phi|^2 - c\phi|\phi|^4 = 0, \tag{16.4}$$

$$\phi^{tr} = (\phi_1, ..., \phi_4),$$

where a,b,c are constants and $\phi_i(x,y,t)$ is the one-particle wave function, $|\phi|^2 = \Sigma_{i=1}^4 |\phi_i|^2$.

The statement of the physical problem and the limits of applicability of this theory proposed are presented in detail in [41] and the references therein cited.

In the framework of eqn. (4) head-on collisions of two identical cs quasi-solitons are studied for various energies ($v_1 = -v_2 = v$) and impact parameters p. The constants a,b and c are chosen from experiments on the measurement of the alpha particle binding energy and mean-square of the α particle.

In the numerical experiments, it has been observed that there is a rather narrow (resonance) region of impact parameters where the inelasticity of the soliton interaction increases appreciably: the excitation energy $E_{exc} = K_{in} - K_{fin}$ (where K_{in} and K_{fin} are the soliton kinetic energies before and after the collision, respectively), the delay time τ_d (the difference of kinetic times of interacting and noninteracting solitons) and the deflection angle θ increase.

When the colliding soliton energy $K_{in} \propto v^2$ decreases, the resonance region is expanded for some value of K_{in} bound states arise in the centre of this region ($\tau_d \to \infty$). In Fig. 16.2, regions are schematically shown where elastic scattering, deep inelastic scattering and bound state production are presented in the K_{in},l plane ($l = 2vp$ is the relative angular mementum). (An example of weak inelastic soliton interaction is demonstrated in Fig. 16.3.)

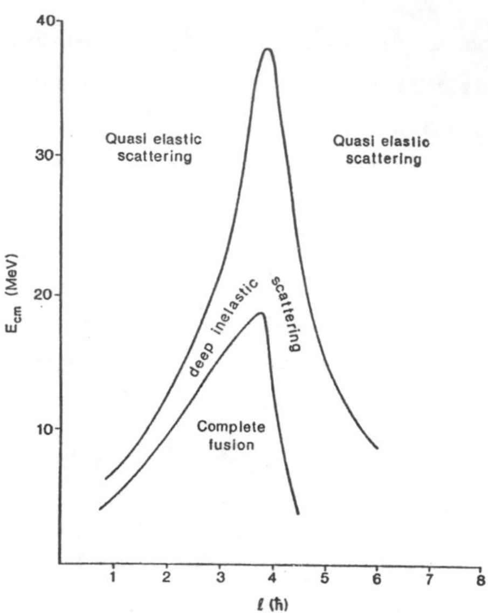

Figure 16.2. The areas of various types of $\alpha-\alpha$ interaction in the plane E_{cm}, l (energy, angular momentum) according to Ref. [41].

2. Relativistic models

Below, we discuss some properties of soliton-like solutions which, from our point of view, are nontrivial. Particular attention will be paid to their dynamic characteristics. We shall see that even in the framework of fairly simple models of nonlinear classical field theory such solutions can have not only nontrivial but, at first glance, striking properties very reminiscent of those of complex real objects.

As we already noted, one possibility for obtaining stable soliton-like solutions is to introduce a certain isotopic symmetry group of the Lagrangian and the conservation laws associated with this. We discuss the properties of models with the simplest $U(1)$ group, which leads to a conservation law for the 'isocharge' Q. The conservation of Q means that there are restrictions on the possible form of perturbations, namely, $\delta Q[\phi] = 0$, which leads to the stability condition $(\omega/Q)dQ/d\omega < 0$ for soliton-like solutions. It is obvious that real stationary field configurations cannot satisfy this condition and will be unstable. These conclusions were confirmed earlier in numerical experiments of various groups.

At the present time, a nonlinear relativistically invariant model is known that is

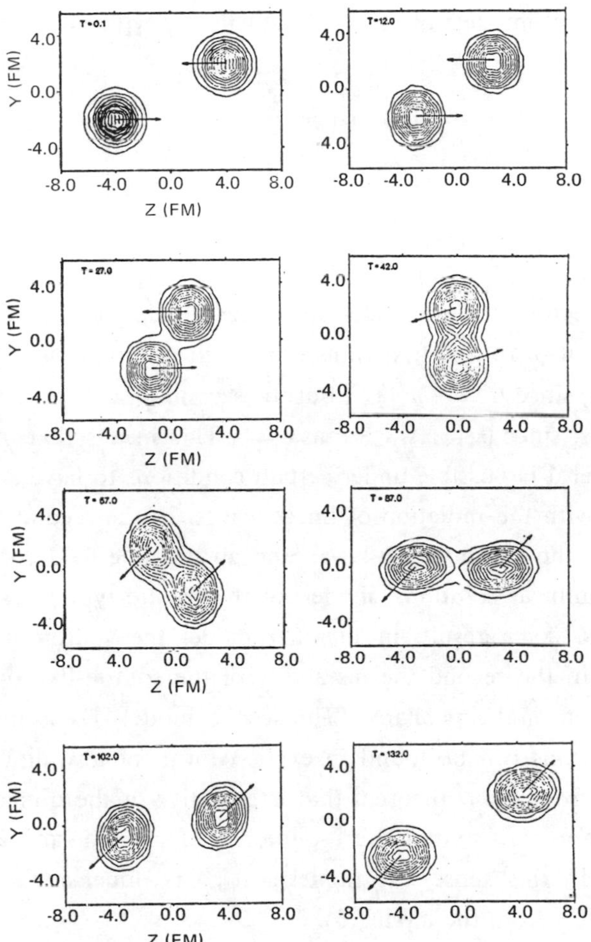

Figure 16.3. An example of $\alpha-\alpha$ inelastic collision from Ref. [41].

integrable in three-dimensional space-time. It is described by a self-duality equation in a four-space with metric (2,2) where the potentials do not depend on one of the coordinates. There is no interaction of solitons. Therefore, particular interest attaches to computer investigations of the dynamical properties of two-dimensional (2,1) and also three-dimensional (3,1) well-localized solutions for different field-theory models. The use of the qualitative properties of these solutions obtained by the computer may suggest ways of further study by analytical (possible approximate) methods.

We consider two models of classical field theory with an interaction potential in the Lagrangian[*)]:

$$U = -\ln(1+|\varphi|^2)-(m^2-1)|\varphi|^2 \qquad (16.5)$$

and

$$U = -|\varphi|^2\ln(|\varphi|^2). \qquad (16.6)$$

It is readily seen that these models are fundamentally different in the following sense: in the limit $\varphi\to 0$, the first transforms with accuracy $O(|\varphi|^2)$ into the ordinary free theory, since $U \simeq -m^2|\varphi|^2$, but the second model contains elements having infinite mass, since $\ln(|\varphi|^2)\to-\infty$ as $\varphi\to 0$. This means that in the framework of the first model it is possible under certain conditions to have a decay of a solution into parts with the radiation of linear waves. In the second case, such decay is forbidden (by the conservation laws), and all possible field configurations consist solely of nonlinear solutions; models of the second type are sometimes called *confining models*. As a result, in the first model the soliton-like solutions can decay, whereas in the second the instability of the soliton-like solutions is manifested in the form of their collapse. The second model is also interesting in that soliton-like solutions can be found in explicit form for any dimension D. Moreover, it follows from the Q theorem that irrespective of the dimension there exist stable $U(1)$-symmetric soliton-like solutions of the form $\varphi = \psi(r)e^{-i\omega t}$ for $\omega>\omega_{cr} = 2^{-1/2}$. In this sense, the model $\ln(|\varphi|^2)$ is dimensionally invariant and differs qualitatively from the model (5).

Hypothesis: The nature of the interaction of solitons in collisions is determined by the dispersion dependence $Q(\omega)$ rather than by the type of the model (independent of the nature of the instability - decay or collapse).

This hypothesis has been checked in a series of numerical experiments [7,8,11]. In these calculations, two parameters were varied - the velocity v of the relative motion of the quasisolitons and their charge Q. In both cases, four types of

[*)] For the ϕ^4_D theory when $D>1$ stable soliton-like solutions do not exist even in systems with an isogroup.

interaction were found[f)];

(1) elastic and quasi-elastic interaction of quasisolitons;

(2) decay (collapse) of quasisolitons after interaction;

(3) decay (collapse) through a short-lived bound state (resonance)[**)];

(4) a long-lived bound state of two central quasisolitons, a bion, which indicates that, in reality, there is a model-independent nature of soliton interaction, at least in the framework of the models considered.

The two central types of interaction are possible only in the region $dQ/d\omega \simeq 0$. This gives grounds for assuming that such forms of interaction will also be present in models whose Lagrangians admit higher symmetry groups, provided the dependence of the corresponding 'isocharge' ('isospin', etc.) has extremal points as in the considered models.

A more detailed investigation of the interaction of quasisolitons showed [11] that it also depends on the impact parameter p or, which is the same, the angular momentum l and the initial phase difference $\Delta\theta$ (similar to Fig. 16.2). The numerical experiments showed that, as in the nonlinear Schrödinger model:

(a) there exists a certain resonance region with respect to the angular momentum l in which the inelasticity of the interaction of the quasisolitons increases sharply (see also [41]);

(b) a pure antisymmetric initial field configuration, $\Delta\theta = \pi \pm 2\pi n$ leads to elastic repulsion of the solitons.

As we have already noted, stationary configurations of real fields cannot be stable, i.e. real quasisolitons do not exist. Moreover, it is not in all systems with internal isosymmetry and not in all cases that stable quasisolitons will exist. Such solutions can arise in systems in which the surfaces of constant energy in the

---

[f)]It is very interesting to note that nearly the same types of interactions were recently claimed to be observed in the experiments on the head-on collisions of vertical Bloch lines clusters moving on the domain walls of garnet films (the uniaxial ferromagnetic system with quite a strong dissipation and a stabilizing external magnetic force) [48].

[**)]We recall the resonances in the (1,1)-dimensional Higgs model ϕ^4_-.

functional space can have conditional or local minima. The following question arises naturally: In such systems, do there exist nonstationary stable configurations of real fields? In such a case, the nonstationarity plays the role of a stabilizing factor like the dependence $\exp(-i\omega t)$ in the case of the $U(1)$ group *).

This conjecture has been verified in a series numerical experiments in the framework of model (5) made in Dubna [8]. The results obtained in these experiments seemed, at first glance, to be paradoxical.

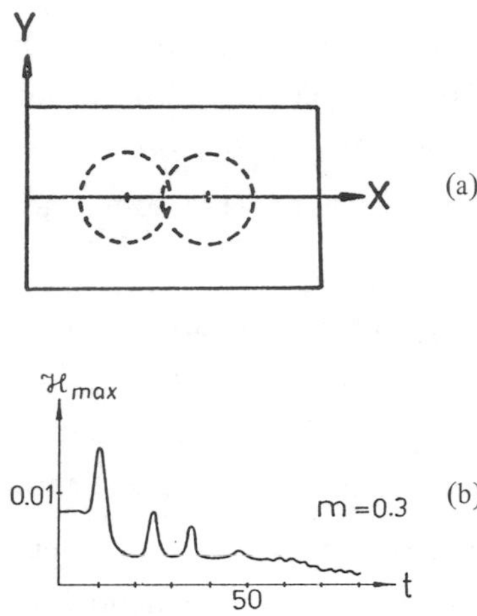

Figure 16.4a. The initial state consisting of two unstable solitons.

Figure 16.4b. The energy density of the system in the origin as a function of time. Inverse growth rate of single soliton instability $\gamma^{-1} \sim 1 \div 2$.

Placing, for example, unstable soliton-like objects sufficiently close to each other - so that the kinematic time of their interaction was shorter than the decay time of each of them (Fig. 16.4a) - we found that if the quasisolitons approached each other at sufficiently low velocities they formed a bound state, i.e. a two-dimensional bion. The amplitude at the centre of the bion oscillated regularly, decreasing only very slightly during the calculation (of several oscillation periods: see Fig. 16.4b). Further study has been shown that similar objects can arise from

*)We recall here Kapitsa's problem of a pendulum with a rocking point of suspension.

a sufficiently heavy single-soliton initial state. The behaviour in time and the form of the discovered bions agree qualitatively with those of the pulsons described above. This has shown that the existence of pulsons is not a privilege of systems with a degenerate vacuum as in the case of Higgs (ϕ^4_-) and Sine-Gordon field equations in which the field function oscillates between two adjacent vacua. We note that similar pulsons also arise naturally in the framework of the system (6).

This confirmed the hypothesis that in models with a sufficiently complicated constant-energy surface that admit the existence of stable solitons there exist long-lived (stable) pulsons, i.e. quasiperiodical solutions. This conjecture was made in [34,42].

This hypothesis enables us to consider from a common point of view all the results so far obtained concerning the interaction of quasisolitons. In the models considered, besides the stable (and unstable) charged ($Q\neq 0$) quasisolitons, there exist stable (and unstable) charged ($Q\neq 0$) and uncharged ($Q=0$) pulsons. This means that for definite parameters of the quasisolitons (both charged and uncharged) the evolution of their interaction either ends with the formation of a corresponding pulson (bound state) or passes through a pulson phase (resonance) before the development of instability (decay).

From this point of view, the appearance of a stable pulson as a result of the collision of two unstable quasisolitons ceases to be paradoxical. Pulsons such as those described above were found in the numerical experiments of [40] in a space of dimension (1,1) in collisions of charged quasisolitons with different total 'iso-charge' ($Q=2Q_1$, $Q_1=Q_2$ and $Q=Q_1+Q_2=0$, $Q_1=-Q_2$). In the first case, we evidently have the formation of a charged pulson, since $Q_p \propto d\theta/dt \neq 0$ (see Fig. 2 of [40]); and in the second case, the formation of an uncharged pulson: $Q_p\to 0$ (see p. 383 of [40]). The region of initial conditions for the formation of an uncharged pulson is much narrower.

Finally, pulsons were also detected in [41], but in the framework of the Schrödinger equation with a nonlinearity of the form $\phi^3-\alpha\phi^5$. Do non-one-dimensional pulsons exist in the framework of the Schrödinger equation with other forms of nonlinearity? In the numerical experiments described in [38] non-one-dimensional pulsons were not found, at least in the case of the gausson model.

We note also that stable bound states formed by unstable constituents have already long been known in nuclear physics (the deuteron). As in our case, this state bears little resemblance to a bound state of two classical objects such as the Earth-Moon, or binary-starts, etc. Upon the formation of a bound state, the constituents that make up such a system lose their individuality.

A few words about the methods used in the calculations is in place. In [41] for two dimensions a 25×25 space grid is used. For the x,y integration a difference scheme of the fourth order (with step = 0.8) is applied. The time integration uses a fifth-order predictor-corrector technique with a time step $\Delta t = 0.15$. All the calculations have been carried out on an IBM 370/165 computer. The average time for a two-dimensional calculation for given K_{in} and l was about 15 minutes. A full three-dimensional calculation with 15 spatial mesh points on the perpendicular axis will take 15 times longer, i.e. nearly 4 hours. However, citing Professor H. Flocard, the authors assert that they could improve the time integration technique and reduce the computation time by a factor 10. Then the study of three-dimensional interactions with the same degree of symmetry, (i.e. head-on collisions of identical solitons with the same initial phase modules) becomes realistic.

In [7,8,11,25] a one-soliton solution $\tilde{\psi}(r)$ is first found with a given accuracy, and then approximated by a set of Gaussian exponents (the mixed method, see Section 11.1)

$$\tilde{\psi}(r) \simeq \sum_i^n \alpha_i \exp\{\beta_i (r-\delta_i)^2\}.$$

By choosing the coefficients α_i, β_i and δ_i appropriately, we can approximate $\tilde{\psi}(r)$ at $n = 3$ so that the condition

$$\max|\tilde{\psi}(r) - \sum_i^3 \alpha_i \exp\{\beta_i (r-\delta_i)^2\}| \leqslant 0.005\tilde{\psi}(0)$$

holds for all ω. Using the Lorentz transformation we obtain the moving soliton

$$\phi(x,y,t) \simeq \sum_i^3 \alpha_i \exp\{\beta_i (\sqrt{\gamma^2 (x-vt)^2 + y^2} - \delta_i)^2\} \exp\{-\omega\gamma(t - vx)\}.$$

Collisions of these solitons are studied using a symmetrical difference scheme of second order with respect to x,y and t. The time step is $\Delta t = 0.1$ and the step $\Delta x = \Delta y$ is taken from the interval (0.1, 0.4) depending on the parameter ω. At moments $t = 0, 1, 2, \ldots$ the field energy density

$$\mathcal{H} = |\phi_t|^2 + |\phi_x|^2 + |\phi_y|^2 + \ln(1+|\phi|^2)$$

is computed. For a controlling the accuracy of the calculation the relation

$$\frac{dE}{dt} = 0, \quad E = \int \mathcal{H} dx dy$$

is employed. In the calculation, the maximum of

$$\epsilon_T = \left| \frac{E_0 - E_T}{E_0} \right|$$

is not to exceed the value 0.01. Here $E_0 = E(t=0)$, $E_T = E(t=T)$.

3. Formfactors and DSF

To complete (in a certain sense) the purpose of this volume (as indicated in the introduction) we discuss two remaining problems: formfactors of solitons in $D \geqslant 2$ dimensions and their structural factors, i.e. their statistical mechanics. We emphasize that exact results in this field are not known to the author and the consideration will be, to a certain extent, of a speculative character.

A variety of soliton-bearing models possess a number of common properties. We demonstrate these by giving an example of the classical field theory with the Lagrangian

$$\mathcal{L} = \bar{\psi}_\mu \psi^\mu + \epsilon U(\bar{\psi}\psi), \tag{16.7}$$

with ψ being the complex or real scalar field in the simplest case. Suppose now that we have the following expansion for $\psi \to 0$

$$U \sim [m^2 - g(\bar{\psi}\psi)](\bar{\psi}\psi) + ... \tag{16.8}$$

There exist two possibilities

$$(1) \quad \epsilon = 1, \tag{16.9a}$$

$$(2) \quad \epsilon = -1. \tag{16.9b}$$

In the first case, a stable vacuum state of the system, for which soliton solutions may be constructed, is trivial: $\psi = 0$. In the latter, a nontrivial stable vacuum state is possible: $|\psi|^2 = m^2 g^{-1}$. Therefore, to avoid the appearance of Goldstone's bosons and gauge fields at this stage, we take the field ψ to be real. In the computer experiments described above [8,25,26,28-33,36] the existence of stable neutral ($Q=0$) oscillating localized solutions (pulsons) has been demonstrated for the models both first- and second-type. Stable charge ($Q\neq 0$) pulsons were also found for the first-type models. Moreover, it follows that the evolution of various initial perturbations over the stable vacuum goes, as a rule, through the pulson phase, i.e. either stable solitons or weakly radiating pulsons of large or small amplitude appear as a result of the decay of the initial perturbations*).

It may be easily verified that the energy and 'charge' (or the appropriate adiabatic invariant) of the system are proportional

$$Q \simeq \frac{E}{\Omega} \sim \int R^2(\kappa r) d^D r = \text{const}\kappa^{2-D}, \quad \omega = \frac{\Omega}{m}, \tag{16.10}$$

where, again, $\kappa^2 = m^2 - \Omega^2$, $\psi \simeq R(r)f(t)$. It follows from (10) that the sign of $dQ/d\omega$ (or $dE/d\omega$) depends, at $\kappa \ll 1$ on the number of space dimensions D:

$D = 1$: $\quad dQ/d\omega < 0$,

$D = 2$: $\quad dQ/d\omega \simeq 0$,

$D = 3$: $\quad dQ/d\omega > 0$.

Small amplitude Q-solitons (and pulsons) are therefore stable only in one-space dimension geometry. The approximate shape of the curves $Q(\omega)$ is depicted in Fig. 16.5. It follows from this figure that when $D = 2,3$ the sign of $dQ/d\omega$ can alter for finite amplitudes, say, at $\omega = \omega_{cr}$ and $E = E_{cr}$, i.e. Q-solitons of mass $M \geqslant E_{cr}$ become stable. The values of ω_{cr} and E_{cr} are model dependent (e.g. in ϕ^4_+ theory $dQ/d\omega > 0$ everywhere in the existence region of soliton-like solutions if $D > 1$); these have been found by computer for various field - [8,25,43] as well as spin (magnetic) - [16] models. Analogous results can also be obtained for the second class of models (9b) with nontrivial vacuum (including models with broken symmetry describing, in particular, structural phase transitions). Here, small amplitude solutions are constructed over one of the vacua ($\varphi = \varphi_v$), and the expansions look less symmetrical. In the region $\kappa \ll 1$, the dependence of the integral

*) Pulsons may be created by soliton interactions.

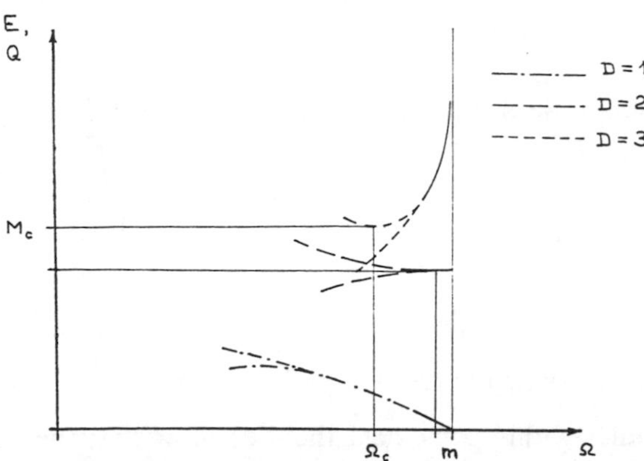

Figure 16.5. Approximate shape of dispersion curves $Q(\Omega)$ for $D = 1,2$ and 3. Possible variants are given.

$\int(\varphi-\varphi_v)^2 d^D x$ on κ (and then on pulson mass) is the same as that obtained above for Q-solitons

$$M_p \propto \kappa^{2-D},$$

i.e. small amplitude stable pulsons exist but in one-dimensional systems. In more-dimensional systems, stable pulsons (if they exist) have a lower critical mass M_{pc} and an upper critical frequency Ω_{pc}, which are also model dependent.

The technique proposed in Chapter 13 for calculating DSF of solitons may be applied to the study of more-dimensional systems. For soliton-like solutions of not too large amplitude we have formula of form (13.29a) and (13.41a)*)

$$S = \frac{\bar{n}_b}{Z_1} |F_b|^2 (e^{-\beta E(v_0)\gamma_0} + \ldots), \tag{16.11}$$

in which

$$\bar{n}_b / Z_1 \simeq (2\pi)^{-(D+1)}$$

and

*)In obtaining (11) we express again $\varphi \propto \cos(kx - \bar{\Omega} t)$ which makes sense for small amplitudes.

$$F_b = \frac{1}{\Delta}\int \psi(-\frac{\mathbf{x}_0}{\Delta})e^{i\mathbf{q}\mathbf{x}_0}d^D x_0.$$

Taking the asymptotic behaviour of the function ψ as given in (14.23) one can calculate F_b as follows:

$$F_b = \int d\mathbf{x}_\perp dx_3(x_\perp^2+x_3^2)^{1/2}e^{-\kappa(x_\perp^2+x_3^2)^{1/2}}e^{-iqx_3} \rightarrow \frac{4\pi}{\kappa^2+q^2}$$

with $\kappa = m(1-\omega^2)^{1/2}$ the effective mass.

Let us summarize briefly the results:

(1) Numerical studies within some field theories show that the appearance of soliton (bion) - type excitations is not a property of integrable two space-time dimensional systems but rather, is a natural behaviour of systems possessing stable soliton-like solutions.

(2) The lower energy limit, for the existence of such objects grows with the number of system dimensions D in the framework of the models considered and vanishes at $D = 1$ (in the classical limit).

(3) Using the hypothesis of the dilute soliton (bion) gas being in thermodynamic equilibrium, a dynamic structure factor $S(q,\omega,\bar{\Omega})$ may be calculated which describes the scattering of light, neutrons, etc.

(4) The functions $S(q,\omega,\bar{\Omega})$ of T are different for different D in the framework of the same model. When $D = 1$, stable solitons (bions) arise at every low temperature T, as a result of which S contains the central peak and satellites. In the $D>1$ case, stable solitons (bions) and hence the central peak and satellites in the dynamic structure factor appear only when the temperature exceeds a certain critical value $T > T_{cr} \simeq E_{cr}$. The satellites move towards the central peak with increasing temperature for any D.

(5) Such a behaviour of the system may be regarded as a phase transition with respect to its clusterization. The temperature of this transition (clusterization) vanishes in the $D = 1$ dimensional systems and is finite in more-dimensional ones increasing with D.

We have simplified as much as possible the picture, so that several specific features of soliton behaviour are ignored. These included the 'spectroscopy' of solitons, i.e. their linear excitation modes such as the localized shape mode in the case of the ϕ^4_- kink. The number of such modes increases considerably for $D \geqslant 2$

solitons to include various radial, angular and rotating modes. Their contribution must be, in principle, considered or estimated in the treatment of experimental results.

The above picture is not limited by the scope of a class of single-field models, it is also valid for other more complicated multi-field models with internal symmetry as well [37].

References Part VI

1. See, e.g., V.E. Zakharov. *The inverse scattering method.* In Solitons, R.Bullough, P. Caudrey eds. Springer and Heidelberg 1980, p. 243-286.

2. I. Krichever and S.P. Novikov. *Holomorphic bundles on Riemann surfaces and KP equation.* Funct. Anal. Priloz. 1978, 12, p. 41-52.

- M. Ablovitz, D. van Yaakov and A. Fokas. *On the inverse scattering transform for KP eq.* Stud. Appl. Math. 1983, 69, p. 135-143.

- S.V. Manakov. *The inverse scattering transform for the time-dependent Schrödinger equation and KP equation.* Physica 1981, 3D, p. 420-427.

- A. Fokas and M. Ablovitz. *On the inverse scattering of the time-dependent Schrödinger equation and the associated KP equation (1).* Stud. Appl. Math. 1983, 69, p. 211-228.

3. L. Nehari. See [24] part IV.

- E.P. Zhidkov and V.P. Shirikov. See [25] part IV.

4. P.E. Zhidkov and V. Makhankov. *On existence of static hole-like solutions in nonlinear Bose-gas model.* JINR Comm. P5-86-341. Dubna 1986.

5. A.M. Kosevich, et al. See [1a] part I.

6. A. Kosevich, B. Ivanov and A. Kovalev. *Magnetic solitons: a new type of collective excitation in magnet ordered crystals.* Sov. Sci. Rev. A. Phys. 1985, 6A, p. 161-260.

7. V. Makhankov, G. Kummer and A. Shavachka. *Many-dimensional U(1) solitons, their interactions, resonances and bound states.* Phys. Scripta 1979, 20, p. 454-461; Phys. Lett. 1979, 70A, p. 171-173.

8. V. Makhankov, G. Kummer and A. Shavachka. *Novel pulsons (or stability*

from instability). Physica 1981, 3D, p. 344-349.

9. Bialynicki-Birula and J. Mycielski. *Gaussons: solitons and the lg Schrödinger equation.* Physica Scripta 1979, 20, p. 539-544.

10. I.L. Bogolubsky. *Bohr-Sommerfeld quantization of n-dimensional neutral and charged pulsons.* ZhETF 1979, 76, p. 422-430.

11. V. Makhankov, I.L. Bogolubsky, G. Kummer and A. Shvachka. *Interaction of relativistic Gaussons.* Phys. Scripta 1981, 23, p. 767-773.

12. L.P. Pitaevskii. *Vortices in non-ideal Bose-gas.* ZhETF 1961, 40, p. 646-654.

13. N. Pereira, R. Sudan and J. Denavit. *Numerical study of 2-D generation and collapse of Langmuir solitons.* Phys. Fluids 1977, 20, p. 936-945, see [16] part V.

14. N. Yajima. *Stability of envelope soliton.* Progr. Theor. Phys. Jpn. 1974, 52, p. 1066-1067.

15. G. Schmidt. *Stability of envelope solitons.* Phys. Rev. Lett. 1975, 34, p. 724-726.

- L. Degtyarev, L. Rudakov and V. Zakharov. *Two examples of collapse of Langmuir waves.* Sov. Phys. JETP, 1976, 41, p. 57-67. ZhETF 1975, 68, p. 115-126.

16. N. Pereira, A. Sen and A. Bers. *Nonlinear development of lower hybrid cones.* Phys. Fluids 1978, 21, p. 117-120.

17. N.C. Freeman and A. Davey. *On the evolution of packets of long surface waves.* Proc. Roy. Soc. Lond. 1975, 344A, p. 427-433.

- D. Benney and G. Roskes. Stud. Appl. Math. 1969, 48, p. 377.

18. D. Anker and N. Freeman. *On the soliton solutions of the Davey-Stewartson equation for long waves.* ibid, 1978, 360A, p. 529-540.

19. A. Sen, C. Karney, G. Johnston and A. Bers. MIT Rep. PLE-PRR 77-16, 1977.

20. V. Makhankov. See [7] part I.

21. N. Vakhitov and A. Kolokolov. See [53b] part I.

22. D. Anderson and G. Derrick. See [39] part IV.

23. V.G. Kartavenko. See [36] part III and [43] part I.

24. I.V. Barashenkov and V.G. Makhankov. *Soliton-like excitations in a one-dimensional nuclear matter.* JINR E2-84-173, Dubna 1984;

- *Soliton-like bubbles in the system of interacting bosons.* JINR E17-87-29, Dubna 1987, Phys. Lett. 1988, 121A, p. 52-56.

25. V. Makhankov and A. Shvachka. *Dynamical properties of many-dimensional U(1) solitons.* Physica 1981, 3D, p. 396-399.

26. I.L. Bogolubsky and V. Makhankov. See [80] part V.

- *On the life-time of pulsating solitons in certain classical models.* Sov. Phys. JETP Lett. 1976, 24, p. 12-14. Pisma ZhETF 1976, 24, p. 15-17.

27. M.S. Kaschiev, et al. *Numerical investigation of stability and bifurcation points of bound static states of fluxons in a circular Josephson junction with microinhomogeneity.* JINR P11-84-832 Dubna 1984 (V. Kaschieva, V. Makhankov, I. Puzynin, T. Puzynina and A.T. Filippov).

28. I.L. Bogolubsky and V. Makhankov. See [80] part V.

29. P.L. Christiansen and P.S. Lomdahl. *Numerical study of 2+1 dimensional SG solitons.* Physica 1981, 2D, p. 482-494 (and references cited).

30. J. Geicke. *Cylindrical pulsons in nonlinear relativistic wave equations.* Phys. Scripta 1984, 29, p. 431-434 (and references cited).

31. T. Belova, N. Voronov, I. Kobzarev and N. Konyukhova. *Particle-like solutions of the scalar Higgs equation.* ZhETF 1977, 73, p. 1611-1622.

32. I.L. Bogolubsky. *Cascade evolution of spherically symmetric pulsons in multivacuum field theory models.* Phys. Lett. 1977, 61A, p. 205-207.

33. I. Bogolubsky, V. Makhankov and A. Shvachka. *Dynamics of the collisions of 2-D pulsons in ϕ^4 field theory.* Phys. Lett. 1977, 63, p. 225-227.

34. S. Manakov. *On pulsating solitons.* Sov. Phys. JETP Lett. 1977, 25, p. 589-593.

35. J. Krumhansl and J. Schrieffer. See [126] part V.

36. See [33], this part.

37. R. Friedberg and T.D. Lee. *Fermion-field non-topological solitons.* Part I. Phys. Rev. 1977, 15D, p. 1694-

    Part II. *Models for hadrons.* ibid, 16D, p. 1096-1118.

38. J. Oficjalski and I. Bialynicki-Birula. See [18] part V.

39. G. Marques and I. Ventura. Preprint IFUSP P-83, 1976.

40. Yu. Simonov and J. Tjon. *Soliton-soliton interaction in confining models.* Phys. Lett. 1979, 85B, p. 380-385.

41. Sandhya Devi, et al. see [37] part III.

42. V. Makhankov. *Report on International Workshop 'Mystery of Soliton',* Yadvizin, Poland 1977; see also Computer and solitons. Physica Scripta 1979, 20, p. 558-562.

43. R. Friedberg, et al. see [23] part IV.

44. J. Geicke. *An approach to radially symmetrical solutions of the SG equation.* Physica, 1982, 4D, p. 197-206.

45. V. Makhankov. See [8] part I.

46. O.H. Olsen and M.R. Samuelsen. *Rotationally symmetric breather-like solutions to the SG equation.* Phys. Lett. 1980, 77A, p. 95-99.

47. R.T. Glassey. *On the blowing up of solutions to the Cauchy problem for nonlinear Schrödinger equations.* J. Math. Phys. 1977, 18, p. 1794-1797.

48. M.V. Chetkin, I.V. Parygina, V.B. Smirnov, S.N. Gadetsky and A.K. Zvezdin. *Soliton-like behaviour of VBL clusters,* Phys. Lett. 1989, 140A, p. 428-430.

Index

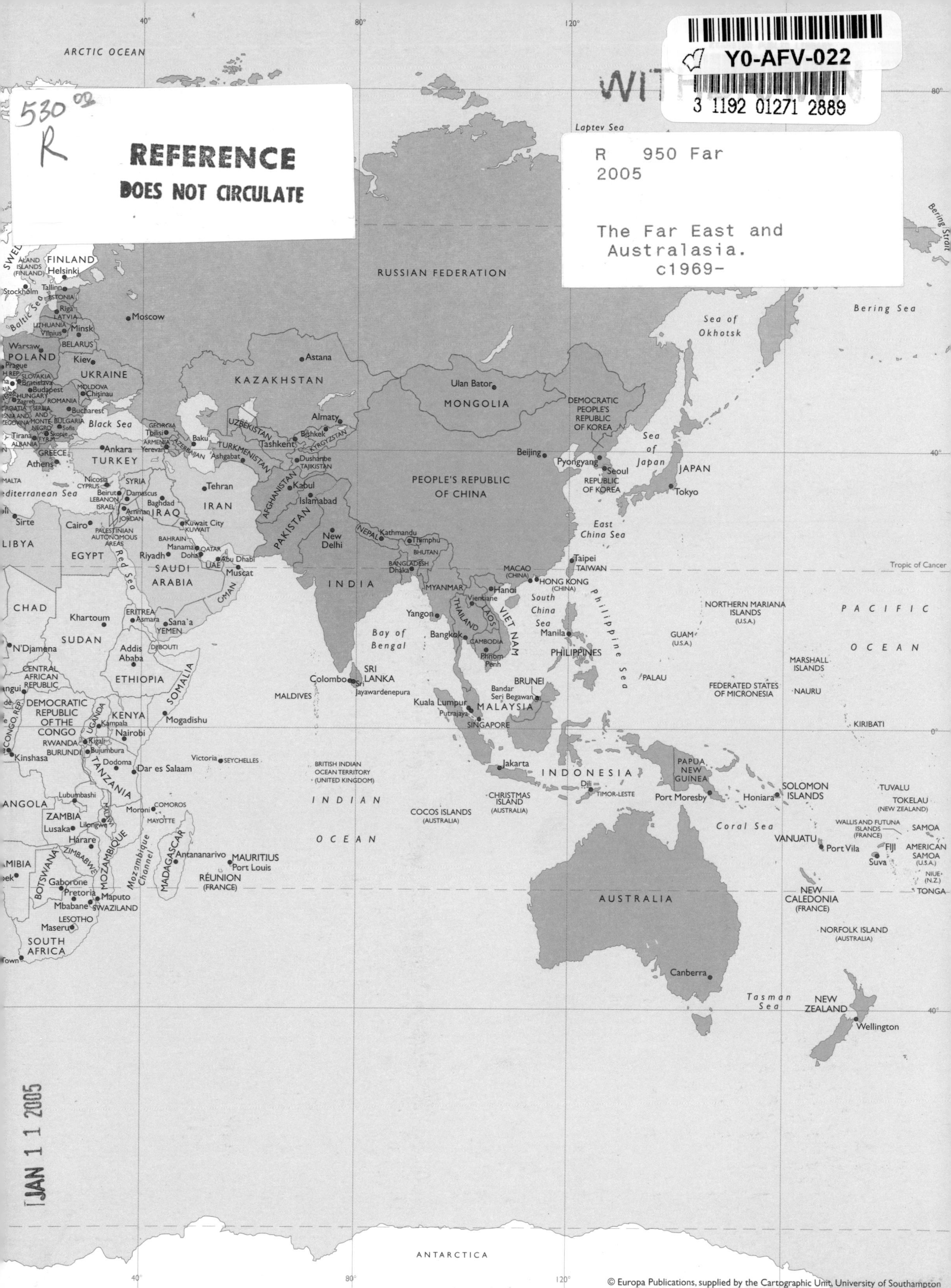

ARCTIC OCEAN
Laptev Sea
Bering Strait
Bering Sea
Sea of Okhotsk
RUSSIAN FEDERATION
FINLAND
Helsinki
Stockholm
Tallinn
ESTONIA
Baltic Sea
Riga
LATVIA
LITHUANIA
Vilnius
Minsk
BELARUS
Moscow
Warsaw
POLAND
Prague
Kiev
UKRAINE
SLOVAKIA
Bratislava
Budapest
HUNGARY
MOLDOVA
Chişinău
ROMANIA
Bucharest
BULGARIA
Sofia
Skopje
Tirana
ALBANIA
GREECE
Athens
Black Sea
GEORGIA
Tbilisi
ARMENIA
Yerevan
AZERBAIJAN
Baku
Ankara
TURKEY
MALTA
Mediterranean Sea
Nicosia
CYPRUS
SYRIA
Beirut
LEBANON
Damascus
ISRAEL
Amman
JORDAN
Baghdad
IRAQ
Tehran
IRAN
Sirte
LIBYA
Cairo
PALESTINIAN AUTONOMOUS AREAS
Kuwait City
KUWAIT
BAHRAIN
Manama
QATAR
Doha
Abu Dhabi
UAE
EGYPT
Riyadh
SAUDI ARABIA
Muscat
OMAN
Red Sea
CHAD
Khartoum
SUDAN
ERITREA
Asmara
Sana'a
YEMEN
N'Djamena
Addis Ababa
DJIBOUTI
ETHIOPIA
SOMALIA
CENTRAL AFRICAN REPUBLIC
DEMOCRATIC REPUBLIC OF THE CONGO
UGANDA
Kampala
KENYA
Nairobi
Mogadishu
RWANDA
Kigali
BURUNDI
Bujumbura
Kinshasa
Dodoma
TANZANIA
Dar es Salaam
Victoria
SEYCHELLES
Lubumbashi
ANGOLA
ZAMBIA
Lusaka
MALAWI
Lilongwe
Harare
ZIMBABWE
MOZAMBIQUE
Moroni
COMOROS
MAYOTTE
Mozambique Channel
MADAGASCAR
Antananarivo
MAURITIUS
Port Louis
RÉUNION (FRANCE)
BOTSWANA
Gaborone
Pretoria
Maputo
Mbabane
SWAZILAND
LESOTHO
Maseru
SOUTH AFRICA
KAZAKHSTAN
Astana
UZBEKISTAN
Tashkent
Almaty
Bishkek
KYRGYZSTAN
TURKMENISTAN
Ashgabat
Dushanbe
TAJIKISTAN
AFGHANISTAN
Kabul
Islamabad
PAKISTAN
New Delhi
INDIA
NEPAL
Kathmandu
Thimphu
BHUTAN
BANGLADESH
Dhaka
Ulan Bator
MONGOLIA
DEMOCRATIC PEOPLE'S REPUBLIC OF KOREA
Beijing
Pyongyang
Seoul
REPUBLIC OF KOREA
Sea of Japan
JAPAN
Tokyo
PEOPLE'S REPUBLIC OF CHINA
East China Sea
Taipei
TAIWAN
Tropic of Cancer
MACAO (CHINA)
HONG KONG (CHINA)
MYANMAR
Hanoi
Vientiane
LAOS
THAILAND
VIET NAM
Yangon
Bay of Bengal
Bangkok
CAMBODIA
Phnom Penh
South China Sea
Philippine Sea
Manila
PHILIPPINES
SRI LANKA
Colombo
Sri Jayawardenepura
MALDIVES
BRUNEI
Bandar Seri Begawan
Kuala Lumpur
Putrajaya
MALAYSIA
SINGAPORE
PALAU
NORTHERN MARIANA ISLANDS (U.S.A.)
GUAM (U.S.A.)
PACIFIC OCEAN
MARSHALL ISLANDS
FEDERATED STATES OF MICRONESIA
NAURU
KIRIBATI
BRITISH INDIAN OCEAN TERRITORY (UNITED KINGDOM)
Jakarta
INDONESIA
Dili
TIMOR-LESTE
PAPUA NEW GUINEA
Port Moresby
Honiara
SOLOMON ISLANDS
TUVALU
TOKELAU (NEW ZEALAND)
CHRISTMAS ISLAND (AUSTRALIA)
COCOS ISLANDS (AUSTRALIA)
INDIAN OCEAN
Coral Sea
WALLIS AND FUTUNA ISLANDS (FRANCE)
SAMOA
VANUATU
Port Vila
FIJI
Suva
AMERICAN SAMOA (U.S.A.)
NIUE (N.Z.)
TONGA
NEW CALEDONIA (FRANCE)
AUSTRALIA
NORFOLK ISLAND (AUSTRALIA)
Canberra
Tasman Sea
NEW ZEALAND
Wellington
ANTARCTICA

The Far East and Australasia 2005

The Far East and Australasia 2005

36th Edition

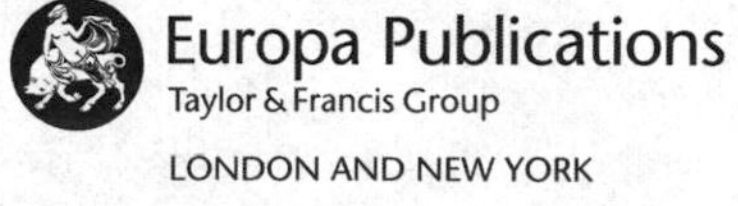

LONDON AND NEW YORK

First published 1969

Thirty-Sixth Edition 2005

Haines House, 21 John Street, London WC1N 2BP, United Kingdom
(A member of the Taylor & Francis Group)

ISBN 1-85743-275-4
ISSN 0071-3791

Library of Congress Catalog Card Number 74-417170

Editor: Lynn Daniel
Associate Editor: Priya Shah
Acting Associate Editor: Elizabeth Kerr
Regional Organizations Editors: Catriona Appeatu Holman, Helen Canton
Statistics Editor: Philip McIntyre
Technology Editor: Ian Preston
Assistant Editors: Mary Bispham, Katie Dawson, Michael Grayer, James Middleton, Patrick Raleigh
Contributing Editor (Commodities): Simon Chapman
Production Co-ordinator: Andreas Gosling
Series Editor: Joanne Maher

Typeset in New Century Schoolbook

Typeset by Unwin Brothers Limited, The Gresham Press, Old Woking, Surrey
Printed and bound by Polestar Wheatons, Exeter

FOREWORD

The 36th edition of THE FAR EAST AND AUSTRALASIA encompasses another year of important events in the region. As the volume went to press in October 2004, King Sihanouk of Cambodia abdicated and was succeeded by his son, Prince Norodom Sihamoni. The Prime Minister of Myanmar, Gen. Khin Nyunt, was dismissed and reportedly placed under house arrest. At legislative elections in Australia, also in October, Prime Minister John Howard was returned to office for a fourth consecutive term. The Philippine electorate voted to return President Gloria Macapagal Arroyo to office at the elections of May. In Indonesia, however, the country's first direct presidential election resulted in the replacement in October of President Megawati Sukarnoputri by Susilo Bambang Yudhoyono. Controversy surrounded the re-election of Taiwanese President Chen Shui-bian in March, compounded by the repercussions of a dramatic shooting incident in which both Chen and his Vice-President had been injured. Following legislative elections in French Polynesia in May, the pro-independence Oscar Temaru replaced the long-standing conservative Gaston Flosse as President, only to be removed from office four months later in a highly controversial motion of censure. During 2004 legislative elections were also held in Hong Kong, Japan, the Republic of Korea, Malaysia, Mongolia, the Cook Islands, Nauru, New Caledonia and Vanuatu.

A calendar of the key political events between November 2003 and October 2004 provides a convenient reference guide to the developments of the last 12 months. Specially commissioned introductory articles on human rights, on population and development issues, on Asian environmental issues, on recent economic trends and on security issues examine subjects of particular relevance to the Asia-Pacific region. A newly commissioned essay on the environmental issues of the Pacific Islands surveys in detail the particular problems of that region. All essays are updated annually by specialist authors and researchers. The statistical surveys, directories and bibliographies have also been extensively revised. Detailed coverage of international organizations and their recent activities in the Far East and Australasia is provided, along with a directory of research institutes and a general bibliography. Information on the region's major commodities, including a new section on cocoa, is also incorporated.

The editors are once again grateful to all the contributors for their articles and advice and to the numerous governments and organizations that have provided statistical and other information.

October 2004

ACKNOWLEDGEMENTS

The editors gratefully acknowledge the co-operation, interest and advice of all the authors who have contributed to this volume. We are also greatly indebted to innumerable organizations connected with the Asia-Pacific region, especially the national statistical and information offices, whose valued co-operation in providing information is greatly appreciated.

We are particularly grateful for the use of material from the following sources: the UN's *Statistical Yearbook, Statistical Yearbook for Asia and the Pacific, Demographic Yearbook, Industrial Commodity Statistics Yearbook, National Accounts Statistics* and *International Trade Statistics Yearbook*; the *UNESCO Statistical Yearbook*; FAO's statistical database; the ILO's statistical database and *Yearbook of Labour Statistics*; the IMF's statistical database, *Government Finance Statistics Yearbook* and monthly *International Financial Statistics*; the World Bank's *World Bank Atlas, Global Development Finance, World Development Report* and *World Development Indicators*; the World Tourism Organization's *Yearbook of Tourism Statistics*; and the Asian Development Bank's *Asian Development Outlook* and *Key Indicators of Developing Asian and Pacific Countries*. We are also grateful to the International Institute for Strategic Studies, Arundel House, 13–15 Arundel Street, London, WC2R 3DX, United Kingdom, for the use of defence statistics from *The Military Balance 2003–2004*.

The following publications have been of special value in providing regular coverage of the affairs of the Asian and Pacific region: *The Far Eastern Economic Review*, Hong Kong; *IMF Survey*, Washington, DC, USA; and *Keesing's Record of World Events*, Cambridge, United Kingdom.

HEALTH AND WELFARE STATISTICS: SOURCES AND DEFINITIONS

Total fertility rate Source: WHO, *The World Health Report* (2003). The number of children that would be born per woman, assuming no female mortality at child-bearing ages and the age-specific fertility rates of a specified country and reference period.

Under-5 mortality rate Source: UNICEF, *The State of the World's Children* (2004). The ratio of registered deaths of children under 5 years to the total number of registered live births over the same period.

HIV/AIDS Source: UNAIDS. Estimated percentage of adults aged 15 to 49 years living with HIV/AIDS. < indicates 'fewer than'.

Health expenditure Source: WHO, *The World Health Report* (2003).

US $ per head (PPP)

International dollar estimates, derived by dividing local currency units by an estimate of their purchasing-power parity (PPP) compared with the US dollar. PPPs are the rates of currency conversion that equalize the purchasing power of different currencies by eliminating the differences in price levels between countries.

% of GDP

GDP levels for OECD countries follow the most recent UN System of National Accounts. For non-OECD countries a value was estimated by utilizing existing UN, IMF and World Bank data.

Public expenditure

Government health-related outlays plus expenditure by social schemes compulsorily affiliated with a sizeable share of the population, and extrabudgetary funds allocated to health services. Figures include grants or loans provided by international agencies, other national authorities, and sometimes commercial banks.

Access to water and sanitation Source: WHO, *Global Water Supply and Sanitation Assessment* (2000 Report). Defined in terms of the type of technology and levels of service afforded. For water, this includes house connections, public standpipes, boreholes with handpumps, protected dug wells, protected spring and rainwater collection; allowance is also made for other locally defined technologies. 'Access' is broadly defined as the availability of at least 20 litres per person per day from a source within 1 km of the user's dwelling. Sanitation is defined to include connection to a sewer or septic tank system, pour-flush latrine, simple pit or ventilated improved pit latrine, again with allowance for acceptable local technologies. Access to water and sanitation does not imply that the level of service or quality of water is 'adequate' or 'safe'.

Human Development Index (HDI) Source: UNDP, *Human Development Report* (2004). A summary of human development measured by three basic dimensions: prospects for a long and healthy life, measured by life expectancy at birth; knowledge, measured by adult literacy rate (two-thirds' weight) and the combined gross enrolment ratio in primary, secondary and tertiary education (one-third weight); and standard of living, measured by GDP per head (PPP US $). The index value obtained lies between zero and one. A value above 0.8 indicates high human development, between 0.5 and 0.8 medium human development, and below 0.5 low human development. Countries with insufficient data were excluded from the HDI. In total, 177 countries were ranked for 2001.

CONTENTS

CONTENTS

PART THREE
Regional Information

List of Maps

THE CONTRIBUTORS

Zachary Abuza. Assistant Professor of Political Science and International Relations, Simmons College, Boston, USA.

Rokiah Alavi. Associate Professor, Head, Department of Economics, International Islamic University, Kuala Lumpur, Malaysia.

Bryant J. Allen. Senior Fellow, Department of Human Geography, Research School of Pacific and Asian Studies, The Australian National University, Canberra, Australia.

Robert Ash. Professor, Department of Economics, School of Oriental and African Studies, University of London, United Kingdom.

Michael Barr. Post-Doctorate Research Fellow in History, The University of Queensland, Brisbane, Australia.

Ian Brown. Professor, Department of History, School of Oriental and African Studies, University of London, United Kingdom.

Jenny J. Bryant. Former Reader, Geography Department, University of the South Pacific, Suva, Fiji.

Kenneth Christie. Professor, Department of Social and Behavioural Sciences, Faculty of Arts and Social Sciences, Zayed University—Dubai Campus, Dubai, United Arab Emirates.

Joseph S. Chung. Professor Emeritus of Economics, Stuart School of Business, Illinois Institute of Technology, Chicago, USA.

Lesley Connors. Former Lecturer, Department of Political Studies, School of Oriental and African Studies, University of London, United Kingdom.

Robert Cribb. Senior Fellow, Division of Pacific and Asian History, Research School of Pacific and Asian Studies, The Australian National University, Canberra, Australia.

Harvey Demaine. Associate Professor of Regional, Rural and Agricultural Development Planning, Asian Institute of Technology, Bangkok, Thailand.

Chris Dixon. Professor of Development Studies, Department of Politics and Modern History, London Metropolitan University, London, United Kingdom.

Tilak Doshi. Fellow, Institute of Southeast Asian Studies, Singapore.

Chris Edwards. Former Senior Lecturer, School of Development Studies, University of East Anglia, Norwich, United Kingdom.

C. P. Fitzgerald. Late Emeritus Professor, The Australian National University, Canberra, Australia.

Aidan Foster-Carter. Honorary Senior Research Fellow in Sociology and Modern Korea, University of Leeds, United Kingdom.

Michael Freeberne. Former Lecturer in Geography, School of Oriental and African Studies, University of London, United Kingdom.

Nick Freeman. Associate Senior Fellow, Institute of Southeast Asian Studies, Singapore.

François Gipouloux. Senior Research Fellow, Centre d'études sur la Chine moderne et contemporaine, Ecole des Hautes Etudes en Sciences Sociales, Paris, France.

Jeanine Graham. Senior Lecturer in History, University of Waikato, Hamilton, New Zealand.

Sok Hach. Economist, Economic Institute of Cambodia, Phnom-Penh, Cambodia.

Hans Hendrischke. Associate Professor and Head, Department of Chinese and Indonesian Studies, The University of New South Wales, Sydney, Australia.

Sarwar O. H. Hobohm. Former Senior Economist/Editor, Asia-Pacific, Country Analysis and Forecasting Division, The Economist Intelligence Unit, London, United Kingdom.

Edith Hodgkinson. Economic journalist specializing in developing countries, London, United Kingdom.

Athar Hussain. Deputy Director, Asia Research Centre, London School of Economics and Political Science, University of London, United Kingdom.

Kenneth E. Jackson. Associate Professor and Director, Centre for Development Studies, Anthropology Department, University of Auckland, New Zealand.

Patrick Jory. Lecturer, School of Asian Studies, The University of Western Australia, Nedlands, Western Australia.

Frank H. H. King. Former Professor of Economic History, The University of Hong Kong, Hong Kong.

Bruce Henry Lambert. Director, Asia Information Centre, Stockholm School of Economics, Stockholm, Sweden.

Roger Lawrey. Faculty of Business, Economics and Policy Studies, Universiti Brunei Darussalam, Brunei.

Philip E. T. Lewis. Director, Centre for Labour Market Research, Murdoch University, Western Australia.

Stuart Macintyre. Professor and Dean of the Faculty of Arts, Department of History, University of Melbourne, Australia.

Adam McCarty. Economic Consultant, Hanoi, Viet Nam.

A. E. McQueen. Former Assistant General Manager, New Zealand Government Railways, New Zealand.

Ruth McVey. Emeritus Reader in Politics with reference to South-East Asia, School of Oriental and African Studies, University of London, United Kingdom.

N. J. Miners. Former Reader in Political Science, Department of Politics and Public Administration, The University of Hong Kong, Hong Kong.

Murni Mohamed. Director, Department of Economic Planning and Development, Ministry of Finance, Bandar Seri Begawan, Brunei.

Andrew C. Nahm. Former Professor Emeritus of History and Consultant to the Office of International Education and Programs, Western Michigan University, USA.

Neville Norman. Professor, Department of Economics, University of Melbourne, Australia.

Gavin Peebles. Former Associate Professor, Department of Economics, The National University of Singapore, Singapore.

Sorpong Peou. Associate Professor of Political Science/International Relations, Faculty of Comparative Culture, Sophia University, Tokyo, Japan.

Michael Pinches. Associate Professor, Discipline of Anthropology and Sociology, School of Social and Cultural Studies, The University of Western Australia, Crawley, Western Australia.

J. W. Rowe. Former Director-General, New Zealand Employers' Federation, New Zealand.

Alan J. K. Sanders. Mongolist, Reading, United Kingdom.

Graham Saunders. Historian, Yorkshire, United Kingdom.

John Sargent. Former Reader in Geography, School of Oriental and African Studies, University of London, United Kingdom.

Josef Silverstein. Former Professor Emeritus of Political Science, Rutgers University, New Brunswick, USA.

Ronald Skeldon. Professorial Fellow in Geography, School of African and Asian Studies, University of Sussex, Brighton, United Kingdom.

Ralph Smith. Former Professor of the History of South-East Asia, School of Oriental and African Studies, University of London, United Kingdom.

THE CONTRIBUTORS

Ross Steele. Lecturer and Director, Graduate Program in Population and Human Resources, School of Geography, Population and Environmental Management, Flinders University, Adelaide, Australia.

Colin Steley. Economic Consultant, Hanoi, Viet Nam.

Richard Storry. Late Director of the Far East Centre, Oxford, United Kingdom.

Martin Stuart-Fox. Professor and Head of Department, Department of History, The University of Queensland, Brisbane, Australia.

Laura Summers. Former Lecturer, Department of Politics and Asian Studies, University of Hull, United Kingdom.

James Tang. Associate Professor and Head, Department of Politics and Public Administration, University of Hong Kong, Hong Kong.

Mya Than. Associate Senior Fellow, Institute of South East Asian Studies, Singapore.

Tran Van Hoa. Professor, Department of Economics, Wollongong University, Wollongong, NSW, Australia.

Hajah Rosni Haji Tungkat. Acting Director, Department of Economic Planning and Development, Ministry of Finance, Bandar Seri Begawan, Brunei.

C. M. Turnbull. Honorary Research Fellow, Centre for Asian Studies, and Former Professor of History, The University of Hong Kong, Hong Kong.

Richard Vokes. Former Lecturer in Economics and South-East Asian Studies, Centre of South-East Asian Studies, University of Kent at Canterbury, United Kingdom.

Peter Wilson. Associate Professor, Department of Economics, National University of Singapore, Singapore.

Michael Yahuda. Sigur Center, Elliott School of International Relations, George Washington University, Washington, DC, USA.

Akira Yamazaki. Former Editorial Writer, *Nihon Keizai Shimbun* (Japan Economic Journal), Tokyo, Japan.

ABBREVIATIONS

Acad.	Academician; Academy
ACP	African, Caribbean and Pacific (States)
ACT	Australian Capital Territory
AD	Anno Domini
ADB	Asian Development Bank
Adm.	Admiral
Admin.	Administration, Administrative, Administrator
AFTA	ASEAN Free Trade Area
AG	Aktiengesellschaft (limited company)
a.i.	ad interim
AIDS	acquired immunodeficiency syndrome
Alt.	Alternate
AM	Amplitude Modulation
ANZUS	Australia, New Zealand and the United States
Apdo	Apartado (Post Box)
APEC	Asia-Pacific Economic Co-operation
approx.	approximately
Apt	Apartment
AR	Autonomous Region
ARF	ASEAN Regional Forum
ASEAN	Association of South East Asian Nations
Asscn	Association
Assoc.	Associate
Asst	Assistant
ATM	automated teller machine
Aug.	August
auth.	authorized
Av.	Avenida (Avenue)
Ave	Avenue
BC	Before Christ
Bd	Board
b/d	barrels per day
Bhd	Berhad (public limited company)
BIS	Bank for International Settlements
Bldg(s)	Building(s)
Blk	Block
Blvd	Boulevard
BOT	build-operate-transfer
BP	Boîte Postale (Post Box)
br.(s)	branch(es)
Brig.	Brigadier
Btu	British thermal unit
C	Centigrade
c.	circa
cap.	capital
Capt.	Captain
CCP	Chinese Communist Party
Cdre	Commodore
Cen.	Central
CEO	Chief Executive Officer
cf.	confer (compare)
CFC	chlorofluorocarbon
CFP	Communauté française du Pacifique
Chair.	Chairman/Chairwoman
CIA	(US) Central Intelligence Agency
Cie	Compagnie
c.i.f.	cost, insurance and freight
C-in-C	Commander-in-Chief
circ.	circulation
CIS	Commonwealth of Independent States
cm	centimetre(s)
CMEA	Council for Mutual Economic Assistance (COMECON)
cnr	corner
c/o	care of
Co	Company
Col	Colonel
Comm.	Commission
Commdr	Commander
Commdt	Commandant
Commr	Commissioner
Conf.	Conference
Confed.	Confederation
COO	Chief Operating Officer
Corpn	Corporation
CP	Case Postale, Caixa Postal (Post Box)
CPO	Central Post Office
Cttee	Committee
cu	cubic
cwt	hundredweight
Dec.	December
Dem.	Democratic
Dep.	Deputy
dep.	deposit(s)
Dept	Department
Devt	Development
Dir	Director
Div.	Division
DPRK	Democratic People's Republic of Korea
Dr	Doctor
Dra	Doutora (Doctor)
DRAM	dynamic random access memory
Drs	Doctorandus (Netherlands doctor's degree)
dwt	dead weight tons
E	East, Eastern
EAEC	East Asia Economic Caucus
EAGA	East ASEAN Growth Area
EC	European Community
Econ.	Economic
ECU	European Currency Unit(s)
Ed.(s)	Editor(s)
Edif.	Edificio (Building)
edn	edition
e.g.	exempli gratia (for example)
Eng.	Engineer; Engineering
EPZ	Export Processing Zone
ESCAP	Economic and Social Commission for Asia and the Pacific
est.	established; estimate, estimated
etc.	etcetera
EU	European Union
excl.	excluding
Exec.	Executive
exhbn(s)	exhibition(s)
Ext.	Extension
f.	founded
FAO	Food and Agriculture Organization
FDI	foreign direct investment
Feb.	February
Fed.	Federal, Federation
Flt	Flight
FM	Frequency Modulation
fmr(ly)	former(ly)
f.o.b.	free on board
Fr	Father
Fri.	Friday
ft	foot (feet)
g	gram(s)
gall(s)	gallon(s)
GATT	General Agreement on Tariffs and Trade
GDP	gross domestic product
Gen.	General
GNI	gross national income
GNP	gross national product
Gov.	Governor
Govt	Government
GPO	General Post Office
grt	gross registered tons
GSP	gross state product
GWh	gigawatt hour(s)
ha	hectare(s)
HDI	human development index
HE	His (or Her) Excellency; His Eminence
HH	His (or Her) Highness; His Holiness
HI	Hawaii
HIV	human immunodeficiency virus
hl	hectolitre(s)
HM	His (or Her) Majesty
HMS	His (or Her) Majesty's Ship
HMSO	Her Majesty's Stationery Office
Hon.	Honorary
HQ	Headquarters
hr(s)	hour(s)
HRH	His (or Her) Royal Highness
HYV	high-yielding variety
ibid.	ibidem (from the same source)
IBRD	International Bank for Reconstruction and Development (World Bank)
IDPs	Internally Displaced Persons
i.e.	id est (that is to say)
ILO	International Labour Organization
IMF	International Monetary Fund
in	inch(es)
Inc	Incorporated

ABBREVIATIONS

incl.	including
Ind.	Independent
Inst.	Institute
Int.	International
IPCC	Intergovernmental Panel on Climate Change
Ir	Insinyur (Engineer)
IRF	International Road Federation
Is	Islands
ISIC	International Standard Industrial Classification
Jan.	January
Jr	Junior
Jt	Joint
Jtly	Jointly
K	kina (Papua New Guinea currency)
kg	kilogram(s)
kHz	kilohertz
KK	Kaien Kaisha (limited company)
Kl	Kilolitre
km	kilometre(s)
kW	kilowatt(s)
kWh	kilowatt hour(s)
lb	pound(s)
LDC	less-developed country
LNG	liquefied natural gas
LPG	liquefied petroleum gas
Lt	Lieutenant
Ltd	Limited
m	metre(s)
m.	million
Maj.	Major
Man.	Manager, Managing
MCPO	Manila Central Post Office
mem.	member
mfg	manufacturing
mfr(s)	manufacturer(s)
mg	milligram
Mgr	Monseigneur, Monsignor
MHz	megahertz
MIA	Missing in Action
Mil.	Military
Mlle	Mademoiselle
mm	millimetre(s)
Mme	Madame
Mon.	Monday
MP	Member of Parliament
Mt	Mount
MV	Motor Vessel
MW	megawatt(s)
MWh	megawatt hour(s)
N	North, Northern
n.a.	not available
NAFTA	North American Free Trade Agreement
Nat.	National
NCO	Non-Commissioned Officer
n.e.s.	not elsewhere specified
NGO	non-governmental organization
NIC	newly-industrializing country
NIE	newly-industrializing economy
No.	number
Nov.	November
NPL	non-performing loan(s)
nr	near
nrt	net registered tons
NSW	New South Wales
NT	Northern Territory
NV	Naamloze Vennootschap (limited company)
NZ	New Zealand
Oct.	October
OECD	Organisation for Economic Co-operation and Development
OER	official exchange rate
OPEC	Organization of the Petroleum Exporting Countries
opp.	opposite
Ord.	ordinary
Org.(s)	Organization(s)
oz	ounce(s)
p(p).	page(s)
p.a.	per annum
Parl.	Parliament(ary)
PB	Private Bag
PC	personal computer
Perm.	Permanent
PICTA	Pacific Island Countries Trade Agreement
PLC	Public Limited Company
PMB	Private Mail Bag
PNG	Papua New Guinea
PO	Post Office
POB	Post Office Box
PPP	purchasing-power parity
Pres.	President
Prin.	Principal
Prof.	Professor
Propr	Proprietor
Prov.	Province, Provincial
PT	Perseroan Terbatas (limited company)
Pt	Point
Pte	Private
Pty	Proprietary
p.u.	paid up
publ.(s)	publication(s); published
Publr	Publisher
Pvt	Private
QIP	Quick Impact Project
Qld	Queensland
q.v.	quod vide (to which refer)
R & D	research and development
Rd	Road
regd	registered
Rep.	Representative
Repub.	Republic
res	reserves
retd	retired
Rev.	Reverend
Rm	Room
RM	Ringgit Malaysia
ro-ro	roll-on roll-off
Rp.	Rupiah (Indonesian currency)
Rt	Right
S	South, Southern
SA	South Australia; Société anonyme, Sociedad Anónima (limited company)
SAARC	South Asian Association for Regional Co-operation
SAR	Special Administrative Region
SARL	Sociedade Anônima de Responsibilidade Limitada (limited company)
SARS	Severe Acute Respiratory Syndrome
Sat.	Saturday
Sdn Bhd	Sendirian Berhad (private limited company)
SDR	Special Drawing Right(s)
SEATO	South-East Asia Treaty Organization
Sec.	Secretary; Section
Secr.	Secretariat
Sept.	September
SEZ	Special Economic Zone
SITC	Standard International Trade Classification
Soc.	Society
SOE	State-Owned Enterprise
Sq.	Square
sq	square (in measurements)
Sr	Senior
St	Street; Saint, San, Santo
subs.	subscribed
Sun.	Sunday
Supt	Superintendent
Tas	Tasmania
tech.	technical
tel.	telephone
TEU	20-foot equivalent unit
Thur.	Thursday
trans.	translator, translated
Treas.	Treasurer
Tue.	Tuesday
TV	television
UHF	ultra-high frequency
UK	United Kingdom
UN	United Nations
UNCTAD	United Nations Conference on Trade and Development
UNDP	United Nations Development Programme
UNEP	United Nations Environment Programme
UNESCO	United Nations Educational, Scientific and Cultural Organization
UNFPA	United Nations Population Fund
UNHCHR	United Nations High Commissioner of Human Rights
UNHCR	United Nations High Commissioner for Refugees
UNIDO	United Nations Industrial Development Organization
UNMISET	United Nations Mission of Support in East Timor
UNODC	United Nations Office on Drugs and Crime
UNTAC	United Nations Transitional Authority in Cambodia
UNTAET	United Nations Transitional Administration in East Timor
USA	United States of America
USAID	United States Agency for International Development
USSR	Union of Soviet Socialist Republics
VAT	value-added tax

Ven.	Venerable
VHF	very high frequency
Vic	Victoria
Vol(s)	Volume(s)
W	West, Western
WA	Western Australia
Wed.	Wednesday
WHO	World Health Organization
WSSD	World Summit on Sustainable Development
WTO	World Trade Organization
yd	yard
yr(s)	year(s)

INTERNATIONAL TELEPHONE CODES

To make international calls to telephone and fax numbers listed in The Far East and Australasia, dial the international code of the country from which you are calling, followed by the appropriate country code for the organization you wish to call (listed below), followed by the area code (if applicable) and telephone or fax number listed in the entry.

	Country code	+ or − GMT*
Australia	61	+8 to +10
Australian Dependencies in the Indian Ocean		
Christmas Island	61	+7
Cocos (Keeling) Islands	61	+6½
Brunei	673	+8
Cambodia	855	+7
China, People's Republic	86	+8
Special Administrative Regions:		
Hong Kong	852	+8
Macao	853	+8
China (Taiwan)	886	+8
Indonesia	62	+7 to +9
Japan	81	+9
Korea, Democratic People's Republic (North Korea)	850	+9
Korea, Republic (South Korea)	82	+9
Laos	856	+7
Malaysia	60	+8
Mongolia	976	+7 to +9
Myanmar	95	+6½
New Zealand	64	+12
The Pacific Islands		
Australian Pacific Territories		
Norfolk Island	672	+11½
Fiji	679	+12
French Pacific Territory		
Wallis and Futuna Islands	681	+12
French Pacific Overseas Countries		
French Polynesia	689	−9 to −10
New Caledonia	687	+11
Kiribati	686	+12 to +13
The Marshall Islands	692	+12
The Federated States of Micronesia	691	+10 to +11
Nauru	674	+12
New Zealand Pacific Territory		
Tokelau	690	−10
New Zealand Pacific: Associated States		
Cook Islands	682	−10
Niue	683	−11
Palau	680	+9
Papua New Guinea	675	+10
Samoa	685	−11
Solomon Islands	677	+11
Tonga	676	+13
Tuvalu	688	+12
United Kingdom Pacific Territory		
Pitcairn Islands	872	−8
US Commonwealth Territory in the Pacific		
Northern Mariana Islands	1 670	+10
US External Territories in the Pacific		
American Samoa	1 684	−11
Guam	1 671	+10
Vanuatu	678	+11
The Philippines	63	+8
Singapore	65	+8
Thailand	66	+7
Timor-Leste (East Timor)	670	+9
Viet Nam	84	+7

* Time difference in hours + or − Greenwich Mean Time (GMT). The times listed compare the standard (winter) times. Some countries adopt Summer (Daylight Saving) Times—i.e. + 1 hour—for part of the year.

EXPLANATORY NOTE ON THE DIRECTORY SECTION

The Directory section of each chapter is arranged under the following headings, where they apply:

THE CONSTITUTION

THE GOVERNMENT

- HEAD OF STATE
- CABINET/COUNCIL OF MINISTERS
- MINISTRIES

LEGISLATURE

STATE GOVERNMENTS

POLITICAL ORGANIZATIONS

DIPLOMATIC REPRESENTATION

JUDICIAL SYSTEM

RELIGION

THE PRESS

PUBLISHERS

BROADCASTING AND COMMUNICATIONS

- TELECOMMUNICATIONS
- RADIO
- TELEVISION

FINANCE

- CENTRAL BANK
- STATE BANKS
- COMMERCIAL BANKS
- DEVELOPMENT BANKS
- INVESTMENT BANKS
- SAVINGS BANKS
- FOREIGN BANKS
- STOCK EXCHANGE
- INSURANCE

TRADE AND INDUSTRY

- GOVERNMENT AGENCIES
- DEVELOPMENT ORGANIZATIONS
- CHAMBERS OF COMMERCE AND INDUSTRY
- INDUSTRIAL AND TRADE ASSOCIATIONS
- EMPLOYERS' ORGANIZATIONS
- UTILITIES
- MAJOR COMPANIES
- CO-OPERATIVES
- TRADE UNIONS

TRANSPORT

- RAILWAYS
- ROADS
- INLAND WATERWAYS
- SHIPPING
- CIVIL AVIATION

TOURISM

DEFENCE

EDUCATION

CALENDAR OF POLITICAL EVENTS IN THE FAR EAST AND AUSTRALASIA, NOVEMBER 2003–OCTOBER 2004

NOVEMBER 2003

2–9 **Myanmar:** UN human rights envoy Paulo Sérgio Pinheiro visited the country; during his time there he visited Aung San Suu Kyi, leader of the opposition National League for Democracy (NLD), who continued to be held under house arrest.

4 **Australia/Indonesia:** in response to the arrival of an Indonesian boat carrying 14 asylum-seekers at Melville Island, the Government of Australia immediately separated the island and 4,000 other small islands from the country's migration zone and ordered an Australian warship to tow the boat to the Indonesian island of Yamdena, provoking strong criticism.

4 **Cambodia:** it was reported that an appeal by former Khmer Rouge leader Chhouk Rin against his conviction in September 2002 for his role in the murder of three Western travellers in 1994 had been rejected.

4 **Thailand:** businessman Chuwit Kamonvisit launched his own political party, the Thon Trakul Thai (First Thai Nation) party, in order to further his campaign against political corruption.

5 **Cambodia:** the Cambodian People's Party (CPP), FUNCINPEC and the Sam Rainsy Party held their first meeting since the 27 July general election, in an attempt to end the political stalemate that had arisen following the CPP's failure to secure a two-thirds' majority in Parliament.

5 **Indonesia:** it was reported that government troops in Papua province had killed 10 members of the separatist organization, Organisasi Papua Merdeka (OPM), during an attack on the Jayawijaya district.

6 **Australia:** the Queensland Court of Appeal overturned the convictions of electoral fraud against Pauline Hanson, the former leader of the One Nation party.

6 **Tuvalu:** an opposition motion of 'no confidence' in the Government was defeated by nine votes to six at the first parliamentary sitting of 2003.

7 **Kiribati/Taiwan:** the Kiribati Government announced the establishment of diplomatic relations with Taiwan, stating that it hoped to be able to retain its links with the People's Republic of China.

7 **Philippines:** charges against the 290 soldiers who had participated in a mutiny in Manila in July were abandoned; however, the 31 officers believed to have led the mutiny continued to await trial.

8 **Philippines:** two armed men seized control of the air traffic control tower at the Ninoy Aquino International Airport in Manila, prompting fears of an attempted coup; both men were killed by the armed forces following police intervention.

8 **Thailand:** a cabinet reorganization took place, resulting in the dismissal of Chart Pattana members from the ruling coalition.

9 **Japan:** in a general election, the Liberal Democratic Party (LDP) led by Prime Minister Junichiro Koizumi secured 237 seats of the 480 seats in the House of Representatives and the Democratic Party of Japan (DPJ) won 177 seats; having failed to secure a majority, the LDP would therefore be obliged henceforth to rely on the parliamentary support of its coalition partners, the New Conservative Party and New Komeito.

10 **Malaysia:** 13 students were detained under the Internal Security Act (ISA), having been deported from Pakistan on suspicion of membership of the regional terrorist organization Jemaah Islamiah; four of the students were later released.

10 **Viet Nam:** Minister of National Defence Lt-Gen. Pham Van Tra paid an historic visit to Washington, DC, where he held talks with US Secretary of Defense Donald Rumsfeld.

11 **Cook Islands:** Prime Minister Robert Woonton effected a reorganization of cabinet portfolios after a failed attempt by Deputy Prime Minister Terepai Maoate to propose a motion of 'no confidence' in the Government, accusing it of 'mismanagement and gross neglect of the public interest'.

11 **Philippines:** following a ruling by the Supreme Court on the previous day, the House of Representatives voted not to proceed with the impeachment of Hilario Davide, Jr, Chief Justice of the Supreme Court, on charges relating to the alleged misuse of judicial funds.

12 **Australia/Timor-Leste:** bilateral negotiations began in Darwin, Australia, intended to resolve a dispute over the location of the maritime boundaries between the two countries.

12 **Papua New Guinea:** Prime Minister Sir Michael Somare effected a reorganization of cabinet portfolios following a change in leadership of one of the governing coalition partners.

13 **Papua New Guinea:** Speaker Bill Skate assumed the role of Governor-General in an acting capacity, pending a new election, following uncertainty over the validity of the recent election to that post.

17 **Indonesia:** further sectarian violence occurred in Poso, Central Sulawesi, resulting in the deaths of three people.

17 **Marshall Islands:** at a legislative election, the United Democratic Party won 20 of the 33 seats in the Nitijela.

17 **Viet Nam:** the trial began in Hanoi of eight government officials, including two former junior agriculture ministers and the director of a state agriculture company, charged with the embezzlement of state funds.

18 **Thailand:** the trials commenced of four men suspected of membership of Jemaah Islamiah and of planning attacks on embassies and tourist destinations in the country.

18 **Vanuatu:** Prime Minister Edward Natapei effected a ministerial reorganization that removed from the Cabinet six UMP members of the governing coalition, including Deputy Prime Minister and leader of the UMP Serge Vohor, reportedly in order to avert a vote of 'no confidence' in the Government.

19 **Indonesia:** President Megawati Sukarnoputri announced the extension of martial law in the province of Aceh for a further six months, following the expiry of the initial period of military rule, which had begun in May 2003.

19 **Viet Nam:** a US frigate spent four days anchored at Ho Chi Minh City, becoming the first US warship to visit the country since the end of the Viet Nam war in 1975.

21 **Papua New Guinea:** the Supreme Court ruled that the election of Sir Albert Kipalan in September to the position of Governor-General was invalid.

21 **Philippines:** Secretary of Finance José Isidro Camacho resigned; Under-Secretary of Finance Juanita Amatong was appointed to succeed him, initially on an acting, but later on a permanent, basis.

23 **Hong Kong:** the pro-Beijing Democratic Alliance for the Betterment of Hong Kong (DAB) lost a substantial number of seats in District Council elections, prompting the resignation of DAB Chairman Tsang Yok-sing.

24 **Malaysia:** 15 members of the Islamic cult al-Ma'unah were released from prison, having expressed remorse following their convictions in December 2001 for participating in an attempted insurrection in July 2000.

26 **Thailand:** King Bhumibol Adulyadej refused to give his assent to an education reform bill that was opposed by most of the country's teachers; it was believed to be the first time that a monarch had refused to sign a major piece of legislation since Thailand became a constitutional democracy in 1932.

27 **Taiwan:** legislation allowing referendums to be held in Taiwan was approved by the Legislative Yuan; however, the legislation prohibited referendums on sovereignty issues.

27 **Vanuatu:** at a by-election for the seat vacated by Barak Sope following the MP's conviction on charges of fraud, Sope himself was re-elected, having received a presidential pardon for his crime.

DECEMBER 2003

1 **Indonesia:** the subversion conviction of the alleged spiritual leader of Jemaah Islamiah, Abu Bakar Bashir, was overturned following an appeal; however, his conviction for holding forged identity documents was upheld, resulting in a reduction of his prison term from four to three years.

2 **Australia:** the republican Mark Latham was elected as the leader of the opposition Australian Labor Party several days after Simon Crean resigned from the post.

2 **Viet Nam:** two former junior agriculture ministers were sentenced to three-year prison terms following the conclusion of the corruption case against them; a former director of a state agriculture company was sentenced to death following the trial, while five other government officials were sentenced to prison terms of between four and 15 years.

3 **Thailand:** Prime Minister Thaksin Shinawatra claimed that the six-month campaign to eliminate drugs-related activity in the country, concluded earlier in the year, had been successful.

4 **Papua New Guinea:** Sir Pato Kakaraya was elected the country's Governor-General by the National Parliament.

4 **Philippines:** Senator Edgardo Angara announced the merger of his faction of the opposition Laban ng Demokratikong Pilipino (LDP) with the Partido Demokratikong Pilipino—Laban (PDP—Laban) and Estrada's Puwers ng Masang Pilipino (PMP) to form the Koalisyon ng Nagkakaisang Pilipino (KNP—Coalition of the United Filipino).

7 **Australia:** the leader of the Australian Democrats Party, Senator Andrew Bartlett, publicly apologized for swearing at and assaulting another senator during a drunken argument and temporarily relinquished his post; the party's deputy leader, Lyn Allison, was appointed acting leader for the duration of Bartlett's leave.

7 **Australia:** the ruling Australian Labor Party was re-elected for a third term at a state election in Queensland.

7 **People's Republic of China:** Premier Wen Jiabao began his first official visit to the USA.

7 **Philippines:** the armed forces were reported to have captured a senior member of Abu Sayyaf, known as 'Commander Robot', in Sulu province.

9 **Japan:** Prime Minister Koizumi's Cabinet approved a plan to dispatch a contingent of Self-Defence Forces (SDF) to Iraq, despite strong popular opposition to the deployment.

10 **People's Republic of China:** in elections to the District People's Congress in Beijing, two seats were secured by independent candidates.

10 **Philippines:** the Secretary of Trade and Industry, Manuel Roxas, II, resigned in order to stand for a seat in the Senate at the May 2004 elections.

11 **Australia/Papua New Guinea:** the Australian Government finalized arrangements to send around 300 police officers and civil servants to Papua New Guinea to assist in the fight against crime and corruption.

14 **Philippines:** Secretary of Foreign Affairs Blas Ople died whilst on a flight from Japan to Thailand; he was later succeeded by Delia Domingo-Albert.

15 **Cambodia:** the newly elected National Assembly held its first session since the general election of 27 July, despite the fact that a new Government had yet to be formed.

15 **People's Republic of China:** the Government named the East Turkestan Islamic Movement and several other Uygur separatist groups based in Xinjiang Province as 'terrorist' organizations.

18 **Laos:** the USA announced that it would accept 15,000 ethnic Hmong currently living in refugee camps in Thailand.

18 **Singapore:** the Government announced that it had detained under the ISA two alleged Singaporean members of Jemaah Islamiah in late October.

22 **Myanmar:** following a visit to the country, the human rights organization Amnesty International reported that the State Peace and Development Council (SPDC) was not implementing its seven part 'road map' to democracy and that there had been a rise in the number of political detentions taking place in recent months.

31 **Indonesia:** a bomb exploded in the town of Peureulak, Aceh, killing 10 people; the attack was believed to have been perpetrated by the Gerakan Aceh Merdeka (GAM—Free Aceh Movement), although the movement denied responsibility for the bombing.

31 **Viet Nam:** Nguyen Vu Binh, a journalist, was sentenced to seven years in prison, having been convicted of espionage charges following a three-hour trial.

JANUARY 2004

1 **Hong Kong:** thousands of people participated in demonstrations demanding direct election of the territory's Chief Executive and Legislative Council.

3–5 **Thailand:** an outbreak of violence in the south of the country, including a spate of arson attacks on schools in the area, resulted in the deaths of six security personnel; as the violence continued, the Government declared martial law in the southern provinces of Yala, Pattani and Narathiwat.

4 **Marshall Islands:** President Kessai Note was re-elected by the Nitijela.

4 **Philippines:** a bomb exploded in the town of Parang on Mindanao, resulting in the deaths of 15 people.

7 **Malaysia:** a minor cabinet reorganization took place, in which the most significant appointment was that of Minister of Defence Najib Tun Razak to the previously vacant post of Deputy Prime Minister.

7 **Niue:** a state of emergency was declared following widespread devastation wreaked by Cyclone Heta; much of the capital, Alofi, and the island's only hospital were destroyed, resulting in speculation regarding the island's viability as a self-governing territory.

10 **Cambodia:** King Norodom Sihanouk announced that, following a request from the President of the Senate, Chea Sim, the mandate of the upper house would be extended by one year in order to avert a constitutional

crisis, owing to the continuation of the political deadlock that had ensued following elections to the House of Representatives in July 2003.

10 **Indonesia:** a bomb exploded in Palopo, South Sulawesi, killing four people.

10 **Philippines:** members of the rebel New People's Army (NPA) attacked a power station near Manila; three government soldiers and three rebels were killed.

13 **Indonesia:** a court in Ambon, South Maluku, convicted nine men of subversion for membership of the banned separatist organization, the Maluku Sovereignty Front.

13 **Philippines:** President Macapagal Arroyo announced that her Government was ready to enter into new peace talks with the NPA.

16–22 **Myanmar:** the SPDC held talks with members of the separatist Karen National Union (KNU), in the hope of reaching an agreement to end the group's ongoing insurgency; it was reported following the talks that an informal cease-fire arrangement had been agreed upon.

17 **Papua New Guinea:** Prime Minister Sir Michael Somare effected a reorganization of cabinet portfolios in which he replaced two ministers who had voted against a government proposal to extend the grace period before which a parliamentary vote of 'no confidence' could take place from 18 months to three years.

21 **Malaysia:** the Court of Appeal refused an application for bail by the imprisoned former Deputy Prime Minister Anwar Ibrahim, pending the outcome of his appeal against his conviction in August 2000 for sodomy.

22 **Cambodia:** the President of the Free Trade Union of Workers of the Kingdom of Cambodia (FTUWKC), Chea Vichea, was shot dead in Phnom-Penh; two men were later charged with his murder, one of whom admitted to having been paid to carry out the shooting.

22 **Samoa:** it was reported that the Fono had voted to abolish the death penalty in the country, although no executions had been carried out since 1952.

23 **Myanmar:** the SPDC announced that it had freed 151 members of the opposition NLD who had been detained since the attack on the motor convoy of NLD leader Aung San Suu Kyi in the previous May.

FEBRUARY 2004

2 **Myanmar:** members of 25 ethnic parties attended the Third Ethnic Nationalities Seminar, at the end of which those who had participated reportedly rejected the SPDC's 'road map' to democracy and called instead for the resolution of the country's problems through a tripartite dialogue involving the Government, the opposition and other political parties, and ethnic minorities.

4–5 **Australia/Indonesia:** the two countries co-hosted a counter-terrorism conference on the Indonesian island of Bali, concluding with an agreement that a regional counter-terrorism centre would be established in Jakarta.

10–11 **Viet Nam:** US Adm. Thomas Fargo, head of the Pacific Command, visited the country, becoming the first foreign official to visit the naval base at Da Nang during his stay.

12 **Indonesia:** the September 2002 corruption conviction of Akbar Tandjung, the Speaker of the House of Representatives, was overturned following an appeal.

14 **Philippines:** the first, four-day round of peace negotiations between the National Democratic Front (NDF), the political wing of the NPA, and the Government concluded in Oslo, Norway, with both sides agreeing jointly to investigate allegations of human rights abuses.

15–16 **Australia:** riots broke out in the predominantly Aboriginal district of Redfern in Sydney in protest against the death of an Aboriginal youth in an apparent cycling accident; three separate inquiries into the death of the youth and into the riots were launched.

15 **Myanmar:** Tin Oo, Vice-Chairman of the NLD, was released from prison and placed under house arrest in Yangon, following his detention by the SPDC in May 2003.

16–17 **Philippines:** the Government held exploratory talks with the separatist group the Moro Islamic Liberation Front (MILF), following which it was announced that formal peace negotiations between the two sides would resume in Malaysia in April.

17 **Malaysia/Thailand:** Thai Prime Minister Thaksin Shinawatra proposed the construction of a security fence on the joint border, in order to prevent insurgents in southern Thailand from taking refuge in Malaysia.

17 **Viet Nam:** an agreement with Russia was signed in Hanoi, concerning the provision of Russian assistance to help build Viet Nam's first nuclear power station.

18 **Malaysia:** Minister of Land and Co-operative Development Kasitah bin Gaddam resigned from his cabinet position, having been charged with corruption six days previously.

20–21 **Thailand/Viet Nam:** the first joint meeting of the two Cabinets was held, during which eight agreements on co-operation in various social and economic areas were signed.

20 **Vanuatu:** Prime Minister Edward Natapei announced a reorganization of cabinet portfolios amid continuing instability within the governing coalition.

24 **New Zealand:** a cabinet reorganization was announced following the resignation of the Minister of Commerce and of Immigration, Lianne Dalziel; the Minister of Transport, of Corrections, of Communications and for Information Technology, Paul Swain, was assigned the immigration portfolio in place of his responsibilities for transport.

24 **Solomon Islands:** the former Minister of Communications, Aviation and Meteorology, Daniel Fa'afunua, was sentenced to two years' imprisonment after being found guilty of assaulting a policewoman; Fa'afunua had received a similar sentence earlier in the month for demanding money with menaces.

25–28 **Democratic People's Republic of Korea:** a second round of six-party talks, involving North and South Korea, the USA, Japan, China and Russia, took place in Beijing to discuss North Korea's nuclear weapons development but ended without agreement.

26 **Indonesia:** Sardono Siliwangi became the first person to be convicted of involvement in the bombing that had taken place at the Marriott Hotel in Jakarta in August 2003; he was sentenced to a 10-year prison term.

26 **Myanmar:** several exiled dissident organizations, associated with the pro-democracy demonstrations that had taken place in the country in August 1988, reportedly formed the Burma Democratic Alliance.

MARCH 2004

1 **Australia:** the report of a parliamentary committee on intelligence on Iraq's weapons of mass destruction was published, largely exonerating the Australian Government of manipulating the intelligence used to justify the case for the removal of the regime of Saddam Hussain in Iraq.

1–4 **Myanmar:** UN envoy Razali Ismail visited the country and held talks with Prime Minister Gen. Khin Nyunt, as well as Aung San Suu Kyi and leaders of ethnic groups.

2 **French Polynesia:** the status of the territory was officially changed to that of an Overseas Country of France, thus according greater autonomy to French Polynesia.

3 **Philippines:** the Supreme Court voted in favour of allowing the actor Fernando Poe, Jr, to contest the forthcoming presidential election; his eligibility for the presidency had been contested owing to doubts concerning his nationality, as his mother was a US citizen.

4 **Brunei:** the Government announced that three men had been detained under the Internal Security Act on treason charges, owing to their alleged participation in subversive activities.

4 **Laos:** the Ministry of Defence claimed that approximately 700 ethnic Hmong rebels, including five senior commanders, had surrendered to the authorities in late February.

8 **Vanuatu:** Prime Minister Edward Natapei effected a further reorganization of cabinet portfolios (the third in three months) following the resignation of the Minister for Finance Sela Molisa.

9 **Indonesia:** after the subversion charges against him were overturned, the revised three-year prison sentence imposed on Abu Bakar Bashir was further reduced to 18 months by the Supreme Court.

10 **Thailand:** a cabinet reorganization took place, in which the defence, interior and finance portfolios were reallocated, owing in part to increasing fears concerning the security situation in the south of the country.

11 **Indonesia:** the Co-ordinating Minister for Political Affairs, Security and Social Welfare, Gen. (retd) Susilo Bambang Yudhoyono, resigned from the Cabinet; it was believed that he intended to contest the forthcoming presidential election.

12 **Republic of Korea:** President Roh Moo-Hyun was impeached following the introduction of a motion in the National Assembly that was supported by more than two-thirds of legislators; the President was accused of violating election law by publicly supporting the pro-Government Uri Party in advance of legislative elections in April; Prime Minister Goh Kun became acting head of state.

13 **Indonesia:** it was reported that Leo Warisman, a leader of the OPM, had been killed in a gunfight with security forces in Papua province.

14 **People's Republic of China:** delegates at the second session of the 10th National People's Congress approved constitutional amendments, including provisions for the protection of private property.

19 **Taiwan:** President Chen Shui-bian and Vice-President Annette Lu were slightly injured in an apparent assassination attempt on the final day of the presidential election campaign.

19 **Thailand:** a spate of arson attacks took place in the increasingly troubled southern area of the country, resulting in the destruction of 36 government buildings.

20 **Taiwan:** President Chen Shui-bian was narrowly re-elected, amid demands for a recount of disputed ballot papers and confusion surrounding the shooting incident of the previous day; in a concurrent referendum concerning defence issues, voter participation was too low to produce a valid result.

21 **Australia:** James Bacon formally resigned as the Premier of Tasmania, owing to ill health; he was replaced by his deputy, Paul Lennon.

21 **Malaysia:** elections were held to the 219-member House of Representatives and to 12 of the 13 State Legislative Assemblies; the Barisan Nasional coalition of Prime Minister Abdullah Ahmad Badawi secured a comprehensive victory, winning 198 seats in the House of Representatives and taking control of 11 of the 12 State Legislative Assemblies.

27 **Malaysia:** Prime Minister Abdullah announced a cabinet reorganization, in which the cabinet was enlarged through the division of three existing ministries and the creation of two new ones.

27 **Thailand:** a bomb exploded in Sungai Kolok, in Narathiwat province, injuring 29 people, including 10 Malaysian tourists, prompting the Government to suspend a US $300m. development aid programme intended for the area.

29 **Papua New Guinea:** it was reported that the Bougainville Interim Provincial Government had passed legislation providing for the establishment of the Bougainville Constituent Assembly, which was to help formulate a constitution for the future autonomous region.

30 **Australia**: the House of Representatives approved a resolution opposing 'artificial deadlines' for the withdrawal of Australian troops from Iraq.

31 **Myanmar:** the SPDC announced that the National Convention, which had been in recess since 1996, would be reconvened on 17 May in order to draw up a new constitution for the country.

APRIL 2004

5 **Indonesia:** elections took place to the 550-member House of Representatives and to the newly created House of Representatives of the Regions.

8 **Philippines:** the armed forces claimed to have killed a leading member of the rebel Abu Sayyaf group following a gun battle on the island of Basilan.

9 **Federated States of Micronesia:** a state of emergency was declared on the island of Yap as a result of the damage caused by Typhoon Sudal.

12 **Vanuatu:** Alfred Maseng Nalo was elected President following four rounds of voting by Parliament in an election contested by more than 30 candidates.

13 **People's Republic of China:** US Vice-President Dick Cheney arrived in China on an official visit.

13 **Myanmar:** the Chairman of the opposition NLD, Aung Shwe, and its Secretary-General, U Lwin, were released from house arrest, having been detained since May 2003 following the attack on the entourage of Aung San Suu Kyi that had led to her renewed detention by the Government.

15 **Republic of Korea:** at elections to the National Assembly, the Uri Party secured a narrow majority, winning 152 of the 299 seats; the opposition Grand National Party (GNP) won 121 seats.

16 **Nauru:** receivers were appointed to take control of the country's assets, including its property portfolio, in order to repay debts of some $A230m.

17 **Myanmar:** the SPDC permitted the headquarters of the NLD, in Yangon, to reopen; they had been closed since May 2003.

19 **Federated States of Micronesia:** Micronesian state governors signed a co-operation pact aimed at ending decades of disunity with regard to policy objectives.

19 **Pitcairn:** the island's Supreme Court ruled that the trial of seven men on 96 charges of sexual offences should be held in Auckland, New Zealand, under British jurisdiction, despite strong opposition from the islanders.

21 **Indonesia:** the Golkar party announced that it had nominated Gen. Wiranto as its presidential candidate for the July election.

21 **Philippines:** the Government announced a break in ongoing peace talks with the NPA owing to the forthcoming elections; talks were expected to resume later in the following month.

22 **Democratic People's Republic of Korea:** at least 161 people were believed to have been killed and a further 1,300 injured by an explosion on a railway line at Ryongchon, only hours after North Korean leader Kim Jong Il had travelled through Ryongchon, returning from a visit to Beijing.

25 **Indonesia:** violence broke out in Ambon following a rally by a predominantly Christian separatist group on the island; the unrest was the worst to have occurred in the area since the signing of a peace agreement in 2002.

26 **Hong Kong:** the Standing Committee of China's National People's Congress issued a ruling stating that direct election of Hong Kong's Chief Executive and

Legislative Council by universal suffrage would not be permitted in elections of 2007 and 2008.

MAY 2004

5 **Indonesia:** the official results of the legislative elections held in the previous month were finally announced; the Golkar party secured 128 of the 550 seats in the House of Representatives, followed by President Megawati's PDI—P with 109.

6 **Indonesia:** President Megawati Sukarnoputri announced that she had chosen Ahmad Hasyim, Chairman of the Nahdlatul Ulama, the country's largest Muslim organization, as her vice-presidential candidate for the forthcoming elections.

6 **New Zealand:** Prime Minister Helen Clark survived a vote of 'no confidence', taken following her dismissal of the Associate Minister of Maori Affairs, Tariana Turia, in the previous month, after Turia had stated that she would vote against the Government's Seabed and Foreshore Bill, which she claimed contravened the rights of indigenous Maori; Turia subsequently resigned from the House of Representatives.

9 **New Caledonia:** legislative elections were held, the results demonstrating an increase in support for the pro-independence movements.

10 **Australia:** the country's first national party for Aboriginal people, Your Voice, was launched; the organization was created in response to the Government's efforts to abolish the Aboriginal and Torres Strait Islander Commission.

10 **Indonesia/Timor-Leste:** a UN-supported tribunal in Timor-Leste issued an arrest warrant for the Indonesian former leader of the armed forces in East Timor, Gen. Wiranto; Wiranto had been formally indicted in February 2003 for human rights abuses relating to the violence that had taken place in the former territory in 1999 but the Indonesian Government had refused to extradite him to face trial.

10 **Japan:** Naoto Kan resigned as leader of the Democratic Party of Japan (DPJ) after admitting that he had failed to pay mandatory pension contributions.

10 **Philippines:** elections were held to select the country's President, Vice-President, 12 members of the Senate, 212 members of the House of Representatives and more than 17,000 local officials.

10 **Vanuatu:** the Supreme Court ruled that President Alfred Maseng Nalo was to be removed from office, following revelations of his criminal convictions.

11 **New Caledonia:** following the announcement of the results of the elections held two days previously, 54 of the 76 elected representatives were chosen in turn to sit in the New Caledonian Congress; Jacques Lafleur, leader of the conservative Rassemblement-UMP, was subsequently replaced as President of the southern province (his party having suffered its first ever defeat), as was the President of the Loyalty Islands.

14 **Republic of Korea:** South Korea's Constitutional Court reinstated President Roh Moo-Hyun as head of state, rescinding the National Assembly's impeachment of the President in March.

17 **Myanmar:** the SPDC convened the National Convention for the first time since 1996 in order to begin drawing up a new constitution; however, the Convention was not attended by Myanmar's main opposition party, the NLD, owing to the SPDC's continued failure to free its leader, Aung San Suu Kyi, from house arrest.

18 **Myanmar:** the USA announced that it would extend economic sanctions against the country for another year, owing to the threat that it believed the country still posed to US interests.

18 **Papua New Guinea:** the Prime Minister, Sir Michael Somare, announced a cabinet reorganization, dismissing all People's National Congress ministers from the coalition Government.

19 **Indonesia:** the Government downgraded the status of martial law in the province of Aceh, first imposed one year previously, to a state of emergency.

20 **Taiwan:** following his recent re-election, on the day of his inauguration President Chen Shui-bian announced the composition of the new Executive Yuan, retaining Premier Yu Shyi-kun.

21 **Nauru:** David Adeang was elected to the position of Speaker, ending some six weeks of political impasse following the resignation of the previous Speaker, during which time Parliament had been unable to be convened.

23 **French Polynesia:** at elections to the newly expanded French Polynesian Assembly, the ruling Tahoeraa Huiraatira party won 28 of the 57 seats in the legislature but lost its overall majority; a coalition led by the pro-independence Tavini Huiraatira won 27 seats; the remaining two seats were taken by pro-autonomy candidates.

24 **Republic of Korea:** Prime Minister Goh Kun, who had been acting head of state during the impeachment of President Roh, resigned from office.

25 **Indonesia:** a bomb exploded on the island of Ambon, killing one person.

27 **Papua New Guinea:** Sir Paulias Matane was narrowly elected Governor-General after four rounds of voting; two previous elections to the post had been nullified by the Supreme Court on procedural grounds.

31 **Papua New Guinea:** the Government declared the position of Governor-General to be vacant pending the resolution of an injunction against the newly elected Matane by Sir Pato Kakaraya, a candidate previously elected to the post.

JUNE 2004

1 **Australia:** Jack Roche, a British-born Australian, was convicted under new anti-terrorism laws of plotting to attack the Israeli embassy in Canberra; he was sentenced to nine years' imprisonment.

1 **Viet Nam:** the National Assembly approved the dismissal of the Minister of Agriculture and Rural Development, Le Huy Ngo, over his suspected involvement in an embezzlement case; he was succeeded by Cao Duc Phat in an acting capacity.

2 **Australia:** the House of Representatives endorsed controversial legislation introduced by the Government to abolish the Aboriginal and Torres Strait Islander Commission; the bill was then referred to the Senate, which two weeks later launched an inquiry into the proposal.

4 **People's Republic of China:** public recognition of the 15th anniversary of the violent suppression of student protests in 1989 was prohibited.

4 **Myanmar/Thailand:** Prime Minister Gen. Khin Nyunt of Myanmar visited Thailand for the first time since assuming office in the previous year and discussed various economic, development and border issues with his Thai counterpart, Thaksin Shinawatra.

15 **French Polynesia:** the French Polynesian Assembly elected Oscar Temaru, leader of the pro-independence Tavini Huiraatira, as the country's President; he announced the composition of the new Government on the following day.

15 **Indonesia:** three leaders of the rebel Gerakan Aceh Merdeka (GAM) were arrested in Sweden.

15 **Democratic People's Republic of Korea/Republic of Korea:** propaganda broadcasts by North and South Korea across their common border were silenced, as loudspeakers were switched off in an attempt to reduce tension between the two sides.

15 **Solomon Islands:** in the 50-member National Parliament Nathaniel Waena was elected Governor-General, securing 27 votes.

20 **Philippines:** the results of the 10 May presidential election were announced; the incumbent President Gloria Macapagal Arroyo was declared the winner, receiving 39.99% of the votes cast, followed by the actor Fernando Poe, Jr, with 36.51%.

22 **Nauru:** President Rene Harris was ousted in a vote of 'no confidence' and replaced by Ludwig Scotty.

23–26 **Democratic People's Republic of Korea:** a third round of six-party talks on North Korea's nuclear weapons programme took place in Beijing, but little progress was achieved.

25 **Australia:** the Council of Australian Governments signed an 'historic' 10-year agreement to preserve the country's water resources.

27 **Mongolia:** elections to the Great Khural took place; preliminary results indicated that the Mongolian People's Revolutionary Party (MPRP) had secured 36 of the 76 seats in the legislature and the Motherland Democracy coalition (incorporating the Mongolian Democratic New Socialist Party, the Democratic Party and the Civil Courage Republican Party) had won 34 seats; two seats remained in dispute following allegations of malpractice.

29 **Republic of Korea:** the National Assembly approved President Roh's nomination of Lee Hae-Chan as the country's new Prime Minister, replacing Goh Kun who had resigned at the end of May.

29 **Papua New Guinea:** Sir Paulias Matane was sworn in as Governor-General, the ceremony having been delayed by a legal injunction.

29 **Solomon Islands:** leader of the opposition, John Garo, crossed the floor to join the Government as Minister of State, assisting the Prime Minister.

29 **New Caledonia:** following a protracted dispute, Marie-Noëlle Thémereau of the anti-independence party, L'Avenir Ensemble, was chosen as the country's President, with the pro-independence Déwé Gorodey being selected as her Vice-President.

30 **Cambodia:** following an 11-month deadlock, the CPP signed a power-sharing agreement with FUNCINPEC enabling the two parties to form a coalition government.

30 **Philippines:** President Macapagal Arroyo was inaugurated for a second term in office following her victory in the 10 May presidential election; Noli de Castro was sworn in as Vice-President.

JULY 2004

5 **Indonesia:** five candidates contested the first round of voting in the country's first direct presidential election.

6 **Vanuatu:** a general election took place, at which 25 new MPs were elected to the 52-member Parliament.

7 **Hong Kong:** the Secretary for Health, Welfare and Food, Yeoh Eng-kiong, resigned following publication of a report criticizing the Government's handling of the outbreak of Severe Acute Respiratory Syndrome (SARS) in 2003.

7 **New Zealand:** a new political organization, the Maori Party, was inaugurated, with Tariana Turia as one of the co-leaders.

10 **New Zealand:** a by-election, necessitated by the resignation of Tariana Turia from the House of Representatives in May, was held for the seat of Te Tai Hauauru on North Island; Turia secured victory in the election.

11 **Japan:** in an election to the House of Councillors, in which 121 of the 242 seats were contested, Prime Minister Koizumi's Liberal Democratic Party (LDP) won only 49 seats in the upper house, whereas the opposition Democratic Party of Japan (DPJ) won 50 seats; the result was regarded as a defeat for the LDP.

12 **Philippines:** the court trying former President Estrada on corruption charges acquitted him of one of the charges against him, that of opening a bank account in a different name.

14 **Australia:** Prime Minister John Howard effected a minor government reorganization.

15 **Cambodia:** the National Assembly sat for the first time since an agreement was reached to end the 11-month political deadlock that had followed the general election of July 2003 and formally ratified the new 207-member Government; Hun Sen had been reappointed as Prime Minister by the King on the previous day; the Sam Rainsy Party boycotted the session.

19 **Norfolk Island:** Deputy Chief Minister Ivens Buffett was shot dead in his office; the killing represented the second murder in the island's recent history.

19 **Philippines:** Philippine troops left Iraq after President Macapagal Arroyo agreed to their withdrawal in order to comply with the demands of militants who had taken a Filipino civilian hostage in the country; the decision jeopardized relations between the Philippines and the USA.

23 **Philippines:** the defeated presidential candidate Fernando Poe, Jr, filed a petition with the Supreme Court demanding a recount of 60% of the votes cast in the May presidential election, owing to allegations of widespread electoral fraud on the part of the victor, Gloria Macapagal Arroyo.

26 **Indonesia:** the results of the presidential election held earlier in the month were announced; Gen. (retd) Susilo Bambang Yudhoyono secured victory in the poll, receiving 33.5% of the votes cast, while the incumbent Megawati Sukarnoputri came second, with 26.6%; as neither candidate secured more than 50% of the votes, the two were to contest a second round of voting on 20 September.

26 **Nauru:** Australian officials arrived in Nauru to assume control of the country's finances.

29 **Vanuatu:** Serge Vohor was elected Prime Minister, defeating Ham Lini by 28 parliamentary votes to 24.

29 **Viet Nam:** the dissident Nguyen Dan Que was sentenced to a 30-month prison term, having been convicted of using the internet to criticize the Government.

AUGUST 2004

6 **Fiji:** Vice-President Ratu Jope Seniloli received a four-year prison sentence, having been found guilty of treason on account of his involvement in the coup of May 2000.

6 **Indonesia/Timor-Leste:** a court in Jakarta overturned the convictions of four men, including former regional commander Maj.-Gen. Adam Damiri, found guilty of human rights abuses in East Timor in 1999.

9 **Australia:** Richard Butler resigned as the Governor of Tasmania, following threats from the state opposition to withdraw its support for him; Chief Justice William Cox was designated acting Governor and was expected to serve until the appointment of a permanent replacement.

12 **Indonesia:** the head of the Kopassus special forces regiment, Maj.-Gen. Sriyanto Muntrasan, was cleared by a court in Jakarta of human rights violations in connection with the shooting of several human rights activists at Tanjung Priok in 1984.

12 **Singapore:** former Deputy Prime Minister Lee Hsien Loong was sworn in as Prime Minister following the resignation of Goh Chok Tong; Goh remained in the Cabinet as Senior Minister, while former Senior Minister Lee Kuan Yew was appointed to the newly created post of Minister Mentor; the new Prime Minister retained the finance portfolio.

13 **Philippines:** a total of 17 members of Abu Sayyaf were sentenced to death, having been found guilty of the kidnapping of four people on the island of Basilan three years previously.

16 **Vanuatu:** Kalkot Mataskelekele was appointed President, having been elected from among 16 candidates by the country's 58-member electoral college.

18 **Philippines:** a cabinet reorganization took place, in which former Executive Secretary Alberto G. Romulo succeeded Delia D. Albert as Secretary of Foreign Affairs.

19 **Vanuatu:** the resignation of Maxime Carlot Korman as Deputy Prime Minister, less than three weeks after his appointment and his party's decision to leave the Government and join the opposition, prompted the formation of a new Government of National Unity and a reorganization of cabinet portfolios.

20 **Mongolia:** Tsakhiagiin Elbegdorj of the Motherland Democracy coalition was appointed Prime Minister; outgoing Prime Minister Nambaryn Enkhbayar of the Mongolian People's Revolutionary Party (MPRP) had been appointed Chairman of the Great Khural a few days previously.

25 **Tonga:** Prime Minister 'Ulukalala Lavaka Ata announced a ministerial reorganization that was to include the reallocation of some of his numerous portfolios to other cabinet members.

25 **Tuvalu:** Prime Minister Saufatu Sopoanga was ousted in a parliamentary vote of 'no confidence' after a member of the Government and the Speaker crossed the floor to vote with the opposition.

26 **Thailand:** a bomb exploded in a market place in Sukhirin, Narathiwat province, killing one person, one day before Prime Minister Thaksin Shinawatra was due to visit the town.

28 **Malaysia:** the Barisan Nasional succeeded in retaining its seat at a state assembly by-election in the Kuala Berang constituency in Terengganu.

29 **Macao:** Chief Executive Edmund Ho Hau Wah was re-elected by an election committee for a second term, in advance of the expiry of his first term of office in December.

29 **Thailand:** at the gubernatorial election in Bangkok, Apirak Kosayodhin, the candidate of the Democrat Party (DP), secured victory; the ruling Thai Rak Thai party had not presented a candidate.

31 **Cambodia:** the National Assembly finally ratified the country's entry into the World Trade Organization, almost a year after its membership had been approved; the bill was then to be referred to the Senate for approval.

SEPTEMBER 2004

2 **Malaysia:** former Deputy Prime Minister Anwar Ibrahim was released from prison when his appeal against his conviction for sodomy was upheld; he had first been imprisoned in 1998.

7 **Cook Islands:** at a general election the Democratic Party secured 14 of the 24 seats in Parliament and incumbent Prime Minister Robert Woonton retained his seat by only four votes; in a concurrent referendum more than 80% of voters registered their approval for the shortening of the parliamentary term from five years to four.

9 **Indonesia:** a bomb exploded outside the Australian embassy in Jakarta, killing at least nine people.

12 **Hong Kong:** at elections to the Legislative Council in which a record number of voters participated, parties supportive of the Government of mainland China secured 34 seats (12 of which were directly elected); 'pro-democracy' parties failed to make significant overall advances, taking a total of 25 seats (of which 18 were directly elected).

16 **Taiwan:** for the 12th consecutive year, the General Committee of the UN General Assembly rejected a proposal for Taiwan's participation in the UN.

18 **Myanmar:** a minor cabinet reorganization took place, in which Maj.-Gen. Nyan Win was appointed Minister of Foreign Affairs following the retirement of U Win Aung.

19 **People's Republic of China:** former President Jiang Zemin relinquished his position as Chairman of the Central Military Commission at the fourth plenary session of the 16th Central Committee of the Chinese Communist Party (CCP); the post was ceded to President Hu Jintao.

20 **Indonesia:** the second round of voting took place in the country's first direct presidential election; Susilo Bambang Yudhoyono defeated the incumbent, Megawati Sukarnoputri, by a significant margin.

27 **Japan:** in a major reallocation of cabinet portfolios, Prime Minister Junichiro Koizumi replaced 11 of his 18 ministers; an important new portfolio for privatization of the postal services was created, which was assigned to the Minister of Economic and Fiscal Policy Heizo Takenaka.

28 **Mongolia:** three months after legislative elections in June, a new Government was appointed under the leadership of incoming Prime Minister Tsakhiagiin Elbegdorj of the Motherland Democracy coalition; following an agreement on the terms of office of the Prime Minister and the newly created post of Deputy Prime Minister; the posts would be shared between the two main parliamentary groups, the Motherland Democracy coalition and the Mongolian People's Revolutionary Party (MPRP); it was expected that Prime Minister Elbegdorj would be required to cede his position to an MPRP Prime Minister after two years.

29 **Brunei:** the Sultan gave his assent to a constitutional amendment that would permit the country to hold its first legislative elections since 1962; the country's Parliament had reconvened four days previously, having been abolished in 1984.

30 **Fiji:** the Minister for National Reconciliation and Multi-Ethnic Affairs, George Shiu Raj, resigned, following allegations that he had overspent government allowances during an official trip to India in 2003.

30 **Nauru:** President Ludwig Scotty dissolved Parliament and declared a state of emergency, following the creation of a deadlock in Parliament caused by the suspension of a cabinet minister which removed the Government's one-seat majority.

OCTOBER 2004

1 **People's Republic of China:** for the first time, the People's Republic of China was represented at a meeting of the Group of Seven (G-7) industrialized countries in Washington, DC.

4 **Cambodia:** the National Assembly finally ratified legislation to establish a UN-supported tribunal to try former members of the Khmer Rouges.

4 **Thailand:** the Government announced the commencement of a second 'war on drugs', which was to last until September 2005.

6 **Thailand:** a minor cabinet reorganization took place following the dismissal of the Minister of Defence, Chettha Thanajaro, and the Minister of Agriculture and Co-operatives, Somsak Thepsuthin; they were replaced by Gen. Sanphan Boonyanan and Wan Muhamad Nor Matha, respectively.

8–9 **ASEM/Myanmar/Viet Nam:** representatives of 39 countries attended the biennial Asia-Europe Meeting; delegates urged Myanmar to pursue a course of democratic reform.

8 **French Polynesia:** President Oscar Temaru was removed from office, following his defeat in a parliamentary motion of censure.

9 **Australia:** at the general election, the Liberal-National coalition increased its majority in both the House of Representatives and the Senate; Prime Minister John Howard was returned to office for a fourth consecutive term.

11 **Tuvalu:** Maatia Toafa, who had been serving as acting Prime Minister, was elected to the position, defeating

the previous Prime Minister, Saufatu Sopoanga, by eight parliamentary votes to seven.

14 **Cambodia:** Norodom Sihamoni was appointed King by the Royal Council of the Throne, following the abdication of his father, Norodom Sihanouk, in the previous week.

16 **French Polynesia:** in the territory's largest-ever demonstration, 15,000 people took part in a rally in support of ousted President Oscar Temaru; the protesters demanded the dissolution of the French Polynesian Assembly and the holding of fresh elections.

19 **Myanmar:** it was reported that Prime Minister Gen. Khin Nyunt had been removed from his position and placed under house arrest, owing to his alleged involvement in corruption; former First Secretary of the SPDC Lt-Gen. Soe Win was subsequently appointed as his replacement.

20 **Indonesia:** Gen. (retd) Susilo Bambang Yudhoyono was inaugurated as President following his victory over Megawati Sukarnoputri in a second round of voting in the previous month; his Vice-President, Yusuf Kalla, was sworn in on the same day; Yudhoyono announced the composition of his Cabinet on the following day.

20 **Norfolk Island:** at a general election, the territory's Chief Minister, Geoffrey Gardner, regained his seat in the nine-member Legislative Assembly, as did Speaker David Buffett; two long-serving members were defeated.

22 **French Polynesia:** in a parliamentary session boycotted by members of the pro-independence coalition, the conservative Gaston Flosse was narrowly elected to replace Oscar Temaru as President of French Polynesia.

23 **Nauru:** at a general election held to end the parliamentary stalemate arising from reformist budgetary proposals, voters expressed their support for the Government of President Ludwig Scotty; seven of the nine opposition members, including former Presidents Kinza Clodumar and Derog Gioura, lost their seats in the 18-member Parliament.

25 **Pitcairn Islands:** the territory's Mayor, Steve Christian, was among six men found guilty of various charges of sexual assault; four of the convicted men subsequently received prison sentences of between two and six years (the Mayor receiving a three-year term), leading some to question the future viability of the 47-member settlement of Pitcairn.

25 **Thailand:** seven people died following clashes with security forces during a protest held outside a police station in the southern province of Narathiwat; more than 1,000 arrests were made; 78 Muslim protesters were subsequently reported to have died, many apparently having suffocated or been crushed inside the trucks used to transport them.

26 **French Polynesia:** amid continuing controversy regarding the validity of his election, President Gaston Flosse announced the composition of his Council of Ministers.

27 **Nauru:** following his success at the recent legislative election, President Ludwig Scotty named his new Cabinet.

28 **Thailand:** two people were killed and almost 20 were injured in a bomb attack in the province of Narathiwat, amid international concern and growing unrest following the deaths of more than 80 protesters earlier in the week, for which Muslim separatists had vowed revenge; further explosions followed.

PART ONE
General Survey

HUMAN RIGHTS IN THE ASIA-PACIFIC REGION

KENNETH CHRISTIE

BACKGROUND

This essay will examine the general and, secondly, the more specific state of human rights in the Asia-Pacific region. To help understand the subject in this vast area, the essay is divided into three sections. The first part provides a general overview of the problems of, and views on, human rights, even though it will be seen that there are not only huge differences in the ways that different parts of the region perceive human rights but also variations among individual states. Secondly, some specific information on the state of human rights in selected countries in the region (particularly on the most pressing and current problems) is given. Lastly, a summary of the prospects and challenges for human rights throughout the Asia-Pacific region will be provided. Overall, an attempt will be made to demonstrate that there is no uniform, homogeneous version of human rights here. The area will always contain numerous, multi-faceted versions of what human rights constitute, how they are implemented and what occurs in practice.

The Context

One of the most dramatic changes affecting human rights agendas, problems and prospects has taken place in the aftermath of the attacks on the USA on 11 September 2001. The subsequent reaction of the US Government against 'terrorism' (the so-called war on terror) has generated a crisis for human rights in many parts of Asia, as neo-conservative, authoritarian regimes seek to pursue long-standing agendas against their domestic oppositions. The notion that 'you're either for us or against us', heavily publicized by US President George W. Bush following the 11 September attacks, has enabled states in Asia that previously were seen to be improving their human rights records to suppress dissent by 'securitizing' the argument. That is, in view of terrorism and its functionaries around the world, the issue of domestic security has acquired a new significance, and if this means emphasis on the restriction and even abolition of civil liberties, this has become the new agenda of many governments in the region, which have intensified repressive internal security measures, some of these having been inherited from the former colonial powers. This is particularly so in countries such as the People's Republic of China, Indonesia, Singapore, Malaysia and others. These 'hard' and 'soft' authoritarian governments have seized much of the political context of the events of 11 September 2001 as a means of increasing the suppression of internal dissent against their rule and a way of bolstering their legitimacy. This context and environment follow the disappearance of the 'Asian values' debate as a result of the Asian financial crisis of 1997/98. This debate is now redundant but the latest trend reflects a far more anxiety-driven notion that human rights can be completely discarded owing to so-called 'internal' and 'external' threats. This, therefore, is one of the basic themes running throughout this examination of human rights in the region.

OVERVIEW

Theoretical Context

A number of observations help to establish a context for the discussion of human rights in the Asia-Pacific region. First, it could be argued that there is no history in Asia of 'human rights' as understood by the inventors of the term. Human rights, in their original Western conception (to which Asians have been obliged to react) not only apply to all groups of people in all societies, regardless of socio-economic status, but are also conceived of as fundamental entitlements that 'trump' all other considerations that may arise from an individual's relationship to social networks or to the state. In Asia and other societies in the developing world, however, there is no tradition or political history of such entitlements. People are believed to have 'basic duties', not 'basic rights', perhaps because such societies did not follow the same liberal trajectory of Western modernization. These duties in part arise from a person's status or group affiliation. From this standpoint, a ruler may have a duty to rule justly and the rich a duty to give to the poor, but this does not imply that the people have a right to be ruled justly or that the poor have a right to receive hand-outs. For many Asian governments, the priority of second-generation (economic and social) rights has always featured heavily in their agendas, much more so than the first-generation civil and political rights.

A 'Growth Industry'

The subject of human rights in Asia became a 'growth industry' in academia from the late 1980s, in part because of the region's dynamic economic expansion and the much-heralded 'East Asian miracle' and that particular miracle's spectacular failure in 1997. Regardless of this, over time, we have witnessed a substantial gap between the levels of economic growth in the region and the dual processes of democratization and human rights; and this is despite a standard hypothesis in political science that an increase in economic growth leads to political liberalization. In fact, the reverse seems to be the case for some states in the region: the richer they become, the more they have to lose; and the more they have to lose then the greater the suppression of essential, individual rights. Some have argued that the region shows itself to be recalcitrant to the ideas and practice of democracy and human rights agendas. While the paths of democratization may appear universal to Western political scientists, something very different was taking place in the Asia-Pacific region. In terms of the Asian values debate used to justify the anti-human rights position of 'soft' authoritarian rule, this disintegrated when the economic crisis came in 1997. These ideas were discarded extraordinarily quickly, as it appeared that the much-heralded Asia-Pacific century might not materialize after all. The whole point of the Asian values movement was to maintain the high rates of economic growth to which the region had become accustomed, without developing the social deviation that had followed affluence in the West; or, to put it another way, to enjoy the positive aspects of modernization without the problems. The neo-conservative debate ceased with the devastating and debilitating effects of the Asian financial crisis in 1997, as the less appealing aspects of modernization, such as unemployment, crime, decline in social welfare and many others, came to the region. This caused erstwhile *nouveaux riches* East Asian states to abandon the moral high ground in the values debate.

Human Rights and Globalization

Although the size and diversity of the region induce caution in making generalizations, it seems defensible to assert that Asian authoritarian states have tended to view human rights as a threat to national security, at least in comparison with the typical outlook of Western states. This is precisely because human rights can be viewed as part of that liberal 'package' of democracy and individualism, that is anathema to them.

In July 1997 an economic and financial crisis emerged in South-East Asia. This broadened and deepened in scope and nature thereafter, affecting the global economy and, of course, neighbouring countries in East Asia and Australasia. The effects of the crisis have been devastating. The argument for community over individual loses much of its credence in the region if economies in the West were growing at a higher rate than those in East Asia; and the fact that these countries were facing economic crisis can be seen in the movement towards political liberalization, as consumers who had been denied continued economic growth and success looked for other political benefits to claim from the paternalistic state. Arguing for an 'Asian Way' and an Asian form of democracy based on illiberal, undemocratic values was a mainstay of South-East Asian politics as long as these countries continued to benefit from the momentum generated by the 'miracle economies'. This, of course, had certain implications for politics in the region. East Asian states have been more inclined to believe, as the West once did, that the manner in which a government treats its own citizens is a matter of sovereign prerogative and should not be subject to external (i.e. 'foreign') interference. Chinese officials speak for many others in the region when they contend that human rights pressure, whether it is from domestic sources or from overseas, is an affront to the authority of the ruling party and is unwelcome. There are tremendous variations on notions of human rights and levels of democratization throughout this vast region. China, for example, is clearly quite different from Thailand, and Viet Nam from Taiwan.

Thus, the promotion of a security agenda in the aftermath of the attacks of September 2001, and the response to those attacks, could not have come at a worse time for human rights defenders in the East Asian region if the belief that human rights are vital for a country's well-being is to be sustained. Asian values may have been discarded but the new 'war on terror' was about to witness the capacity of the state in East Asia once again to sideline and marginalize individual human rights. This time the antagonists were named 'security' and 'civil liberties'. Although Western governments and Asian activists long ago rejected the Asian values argument, in the aftermath of 11 September some Western governments expressed sympathy for, and even encouraged prioritizing security over, human rights. For Asians who have struggled for many years abroad against excessive national security laws and anti-crime campaigns aimed at quelling political dissent, this has been a crushing blow. Few would argue about the importance of security, but the unanswered question in Asia was: at what price? The constant focus on terrorism and security overshadowed human rights developments that in other times might have made headlines.

It is clear that human rights in the region have undergone change in terms of the agendas and associated views. From the neo-conservative emphasis on different values and their political implications in the early 1990s to the hegemonic imposition of the 'war on terror' in the new millennium, it seems clear that these issues will remain contentious for some time to come. However, it does appear that there has been a general degradation of human rights in favour of state security

COUNTRY ANALYSIS

China

The People's Republic of China is constantly cited as one of the worst abusers of human rights by Western Governments and non-governmental organizations (NGOs). One of the reasons for this is its size and the structure of the communist Government; the main concern in China is how to balance any kind of individual rights with the overwhelming need for economic and social stability, and China has always focused on second-generation economic rights at the expense of political and civil liberties. Recently, China has been criticized over two major issues: that of religious freedom and freedom of speech, and that of Tibet (Xizang). The Chinese Government has long conducted campaigns against what it perceives as 'heretical' organizations, but the most recent development, regarded as one of the most dangerous in China, is the emergence of a spiritual movement called Falun Gong, a group that has been banned since 1999 and which the Government sees as a threat to social and political stability. Tens of thousands of Falun Gong members have been arrested, detained and allegedly mistreated in custody by the police for short and long periods. Their persecution by the Chinese authorities has attracted world-wide attention. By early 2003 the human rights group Amnesty International, for instance, had recorded nearly 1,600 cases of detention, arrest or sentencing of members of this group since June 1999, but the real figures appeared to be much higher.

Second, the authorities also continued to restrict and repress minorities within their borders. Tibet has always been the main issue, but the global 'war on terror' announced by the Bush Administration allowed the Chinese Government (like many other authoritarian governments) to repress its own citizenry. The Chinese authorities have deliberately obscured the differences between terrorism and demands for independence by ethnic Uygurs. In Xinjiang, for example, the regime has tightened many restrictions, alleging that the Uygurs were in collaboration with al-Qa'ida, the Islamist fundamentalist organization responsible for the attacks of 11 September 2001. The Xinjiang Uygur Autonomous Region constitutes nearly 17% of the territory of the People's Republic of China, and Amnesty International has reported gross violations of the rights of the Uygurs, who represent the majority ethnic group among the predominantly Muslim population. In China this led to the development of a campaign against Muslims and ethnic separatists. These violations include executions, arbitrary detention and arrest, imprisonment and detention without trial and unfair trials. Part of the problem clearly relates to the lack of transparency in China. In early 2003, for instance, the widespread occurrence of Severe Acute Respiratory Syndrome (SARS), a previously unknown pneumonia-like illness, could clearly be related to the fact that the Chinese Government had deliberately attempted to conceal the extent of the outbreak, in an effort to avoid negative publicity arising from the identification of China as the area where the virus had originated. Saving face is clearly more important to the Chinese regime than the violation of rights of free speech and information. The struggle to control information flows relating to this and other subjects, such as the HIV/AIDS epidemic, has meant that human rights are severely compromised.

With regard to Tibet, there appeared to be some liberalization in 2002–03. A 'private' visit to Tibet by representatives of the Dalai Lama, the region's spiritual leader, was organized, and seven high-profile Tibetan prisoners were released before their terms expired. Cynics argued that this was simply meant to prevent international criticism and to open up new avenues for investment. In their everyday lives, little changed for Tibetans, and the Chinese authorities continued to arrest 'political' dissidents and to place restrictions on religious worship.

China, therefore, remains a serious violator of rights. The number of executions is staggering. In April 2004 Amnesty International reported that at least 726 people had been executed in China in the previous year, but as death sentences and executions are officially kept secret, the number could be much higher.

Myanmar

The situation in Myanmar (formerly Burma) has been a consistent priority in terms of human rights agendas, but little actual progress has been witnessed there. Since 1988 the country has been ruled by a ruthless dictatorship, which initially called itself the Orwellian-sounding State Law and Order Restoration Council (SLORC) but later renamed itself the State Peace and Development Council (SPDC), in a bid to soften the image of the regime. Human rights conditions appeared to have improved slightly in recent years; in 2003 Amnesty International gained access to the country for the first time ever but the renewed detention of opposition leader Aung San Suu Kyi in May 2003 provoked serious concern and the failure to allow her to take part in a National Convention on the Constitution appeared to undermine any attempts, however limited, at reform. There remain the problematic issues of between 1,300 and 1,400 political prisoners detained for all or most of the years since 1988, not to mention the fact that the regime has murdered more than 70 of Suu Kyi's supporters. There is also the difficult question of ethnic minorities (a common theme to most discussions of human rights in the overall region). Ethnic minorities in Myanmar, such as the Karen and the Shan and also Burmese Muslims, are frequent targets of state repression and violations, the most common abuse being their induction into forced labour. Men, women and children are forced systematically to work for the regime, carrying heavy loads for days and weeks on end in order to construct railways, roads, dams and other projects the regime deems necessary for development. Often these labourers endure beatings, torture and widespread malnutrition; many hundreds have died in the process. Child soldiers are prevalent in Burma and there have also been reports of labourers being used as human minesweepers. Moreover, ethnic populations, particularly the Shan, often have to bear the brunt of the rape and sexual violence carried out by the Myanmar military. In some cases women have been held in sexual slavery. This systematic abuse has created thousands of ethnic refugees who have tried to flee to nearby countries such as Thailand and Malaysia, often only to be repatriated by authoritarian governments.

Indonesia

Indonesia is the world's largest Muslim nation and is a country of strategic and political importance to the West. It has only recently begun to progress towards democracy and the respect of human rights. In some ways Indonesia is regarded as a front-line state in the global 'war on terror' and, to some extent, for good reason. In October 2002 two car-bombs exploded in the Balinese resort of Kuta, killing about 200 civilians, including both Indonesian citizens and Australian tourists, among others. In response to Western pressure, the Indonesian Government enacted measures to deal with terrorism, and arrests followed the atrocity. The first trial of the accused concluded in August 2003. Meanwhile, President Megawati issued two presidential decrees to address the issue of terrorism and to investigate the Bali bombings. This has had repercussions in the provinces of Aceh and Papua, which are both engaged in long-standing campaigns for independence. The Indonesian Government has enacted harsh security measures and repressed separatism (easily categorized as part of its 'war on terror'). In addition to the separatist conflicts in the provinces of Aceh and Papua, both of which have witnessed gross human rights violations, religious and ethnic conflict has been precipitated between Christian and Muslim factions in Maluku and Poso.

In Timor-Leste (formerly East Timor), the long-standing issue within Indonesia's human rights record, there remained an overall failure to bring to justice those responsible for human rights abuses, despite efforts by the Indonesian Government to appoint more judges and prosecutors. In the period after September 1999, when the majority of Timorese voted for independence, the Indonesian army embarked upon a campaign of murder, arson and forced expulsion. During this period a total of 1,000–2,000 people were killed and a further 500,000 forced from their homes. The notion of an amnesty for past abuses (based on work conducted in 'truth commission' style) failed to be resolved in any satisfactory manner. The question of justice for the atrocities perpetrated under Indonesian rule (recent and in the larger past) continues to dominate much of the Timorese political scene. On the other hand, mainland human rights commissions, such as the Indonesian National Human Rights Commission (Komisi Nasional Hak Asasi Manusia—Komnas, HAM), became more and more ineffective and marginalized from the process of human rights activities, a pattern reflected in the treatment of human rights advocates across the region.

In Aceh the conflict between the Indonesian military and the Free Aceh Movement (Gerakan Aceh Merdeka—GAM) has intensified. In 2002 an estimated 1,230 people were killed in Aceh. By mid-2003 at least 25,000 military personnel were operating there, despite the provision of a new 'special autonomy' law issued on 1 January 2002. The Indonesian Government declared a state of military emergency in Aceh and a state of martial law in May 2003, after the failure of peace talks. Indonesia has, in effect, launched its largest military operation in Aceh since the late 1970s, in an effort to eradicate the separatist movement there. The situation is uncertain and extremely worrying for human rights in the province. In Papua (formerly Irian Jaya) the security situation has also deteriorated, and civilians and human rights advocates have borne the brunt of this deterioration, facing a heightened level of violence and insecurity, despite the approval of a 'special autonomy' law for Papua in October 2001. US proposals to spend millions of dollars to equip a 'domestic peace-keeping force' have been criticized by human rights advocates as yet another way of repressing human rights and may be reconciled with the global anti-terror agenda. More than 100,000 West Papuans have been killed since Indonesia took control of the province from the Dutch in 1963. Both peripheries continue to be subject to widespread violence, random killings and human rights abuses, corresponding to the general internal security concerns for the Indonesian Government that resulted from the events of 11 September 2001; both would fit into a general pattern of violence and human rights abuse against minorities that exist on the periphery of the state in the new millennium. Furthermore, senior Indonesian officials have used the term 'terrorism' in their rhetoric when referring to domestic groups that are seen to be a threat to the overall unity of Indonesia. With the country regarded as one of the front-line states in the 'war on terror', partly because Indonesia contains the largest Islamic population of any country, these events do not augur well for the positive development of human rights in this South-East Asian archipelago.

Viet Nam

The most recent human rights problems in Viet Nam emanate from the authorities' treatment of ethnic minorities, again a common theme in South-East and East Asia overall. Recent documents show that the Montagnards of Viet Nam are one of the main minorities subjected to such violations. The abuses committed against them include assaults on church leaders by police and other officials, the destruction of churches, the banning of night-time gatherings and of travel outside villages and the large-scale appropriation of farm land by the authorities. This systematic campaign against what are termed 'illegal religious organizations' bears some resemblance to the Chinese authorities' suppression of Falun Gong and other movements within China. Montagnard Christians have been the principal targets of the attacks and persecution, many of them fleeing or going into hiding when asylum proved impossible to obtain in Cambodia. In early 2003, for example, more than 100 Montagnards were forcibly returned to Viet Nam from Cambodia. The status of ethnic minorities and refugees is a recurring theme throughout this article, once again because the situation corresponds to a general renewal of state repression following the attacks of September 2001. Viet Nam has also witnessed a sharp rise in the imposition of the death penalty, mainly for drug and economic-related crimes. A total of 103 people were sentenced to death in 2003, and 64 were executed, many of them in public. Clearly, this is a worrying trend in states such as Viet Nam and Thailand, where there is a determined policy of addressing drug-related crime.

Singapore, Malaysia and Thailand

The major issues in Singapore and Malaysia continue to relate to the lack of freedom of expression. Despite their 'developed' status (Singapore more so than Malaysia), these countries continue to restrict freedom of speech and have effectively curbed and weakened opposition activists by either imprisoning them (as in the case of Malaysia) or issuing defamation suits against opponents (as in the case of Singapore). Malaysia has been accused by human rights groups of failing to give a fair trial to Anwar Ibrahim, the former Deputy Prime Minister, imprisoned in 1998 on politically motivated charges (but released in September 2004). In 2002 members of religious groups in Singapore, such as the Jehovah's Witnesses, were imprisoned for conscientious objection to military service (which is compulsory in Singapore); similarly, members of Falun Gong were arrested for holding a vigil in memory of group members who had died in custody in China. In Malaysia harsh treatment of asylum-seekers and refugees served to reflect the trend on a global scale. From 2001 Malaysia enacted strict laws against illegal immigrants, which prompted an exodus of at least 300,000 workers to Indonesia, while a further 400,000 remained in Malaysia with no financial support. The Malaysian Government at the time agreed to a one-month extension of the departure deadline, but those who remained still faced huge fines, prison sentences in some cases and even caning by the authorities. In May 2003 thousands of Indonesians fled from the brutal conflict in Aceh to Malaysia but did not receive the status or treatment normally given to refugees as a result of conflict and persecution. If they do manage to avoid deportation, many are subject to harsh conditions and poverty. In Thailand in 2003 the Government was waging a 'war' on drugs, which resulted in many extra-judicial killings. More than 2,000 people died during this campaign and more than 50,000 were arrested. In conjunction with this, the Thai Government also resumed executions, by lethal injection. Lastly, there has been widespread violence in southern Thailand, which has a fairly high concentration of the country's Muslim population. There, several Muslim lawyers and human rights defenders have been intimidated and harassed by the security forces.

Australia

Australia is not often viewed as a country that is responsible for gross human rights violations, despite a history in which Aboriginal groups have been treated unfavourably and have formed an impoverished underclass. Australia has often been seen in the light of a Western democracy where human rights are prioritized. Without detailing the treatment of indigenous people in the past, recent indications are that Australia has been affected by the global repercussions of the attacks of 11 September 2001. The reaction against migrants, refugees and asylum-seekers has attracted the attention of human rights groups around the world. Many governments adopted punitive and restrictive measures against these categories of displaced people. In Australia there was a particularly harsh turn of events, as major efforts to prevent and punish 'secondary' movements of asylum-seekers took place. A secondary movement is one in which refugees and asylum-seekers try to move away from the first country they reach (for example, Indonesia) and on to a second (and hopefully, in their view, permanent) country, in this case Australia. Thus, since 2001 the Australian Government has enacted measures such as the interdiction at sea of boats carrying asylum-seekers, diverting them to offshore camps in third countries (such as the Pacific nations of Nauru and Papua New Guinea, and also the Australian territory of Christmas Island), and mandatory detention of persons applying for asylum within Australia, as well as the granting of temporary protection instead of refugee status. Many of these refugees, particularly those from Afghanistan, Iraq and Iran, were subject to repression and human rights abuses within their countries of origin and had little choice but to flee their respective situations and circumstances. Numerous asylum-seekers (including many children) have been detained in very poor conditions for lengthy periods of time; some have attempted to commit suicide. In remote desert detention centres children have been held with adults for up to two years in unsanitary, hot conditions, and there have been reports that some have been sexually assaulted. There have also been reports that children have participated in hunger strikes by stitching their mouths closed.

The option of seeking asylum appears to have been closed. If relocated to a territory like Christmas Island, for instance, the right to apply for protection in Australia is effectively removed. Despite recognition as 'real' refugees under the 1951 UN Convention relating to the Status of Refugees, many of these persons can be denied a secure legal status indefinitely, resulting in a sort of permanent 'Temporary Protection Visa' status. They are, in other words, permanently stateless, hardly an attractive proposition. Australia has been heavily criticized by NGOs for its 'uniqueness' in granting only this temporary status to Afghan refugees.

CONCLUSION

Prospects for Human Rights

The prospects for establishing a solid human rights culture following the September 2001 attacks in the USA appear unclear and problematic in the Asia-Pacific region. The countries of Australasia, notably Australia and New Zealand (the latter not reviewed here), are of course much more firmly rooted in the Western tradition of individual rights. Yet even they have their problems, as evinced by Australia's treatment of refugees arriving from Afghanistan via Indonesia. The US response to the events of September 2001 led to the establishment of a much more unilateral, hard-line stance on terrorism, a position that was to have enormous repercussions for domestic security systems around the world, but one with dramatic implications for local populations in parts of Asia. 'Soft' authoritarian states like Singapore, Malaysia, Indonesia and others could now justify internal repression with reference to the threat of terrorism within their societies. The Philippines could suppress internal dissent with the aid of the US military by fighting the 'war on terror' and had a convenient 'home-grown' Islamist movement on its doorstep, Abu Sayyaf, to justify repressive security measures. China could also justify its repression of Muslim ethnic minorities, in these terms. In other words, human rights were not so much a priority for the USA and the West as fighting the 'war on terror'; therefore, the USA could disregard internal human rights abuses, while states with problematic human rights records no longer needed to acquiesce to demands for rights reforms, with attention so diverted. Indeed, this encouraged many developing states with less than adequate rights records to enhance and 'securitize' their country's reaction to internal dissent.

In some ways it is difficult to discuss the development of human rights in Asia without reference to globalization, which not only played its part in the region's dramatic economic growth patterns but was also a factor in its rapid downfall, with the onset of the Asian economic crisis in mid-1997. Globalization has made national state borders increasingly porous, with the result that economies and populations have become correspondingly more interdependent. Studying the role of multinational corporations in the movement of capital, goods, services and labour around the world has become a small growth industry in itself. One of the many effects of globalization has been the accompanying increase in poverty and in disparities on an economic and social scale; globalization is at once paradoxically an advantage and a disadvantage. It helps to provide economic growth, but much of that growth is economically unbalanced and uneven. Financial crises appear more often and have severe, damaging effects on local populations, causing social disintegration and maldevelopment.

Despite the promotion of liberal market economies and economic development as the remedy for poverty, more often than not the disparities and the tensions between development and individual rights have been accentuated. The 'East Asian miracle' might have been a miracle for the World Bank and for development economists, but it always concealed the underlying disparities and social problems. The financial crisis and the attendant currency turmoil of the late 1990s simply brought these issues into sharp and dramatic focus. There are serious questions involved in globalization: questions of illegal (and legal) migration, labour standards, prostitution, child labour, child prostitution and discrimination against ethnic minorities, to name just a few, and all of these issues have an impact on human rights. However, globalization has also had positive effects in that at least these debates have started to take place, and a consciousness of rights is developing. Indigenous groups have now become more assertive in their claims to political and economic rights. The tensions between economic development, rapid growth and the state of governance is an important one. The long-running conflict in Timor-Leste testifies to this; in part, globalization helped to highlight the case of severe human rights abuses there.

Another debate that dominates the prospects for human rights in the region concerns nation-building. The question remains whether or not authoritarian government is necessary for successful economic and social development. Those who posit this view argue that national development requires discipline, austerity and obedience to measures that may be unpopular in the short term but will yield great dividends in the future. In such circumstances, effectively a national emergency, governments cannot ensure the successful prosecution of their policies without limiting civil liberties such as the freedom of speech and of the press, the right of accused criminals to a fair trial, freedom from physical abuse by the police and so on. As a society becomes wealthier, more orderly and more educated, it is debatable whether civil and political privileges will naturally increase. Yet if the basic goals of development are not achieved, all subsequent aspirations, including human rights, are doomed. It is therefore unfair of the West, where the sense of human rights developed within a particular political and historical context, to demand that poor, politically unstable, underdeveloped countries immediately guarantee as broad a range of individual freedoms as exists in the developed world. The problem with this, in the view of human rights advocates, is the uncertainty over the point at which a society becomes developed enough to embrace human rights; at what stage does it determine that human rights are not expedient and are elevated to a different level in terms of priority? As noted above, the issue of human rights (at least as an individual interpretation of the term) has been downgraded in many of these countries.

In Cambodia the subject of human rights has not appeared on the list of government priorities, as justice has seemed unforthcoming in terms of addressing the legacy of the Khmers Rouges. (In June 2003, however, the Cambodian Government reached a formal agreement with the UN on the establishment of a tribunal to try former Khmer Rouge leaders on charges of genocide.) In Viet Nam ethnic minorities, such as the Montagnards, continue to be suppressed, as they do in China (notably the Uygurs) and in Myanmar (like the Karen). Legitimate ethnic grievances can be disregarded in states that ignore rights and the rule of law. In many parts of Asia the problem of child labour, much sought after in the neo-liberal development model as cheap labour, continues to be an enduring issue, along with persistent encroachments in terms of women's rights. Women in South-East Asia and East Asia have been vulnerable to trafficking. The issue of the 'comfort women' of Korea, who were forced into prostitution by the Japanese army during the Second World War, has never been fully resolved in terms of acknowledgement or compensation.

After September 2001 the global 'war on terror' produced a more difficult and complex world than ever before. Countries in Asia could not be seen to be less than wholly committed in their support of the 'war on terror'. Bombings in Saudi Arabia and Morocco in early 2003, the continuing conflict in Iraq with the resultant tragic loss of life and destruction, have ensured that terrorism, and the challenge of fighting terrorism, will remain a central theme of Western policy. Such global strife will not have positive implications for local human rights. Furthermore, the geopolitical considerations of war in the Middle East, and potential war on the Indian subcontinent, overshadowed much of the previously expressed concern for human rights. There is little doubt that human rights as an issue has become increasingly marginalized. The counter-terrorist measures used by the USA and its 'allies' around the world can definitely be seen to have yielded counter-productive results, as governments often use these measures to perpetuate widespread abuse against people with legitimate grievances. Thus, the Uygurs in China and ethnic groups in the Philippines and Indonesia (among others) can be targeted by governments determined to avert any threat to their country's unity and integrity. More often than not, these efforts are at the expense of people's human rights. Global repression and human rights abuses have increased and will likely assume a more intensive element than previously, with the support of the USA, in a continuing 'war on terror' at the expense of individual rights.

BIBLIOGRAPHY

General

Asia Watch. *Human Rights in the APEC Region*. New York, Asia Watch, 1993.

Bauer, Joanne, and Bell, Daniel A. (Eds). *The East Asian Challenge for Human Rights*. Cambridge, Cambridge University Press, 1999.

Christie, Kenneth. 'Regime Security and Human Rights in Southeast Asia' in *Political Studies (UK)*, 1995, XL 111.

Christie, Kenneth, and Roy, Denny. *The Politics of Human Rights in East Asia*. London, Pluto Press, 2001.

Eide, Asbjorn, and Hagtvet, Bernt (Eds). *Human Rights in Perspective: A Global Assessment*. Nobel Symposium 74, Oxford, Basil Blackwell, 1992.

Espirit, Caesar. *Law and Human Rights in the Development of ASEAN: with Special Reference to the Philippines*. Singapore, Friedrich-Naumann-Stiftung, 1986.

Frank, Thomas M. *Human Rights in Third World Perspective*, 3 vols. New York, Oceana, 1982.

Hsiung, J. *Human Rights in East Asia: A Cultural Perspective*. New York, Paragon, 1985.

Human Rights Watch. Background—Martial Law and Armed Conflict in Aceh and Refugee Flows; internet hrw.org/reports/2004/malaysia0404/2.htm.

Human Rights Watch. *Asia Overview*. Human Rights Watch World Report 2003; internet hrw.org/wr2k3/asia.html.

Oraa, Jaime. *Human Rights in States of Emergency in International Law*. Oxford Monographs in International Law (Ed. Ian Brownlie), Oxford, Clarendon Press, 1992.

Scoble, Harry M., and Wiseberg, Laurie S. *Access to Justice: Human Rights Struggles in Southeast Asia*. London, Zed Books, 1985.

Welch, Claude E., Jr, and Leary, Virginia A. (Eds). *Asian Perspectives on Human Rights*. Boulder, CO, Westview Press, 1990.

Women

Anker, Richard, and Hein, Catherine (Eds). *Sex Inequalities in Urban Employment in the Third World*. New York, St Martin's Press, 1986.

Hazou, Winnie. *The Social and Legal Status of Women: A Global Perspective*. New York, Praeger Press, 1990.

Heyzer, Noeleen. *Working Women in South-East Asia: Development, Subordination, and Emancipation*. Milton Keynes, Open University Press, 1986.

Jahan, Rounaq (Ed.). *Women in Asia* (revised edn). London, Minority Rights Group, 1982.

Kerr, Joana (Ed.). *Ours by Rights: Women's Rights as Human Rights*. London and Ottawa, Zed Books, in association with the North-South Institute, 1993.

Li Yu-ning (Ed.). *Chinese Women: Through Chinese Eyes*. New York, East Gate Books, 1992.

Shiva, Vandana. *Staying Alive: Women, Ecology, and Development*. Atlantic Highlands, NJ, Zed Books, 1988.

Selected Countries

Australia

Amnesty International. *Australia, a Criminal Justice System Weighted against Aboriginal People*. New York, NY, Amnesty International USA, 1993.

Bailey, Peter H. *Human Rights: Australia in an International Context*. Sydney, Butterworths, 1990.

Hocking, Barbara (Ed.). *International Law and Aboriginal Rights*. Sydney, Law Book Company, 1988.

Poynton, Peter. *Aboriginal Australia: Land, Law and Culture*. London, Institute of Race Relations, 1994.

Cambodia

Asia Watch. *Political Control, Human Rights, and the UN Mission in Cambodia*. New York, Asia Watch, Human Rights Watch, 1992.

Duffy, Terence. 'Toward a Culture of Human Rights in Cambodia', in *Human Rights Quarterly*, pp. 82–104, 1994.

Marks, Stephen P. 'Forgetting the Policies and Practices of the Past: Impunity in Cambodia'. The Fletcher Forum of World Affairs, pp. 17–43, 1994.

Pokempner, Dinah. *Cambodia at War*. New York, Human Rights Watch, 1995.

China

Amnesty International. *China: Execution is not a Solution. Amnesty International condemns 46 executions in two days*. ASA 17/05/2002. 04/11/2002; internet www.amnesty.org.

China, Punishment without Crime: Administrative Detention. New York, NY, Amnesty International USA, 1991.

Asia Watch. *Continuing Religious Repression in China*. New York, NY, Asia Watch, 1993.

People's Republic of China: Repression in Tibet, 1987–1992. New York, NY, Asia Watch, 1992.

Punishment Season: Human Rights in China after Martial Law. New York, NY, Asia Watch Committee, 1990.

Cohen, Roberta. *People's Republic of China: the Human Rights Exception*. London, Parliamentary Human Rights Group, 1988.

Davis, Michael C. (Ed.). *Human Rights and Chinese Values: Legal, Philosophical, and Political Perspectives*. Hong Kong and New York, Oxford University Press, 1995.

Edwards, R. Randle, Henkin, Louis, and Nathan, Andrew J. *Human Rights in Contemporary China*. New York, Columbia University Press, 1986.

Human Rights Watch. *Merciless Repression: Human Rights in Tibet*. New York, NY, Human Rights Watch, 1990.

Young, Stephen B. *The Tradition of Human Rights in China and Vietnam*. New Haven, CT, Council on Southeast Asia Studies, Yale Center for International and Area Studies, 1990.

Indonesia

Amnesty International. *Indonesia & East Timor*. New York, Amnesty International, 1994.

Indonesia: 'Shock Therapy': Restoring Order in Aceh, 1989–1993. New York, NY, Amnesty International USA, 1993.

Budiardjo, Carmel. *West Papua: the Obliteration of a People*. Surrey, TAPOL, 1988.

Jones, Sydney. *Injustice, Persecution, Eviction: a Human Rights Update on Indonesia and East Timor*. New York, NY, Asia Watch Committee, Human Rights Watch distributor, 1990.

The Limits of Openness: Human Rights in Indonesia and East Timor. New York, NY, Human Rights Watch, 1994.

Orentlicher, Diane. *Human Rights in Indonesia and East Timor*. New York, NY, Asia Watch Committee, 1989.

Richards, Chris. Bali, Bombs and Human Rights, in *New Internationalist*; internet www.newint.org/features/bali/171002.htm 17 October 2002.

Taylor, J. G. *Indonesia's Forgotten War: The Hidden History of East Timor*. London, Zed Books, 1991.

Malaysia

Eldridge, Philip. 'Human Rights and Democracy in Indonesia and Malaysia' in *Contemporary Southeast Asia*, Vol. 18, No. 3, 1996.

Human Rights Watch Report 1999. *Malaysia: Human Rights Developments*; internet www.hrw.org/hrw/worldreport99/asia/malaysia.html.

Myanmar

Amnesty International. *Myanmar: the climate of fear continues, members of ethnic minorities and political prisoners still targeted*. New York, NY, Amnesty International USA, 1993.

Goldston, James. *Human Rights in Burma (Myanmar)*. New York, NY, Asia Watch Committee, Human Rights Watch, 1990.

Guyon, Rudy. *Violent Repression in Burma: Human Rights and the Global Response*, UCLA Pacific Basin Law Journal, 10, pp. 409–459, 1992.

Smith, Martin. *Burma: Insurgency and the Politics of Ethnicity*. London, Zed Books, 1991.

US Department of State. Fact Sheet, Bureau of Democracy, Human Rights and Labour, *Rape by the Burmese Military in Ethnic Regions,* 17 December 2002.

Singapore

Asia Watch. *Silencing All Critics: Human Rights Violations in Singapore*. New York, Asia Watch, 1990.

Seow, Francis. *To Catch a Tartar*. New Haven, CT, Yale University Press, 1994.

Viet Nam

Amnesty International. *Vietnam: 'Renovation' (doi moi), the Law and Human Rights in the 1980s*. New York, NY, Amnesty International USA, 1990.

Kolko, Gabriel. *Vietnam: Anatomy of a Peace*. London, Routledge, 1997.

Ta Van Tai. *The Vietnamese Tradition of Human Rights*. Berkeley, CA, Institute of East Asian Studies, University of California, 1988.

POPULATION AND HUMAN DEVELOPMENT ISSUES IN THE ASIA-PACIFIC REGION

RONALD SKELDON

BACKGROUND

Diversity appears to characterize the populations of the Asia-Pacific region: diversity in terms of culture, level of economic development, the size of the state and the political system. The most populous country in the world, China, is found in the region, as are some of the world's smallest countries and territories, including the Pacific islands of Niue, Tuvalu and Nauru. Hong Kong and Singapore represent some of the most highly developed economies in the world, but parts of Papua New Guinea incorporate small groups of peoples at quite simple levels of technology. Nevertheless, despite the very real diversity that exists, patterns and trends have emerged that, over the long term, will lead to greater uniformity across the region. Declining mortality and fertility rates and increasing urbanization are producing societies that demographically are becoming more similar. This does not mean that each state inevitably passes through the same sequence of change: states appear to undergo the transition to low fertility and mortality under different conditions and at different rates. However, all appear to be moving towards an increasingly universal demographic pattern that is an integral part of the process of globalization.

The populations of eastern and south-eastern Asia have been characterized by a marked deceleration in their rates of growth over the last 50 years. From average annual rates of population growth of 1.80% in eastern Asia and of 2.08% in south-eastern Asia during the early 1950s, rates had fallen to 0.67% and 1.40% respectively by the early years of the 21st century. That is, population growth rates decreased by some 63% in eastern Asia and by 33% in south-eastern Asia over the period. To put this decline into perspective, the rate of growth for the population of the world as a whole decreased from 1.80% to 1.22%, a decline of 32% over the same period, a fall heavily influenced by the slowing growth of the demographic giant of China. The time that it takes a population to double in size in eastern Asia shifted from 40 years in the early 1950s to around 107 years in the year 2000. Only among most of the small islands of the Pacific and in a very few of the smallest countries in south-eastern Asia did population growth rates remain high at the beginning of the 21st century.

The declining growth rates conceal very real improvements in that basic indicator of human welfare, longevity. In the early 1950s the expectation of life at birth of people in eastern Asia was 42.9 years and that of people in south-eastern Asia was 41.0 years; 50 years later these figures had risen to 72.1 and 66.7 years respectively. A significant difference between male and female survivorship remains: expectation of life at birth for men in eastern Asia in 2000–05 is 69.7 years, compared with 74.7 years for women. In south-eastern Asia, the respective figures are 64.4 and 69.1 years. Infant mortality had decreased by as much as 79% in parts of the region. In terms of fertility, sharp declines have been observed. Whereas women on average were having between five and six children in the early 1950s, they were having fewer than two in eastern Asia and fewer than three in south-eastern Asia by 2000. The Asian region has truly experienced a revolution in its demography, which has clearly been associated with the rapid and sustained economic development of parts of the region since the mid-1950s.

The Pacific islands, too, have experienced great change in the recent past, even if annual population growth rates are generally considerably higher than those in eastern and south-eastern Asia. Mortality rates have declined, but fertility levels in some countries remain high. With the exceptions of Papua New Guinea and Solomon Islands, the small populations can be profoundly affected by migration, and some populations, for example, those of the Cook Islands, Samoa and Tonga, where fertility remains high, record low growth rates as a result of outmigration. The small population of Niue even exhibits marked demographic decline. Thus, the populations of Asia and the Pacific still display considerable variation, although the general direction of change is towards a greater uniformity. That variation is associated with the uneven pattern of economic development across the region, and the broad regional divisions will be briefly described.

ASIAN AND PACIFIC DEMOGRAPHIC TRANSITIONS

Table 1 shows the Asian countries and economies ranked in terms of their gross national product (GNP) at purchasing-power parity (PPP) per caput. A group of wealthy economies with a PPP greater than US $15,000 per caput in 2003 are characterized by a life expectancy at birth of 75 years or more, infant mortality rates (IMRs) of 6 or less per 1,000 live births and, with the single exception of Brunei, below-replacement-level fertility (normally assumed to be 2.1). These economies are located primarily in North-East Asia but also include Singapore and the oil-rich state of Brunei, as well as the two traditional developed countries of Australia and New Zealand. The IMRs of Singapore, Hong Kong and Japan are among the lowest in the world and reflect the well-developed and well-funded health sectors in these economies. Life expectancy for women in Singapore, Hong Kong, Japan and South Korea, as well as for Australia and New Zealand, is very close to, or exceeds, 80 years of age. Fertility in the two settler societies of Australia and New Zealand, while below replacement level, is higher than that of all the other Asian economies in this group except Brunei. This pattern reflects the history of immigration to these countries, with their intake of young families from groups that have higher fertility than the locally born. This is particularly important in New Zealand, with its substantial intake from the Pacific islands.

At the other end of the PPP ranking are those countries with an average per caput income of less than US $2,000 per annum. Among these countries, fertility remains at replacement level or higher, and in the cases of Cambodia and Laos markedly higher. Infant mortality generally ranks among the highest in the region and people can expect to live 20–25 fewer years than in the most advanced economies in the region. Fertility and, to some extent, mortality are, however, lower than one might expect in Mongolia and North Korea given their per caput income, which perhaps shows the relative success of socialist policies in these particular countries in extending health services to their populations. This is not the case with Laos, Cambodia and Myanmar, where high levels of infant mortality and relatively low expectations of life at birth are still to be found.

In between the wealthiest and poorest groups in the Asian region are the larger countries of south-eastern Asia, namely Malaysia, Thailand, the Philippines, Indonesia and Viet Nam, together with China in eastern Asia. In common with the overall trend in Asia, mortality generally declines with per caput income, but fertility among the countries of this group is more variable. It is lower than might be expected given levels of income in Thailand, China and, to some extent, Indonesia, but even where it remains relatively high, as in Malaysia and the Philippines, considerable progress has been made in lowering the number of births per woman in recent decades. Infant mortality is lower, and consequently expectation of life at birth higher, than might be expected in Malaysia, where the indicators are close to those of countries where average per caput income is considerably higher.

Table 1. North-East and South-East Asian Economies: Basic Demographic and Development Data

	Population at mid-2004 ('000)	Annual growth rate (%)*	Per caput GDP, (PPP) in US $, 2003	Expectation of life, years at birth, 2002	Infant mortality rate, 2004	Total fertility rate, 2004 (children per woman)	Contra-ceptive prevalence —modern methods (%)	Urban population, 2004 (%)	Urban growth rate, 2004 (%)
Australia	20,122	1.2†	28,900	80.4	5†	1.8†	72.2†	92	1.3
Hong Kong	6,878	0.9‡	28,700	n.a.	2‡	1.0‡	79.7‡	100	0.9
Japan	127,778	0.1‡	28,000	81.9	3‡	1.3‡	51.0‡	79	0.3
Singapore	4,261	1.7‡	23,700	79.6	3‡	1.4‡	53.0‡	100	1.8
Taiwan	22,640	0.5	23,400	76.3	6	1.2§	n.a.	95	1.3
New Zealand	4,061	1.3‡	21,600	78.9	6‡	1.9‡	72.0	86	0.8
Macao	452	1.1‡	19,400	79.0	4‡	1.1	n.a.	99	0.8
Brunei	366	2.2	18,600‡	76.1	6	2.4	n.a.	74	2.4
South Korea	48,199	0.6‡	17,700	75.5	5‡	1.2‡	66.9‖	84	1.2
Malaysia	25,493	1.8§	9,000	72.0	10	3.0§	29.8§	60	2.8
Thailand	63,763	0.8	7,400	69.3	20	1.7	69.8	31	0.6
China	1,313,309	0.7	5,000	71.1	35	1.8	83.3	40	3.1
Philippines	81,408	1.7	4,600	68.3	28	3.1	28.2	62	3.0
Indonesia	222,611	1.2	3,200	66.4	39	2.3	54.7	45	3.4
Viet Nam	82,481	1.3	2,500	69.6	32	n.a.	56.7	26	3.1
Myanmar	50,101	1.2	1,900	58.9	81	2.8	32.8	30	2.9
Mongolia	2,630	1.3	1,800	62.9	56	2.4	54.3	57	1.4
Cambodia	14,482	2.4	1,700	54.6	71	4.6	18.5	19	5.4
Laos	5,787	2.2	1,000	55.1	86	4.6	28.9	21	4.6
North Korea	22,776	0.5	1,000	65.8	44	2.0	53.0	61	1.2
Timor-Leste	820	3.6	500‖	57.5	120	3.7	n.a.	8	4.6

* Exponential growth rate, including estimates of international migration.
† Figure for 2000.
‡ Figure for 2002.
§ Figure for 2003.
‖ Figure for 2001.

Sources: UN Economic and Social Commission for Asia and the Pacific, *Population Data Sheet 2004*, supplemented by data from UN, *World Population Prospects: The 2002 Revision*, *World Urbanization Prospects: The 2001 Revision*, *Human Development Report*, and *World Contraceptive Use 2003*; WHO, *World Health Report*; *CIA World Factbook 2004*; and Directorate General of Budget, Accounting and Statistics (DGBAS), Bureau of Statistics, Republic of China (Taiwan)—*Statistical Yearbook of the Republic of China, 2003*.

The decline in mortality has clearly preceded significant economic development, as even in the poorest countries of the region today IMRs are less than half the regional levels of the early 1950s when rates in excess of 180 per 1,000 live births were common. A transition to extremely low levels of mortality, nevertheless, must await the emergence of an advanced and wealthy economic system. The transition in fertility is less easy to explain. Clearly, the majority of wealthy economies have shifted to very low levels of fertility, but all societies in the Asian region have witnessed significant declines—particularly in what might be termed some middle-income economies, such as China and Thailand, where fertility is currently significantly below replacement level and equivalent to that of countries at much higher levels of economic development. Viet Nam, Mongolia and North Korea, at much lower levels of economic prosperity, also have lower fertility than might be expected. From the contraceptive prevalence rates, one can infer that family-planning programmes and population policies have been able to accelerate a transition to lower fertility. Not that these were absent in the wealthier economies but, there, the programmes were more likely to be responsive to demands for smaller families consequent upon the massive economic shifts to urban-industrial societies, whereas in the poorer economies the programmes were likely to play a more proactive role in the transition.

The demographic situation in the Pacific islands is less clear, with their small populations exhibiting considerable variation in growth and in fertility and mortality (Table 2). Again, there is a tendency for the wealthier economies to exhibit the lowest infant mortality rates, the highest expectation of life at birth and the lowest fertility although, generally, the island populations have significantly higher fertility than the majority of East Asian societies, with women in several countries having on average four children, and all Pacific island societies but one have above-replacement-level fertility. In the Marshall Islands, a rate of more than five children per woman is found, a level that was last seen in the economies of North-East Asia in the 1950s. Mortality, too, tends to be lower in the wealthier economies, but the pattern is nowhere as clear as in East Asia, and some poorer Pacific economies exhibit higher expectation of life than economies further up the wealth scale. Certainly, the results are distorted by Papua New Guinea, the largest of the Pacific economies, where infant mortality is higher than one would expect given its per caput income. The lack of such a clear relationship between the demographic variables and the economy of the Pacific is suggestive of a very different basis for the generation of wealth compared with East Asia, extractive rather than productive, and with different distributions of wealth within the societies. The small size of Pacific populations and cultural factors are also important.

THE AGEING OF POPULATIONS AND INCREASE IN LONGEVITY

The marked decline in fertility in East Asia has brought about significant shifts in age composition that have resulted in one of the major challenges facing the region: the ageing of populations (Table 3). The proportion of the population 65 years of age and older is highest among the most developed economies in the region, almost always above 7% but 18% in the case of Japan, whereas among the poorer economies the proportion is generally 5% or lower. The annual growth rate of the population 65 years and older is markedly higher than the growth rate of the labour force, or that population 15–64 years old, among all the countries and territories of East Asia, except for Macao (although this is perhaps a short-term statistical anomaly) and for most of those in the poorest economic category apart from North Korea and to some extent Laos. As an indicator of the depth of the transition in fertility, the growth rates for the cohort 0–14 years is zero or negative for all the most developed economies and significantly positive only for Malaysia, the Philippines, Laos and Cambodia. Although the growth rates of the elderly cohort are high in the small island states of the Pacific, the proportions of those 65 years and over are low (Table 4).

The decrease in numbers of young people and increase in numbers of old have profound implications not only for the planning of social services, for example, the need for fewer schools but more old people's homes, but also for the economy in general. Most critical will be the provision of adequate pension programmes. How will a country such as Thailand, with a per

Table 2. Pacific Island Economies: Basic Demographic and Development Data

	Population at mid-2004 ('000)	Annual growth rate (%)*	Per caput GDP, (PPP) in US $, 2003	Expectation of life, years at birth, 2002	Infant mortality rate, 2004	Total fertility rate, 2004 (children per woman)	Contra-ceptive prevalence —modern methods (%)	Urban population, 2004 (%)	Urban growth rate, 2004 (%)
Guam	165	1.5	21,000‡	n.a.	9	2.8	n.a.	41	3.1
French Polynesia	248	1.5	17,500§	n.a.	9	2.4	n.a.	53	1.5
New Caledonia	237	1.9†	15,000§	n.a.	7†	2.4†	n.a.	81	3.0
Northern Marianas	79	3.3‡	12,500‡	n.a.	5	1.6‡	n.a.	n.a.	n.a.
Palau	21	2.1‡	9,000§	68.5	17	2.5‡	n.a.	69	2.0
American Samoa	63	2.0‡	8,000‡	n.a.	73	4.0‡	n.a.	55	4.0
Fiji	847	0.9	5,800	67.3	17	2.8	35.1	52	2.4
Samoa	183	0.9§	5,600†	68.2	n.a.	4.6§	n.a.	23	1.7
Cook Islands	18	0.1§	5,000§	71.6	16	3.1§	60.4‡	68	1.6
Nauru	10	1.6†	5,000§	62.7	47	3.4†	n.a.	100	1.6
Niue	2	n.a.	3,600‡	70.3	29	n.a.	n.a.	34	0.0
Vanuatu	217	2.4	2,900†	67.7	28	4.0	n.a.	23	4.2
Tonga	105	0.9	2,200§	70.7	32	3.6	n.a.	34	1.2
Papua New Guinea	5,836	2.1	2,200	59.8	60	3.9	19.6	18	3.7
Micronesia	113	1.2‡	2,000†	66.5	33‡	4.1‡	n.a.	30	3.6
Solomon Islands	491	2.8	1,700§	65.4	20	4.3	n.a.	22	5.9
Marshall Islands	55	1.6‖	1,600§	62.7	37	5.7‖	n.a.	65	1.6
Tuvalu	10	0.4†	1,100‡	60.6	34	3.8†	n.a.	56	2.9
Kiribati	93	2.3‡	800§	64.1	44	4.3‡	n.a.	43	5.2

* Exponential growth rate, including estimates of international migration.
† Figure for 2002.
‡ Figure for 2000.
§ Figure for 2001.
‖ Figure for 1999.

Sources: UN Economic and Social Commission for Asia and the Pacific, *Population Data Sheet 2004*, supplemented by data from the UN, *World Population Prospects: The 2002 Revision*, *World Urbanization Prospects: The 2001 Revision*, *Human Development Report* and *World Contraceptive Use 2003*; WHO, *World Health Report*; *CIA World Factbook 2004*.

caput income of US $7,400 in 2003, provide adequately for an elderly population projected to grow at over 4% per annum, or, perhaps more critically, China, with an income of $5,000 per caput and an elderly population growing at an annual rate of 2.6%? The absolute number of people 65 years and older in China was estimated to be just under 95m. in 2003 but is projected to rise to 111m. by 2010 and to 167m. by 2020. By 2020 China will have more elderly people than in the whole of Europe and virtually three times as many elderly as the USA.

With decreasing family size, the base for continuing family support and care for the elderly is eroded as the traditional extended family responsibilities decline, forcing more of the costs onto the state. Later marriage, higher proportions of people choosing not to marry (in Japan, for example, 69% of men and 54% of women between the ages of 25 and 29 years remain unmarried), increasing proportions of women entering the labour force and rising divorce rates all contribute not just to lower fertility but to changed family structures: changed structures that will have profound effects for the increasing numbers of elderly in Asia. For example, the annual number of divorces in Hong Kong more than doubled between 1991 and 2001, and the number of single-parent families increased by 70% over the same period. Nevertheless, only in the cases of Japan and South Korea were rates of labour force growth less than those for the populations as a whole, showing the 'demographic dividend' still possessed by the majority of countries in the region that should allow higher rates of saving and greater investment in human capital, with resulting higher productivity growth.

While the issue of ageing appears far from the concerns of some of the poorer countries in south-eastern Asia where fer-

Table 3. North-East and South-East Asian Economies: Simplified Age Structure and Growth 2002–03

	Annual growth rate, 0–14 years	Annual growth rate, 15–64 years	Annual growth rate, 65+ years	Percentage, 0–14 years	Percentage, 15–64 years	Percentage, 65+ years
Singapore	−0.1	1.9	4.9	22	71	7
Hong Kong	−0.8	1.2	2.6	17	72	11
Japan	−0.7	−0.4	2.7	15	68	18
Australia	0.0	1.2	1.7	20	68	12
New Zealand	−0.4	1.0	1.3	23	66	12
Brunei	−0.3	2.6	5.6	31	66	4
Macao	−4.8	2.3	1.5	20	73	7
Taiwan	−0.9	0.7	2.7	21	70	9
South Korea	−0.4	0.5	4.8	22	71	8
Malaysia	0.9	1.9	4.6	33	63	4
Thailand	0.1	1.3	4.1	25	69	6
Philippines	0.6	2.5	3.7	37	60	4
China	−2.2	1.5	2.6	24	69	7
Indonesia	−0.4	1.8	3.7	30	65	5
Viet Nam	−1.4	2.6	1.4	32	63	5
Mongolia	−2.1	2.8	1.0	36	61	4
Laos	1.3	3.0	3.9	42	54	4
Myanmar	−0.2	1.8	1.6	32	63	5
Cambodia	1.4	3.2	2.5	43	54	3
North Korea	−0.4	0.7	4.8	26	68	6
Timor-Leste	n.a.	n.a.	n.a.	43	54	3

Source: UN, Unpublished tabulations.

Table 4. Pacific Island Economies: Simplified Age Structure and Growth 2002–03

	Annual growth rate, 0–14 years	Annual growth rate, 15–64 years	Annual growth rate, 65+ years	Percentage, 0–14 years	Percentage, 15–64 years	Percentage, 65+ years
Guam	2.7	1.8	3.8	35	59	6
New Caledonia	0.8	2.2	4.5	29	65	6
Northern Marianas	n.a.	n.a.	n.a.	27	72	2
French Polynesia	–0.4	2.2	4.4	30	66	5
American Samoa	n.a.	n.a.	n.a.	38	59	3
Fiji	0.3	1.3	4.0	33	64	4
Palau	n.a.	n.a.	n.a.	27	68	5
Cook Islands	n.a.	n.a.	n.a.	35	60	5
Nauru	n.a.	n.a.	n.a.	39	60	2
Samoa	–0.5	0.8	1.7	41	55	5
Niue	n.a.	n.a.	n.a.	33	59	8
Papua New Guinea	2.1	2.3	3.3	40	58	3
Tonga	n.a.	n.a.	n.a.	37	57	6
Micronesia	n.a.	n.a.	n.a.	40	56	4
Solomon Islands	3.2	3.4	3.5	45	53	3
Marshall Islands	n.a.	n.a.	n.a.	44	53	2
Vanuatu	1.4	3.3	2.7	41	56	3
Tuvalu	n.a.	n.a.	n.a.	34	60	7
Kiribati	n.a.	n.a.	n.a.	41	55	4

Source: UN, Unpublished tabulations.

tility remains high, so, too, does it seem of low priority to the island countries of the Pacific. Both the proportions of those 65 years and older and the absolute numbers are generally small. However, the annual growth rate of the elderly is often the fastest rate and family structures, there too, are changing. Emigration is taking many of the able-bodied away to Australasia and North America, and the issue of social and economic support of the elderly left behind is becoming an increasing concern.

INTERNATIONAL MIGRATIONS

The slowing in the growth of the labour forces in the most advanced economies in the region provides the demographic context for what is emerging as one of the critical areas of policy concern in Asia and the Pacific today: the movement of people across international borders. While there has been a long tradition dating back many hundreds of years of the movement of peoples in the region, and particularly of the Chinese into South-East Asia, there were few movements across international borders in the region from the mid-1950s through to the 1980s. Only Australia and New Zealand, traditional countries for European, particularly British, settlement, continued as major destinations for immigration throughout that period. The origins of their migrants, however, changed significantly from the 1970s, as immigration policies were modified to allow the entry of settlers from a wider range of countries. By 2001/02 some 38.7% of the 88,900 settler arrivals in Australia were from Asia, 28% of them coming from north-eastern and south-eastern Asia.

The most noticeable trend since around the late 1980s has been the emergence of significant destinations for migration within Asia itself. The sustained development in the 1960s demonstrated by the emergence of the 'Asian tigers', in association with the fertility declines described above, created demands for labour that could not be satisfied from local sources. Economies such as that of South Korea shifted from being net exporters of labour in the 1970s and 1980s, when fertility was high and domestic rural–urban migration was pronounced, to being net importers of labour in the 1990s. This 'migration transition' has characterized most of the developed East Asian economies to some extent, even if there were considerable variations among them, and all today are characterized by both inward and outward flows of people. For example, 20,742 South Koreans entered the USA as immigrants in 2000/01, while the stock of foreign migrant workers in South Korea increased from 329,555 to 362,529 during calendar year 2002. In that year more than 400,000 Koreans went overseas as skilled workers to support Korean companies, as students, or as some other type of non-permanent migrant. In addition, some 1.6m. Koreans travelled on business trips overseas.

With the exception of Hong Kong, with its policy of allowing 150 people from mainland China a day to enter the Special Administrative Region (SAR) and settle, and, to some extent Singapore, no Asian economy actively pursues open immigration policies to allow permanent settlement. The flow to Hong Kong from China is basically made up of the wives and children of permanent Hong Kong residents. In Asia, unlike migration to Australasia, North America or Europe, international population movement is essentially the movement of labour. That labour is both skilled and unskilled, with greater control over the latter. In the majority of Asian economies, policies are implemented to promote skilled labour migration to fill particular shortages. Singapore, for example, facilitates the acquisition of permanent resident status for those with advanced degrees, a status that can be converted to citizenship at a later date under certain circumstances. Although unskilled migrants are also required across East Asian economies, policy-makers are extremely concerned about temporary entrants becoming more permanent migrants. In some of the economies, unskilled workers are allowed entry under direct labour importation schemes, perhaps the most effective being that of Taiwan, or under a variety of trainee and study programmes that also allow migrants to work legally, as in South Korea and Japan.

The Asian financial crisis of 1997 did have an impact in reducing certain flows, particularly to South Korea, Malaysia and Indonesia, although there was little 'migration effect' in Taiwan, Hong Kong or Japan (Table 5). Labour migration to South Korea quickly resumed, and by 2001 the number of foreign workers was more than twice that of 1998. Some changes during 2002 can perhaps be observed. In Taiwan there was a shift from recruiting foreign workers for the manufacturing sector towards the service sector and, as existing contracts in manufacturing expired, the total number of foreign workers declined. Growth in numbers of foreign workers in Hong Kong and Singapore appears to have decelerated or at least stabilized. Only in South Korea, with its relatively fast economic growth, and in Japan, do the inflows of migrant workers appear to have continued unabated and, as observed above, it is in these two economies that the decline in domestic labour force growth has been most acute.

The whole issue of numbers of foreign workers in Asian economies is complicated by significant numbers of undocumented or 'irregular' migrants. Many of the migrants may either enter legally as tourists, or on some other short-term visitor visa, and then stay on after the expiry of their visas as 'overstayers', or they may simply enter illegally. The former strategy is more common in the economies of North-East Asia, while the latter is more common in South-East Asia, and specifically in Malaysia and Thailand. In a UN survey, these two countries were the only ones in the eastern Asian region to consider that their immigration levels were 'too high'. Accurate figures on the number of irregular migrants are elusive, with numbers constantly changing, and both Thailand and Malaysia took measures in 2001 and 2002 to try to manage the inflows more

Table 5. Official Estimates of Total Number of Foreign Workers in Asian Economies, 1997–2002

	1997	1998	1999	2000	2001	2002
Taiwan	245,697	255,606	278,872	326,515	304,605	303,684
Hong Kong[1]	171,000	180,600	193,700	216,790	235,280	237,110
Japan[2]	630,000	660,000	670,000	710,000	740,000	760,000
South Korea[2]	245,399	157,689	217,384	285,506	329,555	362,597
Singapore	n.a.	n.a.	530,000	612,233	590,000	590,000
Indonesia[3]	35,213	33,295	21,276	14,780	19,789	23,850
Malaysia[2]	1,471,645	1,127,652	818,677	799,685	804,984	n.a.
Philippines[3]	6,055	5,335	5,956	5,576	6,859	10,739
Thailand[2]	1,125,780	1,103,546	1,089,656	1,102,612	655,338*	1,200,000‡
China[3]	82,000	83,000	85,000	n.a.	440,000†	n.a.
Viet Nam[3]	n.a.	n.a.	n.a.	30,000	30,000	n.a.

[1] Includes an estimate of foreign domestic workers only; there are no stock figures for the highly skilled.
[2] Includes estimates of undocumented workers.
[3] Estimate of foreign experts only, primarily professionals, the highly skilled and teachers.
* Excluding an estimated 300,000 undocumented workers.
† Including 190,000 workers from Hong Kong, Macao and Taiwan.
‡ Estimate of 1,000,000–1,200,000.

Source: OECD, *International Migration and the Labour Market in Asia (2003)*, and updated information from OECD sources.

effectively and reduce the number of irregular migrants. In 2001 Thailand instigated a programme to register irregular migrants that resulted in 568,249 persons being registered, the largest number ever to have done so to date. Despite this intensive programme, it was estimated that at least another 300,000 irregular migrants did not register. With more than 300,000 arrests in 1999 and more than 400,000 in 2000, but relatively few deportations, and with a very porous border that allows people to move easily between Myanmar and Thailand, the total number of irregular migrants in Thailand is estimated to fluctuate between 760,000 and 1m. people at any one time. In 2002 Malaysia implemented one of the most extensive changes to its migrant labour programme. During an amnesty from 21 March to 31 July, an estimated 570,000 migrants left the country, primarily for Indonesia. While many were later recruited back into Malaysia on a legal basis, Malaysia also took the opportunity to attempt to diversify the source areas of its migrant labour away from an almost total reliance on Indonesia for unskilled workers to include government-to-government agreements with Thailand, Cambodia, Nepal, Myanmar, Laos, Viet Nam and the Philippines, as well as Uzbekistan, Turkmenistan and Kazakhstan, to supply labour for the manufacturing, service and construction industries. After a period of flux, the programme of regularization appears to have succeeded, at least partially, with some 1,163,194 foreign workers being recognized in Malaysia in 2003.

The lack of adequate legal channels for potential migrants to gain access to labour markets within the East Asian region is one factor in accounting for the importance of irregular migration, and that migration is usually facilitated by the trafficker or smuggler of human beings across international borders. Trafficking is distinguished from smuggling by the presence of deception and exploitation in the migration. The smuggler provides a service that is faster and cheaper than the cumbersome bureaucracies of existing legal channels, and the majority of those who resort to the smugglers do so of their own accord. Much is made by policy-makers and mass media alike of the deception and exploitation brought about by the human trafficker, particularly of young girls into the sex industry in South-East Asia, Japan and beyond. However, deception and exploitation also occur within smuggling networks after migrants have chosen to use their services. For example, Chinese citizens who are smuggled to the USA, particularly from the southern province of Fujian, may be held until final payments are made, and some of these migrants work under virtual debt bondage at destinations. Nevertheless, the majority of Chinese smuggled to the USA appear to pay off their debts fairly quickly. Although cases of exploitation in the trafficking of women within the Asian region unquestionably exist, the majority may not simply be victims but are pro-actively expanding their markets regionally and globally, and a clear distinction between trafficking and smuggling is often difficult to sustain. Trafficking and smuggling networks are complex, constantly changing and highly flexible, and although international criminal groups can be involved at various points along the way, the networks are not monolithic organizations in the same way as are the groups dealing in narcotics.

Overall, the economic downturn and security concerns across the region have prompted increasing attempts to manage migration in order to reduce irregular movements and achieve a diversification in migration flows and perhaps a stabilization in migrant numbers in all but the most labour-scarce economies in the region. However, with renewed signs of economic growth throughout the East Asian region by 2004, international migration to destinations within the region will continue to provide policy-makers with one of their major challenges. Nevertheless, it is important to keep a sense of perspective. With the exceptions of Singapore, Brunei, and to a lesser exten, Malaysia, the current migrant stock in Asian economies represents a very small proportion of the total population, generally less than 2%. As in the case of general trends in fertility and mortality, the Pacific islands exhibit much greater variation in patterns of international migration, with some countries showing substantial proportions of foreign-born. American Samoa, Guam and Nauru all have proportions of foreign-born in excess of 33%, whereas migration to the larger countries of Papua New Guinea, Solomon Islands and even Fiji is negligible. Emigration, too, can have a significant impact on small populations. Unlike eastern Asia, emigration to Australasia and the USA is making an impact on the total population growth of even some of the larger populations of Pacific islands, such as Samoa and Tonga.

URBANIZATION AND POPULATION REDISTRIBUTION

Although international migration is currently one of the main policy concerns in the East Asian region, the numbers of international migrants are tiny, compared with the number of those moving internally within countries. Again, the lack of accurate data is problematic, and differences in how 'urban' is defined make valid comparison difficult. Estimates for the 'floating' population in China, or those away from their place of registration (their *hukou*), can range from 100m. to 200m. In contrast, there are probably just over 400,000 Chinese workers abroad, although there are also several tens of thousands of students and other skilled workers overseas, plus many thousands of settlers in North America and Australasia. Perhaps one-fifth of Thailand's population, or 11m.–12m., migrate internally over any five-year period, compared with a total overseas work-force of 160,000 in 2001. Such comparisons between flows and stocks of migrants are hardly justified except to give a highly generalized impression of the degree of difference in volume between internal and international migrations. These substantial domestic flows have been central to the redistribution of population from rural to urban areas.

The most developed economies are also the most highly urbanized, with three-quarters to four-fifths of their populations living in urban areas. Among the wealthiest group, there is little difference between urban and total population growth rates, showing that urbanization, or the changing balance between urban and rural sectors, has all but stopped. Only in the case of

South Korea is there still a marked difference. Hong Kong, Macao and Singapore are essentially total city economies, but they extend their urban settlement well across their boundaries into mainland China and Malaysia to form integrated mega-urban regions around the Pearl River delta and in the southern Malay peninsula respectively. Within Japan, the Tokyo-Yokohama urban region, with a population of some 30m. people in 2000, forms the largest urban region in the world, and another five of the world's 19 largest cities in 2000 were also to be found in the East Asian region: Shanghai (12.9m.), Jakarta (11.0m.), Osaka (11.0m.), Metro Manila (10.9m.) and Beijing (10.8m.). The great period of urban concentration in Japan took place prior to the 1970s, after which there was a slackening and reversal of migration to metropolitan peripheries, smaller towns and rural areas. With decreasing population growth in Japan, and, more importantly, the decline of property prices in Tokyo in the late 1990s, a renewed if somewhat weaker trend towards concentration has been observed in the early 21st century. Virtually half of the land area of Japan, accounting for only 6% of the population, is classified as 'severely depopulating' (*kaso*).

While the great urban transformation of developed countries like Japan and South Korea has stabilized at around 80%, other countries such as China, Indonesia, Thailand and also Viet Nam are currently experiencing very rapid urban growth and urbanization. It should be noted that the urban proportion and the urban growth rate for Thailand both appear low, given its level of development, but the urban areas of Thailand have for long been 'underbounded' and have not been adjusted outwards to cover the rapid urban expansion that the country has experienced. Thus, the fastest-growing urban areas around the periphery of existing towns and cities were still, in 2003, classified in the rural sector, giving a false impression of both urban growth and urbanization in that country. Rural–urban migration is intense among all countries at this level of development, and there is much circulation, or moving backwards and forwards between towns and villages. Urban growth is also generally fast among the poorest countries in the East Asian region, although among these economies the urban proportion is markedly lower, except in the cases of both Mongolia and North Korea where geographical and political factors account for higher proportions of urban population. Yet again, among the islands of the Pacific there is much variation in levels of urbanization and rates of urban growth, although the majority of island nations are experiencing rapid rates of urban growth. This growth is influenced not just by patterns of internal migration within the country or territory, much of which is circulatory, but also by the prevailing fertility levels that contribute to a rapid natural increase in urban populations.

PUBLIC HEALTH CONCERNS

HIV/AIDS: An Impending Crisis?

Within the East Asia and Pacific region, some 5.3m. people were living with HIV/AIDS at the beginning of the 21st century. Officially, there are some 1m. carriers in China alone, and a total of 10m. carriers is projected for 2010. However, overall prevalence rates are low, compared with those of sub-Saharan Africa. In East Asia rates above 0.1% are unusual, whereas rates above 10% in sub-Saharan Africa are common. The highest rates in the East Asian region in 2003 were to be found in the countries of mainland South-East Asia: Cambodia at 2.6%; Thailand at 1.5%; Myanmar at 1.2%; and Malaysia at 0.4%. The rate of prevalence for China was estimated at 0.1%. The fact that overall rates are generally low should not lead to complacency. Prevalence rates are significantly higher in specific parts of countries with otherwise low rates, and they can be elevated among particular high-risk groups. For example, the prevalence rate among injecting drug-users in Jakarta showed a steep rise to almost 50% in 2001, and the rate among commercial sex workers in Ho Chi Minh City in Viet Nam has recently risen to 20%. The scandal that emerged in 2001 regarding paid blood donors in parts of rural eastern China, in which plasma was extracted from the donated blood and then the blood reinjected into the donors from a common pool, has resulted in local HIV prevalence rates of 12.5%. Given the high rates of spatial population mobility observed in the previous section in all these countries, the potential for the rapid spread of the disease appears to be substantial.

Nevertheless, there are grounds for hope. The implementation of the 100% condom programme in Thailand clearly helped to raise awareness of, and encourage preventive measures against, the disease. HIV prevalence among military recruits decreased from about 4% to less than 1% during the 1990s, and patronage of brothels in Bangkok appears to have declined markedly. There is some evidence to suggest that the epidemic in the most affected country, Cambodia, may have stabilized, with prevalence among sex workers having declined from 42% in 1998 to 29% in 2002, and the incidence among pregnant women attending antenatal clinics in the major urban areas decreasing from 3.2% in 1996 to 2.8% in 2002. Condom use among sex workers in Cambodia appeared to rise to 90% in 2001. In contrast to Cambodia, 85% of sex workers in the two largest urban centres in Papua New Guinea reported inconsistent use of condoms, and HIV prevalence doubled between 2000 and 2001 among those seeking treatment for sexually transmitted diseases in the capital, Port Moresby, which indicates the potential for a rapidly expanding epidemic in that country. Thus, although the development path of the disease in East Asia need not necessarily follow that of sub-Saharan Africa to such high overall prevalence rates, the disease nevertheless has considerable potential for expansion in the region. HIV/AIDS tends to have its greatest impact in some of the poorest countries in the region, and among the poorer parts of richer countries, or on those economies and people least able to afford to take effective protective measures against the disease.

SARS

During the first half of 2003 a new infectious disease swept through much of East Asia, with its epicentre in southern China and Hong Kong. This was Severe Acute Respiratory Syndrome (SARS). With air travel, the disease was quickly spread around the world, and particularly to Canada, with its substantial Asian populations. By the end of 2003 some 8,096 cases had been reported world-wide, with 774 deaths directly attributable to the disease. Minor outbreaks of the disease were reported in China during early 2004 but were quickly contained through public health response. The closure of feral animal meat markets in southern China may also have made some impact on the restriction of the spread of the disease. The damage from the epidemic was likely to be more economic than demographic. Tourist arrivals to Hong Kong in April 2003 were almost 65% lower than in the corresponding month of 2002, and the airline industry was particularly badly affected, for example. The newness of the disease, and a lack of understanding of its nature and means of transmission, contributed to the public fear of an unknown epidemic. In reality, the total world-wide death toll from the outbreak was likely to be equivalent to the number of people killed on the roads of China each day. Traffic accidents, increasing cigarette consumption and more traditional killer diseases such as tuberculosis remain the real, if less visible, public health crises in East Asia today, rather than the dramatic epidemic of SARS, which was short-lived. Nevertheless, SARS is an atypical form of pneumonia that evolved in the area from which new variants of pneumonia have so often emerged, southern China. Its appearance serves as a warning of the constant danger of new and virulent infections that can still threaten modern economies in a globalizing world.

PROSPECTS

The number of people in the East Asian and Pacific region accounts for about one-third of the world's population. By 2050 that proportion is projected to decline to about one in four, as other parts of the world, notably Africa and South and West Asia, experience more rapid growth. Thus, the East Asian region will continue to grow, but slowly. In North-East Asia growth rates are projected to fall below 0.5% per annum from 2015 and will be negative from 2035. Annual growth rates will be faster in South-East Asia but are projected to fall below 1% by 2015 and to be below 0.5% by 2040. The population of the region is projected to reach about 2,500m. in 2050. The demographic projections envisage trends towards a balanced sex ratio even though it is probable that these are currently skewed towards males in certain local areas of China owing to sex-biased abortion of female foetuses. Such a bias, if real, may cause difficulties when the sex-biased cohorts reach marriageable age, and perhaps increased status of women and migration can be expected to adjust for the biases introduced by present practices.

Most importantly, the populations will become older, with North-East Asia projected to have more than 10% of its population older than 65 years by 2015 and South-East Asia to attain the same proportion by 2035. It will also be an urban world. UN projections envisage that 62.7% of the population of North-East Asia and 56.5% of the population of South-East Asia will be living in urban places by 2030. Tokyo will continue to be the largest city in the world, and 19 of 59 cities larger than 5m. will be found in the region. Australia and New Zealand will continue to form the most highly urbanized part of the world. Assuming that economic development continues, even if at a slower pace than in the past, declining labour force growth but rising demand for certain types of labour almost certainly implies continuing and increasing international migration directed primarily at the largest cities in the most dynamic economies. The result will be a growing multiculturalism, with all the benefits and stresses that such mixing can bring. The populations of the East Asian region will begin to take on the characteristics of those of North America, Australasia and Europe, not just in terms of the basic demographic variables of fertility and mortality but also in terms of their increasingly heterogeneous cultural compositions, which is all part of the process of globalization.

BIBLIOGRAPHY

Abella, M. 'Turning Points in Labor Migration', *Asian and Pacific Migration Journal*, 3, 1, special issue, 1994.

ADB. *Key Indicators 2002. Population and Human Resources: Trends and Challenges*. Manila, Asian Development Bank, 2002.

Caldwell, J. C., and Caldwell, B. K. 'Poverty and Mortality in the Context of Economic Growth and Urbanization', *Asia-Pacific Population Journal*, 17, 4, pp. 49–66, 2002.

Chin, K.-L. *Smuggled Chinese: Clandestine Immigration to the United States*. Philadelphia, PA, Temple University Press, 1999.

ESCAP. *Population Data Sheet*. Bangkok, United Nations Economic and Social Commission for Asia and the Pacific, 2002.

Kempadoo, K., and Doezema, J. (Eds). *Global Sex Workers: Rights, Resistance and Redefinition*. New York, Routledge, 1998.

OECD. *Migration and the Labour Market in Asia: Recent Trends and Policies*. Paris, Organisation for Economic Co-operation and Development, 2002.

Seetharam, K. S. 'Half a Century of Unparalleled Demographic Change: the Asia-Pacific Experience', *Asia-Pacific Population Journal*, 17, 4, pp. 13–30, 2002.

Skeldon, R. 'Irregular Migration in the Greater Mekong Subregion: Policy Dimensions of a Growing Issue', in *Labour Migration and Trafficking within the Greater Mekong Subregion*. Bangkok, International Labour Organization, pp. 35–75, 2001.

Population Mobility and HIV Vulnerability in South East Asia: An Assessment and Analysis. Bangkok, United Nations Development Programme, 2000.

'Trends in International Migration in the Asian and Pacific Region', *International Social Science Journal*, 165, pp. 279–382, 2000.

UNAIDS. *AIDS Epidemic Update: December 2002*. Geneva, Joint United Nations Programme on HIV/AIDS, 2002.

United Nations. *International Migration 2002*. New York, Population Division, Department of Economic and Social Affairs, 2003.

World Urbanization Prospects: The 2001 Revision. New York, Population Division, Department of Economic and Social Affairs, 2002.

World Contraceptive Use 2001. New York, Population Division, Department of Economic and Social Affairs, 2002.

World Population Prospects: The 2002 Revision. New York, Population Division, Department of Economic and Social Affairs, 2003.

ENVIRONMENTAL ISSUES OF ASIA

ROSS STEELE

POPULATION, DEVELOPMENT AND THE ENVIRONMENT

The Asia-Pacific region illustrates the stark reality that 'environmental' issues cannot be considered independently of issues of human and economic resource development. In mid-2004 this diverse and immense region, excluding South Asia, was home to more than 2,112m. people, about 33% of the world's population, whose numbers are growing annually by an additional 18m. persons. The region is also home to the industrialized economies of Japan, Australia and New Zealand and to the 'tiger economies' of East and South-East Asia which in recent years have once again recorded impressive rates of economic growth. The Asia-Pacific region also contains some of the world's poorer nations where development capital is scarce, and large minorities lack the most basic human needs, such as sufficient food, adequate supplies of clean drinking water, minimal shelter, and access to health care and education. Governance in these circumstances is weak and undeveloped; immediate concerns dominate the political agenda, with governments and citizens taking the view that economic growth must be the first priority, even if it means serious environmental damage. In this context, 'environmental' strategies not concerned with the economic and political realities of development and with satisfying legitimate human aspirations will prove to be ineffectual.

Paul Ehrlich, the eminent US population biologist, has advanced the I=PAT model in an attempt to predict the impact on the environment (I) of the three major components of development: population growth (P), increased affluence (A) and technology (T). Given that the Asia-Pacific region's population will continue to grow significantly for the next 20–30 years and that the level of affluence, and therefore consumption, of the rapidly growing middle classes will rise even more quickly, then total environmental impact must inevitably increase greatly. In order to achieve the increase in affluence demanded by their growing populations, with minimal environmental damage, the developing nations of the region must receive massive transfers of new, ecologically appropriate technology from the developed nations. Unfortunately, many of the attempts at rapid industrialization over the last 30 years or so have made use of outmoded 'dirty' technology, which has only served to exacerbate environmental damage. This cannot be allowed to continue in the Asia-Pacific region because the environmental challenges that this area will face in the next 15–30 years will be of such magnitude that any failure will have an impact on our global future.

THE REGION'S MAJOR ENVIRONMENTAL ISSUES

The Asia-Pacific region experiences a great variety of environmental problems. This is not unexpected in view of the region's enormous size and extreme diversity of physical and human environments. The problems that are most serious over large geographic areas include deforestation and loss of biodiversity, loss and degradation of agricultural land, ground, water and air pollution, global warming and rising sea-levels. The region is also experiencing rapidly growing urban environmental problems with polluted air, and difficulties of water supply and waste disposal. The environmental consequences of megaprojects, such as the Three Gorges Dam in China, are attracting much international attention, along with concern about the impact of events such as the forest fires in Indonesia in 1997–98 and major flooding in China in 1998 and 1999 and in Jakarta, the Indonesian capital, in 2002. More recently, salinization of water and land and a continuation of prolonged drought in the principal agricultural regions of Australia have become a major cause of concern, along with damage done to coral reefs in South-East Asia and the Pacific, including Australia's Great Barrier Reef, by global warming, coastal development, overfishing and other human activity. An assessment of which of these environmental issues is of most concern is a matter of subjective judgement; however, as the most populous countries of the region still have a majority of their population resident in rural areas, environmental issues that affect rural people will be given most emphasis. Nevertheless, it should be stressed that, although these so-called 'green issues' are currently of greatest concern to informed observers in the developed countries, they will be surpassed in importance by 'brown issues' of the urban environment very shortly, since it is projected that during the first two decades of the 21st century a majority of the region's population will become urban dwellers.

A selection of these major environmental issues will now be discussed. Issues in the most populous developing countries will be emphasized because of their global and life-threatening nature, but this is not to imply that the more industrially advanced countries in the region do not experience serious difficulties. Because of the immense scale of the environmental problems, detailed case studies are included in preference to general discussions, so that the human dimensions of the problems and their inter-relationships may be better understood.

LOSS AND DEGRADATION OF AGRICULTURAL LAND

The most critical environmental issue facing the Asia-Pacific region today is the continual loss and degradation of its most highly productive arable land which, if allowed to continue at the present rates, will threaten the ability of the region's most populous countries to feed their own people. Land is lost to agricultural production by many processes, such as conversion to non-agricultural uses or by erosion and pollution of the topsoil. These processes can convert hitherto productive arable land, capable of producing food crops, into unproductive wasteland, which then requires expensive rehabilitation before production can be resumed. More serious than the absolute loss of land to agricultural production is what two observers, Blaikie and Brookfield, have termed the 'quiet crisis', whereby the productivity of agricultural land is gradually, but insidiously, reduced by the dual processes of degradation and pollution. Another analyst, A. Burnett, defines 'degradation' as the term used to refer to the processes that lead to the loss of productivity in soils and to a decline of surface and underground storages and flows of water. Pollution is described as the contamination of air, water and soil by human activity. These two processes have had their most damaging impact on agricultural productivity in the most populous country in the region, China. However, the ecologically fragile semi-arid environments of Australia, and the steeply sloping uplands of Java, the Philippines, New Guinea and New Zealand have also been seriously eroded and degraded by ecologically unsuited land-management practices driven by economic and population pressures.

Loss and Degradation of Arable Land in China

Faced with the problem of feeding a major portion of the world's population on a disproportionately small share of the earth's arable land, for the last 2,000 years the Chinese people have been exhorted to subjugate nature. From pre-modern times China has converted extensive areas of natural ecosystems to crop fields, but this trend intensified during the 'grain first' policy of the 1960s, when the leadership was belatedly forced to give top priority to increased food production after the famine of 1959–61. The environmental consequences soon became apparent, and the early 1960s witnessed a greatly accelerated degradation of China's farm soils, grasslands, forests and wetlands. Chinese sources put the total losses of arable land for 1957–77 at a conservative 29.3m. ha, equivalent to 26% of the nation's total arable area in 1957. Vaclav Smil, an expert on China's environmental problems, has claimed that about 7m. ha of this arable land was lost through requisitions for construction projects. Desertification, salinization and alkalization claimed another 5m. ha of this loss of farmlands and grasslands between 1957 and 1977, with some 90% blamed on improper land use, involving conversions of semi-arid grasslands to grain fields and to overgrazing. About 10m. ha of sloping grasslands unwisely converted into fields was lost to erosion. During this 20-year period the total area of arable land was boosted by the reclamation of 17.3m. ha of new land coming from the conversion of forests and wetlands, but yields were very often disappointing, with cultivation frequently being abandoned after several years. The remaining 7m. ha of the total 29m. ha was lost when this converted land reverted to non-arable uses. Therefore, the largest single cause of arable land loss would appear to be erosion on grassland or wasteland taken into use since the late 1950s although, as Smil points out, some erosion has a much earlier origin, especially in the Loess Plateau of northern China.

Official estimates of arable land in China seriously underestimate its extent. In the late 1980s, early 1990s and late 1990s official sources estimated the total area at about 96m. ha (Table 1), whereas country-wide sample surveys and satellite images suggest a total of at least 130m. ha. Despite this discrepancy, Chinese experts suggest that the extent of reporting inaccuracies has not changed since the 1960s, and that the large gross losses of farmland revealed by the official estimates (Table 1) are indeed real. These official figures suggest a cumulative loss of about 40m. ha of arable land since the late 1950s, which Smil has estimated is the equivalent of a loss in food production capacity for at least 350m. people. The mean annual loss of more than 400,000 ha during the 1980s was equal to a loss in food production for about 5m. people every year. The mean annual loss for the first four years of the 1990s was even higher, at about 550,000 ha. Comprehensive estimates comparable to those shown in Table 1 were not available in 2004, but estimates of water-eroded land areas alone (increased from 1.53m. sq km in the 1950s to 1.796m. sq km in the 1990s) suggest that the losses have probably accelerated in recent years.

Table 1. Total and Per Caput Availability of Farmland in China and Reported Losses of Farmland, 1949–93

Year	Farmland Total (million ha)	Per caput (ha)	Gross loss of farmland
1949	97.88	0.180	n.a.*
1957	111.83	0.173	n.a.
1965	103.59	0.143	n.a.
1970	101.10	0.122	n.a.
1979	99.50	0.102	n.a.
1985	97.60	0.092	1,000,000
1986	96.85	0.090	640,000
1987	95.89	0.088	470,000
1988	95.72	0.086	510,000
1989	95.65	0.085	198,000
1990	95.67	0.084	733,000
1991	95.65	0.083	580,000
1992	95.60	0.082	243,000
1993	95.50	0.081	620,000

* Note: The farmland loss data are not available for 1949–79; however, the average annual loss for 1957–86 was 526,000 ha.

Source: Smil, V. 'Environmental problems in China: estimates of economic costs', *East-West Center Special Reports*, No. 5, Table 4, p. 34, East-West Center, 1996.

Since 1978 a new household production responsibility system (*baogan*) has provided the incentive for families to intensify crop production and revert to more profitable and environmentally sustainable forms of land use. Much of the subsequent loss of farmland has been a result of restoration of wetlands, grasslands or slopelands unwisely converted to farmland in the two decades of Maoist campaigns. However, deforestation and accelerated erosion continue, despite strict edicts banning excessive timber-logging and cutting of firewood. Between 1980 and 1985 a total of 2.45m. ha of cropland was lost, about one-quarter to frantic house-building by newly private farmers. In 1987 a national limit was introduced, which decreed that no more than 200,000 ha of farmland may be used for non-agricultural construction projects in a year, but the rapid pace of economic expansion, especially the uncontrolled development of rural industries and the establishment of special industrial zones, led to increased losses of farmland in the 1990s. Moreover, these losses will probably include some of the most productive agricultural land on rich alluvial soils close to major towns and cities. Increasing population, rapid industrialization and extension of transport links will make continued losses inevitable in future years, and although some of the causes of the losses in the past such as over-farming on waste land and deforestation have been better controlled in the last few years, in April 2004 a senior engineer with the surveillance centre for water and soil conservation, under the Ministry of Water Resources, claimed that China's severe soil erosion problem was continuing to worsen, with large-scale construction projects being an important factor. Smil estimates that recent losses represent an annual decline of 300,000–750,000 ha of sown farmland when the multi-cropping ratio is considered, with wet rice land accounting for at least one-fifth of the total loss. An extensive programme of creating new agricultural land by labour-intensive practices, such as the terracing of deforested slopes, the manual deposition of alluvial soils behind these embankments and the development of intricate water-control schemes, have added to the area of farmland since the 1950s, but since the mid-1960s losses have surpassed gains. In recent years it has been estimated that there are no more than 10m. ha of land in the whole country that can be converted to farmland at an acceptable cost.

To offset the losses of agricultural land, Chinese farming has become much more intensive. Traditional Chinese agriculture has been seen by the West as a model of sustainable agriculture, where soil fertility was maintained and increased by complex crop rotations, planting of green manures and the recycling of crop residues, human and animal wastes and other organic matter. More recently, China has become the world's largest producer and consumer of synthetic fertilizers. Despite these efforts, Chinese agriculture is now faced with the serious problem of declining soil quality caused by increased soil erosion and a rapid decrease in traditional organic recycling. Between the early 1950s and the late 1990s the extent of all lands affected by soil erosion (caused by both wind and water) more than doubled, from approximately 150m. ha to 367m. ha, equivalent to 39% of China's territory. About 30m. ha of Chinese farm fields are affected and, although the regional rates of damage vary considerably, the Loess Plateau (an area half the size of France) has the highest rates in the world, with the equivalent of 2 cm of topsoil being denuded each year. Very high rates were also reported from the provinces of Heilongjiang and Sichuan, and recent Chinese reports suggest that, overall, China loses some 5,000m. metric tons of topsoil per year from its eroding soils. It is claimed that this loss is equivalent to all macronutrients (nitrogen, phosphorus and potassium) applied in chemical fertilizers.

The economic costs of soil erosion include crop-yield losses caused by the removal of the soil's organic matter, fine clay content, water retention capacity and plant nutrients. The actual yield losses have been estimated as negligible in most of China's wet farmlands, at 5% over a century on fields with tolerable erosion rates, and at 25% on the most seriously affected farmland. Other economic costs of soil erosion include the value of soil nutrients lost which, if yields are to be maintained, have to be replaced by chemical fertilizers; lost irrigation and hydroelectricity-generating capacity of reservoirs owing to silting; additional expenses for filtering silted water for urban and industrial uses; capital costs required to counteract the damage caused by increased siltation of river beds, lakes and reservoirs; and damage done by the increased frequency and severity of flooding. One example will illustrate the scale of these problems. Increased siltation along the lower Huanghe (Yellow River), downstream from the Loess region in the north, has raised the height of the river-bed above the adjacent floodplain by about 1 m per decade, increasing dramatically the risks of catastrophic flooding of the North China plain. If a break in a dike south of Jinan occurred today, it would flood up to 33,000 sq km and affect the livelihoods of some 18m. people. A 1994 decision requiring that all cropland used for construction be offset by land reclaimed elsewhere has exacerbated the extent of soil erosion. The fast-growing coastal provinces, such as Guangdong, Shandong and Jiangsu, that have lost much cropland to urban development and industrial construction are now paying other more ecologically fragile regions in the north-west, such as

Inner Mongolia, Gansu, Qinghai and Xinjiang, to plough ever more marginal land to offset their own losses. Intensified wind erosion, land abandonment and migration eastward have been the ultimate result. In April 2001 a huge dust storm originating in northern China was reported to have covered parts of western Canada and the USA with a layer of dust.

The advent of the 'grain first' policy in the early 1960s discouraged many traditional crop rotations and reduced the cultivation of green manures, which in turn led to a reduction in soil organic matter. Further reduction occurred with the increased use of stalks and straw for household fuel rather than for recycling and with other changes arising from the economic privatization of agriculture in the late 1970s. The increased availability of chemical fertilizers in the early 1980s, the pressures of privatized farming which encouraged the monocropping of cereals, and the construction of modern sewage treatment plants, all worked to discourage the continuation of traditional organic recycling practices. The loss of organic matter in cultivated soils leads to a breakdown in the soil structure with the soil becoming more compacted and less permeable, which increases soil run-off. It also dramatically reduces the number of soil invertebrates such as earthworms, which play a key role in maintaining soil fertility. By the late 1980s some 13m. ha of China's arable land were contaminated by pesticide residues and another 7m. ha were polluted by industrial wastes. In recent years torn plastic sheeting, originally used to conserve soil moisture and accelerate crop ripening, has become mixed with cropland soils. It is believed to be responsible for reduced yields on at least 1m. ha of cropland.

The cumulative loss of about 40m. ha of farmland in China since the late 1950s and the serious degradation of the fertility and quality of the remaining arable soils, owing to increased soil erosion and the spread of less environmentally sustainable management practices, pose doubts about the ability of China to feed itself in future decades. Technological advances will further increase the productivity of agriculture in China over the coming decades, but the above discussion reveals that, unless the environmental degradation caused by present and past farming practices is dramatically reduced, major advances in productivity will be required simply to maintain production at existing levels. This raises the question as to who will be able to provide the increased food that will be demanded by China's expanding population and required by even more rapidly rising living standards.

A report by the environmentalist group, Greenpeace China, delivered in Beijing in August 1999, reiterated these concerns and claimed that China faced 'environmental meltdown' if it waited to get rich before addressing its many serious problems. China's rapid economic growth rates were blamed for most of the damage, which has caused the loss of some of its best farmland to industrial development and urban growth, recurring and worsening floods, the worst dust storms in memory in northern China, acid rain and foul urban water and air. However, the fact that this message could be delivered in Beijing, and the unprecedented candour of reporting on the role of deforestation in the fatal summer floods of 1998, suggest that the authorities are beginning to accept the seriousness of environmental problems. The western interior, which is home to one-quarter of China's 1,300m. people and comprises more than one-half of the country's land area, is the most adversely affected region, with the poorest 10 areas of the west accounting for just 15% of gross domestic product. In March 2000 China's most senior economic planner pledged to reverse the environmental destruction of its western interior by turning steeply terraced fields into forest and exhausted cropland into pasture. He stated that some 70% of that year's fixed asset investments would be directed to the west, in an attempt to bring the region out of poverty and environmental degradation. Official sources claim that this process has already begun, with 4.05m. ha of afforestation and 5.7m. ha of mountain passes converted into forest in 1999. However, whether the more than 35,000m. trees planted in China since 1981 have received proper care and survived is another matter. An official report in 2001 estimated that every year an average of 2,460 sq km of vegetated land in China deteriorates into desert and another 1m. ha of land suffer from serious erosion.

DEFORESTATION

Over the last two decades or so deforestation has received far more publicity than any other environmental issue in the Asia-Pacific region. Much of the initial concern was limited to ecologists, but this is no longer the case, with citizens increasingly aware of the importance of natural forests in protecting soil from erosion, maintaining soil productivity, preventing flooding and providing valuable timbers for national revenue. Moreover, the forests of South-East Asia have been inhabited by local people for at least 39,000 years and today are still the home of some 30m. indigenous people; any damage to their habitat has a critical impact on their way of life and survival.

The populous countries of China, Indonesia and the Philippines historically contained substantial areas of forest, but most of this has long been lost and converted to other uses such as agriculture. Much of the forest that has survived in these countries does so only in a seriously degraded form, with the extent and quality of the forest cover being reduced by indiscriminate logging and the illegal gathering of firewood, inappropriate road-making methods, forest fires and heavy grazing. In

Table 2. Loss of Natural Forest Cover in Selected South-East Asian Countries, 1980–2000

	Land area	Natural forest cover ('000 sq km)*						Loss of natural forest cover					
		1980		1990		2000		1980–1990		1990–2000		1980–2000	
	('000 ha)	Area	%†	Area	%†	Area	%†	Area	%‡	Area	%‡	Area	%‡
Myanmar§	65,755	32,862	50.0	28,856	43.9	33,598	51.1	–4,006	–12.2	4,742	16.4	736	2.2
Thailand	51,089	17,888	35.0	12,735	24.9	9,842	19.3	–5,153	–28.8	–2,893	–22.7	–8,046	–45.0
Cambodia	17,652	13,477	76.3	12,163	68.9	9,245	52.4	–1,314	–9.7	–2,918	–24.0	–4,232	–31.4
Laos	23,080	14,467	62.7	13,173	57.1	12,507	54.2	–1,294	–8.9	–666	–5.1	–1,960	–13.5
Viet Nam	32,550	9,683	29.7	8,312	25.5	8,108	24.9	–1,371	–14.2	–204	–2.5	–1,575	–16.3
Malaysia	32,855	21,546	65.6	17,583	53.5	17,542	53.4	–3,963	–18.4	–41	–0.2	–4,004	–18.6
Indonesia	181,157	121,669	67.2	109,549	60.5	95,115	52.5	–12,120	–10.0	–14,434	–13.2	–26,554	–21.8
Philippines	29,817	10,991	36.9	7,831	26.3	5,036	16.9	–3,160	–28.8	–2,795	–35.7	–5,955	–54.2
Total	433,955	242,583	55.9	210,202	48.4	190,993	44.0	–32,381	–13.3	–19,209	–9.1	–51,590	–21.3

* The Forest Resources Assessment 2000 did not identify natural forest cover separately; therefore this was calculated by subtracting 'Forest Plantations' from 'Total Forest' (Table 2, FAO, 2003), as suggested by Emily Matthews in 'Understanding the FRA 2000', p. 4, *Forest Briefing No. 1*, World Resources Institute, 2001.

† Proportion of total land area covered.

‡ Proportion of forest area lost.

§ The inferred gains in natural forest cover in Myanmar between 1990 and 2000 are not credible. They were derived from the 1990 natural forest baseline data published in Table 19.1, p. 307, *World Resources 1994–95*, World Resources Institute, 1994. However, the FRA 2000 provided higher estimates of forest cover for 1990 than the earlier FRA reports (a 13% upward adjustment in Asia) and it is possible that the 1990 estimates of natural forest cover in Myanmar may have underestimated its extent and resulted in a 'false' gain in natural forest cover during 1990–2000.

Source: Adapted from Tables 1 and 2 (pp. 128–135) of FAO's *State of the World's Forests 2003,* published in 2003 and Table 19.1 (pp. 306–307) of World Resources Institute's *World Resources 1994–95,* published in 1994.

China, despite a programme of reafforestation, annual cutting of forest resources surpasses new growth by more than 40%, and approximately 45% of the uplands suffer soil erosion as a result of deforestation. In 2000 just 12.7% of the total land area remained as natural forest.

It is the deforestation and degradation of the tropical rain forests of South-East Asia, however, that is the current issue of greatest concern to environmentalists. There are a number of reasons for this concern. First, unlike East Asia, South-East Asia still contains very large areas of primary tropical rain forest. Second, if measured in terms of hectares of forest deforested per hectare of national territory, the rate of deforestation in South-East Asia has probably been the highest in the world since 1980. Third, the forests of this area are the habitat of millions of indigenous people, many of whom still practise traditional lifestyles. Fourth, this region, especially the western part of insular South-East Asia (sometimes termed Malesia, and comprising Sumatra, Peninsular Malaysia and the island of Borneo), is renowned for its exceedingly rich biodiversity. There are estimated to be at least 25,000 species of flowering plants in Malesia, along with 350 species of the dipterocarp family. The dense, tall forests have recently provided the tropical timber trade with some of its most valuable hardwood species. Borneo, for example, as well as being a centre of diversity of plant species, contains an immense array of animal and insect species with very high rates of endemism. Finally, the forests of South-East Asia have been regarded by local political leaders as one of their last untamed resources and an appropriate site of mega-projects to dam rivers, construct commercial plantations of rubber and oil palm, and as settlement destinations for landless citizens from more densely populated regions.

Recent research has revealed large variations in estimates of deforestation rates in South-East Asia, but there is little doubt that since the early 1990s most of the South-East Asian countries have at least equalled the highest in the tropical world, at well over 1% of the forest cover per year. Table 2 shows the latest available estimates of natural forest cover in selected countries of South-East Asia and inferred losses of this cover since 1980. The decrease in the percentage of total land area in natural forest cover is marked. Other sources suggest that in the late 1960s about 66% of the total land area of South-East Asia was covered in forest (Bernard, S. and De Koninck, R., Table 1, p. 6, 1996), but that by 1980 this proportion had decreased to 56% and by 2000 it had been reduced further to just 44%. Thus, over the past 30 years or so one-third of the natural forest cover of the region has been lost completely to agriculture or to other uses. Definitions of deforestation, however, are inconsistent. The FAO data do not classify as deforestation temporary clearing of forest that was logged and left to regrow, even if it was clearcut, whereas temporary clearing by shifting cultivators is included. Moreover, the data provide no estimates of degraded forest lands, as secondary forest is not identified, but they are probably very extensive.

The area of forest lost shows great variation by country and by data source, with the largest absolute losses in the period 1980–2000, according to the FAO Forest Resources Assessment (FRA—described by Tole in 1998 as the most consistent and timely data set) occurring in Indonesia, Thailand, the Philippines, Cambodia and Malaysia. The gains in natural forest cover inferred from the FRA data to have occurred in Myanmar since the early 1990s are improbable (see notes on Table 2), and that country has most likely also experienced a continuation of its earlier losses. It is very unfortunate that the countries with the most rapid rate of deforestation over the last decade, such as the Philippines, Cambodia and Thailand (Table 2), were among the countries with the smallest areas of remaining natural forest in 1990, with serious consequences for the quality of runoff from their water catchments and for the local villagers. About 4.8m. ha of the forest loss in Viet Nam and Cambodia occurred during the Viet Nam War, but since the mid-1970s there has also been serious deforestation by loggers and farmers. Thailand lost some 40% of its forest cover in the 1970s and 1980s, leading it to impose a ban on logging, as did Viet Nam, but illegal logging remains a problem and the ban has placed greater pressure on the more plentiful forest resources of their poorer neighbours, Myanmar, Laos and Cambodia. In Thailand and Myanmar much of the forest being lost is open broadleaf forest, which contains valuable timber such as teak. The Philippines has probably the worst record of deforestation in South-East Asia, with the small areas of surviving virgin forest confined to the most inaccessible mountainous areas. However, since the early 1980s the absolute scale of deforestation in Indonesia has overshadowed that in any other South-East Asian country (Table 2). In 2002 a report entitled *The State of the Forest: Indonesia* by Forest Watch Indonesia and Global Forest Watch concluded that 40% of the forests existing in Indonesia in 1950 had been cleared by 2000, forest cover declining from 162m. ha to 98m. ha. It is of even greater concern that the rate of forest loss appears to be accelerating. During the 1980s the amount of forest lost was about 1.2m. ha per year, rising further in the early 1990s, and then increasing to 2m. ha from 1996. Indonesia's lowland tropical forests, the richest in timber resources and biodiversity, are those most at risk. They have almost entirely disappeared in Sulawesi and are predicted to be entirely cleared in Sumatra by 2005 and in Kalimantan by 2010 if current trends continue. Rapid deforestation has also occurred in Malaysia and probably in Myanmar in recent years, with most of the large areas lost being closed broadleaf forests, the most valuable of all forest types in terms of biodiversity.

What has been the environmental impact of this deforestation? Its direct effects are on the forest and its biodiversity (both in terms of its flora and fauna), and on soils and hydrology. Few hard data are available to quantify the species loss rates, as a complete biological inventory of many areas has never been done, but there is little doubt that the forest areas are suffering accelerating and irreversible loss of biodiversity. The Malesian region alone has been a major source of the world's crop plants and fruits, such as sugar cane, bananas and rice on the margins of the region. It has also been the source of some medicinal plants. The house-gardens and plots of shifting cultivators contain a huge variety of plant types, the survival of which is critical to the shrinking crop gene-pools of modern agriculture. The impact of deforestation on wildlife has been acute. In the Republic of Korea, for example, deforestation and the use of pesticides and fertilizers has led to the loss of more than 100 animal species in recent years. In Cambodia deforestation has resulted in the destruction of three-quarters of the country's wildlife habitat. The tiger, elephant and rhinoceros are now reduced to very small numbers in South-East Asia owing to the loss of their habitat, and the forest fires in East Kalimantan in 1997–98, themselves at least partially a result of logging and deforestation, have dramatically illustrated the dangers posed to the survival of the orangutan, the proboscis monkey and other monkey species.

Once the forest canopy is reduced, the exposed soil is eroded and leached of its nutrients and organic material, which then limits forest regrowth and reduces its quality. It becomes degraded secondary forest. This, in turn, by a series of positive feedback mechanisms, alters the carbon cycle and the radiation balance, reduces transpiration and thereby contributes to climate change and desertification. Logging and conversion to agricultural uses fragment the forest areas that remain and increase the length of the borders of these forests relative to their total areal extent. This fragmentation magnifies the impact of forest-clearing and degradation on biodiversity loss. These exposed and degraded zones are more vulnerable to dry spells than inaccessible heavily forested areas; they transfer less moisture into the atmosphere, reduce humidity and are more vulnerable to fire. Because secondary growth is easier to clear and cultivate than primary forest, it is attractive to swidden cultivators and is more likely to be completely deforested. The topsoils that remain, once deprived of their biomass cover (their critical nutrient source), lose nutrients after several cropping cycles and are soon abandoned by shifting cultivators. The exposed soils allow more rapid run-off than forest cover and are more easily eroded. Modern mechanical logging practices, requiring roads and using heavy machinery on steep, previously inaccessible land, have accelerated rates of soil compaction, loss of topsoil, reductions in water percolation and exposure to erosion. Some studies have shown an 18-fold increase in soil loss after logging, which increases siltation, raises river-beds and causes flooding downstream. For example, in November and December 1999 the central coast of Viet Nam was inundated by the worst floods of the century, which cost more than 700 lives and caused extensive damage to houses and farmland. Widespread illegal logging was blamed for much of this damage.

Although the causes of deforestation vary by country, traditional shifting or swidden cultivation, along with agricultural development, have for many years been held responsible for about 50% of the deforestation; another 25% has been blamed on settlers who move into areas after logging operations, with an additional 10% being due to commercial logging and another 10% due to infrastructure development such as road-building or dam construction. Recent research in Indonesia suggests that at least 67% of the deforestation was due to macro-level programmes deliberately encouraged by the Government, such as commercial logging (including the attendant damage caused by machinery, camps, roads and log yards), agricultural resettlement known locally as transmigration (this programme was stopped in 1999) and smallholder cash-crop schemes emphasizing tree crops such as rubber and oil palm. Forest was also cleared for large-scale plantations of industrial timber and estate crops, but 75% of these proposed timber plantations were never planted. In contrast, traditional shifting cultivation accounted for just 27% of deforestation and in reality probably much less than this, for there are now very few genuine examples of these traditional farmers who practise rotational swidden farming. Those who remain are likely to be limited to inaccessible inner areas of the forest remote from the market economy and population pressures. A study published in 1999 by two analysts, A. Angelsen and D. P. Resosudarmo, of areas of Riau, West and East Kalimantan and Central Sulawesi concluded that the monetary crisis and its associated breakdown of governance created opportunities for better-off farmers, immigrants and urban dwellers with capital to convert forests to highly profitable crops. The situation has been made worse by an excess capacity in the wood-processing industry. Much of the gap between the legal supply of wood and demand is being filled by illegal logging, which accounted for approximately 65% of the total timber supply in 2000. The situation appears to be worsening in the short term, as Indonesia moves rapidly towards a system of 'regional autonomy', with the provincial and district governments largely lacking the capacity, the funds and often the will to govern effectively. The ready availability of chain saws and vehicles has increased the impact of illegal logging in recent times.

Some writers explain the causes of deforestation in terms of two contrasting models: the frontier model, which identifies networks of entrepreneurs, companies and small farmers as the chief agents of deforestation, using private capital and with assistance from the state and local élites; and the immiserization model, which sees local shifting cultivators and landless peasants as the key protagonists, forced to clear land for subsistence and smallholder cash-cropping under the pressure of growing populations, devalued local currencies and poverty. Research has shown that both models have some validity, with the immiserization model being more relevant in places with small remnant forests and rapidly growing populations, but the frontier model more relevant where there are large blocks of forests and where frontier conditions prevail. Often the frontier model describes the initial stages of deforestation during periods of rapid economic expansion, whereas the alternative model is more appropriate at a later stage once the forests have been made more accessible and deforestation rises and falls in response to changes in the economic and social conditions of the local rural poor. Economic recession, for example, would be expected to retard frontier deforestation, but promote deforestation in already settled areas. The risks to the forests from modern and traditional sources wax and wane, but generally, as a country becomes more developed, the overall risk of continued deforestation tends to decrease once a threshold level of economic development has been achieved. Development-planners and policy-makers interested in conserving forestry resources and in improving their sustainability must address not only the local pressures that lead to deforestation, but also the macro-development framework that continues to make deforestation profitable to private entrepreneurs and responsible government officials.

WATER AVAILABILITY AND POLLUTION

Obtaining sufficient supplies of unpolluted water is an increasing problem for the rural and urban populations of the region. Much of the contamination of water supplies is caused by the improper disposal of human waste, chemicals and other hazardous materials from both manufacturing and agricultural activities. The quality of surface and groundwater supplies has deteriorated as the pace of industrialization has accelerated and as farmers have attempted to maximize their returns with increased applications of fertilizers and pesticides. Growing populations and improved living standards have at the same time raised the demand for clean water for agricultural, industrial, domestic, and even recreational uses. A brief discussion of the situation in China will highlight current and future problems of water availability.

The pressure placed on China's water resources is the greatest in the region. Between 1950 and 1990 the population of China doubled, urban population increased five-fold, industrial output rose about 80-fold and the area of irrigated land tripled. These trends have continued, with China's cities being required to absorb more than 100m. additional residents in the 1990s. The Worldwatch Institute, a research organization based in Washington, DC, has estimated that by the year 2030 China's total water demand will have increased from 483,000m. tons in 1995 to 1,068,000m. tons. The most dramatic increases in demand will be for industrial (417%) and residential (330%) uses, but the demand for agricultural water is also projected to increase, by 66%. China's total water resources may be enough to meet these quantitative demands, but they are very unevenly distributed. The humid south, with four-fifths of the water and 700m. people, is dominated by the vast Changjiang (Yangtze River) which will probably continue to supply an adequate quantity of water, even if industrial pollution may limit its use. Some Chinese experts have warned that certain stretches of even the mighty Changjiang may be dry by the year 2020. The arid north, comprising the four basins of the Huanghe (Yellow), Liao, Hai, and Huai Rivers, is home to 550m. persons and two-thirds of the nation's cropland. All four of these northern basins currently face acute water shortages, which can only become more desperate as demand increases. In the drought year of 1997 the Huanghe failed to reach Shandong Province, the source of one-fifth of China's corn and one-seventh of its wheat production, much of which is irrigated from the river. As agriculture simply cannot economically compete with industry or residential demands for water, more and more of China's surface water is being diverted to non-farm uses. To survive, farmers are increasingly turning to underground water reserves, many of which are being pumped out more rapidly than the rate at which they can be naturally recharged. The result is that in the North China Plain the water table has fallen by an average of 1.5m annually over the last five years. This is unsustainable in the long term. The eventual outcome will be a cessation of pumping and a reversion to dryland farming, with a 50%–66% decline in yields. In 1998 senior Chinese officials estimated that rural villages nationwide were facing annual water shortages of 30,000m. tons, which had reduced grain production by 20m. tons, enough to feed more than 65m. people. In 1998 two observers, Lester R. Brown and Brian Halweil, predicted that unless China can significantly curtail its projected demand for non-agricultural water, by the year 2030 water shortages will have reduced China's ability to feed itself to such an extent that the country's growing demand for grain imports will pose a threat to the world's food security. Water shortages are already being faced in 300 of China's 617 largest cities, including Beijing and Tianjin, where overexploitation of underground aquifers has caused widespread surface subsidence.

Water pollution in China is serious and further limits the supply of clean water available for drinking and irrigation. Surveys in the 1980s revealed that more than 20% of the 878 major rivers examined had water so badly polluted that it could not be used for irrigation. In urban areas 80% of China's surface water was contaminated, and only six of the country's 27 largest cities had drinking water within state standards. Municipal wastes are commonly released untreated into local water bodies, and the rapid growth of small and medium-sized factories in rural areas has resulted in industrial wastes contaminating local surface water used for drinking. Phenols, heavy metals and organic wastes are the most common pollutants. Because the local farmers use river mud as a fertilizer, residues of chromium, lead and cadmium are now being found in many staple crops. In the Huanghe basin abnormally high rates of stunting, mental retardation and developmental diseases are linked to high concentrations of arsenic and lead in the water and food. To date, pollution of fresh water with nitrates has not been a problem in China, but the experience of the USA suggests that given China's extremely high usage of nitrogen fertilizers, high nitrate loadings will appear in China's drinking water within the next decade and result in increased risks to human health. Continued growth in energy consumption will involve increased use of coal, which in the case of China is high in sulphur content. This will result in heavier acid deposition on water bodies from acid rain, which will degrade water quality even further and pose threats to aquatic life and humans in both China and Japan.

In May 2004 blooms of toxic algae called phytoplankton formed a vast 'red tide' that blanketed an ocean area of 8,000–10,000 sq km in the East China Sea off the island of Zhoushan Dao, south of Shanghai. The algae developed very rapidly and fed on pollution from sewerage, nitrogen fertilizers used in agriculture, fish farms, car emissions and industrial waste discharged into the Changjiang. Similar 'red tides' occurred in October 2000 in a bay off Aotou in southern Guangdong Province, and in 1998 a massive algae bloom near Hong Kong killed up to 75% of the entire stock of Hong Kong's fish farms. In 1998 alone, more than 20 'red tides' were reported along China's coasts. In 1999 toxic algae covered a large area of the Bohai Sea, south-east of Beijing in the north-east of the country. Toxic algae blooms are of course not confined to the coasts of China; they also occur in the bays and coastal waters off Japan and in the Gulf of Thailand, and have been a long-standing problem in the Gulf of Mexico and in the Baltic Sea. However, in the first issue of its *Global Environment Outlook Year Book* (2003) published in March 2004, the United Nations Environment Programme (UNEP) gave new prominence to the issue by claiming that in 2003 there were 150 oxygen-starved or 'dead zones' in the world's oceans and seas, and that the number had been doubling every decade since the 1970s as pollution had increased. The effects of the algae blooms are sometimes mild, but they always cause low levels of oxygen in the water and make it difficult for fish, oysters and other marine creatures to survive along with their most important habitats such as sea-grass beds. At other times, the effects are dramatic, with fish fleeing the oxygen-depleted waters and slow-moving, bottom-dwelling creatures dying in large numbers. In all instances the fish and shellfish contaminated by the 'red tide' are regarded as unsafe to eat, with the annual economic cost to China estimated at US $100m. in 2000. UNEP claims that these 'dead zones' are becoming an even greater threat to the life of the sea than the overexploitation of 72% of the world's fish stocks. The 2003 UNEP report places the major responsibility for 'dead zones' on the massive increase in the amount of reactive nitrogen that humans create, predominantly as fertilizers for food crops. It has been estimated by UNEP that, globally, humans create about 160m. metric tons of reactive nitrogen per year, compared with natural rates of terrestrial biological nitrogen fixation of between 90 and 120m. tons annually. This reactive nitrogen accumulates in the environment, as denitrification processes that convert nitrogen primarily back to non-reactive nitrogen have not been similarly increased. Most of this reactive nitrogen is not used by crops, and it then ends up moving freely through the environment in what UNEP terms a 'nitrogen cascade', where the same atom of nitrogen can contribute to many different negative impacts in sequence on the atmosphere (global warming, air pollution and acid rain), land (acidification of soils), freshwater and ocean (pollution, algae blooms and subsequent oxygen depletion), as well as on human health. UNEP has recommended that actions to reduce algae blooms should focus on reducing the amount of fertilizer and sewerage running off the land. An example of successful intervention has occurred along the River Rhine in Europe where the quantity of nitrogen entering the North Sea was reduced by 37% over the period 1985–2000.

ENVIRONMENTAL PROBLEMS IN URBAN AREAS

Although, in terms of global impact and the number of people immediately affected, rural environmental problems have hitherto been of most significance, it is now recognized that 'brown' environmental issues in the cities of the Asia-Pacific region will be of greater importance to local residents in the 21st century. There are several reasons for this. First, the number and proportion of the region's inhabitants resident in urban areas will increase dramatically over the next 25 years or so. In 2003 the urban population totalled 888m., some 42% of the total population; by 2030 it is projected to total at least 1,504m., an increase of some 69% in just 27 years, and to comprise 61% of the region's total inhabitants. Second, by the year 2015, it is expected that there will be 21 megacities world-wide (cities with more than 10m. inhabitants) and that seven of these will be in East and South-East Asia. Probably as much as 80% of the region's economic growth will be generated by that urban population, with the megacities being in the forefront of the challenge of determining if this growth can be sustained environmentally. Third, the majority of the major cities are sited on or near river estuaries, some of the most productive natural ecosystems in the region. Any serious environmental pollution here has the potential to damage fertile agricultural land, surrounding aquaculture and productive marine environments. Fourth, these large cities are characterized by rapidly growing middle classes who, unlike the environmentalists of the West, are relatively indifferent to deforestation, global warming, wildlife issues and endangered species, but instead are increasingly concerned that their politicians do something about environmental problems that create threats to their health, such as air pollution, traffic accidents, contaminated or insufficient water supplies, and inadequate waste treatment and disposal systems.

Kirk Smith and others have recently developed Abdel Omran's concept of the Epidemiological Transition to help explain the dynamics of environmental health risks in cities in the developing world. This suggests that, as economic development progresses, environmental health risks and the patterns of overall ill health undergo transition. For example, in the early stages of development traditional infectious and parasitic diseases are common, and are associated with rural and urban poverty, poor nutrition and sanitation, and unclean water supplies. These may be termed traditional health risks, which at this stage are considerable and reflected in high infant and child mortality rates. As development progresses, the risks to health decrease as water and sanitation conditions improve, along with better nutrition and higher standards of living. As this occurs, however, modern risks to health arise from new factors such as pesticide run-off, inadequate toxic-waste disposal, air pollution and motor vehicle accidents, which in time increase the risks to health and cause illnesses like cancer and heart disease. The transition point is the level of development at which there is equal risk to health from both modern and traditional environmental and technological factors. In short, there is a risk transition from a condition of less development, when there are many and high risks to health from traditional environmental factors, to a more developed condition where the overall risks to health from environmental factors are few and low, but where the causal factors are modern and post-industrial in nature.

In this article it is argued that the total risks to health (both short-term and long-term) during the transition point may well be greater than the low point predicted in the risk transition model. This is because in the Asia-Pacific region there is a considerable overlap of traditional and modern environmental risk factors and the possibility of synergistic consequences. For example, entirely new kinds of airborne environmental risks are created when crop residues contaminated with modern pesticides are used as traditional cooking fuels inside rudimentary shelters. However, this model provides a useful framework, which will facilitate inter-country comparison of traditional and modern urban environmental risk factors in relation to the level of development. The hypothesis of the environmental risk transition predicts that risks to the health of a city resident may be predicted by the level of economic development of that city. Hence, in very poor cities (Category 1) such as Hanoi, Ho Chi Minh City, Vientiane, Yangon (Rangoon), and also in the poorest slum areas of Manila and Jakarta, traditional problems such as inadequate access to clean drinking water, pollution of drinking water by human waste and poor community sanitation have increased the incidence of water-borne disease. In Manila, untreated or poorly treated domestic water accounts for 40% of water pollution, followed by industrial waste water (38%) and domestic solid waste (22%). Less than 15% of Metro Manila is served by sewers, and there has been an increase in water-borne diseases such as diarrhoea, typhoid, paratyphoid and salmonella infections.

In middle-income cities (Category 2) such as Bangkok, Kuala Lumpur, Beijing, Shanghai, Jakarta and Surabaya, more modern problems such as air pollution are a major concern. This pollution is caused by a rapid growth of motorized vehicles, combined with antiquated road systems and heavy concentrations of industry. Improved living standards have increased the production of solid waste in these cities, and the safe disposal of domestic and industrial waste is now a major problem. The poorer neighbourhoods of these cities are concurrently exposed to high traditional health risks from water-borne disease, which probably means that the overall risk to their health from this overlap and possible synergism of traditional and modern risk factors may be greater than that faced by inhabitants of Category 1 cities. A study prepared by the United Nations Environment Programme (UNEP) and the World Health Organization (WHO) revealed that all of the cities surveyed in the region (Bangkok, Beijing, Jakarta, Manila, Seoul and Shanghai) had their air quality seriously affected by suspended particulate matter; Beijing, Seoul and Shanghai have serious to moderate sulphur dioxide pollution; Bangkok, Manila and Jakarta have a moderate lead problem; Jakarta has a moderate carbon monoxide problem; Beijing and Jakarta have a moderate ozone problem; while none of the cities had excessive concentrations of nitrogen dioxide in their air. Virtually all of the major cities of China, along with Bangkok and Jakarta, experience more than 100 days annually during which suspended particulate matter concentrations exceed the WHO standard of 230 micrograms/cu m. This compares with major cities in Japan and Australia, which do not suffer any days above the standard. Bangkok's estimated losses in 1989 from particulate matter amounted to 51m. restricted-activity days and 26m. lost workdays, together with 1,400 excess mortalities. In China sulphur dioxide concentrations are believed to be largely responsible for lung cancer mortality rates three to seven times higher in the nation's cities than in the country as a whole. In cities where there is a large concentration of heavy industry, people living near the congested industrial areas suffered from a much higher incidence of diseases such as chronic bronchitis, tuberculosis, skin allergies, anaemia and eye irritations than residents in cleaner suburbs.

In high-income cities (Category 3) such as Tokyo, Singapore, Osaka, Sydney and Melbourne, urban environmental health risks are much lower and confined to hazardous wastes (biological and toxic wastes associated with advanced technology), indoor air pollution, and occasional but usually minor problems with localized water contamination. Amenity issues are of more concern to the relatively affluent residents, as they demand high-quality design, landscaping, green space and recreational facilities.

However, an interesting initiative is being attempted in Taiyuan, a grimy industrial city in north-east China, to overcome its problem of air pollution. In a national survey in 1999 and 2000 this city was found to have the nation's worst level of air pollution at nearly nine times the safe level. Most of this was caused by smoke and sulphur dioxide emanating from the city's many coal-fired hearths and industrial boilers. Unlike London and Los Angeles, which adopted classic command-and-control measures to deal with their problems, Taiyuan is letting the market decide. It proposes to use emissions trading, in an attempt to achieve a 50% reduction in sulphur-dioxide output within five years. Agreement was reached with the Asian Development Bank to develop a system whereby permits to pollute will be issued to the city's worst offenders. These permits will be tradable, so that polluters can either retain the right to pollute and pay for this privilege, or restrain their own output of sulphur dioxide and sell the right to another organization. By reducing the total permitted tonnage of gas each year, overall pollution levels can hopefully be reduced in the most economically efficient manner. Nantong, another Chinese city, has already carried out a trial of this approach with the help of a US environmental group. The approach is a good illustration of the mistaken logic of the conventional view that poor cities like Taiyuan cannot afford rich-world environmental standards and that clean air and greenery are luxury goods. Instead, the Chinese leadership appears to be moving towards acceptance of the contrary view that, besides being bad for individuals, pollution is bad for the economy and that improvements in economic growth and welfare will eventually more than compensate for the immediate costs required to reduce it. This change in attitude has been strongly endorsed by the World Bank, which estimated that in the late 1990s China lost between 3.5% and 7.7% of its potential economic output as a result of the detrimental health effects of pollution on the country's work-force.

Most of the above-mentioned urban environmental problems are a result of the failure of governance, which in some cases, but by no means in all, is due to weaknesses in the national economy. Governments fail to control industrial pollution and occupational exposure; to promote environmental health; to ensure that city-dwellers have the basic infrastructure and services essential to health and a decent living environment; and to plan in advance so that sufficient land is made available for low-income housing. These shortcomings are all too often due to the oversight, corruption, vested interests and a lack of commitment to the public good by government officials and politicians. These failures of governance are just as common in the small cities as in the large cities of the region, and they are also emerging even in the cities of the South Pacific. It is obvious from the above discussion that strategies designed to improve environmental conditions in the major urban centres of the region must focus explicitly on the problems of the poor. In particular, they must focus on public health and environmental measures to prevent or limit the transmission of diarrhoeal diseases, typhoid, cholera, and other water-borne, or water-based diseases, or acute respiratory infections and tuberculosis, or vector-borne diseases such as malaria, dengue fever and yellow fever.

GLOBAL WARMING AND RISING SEA-LEVELS

There is now widespread agreement among scientists that the world's climate will warm over the 21st century owing to a more pronounced 'greenhouse' effect caused by the huge amounts of carbon dioxide and other gases, such as methane, being released into the atmosphere as a result of modern human activities. The major contributing activities have been greater energy use, in particular the increased burning of fossil fuels, the expansion of more intensive agricultural land uses, deforestation, and the manufacture of particular chemicals such as the now widely banned chlorofluorocarbons (CFCs). The Asia-Pacific region already contributes more than one-quarter of the global emissions of carbon dioxide. The rapid industrial growth predicted for populous countries like China and India over the 21st century, and the fact that at least for the next 30 years this growth will be driven by a huge rise in the use of fossil fuels, will increase this share dramatically. According to the United Nations' Intergovernmental Panel on Climate Change (IPCC), a doubling of carbon dioxide in the atmosphere will make the climate warmer by approximately 3.0°C. What is not known is how rapidly this doubling will occur and what effect this will have on future temperatures. The World Meteorological Organization suggests that a rise of 3.0°C may occur as early as 2030 if the effects of other greenhouse gases are included, but Mike Hulme, a British climatologist, offers more conservative estimates of temperature change in the range of 1.5°C to 4.5°C by 2100, and a best estimate of transient warming of only 1.2°C by 2050. However, the Third Assessment Report (TAR) of the IPCC released in 2001, has projected globally averaged surface temperature increases of 1.4°C to 5.8°C over the period 1990 to 2100, greater than those predicted by the Second Assessment Report (SAR) of 1996. Already the earth's average surface temperature has risen by approximately 0.6°C in the last century, with new analyses for the Northern Hemisphere indicating that these temperature increases have been the largest of any century during the past 1,000 years.

The major likely effect of global warming is a potential rise in the mean sea-level owing to the thermal expansion of ocean water and the melting of ice caps. The TAR has projected that mean sea-levels will rise by 0.09 to 0.88 m between 1990 and 2100, slightly less than predicted in the SAR. This will seriously threaten the very existence of low-lying developing countries such as the Maldives, Tuvalu, Tonga, Kiribati, and Vanuatu where the vast majority of the population lives along the coasts. Flooding during cyclones and king tides will increase, affecting soils and salinizing freshwater supplies. Water tables will rise, and coastal farmland will be abandoned, forcing inhabitants to cultivate steeper slopes, leading to soil erosion. Tourism, which is the mainstay of many small island economies, will also be adversely affected as beaches and sea walls are destroyed. Coral reefs, which protect tropical islands, will also be severely affected, and there will be fewer fish and shellfish. Coastal vegetation, particularly mangroves, will be lost, along with their protection from wave erosion and their fish-breeding environments. This may well lead to significant local human migration and the loss of cultural heritages.

Natural systems can be particularly vulnerable to climate change in direct relation to the amount and speed of that change because of their limited adaptive capacity. Working Group II of the IPCC (2001) has already observed changes including shrinkage of glaciers, break-up of part of the Antarctic ice shelf, thawing of permafrost, later freezing and earlier break-up of ice on rivers and lakes, lengthening of mid- to high-latitude growing seasons, poleward and altitudinal shifts of plant and animal ranges, declines of some plant and animal populations, and earlier flowering of trees, emergence of insects, and egg-laying in birds. The Working Group warned that natural systems at risk of irreversible damage include glaciers, coral reefs and atolls, mangroves, boreal and tropical forests, polar and alpine ecosystems, prairie wetlands and remnant native grasslands.

Hulme has modelled the likely effects of temperature increases on the vegetation of China and concluded that in the Xinjiang Uygur Autonomous Region of north-west China, the climate would become warmer and drier, rendering sustainable livestock production difficult. In the more arable areas to the east and the north-east a rise in temperatures would increase the diversity of crops that could be grown and may lead to higher production. However, wet rice, the main food crop, requires plentiful water supplies, and Hulme concludes that any rise in temperature would decrease the effectiveness of water available from precipitation and irrigation sources since evapo-transpiration rates would increase. The decrease in water availability would tend to depress yields of rice and also wheat.

Similar effects of global warming have already been observed in the 2002 drought in Australia. In the nine months March–November 2002 the average Australian rainfall was the lowest since reliable records began, but the impact of this drought was more severe than that of any other since 1950 because the temperatures in 2002 were significantly higher. Researchers claim that although it cannot yet be shown that climatic change led to the drought of 2002, global warming has exacerbated its effects by causing increases in potential evaporation. These increased temperatures were not simply the result of the drought, but they appear to be part of a 50-year warming trend in Australia that has exceeded the range of natural variability over the period. They conclude that the warming is likely to be a result of enhanced greenhouse gas concentrations that will probably more than offset any possible, but unlikely, precipitation increases in the future. Intense droughts are therefore likely to become much more common in Australia than in the past. Model simulations also suggest that the frequency of cyclones would increase as a result of global warming and that this could have catastrophic effects for the rural populations located in coastal low-lying areas. Models to predict the effects of global warming have been developed for other countries, but the results are not consistent and few predictions can be made with certainty. However, the 2001 reports of the IPCC project a number of adverse impacts on human systems from global warming. These include a general reduction in crop yields in most tropical, sub-tropical and mid-latitude regions, decreased water availability for populations in many water-scarce regions, particularly in the subtropics, an increase in the number of people exposed to vector-borne and water-borne diseases and to heat-stress mortality, a widespread increase in the risk of flooding for many human settlements and an increased energy demand for space cooling, owing to higher summer temperatures. A study by A. McMichael and others (2003) has attempted to predict changes in the populations at risk of succumbing to malaria, dengue fever and diarrhoeal disease in Oceania if predicted climate changes eventuate in future decades.

Perhaps one of the most underestimated effects of global warming will be that it will bring about shifts in major climatic zones and in vegetation types. Many animal and plant species may be able to follow these shifts in habitat, provided the changes are relatively slow. Rapid changes in habitat, however, combined with artificial barriers such as agricultural land use or urban development, may make it impossible for many species to relocate to more favourable ecological sites, and this may become a major threat to future biodiversity in the most densely settled countries of the region.

EVOLUTION OF A CLIMATE CHANGE REGIME: THE UNFCCC AND KYOTO PROTOCOL

The first formal global attempt to control the increasing concentrations of greenhouse gases such as carbon dioxide and methane, which are trapping heat in the earth's atmosphere and causing global warming, was made in 1992, when most nations of the world joined together to sign the United Nations Framework Convention on Climate Change (UNFCCC). The UNFCCC included a legally non-binding pledge that the industrialized nations would reduce their greenhouse gas emissions to 1990 levels by the year 2000. Within a short time of its signing, scientific research had produced more evidence of climate warming and its serious impacts on sea-level rise, changes in weather patterns and effects on health. It had also become increasingly clear that major nations such as the USA and Japan would not meet the voluntary stabilization target by 2000. Owing to these concerns, parties to the treaty decided in 1995 to enter into negotiations on a protocol to establish legally binding limitations or reductions in greenhouse gas emissions, and in December 1997 the results were formalized in the Kyoto Protocol, which committed the industrialized nations to specified, legally binding reductions in emissions of: carbon dioxide, methane, nitrous oxide, hydrofluorocarbons (HFCs), perfluorocarbons (PFCs) and sulphur hexafluoride (SF6).

The Protocol established limitations only for the so-called developed countries listed in Annex I to the UNFCCC, which by late 2002 included the 24 developed countries that were members of the Organisation for Economic Co-operation and Development (OECD) in 1992, along with the then European Community, Liechtenstein, Monaco and 14 former Communist countries with economies in transition (EITs)—a total of 41 countries. The classification was based on political groupings rather than any objective criteria, and this raised legitimate concerns about equity and the effectiveness of the treaty in ameliorating global warming. Turkey, an OECD member, has successfully argued that its lower historical emissions and less advanced economy did not warrant its inclusion in Annex I, whereas Mexico and South Korea, accepted as members of the OECD since 1992, have not joined Annex I. The 1992 OECD members (except for Turkey) are also listed in the Convention's Annex II, which means they must provide financial assistance to developing countries and technology transfer to both developing countries and EITs.

The application of emission limits only to a selective group of developed countries has been severely criticized by the USA, Japan and Australia as restricting the effectiveness of the controls, particularly since carbon dioxide emissions are expected to grow at much faster rates in China, India, Latin America and other rapidly industrializing developing regions over the next few decades. Moreover, they argue that while they have accepted their responsibilities, all of this has been done without any promise of future action from the developing world. However, the decision to obtain commitments from the developed countries is based on the principles of 'common but differentiated responsibilities' and industrialized country leadership that have precedents in other international environment agreements. It reflects an understanding that while it is the responsibility of all countries to address global climate change, the industrialized countries have a special duty to take the lead, owing to their greater historical contribution to climate change, their generally higher per caput emissions, and their more abundant financial and technological resources to respond to the problem. The special responsibilities of the developed countries are emphasized by the fact that they are collectively responsible for 63% of accumulated atmospheric carbon dioxide emissions, whereas the 80% of the world's population living in developing countries has contributed about 37%.

Under the Kyoto Protocol all remaining countries are grouped together as 'non-Annex I Parties'. The only sub-category within this group consists of the least developed countries as defined by the UN General Assembly, which are to be granted special assistance and leeway in the submission of their national reports. Most of the 145 non-Annex I countries are members of the enlarged Group of 77, which now comprises 133 members and is active throughout the UN system. China is not a member of the G-77, but is usually allied with it. The G-77, along with China, covers a wide spectrum of countries at diverse levels of development and with differing interests relative to climate change. This spectrum ranges from China, which was responsible for approximately 17% of the world's carbon emissions in 2000 and is predicted to more than double its emissions by 2020, through small island states seriously threatened by sea-level rise, to the wealthy oil-exporting countries, which fear that greenhouse gas emission controls will seriously damage their economies.

The Kyoto Protocol was opened for signature in March 1998 and was designed to enter into force when at least 55 nations had ratified it, provided that these nations included Annex I countries that accounted for at least 55% of total carbon emissions in 1990. By June 2003 a total of 110 Parties had ratified the Protocol, but the 31 Annex I countries that had ratified it accounted for only 43.9% of the emissions of all Annex I Parties, still short of the 55% required for the Protocol to come into effect. The USA and Australia, accounting for 36.1% and 2.1% respectively of the Annex I Parties' emissions in 1990, both signed but refused to ratify the Protocol. The Administration of George W. Bush refused to ask the US Senate to ratify the Protocol because of lack of developing-country participation in emission controls and potentially high economic costs. The Australian Government also refused to ratify the Protocol at that time because it would have limited effectiveness without US ratification and because it did not provide for developing-country commitments. Nevertheless, Australia is working towards restricting growth in greenhouse gas emissions to its agreed limit of 8% above 1990 levels by 2008–12. Fears that the Kyoto Protocol would be permanently relegated to international irrelevance grew during 2003 as the Russian Federation showed no interest in ratifying it. This was because Russia accounts for 17.4% of the 1990 emissions of the Annex I countries, and its ratification was required to achieve the 55% 'second trigger' needed to enable the Kyoto Protocol to come into force. In May 2004 the Protocol rapidly returned to the global agenda when President Vladimir Putin committed Russia to signing the climate treaty in return for European concessions on Russian entry to the World Trade Organization (WTO). Some commentators remained sceptical, however, pointing out that the deal related less to decreasing greenhouse emissions as Russia would not have to reduce emissions at all (with the demise of many Soviet-era enterprises, Russia's 2004 output was between one-half and one-third of its 1990 levels) and more to Russia moving to the next stage of economic development as a full member of the global trading system. Others pointed out that President Putin had stopped just short of offering his unequivocal support and that the Russian Academy of Sciences had issued a report disputing the scientific basis of the Protocol and its low effectiveness for reducing the concentration of greenhouse gases in the atmosphere.

Each Annex I country that has ratified the Kyoto Protocol must reduce its emissions or, in some cases, limit its emissions growth from 1990 levels by the 2008–12 commitment period (the EITs may apply to use a different base year). The individual country targets are listed in Annex B of the Protocol and amount to a collective environmental goal of reducing total Annex I Party emissions by at least 5% below 1990 levels. The members of the European Union (EU) each have an 8% reduction target, but have redistributed the target among themselves according to a 'bubble' procedure approved under the Protocol. Despite attempts to establish objective scientific criteria such as emissions per caput, emissions per unit of GDP and projected population growth, countries tended to support whichever criteria granted them the most lenient target. The result was a set of targets derived from political negotiation, which if fully achieved will make only a very modest 1% reduction in the growth of greenhouse emissions, far short of the 60%–80% reduction from 1990 levels required to prevent the rise of between 1.4° and 5.8° in global temperatures by 2100 predicted by the latest IPCC assessment.

Flexibility mechanisms designed to help Annex I Parties meet their targets as cost-effectively as possible, such as emissions trading, joint implementation and the Clean Development Mechanism, have 'watered down' the original reduction targets. They have also opened the door to allow remaining countries to meet their targets by buying hot air and forest sink credits and thereby minimize or eliminate their reductions of greenhouse gas emissions. For example, one expert has estimated that the USA, the world's largest producer of greenhouse gases, with effective use of greenhouse gas sinks and joint implementation, could achieve compliance with its required target of a 7% reduction on 1990 levels by 2008–12 with just a 2%–3% real reduction on its 1990 emission levels. The Bush Administration rejected even this modest reduction, however, and instead in early 2003 announced a programme urging voluntary action by corporate America to meet the President's 2002 greenhouse gas intensity reduction target of 18% by 2012. Greenhouse gas intensity measures the ratio of greenhouse gas emissions to GDP and differs markedly from the Kyoto Protocol, which has mandated reductions in total emissions. The World Resources Institute (WRI) has asserted that this initiative is inherently flawed, as it will not reduce greenhouse gas emissions. Indeed, in the view of the WRI, the initiative lacks any real environmental benefits and, instead of encouraging a reduction in emissions, will allow an increase of 14% by 2012.

Current trends in greenhouse gas emissions in other parts of the developed world are also not encouraging. A report by the European Environment Agency showed that greenhouse gas emissions, far from declining as was confidently predicted in the EU, had risen for the second consecutive year and were 1% greater in 2001 than in 2000. It now appears that the EU has no chance of meeting the 8% reduction demanded by the Kyoto Protocol. Germany and the United Kingdom have experienced modest increases, but Spain and Ireland have witnessed rises of 32% and 31% respectively, instead of their permitted increases of 15% and 13%, while Greece and Portugal have also exceeded their targets. The UNFCCC cites projections provided by governments of developed countries themselves to predict that the combined emissions of Europe, Japan, the USA and other highly industrialized countries could grow by 8% from 2000 to 2010 (to about 17% over the 1990 level), despite current attempts to limit them. The transition countries of Central and Eastern Europe are also starting to increase their emissions again. New Zealand, by ratifying the Protocol, agreed to stabilize emissions at 1990 levels by 2008–12 but, since 1990, with continued economic growth, these have risen by 2%–5% annually. In order to achieve its target, one analyst claims that the country must now reduce emissions by more than 33% from its current business-as-usual level by 2012, an impossible goal for the New Zealand economy. With depletion of its Maui gas reserves, New Zealand is set to become more dependent on coal for power generation, with the resulting increase in carbon emissions. Moreover, as an agricultural exporter, agriculture is responsible for most of the nation's greenhouse gas emissions, and of all non-carbon-dioxide greenhouse gas emissions, 65% are agricultural methane emissions from farm livestock. In view of the economic costs involved, enforcement of the Kyoto Protocol could be very damaging to the economies of the countries of the EU and New Zealand.

Critics of the Protocol claim that, given its inability to establish long-term scientifically based goals, the threats it poses to national economies and its ineffectiveness in addressing climate change because it excludes major parts of the world, enforcement of the Kyoto Protocol could be worse than doing nothing. Its supporters argue that it has shown a way forward, even if that way is still flawed and uncertain, and that it can be adapted to accommodate a variety of new approaches when negotiations begin in 2005 on commitments for Kyoto's second commitment period beyond 2012. In particular, Dr Rajendra Pachauri, Chairman of the UN's IPCC, would like to see a refocusing on the long-term problems of climate change and what should be done to solve it, rather than continued acrimonious debate about the details of the Protocol. According to Pachauri, the world has to agree on target greenhouse gas concentration levels and the way to achieve them, keeping in mind that changes arising from emissions that have already occurred will be evident for hundreds of years, no matter what is done to reduce future emissions. In a recent review of the options available to build on the Kyoto Protocol, the WRI highlighted six key lessons: the current regime provides a solid foundation on which to build; the climate regime needs stronger leadership and US participation; information systems and capacity in developing countries need strengthening; successful implementation requires domestic buy-in and acceptance; simplicity is a virtue; a two-track future approach might best balance interests, simplicity, and fairness.

Certainly, the construction of an internationally accepted climate change regime has proved much more difficult than envisaged at the 1992 Earth Summit. This is because today's national and international institutions are relatively poorly equipped to deal with a problem that is still inadequately understood, and which has the potential to cause irreversible and differential change to the environment of 6,400m. people and nearly 200 countries. Creativity and innovation are needed to overcome issues of economic development, political loyalties and welfare, but perseverance with a system of emission reductions, flawed though it is, may be the best way of building the mutual trust and confidence required to prepare the way for more effective longer-term solutions.

THE IMPACT OF MEGAPROJECTS

Encouraged by economic success and by the immense needs of their burgeoning populations, the countries of East and South-East Asia have launched a disproportionately high number of megaprojects over past decades. These have ranged from massive population redistribution schemes, such as the transmigration project of Indonesia, and the Federal Land Development Authority (FELDA) schemes of Malaysia, to projects aimed at hydroelectric power generation and the provision of increased water storage by the building of huge dams such as the Three Gorges Dam in China, the Mekong Basin dams in Thailand and Laos and the Bakun Dam in Sarawak.

Three Gorges Project, China

Perhaps the project that emphasizes most clearly the environmental and social issues raised by these enormous schemes is the Three Gorges Project in China. By any measure this is a huge project, which will dam the Changjiang (River Yangtze), China's biggest river and the world's third longest, with a massive concrete wall 2 km long and 185 m high. This will be the largest concrete structure in the world which, like the Great Wall before it, will be visible from space. The dam will create a reservoir 660 km long, stretching from the dam wall near Yichang to Chongqing, China's largest municipality with more than 30m. inhabitants. When complete, it will be the biggest electricity-generating facility in the world, with an installed capacity of 18,200 MW (equivalent to 11% of China's total consumption). It will include the largest ship-lift in the world and make the Changjiang navigable to ships of up to 10,000 tons for over 1,600 km inland, all the way to Chongqing, which will thus become one of the world's most important seaport cities. Cost estimates range from US $11,000m., when approved in 1992, to as much as $100,000m. Construction began in 1994, with completion scheduled for 2009. Proponents of the dam claim that it will protect from disastrous floods millions of people living along the middle and lower reaches of the river (which killed some 300,000 people in floods in the 20th century alone); that it will generate up to 18,000 MW of electricity for China's industrial centres and thereby provide a much-awaited economic boost for the long-neglected hinterland; and transform the fast-flowing Changjiang into a smooth, navigable waterway for ocean-going vessels. The dam's power-generation capacity will save the burning of 100m. tons of coal and prevent the emission of 2m. tons of sulphur dioxide into the atmosphere, where it is the major cause of acid rain.

Unfortunately, this massive project is predicted to extract equally huge social, cultural and environmental costs, so much so that its environmental impact has been likened to that of the Soviet draining of the Aral Sea. Because of these fears, international financial support for the project has been withdrawn, and China now has to raise the funds itself. Premier Zhu Rongji, appointed in March 1998, limited the project's access to state funds and, unlike his predecessors, failed publicly to endorse the scheme. The first problem is concern over the ability of China effectively to manage quality control and minimize corruption in such a huge project. These doubts were confirmed with the arrest of the head of the now-defunct Three Gorges Economic Development Corporation for corruption, the scandal also involving other senior officials. Premier Zhu Rongji reinforced these concerns when in 1999 he used the phrase 'tofu dregs' to describe inferior dikes and dams built along the flood-prone Changjiang, and in April 2002 a senior Chinese expert stated that cracks had appeared in the concrete wall of the dam. In June 2002, just as the gates on the dam were closed for the first time, persistent leaks under the dam wall led to renewed concern. The second problem is that the three gorges themselves, the site of some of the most scenic and beautiful landscape anywhere in the world, will be partially flooded and devalued as a wilderness area. Third, the reservoir's 47,280 gigalitres of water will flood six counties and some 100,000 ha of cropland, including 46,000 ha of the most fertile land along the river basin, at a time when the loss of agricultural land is threatening China's ability to feed itself. Fourth, by changing the hydrology of the river for thousands of kilometres, the dam will destroy commercial fish stocks and threaten with extinction endangered animal species unique to the region such as the Siberian white crane, the Yangtze River dolphin, the Chinese alligator, the finless porpoise, as well as the rare Chinese sturgeon. Fifth, the rising dam waters will force at least 1.2m. local residents to abandon their ancestral homes and terraced farmland. Officials have planned for these people to be relocated in stages, 600,000 by 2003, and the remainder by 2009, hopefully to other farming areas. Recent reports acknowledge that there is widespread opposition to resettlement, villagers being fearful that they will suffer a fate similar to most of those previously relocated from other dam sites in China, who have been left destitute, without adequate compensation or shelter, forced to eke out an existence on poor farmland in inhospitable locations far from their homeland and resented by the local people. Sixth, there is increasing concern from environmentalists that by slowing the flow of water the dam will lower the Changjiang's ability to dissipate the huge load of human, industrial and mining waste which it currently receives. In the reservoir area alone there are 3,000 factories and mines, which produce 10,000m. tons of waste per year containing 50 different toxins. Some of these chemicals and heavy metals will end up in the reservoir along with the sewage from Chongqing, and result in increased waste content levels. International medical literature has recently highlighted the public health risks posed by the dam, particularly the threat of schistosomiasis, a parasitic disease that is endemic along the Changjiang below the dam.

Geologists are also concerned about a seventh problem, the high degree of sediment carried by the Changjiang, which is presently transported downstream and deposited as fertile silt on the river's floodplain or as sediment at the river's mouth near Shanghai. Construction of the dam will trap much of this silt in the reservoir where it could shorten the life of the dam, perhaps block effluent pipes, and even prevent the opening of sluice gates, thereby threatening the safety of the dam in a major flood. Similar problems occurred at the 'indestructible' Banqiao dam, which drowned up to 230,000 persons when it collapsed after record rains in 1975. Uncertainty over the effectiveness of methods to control the build-up of sediment, combined with China's past history of dam disasters, heightens concerns about the safety of such a huge structure on China's biggest river. An eighth and related concern is that, by reducing sediment supplies to the coastline, the dam will cause increased coastal erosion at the mouth of the Changjiang and lead to the loss of some of China's most productive farmland, an effect that would be compounded by any future sea-level rises. A ninth effect is that river discharge downstream of the dam will be reduced and expose the water supply of Shanghai to salt-water intrusion, which is already a problem. Finally, the dam will submerge some of China's most outstanding historical sites and cultural relics, including Dachang, a town settled 1,500 years ago during the Three Kingdoms period. Unfortunately, the time and money available to save the archaeological relics is insufficient, and in the absence of a major international effort, something still not permitted by the Government, only a fraction of the relics can be saved.

In 1997 officials decided to flood the dam's 660-km reservoir to its upper operating level of 175 m in the sixth year of operation (2009), instead of during the 10th year as originally planned. This, it was hoped, would quickly maximize electricity output and speed up repayment of the capital costs. A reversal of this decision was requested in a petition to China's senior leaders by 53 eminent academics and engineers, on the grounds that rapidly flooding the reservoir would increase sediment build-up (it would begin before erosion in the tributary streams could be controlled), submerge drainage outlets and back up sewerage from Chongqing city, and force an additional 300,000 people to leave their homes. In May 2004 the Three Gorges reservoir was lowered to prepare for the summer flood season, and some strengthening work was ordered on the temporary cofferdams built during construction of the right-bank powerhouse. However, pollution in the reservoir is emerging as a problem, with the State Environmental Protection Authority (SEPA) issuing an emergency notice to clean up floating matter comprising trees, rubbish and even livestock washed into the water by torrential rain. SEPA is concerned that the build-up of such material could hinder navigation, further damage water quality in the reservoir and even pose a threat to the turbines currently generating electricity at the dam.

RECENT DEVELOPMENTS

Of the many recent events that have caused environmental concern in the Asia-Pacific region, four have been selected for brief discussion. The first is that of the forest fires in Indonesia in 1997–98, which sent a cloud of unwelcome haze across the borders of neighbouring Singapore and Malaysia, emphasizing once again that major environmental problems do not respect national borders. The second is the environmental significance of major flooding in China and Indonesia in 1998 and 2002. Two long-term very difficult environmental problems will then be highlighted. The first is recent evidence that suggests global warming is threatening the very survival of coral reefs in the Asia-Pacific region, and the final topic is the threat of wet and dry land salinization to agricultural land in Australia. Both suggest that the problems facing the region must be robustly addressed by both its developed and less developed countries.

Forest Fires in Indonesia

Between August and November 1997 the Asia-Pacific region experienced its most serious transborder environmental problem when Indonesia, Malaysia, Singapore, Brunei and southern Thailand were smothered by thick smoke haze originating from forest fires in Indonesia. Although geological and historical evidence suggests that similar forest fires have a long history in the region and occurred in recent severe dry seasons in 1982, 1987, 1990 and 1991–95, the smoke generated by the fires of 1997 was more serious. The 1997 haze was so severe that it provoked international consternation and an unprecedented apology from the Indonesian Government. The 1997 forest fires began in July of that year, and by late September were raging out of control in central and South Sumatra, and in central, South and West Kalimantan. Smaller fires were also reported in East Kalimantan, Sulawesi, eastern Indonesia and in the mountain areas of Java and Irian Jaya (West Papua, now Papua). By early December 1997 most of these fires had been doused by north-east monsoon rains. Unfortunately, these rains failed to reach East Kalimantan, and by mid-January 1998 extremely dry conditions had combined with economic crisis and impoverishment to drive people into the forest to clear land by means of fire, thereby creating a wildfire in the whole of the Mahakam basin, west and south of Samarinda, the provincial capital. Kutai National Park, north of Samarinda, one of the island's most significant rainforest and ecological reserves and home to around 2,500 endangered orangutans, had by late March been extensively damaged by fire. By the beginning of May 1998, when rain put out the fires, 5.2m. ha had been burned in East Kalimantan, of which more than 630,000 ha was totally burned out or severely burned dipterocarp forest (most of which had been selectively logged), while another 590,000 ha was peat-swamp forest, plantation and degraded forest. Serious damage had been done to the habitat of many endan-gered animals, and a toxic smoke haze was generated over the province causing respiratory problems for some 5,000 people.

In economic terms, the direct fire-related damage to Indonesia has been estimated to total US $7,900m., and the loss of forest resources has been considerable. Valuable timber and rattan have been destroyed, along with the houses, crops and fruit trees of local residents. The total land area burned by the 1997–98 fires was 9.7m. ha, of which 4.8m. ha was forest, and the damage to biodiversity and to the health of forest ecosystems has been profound. The most serious large-scale and intense fires of 1997–98 were in logged-over forest where the canopy had been broken and logging residue on the forest floor had dried out and became ready fuel. Other areas of disturbed forest, particularly areas of secondary forest favoured by swidden cultivators, were severely damaged, and with regular fires the floristically-rich multi-layered forest vegetation can be converted into a single-layer fire-resistant *alang-alang* (*Imperata cylindrica*) grassland. Damage to the primary forest was less widespread and was concentrated in the more exposed areas near logging roads or other disturbances, and often confined only to the undergrowth. Other serious concerns were expressed about the damage the fires had done to already endangered animal species such as the Sumatran tiger, Asian elephant, Javan rhinoceros, sunbear and the orangutan, particularly when the fires spread into national parks.

Interestingly, the health and economic costs of the smoke haze created by the fires received more attention than the direct damage, probably because of the inter-national implications and the discomfort caused to residents of Singapore and Kuala Lumpur. The immediate dangers of the smoke haze were highlighted on 27 September 1997, when an airbus crashed while attempting to land at Medan, North Sumatra, killing all 234 persons on board, followed about 10 hours later by the collision of two vessels in the Melaka (Malacca) Strait with the loss of 29 sailors. In both accidents, poor visibility caused by the haze was believed to be a contributing factor.

There is little doubt that 'background haze' caused by local air pollution has been getting worse over the last decade in Peninsular Malaysia, where it is particularly serious during the dry season from July to August. In 1997, however, the level of pollution was increased dramatically by the proximity of the fires in Sumatra and the transport of smoke particulates from Kalimantan by a south-easterly wind stream. During September–October 1997 the smoke haze became so dangerous to health that many local populations in Sumatra and Kalimantan had to remain indoors. Analysis of the smoke haze revealed that of the five main haze components, ozone, sulphur dioxide, carbon monoxide, nitrogen dioxide and particulates, it was only the level of particulates that was elevated enough to cause concern in Kuala Lumpur and Singapore, but the high level of particulates in Sarawak quadrupled the incidence of adult asthma and increased significantly the incidence of upper respiratory tract infections and conjunctivitis. An estimated 70m. people in South-East Asia were affected by the haze, of whom 2,000–3,000 may have died as a result of related diseases.

The haze-related damages alone incurred by the 1997 fires have been estimated at US $1,384m. for the whole region, with the greatest burden falling on Indonesia, where estimates exceeded $1,000m., with more than 90% of the losses attributable to short-term health costs and the remainder owing to lost tourist revenues, airline cancellations and airport closures. Costs to Malaysia exceeded $300m., mainly from industrial and agricultural production losses and a sharp decline in tourism revenue, while Singapore lost more than $60m. in tourist receipts. The estimates do not take into account long-term damage to health, which may persist for decades and surpass short-term health costs.

The onset in May 1997 of the second most intense climatic phenomenon of the last 30 years, known as El Niño, exacerbated normal dry-season drought conditions and increased the fire hazard in the months that followed, but almost all of the fires were intentionally lit to assist land-clearing by commercial oil-palm and timber plantations, government-sponsored transmigration projects, 'spontaneous' migrants and local small-scale swidden cultivators. Logging activities, resulting in clearing, road-building, the expansion of secondary forest and increased fragmentation of forest cover, have been the major contributor to an increase in the risk of forest fires over the last decade. Any future attempts to prevent and control the impending risks and hazards of fire must incorporate the enforcement of government regulations, which prohibit the use of fire by large timber and plantation companies and by small farmers. They must also educate swidden cultivators, who have traditionally used small fires to prepare secondary forest for planting, of the increased risks of controlled burns becoming extensive wildfires in times of intense drought and in increasingly degraded forest environments. Attempts must also be made to encourage the application of fire-breaks and the use of tractors to remove the larger trees before controlled burns clear the rest of the land. In general, what Indonesia and its neighbours need is not more and better fire-fighting, but more effective governance that can enforce a fundamental change in land management. Problems posed by the monetary and political crises currently overwhelming Indonesia, and its excessively rapid move to regional autonomy, suggest that the changes in governance to which its citizens aspire will not occur in the near future.

In August 1999, and again in March 2000, smoke haze from forest fires in the Indonesian islands of Sumatra and Kalimantan once more began to shroud parts of neighbouring Malaysia and Singapore. However, this time the Indonesian Government's publicly stated intention to summon logging companies and review their licences appeared to pre-empt another crisis like that of 1997–98.

Flooding in China and Indonesia

Across East, South and South-East Asia the summer monsoon months of June to September of 1998 brought major flooding which resulted in the deaths of some 6,700 people. In China the Changjiang basin, home to some 400m. people, bore most of the damage from floods said to be the worst since 1954, with more than 3,000 people being killed, 14m. persons made homeless and 4.4m. ha of cropland destroyed, but streams in the north-west and north-east of the country also reached record levels. Total damage in central China alone was estimated at US $20,000m. The flooding extended to other parts of East and South-East Asia, hundreds of people being killed in South Korea and in February 1999 southern Mindanao island in the Philippines also experienced death and devastation from severe flooding.

Government officials and meteorologists have placed much of the blame for the floods on heavier-than-usual monsoon rains caused by La Niña, the destructive twin of the weather phenomenon known as El Niño, which caused drought and contributed to forest fires in South-East Asia in 1997–98. Prominent environmentalists, however, while acknowledging the role of La Niña, also point out that, although heavy, rainfall in 1998 was in fact lower than in previous major flood years. Instead, they attribute most of the blame for the record floods to decades of poor land-use management, which has resulted in widespread deforestation, soil erosion, siltation, and more rapid and concentrated run-off. In China the last 20 years have witnessed widespread deforestation from the once heavily forested Qinghai-Tibet Plateau to the Changjiang basin, which has now lost 85% of its original forest cover. The result is much more rapid run-off and serious erosion, which has pushed massive quantities of silt into the river and raised the river bed and flood levels downstream. Reclamation of more than 13,000 sq km of natural lakes in the middle and lower reaches of the river and the silting up of those that have survived has reduced the capacity of these lakes to act as natural flood reservoirs. The storage capacity lost has been estimated at equivalent to about 5.8 times the capacity of the Three Gorges Dam when it is completed. The spontaneous settlement of landless migrants on embankments in the flood-prone lake areas where they build their own dikes has worsened the flood problem by further reducing the storage capacity of the lakes and exposing more people to the risk of annual flooding. The Philippine floods were also attributed to past deforestation.

The positive outcome is that the damage caused by the flooding, at least in China, has forced the authorities to take seriously fears expressed by a new generation of ecologists that China's economic growth is environmentally unsustainable if present policies and practices are continued. In an attempt to prevent further deforestation, in August 1998 China's State Council banned the opening up of new lands by the cutting of forests, suspended all forest land construction projects, and directed that occupation of all forest land required cabinet-level approval. These logging bans initially applied mainly to the Changjiang and Huanghe valleys, but in early 1999 they were extended to 18 provinces with 1.2m. sq km of woodlands. China has also initiated plans to grow an additional 6.7m. ha of young quick-growing trees to offset the shortage of timber in its domestic market caused by the logging ban and obviate the need for imports. The lack of basic research prior to the announcement of these decisions has been criticized, but at least the Chinese authorities have taken the problem seriously and made some courageous decisions. Unfortunately, there is little evidence of a similar political commitment in the other countries that experienced the serious floods of 1998.

In January, February and March 2002 the 16m. residents of Greater Jakarta were reminded of the intimate interrelationship between urban and rural environmental changes and their consequences. At the end of January the city experienced heavy, but not exceptional, rains at the height of the annual wet season. After almost a week of rain to the south of Jakarta in the vicinity of Bogor and Ciawi, in the catchment area of some of the 13 streams that criss-cross the alluvial plain on which Jakarta is sited, residents of the city awoke in the early hours of Saturday 2 February to find major roads in the city and even the grounds of the Presidential Palace under water. In the next three weeks more than 300,000 residents, mostly poor people with houses on low-lying land near the Ciliwung and Pesanggrahan Rivers and their tributaries and many others in North, East and West Jakarta, were evacuated to temporary shelter in public buildings, as their houses were flooded with filthy water to a depth of 2–3 m. At least 20 people lost their lives, and subsequent disease caused by polluted drinking water and piles of garbage caused additional deaths. At the height of the floods it was estimated that some 35% of the city was under water, in what officials acknowledge was the most extensive flooding ever experienced by Jakarta.

It should be emphasized that annual flooding is not a new problem for Jakarta; the Dutch colonial administration built an intricate system of canals that minimized, but did not solve, the problem. However, Indonesia's environmentalists claim that the impact of the major recent floods of 1996 and 2002 has been exacerbated by rapid population growth, unplanned and badly planned urban development, rapid deforestation, ineffective governance and a lack of an appreciation that if environmental constraints are ignored then widespread flooding, landslides and expensive disruptions to the city's economic life and increased hardships for the city's poor, will be the inevitable result. In just four years between 1988 and 1992 some 2,800 ha of agricultural land on the city's outskirts were converted into housing and industrial estates, causing massive disruption to the city's natural drainage system and increasing the rate of run-off. In the 1950s there were some 185 small ponds in and around Jakarta with a total area of 1,304 ha, which created a natural catchment for run-off during the wet season. By the late 1980s there were still six natural ponds in Jakarta, but by 1994 these had been reduced to just two. Suburbanization, golf course construction and ill-planned industrial development has also reduced the number of drainage ponds in the neighbouring local government areas of Bogor, Bekasi and Tangerang in West Java. At the same time no significant progress has been made on construction of the East Flood Canal, first planned in the early 1970s, and on public works to clear silt from the West Flood Canal, built by the Dutch in 1924. Development of an exclusive residential complex on a former protected marsh area and mangrove forest in North Jakarta, which was approved by the former Suharto Government, has been blamed by ecologists for worsening flooding of the toll road linking Jakarta to its international airport and for disruption of the ecological balance of the Jakarta Bay coastline. Serious ecological damage has also occurred in the catchment areas south of Jakarta, where past governments have encouraged excessive deforestation of hilly and mountainous terrain that has increased the rapidity and scale of the run-off following torrential wet season downpours. Official sources downplay the role of corrupt and ineffective governance in creating this ecological tragedy, and instead tend to shift the blame to Jakarta's poor, who are held responsible for clogging the city's drainage system with their garbage and squatter housing. Of course, some of the problems can be attributed to the behaviour of the city's poor, but it is the large conglomerates, estate developers and senior government officials that have had the power and the wealth to prevent these cases of ecological negligence. Unless this behaviour changes in the near future, Jakarta will have little choice but to prepare its poorer residents for more serious annual flooding in future years.

Threats to Coral Reefs

In general, coral reefs are found in shallow waters, between the Tropic of Capricorn and the Tropic of Cancer. Their total extent is unknown, although it probably exceeds 600,000 sq km. They represent less than 0.2% of the total area of oceans and cover an area equivalent to 4% of the world's cropland area. Coral reefs resemble tropical rain forests in that they both thrive in nutrient-poor conditions (where nutrients are largely tied up in living matter), yet support rich communities through incredibly efficient recycling processes. Both exhibit very high levels of species diversity. Coral reefs and other marine ecosystems contain more varied life forms than do land habitats, with all but one of the world's 33 phyla (major kinds of organisms) being found in marine environments. The species richness of coral reefs is unparalleled in the marine environment and the most species-rich reefs are found in the Asia-Pacific region in a swath extending throughout South-East Asia to the Great Barrier Reef, off north-eastern Australia. The Great Barrier Reef (GBR), the world's largest system of coral reefs (covering 349,000 sq km), supports more than 700 species of coral, 1,500 species of fish, and more than 4,000 species of molluscs. In addition, 252 species of birds nest and breed on the coral cays, five species of turtles live on the reef, and several species of whales and dolphins are associated with it. Indeed, coral reefs of the Indo-West-Pacific support more than 16% of the world's estimated 19,000 species of freshwater and marine fish. The GBR alone, occupying only 0.1% of the ocean surface, supports nearly 8% of the world's fish species; in comparison the coastal waters of the Mediterranean Sea support less than 25% as many fish species. The level of species endemism is much lower on coral reefs than in tropical rain forests, however, which means that they are less threatened by species extinction. Nevertheless, for many communities in the Asia-Pacific region coral reefs are a vital source of food, an attraction for much-needed tourist dollars and a protective buffer for vulnerable coastlines. The survival of many of their species also depends on other affiliated ecosystems with coral reefs, mangroves, and sea-grass beds being linked physically and biologically.

In the 1990s, however, it became obvious that the coral reefs of the world were in serious trouble, with large-scale degradation occurring in East Africa, South and South-East Asia, parts of the Pacific and across the Caribbean. Nearly all of this degradation has been caused by human activity. In South-East Asia, which has about 30% of the world's reefs, rapid economic growth and increased population pressure have been concentrated in coastal areas, placing heavy reliance on marine resources. This has produced considerable sediment and nutrient pollution, especially from the major coastal cities and numerous coastal resorts on the islands in the region. The result has been the loss of much of the coral cover in reefs near centres of population, followed by a dramatic decline in fish numbers and catches. The damage is exacerbated by roving bands of fishermen who are damaging reefs by using dynamite, cyanide and muro-ami, a fishing method whereby divers (often children) pound on coral reefs to drive fish out into the net. These are the most threatened reefs of any region of the world, with some 80% of their number now at risk of degradation. In contrast, human stresses to Australian reefs are minimal, except on some reefs close to the land because population density is low, the level of economic development is high, and there is low fishing pressure. However, inner shelf reefs in the GBR have suffered degradation from increased sediment and nutrient run-off caused by cattle-grazing and sugar-cane growing

There were reports of unprecedented widespread bleaching of hard and soft corals in widely separated parts of the world from mid-1997 to the last months of 1998. This bleaching occurred in parallel with big changes in the global climate with a severe El Niño event during late 1997 and early 1998, which switched to a strong La Niña in mid-1998. This has suggested to some researchers that global warming may be posing a much greater danger to the survival of the world's corals than direct human actions. Some bleaching may be a seasonal event in the Pacific, Indian Ocean and the Caribbean. Frequently corals recover from bleaching, but death may result if the stress is prolonged or extreme. Normally fast-growing, branching corals in the Indo-Pacific are more susceptible than slow-growing boulder corals, which, if they are bleached, frequently recover in one to two months. However, in this event, some 700-year-old *Porites* corals on inner reefs of the GBR and Viet Nam were extensively bleached, and some died. Water temperatures over the entire Indian Ocean were exceptionally warm for four to five months in 1998, and corals all the way from Mozambique to Western Australia and Indonesia bleached very heavily, with many dying. Prior to this, localized bleaching had been known to reef scientists for many decades, but only on a small scale as in overheated tidal pools. Mass instances of coral bleaching, of which six have occurred since 1979, are unrecorded in earlier scientific literature, and it is claimed indigenous peoples with long reliance on reefs have no words in their languages to describe this phenomenon.

Bleaching occurs when coral organisms are stressed by excessive heat or other factors. When they are stressed they suddenly expel the microscopic plants or algae, known as zoo-xanthellae, which live in their cells in a normally very successful symbiotic relationship. The coral polyps provide the plants with crucial nutrients such as ammonia and phosphate, while the zooxanthellae deliver to the coral polyps food and essential compounds from photosynthesis. It is the microscopic algae that give the corals their tremendous range of brilliant colours, reds, browns, greens, yellows and blues, so when the corals expel their algae they lose their colour and become bleached. When this happens the coral animal loses much of its food supply and in the nutrient starved waters in which coral thrives they begin to starve and eventually die, unless water temperatures quickly return to normal.

The coral bleaching of 1997–98 was the most geographically widespread ever recorded, and the most important question is whether this was simply a severe isolated occurrence, or whether events like this will happen more frequently as temperatures rise with global warming. It is too soon to answer this question with any certainty. However, sea temperatures in tropical waters, where most corals live, have increased by nearly 1°C in the last 100 years. Being highly temperature-sensitive, many corals are already close to the top of their thermal range and with this in mind Professor Ove Hoegh-Guldberg, the director of the University of Sydney's coral reef research institute, projected that if current rates of warming continue, coral bleaching is set to increase steadily in intensity all over the world until it occurs annually by 2030–70. This would devastate coral reefs globally to such an extent that they would be eliminated from most areas of the world by 2100. Hoegh-Guldberg's projections were regarded by many as unduly pessimistic until late 1998, but have since been taken much more seriously and generally coincide with other major international models of climate change. There is now widespread concern that unless global warming and growth in the emission of greenhouse gases are rapidly curbed, most Caribbean reefs may well be dead by 2020, with southern and central parts of the GBR and many reefs in South-East Asia following by 2030. In February 2004 the World Wildlife Fund of Australia and the Queensland Tourism Industry Council released a study of how climate warming will affect the GBR. The authors of the report concluded that the best-case scenario for the Reef would be a recoverable loss of coral populations if global temperature increases remain below 2°C. Under the worst-case scenario with temperatures increasing by 6°C, they predicted that coral populations will collapse by 2100 and that the re-establishment of coral reefs will be highly unlikely over the following 200–500 years. Current indications are that the world's climate may well experience temperature increases of over 2° well before 2100.

Salinization of Water and Land in the Agricultural Regions of Australia

Located in the relatively fertile well-watered south-east of Australia, the Murray-Darling Basin is Australia's largest and economically most significant river system. It covers over 1m. sq km, equivalent to 14% of the country's total area, and is the source of 41% of the nation's agricultural wealth. The basin grows 71% of Australia's total area of irrigated crops and pastures. Although one of the world's major river systems in terms of length and catchment area, its surface run-off is among the smallest in comparison with that of other systems of similar size. Indeed, owing to the semi-arid climate, some 86% of the basin contributes virtually no surface run-off to the river systems except during floods. The basin is a naturally saline environment in terms of its soils, geology, surface water and groundwater, with the low river flows and limited leaching possible in this dry environment concentrating salinity levels. The natural flora and fauna are adapted to these conditions, but human activities have greatly increased salinity levels until they have begun to cause significant problems for all water and land users, agricultural, domestic and industrial.

In terms of surface water and groundwater salinity, the problems are of more concern in the southern parts of the basin. In the Murray, the major stream of the basin, the salinity levels increase markedly downstream owing to natural processes such as increased evaporation, and the inflow from naturally occurring saline groundwater which can have salinities up to 50,000 electrical conductivity units (EC). Water with levels of more than 700 EC is unsuitable for irrigating most horticultural crops, while 800 EC is the accepted maximum level for domestic supplies in large towns and cities. However, the naturally occurring levels of salinity have been exacerbated by drainage flows of saline water from irrigation areas and from rising groundwater levels, both of which are results of inappropriate forms of land-use management. So far, even in years of very low flow, the Murray and its major tributaries have delivered water of acceptable quality for drinking purposes, although in dry years the lower Murray water is of such high salinity that it may cause crop damage if used for irrigation. Recent research, however, shows an alarming increase in salt loads and salinity in many basin streams even further north in New South Wales, with some predictions estimating that salinity levels may exceed safe maximum levels for human consumption. This finding is of grave concern for irrigators and for residents of Adelaide and other settlements dependent on the Murray and its tributaries for domestic water supplies.

Land salinization occurs naturally in parts of the basin, but the concern here is with secondary or induced salinization resulting from European-style land use practices. In order to maximize the feed for their introduced sheep and to obtain commercial returns from the land, European settlers removed deep-rooted native grasses, shrubs and trees, which were replaced with shallow-rooted annual crops and pastures. This resulted in a major reduction in water use and increased the quantity of water added to groundwaters. The water tables then rose, gradually dissolving naturally occurring salts that had hitherto been harmlessly held in the soils above the natural water tables. As the water table rose, these newly dissolved salts were then brought towards the surface, where the salt was concentrated by evaporation. When the water table is within one metre of the surface, waterlogging and salinization frequently occur and, since the concentrated salts are now within the root zone, substantial losses in agricultural production can result. In the basin's irrigation areas (the most extensive in Australia and the basis of the country's rice, horticultural and viticultural industries) the problem of rising water tables owing to removal of the native vegetation has been compounded by the application of large quantities of water, frequently, at least in the early years, without any drainage facilities to remove surplus water. A decade ago it was estimated that 96,000 ha were affected by salinization, while another 560,000 ha had water tables within 2 m of the surface and rising rapidly. This rise in water tables in one irrigation district of southern New South Wales has been at the rate of 27 cm per year, and in another area in Victoria it has been described as an 'underground flood'. High salinity levels reduce irrigation crop yields and can cause loss of orchard trees. In some areas productivity losses of at least 7% are now common, with losses of 30% predicted within 30 years. By the year 2010 it is expected that all irrigation areas in the southern basin will have water tables within 2 m of the surface, casting severe doubts on the sustainability of agricultural production in areas where the groundwaters are saline. Fortunately, the problem is now being slowly recognized, and a range of measures such as surface and sub-surface drains, groundwater pumping, use of groundwater for irrigation, water harvesting, tree planting, disposal of saline water into evaporation basins, and more efficient water usage and district-wide action plans supported by public funds are now being implemented. The problem is now so serious, however, that a long-term commitment beyond the tenure of several governments is urgently required.

Although induced salinization was once regarded as a problem only in irrigated lands, it is now known that salinization is a far more extensive problem on the basin's drylands and in other parts of Australia (especially in the south-west of Western Australia where some 443,000 ha of land is affected), with both croplands and pastures being damaged by rising water tables. Nationally, it has recently been estimated that dryland salinity threatens one-fifth of Australia's arable land (an area three times the size of Tasmania), and this process cannot be reversed easily or quickly. Excessive removal of native vegetation has been the major cause of these rising water tables. A 1993 study estimated that at least 200,000 ha of land were seriously affected and that more than 1m. ha were at risk of dryland salinity. A 1995 study greatly increased the area at risk, estimating that in New South Wales alone up to 5m. ha were endangered if present land-management practices continued. The land affected may be lost to agricultural production or experience reduced productivity if it is still capable of use. Local communities have become more aware of the problem in the last few years, and are now growing more trees and perennial pastures and making less use of long fallows, in an attempt to reduce accessions of water to the water tables. The abandonment of land and the introduction of more salt-tolerant crops, however, may be the only alternative in the more badly affected areas. In June 2001 all of the state and territory governments responsible for parts of the Murray-Darling Basin joined the Commonwealth of Australia and a Community Advisory Committee in committing themselves to an integrated catchment management programme to rehabilitate the Basin. This is a critical first step in the process; however, difficult political decisions, such as forgoing current water rights in the interests of enhancing environmental stream flows to safeguard long-term agricultural productivity, have yet to be taken. The drought of 2002 again brought the parlous state of the River Murray to the nation's attention. By mid-2003 political parties at the state and Commonwealth level had begun to address the problems of the river and had made their first substantial financial commitments.

Future Concerns

This article has served to emphasize the nexus between environmental issues and the demographic, economic and political realities of development in the Asia-Pacific region. This linkage will become even stronger in the future as the people of the region strive to achieve their legitimate aspirations of a better lifestyle. Looking ahead in this new century, it is clear that environmental issues will increasingly be international, rather than national or local, in their significance. Efforts to minimize the environmental 'footprint' of development must inevitably involve technology transfer, financial support and compensation mechanisms that are at least multinational, if not global in scale, if they are to succeed.

BIBLIOGRAPHY

General

Blaikie, P., and Brookfield, H. (Eds). *Land Degradation and Society*. London, Methuen, 1987.

Brookfield, H., and Byron, Y. (Eds). *South-East Asia's Environmental Future: The Search for Sustainability*. Kuala Lumpur,

United Nations University Press and Oxford University Press, 1993.

Burnett, A. *The Western Pacific: Challenge of Sustainable Growth*. Aldershot, England, Edward Elgar, 1992.

Ehrlich, P. R., and Holdren, J. P. 'Impact of population growth', *Science*, Vol. 171, pp. 1212–17, 1971.

Hardjono, J. (Ed.). *Indonesia: Resources, Ecology, and Environment*. Singapore, Oxford University Press, 1991.

Harrison, P. *The Third Revolution: Population, Environment and A Sustainable World*. London, Penguin Books, 1992.

Howard, M. C. *Asia's Environmental Crisis*. Boulder, CO, Westview Press, 1993.

Meyer, W. B. *Human Impact on the Earth*. Cambridge, Cambridge University Press, 1996.

McDowell, M. A. 'Development and the environment in ASEAN', *Pacific Affairs*, Vol. 62, No. 3, pp. 307–329, 1989.

Parnwell, M. J. G., and Bryant, R. L. (Eds). *Environmental Change in South-East Asia: People, Politics and Sustainable Development*. London, Routledge in association with the ESRC Global Environmental Change Programme, 1996.

Ramphal, S., and Sinding, S. W. (Eds). *Population Growth and Environmental Issues*. Westport, CT, Praeger Publishers, 1996.

World Bank. *World Development Report 1992: Development and the Environment*. New York, Oxford University Press, 1992.

Loss and Degradation of Agricultural Land

Centre for International Earth Science Information Network, Columbia University, USA; internet www.ciesin.org.

Huang, Jikun. 'Land Degradation in China: Erosion and Salinity'. A report submitted to the World Bank. Beijing. Center for Chinese Agricultural Policy, Chinese Academy of Agricultural Sciences, March 2000; internet www.ccap.org.cn/englishtalk/WP-00-E17.pdf.

Smil, V. 'Land degradation in China: an ancient problem getting worse' in P. Blaikie and H. Brookfield. *Land Degradation and Society*. London, Methuen, pp. 214–222, 1987.

'Environmental problems in China: estimates of economic costs', *East-West Center Special Reports*, No. 5, pp. 1–62, Honolulu, HI, East-West Center, 1996.

Deforestation

Angelsen, A. 'Shifting cultivation and 'deforestation': A study from Indonesia', *World Development*, Vol. 23, No. 10, pp. 1713–29, 1995.

Angelsen, A., and D. P. Resosudarmo. 'Krismon, Farmers and Forests: the Effects of the Economic Crisis on Farmers' Livelihoods and Forest Use in the Outer Islands of Indonesia', Bogor, Center for International Forest Research, 1999.

Bernard, S., and De Koninck, R. 'The retreat of the forest in Southeast Asia: A cartographic assessment', *Singapore Journal of Tropical Geography*, Vol. 17, No. 1, pp. 1–14, 1996.

Brookfield, H., Potter, L., and Byron, Y. *In Place of the Forest: Environmental and Socio-economic Transformation in Borneo and the Eastern Malay Peninsula*. Tokyo, United Nations University Press, 1995.

Collins, N. M., Sayer, J. A., and Whitmore, T. C. (Eds). *The Conservation Atlas of Tropical Forests: Asia and the Pacific*. London, Macmillan, 1991.

Food and Agriculture Organization of the United Nations. State of the World's Forests 2001 and 2003. Rome; internet www.fao.org/forestry/index.jsp. Link to Publications: State of the World's Forests.

Forest Watch Indonesia and Global Forest Watch. *The State of the Forest: Indonesia*. Washington, DC, Global Forest Watch, 2002; internet http://pubs.wri.org/index.cfm. Link to Publications: Forests, grasslands, and drylands.

Magdalena, F. V. 'Population growth and the changing ecosystem in Mindanao', *Sojourn*, Vol. 11, No. 1, pp. 105–127, 1996.

Matthews, E. 'Understanding the FRA 2000', World Resources Institute, Forest Briefing No. 1. Washington, DC, March 2001; internet www.wri.org.

Potter, L. 'Environmental and social aspects of timber exploitation in Kalimantan, 1967–1989' in J. Hardjono (Ed.). *Indonesia: Resources, Ecology, and Environment*. Singapore, Oxford University Press, pp. 177–211, 1991.

'The onslaught on the forests in South-East Asia' in H. Brookfield and Y. Byron (Eds). *South-East Asia's Environmental Future: The Search for Sustainability*. Tokyo, United Nations University Press, pp. 103–123, 1993.

Rudel, T., and Roper, J. 'The paths to rainforest destruction: Crossnational patterns of tropical deforestation, 1975–90', *World Development*, Vol. 25, No. 1, pp. 53–65, 1997.

Tole, L. 'Sources of deforestation in tropical developing countries'. *Environmental Management*, Vol. 22, No. 1, pp. 19–33, 1998.

World Bank, Environment and Social Development Unit. *Indonesia: Environment and Natural Resource Management in a Time of Transition*. New York, The World Bank, 2001; internet www.worldbank.org. Documents and Reports Section—Indonesia: Environment. World Resources Institute in collaboration with the United Nations Environment Programme and the United Nations Development Programme. World Resources 1994–95. New York, Oxford University Press, 1994.

Water Availability and Pollution

Brown, L. R., and Halweil, B. 'China's water shortage could shake world food security', *World-Watch*, pp. 10–21, July/August 1998.

Smil, V. 'Environmental problems in China: estimates of economic costs', *East-West Center Special Reports*, No. 5, pp. 1–62, Honolulu, HI, East-West Center, 1996.

United Nations Environment Programme. *GEO Year Book 2003*. UNEP, 2004; internet www.unep.org/geo/yearbook/.

Environmental Problems in Urban Areas

Bartone, C., Bernstein, J., Leitmann, J., and Eigen, J. *Toward Environmental Strategies for Cities: Policy Considerations for Urban Environmental Management in Developing Countries*. Strategic Options for Managing the Urban Environment, No. 18, Washington, DC, The World Bank, 1994.

Douglass, M. 'Planning for environmental stability in the extended Jakarta metropolitan region', in N. Ginsburg, B. Koppel and T. G. McGee (Eds). *The Extended Metropolis: Settlement Transition in Asia*. Honolulu, HI, University of Hawaii Press, pp. 239–273, 1991.

Firman, T. 'Land conversion and urban development in the Northern Region of West Java, Indonesia', *Urban Studies*, Vol. 34, No. 7, pp. 1027–46, 1997.

Firman, T., and I. A. I. Dharmapatni, 'The challenges to sustainable development in Jakarta metropolitan development', *Habitat International*, Vol. 18, No. 3, pp. 79–94, 1994.

Hardoy, J. E., Mitlin, D., and Satterthwaite, D. *Environmental Problems in Third World Cities*. London, Earthscan Publications, 1992.

Lohani, B. N., and Whitington, T. P. 'Environmental management: An intersectoral approach' in Stubbs, J., and Clarke, G. (Eds). *Megacity Management in the Asian and Pacific Region: Policy Issues and Innovative Approaches*. Volume One of the Proceedings of the Regional Seminar on Megacities Management in Asia and the Pacific, Manila, Asian Development Bank, pp. 131–174, 1996.

McKee, D. L. *Urban Environments in Emerging Economies*. Westport, CT, Praeger Publishers, 1994.

Parai, A., Benhart, J. E., and Rense, W. C. 'Water Supply in Selected Mega Cities of Asia' in A. K. Dutt, F. J. Costa, S. Aggarwal and A. G. Noble (Eds). *The Asian City: Processes of Development, Characteristics and Planning*. Dordrecht, Kluwer Academic Publishers, pp. 205–212, 1994.

Serageldin, I., Cohen, M. A., and Sivaramakrishnan, K. C. (Eds). *The Human Face of the Urban Environment*. Proceedings of the Second Annual World Bank Conference on Environmentally Sustainable Development, Environmentally Sustainable Development Proceedings Series No. 6, Washington, DC, The World Bank, 1994.

Smith, K. R., and Lee, Y. F. 'Urbanisation and the Environmental Risk Transition' in J. D. Kasarda and A. M. Parnell (Eds). *Third World Cities: Problems, Policies and Prospects*. London, Sage, pp. 161–179, 1993.

Stubbs, J., and Clarke, G. (Eds). *Megacity Management in the Asian and Pacific Region: Policy Issues and Innovative Approaches*. Volume One of the Proceedings of the Regional Seminar on Megacities Management in Asia and the Pacific, Manila, Asian Development Bank, 1996.

'Economic losses provide spur for cleaner air', *The Weekend Australian*, Melbourne, p. 18, 1–2 June 2002. Reprinted from *The Economist*.

Webster, D. 'The Urban Environment in Southeast Asia: Challenges and Opportunities', *Southeast Asian Affairs 1995*. Singapore, Institute of Southeast Asian Studies, 1995.

Global Warming and Rising Sea-Levels

The Australian Greenhouse Office. *Kyoto Protocol*; internet www.greenhouse.gov.au/international/kyoto/index.html.

Hulme, M. 'Global warming and the implications for Asia and the Pacific', in *The Far East and Australasia 1998*, London, Europa Publications, pp. 42–49, 1997.

Intergovernmental Panel on Climate Change. The Summaries for Policymakers and Technical Summaries, *IPCC Third Assessment Report: Contributions of IPCC Working Groups*; internet www.ipcc.ch. Publications link.

McMichael, A., Woodruff, R., Whetton, P., Hennessy, K., Nicholls, N., Hales, S., Woodward, A. and Kjellstrom, T. *Human Health and Climate Change in Oceania: A Risk Assessment 2002*. Canberra, Australian Department of Health, 2003; internet www.health.gov.au/pubhlth/strateg/envhlth/climate/.

Pyper, W. 'Change in the wind: using a range of climate models, the Climate Impact Group from CSIRO Atmospheric Research has predicted significant changes in temperature, rainfall and evaporation in the next 100 years'. *Ecos*, No. 6, p. 16, July–Sept 2001.

Risbey, J., Karoly, D., Reynolds, A., and Braganza, K. 'Global warming signature in Australia's worst drought', *Bulletin of the Australian Meteorological and Oceanographic Society*, Vol. 16, p. 6, 2003.

United Nations Environment Programme (UNEP) Regional Resource Centre for Asia. *The Asian Brown Cloud: Climate and Other Environmental Impacts* I; internet www.rrcap.unep.org/issues/air/impactstudy/index.cfm.

Watson, R. T., Zinyowera, M. C., and Moss, R. H. *The Regional Impacts of Climate Change: An Assessment of Vulnerability*. New York, Intergovernmental Panel on Climate Change and Cambridge University Press, 1998.

Evolution of a Climate Change Regime: the UNFCCC and Kyoto Protocol

Baumert, K. A., with Blanchard, O., Llosa, S., and Perkaus, J. F. (Eds). *Building on the Kyoto Protocol: Options for Protecting the Climate*. Washington, DC, World Resources Institute, 2002; internet climate.wri.org/publications.cfm.

De Freitas, C. 'Europeans hoist on petard of Kyoto targets', *The National Business Review*, 29 May 2003.

Fallow, B. 'Time to focus beyond Kyoto', *The New Zealand Herald*, 15 May 2003.

Fletcher, S. R. 'Global Climate Change Treaty: The Kyoto Protocol', *CRS Report for Congress*, Vol. 98, No. 2, 2000.

United Nations Framework Convention on Climate Change (UNFCCC). Kyoto Protocol Thermometer; internet unfccc.int/resource/kpthermo_if.html.

Press releases; internet unfccc.int/press/releases/index.html.

The Impact of Megaprojects

Three Gorges Dam Project, China

Dai Qing. *The River Dragon Has Come*. Edited by J. G. Thibodeau and P. Williams, translated by Ming Yi. Probe International and International Rivers Network, 1998.

International Rivers Network. *IRN's Three Gorges Campaign*; internet www.irn.org/programs/threeg/.

Human Rights Dammed off at Three Gorges: An Investigation of Resettlement and Human Rights Problems in the Three Gorges Dam Project. 2003; internet www.irn.org.China. Link to Campaigns.

Probe International, Three Gorges Probe; internet www.probeinternational.org/pi/3g/index.cfm.

Transmigration in Indonesia

Donner, W. *Land Use and Environment in Indonesia*. Honolulu, HI, University of Hawaii Press, 1987.

Fearnside, P. M. 'Transmigration in Indonesia: Lessons from its environmental and social impacts', *Environmental Management*, Vol. 21, No. 4, pp. 553–570, 1997.

Secrett, C. 'The environmental impact of transmigration', *The Ecologist*, Vol. 16, No. 2/3, pp. 77–88, 1986.

Recent Developments

Forest Fires in Indonesia, 1997–98

Integrated Forest Fire Management Project and German Agency for Technical Cooperation; internet www.iffm.org.

Parry, R. L. 'Borneo's forests burn on the bonfire of big business', *The Independent*, London, p. 11, 9 February 1998.

Schindler, L. 'Fire Management in Indonesia-quo vadis?' Report, Integrated Forest Fire Management Project; internet www.iffm.org/ See under 'Report'.

The Straits Times, Singapore, 22 September–4 October 1997.

Wirawan, N. 'The hazard of fire' in H. Brookfield and Y. Byron (Eds). *South-East Asia's Environmental Future. The Search for Sustainability*. Kuala Lumpur, United Nations University Press and Oxford University Press, pp. 242–266, 1993.

Flooding in China and Indonesia

Becker, J. 'China's struggle to hold back floodwaters, a disaster born of progress', *South China Morning Post*, 15 August 1998.

China Daily, 'China—Logging ban scope expanded', 8 February 1999.

Dwi Iswandono. 'Corruption has contributed to flooding', *The Jakarta Post,* 31 January 2002.

Eckert, P. 'Forest loss contributes to Asia's flood woes', Reuters, 27 August 1998.

Threats to Coral Reefs

Birkeland, C. *Life and Death of Coral Reefs*. New York, Chapman and Hall, 1997.

Bryant, D., Burke, L., McManus, J., and Spalding, M. *Reefs at Risk in Southeast Asia*. Cambridge, UK, UNEP-WCMC, 2002; internet http://marine.wri.org/publications.cfm.

Hoegh-Guldberg, H. and Hoegh-Guldberg, O. The Implications of Climate Change for Australia's Great Barrier Reef. World Wildlife Fund of Australia in collaboration with the Queensland Tourism Industry Council, the Queensland Government State Development Environment Protection Agency and Tourism Queensland, 2004; internet www.wwf.org.au.

Wilkinson, C. (Ed.). 'Status of coral reefs of the world: 1998'. Townsville, Global Coral Reef Monitoring Network and Australian Institute of Marine Science, 1998.

Wilkinson, C. 'The 1997–1998 mass bleaching event around the world', Townsville, Global Coral Reef Monitoring Network and Australian Institute of Marine Science, 1998.

Salinization of Water and Land in the Agricultural Regions of Australia

Australia, Department of the Environment, Sport and Territories, State of the Environment Advisory Council. *Australia: State of the Environment 1996*. Collingwood, Melbourne, CSIRO Publishing, 1996.

Crabb, P. *Murray-Darling Basin Resources*. Canberra, Murray-Darling Basin Commission, 1997.

Murray-Darling Basin Ministerial Council. *The Salinity Audit: A 100-year perspective, 1999*. Canberra, Murray-Darling Basin Commission, 1999.

Murray-Darling Basin Ministerial Council. The Integrated Catchment Management Policy Statement under publications, *Integrated Catchment Management in the Murray-Darling Basin 2001–2010: Delivering a sustainable future*. Canberra, Murray-Darling Basin Ministerial Council, 2001; internet publications.mdbc.gov.au/; advanced search.

Walker, G., Gilfedder, M., and Williams, J. *Effectiveness of Current Farming Systems in the Control of Dryland Salinity*. Canberra, CSIRO, 1999.

RECENT ECONOMIC TRENDS IN THE ASIA-PACIFIC REGION

TRAN VAN HOA

In the years leading to 2004 the major countries of the Asia-Pacific region experienced significant economic trends. These deeply reflected a number of serious problems and issues that the countries had not encountered nor had to address in their recent history of economic development and growth and in their international relations. First, there was the continuing effect of the 1997 Asian economic and financial crisis, with severely damaging regional repercussions and 'contagion' beyond Asia. Second, these countries witnessed the sudden decline of high-technology industry around the globe, which adversely affected their important export trade in, and income from, manufactured products. Third, there was a perceived global economic deceleration owing to reduced demand in the US market, the stagnant state of the economies of the European Union (EU) and the seemingly insoluble paralysis of the Japanese economy, the world's second largest. Fourth, the terrorist attacks of 11 September 2001 in the USA, the bombing incidents in the popular tourist resort of Bali in Indonesia on 12 October 2002 and the Madrid bombings of 11 March 2004 seemed to have further exacerbated this economic deceleration. Finally, in early 2003 the Asia-Pacific countries were further troubled by the US-led coalition's military campaign (and its emerging problems one year later) to remove the regime of Saddam Hussain from Iraq and by the possibility of similar attacks on 'rogue' states in the Middle East and East Asia. Furthermore, only a few weeks after the start of the conflict in Iraq, the outbreak of Severe Acute Respiratory Syndrome (SARS), a pneumonia-like illness, affected thousands of people in more than 30 countries around the world but was most serious in Canada, the People's Republic of China, Hong Kong, Singapore, Taiwan, the Philippines and Viet Nam. In early 2004 some of these countries, and even the USA, were affected by an outbreak of avian flu, which had equally damaging consequences on their sectoral economic activities. In addition, the Association of South East Asian Nations (ASEAN) was reassessing its invitation to Australia and New Zealand to join long-awaited free-trade talks; and on 1 May 2004 a new regionalism resulted from the enlargement of the EU from 15 to 25 members, thus boosting and protecting Europe's trade and growth. This essay is an attempt to survey recent economic developments and trends in the major Asia-Pacific countries against a background of regional economic setbacks, management difficulties, persistent global slowdown, financial volatility and uncertainty, amid some good signs of recovery and better economic and trade relations in the Asia-Pacific region. Some discussion on the region's challenges, opportunities and prospects in the medium and long term, however, is also given.

RECENT DEVELOPMENTS

Period prior to the 1997 Asian Crisis

After more than two decades of 'miracle' rapid development and the robust growth that gave rise to what is known as the 'Asian (or East Asian) Economic Development Model' (AEDM), adopted and emulated by other developing and developed countries alike in the relevant contemporary literature, the major countries of the Asia-Pacific region (including Oceania but excluding the subcontinent) experienced, early in 2003, a number of difficulties that were likely to have a profound short- and medium-term impact on their economic growth and development, trade and investment, and also on international economic relations.

During the 10-year period 1988–97, for example, the newly industrialized economies (NIEs) in Asia together recorded an extraordinary average annual growth rate of 7.12% and the countries of developing Asia a rate of 7.94%, while all advanced economies combined and the EU attained only 2.91% and 2.31% respectively. According to the IMF, during the same period Australia and New Zealand achieved respective growth rates of only 3.23% and 2.29%. While many arguments have been put forward by economic and business analysts to explain these achievements in Asia, the main reasons seem to be the 'open door policy' introduced in China in 1978, government and labour reform, industrial restructuring and, most importantly, capital inflows and policy to attract and retain these funds. Some observers have preferred to group all these together in the form of an AEDM, but this position has been challenged by others as an overstatement of facts or policies that were often sequential in time but irregular in systemic policy formulation. Not all the gains above, however, were achieved without difficulty. The average annual inflation rates of the advanced economies and of the EU during the period were only 3.43% and 3.78% respectively, but those of the NIEs and developing Asia were 5.38% and 10.76% respectively. With their relatively low growth levels, Australia and New Zealand recorded inflation rates of 3.48% and 2.34% respectively. In addition, during this period, total debt outstanding and service interest and amortization reached US $451,550m. and $53,940m. respectively for developing Asia but only $526,170m. and $85,470m. for all Western hemisphere economies, according to the IMF. Furthermore, during the time of high development and growth, welfare improvement and political stability, along with social harmony and religious tolerance, appeared generally to prevail in the countries of the Asia-Pacific region.

The Aftermath of the 1997 Asian Crisis

All the above achievements of the Asia-Pacific economies seemed to dissipate in mid-1997 when Thailand suffered a massive capital outflow that resulted in the collapse of its exchange rate and currency, and then the failure of its banking and financial systems and an end to the long-standing economic development and growth. The crisis swiftly spread to other economies in the immediate region, such as Indonesia, Malaysia and the Philippines, and also to North-East Asia, affecting Hong Kong and the Republic of Korea. In 1998, therefore, the countries of the Asian NIEs recorded a contraction of 2.4%, compared with a growth rate of 5.8% in 1997. Developing Asia similarly achieved growth of only 4.0% in 1998, compared with 6.6% in the previous year. Overall, during 1998–2003 Asian NIEs reached an average annual growth rate of only 3.92%, while developing Asia attained 5.90%. Inflationary pressure in Asia-Pacific economies was reduced in the post-crisis period, however, with an average annual inflation rate of 1.70% being recorded in the Asian NIEs and 3.15% in developing Asia. During this period Australia and New Zealand experienced relatively low average annual growth rates, of 3.68% and 2.83% respectively, and low inflation rates, of 2.83% and 2.12% respectively. Total debt outstanding and service interest and amortization, which were needed to implement nationally and internationally inspired policies to rescue crisis economies, increased in developing Asia, reaching US $704,200m. and $103,300m. respectively in 2003, or an average of $69,000m. and $99,030m. during 1998–2003. Net private capital flows to developing Asia, providing the necessary momentum for development and growth, were as high as $123,200m. in 1996, but the countries then suffered massive outflows and such funds subsequently declined in the post-crisis period (to $44,900m. in 1998, for example) and mustered only $18,400m. in 2003, thus yielding an annual average of $7,750m. during 1998–2003, according to the IMF. In the first quarter of 2004, however, in spite of the ongoing problems in Iraq, the avian flu outbreak, the Spanish train bombings, the continuing and growing terrorist threat posed by al-Qa'ida, the geopolitical risks and increasing oil prices, there were signs of a weak recovery in the Asian economies and also globally, owing to stronger domestic demand and a more favourable international (mainly US) outlook. In Asia, real GDP growth was predicted to reach 7.2% in 2004 and 6.8% in 2005 in the countries of emerging Asia, 8.5% and 8.0% respectively for China, 6.5% and 6.0% for ASEAN + 4, and 5.3% and 5.0% in the Asian NIEs. For these same countries, inflationary pressure was expected to be 3.8% and 3.4% in 2004 and 2005 in emerging Asia, 3.5% and 3.0% in China, 3.7% and 3.3% in the countries of ASEAN + 4, and 2.1% and 2.4% in the NIEs. The current-account balance was forecast to be 3.0% and 2.7%, 1.6% and 1.9%, 4.6% and 3.4%, and 6.2% and 5.5% respectively (according to the IMF). In comparison, advanced countries were expected to achieve a modest growth rate of 3.5% in 2004 and 3.1% in 2005 (in contrast to the rates of 1.7% and 2.1% recorded in 2002 and 2003).

THE LONG-TERM IMPACT OF THE ASIAN CRISIS

The long-term impact of the Asian crisis has been serious, but its effects have varied in causal direction and intensity in the major countries of the Asia-Pacific region. More specifically, while all aspects of the Asian economies (except that of Japan, which had experienced a paralysis in its economic activities and performance over a decade or so) have apparently been severely affected by the crisis, other countries in the region, such as Australia and New Zealand, or some major Asian transition economies, such as Cambodia, China, Laos, Myanmar and Viet Nam, seem to have avoided being damaged by the turmoil, at least as far as their growth and trade were concerned. The impact on Australia and New Zealand and on the transition economies of Asia can be assessed from other economic perspectives such as external debts, current accounts and unemployment rates. Overall for developing Asia, the impact as measured by the critical ratio of average post-crisis over pre-crisis growth rates ranges from 7% for Indonesia, 19% for Thailand, 30% for Malaysia, 66% for Viet Nam, 78% for China, 84% for the Philippines, 95% for Laos and to 98% for Cambodia, according to IMF estimates. For advanced Asia, Taiwan suffered a decline in growth of 51%, Hong Kong 59%, and Singapore 64% between the 1988–97 and 1998–2001 periods. Surprisingly, Myanmar, a so-called authoritarian and repressive state in South-East Asia, recorded an apparently impressive ratio of 238%, but the growth rate subsequently declined. In early 2004 available indicators showed that a recovery for ASEAN + 4 (notably Indonesia, Malaysia, Philippines and Thailand) and even the Asian NIEs that were the most badly affected by the Asian crisis had taken place with, according to the IMF, a predicted growth rate of up to 77% between 2003 and 2004 for Asian NIEs.

Economic Effects

The sudden emergence of the crisis of July 1997 in Thailand had a huge economic and financial impact on that country and on other countries in the neighbouring and North-East Asian region. This crisis curbed the decades-long growth pattern of the once 'miracle' economies of Asia and arguably took them from 'being a miracle to needing one'. What caused the crisis to emerge in the first place and to spread subsequently has occupied the minds of economists in Asia and beyond since the late 1990s. Many national and international studies have, as a result, been carried out in an attempt to reach appropriate diagnosis and subsequent suitable prescriptions. The hypotheses put forward in such studies include China's 1994 currency devaluation, Japan's monetary policy of interest rate increases in the mid-1990s in an effort to revitalize its perennially ailing economy, serious nepotism and cronyism in parts of Asia, corruption in the governance of both corporate and government institutions, an alleged international conspiracy to transfer superannuation funds so as to damage national economies in Asia, a lack of transparency in financial management and reporting at the corporate and government level, and, finally, an absence of national and international (for example, the IMF) regulatory controls over international capital inflows and outflows. The consensus now is that massive capital outflows from Thailand, for whatever reasons, created the financial crisis in that country that resulted in a large number of banks and financial institutions finding themselves unable to meet obligations and commitments. The Thai Government, therefore, had no option but to assume responsibility in the midst of the turmoil. These banks and institutions were subsequently shut down, and massive redundancies in all related industries in the country quickly followed. The government budget deficit rose sharply, and advice and emergency loans were urgently sought from the IMF.

Trade and Development Effects

The sudden and substantial devaluation of the currencies of crisis economies in the Asia-Pacific region was initiated as the first line of defence against capital outflows. This was supposed to achieve its aim by expanding exports and export income and thus improving the balance account. However, in the case of some crisis economies, such as Thailand, exports were already beginning to stagnate several years before 1997. The devaluation did not seem to have any significant impact on restricting capital outflows or on increasing capital inflows. The same phenomenon was also observed in other crisis economies in Asia, such as Indonesia, Malaysia, the Philippines, Hong Kong and the Republic of Korea. Since the main impetus of development and growth in the Asia-Pacific region in the past 30 years or so had been provided by capital inflows, along with technology or the transfer of expertise, a substantial decline in these inflows from 1997 damaged the volume and intensity of development and growth. In one member country of the Association of South East Asian Nations (ASEAN), namely Viet Nam, for example, an amount of $40,000m. in foreign direct investment was regarded, after the 1997 crisis, as necessary to support and sustain the country's high average annual growth rate of 7.83% achieved during the pre-crisis period of 1988–97. In 1998, however, Viet Nam experienced a decline of more than 50% in this required level of foreign direct investment and, according to the IMF, the country's rate of growth was reduced to only 3.5% in that year.

Social and Political Effects

An early contribution to the analysis of the Asian turmoil was made in 1998 by two Australian observers, Ross McLeod and Ross Garnaut, in which case studies of 12 Asian economies and related experience from the past (for example, Mexico), as well as the regulatory and prudential functions of various international organizations in crisis management, were investigated. A close examination of the social impact of the crisis, however, was not attempted to any meaningful extent in this work. A more recent detailed analysis was given by Tran Van Hoa in 2000. In this study, the collapse of the banking and financial systems in crisis economies and subsequent redundancies in the Asia-Pacific region were seen as having resulted in massive unemployment and a higher incidence of poverty. A figure of 200m. more people made poor (not poorer) in these economies has been calculated and put forward by various individual economists and international organizations, such as the Asian Development Bank (ADB). Here, the minimum reference standard of poverty was less than one US dollar per day. Unemployment then led to a diversion of education resources, affecting the skills and training of the population for the future (a concern even for the developed countries of Asia such as Singapore) and child labour exploitation or abuses affecting family harmony and stability (especially in Indonesia and Thailand). In Thailand, the crisis had resulted in a change of government. In Indonesia, urban and regional riots, as well as sectarian and religious conflict, occurred. In the Philippines, a consequence of the Asian economic turmoil was a criminal indictment at the highest level of government. In the Republic of Korea, the whole structure of the *chaebol* (large corporations), which for many decades had served as a model of corporate success and growth (in a manner similar to the conglomerates of Japan), was deemed largely responsible for the decline in exports and in growth and was totally reformed.

ECONOMIC CRISIS MANAGEMENT

International Rescue Programmes and Their Effectiveness

While the damaging impact of the benchmark event, the Asian crisis, has been assessed as severe in the economic, financial, social, political and religious context in crisis economies, its causes have not been correctly or appropriately diagnosed either for the whole region or country-by-country, according to many economic analysts. This apparent lack of insight or understanding on the part of the IMF of the causes of the crisis in

Asian economies has given rise to heated debates on the suitability of so-called rescue prescriptions and policies imposed by the IMF as necessary conditionality for securing IMF rescue loans.

One study in 2002 addressed in detail the economic and financial crises and crisis management in a number of major NIEs and developing and transition countries alike in the Asian region. It presented assessments by well-known experts, knowledgeable in the countries under study, on the outcomes of this management as implemented by governments and monetary authorities in the crisis economies. A general evaluation of these outcomes indicates that economic and financial management policies, strongly recommended by international organizations and agencies and adopted voluntarily or involuntarily by crisis countries, have not been effective in crisis management and resolution. The obvious evidence was that, in 2003, seven years after the emergence of the Asian crisis, its impact was still causing havoc at all levels of activity in the crisis countries, with economic, financial, political and social repercussions. The study also discussed plausible and more effective alternative policies for similar crisis issues, and analysed aspects of crisis management that would achieve better short-term and long-term resolutions in any similar future economic turmoil. The relevance of the discussion and analysis to economic and financial crises and management in other regions (such as Latin America) in the early 1990s is straightforward.

Search for a New Crisis-Management Model

A general assessment of the outcomes of economic crisis management in the relevant Asian economies in recent years has been that, on a conceptual level, the simple one-instrument policy of the Keynesian theory (demand management or fiscal expansion and contraction or international loans) or monetarist (money supply or interest rate manipulation) has been inappropriate in dealing with what had emerged as immense issues of diverse economic, regional, political, cultural and social complexity in the Asia-Pacific region.

Among the alternative and more appropriate multi-instrument policies was a macro-economic mix theory first proposed in the mid-1980s and, more importantly, its adoption in conjunction with a deep understanding or knowledge about the long-term and fundamental situation and social fabric of the troubled economies under consideration. An example of the importance of local social factors in crisis management is the underestimation by international organizations of the effect of the removal of price subsidies (resulting from a budget deficit reduction policy) from energy commodities in Indonesia during the height of its economic crisis. The ensuing hardship and unrest caused a delay in the country's major economic and administrative reform programme. Another example is the misdirected focus (or blame) on the public sector in Thailand for having started the initial Asian turmoil (resulting in severe budget reductions and massive unemployment). Later analysis showed that the real cause of the crisis was the uncontrolled and rapid increase in unrestrained lending and borrowings in the private sector itself a year or so prior to 1997.

It is unfortunate that the issue of proper economic crisis management for countries in difficulty in Asia has been recently regarded as obsolescent by a number of influential countries in the world; as a result, less attention and fewer resources have been allocated to finding proper prescriptions. This may have explained why the impact of the crisis was still persisting in the region more than seven years later. There are three reasons for this neglect. First, official reports and assessments by international organizations and agencies seemed to indicate that the impact of the Asian turmoil on the non-Asian economies had been benign. Second, the complexity of crisis management in the once 'miracle' economies of the Asia-Pacific region has been so immense that it has rendered external remedial contributions less effective and immediate. Third, the underlying orthodoxy of crisis-management policies prescribed for crisis economies has been inappropriate both in the short term and in the long term.

The prospects for the Asian economies (including Japan) were rather more dismal and demanded a serious reconsideration of national policy options. The assessment was based on these economies' historical economic and international trade trends and projections and also on new developments in the region and in the US economy in early 2001. The strong persistent effect of the Asian turmoil was found upon deeper analysis to depend crucially on the fundamental situation of crisis economies. This included aspects of economic, political, social, religious, historical and regional and international relations that had both short-term and long-term implications.

The prospects for proper economic crisis management were regarded as slight unless crisis economies in Asia (or even elsewhere) and international organizations took note of the inadequacy or unsuitability of the orthodox policies adopted (which might have been very useful during the Great Depression of the 1930s or, according to the IMF's assessment, to Mexico in the early 1990s, but which might be second or third best in late 2000 or early 2004, for example, in Asia). The prospects would also be slight unless other local and national dimensions of the management were taken into account in designing and implementing economic crisis management.

ASIA'S RECOVERY

Signs of Revival

In the first half of 2000 there were signs that the Asian recovery had started. This could be seen from a number of general aspects, as judged from various reports and published statistics by the national and international organizations involved in the crisis analysis and in management advice. First, any further impact of the crisis in the Asian region had reportedly been halted. Second, emergency loans earmarked by such institutions as the IMF as financial assistance for some crisis economies had been repaid or not even used, owing to improvements in the countries' budgetary situations. Third, growth, along with various economic activities and especially exports, in most of the crisis economies in Asia had resumed, albeit at a hesitant pace and at a much lower rate than that attained by these economies in the 30 years or so before July 1997. From other relevant perspectives (for example, the social and political situations or even a closer examination of economic and financial activities at the sectoral level in detail), however, it would be rather premature to state that the countries affected by the Asian crisis have completely left the turmoil behind and to predict that the repercussions of the crisis will not remain in the months or years to come. There were many reasons for this assessment.

First, at the regional level, the Bank for International Settlements (BIS—which was established in the 1930s to promote co-operation among the major central banks around the world) reported in July 2000 that, in the first three months of that year, international bank lending to Asia had decreased by 2%, as companies repaid foreign debt and as demand for new bank loans remained weak. In fact, lending to Asia declined by $6,000m. in this quarter to $297,000m., with $1,600m. coming from bank loan repayments. Since foreign bank loans have been a main source of finance to support development, growth and technology transfer in the Asia-Pacific region, a shortage of this funding in the region would not bode well for its prospects in expanding trade, improving living standards, enhancing national and individual welfare, and participating in international economic integration and globalization. Second, at the national level, the huge costs (estimated to be about US $1,000m. for the part played by Australia alone) of restoring and maintaining peace in and reconstructing Timor-Leste (formerly East Timor) after the ethnic, religious and racial conflict there in 1999 were seen as a new and additional financial burden on the countries of the region at a time (the post-crisis period) when they were least able to afford it. Timor-Leste was not the only problem affecting a full economic recovery by the crisis countries in Asia in 2000.

In the first quarter of 2004 the World Bank assessed the global growth rate for 2003 at 2.0%. Thus, for the third consecutive year the world economy was growing at a level well below its potential, owing to international events that had undermined confidence, such as the build-up to the Iraq conflict and subsequent developments, trans-Atlantic trade tensions and concerns about SARS. Furthermore, the global economy was expected to grow by only 3.0% in 2004 and by 2.9% in 2005, renewed economic activity in the USA being offset by weaker conditions in the EU and Japan. Improvement in confidence was seen as crucial to a revival in capital spending and growth. In contrast, the East Asia and Pacific region was expected to record higher rates of growth of 6.7% and 6.6% in 2004 and 2005, mainly owing to the strength of the Chinese economy. (Excluding China, the rate of growth for the region was expected to reach only 4.1% in both 2004 and 2005.)

Volatility of Recovery Prospects in the Asia-Pacific Region

In 2001 the Asian crisis economies were still individually beset with problems that had the potential seriously to hinder their recoveries. Indonesia remained a country deeply embroiled in religious and ethnic rioting, violent and widespread unrest arising from independence movements, general mistrust of the Government, power struggles among major political parties and continuing corruption. As an example, in mid-2000 the Indonesian Government's Supreme Audit Agency reported that the country was facing $16,000m. in potential losses, owing to the misuse of emergency loans and financial abuses, and that this involved Bank Indonesia officials and several prominent business groups.

Malaysia in the early 2000s was still affected by internal political turmoil and allegations of cronyism, in addition to a severe shortage of skilled labour for its information-technology-based economic development plan. In the Republic of Korea, the restructuring of the banking and financial system and of the *chaebol* (the main targets for the country's reform programme during the crisis) has not been regarded by international economists as very successful.

The Philippines continued to be troubled by slow growth and by the emergence of zealous Islamist militants and separatists, who have resorted to violence and the kidnapping of foreign tourists to support their causes. Even Viet Nam which, after many years of negotiations, signed an historic trade agreement with the USA in July 2000, continued to be negatively affected by the Asian turmoil and remained uneasy with free-market reforms. The country was quite apprehensive about the possible adverse consequences of this trade agreement and the perceived high costs and uncertain benefits (economic and social) of its accession to the World Trade Organization (WTO) in the near future. China, while being admitted to the WTO in 2001, still faced local and national problems in implementing the organization's numerous conditions and requirements.

In late 1999 Thailand was regarded, in the reports prepared by staff of the international organizations, as a country with fairly successful economic and financial reforms and in which the longer-term impact of the Asian crisis was relatively benign. However, this reported perception (and that of similar reports and assessments for other crisis countries in the Asia-Pacific region) was seen by economic and financial experts in Thailand and overseas as having been superficially investigated, inappropriately analysed and dangerously misleading. In a country with at least 80% of its population living and working in the traditional centuries-old way in the rural areas, the impact of a major modern economic or financial crisis on the country was necessarily centred on the city/urban population. As a result, the weighted average impact on both the city/urban and rural populations of that country must be regarded as negligible. The harsh facts were that Thai business people and their relatives still continued to commit suicide, even publicly, as a result of the hardships and problems brought about by the Asian crisis. Meanwhile, strong growth in some sectors of the economy (such as exports), without equally strong transmissions of the benefits to other major sectors, was regarded as insufficient to rescue the economy from depression or stagnation and to lead it to a full recovery.

Coupled with the global conditions discussed earlier, economic development and recovery in the Asia-Pacific (but excluding China) in early 2004 were conditional upon a number of uncertain factors—local, regional and international. Local uncertain conditions included the recent violent clash between government security forces and Muslim rebels in the Songkla province of southern Thailand and the direction Malaysia might take after the change of its long-standing leadership. Regional factors included the political turmoil in the Republic of Korea, the terrorist bombings in the Philippines, the outbreak of avian flu in Asia and North America, the re-emergence of SARS in Beijing and the renewed ethnic violence in Indonesia. Even the recent strong growth in China, which has boosted development and growth in the Asia-Pacific region as a whole and beyond, has also tended to 'overheat' the economy, with a damaging domestic, regional and global impact on resources and inflation. There were, as a result, regional and global calls for China to curb its development and growth in order to effect an economic 'cooling down' through government directives on bank lending before the May 2004 holiday and in policies to halt investment in 'hot' areas such as steel, cement and property development. Globally, the emerging problems of the Iraq conflict and their possible implications for the forthcoming elections in the USA, the United Kingdom and Australia (as manifested in the election result in Spain in March 2004), the spreading of terrorist attacks to Muslim countries in the Middle East, and the effect of the enlargement of the EU to include new members (mainly from the former Soviet 'satellites'), all contribute to some significant extent to the volatility and uncertainty in the economic recovery of the Asia-Pacific region in the near future.

The above observations for some major countries in the Asia-Pacific region and in a global context seem to indicate that, despite the good signs of an improvement in various economic and financial activities, as reported or expected by national and international organizations, a full recovery in the major economies affected by the crisis had yet, in the early 2000s, to emerge.

CHALLENGES AND PROSPECTS

Economic Recovery and the USA, Japan, the EU and Global Turmoil

The role played by the USA and Japan (the latter remaining the second largest economy in the world after the USA) in the Asian crisis and in the economic recovery in major crisis countries should not be underestimated. In the early 2000s the US economy, the main market for the Asia-Pacific countries' exports, remained beset by weak demand, slow growth and enormous corporate scandals. Repeated interest rate reductions during this period seemed ineffective in stimulating the US economy. It was recognized that, while Japan's often generous foreign aid was important as an interim measure to help Asian development, the region's recovery crucially depended on the Japanese economy, especially with regard to Japan's fiscal and monetary policy. In mid-2000, however, the Governor of the Bank of Japan, Masaru Hayami, stipulated that the country's emergency fiscal and monetary policy measures, which included an interest rate policy of 0%, might have to be adjusted as soon as possible to the normal positive level. The effect of such a shift in policy might result in depressed private demand and therefore reduced growth, not only in Japan but also in the neighbouring countries of the Asia-Pacific region and beyond. Even in Japan, the then Director-General of what was the country's Economic Planning Agency, Taichi Sakaiya, and a number of cabinet ministers had urged caution in this policy shift, since it was too early to be optimistic about the Government's efforts to allow private demand to supersede government spending as the impetus of economic growth. At May 2003 Japan's most senior financial official, Heizo Takenaka, planned to use public funds to boost capital holdings at the nation's still debt-ridden banks, as their bad loans and subsequent decreases in lending to small and medium-size enterprises again curbed growth.

The deceleration in the world's two largest economies, namely the USA and Japan, as observed in early 2001 and again in 2003, would demand that crisis management in troubled economies needs effective, resilient and long-term resolutions. It also should have plausibility and minimal conditionality. Owing to the great effect a crisis can have on the poor (an increase in poverty incidence) and disadvantaged and vulnerable (social and welfare deterioration) of the population, especially in developing and transition economies in the Asia-Pacific region (where 'safety net' and social security systems are not yet well developed or even seriously considered by many governments), a proper crisis management policy should also impart an aspect of urgency to all national governments and international organizations responsible for highly effective crisis resolution.

During 2003 an international assessment of the EU gave a bleak review of the organization's institutions and structural reform and its impact on growth prospects and on the global and the Asian economy. The problems cited by the IMF included generous unemployment compensation, central wage bargaining processes, strict employment protection (for example, substantial dismissal costs) and high taxation of labour income, all of which contributed to regional unemployment rates far higher than those recorded in the USA. Lack of labour market competitiveness in the EU was claimed to be responsible for a 3% increase in its unemployment rate and lack of product market competitiveness for a 5% reduction in its consumption and investment over the medium term. Since countries in the Asia-Pacific region have had strong trade links with the EU (second only to those with the USA), their development and growth will be substantially affected by these stagnant and restrictive conditions in the EU. The discussion earlier on the weak recovery of the global economy as of mid-2004 (with growth rates anticipated by the World Bank, in 2004 and 2005, of 3.4% and 2.8% respectively for the USA, 1.3% and 1.3% for Japan, and 2.5% and 2.3% for the euro area) indicates the limited extent of the economic recovery the Asia-Pacific region can expect to achieve, owing to the well-known linkage between the four regions. This was exacerbated further, in the first months of 2004, by an increase in inflation and by a lower-than-expected rise in consumer spending in the USA; and in Europe by the problems of internal cohesion, complex decision-making, bureaucratic structures and the divisive task of drafting a new constitution for the enlarged EU.

Economic Development and Terrorist Attacks

All the problems associated with the recent economic development and trends in the Asia-Pacific region discussed earlier for some major countries and related to the emergence of the Asian crisis—its impact and 'contagion', its nationally and internationally inspired management schemes and their possible economic recovery—have been the primary interest and concern of academic, business and institutional economists and policy-makers world-wide. These problems, however, were exacerbated by the terrorist attacks on the World Trade Center in New York and on the Pentagon in Washington, DC, on 11 September 2001 and by the bombing on the Indonesian island of Bali, a resort popular with Australian tourists, on 12 October 2002. The impact of the attacks was particularly deep and acute as the whole world, and especially the economies of the Asia-Pacific region, at that time seemed to have been voluntarily or involuntarily involved in, and to be accepting, the process of inevitable and increasing globalization for better or for worse. To many analysts, the terrorist attacks in particular had an impact on development and growth of at least the same magnitude as that of the Asian crisis of the late 1990s and affected not only the economies of the Asia-Pacific region but also the rest of the developed and developing world, through the traditional trade and investment links between these blocs. The impact was almost immediate through the new transmission mechanism of globalization, including electronic commerce, electronic trade and even electronic government, and if previous experience with the Asian crisis is any guide, the impact will continue for years to come.

Some of the damage caused by these terrorist attacks can be evaluated more specifically and numerically. For example, as a result of the attacks in 2001, in that year, according to the IMF, advanced countries attained an average growth rate of only 0.90% (compared with 3.8% in 2000), the EU 1.40% (3.5% in 2000), the Middle East and Turkey 1.4% (5.8% in 2000), the Asian NIEs 0.8% (8.4% in 2000), developing Asia 5.7% (6.8% in 2000), all developing countries 3.9% (5.7% in 2000), and transition economies 5.1% (6.6% in 2000). The 'crowding-out' effects of the terrorist attacks or, similarly worded, the 'costs of hatred', have also been recently studied and calculated by economists. These effects can be measured in terms of resources and resource allocation (for example, the 11 September 2001 terrorist attacks cost the US infrastructure, property, life and business activity an amount of $80,000m., and the costs of the USA waging a war on Iraq would cost it up to $140,000m.), a reduction in the value of goods produced and services provided, and a lower level of well-being owing to changes in the composition of what is produced and what is allocated (for example, more expenditure on defence and terrorism prevention and therefore less expenditure on education and health).

The Effects of the Iraq Conflict, of SARS and of Avian Flu

Early in 2003 the world witnessed the establishment of a 'coalition of the willing' to oust the regime of Saddam Hussain from Iraq and presumably to destroy the country's elusive 'weapons of mass destruction'. The cost of the damage caused by the war operation was estimated at more than $100,000m. in terms of restoration or reconstruction. While it has been claimed that these war expenses will be offset by Iraq's future oil receipts, the 'crowding-out' effect of this revenue for the recouping of war expenses, along with the country's outlay on development and the restoration of trade links with other regions, should not be disregarded by serious economic analysts. In addition, in early 2003 the outbreak of SARS, which originated in China and Hong Kong before spreading to more than 30 countries across the world, was seen as another major crisis for the principal economies of the Asia-Pacific region and beyond, including Australia and India. Calculations by economic analysts indicated that the SARS outbreak would have an economic impact first on immediate tourism and ancillary or supporting industries, and then on the rest of the Asia-Pacific economies, which would be more damaging than that resulting from the Iraq conflict or even from the September 2001 attacks, especially in relation to world aviation, and that economic growth would be reduced by as much as 1%–2% in the countries of Asia or the SARS-infected region. By August 2003, the SARS outbreak appeared to have been contained. In the first months of 2004, however, the Asia-Pacific region was affected not only by the re-emergence of SARS in China but also by an outbreak of avian flu.

Based on the information available at mid-2003, the IMF forecast that growth rates in the world economies and major regions in 2004 would be less than 2.9% for advanced economies, less than 2.4% for the EU, less than 4.5% for Asian NIEs, less than 6.5% for developing Asia and less than 4.1% for countries in transition in general. As discussed above, the World Bank has predicted a worse outcome for these groups in 2004 and 2005, owing to the deteriorating situation in Iraq and other emerging problems such as the avian flu outbreak.

WTO, APEC, Competition Policy, Economic Integration and New Asian Regionalism

The recent economic development and trends and crises in the countries of the Asia-Pacific region, within the process of increasing globalization, accompanied by international competitiveness and the voluntary or involuntary adoption by these countries, have been discussed above. While globalization is a major component or implication of the WTO, a multilateral free-trade agreement and a successor to the General Agreement on Tariffs and Trade (GATT), it cannot operate without underlying and supporting international competitiveness or its subset, competition policy and law, as the main impetus. One study on competition policy and law discussed and critically analysed major aspects and issues of competition policy and anti-trust laws (CPA) and global competitiveness (GCO) as a recent development in the economies in the Asia-Pacific region from a number of perspectives. First, it surveyed the philosophy and the micro-foundation underlying the concepts of CPA and GCO and being adopted in the international literature and institutions that specialized in this field of enquiry. This philosophy and foundation have often been neglected in previous studies on the subject. Second, it discussed the development and implementation of CPA and GCO in a number of major APEC countries in the Asian region, both developed and developing, and critically analysed aspects and issues as well as achievements, obstacles and challenges in the practices of CPA of these countries for possible adoption or adaptation by other economies.

It was also noted that, in many countries under study, CPA and GCO are seen essentially as the outcomes of the philosophy and concepts of a laissez-faire economy or market, strongly advocated by the West, mainly the USA and the EU, with the support of big transnational companies. Just as a laissez-faire economy is known to have optimal benefits in resource allocation and income distribution and maximum welfare generation, it has also been known to have its weaknesses, namely market failures. This may explain why in major market economies with a socialist orientation in Asia, such as China and Viet Nam, CPA and GCO have not been enthusiastically welcomed by government officials and the public alike. This perception, rightly or wrongly, has hindered the development and implementation of CPA and GCO. It is interesting to note, from such studies, that this perception and attitude has, with increasing globalization and the growing influence of WTO membership, slowly changed only in the early 2000s.

To minimize the occurrence and impact of market failures in order to gain optimal benefits of a truly laissez-faire economy model or something close to it (a free-market economy with government or statutory authority regulatory (and anti-trust) controls is another model, being in use in such advanced countries as Australia or even the USA), governments in both developed and developing countries have established their own trade or competition commissions with ministerial power and function to supervise corporate structuring and operation. It seems that no single country from all major trading blocs in the world has not embraced, for its local and international trade and industry policies, the ideas and concepts of CPA and GCO, and has developed or is developing CPA and aspects of GCO.

For these countries, this embracing of ideas is a necessity as trade and investment liberalization can only improve their growth and economic development, enhance their business environment, improve the welfare of their populations, alleviate their poverty and facilitate their integration into the global economy. It is also a necessity as international pressure to adopt and implement CPA and GCO has been intense, coming from international organizations such as the WTO, the World Bank and the IMF and even from regional organizations such as Asia-Pacific Economic Cooperation (APEC). However, the pace of progress of successful negotiations on multilateral free-trade agreements, such as those of the WTO and of economic integration, such as that proposed by APEC, has been notoriously slow. Therefore, alternative conceptual and policy models with more immediate expected outcomes on trade and growth, including the ASEAN and plurilateral free-trade agreements or new Asian regionalism such as ASEAN + 3 (that is, ASEAN + China, the Republic of Korea and Japan) and ASEAN + 5 (ASEAN + 3, along with Australia and New Zealand), have been eagerly promoted and supported by political leaders of the major countries of the Asia-Pacific region since the 1990s. Notable within the ASEAN countries themselves is the ASEAN or Bali Concord II, which was approved in October 2003 and which has the broader objectives of security, economic and sociocultural co-operation for all members.

Growth of Plurilateral and Bilateral Free Trade Agreements

The ASEAN + 3 proposal (also known as the Kim Young-Ho proposal, named after the Republic of Korea's former Minister of Commerce who first presented it) was initially discussed in the mid-1990s. A number of factors can be attributed to its emergence. First, it was the result of decades of fast growth and a number of economic, financial and restructuring developments in North-East Asia and in other major trading blocs in the world. Second, it arose from developments and shifts in focus in North America (through the North American Free Trade Agreement—NAFTA) and the EU in the aftermath of the damaging Asian crisis, which although originating in Thailand in July 1997, ultimately spread to a number of once 'miracle' economies in East and South-East Asia, the former USSR, and to a lesser extent, North and South America and the EU. Third, it was the result of a benign neglect by such international organizations as the IMF or of the economic power of North America and the EU on the plights of crisis countries in Asia and their lack of interest in seriously helping to solve the economic, financial and social problems arising from the Asian crisis.

In 2001 and in early 2002 other new developments in East and South-East Asia (for example, the quick recovery and resumption of growth in the Republic of Korea and the continuing stagnation of the economy of Japan) gained prominence and gave rise to a number of new Asian regionalisms (NARs) or economic integrations and to Asian free-trade agreements (FTAs) either in a plurilateral or bilateral framework.

These NARs and FTAs are currently indeed numerous and are proliferating at an amazing speed at the behest of government leaders, especially in the Asia-Pacific region. They include: ASEAN + 3; ASEAN + 5; ASEAN + 5 + Taiwan; Japan + Singapore; Japan + (the Republic of) Korea; Japan + Mexico; Korea + Mexico + Chile; Singapore + New Zealand; China + Japan + Korea; Hong Kong + New Zealand; and finally, Viet Nam + USA. There was even some discussion on the establishment of a North Asian FTA, in which Japan would play an important part. In May 2002 a protocol was also being negotiated between Washington and Canberra to address key US complaints about the Australian market and to prepare for the setting up of a radical US-Australia FTA (USAFTA), as proposed by the Australian Government, to the dismay of New Zealand which wanted a trilateral 'closer economic relationship' with the USA (along the lines of the CER between Australia and New Zealand—a bilateral agreement reached in 1982). The ratification of the USAFTA by the US Congress was expected in May 2004, with approval by the Australian Senate pending, amidst strong protests by unions and sugar farmers. Also in May 2002, the New Zealand Prime Minister, Helen Clark, suggested the establishment of an Australia-New Zealand Economic Co-operation (ANZEC) to revitalize the CER.

The main focus and objective of these NARs and Asian FTAs (as distinct from currency and customs unions) are to promote trade and, subsequently, growth and development, either among the economies of the Asia-Pacific region itself or with the membership of other economies outside Asia, such as those of the USA, Mexico and Chile in the Americas, and Australia and New Zealand in Oceania. Prominent among these NARs and Asian FTAs is the Kim Young-Ho proposal mentioned above (and part of it, the ASEAN + 1 or ASEAN + China FTA) which would encompass a market of 1,700m. people, a combined GDP of $2,000,000m. and trade to the value of $1,200m.). ASEAN + China was endorsed by the 10 leaders of the ASEAN member countries at a meeting in Brunei in November 2001, and the details were worked out at a negotiating session in Beijing in May 2002.

In view of the establishment of the ASEAN + 3 (and other Asian FTA variations) and the inherent economic issues and problems (historically, socially, politically and religiously), along with other emerging concerns as expected in the future, one may legitimately ask whether the structure or policies of some functions of the existing international organizations, such as APEC, the WTO or even the IMF, can be amended or whether special divisions of them should be established to accommodate the concerns and designs by the ASEAN + 3 trade strategists and policy-makers in dealing with Asia's contemporary economic issues and problems. The perceived effect of the argument often advanced by these organizations is that an ASEAN + 3 FTA is not necessary or will not ultimately be endorsed by the WTO.

The reasons for this proposition are that, first, to some observers, the emergence of the Asian crisis in 1997 and its subsequent 'contagion' were simply the outcome of a volatile international capital market that did not have proper regulatory controls. This can be rectified, however, by appropriate financial reform. Second, the impact of this crisis was still lingering in Asia, more than six years later, because the rescue and reform programmes imposed by the IMF on the crisis economies were based on wrong diagnosis. As a result, incorrect and ineffective prescriptions were recommended. It has been argued that this problem can be solved with more appropriate rescue policies by the IMF and, as a result, not by creating an ASEAN + 3 FTA or economic integration that may have the effect of diminishing some of the principal functions of the IMF, the WTO or even APEC.

The advocates of this solution suggest the establishment of a specialist division at APEC, the WTO or IMF, with a deep and proper understanding of, and expertise in, Asian economies and societies in general and of ASEAN + 3 aspects and issues in particular. To them, this solution may go a long way towards promoting economic growth and development, along with trade and investment, in Asia itself and in relation to the rest of the world.

The debate on what has been known in the contemporary literature on Asian economic development and growth as a repositioning of dialogue between the existing and dominant US- and EU-orientated international institutions (such as the IMF on the one hand and the emerging and powerful economies in South-East and North-East Asia on the other), will increase in the future. Development, growth patterns and trends in trade and economic activities and international relations in the countries of the Asia-Pacific region will be strongly influenced mainly by the outcome of this debate and by the international geopolitical climate of inevitable dynamic change world-wide in the years to come. In early 2004 there was a distinct trend among the countries of the Asia-Pacific region to develop bilateral and plurilateral FTAs and to 'fast-track' the major objectives of the WTO (which currently has 147 members and requires agreement from all of them on trade issues) but without the participation of the WTO. The grouping's 5th Ministerial Conference held in Cancún, Mexico in September 2003, yielded no tangible results; this was due to disagreement on the question of liberalization of agricultural products for WTO members, put forward by the Cairns Group (or similarly composed agricultural trading associations), and the emergence of the 'Singapore Issues'. Recent bilateral FTAs in the Asia-Pacific region include the Australia-Singapore FTA which was signed in July 2003 and the Australia-Thailand FTA signed in October 2003. Japan, which over the years has had a distaste for FTAs, nevertheless concluded a New Age Economic Partnership Agreement with Singapore in November 2002 followed by the Japan-Thailand Economic Co-operation Partnership Agreement in December 2003 and in October 2003 started a joint feasibility study for a Japan-Korea FTA. An ASEAN and India Framework on Comprehensive Economic Co-operation entered into force in July 2004, while a joint feasibility study was being carried out for a proposed Australia-China Trade and Economic Framework. In view of the benign neglect from North America and the EU on Asia and its problems and issues, this proliferation of NARs and FTAs reflects the need of the countries in the Asia-Pacific region to expand trade, to enhance development and growth, to build up economic integration and, at the same time, to promote regional co-operation, political stability and security, and also to improve welfare for mutual benefits. Strategically, the trend is a good way for the countries in the region, led perhaps by the East Asia 3 (China, Japan and the Republic of Korea), to position themselves against the economic and trade power and, consequently, the global influence of the other major trading blocs such as the NAFTA and the enlarged EU.

BIBLIOGRAPHY

(See also Select Bibliography at the end of the volume)

Asian Development Bank. *East Asian Economic Outlook*; internet www.adb.org, January 2001.

Economic and Social Statistics; internet www.adb.org, March 2001.

Australian Financial Review, 5 and 6 April 2001; internet www.afr.com.

Borland, J. 'Counting the Cost of Hatred', *Australian Financial Review*, 29 March 2004, p. 23.

Callick, R. 'Finally facing the truth—but still the yen falls'. *Australian Financial Review*, 9 March 2001.

CHELEM-CEPII. *CHELEM-CEPII International Database*, Paris, CEPII, 2001.

Cornell, A. 'Japan's finances close to collapse'. *Australian Financial Review*, 9 March 2001.

Desai, M. *Financial Crises in Global Governance*. International Conference, London School of Economics and Political Science, 13–14 October 2000.

Harvie, C., and Lee, H. H. 'New Regionalism in East Asia: How Does It Relate to the East Asian Economic Development Model?', *ASEAN Economic Bulletin*, Vol. 19, August 2002.

ICSEAD. 'Recent Trends and Prospects for Major Asian Economies'. *International Centre for the Study of East Asian Development*; internet www.icsead.or.jp/indexe.htm.

International Monetary Fund. *World Economic Outlook*, Opening Remarks by K. Rogoff; internet www.imf.org/external/np/speeches/2002, October 2002.

World Economic Outlook Databases; internet www.imf.org/external/pubs/weo/2003.

Kim, S. 'Seoul Sets Cracking Pace in Asia', *Australian Financial Review*, 23 May 2002, p. 12.

Kojima, K. 'Asian Economic Integration for the 21st Century', *East Asian Economic Perspectives*, Vol. 13, March 2002, pp. 1–38.

Lloyd, P. J. 'The Systemic Effects of Recent Regionalism' in Tran Van Hoa and C. Harvie (Eds). *New Asian Regionalism: Responses to Globalisation and Crises*. London, Palgrave, 2003.

McLeod, R. H., and Garnaut, R. (Eds). *East Asia in Crisis: From Being a Miracle to Needing One?* London, Routledge, 1998.

Pearson, B. 'Little Brother Korea Zooms Ahead', *Australian Financial Review*, 27 May 2002, p. 12.

Rogoff, K. *World Economic Outlook: Foreword*; internet www.imf.org/, May 2003.

Ryan, C. 'Markets Spooked by China's Curbs', *Australian Financial Review*, 3 May 2004, p. 8.

Singapore Trade Development Board. *Review of Year 2000 Trade Performance and Outlook for Year 2001*; internet www.tdb.gov.sg/newsroom/press/pr-00301.shtmlm, March 2001.

Tran Van Hoa. *Sectoral Analysis of Trade, Investment and Business in Vietnam*. London, Macmillan, 1999.

Prospects for Trade, Investment and Business in Vietnam and East Asia. London, Macmillan, 2000.

The Asia Crisis: The Cures, their Effects and the Prospects After. London, Macmillan, 2000.

Social Impact of the Asia Crisis. London, Macmillan, 2000.

China's Trade and Investment After the Asia Crisis. Cheltenham, Edward Elgar, 2000.

'Towards the Formulation and Testing of a New and General Theory of Macroeconomic Policy (1987)', in J. O. N. Perkins, *The Reform of Macroeconomic Policy: From Stagflation to Low or Zero Inflation*, pp. 53–70, London, Macmillan, 2000.

The Asia Recovery: Aspects and Issues in Growth, Development, Trade and Investment. Cheltenham, Edward Elgar, 2001.

Economic Crisis Management. New York, Edward Elgar, 2002.

Competition Policy and Global Competitiveness in Major Asian Economies. New York, Edward Elgar, 2003.

'New Asian Regionalism and ASEAN + 3 Free Trade Agreement: Theoretical and Empirical Foundation, Policy Challenges and Growth Prospects', *Chulalongkorn Journal of Economics* (Thailand), 2003.

Tran Van Hoa (with Ho Trung Thanh). *Co Hoi Dau Tu va Thuong Mai Viet Uc (Australia-Vietnam Investment and Business Opportunities)*. Hanoi, Labour Publishing House, 1998.

Tran Van Hoa (with Pham Quang Thao). *Anh Huong cua Khung Hoang Tai Chinh den Dau Tu va Thuong Mai Vietnam (Impact of the Asian Financial Crisis on Investment and Business in Vietnam)*. Hanoi, Labour Publishing House, 1998.

Tran Van Hoa (with Charles Harvie). *Causes and Impact of the Asian Financial Crisis*. London, Macmillan, 2000.

Tran Van Hoa (with Charles Harvie). *New Asian Regionalism: Responses to Globalisation and Crises*. London, Palgrave, 2003.

Wong, K. Y. *Economic Growth of Selected Asian Countries*; internet faculty.washington.edu/karyiu; March 2001.

World Bank. *Regional Strategy*; internet wbln0018.worldbank.org/eap/eap.nsf, February 2001.

World Development Report 2004: Making Services Work for Poor People; internet econ.worldbank.org/wdr/wdr2004.

CURRENT SECURITY ISSUES IN THE ASIA-PACIFIC REGION

ZACHARY ABUZA

The overall security situation in the Asia-Pacific region was relatively stable in 2003–04. There were no major outbreaks of hostilities, and several countries with long-standing insurgencies were in fact experiencing cease-fires. The major security concerns for the region were related more to the ongoing tensions in the Middle East, and especially to the continued US occupation of Iraq. Australia, the Republic of Korea (South Korea) and Japan all deployed contingents to participate in the US-led stabilization force. The other major security concerns remained the continuing tensions over North Korea's clandestine nuclear weapons programme and the failure to resume the six-party talks (involving North and South Korea, the People's Republic of China, Japan, Russia and the USA). The re-election in March 2004 of the pro-independence leader of Taiwan (see below), heightened tensions with mainland China, while there was a major increase in military spending by both China and Taiwan. Most states in South-East Asia were focused on the continued war against terrorists linked to the al-Qa'ida (Base) organization, especially following the sudden outbreak of violence in southern Thailand in the first quarter of 2004. Likewise, sectarian strife broke out again in Central Sulawesi and in the Maluku (the Moluccas) in Indonesia. Only in the restive Indonesian province of Aceh was a large-scale military campaign being waged.

For many states in the region, 2003–04 was a time of political transition: in China, in 2003, power was transferred to Hu Jintao and the 'fourth generation' of leaders, who sought to maintain a peaceful external environment in which to consolidate their power. In Taiwan, the leader of the pro-independence Democratic People's Party (DPP), Chen Shui-bian, secured re-election in March 2004. During the election campaign, President Chen urged the holding of a popular referendum on the future status of the island, a highly provocative act in Beijing's view; China responded with a significant build-up of forces and continued to reserve the right to invade the island if Taiwan declared independence. Yet Chen's electoral majority emboldened his party to continue to press for Taiwanese independence. In South Korea Roh Moo-Hyun, a former dissident and human-rights lawyer, was elected President in December 2002, taking advantage of the surge of anti-US sentiment that followed the death of two Korean girls in a road accident caused by US soldiers. President Roh had to balance the interests of his electoral constituency with the immediate security concerns of South Korea and with the need to maintain the country's alliance with the USA, especially following North Korea's confirmation in October 2002 of the existence of its nuclear weapons programme.

South-East Asia, too, was mired in political transition during 2003–04. Malaysia witnessed its first major political change in 22 years, when Prime Minister Mahathir Mohamad relinquished his post. Under the leadership of his carefully chosen successor, Abdullah Badawi, at the elections of March 2004, the ruling coalition made substantial gains, at the expense of the Islamic opposition. In the Philippines, the electoral contest was far less decisive and much more controversial. Incumbent President Gloria Macapagal Arroyo defeated the challenge of a popular film star, Fernando Poe, Jr, but only after six weeks of congressional canvassing and political manoeuvring by the opposition. The elections brought the peace process with the Moro Islamic Liberation Front (MILF), formally suspended since November 2002, to a halt. Indonesia, which has the largest Muslim population in the world, held the first direct presidential election in the nation's history. After legislative elections in April 2004, which resulted in heavy losses for President Megawati's Indonesian Democratic Struggle Party (PDI—P) and notable gains for Golkar, the party of former President Suharto, and for the Islamist Prosperous Justice Party, the country entered the first round of a heated presidential election in July. As no candidate received the requisite 50% of the votes cast, a second round of voting took place in September 2004, resulting in President Megawati's replacement by Susilo Bambang Yudhoyono.

THE REPERCUSSIONS OF THE IRAQ CONFLICT

The surge in violence in Iraq against US and coalition forces was disturbing for the Asia-Pacific region, leading to greater concerns about the commitment of US forces to the region and the ability of the USA to maintain its deterrent posture on the Korean Peninsula and in the Taiwan Strait. By mid-2004 more than 900 US soldiers had been killed in Iraq since the end of the campaign to oust Saddam Hussain in March 2003, despite the capture of the former Iraqi President and of many of the leaders of the Baathist regime. Shi'ite militias, as well as terrorist cells comprising both Iraqis and foreigners, continued to create instability. Although the USA transferred formal sovereignty to an Iraqi provisional Government on 28 June 2004, anti-US sentiment remained high; violence continued unabated, especially following the release in April 2004 of shocking photographic evidence of abuse by US soldiers and civilian contractors of Iraqi detainees at Abu Ghraib prison, near Baghdad. The systematic murder of provisional government leaders, coupled with widespread violence, called into question the political future of the country and ensured that 130,000 US and several thousand additional coalition forces would remain deployed in Iraq for the foreseeable future. The violence in Iraq, including terrorist attacks on the country's oil industry, caused anxiety in the region as international petroleum prices surged, thereby precipitating economic deceleration.

The situation in Iraq had another important impact in the Asia-Pacific region. Overstretched, the US Government urged other states to contribute forces to the stabilization force. Several countries from the Asia-Pacific region, including Australia, the Philippines, Thailand, South Korea and Japan, contributed forces. Australia was a leading member of the coalition and participated in the invasion of Iraq in March 2003. Australian naval forces led the blockade of Iraq in preparation for the removal of the regime of Saddam Hussain. Although the war was extremely unpopular in Australia, John Howard, the country's Prime Minister, remained a key ally of the Administration of George W. Bush and the 'war on terror'. Howard supported the Bush Administration's assertion that there were clear links between al-Qa'ida and the regime of Saddam Hussain, although the independent National Commission on Terrorist Attacks upon the United States (known as the 9/11 Commission) concluded that no operational links existed. Australia maintained a force of several hundred personnel during the occupation of Iraq. By mid-2004 Australian forces had not suffered any casualties in Iraq, which averted any mass protest against the Howard Government's policies.

Japan joined the US-led coalition in Iraq in mid-2003. This represented the first overseas deployment of Japanese troops in a combat zone since the Second World War. Although Japan's action aroused some opposition from neighbouring countries, notably China and South Korea, Prime Minister Junichiro Koizumi pledged Japanese involvement and was an early supporter of the US-led war. Iraqi insurgents tried to break the Japanese resolve to deploy troops. In late November 2003 two Japanese diplomats were killed near the town of Tikrit. The April 2004 kidnapping of three Japanese citizens by Iraqi insurgents likewise failed to deter the deployment of the forces, which Prime Minister Koizumi argued was necessary to consolidate the security relationship with the USA. Although the militants threatened to burn alive the three hostages, all were later released unharmed, and Japan insisted that it would maintain its deployment of some 600 troops. Japan also pledged US $1,500m. in support for Iraq in 2004 and allocated a future sum of US $3,500m.

The Japanese public remained divided on the issue of Iraq. While many Japanese regarded the deployment as a necessary step for Japan to assume a greater political and diplomatic role in world affairs, the war itself was unpopular and many Japanese feared becoming embroiled in an extremely difficult situation. Yet in April 2003 an opinion poll by the Japanese daily *Yomiuri Shimbun* found that 42% of the respondents supported amending Article 9 of the Constitution whereby Japan renounces the right to use war as a means of settling international disputes. Some lawmakers in the ruling Liberal-Democratic Party tabled an amendment that would allow Japanese participation not only in UN-sponsored peace-keeping operations but also in non-UN peace-keeping operations. The deployment of troops in Iraq was part of a very gradual progression by Japan towards the enhancement of its military. In mid-2003 Japan's Self-Defence Forces (SDF) totalled almost 240,000, of whom 148,000 were ground troops; with around 480 aircraft and 140 ships, Japan is the second largest naval power in the Pacific, behind the USA. Furthermore, the SDF was to deploy two 13,500-ton helicopter carriers by 2008, greatly increasing their ability to project power.

South Korea deployed 600 military engineers and medical personnel in Iraq. President Roh Moo-Hyun wanted to preserve his country's relationship with the USA, which had come under great strain (see below). South Korea pledged to send an additional 3,000 troops to Iraq in August 2004. This deployment would make the South Korean force the third largest contingent in Iraq after that of the USA and the United Kingdom. In June, however, a Korean employee of a US contractor, was abducted and decapitated by Iraqi militants who tried to force the South Korean Government to reverse its decision to deploy the troops. Despite popular opposition at home, President Roh refused to accede to the demands of the militants, and South Korean troops were to be duly deployed in August 2004.

The token Philippine contingent of 51 personnel was almost withdrawn in May 2004 during an upsurge in fighting yet, under US pressure, the troops remained. Five Philippine civilians had been killed in Iraq since the beginning of the conflict. In July a Filipino was abducted by Islamist militants who threatened to behead him if the Philippines did not withdraw its contingent. Amid much controversy, the Philippine Government complied with the captors' demands; the hostage was released unharmed following the withdrawal, ahead of schedule, of all Philippine troops from Iraq.

THE KOREAN PENINSULA

By the end of 2003 the situation on the Korean Peninsula had reached its lowest point since the conclusion of the 1994 agreement suspending North Korea's clandestine nuclear weapons programme. The complex Agreed Framework had provided for the establishment of the multilateral Korean Energy Development Organization (KEDO), which was to build two light-water nuclear reactors in North Korea. These reactors, which do not produce plutonium and are thus regarded as 'proliferation-proof', were to replace North Korea's aged heavy-water reactors. In the interim, KEDO provided fuel to North Korea to compensate for the closure of the reactors. The South Korean-designed reactors were intended to be readily integrated into the South's electricity grid. In return, North Korea pledged to 'freeze' its nuclear weapons development programme, remain within the Treaty on the Non-Proliferation of Nuclear Weapons (the Non-proliferation Treaty—NPT), accept international inspections and fully comply with the requirements of the International Atomic Energy Agency (IAEA). US intelligence officials had concluded that, at the time, North Korea had sufficient uranium and plutonium to produce two nuclear weapons. The Agreed Framework began to disintegrate in January 2002, when US President Bush designated North Korea (along with Iraq and Iran) as a member of the 'axis of evil'. North Korea's own security concerns heightened as US forces invaded Iraq in March 2003, while the confrontation with Iran over its own nuclear weapons programme escalated in 2003–04.

In October 2002 North Korea shocked the USA by admitting that it had continued to maintain a clandestine programme to enrich uranium for the purposes of nuclear weapons, in violation of the 1994 Agreed Framework. In November, therefore, the USA suspended fuel oil shipments to North Korea, and the agreement came to a halt. In response, North Korea announced that it would reactivate its nuclear facilities because of power shortages. In late December North Korea removed seals and monitoring cameras from its nuclear complex at Yongbyon, north of Pyongyang, which had remained closed since 1994. Seals were also removed from some 8,000 spent nuclear fuel rods, which were capable of producing between six and 10 nuclear warheads. North Korea expelled monitors from the IAEA and stated that it planned to reactivate its reprocessing plant. In January 2003 North Korea formally withdrew from the NPT. In the following month the IAEA declared North Korea to be in violation of non-proliferation accords and referred the crisis to the UN Security Council.

In many ways North Korean provocations were designed to create a division between the South Koreans and their US allies. In December 2002 the liberal Roh Moo-Hyun, who campaigned against the USA's hardline policy on North Korea, was elected as South Korea's next President. Like his predecessor, Roh supported the 'sunshine'policy, which sought to improve relations between the two countries. Based on the premise that South Korea could not afford to absorb North Korea, were the latter to collapse, the 'sunshine' policy was designed to encourage South Korean investment in the North, as well as increased trade, tourism and aid. The US position, however, was very rigid, demanding a 'complete verifiable and irreversible' dismantling of the nuclear weapons programme. The USA refused to negotiate on any issue before North Korea had suspended all enrichment operations, and declined to make any concessions that would merely reward Pyongyang for violating the agreement.

Despite the election of a South Korean President more inclined to be conciliatory towards Pyongyang, the North Korean leadership continued to raise the tension, expecting a rift to develop between South Korea and the USA, as well as hoping for greater concessions by the former. In February 2003 Moody's Investors Service, a leading US credit rating agency, downgraded South Korea's sovereign ratings outlook from positive to negative, citing nuclear security concerns. Just hours before Roh Moo-Hyun's inauguration, North Korea test-fired a missile into the Sea of Japan. The day after the swearing-in ceremony, North Korea restarted its 5-MW nuclear reactor at Yongbyon.

In April 2003 North Korea abandoned its demand for bilateral talks with the USA, stating that it would accept any format for dialogue if Washington were prepared to make a 'bold switch-over' in its Korea policy. No 'bold switch-over' was likely, however, as North Korea claimed that it had begun reprocessing more than 8,000 spent fuel rods to make weapons-grade plutonium the following week. In the same month the USA, North Korea and China met for three days of talks in Beijing, where North Korea admitted to possessing nuclear weapons. The USA announced that it had 'no intention of invading or attempting regime change' in North Korea, but the statement fell far short of the security agreements that Pyongyang was seeking. North Korea announced that it would suspend any further development of its nuclear weapons programme in return for a series of synchronized concessions by the USA, including: the restoration of normal diplomatic relations, the removal of economic sanctions, an increase in food aid, the completion of the construction of the light-water reactors, the resumption of fuel oil shipments and a pledge not to attack the North. The USA refused, as North Korea did not offer to dismantle or relinquish any bomb-grade material already reprocessed. Washington, again, refused to enter into any agreement that would appear to reward Pyongyang's duplicity. The USA quickly turned to its allies to intensify the pressure on North Korea. In May, during a meeting in Washington, President Roh and President Bush agreed that 'further steps' might be needed against North Korea. President Bush and Japanese Prime Minister Junichiro Koizumi denounced North Korea's nuclear weapons campaign at a summit meeting in Texas, referring to the possible need for 'stronger measures'. In June 2003 a US congressional delegation visited Pyongyang and reported that North Korean officials had stated that they had almost completed the reprocessing of 8,000 of its spent nuclear fuel rods for the purposes of making more nuclear bombs, although this could not be verified by any intelligence service.

North Korea agreed to attend six-party talks, encompassing China, Japan, Russia, South Korea and the USA, scheduled to be held in late August 2003 in Beijing. After agreeing, however, North Korea immediately laid down further preconditions: a non-aggression pact with the USA; and the rejection of an early inspection of its nuclear facilities as 'absolutely unacceptable'. The day after these six-party talks began in China, North Korea threatened to conduct a nuclear test and to declare itself a nuclear power. The talks, not surprisingly, ended in acrimony, with North Korea threatening to strengthen its nuclear arsenal if the USA refused its demands to end the nuclear crisis. North Korean officials contended that a nuclear deterrent was necessary to deter a US invasion and that a non-aggression pact had to be signed between the two states. While the negotiators from the six countries agreed that the North's security concerns were legitimate and needed to be addressed at later discussions, they failed to set a timetable for future negotiations. North Korea stated that any further talks on the nuclear crisis would be 'useless'.

On 9 September 2003, the day of the country's national holiday celebrating the 55th anniversary of the establishment of the Stalinist regime, North Korean leaders vowed to press ahead with nuclear weapons development. A few weeks later North Korea rejected IAEA demands to dismantle its nuclear programme and accused the UN body of acting as a 'stooge' for the USA. On 30 September North Korea announced that it was taking 'practical measures' to enhance its nuclear deterrent, while rejecting another round of six-party talks. In early October North Korea declared that it had completed the reprocessing of all 8,000 spent fuel rods and had produced enough weapons-grade plutonium for half-a-dozen atomic bombs, as a step towards boosting its nuclear deterrent, although it did not announce whether the warheads had been constructed. US officials remained sceptical about Pyongyang's assertion regarding the reprocessing of all 8,000 rods and believed the number to be closer to 1,000.

The second round of six-party talks was held in Beijing in February 2004, but achieved little. North Korea demanded the conclusion of a peace treaty and the establishment of full diplomatic relations with the USA, as well as the ending of its 'hostile policy' towards Pyongyang. The USA responded with the offer of a 'permanent peace mechanism' as a framework for future talks. Acrimony between the USA and North Korea increased with the revelations of A. Q. Khan, the founder of Pakistan's nuclear weapons programme. Khan confessed to having transferred nuclear material to North Korea, as well as to Libya and Iran. Evidence from the Khan investigation showed that by 1996, only two years after signing the Agreed Framework, the North Koreans had decided to continue with their clandestine programme, for which Pakistani assistance was required. North Korea shifted from plutonium production to enriched uranium, an operation that the USA, South Korea and the IAEA did not believe the North Koreans had the capability to conduct. In exchange for the technology to work with uranium hexafluoride and blueprints for nuclear bombs, North Korea sold to Pakistan its latest missile technology.

The third round of six-party talks took place in Beijing in June 2004, with very low expectations of a breakthrough. The USA, however, did make an unexpected conciliatory overture at the start of the talks. It offered a three-month test period in which it would pledge not to invade North Korea, to end all economic sanctions and immediately provide aid (especially energy supplies). In return, North Korea would be obliged to provide full disclosure of its nuclear weapons programme. Pyongyang neither accepted nor rejected the agreement. In many ways, the USA's departure from its uncompromising position was seen as a concession to Seoul and to Tokyo, Washington's two key allies. Both South Korea and Japan opposed such a rigid stance and a nuclear confrontation. One official of the Bush Administration acknowledged this pressure from the USA's allies and described the new position as 'alliance management'. South Korea continued to support the 'sunshine' policy, hoping to enmesh the North in economic interdependence, but the leadership was constrained in its relations with the USA, which was the ultimate guarantor of South Korean security.

President Roh Moo-Hyun had been elected in a wave of anti-US sentiment, and public opinion surveys in 2004 showed that the majority of South Koreans considered the USA to be a greater threat to international peace and security than North Korea. Such sentiment clearly angered Washington. In June the USA suddenly announced that it was to withdraw one-third of the 37,000 troops based in South Korea. The timing of the announcement, which was made without any prior consultation with the South Korean Government, caused great concern in South Korea. Some strategic rationale was provided. The US military was clearly overstretched as a result of the continued and unexpectedly long-term deployment of its troops in Iraq. Moreover, US Secretary of Defense Donald Rumsfeld had called for a leaner and more mobile US military, less tied to fixed bases. South Korean officials, however, were now questioning the durability of the alliance, as well as the USA's commitment to the defence of South Korea; many in the Government hoped to negotiate a settlement of the situation on the Korean Peninsula at all costs.

It was clear that North Korea was stalling for time for several reasons. First, the more protracted the negotiations became, the further advanced Pyongyang's weapons programme would be. Past agreements had suspended weapons development programmes, but not required the North either to reveal or to dismantle weapons that had already been developed. Second, North Korea was in the final stages of developing new delivery systems, essential to its deterrent capability. Pyongyang was about to deploy a new version of its intermediate -range ballistic missile (IRBM), the Taepo Dong-1, with a range of 1,800 to 2,500 miles (2,897 to 4,023 km), and capable of striking Japan and all US bases there. Pyongyang was also in the final stages of developing the Taepo Dong-2, the country's first intercontinental ballistic missile (ICBM), which is capable of reaching the west coast of the USA. Finally, Pyongyang was awaiting the outcome of the US presidential election in November 2004, in the hope that a Democratic administration would be more willing to negotiate with North Korea, offering aid and a security guarantee in return for concessions on the weapons programme.

THE TAIWAN STRAIT

The tension in the Taiwan Strait increased in 2004 during the Taiwanese presidential election campaign. After some initial progress, direct discussions between Taiwan and mainland China, albeit conducted by non-official organizations, had been suspended in 1998 when Taiwan insisted that the talks must take place at the state-to-state level. Both sides subsequently increased their weapons capabilities and hardened their stances. In March 2004 Chen Shui-bian, the incumbent President and leader of the pro-independence DPP, was narrowly re-elected. Chen had disappointed many of his constituents after his 2000 inauguration by not pressing for Taiwanese independence, the central aim of the DPP. During the course of the 2004 election campaign, however, he moved the party closer towards a pro-independence stance. Chen had called for the holding of a popular referendum on the future status of the island, which would seek to determine whether the people of Taiwan wanted independence, through the rewriting of the Constitution to 'modernize Taiwan's political institutions and enshrine certain rights'. The DPP sought to take advantage of the pro-independence sentiment that has been growing in Taiwan and has entered the political mainstream, as a younger and affluent generation with no ties to the mainland has matured. Although the referendum would not be binding, the prospect of such a vote infuriated China, which saw it as a first step towards independence.

In January 2004 it was announced that, at the forthcoming presidential election, Taiwanese voters would also be given the opportunity to vote on two issues in a referendum. The first proposed question was 'The people of Taiwan demand that the Taiwan Strait issues be resolved through peaceful means.

Should Mainland China refuse to withdraw the missiles it has targeted at Taiwan and to openly renounce the use of force against us, would you agree that the Government should acquire more advanced anti-missile weapons to strengthen Taiwan's self-defense capabilities?' This question was extremely offensive to China, which warned that the referendum 'heightened tensions and even a crisis' and that Chen had pushed Taiwan to 'the abyss of war'. In China, President Hu Jintao and his fellow 'fourth generation' of leaders, who had yet fully to consolidate their power, were under pressure from conservatives and the People's Liberation Army (PLA) to counter Taiwan's moves towards independence. China not only refused to renounce the use of force to liberate Taiwan, but reiterated its pledge to invade the island, should Taiwan declare independence. China warned that it would 'pay any price to safeguard the unity of the motherland'. Yet President Chen was placed under intense diplomatic pressure from the USA, which intervened and persuaded him to tone down the language of the questions, for fear of provoking a military response from China. President Bush had shifted the USA's traditional stance of 'strategic ambiguity', designed to keep both Beijing and Taipei guessing whether the USA would indeed intervene on Taiwan's behalf should China invade Taiwan, thereby preventing a Taiwanese declaration of independence and deterring a Chinese invasion. The US President announced that the USA would defend Taiwan if it were attacked, but cajoled the island into accepting the political status quo and not declaring independence. Specifically, he warned President Chen 'against any actions which risked altering the status quo'. During US Vice-President Richard Cheney's trip to Beijing, Chinese leaders demanded that the USA 'oppose Taiwanese independence' and 'avoid sending the wrong signals' to Taiwan by selling weapons to the island. To that end, China rapidly increased its military expenditures and built up its forces opposite Taiwan. China deployed approximately 500 nuclear-capable short-range missiles opposite Taiwan. A July 2003 report prepared by the US Congress contended that China was deploying approximately 75 additional missiles a year. All of China's short-range ballistic missiles (SRBMs) are based within 600 km of the island in the Nanjing military region. China announced that it had increased its military expenditure in 2004 by 11.9%, to the equivalent of US $25,000m. US intelligence officials questioned that figure and contended that it was closer to US $65,000m., compared with Taiwanese expenditure of just US $8,000m. China is currently developing two new classes of cruise missiles, the Dong Feng-11, with a range of 300 km, and the Dong Feng-15, which has a range of 600 km. Chinese missiles are increasingly accurate, based on newer Western guidance systems.

In response, there has been a significant increase in Taiwan's own military expenditure. For the first time, Taiwan sent a high-level team of military officials and procurement officers to the USA, where they met senior US officials. The USA renewed its commitment to help Taiwan to defend itself. In response to China's provocative build-up, the USA offered to sell Taiwan an enormous arms 'package' worth US $6,000m. The deal included additional F-16 fighter jets, two batteries of a new generation of *Patriot* missiles, a fleet of anti-submarine aircraft and eight diesel submarines. For the first time, the USA also considered selling Taiwan *Aegis* class destroyers with advanced over-the-horizon radar. In June 2004 a special session of the Taiwanese cabinet approved more than US $18,200m. in funds to purchase advance weaponry over a 15-year period beginning in 2005.

Taiwan has been under pressure from the USA to increase its military strength as a deterrent against China. In early 2001 the Bush Administration approved a massive sale of weapons to Taiwan; yet the island's legislature was reluctant to agree to the high cost. By 2003 the backlog of unfilled Taiwanese weapons purchases had exceeded US $20,000m. Many Taiwanese legislators felt that the USA was trying to reduce its own commitment to defend the island by increasing Taiwan's capacity to defend itself. As Taiwan's military strength is completely overshadowed by that of mainland China, the island needs to maintain a technological advantage in its ability to deter a Chinese attack. Once again, within Taiwan there was some domestic opposition to President Chen's weapons programme. Several lawmakers and analysts argued that Taiwan simply could not afford such costly arrangements. Others said that the time frame of the deal (that is, the amount of time that it would take to procure the weapons and integrate them into the military) would not deter China in the short term. Others considered the purchases to be too provocative and that they would simply result in further purchases of weaponry and deployments of Chinese forces on the mainland. Yet many other lawmakers and analysts argued that it was an important new opportunity for Taiwan to increase its deterrent capability. In the event, the only programme that was not supported was the submarine procurement. Taiwanese legislators who had travelled to the USA on an inspection tour complained that the cost of the eight US-made submarines was too great. Taiwan pressed for the USA to allow a local shipbuilder to make at least two of the craft, but officials at the US Department of Defense objected to the transfer of technology to Taiwan. The USA included Taiwan in six days of advance 'war games' in June 2004. A total of 60 US officers were dispatched to Taiwan, this being the first time that US forces had engaged in such simulations with their counterparts since the USA's withdrawal of its diplomatic recognition of Taiwan in 1979.

In response to the proposed US-Taiwan weapons deal, the PLA effectively lobbied China's leaders for more funds and 'faster war preparation'. Generals also petitioned the senior communist party leadership to accelerate the reunification process. The PLA pressed the National People's Congress in Beijing to pass the National Reunification Law. In July 2004, at a meeting between US National Security Advisor Condoleezza Rice and senior Chinese leaders, the USA refused to suspend weapons sales to Taiwan. This demand has been repeatedly rejected by the USA, which remains committed to the status quo in the Taiwan Strait. Although the USA will continue to place pressure on President Chen to refrain from asserting, as he did after his election, that 'Taiwan is a sovereign state', it will continue to give Taiwan the means to deter a Chinese attack. For now though, the US Government understands that shared strategic interests with China, especially regarding Iraq and North Korea, are far more important.

TERRORISM

The threat of terrorist attacks remained one of the principal security fears in the Asia-Pacific region during 2003–04. Although much of the concern continued to centre on the Philippines and Indonesia, even Japan acted against a suspected five-member al-Qa'ida cell in May 2004. The main threat to the region comes from Jemaah Islamiyah, an organization linked to al-Qa'ida, the stated goals of which are to turn Indonesia into an Islamic state and then to attempt to create a pan-Islamic state, which would also encompass Malaysia, southern Thailand and the Philippines. Jemaah Islamiyah was founded during 1992–93 by two Indonesian clerics living in exile in Malaysia, Abdullah Sungkar and Abu Bakar Bashir. Sungkar met with Osama bin Laden, the Saudi-born leader of al-Qa'ida, around that time and obtained support for the establishment of the organization. The organization itself was assembled and administered by a member of al-Qa'ida and veteran of the Afghanistan conflict, Riduan Isamuddin, better known as Hambali. Hambali spent the remainder of the decade patiently building up a network of cells, establishing *madrassas* (mosque schools) that would serve as centres of recruitment, training new members and dispatching them to al-Qa'ida camps in Afghanistan, and later to MILF camps in the southern Philippines. Hambali divided Jemaah Islamiyah into four different *mantiqis* or regions, each of which seemed to focus on a specific task or function. Each *mantiqi* was divided into *wakalah* (sub-regions) and then into *fiah* (autonomous cells). Throughout this period Jemaah Islamiyah was at al-Qa'ida's disposal and served as an important 'back office' for the terrorist organization, its activities including the establishment of 'front' companies, the opening of bank accounts, the forging of documents, the procurement of weapons, the organization of meetings, and the 'laundering' of money.

In 1999, through his efforts in waging *jihad* (holy war) in the Maluku and later Central Sulawesi where sectarian violence had erupted, Hambali established two paramilitary organizations, the Laskar Mujahidin and the Laskar Jundullah, financed by money diverted from Saudi charities, for the purpose of promoting *jihad*. In 2000 Jemaah Islamiyah cells began their first actual terrorist operations, including the August assassination attempt of the Philippine ambassador to Indonesia, the Christmas church bombings across Indonesia, the light rail transit bombing in Manila and the attacks on a train station and on hotels in southern Thailand. At the time, investigators were unaware of the existence of Jemaah Islamiyah and did not connect the bombings. Jemaah Islamiyah operatives assisted in the 1995 plot to blow up 11 US jetliners over the Pacific. They later conducted a reconnaissance of the US embassy in Jakarta in preparation for a planned al-Qa'ida attack. Jemaah Islamiyah and al-Qa'ida operatives were in the final stages of launching a major attack on US targets in Singapore in late 2001 when their cell was uncovered by Singaporean and Malaysian investigators. Following the US invasion of Afghanistan and the assault on the al-Qa'ida leadership, Hambali was given funding and ordered to execute a major attack in South-East Asia. This resulted, on 12 October 2002, in the attack on the nightclubs on the Indonesian island of Bali, which killed 202 people. The al-Qa'ida leadership was reportedly so pleased with the results of the mission that it transferred another US $100,000 to Hambali to launch further attacks: the first of these was the bombing of the Marriott Hotel in Jakarta on 5 August 2003.

By mid-2004 Jemaah Islamiyah had been severely degraded during more than two-and-a-half years of counter-terrorist operations. The Bali investigations, in particular, led to a far greater understanding of the operational nature of the network and of its command and control structure, resulting in subsequent arrests. Across South-East Asia, more than 250 people had been arrested, including many members of the organization's leadership. Hambali was arrested in Thailand in August 2003. Estimates of the size of Jemaah Islamiyah vary from around 500 to several thousand, but it is not a large organization, and the rate of arrests is not sustainable. Although a number of leading operatives remain at large, including those with operational experience and bomb-making expertise, many of their *madrassas* have been shut down, and they are less able to send their recruits abroad for advanced training. Moreover, there is significantly more inter-state co-operation in terms of police operations and intelligence-sharing. In short, by mid-2004 Jemaah Islamiyah was less able to plan and execute terrorist attacks than it had been a year previously, especially against hardened targets such as US embassies, although it still maintained its capacity to attack soft targets. Jemaah Islamiyah had not been able to execute any attacks since the August 2003 bombing of the Marriott Hotel in Jakarta, in which 12 people were killed. The group appears to be slowly rebuilding its ranks and training a new generation of members; arrests have taken place in Malaysia, Indonesia, Thailand and as far away as Pakistan.

During 2003–04 there were a number of terrorism-related arrests in the region. Most significantly, the Thai security forces and members of the US Central Intelligence Agency (CIA) arrested Hambali, Jemaah Islamiyah's operational chief, in August 2003, in Ayudhya. In March 2004 it was reported that the Malaysian authorities had recently arrested six Indonesians who were trying to enter the country from MILF camps in Mindanao in the southern Philippines. In December 2003, outside Kuala Lumpur, the authorities arrested a cell with a large cache of explosives. An al-Qa'ida suspect was arrested in the remote state of Sarawak. A cell composed mainly of the children and brothers of senior Jemaah Islamiyah members was arrested in Pakistan; all were deported to their respective countries, where they were being investigated and charged. Indonesian police arrested a number of suspects in connection with the bombing of the Marriott Hotel. In Indonesia, however, there was considerable consternation that, in the midst of election campaigning, the Government appeared unwilling to take action against Islamist militants. The Indonesian Supreme Court reduced the sentence of the alleged spiritual leader of Jemaah Islamiyah, Abu Bakar Bashir; based on new evidence provided by Hambali, he was declared a criminal suspect in the Bali bombing and was rearrested.

There has been considerable concern about a new wave of terrorism in the Philippines, atttributed to the resurgence of Abu Sayyaf, a Muslim secessionist group. Abu Sayyaf was founded in 1991 by Abdurrajak Janjalani, who was a leader of the Philippine Muslim contingent of the Afghan *mujahidin* that fought against Soviet forces. An associate of Osama bin Laden, Janjalani received funding from al-Qa'ida to establish a secessionist force. Abu Sayyaf grew, as al-Qa'ida-linked fronts and charities established themselves in the southern Philippines. Very quickly, the grouping made its mark and came to be regarded as a small but lethal organization. Abu Sayyaf began its terrorist attacks in the Philippines in 1991, when it killed two US evangelists in a grenade attack in Zamboanga City. The group's operatives attacked Ipil in April 1995 and assassinated a Catholic bishop in February 1997. Between 1991 and 1996 Abu Sayyaf was responsible for 67 terrorist attacks, more than half of which were indiscriminate bombings. The campaign led to the death of a total of 58 people and to 398 injuries. However, following the enforced departure of Mohammed Jamal Khalifa, the brother-in-law of Osama bin Laden and a financial supporter of the MILF and of Abu Sayyaf in December 1994, the latter began to degenerate into a group of criminal gangs. With the death of Janjalani in 1998, all pretence of being a secessionist group was abandoned, and Abu Sayyaf began a series of high-profile abductions. These included kidnappings of schoolchildren, missionaries, and of tourists from a Malaysian diving resort. Many were freed following the intercession of the Libyan Government.

With the onset of the global 'war on terror', following the attacks of 11 September 2001, the USA gave considerable support to the Government of the Philippines in order to deal with the al-Qa'ida-linked Abu Sayyaf. In the latter part of 2001, US forces began to conduct needs assessments of Philippine troops and to rearm several companies of light-reaction forces. In January 2002 some 1,000 US special operations forces and military engineers were deployed to the Abu Sayyaf stronghold of Basilan, to assist the Philippine military. In that year the US Government provided US $100m. in military aid, more than in the previous 10 years combined, thus greatly restoring the bilateral military relationship, which had deteriorated after the Philippine Senate voted not to renew the leases for the US military installations at Clark Air Base and at Subic Bay. The bases were shut down in 1991–92.

Many on the Philippine left wing saw the deployment of US forces as the first step towards the reopening of the US bases. Under the terms of the bilateral agreement, however, the US forces engaged in Operation Balikatan were withdrawn in July 2002, the six-month joint campaign against Abu Sayyaf being declared a success. US forces were angered that despite the military assistance, training and intelligence, the Philippine armed forces were still not able to defeat the 400–500 member Abu Sayyaf. Indeed, following the withdrawal of US troops, Abu Sayyaf was able to regroup.

In October 2002 the detonation of two bombs in Zamboanga, killing several people and wounding about 160, was attributed to Abu Sayyaf, in retaliation for increased military pressure on the group and the continued presence of 23 US military advisers. Abu Sayyaf also claimed responsibility for a series of bombings in April 2002 in General Santos City, which killed 15 people and wounded almost 70. These attacks and reports of growing linkages between Jemaah Islamiyah and Abu Sayyaf prompted the USA to press for a renewed and expanded role in the Philippines. In February 2003 the US Department of Defense announced that a second Balikatan exercise would be held, but this time some 3,000 troops were to be deployed in a combat role for an indefinite period, until Abu Sayyaf was eliminated. When publicly announced, the proposals created a strong nationalist reaction, as the Constitution explicitly forbids foreign troops from fighting on Philippine soil. Under intense political pressure, President Gloria Macapagal Arroyo was forced to withdraw plans for the operation.

Between mid-2003 and mid-2004 Abu Sayyaf executed, or attempted to execute, a number of terrorist attacks, including an explosion on a superferry that killed more than 100 people. A member of Abu Sayyaf was arrested in Manila in April 2004, with a large cache of explosives. At the same time, most kidnapping incidents by the group appeared to have ceased. Security officials were perplexed by Abu Sayyaf's apparent reversion to terrorism and were concerned that Jemaah Islamiyah had been able to re-establish a working relationship with the group. There was also concern that with the death in a military ambush in

June 2002 of Abu Sabaya and with the capture in December 2003 of Galib Andang, another Abu Sayyaf leader, the loose grouping had been taken over by Khadafi Janjalani, the younger brother of Abu Sayyaf's founder, who seemed more intent on returning the organization to its religious and secessionist roots.

The sudden outbreak of sectarian violence in southern Thailand in the first quarter of 2004 caused considerable consternation in the region. Muslim insurgents raided a number of police and military bases in a series of well-executed attacks, stealing large caches of weapons. In January militants torched 18 schools and killed four government soldiers. Militants then launched a series of simultaneous arson attacks against some 39 government targets in March. During March and April there was a spate of bombings in Narathiwat and Yala provinces. In April Thai military forces ambushed and killed some 100 young militants. There is still little understanding of the groups responsible for these attacks. Although Thailand waged a battle against Muslim separatists from the 1970s to the 1990s, the major groups such as the Pattani United Liberation Organization (PULO) had been defeated or had become defunct by the mid-1990s. Rather than an ethno-nationalist campaign, the new militants are clearly organizing through the *pondok* (Islamic school) network and trying to convert the struggle into a religious war. Some small militant groups, led by former veterans of the Afghan *mujahidin*, are also suspects. Regional security officials expressed concern that Jemaah Islamiyah was behind the attacks, especially after Thai forces killed a number of foreign *jihadis*, although there was no clear evidence to support that suspicion

There has been great concern that future terrorist attacks might be maritime-based. South-East Asia straddles some of the most critical maritime trade routes in the world, including the Strait of Melaka (Malacca). Global shipping remains highly vulnerable. Asia owns 40% of all cargo ships, is home to some of the largest container ports, and more than 25% of the world's cargo and 50% of the world's oil passes through the Strait of Melaka. The International Maritime Bureau reported that more than 80% of the world's pirate attacks take place in South-East Asian waters annually. Whereas an attack on a US naval vessel might currently be beyond the easy reach of al-Qa'ida and Jemaah Islamiyah operatives, an attack on commercial vessels is not. A senior al-Qa'ida operative, Abdullah al-Rahim al-Nishiri, arrested in Yemen in October 2002 while reportedly en route to Malaysia, was found in possession of a 180-page dossier listing maritime targets of opportunity. This information corresponds with the overall al-Qa'ida strategy set out in the same month in a broadcast in which Ayman al-Zawahiri warned that al-Qa'ida 'would target the nodes of your (i.e. the West's) economy'. Any attack on commercial shipping would have a devastating impact on the world's economy. An attack in the congested Strait of Melaka would slow down traffic through that important sea-lane. More than 50,000 large ships annually pass through the Strait, which is more than 965 km (600 miles) long but only 2.4 km (1.5 miles) wide at its narrowest point. In 2002 between 40 and 50 oil tankers (carrying approximately 10m. barrels of oil) and 10–12 liquefied natural gas (LNG) tankers sailed in the Strait each day. These numbers are set to increase greatly as Asian countries, especially China, continue their current rates of economic growth and as energy imports surge. As was the case following the October 2002 attack on a French supertanker off the coast of Yemen, any further attack would lead to a sharp increase in maritime insurance rates as well as in economic closures, with manufacturers depending more on just-in-time deliveries than previously.

As al-Qa'ida stated on 13 October 2002, 'If a boat that did not cost US $1,000 managed to devastate an oil tanker of that magnitude, imagine the extent of the danger that threatens the West's commercial lifeline which is petroleum'. Half the world's oil and much of its LNG passes through the Strait of Melaka, thus creating an attractive target in an area already riddled with piracy and poorly guarded by Indonesia's navy. The appalling condition of the Indonesian navy and the lack of concern for maritime terrorism have created conditions where terrorists might act with near-impunity. The commander of US forces in the Pacific, Adm. Thomas Fargo, proposed the deployment of US naval vessels in the Melaka Strait, hoping to engage in joint patrols with Malaysian, Singaporean and Indonesian navies. This proposal was rebuffed by Indonesia and Malaysia, which regarded the prospect as a violation of their authority and continued to insist that they possessed the capability and means to protect the Strait.

MINDANAO

One of the important factors in the continued existence of Jemaah Islamiyah is the co-operation of the MILF. The latter has been fighting for a homeland since the 1970s, when it broke away from the secular Moro National Liberation Front. In September 1996 the Government of the Philippines and the Moro National Liberation Front signed the Jakarta Peace Accord, ending the 24-year struggle. The former leader of the Moro National Liberation Front, Nur Misuari, became the head of the Southern Philippines Council for Peace and Development and was appointed Governor of the Autonomous Region of Muslim Mindanao (ARMM), which includes the four provinces of Sulu, Tawi-Tawi, Maguindanao (but not the city of Cotabato) and Lanao del Sur, and also the city of Marawi, thus accounting for 12,000 sq km, approximately 4% of Philippine territory. The MILF rejected the peace accord outright and vowed to continue fighting for an independent homeland that would include not only the four provinces of the ARMM, but also Cotabato, Lanao del Norte, Davao del Sur, Sultan Kudarat, South Cotabato, Zamboanga del Sur, Zamboanga del Norte and Palawan. By 1996 the MILF controlled about 10% of Mindanao, although it dominated two provinces. The Front has some 15,000 armed combatants.

The MILF believes (not without reason) that the ARMM has been a failure, refusing to accept a similar proposal, and that autonomy has led to the destruction of the rival MNLF; the MILF demands nothing short of a (qualified) referendum for independence. The Philippine Government believes that the MILF will accept an 'expanded and enhanced' autonomy. Its precondition for talks, however, is that the MILF accept the Constitution and territorial integrity of the Republic of the Philippines. The MILF did enter into formal talks with the Philippine Government in January 1997. The discussions resulted in July in the signing of a cease-fire, which lasted for almost three years. This was the first cease-fire to be negotiated, with the current battlefield status quo being accepted. Importantly, it recognized seven of the MILF's camps and base areas. This agreement was followed by the General Framework of Agreement of Intent, which was signed in August 1998; this more general confidence-building measure led to several government-sponsored development projects in MILF territories. The peace process broke down in 2000. The administration of President Joseph Estrada considered the joint development projects as 'unnecessary' and that they were 'simply coddling the MILF'. Estrada ordered renewed attacks on the MILF after the Government claimed that, in violation of the cease-fire, MILF forces had occupied regions from which the military had withdrawn. By July the Government had retaken much of the lost territory, eventually capturing the MILF's main base, Camp Abu Bakar as-Siddique, in early 2000. Hostilities continued until early 2001, when Estrada was ousted in a popular uprising. The newly inaugurated President Gloria Macapagal Arroyo ceased offensive military action, and discreet discussions resumed.

The formal peace process with the Government was suspended following the Government's violation of the cease-fire and its attack on the rebel stronghold of Pikit, Cotabato. Between March 2003 and mid-2004 there were three rounds of exploratory discussions aiming to resolve the preconditions for the resumption of formal talks. Although three issues were agreed upon in principle, by mid-2004 none had been implemented. Nevertheless, the cease-fire has held relatively well since early 2003: outbreaks have been resolved quickly without being allowed to escalate into a larger conflict.

The talks have been complicated by reports that the MILF continues to give sanctuary to and to train Jemaah Islamiyah operatives. This relationship is long-standing. The MILF began to receive significant amounts of funding (lethal and non-lethal) from al-Qa'ida in the early 1990s. In return for this aid, the MILF opened its facilities to al-Qa'ida trainers, who instructed not only MILF cadres in terrorism but also local Jemaah Islamiyah operatives who were unable to get to Pakistan and Afghanistan in significant numbers.

Since 2000 the MILF has used urban terror tactics whenever it suffers battlefield losses, such as in March 2003 when it carried out a bomb attack on the airport in Davao or after the 1999 offensive when it bombed the light rail transit system in Manila. The MILF has denied every act of terrorism or, when confronted with overwhelming evidence that implicates it, blames the attack on 'lost commands'. The organization is divided over its relationship to Jemaah Islamiyah. While radicals in the MILF regard Jemaah Islamiyah as fellow *jihadis*, many others in the leadership believe that the latter group discredits the legitimate struggle and is a pretext for the Government not to negotiate. Yet, until the MILF is given some encouragement at the negotiating table, there is little incentive for the Front to sever its links to Jemaah Islamiyah. Until that happens, the terrorist problem in the region will continue, as without the MILF camps and a secure base area Jemaah Islamiyah cannot train effectively.

Malaysia, which has been very concerned about the expansion of Jemaah Islamiyah, has put considerable pressure on the MILF to end its links with the group. The Malaysian Government has also sponsored peace talks between the two sides, and in April 2004 dispatched an advance team of peace monitors to Mindanao. A full deployment of monitors from Malaysia and other member countries of the Organization of the Islamic Conference (OIC) was expected to take place in mid-2004.

ACEH

In Indonesia the secessionist movement in Aceh has been waged during 1976–79 and again since 1989. The Acehnese have a history of violent secessionism that dates back to the colonial era when it was the last region to be conquered by the Dutch, and then only with considerable loss of life. The Darul Islam rebellion in support of an Islamic Indonesia won 'special region' status for Aceh in 1959, which gave the territory greater freedom in applying *Shari'a* (Islamic law) at the local level. The Indonesian Government responded to the 1976 rebellion by Gerakan Aceh Merdeka (GAM—Free Aceh Movement) with brutality and systematic abuse of human rights. By 1979 the Government believed that the rebels had been suppressed. Aceh was gradually brought into the mainstream of Indonesian affairs, but continued atrocities by the military and the diversion to Jakarta of Aceh's vast natural resource wealth (mainly from gas fields) led to a resumption of the rebellion. Only 7% of Aceh's revenues remained within the province, and all income from Indonesia's oil and gas production went directly to the central Government. Although Aceh's gas exports in 1998 were worth US $1,300m., the region received almost none of that revenue. GAM renewed its insurgency with the avowed goal of creating an independent Islamic state governed by *Shari'a*. Its leader, Hasan di Tiro, won considerable support from Libya, which armed the movement and provided training for some 600 Acehnese.

In 1991 Aceh was placed under 'operational military status'. For more than a decade the Indonesian military conducted a brutal counter-insurgency campaign and in the process committed egregious violations of human rights. Almost 3,000 deaths are believed to have been caused by the security forces and some 2,000 people are 'missing'. The military's approach drove more people towards GAM; about 90,000 people lived in some 61 refugee camps, a fertile ground for recruitment. GAM, for its part, was also quite brutal. It put pressure on village leaders and clerics to lead mass exoduses to refugee camps in order to help win international sympathy. Brutality in Aceh is also a result of the fact that the central Government gives the armed forces only 25% of its operating costs, thus compelling the military to make up the rest. Therefore, they extort local businessmen, kidnap, kill and engage in marijuana-trafficking.

Under the administration of President Abdurrahman Wahid (1999–2001), there was some progress in reaching a political settlement. Many of the 20,000 troops and 11,000 police officers were withdrawn. In April 1999 the Law on Intergovernmental Fiscal Relations was passed, allowing provinces to retain a greater proportion of the revenue earned from the extraction of natural resources. The law also devolved administrative power to the nation's 306 districts. The central Government offered the provinces 15% of the state share of net oil revenue, 30% of net gas revenue and 80% of income derived from forestry, mining and fisheries. Prior to the implementation of this law, the central Government had collected 85% of gross oil revenue and 70% of gross gas revenue. In 1997/98 the value of oil and gas production amounted to 35,400,000m. rupiah (US $4,100m.), or 32.7% of government revenue; 89% of national wealth flowed to Jakarta. Already the Acehnese provincial government had the right to implement *Shari'a*. Wahid also granted autonomy in education, and in political and economic management.

The result of the independence referendum in the Indonesian province of East Timor (now Timor-Leste), held in August 1999, emboldened the Acehnese to continue their struggle. Following a campaign of intimidation and violence, the Indonesian military withdrew from East Timor and the Commander-in-Chief of the Armed Forces, Gen. Wiranto, apologized for the 'excesses'. The Aceh peace talks broke down in 2000, but resumed somewhat reluctantly under President Megawati Sukarnoputri in 2001. Like the military, no politician is prepared to countenance the possibility of Aceh winning independence from Indonesia. As fighting resumed in 2001, in October Megawati extended a government order to allow the military to continue its operations in the province, where some 1,200 people had been killed in 2001 alone. By July 2002 the Indonesian miltary was requesting an additional 8,000 troops, in addition to the 21,000 soldiers and 15,000 police officers already deployed in Aceh to counter GAM's 1,800 lightly armed guerrillas. The military also lobbied the Government to impose a state of emergency. The Government insisted that GAM abandon its demand for independence and accept the Special Autonomy Act, which granted Aceh a 70% share of revenue from natural resource extraction and greater political autonomy as the basis for negotiations. An accord on the cessation of hostilities was signed by GAM and the Indonesian Government on 9 December 2002. Although this gave the Acehnese greater autonomy, the agreement came to naught. Neither side withdrew to defensive positions, nor was there any progress at the peace talks. In mid-2003 the Indonesian military resumed its offensive against GAM. Threatening 'shock and awe', the armed forces launched a massive invasion of the province, with some 40,000 troops fighting an insurgent army of 1,500–2,000 combatants.

An estimated 2,000 people had been killed in the new offensive, but an accurate assessment was all but impossible as a complete news 'black-out' was imposed in Aceh. No foreign or domestic journalists have been allowed into the province, with the exception of a few meetings in the provincial capital of Banda Aceh. Sporadic reports of egregious human rights abuses by government forces suggested that the military, without any fear of accountability, was operating with impunity. The brutal tactics might prove counter-productive to the Indonesian armed forces. Although GAM was reported to have only limited resources, the armed forces have not been able to defeat the movement. In May 2004 the armed conflict in the province was downgraded from a military to a civilian emergency, but little changed on the ground. Fighting continued, and restrictions on access remained in place. Some 40,000 troops were still stationed in Aceh. Indonesian military officials asserted that GAM rebel forces had been reduced to 40% of their former strength.

Some political campaigning for the parliamentary and presidential elections was permitted in Aceh in mid-2004, but the situation remained very tense. No politicians left the relative safety of the major urban areas to campaign. The Indonesian press reported that in the national election for the House of Representatives in April the turn-out at the polls in Aceh had reached 95%—a figure that would be surprisingly high for even the most peaceful and prosperous of Indonesia's provinces. The percentage was all the more remarkable given that the Acehnese had boycotted the 1999 polls and that GAM rebels often intimidate the electorate into not voting. The armed forces claimed this to be a vindication of their 'hearts and minds' campaign, but it was more likely that this apparent high rate of participation demonstrated either massive electoral fraud or the military's ability to intimidate the population into casting its vote.

In June 2004 Swedish police detained two senior GAM officials, Malik Mahmud and Zaini Abdullah, in Stockholm and raided the house of the 80-year-old titular leader of GAM, Hasan di Tiro. Di Tiro was not taken into custody, owing to his poor health. The Swedish authorities initially declared the three officials to be suspected of 'grave breaches of international law'. Indonesia has accused GAM of being a terrorist organization, and has charged its members with bombing civilian targets and with conducting assassinations and kidnappings. It accused GAM leaders resident in Sweden of ultimate responsibility and for many years had pressed the Swedish Government to arrest them. In June 2003 the Indonesian Government sent a delegation, led by Ali Alatas, the former Minister of Foreign Affairs, to Sweden to deliver evidence to the Prosecutor-General, in an attempt to persuade him to act against the GAM leaders. It linked 'the leadership of the Aceh separatist movement in Sweden and acts of armed rebellion, terrorism and criminal activities of the GAM in Indonesia'. Indonesian officials were very pleased with the arrests, asserting that they represented a major set-back for the rebel group. As all three detainees were Swedish citizens, however, they could not be extradited to Indonesia. Only three days later, a Swedish court ordered the release of the exiled leaders of the rebel movement on the grounds of insufficient evidence. Indonesia vowed to continue to exert pressure on the Swedish Government for the three to be rearrested.

BIBLIOGRAPHY

Abuza, Zachary. *Militant Islam in Southeast Asia: Crucible of Terror.* Boulder, CO, Lynne Rienner Publishers, 2003.

Acharya, Amitav. *Constructing a Security Community in Southeast Asia—ASEAN and the Problem of Regional Order.* London, Routledge, 2000.

Bracken, Paul. *Fire in the East: The Rise of Asian Military Power and the Second Nuclear Age.* London, HarperCollins, 1999.

Buszynski, Leszek. *Asia Pacific Security: Values and Identity (RoutledgeCurzon Security in Asia Studies).* London, Routledge-Curzon, 2004.

Collins, Alan. *The Security Dilemmas of Southeast Asia.* Singapore, Institute of Southeast Asian Studies, 2001.

Security and Southeast Asia: Domestic, Regional and Global Issues. Boulder, CO, Lynne Rienner Publishers, 2003.

Dent, Christopher M. (Ed.). *Asia-Pacific Economic and Security Cooperation: New Regional Agendas.* Basingstoke, Palgrave Macmillan, 2003.

Emmers, Ralf. *Cooperative Security and the Balance of Power in ASEAN and the ARF.* London, RoutledgeCurzon, 2003.

The Threat of Non-Traditional Security in the Asia-Pacific: The Dynamics of Securitisation. Singapore, Times Academic Press, 2004.

Hiro, Dilip. *War Without End: The Rise of Islamist Terrorism and Global Response.* London, Routledge, 2004.

Jeshurun, Chandran (Ed.). *China, India, Japan and the Security of Southeast Asia.* Singapore, Institute of Southeast Asian Studies, 1993.

Kapur, Ashok. *Regional Security Structures in Asia.* London, RoutledgeCurzon, 2002.

Kwak Tae-Hwan and Olsen, Edward A. (Eds). *The Major Powers in Northeast Asia: Seeking Peace and Security.* Boulder, CO, Lynne Rienner, 1996.

Lovell, David W. (Ed.). *Asia-Pacific Security: Policy Challenges.* Singapore, Institute of Southeast Asian Studies, 2003.

Martin Jones, David (Ed.). *Globalization and the New Terror: The Asia Pacific.* Cheltenham, Edward Elgar, 2004.

McNally, Christopher A., and Morrison, Charles E. (Eds). *Asia Pacific Security Outlook: 2002.* Washington, DC, The Brookings Institution, 2002.

Morrison, Charles E. *Asia Pacific Security Outlook: 2003.* Washington, DC, The Brookings Institution, 2003.

Tan See Sang, and Acharya, Amitav (Eds). *Asia-Pacific Security Cooperation: National Interests and Regional Order.* Armonk, NY, M. E. Sharpe, 2004.

Tien Hung-mao and Cheng Tun-jen (Eds). *The Security Environment in the Asia-Pacific.* Armonk, NY, M. E. Sharpe, 2000.

Vicziany, Marika, Wright-Neville, David, and Lentini, Pete (Eds). *Regional Security in the Asia Pacific: 9/11 and After.* Cheltenham, Edward Elgar, 2004.

PART TWO

Country Surveys

AUSTRALIA

Physical and Social Geography

A.E. MCQUEEN

With revisions by the editorial staff

The Commonwealth of Australia covers an area of 7,692,030 sq km (2,969,909 sq miles). Nearly 39% of its land mass lies within the tropics; Cape York, the northernmost point, is only 10° S of the Equator. At the other extreme, the southern limit of the mainland lies at 39° S or, if Tasmania is included, at 43° S, a distance on the mainland alone of 3,134 km from north to south. From east to west, Australia spans a distance of 3,782 km.

CLIMATE AND VEGETATION

The wide latitudinal range as well as the size and compact shape of Australia produce a climate with widely varying effects in different parts of the country. The climatic differences can be assigned generally to latitude, and therefore to its liability to influence rainfall either from tropical rain-bearing air masses or from the westerly wind belt which affects the southern areas of Australia. It is important to note, too, that the average elevation of the land surface is only about 275 m; nearly three-quarters of Australia is a great central plain, almost all of it at 185 m–460 m above sea-level, with few high mountains. The Dividing Range, running parallel to most of the east coast, is the most notable; the highest peak, Mt Kosciusko, reaches 2,228 m. This general lack of mountains, coupled with the moderating effects of the surrounding oceans, means that there are fewer abrupt regional climatic changes than would be found on land masses in comparable latitudes elsewhere.

The northern part of the continent, except the Queensland coast, comes under the influence of summer tropical monsoons. This produces a wet summer as the moist air flows in from the north-west; but winter is dry, with the prevailing wind coming from the south-east across the dry interior. Both the north-east and north-west coasts are liable to experience tropical cyclones between December and April, and these storms, with accompanying heavy rain, will occasionally continue some distance inland.

The southern half of Australia lies in the mid-latitude westerly wind belt for the winter half of the year; consequently winter is the wet season. Winter rainfall in the south-east and south-west corners of Australia and Tasmania is particularly high—at least by Australian standards—with maximum falls occurring on the windward sides of the mountains in each area. Rainfall decreases rapidly inland with distance from the coast, with the result that parts of central Australia record some very low annual average rainfall figures; the area of lowest average annual rainfall is the 460,000 sq km around Lake Eyre in South Australia, which receives an average of some 100 mm per year. At the other extreme lies Tully (17° 55′ S), on the east coast of Queensland, with an annual average of 4,048 mm. Overall, few parts of Australia enjoy abundant rainfall; and even where occasional heavy falls are recorded, the unreliability of its seasonal distribution may well lessen its value in terms of pasture growth. Only south-west Western Australia, western Tasmania and Victoria south of the divide can be classed as areas of reliable precipitation; elsewhere reliability decreases away from the coast, with wide variations being recorded at stations such as Whim Creek (20° 52′ S, 117° 51′ E) where 745 mm once fell in a single day, but only 4.3 mm was recorded in all of 1924.

Very high temperatures, sometimes exceeding 50°C (122°F), are experienced during the summer months over the central parts of the country and for some distance to the south, as well as during the pre-monsoon months in the north. Australia's insular nature and other features tend to hold temperatures at a rather lower general level than other southern hemisphere land areas in the same latitudes, but temperatures are high enough to produce an evaporation rate, especially in inland areas, which in turn is high enough to exert a marked influence on soil and vegetation patterns. In much of the interior, xerophytic plant species, such as spinifex, salt bush, blue bush and dwarf eucalyptus, adapted to very dry and variable conditions, are capable of supporting a limited cattle population. Between these arid areas and the zones of higher rainfall lie the semi-arid plains on which the main vegetation is mulga (*Acacia*) and mallee scrub (*Eucalyptus* spp.) in which several stems rise from a common woody base. It is this type of land that carries most of the sheep in New South Wales and Western Australia; it is also here, as well as in the still drier interior, that the major effects of drought are seen—and drought occurs with sufficient regularity to have a limiting effect upon the long-term stock population. Since the 1860s Australia has suffered from more than 10 major droughts affecting most of the country (most recently in 1999–2003), as well as several others causing severe losses in particular areas.

In the colder south-eastern areas the length of the growing season is mainly dependent upon temperatures, but elsewhere the availability of soil moisture is the major variable. The growing season lasts for nine months or more along the east coast and in south-western Australia; elsewhere, and especially in the interior, there are wide variations according to both the intensity and seasonal distribution of rainfall. Underground water supplies are fairly widespread in the semi-arid parts of Australia, including the resources of the Great Artesian Basin, fed from inland slopes of the mountain range to make one of the largest such catchments in the world. In some areas, notably the Barkly Tableland, stock-raising is largely dependent upon bore water.

SOILS AND LAND USE

Soils are very diverse in both type and origin, and are of low natural fertility over large areas, owing to the great geological age of Australia and the subsequent poor qualities of the parent materials. Climate has a marked effect on soil type, with seasonal desiccation and surface erosion coinciding with the extreme dry and wet seasons which affect much of the continent. Salinity and alkalinity are also problems, especially in arid southern Australia.

The generalized pattern of land use falls into three broad zones. The first comprises some 70% of the land area, and covers all central Australia, reaching the coast along the shores of the Bight and in north-west Australia. About one-third of the area is desert, useless for farming; the rest is of only marginal value for pastoral activities, and then only in the areas close to the rather more favourable conditions of the second zone. This second zone covers only some 17% of Australia, and contains a wide variety of climate and soil types; it is included within a broad belt over 300 km wide extending from the Eyre peninsula paralleling the east coast, and across the northern part of Australia to the Kimberleys. In this zone most farming is practised in the temperate part of the area, including the growing of more than 90% of all wheat sown; the zone also supports some 40% of the nation's sheep, 30% of the beef cattle and 20% of the dairy cattle.

The third general zone comprises a belt of land along the east coast from Cairns southward and then westward to south-east South Australia, all of Tasmania, south-west Western Australia and a small part of the Northern Territory around Darwin. Much of this zone (which covers some 13% of the continent) is of broken relief; in the remaining areas the pattern of land use is quite complex. In the northern parts beef grazing dominates, but in the remainder most forms of cropping and livestock production are found. Nearly all Australia's forests are within this zone, as are almost all of the dairying, sugar, fat-lamb, horticulture and high-producing beef cattle areas. The potential for pasture improvement, especially in the southern areas, is considerable.

POPULATION

Although there has been much immigration from Asia, particularly Viet Nam, in recent years, the majority of Australia's population is still of European stock. At the 1991 census 238,950 people were enumerated as Aboriginal people and 26,902 as Torres Strait Islanders. The total population at the census of August 1996 was 17,892,423 (of whom 386,049 were enumerated as Aboriginal and Torres Strait Islanders). At the census of August 2001 the population was estimated at 18,972,350, giving a density of only about 2.5 inhabitants per sq km, one of the lowest national figures in the world. In terms of population, as well as land use, Australia can be divided into three broad zones, corresponding, in many respects, to those outlined in the last section: one part almost unpopulated, another sparsely populated and the final part containing the great majority of the people. This distribution pattern means that any discussion of 'averages', in terms of population densities, is of only limited value; this is specially so when the proportion of each state's population living in the respective state capitals is revealed (see Statistical Survey). The concentration of population within each state is matched by a concentration, on a national scale, in the south-east of Australia. New South Wales and Victoria contained almost 59% of the nation's population in 1995; if that of Tasmania is added, the figure exceeds 61%. Over the country as a whole, settlement is closely related to the areas of moderate rainfall and less extreme temperatures, a pattern initiated by the early growth of towns and cities based on a predominantly pastoral farming community dependent upon farm exports for a livelihood. The result has been the rapid growth of settlement around major ports on which state railway systems were centred, and a subsequent development of manufacturing industry at these port centres where imported raw materials were available, where a skilled labour force could be found and where distribution facilities to all parts of the respective states were readily available. Only in Queensland and Tasmania did this basic pattern vary to any extent; in Queensland because of the widespread distribution of intensive farming (especially the cultivation of sugar cane) along a coast well serviced with ports, and in Tasmania because the more dispersed distribution of agricultural and other resources called into being a number of moderate-sized towns and commercial centres.

History

STUART MACINTYRE

ABORIGINAL AUSTRALIA

The human settlement of Australia began at least 40,000 years ago, and recent research suggests a longer occupation. The settlers came from the north-west, by means of the chain of islands that extends from Indo-China to the Timor Sea, and they spread rapidly over the island continent. As hunter-gatherers, these settlers met their needs without resort to agriculture or domesticated animals, and adapted their practices to a highly diverse environment. The primary social unit was the extended family group, with links of kinship, language and territory to larger aggregations. At the time of European arrival there were about 250 language groups, and population estimates vary between 200,000 and 750,000. They are now known as Aboriginals, although many of them also use localized names from these indigenous languages.

Aboriginal life was characterized by rhythms of work and leisure, aggregation and dispersal. Sex and age were the main forms of differentiation in a highly egalitarian social structure where kinship defined obligations and elaborate systems of religious belief regulated behaviour. Through many millennia Aboriginals evolved complex relationships with the country in which they lived, expressed in mythologies that explained its creation by spiritual beings and the continuing spiritual presence within it. Through their ceremonies and practices the Aboriginals protected the well-being of their land and their spirit ancestors.

The Aboriginal Australians adapted to changing environmental conditions and also contributed to them. Through periodic burning, for example, they created grasslands for the marsupial animals they hunted, although the larger marsupial species became extinct during their occupation. They acquired dogs some 5,000 years before the present, and improved their technology with nets, hafted axes and specialized stone implements. They also developed extensive trading networks.

BRITISH COLONIZATION

There was contact with Aboriginal Australia from the north before the European arrival. Traders from South-East Asia brought pottery, cloth and metal tools and took timber and sea produce. Spanish navigators sailed through the Torres Strait that separates Australia from Papua New Guinea at the end of the 16th century, and Portuguese vessels may have landed further south. The Dutch trade route to the East Indies took their ships east after rounding the Cape of Good Hope, then north up the western coast of Australia; and in 1642 Abel Tasman proceeded further east to the island of Tasmania, which lies at the south-east corner of the continent. Tasman's return through the Torres Strait demonstrated the limits of what had previously been designated as the Great South Land and was henceforth known as New Holland.

None of these contacts led to settlement, which followed James Cook's Pacific voyage of 1768 to 1771 and his navigation of the eastern coast. Subsequent British and French expeditions, impelled both by scientific investigation and strategic rivalry, completed the mapping of the island continent. Matthew Flinders, the first to make a full circumnavigation, was imprisoned on the French island of Mauritius during his return to the United Kingdom in 1803, and his account of the *Voyage to Terra Australis* did not appear until after his release in 1810. In it he suggested that the name Australia was 'more agreeable to the ear', and it was subsequently adopted.

Cook had declared British sovereignty over the Australian continent in 1770. The British took possession in 1788 when Capt. Arthur Phillip established a colonial settlement at Sydney. Although Cook had been instructed to obtain the consent of the inhabitants, he had not done so; and while Phillip was instructed to 'open an intercourse with the natives, and to conciliate their affections', the British made no recognition of the indigenous peoples' independent rights. The failure to pursue a formal agreement with them distinguished this British settlement from others, with consequences that still exercise the Australian nation state.

The settlement of New South Wales was designed as a penal colony, a place for the transportation of felons, but it was also expected to become a useful acquisition. Influenced by the reports of Cook and his companion Joseph Banks, the British Government intended to establish agriculture in New South Wales and to use the timber and native flax on Norfolk Island (1,700 km to the east of Sydney) for naval supplies. These expectations proved ill-founded: the first crops failed and livestock died, while the trees and flax on Norfolk Island were found to be unsuitable for masts, spars or sails. The main settlement at New South Wales was beset by early hardship and discontent, constrained by the demands of war with France. To forestall

French ambitions, however, an additional settlement was established in Tasmania in 1803.

The conclusion of the Napoleonic Wars in 1815 brought increased numbers of convicts. It coincided with the finding of a passage across the mountain range that had confined the Sydney colony to the coastal region, and the spread of settlement onto inland plains suitable for pastoralism. With the growing demands of British manufacturers for wool, sheep numbers increased rapidly and the Australian colonies found a durable staple export. The new prospects of wealth attracted increasing numbers of free settlers, who were able to acquire land cheaply and to draw on convict labour. European settlement spread through the south-east corner of the continent and exacerbated conflict with Aboriginal landowners who were driven from their hunting grounds. While Aboriginals were officially protected from assault, their resistance was suppressed by military action and unofficial massacres. The numbers killed in frontier violence remain the subject of controversy.

A further convict colony had been established in Tasmania in 1803. New colonies were founded in Western Australia in 1829 and South Australia in 1836, both outside the convict system, while unofficial settlement in Victoria led to official recognition in the same year. The growth of the eastern colonies led to demands for the abolition of convict transportation, which was suspended on the mainland in 1840. The British Government's attempt to revive it in 1849 led to protest movements and demands for greater local consultation. The absolute authority of the governors had been qualified in the 1840s with partly elected legislatures, and in 1852 the British Government invited the colonies to draft constitutions for self-government. New South Wales, Tasmania and Victoria had bicameral legislatures and systems of responsible government by 1855; South Australia followed in 1856, and in 1859 the north-eastern colony of Queensland was separated from New South Wales and similarly endowed. Western Australia still languished and did not become self-governing until 1890.

The advent of self-government coincided with the discovery of gold, which brought a rapid influx of new arrivals: the non-Aboriginal population increased from 430,000 in 1851 to 1.15m. a decade later. The richest finds were in Victoria, and mining laws allowed fossickers to stake a small claim and work it with little capital. With restless newcomers drawn from around the world, including British Chartists and continental Europeans who had participated in the revolutionary movements of 1848, the goldfields were a crucible of vigorous democracy. Declining yields, high charges for the miner's licence and arbitrary administration caused a rebellion at Ballarat in 1854, which was suppressed with the loss of 28 lives, but juries acquitted the leaders who were tried for treason. The egalitarian mood quickly spread to the new parliaments, which adopted manhood suffrage and the secret ballot. A further aspect of this popular radicalism was racial violence directed against the Chinese immigrants on the goldfields.

The Victorian gold rush was followed by subsequent movements resulting from further discoveries in New Zealand and New South Wales during the 1860s; then Queensland in the 1870s and 1880s, and finally Western Australia during the 1890s. The moving frontier of mining settlement was marked by youthful energy, personal ambition and collective endeavour that were channelled into civic buildings, voluntary societies, churches, schools and families. As the initial rush slowed, the colonies turned their attention to consolidating their new wealth. Using their political power, popular movements agitated for reform to land tenure, so that the miners could become farmers, and large pastoral holdings were broken up into agricultural selections. Tariffs were imposed to develop and protect local industry. The colonial Governments raised public loans in London to construct railways and ports, to build and operate public utilities and to create a system of compulsory education.

Australia enjoyed a high standard of living during this phase of sustained economic growth. Improvements in the pastoral industry increased output while export prices for wool, minerals and a growing range of agricultural products were high. Assisted passages, better prospects and comparatively good wages attracted further waves of immigrants. With natural increase they brought the population to 3m. by the end of the 1880s, and provided a further stimulus to the construction and service sectors. The overwhelming majority were from the United Kingdom, with which the colonies retained close ties of kinship; the free movement of people, capital, knowledge and ideas assisted their rapid progress. Ethnic and religious differences were contained with a low degree of friction.

Urban concentration was a marked feature of this phase of rapid growth. Half of the population lived in towns by the 1880s. Each of the colonies was dominated by its seat of government, which also served as the financial, commercial and administrative centre, and used its political influence to augment control over the hinterland. The Victorian capital of Melbourne had expanded to 420,000 inhabitants and Sydney to 360,000. Adelaide, the capital of South Australia had 115,000 residents and Brisbane, the capital of Queensland, 86,000.

The prosperity ended in 1890 as export prices fell and foreign lenders refused further advances. The pressure on production costs led to the formation of unions among workers in the key sectors of pastoralism, mining and transport, matched by the formation of national combinations of employers. A confrontation in late 1890 brought strikes and lock-outs that paralysed the economy. The unions were broken by the use of volunteer labour, supported by the colonial governments and protected by the volunteer militia, but the Australian economy rapidly declined into financial crisis and subsequent depression, which was deepest in Melbourne where the speculative boom had been the most intense. This was followed later in the decade by a severe drought that had halved sheep numbers by the turn of the century.

NATIONAL RECONSTRUCTION

From these misfortunes arose some of the most powerfully formative principles of Australian national life. The industrial conflict gave rise to a labour movement that entered Parliament to make good the losses it had suffered and has remained ever since a force in politics. The colonies came together to create a federal nation state with responsibility for defence and external affairs, trade and national development. The new Commonwealth Government devised institutions designed to insulate Australia from economic disturbance, to safeguard living standards and address social divisions. The national ethos was expressed in art, literature and popular culture.

The movement to federate Australia began in 1890 with a meeting in Melbourne of colonial statesmen, which was called by Sir Henry Parkes, the Premier of New South Wales, and stimulated by a recent report on the country's defence capacity; imperial rivalry had come closer to Australia with German and French activity in the South-West Pacific. A Federal Convention in Sydney in the following year drafted a Constitution, but the colonial Parliaments quibbled over its implications. The federal movement revived in 1893 with an unofficial conference that proposed a new and directly elected convention, which met from 1897 to 1898. Its revised draft was submitted to popular referendums and finally enacted by the Imperial Parliament, so that the new Commonwealth was declared on the first day of the new century. The Constitution combined the principles of responsible government and federalism, with a Parliament consisting of a people's House of Representatives and a Senate made up of equal numbers of representatives of each of the six states (the name now given to the colonies). The rivalry between New South Wales and Victoria was resolved by the provision that Melbourne should be the temporary national capital until a new one was found within New South Wales but at least one hundred miles from Sydney; the federal Parliament moved to Canberra in 1927.

The federal movement was led by middle-class liberals, who saw no inconsistency in their national ardour and Imperial loyalty. Alfred Deakin, the fervent idealist who championed the federal cause in Victoria and prayed for its fortunes, regarded himself as an 'Independent Australian Briton'. While critical of British neglect and condescension, he upheld the Crown as a symbol of Imperial unity. Republican sentiment had receded with the defeat of the unions and the decline of radicalism in the early 1890s. The unions turned to parliamentary activity in response to their defeat and made immediate gains, but failed to win significant representation in the federal conventions; and as they directed their attention to electoral politics they moderated their policies. The first Commonwealth elections in 1901 nevertheless confirmed the influence of organized labour. In 1910 the Labor Party achieved an absolute majority; previous governments had lacked a majority, and office was therefore shared, except for a brief interregnum, between the protectionist Liberals and Labor, each depending upon the parliamentary support of the other. These two parties gave the Commonwealth a popular cast.

The initial Commonwealth elections retained the voting arrangements of the states. The first parliament extended the right to vote to women. South Australia and Western Australia had already granted voting rights to women, and Australia was a pioneer in the suffrage movement as well as in the emancipation of women from other restrictions. Like the labour movement, the feminist movement regarded Australia as a land of freedom and opportunity that had to be safeguarded from the evils of exploitation. It sought to purify domestic life and elevate public life. Campaigns for women's rights of education, employment and public participation ran in parallel with campaigns for temperance and against gambling, prostitution and domestic violence. The maternalist conception of citizenship treated emancipation from masculine tyranny as a necessary condition of women's contribution to national life.

The same Franchise Act of 1902 denied the vote to Aboriginal Australians, despite the fact that they voted in several of the states. This was in keeping with the exclusion of Aboriginals from most rights of citizenship at this time and with the increasing restriction of their lives by the states, which retained legislative power over them. There was a widespread expectation that the Indigenous peoples were doomed to extinction because they were incapable of bridging the gulf that separated them from civilization. Some of mixed parentage might succeed, and to this end Aboriginals were confined to designated settlements as wards of the State, while children of mixed descent were removed and expected to assimilate. In contrast to other white supremacist settler societies, Australia created no absolute barrier between black and white, but its own mixture of segregation and assimilation was designed to eliminate Aboriginality. The motives of those who devised the policy and the consequences for its victims are again subjects of present controversy.

'White Australia', declared Deakin in 1903, when he became the country's second Prime Minister, 'is not a surface but it is a reasoned policy which goes to the roots of national life, and by which the whole of our social, industrial and political organization is governed'. Among the first acts of the new Parliament was the approval of immigration legislation whereby Asians could be denied entry to the country. To meet British concerns, it did not explicitly discriminate against Asians but rather allowed the Commonwealth to impose a test in any European language. Frequently modified and gradually relaxed, this White Australia policy lasted until the 1960s. Additional legislation repatriated Pacific islanders who had been introduced to Queensland as plantation labourers.

The fear of aliens was expressed in the language of national purity. Asians were perceived as a threat to living standards, a cause of miscegenation and moral decay, and a danger to territorial integrity. As the United Kingdom felt the strain of naval competition and withdrew most of its squadron from the Far East, insecurity increased. The British naval treaty with Japan of 1902 only exacerbated this. Thwarted in his attempts to restore the British naval presence in the Pacific, Deakin invited the US fleet to visit Australia in 1908 and in the following year commenced a building programme for the Australian navy. Compulsory military training was introduced in 1910.

Protection of local industry through tariff duties on competing imports was a further policy of the new Commonwealth. In 1906 the Deakin Government linked the price advantage enjoyed by local employers to the payment of 'a fair and reasonable wage' to their employees. It fell to the new Arbitration Court to determine whether the employers complied with this condition. The Court had been established in 1904 to provide a remedy for industrial conflict, and it had compulsory powers to settle disputes and make awards specifying wages and conditions that were binding on the parties. In 1907 the Court considered the meaning of a fair and reasonable wage in a case concerning a large manufacturer of agricultural machinery, and its president determined that such a wage should be sufficient to cover the needs of a man, a wife and three children. This 'basic wage', regularly adjusted for changes in the cost of living, was gradually extended across the work-force as a legal minimum. The basic wage confirmed the role of the male as the family breadwinner: women were restricted in employment opportunities and their minimum wage was designed to support a single person only. Around these arrangements a system of social welfare was created: an old-age pension from 1908, an invalidity pension from 1910 and a maternity benefit from 1912 to assist mothers with the expenses of childbirth.

This period of conscious nation-building was accompanied by an exploration of local forms in art and literature. The artists of the Heidelberg school revealed a landscape of dazzling light. Popular verse and fiction celebrated the independence, egalitarianism and irreverence for authority of the bushman. Australian flora and fauna became common decorative motifs in architecture and advertising. The green of the Australian eucalypt and the gold of the wattle were adopted first by the national cricket team and then by other sporting representatives. The national coat of arms featured a kangaroo and an emu, backed with sprays of wattle. Australians thought of their country as an advanced democracy and took pride in the innovative public policies that earned it the reputation of a social laboratory.

WAR AND DEPENDENCE

Australia entered the First World War (1914–18) along with the other British Dominions in support of the United Kingdom. An Australian Imperial Force was formed by voluntary enlistment, and a first contingent fought alongside New Zealand in the Dardanelles and the Middle East. The unsuccessful attempt of this Australian and New Zealand Army Corps (better known by its acronym as ANZAC) to storm the Turkish peninsula of Gallipoli became a formative legend of heroic endeavour; with Australia Day on 26 January, marking the first white settlement, Anzac Day on 25 April remains the country's principal public anniversary. After evacuation from Gallipoli at the end of 1915 the main Australian force was deployed on the Western Front, and mounting casualties persuaded the Labor Prime Minister William Morris Hughes to propose conscription for overseas service. Referendums on the proposal in 1916 and 1917 failed narrowly and split the Labor Party; the pro-conscriptionists joined with the non-Labor forces to establish the National Party and Hughes remained in office.

His vigorous prosecution of the war and liberal use of his emergency powers caused deep division. The instigation of legal proceedings against industrial militants, the internment of enemy aliens (there was a substantial German community) and vilification of anti-conscriptionists brought a new tone of acrimony to public life. The Prime Minister's conflict with the Irish-born Daniel Mannix, who was Catholic Archbishop of Melbourne, widened sectarian divisions. At the peace conference in Paris at the end of the war Hughes antagonized the US President Woodrow Wilson with his insistence on heavy reparations and offended the Japanese delegates with his determination to retain control of Pacific territory and reject a declaration of racial equality. The war resulted in the loss of 60,000 Australian lives and left a large debt. It also left lingering fears for Australia's security and deep suspicions of national loyalty. Ex-service organizations were formed to safeguard the country from sedition. The Nationalist Party affirmed a more conservative patriotism that was resistant to further experiment in social policy and hostile to alien influences.

A post-war slump was accompanied by renewed industrial unrest. The Hughes Government responded by raising tariffs and extending public enterprises, a policy that offended the farmers who formed their own Country Party. It held the balance of power after the 1922 election, and Hughes lost the leadership of the Nationalist Party to the Melbourne businessman S. M. Bruce (later Lord Bruce), who formed a coalition with the Country Party. The coalition sought to resume the policies of national development by extending government assistance to the farmers. Australia's population increased from 5,412,000 in 1920 to 6,336,000 in 1929. Further loans were raised to settle 200,000 British migrants on the land and to create new farms for ex-servicemen. A wider range of agricultural products, dairy produce, meat, fruit and sugar, as well as wheat and wool, were produced for the British market. Heavy industry followed the establishment of major steelworks at Newcastle during the war, and local manufacturers also expanded to meet the growing domestic market for consumer durables. The construction of the Sydney Harbour Bridge between 1926 and 1932 symbolized this triumph of urban modernity. Yet the city factories and offices remained heavily dependent on tariff protection, and their high cost structure imposed an increasing burden on the rural export sector.

Even before the Wall Street Crash in October 1929, the Nationalist Government attempted to restrict the growing trade imbalance by reducing wage costs. A series of protracted strikes followed in the construction, maritime transport and mining industries. Bruce's attempt to curtail the industrial arbitration system led to a party rebellion, and Labor won the ensuing election at the end of 1929, thus assuming responsibility for the Australian response to the Depression. With the collapse of export earnings and foreign investment, Australia's reliance on the traditional methods of protecting the economy proved futile: gross domestic product declined by more than 10% and the unemployment rate rose to more than 20%. Denied further credit by the banks, the Commonwealth and state Governments were forced to curtail their expenditure. The Arbitration Court reduced the basic wage by 10% in early 1931, while the currency was devalued by 30%. These economy measures caused widespread distress and political unrest. The Labor Government fell at the end of 1931 and the former Labor Treasurer, J. A. Lyons, took office at the head of a newly formed United Australia Party. The Government relied heavily on the resumption of trade with the United Kingdom, and an agreement made at Ottawa in 1932 gave reciprocal preference to British imports. This arrangement caused particular difficulties with Japan, which retaliated in 1937 with damaging sanctions against Australian wool-growers. In foreign as well as trade policy, Australia followed the United Kingdom closely and therefore found itself again at war in 1939.

Australian troops were once more sent overseas, to North Africa, and fought in campaigns there, as well as in the unsuccessful defence of Greece in 1941. The entry of Japan into the Second World War (1939–45) at the end of 1941 and its rapid advance down the South-East Asian archipelago forced a concentration on home defence and a realization that Australian could no longer depend upon British assurances. A Labor Government took office and sought a closer relationship with the USA, which established its regional headquarters in Australia and led the Allied campaign to repulse the Japanese. The same circumstances forced a greater degree of national self-sufficiency: the Government assumed unprecedented controls to manage the war economy and supply the Allied forces; military conscription for overseas service was introduced in 1943. Australian losses in the Second World War, at 37,000, were lower than in the First, although the maltreatment of prisoners of war by the Japanese left bitter resentment (of 22,000 taken into captivity, only 14,000 returned). There was a greater spirit of unity in support of the war effort, encouraged by the Labor Government's insistence on equality of sacrifice, as well as its ambitious programme of post-war reconstruction.

POST-WAR EXPANSION

Again in contrast to the settlement of the First World War, Australia was fully committed to international principles that led to the formation of the UN. It placed special emphasis on the democratic political imperative of full employment, and was quick to adopt the Keynesian techniques of economic management that expanded the government's role. Despite the backlog of consumer demand and shortages of essential materials, it introduced a programme of housing construction and major public works, created a new range of welfare measures and set ambitious immigration targets. Few of the personnel who returned from the armed services after 1945 were placed on the land; many more undertook university education and training, as the aim was to harness science, foster expertise and increase national capacity. As part of this endeavour, a local car-manufacturing industry was established in agreement with General Motors.

While the Labor Government had secured control of income taxation and confirmed the Commonwealth's primacy in welfare provision, it failed in a series of constitutional referendums to extend its wartime economic powers and was particularly concerned that full employment in a period of shortages would lead to inflation. The reluctance to allow wage increases and delays in the hearing by the Arbitration Court of an application for a 40-hour working week strained the patience of the unions, while the onset of the period of mutual hostility between the USA and the Soviet Union known as the Cold War caused an open breach with the Communist Party, which had grown rapidly during the Second World War and held leading positions in major trade unions. A series of strikes in essential industries increased the dissatisfaction of an electorate weary of restrictions and controls. The United Australia Party had been reformed into a Liberal Party during the war and returned to office in coalition with the Country Party in 1949.

Under Robert (later Sir Robert) Menzies, the Coalition established political dominance. Its initial attempt to outlaw the Communist Party failed when the High Court declared the Government's legislation to be unconstitutional, and a referendum in 1951 failed to win popular approval for the necessary constitutional amendment. Yet the new Government was able to link the Cold War with the long-standing fear of invasion from the north. Australia therefore dispatched troops to fight in Korea in 1950, and in 1951 entered into the ANZUS pact with the USA and New Zealand, as well as the South-East Asia Treaty Organization (SEATO), arranged by the USA after communist forces defeated France in Viet Nam in 1954. The Menzies Government emphasized that communist influence in trade unions compromised the Labor Party, a claim that was strengthened when a Soviet diplomat and his wife defected and in 1954 told a royal commission of their links with members of the staff of the Labor leader, Dr Herbert Evatt. An anti-communist organization had been formed in the unions under Catholic leadership, and was extending its influence into the Labor Party. When Evatt denounced this organization, the Labor Party split in 1955, and a breakaway Democratic Labor Party directed its preferences to the Coalition parties at subsequent elections. Labor was destined to languish in opposition in the federal Parliament until 1972.

The Menzies Government benefited from two decades of sustained economic growth, aided by the liberalization of world trade and increased demand for Australian exports. A trade treaty with Japan in 1957 signalled a reorientation of trade from the United Kingdom and Europe towards East Asia as that region began to recover from war and to industrialize; minerals joined wool and agricultural products as major sources of export earnings. Australian manufactures expanded rapidly with the advantage of tariffs and import quotas. The service sector grew even faster as mechanization released blue-collar workers to join the white-collar salariat. An annual economic growth rate of 4% was maintained throughout the 1950s and into the 1960s. While Menzies condemned socialism and continued to proclaim his commitment to private enterprise, his Government maintained controls over foreign exchange and investment, regulated the finance sector and retained the Arbitration Court (restructured in 1956 as the Arbitration Commission) to determine wages. It increased public revenue and provided annual grants to the states for their operation of hospitals, schools and public transport; but relied increasingly on private provision for its own schemes of subsidized medical insurance and assistance to home-buyers.

The rate of home ownership had increased to 70% by 1961, while the rapid spread of car ownership and television supported an increasingly self-sufficient form of suburban domesticity. Women had enjoyed new opportunities during the Second World War, both as members of the auxiliary forces and as workers in war industries earning higher wages; but with the advent of peace they were expected to revert to their roles of wives and mothers. Most did so, and the post-war baby boom caused a rapid expansion of the education sector. Full employment, however, allowed a growing minority to undertake paid employment, and in 1950 the Arbitration Court increased women's pay to 75% of the male rate.

The population rose from 7,579,000 in 1947 to 11,550,000 by 1971, and more than half the increase came from migration. The

Labor Government had set a target of 1% per annum at the end of the war, and tried initially to find migrants from the traditional source of the United Kingdom and Ireland. Unable to fill the quota, it turned to continental Europe and took in 170,000 from the refugee camps there. Agreements were made, initially with the governments of north-west Europe, and later Italy, Greece and other Mediterranean countries, for assisted passage and settlement. Two-thirds of the 1m. settlers during the 1950s were non-British, a reorientation recognized in 1948 with the creation of Australian citizenship (previously Australians had the official status of British subjects). With this came the designation of immigrants as 'new Australians' and procedures designed to assimilate them into 'the Australian way of life' through shipboard instruction, then language classes and educational programmes in the migrant centres, along with 'Good Neighbour' committees and other voluntary organizations that culminated in naturalization ceremonies. There was no immediate relaxation of the White Australia policy, which was enforced by the repatriation of wartime refugees and became a growing embarrassment to Australia in its dealings with Asian neighbours, especially as the Colombo Plan (a post-war scheme for co-operative development of British Commonwealth countries in the region) brought South and South-East Asian students to Australia for university study. The dictation test was quietly abandoned in 1958, naturalization facilitated in 1966 and by the end of the decade 10,000 non-white immigrants were arriving annually.

Menzies gave little recognition of this growing diversity. While he presided over the transfer of Australia's primary alliance from the United Kingdom to the USA, he remained a romantic monarchist and champion of British ties. His attempt to intervene in support of the homeland during the Suez crisis in 1956 only revealed that the United Kingdom was no longer a world power. While his gallantry during the Royal Tour of 1954 was attuned to the immense popularity of the young Queen Elizabeth, the excessive fervour he displayed during her subsequent visit a decade later made listeners uneasy and embarrassed the Queen. Menzies was most confident during visits to London for meetings of the leaders of the white Commonwealth, ill at ease with the non-white leaders of the former colonies, and his defence of the apartheid regime in South Africa damaged Australia's international reputation. It was the United Kingdom's decision to enter the European Common Market in the early 1960s that finally ended the illusion of a special relationship, leaving his own elevation to imperial honours and appointment to the antique office of Lord Warden of the Cinque Ports after he retired in 1966 as nostalgic anachronisms.

Menzies had committed Australia to support the USA in the Viet Nam War. His successors were left to deal with the consequences. Harold Holt, who became Liberal leader and Prime Minister in 1966, told US President Lyndon Baines Johnson that Australia would go 'all the way with LBJ' and sent conscripts to Viet Nam; but Holt's initial popularity waned before he died while swimming at the end of 1967. His successor, John Gorton, pursued a more independent foreign policy but was overthrown two years later and replaced by William McMahon. By this time the issue of Viet Nam had become a huge burden to the Coalition Government. Labor won the election at the end of 1972 and immediately implemented its commitment to withdraw all troops.

The new Labor Prime Minister, Gough Whitlam, came to power with an ambitious programme of reform. A large and magisterial man, he set aside the party's preoccupation with trade union concerns so that it could appeal to the mood for change among the educated middle class. With great speed he signalled a more independent stance for the country by establishing diplomatic relations with the People's Republic of China, announcing independence for Australia's colony of Papua New Guinea, ratifying international conventions on nuclear weapons, labour rights and racism, and abolishing imperial honours. Among his domestic initiatives were the creation of a Department of Aboriginal Affairs and recognition of Aboriginal land rights; the abandonment of migrant assimilation in favour of multiculturalism; equal pay for women along with reform of family law and programmes to promote affirmative action; social welfare reform, a public health system, expansion of educational provision, environmental protection and urban renewal.

It was Whitlam's misfortune to take office just as the circumstances to support such an expansive policy ended. His Government doubled public expenditure in just three years. Prices and wages were increasing rapidly, despite a reduction of 25% in the tariff level, even before the Yom Kippur war of 1973 and the subsequent embargo imposed by the Arab oil producers that disrupted the world economy. While the Labor Government delivered a contractionary budget in early 1975, it was in deep trouble. The non-Labor parties used their majority in the Senate to refuse support to the Government, and on 11 November 1975 the Governor-General took the extraordinary step of dismissing Whitlam from office. The Liberal and Country parties, led by Malcolm Fraser, won a decisive victory in the subsequent election. The Dismissal, as it was known, was the most serious constitutional crisis in Australian history and cast a shadow over the new Prime Minister's legitimacy as he grappled with new problems.

UNCERTAINTY

Malcolm Fraser searched for a solution to the novel combination of high inflation and high unemployment, dual scourges that had been regarded as alternatives in the preceding period of economic management. His Government determined to tackle inflation in order to reduce wage costs and restore profits, and to decrease government expenditure in order to ease interest rates and encourage private investment. While inflation declined, unemployment rose from 250,000 in 1975 to 400,000 by 1978. Australia's difficulties were compounded by a long-term decline in the price of its export commodities, setting at risk its traditional role as a producer of raw materials for overseas manufacturers. The best prospects seemed to be minerals, oil and gas but these were capital-intensive industries that created few jobs and increased the pressure on domestic industries through their effect on the exchange rate.

Fraser won fresh elections in 1978 and 1980, although he lacked control of the Senate and was hampered by ministerial scandals as well as by his own irresolution. Forthright in his anti-communist foreign policy, he was at the same time a strong critic of white supremacy in southern Africa, and a supporter of both multiculturalism and Aboriginal land rights at home. These sympathies were criticized by some of his colleagues, while his abrasive industrial relations policy and insistence on uranium exports aroused protest. As the mineral boom suddenly ended in 1982, the Government relaxed its tight control of public expenditure and the money supply, bringing increases in wages, renewed inflation and a further rise in unemployment. No sooner was Fraser defeated at the 1983 election than those who had served under him dissociated themselves from his legacy.

The new Prime Minister was Bob Hawke, previously the leader of the Australian Council of Trade Unions, and he came to office with an agreement by the unions to co-operate in economic reconstruction. Under the terms of a Prices and Income Accord, workers would accept wage restraint in return for job creation. The Government helped reconstruct key industries such as steel and vehicle construction, and restored health care and other welfare benefits. As an employment strategy, the Accord was successful. A total of 1.5m. jobs were created during the remainder of the decade and the unemployment rate declined from 10% to 6% by 1989. With subsequent revisions that included a national superannuation scheme, the Accord was maintained. Industrial disputes were brief and infrequent.

These arrangements alone, however, did not restore competitiveness and the Labor Treasurer, Paul Keating, implemented a series of further changes. The Australian dollar was floated, exchange controls were removed, the financial sector was deregulated, tariffs were lowered and other forms of protection were dismantled. This exposure of the Australian economy to market forces brought high borrowings, speculative investments, persistent trade deficits and rapid currency fluctuations. Traditional industries went bankrupt, and with them disppeared the job security that Australians had once expected; there was a rapid rise in part-time and casual employment. The rural sector struggled to adjust, and country towns lost the banks, public agencies and professional services that had sustained them. A rapid increase in interest rates at the end of the 1980s brought recession and renewed unemployment. The Government responded with further instalments of economic reform, including the privatization of public enterprises, and moved away from national wage determination by introducing enterprise agreements.

These changes completed the Government's abandonment of the institutions that had guided national life for most of the 20th century. The protection of local industry, the restriction of migration, the regulation of wages, the construction of welfare provision around the entitlements of the male breadwinner, the maintenance of full employment and the recourse to public enterprise were all designed to meet the needs and expectations of the inhabitants of a small, trading nation located on the opposite side of the world from its main sources of trade, finance, technology and culture. The institutions were dismantled because they were no longer effective in the changed circumstances that prevailed by the end of the century. Paul Keating, who drove the economic reforms, warned in 1986 that without them Australia was doomed to become a 'banana republic'. The Government therefore searched for greater competitive efficiency in the new global economy by embracing economic liberalism.

In contrast to similar reforms in the USA and the United Kingdom, those in Australia were undertaken by Labor Governments. The Hawke Ministry's willingness to embrace deregulation outflanked the Liberal and National parties (this change of name by the Country Party signalled the decline of the rural constituency); they found it harder to break free of their attachment to industrial interests that were accustomed to receiving government support. Labor's appeal to the national interest brought victories at federal elections in 1984, 1987 and 1990, while a similar reforming zeal enabled the party to hold office in five of the six states by 1990. This Labor programme retained vestiges of social democracy. While the Treasurer sought a more flexible labour market through enterprise agreements, the Government did retain a lower limit for wages. While income inequality increased, social welfare and employment programmes preserved the living standards of the lowest income earners. It was the rapid increase in the wealth of those who took advantage of financial deregulation to engross corporate assets that increased inequality, the recklessness with which they borrowed and the way they flaunted their spoils that caused resentment.

As trade union membership declined, the Government was more dependent on a coalition of social movements. It attempted to attract the environmentalists with conservation activity, and gained electoral support under the preferential voting system from the new Green Party. It incorporated feminists with affirmative action programmes, as the movement of women into the work-force increased. It recognized ethnic diversity with the encouragement of multiculturalism and it fostered greater Indigenous self-management through the creation of a representative Aboriginal and Torres Strait Islander Commission that assumed responsibility for a range of services and support.

In foreign policy the Hawke Government maintained the alliance with the USA, but moved to realign diplomacy, trade and migration. By 1990 half of Australia's trade was conducted within its region, and the trade policies of the European and North American blocs threatened Australia's foothold in the markets of East and South-East Asia. The Australian Government played a leading role in forming the Cairns Group, an alliance of agricultural producers in preparation for a new round of international trade negotiations in 1986, and also tried to promote the Asia-Pacific Economic Co-operation forum as a regional bloc. Closer Economic Relations (CER) with New Zealand had been formalized in 1983 and brought increasing integration. This relationship became more problematic in 1986 when the New Zealand Government refused access to US warships with nuclear weapons, and the USA suspended its obligations under the ANZUS treaty to New Zealand. Australia adhered firmly to its more powerful ally. Immigration policy allowed for substantial intakes based on economic, humanitarian and family reunion criteria, and increased the numbers that came from the Asian region. More generally, Australia encouraged multilateral strategies and was active in the settlement of the conflict in Cambodia, efforts to encourage nuclear disarmament and the peace-keeping activities of the UN.

These tendencies were intensified after Keating replaced Hawke as Prime Minister at the end of 1991 and the Liberal and National Coalition parties turned to John Hewson, an economist who offered a bold programme of tax reform, further labour market deregulation and reduction of government activity in favour of the principle of 'user pays'. Keating now reinvented himself as a defender of what he called the 'social fabric' and insisted that prosperity was greatest in 'social democracies where the government is involved in making societies tick'. He embraced a landmark decision by the High Court in 1992 that found native title survived on Crown land and in the same year he addressed an Aboriginal audience in inner Sydney with a frank admission of the wrongs that had been committed against them. He argued that Hewson's policies would penalize the poor, marginalize Indigenous and ethnic groups, and imperil the hard-won advances of the previous decade.

Keating was himself an abrasive polemicist and many thought him arrogant, but the electorate was weary of change and reluctant to embark on a further round of radical reform. Yet the Prime Minister interpreted his victory in the March 1993 election as a vindication of what he called his 'big picture' of a 'competitive, outward-looking, phobia-free' society. His republican enthusiasms, his pursuit of closer relations with Asia, his lavish patronage of the arts, his legislation to give effect to the High Court judgment on native title and his own weariness with the complexities of political management gave the impression of a Government that was out of touch. The Coalition parties turned to John Howard, a political veteran of great stamina and discipline, who took advantage of the growing discontent. Howard appealed to the 'battlers', as he called them, the ordinary men and women whose interests had been subordinated to the 'noisy minority groups' of feminists, environmentalists, the ethnic lobby, the Aboriginal 'industry' and the intellectuals. Juxtaposing the practical concerns of his battlers to the indulgences of Labor's pampered 'élites', he emphasized the national interest. Howard's promise at the March 1996 election was that he would govern 'for all of us' and he made major advances into Labor's heartland to win a decisive victory.

The Howard Administration

The new Government pursued a policy of economic liberalism and social conservatism. It embarked on a further round of economic reform with renewed attention to trade liberalization, changes in the labour market to eliminate inefficiencies and uncompetitive practices, the partial sale of the telecommunications corporation Telstra, the largest of the remaining public enterprises, a retrenchment in government services and reduction of the public service in order to restore the budget to surplus. The collapse of stock exchanges in South-East Asia in 1997 and the deep recession that followed in the Asian region provided further justification for these policies. Reduced demand from Asian customers for Australian exports was partially offset by new markets, where the falling value of the Australian dollar gave a competitive advantage. Despite a persistent trade imbalance, interest rates remained low. Productivity gains allowed Australia to enjoy low inflation, low unemployment and sustained growth.

The Howard Government abolished most of its predecessor's training programmes for the unemployed and replaced them with a new scheme that required young unemployed people to undertake community projects. The administration subsequently extended the same principle to other welfare recipients, although the number in receipt of benefits continued to increase. At the beginning of the 1990s 1.5m. Australians of working age were receiving income support; by the end of the decade this number had risen to 2.6m., or 20% of the work force. There were more working poor because of changes to the labour market. The Workplace Relations Act of 1996 encouraged the establishment of individual contracts in addition to enterprise agreements, and restricted the scope of the Industrial Relations Commission's awards.

The Coalition would have liked to extend its policy on industrial relations but lacked a majority in the Senate. The Government therefore had to negotiate a compromise with the smaller parties that held the balance of legislative power. In particular, it wanted to break the power of trade unions. Union membership had declined, to just 31% of the work-force by 1996, but remained strong in key industries such as maritime transport. In 1997 the Government drew up a plan with the National Farmers Federation and one of the principal stevedoring companies to destroy the Maritime Union of Australia. New workers were recruited from the Army and flown to Dubai, in the United Arab Emirates, for training. On the evening of 7 April 1998 security guards, wearing balaclavas and with dogs on leashes, moved onto the nation's ports to dismiss all members of the union who worked for the company. While pickets prevented the movement of cargo, and the courts ordered the reinstatement of the workers, negotiations brought major concessions from the union.

The new Government abolished the principal multicultural agencies and reduced the scope of the office for women's affairs. It decreased funding for Aboriginal programmes, legislated to circumscribe the Aboriginal land rights and refused to accept the findings of an inquiry into the removal from their families of Aboriginal children that had been commissioned by the previous Government. The Prime Minister refused to make a formal apology to the 'stolen generation' or to support the programme for reconciliation that had begun under the previous administration. His Government dismissed reports issued by the UN in 2000 that were critical of Australia's policies.

This rift with the Indigenous community was the most heavily contested of the ideological battles that were fought following the conservative accession to office since, by the closing years of the 20th century, the claims of Aboriginal Australians for recognition and restitution had become inescapable. The census revealed a dramatic increase in Indigenous numbers, from 156,000 in 1976 to 410,000 by 2001, and with the rebuilding of population came a renaissance of both traditional culture and new cultural forms in music, theatre, dance, art and literature. The Aboriginal flag, a striking arrangement of black, gold and ochre, was one of them. It became increasingly common in the 1990s for public meetings to begin with an Aboriginal elder giving a 'welcome to country' and a non-Aboriginal Australian to respond with an acknowledgement of the original inhabitants of the country on which they were meeting. There was a revival of Aboriginal languages, and a return to Aboriginal names.

There were few advantages, however, in being Aboriginal. Those who identified themselves as such in the census had a higher rate of unemployment and lower income levels than the rest of the population. More of them were arrested and imprisoned. Their life expectancy was 15 years below the national average, and this despite a quarter-century of government activity that was meant to overcome their disadvantages. In 1991 the Hawke Government had initiated a formal process of reconciliation led by a statutory body of Indigenous and non-Indigenous Australians, who embarked on a process of consultation and discussion that was to culminate on the centenary of the Australian Commonwealth in 2001. It was at a convention of this Council for Aboriginal Reconciliation in 1997 that Howard dismissed what he called symbolic gestures and over-exaggerated promises. The dashing of Aboriginal aspirations was accompanied by an attack on their moral legitimacy, both by denial of past wrongs and refusal of responsibility for their lasting effects. Aboriginal litigants who sought compensation for being separated from their families and Aboriginal women who claimed that an island marked for development had sacred significance were derided and discredited. Historians who wrote of violence against Aboriginals on the colonial frontier were accused of scholarly malpractice and malevolent denigration of the national honour.

On the other hand, the Coalition Government was on stronger ground in calling for greater accountability in public expenditure and emphasizing health, housing education and employment programmes with practical outcomes. The Aboriginal and Torres Strait Islander Commission (ATSIC) lost much of its credibility and was scheduled to be abolished in 2004. Noel Pearson, one of a younger generation of Aboriginal activists, would speak out on the need to address the problems of alcohol, violence and family breakdown in Aboriginal communities; he condemned welfare dependency as a threat to the very survival of his people and urged them to take greater responsibility for their own circumstances.

The Government's rebuff to Aboriginal aspirations was assisted by an independent member of the Commonwealth parliament, Pauline Hanson. A small business proprietor, she had been endorsed as a Liberal candidate in 1996 for an electorate based on the Queensland industrial town of Ipswich and then lost that endorsement after she alleged that excessive public funding was being provided for Aboriginals 'at the expense of hard-working, white Australians'. In her maiden speech six months later she promised to end the 'reverse racism' of multiculturalism, Asian migration and 'the Aboriginal industry'. Hanson's rapid rise to prominence divided Labor's support base: she affronted its tertiary-educated, cosmopolitan, middle-class progressives while she appealed to older, less mobile, blue-collar workers. Labor condemned her statements, while John Howard insisted she was articulating the feelings of many people.

Assisted by a former Liberal organizer, Hanson formed her own One Nation party. Its threat became apparent in elections for the state of Queensland in June 1998. One Nation candidates won nearly one-quarter of the vote, pushing the Liberals into third place and allowing Labor to gain office. Fear of a similar result in the imminent federal election opened divisions in the Coalition. Although John Howard began criticizing Hanson's views, he resisted Labor's proposal that the major parties direct their preferences (the electoral system requiring voters to order candidates in terms of preference) away from One Nation, until the disastrous outcome for the Coalition in the Queensland election forced his hand. In the event, Hanson ran a poor federal campaign and attracted just 10% of the national vote. Even so, just under 1m. Australians endorsed her views. Howard was therefore under pressure during the election in October 1998. It was a mark of his leadership that he made tax reform the centrepiece of his campaign with an undertaking to introduce a goods-and-services tax. The contest was tight and a narrow majority of voters indicated a preference for Labor; but the distribution of electorates favoured the Coalition and it was returned with a safe parliamentary majority.

The Howard Government pressed on during its second term with the economic measures it had foreshadowed: tax reform, the full sale of Telstra, further deregulation of industrial relations, and further conditions on welfare beneficiaries. The Government's lack of control of the Senate necessitated lengthy negotiations with the smaller parties and independent senators, leading to substantial compromises. The implementation of the goods-and-services tax went ahead, with exemptions and compensation. The sale of Telstra was restricted to just one further tranche, leaving 50.1% still in public ownership. Tony Abbott, the belligerent new Minister for Employment Services, made little progress in his attempt to change legislation that gave protection against unfair dismissal but did tighten eligibility for unemployment benefits.

The second term of the Howard Government also witnessed the defeat of the republican movement. Republican sentiment had been building steadily. The Whitlam Government had abandoned imperial honours in 1975. In 1984 the Hawke Government proclaimed 'Advance Australia Fair' as the national anthem, to the chagrin of many monarchists, and in 1986 the supremacy of the Australian Parliament was formalized. To neutralize the republican issue during the 1996 election, Howard had undertaken to provide an opportunity for the country to decide. He skilfully exploited the prominence of republican celebrities by suggesting that they were part of the 'élites' and he also arranged that the Constitutional Convention that met in Canberra during 1998 to settle on a republican model was composed of equal numbers of appointees and elected representatives. The model proved to be remarkably timid and was not enhanced by the new constitutional preamble that Howard himself drafted to accompany it. Both proposals were rejected by referendum in November 1999.

The country thus hosted the Olympic Games in 2000 and celebrated the centenary of the Commonwealth early in the following year as a constitutional monarchy. The Games confirmed the national capacity for sport, spectacle and irreverence. The Centenary of Federation ceremonies betrayed the seeming exhaustion of the civic tradition. The most memorable speeches were indeed delivered by the Governor-General, Sir William Deane, who observed that 'the founders of our nation were, by and large, good and decent men', but they had 'simply ignored the tragedy of the circumstances of Australia's indigenous peoples'. In celebrating the centenary, he continued, it was well and truly time that the nation acknowledged the past injustice and the continuing problems of Aboriginal disadvantage with an unreserved apology to bring about a 'true and lasting reconciliation'. Deane was a source of continual irritation to the Howard Government until his term ended later in the year; he was replaced by a more compliant Governor-General.

During its first term in office the Howard Government had sought to expunge Keating's legacy at home and abroad. It had been particularly critical of Labor's attempt to forge closer relations in the region and decried the suggestion that Australia's destiny lay in Asia, preferring to emphasize the distinctive character of Australian culture and traditions. It eschewed the former Labor Government's idea of Australia as a 'good international citizen', reiterating instead a 'hard-headed' pursuit of the national interest. While supportive of continued trade liberalization through the new World Trade Organization, the Howard Government refused to accept the greenhouse gas emission targets at the Kyoto conference on global warming in 1997, and was critical of UN forums.

The Asian economic crisis forced a more active engagement. Australia was drawn into the rescue 'packages' organized by the IMF, and also supported the strict conditions imposed by the Fund when it provided financial assistance. Those conditions increased poverty and unrest in Indonesia, forced the resignation of President Suharto in May 1998 and aggravated communal violence and violent repression by the armed forces in Indonesian provinces, including East Timor. Although Indonesia was required to give greater autonomy to East Timor, the military was complicit in the intimidation of the population by local militias until in September 1999 an international force of 10,000 under the authority of the UN landed to restore order; Australia provided the commander and the majority of the force.

The continuing Indonesian resentment had a further consequence. The authorities there did little to curb the passage of refugees who made their way by boat through South-East Asia to Australia. The numbers were small—3,300 in 1999 and 2,900 in 2000—but the Government regarded them as 'queue-jumpers' and treated people-smuggling as a threat to Australian sovereignty and security. A solution required international co-operation but Australia was increasingly at loggerheads with the agencies of the UN over domestic human rights as well as refugee issues. In 2000 the Government announced that it would adopt a more economical and selective approach to requests for information from the UN, and a more robust and strategic approach to Australia's interaction with the treaty system.

The destruction of the World Trade Center on 11 September 2001 brought US President George W. Bush's 'war on terror' and Howard's immediate support for the invasion of Afghanistan. A fortnight earlier Australia had been caught up in its own military drama when a Norwegian container vessel, the *Tampa*, came to the rescue of a drifting Indonesian fishing vessel loaded with 433 passengers, mostly from Afghanistan. The captain of the *Tampa* headed for nearby Christmas Island, an Australian territory to the south of Java, but the Australian Government ordered him not to land. With the distressed passengers outnumbering his crew by 16 to one, the captain dropped anchor off shore. A detachment of Australia's élite SAS force seized control of the *Tampa*. Some of the asylum-seekers were accepted by New Zealand, most were dispatched to specially constructed internment centres on the tiny island of Nauru, and some were subsequently taken to another camp in Papua New Guinea.

The crisis also turned the 2001 election campaign into a debate on border control. Labor initially refused to support the special legislation, then capitulated and lost both ways. The Government took advantage of its position and claimed that another boatload of illegal immigrants had thrown children overboard when approached by a naval vessel. The fears and uncertainties arising from 11 September augmented the issue of border control. Full-page newspaper advertisements appeared on the morning of the November election showing John Howard with his fists clenched and the declaration 'We decide who comes to this country'. He was returned to office with a slightly increased majority.

The war on terrorism began with the strike on Afghanistan and broadened after President Bush denounced the 'axis of evil' (of Iraq, Iran and North Korea) in his State of the Union address at the beginning of 2002. Howard supported the new US doctrine of pre-emptive defence, and ignored regional sensitivities when he announced at the end of the year that Australia also would consider sending troops to a neighbouring country to strike at terrorist cells if they threatened homeland security. By then a terrorist cell had in October attacked tourists in a nightclub on the Indonesian island of Bali: 88 Australians were among the more than 200 killed by the bombing. Australia joined the USA and United Kingdom in their operation to disarm Iraq, and a contingent of Australian forces participated in the three-week war that ousted the regime of Saddam Hussain in March 2003. The Government's close relationship with the USA extended to the negotiation of a Free Trade Agreement in 2004.

There were 13.5m. inhabitants when the long boom ended in the mid-1970s, and 20m. by the early years of the new century. While Tasmania and South Australia stagnated, the large, resource-rich states of Queensland and Western Australia grew more rapidly, but the concentration of population in the south-eastern corner of the continent remained the dominant feature. The vast conurbation of Sydney contained more than 4m. people, while Melbourne approached that number. Queensland's Gold Coast attracted numerous retired southerners and by the end of the 20th century a corridor of 2.4m. people stretched up through Brisbane. More than one in five of Australia's inhabitants was born elsewhere, a higher proportion than Canada, the USA or any other settler society. Sydney and Melbourne were the destinations of most migrants—the overwhelmingly European composition of smaller cities contrasted with the cosmopolitan ambience of the two great entrepôts. Melbourne had attracted larger post-war contingents from southern and eastern Europe, and took in many of the 225,000 Vietnamese who settled here after 1975 as well as other non-European immigrant groups; but Sydney became particularly attractive to new arrivals from East and South-East Asia, the Middle East, Latin America and the Pacific islands. There was also a growing contrast between Melbourne's more mixed distribution of migrants and Sydney's growing enclaves of particular migrant communities.

After his third electoral victory in 2001 (and securing a fourth victory at the election held on 9 October 2004), John Howard was less troubled by ministerial mishaps, and more adept in addressing problems in troublesome policy areas. The Labor Party held power in all six states as well as in the Northern Territory and the Australian Capital Territory, which also now had elected governments. Opinion polls suggested that the Labor Party enjoyed greater confidence in the provision of health services and education, which are largely state responsibilities, but the Coalition was thought to be stronger on economic management and foreign policy. Australians feel the loss of earlier certainties. The removal of earlier forms of protection, the increased emphasis on individual responsibility and the growing divisions between rich and poor strain social cohesion; nevertheless, Australians still enjoy peace and prosperity.

Economy

NEVILLE NORMAN

It is common to measure the performance of economies by their record of economic growth. On this test Australia has clearly led the Western world from about 1999 to 2003. In the year to June 2004 Australia's growth rate of 3.2% (in real terms) had been matched or exceeded by Japan (5.5%), the USA (5.0%) and the United Kingdom (3.2%). With the euro area registering growth of only 1.3% in the year to June 2004, Australia's growth record remains impressive. Only the People's Republic of China and a limited number of developing or recovering economies grew faster than the Australian economy over the 1999–2003 period.

Economic performance, however, is much more than economic growth. In this article a broader dimension of industries, technology, products and related economy-wide and industry-based data is analysed and reviewed. Australia is geographically, and in some ways economically, an Asian country. Its setting against spectacular and recent changes in Asian economies requires careful analysis.

The Australian economy is distinguished from other economies in the Asia-Pacific region, and from other advanced economies. Australia remains reliant on primary and mining activities to a greater degree than most other advanced economies; however, its government sector is relatively small and it remains sparsely populated. Many institutional and legal features of the Australian economy stand firmly in contrast to other nations. Recently, Australia's rate of economic growth has been in advance of almost every other economy in the Western world, and the relative importance of traditional primary and manufacturing activities has been substantially reduced. In the year to July 2004 Australia continued to operate in the face of the longest-ever drought and was one of only a few nations engaged in war against the former regime in Iraq.

Despite the restraints on economic activity arising from the Asian financial crisis that began in 1997, from the terrorist attacks on the USA in September 2001 and from the Iraq conflict of 2003 (where Australia maintained a military presence until at least mid-2004), the Australian economy exhibited a leading position of economic growth in the four years to mid-2004. Indeed, in mid-2004 the country recorded its 12th consecutive year of positive economic growth, accompanied by a low rate of inflation. No period in Australia's history from 1901, or of the colonies before that, has displayed such a sustained span of recession-free economic activity. Few Western-style economies could boast the 3.8% growth rate that Australia had achieved in the fiscal year 2001/02 (July 2001 to June 2002). Since then, Australia's economic performance has been emulated by a number of other developed economies. However, even in the period 1999–2003, when Australia was in the ascendancy, good growth was accompanied by structural weaknesses and many significant corporate failures. Indeed, the years 2001–04 witnessed the demise of some large companies in the airline, telephony and insurance sectors. There were also concerns arising from Australia's external dependence (with the current-account deficit on the balance of payments almost doubling between 2000 and 2004), some ongoing tax tensions, major contentious domestic regulatory policy reviews and, at times, the risk of higher inflation and interest rates. From June 2002 to July 2003 the Australian monetary authorities resisted the temptation to emulate four interest-rate reductions made by the Federal Reserve Board in the USA. By mid-2004 every sign pointed to a rise in world interest rates, placing pressure on the Australian monetary authorities at a time when housing activity and values were starting to decline after a long boom. This brief statement of the economic structure and economic condition suggests the need for a fuller and deeper appreciation of the nature and functioning of the Australian economy.

In the year to March 2004 the national income of Australia, in current prices, totalled $A793,100m. Within this aggregate figure, household income was $A648,000m., which was derived mainly from employment compensation and social security payments. From this national aggregate, Australian households paid $A98,000m. in income tax and consumed more than the balance available, leading to the first recorded year of aggregate negative saving, amounting to $A11,000m.

The revenue from exports provided by Australian producers was $A154,000m. and spending on imports was $A165,000m., over the same period. The widening gap between export and import values reflected both the drought, international restrictions on exports and the impact of rapid Australian economic growth on imports in the latest year (to March 2004). Real gross domestic product (GDP), after allowing for price movements, rose by 3.2% in the year to March 2004. This rate of economic growth was lower than the average Australia achieved in the 1960s, but it remains higher than the trend established from 1970. It remained above the growth performance of most other Western economies in the period 2000–03. In the 1950s and 1960s the average annual rate of economic growth was 5.8%. The 1970s in Australia witnessed a slowdown in both population growth and productivity growth, resulting in an historically low average annual rate of economic growth of just 1.2%. The country's economic growth rate recovered to an annual average of 2.7% in the 1980s and, further, to 3.7% in the 1990s. Australia did not experience the extent of growth of equity values when the US stock markets surged upwards in the 1990s, nor the degree of downturn that took place from mid-2002.

The trends in some overall ratios give a useful impression of the Australian economy. The share of government spending at all levels in national income rose from just under 20% in 1948 to more than 30% in 1984. Since then the government share has stabilized or slightly fallen. The ratio of imports to domestic product sales has increased from 17% in 1948 to 27% in 1990 and then, markedly, to 44% in 2004. This can be taken as confirmation of a substantially open Australian economy. Meanwhile, household saving has systematically fallen, both in absolute terms and as a ratio to household income after tax. This ratio decreased from just over 20% to 8% in 1990, and to 1.9% in the year to March 2003. Since then, the ratio has recorded a negative value, implying that Australian households are spending more than their available incomes. There is no reason why this situation should not prevail. However, sharp upward increases in interest rates would cause difficulties here. In the half-century from 1950 Australia became less dependent on agricultural exports, and it developed an advanced industrial and technological base with world-standard transport and communications networks, despite its low population density. As external perceptions about the structure and performance of the Australian economy can be misled by false impressions, it will be important to document how far Australia has moved from its image, and reality, of the 1950s, by the first decade of the new millennium.

PERCEPTIONS, REALITY AND POLICY

So often the external perception of Australia is formed through its tradition of primary-product exports, its films and its sporting achievements. These visions of Australia can easily depart from the reality of a modern economy, where the majority of its people reside in cities, use computers actively and exhibit one of the highest comparative standards of educational and training achievement in the world.

This is not to say that Australia has not sought to gain advantage from its image as a primary-product exporter. After the Havana Conference of 1947 that established the General Agreement on Tariffs and Trade (GATT), which was subsequently superseded by the World Trade Organization (WTO), there was much pressure from the contracting parties, as they were known, to reduce tariffs and other trade barriers multilaterally. Australia sought dispensation from this process on the grounds that it was a 'midway' country. By this it was meant that Australia was in many ways similar to a developed country, in terms of national income per head (indeed in the later 20th century it led the world on this test). However, in relation to the structure of its exports and its exposure to large inherited fluctuations in economic activity arising from significant dependence on the global economy, it was more akin to the developing world. Images of outback Australia in films and tourism promotions appear to have spread this perception. Yet from the earliest days of Western-style settlement the majority of the population lived in cities relatively close to the coast. Even in the early 21st century about 40% of the national population resided in just two cities—Sydney and Melbourne.

In the high-technology area, it is true that relatively little of the 'innovation' aspect of scientific advancement is performed in Australia, notwithstanding that several major inventions in electronic technology, aircraft navigation, agricultural chemicals, agricultural harvesting, materials processing and logistics have been made in Australia. This is because large companies find it to their comparative advantage to focus the development side of these inventions in huge research establishments in North America and Europe in particular. Yet Australian consumers, businesses and governments remain advanced users of high-technology products. Research and development policy continues to be a major focus of attention in Australia.

Australia has moved a considerable distance from the 'quarry and farm' image it had acquired and which still figures prominently in some international perceptions. Australia remained in the leading 20 countries in the world on the basis of gross national product (GNP) per head of population from 1960 to 2003. Australia's rate of economic growth in real terms, at 3.8% in 2001/02, exceeded that of any other Western economy.

Economics and International Diplomacy

There is a complex connection between Australia's international political stance and material developments in trade and economic activity. For example, Australian merchandise export trade became prominent with Japan, the Republic of Korea (South Korea) and Viet Nam relatively soon after Australia ceased hostilities against these countries in the 1940s, 1950s and 1960s, respectively. In 1973 Australia became one of the first countries to recognize the sovereignty of China and to establish formal diplomatic links with the People's Republic. Very considerable trade and educational links then followed between the two countries. There was some lapse in this connection when Australia became a prominent official critic of the massacre in Tiananmen Square in the Chinese capital, Beijing, in 1989. The economic and cultural links, however, were little affected and had again become extremely strong by mid-2002, when a huge long-term contract for supplies of natural gas to China was won by the Australian north-west-shelf energy supplier, Woodside. Illustrating this theme with a different example, Australia took a lead in international criticism of the Iraqi leadership during mid-2002, causing retaliation from that country which threatened to cancel significant orders of Australian wheat. The shipments were refused entry to Iraqi ports for some weeks. These examples serve to illustrate the ongoing close relationships between international strategic policy and economic conditions. They also bring to prominence the way in which international images of a country may serve to enhance or limit its commercial progress. During March 2003 Australia joined the USA and the United Kingdom in active engagement against the Iraqi leadership, and it maintained a presence in Iraq for some months after the main conflict ended in April.

Planning or the Market?

As regards the philosophy of economic policy, Australia maintains a commitment to national budgets that are balanced or in surplus, to significant growth in living standards and to low inflation. The Australian Government as such does not make or accept formal plans or targets for the population size and structure and for the main national economic variables. There are frequently calls from within Australia for formal population targets and for economic planning agencies to be given official status. During 2002–04 conferences convened by state governments and industrial bodies took place, at which many delegates supported population size, structure and growth targets. The matters were discussed and debated by a wide range of representative groups. However, Australian Governments have not embraced these proposals.

From 1996 to 2004 a target zone was announced for consumer price inflation rates in the range of 2% to 3% per year; this target zone was proposed and maintained by the central bank, the Reserve Bank of Australia, which is based in Sydney. That authority has also, since its creation in 1959, carried formal responsibility for stability of the currency and overall supervision of the financial system. In 1965 a Committee for Economic Inquiry (Vernon Committee) proposed a more formal economic planning mechanism, a Council of Economic Planning. That notion was roundly condemned by the Government of the day, and an aversion to central-planning concepts has been part of the written and unwritten Australian political ethos ever since. During the tenure of the Bob Hawke Government (1983–91) an Economic Planning and Advisory Council did operate, until its functions were absorbed into other bodies; it had effectively disappeared by 1996 when the Government of John Howard took office.

In relation to industrial policy, the Australian approach has been for decades to follow a monitored market course. That is, there are regulations to oversee the workings of the market in relation to mergers, corporate affairs, trade practices, construction proposals, consumer protection, the environment, foreign investment, financial transactions, insurance and banking. In the 1980s and 1990s some of these regulations were eased, especially in relation to the labour market and to international and financial transactions. In early March 2003 the official response to a regulatory-policy inquiry (Dawson Committee) was to affirm the principle of light regulation based on case-by-case inquiries. After considerable debate, the Howard Government introduced legislation in June 2004 to give effect to the liberal approach to abuse of 'market power', as proposed by the Dawson Committee. A fierce and unresolved debate developed by mid-2004, with many smaller businesses arguing that an opportunity for embracing more rights of action against aggressive larger companies was being missed. The traditional Australian approach had been to limit international transactions at the point of entry into Australia: substantial barrier protection, mainly in the form of tariff duties, was erected against foreign supplies of goods; foreign investment was examined and restricted; and immigration was strictly controlled. Through a series of reductions in tariffs and the elimination of most non-tariff means of border protection, Australia has liberalized its trading policy. It has been prominent in the international arena in securing and pressing for the removal of agricultural trade barriers multilaterally. Banking regulations that in the 1950s limited the deposit structure, entity ownership and asset allocation of the main banks and financial institutions have been liberalized. Prudential supervision and control of the financial sector has since 1998 been predominantly the responsibility of a newly created Australian Prudential Regulation Authority (APRA). Competition policy generally prohibits few transactions absolutely. Practices of firms that mislead consumers or are deemed unfair are illegal, as are a number of trade practices, arrangements and mergers that substantially lessen competition.

TRENDS IN POPULATION AND VITAL STATISTICS

In December 2003 the population of Australia had just passed 20m. people (estimated at 20.01m.). While the population growth from both natural increase and migration has been reduced as compared with most periods since 1947, recorded population growth remains at 1.29% annually, which is higher than most advanced economies, although it is lower than the population growth rate found in most countries in the Asia-Pacific region. At June 2004 births occurred on average at just over one birth every two minutes, and deaths on average at just less than one death every four minutes. The natural increase was supplemented by long-term overseas migration, equivalent in 2003 to almost exactly one-half of the population increase itself. The equal contribution of long-term migration and natural increase to population growth represented a return to conditions experienced in the 1960s when natural increase and migration were each relatively higher than they were recorded for 2003. During the 1980s the share of population growth attributed to international migration was just less than 30%. The resumption of approximately equal contributions from both migration and natural increase has more to do with the decline in Australian fertility affecting the latter than with any acceleration of inward international migration. The overseas-born component of the Australian population in March 2004 was 24.8%.

As in most Western-style economies, the rates of marriage and childbirth have been declining significantly, while the average age at marriage has been rising, as has the ratio of aged dependants to the total working population. The median age of the Australian resident population was 36.2 years in December 2002, having risen from 31.2 years in 1986 to 34.2 years in 1996. Reflecting Australia's position as a favoured and rapidly growing tourist destination, some 4.656m. persons arrived in Australia as short-term visitors in the year 2002/03. This figure represents a slight decrease on the previous year, undoubtedly reflecting increased international uncertainties and tensions arising from the Iraq conflict and the perceived health risks associated with Severe Acute Respiratory Syndrome (SARS). With an average stay of just over one month, these temporary residents comprised an average of 2.9% of the Australian population during 2002/03.

The proportion of the Australian population aged 65 years and over at census night on 7 August 2001 increased to 12.6%, as compared with 12.1% in 1996 and 10.2% in 1986. This so-called aged population ratio is similar to the aged shares of populations found in the USA, New Zealand and Canada. It remains considerably less than those aged shares found in France (21%), Japan (23%) and Italy (24%), but well above those found in developing countries, such as China (10%) and Malaysia (7%), with all data as at early 2003. Under official population projections released in Australia in April 2003, this aged population ratio for Australia is expected to be just over 17% in 2020, using plausible assumptions of fertility, mortality and immigration. There are considerable commercial, housing and social implications of this rapidly ageing population, as experienced in several other countries for some time. A total of 50.7% of the population in December 2003 was female. There is a tendency for the femininity of the total Australian population to rise over time as the slight majority of males at birth is overtaken by a tendency for females to live longer in an ageing population. English was the only language spoken in 79.1% of the homes at the census of August 1991, while Italian was the most common other language. By 2001, however, Chinese languages had become the most common non-English spoken language. Some 47.0% of the homes in 1991 contained families without children, a proportion that despite the image of increasing incidence of childless couples has remained stable since the early 1990s. There has been an increasing proportion of adult persons living alone in Australian homes. Some 7.4% of the resident Australian population in 2001 was divorced (and not then remarried), as compared with 1.9% of the Australian population being of this status in 1971. Lone persons living in an Australian dwelling overall accounted for 22.9% of the population in 2001, compared with 18.1% in 1971.

Statistical projections made in 2002 by the Australian Bureau of Statistics indicate that the proportion of adults who will never marry is rising, being 29% for men and 23% for women, as seen at 1999, compared with equivalent demographic conditions at 1986 of 21% and 14%, respectively. Using the same methodology, the likelihood of marriages ending in divorce was 32%. The crude divorce rate at 2001 was 2.6 per 1,000 of population, compared with 4.3 for the USA, 2.9 for the United Kingdom and 2.5 for Canada. The crude marriage rate in Australia in 2001 was 5.9 per 1,000 of population, compared with 8.9 in the USA and 5.4 in the United Kingdom.

According to the Australian Government Actuary, life expectancy in Australia at December 2003 was 77 years at birth for males and 82 years for females. These figures are higher than in the USA (74 and 79 years) and the United Kingdom (74 and 80 years), and they are similar to life expectancy data for Japan, Switzerland and Sweden. The main causes of death in Australia at December 2001 were cancer (27.8% of all deaths), heart disease (20.7%) and stroke (9.6%). At the rates of childbirth prevailing at October 2001, 24% of Australian women would remain childless at the end of their reproductive lives. During the 1990s the proportion of the Australian population aged 25–64 years with a professional, vocational or higher educational qualification rose from 46.5% to 51.0%.

AGRICULTURE, MINING AND MANUFACTURING

Agriculture

Agriculture remains an important source of export earnings for Australia. While farming activities may have accounted for just 3% of Australia's GDP in 2003/04, and only 2% of its employment, the agricultural sector nevertheless contributed 15.5% of export earnings in the year to June 2003. The share of Australian export earnings arising from agriculture has declined from a range of 29%–35% during the 1980s to 12%–16% in 2000–03. This farm-product share of Australia's exports compared with 24.4% in 1990 and as much as 49.0% in 1970. The proportion of the Australian total employment base engaged in agriculture was only 2.0% in 2003, having decreased from 3.8% in 1980 and 7.8% in 1960. There were 110,500 agricultural establishments in Australia in 2000/01, declining from 202,800 in 1960/61.

Agricultural activities occupy a high proportion of the arable land in Australia, although they offer relatively little employment. This is due to the rapid productivity growth achieved in its production processes, and declining relative demand for agricultural products. Within the aggregate value of gross farm production in the year to June 2004, some $A36,000m. of crops accounted for 45% of the total; livestock provided the balance. Some 76% of the crop production by value and 72% of the livestock was exported in the same year. Within the aggregate value of crop and livestock products, the most prominent contributors were beef (15%), wheat (13%), fruit and nuts (10%), milk (10%), wool (8%), and vegetables (6%). In the first half of the 20th century wool alone dominated Australian export earnings, leaving aside some periods of intense gold-mining activity. Beef, lamb and wheat have also been traditionally the main export-earners. The erosion of preferential entry into the United Kingdom in the 1970s, the emergence of mineral and, more recently, manufacturing exports, and competition from new international suppliers, have diminished the relative position of these traditional exports.

From the mid-1990s wool suffered the further specific difficulties of competition from synthetic products, declining demand and prices, compounded by the Asian financial crisis of the late 1990s and the aftermath of an unsuccessful attempt at policy intervention in Australia to maintain wool prices. A huge stockpile of wool had emerged by 1996. It had been completely sold before 2002. In the year to March 2004 the share of wool in all Australia's exports had declined to just 2.4%, with beef and veal at 3.3% and wheat at 2.5%. A privatized Australian Wheat Board carries out the marketing of Australian-grown wheat. The direct role of government in marketing agricultural products in general is very limited. The traditional exports now appear as a stable and efficient, if smaller, component of the economy. These figures confirm that the dependence of Australia on primary or farm products for its exports is much reduced in recent years.

Intensive production of fruit and vegetables is conducted in Australia around urban settlement areas in every state and on land areas that are relatively close to the coastline. However, the major broad-acre developments are found further inland. For example, the sheep-wheat belt is located in an ellipse on the eastern and southern part of the Australian land mass, some 150 km–1,300 km from the coast. Dairy production has moved further from the cities with the development of irrigation and water-storage facilities. These facilities were severely tested in the 1999–2004 drought, until some rains began to fall again in mid-2003. Wine-producing areas are found, in some cases, close to Adelaide, Melbourne and Perth, but also in low-rainfall zones in every state. About two-thirds of the continent is classified as low-rainfall desert, being the central, northern and western zones. Little economic activity, except for mining operations and large-scale grazing activities, is carried on in these inland zones, with some tourism locations interspersed.

On a value basis, crops and livestock each contributed substantially and similarly to Australian national production, with the gross value of crops being $A18,900m., and livestock $A16,750m. in the year to March 2004. The gross value of agricultural production overall in the same year was $A35,860m., a 7% increase on the previous year. Some 5.54m. ha of coarse grains (mainly corn) were planted in 2000/01, with production volume reaching 11.39m. metric tons, of which 40.3% was exported. Within this domestic production figure for coarse grains, corn accounts for 68% of the tonnage produced and barley for 17%. Australia also produced 1.06m. tons of cotton, 10.56m. litres of milk, 0.16m. tons of butter and 0.36m. tons of cheese in 2000/01.

In 2000 there were 27.59m. head of cattle and 118.6m. sheep (mostly in New South Wales and Victoria), yielding production of some 3.11m. metric tons of red meat, of which 1.745m. tons were exported, a figure that increased markedly in 2001 as a result of meat production difficulties and health problems in Europe. Australia produces 5% of the world's pulses and 3% of the world's sugar. There were 12.08m. ha of wheat planted in 2000/01, yielding 21.17m. tons, of which 16.26m. tons (77%) were exported. The rapid productivity growth of wheat farming is reflected in the fact that in 1980 production was almost exactly one-half of this figure. The crop was planted on approximately the same land area allocated to wheat in each period. In 2000/01 Australia exported 15% of the world's internationally traded wheat. The country's wool production totalled 652,000 tons in 2000/01, some 28% of the world aggregate greasy wool clip. In 2000 Australia produced a turnover of $A51,100m. in food-processing and beverage products, of which $A14,900m. represented value added by the food-products processing function.

Some individual agricultural industries face specific difficulties. The dairy industry by mid-2003 was confronting adverse conditions owing to water shortages, long-term competition from margarine, and greatly increased competition from New Zealand, in cheese production especially, since freer market access was granted in the early 1990s. Aggressive competition in wheat markets since 1999 has caused difficulty when Australia, as in its response to falling sugar prices in the 1970s, failed to follow market prices fully downwards. There are also some notable successes, including greatly increased penetration of Australian beef and rice in Japan and other Asian outlets. Australia had also reached considerable prominence as a premium exporter of table wines. By mid-2004 Australian wine producers were observing anxiously as the huge world-wide wine surplus, which had been emerging for some time, began to curb the industry's high-growth prospects. By 2003 some 36% of Australian wine production was exported, mainly to Europe, as compared with 12% in 1980. Some 1.98m. metric tons of grapes were crushed in 2001/02, making 1.61m. tons of wine, with 395m. litres sold domestically. Some 345m. litres were exported, 168.0m. litres to the United Kingdom and 68m. litres to the USA. Imports of wines in that year reached just 12.8m. litres. In the year to June 2003 Australian domestic sales of wine reached 402m. litres, of which 82m. litres was sold as white wine in glass bottles of less than 2 litres capacity each and some 119m. litres as white wine in soft packs. Total sales of table white wines and red wines were equal, at 201m. litres each.

Australia reaps the benefit from its providential endowment of agricultural resources and a reputation for high quality in processing. These benefits are tempered by the long-term trend of declining relative prices of agricultural staples in world markets. Australia has been prominent in seeking freer access for primary products in protected world markets, using overt and other forms of diplomatic pressure to make its position known. In June 2002 Australian Prime Minister John Howard addressed the US Congress, combining unequivocal endorsement for the USA's position on international terrorism with firm disagreement on the protectionist aspects of that country's trade policy. Australia is also subject to the vagaries of climatic conditions. Thus, in December 2002, before the long-running Australian drought had broken (in mid-2003), the office of the Australian Statistician presented estimates to the effect that the drought (since about 1999) had had the effect, through the reduction in potential agricultural production, of taking some 0.7 percentage points from the Australian real GDP between 2001/02 and 2002/03. Even at June 2004 substantial water restrictions remained in place, affecting agricultural producers as well as town dwellers.

Mining

Some $A33,796m., or 4.8% of Australian GDP, was derived directly from mining activities in the year to March 2003. A high proportion of all major minerals (other than brown coal) was exported. In the case of uranium oxide, the entire mineral output was exported because of domestic consumption prohibitions. About 30% of Australian export earnings were derived from mineral exports in the year to March 2004. There was much environmental attention to the conditions of mineral exploration, production and handling. Legal issues arising from land-rights claims with traditional landowners remain controversial in some areas of the industry.

There is a considerable disparity between the location of the major valuable mineral deposits and the areas desired for habitation. Coal and valuable mineral deposits occur mainly inland and in low-rainfall areas. This potential problem has been addressed in three ways. First, some major settlements have developed at mine sites, such as: Broken Hill, about 800 km west of Sydney; Newcastle, 160 km north of Sydney; Wollongong, 130 km south of Sydney; and Ballarat and Bendigo, both around 160 km north-west of Melbourne on slightly different arcs. Second, the development of high-productivity extraction techniques and the availability of concentrate ores and their transport that has limited the need for workers to be at the location of the major deposits or agricultural areas. Third, there is a well-developed system of moving workers to mine sites. There is also the capacity to take workers into temporary locations, for instance in the agricultural areas during harvesting and shearing times. Quite commonly, workers will be engaged for periods of two–three weeks working intensively, followed by two weeks off site.

Historically, Australia has experienced pronounced booms associated with specific mineral products, especially gold. Since the late 1960s, commencing with iron ore, bituminous coal and bauxite, a long-sustained phase of mineral development has been evident. Both Australian and foreign-owned companies have developed huge resources of natural gas, coal, uranium, copper, lead, zinc and petroleum. There was no sign in the early 21st century that this long phase of mineral development had ended. Government financial support for mining development through state and federal budgets has been reduced significantly since 1990. Industrial relations conditions have generally improved, although the 80% reduction in working days lost per employee between 1984 and 2003 conceals continuing tensions over pay and conditions that cause international buyers of coal, particularly, to be ready to substitute alternative suppliers if Australian sources prove unreliable.

Australia has impressive endowments of the world's proven mineral reserves, with over 80% of the world's rutile and beach sands, 27% of the uranium oxide, around 40% of brown coal (lignite) and 22% of black (bituminous) coal. While Australia has substantial petroleum resources, mostly beyond its south-eastern and north-western coastlines, it still needs to import about one-third of its domestic oil and petroleum needs. The main production data recorded that in the year to June 2004 oil and gas exploration accounted for 38% of mining value added in Australia, followed by coal (26%), iron ore (12%) and gold ore (8%).

In 2003 Australia mined 55m. metric tons of bauxite, processing this into 16.3m. tons of alumina, of which 81% was exported; primary aluminium production reached 1.77m. tons of which 1.402m. tons were exported, mostly to Japan and the Republic of Korea. This export trade represented 10% of the world trade in aluminium. In 2000/01 322m. tons of black coal were produced, mostly from Queensland (54%) and New South Wales (43%). After washing the coal, this figure was reduced to a net 257.8m. tons, of which 193.5m. tons (77%) were exported. Australia produced 513,000 tons of refined copper and 301 tons of gold in 2000/01. In the same year Australian output of crude petroleum and condensate reached 43,264m. litres, of which 44% was exported, together with 2,785m. litres of liquefied petroleum gas (LPG), while the country produced 33,800m. cu m of natural gas. Australian rutile production was just over one-half of the world's supply. In 2000/01 Australia produced 7.58m. tons of uranium, some 28.5% of the world's aggregate uranium production (Canada being the leader with 9.92m. tons). Output of zinc in 2000 reached 1.38m. tons, some 15.9% of the world's total supply.

Rationalization and consolidation in Australia's mineral industries was substantial in the 1990s. This was partly caused by the ending of decades of income tax exemption for gold-mining operations, and by the mergers of larger operators. Mergers and take-overs are still closely monitored and are often prevented. In late 2001 the (Royal Dutch) Shell company was blocked by a Treasurer decree on 'public interest' grounds from taking over the operations of the Australian north-west-shelf gas producer, Woodside, although it retained significant interests therein. Seldom are gold producers out of the news concerning financial reconstruction and policy agitation. Some of the policy guide-lines remain unclear or inconsistent as viewed by investors, explorers and producers.

Manufacturing

As in many Western-style economies, there has been a declining trend in the relative position of the manufacturing sector since the early 1970s. This decrease has resulted from a combination of reduced import-duty protection, the emergence of low-cost foreign suppliers in textiles, clothing, footwear and metal manufactures, especially from Asia, and declining household budget shares devoted to manufactured goods. A considerable proportion of manufacturing activity is geared to serving the needs of mining and agricultural sectors, which are relatively larger than in countries with a similar degree of economic development. In the calendar year 2002 manufacturing value added totalled $A77,580m., a growth rate of 4.2% on the previous year and representing 10.9% of the national GDP. Food, beverages and tobacco provided some 17% of employment, 19% of the added value and 24% of the exports within the manufacturing sector. The contribution of machinery and equipment sectors was similar. A further 17% of manufacturing added value was contributed by metal product manufacturing, petroleum and chemical products provided 14%, printing and publishing 9%, wood products 7.2%, and textiles, clothing and footwear 3%.

In the year to June 2002 Australian manufacturing establishments produced 1,744m. litres of beer, 18,367 metric tons of tobacco and cigarette products, 398,000 tons of newsprint, 847,000 tons of wood pulp, 17,652m. litres of automotive gasoline, 1,515m. clay bricks and 6.29m. tons of basic iron.

There are several areas of policy uncertainty that impinge especially on the goods-producing sectors of the Australian economy, notably the large variations in the approach to research and development, export promotion and certain taxation issues. However, the manufacturing sector has arguably gained in relation to other sectors of the Australian economy from the introduction of the New Tax System (NTS) on 1 July 2000. This is because that tax initiative involved the elimination of wholesale sales taxation, which applied only to manufactured goods, and its replacement at the substantially lower rate of 10% by the goods and services tax (GST—see below).

Within the manufacturing sector there is considerable interest in the motor vehicle industry, in which the major international vehicle builders have long been represented. Vehicle registrations were rising at an average annual rate of 2.6% in the period 1999–2002. However, in the year to June 2003 the annual growth rate of motor vehicle registrations had increased to 4.2%, the number of motor vehicles registered (new sales) being 860,000, of which 560,000 were passenger motor vehicles. In the calendar year 2003 total vehicle registrations passed 900,000 units for the first time. It was believed that the release of accumulated savings among younger people for whom residential property prices had become prohibitive had benefited the motor vehicle industry. In January–June 2004 motor vehicle sales remained very strong. The average age of cars on the road in Australia was 10.4 years in 2003, having risen from 6.1 years in 1971. Public policy support for this sector has been reduced, although in April 1998 the Howard Government announced a pause in the programmed reduction of motor vehicle import duties, based on agreement with the leading vehicle builders that cost-reducing plant and equipment investment would take place. Considerable modernization and extension of facilities subsequently followed.

INFRASTRUCTURE, ENERGY AND COMMUNICATIONS

Transport

There are clearly difficulties in covering a land area similar to that of the USA with advanced transport networks, given Australia's small population. Yet through significant government involvement and an active private sector there has been substantial coverage by road, rail, air and shipping facilities. From the late 1980s the domestic and international airline operations controlled by the Commonwealth Government were combined into Qantas, which was then sold to the private sector, a significant share being acquired by British Airways. In addition, the main airport infrastructure was divided into lots for sale by auction. Similarly, port and rail systems were exposed to greater competition by the removal of legislation requiring carriage of selected goods by designated modes, by privatization of some of these operations, and the introduction of corporatization principles into management and performance procedures. In 1999–2003 negotiations and disputes concerning the access arrangements and pricing of these facilities occasioned much activity in competition law.

Since the mid-1990s Australia has followed a more 'open-skies' policy in permitting a wider range of international airline operators to participate in the Australian market. In domestic operations the traditional concept was for a 'two-airline policy' to operate. This involved one carrier being government owned and one (Ansett) being privately owned, with entry by any other potential provider being made impossible through bans on aircraft imports. Heavily monitored 'rationalization agreements' ensured that the two carriers had equal traffic, timetables, aircraft, facilities and profits. From the late 1970s the scope for competition and differential outcomes was markedly increased. The liberalization proceeded rapidly in the 1980s and early 1990s. The entry of third carriers and their failure in the 1990s provided warnings that competition might have become 'excessive'. Just three days after the terrorist attacks of 11 September 2001 the main always-private airline carrier, with its origins in the 1930s, suddenly and completely collapsed. Ansett had previously been sold to international trucking and news operators TNT and News Corp and then was acquired by Air New Zealand. Two further attempts at entry were more successful, the sustained entrant, Virgin Blue, occupying a significant role alongside Qantas by mid-2003. In response, Qantas has introduced a low-cost affiliated airline, which commenced operations in June 2004. Again, the picture was one of rapid change and highly competitive business conditions. The transport sector accounted for almost 6% of Australia's GDP in 2003.

Communications

As in the transport sector, the traditional dominant role in the communications sector was played by a government-owned and -controlled enterprise that has since been privatized. With responsibilities once for both post and telephony, this operation was divided in the 1970s, and in 1993 Telecom was renamed Telstra and partly privatized. Again, as in transport, new entrants have appeared, the main operator being Cable & Wireless Optus. Many smaller operators exist in specialist areas. During 2002 a government review of the five-year process of telecommunications deregulation was carried out. Australia Post is the corporatized principal supplier of postal services. It is also exposed to competition and in turn has widened its functions to provide stationery sales and bill-paying services. The communications sector accounts for 5% of Australia's GDP.

Information Technology (IT)

During the world-wide 'technology boom' of 1995–2001 in particular, Australia was commonly classified as an old-technology economy. The country's continuing substantial dependence on agricultural and mineral exports may encourage this perception. Such an image is not supported by facts relating directly to information technology in Australia. There is almost complete coverage of basic telephony and the adoption and use of advanced IT. There were some 560 internet service providers (ISPs) in Australia at September 2002, this number having declined by 5% over the previous year owing to the difficulties the technology industries were experiencing world-wide in 2001. Six of the larger ISPs each serviced in excess of 100,000 subscribers. The proportion of Australian businesses with a web presence had risen from 6% in 1997/98 to 26% in 2001/02. The percentage of businesses with internet access increased from 29% in 1997/98 to 58% in 2001/02. Some 45% of Australian households had internet access in 2003. The number of internet subscribers was growing at an annual rate of 8% in September 2002, by when the total had reached 3.9m. subscribers. An increased proportion of broadband services was evident by early 2003. The incidence of internet usage increased significantly with household income and where children under 18 years of age were in the household complement (48% compared with 32%); internet usage was greater in metropolitan areas than in rural areas (40% compared with 32%).

Energy Supply and Distribution

Energy and gas supplies had until recently been provided by state-owned and -operated monopolies with considerable fiscal support. Privatization of most of these services had been accompanied by debate and efficiency drives, which increased the accountability of the enterprises concerned and generally reduced or restrained consumer energy prices. Each of the energy-sector enterprises is subject to new 'access' regulations relating both to the conditions on which supplies are made available and to the prices charged. The dominant source of energy is electricity, which in turn is generated from (black and brown) coal and natural gas. Including the domestic energy component of coal and natural gas, the energy sector accounted for 4.6% of Australia's GDP in the year to March 2003.

The responsibility for energy provision and distribution has traditionally rested with state governments. Each tended to permit gas and electricity boards to assume quasi-monopoly positions in the entire production chain from energy extraction, to production and distribution to consumers. For legal and logistical reasons, there was no possibility of trade or exchange of energy products between the separate Australian states until recent times. The emergence of gas pipelines that crossed state boundaries in the 1970s and the 1991 facilitation of electricity sales between the states typify the liberalization that has taken place. In addition, the main electricity and gas providers have been split up, the common arrangement being for distribution systems in each state to be divided into some three to six regional areas sold to private interests that were by 2003 free to compete with each other. Even where government maintained ownership, as in New South Wales, the component businesses were subject to stringent profit-making requirements combined with consumer service monitoring. In 2002 per-head energy use by Australians was 261 gigajoules, within which the per-head electricity use was 75 gigajoules.

OTHER SERVICES

Despite the image of rural and mining production being associated traditionally with the Australian economic and cultural landscape, the majority of residents since British occupation over the last two centuries have lived in towns and cities and engaged in service activities. Australia has an advanced network of personal and business service industries, including legal, finance, education, medical, health and community services, leisure and sporting service activities. Banking, insurance and accounting services are predominantly controlled by large companies with international operations. Retail distribution activities involve both large and small operators. After correction for price movements, the value of retail sales in Australia has remained particularly constant on a per-head basis since 1980. Within the aggregate there has been substitution against some food groups (meat and sugars) and an increasing trend towards processed food not prepared in the home. Shopping hours are very liberal, and internet purchasing of services, especially travel services, is rising rapidly.

The tertiary sector can be usefully divided into the wholesaling, storage and retailing of products; community and charitable services; health and medical services; education; tourism and leisure activities; legal, accounting and business services; and financial services. The retail sector features two prominent chain-store entities: Coles Myer, established in 1985 by a merger of two successful family companies, which were involved in over 40% of the retail transactions in Australia; and Woolworth-Safeway, which now shares the market more evenly. There are also thousands of smaller establishments. Within the retail sector, food and drink retailing comprises 43% of the total, household products 15% and departmental stores 10%. The health services sector features a hospital system shared between government and private ownership, supported by national and private insurance arrangements; independent medical practitioners and a large network of pharmaceutical retailers supplied by production and importing companies featuring most of the world's leading pharmaceutical manufacturers.

In 2003–04 there were many contentious issues and pressures within the tertiary or services sector, including debates about education funding, retrenchments in the finance and accounting sector, and pressures in support and opposition to an informal 'four pillars' policy to prevent mergers among the leading banks. In addition, since July 2000 the services sector has sustained significantly increased tax burdens with the introduction of the NTS (see below). Arguably, the tertiary sector was relatively untaxed under the previous arrangements, so the tax burden has become more uniformly spread.

Education

An advanced internationally-focused education sector exists in Australia. Private (mostly church-based) primary and secondary schools and government (taxpayer-funded) schools operate in competition with each other. Universities and tertiary technical training institutes are mainly government-owned. There are some private tertiary institutions. From the mid-1990s full fee-paying tertiary students, many of them from other countries, became an increasing part of the educational intake. Australian universities began to market their services by website, promotional visits to other (mainly Asian) countries and in some cases by setting up establishments in other countries. For example, Monash University (Melbourne) established a campus in Malaysia in the mid-1990s.

Total public (that is, government) financial support and provision of education in Australia at all levels amounted to the equivalent of 4.5% of GDP in 2001. This relative contribution was similar to that in France, Italy, Japan and the United Kingdom, as compared with 4.9% in the USA and 5.9% in New Zealand. At August 2002 there were 9,632 schools in Australia. Some 3.31m. full-time students attended these schools, 68% in the government-provided (non-fee-paying) sector as compared with 72% in 1990. In 2002 the student teacher ratio was 16.9 in primary schools and 12.4 in secondary schools, each decreasing from 20.8 and 13.1, respectively, in 1982. However, on the supply side, the number of people completing a university qualification in teacher education decreased by 13% between 1991 and 2002. There were 225,353 full-time equivalent teachers in Australian schools in 2002. There were about 180 universities or higher education institutes (mostly government funded) in 2002. There has been a significant increase in the age-specific participation rates in education in Australia. For example, 63% of 17 year-olds were full-time students in 2002, compared with 59% in 1992. A much greater commercial focus came over all aspects of education in Australia from the mid-1980s—schools and universities are run more like businesses, with leaders actively promoting their institutions, and fees have been widely reintroduced into the tertiary sector, while the overseas-born student population has risen rapidly in response to opportunities in Australia and some active marketing campaigns. By the end of 2002 some 18% of all students in Australian universities were foreign-born. That figure was 6.1% in 1992. By 2002 the overseas-born percentage was nearing 40% in many economics–commerce–finance courses.

Legal, Accounting and Business Services

Legal firms went through a process of merger in the 1990s to establish some large firms, mainly serving government and major companies, not least with the work required in association with the privatization of the power and telephony sectors. Small-scale service providers are found in every main town offering legal, accounting or financial services. In the accounting area, all major international firms are prominent in Australia. However, during 2001/02 the firm of Arthur Andersen was merged into other operators, as it was in most parts of the world. In banking there are four major banks, each private or privatized. An unlegislated 'four pillars' policy operates to ensure that the 'big four' cannot merge further, although it is rumoured that moves have been made to test this policy through the courts. Insurance services are provided by major international underwriters, supported by affiliated agents and independent brokers, each subject to increasingly onerous regulations. In every one of these service areas, large and small firms co-exist, serving different parts of the market.

Tourism and Leisure Industries

The number of international visitors to Australia totalled 4.44m. in the year to March 2003, being almost identical to the number arriving in the previous year, a 'stand-still' result for a sector that had demonstrated growth rates of around 10% annually in the 1980s. This temporary decline reflected the global pause in international tourism from late 2001, and was an uncharacteristic interruption in a sustained period of rapid growth—the number of international tourist arrivals rose from 1.1m. in 1984 to 2.1m. in 1988 alone, and the sector then maintained an average annual compound growth of 6.3% between 1988 and 2001. From this point, one of the most rapidly growing sectors of economic activity in Australia began to stall, not least because of international economic uncertainties associated with the serious terrorist incidents in the USA on 11 September 2001 or, in 2003, the outbreak of SARS, as well as economic weakness and war and security threats more generally. During 2003 international visits to Australia returned to the levels experienced in 2001, although the growth in numbers had decreased. The main source of international tourists to Australia in recent years has been New Zealand (15.8% of the total in the year to May 2004), followed by Japan (13.1%), the United Kingdom (8.5%) and the USA (6.1%). Tourism services comprised $A26,284m. value added in 2002, or some 4.3% of the Australian economy's GDP, having risen from 3.0% in 1990. Around 24% of this activity is attributed to serving international tourists. There were 227m. visitor nights occasioned by domestic travellers in 2001. The occupancy rate for hotels and serviced apartments in March 2002 was 63%, while that for motels and guesthouses was 54%. There were 551,000 persons engaged in tourism-generating employment in 2002, or 6.0% of national employment. Findings by the Bureau of Tourism Research show that some 55% of international tourist arrivals land initially in Sydney, although only about 34% of visitor nights are spent by them in New South Wales, Sydney's state. Considerable statistical information is available on the events and theme parks tourists attend and satisfaction surveys derived therefrom.

FOREIGN TRADE AND THE BALANCE OF PAYMENTS

Since December 1983 Australia has operated a freely floating currency regime with little interference either directly or indirectly from Parliament or the monetary authorities. This means that all overseas monetary inflows will match the overseas outflows in aggregate, although at various stopping points in the external accounts imbalances can be, and are, evident. Through Australia's history the gap between exports and imports by value, the merchandise trade balance, has swung significantly from surplus to deficit. However, the overall external accounts exhibit a pronounced and permanent strong long-term capital inflow, alongside persistent current-account outflows of freight payments and interest and debt-servicing payments. Undoubtedly this situation reflects the combination of huge natural resources and development potential with the limited population and savings base within Australia. It also means that Australia is significantly influenced by world interest rates. It has, therefore, gained from the epoch of relatively low interest rates since the mid-1990s. In principle, the adjustments that once took place through movements in Australia's level of international reserves have in the years since 1983, or even since 1977, been absorbed through exchange-rate changes. Owing to its diminished reliance on primary export staples and greater economic flexibility, not least through the exchange-rate mechanism described here, the Australian economy is now less likely to suffer from the cyclical behaviour of world markets that once took it quickly from boom to slump.

Merchandise Trade

Exports and imports of goods alone were each worth almost exactly $A120,000m. in the year to June 2002; in the following year exports declined slightly, to $A115,440m. in value, while imports rose to $A133,800m., opening up a considerable trade deficit, reflecting the strong Australian economy, the supply constraints of drought, and market difficulties associated with international tensions, war and SARS in early 2003. In the year to March 2004 the value of merchandise exports was $A106,032m., while import values stood at $A130,219m.

The main export destinations in the year to March 2004 were Japan (which purchased 20.0% of total Australian exports by value in that period), the USA (9.5%), China (8.1%), the Republic of Korea (7.3%), New Zealand (7.3%), the United Kingdom (6.8%) and Taiwan (4.2%). Collectively, 12.6% was exported by Australia to the countries of the Association of South East Asian Nations (ASEAN) and 14.5% to the European Union in the year to December 2003. Overall, the rural component of exported goods had declined from 30% in 1981 to around 14% in mid-2004. By type of product, metal ores accounted for 12.2% of the total in the year to December 2003 (largely exported to Japan, China and the Republic of Korea), coal at 11.8%, non-ferrous metals and petroleum 7.0%, meat 4.8%, chemicals 4.4%, cereals 3.8% and road vehicles 3.6%. The traditionally dominant beef and wool products were each around 3%. This statistic alone emphasizes the substantial transformation of the Australian export structure away from staple products.

The trade pattern by source and destination is also changing rapidly, with substantial growth in exports to India (up by 30% between 1999 and 2003) and of motor vehicle exports to the Middle East, especially Saudi Arabia. In 2002/03 some 59% of exports were manufactures, particularly non-ferrous metals and meat products, and the mining sector provided 27% of exports by value. The dominant product category in imports is machinery, accounting for 43.9% of total import values in the year to March 2003. In the same period the main sources of imports were the USA (which supplied 17.9%), Japan (13.8%) and China (9.4%).

Invisibles and Internationally Traded Services

Despite the movement to a net surplus in relation to international tourism activity since 1994, the non-merchandise current account between Australia and other countries remained significantly in deficit. This category is dominated by a small number of net deficit items, especially freight and insurance outgoings and the repatriation of interest payments and dividends abroad. In 2002/03 the main destinations for exports of services were the USA (which purchased 16.2%), Japan (11.8%) the United Kingdom (11.0%) and New Zealand (7.2%). Travel and tourism services comprised 47% of the total, with transport and freight services 25% and communications 5%.

Taking trade and other current international financial flows into consideration, the current-account deficit expanded to a record high $A46,726m. in the calendar year 2003. This statistic had stood at just $A18,565m. in the year to March 2002. It is important to remember that this figure is a difference variable and can be subject to huge fluctuations when relatively minor variations in its components take place; furthermore, under flexible exchange-rate regimes there is no need to take corrective policy action when these swings actually occur. At 31 December 2003 Australia's net foreign debt stood at $A374,488m., equivalent to just over 50% of the GDP for that period. The fact that some 80% of this debt was privately incurred and was secured on prime, often export-orientated, assets caused independent international assessments by analysts and credit-rating agencies to offer a confident view of Australia's financial future. Moreover, while GDP is used as a scaling device here, it is not implied that the debt must be funded, or repaid soon out of the GDP. Any significant increase in world interest rates, however, would immediately raise the invisible debit component of the balance of payments. This is perhaps a more relevant risk factor than the traditional concerns about primary product export dependence, which are largely confined to history.

Exchange Rates

Given the structural changes in relation to trade and other international economic dealings and the enhanced role for the exchange-rate mechanism, it is not surprising that significant movements in exchange rates involving the Australian dollar have taken place. Mostly the tendency has been for a long-term decline of the Australian dollar in the cross rates associated with almost every major currency. In the period from 31 May 1980 to 31 July 2002, the Australian dollar decreased in value by 39% on a trade-weighted basis against other currencies, including a depreciation of 53% against the US dollar, 76% against the Japanese yen and 23% against the pound sterling. In September 2001 the Australian dollar reached a low of just over 47 US cents, from which it had recovered to 66 US cents by July 2003. Thereafter, a strong recovery briefly pushed the Australian dollar to more than 80 US cents in February 2004 before the value of the currency declined to around 70 US cents in mid-2004.

From 1949 to 1967 the Australian dollar was fixed to the pound sterling and followed the IMF practice of maintaining fixed rates in the absence of a 'fundamental disequilibrium' in the balance of payments. That rule was tested many times, in both directions, including a strong surplus position in 1972 that moved quickly into deficit in 1974. From 1977 a managed float achieved some flexibility, until the Hawke Government in December 1983 adopted the free float. Significant currency depreciation took place from January 1985. It is not credible to argue that the floating dollar has 'caused' the pronounced long-term fall in the value of the Australian dollar. It is arguable that the adjustments would have taken place under any exchange-rate system, and that under the free-float arrangements they have been smoother and form part of a much better managed Australian economy.

ECONOMIC DEVELOPMENT AND GOVERNMENT POLICY

Labour Market Trends

At May 2004 the Australian labour force totalled 10,238,800 persons, of whom about 570,000 remained unemployed (5.6% of the work-force). Predictably, as has been evident in many countries, the unemployment rates of more fully qualified workers are demonstrably lower. In Australia in 2003 the unemployment rates among those with at least a first university degree was 2.3%, as compared with 10.1% for those who did not complete year 12 of their schooling. The Australian labour force comprised 63.5% of the total population at May 2004. This work-force participation rate has, as in many countries, grown markedly since the 1960s, especially with the increasing propensity of women, particularly married women, to work outside the home. The unemployment rate had in the late 1990s been in the range of 7%–9%, although in the 1960s it was seldom above 2%. In late 1992 the unemployment rate reached a post-war peak of 10.7%. In May 2004 the number of employed males was 4.72m., of whom 3.94m. were employed full time. The remainder—some 16% of all male employees—was employed part time. The employment data for females were 6.64m. and 4.31m., respectively, at May 2004. The proportion of the Australian female work-force engaged in part-time work (35%) is thus far greater than in the case of males, and it has been rising significantly in recent years.

Economic Management

The Australian experience provides a useful case study for debates about the merits of short-term policy intervention to manage the economy. The country's history contains some classic mistakes of policy where demand management was moving to intensify cyclical fluctuations, as in 1961 when boom-curbing tax increases and monetary restraints deepened the recession that was already in train. More recently, problems in reading the movements in the economy and the effects of monetary policy already initiated caused cyclical intensification at the end of the 1980s. Since that time, the demand management record has tended to support the view that active short-term policy intervention can be beneficial.

By the year 2003/04 annual economic growth since 1992 had averaged 3.5%. During that time quarterly growth rates had been negative on only two occasions: September 1993 and December 2000. Adopting the convention of two successive quarters of negative GDP growth to indicate 'recession', Australia has thus been recession-free for over a decade. No other period of its history boasts this record. By mid-2003, however, there remained pressures and tensions, arising from drought, international conflict and SARS, for both the economy and policy-makers. By mid-2004 at least the external factors contributing to this uncertainty had abated considerably. In early 2003 inconsistent signals for short-term policy were emerging. A housing boom in both turnover and price, which had been expected on official forecasts to end in 2002, continued apace, provoking calls for higher interest rates and policy restraint. In the same period Australia had resisted the temptation to follow the USA's Federal Reserve in reducing interest rates on four occasions between mid-2002 and July 2003. At one point, in the early stages of the war in Iraq in March 2003, there were fears of a prolonged conflict that might take both interest rates and inflation into the double-digit range not seen since 1990. The policy-makers survived these tensions to enter a period of lesser policy tensions by July 2003. This resumption of greater confidence was assisted by the ending of the SARS concerns and of major conflict in Iraq, by the relatively subdued level of industrial-relations conflict and by strong economic leadership by an increasingly politically dominant Commonwealth Government.

Australia incurred high inflation rates in the 1970s, both relative to its own past and in comparison with most other Western-style economies. Wage surges, sensitivity to world oil prices, and currency depreciation can be cited as the main causative factors. Between 1990 and the March quarter of 2004, the general index of consumer prices (CPI) rose from 100 to 144. The prices of individual products subject to technological change, such as household appliances (120) and communications (110), rose less quickly, while areas subject to tax and public-policy intervention on pricing rose most rapidly, such as alcohol and tobacco products (219) and education (232). Taking into account these latter categories, in which microeconomic policy dictates caused substantial price increases, the general rise in consumer prices from 1990 reflected centrally on overall economic management. The compound average 2.63% annual CPI increase was the result. These data also incorporated the exceptional 6.1% annual inflation rate of 2000/01 experienced with the fundamental policy switch both among and towards indirect taxes under the NTS. In the year to the March quarter of 2003 the overall CPI rose by 2.7%, with the consumer cost of education rising by 4.9%, food prices going up by 4.4% and transport by 3.7%, while household furnishings, supplies and services prices remained stationary. In the year to March 2004 clothing and footwear prices declined slightly, perhaps reflecting the stronger Australian dollar and the dominance of imported supplies in this sector.

Average weekly earnings for all employees on a national accounts basis were estimated at $A997 (around US $640) in February 2004, compared with $A781 in March 2000. At February 2004 the average rate of increase in earnings was 5.3%. The growth in rates of pay for males and females has been similar in recent years, following the move to more equal pay in the 1970s. The contained and regular growth in these data compared with spates of wage acceleration in the early 1970s, with earnings rising by more than 35% in 1973/74 and, again, when wages rose by 18% in 1981/82. Considerable policy effort has been made by both major political parties in office to prevent the recurrence of these sharp increases.

Lying behind the lower inflation regime that has prevailed in Australia since 1990 is a commitment in policy manifesto and action to budgetary restraint that avoids the excesses of government in the 1970s. Formal budgetary targets were tried and removed in the 1970s and early 1980s. However, the commitment to fiscal discipline is reflected in policy pledges, speeches by the Treasurer and budget documents. When the Labor Government of Bob Hawke assumed office in 1984 it demonstrated a commitment to economic deregulation and financial liberalization, freer trade and budgetary responsibility. That tradition was continued by both political parties: the Labor administration until March 1996 and the Liberal-National coalition subsequently. The Howard Government, since March 1996, turned the headline (federal) budget deficit of $A5,001m. in 1995/96 into a surplus of $A22,100m. in 1999/2000. However, the net operating balances of all levels of government in Australia declined from $A14,365m. in 1999/2000 to $A3,264m. in 2000/01. The federal Government remained a net lender in

2000/01, allocating $A5,459m., while the states were new borrowers of $A2,350m. Although the annual budget surpluses were reduced, their positive sign implies ongoing debt reduction from the public authorities and a tribute to good economic management. With an election due in late 2004, the Howard Government pressed ahead with income-tax reductions and increased welfare support, thereby expending some of the substantial surplus. It is reassuring that both major political parties are committed to strong and disciplined economic management.

Competition Policy Reform

By mid-2002 Australia had introduced significant changes to its personal income tax and company tax systems. However, there were political pressures, from consumer groups and smaller business organizations especially, for major amendments also to be made to the Trade Practices Act (TPA), the principal organ of competition policy. In April 2003 the Australian Government released both the Dawson Committee report into the operation of the TPA and its own response. Both proposed that no major changes should be made. The existing form of the TPA had been in operation since 1974 and was modelled more on the US approach of court-based litigation than on the British system of investigation and examination in a more neutral setting. There are limited exemptions for any specific entity or trade practice, although immunity through authorization is available for potentially illegal agreements between corporations on the grounds of public interest justification. The central provisions are subject to an economic-based test of 'substantial lessening for competition in a market for goods and services', and this is the principal test for assessing proposed mergers and horizontal and vertical arrangements. Among the per se offences not subject to a test of competition as such, the Act contains strictures against misleading consumers and price fixing. These provisions have been clarified by case law and by clear legal drafting. One further area of per se offence concerns corporations that abuse their 'market power' by taking advantage thereof for the purpose of damaging competitors, preventing entry or deterring competitive behaviour. It is this section of the TPA that has generated most controversy and prompted some agitation for it to be strengthened. At July 2003 a further parliamentary process was in operation, under some political pressure, to determine if the TPA was operating to the disadvantage of smaller businesses. After some delays and intense political and lobbying activity, some minor changes in merger evaluation procedures were proposed in legislation introduced by the Howard Government in June 2004. However, it was the Government's decision not to accept smaller business proposals for more aggressive regulation of larger companies.

Tax Reform and Tax Trends

There are distinctive features of both the Australian taxation system and the process that has led to its current structure and administration. The Australian Taxation Office administers the system, collects the dominating federal taxes and handles rulings, objections and appeals. Massive legislation and case law supports the system, especially in relation to income and deductions definitions applying to persons and companies. Controversial areas also include international company transfer pricing, fringe benefits (for which a special tax exists) and anti-avoidance provisions.

With overall (federal, state and local) taxes comprising just under 30% of its GDP, Australia is one of the least taxed countries among the members of the Organisation for Economic Co-operation and Development (OECD), based in Paris, France. With nearly 60% of the federal taxes collected coming from taxes on personal income alone, few other countries have tax jurisdictions that rely so prominently on any one tax type or tax base. This is despite the introduction in July 2000 of a value-added tax, the so-called goods and services tax (GST), at the general rate of 10%. The personal income tax rate schedule is also steeply progressive with high marginal tax rates applying at relatively low taxable incomes. The highest marginal tax rate is just under 50%. For the tax year 2002/03 it applied at taxable incomes at or above $A60,000 (or around US $33,000, using exchange rates as at July 2002). In the Australian annual federal budget of May 2003 there was announced a surprise but minor revision in income-tax bracket limits (taxable income values at which the four main marginal tax rates commence). This policy initiative was presented to the electorate as a 'tax cut'. However, it was in effect the first adjustment in the Australian personal income-tax system since the NTS in July 2000. The May 2003 initiative amounted to delivering only about one-third of the tax relief which automatic inflation-based indexation of these taxable income bracket limits would have provided. The 'surprise' was that there was no apparent political need for this measure, even though regular indexation of the bracket limits is technically justified as a standing policy measure. In the May 2004 federal budget more substantial changes were made to the personal tax rate structure, although entirely by changing the income levels (brackets) at which the relatively high marginal tax rates applied. The adjustments were also confined to the upper end of the scale, and thus to the benefit of higher-income earners. In compensation, lower-income earners were offered welfare concessions. In some ways, the Howard Government was able in this budget to achieve the type of tax restructuring it sought (and could not achieve) with the introduction of the NTS.

In Australia high rates of excise taxes are levied on alcoholic drinks (with wine treated more lightly), petroleum products, and tobacco and cigarettes. That said, the excise component of petroleum products is considerably lower than the tax component of motor fuels found in most European countries—the Australian motorist pays less than half the petrol-pump price found in most of Europe. There is no excise tax on wine, although a specific wine tax to take the place of a former wholesale sales tax has operated since July 2000. The broadly based GST applies widely to all purchases, with zero-rating (called in Australia 'GST-free') treatment of food and exports and exemption ('input-taxed' is the Australian term) is given for financial transactions and rent. As from July 2002 companies were taxed at a standard rate on company profits of 30.0%, with allowances (tax imputation method) in dividend distributions to tax-paying persons for tax already paid by companies. State governments rely on transfers from federal authorities for at least one-half of their revenues. Thus, there exists considerable vertical fiscal imbalance: the federal authority raising far more revenue than it requires for its own expenditures, the states being dependent on the intergovernmental transfers, supplementing them with taxes on share and land transactions and on gambling. Local authorities levy rates geared to property values.

This description of the Australian tax system as at mid-2004 does not reveal the controversies and dynamic aspects of tax reform that lay behind the system. Some seven separate rounds of tax reform debates and proposals commenced in 1972. They predated the introduction of the NTS, which involved as its centrepiece the GST. The NTS was implemented by the federal Government of John Howard as from July 2000. It was accompanied by significant 'price exploitation provisions' to limit price increases, despite which Australia's annual inflation rate increased sharply, from an underlying rate of 2.3% in June 2000 to 6.1% in the year following its introduction. The price exploitation provisions involved price monitoring, heavy fines and public exposure for companies breaching the guide-lines. The provisions ended in June 2002.

Prominent politicians of each major political party had previously promoted the structural shift towards GST or its equivalent, in all cases without success until the Howard initiative put the issue to the federal electorate in October 1998. Economists generally supported the reform proposals. Lying behind them was a wartime initiative in 1942 to pass control of income tax 'as a temporary measure' from the states to the Commonwealth. These powers have not since been returned. In the mean time, the steeply progressive personal tax structure adopted and developed in the high inflation years of the 1950s meant that by the 1970s relatively low-income persons were incurring marginal tax rates designed many years previously for high-income persons. Opponents in each round of reform proposals objected to the administrative burdens resulting from any change on businesses, to the adverse effects of GST-type taxes on low-income households (despite proposals of generous compensation measures at each stage), and to the repercussions for inflation levels and for industry. Even when the Howard Government was returned to office at the 1998 election, with the issue of GST dominating the debates and differences between the parties, political difficulties required it to make many concessions and in several ways to increase the administrative complexity of the system. At the November 2001 federal election, the opposition Labor Party did not succeed with proposals to 'roll back' the GST. While the post-11 September environment favoured the incumbent Government, it was safe to conclude that the GST had become an entrenched part of the Australian tax system. In the period since the November 2001 elections there has been no serious proposal to change the essential structure of the Australian taxation system.

The NTS itself, applying in two stages from July 1999, but mainly from July 2000, involved repealing other indirect taxes, notably a wholesale sales tax that applied at rates of up to 30% on manufactured goods alone. The NTS thus caused prices in general, but especially of services, to rise at rates well above the pre-NTS annual inflation rate of around 2.3%, while some manufactured goods experienced price reductions, for example motor vehicles and jewellery. The short-term dynamics of effecting these changes caused disruption, especially in housing and vehicle sales. One curiosity of the NTS is that the federal Government that created it does not classify any GST revenue as its own. Instead, it collects GST as an 'agency function', passing the proceeds to the states in compensation for their loss of some of their fiscal transfer supports. Students of tax and government trends in Australia thus need to be careful in drawing inferences from the downward trend in the ratio of federal taxes and outlays to GDP from the fiscal year 2000/01 onwards.

In addition to the struggle to change the longer-term shape of the tax system, there remain short-term economic management tensions reflected in annual budgets and other fiscal policies. The date for the delivery of the federal budget and most state budgets has settled on May, the convention until the mid-1990s having been for budgets to be announced in August. After the experience of fiscal neglect in the mid-1970s, during which government outlays rose by 46% in one year alone (1974/75), Australian governments have shown disciplined commitment to fiscal responsibility.

FUTURE CHALLENGES

The challenges for the main decision-makers in the Australian economy are a combination of issues that face most advanced economies and some that are specific to the Australian economic structures and legal and political environment. Foremost is the need to continue economic growth, avoiding any recurrence of recession or unacceptable inflationary pressures. In some ways, the real economy moves disparately from the financial sector, although profit expectations, which drive financial values, cannot for long be separated from the position and prospects in real terms of underlying economic activity. As in the USA, share-price indices rose spectacularly in the late 1990s at a time when the real economy grew only moderately. Between 31 December 1996 and 31 December 1999 the USA's Dow Jones industrial share-price index rose by 78.3%. The ASX200, the Australian stock market price index, also rose substantially, but by the much more modest rate of 29.9%. Subsequently, and until March 2003, the US index fell by 15%, while the Australian index rose slightly. In the short period between March 2003 and July 2003 there was some signs of equity-market price recoveries in both the USA and Australia, though they have not been supported by a stronger US real economy. A challenge is for policy-makers in all countries to see that these are equilibrium corrections in financial markets and not the portents of equity-market plunges from equilibrium values. It was comforting for the Australian financial markets and Australian policy-makers that in the period September 2001–March 2003 the Australian index did not need to undergo the same degree of financial-market correction experienced in the USA. Meanwhile, some domestic markets in Australia were, over the same period, booming in such ways as to concern the policy-makers, notably the housing market. In the year to May 2002 the number of dwelling approvals rose by 23%. The federal budget economic forecasts released in May 2002 predicted a 3% decline in such numbers. A number of credible economic analysts projected sharper declines in housing construction activity in the year to mid-2003. In the year to May 2003, however, housing activity actually rose by over 2%, defying these projections and increasing pressure on the monetary authorities. For the third consecutive year, housing approvals defied predictions of a downturn and rose by a moderate 2% in the year to April 2004. The concurrent rapid growth in motor vehicle sales to record rates has been cited earlier in this essay. In part, and especially in the period before 2003, this rapid activity growth in the two main forward indicators of the economic cycle (housing and car demand) reflected some curious fiscal incentives designed to overcome timing difficulties with the introduction of the GST. It also reflected the low interest rates that had prevailed from about 1996 and which declined further in 2001 and remained relatively stable in the period to mid-2004. In the face of these tensions, the fiscal and monetary authorities maintained a strong and stable policy leadership, which no doubt supported the Australian position in 2002–03 as the leading economic-growth nation among advanced countries. Beyond the challenges for short-term economic management, there is a medium-term imperative for Australia to maintain the trend of national productivity growth above 2% per year and closer to 3%. For long periods Australian underlying productivity growth has fallen short of both expectations and other advanced countries. The higher rate of economic growth has been due to population (more accurately work-force) growth in a country that is one of the few to experience net migration gain. Policy reversals in relation to science and technology have not assisted this objective of maintaining high rates of productivity growth. Even in the decade from 1994 to 2003 uncertainties over the application of property-rights laws, research incentives and research tax supports have been destabilizing.

Overall, Australia can claim to have exhibited an excellent record of economic management between 1984 and mid-2004. The country has achieved a much calmer industrial scene (with radical reductions in working days lost through industrial disputes), a more open and flexible economy and much less reliance on exports of primary products. Australia has a modern economy that has embraced advanced information technology. The country is arguably underpopulated from an economic viewpoint. As the data and descriptions in this article confirm, some of the images of Australia being little more than a primary-product exporter, with huge fluctuations in the external payments balances limiting its growth potential, are long outdated.

Statistical Survey

Source (unless otherwise stated): Australian Bureau of Statistics, POB 10, Belconnen, ACT 2616; tel. (2) 6252-7983; fax (2) 6251-6009; internet www.abs.gov.au.

Area and Population

AREA, POPULATION AND DENSITY

Area (sq km)	7,692,024*
Population (census results)†	
6 August 1996	17,892,423
7 August 2001	
Males	9,362,021
Females	9,610,329
Total	18,972,350
Population (official estimates at mid-year)†	
2001	19,485,300
2002	19,662,800
2003‡	19,881,500
Density (per sq km) at mid-2003	2.6

* 2,969,907 sq miles.

† Census results exclude, and estimates include, an adjustment for under-enumeration, estimated to have been 1.9% in 1991. Estimates also exclude overseas visitors in Australia and include Australian residents temporarily overseas. The estimates shown above have been revised to take account of the August 2001 census result.

‡ Provisional figure.

STATES AND TERRITORIES

(census results, 7 August 2001)

	Area (sq km)	Population	Density (per sq km)
New South Wales (NSW)	800,642	6,371,745	8.0
Victoria	227,416	4,644,950	20.4
Queensland	1,730,648	3,655,139	2.1
South Australia	983,482	1,467,261	1.5
Western Australia	2,529,875	1,851,252	0.7
Tasmania	68,401	456,652	6.7
Northern Territory	1,349,129	210,664	0.2
Australian Capital Territory (ACT)	2,358	311,947	132.3
Jervis Bay Territory	73	611	8.4
Total	7,692,024	18,972,350 *	2.5

* Includes populations of Christmas Island (1,508) and the Cocos (Keeling) Islands (621).

PRINCIPAL TOWNS

(estimated population at 30 June 2002)*

Canberra (national capital)	321,400	Wollongong	272,100
Sydney (capital of NSW)	4,170,900	Hobart (capital of Tasmania)	198,000
Melbourne (capital of Victoria)	3,524,100	Sunshine Coast	191,900
Brisbane (capital of Queensland)	1,689,100	Geelong	161,700
Perth (capital of W Australia)	1,114,300	Townsville	137,400
Adelaide (capital of S Australia)	1,114,300	Cairns	114,500
Newcastle	497,500	Toowoomba	111,400
Gold Coast-Tweed	439,700	Darwin (capital of N Territory)	107,400

* Figures refer to metropolitan areas, each of which normally comprises a municipality and contiguous urban areas.

BIRTHS, MARRIAGES AND DEATHS*

	Registered live births		Registered marriages		Registered deaths	
	Number	Rate (per 1,000)	Number	Rate (per 1,000)	Number	Rate (per 1,000)
1994	258,051	14.5	111,174	6.2	126,692	7.1
1995	256,190	14.2	109,386	6.1	125,133	6.9
1996	253,834	13.9	106,103	5.8	128,719	7.0
1997	251,842	13.6	106,735	5.8	129,350	7.0
1998	249,616	13.3	110,598	5.9	127,202	6.8
1999	248,870	13.1	114,316	6.0	128,102	6.8
2000	249,636	13.0	113,429	5.9	128,291	6.7
2001	246,394	12.6	103,130	5.3	128,544	6.6

* Data are tabulated by year of registration rather than by year of occurrence.

Expectation of life (WHO estimates, years at birth): 80.4 (males 77.9; females 83.0) in 2002 (Source: WHO, *World Health Report*).

IMMIGRATION AND EMIGRATION

(year ending 30 June)*

	2000/01	2001/02	2002/03
Permanent immigrants	107,366	88,900	93,914
Permanent emigrants	46,521	48,241	50,463

* Figures refer to persons intending to settle in Australia, or Australian residents intending to settle abroad.

Source: Department of Immigration and Multicultural and Indigenous Affairs, Belconnen.

ECONOMICALLY ACTIVE POPULATION

(annual averages, '000 persons aged 15 years and over, excluding armed forces)

	2000	2001	2002
Agriculture, hunting and forestry	421.2	414.9	387.3
Fishing	17.5	18.6	17.4
Mining and quarrying	68.4	68.3	70.9
Manufacturing	1,141.1	1,109.1	1,120.0
Electricity, gas and water	64.8	67.4	66.8
Construction	700.7	678.3	711.4
Wholesale and retail trade; repair of motor vehicles, motorcycles and personal and household goods	1,779.4	1,792.4	1,857.0
Hotels and restaurants	457.2	461.8	460.8
Transport, storage and communications	592.0	600.9	574.3
Financial intermediation	335.5	347.9	346.3
Real estate, renting and business activities	1,067.6	1,085.4	1,103.8
Public administration and defence; compulsory social security	456.8	480.4	523.4
Education	612.3	633.9	654.8
Health and social work	849.3	898.4	932.6
Other community, social and personal service activities	437.5	459.5	477.3
Private households with employed persons	6.6	5.0	6.0
Extra-territorial organizations and bodies	1.4	1.8	1.4
Total employed	9,009.4	9,123.9	9,311.4
Unemployed	615.6	666.7	631.3
Total labour force	9,625.0	9,790.6	9,942.7
Males	5,415.7	5,478.5	5,553.9
Females	4,209.3	4,312.1	4,388.8

Source: ILO.

Health and Welfare

KEY INDICATORS

Total fertility rate (children per woman, 2002)	1.7
Under-5 mortality rate (per 1,000 live births, 2002)	6
HIV/AIDS (% of persons aged 15–49, 2003)	0.1
Physicians (per 1,000 head, 2001)	2.5
Hospital beds (per 1,000 head, 1999)	7.9
Health expenditure (2001): US $ per head (PPP)	2,532
Health expenditure (2001): % of GDP	9.2
Health expenditure (2001): public (% of total)	67.9
Access to water (% of persons, 2000)	100
Access to sanitation (% of persons, 2000)	100
Human Development Index (2002): ranking	3
Human Development Index (2002): value	0.946

For sources and definitions, see explanatory note on p. vi.

Agriculture

PRINCIPAL CROPS
('000 metric tons)

	2000	2001	2002
Wheat	22,108	24,854	10,059
Rice (paddy)	1,098	1,760	1,291
Barley	6,743	8,280	3,713
Maize	406	345	457
Oats	1,050	1,434	926
Millet	57	55*	55*
Sorghum	2,116	1,935	2,021
Triticale (wheat-rye hybrid)	841	860	269
Potatoes	1,200	1,250*	1,260*
Sugar cane	38,165	31,228	32,260
Dry beans	37	50	39
Dry broad beans	253	350	108
Dry peas	456	512	160
Chick-peas	162	258	136
Lentils	163	180*	67
Lupins	1,055	1,220	537
Soybeans (Soya beans)	105	62	70
Sunflower seed	147	77	63
Rapeseed	1,775	1,757	841
Cottonseed	1,140	1,054	546
Cabbages	69	60*	60*
Lettuce	152	145*	145*
Tomatoes	414	400*	400*
Cauliflower	116	120*	120*
Pumpkins, squash and gourds	124	125*	125*
Dry onions	247	250*	250*
Green peas	67	65*	60*
Carrots	283	285*	270*
Other fresh vegetables	287	274*	269*
Watermelons	85	86*	80*
Cantaloupes and other melons	87	87*	85*
Bananas	257	250*	250*
Oranges	510	624	437
Tangerines, mandarins, clementines and satsumas	85	75	92
Apples	320	321†	340†
Pears	156	145†	165†
Peaches and nectarines	86	90*	92*
Grapes	1,311	1,546	1,754
Pineapples	139	140*	140*
Cotton (lint)	806	745	386

* FAO estimate.
† Unofficial figure.

Source: FAO.

LIVESTOCK
('000 head at 30 June)

	2000	2001	2002
Horses*	220	220	220
Cattle	27,588	27,721	27,870
Pigs	2,511	2,748	2,940
Sheep	118,552	110,900	106,200
Goats*	260	295	400
Chickens	85,000	90,000*	93,000*
Ducks	517	540*	540*
Turkeys	1,360	1,400*	1,400*

* FAO estimate(s).

Source: FAO.

LIVESTOCK PRODUCTS
('000 metric tons)

	2000	2001	2002
Beef and veal	1,987.9	2,119.0	2,028.0
Mutton and lamb	680	715	644
Goat meat*	10.6	10.6	13.8
Pig meat	364.4	378.5	407.3
Horse meat*	21.3	21.3	21.3
Chicken meat	610.0	619.4	667.5
Duck meat	8.0	9.0*	9.0*
Turkey meat*	25.2	25.9	25.9
Cows' milk	11,183	10,872	11,620
Butter	170	151	164
Cheese	373	376	431
Hen eggs†	143	141	140
Honey	21.4	21.5*	22.0*
Wool: greasy	671	657	607
Wool: scoured†	452	416	395
Cattle hides (fresh)*	238	248	209
Sheepskins (fresh)*	155.8	165.7	149.7

* FAO estimate(s).
† Unofficial figures.
Note: Figures for meat and milk refer to the 12 months ending 30 June of the year stated.

Source: FAO.

Forestry

ROUNDWOOD REMOVALS
('000 cubic metres, excl. bark)

	2000	2001	2002
Sawlogs, veneer logs and logs for sleepers	11,005	10,727	12,224
Pulpwood	12,345	12,691	11,231
Other industrial wood	692	935	867
Fuel wood*	6,333	6,707	7,104
Total	30,375	31,060	31,426

* FAO estimates.

Source: FAO.

SAWNWOOD PRODUCTION
('000 cubic metres, incl. railway sleepers)

	2000	2001	2002
Coniferous (softwood)	2,637	2,351	3,011
Broadleaved (hardwood)	1,346	1,174	1,108
Total	3,983	3,525	4,119

Source: FAO.

Fishing

('000 metric tons, live weight, year ending 30 June)

	1999/2000	2000/01	2001/02
Capture	190.1	194.4	194.2
Blue grenadier	9.5	7.6	9.2
Clupeoids	5.6	8.6	14.6
Australian spiny lobster	14.6	11.6	17.7
Penaeus shrimps	20.3	23.8	22.4
Scallops	12.0	9.2	5.7
Aquaculture	31.7	35.4	38.8
Atlantic salmon	10.9	12.7	14.4
Total catch	221.8	229.8	233.0

Note: Figures exclude aquatic plants ('000 metric tons, capture only): 13.7 in 1999/2000; 13.5 in 2000/01; 9.9 in 2001/02. Also excluded are crocodiles, recorded by number rather than by weight. The number of estuarine crocodiles caught was: 13,296 in 1999/2000; 11,849 in 2000/01; 7,205 in 2001/02. The number of Australian crocodiles caught was: 10 in 1999/2000; 0 in 2000/01; 2 in 2001/02. Also excluded are whales, recorded by number rather than weight. The number of Baleen whales caught was: 2 in 1999/2000; 10 in 2000/01; 1 in 2001/02. The number of toothed whales caught was: 25 in 1999/2000; 24 in 2000/01; 30 in 2001/02. Also excluded are pearl oyster shells (metric tons): 240 in 1999/2000; 190 in 2000/01; 220 in 2001/02.

Source: FAO.

Mining

('000 metric tons, unless otherwise indicated)

	2000	2001	2002
Hard coal	245,500	264,680	347,890
Lignite	67,000	70,000	73,000[1]
Crude petroleum ('000 barrels)	263,500	231,000	240,000[1]
Natural gas (million cu metres)	30,794	30,000[1]	31,000[2]
Iron ore: gross weight	171,508	181,435	182,704[2]
Iron ore: metal content	106,563	112,592	113,548
Copper ore[3]	832	873	883
Nickel ore[3]	166	197	186
Bauxite	53,802	53,799	54,024
Lead ore[3]	739	714	683
Zinc ore[3]	1,420	1,519	1,469
Tin ore (metric tons)[3]	9,146	9,802	6,268
Manganese ore (metallurgical): gross weight	1,613	2,069	2,187
Manganese ore (metallurgical): metal content	787	948	983
Chromite	90	12	133
Ilmenite	2,146	2,017	1,917
Leucoxene	27	30	39
Rutile	208	206	218
Tantalum and niobium (Columbium) concentrates (metric tons)[4]	1,600	2,220	3,100
Zirconium concentrates	374	394	412
Antimony ore (metric tons)[3]	1,511	1,380	1,200
Cobalt ore (metric tons)[3,1]	5,100	6,200	6,600
Silver (metric tons)[3]	2,060	1,970	2,077
Uranium (metric tons)[3,5]	7,609	7,756	6,854
Gold (kilograms)[3]	296,410	285,030	273,010
Limestone[1]	12,000	12,000	20,000
Bentonite and bentonitic clay[1]	180	180	200
Kaolin and ball clay[1]	220	220	230
Brick clay and shale[1]	8,000	8,000	8,000
Magnesite (metric tons)	349,783	605,314	484,314
Phosphate rock[1,6]	977	1,893	2,025
Barite (Barytes)[1]	20	20	20
Salt (unrefined)	8,798	9,536	9,887
Gypsum (crude)[1]	3,800	3,800	4,000
Talc (metric tons)[7]	178,545	173,446	173,741
Pyrophyllite (metric tons)[7]	1,727	1,500[1]	868
Diamonds ('000 carats): gem	14,656	14,397	15,136
Diamonds ('000 carats): industrial	11,992	11,779	18,500

[1] Estimated production.
[2] Provisional figure.
[3] Figures refer to the metal content of ores and concentrates.
[4] The estimated metal content (in metric tons) was: Niobium (Columbium) 160 in 2000, 230 in 2001, 290 in 2002; Tantalum 485 in 2000, 660 in 2001, 940 in 2002.
[5] Data from the World Nuclear Association (London, United Kingdom).
[6] Figures refer to the gross weight of concentrates. The phosphoric acid (P_2O_5) content (in '000 metric tons) was: 225 in 2000; 438 in 2001; 482 in 2002.
[7] Production during 12 months ending 30 June of the year stated.

Source (unless otherwise indicated): US Geological Survey.

Industry

SELECTED PRODUCTS

(year ending 30 June, '000 metric tons, unless otherwise indicated)

	1999/2000	2000/01	2001/02
Pig-iron	6,489	6,096	6,062
Raw steel	8,053	8,003	8,311
Aluminium—unwrought*†	1,742	1,809	1,855
Copper—unwrought*†	477	517	561
Lead—unwrought*†	233	215	275
Zinc—unwrought*†	405	534	572
Tin—unwrought (metric tons)*†	600	1,039	829
Motor spirit (petrol—million litres)†	18,652	17,887	18,000
Fuel oil (million litres)†	1,839	1,951	1,684
Diesel-automotive oil (million litres)†	12,737	13,212	13,064
Aviation turbine fuel (million litres)†	5,538	5,836	5,390
Clay bricks (million)	1,735	1,448	1,514
Woven cotton fabrics (incl. towelling, '000 sq metres)‡	47,000	39,000	n.a.
Woven woollen fabrics (incl. blanketing, '000 sq metres)‡	5,400	4,000	n.a.
Electricity (million kWh)	184,790	188,546	n.a.
Cement	7,937	6,820	7,236
Concrete—ready-mixed ('000 cu m)	20,634	17,250	19,447
Newsprint	381	392	398
Wheat flour‡	2,030	2,077	n.a.
Beer (million litres)	1,768	1,745	1,744
Unfortified wine (million litres)	779	1,016	1,151
Tobacco and cigarettes (metric tons)	20,688	19,125	18,367

* Primary refined metal only.
† Source: Australian Bureau of Agricultural and Research Economics.
‡ Source: UN, *Industrial Commodity Statistics Yearbook*.

2002/03 ('000 metric tons, unless otherwise indicated): Pig iron 6,634; Raw steel 8,467; Aluminium (primary refined metal) 1,855; Copper (primary refined metal) 537; Lead (primary refined metal) 267; Zinc (primary refined metal) 570; Tin (primary refined metal, metric tons) 708; Motor spirit (petrol—million litres) 17,984; Fuel oil (million litres) 1,441; Diesel-automotive oil (million litres) 13,335; Aviation turbine fuel (million litres) 5,149 (Source: Australian Bureau of Agricultural and Research Economics).

Finance

CURRENCY AND EXCHANGE RATES

Monetary Units

100 cents = 1 Australian dollar ($A).

Sterling, US Dollar and Euro Equivalents (31 May 2004)

£1 sterling = $A2.5685;
US $1 = $A1.4000;
€1 = $A1.7144;
$A100 = £38.93 = US $71.43 = €58.29.

Average Exchange Rate (US $ per Australian dollar)

2001 0.5176
2002 0.5439
2003 0.6519

COMMONWEALTH GOVERNMENT BUDGET

($A million, year ending 30 June)

Revenue	2000/01	2001/02	2002/03
Tax revenue	151,156	149,848	163,055
Direct taxes	120,861	119,032	131,278
Individuals	76,599	86,422	91,303
Companies	35,136	27,133	33,365
Indirect taxes, etc.	25,601	25,634	27,255
Non-tax revenue	10,369	12,540	11,958
Total	161,526	162,388	175,014

Expenditure	2000/01	2001/02	2002/03
Defence	11,360	12,017	13,307
Education	10,966	11,761	12,109
Health	25,242	27,614	29,400
Social security and welfare	66,898	69,081	71,263
Economic services	9,940	12,974	12,646
General public services	11,923	12,097	12,956
Public-debt interest	5,836	4,995	4,629
Total (incl. others)	156,783	166,482	169,247

Source: Reserve Bank of Australia, *Bulletin*.

OFFICIAL RESERVES

(US $ million at 31 December)

	2001	2002	2003
Gold*	709	878	1,070
IMF special drawing rights	109	136	170
Reserve position in IMF	1,412	1,934	2,053
Foreign exchange	16,434	18,618	29,966
Total	18,664	21,567	33,259

* Valued at market-related prices.

Source: IMF, *International Financial Statistics*.

MONEY SUPPLY

($A million at 31 December)

	1999	2000	2001
Currency outside banks	24,604	26,928	28,471
Demand deposits at trading and savings banks	101,179	110,660	138,456
Total money (incl. others)	125,945	137,720	167,035

Currency outside banks ($A million at 31 December): 29,703 in 2002; 31,470 in 2003.

Source: IMF, *International Financial Statistics*.

COST OF LIVING

(Consumer Price Index*; base 1990 = 100)

	2000	2001	2002
Food	129.5	138.0	142.9
Electricity, gas and other fuels†	126.2	133.3	138.8
Clothing	106.8	109.3	110.7
Rent‡	101.2	106.0	109.5
All items (incl. others)	124.4	129.8	133.8

* Weighted average of eight capital cities.

† From September 1998 including water and sewerage.

‡ Including expenditure on maintenance and repairs of dwellings; from September 1998 excluding mortgage interest charges and including house purchase and utilities.

Source: ILO.

NATIONAL ACCOUNTS

($A million, current prices, year ending 30 June)

National Income and Product

	1999/2000	2000/01	2001/02
Compensation of employees	302,116	321,024	337,657
Gross operating surplus	196,422	210,279	225,766
Gross mixed income	54,272	57,522	65,365
Total factor incomes	552,810	588,825	628,788
Indirect taxes *Less* Subsidies	73,227	82,295	85,582
GDP in market prices	626,037	671,120	714,370
Net primary incomes from abroad	−18,249	−19,241	−20,273
Gross national income	607,788	651,879	694,097
Current taxes on income, wealth, etc.	1,135	1,100	1,002
Other current transfers (net)	−917	−1,068	−1,019
Gross disposable income	608,006	651,911	694,080

Expenditure on the Gross Domestic Product

	1999/2000	2000/01	2001/02
Government final consumption expenditure	113,305	120,390	127,413
Private final consumption expenditure	374,922	404,271	426,154
Change in inventories	1,791	510	1,136
Gross fixed capital formation	150,609	145,301	160,900
Total domestic expenditure	640,626	670,471	715,603
Exports of goods and services	126,222	153,854	153,340
Less Imports of goods and services	140,811	153,205	154,573
GDP in market prices	626,037	671,120	714,370
GDP at constant 2001/02 prices	673,944	687,720	714,370

Gross Domestic Product by Economic Activity

	1999/2000	2000/01	2001/02
Agriculture, hunting, forestry and fishing	18,527	22,881	27,663
Mining and quarrying	26,700	34,053	33,822
Manufacturing	71,084	73,011	76,686
Electricity, gas and water	14,325	15,319	15,977
Construction	38,550	34,148	39,540
Wholesale and retail trade	65,226	66,251	72,123
Hotels and restaurants	14,334	14,743	14,630
Transport, storage and communications	49,256	51,127	53,881
Financial intermediation (incl. insurance)	41,335	46,927	50,792
Property and business activities	66,608	73,521	75,524
Ownership of dwellings	57,606	60,233	63,326
Public administration and defence	23,795	25,116	27,755
Education	27,683	29,805	31,201
Health and community services	35,055	38,416	41,236
Cultural and recreational services	10,770	11,895	12,470
Personal and other services	14,384	14,800	16,011
Gross value added at basic prices	575,238	612,246	652,637
Taxes on products *Less* subsidies on products	50,799	58,874	61,733
GDP in market prices	626,037	671,120	714,370

2002/03: Agriculture, hunting, forestry and fishing 20,059; Mining and quarrying 34,943; Manufacturing 80,741; Electricity, gas and water 16,906; Construction 45,568; Wholesale and retail trade 78,857; Hotels and restaurants 15,158; Transport, storage and communications 56,438; Financial intermediation (incl. insurance) 55,620; Property and business activities 79,811; Ownership of dwellings 65,613; Public administration and defence 29,743; Education 33,560; Health and community services 44,039; Cultural and recreational services 13,201; Personal and other services 17,553; *Gross value added at basic prices* 687,810; Taxes, *less* subsidies on products 66,693; Statistical discrepancy –1,251; *GDP in market prices* 753,252.

BALANCE OF PAYMENTS
(US $ million)

	2001	2002	2003
Exports of goods f.o.b.	63,676	65,099	70,596
Imports of goods f.o.b.	–61,890	–70,530	–85,850
Trade balance	1,786	–5,431	–15,254
Exports of services	16,698	17,906	21,204
Imports of services	–16,948	–18,107	–21,514
Balance on goods and services	1,536	–5,632	–15,565
Other income received	8,063	8,154	9,579
Other income paid	–18,332	–19,823	–24,490
Balance on goods, services and income	–8,733	–17,301	–30,476
Current transfers received	2,242	2,310	2,767
Current transfers paid	–2,221	–2,373	–2,845
Current balance	–8,712	–17,365	–30,554
Capital account (net)	591	443	764
Direct investment abroad	–12,228	–7,393	–17,115
Direct investment from abroad	4,667	16,141	8,601
Portfolio investment assets	–9,767	–16,400	–8,911
Portfolio investment liabilities	20,735	17,007	50,692
Financial derivatives assets	161	2,243	3,553
Financial derivatives liabilities	336	–2,041	–3,459
Other investment assets	555	–2,955	–6,295
Other investment liabilities	4,141	10,705	9,173
Net errors and omissions	616	–263	430
Overall balance	1,096	122	6,877

Source: IMF, *International Financial Statistics.*

External Trade

PRINCIPAL COMMODITIES
(distribution by SITC, $A million, year ending 30 June)

Imports f.o.b.	2001/02	2002/03
Food and live animals	4,613	5,107
Mineral fuels, lubricants, etc.	9,030	10,598
Petroleum and bituminous oils, crude	6,785	7,812
Chemicals and related products	14,635	15,025
Medicaments (incl. veterinary)	4,002	4,241
Basic manufactures	14,819	16,074
Machinery and transport equipment	53,654	60,640
Automatic data-processing machines and units thereof, magnetic, optical readers; data transcribers and processors	5,028	4,871
Miscellaneous telecommunications equipment and parts, and accessories of radio, television, etc.	4,507	4,239
Motor vehicles principally designed for transport of persons (excl. public transport, incl. racing cars)	8,955	10,283
Aircraft and associated equipment; spacecraft (incl. satellites and launch vehicles) and parts thereof	3,060	5,481
Miscellaneous manufactured articles	17,416	18,716
Total (incl. others)	119,649	133,131

Exports f.o.b.	2001/02	2002/03
Food and live animals	22,380	18,374
Meat of bovine animals, fresh, chilled or frozen	4,333	3,906
Wheat (incl. spelt) and meslin, unmilled	4,527	3,036
Crude materials (inedible) except fuels	22,448	21,429
Iron ore and concentrates	5,160	5,328
Aluminium ores and concentrates (incl. alumina)*		
Mineral fuels, lubricants, etc.	25,130	23,807
Coal, not agglomerated	13,403	11,946
Petroleum and bituminous oils, crude	5,963	5,881
Chemicals and related products	5,293	5,099
Basic manufactures	13,572	12,597
Machinery and transport equipment	14,160	13,528
Miscellaneous manufactured articles	4,483	4,413
Aluminium	4,412	4,059
Non-monetary gold (excl. gold ores and concentrates)	5,129	5,584
Total (incl. others)	121,108	115,442

* Excludes commodities subject to a confidentiality restriction. Such commodities are included in the total.

PRINCIPAL TRADING PARTNERS
($A million, year ending 30 June)

Imports f.o.b.	2000/01	2001/02	2002/03
Canada	1,869	1,607	1,755
China, People's Republic	9,881	11,275	13,792
France	2,478	2,691	3,781
Germany	6,172	6,732	7,953
Hong Kong	1,362	1,410	1,234
Indonesia	3,330	4,010	4,600
Ireland	1,140	1,469	1,607
Italy	3,257	3,410	4,149
Japan	15,371	15,461	16,335
Korea, Republic	4,710	4,722	4,753
Malaysia	4,177	3,857	4,262
New Zealand	4,565	4,740	5,019
Papua New Guinea	1,457	1,124	1,502
Saudi Arabia	1,634	1,026	1,284
Singapore	3,898	3,972	4,370
Sweden	1,624	1,625	1,810
Switzerland	1,171	1,302	1,230
Taiwan	3,327	3,312	3,401
Thailand	2,780	2,886	3,469
United Kingdom	6,321	6,219	5,770
USA	22,351	21,488	22,496
Viet Nam	2,431	1,855	2,503
Total (incl. others)	118,317	119,649	133,131

Exports f.o.b.	2000/01	2001/02	2002/03
Canada	1,768	1,900	1,816
China, People's Republic	6,846	7,816	8,793
France	1,079	1,343	1,171
Germany	1,490	1,349	1,579
Hong Kong	3,904	3,996	3,215
India	2,086	2,524	2,577
Indonesia	3,119	3,194	2,908
Italy	2,100	2,165	1,861
Japan	23,479	22,796	21,738
Korea, Republic	9,209	9,818	9,116
Malaysia	2,506	2,519	2,147
Netherlands	1,738	1,522	1,364
New Zealand	6,872	7,669	8,120
Philippines	1,501	1,252	1,091
Saudi Arabia	2,196	2,598	1,990
Singapore	6,009	4,936	4,655
South Africa	1,301	1,341	1,313
Taiwan	5,894	4,828	4,314
Thailand	2,222	2,291	2,479
United Kingdom	4,653	5,199	7,236
USA	11,654	12,008	10,369
Total (incl. others)	119,539	121,108	114,442

Transport

RAILWAYS
(traffic)*

	1997/98	1998/99	1999/2000
Passengers carried (million)	587.7	595.2	629.2
Freight carried (millon metric tons)	487.5	492.0	508.0
Freight ton-km ('000 million)	125.2	127.4	134.2

* Traffic on government railways only.

ROAD TRAFFIC
('000 vehicles registered at 31 October)

	1997	1998	1999
Passenger vehicles	9,206.2	9,526.7	9,719.9
Light commercial vehicles	1,632.2	1,686.4	1,721.2
Trucks	418.4	426.9	427.8
Buses	61.1	64.0	65.9
Motorcycles	313.1	328.8	333.8

2002 ('000 vehicles registered at 31 March): Passenger vehicles 10,100; Light commercial vehicles 1,800; Trucks 405.4; Buses 70.2; Motorcycles 371.0.

2003 ('000 vehicles registered at 31 March): Passenger vehicles 10,365.9; Light commercial vehicles 1,879.8; Trucks 412.9; Buses 70.1; Motorcycles 377.3.

SHIPPING

Merchant Fleet
(registered at 31 December)

	2001	2002	2003
Number of vessels	622	624	643
Total displacement ('000 grt)	1,887.8	1,861.3	1,905.8

Source: Lloyd's Register-Fairplay, *World Fleet Statistics*.

International Sea-borne Traffic
('000 metric tons, year ending 30 June)

	2000/01	2001/02	2002/03
Goods loaded	496,204	506,317	540,570
Goods unloaded	54,579	58,041	62,459

CIVIL AVIATION
(traffic)*

	1997	1998	1999
International services:			
Passenger arrivals	7,090,979	7,153,514	7,540,535
Passenger departures	7,010,931	7,084,655	7,442,226
Freight carried (metric tons)	649,371	631,908	680,458
Mail carried (metric tons)	21,975	23,437	25,316
Domestic services†:			
Passengers carried	23,375,317	23,574,788	24,375,906
Passenger-km ('000)	26,357,069	26,774,140	27,842,795
Freight and mail carried (metric tons)	190,680	192,770	192,326

* Includes Christmas Island and Norfolk Island.
† Year ending 30 June.

2001/02 ('000, year ending 30 June): International passenger arrivals 8,254.1; International passenger departures 8,194.9.

Tourism

VISITOR ARRIVALS BY COUNTRY OF ORIGIN*

	2000	2001	2002
China, People's Republic	120,300	158,000	190,000
Germany	143,300	147,600	134,800
Hong Kong	154,100	154,200	150,900
Indonesia	98,100	97,900	89,400
Japan	721,000	673,600	715,500
Korea, Republic	157,400	175,600	189,700
Malaysia	152,100	149,400	159,000
New Zealand	817,000	814,900	790,100
Singapore	285,700	296,000	286,900
Taiwan	134,300	110,100	97,400
United Kingdom	580,400	617,200	642,700
USA	488,100	446,400	434,500
Total (incl. others)	4,931,400	4,855,700	4,841,200

* Visitors intending to stay for less than one year.

Receipts from tourism ($A million): 17,000 in 1999; 19,800 in 2000; 15,344 in 2001 (estimate).

Source: Australian Tourist Commission, Sydney.

Communications Media

	2000	2001	2002
Television receivers ('000 in use)	14,129	14,168	n.a.
Telephones ('000 main lines in use)	10,350	10,485	10,905
Mobile cellular telephones ('000 subscribers)	8,562	11,132	12,575
Personal computers ('000 in use)	9,000	10,000	11,100
Internet users ('000)	6,600	7,700	9,472

2003: Telephones ('000 main lines in use) 10,815; Mobile cellular telephones ('000 subscribers) 14,347.

Source: International Telecommunication Union.

Radio receivers ('000 in use, 1997): 25,500.

Facsimile machines ('000 in use, 1997): 900.

Book production (1994): 10,835 titles.

Newspapers (1996): 65 dailies (estimated combined circulation 5,370,000); 98 non-dailies (circulation 383,000).

Source: mainly UNESCO, *Statistical Yearbook*.

Education

(August 2001)

	Institutions	Teaching staff	Students ('000)
Government schools	6,942	152,138*	2,248.2†
Non-government schools	2,654	69,789*	1,019.9
Higher educational institutions	47‡	78,228	726.4

* Full-time teaching staff and full-time equivalent of part-time teaching staff.

† Comprising 1,384,866 primary and 863,353 secondary students.

‡ Public institutions only.

Directory

The Constitution

The Federal Constitution was adopted on 9 July 1900 and came into force on 1 January 1901. Its main provisions are summarized below:

PARLIAMENT

The legislative power of the Commonwealth of Australia is vested in a Federal Parliament, consisting of HM the Queen (represented by the Governor-General), a Senate and a House of Representatives. The Governor-General may appoint such times for holding the sessions of the Parliament as he or she thinks fit, and may also from time to time, by proclamation or otherwise, prorogue the Parliament, and may in like manner dissolve the House of Representatives. By convention, these powers are exercised on the advice of the Prime Minister. After any general election Parliament must be summoned to meet not later than 30 days after the day appointed for the return of the writs.

THE SENATE

The Senate is composed of 12 senators from each state, two senators representing the Australian Capital Territory and two representing the Northern Territory. The senators are directly chosen by the people of the state or territory, voting in each case as one electorate, and are elected by proportional representation. Senators representing a state have a six-year term and retire by rotation, one-half from each state on 30 June of each third year. The term of a senator representing a territory is limited to three years. In the case of a state, if a senator vacates his or her seat before the expiration of the term of service, the houses of parliament of the state for which the senator was chosen shall, in joint session, choose a person to hold the place until the expiration of the term or until the election of a successor. If the state parliament is not in session, the Governor of the state, acting on the advice of the state's executive council, may appoint a senator to hold office until parliament reassembles, or until a new senator is elected.

The Senate may proceed to the dispatch of business notwithstanding the failure of any state to provide for its representation in the Senate.

THE HOUSE OF REPRESENTATIVES

In accordance with the Australian Constitution, the total number of members of the House of Representatives must be as nearly as practicable double that of the Senate. The number in each state is in proportion to population, but under the Constitution must be at least five. The House of Representatives is composed of 150 members, including two members for the Australian Capital Territory and two members for the Northern Territory.

Members are elected by universal adult suffrage and voting is compulsory. Only Australian citizens are eligible to vote in Australian elections. British subjects, if they are not Australian citizens or already on the rolls, have to take out Australian citizenship before thay can enrol and before they can vote.

Members are chosen by the electors of their respective electorates by the preferential voting system.

The duration of the Parliament is limited to three years.

To be nominated for election to the House of Representatives, a candidate must be 18 years of age or over, an Australian citizen, and entitled to vote at the election or qualified to become an elector.

THE EXECUTIVE GOVERNMENT

The executive power of the Federal Government is vested in the Queen, and is exercisable by the Governor-General, advised by an Executive Council of Ministers of State, known as the Federal Executive Council. These ministers are, or must become within three months, members of the Federal Parliament.

The Australian Constitution is construed as subject to the principles of responsible government and the Governor-General acts on the advice of the ministers in relation to most matters.

THE JUDICIAL POWER

See Judicial System, below.

THE STATES

The Australian Constitution safeguards the Constitution of each state by providing that it shall continue as at the establishment of the Commonwealth, except as altered in accordance with its own provisions. The legislative power of the Federal Parliament is limited in the main to those matters that are listed in section 51 of the Constitution, while the states possess, as well as concurrent powers in those matters, residual legislative powers enabling them to legislate in any way for 'the peace, order and good Government' of their respective territories. When a state law is inconsistent with a law of the Commonwealth, the latter prevails, and the former is invalid to the extent of the inconsistency.

The states may not, without the consent of the Commonwealth, raise or maintain naval or military forces, or impose taxes on any property belonging to the Commonwealth of Australia, nor may the Commonwealth tax state property. The states may not coin money.

The Federal Parliament may not enact any law for establishing any religion or for prohibiting the exercise of any religion, and no religious test may be imposed as a qualification for any office under the Commonwealth.

The Commonwealth of Australia is charged with protecting every state against invasion, and, on the application of a state executive government, against domestic violence.

Provision is made under the Constitution for the admission of new states and for the establishment of new states within the Commonwealth of Australia.

ALTERATION OF THE CONSTITUTION

Proposed laws for the amendment of the Constitution must be passed by an absolute majority in both Houses of the Federal Parliament, and not less than two or more than six months after its passage through both Houses the proposed law must be submitted in each state to the qualified electors.

In the event of one House twice refusing to pass a proposed amendment that has already received an absolute majority in the other House, the Governor-General may, notwithstanding such refusal, submit the proposed amendment to the electors. By convention, the Governor-General acts on the advice of the Prime Minister. If in a majority of the states a majority of the electors voting approve the proposed law and if a majority of all the electors voting also approve, it shall be presented to the Governor-General for Royal Assent.

No alteration diminishing the proportionate representation of any state in either House of the Federal Parliament, or the minimum number of representatives of a state in the House of Representatives, or increasing, diminishing or altering the limits of the state, or in any way affecting the provisions of the Constitution in relation thereto, shall become law unless the majority of the electors voting in that state approve the proposed law.

STATES AND TERRITORIES

New South Wales

The state's executive power is vested in the Governor, appointed by the Crown, who is assisted by an Executive Council composed of cabinet ministers.

The state's legislative power is vested in a bicameral Parliament, composed of the Legislative Council and the Legislative Assembly. The Legislative Council consists of 42 members directly elected for the duration of two parliaments (i.e. eight years), 21 members retiring every four years. The Legislative Assembly consists of 93 members and sits for four years.

Victoria

The state's legislative power is vested in a bicameral Parliament: the Upper House, or Legislative Council, of 44 members, elected for two terms of the Legislative Assembly; and the Lower House, or Legislative Assembly, of 88 members, elected for a minimum of three and maximum of four years. One-half of the members of the Council retires every three–four years.

In the exercise of the executive power the Governor is assisted by a cabinet of responsible ministers. Not more than six members of the Council and not more than 17 members of the Assembly may occupy salaried office at any one time.

The state has 88 electoral districts, each returning one member, and 22 electoral provinces, each returning two Council members.

Queensland

The state's executive power is vested in the Governor, appointed by the Crown, who is assisted by an Executive Council composed of Ministers. The state's legislative power is vested in the Parliament comprising the Legislative Assembly (composed of 89 members who are elected at least every three years to represent 89 electoral districts) and the Governor, who assents to bills passed by the Assembly. The state's Constitution anticipates that Ministers are also members of the Legislative Assembly and provides that up to 18 members of the Assembly can be appointed Ministers.

South Australia

The state's Constitution vests the legislative power in a Parliament elected by the people and consisting of a Legislative Council and a House of Assembly. The Council is composed of 22 members, one-half of whom retire every three years. Their places are filled by new members elected under a system of proportional representation, with the whole state as a single electorate. The executive has no authority to dissolve this body, except in circumstances warranting a double dissolution.

The 47 members of the House of Assembly are elected for four years from 45 electoral districts.

The executive power is vested in a Governor, appointed by the Crown, and an Executive Council consisting of 10 responsible ministers.

Western Australia

The state's administration is vested in the Governor, a Legislative Council and a Legislative Assembly.

The Legislative Council consists of 34 members, two of the six electoral regions returning seven members on a proportional representation basis, and four regions returning five members. Election is for a term of four years.

The Legislative Assembly consists of 57 members, elected for four years, each representing one electorate.

Tasmania

The state's executive authority is vested in a Governor, appointed by the Crown, who acts upon the advice of his premier and ministers, who are elected members of either the Legislative Council or the House of Assembly. The Council consists of 15 members who sit for six years, retiring in rotation. The House of Assembly has 25 members elected for four years.

Northern Territory

On 1 July 1978, the Northern Territory was established as a body politic with executive authority for specified functions of government. Most functions of the Federal Government were transferred to the Territory Government in 1978 and 1979, major exceptions being Aboriginal affairs and uranium mining.

The Territory Parliament consists of a single house, the Legislative Assembly, with 25 members. The first Parliament stayed in office for three years. As from the election held in August 1980, members are elected for a term of four years.

The office of Administrator continues. The Northern Territory (Self-Government) Act provides for the appointment of an Administrator by the Governor-General charged with the duty of administering the Territory. In respect of matters transferred to the Territory Government, the Administrator acts with the advice of the Territory Executive Council; in respect of matters retained by the Commonwealth, the Administrator acts on Commonwealth advice.

Australian Capital Territory

On 29 November 1988 the Australian Capital Territory (ACT) was established as a body politic. The ACT Government has executive authority for specified functions, although a number of these were to be retained by the Federal Government for a brief period during which transfer arrangements were to be finalized.

The ACT Parliament consists of a single house, the Legislative Assembly, with 17 members. The first election was held in March 1989. Members are elected for a term of three years.

The Federal Government retains control of some of the land in the ACT for the purpose of maintaining the Seat of Government and the national capital plan.

Jervis Bay Territory

Following the attainment of self-government by the ACT (see above), the Jervis Bay Territory, which had formed part of the ACT since 1915, remained a separate Commonwealth Territory, administered by the then Department of the Arts, Sport, the Environment and Territories. The area is governed in accordance with the Jervis Bay Territory Administration Ordinance, issued by the Governor-General on 17 December 1990.

The Government

Head of State: HM Queen Elizabeth II (succeeded to the throne 6 February 1952).

Governor-General: Maj.-Gen. Michael Jeffery (assumed office 11 August 2003).

THE MINISTRY
(September 2004)

Cabinet Ministers

Prime Minister: John Howard.

Deputy Prime Minister and Minister for Transport and Regional Services: John Anderson.

Minister for Foreign Affairs: Alexander Downer.

Treasurer: Peter Costello.

Minister for Trade: Mark Vaile.

Minister for Defence and Leader of the Government in the Senate: Senator Robert Hill.

Minister for Communications, Information Technology and the Arts: Senator Helen Coonan.

Minister for Employment and Workplace Relations: Kevin Andrews.

Minister for Education, Science and Training: Dr Brendan Nelson.

Minister for Health and Ageing: Tony Abbott.

Minister for Industry, Tourism and Resources: Ian Macfarlane.

Minister for the Environment and Heritage: Senator Ian Campbell.

Minister for Finance and Administration, and Deputy Leader of the Government in the Senate: Senator Nicholas Minchin.

Minister for Family and Community Services, and Minister Assisting the Prime Minister for the Status of Women: Kay Patterson.

Minister for Agriculture, Fisheries and Forestry: Warren Truss.

Minister for Immigration and Multicultural and Indigenous Affairs: Senator AMANDA VANSTONE.

Attorney-General: PHILLIP RUDDOCK.

Other Ministers

Minister for Local Government, Territories and Roads: JIM LLOYD.

Minister for Revenue and Assistant Treasurer: MAL BROUGH.

Minister for Small Business and Tourism: JOE HOCKEY.

Minister for Fisheries, Forestry and Conservation: Senator IAN MACDONALD.

Minister for Science: PETER MCGAURAN.

Minister for Employment Services and Minister assisting the Minister of Defence: FRAN BAILEY.

Minister for Children and Youth Affairs: LAWRENCE ANTHONY.

Minister for Veterans' Affairs: DANNA VALE.

Minister for Ageing: JULIE BISHOP.

Special Minister of State: Senator ERIC ABETZ.

Minister for Justice and Customs: Senator CHRISTOPHER ELLISON.

Minister for the Arts and Sport: Senator ROD KEMP.

Minister for Citizenship and Multicultural Affairs and Minister Assisting the Prime Minister: GARY HARDGRAVE.

DEPARTMENTS

Department of the Prime Minister and Cabinet: 3–5 National Circuit, Barton, ACT 2600; tel. (2) 6271-5111; fax (2) 6271-5414; internet www.dpmc.gov.au.

Aboriginal and Torres Strait Islander Commission: Lovett Tower, Woden Town Centre, Phillip, ACT 2606; tel. (2) 6121-4000; fax (2) 6121-4621; internet www.atsic.gov.au.

Department of Agriculture, Fisheries and Forestry: Edmund Barton Bldg, Broughton St, Barton, GPOB 858, Canberra, ACT 2601; tel. (2) 6272-3933; fax (2) 6272-3008; internet www.affa.gov.au.

Attorney-General's Department: Robert Garran Offices, Barton, ACT 2600; tel. (2) 6250-6666; fax (2) 6250-5900; internet www.law.gov.au.

Department of Communications, Information Technology and the Arts: GPOB 2154, Canberra, ACT 2601; tel. (2) 6271-1000; fax (2) 6271-1901; internet www.dcita.gov.au.

Department of Defence: Russell Offices, Russell Drive, Campbell, Canberra, ACT 2600; tel. (2) 6265-9111; e-mail webmaster@cbr.defence.gov.au; internet www.defence.gov.au.

Department of Education, Science and Training: GPOB 9880, Canberra, ACT 2601; tel. (2) 6240-8111; fax (2) 6240-8571; e-mail library@dest.gov.au; internet www.dest.gov.au.

Department of Employment and Workplace Relations: GPOB 9879, Canberra, ACT 2601; tel. (2) 6121-6000; fax (2) 6121-7542; e-mail webmaster@dewr.gov.au; internet www.dewr.gov.au.

Department of the Environment and Heritage: GPOB 787, Canberra, ACT 2601; tel. (2) 6274-1111; fax (2) 6274-1123; internet www.ea.gov.au.

Department of Family and Community Services: Library and Information Research Service, Box 7788, Canberra Mail Centre, ACT 2610; tel. (2) 6244-6385; fax (2) 6244-7919; e-mail library.research@facs.gov.au; internet www.facs.gov.au.

Department of Finance and Administration: John Gorton Bldg, King Edward Tce, Parkes, ACT 2600; tel. (2) 6263-2222; fax (2) 6273-3021; internet www.dofa.gov.au.

Department of Foreign Affairs and Trade: R. G. Casey Bldg, John McEwan Crescent, Barton, ACT 0221; tel. (2) 6261-1111; fax (2) 6261-3111; internet www.dfat.gov.au.

Department of Health and Ageing: GPOB 9848, Canberra, ACT 2601; tel. (2) 6289-1555; fax (2) 6281-6946; internet www.health.gov.au.

Department of Immigration and Multicultural and Indigenous Affairs: Benjamin Offices, Chan St, Belconnen, ACT 2617; tel. (2) 6264-1111; fax (2) 6264-2670; internet www.immi.gov.au.

Department of Industry, Tourism and Resources: GPOB 9839, Canberra, ACT 2601; tel. (2) 6213-6000; fax (2) 6213-7000; e-mail customerrelation@industry.gov.au; internet www.industry.gov.au.

Department of Transport and Regional Services: GPOB 594, Canberra, ACT 2601; tel. (2) 6274-7111; fax (2) 6257-2505; e-mail publicaffairs@dotrs.gov.au; internet www.dotrs.gov.au.

Department of the Treasury: Langton Crescent, Parkes, ACT 2600; tel. (2) 6263-2111; fax (2) 6273-2614; internet www.treasury.gov.au.

Department of Veterans' Affairs: POB 21, Woden, ACT 2606; tel. (2) 6289-1111; fax (2) 6289-6025; internet www.dva.gov.au.

Legislature

FEDERAL PARLIAMENT

Senate

President: to be appointed.

Election, 9 October 2004 (provisional results)

Party	Seats*
Liberal-National Coalition	39
Australian Labor Party	28
Australian Democrats Party	4
Greens	3
Family First	1
Unallocated	1
Total	76

* The election was for 36 of the 72 seats held by state senators and for all four senators representing the Northern Territory and the Australian Capital Territory (See The Constitution).

House of Representatives

Speaker: to be appointed.

Election, 9 October 2004 (provisional results)

Party	Seats
Liberal Party of Australia	72
Australian Labor Party	57
National Party of Australia	12
Independents	3
Others	1
Unallocated	5
Total	150

State and Territory Governments

(September 2004)

NEW SOUTH WALES

Governor: MARIE BASHIR, Level 3, Chief Secretary's Bldg, 121 Macquarie St, Sydney, NSW 2000; tel. (2) 9242-4200; fax (2) 9242-4266; internet www.nsw.gov.au.

Premier: ROBERT (BOB) J. CARR (Labor), Level 40, Governor Macquarie Tower, 1 Farrer Place, Sydney, NSW 2000; tel. (2) 9228-5239; fax (2) 9228-3935; e-mail bob.carr@nsw.gov.au.

VICTORIA

Governor: JOHN LANDY, Government House, Melbourne, Vic 3004; tel. (3) 9655-4211; fax (3) 9654-8430; internet www.governor.vic.gov.au.

Premier: STEPHEN P. BRACKS (Labor), 1 Treasury Place, Vic 3000; internet www.premier.vic.gov.au.

QUEENSLAND

Governor: QUENTIN BRYCE, Government House, Brisbane, Qld 4001; tel. (7) 3858-5700; fax (7) 3858-5701; e-mail govhouse@govhouse.qld.gov.au; internet www.govhouse.qld.gov.au.

Premier: PETER D. BEATTIE (Labor), Executive Bldg, 100 George St, Brisbane, Qld 4002; tel. (7) 3224-2111; fax (7) 3229-2990; internet www.premiers.qld.gov.au.

SOUTH AUSTRALIA

Governor: MARJORIE JACKSON-NELSON, Government House, North Terrace, Adelaide, SA 5000; tel. (8) 8203-9800; fax (8) 8203-9899; e-mail govthsesa@saugov.sa.gov.au; internet www.sa.gov.au.

Premier: MICHAEL (MIKE) RANN (Labor), 200 Victoria Square, Adelaide, SA 5001; tel. and fax (8) 8463-3166; e-mail premier@saugov.sa.gov.au; internet www.ministers.sa.gov.au.

WESTERN AUSTRALIA

Governor: Lt-Gen. JOHN MURRAY SANDERSON, Government House, Perth, WA 6000; tel. (8) 9429-9199; fax (8) 9325-4476; e-mail enquiries@govhouse.wa.gov.au; internet www.wa.gov.au.

Premier: GEOFFREY GALLOP (Labor), 24th Floor, 197 St George's Terrace, Perth, WA 6000; tel. (8) 9222-9888; fax (8) 9322-1213; e-mail wa-government@mpc.wa.gov.au; internet www.premier.wa.gov.au.

TASMANIA

Governor: WILLIAM JOHN ELLIS COX (acting), Government House, Hobart, Tas 7000; tel. (3) 6234-2611; fax (3) 6234-2556; internet www.tas.gov.au.

Premier: PAUL LENNON (Labor), GPOB 123B, Hobart, Tas 7001; tel. (3) 6233-3464; fax (3) 6234-1572; e-mail premier@dpac.tas.gov.au; internet www.premier.tas.gov.au.

NORTHERN TERRITORY

Administrator: TED EGAN, GPOB 497, Darwin, NT 0801; tel. (8) 8999-7103; fax (8) 8981-5521; e-mail governmenthouse.darwin@nt.gov.au; internet www.nt.gov.au/administrator.

Chief Minister: CLARE MARTIN (Labor), GPOB 3146, Darwin, NT 0801; tel. (8) 8901-4000; fax (8) 8901-4099; e-mail chiefminister.nt@nt.gov.au; internet www.nt.gov.au/ministers.

AUSTRALIAN CAPITAL TERRITORY

Chief Minister: JONATHAN STANHOPE (Labor), Legislative Assembly Bldg, Civic Square, London Circuit, Canberra, ACT 2601; tel. (2) 6205-0104; fax (2) 6205-0399; e-mail canberraconnect@act.gov.au; internet www.act.gov.au.

Political Organizations

Australians for Constitutional Monarchy (ACM): GPOB 9841, Sydney, NSW 2001; tel. (2) 9231-2200; fax (2) 9231-2359; e-mail acmhq@norepublic.com.au; internet www.norepublic.com.au; f. 1992; also known as No Republic; Exec. Dir KERRY JONES.

Australian Democrats Party: Kingston, POB 1059, Canberra, ACT 2602; tel. (2) 6273-1059; e-mail inquiries@democrats.org.au; internet www.democrats.org.au; f. 1977; comprises the fmr Liberal Movement and the Australia Party; Fed. Parl. Leader Senator ANDREW BARTLETT; Nat. Pres. NINA BURRIDGE.

Australian Greens: GPOB 1108, Canberra, ACT 2601; e-mail frontdesk@greens.org.au; internet www.greens.org.au; f. 1992; Leader Senator BOB BROWN.

Australian Labor Party (ALP): Centenary House, 19 National Circuit, Barton, ACT 2600; tel. (2) 6120-0800; fax (2) 6120-0801; e-mail info@alp.com; internet www.alp.com; f. 1891; advocates social democracy; trade unions form part of its structure; Fed. Parl. Leader MARK LATHAM; Nat. Pres. CARMEN LAWRENCE; Nat. Sec. TIM GARTRELL.

Australian Republican Movement (ARM): POB A870, Sydney South, NSW 1235; tel. (2) 9267-8022; fax (2) 9267-8155; e-mail republic@republic.org.au; internet www.republic.org.au; Chair. Prof. JOHN WARHURST; Nat. Dir ALLISON HENRY.

Communist Party of Australia: 65 Campbell St, Surry Hills, NSW 2010; tel. (2) 9212-6855; fax (2) 9281-5795; e-mail cpa@cpa.org.au; internet www.cpa.org.au; f. 1971; fmrly Socialist Party; advocates public ownership of the means of production, working-class political power; Pres. Dr H. MIDDLETON; Gen. Sec. P. SYMON.

Family First: c/o The Senate, Canberra; supports family values.

Liberal Party of Australia: Federal Secretariat, Cnr Blackall and Macquarie Sts, Barton, ACT 2600; tel. (2) 6273-2564; fax (2) 6273-1534; e-mail libadm@liberal.org.au; internet www.liberal.org.au; f. 1944; advocates private enterprise, social justice, individual liberty and initiative; committed to national development, prosperity and security; Fed. Dir BRIAN LOUGHNANE; Fed. Parl. Leader JOHN HOWARD.

National Party of Australia: John McEwen House, National Circuit, Barton, ACT 2600; tel. (2) 6273-3822; fax (2) 6273-1745; e-mail federal@nationalparty.org; internet www.nationalparty.org; f. 1916 as the Country Party of Australia; adopted present name in 1982; advocates balanced national development based on free enterprise, with special emphasis on the needs of people outside the major metropolitan areas; Fed. Pres. HELEN DICKIE; Fed. Parl. Leader JOHN ANDERSON; Fed. Dir ANDREW HALL.

One Nation: GPOB 812, Ipswich, Qld 4304; e-mail pauline@onenation.com.au; internet www.onenation.com.au; f. 1997; opposes globalization, high immigration and special funding for Aboriginal people, advocates public ownership of major services; Pres. (vacant); Vice-Pres. JOHN FISHER.

Your Voice: internet www.yourvoice.org.au; f. 2004; represents the interests of Aboriginal people; Co-founder RICHARD FRANKLAND.

Diplomatic Representation

EMBASSIES AND HIGH COMMISSIONS IN AUSTRALIA

Afghanistan: POB 155, Deakin West, ACT 2600; tel. (2) 6282-7311; fax (2) 6282-7322; e-mail ambassador@afghanembassy.net; internet www.afghanembassy.net; Ambassador MAHMOUD SAIKAL.

Argentina: POB 4835, Kingston, ACT 2604; tel. (2) 6273-9111; fax (2) 6273-0500; e-mail info@argentina.org.au; internet www.argentina.org.au; Ambassador NÉSTOR E. STANCANELLI.

Austria: POB 3375, Manuka, ACT 2603; tel. (2) 6295-1533; fax (2) 6239-6751; e-mail austria@bigpond.net.au; internet www.austriaemb.org.au/; Ambassador Dr OTMAR KOLER.

Bangladesh: POB 5, Red Hill, ACT 2603; tel. (2) 6290-0511; fax (2) 6290-0544; e-mail bdoot.canberra@cyberone.com.au; internet www.bangladoot-canberra.org; High Commissioner Lt-Gen. (retd) M. HARUN-AR-RASHID.

Belgium: 19 Arkana St, Yarralumla, ACT 2600; tel. (2) 6273-2501; fax (2) 6273-3392; e-mail canberra@diplobel.org; Ambassador LUK DARRAS.

Bosnia and Herzegovina: 5 Beale Crescent, Deakin, ACT 2600; tel. (2) 6232-4646; fax (2) 6232-5554; e-mail embaucbr@webone.com.au; internet www.bosnia.webone.com.au; Ambassador Dr ZDRAVKO TODOROVIĆ.

Botswana: Canberra; High Commissioner MOLOSIWA SELEPENG.

Brazil: GPOB 1540, Canberra, ACT 2601; tel. (2) 6273-2372; fax (2) 6273-2375; e-mail brazil@connect.net.au; internet brazil.org.au; Ambassador FREDERICO CEZAR DE ARAUJO.

Brunei: 10 Beale Crescent, Deakin, ACT 2603; tel. (2) 6285-4500; fax (2) 6285-4545; High Commissioner Haji ZAKARIA Haji AHMAD (acting).

Cambodia: 5 Canterbury Crescent, Deakin, ACT 2600; tel. (2) 6273-1259; fax (2) 6273-1053; e-mail cambodianembassy@ozemail.net.au; internet www.embassyofcambodia.org.nz/au.htm; Ambassador HOR NAMBORA.

Canada: Commonwealth Ave, Canberra, ACT 2600; tel. (2) 6270-4000; fax (2) 6273-3285; e-mail cnbra@international.gc.ca; internet www.canada.org.au; High Commissioner (vacant).

Chile: POB 69, Red Hill, ACT 2603; tel. (2) 6286-2430; fax (2) 6286-1289; e-mail chilemb@embachile-australia.com; internet www.embachile-australia.com; Ambassador CRISTÓBAL VALDÉS.

China, People's Republic: 15 Coronation Drive, Yarralumla, ACT 2600; tel. (2) 6273-4780; fax (2) 6273-4878; Ambassador WU TAO.

Colombia: GPOB 2892, Canberra City, ACT 2601; tel. (2) 6257-2027; fax (2) 6257-1448; e-mail emaustralia@iprimus.com.au; internet www.embacol.org.au; Ambassador JUAN SANTIAGO URIBE.

Croatia: 14 Jindalee Crescent, O'Malley, ACT 2606; tel. (2) 6286-6988; fax (2) 6286-3544; e-mail croemb@bigpond.net.au; Ambassador Dr MLADEN IBLER.

Cyprus: 30 Beale Crescent, Deakin, ACT 2600; tel. (2) 6281-0832; fax (2) 6281-0860; e-mail cyphicom@bigpond.net.au; High Commissioner ACHILLEAS ANTONIADES.

Czech Republic: 8 Culgoa Circuit, O'Malley, ACT 2606; tel. (2) 6290-1386; fax (2) 6290-0006; e-mail canberra@embassy.mzv.cz; internet www.mzv.cz/canberra; Chargé d'affaires a. i. ADAM PINOS.

Denmark: 15 Hunter St, Yarralumla, ACT 2600; tel. (2) 6273-2196; fax (2) 6273-3864; e-mail dkembact@dynamite.com.au; Ambassador JENS OSTENFELD.

Ecuador: 6 Pindari Crescent, O'Malley, ACT 2606; tel. (2) 6286-4021; fax (2) 6286-1231; e-mail embecu@bigpond.net.au; Chargé d'affaires Dr CUTY GARCÍA.

Egypt: 1 Darwin Ave, Yarralumla, ACT 2600; tel. (2) 6273-4437; fax (2) 6273-4279; Ambassador ASSEM AHMED MEGAHED.

Fiji: POB 159, Deakin West, ACT 2600; tel. (2) 6260-5115; fax (2) 6260-5105; e-mail fhc@cyberone.com.au; High Commissioner JIOJI KONROTE.

Finland: 12 Darwin Ave, Yarralumla, ACT 2600; tel. (2) 6273-3800; fax (2) 6273-3603; e-mail finland@austarmetro.com.au; internet www.finland.org.au; Ambassador ANNELI PUURA-MÄRKÄLÄ.

France: 6 Perth Ave, Yarralumla, ACT 2600; tel. (2) 6216-0100; fax (2) 6216-0127; e-mail ambassade@ambafrance-au.org; internet www.ambafrance-au.org; Ambassador PATRICK HÉNAULT.

Germany: 119 Empire Circuit, Yarralumla, ACT 2600; tel. (2) 6270-1911; fax (2) 6270-1951; e-mail info@germanembassy.org.au; internet www.germanembassy-canberra.com; Ambassador Dr KLAUS-PETER KLAIBER.

Greece: 9 Turrana St, Yarralumla, ACT 2600; tel. (2) 6273-3011; fax (2) 6273-2620; e-mail greekemb@greekembassy-au.org; Ambassador FOTIOS-JEAN XYDAS.

Holy See: POB 3633, Manuka, ACT 2603 (Apostolic Nunciature); tel. (2) 6295-3876; Apostolic Nuncio Most Rev. FRANCESCO CANALINI (Titular Archbishop of Valeria).

Hungary: 17 Beale Crescent, Deakin, ACT 2600; tel. (2) 6282-3226; fax (2) 6285-3012; e-mail hungcbr@ozemail.com.au; internet www.hunconsydney.com; Ambassador Gen. LAJOS FODOR.

India: 3–5 Moonah Place, Yarralumla, ACT 2600; tel. (2) 6273-3999; fax (2) 6273-1308; e-mail hciisi@bigpond.com.au; internet www.hcindia-au.org; High Commissioner P. P. SHUKLA.

Indonesia: 8 Darwin Ave, Yarralumla, ACT 2600; tel. (2) 6273-8600; fax (2) 6250-6017; e-mail embindo@cyberone.com.au; Ambassador SUDJADNAN PARNOHADININGRAT.

Iran: POB 705, Mawson, ACT 2607; tel. (2) 6290-2427; fax (2) 6290-2431; e-mail general@embassyiran.org.au; internet www.embassyiran.org.au; Ambassador Dr HAMID ABOUTALEBI.

Ireland: 20 Arkana St, Yarralumla, ACT 2600; tel. (2) 6273-3022; fax (2) 6273-3741; e-mail irishemb@cyberone.com.au; Ambassador DECLAN M. KELLY.

Israel: 6 Turrana St, Yarralumla, ACT 2600; tel. (2) 6273-1309; tel. 62224; fax (2) 6273-4273; e-mail israelembassy@israemb.org; internet www.mfa.gov.il; Ambassador GABBY LEVY.

Italy: 12 Grey St, Deakin, ACT 2600; tel. (2) 6273-3333; fax (2) 6273-4223; e-mail embassy@ambitalia.org.au; internet www.ambitalia.org.au; Ambassador DINO VOLPICELLI.

Japan: 112 Empire Circuit, Yarralumla, ACT 2600; tel. (2) 6273-3244; fax (2) 6273-1848; e-mail embofjpn@ozemail.com.au; internet www.japan.org.au; Ambassador ATSUSHI HATAKANAKA.

Jordan: 20 Roebuck St, Red Hill, ACT 2603; tel. (2) 6295-9951; fax (2) 6239-7236; Ambassador SAMIR MASARWEH.

Kenya: GPOB 1990, Canberra, ACT 2601; tel. (2) 6247-4788; fax (2) 6257-6613; e-mail kenrep@austarmetro.com.au; High Commissioner JOHN L. LANYASUNYA.

Korea, Democratic People's Republic: Canberra; Ambassador CHON JAE HONG.

Korea, Republic: 113 Empire Circuit, Yarralumla, ACT 2600; tel. (2) 6270-4100; fax (2) 6273-4839; e-mail embassy-au@mofat.go.kr; internet www.korea.org.au; Ambassador CHO SANG-HOON.

Laos: 1 Dalman Crescent, O'Malley, ACT 2606; tel. (2) 6286-4595; fax (2) 6290-1910; e-mail clao@cyberone.com.au; Ambassador VICHIT XINDAVONG.

Lebanon: 27 Endeavour St, Red Hill, ACT 2603; tel. (2) 6295-7378; fax (2) 6239-7024; e-mail lebanemb@tpg.com.au; internet www.lebanemb.org.au; Ambassador MICHEL BITAR.

Malaysia: 7 Perth Ave, Yarralumla, ACT 2600; tel. (2) 6273-1544; fax (2) 6273-2496; e-mail malcnbera@kln.gov.my; High Commissioner M. H. ARSHAD.

Malta: 38 Culgoa Circuit, O'Malley, ACT 2606; tel. (2) 6290-1724; fax (2) 6290-2453; e-mail maltahc@bigpond.com; High Commissioner Dr IVAN FSADNI.

Mauritius: 2 Beale Crescent, Deakin, ACT 2600; tel. (2) 6281-1203; fax (2) 6282-3235; e-mail mhccan@cyberone.com.au; High Commissioner PATRICE CURÉ.

Mexico: 14 Perth Ave, Yarralumla, ACT 2600; tel. (2) 6273-3963; fax (2) 6273-1190; e-mail embamex@mexico.org.au; internet www.embassyofmexicoinaustralia.org; Ambassador MARTHA ORTHIS DE ROSAS.

Myanmar: 22 Arkana St, Yarralumla, ACT 2600; tel. (2) 6273-3811; fax (2) 6273-4357; Ambassador U SOE WIN.

Netherlands: 120 Empire Circuit, Yarralumla, ACT 2600; tel. (2) 6220-9400; fax (2) 6273-3206; e-mail can@minbuza.nl; internet www.netherlands.org.au; Ambassador H. H. M. SONDAAL.

New Zealand: Commonwealth Ave, Canberra, ACT 2600; tel. (2) 6270-4211; fax (2) 6273-3194; e-mail nzhccba@austarmetro.com.au; internet www.nzembassy.com/australia; High Commissioner KATE LACKEY.

Nigeria: POB 241, Civic Square, ACT 2608; tel. (2) 6282-7411; fax (2) 6282-8471; e-mail chancery@nigeria-can.org.au; internet www.nigeria-can.org.au; High Commissioner Dr RUFAI A. O. SOULE.

Norway: 17 Hunter St, Yarralumla, ACT 2600; tel. (2) 6273-3444; fax (2) 6273-3669; e-mail emb.canberra@mfa.no; internet www.canberra.mfa.no; Ambassador OVE THORSHEIM.

Pakistan: POB 684, Mawson, ACT 2607; tel. (2) 6290-1676; fax (2) 6290-1073; e-mail parepcanberra@actonline.com.au; High Commissioner KHIZAR HAYAT KHAN NIAZI.

Papua New Guinea: POB E432, Queen Victoria Terrace, Kingston, ACT 2604; tel. (2) 6273-3322; fax (2) 6273-3732; High Commissioner RENAGI R. LOHIA.

Peru: POB 106, Red Hill, ACT 2603; tel. (2) 6286-9507; fax (2) 6286-7490; e-mail embassy@embaperu.org.au; internet www.embaperu.org.au; Ambassador MARTHA TOLEDO-OCAMPO.

Philippines: 1 Moonah Place, Yarralumla, Canberra, ACT 2600; tel. (2) 6273-2535; fax (2) 6273-3984; e-mail cbrpe@philembassy.au.com; internet www.philembassy.au.com; Ambassador CRISTINA GARCIA ORTEGA.

Poland: 7 Turrana St, Yarralumla, ACT 2600; tel. (2) 6273-1208; fax (2) 6273-3184; e-mail polish.embassy@tpg.com.au; internet www.poland.org.au; Ambassador JERZY WIECŁAW.

Portugal: 23 Culgoa Circuit, O'Malley, ACT 2606; tel. (2) 6290-1733; fax (2) 6290-1957; e-mail embport@internode.on.net; Ambassador Dr JOSÉ VIEIRA BRANCO.

Romania: 4 Dalman Crescent, O'Malley, ACT 2606; tel. (2) 6286-2343; fax (2) 6286-2433; e-mail roembcbr@cyberone.com.au; internet www.roembau.org.au; Ambassador MANUELA VULPE.

Russia: 78 Canberra Ave, Griffith, ACT 2603; tel. (2) 6295-9033; fax (2) 6295-1847; e-mail rusembassy@lightningpl.net.au; Ambassador LEONID MOISEEV.

Samoa: POB 3274, Manuka, ACT 2603; tel. (2) 6286-5505; fax (2) 6286-5678; e-mail samoahcaussi@netspeed.com.au; High Commissioner Leiataua Dr KILIFOTI S. ETEUATI.

Saudi Arabia: POB 63, Garran, ACT 2605; tel. (2) 6282-6999; fax (2) 6282-8911; e-mail saudiemb@hotmail.com; Chargé d'affaires a.i. ABDULLAH A. AL-QORDI.

Serbia and Montenegro: POB 728, Mawson, ACT 2607; tel. (2) 6290-2630; fax (2) 6290-2631; e-mail yuembau@ozemail.com.au; Ambassador MILIVOJE GLISIC.

Singapore: 17 Forster Crescent, Yarralumla, ACT 2600; tel. (2) 6273-3944; fax (2) 6273-9823; e-mail singapore_hc@bigpond.com; internet www.mfa.gov.sg/canberra; High Commissioner JOSEPH K. H. KOH.

Slovakia: 47 Culgoa Circuit, O'Malley, ACT 2606; tel. (2) 6290-1516; fax (2) 6290-1755; e-mail slovak@cyberone.com.au; internet www.slovakemb-aust.org; Ambassador Dr ANNA TURENIČOVÁ.

Slovenia: POB 284, Civic Square, Canberra, ACT 2608; tel. (2) 6243-4830; fax (2) 6243-4827; e-mail vca@mzz-dkp.gov.si; internet slovenia.webone.com.au; Chargé d'affaires BOJAN BERTONCELJ.

Solomon Islands: POB 256, Deakin West, ACT 2600; tel. (2) 6282-7030; fax (2) 6282-7040; e-mail info@solomon.emb.gov.au; High Commissioner MILNER TOZAKA.

South Africa: cnr State Circle and Rhodes Place, Yarralumla, ACT 2600; tel. (2) 6273-2424; fax (2) 6273-3543; e-mail info@rsa.emb.gov.au; internet www.rsa.emb.gov.au; High Commissioner ANTHONY MONGALO.

Spain: POB 9076, Deakin, ACT 2600; tel. (2) 6273-3555; fax (2) 6273-3918; e-mail embespau@mail.mae.es; internet www.embaspain.com; Ambassador JOSÉ RAMÓN BARANAÑO.

Sri Lanka: 35 Empire Circuit, Forrest, ACT 2603; tel. (2) 6239-7041; fax (2) 6239-6166; e-mail slhc@atrax.net.au; internet slhccanberra.webjump.com; High Commissioner Maj.-Gen. JANAKA PERERA.

Sweden: 5 Turrana St, Yarralumla, ACT 2600; tel. (2) 6270-2700; fax (2) 6270-2755; e-mail sweden@austarmetro.com.au; Ambassador KARIN EHNBOM-PALMQUIST.

Switzerland: 7 Melbourne Ave, Forrest, ACT 2603; tel. (2) 6273-3977; fax (2) 6273-3428; e-mail vertretung@can.rep.admin.ch; Ambassador ANDRÉ FAIVET.

Thailand: 111 Empire Circuit, Yarralumla, ACT 2600; tel. (2) 6273-1149; fax (2) 6273-1518; e-mail rtecanberra@mfa.go.th; Ambassador SAWANIT KONGSIRI.

Timor-Leste: Canberra; Ambassador JORGE TEME.

Turkey: 60 Mugga Way, Red Hill, ACT 2603; tel. (2) 6295-0227; fax (2) 6239-6592; e-mail turkembs@bigpond.net.au; internet www.turkishembassy.org.au; Ambassador TANSU OKANDAN.

United Arab Emirates: 36 Culgoa Circuit, O'Malley, ACT 2606; tel. (2) 6286-8802; fax (2) 6286-8804; e-mail uaeembassy@bigpond.com; internet www.users.bigpond.com/uaeembassy; Ambassador Khalifa Mohammed Bakhit al-Falasi.

United Kingdom: Commonwealth Ave, Canberra, ACT 2600; tel. (2) 6270-6666; fax (2) 6273-3236; e-mail BHC.Canberra@uk.emb.gov.au; internet www.uk.emb.gov.au; High Commissioner Sir Alastair Goodlad.

USA: Moonah Place, Yarralumla, ACT 2600; tel. (2) 6214-5600; fax (2) 6214-5970; internet usembassy-australia.state.gov; Ambassador J. Thomas Schieffer.

Uruguay: POB 5058, Kingston, ACT 2604; tel. (2) 6273-9100; fax (2) 6273-9099; e-mail urucan@austarmetro.com.au; Ambassador Pedro Mó Amaro.

Venezuela: 5 Culgoa Circuit, O'Malley, ACT 2606; tel. (2) 6290-2968; fax (2) 6290-2911; e-mail embaustralia@venezuela-emb-org.au; internet www.venezuela-emb.org.au; Ambassador Lionel Vivas.

Viet Nam: 6 Timbarra Crescent, O'Malley, ACT 2606; tel. (2) 6286-6059; fax (2) 6286-4534; e-mail vnembassy@webone.com.au; internet www.au.vnembassy.org; Ambassador Vu Chi Cong.

Zimbabwe: 11 Culgoa Circuit, O'Malley, ACT 2606; tel. (2) 6286-2700; fax (2) 6290-1680; e-mail zimbabwe1@austarmetro.com.au; Ambassador Florence L. Chitauro.

Judicial System

The judicial power of the Commonwealth of Australia is vested in the High Court of Australia, in such other Federal Courts as the Federal Parliament creates, and in such other courts as it invests with Federal jurisdiction.

The High Court consists of a Chief Justice and six other Justices, each of whom is appointed by the Governor-General in Council, and has both original and appellate jurisdiction.

The High Court's original jurisdiction extends to all matters arising under any treaty, affecting representatives of other countries, in which the Commonwealth of Australia or its representative is a party, between states or between residents of different states or between a state and a resident of another state, and in which a writ of mandamus, or prohibition, or an injunction is sought against an officer of the Commonwealth of Australia. It also extends to matters arising under the Australian Constitution or involving its interpretation, and to many matters arising under Commonwealth laws.

The High Court's appellate jurisdiction has, since June 1984, been discretionary. Appeals from the Federal Court, the Family Court and the Supreme Courts of the states and of the territories may now be brought only if special leave is granted, in the event of a legal question that is of general public importance being involved, or of there being differences of opinion between intermediate appellate courts as to the state of the law.

Legislation enacted by the Federal Parliament in 1976 substantially changed the exercise of Federal and Territory judicial power, and, by creating the Federal Court of Australia in February 1977, enabled the High Court of Australia to give greater attention to its primary function as interpreter of the Australian Constitution. The Federal Court of Australia has assumed, in two divisions, the jurisdiction previously exercised by the Australian Industrial Court and the Federal Court of Bankruptcy and was additionally given jurisdiction in trade practices and in the developing field of administrative law. In 1987 the Federal Court of Australia acquired jurisdiction in federal taxation matters and certain intellectual property matters. In 1991 the Court's jurisdiction was expanded to include civil proceedings arising under Corporations Law. Jurisdiction has also been conferred on the Federal Court of Australia, subject to a number of exceptions, in matters in which a writ of mandamus, or prohibition, or an injunction is sought against an officer of the Commonwealth of Australia. The Court also hears appeals from the Court constituted by a single Judge, from the Supreme Courts of the territories, and in certain specific matters from State Courts, other than a Full Court of the Supreme Court of a state, exercising Federal jurisdiction.

In March 1986 all remaining categories of appeal from Australian courts to the Queen's Privy Council in the United Kingdom were abolished by the Australia Act.

FEDERAL COURTS

High Court of Australia

POB 6309, Kingston, Canberra, ACT 2604; tel. (2) 6270-6811; fax (2) 6270-6868; internet www.hcourt.gov.au.

Chief Justice: Anthony Murray Gleeson.

Justices: Michael Hudson McHugh, William Montague Charles Gummow, Michael Donald Kirby, Kenneth Madison Hayne, Ian David Francis Callinan, John Dyson Heydon.

Federal Court of Australia

Chief Justice: Michael Eric John Black.

In 2003 there were 46 other judges.

Family Court of Australia

Chief Justice: Alistair Bothwick Nicholson.

In 2003 there were 50 other judges.

NEW SOUTH WALES

Supreme Court

Chief Justice: James Jacob Spigelman.

President: Keith Mason.

Chief Judge in Equity: Peter Wolstenhome Young.

Chief Judge at Common Law: James Roland Tomson Wood.

VICTORIA

Supreme Court

Chief Justice: Marilyn Warren.

President of the Court of Appeal: John Spence Winneke.

QUEENSLAND

Supreme Court

Chief Justice: Paul de Jersey.

President of the Court of Appeal: Margaret McMurdo.

Senior Judge Administrator, Trial Division: Martin Patrick Moynihan.

Central District (Rockhampton)

Resident Judge: Peter Richard Dutney.

Northern District (Townsville)

Resident Judge: Keiran Anthony Cullinane.

Far Northern District (Cairns)

Resident Judge: Stanley Graham Jones.

SOUTH AUSTRALIA

Supreme Court

Chief Justice: John Jeremy Doyle.

WESTERN AUSTRALIA

Supreme Court

Chief Justice: David Kingsley Malcolm.

TASMANIA

Supreme Court

Chief Justice: William John Ellis Cox.

AUSTRALIAN CAPITAL TERRITORY

Supreme Court

Chief Justice: John Higgins.

NORTHERN TERRITORY

Supreme Court

Chief Justice: Brian Frank Martin.

Religion

CHRISTIANITY

According to the provisional results of the population census of August 2001, Christians numbered 12,764,342.

National Council of Churches in Australia: Locked Bag 199, Sydney, NSW 1230; tel. (2) 9299-2215; fax (2) 9262-4514; e-mail info@ncca.org.au; internet www.ncca.org.au; f. 1946; 16 mem. churches; Pres. Rev. Prof. JAMES HAIRE; Gen. Sec. Rev. JOHN HENDERSON.

The Anglican Communion

The constitution of the Church of England in Australia, which rendered the church an autonomous member of the Anglican Communion, came into force in January 1962. The body was renamed the Anglican Church of Australia in August 1981. The Church comprises five provinces (together containing 22 dioceses) and the extra-provincial diocese of Tasmania. At the 2001 population census there were an estimated 3,881,162 adherents.

National Office of the Anglican Church

General Synod Office, Box Q190, QVB PO, Sydney, NSW 1230; tel. (2) 9265-1525; fax (2) 9264-6552; e-mail gsoffice@anglican.org.au; internet www.anglican.org.au.

Gen. Sec. Rev. Dr B. N. KAYE.

Archbishop of Adelaide and Metropolitan of South Australia: Ven. JOHN COLLAS (acting), Bishop's Court, 45 Palmer Place, North Adelaide, SA 5006; fax (8) 8305-9399; internet www.anglicare-sa.org.au.

Archbishop of Brisbane and Metropolitan of Queensland: Most Rev. Dr PHILLIP JOHN ASPINALL, Bishopsbourne, GPOB 421, Brisbane, Qld 4001; tel. (7) 3835-2218; fax (7) 3832-5030; e-mail archbishop@anglicanbrisbane.org.au; internet www.anglicanbrisbane.gil.com.au.

Archbishop of Melbourne and Metropolitan of Victoria: Most Rev. PETER R. WATSON, Bishopscourt, 120 Clarendon St, East Melbourne, Vic 3002; tel. (3) 9653-4220; fax (3) 9650-2184; e-mail archbishop@melbourne.anglican.com.au.

Archbishop of Perth and Metropolitan of Western Australia, Primate of Australia: Most Rev. Dr PETER F. CARNLEY, GPOB W2067, Perth, WA 6846; tel. (8) 9325-7455; fax (8) 9325-6741; e-mail abcsuite@perth.anglican.org; internet www.perth.anglican.org.

Also has jurisdiction over Christmas Island and the Cocos (Keeling) Islands.

Archbishop of Sydney and Metropolitan of New South Wales: Most Rev. Dr PETER F. JENSEN, POB Q190, QVB PO, Sydney, NSW 1230; tel. (2) 9265-1521; fax (2) 9265-1504; e-mail archbishop@sydney.anglican.asn.au; internet www.sydney.anglican.asn.au.

The Roman Catholic Church

Australia comprises five metropolitan archdioceses, two archdioceses directly responsible to the Holy See and 24 dioceses, including one diocese each for Catholics of the Maronite and Melkite rites, and one military ordinariate. At 31 December 2002 there were an estimated 5.2m. adherents in the country.

Australian Catholic Bishops' Conference

GPOB 368, Canberra, ACT 2601; tel. (2) 6201-9845; fax (2) 6247-6083; e-mail gensec@catholic.org.au; internet www.catholic.org.au.

f. 1979; Pres. Most Rev. FRANCIS PATRICK CARROLL (Archbishop of Canberra and Goulburn); Sec. Rev. BRIAN LUCAS.

Archbishop of Adelaide: Most Rev. PHILIP WILSON, GPOB 1364, Adelaide, South Australia 5001; tel. (8) 8210-8108; fax (8) 8223-2307; e-mail cco@adelaide.catholic.org.au.

Archbishop of Brisbane: Most Rev. JOHN A. BATHERSBY, Archbishop's House, 790 Brunswick St, New Farm, Brisbane, Qld 4005; tel. (7) 3224-3364; fax (7) 3358-1357; e-mail archbishop@bne.catholic.net.au.

Archbishop of Canberra and Goulburn: Most Rev. FRANCIS PATRICK CARROLL, GPOB 89, Canberra, ACT 2601; tel. (2) 6248-6411; fax (2) 6247-9636; e-mail archbish@cangoul.catholic.org.au.

Archbishop of Hobart: Most Rev. ADRIAN DOYLE, GPOB 62, Hobart, Tas 7001; tel. (3) 6225-1920; fax (3) 6225-3865; e-mail archbishop.hobart@cdftas.com; internet www.hobart.catholic.org.au.

Archbishop of Melbourne: Most Rev. DENIS HART, GPOB 146, East Melbourne, Vic 3002; tel. (3) 9926-5677; fax (3) 9926-5698; e-mail archbishop@melbourne.catholic.org.au.

Archbishop of Perth: Most Rev. BARRY J. HICKEY, Catholic Church Office, 25 Victoria Ave, Perth, WA 6000; tel. (8) 9223-1351; fax (8) 9221-1716; e-mail archsec@perth.catholic.org.au; internet www.perth.catholic.org.au.

Archbishop of Sydney: Cardinal GEORGE PELL, St Mary's Cathedral, Sydney, NSW 2000; tel. (2) 9390-5100; fax (2) 9261-0045; e-mail chancery@ado.syd.catholic.org.au.

Orthodox Churches

Greek Orthodox Archdiocese of Australia: 242 Cleveland St, Redfern, Sydney, NSW 2016; tel. (2) 9698-5066; fax (2) 9698-5368; f. 1924; 700,000 mems; Primate His Eminence Archbishop STYLIANOS.

The Antiochian, Coptic, Romanian, Serbian and Syrian Orthodox Churches are also represented.

Other Christian Churches

Baptist Union of Australia: POB 377, Hawthorn, Vic 3122; tel. (3) 9810-7800; fax (3) 9810-1041; e-mail bua@baptist.org.au; internet www.baptist.org.au; f. 1926; 64,159 mems; 883 churches; Nat. Pres. GWYN MILNE; Nat. Sec. NAOMI McCONNELL.

Churches of Christ in Australia: POB 55, Helensburgh, NSW 2508; tel. (2) 4294-1913; fax (2) 4294-1914; e-mail bobsmith@ozemail.com.au; internet www.churchesofchrist.org.au; 36,000 mems; Pres. Rev. PETER OVERTON; Co-ordinator Rev. ROBERT SMITH.

Lutheran Church of Australia: National Office, 197 Archer St, North Adelaide, SA 5006; tel. (8) 8267-7300; fax (8) 8267-7310; e-mail president@lca.org.au; internet www.lca.org.au; f. 1966; 75,000 mems; Pres. Rev. M. P. SEMMLER.

United Pentecostal Church of Australia: GPOB 1434, Springwood, Qld 4127; tel. (7) 3806-1817; fax (7) 3806-0029; e-mail homemissions@powerup.com.au; internet www.upca.org.au; 174,720 adherents in 1996.

Uniting Church in Australia: POB A2266, Sydney South, NSW 1235; tel. (2) 8267-4202; fax (2) 8267-4222; e-mail assysec@nat.uca.org.au; internet nat.uca.org.au; f. 1977 with the union of Methodist, Presbyterian and Congregational Churches; 1.4m. mems; Pres. Rev. Dr DEAN DRAYTON; Gen. Sec. Rev. TERENCE CORKIN; third largest Christian denomination in Australia.

JUDAISM

The Jewish community numbered an estimated 83,993 at the census of August 2001.

Great Synagogue: 166 Castlereagh St, Sydney, NSW; tel. (2) 9267-2477; fax (2) 9264-8871; e-mail admin@greatsynagogue.org.au; internet www.greatsynagogue.org.au; f. 1828; Sr Minister Sr Rabbi JEREMY LAWRENCE.

OTHER FAITHS

According to the August 2001 census, Buddhists numbered an estimated 357,813, Muslims 281,578 and Hindus 95,473.

The Press

The total circulation of Australia's daily newspapers is very high, but in the remoter parts of the country weekly papers are even more popular. Most of Australia's newspapers are published in sparsely populated rural areas where the demand for local news is strong. The only newspapers that may fairly claim a national circulation are the dailies *The Australian* and *Australian Financial Review*, and the weekly magazines *The Bulletin, Time Australia* and *Business Review Weekly*, the circulation of most newspapers being almost entirely confined to the state in which each is produced.

ACP Publishing Pty Ltd: 54–58 Park St, Sydney, NSW 1028; tel. (2) 9282-8000; fax (2) 9264-4541; internet www.acp.com.au; publishes *Australian Women's Weekly*, *The Bulletin with Newsweek*, *Cleo*, *Cosmopolitan*, *Woman's Day*, *Dolly*, *Belle*, *Wheels*, *Motor* and more than 60 other magazines; Chief Exec. JOHN ALEXANDER.

APN News and Media Ltd: 10th Floor, 300 Ann St, Brisbane, Qld 4000; tel. (7) 3307-0300; fax (7) 3307-0307; Chair. L. P. HEALY; Chief Exec. VINCENT CROWLEY.

John Fairfax Holdings Ltd: POB 506, Sydney, NSW 2001; tel. (2) 9282-2833; fax (2) 9282-3133; e-mail rvictor@mail.fairfax.com.au; internet www.fairfax.com.au; f. 1987; Chair. DEAN WILLS; Chief Exec. FREDERICK G. HILMER; publs include *The Sydney Morning Herald*, *The Australian Financial Review* and *Sun-Herald* (Sydney), *The Age* and *BRW Publications* (Melbourne); also provides online and interactive services.

The Herald and Weekly Times Ltd: POB 14999, Melbourne MC, Vic 8001; tel. (3) 9292-2000; fax (3) 9292-2002; e-mail newspapers@hwt.newsltd.com.au; internet www.heraldsun.com.au; acquired by News Ltd in 1987; Chair. JANET CALVERT-JONES; Man. Dir JULIAN CLARKE; publs include *Herald Sun*, *Sunday Herald Sun*, *The Weekly Times*, *MX*.

News Corporation: 2 Holt St, Surry Hills, Sydney, NSW 2010; tel. (2) 9288-3000; fax (2) 9288-2300; internet www.newscorp.com; Chair. and CEO K. RUPERT MURDOCH; controls *The Australian* and *The Weekend Australian* (national), *Daily Telegraph*, *Sunday Telegraph* (Sydney), *The Herald Sun* and *Sunday Herald Sun* (Victoria),

Northern Territory News (Darwin), *Sunday Times* (Perth), *Townsville Bulletin*, *Courier Mail*, *Sunday Mail* (Queensland), *The Mercury* (Tasmania), *The Advertiser*, *Sunday Mail* (South Australia).

Rural Press Ltd: 159 Bells Line of Road, North Richmond, NSW 2754; tel. (2) 4570-4444; fax (2) 4570-4663; e-mail cosec@rpl.com.au; internet www.rpl.com.au; Chair. John B. Fairfax; Man. Dir B. K. McCarthy.

West Australian Newspapers Holdings Ltd: Newspaper House, 50 Hasler Rd, Osborne Park, WA 6017; tel. (8) 9482-3111; fax (8) 9482-9080; Chair. W. G. Kent; Man. Dir I. F. Law.

Other newspaper publishers include Federal Capital Press (K. Stokes).

NEWSPAPERS

Australian Capital Territory

The Canberra Times: 9 Pirie St, Fyshwick, ACT 2609; POB 7155, Canberra Mail Centre, ACT 2610; tel. (2) 6280-2122; fax (2) 6280-2282; internet www.canberratimes.com; f. 1926; daily and Sun.; morning; Editor-in-Chief Jack Waterford; circ. 38,694 (Mon.–Fri.), 72,080 (Sat.), 39,075 (Sun.).

New South Wales

Dailies

The Australian: News Ltd, 2 Holt St, POB 4245, Surry Hills, NSW 2010; tel. (2) 9288-3000; fax (2) 9288-3077; f. 1964; edited in Sydney, simultaneous edns in Sydney, Melbourne, Perth, Townsville, Adelaide and Brisbane; Editor-in-Chief David Armstrong; Editor Campbell Reid; circ. 122,500 (Mon.–Fri.); *The Weekend Australian* (Sat.) 311,000.

Australian Financial Review: 201 Sussex St, GPOB 506, Sydney, NSW 2000; tel. (2) 9282-2512; fax (2) 9282-3137; f. 1951; Mon.–Fri.; distributed nationally; Publr/Editor-in-Chief Michael Gill; Editor Glen Burge; circ. 92,500 (Mon.–Fri.), 92,000 (Sat.).

The Daily Telegraph: 2 Holt St, Surry Hills, NSW 2010; tel. (2) 9288-3000; fax (2) 9288-2300; f. 1879; merged in 1990 with Daily Mirror (f. 1941); 24-hour tabloid; CEO Lachlan Murdoch; circ. 442,000.

The Manly Daily: 26 Sydney Rd, Manly, NSW 2095; tel. (2) 9977-3333; fax (2) 9977-2831; e-mail manlydailynews@cng.newsltd.com.au; f. 1906; Tue.–Sat.; Editor Steve Stickney; circ. 89,326.

The Newcastle Herald: 28–30 Bolton St, Newcastle, NSW 2300; tel. (2) 4979-5000; fax (2) 4979-5888; e-mail ntlinfo@newcastle.fairfax.com.au; f. 1858; morning; 6 a week; Editor Rod Quinn; circ. 54,006.

The Sydney Morning Herald: 201 Sussex St, GPOB 506, Sydney, NSW 2001; tel. (2) 9282-2822; fax (2) 9282-3253; internet www.smh.com.au; f. 1831; morning; Editor-in-Chief and Publr Alan Revell; circ. 231,508 (Mon.–Fri.), 400,000 (Sat.).

Weeklies

Bankstown Canterbury Torch: 47 Allingham St, Condell Park, NSW 2200; tel. (2) 9795-0000; fax (2) 9795-0096; e-mail trevor@torchpublishing.com.au; internet www.torchpublishing.com.au; f. 1920; Wed.; owned by Trevor Publishing Co; Editor Mark Kirkland; circ. 89,931.

Northern District Times: 79 Rowe St, Eastwood, NSW 2122; tel. (2) 9858-1766; fax (2) 9804-6901; f. 1921; Wed.; Editor D. Bartok; circ. 55,302.

The Parramatta Advertiser: 142 Macquarie St, Parramatta, NSW 2150; tel. (2) 9689-5370; fax (2) 9689-5353; Wed.; Editor Les Pobjie; circ. 96,809.

St George and Sutherland Shire Leader: 182 Forest Rd, Hurstville, NSW 2220; tel. (2) 9598-3999; fax (2) 9598-3987; f. 1960; Tue. and Thur.; Editor Peter Allen; circ. 143,595.

Sun-Herald: GPOB 506, Sydney, NSW 2001; tel. (2) 9282-1679; fax (2) 9282-2151; e-mail shnews@mail.fairfax.com.au; internet www.sunherald.com.au; f. 1953; Sun.; Editor Philip McLean; circ. 580,000.

Sunday Telegraph: 2 Holt St, Surry Hills, NSW 2010; tel. (2) 9288-3000; fax (2) 9288-3311; f. 1938; Editor Jeni Cooper; circ. 720,000.

Northern Territory

Daily

Northern Territory News: Printers Place, POB 1300, Darwin, NT 0801; tel. (8) 8944-9900; fax (8) 8981-6045; f. 1952; Mon.–Sat.; Gen. Man. D. Kennedy; circ. 24,470.

Weekly

Sunday Territorian: Printers Place, GPOB 1300, Darwin, NT 0801; tel. (8) 8944-9900; fax (8) 8981-6045; Sun.; Editor David Coren; circ. 26,437.

Queensland

Daily

Courier-Mail: 41 Campbell St, Bowen Hills, Brisbane, Qld 4006; tel. (7) 3252-6011; fax (7) 3252-6696; f. 1933; morning; Editor-in-Chief C. Mitchell; circ. 250,875.

Weekly

Sunday Mail: 41 Campbell St, Bowen Hills, GPOB 130, Brisbane, Qld 4006; tel. (7) 3666-8000; fax (7) 3666-6787; e-mail smletters@qnp.news; internet www.thesundaymail.com.au; f. 1953; Editor Michael Prain; circ. 611,298.

South Australia

Daily

The Advertiser: 121 King William St, Adelaide, SA 5001; tel. (8) 8206-2220; fax (8) 8206-3669; e-mail tiser@adv.newsltd.com.au; f. 1858; morning; Editor Melvin Mansell; circ. 203,440 (Mon.–Fri.), 274,045 (Sat.).

Weekly

Sunday Mail: 9th Floor, 121 King William St, Adelaide, SA 5000; tel. (8) 8206-2000; fax (8) 8206-3646; e-mail gardnerp@adv.newsltd.com.au; internet www.sundaymail.com.au; f. 1912; Editor Phillip Gardner; circ. 345,036.

Tasmania

Dailies

The Advocate: POB 63, Burnie, Tas 7320; tel. (3) 6440-7409; fax (3) 6440-7345; e-mail letters@theadvocate.com.au; internet www.theadvocate.com.au; f. 1890; morning; Editor Peter Dwyer; circ. 25,623.

Examiner: 71–75 Paterson St, POB 99, Launceston, Tas 7250; tel. (3) 6336-7111; fax (3) 6334-7328; e-mail mail@examiner.com.au; internet www.examiner.com.au/examiner/; f. 1842; morning; independent; Editor D. Southwell; circ. 38,721.

Mercury: 91–93 Macquarie St, Hobart, Tas 7000; tel. (3) 6230-0622; fax (3) 6230-0711; e-mail mercuryedletter@dbl.newsltd.com.au; internet www.news.com.au; f. 1854; morning; Man. Dir Rex Gardner; Editor G. Bailey; circ. 52,998.

Weeklies

Sunday Examiner: 71–75 Paterson St, Launceston, Tas 7250; tel. (3) 6336-7111; fax (3) 6334-7328; e-mail mail@examiner.com.au; f. 1924; Editor D. Southwell; circ. 42,000.

Sunday Tasmanian: 91–93 Macquarie St, Hobart, Tas 7000; tel. (3) 6230-0622; fax (3) 6230-0711; e-mail e-mailsuntas.news@dbl.newsltd.com.au; f. 1984; morning; Man. Dir Rex Gardner; Editor Ian McCausland; circ. 58,325.

Victoria

Dailies

The Age: 250 Spencer St (cnr Lonsdale St), Melbourne, Vic 3000; tel. (3) 9600-4211; fax (3) 9601-2598; e-mail newsdesk@theage.fairfax.com.au; internet www.theage.com.au; f. 1854; independent; morning, incl. Sun.; Editor-in-Chief Michael Gawenda; circ. 197,000 (Mon.–Fri.), 308,000 (Sat.), 191,000 (Sun.).

Herald Sun: HWT Tower, 40 City Rd, Southbank, Vic 3006; tel. (3) 9292-2000; fax (3) 9292-2112; e-mail news@heraldsun.com.au; internet www.heraldsun.com.au; f. 1840; Editor-in-Chief Peter Blunden; circ. 548,764.

Weeklies

Progress Press: 360 Burwood Rd, Hawthorn, Vic 3122; tel. (3) 9818-0555; fax (3) 9818-0029; e-mail editor@ldr.newsltd.com.au; f. 1960; Tue.; Editor Lynne Kinsey; circ. 74,829.

Sunday Herald Sun: HWT Tower, 40 City Rd, Southbank, Vic 3006; tel. (3) 9292-2962; fax (3) 9292-2080; e-mail sundayhs@heraldsun.com.au; internet www.sundayheraldsun.com.au; f. 1991; Editor Alan Howe; circ. 600,000.

Western Australia

Daily

The West Australian: POB D162, Perth, WA 6001; tel. (8) 9482-3111; fax (8) 9482-3399; e-mail editor@wanews.com.au; f. 1833; morning; Editor BRIAN ROGERS; circ. 214,000 (Mon.–Fri.), 386,000 (Sat.).

Weekly

Sunday Times: 34–42 Stirling St, Perth, WA 6000; tel. (8) 9326-8326; fax (8) 9326-8316; e-mail editorial@sundaytimes.newsltd.com.au; f. 1897; Man. Dir DAVID MAGUIRE; Editor BRETT MCCARTHY; circ. 346,014.

PRINCIPAL PERIODICALS

Weeklies and Fortnightlies

The Bulletin: 54 Park St, Sydney, NSW 2000; tel. (2) 9282-8227; fax (2) 9267-4359; e-mail cclegg@acp.com.au; f. 1880; Wed.; Editor-in-Chief GARRY LINNELL; circ. 70,000.

Business Review Weekly: Level 2, 469 La Trobe St, Melbourne, Vic 3000; tel. (3) 9603-3888; fax (3) 9670-4328; f. 1981; Chair. and Editorial Dir ROBERT GOTTLIEBSEN; Editor ROSS GREENWOOD; circ. 75,166.

The Countryman: 219 St George's Terrace, Perth, WA 6000; GPOB D162, Perth 6001; tel. (8) 9482-3322; fax (8) 9482-3324; e-mail countryman@wanews.com.au; f. 1885; Thur.; farming; Editor GARY MCGAY; circ. 13,444.

The Medical Journal of Australia: Locked Bag 3030, Strawberry Hills, NSW 2012; tel. (2) 9562-6666; fax (2) 9562-6699; e-mail medjaust@ampco.com.au; internet www.mja.com.au; f. 1914; fortnightly; Editor Dr MARTIN VAN DER WEYDEN; circ. 27,318.

New Idea: 35–51 Mitchell St, McMahons Point, NSW 2060; tel. (2) 9464-3200; fax (2) 9464-3203; e-mail newidea@pacpubs.com.au; weekly; women's; Editor JENNI GILBERT; circ. 471,000.

News Weekly: POB 186, North Melbourne, Vic 3051; tel. (3) 9326-5757; fax (3) 9328-2877; e-mail freedom@connexus.net.au; f. 1943; publ. by National Civic Council; fortnightly; Sat.; political, social, educational and trade union affairs; Editor PETER WESTMORE; circ. 12,000.

People: 54 Park St, Sydney, NSW 2000; tel. (2) 9282-8743; fax (2) 9267-4365; e-mail sbutler-white@acp.com.au; weekly; Editor SIMON BUTLER-WHITE; circ. 70,000.

Picture: GPOB 5201, Sydney, NSW 2001; tel. (2) 9288-9686; fax (2) 9267-4372; e-mail picture@acp.com.au; weekly; men's; Editor TOM FOSTER; circ. 110,000.

Queensland Country Life: POB 586, Cleveland, Qld 4163; tel. (7) 3826-8200; fax (7) 3821-1236; f. 1935; Thur.; Editor MARK PHELPS; circ. 33,900.

Stock and Land: 10 Sydenham St, Moonee Ponds, Vic 3039; tel. (3) 9287-0900; fax (3) 9370-5622; e-mail stockandland@ruralpress.com; internet www.stockandland.com; f. 1914; weekly; agricultural and rural news; Editor JOHN CARSON; circ. 11,899.

That's Life!: 35–51 Mitchell St, McMahons Point, NSW 2060; tel. (2) 9464-3300; fax (2) 9464-3480; e-mail thatslife@pacpubs.com.au; f. 1994; weekly; features; Editor SANDY MCPHIE; circ. 465,500 (incl. New Zealand).

Time Australia Magazine: GPOB 3873, Sydney, NSW 2001; tel. (2) 9925-2646; fax (2) 9954-0828; e-mail time.letters@time.com.au; internet www.time.com.au; Editor STEVE WATERSON; circ. 111,000.

TV Week: 54 Park St, Sydney, NSW 2000; tel. (2) 9288-9611; fax (2) 9283-4849; e-mail tvweek@acp.com.au; internet www.tvweek.ninemsn.com.au; f. 1957; Wed.; colour national; Editor EMMA NOLAN; circ. 364,044.

The Weekly Times: POB 14999, Melbourne City MC, Vic 8001; tel. (3) 9292-2000; fax (3) 9292-2697; e-mail wtimes@theweeklytimes.com.au; internet www.theweeklytimes.com.au; f. 1869; farming, regional issues, country life; Wed.; Editor PETER FLAHERTY; circ. 78,900.

Woman's Day: 54–58 Park St, POB 5245, Sydney, NSW 1028; tel. (2) 9282-8000; fax (2) 9267-4360; e-mail Womansday@acp.com.au; weekly; circulates throughout Australia and NZ; Editor-in-Chief PHILIP BARKER; circ. 765,170.

Monthlies and Others

Architectural Product News: Architecture Media Pty Ltd, Level 3, 4 Princes St, Port Melbourne, Vic 3207; tel. (3) 9646-4760; fax (3) 9646-4918; e-mail apn@archmedia.com.au; internet www.architecturemedia.com; 6 a year; Editorial Dir SUE HARRIS; circ. 25,063.

Architecture Australia: Architecture Media Pty Ltd, Level 3, 4 Princes St, Port Melbourne, Vic 3207; tel. (3) 9646-4760; fax (3) 9646-4918; e-mail aa@archmedia.com.au; internet www.architecturemedia.com; f. 1904; 6 a year; Editor JUSTINE CLARK; circ. 14,999.

Artichoke: Architecture Media Pty Ltd, Level 3, 4 Princes St, Port Melbourne, Vic 3207; tel. (3) 9646-4760; fax (3) 9646-4918; e-mail artichoke@archmedia.com.au; internet www.architecturemedia.com; 4 a year; Editorial Dir SUE HARRIS; circ. 7,096.

Australian Hi-Fi: POB 5555, St Leonards, NSW 1590; tel. (2) 9901-6100; fax (2) 9901-6198; e-mail hifi@horwitz.com.au; f. 1970; every 2 months; consumer tests on hi-fi and home theatre equipment; Editor GREG BORROWMAN; circ. 11,800.

Australian Home Beautiful: 35–51 Mitchell St, McMahons Point, NSW 2060; tel. (2) 9464-3218; fax (2) 9464-3263; e-mail homebeaut@pacpubs.com.au; internet www.homebeautiful.com.au; f. 1925; monthly; Editor ANDREA JONES; circ. 83,808.

Australian House and Garden: 54 Park St, Sydney, NSW 2000; tel. (2) 9282-8456; fax (2) 9267-4912; e-mail h&g@acp.com.au; f. 1948; monthly; design, decorating, renovating, gardens, food and travel; Editor ANNY FRIIS; circ. 110,000.

Australian Journal of Mining: Informa Australia Pty Ltd, Level 2, 120 Sussex St, Sydney, NSW 2000; e-mail charles.macdonald@informa.com.au; internet www.theajmonline.com; f. 1986; bi-monthly; mining and exploration throughout Australia and South Pacific; Editor CHARLES MCDONALD; circ. 6,771.

Australian Journal of Pharmacy: Suite F2, 1–15 Barr St, Balmain, NSW 2041; tel. (2) 9818-7800; fax (2) 9818-7811; internet appco.com.au; f. 1886; monthly; journal of the associated pharmaceutical orgs; Publ. DAVID WESTON; circ. 6,443.

Australian Law Journal: 100 Harris St, Pyrmont, NSW 2009; tel. (2) 8587-7000; fax (2) 8587-7104; f. 1927; monthly; Editor Justice P. W. YOUNG; circ 4,500.

Australian Photography: POB 606, Sydney, NSW 2001; tel. (2) 9281-2333; fax (2) 9281-2750; e-mail robertkeeley@yaffa.com.au; monthly; Editor ROBERT KEELEY; circ. 9,010.

The Australian Women's Weekly: 54–56 Park St, Sydney, NSW 2000; tel. (2) 9282-8000; fax (2) 9267-4459; e-mail womensweekly@acp.com.au; internet www.ninemsn.com.au/aww; f. 1933; monthly; Editor DEBORAH THOMAS; circ. 701,088.

Belle: 54 Park St, Sydney, NSW 2000; tel. (2) 9282-8000; fax (2) 9267-8037; e-mail belle@acp.com.au; f. 1975; every 2 months; Editor ERIC MATTHEWS; circ. 44,663.

Better Homes and Gardens: GPOB 1203, Sydney, NSW 1045; e-mail philippah@mm.com.au; internet www.bhg.com.au; f. 1978; 13 a year; Editor TONI EATTS; circ. 340,133.

Cleo: 54 Park St, Sydney, NSW 2000; POB 4088, Sydney, NSW 2001; tel. (2) 9282-8617; fax (2) 9267-4368; f. 1972; women's monthly; Editor DEBORAH THOMAS; circ. 263,353.

Commercial Photography: GPOB 606, Sydney, NSW 1041; tel. (2) 9281-2333; fax (2) 9281-2750; e-mail yaffa@flex.com.au; every 2 months; journal of the Professional Photographers Asscn of Australia and Photographic Industry Marketing Asscn of Australia; Editor SAIMA MOREL; circ. 3,835.

Cosmopolitan: 54 Park St, Sydney, NSW 2000; tel. (2) 9282-8039; fax (2) 9267-4457; e-mail cosmo@acp.com.au; internet www.cosmopolitan.com.au; f. 1973; monthly; Editor MIA FREEDMAN; circ. 209,347.

Dolly: 54–58 Park St, Sydney, NSW 1028; tel. (2) 9282-8437; fax (2) 9267-4911; e-mail dolly@ninemsn.com.au; internet www.ninemsn.com.au/dolly; f. 1970; monthly; for young women; Editor VIRGINIA KNIGHT; circ. 177,268.

Ecos: CSIRO, POB 1139, Collingwood, Vic 3066; tel. (3) 9662-7500; fax (3) 9662-7555; internet www.publish.csiro.au; f. 1974; quarterly; reports of CSIRO environmental research findings for the non-specialist reader; Editor BRYONY BENNETT; circ. 8,000.

Electronics Australia: POB 199, Alexandria, NSW 1435; tel. (2) 9353-0620; fax (2) 9353-0613; e-mail electaus@fpc.com.au; internet www.electronicsaustralia.com.au; f. 1922; monthly; technical, radio, television, microcomputers, hi-fi and electronics; Editor GRAHAM CATTLEY; circ. 20,900.

Elle: 54 Park St, Sydney, NSW 2000; tel. (2) 9282-8790; fax (2) 9267-4375; f. 1990; monthly; Editor MARINA GO; circ. 68,154.

Family Circle: Pacific Publications, 35–51 Mitchell St, McMahons Point, NSW 2060; tel. (2) 9464-3471; fax (2) 8220-2111; e-mail familycircle@pacpubs.com.au; 13 a year; circ. 235,860.

Gardening Australia: POB 199, Alexandria, NSW 1435; tel. (2) 9353-6666; fax (2) 9317-4615; f. 1991; monthly; Editor BRODIE MYERS-COOKE; circ. 87,000.

Houses: Architecture Media Pty Ltd, Level 3, 4 Princes St, Port Melbourne, Vic 3207; tel. (3) 9646-4760; fax (3) 9646-4918; e-mail houses@archmedia.com.au; internet www.architecturemedia.com; f. 1989; 4 a year; Editor JULIE DILLAN; Man. Editor SUE HARRIS; circ. 25,246.

HQ: 54 Park St, Sydney, NSW 1028; tel. (2) 9282-8260; fax (2) 9267-3616; e-mail hq@publishing.acp.com.au; internet www.hq.ninemsn.com.au; f. 1989; every 2 months; Publr JOHN ALEXANDER; Editor KATHY BAIL; circ. 32,837.

Manufacturers' Monthly: Level 1/28, Riddell Parade, Elsternwick, Vic 3185; tel. (3) 9245-7777; fax (3) 9245-7750; f. 1961; Editor GREG VIDEON; circ. 15,188.

Modern Boating: The Federal Publishing Co Pty Ltd, Unit 2, 160 Bourke Rd, Alexandria, NSW 2015; tel. (2) 9353-6666; fax (2) 9353-0935; e-mail imacrae@fpc.com.au; internet www.modernboating.com.au; f. 1965; every 2 months; Editor IAN MACRAE; circ. 11,000.

Motor: Locked Bag 12, Oakleigh, Vic 3166; tel. (3) 9567-4200; fax (3) 9563-4554; e-mail motor@acpaction.com.au; f. 1954; monthly; Editor GED BULMER; circ. 41,414.

New Woman: EMAP Australia, Level 6, 187 Thomas St, Haymarket, NSW 2000; tel. (2) 9581-9400; fax (2) 9581-9570; internet www.newwoman.com.au; monthly; Editor SUE WHEELER; circ. 85,539.

The Open Road: L27, 388 George St, Sydney, NSW 2000; tel. (2) 9292-9275; fax (2) 9292-9069; f. 1927; every 2 months; journal of National Roads and Motorists' Assen (NRMA); Editor STEVE FRASER; circ. 1,555,917.

Personal Investor: Level 2, 469 La Trobe St, Melbourne, Vic 3000; tel. (3) 9603-3888; fax (3) 9670-4328; e-mail pieditor@brw.fairfax.com.au; internet www.personalinvestor.com.au; monthly; Editor ROBIN BOWERMAN; circ. 61,016.

Reader's Digest: 26–32 Waterloo St, Surry Hills, NSW 2010; tel. (2) 9690-6111; fax (2) 9690-6211; monthly; Editor-in-Chief BRUCE HEILBUTH; circ. 508,142.

Street Machine: Locked Bag 756, Epping, NSW 2121; tel. (2) 9868-4832; fax (2) 9869-7390; e-mail streetmachine@acpaction.com.au; Editor MARK OASTLER; circ. 55,000.

TV Hits: Private Bag 9900, North Sydney, NSW 2059; tel. (2) 9464-3300; fax (2) 9464-3508; f. 1988; monthly; circ. 114,509.

TV Soap: 55 Chandos St, St Leonards, NSW 2065; tel. (2) 9901-6100; fax (2) 9901-6166; f. 1983; monthly; Editor BEN MITCHELL; circ. 103,000.

Vogue Australia: Level 2, 170 Pacific Highway, Greenwich, NSW 2065; tel. (2) 9964-3888; fax (2) 9964-3879; f. 1959; monthly; fashion; Editor JULIET ASHWORTH; circ. 54,705.

Wheels: GPOB 4088, Sydney, NSW 2001; tel. (2) 9263-9732; fax (2) 9263-9702; internet www.wheelsmag.com.au; f. 1953; monthly; international motoring magazine; circ. 68,155.

Wildlife Research: CSIRO Publishing, 150 Oxford St, POB 1139, Collingwood, Vic 3066; tel. (3) 9662-7629; fax (3) 9662-7611; e-mail publishing.wr@csiro.au; internet www.publish.csiro.au/nid/144.htm; f. 1974; 6 a year; Man. Editor C. M. MYERS; circ. 1,000.

Your Garden: 35–51 Mitchell St, McMahons Point, NSW 2060; tel. (2) 9464-3586; fax (2) 9464-3487; e-mail yg@pacpubs.com.au; internet www.yourgarden.com.au; monthly; Editor MAREE TREDINNICK; circ. 60,246.

NEWS AGENCIES

AAP: Locked Bag 21, Grosvenor Place, Sydney, NSW 2000; tel. (2) 9322-8000; fax (2) 9322-8888; e-mail info@aap.com.au; f. 1983; owned by major daily newspapers of Australia; Chair. MICHAEL GILL; CEO CLIVE MARSHALL.

Foreign Bureaux

Agence France-Presse (AFP): 7th Floor, 259 George St, Sydney, NSW 2000; tel. (2) 9251-1544; fax (2) 9251-5230; e-mail afpsyd@afp.com; internet www.afp.com; Bureau Chief DAVID MILLIKIN.

Agenzia Nazionale Stampa Associata (ANSA) (Italy): Suite 4, 2 Grosvenor St, Bondi Junction, NSW 2022; tel. (2) 9369-1427; fax (2) 9369-4351; e-mail ansasyd@ozemail.com.au; internet www.ansa.it; Bureau Chief CLAUDIO MARCELLO.

Deutsche Presse-Agentur (dpa) (Germany): 36 Heath St, Mona Vale, NSW 2103; tel. (2) 9979-8253; fax (2) 9997-3154; e-mail hofman@zip.com.au; Correspondent ALEXANDER HOFMAN.

Jiji Press (Australia) Pty Ltd (Japan): GPOB 2584, Sydney, NSW 2001; tel. (2) 9230-0020; fax (2) 9230-0024; e-mail jijiaust@bigpond.com; Bureau Chief NOBOTUSHI KOBAYASHI.

Kyodo News Service (Japan): Level 7, 9 Lang St, Sydney, NSW 2000; tel. (2) 9251-5240; fax (2) 9251-4980; e-mail ishii.masato@kyodonews.jp; Bureau Chief MASATO ISHII.

Reuters Australia Pty Ltd: Level 30, 60 Margaret St, Sydney, NSW 2031; e-mail Sydney.newsroom@reuters.com; internet www.reuters.com; Bureau Chief PHIL SMITH.

Xinhua (New China) News Agency (People's Republic of China): 50 Russell St, Hackett, Canberra, ACT 2602; tel. (2) 6248-6369; fax (2) 6257-7633; e-mail xinhua@webone.com.au; Chief Correspondent MI LIGONG.

The Central News Agency (Taiwan) and the New Zealand Press Association are represented in Sydney, and Antara (Indonesia) is represented in Canberra.

PRESS ASSOCIATIONS

Australian Press Council: Suite 10.02, 117 York St, Sydney, NSW 2000; tel. (2) 9261-1930; fax (2) 9267-6826; e-mail info@presscouncil.org.au; internet www.presscouncil.org.au; Chair. Prof. KEN MCKINNON.

Community Newspapers of Australia Pty Ltd: POB 234, Auburn, NSW 1835; tel. (2) 8789-7300; fax (2) 8789-7387; e-mail robyn@printnet.com.au; Fed. Pres. JULIE UPSON; Sec. ROBYN BAKER.

Country Press Association of SA Incorporated: 198 Greenhill Rd, Eastwood, SA 5063; tel. (8) 8373-6533; fax (8) 8373-6544; e-mail sacp@iweb.net.au; internet www.cpasa.ans.au; f. 1912; represents South Australian country newspapers; Pres. T. MCAULIFFE; Exec. Dir M. S. MANUEL.

Country Press Australia: POB Q182, QVB PO, Sydney, NSW 1230; tel. (2) 9299-4658; fax (2) 9299-1892; f. 1906; Exec. Dir D. J. SOMMERLAD; 420 mems.

Queensland Country Press Association: POB 103, Paddington, Qld 4064; tel. (7) 3356-0033; Pres. M. HODGSON; Sec. N. D. MCLARY.

Tasmanian Press Association Pty Ltd: 71–75 Paterson St, Launceston, Tas 7250; tel. (3) 6336-7111; Sec. TOM O'MEARA.

Victorian Country Press Association Ltd: 113 Rosslyn St, West Melbourne, Vic 3003; tel. (3) 9320-8900; fax (3) 9328-1916; e-mail vcpa@vcpa.com.au; internet www.vcpa.com.au; f. 1910; Pres. G. KELLY; Exec. Dir J. E. RAY; 110 mems.

Publishers

Allen and Unwin Pty Ltd: 83 Alexander St, Crows Nest, NSW 2065; tel. (2) 8425-0100; fax (2) 9906-2218; e-mail info@allenandunwin.com; internet www.allenandunwin.com; fiction, trade, academic, children's; Man. Dir PATRICK A. GALLAGHER.

Australasian Medical Publishing Co Pty Ltd: Level 2, 26-32 Pyrmont Bridge Road, Pyrmont, NSW 2009; tel. (2) 9562-6666; fax (2) 9562-6600; e-mail ampco@ampco.com.au; internet www.ampco.com.au; f. 1913; scientific, medical and educational; CEO Dr MARTIN VAN DER WEYDEN.

Cambridge University Press (Australia): 10 477 Williamstown Rd, PB 31, Port Melbourne, Vic 3207; tel. (3) 8671-1400; fax (3) 9676-9966; e-mail customerservice@cambridge.edu.au; internet www.cambridge.edu.au; scholarly and educational; Dir SANDRA MCCOMB.

Commonwealth Scientific and Industrial Research Organisation (CSIRO Publishing): 150 Oxford St, POB 1139, Collingwood, Vic 3066; tel. (3) 9662-7500; fax (3) 9662-7555; e-mail publishing@csiro.au; internet www.publish.csiro.au; f. 1926; scientific and technical journals, books, magazines, videos, CD-ROMs; Gen. Man. P. W. REEKIE.

Doubleday Australia Pty Ltd: 91 Mars Rd, Lane Cove, NSW 2066; tel. (2) 9427-0377; fax (2) 9427-6973; educational, trade, non-fiction, Australiana; Man. Dir BARRY MACMULLEN.

Elsevier Australia (a division of Reed International Books Australia Pty Ltd): 30–52 Smidmore St, Marrickville, NSW 2204; tel. (2) 9517-8999; fax (2) 9517-2249; e-mail service@elsevier.com.au; internet www.elsevier.com.au; health sciences, science and medicine; Man. Dir FERGUS HALL.

Encyclopaedia Britannica Australia Ltd: Level 1, 90 Mount St, North Sydney, NSW 2060; tel. (2) 9923-5600; fax (2) 9929-3758; e-mail sales@Britannica.com.au; internet www.britannica.com.au; reference, education, art, science and commerce; Man. Dir DAVID CAMPBELL.

Harcourt Education Australia: POB 460, Port Melbourne, Vic 3207; tel. (3) 9245-7111; fax (3) 9245-7333; e-mail admin@harcourteducation.com.au; internet www.hi.com.au; primary, secondary and tertiary educational; division of Reed International; Man. Dir DAVID O'BRIEN.

Harlequin Enterprises (Australia) Pty Ltd: Unit 2, 3 Gibbes St, Chatswood, NSW 2067; tel. (2) 9415-9200; fax (2) 9415-9292; internet www.eHarlequin.com.au; Man. Dir MICHELLE LAFOREST.

Hyland House Publishing Pty Ltd: POB 122, Flemington, Vic 3031; tel. and fax (3) 9376-4461; e-mail hyland3@netspace.net.au; f. 1976; trade, general, gardening, pet care, Aboriginal, Asian-Pacific public policy, fiction; Rep. MICHAEL SCHOO.

Lansdowne Publishing: Level 1, 18 Argyle St, The Rocks, NSW 2000; tel. (2) 9240-9222; fax (2) 9241-4818; e-mail steven@lanspub.com.au; cookery, gardening, health, history, pet care; Chief Exec. STEVEN MORRIS.

LexisNexis: Tower 2, 475 Victoria Ave, Chatswood, NSW 2067; tel. (2) 9422-2222; fax (2) 9422-2444; internet www.lexisnexis.com.au; f. 1910; div. of Reed International Books Australia Pty Ltd; legal and commercial; Man. Dir MAX PIPER.

Lothian Books: Level 5, 132 Albert Rd, South Melbourne, Vic 3205; tel. (3) 9694-4900; fax (3) 9645-0705; e-mail books@lothian.com.au; internet www.lothian.com.au; f. 1888; gardening, health, sport, business, New Age, Buddhism, self-help, general non-fiction, young adult fiction and children's picture books; Man. Dir PETER LOTHIAN.

McGraw-Hill Australia Pty Ltd: Level 2, 82 Waterloo Rd, North Ryde, NSW 2113; tel. (2) 9900-1800; fax (2) 9878-8640; e-mail cservice_sydney@mcgraw-hill.com; internet www.mcgraw-hill.com.au; educational, professional and technical; Man. Dir JOHN BLACK.

Melbourne University Press: 268 Drummond St, Carlton South, Vic 3053; tel. (3) 9347-0300; fax (3) 9342-0399; e-mail info@mup.unimelb.edu.au; internet www.mup.com.au; f. 1922; scholarly non-fiction, Australian history and biography; Chair. Prof. BARRY SHEEHAN; Dir JOHN MECKAN.

Murdoch Books: GPOB 1203, Sydney, NSW 2001; tel. (2) 8220-2000; fax (2) 8220-2558; e-mail incaw@mm.com.au; cooking, gardening, DIY, craft, gift, general leisure and lifestyle; CEO JULIET ROGERS; Publr KAY SCARLETT.

National Library of Australia: Parkes Place, Canberra, ACT 2600; tel. (2) 6262-1111; fax (2) 6273-4493; e-mail serialorders@nla.gov.au; internet www.nla.gov.au; f. 1960; national bibliographic service, etc.; Publications Dir PAUL HETHERINGTON.

Nelson Thomson Learning: 102 Dodds St, South Melbourne, Vic 3205; tel. (3) 9685-4111; fax (3) 9685-4199; e-mail customerservice@nelson.com.au; internet www.nelsonitp.com; educational; Man. Dir G. J. BROWNE.

Oxford University Press: 253 Normanby Rd, South Melbourne, Vic 3205; tel. (3) 9934-9123; fax (3) 9934-9100; f. 1908; general non-fiction and educational; Man. Dir MAREK PALKA.

Pan Macmillan Australia Pty Ltd: Level 18, St Martin's Tower, 31 Market St, Sydney, NSW 2000; tel. (2) 9285-9100; fax (2) 9285-9190; e-mail pansyd@macmillan.com.au; general, reference, children's, fiction, non-fiction; Chair. R. GIBB.

Pearson Education Australia Pty Ltd: 95 Coventry St, South Melbourne, Vic 3205; tel. (3) 9697-0666; fax (3) 9699-2041; e-mail longman.sales@pearsoned.com.au; internet www.pearsoned.com.au; f. 1957; mainly educational, academic, computer, some general; Man. Dir DAVID BARNETT.

Penguin Group (Australia): POB 701, Hawthorn, Vic 3122; tel. (3) 9811-2400; fax (3) 9811-2620; internet www.penguin.com.au; f. 1946; general; Man. Dir GABRIELLE COYNE; Publishing Dir ROBERT SESSIONS.

Random House Australia Pty Ltd: 20 Alfred St, Milsons Point, NSW 2061; tel. (2) 9954-9966; fax (2) 9954-4562; e-mail random@randomhouse.com.au; internet www.randomhouse.com.au; fiction, non-fiction and children's; Man. Dir MARGARET SEALE.

Reader's Digest (Australia) Pty Ltd: POB 4353, Sydney, NSW 2000; tel. (2) 9690-6111; fax (2) 9699-8165; general; Man. Dir WILLIAM B. TOOHEY.

Scholastic Australia Pty Ltd: Railway Crescent, Lisarow, POB 579, Gosford, NSW 2250; tel. (2) 4328-3555; fax (2) 4323-3827; internet www.scholastic.com.au; f. 1968; educational and children's; Man. Dir KEN JOLLY.

Schwartz Publishing (Victoria) Pty Ltd: 45 Flinders Lane, Melbourne, Vic 3000; tel. (3) 9654-2000; fax (3) 9650-5418; fiction, non-fiction; Dir MORRY SCHWARTZ.

Simon and Schuster Australia: 20 Barcoo St, POB 507, East Roseville, NSW 2069; tel. (2) 9415-9900; fax (2) 9417-4292; general fiction, non-fiction, cooking, gardening, craft, parenting, health, history, travel and biography; Man. Dir JON ATTENBOROUGH.

Thames and Hudson (Australia) Pty Ltd: 11 Central Boulevard, Portside Business Park, Fishermans Bend, Vic 3207; tel. (3) 9646-7788, fax (3) 9646-8790; e-mail enquirics@thaust.com.au; internet www.thamesandhudson.com; art, history, archaeology, architecture, photography, design, fashion, textiles, lifestyle; Man. Dir PETER SHAW.

Thomson Legal & Regulatory, Australia: Level 5, 100 Harris St, Pyrmont, NSW 2009; tel. (2) 8587-7000; fax (2) 8587-7100; e-mail support@thomson.com.au; internet www.thomson.com.au; legal, professional, tax and accounting; Man. Dir JACKIE RHODES.

Thorpe-Bowker: C3, 85 Turner St, POB 101, Port Melbourne, Vic 3207; tel. (3) 8645-0300; fax (3) 8645-0333; e-mail customer.service@thorpe.com.au; internet www.thorpe.com.au; bibliographic, library and book trade reference; Gen. Man. RICHARD SIEGERSMA.

Time Life Australia Pty Ltd: Level 12, 33 Berry St, North Sydney, NSW 2060; tel. (2) 8925-3800; fax (2) 9957-4227; general and educational; Man. Dir ROBERT HARDY.

UNSW Press Ltd: University of New South Wales, Sydney, NSW 2052; tel. (2) 9664-0999; fax (2) 9664-5420; e-mail info.press@unsw.edu.au; f. 1961; scholarly, general and tertiary texts; Man. Dir Dr ROBIN DERRICOURT.

University of Queensland Press: POB 6042, St Lucia, Qld 4067; tel. (7) 3365-2127; fax (7) 3365-7579; e-mail uqp@uqp.uq.edu.au; internet www.uqp.uq.edu.au; f. 1948; scholarly and general cultural interest, incl. Black Australian writers, adult and children's fiction; Gen. Man. GREG BAIN.

University of Western Australia Press: 35 Stirling Highway, Crawley, WA 6009; tel. (8) 6488-3670; fax (8) 6488-1027; e-mail uwap@cyllene.uwa.edu.au; internet www.uwapress.uwa.edu.au; f. 1935; natural history, history, literary studies, Australiana, children's, general non-fiction; Dir Dr JENNY GREGORY.

John Wiley & Sons Australia, Ltd: POB 1226, Milton, Qld 4064; tel. (7) 3859-9755; fax (7) 3859-9715; e-mail brisbane@johnwiley.com.au; internet www.johnwiley.com.au; f. 1954; educational, reference and trade; Man. Dir PETER DONOUGHUE.

Government Publishing House

AusInfo: GPOB 1920, Canberra, ACT 2601; tel. (2) 6275-3442; fax (2) 6275-3682; internet www.ausinfo.gov.au; f. 1970; fmrly Australian Govt Publishing Service; Assistant Sec. MICHELLE KINNANE.

PUBLISHERS' ASSOCIATION

Australian Publishers Association Ltd: 60/89 Jones St, Ultimo, NSW 2007; tel. (2) 9281-9788; fax (2) 9281-1073; e-mail apa@publishers.asn.au; internet www.publishers.asn.au; f. 1948; 115 mems; Pres. GREG BROWNE; Chief Exec. SUSAN BRIDGE.

Broadcasting and Communications

TELECOMMUNICATIONS

By July 2004, 101 licensed telecommunications carriers were in operation.

AAPT Ltd: AAPT Centre, 9 Lang St, Sydney, NSW 2000; tel. (2) 9377-7000; fax (2) 9377-7133; internet www.aapt.com.au; f. 1991; long-distance telecommunications carrier; Chair C. L. CASEY; CEO and Man. Dir L. WILLIAMS.

Matrix Telecommunications Services: 1st Floor, 24 Artamon Rd, Willoughby, NSW 2068; tel. (2) 9290-4111; fax (2) 9262-2574; f. 1985; mobile communication services; Chair. MARK CARNEGIE; Pres. and CEO JOSEPH YANG.

One.Tel Ltd: Level 28, 9 Castlereagh St, Sydney, NSW 2000; tel. (2) 9777-8111; fax (2) 9777-8199; internet www.onetel.com.au; telecommunication services; Chair. JOHN GREAVES.

Optus Ltd: POB 1, North Sydney, NSW 2059; tel. (2) 9342-7800; fax (2) 9342-7100; internet www.optus.com.au; f. 1992; division of Singapore Telecommunications Ltd; general and mobile telecommunications, data and internet services, pay-TV; Chair. Sir RALPH ROBINS; Chief Exec. PAUL O'SULLIVAN.

Telstra Corpn Ltd: Level 14, 231 Elizabeth St, Sydney, NSW 2000; tel. (2) 9287-4677; fax (2) 9287-5869; internet www.telstra.com.au; general and mobile telecommunication services; Man. Dir and Chief Exec. ZIGGY SWITKOWSKI.

Vodafone Australia Ltd: Tower A, 799 Pacific Highway, Chatswood, NSW 2067; tel. (2) 9415-7000; fax (2) 9878-7788; internet www.vodafone.com.au; mobile telecommunication services.

Regulatory Authority

Australian Communications Authority (ACA): POB 13112, Law Courts, Vic 8010; tel. (3) 9963-6800, fax (3) 9963-6899; e-mail candinfo@aca.gov.au; internet www.aca.gov.au; f. 1997; Commonwealth regulator for telecommunications and radiocommunications; Chair. BOB HORTON (acting).

BROADCASTING

Many programmes are provided by the non-commercial statutory corporation, the Australian Broadcasting Corporation (ABC). Commercial radio and television services are provided by stations operated by companies under licences granted and renewed by the Australian Broadcasting Authority (ABA). They rely for their income on the broadcasting of advertisements. In mid-2004 there were about 256 commercial radio stations in operation, and 48 commercial television stations.

In 1997 there were an estimated 25.5m. radio receivers. The number of television receivers in use totalled 14,168,000 in 2001.

Australian Broadcasting Corporation (ABC): 700 Harris St, Ultimo, POB 9994, Sydney, NSW 2001; tel. (2) 9333-1500 (radio); fax (2) 8333-2603 (radio), (2) 9950-3050 (television); e-mail comments@your.abc.net.au; internet www.abc.net.au; f. 1932 as Australian Broadcasting Commission; became corporation in 1983; one national television network operating on about 700 transmitters, one international television service broadcasting via satellite to Asia and the Pacific and six radio networks operating on more than 6,000 transmitters; Chair. DONALD MCDONALD; Man. Dir RUSSELL BALDING.

Radio Australia: international service broadcast by short wave and satellite in English, Indonesian, Standard Chinese, Khmer, Tok Pisin and Vietnamese.

Radio

Commercial Radio Australia Ltd: Level 5, 88 Foveaux St, Surry Hills, NSW 2010; tel. (2) 9281-6577; fax (2) 9281-6599; e-mail mail@commercialradio.com.au; internet www.commercialradio.com; f. 1930; represents the interests of Australia's commercial radio broadcasters; CEO JOAN WARNER.

Major Commercial Broadcasting Station Licensees

5AD Broadcasting Co Pty Ltd: 201 Tynte St, Nth Adelaide, SA 5006; tel. (8) 8300-1000; fax (8) 8300-1020; internet www.5adfm.com.au; also operates 5DN and Mix102.3; Gen. Man. GRAEME TUCKER.

Associated Communications Enterprises (ACE) Radio Broadcasters Pty Ltd: POB 7515, Melbourne, Vic 3004; tel. (3) 9645-9877; fax (3) 9645-9886; operates six stations; Man. Dir S. EVERETT.

Austereo Pty Ltd: Ground Level, 180 St Kilda Rd, St Kilda, Vic 3182; tel. (3) 9230-1051; fax (3) 9593-9007; e-mail pharvie@austereo.com.au; internet www.austereo.com.au; operates 14 stations; Exec. Chair. PETER HARVIE.

Australian Radio Network Pty Ltd: Level 8, 99 Mount St, North Sydney, NSW 2060; tel. (2) 9464-1000; fax (2) 9464-1010; operates nine stations; CEO NEIL MOUNT.

Australian Regional Broadcasters: 1 June Rd, Gooseberry Hill, WA 6076; tel. (8) 9472-8900; fax (8) 9472-8911; operates three stations; Man. Dir NICK RINGROSE.

Bass Radio: 109 York St, Launceston, Tas 7250; tel. (3) 6331-4844; fax (3) 6334-5858; operates five radio stations and part of RG Capital Radio; Man. DAVE HILL; Contact MATT RUSSELL.

Capital Radio: 28 Sharp St, Cooma, NSW 2630; tel. (2) 6452-1521; fax (2) 6452-1006; operates four stations; Man. Dir KEVIN BLYTON.

DMG Regional Radio Pty Ltd: Level 5, 33 Saunders St, Pyrmont, NSW 2009; tel. (2) 9564 9888; fax (2) 9564 9867; e-mail sydoff@dmgradio.com.au; internet www.dmgradio.com.au; operates 59 stations; Group Mans STEVE HIRST (Western Group), SEAN RYAN (Northern Group), GARRY LEDDIN (Southern Group); Gen. Man. ROB GAMBLE.

Grant Broadcasting: 63 Minimbah Rd, Northbridge, NSW 2063; tel. (2) 9958-7301; fax (2) 9958-6906; operates seven stations; Gen. Man. JANET CAMERON.

Greater Cairns Radio Ltd: Virginia House, Abbott St, Cairns, Qld 4870; tel. (7) 4050-0800; fax (7) 4051-8060; e-mail cnssales@dmgradio.com.au; Gen. Man. J. ELLER.

Macquarie Radio Network Pty Ltd: POB 4290, Sydney, NSW 2001; tel. (2) 9269-0646; fax (2) 9287-2772; operates 2GB and 2CH; CEO GEORGE BUSCHMAN.

Moree Broadcasting and Development Company Ltd: 87–89 Balo St, Moree, NSW 2400; tel. (2) 6752-1155; fax (2) 6752-2601; operates two stations; Man. KEN BIRCH.

Radio 2SM Gold 1269: 8 Jones Bay Rd, Pyrmont, NSW 2009; tel. (2) 9922-1269; e-mail info@2sm.com.au; internet www.2sm.com.au; f. 1931; CEO and Chair. C. M. MURPHY.

RadioWest Hot FM: POB 10067, Kalgoorlie, WA 6430; tel. (8) 9021-2666; fax (8) 9091-2209; e-mail radio6KG@gold.net.au; f. 1931.

Regional Broadcasters (Australia) Pty: McDowal St, Roma, Qld 4455; tel. (7) 4622-1800; fax (7) 4622-3697; Chair. G. MCVEAN.

RG Capital Radio Pty Ltd: Level 2, Seabank Bldg, 12–14 Marine Parade, Southport, Qld 4215; tel. (7) 5591-5000; fax (7) 5591-2869; operates 34 stations; Man. Dir RHYS HOLLERAN.

Rural Press Ltd: Cnr Pine Mt Rd and Hill St, Raymonds Hill, Qld 4305; tel. (7) 3201-6000; fax (7) 3812-3060; internet www.rpl.com.au; f. 1911; operates five stations; Gen. Man. RICHARD BURNS.

SEA FM Pty Ltd: POB 5910, Gold Coast Mail Centre, Bundall, Qld 4217; tel. (7) 5591-5000; fax (7) 5591-6080; operates 28 stations; Man. Dir RHYS HOLLERAN.

Southern Cross Broadcasting (Australia) Ltd: see under Television.

Supernetwork Radio Pty Ltd: POB 97, Coolangatta, Qld 4225; tel. (7) 5524-4497; fax (7) 5554-3970; operates 15 stations; Chair. W. CARALIS.

Tamworth Radio Development Company Pty Ltd: POB 497, Tamworth, NSW 2340; tel. (2) 6765-7055; fax (2) 6765-2762; operates five stations; Man. W. A. MORRISON.

Tasmanian Broadcasting Network (TBN): POB 665G, Launceston, Tas 7250; tel. (3) 6431-2555; fax (3) 6431-3188; operates three stations; Chair. K. FINDLAY.

Wesgo Ltd: POB 234, Seven Hills, NSW 2147; tel. (2) 9831-7611; fax (2) 9831-2001; operates eight stations; CEO G. W. RICE.

Television

Free TV Australia (CTVA): 44 Avenue Rd, Mosman, NSW 2088; tel. (2) 8968-7100; fax (2) 9969-3520; e-mail ctva@ctva.com.au; internet freetvaust.com.au; f. 1960; fmrly Commercial Television Australia; represents all commercial free-to-air television stations; Chair. DAVID GYNGELL; CEO JULIE FLYNN.

Commercial Television Station Licensees

Amalgamated Television Services Pty Ltd: Mobbs Lane, Epping, NSW 2121; tel. (2) 9877-7777; fax (2) 9877-7888; f. 1956; originating station for Seven Network TV programming; Exec. Chair. KERRY STOKES.

Australian Capital Television Pty Ltd (Southern Cross Ten): Private Bag 10, Dickson, ACT 2602; tel. (2) 6242-2400; fax (2) 6241-7230; e-mail reception@southerncrossnsw.com.au; f. 1962; Gen. Man. ERIC PASCOE.

Broken Hill Television Ltd: POB 472, Rocky Hill, Broken Hill, NSW 2880; tel. (8) 8087-6013; fax (8) 8087-8492; internet www.centralonline.com.au; f. 1968; operates one station; Chair. PETER STORROCK; Chief Exec. D. WESTON.

Channel 9 South Australia Pty Ltd: 202 Tynte St, North Adelaide 5006; tel. (8) 8267-0111; fax (8) 8267-3996; f. 1959; Gen. Man. M. COLSON.

Channel Seven Adelaide Pty Ltd: 45–49 Park Terrace, Gilberton, SA 5081; tel. (8) 8342-7777; fax (8) 8342-7717; f. 1965; mem. of Seven Network; Man. Dir MAX WALTERS.

Channel Seven Brisbane Pty Ltd: GPOB 604, Brisbane, Qld 4001; tel. (7) 3369-7777; fax (7) 3368-2970; f. 1959; operates one station; mem. of Seven Network; Man. Dir L. M. RILEY.

Channel Seven Melbourne Pty Ltd: 119 Wells St, Southbank, Vic 3006; tel. (3) 9697-7777; fax (3) 9697-7888; e-mail daspinall@seven.com.au; f. 1956; operates one station; Chair. KERRY STOKES; Man. Dir DAVID ASPINALL.

Channel Seven Perth Pty Ltd: POB 77, Tuart Hill, WA 6939; tel. (8) 9344-0777; fax (8) 9344-0670; f. 1959; Chair. C. S. WHARTON.

General Television Corporation Pty Ltd: 22–46 Bendigo St, POB 100, Richmond, Vic 3121; tel. (3) 9429-0201; fax (3) 9429-3670; internet www.nine.msn.com.au; f. 1957; operates one station; Man. Dir GRAEME YARWOOD.

Golden West Network: POB 5090, Geraldton, WA 6531; tel. (8) 9921-4422; fax (8) 9921-8096.

Golden West Network Pty Ltd: POB 1062, West Perth, WA 6872; tel. (8) 9481-0050; fax (8) 9321-2470; f. 1967; operates three stations (SSW10, VEW and WAW); Gen. Man. W. FENWICK.

Imparja Television Pty Ltd: POB 52, Alice Springs, NT 0871; tel. (8) 8950-1411; fax (8) 8953-0322; e-mail imparja@imparja.com.au; internet www.imparja.com.au; CEO ALISTAIR FEEHAN.

Independent Broadcasters of Australia Pty Ltd: POB 285, Sydney, NSW 2001; tel. (2) 9264-9144; fax (2) 9264-6334; fmrly Regional Television Australia Pty Ltd; Chair. GRAEME J. GILBERTSON; Sec. JEFF EATHER.

Mt Isa Television Pty Ltd: 110 Canooweal St, Mt Isa, Qld 4825; tel. (7) 4743-8888; fax (7) 4743-9803; f. 1971; operates one station; Station Man. LYALL GREY.

NBN Ltd: Mosbri Crescent, POB 750L, Newcastle, NSW 2300; tel. (2) 4929-2933; fax (2) 4926-2936; f. 1962; operates one station; Man. Dir DENIS LEDBURY.

Network Ten Ltd: GPOB 10, Sydney, NSW 2001; tel. (2) 9650-1010; fax (2) 9650-1170; operates Australian TV network and commercial stations in Sydney, Melbourne, Brisbane, Perth and Adelaide; CEO JOHN MCALPINE.

Nine Network Australia Pty Ltd: POB 27, Willoughby, NSW 2068; tel. (2) 9906-9999; fax (2) 9958-2279; internet www.ninemsn.com.au; f. 1956; division of Publishing and Broadcasting Ltd; operates three stations: TCN Channel Nine Pty Ltd (Sydney), Queensland Television Ltd (Brisbane) and General Television Corporation Ltd (Melbourne); CEO DAVID LECKIE.

Northern Rivers Television Pty Ltd: Peterson Rd, Locked Bag 1000, Coffs Harbour, NSW 2450; tel. (2) 6652-2777; fax (2) 6652-3034; f. 1965; CEO GARRY DRAFFIN.

Prime Television Group: Level 6, 1 Pacific Highway, North Sydney, NSW 2060; tel. (2) 9965-7700; fax (2) 9965-7729; e-mail primetv@primetv.com.au; internet www.primetv.com.au; Chair. PAUL RAMSAY; CEO BRENT HARMAN.

Prime Television (Northern) Pty Ltd: POB 2077, Elermore Vale, NSW 2287; tel. (2) 4952-0500; fax (2) 4952-0502; internet www.primetv.com.au; Gen. Man. BRAD JONES.

Prime Television (Southern) Pty Ltd: POB 465, Orange, NSW 2800; tel. (2) 6361-6888; fax (2) 6363-1889; Gen. Man. D. EDWARDS.

Prime Television (Victoria) Pty Ltd: Sunraysia Highway, Ballarat, Vic 3350; tel. (3) 5337-1777; fax (3) 5337-1700; e-mail primetv.ballarat@primetv.com.au; Gen. Man. CRAIG WHITFIELD.

Queensland Television Ltd: GPOB 72, Brisbane, Qld 4001; tel. (7) 3214-9999; fax (7) 3369-3512; f. 1959; operates one station; Gen. Man. IAN R. MÜLLER.

Seven Network Ltd: Level 13, 1 Pacific Highway, North Sydney, NSW 2060; tel. (2) 9967-7903; fax (2) 9967-7972; internet www.seven.com.au; owns Amalgamated Television Services Pty Ltd (Sydney), Brisbane TV Ltd (Brisbane), HSV Channel 7 Pty Ltd (Melbourne), South Australian Telecasters Ltd (Adelaide) and TVW Enterprises Ltd (Perth); Exec. Chair. KERRY STOKES.

Seven Queensland: 140–142 Horton Parade, Maroochydore, Qld 4558; tel. (7) 5430-1777; fax (7) 5430-1767; f. 1965; fmrly Sunshine Television Network Ltd.

Southern Cross Broadcasting (Australia) Ltd: 41–49 Bank St, South Melbourne, Vic 3205; tel. (3) 9243-2100; fax (3) 9690-0937; internet www.scbnetwork.com.au; f. 1932; operates four TV and four radio stations; Man. Dir A. E. BELL.

Southern Cross Television (TNT9) Pty Ltd: Watchorn St, Launceston, Tas 7250; tel. (3) 6344-0202; fax (3) 6343-0340; f. 1962; operates one station; Gen. Man. BRUCE ABRAHAM.

Special Broadcasting Service (SBS): Locked Bag 028, Crows Nest, NSW 1585; tel. (2) 9430-2828; fax (2) 9430-3700; e-mail sbs.com.au; internet www.sbs.com.au; f. 1980; national multi-cultural broadcaster of TV and radio; Man. Dir NIGEL MILAN.

Spencer Gulf Telecasters Ltd: POB 305, Port Pirie, SA 5540; tel. (8) 8632-2555; fax (8) 8633-0984; e-mail dweston@centralonline.com.au; internet www.centralonline.com.au; f. 1968; operates two stations; Chair. P. M. STURROCK; Chief Exec. D. WESTON.

Swan Television & Radio Broadcasters Pty Ltd: POB 99, Tuart Hill, WA 6939; tel. (8) 9449-9999; fax (8) 9449-9900; Gen. Man. P. BOWEN.

Telecasters Australia Ltd: Level 8, 1 Elizabeth Plaza, North Sydney, NSW 2060; tel. (2) 9922-1011; fax (2) 9922-1033; internet www.telecasters.com.au; operates commercial TV services of TEN Queensland, TEN Northern NSW, Seven Central and Seven Darwin.

Territory Television Pty Ltd: POB 1764, Darwin, NT 0801; tel. (8) 8981-8888; fax (8) 8981-6802; f. 1971; operates one station; Gen. Man. A. G. BRUYN.

WIN Television Griffith Pty Ltd: 161 Remembrance Driveway, Griffith, NSW 2680; tel. (2) 6962-4500; fax (2) 6962-0979; e-mail mtntv@ozemail.com.au; fmrly MTN Television; Man. Dir RAY GAMBLE.

WIN Television Loxton SA Pty Ltd: Murray Bridge Rd, POB 471, Loxton, SA 5333; tel. (8) 8584-6891; fax (8) 8584-5062; f. 1976; operates one station; Exec. Chair. E. H. URLWIN; Gen. Man. W. L. MUDGE.

WIN Television Mildura Pty Ltd: 18 Deakin Ave, Mildura, Vic 3500; tel. (3) 5023-0204; fax (3) 5022-1179; f. 1965; Chair. JOHN RUSHTON; Man. NOEL W. HISCOCK.

WIN Television NSW Network: Television Ave, Mt St Thomas, Locked Bag 8800, South Coast Mail Centre, NSW 2521; tel. (2) 4223-4199; fax (2) 4227-3682; internet www.wintv.com.au; f. 1962; Man. Dir K. KINGSTON; CEO JOHN RUSHTON.

WIN Television Qld Pty Ltd: POB 568 Rockhampton, Qld 4700; tel. (7) 4930-4499; fax (7) 4930-4490; Station Man. R. HOCKEY.

WIN Television Tas Pty Ltd: 52 New Town Rd, Hobart, Tas 7008; tel. (3) 6228-8999; fax (3) 6228-8991; e-mail wintas.com.au; internet www.wintv.com.au; f. 1959; Gen. Man. GREG RAYMENT.

WIN Television Vic Pty Ltd: POB 464, Ballarat, Vic 3353; tel. (3) 5320-1366; fax (3) 5333-1598; internet www.winnet.com.au; f. 1961; operates five stations; Gen. Man. MICHAEL TAYLOR.

Satellite, Cable and Digital Television

Digital television became available in metropolitan areas in January 2001 and was to be available in all major regional areas by 2004.

Austar United Communications: Level 29, AAP Centre, 259 George St, Sydney, NSW 2000; tel. (2) 9251-6999; fax (2) 9251-6136; e-mail austar@austarunited.com.au; internet www.austarunited.com.au; began operations in 1995; 459,819 subscribers (June 2004); CEO JOHN C. PORTER.

Australia Television: international satellite service; broadcasts to more than 30 countries and territories in Asia and the Pacific.

Foxtel: Foxtel Television Centre, Pyrmont, Sydney; internet www.foxtel.com.au; owned by the News Corpn, Telstra Corpn and PBL; 800,100 subscribers (Aug. 2002).

Optus Vision: Tower B, Level 15, 16 Zenith Centre, 821–841 Pacific Highway, Chatswood, NSW 2067; commenced cable services on 11 channels in 1995; 210,000 subscribers (March 1999).

Regulatory Authority

Australian Broadcasting Authority: POB Q500, QVB PO, NSW 1230; tel. (2) 9334-7700; fax (2) 9334-7799; e-mail info@aba.gov.au; internet www.aba.gov.au; regulates radio and TV broadcasting, and internet content; Chair. LYN MADDOCK (acting).

Finance

(cap. = capital; p.u. = paid up; res = reserves; dep. = deposits; m. = million; brs = branches; amounts in Australian dollars)

Radical reforms of the financial sector were introduced in 1998. The banking system was opened up to greater competition. The licensing and regulation of deposit-taking institutions was supervised by the new Australian Prudential Regulation Authority, while consumer protection was the responsibility of the Australian Corporations and Financial Services Commission.

Australian Prudential Regulation Authority (APRA): GPOB 9836, Sydney, NSW 2000; tel. (2) 9210-3000; fax (2) 9210-3411; e-mail aprainfo@apra.gov.au; internet www.apra.gov.au; f. 1998; responsible for regulation of banks, insurance cos, superannuation funds, credit unions, building societies and friendly societies; Chair. Dr JOHN LAKER.

BANKING

Central Bank

Reserve Bank of Australia: GPOB 3947, Sydney, NSW 2001; tel. (2) 9551-8111; fax (2) 9551-8000; e-mail rbainfo@rba.gov.au; internet www.rba.gov.au; f. 1911; responsible for monetary policy, financial system stability, payment system development; cap. 40m., res 9,374m., dep. 14,736m., total assets 66,593m., notes on issue 32,172m. (June 2003); Gov. IAN MACFARLANE.

Development Bank

Primary Industry Bank of Australia Ltd: GPOB 4577, Sydney, NSW 1042; tel. (2) 9234-4200; fax (2) 9221-6218; internet www.piba.com.au; f. 1978; cap. 123.2m., res 2.5m.; Chair. H. G. GENTIS; 21 brs.

Trading Banks

ABN AMRO Australia Ltd: Level 29, ABN AMRO Tower, 88 Phillip St, Sydney, NSW 2000; tel. (2) 8259-5000; fax (2) 8259-5444; internet www.abnamro.com.au; f. 1971; cap. 70m., res 590,000 (Dec. 2000); CEO STEVE CRANE.

Arab Bank Australia Ltd: GPOB N645, Grosvenor Place, 200 George St, Sydney, NSW 2000; tel. (2) 9377-8900; fax (2) 9221-5428; internet www.arabbank.com.au; cap. 55.0m., dep. 464.9m. (Dec. 2000); Chair. KHALID SHOMAN; Man. Dir JACK BEIGHTON.

Australia and New Zealand Banking Group Ltd: 100 Queen St, GPOB 537E, Melbourne, Vic 3000; tel. (3) 9273-5555; fax (3) 9658-2484; internet www.anz.com; f. 1835; present name adopted in 1970; cap. 5,402.0m., res 786.0m., dep. 112,849.0m. (2000); 871 brs; Chair. C. B. GOODE; CEO JOHN MCFARLANE.

Bank of America Australia Ltd: Level 63, MLC Centre, 19–29 Martin Place, Sydney, NSW 2000; tel. (2) 9931-4200; fax (2) 9221-1023; f. 1964; cap. 150.3m. (Dec. 1998); Man. Dir JOHN LILES.

Bank of Melbourne: 360 Collins St, Melbourne, Vic 3000; tel. (3) 9608-3222; fax (3) 9608-3700; division of Westpac Banking Corpn; f. 1989; cap. 752m., dep. 8,706m. (1997); Chair. CHRIS STEWART; CEO MATTHEW SLATTER; 129 brs.

Bank of Queensland Ltd: 229 Elizabeth St, POB 898, Brisbane, Qld 4001; tel. (7) 3212-3333; fax (7) 3212-3399; internet www.boq.com.au; f. 1874; cap. 321.9m., res 0.2m., dep. 6,114.2m. (Aug. 2004); Chair. NEIL ROBERTS; Man Dir. DAVID P. LIDDY; 137 brs.

Bank of Tokyo-Mitsubishi (Australia) Ltd: Level 26, Gateway, 1 Macquarie Place, Sydney, NSW 2000; tel. (2) 9296-1111; fax (2) 9247-4266; e-mail btmacorp@btma.com.au; f. 1985; cap. 152.9m., res 33,000m., dep. 1,910.8m. (Dec. 2000); Chair. R. NICOLSON; Man. Dir H. KOJIMA.

Bank of Western Australia Ltd (BankWest): Level 7, BankWest Tower, 108 St George's Terrace, POB E237, Perth, WA 6001; tel. (8) 9449-7000; fax (8) 9449-7050; e-mail finmkts@bankwest.com.au; internet www.bankwest.com.au; f. 1895 as Agricultural Bank of Western Australia, 1945 as Rural and Industries Bank of Western Australia; present name adopted in 1994; cap. 715.9m., res 6m., dep. 12,753.2m. (Feb. 2001); Chair. IAN C. R. MACKENZIE; Man. Dir TERRY C. BUDGE; 109 brs.

Bankers' Trust Australia Ltd: GPOB H4, Australia Sq., Sydney, NSW 2000; tel. (2) 9259-3555; fax (2) 9259-9800; internet www.btal.com.au; f. 1986; cap. 273.3m., dep. 6,266.8m. (Dec. 1997); Man. Dir R. A. FERGUSON; 5 brs.

The Chase Manhattan Bank: GPOB 9816, Sydney, NSW 2001; tel. (2) 9250-4111; fax (2) 9250-4554; internet www.chase.com; Man. Dir W. SCOTT REID.

Citibank Ltd: GPOB 40, Sydney, NSW 2000; tel. (2) 8225-0000; fax (2) 9239-9110; internet www.citibank.com.au; f. 1954; cap. 457m., res 1m., dep. 5,619m. (Dec. 1998); CEO (Australasia) AHMED FAHOUR; Man. Dir LES MATHESON.

Commonwealth Bank of Australia: Level 7, 48 Martin Place, Sydney, NSW 1155; tel. (2) 9378-2000; fax (2) 9378-3317; internet www.commbank.com.au; f. 1912; merged with Colonial Ltd in 2000; cap. 13,365.0m., res 3,850.0m., dep. 172,373.0m. (June 2003); Chair. J. T. RALPH; CEO and Man. Dir D. V. MURRAY; more than 1,200 brs world-wide.

HSBC Bank Australia Ltd: Level 10, 1 O'Connell Street, Sydney, NSW 2000; tel. (2) 9255-2888; fax (2) 9255-2332; internet www.hsbc.com.au; f. 1985; fmrly Hongkong Bank of Australia; cap. 560.0m., res 100,000m., dep. 5,052.8m. (Dec. 2000); Chair D. J. SAY; CEO STUART DAVIS; 18 brs.

ING Bank (Australia) Ltd: Level 13, 140 Sussex St, Sydney, NSW 2000; tel. (2) 9028-4000; fax (2) 9028-4708; f. 1994; cap. 60m., res 1.3m., dep. 1,015.4m. (Dec. 1998); Gen. Man. JULIE BROWN.

Macquarie Bank Ltd: 1 Martin Place, Sydney, NSW 2000; tel. (2) 8232-3333; fax (2) 8232-3350; internet www.macquarie.com.au; f. 1969 as Hill Samuel Australia Ltd; present name adopted in 1985; cap. 2,020m., dep. 13,954m. (March 2002); Chair. DAVID S. CLARKE; Man. Dir ALLAN E. MOSS; 2 brs.

National Australia Bank Ltd: 500 Bourke St, Melbourne, Vic 3000; tel. (3) 8641-3500; fax (3) 8641-4912; internet www.national.com.au/; f. 1858; cap. 9,855m., res 2,006m., dep. 244,661m. (Sept. 2000); Chair. GRAHAM KRAEHE; Chief Exec. JOHN STEWART; 2,349 brs.

N. M. Rothschild & Sons (Australia) Ltd: Level 16, 1 O'Connell St, Sydney, NSW 2000; tel. (2) 9323-2000; fax (2) 9323-2323; internet www.rothschild.com.au; f. 1967 as International Pacific Corpn; cap. 130.0m., dep. 978.1m. (March 1999); Chair. PHILIP BRASS; Chief Exec. RICHARD LEE.

SG Australia Ltd: Level 21, 400 George St, Sydney, NSW 2000; tel. (2) 9210-8000; fax (2) 9235-3941; internet www.au.sg-ib.com; f. 1981; fmrly Société Générale Australia Ltd; cap. 21.5m., res 208m., dep. 5,656m. (Dec. 1999); CEO MICHEL L. MACAGNO.

St George Bank Ltd: Locked Bag 1, Kogarah, NSW 2217; tel. (2) 9952-1311; fax (2) 9952-1000; e-mail stgeorge@stgeorge.com.au; internet www.stgeorge.com.au; f. 1937 as building society; cap. 3,174m., res 130m., dep. 35,047m. (Sept. 2000); Chair. F. J. CONROY; CEO and Man. Dir GAIL KELLY; 421 brs.

Standard Chartered Bank Australia Ltd: Level 11, 345 George St, Sydney, NSW 2000; tel. (2) 9232-9333; fax (2) 9232-9345; f. 1986; cap. 226.2m., dep. 667.8m. (Dec. 1999); Chair. RICHARD NETTLETON; CEO EUGENE ELLIS.

Toronto Dominion Australia Ltd: Level 34, Rialto South Tower, 525 Collins St, Melbourne, Vic 3000; tel. (3) 9993-1344; fax (3) 9614-0083; internet www.tdbank.ca; f. 1970; cap. 191.5m., res 6.0m., dep. 3,435.4m. (Oct. 1997); Man. Dir STEVE FRYER.

Westpac Banking Corporation: 60 Martin Place, Sydney, NSW 2000; tel. (2) 9226-3311; fax (2) 9226-4128; e-mail westpac@westpac.com.au; internet www.westpac.com.au; f. 1817; cap. 2,258m., res 4,892m., dep. 113,169m. (Sept. 2000); Chair. L. A. DAVIS; Man. Dir DAVID MORGAN.

Foreign Banks

Bank of China (People's Republic of China): 39–41 York St, Sydney, NSW 2000; tel. (2) 9267-5188; fax (2) 9262-1794; e-mail bocsyd@bigpond.com.au; Gen. Man. GAO JI LU.

Bank of New Zealand: 9th Floor, BNZ House, 333–339 George St, Sydney, NSW 2000; tel. (2) 9290-6666; fax (2) 9290-3414; Chief Operating Officer G. ARMBRUSTER.

BNP Paribas (France): 60 Castlereagh St, Sydney, NSW 2000; POB 269, Sydney, NSW 2001; tel. (2) 9216-8633; fax (2) 9619-6104; e-mail finance@au.bnpparibas.com; internet www.bnpparibas.com.au; CEO FRANÇOIS CRISTOFARI; 4 brs.

Deutsche Bank AG (Germany): GPOB 7033, Sydney, NSW 1170; tel. (2) 9258-1234; fax (2) 9241-2565; internet www.australia.db.com; CEO CHUM DARVALL.

STOCK EXCHANGE

Australian Stock Exchange Ltd (ASX): Level 9, 20 Bridge St, Sydney, NSW 2000; tel. (2) 9227-0000; fax (2) 9227-0885; e-mail info@asx.com.au; internet www.asx.com.au; f. 1987 by merger of the stock exchanges in Sydney, Adelaide, Brisbane, Hobart, Melbourne and Perth, to replace the fmr Australian Associated Stock Exchanges; demutualized and listed Oct. 1998; Chair. MAURICE NEWMAN; Man. Dir and CEO RICHARD HUMPHRY.

Supervisory Body

Australian Securities and Investments Commission (ASIC): GPOB 9827, Sydney, NSW 2001; tel. (2) 9111-2000; fax (2) 9911-2333; e-mail infoline@asic.gov.au; internet www.asic.gov.au; f. 1990; corporations and financial products regulator; Chair. JEFFREY LUCY.

PRINCIPAL INSURANCE COMPANIES

Allianz Australia Ltd: 2 Market St, Sydney, NSW 2000; tel. (2) 9390-6222; fax (2) 9390-6425; internet www.allianz.com.au; f. 1914; workers' compensation; fire, general accident, motor and marine; Chair. J. S. CURTIS; Man. Dir T. TOWELL.

AMP Ltd: AMP Bldg, 33 Alfred St, Sydney, NSW 2000; tel. (2) 9257-5000; fax (2) 9257-7886; internet www.amplimited.com.au; f. 1849; fmrly Australian Mutual Provident Society; life insurance; Chair. STAN WALLIS; Man. Dir (vacant).

AMP General Insurance Ltd: 10 Loftus St, Sydney Cove, NSW 2000; tel. (2) 9257-2500; fax (2) 9257-2199; internet www.amp.com.au; f. 1958; Chair. GREG COX; Man. Dir GAVIN PEACE.

Australian Guarantee Corpn Ltd: 130 Phillip St, Sydney, NSW 2000; tel. (2) 9234-1122; fax (2) 9234-1225; f. 1925; Chair. J. A. UHRIG; Man. Dir R. THOMAS.

Australian Unity General Insurance Ltd: 114 Albert Rd, South Melbourne, Vic 3205; tel. (3) 9697-0219; fax (3) 9690-5556; e-mail webmaster@austunity.com.au; internet www.austunity.com.au; f. 1948; Chair. LEON HICKEY; Chief Exec. M. W. SIBREE.

Aviva Australia Holdings Ltd: GPOB 2567W, 509 St Kilda Rd, Melbourne 3001; tel. (3) 9829-8989; fax (3) 9820-1534; internet www.avivagroup.com.au; f. 2000 as CGNU following merger of CGU and Norwich Union, renamed as above in 2003; CEO ALLAN GRIFFITHS.

Catholic Church Insurances Ltd: 324 St Kilda Rd, Melbourne, Vic 3004; tel. (3) 9934-3000; fax (3) 9934-3460; f. 1911; Chair. Most Rev. KEVIN MANNING (Bishop of Parramatta); Gen. Man. PETER RUSH.

The Copenhagen Reinsurance Co Ltd: 60 Margaret St, Sydney, NSW 2000; tel. (2) 9247-7266; fax (2) 9235-3320; e-mail david.kennedy@copre.com; internet www.copre.com; reinsurance; Gen. Man. DAVID KENNEDY.

FAI Insurances Ltd: FAI Insurance Group, 333 Kent St, Sydney, NSW 1026; tel. (2) 9274-9000; fax (2) 9274-9900; internet www.fai.com.au; f. 1953; Chair. JOHN LANDERER; CEO RODNEY ADLER.

General Reinsurance Australia Ltd: Level 24, 123 Pitt St, Sydney, NSW 2000; tel. (2) 8236-6100; fax (2) 9222-1500; f. 1961; reinsurance, fire, accident, marine; Chair. F. A. MCDONALD; Man. Dir G. C. BARNUM.

GIO Australia Holdings Ltd: Level 39, Governor Phillip Tower, 1 Farrer Place, Sydney, NSW 2000; tel. (2) 9255-8090; fax (2) 9251-2079; e-mail emailus@gio.com.au; internet www.gio.com.au; f. 1926; CEO PETER CORRIGAN.

Guild Insurance Ltd: Guild House, 40 Burwood Rd, Hawthorn, Vic 3122; tel. (3) 9810-9820; fax (3) 9819-5670; f. 1963; Man. Dir W. K. Bastian.

HIH Insurance Ltd: AMP Centre, 50 Bridge St, Sydney, NSW 2000; tel. (2) 9650-2000; fax (2) 9650-2030; internet www.hih.com.au; f. 1968; Chair. G. A. Cohen; CEO R. R. Williams.

Lumley General Insurance Ltd: Lumley House, 309 Kent St, Sydney, NSW 1230; tel. (2) 9248-1111; fax (2) 9248-1122; e-mail general@lumley.com.au; Man. Dir D. M. Matcham.

The National Mutual Life Association of Australasia Ltd: 447 Collins St, Melbourne, Vic 3000; tel. (3) 9618-4920; fax (3) 9616-3445; e-mail investor.relations@axa.com.au; internet www.axa.com.au; f. 1869; life insurance, superannuation, income protection; Chair. R. H. Allert; Group CEO A. L. Owen.

NRMA Insurance Ltd: 151 Clarence St, Sydney, NSW 2000; tel. (2) 9292-9222; fax (2) 9292-8472; f. 1926; CEO Ian Brown (acting).

NZI Insurance Australia Ltd: 9th Floor, 10 Spring St, Sydney, NSW 2000; tel. (2) 9551-5000; fax (2) 9551-5865; Man. Dir H. D. Smith.

QBE Insurance Group Ltd: 82 Pitt St, Sydney, NSW 2000; tel. (2) 9375-4444; fax (2) 9235-3166; internet www.qbe.com; f. 1886; general insurance; Chair. E. J. Cloney; Man. Dir F. M. O'Halloran.

RAC Insurance Pty Ltd: 228 Adelaide Terrace, Perth, WA 6000; tel. (8) 9421-4444; fax (8) 9421-4593; f. 1947; Gen. Man. Tony Carter.

RACQ Insurance: POB 4, Springwood, Qld 4127; tel. (7) 3361-2444; fax (7) 3841-2995; e-mail inorris@racqi.com.au; internet www.racqinsurance.com.au; f. 1971; CEO I. W. Norris.

RACV: 550 Princes Highway, Noble Park, Vic 3174; tel. (3) 9790-2211; fax (3) 9790-3091.

RSA Insurance Australia Ltd: 465 Victoria Ave, Chatswood, NSW 2067; tel. (2) 9978-9000; fax (2) 9978-9807; fire, accident and marine insurance; Gen. Man. E. Kulk.

Suncorp-Metway Ltd: Level 18, 36 Wickham Tce, Brisbane, Qld 4000; tel. (7) 3835-5355; fax (7) 3836-1190; e-mail direct@suncorp.co.au; internet www.suncorp.com.au; f. 1996; Man. Dir Steve Jones.

Swiss Re Australia Ltd: 363 George St, Sydney, NSW 2000; tel. (2) 8295-9500; fax (2) 8295-9804; f. 1962; Man. Dir R. G. Watts.

Wesfarmers Federation Insurance Ltd: 184 Railway Parade, Bassendean, WA 6054; tel. (8) 9273-5770; fax (8) 9273-5290; e-mail mel.rom@wfi.wesfarmers.com.au; internet www.wfi.com.au; Gen. Man. R. J. Buckley.

Westpac Life Ltd: 35 Pitt St, Sydney, NSW 2000; tel. (2) 9220-4768; f. 1986; CEO David White.

World Marine & General Insurances Ltd: 600 Bourke St, Melbourne, Vic 3000; tel. (3) 9609-3333; fax (3) 9609-3634; f. 1961; Chair. G. W. McGregor; Man. Dir A. E. Reynolds.

Zurich Financial Services Australia Ltd: 5 Blue St, North Sydney, NSW 2060; tel. (2) 9391-1111; fax (2) 9922-4630; CEO Malcolm M. Jones.

Insurance Associations

Australian Insurance Association: 54 Marcus Clarke St, Canberra City ACT; tel. (2) 6249-7666; f. 1968; Pres. Raymond Jones; Exec. Sec. P. M. Murphy.

Australian and New Zealand Institute of Insurance and Finance: Level 17, 31 Queen St, Melbourne, Vic 3000; tel. (3) 9629-4021; fax (3) 9629-4204; e-mail ceo@theinstitute.com.au; internet www.theinstitute.com.au; f. 1919; Pres. John Richardson; CEO Joan Fitzpatrick; 11,984 mems.

Insurance Council of Australia Ltd: Level 3, 56 Pitt St, Sydney, NSW 2000; tel. (2) 9253-5100; fax (2) 9253-5111; internet www.ica.com.au; f. 1975; CEO Alan Mason.

Investment and Financial Services Association (IFSA): Level 24, 44 Market St, Sydney, NSW 2000; tel. (2) 9299-3022; fax (2) 9299-3198; e-mail ifsa@ifsa.com.au; internet www.ifsa.com.au; f. 1997 following merger of Australian Investment Managers' Association, Investment Funds Association and Life, Investment and Superannuation Asscn of Australia Inc; non-profit organization; Chair. Dr Doug McTaggart; CEO Richard Gilbert.

Trade and Industry

GOVERNMENT AGENCY

Austrade: GPOB 5301, Sydney, NSW 2001; tel. (2) 9390-2000; fax (2) 9390-2024; e-mail info@austrade.gov.au; internet www.austrade.gov.au; f. 1931; export promotion agency; Chair. Ross Adler; Man. Dir Peter O'Byrne.

CHAMBERS OF COMMERCE

Australian Chamber of Commerce and Industry (ACCI): POB E14, Kingston, ACT 2604; tel. (2) 6273-2311; fax (2) 6273-3286; e-mail acci@acci.asn.au; internet www.acci.asn.au; Pres. Dr John Keniry; CEO Mark Paterson.

Chamber of Commerce and Industry of Western Australia (CCIWA): POB 6209, East Perth, WA 6892; tel. (8) 9365-7555; fax (8) 9365-7550; e-mail info@cciwa.com; internet www.cciwa.com; f. 1890; 6,000 mems; Chief Exec. Ross McLean; Pres. Andrew Thompson.

Commerce Queensland: Industry House, 375 Wickham Terrace, Brisbane, Qld 4000; tel. (7) 3842-2244; fax (7) 3832-3195; e-mail info@commerceqld.com.au; internet www.commerceqld.com.au; f. 1868; operates World Trade Centre, Brisbane; 5,500 mems; CEO Andrew Craig.

South Australian Employers' Chamber of Commerce and Industry Inc: Enterprise House, 136 Greenhill Road, Unley, SA 5061; tel. (8) 8300-0000; fax (8) 8300-0001; e-mail enquiries@business-sa.com; internet www.business-sa.com; f. 1839; 4,700 mems; CEO P. Vaughan.

State Chamber of Commerce (New South Wales): Level 12, 83 Clarence St, Sydney, NSW 2000; tel. (2) 9350-8100; fax (2) 9350-8197; e-mail trade@thechamber.com.au; internet www.thechamber.com.au; f. 1825; offers advice to and represents over 70,000 businesses; Chief Exec. Margy Osmond.

Tasmanian Chamber of Commerce and Industry: GPOB 793H, Hobart, Tas 7001; tel. (3) 6234-5933; fax (3) 6231-1278; CEO Tim Abey.

Victorian Employers' Chamber of Commerce and Industry: Industry House, 486 Albert St, Melbourne, Vic 3002; tel. (3) 8662-5333; fax (3) 8662-5462; e-mail vecci@vecci.org.au; internet www.vecci.org.au; f. 1885; CEO Neil Coulson.

AGRICULTURAL, INDUSTRIAL AND TRADE ASSOCIATIONS

Australian Business Ltd: Locked Bag 938, North Sydney, NSW 2059; tel. (2) 9458-7500; fax (2) 9923-1166; e-mail member.services@australianbusiness.com.au; internet www.australianbusiness.com.au; f. 1885; fmrly Chamber of Manufactures of NSW; CEO Mark Bethwaite.

Australian Manufacturers' Export Council: POB E14, Queen Victoria Terrace, ACT 2600; tel. (2) 6273-2311; fax (2) 6273-3196; f. 1955; Exec. Dir G. Chalker.

Australian Wine and Brandy Corporation (AWBC): POB 2733, Kent Town Business Centre, Kent Town, SA 5071; tel. (8) 8228-2000; fax (8) 8228-2022; e-mail awbc@awbc.com.au; internet www.awbc.com.au; Chief Exec. Sam Totley.

Australian Wool Services Ltd: Wool House, 369 Royal Parade, Parkville, Vic 3052; tel. (3) 9341-9111; fax (3) 9341-9273; internet www.woolmark.com; f. 2001 following the privatization of the Australian Wool Research and Promotion Organisation; operates two subsidiaries: Australian Wool Innovation and The Woolmark Company, which is responsible for commercial development; Chair. Trevor Flugge.

AWB Ltd: Ceres House, 528 Lonsdale St, Melbourne, Vic 3000; tel. (3) 9209-2000; fax (3) 9670-2782; e-mail awb@awb.com.au; internet www.awb.com.au; f. 1939; fmrly Australian Wheat Board; national and international marketing of grain, financing and marketing of wheat and other grains for growers; 12 mems; Chair. Brendan Stewart; CEO Andrew Lindberg.

Business Council of Australia: GPOB 1472N, Melbourne, Vic 3001; tel. (3) 8664-2664; fax (3) 8664-2666; e-mail info@bca.com.au; internet www.bca.com.au; public policy research and advocacy; governing council comprises chief execs of Australia's major cos; Pres. Stan Wallis; Chair. Hugh Morgan.

Cotton Australia: Level 2, 490 Crown St, Surry Hills, NSW 2010; tel. (2) 9360-8500; fax (2) 9360-8555; e-mail talktous@cottonaustralia.com.au; internet www.cottonaustralia.com.au; Chair. Charles Wilson.

Meat and Livestock Australia: Level 1, 165 Walker St, North Sydney, NSW 2060; tel. 1800-023-100; fax (2) 9463-9393; internet www.mla.com.au; producer-owned co; represents, promotes, protects and furthers interests of industry in both the marketing of meat and livestock and industry-based research and devt activities; Chair. David Crombie.

National Farmers' Federation: POB E10, Kingston, ACT 2604; tel. (2) 6273-3855; fax (2) 6273-2331; e-mail nff@nff.org.au; internet www.nff.org.au; Pres. PETER CORISH; CEO ANNA CRONIN.

Natural Resources Management Ministerial Council (NRMC): Dept of Agriculture, Fisheries and Forestry—Australia, Barton, Canberra, ACT 2600; tel. (2) 6272-5216; fax (2) 6272-4772; internet www.mincos.gov.au; f. 2002 to replace the Agricultural and Resource Management Council of Australia and New Zealand to promote the conservation and sustainable use of Australia's natural resources; mems comprising the Commonwealth/state/territory and New Zealand ministers responsible for environment, water and natural resources.

Primary Industries Ministerial Council (PIMC): Dept of Agriculture, Fisheries and Forestry, Barton, Canberra, ACT 2600; tel. (2) 6272-5216; fax (2) 6272-4772; e-mail pimc@mincos.gov.au; internet www.mincos.gov.au; f. 2002 to develop and promote sustainable, innovative and profitable agriculture, fisheries, food and forestry industries; mems comprising the state/territory and New Zealand ministers responsible for agriculture, fisheries, food and forestry.

Trade Policy Advisory Council (TPAC): c/o Dept of Foreign Affairs and Trade, R. G. Casey Bldg, John McEwen Cres., Barton, ACT 0221; tel. (2) 6261-2910; fax (2) 6261-1385; e-mail tpac@dfat.gov.au; internet www.dfat.gov.au/trade/opening_doors/tpac.html; advises the Minister for Trade on business and investment issues; Chair. GEOFF ALLEN.

WoolProducers: POB E10, Kingston, Canberra, ACT 2604; tel. (2) 6273-2531; fax (2) 6273-1120; e-mail woolproducers@nff.org.au; internet www.woolproducers.com; fmrly Wool Council Australia; comprises 20 mems; represents wool-growers in dealings with the Federal Govt and industry; Pres. SIMON CAMPBELL.

EMPLOYERS' ORGANIZATIONS

Australian Co-operative Foods Ltd: 433 Victoria St, Wetherill Park, NSW 2164; tel. (2) 4821-1391; e-mail exports@dairyfarmers.com.au; f. 1900; Man. Dir A. R. TOOTH.

Australian Industry Group: 51 Walker St, North Sydney, NSW 2060; tel. (2) 9466-5566; fax (2) 9466-5599; e-mail louisep@aignsw.aigroup.asn.au; internet www.aigroup.asn.au; f. 1998 through merger of MTIA and ACM; 11,500 mems; Nat. Pres. G. J. ASHTON; CEO ROBERT N. HERBERT.

Australian Meat Industry Council: 25–27 Albany St, Crows Nest, NSW 2065; POB 1208, Crows Nest, NSW 1585; tel. (2) 9086-2200; fax (2) 9086-2201; e-mail admin@amic.org.au; internet www.amic.org.au; f. 1928; Chair. GARY HARDWICK; CEO KEVIN COTTRILL.

NSW Farmers' Association: GPOB 1068, Sydney, NSW 2001; tel. (2) 8251-1700; fax (2) 8251-1750; e-mail emailus@nswfarmers.org.au; internet www.nswfarmers.org.au; f. 1978; CEO JONATHAN MCKEOWN; Pres. MAL PETERS.

UTILITIES

Australian Institute of Energy: POB 268, Toukley, NSW 2263; tel. 1800-629-945; fax (2) 4393-1114; e-mail aie@tpgi.com.au; internet www.aie.org.au.

Australian Water Association: POB 388, Artarmon, NSW 1570; tel. (2) 9413-1288; fax (2) 9413-1047; e-mail info@awa.asn.au; internet www.awa.asn.au; f. 1962; c. 4,000 mems; CEO CHRIS DAVIS.

Electricity Supply Association of Australia: GPOB 1823Q, Melbourne, Vic 3001; tel. (2) 9670-0188; fax (2) 9670-1069; internet www.esaa.com.au; CEO BRAD PAGE.

Electricity Companies

Actew AGL: GPOB 366, Canberra, ACT 2601; tel. (2) 6248-3111; e-mail webmaster@actewagl.com.au; internet www.actewagl.com.au; f. 2000 by amalgamation of ACTEW Corpn Ltd and AGL; supplier of electricity, gas, water and wastewater services; Chief Exec. PAUL PERKINS.

Delta Electricity: POB Q863, QVB PO, NSW 1230; tel. (2) 9285-2700; fax (2) 9285-2777; internet www.de.com.au; f. 1996; Chief Exec. JIM HENNESS.

ENERGEX: GPOB 1461, Brisbane, Qld 4001; tel. (7) 3407-4000; fax (7) 3407-4609; e-mail enquiries@energex.com.au; internet www.energex.com.au; spans Queensland and New South Wales; CEO GREG MADDOCK.

EnergyAustralia: 145 Newcastle Rd, Wallsend, NSW 2287; tel. (2) 4951-9346; fax (2) 4951-9351; e-mail energy@energy.com.au; internet www.energy.com.au; supplies customers in NSW; CEO PETER HEADLEY; Man. Dir PAUL BROAD.

Ergon Energy: POB 107, Albert St, Brisbane, Qld 4002; tel. (7) 3228-8222; fax (7) 3228-8118; internet www.ergon.com.au; national retailer of electricity.

Generation Victoria Corpn—Ecogen Energy: 5th Floor, 416 Collins St, Melbourne, Vic 3000; tel. (3) 9679-4600; fax (3) 9679-4619; internet www.ecogen-energy.com.au/; f. 1994; CEO GERRY BASTEN.

Great Southern Energy: Level 1, Citilink Plaza, Morriset St, Queanbeyan, NSW 2620; tel. (2) 6214-9600; fax (2) 6214-9860; e-mail mail@gsenergy.com.au; internet www.gsenergy.com.au; state-owned electricity and gas distributor; Chair. BRUCE RODELY.

Power and Water Corpn: Energy House, 18–20 Cavenagh St, Darwin, NT 0800; tel. 1800-245-092; fax (8) 8924-7730; e-mail customerservice@powerwater.com.au; internet www.powerwater.com.au; state-owned; supplier of electricity, water and sewerage services in NT; Man. Dir KIM WOOD; Chair. NEIL PHILIP.

Powercor Australia Ltd: 40 Market St, Melbourne, Vic 3000; tel. (3) 9683-4444; fax (3) 9683-4499; e-mail info@powercor.com.au; internet www.powercor.com.au; Chair. WILLIAM SHURNIAK; CEO C. T. WAN.

Snowy Mountains Hydro-electric Authority: POB 332, Cooma, NSW 2630; tel. (2) 6452-1777; fax (2) 6452-3794; e-mail info@snowyhydro.com.au; internet www.snowyhydro.com.au.

United Energy Ltd: Level 13, 101 Collins St, Melbourne, Vic 3000; fax (3) 9222-8588; e-mail info@mail.ue.com.au; internet www.ue.com.au; f. 1994 following division of State Electricity Commission of Victoria; transferred to private sector; distributor of electricity and gas.

Western Power Corpn: GPOB L921, Perth, WA 6842; tel. (8) 9326-4911; fax (8) 9326-4595; e-mail info@wpcorp.com.au; internet www.wpcorp.com.au; f. 1995; principal supplier of electricity in WA; Chair. HECTOR STEBBINS (acting); Man. Dir DAVID EISZELE.

Gas Companies

AlintaGas: GPOB W2030, Perth, WA 6846; internet www.alintagas.com.au; f. 1995; CEO ROBERT BROWNING.

Allgas Energy Ltd: 150 Charlotte St, Brisbane, Qld 4000; tel. (7) 3404-1822; fax (7) 3404-1821; e-mail corporate@allgas.com.au; internet www.allgas.com.au; f. 1885; Chief Exec. TOM BLOXSOM.

Australian Gas Light Co: AGL Centre, Corner Pacific Highway and Walker St, North Sydney, NSW 2060; tel. (2) 9922-0101; fax (2) 9957-3671; e-mail aglmail@agl.com.au; internet www.agl.com.au; f. 1837; Chair. M. J. PHILLIPS; Man. Dir. GREG MARTIN.

Envestra: 10th Floor, 81 Flinders St, Adelaide, SA 5000; tel. (8) 8227-1500; fax (8) 8277-1511; e-mail des.petherick@envestra.com.au; internet www.envestra.com.au; f. 1997 by merger of South Australian Gas Co, Gas Corpn of Queensland and Centre Gas Pty Ltd; purchased Victorian Gas Network in 1999; Chair. J. G. ALLPASS; Man. Dir I. B. LITTLE.

Epic Energy: GPOB 657, Brisbane, Qld 4001; tel. (7) 6200-1600; fax (7) 3218-1650; internet www.epicenergy.com.au; f. 1996; privately-owned gas transmission co; CEO SUE ORTENSTONE.

Origin Energy: GPOB 5376, Sydney, NSW 2001; tel. (2) 9220-6400; fax (2) 9235-1661; internet www.origin.energy@originenergy.com.au; Man. Dir GRANT KING.

TXU: Locked Bag 14060, Melbourne, Vic 8001; tel. (3) 133-466; fax (3) 9299-2777; e-mail enq@txu.com.au; internet www.txu.com.au.

Water Companies

Actew AGL: see Electricity, above.

Melbourne Water Corpn: 100 Wellington Parade, East Melbourne, Vic 3002; tel. (3) 9235-7100; fax (3) 9235-7200; state-owned; Man. Dir BRIAN BAYLEY.

Power and Water Corpn: see Electricity, above.

South Australian Water Corpn: 77 Grenfell St, Adelaide, SA 5000; tel. (8) 8204-1000; fax (8) 8204-1048; internet www.sawater.com.au; state-owned; Chief Exec. ANNE HOWE.

South East Water Ltd: POB 1382, Moorabbin, Vic 3189; tel. (3) 9552-3000; fax (3) 9552-3001; e-mail info@sewl.com.au; internet www.sewl.com.au; f. 1995; state-owned; Man. Dir DENNIS CAVAGNA.

Sydney Water Corpn: Ground Floor, 115–23 Bathurst St, Sydney NSW 2000; internet www.sydneywater.com.au; state-owned; Man. Dir GREG ROBINSON; Chair. GABRIELLE KIBBLE.

Water Corpn: 629 Newcastle St, Leederville, WA 6007; tel. (8) 9420-2420; fax (8) 9420-3200; e-mail webmaster@watercorporation.com.au; internet www.watercorporation.com.au; state-owned; Man. Dir Dr JIM GILL.

Yarra Valley Water Ltd: Lucknow St, Mitcham, Vic 3132; tel. (3) 9874-2122; fax (3) 9872-1353; e-mail enquiry@yvw.com.au; internet www.yvw.com.au; f. 1995; state-owned; Man. Dir TONY KELLY.

MAJOR COMPANIES

Mining and Metals

Alcoa World Alumina Australia: POB 252, Applecross, WA 6153; tel. (8) 9316-5111; fax (8) 9316-5228; internet www.alcoa.com.au; f. 1961; cap. and res $A2,317.1m., sales $A3,441.2m. (2001); producer of aluminium, bauxite, etc.; Chair. G. John Pizzey; Pres. Wayne Osborn; 5,379 employees.

BHP Billiton Ltd: 48th Floor, BHP Tower-Bourke Place, 600 Bourke St, Melbourne, Vic 3000; tel. (3) 9609-3333; fax (3) 9609-3015; internet www.bhpbilliton.com.au; f. 2001 by merger of BHP (f. 1885) and Billiton (UK); cap. and res US $12,356m., sales US $17,778m. (2002); mining of iron ore, coal, copper, silver and diamonds; iron and steelmaking; oil and natural gas production; operates in every state, the NT and in many foreign countries; has 165 subsidiaries and nine major associated cos; Chair. Don Argus; CEO Charles Goodyear; 61,000 employees.

Comalco Ltd: Level 25, Comalco Place, 12 Creek St, Brisbane, Qld 4000; tel. (7) 3867-1711; fax (3) 3867-1775; internet www.comalco.com.au; f. 1960; cap. and res $A1,896m., sales $A2,404m. (1999); aluminium production; Chair. John P. Morschel; Chief Exec. W. Terry Palmer; 4,747 employees.

Goldfields Ltd: Level 16, 1 Castlereagh St, Sydney, NSW 2000; tel. (2) 8223-2400; fax (2) 8223-2444; f. 1995; fmrly Goldfields Kalgoorlie; cap. and res $A89.3m., sales $301.7m. (1999/2000); mining; Chair. R. F. E. Warburton; Man. Dir P. W. Cassidy; 405 employees.

MIM Holdings Ltd: 410 Ann St, Brisbane, Qld 4000; tel. (7) 3833-8000; fax (7) 3832-2426; e-mail corpaff@mimholdings.com.au; internet www.mimholdings.com.au; f. 1970; cap. and res $A2,548m., sales $A4,011m. (2001/02); exploration for precious and base metals; mining, processing and marketing of copper, gold, zinc, lead and silver; mining and marketing of coal; recycling of metals; Chair. L. E. Tutt; Man Dir. V. P. Gauci; 8,434 employees.

Newcrest Mining Ltd: Level 9, 600 St Kilda Rd, Melbourne, Vic 3004; tel. (3) 9522-5333; fax (3) 9525-2996; e-mail invrelations@newcrest.com.au; internet www.newcrest.com.au; f. 1990; mining of gold and other minerals; cap. and res $A448.2m., sales $A581.1m. (2000/01); Chair. Ian R. Johnson; 775 employees.

Normandy Mining Ltd: 100 Hutt St, Adelaide, SA 5000; tel. (8) 8303-1700; fax (8) 8303-1900; internet www.normandy.com.au; f. 1991; cap. and res $A1,398m., sales $A1,543.7m. (2000/01); mining of gold, diamonds and base metals; Chair. and CEO Robert J. Champion de Crespigny; 3,000 employees.

Placer Dome Asia Pacific Ltd: RAMS House, Level 2, 189 Corporation Drive, Milton, Qld 4064; tel. (7) 3510-6700; fax (7) 3510-6740; e-mail placer_pacific@placerdome.com; internet www.placerdome.com; f. 1986; mineral exploration, mining of gold, silver and copper ores; Chair. J. M. Willson; Man. Dir D. W. Zandee; 3,260 employees.

Rio Tinto Ltd: 33rd Floor, 55 Collins St, Melbourne, Vic 3001; tel. (3) 9283-3333; fax (3) 9283-3707; internet www.north.com.au; f. 1962; fmrly CRA; cap. and res $A14,029m., sales $A20,187m. (2001); exploration and mining group, principally iron ore, aluminium, coal, salt, gold, silver and diamonds; Chair. Sir Robert P. Wilson; Chief Exec. Leigh Clifford; 36,000 employees.

Santos Ltd: Santos House, 91 King William St, Adelaide, SA 5000; tel. (8) 8218-5111; fax (8) 8218-5274; e-mail investor.relations@santos.com; internet www.santos.com; market cap. $A4,016m., sales $A1465m. (2003); gas and petroleum exploration and production; Chair. Stephen Gerlach; Man. Dir J. C. Ellice-Flint; 1,600 employees.

Tubemakers of Australia Ltd: 1 York St, Sydney, NSW 2000; tel. (2) 9239-6666; fax (2) 9251-3042; f. 1946; cap. $A141.4m., sales $A1,400m. (1999); mfrs of steel pipes and tubes; merchandiser of steel and aluminium products; wholly-owned subsidiary of BHP Steel; 4,000 employees.

WMC Ltd: Level 16, IBM Centre, 60 City Road, Southbank, Vic 3006; tel. (3) 9685-6000; fax (3) 9670-3569; internet www.wmc.com.au; f. 1933; fmrly Western Mining Corpn Holdings Ltd; cap. and res $A4,853.4m., sales $A2,796.9m. (2001); mining and processing of nickel, gold, copper, uranium, talc and petroleum; exploration for and development of mineral resources; investment in alumina and alumina chemicals; Chair. Ian Burgess; CEO H. M. Morgan; 4,634 employees.

Motor Vehicles

Ford Motor Company of Australia Ltd: 1735 Sydney Rd, Campbellfield, Vic; tel. (3) 9359-8211; fax (3) 9359-8200; internet www.ford.com.au; f. 1925; total assets $A1,670m., sales $A3,261m. (2002); mfrs of passenger and commercial motor vehicles and parts and accessories; Pres. Geoff Polites; 5,000 employees.

Holden Ltd: 241 Salmon St, Port Melbourne, Vic 3207; tel. (3) 9647-1111; fax (3) 9647-2550; internet www.holden.com.au; f. 1986; subsidiary of General Motors Corpn; sales $A4,700m. (2000); mfrs of passenger and commercial vehicles; Chair. and Man. Dir Peter H. Hanenberger; 7,858 employees.

Iveco Trucks Australia Ltd: Princes Hwy, Dandenong, Vic 3175; POB 117, Dandenong, Vic 3175; tel. (3) 9238-2200; fax (3) 9238-2387; e-mail iveco@iveco.com.au; internet www.iveco.com.au; f. 1912; cap. and res $A81.3m. (2001), sales $A377.5m. (2003); designers, mfrs and marketers of trucks; Man. Dir Alain Gajnik; 800 employees.

Mitsubishi Motors Australia Ltd: 1284 South Rd, Clovelly Park, SA 5042; tel. (8) 8275-7111; fax (8) 8275-6841; internet www.mitsubishi-motors.com.au; issued cap. $A279.3m., sales $A2,694m. (2001); mfrs of cars, service parts, accessories, automotive components, engines; Pres. and CEO T. R. Phillips; 4,100 employees.

Toyota Motor Corpn Australia Ltd: 155 Bertie St, Port Melbourne, Vic 3207; tel. (3) 9647-4444; fax (3) 9645-1311; internet www.toyota.com.au; cap. and res $A1,830m., sales $A5,660m. (2001); Pres. K. Asano; 4,200 employees.

Petroleum

BP Australia Holdings Ltd: Level 30, The Tower, Melbourne Central, 360 Elizabeth St, Melbourne, Vic 3000; tel. (3) 9268-4111; fax (3) 9268-3321; f. 1952; cap. and res $A1,379.6m. (2000); refining, marketing, exploration, transportation of petroleum products; 30 subsidiaries; Chair. and Man. Dir Gregory D. Bourne; 1,800 employees.

ChevronTexaco Australia Pty Ltd: L 24, 250 St George's, GPOB S1580, Perth, WA 6845; tel. (8) 9216-4000; fax (8) 9216-4444; e-mail ctapl@chevrontexaco.com; fmrly West Australian Petroleum; exploration and production of petroleum and natural gas; Man. Dir Jay Johnson.

Esso Australia Ltd: 12 Riverside Quay, Southbank, Vic 3000; tel. (3) 9270-3333; fax (3) 9270-3995; internet www.exxon.com; revenue $A1.5m. (1998); subsidiary of Exxon Corpn; active in the upstream oil and gas business; Chair. and Man. Dir Robert C. Olsen; 1,002 employees.

Mobil Exploration & Producing Australia Pty Ltd: GPOB 4507, Melbourne, Vic 3001; tel. (3) 9252-3111; internet www.mobil.com.au; f. 1996 through acquisition of Ampolex Ltd; upstream exploration activities and producing activities in Australia and Papua New Guinea; Man. Dir D. P. Haworth.

Shell Australia Ltd: POB 872K, GPO, Melbourne, Vic 3001; tel. (3) 9666-5444; fax (3) 9666-5008; internet www.shell.com.au; inc. 1958; cap. $A900m., total assets $A4,879.1m. (2000); group manufactures and markets petroleum and petroleum products; exploration and production of oil, gas and coal; 50 group cos; Chair. and CEO P. Duncan; 5,900 employees.

Woodside Petroleum Ltd: 1 Adelaide Terrace, Perth, WA 6000; tel. (8) 9348-4000; fax (8) 9325-8178; internet www.woodside.com.au; cap. $A708.3m., total assets $A2,554.2m. (2001); exploration and production of petroleum and natural gas; Chair. Charles Goode; Man. Dir John H. Akehurst; 2,198 employees.

Rubber and Textiles

Pacific Dunlop Ltd: Level 41, 101 Collins St, Melbourne, Vic 3000; tel. (3) 9270-7270; fax (3) 9270-7300; internet www.pacdun.com; cap. and res $A855m., sales $A5,726m. (1999/2000); marketing, manufacturing and importing of batteries, tyres, plastic and industrial rubber products; latex rubber products; surgical gloves; household gloves; sports equipment; foams and bedding; industrial and thermal insulation; textiles and clothing; footwear; electrical products and telecommunications cables; Chair. John Ralph; Man. Dir Rod Chadwick; 37,836 employees.

Paper and Pulp

Amcor Ltd: 679 Victoria St, Abbotsford, Vic 3067; tel. (3) 9226-9000; fax (3) 9226-9050; internet www.amcor.com.au; f. 1926 as Australian Paper Manufacturers (APM) Ltd; cap. and res $A2,569.7m., total revenue $A5,891.7m. (2000/01); afforestation; production of woodpulp, paper and paperboard and associated goods; manufacture and sale of metal, paper, plastic and corrugated packaging products; trading of industrial and consumer products; 300 locations worldwide, operating in 25 countries; Chair. Chris Roberts; Man. Dir Russell H. Jones; 23,300 employees.

Food and Drink, etc.

Arnotts Ltd: 11 George St, Homebush, NSW 2160; tel. (2) 9394-3555; fax (2) 9394-3500; e-mail croberts@arnotts.com; internet www.arnotts.com.au; cap. and res $A330.6m., sales $A723.5m. (1996/97); manufacture of biscuits and snacks; Chair. D. M. McDonald; Man. Dir and CEO C. I. Roberts; 4,900 employees.

British American Tobacco Australasia Holdings Ltd: Virginia Park, Westfield Drive, Eastgardens 2036, NSW; tel. (2) 9370-1500; internet www.bata.com.au; Chair. Nick Greiner; 1,200 employees.

Burns Philp & Co Ltd: Level 23, 56 Pitt St, Sydney, NSW 2000; tel. (2) 9259-1111; fax (2) 9247-3272; e-mail shareholder.enquiries@burnsphilp.com; internet www.burnsphilp.com; f. 1883; cap. and res $A733m., sales $A1,880.2m. (2002/03); global food mfr, mainly yeast, bread, breakfast cereals, snack foods, edible oils, herbs and spices; Chair. Alan McGregor; Man. Dir and Chief Exec. Tom Degnan; 14,000 employees.

Cadbury Schweppes Australia Ltd: 636 St Kilda Rd, Melbourne, Vic 3004; tel. (3) 9520-7444; fax (3) 9520-7400; internet www.cadbury.com.au; f. 1971; cap. and res $A634.2m., sales $A1,529m. (2001); mfrs and distributors of chocolate and sugar confectionery, jams, soft drinks, post mix syrups, fruit juices; Chief Exec. G. M. Cassagne (Asia-Pacific); 4,000 employees.

Coca-Cola Amatil Ltd: GPOB 145, GPO Sydney, NSW 2001; tel. (2) 9259-6666; fax (2) 9259-6623; internet www.ccamatil.com; f. 1904; cap. and res $A3,032m., total revenue $A5,904.4m. (2001); manufacturing and distribution of beverages in Asia-Pacific; 24 plants, of which 18 are overseas; Chair. David M. Gonski; 31,800 employees.

Foster's Brewing Group: 77 Southbank Boulevard, Southbank, Vic 3006; tel. (3) 9633-2000; fax (3) 9633-2002; e-mail info@forestplace.com.au; internet www.fosters.com.au; f. 1962; cap. and res $A3,779m., total revenue $A4,176.7m. (2000/01); production and distribution of beer and wine; operates in Australia, Asia-Pacific and the UK; Chair. Frank Swan; CEO E. T. Kunkel; 13,700 employees.

Goodman Fielder Ltd: 75 Talavera Rd, Macquarie Park, North Ryde, NSW 2113; tel. (2) 8874-6000; fax (2) 8874-6099; e-mail corporate.affairs@goodmanfielder.com.au; internet www.goodmanfielder.com.au; cap. and res $A1,136.7m., sales $A3,062.5m. (2000/01); manufacture, sale and export of foods; Chair. Dr Keith Barton; CEO Thomas P. Park; 14,299 employees.

National Foods Ltd: Level 10, 5 Queens Rd, Melbourne, Vic 3004; tel. (3) 9234-4000; fax (3) 9234-4100; e-mail ian.greenshields@natfoods.com.au; internet www.natfoods.com.au; cap. and res. $A370.6m., sales $A970.9m. (2000/01); Chair. W. B. Capp; Man. Dir M. G. Ould; 1,712 employees.

Philip Morris Ltd: 252 Chesterville Rd, POB 1093, Moorabbin, Vic 3189; tel. (3) 8531-1000; fax (3) 8531-1900; internet www.philipmorris.com; mfrs of tobacco products; total revenue US $80.4m. (2000); CEO Geoffrey Bible; 1,064 employees.

Southcorp Ltd: 403 Pacific Highway, Artarmon, NSW 2064; tel. and fax (2) 9465-1000; internet www.southcorp.com.au; cap. and res $A2,375m., sales $A2,667m. (2002/03); wine; Chair. T. Brian Finn; Man. Dir and Chief Exec. John C. Ballard; 3,000 employees.

Unilever Australia Export: Private Bag 2, Epping, NSW 2121; tel. (2) 9869-6450; fax (2) 9869-6480; e-mail mike.a.newman@unilever.com; internet www.uae.com.au; sales $A1,288.3m. (2000); marketing ice cream, food, edible oils, toiletries, soap and detergents; Chair. Enzo Allara.

George Weston Foods Ltd: Level 20, Tower A, Zenith Centre, 821 Pacific Highway, Chatswood, NSW 2067; tel. (2) 9415-1411; fax (2) 9419-2907; internet www.gwf.com.au; cap. and res $A673.5m., sales $A1,607.5m. (2000/01); Chair. John Henry Pascoe; CEO Marvin Weinman; 6,801 employees.

Miscellaneous

Amcor Containers Packaging Australasia: 971 Burke Rd, Camberwell, Vic 3124; tel. (3) 9811-7444; fax (3) 9811-7474; f. 1950; mfrs of aluminium, steel and aerosol cans for food, beverage and other consumer goods, metal and plastic caps and closures, flexible, foil and film packaging, rigid plastic containers, composite cans for food and industrial products, multi-wall paper sacks, and packaging machinery; Man. Dir D. Goldthorp; 3,600 employees.

Boral Ltd: AMP Centre, Level 39, 50 Bridge St, Sydney, NSW 2000; tel. (2) 9220-6300; fax (2) 9233-6605; e-mail info@boral.com.au; internet www.boral.com.au; cap. and res $A3,112.1m., sales $A4,898.6m. (1999); quarrying, sand extraction, premixed concrete, fly ash, lightweight aggregate producer, clay and concrete products, road surfacing, road transport, marketing and distribution of cements and industrial lime, windows, doors, timber, plasterboard, bricks, etc.; cap. and res $A1,854.5m., sales $A3,280.2m. (2000/01); Chair. Dr Kenneth J. Moss; Man. Dir Rod T. Pearse; 11,411 employees.

Brambles Industries Ltd: Level 40, Gateway, 1 Macquarie Place, Sydney, NSW 2000; tel. (2) 9256-5222; fax (2) 9256-5299; e-mail info@brambles.com; internet www.brambles.com.au; f. 1875; cap. and res $A2,200.7m., sales $A3,493.8m. (2000/01); materials movement and distribution, including industrial plant hire, equipment pools, scheduled freight forwarding by road, rail, sea and air, heavy haulage, logistical support programmes for major projects, marine towage and transportation, pollution control services, security services, records management, shipping and travel agencies; Chair. D. R. Argus; CEO David Turner; 13,283 employees.

CSR Ltd: Level 1, 9 Help St, Chatswood, NSW 2057; tel. (2) 9235-8000; fax (2) 9235-8044; e-mail info@csr.com.au; internet www.csr.com.au; f. 1855; cap. and res $A4,022m., total revenue $A7,195m. (2001/02); manufacture of building materials, sugar milling, and investments in aluminium; Chair. John Morschel; 16,057 employees.

Futuris Corpn Ltd: 26/27 Currie St, Adelaide, SA 5000; tel. (8) 8425-4999; fax (8) 8410-1597; e-mail information@futuris.com.au; internet www.futuris.com.au; cap. and res $A384.3m., sales $A6,776.1m. (2001/02); provision of services to rural sector, automotive components, etc.; Chair. William Beischer; Man. Dir Alan Newman.

James Hardie Industries Ltd: 10 Colquhoun St, Rosehill, NSW 2142; tel. (2) 8837-4709; e-mail info@jameshardie.com.au; internet www.jameshardie.com; cap. and res $A510m., sales $A1,549m. (2000/01); mfrs of fibre cement building products and building systems, supply and installation of insulated panel systems; Chair. A. G. McGregor; CEO Dr R. K. Barton; 3,729 employees.

Kodak (Australasia) Pty Ltd: 173 Elizabeth St, Coburg, Vic 3058; tel. (3) 9350-1222; fax (3) 9350-2416; internet www.kodak.com.au; cap. and res $A190.3m., sales $A1,095.2m. (2000/01); mfrs of sensitized photographic materials, photographic chemicals and equipment; distributors and retailers; Chair. and Man. Dir John Allen; 2,200 employees.

Leighton Holdings Ltd: 472 Pacific Highway, St Leonards, NSW 2065; tel. (2) 9925-6666; fax (2) 9925-6005; internet www.leighton.com.au; cap. and res $A740.3m., total revenue $A4,393.3m. (2000/01); construction, infrastructure development, contract mining, property development, telecommunications, waste management; Chair. John Morschel; CEO W. M. King; 12,688 employees.

Lend Lease Corpn Ltd: Level 46, Tower Bldg, Australia Square, George St, Sydney, NSW 2000; tel. (2) 9236-6111; fax (2) 9252-2192; internet www.lendlease.com.au; cap. and res. $A4,580m. (2000/01); property management, etc.; Chair. Jill Conway; Man. Dir David Higgins; 4,627 employees.

Mayne Group Ltd: Level 21, 390 St Kilda Rd, Melbourne, Vic 3004; tel. (3) 9868-0700; fax (3) 9867-1179; e-mail webrequests@maynegroup.com; internet www.maynegroup.com.au; sales $A3,158.7m. (2000/01); transport, security and health-care services; Chair. M. R. Rayner; Man. Dir P. Smedley; 29,000 employees.

Orica Ltd: 1 Nicholson St, Melbourne, Vic 3000; tel. (3) 9665-7111; fax (3) 9665-7937; e-mail companyinfo@orica.com; internet www.orica.com; cap. and res $A1,402.8m., sales $A4,187.4m. (2000/01); mfrs of fertilizers, chemicals, plastics, etc.; Chair Donald Mercer; Man. Dir Malcolm Broomhead; 9,000 employees.

Wesfarmers Ltd: 11th Floor, Wesfarmers House, 40 The Esplanade, Perth, WA 6000; tel. (8) 9327-4211; fax (8) 9327-4216; internet www.wesfarmers.com.au; cap. and res $A3,572m., total revenue $A7,800m. (2003); mfrs of fertilizers and chemicals; gas-processing and distribution; coal mining; building materials and hardware; insurance; Chair. Trevor Eastwood; Man. Dir and Chief Exec. Michael A. Chaney; 30,000 employees.

TRADE UNIONS

Australian Council of Trade Unions (ACTU): Level 2, 393 Swanston St, Melbourne, Vic 3000; tel. (3) 9663-5266; fax (3) 9663-4051; e-mail mailbox@actu.asn.au; internet www.actu.asn.au; f. 1927; br. in each state, generally known as a Trades and Labour Council; 46 affiliated trade unions; Pres. Sharan Burrow; Sec. Gregory Combet.

Principal Affiliated Unions

Ansett Pilots Association (APA): POB 415, Essendon, Vic 3040; tel. (3) 9325-4233; fax (3) 9375-7405; e-mail secretary@apa.asn.au; internet www.apa.asn.au; Pres. Henry Otto; Sec. John Doggett.

Association of Professional Engineers, Scientists & Managers, Australia (APESMA): POB 1272L, Melbourne, Vic 3001; tel. (3) 9695-8800; fax (3) 9696-9312; e-mail info@apesma.asn.au; internet www.apesma.asn.au; Nat. Pres. Barry Tonkin; Nat. Sec. Tim Griffin; 24,000 mems.

Australasian Meat Industry Employees' Union (AMIEU): Level 1, 39 Lytton Rd, East Brisbane, Qld 4169; tel. (7) 3217-3766; fax (7) 3217-4462; e-mail admin@amieuqld.asn.au; internet www.amieu.asn.au; Fed. Pres. Ross Richardson; Fed. Sec. T. R. Hannan; 20,484 mems.

Australian Airline Flight Engineers Association (AAEA): Essendon, Vic 3040; f. 1984; Fed. Pres. JEFF SEABURN; Fed. Sec. RON HARE.

Australian Collieries Staff Association (ACSA): POB 21, Merewether, NSW 2291; tel. (2) 4963-5656; fax (2) 4963-3425; e-mail acsa@acsa.org.au; internet www.acsa.org.au; Gen. Pres. MICK BURGESS; Gen. Sec. WENDY CLEWS.

Australian Education Union (AEU): 120 Clarendon St, Southbank, Vic 3006; tel. (3) 9693-1805; fax (3) 9254-1805; e-mail aeu@aeufederal.org.au; internet www.aeufederal.org.au; f. 1984; Fed. Pres. PATRICIA BYRNE; Fed. Sec. ROBERT DURBRIDGE; 164,642 mems.

Australian Manufacturing Workers' Union (AMWU): POB 160, Granville, NSW 2142; tel. (2) 9897-9133; fax (2) 9897-9274; e-mail amwu2@amwu.asn.au; internet www.amwu.asn.au; Nat. Pres. JULIUS ROE; Nat. Sec. DOUG CAMERON; 170,000 mems.

Australian Services Union (ASU): Ground Floor, 116 Queensberry St, Carlton South, Vic 3053; tel. (3) 9342-1400; fax (3) 9342-1499; e-mail asunatm@asu.asn.au; internet www.asu.asn.au; f. 1885; amalgamated, in present form in 1993; Nat. Sec. PAUL SLAPE; 140,000 mems.

Australian Workers' Union (AWU): 685 Spencer St, West Melbourne, Vic 3003; tel. (3) 8327-0888; fax (3) 8327-0899; e-mail awu@alphalink.net.au; internet www.awu.net.au; f. 1886; Nat. Pres. BILL LUDWIG; Nat. Sec. BILL SHORTEN; 135,000 mems.

Communications, Electrical, Electronic, Energy, Information, Postal, Plumbing and Allied Services Union of Australia (CEPU): Suite 701, Level 7, 1 Rosebery Ave, Rosebery, NSW 2018; tel. (2) 9663-3699; fax (2) 9663-5599; e-mail edno@nat.cepu.asn.au; internet www.cepu.asn.au; Nat Pres. BRIAN BAULK; Nat. Sec. PETER TIGHE; 180,000 mems.

Community and Public Sector Union (CPSU): Level 5, 191–199 Thomas St, Haymarket, NSW 2000; tel. (2) 9334-9200; fax (2) 8204-6902; e-mail cpsu@cpsu.org; internet www.cpsu.org; Nat. Pres. MATTHEW REYNOLDS; Nat. Sec. WENDY CAIRD; 200,000 mems.

Construction, Forestry, Mining and Energy Union (CFMEU): Box Q235, QVB PO, Sydney, NSW 1230; tel. (2) 9267-3393; fax (2) 9267-2460; internet www.cfmeu.asn.au; f. 1992 by amalgamation; Pres. TREVOR SMITH; Sec. JOHN MAITLAND; 120,000 mems.

Finance Sector Union of Australia (FSU): GPOB 9893, 341 Queen St, Melbourne, Vic 3001; tel. (3) 9261-5300; fax (3) 9670-2950; e-mail fsuinfo@fsunion.org.au; internet www.fsunion.org.au; f. 1991; Nat. Pres. JOY BUCKLAND; Nat. Sec. ANTHONY BECK; 61,000 mems.

Health Services Union of Australia (HSUA): Level 2, 106–108 Victoria St, Carlton, Vic 3053; tel. (3) 9341-3328; fax (3) 9341-3329; e-mail hsu@hsu.net.au; internet www.hsu.net.au; Nat. Pres. MICHAEL WILLIAMSON; Nat. Sec. CRAIG THOMSON; 90,000 mems.

Independent Education Union of Australia (IEU): POB 1301, South Melbourne, Vic 3205; tel. (3) 9254-1830; fax (3) 9254-1835; e-mail ieu@ieu.org.au; internet www.ieu.org.au; Fed. Sec. LYNNE ROLLEY; Fed. Pres. RICHARD SHEARMAN; 45,000 mems.

Liquor, Hospitality and Miscellaneous Workers Union (LHMU): Locked Bag 9, Haymarket, NSW 1240; tel. (2) 8204-7200; fax (2) 9281-4480; e-mail lhmu@lhmu.org.au; internet www.lhmu.org.au; f. 1992; Nat. Pres. HELEN CREED; Nat. Sec. JEFF LAWRENCE; 143,800 mems.

Maritime Union of Australia (MUA): 2nd Floor, 365 Sussex St, Sydney, NSW 2000; tel. (2) 9267-9134; fax (2) 9261-3481; e-mail muano@mua.org.au; internet www.mua.org.au; f. 1993; Nat. Sec. PADDY CRUMLIN; 10,012 mems.

Media, Entertainment & Arts Alliance (MEAA): POB 723, Strawberry Hills, NSW 2012; tel. (2) 9333-0999; fax (2) 9333-0933; e-mail mail@alliance.org.au; internet www.alliance.org.au; Fed. Sec. CHRISTOPHER WARREN; 30,000 mems.

National Union of Workers (NUW): POB 343, North Melbourne, Vic 3051; tel. (3) 9287-1850; fax (3) 9287-1818; e-mail nuwnat@nuw.org.au; internet www.nuw.org.au; Gen. Sec. CHARLES DONNELLY; Gen. Pres. DOUG STEVENS; 100,000 mems.

Rail, Tram and Bus Union (RTBU): 83–89 Renwick St, Redfern, NSW 2016; tel. (2) 9310-3966; fax (2) 9319-2096; e-mail rtbu@rtbu-nat.abn.au; internet www.rtbu-nat.asn.au; Nat. Pres. R. HAYDEN; Nat. Sec. ROGER JOWETT; 35,000 mems.

Shop, Distributive & Allied Employees Association (SDA): 5th Floor, 53 Queen St, Melbourne, Vic 3000; tel. (3) 9629-2299; fax (3) 9629-2646; e-mail sdanat@c031.aone.net.au; internet www.sda.org.au; f. 1908; Nat. Pres. DON FARRELL; Nat. Sec. JOE DE BRUYN; 214,029 mems.

Textile, Clothing and Footwear Union of Australia (TCFUA): Ground Floor, 28 Anglo Rd, Campsie, NSW 2194; tel. (2) 9789-4188; fax (2) 9789-6510; e-mail tcfua@tcfua.org.au; f. 1919; Pres. BARRY TUBNER; Nat. Sec. TONY WOOLGAR; 21,354 mems.

Transport Workers' Union of Australia (TWU): St Kilda Rd, POB 7419, Vic 8004; tel. (3) 8645-3333; fax (3) 9676-2669; e-mail twu@twu.com.au; internet www.twu.com.au; Fed. Pres. HUGHIE WILLIAMS; Fed. Sec. JOHN ALLAN; 82,000 mems.

Transport

Australian Transport Council: POB 594, Canberra, ACT 2601; tel. (2) 6274-7851; fax (2) 6274-7703; e-mail atc@dotars.gov.au; internet www.atcouncil.gov.au; f. 1993; mems include: Federal Minister for Transport and Regional Services, State, Territory and New Zealand Ministers responsible for transport; Sec. D. JONES.

State Transit Authority of New South Wales: 219–241 Cleveland St, Strawberry Hills, NSW 2010; tel. (2) 9245-5777; e-mail info@sydneybuses.nsw.gov.au; internet www.sta.nsw.gov.au; operates government buses and ferries in Sydney and Newcastle metropolitan areas; Chair. DAVID HERLIHY; CEO JOHN STOTT.

TransAdelaide (South Australia): GPOB 2351, Adelaide, SA 5001; tel. (8) 8218-2200; fax (8) 8218-4399; e-mail info@transadelaide.sa.gov.au; internet www.transadelaide.com.au; f. 1994; fmrly State Transport Authority; operates metropolitan train, bus, tram and Busway services; Gen. Man. SUE FILBY.

RAILWAYS

In June 2001 there were 39,844 km of railways in Australia (including tram and light rail track). In September 2003 the construction of a 1,400-km railway between Alice Springs and Darwin was completed. The rail link was to be used principally for transporting freight.

Pacific National: Locked Bag 90, Parramatta, NSW 2124; tel. (2) 9893-2500; fax (2) 9893-2501; e-mail communication@pacificnational.com.au; internet www.pacificnational.com.au; freight; fmrly National Rail Corporation Ltd; CEO STEPHEN O'DONNELL.

Public Transport Corporation (Victoria): Level 15, 589 Collins St, Melbourne, Vic 3000; tel. (3) 9619-4222; fax (3) 9619-4911; e-mail j.barry@ptc.vic.gov.au; f. 1989; Exec. Dir JOHN R. BARRY.

QR (Queensland Rail): POB 1429, Brisbane, Qld 4001; tel. (7) 3235-2180; fax (7) 3235-1799; internet www.qr.com.au; f. 1863; passenger commuter and long-distance services, freight and logistic services, track access and rail-specific expert services; CEO BOB SCHEUBER.

RailCorp: POB K349, Haymarket, NSW 1238; tel. (2) 8202-2000; fax (2) 8202-2111; internet www.railcorp.info; f. 1980; responsible for passenger rail and associated coach services in NSW; Chief Exec. VINCE GRAHAM.

Western Australian Government Railways (Westrail): POB 8125, Perth 6849, WA; tel. (8) 9326-2000; fax (8) 9326-2500; internet www.wagr.wa.gov.au; statutory authority competing in the freight, passenger and related transport markets in southern WA; operates 1,029 main line route-km of track; Commr REECE WALDOCK (acting).

ROADS

In June 2001 there were 808,294 km of roads open for general traffic. In 1996 this included 1,000 km of freeways, a further 103 km of toll roads, 45,889 km of highways, 77,045 km of arterial and major roads and 30,596 of secondary tourist and other roads. Local roads in urban areas account for 93,677 km of the network and those in rural localities for 537,278 km.

Austroads Inc: POB K659, Haymarket, NSW 2000; tel. (2) 9264-7088; fax (2) 9264-1657; e-mail austroads@austroads.com.au; internet www.austroads.com.au; f. 1989; asscn of road transport and traffic authorities.

SHIPPING

In December 2003 the Australian merchant fleet comprised 643 vessels, with a total displacement of 1,905,778 grt.

Adsteam Marine Ltd: Level 22, Plaza 2, 500 Oxford St, Bondi Junction, NSW 2022; tel. (2) 9369-9200; fax (2) 9369-9288; e-mail info@adsteam.com.au; internet www.adsteam.com.au; f. 1875; fmrly Adelaide Steamship Co; Man. Dir and Chief Exec. JOHN MOLLER.

ANL Ltd (Australian National Line): GPOB 2238T, Melbourne, Vic 3004; tel. (3) 9257-0613; fax (3) 9257-0517; e-mail anl@anl.com.au; f. 1956; shipping agents; coastal and overseas container shipping and coastal bulk shipping; container management services; overseas container services to Hong Kong, Taiwan, the Philippines, Korea, Singapore, Malaysia, Thailand, Indonesia and Japan; extensive transhipment services; Chair. E. G. ANSON; Man. Dir MALCOLM TURNBULL.

BHP Transport Pty Ltd: 27th Level, 600 Bourke St, POB 86A, Melbourne, Vic 3001; tel. (3) 9609-3333; fax (3) 9609-2400; Chair. D. Argus; Man. Dir P. Andersen.

William Holyman and Sons Pty Ltd: No. 3 Berth, Bell Bay, Tas 7253; tel. (3) 6382-2383; fax (3) 6382-3391; coastal services; Chair. R. J. Hoy.

CIVIL AVIATION

Eastern Australia Airlines: POB 538, Mascot, Sydney, NSW 2020; tel. (2) 9691-2333; fax (2) 9693-2715; internet www.qantas.com.au; subsidiary of Qantas; domestic flights; Gen. Man. Ashley Kilroy.

Impulse Airlines: Eleventh St, Sydney Kingsford-Smith Airport, Mascot, NSW 2020; tel. (2) 9317-5400; fax (2) 9317-3440; e-mail info@ImpulseAirlines.com.au; internet www.impulseairlines.com.au; f. 1992; domestic services; Chief Exec./Chair. Gerry McGowan; Exec. Dir Sue McGowan.

Jetstar Airways Pty Ltd: Tullamarine, POB 1503, Melbourne, Vic 3043; tel. (3) 8341-4901; internet www.jetstar.com.au; f. 2004; owned by Qantas Airways Ltd; low-cost domestic passenger service; Chief Exec. Alan Joyce.

National Jet Systems: Adelaide Airport, 28 James Schofield Drive SA 5000; tel. (8) 8238-7200; fax (8) 8238-7238; internet www.nationaljet.com.au; f. 1989; domestic services; Man. Dir Daniela Marsilli; Group Gen. Man. Commdr Robert Birks.

Qantas Airways Ltd: Qantas Centre, 203 Coward St, Mascot, NSW 2020; tel. (2) 9691-3636; fax (2) 9691-3339; internet www.qantas.com.au; f. 1920 as Queensland and Northern Territory Aerial Services; Australian Govt became sole owner in 1947; merged with Australian Airlines in Sept. 1992; British Airways purchased 25% in March 1993; remaining 75% transferred to private sector in 1995; services throughout Australia and to 34 countries, including destinations in Europe, Africa, the USA, Canada, South America, Asia, the Pacific and New Zealand; Chair. Margaret Jackson; CEO Geoff Dixon.

Spirit Airlines: Level 9, 580 St Kilda Rd, Melbourne, Vic 3004; fax (3) 9510-4095; e-mail info@spiritairlines.com.au; internet www.spiritairlines.com.au/; f. 1998; domestic services; CEO Mike Dixon.

Sunstate Airlines: Lobby 3, Level 3, 153 Campbell St, Bowen Hills, Qld 4006; tel. (7) 3308-9022; fax (7) 3308-9088; e-mail edalessio@qantas.com.au; f. 1982; wholly owned by Qantas; operates passenger services within Queensland and to Newcastle (NSW) and Lord Howe Island; Gen. Man. Elsa D'Alessio.

Virgin Blue: Centenary Place, Level 7, 100 Wickham St, Brisbane, Qld; tel. (7) 3295-3000; e-mail customercare@virginblue.com.au; internet www.virginblue.com.au; domestic services; CEO Brett Godfrey.

Tourism

The main attractions are the cosmopolitan cities, the Great Barrier Reef, the Blue Mountains, water sports and also winter sports in the Australian Alps, notably the Snowy Mountains. The town of Alice Springs, the Aboriginal culture and the sandstone monolith of Ayers Rock (Uluru) are among the attractions of the desert interior. Much of Australia's wildlife is unique to the country. Australia received 4,745,900 foreign visitors in 2003. The majority of visitors came from New Zealand, Japan, the United Kingdom and the USA. Receipts totalled US $8,087m. in 2002. The 2000 Olympic Games were held in Sydney in September.

Australian Tourist Commission: GPOB 2721, Sydney, NSW 1006; Level 4, 80 William St, Woolloomooloo, Sydney, NSW 2011; tel. (2) 9360-1111; fax (2) 9331-6469; internet www.atc.australia.com; f. 1967 for promotion of international inbound tourism; 13 offices overseas; Chair. Nick Evers; Man. Dir Ken Boundy.

Defence

In August 2003 Australia's active armed forces numbered 53,650: army 26,600, navy 12,850, air force 14,200. There were also reserve forces of 20,300. Military service is voluntary. Australia's defence policy is based on collective security and it is a member of the ANZUS Security Treaty (with New Zealand and the USA), and of the Five Power Defence Arrangements (with Singapore, Malaysia, the United Kingdom and New Zealand).

Defence Expenditure: Budgeted at $A13,307m. for 2002/03.

Chief of the Defence Force: Gen. Peter J. Cosgrove.

Chief of Navy: Vice-Adm. Christopher Ritchie.

Chief of Army: Lt-Gen. Peter Leahy.

Chief of Air Force: Air Marshal Allan Grant Houston.

Education

Under the federal system of government in Australia, the six states and two territories are responsible for providing education services for their own residents. The Australian Constitution, however, empowers the Federal Government to make special-purpose financial grants to the states for education in both government and non-government schools. Expenditure on education by all levels of government in 2001/02 was $A37,546m. (equivalent to 5.3% of GDP), while private education accounted for a further $A11,445m. Responsibility for educational policy rests with the Minister for Education, Science and Training, and an education department headed by a Director-General deals with all aspects of education within each state. A small proportion of Australia's school-age children live in remote districts, thus requiring distance education. School attendance in Australia is compulsory and free between the ages of six and 15 years (16 in Tasmania) for all children except those exempted on account of distance, for whom a special form of education is provided. The Federal Government is the most important source of funding for universities.

GOVERNMENT SCHOOLS

In 2002 there were 6,969 government primary and secondary schools, with a total enrolment of 2,268,800 full-time pupils and 152,982 teachers. In 2000/01 some 95.7% of children of the relevant age-group (males 95.3%; females 96.1%) were enrolled in primary schools, while the equivalent rate for secondary enrolment was estimated at 89.7% (males 88.5%; females 91.0%). Schooling is not compulsory in Australia until the age of six but most children start earlier. Primary schools are generally mixed and cater for children up to the age of 11 or 12.

The co-educational high school offers a wide range of subjects. Most have modern facilities for teaching sciences, information technology and general technology, including woodwork and metal fabrication. Many schools are beginning to focus on vocational education, which prepares students for work. Some states also have separate high schools and colleges specializing in the agricultural, commercial and vocational/technical fields. Some agricultural high schools are residential. The curriculum in these specialist schools consists of general academic subjects and training in the particular area of focus.

INDEPENDENT SCHOOLS

A total of 1,046,200 full-time students were enrolled in the 2,663 non-government or independent schools in 2002. Almost 68% of primary and secondary students at independent schools attended Roman Catholic schools. Most other independent schools are under the auspices of, or are actually administered by, other religious denominations. The teaching staff at independent schools in 2002 totalled 72,371.

The organization of Roman Catholic primary schools is largely through a diocesan Education Office. Many secondary schools are administered by religious orders; however, a growing number now come within general administrative structures in a similar way to primary schools.

EDUCATION FOR CHILDREN IN ISOLATED AREAS

Some children in Australia live in remote areas, which are geographically isolated and do not allow access to the resources usually offered by schools. Australia has developed some innovative approaches to delivering education to overcome these disadvantages. A long-standing example is that of the Schools of the Air, which have been broadcasting lessons for remote primary and secondary school children, using two-way radio equipment, since 1950. Rapid changes in technology have improved the type of educational support offered for remote students. Schools can make use of satellite and television broadcasting and use computer networks to provide an up-to-date curriculum for their students.

EDUCATION NETWORK AUSTRALIA (EdNA)

One Australian initiative that aims to facilitate the provision of cost-effective education to all parts of the community is Education Network Australia (EdNA). EdNA is a multi-faceted process of co-operation and consultation between representatives of all sectors of the education community, including Federal, State and Territory governments, non-government schools, the vocational education and training sector, the higher education sector and the adult and community education sectors. Its aim is to maximize the benefits of information technology for all sectors in Australian education and to avoid duplication between the various sectors and systems. The most visible part of the EdNA initiative is the directory of services, which is available on the Internet. Each sector of the education community is developing its part to meet the needs of its particular client group. The directory is highly innovative and flexible. EdNA is available free of charge on the Internet. Its address is: www .edna.edu.au.

HIGHER EDUCATION

There are 47 public universities and several additional public institutions of higher education, which enrolled a total of 896,621 students in 2002. There are also two small private universities. Most courses last from three to six years. The majority of higher education institutions offer courses to both internal and external students and have postgraduate research facilities.

Most Australian students contribute to the cost of their courses under the Higher Education Contribution Scheme (HECS). Students commencing a new course of study are subject to a three-tiered contribution based on the cost of the course undertaken and its likely future benefits to the individual. Some postgraduate courses are offered on a fee-paying rather than on an HECS-liable basis. From 1998 Australian undergraduate students may also choose to enrol on a fee-paying basis. Financial assistance is available to certain students subject to a means test. In addition, there are a limited number of equity and merit scholarships, which exempt students from the HECS charge.

VOCATIONAL EDUCATION AND TRAINING

Vocational education and training is aimed at assisting people seeking to upgrade their skills, join the work-force or change career paths. Off-the-job training for apprentices and trainees is provided by this sector. In 2002 an estimated total of 2,276,600 students were enrolled in vocational education and training, provided by organizations such as Technical and Further Education (TAFE) institutes, Skills Centres, business colleges, registered private training providers, computer-training colleges and Adult and Community Education centres, which offer a wide range of courses and subjects.

TAFE

Technical and Further Education (TAFE) is the major provider of post-secondary vocational education and training. There are more than 7,000 courses (including non-vocational courses) available at almost 1,500 locations throughout Australia. Most TAFE courses are developed in consultation with the industry concerned. TAFE award courses can be studied both full time and part time, by internal study at campuses, by external studies and also through flexible delivery (open learning).

SKILLS CENTRES

Skills Centres are industry- and enterprise-based training facilities that offer a wide range of accredited training to enterprise employees, industry groups and individuals. Some Skills Centres are located in-plant, some are 'stand alone' and others are linked to a TAFE college. There are approximately 100 Skills Centres around Australia.

REGISTERED PRIVATE TRAINING PROVIDERS

There are many private training institutions, including business colleges, secretarial colleges, computer-training centres and industry-based training centres. They provide training in a wide range of fields such as beauty therapy, hospitality, travel, business, computing, music, aviation, child care, art and design, naturopathy and languages.

ADULT AND COMMUNITY EDUCATION

Adult and Community Education (ACE) centres include neighbourhood centres, non-government adult education providers, community centres and SkillShare centres. They provide recreational, pre-vocational and vocational training to local communities focusing particularly on mature students and long-term unemployed people. Much of the training is in subjects such as arts and humanities, health and community services, office administration and English as a second language. SkillShare centres provide access to training courses and employment-related services to disadvantaged jobseekers. There are more than 1,500 ACE centres around Australia.

Bibliography

GENERAL

Aitken, J. *Land of Fortune: A Study of the New Australia*. London, Secker and Warburg, 1976.

Anderson, W. *The Cultivation of Whiteness: Science, Health and Racial Destiny in Australia*. New York, NY, Basic Books, 2003.

Australian Bureau of Statistics. *Year Book Australia*. Canberra.

Berndt, R. M. and C. H. *The World of the First Australian*. Sydney, Ure Smith, 1964.

Dalrymple, R. *Continental Drift: Australia's Search for a Regional Identity*. Burlington, VT, Ashgate, 2003.

Davies, A. F. (Ed.). *Australian Society: A Sociological Introduction*. 2nd Edn, Melbourne, Cheshire, 1970.

Encel, S. *Equality and Authority: A Study of Class, Status and Power*. Melbourne, Cheshire, 1970.

Germov, R. and Motta, M. *Refugee Law in Australia*. South Melbourne, Vic, Oxford University Press, 2003.

Gothard, J., Jayasuriya, L. and Walker, D. (Eds). *Legacies of White Australia: Race, Culture and Nation*. WA, University of Western Australia Press, 2003.

Greig, A., Lewins, F. and White, K. *Inequality in Australia*. Melbourne, Vic, Cambridge University Press, 2003.

Hillstrom, K. and L. C. (Eds). *Australia, Oceania, and Antarctica: a Continental Overview of Environmental Issues*. Oxford, ABC-CLIO, 2003.

Horne, D. *The Australian People: Biography of a Nation*. Sydney, Angus and Robertson, 1976.

Howe, B. and Hughes, P. (Eds). *Spirit of Australia II: Religion in Citizenship and National Life*. London, Sheldon Press, 2003.

Inglis, K. *Sacred Places: War Memorials in the Australian Landscape*. Carlton, Vic, Melbourne University Press, 1998.

Jeans, D. N. (Ed.). *Australia: A Geography*. Sydney University Press, 1977.

Jupp, J. *From White Australia to Woomera: The Story of Australian Immigration*. Melbourne, Vic, Cambridge University Press, 2002.

Kewley, T. *Social Security in Australia*. Sydney University Press, 1972.

Maddock, K. J. *The Australian Aborigines*. London, Penguin, 1973.

Mares, P. *Borderline: Australia's Treatment of Refugees and Asylum Seekers in the Wake of the Tampa*. Sydney, NSW, New South Wales University Press, 2003.

Pilger, J. *A Secret Country*. London, Jonathan Cape, 1989.

Saeed, A. *Islam in Australia*. St Leonards, NSW, Allen and Unwin, 2003.

Simpson, C. *The New Australia*. Sydney, Angus and Robertson, 1971.

Spate, O. H. K. *Australia*. Nations of the Modern World Series, London, Benn, 1968.

Stone, Sharman. *Aborigines in White Australia*. Adelaide, Heinemann Educational Books, 1974.

Sutton, P. *Native Title in Australia: An Ethnographic Perspective*. Melbourne, Vic, Cambridge University Press, 2003.

Venturini, V. G. (Ed.). *Australia: A Survey*. Wiesbaden, Otto Harrassowitz, 1970.

HISTORY AND POLITICS

Aitkin, D. *Stability and Change in Australian Politics*. Canberra, 1977.

Alexander, F. *From Curtin to Menzies and After*. Melbourne, Nelson, 1973.

Atkins B. *Governing Australia*. Sydney, Wiley, 1972.

Attwood, Bain, and Markus, Andrew (Eds). *The Struggle for Aboriginal Rights: A Documentary History*. Sydney, Allen and Unwin, 2000.

Australian Dictionary of Biography (12 vols). Melbourne University Press, 1966–90.

Bennett, S. C. *The Making of the Commonwealth*. Melbourne, Cassell, 1971.

Benwell, P. *In Defense of Australia's Constitutional Monarchy*. New York, NY, Edwin Mellen Press, 2003.

Blainey, Geoffrey. *The Tyranny of Distance*. 2nd Edn, Melbourne, Sun Books, 1976.

Brett, J. *Australian Liberals and the Moral Middle Class: From Alfred Deakin to John Howard*. Melbourne, Vic, Cambridge University Press, 2003.

Broome, R. *Aboriginal Australians—the Black Response to White Dominance, 1788–1980*. Sydney, 1982.

Clark, M. *A History of Australia* (6 vols). Melbourne University Press, 1962–87. (Abridgement by Michael Cathcart, 1994.).

A Short History of Australia. Melbourne, Macmillan, 1986.

Crisp, L. F. *Australian National Government*. Melbourne, Longman, 1975.

Crowley, F. K. (Ed.). *A New History of Australia*. Melbourne, Heinemann, 1974.

Davies, S. R. (Ed.). *The Government of the Australian States*. London, Longman, 1960.

Davison, Graeme, et al. (Eds). *The Oxford Companion to Australian History*. Oxford, Oxford University Press, 2002.

Day, David. *Claiming a Continent: A History of Australia*. Pymble, NSW, Angus & Robertson, 1996.

Edwards, John. *Keating: The Inside Story*. Ringwood, Vic, Penguin Books, 1997.

FitzGerald, Stephen. *Is Australia an Asian Country?* St Leonards, NSW, Allen and Unwin, 1997.

Goot, Murray, and Rowse, Tim (Eds). *Make a Better Offer: The Politics of Mabo*. Leichhardt, NSW, Plato Press, 1994.

Gordon, Michael. *Paul Keating: True Believer*. St Lucia, Qld, University of Queensland Press, 1997.

Greer, Germaine. *Whitefella Jump Up*. London, Profile, 2004.

Hawke, Bob. *The Hawke Memoirs*. Port Melbourne, William Heinemann Australia, 1994.

Headon, D., Warden, J., and Gammadge, B. (Eds). *The Traditions of Australian Republicanism*. St Leonards, NSW, Allen and Unwin, 1994.

Hearn, Mark, and Knowles, Harry. *One Big Union: A History of the Australian Workers' Union, 1886–1994*. Cambridge University Press, 1997.

Hough, Richard. *Captain James Cook: A Biography*. London, Hodder and Stoughton, 1994.

Hughes, C. A. *A Handbook of Australian Government and Politics, 1965–74*.

Hughes, R. *The Fatal Shore*. London, Collins Harvill, 1987.

Irving, Helen. *To Constitute A Nation: A Cultural History of Australia's Constitution*. Cambridge, Cambridge University Press, 1997.

Jack, Ian (Ed.). *Australia, the New New World*. London, Granta Books, 2000.

Keating, Paul. *Advancing Australia: The Speeches of Paul Keating, Prime Minister*. Cremorne Pt, NSW, Big Picture Publications, 1996.

Engagement: Australia Faces the Asia-Pacific. Sydney, Pan Macmillan, 2000.

Kelly, Paul. *The End of Certainty: The Story of the 1980s*. Sydney, Allen and Unwin, 1993.

One Hundred Years. Sydney, Allen and Unwin, 2001.

Knightley, Phillip. *Australia –A Biography of a Nation*. London, Jonathan Cape, 2000.

Lacour-Gayet, R. *A Concise History of Australia*. London, Penguin, 1976.

La Nauze, J. *The Making of the Australian Constitution*. Melbourne University Press, 1972.

Loveday, P., Martin, A. W., and Parker, R. S. *The Emergence of the Australian Party System*. Sydney, 1977.

Marr, D. and Wilkinson, M. *Dark Victory*. St Leonards, NSW, Allen and Unwin, 2003.

McGillivray, Mark, and Smith, Gary (Eds). *Australia and Asia*. Oxford, Oxford University Press, 1998.

McIntyre, K. G. *The Secret Discovery of Australia*. London, Souvenir Press, 1977.

MacIntyre, Stuart. *A Concise History of Australia*. Cambridge, Cambridge University Press, 2000.

McKenna, Mark. *The Captive Republic: A History of Republicanism in Australia 1788–1996*. Cambridge, Cambridge University Press, 1997.

Meyer, H. *Australian Politics*. Melbourne, Cheshire, 1971.

Miller, J. D. B. *Australian Government and Politics*. London, Duckworth, 1971.

Moon, J. and Sharman, C. (Eds). *Australian Politics and Government: The Commonwealth, The States and The Territories*. Melbourne, Vic, Cambridge University Press, 2003.

Oxford History of Australia (5 vols). Melbourne, Oxford University Press, 1986–91.

Reid, G. *Australia's Commonwealth Parliament*. Canberra, Australian Government Publishing Service, 1988.

Richardson, Graham. *Whatever It Takes*. Neutral Bay Junction, NSW, Bantam Books, 1994.

Sawer, G. *The Australian Constitution*. Canberra, Australian Government Publishing Service, 1975.

Sharp, C. A. *The Discovery of Australia*. London, Oxford University Press, 1963.

Shaw, A. G. L. *The Story of Australia*. 5th Edn, London, Faber and Faber, 1983.

Shaw, A. G. L., and Nicolson, H. D. *Australia in the Twentieth Century*. Sydney, Angus and Robertson, 1967.

Sheridan, Greg. *Living with Dragons: Australia Confronts its Asian Destiny*. St Leonards, NSW, Allen and Unwin, 1995.

Solomon, D. *The People's Palace: Parliament in Modern Australia*. Canberra, Australian Government Publishing Service, 1986.

Souter, G. *Acts of Parliament*. Canberra, Australian Government Publishing Service, 1988.

Statham-Drew, P. *James Stirling: Admiral and Founding Governor of Western Australia*. WA, University of Western Australia Press, 2003.

Theophanous, Andrew. *Australian Democracy in Crisis: A New Theoretical Introduction to Australian Politics*. Oxford University Press, 1980.

Ward, R. *A History of Australia: The Twentieth Century 1901–1975*. London, Heinemann, 1978.

Watt, A. *The Evolution of Australian Foreign Policy*. Cambridge University Press, 1967.

Welsh, Frank. *Great Southern Land: A New History of Australia*. London, Allen Lane, 2004.

Wesley, M. and Gyngell, A. *Making Australian Foreign Policy*. Melbourne, Vic, Cambridge University Press, 2003.

Whitington, B. L. *The Menzies Era and After*. Melbourne, Cheshire, 1972.

ECONOMY

Andrews, J. A. *Australia's Resources and their Utilisation*. Sydney, University Press, 1970.

Bell, Stephen, and Head, Brian (Eds). *State Economy and Public Policy in Australia*. Oxford University Press, 1994.

Blainey, G. *The Rush That Never Ended: A History of Australian Mining*. Carlton South, Vic, Melbourne University Press, 2003.

Brown, D. *Market Rules: Economic Union Reform and Intergovernmental Policy-Making in Australia and Canada*. Montréal, QC, McGill-Queen's University Press, 2003.

Byrnes, Michael. *Australia and the Asia Game*. St Leonards, NSW, Allen and Unwin, 1994.

Bryan, Dick, and Rafferty, Michael. *The Global Economy in Australia: Global Integration and National Economic Policy*. Allen and Unwin, 1999.

Catley, Bob. *Globalising Australian Capitalism*. Cambridge, Cambridge University Press, 1997.

Caves, R., and Kraus, L. (Eds). *The Australian Economy*. Sydney, George Allen and Unwin, 1983.

Chapman, B., Isaacs, J. E., and Niland, J. R. *Australian Labour Economics Readings*. 3rd Edn, Melbourne, Macmillan, 1985.

Coghill, I. G. *Australia's Mineral Wealth*. Melbourne, Sorrett Publishing, 1971.

Crawford, J. G., and others. *Study Group on Structural Adjustment, March 1979*. Canberra, Australian Government Publishing Service, 1979.

Crawford, J. G., and Okita, S. (Eds). *Raw Materials and Pacific Economic Integration*. London, Croom Helm, 1978.

Downing, R. I. *National Income and Social Accounts: An Australian Study*. 12th Edn, Melbourne University Press, 1971.

Fenna, Alan. *Introduction to Australian Public Policy*. Melbourne, Longman, 1998.

Fitzpatrick, B. C. *British Imperialism and Australia 1783–1833: An Economic History of Australasia*. Sydney University Press, 1971.

The British Empire in Australia: An Economic History, 1834–1939. Melbourne, Macmillan, 1969.

Higgins, V. J. G. *Constructing Reform: Economic Expertise and the Governing of Agricultural Change in Australia*. Canberra, ACT, Nova Science Publishers, 2003.

INDECS ECONOMICS. *State of Play 6*. Sydney, George Allen and Unwin, 1990.

Isaac, J. E., and Ford, G. *Australian Labour Relations: Readings*. 2nd Edn, Melbourne, Sun Books, 1971.

Kenwood, George, and Kenwood, A. G. *Australian Economic Institutions Since Federation : An Introduction.* Oxford University Press, 1996.

Kreisler, Peter (Ed.). *The Australian Economy.* Allen and Unwin, 1999.

Lewis, M. K., and Wallace, R. H. *Australia's Financial Institutions and Markets.* Melbourne, Longman Cheshire, 1985.

Marginson, Simon. *Educating Australia: Government, Economy and Citizenship since 1960.* Cambridge, Cambridge University Press, 1997.

McMichael, P. *Settlers and the Agrarian Question: Capitalism in Colonial Australia.* Melbourne, Vic, Cambridge University Press, 2003.

Meredith, David, and Dyster, Barrie. *Australia in the Global Economy: Continuity and Change.* Cambridge, Cambridge University Press, 2000.

Painter, Martin. *Collaborative Federalism: Economic Reform in Australia in the 1990s.* Cambridge, Cambridge University Press, 1998.

Peter, P., Dawkins, P. and Kelly, P. (Eds). *Hard Heads, Soft Hearts: A New Reform Agenda for Australia.* St Leonards, NSW, Allen and Unwin, 2003.

Pinkstone, Brian, and Meredith, David. *Global Connections: A History of Exports and the Australian Economy.* International Specialized Book Services, 1997.

Pusey, M. *The Experience of Middle Australia: The Dark Side of Economic Reform.* Melbourne, Vic, Cambridge University Press, 2003.

Saunders, P. *The Ends and Means of Welfare: Coping with Economic and Social Change in Australia.* Melbourne, Vic, Cambridge University Press, 2003.

Shaw, A. G. L. *The Economic Development of Australia.* 7th Edn, Melbourne, 1980.

Sheehan, Peter, et al. *Dialogues on Australia's Future.* Melbourne, Victoria University of Technology, 1996.

Sinclair, W. A. *The Process of Economic Development in Australia.* Melbourne, Cheshire, 1976.

Sinden, J. A. (Ed.). *The Natural Resources of Australia: Prospects and Problems for Development.* Sydney, Angus and Robertson, 1972.

Singh, Anoop, et al. *Australia: Benefiting from Economic Reform.* Washington, DC, IMF, 1998.

Sykes, Trevor. *The Bold Riders: Behind Australia's Corporate Collapses.* St Leonards, NSW, Allen and Unwin, 1994.

AUSTRALIAN DEPENDENCIES IN THE INDIAN OCEAN

Australia's 'non-self-governing' territories in the Indian Ocean comprise the Ashmore and Cartier Islands, Christmas Island and the Cocos (Keeling) Islands. Heard Island and the McDonald Islands are not permanently inhabited.

TERRITORY OF ASHMORE AND CARTIER ISLANDS

The Ashmore Islands (known as West, Middle and East Islands) and Cartier Island are situated in the Timor Sea, about 850 km and 790 km west of Darwin respectively. The Ashmore Islands cover some 93 ha of land and Cartier Island covers 0.4 ha. The islands are small and uninhabited, consisting of sand and coral, surrounded by shoals and reefs. Grass is the main vegetation. Maximum elevation is about 2.5 m above sea-level. The islands abound in birdlife, sea-cucumbers (*bêches-de-mer*) and, seasonally, turtles.

The United Kingdom took formal possession of the Ashmore Islands in 1878, and Cartier Island was annexed in 1909. The islands were placed under the authority of the Commonwealth of Australia in 1931. They were annexed to, and deemed to form part of, the Northern Territory of Australia in 1938. On 1 July 1978 the Australian Government assumed direct responsibility for the administration of the islands, which rests with a parliamentary secretary appointed by the Minister for Regional Services, Territories and Local Government. Periodic visits are made to the islands by the Royal Australian Navy and aircraft of the Royal Australian Air Force, and the Civil Coastal Surveillance Service makes aerial surveys of the islands and neighbouring waters. The oilfields of Jabiru and Challis are located in waters adjacent to the Territory.

In August 1983 Ashmore Reef was declared a national nature reserve. An agreement between Australia and Indonesia permits Indonesian traditional fishermen to continue fishing in the territorial waters and to land on West Island to obtain supplies of fresh water. In 1985 the Australian Government extended the laws of the Northern Territory to apply in Ashmore and Cartier, and decided to contract a vessel to be stationed at Ashmore Reef during the Indonesian fishing season (March–November) to monitor the fishermen.

During 2000–01 increasing numbers of refugees and asylum-seekers attempted to land at Ashmore Reef, hoping to gain residency in Australia. The majority had travelled from the Middle East via Indonesia, where the illegal transport of people was widespread. Consequently, in late 2000 a vessel with the capacity to transport up to 150 people was chartered to ferry unauthorized arrivals to the Australian mainland. In September 2001 the Australian Government introduced an item of legislation to Parliament excising Ashmore Reef and other outlying territories from Australia's migration zone. However, in March 2004 a group of nine women and six men, believed to be seeking asylum in Australia, were discovered on Ashmore Reef. A government spokesperson reiterated the Territory's exclusion from Australia's migration zone, stating that this would preclude the group from seeking any form of residency in the country.

CHRISTMAS ISLAND

Introduction

Christmas Island lies 360 km south of Java Head (Indonesia) in the Indian Ocean. The nearest point on the Australian coast is North West Cape, 1,408 km to the south-east. Christmas Island has no indigenous population. The population was 1,508 at the 2001 census (compared with 1,906 in 1996), comprising mainly ethnic Chinese (some 70%), but there were large minorities of Malays (about 10%) and Europeans (about 20%). A variety of languages are spoken (more than 60% of the population spoke a language other than English in 2001), but English is the official language. The predominant religious affiliation is Buddhist (55% in 1991). The principal settlement and only anchorage is Flying Fish Cove.

Following annexation by the United Kingdom in 1888, Christmas Island was incorporated for administrative purposes with the Straits Settlements (now Singapore and part of Malaysia) in 1900. Japanese forces occupied the island from March 1942 until the end of the Second World War, and in 1946 Christmas Island became a dependency of Singapore. Administration was transferred to the United Kingdom on 1 January 1958, pending final transfer to Australia, effected on 1 October 1958. The Australian Government appointed Official Representatives to the Territory until 1968, when new legislation provided for an Administrator, appointed by the Governor-General. Responsibility for administration lies with the Minister for Regional Services, Territories and Local Government. In 1980 an Advisory Council was established for the Administrator to consult. In 1984 the Christmas Island Services Corporation was created to perform those functions that are normally the responsibility of municipal government. This body was placed under the direction of the Christmas Island Assembly, the first elections to which took place in September 1985. Nine members were elected for one-year terms. In November 1987 the Assembly was dissolved, and the Administrator empowered to perform its functions. The Corporation was superseded by the Christmas Island Shire Council in 1992.

In May 1994 an unofficial referendum on the island's status was held concurrently with local government elections. At the poll, sponsored by the Union of Christmas Island Workers, the islanders rejected an option to secede from Australia, but more than 85% of voters favoured increased local government control. The referendum was prompted, in part, by the Australian Government's plans to abolish the island's duty-free status (which had become a considerable source of revenue).

Since 1981 all residents of the island have been eligible to acquire Australian citizenship. In 1984 the Australian Government extended social security, health and education benefits to the island, and enfranchised Australian citizens resident there. Full income-tax liability was introduced in the late 1980s.

During the late 1990s an increasing number of illegal immigrants travelling to Australia landed on Christmas Island. In January 2001 Australian government officials denied claims by Christmas Islanders that some 86 illegal immigrants who had arrived at the island from the Middle East via Indonesia were being detained in inhumane conditions. Local people claimed that the detainees were sleeping on concrete floors and were being denied adequate food and medical care.

International attention was focused on Christmas Island in August 2001 when the *MV Tampa*, a Norwegian container ship carrying 433 refugees whom it had rescued from a sinking Indonesian fishing boat, was refused permission to land on the island. As the humanitarian crisis escalated, the Australian Government's steadfast refusal to admit the mostly Afghan refugees prompted international condemnation and led to a serious diplomatic dispute between Australia and Norway. The office of the United Nations High Commissioner for Refugees (UNHCR) and the International Organization for Migration (IOM) expressed grave concern at the situation. Hundreds of Christmas Island residents attended a rally urging the Australian Government to reconsider its uncompromising stance. In September the refugees were transferred (via Papua New Guinea and New Zealand) to Nauru, where their applications for asylum were to be processed. In the same month the Senate in Canberra approved new legislation, which excised Christmas Island and other outlying territories from Australia's official migration zone. The new legislation also imposed stricter criteria for the processing of asylum-seekers and the removal of their right to recourse to the Australian court system. Meanwhile, increasing numbers of asylum-seekers continued to attempt to reach Christmas Island via Indonesia. Among the many controversial incidents that occurred in the waters of Christmas Island in September–December 2001 was that of 186 Iraqis who jumped into the sea when ordered to leave Australian waters in October. They were temporarily housed on Christmas Island before being transferred to Nauru. According to Australian Immigration Department figures, 146 asylum-seekers were turned away from Christmas Island and Ashmore Reef in December 2001. In January 2002 211 asylum-seekers remained on Christmas Island awaiting transferral. In March the Government announced plans to establish a permanent detention centre on the island, the construction of which was expected to cost more than $A150m. The Government was thus preparing to accommodate an anticipated total of 18,000 illegal immigrants who were expected to arrive at Christmas Island during 2002–06. However, plans to scale down the project were announced in February 2003. In July of that year a boat carrying 53 asylum-seekers from Viet Nam was intercepted in Australian waters (the first such vessel to be found in the region since 2001), and its passengers were transferred to Christmas Island. The construction of the detention centre was expected to begin during 2004.

The economy has been based on the recovery of phosphates. During the year ending 30 June 1984 about 463,000 metric tons were exported to Australia, 332,000 tons to New Zealand and 341,000 tons to other countries. Reserves were estimated to be sufficient to enable production to be maintained until the mid-1990s. In November 1987 the Australian Government announced the closure of the phosphate mine, owing to industrial unrest, and mining activity ceased in December. In 1990, however, the Government allowed private operators to recommence phosphate extraction, subject to certain conditions such as the preservation of the rainforest. A total of 220,000 metric tons of phosphates were produced in 1995. In 2001 the mine employed some 180 workers. The owner of the mine, Phosphate Resources Ltd (trading as Christmas Island Phosphates), was thus the island's largest employer. A total of 570,000 metric tons of phosphate dust were exported in 2001/02, and phosphate exports were worth $A43.8m. in 1997/98. A new 21-year lease, drawn up by the Government and Phosphate Resources Ltd, took effect in February 1998. The agreement incorporated environmental safeguards and provided for a conservation levy, based on the tonnage of phosphate shipped, which was to finance a programme of rainforest rehabilitation.

Efforts have been made to develop the island's considerable potential for tourism. In 1989, in an attempt to protect the natural environment and many rare species of flora and fauna (including the Abbott's Booby and the Christmas frigate bird), the National Park was extended to cover some 70% of the island. A hotel and casino complex, covering 47 ha of land, was opened in November 1993. In 1994 revenue from the development totalled $A500m. A 50-room extension to the complex was constructed in 1995. In early 1997, however, fears for the nascent industry were expressed, following the decision by Ansett Australia to discontinue its twice-weekly air service to the island from September of that year. Despite the subsequent commencement of a weekly flight from Perth, Australia, operated by National Jet Systems, the complex was closed down in April 1998 and some 350 employees were made redundant. In July the resort's casino licence was cancelled and a liquidator was appointed to realize the complex's assets. The closure of the resort had serious economic and social repercussions for the island. An audit of tourist accommodation, conducted in May 2000, counted approximately 140 beds on the island. Only 60 visitors arrived on the island from overseas in 2001, according to official Australian figures.

Between 1992 and 1999 the Australian Government invested an estimated $A110m. in the development of Christmas Island's infrastructure as part of the Christmas Island Rebuilding Programme. The main areas of expenditure under this programme were a new hospital, the upgrading of ports facilities, school extensions, the construction of housing, power, water supply and sewerage, and the repair and construction of roads. In 2000 further improvements to marine facilities and water supply were carried out, in addition to the construction of new housing to relocate islanders away from a major rockfall risk area. In mid-2001 the Australian Government pledged a total of more than $A50m. for further developments including improvements to the airport and the road network, as well as an alternative port. The proposed additional port was to be constructed on the east coast of the island, allowing for the handling of sea freight and the launching of emergency vessels at times when the existing port at Flying Fish Cove is closed owing to north-west swells. These closures, which most commonly occur between December and March, had resulted in inflated costs for shipping companies and inconvenience for Christmas Islanders. The new port was expected to be built by 2005. The Australian Government announced in April 2004 that it would provide $A2.5m. to fund a new mobile telephone network for the Territory. It was hoped that enhanced communications would facilitate new business prospects for the island. The cost of the island's imports from Australia increased from $A13m. in 1998/99 to $A17m. in 1999/2000 and declined to $A16m. in 2000/01, when the Territory's exports to that country earned $A7m. In 1998/99 imports from New Zealand cost $NZ9m. Christmas Island purchased no imports from New Zealand in 2000/01, but exports to that country in the same year were worth $NZ4m. An estimated 7.5% of the island's population were unemployed in 1996.

Proposals for the development of a communications satellite launching facility on the island were under consideration in 1998. An assessment of the environmental impact was subsequently undertaken, and the scheme received government approval in May 2000. Following an agreement between the Governments of Australia and Russia in May 2001, preparations for the establishment of a space control centre on Christmas Island commenced. Australia and Russia were to contribute up to US $100m. towards the project, the total cost of which was expected to be US $425m. The Christmas Island site was to be developed by the Asia Pacific Space Centre, an Australian company. Construction began in November 2001, with the first rocket launch being planned for 2004. The space centre was expected to become the world's first wholly privately-owned land-based launch facility, with an operational lifespan of 15–20 years. Asia Pacific planned to launch 10–12 satellites annually by 2005. Although the project would create many employment opportunities on the island (an estimated 400 new jobs during construction and 550 during operation) and lead to substantial infrastructural improvements, environmentalists continued to express concern.

Statistical Survey

AREA AND POPULATION

Area: 135 sq km (52 sq miles).

Population: 1,906 (males 1,023, females 883) at census of 30 June 1996; 1,508 (males 850, females 658) at census of 7 August 2001; 1,506 (official estimate) at August 2002. *Ethnic Groups* (census of 30 June 1981): Chinese 1,587; Malay 693; European 336; Total (incl. others) 2,871. Source: mainly UN, *Demographic Yearbook*.

Density (2002): 11.2 per sq km.

Births and Deaths (1985): Registered live births 36 (birth rate 15.8 per 1,000); Registered deaths 2.

Labour Force (1996): 971 (employed 898, unemployed 73).

MINING

Natural Phosphates (official estimates, '000 metric tons): 285 in 1994; 220 in 1995.

FINANCE

Currency and Exchange Rates: Australian currency is used.

EXTERNAL TRADE

Principal Trading Partners (phosphate exports, '000 metric tons, year ending 30 June 1984)): Australia 463; New Zealand 332; Total (incl. others) 1,136.

2000/01 ($A million): *Imports:* Australia 16. *Exports:* Australia 7. Source: Australian Bureau of Statistics, *Year Book Australia*.

2000/01 ($NZ million): *Exports:* New Zealand 4. Source: Ministry of Foreign Affairs and Trade, New Zealand.

TRANSPORT

International Sea-borne Shipping (estimated freight traffic, '000 metric tons, 1990): Goods loaded 1,290; Goods unloaded 68. Source: UN, *Monthly Bulletin of Statistics*.

TOURISM

Visitor Arrivals and Departures by Air: 14,513 in 1996; 3,895 in 1997; 2,712 in 1998. Source: *Year Book Australia*.

COMMUNICATIONS MEDIA

Radio Receivers (1997): 1,000 in use.

Television Receivers (1997): 600 in use.

Personal Computers (home users, 2001): 506.

Internet Users (2001): 1,450.

EDUCATION

Pre-primary (2002): 37 pupils.

Primary (2002): 217 pupils.

Secondary (2002): 124 pupils.

Source: Education Department of Western Australia.

Directory

The Government

The Administrator, appointed by the Governor-General of Australia and responsible to the Minister for Regional Services, Territories and Local Government, is the senior government representative on the island.

Administrator: EVAN WILLIAMS.

Administration Headquarters: POB 868, Christmas Island 6798, Indian Ocean; tel. (8) 9164-7901; fax (8) 9164-8524.

Shire of Christmas Island: George Fam Centre, POB 863, Christmas Island 6798, Indian Ocean; tel. (8) 9164-8300; fax (8) 9164-8304; e-mail soci@pulau.cx.

Judicial System

The judicial system comprises the Supreme Court, District Court, Magistrate's Court and Children's Court.

Supreme Court

c/o Govt Offices, Christmas Island 6798, Indian Ocean; tel. (8) 9164-7911; fax (8) 9164-8530.

Judges (non-resident): ROBERT SHERATON FRENCH, MALCOLM CAMERON LEE.

Managing Registrar: JEFFERY LOW, Govt Offices, Christmas Island 6798, Indian Ocean; tel. (8) 9164-7911; fax (8) 9164-8530.

Religion

According to the census of 1991, of the 1,275 residents of Christmas Island, some 55% were Buddhists, 10% were Muslims, and 15% were Christians. Within the Christian churches, Christmas Island lies in the jurisdiction of both the Anglican and Roman Catholic Archbishops of Perth, in Western Australia.

The Press

The Islander: Shire of Christmas Island, George Fam Centre, POB 863, Christmas Island 6798, Indian Ocean; tel. (8) 9164-8300; fax (8) 9164-8304; e-mail chong@shire.gov.cx; newsletter; fortnightly; Editor M.G. CHONG.

Broadcasting and Communications

BROADCASTING

Radio

Christmas Island Community Radio Service: f. 1967; operated by the Administration since 1991; daily broadcasting service by Radio VLU-2 on 1422 KHz and 102 MHz FM, in English, Malay, Cantonese and Mandarin; Station Man. WILLIAM TAYLOR.

Christmas Island Radio VLU2–FM: POB 474, Christmas Island 6798, Indian Ocean; tel. (8) 9164-8316; fax (8) 9164-8315; daily broadcasts on 102.1FM and 105.3FM in English, Malay, Cantonese and Mandarin; Chair. and Station Man. TONY SMITH.

Television

Christmas Island Television: POB AAA, Christmas Island 6798, Indian Ocean.

Finance

BANKING

Commercial Bank

Westpac Banking Corpn (Australia): Flying Fish Cove, Christmas Island, Indian Ocean; tel. (8) 9164-8221; fax (8) 9164-8241.

Trade and Industry

In April 2001 there were 67 small businesses in operation.

Administration of Christmas Island: POB 868, Christmas Island 6798, Indian Ocean; tel. (8) 9164-7901; fax (8) 9164-8524; operates power, airport, seaport, health services, public housing, local courts; Dir of Finance JEFFERY TAN.

Christmas Island Chamber of Commerce: Christmas Island 6798, Indian Ocean; tel. (8) 9164-8249; Pres. DON O'DONNELL; Vice-Pres. PHILLIP OAKLEY.

Shire of Christmas Island: George Fam Centre, POB 863, Christmas Island 6798, Indian Ocean; tel. (8) 9164-8300; fax (8) 9164-8304; e-mail soci@iocomm.com.au; f. 1992 by Territories Law Reform Act to replace Christmas Island Services Corpn; provides local govt services; manages tourism and economic development; Pres. DAVID MCLANE; CEO DAVID PRICE.

Union of Christmas Island Workers (UCIW): Poon Saan Rd, POB 84, Christmas Island 6798, Indian Ocean; tel. (8) 9164-8471; fax (8) 9164-8470; e-mail uciw@pulau.cx; fmrly represented phosphate workers; Pres. FOO KEE HENG; Gen. Sec. GORDON THOMSON; 800 mems.

Transport

There are good roads in the developed areas. National Jet Systems operate a twice-weekly flight from Perth, via the Cocos (Keeling) Islands, and a private Christmas Island-based charter company operates services to Jakarta, Indonesia. In 1998 arrivals and departures by air in Christmas Island totalled 2,712 (compared with 3,895 in the previous year). The Australian National Line (ANL) operates ships to the Australian mainland. Cargo vessels from Fremantle deliver supplies to the island every four to six weeks. The Joint Island Supply System, established in 1989, provides a shipping service for Christmas Island and the Cocos Islands. The only anchorage is at Flying Fish Cove.

Tourism

Tourism is a growing sector of the island's economy. Visitors are attracted by the unique flora and fauna, as well as the excellent conditions for scuba-diving and game-fishing. In 2000 there were approximately 90 hotel rooms on the island.

Christmas Island Tourism Association/Christmas Island Visitor Information Centre: POB 63, Christmas Island 6798, Indian Ocean; tel. (8) 9164-8382; fax (8) 9164-8080; e-mail cita@christmas.net.au; internet www.christmas.net.au.

Christmas Island Tours and Travel: Christmas Island 6798, Indian Ocean; tel. (8) 9164-7168; fax (8) 9164-7169; e-mail xch@citravel.com.au; Dir TAN SIM KIAT.

Island Bound Travel: tel. (8) 9381-3644; fax (8) 9381 2030; e-mail info@islandbound.com.au.

Parks Australia: Christmas Island 6798, Indian Ocean; tel. (8) 9164-8700; fax (8) 9164-8755.

Education

The Christmas Island District High School, operated by the Western Australia Ministry of Education, provides education from pre-school level up to Australian 'year 10'. In 2002 enrolment totalled 37 pre-primary, 217 primary and 124 secondary pupils.

COCOS (KEELING) ISLANDS

Introduction

The Cocos (Keeling) Islands are 27 in number and lie 2,768 km north-west of Perth, in the Indian Ocean. The islands, with a combined area of 14 sq km (5.4 sq miles), form two low-lying coral atolls, densely covered with coconut palms. The climate is equable, with temperatures varying from 21°C (69°F) to 32°C (88°F), and rainfall of 2,000 mm per year. In 1981 some 58% of the population were of the Cocos Malay community, and 26% were Europeans. The Cocos Malays are descendants of the people brought to the islands by Alexander Hare and of labourers who were subsequently introduced by the Clunies-Ross family (see below). English is the official language, but Cocos Malay and Malay are also widely spoken. Most of the inhabitants are Muslims (56.8% in 1981). Home Island, which had a population of 446 in mid-1992, is where the Cocos Malay community is based. The only other inhabited island is West Island, with a population of 147 in mid-1992 and where most of the European community lives, the administration is based and the airport is located. The West Island population is comprised mainly of the mainland-based employees of various government departments and their families. The total population of the islands was 621 at the census of both 1998 and August 2001, although official estimates had registered slight increases in the intervening years.

The islands were uninhabited when discovered by Capt. William Keeling, of the British East India Company, in 1609, and the first settlement was not established until 1826, by Alexander Hare. The islands were declared a British possession in 1857 and came successively under the authority of the Governors of Ceylon (now Sri Lanka), from 1878, and the Straits Settlements (now Singapore and part of Malaysia), from 1886. Also in 1886 the British Crown granted all land on the islands above the high-water mark to John Clunies-Ross and his heirs and successors in perpetuity. In 1946, when the islands became a dependency of the Colony of Singapore, a resident administrator, responsible to the Governor of Singapore, was appointed. Administration of the islands was transferred to the Commonwealth of Australia on 23 November 1955. The agent of the Australian Government was known as the Official Representative until 1975, when an Administrator was appointed. The Minister for Regional Services, Territories and Local Government is responsible for the governance of the islands. The Territory is part of the Northern Territory Electoral District.

In June 1977 the Australian Government announced new policies concerning the islands, which resulted in its purchase from John Clunies-Ross of the whole of his interests in the islands, with the exception of his residence and associated buildings. The purchase for $A6.5m. took effect on 1 September 1978. An attempt by the Australian Government to acquire Clunies-Ross' remaining property was deemed by the Australian High Court in October 1984 to be unconstitutional.

In July 1979 the Cocos (Keeling) Islands Council was established, with a wide range of functions in the Home Island village area (which the Government transferred to the Council on trust for the benefit of the Cocos Malay community) and, from September 1984, in the greater part of the rest of the Territory.

On 6 April 1984 a referendum to decide the future political status of the islands was held by the Australian Government, with UN observers present. A large majority voted in favour of integration with Australia. As a result, the islanders were to acquire the rights, privileges and obligations of all Australian citizens. In July 1992 the Cocos (Keeling) Islands Council was replaced by the Cocos (Keeling) Islands Shire Council, comprised of seven members and modelled on the local government and state law of Western Australia. The first Shire Council was elected in 1993. The Clunies-Ross family was declared bankrupt in mid-1993, following unsuccessful investment in a shipping venture, and the Australian Government took possession of its property.

In September 2001, following an increase in the numbers of illegal immigrants reaching Australian waters (see the section on Christmas Island), legislation was enacted removing the Cocos Islands and other territories from Australia's official migration zone. In October of that year the Australian Government sent contingency supplies to the islands as a precaution, should it be necessary to accommodate more asylum-seekers. This development provoked concern among many Cocos residents that the former quarantine station used as a detention centre might become a permanent asylum-processing facility under the order of the Australian Government. In December 123 Sri Lankan and Vietnamese asylum-seekers were housed at the station, which was built to accommodate only 40. They were transferred to Christmas Island in February 2002. The Government admitted that the establishment of an asylum-processing facility on Cocos might become necessary.

Although local fishing is good, some livestock is kept, and domestic gardens provide vegetables, bananas and papayas (pawpaws), the islands are not self-sufficient, and other foodstuffs, fuels and consumer items are imported from mainland Australia. Coconuts, grown throughout the islands, are the sole cash crop: total output was an estimated 7,625 metric tons in 2002. In 2001/02 the Administration and the Shire provided support for a study investigating the potential economic benefits of various coconut products, including the production of high-quality carbon from coconut kernels and the manufacture of furniture from coconut palm wood. A clam farm was established in 2000. A Cocos postal service (including a philatelic bureau) came into operation in September 1979, and revenue from the service is used for the benefit of the community. In early 2000 the islands' internet domain name suffix, '.cc', was sold to Clear Channel, a US radio group, thus providing additional revenue. The conversion of a disused quarantine station into a business centre was under consideration in 2001. In October 2003 a ship carrying 52,000 Australian sheep was diverted to the Cocos Islands, having been refused entry in Saudi Arabia, where inspectors claimed to have found high levels of disease among the animals. Many islanders expressed concern over the arrangement and the Territory's lack of facilities to deal with the situation. The islands have a small tourist industry. During August, September and October 2003 there were a total of 76 non-resident arrivals, of whom 28 were travelling on business and 19 were visiting relatives.

The cost of the islands' imports from Australia increased from $A5m. in 1999/2000 to $A11m. in 2000/01. Exports to Australia totalled $A2m. in 1996/97, but by 2002 the Cocos Islands no longer exported any goods or produce. The Grants Commission estimated that the Australian Government's net funding of the Territory amounted to some $A18m. in 1999. In 2000/01 the Cocos Administration expended approximately $A9m. An estimated 19% of the total labour force was unemployed in 2001, 60% of whom were under the age of 30. The Shire Council and the Co-operative Society were the principal employers.

Primary education is provided at the schools on Home and West Islands. Secondary education is provided to the age of 16 years on West Island. A bursary scheme enables Cocos Malay children to continue their education on the Australian mainland.

Statistical Survey

AREA AND POPULATION

Area: 14.2 sq km (5.5 sq miles).

Population: 655 (males 338, females 317) at census of 30 June 1996; 621 (males 334, females 287) at census of 7 August 2001. *Ethnic Groups* (census of 30 June 1981): Cocos Malay 320; European 143; Total (incl. others) 555. Source: mainly UN, *Demographic Yearbook*.

Density (2001): 43.7 per sq km.

Births and Deaths (1986): Registered live births 12 (birth rate 19.8 per 1,000); Registered deaths 2.

Labour Force (2001): 270 (employed 218, unemployed 52). Source: Indian Ocean Group Training.

AGRICULTURE

Production (FAO estimates, metric tons, 2002): Coconuts 7,600; Copra 1,000. Source: FAO.

FINANCE

Currency and Exchange Rates: Australian currency is used.

EXTERNAL TRADE

Principal Commodities (metric tons, year ending 30 June 1985): *Exports*: Coconuts 202. *Imports*: Most requirements come from Australia. The trade deficit is offset by philatelic sales and Australian federal grants and subsidies. *2000/01* ($A '000, imports from Australia): 11,000.
Source: *Year Book Australia*.

COMMUNICATIONS MEDIA

Radio Receivers (1992): 300 in use.

Personal Computers (home users, 2001): 142.

Internet Users (2001): 618.

EDUCATION

Pre-primary (2002): 23 pupils.

Primary (2002): 84 pupils.

Secondary (2002): 31 pupils.

Teaching Staff (2002): 18.

Source: Education Department of Western Australia.

Directory

The Government

The Administrator, appointed by the Governor-General of Australia and responsible to the Minister for Regional Services, Territories and Local Government, is the senior government representative in the islands.

Administrator: EVAN WILLIAMS (non-resident).

Administrative Offices: POB 1093, West Island, Cocos (Keeling) Islands 6799, Indian Ocean; tel. (8) 9162-6615; fax (8) 9162-6697.

Cocos (Keeling) Islands Shire Council: POB 94, Home Island, Cocos (Keeling) Islands 6799, Indian Ocean; tel. (8) 9162-6649; fax (8) 9162-6668; e-mail info@shire.cc; f. 1992 by Territories Law Reform Act; Pres. MOHAMMED SAID CHONGKIN; CEO BOB JARVIS.

Judicial System

Supreme Court, Cocos (Keeling) Islands: West Island Police Station, Cocos (Keeling) Islands 6799, Indian Ocean; tel. (8) 9162-6615; fax (8) 9162-6697; Judge ROBERT SHERATON FRENCH; Additional Judge MALCOLM CAMERON LEE.

Magistrates' Court, Cocos (Keeling) Islands

Special Magistrate (vacant).

Managing Registrar: ROBYN JENKINS, Cocos (Keeling) Islands 6799, Indian Ocean; tel. (8) 9162-6615; fax (8) 9162-6697.

Religion

According to the census of 1981, of the 555 residents, 314 (some 57%) were Muslims and 124 (22%) Christians. The majority of Muslims live on Home Island, while most Christians are West Island residents. The Cocos Islands lie within both the Anglican and the Roman Catholic archdioceses of Perth (Western Australia).

Broadcasting and Communications

BROADCASTING

Radio

Radio VKW Cocos: POB 33, Cocos (Keeling) Islands 6799, Indian Ocean; tel. (8) 9162-6666; e-mail vkw@kampong.cc; non-commercial; daily broadcasting service in Cocos Malay and English; 200 listeners; Station Man. CATHERINE CLUNIES-ROSS.

Television

A television service, broadcasting Indonesian, Malaysian and Australian satellite television programmes and videotapes of Australian television programmes, began operating on an intermittent basis in September 1992.

Industry

Cocos (Keeling) Islands Co-operative Society Ltd: Home Island, Cocos (Keeling) Islands 6799, Indian Ocean; tel. (8) 9162-6702; fax (8) 9162-6764; f. 1979; conducts the business enterprises of the Cocos Islanders; activities include boat construction and repairs, copra and coconut production, sail-making, stevedoring and airport operation; owns and operates a supermarket and tourist accommodation; Chair. MOHAMMED SAID BIN CHONGKIN; Gen. Man. RONALD TAYLOR.

Transport

National Jet Systems operate a twice-weekly service from Perth, via Christmas Island, for passengers, supplies and mail to and from the airport on West Island. Cargo vessels from Singapore and Perth deliver supplies, at intervals of four to six weeks.

Cocos Trader: Cocos (Keeling) Islands 6799, Indian Ocean; tel. (8) 9162-6612; fax (8) 9162-6568; e-mail manpower@kampong.cc; shipping agent.

Tourism

Cocos Island Tourism Association: POB 30, Cocos (Keeling) Islands 6799, Indian Ocean; tel. (8) 9162-6790; fax (8) 9162-6696; e-mail info@cocos-tourism.cc; internet www.cocos-tourism.cc.

Education

Primary education is provided at the schools on Home and West Islands. Secondary education is provided to the age of 16 years on West Island. A bursary scheme enables Cocos Malay children to continue their education on the Australian mainland.

TERRITORY OF HEARD ISLAND AND THE MCDONALD ISLANDS

These islands are situated about 4,000 km (2,500 miles) south-west of Perth, Western Australia. The Territory, consisting of Heard Island, Shag Island (8 km north of Heard) and the McDonald Islands, is almost entirely covered in ice and has a total area of 369 sq km (142 sq miles). Sovereignty was transferred from the United Kingdom to the Commonwealth of Australia on 26 December 1947, following the establishment of a scientific research station on Heard Island (which functioned until March 1955). The islands are administered by the Antarctic Division of the Australian Department of the Environment and Heritage. There are no permanent inhabitants. However, in 1991 evidence emerged of a Polynesian community on Heard Island some 700 years before the territory's discovery by European explorers. The island is of considerable scientific interest, as it is believed to be one of the few Antarctic habitats uncontaminated by introduced organisms. Heard Island is about 44 km long and 20 km wide and possesses an active volcano, named Big Ben. In January 1991 an international team of scientists travelled to Heard Island to conduct research involving the transmission of sound waves, beneath the surface of the ocean, in order to monitor any evidence of the 'greenhouse effect' (melting of polar ice and the rise in sea-level as a consequence of pollution). The pulses of sound, which travel at a speed largely influenced by temperature, were to be received at various places around the world, with international co-operation. Heard Island was chosen for the experiment because of its unique location, from which direct paths to the five principal oceans extend. The McDonald Islands, with an area of about 1 sq km (0.4 sq miles), lie some 42 km west of Heard Island. Only two successful landings by boat have been recorded since the discovery of the McDonald Islands in the late 19th century. In late 1997 Heard Island and the McDonald Islands were accorded World Heritage status by UNESCO in recognition of their outstanding universal significance as a natural landmark.

In 1999 concern was expressed that stocks of the Patagonian toothfish in the waters around the islands were becoming depleted as a result of over-exploitation, mainly by illegal operators. (The popularity of the fish in Japan and the USA, where it is known as

Chilean sea bass, increased significantly during the early 2000s.) This problem was highlighted in August 2003 when a Uruguayan fishing vessel was seized, following a 20-day pursuit over 7,400 km by Australian, South African and British patrol boats. The trawler had been fishing illegally in waters near Heard and McDonald and had a full cargo of Patagonian toothfish worth some US $1.5m. Experts feared that, if poaching continued at current rates, the species would become extinct by 2007. In response to this situation, the Australian Government announced in December 2003 that it was to send an ice-breaking patrol vessel with deck-mounted machine guns to police the waters around Heard and McDonald Islands. It was hoped that this action might serve to deter illegal fishing activity in the area, much of which was believed to involve international criminal organizations.

In 2001 the Australian Government's Antarctic Division conducted a five-month scientific expedition to Heard Island. It claimed that glacial cover had retreated by 12% since 1947 as a result of global warming. In October 2002 the Australian Government declared the establishment of the Heard Island and McDonald Islands Marine Reserve. Covering 6.5m. ha, the marine reserve was to be the largest in the world, strengthening existing conservation measures and imposing an official ban on all fishing and petroleum and mineral exploitation. Among the many species of plant, bird and mammal to be protected by the reserve were the southern elephant seal, the sub-Antarctic fur seal and two species of albatross. Limited scientific research and environmental monitoring were to be allowed in the Marine Reserve.

BRUNEI

Physical and Social Geography

HARVEY DEMAINE

PHYSICAL FEATURES AND CLIMATE

The Sultanate of Brunei (Negara Brunei Darussalam) covers an area of 5,765 sq km (2,226 sq miles) and faces the South China Sea along the north-west coast of the island of Borneo, most of which comprises the Indonesian territory of Kalimantan. On its landward side, Brunei is both surrounded and split into two separate units by Sarawak, part of Malaysia. Brunei is divided into four districts: Brunei/Muara, Tutong and Seria/Belait, in the western section of Brunei, and Temburong, forming the eastern section.

The greater part of Brunei's small territory consists of a low coastal plain, and only on its southern margins does it attain heights of more than 300 m above sea-level. Brunei's highest point is Bukit Pagon, in the east of the country, which reaches 1,841 m above sea-level. Situated only 4°–5° N of the Equator, Brunei has a consistently hot and humid climate, with mean monthly temperatures of around 27°C and a heavy annual rainfall, well distributed throughout the year, of more than 2,500 mm. Except for those areas that have been cleared for permanent cultivation in the coastal zone, about three-quarters of the country is covered by dense equatorial forest, although this has deteriorated in places as a result of shifting cultivation.

POPULATION AND RESOURCES

At the census of August 2001 the population of Brunei was 332,844, compared with 260,482 at the August 1991 census. In 2003, according to official estimates, the population was 348,800. In that year 66.6% of the population were Malay and 11.0% Chinese. Indigenous races, which comprised 3.6% of the population, are mainly Muruts, Kedayans and Dusuns. More than 50% of the population were less than 20 years of age in 2000. The Chinese reside mainly in Bandar Seri Begawan (formerly Brunei Town), the capital, and Seria. Bandar Seri Begawan, which occupies an impressive site overlooking the large natural inlet of Brunei Bay, had an estimated population of 27,285 in 2001.

Brunei's natural resources consist almost exclusively of petroleum and natural gas. In the 1980s new petroleum reserves were discovered at Seria, the first significant onshore oilfield in Brunei. Further reserves were discovered in new and existing offshore fields in the early 1990s. At the end of 2003 proven petroleum reserves amounted to 1,100m. barrels, sufficient to maintain output at that year's levels (averaging 214,000 barrels per day) for approximately 14 years.

The major source of future development is expected to be natural gas, of which Brunei had proven reserves of 350,000m. cu m at the end of 2003. Natural gas reserves in 2003 were estimated to be sufficient to maintain output for approximately 28 years, assuming current rates of production. Production of natural gas, which in 2003 totalled 12,400m. cu m, is mostly destined for export to Japan under long-term contracts.

History

C. M. TURNBULL

Revised by GRAHAM SAUNDERS

EARLY HISTORY

One of the world's smallest states, Brunei was once the centre of a great maritime empire. Its origins are obscure, but it was probably founded in the late seventh century by a refugee prince fleeing from the Khmer conquest of Funan. From that time until the 16th century it was the centre of three successive empires, which at their height held sway over much of coastal Borneo and the Philippines. Known to the Chinese as Po-ni, the port lay on the main trading route between China, the western part of the Indonesian archipelago and the Indian Ocean. There was extensive trade between the area and China during the period of the Tang and Song dynasties.

About the middle of the 15th century Brunei became an independent sultanate. The first Sultan married a Malay princess from the powerful Muslim sultanate of Melaka (Malacca), adopted Islam as the court religion and introduced an efficient administration, modelled on Melaka. Brunei profited from trade with Melaka but achieved even greater prosperity after the great Malay port was conquered by the Portuguese in 1511, when many Muslim traders diverted their custom to Brunei. The first European visitors were members of the Spanish expedition which had been led by the Portuguese navigator, Fernão de Magalhães (Magellan). They arrived in 1521, during the reign of Bolkiah, the most illustrious of Brunei's sultans, and described the capital as a large and wealthy city of some 25,000 households, with an impressive and cultivated royal court. Brunei established friendly relations with Portugal. In 1526 the two countries concluded a commercial arrangement, and the Portuguese established a trading post at Brunei. As Portuguese trade with China and Japan expanded, Brunei became a regular port of call on the route between Melaka and Macao, and the Chinese community expanded greatly. Brunei already had quite a large Chinese community, and a 15th-century Chinese trader had married into the royal family and become Sultan himself.

The first half of the 16th century was Brunei's 'golden age', when it claimed suzerainty over the whole coast of Borneo, the Sulu archipelago and Mindanao, and forced Manila to pay tribute. The empire was not a centralized polity but comprised a group of individual river states ruled by vassals, who paid obeisance to the Sultan and secured a revenue through river tolls and poll taxes. The empire depended almost entirely on its ability to control regional trade.

Brunei came into conflict with the Spaniards, who established themselves in the northern Philippines in the latter part of the 16th century. The rivals clashed over trade and religion, and in 1578 the Spaniards seized Brunei for a short period and attempted unsuccessfully to impose Christianity. Eventually the Spaniards abandoned attempts to subdue the southern Philippines, but they raided Brunei periodically, and vassal chiefs in Mindanao and the Sulu archipelago took advantage of Brunei's weakness to break away and establish their own independence. Sulu was an aggressive state, the home of the infamous Balanini pirates, who ravaged the Borneo coasts and even ventured as far as the Straits of Melaka. As recompense for its intervention in a civil war that racked Brunei for 12 years in the mid-17th century, Sulu claimed the whole of North Borneo (present-day Sabah), a claim that remains unresolved.

Brunei continued to decline, and by the early 19th century the Sultan could lay claim only to the district centring on Brunei Town itself, the Sarawak River and the western coast of northern Borneo. Even here he exerted only weak control along the coastal strip and the lower reaches of the main rivers. Pirates operated along the coast of North Borneo, and Brunei Town became little more than a trading centre for their plunder and slaves. The stability of the sultanate was also threatened by disputed successions and rebellious chiefs. When antimony was discovered in the Sarawak valley in 1824, Brunei officials were sent to organize local Dayaks to mine the ore. In 1835 the corruption of the Brunei Governor drove the local chiefs and Dayaks to revolt. The province was still in a state of armed rebellion when an English gentleman adventurer, James (later Sir James) Brooke, arrived four years later. In 1841, in return for his help in settling the revolt, Brooke was granted the Sarawak River district and the title of Raja, which was confirmed by the Sultan the following year when Brooke paid the customary tribute at the Brunei court.

BRITISH INTERVENTION

The British connection was to lead to the dissolution of most of Brunei's empire, but was arguably responsible for the continued existence of the sultanate. When Brooke's position at the Brunei court was threatened, the British Navy intervened, and the Sultan was compelled to confirm Brooke's tenure at Sarawak for himself and his successors in perpetuity, to give the island of Labuan to the United Kingdom, and to sign a treaty, in 1847, undertaking not to cede any further territory without British approval. In 1853, as a result of conflict with pirates, the Sultan ceded to Brooke the troublesome Saribas and Skarang districts, which later constituted the Second Division of Sarawak. Eight years later further piracy, which threatened the profitable sago trade of the Mukah and Oya valleys, compelled the Sultan to cede to Brooke the vast Rejang River basin, which became the Third Division of Sarawak.

James Brooke's successor, his nephew Charles (later Sir Charles) Brooke, wished to extend Sarawak's rule over the lawless upper Baram area, the scene of recurrent friction between the people of Sarawak and Brunei, but for many years the British Government opposed this. In 1874 the British Government rejected Charles Brooke's proposal to place Brunei under the protection of either the United Kingdom or Sarawak. In 1877, however, the Sultan granted the northern part of Borneo (present-day Sabah) to a Hong Kong-based company owned by a British businessman, Alfred Dent, and the Austrian Consul, Von Overbeck. Four years later, Dent purchased the Austrian share, and in that same year the British Government granted him a royal charter to form the British North Borneo Company.

Fears of rival foreign ambitions in the region, inspired by Sultan Abdul Mumin's advancing years and uncertainty concerning the succession, prompted the United Kingdom to give approval in 1882 for Charles Brooke to obtain cession of the Baram basin, which became the Fourth Division of Sarawak. In 1884 he also acquired the Trusan valley. Now only the heartland of Brunei remained, under growing pressure from both Sarawak and the British North Borneo Company. In 1885 the new Sultan, Hashim, promised the Brunei chiefs that he would not alienate any further territory, and in 1888 the United Kingdom made Brunei, Sarawak and North Borneo protectorates, thus assuming paramountcy over the whole of North-West Borneo. The Governor of the Straits Settlements was appointed High Commissioner for Brunei, but the British Government did not appoint a Resident at the Brunei court, and Sarawak continued to present a threat to the sultanate. In 1890 the chiefs of Limbang, which had been in a state of rebellion for some years, asked Raja Brooke to take over their district, which was then joined with Trusan to form the Fifth Division of Sarawak. This deprived Brunei of a valuable food-producing area and divided it into two parts. The British Government offered to pay compensation to the Sultan but the offer was never formally accepted, so that this came to be regarded as a cession by default. In 1916 the United Kingdom formally recognized Limbang as part of Sarawak, but the Sultans of Brunei never acknowledged the cession, and the status of Limbang remained controversial.

By the early 20th century the Tutong and Belait districts were in revolt, many people were migrating into adjoining Sarawak, and the sultanate had contracted to little more than 2,000 sq miles. The United Kingdom considered dividing what remained of Brunei between Sarawak and the British North Borneo Company, but decided against this after considering a detailed report by Stewart McArthur, an officer of the Malayan civil service, who spent six months investigating the situation in Brunei in 1904. In the following year the British Government signed an agreement with the Sultan and senior chiefs, establishing Brunei as a full protectorate, where all matters relating to administration, legislation and taxation were to be conducted on the advice of a British Resident. McArthur was appointed as first incumbent of this post in 1906, and, apart from the period of Japanese occupation, the British Resident remained the effective ruler of Brunei until 1959. The administration was modelled on that of the (British) Federated Malay States, and the Resident was always an official of the Malayan civil service seconded from the peninsula. A modern civil service was created, a land code introduced, and the state revenues organized. The traditional State Council was formalized, with the Sultan presiding but with the Resident as the dominant influence. Despite some dissatisfaction, quasi-colonial rule brought peace and stability to the country and guaranteed Brunei's survival. Brunei's future economic prosperity was also foreshadowed in this period, when petroleum was first discovered in 1903, and a major oilfield was eventually located at Seria in 1929. By the 1930s Brunei's debts had been discharged, and revenues from petroleum exports helped to finance modest programmes in education and social services.

THE POST-WAR PERIOD

Brunei was occupied by Japanese forces from December 1941, shortly after Japan entered the Second World War, until it was liberated by Australian troops in July 1945. The immediate post-war years were devoted to rehabilitation and the resumption of petroleum production, which had been interrupted by the war.

The United Kingdom envisaged an eventual self-governing confederation embracing all the British dependencies in South-East Asia, namely the Malay Peninsula, Singapore and Borneo. As a first step, it planned to abolish protectorate status, bringing all the territories under the direct rule of the British crown. Sarawak and British North Borneo did become crown colonies, but the transition in Brunei was deferred, following difficulties encountered in abrogating the Malay States' treaties. In 1948 the Governor of Sarawak was appointed High Commissioner for Brunei, in place of the Governor of the now defunct Straits Settlements, but this did not alter the protectorate's status or the Resident's powers. Nevertheless, political changes in the region and the prospect of ultimate British withdrawal presented new threats to Brunei's security. These problems dominated Brunei politics for the next 20 years, producing two rival forms of nationalism: an enlightened paternalism, propounded by the Sultan and his supporters, and a form of popular democracy.

In 1953 Sultan Omar Ali Saifuddin III established a commission to help to formulate a written constitution for Brunei. District Councils, nominated by the Sultan, were created in 1954, but little progress was made in devising a constitution. Meanwhile, Brunei's first political party, the Parti Rakyat Brunei (PRB—Brunei People's Party), was formed in 1956, modelled on the left-wing Malayan Parti Rakyat. Its charismatic President, Sheikh Ahmad Azahari (born in Labuan of Arab-Malay parentage), had studied in Java during the Japanese occupation and taken part in the Indonesian struggle for independence against the Dutch. He spent some time in Singapore in the early 1950s, and was imprisoned briefly for organizing the first political demonstration in Brunei. While remaining loyal to the sultanate, the party advocated democratic self-government for Brunei as part of a federation of the three Borneo states. The PRB attracted considerable popular support and in 1957 petitioned the Sultan and the Colonial Office for independence. Leaders of the PRB were angered by their exclusion from the delegation that the Sultan led to London in 1958 for constitutional talks. Under the Brunei Constitution, which was promulgated in 1959, the United Kingdom retained responsibility for Brunei's defence and foreign relations but transferred internal government to the Sultan, who was to preside over an Executive Council and rule with the help of a Legislative Council and District Councils, the latter being elected by universal adult suffrage. The PRB pressed for immediate elections, for independence in Brunei by 1963 and for a merger with the other Borneo states. The Sultan preferred a closer association with the Federation of Malaya, which had gained its independence in 1957. He welcomed the proposal made by Tunku Abdul Rahman, the Malayan Prime Minister, in May 1961, mooting a Malaysian federation to include Malaya, Singapore and the three Borneo territories. This led to heated confrontation between the Sultan's supporters and the PRB.

At district council elections, which were eventually held in August 1962 after many delays, the PRB won all but one of the 55 seats, which also gave its candidates all the indirectly elective legislative council seats. The PRB's campaign had advocated internal democratic reform and rejection of the Malaysia proposal, in favour of a Borneo federation. In September the party united with politicians in Sarawak and North Borneo to form an Anti-Malaysia Alliance, for which Azahari tried to rouse international support. In December 1962, after failing in an attempt to present to the Legislative Council a proposal in favour of independence, separate from Malaysia, Azahari resorted to force and staged a rebellion through the North Borneo Liberation Army, which had strong links with the Indonesian Communist Party and with left-wing extremists in Singapore. The rebels proclaimed a Revolutionary State of North Kalimantan, with Azahari as Prime Minister. The Sultan quickly suppressed the revolt, however, with the aid of British forces from Singapore, and most of Azahari's former supporters in Sarawak and North Borneo disowned his use of force. A state of emergency was declared, the PRB was banned, Azahari went into exile in Indonesia, and his supporters fled or were imprisoned. From that time the Sultan ruled by decree, and the emergency laws have been renewed every two years.

The rebellion initially strengthened Sultan Omar's resolve to join Malaysia, as a means of ensuring Brunei's permanent security. In June 1963, however, negotiations collapsed, owing mainly to disputes concerning petroleum revenues, but also to disagreement regarding the Sultan's precedence among Malay rulers. Two meetings between the Tunku and Sultan Omar failed to resolve the issue, and Brunei withdrew from the final negotiations which led to the establishment of Malaysia in September 1963. Brunei's decision not to join Malaysia soured its relations with Kuala Lumpur, and allegations about Indonesia's involvement in the December 1962 revolt strained relations with Jakarta. Brunei was isolated, more than ever dependent on the United Kingdom, and its autocratic Government and semi-colonial status exposed it to international criticism.

Following the 1962 revolt, the ban on the PRB had removed the most articulate opposition, and most other parties united to form a Brunei Alliance Party (BAP), which supported entry to Malaysia. The Sultan ignored the BAP's demand for a fully elected legislature, although he agreed to elections in 1965 for some legislative council seats and replaced the Executive Council with a Council of Ministers. The British Government exerted pressure on Sultan Omar to quicken the pace of constitutional reform, but he insisted on an appointed cabinet. In August 1966 all existing political groups united to form a Brunei People's Independence Party (BPIP), which demanded responsible government, a full ministerial system and a fully elected legislature. The British Government was impatient at the slow progress of constitutional reform, at a time when it was preparing to withdraw most of its forces 'east of Suez'. In 1967 Sultan Omar abdicated in favour of his 21-year-old son, Hassanal Bolkiah, but the ex-Sultan retained effective power. The BPIP failed to gain support at district council elections held in 1968, which were overshadowed by the young Sultan's forthcoming talks in London about the future defence of Brunei. Under a new treaty, which was signed with the United Kingdom in 1971, the 1959 Constitution was amended to give the Sultan full control of all internal matters, with the United Kingdom retaining responsibility for foreign affairs. A separate agreement provided for the stationing of a British battalion of Gurkhas in Brunei.

Brunei developed close ties with Singapore, after the latter's secession from the Federation of Malaysia in 1965, but relations with Malaysia continued to be strained for many years. Malaysia offered political asylum to PRB leaders, permitted the illegal party to open an office in Kuala Lumpur, and in 1975 sponsored a PRB delegation which presented a case for independence to the UN Committee on Decolonization. Brunei recalled all its students from Malaysia for fear that they might become a focus for dissidence, and it officially revived the Limbang claim, with former Sultan Omar crossing the border to incite Limbang villagers against Malaysia.

In 1977 the UN General Assembly adopted a Malaysian-sponsored resolution proposing free elections in Brunei, the end of the ban on political parties and the return of all political exiles to Brunei. While the British Labour Government (whose representative abstained from voting on the UN resolution) was prepared to sever its links with Brunei, the Sultan regarded the association as a protection against the possible encroachment of neighbouring governments, secessionists and political opponents within Brunei itself. The sultanate was reluctant to revise the terms of its 1971 treaty with the United Kingdom until it received assurances that Malaysia and Indonesia would respect Brunei's independence. In June 1978 Sultan Sir Hassanal Bolkiah and his father visited London in an unsuccessful attempt to resist separation from the United Kingdom. In 1979, however, they were compelled to sign an agreement whereby Brunei became a sovereign independent state on 1 January 1984. In September 1983 Brunei concluded a new defence agreement with the United Kingdom, whereby Brunei would continue, at its own expense, to employ the battalion of Gurkhas under British command.

AFTER INDEPENDENCE

When Brunei became independent the Council of Ministers was abolished in favour of a seven-member Cabinet, headed by the Sultan and including his father and two brothers. *Melayu Islam Beraja* (MIB—Malay Islamic Monarchy) was proclaimed as the state ideology promoting Islamic values, an emphasis on the unique nature of Brunei-Malay culture and the importance of the role of the monarchy.

In the 1980s new laws were adopted to increase the share of petroleum revenue accruing to the State. While the demarcation between state revenue and the Sultan's personal wealth was not clearly defined and much was spent on royal prestige projects, all citizens enjoyed free medical care and education, and government housing loans. Indigenous, mainly Malay, inhabitants, *bumiputras* (sons of the soil), received preferential treatment. Even Brunei-born non-*bumiputras* were subject to stringent requirements with respect to residence and language when applications for citizenship were considered. In 1985 90% of the ethnic Chinese, who at that time constituted about one-third of the population, were classified as non-citizens excluded from state benefits. Although they still dominated the private sector, many Chinese began to emigrate.

Increasingly, Brunei's modernization and exposure to the rest of the world were regarded as a potential threat to its moral, cultural and religious traditions. At the same time rising unemployment and, more particularly, a shortage of non-manual jobs (menial work was generally undertaken by immigrant labourers) led to the emergence in the early 1990s of social problems, including drug and alcohol abuse. There was concern at the prospect of the situation deteriorating since more than one-half of the Bruneian population were aged under 20, and educational standards and employment expectations were rising. Partly in response to these incipient problems, from 1990 the state ideology (MIB) was promoted more vigorously. Muslims were encouraged to adhere more closely to the tenets of Islam, with greater emphasis on Islamic holiday celebrations, and in January 1991 the import of alcohol was banned. In December of that year the public celebration of Christmas, the Christian festival, was forbidden. The state Mufti was brought under the direct control of the Sultan rather than the Ministry of Religious Affairs and in that same year the first Islamic bank was established. Under the Seventh National Development Plan (1996–2000) more resources were devoted to building mosques, religious schools and an Islamic college. An Islamic radio station began broadcasting in 1997.

After the former Sultan's death in 1986, the Cabinet was enlarged to 11 ministers, incorporating members of the educated élite. The royal family, however, remained the dominant force in the Government, with the Sultan as Prime Minister and Minister of Defence, and his brothers, Mohamed and Jefri, responsible for foreign affairs and finance respectively. In 1985 the Sultan considered permitting the introduction of a party political system and agreed to the formation of the Parti Kebangsaan Demokratik Brunei (PKDB—Brunei National Democratic Party). The PKDB advocated greater participation in the administration of the Government, democratization and a more equitable distribution of wealth. It attracted a membership of some 3,000, comprising mainly Malay professionals and business executives, but it aroused little public support. Within a year a breakaway faction formed a new party, the Parti Perpaduan Kebangsaan Brunei (PPKB—Brunei National Solidarity Party), which emphasized greater co-operation with the Government. In 1988, however, the President and the Secretary-General of the PKDB, Abdul Latif Hamid and Abdul Latif Chuchu, were arrested and detained for two years under the provisions of the Internal Security Act. The party was dissolved after it had demanded the Sultan's resignation as Head of Government (although not as Head of State), the holding of democratic elections and the ending of the 26-year state of emergency. Meanwhile, at the beginning of 1988 a number of political detainees of the former PRB were released in a general amnesty. In 1990 all political prisoners were released, and in 1996 the few remaining PRB members were permitted to return from exile, providing that they refrained from political activity. Azahari himself died in Bogor, Indonesia, in May 2002. In any event, radical politicians found it difficult to attract popular support while most Bruneians continued to enjoy a high standard of living.

In October 1992 Brunei celebrated Sultan Hassanal's Silver Jubilee, but hopes of liberal political concessions were dispelled when the Sultan marked the occasion by reaffirming the central role of the monarchy in a Malay Islamic nation. He assumed a more paternal stance, as 'the People's Sultan', distancing himself from the extravagant lifestyle of earlier days to stress diligence and mutual responsibility to Brunei as a country with 'its own firm identity and image among the non-secular nations of the world'. District and village councils were established in 1993 and held their first general assembly in May 1996. In 1994 a constitutional committee, chaired by Prince Mohamed, which had been appointed by the Sultan to review the 1959 constitutional arrangements, submitted a new draft constitution to the Sultan for consideration. In February 1995 the PPKB was permitted to hold its first national assembly, at which its newly elected President, Abdul Latif Chuchu, reaffirmed support for the monarchy and the national ideology. However, he also called for democratic elections, and was soon forced to resign by the Government, which renewed the emergency laws. Following an inactive period of three years, in May 1998 the PPKB held an annual general meeting, at which a business executive, Hatta Zainal Abidin, was elected President. The party briefly voiced concern at allegations of official corruption during the Amedeo court case in May 2000 (see below), but soon lapsed into infighting in attempts to oust its President.

The anachronistic political system remained firmly entrenched, and the dominance of the monarchy was reaffirmed in August 1998 when the Sultan's eldest son, Al-Muhtadee Billah Bolkiah, was formally installed as the heir to the throne in a lavish ceremony. The royal family enjoyed a monopoly of power, but maintained its popularity by enabling all citizens to share to some extent in the wealth of the State. There was no income tax, while housing, fuel and other essentials were subsidized and until 1995, when nominal charges were introduced for medical and dental treatment, both health services and education were free for citizens. The populace, as well as foreign dignitaries, were involved in extravagant royal festivities, and at his birthday celebrations in 1998 the Sultan pledged salary increases for lower-paid civil servants and greater state support for pensioners and the destitute. He frequently exhorted the wealthy to contribute generously to the less privileged, and warned of the dangers of putting economic progress ahead of social development.

The economic crisis that beset South-East Asia from late 1997, together with scandals in Brunei itself, provided the catalyst for economic and, potentially, political change. Initially, Brunei provided assistance to other parts of the region; the sultanate contributed to an IMF emergency programme for Thailand, the state-owned Brunei Investment Agency (BIA) helped to stabilize currencies by buying Singapore dollars and Malaysian ringgit, and Brunei promised to invest in Malaysia to assist its recovery. Brunei itself, however, was adversely affected by a sharp decline in the international price of petroleum and by haze pollution from forest fires in Indonesia and Malaysia in 1997 and 1998, which threatened the development of tourism. The downturn was compounded in July 1998 with the collapse of the Amedeo Development Corporation, Brunei's largest investment and construction firm, which Prince Jefri controlled through his son and which had benefited from numerous lucrative government contracts.

Relations between the Sultan and his younger brother, Jefri, were already severely strained. Lawsuits were brought against Prince Jefri in the USA in 1997 and 1998 by US and British beauty queens alleging sexual misconduct, and in February 1998 two former business associates sued the prince for £80m. in London, claiming that he had reneged on property agreements. The first case was withdrawn when Prince Jefri was granted diplomatic immunity, and the other two lawsuits were settled out of court for undisclosed sums. The cases were unreported in Brunei but attracted wide international publicity and brought Prince Jefri's extravagant and profligate lifestyle into disrepute. In February 1998 the Sultan removed the finance portfolio from Prince Jefri, assuming responsibility for it himself, in the first major cabinet change for 10 years. In the following month the Minister of Health was dismissed, reportedly owing to his inadequate response to the haze over Brunei caused by forest fires in Indonesia and Malaysia, and in June the Sultan accepted the resignation of the Attorney-General and of the Solicitor-General. In July 1998 the Sultan appointed international accountants as executive managers of Amedeo, removed Prince Jefri from the boards of seven telecommunications companies and dismissed him as Chairman of the BIA, which controlled much of Brunei's overseas investment.

For a long time the Brunei Government made no official comment about the collapse of the Amedeo group, the misappropriation of BIA funds, or the role played by Prince Jefri. In August 1999, however, Abdul Aziz Umar, the Minister of Education, who was Chairman of the government task force appointed to investigate the missing BIA funds, admitted that there had been mismanagement. After the failure of private negotiations, the Government and the BIA began civil proceedings against Prince Jefri in Brunei and England, alleging improper withdrawal and use of more than US $28,000m. while he was Minister of Finance and Chairman of the BIA. A total of 71 others were named in the action, including Prince Jefri's eldest son, Prince Muda Abdul Hakeem, along with his private secretary, and more than 60 overseas companies believed to be controlled by Prince Jefri. In April the Brunei court dismissed an appeal against disclosure of his assets, which were 'frozen' world-wide, and rejected his plea for an independent judge from outside Brunei. The court case, which opened in May, dominated the local press, but Prince Jefri and his son reached confidential out-of-court settlements, in which the case against them was abandoned when they agreed to return all assets purchased with BIA funds. In October 2000, however, Haji Awang Kassim, Jefri's former confidential secretary, who had virtually run Amedeo and was deputy managing director of the BIA, was arrested after being extradited from Manila. The State and the BIA also began civil proceedings to recover funds from six other former associates of Prince Jefri, including the former managing directors of two state-owned corporations. Meanwhile, there were angry scenes at meetings of Amedeo creditors in November 2000, when they refused the BIA's settlement terms, and in May 2001 the liquidators sued a former senior official.

In September 1998 the Brunei Darussalam Economic Council (BDEC), chaired by Prince Mohamed, was established to seek ways of improving the economy. The Sultan approved the

BDEC's report, released in February 2000, which warned that Brunei's economy was becoming increasingly unsustainable, and appointed Prince Mohamed to oversee the recommended economic recovery plan. There were no proposals to match economic modernization with political reform. The aim was to create a corporate system of government presided over by a traditional monarchy and to transform the bureaucracy into a 'technocracy'. To this end, in February 2004, during the celebrations of the 20th anniversary of independence, the Sultan announced the establishment of a special Task Force to update and streamline the 'National Vision', along with the formation of a long-term development planning body, which would prepare the framework and formulate plans for a 30-year period, beginning in financial year 2006. The aim was to involve the people as 'stakeholders' in the country's development. These themes of development, adaptation and involvement recurred in the Sultan's speeches. In mid-1999 Prince Mohamed announced that the report on the Constitution had been completed, and in March 2000 seminars on modern management were organized for village headmen, but the Sultan, in his capacity as Prime Minister and Minister of Defence and of Finance, continued to rule by decree under the state of emergency. Modernization was led by the Sultan, who carried out random checks on government departments. At the same time he continued to play the role of a good Muslim ruler, visiting rural kampongs and making himself accessible to his people, and holding open days every year during the three-day Hari Raya festivities, when thousands of people would visit the palace. Crown Prince Al-Muhtadee Billah played an increasing role in public life at home and also visited the People's Republic of China and Japan in March 2002. The Prince's status was raised further in March 2004, when he was bestowed with the rank of Four-Star General. In this capacity he made visits to installations of the Royal Brunei Armed Forces. In May it was announced that he was to marry Dayangku Sarah binte Pengiran Haji Salleh Abdul Rahaman in September. The Crown Prince's activities became more closely identified with those of the Sultan and, at the end of May, both attended the wedding of the Crown Prince of Jordan. In early June Prince Billah spent a full day visiting factories at Muara while the Sultan was distributing land titles and house keys to the landless as part of the Kampong Ayer Resettlement Scheme.

In a speech celebrating his 54th birthday in July 2000 the Sultan, while calling for a new mindset to counter antiquated regulations, with a view to making Brunei a financial hub for the region, also insisted that the economy must be strengthened in line with Islamic teaching in order to preserve social and moral values; repeatedly he urged Muslim devotion, especially at a time of economic difficulty. In September Brunei declared its aim to be a regional safe haven for international financial services compatible with Islamic principles. In the following month the Sultan, as Chancellor of the University of Brunei Darussalam, called for Islamic studies to be upheld as the most important field of education, and in November he officially launched the Islamic Development Bank of Brunei, seven years after the opening of the country's first Islamic bank. In opening the first International Islamic Exposition, which was held in Brunei in August 2001 with participants from 25 countries, Sultan Hassanal urged all Muslims to unite in order to regain the glory of Islam. Brunei reacted with horror to the suicide attacks on the USA on 11 September 2001, repudiating any link between religion and terrorism. While Brunei itself was not threatened, the country aligned itself with the other member states of the Association of South East Asian Nations (ASEAN) and Asia-Pacific Economic Co-operation (APEC), closing ranks against terrorism. The Sultan continued to exhort young people to follow the teachings of the Koran, and he himself went regularly on pilgrimage to Mecca. Meanwhile, the State was determined to eradicate so-called deviant or extremist teachings among Muslims. In 2000 a Malay martial arts group, which planned armed attacks on what it held to be un-Islamic institutions, was uncovered, and in December three men, including one retired senior police officer and two Malay businessmen, were arrested for allegedly supporting subversive Christian practices, with close links to groups in Sabah and Sarawak.

The broadcast media and the press remained closely controlled. Legislation passed in 1997 required all journalists to register and prohibited undesirable foreign broadcasts, criticism of the royal family, and objectionable religious or cultural material. Until 1999 television was state-owned, but the first commercial television channel was introduced in that year. The Government began to adopt a slightly more relaxed attitude to the press, however, paying heed to mildly critical letters about administrative shortcomings, which began to appear from 1999 in the correspondence pages of the English-language newspaper *Borneo Bulletin*, which was owned by Prince Mohamed. A second English-language newspaper, *News Express*, which was partly owned by the former Attorney-General, was established in August, and was later permitted to publish Brunei's first Chinese-language newspaper and a Malay daily. While official willingness to accept restrained criticism marked a substantial advance, the press exerted self-censorship, avoiding any questioning of the sultanate or national philosophy, and in March 2000 the Government warned newspapers to focus on national development, social well-being and character-building, instead of 'negative news'. In October 2001 a Local Newspapers (Amendment) Order introduced more stringent measures, requiring newspapers to obtain an annual permit, increasing deposit fees and fines and permitting the authorities to suspend any local newspaper without appeal and to ban foreign newspapers from entering the country. *News Express* had to suspend publication temporarily in order to meet the new financial requirements; it finally closed down in September 2002.

The Eighth National Development Plan, encompassing the years 2001–05, repeated aims for diversification, although the discovery of a substantial new oil and gas field by Brunei Shell Petroleum (BSP) in November 2001 raised the prospect of a renewal of intense activity in the sector. Petroleum prices continued to recover, despite experiencing a sharp decline in the aftermath of the terrorist attacks of September 2001, but Brunei was affected by the continuing regional economic malaise, which brought considerable hardship to many sectors of the population. Unemployment remained a pressing problem, particularly among young university graduates and school-leavers, who constituted 90% of the 6,812 people unemployed in February 2002. The Government wanted to extend corporatization of government departments and agencies, to divert employees from public service into the private sector, and to make the bureaucracy more efficient. However, in view of deeply entrenched practices and prejudices, in November 2000 the Government abandoned plans to substitute performance-related bonuses for the system of automatic salary increments in the civil service.

Meanwhile, the ongoing Amedeo crisis threatened to impede the nation's recovery. In September 2001 about 300 creditors remained unpaid, with no settlement in prospect. The Government was engaged in disputes with Prince Jefri concerning the assets covered by the out-of-court settlement that he had made, while the case was generating huge costs in fee payments to foreign liquidators, lawyers and accountants. In October the Government intervened and established Global Evergreen Sdn Bhd—a government-owned corporation under the chairmanship of the Minister of Education, the former Chairman of the BIA. Within two months Global Evergreen had settled 97% of all claims and dispensed with the services of overseas consultants. However, in May 2002 it emerged that several of the overseas consultants employed by Global Evergreen had allegedly been prevented from leaving Brunei owing to visa irregularities. The consultants had been investigating the reported embezzlement of an estimated B $10,000m. from Sultan Hassanal and had made inquiries into the conduct of the Minister of Home Affairs, Isa Ibrahim, following which the immigration authorities had raided the company's offices and uncovered the alleged visa errors. The Government denied that it had prevented the consultants from leaving the country, claiming that they had been subjected only to normal immigration procedures, and those involved were finally permitted to leave Brunei. By August 2002 litigation over Amedeo had almost been completed, and the country hoped to be able to put the crisis behind it and begin to attract the external investment that was vital to the success of the Eighth National Development Plan. After a decent interval, Prince Jefri was publicly rehabilitated, appearing at the end of July 2003 along with the Sultan and other members of the royal family at a concert sponsored by the Johor royal family. Within a week he made his second public appearance at a dinner given by the Royal Brunei Police to mark the Sultan's 57th birthday and, a few days later, attended the annual National Koran Reading Competition. These appearances provoked much comment in the Brunei media, which felt free enough to refer to his past disgrace whilst speculating about his future.

Nevertheless, financial scandal in the country did not disappear. In May 2004 the former Minister of Development, Pengiran Haji Ismail bin Pengiran Haji Damit, who had been replaced in May 2001, was charged with corruption, along with a Chinese businessman, Wong Tom Kai.

Economic change, rising expectations of accountability and transparency, the exposure to outside influences through the internet and satellite television, and the expansion of education in information technology would, it was hoped, eventually bring changed attitudes to government. This hope found cautious expression in April 2004 with the election of a new village chief in Kampong Lambak 'A', near the capital, leading to speculation that elected village chiefs might be the precursors of a more representative government. Such speculation intensified in July when the Sultan announced that he intended to reinstate the country's Legislative Council, suspended in 1962, by the end of 2004. However, its members were to be elected by a governmental committee, and the Sultan confirmed his intention to continue ruling by decree.

Recent Foreign Relations

After independence, Brunei began to develop extensive international links. It became a member of the UN, ASEAN, the Commonwealth and the Organization of the Islamic Conference. In 1992 it joined the Non-aligned Movement and established diplomatic ties with Russia and the People's Republic of China. In the same year diplomatic links were formally established with Viet Nam, and in 1993 with Myanmar. In 1995 Brunei entered the World Bank and the IMF. Brunei remained heavily dependent on the United Kingdom and Singapore for defence but the sultanate also forged defence links with Australia and the USA. Sultan Hassanal played an active part, presiding over regional and international meetings in Brunei and travelling widely to consolidate friendly relations with other countries.

The most immediate foreign policy objective was to establish cordial relations with ASEAN partners. Brunei was closest to Singapore, which remains an important trading partner, a major source of skilled labour and a repository of Brunei's petroleum revenues, with the two currencies 'pegged' under an inter-changeability agreement dating from 1967. As small states, Brunei and Singapore share a concern to promote peace and stability in the region, working closely together in bodies such as the ASEAN Regional Forum (ARF—established to address security issues) and APEC. They co-operate in offering training facilities for their armed forces, officials exchange regular visits, and in November 2002 the Port of Singapore Authority Corporation and Brunei Archipelago Development Corporation signed a joint-venture contract to manage and develop Muara Container Terminal for the next 25 years.

From the late 1980s Brunei provided generous aid and investment to promote economic development in Indonesia and the Philippines, and in 1994 the ministers responsible for foreign affairs of Brunei, Indonesia, Malaysia and the Philippines (BIMP) agreed to establish an East ASEAN Growth Area (EAGA). Affected by the economic crisis of the late 1990s and more recently by the consequences of terrorism and the Severe Acute Respiratory Syndrome (SARS) epidemic, the BIMP-EAGA project faltered. However, at the first-ever meeting of BIMP-EAGA leaders, held on the Indonesian island of Bali in October 2003, the Sultan joined with the Presidents of Indonesia and the Philippines and Prime Minister Mahathir Mohamad of Malaysia in resolving to revitalize business activities in the region.

Meanwhile, relations with Malaysia showed the most dramatic improvement. In 1993 Mahathir Mohamad headed a delegation to Brunei, and the two countries agreed to resolve all border disputes, including the Limbang question, through bilateral negotiations. In 1994 Brunei signed an agreement with Malaysia establishing a joint commission to promote co-operation in trade, industry, finance, education, culture and religion. Despite occasional friction, relations remained cordial. Senior officials exchanged frequent visits, Sultan Hassanal held annual consultations with Mahathir in Brunei or Kuala Lumpur, and in August 2002 the King of Malaysia paid a state visit to Brunei. In May 2003, however, a dispute began concerning the offshore boundary between the two countries, when both Malaysia and Brunei granted conflicting oil exploration licences. The Sultan immediately flew to the Malaysian island of Penang, to meet Mahathir, and both countries suspended their drilling operations pending a compromise.

From the mid-1990s Brunei became more assertive in the region, urging economic co-operation and trade liberalization; the sultanate expressed support for free trade within APEC by the year 2020 and wished to accelerate implementation of the ASEAN Free Trade Area (AFTA).

In August 1999 Brunei hosted the South-East Asian Games. At the third unofficial ASEAN summit meeting, held in Manila in November, the Sultan urged greater openness to achieve economic integration, strengthen economies and attract foreign investment. At the same time he argued for closer but more gradual co-operation with Japan, China and the Republic of Korea. In February 2000 Brunei hosted its first APEC Senior Officials' Meeting, with 3,000 delegates from 21 countries, and ASEAN ministers responsible for finance met in Brunei in the following month to discuss a proposal for an Asian monetary fund. In November 2000 Brunei played host to the annual APEC meeting, the largest international event ever staged in Brunei: 6,000 delegates attended, including Presidents Clinton of the USA, Jiang Zemin of China and Putin of Russia. Brunei lobbied intensively for APEC unity, but its hopes of achieving a major breakthrough in obtaining consensus for a new round of World Trade Organization (WTO) talks encountered resistance from some delegates, notably Malaysia, and a compromise was reached.

Three ASEAN heads of state visited Brunei in swift succession in August 2001—the Prime Minister of Thailand, the President of the Philippines, and the new President of Indonesia, Megawati Sukarnoputri. In the same month Brunei and the Philippines agreed to establish a Bilateral Commission for Co-operation. In view of its ambitions to promote tourism, Brunei was particularly concerned to improve regional co-operation to protect the environment. The sultanate celebrated World Forestry Day in 2000, introduced a total ban on open burning and illegal logging as agreed by ASEAN, and inaugurated 'ASEAN Environment Year 2000'. In January 2001 the Sultan launched both 'Visit Brunei Year 2001' and 'Visit ASEAN Year 2002', but the events of 11 September 2001 caused a depression in the tourism sector after what had been a promising start. Brunei suffered from the downturn in tourism in the region that resulted from the outbreak of SARS in early 2003, although the state itself emerged largely unaffected by the epidemic. By May 2004 greater confidence in Brunei as a tourist destination linked to Labuan and East Malaysia was being expressed, but the rise in petroleum prices in mid-2004 was viewed with caution. While higher revenues from oil would bring greater financial benefits, the tourism sector might be adversely affected.

Brunei staged the seventh ASEAN summit meeting in November 2001, with representatives from China, Japan and the Republic of Korea also in attendance. At the meeting of the so-called 'ASEAN + 3' the national leaders declared war on terrorism and agreed to form an ASEAN + 3 secretariat. The Sultan envisaged that this would lead to the creation of the largest free trade area in the world. In July 2002 Brunei hosted a meeting of ASEAN ministers responsible for foreign affairs and was also the venue for the ninth meeting of the ARF, which was attended by delegates from the 10 member nations, together with representatives of 13 dialogue partners. These included US Secretary of State Colin Powell, who signed a US-ASEAN Joint Declaration for Co-operation to Counter Terrorism with the Sultan; this provided for the exchange of information and intelligence and made a commitment to suppress all forms of terrorism, but without creating any new US bases in the region. The issue of terrorism dominated the meeting of ASEAN ministers of foreign affairs and the ARF. However, in opening the 34th meeting of ASEAN ministers responsible for economic affairs, which was held in Brunei in September 2002 with representatives from Australia, New Zealand and India, the Sultan stressed economic development as being integral to ASEAN's progress. In October 2003 the annual summit meeting of ASEAN leaders, held in Bali, endorsed plans to accelerate the creation of an ASEAN Economic Community. While in Bali, the ASEAN leaders met representatives from China, Japan, the Republic of Korea and India for bilateral talks. In December the Sultan joined other ASEAN leaders in Tokyo, Japan, to sign a declaration for an enduring ASEAN-Japan partnership. In his speech the Sultan endorsed the ASEAN + 3 policy as a mechanism for the promotion of co-operation between ASEAN and Japan, China and the Republic of Korea.

Brunei was vigilant in excluding terrorism from its own territory and denied international media reports in February 2003 that the regional Islamist fundamentalist organization, Jemaah Islamiah, was planning to set up a network in the state. The Director of the Internal Security Department was replaced in May 2003 by the Deputy Minister of Home Affairs. The Sultan made a statement in February 2003, on the eve of the commencement of the US-led campaign to remove the regime of Saddam Hussain in Iraq, declaring his hope that the UN Security Council would settle the matter. Following the conclusion of the campaign, in April the Government of Brunei established a humanitarian fund for Iraq, and in May Prince Mohamed publicly expressed concern at the post-war situation in the country. Brunei otherwise kept a low profile concerning the war, a position it maintained into 2004. Nevertheless, Brunei made clear its support for reform of the UN Security Council in a speech by Prince Mohamed Bolkiah, Minister of Foreign Affairs, to the UN General Assembly in September 2003. APEC was another forum in which Brunei actively participated. In October 2003 the Sultan met APEC leaders at the forum's annual summit meeting in Bangkok, Thailand. Immediately prior to this he had attended the 10th session of the OIC in Kuala Lumpur, illustrating Brunei's determination to voice its concerns and to confirm its role as an independent player on the world stage.

Meanwhile, the country continued to strengthen relations both with its neighbours and further afield. Sultan Hassanal made state visits to Thailand in August 2002 and the Philippines in January 2003, and in May 2003 Brunei and the Philippines held their first joint naval exercise. In February 2004 the first Commission for Bilateral Co-operation between Brunei and the Philippines met in Manila. Agreements were signed relating to defence co-operation and technical and trade co-operation in fisheries and air services. In December 2003 three Indonesian warships visited Brunei. In April 2004 the Vietnamese Minister of National Defence, Lt-Gen. Pham Van Tra, led a delegation of officials from his ministry on a four-day visit, during which the strengthening of military ties was discussed. Relations with China became closer in 2003 with the signing in October of an agreement for the sale of Brunei's crude petroleum to the People's Republic and the first goodwill visit of two Chinese warships in November. In the same month a French navy vessel visited. In September 2002 the Sultan paid a three-day working visit to Syria, opening the way for the establishment of bilateral diplomatic links in April 2003. In November 2002 the Sultan made a two-day goodwill trip to Bahrain, and relations continued to improve in 2004 as Brunei drew upon Bahrain's expertise in Islamic banking and international financial and economic matters. In May 2004 the Sultan paid a three-day visit to Pakistan and the two Governments signed memoranda of understanding relating to counter-terrorism, defence and co-operation. In December 2002 he had met US President George W. Bush in Washington, DC. Relations with the USA remained cordial, but during 2003–04 Brunei distanced itself from events in Iraq.

In his customary birthday *titah* (royal address), delivered in July 2002, Sultan Hassanal stressed the importance of the quest for peace, stability, good regional relations and firmness in condemning terrorism in co-operation with other countries, while emphasizing in a domestic context the need to increase the involvement of young people in society and the private sector in economic development, themes returned to in 2003–04. By June 2004 he had become more concerned with Brunei's security. In March of that year three people—a high-ranking former soldier, a senior police officer and a leading businessman—were detained for subversion and treason for allegedly leaking official government secrets on the internet. The Sultan addressed broader regional security concerns in a national broadcast on 31 May, the 43rd anniversary of the foundation of the Royal Brunei Armed Forces, when he outlined the strategic and economic importance of the South China Sea and the threat from international terrorism, and announced that a Defence 'White Paper' would soon be issued.

The Sultan continued to be concerned about the effects of the outside world upon the youth of Brunei, causing him, in his speeches on religious and national anniversaries, to stress the importance of Islamic principles and observance. In March 2004, addressing the opening session of the 39th South-East Asian Ministers of Education Organization (SEAMEO), held in Brunei, the Sultan revealed his desire that children acquiring an ever-increasing amount of technological, scientific and electronic information would also acquire tolerance and respect for others. His speech reflected Brunei's underlying concerns as a small state in a global community, as it tried to balance the demands of modernization and tradition while maintaining harmonious relations with its ASEAN neighbours and with other countries with interests in the region.

Economy

MURNI MOHAMED

Revised by HAJAH ROSNI HAJI TUNGKAT and for this edition by ROGER LAWREY

INTRODUCTION

Brunei Darussalam—the Abode of Peace—acceded to full independence in January 1984, following nearly 80 years as a British protectorate. The population in 2003 was estimated at 348,800, with Malays forming the largest ethnic group, accounting for approximately 66.6% of the population. Chinese and indigenous people account for about 11.0% and 3.6% of the population, respectively. The Brunei/Muara district is relatively densely populated (426 people per sq km in 2003), while the Tutong and Seria/Belait districts (34 and 21 people per sq km respectively), and especially the Temburong district (7 people per sq km), are sparsely inhabited. During 1990–2002 the population of Brunei increased at an average annual rate of 2.46%.

In 2002 gross domestic product (GDP) was estimated at B $7,351m.,or B $21,570 per caput. Brunei has no personal income tax, sales or export tax, although there is a flat-rate corporate tax of 30%. Brunei citizens receive education and health-care at nominal charges, and citizens who are government employees are eligible for interest-free loans for the construction of houses and the purchase of motor cars.

The economy relies largely on revenues from oil and natural gas. The Brunei Investment Agency (BIA) is responsible for investing net revenues from this sector and net income from overseas from these investments also plays a large role in supporting the economy, although these figures are not reported in the national accounts. Brunei is the fourth largest crude petroleum producer in South-East Asia, after Indonesia, Malaysia and Viet Nam, and is the fourth largest producer of liquefied natural gas (LNG) in the world. Japan and the Republic of Korea are the major customers for Brunei's natural gas and crude petroleum exports.

The labour force in 2001 was estimated at 154,200, of whom approximately 98,400 were employed in the private sector,

47,200 in the public sector and 8,600 were unemployed. Approximately 72,000 private-sector employees are foreign workers (temporary residents). Bruneian citizens and permanent residents have typically preferred the public sector because salaries and employment conditions are generally perceived to be better than in the private sector. The Brunei Shell Petroleum Co Sdn Bhd (BSP) is the largest private-sector employer. However, the oil and gas sector is relatively capital-intensive and there is limited scope for it to absorb a growing labour force. The industry that accounted for the largest share of private-sector employment in 2001 was construction, followed by wholesale and retail trade and manufacturing. Both of these sectors were ultimately dependent on oil and gas revenues for their livelihoods. The oil and gas sector provided a mere 3,700 jobs.

Substantial oil revenues began to flow in the 1960s with the development of offshore production and the oil price increases of the 1970s. Subsequently, Brunei's economy remained largely dependent on these revenues for its wealth and on the Government for employment. Economic development was organized through a series of formal five-year development plans, and infrastructure built in the form of roads, telecommunications networks, educational establishments and an international airport and airline.

Oil and gas create substantial revenues but little direct employment. Moreover, international petroleum prices are extremely volatile and, although the prospects for the discovery and development of substantially more oil and gas appeared favourable, oil remained a finite resource. Accordingly, the major focus of economic policy in Brunei in the early 21st century was the diversification of the economy into the downstream oil and gas sector and non-petroleum-related activities.

GDP AND THE STRUCTURE OF THE ECONOMY

As annual GDP estimates usually include elements of both increased output and increased prices, it is normal to report GDP growth in constant dollar (inflation adjusted) terms by using a GDP deflator. This adjustment removes the element of inflation so that any changes in real GDP reflect actual physical changes in output. Data is available in Brunei for GDP in both nominal and constant dollar terms. However, deflating nominal dollar GDP in the case of Brunei can be misleading. Oil and gas are sold in the international market in US dollars. In recent years, the average annual price received for Brunei's oil varied between US $13.46 per barrel in 1998 and US $30.17 in 2003, and the international price of light crude petroleum exceeded US $40 in mid-2004. To deflate these prices would be misleading, since the change in price was not due to inflation in Brunei. Higher US dollar prices do translate into greater real value of production. Deflating nominal GDP through the consumer price index (CPI) would give some measure of changes in the purchasing power associated with GDP, but it might be most appropriate to use nominal dollar GDP when comparing figures over time.

GDP is calculated in Brunei using the production approach. All figures are subject to revision. The data given here are provisional but are the latest complete, consistent accounts. In 2002 the largest contribution to the total nominal GDP of an estimated B $7,351m. came from the mining, quarrying and manufacturing sector (including the oil and gas sector), at B $2,739m., or 37%. The second largest contribution came from community, social and personal services, at B $2,262m., or 31%. Other important sectors were retail trade (5.6%), construction (5.5%), transport, storage and communications (5.3%) and banking and finance (5.1%).

Nominal GDP in 2002 was 2.6% higher than in 2001. It was estimated that, in 2003, real GDP would increase by 3.2% compared with 2002, a figure that would be substantially higher in nominal terms. Per caput nominal GDP has demonstrated a general trend of decline since 1990, when it was estimated to be B $25,685. Growth in nominal GDP has failed to keep pace with a rate of population growth of 2.46% per annum over this period.

Statistics are also available for government and private-sector production and for oil and non-oil sectors, although the distinctions are sometimes unclear. According to the latest available figures, the oil and gas sector accounted for 37% of GDP in 2002 compared with 83.7% in 1980. The non-oil sector contributed some 63% of GDP. Within the non-oil sector, the private sector accounted for 62% and the government sector for 38% of production.

Although the proportion of GDP contributed by the petroleum sector has declined steadily since 1980, its influence on the health of the economy is still very substantial. The petroleum sector accounted for 88% of the value of exports in 2002. The only other major export earner is the garment industry, which has been established to take advantage of the international quota system in this area. The garment sector accounted for 5.8% of exports in 2002. Exports of crude petroleum, natural gas and casing head petroleum spirit totalled B $5,743m. in 2002, compared with reported GDP of B $7,351m. Oil production in 2002 averaged 203,000 barrels per day (b/d), of which 199,000 b/d (98%) were exported, at an average price of US $25.33 per barrel. Natural gas production averaged 1,124m. cu ft per day (MMcf/d), which was exported as liquefied natural gas (LNG) at an average rate of 965,799m. British thermal units per day (MMBtu/d) with a price of US $4.1608 per MMBtu.

Imports in 2003 totalled B $2,312m. The largest category of imports was machinery and transport equipment (B $815m.), followed by manufactured goods (B $597m.) and food (B $348m.). Brunei's reliance on imports exposes it to external inflationary pressures related to changes in the international value of the Brunei dollar. The Brunei dollar is pegged to the Singapore dollar, and there is a currency interchangeability arrangement between the two countries. The concept of a GDP deflator is somewhat irrelevant for Brunei because of the importance of the petroleum sector, the output of which is traded internationally in US dollars. A CPI is estimated monthly, quarterly and annually. The CPI in 2000 increased by 1.2% compared with the previous year, and in 2001 it increased by 0.6%. It was estimated that the CPI for the fourth quarter of 2003 decreased by 0.1% compared with the fourth quarter of 2002.

PETROLEUM AND NATURAL GAS

In 1929 Shell discovered an onshore petroleum deposit in Seria, from which production was gradually expanded to 17,000 b/d by 1940. The Seria oilfield was extensively damaged during the Second World War, but production was resumed in 1945 and had reached 114,700 b/d by 1956.

BSP began offshore exploration in 1954, and started exploratory drilling on the continental shelf in 1956. In 1963 the South West Ampa field was found to contain large quantities of both petroleum and natural gas. Since 1964 numerous oilfields, including Fairley, Fairley-Baram, Champion, Magpie, Iron Duke, Jurajan, Perdana, Gannet and, most recently, Egret, have been discovered, in addition to further onshore reserves at Seria and Rasau. Production declined, however, from a record 254,000 b/d in 1979 to 150,000 b/d in 1988 and 1989, in an effort to ensure that extraction rates were suited to economic needs. Following an increase to its highest levels in more than a decade, to 182,000 b/d in 1992, production declined to 157,000 b/d in 1998, but increased again to 193,000 b/d in 2000 and to 203,000 b/d in 2001. In 2003 production averaged 207,240 b/d, compared with 203,021 b/d in 2002.

Most of Brunei's proven reserves are in shallow water. The most productive offshore field is Champion, in 30 m of water, about 70 km north-east of Seria. It holds 40% of the country's known reserves and produces more than 50,000 b/d. The field already has more than 260 wells drilled from 40 platforms. The oldest field offshore is South West Ampa, 13 km from Kuala Belait. Its reservoirs hold more than half of Brunei's total gas reserves, and gas production from the field accounts for 60% of Brunei Shell's total production. Gas from its 56 gas wells is piped 39 km to the Brunei LNG plant in Lumut. South West Ampa also has substantial oil reserves, with 164 oil-producing wells.

In 1998 the average price for crude petroleum was US $13.46 per barrel, its lowest level since 1984. Although partially offset by reduced operating costs and a weaker exchange rate, the decline in prices had an adverse effect on Brunei's financial position, prompting the Government to place emphasis upon maximizing output while reducing production costs. Prices subsequently recovered, averaging US $25.33 per barrel in 2002 and US $30.17 in 2003. In June 2004 international petroleum prices averaged between US $38–US $40 per barrel. This volatility in prices, and Brunei's dependence on the petroleum sector, highlighted the need for diversification of the economy.

In 2002 the major export destinations for crude petroleum were Thailand (accounting for 28% of petroleum exports), Australia (18.5%), the Republic of Korea (14%), the People's Republic of China (13%) and Japan (12%). Crude petroleum exports in 2002 were valued at B $3,139.6m. Japan is by far the largest export destination for Brunei's LNG, accounting for sales valued at B $2,274m. in 2002 (87% of total LNG exports of B $2,603.5m.), with the Republic of Korea taking the majority of the balance (11%).

Export volumes of both crude petroleum and LNG were higher in 2002 than in 2001. Crude petroleum exports averaged 199,000 b/d in 2002, compared with 189,000 b/d in 2001, while LNG exports averaged 965,799 MMBtu/d in 2002, compared with 948,803 MMBtu/d in 2001. Prices were also higher for crude petroleum, although 2002 LNG prices were lower than they had been in 2001.

In 2002 Brunei was the world's sixth largest exporter of LNG. A 20-year LNG contract with Japan expired in 1993 and was replaced by a new 20-year contract, under which the volume of LNG exported increased to 5.54m. metric tons (from 5.14m. previously) and the price of gas increased by about 4%. A sales and purchase agreement was signed in October 1997 between Brunei LNG and Korea Gas Corporation (KOGAS). Under the agreement, Brunei LNG is to supply KOGAS with 0.7m. tons per annum of LNG from 1997 to 2013. The agreement confirmed KOGAS as the fourth long-term customer of Brunei LNG, in addition to the three Japanese companies (Tokyo Electric Power, Tokyo Gas and Osaka Gas). Brunei LNG will supply the four customers with a total of 6.24m. tons of LNG until the year 2013.

BSP's operations are conducted in accordance with the Government's political objectives. Legislation was introduced in the early 1980s to ensure that Brunei would derive maximum benefit from any petroleum exploration by foreign companies. The State was to be an equal partner in all petroleum ventures with foreign interests. Any company wishing to apply for a concession was required to offer the Government a percentage of its annual net profit from petroleum or gas production, at a rate to be fixed before contracts were signed. The aim was to increase competition among prospective concession-holders. A state-controlled petroleum company, similar to those in Malaysia and Indonesia, however, was not envisaged at that point. In January 1993 the Brunei Oil and Gas Authority (BOGA) was formed, its main function being to advise and recommend on policies in all matters pertaining to oil and gas production. Brunei is not a member of the Organisation of the Petroleum Exporting Countries (OPEC).

There are four principal companies involved in the petroleum and natural gas sector, of which BSP is the largest. BSP is jointly owned by the Brunei Government (50%) and a company within the Royal Dutch /Shell Group (50%). In 1969 Brunei LNG was established by the Government, Shell and the Mitsubishi Corporation of Japan, to purchase gas brought on shore by BSP. In 1987 Brunei Coldgas was owned by these three partners, and shipped gas to Japan, after it had been cooled and liquefied by Brunei LNG. In 1995 Brunei Coldgas merged with Brunei LNG under the name of Brunei LNG. Brunei Shell Marketing Co, jointly owned by the Government and Shell, was established to service the domestic market. Brunei Shell Tankers is the fifth company of the group. A few other companies, namely Jasra-Elf (or Jasra International), a joint venture formed in 1986 between Elf Aquitaine and the locally-owned Jasra Jackson, were given concessions, mainly on shore.

Jasra-Elf began to produce petroleum and natural gas in March 1999 from the Maharaja Lela Jamalulam discovery made in 1990. Jasra-Elf-Fletcher, a joint venture of three companies (Jasra Jackson, Elf Aquitaine and Jaspet Fletcher Challenge), began to produce petroleum and natural gas in April 1999. Exploration activities continued, and in 1999 a new seismic survey called the Seria High Resolution 3D Seismic Survey was initiated. Early studies have revealed great potential. In November 2001 BSP announced that it intended to develop a new offshore oil and gas field at Egret, north-east of the Fairley field. The development produced its first natural gas in August 2003. Also in November 2001 the Government announced that two international consortia had advanced bids for two deep-sea areas in Brunei's offshore exclusive economic zone (EEZ)—Block J (with a total area of 5,020 sq km) and Block K (4,944 sq km). In February 2002 the French company TotalFinaElf (now known as Total), in partnership with BHP Billiton of Australia and Amerada Hess of the USA, won exploration rights to Block J, while a consortium comprising BSP, Japan's Mitsubishi Corporation and Conoco Brunei Ltd won the right to explore for natural gas in Block K. In early 2003 TotalFinaElf announced that it had made a significant discovery of natural gas and condensate in Block B, below the existing Maharaja Lela Jamalulam field.

In November 2001 a national oil company—Brunei National Petroleum Company Sdn Bhd (PetroleumBrunei) was created. BOGA was dissolved following the incorporation of the new company in early 2002. PetroleumBrunei initiated the use of the Production Sharing Contract (PSC) as a new model for the future exploration and production of petroleum and natural gas in Brunei, using it for the first time for the new areas to which rights had been offered in the EEZ. The PSC replaced the granting of concession rights to companies operative in the region, allowing the Government to exercise greater control over national resources. Production-sharing arrangements were not subject to the tax and royalty payments required by concession contracts.

In April 2003 it was announced that the consortium led by TotalFinaElf had signed a production-sharing agreement with PetroleumBrunei concerning exploration of Block J. The agreement included a commitment by the consortium to use local goods and services during its operations and to ensure that local people comprised a significant proportion of the work-force. Meanwhile, negotiations over Block K were also approaching completion. However, in mid-2003 a dispute with neighbouring Malaysia over the ownership of the EEZ (following Malaysia's discovery of a potential new oilfield—Kikeh—in the area in 2002) was initiated by fears that the new oilfield might extend into Block J. In June Total was compelled to suspend temporarily its drilling operations in the area when a Malaysian patrol boat forced its vessels to leave. Following a meeting between officials from the two Governments, it was hoped that the dispute could be resolved amicably, possibly through the creation of a joint development area. No further publicly available information was obtainable regarding the situation.

Discoveries of petroleum and gas in the 1990s resulted in the continuous upward revision of estimates of the country's hydrocarbons base. BSP increased its reserves estimates substantially, especially for natural gas. In 1993 it was estimated that current rates of gas production could be maintained for another 40 years. In 2002 proven recoverable reserves of natural gas were estimated to be 391,000m. cu m. BSP's reserves were restated in 2004 by the Royal Dutch /Shell Group, but no public statements had been made as to the extent of the effect on proven reserves in Brunei.

In early 1998 a new company, Brunei Gas Carriers Sdn Bhd, was formed to build and manage a new LNG vessel with a capacity of 135,000 cu m, which had become operational by July 2002. The shareholders were the Brunei Government (which controlled 80% of the company), Shell International Gas (10%) and the Mitsubishi Corporation (10%).

INTERNATIONAL TRADE

Brunei joined the Association of South East Asian Nations (ASEAN) on the attainment of independence in 1984. Participation in ASEAN projects has also given Brunei an interest in the economic development of the region. In October 1991 the member states of ASEAN formally announced the establishment of the ASEAN Free Trade Area (AFTA), of which Brunei was one of the principal proponents. Brunei is also a member of the Asia-Pacific Economic Co-operation (APEC) and of the Organization of the Islamic Conference (OIC). In October 1993 the idea of a 'growth quadrangle', encompassing Mindanao and Palawan (Philippines), Sarawak, Labuan and Sabah (Malaysia), East and West Kalimantan and Sulawesi (Indonesia) and Brunei was mooted, aiming to emulate the Singapore-Johor-

Riau 'growth triangle'. At a meeting in Mindanao in November 1994, it was agreed to establish the 'growth quadrangle' as the Brunei-Indonesia-Malaysia-Philippines-East ASEAN Growth Area (BIMP-EAGA). The area has since been expanded with the announcement of the incorporation of additional provinces in Indonesia, including North and South Kalimantan, the Maluku Islands and Irian Jaya (now known as Papua), in July 1996. It was also decided to locate the secretariat of the East ASEAN Business Council (EABC) in Brunei.

Since the formation of BIMP-EAGA, a number of successful working groups have been established through which various projects have been implemented: a construction consortium and a Pan-EAGA Multi-Capital Transportation Network System were formed; the first phase of a programme to reduce telephone tariffs for the four countries comprising BIMP-EAGA was implemented; and a shipping agreement was established between Brunei and Mindanao. A number of joint ventures have also been formed and memorandums of understanding signed that involved private-sector participation, including the establishment of a Regional Ship Operator Association in BIMP-EAGA, a bark-medium processing plant, and co-operation and exchange in education and training.

In August 2001 representatives from BIMP-EAGA and the ASEAN Centre for Energy (ACE) met for the first time to discuss the development of several trans-Borneo power grid interconnection projects, one of which would link Brunei's grid with those of Sarawak and Sabah. In 2002 the feasibility of a Trans-Borneo railway connecting Brunei with several destinations within BIMP-EAGA was also being considered; it was to constitute part of the Pan-EAGA Multi-Capital Transportation Network. In July 2002 the first BIMP-EAGA Customs, Immigration, Quarantine and Security (CIQS) conference took place, as a result of which the member nations agreed to adopt common CIQS policies in an attempt to accelerate regional integration and growth. BIMP-EAGA had been subjected to criticism for its failure to establish adequate institutional mechanisms to facilitate its development, particularly following the formal implementation of AFTA in January 2002.

Brunei has had large surpluses on its balance of trade from 1972, owing to the country's exports of petroleum and natural gas. Its trade surplus reached a peak of B $8,622.3m. in 1980 and a low point of B $153.3m. in 1996. The balance of trade was B $4,582.7m. in 2001, B $3,987.4m. in 2002 and B $5,513.0m. for the first quarter of 2003. In addition to export revenue, substantial income accrued from the investment of Brunei's reserves, contributing to the current account of the balance of payments.

For the first six months of 2002 Japan continued to be Brunei's major trading partner, accounting for 42.0% of total exports. Other major export markets included Thailand (16.5%) and the Republic of Korea (12%). Singapore was the principal source of imports in 2002, accounting for 18.1% of the total. Trade with Malaysia increased rapidly during the mid-1990s, accounting for 17.4% of total trade in 2002. Brunei has an unweighted average tariff rate of 5%.

AGRICULTURE, FORESTRY AND FISHING

Agriculture and forestry were the basis of Brunei's economy before the discovery of petroleum in the late 1920s. In 1947 more than 50% of the population was employed in the primary sector. By 1971 this had been reduced to 10%. Farming had become very much a part-time business for most rural families, owing to the availability of more lucrative forms of employment.

Agriculture accounted for just 3.3% of GDP in 2002, and Brunei imports more than 80% of its food requirements. In 2001 imports of food, beverages and tobacco amounted to B $354.4m. (17.0% of total imports). Government policy is to reduce Brunei's dependence on food imports and to aim to achieve a greater degree of self-sufficiency in agriculture. From the early 1980s the Government began a series of measures intended to contribute towards the development of Brunei's agricultural industry.

The enhancement of long-term food security was the major agricultural objective of the Seventh National Development Plan (1996–2000). For this purpose, attention was focused on the development of basic infrastructure, materials inputs, marketing and regulatory service, with the aims of, firstly, meeting domestic demand and, secondly, developing export potential. The Seventh Plan allocated B $173m., or 2.4% of the overall allocation and triple that of the previous Plan, to the development of agriculture.

Under the plan for enhancing national food security, the priority product groups were poultry and livestock, vegetables and ornamentals, fruits and plantation crops, rice and root crops. The plan for the development of the poultry industry had two aspects: the first was the encouragement of integrated medium-sized enterprises; the second was the development of the small-scale farm producers linked to processing and marketing centres, such as the Mulaut Abattoir. Such centres act as market outlets for the small-scale poultry producers. The programme emphasized the participation of private enterprise in the development of the industry, government involvement being limited to the provision of basic infrastructure and production-support services. The country is now self-sufficient in egg production and approximately 90% self-sufficient in the production of chicken meat.

The programme for vegetable production was orientated towards the development of high-value products and high-technology farming, such as protected cultivation and hydroponics. At the same time, the technical assistance required by small-scale local farmers was to continue to be provided. The production target for vegetables was 8,500 metric tons, or 75% of local requirements. In 2003 vegetable production was estimated at 10,360 tons.

The Plan also envisaged the collection of specimens of local fruit trees as the basis for developing small plantations and for the production of seedlings. It was hoped that this would support the development of large-scale fruit plantations in the future, and also encourage the planting of more local fruit trees. The aim was to increase the level of self-sufficiency from 10% to 70% by the end of the Seventh Plan period. In 1997 the achievement was only 12%. In 2001 fruit and arable crop production was reported at 4,140 metric tons.

With regard to the production of rice, the intention was to develop a production capacity equivalent to 7% of local needs. In 2000, however, the achievement was only 1%. The Eighth National Development Plan (2001–05) set a target to increase rice production to 1,300 metric tons, or 3% of total requirements, by 2005. Rice production was 600 tons in 2003, an increase of 71% compared with the 2001 figure. The Government encouraged the planting of rice through various schemes such as price support and the provision of improved infrastructure, irrigation and drainage facilities.

Live cattle are imported from Australia; there were also plans to develop goat and deer farming, and to increase the supply (and variety) of red meat in the country from 20 metric tons to 50 during the Seventh Plan. Production of cattle rose by 13% in 2001, while that of goats increased by 37%. In June 1997 the Government bought the Brunei Meat Export Company to ensure that consumers in Brunei would continue to obtain reasonably priced genuine *halal* products. It also bought shares in a *halal* abattoir in Tennant Creek, owned by the Meat Processing Company in Australia. The Eighth National Development Plan identified *halal* food processing for export as a prospective growth industry.

The Government, in its effort to encourage greater private-sector participation, has also identified a number of projects for possible private involvement. These include rice production, meat processing, day-old chick production, supply of fruit-planting materials and ornamental potted plants and tissue-cultured planting materials. The vast BIMP-EAGA market promised better prospects for the agricultural development of Brunei, particularly owing to its potential as a reliable source of *halal* food products.

There is a long tradition of fishing in Brunei, and fish consumption per caput is one of the highest in the world. In the first six months of 2002 3.1m. kg of fish, with a value of B $16.9m., was marketed; 1.8m. kg (B $10m.) of imported fish and 1.3m. kg (B $6.9m.) of locally-caught fish was sold. Imported fish volumes have steadily declined since peaking in 1995 at 5.2m. kg. During the same period local production has increased from 956,000 kg to its current level. To increase fish production, the Government encouraged the establishment of a small, modern, offshore fishing industry to meet projected demand and, under the Seventh Plan, B $72.9m. was allocated for the continued development of the country's fisheries sector. The Government emphasized the sustainable exploitation of the fishery resources. The current permitted exploitation level is self-imposed at the estimated maximum economic yield, and is managed through limiting the number of fishing licences issued for each fishing method. The Brunei Fisheries Limits Act (1982) delimits a maritime area of about 57,000 sq km extending to about 200 nautical miles (370 km) from the coast.

Brunei's relatively clean environment and freedom from extreme natural phenomena, such as typhoons, is conducive to the development of aquaculture. Once identified, potential areas are prepared with the required basic infrastructure such as access roads, water and power supplies. In line with the policy of sustainability, a buffer zone of 50 m is imposed in all aquaculture areas from other land use.

Forestry is not economically significant in Brunei, but is important in the conservation of soils, water, wildlife and the environment. Primary (60%) and secondary (20%) rain forest covers about 80% of total land area in Brunei. Forests in Brunei contain about 5,000 species of plants, including some 2,000 species of trees. These forests are of five main types: mangrove swamps, heath forests, peat swamp forests, mixed dipterocarp forests and montane forests. The Forestry Act of Brunei (1934, revised 1984) provides the legal framework for the protection and conservation of forestry resources of the country. In addition, the National Forestry Policy (NFP), issued in 1989, guides and governs all future forestry activities throughout Brunei. The NFP emphasizes the need to protect and conserve the forest resources of the country for both socio-economic and environmental purposes, and institutes sound and sustainable management of the national forest resources. Accordingly, these forests have been functionally categorized into protection forest, conservation forest, production forest, forest recreational areas and National Park, which covers a total area of 50,000 ha.

Logging in Brunei is very limited and is being monitored closely. The Reduced Cut Policy, introduced in 1990, restricted the total log production of the country, ensuring the sustainable production of local timbers for domestic use. In line with the strong conservation policy, the Forestry Department proposed to incorporate an additional area of 86,411 ha into the forest reserves in an effort to increase reserves from 39% to 55% of existing forest and unallocated state lands. In accordance with the NFP, many areas have been developed and promoted as 'eco-tourism' attractions. By 1998 11 forest recreation parks (FRP) had been developed, namely Sungai Liang, Luangan Lalak, Bukit Shahbandart, Berakas, Peradayan, Selirong, Bukit Subok, Sungai Basong, Kuala Belalong, Batang Duri and Ulu Temburong; these have become very popular, particularly among domestic tourists. Under the Seventh National Development Plan, various programmes were introduced for the conservation and protection of forests for both environmental and economic reasons.

The forestry sector contributed B $27.9m. (0.4%) of GDP in 2002. The Reduced Cut Policy limited logging to 100,000 cu m per annum, accounting for only 30% of domestic demand, while the rest had to be imported. Within the framework of the forest conservation policy, B $104m. was allocated to develop the forestry sector, with plantation activity, sustainable timber production, the creation of National Park and forestry recreation areas, the rehabilitation of wasteland and rattan production for the furniture industry all being areas to receive particular emphasis.

The theme of conservation is expected to continue to form the basis of the forestry policy. It is anticipated that the long-term plantation programme will ensure that the country becomes self-sufficient in sawn timber and will therefore minimize unnecessary pressures on the forest reserves. It is hoped that this strategy will help maintain the 80% forest cover, and also preserve the rich diversity of the tropical rain forest, making Brunei an ideal centre for eco-tourism and also providing an 'international laboratory' in which scientists can conduct research and studies. With this in mind, a new national herbarium complex in Sungai Liang was completed in 1998. This complex serves as an important referral institution for plant collection in the country. An Andulau Forest Reserve has also been developed as a centre for the conservation of plant biodiversity.

FINANCE AND BANKING

The investment of the surplus revenue from oil and gas production is managed jointly by the government-controlled BIA and a consortium of US, British and Japanese financial institutions. The BIA, formed in July 1983, was initially entrusted with the management of B $12,000m. transferred from the direct management of the British Crown Agents. The BIA established offices in London, Paris, Brussels, Boston, Tokyo, Hong Kong, Singapore, Kuala Lumpur, Bali and Yangon, managing its foreign reserves in European, North American and Asian markets. The Asian markets are also covered by the BIA's headquarters in Brunei. The reserves are invested in government bonds, equities, properties, precious metals and short-term deposits. BIA funds were seriously depleted in the 1990s by the effects of the regional economic crisis and a significant decline in the price of petroleum in the late 1990s. With oil prices at their current levels it will now be possible for these reserves to be replenished. There is no external public debt.

There is no central bank in Brunei, with most of the central banking functions being performed by the Financial Institution Division (FID) and the Brunei Currency Board, both under the Ministry of Finance. The FID is the regulator of the finance, banking and insurance sector. It issues licences and regulates financial institutions, including the enforcement of a statutory reserve requirement. The Brunei dollar is issued by the Brunei Currency Board

There are no restrictions on the import of capital from any country, nor are there restrictions on overseas remittances of capital or profits. Non-resident accounts can be maintained, and there is no restriction on borrowing by non-residents. Interest rates tend to move in line with those of Singapore.

In 2000 the financial sector comprised 12 banks, of which nine were branches of foreign institutions, and three finance houses. In January 1993 the Islamic Bank of Brunei (IBB) was established, replacing the International Bank of Brunei. The potential of foreign banks in Brunei, however, is limited by the small domestic market and an insufficient demand for traditional banking services, owing to the Government's provision of loans for the purchase of housing and cars. In 1991 the Brunei Islamic Trust Fund (Tabung Amanah Islam Brunei—TAIB) was established to promote trade and industry and to further the country's economic development. In early 1995 the Development Bank of Brunei (DBB) was established. In July 2000 the DBB was transformed into the Islamic Development Bank of Brunei (IDBB). The IDBB became the third Islamic Bank in Brunei and was to provide financial assistance to small and medium enterprises, in line with the teachings of Islam.

By 1999 there were 125 insurance companies registered to operate in Brunei, of which 23 were active. Most of these are branches of foreign companies but some are locally incorporated, namely the Takaful IBB Bhd and Syarikat Insurans Islam TAIB Sdn Bhd. The insurance companies currently in operation are predominantly small companies. All insurance companies transacting business in Brunei can join the General Insurance Association of Brunei, formed in 1985.

The Brunei International Financial Centre (BIFC) was established in July 2000 with the specific aim of diversifying and stimulating the service sector of the economy of Brunei and also attracting overseas professionals and qualified Bruneians to the international business sector. The Eighth National Development Plan included plans for the establishment of a Brunei International Offshore Financial Centre (BIOFC), in an attempt to encourage economic diversification and the development of Brunei's technological and service sectors. In March 2002 an amendment to the existing Bank Act was enforced in an attempt to eliminate illegal deposit taking and discourage Bruneians from becoming involved in 'get rich quick' schemes. In 2002 the Government also issued a licence to the International Brunei Exchange Ltd, enabling it to operate an International Securities Exchange in Brunei. It was hoped that the opening of Brunei's first exchange would enlarge the country's role within the global financial market.

Brunei joined the IMF and the World Bank in October 1995, which qualifies the country for technical assistance and consultative advice from the array of experts from both institutions, particularly with regard to the diversification of the economy. In May 2000 attempts to establish a regional framework for financial co-operation, in order to prevent a recurrence of the regional financial crisis of 1997, advanced with an agreement among member nations of ASEAN and Japan, the Republic of Korea and the People's Republic of China (the Chiang Mai initiative) to swap foreign-exchange reserves to avert potential crises.

LABOUR FORCE AND EMPLOYMENT

In 2002 the labour force was estimated at 158,000, but no more detailed breakdown of this figure was available. The labour force in 2001 was estimated at 154,200, of whom approximately 98,400 were employed in the private sector, 47,200 in the public sector and 8,600 were unemployed. Approximately 72,000 private-sector employees were foreign workers (temporary residents). It was estimated that another 5,000 or so temporary residents might be employed in the government sector, bringing the proportion of foreign workers to approximately 49% of the labour force.

Despite the persistent shortage of manual workers and of skilled labour, unemployment among indigenous citizens began to rise in the late 1980s. Increasing youth unemployment is a matter of particular concern, since it is believed to be a major reason for nascent social problems in Brunei. The more general employment issue is, however, a crucial one for the long-term viability of the economy: the strong preference for public-sector jobs among Brunei Malays is a major problem; also, there are only a relatively small number of non-manual jobs in the private sector for increasingly well-educated, but not necessarily vocationally and technically skilled, Bruneians. In response, the Government has enforced stricter implementation of the 'Bruneianization' employment policy, whereby Malays receive positive discrimination, and has invested heavily in vocational and technical education. As a result of the Bruneianization programme, Brunei nationals fill an increasing number of senior, supervisory and specialist positions. In 2001 unemployment was estimated at 8,600 persons, and in 2002 at 5,500. This translated to unemployment rates of 5.6% and 3.5%, respectively. However, unemployment is a more serious problem than indicated by these figures. All foreign workers in Brunei hold work permits for specific jobs, and therefore there is no unemployment among legally-resident foreign workers. Accordingly, unemployment among Bruneians in 2001 was 8,600 from a Bruneian (citizen and permanent resident) work-force of 77,200, or 11%. Moreover, most of those unemployed were young Bruneians looking for their first jobs. It is unusual for established government workers to lose their jobs.

Women play a significant role in both the government and private sectors, having taken advantage of the equal educational opportunities in Brunei. The labour force participation rate of women was 20% in 1971, 54% in 1995 and 59.1% in 2002. The male participation rate has been reasonably stable over the last few years, at approximately 79%, and was reported as being 78.9% in 2002, in which year the overall participation rate was 69%.

The establishment of a Government Workers' Provident Fund in 1993, which organized a pension system for non-pensionable government employees and subsequently for employees in the private sector, helped make the private sector more attractive to Bruneians by giving employees new financial security. Of the 98,400 private-sector workers in 2001, the construction sector was the largest employer with 27,600 workers. Of this figure, a mere 2,980 were Brunei citizens or permanent residents. BSP is the largest private employer. A total of 3,041 Bruneians were reported to be working in the oil and gas sector in 2001, but the employment capacity of the hydrocarbon sector is limited owing to the highly capital-intensive nature of the industry.

The private sector remains highly dependent on immigrant labour, primarily from the ASEAN countries, but also from countries such as India, Bangladesh and Pakistan. Professional staff, including doctors, medical specialists and teachers, are mostly recruited from ASEAN, the Indian subcontinent, Australia, New Zealand and the United Kingdom. Union activity is low and of very limited scope; collective agreements exist only in the oil sector.

DEVELOPMENT PLANS AND CURRENT ECONOMIC OUTLOOK

There are a number of agencies involved in efforts to diversify the economy and to reduce the size of the public sector, and a variety of focal points for development. The Ministries of Development, Industry and Primary Resources and Finance and the Brunei Economic Development Board (BEDB) are all involved in development planning. In addition to in-house planning, reports from international consultants are often used to investigate development scenarios.

There have been seven National Development Plans, covering the periods 1953–58, 1962–66, 1975–79, 1980–84, 1986–90, 1991–95 and 1996–2000, respectively, with the Eighth National Development Plan encompassing the period 2001–05. These Plans, although far from comprehensive, have delineated proposals for government investment in infrastructure, services and incentives, all aimed at diversifying the economy and at increasing private-sector participation. In each of its Development Plans, the Government of Brunei has emphasized the need to diversify the economy through the expansion of agriculture and industry and the development of financial services and tourism.

In 1995 the Brunei Industrial Development Plan (IDP) was commissioned to reactivate the non-oil sector; the IDP encompassed several policy recommendations, including the development of a 'niche strategy' for industrial activities and the creation of an environment more conducive to investment promotion.

The Eighth National Development Plan (2001–05) was announced in September 2001. Total expenditure of B $7,300m. was approved for its implementation, of which the Government intended to provide only B $2,900m. For the first time, foreign and local investors were to supply the remaining funding. In November 2001 the Sultan announced that an additional B $1,000m. would be allocated specifically to economic development projects in 2002. The Plan continued to emphasize the importance of diversification to Brunei's long-term economic prospects, and to encourage industrial expansion. The development of the private sector and human resources was accorded priority, particularly in service industries such as finance, tourism, and information and communications technology.

In 1996 the Government announced plans to develop Brunei as a 'Service Hub for Trade and Tourism' (or SHuTT) by the year 2003. Brunei wants to see itself as a bridge for the EAGA member countries to the regional and global markets. At the same time it aspires to be the gateway to EAGA markets for the rest of the world. Ports, airport facilities and tourism services are being upgraded as part of plans to build on Brunei's telecommunications network both regionally and internationally. Tourism is considered to be a key element of the SHuTT plan and a new tourism unit was established in 1995 under the Ministry of Industry and Primary Resources. Visa requirements were relaxed and border check-points were upgraded. The SHuTT plan aims to exploit and maximize the economic potential of Brunei as a centre for trade and tourism, and forms part of the Government's overall scheme to develop the non-petroleum-based sector of the economy.

Under the SHuTT 2003 plan, Brunei International Airport was to become an airline hub through its promotion as an International Aviation Centre, an Air Freight Services Centre and an Airport City. The widening of the airport's operation area (apron) to accommodate up to eight Boeing-767 aircraft was completed in March 1999. The SHuTT 2003 programme also identified Muara Port as a gateway for international trade. The Ports Department has developed extensive plans for the improvement of infrastructure and operational equipment.

The Brunei Darussalam Economic Council (BDEC) was established in September 1998 under the chairmanship of Prince Mohamed Bolkiah to examine the situation in Brunei in the light of the regional economic crisis and to recommend measures to revitalize the economy. In its report, published in February 2000, the BDEC warned that the Bruneian economy was unsustainable, citing the inability of income growth to keep pace with the rise in the population (since 1984 real GDP had increased at less than half the rate of population growth), and rising unemployment. More than 75% of Brunei nationals in the labour force were employed by the Government, and government contracts accounted for most of domestic economic activity. As a result of reduced revenues from petroleum and gas, owing to a significant decline in international prices from mid-1997, the Government's capacity to provide employment and to supply the impetus for economic development was impaired. However, the private sector was too small and too dependent on government expenditure to create much-needed employment opportunities.

The implementation of the Strategy for Sustainable Growth commenced within one year, following the receipt of detailed expert advice to develop the policies. Under the strategy, government finances were to be improved by prioritizing government expenditure and eliminating waste, by reducing the size of the public sector through privatization and by expanding government revenue through the broadening of the tax base and the gradual removal of subsidies. The private sector was to be encouraged to expand in order to reduce the country's dependence on government expenditure. This expansion was to be achieved through privatization, the promotion of local and foreign investment through the simplification of bureaucratic procedures and the encouragement of local business creation through easier access to financing. The restrictions on ownership and development of land were to be reviewed to facilitate investment, and private sector liquidity was to be improved by ensuring prompt government payment. In addition to these measures, the regulatory and legal framework of the country was to be modernized, the transparency of government policies and regulations was to be improved, and emphasis was to be placed upon the development of a superior communications and information technology infrastructure. The competitiveness of the petroleum and gas industry was to be improved and local small- and medium-scale enterprises were to be encouraged through the expansion of existing financial assistance programmes.

The BEDB was founded in 2001 to promote Brunei as a foreign investment destination. In 2003 it announced a two-part, five-year strategy to encourage economic development and diversification in Brunei. Plans were formalized to investigate the construction of an aluminium smelter at Sungai Liang and additional port facilities at Pulau Muara Besar. In September the BEDB and US company Alcoa Incorporated announced that they had signed a Memorandum of Understanding (MOU) to undertake a feasibility study on the establishment of an aluminium smelting plant and its associated infrastructure in Brunei, which would utilize the country's natural gas reserves.

Emphasis was also placed upon the identification and prioritization of certain industry 'clusters' with the potential to achieve growth in the country. Consultants' reports subsequently identified four priority areas: business services, hospitality and tourism, financial services and transport and logistics.

In July 2001 the Government enacted the Investment Incentives Order 2001, together with Pioneer Status and Income Tax Relief regulations. These issued guide-lines under which certain industries could achieve pioneer status and tax relief could be granted for foreign and local investment, as well as extending the existing period of tax relief. Pioneer service companies were specifically designated as any engaged in the provision of engineering, technical, educational, medical or agricultural technology services. Companies engaged in the manufacture of goods for export were also eligible for pioneer status. The Government hoped that the new law and regulations would encourage the development of priority industries within the private sector.

In July 2003 a memorandum of understanding was signed between the BEDB and ACI Corporation of the United Kingdom regarding the establishment of a rubber-recycling plant in the country. The project was to involve the construction of up to four plants over a five-year period and would require total investment of US $1,820m., while creating 1,220 jobs. The plant would use cryogenic technology to convert used tyres into pure rubber powder and would be the first of its kind in the Asian region.

Between 1989 and 2001 nine textile manufacturing companies from Malaysia, Singapore and Indonesia established operations in Brunei. Garment manufacturing is now the second largest contributor to export earnings, after oil and gas: the total value of garments exported in 2002 was B $383m., compared with B $159.8m. in 1999 and B $10m. in 1989. The increase was due to the establishment of new factories and the expansion of existing ones. A total of 99% of the ready-made garments was exported to the USA, Canada and the European Union (EU). The manufacture of traditional handicrafts has also been encouraged by the Government, though on a very small scale.

The Ministry of Industry and Primary Resources was established in January 1989 to facilitate foreign investment and to enhance the diversification process. The Ministry functions as a 'one-stop' agency for foreign investors and is responsible for co-ordinating the licensing and approval process. It acts as a focal point for all industrial development. The Government has established a trading and investment division under the Ministry, called Semaun Holdings, to help accelerate industrial development through direct investment. The main sphere of operation is in trading and commercial ventures, particularly in the food, manufacturing and service sectors. Semaun Holdings has since established its own subsidiaries, namely Semaun Seafood and Semaun Prim, and has signed several agreements relating to joint-venture projects. The Ministry also arranges consultative meetings with representatives from the textile industries and industrial complexes, as well as from co-operative societies.

In January 2001 the Brunei Information Technology (BIT) Council was established, following the implementation of the country's first IT Strategic Plan (ISP)—'IT 2000 and beyond'. The ISP's goals included: raising IT literacy within society; promoting its application in both the public and private sectors; and ensuring that the sector was provided with adequate IT-skilled personnel to facilitate growth within it. Under the Eighth National Development Plan, B $526m. was allocated specifically to national IT development for the first time. The Department of Telecommunications was to be corporatized and transformed into a government-administered private company as part of an ongoing programme to corporatize many government services. Syarikat Telekom Brunei Bhd (TelBru) was registered in May 2002. The corporatization programme was intended to reduce the Government's direct involvement in the provision of public services and to enable it to become a more effective regulator and facilitator. However, in mid-2004 the programme remained incomplete.

Structural constraints remain in trying to develop and widen the industrial base. Firms face high labour costs, a very small domestic market, a narrow resource base and a lack of indigenous skilled labour and entrepreneurs. In addition, long-term titles to land ownership are difficult to acquire, and the government bureaucracy has a reputation for being slow-moving although benevolent. These combined factors have hindered significant foreign investment. The sound oil sector (and the fact that the currency is tied to the Singapore dollar) maintains the exchange rate at a high level, thus hampering export possibilities further.

Statistical Survey

Source (unless otherwise stated): Department of Economic Planning and Development, Prime Minister's Office, Block 2A, Jalan Ong Sum Ping, Bandar Seri Begawan BA 1311; tel. 2244433; fax 2230236; e-mail info@jpke.gov.bn; internet www.depd.gov.bn.

AREA AND POPULATION

Area: 5,765 sq km (2,226 sq miles); *By District:* Brunei/Muara 570 sq km (220 sq miles), Seria/Belait 2,725 sq km (1,052 sq miles), Tutong 1,165 sq km (450 sq miles), Temburong 1,305 sq km (504 sq miles).

Population (excluding transients afloat): 260,482 at census of 7 August 1991; 332,844 (males 168,925, females 163,919) at census of 21 August 2001; 348,800 (males 176,300, females 172,500) in 2003 (official estimate). *By District* (official estimates, 2003): Brunei/Muara 242,600; Seria/Belait 58,000; Tutong 39,600; Temburong 8,600.

Density (2003): 60.5 per sq km.

Ethnic Groups (2003, official estimates): Malay 232,200, Chinese 38,200, Other indigenous 12,500, Others 65,900, Total 348,800.

Principal Towns: Bandar Seri Begawan (capital): population 27,285 at 2001 census, 45,900 at 1991 census; Kuala Belait: population 21,200 at 1991 census; Seria: population 21,100 at 1991 census; Tutong: population 13,000 at 1991 census.

Births, Marriages and Deaths (registrations, 2002): Live births 7,464 (birth rate 21.9 per 1,000); Deaths 1,041 (death rate 3.1 per 1,000); Marriages 2,288 (marriage rate 6.3 per 1,000 in 1999).

Expectation of Life (WHO estimates, years at birth): 76.1 (males 74.8; females 77.4) in 2002. Source: WHO, *World Health Report*.

Economically Active Population (persons aged 15 years and over, 2001 census, provisional): Agriculture, hunting, forestry and fishing 1,994; Mining and quarrying 3,954; Manufacturing 12,455; Electricity, gas and water 2,639; Construction 12,301; Trade, restaurants and hotels 20,038; Transport, storage and communications 4,803; Financing, insurance, real estate and business services 8,190; Community, social and personal services 79,880; *Total employed* 146,254 (males 85,820, females 60,434); Unemployed 11,340 (males 6,734, females 4,606); *Total labour force* 157,594 (males 92,554, females 65,040).

HEALTH AND WELFARE

Key Indicators

Total Fertility Rate (children per woman, 2002): 2.5.

Under-5 Mortality Rate (per 1,000 live births, 2002): 6.

HIV/AIDS (% of persons aged 15–49, 2003): <0.1.

Physicians (per 1,000 head, 2002): 0.87.

Health Expenditure (2001): US $ per head (PPP): 638.

Health Expenditure (2001): % of GDP: 3.1.

Health Expenditure (2001): public (% of total): 79.4.

Human Development Index (2002): ranking: 33.

Human Development Index (2002): value: 0.867.

For sources and definitions, see explanatory note on p. vi.

AGRICULTURE, ETC.

Principal Crops ('000 metric tons, 2003): Rice (paddy) 0.6, Vegetables 10.4, Fruit 4.7.

Livestock ('000 head, 2003): Cattle 1.4, Buffaloes 5.9, Goats 3.1, Poultry 12,093.

Livestock Products (FAO estimates, '000 metric tons, 2002): Beef and veal 3.3, Poultry meat 12.8, Poultry eggs 5.9, Cattle hides (fresh) 0.6. Source: FAO.

Forestry ('000 cubic metres, 2003): Round timber 101.1; Sawn timber 50.9; Firewood 0.1; Bakau poles 89.9; Charcoal ('000 kg) 28.4; Poles ('000 pieces) 3.1.

Fishing (metric tons, live weight, 2003): Capture 14,546.4; Aquaculture 583.3; *Total catch* 15,129.7.

MINING

Production (2003): Crude petroleum ('000 barrels, incl. condensate) 75,642; Natural gas (million cu m, gross) 12,702. Source: US Geological Survey.

INDUSTRY

Production ('000 barrels, unless otherwise indicated, 2003): Motor spirit (petrol) 1,717; Distillate fuel oils 1,195; Residual fuel oil 511; Cement ('000 metric tons) 235; Electric energy (million kWh, 2000, estimate) 2,579. Source: mainly US Geological Survey.

FINANCE

Currency and Exchange Rates: 100 sen (cents) = 1 Brunei dollar (B $). *Sterling, US Dollar and Euro Equivalents* (31 May 2004): £1 sterling = B $3.1155; US $1 = B $1.6981; €1 = B $2.0795; B $100 = £32.10 = US $58.89 = €48.09. *Average Exchange Rate* (Brunei dollars per US $): 1.7917 in 2001; 1.7906 in 2002; 1.7422 in 2003. Note: The Brunei dollar is at par with the Singapore dollar.

Budget (B $ million, 2002): *Revenue:* Tax revenue 2,331.7 (Import duty 111.2; Corporate income tax 2,204.1); Non-tax revenue 1,936.1 (Commercial receipts 301.9; Property income 1,631.0); Total 4,267.8. *Expenditure:* Personal emoluments 1,337.4; Other annual recurrent charges 1,204.3; Other special expenditure charges 497.6; Development expenditure 419.8; Charged expenditure 1,137.2; Investment in public sector 50.1; Total 4,646.3.

Money Supply (B $ million, 2002): Currency outside banks 654.8; Demand deposits at banks 1,617.8; Total money 2,272.6.

Cost of Living (Consumer Price Index; base: 2002 = 100): *2003:* All items 100.3 (Food and non-alcoholic beverages 99.2; Clothing and footwear 97.1; Housing, water, electricity and maintenance 99.2; Transport goods, services and operations 98.5; Communications 95.5; Education 99.6; Medical and health 98.1; Recreation and entertainment 101.3; Miscellaneous goods and services 100.6).

Gross Domestic Product by Economic Activity (B $ million in current prices, 2003, provisional): Agriculture, hunting, forestry and fishing 294.5; Mining, quarrying and manufacturing 3,402.3; Electricity, gas and water 62.1; Construction 334.0; Trade, restaurants and hotels 803.7; Transport, storage and communications 431.9; Finance, insurance, real estate and business services 722.2; Community, social and personal services 2,441.1; *Sub-total* 8,491.8; *Less* Imputed bank service charge 254.8; *GDP in purchasers' values* 8,236.9.

Balance of Payments (B $ million, 2003): Exports of goods 7,704.3; Imports of goods –2,191.3; *Trade balance* 5,513.0; Exports of services 760.5; Imports of services –1,797.4; *Balance on goods and services* 4,476.1; Other income received 3,917.0; Other income paid –651.4; *Balance on goods, services and income* 7,741.7; Current transfers received 0; Current transfers paid –241.9; *Current balance* 7,499.8; Capital balance –16.6; Financial balance –4,613.5; *Overall balance* 2,869.7.

EXTERNAL TRADE

Principal Commodities (B $ million, 2003): *Imports c.i.f.:* Food and live animals 348.02; Chemicals 175.87; Basic manufactures 597.47; Machinery and transport equipment 815.20; Miscellaneous manufactured articles 264.03; Total (incl. others) 2,311.82. *Exports f.o.b.:* Mineral fuels, lubricants, etc. 6,754.15; Machinery and transport equipment 411.63; Miscellaneous manufactured articles 418.78; Total (incl. others) 7,704.33.

Principal Trading Partners (B $ million, 2003): *Imports:* Australia 68.9; People's Republic of China 112.5; France 38.6; Germany 74.6; Hong Kong 151.8; Indonesia 61.7; Italy 16.1; Japan 231.2; Republic of Korea 54.4; Malaysia 468.1; Netherlands 10.9; Singapore 465.2; United Kingdom 85.0; USA 266.0; Total (incl. others) 2,311.8. *Exports:* Japan 3,155.9; Republic of Korea 862.1; Malaysia 284.0; Singapore 344.4; Thailand 726.2; USA 597.1; Total (incl. others) 7,704.3.

TRANSPORT

Road Traffic (registered vehicles, 2001): Private cars 188,720, Goods vehicles 17,828, Motorcycles and scooters 7,162, Buses and taxis 2,267, Others 4,470.

Merchant Fleet (displacement, '000 grt at 31 December): 362.7 in 2001; 482.6 in 2002; 480.0 in 2003. Source: Lloyd's Register-Fairplay, *World Fleet Statistics*.

International Sea-borne Shipping (freight traffic, freight tons, 2003): Goods loaded 127,547; Goods unloaded 1,260,507. Note: One freight ton equals 40 cubic feet (1.133 cubic metres) of cargo.

Civil Aviation (2003): Passenger arrivals 487,768, passenger departures 487,701; freight loaded 16,038 metric tons, freight unloaded 11,265 metric tons; mail loaded 440 metric tons, mail unloaded 124 metric tons.

TOURISM

Visitor Arrivals by Nationality (incl. excursionists, 2000): Indonesia 68,527; Malaysia 974,132; Philippines 65,842; Singapore 27,995; United Kingdom 43,865; Total (incl. others) 1,306,764.

Tourism Receipts (US $ million): 38 in 1996; 39 in 1997; 37 in 1998. Source: World Bank.

COMMUNICATIONS MEDIA

Radio Receivers (2000, estimate): 362,712 in use.

Television Receivers (2000, estimate): 216,223 in use.

Telephones (2003): 81,900 direct exchange lines in use.

Facsimile Machines (1996, estimate): 2,000 in use. Source: UN, *Statistical Yearbook*.

Mobile Cellular Telephones (2003): 177,372 subscribers.

Personal Computers ('000 in use, 2001 estimate): 25. Source: International Telecommunication Union.

Internet Users ('000, 2001): 35.0. Source: International Telecommunication Union.

Book Production (1992): 25 titles; 56,000 copies. Source: UNESCO, *Statistical Yearbook*.

Newspapers (2002): Daily 4; Non-daily 3 (English 2, with circulation of 22,000 copies; Malay 3, with circulation of 39,500 copies; Malay and English 1, with circulation of 10,000 copies).

Other Periodicals (1998): 15 (estimated combined circulation 132,000 copies per issue).

EDUCATION

Pre-primary and Primary: 207 schools; 4,828 teachers; 58,837 pupils in 2003.

General Secondary: 33 schools; 2,913 teachers; 37,743 pupils in 2003.

Teacher Training: 1 college; 37 teachers; 244 students in 2003.

Vocational: 7 colleges; 501 teachers; 2,780 students in 2003.

Higher Education: 2 institutes (incl. 1 university); 392 teachers; 3,805 students in 2003.

Adult Literacy Rate (UNESCO estimates): 93.9% (males 96.3%; females 91.4%) in 2002. Source: UN Development Programme, *Human Development Report*.

Directory

The Constitution

Note: Certain sections of the Constitution relating to elections and the Legislative Council have been in abeyance since 1962.

A new Constitution was promulgated on 29 September 1959 (and amended significantly in 1971 and 1984). Under its provisions, sovereign authority is vested in the Sultan and Yang Di-Pertuan, who is assisted and advised by five Councils: the Religious Council, the Privy Council, the Council of Cabinet Ministers, the (inactive) Legislative Council and the Council of Succession. Power of appointment to the Councils is exercised by the Sultan.

The 1959 Constitution established the Chief Minister as the most senior official, with the British High Commissioner as adviser to the Government on all matters except those relating to Muslim and Malay customs.

In 1971 amendments were introduced reducing the power of the British Government, which retained responsibility for foreign affairs, while defence became the joint responsibility of both countries.

In 1984 further amendments were adopted as Brunei acceded to full independence and assumed responsibility for defence and foreign affairs.

THE RELIGIOUS COUNCIL

In his capacity as head of the Islamic faith in Brunei, the Sultan and Yang Di-Pertuan is advised on all Islamic matters by the Religious Council, whose members are appointed by the Sultan and Yang Di-Pertuan.

THE PRIVY COUNCIL

This Council, presided over by the Sultan and Yang Di-Pertuan, is to advise the Sultan on matters concerning the Royal prerogative of mercy, the amendment of the Constitution and the conferment of ranks, titles and honours.

THE COUNCIL OF MINISTERS

Presided over by the Sultan and Yang Di-Pertuan, the Council of Cabinet Ministers considers all executive matters.

THE LEGISLATIVE COUNCIL

The role of the Legislative Council is to scrutinize legislation. However, following political unrest in 1962, provisions of the Constitution relating, *inter alia*, to the Legislative Council were amended, and the Legislative Council has not met since 1984. In the absence of the Legislative Council, legislation is enacted by royal proclamation.

In July 2004 the Sultan announced that the Legislative Council would be reinstated by the end of that year. Its members were to be appointed by a governmental committee. However, the state of emergency that had been declared in 1962 would remain in force and the Sultan would continue to rule by decree.

THE COUNCIL OF SUCCESSION

Subject to the Constitution, this Council is to determine the succession to the throne, should the need arise.

The State is divided into four administrative districts, in each of which is a District Officer responsible to the Prime Minister and Minister of Home Affairs.

The Government

HEAD OF STATE

Sultan and Yang Di-Pertuan: HM Sultan Haji HASSANAL BOLKIAH (succeeded 4 October 1967; crowned 1 August 1968).

COUNCIL OF CABINET MINISTERS

(October 2004)

Prime Minister, Minister of Defence and of Finance: HM Sultan Haji HASSANAL BOLKIAH.

Minister of Foreign Affairs: HRH Prince MOHAMED BOLKIAH.

Minister of Home Affairs and Special Adviser to the Prime Minister: Pehin Dato' Haji ISA BIN Pehin Haji IBRAHIM.

Minister of Education: Pehin Dato' Haji ABDUL AZIZ BIN Pehin Haji UMAR.

Minister of Industry and Primary Resources: Dato' Haji ABDUL RAHMAN TAIB.

Minister of Religious Affairs: Pehin Dato' Dr Haji MOHAMAD ZAIN BIN Haji SERUDIN.

Minister of Development: Dato' Seri Paduka Dr Haji AHMAD BIN Haji JUMAT.

Minister of Health: Pehin Dato' Haji ABU BAKAR BIN Haji APONG.

Minister of Culture, Youth and Sports: Pehin Dato' Haji HUSSEIN BIN Pehin Haji MOHAMAD YOSOF.

Minister of Communications: Pehin Dato' Haji ZAKARIA BIN Haji SULEIMAN.

There are, in addition, eight deputy ministers.

MINISTRIES

Office of the Prime Minister (Jabatan Perdana Menteri): Istana Nurul Iman, Bandar Seri Begawan BA 1000; tel. 2229988; fax 2241717; e-mail PRO@jpm.gov.bn; internet www.pmo.gov.bn.

Ministry of Communications (Kementerian Perhubungan): Jalan Menteri Besar, Bandar Seri Begawan BB 3910; tel. 2383838; fax 2380127; e-mail info@mincom.gov.bn; internet www.mincom.gov.bn.

Ministry of Culture, Youth and Sports (Kementerian Kebudayaan, Belia dan Sukan): Simpang 336, Jalan Kebangsaan, Bandar Seri Begawan BC 4415; tel. 2380911; fax 2380653; e-mail info@kkbs.gov.bn; internet www.kkbs.gov.bn.

Ministry of Defence (Kementerian Pertahanan): Bolkiah Garrison, Bandar Seri Begawan BB 3510; tel. 2386371; fax 2331615; e-mail info@mindef.gov.bn; internet www.mindef.gov.bn.

Ministry of Development (Kementerian Pembangunan): Old Airport, Jalan Berakas, Bandar Seri Begawan BB 3510; tel. 2241911; e-mail info@mod.gov.bn; internet www.mod.gov.bn.

Ministry of Education (Kementerian Pendidikan): Old Airport, Jalan Berakas, Bandar Seri Begawan BB 3510; tel. 2382233; fax 2380050; e-mail sutmoe@brunet.bn; internet www.moe.gov.bn.

Ministry of Finance (Kementerian Kewangan): Tingkat 5, Bangunan Kementerian Kewangan, Commonwealth Dr., Jalan Kebangsaan, Bandar Seri Begawan BB 3910; tel. 2241991; fax 2226132; e-mail info@finance.gov.bn; internet www.finance.gov.bn.

Ministry of Foreign Affairs (Kementerian Hal Ehwal Luar Negeri): Jalan Subok, Bandar Seri Begawan BD 2710; tel. 2261177; fax 2262904; e-mail info@mfa.gov.bn; internet www.mfa.gov.bn.

Ministry of Health (Kementerian Kesihatan): Jalan Menteri Besar, Bandar Seri Begawan BB 3910; tel. 2226640; fax 2240980; e-mail moh2@brunet.bn; internet www.moh.gov.bn.

Ministry of Home Affairs (Kementerian Hal Ehwal Dalam Negeri): Jalan Menteri Besar, Bandar Seri Begawan BB 3910; tel. 2223225; e-mail info@home-affairs.gov.bn; internet www.home-affairs.gov.bn.

Ministry of Industry and Primary Resources (Kementerian Perindustrian dan Sumber-sumber Utama): Jalan Menteri Besar, Bandar Seri Begawan BB 3910; tel. 2382822; fax 2382807; e-mail MIPRS2@brunet.bn; internet www.industry.gov.bn.

Ministry of Religious Affairs (Kementerian Hal Ehwal Ugama): Jalan Menteri Besar, Jalan Berakas, Bandar Seri Begawan BB 3910; tel. 2382525; fax 2382330; e-mail info@religious-affairs.gov.bn; internet www.religious-affairs.gov.bn.

Political Organizations

Parti Perpaduan Kebangsaan Brunei (PPKB) (Brunei National Solidarity Party—BNSP): Bandar Seri Begawan; f. 1986 after split in PKDB (see below); ceased political activity in 1988, but re-emerged in 1995; Pres. HATTA ZAINAL ABIDIN.

Former political organizations included: **Parti Rakyat Brunei—PRB** (Brunei People's Party), banned in 1962, leaders are all in exile; **Barisan Kemerdeka'an Rakyat—BAKER** (People's Independence Front), f. 1966 but no longer active; **Parti Perpaduan Kebangsaan Rakyat Brunei—PERKARA** (Brunei People's National United Party), f. 1968 but no longer active; and **Parti Kebangsaan Demokratik Brunei—PKDB** (Brunei National Democratic Party—BNDP), f. 1985 and dissolved by government order in 1988.

Diplomatic Representation

EMBASSIES AND HIGH COMMISSIONS IN BRUNEI

Australia: Teck Guan Plaza, 4th Floor, Jalan Sultan, Bandar Seri Begawan BS 8811; tel. 2229435; fax 2221652; e-mail austhicom.brunei@dfat.gov.au; internet www.australia.org.bn; High Commissioner CHRISTIAN BENNETT.

Bangladesh: 125 Kampong Kiulap, Bandar Seri Begawan BE 1518; tel. 2238420; fax 2238421; e-mail bdoot@brunet.bn; High Commissioner M. A. HAKIM.

Cambodia: 8 Simpang 845, Kampong Tasek, Meradun, Jalan Tutong, Bandar Seri Begawan BF 1520; tel. 2654046; fax 2650646; Ambassador ITH DETTOLA.

Canada: 5th Floor, Jalan McArthur Bldg, 1 Jalan McArthur, Bandar Seri Begawan BS 8711; tel. 2220043; fax 2220040; e-mail hicomcda@brunet.bn; internet www.dfait-maeci.gc.ca/Brunei/; High Commissioner PAUL S. H. LAU.

China, People's Republic: 1, 3 & 5 Simpang 462, Kampong Sungai Hanching, Jalan Muara, Bandar Seri Begawan BC 2115; tel. 2334163; fax 2335710; e-mail embproc@brunet.bn; Ambassador WEI WEI.

France: Kompleks Jalan Sultan, Units 301–306, 3rd Floor, 51–55 Jalan Sultan, Bandar Seri Begawan BS 8811; tel. 2220960; fax 2243373; e-mail france@brunet.bn; internet www.france.org.bn; Ambassador THIERRY BORJA DE MOZOTA.

Germany: Kompleks Bangunan Yayasan Sultan Haji Hassanal Bolkiah, Unit 2.01, Block A, 2nd Floor, Jalan Pretty, Bandar Seri Begawan BS 8711; tel. 2225547; fax 2225583; e-mail prgerman@brunet.bn; Ambassador CONRAD KARL CAPPELL.

India: 'Baitussyifaa', Simpang 40–22, Jalan Sungai Akar, Bandar Seri Begawan BC 3915; tel. 2339947; fax 2339783; e-mail hicomind@brunet.bn; internet www.brunet.bn/gov/emb/india; High Commissioner AJAI CHOUDHRY.

Indonesia: Simpang 528, Lot 4498, Kampong Sungai Hanching Baru, Jalan Muara, Bandar Seri Begawan BC 2115; tel. 2330180; fax 2330646; e-mail kbribsb@brunet.bn; internet www.indonesia.org.bn; Ambassador YUSBAR DJAMIL.

Iran: 19 Simpang 477, Kampong Sungai Hanching, Jalan Muara, Bandar Seri Begawan BC 2115; tel. 2330021; fax 2331744; Ambassador ABD AL-FAZL MUHAMMAD ALIKHANI.

Japan: 1 and 3 Jalan Jawatan Dalam, 33 Simpang 122, Kampong Kiulap, Bandar Seri Begawan BE 1518; tel. 2229265; fax 2229481; e-mail embassy@japan.com.bn; internet www.bn.emb-japan.go.jp; Ambassador YOSHINOBU NISAKA.

Korea, Republic: POB 2169, Bandar Seri Begawan BS 8674; tel. 2426038; fax 2426041; e-mail koreaemb@brunet.bn; Ambassador KIM WOONG-NAM.

Laos: Lot 19824, 11 Simpang 480, Jalan Kebangsaan Lama, off Jalan Muara, Bandar Seri Begawan BC 4115; tel. 2345666; fax 2345888; e-mail LAOSEMBA@brunet.bn; Ambassador BOUNTHONG VONGSALY.

Malaysia: 61 Simpang 336, Jalan Kebangsaan, Kampong Sungai Akar, Bandar Seri Begawan BC 4115; tel. 2381095; fax 2381278; e-mail mwbrunei@brunet.bn; High Commissioner Tan Sri MOHAMED JAMIL JOHARI.

Myanmar: 14 Lot 2185/46292, Simpang 212, Jalan Kampong Rimba, Gadong, Bandar Seri Begawan BE 3119; tel. 2450506; fax 2451008; e-mail myanmar@brunet.bn; Ambassador U THET WIN.

Oman: 35 Simpang 100, Kampong Pengkalan, Jalan Tungku Link, Gadong, Bandar Seri Begawan BE 3719; tel. 2446953; fax 2449646; e-mail omnembsb@brunet.bn; Ambassador Haji MOHAMMAD BIN OMAR AHMAD AIDID.

Pakistan: 10A Simpang 350, Kampong Sungai Akar, Jalan Subok, Bandar Seri Begawan; tel. 2261354; fax 2261353; e-mail hepak@brunet.bn; internet www.brunet.bn/gov/emb/Pakistan; High Commissioner BADR-UD-DEEN.

Philippines: 17 Simpang 126, Km 2, Jalan Tutong, Bandar Seri Begawan BA 2111; tel. 2241465; fax 2237707; e-mail bruneipe@brunet.bn; Ambassador VIRGINIA H. BENDAVIDEZ.

Saudi Arabia: 1 Simpang 570, Kampong Salar, Jalan Muara, Bandar Seri Begawan BT 2528; tel. 2792821; fax 2792826; e-mail bnemb@mofa.gov.sa; Ambassador ESAM BIN AHMED BIN JAMAL ABID AL-THAGAFI.

Singapore: 8 Simpang 74, Jalan Subok, Bandar Seri Begawan; tel. 2262741; fax 2262743; e-mail singa@brunet.bn; internet www.gov.sg/mfa/brunei/; High Commissioner V. P. HIRUBALAN.

Thailand: 2 Simpang 682, Jalan Tutong, Kampong Bunut, Bandar Seri Begawan BF 1320; tel. 2653108; fax 2653032; e-mail thaiemb@brunet.bn; Ambassador Lt SUNTHON NIYOMYATH.

United Kingdom: POB 2197, Bandar Seri Begawan BS 8674; tel. 2222231; fax 2234315; e-mail brithc@brunet.bn; internet www.britishhighcommission.gov.uk/brunei; High Commissioner ANDREW CAIE.

USA: Teck Guan Plaza, 3rd Floor, Jalan Sultan, Bandar Seri Begawan BS 8811; tel. 2220384; fax 2225293; e-mail ConsularBrunei@state.gov; internet bandar.usembassy.gov; Ambassador GENE B. CHRISTY.

Viet Nam: 7 Simpang 5383719, Jalan (Duong) Kebangsaan Lama, Bandar Seri Begawan BC 2115; tel. 2343167; fax 2343169; e-mail vnembassy@hotmail.com; Ambassador HA HONG HAI.

Judicial System

SUPREME COURT

The Supreme Court consists of the Court of Appeal and the High Court. Syariah (*Shari'a*) courts coexist with the Supreme Court and deal with Islamic laws.

Supreme Court

Km 11/2, Jalan Tutong, Bandar Seri Begawan BA 1910; tel. 2225853; fax 2241984; e-mail judiciarybn@hotmail.com; internet www.judicial.gov.bn/supr_court.htm.

Chief Registrar: Haji HAIROLARNI ABD MAJID.

The Court of Appeal

Composed of the President and two Commissioners appointed by the Sultan. The Court of Appeal considers criminal and civil appeals against the decisions of the High Court and the Intermediate Court. The Court of Appeal is the highest appellate court for criminal cases. In civil cases an appeal may be referred to the Judicial Committee of Her Majesty's Privy Council in London if all parties agree to do so before the hearing of the appeal in the Brunei Court of Appeal.

President: Sir DEREK CONS.

The High Court

Composed of the Chief Justice and judges sworn in by the Sultan as Commissioners of the Supreme Court. In its appellate jurisdiction, the High Court considers appeals in criminal and civil matters against the decisions of the Subordinate Courts. The High Court has unlimited original jurisdiction in criminal and civil matters.

Chief Justice: Dato' Seri Paduka MOHAMMED SAIED.

OTHER COURTS

Intermediate Courts: have jurisdiction to try all offences other than those punishable by the death sentence and civil jurisdiction to try all actions and suits of a civil nature where the amount in dispute or value of the subject/matter does not exceed B $100,000.

The Subordinate Courts

Presided over by the Chief Magistrate and magistrates, with limited original jurisdiction in civil and criminal matters and civil jurisdiction to try all actions and suits of a civil nature where the amount in dispute does not exceed B $50,000 (for Chief Magistrate) and B $30,000 (for magistrates).

Chief Magistrate: ROSTAINA BINTI Pengiran Haji DURAMAN.

The Courts of Kathis

Deal solely with questions concerning Islamic religion, marriage and divorce. Appeals lie from these courts to the Sultan in the Religious Council.

Chief Kathi: Dato' Seri SETIA Haji SALIM BIN Haji BESAR.

Attorney-General: Dato' Paduka Haji KIFRAWI BIN Dato' Paduka Haji KIFLI, Attorney-General's Chambers, The Law Bldg, Km 1, Jalan Tutong, Bandar Seri Begawan BA 1910; tel. 2244872; fax 2223100; e-mail info@agc.gov.bn; internet www.agc.gov.bn.

Solicitor-General: Datin Paduka MAGDALENE CHONG.

Religion

The official religion of Brunei is Islam, and the Sultan is head of the Islamic community. The majority of the Malay population are Muslims of the Shafi'is school of the Sunni sect; at the 1991 census Muslims accounted for 67.2% of the total population. The Chinese population is either Buddhist (accounting for 12.8% of the total population at the 1991 census), Confucianist, Daoist or Christian. Large numbers of the indigenous ethnic groups practise traditional animist forms of religion. The remainder of the population are mostly Christians, generally Roman Catholics, Anglicans or members of the American Methodist Church of Southern Asia. At the 1991 census Christians accounted for 10.0% of the total population.

ISLAM

Supreme Head of Islam: HM Sultan Haji HASSANAL BOLKIAH (Sultan and Yang Di-Pertuan).

CHRISTIANITY

The Anglican Communion

Within the Church of the Province of South East Asia, Brunei forms part of the diocese of Kuching (Malaysia).

The Roman Catholic Church

Brunei comprises a single apostolic prefecture. At December 2002 an estimated 6.3% of the population were adherents.

Prefect Apostolic: Rev. CORNELIUS SIM, Church of Our Lady of the Assumption, POB 527, Bandar Seri Begawan BS 8670; tel. 2222261; fax 2238938; e-mail frcsim@brunet.bn; internet www.catholicbrunei.org.

The Press

NEWSPAPERS

Borneo Bulletin: Locked Bag No. 2, MPC (Old Airport, Berakas), Bandar Seri Begawan 3799; tel. 2451468; fax 2451461; e-mail brupress@brunet.bn; internet www.brunet.bn/news/bb; f. 1953; daily; English; independent; owned by QAF Group; Editor CHARLES REX DE SILVA; circ. 25,000.

Brunei Darussalam Newsletter: Dept of Information, Prime Minister's Office, Istana Nurul Iman, Bandar Seri Begawan BA 1000; tel. 2229988; fortnightly; English; govt newspaper; distributed free; circ. 14,000.

Daily News Digest: Dept of Information, Prime Minister's Office, Istana Nurul Iman, Bandar Seri Begawan BA 1000; internet www.brunet.bn/news/dndbd/digest.htm; English; govt newspaper.

Media Permata: Locked Bag No. 2, MPC (Old Airport, Berakas), Bandar Seri Begawan BB 3510; tel. 2451468; fax 2451461; e-mail mediapermata@brunet.bn; internet www.brunei-online.com/mp; f. 1995; daily (not Sun.); Malay; owned by QAF Group; Editor ABDUL LATIF; circ. 10,000.

Pelita Brunei: Dept of Information, Prime Minister's Office, Old Airport, Berakas BB 3510; tel. 2383941; fax 2381004; e-mail pelita@brunet.bn; internet www.brunet.bn/news/pelita/pelita1.htm; f. 1956; weekly (Wed.); Malay; govt newspaper; distributed free; Editor TIMBANG BIN BAKAR; circ. 27,500.

Salam: c/o Brunei Shell Petroleum Co Sdn Bhd, Seria 7082; tel. 34184; fax 34189; internet www.shell.com.bn/salam; f. 1953; monthly; Malay and English; distributed free to employees of the Brunei Shell Petroleum Co Sdn Bhd; Editor SAIFUL MAHADHIR NORDIN; circ. 46,000.

Publishers

Borneo Printers & Trading Sdn Bhd: POB 2211, Bandar Seri Begawan BS 8674; tel. 2651387; fax 2654342; e-mail bptl@brunet.bn.

Brunei Press Sdn Bhd: Lots 8 and 11, Perindustrian Beribi II, Jalan Gadong, Bandar Seri Begawan BE 1118; tel. 2451468; fax 2451462; e-mail brupress@brunet.bn; internet www.bruneipress.com.bn; f. 1953; Gen. Man. REGGIE SEE.

Capital Trading & Printing Pte Ltd: POB 1089, Bandar Seri Begawan; tel. 2244541.

Leong Bros: 52 Jalan Bunga Kuning, POB 164, Seria; tel. 322381.

Offset Printing House: POB 1111, Bandar Seri Begawan; tel. 2224477.

The Star Press: Bandar Seri Begawan; f. 1963; Man. F. W. ZIMMERMAN.

Government Publishing House

Government Printer: Government Printing Department, Office of the Prime Minister, Bandar Seri Begawan BB 3510; tel. 2382541; fax 2381141; e-mail jpkkuu@brunet.bn; internet www.printing.gov.bn; Dir MOHIDIN BIN Haji DAUD (acting).

Broadcasting and Communications

TELECOMMUNICATIONS

Authority for the Information Communication Technology Industry (AiTi): Block B14, Simpang 32-5, Kampong Anggrek Desa, Jalan Berakas, Bandar Seri Begawan BB 3713; tel. 2323232; fax 2382445; e-mail info@aiti.gov.bn; internet www.aiti.gov.bn; f. 2003; regulatory body for telecommunications industry; responsible for development of information communication technology industry; Chair. Dato' Paduka Awang Haji ABDULLAH BIN Haji BAKAR.

DST Communications Sdn Bhd: Block D, Yayasan Sultan Haji Hassanal Bolkiah Kompleks, Bandar Seri Begawan; tel. 2232323; e-mail dst-group@simpur.net.bn; internet www.dst-group.com; mobile service provider; signed agreement with Alcatel in 2003 to improve provision of mobile services in Brunei; Man. Dir Pengiran Haji MOHAMED ZIN BIN Pengiran DAMIT.

Jabatan Telecom Brunei (Department of Telecommunications of Brunei): Ministry of Communications, Jalan Berakas, Bandar Seri Begawan BB 3510; tel. 2382382; fax 2382445; e-mail info@telecom.gov.bn; internet www.telecom.gov.bn; telecommunications services provider; originally scheduled to become Telekom Brunei Bhd (govt-owned company) in 2003; corporatization later postponed indefinitely; Dir of Telecommunications Awang BUNTAR BIN OSMAN.

BROADCASTING

Radio

Radio Televisyen Brunei (RTB): Prime Minister's Office, Jalan Elizabeth II, Bandar Seri Begawan BS 8610; tel. 2243111; fax 2241882; e-mail rtbpits@brunet.bn; internet www.rtb.gov.bn; f. 1957; five radio networks: four broadcasting in Malay, the other in English, Chinese (Mandarin) and Gurkhali; also broadcasts on the internet; Dir Pengiran Dato' Paduka ISMAIL MOHAMED.

The British Forces Broadcasting Service (Military) broadcasts a 24-hour radio service to a limited area.

Television

Radio Televisyen Brunei (RTB): Prime Minister's Office, Jalan Elizabeth II, Bandar Seri Begawan BS 8610; tel. 2243111; fax 2241882; e-mail rtbpits@brunet.bn; internet www.rtb.gov.bn; f. 1957; three television channels with one satellite channel; Dir Pengiran Dato' Paduka ISMAIL MOHAMED.

Finance

(cap. = capital; res = reserves; dep. = deposits; brs = branches; amounts in Brunei dollars unless otherwise stated)

BANKING

The Department of Financial Services (Treasury), the Brunei Currency Board and the Brunei Investment Agency (see Government Agencies below), under the Ministry of Finance, perform most of the functions of a central bank.

Commercial Banks

Baiduri Bank Bhd: Block A, Units 1–4, Kiarong Complex, Lebuhraya Sultan Hassanal Bolkiah, Bandar Seri Begawan BE 1318; tel. 2455111; fax 2455599; e-mail bank@baiduri.com; internet www.baiduri.com; f. 1994; cap. 30m., res 32.0m., dep. 1,329.7m. (Dec. 2003); Chair. Pengiran ANAK ISTERI Pengiran ANAK HAJJAH ZARIAH; Gen. Man. PIERRE IMHOF; 10 brs.

Islamic Bank of Brunei Bhd: Lot 159, Bangunan IBB, Jalan Pemancha, POB 2725, Bandar Seri Begawan BS 8711; tel. 2235687; fax 2235722; e-mail ibb@brunet.bn; internet www.ibb.com.bn; f. 1981 as Island Development Bank; name changed from International Bank of Brunei Bhd to present name in Jan. 1993; practises Islamic banking principles; Chair. Haji ABDUL RAHMAN BIN Haji ABDUL KARIM; Man. Dir Haji ZAINASALLEHEN BIN Haji MOHAMED TAHIR; 13 brs.

Islamic Development Bank of Brunei Bhd: Ground–4th Floors, Kompleks Setia Kenangan, Kampong Kiulap, Jalan Gadong, Bandar Seri Begawan BE 1518; tel. 2232547; fax 2233540; e-mail dbbcom@brunet.bn; internet www.idbb-bank.com.bn; f. 1995 as Development Bank of Brunei Bhd; name changed to present in July 2000; practises Islamic banking principles; Man. Dir Pengiran Datin Paduka Hajah URAI Pengiran ALI; 3 brs.

Foreign Banks

Citibank NA (USA): Darussalam Complex, 12–15 Jalan Sultan, Bandar Seri Begawan BS 8811; tel. 2243983; fax 2237344; e-mail glen.rase@citicorp.com; Vice-Pres. and Country Head PAGE STOCKWELL; 2 brs.

The Hongkong and Shanghai Banking Corpn Ltd (Hong Kong): Jalan Sultan, cnr Jalan Pemancha, Bandar Seri Begawan BS 8811; tel. 2252252; fax 2241316; e-mail hsbc@hsbc.com.bn; internet www.hsbc.com.bn; f. 1947; acquired assets of National Bank of Brunei in 1986; CEO MARCUS HURRY; 10 brs.

Maybank (Malaysia): 1 Jalan McArthur, Bandar Seri Begawan BS 8711; tel. 2226462; fax 2226404; e-mail maybank@brunet.bn; f. 1960; Country Man. AZIZUL ABDUL RASHID; 3 brs.

RHB Bank Bhd (Malaysia): Unit G. 02, Block D, Kompleks Bangunan Yayasan Sultan Haji Hassanal Bolkiah, Ground Floor, Jalan Pretty, Bandar Seri Begawan BS 8711; tel. 2222526; fax 2237487; e-mail rhbbsb@brunet.bn; fmrly Sime Bank Bhd; Country Man. APANDI BIN KLOMPOT; 1 br.

Standard Chartered Bank (United Kingdom): 1st Floor, 51–55 Jalan Sultan, POB 186, Bandar Seri Begawan BS 8811; tel. 2242386; fax 2220103; e-mail scb.brunei@bn.standardchartered.com; internet www.standardchartered.com/bn; f. 1958; CEO HANS THEILKUHL; 8 brs.

United Overseas Bank Ltd (Singapore): Unit G5, RBA Plaza, Jalan Sultan, Bandar Seri Begawan BS 8811; tel. 2225477; fax 2240792; f. 1973; Gen. Man. SIA KEE HENG; 2 brs.

'Offshore' Bank

The Brunei International Financial Centre (BIFC), under the Ministry of Finance, supervises the activities of the offshore banking sector in Brunei.

Royal Bank of Canada: 1 Jalan McArthur, 4A, 4th Floor, Bandar Seri Begawan BS 8711; tel. 2224366; fax 2224368; Gen. Man. MATTHEW YONG.

STOCK EXCHANGE

In May 2002 the International Brunei Exchange Ltd (IBX) was granted an exclusive licence to establish an international securities exchange in Brunei.

International Brunei Exchange Ltd (IBX): The Empire, Muara-Tutong Highway, Jerudong BG 3122; tel. 2611222; fax 2611020; e-mail info@ibx.com.bn; internet www.ibx.com.bn; f. 2001; CEO B. C. YONG.

INSURANCE

In 1998 there were 18 general, three life and two composite (takaful) insurance companies operating in Brunei.

General Companies

AGF Insurance (Singapore) Pte Ltd: c/o A&S Associates Sdn Bhd, Bangunan Gadong Properties, 03-01, Jalan Gadong, Bandar Seri Begawan BE 4119; tel. 2420766; fax 2440279; Gen. Man. JEAN NOEL ROUSSELLE.

The Asia Insurance Co Ltd: Unit A1 & A2, 1st Floor, Block A, Bangunan Hau Man Yong, Simpang 88, Kampong Kiulap, Bandar Seri Begawan BE 3919; tel. 2236100; fax 2236102; e-mail asiains_d.wong@brunet.bn; Man. DAVID WONG KOK MIN.

Aviva Insurance Bhd: Unit 311, 3rd Floor, Kompleks Mohd Yussof, Mile 1½, Jalan Tutong, Bandar Seri Begawan BA 1714; tel. 2223632; fax 2220965; e-mail aviva@brunet.bn; fmrly CGU Insurance Bhd; Man. LIANG VOON CHIANG.

AXA Insurance (B) Sdn Bhd: Unit 608, 6th Floor, Jalan Sultan Complex, 51–55 Jalan Sultan, Bandar Seri Begawan BS 8811; tel. 2226138; fax 2243474; fmrly GRE Insurance (B) Sdn Bhd; Man. MOK HAI TONG.

BALGI Insurance (B) Sdn Bhd: Unit 7, Simpang 88, Bangunan Haji Ahmad bin Hassan and Anak-Anak, Kampong Kiulap, Bandar Seri Begawan BE 1518; tel. 2234020; fax 2233981; Man. Dir PATRICK SIM SONG JUAY.

Borneo Insurance Sdn Bhd: Unit 103, Bangunan Kambang Pasang, Km 2, Jalan Gadong, Bandar Seri Begawan BE 4119; tel. 2420550; fax 2428550; Man. LIM TECK LEE.

Commercial Union Assurance (M) Sdn Bhd: c/o Jasra Harrisons Sdn Bhd, Jalan McArthur, cnr Jalan Kianggeh, Bandar Seri Begawan BS 8711; tel. 2242361; fax 2226203; Man. WHITTY LIM.

Cosmic Insurance Corpn Sdn Bhd: Block J, Unit 11, Abdul Razak Complex, 1st Floor, Jalan Gadong, Bandar Seri Begawan BE 3919; tel. 2427112; fax 2427114; Man. RONNIE WONG.

ING General Insurance International NV: Shop Lot 86, 2nd Floor, Jalan Bunga Raya, Kuala Belait KA 1131; tel. 3335338; fax 3335338; Man. SHERRY SOON PECK ENG.

Liberty Citystate Insurance Pte Ltd: 1st Floor, Unit 25, Block C, Bangunan Hau Man Yong Complex, Simpang 88, Kampong Kiulap, Bandar Seri Begawan BE 1518; tel. 2238282; fax 2236848; e-mail libertyintl@brunet.bn; Man. ROBERT LAI CHIN YIN.

Malaysia National Insurance Bhd: 9 Bangunan Haji Mohd Salleh Simpang 103, 1st Floor, Jalan Gadong, Bandar Seri Begawan BE 4119; tel. 2443393; fax 2427451; e-mail tsang.py@brunet.bn; Man. ANDREW AK NYAGORN.

MBA Insurance Sdn Bhd: 7 Bangunan Hasbullah I, 1st Floor, Km 4, Jalan Gadong, Bandar Seri Begawan BE 3519; tel. 2441535; fax 2441534; e-mail cheahmbabrunei@brunet.bn; Man. CHEAH LYE CHONG.

Motor and General Insurance Sdn Bhd: 6 Bangunan Hasbullah II, Km 4, Jalan Gadong, Bandar Seri Begawan BE 3919; tel. 2440797; fax 2445342; Man. Dir Haji ABDUL AZIZ BIN ABDUL LATIF.

National Insurance Co Bhd: Scouts' Headquarters Bldg, Jalan Gadong, Bandar Seri Begawan BE 1118; tel. 2426888; fax 2429888; e-mail insurance@brunet.bn; internet www.national.com.bn; Gen. Man. CHAN LEK WEI.

Royal and Sun Alliance Insurance (Global) Ltd: Unit 7, 1st Floor, Block B, Kiarong Complex, Lebuhraya Sultan Hassanal Bolkiah, Bandar Seri Begawan BE 1318; tel. 2423233; fax 2423325; Gen. Man. TOMMY LEONG TONG KAW.

South East Asia Insurance (B) Sdn Bhd: Unit 2, Block A, Abdul Razak Complex, 1st Floor, Jalan Gadong, Bandar Seri Begawan BE 3919; tel. 2443842; fax 2420860; Gen. Man. SHIM WEI HSIUNG.

Standard Insurance (B) Sdn Bhd: 2 Bangunan Hasbullah I, Ground Floor, Bandar Seri Begawan BE 3919; tel. 2450077; fax 2450076; e-mail feedback@standard-ins.com; internet www.standard-ins.com; Man. PAUL KONG.

Winterthur Insurance (Far East) Pte Ltd: c/o Borneo Co (B) Sdn Bhd, Lot 9771, Km 3½, Jalan Gadong, Bandar Seri Begawan BE 4119; tel. 2422561; fax 2424352; Gen. Man. ANNA CHONG.

Life Companies

American International Assurance Co Ltd: Unit 509A, Wisma Jaya Building, 5th Floor, 85–94 Jalan Pemancha, Bandar Seri Begawan BS 8811; tel. 2239112; fax 2221667; e-mail Kenneth-WC.Ling@AIG.com; Man. PHILIP TAN.

The Asia Life Assurance Society Ltd: Unit 2, 1st Floor, Block D, Abdul Razak Complex, Jalan Gadong, Bandar Seri Begawan BE 4119; tel. 2423755; fax 2423754; e-mail asialife@simpur.net.bn; Br. Man. JOSEPH WONG SIONG LION.

The Great Eastern Life Assurance Co Ltd: Suite 1, Badi'ah Complex, 2nd Floor, Jalan Tutong, Bandar Seri Begawan BA 2111; tel. 2243792; fax 2225754; e-mail gelife@brunet.bn; Man. HELEN YEO.

Takaful (Composite Insurance) Companies

Syarikat Insurans Islam TAIB Sdn Bhd: Bangunan Pusat Komersil dan Perdagangan Bumiputera, Ground Floor, Jalan Cator, Bandar Seri Begawan BS 8811; tel. 2237724; fax 2237729; e-mail insuranstaib@brunet.bn; internet www.insuranstaib.com.bn; f. 1993; provides Islamic insurance products and services; Man. Dir Haji MOHAMED ROSELAN BIN Haji MOHAMED DAUD.

Takaful Bank Pembangunan Islam Sdn Bhd (TBPISB): Bangunan Seri Kiulap, Kampong Kiulap, Gadong, Bandar Seri Begawan; internet www.tbpisb.com; f. 2001; fmrly Takaful IDBB Sdn Bhd; name changed as above in 2003; Islamic life and non-life insurance products; Chair. Pehin Dato' Paduka Seri LAILA JASA Haji Awang AHMAD WALLY SKINNER; Man. Dir Haji AISHATUL AKMAR SIDEK.

Takaful IBB Bhd: Unit 5, Block A, Kiarong Complex, Lebuhraya Sultan Hassanal Bolkiah, Bandar Seri Begawan BE 1318; tel. 2451804; fax 2451808; e-mail takaful@brunet.bn; internet www.takaful-ibb.com; f. 1993; Chair. Pehin Dato' Haji ABU BAKAR BIN Haji APONG DAUD.

Insurance Association

General Insurance Association of Negara Brunei Darussalam (GIAB): Unit C2-2, Block C, Shakirin Complex, Kampong Kiulap, Bandar Seri Begawan BE 1318; tel. 2237898; fax 2237858; e-mail giab@brunet.bn; internet www.giab.com.bn; f. 1986; Chair. DOROTHY NEWN.

Trade and Industry

GOVERNMENT AGENCIES

Brunei Currency Board: Tingkat 5, Bangunan Kementerian Kewangan, Commonwealth Dr., Jalan Kebangsaan, Bandar Seri Begawan BB 3910; tel. 2383999; fax 2382232; e-mail bcb@brunet.bn; internet www.finance.gov.bn/bcb/bcb_index.htm; f. 1967; maintains control of currency circulation; Sec. Haji MAHADI Haji IBRAHIM; Chair. HM Sultan Haji HASSANAL BOLKIAH (Minister of Finance).

Brunei Darussalam Economic Council (BDEC): Bandar Seri Begawan; internet www.brudirect.com/BruneiInfo/info/BD_EconomicCouncil.htm; f. 1998; convened to examine the economic situation in Brunei and to recommend short- and long-term measures designed to revitalize the economy; Chair. HRH Prince MOHAMED BOLKIAH.

Brunei International Financial Centre (BIFC): Level 14, Ministry of Finance, Commonwealth Dr., Bandar Seri Begawan BB 3910; tel. 2383750; fax 2383787; e-mail bifc@finance.gov.bn; internet www.bifc.finance.gov.bn; f. 2000; regulates international financial sector and encourages development of Brunei as investment destination; CEO ROBERT MILLER.

Brunei Investment Agency: Tingkat 5, Bangunan Kementerian Kewangan, Commonwealth Dr., Jalan Kebangsaan, Bandar Seri Begawan BB 3910; f. 1973; Man. Dir Awang Haji ALI BIN Haji APONG (acting).

DEVELOPMENT ORGANIZATIONS

Brunei Economic Development Board (BEDB): Block 2K, Jalan Ong Sum Ping, Bandar Seri Begawan BA 1311; tel. 2230111; fax 2230053; f. 2001; promotes Brunei as an investment destination; facilitates and assists industrial development; Chair. Pehin Dato' Haji MOHAMMAD BIN Haji DAUD; CEO JOHN A. PERRY.

Brunei Industrial Development Authority (BINA): Ministry of Industry and Primary Resources, Km 8, Jalan Gadong, Bandar Seri Begawan BE 1118; tel. 2444100; fax 2423300; e-mail bruneibina@brunet.bn; internet www.bina.gov.bn; f. 1996; Dir Haji MOHD ZAIN BIN Haji GAFFAR.

Brunei Islamic Trust Fund (Tabung Amanah Islam Brunei): Block A, Unit 2, Ground Floor, Kiarong Complex, Lebuhraya Sultan Hj Hassanal Bolkiah, Bandar Seri Begawan BE 1318; tel. 2452666; fax 2450877; f. 1991; promotes trade and industry; Chair. Haji Awang YAHYA BIN Haji IBRAHIM.

Industrial and Trade Development Council: Bandar Seri Begawan; facilitates the industrialization of Brunei; Chair. Dato' Haji ABDUL RAHMAN TAIB (Minister of Industry and Primary Resources).

Semaun Holdings Sdn Bhd: Unit 2.02, Block D, 2nd Floor, Yayasan Sultan Haji Hassanal Bolkiah Complex, Jalan Pretty, Bandar Seri Begawan BS 8711; tel. 2232950; e-mail semaun@brunet.bn; internet www.semaun.gov.bn; promotes industrial and commercial development through direct investment in key industrial sectors; 100% govt-owned; the board of directors is composed of ministers and senior govt officials; Chair. Dato' Haji ABDUL RAHMAN TAIB (Minister of Industry and Primary Resources); Man. Dir Pengiran Haji MARIANA P. D. N. Pengiran Haji ABDUL MOMIN.

CHAMBERS OF COMMERCE

Brunei Darussalam International Chamber of Commerce and Industry: Unit 402–403A, 4th Floor, Wisma Jaya, Jalan Pemancha, Bandar Seri Begawan BS 8811; tel. 2228382; fax 2228389; Chair. SULAIMAN Haji AHAI; Sec. Haji SHAZALI BIN Dato' Haji SULAIMAN; 108 mems.

Brunei Malay Chamber of Commerce and Industry: POB 1099, Bandar Seri Begawan BS 8672; tel. 2422752; fax 2422753; f. 1964; Pres. Dato' A. A. HAPIDZ; 160 mems.

Chinese Chamber of Commerce: Chinese Chamber of Commerce Bldg, 72 Jalan Robert, 2nd–4th Floors, Bandar Seri Begawan BS 8811; tel. 2235494; fax 2235492; e-mail ccc@brunet.bn; internet www.bruneichinesechamber.com; Pres. ROBERT KOH HOE KIAT.

Indian Chamber of Commerce: Lot 20021, Taman Aman Complex, Jalan Gadong-Tutong, Kampong Beribi, Bandar Seri Begawan BE 3188; tel. 2650793; fax 2650793.

National Chamber of Commerce and Industry of Brunei Darussalam: 2nd Floor, 144 Jalan Pemancha, Bandar Seri Begawan BS 8711; tel. 2243321; fax 2228737; e-mail abas@nccibd.com; internet www.bruneichamber.com; Pres. Sheikh ABAS Sheikh MOHAMAD.

STATE HYDROCARBON COMPANIES

Brunei Gas Carriers Sdn Bhd (BGC): 7th Floor, Setia Kenangan Mall, Kampong Kiulap, Bandar Seri Begawan; internet www.syarikatbgc.com; f. 1998; LNG shipping co; owned jtly by Brunei Govt, Shell International Gas and Diamond Gas Carriers.

Brunei LNG Sdn Bhd: Lumut KC 2935, Seria; tel. 3236901; fax 3236749; e-mail GP-External-Affairs@BruneiLNG.com; internet www.blng.com.bn; f. 1969; natural gas liquefaction; owned jointly by the Brunei Govt, Shell and Mitsubishi Corpn; operates LNG plant at Lumut, which has a capacity of 7.2m. tons per year; Man. Dir Dato' HAMDILLAH H. A. WAHAB.

Brunei National Petroleum Co Sdn Bhd (PetroleumBrunei): 5th Floor, Bangunan Bahirah, Jalan Menteri Besar, Bandar Seri Begawan BB 3910; tel. 2383022; fax 2383004; e-mail bnpc@brunet.bn; f. 2001; wholly govt-owned; CEO ALIAS Haji MOHD YUSOF (acting).

Brunei Shell Marketing Co Bhd: Maya Puri Bldg, 36/37 Jalan Sultan, POB 385, Bandar Seri Begawan; tel. 2229304; fax 2240470; internet www.bsm.com.bn; f. 1978 (from the Shell Marketing Co of Brunei Ltd), when the Govt became equal partner with Shell; markets petroleum and chemical products throughout Brunei; Man. Dir MAT SUNY Haji MOHD HUSSEIN.

Brunei Shell Petroleum Co Sdn Bhd (BSP): Jalan Utara, Panagia, Seria KB 3534; tel. 3373999; fax 3372040; internet www.shell.com.bn; f. 1957; the largest industrial concern in the country; 50% state holding; Man. Dir MARK CARNE; Dep. Man. Dir Haji ZAINAL ABIDIN Haji MOHAMED ALI.

Jasra International Petroleum Sdn Bhd: RBA Plaza, 2nd Floor, Jalan Sultan, Bandar Seri Begawan; tel. 2228968; fax 2228929; petroleum exploration and production; Man. Dir ROBERT A. HARRISON.

MAJOR COMPANIES

Ath Garments Sdn Bhd: Simpang 245, Plot 18, Serambangan Industrial Estate, Tutong TA 1141; tel. 4261383; fax 4261390; e-mail athgmt_gm@brunet.bn; f. 1998; mfr of textiles and garments; Man. Dir Hajah SAADATENA A. BAKAR; 879 employees.

BlueScope Lysaght (B) Sdn Bhd: Industrial Complex, Beribi, Phase 1, 6 Km, Jalan Gadong, Bandar Seri Begawan BE 1118; tel. 2447155; fax 2447154; e-mail bluescopelysaght@brunet.bn; internet www1.bluescopesteelasia.com/BlueScopeSteel/country/brunei/lysaght/en/index.cfm; f. 1993; fmrly BHP Steel Lysaght (B) Sdn Bhd; supplier of steel building solutions; Pres. DORAIRAJ MARAN.

Brunei Oxygen Sdn Bhd: Lot 5761, Tapak Perindustrian, Pekan Belait, Kuala Belait KA 1931; tel. 3332861; fax 3333466; owned by QAF Brunei Sdn Bhd; mfr of industrial gases; Gen. Man. LEE CHUN WAH.

Brutex Manufacturing Co Sdn Bhd: 4090 Jalan Perusahaan, Kampong Serasa, POB 106, Jejabat Pos Muara, Muara BP 1128; tel. 2771584; e-mail brutex@pso.brunet.bn; f. 1988; mfr of garments; Man. Dir KIM K. HOO.

Butra Heidelberger Zement Sdn Bhd (Brunei Cement): Lot 3, Serasa Industrial Area, POB 153, Muara BT 1128; tel. 2771395; fax 2771404; e-mail bhz@bruneicement.com; internet www.bruneicement.com; f. 1993; mfr and distributor of cement; cap. and res B $30m. (1999); Man. Dir YAVUZ ERMIS; 115 employees.

Fraser and Neave (B) Bhd: 80 Jalan Perusahaan, Kampong Serasa, Muara; tel. 2770839; fax 2770151; f. 1981; production and bottling of soft drinks, incl. Coca-Cola; Chair. Dato' Haji ALI BIN Haji MOHD DAUD; 35 employees.

Hunt Concrete Industries Sdn Bhd: Room 302, 1st Floor, Block C, Chandrawaseh Complex, Mile 1, Jalan Tutong, Bandar Seri Begawan; tel. 2229249; fax 2226596; e-mail hunt@brunet.bn; f. 1983; Chair LIM MING SIONG; Man. Dir KOH MING SHAM; 180 employees.

Jati Freedom Textile Sdn Bhd: Jalan Perusahaan, Simpang 15, Pekan Muara, Muara Negara BT 1728; tel. 2770010; fax 2770017; e-mail jatiinvest@brunet.bn; f. 1988; mfr of garments; Station Man. ALICE LEE; 420 employees.

Mulaut Abattoir Sdn Bhd: POB 28, Sengkurong BG 1121; tel. 2670289; fax 2670800; f. 1990; supply and marketing of livestock and meat products; Chair. Dato' Paduka Awang Haji ALIMIN Haji ABDUL WAHAB; 100 employees.

O'Connor's (B) Sdn Bhd: Sufri Complex, Block D, Km 1, Jalan Tutong, Bandar Seri Begawan; tel. 2223109; fax 2220391; mfr of telecommunication and electrical equipment.

QAF Brunei Sdn Bhd: QAF Centre, Lot 65–66, Perindustrial Beribi II, Bandar Seri Begawan BE 1118; tel. 2453388; fax 2452152; e-mail qaf@brunet.bn; internet www.qaf.com; f. 1982; investment holding co with interests in wholesale and retail trade, investment, engineering, offshore services and publishing; subsidiary of Baiduri Holdings Bhd; Chair. HRH Pengiran ANAK ISTERI Pengiran ANAK Haji ZARIAH; CEO HRH Prince ABDUL QAWI.

Supercrete Sdn Bhd: Supercrete Complex, LTS 42/92, Simpang 99, Bengkurong, Bandar Seri Begawan BF 1920; tel. 2654744; fax 2654731; e-mail firstmix@brunet.bn; internet www.supercretebrunei.com; f. 1978; mfr and supplier of pre-mixed concrete.

VSL Systems (N) Sdn Bhd: POB 291, MPC-Old Airport, Bandar Seri Begawan BB 3577; tel. 2380153; fax 2381954; f. 1984; mfr of ready-mixed concrete and concrete products; Man. Dir CHUAH MENG HU; 60 employees.

TRADE UNIONS

Trade unions are legal in Brunei, but must be registered with the Government. In 2003 there were three officially registered trade unions, all in the petroleum sector, with a total membership constituting less than 5% of that sector's work-force. These included:

Brunei Oilfield Workers' Union: XDR/11, BSP Co Sdn Bhd, Seria KB 3534; f. 1964; 470 mems; Pres. SUHAINI Haji OTHMAN; Sec.-Gen. ABU TALIB BIN Haji MOHAMAD.

Transport

RAILWAYS

There are no public railways in Brunei. The Brunei Shell Petroleum Co Sdn Bhd maintains a 19.3-km section of light railway between Seria and Badas.

ROADS

In 1999 there were 1,150 km of roads in Brunei, of which almost 400 km were paved. The main highway connects Bandar Seri Begawan, Tutong and Kuala Belait. A 59-km coastal road links Muara and Tutong. The Eighth National Development Plan (2001–05) prioritized the development of Brunei's roads and, in particular, the construction of a network of main roads that would connect Brunei Muara, Tutong, Kuala Belait and Temburong. In September 2002 construction of the 15-km Jalan Lumut Bypass was completed.

Land Transport Department: 451979 Km 6, Jalan Gadong, Beribi, Bandar Seri Begawan BE 1110; tel. 2451979; fax 2424775; e-mail latis@brunet.bn; internet www.land-transport.gov.bn; Dir Awang Haji OTHMAN BIN Haji MOMIN.

SHIPPING

Most sea traffic is handled by a deep-water port at Muara, 28 km from the capital, which has a 611-m wharf and a draught of 8 m. The port has a container terminal, warehousing, freezer facilities and cement silos. In October 2000 a plan to deepen the port to enable its accommodation of larger vessels was announced. The original, smaller port at Bandar Seri Begawan itself is mainly used for local river-going vessels, for vessels to Malaysian ports in Sabah and Sarawak and for vessels under 30 m in length. There is a port at Kuala Belait, which takes shallow-draught vessels and serves mainly the Shell petroleum field and Seria. Owing to the shallow waters at Seria, tankers are unable to come up to the shore to load, and crude petroleum from the oil terminal is pumped through an underwater loading line to a single buoy mooring, to which the tankers are moored. At Lumut there is a 4.5-km jetty for liquefied natural gas (LNG) carriers.

Four main rivers, with numerous tributaries, are an important means of communication in the interior, and boats or water taxis are the main form of transport for most residents of the water villages. Larger water taxis operate daily to the Temburong district.

Bee Seng Shipping Co: 7 Block D, Sufri Complex, Km 2, Jalan Tutong, POB 1777, Bandar Seri Begawan; tel. 2220033; fax 2224495; e-mail beeseng@brunet.bn.

Belait Shipping Co (B) Sdn Bhd: B1, 2nd Floor, 94 Jalan McKerron, Kuala Belait 6081; POB 632, Kuala Belait; tel. 3335418; fax 3330239; f. 1977; Man. Dir Haji FATIMAH BINTE Haji ABDUL AZIZ.

Brunei Shell Tankers Sdn Bhd: Seria KB 3534; tel. 3373999; f. 1986; owned jtly by Brunei Govt (50%), a co in the Royal Dutch/Shell Group of Cos (25%) and Mitsubishi Corpn of Japan (25%); seven vessels operated by Shell International Trading and Shipping Co Ltd; delivers LNG to overseas markets; Man. Dir CHRIS FINLAYSON.

Harper Wira Sdn Bhd: B2 Bangunan Haji Mohd Yussof, Jalan Gadong, Bandar Seri Begawan 3188; tel. 2448529; fax 2448529.

Inchcape Borneo: Bangunan Inchcape Borneo, Km 4, Jalan Gadong, Bandar Seri Begawan; tel. 2422561; fax 2424352; f. 1856; Gen. Man. LO FAN KEE.

New Island Shipping: POB 850, Bandar Seri Begawan 1908; tel. 2451800; fax 2451480; f. 1975; Chair. TAN KOK VOON; Man. JIMMY VOON.

Pansar Co Sdn Bhd: First Floor, 27 Kompleks Mubibbah 3, Jalan Gadong, Bandar Seri Begawan 3180; tel. 2445246; fax 2445247.

Seatrade Shipping Co: POB 476, Bandar Seri Begawan 1904; tel. 2421457; fax 2421453.

Silver Line (B) Sdn Bhd: Muara Port; tel. 2445069; fax 2430276.

Wei Tat Shipping and Trading Co: Mile 41, Jalan Tutong, POB 103, Bandar Seri Begawan; tel. 265215.

CIVIL AVIATION

There is an international airport near Bandar Seri Begawan, which can handle up to 1.5m. passengers and 50,000 metric tons of cargo per year. In 2004 expansion of the airport was ongoing. The Brunei Shell Petroleum Co Sdn Bhd operates a private airfield at Anduki for helicopter services.

Department of Civil Aviation: Brunei International Airport, Bandar Seri Begawan BB 2513; tel. 2330142; fax 2331706; e-mail dea@brunet.bn; internet www.civil-aviation.gov.bn; Dir Haji KASIM BIN Haji LATIP.

Royal Brunei Airlines Ltd: RBA Plaza, Jalan Sultan, POB 737, Bandar Seri Begawan BS 8671; tel. 2240500; fax 2244737; e-mail feedback@rba.com.bn; internet www.bruneiair.com; f. 1974; operates services within the Far East and to the Middle East, Australia and Europe; Chair. Dato' Haji HAZAIR ABDULLAH; CEO PETER WILLIAM FOSTER.

Tourism

Tourist attractions in Brunei include the flora and fauna of the rain forest and the national parks, as well as mosques and water villages. There were 964,080 foreign visitor arrivals (including same-day visitors) in 1998. Foreign visitor arrivals increased to 1,306,764 in 2000. In 1998 international tourist receipts totalled US $37m. The year 2001 was designated 'Visit Brunei Year'.

Brunei Tourism: c/o Ministry of Industry and Primary Resources, Jalan Menteri Besar, Bandar Seri Begawan BB 3910; tel. 2382822; fax 2382824; e-mail info@industry.gov.bn; internet www.tourismbrunei.com; Dir-Gen. Sheikh Jamaluddin bin Sheikh Mohamed.

Defence

At August 2003 the total strength of the Royal Brunei Malay Regiment was 7,000 (including 700 women): army 4,900 (including 250 women); navy 1,000; air force 1,100 (including 75 women). Military service (for which only ethnic Malays are eligible) is voluntary. Paramilitary forces comprise an estimated 3,750, of which 1,750 are members of the Royal Brunei Police. Since 1971 national defence has been the responsibility of the Brunei Government. A Gurkha battalion of the British army, comprising about 1,120 men, has been stationed in Brunei since 1971, but is not responsible for internal security (its duty now being to guard the petroleum and gas fields). About 500 Singapore troops, operating a training school, are also stationed in Brunei.

Defence Expenditure: B $450m. in 2003.

Commander of the Royal Brunei Armed Forces: Pehin Datu Lailaraja Maj.-Gen. Awang Haji Halbi bin Haji Mohamed Yussof.

Education

There are three official languages of instruction, Malay, English and Chinese, and schools are divided accordingly. There are also religious schools. Education is free and is compulsory for 12 years between the ages of five and 16.

All Malay schools are state-administered and are, in general, co-educational. Pre-primary education begins at the age of five years. Primary education in the Malay schools lasts for six years from the age of six years; it is divided into two cycles of three years, lower primary and upper primary. At lower primary level all instruction is in Malay but at upper primary certain subjects, e.g. maths, geography and science, are taught in English. Pupils sit for the Primary Certificate of Education (PCE) examination at the end of primary education.

Secondary education lasts for seven years. The first five years are divided into lower secondary, lasting for three years at the end of which pupils sit the Penilaian Menengah Bawah (PMB—Lower Secondary Assessment) examination, and upper secondary which lasts for two or three years. After the PMB some pupils enrol in vocational schools where courses last for 18 months leading to the National Trade Certificate. Other pupils proceed to upper secondary level where abler students follow a two-year course leading to the Brunei-Cambridge General Certificate of Education Ordinary Level (BC-GCE 'O' Level) examinations and others follow a two-year course leading to the BC-GCE 'N' Level (which was introduced in 1997). At 'N' Level those gaining enough credits follow a further one-year course to 'O' Level. The technical colleges then offer two-year courses to those with adequate 'O' Level passes. Students with the requisite 'O' Level results proceed to the pre-university level to pursue a two-year course leading to the BC-GCE Advanced Level ('A' Level) examination. Students with adequate 'A' Level passes may be eligible for entry to the University of Brunei Darussalam (UBD) or other tertiary institutions or be awarded scholarships to study abroad. The Institut Teknologi Brunei provides courses leading to a Higher National certificate (part-time) or a National Diploma (full-time).

Chinese schools are privately run and not assisted by the Government. They cater for pupils at both primary and secondary levels.

In 1994 enrolment at primary level included 91% of those in the relevant age-group (males 90%; females 91%) and enrolment at secondary level included 68% of pupils (males 64%; females 71%). Enrolment at tertiary level in 1996 was equivalent to 6.6% of the relevant age-group (males 5.3%; females 8.0%).

Public expenditure on education totalled an estimated B $347m. in 1997 (8.6% of total expenditure). Under the Eighth National Development Plan (2001–05) the education sector was allocated B $300m. of the total development budget (4.19%).

Bibliography

See also Malaysia Bibliography

Asian Development Bank Technical Assistance Study. *Brunei Darussalam-Indonesia-Malaysia-Philippines East ASEAN Growth Area (BIMP-EAGA), Final Report Volume 1A, The Overall Integrative Report,* 1996.

Asian Development Bank Technical Assistance Study. *Brunei Darussalam-Indonesia Malaysia-Philippines East ASEAN Growth Area (BIMP-EAGA), Final Report Volume 2A, Brunei Darussalam Country Profile,* 1996.

Badan Kemajuan Industri Brunei (BINA), Kementerian Perindustrian dan Sumber-Sumber Utama. *Peranan BINA Dalam Mengembangkan Perusahaan Kecil dan Sederhana,* 1997.

Bartholomew, James. *The Richest Man in the World: The Sultan of Brunei.* London, Viking, 1989.

Blomqvist, Hans C. 'Brunei's Strategic Dilemmas'. *The Pacific Review,* Vol. 6, No 2, 1993.

Bolkiah, Prince Mohamed. *Time and the River.* 2000.

Borneo Bulletin. *Brunei Yearbook Key Information on Brunei.* Bandar Seri Begawan.

Braighlinn, G. *Ideological Innovation under Monarchy: Aspects of Legitimisation Activity in Contemporary Brunei.* Amsterdam, V U University Press, 1992.

Brown, D. E. *Brunei, The Structure and History of a Bornean Malay Sultanate.* Brunei, Brunei Museum, 1970.

Brunei Currency Board. *Brunei Darussalam Financial Structure, Functions and Policies.* Bandar Seri Begawan, 1996.

Chalfont, Lord. *By God's Will: A Portrait of the Sultan of Brunei.* London, Weidenfeld and Nicolson, 1989.

Cleary, Mark, and Shuang Yann Wong. *Oil, Economic Development and Diversification in Brunei Darussalam.* London and New York, Macmillan and St Martin's Press, 1994.

Department of Agriculture. *Investment Opportunities in Agro Industry.* Bandar Seri Begawan, 1997.

Department of Telecommunications. *Corporate Mission and Vision.* Bandar Seri Begawan, 1997.

Doshi, Tilak. 'Brunei: the Steady State'. *Southeast Asian Affairs 1991.* Singapore, Institute of Southeast Asian Studies/Heinemann, 1991.

Economic Development Board, Ministry of Finance. *Incentives for Investment in Brunei Darussalam,* Fifth Edn. Bandar Seri Begawan, 1996.

Economic Planning Unit, Ministry of Finance. *Sixth National Development Plan, 1991–1995.* Bandar Seri Begawan.

Economic Planning Unit, Ministry of Finance. *Seventh National Development Plan, 1996–2000.* Bandar Seri Begawan.

Economist Intelligence Unit. *Country Profile: Malaysia, Brunei.* London, 1995–96.

Government of Brunei Darussalam. *Brunei Darussalam in Profile.* London, Shandwick, 1988.

Horton, A. V. M. *The British Residency in Brunei, 1906–1959.* Hull, Centre for South-East Asian Studies, 1984.

Hussainmiya, B. A. *Sultan Omar Ali Saifuddin III and Britain.* Oxford, Oxford University Press, 1995.

Information Department, Prime Minister's Office. *Brunei Darussalam In Brief 1995,* Sixth Edn. Bandar Seri Begawan.

Krausse, Sylvia C. Engelen, and Gerald, H. *Brunei* (An annotated bibliography.) Oxford, Clio Press, 1988.

Leake, David, Jr. *Brunei: The Modern Southeast Asian Islamic Sultanate.* Jefferson, McFarland and Co, 1990.

McArthur, M. H. S. *Report on Brunei in 1904.* Athens, Ohio, 1987.

Mani, A. 'Negara Brunei Darussalam in 1992. Celebrating the Silver Jubilee'. *Southeast Asian Affairs 1993*. Singapore, Institute of Southeast Asian Studies, 1993.

Metra Consulting. *Handbook of National Development Plans*. London, Graham and Trotman, 1986.

Ministry of Communications. *Brunei Darussalam: Service Hub for Trade and Tourism (SHuTT) 2000 and Beyond*. Bandar Seri Begawan.

Ministry of Industry and Primary Resources. *Brunei Darussalam Investment Guide*. Bandar Seri Begawan.

Ports Department. *Your Regional Business Base, Investors' Handbook*. Bandar Seri Begawan.

Muara Port. Bandar Seri Begawan, 1997.

Ranjit Singh, D. S. *Brunei, 1839-1983, The Problems of Political Survival*. Singapore, Oxford University Press, 1984.

Saunders, Graham. *Bishops and Brookes*. Singapore, Oxford University Press, 1992.

'Brunei: A Vision for the the Future?' *Southeast Asian Affairs 1997*. Singapore, Institute of Southeast Asian Studies/Heinemann, 1997.

A History of Brunei. Second Edn. London, RoutledgeCurzon, 2002.

Siddique, Sharon. 'Brunei Darussalam 1991: the Non-secular State'. *Southeast Asian Affairs 1992*. Singapore, Institute of Southeast Asian Studies/Heinemann, 1992.

Singh, D. S. Ranjit, and Sidhu, Jatswan S. *Historical Dictionary of Brunei Darussalam*. London, Scarecrow Press, 1997.

Statistics Division, Economic Planning Unit, Ministry of Finance. *Brunei Darussalam Statistical Yearbook*. Bandar Seri Begawan.

Tarling, Nicholas. *Britain, the Brookes and Brunei*. Kuala Lumpur, Oxford University Press, 1971.

Tourism Development Division, Ministry of Industry and Primary Resources. *Explore Brunei: A Visitor's Guide*. Third Edn. Bandar Seri Begawan, 1997.

Turnbull, C. M. *A History of Malaysia, Singapore and Brunei*. Sydney, Allen and Unwin, 1989.

Zaini Haji Ahmad, Haji. *The People's Party of Brunei, Selected Documents*. Kuala Lumpur, Insan, 1988.

CAMBODIA

Physical and Social Geography

HARVEY DEMAINE

Cambodia comprises a relatively small and compact territory on the Indo-Chinese peninsula and covers an area of 181,035 sq km (69,898 sq miles), bordered by Thailand to the west, by Laos to the north, and by Viet Nam to the east.

PHYSICAL FEATURES

Apart from the Cardamom and related mountains in the south, which divide the country's interior from its short southern coastline, the greater part of Cambodia consists of a shallow lacustrine basin, centred on Tonlé Sap ('the Great Lake'), which was historically of far greater extent than it is today. This lowland drains eastwards, via the Tonlé Sap River, to the Mekong, which flows through the eastern part of the lowlands from north to south before turning eastwards into Viet Nam and to the South China Sea.

Throughout its course across Cambodia, the Mekong River averages about 2 km in width, but it is interrupted by precipitous rapids at Kratié, and by falls at Khone along the Laotian border. Moreover, its flow fluctuates widely from season to season, and, during the period of greatest volume between June and October, a substantial portion of its flood-waters is diverted up the Tonlé Sap River (the flow of which is thus reversed) into the Great Lake itself, which comes to occupy an area at least twice as great as it does during the dry season in the early months of the year. The temperature is generally between 20°C and 36°C (68°F to 97°F), and the annual average in Phnom-Penh is 27°C (81°F).

RESOURCES AND POPULATION

With some good alluvial soils, abundant water for irrigation, and a tropical monsoon climate without excessive rainfall, Cambodia has considerable agricultural potential and could undoubtedly support both a wider area and a greater intensity of cultivation than it does at present.

In 1975 Cambodia had an estimated 7.1m. inhabitants and an average population density of 39 per sq km. By 1981, according to a census, the country's population had fallen to 6.7m., owing to warfare, famine and migration, giving a density of only 36.9 per sq km. At the census of 3 March 1998 the population was 11.44m., and density had increased to 63.18 per sq km. In mid-2003 the population was officially estimated at 13.8m. The capital city, Phnom-Penh, had an estimated population of only 20,000 in 1978. However, in mid-2003, with the resurgence of internal trade, the city's population was estimated to have risen to 1,157,066.

History

LAURA SUMMERS

Revised by SORPONG PEOU

Early Khmer civilization, owing to its situation on major Chinese and Indian trade routes, was greatly affected by foreign cultural influences. The assimilation of Indian Brahmans into Khmer society encouraged the adoption of Hindu cults, including the recognition of the supremacy of the god Siva (Shiva) in the 'Funan' period (which extended from the first to the sixth century AD). Archaeological discoveries indicate the existence, at that time, of a highly pluralistic, peninsular political system containing a number of Khmer princely families, each of which supported and promoted family cults. Religious syncretism in the sixth century signalled the beginning of military competition for ascendancy in the Mekong delta. The ensuing wars between the cult-based principalities eventually gave rise to the highly centralized Angkorian empires during the ninth–14th centuries. The rejection of established religious cults and the rise of a moral tradition which united images of royalty, divinity and fertility corresponded to a massive intensification of rice production in the area surrounding the modern city of Siem Reap. A complex system of hydraulics, which diverted the water of highland streams and retained flood waters from the Tonlé Sap (a natural reservoir of the Mekong), permitted year-round agriculture and hitherto unthinkable concentrations of population. Much of the labour required for the building of the irrigation system and its associated, extraordinary temple complexes was provided by slaves, most of whom were prisoners of war or captured tribespeople from the highlands. With each succeeding century, the political influence of Angkor expanded, as did its agrarian economy and artistic achievements. At its greatest extent in the 12th century, during the reign of Jayavarman VII (1181–1218), the Angkorian empire embraced the Chao Phraya plain and parts of the Malay peninsula as well as all principalities and populations south of the Annamite chain in present-day Viet Nam, including the powerful Cham state of Champa. Jayavarman VII was also the first Buddhist King of Angkor. By the late 13th century, the Angkorian civilization displayed unmistakable signs of decadence. In 1431 the declining economic and military capacity of Cambodia were further eroded when the ascendant Thai civilization, based at Sukothai, sacked Angkor and its surrounding sites. Succeeding monarchs abandoned all efforts to continue state-controlled rice cultivation, permanently renounced Hindu cults and shifted their capitals southward to Lovek and Oudong, to the north of the riparian crossroads of Phnom-Penh.

PRE-COLONIAL CAMBODIA

Historians differ significantly in their assessment of society and politics from the 16th to the 19th century. Most French historians, focusing on the dramatic decrease in royal military power after the collapse of the empire, argue that Khmer civilization went into decline. By contrast, US scholars, reacting critically to the socially oppressive features of Angkorian centralization and the construction of monuments (especially since the failed socialist revolution in the 1970s), have argued that the 16th century gave rise to a pluralistic, dispersed, village-centred political order, in which communities, organized around Buddhist temples (*wat*), were able to avoid some of the earlier tyrannical excesses. Nevertheless, it is recognized that these communities were under the control of royally appointed governors who collected taxes and demanded labour service (*corvée*) in the name of the still revered monarch.

By the end of the 18th century, the Khmer kingdom had contracted in geographical size to approximately two-thirds of its present area. The rise of the powerful Chakri dynasty, to the west, and the southward expansion of the demographically buoyant Vietnamese nation resulted in the need to pay tribute to both foreign courts so as to solicit royal respect and paternalist protection. This period of dual Siamese-Vietnamese suzerainty came to an abrupt end after an aborted Vietnamese attempt to annex Cambodia. An indigenous, Buddhist-led rebellion, encouraged (although discreetly) by the monarch and supported by timely military intervention from Siam (Thailand), ultimately defeated the Vietnamese plan, but inevitably led to near-total subordination to the Siamese court. Although, unlike the Vietnamese, the Siamese did not insist on cultural assimilation or commence settling on Khmer land, King Ang Duang approached the French in the hope of signing a treaty of protection. The treaty, signed in 1863 by his son King Norodom, gave France complete control over Cambodia's foreign policy and required the royal court to accept a permanent Resident-General. France proceeded almost immediately to assume colonial control.

COLONIAL CAMBODIA

Inevitably, colonialism had a profound impact on the development of the Cambodian state and its politics. By 1884 a second, far-reaching treaty, imposed on Norodom, established French control over the royal administration, royal treasury and foreign trading. Cambodia was incorporated into the Indo-Chinese Union in 1887, and Union budgetary resources were used to reinforce the prestige and near-sacred regard for the Khmer monarch, as a shield against popular or élite discontent (following the successful suppression of a national uprising in 1884–86).

The colonial order was undermined only by the war in the Pacific. Vichy French collaboration with Japan after 1940, combined with Thailand's alliance with the Japanese, resulted in the Japanese-approved Thai annexation of Bat Dambang province in 1941. King Monivong died and was replaced soon afterwards by King Norodom Sihanouk, who was only 18 years of age. King Sihanouk initially displayed sympathy for the emergent nationalist sentiments forcefully articulated by Son Ngoc Thanh and Buddhist monks, among others. (Thanh was appointed Prime Minister in an 'independent' Cabinet hastily established by the Japanese in 1945.) Sihanouk, however, alarmed by the increasingly anti-royalist tendencies of the nationalist movement, subsequently initiated secret discussions with France, arranging the arrest and exile of Thanh, the resumption of French colonial rule in 1946 and the promulgation of a new Constitution in 1947. The post-war Constitution permitted the formation of political parties and provided for the holding of legislative elections. These reforms satisfied the aspirations of some of the élite nationalists. France, meanwhile, used Cambodia as a rear military staging-area in its war against the Viet Minh (see the chapter on Viet Nam), a strategic manoeuvre which was also attempted by the USA after 1970. The resumption of colonial rule and the exile of Thanh divided the nascent nationalist movement. Some nationalists, responding partly to Viet Minh urgings, formed resistance groups seeking to overthrow both colonialism and monarchy, which was finally seen as a French instrument of repression. These *Issaraks*, or freedom fighters, displayed little internal unity. The mainstream of the nationalist movement (represented by the Democrat Party of Prince Yutevong), although anti-royalist, eschewed violence and opted to remain in Phnom-Penh, seeking to gain power through Parliament. The Democrats were highly successful in the legislative elections, but encountered strong opposition from the powerful Liberal Party, which represented royalist and landowning families, and the smaller, ultra-conservative Khmer Renovation Party, a movement led by Lt-Gen. (later Marshal) Lon Nol and other high-ranking state functionaries and their families. The nationalist threat prompted Sihanouk to suspend the Constitution in June 1952 and to assume governmental powers. Fearing French defeat in Viet Nam and Democrat and *Issarak* ascendency, Sihanouk undertook a diplomatic mission to France and the USA, pleading for his country's right to independence. This 'Royal Crusade for Independence', as it was later called, succeeded. France conceded independence on 9 November 1953, and thus indirectly awarded to Sihanouk the title of 'father' of independence. Cambodia's independence was ratified by the Geneva Conference on Indo-China in July 1954.

THE SIHANOUK, LON NOL AND POL POT YEARS

The Sihanouk years of 1954–70 brought the restoration of limited constitutional rights to Cambodians, but were also characterized by Sihanouk's efforts to achieve stability, while being faced by renewed challenges from the socialist left, the liberal democratic and reformist centre and extreme right-wing parties and interests. In foreign policy, Sihanouk constantly renewed his nationalist credentials by steadily opposing US imperialism in Viet Nam. In 1955 he abdicated in favour of his father, Norodom Suramarit, to avoid the limitations of his role as constitutional monarch and thus to play a more direct role in politics and government. He founded the Sangkum Reastr Niyum (Popular Socialist Community), which decisively defeated the Democrats at the polls. Prince Sihanouk also created a biannual National Congress, effectively a mass meeting at which the public was invited to present its complaints. Thus, the power of the elected National Assembly was steadily eroded, and the mass media were also increasingly subject to state control and repression. Confronted by severe economic problems after 1966, Sihanouk resorted to arbitrary arrests and to some public executions, especially of 'pro-American' Thanists, who were known in this period as *Khmer Serei*. The political repression and economic disorder that marred the late 1960s affected nearly every well-placed family. The appointment of an emergency Government of National Salvation in 1969, under the leadership of Lon Nol, resulted in a carefully orchestrated *coup d'état* in March 1970. The ostensible motive for the coup was Sihanouk's alleged collaboration with the Vietnamese communist revolutionaries who used Cambodian territory for sanctuary and who seemingly presented a new threat to Cambodia's independence. The organizers of the coup comprised remnants of the old Democrat and Khmer Renovation Parties. In October 1970 Lon Nol proclaimed the Khmer Republic, of which he was elected President in 1972. Thanh became Prime Minister once again briefly in 1972. The Khmer Republic administration was, however, rapidly overwhelmed by the corruption attendant upon a war economy.

Informed of the coup while on a diplomatic mission and convinced of the involvement of the US Central Intelligence Agency, Sihanouk formed an alliance with North Viet Nam and with an underground Marxist insurgency group, the Khmers Rouges, led by Saloth Sar, who later became known as Pol Pot. The Khmers Rouges had initiated an armed struggle against Sihanouk in 1968 and already had a guerrilla force of 3,000 by the time of the coup. Following organized mass demonstrations opposing the coup, there were fears in Saigon and Washington that Lon Nol's administration might rapidly be overthrown. A joint US-South Vietnamese invasion of some 50,000 troops, officially for the purpose of clearing communist Vietnamese forces out of their Cambodian sanctuaries, served only to drive an estimated 30,000 Vietnamese revolutionary troops deeper into the country, where they systematically assisted the Khmers Rouges in raising support and troops for their new, Sihanouk-led United National Front of Cambodia—FUNC. Despite massive US military and economic assistance, the armed forces of Lon Nol were effectively defeated by the end of 1972, perhaps partly because of the unpopularity of the South Vietnamese occupation of parts of eastern Cambodia until 1973. In a desperate attempt to redress the military balance in favour of Lon Nol, the US Air Force engaged in nine months of round-the-clock saturation bombings in 1973 until the US Congress terminated funding. Between 1973 and 1975, confident of victory, the Khmers Rouges gradually assumed control of ministerial portfolios in Sihanouk's Royal Government of National Union of Cambodia (in exile) and put pressure on Vietnamese armed forces and advisers to leave Cambodia. Party cadres who were judged to be too loyal to the Vietnamese or to their revolutionary traditions and ideology were secretly purged; many were killed rather than demoted. By early 1975 Phnom-Penh was completely isolated from all overland and river communications and was dependent on US airlifts. Lon Nol was flown to Hawaii on 1 April 1975, just in advance of a revolutionary occupation of the capital on 17 April. In the following weeks, the entire populations of Phnom-Penh and other refugee-swollen cities were evacuated and resettled in rural areas in agricultural collectives under Khmer Rouge control. Over the next three years an estimated 1.7m. people died as a result of hard labour, inadequate food and medical supplies, harsh treatment and executions. The Khmers Rouges' campaign to transform Cambodia so rapidly was partly stimulated by fears that Viet Nam and the USA would not respect Cambodia's right to independence, especially its right to an 'independent' socialist revolution.

Although he returned to Phnom-Penh, Sihanouk was rapidly eclipsed in the post-war revolutionary turmoil. A new Constitution renamed the state 'Democratic Kampuchea' (DK). National elections were held in March 1976 for a legislative body, the People's Representative Assembly, with the franchise restricted to full member-supporters of the state collectives (approximately one-half of the adult population). In a typical gesture, calculated to initiate bargaining over terms and conditions, Sihanouk declined to serve as Head of State in April. The Assembly, however, elected Khieu Samphan to the Chairmanship of the State Presidium, while the relatively unknown Pol Pot was named Prime Minister. In September 1977, as border conflicts with Viet Nam increased, Pol Pot revealed that the ruling organization was the Communist Party of Kampuchea (CPK). His revised version of party history eliminated all reference to early Vietnamese involvement in the Cambodian communist movement, a clear indication that international solidarity had been permanently ruptured. Social tensions, arising from catastrophic shortfalls in production and the outbreak of full-scale border war with Viet Nam at the end of 1977, provoked extensive purges inside the CPK and emergency attempts to reorganize rural collectives, in an attempt to support the armed forces. Diplomatic relations with Viet Nam were severed on 31 December 1977, after an aborted invasion attempt appeared to confirm CPK fears that Viet Nam intended to incorporate the Kampuchean revolution into a Vietnamese-dominated, communist federation of Indo-China. During the following year relations between the two countries continued to deteriorate. Viet Nam feared that the harsh conditions and instability in Cambodia would make the fiercely independent but increasingly beleaguered CPK dependent upon and vulnerable to the will of the People's Republic of China. On 25 December 1978 Viet Nam invaded Democratic Kampuchea, with a force estimated at more than 200,000 men, supported by the newly-formed Kampuchean National United Front for National Salvation (KNUFNS, now United Front for the Construction and Defence of the Kampuchean Fatherland—UFCDKF) led by Heng Samrin and comprising CPK dissidents. The rebel forces occupied Phnom-Penh on 7 January 1979. By the beginning of 1980, the defeated Democratic Kampuchean army, which numbered 70,000 on the eve of the assault, had been forced to retreat into remote mountain redoubts along the frontier, with troops estimated at fewer than 30,000.

ESTABLISHMENT OF THE PEOPLE'S REPUBLIC OF KAMPUCHEA

The Vietnamese-supported People's Republic of Kampuchea (PRK) failed to secure widespread international recognition, despite its efforts to portray its installation as the result of an indigenously supported revolutionary uprising against the horrendous violations of human rights by the Democratic Kampuchean Government. The People's Republic of China and members of the Association of South East Asian Nations (ASEAN) viewed Viet Nam's intervention in Kampuchea as another manifestation of traditional Vietnamese expansionism, which constituted a threat to their own security. The USA responded to the invasion by strengthening its economic embargoes on aid and trade with Indo-China, persuading Japan and most member countries of the European Community (now European Union—EU) to join the embargo. Between 1979 and 1981 the dislodged Pol Pot Government continued to be recognized by the UN in view of Viet Nam's open violation of the UN Charter. From 1982 the Government of Democratic Kampuchea had taken the form of a coalition Government-in-exile comprising the Party of Democratic Kampuchea (PDK—the CPK was officially disbanded in 1981), a royalist movement known as FUNCINPEC (a French acronym for United National Front for an Independent, Neutral, Peaceful and Co-operative Cambodia), led by Sihanouk, and the Khmer People's National Liberation Front (KPNLF), led by a former Prime Minister under Sihanouk, Son Sann. An anti-communist republican movement, the KPNLF embraced many important personalities from the Khmer Republic regime and the old Democrat Party. Sihanouk agreed to serve as President in the new Coalition Government of Democratic Kampuchea (CGDK). Although united by their opposition to the Vietnamese occupation of Cambodia and to the communist Government headed by Heng Samrin, which had been installed in Phnom-Penh, the parties that formed the CGDK were unable to function as a political alliance in view of their mutually hostile political visions and aspirations. Each of the three parties fielded its own army, and periodic attempts by the foreign supporters of the CGDK to encourage more than sporadic military co-operation were largely unsuccessful. China supplied nearly all of the weapons required by each of the three armies, while smaller amounts of military, humanitarian and 'non-lethal' aid were supplied to the non-communist FUNCINPEC and the KPNLF by the ASEAN countries, the USA, France and the United Kingdom. The USSR, its allies in Eastern Europe and Cuba supplied and financed the PRK.

By 1988 fighting between the CGDK and the PRK armed forces was of low intensity. The stalemate extended to the political arena. Although Viet Nam formally ignored UN resolutions appealing for a full and unconditional withdrawal and steadfastly rejected appeals from the CGDK and ASEAN for a negotiated end to the occupation, it tacitly responded to international criticism after 1983 by reducing its troop levels during annual rotations. Viet Nam's determination to 'rescue' the revolution in Cambodia and forcibly to restore solidarity between the Cambodian and Vietnamese revolutions was disrupted only by developments in other parts of the communist world. Following reductions in military and economic aid from the USSR in 1987, rapid Sino-Soviet *rapprochement* in 1988 and insistent Soviet pressure to seek a settlement in Cambodia based on 'national reconciliation' and the restoration of the monarchy, Viet Nam made increasingly firm unilateral pledges to withdraw all of its forces from Cambodia by the end of September 1989. Initially, the pledges were made conditional upon Thailand, China and other powers agreeing to withhold their military aid to the Cambodian resistance armies (for Viet Nam refused to accept that it had any unique responsibility to withdraw). Adding its terms to the effort to obtain a 'partial' or 'external' solution, as it was known, the leadership of the ruling Kampuchean People's Revolutionary Party (KPRP) in Phnom-Penh issued an appeal for 'national reconciliation', calling for the formation of a broad coalition government embracing all nationalist forces. The appeal made clear that the envisaged coalition would be guided by the KPRP communists and be based on the legal and administrative framework that had been established in Phnom-Penh in 1979. The KPRP also proposed to offer a senior position in the PRK to Sihanouk, with whom informal talks were arranged. Both Viet Nam and the KPRP continued to oppose any political role for the PDK, and to reject the involvement of the UN in the monitoring of the Vietnamese troop withdrawal.

Mindful of the need to widen its social base, the KPRP announced in 1988 that it would reform its state-controlled economy and proceeded to award managerial autonomy to nationalized industries. In 1989 peasants were informed that traditional usufruct titles to land would be reintroduced and tenants in state-supplied housing were promised property deeds, an act which simultaneously dispossessed pre-1975 owners of urban housing stock. In April 1989 the KPRP-controlled National Assembly voted to change the official name of the PRK to the State of Cambodia (SOC), thus removing the communist designation for a democratic republic. Sihanouk, who had requested the change as a concessionary gesture, refused nevertheless to join the reformed SOC, and continued to urge the creation of an interim, quadripartite coalition government to replace both the CGDK and SOC governmental frameworks, UN supervision of the Vietnamese troop withdrawal, a cease-fire and UN-supervised, national elections. In spite of SOC reluctance to form a coalition, the collapse of communist power elsewhere in the world encouraged Sihanouk as well as France to believe that the situation in Cambodia was evolving towards a settlement. Thus France, aided by Indonesia (acting on behalf of the ASEAN countries), convened an international conference on Cambodia in Paris on 30 July. By this time, the USSR had lent its approval to proposals for an interim, quadripartite coalition government to be led by Sihanouk, but in Paris the delegations representing Viet Nam, Laos and the KPRP continued to oppose an all-party interim government, advocating instead a coalition of 'national reconciliation' excluding the PDK, which was denounced as 'genocidal' and 'anti-national'. They also rejected all proposals for UN intervention in Cambodia, accusing the UN of being politically biased. The communist states also insisted that the SOC deserved legal recognition because it controlled most of the population and administered most of Cambodia's territory. It was further asserted that the SOC alone possessed the sovereign rights to negotiate a solution to Cambodia's internal conflict and to organize national elections.

It rapidly became clear that the allied Indo-Chinese communist parties sought a 'partial' solution focusing on the cessation of external weapons supplies to Cambodia and the establishment of a coalition government that would further weaken and divide the resistance coalition while enhancing the nationalist appeal and legitimacy of the KPRP. Such proposals failed to address the security concerns of China or of the ASEAN countries. The regional supporters of the CGDK sought, if possible, to eradicate Vietnamese influence in Cambodia or, at least, to ensure the restoration of a neutral, Sihanoukist government in which the PDK exercised a clear role as a form of insurance against renewed Vietnamese intervention. Unable to bridge such enormous differences, the Paris Conference deliberations were suspended after one month. One obstacle to the success of the Conference was the failure of the major powers to present a unified view to the Cambodian parties. US delegates reinforced the tendency of each of the two Cambodian sides to resist compromise, for while the USA accepted that peace required awarding political roles to all four parties, its diplomats made no secret of their personal preference for a solution excluding the PDK, as demanded by the SOC. Despite the diplomatic stalemate and encouraged by the open US hostility towards the PDK, Viet Nam proceeded unilaterally to withdraw its regular armed forces from the country at the end of September. Its appeals for reciprocal withdrawals of foreign military support for other Cambodian parties were discounted by foreign intelligence reports noting that hundreds and perhaps thousands of Vietnamese cadres remained in Cambodia in advisory roles and that as many as 3,000 rapid-intervention troops were reintroduced to Cambodian battlefields within a matter of weeks, when fighting between the two Cambodian sides intensified.

As anticipated by Viet Nam, international concern to resolve the Cambodian crisis intensified following the withdrawal of its regular forces. The retreat also prompted a dramatic increase in fighting in several parts of Cambodia, but especially in the west, with the National Army of Democratic Kampuchea (NADK), the army of the PDK, making the most significant advances. This raised the spectre of a military outcome, which no one other than the PDK wanted. Responding to the Cambodian failure to agree on how to form an interim coalition government, a US Congressman, Stephen Solarz, proposed the establishment of an interim UN administration in Cambodia, an idea that circumvented debate about which Government, the SOC or the CGDK, should form the basis for an interim state authority. Solarz also favoured a UN role as a means of promoting human rights education in Cambodia. The Australian Government, acting on Solarz's proposals, undertook the difficult task of persuading Viet Nam, as well as the KPRP, that support for a UN administrative role, UN monitoring of a cease-fire and UN-supervised elections were not incompatible with KPRP desires to retain power and were perhaps the best options for the SOC and Viet Nam to bring an end to their international isolation, which was increasing as a result of the loss of Soviet support. Cautiously, Viet Nam in January 1990 and the KPRP in February agreed to a 'limited' UN role. Concurrently, at the initiative of the USA, the five permanent members of the UN Security Council began monthly meetings to establish a mutually acceptable and practical framework for a settlement which could then be recommended to the Cambodian parties. Since the USA and the USSR had already agreed on the desirability of a non-communist, neutral administration led by Sihanouk, the concern in the first half of 1990 was to persuade China to abandon its continuing diplomatic and military support for the PDK, which remained the principal obstacle to a settlement of the conflict.

The comprehensive political settlement negotiated by the permanent members of the Security Council, which was finally presented in September 1990, envisaged: free and fair elections to be conducted under direct UN administration; the verified withdrawal of foreign forces; the cessation of all military assistance to Cambodia; the repatriation of refugees and displaced persons from Thailand under the auspices of the office of the UN High Commissioner for Refugees (UNHCR); the rehabilitation and reconstruction of Cambodia's economy; the formation of a Supreme National Council (SNC) by the four Cambodian parties (none being treated as a government); and the creation of a UN Transitional Authority in Cambodia (UNTAC), which would have special powers of administration and supervision during a transitional period and which would be headed by a Special Representative of the UN Secretary-General. The SNC would represent Cambodian sovereignty externally during a transitional period and would occupy Cambodia's seat in the UN General Assembly; UNTAC was to have powers of control or supervision over wide areas of national government (especially all agencies responsible for defence, internal security, finance and public information), primarily for the purpose of creating a neutral political environment for the holding of free elections. The UN plan envisaged a substantial reduction in the power wielded by the existing administrative structures, especially the SOC ministries in Phnom-Penh. Sihanouk, the President of the resistance coalition, and Hun Sen, the Chairman of the Council of Ministers of the SOC, had informally agreed in June to an equal division of the SNC seats between the two rival Governments (rather than among the four parties). Although this was initially unacceptable to the PDK, which preferred equal representation for each party, as was in fact proposed by the UN, the four parties agreed in September to a 12-member SNC comprising six representatives from the KPRP, two from FUNCINPEC, two from the KPNLF and two from the PDK. It was further agreed that SNC decisions would be taken by consensus, a procedure that awarded each party the power of veto. These compromises were influenced by Chinese diplomats, who quietly informed the PDK, the KPNLF and FUNCINPEC that all military aid to their armies would be gradually reduced and then cease altogether perhaps from the beginning of 1991, and by Soviet diplomats, who reciprocated by advising Viet Nam and Phnom-Penh, to which aid had effectively ceased, to abandon demands for the exclusion of the PDK from a settlement.

With each side none the less fiercely determined to secure the political advantages of holding office, the first meeting of the SNC in September 1990 finished acrimoniously with no agreement on who should assume the chairmanship of the SNC or who would represent it in the UN General Assembly. The permanent members of the UN Security Council intervened by proposing that Sihanouk head the SNC, a proposal initially unacceptable to the SOC side, which had been seeking joint chairmanship, rotating chairmanship or the compensatory appointment of Hun Sun as Sihanouk's deputy, together with an additional, compensatory seat for the SOC to rebalance the numbers at seven members each. The SOC also raised objections to proposals for the full disarmament of all four Cambodian armies, insisting upon the need to have access to weapons both during and after the proposed transitional period as a guarantee against attempts by the PDK (whose troops and weapons caches might escape UN monitoring) to seize power. Still seeking to portray the UN role as indirect diplomatic recognition, the KPRP also objected to UN plans to exercise control over its ministries. Neither the SOC nor the resistance coalition was reassured when in December the UN Secretary-General's Special Representative for Cambodia explained that the UN's draft agreement for a settlement was based on 'full respect' for Cambodia's existing administrative structures and that the UN accepted that there would be 'three categories of entities exercising powers' during the transitional period: the SNC; UNTAC, which would assume control of all administrative agencies concerned with foreign affairs, national defence, finance, public security and information; and the existing administrative structures in the zones controlled by the four parties, which would continue to function either under UNTAC control or supervision or with no control or supervision. With each side seeking to promote its claim for leadership of the SNC and both anxious (for opposed reasons) about the administrative influence of the KPRP state apparatus on the population, diplomatic progress ceased in early 1991. Military activity intensified during the dry season, rapidly increasing the number of refugees housed in holding centres in Thailand. The UN Secretary-General, Javier Pérez de Cuéllar, appealed in April for a cease-fire.

The diplomatic impasse was broken by Sihanouk, who was irritated by the unproductive informal talks that had resumed in Jakarta; fearful that neglect of the UN plan and pleas for a cease-fire would result in the abandonment of efforts to find a solution to the Cambodian problem, he was also prompted by 'rumours' that China would terminate all supplies of weapons to the PDK, KPNLF and FUNCINPEC armies as early as September 1991. In June Sihanouk announced the resignation of one of the FUNCINPEC representatives in the newly formed, but inoperative, SNC, and appointed himself to the empty seat as a 'simple member'. He then convened and presided, unappointed, over a meeting of the SNC held in Pattaya, Thailand—an effective and logical act of usurpation when it is recalled that each side sought prestige from association with the monarchy and could ill afford to oppose the Prince. At the June SNC meeting in Pattaya, and two others subsequently held in Beijing (in July) and in Pattaya (in August), delegates representing the four Cambodian parties finally agreed that Sihanouk would assume the chairmanship of the SNC, that the SNC would be based in Phnom-Penh and that it would commence functioning in November. The accession of the Prince to the chairmanship of the SNC was achieved without compensation to the SOC because China, in a clear change of foreign policy, concurrently ceased to extend diplomatic recognition to the resistance Government and began to accord equal recognition to all Cambodian parties, and their representatives, specifically to Hun Sen, head of the SOC delegation in the SNC. Adding to the momentum, China also reportedly conceded secretly to Viet Nam that military demobilization, fixed at 100% in the UN draft plan, did not have to be total. The Cambodian parties then agreed at their second meeting in Pattaya to a mutual reduction of 70% in their force levels and weapons stocks, a compromise forced on the PDK by Thailand. The guerrilla-based PDK favoured 100% disarmament, as originally recommended in the UN draft, for this would have effectively destroyed the mechanized SOC army; the NADK would not be so radically affected and remobilization of their forces posed comparatively fewer problems. Whatever PDK intentions were, the Thai Premier informed Khieu Samphan, the leader of the PDK and a PDK member of the SNC, that Thailand might cease selling and transporting food and other supplies to guerrilla bases if there were no compromise. The final significant dispute among the parties involved the modalities of the free and fair elections envisaged by the UN. Under a compromise negotiated within the SNC in September, the SOC abandoned its demands for single-member constituencies with simple plurality elections, a formula favouring the largest, nationally-organized parties, and agreed to a system of proportional representation for each of 21 constituencies, these being the existing SOC provinces (19) and municipalities (two). These issues having been resolved and the UN draft amended accordingly, comprehensive political agreements and treaties were formally signed by the UN Secretary-General, the four Cambodian parties, Viet Nam and 17 other states at a reconvened Paris Conference on 23 October. Although no announcements were made, China halted weapons shipments to its three former allies; Viet Nam, equally, recalled its advisers and ceased to intervene militarily in Cambodia.

IMPLEMENTATION OF THE 1991 PEACE ACCORDS

UN intervention in Phnom-Penh accelerated the liberalizing trends which the KPRP had set in motion in 1988, but also exposed, albeit inadvertently, the profoundly illiberal and authoritarian character of Cambodian politics. Anticipating the arrival of the UN, at an extraordinary party congress the KPRP formally abandoned its one-party state on 18 October 1991, while announcing its support for the creation of a multi-party democracy as the means for securing national reconciliation. The need for an electoral vehicle, rather than a vanguard party, was met by two additional significant developments: the Congress changed the name of the KPRP to the Cambodian People's Party (CPP) and abandoned its Marxist-Leninist ideology. The veteran leader of the KPRP, Heng Samrin, was retired to an honorary role in the party, and replaced as Chairman of the Central Committee by the lesser-known Chea Sim, who named the youthful Hun Sen as his deputy and principal party spokesman. Revitalizing the party's invitation to Sihanouk, Hun Sen declared his support for Sihanouk as an elected President in a new constitutional order to be elaborated following the planned national elections. Sihanouk returned to Phnom-Penh in November to establish the SNC in its national headquarters and negotiated a plan for co-operation between the CPP and FUNCINPEC with Prince Norodom Ranariddh, Sihanouk's son and the leader of FUNCINPEC. The vaguely worded agreement was immediately widely criticized as the basis for a power-sharing arrangement following the elections, and was therefore renounced in December. The friendly approach to FUNCINPEC from the CPP served to emphasize its continuing opposition to a political role for the PDK. Khieu Samphan's return to Phnom-Penh in November provoked a violent demonstration (widely believed to have been orchestrated by the CPP). The villa that housed the PDK delegation to the SNC was besieged and forcibly entered by the demonstrators, and Khieu Samphan and Son Sen were forced to flee to Thailand. A third SNC meeting in Pattaya, at which security arrangements for the PDK were discussed, finally permitted the SNC to begin functioning in Phnom-Penh at the end of December.

In the mean time, civic order in Phnom-Penh had collapsed. Former state employees, dispossessed of their jobs as a result of SOC 'privatizations', began to picket their former workplaces; people dispossessed of their assigned, cheaply leased housing as the result of the now lawful, but often corrupt, sale of state-owned buildings, or evicted from squatter settlements, demonstrated, demanding compensation and new homes. Students demonstrated against corruption amongst senior officials. Public order was only slowly restored by the imposition of a curfew, the assassination of one prominent critic, the use of armed police and security services and the arrest and detention of a large number of demonstrators.

Cambodian expectations of help and protection from the UN could not be satisfied by UNTAC. For its time, UNTAC was the largest multi-functional mission ever attempted, involving 16,000 troops, 3,600 civilian police, 2,400 civilian administrators and approximately 5,000 local employees, giving rise to delays linked to fund-raising (US $1,800m., excluding the costs of repatriation, which were raised separately by the UNHCR) and the recruitment of appropriate personnel. Yasushi Akashi, the most senior Japanese diplomat at the UN, was named Special Representative of the UN Secretary-General in Cambodia and Head of UNTAC, partly to encourage important financial and personnel contributions from Japan. Maj.-Gen. John Sanderson, an Australian, was chosen to head UNTAC's crucially important military component. Unfortunately, UNTAC was not formally established in Phnom-Penh until mid-March 1992, and then with only limited staff. By this time the cease-fire agreed in October had collapsed, for several reasons. Realizing that the arrival of the UN troops would be delayed, the four factions were determined to secure further territorial gains. The PDK refused to comply with the peace process, and in April the UN condemned the party for its failure to co-operate and particularly for its refusal to allow UNTAC officials free access to PDK-controlled territory. The PDK army, the NADK, engaged the CPP army in Kampong Thum, accusing the 'Vietnamese' forces of initiating the offensive. Concurrently, PDK spokesmen accused UNTAC of ignoring its responsibilities under the Paris agreement to control and supervise the withdrawal of all 'foreign forces', claiming the existence of thousands of concealed Vietnamese troops in Cambodia. By June the PDK announced that its army would not regroup or disarm, in compliance with the demands of the incomplete military contingent of UNTAC. The PDK campaigned for more powers of government, supervision and control to be given to the SNC. Sihanouk, Akashi and all other members of the SNC opposed the PDK proposals on the grounds that the Paris agreement could not be renegotiated. The PDK strongly criticized provisions of the election law introduced by Akashi, which permitted Vietnamese residents to vote providing that the intending voter was born in Cambodia with at least one parent who was also born in Cambodia, or, wherever born, able to prove that at least one parent was a Cambodian person by the place of birth principle.

As the election campaign advanced, a total of 20 parties met UNTAC requirements for a place on the ballot paper. Most of the parties were poorly organized; the majority were vehicles for prominent individuals who in some cases had returned from long periods of exile in the USA, France or Switzerland. Excluding the parties forged during the 1978–91 war, namely the CPP, FUNCINPEC (which altered its title to the FUNCINPEC Party when adopting political status in February 1992) and the political party formed by the KPNLF, the Buddhist Liberal Democratic Party (BLDP) led by Son Sann, the parties lacked the organizational capacity as well as the material means essential for campaigning effectively in every province. Difficulties of access to voters were further compounded by the SOC's obstruction of physical displacement, including private air flights, and SOC refusal to allow access to state-controlled radio and television stations or printing facilities to any party but the CPP, in clear violation of the Paris agreement. UNTAC quickly installed its own radio station in order to educate the Cambodian public about free elections.

The election campaign focused increasingly on issues of war, peace and national survival. The CPP leader, Hun Sen, questioned whether any other party had sufficient 'forces' to govern the country or to oppose Pol Pot. He pledged that the CPP would outlaw the PDK and defeat its forces militarily. The BLDP, together with the majority of parties formed on the basis of bonds of personal loyalty, judged the major issues to be the need to defend Cambodia from Vietnamese immigration and annexation and from parties who worked with and, allegedly, for the Vietnamese, specifically the CPP. To achieve this, it was argued, unity rather than conflict among Cambodians was essential. With their claims to office under open challenge, provincial CPP leaders systematically bullied and threatened anti-CPP party workers and candidates, discouraging many from campaigning; among those who persisted, a large number were murdered. The climate of fear and violence was neutralized in more populous provinces and cities by greater security and transparency in the campaigning and a rising sense of anticipation. Even though they were not competing for votes, the PDK brutally encouraged the rising tides of anti-Vietnamese nationalism by massacring Vietnamese civilians in long-established fishing communities on the Tonlé Sap, provoking a mass exodus of more than 20,000 Cambodian Vietnamese to Viet Nam in April–May 1993. Ranariddh, leading the FUNCINPEC Party's campaign, warned that a CPP election victory would result in renewed civil war between the CPP and the PDK. A FUNCINPEC victory, he stressed, would not return the PDK to power, but would end the war, since a government chosen by the Cambodian people and composed entirely of Cambodians would eliminate the PDK pretext for continuing to fight the Vietnamese. Ranariddh also promised that his father, Sihanouk, would deal with the problem of Vietnamese immigration; the FUNCINPEC Party's policy on immigration was described summarily as 'non-racist'.

Despite an increase in fighting between Phnom-Penh forces and the PDK, UNTAC's voter registration campaign was extremely successful; by the end of the process in February 1993 4.7m. Cambodians (constituting 97% of the estimated eligible electorate) had been registered. The repatriation of refugees from camps on the Thai border also proceeded on schedule; all 360,000 had been returned to Cambodia by the end of April. Despite fears of PDK assaults against polling stations, 89.6% of all eligible registered voters participated in the elections, which took place on 23–28 May. The PDK, which apparently lacked the capability to disrupt the polls systematically, instead offered its support to the FUNCINPEC Party in the hope of securing a role in government following the election. Early indications of a FUNCINPEC victory in the election prompted allegations of electoral malpractice from the CPP. UNTAC, however, rejected CPP requests for fresh elections in at least four provinces. Hun Sen indicated that the CPP might not recognize the results of the elections, and SOC National Security Ministry officials allegedly prepared a putsch. Sihanouk, supposedly advised of the imminent coup as troops surrounded the Royal Palace, agreed to an interim FUNCINPEC-CPP coalition Government in talks with the CPP Chairman, Chea Sim. The proposal was vetoed, unexpectedly, by Ranariddh, who was out of the country. On 5 June the official results of the election were released; the FUNCINPEC Party secured 58 seats and the CPP 51. The CPP carried 11 of the 21 constituencies, with most of its votes coming from the smaller, rural provinces in which many opposition parties had failed to campaign. The BLDP finished a distant third, securing 10 seats, while the only other party to gain representation in the Constituent Assembly was MOLINAKA (National Liberation Movement of Cambodia, a breakaway faction from FUNCINPEC), which secured one seat. In June CPP militants in several eastern provinces bordering Viet Nam announced the creation of an 'autonomous' zone. The secessionists, who were nominally led by Prince Norodom Chakrapong, a son of Sihanouk's and a Vice-Chairman of the Council of Ministers in the SOC administration, expelled UNTAC officials and FUNCINPEC supporters from three provinces in their zone, a partition which Hun Sen and other CPP officials in Phnom-Penh publicly opposed, but secretly endorsed as a means to force Sihanouk and Ranariddh into a power-sharing coalition. Hun Sen arranged for the Constituent Assembly to vote special powers to Sihanouk at its inaugural meeting on 14 July 1993, prefiguring the award of a senior position to the Prince at a later date or, as the FUNCINPEC Party desired, the restoration of the monarchy. Sihanouk's spokesmen announced in July 1993 that Ranariddh and Hun Sen would be co-chairmen of the Provisional National Government of Cambodia, pending a new constitution. The secessionist movement collapsed on 17 July. Although the formation of an interim government and the resumption of powers by Sihanouk were outside the terms of the Paris agreement, UNTAC was powerless to intervene in these developments. In July and August, as Sihanouk attempted to arrange 'round-table' talks with the PDK, so as to end the partition of Cambodia and to complete the process of national reconciliation as understood in royalist terms, the USA continued to object to a role, advisory or otherwise, for the PDK in the coalition. US objections to any involvement of the PDK lent support to the CPP, which wished, not only to exclude the PDK from the coalition, but to secure international support for a resumption of war.

While Sihanouk's attempts to arrange a settlement between the PDK and CPP at meetings in 1993 and 1994 failed, the Prince exerted considerable influence in determining his role—and the role of future kings—in the new constitutional order. On 21 September 1993, the FUNCINPEC-CPP-controlled Constituent Assembly adopted a new Constitution with 139 articles. It was signed and promulgated by Sihanouk on 24 September (in his newly resumed, extra-legal role of 'Head of State' and as Chairman of the SNC, a legal but non-functioning entity). Under the provisions of the Paris peace agreement the Constituent Assembly became the National Assembly, and on the same day Sihanouk acceded to the throne of the new Kingdom of Cambodia. Articles 1–30 of the Constitution proclaimed the country a Kingdom and a 'multi-party liberal democracy'. The duties and responsibilities awarded to the King were those of a constitutional monarch, but Article 9 also specified that the King should serve in the role of 'referee to guarantee the normal functioning of public authorities'. The reigning King, according to the Constitution, should 'not have the power to appoint his successor' who was to be elected by a Royal Council of the Throne. The members of this Council were fixed by the Constitution; they were the Chairman of the National Assembly, the Prime Minister, the Supreme Patriarchs of the Mohanikay and Thammayut Buddhist sects and the First and Second Vice-Chairmen of the National Assembly. The promulgation of the Constitution coincided with the first public acknowledgements that Sihanouk was seriously ill with cancer. Following his enthronement, Sihanouk underwent two major courses of chemotherapy. Although prevented by law from identifying a crown prince, Sihanouk awarded royal titles to three sons, a half-brother and one relative from the Sisowath line, which indicated his preference in 1994 for Ranariddh to succeed him. Ranariddh's ability to do so remained dependent upon support from the CPP, and specifically from Chea Sim, who was re-elected Chairman of the National Assembly.

THE ROYAL GOVERNMENT OF CAMBODIA

At the end of October 1993 the National Assembly approved the composition of the new Royal Government of Cambodia (RGC), which had been endorsed by Sihanouk. Ranariddh was named First Prime Minister and Hun Sen Second Prime Minister. Paradoxically, the restoration of constitutional government to Cambodia in September 1993, which brought an end to UNTAC's mission, unleashed serious factional disputes within the ruling CPP and FUNCINPEC Party. One reason for this, as indicated above, was that the restoration of Sihanouk to the throne placed the King's personal agenda for the future in confrontation with the predominantly bureaucratic and militarist impulses within the CPP as well as anti-communist, technocratic tendencies within the royalist movement. After 25 years of political turmoil, Cambodia was no longer self-sufficient in cereal production, had little modern infrastructure, and supported excessively large and ill-disciplined security forces. The use of patronage to secure political support during the PRK era had also led to an oversized bureaucracy staffed by a highly politicized corps of civil servants. The parlous state of the national budget (which was supported by international aid donations amounting to approximately US $10m. a month in 1993) lent separate momentum to disagreements both within and between the two ruling parties over their response to the international donor community.

Disputes relating to policy and supremacy within the CPP revolved around Chea Sim, who represented the deeply authoritarian and traditional element of the party. Hun Sen, by contrast, spoke for a younger generation of cadres who had been recruited to the revolutionary cause during the 1970s. This younger generation was widely regarded as more pragmatic, liberal and at greater ease with market-led economic development.

In September 1993 Chea Sim indicated a desire to remove Hun Sen from the Government in advance of the promulgation of the new Constitution by criticizing him for not having made 'sacrifices' for the party. With FUNCINPEC support, Chea Sim succeeded in being elected Chairman of the National Assembly, thereby dislodging the BLDP leader, Son Sann, who had held the post during the period of provisional national government (which extended from June to October 1993). In this new role Chea Sim ignored Sihanouk's initiatives, defended the corporate interests of the army and of the state administration (which were broadly indistinguishable from the organizational interests of the CPP) and urged uncompromising policies towards the PDK. The most significant such policy was the adoption in July 1994 of a law that declared the PDK to be 'an illegal and criminal group', thereby proscribing the party.

Despite its impressive electoral success, the FUNCINPEC Party, in contrast to the CPP, lacked a nation-wide organization of disciplined cadres and supporters. At a ministerial level, however, FUNCINPEC was stronger, and during 1993–94 displayed a pronounced technocratic orientation. Serious differences of opinion on how best to resolve the PDK issue continued to divide party intellectuals; with the PDK being in control of 5%–10% of the population and around 10% of national territory, the issue had become one of partitioning the country. The conciliatory position of Sihanouk towards the PDK commanded most support, and there was a belief that if he failed to persuade the PDK to surrender, they could be ignored, on the assumption that they posed no military threat to the cities or to the RGC. The Minister of Finance, Sam Rainsy, a FUNCINPEC member of the National Assembly, judged the traditional, incorporative position of the King as desirable if not essential, but regarded economic growth, rather than political unity, as the key to the Government's stability. As a former banker and advocate of free-market development, Rainsy believed in co-operation with foreign donors in the drafting of reconstruction and stabilization plans. Rainsy attempted to review business contracts negotiated previously by SOC officials and tried, unsuccessfully, to introduce an independent system of assessing customs duties on goods. In March 1994 Ranariddh and Hun Sen jointly proposed that Sihanouk remove Rainsy from the Ministry of Finance. The two were drawn together through their common interest in protecting party patronage from the challenges posed by Rainsy. Sihanouk withheld his consent, however, due to Rainsy's evident national popularity (corruption, against which he campaigned, had become a major public issue) and his success in gaining the confidence of donor countries and multilateral agencies upon whom the country depended. The second session of the National Assembly, which convened shortly thereafter, was quickly suspended following disputes over vacancies in the Assembly caused by requests from leaders of the 1993 secessionist movement, and because certain pieces of draft legislation were unfinished.

In June 1994 the two Prime Ministers transferred control of timber exports (then legally proscribed) to the Ministry of National Defence in order to supplement its 1994 budget allocation. The decision, which was undertaken without consultation, undermined IMF attempts to restore fiscal rationality to the budget (see Economy). By this time Ranariddh had begun openly to support the CPP election promise to ban the PDK. Hun Sen publicly challenged Sihanouk's renewed willingness to accommodate PDK demands for a role in government, and his stated desire to assume power if the political system collapsed, by asking him not to denigrate the Constitution. Once introduced in the National Assembly, the law to ban the PDK was initially opposed by a group of approximately 15 legislators, led by Rainsy, on the grounds that it would engender more civil strife and threaten civil liberties. Most members of the National Assembly, including Rainsy, subsequently voted for the law, however, after amendments were agreed safeguarding citizens from possibly false accusations and intimidation. In spite of Sihanouk's indication that he would not sign the measure, as it amounted to a rebuff to his hopes for a government of national unity (he was undergoing chemotherapy in China during the vote), the proposed legislation was quickly signed into law, eventually with Sihanouk's assent, by the acting Head of State, Chea Sim.

Political co-operation between Ranariddh and Hun Sen grew considerably from July 1994, following a coup attempt allegedly instigated by Chakrapong. With the support of ex-FUNCINPEC army generals and units loyal to Ranariddh, Hun Sen succeeded in arresting Chakrapong, who protested his innocence and appealed successfully to his father, Sihanouk, to be allowed to go into exile. Sihanouk announced that he would no longer intervene in the affairs of the RGC, adding that he believed reconciliation between the Government and the outlawed PDK (which he had been attempting to promote) had become impossible. Former CPP general Sin Song was also arrested, but escaped to Thailand in September before he could be brought to trial. The Government used the incident to stifle criticism in the national press. A new press law, adopted by the National Assembly in July 1995, made defamation a criminal offence and codified governmental rights to suspend publication of newspapers that carried articles deemed disruptive of 'national security' and 'political stability'.

The adoption of the law proscribing the PDK, the political eclipse of Sihanouk, and the clamp-down on the press signalled an end to the era of political transition and realignment promoted by the UN in 1992 and 1993. Having neutralized the role of Sihanouk, the two Prime Ministers finally succeeded in removing Rainsy from the Ministry of Finance and from the Cabinet during a government reshuffle in October 1994. However, Rainsy continued to criticize government policies from the floor of the Assembly and, in retaliation, he was expelled from the FUNCINPEC Party in May 1995 and excluded from his FUNCINPEC seat in the National Assembly in June. Ranariddh, who had made a point of refraining from public disagreements with his CPP and BLDP coalition partners throughout 1994, claimed that it was 'discipline' rather than democracy that was lacking in Cambodia.

The fissiparous and fractious tendencies within the two major parties were reflected in the demise of most other political parties in 1995, the direct consequence of the centralizing, undemocratic consolidation of power in the hands of the leaders of the CPP-FUNCINPEC coalition. In July 1995 Son Sann, the President of the BLDP, and Ieng Muli, the party's Secretary-General and the Minister of Information and the Press in the coalition Government (and a long-standing rival of Son Sann), convened an extraordinary, unofficial congress of the party. Although the two leaders were ostensibly in dispute over the allocation of public appointments offered to the party (one list prepared by Son Sann had been disregarded by the Government and a shorter list prepared by Muli had been accepted), Muli claimed at the special congress that his strategic concern was party political: to determine decisively whether the BLDP was part of the RGC of which he was a member, or whether, as Son Sann believed, the party was part of the parliamentary opposition. The congress elected Muli's candidates as new party officers and passed a vote of 'no confidence' in four of the BLDP members of the National Assembly, including Son Sann. Contrary to expectation, the four were not immediately expelled from the Assembly, and retained their seats as of mid-1996, but a rival congress convened by Son Sann in September 1995 was initially banned and then disrupted by grenade attacks.

The formation in November 1995 of the Khmer Nation Party (KNP) by Rainsy, who renewed appeals for peace, social justice and the protection of national land and forests, was highly significant, most of all for challenging the organizational viability of Rainsy's former party. In rapid succession a planned FUNCINPEC party congress was postponed until March 1996, Prince Norodom Sirivudh, the party's General Secretary and a personal friend of Rainsy's, was arrested for allegedly expressing a wish to assassinate Hun Sen, and in December 1995, following international and domestic protests over political abuse of the legal system, another intervention from Sihanouk secured Sirivudh's release from prison and exile to France. The violent turn of events clearly alarmed the general public.

The arrest of Sirivudh, organized by Hun Sen, which ultimately deprived FUNCINPEC of its leading organizational personality at the same moment that it faced major defections to the KNP, disrupted the CPP-FUNCINPEC alliance, decisively, if not permanently. However, although the Second Prime Minister appeared to emerge from these events as accountable to no-one, his dominance and the leading role of the CPP did not remain unchallenged. In January 1996 Hun Sen insisted on the reintroduction of 7 January as a public holiday in commemoration of the Vietnamese 'liberation' of 1979. For supporters of the former non-communist resistance of 1979–91, which included FUNCINPEC, this holiday was an affront, representing a denial of Khmer rights to dignity, independence and autonomy, even though the regime installed by the Vietnamese in Phnom-Penh on 7 January 1979 replaced that of Pol Pot. Ranariddh retaliated in January 1996 by denouncing Vietnamese encroachments on Cambodian border territory as a 'full invasion'. The CPP-controlled judiciary proceeded in February with the trial *in absentia* of Sirivudh, who was found guilty of criminal conspiracy and of possession of unlicensed firearms and sentenced to 10 years' imprisonment. Recriminations were openly aired in March when Ranariddh accused the CPP of reneging on power-sharing agreements at district (*srok*) level. He also threatened to force early elections by leaving the ruling coalition if the CPP failed to hand over promised positions, which amounted to at least one-half of the headships and one-half of the first deputy headships. Hun Sen demanded a public apology from Ranariddh and in June 1996 instructed CPP provincial governors to ignore normal protocols and not to facilitate visits of the First Prime Minister to their provinces.

The dispute between the two Prime Ministers was communicated through the ranks of their parties, paralysing public administration, obstructing decisions relating to foreign investment and making it impossible to set an agenda for meetings of the National Assembly. Political and ideological tensions were exacerbated by an economic downturn, and especially by the suspension by the IMF of aid to the budget, the devaluation of the national currency in dollar exchanges and labour unrest (see Economy). Being heavily dependent upon informally secured revenues, the two leaders agreed upon a division of spoils in Pailin in anticipation of a government victory in an offensive against the PDK base there. The offensive failed, the morale of poorly paid government soldiers being sapped by the absence of unity and purpose in Phnom-Penh. Military reverses, together with the stalemate over the sharing out of district headships, heightened fears within FUNCINPEC about its prospects for success in both the approaching *khum* (sub-district or commune) headship elections and the legislative elections, the latter scheduled by the Constitution to be held no later than November 1998.

International consultants recommended that both elections be held concurrently in 1998 and that a National Election Commission to include NGOs be established. Confident of their administrative control at the basic level, CPP spokesmen were steadily seeking to limit international expectations of involvement in the elections, beyond that of supplying funding, technical assistance and a modest number of observers. CPP officials and local CPP-controlled security forces also forced the closure of many provincial party offices opened by the KNP in the second half of 1996, and refused KNP requests for permission to establish a radio station. Police harassment, and the assassination of several party officials, signalled a refusal to allow opposition parties or leaders to challenge the existing political configuration. Sam Rainsy, with tacit support from FUNCINPEC, emphasized the need for the widest possible international involvement in the forthcoming elections. In June Hun Sen protested his support for political pluralism and the formation of new parties, while denying that the CPP had any aspirations to integrate all parties into one and thereby restore the one-party state of the 1980s. From mid-1996 it was the official policy of the CPP to co-operate with other parties, including FUNCINPEC, on a 'no power-sharing' basis. While conflict between the CPP and FUNCINPEC increased, and the repression of the KNP continued, the CPP lent support to small or newly formed parties and also encouraged defections from the KNP and FUNCINPEC.

In the wake of military set-backs in 1995 and the substantial losses suffered as a result of defections and self-demobilizations in 1995–96, the PDK leadership was shaken by the near-loss of its economic capital at Pailin in April–May 1996. Deprived of Chinese military assistance in 1991, the Khmers Rouges had financed their armed struggle by selling logging and gem concessions to entrepreneurs in neighbouring Thailand and by purchasing weapons and ammunition in private markets. Inadequate or poor agricultural land in zones under the movement's control pushed many communities in the interior to compensate for revenue shortfalls by resorting to banditry, extortion and theft. The appropriation of goods and wealth and the coercive treatment of civilians undermined the historically good social relations between the PDK/NADK and the peasantry. War-weariness was compounded by Phnom-Penh's adoption of the law proscribing the movement in 1994; by January 1995 more than 7,000 fighters had taken advantage of an offer of amnesty made by the RGC. The PDK formed a Provisional Government of National Union and National Salvation of Cambodia (PGNUNSC) in July 1994 and, in early 1995, launched an assault on villages in the north-west, leaving more than 40,000 civilians temporarily displaced. Evidence from defectors and captured archives, however, revealed schisms within the movement arising from policy disputes focusing on various issues, including the Paris peace process in 1991–92, the movement's decision not to participate in the UNTAC election process, its desire to resume armed struggle for the purpose of restoring Democratic Kampuchea, and its decision to promote family production, reliance on free markets and regional autonomy in economic life. Appeals broadcast to the Government to engage in political talks received no public acknowledgement, but negotiators from FUNCINPEC, led by royalist Gen. Nhek Bun Chhay and assisted by Thai authorities, are known to have initiated secret contacts with several DK commanders based near the border with Thailand by no later than mid-1996.

Independent of the FUNCINPEC initiative, policy disputes within the PDK led younger military leaders in the movement's commercial regions of Pailin and Phnom Malai into open confrontation with senior civilian leaders in Anlong Veng. In early August 1996 the Pol Pot-controlled, clandestine radio denounced several field commanders and Ieng Sary, the former DK Minister of Foreign Affairs, as 'traitors' and ordered their immediate arrest. Rejecting the charges, Commanders Y Chhien at Pailin and Sok Pheap at Phnom Malai revealed that they had been unwilling to carry out instructions to recollectivize the economy, starting with the confiscations of means of transport. Once effectively expelled from the movement, the dissident Khmers Rouges indicated their willingness to recognize the authority of the Royal Government and their respect for the Constitution, but made clear their refusal either to surrender or to defect to Phnom-Penh. The Democratic National Union Movement (DNUM), founded by Ieng Sary, was quickly established as a vehicle for negotiating a union with the Government while resisting integration by either side and carefully asserting the political and territorial autonomy of the breakaway region. Subsequently, the DNUM and generals from FUNCINPEC and the CPP competed for brokering opportunities and political influence. Together, they ultimately garnered support from 11 other DK divisions and fronts during late 1996. At the joint request of the two Prime Ministers, a royal amnesty was granted to Ieng Sary for the death sentence given *in absentia* by a PRK tribunal in 1979 and also for criminal penalties arising from the 1994 law that outlawed the DK group. Breakaway troops associated with the DNUM were formally inducted into the command structure of the Royal Cambodian Armed Forces (RCAF) in November, although most refused reassignment. In official appointments and commissions announced in January 1997, Y Chhien was appointed Governor of Pailin, while retaining control of the lucrative gem and logging activities, the principal source of income for the DK movement after 1991. Although the CPP initially welcomed the dissolution of the DK movement, and acquired the political loyalty of some breakaway DK military commanders, the rekindling of resistance era comradeship among many dissident Khmers Rouges and pro-BLDP and FUNCINPEC military leaders, together with military confrontations between pro-FUNCINPEC and pro-CPP forces in Bat Dambang, unleashed suspicions that FUNCINPEC was aiming for a new political alliance and was even planning to seize control of the provinces of Bat Dambang, Banteay Mean Chey and Siem Reab by military force.

The competition for forces, land rights, and positions quickly renewed power struggles between the two Prime Ministers and their parties, and also fuelled constitutional tensions dividing Hun Sen from the King, whose desire to pardon Chakrapong, Sirivudh and other civil criminals was rebuffed. Stopping short of full repudiation of the coalition, senior CPP officials announced that the governing alliance with FUNCINPEC existed in theory only, and asserted that their party was in full control and would win most or all of the approaching 1,453 *khum* headship elections. In an attempt further to marginalize their royalist allies and opponents, many of whom, like Ranariddh and Sam Rainsy, held dual French and Cambodian nationality, the quinquennial congress of the CPP proposed to require candidates for all public offices in Cambodia, including *khum* headships, to hold Cambodian nationality only. FUNCINPEC officials responded by accusing the CPP of attempting to arrange a coup. FUNCINPEC also criticized the predominantly CPP-controlled media and the Ministry of Information for giving greater broadcast coverage to Hun Sen than to Prince Ranariddh. Royalist officials began overtly to supply military protection and political assistance to the KNP at the opening of its party offices in Bat Dambang, and to seek negotiations with the last of the insurgent, Pol Pot-led PDK forces in Anlong Veng.

Alongside the KNP, a new electoral National United Front (NUF) was established in February 1997, to which the BLDP Son Sann faction and the small Khmer Neutral Party quickly rallied. This revitalization of the historic nationalist front, which had propelled the royalists to power in the 1993 elections, coincided with the publication of an interview with King Sihanouk, in which he revealed his unhappiness with the way in which the governing parties had restored some respectability to the Khmers Rouges, his concerns about the country's future in view of the suspension of IMF assistance, and about deforestation. He also suggested that he might abdicate the throne, as he had done in 1955. Clearly alarmed by the prospect of the King usurping CPP state power, Hun Sen announced that he would cancel local and national elections if Sihanouk should abdicate. He added that if the King did not refrain from interfering in politics, he would seek to amend the Constitution to prohibit all members of the Royal family from participating in politics, thereby ensuring the neutrality of the constitutional monarchy. Prince Ranariddh, in response, stressed that the Constitution allowed no-one to cancel elections, reserving to the National Assembly alone the right to postpone or prolong electoral mandates by a two-thirds' majority. The King, though silenced, had successfully exposed the gradual realignment of the royalist movement as well as the authoritarian orientations of the Second Prime Minister. The FUNCINPEC-CPP talks of March 1997 produced little reconciliation and resulted in no renewal of their alliance.

The decline in public order in the second quarter of 1997 was rapid and seemingly irreversible, despite attempts made by a bipartisan FUNCINPEC-CPP Commission for Abnormal Conflict Resolution to uphold the neutrality of national policing and of the army. As the CPP began in earnest to form a new ruling alliance along Malaysian-style lines, it secured support from Ieng Muli of the BLDP, the Democrat Party (In Tam), the Free Development Republican Party (Ted Ngoy), the Khmer Citizens' Party (Nguon Soeur), and the LDP (Chhim Om Yon), and abandoned its demand for electoral candidates to be Cambodian nationals. Despite attempts to negotiate with FUNCINPEC over dates for the forthcoming elections, no agreement was reached. Meanwhile, Hun Sen lent support to a rebellion against Ranariddh's leadership of the FUNCINPEC party promoted by eight FUNCINPEC members of the National Assembly, thus beginning the process of the accumulation of the necessary parliamentary votes for the deposition of the First Prime Minister on a constitutional basis. A party congress organized by the rebels in June resulted in the formation of FUNCINPEC II and the election of Toan Chhay, a former resistance commander and the FUNCINPEC Governor of Siem Reap, as Chairman. At the same time, the attempt by the exiled Prince Sirivudh (still a seated FUNCINPEC legislator) to return to Phnom-Penh was blocked in Hong Kong. Equally concerned about political realignments and the possible disintegration of FUNCINPEC, PDK-Anlong Veng radio broadcasts urged public support for the Ranariddh-led NUF, even though the movement continued to detain 15 FUNCINPEC party negotiators who had been taken hostage in mid-February. DNUM leaders expressed fears that inter-party disputes were undermining aspirations for national reconciliation and peace: their former leader, Pol Pot, was held responsible. While indicating sympathy towards the NUF, however, DNUM leaders declined to join, reiterating their intentions to remain neutral (and autonomous). Ieng Sary nevertheless promised that his party would respect the outcome of future elections, saying that he would employ his forces to make peace if, as in 1993, violence should follow.

The Removal of Prince Ranariddh from Power

One final attempt at reconciling the personal and political disputes dividing the two Prime Ministers occurred in May 1997; within 24 hours, however, Prince Ranariddh had accused Hun Sen of planning to restore a communist dictatorship if the CPP won the elections. The Prince urged the dissolution of the Assembly and the holding of early elections but, as a means of delaying a confrontation in the Assembly, the FUNCINPEC Party General Secretary and acting Chairman of the National Assembly, Loy Simchheang, postponed conflict on the issue of the FUNCINPEC-Ranariddh proposal to expel renegade FUNCINPEC deputies from the Assembly via the procedural device of suspending steering committee meetings. Polarization of party politics was further accentuated at the end of May, when containers of weapons, destined for the First Prime Minister's 1,500-strong bodyguard unit, were seized by CPP officials, and when Ranariddh revealed that Khieu Samphan, the nominal leader of the PDK, had communicated to him a desire to return to mainstream Cambodian politics. Hun Sen sternly warned the Prince against entering into any alliance that would allow the return of the genocidal regime. Controversy over the treatment to be accorded to Pol Pot and the last of his close associates accelerated in early June, when speculation that the FUNCINPEC Party was on the verge of reaching an agreement with Khieu Samphan intensified. FUNCINPEC's senior military adviser, Gen. Nhek Bun Chhay, announced on 2 June that Pol Pot, Ta Mok (Pol Pot's Chief of Staff) and Son Sen (the former DK Commander-in-Chief) would go into voluntary exile in exchange for immunity from prosecution; it was indicated by FUNCINPEC that the exile of these individuals would constitute acceptable grounds for the return to the mainstream political arena of Khieu Samphan, with reports suggesting that there were plans for him to form a new political alliance with FUNCINPEC and Sam Rainsy's KNP. Within days, however, the likelihood of any such arrangement being made disappeared: Pol Pot—who, according to reports, had initially supported the deal with FUNCINPEC—apparently vetoed the agreement at the last moment, and then reportedly ordered the assassination of Son Sen, his wife and nine relatives in retaliation for Son Sen's suspected secret dealings with Hun Sen and a CPP spy network. The news of Son Sen's death was later confirmed by Ranariddh. Pol Pot then reportedly fled in a 10-vehicle convoy, which included Khieu Samphan. The convoy, however, was intercepted by troops wanting to defect to the Government, led by Ta Mok. The surrounding of Pol Pot by mutinous NADK soldiers gave rise to international, and some national, appeals for him to be handed over to the Cambodian Government and brought to justice. These appeals were supported by the two Prime Ministers, but produced no lessening of strife between the two coalition parties. Ranariddh reaffirmed his willingness, in principle, to accept the defection of nearly all Anlong Veng guerrillas, barring only Pol Pot and Ta Mok, and to welcome Khieu Samphan and his National Solidarity Party into the NUF provided that he received the necessary royal amnesty. Hun Sen, however, disregarding his previous negotiations with Ieng Sary, began insisting, *inter alia*, that negotiations with the DK were illegal.

In mid-June 1997, after denouncing the fleeing Pol Pot for acts of treason in an extraordinary public criticism of the former leader, a radio broadcast made on Anlong Veng radio in Khieu Samphan's name pledged the loyalty of the latter's National Solidarity Party to the FUNCINPEC-led NUF and urged all national forces to unite in a struggle against Hun Sen, stigmatized as a 'lackey' of Viet Nam. Within hours, the military bodyguard units of the two Prime Ministers and other high officials clashed on the streets of Phnom-Penh. Characterizing a personal meeting between Ranariddh and Khieu Samphan as an intolerable betrayal, Hun Sen then issued an ultimatum to the Prince giving him a few days in which to decide whether he wished to work with the coalition Government or with Khieu Samphan. The US State Department released a statement at this point, warning that it would be gravely concerned if senior PDK leaders were to be awarded roles in national politics or allowed to retain effective control over any territorial domain, but diplomatically avoided reference to the situation in Pailin. On 20 June Anlong Veng radio announced triumphantly that Pol Pot had surrendered. As the Consultative Group on Cambodia (CGC) met in Paris on 1–2 July, Hun Nheng, the CPP Governor of Kampong Thum province and a brother of Hun Sen, forcibly disarmed 70 of Prince Ranariddh's security guards as the Prince completed a tour of the province. As the DK defector Keo Pong, allegedly acting upon orders from the Second Prime Minister, positioned his troops for an assault on Gen. Nhek Bun Chhay's garrison, the Prince boarded a flight to France.

The airport and large parts of Phnom-Penh were subsequently cordoned off and looted by marauding troops during 4–6 July 1997, and the FUNCINPEC and KNP party offices were ransacked. Hun Sen denied that he was staging a coup or aiming to govern on his own; he insisted, however, that the First Prime Minister had to be replaced. He blamed the outbreak of violence in the capital on a criminal conspiracy mounted by Ranariddh together with the outlawed Khmers Rouges and accused Ranariddh of having broken the law by negotiating with the DK, by unlawfully smuggling ex-DK troops into Phnom-Penh to strengthen his own forces and by secretly importing weapons to arm those forces. Denouncing Ranariddh's real and alleged actions as criminal and unacceptable, Hun Sen ordered the two factions in FUNCINPEC to replace the Prince, and thereby to restore stability to the ruling FUNCINPEC-CPP coalition, protecting the 1993 Constitution and laying the basis for democratic elections in May 1998. Furthermore, Hun Sen revealed that he had already asked both the Co-Minister of Defence, Tie Chamrath, and the leader of FUNCINPEC II, Toan Chhay, if they would serve as First Prime Minister, but each had declined. Only three FUNCINPEC leaders—Nhek Bun Chhay, Chau Sambath and Serey Kosal—were unacceptable to Hun Sen. Troops loyal to the CPP and Keo Pong were ordered to locate and eliminate these three; other leading FUNCINPEC figures were also similarly named as targets. In total, approximately 40 people, including Sambath and Sok, were murdered or assassinated during the week beginning 4 July; tens of thousands of civilians associated with the royalist, democratic or human-rights movements and parties fled into temporary hiding, and approximately one-half of the FUNCINPEC party members in the legislature fled overseas, fearing for their lives. Hun Sen pledged that all FUNCINPEC ministers, legislators and cabinet officials who agreed to withdraw their support from Ranariddh as party leader and Prime Minister would be allowed to retain their positions; he also promised to amend the Constitution to create more positions, allowing individuals belonging to parties not represented in the National Assembly to hold government portfolios.

From France, Prince Ranariddh announced his intention to resist his expulsion from the Government, vowing to employ military force if necessary. Although five FUNCINPEC generals had been killed, Gen. Nhek Bun Chhay and others escaped, and proceeded to establish resistance bases in the north-west. At the UN and in Washington, Ranariddh continued to be recognized as Prime Minister. The US Government expressed strong opposition to the use of force to change the results of the 1993 election and effectively to rupture the Paris accords of 1991. Sam Rainsy, leader of the KNP but acting on behalf of the NUF, issued an appeal to the international community to suspend economic assistance to Cambodia, excluding essential humanitarian aid; Germany and the USA obliged, while Australia suspended military assistance. Both Sam Rainsy and Prince Ranariddh rushed to the Thai-Cambodian border to make contact with the more than 20,000 people fleeing the country, many of whom were their supporters, and quickly agreed with more than 20 temporarily exiled BLDP and FUNCINPEC legislators to establish a Union of Cambodian Democrats (UCD) for the purpose of restoring the legitimate Royal Cambodian Government by peaceful means. ASEAN members agreed at a ministerial conference on 10 July to postpone indefinitely Cambodia's accession to the organization, originally scheduled for 24 July, and formally requested Cambodia to take steps to preserve until the forthcoming elections the power-sharing arrangement agreed following the elections of 1993. ASEAN also agreed to send mediators to Beijing, Bangkok and Phnom-Penh in an attempt to facilitate a peaceful solution to the governmental and political crisis, despite Hun Sen's initial rejection of previous offers of mediation. Hun Sen's decision in mid-July to ask the FUNCINPEC Minister of Foreign Affairs, Ung Huot, to serve concurrently as First Prime Minister provoked international criticism and was widely rejected. In the first DNUM comment on the events of 5–6 July, and as concern grew over the impact of sustained civil strife on the RCAF, Gen. Y Chhien of Pailin stated that his party opposed the ousting of Ranariddh and regarded the nomination of Ung Huot as inappropriate. He appealed for the return of the UN to Cambodia to ensure the restoration of peace and the holding of elections.

In August 1997 the National Assembly voted on the nomination of Ung Huot as First Prime Minister, replacing Prince Norodom Ranariddh: 86 members of the 120-seat National Assembly voted in favour of his appointment. Ung Huot was formally elected when acting Head of State Chea Sim signed a royal decree approving the appointment after King Norodom Sihanouk reportedly gave his authorization; Ung Huot and Hun Sen agreed that a legislative election would be held in May 1998 as planned, and that the winner of that election would become Cambodia's sole Prime Minister. Meanwhile, Hun Sen gave National Assembly members who had left the country three months in which to return before being replaced. Hun Sen insisted that, should Prince Norodom Ranariddh return, however, he would face trial for attempting to negotiate an alliance with remaining Khmer Rouge rebels.

A further significant development in Cambodia in July 1997, meanwhile, was the denunciation and trial of former Khmer Rouge leader, Pol Pot, by his own comrades: an announcement broadcast on the PDK radio station stated that Pol Pot had been sentenced to life imprisonment at the Anlong Veng guerrilla base in north-west Cambodia for 'betraying the Khmer Rouge movement'. It was thought unlikely, however, that Pol Pot would be handed over to the Cambodian Government or to the international authorities to be tried for genocide.

The events of 5–6 July 1997 resulted, in the first instance, in the removal of Prince Ranariddh as First Prime Minister. Yet, more fundamentally, it was an attempt by Hun Sen to re-establish CPP control of the State and to put an end to the parallel FUNCINPEC structure in the armed forces, police and bureaucracy which had developed since 1993. The killing of several of Ranariddh's senior army and police commanders (a sixth general, Thach Kim Sang, was assassinated in March

1998) and the sentencing *in absentia* of two others (Nhek Bun Chhay and Serey Kosal) to long prison terms, severely weakened Ranariddh's capacity to mount a military challenge. In the immediate aftermath of the fighting, FUNCINPEC forces either agreed to be disarmed or retreated rapidly to their pre-1993 bases on the Thai border, from where they sought to link up with the remaining Khmer Rouge insurgents. Within a month, Nhek Bun Chhay's forces retained only one stronghold, the border village of O Smach in Otdar Mean Chey province. Total military control of the country evaded Hun Sen, owing to the failure of repeated government offensives against O Smach and the persistence of FUNCINPEC and PDK activity in other remote areas of the north and west; an internationally sponsored cease-fire was implemented on 27 February 1998.

Achieving total control of the state apparatus proved less difficult. Ung Huot, Ranariddh's replacement as First Prime Minister, served Hun Sen's purposes faithfully: the principle of 'consensus' and equality between the two parties, on which the coalition Government had been founded, was retained in form but not in substance. Except for Ranariddh, Hun Sen left almost all FUNCINPEC appointees in place: his priority was to retain international legitimacy by preserving an unchanged façade of the Royal Government, and it proved as easy for the CPP to work around FUNCINPEC officials as to dismiss them. Hun Sen thus invited FUNCINPEC and BLDP ministers and members of the National Assembly to return from exile to their former positions, and even permitted a small UN team to monitor their safety. Hun Sen did attempt to reorganize the Cabinet in September 1997, with a view to rewarding Ieng Muli, Toan Chhay and others, but this initiative was unexpectedly obstructed by internal disagreements within the remnants of FUNCINPEC and by opposition from the Chea Sim faction of the CPP, whose suspicion of Hun Sen's intentions reached new heights in the period immediately following the events of 5–6 July.

Hun Sen did successfully exploit his newly obtained parliamentary majority to ensure that the CPP would control the preparations for the elections due in 1998 and would retain judicial, as well as bureaucratic and military, power during and after the polls. The National Election Committee (NEC), the Supreme Council of Magistracy and the Constitutional Council (the bodies mandated under the Constitution to organize elections, to appoint and supervise judges, and to monitor the constitutionality of laws and judge electoral disputes) were finally established with clear CPP majorities in each. The elections thus took place in a framework determined and managed by Hun Sen.

None the less, matters did not proceed completely as planned. The main obstacle Hun Sen faced was that of international opposition: in particular, ASEAN's refusal to admit Cambodia in July 1997 and the decision of a committee of the UN General Assembly in September to leave Cambodia's seat in the UN vacant until the elections. This latter decision was effectively taken by the USA, Hun Sen's principal critic, and Washington was also responsible for the IMF's decision to suspend aid. Cambodia, meanwhile, suffered from a collapse in investment following fighting in Phnom-Penh and the disastrous decline in the regional economy. Whilst Hun Sen consistently rejected offers from ASEAN, Thailand, the USA and King Sihanouk to mediate between him and Ranariddh, insisting that this was an internal criminal affair to be dealt with by Cambodia's Government and courts, Hun Sen's need for international recognition and aid ultimately proved too great. In February 1998 a 'four-pillar' peace plan, proposed by Japan and strongly supported by the 'Friends of Cambodia', was accepted by both sides. The 'four pillars' were: a cease-fire, to be followed by the reintegration of Ranariddh's forces into the RCAF; the end of ties between the Prince's forces and the PDK; a prompt trial and pardoning of Ranariddh; and his participation in the elections. A cease-fire was declared and was largely adhered to, but reintegration failed to take place and both sides continued to exchange accusations of links with the Khmers Rouges. Two 'show' trials were held (in March) at which Ranariddh was found guilty *in absentia* of smuggling weapons, causing instability, disobeying the orders of superiors and complicity with the Khmers Rouges. He was sentenced to a total of 35 years' imprisonment and ordered, together with his senior military commanders, to pay compensation of more than US $54m. to cover the damage caused in the fighting. After a tense battle of wills, and faced with the real prospect of Japan, ASEAN and the EU withdrawing support for the elections, Hun Sen wrote to the King requesting a full pardon for Ranariddh.

The National Election of 1998

Ranariddh thus returned to Cambodia on 30 March 1998, the last of the opposition politicians to do so. None the less, the opposition's ability to compete in the elections (which had been delayed from 23 May to 26 July to allow sufficient time for the NEC to carry out preparations) had been severely weakened. As well as losing equipment, money and their headquarters during the events of July 1997, the opposition had lost all access to the electronic media: the FUNCINPEC-aligned television and radio stations and the radio station run by the BLDP had been taken over by the Government. Opposition and UN demands that past political violence be investigated and punished went unheeded. Both the KNP and the BLDP were forced to change their names (to the Sam Rainsy Party (SRP) and Son Sann Party, respectively) owing to ongoing court disputes. The uncertainty over Ranariddh's participation also exacerbated the split within FUNCINPEC: three breakaway parties with quite close ties to the CPP registered for the elections—they were led by Toan Chhay, Ung Huot and Loy Simchheang. Meanwhile, the opposition was adversely affected by internal tensions within the UCD (particularly between Ranariddh and Rainsy), which put an end to the possibility of contesting the election as a coalition. None the less, under strong international pressure and despite grave reservations about the entire electoral process, the opposition did compete.

The election proceeded unexpectedly smoothly. In spite of widespread low-level intimidation and an estimated 12 killings of their members, the FUNCINPEC Party and the SRP managed to attract significant crowds to their campaign rallies. Officially, 93.7% of the 5,395,024 registered went to vote on 26 July 1998, and on the next day the UN-co-ordinated Joint International Observation Group expressed its confidence that the elections had been 'free and fair'. Unsurprisingly, however, the situation soon deteriorated. Official preliminary results awarded the CPP 64 seats, the FUNCINPEC Party 43 and the SRP 15, under a complicated system of proportional representation, although combined support for the FUNCINPEC Party and the SRP, which had gained 31.7% and 14.3% of the vote, respectively, represented a majority of the popular vote; the CPP secured 41.4% of the vote. The 36 other parties failed to obtain a single seat. The results gave the CPP a parliamentary majority but not the two-thirds' majority needed to form a government; Hun Sen thus offered to form a 60:40 coalition government with FUNCINPEC, with the provision that he be premier and that the CPP retain all the key ministries. He also suggested the formation of a tripartite coalition, with the SRP being given 10% of ministerial positions and the FUNCINPEC Party 30%. These offers were rejected as premature by FUNCINPEC and the SRP, which alleged massive electoral irregularities (in the voting, counting and seat allocation process). Both parties' leaders refused to recognize the election results until their complaints were investigated by the NEC. They also announced their intention to boycott the new National Assembly, which was scheduled to convene in September. FUNCINPEC and the SRP lodged nearly 900 complaints of election irregularities. In mid-August the Constitutional Council rejected all but one of the complaints lodged by the SRP, and this concerned a vote recount. In October the NEC issued a report, which concluded that there were no discrepancies in the vote count and that the results would stand.

Coalition Government and Democratic Stagnation

In September 1998, amidst steadily rising politically-motivated violence, King Sihanouk convened unsuccessful talks in Siem Reap, involving representatives from the CPP, FUNCINPEC, the SRP and the Constitutional Council. While these discussions were in progress two grenades were thrown into the residence of Hun Sen. Hun Sen promptly returned to Phnom-Penh and ordered that 'Democracy Square'—an area near the National Assembly that had been occupied by opposition supporters, encouraged by Sam Rainsy—be cleared and that the leaders of the demonstrators be arrested. Sam Rainsy, whom Hun Sen accused of responsibility for the grenade attack, immediately sought refuge in a nearby UN office. The police subsequently cleared Democracy Square, precipitating street violence in the capital as groups loyal to Sam Rainsy clashed with supporters of the CPP. Two monks were shot dead, and the Hun Sen Government banned members of the National Assembly from leaving the country. In mid-September Thomas Hammarberg, the UN Secretary-General's Special Representative for Human Rights in Cambodia, announced that, since the police action, 16 bodies (including those of the two monks) had been found. This figure was later revised to 26.

It was in this context that external diplomatic pressures resulted in the resumption of political talks. Ranariddh announced that he was calling off street demonstrations and would meet with the King. He also stated that FUNCINPEC would attend the first session of the National Assembly on 24 September 1998. Sam Rainsy made a similar public announcement. The King convened a summit meeting in Siem Reap involving Chea Sim, Hun Sen, Ranariddh and Sam Rainsy. The ban on foreign travel was lifted. None the less, the tripartite talks broke down in late September, and Ranariddh and Rainsy fled abroad in October.

In early October 1998 the King travelled to Phnom-Penh to renew his efforts to break the political impasse. In November the King hosted discussions involving Hun Sen, Chea Sim and Ranariddh, which resulted in a protocol on power-sharing. Under the terms of this agreement, Hun Sen would remain in office as Prime Minister, while Ranariddh would become the President of the National Assembly, replacing Chea Sim. A full amnesty was granted to Princes Norodom Sirivuth and Norodom Chakrapong and Generals Nhek Bun Chhay, Serey Kosal and Sin Song. A second parliamentary chamber, a senate, was to be created under the chairmanship of Chea Sim, who would also serve as acting Head of State when the King was out of the country. Sam Rainsy, who had not been present at these discussions, was highly critical of them. He nevertheless returned to Cambodia from Paris; Prince Sirivuth also returned to Cambodia in early 1999.

On 4 March 1999 the National Assembly finally voted by 106 to five to amend the Constitution and create a Senate with the power to scrutinize and amend bills before sending them back to the National Assembly for affirmation and final royal assent. Chea Sim was duly elected Chairman of the Senate. Prince Sisowath, Chivoun Monirak and Nhek Bun Chhay were chosen as Deputy Chairmen. Representation in the 61-member Senate was proportional to party strength in the lower house: 31 seats were allocated to the CPP; 21 seats to FUNCINPEC; seven seats to the SRP; and two members were nominated by the King. The inaugural session of the Senate was held on 25 March.

On 30 November 1998 Hun Sen was approved as Prime Minister by a vote of 99 to 13 in the National Assembly. Of the 29 cabinet posts, 15 were accorded to the CPP and the remainder to FUNCINPEC. Both parties shared control of the Ministry of Interior and the Ministry of Defence. The new Government began drafting legislation for the conduct of 'free and fair' elections in Cambodia's *khum*, which were not held until February 2002.

As a result of the November protocol on power-sharing, Hun Sen and Ranariddh agreed to rationalize and reform the RCAF. Cambodia's military was widely viewed as overstaffed, underpaid, unruly and heavily engaged in illegal logging. The exact size of the armed forces was unknown, owing to the large numbers of 'ghost soldiers' on the payroll (soldiers who had either been killed or had returned to their villages but whose pay continued to be collected by senior officers). In late 1999 it was estimated that these soldiers constituted at least one-third of the military's approximately 155,000 personnel. The military was regularly implicated in armed robberies, kidnapping for ransom and drugs-trafficking. In late January 1999, as agreed, Hun Sen resigned as Commander-in-Chief of the RCAF; Gen. Ke Kimyan, the former Chief of the General Staff, was elevated to this position. Of 26 senior positions within the armed forces, FUNCINPEC representatives were given only three posts. At the change of command ceremony Hun Sen urged that the armed forces be reduced by 55,000 and the police by 24,000 over the next few years. Germany and the Asian Development Bank (ADB) pledged financial assistance for this programme.

By 2000 there were signs that Cambodia was progressing towards the development of a more mature political situation and was beginning to experience greater stability. Within the CPP, Hun Sen's power base appeared to have been consolidated, despite rumours of intra-party struggles. Hun Sen, however, remained guarded against any potential political challengers. Deputy Prime Minister and Co-Minister of the Interior, Sar Kheng, was perceived by observers to represent the greatest threat to Hun Sen's leadership; it was also alleged that a rift existed between Sar Kheng and Hun Sen's closest ally, Sok An, Minister in charge of the Council of Ministers and perceived by many as the CPP's alternative potential successor to Hun Sen. FUNCINPEC, meanwhile, remained under the leadership of Ranariddh. At the Party Congress held in March 2000 it was acknowledged that the Party would need to regroup if it were to become an effective political force at the forthcoming *khum* headship elections and beyond. Ranariddh was subject to criticism for his ambivalence regarding the return to FUNCINPEC of former party members who were alleged to have abandoned the Party after the unrest of July 1997. Hun Sen, while retaining FUNCINPEC as his coalition partner in the Government, sought to improve domestic security by adopting severe measures to control weapons and by seeking to reform the armed forces. By late 1999 the Government had reportedly destroyed some 60,000 illegal weapons; however, an estimated 450,000 guns remained in the country, many in the hands of the police and military.

In 1999/2000 opposition parties received better treatment from the Government. Repressive violence against members of the SRP by elements allied with the Government appeared to have declined in comparison with 1997 and 1998. Rainsy continued to challenge the Government, however, voicing the concerns of groups such as protesting garment workers and the landless. In October 1999 a member of the legislature belonging to the SRP was abducted. Rainsy himself considered this incident to have been politically motivated and asserted that Cambodia continued to be ruled by dictatorial leaders, whom he described as 'crooks, criminals and clowns'. His comments resulted in the issue of a series of threats (apparently from the CPP) against Rainsy and members of his party.

Relations between the two coalition partners in the incumbent Government in 2000–02 remained officially cordial. Ranariddh, as leader of FUNCINPEC and Chairman of the National Assembly, did not want his ministers and members of the legislature to be in conflict with the Hun Sen-dominated CPP; he thus assumed a secondary role in the Government. During its annual congress in March 2001, which marked the party's 20th anniversary, FUNCINPEC reaffirmed its commitment to working closely with the CPP in the interests of peace and reconciliation. After he had taken up the position of FUNCINPEC Secretary-General in July, Prince Norodom Sirivudh also advocated the politics of non-confrontation. This accommodating tone was motivated more by practical considerations than by goodwill: the royalists could no longer afford to risk another violent confrontation with Hun Sen, who had exiled Sirivudh after the latter had been accused of plotting a coup against him and who had also ousted Ranariddh in 1997. Positive relations between the two leaders were based on a mutual understanding that each required the presence of the other for their own political interests. However, preparations for *khum* headship elections, scheduled to take place in February 2002, appeared to prompt the royalists to think strategically about competing with the CPP.

The *Khum* Headship Elections of 2002 and the National Election of 2003

Although the *khum* headship elections that took place on 3 February 2002 appeared to advance the democratization process, it was uncertain whether they brought the country closer to the attainment of what some people referred to as 'grassroots' democracy. While the opposition political parties did participate, they failed to make serious inroads into gaining control of the 1,621 *khum*. The CPP retained approximately 98% of *khum* seats, leaving only 11 for the SRP and 10 for FUNCINPEC. Although the monopoly of power exercised by the CPP over the *khum* was broken, the participation of the SRP and FUNCINPEC appeared to contribute to the legitimacy of elections that were regarded by many as being far from 'free and fair', contrary to the testimony of the EU Election Observation Mission. Moreover, about 1.75m. of the 6.2m. eligible voters did not turn out to vote. Although the cause of the lower than expected turn-out was unclear, the campaign period had been characterized by political intimidation and violence directed at members of the opposition parties. At least 22 *khum* candidates and political activists belonging to FUNCINPEC and the SRP were reported to have been murdered. The Election Monitoring Organizations—comprising non-governmental groups—issued a joint statement on 12 February rejecting claims that the elections had been 'free and fair' and citing several instances of electoral irregularities to support this claim. Overall, the elections allowed the two major opposition parties to claim some victory without weakening the CPP, which had now become more dominant than ever.

Before and after the *khum* elections took place, relations between the CPP and FUNCINPEC remained stable, as the CPP continued to consolidate its control over the country. Political stability was promoted both by FUNCINPEC's new commitment to a non-confrontational style of politics and the weakened ability of the royalists to challenge the CPP. Their party was beleaguered by rifts and new internal frictions. Internal conflicts intensified when the royalists were forced to take sides between those who supported the Deputy Commander-in-Chief of the RCAF, Khan Savoeun (a FUNCINPEC member), and supporters of the Co-Minister of the Interior, You Hockry (another FUNCINPEC member), who was accused of nepotism and corruption. Under growing pressure from the party to resign from his cabinet post, You Hockry finally announced in May 2002 that he would accept the party's decision. (The National Assembly later rejected the efforts of Savoeun's supporters to remove You Hockry from his post.) At the same time Hang Dara, a senior member of FUNCINPEC, announced that he intended to create a new party. In late May he registered the Hang Dara Movement Democratic Party with the Ministry of the Interior, citing his dissatisfaction with FUNCINPEC's performance as his motivation. Another set-back to FUNCINPEC came when Prince Chakrapong (the half-brother of Prince Ranariddh) chose to create a new royalist party, the Prince Norodom Chakrapong Khmer Soul Party; its members included Toan Chhay. Chakrapong sought permission to contest the general election, scheduled for July 2003, and offered to form a political alliance with the SRP. With the royalist camp fragmenting, the CPP gained strength. Prince Ranariddh seemed disheartened by these developments; he warned that political divisions within FUNCINPEC would only work to Hun Sen's advantage.

As campaigning for the national election of 2003 got under way, the CPP continued to maintain its dominance over the electoral process. In August 2002 the National Assembly approved amendments to the Election Law which allowed the Ministry of the Interior to nominate five 'dignitaries' to the NEC, allowing the CPP to exercise greater control over the electoral process. The amendments were passed without any debate in the Senate. Three of the five new NEC members reportedly enjoyed close links with the CPP, while the other two belonged to FUNCINPEC. Some CPP officials became more confident of electoral victory and entertained the idea that their party might not need to form a coalition government with FUNCINPEC after the election.

Tensions between King Sihanouk and the CPP did not help FUNCINPEC. The King's position on national issues was challenged by CPP officials, in particular Hun Sen. On several occasions the King threatened to abdicate unless the Government convened the Royal Council of the Throne to be entrusted with the power to elect a new king, but this threat was disregarded by Hun Sen, who declared that the 2003 election was of more importance. In April Hun Sen took issue with the King by demanding that the latter's unidentified overseas correspondent, 'Ruom Ritt' (suspected to be an alias for the King himself), be brought to justice for insulting the Government. The King promised to stop publishing Ruom Ritt's letters in his public bulletin, but expressed his preference that France, rather than China, be the country of his possible exile. In May his prediction that his death was 'coming soon' raised uncertainty concerning the future of the monarchy.

Meanwhile, the royalists still found themselves unable to co-operate with the SRP. Members of the two parties engaged in a campaign of defections. In March it was reported that about 20 royalist generals intended to defect to the SRP. By May 2003 six royalists had defected to the SRP, and seven SRP members had joined FUNCINPEC.

On 27 July 2003 23 parties contested elections to the 123-seat National Assembly. The CPP secured victory in the poll, obtaining 73 seats and 47.35% of all votes cast, an improvement on its performance at the 1998 election, followed by FUNCINPEC with 26 seats (20.75% of votes) and the SRP with 24 (21.87%). However, the CPP failed to obtain the two-thirds' majority of National Assembly seats that would enable it to form a single-party government, raising the prospect of further political instability in the country.

When assessed in terms of its overall level of freeness and fairness, the election bore testament to the fact that the country had made further democratic progress. Prior to polling day, the pre-election period had proceeded more smoothly than in the past. Although, in the period between the *khum* elections and the commencement of the 2003 election campaign, 12 acts of lethal violence (including the murder of Om Rasady, Prince Ranariddh's senior adviser) were committed against political activists, the campaign period itself was relatively free from violence. On polling day no major incidents were reported. The level of violence after polling day was also much lower than at previous elections.

However, democracy in Cambodia after the 2003 election could still not be termed mature. The regime's political legitimacy continued to be threatened by the effects of corruption, the extent of which was unclear, although critics remained pessimistic. It was estimated that corruption cost Cambodia between US $120m. and $1,000m. per year. Politicking also continued. FUNCINPEC and the SRP denied that the election was free and fair, alleging that 1.5m. Cambodians had been prevented from exercising their right to vote, and consequently formed the Alliance of Democrats in August, demanding that Hun Sen step down as Prime Minister. By late June 2004, 11 months after the election, the transfer of power had not taken place and the interim Government remained in place, as the CPP, FUNCINPEC and the SRP failed to agree on how to form the next administration. The CPP had won more seats than the two opposition parties combined, but the number fell short of the two-thirds' majority required for a single-party government. The Alliance of Democrats made several demands, but the CPP was unprepared to meet all of them. They included the formation of a tripartite government incorporating FUNCINPEC and the SRP, an increase in salary for public servants from $30 to $100 per month, and a commitment to combating corruption through the promotion of good governance. Hun Sen rejected the idea of any government that included the SRP. The idea of a 'two-and-a-half party' government was entertained, but the post-election crisis remained unresolved.

Meanwhile, the CPP continued to consolidate its power at the expense of the opposition. In September 2003 Hun Sen reportedly dismissed 17 senior FUNCINPEC government officials, accusing them of failing to carry out their duties. On 14 June 2004 the political stalemate was exacerbated by FUNCINPEC's unreasonable demand that it be given half of all senior government-appointed posts; the CPP was reportedly prepared to give the royalist party only 40% of them. On 26 June, however, the two parties finally signed a deal that would increase the number of cabinet posts by 150%: from only 89 in the last Government to 207 (with 136 to be given to the CPP and 71 to FUNCINPEC). By the end of June, the SRP's political role remained unclear.

In October 2004 King Norodom Sihanouk prompted considerable political speculation by announcing that he intended to abdicate, citing reasons of ill health. The Royal Council of the Throne was hastily convened, as required by the Constitution, and appointed Sihanouk's son, Prince Norodom Sihamoni, as his successor in the following week.

The Recent Security Situation

In recent years the overall security environment has improved significantly. Between 1999 and 2001 Cambodia destroyed approximately 50,604 weapons. In November 2000, however, a group of Cambodian Freedom Fighters (CFF) staged an armed attack on government buildings in Phnom-Penh, leaving eight people dead. In total, 47 people, including three US citizens, were charged in connection with the coup attempt. In July 2001 two bombs exploded in separate hotels in Phnom-Penh, killing three people. A Cambodian man later admitted to the bombings, claiming that he had acted to extort money. In August two explosions in the main offices of FUNCINPEC in Phnom-Penh injured three people. By mid-2002 the security situation had become somewhat precarious, as politically motivated violence against members of the opposition parties increased and an average of two 'mob' killings a month were reported. In an attempt to prevent the rise of nocturnal crime, Hun Sen closed down night-clubs.

Increasing political stability in Cambodia, however, was still being gained at the expense of genuine democracy. Hun Sen acted as a temporary stabilizing force within the Government, although the death of the Minister of Agriculture, Forestry, Hunting and Fisheries, Chhea Song, in April 2001 precipitated the need for a cabinet reshuffle. Hun Sen continued to describe the inspiration he received from two anti-democratic leaders: the late Prime Minister of Laos, Kaysone Phomvihane, and the anti-Western then Prime Minister of Malaysia, Mahathir Mohamad. Although Hun Sen referred to the President of the Republic of Korea, Kim Dae-Jung, as a third role model in his treatment of domestic enemies, he did little to accelerate democratic reform at his party's expense.

Major symptoms of the democratic stagnation in the country included the Government's willingness to pre-empt and suppress domestic challenges to its political authority. Human Rights Watch, an international organization based in New York, USA, reported that during 2000 the most serious incidents of violence since 1997 had occurred in Phnom-Penh, including numerous attacks against *khum* leaders, mostly directed at members of the SRP. In June 2001 a Cambodian court tried 32 coup plotters arrested after the failed coup attempt in November 2000. Twelve of the 15 defence lawyers left the courtroom after the start of the trial because the court was perceived to be subject to government pressure. In March 2002 a further 18 men were sentenced to prison terms of between seven and 18 years. Amongst those convicted were members of FUNCINPEC. The judiciary remained subject to criticism.

Cambodia's record on human rights remained questionable. In June 1998 Cambodia established a Human Rights Committee, to respond to concerns expressed by the Office of the UN High Commissioner for Human Rights regarding 80 unresolved deaths following the events of July 1997. More than one year after its formation the committee had yet to acknowledge that any politically motivated killings had taken place, and there had been no arrests or prosecutions in connection with these murders. Allegations of human rights violations continued. In early 2002 Human Rights Watch issued a review of developments in Cambodia. The report criticized the rise of political violence prior to the *khum* headship elections, the lack of improvement in prison conditions and the continuation of torture by police and prison officials, who continued to act with impunity. Human Rights Watch also accused the Government of violating the fundamental principle of non-refoulment (under which refugees are protected from returning to any country where they might be subjected to inhumane treatment) when deporting more than 100 asylum-seekers (who complained of persecution and repression by Hanoi) back to Viet Nam. In March a senior UN official stated that Cambodia had violated international law by deporting 63 Vietnamese refugees against their will. In June 2002 a UN human rights agency claimed that Cambodians had lost faith in the judiciary, citing 'mob' killings and the failure of the police to intervene, as well as its role in instigating the attacks.

Despite the holding of elections, socio-economic conditions continued to obstruct the process of democratic consolidation. Cambodia remained one of the poorest countries in the world. Corruption continued to be a serious issue, further weakening the Government's political legitimacy; even Hun Sen admitted the seriousness of corruption within the military and security apparatus. In recent years the economy performed better than in the immediate post-coup period, but the gap between rich and poor continued to widen. Ordinary people still suffered from a lack of health care and could not afford to live comfortably in Phnom-Penh, where the average monthly cost of living was about US $250. The problem of landlessness remained largely unresolved, and was so serious an issue that the UN Secretary-General's Special Representative for Human Rights in Cambodia, Peter Leuprecht (who replaced Thomas Hammarberg in 2000), declared it to be a threat to social stability.

The CPP-dominated regime, however, became even more determined to take repressive action against any protesters or demonstrators accused of disrupting law and order. In December 2002 the Government formed the Central Bureau for Security (CBS), most of the members of which were powerful CPP leaders. The decision to pursue more repressive measures against potential trouble-makers came in January 2003, after demonstrators protesting against comments made by a Thai actress concerning the ownership of the temples at Angkor Wat became violent, burning down the Thai embassy and looting other Thai business interests in Phnom-Penh. The total damage was estimated at US $50m. In February the Chief of the National Police, Hok Lundy, reportedly claimed that the police force was now prepared to demonstrate its strength by taking action against any demonstrators who refused to accept the election results. In March about 200 riot police, armed with AK-47 rifles and plastic batons, dispersed a group of demonstrators loyal to the SRP, injuring 13 of them. In June the police opened fire on around 300 workers staging a protest at a garment factory, leaving two protesters dead and more than 20 policemen injured.

Following the 2003 national election, restrictions on political freedoms continued. Post-election demonstrations were prohibited. In August 2003 the small Khmer Front Party sought to hold a demonstration in Phnom-Penh but was not granted permission. When the demonstration proceeded regardless, police dispersed it and arrested 21 party members, who were forced to sign statements promising that there would be no repeat of the protest. In November garment-factory workers staged a demonstration but were confronted by armed riot police, who used excessive force to disperse them.

Any serious opposition to the CPP regime continued to risk violent repression. In October 2003 Ta Prohm Radio editor and journalist Chour Chetharith was shot dead outside his office, the first time that a journalist had been killed in the country since 1996. In January 2004 Chea Vichea—the leader of a major trade union, a founding member of the SRP and a member of the Alliance of Democrats—was also shot dead, in broad daylight. Even monks remained subject to control. In October 2003 12 Buddhist monks were threatened with expulsion from their pagoda because they allegedly supported the SRP; their leader had been disrobed and expelled. One year later Cambodia's most senior monk and head of the Mohanikay Buddhist majority, Ven. Patriarch Tep Vong, ordered all monks who had arrived in Phnom-Penh after 2003 to return to their home pagodas. He was reported by the *Phnom Penh Post* to be 'aligned with the CPP'.

The Demise of the Khmers Rouges and the Issue of Their Trial

The dissolution of the Khmers Rouges entered its final stage in late March 1998, when five divisions rebelled against the leadership of Ta Mok and defected to Hun Sen. Within weeks Anlong Veng was in government possession, and the ever-dwindling Khmer Rouge forces had been almost entirely forced into Thailand. Although Hun Sen embraced Keo Pok (a 68-year-old former DK zonal secretary during the Pol Pot period) as the leader of the rebels, in fact some were Pol Pot and Son Sen loyalists, others were resentful that Ta Mok had reneged on his promises of liberalization, while almost all were weary of the war and realized that defeat was inevitable. Pol Pot himself died on 14 April, only a day after a desperate Ta Mok had offered to hand Pol Pot over to the international community. Pol Pot's death (which was later reliably reported to have been suicide, following a radio broadcast describing Ta Mok's plans) deprived the PDK of its last opportunity to negotiate.

Throughout the last quarter of 1998 the remaining Khmer Rouge leaders negotiated an end to their armed resistance and their re-entry into Cambodian society. A ceremony was held in February 1999, at which time more than 1,500 troops were reintegrated into the ranks of the RCAF. The Khmer Rouge officials, Nuon Chea (the former Chairman of the National

Assembly and the second most senior member of the PDK) and Khieu Samphan, left the jungle in December 1998 under an agreement with the Government. They were received warmly by Hun Sen in Phnom-Penh. In March 1999 Ta Mok was captured along the Thai border and placed in detention. In the following month Duch (Kang Khek Ieu), the former director of Tuol Sleng prison (where at least 17,000 prisoners were tortured and executed), was discovered working in Bat Dambang province; he was arrested in May and charged with violating a 1994 law outlawing the Khmers Rouges. The presence of senior Khmer Rouge leaders in Cambodian society prompted calls from within Cambodia, and especially from the international community, for their trial and punishment for crimes committed under the PDK regime. In May 2002 Gen. Sam Bith, a former Khmer Rouge commander, was arrested, ostensibly for his participation in a train ambush in 1994 in which 16 people, including three foreign tourists, died. The most senior of the three former Khmer Rouge commanders (the others being Gen. Nuon Paet and Col Chhouk Rin) to stand trial for the killings, he was convicted of the charges against him in December 2002 and sentenced to life imprisonment, pending an appeal. The appearance of former Khmer Rouge leader Nuon Chea as a witness at his trial provoked criticism that the Government had made no effort to detain him. Meanwhile, pressure from Australia and the United Kingdom had resulted in the arrest of Gen. Nuon Paet in August 1998. Paet was also implicated in the ambush of 1994. In June 1999 he was convicted and sentenced to life imprisonment and in September 2002 the Supreme Court confirmed his life sentence. Ten other Khmer Rouge leaders, including two former commanders and eight subordinates, were also charged. In July 2000 Col Chhouk Rin was freed after it was decreed that he was covered by the amnesty granted to Khmer Rouge cadres who surrendered to the Cambodian Government. (He had been appointed a colonel in the Cambodian army after his surrender; however, following strong formal protests at the acquittal from the British, French and Australian Governments, the Cambodian Government subsequently announced that it was to appeal against the court's decision, and in September 2002 Chhouk Rin received a sentence of life imprisonment.) Hun Sen resisted external pressures to try high-level Khmer Rouge leaders, arguing that such action would undermine his attempts to reach national reconciliation with the Khmer Rouge rank and file and prompt a renewal of insurgency.

None the less, momentum continued towards some sort of international accounting. A US-drafted UN Security Council resolution to extend the mandate of the Hague tribunal on war crimes in Yugoslavia and Rwanda to cover senior Khmer Rouge leaders made little progress, but a Commission of Experts, appointed by the UN Secretary-General, was established in August 1998 to examine the issue. In March 1999 a team of three legal experts commissioned by the UN recommended that 20–30 Khmer Rouge leaders be brought to trial and reparations be made to their victims. The international community and international observers remained consistently sceptical of Cambodia's ability to establish an independent tribunal.

Hun Sen was reported to have considered the idea of a 'truth and reconciliation commission' modelled on that of South Africa. The UN Secretary-General, Kofi Annan, however, argued in a letter to Hun Sen in March 1999 that a tribunal for the Khmers Rouges should be international in character. After several leaders were captured, Hun Sen announced they would be tried under the jurisdiction of the Kingdom of Cambodia, strenuously arguing that Cambodian sovereignty should be respected. Both the People's Republic of China and Thailand endorsed his stance. In May, however, when Hun Sen met Thomas Hammarberg he accepted the UN envoy's offer of assistance in refining draft legislation to create a domestic tribunal using international standards, prior to its submission to the National Assembly. This tribunal would include both Cambodian and foreign judges and prosecutors. However, in August it was clear that Hun Sen was still resisting pressures to form a tribunal for this purpose. The National Assembly approved a law extending the period of detention without trial for those who were accused of 'genocide, crimes against humanity or war crimes' from six months to three years. The effect of this law was to postpone the trials of Ta Mok and Duch, who were both, however, charged with genocide (under a 1979 decree) in early September.

International concern over the culture of judicial impunity that pervaded the country's legal system remained high in 2000. More than 20 years after the collapse of the PDK regime, none of the leaders of the Khmers Rouges had been brought to trial for the atrocities committed under the regime, and several principal Khmer Rouge figures continued to live openly in Pailin. At mid-2004 Ta Mok and Duch remained in government custody, awaiting trial. Although, in the late 1990s, the CPP was believed to have largely opposed any compromise with the UN on the issue of a Khmer Rouge trial, external pressure eventually forced the Government to take action. In April 2000 Hun Sen agreed to co-operate with the UN over a US proposal for the establishment of a UN-sponsored court, which would include both Cambodian and foreign judges and would uphold international standards of justice, as well as maintain Cambodia's national sovereignty. It remained unclear, however, how many Khmer Rouge leaders would stand trial before such a court. In July the Cambodian Government and the UN finalized the details of a draft accord on the establishment of a special tribunal to try former Khmer Rouge leaders implicated in atrocities carried out during the regime's rule.

Legislation to establish the tribunal—unanimously approved by the legislature in January 2001—gave rise to serious controversy, domestically and internationally. The draft law allegedly failed to meet international standards of justice. In June 2001, however, Hun Sen accused the UN of violating Cambodia's sovereignty, reaffirming his desire to conduct a trial without UN participation. His attack on the UN followed the departure of Peter Leuprecht, the UN Special Representative, who expressed misgivings about the way in which the tribunal law had been formulated and stated that the legislation would benefit from further scrutiny. Reiterating the views of Hun Sen, Ranariddh declared that he was no longer prepared to conform to UN stipulations.

Until late 2002 progress on issues related to the trials of the Khmers Rouges remained indeterminate. On 11 July 2001 86 of the 88 members of the National Assembly who supported the idea of bringing Khmer Rouge leaders to justice approved the legislation. It specified that life imprisonment would be the heaviest penalty that the tribunal could mete out to convicted Khmer Rouge leaders. A general desire for peace and stability, rather than justice, still seemed to have served as the guiding principle for the formulation of the tribunal law. In August King Sihanouk endorsed the legislation. (Meanwhile, Hun Sen said that he expected only about 10 former Khmer Rouge commanders to stand trial and reassured the others that they should have nothing to fear.) Cambodia still failed to satisfy the UN. In February 2002 the UN Secretary-General decided to terminate negotiations with the Cambodian Government, having failed to gain support from the latter for the establishment of a court that would conform to international standards of independence, impartiality and objectivity. The Hun Sen Government responded by stating that it would proceed with the planned trials with or without UN support, blaming the UN for creating obstacles, while defending the position that it had done everything possible to co-operate. It justified its decision by stating that it had not invited the UN to dictate terms to Cambodia.

In 2003, however, significant progress was made on the trials of the Khmers Rouges. Following the issue of a UN press release in Phnom-Penh in August 2002 (which indicated that the UN was prepared to restart talks on the trials, that it would allow the Cambodian Government to conduct them and that it would only provide assistance to help ensure that the future tribunal would observe international standards of justice), steps were taken to resume negotiations between the UN and Cambodia. On 6 June 2003 the UN and the Cambodian Government finally signed an agreement providing for the establishment of one Pre-Trial Chamber of five judges (three appointed by the Supreme Council of the Magistracy and two appointed by the Council upon nomination by the UN Secretary-General) and two Extraordinary Chambers (the Trial Chamber made up of three Cambodian judges and two international judges, and the Supreme Court Chamber made up of four Cambodian judges and three international judges). The Chambers would have jurisdiction over those 'senior (Khmer Rouge) leaders of Democratic Kampuchea' who were considered to be 'most responsible for the crimes and serious violations of Cambodian penal law, international humanitarian law and custom and international conventions recognized by Cambodia that they had committed during the period from 17 April 1975 to 6 January 1979'. In addition to their agreement on the Chambers, which would operate within the existing Cambodian court structure, the UN and the Government of Cambodia also agreed to create two co-investigating judges (one Cambodian and one international) and a Prosecutors' Office (one Cambodian and one international). An Office of Administration would service them. The total cost of the entire judicial process was expected to be an estimated US $60m. over a three-year period. However, owing to the prolonged political stalemate that had resulted from the inconclusive national election of July 2003, legislation to approve the establishment of the tribunal continued to await ratification by the National Assembly in mid-2004.

Foreign Relations Restored and Strengthened

Cambodia's relations with multilateral organizations have generally been strengthened. As a result of the formation of a second coalition Government in November 1998, Cambodia regained its seat at the UN General Assembly in December, following an absence of 15 months. However, relations between Cambodia and the UN were somewhat strained from 1999. In December of that year the office of the UN Secretary-General's special envoy to Cambodia was closed at the request of the Cambodian Government, despite UN pleas for it to remain open for a further year. The Cambodian Government had, however, agreed in August to extend the mandate of the UN Special Representative for Human Rights in Cambodia, Thomas Hammarberg. Hun Sen, meanwhile, contended that the UN should focus more on providing assistance in areas such as the drafting of laws and judicial reform. Relations continued to be uneasy in 2001 and 2002, largely because the Hun Sen Government felt that the UN had forced the country to yield to its demands regarding a Khmer Rouge war crimes tribunal. Relations deteriorated further when the UN Drug Control Programme (UNDCP) concluded in February 2001 that the country was fast becoming the most attractive in South-East Asia for transnational criminals. The issue of the Khmer Rouge trial constituted a persistent source of tension in the relationship, as the UN continued to urge Hun Sen's Government to bring Khmer Rouge leaders to justice (see above). During her visit to Cambodia in August 2002 the UN High Commissioner for Human Rights, Mary Robinson, attacked the country for its pervasive human rights abuses and urged the Government to promote judicial independence and to reform the Supreme Council of Magistracy.

Since the coup in 1997 Cambodia's relations with the international donor community have been generally positive, as the latter continued to pledge financial support. In January 1999 Japan announced the resumption of aid by pledging to fund an electricity project in Phnom-Penh. The CGC agreed in February to provide Cambodia with US $470m. in assistance in the following year, conditional upon the implementation of constitutional reforms and a quarterly review. In June the CGC carried out its first review, identifying as priority areas the end of illegal logging and the reform of the armed forces, the civil service and the financial sector. In late 1999 the IMF (which had provided $85m. in budgetary support to Cambodia since 1994 but had suspended its assistance after the political events of July 1997) reached an agreement with Cambodia on a new three-year $81m. loan. In early 2000 the World Bank, which had previously cancelled a three-year $120m. loan arrangement, also agreed to grant Cambodia a new $30m. structural adjustment credit facility. At a meeting of the CGC in Paris in May, donors pledged to provide the country with $548m. in 2001; this aid, however, was to be dependent upon the Cambodian Government's honouring of its commitments to reduce military spending, suppress trafficking in illegal drugs and trim the civil-service payroll. In June 2001 14 donor nations met in Tokyo for a World Bank-sponsored conference on funding for Cambodia. The country had requested $500m. The donors, however, pledged a total of $615m., although with the conditions that Cambodia would take appropriate measures to combat corruption, adopt the tribunal law and promote human rights. Despite the fact that relations between Cambodia and the UN had deteriorated, the IMF remained optimistic regarding Cambodia. In February 2002 it released a $10.4m. credit under the Poverty Reduction and Growth Facility, bringing disbursements under the facility to a total of $52m. In March it issued a report praising Cambodia for the significant progress that it had made on reform; taxation and customs administration had been improved and the Ministry of Economy and Finance had exerted more stringent control over financial management. In June donors from 22 countries and seven international organizations pledged an annual donation of $635m. to Cambodia; the sum exceeded that pledged in the previous year and was approximately $116m. more than the Government had requested. However, the donor community adopted a more severe policy towards Cambodia by insisting that its support depended upon the country making substantive progress in policy reforms, particularly efforts to combat corruption and to improve the country's legal and judicial institutions.

Meanwhile, Cambodia's relations with ASEAN and individual states in South-East Asia remained positive. The improvement in political stability and domestic security prompted ASEAN officials to agree at their informal summit meeting in Hanoi in late 1998 to admit Cambodia as their 10th member; Cambodia formally acceded to the organization on 30 April 1999. The prospect of ASEAN membership effectively ended Cambodia's diplomatic isolation, and normal foreign relations were restored. In July 1999 Cambodia attended a meeting of the ASEAN Ministers of Foreign Affairs, its first attendance as a full member of the association. The Cambodian Minister of Foreign Affairs, Hor Nam Hong, pledged the Government's commitment to political and economic reforms as Cambodia sought to integrate itself with its South-East Asian neighbours. Hun Sen, meanwhile, maintained that the main reasons for his Government's decision to join ASEAN in 1999 were related to Cambodia's interest in regional security and stability, the ASEAN norm of consensus and the principle of non-interference, the economic growth experienced by ASEAN states over the previous 30 years, and the possibility of ASEAN serving as a gateway for trading relations with the outside world. As a new member of ASEAN, Cambodia was perceived to present a challenge to the unity of the regional group. In October 1999, for instance, Cambodia and the other two most recent entrants to ASEAN—Laos and Viet Nam—held their first unofficial Indo-China 'summit' meeting, which prompted concern across the region. At the meeting, Hun Sen and his Laotian and Vietnamese counterparts expressed their joint opposition to outside intervention in the newly-independent territory of East Timor (now Timor-Leste), hitherto part of Indonesia. Cambodia was perceived by some observers to be part of the 'Indo-China enclave' aligned against some other ASEAN members (particularly Thailand and the Philippines) which had contributed troops to the international peace-keeping force in East Timor.

Cambodia's relations with Viet Nam were periodically strained. In March 1999 Cambodia formally established a Cambodia-Viet Nam border joint committee, the first meeting of which was held in March. In May Prince Ranariddh, in his capacity as Chairman of the National Assembly, met his Vietnamese counterpart, Nong Duc Manh, to draw up a programme of co-operation between the two legislatures. In July Chea Sim, the new Chairman of the Senate, also visited Viet Nam. Meanwhile, in June 1999 Viet Nam's party Secretary-General, Le Kha Phieu, made an official visit to Phnom-Penh. He met King Sihanouk, and held discussions with Hun Sen, Ranariddh and Chea Sim. Although this visit provoked anti-Vietnamese demonstrations, the leaders of the two countries consolidated bilateral relations, agreeing to resolve outstanding border demarcation issues before the end of 2000, and signing co-operation agreements for education and energy. In February 2000 the Chairman of the Vietnamese National Assembly visited Cambodia; the Vietnamese Minister of Foreign Affairs, Nguyen Dy Nien, visited the country in the following month. In May 2000 Cambodia and Viet Nam signed an agreement allowing the latter to search for the remains of Vietnamese soldiers listed as missing in action during the Viet Nam War. While bilateral ties improved, elements of tension remained. Disputes over border issues and illegal migration featured prominently in the relationship. The Cambodian Government's decision to allow a number of Montagnard refugees from Viet Nam to enter Cambodia caused strain, and border demarcation issues also remained a major obstacle to cordial relations. It was reported in

August 2000 that Prince Ranariddh had urged Hun Sen to negotiate with Viet Nam owing to the latter's violation of a border agreement signed in 1995. In June 2001 King Sihanouk urged Hun Sen to 'save the sovereignty' of Cambodia. This was in response to Buth Rasmei Kongkea, the director of the Khmer Border Protection Organization, who accused both Viet Nam and Thailand of having illegally entered all 15 Cambodian provinces that border the two neighbouring countries. In May 2002 a group of Cambodian protesters travelled to Phnom-Penh to complain about land encroachment by Vietnamese authorities. Cambodian leaders, including the Governor of Svay Rieng (the brother of Hun Sen), as well as officials and lawmakers from FUNCINPEC and the SRP, voiced their concerns regarding the problem. In late May the King requested again that the Government investigate this matter and employ harsher measures to prevent border encroachment by neighbouring countries. Overall, however, Cambodia's relations with Viet Nam remained manageable, even though the border issues had yet to be resolved.

Cambodia also continued to maintain stable relations with Thailand. In February 1999 the Supreme Commander of the Royal Thai Armed Forces visited Phnom-Penh to discuss co-operation on the repatriation of 30,000 refugees from Thailand to Cambodia, which constituted the major issue in bilateral relations between the two countries. Later that month the Thai Deputy Minister of Foreign Affairs, Sukhumbhand Paribatra, visited Phnom-Penh to attend the tripartite meeting of Cambodia, Thailand and the UNHCR concerning the final repatriation of Cambodian refugees. In his meeting with Hun Sen, Sukhumbhand also discussed border issues and Cambodia's impending membership of ASEAN. In June the Prime Minister of Thailand, Chuan Leekpai, visited Phnom-Penh, where representatives of the Thai and Cambodian Governments pledged to resolve the territorial disputes between their respective countries 'in the spirit of friendship and neighbourliness'. In March a senatorial delegation from Thailand visited Cambodia, seeking ways to settle land- and sea-border disputes between the two countries. In June 2001 the new Thai Prime Minister, Thaksin Shinawatra, also visited Cambodia. However, border issues between the two countries remained unresolved.

In January 2003 Thai-Cambodian relations were tested when Cambodian demonstrators burned down the Thai embassy and looted other Thai-owned business properties (see above). The Thai Government evacuated its citizens resident in Cambodia, recalled its ambassador, ordered the Cambodian ambassador in Bangkok to leave and demanded that the Cambodian Government pay an estimated US $47m. in compensation for the damage inflicted on Thai interests. However, normal relations between the two countries were restored following the reopening of the border on 21 March; the Cambodian ambassador returned to Bangkok on 13 April and the Thai ambassador came back to Phnom-Penh on 24 April, where he declared that the two countries should 'let bygones be bygones'. Cambodia had already paid Thailand $5.9m. in compensation.

Cambodia appeared to enjoy better relations with the other members of ASEAN. After the ASEAN summit meeting in Hanoi in December 1998, the Government made a concerted attempt to improve relations with Cambodia's regional neighbours. Malaysia was an advocate of Cambodia's membership of ASEAN long before the country was formally admitted to the association. Hun Sen paid an official three-day visit to Kuala Lumpur in February 1999 when he discussed Cambodia's proposed new Senate with the Malaysian Prime Minister, Dato' Seri Dr Mahathir Mohamad. At the conclusion of the visit both countries signed an agreement further to expand trade, economic and industrial ties. Cambodia and Indonesia also improved their bilateral ties. In March 2001 Hun Sen paid an official visit to Indonesia, where he held discussions with the President, B. J. Habibie. At the end of the visit an agreement on the protection of investments, trade and tourism was signed. Indonesia also agreed to provide training assistance to the Cambodian police. In August 2001 Indonesia's new President, Megawati Sukarnoputri, arrived in Phnom-Penh for an official visit. Cambodia also sought to improve ties with Singapore. In June 1999 Hun Sen visited Singapore, where he asserted that peace and political stability had returned to his country, which would strive to catch up with other ASEAN members and fulfil its commitments to achieving ASEAN's 2020 vision and Hanoi's Plan of Action. Cambodia-Laos relations were generally positive. The Prime Minister of Laos, Sisavat Keobounphan, visited Cambodia in April 2000 and the new Prime Minister, Boungnang Volachit, also visited in August 2001.

Bilateral ties between Cambodia and other extra-regional states have also been strengthened. During the period from July 1997 until April 1999 Cambodia's relations with most external states were strained. The USA, Germany and Japan suspended all but humanitarian assistance. Hun Sen's Government was repeatedly criticized for refusing to take steps to identify and punish those responsible for politically motivated killings and human rights abuses in 1997–98. In October 1998 the US House of Representatives passed a resolution accusing Hun Sen of genocide and other crimes. In March 1999 it was indicated that a resumption of US assistance was dependent on democratic reforms and Cambodia's agreement to the establishment of an international tribunal to try Khmer Rouge leaders. In July a US official paid a brief visit to Phnom-Penh for discussions with Hun Sen and Ranariddh regarding the trial of Khmer Rouge leaders, democratic reforms and human rights. While little progress was made on these matters, Cambodia had earlier agreed to host a regional conference on the issue of US soldiers listed as missing in action during the Viet Nam War. Hun Sen continued to welcome US officials to his country, despite lingering resentment. At the CGC meeting in Tokyo in June 2001 the USA finally announced that it would resume direct aid to Cambodia. At the donor conference in June 2002 the USA pledged US $45m. Overall, the US strategy of combining pressure with reward was aimed at prompting Cambodia to accelerate the process of bringing Khmer Rouge leaders to justice and at curbing growing Chinese influence in Cambodia. However, there was no significant improvement in bilateral relations in 2002. In June the US State Department released its second annual Trafficking in Persons Report, which described Cambodia as a 'Tier 3' country (owing to its failure to comply fully with the minimum standards for the elimination of trafficking and to make sufficient efforts to do so) and warned that it could be subjected to aid sanctions by the USA. In June 2003 the State Department upgraded Cambodia's ranking to Tier 2, which meant that the country would not face the imposition of US sanctions. Up to this point, however, US politicians had remained highly critical of the Hun Sen Government and, in September 2002, US Senator Mitch McConnell had even called for 'regime change' in Cambodia. A US State Department report described the Cambodian Government's handling of the anti-Thai riots in January 2003 as 'irresponsible' and 'incompetent'. Although the US Administration pledged $8m. to Cambodia for its national elections, leading US lawmakers continued to criticize the country's electoral process and urged Washington to put pressure on Phnom-Penh. While attending the 2003 ASEAN Regional Forum (ARF), held in Phnom-Penh, US Secretary of State Colin Powell expressed his Government's desire to see peaceful elections conducted in Cambodia. Overall, however, it appeared that US fears of Islamist militancy in Cambodia had tempered its criticism of the Cambodian Government. The arrest of three foreign Muslim men in Cambodia, who were later charged with terrorist offences, and Cambodia's plan to expel Islamic foreigners before the holding of the ARF were thought to have pleased the US Administration.

In the mean time the People's Republic of China continued to provide political and financial support to the Hun Sen Government and vehemently criticized what it perceived as outside attempts to interfere in Cambodia's internal affairs. In January 1999 it was announced that China would supply agricultural equipment to Cambodia and construct a building to house the new Senate. Hun Sen visited Beijing in February, when he held discussions with the Chinese Premier, Zhu Rongji, and met President Jiang Zemin. Two months later the Cambodian Co-Ministers of Defence, Tea Banh and Prince Sisowath Sireirath, travelled to Beijing for discussions with their counterpart, Gen. Chi Haotian. Cambodia, which had been the recipient of US $3m. of military vehicles from China, secured additional funding of $1.5m. for its armed forces reform programme. Following Hun Sen's reiteration of an earlier pledge by the Cambodian Government that Cambodia would not engage in political relations with Taiwan, Cambodia's 'one-China' policy was further reaffirmed in March 2000 during Hun Sen's meeting with a visiting Chinese government delegation. In November 2000 President Jiang Zemin visited Cambodia (the first visit by a Chinese head of state since 1966). China offered Cambodia commercial credit worth $200m., including an allocation of $2.7m. for military training. In September 2002 the Chief of General Staff of the Chinese armed forces, Gen. Fu Quanyou, visited Phnom-Penh and reportedly pledged to triple China's military aid to Cambodia. Afterwards a Cambodian military spokesman declared that Sino-Cambodian ties were becoming 'stronger and stronger'. In November Hun Sen also met with his Chinese counterpart, Zhu Rongji, in Phnom-Penh and announced subsequently that the Chinese Government had agreed to relieve Cambodia of its debt to China, which amounted to approximately $200m. In 2003–04 Sino-Cambodian ties remained strong. High-ranking Chinese officials reportedly paid several visits to Cambodia, pledging $50m. to help the Cambodian Government address its financial problems. In April 2004 Hun Sen led a large group of CPP officials on a trip to China, where they met with Chinese President Hu Jintao and visited several provinces.

Relations between Cambodia and Japan remained excellent. In January 2000 the Prime Minister of Japan, Keizo Obuchi, made an official visit to Cambodia, the first by a Japanese head of government in over 40 years. During his visit, Obuchi met with King Norodom Sihanouk and Hun Sen. Obuchi declared the Japanese Government's intention to assist Cambodia's development over the forthcoming decade, having pledged US $140m. in assistance to Cambodia; the Japanese leader also expressed his readiness to send Japanese financial experts to Cambodia to assist in the areas of taxation and debt management, and was reported to be considering providing the Cambodian Government with non-project grants valued at $2,000m. yen. In return, Hun Sen pledged his Government's support for Japan's bid for permanent membership of the UN Security Council. In June 2001 Tokyo pledged to provide annual aid of $118m., thus maintaining its position as Cambodia's largest donor. During the same month Japan's Prince Akishino and his wife, Princess Kiko, made the first visit to Cambodia by members of the Japanese royal family. In early November 2001 the Japanese Government was reported to have planned to nominate a university professor of law as a judge for the Khmer Rouge trials, illustrating its continued interest in Cambodia.

Cambodia also continued to strengthen its bilateral ties with India. In April 2002 Atal Bihari Vajpayee paid a three-day visit to Cambodia—the first visit by an Indian Prime Minister since Jawaharlal Nehru's visit in 1954. Vajpayee reportedly supported Cambodia's apparent intention to proceed with the Khmer Rouge trials without any UN involvement and indicated that his Government would be willing to send an Indian judge to assist in the Cambodian judicial process.

There were several other positive developments in Cambodia's foreign relations. The Government hosted three major international meetings in 2002–03. In November 2002 Cambodia was the venue for the eighth ASEAN Summit meeting, which was attended by the leaders of the 10 ASEAN members, as well as those of China, Japan, India, the Republic of Korea and South Africa. Hun Sen reaffirmed his support for ASEAN's commitment to the 'Initiative for ASEAN Integration'. In June 2003 Cambodia, as Chair of ASEAN, organized both the 36th ASEAN meeting of ministers responsible for foreign affairs and the 10th ARF, attended by ministers responsible for foreign affairs and senior officials from its 22 members, including major powers such as the USA, China, Russia, Japan and India.

One major set-back for Cambodia in 2003 was that, due to the ongoing political stalemate, the donor community did not hold its annual CGC meeting to make the usual pledges. Only a few donors, in particular China and Japan, provided financial assistance, while other countries were prepared to wait until a new government was formed. However, relations between Cambodia and international organizations continued to improve. The country's relations with the UN from 2003 took a more positive turn. In response to the signing of an agreement on Khmer Rouge trial issues, the UN became less confrontational. At the end of 2003 the General Assembly removed from its agenda its annual discussion of the human rights situation in Cambodia. On 20 April 2004 the UN Commission on Human Rights also passed a resolution regarding Cambodia, which was viewed by critics as being 'weak' and 'different' from the real situation inside the country. In addition, in September 2003 Cambodia received permission to become the 147th member of the World Trade Organization (WTO)—the first 'least-developed' country to be invited to join. However, it was required to pass 46 laws by 2006 if it was to comply with WTO regulations.

Overall, during 2002–04 Cambodia remained internationally active. In September 2003 it hosted the 35th ASEAN meeting of ministers responsible for economic affairs, discussing regional economic integration among member states in South-East Asia. Minister of Foreign Affairs and International Co-operation Hor Nam Hong also led the Cambodian delegation to the third Asia Cooperation Dialogue (ACD) Foreign Ministers' Meeting in China in June 2004.

Economy

LAURA SUMMERS

Revised by SOK HACH

ECONOMIC CONDITIONS PRIOR TO 1979

At the time of independence from France the prospects for economic development were promising, not least because King Sihanouk's 'Royal Crusade for Independence' averted the economic devastation of war. Equally important, and in contrast to Viet Nam or Laos, Cambodia possessed ample food supplies and a satisfactory balance of trade, with most foreign exchange being derived from exported agricultural surpluses such as rice, freshwater fish or rubber latex. Nevertheless, the French were frequently criticized for leaving the country with inadequate infrastructure, little industry and high rates of illiteracy. The many formidable constraints on agriculture included low national yields (on average one metric ton or less of paddy rice per ha), mediocre soils, unreliable rain-fed irrigation systems, low mechanization and productivity, inadequate rural credit mechanisms, and inheritance traditions that resulted in the steady division of land holdings. Population growth rates as high as 3%–4% a year further contributed to pressure on the available arable land. Budget revenues, which came primarily from taxation of external trade, augmented by aid from the USA and France, were increasingly threatened during the 1960s.

Public investment in economic development, and specifically in import-substitution industrialization, began in 1960 in response to the rising costs of imports and the receipt of state-sector aid from the People's Republic of China and (from 1964) Czechoslovakia. State-owned industries producing plywood, cement, textiles, rubber tyres, paper, sugar, glass and tractors (assembled from parts supplied by Czechoslovakia) were established and created competition for private sector importers and retailers; many of the state-sector industries also benefited from US-financed commodity import programmes. Although alarmed by Cambodia's close relations with China, and the combining of East-West aid programmes, the USA continued to be Cambodia's principal aid donor until 1963. Despite the prohibition of defence alliances under the Geneva Agreements of 1954, the USA also supplied 'defence support' aid to the royal constabulary, and granted riel revenues from the commodity imports counterpart fund, directly to the Ministry of Defence, for the payment of army salaries. From Sihanouk's nationalist perspective, which many shared, US aid was meagre, amounting to less than the budget required and less than that awarded to neighbouring states; neither did it respond to Cambodia's mounting deficit nor to its developmental needs as perceived by the Government. Excessively conservative controls on monetary policies, supported by a belief in the need for a 'strong' riel (i.e. an overvalued rate of foreign exchange), discouraged both state and private-sector investment, especially in the faltering agricultural sector. In November 1963, following concealed moves to encourage the USA to increase its aid, the Cambodian Government requested an end to US aid programmes. When this decision, accompanied by sharp denunciations of US policies in southern Viet Nam, failed to secure the expected massive aid increases, Sihanouk proceeded with the nationalization of the financial sector (banking and insurance) and of international trade in January 1964. A state monopoly was also imposed on the production, sale and import of alcohol. (Hitherto informal trading and local brewing had lost the State valuable tax revenues.) Nine private banks, including five foreign banks, were closed and their accounts taken over by the National Bank and two hastily established subsidiaries.

The *étatist* reforms, however, failed significantly to ease the crisis caused by agricultural stagnation and population growth. Neither did the reforms succeed in solving investment and credit problems, despite the imposition of state controls on the import of non-essential luxuries and encouragement of the import of capital goods such as fertilizers. The army and the private sector opposed the reforms, which had become associated with a general foreign policy shift towards the Eastern Bloc, and created their own, informal 'black' markets in trade with neighbouring Thailand and Viet Nam. The state industries, already troubled by inefficiencies arising from poor equipment, shortages of spare parts and ineffective and often corrupt management, were unable to compete for local market shares. The eclipse of the formal market by the informal and the deceleration in agricultural production were represented by a decline in annual rates of economic growth, which averaged 7% per annum in 1953–63, but decreased to around 3% in 1963–70, thus barely keeping pace with the official rate of population growth. A state casino, established in Phnom-Penh in 1969 in a desperate attempt to increase budget revenue, failed to generate sufficient income and greatly alienated Sihanouk from important urban political groups. The group of officers led by Gen. Lon Nol, which instigated the military coup and which was supported by important technocrats and business groups, subsequently restored free-market policies, US aid and links to the IMF, the World Bank and the Asian Development Bank (ADB). Expectations of a rapid economic recovery following political realignment with the West, however, were unrealistic. The mobilization of an army numbering 200,000 (in place of a peacetime force of 35,000), intensive US bombing of the countryside, and the flow of rural dwellers into urban areas, exacerbated the already acute economic problems.

In retrospect, the material destruction and human displacements arising from the 1970–75 war can be seen to have prepared the way for the subsequent economic radicalism. Capitalizing on widespread sympathy amongst the Cambodian peasantry for the deposed Sihanouk and, as normal subsistence patterns were disrupted, their alienation from the Phnom-Penh Government, the clandestine Communist Party of Kampuchea (CPK, later known as the Khmers Rouges) began introducing collective forms of rice production in 1973. Initially achieving important economies of scale and gains in productivity, the collectives pooled family land, made use of large-scale irrigation systems and introduced high-yield seed varieties. By 1974 the revolutionary movement had secured control of much of the country's dwindling rice output as well as foreign-owned rubber plantations, where reduced operations were taxed. Although the work regime was harsh and food rations minimal, the system was perceived by those it served as a means of survival during the wartime emergency. In total the 1970–75 war reduced the labour force by an estimated 500,000 to 1m., including the departure for overseas exile of tens of thousands of professional people; around 3m. other Cambodians (nearly half the estimated population of about 7m.) became internal refugees, most of whom were located in Phnom-Penh. In the final months of the fighting US food airlifts supplied the capital city, which was finally occupied on 17 April 1975.

The National United Front of Cambodia (NUFC) ordered the inhabitants of Phnom-Penh and other urban centres to leave the cities for the countryside, where they were gradually incorporated into co-operative production units. In economic terms, the NUFC portrayed the evacuation as the only alternative to the massive suffering caused by malnutrition and starvation and the threat of US bombardments. In May 1975 the USA imposed an embargo on aid and trade. Economic realities and the US hostility heightened revolutionary communist suspicions of urban 'feudalists' and the middle classes, both groups being portrayed as exploiters and oppressors of the poor, unproductive 'parasites' on the economy, and allies of US capitalism. The revolutionary aim was to establish a new and classless society in which the unproductive 'new people' would work together with wartime 'base' (or 'old') people in large-scale agricultural co-operatives. Both markets and money were viewed as mechanisms of exploitation, rather than of production, and were abolished. Each co-operative was to be a self-reliant, autonomously managed and mixed production unit. A few state industries, producing goods essential for the modernization of agriculture, were created. New workers, consisting mainly of demobilized soldiers of peasant origin, were ordered to produce necessities such as clothing and blankets, farm machinery, boats for river transport, medicines, tyres and printed materials, often using inadequate machinery and materials. Most water pumps for agricultural use were donated by the Democratic People's Republic of Korea, while the People's Republic of China supplied technical assistance (and from 1978, significant amounts of military aid).

The leaders of Democratic Kampuchea, as the country became known, planned to reconstruct and to modernize Cambodia's backward agricultural sector over a period of 10–20 years and to promote industrialization, based primarily on the use of accumulated national capital, over 15–20 years. A five-year plan, introduced in 1976, set overly ambitious annual targets of three tons per ha for paddy, or six or seven tons per ha in regions where a second, dry-season crop was possible. The plan gave priority to rural infrastructure, particularly the construction of dams and industrial-scale irrigation systems. Labour teams, often numbering tens of thousands of manual workers, completed dozens of major projects, but, because engineering skills were denigrated and peasant revolutionary cadres worked to formulas (such as the construction of perfectly square canals regardless of topography), many of the new irrigation systems were technically flawed; some collapsed, disrupting production over wide areas. Output was also affected by a shortage of draught animals or tractors, inadequate seed supplies and high-yield seed stocks, shortages of fertilizers and pesticides, war-damaged (impacted) paddy fields, soil erosion and the rapid exhaustion of newly cleared land. As the labour force became increasingly demoralized and exhausted, production targets were rarely attained. Although rice output in 1976 may have satisfied national needs, there were acute food shortages in 1977 and in 1978, when adverse weather compounded the profound disorder. With popular hostility growing, the country's leaders blamed foreign powers and their allegedly concealed 'agents' for sabotaging the revolution. More than 20,000 party and administration cadres were secretly detained and executed by the state security service, while, in addition, tens of thousands of alleged enemies, often, but not exclusively 'new people', were detained and then executed by local authorities. In total, some 1.7m. people were believed to have died as a result of malnutrition, disease, exhaustion or execution. A border war with Viet Nam is estimated to have resulted in the deaths of thousands of others.

ECONOMIC DEVELOPMENT FROM 1979 TO 1991

The Vietnamese invasion of Cambodia at the end of 1978 brought most production to an abrupt halt. Vietnamese forces advanced through the countryside, pursuing remnants of the retreating army towards the western and northern frontiers and causing massive population displacement. Many people fled to avoid the fighting or to escape from the oppressive collectives and returned to their home villages in search of relatives or lost property. Following the looting of food stocks and the abandoning of agricultural production, only an international relief effort that cost more than US $300m. prevented a major famine from ensuing. With the inauguration of the People's Republic of Kampuchea (PRK) in 1979, peasant farmers were encouraged to form small production groups known as *krom samaki* ('solidarity groups'), each consisting of approximately 10–15 families. Land remained public property, to be farmed on a collective basis, but each family was also allotted a small private plot. These efforts to reorganize agricultural production were hampered by armed opposition to the new communist regime and by the refusal of most states to recognize a government imposed by Vietnamese invasion forces. Existing embargoes on aid, trade and investment in Cambodia were joined by the European Community (EC—later European Union—EU), Japan and other states in an effort to oblige Viet Nam to withdraw its military forces. Although generous economic assistance to other sectors of the economy was supplied by the USSR and its allies in the Council for Mutual Economic Assistance (CMEA), little aid was available for agriculture.

Another factor that constrained agrarian reconstruction and development was popular resistance to renewed attempts to promote socialism or any collective form of ownership and production. In most rural regions peasants initially agreed to form *krom samaki* and to work rice paddies on a collective basis for as long as shortages of land, draught animals, hand tools, water pumps and petrol made this necessary. Where possible, however, villagers persuaded local party and government officials to distribute farming land and equipment to individual families, who managed them as they chose and who marketed their own produce after paying taxes and collective, village dues owed to local schoolteachers, disabled people or war veterans and the unsupported elderly. Local cadres sometimes continued to organize work such as ploughing and harvesting on a mutual aid basis and as a labour-saving device. *Krom samaki* engaging in fully collectivized production with no family management of rice paddies or inputs were classified as Level One collectives; those *krom samaki* engaging in only limited collective work, with families retaining effective managerial control over inputs, were known as Level Two administrative units. Level Three units comprised those in which families engaged wholly in private production on land claimed as family property. By 1989 almost 90% of all organized solidarity groups were classified as Level Three. The new administration proved too weak to reimpose socialist forms of agriculture. The promotion of production in any way possible, securing tax revenues, military conscription and political support from the peasantry became the administration's primary concerns.

Between 1981 and 1984 the PRK permitted or recognized three economic sectors: the state sector, which comprised 57 state industries in 1984; a co-operative sector, composed of rural production groups; and a family sector, consisting of rural or urban families engaged in the production of handicrafts or garden crops for exchange in state-controlled markets. Capital for family enterprises was usually derived from hoarded or retrieved savings, or was accumulated in 'black market' trading in refugee camps along the border with Thailand. By 1985 the ruling party had recognized the existence of widespread street vending and petty marketing, and legalized a fourth, private, sector to promote the manufacture of light consumer goods. Small factories, producing everyday necessities such as household utensils or containers, were quickly established. In 1988 a fifth, mixed state and private, sector was created which allowed the establishment of joint ventures with the participation of foreign as well as national entrepreneurs. In the late 1980s the Ministry of Planning acknowledged the existence of about 2,000 private businesses. Concern, however, for the survival of state enterprises, following the reduction in supplies of raw materials from the USSR, prompted their leasing to national and foreign investors.

The pace of economic reform accelerated from 1989 as the Vietnamese army withdrew from Cambodia. Traditional usufruct rights to land and rights to inherit inhabitable property were restored in order to encourage more investment in agriculture and in construction. Transport, health care and education were also partly privatized, as were most state markets. The relaxation of many of the remaining controls on the economy dramatically boosted economic activity, especially in the capital, Phnom-Penh, where, by mid-1989, there were an estimated 13,000 small businesses. The value of private-sector output quickly exceeded that of the State, but with most private investment concentrated in service industries, and in particular import-export services, and with the state sector in decline, real growth for 1989 was only 2.4% compared with the previous year. In 1990, measured at constant 1989 prices, economic growth was negligible.

Economic growth was also constrained by an intensification of the civil war in 1990, resulting in a significant decline in state revenue. Sharp increases in budget expenditure for defence, arising from the expansion of the armed forces and the first ever purchases of military equipment, led to rapid increases in the money supply and in the rate of inflation, which stood at around 200% in 1990 and 1991. Foreign aid, excluding military assistance, averaged around US $150m. a year during the five-year plan period (1986–90), most of which the USSR provided. Soviet aid ceased in 1991, and from January of that year the State of Cambodia (SOC), the name of the administration in Phnom-Penh between 1989 and 1993, paid for all its imports from the USSR at international prices and in convertible currency. This represented a double crisis for the budget. Previously, commodities imported from the USSR, on a free or cheap, long-term credit basis, had been sold to finance the budget, and had represented as much as 40% of revenue in some years. From 1991, however, the same commodities, and in particular petroleum and its products, had to be paid for with foreign-currency earnings from Cambodian exports, principally timber. In 1991 30% of the budget was deficit-financed by the printing of new bank notes.

THE ECONOMIC EFFECTS OF THE IMPLEMENTATION OF THE UN PEACE AGREEMENT (1991–93)

Following the signing of the Paris Agreements on a Comprehensive Political Settlement of the Cambodian Conflict in October 1991, international embargoes on aid and trade were removed. The signatories to the Agreements, led by Japan and France, also pledged to assist the Cambodian people in the rehabilitation and reconstruction of the national economy. The first pledge of US $880m., made at the inaugural conference of the International Committee on the Reconstruction of Cambodia (ICORC) held in Tokyo, Japan, in 1992, was committed to reconstruction projects extending over several years, and was to commence after 1993 once a lawfully elected and internationally recognized government had been installed. In 1992 the urban-based, private-sector economy continued to thrive, and activity in housing construction and repair barely kept pace with the demand from the UN Transitional Authority in Cambodia (UNTAC), whose personnel numbered about 21,000. Inflation also increased rapidly as the influx of affluent employees added to existing pressures on the economy. By October 1992 UN contributions to the SOC budget, mostly for the payment of civil servants, had helped to ease overall inflationary pressure, decreasing the annual rate of increase in the consumer price index to around 175% by the end of 1992. In total, UN funds and spending accounted for an estimated one-tenth of Cambodia's GDP in 1992, and for much of the growth in GDP, which was estimated at 7% by the IMF in that year.

The economic recovery was largely confined to urban centres, however, and specifically to import-consuming, property-owning or retailing families. Little filtered through to rural areas, and poor urban consumers complained of price rises for food. Some earnings were repatriated to Viet Nam by the tens of thousands of migrant Vietnamese workers who, encouraged by the UN presence, took most construction jobs. Uncertainty about the economy and about who would be in government after 1993 undermined efforts to collect taxes or customs duties. Senior government officials were witnessed selling off state assets for personal gain, which led to protests from students, workers or the families who found themselves evicted following the auction of government housing. The signing of timber agreements between Thai companies and the political parties also spurred some of the national growth in 1992, but, with hardwoods being irreplaceable and with no reafforestation programme, the rapacious expansion in timber-logging served to undermine confidence in the country's economic future.

Foreign investors and Cambodians returning from abroad after 1991 were wary of investing large sums in an economy that lacked both the legal infrastructure and the tax base essential for stability. The economy's vulnerability to political events was clearly demonstrated in 1993 when growth in GDP slumped to 3.9% because of the civil disorder surrounding the UN-organized elections held in May. Output also declined in agriculture, having been disrupted by continuing, low-intensity warfare, and by sharp fluctuations in the value of the riel, which halved when the inflation rate reached 450%–500% (in yearly terms); this was followed by a recessionary loop in the latter half of 1993. (The annual rate of inflation was finally estimated at 55%.) A measure of stability and confidence in the economy was restored following the successful completion of the elections, the prevention of a coup attempt in June and the formation of a coalition Government.

ECONOMIC DEVELOPMENTS UNDER THE ROYAL GOVERNMENT OF CAMBODIA (1993–)

The withdrawal of UNTAC in the final quarter of 1993 did not provoke the collapse in economic confidence that many, primarily in the expanding service sector, feared. Economic growth reached 5.7% in 1994, 7.6% in 1995 and 7.0% in 1996. The emergency rehabilitation, stabilization and structural adjustment agreements concluded with the World Bank, IMF and ADB helped stabilize the riel. Contrary to public expectations, the riel, in fact, appreciated against a falling US dollar, making the prices of many imports cheaper and allowing structural improvements in the organization of the national budget.

At the end of December 1993 the National Assembly approved new financial laws and a national budget for 1994, which met with widespread approval within the international donor community. Just over one-half (52%) of the proposed budget of US $342.2m. was to have been funded by domestically generated revenue, with the remaining expenditure ($165.2m.—mostly for capital outlays of $120m.) being covered by international assistance. In common with previous policy, most domestically generated income was to come from customs duties and consumption taxes levied by the customs authorities. In a significant departure, however, the budget was centralized and placed under the control of the Ministry of Finance, thus depriving individual ministries, state agencies and provincial authorities of their previous financial autonomy, and, specifically, their right to dispose of central budget allocations without accounting, to raise local tax revenues and even to accumulate foreign-exchange earnings. Central control over all revenue collection theoretically permitted government officials to restrain expenditure in some areas (such as defence), to monitor the use of aid more effectively and to direct investment into areas identified as of national importance (for example, production for export and agriculture). A centralized budget also represented a countermeasure to inflation, for which annual benchmarks of 5%–10% are required by donor agencies.

In 1993–94 important progress was made in the area of tax collection, which was a sector that traditionally yielded less than 5% of national output and which was constrained by inadequate revenue laws and corrupt practice. With a combination of new controls and political pressure from the Ministry of Trade, revenue from taxes increased from around US $5m. per month in June 1993 to $10m. per month by the end of the year. As the Government's fiscal position noticeably improved, the Ministry of Finance drafted new legislation designed to attract foreign capital investment, especially long-term inflows, which would encourage the introduction of new technology, export-based industry and tourism, agro-industry and rural development. The new Law on Investment in Cambodia, approved by the National Assembly in August 1994, provided for a low corporate taxation rate of 9%, tax exemptions of up to eight years, five-year loss carried forward, no taxation of reinvested profits, the free repatriation of profits, the leasing of land for up to 70 years and tariff exemptions on imported capital goods destined for use in export-orientated production. In 1994 investments worth $465m. (of which $285m. were domestic) were approved by a newly created Cambodian Development Council (CDC), more than one-half of which was for tourism and other service industries.

Cambodia's attempts to secure foreign investment in the following years were hindered by: investor anxiety about political and budgetary stability; a lack of banking regulations and legal guarantees; the slow implementation of projects as a result of ministerial or local government agencies demanding bribes; and expanding criminal activity linked to poverty, unemployment and other social grievances. The CDC approved investments totalling about US $2,250m. in 1995, and more than $1,000m. in 1996. According to the IMF, however, actual investments remained substantially lower; net foreign investment was estimated at $151m. in 1995 and at $294m. in 1996. Investor confidence was reinforced by other developments: the USA granted Most Favoured Nation (MFN—now Normal Trade Relations) trading status in September 1996, and admitted Cambodia to its Generalized System of Preferences (GSP) in August 1997; Cambodia also joined the International Finance Corporation (IFC) of the World Bank in December 1996, thereby gaining access to multilateral sources of finance for private-sector interests. However, following domestic political unrest in July 1997, which resulted in the removal of Prince Ranariddh as First Prime Minister, and the onset of the regional economic crisis in the same year, foreign direct investment decelerated, and new commitments declined to $759m. for the year. Although the domestic political situation improved following the formation of a potentially stable coalition Government in November 1998 and the cessation of any effective armed resistance in the country, with the surrender or capture of many of the remaining Khmer Rouge leaders, investment approved by the CDC continued to decline sharply during 1999–2001. Such investment reached only $218m. in 2001. Registered investment increased to $235m. in 2002 and to $295m. in 2003. However, actual foreign direct investment was estimated to be only $66m and $59m., respectively, in the same years. One effect of the low level of investment on the country's economy was that it became increasingly difficult for new entrants into the labour market (numbering about 200,000 people per year) to be absorbed into the work-force.

A major factor governing future prospects for budgetary control of the economy concerned the decision in June 1994 to transfer control of timber extraction and export to the Ministry of National Defence, which had insisted on higher budgetary allocations. To the dismay of environmentalists and international donors, revisions to the 1994 budget in September of that year allocated the security forces a total of 48.3% of recorded, current expenditure; revenues from logging, theoretically controlled and originally set at US $2.2m., rose to $33.2m. Since the transfer of rights of revenue collection to a ministry was in contravention of agreements concluded with international donors, the IMF withheld a disbursement of aid funds at the end of 1994, and resumed payments only in 1995. Although donors publicly renewed their pledges of support, these were conditional upon private government assurances related to budgetary procedures and corruption. The 1995 budget aimed to reduce current expenditure by about 10%, (mainly through reductions in defence and salary outlays) and to increase capital or developmental expenditure. These objectives were to be achieved without resort to environmentally destructive, windfall forestry revenues (set at only $1.5m. for the fiscal year 1995). A planned deficit equal to around 6% of GDP was to be covered by concessional financing, thereby providing a measure of continuing stability for the currency. In the second half of the year the 1995 budget was revised upwards by 20%, owing to higher than expected revenue collections, mostly from timber royalties and exports (recorded at $34m.). Reports of new logging deals in January 1996 (see Forestry, below) prompted the IMF, as well as the World Bank, to withhold a total of $47m. in aid, in support of reserves and agriculture respectively. The Government secured $501m. in new aid pledges at the annual meeting of the Consultative Group for Cambodia (CGC) in July 1996, but was advised by donors of the need for 'prior actions' before structural adjustment projects could proceed. In particular, the Government was expected to honour 1994 pledges to reduce the size of the civil service by 20% (from an October 1994 level of 143,855 employees) and to proceed with military demobilization (reducing the armed forces from 130,000 to 90,000);

public expenditure was also to accord higher priority to health, education, agriculture and rural development. Donor advice was taken seriously in parts of the Ministry of Finance, but disorder in the budgetary process worsened: up to 20,000 defecting Khmer Rouge soldiers were inducted into the Royal Cambodian Armed Forces (RCAF); ministries continued to provide jobs as a means of securing political support; and the value of the riel declined sharply against the US dollar in local exchanges, as the effects of the increase in the money supply and pressures on the country's limited reserves were felt.

The violence and political unrest of July 1997 rendered prospects for the future of Cambodia's economic development uncertain. Following Prince Ranariddh's removal from office, several international lenders and donors (including the USA, the IMF and the World Bank) withheld approximately US $100m. in aid for 1997, and made resumption of aid dependent on the successful completion of free and fair national elections. The withdrawal of aid and investment resulted in the estimated loss of 40,000 jobs, mostly in the service sector, and damage in Phnom-Penh from the fighting and ensuing looting was estimated at $100m. The regional economic crisis also affected Cambodia's development, leading to a decrease in investment from the region and to a decline in demand for Cambodian exports. Economic growth thus decelerated in 1997 to about 1%, down from 7% in 1996. Government revenue collection fell by 6% for 1997 compared with the planned budget, owing to decreased revenue from customs duties (caused by a decline in imports) and income taxes. Threatened with a large budget deficit, the Government attempted to reduce spending, but the continued fighting in the north-west and the need to incorporate defecting Khmer Rouge soldiers into the RCAF necessitated increased military spending, and the reductions were thus made in other areas. Inflation increased in 1997, owing to large-scale purchasing prompted by public anxiety in the wake of shortages of certain goods following the political turmoil.

There was no discernible GDP growth in 1998, owing to a poor agricultural performance following a drought, the continued effects of the regional financial crisis, reduced aid levels, and the political uncertainty surrounding the election in July 1998 and its aftermath. Per caput GDP actually declined in real terms by 1.8%. Consumer spending remained depressed following the election. The riel depreciated against the US dollar and some regional currencies, causing an increase in inflation to 12% in 1998. However, the economy recovered in 1999–2002, with growth rates estimated at an annual average of approximately 6.0%, as a result of improved consumer confidence and renewed aid pledges. GDP growth in 2003 was estimated at 5.0%, driven by significant expansion in the agricultural sector and the strong performance of the garment industry.

Cambodia's long-term development requires major investment in infrastructural improvements to attract substantial foreign investments. The country also needs significant resources to be allocated to education, to provide skilled personnel in order to improve its administrative, legal, educational and medical institutions, and also to the health sector in order to improve general health care and to avert an impending HIV/AIDS crisis.

AGRICULTURE

After many years of political instability and the neglect of economic development, agriculture (including forestry and fishing) continued to provide employment for approximately 70% of the population, representing an estimated labour force of 4.5m. in 2002. The declining contribution of agriculture to GDP, from about 60% in 1988 to about 36% in 2002, is indicative of the continuing poverty in rural areas. Agricultural GDP averaged 0.5% growth per annum, in real terms, during 1995–2002, about two percentage points lower than the rate of growth of the population during the same period. In 2003 the agriculture sector, having recovered from a severe drought in the previous year, grew by 9.4%, its best performance in seven years. To the perennial problems of poor soils, uncertain rainfall and irregular flooding, inadequate fertilization (by silt or appropriate chemicals) or use of pesticides, may be added shortages of male labour power. As a result of war losses, and the greater political vulnerability of men during the period of Democratic Kampuchea, in the mid-1990s women constituted 54% of the adult population over 15, headed 20% of rural households and held title to substantial paddy land. Widowed women who received land after 1989 experienced difficulty working it, with hired help or the assistance of relatives, and without labour support from *krom samaki*. Many women were forced to sell land or to neglect agriculture in pursuit of other means of income generation. An additional, and serious, problem is the presence of land-mines over an estimated 300,000 ha, including 55,000 ha of prime rice-growing land in Siem Reab, Banteay Mean Chey and Bat Dambang Provinces, the country's traditional grain basket. In 1997 total cultivated area increased to 2.1m. ha, from 1.8m. ha in 1994, as some land previously laid with land-mines was reclaimed. Following the establishment of the new coalition Government in November 1998, however, plans were announced to improve the agricultural sector through the further clearance of land-mines to augment cultivable land and to consolidate the implementation of the land ownership law to guarantee the safe occupation of land, thus encouraging investment and long-term development.

The retreat from commercial production to subsistence rice production, which characterized the 1980s, can be attributed to the lack of price incentives offered by the successive PRK and SOC Governments. Official purchasing prices for the state markets were low, and only slowly increased when procurement difficulties created food problems in urban areas. As informal or free markets for rice expanded in 1989, prices reportedly reached 40 riels per kg, compared with the state purchasing price of 15 riels. Compulsory deliveries of rice—either in the form of taxation, which was slowly reintroduced in 1983 as voluntary 'patriotic contributions', and later, in the form of levies on output (measured at around 10% of output above one metric ton per ha)—were abandoned in 1989 when progressive rates of taxation were introduced (e.g. 25 kg per ha on land yielding less than one ton per ha, up to 60 kg per ha on land yielding more than 2.5 tons per ha). These rates were abandoned in 1992 when the authorities, seeking peasant support in the 1993 elections, declared a 15-year moratorium on the taxation of agriculture. In the early 1990s the introduction of other protectionist policies, including a ban on the export of paddy (which was rescinded in 1995) and exemptions from duties on rice imports, were designed to keep rice prices as low as possible, although, ultimately, they served to undermine producer incentives.

Draught animal stocks were severely depleted during the conflicts of the 1970s, and by 1979 there were only about 1.5m. head of cattle and buffaloes, compared with 3.1m. in 1970. Herds have subsequently recovered and by early 1991 there were 2.9m. head of cattle, including 1.5m. draught animals, representing a sufficient number to plough 2.1m. ha of paddy by traditional, manual means. During the 1990s livestock numbers increased further, owing to a rising consumption of meat rather than expansion of cereal production. The UN Food and Agriculture Organization (FAO) estimated total cattle head at 2.9m. in 2002. The livestock sector's contribution to GDP increased by about 4% per annum during the latter half of the 1990s.

Despite the obstacles and constraints, agricultural production improved significantly between the early 1980s and the early 1990s. According to official figures, the annual harvest of paddy rice increased from 565,000 metric tons in 1979 to about 2.55m. tons in 1991, representing the highest level achieved since 1967. Paddy production attained self-sufficiency levels in 1995, rising to 3.5m. tons. This increase was due both to the expansion of the area under cultivation and to improvements in productivity. Paddy production continued to increase steadily thereafter, reaching 4m. tons in 1999. About 300,000 tons of milled rice were exported in that year. Productivity rose from 1.3 tons per ha in the early 1980s to 1.8 tons per ha in 1995, and then to 2.0 tons per ha in 1999, although yields remained among the lowest in Asia. Despite efforts to improve farming techniques and promote new varieties of seeds and fertilizers, little perceptible progress has been made since 1999, owing to successive floods and droughts, and to a lack of development in the irrigation system. In 2001–03 less than 10% of rice fields were irrigated and irregular rainfall and serious flooding in the Mekong River basin damaged approximately 250,000 ha of rice crops. Extension of irrigation systems, better distribution of new land brought into cultivation and improvement in marketing systems would significantly increase crop production in Cambodia.

Many problems face the agricultural sector, apart from the inability of peasants to bring available land back into production. Among the most important are problems of water control, which have worsened as a direct result of deforestation. Of the 841 irrigation systems surveyed in 1994 by the UN Development Programme (UNDP), some 80% had malfunctioned since 1991 owing to damage from recent environmental changes. In 1999 the Government announced plans to expand irrigation networks from 16.6% of cultivable land to 20%. A flat topography prohibits the construction of large dams, except on the Mekong River, while smaller systems were designed for flood patterns that no longer exist in the Mekong-Tonlé Sap basin in Cambodia.

Vegetables and other secondary food crops feature prominently in household consumption and income generation in Cambodia. Vegetables are grown largely on homestead lands and their cultivation is considered to constitute a small-scale production practice. The size of the area under cultivation and levels of production have fluctuated from year to year owing to pest and disease attacks, natural disasters and price fluctuations. Daily per caput vegetable consumption in Cambodia averages about one-half of the standard 200g recommended by FAO. It is estimated that the total annual consumption of vegetables is about 600,000 metric tons. Forestry and lakes are the source of about 25% of all vegetables consumed, while local cultivation accounts for another 25%. The remaining 50% is imported from neighbouring countries. There are more than 40 local and international NGOs who work to improve food security for rural people through the promotion of vegetable growing for both self-consumption and income generation. As a consequence of these activities, the productivity of some areas increased significantly, especially since 2001. Moreover, the horticulture and livestock sectors should benefit from the increased demands of the tourism sector (hotels and restaurants). However, poor infrastructure in rural areas and the inability of local products to compete with imported vegetables in terms of either quality or price cause problems for the growers. In addition, improved distribution of land resources, and the definition of rights and responsibilities over them, are of fundamental importance to Cambodia's development. The country's lack of clear land tenure rights has inhibited business people from investing in the agricultural sector. Nevertheless, the passage of a new Land Law in August 2001 marked a watershed in creating a legal framework for the management of land in Cambodia.

Exports of rubber were once an important source of foreign exchange. Output is generally lower than in the 1960s, however, owing to the neglect of the six old plantations (comprising 52,439 ha in 1969) and their slow rehabilitation during the 1980s. Rubber production was adversely affected by the ageing of trees. While rubber trees began to be replanted in the early 1980s, they took time to mature. Rubber production increased during the mid-1990s, reaching 46,000 metric tons in 1996, thus approaching the annual outputs achieved in the 1960s when the plantations were mature and routinely restocked. Production declined to 38,600 tons in 1997 and 1998, but significantly recovered to about 47,000 tons annually in 1999–2001. In 1996, meanwhile, world production of rubber began to exceed demand, causing a decline in international prices of 27%. In an attempt to remain competitive, the Government suspended a 10% export tax on rubber in October 1997. This did not, however, halt illegal exports to Viet Nam, which had become prevalent in the early 1990s. Owing to a sharp fall in international rubber prices following the onset of the Asian economic crisis in mid-1997, Cambodia's unit value declined from US $1,108 per ton in 1995 to $450 in 2001, resulting in export values of $45m. in 1995 and an estimated $21m. in 2001. Exports improved to about $29m. and $35m. in 2002 and 2003, respectively, owing to the higher international price of rubber. Plans for the eventual privatization of the six state-owned plantations (five in Kampong Cham Province and one in Kracheh Province) were agreed with the Caisse Française de Développement (now Agence Française de Développement) in January 1995, and involved the introduction of foreign managerial control. By the end of 2002 the six plantations were still in the process of being transformed into 'public enterprises' in preparation for privatization.

FORESTRY

In 1969 the area under forest cover was 13.2m. ha, which accounted for 73% of the national land area. By 1997, however, forest cover had been reduced to about 10.6m. ha, representing only 58% of total land area. Provincial authorities were engaged in widespread illegal logging activities, despite bans on felling and efforts to centralize revenue collection. Much of the exploitation of Cambodia's forests was carried out by Thai companies following the granting of concessionary areas, with rights of exploitation, either directly by the Phnom-Penh authorities, or indirectly by local provincial authorities or opposition party leaders and military officials. Timber concessions incorporating transport across disputed zones often involved payments to more than one party.

On the advice of international environmental agencies, UNTAC imposed a moratorium on round timber exports at the end of 1992. The ban was suspended by the Royal Government in October 1993, ostensibly to permit the export of old timber felled before the UN moratorium and to prevent it from rotting. A system requiring certificates of origin was established, which facilitated taxation and generated immediate revenues for the budget. The suspension was abused, however, and, when evidence of fresh logging activity emerged, the Government re-imposed a ban in March 1994. This ban remained in force until June, when the two Prime Ministers decided independently to revise central government controls and transferred major responsibilities for decisions concerning export licences, concessions and revenue gathering to the Ministry of National Defence. Protests against the resumption of ministerial 'tax farming' in August and September 1994 coincided with severe drought and then flooding in rural areas, thus renewing awareness of the ecological damage arising from deforestation. By November Prince Ranariddh was obliged to pledge the restoration of the ban on logging and to resume attempts to impose fiscal responsibility on national ministries. The new ban on logging came into effect in April 1995, but a timber concession granted to Malaysia's Samling Corporation in August 1994, which awarded rights to exploit 805,509 ha (about 12% of the remaining forest area) over a 60-year period, was not rescinded. Thai companies received renewed authorization to remove previously felled logs from May 1995, but this concession was rapidly and openly abused. In March 1996 it was revealed that the two Prime Ministers had personally authorized 20 Thai companies to remove up to 1,079,300 cu m of logs, when accounts showed that felled, lying timber inside Cambodia following the ban on cutting amounted to only 330,648 cu m. The revelations precipitated the suspension of IMF loans and protests from the co-premiers that the contracts were being misinterpreted.

Seeking to avert criticism at the July 1996 donors' meeting in Tokyo, the Government informed the Thai authorities that the rights of Thai firms to remove felled timber would expire, definitively, on 31 December 1996. By June 1997 environmentalist groups revealed that illegal logging continued without disruption, with companies often securing protection and labour from regional military commanders. Official revenue yields, however, were appallingly low: approximately US $20m. in 1995, $10m. in 1996 and $12.7m. in 1997. Revenue continued to decline sharply from 2001, reaching a low of $2m. in 2003. Reafforestation projects were being implemented, but would not immediately affect disruptions to seasonal rainfall patterns and falling water levels in the Tonlé Sap.

An intensified campaign against illegal logging was ordered in January 1999 prior to the meeting of multilateral and bilateral aid donors in February. The donors reiterated their stance that aid was to be dependent, amongst other conditions, on Cambodia's implementation of a sustainable forestry management programme to halt the rapid environmental degradation caused by deforestation. By mid-1999 the Government had revoked logging concessions totalling 2,173,041 ha and made this area into forest reserves. In mid-1999, according to official sources, 4,739,153 ha of land was being worked by 14 companies. The World Bank estimated that 4.2m. cu m of commercial timber were illegally felled in 1997 (costing the Government US $100m. in lost revenue), compared with an estimated sustainable yield of 1.5 cu m, and that at this rate Cambodia's timber resources would be depleted within five years. To protect the forests from further deterioration, reforms have been initiated with the support of donors, and remedial measures have been implemented. As a result, production of logs has declined steeply since 2000. To ease access to the country's common property, and as part of the reforms, the Government has allowed local populations access to some forest land (about 5% of the total forest land). Assisted by NGOs, about 200 community forests were established in over 16 provinces at the end of 2001. However, the role played by the village communities in the management of forestry remained in need of enhancement.

FISHING

Fish is the primary source of protein in the national diet. Historically, freshwater catches from the Mekong-Tonlé Sap basin satisfied most of the national requirement, with marine products accounting for less than 20%. Throughout the 1980s total recorded catches increased steadily, reaching 117,800 metric tons (including 36,400 tons of marine products) in 1991, compared with only 20,000 tons during the famine year of 1980. However, the total catch declined to 111,150 tons in 1992 and 108,900 tons in 1993. Fish farming, introduced by foreign aid agencies and the Vietnamese, supplied about 10% of the freshwater catch in 1990. Ecological changes, such as falling water levels, heavy and rapid flooding, and loss of the forested regions once completely inundated by flooding in the Tonlé Sap, have disrupted fish migrations, breeding and feeding patterns. The forested area around the Tonlé Sap was the main breeding ground for the lake's fish. By 1998 only 39% of the original 10,000 sq km of flooded forest remained under natural vegetation. Moreover, substantial shares of the marine catch are lost to the economy as a result of illegal, offshore fishing by Thai companies which receive unofficial licences, and protection from the coastal marine and rivers police and provincial authorities, to continue their activities. The officially recorded freshwater catch in 1995 was 72,500 tons, an increase of 7,500 tons on the 1994 catch of 65,000 tons. The appearance of improvement is misleading, concealing the over-harvesting of baby *pra kchao* fish which fetch higher prices than mature fish, reductions in the catch of other varieties, owing to ecological changes disrupting breeding and migration patterns, and illegal fishing. Laws prohibiting grenade and electric-shock fishing are enforced in some regions, with aid from the army, but these illegal techniques, and others, are still in common use. Deforestation presents the most serious challenge to inland fishing: according to an Institute of Technology study, the Tonlé Sap will be silted over by the year 2023 if logging continues at the pace set in the early 1990s. The mean depth of the lake declined by 0.5 m to 1 m over the 15 years between January 1979 and January 1994.

Fishing was transferred to the private sector in late 1993, following the creation of the coalition Government. Leases, or 'concessions' as they were called in colonial times, to lots on the Tonlé Sap and inland rivers are sold on a two-yearly basis to the highest bidders. Sales, however, are organized by provincial governors, with arrangements being open to manipulation and abuse at several stages and levels. Successful bidders for one or more of the 280 leases available must pay 40% of their bid as a deposit for exploitation rights, the balance falling due at the end of the lease period. Public revenues from sales of leases increased sharply to US $3.7m. in 1995, from $1.2m. in 1993 and $1.9m. in 1994, but subsequently declined steadily to less than $2.0m. in 2002 and to approximately $1.4m. in 2003, owing to high levels of corruption within the sector. In 2003 the commercial freshwater catch was estimated at 90,000 metric tons, and the sector contributed about 3.6% of total GDP.

INDUSTRY

Following the installation of the socialist PRK regime in 1979, many state industries were rehabilitated. In contrast to the patterns of industrial concentration in earlier periods, however, the PRK revival plan involved substantial promotion of industrial activity at the provincial level and significant encouragement of small-scale, family-based handicraft activity.

Without heavy industry in Cambodia, until the mid-1990s the industrial sector was dominated by rice mills (of which there were approximately 1,500) and by 80–100 state industries or factories, including the latex-processing plant in Kampong Cham, which remains wholly state-owned. However, in subsequent years there was significant industrial growth in such areas as textiles, garments, beer and soft drinks, food-processing and construction materials, and in early 1998 a consumer electronics plant started production. Many factories have operated at low levels of efficiency in recent years, being constrained by outdated and virtually unserviceable equipment, frequent interruptions to the power supply and difficulties in securing supplies. The total contribution of industry to GDP was estimated at 15% in 1994, around the same level as the estimated contribution to GDP of the sector in the 1960s. However, owing to the rapid expansion of the garment industry in the latter half of the 1990s, the contribution to GDP of the industrial sector as a whole increased to 27% in 2003.

Many difficulties in privatizing, encouraging and reorganizing industry can be attributed to policies introduced in the socialist era of the 1980s when state companies were overstaffed and depended upon the command economy for administrative support and subsidies. After 1988 managers were awarded local autonomy: they could retain profits earned from the disposal of products surplus to plan targets; they could reinvest such profits in new product lines; and they could pay piece-rates, or bonuses, to productive workers. Many state enterprises received full financial autonomy from the budget after June 1990, which meant that managers had to procure all their supplies from the private sector, to keep accurate accounts and to finance deficits on operations through acquiring loans. As a result many industries were unable to record profits, and by July 1991 about 25 state factories, in economic difficulty, had been leased to private entrepreneurs. Part or all of the labour force were dismissed in some enterprises creating political disruptions and drawing attention, *inter alia*, to the absence of labour laws. At the end of 1998 the Government had privatized 73 industrial enterprises, of which 68 were leased to the private sector and five were joint ventures. There remained only two enterprises to be privatized, whilst one company was to be retained in public ownership.

In the early 1990s the largest industries were engaged in agro-food processing or in the production of construction materials. Between 1995 and 1998, because Cambodia was not yet subject to international quotas, considerable foreign direct investment from Malaysia, Hong Kong, Taiwan, the Republic of Korea, the People's Republic of China and Singapore resulted in the rapid development of the garment sector. By the end of 2000 there were about 220 garment factories, employing 160,000 workers, in operation in Cambodia, compared with 13 in 1995. The garment sector was characterized by poor working conditions and low wages, a situation that provoked labour unrest in the late 1990s. About 70% of the garments were exported to the USA, which granted Cambodia MFN status (now styled Normal Trade Relations) in 1997, and the rest to Europe. Garment exports, which were negligible in 1994, increased to US $378m. in 1998. In January 1999 the USA introduced quantitative restrictions on imports of 12 categories of garment products from Cambodia, necessitating an increased focus on European markets. Nevertheless, in 1999 garment exports increased by 46%, and were valued at $553m. Despite the US quota and global economic slowdown, expansion of the sector remained robust in 2000, with an increase in production of 78% compared with the previous year. However, owing both to the normalization of trade relations between Viet Nam and the USA and to the accession of the People's Republic of China to the World Trade Organization (WTO), the number of factories operating in Cambodia had declined to 184 by the end of 2002. Meanwhile, Cambodian garment exports continued to increase, but at a slower rate of 18% per annum, during 2001–03. Exports were valued at $1,580m. in the latter year.

Investment in industry decelerated following the events of July 1997. Some factories were damaged or looted during the fighting, and many expatriate staff left the country. The Government pledged to settle claims with companies that suffered damage, but owing to a lack of funds it hoped to achieve this through tax rebates. The industrial sector was also adversely affected by the regional financial crisis, which caused a decline in investment from countries in the region and increased competition in the domestic market in some industries, as the weakening of the baht led to an influx of cheap Thai imports.

The Royal Government announced plans in December 1998 to promote the establishment of agro-industrial enterprises (for products like palm oil, rubber and tapioca) and to encourage the development of industrial zones, especially in coastal areas, to prepare goods for export.

Industrial GDP increased by 17.9% in 1996, but the growth rate decelerated to 2.4% in 1997 before registering a strong recovery to about 15% per year during 1998–2002. The GDP of the construction sector followed the same pattern until 1998, increasing by 8.6% and 20.9% in 1995 and 1996 respectively, but then, owing to the loss of external funding in 1997, declining by 8.9% in that year, before growing by 13.4% in 1998. Construction activity remained depressed in 1999, with the sector registering a GDP growth rate of only 1.5%, and then sharply declined in 2000 and 2001, owing to a lack of new investment during that period. In 2002 construction activity recovered significantly (registering a 12% increase in real terms), as a result of significant inflows of foreign aid. However, the GDP of the construction sector contracted by 4% in 2003, affected by the postponement of building in the hotel sector owing to a decrease in tourist arrivals, as a result of the epidemic of Severe Acute Respiratory Syndrome (SARS) in the region in that year.

Mining

Mining contributed an estimated 0.3% of GDP in 2002. The sector expanded by 20.1% in 1996 but declined by 1.0% in 1998, before recovering by an estimated 14.4% in 1999. Cambodia possesses few commercially exploitable mineral reserves. Phosphate deposits are processed for use as fertilizer, but the exploitation of deposits of iron ore, silver, bauxite, tin, silicon and manganese is judged unviable. The commercial mining of sapphires and rubies in the Pailin district of Bat Dambang increased dramatically after 1991 when the Khmers Rouges used revenues from this source to offset lost aid from China. In common with the timber trade, Thai companies profited the most. Although the Khmers Rouges were effectively defunct by 1999, many of the gem-mining areas remained under the control of former Khmer Rouge leaders, and exploitation of these resources was largely uncontrolled by the central Government. Use of the Bavel River for the industrial sifting of excavated soil for gems has contaminated water supplies in Bat Dambang Province. There is some gold mining—at a household level—in the eastern provinces. A number of Australian, Canadian and Malaysian companies have signed agreements to prospect in this region, but by 1999 the Government had yet to pass a draft mining law.

In late 1991 international oil companies began offshore exploration for deposits of petroleum and natural gas. Although initial seismic surveys were promising, and Thai geological assessments suggested reserves worth as much as US $15,000m. in disputed parts of the Pattani trough, the extent and accessibility of the reserves have yet to be determined. The most successful tests in 1994 produced flows of 4.7m. cu ft of gas (Enterprise Oil consortium) or 224 barrels of oil per day (Campex). The area of overlapping claims is estimated at 5,570 sq km; Thailand awarded concessions in the area in the 1970s. Proposals for a joint development zone have come to nothing so far, but a memorandum of understanding, issued in June 1996 following talks at prime ministerial level, indicated that negotiations concerning the competing claims would continue. Enterprise Oil halted all exploration in December 1996, but expressed an interest in securing new claims in Cambodian waters. Cambodia signed agreements in November 1997 with five foreign oil companies (two from Japan, and others from the USA, the United Kingdom and Australia) allowing them to explore in four offshore concessions in the disputed area for an initial payment of $9.6m., followed by an additional $36.6m. over the next 10 years.

Energy

All commercial energy used in Cambodia is imported. Domestic energy is principally derived from timber (6m. cu m of logs were felled in 1997 for domestic consumption). The country has an installed capacity of 28.7 MW, of which 15 MW is accounted for by an oil-fired thermal power plant and the rest by diesel generating units. Only a small proportion of the generating capacity can be utilized, however, owing to a lack of spare parts and a shortage of fuel. Installed generating capacity was 51.97 MW in 1984. Cambodia has considerable hydropower potential. Two hydroelectric plants were under construction in the late 1990s, with numerous other projects under discussion, owing to the potential importance of revenue from sales of electricity to Thailand and Viet Nam. In 1999 the German company Siemens AG proposed a US $500m. project to build two power plants within two years, one with an installed capacity of 320 MW in Sihanoukville and the other of 180 MW in Kampong Spueu Province. According to a preliminary strategy drafted by Electricité du Cambodge (ECD), with assistance from the World Bank, almost all of Cambodia was to be electrified by 2015. Meanwhile, production of electricity in Cambodia increased by about 12% a year.

FINANCE

Total state revenue in 1989 amounted to only 5.5% of estimated gross national output, with revenue from taxation equal to only 2%. As budget support from the USSR was also falling, the SOC had to seek alternative sources of revenue and to reduce expenditure wherever possible. Steps were taken in 1989–90 to reduce the number of state employees—both civil servants and other workers—and to bring an end to state subsidies in accordance with practices already adopted in Viet Nam and in Laos. As noted above, state factories in deficit were leased to private entrepreneurs, a commercial banking system was introduced and attempts were made to secure improved revenues from taxation. The continued military expenditure, however, (see above) was funded mostly by timber sales and by budget deficit financing.

Until 1989 the new riel, introduced in March 1980, had been relatively stable. However, in preparation for the new terms of trade being arranged with the USSR and the CMEA, in October 1988 a policy of floating rates of exchange was adopted, which closely shadowed changes in 'black market' rates fixed to the value of gold in the Hong Kong market and to the value of the US dollar. The rate of inflation increased throughout 1989, and reached very high levels in 1990–92 in response to the lack of foreign exchange and the heavy government reliance on the printing of new currency. When it appeared that gold and other currencies might replace the national currency in urban market exchanges, in September 1991 the Government banned the use of gold and other currencies and devalued the Cambodian currency to 750 riels per US $1. By the second quarter of 1993, when the rising inflation rate affecting the riel had reached its peak, more than 4,000 riels were required to purchase one dollar. Following the elections, and with stabilization programmes in place, the riel recovered and maintained its position on foreign exchanges at 2,200–2,600 riels to US $1. Since early 1994 official market exchange rates have been kept in line with free market exchanges and adjustments have been made on a daily basis. The Law on Investment, introduced in August 1994, removed all foreign-exchange restrictions that had previously applied to investors. New bank notes, in larger denominations and with portraits of King Sihanouk replacing the socialist images favoured by the PRK/SOC Government, were issued in April 1995.

A principal problem for Cambodia is its failure to improve revenue generation; fiscal revenue has remained equivalent to about 9.5% of GDP since 1994. Meanwhile, budget revenue declined in 1998, reaching only 8.6% of GDP, reflecting the acceleration of problems caused by weak governance and lack of a clear fiscal policy. The implementation of a very generous investment law and various *ad hoc* tax exemptions cost the country millions of US dollars per year in revenue. Following the formation of a second coalition Government, some additional fiscal measures were implemented. The introduction of a value added tax (VAT) and competitive bidding for garment export quotas in 1999 sharply boosted government revenue (a 40% increase compared with the previous year). Although significantly lower than forecast, revenue collection in 2000–02 continued to improve slightly, largely as a result of the enforcement of VAT regulations and non-tax revenue collection. Revenue reached about 12% of GDP in 2002, but again declined, to 11%, in 2003, owing to lax management during the political deadlock that arose after the July national election. Gross official reserves increased from the equivalent of 2.2 months of imports in 1996 to 3.3 months at the end of 2003, owing to the release by the Bank for International Settlements of gold reserves which had been 'frozen' since the 1970s. This increase in reserves would permit the authorities greater flexibility in exchange-rate management. A new foreign-exchange law had been approved in 1997, which formalized the liberal exchange structure.

The riel depreciated less than other regional currencies in the initial period of the financial crisis beginning in mid-1997, owing to the widespread use of the US dollar in the Cambodian economy. This dependence on the US dollar also mitigated the impact of the depreciation by 27% against the US dollar and the appreciation by 4% against the Thai baht in 1998, which, despite the use of dollars, resulted in an increase in inflation to 12.6% in that year, from 9.1% in 1997. However, inflation slowed from mid-1998, reaching zero in 1999, despite the introduction of the 10% VAT and a 30% increase in the salaries of public-sector workers in that year. On average, inflation was slightly negative during 2000–02, mainly as a result of the weakness of the prices of food and some manufactured goods. In 2003 inflation and the exchange rate remained basically stable, despite the political uncertainty arising from the July election.

BANKING

The banking sector currently plays only a very modest role in public or private finance. A Law on the Management of Foreign Exchange, adopted in August 1991, and a Law on the State Bank of Cambodia, adopted in March 1992, gave the former socialist State Bank the powers of a central bank, including the authority to issue money. The ability of the National Bank of Cambodia (NBC, as it was restyled in 1992) to implement monetary policy is severely limited by the widespread use of the US dollar. Although the riel is not convertible in international exchanges, it is freely exchanged. Figures for the end of 2003 showed that total liquidity recorded by the NBC was approximately 21% of GDP, a sharp increase from the end of 2001 (when it stood at about 16%). This rise was due to a rapid increase in foreign currency deposits, as a result of the significant improvement in the business environment and the regional economic recovery. Liquidity in riels increased to about 5.8% of GDP in 2003 (from 5% at the end of 2002), while the amount of US dollars circulating through the banking system rose to about 21% of GDP (from 19% at the end of 2002). Foreign currencies circulating outside the banking system are estimated to be at least twice as high as the total liquidity recorded by the NBC. The use of riels is confined to small transactions and low-wage payments, while larger transactions are conducted in foreign currencies such as the US dollar and the Thai baht. Despite this improvement, Cambodian bankers still appear to have little confidence in domestic developments. According to a monetary survey conducted at the end of 2003, net domestic credit remains largely negative, suggesting that Cambodian bankers prefer to place money outside the country rather than to lend it to local investors. Most of the population, meanwhile, make no use of banking services, preferring to keep their limited savings in the form of hoarded gold or US dollars. While the credit services provided by banks are competitive, small-scale business people prefer to borrow from relatives, and small amounts of credit are routinely raised via the organization of tontines. Peasants also rely on relatives for loans, or on shopkeepers and mill-owners who advance cash or goods at exorbitant rates of interest. By the end of 2003 there were 17 commercial banks, most of them small, private ventures, many of which were alleged to be engaged in the 'laundering' of regional profits from drugs-trafficking. The NBC increased reserve requirements for banks from 5% to 8% and the capital guarantee deposit from 5% to 10% with effect from 1 January 1998, to ensure liquidity and solvency in the banking system. In order further to strengthen the credibility of the sector, the NBC also requested in May 2000 that all commercial banks apply for a new licence. As a result, several non-viable banks have shut down, having been unable to comply with the required increase of registered capital to US $13m. from $5m.

TOURISM

Tourism is regarded as one of the most important immediate and long-term sources of foreign exchange. Attempts were made to increase tourism from 1986 as the economy began to experience substantial deficits on its balance of trade with socialist countries (approximately 90m. roubles per annum on average for 1987–90); in 1990 the deficit incurred in trading in the convertible-currency zone exceeded US $20m. Apart from the creation of tourism services, with technical assistance from Viet Nam, in 1988 the Government authorized the establishment of free-trading zones with Thailand and Singapore in the two coastal provinces of Kampot and Kaoh Kong. Their subsequent success led to the expansion and legalization of cross-border trade with Thailand along the western frontier, with Cambodia exporting rubber, timber and agricultural produce and importing petroleum, machinery and consumer goods, including textiles.

Although tours of Phnom-Penh and Angkor Wat, via Ho Chi Minh City, were made available from 1986, Cambodia attracted fewer than 1,000 visitors in 1987. However, foreign visitor arrivals increased from 118,200 in 1993 to 260,500 in 1996. Visitor arrivals declined substantially following the events of mid-1997, and the year ended with a total of 255,124 arrivals. In December the 131-room Grand Hotel d'Angkor in Siem Reap, which had been refurbished by Raffles Holdings of Singapore at a cost of US $30m., was opened, but vacancy rates in hotels remained high. The Government hoped for an increase in visitors in 1998, and developed a plan for the tourism industry that focused on the Angkor temples, the beaches of Sihanoukville, and 'eco-tourism' in Ratanakiri. However, arrivals (by air) declined to 186,333 in 1998, owing largely to the continued political instability and the violence surrounding the election in July, combined with the effect of the regional financial crisis. The tourism sector recovered significantly following the establishment of a potentially stable coalition Government in November 1998, with the total number of visitors reaching 367,743 in 1999, of whom 262,907 had arrived by air. The significant decline in the number of visitors to Cambodia in 1997 prompted the Government to initiate an 'open-skies' policy, in accordance with which direct flights from overseas would be permitted to land in Siem Reap. Many tour and hotel operators in Phnom-Penh opposed this policy, arguing that it would directly affect their business and other enterprises in the capital. The number of foreign visitors arriving on direct flights to Siem Reap almost tripled in 1999 relative to 1998, and continued to increase significantly during 2000–02, representing 35% of the total number of arrivals by air in Cambodia in 2002. Total arrivals (including arrivals by land and sea) rose to 786,546 in 2002, when tourist receipts totalled $379m. However, the outbreak of SARS in the region in 2003 contributed to a decline in tourist arrivals to 701,014; receipts were reduced to $179m. Tourists tend to stay only for short periods, usually no more than two days, and it remained unclear whether the 'open-skies' policy would benefit the entire economy in the longer term. According to a study prepared by a group of NGOs, only 30% of the income generated by foreign tourists in Siem Reap benefited business inside Cambodia, the remaining 70% returning to foreign countries, as most of the assets in the tourism sector (airlines, hotels and even airports) are foreign-owned.

Since 1994 the tourism sector has attracted a very high proportion of foreign capital investment. However, the rate of implementation of approved projects is very poor. In 1995 approvals totalled US $1,572m., of which $1,350m. was accounted for by an approved tourism centre project proposed by Ariston of Malaysia; at mid-2004, however, this project had yet to be realized.

TRADE

From 1997 traditional exports like logs, sawn timber and rubber were superseded by garments, which accounted for 44% of domestic exports (excluding re-exports) in that year and for about 92% in 2003. Logs, sawn timber and rubber accounted for only 3% of domestic exports in the latter year. Imports were dominated by investment-related products, durable consumer goods and petroleum products. Re-exports represented a significant feature of external trade, prompted by the differences in import tariffs between Cambodia and its neighbours, and included cigarettes, motorcycles, beer and electric equipment (which were re-exported to Viet Nam) and gold (which was principally re-exported to Thailand). In the late 1990s, however, re-exports declined sharply. Re-exports accounted for 62% of total exports in 1995, but declined to an estimated 12% of total exports in 2003.

Despite the slowdown in garment and timber exports, the volume of Cambodia's external trade continued to grow steadily, reaching US $3,343m. in 2001, exceeding the nominal GDP for the first time in Cambodia's history. In 2002 and 2003 exports of goods increased by an average annual rate of about 14%, in contrast to an expansion of 20% in 2000; similarly, imports of goods rose by 10% per year on average, compared with a 20% increase in 2000. In terms of the balance of trade, the trade surplus in agriculture was very significant owing to the huge increase in production of paddy and higher prices for rubber exports. The trade surplus in the garment sector remained strong. However, the trade deficit in the energy and non-garment manufacturing sectors increased significantly owing to the expansion in economic activity. As a result of the improved performance of the export sector, in 2003 the trade deficit narrowed to $385m., the equivalent of 9.2% of GDP, compared with $517m., or 14% of GDP, in 2002. The balance in the service sector remained in surplus, although it was adversely affected by reduced tourism revenues. The outflow of income owed by expatriates (including salaries and profits generated by foreign direct investment) was broadly balanced by the inflow of money repatriated by Cambodians residing overseas. Overall, Cambodia's current-account deficit was reduced to the equivalent of 6.2% of GDP in 2003, compared with 9.1% in 2002.

Cambodia's relations with the Association of South East Asian Nations (ASEAN) improved significantly following the 1993 elections, although Cambodia was not formally admitted into the organization until April 1999. ASEAN membership requires the implementation of the Common Effective Preferential Tariff (CEPT), a system of reciprocal tariff reductions. Cambodia, however, will find it difficult to reduce tariffs for customs duties, which provide a substantial proportion of non-aid revenues to the national budget. The Cambodian Government expected to increase domestic taxes, principally excise duties, to offset the decline in customs duties.

FOREIGN AID

The exact level of Cambodia's outstanding external debt remained unclear at mid-2003, although World Bank documents showed that at the end of 2002 Cambodia's overall foreign debt totalled more than US $2,400m., most of which was owed to Russia. These debts were contracted during four different periods: at the end of the 1960s Cambodia's outstanding external debt amounted to about $50m., all of which was already rescheduled or partly cancelled in 1995, according to the procedure of the 'Paris Club' of creditor nations, and which by mid-2002 had been reduced to about $30m.; however, it appears that Cambodia also owed at least $300m. to the USA during 1970–75, and about $800m. of convertible rouble (equivalent to $1,400m.) to Russia during 1980–91. (By mid-2003 the incumbent Cambodian Government had not yet recognized these two debts, although discussions with the USA and Russia on the issue were under way.) After 1993 the Cambodian Government began to borrow again from the World Bank and ADB in order to finance the rehabilitation of the country's public infrastructure; this debt is estimated to amount to $710m. In sum, the total outstanding external debt recognized by the Cambodian Government at the end of 2002 amounted to $740m., while pending debt was estimated at $1,700m. At a conference in Tokyo in June 1992 aid for economic rehabilitation and infrastructural development amounting to $880m. was pledged by approximately 30 donors, including ASEAN. At subsequent donor meetings between 1992 and 2002, pledges rose to more than $3,000m. Disbursements for the repatriation and resettlement of refugees, community development, agriculture, health and sanitation, education and training, public utilities, industry and public administration were spread over several years, and by 1996 were behind schedule, owing to problems of absorption. However, the removal of the first Prime Minister in July 1997 prompted several lenders and donors to cut their aid programmes. The USA, Germany, the IMF and the World Bank suspended all non-humanitarian aid. Other donors made more modest reductions, or continued providing assistance—Australia only ceased providing military assistance, and Japan briefly delayed the release of some funds, while the People's Republic of China made no decreases in aid. Nevertheless, the aid cuts forced the Government to reduce its budget, and to take measures to increase revenue collection, steps that mitigated but did not prevent a growth in the budget deficit. In 1998 foreign assistance was provided for the elections held in July; the EU donated $11.5m., Japan $3m. and the USA an additional $2.3m. (the latter to be administered by NGOs). Donors also helped fund a national census conducted in March 1998, the first since 1962.

In the past donors have regularly requested that budgetary procedures and transparency be improved, and that greater priority be given to basic needs via the expansion of health and education services and the promotion of rural development (the latter of which necessarily requires protection of the environment by means of effective regulation of logging and reafforestation). Among the most important elements of structural reform is the planned reduction in the number of state employees, both civil and military. The fiscal gains have been undermined to some extent by approximately 35,000 patronage appointments offered to parties in the ruling coalition, the obstruction of the privatization of state-owned enterprises, continuing problems with the collection of tax and customs revenues and recourse to parallel budgets or sources of revenue. In pursuing their reform agenda, lending institutions and countries have increasingly attempted to make aid conditional upon improved performance and accountability from the Government, and to transfer aid directly to local areas, thereby circumventing central ministries. A significant number of functions normally assumed by central government, such as the organization and management of credit associations, have been assumed by NGOs. (In June 2002 there were about 1,000 national and international NGOs operating in Cambodia.)

Following the successful formation of a coalition Government in November 1998, the CGC, which comprises 16 countries and six international financial institutions, pledged an assistance grant and loan of US $470m. for Cambodia, at a meeting held in Tokyo in February 1999. The CGC promised a further $548m. in May 2000 in Paris and another $615m. in June 2001 in Tokyo. In June 2002 in Phnom-Penh, it pledged further assistance of $635m. The funds were dependent on the Cambodian Government's commitment to implement reforms, including: military demobilization; administrative and fiscal reforms; the reduction of corruption; and the elimination of illegal logging, in order to prevent further environmental damage. In 2003 the scheduled meeting of the CGC was delayed because of the political stalemate that had occurred after the national election in July. It was to be rescheduled soon after the establishment of the new National Assembly and Government, which finally took place in July 2004. In 1999, following Cambodia's achievement of political progress, Japan (Cambodia's leading aid donor) resumed yen-denominated loans for the first time in 30 years. The IMF resumed assistance to Cambodia in late 1999, following its approval in October of a three-year arrangement under the enhanced structural adjustment facility (ESAF) for Cambodia to support the Government's economic programme for 1999–2000; however, the organization remained critical of Cambodia's fiscal performance. Relative political stability in Cambodia, and the cessation of any formal armed resistance with the collapse of the Khmers Rouges, augurs well for Cambodia to achieve rapid economic development in the coming years. Donors hoped that the importance of aid in the Government's budget would provide impetus for political and financial reform.

Statistical Survey

Source (unless otherwise stated): National Institute of Statistics, Ministry of Planning, Sangkat Boeung Keng Kong 2, blvd Monivong, Phnom-Penh; tel. (23) 216538; fax (23) 213650; e-mail census@camnet.com.kh; internet www.nis.gov.kh .

Note: Some of the statistics below represent only sectors of the economy controlled by the Government of the former Khmer Republic. During the years 1970–75 no figures were available for areas controlled by the Khmers Rouges.

Area and Population

AREA, POPULATION AND DENSITY

Area (sq km)	181,035*
Population (census results)†	
17 April 1962	5,728,771
Prior to elections of 1 May 1981	6,682,000
3 March 1998	
Males	5,511,408
Females	5,926,248
Total	11,437,656
Population (official estimates at mid-year)	
2001	13,148,363
2002	13,473,352
2003	13,798,237
Density (per sq km) at mid-2003	76.2

* 69,898 sq miles.

† Excluding adjustments for underenumeration.

PROVINCES
(1998 census)

	Area (sq km)*	Population	Density (per sq km)
Banteay Mean Chey	6,679	577,772	86.5
Bat Dambang	11,702	793,129	67.8
Kampong Cham	9,799	1,608,914	164.2
Kampong Chhnang	5,521	417,693	75.7
Kampong Spueu	7,017	598,882	85.3
Kampong Thum	13,814	569,060	41.2
Kampot	4,873	528,405	108.4
Kandal	3,568	1,075,125	301.3
Kaoh Kong	11,160	132,106	11.8
Kracheh	11,094	263,175	23.7
Mondol Kiri	14,288	32,407	2.3
Phnom Penh	290	999,804	3,447.6
Preah Vihear	13,788	119,261	8.6
Prcy Veaeng	4,883	946,042	193.7
Pousat	12,692	360,445	28.4
Rotanak Kiri	10,782	94,243	8.7
Siem Reab	10,299	696,164	67.6
Krong Preah Sihanouk	868	155,690	179.4
Stueng Traeng	11,092	81,074	7.3
Svay Rieng	2,966	478,252	161.2
Takaev	3,563	790,168	221.8
Otdar Mean Chey	6,158	68,279	11.1
Krong Kaeb	336	28,660	85.3
Krong Pailin	803	22,906	28.5
Total	178,035	11,437,656	64.2

* Excluding Tonlé Sap lake (3,000 sq km).

PRINCIPAL TOWNS
(population at 1998 census)

Phnom-Penh (capital)	999,804	Bat Dambang (Battambang)	139,964
Preah Sihanouk (Sihanoukville)*	155,690	Siem Reab (Siem Reap)	119,528

* Also known as Kampong Saom (Kompong Som).

Mid-2003 (UN estimate, incl. suburbs): Phnom-Penh 1,157,066 (Source: UN, *World Urbanization Prospects: The 2003 Revision*).

BIRTHS AND DEATHS
(annual averages)

	1985–90	1990–95	1995–2000*
Birth rate (per 1,000)	44.2	38.2	36.8
Death rate (per 1,000)	16.5	14.1	10.4

* UN estimates (Source: UN, *World Population Prospects: The 2002 Revision*).

Source (unless otherwise indicated): Ministry of Economy and Finance, Phnom-Penh.

1997: Death rate (per 1,000) 12.0 (Source: WHO).

1998: Birth rate (per 1,000) 38.0 (Source: WHO).

Expectation of life (WHO estimates, years at birth): 54.6 (males 51.9; females 57.1) in 2002 (Source: WHO, *World Health Report*).

EMPLOYMENT

	2000	2001	2002
Agriculture, forestry and fishing	3,889,048	4,384,250	4,479,773
Mining and quarrying	3,328	13,525	10,751
Manufacturing	367,286	544,832	556,388
Electricity, gas and water	3,799	3,795	4,704
Construction	69,773	94,077	100,123
Wholesale and retail trade	436,308	644,307	628,960
Restaurants and hotels	18,794	10,412	32,446
Transport and communications	119,596	169,307	174,711
Financial intermediation, real estate and renting	16,636	11,854	16,224
Public administration	146,986	149,382	143,513
Education	87,385	88,446	102,331
Health and social work	30,235	24,810	36,190
Other social services	40,098	35,153	44,926
Other services	45,905	69,179	68,637
Total employed	5,275,177	6,243,329	6,399,677

Health and Welfare

KEY INDICATORS

Total fertility rate (children per woman, 2002)	4.8
Under-5 mortality rate (per 1,000 live births, 2002)	138
HIV/AIDS (% of persons aged 15–49, 2003)	2.60
Physicians (per 1,000 head, 1998)	0.30
Hospital beds (per 1,000 head, 1990)	2.07
Health expenditure (2001): US $ per head (PPP)	184
Health expenditure (2001): % of GDP	11.8
Health expenditure (2001): public (% of total)	14.9
Access to water (% of persons, 2000)	30
Access to sanitation (% of persons, 2000)	18
Human Development Index (2002): ranking	130
Human Development Index (2002): value	0.568

For sources and definitions, see explanatory note on p. vi.

Agriculture

PRINCIPAL CROPS

('000 metric tons)

	2000	2001	2002
Rice (paddy)	4,026.1	4,099.0	3,740.0
Maize	157.0	185.6	168.1
Sweet potatoes	28.2	26.3	25.5
Cassava (Manioc)	147.8	142.3	186.8
Other roots and tubers*	19.0	20.0	20.0
Dry beans	15.1	17.2	19.6
Soybeans (Soya beans)	28.1	24.7	21.3
Groundnuts (in shell)	7.5	8.9	8.7
Sesame seed	9.9	9.0	8.7
Coconuts	70.0†	70.0†	70.0*
Sugar cane	164.2	169.3	168.9
Tobacco (leaves)	7.7	4.7	4.7
Natural rubber	42.4	38.7	32.4
Vegetables*	470.0	473.0	473.0
Oranges	63.0	63.0	63.0
Mangoes	35.0	35.0	35.0*
Pineapples	16.0	16.0	16.0*
Bananas	146.0	146.0	146.0
Other fruits and berries*	61.7	62.2	62.2

* FAO estimate(s).

† Unofficial figure.

Source: FAO.

LIVESTOCK

('000 head, year ending September)

	2000	2001	2002
Horses*	26	26	27
Cattle	2,993	2,869	2,924
Buffaloes	694	626	626
Pigs	1,934	2,115	2,105
Chickens	15,249	15,248	16,678
Ducks*	4,600	4,600	6,500

* FAO estimates.

Source: FAO.

LIVESTOCK PRODUCTS

('000 metric tons)

	2000	2001	2002
Beef and veal*	56.7	58.2	53.4
Buffalo meat*	13.1	13.4	13.4
Pig meat*	105.0	107.8	87.5
Poultry meat*	25.1	25.6	26.5
Cows' milk*	20.4	20.4	20.4
Hen eggs	11.7	11.7	12.8*
Other poultry eggs*	3.3	3.3	3.5
Cattle hides (fresh)*	14.2	14.6	13.4
Buffalo hides (fresh)*	2.7	2.8	2.8

* FAO estimate(s).

Source: FAO.

Forestry

ROUNDWOOD REMOVALS

('000 cubic metres, excl. bark)

	2000	2001	2002*
Sawlogs, veneer logs and logs for sleepers	143	100	100
Other industrial wood	36	21	21
Fuel wood*	10,119	9,924	9,737
Total	10,298	10,045	9,858

* FAO estimates.
Source: FAO.

SAWNWOOD PRODUCTION

('000 cubic metres, incl. railway sleepers)

	2000	2001	2002
Total (all broadleaved)	20	5	5*

* FAO estimate.

Source: FAO.

Fishing

('000 metric tons, live weight)

	2000	2001	2002
Capture	284.4	428.2	406.2
Freshwater fishes	245.3	384.5	359.8
Marine fishes	26.6	31.0	33.8
Aquaculture*	14.4	14.0	14.6
Total catch*	298.8	442.2	420.8

* FAO estimates.

Note: Figures exclude crocodiles, recorded by number rather than by weight. The total number of estuarine crocodiles caught was: 26,300 in 2000; 35,970 in 2001; 50,850 in 2002.

Source: FAO.

Mining

('000 metric tons)

	2001	2002	2003
Salt (unrefined)	11.0	72.5	50.0*

* Estimate.

Source: US Geological Survey.

Industry

SELECTED PRODUCTS

('000 metric tons, unless otherwise indicated)

	1971	1972	1973
Distilled alcoholic beverages ('000 hectolitres)	45	55	36
Beer ('000 hectolitres)	26	23	18
Soft drinks ('000 hectolitres)	25	25*	25*
Cigarettes (million)	3,413	2,510	2,622
Cotton yarn—pure and mixed (metric tons)	1,068	1,094	415
Bicycle tyres and tubes ('000)	208	200*	200*
Rubber footwear ('000 pairs)	1,292	1,000*	1,000*
Soap (metric tons)	469	400*	400*
Motor spirit (petrol)	2	—	—
Distillate fuel oils	11	—	—
Residual fuel oils	14	—	—
Cement	44	53	78
Electric energy (million kWh)†	148	166	150

* Estimate.

† Production by public utilities only.

Cigarettes (million): 4,175 in 1987; 4,200 annually in 1988–92 (estimates by US Department of Agriculture).

Cement ('000 metric tons): 50 in 2001 (estimate by the US Geological Survey).

Electric energy (estimates, million kWh): 194 in 1995; 201 in 1996; 208 in 1997; 215 in 1998; 222 in 1999; 229 in 2000.

Plywood ('000 cu m): 15 in 1999; 18 in 2000; 14 in 2001; 14 in 2002 (FAO estimate) (Source: FAO).

Source: partly UN, *Industrial Commodity Statistics Yearbook*.

Finance

CURRENCY AND EXCHANGE RATES

Monetary Units

100 sen = 1 riel.

Sterling, Dollar and Euro Equivalents (31 May 2004)

£1 sterling = 7,384.7 riels;
US $1 = 4,025.0 riels;
€1 = 4,929.0 riels;
10,000 riels = £1.354 = $2.484 = €2.029.

Average Exchange Rate (riels per US $)

2001 3,916.3
2002 3,912.1
2003 3,973.3

BUDGET

('000 million riels)

Revenue*	1999	2000	2001
Tax revenue	947.7	1,026.0	1,087.4
Direct taxes	82.7	135.6	140.5
Profit tax	63.8	100.9	112.8
Indirect taxes	431.6	500.0	571.2
Turnover tax	21.8	12.6	9.9
Value-added tax	313.6	371.6	402.8
Excise duties	91.8	112.6	154.8
Taxes on international trade	433.4	390.4	375.7
Import duties	415.3	372.8	364.1
Other current revenue	354.8	353.3	423.8
Forestry	36.3	41.0	29.1
Receipts from public enterprises	30.5	51.4	74.5
Posts and telecommunications	108.9	91.9	122.3
Public services	141.9	101.4	119.6
Capital revenue	13.7	29.3	9.1
Total	1,316.3	1,408.5	1,520.3

Expenditure†	1999	2000	2001
Council of Ministers	54	85	101
Ministry of Defence	336	309	277
Ministry of the Interior	147	142	140
Ministry of Economy and Finance	134	96	19
Ministry of Public Works and Transport	29	20	21
Ministry of Agriculture, Forestry, Hunting and Fisheries	21	23	30
Ministry of Education, Youth and Sport	156	166	209
Ministry of Industry, Mines and Energy	4	5	6
Ministry of Health	126	102	130
Ministry of Posts and Telecommunications	114	29	41
Ministry of Social Affairs, Labour, Professional Training and Youth Rehabilitation	19	26	28
Total (incl. others)	1,825	2,085	2,329
Current expenditure	1,097	1,189	1,354
Capital expenditure	728	896	975

* Excluding grants received.

† Figures for individual ministries exclude externally financed capital expenditure ('000 million riels): 504 in 1999; 593 in 2000; 692 in 2001.

Source: IMF, *Cambodia: Selected Issues and Statistical Appendix* (March 2003).

INTERNATIONAL RESERVES

(US $ million at 31 December)

	2001	2002	2003
IMF special drawing rights	0.51	0.55	0.20
Foreign exchange	586.30	775.60	815.33
Total	586.81	776.15	815.53

Source: IMF, *International Financial Statistics*.

MONEY SUPPLY

(million riels at 31 December)

	2001	2002	2003
Currency outside banks	577,780	765,980	906,390
Demand deposits at deposit money banks	31,940	47,300	29,470
Total money	609,720	813,280	935,860

Source: IMF, *International Financial Statistics*.

COST OF LIVING

(Consumer Price Index for Phnom-Penh; base: December 2000 = 100)

	2000	2001	2002
Food, beverages and tobacco	100.37	97.86	99.58
Clothing and footwear	102.38	95.22	89.18
Housing and utilities	99.96	103.35	110.80
Household furnishings and operations	99.23	98.78	95.85
Medical care and health expenses	98.63	103.24	105.01
Transport and communication	99.74	95.59	95.88
Recreation and education	100.05	107.49	109.55
Personal care and effects	99.70	101.36	103.09
All items	100.11	100.34	103.65

NATIONAL ACCOUNTS

('000 million riels at current prices)

Expenditure on the Gross Domestic Product

	2001	2002	2003*
Government final consumption expenditure	827.9	913.2	1,012.0
Private final consumption expenditure	12,337.7	12,860.3	13,891.9
Increase in stocks	299.0	–68.8	32.0
Gross fixed capital formation	2,786.7	3,549.9	3,692.0
Statistical discrepancy	–247.3	–304.9	–880.0
Total domestic expenditure	16,004.0	16,949.7	17,747.9
Exports of goods and services	7,914.7	9,275.3	9,854.1
Less Imports of goods and services	9,374.9	10,557.8	10,952.0
GDP in purchasers' values	14,543.9	15,667.2	16,650.0
GDP at constant 2000 prices	14,952.6	15,392.4	16,177.4

* Preliminary figures.

Source: Asian Development Bank, *Key Indicators of Developing Asian and Pacific Countries*.

Gross Domestic Product by Economic Activity

	2001	2002	2003*
Agriculture, hunting, forestry and fishing	5,161.7	5,231.8	5,859.6
Mining and quarrying	39.6	46.6	85.0
Manufacturing	2,556.4	2,969.5	3,044.0
Electricity, gas and water	56.8	75.8	84.6
Construction	867.1	1,023.1	1,014.4
Trade, restaurants and hotels	2,021.3	2,140.2	2,247.0
Transport, storage and communications	947.2	960.0	1,046.0
Financing, real estate and business services	980.8	964.8	950.0
Public administration	359.2	390.5	448.0
Other services	751.2	902.9	992.2
Sub-total	13,741.2	14,705.0	15,770.8
Less Imputed bank service charge	118.1	78.3	69.8
GDP at factor cost	13,623.1	14,626.7	15,701.0
Indirect taxes, *less* subsidies	920.7	1,040.4	879.8
Statistical discrepancy	—	—	69.2
GDP in purchasers' values	14,543.9	15,667.2	16,650.0

* Preliminary figures.

Source: Asian Development Bank, *Key Indicators of Developing Asian and Pacific Countries*.

BALANCE OF PAYMENTS
(US $ million)

	2000	2001	2002
Exports of goods f.o.b.	1,401.1	1,571.2	1,750.1
Imports of goods f.o.b.	–1,939.3	–2,094.0	–2,313.5
Trade balance	–538.2	–522.8	–563.5
Exports of services	428.4	524.6	600.2
Imports of services	–327.9	–347.3	–379.6
Balance on goods and services	–437.7	–345.5	–342.8
Other income received	67.1	57.5	50.7
Other income paid	–189.6	–193.4	–219.2
Balance on goods, services and income	–560.2	–481.4	–511.2
Current transfers received	432.0	403.8	456.5
Current transfers paid	–7.3	–8.3	–9.3
Current balance	–135.4	–85.8	–64.0
Capital account (net)	79.8	102.8	76.7
Direct investment from abroad	148.5	148.1	53.8
Other investment assets	–183.7	–118.2	–11.9
Other investment liabilities	224.2	118.4	238.1
Net errors and omissions	16.6	–24.2	–50.2
Overall balance	92.0	68.1	165.7

Source: IMF, *International Financial Statistics*.

External Trade

PRINCIPAL COMMODITIES
(US $ million)

Imports c.i.f.	1999	2000	2001*
Cigarettes	119	70	70
Petroleum products	131	176	160
Motorcycles	36	30	21
Total (incl. others)	1,392	1,846	1,951

Exports f.o.b.	1999	2000	2001*
Crude rubber†	56	63	50
Logs and sawn timber†	111	59	33
Clothing	679	1,013	1,147
Total (incl. others)‡	1,100	1,394	1,475

* Estimates.
† Including estimates for illegal exports.
‡ Including re-exports (US $ million): 178 in 1999; 188 in 2000; 180 (estimate) in 2001.

Source: IMF, *Cambodia: Selected Issues and Statistical Appendix* (March 2003).

PRINCIPAL TRADING PARTNERS
(US $ million)

Imports c.i.f.	2001	2002	2003
China, People's Republic	86.9	276.8	324.1
France	12.6	63.1	53.6
Hong Kong	116.9	372.8	411.2
Indonesia	9.9	75.7	84.5
Japan	19.7	76.8	67.3
Korea, Republic	49.6	126.8	144.6
Malaysia	19.3	60.4	68.9
Singapore	399.5	387.7	338.2
Thailand	503.9	567.0	756.5
Viet Nam	109.5	118.9	135.5
Total (incl. others)	1,600.3	2,311.0	2,469.0

Exports f.o.b.	2001	2002	2003
China, People's Republic	16.7	22.3	23.6
France	35.0	39.4	40.5
Germany	98.7	159.8	211.3
Netherlands	25.7	29.2	25.6
Singapore	28.0	76.8	67.8
Thailand	7.6	10.2	11.3
United Kingdom	126.3	122.1	150.3
USA	832.2	1,041.7	1,214.3
Viet Nam	24.5	26.6	30.3
Total (incl. others)	1,374.4	1,766.0	1,917.0

Source: Asian Development Bank, *Key Indicators of Developing Asian and Pacific Countries*.

Transport

RAILWAYS
(traffic)

	1997	1998	1999
Freight carried ('000 metric tons)	16	294	259
Freight ton-km ('000)	36,514	75,721	76,171
Passengers ('000)	553	438	431
Passenger-km ('000)	50,992	43,847	49,894

Source: Ministry of Economy and Finance, Phnom-Penh.

2000: Passenger-km ('000) 15; Freight ton-km ('000) 91 (Source: UN, *Statistical Yearbook*).

ROAD TRAFFIC
(estimated number of motor vehicles in use)

	2000	2001	2002
Passenger cars	193,851	200,561	209,128
Buses and coaches	2,918	2,996	3,196
Trucks	27,819	28,533	29,968
Other vehicles	371	396	421
Motorcycles and mopeds	513,503	565,931	586,278

Sources: Ministry of Public Works and Transport, Phnom-Penh, and Phnom-Penh Municipal Traffic Police.

SHIPPING

Merchant Fleet
(registered at 31 December)

	2001	2002	2003
Number of vessels	564	727	663
Displacement ('000 grt)	1,996.7	2,425.8	2,048.3

Source: Lloyd's Register-Fairplay, *World Fleet Statistics*.

International Sea-borne Freight Traffic
(estimates, '000 metric tons)

	1988	1989	1990
Goods loaded	10	10	11
Goods unloaded	100	100	95

Source: UN, *Monthly Bulletin of Statistics*.

CIVIL AVIATION
(traffic on scheduled services)

	1975	1976	1977
Passenger-kilometres (million)	42	42	42
Freight ton-kilometres ('000)	400	400	400

Source: Statistisches Bundesamt, Wiesbaden, Germany.

Tourism

FOREIGN TOURIST ARRIVALS
(by air)*

Country of residence	1999	2000	2001
Australia	9,471	11,350	13,078
Canada	5,415	5,646	6,191
China, People's Repub.	26,805	30,586	32,002
France	23,754	24,883	23,328
Germany	6,490	7,298	6,861
Japan	17,885	19,906	17,952
Korea, Repub.	6,377	7,536	9,579
Malaysia	12,541	14,701	15,994
Singapore	10,634	10,734	10,982
Taiwan	20,607	21,626	23,098
Thailand	15,272	16,550	17,496
United Kingdom	13,843	15,912	17,686
USA	30,301	35,814	37,033
Viet Nam	5,217	8,333	7,828
Total (incl. others†)	262,907	351,661	408,377

* Figures for individual countries refer to arrivals at Pochentong (Phnom-Penh) airport only.

† Including arrivals at Siem Reap airport (28,525 in 1999; 87,012 in 2000; 133,688 in 2001).

Total arrivals (incl. arrivals by land and sea): 604,919 in 2001; 786,546 in 2002; 701,014 in 2003.

Source: Ministry of Tourism, Phnom-Penh.

Tourism receipts (US $ million): 228 in 2000; 304 in 2001; 379 in 2002 (Source: World Tourism Organization).

Communications Media

	1999	2000	2001
Television receivers ('000 in use)	98	99	n.a.
Telephones ('000 main lines in use)	27.7	30.9	33.5
Mobile cellular telephones ('000 subscribers)	89.1	130.5	223.5
Personal computers ('000 in use)	13	15	20
Internet users ('000)	4.0	6.0	10.0

Facsimile machines (number in use): 884 in 1995; 1,470 in 1996; 2,995 in 1997.

Internet users ('000): 30.0 in 2002.

Source: International Telecommunication Union.

Radio receivers ('000 in use): 1,120 in 1995; 1,300 in 1996; 1,340 in 1997 (Source: UNESCO, *Statistical Yearbook*).

Education

(2001/02*)

	Institutions	Teachers	Students
Primary	5,741	54,519	2,705,453
Secondary	542	24,884	465,039
Junior high school	379	19,650	357,635
Senior high school	163	5,234	113,404

* Excluding technical and vocational education and higher education.

Source: IMF, *Cambodia: Selected Issues and Statistical Appendix* (March 2003).

Adult literacy rate (UNESCO estimates): 69.4% (males 80.8%; females 59.3%) in 2002 (Source: UN Development Programme, *Human Development Report*).

Directory

The Constitution

The Constitution was promulgated on 21 September 1993; a number of amendments were passed on 4 March 1999. The main provisions are summarized below:

GENERAL PROVISIONS

The Kingdom of Cambodia is a unitary state in which the King abides by the Constitution and multi-party liberal democracy. Cambodian citizens have full right of freedom of belief; Buddhism is the state religion. The Kingdom of Cambodia has a market economy system.

THE KING

The King is Head of State and the Supreme Commander of the Khmer Royal Armed Forces. The monarchist regime is based on a system of selection: within seven days of the King's death the Royal Council of the Throne (comprising the Chairman of the Senate, the Chairman of the National Assembly, the Prime Minister, the Supreme Patriarchs of the Mohanikay and Thoammayutikanikay sects, the First and Second Vice-Chairmen of the Senate and the First and Second Vice-Chairmen of the National Assembly) must select a King. The King must be at least 30 years of age and be a descendant of King Ang Duong, King Norodom or King Sisowath. The King appoints the Prime Minister and the Cabinet. In the absence of the King, the Chairman of the Senate assumes the duty of acting Head of State.

THE LEGISLATURE

Legislative power is vested in the National Assembly (the lower chamber) and the Senate (the upper chamber). The National Assembly has 123 members who are elected by universal adult suffrage. A member of the National Assembly must be a Cambodian citizen by birth over the age of 25 years and has a term of office of five years, the term of the National Assembly. The National Assembly may not be dissolved except in the case where the Royal Government (Cabinet) has been dismissed twice in 12 months. The National Assembly may dismiss cabinet members or remove the Royal Government from office by passing a censure motion through a two-thirds' majority vote of all the representatives in the National Assembly. The Senate comprises nominated members, the number of which does not exceed one-half of all of the members of the National Assembly; two are nominated by the King, two are elected by the National Assembly and the remainder are elected by universal adult suffrage. A member of the Senate has a term of office of six years. The Senate reviews legislation passed by the National Assembly and acts as a co-ordinator between the National Assembly and the Royal Government. In special cases, the National Assembly and the Senate can assemble as the Congress to resolve issues of national importance.

CABINET

The Cabinet is the Royal Government of the Kingdom of Cambodia, which is led by a Prime Minister, assisted by Deputy Prime Ministers, with state ministers, ministers and state secretaries as members. The Prime Minister is designated by the King at the recommendation of the Chairman of the National Assembly from among the representatives of the winning party. The Prime Minister appoints the members of the Cabinet, who must be representatives in the National Assembly or members of parties represented in the National Assembly.

THE CONSTITUTIONAL COUNCIL

The Constitutional Council's competence is to interpret the Constitution and laws passed by the National Assembly and reviewed completely by the Senate. It has the right to examine and settle disputes relating to the election of members of the National Assembly and the Senate. The Constitutional Council consists of nine members with a nine-year mandate. One-third of the members are replaced every three years. Three members are appointed by the King, three elected by the National Assembly and three appointed by the Supreme Council of the Magistracy.

The Government

HEAD OF STATE

King: HM King NORODOM SIHAMONI (appointed by the Royal Council of the Throne on 14 October 2004).

ROYAL GOVERNMENT OF CAMBODIA
(October 2004)

A coalition of the Cambodian People's Party (CPP) and the FUNCINPEC Party.

Prime Minister: HUN SEN (CPP).

Deputy Prime Ministers and Co-Ministers of the Interior: SAR KHENG (CPP), Prince NORODOM SIRIVUDH (FUNCINPEC).

Deputy Prime Ministers and Co-Ministers of National Defence: TEA BANH (CPP), NHEIK BUN CHHAY (FUNCINPEC).

Deputy Prime Minister and Minister of Foreign Affairs and International Co-operation: HOR NAM HONG (CPP).

Deputy Prime Minister and Minister of Rural Development: LU LAY SRENG (FUNCINPEC).

Deputy Prime Minister and Minister in Charge of the Council of Ministers: SOK AN (CPP).

Senior Ministers: CHAM PRASIDH (CPP), CHHAY THON (CPP), KEAT CHHON (CPP), IM CHHUN LIM (CPP), MEN SAM ON (CPP), Dr MOK MARETH (CPP), NHIM VANNDA (CPP), TAV SENGHUO (CPP), HONG SUN HUOT (FUNCINPEC), KHUN HAING (FUNCINPEC), KHY TAING LIM (FUNCINPEC), KOL PHENG (FUNCINPEC), SIREY KOSAL (FUNCINPEC), VENG SEREIVUTH (FUNCINPEC), YOU HOCKRY (FUNCINPEC).

Minister of Agriculture, Forestry, Hunting and Fisheries: CHAN SARUN (CPP).

Minister of Commerce: CHAM PRASIDH (CPP).

Minister of Cults and Religions: KHUN HAING (FUNCINPEC).

Minister of Culture and Fine Arts: SISOVAT PANARA SIRIVUDH (FUNCINPEC).

Minister of Economy and Finance: KEAT CHHON (CPP).

Minister of Education, Youth and Sport: KOL PHENG (FUNCINPEC).

Minister of Environment: Dr MOK MARETH (CPP).

Minister of Health: NUTH SOKHOM (FUNCINPEC).

Minister of Industry, Mines and Energy: SUY SEM (CPP).

Minister of Information: KHIEU KANHARITH (CPP).

Minister of Justice: ANG VONG VATTANA (CPP).

Minister of Parliamentary Affairs and Inspection: MEN SAM ON (CPP).

Minister of Planning: CHHAY THON (CPP).

Minister of Posts and Telecommunications: SO KHUN (CPP).

Minister of Public Works and Transport: SUN CHAN THOL (FUNCINPEC).

Minister of Rural Development: LY THUCH (FUNCINPEC).

Minister of Social Affairs and Youth Rehabilitation: ITH SAM HENG (CPP).

Minister of Territorial Organization, Urbanization and Construction: IM CHHUN LIM (CPP).

Minister of Tourism: LAY PROHORS (FUNCINPEC).

Minister of Vocational Training and Labour: NHEM BUN CHIN (FUNCINPEC).

Minister of Water Resources and Meteorology: LIM KEAN HOR (CPP).

Minister of Women's Affairs and Veterans: UNG KANTHA PHAVY (FUNCINPEC).

Secretary of State for Public Functions: PICH BUN THIN (CPP).

Secretary of State for Civil Aviation: MAO HASVANNAL (FUNCINPEC).

There are also 133 further Secretaries of State.

MINISTRIES

Ministry of Agriculture, Forestry, Hunting and Fisheries: 200 blvd Norodom, Phnom-Penh; tel. (23) 211351; fax (23) 217320; e-mail info@maff.gov.kh; internet www.maff.gov.kh.

Ministry of Commerce: 20A–B blvd Norodom, Phnom-Penh; tel. (23) 427358; fax (23) 426396; e-mail kunkoet@moc.gov.kh; internet www.moc.gov.kh.

Ministry of Cults and Religions: Preah Sisowath Quay, cnr rue 240, Phnom-Penh; tel. (23) 725099; fax (23) 725699; e-mail morac@cambodia.gov.kh; internet www.morac.gov.kh.

Ministry of Culture and Fine Arts: 227 blvd Monivong, cnr rue Red Cross, Phnom-Penh; tel. (23) 217645; fax (23) 725749; e-mail mcfa@cambodia.gov.kh; internet www.mcfa.gov.kh.

Ministry of National Defence: blvd Confederation de la Russie, cnr rue 175, Phnom-Penh; tel. (23) 883184; fax (23) 366169; e-mail info@mond.gov.kh; internet www.mond.gov.kh.

Ministry of Economy and Finance: 60 rue 92, Phnom-Penh; tel. (23) 722863; fax (23) 427798; e-mail mefcg@hotmail.com; internet www.mef.gov.kh.

Ministry of Education, Youth and Sport: 80 blvd Norodom, Phnom-Penh; tel. (23) 210705; fax (23) 215096; e-mail crsmeys@camnet.com.kh; internet www.moeys.gov.kh.

Ministry of Environment: 48 blvd Sihanouk Tonle Bassac, Khan Chamkarmon, Phnom-Penh; tel. (23) 427894; fax (23) 427844; e-mail moe-cabinet@camnet.com.kh; internet www.moe.gov.kh.

Ministry of Foreign Affairs and International Co-operation: 161 Preah Sisowath Quay, cnr rue 240, Phnom-Penh; tel. (23) 216141; fax (23) 216144; e-mail mfaicasean@bigpond.com.kh; internet www.mfaic.gov.kh.

Ministry of Health: 151–153 blvd Kampuchea Krom, Phnom-Penh; tel. (23) 722873; fax (23) 426841; e-mail procure.pcu@bigpond.com.kh; internet www.moh.gov.kh.

Ministry of Industry, Mines and Energy: 45 blvd Preah Norodom, Phnom-Penh; tel. (23) 723077; fax (23) 428263; e-mail mine@cambodia.gov.kh; internet www.mine.gov.kh.

Ministry of Information: 62 blvd Monivong, Phnom-Penh; tel. (23) 724159; fax (23) 427475; e-mail information@cambodia.gov.kh; internet www.information.gov.kh.

Ministry of the Interior: 275 blvd Norodom, Khan Chamkarmon, Phnom-Penh; tel. (23) 726148; fax (23) 212708; e-mail moi@interior.gov.kh; internet www.interior.gov.kh.

Ministry of Justice: 14 blvd Sothearos, Phnom-Penh; tel. (23) 360327; fax (23) 364119; e-mail moj@cambodia.gov.kh; internet www.moj.gov.kh.

Ministry of Parliamentary Affairs and Inspection: rue Jawaharlal Nehru, Phnom-Penh; tel. (23) 884261; fax (23) 884264; e-mail mnasrl@cambodia.gov.kh; internet www.mnasrl.gov.kh.

Ministry of Planning: 386 blvd Monivong, Sangkat Boeung Keng Kong 2, Phnom-Penh; tel. (23) 212049; fax (23) 210698; e-mail mop@cambodia.gov.kh; internet www.mop.gov.kh.

Ministry of Posts and Telecommunications: cnr rue Preah Ang Eng and rue Ang Non, Phnom-Penh; tel. (23) 426510; fax (23) 426011; e-mail mptc@cambodia.gov.kh; internet www.mptc.gov.kh.

Ministry of Public Works and Transport: 106 blvd Norodom, Phnom-Penh; tel. (23) 427845; fax (23) 427852; e-mail mpwt@mpwt.gov.kh; internet www.mpwt.gov.kh.

Ministry of Rural Development: Jok Dimitrov, cnr rue 169, Phnom-Penh; tel. (23) 426850; fax (23) 366790; e-mail mrd@cambodia.gov.kh; internet www.mrd.gov.kh.

Ministry of Social Affairs, Labour, Vocational Training and Youth Rehabilitation: 788B blvd Monivong, Phnom-Penh; tel. (23) 218437; fax (23) 726086; e-mail mosalvy-gdlv@camnet.com.kh; internet www.mosalvy.gov.kh.

Ministry of Territorial Organization, Urbanization and Construction: 771–773 blvd Monivong, Phnom-Penh; tel. (23) 215660; fax (23) 217035; e-mail gdlmup-mlmupc@camnet.com.kh; internet www.mlmupc.gov.kh.

Ministry of Tourism: 3 blvd Monivong, Phnom-Penh 12258; tel. (23) 212837; fax (23) 426877; e-mail info@mot.gov.kh; internet www.mot.gov.kh.

Ministry of Water Resources and Meteorology: 47 blvd Norodom, Phnom-Penh; tel. (23) 724289; fax (23) 426345; e-mail mowram@cambodia.gov.kh; internet www.mowram.gov.kh.

Ministry of Women's Affairs and Veterans: 3 blvd Norodom, Phnom-Penh; tel. and fax (23) 428965; e-mail mwva.cabinet@bigpond.com.kh; internet www.mwva.gov.kh.

Legislature

PARLIAMENT

NATIONAL ASSEMBLY

National Assembly, blvd Samdech Sothearos, cnr rue 240, Phnom-Penh; tel. (23) 214136; fax (23) 217769; e-mail kimhenglong@cambodian-parliament.org; internet www.cambodian-parliament.org.

Chairman: NORODOM RANARIDDH (FUNCINPEC).

Election, 27 July 2003

	% of Votes	Seats
Cambodian People's Party	47.35	73
FUNCINPEC Party	20.75	26
Sam Rainsy Party	21.87	24
Others	10.03	—
Total	100.00	123

SENATE

Senate, Chamkarmon Palace, blvd Norodom, Phnom-Penh; tel. (23) 211446; fax (23) 211441; e-mail oum_sarith@camnet.com.kh; internet www.khmersenate.org.

Chairman: CHEA SIM (CPP).

First Vice-Chairman: CHIVAN MONIRAK (FUNCINPEC).

Second Vice-Chairman: NHIEK BUNCHHAY (FUNCINPEC).

Inauguration, 25 March 1999

	Seats
Cambodian People's Party	31
FUNCINPEC Party	21
Sam Rainsy Party	7
King's appointees	2
Total	61

Political Organizations

Alliance of Democrats (AD): Phnom-Penh; f. 2003; alliance of FUNCINPEC Party and Sam Rainsy Party; Pres. Prince NORODOM RANARIDDH; Vice-Pres. SAM RAINSY.

Buddhist Liberal Party (Kanakpak Serei Niyum Preah Put Sasna): Phnom-Penh; internet www.blp.org; f. 1998; Chair. IENG MULI; Gen. Sec. SIENG LAPRESSE.

Cambodia National Sustaining Party: 56 rue 598, Krom 25, Mondul 3, Sangkat Bang Kork II, Khan Toul Kork, Phnom-Penh; tel. (12) 655442; e-mail cnsp_cambodia@hotmail.com; internet www.pensovann.com; f. 1997 by Pen Sovann, fmr Sec.-Gen. of the Cen. Cttee of the CPP, to contest the 1998 legislative elections; boycotted 2003 legislative elections; Pres. PEN SOVANN.

Cambodian Freedom Fighters: 2728 E 10th Street, Long Beach, CA 90804, USA; tel. (562) 433-9930; fax (562) 433-7490; internet www.cffighters.org; f. 1998 in opposition to Hun Sen's leadership; Leader CHHUN YASITH; Sec.-Gen. RICHARD KIRI KIM.

Cambodian People's Party (CPP) (Kanakpak Pracheachon Kampuchea): 203 blvd Norodom, Sangkat Tonle Bassac, Khan Chamkarmon, Phnom-Penh; tel. and fax (23) 2158801; e-mail cpp@thecpp.org; internet www.thecpp.org; known as the Kampuchean People's Revolutionary Party 1979–91; 21-mem. Standing Cttee of the Cen. Cttee; Cen. Cttee of 153 full mems; Hon. Chair. of Cen. Cttee HENG SAMRIN; Chair. of Cen. Cttee CHEA SIM; Vice-Chair. HUN SEN; Chair. of Permanent Cttee SAY CHHUM.

Cambodian Women's Party: 22 rue 384, Sangkat Toul Svay Prey 1, Khan Chamkarmon, Phnom-Penh; tel. (23) 993297; f. 1998; Pres. NUON BUNNA.

Democratic National United Movement (DNUM): Pailin; f. 1996 by Ieng Sary, following his defection from the PDK; not a national political party, did not contest 1998 election; DNUM members are also free to join other political parties.

Farmers' Party: 21 rue 528, Sangkat Boeung Kak 1, Khan Chamkarmon, Phnom-Penh; tel. (16) 333200; Pres. PON PISITH.

FUNCINPEC Party (United National Front for an Independent, Neutral, Peaceful and Co-operative Cambodia Party): 11 blvd Monivong (93), Sangkat Sras Chak, Khan Daun Penh, Phnom Penh; tel. (23) 428864; fax (23) 426521; e-mail funcinpec@funcinpec.org; internet www.funcinpec.org; FUNCINPEC altered its title to the FUNCINPEC Party when it adopted political status in 1992; the party's military wing was the National Army of Independent Cambodia (fmrly the Armée Nationale Sihanoukiste—ANS); merged with the Son Sann Party in Jan. 1999; Pres. Prince NORODOM RANARIDDH; Sec.-Gen. Prince NORODOM SIRIVUDH.

Hang Dara Movement Democratic Party: 16 rue 430, Sangkat Phsardoeum Thkov, Khan Chamkarmon, Phnom-Penh; tel. (12) 672007; f. 2002 to contest 2003 general election; breakaway faction of the FUNCINPEC Party; Pres. HANG DARA.

Indra Buddra City Party: Commune, Chbarmon District, Kampong Spueu; tel. (12) 710331; Pres. NOREAK RATANAVATHANO.

Khmer Angkor Party: 27C rue 390, Sangkat Boeung Keng Kang 3, Khan Chamkarmon, Phnom-Penh; tel. (12) 865269; f. 1998; breakaway faction of Sam Rainsy Party; Pres. KUNG MUNI.

Khmer Citizens' Party (Kanakpak Pulroat Khmer): 94B rue 146, Sangkat Phsar Depo 2, Khan Toul Kork, Phnom-Penh; tel. (12) 827071; f. 1996; breakaway faction of Khmer Nation Party (now Sam Rainsy Party); Pres. KHIEU SENGKIM.

Khmer Democratic Party (Kanakpak Pracheathippatei Khmer): 79A rue 186, Sangkat Touek Laak 3, Khan Toul Kork, Phnom-Penh; tel. (12) 842947; Pres. OUK PHURIK.

Khmer Front Party (Ronakse Chuncheat Khmer): 34 rue 5, Sangkat Km 6, Kham Roeusey Keo, Phnom-Penh; tel. (12) 762207; f. 2002 to contest 2003 general election; Pres. SUTH DINA.

Khmer Help Khmer Party (Khmer Chuoy Khmer): 21 rue 281, Sangkat Boeung Kak 1, Khan Toul Kork, Phnom-Penh; tel. (23) 885017; f. 1998 as Khmer New Life Party; name changed as above in 2003; Pres. KING SOVANN; Gen. Sec. TIT SARUN.

Khmer Neutral Party (Kanakpak Kampuchea Appyeakroet): 14A rue Keo Chea, Phnom-Penh; tel. (23) 62365; fax (23) 27340; e-mail Masavang@datagraphic.fr; internet www.datagraphic.fr/knp/; Pres. BUO HEL.

Khmer Soul Party: 7 blvd Makara, Sangkat No. 4, Khan Mitta Pheap, Sihanoukville; tel. (12) 573345; Pres. MOEUNG MLOB.

Khmer Unity Party: 54 blvd Russia, Sangkat Phsar Depo 3, Khan Toul Kork, Phnom-Penh; tel. (12) 829092; e-mail kup@free.fr; internet kup.free.fr; Pres. BOU SARIN.

Liberal Democratic Party: 21 rue 614, Sangkat Boeung Kak 2, Khan Toul Kork, Phnom-Penh; tel. (12) 866576; f. 1993; receives support from members of the armed forces; pro-Government; Chair. Gen. CHHIM OM YON.

MOLINAKA (National Liberation Movement of Cambodia): 8 rue 292, Sangkat Boeung Keng Kang 2, Khan Chamkarmon, Phnom-Penh; tel. (12) 941029; f. 1992; a breakaway faction of FUNCINPEC; Pres. KHLOK PRITHY.

National Union Party (Kanakpak Ruop Ruom Cheat): Phnom-Penh; established by rebel mems of FUNCINPEC Party; Chair. (vacant); Sec.-Gen. UNG PHAN.

Norodom Chakrapong Khmer Soul Party (NCPPK): 61 rue 294, Sangkat Boeung Keng Kang 1, Khan Chamkarmon, Phnom-Penh; tel. (23) 212597; e-mail ncppk@bigpond.com.kh; internet www.ncppk.org; f. 2002 by Prince Norodom Chakrapong; breakaway faction of FUNCINPEC Party; Pres. Prince NORODOM CHAKRAPONG.

Reastr Niyum (Nationalist Party): blvd Norodom, Phnom-Penh; tel. (23) 215659; fax (23) 215279; f. 1998; breakaway faction of the FUNCINPEC Party; Pres. UNG HUOT; Sec.-Gen. PU SOTHIRAK.

Rice Party (Svor): 69 blvd Sothearos, Sangkat Tonle Bassac, Khan Chamkarmon, Phnom-Penh; tel. (12) 860060; e-mail riceparty_kh@yahoo.com; internet www.riceparty.web1000.com; f. 1992; Pres. NHOUNG SEAP.

Sam Rainsy Party (SRP): 71 blvd Sothearos, Sangkat Tonle Bassac, Khan Chamkarmon, Phnom-Penh; tel. and fax (23) 217452; e-mail samrainsy@samrainsyparty.org; internet www.samrainsyparty.org; f. 1995 as the Khmer Nation Party; name changed as above in 1998; 444,544 mems (Aug. 2001); Pres. SAM RAINSY; Sec.-Gen. MENG RITA (acting).

Union of National Solidarity Party (UNS): 75 rue 598, Sangkat Phnom Penh Thmei, Khan Roeusey Keo, Phnom-Penh; tel. (12) 829003; Pres. SIM SOKHOM.

United Front for the Construction and Defence of the Kampuchean Fatherland (UFCDKF): Phnom-Penh; f. 1978 as the Kampuchean National United Front for National Salvation (KNUFNS), renamed Kampuchean United Front for National Construction and Defence (KUFNCD) in 1981, present name adopted in 1989; mass organization supporting policies of the CPP; an 89-mem. Nat. Council and a seven-mem. hon. Presidium; Chair. of Nat. Council CHEA SIM; Sec.-Gen. ROS CHHUN.

Diplomatic Representation

EMBASSIES IN CAMBODIA

Australia: Villa 11, R. V. Senei Vinnavaut Oum (rue 254), Sangkat Chaktomouk, Khan Daun Penh, Phnom-Penh; tel. (23) 213470; fax (23) 213413; e-mail australian.embassy.cambodia@dfat.gov.au; internet www.embassy.gov.au/kh.html; Ambassador LISA KIM FILIPETTO.

Brunei: 237 rue Pasteur 51, Sangkat Boeung Keng Kang 1, Khan Chamkarmon, Phnom-Penh; tel. (23) 211457; fax (23) 211455; e-mail brunei@bigpond.com.kh; Ambassador Awang Haji EMRAN B. BAHAR.

Bulgaria: 227/229 blvd Norodom, Phnom-Penh; tel. (23) 217504; fax (23) 212792; e-mail bulgembpnp@camnet.com.kh; Chargé d'affaires a.i. ROUMEN DONTCHEV.

Canada: Villa 9, R. V. Senei Vinnavaut Oum, Sangkat Chaktomouk, Khan Daun Penh, Phnom-Penh; tel. (23) 213470; fax (23) 211389; e-mail pnmpn@dfait-maeci.gc.ca; internet www.dfait-maeci.gc.ca/cambodia; Ambassador STEFANIE BECK.

China, People's Republic: 156 blvd Mao Tse Toung, Phnom-Penh; tel. (23) 720920; fax (23) 364738; e-mail chinaemb_kh@mfa.gov.cn; Ambassador HU QIANWEN.

Cuba: 96/98 rue 214, Sangkat Veal Vong, Khan 7 Makara, Phnom-Penh; tel. (23) 213212; fax (23) 217428; e-mail embacuba@camnet.com.kh; Ambassador NIRSIA CASTRO GUEVARA.

France: 1 blvd Monivong, Phnom-Penh; tel. (23) 430020; fax (23) 430041; e-mail ambafrance@online.com.kh; internet www.ambafrance.gov.kh; Ambassador YVON ROÉ D'ALBERT.

Germany: 76–78 rue Yougoslavie, BP 60, Phnom-Penh; tel. (23) 216381; fax (23) 427746; e-mail germanembassy@everyday.com.kh; internet www.germanembassy-cambodia.org; Ambassador PIUS FISCHER.

India: 777 blvd Monivong, Phnom-Penh; tel. (23) 210912; fax (23) 213640; e-mail hocembindia@online.com.kh; Ambassador PRADEEP KUMAR KAPUR.

Indonesia: 90 blvd Norodom, Phnom-Penh; tel. (23) 216148; fax (23) 216571; e-mail kukppenh@bigpond.com.kh; internet www.indonesia-phnompenh.org; Ambassador NURRACHMAN OERIP.

Japan: 194 blvd Norodom, Sangkat Tonle Bassac, Khan Chamkarmon, Phnom-Penh; tel. (23) 217161; fax (23) 216162; e-mail eojc@bigpond.com.kh; internet www.kh.emb-japan.go.jp; Ambassador FUMIAKI TAKAHASHI.

Korea, Democratic People's Republic: 39 rue 268, Phnom-Penh; tel. (15) 912567; fax (23) 426230; Ambassador CHOE HAN-CHUN.

Korea, Republic: 50 rue 214, Sangkar Beung Rain, Khan Daun Penh, Phnom-Penh; tel. (23) 211900; fax (23) 219200; e-mail cambodia@mofat.go.kr; internet www.koreanembcam.go.kr; Ambassador LEE HAN-GON.

Laos: 15–17 blvd Mao Tse Toung, POB 19, Phnom-Penh; tel. (23) 982632; fax (23) 720907; e-mail laoembpp@camintel.com; Ambassador THOUANE VORASARN.

Malaysia: 5 rue 242, Sangkat Chaktomouk, Khan Daun Penh, Phnom-Penh; tel. (23) 216176; fax (23) 216004; e-mail mwppenh@online.com.kh; Ambassador Datin Paduka MELANIE LEONG SOOK LEI.

Myanmar: 181 blvd Norodom, Phnom-Penh; tel. (23) 213664; fax (23) 213665; e-mail M.E.PHNOMPENH@bigpond.com.kh; Ambassador Dr AUNG NAING.

Philippines: 33 rue 294, Khan Chamkarmon, Sangkat Tonle Bassac, Phnom-Penh; tel. (23) 215145; fax (23) 215143; e-mail phnompenhpe@online.com.kh; Ambassador (vacant).

Poland: 767 blvd Monivong, BP 58, Phnom-Penh; tel. (23) 217782; fax (23) 217781; e-mail emb.pol.pp@online.com.kh; internet www.polishembassy-cambodia.org; Ambassador KAZIMIERZ A. DUCHOWSKI.

Russia: 213 blvd Sothearos, Phnom-Penh; tel. (23) 210931; fax (23) 216776; e-mail russemba@online.com.kh; internet www.embrusscambodia.mid.ru; Ambassador VALERII TERESHCHENKO.

Singapore: 92 blvd Norodom, Phnom-Penh; tel. (23) 360855; fax (23) 210862; e-mail singemb@bigpond.com.kh; internet www.mfa.gov.sg/phnompenh; Ambassador LAWRENCE ANDERSON.

Thailand: 196 Preah Norodom Blvd, Sangkat Tonle Bassac, Khan Chamkarmon, Phnom-Penh; tel. (23) 726306; fax (23) 726309; e-mail thaipnp@mfa.go.th; Ambassador PIYAWAT NIYOMRERKS.

United Kingdom: 27–29 Sras Chak, Khan Daun Penh, Phnom-Penh; tel. (23) 427124; fax (23) 427125; e-mail BRITEMB@online.com.kh; Ambassador STEPHEN BRIDGES.

USA: 16 rue 228, Phnom-Penh; tel. (23) 216436; fax (23) 216437; internet phnompenh.usembassy.gov; Ambassador CHARLES AARON RAY.

Viet Nam: 436 blvd Monivong, Phnom-Penh; tel. (23) 362741; fax (23) 427385; e-mail embbvnpp@camnet.com.kh; Ambassador NGUYEN DU HONG.

Judicial System

An independent judiciary was established under the 1993 Constitution.

Supreme Court: rue 134, cnr rue 63, Phnom-Penh; tel. 17816663; Chair. DID MONTY.

Religion

BUDDHISM

The principal religion of Cambodia is Theravada Buddhism (Buddhism of the 'Tradition of the Elders'), the sacred language of which is Pali. A ban was imposed on all religious activity in 1975. By a constitutional amendment, which was adopted in April 1989, Buddhism was reinstated as the national religion and was retained as such under the 1993 Constitution. By 1992 2,800 monasteries (of a total of 3,369) had been restored and there were 21,800 Buddhist monks. In 1992 about 90% of the population were Buddhists.

Supreme Patriarchs: Ven. Patriarch TEP VONG, Ven. Patriarch BOU KRI.

Patriotic Kampuchean Buddhists' Association: Phnom-Penh; mem. of UFCDKF; Pres. LONG SIM.

CHRISTIANITY

The Roman Catholic Church

Cambodia comprises the Apostolic Vicariate of Phnom-Penh and the Apostolic Prefectures of Battambang and Kompong-Cham. At 31 December 2002 there were an estimated 15,301 adherents in the country, equivalent to about 0.1% of the population. An Episcopal Conference of Laos and Kampuchea was established in 1971. In 1975 the Government of Democratic Kampuchea banned all religious practice in Cambodia, and the right of Christians to meet to worship was not restored until 1990.

Vicar Apostolic of Phnom-Penh: Rt Rev. EMILE DESTOMBES (Titular Bishop of Altava), 787 blvd Monivong (rue 93), BP 123, Phnom-Penh; tel. and fax (23) 212462; e-mail evecam@camnet.com.kh.

ISLAM

Islam is practised by a minority in Cambodia. Islamic worship was also banned in 1975, but it was legalized in 1979, following the defeat of the Democratic Kampuchean regime.

The Press

According to Cambodia's Press Law, newspapers, magazines and foreign press agencies are required to register with the Department of Media at the Ministry of Information. In October 2002 160 Khmer language newspapers, 36 foreign language newspapers, 42 magazines, 19 bulletins and 12 foreign news agencies were registered.

NEWSPAPERS AND MAGAZINES

Newspapers are not widely available outside Phnom-Penh. In June 2004 Cambodia's first regional newspaper, *Somne Thmey*, was launched. It was to circulate in the country's four main provinces.

Areyathor (Civilization): 52 rue Lyuk Lay, Sangkat Chey, Chummneah, Phnom-Penh; tel. (23) 913662; Editor CHIN CHAN MONTY.

Bayon Pearnik: 3 rue 174, BP 2279, Phnom-Penh; tel. (12) 803968; fax (23) 211921; e-mail bp@forum.org.kh; internet www.bayonpearnik.com; f. 1995; English; monthly; Publr and Editor ADAM PARKER; circ. 10,000.

Cambodia Daily: 50B rue 240, Phnom-Penh; tel. (23) 426602; fax (23) 426573; e-mail aafc@forum.org.kh; internet www.cambodiadaily.com; f. 1993; in English and Khmer; Mon.–Sat.; Editor CHRIS DECHERD; Publr BERNARD KRISHER; circ. 3,500.

Cambodia New Vision: BP 158, Phnom-Penh; tel. (23) 219898; fax (23) 360666; e-mail cabinet1b@camnet.com.kh; internet www.cnv.org.kh; f. 1998; official newsletter of the Cambodian Govt.

Chakraval: 3 rue 181, Sangkat Tumnop Teuk, Khan Chamkarmon, Phnom-Penh; tel. (23) 913667; fax (23) 720141; Khmer; daily; Publr KEO SOPHORN; Editor SO SOVAN RITH.

Commercial News: 394 blvd Preah Sihanouk, Phnom-Penh; tel. (23) 721665; fax (23) 721709; e-mail tcnews@camnet.com.kh; f. 1993; Chinese; Chief Editor LIU XIAO GUANG; circ. 6,000.

Construction (Kasang): 126 rue 336, Sangkat Phsar Deum Kor, Khan Tuol Kok, Phnom-Penh; tel. (18) 818292; Khmer; Editor CHHEA VARY.

Equality Voice: 470 rue 163, Sangkat Boeung Keng Kang, Khan Chamkarmon, Phnom-Penh; tel. (12) 842471; Khmer; Publr HUON MARA.

Indradevi: 256 blvd Mao Tse Toung, Phnom-Penh; tel. (23) 215808; e-mail indradevi@camnet.com.kh; internet www.indradevi.com.kh; f. 1998; general interest magazine.

Kampuchea: 158 blvd Norodom, Phnom-Penh; tel. (23) 725559; f. 1979; weekly; Chief Editor KEO PRASAT; circ. 55,000.

Kampuchea Thmey (New Cambodia): 805 rue Kampuchea Krom, Phnom-Penh; tel. (23) 882535; fax (23) 882656; e-mail kampucheathmey@camnet.com.kh; internet www.kampucheathmey.com; daily.

Khmer Wisdom: 1588 Khan Russei Keo, Phnom-Penh; tel. (12) 841377; Khmer; Publr CHEA CHAN THON.

Khmer Youth Voice: 240 rue 374, Sangkat Toul Prey 2, Khan Chamkarmon, Phnom-Penh; tel. (23) 211336; fax (23) 210137; e-mail sovann@camnet.com.kh; Khmer; twice weekly; Editor UO SOVANN.

Koh Santepheap (Island of Peace): 41 rue 338, Sangkat Toul Svay Prey 1, Khan Chamkarmon, Phnom-Penh; tel. (23) 211818; fax (23) 220155; e-mail kohsantepheap@online.com.kh; internet www.kohsantepheapdaily.com.kh; Khmer; daily; Dir THONG UY PANG.

Moneaksekar Khmer: 27 rue 318, Sangkat Toul Svay Prey 1, Khan Chamkarmon, Phnom-Penh; tel. (23) 990777; Editor DAM SITHIK.

Neak Chea: 1 rue 158, Khan Daun Penh, Phnom-Penh; tel. (23) 428653; fax (23) 427229; e-mail adhoc@forum.org.kh.

Phnom Penh Daily: 5 rue 84, Corner 61, Sangkat Srah Chak, Khan Daun Penh, Phnom-Penh; tel. (15) 917682; e-mail ppenhdaily@camnet.com.kn; internet www.phnompenhdaily.com.kh; Khmer; available online in English; Editor VA DANE.

Phnom Penh Post: 10A rue 264, Phnom-Penh; tel. (23) 210309; fax (23) 426568; e-mail michael.pppost@online.com.kh; internet www.phnompenhpost.com; f. 1992; English; fortnightly; Editor-in-Chief and Publr MICHAEL HAYES.

Pracheachon (The People): 101 blvd Norodom, Phnom-Penh; tel. (23) 723665; f. 1985; 2 a week; organ of the CPP; Editor-in-Chief SOM KIMSUOR; circ. 50,000.

Rasmei Kampuchea: 476 blvd Monivong, Phnom-Penh; tel. (23) 362881; fax (23) 362472; e-mail rasmei_kampuchea@yahoo.com; daily; f. 1993; local newspaper in northern Cambodia; Editor PEN SAMITHY.

Samleng Thmei (New Voice): 91 rue 139, Sangkat Veal Vong, Khan 7, Phnom-Penh; tel. (15) 920589; Khmer; Editor KHUN NGOR.

Somne Thmey (New Writing): tel. (23) 224303; e-mail mcd.drans@online.com.kh; f. 2004; Khmer; weekly; publ. in Siem Reab, Preah Sihanouk, Kampong Cham and Bat Dambang; Editor-in-Chief PEN BONA.

NEWS AGENCIES

Agence Kampuchia de Presse (AKP): 62 blvd Monivong, Phnom-Penh; tel. (23) 430564; e-mail akp@camnet.com.kh; internet www.camnet.com.kh/akp; f. 1978; Dir-Gen. KIT-KIM HUON.

Foreign Bureaux

Agence France-Presse (AFP) (France): 8 rue 214, BP 822, Phnom-Penh; tel. (23) 426227; fax (23) 426226; Bureau Chief LUKE HUNT.

Associated Press (AP) (USA): 18C rue 19, BP 870, Phnom-Penh; tel. (23) 426607; e-mail ap@bigpond.com.kh; Correspondent CHRIS FONTAINE.

Deutsche Presse-Agentur (dpa): 5E rue 178, Phnom-Penh; tel. (23) 427846; fax (23) 427846; Correspondent JOE COCHRANE.

Reuters (UK): 201, 2nd Floor, Hong Kong Centre, 108–112 blvd Sothearos, Phnom-Penh; tel. (23) 216977; fax (23) 216970; Bureau Chief ROBERT BIRSEL.

Xinhua (New China) News Agency (People's Republic of China): 19 rue 294, Phnom-Penh; tel. (23) 211608; fax (23) 426613; Correspondent LEI BOSONG.

ASSOCIATIONS

Cambodian Association for the Protection of Journalists (CAPJ): BP 816, 58 rue 336, Sangkat Psadoeum Khor, Khan Tuol Kok, Phnom-Penh; tel. (15) 997004; fax (23) 215834; e-mail umsarin@hotmail.com; Pres. UM SARIN.

Cambodian Club of Journalists: Phnom-Penh; Pres. PEN SAMITHY; Sec.-Gen. PRACH SIM.

Khmer Journalists' Association: 101 blvd Preah Norodom, Phnom-Penh; tel. (23) 725459; f. 1979; mem. of UFCDKF; Pres. PIN SAMKHON.

League of Cambodian Journalists (LCJ): 74 rue 205, Sangkat Toulsvayprey, Khan Chamkarmon, Phnom-Penh; tel. and fax (23) 360612; Pres. OM CHANDARA.

Broadcasting and Communications

TELECOMMUNICATIONS

Cambodian Samart Communication: 56 blvd Norodom, Sangkat Chey Chumneah, Khan Daun Penh, Phnom-Penh; tel. (16) 810001; fax (16) 810004; e-mail somchai.an@hello016-gsm.com; internet www.hello016-gsm.com; f. 1992; operates a national mobile telephone network; CEO SOMCHAI (AN) LERTWISETTHEERAKUL.

CamGSM Co Ltd: 33 blvd Preah Sihanouk, Phnom-Penh; tel. (12) 800800; fax (12) 801801; e-mail helpline@mobitel.com.kh; internet www.mobitel.com.kh; f. 1996; operates national GSM 900 mobile telephone network under trade name MobiTel; CEO DAVID SPRIGGS.

Camintel: 1 cnr Terak Vithei Sisowath and Vithei Phsar Dek, Phnom-Penh; tel. (23) 986789; fax (23) 986277; e-mail victor@camintel.com; internet www.camintel.com; a jt venture between the Ministry of Posts and Telecommunications and the Indonesian co, Indosat; operates domestic telephone network; Chair. NHEK KORSOL VYTHYEA.

Camshin Corporation: Suite 6B–7B, Regency Sq., 294 blvd Mao Tse Toung, Phnom-Penh; tel. (23) 367801; fax (23) 367805; e-mail sales@camshin.com; internet www.camshin.com; a jt venture between the Ministry of Posts and Telecommunications and the Thai co, Shinawatra International Co Ltd; telephone communications co.

BROADCASTING

Radio

Apsara: 69 rue 57, Sangkat Boeung Keng Kang 1, Khan Chamkarmon, Phnom-Penh; tel. (23) 303002; fax (23) 214302; internet www.apsaratv.com.kh; f. 1996; Head of Admin. KEO SOPHEAP; News Editor SIN SO CHEAT.

Bayon: c/o Bayon Media Group, 954 rue 2, Takhmau, Kandal Province; tel. (23) 363695; fax (23) 363795; e-mail bayontv@camnet.com.kh; internet www.bayontv.com.kh; Dir-Gen. KEM KUNNAVATH.

Bee Hive Radio: 949 rue 360, Boeung Keng Kang, Khan Chamkarmon, Phnom-Penh; tel. (23) 992939; fax 240439; e-mail sok@online.com.kh; Dir-Gen. MAM SONANDO.

FM 90 MHZ: 65 rue 178, Phnom-Penh; tel. (23) 363699; fax (23) 368623; Dir-Gen. NHIM BUN THON; Dep. Dir-Gen. TUM VANN DET.

FM 99 MHZ: 41 rue 360, Phnom-Penh; tel. (23) 426794; Gen. Man. SOM CHHAYA.

FM 107 MHZ: 18 rue 562, Phnom-Penh; tel. (23) 880874; fax (23) 368212; e-mail tv9@camnet.com.kh; internet www.tv9.com.kh; Dir-Gen. KHUN HANG.

Krusa FM: Phnom-Penh; e-mail febcam@bigpond.com.kh; f. 2002; controlled by Far East Broadcasting; religious programmes; Dir SAMOEUN INTAL.

Phnom-Penh Municipality Radio: 131–132 blvd Pochentong, Phnom-Penh; tel. (23) 725205; fax (23) 360800; Gen. Man. KHAMPUN KEOMONY.

Radio WMC: 30 rue 488, Sangkat Phsar Dem Thkov, Khan Chamkarmon, Phnom-Penh; tel. and fax (12) 847854; e-mail wmc@forum.org.kh; independent; radio station of Women's Media Centre of Cambodia; Dir CHEA SUDANETH.

RCAF Radio: c/o Borei Keila, rue 169, Phnom-Penh; tel. (23) 366061; fax (23) 366063; f. 1994; Royal Cambodian Armed Forces radio station; Dir THA TANA; News Editor SENG KATEKA.

Ta Prohm Radio: 27B rue 472, Phnom-Penh; tel. (23) 213054; e-mail taprohm@yahoo.com; f. 2003; launched by FUNCINPEC Party as opposition radio station; broadcasts news programmes in Khmer to Phnom-Penh and surrounding area.

Vithyu Cheat Kampuchea (National Radio of Cambodia): rue Preah Kossamak, Phnom-Penh; tel. (12) 852741; fax (11) 745074; e-mail chaosang2002@yahoo.com; f. 1978; fmrly Vithyu Samleng Pracheachon Kampuchea (Voice of the Cambodian People); controlled by the Ministry of Information; home service in Khmer; daily external services in English, French, Lao, Vietnamese and Thai; Dir-Gen. VANN SENG LY; Dep. Dir-Gen. TAN YAN.

Voice of Cambodia: Phnom-Penh; e-mail vocri@vocri.org; internet www.vocri.org; Cambodia's first international internet radio station.

There are also several private local radio stations.

Television

Apsara Television (TV11): 69 rue 57, Sangkat Boeung Keng Kang 1, Khan Chamkarmon, Phnom-Penh; tel. (23) 303002; fax (23) 214302; internet www.apsaratv.com.kh; Dir-Gen. SOK EISAN.

Bayon Television (TV27): 954 rue 2, Takhmau, Kandal Province; tel. (23) 363695; fax (23) 363795; e-mail bayontv@camnet.com.kh; internet www.bayontv.com.kh; Dir-Gen. KEM KUNNAVATH.

National Television of Cambodia (TVK): 62 blvd Preah Monivong, Phnom-Penh; tel. (23) 722983; fax (23) 426407; e-mail tvk@camnet.gov.kh; internet www.tvk.gov.kh; f. 1983; broadcasts for 10 hours per day in Khmer; Dir-Gen. (Head of Television) MAO AYUTH.

Phnom-Penh Television (TV3): 2 blvd Russia, Phnom-Penh; tel. (12) 814323; fax (23) 360800; e-mail tv3@camnet.com.kh; internet www.tv3.com.kh; Dir-Gen. KHAMPHUN KEOMONY.

RCAF Television (TV5): 165 rue 169, Borei Keila, Phnom-Penh; tel. (23) 366061; fax (23) 366063; e-mail mica.t.v.5@bigpond.com.kh; Editor-in-Chief PRUM KIM.

TV Khmer (TV9): 18 rue 562, Phnom-Penh; tel. (23) 880874; fax (23) 368212; e-mail tv9@camnet.com.kh; internet www.tv9.com.kh; f. 1992; Dir-Gen. KHOUN ELYNA; News Editor PHAN TITH.

Finance

(cap. = capital; res = reserves; dep. = deposits; brs = branches)

BANKING

The National Bank of Cambodia, which was established as the sole authorized bank in 1980 (following the abolition of the monetary system by the Government of Democratic Kampuchea in 1975), is the central bank, and assumed its present name in February 1992. The adoption of a market economy led to the licensing of privately owned and joint-venture banks from July 1991. Following the implementation of the 1999 Financial Institutions Law 29 of the banks operative in Cambodia were considered for re-licensing. Between November 1999 and November 2002 17 banks were closed, many having failed to meet new capital requirements. At the end of December 2002 there were 17 banks (excluding the central bank) operating in Cambodia, including: one state-owned bank; four specialized banks; nine locally-incorporated private banks; and three branches of foreign banks.

Central Bank

National Bank of Cambodia: 22–24 blvd Preah Norodom, BP 25, Phnom-Penh; tel. (23) 722563; fax (23) 426117; e-mail nbc@bigpond.com.kh; f. 1980; cap. 100,000m. riels, res 939,154m. riels, dep. 1,777,673m. riels (Dec. 2003); Gov. CHEA CHANTO; Dep. Gov. ENG THAYSAN.

State Bank

Foreign Trade Bank: 3 rue Kramoun Sar, Khan Daun Penh, Phnom-Penh; tel. (23) 724466; fax (23) 426108; e-mail info@ftbbank.com; internet www.ftbbank.com; f. 1979; removed from direct management of National Bank of Cambodia in 2000; scheduled for privatization; Man. TIM BO PHOL.

Specialized Banks

ACLEDA Bank Ltd: 28 blvd Mao Tse Toung, Sangkat Beung Trabek, Khan Chamkarmon, Phnom-Penh; tel. (23) 364619; fax (23) 364914; e-mail acledabank@acledabank.com.kh; internet www.acledabank.com.kh; f. 1993; became specialized bank in Oct. 2000; provides financial services to all sectors; Gen. Man. IN CHANNY; 94 brs and offices.

Cambodia Agriculture Industrial Specialized Bank: 87 blvd Preah Norodom, Sangkat Phsar Thmey III, Khan Daun Penh, Phnom-Penh; tel. (23) 218667; fax (23) 217751; e-mail kien@bigpond.com.kh; Man. CHHOR SANG.

Peng Heng SMI Bank: 72 blvd Norodom, Phnom-Penh; tel. (23) 219243; fax (23) 219185; e-mail pengheng@camnet.com.kh; f. 2001.

Rural Development Bank: 9–13 rue 7, Sangkat Chaktomouk, Khan Daun Penh, Phnom-Penh; tel. (23) 220810; fax (23) 722388; e-mail rdb@online.com.kh; internet www.rdb.com.kh; f. 1998; provides credit to rural enterprises; Chair. and CEO SON KOUN THOR.

Private Banks

Advanced Bank of Asia Ltd: 97–99 blvd Preah Norodom, Sangkat Boeung Raing, Khan Daun Penh, Phnom-Penh; tel. (23) 720434; fax (23) 720435; e-mail aba@ababank.com.kh; Dir CHAE WAN CHO.

Cambodia Asia Bank Ltd: 252 blvd Preah Monivong, Sangkat Phsar Thmey II, Khan Daun Penh, Phnom-Penh; tel. (23) 722105; fax (23) 426628; e-mail cab@camnet.com.kh; Man. WONG TOW FOCK.

Cambodia Mekong Bank: 1 rue Kramoun Sar, Sangkat Phsar Thmey I, Khan Daun Penh, Phnom-Penh; tel. (23) 217114; fax (23) 217122; e-mail ho.mailbox@mekongbank.com; cap. US $13m., dep. US $5.4m. (Dec. 2002); Chair. MICHAEL C. STEPHEN; Pres. and CEO KHOV BOUN CHHAY.

Cambodian Commercial Bank Ltd: 26 blvd Preah Monivong, Sangkat Phsar Thmey II, Khan Daun Penh, Phnom-Penh; tel. (23) 426145; fax (23) 426116; e-mail ccbpp@online.com.kh; internet www.cbb-cambodia.com; f. 1991; cap. US $13m., dep. US $76.6m. (Dec. 2002); Chair MALEERATNA PLUMCHITCHOM; Dir and Gen. Man. SAHASIN YUTTARAT; 4 brs.

Cambodian Public Bank (Campu Bank): Villa 23, rue Kramoun Sar, Sangkat Phsar Thmey II, Khan Daun Penh, Phnom-Penh; tel. (23) 214111; fax (23) 217655; e-mail campu@online.com.kh; f. 1992; cap. US $15m., dep. US $80.9m. (Dec. 2002); Gen. Man. PHAN YING TONG.

Canadia Bank Ltd: 265–269 rue Preah Ang Duong, Sangkat Wat Phnom, Khan Daun Penh, Phnom-Penh; tel. (23) 215286; fax (23) 427064; e-mail canadia@camnet.com.kh; internet www.canadiabank.com; f. 1991; cap. 50.6m. riels, res 32.3m. riels, dep. 411.6m. riels (2001); Man. PUNG KHEAV SE; 10 brs.

Singapore Banking Corporation Ltd: 68 rue Samdech Pan, Sangkat Boeung Reang, Khan Daun Penh, Phnom-Penh; tel. (23) 211211; fax (23) 212121; e-mail info@sbc-bank.com; internet www.sbc-bank.com; f. 1993; cap. US $13m. (2003), dep. US $6.8m. (Dec. 2000); Pres. ANDY KUN SWEE TIONG; Chair. KUN KAY HONG.

Union Commercial Bank Ltd: UCB Bldg, 61 rue 130, Sangkat Phsar Chas, Khan Daun Penh, Phnom-Penh; tel. (23) 427995; fax (23) 427997; e-mail ucb@bigpond.com.kh; internet www.ucb.com.kh; f. 1994; cap. US $13m., dep. US $46m. (Dec. 2002), res US $4m. (Dec. 2001); CEO YUM SUI SANG; Chair. and Pres. YIU KAI KWONG; 3 brs.

Vattanac Bank: 89 blvd Preah Norodom, Sangkat Boeung Raing, Khan Daun Penh, Phnom-Penh; tel. (23) 212727; fax (23) 216687; e-mail csc@vattanacbank.com; internet www.vattanacbank.com.

Foreign Banks

First Commercial Bank (Taiwan): 263 rue Preah Ang Duong, Sangkat Wat Phnom, Khan Daun Penh, Phnom-Penh; tel. (23) 210027; fax (23) 210029; e-mail fcbpp@bigpond.com.kh.

Krung Thai Bank PLC (Thailand): 149 rue Jawaharlal Nehru, Depot Market 1, Khan Tuolkok Division, Phnom-Penh; tel. (23) 366005; fax (23) 428737; e-mail ktbpmp@bigpond.com.kh; Man. NAKROB U-SETTHASAKDI.

Maybank Bhd (Malaysia): 4 rue Kramoun Sar, Sangkat Boeung Raing, Khan Daun Penh, Phnom-Penh; tel. (23) 210123; fax (23) 210099; e-mail mbb@camnet.com.kh; internet www.maybank2u.com.my; Man. ABDUL MALEK MOHD KHAIR.

INSURANCE

Asia Insurance (Cambodia) Ltd: 91 blvd Preah Norodom, Sangkat Boeung Raing, Khan Daun Penh, Phnom-Penh; tel. (23) 427981; fax (23) 216969; e-mail email@asiainsurance.com.kh; internet www.asiainsurance.com.kh; Gen. Man. PASCAL BRANDT-GAGNON.

Cambodia National Insurance Company (CAMINCO): cnr rue 106 and rue 19, Phnom-Penh; tel. (23) 772043; fax (23) 427810; e-mail caminco@caminco.com.kh; state-owned insurance co.

Commercial Union: 28 rue 47, Phnom-Penh; tel. (23) 426694; fax (23) 427171; general insurance; Gen. Man. PAUL CABLE.

Forte Insurance (Cambodia) Ltd: 325 blvd Mao Tse Toung, BP 565, Phnom-Penh; tel. (23) 885077; fax (23) 982907; e-mail forte@forte.com.kh; internet www.forteinsurance.com; f. 1996; Man. Dir CARLO CHEO.

Indochine Insurance Ltd: 55 rue 178, BP 808, Phnom-Penh; tel. (23) 210701; fax (23) 210501; e-mail info@indochine.com.kh; internet www.indochine.net; f. 1994; Man. Dir PHILIPPE LENAIN.

Trade and Industry

DEVELOPMENT ORGANIZATIONS

Council for the Development of Cambodia (CDC): Government Palace, quai Sisowath, Wat Phnom, Phnom-Penh; tel. (23) 981156; fax (23) 428426; internet www.cdc-crdb.gov.kh; f. 1993; Chair. HUN SEN; Sec.-Gen. SOK CHENDA.

Cambodian Investment Board (CIB): Government Palace, quai Sisowath, Wat Phnom, Phnom-Penh; tel. (23) 981156; fax (23) 428426; e-mail CDC.CIB@bigpond.com.kh; internet www.cambodiainvestment.gov.kh; f. 1993; part of CDC; sole body responsible for approving foreign investment in Cambodia, also grants exemptions from customs duties and other taxes, and provides other facilities for investors; Chair. HUN SEN; Sec.-Gen. SOK CHENDA.

National Information Communications Technology Development Authority (NiDA): 3rd Floor, Satellite Bldg, Office of the Council of Ministers, blvd Confederation de la Russie, Phnom-Penh; tel. (23) 880635; fax (23) 880637; e-mail info@nida.gov.kh; internet www.nida.gov.kh; f. 2000; promotes information technology and formulates policy for its development; Chair. HUN SEN; Sec.-Gen. PHU LEEWOOD.

CHAMBER OF COMMERCE

Cambodia Chamber of Commerce: Office Vila 7B, cnr rue 81 and rue 109, Sangkat Boeung Reang, Khan Daun Penh, Phnom-Penh; tel. (23) 212265; fax (23) 212270; e-mail ppcc@camnet.com.kh; internet www.ppcc.org.kh; f. 1995; Pres. SOK KONG; Dir-Gen. Dr NANG SOTHY.

INDUSTRIAL AND TRADE ASSOCIATIONS

Export Promotion Department: Ministry of Commerce, 65–67–69 rue 136, Sangkat Phsar Kandal II, Khan Daun Penh, Phnom-Penh; tel. (23) 216948; fax (23) 217353; e-mail praknork@everyday.com.kh; internet www.moc.gov.kh; f. 1997; Dir PRAK NORK.

Garment Manufacturers' Association in Cambodia (GMAC): rue Jawaharlal Nehru, Phnom-Penh; tel. (23) 723796; fax (23) 369398; e-mail GMAC@bigpond.com.kh; Pres. VAN SOU IENG; Sec.-Gen. KEN LOO.

UTILITIES

Electricity

Electricité du Cambodge: EDC Bldg, rue 19, Wat Phnom, Khan Daun Penh, Phnom-Penh; tel. (23) 724771; fax (23) 426938; e-mail yim_nolson@bigpond.com.kh; f. 1996; state-owned; Man. Gen. TAN KIM VIN.

Electricity Authority of Cambodia (EAC): 2 rue 282, Boeng Keng Kang 1, Phnom-Penh; tel. (23) 217654; fax (23) 214144; e-mail admin@eac.gov.kh; internet www.eac.gov.kh; f. 2001; regulatory authority; Chair. Dr TY NORIN.

Water

Phnom-Penh Water Supply Authority: rue 108, 12201 Phnom-Penh; tel. (23) 724046; fax (23) 428969; e-mail eksonnchan@ppwsa.com.kh; f. 1996 as an autonomous public enterprise; Dir-Gen. EK SONN CHAN.

MAJOR COMPANIES

Asia Flour Mill Corpn: 228 blvd Preah Norodom, Sangkat Tonle Bassac, Khan Chamkarmon, Phnom-Penh; tel. (23) 301228; fax (23) 366928; e-mail AFMCorp@mobitel.com.kh; flour mfr; Man. IGOR HENRI.

Caltex Cambodia Ltd: Olympic Motor Bldg, 173 blvd Jawaharlal Nehru, Sangkat Phsar Deumkor, Phnom-Penh; tel. (23) 880570; fax (23) 880691; e-mail caltex@online.com.kh; f. 1995; retail service station operator; distribution and marketing of fuels and lubricants.

Cambodia Beverage Co Ltd: 287 rue Nationale 5, Khan Russei Keo, Phnom-Penh; tel. (23) 428995; fax (23) 428992; mfr of soft drinks.

Cambodia Cement Co: 3 rue 598, Khum Tuk Thla, Khan Russei Keo, Phnom-Penh; tel. (23) 27540; fax (23) 27815; production of cement.

Continental Indochine Import/Export Co Ltd: 139 blvd Monivong, Phnom-Penh; tel. (23) 366602; fax (23) 366604; e-mail cil@bigpond.com.kh; f. 1992; aviation fuel supply and bulk fuel distribution; mfr of knitwear and woven garments; bar, restaurant and hotel management; Group Chair. CLIVE MCLEOD FAIRFIELD.

Goodhill Enterprise (Cambodia) Ltd: 214/218 Goodhill Bldg, blvd Preah Sihanouk, Phnom-Penh; tel. (23) 217888; fax (23) 213688; e-mail goodhill_rt@bigpond.com.kh; internet www.goodhill.com.kh; f. 1989; distributor of consumer products, stationery, office equipment, lubricants, beverages and fertilizer.

Hung Hiep (Cambodia) Co Ltd: 230A blvd Preah Norodom, Khan Chamkarmon, Phnom-Penh; tel. (23) 213527; fax (23) 216659; e-mail hunghiep@bigpond.com.kh; dealers in passenger and commercial vehicles.

JIT Service Ltd: 50 rue 139, Sangkat Veal Vong, Khan Makara 7, Phnom-Penh; tel. (15) 830031; fax (23) 721197; e-mail jitcambodia@bigpond.com.kh; f. 1996; jt venture with Singapore; export of frozen seafood; freight forwarding services; automotive repairs; Chair. RAMADY MOUN.

JMK Group: 6 blvd Monivong, BP 112, Phnom-Penh; tel. (23) 217366; fax (23) 218884; e-mail jmkt@camnet.com.kh; f. 1992; architectural services; computer sales and maintenance.

Muhibbah Engineering (Cambodia) Ltd: 313–315 blvd Mao Tse Toung, Sangkat Phsar Dept III, Khan Toul Kork, Phnom-Penh; tel. (23) 367988; fax (23) 366888; e-mail admin@muhibbah.com.kh; internet www.muhibbah.com; civil engineering, general contracting, real estate development, architectural services; Man. CHEAH SOON LYE.

Tela Petroleum Group Investment Co: 9 rue Jawaharlal Nehru, Phnom-Penh; tel. (23) 428739; fax (23) 725559; e-mail tela.petroleum@camnet.com.kh; supplies petroleum products to Cambodian armed forces and civil administration; Chair. CHHUN ON; Dir-Gen. MUONG KOMPHEAK.

Total Cambodia: 108–112 blvd Preah Sothearos, BP 600, Phnom-Penh; tel. (23) 218630; fax (23) 217662; e-mail total.cambodge@camnet.com.kh; sales US $30m. (Dec. 2001); international petroleum co; Chair. PIERRE YVES LOISEAU.

TRADE UNIONS

Association of Independent Cambodian Teachers: 33 rue 432, Sangkat Boeng Trabaek, Khan Chamkarmon, Phnom Penh; Pres. RUNG CHHUN; Gen.-Sec. CHEA MUNI.

Cambodia Federation of Independent Trade Unions (CFITU): 45 rue 63, Boeng Keng Kang 1, Khan Chamkarmon, Phnom-Penh; tel. (23) 213356; e-mail CFITU@bigpond.com.kh; f. 1979 as Cambodia Federation of Trade Unions; changed name as above in 1999; Chair. ROS SOK; Vice-Chair. TEP KIM VANNARY, KIENG THISOTHA.

Cambodia Labour Union Federation (CLUF): 78 rue 474, Sangkat Boeung Trabek, Khan Chamkarmon, Phnom-Penh; tel. (23) 866682; f. 1999; Pres. SOM AUN.

Cambodian Union Federation (CUF): 18 rue 112, Sangkat Phsar Depo III, Khan Toul Kok, Phnom-Penh; tel. (23) 882453; fax (23) 427632; e-mail CUF@bigpond.com.kh; f. 1997 with the support of the CPP in response to the formation of the FTUWKC; Pres. CHUON MOM THOL.

Cambodian Union Federation of Building and Wood Workers: 18A rue 112, Sangkat Phsar Depo III, Khan Tuol Kok, Phnom-Penh; tel. (23) 842382; fax (23) 882453; f. 2001; Pres. SAY SAM ON.

Coalition of Cambodia Apparel Workers' Democratic Union: 6C rue 476, Sangkat Tuol Tum Pong I, Khan Chamkarmon, Phnom-Penh; tel. (23) 210481; f. 2001; Pres. CHHORN SOKHA.

Free Trade Union of Workers of the Kingdom of Cambodia (FTUWKC): 222B Sangkat Boeung Reang, Khan Daun Penh, Phnom-Penh; tel. and fax (23) 216870; e-mail ftuwkc@cambodiaworkers.org; internet www.cambodiaworkers.org; fmrly Free Trade Union of Khmer Workers; f. 1996 by Mary Ou with the assistance of Sam Rainsy; Pres. SAM SREY MOM (acting); Gen. Sec. SUM SAMNEANG.

National Independent Federation Textile Union of Cambodia (NIFTUC): 29B rue 432, Sangkat Toul Tompoung II, Khan Chamkarmon , Phnom-Penh; tel. and fax (23) 219239; e-mail niftuc@forum.org.kh; f. 1999; Pres. MORM NHIM.

Transport

RAILWAYS

Royal Railway of Cambodia: Central Railway Station, Railway Sq., Sangkat Srach Chak, Khan Daun Penh, Phnom-Penh; tel. (12) 994168; fax (23) 430815; e-mail RRCcambodia@bigpond.com.kh; comprises two 1,000 mm-gauge single-track main lines with a total length of 650 km: the old 385-km Phnom-Penh to Poipet line (of which the 48-km Sisophon to Poipet link is awaiting restoration), the new 264-km Phnom-Penh to Sihanoukville line and branch lines and special purpose sidings 100 km; the condition of the tracks and bridges on both lines is very poor, with many temporary repairs, owing to mine damage, and the service also suffers from other operational difficulties, such as a shortage of rolling stock; there are 14 'Gares' (main stations), 19 stations and 38 halts; Pres. and Dir-Gen. Sokhom Pheakavanmony.

ROADS

In 1997 the total road network was 35,769 km in length, of which 4,165 km were highways and 3,604 km were secondary roads. In the same year about 7.5% of the road network was paved, but this figure rose to an estimated 11.6% in 1999. West and East Cambodia were linked by road for the first time in December 2001, with the opening of a bridge across the Mekong River.

INLAND WATERWAYS

The major routes are along the Mekong river, and up the Tonlé Sap river into the Tonlé Sap (Great Lake), covering, in all, about 2,400 km. The inland ports of Neak Luong, Kompong Cham and Prek Kdam have been supplied with motor ferries, and the ferry crossings have been improved.

SHIPPING

The main port is Sihanoukville, on the Gulf of Thailand, which has 11 berths and can accommodate vessels of 10,000–15,000 tons. Phnom-Penh port lies some distance inland. Steamers of up to 4,000 tons can be accommodated.

CIVIL AVIATION

There is an international airport at Pochentong, near Phnom-Penh. A further international airport, at Siem Reap, was inaugurated in December 2002. In mid-2003 construction of a new terminal at Pochentong International Airport was completed.

State Secretariat of Civil Aviation (SSCA): 62 blvd Norodom, Phnom-Penh; tel. (23) 360617; fax (23) 426169; e-mail civilaviation@cambodia.gov.kh; internet www.civilaviation.gov.kh; Dir-Gen. Keo Saphal.

Mekong Airlines: Hong Kong Centre, blvd Sothearos, Phnom-Penh; tel. (23) 217299; fax (23) 217277; e-mail bookm8@everyday.com.kh; internet www.marveltour.com/mekong; f. 2002; joint venture between Hun Kim Leng Investment (51%) and Australian co Via Aviation (49%); domestic and international flights to eight destinations.

President Airlines: 298 blvd Mao Tse Toung, Phnom-Penh; tel. (23) 993088; fax (23) 212992; e-mail pnhrr@presidentairlines.com; internet www.presidentairlines.com; f. 1998; domestic and, from Aug. 2002, international passenger services.

Royal Khmer Airlines: 19 Unit 12, rue Preah Kossomak, Phnom-Penh; tel. (23) 216899; fax (23) 428279; f. 2000; domestic and international services.

Royal Phnom-Penh Airways: 209 rue 19, Sangkat Chey Chumneah, Khan Daun Penh, Phnom-Penh; tel. (23) 216487; fax (23) 217420; e-mail pnhairways@rlppairways.com; internet www.rlppairways.com; f. 1999; scheduled and charter passenger flights to domestic and regional destinations; Chair. Prince Norodom Chakrapong.

Siem Reap Airways International: 65 rue 214, Sangkat Beoung Rang, Khan Daun Penh, Phnom-Penh; tel. (23) 720022; fax (23) 720522; e-mail reservation@siemreapairways.com; internet www.siemreapairways.com; f. 2000; scheduled international and domestic passenger services; CEO Prasert Prasarttong-Osoth.

Tourism

Tourist arrivals increased to an estimated 466,365 in 2000, owing to the improvement in the security situation; in that year tourist receipts reached US $178m. Despite the repercussions of the terrorist attacks on the USA in September 2001, an estimated 604,919 tourists visited in that year, an increase of 29.7% compared with 2000. In 2002 tourist arrivals totalled 786,546 and receipts an estimated $379m. In an attempt to encourage the development of the tourism sector, 2003 was designated 'Visit Cambodia Year'. However, in that year tourist arrivals decreased by 10.9%, to 701,014.

Directorate-General of Tourism: 3 blvd Monivong, Phnom-Penh; tel. (23) 427130; fax (23) 426107; e-mail tourism@camnet.com.kh; f. 1988; Dir So Mara.

Defence

In August 2003 the total strength of the Royal Cambodian Armed Forces was estimated to be 125,000 (including provincial forces): army 75,000, navy 3,000, air force 2,000 and provincial forces about 45,000. There was a system of conscription in force, for those aged between 18 and 35, for five years, although this had not been implemented since 1993. Paramilitary forces are organized at village level. The defence budget for 2003 was estimated at 300,000m. riels.

Supreme Commander of the Royal Cambodian Armed Forces: King Norodom Sihamoni.

Commander-in-Chief: Gen. Ke Kimyan.

Education

Primary education is compulsory for six years between the ages of six and 12. In 1997 enrolment in primary schools was equivalent to 113% of school-age children (males 123%; females 104%). Secondary education comprises two cycles, each lasting three years. In 1996 enrolment at secondary level was equivalent to 24% of those in the relevant age-group (males 31%; females 17%).

In 1996 enrolment at tertiary level was equivalent to 1.0% of those in the relevant age-group (males 2.0%; females 0.5%). Institutions of higher education included Phnom-Penh University, an arts college, a technical college, a teacher-training college, a number of secondary vocational schools and an agricultural college. In 1993, according to a sample survey, the average rate of adult illiteracy was 34.7% (males 20.3%, females 46.6%). Rates varied between urban and rural areas; adult illiteracy in Phnom-Penh was 18.0% (males 8.1%; females 36.7%), whereas in rural areas it was 36.5% (males 21.4%; females 49.0%).

The budget for 2001 allocated 209,000m. riels (9.0% of total expenditure) to the Ministry of Education, Youth and Sport.

Bibliography

See also Laos and Viet Nam

Acharya, A., Lizée, P., and Peou, S. (Eds). *Cambodia—The 1989 Paris Peace Conference.* New York, Kraus International, 1991.

Ann, Porn Moniroth. *Democracy in Cambodia: Theories and Realities.* Translated by Khieu Mealy and edited by Sok Siphana, Phnom-Penh, Cambodian Institute for Co-operation and Peace, 1996.

Ayres, David M. *Anatomy of a Crisis: Education, Development and the State in Cambodia, 1953–1998.* Honolulu, HI, University of Hawaii Press, 2001.

Bizot, François. *The Gate.* London, Harvill Press, 2003.

Brown, Frederick Z., and Timberman, David G. *Cambodia and the International Community: The Quest for Peace, Development and Democracy.* Singapore, Institute of Southeast Asian Studies, 1998.

Carney, Timothy, and Tan Lian Choo. *Whither Cambodia? Beyond the Election.* Singapore, Institute of Southeast Asian Studies, 1993.

Chandler, David P. *A History of Cambodia.* Boulder, CO, Westview Press, 2nd Edn, 1991.

The Tragedy of Cambodian History: Politics, War and Revolution since 1945. New Haven, CT, Yale University Press, 1992.

Brother Number One: A Political Biography of Pol Pot. Boulder, CO, Westview Press, 1992.

Voices from S-21: Terror and History in Pol Pot's Secret Prison. Berkeley, CA, University of California Press, 2000.

Chandler, David P., Kiernan, B., and Boua, C. *Pol Pot Plans the Future: Confidential Leadership Documents from Democratic Kampuchea.* New Haven, CT, Yale University Press, 1988.

Clymer, Kenton. *The United States and Cambodia, 1969–2000: A Troubled Relationship.* London, RoutledgeCurzon, 2004.

Corfield, Justin J. *The Royal Family of Cambodia.* Melbourne, Khmer Language and Culture Centre, 2nd revised Edn, 1993.

Khmers Stand Up! Clayton, Vic, Monash University Centre for Southeast Asian Studies, 1994.

Corfield, Justin J., and Summers, Laura. *Historical Dictionary of Cambodia*. Oxford, Rowman and Littlefield Publrs, 2003.

Deedrick, Tami. *Khmer Empires (Ancient Civilisations)*. London, Raintree Steck-Vaughn, 2001.

DePaul, Kim (Ed.). *Children of Cambodia's Killing Fields: Memoirs by Survivors*. New Haven, CT, Yale University Southeast Asia Studies, 1997.

Doyle, Michael W. *UN Peacekeeping in Cambodia: UNTAC's Civilian Mandate*. Boulder, CO, Lynne Rienner Publishers, 1995.

Ebihara, May M., Mortland, Carol, and Ledgerwood, Judy (Eds). *Cambodian Culture Since 1975*. London, Cornell, 1994.

Engelbert, Thomas, and Goscha, Christopher. *Falling Out of Touch: A Study on Vietnamese Policy Towards an Emerging Cambodian Communist Movement, 1930–1975*. Clayton, Vic, Monash University Asia Institute, 1995.

Etcheson, Craig. *The Rise and Demise of Democratic Kampuchea*. Boulder, CO, Westview Press, 1984.

Evans, Grant, and Rowley, Kelvin. *Red Brotherhood at War*. London, Verso, 1984, revised Edn 1990.

Fawthrop, Tom, and Jarvis, Helen. *Getting Away with Genocide: Cambodia's Long Struggle Against the Khmer Rouge*. London, Pluto Press, 2004.

Findlay, Trevor. *Cambodia: The Legacy and Lessons of UNTAC*. Oxford, Oxford University Press, 1995.

Gottesman, Evan R. *Cambodia After the Khmer Rouge: Inside the Politics of Nation Building*. New Haven, CT, Yale University Press, 2002.

Guy, John. *Sanctuary: The Temples of Angkor*. London, Phaidon Press, 2002.

Hamel, Bernard. *Sihanouk et le Drame Cambodgien*. Paris, L'Harmattan, 1993.

Heder, Steve, and Ledgerwood, Judy (Eds). *Propaganda, Politics and Violence in Cambodia*. London, M. E. Sharpe, 1996.

Higham, Charles. *The Civilisation of Angkor*. Weidenfeld, 2001.

Hughes, Caroline. *UNTAC in Cambodia: The Impact on Human Rights*. Singapore, Institute of Southeast Asian Studies, 1996.

The Political Economy of the Cambodian Transition. London, RoutledgeCurzon, 2002.

Human Rights Watch/Asia. *Cambodia at War*. London, Human Rights Watch, 1995.

Jackson, Karl (Ed.). *Cambodia, 1975–1978: Rendezvous with Death*. Princeton, NJ, Princeton University Press, 1990.

Jennar, Raoul M. (Ed.). *The Cambodian Constitutions (1953–1993)*. Bangkok, White Lotus Co, 1995.

Kamm, Henry. *Cambodia: Report from a Stricken Land*. London/New York, Arcade, 1998.

Kèn, Khun. *De la Dictature des Khmers Rouges à l'Occupation Vietnamienne: Cambodge, 1975–1979*. Paris, L'Harmattan, 1994.

Kiernan, Ben. *How Pol Pot Came to Power*. London, Verso/New Left Books, 1985.

(Ed.). *Genocide and Democracy in Cambodia*. New Haven, CT, Yale University Southeast Asia Studies, 1993.

The Pol Pot Regime: Race, Power and Genocide in Cambodia under the Khmer Rouge, 1975–1979. New Haven, CT, Yale University Press, 1996.

Kiljuen, Kimmo (Ed.). *Kampuchea—Decade of the Genocide*. London, Zed Press, 1984.

Lafreniere, Bree. *Music Through The Dark: A Tale of Survival in Cambodia*. Honolulu, HI, University of Hawaii Press, 2000.

Mabbett, I. W., and Chandler, D. *The Khmers*. Oxford, Blackwell, 1995.

Mannika, E. *Angkor Wat: Time, Space and Kingship*. Honolulu, HI, University of Hawaii Press, 1996.

Marston, John, and Guthrie, Elizabeth (Eds). *History, Buddhism and New Religious Movements in Cambodia*. Honolulu, HI, University of Hawaii Press, 2004.

Martin, Marie A. *Le Mal Cambodgien*. Paris, Hachette, 1989.

Cambodia: A Shattered Society. Berkeley, CA, University of California Press, 1994.

Mehta, Harish. *Warrior Prince: Norodom Ranariddh, Son of Sihanouk of Cambodia*. Singapore, Graham Brash, 2001.

Mehta, Harish, and Mehta, Julie. *Hun Sen—Strongman of Cambodia*. Singapore, Graham Brash, 2000.

Morris, Stephen J. *Why Vietnam Invaded Cambodia*. London, Cambridge University Press, 2001.

Népote, Jacques. *Parenté et organisation sociale dans le Cambodge moderne et contemporain*. Geneva, Ouzane, 1992.

Népote, Jacques, and Vienne, Marie-Sybille. *Cambodge, Laboratoire d'une Crise*. Paris, Centre des Hautes Etudes sur l'Afrique et l'Asie Modernes, 1993.

Ngor, Haing, and Warner, Roger. *Survival in the Killing Fields*. London, Constable and Robinson, 2003.

Norodom Sihanouk. *War and Hope: the Case for Cambodia*. London, Sidgwick and Jackson, 1980.

Norodom Sihanouk, and Krisher, Bernard. *Sihanouk Reminisces: World Leaders I Have Known*. Bangkok, Editions Duang Kamol, 1990.

Osborne, Milton E. *Sihanouk: Prince of Light, Prince of Darkness*. St Leonards, NSW, Allen and Unwin, 1994.

Before Kampuchea: Preludes to Tragedy. Bangkok, Orchid Press, 2003.

Ovesen, Jan, Trankell, Ing-Britt, and Öjendal, Joakim. *When Every Household is an Island: Social Organization and Power Structures in Rural Cambodia*. Uppsala, Uppsala Research Reports in Cultural Anthropology, No. 15, 1996.

Peou, Sorpong. *Conflict Neutralization in the Cambodia War: From Battlefield to Ballot-Box*. Kuala Lumpur, New York & Singapore, Oxford University Press, 1997.

Intervention and Change in Cambodia: Towards Democracy? Singapore, Institute of Southeast Asian Studies, 2000.

(Ed.). *Cambodia: Change and Continuity in Contemporary Politics*. Aldershot, Hampshire, Ashgate Publishing Ltd, 2001.

Picq, Laurence. *Beyond the Horizon: Five Years with the Khmer Rouge*. New York, St Martin's Press, 1989.

Pradham, P. C. *Foreign Policy of Kampuchea*. London, Sangam Books, 1988.

Pradhan, M., and Prescott, N. *A Poverty Profile of Cambodia*. World Bank Discussion Papers, No 373, Washington, DC, World Bank, 1997.

Pran, D. (Ed.). *Children of Cambodia's Killing Fields: Memoirs by Survivors*. New Haven, CT, Yale University Press, 1997.

Quigley, John, and Robinson, Kenneth. (Eds). *Documents from the Trial of Pol Pot and Ieng Sary*. Philadelphia, PA, University of Pennsylvania Press, 2000.

Roberts, David. *Political Transition in Cambodia 1991–1999: Power, Elitism and Democracy*. Richmond, Surrey, Curzon Press, 2001.

Ros, Chantrabot. *La République Khmère*. Paris, L'Harmattan, 1993.

Shawcross, William. *Sideshow: Kissinger, Nixon and the Destruction of Cambodia*. London, André Deutsch, 1979.

The Quality of Mercy: Cambodia, Holocaust and Modern Conscience. London, André Deutsch, 1984.

Cambodia's New Deal. Washington, Carnegie Endowment for International Peace, 1994.

Short, Philip. *Pol Pot: The History of a Nightmare*. London, John Murray, 2004.

Slocomb, Margaret. *People's Republic of Kampuchea, 1979–1989: The Revolution after Pol Pot*. Seattle, WA, University of Washington Press, 2004.

Sola, Richard. *Le Cambodge de Sihanouk: Espoir, Désillusions et Amertume, 1982–1993*. Paris, Sudestasie, 1994.

Stuart-Fox, Martin, and Ung, Bunhaeng. *The Murderous Revolution*. Bangkok, Tamarind Press, 1986.

Thion, Serge. *Watching Cambodia*. Bangkok, White Lotus Co, 1993.

Tully, John A. *France on the Mekong: A History of the Protectorate in Cambodia, 1863–1953*. Oxford, Rowman and Littlefield Publrs, 2003.

United Nations. *The United Nations and Cambodia, 1991–1995*. New York, United Nations, 1995.

Utting, Peter (Ed.). *Between Hope and Insecurity: The Social Consequences of the Cambodian Peace Process*. Geneva, UNRISD, 1994.

Vandy, Kaonn. *Cambodge: 1940–1991 ou la Politique sans les Cambodgiens*. Paris, L'Harmattan, 1993.

Vickery, Michael. *Cambodia: 1975–1982*. Sydney, Allen and Unwin, 1984.

Kampuchea: Politics, Economics and Society. Sydney, Allen and Unwin, 1987.

Zhou Mei. *Radio UNTAC of Cambodia: Winning Ears, Hearts and Minds*. Bangkok, White Lotus Co, 1994.

THE PEOPLE'S REPUBLIC OF CHINA

Physical and Social Geography

MICHAEL FREEBERNE

With additions by the editorial staff

The People's Republic of China covers an area of 9,572,900 sq km (almost 3.7m. sq miles) and extends about 4,000 km from north to south and 4,800 km from east to west. Owing to China's mountainous relief and the comparatively undeveloped state of transport, distance creates major economic and political problems. For example, not only is it costly and technically difficult to build a dense communications network, but also repeated attempts to move industry inland and away from the established centres in the east have been seriously hindered by such factors as the long haul for raw materials and markets. Similarly, the vastness of China has made it very hard to provide strong central government from Beijing.

China's land frontiers extend for a total of 20,000 km, and have been the source of some tension. China shares frontiers with the Democratic People's Republic of Korea (North Korea), Mongolia, Russia, Kazakhstan, Kyrgyzstan, Tajikistan, Afghanistan, Pakistan, India, Nepal, Bhutan, Myanmar (formerly Burma), Laos and Viet Nam. The dispute over the boundary between China and India resulted in the border war of 1962. Negotiations regarding the settlement of the dispute by peaceful means continue. The two sections of the Sino-Russian border (in the north-east and the north-west of China) total more than 4,300 km. Various boundary incidents occurred after 1960, but subsequent negotiations resulted in the demarcation of large parts of both sections in the mid-1990s. The delimitation of the eastern section (some 4,200 km) was finally concluded in November 1997. Discussions concerning the shorter western section and the status of two islands continue. China's eastern seaboard is 14,000 km in length. Its territorial waters are dotted with some 5,000 islands, ranging from provincial-sized Hainan down to minute atolls, which include the strategically significant but disputed Xisha (Paracel) and Nansha (Spratly) Islands. Rich in fish (and also petroleum reserves), these waters make a significant contribution to the output of marine and fresh water aquatic products. China lacks an important seafaring tradition, however, partly because the relatively smooth coastline is largely without good natural harbours.

Administratively, the People's Republic of China is divided into 22 provinces, five autonomous regions, and four municipalities, all of which are directly under the central Government. There are more than 2,000 counties, which until the early 1980s were subdivided into more than 50,000 people's communes. As the communes underwent striking changes after their introduction in 1958, much of the effective economic and political organization in China was at production brigade and production team level, which frequently coincided with the natural village. The communes have been superseded by the household contract responsibility system, commonly centred upon the family unit. Other organizational structures, such as macroeconomic and military regions, may embrace various provinces, whilst, as part of Deng Xiaoping's economic reforms and the 'open door' policy, several Special Economic Zones have been established (including Shenzhen, Zhuhai, Shantou, Xiamen and Hainan). In the urban areas, tiny neighbourhood street committees keep a watchful eye on day-to-day activities.

PHYSICAL FEATURES

Physical size alone cannot automatically raise China to the rank of a first-class world power. The West regarded China as a land of fabulous wealth at the height of the Qing empire, but in fact the geographical environment presents considerable obstacles to modernization. According to official figures, in the late 1990s more than 13% of China's surface was cultivated, less than 17% was forest and almost 42% was grassland. Less than 33% of the area was classified as usable. In practice, China must feed more than 20% of the world's population on just 7% of its arable land and with only 8% of its fresh water, and each year farming land is lost while the population continues to expand. Between 1990 and 1994 a total of 1.5m. ha, or 1% of the area sown to grain crops, was lost, not only through increasing urbanization and industrial development, but also through drought and flooding. In May 1997 the Director of the State Land Administration announced the enforcement of stricter curbs on the use of land for non-agricultural development. In 2000 it was calculated that 2.62m. ha of China's territory (or 27% of the total land area) was affected by desertification, with this area expanding by 2,460 sq km annually. The total area affected by soil erosion, meanwhile, was estimated at 3.7m. sq km in 1997. The felling of trees in the remote north-western region of Xinjiang, China's driest area, was banned in mid-1997, in an attempt to halt the process of desertification.

Relief, configuration and climate are critical in suggesting possible settlement areas and zones suitable for economic development. For the most part high in the west and relatively low in the east, China has been compared to a three-section staircase. The Qinghai-Tibet (Xizang) plateau, at over 4,000 m, is the highest flight; next is an arc of plateaux and basins lying at 1,000 m–2,000 m, extending eastwards from the Tarim Basin, across Nei Monggol and the loess lands, then turning south to include the immensely fertile Sichuan Basin and the Yunnan-Guizhou plateau; much of the land that constitutes the lowest flight lies below 500 m and covers the most densely-settled areas, such as the middle and lower Changjiang (Yangtze) basin, the North China plain and the north-eastern plain. About 33% of China's total area comprises mountains; 26% is plateau land; 10% is hill country; 19% is occupied by basins; but only 12% of the surface is composed of plains.

Watering these plains are rivers which in some years bring rich harvests, while in other years they may cause flooding, or dry up altogether with resulting drought famines, which were frequent before 1949. Indeed, China has been characterized as a land that suffers from having either too much water or not enough water, both in terms of regional and seasonal distribution. In the north, the Huanghe (Yellow River) is 5,464 km in length and has a drainage basin of 752,443 sq km. In central China, the Changjiang is 6,300 km long with an annual flow of 9,513,000m. cu m, and a massive drainage basin of 1.8m. sq km, covering one-fifth of the country. The Zhujiang (2,214 km) flows through southern China. Water conservation is actively pursued, and flood control, irrigation, navigation and power generation are all emphasized in numerous multi-purpose projects.

China experienced exceptionally severe flooding of the Changjiang in mid-1998: more than 240m. people were affected, some 14m. people were forced to leave their homes, more than 3,000 people lost their lives and damage costing in excess of US $20,000m. was caused by the worst floods since 1954 (when 30,000 were killed). The authorities conceded that soil erosion, resulting from extensive deforestation, was largely responsible, and acknowledged the need for an accelerated tree-planting programme. The floods of mid-1999 led to over 1,000 deaths and the evacuation of 5.5m. citizens from their homes.

Hydroelectric power provided 7.8% of total energy output in 1999. In 1994 work commenced on the Three Gorges project in Hubei Province. To comprise a 1,983-m dam across the Changjiang and scheduled for completion in 2009, the plant will have an annual generating capacity of 84,700m. kWh of electricity, but involves the submerging of 17,000 ha of farmland and the displacement of more than 1m. people (see Environmental

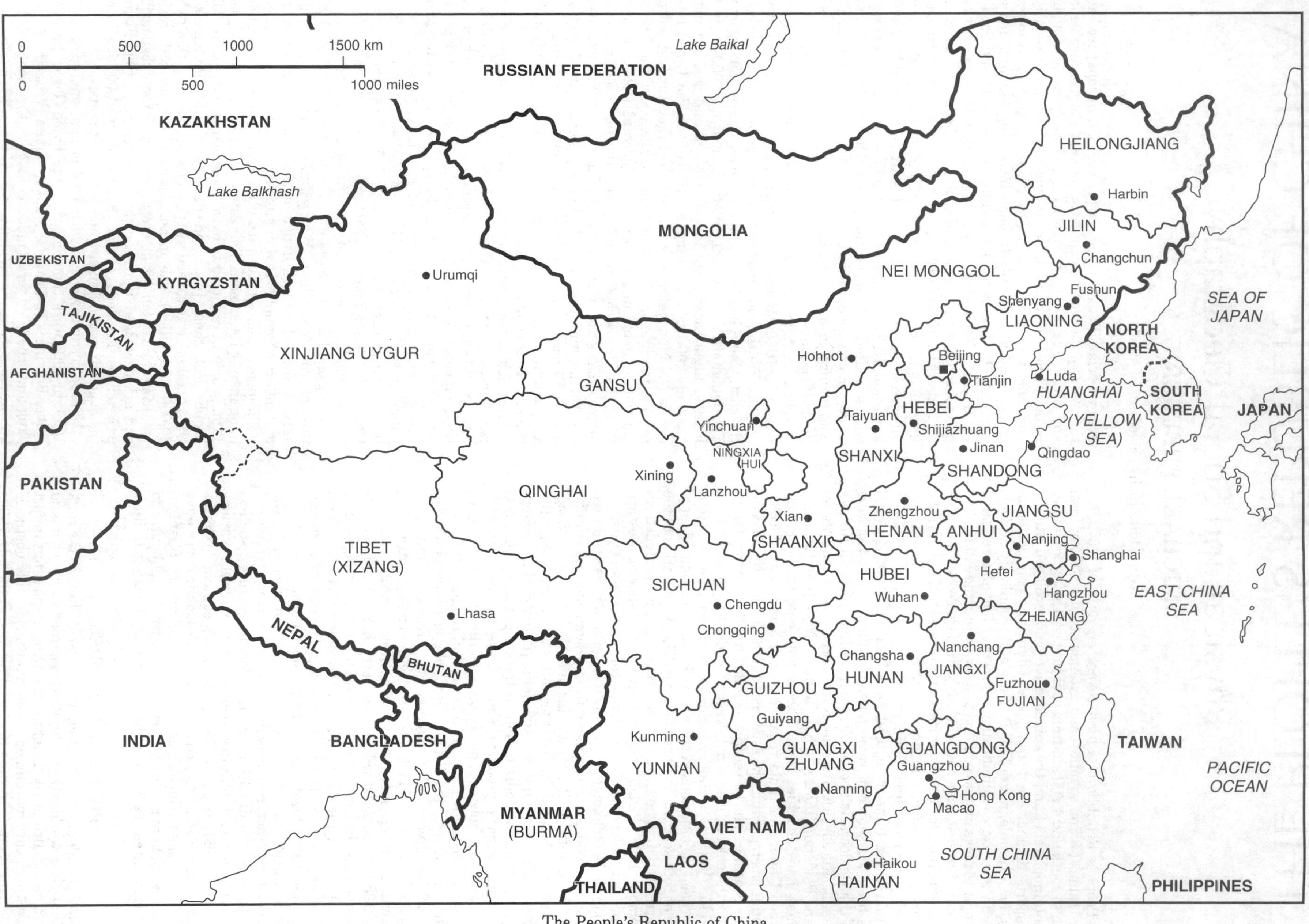

The People's Republic of China

Issues of Asia). The project will incorporate a massive water-conservancy programme. In November 1997 the damming of the Changjiang, and the diversion of the river into a man-made channel, marked the completion of the first phase of the project. The Xiaolangdi dam, part of a major water-conservancy project on the Huanghe, was completed in October 1997; power generation commenced in 2000, with annual capacity reaching 5,100m. kWh.

Increasingly, many Chinese cities, including Beijing and Tianjin, are suffering from acute water shortages. In the 1990s scientists began to warn of a serious shortage of drinking water by the early 21st century. Furthermore, the rapid degradation of the quality of water supplies through industrial pollution has become a cause of grave concern, leading the Government to implement stricter regulations. In mid-2001 the most severe drought for more than a decade was reported to have left millions of Chinese citizens without adequate drinking water, the most seriously affected areas being Shandong and Liaoning Provinces. By 2004 it was being reported that 400 of China's 670 largest cities were suffering serious water deficits. In March 2001, meanwhile, the Government announced details of a major project to transport water, via three channels, from the south of the country to the drier northern regions. Construction began in late 2002.

CLIMATE

Climatically, China is dominated by a monsoonal regime. Cold air masses build up over the Asian land mass in winter, and the prevailing winds are offshore and dry. In summer there is a reversal of this pattern, and the rainy season is concentrated in the summer months over the most densely-settled parts of the country in the east and the south. Running from south to north there are six broad temperature zones: tropical and sub-tropical, warm-temperate and temperate, cold-temperate and the Qinghai-Tibet plateau area, which has its own characteristic regime. January is generally the coldest month and July the hottest. There is a great range in winter temperatures—as much as 15°C between the average for Guangzhou in the south and Harbin in the north. South of the Nanling mountains January temperatures average around 8°C, but they drop to between −8°C and −15°C over much of the north-east, Nei Monggol and the north-west. In summer the temperature difference between Guangzhou and Harbin narrows to 12°C and summer temperatures over much of the country average above 20°C.

The summer monsoon brings abundant rain to coastal China, especially in the south and east, but amounts decrease drastically to the north and west. A humid zone covers much of south-eastern China and the average annual rainfall is above 750 mm. In the semi-humid zone, extending across the north-east, the North China plain and the south-eastern region of the Qinghai-Tibet plateau, the average falls to less than 500 mm. The remainder of the Qinghai-Tibet, the loess, and the Nei Monggol plateaux receive only about 300 mm, while western Nei Monggol and Xinjiang, where there are extensive deserts, get less than 250 mm.

About 80% of the precipitation falls between May and October, with July and August the wettest months. Not infrequently the rain turns the rivers into raging torrents and disastrous floods occur; alternatively, not enough rain falls. To flood and drought can be added other calamities: typhoons, earthquakes, frosts, hailstorms, plant and animal pests, and diseases. The Chinese attributed the grave economic difficulties of the early 1960s, when millions died from starvation, to three factors: the withdrawal of Soviet aid, policy mistakes and natural calamities.

VEGETATION AND NATURAL RESOURCES

Over hundreds of years a great deal of China's natural vegetation has been stripped. The basic contrast is between the forests and woodlands of the eastern half of the country and the grassland-desert complex of the western half. Tree types vary from the tropical rain forests in the south, through evergreen broadleaved forests, mixed mesophytic forests, temperate deciduous broadleaved forests, and mixed northern hardwood and boreal coniferous forests in the north. The eastern Mongolian plateau, the Xiao Hinggan Ling and Da Hinggan Ling (the Lesser and Greater Khingan mountains), and the Changbaishan massif contain 60% of China's forest reserves. Other natural forests are located in Yunnan, Jiangxi, Fujian, Guizhou, Sichuan, Hainan and in the Qinling mountains and along the eastern edge of the Qinghai-Tibet plateau. Most of China's forests are largely inaccessible, however, and there is a serious shortage of workable timber. In 1987 there was a major ecological disaster when a forest fire laid waste vast stretches near to the north-eastern border with the USSR.

Owing to the widespread destruction of natural vegetation, soil erosion is a major problem. Sheet and gully erosion are common; water and wind erosion do great damage in the north, while water erosion is the chief enemy in the south; also, farming malpractices, such as deep ploughing, have aggravated the situation. It has been estimated that about 40% of the total cultivated area comprises 'poor' soils: red loams, saline-alkaline soil and some of the rice paddy soils.

China's total petroleum reserves are estimated at around 140,000m. metric tons and natural gas reserves at more than 33,000,000m. cu m. The Daqing oilfield, China's largest, has provided 47% of China's petroleum output since 1960. The largest reserves are believed to lie in the Tarim Basin, in Xinjiang. China is extremely rich in coal and iron ore. There are abundant reserves of manganese, tungsten and molybdenum, but China is relatively poor in copper, lead and zinc, and nickel supplies are meagre. China has rich resources of salt, moderate reserves of sulphur, while phosphates require development; supplies of gold, tin, fluorite-graphite, magnesite, talc, asbestos and barytes are also comparatively good.

POPULATION

China's fifth national census of 1 November 2000 revealed that the population had grown to a total 1,242,612,226, compared with 1,130,510,638 at the time of the fourth census in 1990, 1,008,180,738 at the third census in 1982 and with 582,603,417 at the first census in 1953. These figures, which represent more than one-fifth of the world's population, are formidable in view of both the pressures that population growth has exerted historically and the contemporary problems in the physical environment already outlined. There is, for instance, a striking imbalance in the distribution of population, which is heavily concentrated in the plain and riverine lands of the south-eastern half of the country, while most of the north-western half is, by comparison, thinly populated. This results in very high densities of population in the richest areas for settlement, such as the Changjiang Delta or the Red Basin of Sichuan. Indeed, 90% of the population inhabit little more than 15% of the country's surface area.

Some 91.53% of the population are Han Chinese. The remaining 8.47% belong to one of the national minority groups. Altogether there are over 106m. non-Chinese living within China, chiefly in the peripheral areas beyond the Great Wall, in the north, the north-west and the south-west. There are 55 different minorities scattered throughout 60% of the country. According to the 2000 census, 18 minorities numbered more than 1m. each (see Statistical Survey). Between 1982 and 1990, whilst the Han Chinese population increased by 102m., or 10.80% (1.29% annually), the national minorities grew by 24m., or 35.52% (3.87% per year). However, between 1990 and 2000 the Han Chinese population increased by 116.92m., or 11.22% (1.07% annually), while the national minorities grew by 15.23m., or 16.7% (1.56% annually), the latter thus recording a faster rate of growth than the Han, but significantly slower than in the previous decade.

Although so-called autonomous regions (and also districts and counties) have been established, the larger minority groups have presented the central Government with serious administrative difficulties. Racial, religious and linguistic problems, as in Muslim Xinjiang and Buddhist Tibet, have resulted in several anti-Chinese uprisings since 1949; these have been forcibly suppressed.

Linguistic differences among the seven main Chinese dialects, as well as between Chinese and minority languages, have proved an intractable issue, despite the adoption of Mandarin (Putonghua in Chinese) as the national language, despite attempts at the simplification of the written language by reducing the number of strokes in individual characters and by romanization, and despite campaigns to increase literacy.

In 2000 455.94m. people in China lived in cities or towns, but this is still predominantly a rural country, with almost 64% of the population living in the countryside (compared with almost 88% in 1952). The inequalities in living standards between urban and rural areas present the Chinese with some of their most urgent ideological and practical problems.

In 2002 China recorded a birth rate of 12.86 per thousand, a death rate of 6.41 per thousand and a natural growth rate of 6.45 per thousand. Family-planning programmes from the mid-1950s failed to make any pronounced inroads in the increase in Chinese numbers, and in 1979 the Government issued further directives favouring couples with only one child and penalizing those who practise 'anarchism in parenthood'. In 1988 the Government conceded that it was unlikely to achieve its declared goal of limiting China's population to 1,200m. by the year 2000. In 1998 the State Family Planning Commission announced the family-planning policy for the end of the 20th century and the first half of the 21st century. It aimed to limit the population to 1,300m. by the year 2000, and to keep the population below 1,400m. in 2010. It estimated that the population would reach a peak of 1,600m. in the mid-21st century, before declining gradually. Peasants are now allowed to have two children, and even larger families are more and more common. Internal migration offers no solution to the population problem.

Despite improved grain harvests in the 1970s and the 1980s, with a harvest of 462.5m. metric tons being recorded in 2000, and because the agricultural sector has to provide not only food for a large population but also investment for industrial growth, population pressure must remain central to all domestic and external issues within the foreseeable future.

History up to 1966

C. P. FITZGERALD

PREHISTORY AND CLASSICAL PERIOD

The earliest Chinese written records, recovered by archaeological excavation at the site of what was the capital of the Shang kingdom in Henan Province, date from approximately 1500 BC. Legendary history, for which there is as yet no archaeological evidence, records a previous kingdom, Xia, and a golden age of the rule of Sages, for at least 1,000 years earlier. The agreement between the king list of Shang, as found on the oracle bones discovered at Anyang, and the list preserved in Chinese official history, which are wholly independent of each other, shows that the official history must be treated with some respect. Shang culture included the making of bronze vessels of great beauty, some of which are briefly inscribed. The succeeding period, the Zhou dynasty, from 1100 to 221 BC, continued the Shang culture, but a more elaborate literature appeared, and in the second part of the Zhou rule, from about 800 BC onwards, the feudal system instituted at the foundation broke down. China became a land of contending kingdoms. At the same time there arose the various schools of philosophy (Confucian, Daoist, Moist and Legalist), whose contention matched the military conflict between the kingdoms.

UNIFICATION OF CHINA

From about the date of the death of Confucius (479 BC) until 221 BC China was constantly subjected to the wars of the contending kingdoms, of which the western state of Qin and the southern state of Chu, in the Yangtze (Changjiang) valley, were the chief protagonists. The history of Chu provides all that is known of south China in the earlier period. The struggle was won by Qin in 221 BC. The ruler of that kingdom then assumed the new title of *Huang Di*, translated as 'emperor', and imposed on his new dominions the harsh Legalist code of laws and administration, which had been in force in his country. Qin law despised art and literature, glorifying war and promoting agriculture as the foundation of military power. To suppress the opposition of the literate class in the new empire, the emperor ordered the burning of all books on history and philosophy not included in his own library. Although much was hidden and preserved, this policy did great damage to the recorded literature of China. A few years after his death, a general revolt overthrew the Qin and they were replaced by the Han dynasty, which ruled from 206 BC to AD 221.

THE FIRST EMPIRE AND PERIOD OF PARTITION

The Han dynasty consolidated the new empire, but ruled with moderation. A civil service, filled with educated men recommended by patrons, replaced the feudal system. Free tenure of land created both a landlord and a tenant system, which endured until recent times. Confucianism became the established orthodoxy. The empire was expanded to include the Guangzhou (Canton) region, and later in central Asia as far as the Caspian Sea. Contact with the Roman empire, although slight, is recorded. In art the Han developed mural painting and bas-reliefs, and in literature history was highly esteemed and developed in a systematic and accurate form. Paper and ink replaced bamboo strips and the stylus for writing. After 400 years the Han empire fell in confusion. Contending states were briefly suppressed by a reunion under the Jin dynasty, which itself soon lost North China to the invading Tartar tribes. China was divided from AD 316 to 589. Tartar rulers held the north, and Chinese dynasties the south (Yangtze valley). Both in north and south the dynasties were brief, and internal conflict frequent. Yet the period, although an 'Age of Confusion', was not a 'Dark Age'. Literature flourished in both north and south. Buddhism was introduced and spread widely. The majority of the population being Chinese, the Tartar invaders were soon absorbed, both ethnically and culturally.

THE SECOND EMPIRE, TANG AND SONG DYNASTIES

The Tang dynasty, founded in AD 618, reunited the empire on a lasting basis. The aristocratic military class gave way to a bureaucracy recruited by public examination open to all literates. The administration of government was developed to a degree unknown elsewhere for several centuries to come. Art and literature, especially poetry, flourished. The population recorded by an accurate census in AD 754 was 52,880,488. Archaeological discovery of a census return shows that this figure included women and children. The Song dynasty, which succeeded the Tang after a brief interval, continued the civil service system which now controlled the government. The Song were unassertive, and failed to recover north-eastern territory lost to nomad invasion at the fall of Tang. In AD 1127 the Song lost North China to an invasion of the Jin Tartars. A century later the Song were conquered, after a long war, by the Mongols. The new Confucian philosophy was the main development in literature; in art Song painting is still the most esteemed. Technical developments include printing, porcelain, the maritime compass, gunpowder and primitive cannon, advances in silk spinning and the development of maritime trade with South and West Asia.

THE LATE EMPIRE AND SUBSEQUENT DYNASTIES

The Mongol conquest of China (1280) was most destructive, especially in the northern half of the country. Huge depopulation occurred, and great areas of fertile land became wilderness. Song culture, though damaged, survived. Mongol rule was oppressive and largely exercised through foreign officials from

West Asia and even Europe (Marco Polo). It was also brief. After barely a century the Chinese revolt ended in the foundation of the Ming dynasty and the expulsion of the Mongols (1368). The development of drama, written by unemployed Chinese scholars, is the significant cultural development of the Mongol period.

The Ming dynasty, however, not only restored Chinese rule, but expanded the limits of the empire. South Manzhou (Manchuria) was settled and incorporated, as was Yunnan, at the opposite extremity of the empire. The Ming aimed to restore the style of Tang and Song government, but their rule was much more autocratic. In early Ming (1405–33) expeditions were sent by sea to South-East Asia, the Indian coast, Persian Gulf, Red Sea and East Africa, but this naval activity was abandoned only a few years before the Portuguese first appeared in Far Eastern waters. In late Ming contact with Europeans increased, and the first Roman Catholic missionaries reached China.

From the middle of the 15th century China was threatened, to an increasing extent, by the growth of a new power in what is called Manzhou (Manchuria), or the Three Eastern Provinces. The Manzu tribes, kindred of the Jin Tartars who had ruled northern China in the late Song period, were at first tributary to the Ming. From China, through this contact, they acquired a knowledge of governing techniques, literacy and organization. Late in the 16th century they coalesced into a new kingdom, which threw off allegiance to the Ming, and before long began to encroach on the Ming territory of South Manzhou, or modern Liaoning Province. By the middle of the 17th century they had seized this region and were raiding the Great Wall frontier of China proper. The Ming fell in 1644 to an internal rebellion, which gave the Manzhous (Manchus) the opportunity to enter China and, after nearly 40 years, control the whole empire.

The Qing (Manzhou) dynasty ruled until 1912. The first 150 years, under the three very competent emperors Kang Xi, Yong Zheng and Qian Long, were prosperous and peaceful. At the end of the 18th century the dynasty began to decline. The growth of trade with Europe was at first very profitable to China, but the Manzhous distrusted the foreign traders and imposed restrictions on their activities. The discovery of opium smuggling by British traders brought about the Opium War of 1842; China, unable to match the strength of the British fleet, was defeated. The Treaty of Nanjing, the first of the 'Unequal Treaties' as they came to be called, which followed this defeat, established the system of Treaty Ports, concession areas and the right of extraterritorial jurisdiction. Shanghai was claimed as a British Treaty Port, and by the end of the century had become an International Settlement, guarded by its own multinational troops.

The unsuccessful war, and the internal rebellions that swept the country in the 1850s, weakened the authority of a dynasty always considered alien in the south. Moreover, in the 1870s China began to suffer the encroachment of the European powers. Russia took advantage of the rebellions in China to obtain territory in the north, while in the 1880s France seized Indo-China and forced Beijing to renounce its suzerainty. The United Kingdom and France together waged war with China in 1858 and actually occupied Beijing, exacting a further 'Unequal Treaty'. Towards the end of the century Japan became involved, and in the war of 1894–95 drove the Chinese out of Korea. At home the young intellectuals, inspired by Western education and thinking, adopted revolutionary ideas under the leadership of Dr Sun Yat-sen. In 1911, three years after the death of the empress dowager Ci Xi (who had ruled from 1862), a revolt of the army at Wuhan led to the fall of the dynasty and the abolition of the monarchy.

REPUBLICAN CHINA

Sun Yat-sen was a native of Guangzhou, who had been educated from childhood in Hawaii and then taken a degree at the medical school of Hong Kong. His formation was thus largely foreign and Western. Finding that radical reform was unacceptable to the official world of China, he turned revolutionary and republican, and for more than 10 years maintained an unceasing effort to stir up rebellion in China. He was for long unsuccessful; but his influence grew steadily among the young Chinese studying abroad, particularly in Japan, where the majority of them went. He built up a nationalist party and a secret organization, obtained funds from the overseas Chinese of South-East Asia, and finally his followers were able to infiltrate the army—the new model army whose officers had also studied abroad.

Thus, when the revolution broke out in 1911, it was from the first dominated by the army, a servitude from which it was not to escape for many years. The court had lost further prestige in 1900 by supporting the peasant anti-foreign movement known as the Boxer Rebellion, which for a time threatened to massacre the diplomatic corps in Beijing; the Rebellion was finally crushed by an international expedition, which took Beijing and drove the court to retreat to the west of China. The southern provinces under their great viceroys refused to follow court policy over the Boxers, and virtually concluded a separate peace with the foreign powers. This was a sign of coming disruption. After signing a further humiliating peace, the court returned to Beijing, and in its last years attempted to put through reforms which might have saved it 50 years earlier. It was too late. When Ci Xi died in 1908, there was no competent successor to continue the regency in the name of the next infant emperor, Xuan Tong or Pu Yi. Within three years the revolution had broken out and the dynasty was doomed.

In its last extremity, the imperial regime appealed to the former commander-in-chief of the imperial army, Yuan Shikai, who was out of favour with the new regent, to save it. The northern troops would obey only their old commander; the southern army had transferred its allegiance to the revolution. Yuan took command, but he did not intend to save the dynasty; he hoped to establish his own. First he showed by a brief campaign that he was a serious contender, then began to negotiate with the republicans. A plan was soon arranged. Yuan would bring about the peaceful abdication of the dynasty, which would, in return, be granted very favourable terms, and the republic would elect Yuan to be president. When the first parliament was convened (under conditions of flagrant corruption), Yuan had some of the more able members assassinated, and soon, having obtained a loan from the foreign powers without the assent of parliament, dissolved that body and ruled by decree. Futile and ineffective resistance in the south was speedily crushed. In 1914 Yuan moved to obtain support for a new dynasty with himself as emperor.

The outbreak of the First World War was a factor that worked against this programme. It divided the foreign powers, and left Japan a comparatively free hand in Asia. Japan bribed and armed Yuan's secret opponents, his own generals, who resented his pretensions to the throne. On 25 December 1915 a revolt broke out, and within a few months it was evident that the generals had turned against him, and the projected monarchy was impossible. He renounced his plans, tried to cling to the presidency, but died in June 1916. His death was soon followed by the contests among his former generals who controlled the provinces. The 'war-lord era' from 1917 to 1927 was marked by a series of short civil wars fought entirely between rival militarists to gain control of revenues, and above all of the impotent Government in Beijing, which could dispense the custom revenue collected under foreign supervision to service the external loans, but which still left a valuable revenue for whichever general could dominate Beijing. Within the country there was an increasing breakdown of law and order, banditry, and rural distress.

Nationalism and Communism

In May 1919 the students of Beijing had rioted against the Government's acceptance of the secret arrangement whereby Japan was to acquire the former German-leased port of Qingdao in Shandong. It was generally known that the corrupt politicians and their militarist master had received large sums from Japan for this virtually treasonable decision. The 'May Fourth Movement', as it became known, spread widely; it was the first sign of a new phase of the revolution, a revolt against Western dictation of China's affairs and fate, the first overt reaction of the generation who had grown up since the fall of the empire. Today the Communist Government commemorates it as the opening of a new era.

In May 1925 another violent outbreak followed upon the shooting by International Settlement police of student demonstrators in Shanghai. This time the wave of anger, directed against the United Kingdom and Japan, was nation-wide. There was a total boycott of British and Japanese trade and enterprise. Hong Kong's labour was withdrawn and its life all but paralysed. Further riots and shootings occurred in Guangzhou, and missionaries were compelled to leave the interior of China. Boycott pickets were established in the Treaty Ports and became an extra-legal militia.

Dr Sun Yat-sen, having failed to obtain any help from the Western powers to reinstate his Government—which he and his followers regarded as the only legal one—had turned to the USSR, which gave him the necessary support in arms, advisers and possibly finance. He regained control of Guangzhou in 1923 and swiftly set about the organization of an efficient government and a new model army. In 1921 the Chinese Communist Party (CCP) had been formally established at a meeting attended by 11 members, one of whom was Mao Zedong. At almost the same time a CCP had been formed in France by students living in Paris. One of its founders was Zhou Enlai. The two parties in China, the CCP still very small, and the Nationalists (Kuomintang—KMT) already gaining wide support, co-operated on the basis that members of the CCP might join the KMT as individuals, but there was no affiliation of the two parties. Aided by the repercussions of May 1925, revolutionary agitation increased rapidly.

After Dr Sun's death in 1925, all hope of peaceful reunion ended, and the KMT Government in Guangzhou prepared for war, which was launched in 1926 against the southern war-lords. Success was rapid, and by early 1927 the whole of the middle Changjiang region had fallen to the KMT, whose armies, commanded by Jiang Jieshi (Chiang Kai-shek), were approaching Shanghai. Alarmed, the Treaty Powers sent troops to defend the International Settlement. The Shanghai workers and boycott pickets, organized by the Communists, rose and seized the Chinese-governed part of Shanghai, expelling the war-lord army. When Jiang's forces arrived, they found Chinese Shanghai already in the hands of the revolutionaries, and facing the acute danger of war with the foreign powers. Jiang had close connections with Chinese big business and finance in Shanghai. These people, good Nationalists, and no friends of the plundering war-lords, were equally very frightened of social revolution and the Communist-controlled workers. Jiang, knowing he had their support, carried out a sudden coup and massacre of the Communists (from which Zhou Enlai narrowly escaped) and broke with the CCP. For several months the situation was confused. Jiang formed a right-wing KMT Government at Nanjing; the former Guangzhou Government was now established at Wuhan, further up the Changjiang, and did not at first break with the CCP. In much of South China, particularly Hunan Province, social revolution, inspired by rural agitators led by Mao Zedong, was sweeping the country.

Before long the two Nationalist Governments coalesced at Nanjing, and Jiang could turn his attention to combating the Communists. From 1929 to 1935 Jiang launched successive extermination campaigns against the Communists, who had now, under the leadership of Mao Zedong and Zhu De, established a Soviet area in the hill country on the Hunan-Jiangxi border. Jiang's campaigns failed until he devised, on the advice of his German staff officers, the plan of blockading the Jiangxi Soviet and thus forcing the Communists to break out or be starved into submission.

In 1935 the Communists set out on what came to be known as the 'Long March', with about 100,000 men and many of their dependants. A year later they reached Yanan, in north Shaanxi, after marching and counter-marching for more than 9,500 km, with 30,000 fighting men. However, they had not been defeated and, during that epic march, Mao Zedong had emerged as the unquestioned leader of the CCP, a position he retained until his death in 1976. The CCP, also, was fully emancipated from long-distance control by Moscow, which had proved uniformly disastrous for several years. Yanan, in the far north-west, was difficult to attack, almost impossible to blockade, and close to the areas soon to be threatened by the impending Japanese invasion.

The Japanese Invasion

In China it was widely recognized that the Japanese had embarked upon an all-out effort to conquer the country. Japan feared that, if it waited, China would grow strong, and it also feared the rise of Communist influence. However, the Nanjing Government was still determined to destroy the CCP before resisting the Japanese. It was not until December 1936, when Jiang's own army, facing the Communists at Xian in Shaanxi, mutinied and held him prisoner until he agreed to cease the civil war, that he was forced to concede to the slogans 'Chinese do not fight Chinese' and 'unite to resist Japanese aggression'. The Japanese did not wait: in July 1937 they struck near Beijing, and the fighting soon escalated into a large-scale, but still undeclared, war.

In the early stages KMT resistance, as at Shanghai and the battle of Taierzhuang in Shandong, had been, at times, effective. The weight of Japanese armament, however, was far superior and they had almost unchallenged air power and complete control of the sea. The KMT forces were driven back from the coast to the mountainous interior of western China, losing nearly two-thirds of the provinces. The difficulties of forcing the Changjiang gorges halted the Japanese at that point, and the added challenge of holding vast conquered territories prevented any further advance. In those conquered territories, particularly North China, the CCP was organizing the guerrilla resistance, which was soon to shake Japanese authority.

The hope of a Chinese military collapse had faded; by early 1942 Japan was involved in the Second World War in the Pacific, and here, too, early victories were proving inconclusive and presaging Japan's eventual defeat. The CCP steadily expanded its guerrilla war until large areas were liberated, in which the CCP set up its own administration, gaining essential experience in social, economic and political reform, including the major problems of land reform. Japanese retaliation was brutal and ruthless, forcing the Chinese peasantry to rely on guerrilla groups for their protection. It roused the national consciousness of an indifferent apolitical peasantry, and was the main factor in building the power of the CCP to national level.

COMMUNIST CHINA

The war was ended neither by the still-passive resistance of the KMT in western China, nor by the activity of the guerrillas, but by the Japanese surrender in August 1945. The termination of the Japanese occupation left China deeply divided. The KMT took over from the Japanese in the southern and eastern provinces. The CCP controlled the rural north, and cut the communications when the KMT flew in men to take over the Japanese-held cities. Civil war loomed close. The USA sent Gen. George C. Marshall to mediate, and to build, if possible, a coalition government. He failed; neither side trusted the other, and the demands made by the KMT would have been a death warrant for the CCP. Early in 1946 the foreshadowed civil war began, but was neither as long nor as destructive as most Chinese had feared it would be. It soon became evident that the CCP was going to win. Its troops fought well under firm discipline; the KMT forces had no will to war, and plundered wherever they went. Gross inflation was ruining the economy and alienating those to whom the KMT looked for support; corruption was rife in the KMT Government and army, and business was almost paralysed. There was nothing that the KMT could offer to enlist the support of any social class, not even the capitalists of Shanghai.

Therefore, despite massive US arms supplies, full control of the air, and vastly superior numbers, the KMT armies were wholly routed in less than three years. Vast numbers surrendered; relatively few were killed in battle. By the end of 1948 the Communists already held all northern China, including the north-eastern provinces that formed Manzhou; they were on the banks of the Changjiang opposite Nanjing. The KMT was no longer united. A large group which favoured peace and negotiation compelled Jiang to renounce his presidency, but was unable to shake his covert control over many units of the army. The Nanjing Government tried to secure peace, and nearly did so, but this effort was sabotaged by the agents of Jiang at the last moment, and the acting President, Li Zongren (Li Tsung-jen), was forced into exile. The war resumed, the Communists crossed the Changjiang, took Nanjing, then Shanghai, and swept on into the south and west. By the middle of 1949, when the People's Republic of China was proclaimed on 1 October in Beijing, the Communists were the masters of China, and Jiang and his remnant forces retreated to Taiwan, where they and their successors have since remained (see the chapter on Taiwan for subsequent history of the island and reunification initiatives).

Yet the failure to end the war by negotiation did China, and the Communists, one serious piece of harm. It destroyed the continuity of the legitimate internationally recognized Government. If the Nanjing regime had made peace—any sort of peace—it would have remained the legal Government, even if it were now run by the CCP. By failing to win this diplomatic victory, the Communists found their new regime subject to recognition, or non-recognition, at the will of foreign states, and their claim to China's seat at the UN disputed by the KMT protégés of the USA.

This situation continued to be one of the main causes of friction between the People's Republic and the Western powers, who, in their attitudes to the new China, were also deeply divided. To many of the Western, in particular the European, countries the fate of China was settled; the KMT on Taiwan was no longer significant. To the USA, on the other hand, its regime was the 'real China' and the CCP considered to be Soviet 'puppets'. Thus, the CCP regime started out with the open ill will of the USA, the doubtful and wary acceptance of the United Kingdom and other smaller Western powers, and the half-hearted and cautious approval of the USSR. Only two years earlier Stalin had assured the USA that he recognized only Jiang Jieshi as the legitimate ruler of China.

The leading figure in China's political affairs was now Mao Zedong, who was Chairman of the CCP from 1935 until his death in 1976. Chairman Mao, as he was known, also became Head of State in October 1949 but he relinquished this post in December 1958. His successor was Liu Shaoqi, First Vice-Chairman of the CCP, who was elected Head of State in April 1959. The first Premier (head of government) was Zhou Enlai, who held this office from October 1949 until his death in 1976. Zhou was also Minister of Foreign Affairs from 1949 to 1958.

Economic and Social Reform

The early policy of the new regime in Beijing was necessarily one of national renewal. The economy was at a standstill, communications almost wholly interrupted, inflation rampant, public utilities run down by years of neglect. Even foreign trade was deflected into the supply of quick-selling consumer goods, largely useless to the economy, while valuable exports could not be moved and necessary imports could not be paid for. Nevertheless, the new regime, headed by men who had had no experience of urban life for more than 25 years, tackled these tasks with skill and expedition. Within weeks, the railways were running, and supplying coal to Shanghai in place of the normal seaborne supplies which were under a KMT naval blockade. Inflation was brought under steady control and ended, with a new currency, in the following year. Thereafter, the Chinese currency, subjected to violent fluctuations for longer than living memory, remained stable. Foreign trade began to revive, cautiously, being limited to imports of essential goods, and to exports that would earn foreign exchange. The restoration of the cities, some of which were still in partial ruins from wartime bombing, and all neglected, insanitary and decaying, was made a high priority. This improvement brought widespread popular support for the CCP regime, and served to offset other policies that were less immediately appealing to many people. Land reform was the first major socialist, or communist, policy implemented. It was at first a simple redistribution of land in equal lots to all cultivators, including the families of former landlords.

The CCP did not intend to leave the matter at the level of peasant proprietorship of tiny plots. From the first, mutual aid teams were organized to manage the busy agricultural period. Later these were developed into the two stages of co-operative farming, and still later the co-operatives were grouped together into communes. In this way, private ownership of agricultural land was abolished and replaced by the communal system under which each former owner had a share of the commune's revenue allotted by 'work points', based on hours worked. State-owned collective farms were confined to newly-opened lands or reclaimed land not previously privately owned. Whatever other defects and difficulties the new land system encountered, owing to bad weather or administrative over-centralization, it can be said with certainty to have made two major advances: the constant threat of starvation in bad times receded; and water control and supply were made more possible by the new institutional units that replaced the smallholdings. These factors helped the commune system to withstand the great drought years 1960–62 without large-scale famine, although not without stern rationing and some malnutrition. In earlier, less severe, droughts the victims were often numbered by the million.

The Korean War and Relations with the USSR

The Korean War (1950–53) gave rise to a large literature, and its origin and the responsibility for its outbreak have remained in dispute. Chinese intervention, after UN forces began to move northwards into the Democratic People's Republic of Korea (DPRK), was forewarned, but the warnings were not heeded. To the Chinese this movement was a direct threat to their vital industrial area of south Manzhou (Liaoning Province), bordering on the DPRK. It was also widely feared in China to be the preliminary move to an invasion of China itself. The extent to which the Chinese intervention was intended to reassert Chinese authority, rather than Soviet influence, in Manzhou and in Korea, remains conjectural. Later developments seem to indicate that this consideration was important. It was certainly a consequence of the war, because, after the cease-fire, the USSR soon renounced the special position that the Chinese had conceded to it in the port of Dalian and over the railways across Manzhou.

In China the effect of the war was to strengthen the prestige of the Government which had, for the first time for more than a century (if ever), shown itself able to meet and match a large-scale Western army. In the years since the truce ending the war was signed at Panmunjom, Chinese relations with the DPRK have not always been smooth. The pretensions of President Kim Il Sung to be a major ideological leader were not appreciated in Beijing. The DPRK's attitude of neutrality in the Sino-Soviet dispute, although undoubtedly very wise, cooled relations with China. On the other hand, foreign observers drew the conclusion that China exercised a restraining influence on the adventurism of President Kim in respect of the Republic of Korea. However, the DPRK was certainly not a Chinese satellite.

In 1957, in the 'Hundred Flowers' movement, the Government permitted open criticism of its methods, if not its basic policies. The extent of the resulting criticisms was probably disconcerting to the authorities, yet much of what was said made its mark and led to some change of style in the CCP. The 'Hundred Flowers' movement was almost contemporary with the first phase of the Sino-Soviet dispute, which grew over the years until the two countries became completely estranged. The original quarrel over ideology developed into a dispute more concerned with national interests, especially after the USSR withdrew its technical aid and experts from China in 1960. This was a severe blow to the developing Chinese industrialization, but the setback was overcome. After a series of border clashes in 1969, negotiations for a settlement of Sino-Soviet differences concerning the border regions opened in Beijing in October 1969. Subsequent relations between the two countries long remained under strain. The fear of a possible Soviet attack, either using conventional or nuclear weapons, strongly influenced China's military and diplomatic planning. The expectation entertained by the Soviet leaders that, after the death of Mao Zedong, China would prove willing to renew the former friendship, or at least to modify its criticisms of the USSR, was not realized until mid-1989, when relations between the two countries were officially normalized.

The Tibet Issue and Relations with India

All Chinese Governments since the fall of the Qing dynasty have continued to assert rights of sovereignty over Tibet (Xizang), although the western two-thirds of the territory had been, in practice, independent since 1912. Tibet was occupied in 1950 by Chinese Communist forces. In March 1959 there was an unsuccessful armed uprising by Tibetans opposed to Chinese rule. As a result, the Dalai Lama, the head of Tibet's Buddhist clergy and thus the region's spiritual leader, fled with some 100,000 supporters to northern India, where a government-in-exile was established. The Chinese ended the former dominance of the lamas (Buddhist monks) and destroyed many monasteries. Tibet became an 'Autonomous Region' of China in September 1965, but the majority of Tibetans have continued to regard the

Dalai Lama as their 'god-king', and to resent the Chinese presence, leading to intermittent unrest.

In 1962 the establishment of Chinese forces on the Indian border with Tibet led to disputes about the position of the undefined and unmarked boundary. China proposed negotiations, but the Indian side rejected them, asserting that the frontier had been established by the United Kingdom before Indian independence. The tension escalated into a border war when Indian forces attempted to expel Chinese troops from some disputed positions. The clash resulted in a Chinese victory, which could have led to an invasion of Indian Assam. China unilaterally terminated the operations and withdrew to the positions already established before the clash. Soviet verbal support for the Indian claim considerably embittered relations between China and the USSR. The frontier dispute remains unsettled, with the Chinese holding what they claim to be the correct frontier line.

Recent History

MICHAEL YAHUDA

THE 'CULTURAL REVOLUTION'

Political Background

There is much about the 'Cultural Revolution' that still awaits historical evaluation, but its origins may be traced to political developments that began in the late 1950s. The upheavals in Eastern Europe in 1956, and Mao's growing doubts about the Soviet leadership, led him to the view that, even after the revolution and the establishment of a socialist society on the Leninist model, it was possible that a restoration of capitalism could take place. By 1965 he had convinced himself that this had indeed happened in the USSR, and that, unless action were taken, it could occur in China too. His theory stressed the importance of leadership and of the values that were espoused and institutionalized by those leaders, rather than the character of economic development as such. In Marxist terms, he emphasized the significance of the superstructure rather than the materialistic economic base. Mao's tendency to eulogize the significance of inspired leadership allied to mass activism, as against an institutionalized functional division of labour through orderly bureaucracies, underlay many of the political conflicts in the Chinese leadership from the 'Hundred Flowers' movement onwards. That campaign, in the spring of 1957, called on professionals and intellectuals to criticize the closed-door autocracy of the Chinese Communist Party (CCP), but it was resisted by some of the other, more orthodox, Leninist leaders. It was soon followed by an anti-rightist movement which was to plunge hundreds of thousands of non-CCP intellectuals into a disgrace from which they were not to be rehabilitated for another 20 years.

The 'Great Leap Forward' of 1958 may be seen as illustrative of Mao's preferred approach to economic development. The intention was to unleash the enthusiasm of the masses, under the leadership of politically inspired CCP cadres, rather than to follow the heavily-bureaucratized system associated with the Soviet model. It turned out to be an economic disaster, which led to the death of 23m. (or possibly as many as 37m.) people through famine. As 70m. peasants laboured to produce worthless steel from back-yard furnaces, the fields remained ill-attended. The cadres vied with each other to provide falsely inflated production figures, so that the scale of the débâcle did not become apparent to the leadership until well into the following year. Although most of the leaders had initially supported the 'Great Leap' programme, the leadership that met to review the situation in August 1959 was divided. The Minister of Defence, Marshal Peng Dehuai, severely criticized the 'Great Leap', and, at Mao's insistence, he was dismissed. Mao had construed the criticism as a personal attack upon himself. The episode may be seen as a break with those norms of CCP procedure that allowed for open debate (at the highest levels at least), and as the beginning of the elevation of Mao to a higher authority than the collective wisdom of the CCP, as expressed through its organization. Peng was replaced by Marshal Lin Biao, who soon ingratiated himself with Mao by elevating the study of 'Mao Zedong thought' in the armed forces and by emphasizing the army's role as a 'people's army', rather than fulfilling the more professionalist role inspired by the Soviet example.

Much of this was not immediately apparent, as Mao retreated from the daily management of domestic affairs, relinquishing responsibility for this to senior leaders such as Liu Shaoqi, Deng Xiaoping (who had become Secretary-General of the CCP Central Committee in 1954) and Zhou Enlai. While these men directed China's recovery from the three difficult years of 1959–61, Mao apparently confined himself to foreign affairs and ostensibly ideological questions related to the Sino-Soviet conflict.

The Initial Phase, 1966–69

Concerned by what he saw as the decline of the revolutionary ethos in China, and thwarted in his attempts to reassert a more dominant role, Mao launched the 'Great Proletarian Cultural Revolution' in 1966. The movement was directed against those 'party leaders in authority taking the capitalist road'. Mao's shock troops were to be the Red Guards, principally middle-school and university students who were to draw inspiration from Mao himself—now the supreme authority in all matters. They were to be guided by the newly established 'Cultural Revolution Group' of the Central Committee, a prominent member of which was Mao's wife, Jiang Qing. Having been denied a political role for more than 25 years, she sought to gain revenge against those who had slighted her, and at the same time she imposed a highly limited orthodoxy on culture and the arts. The Red Guards initially focused their attacks on teachers and CCP leaders in schools and universities. However, at the often clandestine instigation of members of the 'Cultural Revolution Group', they soon toppled government ministers and even members of the CCP's Politburo. Mao himself had singled out Liu Shaoqi (the man whom he had once described as his eventual successor) as the leading 'capitalist-roader' (he was dismissed as Head of State in October 1968 and died in prison in 1969), but Deng Xiaoping was also forced to undergo self-criticism and disgrace (in 1966) and few of the other senior leaders survived unscathed. Moreover, by the end of 1966 the CCP itself had virtually ceased to operate as a national organization, and the state administration was barely functioning. The one truly national organization, with an effective chain of command from the centre down to the localities, was the army.

Chinese politics became highly polarized and factionalized. This was to be its predominant character for the ensuing 10 years. The 'Cultural Revolution' itself lacked coherence. Mao's theory of how to avoid what he regarded as a counter-revolution was seriously flawed, as he had not articulated a clear vision of the political system to replace the one that was seen to have gone wrong. All groups and factions in China claimed to be the true followers of 'Mao Zedong thought', and each one stigmatized its opponents as revisionists. When the two radicals, Yao Wenyuan and Zhang Qunqiao (who had been so useful to Mao in launching the 'Revolution' and who were later to be condemned as members of the notorious 'Gang of Four'), reported to Mao in January 1967 about the establishment of a Shanghai Commune, modelled on the Paris Commune of 1871, Mao turned on them, dismissing the new institution as anarchist and insisting on the need for leaders. Meanwhile, the army had been called in to restore order, and it soon began to play a greater role in public life. Rejecting the Commune, Mao turned to the so-called 'three-way revolutionary committees' as the answer to China's absence of organization. These were largely dominated by the army, and included also representatives of surviving older cadres and of the 'revolutionary masses'.

By 1968 Red Guard factions, reduced to their hard-core members, were engaged in internecine armed struggles. Finally, on Mao's insistence, the organizations were disbanded, and, alongside millions of educated young Chinese, their members were sent to the countryside. Mao called for a reconstruction of the CCP. In early 1969, as fighting took place on the riverine border with the USSR, the Ninth Party Congress was convened. The newly elected Central Committee was dominated by men in military uniform, but it also included people who had been promoted during the 'Cultural Revolution', as well as a small proportion of surviving senior leaders. A new Constitution was proclaimed, in which Lin Biao was designated as Mao's successor—an unusual and irregular constitutional provision. The Congress also reviewed the 'Cultural Revolution'.

Factional Conflicts, 1969–76

Many in the West took the Congress to signal the end of the 'Cultural Revolution'. The events of mid-1968 certainly brought its end in the streets. Henceforth the 'Cultural Revolution' took the form of increasingly ruthless power struggles at the highest level of the leadership, and clashes between those, on the one side, who tried to institutionalize the so-called 'new-born things' of the 'Revolution' (principally in education and culture, but also in resisting allegedly revisionist economic practices) and those, on the other, who stressed the need for modernization. Underlying all these disputes were the questions of the succession to Mao and the future direction of China.

After the 1969 Congress, two main struggles developed. The first centred on Lin Biao and the second was between the 'Cultural Revolutionaries' (led by Jiang Qing), on the one side, and the modernizers (headed by Zhou Enlai and, from 1973, by the rehabilitated Deng Xiaoping), on the other. The official account of the episode leading to Lin Biao's death describes him as plotting the assassination of Mao after having been denied the position of Head of State. Having failed to kill Mao in August–September 1971, Lin allegedly escaped in a requisitioned aircraft, but died as it crashed in neighbouring Mongolia. Whatever the facts may have been, the episode reflected badly on the Chinese political system.

Following Lin's death, many of the leaders who had been disgraced earlier in the 'Cultural Revolution' were brought back to office. The most notable of these was the man who had been reviled in 1966 as China's 'No. 2 capitalist-roader', Deng Xiaoping, in April 1973. Unbeknown to the outside world, Premier Zhou Enlai was terminally ill, and Deng Xiaoping rapidly assumed many of Zhou's responsibilities. This brought about the second major series of political struggles. Lin Biao, who had been reviled as an 'ultra-leftist' in 1972, was now labelled an 'ultra-rightist'—the point being that the previous designation hampered the radical 'Cultural Revolutionaries'. They initiated a highly confusing 'Criticize Lin Biao, criticize Confucius' campaign, which took the form of debates on abstruse historical allegories, and was aimed at discrediting Zhou Enlai. This was followed by equally confusing and inconclusive campaigns on the importance of 'the dictatorship of the proletariat' and on the lessons to be drawn from a medieval novel, *The Water Margin*. For his part, Deng Xiaoping presided over the preparation of three blueprints to modernize education, science and the economy.

The situation reached a climax when Zhou Enlai died in January 1976. Deng Xiaoping, who had re-emerged in the previous year as First Vice-Premier and Chief of the General Staff of the Armed Forces, delivered the funeral address and promptly disappeared from public view. At the Cheng Ming Festival—the traditional time for honouring the dead—in April, public demonstrations in memory of Zhou took place in the major cities. These were the first truly spontaneous demonstrations to have taken place in the history of the People's Republic. The principal demonstration was in Beijing, where hundreds of thousands of people laid wreaths which pointedly eulogized the late Premier, and some went so far as to attack Jiang Qing and to criticize Mao's rule. On 5 April this escalated into a serious public disturbance, known as the 'Tiananmen incident', which led to hundreds of arrests. The episode was condemned as counter-revolutionary by a tense meeting of the CCP's Central Committee. Deng Xiaoping was condemned as the instigator, and was stripped of his official posts, but he was allowed to retain Party membership. The little-known Hua Guofeng, hitherto the Minister of Public Security, was appointed Premier and First Vice-Chairman of the CCP. Although Deng had lost the contest for power (and was dismissed from his posts), the 'Cultural Revolutionary' radicals had not won either. As Mao was becoming increasingly incapacitated, the radicals frantically sought to bolster their position in a final bid to denigrate Deng and his supporters. The country was in a chaotic state, and was later said to have been on the verge of civil war.

Mao died, aged 82, on 9 September 1976. Jiang Qing, her three associates and others of their faction were arrested in October, after manoeuvring to oust Hua Guofeng. They were quickly pilloried as the 'Gang of Four', and the 'Cultural Revolution' was declared to be at an end.

TOWARDS A NEW POLITICAL ORDER

Hua Guofeng and Deng Xiaoping

Hua Guofeng, as Mao's chosen successor and as the man who instigated the arrest of the 'Gang of Four', assumed the post of Chairman of the CCP, alongside his premiership of the State Council (cabinet), thus notionally succeeding both Mao Zedong and Zhou Enlai, but he was, in truth, a newcomer to the central political stage. He and a few undistinguished members of the Politburo were beneficiaries of the 'Cultural Revolution'. They had not the prestige, the seniority, the experience and the wide range of contacts of the pre-'Cultural Revolution' senior leaders who had either survived as leaders or had been rehabilitated. Moreover, neither Hua nor his fellow beneficiaries could fully negate the 'Cultural Revolution' or reverse the verdict of the 'Tiananmen incident' without undermining their own position. They were committed to modernizing China, yet the programme was not theirs. Having been initiated by Zhou Enlai, the modernization strategy had taken shape under Deng Xiaoping and his group.

By 1977 Deng had returned to prominence again, and in August he was restored to his former posts. Hua and his fellow beneficiaries from the 'Cultural Revolution' sought in vain to retain power by clinging to Mao's fading prestige. They advanced the slogan of following whatever Mao said and whatever he did, and became known as the 'whateverist faction'. In early 1978 Hua Guofeng announced a grandiose plan to modernize the country by 1985 (an extension from the 10-Year Plan advocated by Deng in 1975). The economy, already seriously out of balance, was soon dangerously overextended. The situation culminated at the third plenary session of the 11th CCP Central Committee in December 1978, which marked Deng's return to power.

The Significance of the Third Plenum

The third plenum of late 1978 is rightly regarded as one of the major turning-points in the history of the People's Republic. It set in motion a series of reforms that brought about a fundamentally new political and economic order. It also brought new people into the Politburo, notably Hu Yaobang, hitherto director of the CCP's organization department. Above all, it set the stage for a soundly based programme of economic reforms which could take place against a background of greater institutional regularity, legality and a freedom from the fear of chaotic political campaigns. This also entailed a re-examination of CCP history, a rehabilitation of past leaders and, indeed, a reconsideration of hundreds of thousands of unjust verdicts on people of humbler status.

Leading up to the third plenum, and carrying on into the early months of 1979, there took place a movement for democracy, sometimes known as the 5 April Movement (after the 'Tiananmen incident' in 1976). Groups of young people, many of whom were former Red Guards, displayed wall posters in the centre of Beijing and circulated their own unofficial magazines, demanding the restitution of wrongs by officials and, more defiantly, calling for the establishment of democracy and legality along Western lines. The movement had proved useful to Deng and his reformers in the period prior to the third plenum. Thereafter it became an embarrassment, to be crushed in the name of order. Deng issued a statement which declared that the limits of tolerance were to be judged according to four basic principles. These were the primacy of the socialist road, the dictatorship of the proletariat, the leadership of the CCP and Marxism-Leninism, and 'Mao Zedong thought'. The third was obviously the most crucial. At the fourth plenary session of the 11th Central Committee, held in September 1979, Zhao Ziyang was promoted from an alternate to a full member of the Politburo.

By 1981 the new leadership had produced its own verdict on the history of the CCP. Granted that the 'history' was a product of political compromises and that it had to serve current political aims, it contained far fewer factual distortions than had been the case with such 'histories' during Mao's lifetime. Not surprisingly, the 'history' made a positive assessment of Mao, but it blamed him for assuming excessive personal power and for being too leftist in his last 20 years. The 'Cultural Revolution' was condemned as an unmitigated disaster, based on a theory by Mao that bore no relation to reality. However, the terrible excesses of the 'Cultural Revolution' were blamed on alleged schemers such as the 'Lin Biao' and 'Gang of Four' groups, members of which had been put on trial in the previous year, and sentenced to long periods of imprisonment. Jiang Qing and Zhang Qunqiao had been sentenced to death, unless they repented within two years. Although no evidence was presented suggesting their repentance, their sentences were, in the end, commuted to life imprisonment. Meanwhile, some of those of lesser rank who had been found responsible for the persecution and, indeed, for the deaths of fewer people during the 'Cultural Revolution' were executed.

In December 1982 a new Constitution was promulgated, containing more detailed provisions than ever before on citizens' rights and the specific functions of organizations. In 1980, meanwhile, when Hu Yaobang, newly appointed to the restored post of General Secretary, and Zhao Ziyang, who replaced Hua Guofeng as Premier of the State Council, were elected to the Standing Committee of the Politburo, the old Party Secretariat was revived. By the 12th CCP Congress in September 1982, Deng had been able to bring to an end the practice of life-long tenure of senior CCP positions. A new Central Advisory Commission was established for people of at least 40 years' standing as CCP members. Much emphasis was put on the need to promote more professional, young and middle-aged party members to positions of high responsibility.

In short, the third plenum ushered in a period of political change aimed at promoting China's economic development without sacrificing the Party's monopoly of power. Despite the continual tension between the ensuing economic transformations and the unyielding political dominance by the Party, the system that was begun by the plenary session in December 1978 was to endure for longer than any previous period in the history of the People's Republic.

Modernization Under Deng Xiaoping: Aims and Achievements

The purpose of the economic modernization of the new order was defined as the development of the productive forces in order to raise the standard of living of the Chinese people; and to ensure the long-term security of China in accordance with its status as a world power. The ultimate purpose of the reforms was to transform the country into a 'modern socialist state with Chinese characteristics'. The process of reform, described by Deng Xiaoping in 1979 as China's 'second revolution', lacked any coherent theoretical framework. It proceeded in fits and starts in a pragmatic accommodation to changing circumstances. Nevertheless, the economic achievements were remarkable, with an average annual growth rate from 1978 of nearly 10%. Foreign trade expanded rapidly, as did China's engagement with the international economy. This success, however, has been uneven. The state-owned enterprises that employ the bulk of the urban labour force stagnated, while the rural-based industries and the foreign-orientated sectors showed strong growth.

The reforms began in the agricultural sector, as peasants were allowed to revert to family farming at the expense of the collective. Starting in the poorer areas, peasant households were permitted to contract for the use of land for 15 years or more. After fulfilling certain production quotas for the State, they were allowed to cultivate almost any crop they wished, and to sell them either to state agencies at above-quota prices or to market the produce themselves. Up to one-fifth of peasant households have specialized in single commodities or in providing various services or indeed developing local industries. Prominent among them have been the households of CCP members. These reforms led to a phenomenal growth in agricultural output. As the ratio of land to population will necessarily deteriorate, however, there is a need to combine greater efficiency in farming with the provision of employment prospects for the millions of unemployed or underemployed peasants in the countryside.

Reform of the industrial urban sector has proved more difficult by far. Following a three-year period of readjusting structural imbalances between the different sectors of the economy, in October 1984 the Government announced an ambitious programme to revitalize the planning system and the state-run industries. With the exception of a few strategic economic categories, mandatory planning was to be replaced for the bulk of industry by 'guidance planning' or by the macroeconomic controls of market forces. The state-owned enterprises were to be responsible for marketing their products and for their profitability. Accordingly, the programme envisaged that enterprise managers would have greater authority to reward and, if necessary, to dismiss workers. It also anticipated that loss-making enterprises could even face bankruptcy. Managers were given the right to sell above-quota production at higher prices. Unlike other parts of the economy, however, the state-owned enterprises proved very difficult to reform; they were as much social as economic units, providing life-long employment, social services and housing.

These economic and political developments were accompanied by social changes with far-reaching implications. In 1979 Deng Xiaoping finally removed the stigma attendant upon intellectuals for the previous 30 years, by declaring them to be members of the working class. Former business executives and industrialists, and even former landlords and rich peasants, were also, in the main, redefined as members of the working populace. Positive encouragement was extended to overseas Chinese communities to contribute to China's modernization. The 'compatriots' of Hong Kong were the most active element in the development of the various Special Economic Zones (SEZs) located on the southern coast, accounting for up to 80% of the joint ventures and other forms of economic co-operation. The adjoining province of Guangdong also benefited from the association with Hong Kong. Indeed, the 'open door' policy was deemed so successful that, beginning with the seventh Five-Year Plan (1986–92), China's economic strategy was orientated to give priority to the development of its southern and eastern coastal regions. At the same time, laws were enacted to render the SEZs attractive to foreign investors. In April 1988 it was announced that Hainan Island would be developed into China's largest SEZ, with the status of a separate province. Although the 1992–96 Five-Year Plan reflected a more conservative approach, seeking to 'confine' growth to 6% a year, this was soon challenged by Deng Xiaoping in early 1992, when he called for accelerated economic expansion and emphasized still further a pattern of growth modelled upon that of Hong Kong.

Thus, rural society was transformed, as the family unit gained at the expense of collectivism and as the significance of the market and rural-based industries increased sharply. The political role of the CCP declined, but many Party members drew on the administrative experience and extensive personal networks to establish themselves as 'specialized households', so as to generate sizeable incomes. At the same time rural unemployment, or underemployment, became more evident while the long-standing provisions against the mobility of the population remained. In the cities tension grew between the traditionally favoured state enterprises and the numerous bureaucracies on the one hand, and the newer collective and private operators on the other. The impetus for money-making and consumerism replaced Maoist orthodoxy, while corruption, nepotism and official racketeering became more visible and widespread. Most urban dwellers and a growing number of rural people gained access to the 'global village' of international television and radio transmissions. Traditional pre-Communist beliefs and customs re-emerged, particularly in the countryside. Official injunctions, designed to inculcate the values of 'socialist spiritual civilization', were less than successful. The more conservative, veteran leaders initiated short-lived campaigns against the more reformist CCP leaders and intellectuals. The first, against 'spiritual pollution' in 1983–84, rapidly went to excess and was brought to a premature end. The others, against 'bourgeois liberalization' in early 1987 and mid-1989, involved the dismissals of the General Secretaries of the CCP, Hu Yaobang and his replacement, Zhao Ziyang. These could all be seen as related to the social tensions engendered by the reforms, as they involved the problem of upholding the identity of China as a socialist country under CCP rule, while simultaneously reforming the economy along quasi-capitalist lines and opening it up to foreign influence.

PROBLEMS OF POLITICAL REFORM

By the mid-1980s there was evidence of major divisions within the Chinese élite about the direction and pace of economic reform and, above all, about the extent to which political reform should be a necessary component of economic change. Traditional Communist values no longer held sway. Indeed, the lack of belief in socialism and in the CCP itself, which was a legacy of the 'Cultural Revolution', became a crisis for the system. This crisis was intensified by the corruption and nepotism that was prevalent in the Party at all levels. Intellectuals who had hitherto been relatively quiescent under Party rule became increasingly critical and independent in their thinking.

Although it was not immediately apparent, the sixth plenary session of the 12th CCP Central Committee, held in September 1986, signalled the onset of an ideological struggle between contending factions within the Chinese leadership, which continued until mid-1987 and which finally culminated in the Beijing massacre of 4 June 1989. On the one side were those who favoured greater ideological diversity, democratization and structural reform as necessary for long-term economic modernization. On the other was a more conservative group supported by many of the long-serving revolutionary veterans, who held a more doctrinaire concept of the political role of the CCP, and who feared the potentially disruptive effects of some of the proposed reforms. Underlying the struggle was the question of electing a successor to Deng Xiaoping, and, indeed, successors to the generation of revolutionary veterans in their eighties.

In December 1986 demonstrations by students, demanding a greater measure of democracy, took place in Hefei, and then in Shanghai, Beijing and other major cities. By January 1987 the CCP had condemned these demonstrations as a threat to public order, and new municipal laws were enacted to prohibit them. Conservative leaders of the CCP and the army initiated a campaign against 'bourgeois liberalization' (defined, at the sixth plenary session of the 12th CCP Central Committee, as 'the negation of the socialist system in favour of capitalism'). On 16 January it was announced that Hu Yaobang had submitted his resignation as General Secretary of the CCP, having been compelled to undergo 'self-criticism', following accusations that he had allowed 'bourgeois liberalization' to spread. Following the departure of Hu Yaobang, Premier Zhao Ziyang assumed the additional duties of acting General Secretary of the CCP. Although he, too, had been a protégé of Deng Xiaoping and was a leading reformer, Zhao had not taken a public stand in the ideological quarrel, and enjoyed wide respect as an effective administrator and economic manager.

The 13th National Congress of the CCP, held in the period 25 October–1 November 1987, provided a major platform for Zhao, who still enjoyed the support of Deng. His report, which had undergone some drafting and redrafting within the Party, put forward the idea that China was at 'the initial stage of socialism'. Accordingly, there was no problem in abandoning the command economy (which by implication had been introduced prematurely) in favour of the market and of co-operation with Western countries provided, of course, that the Party's leading role remained in place. The slogan was that the State would regulate the market and the market would regulate the enterprises. Zhao went on to argue that the Party and state functions should be separated. That is, the Party should withdraw from the direct administration of the State and from the direct administration of enterprises, leaving those tasks to professional civil servants and trained managers respectively.

A new, younger and better-educated Central Committee was chosen. The 175 full and 110 alternate members had an average age of 55 years, and more than two-thirds had received university-level education. Many of the aged revolutionaries formally retired, although Deng Xiaoping retained the chairmanship of the Central Military Commission. The trend towards reform was confirmed at the 7th National People's Congress (NPC), which took place in March–April 1988. The long-standing Chairman of the NPC Standing Committee, Peng Zhen (who had presided over the delay of crucial legislation on enterprises), was replaced by the more reform-minded Wan Li.

In early 1988 Premier Zhao, with the public support of Deng Xiaoping, pursued a policy of rapid price reform, which had immediate inflationary consequences. The official annual inflation rate (considered to be a gross underestimate) rose from 7% in 1987 to more than 13% in the first quarter of 1988 (in urban areas 22%, including 44% for fresh vegetables). By June the rate had risen to 19%, and by July it had reached 24%. After highly contentious meetings of the senior leaders throughout mid-1988, the State Council announced its decision to defer any further reform of prices until 1990. Meanwhile, social tensions had been aggravated, and the economy was showing alarming signs of 'overheating'. A policy of retrenchment was announced.

The Tiananmen Massacre

The political struggle between the reformist and conservative factions in the higher reaches of the Party continued. On 15 April 1989 the former Party General Secretary, Hu Yaobang, died. As he was regarded as a symbol of political reform, student demonstrations erupted once again. Initially, the students were restrained, congregating around the memorial for revolutionary martyrs in Tiananmen Square in Beijing. Although their demand for Hu's posthumous rehabilitation over his forced resignation in January 1987 embarrassed the Party elders, they were nevertheless tolerant of the students. As protests persisted beyond Hu's funeral, Deng Xiaoping's patience expired, and at the instigation of Li Peng he authorized the issuing of a condemnatory *People's Daily* editorial. Unlike the student demonstrations of 1986/87, these protests were relatively restrained in demanding only a dialogue with state leaders and the ending of party corruption. However, they also demanded a retraction of the editorial and the truthful reporting of their demonstrations. They were soon joined by the ordinary citizens of the capital, by workers, intellectuals, teachers and even civil servants, journalists, members of ministries and the security forces. At one stage, more than 1m. people congregated in the Square and its main approaches. Similar demonstrations took place in 81 other cities in China. The official visit of the Soviet President, Mikhail Gorbachev, from 15 to 18 May, was disrupted and upstaged. Two days earlier, some 3,000 students began a hunger strike to give further impetus to their demands. The General Secretary of the Party, Zhao Ziyang, appeared to show a degree of sympathy to the demonstrators. After political struggles behind the scenes, which Zhao lost, martial law was declared on 20 May by Premier Li Peng and President Yang Shangkun. However, the soldiers sent to the centre of Beijing were stopped in their tracks by crowds of people. The demonstrators raised their demands by calling for the resignations of Li Peng, Yang Shangkun and Deng Xiaoping. On 30 May they erected a 30-m replica of the Statue of Liberty, called the Goddess of Democracy. On the night of 3–4 June 1989 heavily-armed troops, accompanied by tanks and armoured personnel carriers, shot their way into the main square, killing (according to eye-witness accounts) more than 1,000 students, workers and innocent bystanders. Some of the troops were massacred by the crowd. Both the demonstrations and the killings were witnessed by a world-wide television audience.

A conservative backlash ensued. It was alleged that a counter-revolutionary rebellion had been taking place and that the main victims had been soldiers. The principal decision-makers were the revolutionary veterans, now in their eighties and late seventies. On 23–24 June 1989 a plenary session of the CCP's Central Committee was held. It dismissed Zhao Ziyang from all his Party posts, as well as other prominent reformers, and it established an inquiry to examine his case further. The plenary confirmed Deng's version of events but, at the same time, it asserted its continued commitment to economic reform and the policies of the 'open door'. The Party Secretary from Shanghai, Jiang Zemin, was chosen as the new General Secretary of the CCP, and the Standing Committee of the Politburo was enlarged to six members. However, it was clear from the official accounts of the proceedings, and of the enlarged Politburo meeting that preceded it, that the effective rulers of China were the old revolutionary leaders.

The fifth plenary session of the CCP's Central Committee was held from 30 October to 3 November 1989. An attempt was made to prepare for the impending political succession with the resignation of Deng Xiaoping from the chairmanship of the Party's Central Military Commission in favour of Jiang Zemin. Although this meant that Deng no longer held any official posts, he was more than simply the ordinary retired citizen and CCP member that he claimed to be. He remained the country's effective paramount leader.

Deng Xiaoping's Final Years

In his twilight years Deng carried out two major endeavours of lasting significance for the development of China. He arranged for his succession in such a way as to provide for a smooth transition and to strengthen the institutionalization of Chinese political life. Secondly, he propelled China fully down the road of 'marketization' and rapid economic growth so as to break the hold that the remnant command sector had over the economy.

In the absence of an effective legal system, let alone one that could constrain the senior political leaders, China could still be characterized by what was called there 'rule by men instead of rule by law', but one of the striking legacies of Deng Xiaoping was the attempt to institutionalize key areas of political life. Thus, the functions of the main central political organizations were more clearly specified, and meetings were henceforth held at regular intervals according to proper procedures. More importantly, Deng established a system by which office holders must retire once they reach particular ages. Even the senior leaders were technically not allowed to put their names forward for re-election at national congresses once they had reached the age of 70. This development was strengthened by the transition of leadership from the old guard of the revolutionary veterans to the bureaucratic technocrats who were trained in the Soviet-influenced era of the 1950s. If the former drew their authority from their personal stature as founding fathers who had achieved so much, the latter's standing stemmed from their institutional positions. The link between the two sources of authority was that the technocrats had been selected by the founders as their successors. Although political succession might be seen as a fundamental weakness in the Chinese political system, the actual transition from Deng to Jiang proceeded remarkably smoothly. Jiang Zemin emerged as a political leader in his own right, who, none the less, was constrained by the need to compromise with other major leaders and to cultivate powerful constituencies such as the military.

Deng Xiaoping had technically retired in 1989, yet he continued to dominate Chinese politics until, in late 1994, he declined into inactivity, eventually dying on 19 February 1997, at the age of 92. By living to such an age, he succeeded in surviving most of the other revolutionary veterans, many of whom were less enthusiastic about his reforms and who may well have been able to upstage the younger Jiang Zemin. Notable among these was the austere Chen Yun, the most important of the conservative reformers, who died in April 1995. By the time of Deng's death, his successors, the seven-member Standing Committee of the Politburo headed by Jiang Zemin, whom Deng had designated the 'core leader', had been in charge of daily affairs for a considerable time, so that Deng's passing was marked by a singular calmness.

Deng Xiaoping left behind him a country that had largely recovered from the traumas of the Tiananmen killings, the collapse of the European Communist regimes in 1989, and the demise of the USSR, 'the motherland of socialism', two years later. His response to these critical events, which deepened the crisis of Communism in China, was to emphasize rapid economic growth, in order to increase prosperity and elicit popular support for the stability provided by Communist rule. Influenced by the example of Singapore, Deng saw no contradiction between maintaining authoritarian rule and encouraging rapid economic expansion and the attendant socio-economic changes. The alternative to Communist rule, according to Deng, was the chaos evident in the former USSR. Thus, he strengthened the forces of order while simultaneously stimulating economic reforms and the opening-up of China to the outside world.

In early 1992, in the course of an imperial-like tour of southern China, Deng challenged the prevailing conservative, or 'leftist', drift, by calling for a dramatic increase in the rate of economic growth and for the intensification of the policies of economic reform and openness. He urged Guangdong Province to catch up with the newly industrialized economies of East Asia, so as to be a model for the rest of China. He promoted the cause of the market, overcoming resistance from Beijing, by denying the significance in this regard of the distinction between capitalist and socialist practices. The 14th Congress of the CCP, held in October 1992, formally endorsed his concept of the 'socialist market economy.' At the Congress Deng oversaw the succession of a younger generation, with the average age of the Politburo being reduced from 69 to under 63, and the 'election' of a Central Committee with more professionally-qualified younger members, 61% of whom were under 55. Just as Deng found that he had to use his personal authority to overcome resistance to economic advance from members of those institutions whose authority he was trying to enhance, so he realized that he had to carry out a purge of the leadership of the Central Military Commission in the traditional secretive way, rather than in the more modern style he avowedly espoused. Moreover, the only exception to the rejuvenation of the central leaders was in the military, whose two leading figures were 76 and 78 years old. Clearly, Deng had trouble in finding younger men whom both he and the military could trust.

Jiang Zemin as the Core Leader

At the time of his nomination in 1989 many observers inside and outside China regarded Jiang Zemin as a political 'lightweight' and as something of an interim appointment. In the event, Jiang had more than seven years to ease himself into the most senior post. By the time that Deng died in 1997, Jiang Zemin had cleverly manoeuvred some members of his so-called Shanghai faction into leading positions in both the state and party organizations. Perhaps even more significantly, he also made extensive new appointments to the senior military positions. Although some of these were no more than reshuffles of regional leaders, it meant that key military members owed their appointment to Jiang, who has been able to retain the loyalty of the army, despite becoming its first civilian leader without any experience of military service. Thus, by the time of the 15th Party Congress in September 1997, Jiang Zemin was well placed to consolidate his position as the 'core leader'. He successfully removed troublesome opponents, Qiao Shi, the former minister and Chairman of the NPC, who had challenged some of Jiang's positions from an apparently more liberal perspective, and the 81-year-old Gen. Liu Huaqing. Although Jiang did not entirely succeed in determining the new appointments to the Politburo and its Standing Committee, he emerged from the Congress with his personal authority considerably enhanced. The new Central Committee of 193 full and 151 alternate members could be said to reflect Jiang's experience and outlook. Only 43% of the previous Central Committee's members were re-elected, with most of the remainder being excluded owing to their age. Yet the new Committee's average age of 55.9 years was not even six months lower than the previous average at the time of selection. Similarly, the membership remained largely technocratic, with more than 90% educated to college standard, and the vast majority having bureaucratic experience. These developments were taken a stage further at the Ninth NPC, which convened in March 1998, when Li Peng displaced Qiao Shi as Chairman of the NPC Standing Committee and Zhu Rongji, the former Mayor of Shanghai, became the new Premier of the State Council. He announced that the ministries of the State Council would be reduced in number from 40 to 29, with about one-half of these to be headed by ministers aged in their fifties.

Unfortunately, Jiang and Zhu inherited an economy that was already decelerating after the phenomenal rates of double-digit growth achieved in 1993–95. Indeed, Zhu Rongji had been widely praised for having engineered a successful 'soft landing' for the economy. At this stage it became possible for the senior leadership to encourage village democracy as a means to constrain corrupt local officials. The idea was that villagers would hardly choose those who had cheated them. The outcome was mixed as far as democratic accountability was concerned, since the election results needed confirmation by authorities at the next higher level. Nevertheless, this was a sign of progress in itself, and it attracted much interest in the West. The broader difficulty was that, by the time Jiang and Zhu took office, economic deceleration was becoming a problem, partly because of the impact of the Asian economic crisis that began in July 1997. China lost the Asian market that accounted for nearly one-fifth of the value of its exports, and the amount of foreign direct investment coming into the country began to decline sharply. Furthermore, the long-standing difficulties of the state-owned enterprises (SOEs) suddenly became more acute, leading to a precipitate fall in central government revenues. The post-Deng leadership decided to address the situation by undertaking a fundamental restructuring of the SOEs, despite the rapid rise in unemployment that would ensue. Economically, the SOEs received huge subsidies that in one form or another accounted for one-third of central budgetary expenditure. Yet they were also massively in debt to each other and to the banking sector, which meant that reform of the latter had to be postponed. Accordingly, it was announced at the 15th Party Congress that apart from some 500–1,000 large strategic SOEs, the remaining 150,000 would be restructured by means of amalgamations, share flotations and even bankruptcies. The decision was reaffirmed at the following NPC in March 1998, when the new Premier, Zhu Rongji, also announced that the number of civil servants in the central bureaucracies would be reduced by one-half.

Zhu promised to complete the reforms within three years, thereby suggesting that he did not fully appreciate the severity of the deceleration of domestic demand, which continued to decline during the following two years. Official figures claiming growth rates of 7%–8% took no account of the vast unsold inventories accumulated by SOEs, in particular. The growth of exports declined, as did that of foreign direct investment. Consequently, the pace of reform slowed as the Government attempted to promote expansion through massive state investment in infrastructure (principally transport, communications, energy and water conservation). Nevertheless, the authorities continued to press SOEs to restructure in order to reduce losses; this inevitably entailed laying off workers. In 1999, for example, according to official figures, 11.74m. SOE employees were made redundant, some 40% of whom reportedly found new jobs.

It was against this background of failure to stimulate domestic demand that Zhu Rongji visited Washington, DC, in April 1999, prepared to make unprecedented concessions in the hope of gaining US approval for China's bid to enter the World Trade Organization (WTO). The calculation was that the immediate effect of accession would be to stimulate the economy by a rapid inflow of new foreign direct investment and by overcoming obstacles to increasing Chinese exports. However, against the advice of his executive officers, President Bill Clinton refused to endorse the bid for reasons pertaining to US domestic politics. Although Clinton soon changed his mind, the damage was done, and the unfortunate Premier returned home to a storm of criticism from China's telecommunications, finance and agricultural sectors for his readiness to make concessions at their expense. The wave of anti-US feeling provoked by the bombing of the Chinese embassy in Belgrade, Yugoslavia, in May made Zhu's position even worse. Jiang Zemin, who supposedly represented the mainstream approach, continued to call for reform of the SOEs, but qualified this by conveying opposition to privatization. No support was expressed for the Premier's original offer to the USA, and by the end of June the normally assertive and autocratic Zhu seemed in danger of being marginalized. However, in the absence of any alternative economic programme, Zhu survived. As the sentiments aroused by the bombing began to dissipate, in November the USA finally negotiated an agreement on substantially the same terms. The European Union (EU) followed suit in May 2000, and the way was finally clear for China to join the WTO in late 2001—a development that would not only significantly open China still further to international commerce, but that would also result in the transformation of many features of the Chinese economy. When the second session of the Ninth NPC convened in March 1999 it was evident that the position of the reformers had been strengthened, as the Constitution was amended to allow greater scope to private enterprise. Meanwhile, the authorities took steps to increase their control of access to the internet within China, and Jiang Zemin encouraged the CCP to establish party cells in private companies.

As far as the President was concerned, a new and entirely unexpected challenge to the status quo emerged with a demonstration in April 1999 by an ostensibly quietist sect encompassing Buddhist and Daoist elements and composed largely of the middle-aged: the Falun Gong. Shocked to find that the sect had adherents in the highest circles of the élite, Jiang hastily had it condemned as a 'cult' and prosecuted it with the full coercive force of the regime. These alleged subversives continued their quiet demonstrations, and they appeared not to be intimidated by harassment and the arrest of thousands of practitioners, including government and party officials. The President's inability to control the spiritual movement entirely was indicative of how much authority, as opposed to power, the CCP had lost. Perhaps with this in mind, Jiang Zemin personally launched a campaign in February 2000 to revitalize the Party, by urging it to stand for the 'three represents', stressing that it would 'always represent the development needs of China's advanced social productive forces, always represent the onward direction of China's advanced culture, and always represent the fundamental interests of the largest number of Chinese people'. It was not until mid-2001 that the Chinese authorities began to overcome the challenge of the Falun Gong, at least to the extent that its followers were no longer able to mount public demonstrations. Yet adherents continually were reported by human rights organizations to be subject to horrendous treatment in prisons and detention camps, which sometimes resulted in mass suicides. During early 2002 members of the Falun Gong were able to intercept television broadcasts in several cities, and broadcast short messages contradicting the Government's anti-cult propaganda. The incidents raised new questions regarding the authorities' ability to control the activities of Falun Gong.

Meanwhile, during 1999–2001 the Chinese authorities were especially active in suppressing open dissent. Religious groups, indigenous human rights campaigners and democracy activists were subject to direct attacks by the security organs. Better-known activists were among those imprisoned without regard to due process, even by the limited terms of Chinese judicial procedures. The regime was particularly alert to the alleged threat posed by Muslim Uygur groups to Chinese rule in the region of Xinjiang. In the course of these three years various highly publicized violent incidents were associated with them. The regime used these threats as an opportunity to target other religious groups who had not registered with the authorities, thereby resisting their control. Yet during this period of egregious infringements of basic human rights, the Government signed the two main UN Conventions on Human Rights and even ratified the one concerned with economic and social rights. The regime, however, was remorseless when it came to dealing with matters it judged to threaten its ability to control order and social stability. By 2001 the regime had detained a number of mainly US scholars of Chinese origin and thereby received massive adverse publicity. Yet such was the enhanced prestige of China arising, for example, from its new standing in the world as a major trading nation (with its total trade reaching US $475,000m. in 2000, it now ranked as eighth), that Beijing was awarded the Olympic Games for 2008. Many in the West claimed that this would have the effect of encouraging political reform and liberalization. A more likely cause of change might be membership of the WTO, which would require greater adherence to the rule of law and greater respect for the autonomy of professions such as the law, accountancy, banking, etc.

Jiang's attempt to transform the CCP from a proletarian party to one that represented the people as a whole did not command full support within the party. Objections were raised to the idea that businesspeople who had hitherto been regarded as exploiting capitalists could now become party members. Nevertheless, in mid-2001 a plenary session of the Central Committee agreed to admit them. By this stage party leaders had become reconciled to the implications of the fact that the once all-dominant state-owned enterprises contributed less than 30% of the value of industrial production.

The Leadership of Hu Jintao

In November 2002 the 16th Party Congress established a new leadership, but it was circumscribed by Jiang Zemin's determination to retain power and influence. He successfully ensured that his loyalists would command a majority on the Standing Committee of the Political Bureau, which had been enlarged from seven to nine members. Moreover, being roughly of the same age as Hu Jintao (the new General Secretary of the CCP), only one member would have to retire because of having passed the age of 70 at the next Congress in five years' time. As a result, Hu would not be able to look forward to assembling his own team in his second and last term. Despite no longer being a member of the Central Committee, Jiang also secured a place for himself in one of the pinnacles of power by retaining the chairmanship of the Central Military Commission, which exercises operational command of the armed forces. Interestingly, Jiang's influence was less evident in the new Central Committee. More than 50% of the members were newly 'elected', and only one of the 381 (including the alternate members) was a recognized businessman. The pattern was repeated at the NPC held in March 2003, when Jiang's stalwarts were allocated significant positions on the State Council. Zeng Qinghong, Jiang's most able and ambitious associate, was even named the country's Vice-President. Hu, as the new President, together with Wen Jiabao, the new Premier, lost little time in trying to make his mark. Signifying a concern for the poor rather than for the entrepreneurs, they spent time with poor miners and indigenous people in Inner Mongolia during highly publicized visits over the New Year.

The emphasis on a simpler style of leadership was continued into mid-2003 with the abandonment of welcoming ceremonies at airports for departing and returning leaders and with the decision to forgo the annual leadership retreat to the seaside resort of Beidaihe. Before the new approach could be transformed into specific policies, however, the incoming administration was confronted with the crisis arising from the outbreak of the highly infectious Severe Acute Respiratory Syndrome (SARS). This hitherto-unknown pneumonia-like illness first appeared in the southern province of Guangdong in November 2002, but the news was suppressed by local and national officials lest it disrupt New Year travel and thereby damage the economy. Consequently, the disease spread more rapidly both within China and to neighbouring Hong Kong, Taiwan and Viet Nam. Signifying the advent of the more globalized era, the virus was carried by air travellers to Toronto, in Canada. It was that international exposure that eventually ended what had become an official attempt to conceal the extent of the disease. The Minister of Public Health and the Mayor of Beijing were dismissed. Under a strict quarantine regime, monitored by the World Health Organization, the affected areas of the country, including Beijing, were declared in June to be clear of SARS, which had killed more than 800 people across the world. The crisis led to demands that the Government become more open and allow the media greater freedom, so as to provide better and earlier warnings of future threats to public health. However, once the worst was over, it became evident that the Government was reverting to its previous policy of discouraging informants, dismissing editors of investigative newspapers and even closing down offending journals. If there were to be an improvement in accountability, it would be made by inferiors to superiors within the Party and hence would be beyond public scrutiny. Having first exposed some of the deep problems of the Chinese political system, the SARS episode ended by showing how limited the scope still was for fundamental political reform.

By 2004, moreover, it was evident that the scope for reform was further limited by the need to restrict investment in order to curb growth in an economy that was showing signs of 'overheating'. In addition to the economic and systemic constraints, political reform was effectively suspended for the time being by the effects of the power struggles at the highest level. Jiang Zemin was determined to retain his power and influence as Chairman of the Central Military Commission. Fearing that any change might reflect badly on the period of his stewardship of the CCP, he was able to keep Hu and Wen relatively isolated through the majority he commanded in the Standing Committee of the Politburo and through his ability to continue to make senior appointments in the military. The heightened tensions with Hong Kong and especially Taiwan from the end of 2003 also helped to sustain Jiang's position. As the most experienced leader and as head of the military, Jiang was able to insist on a relatively uncompromising approach—which of course militated against the prospect for reform at home.However, Jiang did eventually relinquish his position as Chairman of the Central Military Commission at the fourth plenary session of the 16th Central Committee of the CCP in September 2004, ceding the post to Hu, who was thus seen as having consolidated his hold on power. Hu was not expected to introduce major political reforms, having stated in a speech prior to the plenary session that China would not benefit from indiscriminately copying Western political systems. However, he was believed to favour reforms within the CCP, and was expected to show a less aggressive attitude than Jiang on foreign policy issues such as Taiwan.

The Social Impact of the Reforms

Chinese society has undergone massive changes since the reforms began more than 25 years ago and, if anything, the pace of change has accelerated since the early 1990s. Average incomes have more than doubled in real terms, and the number of those classified in dire poverty has declined from over 300m. to under 60m. The Chinese people may be said to have experienced a revolution even more profound than that accompanying the Communist triumph in 1949. The gradual establishment of a middle class is evident in the prosperous coastal belt, but much of the interior is still held back by the legacy of the Stalinist economic system and by a sense of having fallen behind. Unemployment is increasing, and China's cities are filled with 100m.–150m. migrants from the countryside. After an initial rise, peasant real income has fallen in the last few years. There are many sources for social unrest. Corruption is widespread, and leaders worry that it could lead to the downfall of the Party. Although the regime has strengthened its powers of repression since the Tiananmen killings, and despite its undoubted effectiveness in eliminating all overt forms of political opposition, its actual control over the daily lives of its people has diminished. Communist ideology has few adherents, and Chinese society is pervaded by what the official press describes as 'money worship'. The determinant of schooling, work, type of housing, nature of consumer goods purchased, material quality of life and even place of residence has become money and not the dictates of CCP officials.

In May 1994 the authorities announced the disbandment of the housing registration system (which had traditionally been the means of tying people to one place and subjecting them to easy control by their local or work units), because it had become unsustainable. In the late 1990s the main sentiment holding the Chinese people together was state-inspired 'patriotism'. Yet this, combined with looser forms of social control, also contributed to greater ethnic unrest in Xinjiang and Tibet where the local peoples found themselves in danger of being displaced by ethnic Chinese who were the principal beneficiaries of the economic reforms. Significantly, a report released in May 2001 under the auspices of the CCP's Organization Department gave a grim impression of an unsettled and sharply divided society, which was seething with discontent about official corruption, economic inequalities and the failure of existing social safety nets. It detailed widespread riots and disturbances. Police reports showed that the number of 'mass group incidents' were growing rapidly; 8,700 were recorded in 1993, and by 1999 the number had reached 32,000. Clearly, social stability had become a major source of concern. As seen by the leadership, the answer lay in ensuring that the economy would be able to absorb most of the unemployed by growing at an annual rate of at least 7%. Another issue that arose was that of the 100m.–150m. migrant workers, most of whom lacked the rights available to registered city dwellers and were forced to live in effect as second-class citizens. Pressure grew to ameliorate some of the discriminatory rules that applied to them. Thus, the 1982 ruling on the detention and repatriation of vagabonds that had led to the abuse of migrants by corrupt police was abolished after details of one particularly gruesome case were published in June 2003. In October 2003 the Party finally removed the constraint that all marriages had to be approved by the leader at the workplace. From a longer-term perspective, the key question was whether the new leadership had sufficient confidence and the necessary support within the CCP to effect the required political reforms to survive the social changes that lay ahead.

CHINA AND THE WORLD

During the 'Cultural Revolution', China chose the policy of exclusion—a policy of revolutionary isolationism. In 1965, when the USA first bombed North Viet Nam and then introduced combat troops in the South, Mao decided, after much debate among his colleagues, that China would not join in united action with the USSR to help North Viet Nam, but that it would continue to aid North Viet Nam independently. Meanwhile, he also urged the Vietnamese communists to adopt a low-key guerrilla strategy in the South. His advice was not followed. By 1966 China had suffered several reverses in the developing world, and, having presumably calculated that the USA would neither invade North Viet Nam nor bomb China, Mao was able to focus on the 'Cultural Revolution'. The underlying strategic view, that the USA was the main enemy and that the USSR sought to collude with it (at the expense of China, the developing countries and revolution in general), was shattered in 1968. The USA was seen to have reached the limits of its Viet Nam adventure as a result of the communist offensive during the Tet (lunar new year) festival of that year, while the USSR was regarded as a potential threat because of its invasion of Czechoslovakia (in August 1968) and its major military deployments to the north of China. Beijing promptly abandoned its diplomatic isolation and sought to restore relations with countries bordering the USSR and even with 'heretic' Yugoslavia. This expansion of outside contacts was soon extended to developing countries and to the small and medium-sized capitalist powers.

The military tension with the USSR led to clashes on the Ussuri river border in March 1969. These conflicts were not resolved until a Soviet armoured column penetrated into Xinjiang, and Aleksei Kosygin, the Chairman of the USSR Council of Ministers, held a meeting with Zhou and Mao at Beijing airport in September. Following fears of a Soviet nuclear strike, the US President, Richard Nixon, and his national security adviser, Dr Henry Kissinger, began moves that culminated in the latter's surprise visit to Beijing in July 1971. It was then announced that Nixon would visit China in the early part of the following year. In October the UN General Assembly expelled representatives of Taiwan and invited the People's Republic to take up China's seat. In February 1972 Nixon duly visited China, to be received by Mao and later to sign the famous Shanghai communiqué with Zhou Enlai. In effect, China had helped to change the central balance between the two superpowers by its shift from alliance with the USSR in the 1950s to alignment with the USA in the 1970s.

For the remainder of the 1970s China sought to build an anti-Soviet coalition against what it regarded as the major expansionist power in the world. During this time the closer relations that China forged with the USA were tempered by concern that, in the era of *détente*, the USA was insufficiently vigilant in confronting the alleged Soviet threat, and by anxiety that the US Government was too dilatory over the Taiwan issue.

China's new Western-orientated foreign policy included the normalization of relations with Japan; moreover, in sharp contrast with the earlier period of the 'Cultural Revolution', not only did trade rapidly expand, but, in the period up to Mao's death and the official end of the 'Cultural Revolution' in October 1976, China imported complete industrial plants to the value of more than US $3,000m.

In the aftermath of the final victory of the revolutionary forces in Indo-China in 1975, Chinese fears of Soviet encirclement, following the US withdrawal, increased. These fears were exacerbated by what were perceived as Soviet advances in Africa and in the People's Democratic Republic of Yemen. Events in Indo-China brought matters to a culmination. The rapid deterioration of Sino-Vietnamese relations, accompanied by the emotionally charged exodus of Hoa Chinese from Viet Nam, was centred mainly on the issue of Cambodia, where China was aligned with the infamous Pol Pot as part of a long-standing policy to deny to Viet Nam dominance over the whole of Indo-China. As Viet Nam consolidated its ties with the USSR in 1978, China signed a treaty of peace and friendship with Japan and, in December (during the course of the vital third plenum), normalized relations with the USA. Towards the end of December Viet Nam duly invaded Cambodia and, within a fortnight, had occupied most of the country and established a subservient regime in Phnom-Penh. In February–March 1979 China attacked Viet Nam in an exercise that was announced in advance to be of limited duration and penetration. The campaign did not go well for China's forces. Nevertheless, after the capture of the provincial capital of Long Son, the Chinese army withdrew. Perhaps both China and Viet Nam had been taught a lesson, but, importantly from the Chinese point of view, the USSR did not directly intervene.

For the next 10 years the result was deadlock in Indo-China, which left Viet Nam in a parlous economic condition, politically isolated and dependent upon the USSR. China, meanwhile, acted along parallel lines with the USA, forged an alliance with Thailand and a diplomatic partnership with the countries of the Association of South East Asian Nations (ASEAN), despite misgivings by at least two of its members. The collapse of the USSR left Viet Nam economically bereft and totally incapable of sustaining its occupation of Cambodia. China, therefore, was able to impose its terms for an accommodation with Viet Nam that was achieved in 1991, and at the same time to co-sponsor a UN Security Council Resolution for settling the Cambodian problem. Accordingly, China was able to attain its main objectives in Indo-China through the UN. In this way Chinese leaders could portray themselves as important and responsible members of the international community—a significant consideration in view of the international disapprobation still attached to China after the Tiananmen killings.

Since embarking upon the policies of reform and opening-up in 1978, economic considerations have become an increasingly important component of China's foreign relations. As China's leaders put it, they sought a tranquil international environment in which to concentrate on domestic economic tasks. To this end, they were quick to seize upon the perceived decline of the Soviet threat, and in 1982, at the 12th Congress of the CCP, Hu Yaobang announced that China was to pursue a foreign policy of independence, which would chart a more balanced path between the two superpowers. Indeed, China still tilted towards the USA, but Chinese leaders sought to manoeuvre more freely. China's problems with the USSR were epitomized by the reiterated demands that it remove the so-called 'three obstacles' (withdraw from Afghanistan, reduce the military threat from the north, and end the support to Viet Nam that enabled it to occupy Cambodia). Following the accession of Mikhail Gorbachev to the Soviet presidency, the USSR developed new approaches to foreign policy which met Chinese demands and paved the way for a summit meeting between Gorbachev and Deng Xiaoping in May 1989. The meeting was overshadowed by the student-led demonstrations in Beijing. Hopes that the two reforming Communist powers might build new relations were dashed in part because of the Tiananmen bloodshed, but in the main because of the collapse of Communism in Eastern Europe, for which the Chinese leadership privately held Gorbachev responsible to a considerable degree. The failure of the Soviet coup of August 1991 disappointed China's leaders. Yet they reacted with unusual aplomb to the collapse of the USSR itself. They very rapidly moved to establish correct relations with the successor states, including those of Central Asia.

China's relations with the West in general, and the USA in particular, have been marked by a curious ambivalence. On the one hand, China's leaders recognize the importance of the West for China's modernization as a supplier of technology, managerial expertise, capital, etc.; and on the other, they fear the possibility of Westernization and the erosion of the Chinese Communist system that may follow. These concerns have focused especially upon relations with the USA. For much of the 1980s China's leaders were able to manoeuvre within the so-called strategic triangle involving both the USSR and the USA. As a result, they were able to reap many of the benefits of a quasi-alignment with the USA without paying what they would have regarded as excessive costs in terms of dependency and loss of independence. Thus, China and the USA pursued parallel policies regarding the Cambodian and Afghan conflicts. They were separately allied with Thailand and Pakistan, and they both gave material and diplomatic support to the armed resistance to the Vietnamese and Russian invaders respectively. The Chinese were also able to gain access to advanced technology of military significance. At the same time the existence of the USSR as a counterweight to the USA enabled China's leaders to pursue in principle an independent foreign policy. Despite the continuation of the Taiwan issue as a problem between the two sides, China won favour within the USA because of its economic reforms and its open-door policies. Its poor record on human rights proved to be no obstacle to gaining US support for accession to international economic organizations, such as the World Bank and the IMF, from which China gained greatly. Unlike the USSR, China was also allowed entry into the domestic US market on the basis of Most Favoured Nation (MFN) trading status (now Normal Trade Relations—NTR).

Much of this changed with the end of the Cold War and the ramifications of the Tiananmen killings. At a stroke, the perhaps unduly favourable image of China's leaders in the West changed fundamentally. Moreover, the ending of the Cold War also brought to a close a period in world history in which China was regarded as an important player in global strategy. Thus marginalized strategically and despised politically because of the Tiananmen massacre, China's leaders found themselves subject to a series of Western sanctions. Sanctions began to be withdrawn within about 18 months, and China's international prestige started to recover. Through skilful diplomacy and by virtue of its size and weight in international affairs, as symbolized by its position as one of the five permanent members of the UN Security Council, China has shown that it cannot be ignored.

Owing to its rapid rate of economic growth, China has come to be regarded as the next rising world power. That view contributed to the US-led insistence that China conform better to what were presented as universal norms of state conduct. In particular, the USA emphasized observance of human rights, agreements about the non-proliferation of weapons of mass destruction and medium-range missiles, greater transparency in military matters and various trade practices, including intellectual property rights (IPR). In 1994 US President Bill Clinton gave way on the MFN issue by dissociating it from human rights. In February 1995, after tense negotiations, the Chinese eventually acceded to US demands on IPR. By this stage, however, Chinese strategists had come to the conclusion that the USA was imposing various constraints upon China in a concerted attempt to maintain its weakness. Matters came to a head when President Lee Teng-hui of Taiwan was granted a visa to visit the USA in a private capacity in April 1995. The USA was then accused of seeking to divide China. Beijing denounced President Lee in the most virulent terms for allegedly seeking independence, and in late 1995 initiated a series of intimidatory military manoeuvres. These culminated in March 1996 with the firing of missiles into the sea within less than 100 km of Taiwan's two major ports. If anything, these tactics produced the opposite of the effect intended, as Lee was overwhelmingly returned to office in the island's first direct presidential election. Moreover, the USA responded by sending two aircraft carrier battle groups. The crisis of March 1996 ironically paved the way for an improvement in Sino-US relations, as both sides sought to avoid similar confrontations in the future. Meanwhile, the USA upgraded its strategic alliance with Japan by agreeing new guide-lines that widened the scope of the support that Japan might give to US forces engaged in a conflict in the region. A closer Sino-US dialogue developed, in part to assuage Chinese concerns about the possible implications of those guide-lines for the Taiwan Strait. These developments culminated in an important exchange of presidential visits, by Jiang Zemin to the USA, in late October–early November 1997, and by Bill Clinton to China, in late June–early July 1998. The two Presidents agreed to establish a 'strategic partnership', and Clinton's visit was notable for his address to the Chinese people, broadcast live on television and radio, in which he denounced the Tiananmen killings and called for greater democracy. However, Clinton also praised Jiang's personal qualities and the 'moral' worth of China's reform programmes.

This improvement in relations was tested by the election of George W. Bush to the US presidency. After his assumption of office in January 2001, there was no more talk of strategic partnerships; instead China was viewed as a potential 'strategic competitor' in some respects and as a country with which the USA had important economic links in other respects. In practice, relations were strained initially by an accident on 1 April between a US surveillance aircraft and a Chinese jet that was monitoring it. The collision, which killed the pilot of the Chinese aircraft and forced the US aeroplane to make an emergency landing on Hainan Island, led to a crisis that was not resolved until the 24 US crew members were released 12 days later. Meanwhile, the atmosphere had not been improved by a decision by the new US Administration to accede to requests from Taiwan for most of the modern weaponry required by the island. However, both Beijing and Washington recognized that it was not in their interests to allow their relationship to deteriorate unduly and thus ensured that their economic ties were unaffected. The Chinese leadership, nevertheless, objected to what it saw as US hegemonic attempts to prevent China's rise and favoured the development of multi-polarity among the world's great powers as a means to constrain the unilateralism of the single superpower. The Chinese objected especially to aspects of the anti-ballistic-missile systems supported by the Bush Administration, as they feared that these would undermine the Chinese deterrent force and promote Taiwanese independence by providing protection for Taiwan against China.

The terrorist attacks on the USA on 11 September 2001 led to a major improvement in the all-important relationship with that country. China was quick to join the US-led coalition against international terrorism. Although the Chinese leadership did not contribute directly to the campaign against al-Qa'ida and the Taliban of Afghanistan, it did begin to share intelligence and undertook to deny financial assets and services to international terrorists. China managed to elicit formal US support against alleged Uygur terrorists in Xinjiang in return for greater legal explicitness by China in restricting the export of weapons of mass destruction and related technologies. The Administration of George W. Bush reduced its criticism of Chinese human rights violations. Jiang was pleased by the fact that President Bush visited China twice within six months (in October 2001 and again in February 2002). Bush also appeared to get on well with Hu Jintao, the new President, at the first appearance of a Chinese leader at a G-8 meeting of advanced industrialized nations in Evian, France, in June 2003. Although outstanding differences remained, neither pressed the other hard on any particular issue. From a Chinese perspective this meant having to tolerate a new and potentially prolonged US military presence in Central Asia as well as a revival of the USA's strategic interest in South and South-East Asia.

Nevertheless, the end of the Cold War continued to be strategically beneficial to China, as it ensured that an international setting in which for the first time since its establishment in 1949 the People's Republic was no longer subject to military threat from a superior power. Consequently, the requirements of economic modernization have become even more important in the country's foreign policy. Better relations have been forged with China's neighbours, including the newly established states of Central Asia and Russia itself. Indeed, relations between Moscow and Beijing have been described by both sets of leaders as better than ever before. In the absence of the Soviet factor, its former allies, Viet Nam and India, have accommodated themselves more to Chinese interests.

The Democratic People's Republic of Korea (North Korea) too found itself more dependent upon China and, under Chinese pressure, along with the Republic of Korea (South Korea) it joined the UN in 1991. Yet it did not follow China in reforming its troubled economy and in opening up to the outside world, and the North Korean leaders were greatly angered by China's formal recognition of the South in 1992. North Korea then began to use its perceived nuclear potential in a desperate diplomatic bid to ensure its survival. China skilfully manoeuvred between its own conflicting interests of seeking to avoid the collapse of North Korea, while at the same time striving to prevent the proliferation of nuclear weapons in North-East Asia. In reaching the Framework Agreement with North Korea in 1994, the US Government acknowledged that the Chinese had been crucially helpful behind the scenes. China then continued to play a role in facilitating US diplomacy with North Korea and in encouraging

North-South interactions. China was pleased with the first North-South summit meeting, which was held in Pyongyang in mid-2000, and with the exchange of senior officials between North Korea and the outgoing Clinton Administration. From 2001, however, the Administration of George W. Bush, took an unaccommodating approach to North Korea and ended this phase of active diplomatic engagement. Chinese interests were not best served by an escalation of tension, and China favoured the resumption of US-DPRK negotiations and the establishment of a kind of co-existence between the two Koreas, as that would allow the North to remain as an unofficial buffer state. Pre-occupied with Iraq, the Bush Administration, like its predecessor, soon come to realize the advantage of using China as an interlocutor with North Korea. Tension increased after North Korea's admission in October 2002 that it was engaged in the use of enriched uranium. China began to play a more active diplomatic role, particularly as North Korea began to carry out its threat of building nuclear weapons amid fears of a possible US attack, especially following the conflict in Iraq in early 2003. In early August it seemed as if the US demand for discussions to be held in Beijing on a multilateral basis might not be met, when North Korea renewed its insistence that a bilateral non-aggression pact first be reached with the USA. Nevertheless, China confirmed that talks would open in Beijing in late August, encompassing China, North and South Korea, the USA, Russia and Japan. The meeting was duly held but ended without agreement.

Elsewhere in the Asia-Pacific region China has made significant gains, amid uncertainties as to whether the country is a stabilizing or a destabilizing influence. On the positive side, despite the Asian financial crisis of the late 1990s, China is acknowledged to have assumed a significant role in enhancing the economy of the region as a whole. It has forged closer ties with Japan, which has become a major trading partner, and economic relations have deepened with the Republic of Korea, Taiwan and the ASEAN countries. China's adjacent provinces of Fujian and Guangdong have become increasingly integrated with the economies of Taiwan and Hong Kong; collectively they are recognized as 'Greater China', which represents one of the USA's largest trading partners. On the more negative side, China has territorial claims involving nearly all of its maritime neighbours. No means have been found of settling the competing claims, and China is torn between the nationalist urge to assert control over these islands and disputed territorial seas, on the one hand, and the economic necessity of establishing stable peaceable relations with the same disputants on the other. However, following their ameliorations with the Americans, China's leaders have become less troubled by what they tended to regard as US hegemonic attempts to constrain China's progress, especially by preventing it from recovering Taiwan. Consequently, China's leaders put less pressure on the South-East Asian nations to disavow the US system of alliances as relics from the Cold War. From the viewpoint of the governments of these countries, the US military presence in the Western Pacific provides reassurance about the stability of the region and indeed it is precisely because of these alliances that they feel able to engage China in constructive ways. The problem has been even more evident in the case of Japan. The Chinese authorities still complained that, as the Japanese had not come to terms with their aggression in the recent past (demonstrated by their failure to make a full apology for atrocities committed during 1937–45), the tendency toward militarism was not far below the surface. Consequently, they have quietly appreciated that the US security alliance has kept the Japanese from returning to the military path. However, China has been concerned that perhaps the USA is preparing Japan to play a more active military role in the region. The visits in 2001, 2002 and 2003 by Japanese Prime Minister Junichiro Koizumi to a controversial memorial honouring Japan's war dead angered China, which feared a resurgence of nationalist feeling in Japan.

In April 1996 an unprecedented multilateral treaty was signed with Russia and the three Central Asian Republics of Kazakhstan, Kyrgyzstan and Tajikistan, the countries becoming known as the Shanghai Five. This went beyond agreement about the demarcation of the long-disputed borders to include commitments to engage in military confidence-building measures and a pledge to refrain from exacerbating ethnic or religious tensions in each other's countries. In June 2001 these links were reaffirmed with the establishment of the Shanghai Co-operation Organization, and the number of participating states was increased to six with the accession of Uzbekistan. However, the importance of this grouping seemed to diminish after September 2001, when the USA began establishing a military presence in the region in connection with its war against the ruling Taliban in Afghanistan. Meanwhile, China's leaders had continued a series of exchanges with their Russian counterparts, culminating in the signing of a new treaty of partnership and friendship in July 2001. The new multilateral character of Chinese diplomacy was also evident in dealings with the regional organizations of South-East Asia where it was becoming an ever more important factor in the regional economy, and in February 2002 China reached a landmark agreement with the 10 member states of ASEAN to work towards a free-trade agreement within 10 years.

In 1997 the Chinese Government formally signed the Comprehensive Test Ban Treaty, which prohibits nuclear testing. China's integration with the international community was indeed gathering pace. China's membership of international organizations increased from 71 in 1977 to 677 in 1989, including the World Bank and the IMF. Fifteen years after first applying to rejoin the General Agreement on Tariffs and Trade (GATT), China entered the successor WTO in December 2001. China is a member of the main Asia-Pacific regional organizations, such as APEC and the ASEAN Regional Forum (ARF), but these are mainly of a consultative nature. However, as China emerged as the main source of economic growth in the region it was able to translate its new economic significance into political influence. To this end, China tried to show that it took its neighbours' interests into account. At China's suggestion, agreement was reached with ASEAN countries to establish a China-ASEAN Free Trade Area by 2010. China was also the first non-ASEAN country to sign the Association's Treaty of Amity and Cooperation in South-East Asia, which committed members to settle conflicts by non-military means. It also signed a declaration of a code of conduct in the South China Sea that required it to desist from making unilateral bids to expand its presence in the contested Spratly chain of islands. China was also active in the ASEAN + 3 (also including Japan and South Korea). In short, it seemed as if China was responding positively to attempts to engage it in diverse patterns of co-operation.

On 1 July 1997 Hong Kong was returned to China under the terms of the Sino-British Joint Declaration that was agreed in September 1984. The Chinese Government undertook to allow Hong Kong to enjoy a high degree of autonomy as a Special Administrative Region (SAR), so as to maintain its economic system and way of life for a further 50 years. This unique arrangement of 'one country, two systems' is regarded in China as a potential model for the eventual reunion of Taiwan with the mainland. The negotiations between the United Kingdom and China were often acrimonious, and Chinese leaders were particularly displeased with the last British Governor, Christopher Patten, whom they vilified for his attempts to broaden democracy in the territory without first securing their consent. Consequently, in 1997 they replaced the legislature elected in 1995 with a provisional body. Elections to a new legislative body were held in May 1998, under a mixed system of voting, in which the number of people eligible to vote was severely reduced. Yet, despite apprehensions on all sides, the transition to Chinese sovereignty proceeded smoothly. However, Hong Kong's failure to recover fully from the downturn resulting from the 1997 Asian economic crisis, as well as the political shortcomings of the SAR's Chief Executive, Tung Chee-hwa, compounded popular discontent in the territory. On 1 July 2003 huge demonstrations took place in protest at proposals by the Hong Kong Government to introduce new anti-subversion legislation at Beijing's request. Tung Chee-hwa subsequently modified and finally abandoned the controversial proposals, his action representing a challenge to China's role as the arbiter of Hong Kong affairs. By April 2004 the Chinese Government had 'interpreted' the Basic Law (Hong Kong's Constitution) to allow it to predetermine what proposals for democratic change would be allowed, and declared that universal suffrage would not be permitted either to elect the Chief Executive or in legislative elections in 2007/08, as had been deemed possible under the provisions of the Basic Law itself. Further demonstrations followed on 1 July 2004, and attention was focused on the legislative elections due in September (see the chapter on Hong Kong). Meanwhile Macao, the first and last Western possession in Asia, returned from Portuguese to Chinese sovereignty on a similar basis to that of Hong Kong on 19 December 1999. However, being a much smaller territory, Macao did not experience problems comparable to those that emerged in Hong Kong.

Although China's leaders are reconciled to the global predominance of the USA as the world's sole superpower, they still seek to encourage movement towards a more multi-polar world that would enable China better to balance US power. In practice, however, China wishes to maintain good relations with the USA largely because it is a major export market and because it is a guarantor of security in the Asia-Pacific region. The conditions of US predominance have served Chinese interests well by providing a peaceful environment in which, since the late 1970s, China has been able to develop its economy so well. The US strategic presence has ensured the absence of war in Korea and the containment of possible Japanese militarism. The main practical issue is the question of Taiwan. Beyond that, the Chinese appear to have learned how to balance their practical need to work with the USA against their principled opposition to much of what the USA stands for. The Iraq conflict in 2003 showed how China could maintain a principled opposition without antagonizing the USA in practice. However, China's foreign policy has become increasingly a product of the primary domestic concern to maintain economic growth and social stability as the prerequisites for maintaining CCP rule.

The one exception to China's non-confrontational stance in its foreign relations is the problem of Taiwan. Seen as the last obstacle to attaining the complete unity of the Chinese state and as a remnant of the civil war that has been kept apart from the mainland solely because of US intervention, China's leaders are determined to use all means to prevent the island from declaring itself independent. In mid-2003 China's leaders objected strenuously to a proposal by Chen Shui-bian, the elected President of Taiwan, to change the Constitution by a referendum, as this would have located the sovereignty of the island in the people of Taiwan rather than in the people of China as a whole—a move that would be seen by Beijing as tantamount to a declaration of Taiwanese independence. Following threats of military action by China if Chen's reform proposals were to be realized, US President Bush in December 2003 warned Chen against seeking to change the status quo unilaterally. Chen nevertheless won the presidential election in March 2004, to the consternation of Beijing. China continued to press Washington about weapons sales to the island, which in its view only encouraged the untrustworthy Chen to pursue his unilateralist path towards formal independence. China's leaders claim that they would pay any price to prevent this from happening, and they have mustered sufficient military force to make their threats credible. Taiwan, therefore, is the one issue that threatens to undermine China's goal of fostering a peaceful international environment so as to concentrate on domestic economic development.

Economy

ROBERT F. ASH

Based on an earlier article by HANS HENDRISCHKE

INTRODUCTION

The most recent developments in the Chinese economy have occurred against the background of the major political changes associated with the succession, during the 16th National Congress of the Chinese Communist Party (CCP) in November 2002, from the 'Third Generation' of party and political leaders, under Jiang Zemin, to the 'Fourth Generation', under Hu Jintao, the new General Secretary of the CCP Central Committee and its Political Bureau (Politburo). The importance of this change was underlined in March 2003 by the election, during the First Session of the 10th National People's Congress (NPC), of a new Government, under incoming Premier Wen Jiabao.

The economic impact of these political changes has, however, so far remained quite slight, although Wen Jiabao's evolving vision of China's growth strategy seems destined to place renewed emphasis on sustainability as well as on the mere maximization of growth in gross domestic product (GDP). Since the 1980s pursuit of the economic imperative has been the key to maintaining economic and social stability and, thereby, to upholding the legitimacy and authority of the CCP. This holds true for China's 'Fourth Generation' of leaders no less so than it did for its predecessors. To this extent, the broad economic agenda—one dominated by the programme of 'reform and opening', initiated by Deng Xiaoping—is already in place. At the same time, Wen has stressed the need for a new 'scientific' concept of development that will facilitate the emergence of a more equitable pattern of economic growth than that of recent years. In a very real sense, China's future social and economic trajectory will be determined by the extent to which the existing trend towards growth with inequality can be replaced by growth with equity.

The change in emphasis is captured in official pronouncements. For example, in November 2002 the 'Resolution' passed at the 16th National Congress of the CCP stated that 'development is our Party's top priority in governing and rejuvenating the country and ...it is imperative to take economic development as the central task,...accelerate modernization, maintain a sustained, rapid and sound development of the national economy and steadily uplift the people's living standards'. By contrast, the Government Work Report presented by Wen Jiabao to the Seventh Session of the 16th NPC in March 2004 highlighted the importance of achieving more stable growth through the fulfilment of the 'five balances': the maintenance of an appropriate balance among economic sectors, geographical regions, human and natural resource constraints, economic and social priorities, and domestic and overseas goals. Although China's economic performance under the impact of post-1978 reforms is remarkable by world standards, the momentum of its GDP growth showed a declining trend during the 1990s. Interpreting China's official growth statistics is a hazardous exercise, and while there is a consensus that the growth figures issued by the National Bureau of Statistics (NBS) in Beijing exaggerate reality, adjusting them for this bias is extremely difficult. Suffice to say that until 2003, when there were suggestions that official growth estimates might be underestimates, many would remove around two percentage points from existing official claims in an attempt more closely to approximate to the truth. In any case, taking NBS estimates at face value, the evidence points to a halving of annual GDP growth between 1992, when it reached its peak of 14.2%, and 1999, by which time it had contracted to 7.1%. Since 1999 growth has once more followed a rising trend, and, in particular, has accelerated from 7.5% to 9.1% between 2001 and 2003.

All three major sectors shared in China's buoyant performance in 2003, although industry (including construction) and services (up by 12.5% and 6.7% respectively) easily outpaced the more modest expansion of agriculture (2.5%). As in previous recent years, the implementation of a pro-active fiscal policy, with bond issues substituting for weak domestic demand and large-scale deficit financing of infrastructural projects undertaken by the State, contributed significantly to China's economic success. The positive role of external economic relations (the continued expansion of foreign trade and inflows of foreign direct investment—FDI) has also continued to be a cornerstone of China's economic success. Contrary to expectations, the outbreak of Severe Acute Respiratory Syndrome (SARS) in the first half of 2003 was no more than a temporary constraint on GDP growth, confounding earlier predictions that the disruption of markets would have a more severe and lasting macroeconomic impact.

China's economic performance in the first half of 2004 remained buoyant. Indeed, GDP growth continued to accelerate (rising by 9.7% during the first quarter), with fixed asset investment (increasing by 28.6% year-on-year) reaching a record level. At mid-year, there was widespread concern that the economy was overheating, raising expectations that interest rates might have to be increased for the first time since 1995 in order to direct the economy towards a 'soft landing'. The challenge facing the Government was a difficult one, comprising the need to curb excessive capacity expansion in overheated industries whilst simultaneously supporting private consumption. Against the background of rising consumer price inflation, the consensus view was that reliance on market mechanisms would enable the central authorities to avoid a 'hard landing' and that GDP would expand by around 9.5% in 2004.

The future development of the Chinese economy has both regional and global significance, not least now that the country has joined the World Trade Organization (WTO). However, the future is beset, as Chinese sources themselves acknowledge, by serious problems. The economic and social consequences of continued restructuring, including large-scale redundancies, of state-owned enterprises (SOEs), the associated problem of a huge 'overhang' of non-performing loans (NPLs), increases in urban unemployment alongside massive over-supply of labour in agriculture, widening inter-regional and inter-sector growth and welfare differentials, a growing fiscal deficit—such domestic problems present major challenges to China's Government and policy-makers. Internationally, China's economic growth depends increasingly on mutually-reinforcing trade and investment relations with developed countries, including the USA, European Union (EU) member states and Japan (but also Taiwan and Hong Kong). China's membership of the WTO underlines its commitment to globalization and to the strengthening of domestic market reforms that are expected to result from its increased economic internationalization. One of the effects of the Asian financial crisis of the late 1990s was to highlight the interdependence of economies within the Asia-Pacific region. During the crisis the People's Republic succeeded in presenting itself as a responsible power and a force for stability in the region. Its potential to destabilize neighbouring economies, whether through enhanced competitiveness or the exertion of political pressure (for example with regard to Taiwan) should not, however, be overlooked.

After more than two decades of rapid growth, China's economy has become one of the largest in the world, ranking second only to the USA in terms of purchasing-power parity (PPP). During 2001–03 China accounted for one-third of global economic growth, measured at PPP, compared with a share of one-sixth from the USA. However, in terms of average per caput GDP, China remains firmly in the ranks of less developed countries. On this basis, the Human Development Report of the UN Development Programme ranked China 104th out of 175 countries in 2001, and although its average GDP per head had risen to around US $5,000 (on a PPP basis) by 2003, the gap between China and the first rank of Asian newly-industrializing economies (NIEs)—let alone the USA and countries of Western Europe—remains huge. Measured according to the World Bank Atlas approach, China's average per caput GDP rose from US $758 to US $1,029 between 1998 and 2003. Nevertheless, China has become one of the largest markets for investment goods, and it has demonstrated good potential for growth in new markets, such as telecommunications, automobiles and financial services. It is a leading supplier of cheap consumer items and is often described as the manufacturing base of the entire world. Increasingly, China has emerged as a global economic power, the decisions of which have a significant bearing on other countries. Even allowing for the negative, as well as positive, impact of China's accession to the WTO, the gradual integration of its domestic markets into the international economy will only increase such influence. Meanwhile, the domestic effect of WTO membership is likely to include a short-term increase in unemployment resulting from greater competition among domestic enterprises, as well as between domestic and foreign firms. Inter-regional strains will also accompany the necessary restructuring of domestic markets, as provincial protectionism and trade barriers are abolished. The employment-creation effects of WTO membership will emerge only as and when projected increases in GDP growth are forthcoming in the longer run.

ECONOMIC STRATEGY, GROWTH AND STRUCTURAL CHANGE UNDER MAO ZEDONG

When it assumed power in October 1949, the Chinese Government under the CCP's leadership instituted policies designed to fulfil a broad objective, which was shared by all post-war developing countries. It was to transform a poor, backward, traditional and overwhelmingly agricultural economy into a modern, industrial, high-income society. Characteristic of the Chinese approach, however, have been the specific policies formulated to achieve this goal. Under Mao Zedong, these reflected the aspirations of a country pursuing a socialist-orientated development strategy that borrowed heavily from the CCP's Marxist-Leninist roots, and from the development experience of the former Soviet Union. Since 1978—first under Deng Xiaoping, subsequently under Jiang Zemin and from 2003 under Hu Jintao—the broad development goal has become increasingly market-orientated (the establishment, in Chinese parlance, of a 'socialist market economy'). At the end of 2002 a senior official claimed that China had completed the transition from a planned economy to a socialist market economic system, with the prices of 90% of goods and services throughout the economy being determined by the market. Its status as a 'true' market system has yet, however, to be universally recognized, as indicated in the conclusion of an EU investigation, which in June 2004 argued that China had yet to qualify as a market economy—a decision that elicited protests from Chinese officials anxious to avoid the penalties for export 'dumping' to which non-market economies are more susceptible.

Prior to 1978 China's economic development was shaped by Maoist ideological imperatives of egalitarianism and self-sufficiency. Until the late 1950s China pursued a Soviet-style, even Stalinist, strategy that sought to maximize industrialization (especially heavy industrial growth). At the heart of this strategy was the construction of 156 modern, capital-intensive Soviet 'aid' projects, designed to provide a balanced programme of basic industrial facilities (including iron and steel, energy, machine-building and armaments). In support of this programme, domestic investment, funded by a high savings ratio and low farm wages, was directed disproportionately towards heavy industry and associated infrastructural construction. By contrast, agriculture was largely deprived of resources, its growth left dependent on institutional change (that is, farm collectivization). Meanwhile, foreign trade was dominated by the USSR, with Chinese primary goods and raw materials exchanged for Soviet industrial equipment.

The highly unbalanced nature of investment was reflected in rapid industrial growth, but at the expense of much slower farm growth, as a result of which congestion in agricultural supply of food, raw materials and exports began to emerge, threatening to impede the overall growth momentum. This was the background against which the leadership began to question the suitability of existing economic policies—a process out of which emerged a more indigenous radical strategy (the 'Great Leap Forward' of 1958), considered more appropriate to Chinese conditions. More a vision than a plan, the new strategy advocated a policy of technological dualism ('walking on two legs'), whereby budgetary investment would continue to be focused on heavy industry (the modern leg), while unprecedented mass mobilization of the supposedly under-utilized rural labour force within large-scale communes (the traditional leg) would help promote more rapid agricultural growth, as well as supporting industry through the establishment of rural factories (especially back-yard steel furnaces).

The origins of the catastrophic failure of the 'Leap' lay in economic mismanagement in agriculture, fatally exacerbated by a breakdown of the statistical reporting system, which led the Government grossly to exaggerate the output of food grains. The reality was that by the end of 1960 rural per caput food availability had declined by 25%, causing unprecedented starvation and leading to the deaths of up to 30m. people during 1959–61. The situation was made worse by growing ideological differences between China and the USSR, which, in mid-1960, reached breaking point, prompting Moscow to withdraw aid to China. What started as an agricultural crisis became a deep depression affecting the entire economy. Only by reversing economic priorities in favour of agriculture and adopting pragmatic policies (including the delegation of decision-making responsibility and an emphasis on material incentives, which bore a marked resemblance to those that were to be pursued after 1978), was economic recovery eventually (in 1965) secured.

Such, however, was Mao's antipathy towards what he regarded as the capitalistic nature of these recovery policies that at the end of 1965, he launched an 'anti-rightist socialist education campaign', which in the following year was transformed into a new national experiment in the form of the 'Great Proletarian Cultural Revolution'. Although this movement was not primarily an economic phenomenon, it nevertheless generated temporary periods of serious economic disruption, especially in China's cities during 1967–69, when revolutionary 'Red Guards' entered planning offices and factories, and disrupted production and policy formulation. In addition, throughout the decade of the Cultural Revolution (1966–76), a disproportionate emphasis on egalitarianism and self-sufficiency caused serious structural distortions within the economy. The most extraordinary manifestation of this was a gigantic industrial initiative (the 'Third Front'), designed to create a self-sufficient industrial base in the deep interior of China. It involved the construction of large-scale industrial facilities and associated infrastructure in western provinces of China, supposedly safe from attack from the USA and/or the USSR. About half of the entire central budget was committed to this programme during the late 1960s and early 1970s.

China's economic record under Mao was uneven. Heavy industry expanded rapidly, but at a high cost in terms of agricultural and light industrial growth. Between 1952 and 1978 nominal GDP growth averaged 6.7% annually, but concealed within this figure were rates of expansion of 4.3% for the farm sector, 5.9%. for services and 10.1% for industry (manufacturing and construction). The outcome was that the share of industry in GDP increased from 21% to 48%. The evidence suggests, however, that this growth record was driven by massive inputs of physical capital, unaccompanied by efficiency improvements. As a result, side by side with modest improvements in labour productivity, capital productivity probably declined. China's high degree of isolation within the global economic community also resulted in a widening gap between the technology embodied in its own industrial technology and best-practice techniques available in the USA, Japan and Western Europe. Finally, the imperative of heavy industrial growth was reflected in a extraordinarily high rate of accumulation, which implied severe constraints on welfare levels. With economic growth outstripping consumption growth, the result was that living standards remained largely unchanged between the mid-1950s and the mid-1970s.

ECONOMIC DEVELOPMENT UNDER THE IMPACT OF REFORM

The political victory, in 1978 (two years after the death of Mao Zedong), of a group of pragmatic leaders centred around Deng Xiaoping marked a watershed in China's economic development. It signalled the abandonment of the Maoist system of central control over resource use and allocation in favour of a strategy of reform and experimentation. The new strategy sought to decentralize economic decision-making, increase departmental and regional autonomy, encourage competitiveness, and facilitate the establishment of an increasingly market-driven economic system (later to be defined as a 'socialist market economy' in order to highlight the significant residual role of the State in managing the economy). A cornerstone of the new approach was the implementation of an 'open door' policy, the radical nature of which was most clearly demonstrated in China's willingness to accept foreign capital (including increasing inflows of FDI), in order to help fund investment and upgrade its obsolete industrial technology.

China's domestic economic reform began in the countryside, where it embraced the decollectivization of agricultural production (by 1983 the previous collectivist thrust of farming had disappeared), the raising of prices for farm products and increasing economic diversification (including, most notably, the creation of rural industries). In addition, a path-breaking initiative was the establishment, in 1980, of four Special Economic Zones (SEZs) in two southern provinces (Guangdong and Fujian) located next to Hong Kong and Taiwan. Intended to be windows on the global economy, the SEZs were bases for economic experimentation (technical and institutional), as well as gateways for inflows of foreign investment.

The immediate success of these policies laid the foundation for a gradual transformation of China's planned economy during the following two decades. Price reform was the first stage of macroeconomic reform and addressed the irrationality of a system in which prices had previously been set administratively simply to accommodate plan priorities and which failed to reflect the reality of market-orientated demand-supply relations. It was, moreover, a system that favoured industrial producers at the expense of farmers and other primary producers. In the first stage of price reform, from 1979 onwards, purchase prices for agricultural and rural sideline products were raised to increase incentives for farmers and to create markets for light industrial goods. The price differential between the mining and processing industries was also reduced. In 1985 a dual ('two-track') pricing system was introduced, which allowed planned prices and market prices to coexist. The underlying intention was to facilitate the retention of some control by the centre, as the economy moved towards a free-market system and the proportion of goods produced and sold at market prices expanded. It was believed that in this way, mandatory prices would gradually give way to a market-based pricing system without generating any significant inflationary impact. On the debit side, two-track pricing gave rise to corruption, by enabling officials to make windfall profits by selling goods, which they had received under lower planned prices, at higher market prices. A more important, positive effect of the system, however, was that it encouraged enterprises to adapt their production to market demand. The overall aim of deregulating the prices of most consumer goods, as well as of capital goods, had been achieved by the late 1980s; and by the end of the 1990s more than 90% of retail prices, and about 80% of prices for raw materials and agricultural produce, had been liberalized. Between 1993 and 2003, the share of industrial value added accounted for by shareholding and foreign-owned firms (including those of investors in Hong Kong, Macao and Taiwan) rose from about 10% to more than 50%. As a result of such developments, the degree of central control over the national economy, as measured by the share of GDP produced by state-owned firms, has fallen below that of France, Italy and Singapore. Price controls do still apply in strategic areas, such as in the energy sector, although even here (for example in the electricity sector) the first regional markets have begun to appear in an attempt to break national monopolies. China's WTO membership has also already led to a further reduction in government control over prices and tariffs.

China's dual-track transition strategy resulted in the coexistence of the planned economy with an increasing number of unregulated sectors, such as rural collective enterprises. The outcome was continued economic growth within a framework in which the scope of price controls and economic planning was gradually reduced. Owing to the faster development of the private and non-state sectors, the Chinese economy was literally 'growing out of the Plan'. Yet in the absence of a specific model for the transition to a market economy, policy tensions generated frequent macroeconomic crises, characterized by high inflation and social unrest. The gradual, incremental and experimental nature of China's economic reforms (likened, in Chinese parlance, to 'crossing the river by feeling for the stones') reflected not only the continuing influential role of the CCP, but also the long historical tradition of state interference in the economy. Unlike the Soviet Communist Party in the late 1980s and early 1990s, which lost much of its authority under the impact of economic and political reforms taking place in the USSR, the CCP refused to sanction political change and retained significant control over macroeconomic policies and reform ideology. In general, it maintained social stability by avoiding the abrupt dismantling of the institutions of the command economy, particularly of the dominant state-enterprise sector, which traditionally afforded substantial social service benefits to its workers and their families. The disadvantage of this approach, observable in policy dilemmas that have lasted to the present day, was that in order to keep these inefficient sectors afloat huge state subsidies had to be made available to them. The protection of political structures and the state bureaucracy also led to widespread corruption and stifled public debate. The gap between the reality of economic change and public aspirations for political reform was a major contributory factor to the tragic mishandling of the massive demonstrations in Tiananmen Square (in April–June 1989), which damaged both the domestic and the international standing of the Chinese leadership. The demands of the protesters were varied, but there is no doubt that they embodied concerns about the erosion of urban living standards by high inflation, as well as anger against official corruption.

The Government's reaction to the events of 1989 was to re-emphasize central control and to slow the pace of reform until the next breakthrough in the transition towards a market economy occurred in 1992. The intervening years of retrenchment were marked by a leadership struggle for power over the direction of economic policy. The collapse of the USSR in 1991 had come as a shock to the Chinese leadership, forcing it to reconsider its economic strategy. A conservative faction within the leadership, led by Chen Yun (one of the architects of economic construction in the 1950s and 1960s, and an advocate of economic pragmatism after 1978), advocated maintaining administrative and party control over the economy. This approach, however, was rejected by Deng Xiaoping, who reaffirmed his commitment to the establishment of a market economy with a strong private sector, in which the role of the CCP would be restricted to political and social control. In a manoeuvre typical of traditional Chinese politics, Deng Xiaoping rallied support from provincial leaders during a visit, in early 1992, to southern China. Later that year, the 14th National Congress of the CCP gave official sanction to Deng's calls for accelerated market reforms and economic liberalization through its formal advocacy of the creation of a 'socialist market economy'.

The response to Deng Xiaoping's reaffirmation of economic reform was astonishing. Domestic and foreign investment surged, and average annual GDP growth rose from 5.7% (in 1989–91) to 13.4% (in 1992–94). At the same time, however, extra-budgetary finance and uncontrolled monetary expansion generated renewed inflationary pressures, reflected in the annual growth of consumer prices, which rose from 3.4% in 1991 to 24.1% in 1994. From mid-1993 Zhu Rongji, as Vice-Premier in charge of economic affairs, assumed responsibility for combating a rapidly overheating economy, and his success in effecting an economic 'soft landing' in 1996 was instrumental in his promotion to the premiership in 1998. By 1997 the annual increase in the consumer price index (CPI) had fallen back to 2.8%, while GDP growth remained buoyant at 8.8%. In the following year consumer prices actually declined by 0.8%, alongside an economic growth rate of 7.8%. Meanwhile, the 15th National Congress of the CCP in 1997 established new guide-lines for the corporatization of government functions, redefined the notion of public ownership and called for accelerated restructuring of the SOEs and complementary reforms of the social security and financial systems. Favourable international economic conditions (at least until 1997), huge foreign-exchange reserves and a strong export performance provided a firm foundation for these decisions. However, the impact of the Asian financial crisis and declining domestic demand at the end of the 1990s necessitated an extension of the original timetable for the completion of these reforms. Decelerating economic growth, averaging 7.7% annually during 1999–2002, was also a source of concern in the face of increasing urban and rural unemployment and underemployment, and associated widespread social unrest. Between 1998 and 2002 China experienced virtually zero inflation (on average, the CPI declined by 0.4% annually during this period). In the face of these deflationary pressures and against the background of quite weak consumer demand, the Chinese Government resorted to a strong fiscal stimulus, relying on large-scale bond issues to maintain the growth momentum. Towards the end of 2003, however, the CPI began once again to increase. The rise of 1.8% in October 2003 was the highest since 1997, and by the end of the year the monthly rate of increase had accelerated to 3.2% This trend persisted into the first half of 2004, during which the increase in the CPI reached 3.6% year-on-year. The re-emergence, after so many years, of inflationary pressures generated worries that renewed GDP growth was causing the economy dangerously to overheat. In fact, such extreme concern was an exaggerated response to these latest developments. The rate of consumer price inflation in 2003 (1.2%) or in the first half of 2004 hardly compared with the peak of 24.1% reached in 1994. Moreover, the rising CPI mainly reflected increases in the prices of food (both grain and non-staple products) and residential housing—a pattern that contrasted with continuing deflationary pressures in the manufacturing sector, associated with industrial overcapacity. Given the countervailing tendencies of such domestic forces, the emergence of macro-inflation reflected the impact of an additional external factor: the existence of large-scale speculative inflows of capital. From the perspective of mid-2004, it seemed likely that intensified inflationary pressures would elicit a policy response embodying a familiar combination of tighter control over both domestic bank lending and foreign capital flows.

Associated with accelerated GDP growth since mid-2003 has been excessive investment, especially in heavy industry. By the end of 2003 fixed asset investment constituted a record 43% of GDP, and in January–June 2004 it continued to expand rapidly. This rising trend has exacerbated already serious resource shortages. In 2003 there were power shortages in 22 provinces, and starting late that same year, a number of provinces were forced to impose power cuts as a result of electricity shortages. In mid-2004 China was reportedly experiencing the worst energy shortages since the 1980s, with street lighting being switched off and some factories being forced to halt production. Meanwhile, imports of oil, iron ore and other raw materials have risen sharply. Given that China's consumption of iron and steel, coal and cement accounts for 25%, 30% and 50% of global consumption, it is not surprising that such developments threaten to affect international, as well as domestic, raw material prices. At home, for example, the prices of steel, aluminium, copper and coal all increased sharply during 2003.

Although China's economy has changed dramatically since 1978 under the impact of radical reforms, it remains an economy in transition, in which many important decisions are still taken by state officials, whether at the local or central level. Difficult reforms—especially those of providing the legal, administrative and regulatory framework of a modern economy—still lie ahead, the successful implementation of which presents major challenges. Economic issues are meanwhile closely linked to political developments and power struggles within a system that is far from transparent to outside observers. Even allowing for the smooth transition to the 'Fourth Generation' of leaders under Hu Jintao, and the strong technocratic credentials of the new leadership, the Chinese economy remains heavily influenced by politics. In the absence of a strong institutional and legal framework, new protagonists (including bureaucratic and entrepreneurial élites at central and provincial levels) will continue to emerge and exert considerable power. In particular, as control over industries and enterprises continues to be devolved from central to regional and local authorities, protectionist and expansionist interests will no doubt still prove to be potent driving forces. Meanwhile, the central Government has asserted its authority through several campaigns directed against corruption by local officials. In a move that is indicative of a more sophisticated attitude towards control and distribution of information, Beijing has also actively embraced internet-based technology to disseminate information for domestic and foreign consumption with an extensive range of government-sponsored websites.

In recent historical terms, China's growth performance since 1978 is unprecedented, GDP having expanded, on average, by 9.4% annually in 1978–2003. Compared with that of other reforming countries (most notably the former USSR), an outstanding feature of China's market-orientated economic reforms is that having started earlier than in most other former socialist economies, they have achieved greater success despite (some would argue because of) being implemented without accompanying profound political changes. World Bank data show China to have been the fastest-growing low-income country in the world since the early 1980s. The Chinese economy is now second only to that of the USA, in terms of domestic purchasing power, and is projected to overtake the USA in about 2020. In sharp contrast to the Mao era, post-1978 GDP growth has been accompanied by major improvements in living standards, in which both the urban and rural population have shared.

Nevertheless, there are qualifications to this generally buoyant picture. One is that average per caput income in China remains low. Another is that even allowing for major progress in poverty alleviation, up to one-third of the population still lives beneath the poverty line, as defined by the World Bank. The same source indicates that labour productivity is among the lowest in the world, being only slightly higher than in India and a mere 10% of the US level. Finally, although the impact of economic reforms on absolute material consumption has been positive, the distribution of such gains between regions and sectors has been very uneven. Indeed, investigations of national income distribution show China to be one of the most unequal societies in Asia, and the gap between urban and rural incomes is extremely high, as measured by international comparative standards. Nevertheless, China's post-1978 growth record is, in general, very impressive. Having overtaken its former mentor (the former USSR), China has become a major industrial power, standing alongside Japan within the Asia-Pacific Economic Co-operation (APEC) forum. In addition, after more than four decades of forced-draft industrialization under a highly protective trade regime, China now possesses an independent, comprehensive and largely self-sufficient industrial system of considerable technological sophistication. In this respect, China's recent record is unmatched by any other country in the Asia-Pacific region, except Japan.

Under the impact of reform, major structural alterations have also taken place, as shown in the changing distribution of GDP and employment between primary, secondary and tertiary sectors. From a high of 33% in 1983, agriculture's contribution to GDP had contracted to 14.8% by 2003. The most significant expansion has been that of the service sector, its output share having risen from a low of 21.4% in 1980 to 32.3% in 2003. As for secondary sector activities, their share of GDP initially contracted from 48.2% to 43.1% in 1985, before recovering and increasing to 52.9% (around 45% from manufacturing alone) in 2003. Such changes are reflected in employment trends: since 1978 a contraction in the primary sector's share of total employment from 70.5% to 49.1% has been offset by rises in the corresponding figures for industry (from 17.3% to 21.6%) and, above all, for services (from 12.2% to 29.3%). Perhaps the most radical structural change of all has been the establishment of the 'non-state' sector of the economy (comprising private enterprises as well as enterprises of various intermediate ownership forms). Its size and characteristics vary locally. Unclear ownership categories and the persistence of administrative links between local authorities and enterprises make it impossible to measure the exact size of this sector. What is clear is that it has overtaken the SOE sector in terms of industrial output value (accounting for 58% in 2002), although not in terms of fixed investment (28%) or of employment (two-thirds of all urban employment was in the state-owned sector in 2002). In short, the non-state sector is playing an increasingly important role in creating a private market economy and in absorbing excess labour from the agricultural sector and from urban areas. The shape of China's future market economy will be characterized by locally varying forms of co-existence of an expanding private sector and a reformed SOE sector.

Under the Tenth Five-Year Plan (2001–05), government spending has been primarily directed towards infrastructure and building construction. In particular, a large proportion of state investment has been allocated to infrastructure projects in the western regions of China. This major strategic initiative—the 'Opening of the West'—seeks to transform the economic backwardness of 12 provincial economies, stretching from Inner Mongolia (Nei Mongol) in the north-west to Guangxi Zhuang in the south-west, that have benefited least from post-1978 reforms. Whether this policy will succeed in fulfilling its goal, or whether national economic and social development would have been better served by following the dictates of comparative advantage and directing expenditure towards coastal-based consumer-orientated industries, remains to be seen. What the western initiative demonstrates beyond doubt, however, is the central Government's commitment to trying to reduce potentially destabilizing regional differentials. Meanwhile, the new 'Fourth Generation' leadership faces the challenge of how to co-ordinate the implementation of enterprise, financial reforms and social improvements begun in the 1980s. Considerable success has been achieved towards reducing the major economic burden associated with a preponderance of loss-making SOEs that have traditionally been dependent on state investment and supported by loans from an increasingly heavily indebted banking sector. Thus, comprehensive reform of the SOEs is inextricably linked to reform of the financial sector, which is burdened with the accumulated debt of SOEs. Without reform of the banking and financial sector, China faces a potentially serious economic crisis. Social reform is a further precondition of SOE reform, highlighted in the need to put in place an effective urban social security network in order to provide workers and their families with basic social services (including pensions, unemployment insurance, health care and housing), formerly provided by the state enterprise sector.

One of the last major policy initiatives taken by the 'Third Generation' leadership under Jiang Zemin was to endorse the admission of private entrepreneurs to membership of the CCP. It is too early to judge the extent to which the Party leadership under Hu Jintao and the Government under Wen Jiabao will further encourage and facilitate radical departures from the policies pursued by Jiang and Zhu Rongji. One important development was the change to the wording of the Constitution, announced in March 2004, which gave formal endorsement to state protection of the 'lawful rights and interests of individual and private economies', representing the first legal protection of private property rights in China since 1949. That the new leaders are committed to the centrality of rapid economic growth to China's continued development and stability is not in doubt. However, there is some evidence that the new leadership may be prepared to accept slightly slower GDP growth in the interests of a more economically and socially sustainable pattern of development. From this perspective, it is significant that as someone who belongs to the reformist end of the policy spectrum, Wen Jiabao favours accelerated privatization in addressing the problems of SOEs, has a strong sympathy with farmers and the need to increase their incomes, and advocates an expansion of domestic demand in order to stimulate further economic growth and reduce income inequalities. None of these policy emphases implies, however, any reduction in the commitment to China's increased role in the international economy (not least, through its membership of the WTO).

The first serious—and unexpected—challenge faced by the new leadership was that posed by the outbreak of Severe Acute Respiratory Syndrome (SARS) in the first half of 2003. Quite apart from its major social impact (including the deaths of more than 340 Chinese citizens), the disease also caused serious economic dislocation. The short-term impact manifested itself in a variety of ways, including a contraction in domestic trade, a reduction in domestic consumer spending and a decline in income from tourism. The disruption of travel to China also caused the postponement, although not the cancellation, of some FDI agreements. Employment too was affected: the potential for job creation in services, small-scale enterprises and the private economy—all of which were badly affected by SARS—was seriously constrained; the appointment of university graduates to new jobs was delayed; and up to 10% of the 100m. farm migrants working in cities and economically advanced regions returned to rural areas as a result of the epidemic. The longer-term impact was, however, less severe than anticipated, and although growth fell back sharply in the second quarter of 2003, subsequent recovery enabled the rate of GDP expansion for the year as a whole to reach its highest level since 1997.

POLITICAL AND ECONOMIC REFORMS

To argue that China's economic reforms have been unaccompanied by systemic political reform is not to suggest the complete absence of political change. The CCP still defines the public political agenda, but its role in shaping the gradualist and experimental approach to economic reform in China has encouraged a shift away from communist ideology towards economic and social policies that share some commonalities with a system of social democracy. In the transition from administrative interventionism to more indirect macroeconomic control by the central Government, the Party has relinquished much of its former influence over the economy. What remains of the Party's orthodox tradition is its opposition to open public debate, individual rights and a democratic, parliamentary system, as well as its control over public security and strategic ('pillar') sectors of the economy and State. Since electronic media, such as the internet, threaten the state monopoly on information, the official response has been actively to embrace the new technologies and make government institutions the leading providers of web-based information, while taking selective action against the new media. At the same time, in its efforts to combat widespread and endemic corruption, both Party and Government need public media to project and maintain credibility. The tension between pluralistic tendencies engendered by the market economy and the Party's tradition of ideological control of all political discourse remains unresolved. Economic reforms were introduced in the face of political opposition, both from within the CCP and from other constituencies (for example, the military, state bureaucracies, large SOEs and some provincial governments) with vested interests in preserving a strong role for the central State. The experimental nature of the economic reform process derives, in part, from the need for reformers within the Party to convince their conservative opponents of the merits of the various reforms. When opposition to reform has made this impossible, ideological compromises have emerged to generate formulations such as 'primary stage of socialism' or 'socialist market economy'. Such uncertainties notwithstanding, the core issue throughout has been the extent to which the State might be prepared to relax its control over the economy and society. All of the major macroeconomic reform initiatives, affecting central planning, prices, fiscal and monetary policies, enterprise behaviour and administration, began in the early 1980s, but gave rise to heated debate that has still not been finally resolved, owing to continuing opposition from the proponents of a strong interventionist state.

The 'Third Generation' leadership that took office in 1997–98 after the death of Deng Xiaoping had witnessed the inadequacy of administrative controls to manage the economy during the recession that followed the Tiananmen Square massacre. These leaders were shocked at the collapse of the USSR and had seen at first hand how Deng Xiaoping's reaffirmation of reform in 1992 had precipitated economic boom conditions that strengthened the movement towards marketization. As the economy became increasingly 'overheated', they proved themselves adept in relying more on macroeconomic policies than on administrative intervention to bring about a 'soft landing' and promote a more stable path of development. Even so, Party and State have maintained a degree of administrative and corporate control over the economy through corporatization of state institutions and SOEs, including the listing of majority state-owned corporations on domestic and international share markets. Another important initiative took place in 1998, when China's economic leaders, headed by Premier Zhu Rongji, a trained economist, abolished the central ministries that had been in charge of whole industries and replaced them with state-controlled corporate structures. For the foreseeable future, the State's retention of a strong role in preserving social stability and public confidence in the economy is likely to remain.

Legal and Administrative Reform

The introduction of the rule of law is another area of reform that has transformed the Chinese economy, as well as social and political life. Legal reform was driven by domestic and external pressures on China to establish procedural and other norms that accorded with standard international practice and that served the needs of China's market-orientated path of development. Economic activity is now regulated by a vast body of economic legislation, although China's legal system is better described as 'rule by law' than 'rule of law'. Among the earliest economic legislation passed in the reform period was the Joint Venture Law (1979), which established the legal basis for foreign investment in China and was later used as a model for the development of parallel domestic legislation (e.g., the Contract Law, and the Company Law (1994), which for the first time introduced limited liability and share-issuing companies). There now exists a substantial body of codified commercial law, regulating the activities of domestic and foreign-invested enterprises. Some progress has been made towards the legal protection of intellectual property rights, although there remain weaknesses in its implementation. There is also a Civil Procedure Law, and litigation has become an increasingly common (if not always effective) means of conflict resolution, in addition to the still important methods of arbitration or informal conflict resolution.

Recent reforms of the central ministries and commissions have changed government administration. In 1998 restructuring of the State Council resulted in a contraction in the number of central ministries and commissions from 40 to 29, and in the central government civil service establishment by one-third (in some individual ministries, by more than one-half). Industries such as coal, power, metallurgy, machine-building, electronics and chemicals were removed from central bureaucratic control and placed under the supervision of more general supervisory organs, with their operational activities being administered by newly-corporatized structures. Such changes, which are being extended to provincial and local levels, are intended to reduce remaining government monopolistic powers and introduce industries to competition.

Regionalism

The People's Republic has a unitary system of government that co-exists with strong and mostly informal federalist features. Under the central Government, there are four more levels of government, embracing 31 units at provincial level (22 provinces, five autonomous regions and four municipalities), 333 at prefectural level, 2,055 at county level, and over 44,000 at township level. They are structured in the form of a 'nested' hierarchy in the sense that higher levels of government deal only with directly subordinate levels and that units at every level have independent budget authority. Under the impact of reform, provincial autonomy and regional differentiation have expanded, with coastal provinces—especially those constituting the three major growth hubs of the Pearl and Yangtze River Deltas and the Bohai Gulf (centred on Beijing and Tianjin)—benefiting disproportionately from their better industrial infrastructure and access to markets. GDP in the more developed, highly industrialized coastal areas is much higher than in central or western regions. In 2003 the combined GDP of the four largest provincial economies (Guangdong, Jiangsu, Shandong and Zhejiang) accounted for almost 41% of national GDP, and for more than 45% if including Shanghai. The contribution of Guangdong alone was 11.4%. Provinces not only differ in terms of comparative and competitive advantages, but they have also pursued different privatization policies that have generated distinctive regulatory and business environments (for example, coastal provinces like Guangdong and Zhejiang have a much higher proportion of private enterprises than other provinces). Competition for foreign investment has disproportionately favoured coastal regions; in 2003 just six areas (the provinces of Guangdong, Jiangsu, Shandong, Fujian and Zhejiang, along with the municipality of Shanghai) absorbed 70% of all utilized FDI. Meanwhile, formal and informal trade barriers still impede interprovincial economic exchange, although China's WTO membership will help reduce internal trade restrictions and expand domestic trade and competition. In this respect, it is significant that there is evidence that the immediate interior neighbours of some coastal provinces—Anhui is one example—have begun to benefit from increased FDI flows. In some cases, improvements in transport links have enabled industrial plants to be shifted out of coastal boom areas, such as Guangdong, Zhejiang, Jiangsu and Shandong, to their inland neighbours in order to take advantage of lower labour costs.

Growing social and economic disparities between China's coastal regions and the interior have become a source of great concern to the central Government, not least because of their potential impact on social and political stability. The central Government's declining fiscal authority has limited its ability to enforce equalizing inter-regional revenue transfers between rich and poor regions. Under the Ninth Five-Year Plan (1996–2000), voluntary partnership schemes, involving capital transfers and cross-regional co-operation, sought to combat widening differentials, but to little effect. Hence, the formal adoption, during the Tenth Plan, of the strategic initiative to 'open up the West' in an attempt to strengthen the economic integration, improve the ecological environment and raise the development level of China's 12 western provinces. The scope of this programme is reflected in the huge scale of projected infrastructural investment in developing road and rail links, natural gas pipelines and communication links between less-developed western regions and prosperous coastal provinces. Major funding is also being made available for environmental development, through large-scale afforestation and water diversion projects (especially the massive 'South–North Water Transfer Scheme', designed to obviate critical water shortages in North China).

Fiscal Reform

Fiscal reform, like price reform, was necessary to introduce market incentives into the enterprise sector. Until the first tax reform in 1983, government revenue consisted primarily of the financial surplus of the SOEs, which had to be remitted in full to the State. Tax reform in 1983, under the slogan 'tax for profit', introduced enterprise taxation and allowed enterprises to retain part of their profits as an economic incentive. In a major reform in 1984 value-added tax (VAT) was introduced in an attempt to simplify the tax regime. Taxation was, however, not unified, and different tax rates applied to SOEs, collective enterprises and foreign-funded enterprises. Lower tax rates were used as an incentive to attract foreign investment in the 1980s. A further series of tax reforms in 1994 introduced personal income tax and a uniform corporate taxation rate of 33% for all enterprises, including foreign-invested enterprises. Tax incentives already granted to foreign investors were, however, excluded and foreign-invested enterprises continue to benefit from tax incentives. VAT, at a rate of 17%, was also extended to cover all foreign and domestic enterprises. This tax reform was a response to the decline in total state income and a shift of income from central to local governments. Government revenue had declined from 31% of GDP in 1978 to less than 11% in 1995. Subsequently, in the second half of the 1990s, government revenue as a percentage of GDP once more increased, reaching 15% in 2000 and over 18.5% at the end of 2003. In 2003, revenue rose sharply, thanks to a 20% rise in tax receipts.

The 1994 tax reforms also addressed the decline in the central Government's share of tax revenue, which had been strongly in evidence since the mid-1980s (falling from 38.4% in 1985 to 22% in 1993). Under the new system of revenue-sharing, central and local governments were each given their own separate administration for the collection of central and local taxes. The impact of the initiative was dramatic, and in a single year (1993/94), the central Government's share of budgetary revenue increased from 22% to 55.7%. Although this figure thereafter fell back slightly, it subsequently recovered and reached 57.5% in 2003. Under the tax-sharing system, the central Government's main sources of revenue in 1999 were, in order of importance, a 75% share of VAT revenue, central consumption tax receipts, tariffs, custom tariffs and VAT on import goods and corporate income tax. Local governments were entitled to the proceeds of local personal and corporate income tax, 25% of VAT revenue, the receipts from taxes on urban land and other miscellaneous taxes, although the majority of their revenue derived from extrabudgetary items and administrative charges that were not part of regular government revenue. China's tax-sharing system is complicated by rebate schemes from the central Government to local governments and by inter-governmental transfer mechanisms.

A report by the World Bank in 2002 noted that the heavy expenditure responsibilities of local governments in China were out of line with international practice, and drew special attention to its highly decentralized fiscal system, under which sub-national governments accounted for more than 70% of total budgetary expenditure and over two-thirds of extra-budgetary funds. Erosion of state revenues, the limited ability of the central Government to exercise control over the economy through budgetary tools and the unequalizing effects of inter-regional transfers remain unresolved problems of financial reform. Increasingly, fiscal policies have been used to raise domestic demand and exports in order to stimulate the economy. From 1999 such measures included reduced tax rates on capital and real-estate investments, tax exemptions for investment in western China, increased tax reimbursement for exports, and higher taxation of private savings deposits to encourage consumption. Such forms of deficit spending continued into 2003.

In general, expenditure has grown faster than revenue since the beginning of the 1990s, as a result of which China's overall budget deficit increased from 15,890m. yuan in 1989 to a record 314,900m. yuan in 2002, before falling back to 291,600m. yuan in 2003. Between 1997 and 2002 the budget deficit rose almost five-and-a-half fold, from 0.8% to 3.0% of GDP, giving rise to warnings that government liabilities had reached a level that threatened to precipitate a budgetary crisis. In 2003, however, fiscal balance fell to 2.7% of GDP, a figure that was projected to decline to 2.5% in 2004. Meanwhile, estimates of government debts—including bonds issued by the Treasury, the policy banks, asset management corporations and costs for refinancing the heavily indebted state banks (see below)—have ranged from 100% to 150% of GDP, roughly comparable to Japan and significantly higher than the international debt-to-GDP ratio of 60%, which is generally seen as the highest acceptable 'ceiling'. In the longer term, the burden on the Government to provide funding for pensions and other social security payments for an increasingly ageing population will rise sharply.

POPULATION, LABOUR AND SOCIAL SECURITY

The history of China's population growth since 1949 may be divided into two main periods. Between the early 1950s and the early 1970s total population expanded rapidly, with annual growth averaging more than 2%. By contrast, subsequent years, until the present day, witnessed unprecedented success by the Government in limiting such growth. Such trends highlight the impact of revisions in official policy on demographic change. Rapid population growth after 1949 reflected a sharp reduction in the death rate, associated with major improvements in basic health and hygiene, as well as an increase in the birth rate that was facilitated by a deliberately pro-natalist stance. The radicalism of the Great Leap Forward and the Cultural Revolution effectively prevented the implementation of population control policies, and it was only in 1972 that the Government began to formulate and implement effective family-planning policies, designed to control the number of births. Early efforts sought to restrict family size through late marriages, less frequent births and fewer births, but these were overtaken, in 1979, by advocacy of the one-child family. The new policy attracted strong resistance—especially in the countryside, where a marked cultural preference for boys persisted—which took various forms (including the under-reporting of female births, the use of ultrasound scans to facilitate abortion of female foetuses and female infanticide). China's sex ratio remains abnormally high, the 2000 census showing 106.74 males for every 100 females nationwide (but 120.17 for 0–4 year olds). In recent years there has been a relaxation of the one-child rule, although it remains the norm in urban areas. Meanwhile, official orthodoxy insists on the validity of China's draconian demographic policies, arguing that in the absence of such policies, some 250m. more births would have occurred in the last quarter of the 20th century.

Projections suggest that even without any further decline in fertility, China's total population will have reached 1,350m. by 2010. Meanwhile, in 2003 the total population was 1,292.27m., of whom 40.5% were officially classified as urban. Vital rates for 2003 show crude birth and death rates of 12.41 and 6.40 per thousand, giving a rate of natural increase of 6.01. China's total population is expected to peak in around 2040–50, when it will reach between 1,500m. and 1,600m. people. China's international success in controlling births is highlighted in the finding that around the middle of the 21st century, India's population will overtake that of China to become the largest in the world. Since the 1950s the share of children in total population has steadily declined, while the problem of an ageing population has yet to emerge. As a result, the proportion of working-age people had risen to reach 70.4% by 2003—a figure that emphasizes the huge burden faced by the Government in seeking to achieve full employment. In general, employment generation has declined quite sharply since the early 1990s: between 1980 and 1992 (the year in which GDP growth peaked), the rate of job creation increased annually by 3.8%; thereafter, until 2003, the corresponding figure was a mere 1.4%. Nevertheless, it seems that China's momentum of GDP growth since 1978 has made it relatively easy to provide jobs for new entrants to the labour force. The real, and very formidable, challenge facing the Government lies in the pressure on urban employment resulting from the loss of millions of jobs caused by SOE restructuring (in 2003 the official rate of urban unemployment was 4.3%, although the real figure was closer to 10% and may have been even higher); and on rural employment from the existence of at least 130m. surplus farm labourers. It is against this background that rural–urban migration on a huge scale in recent years should be interpreted. One of the ironies of post-1978 reforms is that despite having created many tens of millions of new jobs (more than 100m. in rural industrial enterprises alone), such is the scale of the problem that as many jobs again are needed, if the economic and social consequences of large-scale urban unemployment, rural underemployment and rural–urban migration are not to worsen.

Resolving China's employment problems is closely linked with the continuing reform of the social security system (including old-age insurance, medical insurance and unemployment insurance), all of which have been established nationally but are based and managed at provincial level. Social security provision has improved significantly in the urban sector, but is still almost totally lacking in the countryside. The transfer of former central industry-based pension funds and responsibilities to provinces has been completed, but the coverage of the system remains limited (although participation in basic health insurance, based on shared contributions by government, enterprise and individual, has risen from 18.78m. in 1998 to 108.95m. in 2003—equivalent to 18% of the urban population). Unemployment insurance has traditionally been the State's responsibility (exercised through SOEs), althoughthe financial burden is increasingly being shared by government, company and employee. At the end of 2003 103.73m. employees and workers participated in unemployment insurance programmes, of whom 4.15m. were in receipt of payments (250,000 fewer than in 2002). In the same year 154.9m. people (116.38m. employees and 38.52m. retirees) were enrolled in basic pension programmes (38% more than in 1998). Some 22.35m. urban residents were in receipt of minimum income relief from the Government—1.7m. more than in 2002.

AGRICULTURE

In line with the changing nature of the Chinese economy since 1978, the role of the farming sector has contracted relative to industry and services, and agriculture's contribution to GDP growth has become more marginalized. Farming itself has become more diversified, and as a result of growing affluence (especially in cities and coastal provinces) non-cropping activities, such as fishing, husbandry and fruit-farming, have grown much more rapidly than grain-farming, which previously had overwhelmingly dominated agriculture. In terms of employment, however, the changes have been less profound: even today, crop cultivators still account for well over 95% of all farmers. The changes in China's agricultural economy are largely, but not solely, attributable to the implementation of reforms, which have freed farmers from the previous constraints of a rigid planning regime. Between the late 1950s and the end of the 1970s, the three-tier collective institutional framework of communes, brigades and production teams had regulated agricultural production to the detriment of individual incentives and efficiency in resource use. As a result, food consumption levels remained low, and the margin of subsistence narrow. Post-1978 reforms addressed such problems by sanctioning *de facto* decollectivization through the introduction of a contract responsibility system, which required producers to deliver fixed-price quotas to the State, but enabled them to sell their surplus to the market at a higher price. Not only did farmers acquire increasingly long-term land use—though not ownership—rights, but by being given greater decision-making powers, they were also encouraged to diversify their operations into cash crop farming, animal husbandry, fishing or forestry. Between 1979 and 1984 these institutional changes, accompanied by increases in purchase prices, generated unprecedented agricultural growth and resulted in a doubling of average per caput farm income. Farmers, who had long been disadvantaged by unfavourable inter-sectoral price terms of trade, were the beneficiaries of these measures, while urban consumers received state subsidies to reduce the negative effect of the rising prices. Meanwhile, private local and inter-regional markets for vegetables and other produce opened. The prices of many products, such as meat and fish, which had either been rationed or kept under state control, were also gradually liberalized. In recent years, under the impact of reforms designed to separate government administration from enterprise functions in the purchase, sale and storage of grain, the responsibility for grain procurement has also been decentralized. In addition, reforms of the grain market (for example, in 1999–2000) have sought to reduce the losses of grain-handling enterprises through greater emphasis on commercialization. Thus, by 2000 even a basic foodstuff like rice was beginning to be 'marketized'.

Since the mid-1980s animal husbandry and aquaculture have registered consistently rapid and sustained growth, although the performance of the crucial grain sector (including cereals, pulses, soybeans and tubers) has been more mixed. Total grain output rose from 305m. metric tons in 1978 to a then-record level of 407m. tons in 1984 (this level was not, however, reattained until 1989). The first half of the 1990s showed no clear trend, leading some to argue that China would be forced to import increasingly large amounts of grain to feed its growing population. Such predictions appeared to be confounded by a series of fine harvests, which took average annual output to 505m. tons during 1996–99 (significantly above the critical 400 kg per head level that is considered sufficient to meet the dietary demands of China's increasingly affluent population). This surge, however, was not sustained, and grain production once again declined and stagnated, averaging only 451m. tons during 2000–03. The 2003 harvest was just 430.7m. tons—almost 78m. tons less than in 1999 (the difference between 505m. and 431m. tons of grain would feed around 185m. Chinese, at current rates of consumption). As a result, average grain output per head in 2003 (333.3 kg) was well below the 400 kg benchmark and lower than in any year since 1981. The sequence of poor harvests since 2000 has affected both China's domestic economy and its foreign economic relations. In late 2003 grain shortages were finally translated into rising prices, especially of maize and wheat. In November wheat prices increased by 32%, year-on-year, in Heilongjiang; in Hebei Province, the corresponding figure for maize was 40%. Such increases set off a chain reaction in farm product industries. Above all, with expenditure on food accounting for 46% of average total consumption spending by rural residents (compared with 38% for their urban counterparts), the grain price rise had a major impact on the CPI, reversing five years of zero inflation. Moreover, with planned grain output for 2004 set at a modest level (455m. tons), it seemed unlikely that the rising trend would be reversed. Given that this fall in China's grain production began in 2000, it was perhaps surprising that upward pressure on prices did not occur earlier. Interestingly too, as of the end of 2003, China had not yet resorted to increasing its cereal imports (having consistently remained a net exporter of cereals—rice, wheat and maize—since 1996). Even in 2003, with per caput grain output at the lowest level since the early 1980s, China's net exports exceeded 10.5m. tons, from which it earned over US $1,000m. Such apparent contradictions are resolved by taking account of changes in domestic grain reserves that had meanwhile taken place. This running down of stocks could not, however, continue indefinitely, and in 2004 the signing of large-scale cereal import deals with the USA indicated that China would revert to becoming a net cereal importer. Indeed, it is possible that the Government may adjust its long-standing policy imperative and reduce its grain self-sufficiency criterion from 95% to 85%–90%. To do so would confer major economic and welfare benefits on the rural population without exposing the country to any significant danger resulting from other countries—notably the USA—seeking to use the grain trade as a means to bring political pressure to bear on China. Accession to the WTO is also likely to underline the shift towards increasing dependence on grain imports, as farmers are forced to face growing competition from more efficient foreign producers.

In 1978 farming was overwhelmingly dominated by grain production (the area under grain accounted for 80% of the total sown area, while grain farming contributed more than 70% of the total value of farm output). The 1980s and 1990s witnessed a rapid diversification of farming, exemplified above all by the rapid growth of animal husbandry and fishing. By 2002 the contribution of grain to agriculture's gross value output (GVO) had contracted to 54%, while the corresponding figures for livestock farming, aquaculture and forestry were 31%, 11% and 4% respectively. Such trends reflect the demand for a more diversified diet by increasingly affluent consumers (especially in cities) and highlight the growing emphasis on grain for indirect uses (feed, inputs for processing, etc.). Concealed within these figures is also the declining proportion of agriculture in overall rural production, as well as in national GDP, although it deserves emphasizing that in 2002, although 133m. workers were employed in rural township and village enterprises (TVEs—see also below), farming still absorbed half of all employment throughout the economy.

Against the background of official government recognition of lagging economic and social development in many parts of the countryside, the role of agricultural and rural development in generating a prosperous society emerged as an important policy theme in the first half of 2004. Indeed, for the first time since 1986, agricultural and rural issues were the subject of a major policy document, which sought to place the 'three rural issues' (namely agriculture, peasants and the rural sector) at the centre of the Party's work agenda. To this end, new policy measures, including farm tax reductions, were announced, designed to address lagging farm income growth by extending more support to the agricultural sector (especially to grain farmers). The document also demanded accelerated reform of the land requisition system. Against the background of farm incomes having grown at half the rate of urban incomes during 1997–2003, it highlighted agriculture's central importance in the stark statement that ' if the peasants cannot increase their income and the rural areas are unstable, the development of the national economy and social stability will lose their essential meaning and motive force'—a statement that underlined the political, as well as economic and social, ramifications of slowing farm growth.

INDUSTRY

The sources of China's modern industrialization are to be found in developments along the coast and, later, in the north-east, driven by western imperialist powers and Japan. Starting in the second half of the 19th century, mainly light industries were established in the Treaty Ports, notably in Shanghai and Tianjin. One-third of the total of China's manufacturing output was produced by foreign-owned factories in these coastal areas, although linkages between modern industry and the rest of the economy remained quite limited. Before the outbreak of war in 1937, modern industries produced less than 4% of net domestic product and employed fewer than 2m. workers. Two-thirds of the gross value of national industrial output was produced by handicraft industries and only one-third by industries that relied on machinery for their main production. Heavy industry on a larger scale developed in Manzhou (Manchuria) during the Japanese occupation after 1931. Such initiatives notwithstanding, when the CCP came to power in 1949, only 5% of all factories throughout China employed more than 500 people, and the weak industrial base remained concentrated in the north-east and a few coastal centres.

This low industrial base facilitated the achievement of high growth rates after 1949. With crucial aid from the USSR, in 1953 (the start of its First Five-Year Plan—FFYP) China embarked on a large-scale programme of industrialization, the intended focus of which, for both economic and defence reasons, was on investment projects in the interior regions. The new industrial infrastructure was overwhelmingly orientated towards heavy industries, including mining, chemicals, machine-building and metallurgy. The FFYP's impact was dramatic, with industrial GDP growing at an unprecedented rate, by almost 20% annually. Inland industrialization continued, though at a slower pace, after China's break with the USSR in 1960, when the latter withdrew its experts and advisers. The shift away from the coast intensified after the mid-1960s, when military and strategic reasons encouraged the establishment of a self-sufficient industrial economy in inaccessible western regions (the so-called 'Third Front') in order to protect China against a possible US and/or Soviet nuclear strike. Such activities, and associated economic and social infrastructural construction, were hugely wasteful, absorbing around half of the central budget during the decade of the Cultural Revolution (1966–76). In addition, as a result of the isolationist policies pursued during this period, China's industrial infrastructure became obsolete, and the country's economic growth fell increasingly behind that of the Asian 'tiger' economies. As post-Mao initiatives were to demonstrate, systemic industrial reform was needed, involving the opening of China's economy to the outside world, implementing technical modernization, relocating industries closer to markets, reorientating production towards consumer demand, and reforming enterprise structures and macroeconomic policies.

The first step towards export-orientated, reform-driven industrial development—and the beginning of China's integration into the international economy—was designed to attract foreign investors (especially those living in the 'Greater Chinese diaspora') to establish joint ventures in China. The Joint Venture Law (1979), underlined by other legislation, created the basis for inflows of FDI, primarily in the SEZs. Three of the first four SEZs were set up in the vicinity of Hong Kong (the fourth was in Xiamen, across the Taiwan Straits from Taiwan), the largest being at Shenzhen, adjacent to the Hong Kong border. Thereafter, but especially in the years following Deng Xiaoping's 'southern tour' of 1992, FDI proved to be a major stimulus to industrial reform. By the late 1990s China had become the second largest recipient of FDI after the USA, and by far the largest among developing countries (see below). In 2003 foreign-invested enterprises—over 400,000 of them, with cumulative utilized investment of US $473,300m.—contributed about 30% of China's industrial GVO and over one-half of the value of its merchandise trade. Foreign investment has made available advanced production technologies and management expertise to China and has introduced competition into hitherto-protected domestic markets. It has also shifted China's industrial focus back to the coastal regions and to the centres of consumer demand. Finally, joint-venture legislation has provided guidelines for the formulation of other kinds of domestic economic legislation.

An important element of China's industrial reform strategy has been the establishment of market segments within the state-owned industrial monopolies, for example by stimulating the growth and diversification of consumer-orientated industries that would have to compete for markets. The development of light consumer industries during the 1980s required less investment than heavy industries and offered lower entry barriers for small-scale collective enterprises. These new forms of regionally-based, non-state enterprises successfully competed for raw materials and eventually forced existing SOEs out of whole industries. Nevertheless, even after two decades of reform, China's industrial structure is still dominated by heavy industry, a reflection of earlier policies of Soviet-style industrialization. In 2003, for example, heavy industry accounted for 64% of annual industrial value added. In 1952 industry (including construction) accounted for less than 21% of GDP, compared with over 50% for agriculture and 29% for services. Subsequently, heavy industry grew twice as fast as light industry, while sustained rapid industrialization raised the industrial share of GDP to 30% in 1957, 40% in 1970 and 48% by 1978—by which time heavy industry accounted for 57% of industrial GVO, and light industry 43%. In 2003 the industrial share of GDP had reached 52.3% (about 45% for manufacturing alone). State-owned industrial activities accounted for 47% of industrial value added (but more than half of that of heavy industry). Projections suggest that by 2005, industry's share of GDP will have risen to 55.5%, with agriculture and services accounting for about 12.5% and 32% respectively.

Non-agricultural, rural industrialization has been a defining characteristic of the Chinese development 'model' since the late 1950s, when small-scale enterprises were set up in the countryside in order to produce simple farm tools, agricultural inputs (especially chemical fertilizers) and raw materials, such as cement. Since the 1980s the scope of rural industrial activities has broadened immeasurably and, in particular, TVEs have made a major contribution to industrial growth. By 2003 TVEs had created almost 60m. jobs in industry, more than 12m. in construction and 37m. in services. They also produced almost 60% of industrial value added throughout the economy and generated about one-third of export earnings.

Pillar Industries

China's planning has traditionally been based on 'pillar industries', such as machine-building, petrochemicals, and construction and building materials, which benefited disproportionately from investment allocations. Recently, however, 'new' industries, such as information technology (IT), electronics and automobiles, have become the fastest-growing branches and have replaced traditional heavy industries to become the principal industrial pillars. The GVO of China's IT sector has grown spectacularly in recent years and the output of IT hardware reached US $49,100m. in 2003 (compared with US $62,500m. in the USA). In 2003 output of integrated circuits and microcomputers rose by 54% and 119.8%, respectively. The IT workforce numbers well over 6m. and has more than quadrupled since 1997. If employment in related fields is included, the figure rises to more than 16m., or about 7% of the urban labour force. In 2003 foreign sales of computers and telecommunications equipment increased by 68.6%, while exports of automatic data-processing machines more than doubled to reach US $41,100m. Chinese sources predict that on the basis of annual IT growth of 30%, by 2005 associated sales revenue will have reached some US $7,250m., to account for 3.0% of the global software market (compared with 1.2% in 2002). By the same year it is estimated that the IT sector will contribute about 7% of GDP. IT production is seen as a difficult environment for SOEs, and high-technology IT industries have particularly benefited from FDI and associated foreign technological expertise, especially from the USA and Taiwan. In 2003, for example, output by Taiwanese firms operating on the mainland is estimated to have accounted for more than 70% of China's IT production.

Around 80% of the world's top 500 companies have now established research and development centres in China, and it is estimated that 85% of new high-technology products are produced by foreign-funded enterprises. US products dominate the software market for word-processing, spreadsheets and graphical user interfaces. China's own software industry has been most successful in adapting software written in English for the Chinese market. Pirated software remains a major problem for the future development of this industry. The output of the machine-building industry almost halved between 1993 and 1998, although thereafter the downward trend was reversed, facilitating the attainment of a new peak level of production in 2002. Within this sector, growth has been strongest in agricultural machinery, especially combine harvesters and tractors, but has also been quite buoyant in the production of numerically controlled machine tools and hydroelectric equipment.

Following the restructuring of China's state-owned petroleum industry, there now exist three corporations. The largest of these is China National Petroleum Corporation (CNPC)—better known under the name of its listed subsidiary, PetroChina, which originally specialized in petroleum and gas exploration and production; its market capitalization was estimated at US $90,700m. in February 2004. Second in importance is China Petrochemical Corporation (Sinopec), the original focus of which was refining and distribution, with a market capitalization of US $51,900m. PetroChina and Sinopec have been transformed into all-embracing, regionally-based oil companies, CNPC operating mainly in the north and west, and Sinopec in the south. The third, and smallest, producer is China National Offshore Oil Corporation (CNOOC), which is responsible for China's offshore exploration and production, and the low cost structure of which has facilitated its major commercial success. Owing to high world oil prices, the petroleum industry was one of the profitable state-owned industries in 2001. In the latter part of 2002 a 'price war' between CNPC and Sinopec resulted in a decline in profitability in the sector, although mutual accommodation, underscored by upward pressure on world oil prices associated with tension in the Middle East, helped to reverse this trend in early 2003. Meanwhile, China's domestic market is expected to change as a result of WTO membership, as international oil companies enter the market. China's own oil companies are seeking to expand their co-operation with foreign partners, although there was a major set-back in May 2003 when CNOOC and Sinopec failed to secure a 16.7% stake, worth US $1,200m., in Kazakhstan's North Caspian Sea Production Sharing Agreement (NCSPSA). A successful conclusion of this agreement would have helped China in its aim of diversifying energy sources, as well as giving it a stake in one of the world's biggest oil (and gas) projects. PetroChina achieved greater success during 2003: in April it increased its investment in Indonesia, as well as securing permission from Russia to build a pipeline from Siberia to China; and in the following month it announced that it had reached agreement to double its output at a refinery in Sudan. CNPC has also signed agreements with companies in Venezuela, Peru, Canada, Thailand, Myanmar, Turkmenistan, Azerbaijan and Oman. A proportion of the shares of subsidiaries of CNPC, Sinopec and CNOOC are listed on stock exchanges in Hong Kong and New York.

The motor vehicle (especially car) industry is another 'pillar sector' that has experienced very rapid growth under the impact of reform. Car production expanded by more than 27% annually between 1980 and 2002. Following 55% growth in 2002, the rate of expansion in 2003 was an astonishing 85% (from 1.092m. to 2.02m. units). Through the first half of 2004, domestic demand remained buoyant, although there were signs of weakening sales in the face of credit restrictions. Even so, with levels of car ownership among the lowest in Asia (a mere 0.88 per 100 urban households in 2003), the potential for future growth is enormous, and projections indicate that car production will increase to 5m.–6m. units by 2010. If China is to benefit fully from this expanding market, it will need to consolidate its domestic industry, which currently comprises 120 car producers. Foreign companies (such as Volkswagen, General Motors (GM), Toyota, Nissan, DC-Hyundai and, most recently, BMW) are also active in China, and in the first quarter of 2003 Volkswagen reported that it had sold more cars in China than in its domestic market; its planned production capacity in China for 2005 is 1.6m. units. Profits made by GM from its activities in China were almost US $450m. in 2003.

Construction activity has benefited from state spending and from the government policy to privatize residential housing. GVO of construction grew consistently at between 6.5% and 6.9% between 1993 and 2002. In 2003 the value added of construction enterprises increased by 11.9%. In the same year real estate development grew by 29.7%. In recent years there have been warnings that a dangerous property market 'bubble' might be emerging. In particular, the price of land for residential use rose sharply during 2002–03. Real estate figures for the first half of 2004 indicated some success in curbing the increase in property prices through credit tightening. However, the picture remains quite mixed: in some cities (such as Beijing and Guangzhou) property prices were stable, or even declining, during 2003; but elsewhere (Shanghai, Ningbo, Hangzhou) property price inflation remained strong.

Enterprise Reform and Privatization

China has two categories of enterprises: state-owned and non-state—the latter comprising rural and urban collective enterprises, private enterprises and foreign-funded enterprises (including those funded by 'compatriots' in Hong Kong, Macao and Taiwan). Non-state enterprises operate in a market environment, while SOEs are still protected by the State through their access to bank loans (regardless of considerations of efficiency and economic performance). SOEs are the main source of the high level of NPLs in China's banking sector. Measuring the level of China's NPLs is a difficult exercise, but some have argued that, at their worst, they may have reached around 25% of GDP. In absolute terms, it is estimated that China's banking system has bad debts of up to 4,000,000m. yuan. Recent evidence does, however, indicate that conditions have improved: between 2001 and 2003, cancellations of debt and the rapid expansion of bank credit resulted in the share of NPLs in total bank loans falling from over 30% to less than 15%; during the same period, NPLs as a share of GDP declined from 23% to 17%. In January 2004 the Government announced a recapitalization of the banking sector, with an injection of US $45,000m. being divided equally between the Bank of China and the China Construction Bank. Similar support is expected to be given to the other two state-owned commercial banks (SOCBs)—the Agricultural Bank, along with the Industrial and Commercial Bank. This latest initiative followed two previous recapitalization exercises in 1998 and 2000. Meanwhile, also in 2003, the Government established the China Banking Regulatory Commission (CBRC) in an attempt to improve the regulation and supervision of the four SOCBs. This important institutional initiative was given legislative support through the enactment (in December 2003) of a Law on the Supervision of the Banking Industry.

Distinguishing state and non-state sectors is not easy, the dividing line between them having become blurred, as local governments have transformed and sold off enterprises under their control without always changing their administrative status. The origins of China's SOE reform programme are found in experimental changes that began in the 1980s, when measures were introduced to encourage incentives for workers and management, and a bankruptcy law was promulgated on a trial basis. In 1998 the Government initiated a three-year programme to reform the then 986,000 SOEs—at one time the mainstay of China's economy—two-thirds of which had reported losses in the previous year. This programme's targets were reportedly successfully fulfilled. In general, development of the emerging non-state enterprise sector has been a fundamental theme of China's market-orientated reform strategy. It is a process that has been strengthened by WTO accession, and by 2003 the share of the non-state sector in China's GDP had reached about 35%.

State enterprises, which in 2003 accounted for 47% of industrial value added and two-thirds of the urban work-force, are concentrated in heavy industry and sectors over which the State seeks to retain control for strategic reasons, as well as in areas that are not served by the growing non-state enterprise sector. Since the 1980s SOE reform has been delayed not only for strategic and ideological reasons, but also because of the critically important social role of such enterprises in providing employment and essential services (such as housing, health care, pensions, etc.) for urban workers and their families. The absence of a comprehensive social security system in the urban sector has made it difficult to downsize, let alone close down, SOEs, for fears—already realized in some cities—of undermining social order. In short, further reforms, including privatization of the housing sector, education and social welfare, are inextricably linked to SOE reform. 'Grasping the big and letting go the small' is the slogan that has guided recent efforts to limit state control to a relatively small number of large corporations concentrated in industrial sectors of strategic importance. In 1999 enterprise reform focused on 6,600 large and medium-scale SOEs, 70% of which were said to have become profitable or to be undergoing restructuring in 2000. Further improvements have since been reported, as a result of the implementation of large-scale government-funded infrastructure and construction projects that offered temporary relief for SOEs. SOE profitability also improved: in 2003 SOEs' profits were estimated to be 378,400m. yuan (46.4% of all industrial profits), a rise of 45.2% above the level of 2002. The broad goal is that some 1,000 large enterprises will eventually be managed on a commercial basis, but under state control, organized in enterprise groups loosely modelled on the Korean *chaebol*. Enterprises selected for inclusion in these groups will receive state support in settling their debt with the banking sector, if necessary through equity swaps, which started in 1999. As enterprises fulfil their financial and organizational requirements, so their shares will be listed on the stock market, a process that has been under way since the late 1990s. Medium- and small-scale SOEs have been privatized or had their status otherwise changed in large numbers, in order to reduce the State's economic and welfare burden in the industrial sector. This has been achieved by various means, including sale, conversion to joint ownership (with employees receiving shares in the enterprise) or leasing arrangements. The result of such restructuring was to more than halve the number of medium and small-scale enterprises to around 34,000 by 2001.

In the 1980s urban collective enterprises were a means for both local government institutions and private entrepreneurs to engage in commercial business activity. Under the label of collective enterprises, SOEs could combine some of their assets with those of other enterprises, or operate profitable sections as independent enterprises. This afforded them tax benefits and commercial flexibility, without endangering their SOE status and associated benefits of that status. Private entrepreneurs meanwhile used the collective nomenclature to minimize the political risks associated with the operation of private enterprises. In rural areas almost 20m. TVEs have been established since 1979. About 30% of these are engaged in manufacturing, and a further two-thirds in service sector activities. They employ a work-force of 132m. (27% of the rural labour force); between 1980 and 1995 their output value increased, on average, by 36% annually, although subsequent growth has slowed to less than 11%. Their contribution to exports has also become increasingly important, generating around US $140m. in export earnings (over one-third of the total). In recent years competition from the urban consumer industries has presented a major challenge to TVEs, to which they have responded unevenly: some have raised their productivity and efficiency successfully to meet the challenge; others have been forced to close.

The 1994 Company Law provided the legal basis for incorporation of these enterprises across the urban–rural divide. This new enterprise sector has combined private shareholding with involvement by official stakeholders, thereby enabling problems posed by weak institutional support and a still rudimentary commercial legal system to be overcome. According to a 1999 report by the Chinese Academy of Social Sciences, the collective and private sectors together accounted for three-quarters of industrial GVO in 1997, a figure that has subsequently increased. Under the influence of provincial policies, the geographical distribution of the non-state sector has been uneven: for example, in a number of wealthy coastal provinces (most notably, Jiangsu, Zhejiang and Guangdong), the weight, in terms of industrial GVO, of the collective sector is considerably greater than that of the state sector. A major contribution of the non-state sector has been its absorption of surplus urban and rural labour, thereby facilitating the downsizing of the state sector. Official figures show that in 2000 1.76m. registered private enterprises employed more than 200m. workers. Alongside the 32,500 foreign-funded enterprises (including more than almost 19,000 funded from Hong Kong, Macao and Taiwan), these enterprises form the basis of China's market economy. The status of the private sector has been continually enhanced (not least, because of its ability to absorb labour), and in 1999 a constitutional amendment strengthened the legal status of private enterprises by giving them equal rights with other businesses in their commercial access to capital and funding.

ENVIRONMENT

China's rapid industrialization in recent decades has led to serious environmental degradation. According to the World Bank, in the late 1990s particulate pollution in northern China, sulphur dioxide pollution in the south (reflecting the high sulphur content of coal in the region) accounted for some 300,000 premature deaths annually as a result of chronic pulmonary disease. Ongoing concerns of the Chinese authorities are air pollution, acid rainfall, worsening water quality and land degradation. China's pollution is mainly energy-related. Emissions of greenhouse gases, sulphur dioxide and particulate matter are very high owing to the widespread use of unwashed coal. Conversion from coal to natural gas for domestic fuel is one of the major measures taken by the Government to reduce urban pollution. Northern China suffers from a chronic water shortage and desertification. Public disclosure of environmental issues has greatly improved, and market-based anti-pollution measures such as emissions charges, 'cap and trade' systems and higher water prices have been introduced. However, although China's commitment to the promotion of sustainable development has strengthened in recent years, it has still not agreed to binding targets for the reduction of carbon dioxide emissions under the Kyoto Protocol. Environmental protection expenditure is expected to account for 1.4% of GDP during the Tenth Five-Year Plan (2001–05).

In 2003 41.5% of the 340 cities under the environmental monitoring programme met high national standards for air quality (7.7% more than in 2002), while 26.7% failed to meet even the lowest standards (4.5% fewer than in the previous year). The improvement is partly attributable to the ban on leaded petrol (gasoline), introduced in mid-2000. In the urban sector, there were improvements in the processing of solid and liquid waste: 42.1% of waste water was processed within centralized systems, 58% of solid waste was processed by non-toxic means, and 53.5% of such waste was recycled. In terms of its economic and social implications, an acute water shortage in the northern half of the country is a critical issue. Water quality in rivers and lakes may have improved in recent years, but the qualitative dimension of the water problem too remains a source of serious concern. Red algae tides have also caused losses to fisheries along the coast. Per caput water resource availability decreased by 5.6% in 2003 to 2,076 cu m—about one-quarter of the world average. The fact that only Egypt, Pakistan and Japan surpass China in their dependence on irrigation highlights the seriousness of this low figure. Grasslands in China's western regions are severely affected by desertification and salination, and the affected area is expanding by 2m. ha each year (possibly causing climatic changes in northern China).

ENERGY

China's energy sector remains a weak link in its industrial development. China is the world's second largest producer and consumer of energy, but energy supply has not kept pace with industrial growth since 1979. During the 1980s and 1990s China's demand for primary energy grew significantly faster than that of the global average, and between 1980 and 1995 its share of global energy consumption rose from 6.5% to 10.3%. Although China's share in global energy exceeds 11%, inadequate energy supply remains a constraint on further industrial development, with some 10%–15% of demand unable to be met. The centres of energy production and consumption are situated more than 1,000 km apart, and long-distance shipments of coal take up 40% of railway freight capacity. Nor was this situation expected to improve in the short term. Projections show energy demand quadrupling between 2000 and 2020, with most of the increase occurring in the coastal provinces of Shandong, Jiangsu, Zhejiang, Fujian and Guangdong, along with Shanghai, which are far from the centres of coal production in the north-west of China. China's unusually high dependence on coal will continue into the foreseeable future, despite the huge environmental cost which it entails. The Government plans to establish a small number of large-scale nation-wide coal companies that will facilitate long-term guaranteed supplies at relatively low cost. Coal distribution remains, however, a problem, and local shortages have caused some power plants to shut down.

For many years dependence on coal has been a striking feature of China's energy industry. The share of coal in total energy consumption reached its peak in 1994, at 75.3%; it subsequently declined, to 66.1% in 2002. This decline has been offset by rises in the shares of crude petroleum (to 23.4%), hydroelectricity (7.8%) and natural gas (2.7%). The contribution of nuclear energy remains negligible, while alternative and sustainable forms of energy are rarely used.

Electricity

China's electricity production in 2003 was 1,910,672m. kWh. From a base of 294,000 MW in 2000, the Tenth Five-Year Plan aimed to raise capacity to 370,000 MW by 2005, and to 500,000 MW–550,000 MW by 2010 (by which time, output is targeted to be 2,500,000m. kWh. Hydropower accounts for about 80% of total electricity output, compared with 17.5% from thermal power, and a mere 1.5% from nuclear power. Hydroelectric power is relatively abundant in China's southern provinces, but the sources are often located in remote areas. Currently, China's largest hydroelectric power station is that at Ertan, which began full-scale operations in December 1999. However, potentially the largest-and certainly the most controversial—hydroelectric project is the Three Gorges Dam, which will add approximately 18,200 MW when construction is completed in 2009 (at a total cost of 200,000m. yuan). Another large-scale hydroelectric scheme is planned for the Huanghe (Yellow River) under the newly established Yellow River Hydroelectric Development Corporation, which plans to achieve an installed capacity of 15.8 GW. In 2002 seven of 25 planned stations were under construction in provinces along the Huanghe.

Small-scale power plants with a capacity of 125 MW or less account for 50% of installed capacity of coal-fired stations. These plants are relatively cheap to build, but they incur significant transport costs for coal and emit high levels of carbon dioxide, sulphur dioxide and particles. For such reasons, where possible, the use of larger and more efficient plants is now being encouraged. Local governments can force factories to use the electricity produced by the smaller plants, even if it costs more than that generated by the larger stations funded by the central Government. The development of larger grids is expected to facilitate the long-distance transport of electricity, to reduce the need for coal.

The structure of the power sector was changed in 1997, when, in an attempt to separate power production and distribution, and to make the industry more competitive, the operation of the State's power plants was transferred from the former Ministry of Power Industry to the State Power Corporation (SPC). The SPC was made responsible for the funding of energy development (state subsidies to the power industry being discontinued in favour of commercial funding from domestic and foreign sources) and for power distribution, through its control of national electricity grids. A further major watershed was passed in October 2002, when the SPC itself underwent restructuring, involving its transformation into five generating and two distributing companies.

Coal

China is the world's largest producer and consumer of coal, and at current levels of consumption, known coal reserves are expected to last for 250 years (compared with only 50–70 years for proven oil and natural gas reserves). However, official figures show that total production decreased by 30% between 1996 and 2000, as the Government sought to rationalize production and, in particular, to address problems of over-staffing and inefficiency associated with the continuing operation of so many small-scale mines (not least, because of the extremely high incidence of accidents with which they were associated). Those same sources indicate that the previous peak level of production has now been reattained and surpassed, with total output reaching 1,667m. metric tons in 2003 (15% higher than in the previous year). Interpreting such figures in difficult, but what is not in doubt is that coal remains China's major energy source (accounting for about two-thirds of total energy consumption). Coal utilization is divided roughly equally among electricity generation, industrial activities, and domestic heating and cooking. Meanwhile, cities have started to prohibit the burning of coal for heating and cooking, because of increasing levels of pollution. Problems arising from the use of coal include greenhouse-gas emissions, air pollution and acid rain. Environmental concerns have encouraged technical advances, designed to make coal cleaner and easier to transport through liquefaction or gasification.

Coal deposits are concentrated in the north-west of China, in the provinces of Shanxi, Shaanxi and Inner Mongolia. Hardly any coal is produced south of the Changjiang (Yangtze River), requiring the fuel to be transported mainly by rail to eastern coastal and southern provinces. Limited rail capacity has added to the energy shortages along the Changjiang valley and in southern regions, while hundreds of trucks make daily journeys from the coalfields of Shanxi Province in order to supply coal to Beijing.

From the early 1980s demand for coal led central government to encourage the growth of small, decentralized coal mines, managed either by provincial or by local authorities, to supplement the bulk of output generated from fewer than 100 large-scale units. By the late 1990s small-scale mines (run privately or by TVEs), each with an annual output of 30,000 tons or less, generated 44% of national production. Many of these were, however, highly inefficient and in 1998 the Government announced the planned closure of some 26,000 small mines, with the expected loss of almost 1m. jobs. To what extent subsequent reported closures really have taken place is difficult to determine: the total number of deaths in small-scale mines seems not to have declined significantly, despite supposedly large-scale closures, and it may be that output from TVE mines has simply been concealed.

Petroleum and Natural Gas

China has proven petroleum reserves of 24,000m. barrels. In 2002 crude petroleum production totalled 167m. metric tons, but had to be supplemented by additional net imports of 81.3m. tons in order to help meet domestic demand.In other words, net imports accounted for about one-third of China's total energy consumption. China's transformation from a major oil exporter in the 1980s to an increasing importer since 1993 emphasizes the energy constraint which high GDP growth has imposed. China's annual petroleum production is not expected to increase significantly in the near future and oil imports will continue to rise. One official Chinese projection shows oil dependence increasing to 42% by 2010, and to 55% by 2020. The impact of WTO membership on the oil sector will also be significant, its rules requiring China to allow foreign companies to import, distribute and sell oil products. For the time being, about three-quarters of the total are extracted from older oilfields in eastern and north-eastern China. Since beginning production in 1963, Daqing in north-eastern China has remained the country's largest oilfield and still accounts for approximately one-third of total output. The second largest oilfield (also located in north-eastern China) is Liaohe, where improved recovery rates have generated more stable production. Expansion of production is expected from new oilfields in Xinjiang (north-west China) where, contingent on infrastructural improvements, total production could reach 1m. barrels per day (b/d) by 2008.

Offshore petroleum contributes 10% of domestic production. Exploitation and new exploration are centred in the Bohai Gulf, near Tianjin (where reserves are estimated at 1,500m. barrels), and the Zhujiang (Pearl River) Delta region in South China. China has also reached agreement with Viet Nam in preparing the way for joint oil and gas production in the Beibu Gulf. Most offshore exploration is based on consortia with international partners. China has participated in international petroleum and gas prospecting and exploitation, involving oil projects in Peru, Canada, Sudan, Thailand, Venezuela and Kazakhstan.

Natural gas consumption remains at a relatively low level, accounting for only 2.7% of total primary energy consumption in 2002. Estimates of China's proven reserves of natural gas range between 1,200,000m. and 5,300,000m. cu m. In larger cities, gas has already replaced coal as a household fuel. Intensified petroleum exploration is expected to generate further discoveries of natural gas. Against the background of China's search for more diversified, as well as clean and efficient, energy sources, natural gas features strongly in national energy planning, the intention being that it should contribute 6% of primary energy consumption by 2010 and 8% by 2020. According to Chinese estimates, an increase in annual output of up to 100,000m. cu m could save as much as 400m. tons of coal. A new gas pipeline from Xinjiang to Shanghai is a core project in the development of western China, based on the expectation that it will create a development corridor along its path. Guangdong Province and Shanghai have been targeted as the first trial areas for conversion from coal to natural gas as their main energy supply. In 2002 China signed a 25-year US $12,000m. agreement with Australia for the supply of natural gas.

Nuclear Power Industry

In 2002 nuclear power supplied less than 1% of total energy, although China National Nuclear Corporation plans to raise this level to 5% by 2020. In 1998 China had three operational nuclear plants: a Chinese-designed 300-MW pressurized water reactor at Qinshan, in Zhejiang Province; and two 900-MW pressurized water reactors supplied and constructed by a French company at Daya Bay (Guangdong). In 2002 the first generation unit of a nuclear power plant in Guangdong Province began operation; a further 1-GW generating unit became fully operational in 2003. In Zhejiang Province two 600-MW generating units have also begun operations. There are also plans for the installation of four French reactors at a 3,600-MW project near Daya Bay, and two Russian 1,000-MW reactors at Liangyungang, in Jiangsu Province. As a result of its dependence on finance from the supplying countries, China has acquired a range of different nuclear technologies. Under the Tenth Five-Year Plan (2001–05), China does not intend to construct any new nuclear power plants, although existing facilities may be expanded and upgraded.

TRANSPORT AND COMMUNICATIONS

Energy supply, transport and telecommunications have long been, and remain, weak links in China's infrastructure. Transport was designated a priority area under the Tenth Five-Year Plan (2001–05), associated projects including the Beijing–Shanghai Express rail project, the Shanghai International Shipping Center, and major transport infrastructure projects in China's western regions. In order to meet infrastructural needs, roads, railways, airports, and ports and harbours are being upgraded, or new facilities built. Meanwhile, freight volume rose to 5,715,200m. freight ton-km in 2003 (an annual increase of 13.1%). Of this, 29.9% was carried by rail, 12.3% by road and 56.5% by water. However, as a result of the SARS crisis, the overall volume of passenger traffic decreased by 2.3% to 1,379,500m. person-km—34.7% of whom were carried by rail, 55.7% by road, 9.2% by air and a mere 0.5% by water.

The railway system, which remains a state monopoly, plays a strategic role in China's infrastructure, since it transports coal from north-western provinces to the eastern seaboard and to the southern provinces where demand is concentrated. New railway corridors are expected to invigorate the economies of adjoining regions, including provinces along the new line between Kowloon (Hong Kong) and Beijing, and the south-western line connecting Guangxi Zhuang Autonomous Region and Yunnan Province. The main trunk lines of Beijing–Guangzhou, Harbin–Dalian, Beijing–Shenyang and Lianyungang–Lanzhou carry a freight volume three times their official capacity. High-speed trains, capable of speeds of up to 140 km per hour, have been introduced on routes between Beijing, Shanghai, Harbin and Guangzhou. This is a considerable improvement over the average speed of 60 km per hour on long-distance routes. The target of expanding the length of rail track to 70,000 km by 2002 (with 21,000 km of double-track lines and 15,000 km of electrified lines) was fulfilled a year ahead of schedule. Yet despite such improvements, China's rail system is still able to meet only about one-third of demand and delays in freight shipments remain common, a situation that reflects under-investment in earlier years. Track length is planned to reach 75,000 km by 2005, and 100,000 km by 2010.

Since the 1980s highway construction has helped to create a long-distance network facilitating interregional road transport (previously, road networks were much more localized and province-orientated). Between 1990 and 1995 China completed 130,000 km of highways, including 1,619 km of expressways and 9,328 km of wider and dual-carriage roads. By 2002 the total length of China's highways was 1.77m. km (31% above the 1999 figure), the total length of expressways having reached 25,100. However, since the 1980s the number of vehicles (especially, in the most recent past, cars) has increased 10 times faster than the expansion of the road system. Long-term plans seek to fulfil the goal of constructing a 35,000-km network of 12 national highways that will link Beijing to all provincial capitals and larger cities.

China has 13 major ocean shipping ports, each with an annual capacity of more than 5m. metric tons. The volume of cargo handled at major coastal ports increased by 17.9% to reach 3,300m. tons in 2003 (foreign trade accounting for 940m. tons of this). The four largest ports are Shanghai, Ningbo, Dalian and Tianjin. In technical terms, however, Chinese ports do not match international standards; only about 20% of trade is containerized, compared with a world average of 50%. Turnover times in Chinese ports are another major concern: for example, deliveries of coal and iron ore have been delayed for up to three weeks because of the lack of unloading berths. Resolving such problems will take time, not least because it takes three–four years to construct a port.

Between 1978 and 1980 there was a sharp contraction in the length of inland shipping waterways, from 136,000 km to 108,500 km. Since then, recovery has taken place, and in 2002 the corresponding figure was 121,600 km. In 1998 some 6,000 km were navigable for vessels with a displacement of 1,000 gross tons. In 2002 freight carried by inland waterways totalled 1,418m. tons (25% more than in 1995). Inland river navigation is particularly developed in southern and south-west China. Since 1997 upgrading of facilities on the Xijiang to accommodate 1,000-ton vessels has given Nanning, the capital of Guangxi, access to Hong Kong and Macao. Subsequently, new navigation networks have been constructed and existing ones improved in the Zhujiang (Pearl River) delta, along the Changjiang and on the Beijing–Hangzhou Grand Canal. Inland river navigation is managed by diverse operators, including household enterprises.

China's aviation industry grew at an annual rate of 20% during the 1990s and continues to expand and upgrade its fleet of aircraft. In 2001, however, 90% of China's 143 airports operated at a loss. Some 95% of passenger traffic is handled by 40 airports, of which only the larger airports, such as Beijing, Shanghai and Guangzhou, are believed to be profitable. This sector has been slow to absorb foreign investment. The first projects started in 1998 with Hong Kong and German investors in Wuhan and Pudong, Shanghai. In 2002 foreign investment was actively sought as part of industrial restructuring designed to merge China's 10 major airlines into three groups, Air China, China Eastern Airlines and China Southern Airlines. Associated with this process of corporatization has been the establishment of three service groups China Civil Aviation Information (Group), China Aviation Supplies Import and Export (Group) and China Aviation Fuel Group (their functions having previously been the responsibility of the government-based General Administration of Civil Aviation of China).

Since the early 1990s China's telecommunications industry has experienced boom conditions, fuelled by huge demand for new technology. However, despite major initiatives undertaken by the Government, such as the accelerated development and expansion of a digital network, fibre-optic cables and satellite systems, and mobile communication facilities, the industry remains quite immature and the potential for growth therefore huge. In 2003 the value of posts and telecommunications transactions grew by 27.8%. There were 49.08m. new fixed telephone subscribers, taking the total number of customers to 268.69m. (171.292m. in cities, 92.013m. in rural areas). The completion of investment projects already under way is expected to take the national fixed-line penetration rate to 33% by 2007—still well below the current 50% rate achieved in South Korea, for example. Meanwhile, China's mobile-phone market is already the largest in the world: in 2003 there were 268.69m. subscribers—an astonishing 30% more than in the previous year. One estimate suggests that by 2005 the corresponding figure could be almost 540m., with major urban centres and much of the coastal region approaching saturation. The length of long-distance fibre-optic cable and of long-distance microwave lines has risen dramatically in recent years: by 274% to 399,082 km, and by 106% to 164,052 km, respectively, between 1995 and 2001). By the end of 2003 the total capacity of telephone exchanges was 350m. lines ('gates'). High fees and sophisticated efforts by the Government to control content continue to restrict access to data communication in China. In 1998, however, government institutions and corporations, including SOEs, in co-operation with US software companies, started to provide their own information sites on the internet and even allowed experiments in electronic commerce. By the end of 2002 there were about 58m. internet users in China, and usage continued to expand rapidly in 2003 and into 2004. Internet use is disproportionately orientated towards commercial consumers in major cities, such as Beijing, Guangzhou and Shanghai. By the end of 2000 China had 620 internet service providers (ISPs) and 1,600 internet content providers (ICPs). In the same year, the value of e-commerce was estimated to be between US $20m. and US $40m.

Major structural changes have affected the communications industry in recent years. Until 1998 service operations under China's telecommunications industry were closed to foreign investment, although there was foreign involvement in the provision of equipment and the construction of network markets. Under the terms of China's WTO membership, foreign firms are now permitted to own up to 35% of the equity in companies located in the coastal areas around Beijing, Shanghai and Guangzhou. The schedule for market access will eventually allow overseas investors to achieve 49% ownership in mobile and fixed-line services, and 50% in value-added services.

THE FINANCIAL SECTOR

Banking and Finance Reform

The strategic role of China's banking sector is highlighted by the fact that bank assets constitute more than 150% of GDP (compared with about 50% in the USA), whereas stock-market capitalization and bond issues account for only 20%–30% (compared with well over 100% in the USA). China's financial and banking system is still dominated by the State. From 1949 to 1983 the People's Bank of China (PBC) was China's only bank. Its remit was all-embracing, ranging from the issue of currency to account settlement and foreign-currency transactions. In 1984 PBC became the central bank, and its commercial functions were taken over by four new policy banks: the Industrial and Commercial Bank of China (ICBC), the Agricultural Bank of China (ABC), the People's Construction Bank of China—subsequently China Construction Bank (CCB)—and the Bank of China (BOC). Under the supervision of the PBC, these four banks have since dominated the banking sector, and they currently hold 60% of total financial assets and 57% of market share. The China International Trust and Investment Corporation (CITIC) was established in 1979: its original *raison d'être* was to attract Overseas Chinese capital, although its remit subsequently came to include the provision of a wider range of financial services (including leasing and foreign investment). Most provinces have since established their own International Trust and Investment Corporations (ITICs), one of which, Guangdong ITIC, spectacularly defaulted and was officially declared bankrupt in 1999. In the 1980s smaller banks, non-banking financial institutions, and urban and rural credit co-operatives were also established. Urban credit co-operatives have been merged into 80 urban commercial banks to provide services for local small and medium-scale enterprises. The private banking sector has yet to be developed as a significant force, although there does exist one private banking institution (Minsheng), which hopes to list internationally. Foreign banks are allowed to provide financial services to foreign-invested enterprises and in foreign currency, although their operations in local currency have mainly been restricted to small enclaves in some SEZs. China's WTO membership is gradually allowing a more visible and significant presence by foreign financial institutions (including HSBC, Citibank, Hang Seng Bank, IFC and Newbridge) not only in banking, but also in other financial activities (such as securities). A number of overseas financial companies have already tried, with varying degrees of success, to buy into China's domestic market.

In the early 1990s as part of state enterprise reform, the enterprise debt of the SOE sector was transferred from government ministries and their state budget to the banking system. The transfer meant that the large commercial banks had to convert this debt into loans to the SOEs, but without being able to cancel bad debts at the required level. Indeed, by the terms of the existing regulatory framework, banks had to expand their loans to SOEs, irrespective of the latter's financial situation. Some estimates indicate that in the absence of effective criteria for extending loans, China's banking system has bad loans totalling up to 4,000,000m. yuan, or 35% of 2003 GDP (although official estimates put the figure much lower, at less than 2,000,000m. yuan). The Government's target is that the NPL ratio should be reduced to no more than 15% by 2005.

The existence of high NPL ratios has been a major obstacle to the commercialization of the banking sector in China. The original intention in establishing the policy banks was to enable them to assume the burden of financing loss-making SOEs, while allowing other banks to take the lead in pursuing genuinely commercial operations. In the event, restrictive regulations, political interference and other problems allowed only limited progress to be made. Accordingly, starting in 1998, in a further effort to reduce the NPL ratio, a second major institutional initiative was undertaken. This took the form of a programme of debt-for-equity swaps for the SOE sector, implemented by newly established asset management corporations (AMCs) under the four big commercial banks. Underlining this initiative was the decision, announced in March 2003, to establish a state asset management commission. By 2003 AMCs had purchased an estimated RMB 1,400,000m. of bad loans since 1998. The first of the AMCs was Cinda, established in 1998 in order to acquire NPLs for CCB. The following year witnessed the establishment of China Orient (for BOC), China Great Wall (for CCB) and Huarong (for ICBC). Under the impact of the AMC programme and the cancellation of bad loans by banks, there is evidence that progress is at last being achieved and that NPL ratios are falling. Official data suggest that between 2002 and 2003 NPLs as a share of total loans declined from over 25% to about 12%. Such success reflects aggressive debt 'write-offs', but also accelerated loans growth through 2003. Interestingly, banks have also diversified their customer base from SOEs towards private households. Of the four major banks, the China Construction Bank has been the most successful in reducing NPLs, whereas the Agricultural Bank has been the least successful.

The Chinese Stock Market

Stock exchanges were established in Shanghai and Shenzhen in 1990, since when they have shown spectacular growth. Even so, China's stock markets are still immature, weak and highly speculative. The World Bank has indicated that China's stock market capitalization of around 1,300,000m. yuan in 2003 (substantially below the peak level of more than 1,600,000m. yuan in 2000) is below the average for low-income countries and less, for example, than that of India. The relative capital value of shares that can be traded freely is even smaller. In 2003 funds raised through the issue of stocks and share rights were 135,800m. yuan, 41% more than in 2002. The number of listed companies increased from 1,224 in 2002 to 1,287 in 2003. In both Shanghai and Shenzhen, control of these companies is dominated by SOEs.

The Chinese stock market continues to be tarnished by problems of mismanagement, although regulations have been tightened and practices have recently improved. A Code of Corporate Governance for Listed Companies was enacted by the China Securities Regulatory Commission in 2002 as part of a move to improve the reporting of inflated profits and the delisting of loss-making companies. Listed companies are required to undergo supplementary audits by the major international auditing firms. In general, opportunities for domestic firms and individuals to hold foreign-currency denominated stocks and bonds (the 'B' share market) remain heavily circumscribed, just as they are for foreign firms and individuals to purchase Chinese-currency denominated assets (the 'A' share market). From this perspective, an important initiative was the Chinese Government's decision, in May 2003, to allow two overseas firms (one Swiss, one Japanese) to trade on the renminbi-denominated 'A' share market, an initiative likely to embrace an increasing number of foreign firms in the coming years. More than 100 of China's 500 largest companies, including China Telecom, Anshan Iron and Steel Company and Handan Iron and Steel Company, have been publicly listed, although state institutions still hold around 60% of shares in such enterprises. China's big state corporations have also increasingly listed their shares in international markets. China's other capital markets remain underdeveloped, although the country's WTO accession is one factor that will help enhance bond, insurance and securities markets. Diversification within the interbank market has strengthened its role as a source of capital for domestic banks. Meanwhile, the corporate bond market remains small: in 2002, for example, a mere RMB 32,000m. of corporate bonds were issued, compared with RMB 568,000m. of government-issued bonds.

Fiscal Policy

Since the late 1990s China's budget deficit has steadily widened. Between 1998 and 2003 the share of central and local government revenue in GDP rose quite impressively, from 12.4% to 18.6%, but during the same period the share of government spending increased from 14.0% to 21.1%. The outcome was a deterioration in budgetary balance, from –1.6% to –2.5%. Forecasts indicated that the fiscal deficit would average 2.6% of GDP in 2004–05, and by 2005 one authoritative projection suggested that government revenue and expenditure would constitute 19.3% and 22.0% of GDP. A major problem faced by the Chinese Government in implementing its fiscal policy is that its revenue system remains tied to pre-reform practices of revenue raising. The decline in budgetary revenue from 35% to 11% of GDP between 1978 and 1995 reflects the extent to which the government budget continued to rely on SOEs that were unable to pay taxes because of their increasing debt burden. The situation has since improved, but further major changes in the government revenue system will be required in order to accommodate ongoing structural changes within the economy. The most important challenges facing the Government in this respect are: to develop new taxes in order to take advantage of the fiscal potential of the growing non-state sector (as in the introduction during 2003–04 of a unified corporate tax rate); to enhance the efficiency of tax management in order to secure greater compliance with existing tax laws; and to devise mechanisms that will allow fiscal resources to be transferred from rich to poorer regions of the country.

FOREIGN TRADE AND INVESTMENT

China's foreign trade and its absorption of foreign investment have played an important part in the reform of its domestic economy and in strengthening its status as a regional and international economic power. China's increased international standing became evident during the Asian economic crisis of 1997/98, when the country won international praise for having resisted pressure to devalue its currency.

Ideological and historical factors had long made foreign trade and China's involvement in the international economy controversial issues. China's reluctance to participate in international exchange manifested itself in a foreign trade regime that was modelled on the USSR and that persisted from the early 1950s until well into the 1980s. Under this structure, some 10 large foreign trade corporations under the Ministry of Foreign Economic Relations and Trade (later the Ministry of Foreign Trade and Economic Co-operation—itself reorganized under a new State Economic and Trade Commission in 2003), and some technical ministries, monopolized all foreign trade. By acting as an independent intermediary between foreign clients and domestic producers or buyers, they isolated the domestic economy from competition, but also from technical innovation. Under this system, foreign trade fulfilled no more than a 'gap-filling' role, exports serving merely to secure the foreign exchange needed to provide the means for importing developmental goods that could not be produced domestically. At some points during the Mao era (especially during the most radical early phase of the Cultural Revolution), China turned increasingly inward, adopting a quasi-autarchic strategy. At the end of the 1970s foreign trade accounted for about 13% of GDP and China ranked 32nd among world trading nations. From this perspective, Deng Xiaoping's adoption of the 'open door' policy was a major watershed, marking a reorientation of China's official view of its role in the international economic community. The most radical manifestation of the change in the Government's mindset was its active encouragement of overseas involvement in domestic economic construction and solicitation of foreign capital (above all, FDI). The impact was profound. By the end of the 1990s China had become the 10th largest trading nation in the world and the second largest recipient—easily the largest among developing countries—of overseas capital. It had also become a pivotal player in the Asian economy and a major player within the global system. This process began in the early 1980s, when the previously passive system of foreign trade began to be replaced by a much more proactive stance that owed a great deal to the export-led growth model of the first-échelon Asian NIEs. Thereafter, foreign trade monopolies were gradually reduced and more than 5,000 Chinese corporations, as well as Sino-foreign joint-venture companies, received foreign trade rights. Significantly, since 1999 more than 40,000 private enterprises have also received the same rights.

Alongside institutional decentralization of foreign trade decision-making, other initiatives took place, including tariff reductions. Not least important, a succession of renminbi devaluations took the official exchange rate to a more rational level of 8.28 RMB to one US dollar , at which it has stabilized since 1998. Initially, a dual exchange rate was introduced (a lower rate being used for commercial transactions), but in 1994 this was replaced by a unified rate. The implicit 'peg' with the US dollar that had existed since the late 1990s remained in place in 2003, despite pressure from overseas trading partners to revalue in order to restrain China's exports and stimulate its purchases overseas. In 1996 China's currency was made fully convertible for current-account transactions, although the Chinese Government has been much more resistant to capital account convertibility in order to protect itself against destabilizing speculative activities. In the second half of 2003, in the face of high current-account surpluses, the Chinese Government came under increasing international pressure to revalue the renminbi. Continuing large-scale inflows of FDI also put pressure on the currency. However, the Government consistently resisted calls from its trading partners to adjust the renminbi in order to contain its booming exports and to stimulate imports. The official Chinese view was that that the existing exchange rate *vis-à-vis* the US dollar was appropriate and that allegations that China was 'exporting deflation' were groundless. It seemed unlikely, as of mid-2004, that any significant exchange rate change would take place in the foreseeable future, although further moves towards a more flexible exchange rate regime were considered possible.

China's foreign trade increased, in nominal terms, from US $20,640m. in 1978 to US $165,500m. in 1992, and reached US $851,200m. in 2003—a remarkable 37.1% above the level of 2002. In the first half of 2004 the value of China's merchandise trade rose by 39.1% year-on-year, to reach US $523,000m. The implication is that merchandise trade grew, on average, by 16% annually between 1978 and 2003 (16.4% for exports and 15.7% for imports). Concealed within these figures is the transformation from a merchandise trade deficit (totalling US $14,900m. in 1985) to a robust surplus (rising from US $87,400m. in 1990 to a peak of US $42,854m. in 1998). As a result of more sharply rising import growth from 1998, this surplus subsequently contracted to US $25,600m. in 2003 and in the first half of 2004 the more rapid increase in import growth generated a merchandise trade deficit of US $6,800m. In 2003 China's current-account balance was US $31,400m. (equivalent to 2.2% of GDP). This last figure represents a significant decline on 2002, and projections indicate that a worsening deficit on the services account may reduce the current account to less than 0.5% of GDP by 2007. Under the impact of its buoyant foreign trade performance, China's foreign-exchange reserves grew steadily to reach US $444,000m. in March 2004, thus significantly greater than those of Taiwan (US $230,100m.), South Korea (US $166,500m. at May 2004), Hong Kong (US $120,800m. at June 2004) and Singapore (US $101,600m. at June 2004).

A notable feature of China's foreign trade performance from the late 1970s was the contribution of crude petroleum and refined oil products to its export expansion. As early as 1975, Deng Xiaoping had identified oil as possessing the potential to finance a major trade expansion. Between 1977 and 1985 physical exports of crude petroleum and petroleum products rose threefold, but owing to the simultaneous rise in international oil prices, the monetary value of the increase was almost sevenfold. As a result, this single product contributed one-third of China's incremental export earnings. It did so, however, at a heavy cost, for almost all of the additional oil produced domestically during these years was diverted to export markets, leaving China's rapidly growing economy facing an increasingly severe energy constraint. The high level of oil exports was not sustainable, and as incremental production was redirected towards domestic needs, so oil exports contracted, to the extent that by 1993 China had reverted to becoming a net oil importer, a position it continued to retain in 2003. In 2003 China imported 91.12m. tons of crude petroleum (31.3% more than in 2002), and spent US $26,700m. on overseas purchases of oil and associated products. This shift in emphasis reflected a move away from dependence on capital-intensive raw material production towards production of labour-intensive exports (for example, textiles, shoes, toys, electrical appliances, etc.), a development that, given the abundance of labour, accorded with the dictates of comparative advantage. Between 1980 and 1990 the share of primary goods in total exports decreased by almost one-half (from 50.3% to 25.6%), and has since fallen to under 10%; in other words, manufactured goods' share has risen to over 90%. In 2003 Chinese exports were dominated by machinery and electrical appliances, of which the earnings of US $172,000m. accounted for almost 40% of the value of all exports. The annual increase of such sales was almost 50%—and 68.6% for computers and telecommunications equipment. In contrast to the buoyant export performance of these 'new economy' products, that of 'old economy' industries was much less impressive (for example, exports of garments and clothing accessories increased by 26.1%, and of footwear by a mere 16.6%). Purchases of machinery and electrical appliances, worth $175,426m., accounted for 42.5% of all imports in 2003. Next in importance were mineral products (9.1%, with 7.7% for oil alone) and chemicals (7.7%). Simultaneously high exports and imports of 'new economy' goods underlines the reality that although China has become a global centre for the production of such goods (especially in the IT sector), many components needed in their production cannot be supplied from domestic sources. Meanwhile, trade in services has also grown rapidly: from a negligible base in 1989, services exports rose to well in excess of US $30,000m. However, the onset of SARS significantly reduced receipts from services in 2003. A notable structural aspect of China's foreign trade development under the impact of reform has been the increasingly important contribution of foreign-funded enterprises (FFEs) as an engine of foreign trade and, by extension, overall economic growth. By 2002 China's officially registered FFEs accounted for 54.3% of Chinese imports and 52.2% of total exports.

Although China's foreign trade remains concentrated within Asia (with over 50% of exports being transported to Asian destinations, and almost 65% of imports coming from Asian origins), China's single largest export market is the USA, which in 2003 purchased 21.1% of all Chinese exports. Next in importance were Hong Kong (17.4%), Japan (13.6%) and South Korea (4.6%). The EU took 16.5% of China's exports, with Germany (4%), Netherlands (3.1%), the United Kingdom (2.5%) as the most important national markets within Europe. The most important source of imports was Japan (18.0%), followed by Taiwan (12.0%) and South Korea (10.4%); the share of purchases from the EU was 12.9% (5.9% from Germany alone), and from the USA, 8.2%.

From its introduction in 1979, the use of foreign capital was intended to help raise China's technical and managerial standards to the level of China's Asian neighbours and, subsequently, to that of major Western economies. Until the mid-1980s the importance of foreign loans outstripped that of FDI, but by 2003 levels of FDI had increased sharply to exceed 95% of all capital inflows. According to official Chinese classification, FDI to China includes investment made by Hong Kong, Macao and Taiwan 'compatriots', as well as Overseas Chinese, and receipts from the 'Chinese diaspora' have dominated FDI inflows, accounting for close to 60% of the total. In order to encourage FDI, China has, since the early 1980s, enacted legislation designed to create a positive investment environment. Non-Chinese foreign investors have faced considerable obstacles and resistance, which their Overseas Chinese counterparts (for example, those from Hong Kong, Taiwan and Singapore), with their more flexible modes of operation, as well as their cultural and linguistic advantages, have been more readily able to overcome. Although the importance of technology transfer was highly publicized in the 1980s, investment in joint-venture enterprises, as the only vehicles for foreign investment, was concentrated in tourism and the service industries. From the mid-1980s to the mid-1990s Asian investors contributed 80% of FDI in China, with Hong Kong alone holding a share of 60%. Hong Kong served as a conduit for Overseas Chinese and Taiwanese investment, but also for Chinese capital that had been transferred overseas and then redirected as foreign investment back into China (the phenomenon of 'round-tripping'). The increasing FDI involvement, in and after the early 1990s, of the USA, Japan and Western European countries, was increasingly directed towards major 'pillar industries', such as the automobile and telecommunications industries, in which investors sought to secure a significant, even dominant market share, a process that was made possible by the opening of domestic markets for the products of FFEs. As foreign investors were allowed to establish wholly owned subsidiaries in China, so they were able to bring in more expertise and to compete directly with Chinese enterprises.

The cumulative value of inflows of all forms of utilized foreign capital during 1979–2002 was US $623,420m., of which FDI was US $446,250m. (72%). From a base of US $1,170m. (1979–82), annual utilized FDI inflows into China exceeded US $11,000m. in 1992 and reached an astonishing US $53,500m. in 2003 (implying average annual growth of 23.4% since 1990). In the first half of 2004 utilized FDI increased by 12% compared with the corresponding period of the previous year, to reach US $33,900m. Annual inflows rose consistently year-on-year between the early 1980s and 1998. Thereafter, however, levels fell sharply (FDI averaged a mere US $4,051.5m. in 1999–2000), before recovering. In response to these fluctuations, the Chinese Government sought to attract more investment from Japan, the USA and Europe, as well as to open up new areas of investment, such as infrastructure, telecommunications and services. The sharp deceleration in annual FDI growth between 2002 and 2003 (from 15.1% to just 1.4%) reflected the dual impact of SARS and the weakness of the global economy during the first half of 2003. However, with about two-thirds of Chinese-foreign joint ventures in profit (often to a greater extent than other overseas ventures) overseas confidence in China as an investment environment remains strong.

The coastal provinces have benefited disproportionately from inflows of FDI, whereas the share directed towards interior provinces—above all, those in western China—has been very much smaller. It is some years now since China sought explicitly to attract more investment into interior regions, an aspiration that has been underlined in the strategy to 'open up the West'. However, the evidence suggests that narrowing the gap between coast and interior, let alone between coastal provinces and the far West, is likely to be a very long-term process. Meanwhile, in 2002 Guangdong was the single largest provincial recipient of FDI inflows (21.5%), followed by Jiangsu (19.3%), Shandong (9%), Shanghai (8.1%), Fujian (7.3%), Liaoning (6.5%) and Zhejiang (5.8%); these seven provincial-level units accounted for more than one-quarter of the national total.

China and the WTO

China's major economic challenge in the coming years, apart from its macroeconomic problems, will be how to maximize the potential benefits associated with its accession to the WTO. China has undertaken to make major market access concessions, in particular, committing itself to a general reduction of tariff levels and other quantitative restrictions, opening up its services sectors (for example, insurance, banking, telecommunications, and wholesale and retail distribution). Between 2001 and 2003 the average tariff was reduced from 15.3% to 11.0%, and all non-tariff protectionist measures are scheduled to be phased out by 2005, when the average level of industrial tariffs will have fallen to 9.4%. Although tariff reduction and increased rivalry by foreign-invested enterprises will place additional competitive pressure on domestic producers, overseas companies will face an improved business environment, in which they will enjoy enhanced rights to export directly to Chinese customers, to engage in distribution and wholesale operations, and to invest without having to accommodate trade-distorting requirements. Increased competition seems likely to have a quite damaging impact on a number of important economic sectors, such as aerospace, agriculture, the automobile industry, electronics and IT, entertainment and communications, financial services and steel. The main beneficiaries will be the textiles sector and machine-building—both major export industries for China. The official Chinese orthodoxy is that WTO membership will have a major and positive impact on the Chinese economy, above all by facilitating the restructuring of large enterprises and enabling them to compete with global firms. Beijing's successful bid for the 2008 Olympic Games will have given a further boost to fulfilling China's global aspirations by attracting accelerated investment from overseas, as foreign firms compete for the huge construction projects associated with the Games. There is, however, an alternative view, held quite widely outside China, which argues that the terms on which the Chinese Government has entered the WTO will generate major difficulties, especially for ailing large enterprises. All that can be said with certainty at this stage is that there will indeed be winners and losers within China, but whether the net impact is positive or negative remains, especially in the short term, uncertain.

Statistical Survey

Source (unless otherwise stated): National Bureau of Statistics of China, 38 Yuetan Nan Jie, Sanlihe, Beijing 100826; tel. (10) 68515074; fax (10) 68515078; e-mail service@stats.gov.cn; internet www.stats.gov.cn.

Note: Wherever possible, figures in this Survey exclude Taiwan. In the case of unofficial estimates for China, it is not always clear if Taiwan is included or excluded. Where a Taiwan component is known, either it has been deducted from the all-China figure or its inclusion is noted. Figures for the Hong Kong Special Administrative Region (SAR) and for the Macao SAR are listed separately. Transactions between the SARs and the rest of the People's Republic continue to be treated as external transactions.

Area and Population

AREA, POPULATION AND DENSITY

Area (sq km)	9,572,900*
Population (census results)	
1 July 1990	1,130,510,638
1 November 2000	
Males	640,275,969
Females	602,336,257
Total	1,242,612,226
Population (official estimates at 31 December)	
2001	1,276,270,000
2002	1,284,530,000
2003	1,292,270,000
Density (per sq km) at 31 December 2003	134.9

* 3,696,100 sq miles.

PRINCIPAL ETHNIC GROUPS
(at census of 1 November 2000)

	Number	%
Han (Chinese)	1,137,386,112	91.53
Zhuang	16,178,811	1.30
Manchu	10,682,262	0.86
Hui	9,816,805	0.79
Miao	8,940,116	0.72
Uygur (Uigur)	8,399,393	0.68
Tujia	8,028,133	0.65
Yi	7,762,272	0.63
Mongolian	5,813,947	0.47
Tibetan	5,416,021	0.44
Bouyei	2,971,460	0.24
Dong	2,960,293	0.24
Yao	2,637,421	0.21
Korean	1,923,842	0.16
Bai	1,858,063	0.09
Hani	1,439,673	0.12
Kazakh	1,250,458	0.10
Li	1,247,814	0.10
Dai	1,158,989	0.09
She	709,592	0.06
Lisu	634,912	0.05
Gelao	579,357	0.05
Dongxiang	513,805	0.04
Others	3,568,237	0.29
Unknown	734,438	0.06
Total	1,242,612,226	100.00

ADMINISTRATIVE DIVISIONS
(previous or other spelling given in brackets)

	Area ('000 sq km)	Population at 1 November 2000: Total	Density (per sq km)	Capital of province or region	Estimated population ('000) at mid-2000*
Provinces					
Sichuan (Szechwan)	487.0	82,348,296	169	Chengdu (Chengtu)	3,294
Henan (Honan)	167.0	91,236,854	546	Zhengzhou (Chengchow)	2,070
Shandong (Shantung)	153.3	89,971,789	587	Jinan (Tsinan)	2,568
Jiangsu (Kiangsu)	102.6	73,043,577	712	Nanjing (Nanking)	2,740
Guangdong (Kwangtung)	197.1	85,225,007	432	Guangzhou (Canton)	3,893
Hebei (Hopei)	202.7	66,684,419	329	Shijiazhuang (Shihkiachwang)	1,603
Hunan (Hunan)	210.5	63,274,173	301	Changsha (Changsha)	1,775
Anhui (Anhwei)	139.9	58,999,948	422	Hefei (Hofei)	1,242
Hubei (Hupeh)	187.5	59,508,870	317	Wuhan (Wuhan)	5,169
Zhejiang (Chekiang)	101.8	45,930,651	451	Hangzhou (Hangchow)	1,780
Liaoning (Liaoning)	151.0	41,824,412	277	Shenyang (Shenyang)	4,828
Jiangxi (Kiangsi)	164.8	40,397,598	245	Nanchang (Nanchang)	1,722
Yunnan (Yunnan)	436.2	42,360,089	97	Kunming (Kunming)	1,701
Heilongjiang (Heilungkiang)	463.6	36,237,576	78	Harbin (Harbin)	2,928
Guizhou (Kweichow)	174.0	35,247,695	203	Guiyang (Kweiyang)	2,533
Shaanxi (Shensi)	195.8	35,365,072	171	Xian (Sian)	3,123
Fujian (Fukien)	123.1	34,097,947†	277	Fuzhou (Foochow)	1,397
Shanxi (Shansi)	157.1	32,471,242	207	Taiyuan (Taiyuan)	2,415
Jilin (Kirin)	187.0	26,802,191	143	Changchun (Changchun)	3,093
Gansu (Kansu)	366.5	25,124,282	69	Lanzhou (Lanchow)	1,730
Hainan	34.3	7,559,035	220	Haikou	438‡
Qinghai (Tsinghai)	721.0	4,822,963	7	Xining (Hsining)	691‡
Autonomous regions					
Guangxi Zhuang (Kwangsi Chuang)	220.4	43,854,538	199	Nanning (Nanning)	1,311
Nei Mongol (Inner Mongolia)	1,177.5	23,323,347	20	Hohhot (Huhehot)	978
Xinjiang Uygur (Sinkiang Uighur)	1,646.9	18,459,511	11	Urumqi (Urumchi)	1,415
Ningxia Hui (Ninghsia Hui)	66.4	5,486,393	83	Yinchuan (Yinchuen)	530‡
Tibet (Xizang)	1,221.6	2,616,329	2	Lhasa (Lhasa)	105§
Municipalities					
Shanghai	6.2	16,407,734	2,646	—	12,887
Beijing (Peking)	16.8	13,569,194	808	—	10,839
Tianjin (Tientsin)	11.3	9,848,731	872	—	9,156
Chongqing (Chungking)	82.0	30,512,763	372	—	4,900
Total	9,572.9	1,242,612,226‖	130		

* UN estimates, excluding population in counties under cities' administration.
† Excluding islands administered by Taiwan, mainly Jinmen (Quemoy) and Mazu (Matsu), with 49,050 inhabitants according to figures released by the Taiwan authorities at the end of March 1990.
‡ December 1998 figure.
§ 1982 figure.
‖ Including 2,500,000 military personnel and 1,050,000 persons with unregistered households.

PRINCIPAL TOWNS
(Wade-Giles or other spellings in brackets)

Population at mid-2000
(UN estimates, incl. suburbs, in '000)

Shanghai (Shanghai)	12,887	Liupanshui	2,023
Beijing (Pei-ching or Peking, the capital)	10,839	Handan	1,996
Tianjin (T'ien-chin or Tientsin)	9,156	Jinxi	1,821
Wuhan (Wu-han or Hankow)	5,169	Liuan	1,818
Chongqing (Ch'ung-ch'ing or Chungking)	4,900	Hangzhou (Hang-chou or Hangchow)	1,780
Shenyang (Shen-yang or Mukden)	4,828	Tianmen	1,779
Guangzhou (Kuang-chou or Canton)	3,893	Changsha (Chang-sha)	1,775
Chengdu (Ch'eng-tu)	3,294	Wanxian	1,759
Xian (Hsi-an or Sian)	3,123	Lanzhou (Lan-chou or Lanchow)	1,730
Changchun (Ch'ang-ch'un)	3,093	Nanchang (Nan-ch'ang)	1,722
Harbin (Ha-erh-pin)	2,928	Kunming (K'un-ming)	1,701
Nanjing (Nan-ching or Nanking)	2,740	Yantai	1,681
Zibo	2,675	Xuzhou	1,636
Dalian (Ta-lien or Dairen)	2,628	Xiantao	1,614
Jinan (Chi-nan or Tsinan)	2,568	Shijiazhuang (Shih-chia-chuang or Shihkiachwang)	1,603
Guiyang	2,533	Heze	1,600
Linyi	2,498	Yancheng	1,562
Taiyuan (T'ai-yüan)	2,415	Yulin	1,558
Qingdao (Ch'ing-tao or Tsingtao)	2,316	Xinghua	1,556
Zhengzhou (Cheng-chou or Chengchow)	2,070	Taian	1,503
Zaozhuang	2,048	Pingxiang	1,502

Source: UN, *World Urbanization Prospects: The 2001 Revision.*

BIRTHS AND DEATHS
(sample surveys)

	2000	2001	2002
Birth rate (per 1,000)	14.03	13.38	12.86
Death rate (per 1,000)	6.45	6.43	6.41

Marriages (number registered): 8,420,044 in 2000; 7,971,144 in 2001; 7,788,000 in 2002.

Expectation of life (WHO estimates, years at birth): 71.1 (males 69.6; females 72.7) in 2002 (Source: WHO, *World Health Report*).

EMPLOYMENT*
(official estimates, '000 persons at 31 December)

	2000	2001	2002
Agriculture, forestry and fishing	333,550	329,740	324,870
Mining	5,970	5,610	5,580
Manufacturing	80,430	80,830	83,070
Electricity, gas and water	2,840	2,880	2,900
Construction	35,520	36,690	38,930
Transport, storage and communications	20,290	20,370	20,840
Wholesale and retail trade and catering	46,860	47,370	49,690
Banking and insurance	3,270	3,360	3,400
Social services	9,210	9,760	10,940
Health care, sports and social welfare	4,880	4,930	4,930
Education, culture, art, radio, film and television broadcasting	15,650	15,680	15,650
Government agencies, etc.	11,040	11,010	10,750
Others	56,430	58,520	62,450
Total	625,940	626,750	634,000

* In addition to employment statistics, sample surveys of the economically active population are conducted. On the basis of these surveys, the total labour force ('000 persons at 31 December) was: 739,920 in 2000; 744,320 in 2001; 753,600 in 2002. Of these totals, the number of employed persons ('000 at 31 December) was: 720,850 (agriculture, etc. 360,430; industry 162,190; services 198,230) in 2000; 730,250 (agriculture, etc. 365,130; industry 162,840; services 202,280) in 2001; 737,400 (agriculture, etc. 368,700; industry 157,800; services 210,900) in 2002.

Health and Welfare

KEY INDICATORS

Total fertility rate (children per woman, 2002)	1.8
Under-5 mortality rate (per 1,000 live births, 2001)	39
HIV/AIDS (% of persons aged 15–49, 2003)	0.1
Physicians (per 1,000 head, 2002)	1.67
Hospital beds (per 1,000 head, 2000)	2.38
Health expenditure (2001): US $ per head (PPP)	224
Health expenditure (2001): % of GDP	5.5
Health expenditure (2001): public (% of total)	37.2
Access to water (% of persons, 2000)	75
Access to sanitation (% of persons, 2000)	38
Human Development Index (2002): ranking	94
Human Development Index (2002): value	0.745

For sources and definitions, see explanatory note on p. vi.

Agriculture

PRINCIPAL CROPS
('000 metric tons)

	2000	2001	2002
Wheat	99,636	93,873	90,290
Rice (paddy)	187,908	177,581	174,739
Barley*	2,646	2,893	3,324
Maize	106,000	114,088	121,310
Rye†	1,000	701	580
Oats†	1,012	790	490
Millet	2,125	1,966	2,176
Sorghum	2,582	2,696	3,329
Buckwheat†	1,950	1,245	968
Triticale (wheat-rye hybrid)	365	644	980
Potatoes	66,282	64,564	75,230
Sweet potatoes	117,978	113,591	108,075
Cassava (Manioc)†	3,800	3,850	3,900
Taro (Coco yam)†	1,500	1,500	1,550
Sugar cane	66,280	75,663	90,107
Sugar beet	8,073	10,889	12,820
Dry beans*	1,650	1,800	2,150
Dry broad beans*	1,788	1,950	2,100
Dry peas*	1,020	1,120	1,500
Other pulses*	230	246	262
Soybeans (Soya beans)	15,411	15,407	16,507
Groundnuts (in shell)	14,437	14,416	14,818
Oil palm fruit†	640	650	660
Castor beans*	513	300	370
Sunflower seed	1,954	1,478	1,946*
Rapeseed	11,381	11,331	10,552
Tung nuts	453	407	388
Sesame seed	811	804	895
Tallowtree seeds†	820	835	850
Linseed	344	253	420*
Cottonseed	8,834	10,647	9,832
Other oilseeds	849†	845	875†
Cabbages†	22,500	24,650	27,400
Asparagus†	3,900	4,200	5,000
Lettuce†	7,250	7,600	9,000
Spinach†	6,950	7,400	9,000
Tomatoes†	22,200	24,000	27,000
Cauliflower†	5,700	6,100	6,500
Pumpkins, squash and gourds†	3,500	3,700	5,000
Cucumbers and gherkins†	19,800	21,600	24,000
Aubergines (Eggplants)†	13,750	14,000	15,400
Green chillies and peppers†	9,400	9,850	10,500
Green onions and shallots†	280	300	400
Dry onions†	14,070	15,000	16,500
Garlic†	7,380	7,800	9,000
Green beans†	1,680	1,800	2,000
Green peas†	1,450	1,530	1,700
Carrots†	5,600	6,000	7,000
Mushrooms†	800	960	1,050
Other vegetables†	120,613	128,126	135,150
Watermelons	51,466	57,178	61,608†
Cantaloupes and other melons	7,244	11,670	12,618*
Grapes	3,282	3,680	4,479
Apples	20,431	20,015	19,241
Pears	8,412	8,796	9,309
Peaches and nectarines	3,827*	4,562	5,230
Plums*	3,900	4,026	4,363
Oranges	1,054*	1,352	1,501
Tangerines, mandarins, clementines and satsumas	6,587*	8,749	8,824
Lemons and limes*	279	376	505
Grapefruit and pomelos*	187	233	261
Other citrus fruit*	676	896	920
Mangoes*	3,000	3,060	3,300
Pineapples	857	869	827
Persimmons	1,592	1,585	1,741
Bananas	4,941	5,272	5,557
— continued			
Other fruits and berries*	3,226	3,089	3,462
Walnuts	310	252	343
Other treenuts†	684	689	803
Tea (made)	683	702	745
Pimento and allspice†	212	215	220
Other spices†	321	347	365
Tobacco (leaves)	2,552	2,350	2,447
Jute and jute-like fibres*	126	106	159
Other fibre crops	426†	575	809*

* Unofficial figure(s).
† FAO estimate(s).

Source: mainly FAO.

LIVESTOCK
('000 head at 31 December)

	2000	2001	2002
Horses	8,914	8,766	8,260
Mules	4,673	4,530	4,362
Asses	9,348	9,227	8,815
Cattle	104,396	105,905	100,959
Buffaloes	22,587	22,758	22,684
Camels	330	326	279
Pigs	430,198	446,815	457,430
Sheep	131,095	133,160	135,893
Goats	148,163	157,159	161,292
Rabbits	185,000*	190,000*	191,247
Chickens	3,500,000*	3,650,000*	3,980,000†
Ducks	600,000*	625,000*	676,000†
Geese	200,000*	205,000*	232,506†

* FAO estimate.
† Unofficial figure.

Source: FAO.

LIVESTOCK PRODUCTS
('000 metric tons)

	2000	2001	2002
Beef and veal*	4,968	5,110	5,460
Buffalo meat*	360	378	386
Mutton and lamb*	1,440	1,540	1,680
Goat meat*	1,300	1,387	1,487
Pig meat	40,314	41,845	43,266
Horse meat†	166	156	156
Rabbit meat	370	406	423
Poultry meat*	12,075	12,103	12,498
Other meat†	261	414	510
Cows' milk	8,274	10,255	12,998
Buffaloes' milk†	2,650	2,680	2,700
Sheep's milk	847	974	1,006
Goats' milk†	200	225	220
Butter	82	84	88
Cheese†	206	217	225
Hen eggs	19,068	19,862	20,933
Other poultry eggs	3,365	3,506	3,694
Honey	246	252	265
Raw silk (incl. waste)*	78	94	100
Wool: greasy	293	298	308
Wool: scoured	146	149	154
Cattle hides (fresh)†	1,226	1,272	1,365
Buffalo hides (fresh)†	108	113	116
Sheepskins (fresh)†	281	300	321
Goatskins (fresh)†	282	297	319

* Unofficial figures.
† FAO estimates.

Source: FAO.

Forestry

ROUNDWOOD REMOVALS
('000 cubic metres, excl. bark)

	2000	2001	2002
Sawlogs, veneer logs and logs for sleepers	53,623	51,923*	51,583*
Pulpwood*	6,718	6,578	6,478
Other industrial wood*	36,080	35,360	35,060
Fuel wood*	191,051	191,049	191,047
Total	287,472	284,910	284,168

* FAO estimate(s).
Source: FAO.

Timber production (official figures, '000 cubic metres): 47,240 in 2000; 45,520 in 2001; 44,361 in 2002.

SAWNWOOD PRODUCTION
('000 cubic metres, incl. railway sleepers)

	2000*	2001	2002
Coniferous (softwood)	3,930	4,923	5,182
Broadleaved (hardwood)	3,415	3,626*	4,249
Total	7,345	8,549	9,431

* FAO estimate(s).
Source: FAO.

Fishing

('000 metric tons, live weight)

	2000	2001	2002
Capture	16,987.3	16,529.4	16,553.1
Freshwater fishes	1,223.0	1,033.3	924.2
Japanese anchovy	1,142.9	1,260.7	1,173.2
Largehead hairtail	1,285.5	1,282.7	1,287.8
Aquaculture	24,580.7	26,050.1	27,767.2
Common carp	2,119.8	2,193.2	2,235.6
Crucian carp	1,375.4	1,523.4	1,697.2
Bighead carp	1,614.0	1,637.9	1,701.0
Grass carp (White amur)	3,162.6	3,310.9	3,419.6
Silver carp	3,227.9	3,275.9	3,401.9
Pacific cupped oyster	3,291.9	3,491.0	3,625.5
Japanese carpet shell	1,616.4	2,014.4	2,300.9
Total catch	41,568.0	42,579.5	44,320.4

Note: Figures exclude aquatic plants ('000 metric tons, wet weight): 8,067.8 (capture 204.3, aquaculture 7,863.5) in 2000; 8,426.3 (capture 266.9, aquaculture 8,159.5) in 2001; 9,106.3 (capture 297.2, aquaculture 8,809.1) in 2002.

Source: FAO.

Aquatic products (official figures, '000 metric tons): 41,224.1 (marine 24,719.2, freshwater 16,504.9) in 1999; 42,784.8 (marine 25,387.4, freshwater 17,397.5) in 2000; 43,813.4 (marine 25,717.0, freshwater 18,096.4) in 2001; 45,645 (marine 26,463, freshwater 19,182) in 2002. The totals include artificially cultured products ('000 metric tons): 23,970.1 (marine 9,743.0, freshwater 14,227.1) in 1999; 25,746.7 (marine 10,612.9, freshwater 15,133.8) in 2000; 27,261.3 (marine 11,310.8, freshwater 15,950.5) in 2001; 29,058 (marine 12,128, freshwater 16,930) in 2002. Figures include aquatic plants on a dry-weight basis ('000 metric tons): 1,194.4 in 1999; 1,222.0 in 2000; 1,249.5 in 2001; 1,333 in 2002. Freshwater plants are not included.

Mining

(estimates, '000 metric tons, unless otherwise indicated)

	1999	2000	2001
Coal*	1,050,000	999,000	1,160,000
Crude petroleum*	160,000	163,000	163,959
Natural gas (million cu m)*	25,198	27,200	30,329
Iron ore: gross weight	237,000	223,000	220,000
Iron ore: metal content	71,000	67,200	n.a
Copper ore†	520	593	587
Nickel ore (metric tons)†	49,500	50,300	51,500
Bauxite	8,500	9,000	9,800
Lead ore†	549	660	676
Zinc ore†	1,476	1,780	1,700
Tin concentrates (metric tons)†	80,100	99,400	95,000
Manganese ore: gross weight	3,190	2,640	2,500
Manganese ore: metal content	630	800	n.a.
Tungsten concentrates (metric tons)†	31,100	37,000	38,500
Ilmenite	180	185	n.a.
Molybdenum ore (metric tons)†	29,700	28,800	28,200
Vanadium (metric tons)†	26,000	30,000	30,000
Zirconium concentrates (metric tons)	15,000	15,000	n.a.
Antimony ore (metric tons)†	89,600	110,000	150,000
Cobalt ore (metric tons)†	250	90	150
Mercury (metric tons)†	200	200	190
Silver (metric tons)†	1,360	1,600	1,910
Uranium (metric tons)†‡	500	500	655
Gold (metric tons)†	173	180	185
Magnesite	2,450	4,070	3,580
Phosphate rock and apatite§	20,000	19,400	n.a.
Potash‖	260	380	385
Native sulphur	280	290	290
Fluorspar	2,400	2,450	2,450
Barite (Barytes)	2,800	3,500	3,600
Arsenic trioxide (metric tons)	16,000	16,000	n.a.
Salt (unrefined)*	28,124	31,280	34,105
Gypsum (crude)	6,700	6,800	6,800
Graphite (natural)	300	430	450
Asbestos	247	320	360
Talc and related materials	3,900	3,500	3,500
Diamonds ('000 carats):			
gem	230	230	235
industrial	920	920	950

* Official figures. Figures for coal include brown coal and waste. Figures for petroleum include oil from shale and coal. Figures for natural gas refer to gross volume of output.
† Figures refer to the metal content of ores, concentrates or (in the case of vanadium) slag.
‡ Data from the World Nuclear Association (London, United Kingdom).
§ Figures refer to gross weight. The estimated phosphoric acid content was 30%.
‖ Potassium oxide (K_2O) content of potash salts mined.

Source: mainly US Geological Survey.

Industry

SELECTED PRODUCTS

Unofficial Figures

('000 metric tons, unless otherwise indicated)

	1999	2000	2001
Plywood ('000 cu m)*†	8,103	10,735	9,856
Mechanical wood pulp*†	455	480	565
Chemical wood pulp*†	1,695	1,745	1,785
Other fibre pulp*†	13,646	14,276	14,321
Sulphur‡§‖(a)	1,630	1,900	2,000
Sulphur‡§‖(b)	3,860	3,370	3,090
Kerosene	7,438	8,723	n.a.
Residual fuel oil	19,594	20,537	n.a.
Lubricating oils	2,658	2,887	n.a.
Paraffin wax	1,276	1,386	n.a.
Petroleum coke	4,253	4,619	n.a.
Petroleum bitumen (asphalt)	3,545	3,849	n.a.
Liquefied petroleum gas	8,166	9,166	n.a.
Aluminium (unwrought)	2,808.9	2,989.2	3,575.8
Refined copper (unwrought)	1,170.0	1,370.0	1,440.0
Lead (unwrought)	945.1	1,113.4	1,195.4
Tin (unwrought)	90.5	101.3	105.1
Zinc (unwrought)	1,669.8	1,931.7	2,037.6

* Data from the FAO.

† Including Taiwan.

‡ Data from the US Geological Survey.

§ Figures refer to (a) sulphur recovered as a by-product in the purification of coal-gas, in petroleum refineries, gas plants and from copper, lead and zinc sulphide ores; and (b) the sulphur content of iron and copper pyrites, including pyrite concentrates obtained from copper, lead and zinc ores.

‖ Provisional or estimated figure(s).

Source: UN, *Industrial Commodity Statistics Yearbook*.

Official Figures

('000 metric tons, unless otherwise indicated)

	2000	2001	2002
Edible vegetable oils	8,353.2	13,831.7	15,312.2
Raw sugar	7,000	6,531	9,260.0
Beer	22,313.2	22,889.3	24,027.0
Cigarettes ('000 cases)	33,970	34,021	34,670.8
Cotton yarn (pure and mixed)	6,570	7,607	8,500
Woven cotton fabrics—pure and mixed (million metres)	27,700	29,000	32,239
Woollen fabrics ('000 metres)	278,323.7	343,033.9	326,912.1
Silk fabrics (metric tons)	73,300	87,300	98,200
Chemical fibres	6,940	8,414	9,912
Paper and paperboard	24,869.4	37,770.7	46,669.9
Rubber tyres ('000)	121,578.7	135,730.0	163,065.9
Sulphuric acid	24,270	26,963.2	30,504.0
Caustic soda (Sodium hydroxide)	6,678.8	7,879.6	8,779.7
Soda ash (Sodium carbonate)	8,340	9,143.7	10,331.5
Insecticides	607	787	929
Nitrogenous fertilizers (a)*	23,981.1	25,273.7	28,084.8
Phosphate fertilizers (b)*	6,630.3	7,525.6	8,010.3
Potash fertilizers (c)*	1,248.6	1,030.8	1,814.9
Synthetic rubber	865.2	1,219.8	1,362.1
Plastics	10,875.1	12,887.1	14,556.7
Motor spirit (gasoline)	41,346.7	41,546.6	43,207.6
Distillate fuel oil (diesel oil)	70,796.2	74,856.6	77,061.0
Coke	121,840.2	131,307.7	142,798.1
Cement	597,000	661,040	725,000
Pig-iron	131,014.8	155,542.5	170,846
Crude steel	128,500	151,634	182,366
Internal combustion engines ('000 horse-power)†	188,573.0	205,311.7	285,057.5
Tractors—over 20 horse-power (number)	41,000	38,200	45,400

— continued	2000	2001	2002
Railway freight wagons (number)	27,300	30,700	31,300
Road motor vehicles ('000)	2,070	2,342	3,251
Bicycles ('000)	29,067.9	29,022.6	39,575.2
Electric fans ('000)	76,616.1	96,161.0	107,613.3
Mobile telephones ('000 units)	n.a.	80,316.6	121,463.5
Floppy disks ('000)	473,000	488,000	718,000
Microcomputers ('000)	6,720	8,777	14,635
Large semiconductor integrated circuits ('000)	2,392,000	2,226,000	4,132,000
Colour television receivers ('000)	39,360	40,937	51,550
Cameras ('000)	55,145.2	59,620.9	53,096.1
Electric energy (million kWh)	1,355,600	1,480,802	1,654,000

* Production in terms of (a) nitrogen; (b) phosphoric acid; or (c) potassium oxide.

† Sales.

Finance

CURRENCY AND EXCHANGE RATES

Monetary Units

100 fen (cents) = 10 jiao (chiao) = 1 renminbiao (People's Bank Dollar), usually called a yuan.

Sterling, Dollar and Euro Equivalents (31 May 2004)

£1 sterling = 15.186 yuan;
US $1 = 8.277 yuan;
€1 = 10.136 yuan;
1,000 yuan = £65.85 = $120.82 = €98.66.

Average Exchange Rate (yuan per US $)

2001 8.2771
2002 8.2770
2003 8.2770

Note: Since 1 January 1994 the official rate has been based on the prevailing rate in the interbank market for foreign exchange.

STATE BUDGET

(million yuan)*

Revenue	2000	2001	2002
Taxes	1,258,151	1,530,138	1,763,645
Industrial and commercial taxes	1,036,609	n.a.	n.a.
Tariffs	75,048	84,052	70,427
Agricultural and animal husbandry taxes	46,531	48,170	71,785
Taxes on income of state-owned enterprises	82,741	n.a.	n.a.
Taxes on income of collectively-owned enterprises	17,222	n.a.	n.a.
Other receipts	109,250	138,470	152,679
Sub-total	1,367,401	1,668,600	1,916,324
Less Subsidies for losses by enterprises	27,878	30,004	25,960
Total	1,339,523	1,638,604	1,890,364
Central Government	698,917	858,274	1,038,864
Local authorities	640,606	780,330	851,500

Expenditure†	2000	2001	2002
Capital construction	209,489	251,064	314,298
Agriculture, forestry and water conservancy	76,689	91,796	110,270
Culture, education, science and health care‡	273,688	336,102	397,908
National defence	120,754	144,204	170,778
Administration	178,758	219,752	297,942
Pensions and social welfare	21,303	26,668	37,297
Subsidies to compensate price increases	104,228	74,151	64,507
Development of enterprises	86,524	99,156	96,838
Other purposes	517,217	647,365	715,477
Total	1,588,650	1,890,258	2,205,315
Central Government	551,985	576,802	677,170
Local authorities	1,036,665	1,313,456	1,528,145

* Figures represent a consolidation of the regular (current) and construction (capital) budgets of the central Government and local administrative organs. The data exclude extrabudgetary transactions, totalling (in million yuan): Revenue 338,517 (central 23,045, local 315,472) in 1999; 382,643 (central 24,763, local 357,879) in 2000; Expenditure 313,914 (central 16,482, local 297,432) in 1999; 352,901 (central 21,074, local 331,828) in 2000.

† Excluding payments of debt interest.

‡ Current expenditure only.

INTERNATIONAL RESERVES
(US $ million at 31 December)

	2001	2002	2003
Gold*	3,093	4,074	4,074
IMF special drawing rights	851	998	1,102
Reserve position in IMF	2,590	3,723	3,798
Foreign exchange†	212,165	286,407	403,251
Total†	218,698	295,202	412,225

* Valued at SDR 35 per troy ounce.

† Excluding the Bank of China's holdings of foreign exchange.

Source: IMF, *International Financial Statistics*.

MONEY SUPPLY
(million yuan at 31 December)*

	2001	2002	2003
Currency outside banking institutions	1,568,730	1,727,800	1,974,600
Demand deposits at banking institutions	4,414,010	5,356,210	6,431,420
Total money (incl. others)	6,168,850	7,266,540	8,644,890

* Figures are rounded to the nearest 10 million yuan.

Source: IMF, *International Financial Statistics*.

COST OF LIVING
(General Consumer Price Index; base: previous year = 100)

	2000	2001	2002
Food	97.4	100.0	99.4
Clothing	99.1	98.1	97.6
Housing*	104.8	101.2	99.9
All items (incl. others)	100.4	100.7	99.2

* Including water, electricity and fuels.

NATIONAL ACCOUNTS
(million yuan at current prices)

Expenditure on the Gross Domestic Product*

	2000	2001	2002
Government final consumption expenditure	1,170,530	1,302,930	1,383,010
Private final consumption expenditure	4,289,560	4,589,810	4,853,450
Increase in stocks	−12,400	64,750	18,710
Gross fixed capital formation	3,262,380	3,681,330	4,216,830
Total domestic expenditure	8,710,070	9,638,820	10,472,000
Exports of goods and services; *Less* Imports of goods and services	224,020	220,470	279,420
Sub-total	8,934,090	9,859,290	10,751,420
Statistical discrepancy†	12,720	−127,810	−272,360
GDP in purchasers' values	8,946,810	9,731,480	10,479,060

* Figures are rounded to the nearest 10 million yuan.

† Referring to the difference between the sum of the expenditure components and official estimates of GDP, compiled from the production approach.

Gross Domestic Product by Economic Activity*

	2000	2001	2002
Agriculture, forestry and fishing	1,462,820	1,541,180	1,611,730
Industry†	3,904,730	4,237,460	4,653,570
Construction	588,800	637,540	700,500
Transport, storage and communications	540,860	596,830	624,090
Wholesale and retail trade and catering	731,600	791,880	821,530
Other services	1,718,000	1,926,590	2,067,640
Total	8,946,810	9,731,480	10,479,060

* Figures are rounded to the nearest 10 million yuan.

† Includes mining, manufacturing, electricity, gas and water.

BALANCE OF PAYMENTS
(US $ million)

	2000	2001	2002
Exports of goods f.o.b.	249,131	266,075	325,651
Imports of goods f.o.b.	−214,657	−232,058	−281,484
Trade balance	34,474	34,017	44,167
Exports of services	30,430	33,334	39,745
Imports of services	−36,031	−39,267	−46,528
Balance on goods and services	28,874	28,084	37,383
Other income received	12,550	9,338	8,344
Other income paid	−27,216	−28,563	−23,289
Balance on goods, services and income	14,207	8,909	22,438
Current transfers received	6,861	9,125	13,795
Current transfers paid	−550	−633	−811
Current balance	20,518	17,401	35,422
Capital account (net)	−35	−54	−50
Direct investment abroad	−916	−6,884	−2,518
Direct investment from abroad	38,399	44,241	49,308
Portfolio investment assets	−11,307	−20,654	−12,095
Portfolio investment liabilities	7,317	1,249	1,752
Other investment assets	−43,864	20,813	−3,077
Other investment liabilities	12,329	−3,933	−1,029
Net errors and omissions	−11,748	−4,732	7,504
Overall balance	10,693	47,447	75,217

Source: IMF, *International Financial Statistics*.

External Trade

PRINCIPAL COMMODITIES
(distribution by SITC, US $ million)

Imports c.i.f.	2000	2001	2002
Food and live animals	4,742.3	4,967.3	5,224.1
Crude materials (inedible) except fuels	19,685.8	21,828.8	22,506.9
Metalliferous ores and metal scrap	5,801.6	7,326.6	7,265.0
Mineral fuels, lubricants, etc.	20,756.6	17,618.5	19,490.0
Petroleum, petroleum products, etc.	19,049.3	16,049.7	17,430.2
Crude petroleum oils, etc.	14,860.7	11,661.3	12,757.3
Chemicals and related products	29,768.7	31,603.0	38,370.9
Organic chemicals	8,301.5	8,867.8	10,979.1
Artificial resins, plastic materials, etc.	13,012.0	13,822.7	15,650.6
Products of polymerization, etc.	10,273.4	11,142.7	12,138.8
Basic manufactures	42,493.0	42,667.6	49,085.9
Textile yarn, fabrics, etc.	13,108.4	12,819.1	13,270.3
Iron and steel	9,795.2	10,870.8	13,680.5
Machinery and transport equipment	91,866.0	106,884.7	136,904.1
Machinery specialized for particular industries	10,556.8	12,397.4	15,252.2
General industrial machinery, equipment and parts	7,995.0	10,062.7	12,607.2
Office machines and automatic data-processing equipment	10,858.4	12,659.8	17,094.0
Telecommunications and sound equipment	12,412.8	13,293.2	14,149.9
Other electrical machinery, apparatus, etc.	35,641.7	39,903.5	55,488.8
Thermionic valves, tubes, etc.	21,155.6	23,611.9	35,167.4
Electronic microcircuits	13,300.0	16,591.4	25,644.1
Road vehicles and transport equipment *	6,348.3	9,988.0	11,482.6
Miscellaneous manufactured articles	12,699.8	15,119.3	20,000.3
Total (incl. others)	225,093.7	243,552.9	295,170.1

Exports f.o.b.	2000	2001	2002
Food and live animals	12,270.8	12,764.8	14,601.7
Mineral fuels, lubricants, etc.	7,862.3	8,415.2	8,449.3
Chemicals and related products	11,917.5	13,145.7	15,062.0
Basic manufactures	43,305.0	44,604.6	53,744.4
Textile yarn, fabrics, etc.	16,316.4	17,031.4	20,769.8
Machinery and transport equipment	82,443.7	94,774.4	126,853.2
Office machines and automatic data-processing equipment	18,637.9	23,572.4	36,227.8
Automatic data-processing machines and units	10,994.1	13,093.8	20,132.3
Telecommunications and sound equipment	19,508.3	23,758.7	32,016.8
Other electrical machinery, apparatus, etc.	24,655.4	25,986.2	32,845.6
Road vehicles and transport equipment*	8,918.1	9,026.3	10,137.3
Miscellaneous manufactured articles	85,582.8	86,682.1	100,760.8
Clothing and accessories (excl. footwear)	36,147.3	36,723.8	41,384.9
Footwear	9,466.6	9,676.4	10,680.6
Baby carriages, toys, games and sporting goods	10,112.4	9,886.2	12,561.2
Children's toys, indoor games, etc.	8,555.3	8,318.2	10,713.1
Total (incl. others)	249,202.6	266,098.2	325,596.0

* Data on parts exclude tyres, engines and electrical parts.

Source: UN, *International Trade Statistics Yearbook*.

PRINCIPAL TRADING PARTNERS
(US $ million)*

Imports c.i.f.	2000	2001	2002
Australia	5,024.0	5,425.9	5,850.6
Brazil	1,621.4	2,347.3	3,003.0
Canada	3,751.1	4,027.8	3,626.9
Finland	2,353.1	2,376.3	1,512.8
France	3,949.8	4,104.8	4,253.1
Germany	10,408.7	13,772.1	16,416.4
Hong Kong	9,429.0	9,423.0	10,726.2
Indonesia	4,402.0	3,887.9	4,508.4
Italy	3,078.4	3,784.3	4,319.5
Japan	41,509.7	42,787.3	53,466.0
Korea, Republic	23,207.4	23,377.0	28,568.0
Malaysia	5,480.0	6,204.0	9,296.3
Oman	3,261.8	1,609.6	1,446.5
Russia	5,769.9	7,958.8	8,406.7
Singapore	5,059.6	5,128.3	7,046.6
Sweden	2,674.7	2,167.4	1,791.0
Taiwan	25,493.6	27,338.8	38,061.4
Thailand	4,380.8	4,713.9	5,599.6
United Kingdom	3,592.5	3,527.0	3,336.0
USA	22,363.2	26,199.9	27,237.6
Total (incl. others)	225,093.7	243,557.2	295,170.1

Exports f.o.b.	2000	2001	2002
Australia	3,428.9	3,569.5	4,585.0
Canada	3,157.8	3,345.9	4,303.5
France	3,705.2	3,685.7	4,071.9
Germany	9,277.8	9,755.1	11,371.9
Hong Kong†	44,518.3	46,541.2	58,463.2
Indonesia	3,061.8	2,835.7	3,426.5
Italy	3,802.0	3,991.6	4,827.4
Japan	41,654.3	44,940.5	48,433.8
Korea, Republic	11,292.4	12,518.8	15,534.6
Malaysia	2,564.9	3,221.1	4,974.2
Netherlands	6,687.2	7,278.4	9,107.6
Russia	2,233.4	2,710.5	3,520.7
Singapore	5,761.0	5,790.7	6,984.2
Taiwan	5,039.0	4,999.6	6,585.7
United Kingdom	6,310.1	6,780.5	8,059.4
USA	52,099.2	54,279.5	69,945.8
Total (incl. others)	249,202.6	266,098.2	325,596.0

* Imports by country of origin; exports by country of consumption.

† The majority of China's exports to Hong Kong are re-exported.

Transport

	2000	2001	2002
Freight (million ton-km):			
Railways	1,366,260	1,457,510	1,551,560
Roads	612,940	633,040	678,250
Waterways	2,373,420	2,598,890	2,751,060
Air	5,027	4,372	5,155
Passenger-km (million):			
Railways	453,260	476,680	496,940
Roads	665,740	720,710	780,580
Waterways	10,050	8,990	8,180
Air	97,050	109,140	126,870

ROAD TRAFFIC
('000 motor vehicles in use)*

	2000	2001	2002
Passenger cars and buses	8,537.3	9,939.6	12,023.7
Goods vehicles	7,163.2	7,652.4	8,122.2
Total (incl. others)	16,089.1	18,020.4	20,531.7

* Excluding military vehicles.

SHIPPING

Merchant Fleet
(registered at 31 December)

	2001	2002	2003
Number of vessels	3,280	3,326	3,376
Total displacement ('000 grt)	16,646.1	17,315.5	18,428.0

Source: Lloyd's Register-Fairplay, *World Fleet Statistics*.

Sea-borne Shipping
(freight traffic, '000 metric tons)

	2000	2001	2002
Goods loaded and unloaded	1,256,030	1,426,340	1,666,280

Tourism

FOREIGN VISITORS
(arrivals, '000)

Country of origin	2000	2001	2002
Hong Kong and Macao	70,099.4	74,344.5	80,808.2
Taiwan	3,108.6	3,442.0	3,660.6
Australia	234.1	255.1	291.3
Canada	236.6	253.9	291.3
France	185.0	199.5	222.1
Germany	239.1	253.4	281.8
Indonesia	220.6	224.2	274.7
Japan	2,201.5	2,385.7	2,925.6
Korea, Republic	1,344.7	1,678.8	2,124.3
Malaysia	441.0	468.6	592.4
Mongolia	399.1	387.1	453.1
Philippines	363.9	408.0	508.6
Russia	1,080.2	1,196.2	1,271.6
Singapore	399.4	415.0	497.1
Thailand	241.1	298.4	386.3
United Kingdom	283.9	302.5	343.0
USA	896.2	949.2	1,121.2
Total (incl. others)	83,443.9	89,012.9	97,908.3

Total tourism receipts (US $ million): 16,224 in 2000; 17,792 in 2001; 20,385 in 2002.

Communications Media

	2000	2001	2002
Television receivers ('000 in use)*	380,000	n.a.	n.a.
Telephones ('000 main lines in use)*	144,829	180,368	214,420
Mobile cellular telephones ('000 subscribers)*	85,260	144,820	206,620
Personal computers ('000 in use)*	20,600	25,000	35,500
Internet users ('000)*	22,500	33,700	59,100
Book production:			
titles	143,376	154,526	170,962
copies (million)	6,270.0	6,310.0	6,870.0
Newspapers:			
number	2,007	2,111	2,137
average circulation ('000 copies)	179,140	181,300	187,210
Magazines:			
number	8,725	8,889	9,029
average circulation ('000 copies)	215,440	206,970	204,060

* Source: International Telecommunication Union.

1997 ('000 in use): Radio receivers 417,000; Facsimile machines 2,000 (Sources: UNESCO, *Statistical Yearbook*, and UN, *Statistical Yearbook*).

2003: 263,000,000 main telephone lines in use; 79,500,000 internet users; 269,000,000 mobile cellular telephone subscribers (Source: International Telecommunication Union).

Education

(2002)

	Institutions	Full-time teachers ('000)	Students ('000)
Kindergartens	111,752	571	20,360
Primary schools	456,903	5,779	121,567
General secondary schools	80,067	4,376	82,879
Secondary technical schools	2,523	170	3,962
Teacher-training schools	430	38	601
Agricultural and vocational schools	7,402	310	5,115
Special schools	1,540	30	375
Higher education	1,396	618	9,034

Adult literacy rate (based on census data): 90.9% (males 95.1%; females 86.5%) in 2002 (Source: UN Development Programme, *Human Development Report*).

Directory

The Constitution

A new Constitution was adopted on 4 December 1982 by the Fifth Session of the Fifth National People's Congress. Its principal provisions, including amendments made in 1993, 1999 and 2004, are detailed below. The Preamble, which is not included here, states that 'Taiwan is part of the sacred territory of the People's Republic of China'. The seventh paragraph of the Preamble was amended in 1993 and 1999 to state: 'The basic task of the nation is to concentrate its efforts on socialist modernization by following the road of building socialism with Chinese characteristics. Under the leadership of the Communist Party of China and the guidance of Marxism-Leninism, Mao Zedong Thought and Deng Xiaoping Theory, the Chinese people of all nationalities will continue to adhere to the people's democratic dictatorship-'. The paragraph was further amended in 2004 to refer to 'Chinese-style socialism' and to 'the important thought of the Three Represents'.

GENERAL PRINCIPLES

Article 1: The People's Republic of China is a socialist state under the people's democratic dictatorship led by the working class and based on the alliance of workers and peasants.

The socialist system is the basic system of the People's Republic of China. Sabotage of the socialist system by any organization or individual is prohibited.

Article 2: All power in the People's Republic of China belongs to the people.

The organs through which the people exercise state power are the National People's Congress and the local people's congresses at different levels.

The people administer state affairs and manage economic, cultural and social affairs through various channels and in various ways in accordance with the law.

Article 3: The state organs of the People's Republic of China apply the principle of democratic centralism.

The National People's Congress and the local people's congresses at different levels are instituted through democratic election. They are responsible to the people and subject to their supervision.

All administrative, judicial and procuratorial organs of the State are created by the people's congresses to which they are responsible and under whose supervision they operate.

The division of functions and powers between the central and local state organs is guided by the principle of giving full play to the initiative and enthusiasm of the local authorities under the unified leadership of the central authorities.

Article 4: All nationalities in the People's Republic of China are equal. The State protects the lawful rights and interests of the minority nationalities and upholds and develops the relationship of equality, unity and mutual assistance among all of China's nationalities. Discrimination against and oppression of any nationality are prohibited; any acts that undermine the unity of the nationalities or instigate their secession are prohibited.

The State helps the areas inhabited by minority nationalities speed up their economic and cultural development in accordance with the peculiarities and needs of the different minority nationalities.

Regional autonomy is practised in areas where people of minority nationalities live in compact communities; in these areas organs of self-government are established for the exercise of the right of autonomy. All the national autonomous areas are inalienable parts of the People's Republic of China.

The people of all nationalities have the freedom to use and develop their own spoken and written languages, and to preserve or reform their own ways and customs.

Article 5: The People's Republic of China shall be governed according to law and shall be built into a socialist country based on the rule of law.

The State upholds the uniformity and dignity of the socialist legal system.

No law or administrative or local rules and regulations shall contravene the Constitution.

All state organs, the armed forces, all political parties and public organizations and all enterprises and undertakings must abide by the Constitution and the law. All acts in violation of the Constitution and the law must be looked into.

No organization or individual may enjoy the privilege of being above the Constitution and the law.

Article 6: The basis of the socialist economic system of the People's Republic of China is socialist public ownership of the means of production, namely, ownership by the whole people and collective ownership by the working people.

The system of socialist public ownership supersedes the system of exploitation of man by man; it applies the principle of 'from each according to his ability, to each according to his work.'

In the initial stage of socialism, the country shall uphold the basic economic system in which the public ownership is dominant and diverse forms of ownership develop side by side, and it shall uphold the distribution system with distribution according to work remaining dominant and a variety of modes of distribution coexisting.

Article 7: The state-owned economy, namely the socialist economy under the ownership of the whole people, is the leading force in the national economy. The State ensures the consolidation and growth of the state-owned economy.

Article 8: The rural collective economic organizations shall implement a two-tier operations system that combines unified operations with independent operations on the basis of household contract operations and different co-operative economic forms in the rural areas—the producers', supply and marketing, credit, and consumers' co-operatives—are part of the socialist economy collectively owned by the working people. Working people who are all members of rural economic collectives have the right, within the limits prescribed by law, to farm plots of cropland and hilly land allotted for their private use, engage in household sideline production and raise privately-owned livestock.

The various forms of co-operative economy in the cities and towns, such as those in the handicraft, industrial, building, transport, commercial and service trades, all belong to the sector of socialist economy under collective ownership by the working people.

The State protects the lawful rights and interests of the urban and rural economic collectives and encourages, guides and helps the growth of the collective economy.

Article 9: Mineral resources, waters, forests, mountains, grassland, unreclaimed land, beaches and other natural resources are owned by the State, that is, by the whole people, with the exception of the forests, mountains, grassland, unreclaimed land and beaches that are owned by collectives in accordance with the law.

The State ensures the rational use of natural resources and protects rare animals and plants. The appropriation or damage of natural resources by any organization or individual by whatever means is prohibited.

Article 10: Land in the cities is owned by the State.

Land in the rural and suburban areas is owned by collectives except for those portions which belong to the State in accordance with the law; house sites and private plots of cropland and hilly land are also owned by collectives.

The State may, in the public interest and in accordance with the provisions of law, expropriate or requisition land for its use and shall make compensation for the land expropriated or requisitioned.

No organization or individual may appropriate, buy, sell or lease land, or unlawfully transfer land in other ways.

All organizations and individuals who use land must make rational use of the land.

Article 11: The non-public sector of the economy comprising the individual and private sectors, operating within the limits prescribed by law, is an important component of the socialist market economy.

The State protects the lawful rights and interests of the non-public sectors of the economy such as the individual and private sectors of the economy. The State encourages, supports and guides the development of the non-public sectors of the economy and, in accordance with the law, exercises supervision and control over the non-public sectors of the economy.

Article 12: Socialist public property is sacred and inviolable.

The State protects socialist public property. Appropriation or damage of state or collective property by any organization or individual by whatever means is prohibited.

Article 13: Citizens' lawful private property is inviolable.

The State, in accordance with the law, protects the rights of citizens to private property and to its inheritance.

The State may, in the public interest and in accordance with the law, expropriate or requisition private property for its use and shall make compensation for private property expropriated or requisitioned.

Article 14: The State continuously raises labour productivity, improves economic results and develops the productive forces by enhancing the enthusiasm of the working people, raising the level of their technical skill, disseminating advanced science and technology, improving the systems of economic administration and enterprise operation and management, instituting the socialist system of responsibility in various forms and improving organization of work.

The State practises strict economy and combats waste.

The State properly apportions accumulation and consumption, pays attention to the interests of the collective and the individual as well as of the State and, on the basis of expanded production, gradually improves the material and cultural life of the people.

The State establishes a sound social security system compatible with the level of economic development.

Article 15: The State practises a socialist market economy. The State strengthens economic legislation and perfects macro-control. The State prohibits, according to the law, disturbance of society's economic order by any organization or individual.

Article 16: State-owned enterprises have decision-making power in operations within the limits prescribed by law.

State-owned enterprises practise democratic management through congresses of workers and staff and in other ways in accordance with the law.

Article 17: Collective economic organizations have decision-making power in conducting economic activities on the condition that they abide by the relevant laws. Collective economic organizations practise democratic management, elect and remove managerial personnel, and decide on major issues in accordance with the law.

Article 18: The People's Republic of China permits foreign enterprises, other foreign economic organizations and individual foreigners to invest in China and to enter into various forms of economic co-operation with Chinese enterprises and other economic organizations in accordance with the law of the People's Republic of China.

All foreign enterprises and other foreign economic organizations in China, as well as joint ventures with Chinese and foreign investment located in China, shall abide by the law of the People's Republic of China. Their lawful rights and interests are protected by the law of the People's Republic of China.

Article 19: The State develops socialist educational undertakings and works to raise the scientific and cultural level of the whole nation.

The State runs schools of various types, makes primary education compulsory and universal, develops secondary, vocational and higher education and promotes pre-school education.

The State develops educational facilities of various types in order to wipe out illiteracy and provide political, cultural, scientific, technical and professional education for workers, peasants, state functionaries and other working people. It encourages people to become educated through self-study.

The State encourages the collective economic organizations, state enterprises and undertakings and other social forces to set up educational institutions of various types in accordance with the law.

The State promotes the nation-wide use of Putonghua (common speech based on Beijing pronunciation).

Article 20: The State promotes the development of the natural and social sciences, disseminates scientific and technical knowledge, and commends and rewards achievements in scientific research as well as technological discoveries and inventions.

Article 21: The State develops medical and health services, promotes modern medicine and traditional Chinese medicine, encourages and supports the setting up of various medical and health facilities by the rural economic collectives, state enterprises and undertakings and neighbourhood organizations, and promotes sanitation activities of a mass character, all to protect the people's health.

The State develops physical culture and promotes mass sports activities to build up the people's physique.

Article 22: The State promotes the development of literature and art, the press, broadcasting and television undertakings, publishing and distribution services, libraries, museums, cultural centres and other cultural undertakings, that serve the people and socialism, and sponsors mass cultural activities.

The State protects places of scenic and historical interest, valuable cultural monuments and relics and other important items of China's historical and cultural heritage.

Article 23: The State trains specialized personnel in all fields who serve socialism, increases the number of intellectuals and creates conditions to give full scope to their role in socialist modernization.

Article 24: The State strengthens the building of socialist spiritual civilization through spreading education in high ideals and morality, general education and education in discipline and the legal system, and through promoting the formulation and observance of rules of conduct and common pledges by different sections of the people in urban and rural areas.

The State advocates the civic virtues of love for the motherland, for the people, for labour, for science and for socialism; it educates the people in patriotism, collectivism, internationalism and communism and in dialectical and historical materialism; it combats capitalist, feudalist and other decadent ideas.

Article 25: The State promotes family planning so that population growth may fit the plans for economic and social development.

Article 26: The State protects and improves the living environment and the ecological environment, and prevents and remedies pollution and other public hazards.

The State organizes and encourages afforestation and the protection of forests.

Article 27: All state organs carry out the principle of simple and efficient administration, the system of responsibility for work and the system of training functionaries and appraising their work in order constantly to improve quality of work and efficiency and combat bureaucratism.

All state organs and functionaries must rely on the support of the people, keep in close touch with them, heed their opinions and suggestions, accept their supervision and work hard to serve them.

Article 28: The State maintains public order and suppresses treasonable and other criminal activities that endanger national security; it penalizes activities that endanger public security and disrupt the socialist economy as well as other criminal activities; and it punishes and reforms criminals.

Article 29: The armed forces of the People's Republic of China belong to the people. Their tasks are to strengthen national defence, resist aggression, defend the motherland, safeguard the people's peaceful labour, participate in national reconstruction, and work hard to serve the people.

The State strengthens the revolutionization, modernization and regularization of the armed forces in order to increase the national defence capability.

Article 30: The administrative division of the People's Republic of China is as follows:

(1) The country is divided into provinces, autonomous regions and municipalities directly under the central government;

(2) Provinces and autonomous regions are divided into autonomous prefectures, counties, autonomous counties and cities;

(3) Counties and autonomous counties are divided into townships, nationality townships and towns.

Municipalities directly under the central government and other large cities are divided into districts and counties. Autonomous prefectures are divided into counties, autonomous counties, and cities.

All autonomous regions, autonomous prefectures and autonomous counties are national autonomous areas.

Article 31: The State may establish special administrative regions when necessary. The systems to be instituted in special administrative regions shall be prescribed by law enacted by the National People's Congress in the light of the specific conditions.

Article 32: The People's Republic of China protects the lawful rights and interests of foreigners within Chinese territory, and while on Chinese territory foreigners must abide by the law of the People's Republic of China.

The People's Republic of China may grant asylum to foreigners who request it for political reasons.

FUNDAMENTAL RIGHTS AND DUTIES OF CITIZENS

Article 33: All persons holding the nationality of the People's Republic of China are citizens of the People's Republic of China.

All citizens of the People's Republic of China are equal before the law.

Every citizen enjoys the rights and at the same time must perform the duties prescribed by the Constitution and the law.

The State respects and preserves human rights.

Article 34: All citizens of the People's Republic of China who have reached the age of 18 have the right to vote and stand for election, regardless of nationality, race, sex, occupation, family background, religious belief, education, property status, or length of residence, except persons deprived of political rights according to law.

Article 35: Citizens of the People's Republic of China enjoy freedom of speech, of the press, of assembly, of association, of procession and of demonstration.

Article 36: Citizens of the People's Republic of China enjoy freedom of religious belief.

No state organ, public organization or individual may compel citizens to believe in, or not to believe in, any religion; nor may they discriminate against citizens who believe in, or do not believe in, any religion.

The State protects normal religious activities. No one may make use of religion to engage in activities that disrupt public order, impair the health of citizens or interfere with the educational system of the state.

Religious bodies and religious affairs are not subject to any foreign domination.

Article 37: The freedom of person of citizens of the People's Republic of China is inviolable.

No citizen may be arrested except with the approval or by decision of a people's procuratorate or by decision of a people's court, and arrests must be made by a public security organ.

Unlawful deprivation or restriction of citizens' freedom of person by detention or other means is prohibited; and unlawful search of the person of citizens is prohibited.

Article 38: The personal dignity of citizens of the People's Republic of China is inviolable. Insult, libel, false charge or frame-up directed against citizens by any means is prohibited.

Article 39: The home of citizens of the People's Republic of China is inviolable. Unlawful search of, or intrusion into, a citizen's home is prohibited.

Article 40: The freedom and privacy of correspondence of citizens of the People's Republic of China are protected by law. No organization or individual may, on any ground, infringe upon the freedom and privacy of citizens' correspondence except in cases where, to meet the needs of state security or of investigation into criminal offences, public security or procuratorial organs are permitted to censor correspondence in accordance with procedures prescribed by law.

Article 41: Citizens of the People's Republic of China have the right to criticize and make suggestions to any state organ or functionary. Citizens have the right to make to relevant state organs complaints and charges against, or exposures of, violation of the law or dereliction of duty by any state organ or functionary; but fabrication or distortion of facts with the intention of libel or frame-up is prohibited.

In case of complaints, charges or exposures made by citizens, the state organ concerned must deal with them in a responsible manner after ascertaining the facts. No one may suppress such complaints, charges and exposures, or retaliate against the citizen making them.

Citizens who have suffered losses through infringement of their civic rights by any state organ or functionary have the right to compensation in accordance with the law.

Article 42: Citizens of the People's Republic of China have the right as well as the duty to work.

Using various channels, the State creates conditions for employment, strengthens labour protection, improves working conditions and, on the basis of expanded production, increases remuneration for work and social benefits.

Work is the glorious duty of every able-bodied citizen. All working people in state-owned enterprises and in urban and rural economic collectives should perform their tasks with an attitude consonant with their status as masters of the country. The State promotes socialist labour emulation, and commends and rewards model and advanced workers. The State encourages citizens to take part in voluntary labour.

The State provides necessary vocational training to citizens before they are employed.

Article 43: Working people in the People's Republic of China have the right to rest.

The State expands facilities for rest and recuperation of working people, and prescribes working hours and vacations for workers and staff.

Article 44: The State prescribes by law the system of retirement for workers and staff in enterprises and undertakings and for functionaries of organs of state. The livelihood of retired personnel is ensured by the State and society.

Article 45: Citizens of the People's Republic of China have the right to material assistance from the State and society when they are old, ill or disabled. The State develops the social insurance, social relief and medical and health services that are required to enable citizens to enjoy this right.

The State and society ensure the livelihood of disabled members of the armed forces, provide pensions to the families of martyrs and give preferential treatment to the families of military personnel.

The State and society help make arrangements for the work, livelihood and education of the blind, deaf-mute and other handicapped citizens.

Article 46: Citizens of the People's Republic of China have the duty as well as the right to receive education.

The State promotes the all-round moral, intellectual and physical development of children and young people.

Article 47: Citizens of the People's Republic of China have the freedom to engage in scientific research, literary and artistic creation and other cultural pursuits. The State encourages and assists creative endeavours conducive to the interests of the people that are made by citizens engaged in education, science, technology, literature, art and other cultural work.

Article 48: Women in the People's Republic of China enjoy equal rights with men in all spheres of life, political, economic, cultural and social, including family life.

The State protects the rights and interests of women, applies the principle of equal pay for equal work for men and women alike and trains and selects cadres from among women.

Article 49: Marriage, the family and mother and child are protected by the State.

Both husband and wife have the duty to practise family planning.

Parents have the duty to rear and educate their minor children, and children who have come of age have the duty to support and assist their parents.

Violation of the freedom of marriage is prohibited. Maltreatment of old people, women and children is prohibited.

Article 50: The People's Republic of China protects the legitimate rights and interests of Chinese nationals residing abroad and protects the lawful rights and interests of returned overseas Chinese and of the family members of Chinese nationals residing abroad.

Article 51: The exercise by citizens of the People's Republic of China of their freedoms and rights may not infringe upon the interests of the State, of society and of the collective, or upon the lawful freedoms and rights of other citizens.

Article 52: It is the duty of citizens of the People's Republic of China to safeguard the unity of the country and the unity of all its nationalities.

Article 53: Citizens of the People's Republic of China must abide by the Constitution and the law, keep state secrets, protect public property and observe labour discipline and public order and respect social ethics.

Article 54: It is the duty of citizens of the People's Republic of China to safeguard the security, honour and interests of the motherland; they must not commit acts detrimental to the security, honour and interests of the motherland.

Article 55: It is the sacred obligation of every citizen of the People's Republic of China to defend the motherland and resist aggression.

It is the honourable duty of citizens of the People's Republic of China to perform military service and join the militia in accordance with the law.

Article 56: It is the duty of citizens of the People's Republic of China to pay taxes in accordance with the law.

STRUCTURE OF THE STATE

The National People's Congress

Article 57: The National People's Congress of the People's Republic of China is the highest organ of state power. Its permanent body is the Standing Committee of the National People's Congress.

Article 58: The National People's Congress and its Standing Committee exercise the legislative power of the State.

Article 59: The National People's Congress is composed of deputies elected from the provinces, autonomous regions, municipalities directly under the Central Government, and the special administrative regions, and of deputies elected from the armed forces. All the minority nationalities are entitled to appropriate representation.

Election of deputies to the National People's Congress is conducted by the Standing Committee of the National People's Congress.

The number of deputies to the National People's Congress and the manner of their election are prescribed by law.

Article 60: The National People's Congress is elected for a term of five years.

Two months before the expiration of the term of office of a National People's Congress, its Standing Committee must ensure that the election of deputies to the succeeding National People's Congress is completed. Should exceptional circumstances prevent such an election, it may be postponed by decision of a majority vote of more than two-thirds of all those on the Standing Committee of the incumbent National People's Congress, and the term of office of the incumbent National People's Congress may be extended. The election of deputies to the succeeding National People's Congress must be completed within one year after the termination of such exceptional circumstances.

Article 61: The National People's Congress meets in session once a year and is convened by its Standing Committee. A session of the National People's Congress may be convened at any time the Standing Committee deems this necessary, or when more than one-fifth of the deputies to the National People's Congress so propose.

When the National People's Congress meets, it elects a presidium to conduct its session.

Article 62: The National People's Congress exercises the following functions and powers:

(1) to amend the Constitution;

(2) to supervise the enforcement of the Constitution;

(3) to enact and amend basic statutes concerning criminal offences, civil affairs, the state organs and other matters;

(4) to elect the President and the Vice-President of the People's Republic of China;

(5) to decide on the choice of the Premier of the State Council upon nomination by the President of the People's Republic of China, and to decide on the choice of the Vice-Premiers, State Councillors, Ministers in charge of Ministries or Commissions and the Auditor-General and the Secretary-General of the State Council upon nomination by the Premier;

(6) to elect the Chairman of the Central Military Commission and, upon his nomination, to decide on the choice of all the others on the Central Military Commission;

(7) to elect the President of the Supreme People's Court;

(8) to elect the Procurator-General of the Supreme People's Procuratorate;

(9) to examine and approve the plan for national economic and social development and the reports on its implementation;

(10) to examine and approve the state budget and the report on its implementation;

(11) to alter or annul inappropriate decisions of the Standing Committee of the National People's Congress;

(12) to approve the establishment of provinces, autonomous regions, and municipalities directly under the Central Government;

(13) to decide on the establishment of special administrative regions and the systems to be instituted there;

(14) to decide on questions of war and peace; and

(15) to exercise such other functions and powers as the highest organ of state power should exercise.

Article 63: The National People's Congress has the power to recall or remove from office the following persons:

(1) the President and the Vice-President of the People's Republic of China;

(2) the Premier, Vice-Premiers, State Councillors, Ministers in charge of Ministries or Commissions and the Auditor-General and the Secretary-General of the State Council;

(3) the Chairman of the Central Military Commission and others on the Commission;

(4) the President of the Supreme People's Court; and

(5) the Procurator-General of the Supreme People's Procuratorate.

Article 64: Amendments to the Constitution are to be proposed by the Standing Committee of the National People's Congress or by more than one-fifth of the deputies to the National People's Congress and adopted by a majority vote of more than two-thirds of all the deputies to the Congress.

Statutes and resolutions are adopted by a majority vote of more than one-half of all the deputies to the National People's Congress.

Article 65: The Standing Committee of the National People's Congress is composed of the following:

the Chairman;
the Vice-Chairmen;
the Secretary-General; and
members.

Minority nationalities are entitled to appropriate representation on the Standing Committee of the National People's Congress.

The National People's Congress elects, and has the power to recall, all those on its Standing Committee.

No one on the Standing Committee of the National People's Congress shall hold any post in any of the administrative, judicial or procuratorial organs of the State.

Article 66: The Standing Committee of the National People's Congress is elected for the same term as the National People's Congress; it exercises its functions and powers until a new Standing Committee is elected by the succeeding National People's Congress.

The Chairman and Vice-Chairmen of the Standing Committee shall serve no more than two consecutive terms.

Article 67: The Standing Committee of the National People's Congress exercises the following functions and powers:

(1) to interpret the Constitution and supervise its enforcement;

(2) to enact and amend statutes with the exception of those which should be enacted by the National People's Congress;

(3) to enact, when the National People's Congress is not in session, partial supplements and amendments to statutes enacted by the National People's Congress provided that they do not contravene the basic principles of these statutes;

(4) to interpret statutes;

(5) to examine and approve, when the National People's Congress is not in session, partial adjustments to the plan for national economic and social development and to the state budget that prove necessary in the course of their implementation;

(6) to supervise the work of the State Council, the Central Military Commission, the Supreme People's Court and the Supreme People's Procuratorate;

(7) to annul those administrative rules and regulations, decisions or orders of the State Council that contravene the Constitution or the statutes;

(8) to annul those local regulations or decisions of the organs of state power of provinces, autonomous regions and municipalities directly under the Central Government that contravene the Constitution, the statutes or the administrative rules and regulations;

(9) to decide, when the National People's Congress is not in session, on the choice of Ministers in charge of Ministries or Commissions or the Auditor-General and the Secretary-General of the State Council upon nomination by the Premier of the State Council;

(10) to decide, upon nomination by the Chairman of the Central Military Commission, on the choice of others on the Commission, when the National People's Congress is not in session.

(11) to appoint and remove the Vice-Presidents and judges of the Supreme People's Court, members of its Judicial Committee and the President of the Military Court at the suggestion of the President of the Supreme People's Court;

(12) to appoint and remove the Deputy Procurators-General and Procurators of the Supreme People's Procuratorate, members of its Procuratorial Committee and the Chief Procurator of the Military Procuratorate at the request of the Procurator-General of the Supreme People's Procuratorate, and to approve the appointment and removal of the Chief Procurators of the People's Procuratorates of provinces, autonomous regions and municipalities directly under the Central Government;

(13) to decide on the appointment and recall of plenipotentiary representatives abroad;

(14) to decide on the ratification and abrogation of treaties and important agreements concluded with foreign states;

(15) to institute systems of titles and ranks for military and diplomatic personnel and of other specific titles and ranks;

(16) to institute state medals and titles of honour and decide on their conferment;

(17) to decide on the granting of special pardons;

(18) to decide, when the National People's Congress is not in session, on the proclamation of a state of war in the event of an armed attack on the country or in fulfilment of international treaty obligations concerning common defence against aggression;

(19) to decide on general mobilization or partial mobilization;

(20) to decide on entering the state of emergency throughout the country or in particular provinces, autonomous regions, or municipalities directly under the Central Government; and

(21) to exercise such other functions and powers as the National People's Congress may assign to it.

Article 68: The Chairman of the Standing Committee of the National People's Congress presides over the work of the Standing Committee and convenes its meetings. The Vice-Chairmen and the Secretary-General assist the Chairman in his work.

Chairmanship meetings with the participation of the Chairman, Vice-Chairmen and Secretary-General handle the important day-to-day work of the Standing Committee of the National People's Congress.

Article 69: The Standing Committee of the National People's Congress is responsible to the National People's Congress and reports on its work to the Congress.

Article 70: The National People's Congress establishes a Nationalities Committee, a Law Committee, a Finance and Economic Committee, an Education, Science, Culture and Public Health Committee, a Foreign Affairs Committee, an Overseas Chinese Committee and such other special committees as are necessary. These special committees work under the direction of the Standing Committee of the National People's Congress when the Congress is not in session.

The special committees examine, discuss and draw up relevant bills and draft resolutions under the direction of the National People's Congress and its Standing Committee.

Article 71: The National People's Congress and its Standing Committee may, when they deem it necessary, appoint committees of inquiry into specific questions and adopt relevant resolutions in the light of their reports.

All organs of State, public organizations and citizens concerned are obliged to supply the necessary information to those committees of inquiry when they conduct investigations.

Article 72: Deputies to the National People's Congress and all those on its Standing Committee have the right, in accordance with procedures prescribed by law, to submit bills and proposals within the scope of the respective functions and powers of the National People's Congress and its Standing Committee.

Article 73: Deputies to the National People's Congress during its sessions, and all those on its Standing Committee during its meetings, have the right to address questions, in accordance with procedures prescribed by law, to the State Council or the Ministries and Commissions under the State Council, which must answer the questions in a responsible manner.

Article 74: No deputy to the National People's Congress may be arrested or placed on criminal trial without the consent of the presidium of the current session of the National People's Congress or, when the National People's Congress is not in session, without the consent of its Standing Committee.

Article 75: Deputies to the National People's Congress may not be called to legal account for their speeches or votes at its meetings.

Article 76: Deputies to the National People's Congress must play an exemplary role in abiding by the Constitution and the law and keeping state secrets and, in production and other work and their public activities, assist in the enforcement of the Constitution and the law.

Deputies to the National People's Congress should maintain close contact with the units which elected them and with the people, listen to and convey the opinions and demands of the people and work hard to serve them.

Article 77: Deputies to the National People's Congress are subject to the supervision of the units which elected them. The electoral units have the power, through procedures prescribed by law, to recall the deputies whom they elected.

Article 78: The organization and working procedures of the National People's Congress and its Standing Committee are prescribed by law.

The President of the People's Republic of China

Article 79: The President and Vice-President of the People's Republic of China are elected by the National People's Congress.

Citizens of the People's Republic of China who have the right to vote and to stand for election and who have reached the age of 45 are eligible for election as President or Vice-President of the People's Republic of China.

The term of office of the President and Vice-President of the People's Republic of China is the same as that of the National People's Congress, and they shall serve no more than two consecutive terms.

Article 80: The President of the People's Republic of China, in pursuance of decisions of the National People's Congress and its Standing Committee, promulgates statutes; appoints and removes the Premier, Vice-Premiers, State Councillors, Ministers in charge of Ministries or Commissions, and the Auditor-General and the Secretary-General of the State Council; confers state medals and titles of honour; issues orders of special pardons; proclaims entering the state of emergency; proclaims a state of war; and issues mobilization orders.

Article 81: The President of the People's Republic of China, on behalf of the People's Republic of China, engages in activities involving State affairs and receives foreign diplomatic representatives and, in pursuance of decisions of the Standing Committee of the National People's Congress, appoints and recalls plenipotentiary representatives abroad, and ratifies and abrogates treaties and important agreements concluded with foreign states.

Article 82: The Vice-President of the People's Republic of China assists the President in his work.

The Vice-President of the People's Republic of China may exercise such parts of the functions and powers of the President as the President may entrust to him.

Article 83: The President and Vice-President of the People's Republic of China exercise their functions and powers until the new President and Vice-President elected by the succeeding National People's Congress assume office.

Article 84: In case the office of the President of the People's Republic of China falls vacant, the Vice-President succeeds to the office of President.

In case the office of the Vice-President of the People's Republic of China falls vacant, the National People's Congress shall elect a new Vice-President to fill the vacancy.

In the event that the offices of both the President and the Vice-President of the People's Republic of China fall vacant, the National People's Congress shall elect a new President and a new Vice-

President. Prior to such election, the Chairman of the Standing Committee of the National People's Congress shall temporarily act as the President of the People's Republic of China.

The State Council

Article 85: The State Council, that is, the Central People's Government, of the People's Republic of China is the executive body of the highest organ of state power; it is the highest organ of state administration.

Article 86: The State Council is composed of the following: the Premier; the Vice-Premiers; the State Councillors; the Ministers in charge of ministries; the Ministers in charge of commissions; the Auditor-General; and the Secretary-General.

The Premier has overall responsibility for the State Council. The Ministers have overall responsibility for the respective ministries or commissions under their charge.

The organization of the State Council is prescribed by law.

Article 87: The term of office of the State Council is the same as that of the National People's Congress.

The Premier, Vice-Premiers and State Councillors shall serve no more than two consecutive terms.

Article 88: The Premier directs the work of the State Council. The Vice-Premiers and State Councillors assist the Premier in his work.

Executive meetings of the State Council are composed of the Premier, the Vice-Premiers, the State Councillors and the Secretary-General of the State Council.

The Premier convenes and presides over the executive meetings and plenary meetings of the State Council.

Article 89: The State Council exercises the following functions and powers:

(1) to adopt administrative measures, enact administrative rules and regulations and issue decisions and orders in accordance with the Constitution and the statutes;

(2) to submit proposals to the National People's Congress or its Standing Committee;

(3) to lay down the tasks and responsibilities of the ministries and commissions of the State Council, to exercise unified leadership over the work of the ministries and commissions and to direct all other administrative work of a national character that does not fall within the jurisdiction of the ministries and commissions;

(4) to exercise unified leadership over the work of local organs of state administration at different levels throughout the country, and to lay down the detailed division of functions and powers between the Central Government and the organs of state administration of provinces, autonomous regions and municipalities directly under the Central Government;

(5) to draw up and implement the plan for national economic and social development and the state budget;

(6) to direct and administer economic work and urban and rural development;

(7) to direct and administer the work concerning education, science, culture, public health, physical culture and family planning;

(8) to direct and administer the work concerning civil affairs, public security, judicial administration, supervision and other related matters;

(9) to conduct foreign affairs and conclude treaties and agreements with foreign states;

(10) to direct and administer the building of national defence;

(11) to direct and administer affairs concerning the nationalities, and to safeguard the equal rights of minority nationalities and the right of autonomy of the national autonomous areas;

(12) to protect the legitimate rights and interests of Chinese nationals residing abroad and protect the lawful rights and interests of returned overseas Chinese and of the family members of Chinese nationals residing abroad;

(13) to alter or annul inappropriate orders, directives and regulations issued by the ministries or commissions;

(14) to alter or annul inappropriate decisions and orders issued by local organs of state administration at different levels;

(15) to approve the geographic division of provinces, autonomous regions and municipalities directly under the Central Government, and to approve the establishment and geographic division of autonomous prefectures, counties, autonomous counties and cities;

(16) in accordance with the provisions of law, to decide on entering the state of emergency in parts of provinces, autonomous regions, and municipalities directly under the Central Government;

(17) to examine and decide on the size of administrative organs and, in accordance with the law, to appoint, remove and train administrative officers, appraise their work and reward or punish them; and

(18) to exercise such other functions and powers as the National People's Congress or its Standing Committee may assign it.

Article 90: The Ministers in charge of ministries or commissions of the State Council are responsible for the work of their respective departments and convene and preside over their ministerial meetings or commission meetings that discuss and decide on major issues in the work of their respective departments.

The ministries and commissions issue orders, directives and regulations within the jurisdiction of their respective departments and in accordance with the statutes and the administrative rules and regulations, decisions and orders issued by the State Council.

Article 91: The State Council establishes an auditing body to supervise through auditing the revenue and expenditure of all departments under the State Council and of the local government at different levels, and those of the state financial and monetary organizations and of enterprises and undertakings.

Under the direction of the Premier of the State Council, the auditing body independently exercises its power to supervise through auditing in accordance with the law, subject to no interference by any other administrative organ or any public organization or individual.

Article 92: The State Council is responsible, and reports on its work, to the National People's Congress or, when the National People's Congress is not in session, to its Standing Committee.

The Central Military Commission

Article 93: The Central Military Commission of the People's Republic of China directs the armed forces of the country.

The Central Military Commission is composed of the following: the Chairman; the Vice-Chairmen; and members.

The Chairman of the Central Military Commission has overall responsibility for the Commission.

The term of office of the Central Military Commission is the same as that of the National People's Congress.

Article 94: The Chairman of the Central Military Commission is responsible to the National People's Congress and its Standing Committee.

Two further sections, not included here, deal with the Local People's Congresses and Government and with the Organs of Self-Government of National Autonomous Areas, respectively.

The People's Courts and the People's Procuratorates

Article 123: The people's courts in the People's Republic of China are the judicial organs of the State.

Article 124: The People's Republic of China establishes the Supreme People's Court and the local people's courts at different levels, military courts and other special people's courts.

The term of office of the President of the Supreme People's Court is the same as that of the National People's Congress; he shall serve no more than two consecutive terms.

The organization of people's courts is prescribed by law.

Article 125: All cases handled by the people's courts, except for those involving special circumstances as specified by law, shall be heard in public. The accused has the right of defence.

Article 126: The people's courts shall, in accordance with the law, exercise judicial power independently and are not subject to interference by administrative organs, public organizations or individuals.

Article 127: The Supreme People's Court is the highest judicial organ.

The Supreme People's Court supervises the administration of justice by the local people's courts at different levels and by the special people's courts; people's courts at higher levels supervise the administration of justice by those at lower levels.

Article 128: The Supreme People's Court is responsible to the National People's Congress and its Standing Committee. Local people's courts at different levels are responsible to the organs of state power which created them.

Article 129: The people's procuratorates of the People's Republic of China are state organs for legal supervision.

Article 130: The People's Republic of China establishes the Supreme People's Procuratorate and the local people's procuratorates at different levels, military procuratorates and other special people's procuratorates.

The term of office of the Procurator-General of the Supreme People's Procuratorate is the same as that of the National People's Congress; he shall serve no more than two consecutive terms.

The organization of people's procuratorates is prescribed by law.

Article 131: People's procuratorates shall, in accordance with the law, exercise procuratorial power independently and are not subject to interference by administrative organs, public organizations or individuals.

Article 132: The Supreme People's Procuratorate is the highest procuratorial organ.

The Supreme People's Procuratorate directs the work of the local people's procuratorates at different levels and of the special people's procuratorates; people's procuratorates at higher levels direct the work of those at lower levels.

Article 133: The Supreme People's Procuratorate is responsible to the National People's Congress and its Standing Committee. Local people's procuratorates at different levels are responsible to the organs of state power at the corresponding levels which created them and to the people's procuratorates at the higher level.

Article 134: Citizens of all nationalities have the right to use the spoken and written languages of their own nationalities in court proceedings. The people's courts and people's procuratorates should provide translation for any party to the court proceedings who is not familiar with the spoken or written languages in common use in the locality.

In an area where people of a minority nationality live in a compact community or where a number of nationalities live together, hearings should be conducted in the language or languages in common use in the locality; indictments, judgments, notices and other documents should be written, according to actual needs, in the language or languages in common use in the locality.

Article 135: The people's courts, people's procuratorates and public security organs shall, in handling criminal cases, divide their functions, each taking responsibility for its own work, and they shall co-ordinate their efforts and check each other to ensure correct and effective enforcement of law.

THE NATIONAL FLAG, THE NATIONAL ANTHEM, THE NATIONAL EMBLEM AND THE CAPITAL

Article 136: The national flag of the People's Republic of China is a red flag with five stars.

The National Anthem of the People's Republic of China is the March of the Volunteers.

Article 137: The national emblem of the People's Republic of China is the Tiananmen (Gate of Heavenly Peace) in the centre, illuminated by five stars and encircled by ears of grain and a cogwheel.

Article 138: The capital of the People's Republic of China is Beijing (Peking).

The Government

HEAD OF STATE

President: HU JINTAO (elected by the 10th National People's Congress on 15 March 2003).

Vice-President: ZENG QINGHONG.

STATE COUNCIL
(September 2004)

Premier: WEN JIABAO.

Vice-Premiers: HUANG JU, WU YI, ZENG PEIYAN, HUI LIANGYU.

State Councillors: ZHOU YONGKANG, Gen. CAO GANGCHUAN, TANG JIAXUAN, HUA JIANMIN, CHEN ZHILI.

Secretary-General: HUA JIANMIN.

Minister of Foreign Affairs: LI ZHAOXING.

Minister of National Defence: Gen. CAO GANGCHUAN.

Minister of State Development and Reform Commission: MA KAI.

Minister of Education: ZHOU JI.

Minister of Science and Technology: XU GUANHUA.

Minister of State Commission of Science, Technology and Industry for National Defence: ZHANG YUNCHUAN.

Minister of State Nationalities Affairs Commission: LI DEZHU.

Minister of Public Security: ZHOU YONGKANG.

Minister of State Security: XU YONGYUE.

Minister of Supervision: LI ZHILUN.

Minister of Civil Affairs: LI XUEJU.

Minister of Justice: ZHANG FUSEN.

Minister of Finance: JIN RENQING.

Minister of Personnel: ZHANG BAILIN.

Minister of Labour and Social Security: ZHENG SILIN.

Minister of Land and Natural Resources: SUN WENSHENG.

Minister of Construction: WANG GUANGTAO.

Minister of Railways: LIU ZHIJUN.

Minister of Communications: ZHANG CHUNXIAN.

Minister of Information Industry: WANG XUDONG.

Minister of Water Resources: WANG SHUCHENG.

Minister of Agriculture: DU QINGLIN.

Minister of Commerce: BO XILAI.

Minister of Culture: SUN JIAZHENG.

Minister of Public Health: WU YI.

Minister of Population and State Family Planning Commission: ZHANG WEIQING.

Governor of the People's Bank of China: ZHOU XIAOCHUAN.

Auditor-General of the National Audit Office: LI JINHUA.

MINISTRIES

Ministry of Agriculture: 11 Nongzhanguan Nanli, Chao Yang Qu, Beijing 100026; tel. (10) 64192293; fax (10) 64192468; e-mail webmaster@agri.gov.cn; internet www.agri.gov.cn.

Ministry of Civil Affairs: 147 Beiheyan Dajie, Dongcheng Qu, Beijing 100721; tel. (10) 65135333; fax (10) 65135332.

Ministry of Communications: 11 Jianguomennei Dajie, Dongcheng Qu. Beijing 100736; tel. (10) 65292114; fax (10) 65292345; internet www.moc.gov.cn.

Ministry of Construction: 9 Sanlihe Dajie, Xicheng Qu, Beijing 100835; tel. (10) 68394215; fax (10) 68393333; e-mail webmaster@mail.cin.gov.cn; internet www.cin.gov.cn.

Ministry of Culture: 10 Chaoyangmen Bei Jie, Dongcheng Qu, Beijing 100020; tel. (10) 65551432; fax (10) 65551433; e-mail webmaster@whb1.ccnt.com.cn; internet www.ccnt.com.cn.

Ministry of Education: 37 Damucang Hutong, Xicheng Qu, Beijing 100816; tel. (10) 66096114; fax (10) 66011049; e-mail webmaster@moe.edu.cn; internet www.moe.edu.cn.

Ministry of Finance: 3 Nansanxiang, Sanlihe, Xicheng Qu, Beijing 100820; tel. (10) 68551888; fax (10) 68533635; e-mail webmaster@mof.gov.cn; internet www.mof.gov.cn.

Ministry of Foreign Affairs: 225 Chaoyangmennei Dajie, Dongsi, Beijing 100701; tel. (10) 65961114; fax (10) 65962146; e-mail webmaster@fmprc.gov.cn; internet www.fmprc.gov.cn.

Ministry of Foreign Trade and Economic Co-operation: 2 Dongchangan Jie, Dongcheng Qu, Beijing 100731; tel. (10) 67081526; fax (10) 67081513; e-mail webmaster@moftec.gov.cn; internet www.moftec.gov.cn.

Ministry of Information Industry: 13 Xichangan Jie, Beijing 100804; tel. (10) 66014249; fax (10) 66034248; e-mail webmaster@mii.gov.cn; internet www.mii.gov.cn.

Ministry of Justice: 10 Chaoyangmennan Dajie, Chao Yang Qu, Beijing 100020; tel. (10) 65205114; fax (10) 65205316.

Ministry of Labour and Social Security: 12 Hepinglizhong Jie, Dongcheng Qu, Beijing 100716; tel. (10) 84201235; fax (10) 64218350.

Ministry of Land and Natural Resources: 3 Guanyingyuanxiqu, Xicheng Qu, Beijing 100035; tel. (10) 66127001; fax (10) 66175348; internet www.mlr.gov.cn.

Ministry of National Defence: 20 Jingshanqian Jie, Beijing 100009; tel. (10) 66730000; fax (10) 65962146.

Ministry of Personnel: 12 Hepinglizhong Jie, Dongcheng Qu, Beijing 100716; tel. (10) 84223240; fax (10) 64211417.

Ministry of Public Health: 1 Xizhimenwai Bei Lu, Xicheng Qu, Beijing 100044; tel. (10) 68792114; fax (10) 64012369; e-mail zhou@chsi.moh.gov.cn; internet www.moh.gov.cn.

Ministry of Public Security: 14 Dongchangan Jie, Dongcheng Qu, Beijing 100741; tel. (10) 65122831; fax (10) 65136577.

Ministry of Railways: 10 Fuxing Lu, Haidian Qu, Beijing 100844; tel. (10) 63244150; fax (10) 63242150; e-mail webmaster@ns.chinamor.cn.net; internet www.chinamor.cn.net.

Ministry of Science and Technology: 15B Fuxing Lu, Haidian Qu, Beijing 100862; tel. (10) 68515050; fax (10) 68515006; e-mail officemail@mail.most.gov.cn; internet www.most.gov.cn.

Ministry of State Security: 14 Dongchangan Jie, Dongcheng Qu, Beijing 100741; tel. (10) 65244702.

Ministry of Supervision: 4 Zaojunmiao, Haidian Qu, Beijing 100081; tel. (10) 62256677; fax (10) 62254181.

Ministry of Water Resources: 2 Baiguang Lu, Ertiao, Xuanwu Qu, Beijing 100053; tel. (10) 63203069; fax (10) 63202650; internet www.mwr.gov.cn.

STATE COMMISSIONS

State Commission of Science, Technology and Industry for National Defence: 2A Guang'anmennan Jie, Xuanwu Qu, Beijing

100053; tel. (10) 63571397; fax (10) 63571398; internet www.costind.gov.cn.

State Development and Reform Commission: 38 Yuetannan Jie, Xicheng Qu, Beijing 100824; tel. (10) 68504409; fax (10) 68512929; e-mail news@sdpc.gov.cn; internet www.sdpc.gov.cn.

State Economic and Trade Commission: 26 Xuanwumenxi Dajie, Xuanwumen Qu, Beijing 100053; tel. (10) 63192334; fax (10) 63192348; e-mail webmaster@setc.gov.cn; internet www.setc.gov.cn.

State Nationalities Affairs Commission: 252 Taipingqiao Dajie, Xicheng Qu, Beijing 100800; tel. and fax (10) 66017375.

State Population and Family Planning Commission: 14 Zhichun Lu, Haidian Qu, Beijing 100088; tel. (10) 62046622; fax (10) 62051865; e-mail sfpcdfa@public.bta.net.cn; internet www.sfpc.gov.cn.

Legislature

QUANGUO RENMIN DAIBIAO DAHUI
(National People's Congress)

The National People's Congress (NPC) is the highest organ of state power, and is indirectly elected for a five-year term. The first plenary session of the 10th NPC was convened in Beijing in March 2003, and was attended by 2,916 deputies. The first session of the 10th National Committee of the Chinese People's Political Consultative Conference (CPPCC, Chair. Jia Qinglin), a revolutionary united front organization led by the Communist Party, took place simultaneously. The CPPCC holds discussions and consultations on the important affairs in the nation's political life. Members of the CPPCC National Committee or of its Standing Committee may be invited to attend the NPC or its Standing Committee as observers.

Standing Committee

In March 2003 158 members were elected to the Standing Committee, in addition to the following:

Chairman: Wu Bangguo.

Vice-Chairmen: Wang Zhaoguo, Li Tieying, Ismail Amat, He Luli, Ding Shisun, Cheng Siwei, Xu Jialu, Jiang Zhenghua, Gu Xiulian, Raidi, Sheng Huaren, Lu Yongxiang, Uyunqimg, Han Qide, Fu Tieshan.

Secretary-General: Sheng Huaren.

Provincial People's Congresses

Chairmen of Standing Committees of People's Congresses:

Provinces: Wang Taihua (Anhui), Song Defu (Fujian), (vacant) (Gansu), Lu Zhonghe (Guangdong), Qian Yunlu (Guizhou), (vacant) (Hainan), Bai Keming (Hebei), Song Fatan (Heilongjiang), Li Keqiang (Henan), Yang Yongliang (Hubei), Yang Zhengwu (Hunan), Li Yuanchao (Jiangsu), Meng Jianzhu (Jiangxi), Wang Yunkun (Jilin), Wen Shizhen (Liaoning), (vacant) (Qinghai), Li Jianguo (Shaanxi), Zhang Gaoli (Shandong), Tian Chengping (Shanxi), Zhang Xuezhong (Sichuan), Bai Enpei (Yunnan), Xi Jinping (Zhejiang).

Special Municipalities: Yu Junbo (Beijing), Huang Zhendong (Chongqing), Gong Xueping (Shanghai), Fang Fengyou (Tianjin).

Autonomous Regions: Cao Bochun (Guangxi Zhuang), Chu Bo (Nei Monggol), Chen Jianguo (Ningxia Hui), Legqog (Tibet—Xizang), Abudureyimu Amiti (acting—Xinjiang Uygur).

People's Governments

Provinces

Governors: Wang Jinshan (acting—Anhui), Lu Zhangong (Fujian), Lu Hao (Gansu), Huang Huahua (Guangdong), Shi Xiushi (Guizhou), Wei Liucheng (Hainan), Ji Yunshi (Hebei), Zhang Zuoji (Heilongjiang), Li Chengyu (Henan), Luo Qingquan (Hubei), Zhou Bohua (acting—Hunan), Liang Baohua (Jiangsu), Huang Zhiquan (Jiangxi), Hong Hu (Jilin), Bo Xilai (Liaoning), Yang Chuantang (acting—Qinghai), Jia Zhibang (Shaanxi), Han Yuqun (Shandong), Liu Zhenhua (Shanxi), Zhang Zhongwei (Sichuan), Xu Rongkai (Yunnan), Lu Zushan (Zhejiang).

Special Municipalities

Mayors: Wang Qishan (acting—Beijing), Wang Hongju (Chongqing), Han Zheng (Shanghai), Dai Xianglong (Tianjin).

Autonomous Regions

Chairmen: Lu Bing (acting—Guangxi Zhuang), Yang Jing (acting—Nei Monggol), Ma Qizhi (Ningxia Hui), Qiangba Puncog (Tibet—Xizang), Ismail Tiliwaldi (Xinjiang Uygur).

Political Organizations

COMMUNIST PARTY

Zhongguo Gongchan Dang (Chinese Communist Party—CCP): Beijing; f. 1921; 61m. mems in Dec. 1998; at the 16th Nat. Congress of the CCP in Nov. 2002, a new Cen. Cttee of 198 full mems and 158 alternate mems was elected; at its first plenary session the 16th Cen. Cttee appointed a new Politburo.

Sixteenth Central Committee

General Secretary: Hu Jintao.

Politburo

Members of the Standing Committee: Hu Jintao, Wu Bangguo, Wen Jiabao, Jia Qinglin, Zeng Qinghong, Huang Ju, Wu Guanzheng, Li Changchun, Luo Gan.

Other Full Members: Wang Lequan, Wang Zhaoguo, Hui Liangyu, Liu Qi, Liu Yunshan, Li Changchun, Wu Yi, Wu Bangguo, Wu Guanzheng, Zhang Lichang, Zhang Dejiang, Chen Liangyu, Luo Gan, Zhou Yongkang, Hu Jintao, Yu Zhengsheng, He Guoqiang, Jia Qinglin, Gen. Guo Boxiong, Huang Ju, Gen. Cao Gangchuan, Zeng Qinghong, Zeng Peiyan, Wen Jiabao.

Alternate Member: Wang Gang.

Secretariat: Zeng Qinghong, Liu Yunshan, Zhou Yongkang, He Guoqiang, Wang Gang, Gen. Xu Caihou, He Yong.

OTHER POLITICAL ORGANIZATIONS

China Association for Promoting Democracy: 98 Xinanli Guloufangzhuangchang, Beijing 100009; tel. (10) 64033452; f. 1945; mems drawn mainly from literary, cultural and educational circles; Chair. Xu Jialu; Sec.-Gen. Zhao Guanghua.

China Democratic League: 1 Beixing Dongchang Hutong, Beijing 100006; tel. (10) 65137983; fax (10) 65125090; f. 1941; formed from reorganization of League of Democratic Parties and Organizations of China; 131,300 mems, mainly intellectuals active in education, science and culture; Chair. Ding Shisun; Sec.-Gen. Zhang Baowen.

China National Democratic Construction Association: 208 Jixiangli, Chaowai Lu, Beijing 100020; tel. (10) 65523229; fax (10) 65523518; e-mail bgt@cndca.org.cn; internet www.cndca.org.cn; f. 1945; 94,544 mems, mainly industrialists and business executives; Chair. Cheng Siwei; Sec.-Gen. Zhang Jiao.

China Zhi Gong Dang (Party for Public Interests): Beijing; e-mail zhigong@public2.east.net.cn; f. 1925; reorg. 1947; mems are mainly returned overseas Chinese and scholars; Chair. Luo Haocai; Sec.-Gen. Qiu Guoyi.

Chinese Communist Youth League: 10 Qianmen Dongdajie, Beijing 100051; tel. (10) 67018132; fax (10) 67018131; e-mail guoji3acyt@yahoo.com; f. 1922; 68.5m. mems; First Sec. of Cen. Cttee Zhou Qiang.

Chinese Peasants' and Workers' Democratic Party: f. 1930 as the Provisional Action Cttee of the Kuomintang; took present name in 1947; more than 65,000 mems, active mainly in public health and medicine; Chair. Jiang Zhenghua; Sec.-Gen. Jiao Pingsheng (acting).

Jiu San (3 September) Society: f. 1946; fmrly Democratic and Science Soc.; 68,400 mems, mainly scientists and technologists; Chair. Han Qide; Sec.-Gen. Liu Ronghan.

Revolutionary Committee of the Chinese Kuomintang: tel. (10) 6550388; f. 1948; mainly fmr Kuomintang mems, and those in cultural, educational, health and financial fields; Chair. He Luli; Sec.-Gen. Liu Minfu.

Taiwan Democratic Self-Government League: f. 1947; recruits Taiwanese living on the mainland; Chair. Zhang Kehui; Sec.-Gen. Zhang Huajun.

During 1998 there were repeated failed attempts by pro-democracy activists to register an opposition party, the Chinese Democratic Party. The leaders of the party (Wang Youcai, Xu Wenli and Qin Yongmin) were sentenced to lengthy terms of imprisonment, and many other members of the party were detained.

Diplomatic Representation

EMBASSIES IN THE PEOPLE'S REPUBLIC OF CHINA

Afghanistan: 8 Dong Zhi Men Wai Dajie, Chao Yang Qu, Beijing 100600; tel. (10) 65321532; fax (10) 653226603; e-mail afgemb_beijing@yahoo.com; Ambassador Dr Qyamuddin Rahi Barlas.

Albania: 28 Guang Hua Lu, Jian Guo Men Wai, Beijing 100600; tel. (10) 65321120; fax (10) 65325451; Ambassador Kujtim Xhani.

Algeria: 2 Dong Zhi Men Wai Dajie, Chao Yang Qu, Beijing 100600; tel. (10) 65321231; fax (10) 65321648; Ambassador Madjid Bouguerra.

Angola: 1-13-1 Tayuan Diplomatic Office Bldg, Beijing 100600; tel. (10) 65326968; Ambassador João Manuel Bernardo.

Antigua and Barbuda: Guomen Bldg, Rm 1N, 1 Zuo Jia Zhuang, Chao Yang Qu, Beijing; tel. (10) 65326518; fax (10) 65326520; Ambassador James Thomas.

Argentina: Bldg 11, 5 Dong Wu Jie, San Li Tun, Beijing 100600; tel. (10) 65322090; fax (10) 65322319; e-mail echin@public.bta.net.cn; Ambassador Juan Carlos Morelli.

Armenia: 9-2-62 Tayuan Diplomatic Office Bldg, Beijing 100600; tel. (10) 65325677; fax (10) 65325654; Ambassador Vasili Ghazaryan.

Australia: 21 Dong Zhi Men Wai Dajie, San Li Tun, Beijing 100600; tel. (10) 65322331; fax (10) 65326718; e-mail pubaff.beijing@dfat.gov.au; internet www.austemb.org.cn; Ambassador Dr Alan William Thomas.

Austria: 5 Xiu Shui Nan Jie, Jian Guo Men Wai, Beijing 100600; tel. (10) 65322726; fax (10) 65321505; e-mail oebpekin@public.bta.net.cn; Ambassador Hans Dietmar Schweisgut.

Azerbaijan: 3-2-31 San Li Tun Diplomatic Compound, Beijing 100600; tel. (10) 65324614; fax (10) 65324615; e-mail safirprc@public.fhnet.cn.net; Ambassador Yashar Tofigi Aliyev.

Bahrain: 2-9-1 Tayuan Diplomatic Office Bldg, Beijing 100600; tel. (10) 65325025; fax (10) 65325016; Ambassador Karim Ebrahim al-Shakar.

Bangladesh: 42 Guang Hua Lu, Beijing 100600; tel. (10) 65321819; fax (10) 65324346; e-mail embbd@public.intercom.com.cn; Ambassador Ashfaqur Rahman.

Belarus: 2-10-1 Tayuan Diplomatic Office Bldg, Xin Dong Lu, Chao Yang Qu, Beijing 100600; tel. (10) 65326426; fax (10) 65326417; Ambassador Anatol Kharlap.

Belgium: 6 San Li Tun Lu, Beijing 100600; tel. (10) 65321736; fax (10) 65325097; e-mail Beijing@diplobel.org; Ambassador Gaston van Duyse-Adam.

Benin: 38 Guang Hua Lu, Jian Guo Men Wai, Beijing 100600; tel. (10) 65323054; fax (10) 65325103; Ambassador Pierre Ago Dossou.

Bolivia: 2-3-2 Tayuan Diplomatic Office Bldg, Beijing 100600; tel. (10) 65323074; fax (10) 65324686; e-mail embolch@public3.bta.net.cn; Ambassador Augusto Arguedas Del Carpio.

Bosnia and Herzegovina: 1-5-1 Tayuan Diplomatic Office Bldg; tel. (10) 65326587; fax (10) 65326418; Ambassador Borislav Marić.

Botswana: 1-8-1/2 Tayuan Diplomatic Office Bldg, Beijing 100600; tel. (10) 65325751; fax (10) 65325713; Ambassador Kgosi Seepapitso IV.

Brazil: 27 Guang Hua Lu, Jian Guo Men Wai, Beijing 100600; tel. (10) 65322881; fax (10) 65322751; e-mail empequim@public.bta.net.cn; Ambassador Luiz Castro Neves (designate).

Brunei: Villa No. 3, Qijiayuan Diplomatic Compound, Jian Guo Men Wai Dajie, Chao Yang Qu, Beijing 100600; tel. (10) 65324094; fax (10) 65324097; Ambassador Haji Abd. Hamid Abd. Halid.

Bulgaria: 4 Xiu Shui Bei Jie, Jian Guo Men Wai, Beijing 100600; tel. (10) 65321946; fax (10) 65324502; e-mail bulemb@public.bta.net.cn; Ambassador Angel Orbetsov.

Burundi: 25 Guang Hua Lu, Jian Guo Men Wai, Beijing 100600; tel. (10) 65321801; fax (10) 65322381; e-mail ambbubei@yahoo.fr; Ambassador Alfred Nkurunziza.

Cambodia: 9 Dong Zhi Men Wai Dajie, Beijing 100600; tel. (10) 65321889; fax (10) 65323507; Ambassador Khek Lerang.

Cameroon: 7 San Li Tun, Dong Wu Jie, Beijing 100600; tel. (10) 65321771; fax (10) 65321761; Ambassador Eleih-Elle Etian.

Canada: 19 Dong Zhi Men Wai Dajie, Chao Yang Qu, Beijing 100600; tel. (10) 65323536; fax (10) 65324311; internet www.beijing.gc.ca; Ambassador Joseph Caron.

Cape Verde: 6-2-121, Tayuan Diplomatic Office Bldg, Beijing; tel. (10) 65327547; fax (10) 65327546.

Chile: 1 Dong Si Jie, San Li Tun, Beijing 100600; tel. (10) 65321591; fax (10) 65323170; e-mail echilecn@public3.bta.net.cn; Ambassador Benny Pollack Eskenazi.

Colombia: 34 Guang Hua Lu, Jian Guo Men Wai, Beijing 100600; tel. (10) 65321713; fax (10) 65321969; Ambassador Guillermo Ricardo Velez Londono.

Congo, Democratic Republic: 6 Dong Wu Jie, San Li Tun, Beijing 100600; tel. (10) 65321995; fax (10) 65321360; Ambassador Johnson Baclongandi Wa Binana.

Congo, Republic: 7 Dong Si Jie, San Li Tun, Beijing 100600; tel. (10) 65321658; Ambassador Pierre Passi.

Côte d'Ivoire: 9 San Li Tun, Bei Xiao Jie, Beijing 100600; tel. (10) 65321223; fax (10) 65322407; Ambassador Konan Kramo.

Croatia: 2-72 San Li Tun Diplomatic Office Bldg, Beijing 100600; tel. (10) 65326241; fax (10) 65326257; e-mail vrhpek@public.bta.net.cn; Ambassador Boris Velić.

Cuba: 1 Xiu Shui Nan Jie, Jian Guo Men Wai, Beijing 100600; tel. (10) 65321714; fax (10) 65322870; Ambassador Alberto Rodríguez Arufe.

Cyprus: 2-13-2 Tayuan Diplomatic Office Bldg, Liang Ma He Nan Lu, Chao Yang Qu, Beijing 100600; tel. (10) 65325057; fax (10) 65324244; e-mail cyembpek@public3.bta.net.cn; Ambassador Petros Kestoras.

Czech Republic: Ri Tan Lu, Jian Guo Men Wai, Beijing 100600; tel. (10) 65326902; fax (10) 65325653; e-mail beijing@embassy.mzv.cz; internet www.mzv.cz/beijing; Ambassador Dr Tomáš Smetánka.

Denmark: 1 Dong Wu Jie, San Li Tun, Beijing 100600; tel. (10) 65322431; fax (10) 65322439; e-mail bjsamb@um.dk; internet www.dk-embassy-cn.org/; Ambassador Ole Loensmann Poulsen.

Djibouti: 2-2-102 Tayuan Diplomatic Office Bldg, Beijing; tel. (10) 65327857; fax (10) 65327858; Ambassador Moussa Bouh Odowa.

Ecuador: 2-62, San Li Tun Office Bldg, Chaoyang Qu, Beijing 100600; tel. (10) 65323158; fax (10) 65324371; e-mail embecuch@public3bta.net.cn; Ambassador Rodrigo Yepes Enriquez Herrera.

Egypt: 2 Ri Tan Dong Lu, Jian Guo Men Wai, Beijing 100600; tel. (10) 65321825; fax (10) 65325365; Ambassador Ali Houssam El Din Mahmoud Elhefny.

Equatorial Guinea: 2 Dong Si Jie, San Li Tun, Beijing; tel. (10) 65323709; fax (10) 65323805; e-mail ntugabeso@hotmail.com; Ambassador Narciso Ntugu Abeso Oyana.

Eritrea: Tayuan Diplomatic Office Bldg, Beijing 100600; tel. (10) 56326534; fax (10) 65326532; Ambassador Tseggai Tesfatsion Sereke.

Ethiopia: 3 Xiu Shui Nan Jie, Jian Guo Men Wai, Beijing 100600; tel. (10) 65325258; fax (10) 65325591; e-mail ethembcn@public.bta.net.cn; Ambassador Mait Martinson.

Fiji: 1-15-2 Tayuan Diplomatic Office Bldg, Beijing 100600; tel. (10) 65327305; fax (10) 6532 7253; e-mail lvratuvuki@hotmail.com; Ambassador Jeremaia Waqanisau.

Finland: Beijing Kerry Centre, 26/F South Tower, 1 Guanghua Lu, Beijing 100020; tel. (10) 85298541; fax (10) 85298547; e-mail sanomat.pek@formin.fi; internet www.finland-in-china.com; Ambassador Benjamin Basin.

France: 3 Dong San Jie, San Li Tun, Chao Yang Qu, Beijing 100600; tel. (10) 65321331; fax (10) 65324841; internet www.ambafrance-cn.org; Ambassador Philippe Guelluy.

Gabon: 36 Guang Hua Lu, Jian Guo Men Wai, Beijing 100600; tel. (10) 65322810; fax (10) 65322621; Ambassador M. Obiang-Ndoudum.

Germany: 17 Dong Zhi Men Wai Dajie, San Li Tun, Beijing 100600; tel. (10) 65322161; fax (10) 65325336; e-mail embassy@peki.diplo.de; internet www.beijing.diplo.de; Ambassador Joachim Broudré-Gröger.

Ghana: 8 San Li Tun Lu, Beijing 100600; tel. (10) 65321319; fax (10) 65323602; Ambassador Afare Apeadu Donkor.

Greece: 19 Guang Hua Lu, Jian Guo Men Wai, Beijing 100600; tel. (10) 65321588; fax (10) 65321277; Ambassador Charalambos Rocanas.

Guinea: 2 Xi Liu Jie, San Li Tun, Beijing 100600; tel. (10) 65323649; fax (10) 65324957; Ambassador El Hadji Djigui Camara.

Guinea-Bissau: Beijing; Ambassador Nicolau Dos Santos.

Guyana: 1 Xiu Shui Dong Jie, Jian Guo Men Wai, Beijing 100600; tel. (10) 65321601; fax (10) 65325741; Ambassador Ronald Mortimer Austin.

Hungary: 10 Dong Zhi Men Wai Dajie, San Li Tun, Beijing 100600; tel. (10) 65321431; fax (10) 65325053; Ambassador Mihály Bayer.

Iceland: Landmark Tower 1, 802, 8 North Dongsanhuan Lu, Beijing 100004; tel. (10) 65907795; fax (10) 65907801; e-mail icemb.beijing@utn.stjr.is; internet www.iceland.org/cn; Ambassador EIDUR GUDNASON.

India: 1 Ri Tan Dong Lu, Jian Guo Men Wai, Beijing 100600; tel. (10) 65321927; fax (10) 65324684; internet www.indianembassy.org.cn; Ambassador NALIN SURIE.

Indonesia: Diplomatic Office Bldg B, San Li Tun, Beijing 100600; tel. (10) 65325486; fax (10) 65325368; e-mail kombei@public3.bta.net.cn; Ambassador AA KUSTIA.

Iran: 13 Dong Liu Jie, San Li Tun, Beijing 100600; tel. (10) 65322040; fax (10) 65321403; Ambassador FEREYDOUN VERDINEJAD.

Iraq: 25 Xiu Shui Bei Jie, Jian Guo Men Wai, Beijing 100600; tel. (10) 65324355; fax (10) 65321599; e-mail iraqbeijing@yahoo.com; Chargé d'affaires a.i. RAHMAN LOUAN MOHSEN.

Ireland: 3 Ri Tan Dong Lu, Jian Guo Men Wai, Beijing 100600; tel. (10) 65322691; fax (10) 65326857; Ambassador DECLAN KELLEHER.

Israel: Room 405, West Wing Office, 1 Jian Guo Men Wai Dajie, Beijing 100004; tel. (10) 65052970; fax (10) 65050328; e-mail israemb@public.bta.net.cn; Ambassador YEHOYADA HAIM.

Italy: 2 Dong Er Jie, San Li Tun, Beijing 100600; tel. (10) 65322131; fax (10) 65324676; e-mail ambpech@ambpech.org.cn; internet www.italianembassy.org.cn; Ambassador GABRIELE MENEGATTI.

Japan: 7 Ri Tan Lu, Jian Guo Men Wai, Beijing 100600; tel. (10) 65322361; fax (10) 65324625; Ambassador FUMIYO ANAMIA.

Jordan: 5 Dong Liu Jie, San Li Tun, Beijing 100600; tel. (10) 65323906; fax (10) 65323283; Ambassador SAMIR I. AL-NAOURI.

Kazakhstan: 9 Dong Liu Jie, San Li Tun, Beijing 100600; tel. (10) 65326182; fax (10) 65326183; e-mail kazconscan@on.aibn.com; Ambassador ZHANYBEK SALIMOVICH KARIBZHANOV.

Kenya: 4 Xi Liu Jie, San Li Tun, Beijing 100600; tel. (10) 65323381; fax (10) 65321770; Ambassador RUTH SERETI SOLITEI.

Korea, Democratic People's Republic: Ri Tan Bei Lu, Jian Guo Men Wai, Beijing 100600; tel. (10) 65321186; fax (10) 65326056; Ambassador CHOE JIN SU.

Korea, Republic: 3rd–4th Floors, China World Trade Centre, 1 Jian Guo Men Wai Dajie, Beijing 100600; tel. (10) 65053171; fax (10) 65053458; Ambassador KIM HA-JOONG.

Kuwait: 23 Guang Hua Lu, Jian Guo Men Wai, Beijing 100600; tel. (10) 65322216; fax (10) 65321607; Ambassador FAISAL RASHED AL-GHAIS.

Kyrgyzstan: 2-4-1 Tayuan Diplomatic Office Bldg, Beijing 100600; tel. (10) 65326458; fax (10) 65326459; e-mail kyrgyzch@public2.east.net.cn; Ambassador ERLAN ABDYLDAEV.

Laos: 11 Dong Si Jie, San Li Tun, Chao Yang Qu, Beijing 100600; tel. (10) 65321224; fax (10) 65326748; e-mail laoemcn@public.east.cn.net; Ambassador THONGSAY BODHISANE.

Latvia: Unit 71, Green Land Garden, No. 1A Green Land Road, Chao Yang Qu, Beijing 100016; tel. (10) 64333863; fax (10) 64333810; e-mail kinas@163bj.com; Ambassador JANIS LOVNIKS.

Lebanon: 10 Dong Liu Jie, San Li Tun, Beijing 100600; tel. (10) 65322197; fax (10) 65322770; Ambassador SLEIMAN RASSI.

Lesotho: 2-3-13 San Li Tun Diplomatic Apartment, Beijing 100600; tel. (10) 65326842; fax (10) 65326845; e-mail doemli@public.bta.net.cn; Ambassador LEBOHANG K. MOLEKO.

Libya: 3 Dong Liu Jie, San Li Tun, Beijing 100600; tel. (10) 65323666; fax (10) 65323391; Secretary of the People's Bureau MUFTAH OTMAN MADI.

Lithuania: B30 King's Garden , 18 Xiaoyun Lu, Chaoyang Qu, Beijing 100016; tel. (10) 84518520; fax (10) 84514442; e-mail emlituan@public3.bta.net.cn; Ambassador ARTURAS ZURAUSKAS.

Luxembourg: 21 Nei Wu Bu Jie, Beijing 100600; tel. (10) 65135937; fax (10) 65137268; e-mail ambluxcn@public.bta.net.cn; Ambassador MARC UNGEHEUER.

Macedonia, former Yugoslav republic: 3-2-21 San Li Tun Diplomatic Office Bldg, Beijing; tel. (10) 65327846; fax (10) 65327847; e-mail macdebas@public3.bta.net.cn; Ambassador GEORGI EFREMOV.

Madagascar: 3 Dong Jie, San Li Tun, Beijing 100600; tel. (10) 65321353; fax (10) 65322102; e-mail ambpek@public2.bta.net.cn; Ambassador VICTOR SIKONINA.

Malaysia: 2 Liang Ma Qiao Bei Jie, Chaoyang Qu, San Li Tun, 100600 Beijing; tel. (10) 65322531; fax (10) 65325032; e-mail mwbjing@kln.gov.my; Ambassador Dato' ABDUL MAJID.

Mali: 8 Dong Si Jie, San Li Tun, Beijing 100600; tel. (10) 65321704; fax (10) 65321618; Ambassador MODIBO TIEMOKO TRAORE.

Malta: 1-52 San Li Tun Diplomatic Compound, Beijing 100600; tel. (10) 65323114; fax (10) 65326125; e-mail maltaemb@public3.bta.net.cn; Ambassador Dr SAVIOUR GAUCI.

Mauritania: 9 Dong San Jie, San Li Tun, Beijing 100600; tel. (10) 65321346; fax (10) 65321685; Ambassador N'GAIDE LAMINE KAYOU.

Mauritius: 202 Dong Wai Diplomatic Office Bldg, 23 Dong Zhi Men Wai Dajie, Chao Yang Qu, Beijing; tel. (10) 65325695; fax (10) 65325706; Ambassador L. K. C. LAM PO TANG.

Mexico: 5 Dong Wu Jie, San Li Tun, Beijing 100600; tel. (10) 65321717; fax (10) 65323744; e-mail embmxchn@public.bta.net.cn; Ambassador SERGIO LEY-LÓPEZ.

Moldova: 3-1-152 Tayuan Diplomatic Office Bldg, Beijing 100600; tel. (10) 65325379; Ambassador VICTOR BORSEVICI.

Mongolia: 2 Xiu Shui Bei Jie, Jian Guo Men Wai, Beijing 100600; tel. (10) 65321203; fax (10) 65325045; e-mail monembbj@public3.bta.net.cn; Ambassador L. AMARSANAA.

Morocco: 16 San Li Tun Lu, Beijing 100600; tel. (10) 65321489; fax (10) 65321453; e-mail embmor@public.bta.net.cn; Ambassador MOHAMED CHERTI.

Mozambique: 1-7-2 Tayuan Diplomatic Office Bldg, Beijing 100600; tel. (10) 65323664; fax (10) 65325189; e-mail embamoc@public.bta.net.cn; Ambassador ANTÓNIO INÁCIO JÚNIOR.

Myanmar: 6 Dong Zhi Men Wai Dajie, Chao Yang Qu, Beijing 100600; tel. (10) 65321584; fax (10) 65321344; Ambassador U THEIN LWIN.

Namibia: 2-9-2 Tayuan Diplomatic Office Bldg, Beijing 100600; tel. (10) 65324810; fax (10) 65324549; e-mail namemb@eastnet.com.cn; Ambassador H. U. IPINGE.

Nepal: 1 Xi Liu Jie, San Li Tun Lu, Beijing 100600; tel. (10) 65322739; fax (10) 65323251; Ambassador NARENDRA RAJ PANDAY.

Netherlands: 4 Liang Ma He Nan Lu, Beijing 100600; tel. (10) 65321131; fax (10) 65324689; Ambassador PHILIP DE HEER.

New Zealand: 1 Ri Tan, Dong Er Jie, Chao Yang Qu, Beijing 100600; tel. (10) 65322731; fax (10) 65324317; e-mail nzemb@eastnet.com.cn; Ambassador JOHN MCKINNON.

Niger: 3-2-12 San Li Tun, Beijing 100600; tel. (10) 65324279; e-mail nigerbj@public.bta.net.cn; Ambassador BOZARI SEYDOU.

Nigeria: 2 Dong Wu Jie, San Li Tun, Beijing; tel. (10) 65323631; fax (10) 65321650; Ambassador JONATHAN OLUWOLE COKER.

Norway: 1 Dong Yi Jie, San Li Tun, Beijing 100600; tel. (10) 65322261; fax (10) 65322392; e-mail emb.beijing@mfa.no; internet www.norway.cn; Ambassador TOR CHR. HILDAN.

Oman: 6 Liang Ma He Nan Lu, San Li Tun, Beijing 100600; tel. (10) 65323956; fax (10) 65325030; Ambassador ABDULLAH HOSNY.

Pakistan: 1 Dong Zhi Men Wai Dajie, San Li Tun, Beijing 100600; tel. (10) 65322504; fax (10) 65322715; e-mail pak@public.bta.net.cn; Ambassador RIAZ MOHAMMAD KHAN.

Papua New Guinea: 2-11-2 Tayuan Diplomatic Office Bldg, Beijing 100600; tel. (10) 65324312; fax (10) 65325483; Ambassador MAX RAI.

Peru: 1-91 San Li Tun, Bangonglou, Beijing 100600; tel. (10) 65323477; fax (10) 65322178; e-mail 1pekin@public.bta.net.cn; internet www.embperu.cn.net; Ambassador LUIS V. CHANG REYES.

Philippines: 23 Xiu Shui Bei Jie, Jian Guo Men Wai, Beijing 100600; tel. (10) 65321872; fax (10) 65323761; e-mail beijingpe@cinet.com.cn; internet www.philembassy-china.org; Ambassador WILLY C. GAA.

Poland: 1 Ri Tan Lu, Jian Guo Men Wai, Chaoyang Qu, Beijing 100600; tel. (10) 65321235; fax (10) 65321745; e-mail polska@public2.bta.net.cn; internet www.polandembassychina.net; Ambassador KSAWERY BURSKI.

Portugal: 8 San Li Tun Dong Wu Jie, Beijing 100600; tel. (10) 65323497; fax (10) 65324637; Ambassador ANTONIO NUNES DE CARVALHO SANTANA.

Qatar: 2-9-2 Tayuan Diplomatic Office Bldg, 14 Liang Ma He Nan Lu, Beijing 100600; tel. (10) 65322231; fax (10) 65325274; Ambassador SALEH ABDULLA AL-BOUANIN.

Romania: Ri Tan Lu, Dong Er Jie, Beijing 100600; tel. (10) 65323442; fax (10) 65325728; e-mail roamb@ht.rol.cn.net; Ambassador VIOREL ISTICIOAIA-BUDURA.

Russia: 4 Dong Zhi Men Nei, Bei Zhong Jie, Beijing 100600; tel. (10) 65322051; fax (10) 65324851; e-mail embassy@russia.org.cn; internet www.russia.org.cn; Ambassador SERGEI RAZOV.

Rwanda: 30 Xiu Shui Bei Jie, Jian Guo Men Wai, Beijing 100600; tel. (10) 65322193; fax (10) 65322006; e-mail ambarwda@public3.bta.net.cn; internet www.embarwanda-china.com; Ambassador JOSEPH BONESHA.

Saudi Arabia: 1 Bei Xiao Jie, San Li Tun, Beijing 100600; tel. (10) 65324825; fax (10) 65325324; Ambassador MOHAMMED A. AL-BESHIR.

Serbia and Montenegro: 1 Dong Liu Jie, San Li Tun, Beijing 100600; tel. (10) 65323516; fax (10) 65321207; e-mail ambjug@netchina.com.cn; internet www.emb-serbia-montenegro.org.cn; Ambassador DRAGAN MOMCILOVIĆ.

Sierra Leone: 7 Dong Zhi Men Wai Dajie, Beijing 100600; tel. (10) 65321222; fax (10) 65323752; e-mail adeen@163bj.com; Ambassador ALHUSINE DEEN.

Singapore: 1 Xiu Shui Bei Jie, Jian Guo Men Wai, Beijing 100600; tel. (10) 65323926; fax (10) 65322215; Ambassador CHIN SIAT-YOON.

Slovakia: Ri Tan Lu, Jian Guo Men Wai, Beijing 100600; tel. (10) 65321531; fax (10) 65324814; Ambassador PETER PAULEN.

Slovenia: Block F, 57 Ya Qu Yuan, King's Garden Villas, 18 Xiao Yun Lu, Chao Yang Qu, Beijing 100016; tel. (10) 64681030; fax (10) 64681040; Ambassador VLADIMIR GASPARIĆ.

Somalia: 2 San Li Tun Lu, Beijing 100600; tel. (10) 65321752; Ambassador MOHAMED HASSAN SAID.

South Africa: 5 Dongzhimen Wai Dajie, Chao Yang Qu, Beijing 100600; tel. (10) 65320171; fax (10) 65327319; e-mail safrican@163bj.com; Ambassador THEMBA M. N. KUBHEKA.

Spain: 9 San Li Tun Lu, Beijing 100600; tel. (10) 65321986; fax (10) 65323401; Ambassador JOSÉ PEDRO SEBASTIÁN DE ERICE Y GÓMEZ-ACEBO.

Sri Lanka: 3 Jian Hua Lu, Jian Guo Men Wai, Beijing 100600; tel. (10) 65321861; fax (10) 65325426; e-mail lkembj@public.east.cn.net; Ambassador B. A. B. GOONETILLEKE.

Sudan: Bldg 27, San Li Tun, Beijing 100600; tel. (10) 65323715; fax (10) 65321280; e-mail mission.sudan@itu.cn; Ambassador MIRHGANI MOHAMED SALIH.

Sweden: 3 Dong Zhi Men Wai Dajie, San Li Tun, Beijing 100600; tel. (10) 65329790; fax (10) 65325008; e-mail ambassaden.peking@foreign.ministry.se; internet www.swedemb-cn.org.cn; Ambassador BORJE LJUNGGREN.

Switzerland: 3 Dong Wu Jie, San Li Tun, Beijing 100600; tel. (10) 65322736; fax (10) 65324353; Ambassador DOMINIQUE DREYER.

Syria: 6 Dong Si Jie, San Li Tun, Beijing 100600; tel. (10) 65321563; fax (10) 65321575; Ambassador MOHAMMED KHEIR AL-WADI.

Tajikistan: 5-1-41 Tayuan Diplomatic Office Bldg, Beijing 100600; tel. (10) 65322598; internet (10) 65323039; Ambassador BOKHADYR NAJMIDINOVICH ABDOULLAYEV.

Tanzania: 8 Liang Ma He Nan Lu, San Li Tun, Beijing 100600; tel. (10) 65321408; fax (10) 65324985; Ambassador CHARLES ASILIA SANGA.

Thailand: 40 Guang Hua Lu, Jian Guo Men Wai, Beijing 100600; tel. (10) 65321903; fax (10) 65321748; Ambassador JULLAPONG NONSRICHAI.

Togo: 11 Dong Zhi Men Wai Dajie, Beijing 100600; tel. (10) 65322202; fax (10) 65325884; Ambassador NOLANA TA-AMA.

Tunisia: 1 Dong Jie, San Li Tun, Beijing 100600; tel. (10) 65322435; fax (10) 65325818; e-mail ambtun@public.netchina.com.cn; Ambassador SALAH HAMDI.

Turkey: 9 Dong Wu Jie, San Li Tun, Beijing 100600; tel. (10) 65322490; fax (10) 65325480; e-mail trkelcn@public.bta.net.cn; Ambassador RAFET AKGUNAY.

Turkmenistan: King's Garden, Villa D-26, 18 Xiao Yuan Rd, Beijing; tel. (10) 65326975; fax (10) 65326976; e-mail China@a-1.net.cn; Ambassador GURBANMUKHAMMET KASYMOV.

Uganda: 5 Dong Jie, San Li Tun, Beijing 100600; tel. (10) 65322370; fax (10) 65322242; e-mail ugembssy@public.bta.net.cn; internet www.uganda.cn777.com.cn; Ambassador PHILIP IDRO.

Ukraine: 11 Dong Liu Jie, San Li Tun, Beijing 100600; tel. (10) 65324013; fax (10) 65326359; e-mail ukrembcn@public3.bta.net.cn; internet www.ukrembcn.org; Ambassador SERGEY ALEKSEEVICH KAMYSHEV.

United Arab Emirates: C801 Lufthansa Center, Office Building 50, Liangmaqiao Lu, Chao Yang Qu, Beijing 100016; tel. (10) 84514416; fax (10) 84514451; Ambassador JUMA RASHED JASSIM.

United Kingdom: 11 Guang Hua Lu, Jian Guo Men Wai, Beijing 100600; tel. (10) 65321961; fax (10) 65321937; internet www.britishembassy.org.cn; Ambassador Sir CHRISTOPHER HUM.

USA: 3 Xiu Shui Bei Jie, Beijing 100600; tel. (10) 65323831; fax (10) 65323178; internet www.usembassy-china.org.cn; Ambassador CLARK T. RANDT.

Uruguay: 1-11-2 Tayuan Diplomatic Office Bldg, Beijing 100600; tel. (10) 65324445; fax (10) 65327375; e-mail urubei@public.bta.net.cn; Ambassador PELAYO JOAQUÍN DÍAZ MUGUERZA.

Uzbekistan: 11 Bei Xiao Jie, San Li Tun, Beijing 100600; tel. (10) 65326305; fax (10) 65326304; Ambassador NOSIRZHON YUSUPOV.

Venezuela: 14 San Li Tun Lu, Beijing 100600; tel. (10) 65321295; fax (10) 65323817; e-mail embvenez@public.bta.net.cn; Ambassador JUAN DE JESÚS MONTILLA SALDIVIA.

Viet Nam: 32 Guang Hua Lu, Jian Guo Men Wai, Beijing 100600; tel. (10) 65321155; fax (10) 65325720; Ambassador TRAN VAN LUAT.

Yemen: 5 Dong San Jie, San Li Tun, Beijing 100600; tel. (10) 65321558; fax (10) 65324305; Ambassador MARWAN ABDULLAH NU'MAN.

Zambia: 5 Dong Si Jie, San Li Tun, Beijing 100600; tel. (10) 65321554; fax (10) 65321891; Ambassador MWENYA LWATULA.

Zimbabwe: 7 Dong San Jie, San Li Tun, Beijing 100600; tel. (10) 65323795; fax (10) 65325383; Ambassador CHRISTOPHER HATIKURI MUTSVANGWA.

Judicial System

The general principles of the Chinese judicial system are laid down in Articles 123–135 of the December 1982 Constitution (q.v.).

PEOPLE'S COURTS

Supreme People's Court: 27 Dongjiaomin Xiang, Beijing 100745; tel. (10) 65136195; f. 1949; the highest judicial organ of the State; handles first instance cases of national importance; handles cases of appeals and protests lodged against judgments and orders of higher people's courts and special people's courts, and cases of protests lodged by the Supreme People's Procuratorate in accordance with the procedures of judicial supervision; reviews death sentences meted out by local courts, supervises the administration of justice by local people's courts; interprets issues concerning specific applications of laws in judicial proceedings; its judgments and rulings are final; Pres. XIAO YANG (five-year term of office coincides with that of National People's Congress, by which the President is elected).

Local People's Courts: comprise higher courts, intermediate courts and basic courts.

Special People's Courts: include military courts, maritime courts and railway transport courts.

PEOPLE'S PROCURATORATES

Supreme People's Procuratorate: 147 Beiheyan Dajie, Beijing 100726; tel. (10) 65126655; acts for the National People's Congress in examining govt depts, civil servants and citizens, to ensure observance of the law; prosecutes in criminal cases; Procurator-Gen. JIA CHUNWANG (elected by the National People's Congress for five years).

Local People's Procuratorates: undertake the same duties at the local level. Ensure that the judicial activities of the people's courts, the execution of sentences in criminal cases and the activities of departments in charge of reform through labour conform to the law; institute, or intervene in, important civil cases that affect the interest of the State and the people.

Religion

During the 'Cultural Revolution' places of worship were closed. After 1977 the Government adopted a policy of religious tolerance, and the 1982 Constitution states that citizens enjoy freedom of religious belief and that legitimate religious activities are protected. Many temples, churches and mosques subsequently reopened. Since 1994 all religious organizations have been required to register with the Bureau of Religious Affairs. In the late 1990s a new religious sect, the Falun Gong (also known as Falun Dafa), emerged and quickly gained new adherents. However, the authorities banned the group, describing it as an 'evil cult'.

Bureau of Religious Affairs: Beijing; tel. (10) 652625; Dir YE XIAOWEN.

ANCESTOR WORSHIP

Ancestor worship is believed to have originated with the deification and worship of all important natural phenomena. The divine and human were not clearly defined; all the dead became gods and were worshipped by their descendants. The practice has no code or dogma and the ritual is limited to sacrifices made during festivals and on birth and death anniversaries.

BUDDHISM

Buddhism was introduced into China from India in AD 67, and flourished during the Sui and Tang dynasties (6th–8th century), when eight sects were established. The Chan and Pure Land sects are the most popular. According to official sources, in 1998 there were 9,500 Buddhist temples in China. There were 100m. believers in 1997.

Buddhist Association of China (BAC): f. 1953; Pres. YI CHENG; Sec.-Gen. XUE CHENG.

Tibetan Institute of Lamaism

Pres. BUMI JANGBALUOZHU; Vice-Pres. CEMOLIN DANZENGCHILIE.

14th Dalai Lama: His Holiness the Dalai Lama TENZIN GYATSO, Thekchen Choeling, McLeod Ganj, Dharamsala 176 219, Himachal Pradesh, India; tel. (91) 1892-21343; fax (91) 1892-21813; e-mail ohhdl@cta.unv.ernet.ind.

Spiritual and temporal leader of Tibet; fled to India after failure of Tibetan national uprising in 1959.

CHRISTIANITY

During the 19th century and the first half of the 20th century large numbers of foreign Christian missionaries worked in China. According to official sources, there were 10m. Protestants and more than 4m. Catholics in China in 2000, although unofficial sources estimate that the Christian total could be as high as 90m. The Catholic Church in China operates independently of the Vatican. In addition, there is an increasing number of Christian sects in China.

Three-Self Patriotic Movement Committee of Protestant Churches of China: Pres. JI JIANHONG; Sec.-Gen. DENG FUCUN.

China Christian Council: 169 Yuan Ming Yuan Lu, Shanghai 200002; tel. (21) 63210806; fax (21) 63232605; e-mail tspmccc@online.sh.cn; f. 1980; comprises provincial Christian councils; Pres. and acting Sec.-Gen. Rev. CAO SHENGJIE.

The Roman Catholic Church: Catholic Mission, Si-She-Ku, Beijing; Bishop of Beijing MICHAEL FU TIESHAN (not recognized by the Vatican).

Chinese Patriotic Catholic Association: Pres. MICHAEL FU TIESHAN; Sec.-Gen. LIU BAINIAN; c. 3m. mems (1988).

CONFUCIANISM

Confucianism is a philosophy and a system of ethics, without ritual or priesthood. The respects that adherents accord to Confucius are not bestowed on a prophet or god, but on a great sage whose teachings promote peace and good order in society and whose philosophy encourages moral living.

DAOISM

Daoism was founded by Zhang Daoling during the Eastern Han dynasty (AD 125–144). Lao Zi, a philosopher of the Zhou dynasty (born 604 BC), is its principal inspiration, and is honoured as Lord the Most High by Daoists. According to official sources, there were 600 Daoist temples in China in 1998.

China Daoist Association: Temple of the White Cloud, Xi Bian Men, Beijing 100045; tel. (10) 6367179; f. 1957; Pres. MIN ZHITING; Sec.-Gen. YUAN BINGDONG.

ISLAM

According to Muslim history, Islam was introduced into China in AD 651. There were some 18m. adherents in China in 1997, chiefly among the Wei Wuer (Uygur) and Hui people, although unofficial sources estimate that the total is far higher, in the tens of millions.

Beijing Islamic Association: Dongsi Mosque, Beijing; f. 1979; Chair. Imam Al-Hadji CHEN GUANGYUAN.

China Islamic Association: Beijing 100053; tel. (10) 63546384; fax (10) 63529483; f. 1953; Chair. Imam Al-Hadji CHEN GUANGYUAN; Sec.-Gen. YU ZHENGUI.

The Press

In December 2000 China had 2,007 newspaper titles (including those below provincial level) and 8,187 periodicals. Each province publishes its own daily. Only the major newspapers and periodicals are listed below. In late 1999 the Government announced its intention to merge or close down a number of newspapers, leaving a single publication in each province. Foreign investment in some areas of the print media was allowed for the first time in December 2002. In June 2003 the Government barred all newspapers and periodicals from taking subscriptions for 2004, in an attempt to prevent publications from coercing readers into buying subscriptions. In November 2003 the Government suspended 673 unprofitable state-run newspapers under new regulations requiring the large majority of newspapers to become financially independent from the central Government.

PRINCIPAL NEWSPAPERS

Anhui Ribao (Anhui Daily): 206 Jinzhai Lu, Hefei, Anhui 230061; tel. (551) 2827842; fax (551) 2847302; Editor-in-Chief ZHANG YUXUAN.

Beijing Ribao (Beijing Daily): 34 Xi Biaobei Hutong, Dongdan, Beijing 100743; tel. (10) 65131071; fax (10) 65136522; f. 1952; organ of the Beijing municipal cttee of the CCP; Dir WAN YUNLAI; Editor-in-Chief LIU ZONGMING; circ. 700,000.

Beijing Wanbao (Beijing Evening News): 34 Xi Biaobei Hutong, Dongdan, Beijing 100743; tel. (10) 65132233; fax (10) 65126581; f. 1958; Editor-in-Chief REN HUANYING; circ. 800,000.

Beijing Youth Daily: Beijing; national and local news; promotes ethics and social service; circ. 3m.–4m.

Changsha Wanbao (Changsha Evening News): 161 Caie Zhong Lu, Changsha, Hunan 410005; tel. (731) 4424457; fax (731) 4445167.

Chengdu Wanbao (Chengdu Evening News): Qingyun Nan Jie, Chengdu 610017; tel. (28) 664501; fax (28) 666597; circ. 700,000.

China Business Times: Beijing; f. 1989; Editor HUANG WENFU; circ. 500,000.

Chongqing Ribao (Chongqing Daily): Chongqing; Dir and Editor-in-Chief LI HUANIAN.

Chungcheng Wanbao (Chungcheng Evening News): 51 Xinwen Lu, Kunming, Yunnan 650032; tel. (871) 4144642; fax (871) 4154192.

Dazhong Ribao (Dazhong Daily): 46 Jinshi Lu, Jinan, Shandong 250014; tel. (531) 2968989; fax (531) 2962450; internet www.dzdaily.com.cn; f. 1939; Dir XU XIYU; Editor-in-Chief LIU GUANGDONG; circ. 2,100,000.

Economic News: Editor-in-Chief DU ZULIANG.

Fujian Ribao (Fujian Daily): Hualin Lu, Fuzhou, Fujian; tel. (591) 57756; daily; Dir HUANG SHIYUN; Editor-in-Chief HUANG ZHONGSHENG.

Gongren Ribao (Workers' Daily): Liupukang, Andingmen Wai, Beijing 100718; tel. (10) 64211561; fax (10) 64214890; f. 1949; trade union activities and workers' lives; also major home and overseas news; Dir LIU YUMING; Editor-in-Chief SHENG MINGFU; circ. 2.5m.

Guangming Ribao (Guangming Daily): 106 Yongan Lu, Beijing 100050; tel. (10) 63017788; fax (10) 63039387; f. 1949; literature, art, science, education, history, economics, philosophy; Editor-in-Chief YUAN ZHIFA; circ. 920,000.

Guangxi Ribao (Guangxi Daily): Guangxi Region; Dir and Editor-in-Chief CHENG ZHENSHENG.

Guangzhou Ribao (Canton Daily): 10 Dongle Lu, Renmin Zhonglu, Guangzhou, Guangdong; tel. (20) 81887294; fax (20) 81862022; f. 1952; daily; social, economic and current affairs; Editor-in-Chief LI YUANJIANG; circ. 600,000.

Guizhou Ribao (Guizhou Daily): Guiyang, Guizhou; tel. (851) 627779; f. 1949; Dir GAO ZONGWEN; Editor-in-Chief GAN ZHENGSHU; circ. 300,000.

Hainan Ribao (Hainan Daily): 7 Xinhua Nan Lu, Haikou, Hainan 570001; tel. (898) 6222021; Dir ZHOU WENZHANG; Editor-in-Chief CHANG FUTANG.

Hebei Ribao (Hebei Daily): 210 Yuhuazhong Lu, Shijiazhuang, Hebei 050013; tel. (311) 6048901; fax (311) 6046969; f. 1949; Dir GUO ZENGPEI; Editor-in-Chief PAN GUILIANG; circ. 500,000.

Heilongjiang Ribao (Heilongjiang Daily): Heilongjiang Province; Dir JIA HONGTU; Editor-in-Chief AI HE.

Henan Ribao (Henan Daily): 1 Weiyi Lu, Zhengzhou, Henan; tel. (371) 5958319; fax (371) 5955636; f. 1949; Dir YANG YONGDE; Editor-in-Chief GUO ZHENGLING; circ. 390,000.

Huadong Xinwen (Eastern China News): f. 1995; published by Renmin Ribao.

Huanan Xinwen (South China News): Guangzhou; f. 1997; published by Renmin Ribao.

Hubei Ribao (Hubei Daily): 65 Huangli Lu, Wuhan, Hubei 430077; tel. (27) 6833522; fax (27) 6813989; f. 1949; Dir ZHOU NIANFENG; Editor-in-Chief SONG HANYAN; circ. 800,000.

Hunan Ribao (Hunan Daily): 18 Furong Zhong Lu, Changsha, Hunan 410071; tel. (731) 4312999; fax (731) 4314029; Dir JIANG XIANLI; Editor-in-Chief WAN MAOHUA.

Jiangxi Ribao (Jiangxi Daily): 175 Yangming Jie, Nanchang, Jiangxi; tel. (791) 6849888; fax (791) 6772590; f. 1949; Dir ZHOU JINGUANG; circ. 300,000.

Jiefang Ribao (Liberation Daily): 300 Han Kou Lu, Shanghai 200001; tel. (21) 63521111; fax (21) 63516517; f. 1949; Editor-in-Chief JIA SHUMEI; circ. 1m.

Jiefangjun Bao (Liberation Army Daily): Beijing; f. 1956; official organ of the Central Military Comm.; Editor-in-Chief Maj.-Gen. ZHANG SHIGANG; circ. 800,000.

Jilin Ribao (Jilin Daily): Jilin Province; Dir and Editor-in-Chief YI HONGBIN.

Jingji Ribao (Economic Daily): 2 Bai Zhi Fang Dong Jie, Beijing 100054; tel. (10) 63559988; fax (10) 63539408; f. 1983; financial affairs, domestic and foreign trade; administered by the State Council; Editor-in-Chief FENG BING; circ. 1.2m.

Jinrong Shibao (Financial News): 44 Taipingqiao Fengtaiqu, Beijing 100073; tel. (10) 63269233; fax (10) 68424931.

Liaoning Ribao (Liaoning Daily): Liaoning Province; Dir XIE ZHENGQIAN.

Nanfang Ribao (Nanfang Daily): 289 Guangzhou Da Lu, Guangzhou, Guangdong 510601; tel. (20) 87373998; fax (20) 87375203; f. 1949; Nanfang Daily Group also publishes *Nanfang Dushi Bao* (Southern Metropolis Daily), *Ershiyi Shiji Jingji Baodao* (21st Century Economic Herald), and weekly edn *Nanfang Zhoumou* (Southern Weekend); Editor-in-Chief, *Nanfang Dushi Bao* (vacant); Gen. Man., *Nanfang Dushi Bao* (vacant); circ. 1m.

Nanjing Ribao (Nanjing Daily): 53 Jiefang Lu, Nanjing, Jiangsu 210016; tel. (25) 4496564; fax (25) 4496544.

Nongmin Ribao (Peasants' Daily): Shilipu Beili, Chao Yang Qu, Beijing 100025; tel. (10) 65005522; fax (10) 65071154; f. 1980; 6 a week; circulates in rural areas nation-wide; Dir ZHANG DEXIU; Editor-in-Chief ZHANG WENBAO; circ. 1m.

Renmin Ribao (People's Daily): 2 Jin Tai Xi Lu, Chao Yang Qu, Beijing 100733; tel. (10) 65368971; fax (10) 65368974; internet www.people.com.cn; f. 1948; organ of the CCP; also publishes overseas edn; Pres. WANG CHEN; Editor-in-Chief ZHANG YANNONG; circ. 2.15m.

Shaanxi Ribao (Shaanxi Daily): Shaanxi Province; Dir LI DONGSHENG; Editor-in-Chief DU YAOFENG.

Shanxi Ribao (Shanxi Daily): 24 Shuangtasi Jie, Taiyuan, Shanxi; tel. (351) 446561; fax (351) 441771; Dir ZHAO WENBIN; Editor-in-Chief LI DONGXI; circ. 300,000.

Shenzhen Commercial Press: Shenzhen; Editor-in-Chief GAO XINGLIE.

Shenzhen Tequ Bao (Shenzhen Special Economic Zone Daily): 4 Shennan Zhonglu, Shenzhen 518009; tel. (755) 3902688; fax (755) 3906900; f. 1982; reports on special economic zones, as well as mainland, Hong Kong and Macao; Editor-in-Chief DONG YANGLUE.

Sichuan Ribao (Sichuan Daily): Sichuan Daily Press Group, 70 Hongxing Zhong Lu, Erduan, Chengdu, Sichuan 610012; tel. and fax (28) 86968000; internet www.sconline.com.cn; f. 1952; Chair. of Bd LI ZHIXIA; Editor-in-Chief TANG XIAOQIANG; circ. 8m.

Tianjin Ribao (Tianjin Daily): 873 Dagu Nan Lu, Heri Qu, Tianjin 300211; tel. (22) 7301024; fax (22) 7305803; f. 1949; Dir and Editor-in-Chief ZHANG JIANXING; circ. 600,000.

Wenhui Bao (Wenhui Daily): 50 Huqiu Lu, Shanghai 200002; tel. (21) 63211410; fax (21) 63230198; f. 1938; Editor-in-Chief WU ZHENBIAO; circ. 500,000.

Xin Min Wan Bao (Xin Min Evening News): 839 Yan An Zhong Lu, Shanghai 200040; tel. (21) 62791234; fax (21) 62473220; f. 1929; specializes in public policy, education and social affairs; Editor-in-Chief JIN FUAN; circ. 1.8m.

Xinhua Ribao (New China Daily): 55 Zhongshan Lu, Nanjing, Jiangsu 210005; tel. (21) 741757; fax (21) 741023; Editor-in-Chief ZHOU ZHENGRONG; circ. 900,000.

Xinjiang Ribao (Xinjiang Daily): Xinjiang Region; Editor-in-Chief HUANG YANCAI.

Xizang Ribao (Tibet Daily): Tibet; Editor-in-Chief LI ERLIANG.

Yangcheng Wanbao (Yangcheng Evening News): 733 Dongfeng Dong Lu, Guangzhou, Guangdong 510085; tel. (20) 87776211; fax (20) 87765103; e-mail ycwbic@ycwb.com.cn; internet www.ycwb.com.cn; f. 1957; Editor-in-Chief PAN WEIWEN; circ. 1.3m.

Yunnan Ribao (Yunnan Daily): Yunnan Province; Editor-in-Chief SUN GUANSHENG.

Zhejiang Ribao (Zhejiang Daily): Zhejiang Province; Dir CHEN MINER; Editor-in-Chief YANG DAJIN.

Zhongguo Qingnian Bao (China Youth News): 2 Haiyuncang, Dong Zhi Men Nei, Beijing 100702; tel. (10) 64032233; fax (10) 64033792; f. 1951; daily; aimed at 14–40 age-group; Dir XU ZHUQING; Editor-in Chief LI XUEQIAN; circ. 1.0m.

Zhongguo Ribao (China Daily): 15 Huixin Dongjie, Chao Yang Qu, Beijing 100029; tel. (10) 64918633; fax (10) 64918377; internet www.chinadaily.com.cn; f. 1981; English; China's political, economic and cultural developments; world, financial and sports news; also publishes *Business Weekly* (f. 1985), *Beijing Weekend* (f. 1991), *Shanghai Star* (f. 1992), *Reports from China* (f. 1992), *21st Century* (f. 1993); Editor-in-Chief ZHU YINGHUANG; circ. 300,000.

Zhongguo Xinwen (China News): 12 Baiwanzhuang Nanjie, Beijing; tel. (10) 68315012; f. 1952; daily; Editor-in-Chief WANG XIJIN; current affairs.

SELECTED PERIODICALS

Ban Yue Tan (China Comment): Beijing; tel. (10) 6668521; f. 1980; in Chinese and Wei Wuer (Uygur); Editor-in-Chief DONG RUISHENG; circ. 6m.

Beijing Review: 24 Baiwanzhuang Lu, Beijing 100037; tel. (10) 68326085; fax (10) 68326628; e-mail bjreview@public3.bta.net.cn; internet www.bjreview.com.cn; f. 1958; weekly; edns in English, French, Spanish, Japanese and German; also **Chinafrica** (monthly in English and French); Publr WANG GANGYI; Editor-in-Chief LII HAIBO.

BJ TV Weekly: 2 Fu Xing Men Wai Zhenwumiao Jie, Beijing 100045; tel. (10) 6366036; fax (10) 63262388; circ. 1m.

Caijing: 10th Floor, Prime Tower 22 Chaoyangmenwai Lu, Beijing 100020 ; tel. (10) 65885047; fax (10) 65885046; internet www.caijing.com.cn/english/; f. 1998; business and finance, 2 a month; Editor HU SHULI.

China TV Weekly: 15 Huixin Dong Jie, Chao Yang Qu, Beijing 100013; tel. (10) 64214197; circ. 1.7m.

Chinese Literature Press: 24 Baiwanzhuang Lu, Beijing 100037; tel. (10) 68326010; fax (10) 68326678; e-mail chinalit@public.east.cn.net; f. 1951; monthly (bilingual in English); quarterly (bilingual in French); contemporary and classical writing, poetry, literary criticism and arts; Exec. Editor LING YUAN.

Dianying Xinzuo (New Films): 796 Huaihai Zhong Lu, Shanghai; tel. (21) 64379710; f. 1979; bi-monthly; introduces new films.

Dianzi yu Diannao (Electronics and Computers): Beijing; f. 1985; popularized information on computers and microcomputers.

Dushu (Reading): Beijing; internet www.dushu.net; academic journal on social science and humanities; Editor WANG HUI.

Elle (China): 14 Lane 955, Yan'an Zhong Lu, Shanghai; tel. (21) 62790974; fax (21) 62479056; f. 1988; monthly; fashion; Pres. YANG XINCI; Chief Editor WU YING; circ. 300,000.

Family Magazine: 14 Siheng Lu, Xinhepu, Dongshan Qu, Guangzhou 510080; tel. (20) 7777718; fax (20) 7185670; monthly; circ. 2.5m.

Feitian (Fly Skywards): 50 Donggan Xilu, Lanzhou, Gansu; tel. (931) 25803; f. 1961; monthly.

Guoji Xin Jishu (New International Technology): Zhanwang Publishing House, Beijing; f. 1984; also publ. in Hong Kong; international technology, scientific and technical information.

Guowai Keji Dongtai (Recent Developments in Science and Technology Abroad): Institute of Scientific and Technical Information of China, 54 San Li He Lu, Beijing 100045; tel. (10) 68570713; fax (10) 68511839; e-mail baiyr@istic.ac.cn; internet www.wanfang.com.cn; f. 1962; monthly; scientific journal; Editor-in-Chief GUO YUEHUA; circ. 40,000.

Hai Xia (The Strait): 27 De Gui Xiang, Fuzhou, Fujian; tel. (10) 33656; f. 1981; quarterly; literary journal; CEOs YANG YU, JWO JONG LIN.

Huasheng Monthly (Voice for Overseas Chinese): 12 Bai Wan Zhuang Nan Jie, Beijing 100037; tel. (10) 68311578; fax (10) 68315039; f. 1995; monthly; intended mainly for overseas Chinese and Chinese nationals resident abroad; Editor-in-Chief FAN DONGSHENG.

Jianzhu (Construction): Baiwanzhuang, Beijing; tel. (10) 68992849; f. 1956; monthly; Editor FANG YUEGUANG; circ. 500,000.

Jinri Zhongguo (China Today): 24 Baiwanzhuang Lu, Beijing 100037; tel. (10) 68326037; fax (10) 68328338; internet www.chinatoday.com.cn; f. 1952; fmrly *China Reconstructs*; monthly; edns in English, Spanish, French, Arabic, German, and Chinese; economic, social and cultural affairs; illustrated; Pres. and Editor-in-Chief HUANG ZU'AN.

Liaowang (Outlook): 57 Xuanwumen Xijie, Beijing; tel. (10) 63073049; f. 1981; weekly; current affairs; Gen. Man. ZHOU YICHANG; Editor-in-Chief JI BIN; circ. 500,000.

Luxingjia (Traveller): Beijing; tel. (10) 6552631; f. 1955; monthly; Chinese scenery, customs, culture.

Meishu Zhi You (Chinese Art Digest): 32 Beizongbu Hutong, East City Region, Beijing; tel. (10) 65591404; f. 1982; every 2 months; art review journal, also providing information on fine arts publs in China and abroad; Editors ZONGYUAN GAO, PEI CHENG.

Nianqingren (Young People): 169 Mayuanlin, Changsha, Hunan; tel. (731) 23610; f. 1981; monthly; general interest for young people.

Nongye Zhishi (Agricultural Knowledge): 21 Ming Zi Qian Lu, Jinan, Shandong 250100; tel. (531) 8932238; e-mail sdnyzs@jn-public.sd.cninfo.net; internet www.sdny.com.cn; f. 1950; fortnightly; popular agricultural science; Dir YANG LIJIAN; circ. 410,000.

Qiushi (Seeking Truth): 2 Shatan Beijie, Beijing 100727; tel. (10) 64037005; fax (10) 64018174; f. 1988 to succeed *Hong Qi*(Red Flag); 2 a month; theoretical journal of the CCP; Editor-in-Chief WANG TIANXI; circ. 1.83m.

Renmin Huabao (China Pictorial): Huayuancun, West Suburbs, Beijing 100044; tel. (10) 68411144; fax (10) 68413023; f. 1950; monthly; edns: two in Chinese, one in Tibetan and 12 in foreign languages; Dir and Editor-in-Chief ZHANG JIAHUA.

Shichang Zhoubao (Market Weekly): 2 Duan, Sanhao Jie, Heping Qu, Shenyang, Liaoning; tel. (24) 482983; f. 1979; weekly in Chinese; trade, commodities, and financial and economic affairs; circ. 1m.

Shufa (Calligraphy): 81 Qingzhou Nan Lu, Shanghai 200233; tel. (21) 64519008; fax (21) 64519015; f. 1977; every 2 months; journal on ancient and modern calligraphy; Chief Editor LU FUSHENG.

Tiyu Kexue (Sports Science): 8 Tiyuguan Lu, Beijing 100763; tel. (10) 67112233; f. 1981; sponsored by the China Sports Science Soc.; every 2 months; summary in English; Chief Officer YUAN WEIMIN; in Chinese; circ. 20,000.

Wenxue Qingnian (Youth Literature Journal): 27 Mu Tse Fang, Wenzhou, Zhejiang; tel. (577) 3578; f. 1981; monthly; Editor-in-Chief CHEN YUSHEN; circ. 80,000.

Women of China English Monthly: 15 Jian Guo Men Dajie, Beijing 100730; tel. (10) 65134616; fax (10) 65225380; e-mail geo@womenofchina.com.cn; internet www.womenofchina.com.cn; f. 1956; monthly; in English; administered by All-China Women's Federation; women's rights and status, views and lifestyle, education and arts, etc.; Editor-in-Chief YUN PENGJU.

Xian Dai Faxue (Modern Law Science): Southwest University of Political Science and Law, Chongqing, Sichuan 400031; tel. (23) 65382527; e-mail MLS@swupl.edu.cn; f. 1979; bi-monthly; with summaries in English; Dirs CAO MINGDE, LI YUPING.

Yinyue Aihaozhe (Music Lovers): 74 Shaoxing Lu, Shanghai 200020; tel. (21) 64372608; fax (21) 64332019; f. 1979; every two months; music knowledge; illustrated; Editor-in-Chief CHEN XUEYA; circ. 50,000.

Zhongguo Duiwai Maoyi Ming Lu (Directory of China's Foreign Trade): CCPIT Bldg, 1 Fuxingmen Wai Da Jie, Beijing 100860; tel. (10) 68022948; fax (10) 68510201; e-mail inform@press-media.com; f. 1974; monthly; edns in Chinese and English; information on Chinese imports and exports, foreign trade and economic policies; Editor-in-Chief YANG HAIQING.

Zhongguo Ertong (Chinese Children): 21 Xiang 12, Dongsi, Beijing; tel. (10) 6444761; f. 1980; monthly; illustrated journal for elementary school pupils.

Zhongguo Guangbo Dianshi (China Radio and Television): 12 Fucheng Lu, Beijing; tel. (10) 6896217; f. 1982; monthly; reports and comments.

Zhongguo Jin Rong Xin Xi: Beijing; f. 1991; monthly; economic news.

Zhongguo Sheying (Chinese Photography): 61 Hongxing Hutong, Dongdan, Beijing 100005; tel. (10) 65252277; fax (10) 65257623; e-mail cphoto@public.bta.net.cn; internet www.cphoto.com.cn; f. 1957; monthly; photographs and comments; Editor WU CHANGYUN.

Zhongguo Zhenjiu (Chinese Acupuncture and Moxibustion): China Academy of Traditional Chinese Medicine, Dongzhimen Nei, Beijing 100700; tel. (10) 84014607; fax (10) 64013968; e-mail weihongliu@263.net; f. 1981; monthly; publ. by Chinese Soc. of Acupuncture and Moxibustion; abstract in English; Editor-in-Chief Prof. DENG LIANGYUE.

Zijing (Bauhinia): Pres. and Editor-in-Chief CHEN HONG.

Other popular magazines include **Gongchandang Yuan** (Communists, circ. 1.63m.) and **Nongmin Wenzhai** (Peasants' Digest, circ. 3.54m.).

NEWS AGENCIES

Xinhua (New China) News Agency: 57 Xuanwumen Xidajie, Beijing 100803; tel. (10) 63071114; fax (10) 63071210; internet www.xinhuanet.com; f. 1931; offices in all Chinese provincial capitals, and about 100 overseas bureaux; news service in Chinese, English, French, Spanish, Portuguese, Arabic and Russian, feature and photographic services; Pres. TIAN CONGMING; Editor-in-Chief NAN ZHENZHONG.

Zhongguo Xinwen She (China News Agency): POB 1114, Beijing; f. 1952; office in Hong Kong; supplies news features, special articles and photographs for newspapers and magazines in Chinese printed overseas; services in Chinese; Dir WANG SHIGU.

Foreign Bureaux

Agence France-Presse (AFP) (France): 11-11 Jian Guo Men Wai, Diplomatic Apts, Beijing 100600; tel. (10) 65321409; fax (10) 65322371; e-mail afppek@afp.com; Bureau Chief ELIZABETH ZINGO.

Agencia EFE (Spain): 2-2-132 Jian Guo Men Wai, Beijing 100600; tel. (10) 65323449; fax (10) 65323688; Rep. CARLOS REDONDO.

Agenzia Nazionale Stampa Associata (ANSA) (Italy): 1-11 Ban Gong Lu, San Li Tun, Beijing 100600; tel. (10) 65323651; fax (10) 65321954; e-mail barbara@public3.bta.net.cn; Bureau Chief BARBARA ALIGHIERO.

Allgemeiner Deutscher Nachrichtendienst (ADN) (Germany): 7-2-61, Jian Guo Men Wai, Qi Jia Yuan Gong Yu, Beijing 100600; tel. and fax (10) 65321115; Correspondent Dr LUTZ POHLE.

Associated Press (AP) (USA): 6-2-22 Jian Guo Men Wai, Diplomatic Quarters, Beijing 100600; tel. (10) 65326650; fax (10) 65323419; Bureau Chief ELAINE KURTENBACH.

Deutsche Presse-Agentur (dpa) (Germany): Ban Gong Lou, Apt 1-31, San Li Tun, Beijing 100600; tel. (10) 65321473; fax (10) 65321615; e-mail dpa@public3.bta.net.cn; Bureau Chief ANDREAS LANDWEHR.

Informatsionnoye Telegrafnoye Agentstvo Rossii (ITAR—TASS) (Russia): 6-1-41 Tayuan Diplomatic Office Bldg, Beijing 100600; tel. (10) 65324821; fax (10) 65324820; e-mail tassbj@itar-tass.com; Bureau Chief ANDREY KIRILLOV.

Inter Press Service (TIPS) (Italy): 15 Fuxing Lu, POB 3811, Beijing 100038; tel. (10) 68514046; fax (10) 68518210; e-mail tipscn@istic.ac.cn; internet www.tips.org.cn; Dir WANG XIAOYING.

Jiji Tsushin (Japan): 9-1-13 Jian Guo Men Wai, Waijiao, Beijing; tel. (10) 65322924; fax (10) 65323413; Correspondents YOSHIHISA MURAYAMA, TETSUYA NISHIMURA.

Korean Central News Agency (Democratic People's Republic of Korea): Beijing; e-mail kcnab@xinhuanet.com; Bureau Chief RI YONG.

Kyodo News Service (Japan): 3-91 Jian Guo Men Wai, Beijing; tel. (10) 6532680; fax (10) 65322273; e-mail kyodob@ccnet.cn.net; Bureau Chief YASUHIRO MORI.

Magyar Távirati Iroda (MTI) (Hungary): 1-42 Ban Gong Lu, San Li Tun, Beijing 100600; tel. (10) 65321744; Correspondent GYÖRGY BARTA.

Prensa Latina (Cuba): 4-1-23 Jianguomenwai, Beijing 100600; tel. and fax (10) 65321914; e-mail prelatin@public.bta.net.cn; Correspondent ILSA RODRÍGUEZ SANTANA.

Press Trust of India: 5-131 Diplomatic Apts, Jian Guo Men Wai, Beijing 100600; tel. and fax (10) 65322221.

Reuters (UK): Hilton Beijing, 1 Dong Fang Lu/Bei Dong Sanhuan Lu, Chao Yang Qu, Beijing; tel. (10) 64662288; fax (10) 64653052; e-mail hilton@hiltonbeijing.com.cn; internet www.hilton.com; Bureau Man. RICHARD PASCOE.

Tanjug News Agency (Serbia and Montenegro): Qijayuan Diplomatic Apt, Beijing 100600; tel. (10) 65324821.

United Press International (UPI) (USA): 7-1-11 Qi Jia Yuan, Beijing; tel. (10) 65323271; Bureau Chief CHRISTIAAN VIRANT.

The following are also represented: Rompres (Romania) and VNA (Viet Nam).

PRESS ORGANIZATIONS

All China Journalists' Association: Xijiaominxiang, Beijing 100031; tel. (10) 66023981; fax (10) 66014658; Chair. SHAO HUAZE.

China Newspapers Association: Beijing; Chair. WANG CHEN.

The Press and Publication Administration of the People's Republic of China (State Copyright Bureau): 85 Dongsi Nan Dajie, East District, Beijing 100703; tel. (10) 65124433; fax (10) 65127875; Dir SHI ZONGYUAN.

Publishers

In 2000 there were 565 publishing houses in China. A total of 143,376 titles (and 6,270m. copies) were published in that year.

Beijing Chubanshe Chuban Jituan (Beijing Publishing House Group): 6 Bei Sanhuan Zhong Lu, Beijing 100011; tel. (10) 62016699; fax (10) 62012339; e-mail geo@bph.com.cn; internet www.bph

.cn; f. 1956; politics, history, law, economics, geography, science, literature, art, etc.; Dir ZHU SHUXIN; Editor-in-Chief TAO XINCHENG.

Beijing Daxue Chubanshe (Beijing University Press): 205 Chengfu Lu, Zhongguancun, Haidian Qu, Beijing 100871; tel. (10) 62752024; fax (10) 62556201; f. 1979; academic and general.

China International Book Trading Corpn: POB 399, 35 Chegongzhuang Xilu, Beijing 100044; tel. (10) 68433113; fax (10) 68420340; e-mail bk@mail.cibtc.co.cn; internet www.cibtc.com.cn; f. 1949; foreign trade org. specializing in publs, including books, periodicals, art and crafts, microfilms, etc.; import and export distributors; Pres. LIU ZHIBIN.

China Publishing Group: Beijing; f. 2002; aims to restructure and consolidate publishing sector; comprises 12 major publishing houses; Pres. of the Bd YANG MUZHI.

CITIC Publishing House: Ta Yuan Diplomatic Office Bldg, 14 Liangmahe Lu, Chao Yang Qu, Beijing 100600; tel. (10) 85323366; fax (10) 85322505; e-mail liyinghong@citicpub.com; internet www.publish.citic.com; f. 1988; finance, investment, economics and business; Pres. WANG BIN.

Dianzi Gongye Chubanshe (Publishing House of the Electronics Industry—PHEI): POB 173, Wan Shou Lu, Beijing 100036; tel. (10) 68159028; fax (10) 68159025; f. 1982; electronic sciences and technology; Pres. LIANG XIANGFENG; Vice-Pres. WANG MINGJUN.

Dolphin Books: 24 Baiwanzhuang Lu, Beijing 100037; tel. (10) 68326332; fax (10) 68326642; f. 1986; children's books in Chinese and foreign languages; Dir WANG YANRONG.

Falü Chubanshe (Law Publishing House): POB 111, Beijing 100036; tel. (10) 6815325; f. 1980; current laws and decrees, legal textbooks, translations of important foreign legal works; Dir LAN MINGLIANG.

Foreign Languages Press: 19 Chegongzhuang Xi Lu, Fu Xing Men Wai, Beijing 100044; tel. (10) 68413344; fax (10) 68424931; e-mail info@flp.com.cn; internet www.flp.com.cn; f. 1952; books in 20 foreign languages reflecting political and economic developments in People's Republic of China and features of Chinese culture; Dir GUO JIEXIN; Editor-in-Chief XU MINGQIANG.

Gaodeng Jiaoyu Chubanshe (Higher Education Press): 55 Shatan Houjie, Beijing 100009; tel. (10) 64014043; fax (10) 64054602; e-mail linm@public.bta.net.cn; internet www.hep.edu.cn; f. 1954; academic, textbooks; Pres. LIU ZHIPENG; Editor-in-Chief ZHANG ZENGSHUN.

Gongren Chubanshe (Workers' Publishing House): Liupukeng, Andingmen Wai, Beijing; tel. (10) 64215278; f. 1949; labour movement, trade unions, science and technology related to industrial production.

Guangdong Keji Chubanshe (Guangdong Science and Technology Press): 11 Shuiyin Lu, Huanshidong Lu, Guangzhou, Guangdong 510075; tel. (20) 37607770; fax (20) 37607770; e-mail gdkjzbb@21cn.com; internet www.gdstp.com.cn; f. 1978; natural sciences, technology, agriculture, medicine, computing, English language teaching; Dir HUANG DAQUAN.

Heilongjiang Kexue Jishu Chubanshe (Heilongjiang Science and Technology Press): 41 Jianshe Jie, Nangang Qu, Harbin 150001, Heilongjiang; tel. and fax (451) 3642127; f. 1979; industrial and agricultural technology, natural sciences, economics and management, popular science, children's and general.

Huashan Wenyi Chubanshe (Huashan Literature and Art Publishing House): 45 Bei Malu, Shijiazhuang, Hebei; tel. 22501; f. 1982; novels, poetry, drama, etc.

Kexue Chubanshe (Science Press): 16 Donghuangchenggen Beijie, Beijing 100717; tel. (10) 64034313; fax (10) 64020094; e-mail icd@cspg.net; f. 1954; books and journals on science and technology.

Lingnan Meishu Chubanshe (Lingnan Art Publishing House): 11 Shuiyin Lu, Guangzhou, Guangdong 510075; tel. (20) 87771044; fax (20) 87771049; f. 1981; works on classical and modern painting, picture albums, photographic, painting techniques; Pres. CAO LIXIANG.

Minzu Chubanshe (The Ethnic Publishing House): 14 Hepingli Beijie, Beijing 100013; tel. (10) 64211126; e-mail nova126@sina.com; f. 1953; books and periodicals in minority languages, e.g. Mongolian, Tibetan, Uygur, Korean, Kazakh, etc.; Editor-in-Chief HUANG ZHONGCAI.

Qunzhong Chubanshe (Masses Publishing House): Bldg 15, Part 3, Fangxingyuan, Fangzhuan Lu, Beijing 100078; tel. (10) 67633344; f. 1956; politics, law, judicial affairs, criminology, public security, etc.

Renmin Chubanshe (People's Publishing House): 8 Hepinglidongjie, Andingmenwai, Beijing; tel. (10) 4213713; managed by the Ministry of Communications; science and technology, textbooks, laws and specifications of communications; Dir and Editor-in-Chief XUE DEZHEN.

Renmin Jiaoyu Chubanshe (People's Education Press): 55 Sha Tan Hou Jie, Beijing 100009; tel. (10) 64035745; fax (10) 64010370; f. 1950; school textbooks, guidebooks, teaching materials, etc.

Renmin Meishu Chubanshe (People's Fine Arts Publishing House): Beijing; tel. (10) 65122371; fax (10) 65122370; f. 1951; works by Chinese and foreign painters, sculptors and other artists, picture albums, photographic, painting techniques; Dir GAO ZONGYUAN; Editor-in-Chief CHENG DALI.

Renmin Weisheng Chubanshe (People's Medical Publishing House): Beijing; tel. (10) 67617283; fax (10) 645143; f. 1953; medicine (Western and traditional Chinese), pharmacology, dentistry, public health; Pres. LIU YIQING.

Renmin Wenxue Chubanshe (People's Literature Publishing House): 166 Chaoyangmen Nei Dajie, Beijing 100705; tel. and fax (10) 65138394; e-mail rwzbs@sina.com; internet www.rw-cn.com; f. 1951; largest publr of literary works and translations into Chinese; Dir and Editor-in-Chief NIE ZHENNING.

Shanghai Guji Chubanshe (Shanghai Classics Publishing House): 272 Ruijin Erlu, Shanghai 200020; tel. (21) 64370011; fax (21) 64339287; e-mail guji1@guji.com.cn; internet www.guji.com.cn; f. 1956; classical Chinese literature, history, art, philosophy, geography, linguistics, science and technology.

Shanghai Jiaoyu Chubanshe (Shanghai Educational Publishing House): 123 Yongfu Lu, Shanghai 200031; tel. (21) 64377165; fax (21) 64339995; f. 1958; academic; Dir and Editor-in-Chief CHEN HE.

Shanghai Yiwen Chubanshe (Shanghai Translation Publishing House): 14 Xiang 955, Yanan Zhonglu, Shanghai 200040; tel. (21) 62472890; fax (21) 62475100; e-mail cpbq@bj.cal.com.cn; internet www.cp.com.cn; f. 1978; translations of foreign classic and modern literature; philosophy, social sciences, dictionaries, etc.

Shangwu Yinshuguan (The Commercial Press): 36 Wangfujing Dajie, Beijing; tel. (10) 65252026; fax (10) 65135899; e-mail comprs@public.gb.com.cn; internet www.cp.com.cn; f. 1897; dictionaries and reference books in Chinese and foreign languages, translations of foreign works on social sciences; Pres. YANG DEYAN.

Shaonian Ertong Chubanshe (Juvenile and Children's Publishing House): 1538 Yan An Xi Lu, Shanghai 200052; tel. (21) 62823025; fax (21) 62821726; e-mail forwardz@public4.sta.net.cn; f. 1952; children's educational and literary works, teaching aids and periodicals; Gen. Man. ZHOU SHUNPEI.

Shijie Wenhua Chubanshe (World Culture Publishing House): Dir ZHU LIE.

Wenwu Chubanshe (Cultural Relics Publishing House): 29 Wusi Dajie, Beijing 100009; tel. (10) 64048057; fax (10) 64010698; e-mail web@wenwu.com; internet www.wenwu.com; f. 1956; books and catalogues of Chinese relics in museums and those recently discovered; Dir SU SHISHU.

Wuhan Daxue Chubanshe (Wuhan University Press): Luojia Hill, Wuhan, Hubei; tel. (27) 87218069; fax (27) 87218069; e-mail wdp3@whu.edu.cn; internet www.wdp.whu.edu.cn; f. 1981; reference books, academic works, maps, audio-visual works etc.; Pres. and Editor-in-Chief Prof. JIANG JIANQIN.

Xiandai Chubanshe (Modern Press): 504 Anhua Li, Andingmenwai, Beijing 100011; tel. (10) 64263515; fax (10) 64214540; f. 1981; directories, reference books, etc.; Dir ZHOU HONGLI.

Xinhua Chubanshe (Xinhua Publishing House): 57 Xuanwumen Xidajie, Beijing 100803; tel. (10) 63074022; fax (10) 63073880; e-mail xhpub@xinhua.org; f. 1979; social sciences, economy, politics, history, geography, directories, dictionaries, etc.; Dir WANG CHUNRONG; Editor-in-Chief ZHANG SHOUDI.

Xuelin Chubanshe (Scholar Books Publishing House): 120 Wenmiao Lu, Shanghai 200010; tel. and fax (21) 63768540; f. 1981; academic, including personal academic works at authors' own expense; Dir LEI QUNMING.

Zhongguo Caizheng Jingji Chubanshe (China Financial and Economic Publishing House): 8 Dafosi Dongjie, Dongcheng Qu, Beijing; tel. (10) 64011805; f. 1961; finance, economics, commerce and accounting.

Zhongguo Dabaike Quanshu Chubanshe (Encyclopaedia of China Publishing House): 17 Fu Cheng Men Bei Dajie, Beijing 100037; tel. (10) 68315610; fax (10) 68316510; e-mail ygh@bj.col.com.cn; f. 1978; specializes in encyclopaedias; Dir SHAN JIFU.

Zhongguo Ditu Chubanshe (SinoMaps Press): 3 Baizhifang Xijie, Beijing 100054; tel. (10) 63530808; fax (10) 63531961; e-mail wuqinjie@sinomaps.com; internet www.sinomaps.com; f. 1954; cartographic publr; Dir Dr BAI BO.

Zhongguo Funü Chubanshe (China Women Publishing House): 24A Shijia Hutong, 100010 Beijing; tel. (10) 65126986; f. 1981; women's movement, marriage and family, child-care, etc.; Dir Li Zhongxiu.

Zhongguo Qingnian Chubanshe (China Youth Press): 21 Dongsi Shiertiao, Beijing 100708; tel. (10) 84015396; fax (10) 64031803; e-mail cyph@eastnet.com.cn; internet www.cyp.com.cn; f. 1950; literature, social and natural sciences, youth work, autobiography; also periodicals; Dir Hu Shouwen; Editor-in-Chief Xu Wenxin.

Zhongguo Shehui Kexue Chubanshe (China Social Sciences Publishing House): 158A Gulou Xidajie, Beijing 100720; tel. (10) 64073837; fax (10) 64074509; f. 1978; Dir Zheng Wenlin.

Zhongguo Xiju Chubanshe (China Theatrical Publishing House): 52 Dongsi Batiao Hutong, Beijing; tel. (10) 64015815; f. 1957; traditional and modern Chinese drama.

Zhongguo Youyi Chuban Gongsi (China Friendship Publishing Corpn): e-mail tmdoxu@public.east.cn.net; Dir Yang Wei.

Zhonghua Shuju (Zhonghua Book Co): 38 Taipingqiao Xili, Fenglai Qu, Beijing; tel. (10) 63458226; f. 1912; general; Pres. Song Yifu.

PUBLISHERS' ASSOCIATION

Publishers' Association of China: Beijing; f. 1979; arranges academic exchanges with foreign publrs; Hon. Chair. Song Muwen; Chair. Yu Youxian.

Broadcasting and Communications

TELECOMMUNICATIONS

Ministry of Information Industry: 13 Xichangan Jie, Beijing 100804; tel. (10) 66014249; fax (10) 66034248; e-mail webmaster@mii.gov.cn; internet www.mii.gov.cn; regulates all issues concerning the telecommunications sector.

China Mobile (Hong Kong) Ltd: 60th Floor, The Center, 99 Queen's Rd, Central, Hong Kong; tel. (852) 31218888; fax (852) 25119092; e-mail ca@chinamobilehk.com; internet www.chinamobilehk.com; f. 1997; provides mobile telecommunications services in 13 provinces, municipalities, and autonomous regions of China; world's biggest cellular carrier (2002); Chair. and CEO Wang Xiaochu.

China Netcom Corpn: Beijing 100032; 9–15/F, Building A, 15/F, Bldg C, Corporate Sq., 35 Financial St, Xicheng Qu; tel. (10) 8809-3588; fax (10) 8809-1446; e-mail cnc@china-netcom.com; internet www.cnc.net.cn; f. 1999; internet telephone service provider; merged with China Telecom northern operations (10 provinces) in May 2002; CEO Edward Tian.

China Telecom: 5th Floor, North Wing, Xibianmennei Jie, Xuanwu Qu, Beijing 100053; e-mail info@chinatelecom.com.cn; internet www.chinatelecom.com.cn; f. 1997 as a vehicle for foreign investment in telecommunications sector; operates 'Xiao Ling Tong' mobile phone services; restructured in May 2002 with responsibility for fixed-line network in 21 southern and western provinces; Pres. Zhou Deqiang.

China Telecommunications Satellite Group Corpn: Beijing; f. 2001 to provide internet, telephone and related services; Gen. Man. Zhou Zehe.

China United Telecommunications Corpn (UNICOM): 1/F, Hongji Centre Office Bldg, 18 Jianguomenei Dajie, Beijing; tel. (10) 65181800; fax (10) 65183405; e-mail webmaster@chinaunicom.com.cn; internet www.chinaunicom.com.cn; f. 1994; cellular telecommunications; Chair. and Pres. Yang Xianzu.

Netease.com: Rm 1901, Tower E3, The Towers, Oriental Plaza No.1, East Chang An Jie, Dong Cheng Qu, Beijing 100738; tel. (10) 85180163; fax (10) 85183618; internet www.netease.com; Nasdaq-listed internet portal; CEO Ted Sun.

Sina.com: Soho New Town, 16F Bldg C, 88 Jianguo Lu, Chaoyang Qu, Beijing 100022; tel. (10) 65665009; fax (10) 85801740; internet www.sina.com.cn; Nasdaq-listed internet portal; Pres. Wang Yan.

Sohu.com: 15th Floor, Tower 2, Bright China Chang An Bldg, 7 Jianguomen Nei Jie, Beijing 100005; tel. (10) 65102160; fax (10) 65101377; internet www.sohu.com; ; Nasdaq-listed internet portal; Chair. Charles Zhang.

BROADCASTING

In 2002 there were 306 radio broadcasting stations, 757 radio transmitting and relay stations (covering 93.3% of the population), 368 television stations and 53,668 television transmitting and relay stations (covering 94.6% of the population).

Regulatory Authorities

State Administration of Radio, Film and Television (SARFT): 2 Fu Xing Men Wai Dajie, POB 4501, Beijing 100866; tel. (10) 68513409; fax (10) 68512174; internet www.dns.incmrft.gov.cn; controls the Central People's Broadcasting Station, the Central TV Station, Radio Beijing, China Record Co, Beijing Broadcasting Institute, Broadcasting Research Institute, the China Broadcasting Art Troupe, etc.; Dir Xu Guangchun.

State Radio Regulatory Authority: Beijing; operates under the State Council; Chair. Zou Jiahua.

Radio

China National Radio (CNR): 2 Fu Xing Men Wai Dajie, Beijing 100866; tel. (10) 68045630; fax (10) 68045631; internet www.cnradio.com; f. 1945; domestic service in Chinese, Zang Wen (Tibetan), Min Nan Hua (Amoy), Ke Jia (Hakka), Hasaka (Kazakh), Wei Wuer (Uygur), Menggu Hua (Mongolian) and Chaoxian (Korean); Dir-Gen. Yang Bo.

Zhongguo Guoji Guangbo Diantai (China Radio International): 16A Shijingshan Lu, Beijing 100039; tel. (10) 68891001; fax (10) 68891582; e-mail crieng@public.bta.net.cn; internet www.cri.com.cn; f. 1941; fmrly Radio Beijing; foreign service in 38 languages incl. Arabic, Burmese, Czech, English, Esperanto, French, German, Indonesian, Italian, Japanese, Lao, Polish, Portuguese, Russian, Spanish, Turkish and Vietnamese; Dir Li Dan.

Television

China Central Television (CCTV): 11 Fuxing Lu, Haidian, Beijing 100859; tel. (10) 8500000; fax (10) 8513025; internet www.wtdb.com/CCTV/about.htm; operates under Bureau of Broadcasting Affairs of the State Council, Beijing; f. 1958; operates eight networks; 24-hour global satellite service commenced in 1996; Pres. Zhao Huayong.

In April 1994 foreign companies were prohibited from establishing or operating cable TV stations in China. By mid-1996 there were more than 3,000 cable television stations in operation, with networks covering 45m. households. The largest subscriber service is Beijing Cable TV (Dir Guo Junjin). Satellite services are available in some areas: millions of satellite receivers are in use. In October 1993 the Government approved new regulations, attempting to restrict access to foreign satellite broadcasts. In September 2001 the Government signed a deal that would allow News Corpn and AOL Time Warner to become the first foreign broadcasters to have direct access to China's markets, although broadcasts would be restricted to Guangdong Province.

Finance

(cap. = capital; auth. = authorized; p.u. = paid up; res = reserves; dep. = deposits; m. = million; amounts in yuan unless otherwise stated)

BANKING

Radical economic reforms, introduced in 1994, included the strengthening of the role of the central bank and the establishment of new commercial banks. The Commercial Bank Law took effect in July 1995. The establishment of private banks was to be permitted.

Regulatory Authority

China Banking Regulatory Commission: 33 Cheng Fang Lu, Xi Qu, Beijing; internet www.cbrc.gov.cn; f. 2003; Chair. Liu Mingkang.

Central Bank

People's Bank of China: 32 Chengfang Jie, Xicheng Qu, Beijing 100800; tel. (10) 66194114; fax (10) 66015346; e-mail master@pbc.gov.cn; internet www.pbc.gov.cn; f. 1948; bank of issue; decides and implements China's monetary policies; Gov. Zhou Xiaochuan; 2,204 brs.

Other Banks

Agricultural Bank of China: 23A Fuxing Lu, Haidian Qu, Beijing 100036; tel. (10) 68424388; fax (10) 68424437; e-mail webmaster@intl.abocn.com; internet www.abchina.com; f. 1951; serves mainly China's rural financial operations, providing services for agriculture, industry, commerce, transport, etc. in rural areas; cap. 129,252m., res 3,183m., dep. 2,620,746m. (Dec. 2002); Pres. Yang Mingsheng; 37,000 brs (domestic).

Agricultural Development Bank of China: 2A Yuetanbei Jie, Xicheng Qu, Beijing 100045; tel. (10) 68081557; fax (10) 68081773; f. 1994; cap. 20,000m.; Pres. He Linxiang.

Bank of China Ltd: 1 Fu Xing Men Nei Dajie, Beijing 100818; tel. (10) 66016688; fax (10) 66016869; e-mail webmaster@bank-of-china.com; internet www.bank-of-china.com; f. 1912; handles foreign exchange and international settlements; operates Orient AMC (asset management corporation) since 1999; fmrly Bank of China, became shareholding company in Aug. 2004, initial public offering expected in 2005; cap. 142,100m., res 66,995m., dep. 2,602,384m. (Dec. 2002); Chair. XIAO GANG; Pres. LI LIHUI; Dir ANTHONY NEOH; 121 brs.

Bank of Communications Ltd: 188 Yin Cheng Lu, Shanghai 200120; tel. (21) 58781234; fax (21) 58880559; e-mail webmaster@hq.bankcomm.com; internet www.bankcomm.com; f. 1908; commercial bank; cap. 15,909.5m., res 15,369.7m., dep. 511,021.8m. (Dec. 2002); a 19.9% stake was acquired by HSBC in Aug. 2004; Chair. YIN JIEYAN; Pres. FANG CHENGGUO; 90 brs.

Bank of Shanghai Co Ltd: 585 Zhongshan Lu (E2), Shanghai 200010; tel. (21) 63370888; fax (21) 63370777; e-mail shenjie@bankofshanghai.com.cn; internet www.bankofshanghai.com; f. 1995 as Shanghai City United Bank, assumed present name in 1998; cap. 2,600m., res 3,883.9m., dep 149,377.2m. (Dec. 2002); Chair. FU JIANHUA; Pres. XIN CHEN.

Beijing City Commercial Bank Corpn Ltd: 2nd Floor, Tower B, Beijing International Financial Bldg, 156 Fu Xing Men Nei Jie, Beijing 100031; tel. (10) 66426928; fax (10) 66426691; e-mail bccbibd@sina.com; internet www.bccb.com.cn; f. 1996 as Beijing City United Bank Corpn, assumed present name in 1998; cap 2,009.9m., res 1,448.4m., dep 89,773.6m. (Dec. 2001); Chair. YAN BINGZHU; Pres. YAN XIAOYAN.

Beijing City Co-op Bank: 65 You An Men Nei Lu, Xuanwu Qu, Beijing 100054; tel. and fax (10) 63520159; f. 1996; cap. 1,000m., res 3,466m., dep. 26,660m. (Dec. 1996); 90 brs.

Bengbu House Saving Bank: 85 Zhong Rong Jie, Bengbu 233000; tel. (552) 2042069.

Changsha City Commercial Bank: 1 Furong Zhong Lu, Changsha, Hunan; tel. (73) 14305570; fax (73) 14305560; internet www.hncccb.com.cn; f. 1997; cap. 312.0m., res 37.5m., dep. 16,785.4m. (Dec. 2002); Chair. YE ZHANG; Pres. XIANG LILI.

China and South Sea Bank Ltd: 410 Fu Cheng Men Nei Dajie, Beijing; tel. (10) 66016688; fax (10) 66016869; f. 1921; cap. 1,200.0m., res 3,735.1m., dep. 38,850.2m. (Dec. 1999); Chair. CHUN PING.

China Construction Bank (CCB): 25 Jinrong Jie, Beijing 100032; tel. (10) 67598628; fax (10) 67598544; e-mail ccb@bj.china.com; internet www.ccb.com.cn; f. 1954; fmrly People's Construction Bank of China; makes payments for capital construction projects in accordance with state plans and budgets; issues medium- and long-term loans to enterprises and short-term loans to construction enterprises and others; also handles foreign-exchange business; housing loans; operates Cinda AMC (asset management corporation) since 1998 and China Great Wall AMC since 1999; cap. 85,115m., res 22,121m., dep. 2,848,248m. (Dec. 2002); 49 brs; Pres. ZHANG ENZHAO.

China Everbright Bank: Everbright Tower, 6 Fu Xing Men Wai Lu, Beijing 100045; tel. (10) 68565577; fax (10) 68561260; e-mail eb@cebbank.com; internet www.cebbank.com; f. 1992 as Everbright Bank of China; acquired China Investment Bank and assumed present name in 1999; cap. 8,216.9m., res 4,104.3m., dep. 309,537.6m. (Dec. 2002); Pres. WANG CHUAN; Chair. WANG MINGQUAN; 10 brs.

China International Capital Corporation (CICC): 28th Floor, China World Tower 2, 1 Jian Guo Men Wai Dajie, Beijing 100004; tel. (10) 65051166; fax (10) 65051156; e-mail info@cicc.com.cn; internet www.cicc.com.cn; f. 1995; international investment bank; 42.5% owned by China Construction Bank, 34.3% owned by Morgan Stanley; registered cap. US $100m.; CEO LEVIN ZHU.

China International Trust and Investment Corporation (CITIC): Capital Mansion, 6 Xianyuannan Lu, Chao Yang Qu, Beijing 100004; tel. (10) 64661105; fax (10) 64662137; e-mail g-office@citic.com.cn; internet www.citic.com.cn; f. 1979; economic and technological co-operation; finance, banking, investment and trade; registered cap. 3,000m.; sales US $3,462.7m. (1999/2000); Chair. WANG JUN; Pres. KONG DAN.

China Merchants Bank: China Merchants Bank Tower, 7088 Shennan Blvd, Shenzhen 518040; tel. (755) 83198888; fax (755) 83195061; e-mail 00430@oa.cmbchina.com; internet www.cmbchina.com; f. 1987; cap. 5,706.8m., res 9,785.9m., dep. 346,127m. (Dec. 2002); Pres. and CEO MA WEIHUA; 19 brs.

China Minsheng Banking Corporation: 4 Zhengyi Lu, Dongcheng Qu, Beijing 100006; tel. (10) 65269578; fax (10) 65269593; e-mail cmbc@public.bta.net.cn; internet www.cmbc.com.cn; first non-state national commercial bank, opened Jan. 1996; cap. 2,586.7m., res 2,792.2m., dep. 235,469.0m. (Dec. 2002); Chair. JING SHUPING; Pres. DONG WENBIAO; 5 brs.

Chinese Mercantile Bank: Ground and 23rd Floors, Dongfeng Bldg, 2 Yannan Lu, Futian Qu, Shenzhen 518031; tel. (755) 3257880; fax (755) 3257801; e-mail szcmbank@public.szptt.net.cn; f. 1993; owned by Hong Kong Chinese Bank 40%, CTS Hong Kong 30%, Industrial and Commercial Bank of China 30%; cap. US $85.3m., res US $3.3m., dep. US $183.8m. (Dec. 1999); Pres. HUANG MINGXIANG.

CITIC Industrial Bank: Block C, Fuhua Bldg, 8 Chao Yang Men Bei Dajie, Dongcheng Qu, Beijing 100027; tel. (10) 65541658; fax (10) 65541671; e-mail webmaster@citicb.com.cn; internet www.citicib.com.cn; f. 1987; cap. 6,809.3m., res 1,111.2m., dep. 290,172.5m. (Dec. 2002); Chair. WANG JUN; Pres. DOU JIANZHONG; 26 brs.

Export and Import Bank of China (China Exim Bank): Jinyun Tower B, No. 43-A, Xizhimenbei Lu, Beijing 100044; tel. (10) 62231626; fax (10) 62231236; internet www.eximbank.gov.cn; f. 1994; provides trade credits for export of large machinery, electronics, ships, etc.; Chair. and Pres. YANG ZILIN.

Fujian Asia Bank Ltd: 2nd Floor, Yuan Hong Bldg, 32 Wuyi Lu, Fuzhou, Fujian 350005; tel. (591) 3330788; fax (591) 3330843; e-mail fablls@pub2.fz.fj.cn; f. 1993; cap. US $27.0m., res US $2.1m., dep. US $0.2m. (Dec. 2002); Chair. MA HONG; Gen. Man. SONG JIANXIN.

Fujian Industrial Bank: Zhong Shang Bldg, 154 Hudong Lu, Hualin, Fuzhou, Fujian 350003; tel. (591) 7839338; fax (591) 7841932; e-mail fjib@pub3.fz.fj.cn; internet www.fib.com.cn; f. 1982; cap. 3,000m., res 2,867.8m., dep. 117,935.2m. (Dec. 2001); Pres. GAO JIANPING; 19 brs.

Guangdong Development Bank: 83 Nonglinxia Lu, Dongshan Qu, Guangzhou, Guangdong 510080; tel. (20) 87310888; fax (20) 87310779; internet www.gdb.com.cn; f. 1988; cap. 3,585.7m., res 1,388.7m., dep. 166,079.3m. (Dec. 2001); Chair. LI RUOHONG; Pres. ZHANG GUANGHUA; 24 brs.

Hua Xia Bank: 9th–12th Floors, Xidan International Mansion, 111 Xidan Bei Dajie, Beijing 100032; tel. (10) 66151199; fax (10) 66188333; e-mail hxbk@public.bta.net.cn; internet www.hxb.cc; f. 1992 as part of Shougang Corpn; cap. 2,500m., res 165.2m., dep. 171,235.7m. (Dec. 2002); Chair. LIU HAIYAN; Pres. WU JIAN.

Industrial and Commercial Bank of China: 55 Fuxingmennan Dajie, Xicheng Qu, Beijing 100031; tel. (10) 66106071; fax (10) 66106053; e-mail webmaster@icbc.com.cn; internet www.icbc.com.cn; f. 1984; handles industrial and commercial credits and international business; operates Huarong AMC (asset management corporation) since 1999; cap. 167,417m., res 15,908m., dep. 3,873,732m. (Dec. 2001); Chair. and Pres. JIANG JIANQING; Gen. Man. CHEN AIPING.

International Bank of Paris and Shanghai: 13th Floor, North Tower, Shanghai Stock Exchange Bldg, 528 Pudong Nan Lu, Shanghai 200120; tel. (21) 58405500; fax (21) 58889232; f. 1992; Industrial and Commercial Bank of China 50%, Banque Nationale de Paris 50%; cap. US $33.6m., res US $1.6m., dep. US $44.5m. (Dec. 2002); Chair. and Dir JI XIAOHUI.

Kincheng Banking Corporation: 410 Fu Cheng Men Nei Dajie, Beijing; internet kincheng.bocgroup.com; f. 1917; cap. 2,200.0m., res 6,768.8m., dep. 51,873.2m. (Dec. 2000); Chair. SUNG HUNGKAY.

Kwangtung (Gwangdong) Provincial Bank: 410 Fu Cheng Men Nei Dajie, Beijing 100818; internet www.kpb-hk.com; f. 1924; cap. 1,500.0m., res 6,240.0m., dep. 71,535.2m. (Dec. 2000).

National Commercial Bank Ltd: 410 Fu Cheng Men Nei Dajie, Beijing; e-mail hkbrmain@natcombank.bocgroup.com; internet www.natcombank.bocgroup.com; f. 1907; cap. 1,200.0m., res 4,927.6m., dep. 45,291.5m. (Dec. 2000); 26 brs.

Nantong City Commercial Bank Co Ltd: 300 Nanda Lu, Nantong, Jiangsu 226006; tel. (513) 5123040; fax (513) 5123039; e-mail ntccb.id@pub.nt.jsinfo.net; internet www.ntccb.com; f. 1997 as Nantong City United Bank; assumed present name in 1998; cap. 215.7m., res 25.4m., dep. 7,444.1m. (Dec. 2003); Chair. and Pres. LIU CHANGJI.

Qingdao International Bank: Full Hope Mansion C, 12 Hong Kong Middle Rd, Qingdao, Shandong 266071; tel. (532) 5026230; fax (532) 5026222; e-mail qibankc@public.qd.sd.cn; internet www.qibank.net; f. 1996; joint venture between Industrial and Commercial Bank of China and Korea First Bank; cap. 165.5m., res 11.8m., dep. 167.3m. (Dec. 2003); Chair. KYO JOONG YOON; Pres. SUNG JAE CHUNG.

Shanghai Pudong Development Bank: 12 Zhongshan Dong Yi Lu, Shanghai 200002; tel. (21) 63296188; fax (21) 63232036; internet www.spdb.com.cn; f. 1993; cap. 3,915.0m., res 7,612.0m., dep. 350,744.3m. (Dec. 2003); Chair. ZHANG GUANGSHENG; Pres. JIN YUN.

Shenzhen Commercial Bank: Shenzhen Commercial Bank Bldg, 1099 Shennan Lu, Central, Shenzhen 518031; tel. (755) 25878092; fax (755) 25878212; e-mail ibd@bankofshenzhen.com; internet www

.18ebank.com; cap. US $193m., res US $15m., dep. US $4,044m. (Dec. 2002); f. 1995; Chair. CHEN ZEMING; Pres. WANG JI; 45 brs.

Shenzhen Development Bank Co Ltd: 5047 Shennan Dong Lu, Shenzhen 518001; tel. (755) 2088888; fax (755) 2081018; e-mail shudi@sdb.com.cn; internet www.sdb.com.cn; f. 1987; cap. 1,945.8m., res 2,370.0m., dep. 158,946.2m. (Dec. 2002); Chair. CHEN ZHAOMIN; Pres. ZHOU LIN.

Sin Hua Bank Ltd: 17 Xi Jiao Min Xiang, Beijing 100031; subsidiary of Bank of China; cap. 2,200.0m., res 7,723.2m., dep. 90,021.3m. (Dec. 1998); Chair. JIANG ZUQI.

State Development Bank (SDB): 29 Fuchengmenwai Lu, Xicheng Qu, Beijing 100037; tel. (10) 68306557; fax (10) 68306541; f. 1994; merged with China Investment Bank 1998; handles low-interest loans for infrastructural projects and basic industries; Gov. CHEN YUAN.

Xiamen International Bank: 8–10 Jiang Lu, Xiamen, Fujian 361001; tel. (592) 2078888; fax (592) 2988788; e-mail xib@public.xm .fj.cn; internet www.xib.com.cn; f. 1985; cap. HK $800m., res HK $370.2m., dep. HK $8,075.8m. (Dec. 2002); Chair. CHEN GUI ZONG; Pres. LU YAO MING; 3 brs.

Yantai House Saving Bank: 248 Nan Da Jie, Yantai 264001; tel. (535) 6207047.

Yien Yieh Commercial Bank Ltd: 17 Xi Jiao Min Xiang, Beijing 100031; f. 1915; cap. 800m., res 4,425m., dep. 39,218m. (Dec. 2000); Chair. ZHAO ANGE; Gen. Man. WU GUORUI; 27 brs.

Zhejiang Commercial Bank Ltd: 88 Xi Zhongshan Lu, Ningbo 315010; tel. (574) 87252668; fax (574) 87245409; e-mail zcbho@mail .nbptt.zj.cn; internet www.zcbl.com; f. 1993; cap. US $49.2m. (1998), dep. US $54.8m. (Dec. 2002); Pres. and Chair. DUAN YONGKUAN.

Zhongxin Shiye Bank is a nation-wide commercial bank. Other commercial banks include the Fujian Commercial Bank and Zhaoshang Bank.

Foreign Banks

Before mid-1995 foreign banks were permitted only to open representative offices in China. The first foreign bank established a full branch in Beijing in mid-1995, and by March 1998 there were 51 foreign banks in China. In March 1997 foreign banks were allowed for the first time to conduct business in yuan. However, they are only entitled to accept yuan deposits from joint-venture companies. Representative offices totalled 519 in December 1996. In March 1999 the Government announced that foreign banks, hitherto restricted to 23 cities and Hainan Province, were to be permitted to open branches in all major cities.

STOCK EXCHANGES

Several stock exchanges were in the process of development in the mid-1990s, and by early 1995 the number of shareholders had reached 38m. By 1995 a total of 15 futures exchanges were in operation, dealing in various commodities, building materials and currencies. The number of companies listed on the Shanghai and Shenzhen Stock Exchanges rose from 323 in 1995 to 1,224 in 2002. In August 1997, in response to unruly conditions, the Government ordered the China Securities Regulatory Commission (see below) to assume direct control of the Shanghai and Shenzhen exchanges.

Stock Exchange Executive Council (SEEC): Beijing; tel. (10) 64935210; f. 1989 to oversee the development of financial markets in China; mems comprise leading non-bank financial institutions authorized to handle securities; Vice-Pres. WANG BOMING.

Securities Association of China (SAC): Olympic Hotel, 52 Baishiqiao Lu, Beijing 100081; tel. (10) 68316688; fax (10) 68318390; f. 1991; non-governmental organization comprising 122 mems (stock exchanges and securities cos) and 35 individual mems; Pres. ZHUANG XINYI.

Beijing Securities Exchange: 5 Anding Lu, Chao Yang Qu, Beijing 100029; tel. (10) 64939366; fax (10) 64936233.

Shanghai Stock Exchange: 528 Pudong Nan Lu, Shanghai 200120; tel. (21) 68808888; fax (21) 68807813; e-mail webmaster@ sse.com.cn; internet www.sse.com.cn; f. 1990; Chair. GENG LIANG; Pres. ZHU CONGJIU.

Shenzhen Stock Exchange: 5045 Shennan Dong Lu, Shenzhen, Guangdong 518010; tel. (755) 20833333; fax (755) 2083117; internet www.sse.org.cn; f. 1991; Chair. ZHENG KELIN; Pres. GUI MINJIE.

Regulatory Authorities

Operations are regulated by the State Council Securities Policy Committee and by the following:

China Securities Regulatory Commission (CSRC): Bldg 3, Area 3, Fangqunyuan, Fangzhuang, Beijing 100078; tel. (10) 67617343; fax (10) 67653117; e-mail csrcweb@publicf.bta.net.cn; internet www.csrc.gov.cn; f. 1993; Chair. SHANG FULIN.

INSURANCE

A new Insurance Law, formulated to standardize activities and to strengthen the supervision and administration of the industry, took effect in October 1995. Changes included the separation of life insurance and property insurance businesses. By the end of 2002 the number of insurance institutions operating in China totalled 44, of which 11 were joint-venture corporations and 17 were foreign-funded insurance branches. Total premiums rose from 44,000m. yuan in 1994 to some 305,400m. yuan in 2002. Of the latter figure, property insurance accounted for 78,000m. yuan, life insurance for 207,400m. yuan and health and accident insurance for 20,000m. yuan.

AXA-Minmetals Assurance Co: f. 1999; joint venture by Groupe AXA (France) and China Minmetals Group; Gen. Man. JOSEPH SIN.

China Insurance (Holdings) Co Ltd: 28 Xuanwumen Xi Lu, Xuanwu Qu, Beijing; tel. (10) 63600601; fax (10) 63600605; internet www.chinainsurance.com; fmrly China Insurance Co, renamed Aug. 2002; f. 1931; cargo, hull, freight, fire, life, personal accident, industrial injury, motor insurance, reinsurance, etc.; Chair. YANG CHAO.

China Insurance Group: 410 Fu Cheng Men Nei Dajie, Beijing; tel. (10) 66016688; fax (10) 66011869; f. 1996; fmrly People's Insurance Co of China (PICC), f. 1949; hull, marine cargo, aviation, motor, life, fire, accident, liability and reinsurance, etc.; 300m. policyholders (1996); Chair. and Pres. MA YONGWEI.

China Life Insurance Co: 16 Chaowai Dajie, Chaoyang Qu, Beijing 10020; tel. (10) 85659999; internet www.chinalife.com.cn; f. 1999; formed from People's Insurance (Life) Co, division of fmr People's Insurance Co of China—PICC; restructured into a parent company and a shareholding company Aug. 2003; initial public offering Dec. 2003; Chair. and Pres. WANG XIANZHANG.

China Pacific Insurance Co Ltd (CPIC): 12 Zhongshan Lu (Dong 1), Shanghai 200001; tel. (21) 63232488; fax (21) 63218398; internet www.cpic.com.cn; f. 1991; joint-stock co; Chair. WANG MINGQUAN; Pres. HUO LIANHONG.

China Ping An Insurance Co: Ping An Bldg, Bagua San Lu, Bagualing, Shenzhen 518029; tel. (755) 82262888; fax (755) 82431019; internet www.pa18.com; f. 1988; Chair. and CEO MA MINGZHE.

Hua Tai Insurance Co of China Ltd: Beijing; tel. (10) 68565588; fax (10) 68561750; f. 1996 by 63 industrial cos; Pres. LIU XUE.

Pacific-Aetna Life Insurance Co: Shanghai; f. 1998 by CPIC and Aetna Life Insurance Co; China's first Sino-US insurance co.

Tai Ping Insurance Co Ltd: 410 Fu Cheng Men Nei Dajie, Beijing 100034; tel. (10) 66016688; fax (10) 66011869; marine freight, hull, cargo, fire, personal accident, industrial injury, motor insurance, reinsurance, etc.; Pres. SUN XIYUE.

Taikang Life Insurance Co Ltd: Beijing; f. 1996; Chair. CHEN DONGSHENG.

Joint-stock companies include the Xinhua (New China) Life Insurance Co Ltd (Gen. Man. Sun Bing).

Regulatory Authority

China Insurance Regulatory Commission (CIRC): 410 Fu Cheng Men Nei Dajie, Beijing 100034; tel. (10) 66016688; fax (10) 66018871; internet www.circ.gov.cn; f. 1998; under direct authority of the State Council; Chair. WU DINGFU.

Trade and Industry

GOVERNMENT AGENCIES

China Council for the Promotion of International Trade (CCPIT): 1 Fuxingmenwai Dajie, Beijing 100860; tel. (10) 68013344; fax (10) 68011370; e-mail ccpitweb@public.bta.net.cn; internet www.ccpit.org; f. 1952; encourages foreign trade and economic co-operation; sponsors and arranges Chinese exhbns abroad and foreign exhbns in China; helps foreigners to apply for patent rights and trade-mark registration in China; promotes foreign investment and organizes tech. exchanges with other countries; provides legal services; publishes trade periodicals; Chair. WAN JIFEI.

Chinese General Association of Light Industry: 22B Fuwai Dajie, Beijing 100833; tel. (10) 68396114; under supervision of State Council; Chair. YU CHEN.

Chinese General Association of Textile Industry: 12 Dong Chang An Jie, Beijing 100742; tel. (10) 65129545; under supervision of State Council; Chair. SHI WANPENG.

Ministry of Foreign Trade and Economic Co-operation: see under Ministries.

State Administration for Industry and Commerce: 8 San Li He Dong Lu, Xicheng Qu, Beijing 100820; tel. (10) 68010463; fax (10) 68020848; responsible for market supervision and administrative execution of industrial and commercial laws; functions under the direct supervision of the State Council; Dir WANG ZHONGFU.

State-owned Assets Supervision and Administration Commission: Beijing; e-mail iecc@sasac.gov.cn ; internet www.sasac.gov.cn; f. 2003; supervision and administration of state-owned assets, regulation of ownership transfers of state-owned enterprises; Chair. LI RONGRONG.

Takeover Office for Military, Armed Police, Government and Judiciary Businesses: Beijing; f. 1998 to assume control of enterprises formerly operated by the People's Liberation Army.

CHAMBERS OF COMMERCE

All-China Federation of Industry and Commerce: 93 Beiheyan Dajie, Beijing 100006; tel. (10) 65136677; fax (10) 65122631; f. 1953; promotes overseas trade relations; Chair. HUANG MENGFU.

China Chamber of International Commerce—Shanghai: Jinling Mansions, 28 Jinling Lu, Shanghai 200021; tel. (21) 53060228; fax (21) 63869915; e-mail ccpitllb@online.sh.cn; Chair. YANG ZHIHUA.

China Chamber of International Commerce (CCOIC)—Zhuhai Chamber of Commerce: Fa Zhan Bldg, Rm 1702, 131 Shui Wan Rd, Gong Bei, Zhuhai, Guangdong 519020; tel. (756) 8890808; fax (756) 8280888; e-mail zhh@ccpit.org.

TRADE AND INDUSTRIAL ORGANIZATIONS

Anshan Iron and Steel Co: Huangang Lu, Tiexi Qu, Anshan 114021; tel. and fax (412) 6723090; e-mail info@ansteel.com.cn; internet www.ansteel.com.cn; Pres. LIU JIE.

Baotou Iron and Steel Co: Gangtie Dajie, Kundulun Qu, Baotou 014010, Inner Mongolia; tel. (472) 2125619; fax (472) 2183708; Pres. ZENG GUOAN.

Beijing Urban Construction Group Co Ltd: 62 Xueyuannan Lu, Haidian, Beijing 100081; tel. (10) 62255511; fax (10) 62256027; e-mail cjp@mail.bucg.com; internet www.bucg.com; construction of civil and industrial buildings and infrastructure.

China Aviation Industry Corporation II: 67 Jiao Nan Street, Beijing 100712; tel. (10) 64094013; fax (10) 64032109; e-mail avic@public3.bta.net.cn; Pres. ZHANG YANZHONG.

China Aviation Supplies Corpn: 155 Xi Dongsi Jie, Beijing 100013; tel. (10) 64012233; fax (10) 64016392; f. 1980; Pres. LI HAI.

China Civil Engineering Construction Corpn (CCECC): 4 Beifeng Wo, Haidian Qu, Beijing 100038; tel. (10) 63263392; fax (10) 63263864; e-mail zongban@ccecc.com.cn; f. 1953; general contracting, provision of technical and labour services, consulting and design, etc.; Pres. DING YUANCHEN.

China Construction International Inc: 9 Sanlihe Lu, Haidian Qu, Beijing; tel. (10) 68394086; fax (10) 68394097; Pres. FU RENZHANG.

China Electronics Corpn: 27 Wanshou Lu, Haidian Qu, Beijing 100846; tel. (10) 68218529; fax (10) 68213745; e-mail cec@public.gb.com.cn; internet www.cec.com.cn; Pres. WANG JINCHENG.

China Garment Industry Corpn: 9A Taiyanggong Beisanhuandong Lu, Chao Yang Qu, Beijing 100028; tel. (10) 64216660; fax (10) 64239134; Pres. DONG BINGGEN.

China General Technology (Group) Holding Ltd: f. 1998; through merger of China National Technical Import and Export Corpn, China National Machinery Import and Export Corpn, China National Instruments Import and Export Corpn and China National Corpn for Overseas Economic Co-operation; total assets 16,000m. yuan; Chair. and Pres. TONG CHANGYIN.

China Gold Co: 1 Bei Jie, Qingnianhu, Andingmenwai, Beijing; tel. (10) 64214831; Pres. CHENG FUMIN.

China Great Wall Computer Group: 38A Xueyuan Lu, Haidian Qu, Beijing 100083; tel. (10) 68342714; fax (10) 62011240; internet www.gwssi.com.cn; f. 1988; Chair. ZHANG ZHIKAI; Gen. Man. GAO KEQIN.

China Great Wall Industry Corpn: Hangtian Changcheng Bldg, 30 Haidian Nanlu, Haidian Qu, Beijing 100080; tel. (10) 68748737; fax (10) 68748865; e-mail cgwic@cgwic.com; internet www.cgwic.com.cn; registered cap. 200m. yuan; Pres. ZHANG XINXIA.

China International Book Trading Corpn: see under Publishers.

China International Contractors Association: 28 Donghouxiang, Andingmenwai, Beijing 100710; tel. (10) 64211159; fax (10) 64213959; Chair. LI RONGMIN.

China International Futures Trading Corpn: 24th Floor, Capital Mansion, 6 Xinyuan Nan Lu, Chao Yang Qu, Beijing 100004; tel. (10) 64665388; fax (10) 64665140; Chair. TIAN YUAN; Pres. LU JIAN.

China International Telecommunications Construction Corpn (CITCC): 22 Yuyou Lane, Xicheng Qu, Beijing 100035; tel. (10) 66012244; fax (10) 66024103; Pres. QI FUSHENG.

China International Water and Electric Corpn: 3 Liupukang Yiqu Zhongjie, Xicheng Qu, Beijing 100011; tel. (10) 64015511; fax (10) 64014075; e-mail cwe@mx.cei.go.cn; f. 1956 as China Water and Electric International Corpn, name changed 1983; imports and exports equipment for projects in the field of water and electrical engineering; undertakes such projects; provides technical and labour services; Pres. WANG SHUOHAO.

China Iron and Steel Industry and Trade Group Corpn: 17B Xichangan Jie, Beijing 100031; tel. (10) 66067733; fax (10) 66078450; e-mail support@sinosteel.com.cn; internet www.sinosteel.com; f. 1999 by merger of China National Metallurgical Import and Export Corpn, China Metallurgical Raw Materials Corpn and China Metallurgical Steel Products Processing Corpn; Pres. BAI BAOHUA.

China National Aerotechnology Import and Export Corpn: 5 Liangguochang, Dongcheng Qu, Beijing 100010; tel. (10) 64017722; fax (10) 64015381; f. 1952; exports signal flares, electric detonators, tachometers, parachutes, general purpose aircraft, etc.; Pres. YANG CHUNSHU; Gen. Man. LIU GUOMIN.

China National Animal Breeding Stock Import and Export Corpn (CABS): 10 Yangyi Hutong Jia, Dongdan, Beijing 100005; tel. (10) 65131107; fax (10) 65128694; sole agency for import and export of stud animals including cattle, sheep, goats, swine, horses, donkeys, camels, rabbits, poultry, etc., as well as pasture and turf grass seeds, feed additives, medicines, etc.; Pres. YANG CHENGSHAN.

China National Arts and Crafts Import and Export Corpn: Arts and Crafts Bldg, 103 Jixiangli, Chao Yang Men Wai, Chao Yang Qu, Beijing 100020; tel. (10) 65931075; fax (10) 65931036; e-mail po@mbox.cnart.com.cn; internet www.cnart-group.com; deals in jewellery, ceramics, handicrafts, embroidery, pottery, wicker, bamboo, etc.; Pres. CHEN KUN.

China National Automotive Industry Corpn (CNAIC): 46 Fucheng Lu, Haidian Qu, Beijing 100036; tel. (10) 88123968; fax (10) 68125556; Pres. GU YAOTIAN.

China National Automotive Industry Import and Export Corpn (CAIEC): 5 Beisihuan Xi Lu, Beijing 100083; tel. (10) 62310650; fax (10) 62310688; e-mail info@chinacaiec.com; internet www.chinacaiec.com; sales US $540m. (1995); Pres. ZHANG FUSHENG; 1,100 employees.

China National Cereals, Oils and Foodstuffs Import and Export Corpn (COFCO): 7th–13th Floors, Tower A, COFCO Plaza, Jian Guo Men Nei Dajie, Beijing 100005; tel. (10) 65268888; fax (10) 65278612; e-mail minnie@cofco.com.cn; internet www.cofco.com.cn; f. 1952; imports, exports and processes grains, oils, foodstuffs, etc.; also hotel management and property development; sales US $12,099.2m. (1999/2000); Chair. ZHOU MINGCHEN.

China National Chartering Corpn (SINOCHART): Rm 1601/1602, 1607/1608, Jiu Ling Bldg, 21 Xisanhuan Bei Lu, Beijing 100081; tel. (10) 68405601; fax (10) 68405628; e-mail sinochrt@public.intercom.co.cn; f. 1950; functions under Ministry of Foreign Trade and Economic Co-operation; subsidiary of SINOTRANS (see below); arranges chartering of ships, reservation of space, managing and operating chartered vessels; Pres. LIU SHUNLONG; Gen. Man. ZHANG JIANWEI.

China National Chemical Construction Corpn: Bldg No. 15, Songu, Anzhenxili, Chao Yang Qu, Beijing 100029; tel. (10) 64429966; fax (10) 64419698; e-mail cnccc@cnccc.com.cn; internet www.cnccc.com.cn; registered cap. 50m.; Pres. CHEN LIHUA.

China National Chemicals Import and Export Corporation (SINOCHEM): SINOCHEM Tower, A2 Fuxingmenwai Dajie, Beijing 100045; tel. (10) 68568888; fax (10) 68568890; internet www.sinochem.com; f. 1950; import and export, domestic trade and entrepôt trade of oil, fertilizer, rubber, plastics and chemicals; it has made notable development in other areas like industry, finance, insurance, transportation and warehousing; sales US $15,066.2m. (1999/2000); Pres. LIU DESHU.

China National Coal Industry Import and Export Corpn (CNCIEC): 88B Andingmenwai, Dongcheng Qu, Beijing 100011; tel. (10) 64287188; fax (10) 64287166; e-mail cnciec@chinacoal.com; internet www.chinacoal.com; f. 1982; sales US $800m. (1992); imports and exports coal and tech. equipment for coal industry, joint coal development and compensation trade; Chair. and Pres. WANG CHANGCHUN.

China National Coal Mine Corpn: 21 Bei Jie, Heipingli, Beijing 100013; tel. (10) 64217766; Pres. WANG SENHAO.

China National Complete Plant Import and Export Corpn (Group): 9 Xi Bin He Lu, An Ding Men, Beijing; tel. (10) 64253388; fax (10) 64211382; Chair. HU ZHAOQING; Pres. LI ZHIMIN.

China National Electronics Import and Export Corpn: 8th Floor, Electronics Bldg, 23A Fuxing Lu, Beijing 100036; tel. (10) 68219550; fax (10) 68212352; e-mail ceiec@ceiec.com.cn; internet www.ceiec.com.cn; imports and exports electronics equipment, light industrial products, ferrous and non-ferrous metals; advertising; consultancy; Chair. and Pres. QIAN BENYUAN.

China National Export Bases Development Corpn: Bldg 16–17, District 3, Fang Xing Yuan, Fang Zhuang Xiaoqu, Fengtai Qu, Beijing 100078; tel. (10) 67628899; fax (10) 67628803; Pres. XUE ZHAO.

China National Foreign Trade Transportation Corpn (Group) (SINOTRANS): Sinotrans Plaza, A43, Xizhimen Beidajie, Beijing 100044; tel. (10) 62295900; fax (10) 62295901; e-mail office@sinotrans.com; internet www.sinotrans.com; f. 1950; agents for Ministry's import and export corpns; arranges customs clearance, deliveries, forwarding and insurance for sea, land and air transportation; registered cap. 150m. yuan; Chair. and Pres. LUO KAIFU.

China National Import and Export Commodities Inspection Corpn: 15 Fanghuadi Xi Jie, Chao Yang Qu, Beijing 100020; tel. (10) 65013951; fax (10) 65004625; internet www.ccic.com; inspects, tests and surveys import and export commodities for overseas trade, transport, insurance and manufacturing firms; Pres. ZHOU WENHUI.

China National Instruments Import and Export Corpn (Instrimpex): Instrimpex Bldg, 6 Xizhimenwai Jie, Beijing 100044; tel. (10) 68330618; fax (10) 68330528; e-mail zcb@instrimpex.com.cn; internet www.instrimpex.com.cn; f. 1955; imports and exports; technical service, real estate, manufacturing, information service, etc.; Pres. ZHANG RUEN.

China National Light Industrial Products Import and Export Corpn: 910, 9th Section, Jin Song, Chao Yang Qu, Beijing 100021; tel. (10) 67766688; fax (10) 67747246; e-mail info@chinalight.com.cn; internet www.chinalight.com.cn; imports and exports household electrical appliances, audio equipment, photographic equipment, films, paper goods, building materials, bicycles, sewing machines, enamelware, glassware, stainless steel goods, footwear, leather goods, watches and clocks, cosmetics, stationery, sporting goods, etc.; Pres. XU LIEJUN.

China National Machine Tool Corpn: 19 Fang Jia Xiaoxiang, An Nei, Beijing 100007; tel. (10) 64033767; fax (10) 64015657; f. 1979; imports and exports machine tools and tool products, components and equipment; supplies apparatus for machine-building industry; Pres. QUAN YILU.

China National Machinery and Equipment Import and Export Corpn (Group): 6 Xisanhuannan Lu, Liuliqiao, Beijing 100073; tel. (10) 63271392; fax (10) 63261865; f. 1978; imports and exports machine tools, all kinds of machinery, automobiles, hoisting and transport equipment, electric motors, photographic equipment, etc.; Pres. HU GUIXIANG.

China National Machinery Import and Export Corpn: Sichuan Mansion, West Wing, 1 Fu Xing Men Wai Jie, Xicheng Qu, Beijing 100037; tel. (10) 68991188; fax (10) 68991000; e-mail cmc@cmc.com.cn; internet www.cmc.com.cn; f. 1950; imports and exports machine tools, diesel engines and boilers and all kinds of machinery; imports aeroplanes, ships, etc.; Chair. and Pres. CHEN WEIGUN.

China National Medicine and Health Products Import and Export Corpn: Meheco Plaza, 18 Guangming Zhong Jie, Chongwen Qu, Beijing 100061; tel. (10) 67116688; fax (10) 67021579; e-mail webmaster@meheco.com.cn; internet www.meheco.com.cn; Pres. LIU GUOSHENG.

China National Metals and Minerals Import and Export Corpn: Bldg 15, Block 4, Anhuili, Chao Yang Qu, Beijing 100101; tel. (10) 64916666; fax (10) 64916421; e-mail support@minmetals.com.cn; internet www.minmetals.com.cn; f. 1950; principal imports and exports include steel, antimony, tungsten concentrates and ferrotungsten, zinc ingots, tin, mercury, pig-iron, cement, etc.; Pres. MIAO GENGSHU.

China National Native Produce and Animal By-Products Import and Export Corpn (TUHSU): Sanli Bldg, 208 Andingmenwai Jie, Beijing 100011; tel. (10) 64248899; fax (10) 64204099; e-mail info@china-tuhsu.com; internet www.china-tuhsu.com; f. 1949; imports and exports include tea, coffee, cocoa, fibres, etc.; 23 subsidiary enterprises; 9 tea brs; 23 overseas subsidiaries; Pres. ZHANG ZHENMING.

China National Non-Ferrous Metals Import and Export Corpn (CNIEC): 12B Fuxing Lu, Beijing 100814; tel. (10) 63975588; fax (10) 63964424; Chair. WU JIANCHANG; Pres. XIAO JUNQING.

China National Nuclear Corpn: 1 Nansanxiang, Sanlihe, Beijing; tel. (10) 68512211; fax (10) 68533989; internet www.cnnc.com.cn; Pres. LI DINGFAN.

China National Offshore Oil Corpn (CNOOC): PO Box 4705, No. 6 Dongzhimenwai, Xiaojie, Beijing 100027; tel. (10) 84521010; fax (10) 84521044; e-mail webmaster@cnooc.com.cn; internet www.cnooc.com.cn; f. 1982; operates offshore exploration and production of petroleum; sales US $1,341.5m. (1999/2000); Pres. FU CHENGYU.

China National Oil Development Corpn: Liupukang, Beijing 100006; tel. (10) 6444313; Pres. CHENG SHOULI.

China National Packaging Import and Export Corpn: Xinfu Bldg B, 3 Dong San Huan Bei Lu, Chao Yang Qu, Beijing 100027; tel. (10) 64611166; fax (10) 64616437; e-mail info@chinapack.net; internet www.chinapack.net; handles import and export of packaging materials, containers, machines and tools; contracts for the processing and converting of packaging machines and materials supplied by foreign customers; registered cap. US $30m.; Pres. ZHENG CHONGXIANG.

China National Petroleum Corpn (CNPC) (PetroChina): 6 Liupukang Jie, Xicheng Qu, Beijing 100724; tel. (10) 62094538; fax (10) 62094806; e-mail admin@hq.cnpc.com.cn; internet www.cnpc.com.cn; restructured mid-1998; responsible for petroleum extraction and refining in northern and western China, and for setting retail prices of petroleum products; Pres. CHEN GENG.

China National Publications Import and Export Corpn: 16 Gongrentiyuguandong Lu, Chao Yang Qu, Beijing; tel. (10) 65066688; fax (10) 65063101; e-mail cnpiec@cnpiec.com.cn; internet www.cnpiec.com.cn; imports and exports books, newspapers and periodicals, records, CD-ROMs, etc.; Pres. SONG XIAOHONG.

China National Publishing Industry Trading Corpn: POB 782, 504 An Hua Li, Andingmenwai, Beijing 100011; tel. (10) 64215031; fax (10) 64214540; f. 1981; imports and exports publications, printing equipment technology; holds book fairs abroad; undertakes joint publication; Pres. ZHOU HONGLI.

China National Seed Group Corpn: 16A Xibahe, Chao Yang Qu, Beijing 100028; tel. (10) 64201817; fax (10) 64201820; imports and exports crop seeds, including cereals, cotton, oil-bearing crops, teas, flowers and vegetables; seed production for foreign seed companies etc.; Pres. HE ZHONGHUA.

China National Silk Import and Export Corpn: 105 Bei He Yan Jie, Dongcheng Qu, Beijing 100006; tel. (10) 65123338; fax (10) 65125125; e-mail cnsiec@public.bta.net.cn; internet www.chinasilk.com; Pres. XU HONGXIN.

China National Star Petroleum Corpn: 1 Bei Si Huan Xi Lu, Beijing; e-mail jf@mail.cnspc.com.cn; internet www.cnspc.com.cn; f. 1997; petroleum and gas exploration, development and production; Pres. ZHU JIAZHEN.

China National Technical Import and Export Corpn: Jiuling Bldg, 21 Xisanhuan Beilu, Beijing 100081; tel. (10) 68404000; fax (10) 68414877; e-mail info@cntic.com.cn; internet www.cntic.com.cn; f. 1952; imports all kinds of complete plant and equipment, acquires modern technology and expertise from abroad, undertakes co-production and jt ventures, and technical consultation and updating of existing enterprises; registered cap. 200m.; Pres. WANG HUIHENG.

China National Textiles Import and Export Corpn: 82 Donganmen Jie, Beijing 100747; tel. (10) 65123844; fax (10) 65124711; e-mail webmaster@chinatex.com; internet www.chinatex-group.com; imports synthetic fibres, raw cotton, wool, garment accessories, etc.; exports cotton yarn, cotton fabric, knitwear, woven garments, etc.; Pres. ZHAO BOYA.

China National Tobacco Import and Export Corpn: 11 Hufang Lu, Xuanwu Qu, Beijing 100052; tel. (10) 63533399; fax (10) 63015331; Pres. XUN XINGHUA.

China National United Oil Corpn: 57 Wangfujing Jie, Dongcheng Qu, Beijing 100006; tel. (10) 65223828; fax (10) 65223817; Chair. ZHANG JIAREN; Pres. ZHU YAOBIN.

China No. 1 Automobile Group: 63 Dongfeng Jie, Chao Yang Qu, Changchun, Jilin; tel. (431) 5003030; fax (431) 5001309; f. 1953; mfr of passenger cars; Gen. Man. GENG ZHAOJIE.

China North Industries Group: 46 Sanlihe Lu, Beijing 100821; tel. (10) 68594210; fax (10) 68594232; internet www.corincogroup.com.cn; exports vehicles and mechanical products, light industrial products, chemical products, opto-electronic products, building materials, military products, etc.; Pres. MA ZHIGENG.

China Nuclear Energy Industry Corpn (CNEIC): 1A Yuetan Bei Jie, Xicheng Qu, Beijing 100037; tel. (10) 68013395; fax (10) 68512393; internet www.cnnc.com.cn; exports air filters, vacuum valves, dosimeters, radioactive detection elements and optical instruments; Pres. ZHANG ZHIFENG.

China Petroleum and Chemical Corporation (SINOPEC): No. A6 Huixin Dong Jie, Chao Yang Qu, Beijing 100029; tel. (10) 64225533; fax (10) 64212429; e-mail webmaster@sinopec.com.cn; internet www.sinopec.com.cn; f. 2000; exploration, development, production, refining, transport and marketing of petroleum and natural gas, production and sales of petrochemicals, chemical fibres, chemical fertilizers, and other chemicals, research and development and application of technology and information; sales US $29m. (1999/2000); Pres. WANG JIMING; Chair. of Bd CHEN TONGHAI.

SINOPEC Shanghai Petrochemical Co Ltd: Wei Er Lu, Jinshanwei, Shanghai 200540; tel. (21) 57943143; fax (21) 57940050; e-mail spc@spc.com.cn; internet www.spc.com.cn; processing of crude petroleum into synthetic fibres, resins and plastics, and other petroleum products; cap. and res 12,294.8m. (1997), sales US $1,695.6m. (1999/2000); Chair. LU YIPING.

SINOPEC Yizheng Chemical Fibre Co Ltd: Zhenzhou, Yizheng, Jiangsu 211900; tel. (514) 3232235; fax (514) 3233880; internet www.sinopec.com.cn; f. 1993; chemical products; China's largest manufacturer of chemical fibre and world's fifth largest polyester maker (2001); revenue US $1,088.9m. (2000); Dir and Gen. Man. XU ZHENGNING.

Sinopec Zhenhai Refining and Chemical Co Ltd: Zhenhai Qu, Ningbo, Zhejiang 315207; tel. (574) 6456425; fax (574) 6456155; internet www.zrcc.com.cn; f. 1974; petroleum refining and production of related chemicals; sales US $1,330.6m. (1999/2000); Chair. SUN WEIJUN.

China Road and Bridge Corpn: Zhonglu Bldg, 88C, An Ding Men Wai Dajie, Beijing 100011; tel. (10) 64285616; fax (10) 64285686; e-mail crbc@crbc.com; internet www.crbc.com; overseas and domestic building of highways, urban roads, bridges, tunnels, industrial and residential buildings, airport runways and parking areas; contracts to do surveying, designing, pipe-laying, water supply and sewerage, building, etc., and/or to provide technical or labour services; Chair. ZHOU JICHANG.

China Shipbuilding Trading Corpn Ltd: 56 Zhongguancun Nan Dajie, Beijing 100044; tel. (10) 88026037; fax (10) 88026000; e-mail webmaster@cstc.com.cn; internet www.chinaships.com; Pres. LI ZHUSHI.

China State Construction Engineering Corpn: Baiwanzhuang, Xicheng Qu, Beijing 100835; tel. (10) 68347766; fax (10) 68314326; e-mail cscec-us@worldnet.att.net; internet www.cscec.com; sales US $4,726.8m. (1999/2000); Pres. MA TINGGUI.

China State Shipbuilding Corpn: 5 Yuetan Beijie, Beijing; tel. (10) 68030208; fax (10) 68031579; Pres. CHEN XIAOJIN; Gen. Man. XU PENGHANG.

China Tea Import and Export Corpn: Zhongtuchu Bldg, 208 Andingmenwai Jie, Beijing 100011; tel. (10) 64204123; fax (10) 64204101; e-mail info@teachina.com; internet www.chinatea.com .cn; Pres. LI JIAZHI.

China Xinshidai (New Era) Corpn: 40 Xie Zuo Hu Tong, Dongcheng Qu, Beijing 100007; tel. (10) 64017384; fax (10) 64032935; Pres. QIN ZHONGXING.

Chinese General Co of Astronautics Industry (State Aerospace Bureau): 8 Fucheng Lu, Haidian Qu, Beijing 100712; tel. (10) 68586047; fax (10) 68370080; Pres. LIU JIYUAN.

Daqing Petroleum Administration Bureau: Sartu Qu, Daqing, Heilongjiang; tel. (459) 814649; fax (459) 322845; Gen. Man. WANG ZHIWU.

Handan Iron and Steel Co Ltd: Handan; e-mail hdgt@mail.hgjt .com.cn; internet www.hdgt.com.cn; f. 1958; Chair. LIU RUJUN.

Ma'anshan Iron and Steel Co: 8 Hongqibei Lu, Maanshan 243003, Anhui; tel. (555) 2883492; fax (555) 2324350; Chair. HANG YONGYI; Pres. LI ZONGBI.

Shanghai Automotive Industry Sales Corpn: 548 Caoyang Lu, Shanghai 200063; tel. and fax (21) 62443223; Gen. Man. XU JIANYU.

Shanghai Baosteel Group Corpn: Baosteel Tower, 370 Pudian Lu, Pudong New District, Shanghai; tel. (21) 58358888; fax (21) 68404832; e-mail webman@baosteel.com; internet www.bstl.sh.cn; f. 1998; incorporating Baoshan Iron and Steel Corpn, and absorption of Shanghai Metallurgical Holding Group Corpn, and Shanghai Meishan Group Corpn Ltd; produces steel and steel products; sales US $14,548.0m. (2003); Pres. and Chair. XIE QIHUA.

Shanghai Foreign Trade Corpn: 27 Zhongshan Dong Yi Lu, Shanghai 200002; tel. (21) 63217350; fax (21) 63290044; f. 1988; handles import-export trade, foreign trade transportation, chartering, export commodity packaging, storage and advertising for Shanghai municipality; Gen. Man. WANG MEIJUN.

Shanghai International Trust Trading Corpn: 201 Zhaojiabang Lu, Shanghai 200032; tel. (21) 64033866; fax (21) 64034722; f. 1979; present name adopted 1988; handles import and export business, international mail orders, processing, assembling, compensation trade, etc.

Shougang Group: Shijingshan, Beijing 100041; tel. (10) 88294166; fax (10) 88295578; e-mail sgjtglb01@shougang.com.cn; internet www.shougang.com.cn; f. 1919; produces iron and steel; sales US $4,396.8m. (1999/2000); Chair. BI QUN; Gen. Man. LUO BINGSHENG.

State Bureau of Non-Ferrous Metals Industry: 12B Fuxing Lu, Beijing 100814; tel. (10) 68514477; fax (10) 68515360; under supervision of State Economic and Trade Commission; Dir ZHANG WULE.

Wuhan Iron and Steel (Group) Co: Qingshan Qu, Wuhan, Hubei Province; tel. (27) 6892004; fax (27) 6862325; proposals for merger with two other steel producers in Hubei announced late 1997; Pres. LIU BENREN.

Xinxing Oil Co (XOC): Beijing; f. 1997; exploration, development and production of domestic and overseas petroleum and gas resources; Gen. Man. ZHU JIAZHEN.

Yuxi Cigarette Factory: Yujiang Lu, Yuxi, Yunnan Province; tel. and fax (877) 2052343; Gen. Man. CHU SHIJIAN.

Zhongjiang Group: Nanjing, Jiansu; f. 1998; multinational operation mainly in imports and exports, contract projects and real estate; group consists of 126 subsidiaries, incl. 25 foreign ventures.

UTILITIES

Regulatory Authority

State Electricity Regulatory Commission: 86 Xichangan Dajie, Beijing 100031; internet www.serc.gov.cn; f. 2003; Chair. CHAI SONGYUE.

Electricity

Beijing Power Supply Co: Qianmen Xidajie, Beijing 100031; tel. (10) 63129201.

Central China Electric Power Group Co: 47 Xudong Lu, Wuchang, Wuhan 430077; tel. (27) 6813398.

Changsha Electric Power Bureau: 162 Jiefang Sicun, Changsha 410002; tel. (731) 5912121; fax (731) 5523240.

China Atomic Energy Authority: Chair. ZHANG HUAZHU.

China Northwest Electric Power Group Co: 57 Shangde Lu, Xian 710004; tel. (29) 7275061; fax (29) 7212451; Chair. LIU HONG.

China Power Grid Development (CPG): f. to manage transmission and transformation lines for the Three Gorges hydroelectric scheme; Pres. ZHOU XIAOQIAN.

China Power Investment Corpn: e-mail engweb@cpicorp.com.cn; internet www.zdt.com.cn; f. 2002; from part of the constituent businesses of fmr State Power Corpn; Pres. WANG BINGHUA.

China Southern Power Grid Co: f. 2002 from power grids in southern provinces of fmr State Power Corpn.

China Yangtze Electric Power Corpn: subsidiary of China Yangtze River Three Gorges Project Development Corpn; initial public offering on the Shanghai Stock Exchange Nov. 2003; Gen. Man. LI YONGAN.

China Yangtze Three Gorges Project Development Corpn: 1 Jianshe Dajie, Yichang, Hubei Province; tel. (717) 6762212; fax (717) 6731787; Pres. LU YOUMEI.

Datang International Power Generation Co Ltd: 8/F, 482 Guanganmennei Dajie, Xuanwu Qu, Beijing 100053; internet www .dtpower.com; one of China's largest independent power producers; Chair. ZHAI RUOYU.

Dalian Power Supply Co: 102 Zhongshan Lu, Dalian 116001; tel. (411) 2637560; fax (411) 2634430; Chief Gen. Man. LIU ZONGXIANG.

Fujian Electric Industry Bureau: 4 Xingang Dao, Taijrang Qu, Fuzhou 350009; tel. and fax (591) 3268514; Dir WANG CHAOXU.

Gansu Bureau of Electric Power: 306 Xijin Dong Lu, Qilihe Qu, Lanzhou 730050; tel. (931) 2334311; fax (93) 2331042; Dir ZHANG MINGXI.

Guangdong Electric Power Bureau: 757 Dongfeng Dong Lu, Guangzhou 510600; tel. (20) 87767888; fax (20) 87770307.

Guangdong Shantou Electric Power Bureau: Jinsha Zhong Lu, Shantou 515041; tel. (754) 8257606.

Guangxi Electric Power Bureau: 6 Minzhu Lu, Nanning 530023; tel. (771) 2801123; fax (771) 2803414.

Guangzhou Electric Power Co: 9th Floor, Huale Bldg, 53 Huale Lu, Guangzhou 510060; tel. (20) 83821111; fax (20) 83808559.

Guodian Power Group: transfer of Jianbi power plant from fmr State Power Corpn completed Sept. 2003.

Hainan Electric Power Industry Bureau: 34 Haifu Dadao, Haikou 570203; tel. (898) 5334777; fax (898) 5333230.

Heilongjiang Electric Power Co: B12Fl High Tech Development Zone, Harbin 150001; tel. (451) 2308810; fax (451) 2525878; Chair. XUE YANG.

Huadian Power Intenational Corpn Ltd: 14 Jingsan Lu, Jinan, Shandong 250001; f. 1994; fmrly Shandong International Power Development, renamed as above 2003; Chair. HE GONG; Dir CHEN JIANHUA.

Huadong Electric Power Group Corpn: 201 Nanjing Dong Lu, Shanghai; tel. (21) 63290000; fax (21) 63290727; power supply.

Huaneng Power International: West Wing, Building C, Tianyin Mansion, 2C Fuxingmennan Lu, Xicheng, Beijing; tel. (10) 66491999; fax (10) 66491888; e-mail ir@hpi.com.cn; internet www.hpi.com.cn; f. 1998; transfer of generating assets from fmr State Power Corpn completed Sept. 2003; Chair. and Pres. LI XIAOPENG.

Huazhong Electric Power Group Corpn: Liyuan, Donghu, Wuhan, Hubei Province; tel. (27) 6813398; fax (27) 6813143; electrical engineering; Gen. Man. LIN KONGXING.

Inner Mongolia Electric Power Co: 28 Xilin Nan Lu, Huhehaose 010021; tel. (471) 6942222; fax (471) 6924863.

Jiangmen Electric Power Supply Bureau: 87 Gangkou Lu, Jiangmen 529030; tel. and fax (750) 3360133.

Jiangxi Electric Power Bureau: 13 Yongwai Zheng Jie, Nanchang 330006; tel. (791) 6224701; fax (791) 6224830.

National Grid Construction Co: established to oversee completion of the National Grid by 2009.

North China Electric Power Group Corpn: 32 Zulinqianjie, Xuanwu Qu, Beijing 100053; tel. and fax (10) 63263377; Pres. JIAO YIAN.

Northeast China Electric Power Group: 11 Shiyiwei Lu, Heping Qu, Shenyang 110003; tel. (24) 3114382; fax (24) 3872665.

Shandong Electric Power Group Corpn: 150 Jinger Lu, Jinan 250001; tel. (531) 6911919.

Shandong Rizhao Power Co Ltd: 1st Floor, Bldg 29, 30 Northern Section, Shunyu Xiaoqu, Jinan 250002; tel. (531) 2952462; fax (531) 2942561.

Shanghai Electric Power Co: 181 Nanjing Dong Lu, Huangpu Qu, Shanghai 200002; tel. (21) 63291010; fax (21) 63291440; e-mail smepc@smepc.com; internet www.smepc.com; total assets 61,946m. yuan, sales 28,183m. yuan (2003); Dir SHUAI JUNQING; 18,531 employees.

Shenzhen Power Supply Co: 2 Yanhe Xi Lu, Luohu Qu, Shenzhen 518000; tel. (755) 5561920.

Sichuan Electric Power Co: Room 1, Waishi Bldg, Dongfeng Lu, Chengdu 610061; tel. (28) 444321; fax (28) 6661888.

State Power Grid Corpn: No. 1 Lane 2, Baiguang Lu, Beijing 100761; tel. (10) 63416475; fax (10) 63548152; e-mail webmaster@sp.com.cn; internet www.cep.gov.cn; f. 1997; from holdings of Ministry of Electric Power; fmrly State Power Corpn of China; became a grid company following division of State Power Corpn into 11 independent companies (five generating companies, four construction companies and two transmission companies) in Dec. 2002; generating assets to be transferred to Huaneng Group, Huadian Group, Guodian Group, China Power Investment Group and Datang Group; Gen. Man. ZHAO XICHENG.

Tianjin Electric Power Industry Bureau: 29 Jinbu Dao, Hebei Qu, Tianjin 300010; tel. (22) 24406326; fax (22) 22346965.

Wenergy Co Ltd: 81 Wuhu Lu, Hefei 230001; tel. (551) 2626906; fax (551) 2648061.

Wuhan Power Supply Bureau: 981 Jiefang Dadao, Hankou, Wuhan 430013; tel. (27) 2426455; fax (27) 2415605.

Wuxi Power Supply Bureau: 8 Houxixi, Wuxi 214001; tel. (510) 2717678; fax (510) 2719182.

Xiamen Power Transformation and Transmission Engineering Co: 67 Wenyuan Lu, Xiamen 361004; tel. (592) 2046763.

Xian Power Supply Bureau: Huancheng Dong Lu, Xian 710032; tel. (29) 7271483.

Gas

Beijing Gas Co: 30 Dongsanhuan Zhong Lu, Beijing 100020; tel. (10) 65024131; fax (10) 65023815; Dir LIU BINGIUN.

Beijing Natural Gas Co: Bldg 5, Dixingju, An Ding Men Wai, Beijing 100011; tel. (10) 64262244.

Changchun Gas Co: 30 Tongzhi Jie, Changchun 130021; tel. (431) 8926479.

Changsha Gas Co: 18 Shoshan Lu, Changsha 410011; tel. (731) 4427246.

Qingdao Gas Co: 399A Renmin Lu, Qingdao 266032; tel. (532) 4851461; fax (532) 4858653.

Shanghai Gas Supply Co: 656 Xizang Zhong Lu, Shanghai 200003; tel. (21) 63222333; fax (21) 63528600; Gen. Man. LI LONGLING.

Wuhan Gas Co: Qingnian Lu, Hankou, Wuhan 430015; tel. (27) 5866223.

Xiamen Gas Corpn: Ming Gong Bldg, Douxi Lukou, Hubin Nan Lu, Xiamen 361004; tel. (592) 2025937; fax (592) 2033290.

Water

Beijing District Heating Co: 1 Xidawang Lu, Hongmiao, Chao Yang Qu, Beijing 100026; tel. (10) 65060066; fax (10) 65678891.

Beijing Municipal Water Works Co: 19 Yangrou Hutong, Xicheng Qu, Beijing 100034; tel. (10) 66167744; fax (10) 66168028.

Changchun Water Co: 53 Dajing Lu, Changchun 130000; tel. (431) 8968366.

Chengdu Water Co: 16 Shierqiao Jie, Shudu Dadao, Chengdu 610072; tel. (28) 77663122; fax (28) 7776876.

The China Water Company: f. to develop investment opportunities for water projects.

Guangzhou Water Supply Co: 5 Huanshi Xi Lu, Guangzhou 510010; tel. (20) 81816951.

Haikou Water Co: 31 Datong Lu, Haikou 570001; tel. (898) 6774412.

Harbin Water Co: 49 Xi Shidao Jie, Daoli Qu, Harbin 150010; tel. (451) 4610522; fax (451) 4611726.

Jiangmen Water Co: 44 Jianshe Lu, Jiangmen 529000; tel. (750) 3300138; fax (750) 3353704.

Qinhuangdao Pacific Water Co: Hebei; Sino-US water supply project; f. 1998.

Shanghai Municipal Waterworks Co: 484 Jiangxi Zhong Lu, Shanghai 200002; tel. (21) 63215577; fax (21) 63231346; service provider for municipality of Shanghai.

Shenzhen Water Supply Group Co: Water Bldg, 1019 Shennan Zhong Lu, Shenzhen 518031; tel. (755) 2137836; fax (755) 2137888; e-mail webmaster@waterchina.com; internet www.waterchina.com.

Tianjin Waterworks Group: 54 Jianshe Lu, Heping Qu, Tianjin 300040; tel. (22) 3393887; fax (22) 3306720.

Xian Water Co: Huancheng Xi Lu, Xian 710082; tel. (29) 4244881.

Zhanjiang Water Co: 20 Renmin Dadaonan, Zhanjiang 524001; tel. (759) 2286394.

Zhongshan Water Supply Co: 23 Yinzhu Jie, Zhuyuan Lu, Zhongshan 528403; tel. (760) 8312969; fax (760) 6326429.

Zhuhai Water Supply General Corpn: Yuehai Zhong Lu, Gongbei, Zhuhai 519020; tel. (756) 8881160; fax (756) 8884405.

MAJOR COMPANIES

(cap. = capital; res = reserves; m. = million; amounts in yuan, unless otherwise stated)

Beijing Changning Group: Changping County, Beijing 102206; tel. (10) 83607016; fax (10) 83607022; e-mail cngroup@public.bta.net.cn; internet www.changning.com.cn; f. 1985; cap. and res 280m., sales 448m. (1995); 18 corporate mems; machinery, electronics, foods, light industry, building, trade; Pres. SHI SHANLIN; 3,000 employees.

Bengang Steel Plates Co Ltd: Gantie Lu, Pingshan Qu, Benxi, Liaoning 117000; tel. (414) 7827344; fax (414) 7827004; e-mail bgbctwg@mail.bxptt.ln.cn; f. 1997; cap. and res 4,027m., sales 5,243m. (2001); ferrous metals processing; Chair ZHANG YINGFU.

Brilliance China Automotive Holdings Ltd: Suites 2303–2306, 23rd Floor, Great Eagle Centre, 23 Harbour Rd, Wanchai, Hong Kong ; tel. (852) 25237227; fax (852) 25268472; internet www.brillianceauto.com; f. 1992; China's largest light truck manufacturer; revenue 6,218m. (2001); Chair. WU XIAO AN.

Broad Air Conditioning: Broad Town, Changsha, Hunan; tel. (10) 82514688; fax (10) 82515208; e-mail international@broad.net; internet www.broad.com; sales US $145.0m. (2001); air conditioning products manufacturer; Chair. ZHANG YUE; 1,600 employees.

Changchai Co Ltd: 123 Huaide Jie, Changzhou, Jiangsu 213002; tel. (519) 6600341; fax (519) 6670765; e-mail ccstp@public.cz.js.cn; internet www.changchai.com.cn; f. 1913; production of diesel engines and agricultural product-processing machinery; cap. and res 1,324.7m. (1997), sales 1,743.5m. (Dec. 2001); Chair. ZHANG JUNYUAN.

Chaoda Modern Agriculture Group: 32-36F, Zhongshan Tower, No.154 Hudong Lu, Fuzhou, Fujian; tel. (591) 7835933; fax (591) 7833208; e-mail chaoda@pub2.fz.fj.cn; internet www.chaoda.com; sales US $87.0m. (2001); organic agriculture; Chair. GUO HAO; 6,000 employees.

China Harbour Engineering Company: 9 Chunxiu Lu, Dong Zhi Men Wai, Beijing 100027; tel. (10) 64154455; fax (10) 64168276; e-mail chechw@homeway.com.cn; civil engineering; sales US $1,333.8m. (1999/2000); Pres. LI HUAIYUAN.

China Orient Group Co Ltd (Dongfang Group): 2nd Bldg of Science Industry and Trade, Nangang Qu, Harbin, Heilongjiang 150001; tel. (451) 3666036; fax (451) 3666030; e-mail master@china-orient.com; internet www.china-orient.com; f. 1988; sales 5,500.0m. (2001); finance, transport, construction, industry, real estate management; Chair. ZHANG HONGWEI.

Chint Group: Chint High-tech Industrial Park, North Baixiang, Wenzhou, Zhejiang 325603; tel. (577) 62777777 ; fax (577) 62775769; e-mail info@chint.com; internet www.chint.com, www.chint.com/foreign; f. 1984; specialized manufacturer of electrical appliances; sales US $970m. (2002); Chair. NAN CUNHUI; 13,000 employees.

Chongqing Chang'an Automobile Co Ltd: 260 Jianxin Dong Lu, Jiangbei Qu, Chongqing 400060; tel. (23) 67591349; fax (23) 67866055; e-mail cazqc@mail.changan.com.cn; internet www.changan.com.cn; sales 6,703.2m. (2000); manufactures and exports road vehicles and vehicle components; Chair. YIN JIA XU.

Chongqing Iron and Steel Co Ltd: 30 Gangtie Lu, Dadukou Qu, Chongqing 630081; tel. (23) 8875551; fax (23) 68831444; internet www.china-metal.com; manufacture and sale of iron and steel products; cap. and res 1,740.6m. (1998), sales 3,722.8m. (2000); Chair. TANG MIN WEI.

Delixi Group: Delixi Industrial Zone, Liushi, Wenzhou 325604; tel. (577) 62723888; fax (577) 62725559; e-mail info@delixi.com; internet www.delixi.com; www.delixi.com/english; sales US $790.0m. (2001); manufacturer of electric power transmission and distribution appliances; Chair. HU CHENGZHONG; 10,000 employees.

Dong Fang Electrical Machinery Co Ltd: 13 Huanghe Xi Jie, Deyang, Sichuan 618000; tel. (838) 2412144; fax (838) 2203305; e-mail info@dfem.com.cn; internet www.dfem.com.cn; production and sale of power-generating equipment; cap. and res 1,204.2m. (1998), sales 757.1m. (1999); Chair. SI ZE FU; 7,804 employees.

Dongfeng Motor Corpn: 1 Checheng Lu, Zhangwan Qu, Shiyan 442001; tel. (719) 226987; fax (719) 226815; internet www.chinacars.com; f. 1969; sales 11,200m. (1998); trucks and automobiles; Gen. Man. MIAO WEI; 90,000 employees.

East Hope Group: PO Box 134-111 Pudong, Shanghai 200134; tel. (21) 58312099 ; fax (21) 68670083; e-mail master@easthope.com.cn; internet www.easthope.com.cn; f. 1995; created from restructuring of Hope Group; sales income 2,800.0m. (1999); feed production; Chair. LIU YONGXING; 6,000 employees.

First Tractor Co Ltd: 154 Jian She Jie, Luoyang City, Henan; tel. (379) 4970038; fax (379) 4978838; e-mail office@first-tractor.com.cn; internet www.first-tractor.com.cn; production and sale of agricultural tractors; cap. and res 2,358.9m., sales 2,703.6m. (1998); Sec. JIANG GUOLIANG.

Gree Group Corp: Beijing Industrial Zone, Zhuhai, Guangdong 519020; tel. (756) 8131888; fax (756) 8885701; e-mail info@gree.com; internet www.gree.com; f. 1985; revenue US $761.8m. (2000); manufactures electronic and electrical goods and machinery; property and transport.

Guangdong Kelon Electrical Appliance Co (Kelon Group): 12 Qiaodong Lu, Ronggui, Shunde 528303; tel. (765) 8361163; fax (765) 8361060; e-mail overseas@kelon.com; internet www.kelon.com; f. 1993; domestic electrical appliances, incl. refrigerators; cap. and res 3,319m. (1998), sales 3,869.5m. (2000); Chair. GU CHU JUN; 20.6% owned by Greencool Technology; 10,210 employees.

Guangdong Wanli Architectural Machinery Equipment Group Co (Guangdong Wanli Scaffolding Company): 102 He Nan Lu, Chikan Qu, Kai Ping, Guangdong; tel. (750) 2618888; fax (750) 2613288; e-mail wanliok@21cn.net; internet www.wanliok.com; f. 1972; construction machinery.

Guanghui Industry Co Ltd (Xinjiang Guanghui Stone Joint-Stock Co Ltd): No. 6 Shanghai Rd, Economic & Technology Development Zone, Urumqi, Xinjiang Province; tel. (991) 3715828; fax (991) 3735502; internet www.guanghui.com (Chinese website); f. 1988; stone excavation, real estate; sales 6,100.0m. (2001); Chair. SUN GUANGXIN; 20,000 employees.

Haci Group Co Ltd: 169 Tongxiang Jie, Harbin, Heilongjiang 150046; tel. (451) 2688688; fax (451) 2686254; internet www.hacico.com; f. 1992; 12 corporate mems; magnetic healthcare products, food engineering; Pres. GUO LIWEN; 480 employees.

Haier Group: Inside Haier Garden, Haier Industrial Park, Haier Lu, Qingdao 266101; tel. (532) 8938888; fax (532) 8938666; e-mail info@haier.com; internet www.haier.com; f. 1991; sales 71,100.0m. (2002); household appliances, pharmaceuticals, air conditioners, refrigerators; Pres. ZHANG RUIMIN; Vice-Pres. SHAO MINGJIN; 18,901 employees.

Harbin Pharmaceutical Group Co Ltd: 94 Gongchang Lu, Harbin, Heilongjiang 150018; tel. and fax (451) 4604688; f. 1991; revenue US $778.4m. (2000); China's biggest pharmaceuticals company (2000).

Heibei Qifa Textiles Co Ltd: New Developing Industry Zone, Baoding, Hebei 071400; tel. (312) 6561030; fax (312) 6561590; internet www.qifagroup.com; f. 1988; cap. 200m., fixed assets 160m. (2002); pure knitting wool, chemical fibres, worsted woollen products; Pres. WANG QIFA; 3,600 employees.

Henan Star Hi-Tech Co Ltd: 38 Jinsuo Lu, Zhengzhou, Henan 450001; tel. (371) 7982920; fax (371) 7982920; e-mail star@hnstar.com; internet www.hnstar.com; www.starhi-tech.com; f. 1988; electric measurement equipment, telecommunication systems ; 66.72% owned by Henan Star Science and Technology Group Ltd; Gen. Man. GENG ZHI; 1,169 employees.

Hengdian Group (Zhejiang Hengdian Imp. & Exp. Co Ltd): 3/F, World Trade Office Plaza, World Trade Centre Zhejiang,15 Shuguang Lu, Hangzhou, Zhejiang; tel. (571) 87950110; fax (571) 87950086; e-mail hdhz@mail.hz.zj.cn; internet www.hengdian.com; www.hengdian-group.com; f. 1975; sales 8,000.0m. (2001); magnetic materials, electronics, chemicals and pharmaceuticals; Chair. XU WENRONG; 35,944 employees (2001).

Huawei Technologies Co Ltd: Banxuegang Industrial Park, Buji Longgang, Shenzhen 518129; tel. (755) 28780808; fax (755) 28780808; e-mail hpeng@huawei.com; internet www.huawei.com.cn; f. 1988; international revenues US $552.0m. (2002); communications technology; Chair. YAFANG SUN.

Jilin Chemical Industrial Co Ltd: 31 Zunyi Dong Jie, Jilin 132021; tel. (432) 3976445; fax (432) 3028146; e-mail webmaster@jcic.com.cn; internet www.jcic.com; production and sale of petroleum and chemical products; revenue US $1,618m. (2000); Chair. JIAO HAIKUN; 6,500 employees.

Jingwei Textile Machinery Co Ltd: 150 Jingwei Jie, Yuci, Shanxi; tel. (354) 2422878; fax (354) 2425428; internet www.jwme.com; manufacture and sale of textile machinery; distribution of computers; cap. and res 658.5m. (1998), sales 803.6m. (2000); Chair. YIN SHOUEN; 5,505 employees.

Jinzhou Petrochemical: 2 Chongqing Lu, Guta, Jinzhou, Liaoning 121001; tel. and fax (416) 4152316; fax (416) 416 9024; e-mail jzpcmaster@public.jzpc.com.sn; internet www.jzpc.com; f. 1997; petroleum processing and coking; revenue US $1,094.8m. (2000); Dir YANG XUEZHUANG; 10,500 employees.

Konka Group Co Ltd: East Industrial Zone, Overseas Chinese Town, 518053 Shenzhen; tel. (755) 86608866; fax (755) 86600082; e-mail chenxuri@konka.com; internet www.konka.com; f. 1979; sales 6,748.1m. (2001); consumer electronics; Chair. REN KELEI.

Kunming Machine Tool Co Ltd: 23 Ciba Jie, Kunming, Yunnan 650203; tel. (871) 5212411; fax (871) 5150317; design, development and production of machine tools, precision-measuring equipment, etc.; cap. and res 521.4m. (1998), sales 46.2m. (1999); Chair. YE XIANGYU; 3,077 employees.

Lenovo Group: 6 Chuang Ye Lu, Haidian Qu, Beijing 100085; tel. (10) 82878888; fax (10) 62570209; e-mail cmk@lenovo.com; internet www.lenovo.com; fmrly Legend Group Corpn; f. 1984; mfr of personal computers; China's largest private company (2001); sales HK $19,267.0m. (2002); Pres. and CEO YANG YUANQING; Chair. LIU CHUANZHI; 10,000 employees.

New Hope Group Co Ltd: No. 45, Sec. 4, Ren Ming Nan Lu, Chengdu, Sichuan 610041; tel. (28) 5225052; fax (28) 5233678; e-mail nhg@mail.sc.cninfo.net; internet www.newhopegroup.com; f. 1983; cap. and res 2,500m. (1997), sales 11,300m. (2001); 76 factories; animal feed, food, trade; Pres. LIU YONGHAO; 10,000 employees.

Panzhihua New Steel and Vanadium Co Ltd: 55 Dadukou Lu, Dong Qu, Panzhihua, Sichuan 617067; tel. (812) 3394123; fax (812) 2226014; f. 1993; metal (incl. high vanadium steel) manufacturing and exports; Rep. HONG JIBI; 20,169 employees.

People's Food Holding Ltd: Bancheng Town, Linyi, Shangdong 276036; tel. (539) 8692888; fax (539) 8692875; internet www.peoplesfood.com.sg; f. 1994; sales 2,400.0m. (2001/02); meat products; Chair. MING KAM SING.

Petrochina Co. Ltd: 16 Andelu Xicheng District, Beijing; tel. (10) 84886034; fax (10) 84886039; e-mail webmaster@petrochina.com.cn; internet www.petrochina.com.cn; exploration, development, and production of crude oil and natural gas, and refining, transport, storage, and trade of crude petroleum and its products; sales US $21,256.7m. (1999/2000); Chair. MA FUCAI.

Qingqi Motorcycle: 34 Heping Lu, Jinan; tel. (531) 6953325; fax (531) 6954219; e-mail jnqqxx@public.jn.sd.cn; internet www.china-qingqi.com/; China's largest producer of motor cycles; Pres. ZHANG JIALING.

Shandong Xinhua Pharmaceutical Co Ltd: 14 Dongyi Jie, Zhangdian Qu, Zibo, Shandong 255005; tel. (533) 2184223; fax (533) 2184991; f. 1943; manufacture and sale of pharmaceuticals; cap. and res 863.8m. (1998), sales 1,044.0m. (2000); Chair. HE DUANSHI; Pres. GUO QIN; 5,650 employees.

Shanggong Co Ltd: 12th Floor, Orient Mansion, No.1500 Shiji Da Dao Pudong New Area Shanghai 200122 China; tel. (21) 68407700; fax (21) 63302939; e-mail shggg@public3.sta.net.cn; internet www.shanggonggroup.com; manufacture and sale of sewing machines; cap. and res 430.4m., sales 481.1m. (1997), export income US $46m. (2001); Chair. NI YONGGONG.

Shanghai Automation Instrumentation Co Ltd (SAIC): 1599 Yan An Xi Lu, Shanghai 200050; tel. (21) 62521870; fax (21) 62801680; f. 1993; manufacture of control systems and meters for industrial use; cap. and res 547.9m. (1998), sales 748.8m. (1999); Chair. ZHOU YONGQING; 14,167 employees.

Shanghai Chlor-Alkali Chemical Co Ltd: 47 Longwu Jie, Shanghai 200241; tel. (21) 64340000; fax (21) 64341341; e-mail public@styc.com; internet www.scacc.com; f. 1992; production of industrial chemicals; cap. and res 2,739.4m. (1997), sales 2,939.3m. (2000); Chair. ZHOU BO; 6,062 employees.

Shanghai Dajiang (Group) Stock Co Ltd: 26 Yuzai Nan Lu, Songjiang, Shanghai 201600; tel. (21) 57822480; fax (21) 57820072; e-mail dajiang@dajiang.com; internet www.dajiang.com; f. 1985; production and sale of animal feed and feed machinery; sales 1,618.8m. (2000); Pres. THANAKOM SERIBURI; Dir THAWEEP LERTDAMRONGCHAI; 1,277 employees.

Shanghai Erfangji Co Ltd: 265 Chang Zhong Jie, Shanghai 200434; tel. (21) 65318888; fax (21) 65421963; e-mail master@shej.com; internet www.shej.com; f. 1944; manufacture and sale of textile machinery; cap. and res 746.8m. (1997), sales 465.3m. (2000); Dir. CHEN XINGQIANG; 5,000 employees.

Shanghai Forever Bicycle Co Ltd (Shanghai Forever Import & Export Co Ltd): 8/F,380 Tian Mu Zhong Lu, Shanghai 200070; tel. (21) 63541511; fax (21) 63171995; e-mail info@cnforever.com; internet www.forever-bicycle.com; design, manufacture and distribution of bicycles; cap. and res 502.0m. (1997), sales 279.7.4m. (2000); one of the world's largest bicycle manufacturers; Chair. TAO GUOQIANG.

Shanghai Fosun Industrial Co Ltd (Shanghai Fortune High Technology Co): No. 2 Fuxing Dong Lu, Shanghai China 200010; tel. (21) 63323318; fax (21) 63325080; e-mail 600196@fosun.com.cn; internet www.fosun.com.cn; sales 5,700m. (2001); biotechnology company; Chair. GUO GUANG CHANG.

Shanghai Haixin Group Co Ltd: 688 Changxing Lu, Dongjing, Song Jiang Qu, Shanghai 201619; tel. (21) 56798100; fax (21) 57698025; e-mail yph@haixin.com; internet www.haixin.com; f. 1986; manufacture and sale of plush and flannel materials; cap. and res 683.4m. (1997), sales 964.9m. (2000); Chair. YAN ZHENBO; 7,500 employees.

Shanghai Hongyuan Lighting & Electric Equipment Co Ltd: 5028 Zhennan Lu, Shanghai 201802; tel. (21) 39120199; fax (21) 59128661; e-mail webmaster@shhongyuan.com; internet www.shhongyuan.com; f. 1989; high-voltage sodium lamps, lighting equipment and components; Pres. LI WEIDE; 500 employees.

Shanghai Huili Building Materials Co Ltd (Shanghai Huili (Group) Ltd): 100 Xinma Lu, Zhoupu, Nanhui, Shanghai 201318; tel. (21) 58113390; fax (21) 58113449; internet www.hiuligp.com; development, production and sale of wall coatings and PVC flooring; cap. and res 208.5m. (1997), sales 212.8m. (1999); Chair. ZHANG YONGDING; 970 employees.

Shanghai Tyre and Rubber Co Ltd: 63 Sichuan Lu, Shanghai 200002; tel. (21) 63290433; fax (21) 63299609; e-mail company@cstarc.com; internet www.cstarc.com; tyres and machinery for production of rubber products; cap. and res 1,827.4m. (1997), sales 3,007.8m. (1999); Chair. ZHANG YINGSHI; 12,000 employees.

Shanghai Volkswagen Automotive Company Ltd: 63 Antingluopu Lu, Shanghai 201805; tel. (21) 59561888; fax (21) 59572815; internet www.csvw.com; jt venture between Volkswagen (Germany) and Shanghai Automotive Industry Corpn to manufacture cars; sales US $3,230.2m. (1999/2000); Chair. LU JIAN.

Shanghai Xin Gao Chao Group Co: 787 Kang Qiao Lu, Kangqiao Industrial Development Area, Shanghai 201315; tel. (21) 58120888; fax (21) 58121988; e-mail director@xingaochao.com; internet www.xingaochao.com.cn; sales 5,600.0m. (2001); flooring and furniture manufacturer; Pres. TAO XIN KANG; 20,000 employees.

Shenyang Xiehe Group Co: 65 Hunnan New & High Tech Industrial Development Zone, Shenyang 110179; tel. (24) 23936916; fax (24) 23936741; e-mail xiehe@xiehegroup.com.cn; internet www.xiehegroup.com.cn; f. 1988; 3 corporate mems; pharmaceutical products, healthcare, software; Pres. CHEN JUYU; 350 employees.

Shenzhen Taita Pharamceutical Industry Co Ltd: 23/F Office Tower, Xin Xing Sq., Di Wang Commercial Centre, 5002 Shen Nan Dong Lu, Shenzhen 518008; tel. (755) 2463888; fax (755) 2461436; e-mail mr@taitai.com; internet www.taita.com; f. 1992; sales US $82.0m. (2001); pharmaceutical products, Chinese herbal medicine; Chair. ZHU BAOGUO; 3,000 employees.

Shenzhen Textile (Holdings) Co Ltd: 6th Floor, Shen Fang Bldg, 3 Hua Qiang Bei Lu, Shenzhen 518031; tel. (755) 23776043; fax (755) 23360139; internet www.sz-textile.com.cn; manufacture of textiles, garments and related products; f. 1994; cap. and res 342.9m. (1997), sales 292.9m. (2000); Chair. GUAN TONGKE; 1,000 employees.

Shide Group: Dalian Shide Group, 38 Gao Er Ji Lu, XiGang Qu, Dalian, Liaoning 116011; tel. (411) 3622218; e-mail webmaster@shide.com; internet www.shide.com, www.shide-global.com; f. 1992; sales US $460.0m. (2001); home electrics, chemical building materials, automobile manufacturing; Pres. XU MING; 5,000 employees.

Sichuan Changhong Electric Co Ltd: 4 Yuejin Lu, Mianyang 621000; tel. (816) 2411114; fax (816) 2337518; f. 1988; manufacture of electrical appliances, electronic components and computers; China's largest TV manufacturer; sales US $1,401.6m. (1999/2000); Chair. NI FUNRENG; 6,500 employees.

Tangshan Iron and Steel Co Ltd: 9 Binhe Lu, Tangshan, Hebei 063016; tel. (315) 2702941; fax (315) 2702198; e-mail tgyang@ts-user.he.cninfo.net; f. 1995; revenue US $839.1m. (2000); manufacturing of ferrous metals; Rep. WANG TIANYI.

Tengen Group: Tengen Mansion, Dongfeng Industrial Zone, Liushi, Wenzhou 325604; tel. (577) 62775688; fax (577) 62776888; e-mail ibd@tengen.com.cn; internet www.tengen.com.cn; sales 5,100.0m. (2001); manufacturer of low voltage electrical appliances; Man. Dir WILL ZHANG.

Tianjin Da Heng (Group) Co: 588 Jintang Lu, Tianjin 300300; tel. (22) 24993988; fax (22) 24993986; e-mail dah@daheng.com.cn; internet www.daheng.com.cn; f. 1988; 8 corporate mems; foods and beverages; Pres. WU ZHENGHAI; 1,800 employees.

Tongwei Group: PO B 618, Hi-tech Development Zone, Chengdu, Sichuan 610041; tel. (28) 85188888; fax (28) 85199999; e-mail atw@tongwei.com; internet www.tongwei.com; sales US $735.0m. (2000); feed production, bio-engineering, construction; Pres. LIU HANYUAN.

Wanxiang Group Corporation: Ningwei Town, Xiaoshan City, Zhejiang Province; tel. (571) 2716888; fax (571) 2602358; internet www.wanxiang.com/group, www.wanxiang.com.cn; sales 8,600.0m. (2001); China's largest manufacturer of car parts; Pres. LIU GUANQIU; 6,500 employees.

Wuhan Steel Processing Co Ltd: 3 Yangang Lu, Qingshan Qu, Wuhan Hubei 430080; tel. (27) 86306023; fax (27) 86807873; internet www.wisco.com.cn; f. 1997; revenue US $836.5m. (2000); metals manufacturing; Chair. LIU BENREN.

Wuxi Little Swan Co Ltd: 1 Hanjiang Lu, Wuxi 214028; tel. (510) 3704003; fax (510) 3704031; e-mail info@littleswan.com.cn; internet www.littleswan.com; f. 1979; design, development, manufacture and sale of domestic washing machines; cap. and res 365.1m., sales 2,723.8m. (2000); Chair. ZHU DEKUN.

Zhejiang Zhongda Group Co Ltd: 21/F, Tower A, Zhongda Plaza, Hangzhou 310003; tel. (571) 85155000; fax (571) 85777050; e-mail gufen@zhongda.com; internet zhongda.zj-zhongda.com; revenue US $787.8m. (2000); garments and textiles; diversifying into other consumer industries; Chair. ZHONG SHAN; 600 employees.

Zongshen Group: Chongqing Zongshen Motorcycles Group, Zongshen Industrial Garden, Banan Qu, Chongqing 400054; tel. (23) 66372880; e-mail zsgroup@zongshenmotor.com; internet www.zongshenmotor.com; f. 1992; assets 4,000.0m. (2002); motorcycle and engines manufacturer; Chair. ZUO ZONGSHEN; 18,000 employees.

TRADE UNIONS

All-China Federation of Trade Unions (ACFTU): 10 Fu Xing Men Wai Jie, Beijing 100865; tel. (10) 68592114; fax (10) 68562030; f. 1925; organized on an industrial basis; 15 affiliated national industrial unions, 30 affiliated local trade union councils; membership is voluntary; trade unionists enjoy extensive benefits; 103,996,000 mems (1995); Chair. WANG ZHAOGUO; First Sec. ZHANG JUNJIU.

Principal affiliated unions:

All-China Federation of Railway Workers' Unions: Chair. HUANG SICHUAN.

Architectural Workers' Trade Union: Sec. SONG ANRU.

China Self-Employed Workers' Association: Pres. REN ZHONG-LIN.

Educational Workers' Trade Union: Chair. JIANG WENLIANG.

Light Industrial Workers' Trade Union: Chair. LI SHUYING.

Machinery Metallurgical Workers' Union: Chair. ZHANG CUNEN.

National Defence Workers' Union: Chair. GUAN HENGCAI.

Postal and Telecommunications Workers' Trade Union of China: Chair. LUO SHUZHEN.

Seamen's Trade Union of China: Chair. ZHANG SHIHUI.

Water Resources and Electric Power Workers' Trade Union: Chair. DONG YUNQI.

Workers' Autonomous Federation (WAF): f. 1989; aims to create new trade union movement in China, independent of the All-China Federation of Trade Unions.

Transport

RAILWAYS

Ministry of Railways: 10 Fuxing Lu, Haidian Qu, Beijing 100844; tel. (10) 63244150; fax (10) 63242150; e-mail webmaster@ns.chinamor.cn.net; internet www.chinamor.cn.net; controls all railways through regional divisions. The railway network has been extended to all provinces and regions except Tibet (Xizang), where construction is in progress. Total length in operation in December 2002 was 71,897.5 km, of which about 17,400 km were electrified. The major routes include Beijing–Guangzhou, Tianjin–Shanghai, Manzhouli–Vladivostok, Jiaozuo–Zhicheng and Lanzhou–Badou. In addition, special railways serve factories and mines. A new 2,536-km line from Beijing to Kowloon (Hong Kong) was completed in late 1995. Plans for a 1,450-km high-speed link between Beijing and Shanghai were announced in 1994, and construction was scheduled to begin by 2005. A high-speed link between Beijing and Guangzhou was also planned. China's first high-speed service, linking Guangzhou and Shenzhen, commenced in December 1994. A direct service between Shanghai and Hong Kong commenced in 1997.

An extensive programme to develop the rail network was announced in early 1998, which aimed to increase the total network to 68,000 km by the year 2000, and to more than 75,000 km by 2005. Railways were to be constructed along the Changjiang valley, starting at Sichuan, and along China's east coast, originating at Harbin. In December 1999 plans were announced for a railway to Kazakhstan. In June 2001 construction began on a new 1,118-km railway linking Tibet with the rest of China, to be completed after 10 years. A new magnetic-levitation ('maglev') railway linking Shanghai to Pudong International airport was built in co-operation with a German consortium in 2002. The first journey on this railway line was made at the beginnning of 2003. The 'maglev' train was expected to carry 10m. passengers a year by 2005.

City Underground Railways

Beijing Metro Corpn: 2 Beiheyan Lu, Xicheng, Beijing 100044; tel. (10) 68024566; f. 1969; total length 54 km, with 98 km of further lines to be built by the year 2010; Gen. Man. FENG SHUANGSHENG.

Guangzhou Metro: 204 Huanshi Lu, Guangzhou 510010; tel. (20) 6665287; fax (20) 6678232; opened June 1997; total length of 18.5 km, with a further 133 km planned; Gen. Man. CHEN QINGQUAN.

Shanghai Metro Corpn: 12 Heng Shan Lu, Shanghai 200031; tel. (21) 64312460; fax (21) 64339598; f. 1995; 65.8 km open, with at least a further 181.5 km under construction or planned; Pres. SHI LIAN.

Tianjin Metro: 97 Jiefangbei Lu, Heping, Tianjin 300041; tel. (22) 23395410; fax (22) 23396194; f. 1984; total planned network 154 km; Gen. Man. WANG YUJI.

Underground systems were under construction in Chongqing, Nanjing, and Shenzhen, and planned for Chengdu and Qingdao.

ROADS

At the end of 2002 China had 1,765,222 km of highways. Four major highways link Lhasa (Tibet) with Sichuan, Xinjiang, Qinghai Hu and Kathmandu (Nepal). A programme of expressway construction began in the mid-1980s. By 2002 there were 25,130 km of expressways, routes including the following: Shenyang–Dalian, Beijing–Tanggu, Shanghai–Jiading, Guangzhou–Foshan and Xian–Lintong. Expressway construction was to continue, linking all main cities and totalling 55,000 km by 2020. A new 123-km highway linking Shenzhen (near the border with Hong Kong) to Guangzhou opened in 1994. A 58-km road between Guangzhou and Zhongshan connects with Zhuhai, near the border with Macao. Construction of a bridge, linking Zhuhai with Macao, began in June 1998 and was completed in late 1999. A bridge connecting the mainland with Hong Kong was to be built, with completion scheduled for 2004.

INLAND WATERWAYS

At the end of 2002 there were some 121,557 km of navigable inland waterways in China. The main navigable rivers are the Changjiang (Yangtze River), the Zhujiang (Pearl River), the Heilongjiang, the Grand Canal and the Xiangjiang. The Changjiang is navigable by vessels of 10,000 tons as far as Wuhan, more than 1,000 km from the coast. Vessels of 1,000 tons can continue to Chongqing upstream.

There were 5,142 river ports at the end of 1996. In 1997 there were some 5,100 companies involved in inland waterway shipping.

SHIPPING

China has a network of more than 2,000 ports, of which more than 130 are open to foreign vessels. In May 2001 plans were announced for the biggest container port in the world to be built on the Yangshan Islands, off shore from Shanghai. The main ports include Dalian, Qinhuangdao, Tianjin, Yantai, Qingdao, Rizhao, Lianyungang, Shanghai, Ningbo, Guangzhou and Zhanjiang. In 2002 the main coastal ports handled 1,666.3m. metric tons of cargo. In December 2002 China's merchant fleet comprised 3,326 ships, totalling 17.3m. grt.

Bureau of Water Transportation: Beijing; controls rivers and coastal traffic.

China International Marine Containers Group Co Ltd: 5/F, Finance Centre, Shekou, Shenzhen 518067; tel. (755) 26691130; fax (755) 26692707; internet www.cimc.com; f. 1980; container-manufacturing, supply and storage; revenue US $1,081.6m. (2000); Chair. and Dir LI JIANHONG.

China National Chartering Corpn (SINOCHART): see Trade and Industrial Organizations.

China Ocean Shipping (Group) Co (COSCO): 11th and 12th Floors, Ocean Plaza, 158 Fu Xing Men Nei, Xi Cheng Qu Chao Yang Qu, Beijing 100031; tel. (10) 66493388; fax (10) 66492288; internet www.cosco.com.cn; reorg. 1993, re-established 1997; head office transferred to Tianjin late 1997; br. offices: Shanghai, Guangzhou, Tianjin, Qingdao, Dalian; 200 subsidiaries (incl. China Ocean Shipping Agency—PENAVIC) and joint ventures in China and abroad, engaged in ship-repair, container-manufacturing, warehousing, insurance, etc.; merchant fleet of 600 vessels; 47 routes; Pres. WEI JIAFU.

China Shipping (Group) Co: Shanghai; f. 1997; Pres. LI KELIN.

China Shipping Container Lines Co Ltd: 5th Floor, Shipping Tower, 700 Dong Da Ming Lu, Shanghai 200080; tel. (21) 65966978; fax (21) 65966498; Chair. LI SHAODE.

China Shipping Development Co Ltd Tanker Co: 168 Yuanshen Lu, Pudong New Area, Shanghai 200120; tel. (21) 68757170; fax (21) 68757929.

Fujian Shipping Co: 151 Zhong Ping Lu, Fuzhou 350009; tel. (591) 3259900; fax (591) 3259716; e-mail fusco@pub2.fz.fj.cn; internet www.fusco-cn.com; f. 1950; transport of bulk cargo, crude petroleum products, container and related services; Gen. Man. LIU QIMIN.

Guangzhou Maritime Transport (Group) Co: 22 Shamian Nan Jie, Guangzhou; tel. (20) 84104673; fax (20) 84103074.

CIVIL AVIATION

Air travel is expanding very rapidly. In 2002 a total of 141 civil airports were in operation. Chinese airlines carried a total of 85.9m. passengers in 2002. By 1998 there were 34 airlines, including numerous private companies, operating in China. In October 2002 the Government merged the nine largest airlines into three groups, based in Guangzhou, Shanghai and Beijing. Air China incorporated China National Aviation Corpn (which also controls Zhejiang Airlines and 43% of Hong Kong's Dragonair) and China Southwest Airlines; China Eastern Airlines incorporated China Northwest Airlines and Yunnan Airlines; and China Southern Airlines incorporated Air Xinjiang and China Northern Airlines.

General Administration of Civil Aviation of China (CAAC): POB 644, 155 Dongsixi Jie, Beijing 100710; tel. (10) 64014104; fax (10) 64016918; f. 1949 as Civil Aviation Administration of China; restructured in 1988 as a purely supervisory agency, its operational functions being transferred to new, semi-autonomous airlines (see below); also China United Airlines (division of the Air Force) and China Capital Helicopter Service); domestic flights throughout

China; external services are mostly operated by **Air China, China Eastern** and **China Southern Airlines**; Dir LIU JIANFENG.

Air China: Beijing International Airport, POB 644, Beijing 100621; tel. (10) 64599068; fax (10) 64599064; e-mail webmaster@mail.airchina.com.cn; internet www.airchina.com.cn; international and domestic scheduled passenger and cargo services; Pres. WANG KAIYUAN.

Changan Airlines: 16/F, Jierui Bldg, 5 South Er Huan Rd, Xian, Shaanxi 710068; tel. (29) 8707412; fax (29) 8707911; e-mail liulei@hnair.com; internet www.changanair.com; f. 1992; local passenger and cargo services; Pres. SHE YINING.

China Eastern Airlines: 2550 Hongqiao Rd, Hongqiao Airport, Shanghai 200335; tel. (21) 62686268; fax (21) 62686116; e-mail webmaster@cea.online.sh.cn; internet www.cea.online.sh.cn; f. 1987; domestic services; overseas destinations include USA, Europe, Japan, Sydney, Singapore, Seoul and Bangkok; Pres. LIU SHAOYONG.

China General Aviation Corpn: Wusu Airport, Taiyuan, Shanxi 030031; tel. (351) 7040600; fax (351) 7040094; f. 1989; 34 domestic routes; Pres. ZHANG CHANGJING.

China National Aviation Corpn: 23 Qianmen Dong Dajie, Dongcheng Qu, Beijing 100006; tel. (10) 5133060; fax (10) 5133060; controls Zhejiang Airlines and 43% of Hong Kong's Dragonair; merged with Air China Oct. 2002.

China Northern Airlines: 3-1 Xiaoheyan Lu, Dadong Qu, Shenyang, Liaoning 110043; tel. (24) 88294432; fax (24) 88294037; e-mail northern_air@163.net; internet www.cna.com.cn; f. 1990; scheduled flights to the Republic of Korea, Russia, Hong Kong, Macao and Japan; merged with China Southern Airlines Oct. 2002; Pres. JIANG LIANYING.

China Northwest Airlines: Laodong Nan Lu, Xian, Shaanxi 710082; tel. (29) 7298000; fax (29) 8624068; e-mail cnwadzz@pub.xa-online.sn.cn; internet www.cnwa.com; f. 1992; domestic services and flights to Macao, Singapore and Japan; merged with China Eastern Airlines in Oct. 2002; Pres. GAO JUNQUI.

China Southern Airlines: Baiyuan International Airport, Guangzhou, Guangdong 510406; tel. (20) 86128473; fax (20) 86658989; e-mail webmaster@cs-air.com; internet www.cs-air.com; f. 1991; merged with Zhong Yuan Airlines, 2000; domestic services; overseas destinations include Bangkok, Fukuoka, Hanoi, Ho Chi Minh City, Kuala Lumpur, Penang, Singapore, Manila, Vientiane, Jakarta and Surabaya; Chair. LIANG HUANFU; Pres. WANG CHANGSHUN.

China Southwest Airlines: Shuangliu Airport, Chengdu, Sichuan 610202; tel. (28) 5814466; fax (28) 5582630; e-mail szmaster@cswa.com; internet www.cswa.com; f. 1987; 70 domestic routes; international services to Singapore, Bangkok, Japan, the Republic of Korea and Kathmandu (Nepal); merged with Air China Oct. 2002; Pres. ZHOU ZHENGQUAN.

China Xinhua Airlines: 1 Jinsong Nan Lu, Chao Yang Qu, Beijing 100021; tel. (10) 67740116; fax (10) 67740126; e-mail infocxh@homeway.com.cn; internet www.chinaxinhuaair.com; f. 1992; Pres. ZHAO ZHONGYING.

China Xinjiang Airlines: Diwopu International Airport, Urumqi 830016; tel. (991) 3801703; fax (991) 3711084; f. 1985; 30 domestic routes; international services to Kazakhstan, Russia, Pakistan, and Uzbekistan; merged with China Southern Airlines Oct. 2002; Pres. ZHANG RUIFU.

Hainan Airlines: Haihang Devt Bldg, 29 Haixiu Lu, Haikou, Hainan 570206; tel. (898) 6711524; fax (898) 6798976; e-mail webmaster@hnair.com; internet www.hnair.com; f. 1989; undergoing major expansion in 2001–02; 300 domestic services; international services to Korea; 14.8% owned by financier George Soros; Chair. FENG CHEN.

Shandong Airlines: Jinan International Airport, Jinan, Shandong 250107; tel. (531) 8734625; fax (531) 8734616; e-mail webmaster@shandongair.com.cn; internet www.shandongair.com; f. 1994; domestic services; Pres. SUN DEHAN.

Shanghai Air Lines: 212 Jiangming Lu, Shanghai 200040; tel. (21) 62558888; fax (21) 62558885; e-mail liw@shanghai-air.com; internet www.shanghai-air.com; f. 1985; domestic services; also serves Phnom Penh (Cambodia); Pres. ZHOU CHI.

Shenzhen Airlines: Lingtian Tian, Lingxiao Garden, Shenzhen Airport, Shenzhen, Guangdong 518128; tel. (755) 7771999; fax (755) 7777242; internet www.shenzhenair.com; f. 1993; domestic services; Pres. DUAN DAYANG.

Sichuan Airlines: Chengdu Shuangliu International Airport, Chengdu, Sichuan 610202; tel. (28) 5393001; fax (28) 5393888; e-mail scaloi@public.cd.sc.cn; internet www.hpis.com/sichuan/sichuan.htm; f. 1986; domestic services; Pres. LAN XINGGUO.

Wuhan Air Lines: 435 Jianshe Dajie, Wuhan 430030; tel. (87) 63603888; fax (87) 83625693; e-mail wuhanair@public.wh.hb.cn; f. 1986; domestic services; Pres. CHENG YAOKUN.

Xiamen Airlines: Gaoqi International Airport, Xiamen, Fujian 361009; tel. (592) 5739888; fax (592) 5739777; internet www.xiamenair.com.cn; f. 1992; domestic services; also serves Bangkok (Thailand); Pres. WU RONGNAN.

Yunnan Airlines: Wujaba Airport, Kunming 650200; tel. (871) 7112999; fax (871) 7151509; internet www.chinayunnanair.com; f. 1992; 49 domestic services; also serves Bangkok, Singapore, and Vientiane (Laos); merged with China Eastern Airlines Oct. 2002; Pres. XUE XIAOMING.

Zhejiang Airlines: Jian Qiao Airport, 78 Shiqiao Lu, Hangzhou, Zhejiang 310021; tel. (571) 8082490; fax (571) 5173015; e-mail zjair@public.hz.zj.cn; internet www.zjair.com; f. 1990; domestic services; owned by China National Aviation Corpn; Pres. LUO QIANG.

Tourism

China has enormous potential for tourism, and the sector is developing rapidly. Attractions include dramatic scenery and places of historical interest such as the Temple of Heaven and the Forbidden City in Beijing, the Great Wall, the Ming Tombs, and also the terracotta warriors at Xian. Tibet (Xizang), with its monasteries and temples, has also been opened to tourists. Tours of China are organized for groups of visitors, and Western-style hotels have been built as joint ventures in many areas. In 2002 8,880 tourist hotels were in operation. A total of 97.9m. tourists visited China in 2002. In that year receipts from tourism totalled US $20,385m. The 2008 Olympic Games were to be held in Beijing.

China International Travel Service (CITS): 103 Fu Xing Men Nei Dajie, Beijing 100800; tel. (10) 66011122; fax (10) 66039331; e-mail mktng@cits.com.cn; internet www.cits.net; f. 1954; makes travel arrangements for foreign tourists; subsidiary overseas companies in 10 countries and regions; Pres. LI LUAN.

China National Tourism Administration (CNTA): 9A Jian Guo Men Nei Dajie, Beijing 100740; tel. (10) 65138866; fax (10) 65122096; Dir HE GUANGWEI.

Chinese People's Association for Friendship with Foreign Countries: 1 Tai Ji Chang Dajie, Beijing 100740; tel. (10) 65122474; fax (10) 65128354; f. 1954; Pres. QI HUAIYUAN; Sec.-Gen. BIAN QINGZU.

State Bureau of Tourism: Jie 3, Jian Guo Men Nei Dajie, Beijing 100740; tel. (10) 65122847; fax (10) 65122095; Dir LIU YI.

Defence

China is divided into seven major military administrative units. All armed services are grouped in the People's Liberation Army (PLA). In August 2003, according to Western estimates, the regular forces totalled 2,250,000, of whom 1,000,000 were conscripts: the army numbered 1,700,000, the navy 250,000 (including a naval air force of 26,000), and the air force 400,000 (including 210,000 air defence personnel). Reserves number some 500,000–600,000, and the People's Armed Police comprises an estimated 1.5m. Military service is usually by selective conscription, and is for two years in all services.

Defence Expenditure: Budgeted at 185,000m. yuan for 2003.

Chairman of the CCP Central Military Commission (Commander-in-Chief): HU JINTAO.

Vice-Chairmen: Gen. XU CAIHUO, Gen. GUO BOXIONG, Gen. CAO GANGCHUAN.

Director of the General Political Department (Chief Political Commissar): XU CAIHOU.

Chief of General Staff: Gen. LIANG GUANGLIE.

Commander, PLA Navy: Adm. ZHANG DINGFA.

Commander, PLA Air Force: Gen. QIAO QINGCHEN.

Director, General Logistics Department: Gen. LIAO XILONG.

Commanders of Military Regions: Lt-Gen. ZHU QI (Beijing), Gen. WANG JIANMIN (Chengdu), Lt-Gen. LIU ZHENWU (Guangzhou), Gen. CHEN BINGDE (Jinan), Lt-Gen. LI QIANYUAN (Lanzhou), Lt-Gen. ZHU WENGQUAN (Nanjing), Gen. QIAN GUOLIANG (Shenyang).

Education

In May 1985 the CCP Central Committee adopted a decision to reform the country's whole educational structure. This stipulated that nine-year compulsory education was to be implemented in stages, with the date of attaining that goal varying according to regional disparities in development. By the year 2000 it was envisaged that nine-year education would be compulsory for 85% of the population. Fees are payable at all levels of education. The 1997 budget allocated an estimated 118,907m. yuan to education. Private education is expanding rapidly, the number of private schools being estimated at 2,000 in early 1995.

According to the census of July 1990, there were 182,246,000 illiterates and semi-illiterates among the population aged 15 years and over, representing 22.3% of the population. In 2000 a total of 2.58m. adults completed basic literacy courses.

Pre-School Education

Kindergartens are regarded as important in introducing children to ideas that will shape their thought in later life; the education takes the form of games, sports and music. The kindergartens also have economic significance in releasing women for productive work. In 2002 a total of 20.4m. children were enrolled in the 111,752 kindergartens.

Primary Education

Most primary schools have adopted a five- or six-year course. The curriculum includes Chinese language, mathematics, natural science, geography, history, music and art. Senior pupils are required to spend some time engaged in manual labour. In 2002 there were 121.6m. primary school pupils enrolled in 456,903 schools.

Secondary Education

Junior secondary education (in which the enrolment rate was 82.4% in 1996) lasts for three years, followed by upper secondary education for two to three years. The curriculum includes Chinese, mathematics, foreign languages, politics, history, geography, science, music and art. There are also specialized secondary schools which offer subjects such as engineering, medicine, agriculture, business administration and law. Four weeks are set aside for physical labour and technical training, although in rural areas many schools allow pupils to spend time instead engaged in agricultural work, while some schools run their own factories. In 2002 there were 82.9m. students at 80,067 general secondary schools and 4.0m. students at 2,523 secondary technical schools.

Higher Education

University courses last for four or five years, but, in line with the general reform of the whole system, emphasis is being shifted to the establishment of shorter vocational courses (lasting for two to three years), while the proportion of students in the fields of finance and economics, political science and law, management and liberal arts is being increased. Entrance to most courses is by state examination. College graduates may be required to spend some time engaged in factory or farm work. College and university students compete for scholarships, which are awarded according to academic ability. Prior to entering college, however, students are required to complete one year of political education. Since 1981 three types of degree have been awarded: bachelor, master and doctoral. In November 1989 it was announced that post-graduate students were to be selected on the basis of assessments of moral and physical fitness, as well as academic ability. In 2002 a total of 9.0m. students were enrolled at the 1,396 institutes of higher education. In recent years exchanges with foreign universities have been encouraged. In February 2000 a new higher education reform project was launched, with the aim of creating a series of modernized courses, textbooks and software, establishing several education demonstration bases and training a group of able teachers.

Teacher Training

The rapid expansion of primary and secondary school education produced a serious shortage of qualified teachers. Teacher training is mostly undertaken in specialist establishments. Graduates of junior secondary schools may enrol for three years to train as primary school teachers, while teachers for secondary schools must have an educational level equivalent to that of a university graduate. Efforts are also being made to raise the standards of teachers, through the use of short training courses and correspondence courses. In 2002 there were 430 teacher-training establishments, with 601,000 students.

Other Institutions

Much of China's educational effort is devoted to part-time and spare-time systems, both for vocational training and ideological dissemination. The education of peasants and cadres in rural areas takes the form of literacy classes, part-farming and part-study classes, and spare-time study classes. Elementary classes are conducted by factories and workshops, while regular colleges provide tuition by means of correspondence courses and night universities. The inauguration in 1979 of a central radio and television university, which offers basic general courses in a variety of subjects, was followed by the establishment of numerous provincial centres. By 1999 the radio and television universities totalled 45, full-time teachers numbering 25,700. In the same year they had an enrolment of some 48,680 students, who would eventually receive a qualification equivalent to that of a regular higher education.

Bibliography

Adshead, S. A. M. *China in World History*. 3rd Edn, Basingstoke, Macmillan Press, 1999.

Alexandroff, Alan S., Ostry, Sylvia, and Gomez, Rafael (Eds). *China and the Long March to Global Trade—The Accession of China to the World Trade Organization*. London, Routledge, 2003.

Amnesty International. *Death in Beijing*. London, 1989.

An Chen. *Restructuring Political Power in China: Alliances and Opposition, 1978–1998*. Boulder, CO, Lynne Rienner, 1999.

Ash, Robert, Howe, Christopher, and Kueh, Y. Y. (Eds). *China's Economic Reform—A Study with Documents*. London, RoutledgeCurzon, 2002.

Barnett, Robert, and Akiner, Shirin (Eds). *Resistance and Reform in Tibet*. London, Hurst and Co, 1995.

Baum, Richard. *Burying Mao: Chinese Politics in the Age of Deng Xiaoping*. Princeton, NJ, Princeton University Press, 1995.

Becker, Jasper. *Hungry Ghosts: China's Secret Famine*. London, John Murray, 1996.

The Chinese. London, John Murray, 2000.

Benewick, Robert, and Wingrove, Paul (Eds). *China in the 1990s*. Basingstoke, Macmillan, 1995.

Bernstein, Richard, and Munro, Ross H. *The Coming Conflict with China*. New York, Alfred A. Knopf, 1997.

Blackman, Carolyn. *Negotiating China: Case Studies and Strategies*. St Leonards, NSW, Allen and Unwin, 1998.

Bonavia, David. *The Chinese: A Portrait*. London, Allen Lane, 1980.

Deng. Harlow, Longman, 1990.

Bowring, Philip, with Blaisdell, David, and Parry, Jane. *The China Investor*. AsiaPacific Financial Publishing, 1993.

Brahm, Laurence J. *China's Century: The Awakening of the Next Economic Powerhouse*. New York, John Wiley and Sons, 2001.

Zhu Rongji and the Transformation of Modern China. Singapore, John Wiley and Sons, 2002.

Breslin, Shaun. *China in the 1980s: Centre-Province Relations in a Reforming Socialist State*. Basingstoke, Macmillan, 1996.

Brodsgaard, Kjeld Erik, and Heurlin, Bertel (Eds). *China's Place in Global Geopolitics*. Richmond, Surrey, Curzon Press, 2001.

Brodsgaard, Kjeld Erik, and Strand, David (Eds). *Reconstructing Twentieth Century China*. Oxford, Oxford University Press, 1998.

Brown, Lester R. *Who Will Feed China?* New York, W. W. Norton, 1995.

Buckley Ebrey, Patricia. *The Cambridge Illustrated History of China*. Cambridge, Cambridge University Press, 1997.

Burstein, Daniel, and de Keijzer, Arne. *Big Dragon: China's Future*. New York, Simon and Schuster, 1998.

Buruma, Ian. *Bad Elements: Chinese Rebels from Los Angeles to Beijing*. London, Random House, 2002.

Buzan, Barry, and Foot, Rosemary (Eds). *Does China Matter? A Reassessment: Essays in Memory of Gerald Segal*. London, Routledge, 2004.

Cao Siyuan. *The ABCs of Political Civilization: an Outline of Political Reform in China*. New York, NY, Cozygraphics, 2003.

Chai, Joseph C. H. (Ed.). *The Economic Development of Modern China*. Cheltenham, Edward Elgar, 2000.

Chang, Gordon G. *The Coming Collapse of China*. London, Random House, 2001.

Chang, Iris. *The Rape of Nanking*. New York, Basic Books, 1998.

Chang Jung. *Wild Swans*. London, HarperCollins, 1992.

Chang, Parris H. *Power and Policy in China*. University Park, PA, Pennsylvania State University Press, 1978.

Cheek, Timothy, and Saich, Tony. *New Perspectives on State Socialism in China*. Armonk, NY, M. E. Sharpe, 1999.

Chen, Chih-Jou Jay. *Transforming Rural China: How Local Institutions Shape Property Rights in China*. London, RoutledgeCurzon, 2004.

Cheng Li. *Rediscovering China: Dynamics and Dilemmas of Reform*. Lanham, MD, Rowman & Littlefield, 1997.

Chi Lo. *Misunderstood China: Uncovering the Truth Behind the Bamboo Curtain*. Upper Saddle River, NJ, Prentice Hall, 2004.

China Publications Centre. *Resolution on CPC History (1949–1981)*. Beijing, Guoji Shudian, 1981.

Ching, Frank (Ed.). *China in Transition*. Hong Kong, Review Publishing Co, 1994.

Ching, Julia. *Probing China's Soul: Religion, Politics and Protest in the People's Republic*. San Francisco, Harper and Row, 1990.

Cohen, Paul A. *China Unbound: Evolving Perspectives on the Chinese Past*. London, RoutledgeCurzon, 2003.

Conboy, Kenneth, and Morrison, James. *The CIA's Secret War in Tibet*. Lawrence, University of Kansas Press, 2002.

Cook, Ian G., and Murray, Geoffrey. *China's Third Revolution: Tensions in the Transition Towards a Post-Communist China*. Richmond, Surrey, Curzon Press, 2001.

Cook, Sarah, Shuje Yao and Juzhong Zhuang (Eds). *The Chinese Economy under Transition*. Basingstoke, Macmillan Press, 2000.

Dai Qing. *The Three Gorges Dam and the Fate of China's Yangtze River and Its People*. Armonk, NY, M. E. Sharpe, 1997.

Dassu, M., and Saich, T. (Eds). *The Reform Decade in China: From Hope to Dismay*. Kegan Paul International, 1993.

Davis, Deborah, and Vogel, Ezra (Eds). *Chinese Society on the Eve of Tiananmen*. Cambridge, MA, Harvard/East Asia, 1991.

Deng Maomao. *Deng Xiaoping, My Father*. New York, Basic Books, 1995.

Department of Foreign Affairs and Trade, Canberra, Australia. *China Embraces the Market: Achievements, Constraints and Opportunities*. 1997.

Dickson, Bruce, and Chao Chien-min. *Remaking the Chinese State: Strategies, Society, and Security*. London, Routledge, 2001.

Ding Yijiang. *Chinese Democracy After Tiananmen*. New York, Columbia University Press, 2002.

Dirlik, Arif. *The Origins of Chinese Communism*. New York, Oxford University Press, 1990.

Donnet, Pierre-Antoine. *Tibet: Survival in Question*. London, Zed Books, 1994.

Dooling, Amy D., and Torgeson, Kristina M. (Eds). *Writing Women in Modern China: An Anthology of Women's Literature from the Early Twentieth-Century*. New York, Columbia University Press, 1998.

Drysdale, Peter, and Song Liang (Eds). *China's Entry into the World Trade Organization*. London, Routledge, 2000.

Economy, Elizabeth C. *The River Runs Black: The Environmental Challenge to China's Future*. Ithaca, NY, Cornell University Press, 2004.

Edmonds, Richard Louis (Ed.). *The People's Republic of China After 50 Years*. Oxford, Oxford University Press, 2000.

Managing the Chinese Environment. Oxford, Oxford University Press, 2000.

Edmonds, Richard Louis, and Wakeman Jr, Frederic (Eds). *Reappraising Republican China*. Oxford, Oxford University Press, 2000.

Elvin, Mark. *The Retreat of Elephants: An Environmental History of China*. New Haven, CT, Yale University Press, 2004.

Ethridge, James M. *Changing China: The New Revolution's First Decade, 1979–88*. Beijing, New World Press, 1989.

Europa Publications. *The Territories of the People's Republic of China*. London, Europa Publications, 2002.

Evans, Grant, Hutton, Christopher, and Kuah Khun Eng (Eds).*Where China Meets Southeast Asia: Social and Cultural Change in the Border Regions*. Basingstoke, Palgrave Macmillan, 2000.

Evans, Richard. *Deng Xiaoping and the Making of Modern China*. London, Hamish Hamilton, 1993.

Fairbank, John K., et al. (Eds). *The Cambridge History of China*. 15 vols. Cambridge, Cambridge University Press, 1987–92.

Fairbank, John K. *The Great Chinese Revolution: 1800–1985*. New York, Harper and Row, 1986.

China: A New History. Cambridge, MA, Harvard University Press, 1992.

Faust, John R., and Kornberg, Judith F. *China in World Politics*. Boulder, CO, Lynne Rienner, 1995.

Feignon, Lee. *Mao: A Reinterpretation*. Chicago, IL, Ivan R. Dee, 2003.

Fenby, Jonathan. *Generalissimo: Chiang Kai-shek and the China He Lost*. New York, NY, Free Press, 2003.

Feuchtwang, Stephan, and Hussain, A. *The Chinese Economic Reforms*. London, Croom Helm, 1983.

Fewsmith, Joseph. *Elite Politics in Contemporary China*. Armonk, NY, M. E. Sharpe, 2000.

Fitzgerald, C. P. *Revolution in China*. London, Cresset Press, 1952.

Foreign Languages Press. *Birth of Communist China*. London, Penguin Books, 1964.

Mao Tse-Tung and China. London, Hodder and Stoughton, 1976.

French, Patrick. *Tibet, Tibet: Dreams and Memories of a Lost Land*. London, HarperCollins, 2003.

Friedman, Edward, and McCormick, Barrett L. *What if China Doesn't Democratize? Implications for War and Peace*. Armonk, NY, M. E. Sharpe, 2000.

Gao Shangquan. *China's Economic Reform*. Basingstoke, Macmillan, 1996.

Gargan, Edward A. *China's Fate: A People's Turbulent Struggle with Reform and Repression, 1980–1990*. New York, Doubleday, 1991.

Garnaut, Ross G., Guo Shutian and Ma Guonan (Eds). *The Third Revolution in the Chinese Countryside*. Cambridge, Cambridge University Press, 1996.

Garnaut, Ross G., and Ligang Song (Eds). *Private Enterprise in China*. London, RoutledgeCurzon, 2003.

Garnaut, Ross G., and Yiping Huang (Eds). *Growth Without Miracles: Readings on the Chinese Economy in the Era of Reform*. Oxford, Oxford University Press, 2000.

Garside, Roger. *Coming Alive: China after Mao*. London, André Deutsch, 1981.

Gilley, Bruce. *China's Democratic Future: How It Will Happen and Where It Will Lead*. New York, Columbia University Press, 2004.

Tiger on the Brink: Jiang Zemin and China's New Elite. Berkeley, CA, University of California Press, 1998.

Gilley, Bruce, and Nathan, Andrew J. *China's New Rulers: The Secret Files*. New York, New York Review Books, 2002.

Gilmore, Fiona, and Dumont, Serge. *Brand Warriors China: Creating Sustainable Brand Capital*. London, Profile Books, 2003.

Gittings, John. *China Changes Face: The Road from Revolution 1949–1989*. Oxford, Oxford University Press, 1989.

Real China: From Cannibalism to Karaoke. London, Simon and Schuster, 1996.

Goldman, Merle. *Sowing the Seeds of Democracy in China: Political Reform in the Deng Xiaoping Era*. Cambridge, MA, Harvard University Press, 1994.

Goldman, Merle, Gu, Edward (Eds). *Chinese Intellectuals between State and Market*. London, RoutledgeCurzon, 2004.

Goldstein, Melvyn C., and Beall, Cynthia M. *Nomads of Western Tibet: The Survival of a Way of Life*. Hong Kong, Odyssey Productions, 1990.

Goldstein, Melvyn, Siebenschuh, William, and Tsering, Tashi. *The Struggle for Modern Tibet: The Autobiography of Tashi Tsering*. Armonk, NY, M. E. Sharpe, 1998.

Goodman, David S. G. *Deng Xiaoping and the Chinese Revolution: A Political Biography*. London, Routledge, 1995.

China's Provinces in Reform: Class, Community and Political Culture. London, Routledge, 1997.

Goodman, David S. G. (Ed.). *China's Regional Development*. London, Routledge, 1989.

Goodman, David S. G., and Segal, Gerald. *China in the Nineties: Crisis Management and Beyond*. Oxford, Oxford University Press, 1991.

Goodman, David S. G., and Segal, Gerald (Eds). *China Rising: Nationalism and Interdependence*. London, Routledge, 1997.

Gray, Jack. *Rebellions and Revolutions*. Oxford, Oxford University Press, 1990.

Gries, Peter. *China's New Nationalism: Pride, Politics and Diplomacy*. Berkeley, CA, University of California Press, 2004.

Gries, Peter, and Rosen, Stanley (Eds). *State and Society in 21st-century China*. London, RoutledgeCurzon, 2004.

Guo Yingjie. *Cultural Nationalism in Contemporary China*. London, RoutledgeCurzon, 2004.

Han Suyin. *Eldest Son: Zhou Enlai and the Making of Modern China, 1898–1976*. London, Jonathan Cape, 1994.

Hannan, Kate. *Industrial Change in China*. London, Routledge, 1998.

Harding, H. (Ed.). *China's Foreign Relations in the 1980s*. New Haven, CT, Yale University Press, 1986.

He Baogang. *The Democratic Implications of Civil Society in China*. Basingstoke, Macmillan, 1997.

Hendrischke, Hans, and Feng Chongyi (Eds). *The Political Economy of China's Provinces: Competitive and Comparative Advantage*. London, Routledge, 1999.

Hicks, George (Ed.). *The Broken Mirror: China After Tiananmen*. Harlow, Longman, 1990.

Hoa, Tran Van. *China's Trade and Investment After the Asia Crisis*. Cheltenham, Edward Elgar, 2000.

Hodder, Rupert. *In China's Image: Chinese Self-perception in Western Thought*. Basingstoke, Macmillan Press, 2000.

Holbig, Heike, and Ash, Robert (Eds). *China's Accession to the World Trade Organization—National and International Perspectives*. London, RoutledgeCurzon, 2002.

Hollingworth, Clare. *Mao and the Men Against Him*. London, Jonathan Cape, 1985.

Hook, Brian (Ed.). *The Cambridge Encyclopedia of China*. Cambridge University Press.

Hsü, Immanuel C. Y. *The Rise of Modern China*. 6th Edn, Oxford University Press, 2000.

Hu Yebi. *China's Capital Market*. Hong Kong, Chinese University Press, 1993.

Hughes, Christopher R. *Chinese Nationalism in a Global Era*. London, RoutledgeCurzon, 2004.

Hughes, Christopher R., and Wacker, Gudrun (Eds). *China and the Internet*. London, RoutledgeCurzon, 2003.

Hughes, Neil. C. *China's Economic Challenge: Smashing the Iron Rice Bowl*. Armonk, NY, M. E. Sharpe, 2002.

Hutchings, Graham. *Modern China: A Guide to a Century of Change*. Cambridge, MA, Harvard University Press, 2001.

International Food Policy Research Institute (IFPRI). *China's Food Economy to the 21st Century*. Washington, DC, 1997.

Jae Ho Chung. *Central Control and Local Discretion in China: Leadership and Implementation during Post-Mao Decollectivization*. Oxford, Oxford University Press, 2000.

Jefferson, Gary H., and Singh, Inderjit (Eds). *Enterprise Reform in China: Ownership, Transition and Performance*. Oxford, Oxford University Press, 1999.

Jenner, W. J. F. *The Tyranny of History: The Roots of China's Crisis*. London, Allen Lane, 1992.

Joffe, E. *The Chinese Army After Mao*. London, Weidenfeld and Nicolson, 1987.

Johnston, Alastair Iain, and Ross, Robert S. (Eds). *Engaging China—The Management of an Emerging Power*. London, Routledge, 1999.

Joint Economic Committee (Congress of the United States). *China under the Four Modernizations*. Parts 1 and 2. Washington, DC, US Government Printing Office, 1981, 1982.

China's Economic Future: Challenges to US Policy. Armonk, NY, M. E. Sharpe, 1997.

Kaplan, Frederic M., Sobin, Julian M., and Andors, Stephen. *Encyclopaedia of China Today*. New York, Eurasia Press/Harper and Row, 1979.

Karl, Rebecca E. *Staging the World: Chinese Nationalism at the Turn of the Twentieth Century*. Durham, NC, Duke University Press, 2002.

Karmel, Solomon M. *China and the People's Liberation Army: Great Power or Struggling Developing State?* Basingstoke, Macmillan Press, 2000.

Kim, Samuel S. *China and the World: Chinese Foreign Policy Faces the New Millennium*. Boulder, CO, Westview Press, 1998.

Kraus, Willy. *Private Business in China*. London, Hurst and Co, 1992.

Kristof, Nicholas D., and Wu Dunn, Sheryl. *China Wakes: The Struggle for the Soul of a Rising Power*. New York, Times Books, 1994.

Krug, Barbara. *China's Rational Entrepreneurs: The Development of the New Private Sector*. London, RoutledgeCurzon, 2004.

Kueh Yak-yeow, Chai, Joseph C. H. and Gang Fan (Eds). *Industrial Reforms and Macroeconomic Instability In China*. Oxford, Oxford University Press, 1999.

Kwong, Julia. *The Political Economy of Corruption in China*. Armonk, NY, M. E. Sharpe, 1997.

Lam, Willy Wo-Lap. *The Era of Zhao Ziyang: Power Struggle in China, 1986–88*. Hong Kong, A. B. Books and Stationery (International) Ltd, 1989.

China after Deng Xiaoping: The Power Struggle in Beijing since Tiananmen. Singapore, John Wiley, 1995.

The Era of Jiang Zemin. Singapore, Prentice Hall, 1999.

Lampton, David M. *The Making of Chinese Foreign and Security Policy in the Era of Reform*. Stanford, CA, Stanford University Press, 2001.

Lardy, Nicholas R. *China in the World Economy*. Washington, DC, Institute for International Economics, 1994.

China's Unfinished Economic Revolution. Washington, DC, Brookings Institution Press, 1998.

Integrating China into the Global Economy. Washington, DC, Brookings Institution Press, 2002.

Lattimore, Owen. *China Memoirs: Chiang Kai-shek and the War Against Japan*. Tokyo, University of Tokyo Press, 1991.

Laurenceson, James, and Chai, Joseph C. H. *Financial Reform and Economic Development in China*. Cheltenham, Edward Elgar Publishing, 2003.

Lawrance, Alan. *China Under Communism*. London, Routledge, 1998.

Lee, Chin-Chuan. *Chinese Media*. London, RoutledgeCurzon, 2003.

Lee, Hong Yung. *The Politics of the Chinese Cultural Revolution*. Berkeley, CA, University of California Press, 1980.

Lee, Ngok. *China's Defence Modernization and Military Leadership*. Sydney, Australian National University Press, 1991.

Lees, Francis A. *China Superpower: Requisites for High Growth*. Basingstoke, Macmillan, 1996.

Li Jun. *Financing China's Rural Enterprises*. London, RoutledgeCurzon, 2002.

Li Shaomin and Tse, David K. *China Markets Yearbook, 1999*. Armonk, NY, M. E. Sharpe, 2000.

Li Si-ming and Tang Wing-shing (Eds). *China's Regions, Polity and Economy*. Hong Kong, Chinese University Press, 2000.

Li Xueqin (trans. Chang, K. C.) *Eastern Zhou and Qin Civilizations*. New Haven, CT, Yale University Press, 1986.

Li Zhensheng. *Red-Color News Soldier*. New York, NY, Phaidon Press, 2003.

Lieberthal, Kenneth. *Governing China: From Revolution Through Reform*. New York, W. W. Norton, 1995.

Liew, Yeong H., and Wang Shaoguang (Eds). *Nationalism, Democracy and National Integration in China*. London, RoutledgeCurzon, 2003.

Lilley, James, with Lilley, Jeffrey. *China Hands: Nine Decades of Adventure, Espionage and Diplomacy in Asia*. New York, NY, Public Affairs, 2004.

Lin Chong-pin. *China's Nuclear Weapons Strategy: Tradition within Evolution*. London, Lexington Books, 1989.

Liu Guoguang, Wang Luolin, Li Jingwen, Liu Shucheng, and Wang Tongsan (Eds). *Economics Blue Book of the People's Republic of China, 1999*. Armonk, NY, M. E. Sharpe, 2000.

Lloyd, P. J., and Zhang Xiao-guang (Eds). *China in the Global Economy*. Cheltenham, Edward Elgar, 2000.

Lo, Dic. *Market and Institutional Regulation in Chinese Industrialization, 1978–94*. Basingstoke, Macmillan, 1997.

Logan, Pamela. *Among Warriors: A Martial Artist in Tibet*. New York, Overlook Press, 1997.

Ma Bo. *Blood Red Sunset: A Memoir of the Chinese Cultural Revolution*. New York, Penguin Books, 1995.

MacFarquhar, Roderick. *The Origins of the Cultural Revolution*. 2 vols. London, Oxford University Press, 1983, and New York, Columbia University Press, 1984.

MacFarquhar, Roderick (Ed.). *The Politics of China: The Eras of Mao and Deng*. Cambridge, Cambridge University Press, 1997.

Mackerras, Colin. *Modern China: A Chronology*. London, Thames and Hudson, 1983.

Western Images of China. Hong Kong, Oxford University Press, 1989.

China's Ethnic Minorities and Globalisation. London, RoutledgeCurzon, 2003.

Mackerras, Colin, McMillen, Donald H., and Watson, Andrew (Eds). *Dictionary of the Politics of the People's Republic of China*. London, Routledge, 1998.

Marton, Andrew M. *China's Spatial Economic Development—Regional Transformation in the Lower Yangzi Delta*. London, Routledge, 2000.

Mastel, Greg. *The Rise of the Chinese Economy: The Middle Kingdom Emerges*. Armonk, NY, M. E. Sharpe, 1997.

Menzies, Gavin. *1421: The Year China Discovered the World*. London, Bantam Press, 2002.

Miles, James. *The Legacy of Tiananmen: China in Disarray*. Ann Arbor, University of Michigan Press, 1996.

Mulvenon, James C. *Soldiers of Fortune: The Rise and Fall of the Chinese Military-Business Complex*. Armonk, NY, M. E. Sharpe, 2000.

Nathan, Andrew J. *China's Crisis*. New York, Columbia University Press, 1990.

China's Transition. New York, Columbia University Press, 1998.

Nathan, Andrew J., Hong Zhaohui and Smith, Steven R. (Eds). *Dilemmas of Reform in Jiang Zemin's China*. Boulder, CO, Lynne Rienner, 1999.

Nathan, Andrew J., and Link, Perry (Eds). *The Tiananmen Papers: The Chinese Leadership's Decision to Use Force Against Their Own People—In Their Own Words*. New York, Public Affairs, 2001.

Nathan, Andrew J., and Ross, Robert S. *The Great Wall and the Empty Fortress: China's Search for Security*. New York, W. W. Norton, 1997.

Naughton, Barry. *Growing Out of the Plan: Chinese Economic Reform 1978–1993*. Cambridge, Cambridge University Press, 1995.

Nolan, Peter. *China's Rise; Russia's Fall*. New York, St Martin's Press, 1995.

OECD, Publications Service. *China in the 21st Century: Long-Term Global Implications*. Paris, 1996.

Ogden, Suzanne. *Inklings of Democracy in China*. Cambridge, MA, Harvard University Press, 2002.

Ogilvy, James A., Schwartz, Peter, and Flower, Joe. *China's Futures: Scenarios for the World's Fastest Growing Economy, Ecology, and Society*. San Francisco, CA, Jossey-Bass, 2000.

Ong, Russell. *China's Security Interests in the Post-Cold War Era*. Richmond, Surrey, Curzon Press, 2001.

Overholt, William H. *China: The Next Economic Superpower*. London, Weidenfeld and Nicolson, 1993.

Panitchpakdi, S., and Clifford, Mark. L. *China and the WTO: Changing China, Changing World Trade*. New York, John Wiley and Sons, 2002.

Perkins, D. W., and Yusuf, S. *Rural Development in China*. World Bank/Johns Hopkins, 1986.

Perry, Elizabeth J., and Selden, Mark (Eds). *Chinese Society: Change, Conflict, and Resistance*. London, Routledge, 1999.

Preston, Peter, and Haacke, Jürgen. *Contemporary China—The Dynamics of Change at the Start of the New Millennium*. London, RoutledgeCurzon, 2002.

Pomfret, Richard. *Investing in China*. London and New York, Harvester Wheatsheaf, 1991.

Rawski, Thomas G. *China's Transition to Industrialization: Producer Goods and Economic Development in the Twentieth Century*. Ann Arbor, MI, University of Michigan Press, 1980.

Economic Growth and Development in China. New York, for World Bank, Oxford University Press, 1980.

Renard, Mary-Françoise. *China and its Regions: Economic Growth and Reform in the Chinese Provinces*. Cheltenham, Edward Elgar, 2001.

Riskin, Carl, Zhao Renwei, and Li Shih. *China's Retreat From Equality: Income Distribution and Economic Transition*. Armonk, NY, M. E. Sharpe, 2000.

Roberts, J. A. G. *A History of China*. London, Macmillan, 1999.

Rui Huaichuan. *Globalisation, Transition and Development in China*. London, RoutledgeCurzon, 2004.

Saez, Lawrence. *Banking Reform in India and China*. New York, NY, Macmillan, 2004.

Saich, Tony (Ed.). *The Chinese People's Movement*. Armonk, NY, M. E. Sharpe, 1991.

Salisbury, Harrison E. *The Long March: The Untold Story*. New York and London, Macmillan, 1985.

Tiananmen Diary: Thirteen Days in June. Boston, MA, Little, Brown and Co, 1989.

The New Emperors—Mao and Deng. London, HarperCollins, 1992.

Schechter, Danny. *Falun Gong's Challenge to China: Spiritual Practice or 'Evil Cult'?* Askashic Books, 2001.

Schell, Orville, and Shambaugh, David (Eds). *The China Reader: The Reform Era*. New York, NY, Vintage Books, 1999.

Schoenfels, Michael (Ed.). *China's Cultural Revolution*. Armonk, NY, M. E. Sharpe, 1997.

Schwartz, Ronald D. *Circle of Protest: Political Ritual in the Tibetan Uprising*. London, Hurst and Co, 1995.

Segal, Gerald. *Defending China*. Oxford University Press, 1985.

Segal, Gerald, and Yang, Richard. *Chinese Economic Reform: The Impact on Security*. London, Routledge, 1996.

Seligman, Scott D. *Dealing with the Chinese*. New York, Warner Books, 1989.

Seymour, James, and Anderson, Richard. *New Ghosts, Old Ghosts: Prisons and Labor Reform Camps in China*. Armonk, NY, M. E. Sharpe, 1998.

Shakya, Tsering. *The Dragon in the Land of Snows: A History of Modern Tibet Since 1947*. London, Pimlico, 1999.

Shambaugh, David. *Is China Unstable? Assessing the Factors*. Armonk, NY, M. E. Sharpe, 2000.

The Modern Chinese State. Cambridge University Press, 2000.

Modernizing China's Military: Progress, Problems, and Prospects. Berkeley, CA, University of California Press, 2003.

Shambaugh, David, and Lilley, James. *China's Military Faces the Future*. Armonk, NY, M. E. Sharpe, 1999.

Shao Kuo-Kang. *Zhou Enlai and the Foundations of Chinese Foreign Policy*. Basingstoke, Macmillan, 1996.

Shapiro, James E., et al. *Direct Investment and Joint Ventures in China: A Handbook for Corporate Negotiators*. New York, Quorum Books, 1992.

Shapiro, Judith. *Mao's War Against Nature: Politics and the Environment in Revolutionary China*. New York, Cambridge University Press, 2001.

Sheel, Kamal. *Peasant Society and Marxist Intellectuals in China: Fang Zhimin and the Origin of a Revolutionary Movement in the Xinjiang Region*. Princeton, NJ, Princeton University Press, 1990.

Sheng Lijun. *China's Dilemma: The Taiwan Issue*. Singapore, Institute of Southeast Asian Studies, 2001.

Shirk, Susan. *How China Opened its Door*. Washington, DC, Brookings Institute, 1994.

Short, Philip. *Mao: A Life*. Henry Holt and Co, 2000.

Simmie, Scott, and Nixon, Bob. *Tiananmen Square*. Vancouver, Douglas and McIntyre, 1989.

Smil, Vaclav. *China's Environmental Crisis: An Inquiry into the Limits of National Development*. Armonk, NY, M. E. Sharpe, 1993.

Smil, Vaclav. *China's Past, China's Future*. London, RoutledgeCurzon, 2003.

Spence, Jonathan. *The Gate of Heavenly Peace: The Chinese and Their Revolution, 1895–1980*. Harmondsworth, Penguin Books, 1982.

The Search for Modern China. London, Century, 2nd Edn, 2000.

God's Chinese Son: The Taiping Heavenly Kingdom of Hong Xiuquan. New York, W. W. Norton, 1996.

The Chan's Great Continent: China in Western Minds. New York, W. W. Norton, 1998.

Starr, John Bryan. *Understanding China: A Guide to China's Economy, History, and Political Structure*. New York, NY, Hill and Wang, 1997.

Steinfeld, Edward S. *Forging Reform in China: The Fate of State-Owned Industry*. Cambridge, Cambridge University Press, 1998.

Stockman, Norman. *Understanding Chinese Society*. Polity, 2002.

Story, Jonathan. *China: The Race to Market*. London, Financial Times Prentice Hall, 2003.

Strange, Roger, Slater, Jim, and Wang Limin (Eds). *Trade and Investment in China—The European Experience*. London, Routledge, 1998.

Studwell, Joe. *The China Dream: The Elusive Quest for the Greatest Untapped Market on Earth*. New York, Atlantic Monthly Press, 2002.

Stuttard, John. B. *The New Silk Road: Secrets of Doing Business in China Today*. New York, NY, John Wiley & Sons, 2000.

Suettinger, Robert L. *Beyond Tiananmen: The Politics of US-China Relations, 1989—2000*. Washington, DC, Brookings Institution Press, 2003.

Sullivan, Lawrence R., with Hearst, Nancy R. *Historical Dictionary of the People's Republic of China, 1949–1997*. Lanham, MD, Scarecrow Press, 1997.

Sulter, Robert G. *Shaping China's Future in World Affairs: The Role of the United States*. Boulder, CO, Westview Press, 1998.

Sun Shuyun. *Ten Thousand Miles Without a Cloud*. London, HarperCollins, 2003.

Swaine, Michael D., and Tellis, Ashley J. *Interpreting China's Grand Strategy: Past, Present, and Future*. Santa Monica, CA, Rand, 2000.

Tam On Kit. *The Development of Corporate Governance in China*. Cheltenham, Edward Elgar, 2000.

Teiwes, Frederick C., and Sun, Warren. *The Tragedy of Lin Biao: Riding the Tiger during the Cultural Revolution*. Honolulu, HI, University of Hawaii Press, 1996.

Terrill, Ross. *Mao: A Biography*. Cambridge, Cambridge University Press, 1998.

The New Chinese Empire and What it Means for the United States. New York, NY, Basic Books, 2003.

Teufel Dreyer, June. *China's Political System: Modernization and Tradition*. Basingstoke, Macmillan, 1996.

Tien Hung-mao and Chu Yun-han (Eds). *China under Jiang Zemin*. Boulder, CO, Lynne Rienner, 1999.

Tregear, T. R. *A Geography of China*. University of London Press, 1965.

China: A Geographical Survey. London, Hodder and Stoughton, 1980.

Tyler, Christian. *Wild West China: The Taming of Xinjiang*. London, John Murray, 2003.

Unger, Jonathan. *Chinese Nationalism*. Armonk, NY, M. E. Sharpe, 1996.

Unger, Jonathan (Ed.). *The Pro-Democracy Protests in China*. Armonk, NY, M. E. Sharpe, 1992.

Vogel, Ezra F. *One Step Ahead in China: Guangdong Under Reform*. Cambridge, MA, Harvard University Press, 1989.

Vogel, Ezra (Ed.). *Living with China: US-China Relations in the Twenty-First Century*. New York, W. W. Norton, 1997.

Wakeman, Carolyn, and Light, Ken (Eds). *Assignment Shanghai: Photographs on the Eve of Revolution*. Berkeley, CA, University of California Press, 2003.

Waldron, Arthur. *The Great Wall of China: From History to Myth*. Cambridge, Cambridge University Press, 1991.

Walter, Carl E., and Howie, Fraser J. T. *Privatizing China*. Singapore, John Wiley, 2003.

Wang Gungwu. *The Chinese Way: China's Position in International Relations*. Oslo, Scandinavian University Press, 1995.

Wang Hui. *China's New Order: Society, Politics and Economy in Transition*. Cambridge, MA, Harvard University Press, 2003.

Wang Jun. *An Evolutionary Record of Beijing City*. Beijing, San Lian Press, 2003.

Wang Ke-wen (Ed.). *Modern China—An Encyclopedia of History, Culture and Nationalism*. New York, Garland Publishing, 1997.

Wang Shaoguang. *The Chinese Economy in Crisis: State Capacity and Tax Reform*. Armonk, NY, M. E. Sharpe, 2001.

Wasserstrom, Jeffrey, and Perry, Elizabeth. *Popular Protest and Political Culture in Modern China*. Boulder, CO, Westview Press, 1992.

Wei, C. X. George, and Liu Xiaoyuan (Eds). *Chinese Nationalism in Perspective*. Westport, CT, Greenwood Press, 2001.

Wei Jingsheng. *The Courage To Stand Alone: Letters from Prison and Other Writings*. London, Viking, 1997.

Wei, Yehua Dennis. *Regional Development in China—State, Globalization and Inequality*. London, Routledge, 2000.

Wen, G., and Xu, D. (Eds). *The Reformability of China's State Sector*. River Edge, NJ, World Scientific Press, 1997.

White, Lynn T. *Politics of Chaos: The Organizational Causes of Violence in China's Cultural Revolution*. Princeton, NJ, Princeton University Press, 1989.

Wilson, Dick. *China, The Big Tiger: A Nation Awakes*. London, Little, Brown and Co, 1996.

Woetzel, Jonathan R. *Capitalist China*. Singapore, John Wiley & Sons, 2003.

Wong, John, and Nah Seok Ling. *China's Emerging New Economy; The Internet and E-Commerce*. Singapore, Singapore University Press, 2001.

Wong, John, and Zheng Yongnian (Eds). *China's Post-Jiang Leadership Succession; Problems and Perspectives*. Singapore, Singapore University Press, 2002.

Wood, Frances. *Did Marco Polo Go to China?* London, Secker and Warburg, 1996.

World Bank. *China: Long-term Development Issues and Options*. Washington, DC, 1985.

China 2020: Development Challenges in the New Century. Washington, DC, 1997.

China's Management of Enterprise Assets: The State as a Shareholder. Washington, DC, 1997.

The Chinese Economy: Fighting Inflation, Deepening Reforms. Washington, DC, 1996.

Wortzel, Larry M. (Ed.). *The Chinese Armed Forces in the 21st Century*. Carlisle, PA, Strategic Studies Institute, 2000.

Wu, Hongda Harry. *Laogai: The Chinese Gulag*. Boulder, CO, Westview Press, 1993.

Wu, Harry. *Troublemaker: One Man's Crusade Against China's Cruelty*. London, Chatto and Windus, 1996.

Wu, Harry, and Wakeman, Carolyn. *Bitter Winds: A Memoir of My Years in China's Gulag*. London, John Wiley and Sons, 1994.

Wu Yanrui. *China's Consumer Revolution—The Emerging Patterns of Wealth and Expenditure*. Cheltenham, Edward Elgar, 1999.

China's Economic Growth. London, RoutledgeCurzon, 2002.

Wu Yanrui (Ed.). *Foreign Direct Investment and Economic Growth in China*. Cheltenham, Edward Elgar, 1999.

Yabuki, Susumu. *China's New Political Economy: The Giant Awakes*. Boulder, CO, Westview Press, 1995.

Yahuda, Michael B. *China's Role in World Affairs*. London, Croom Helm, 1979.

Hong Kong—China's Challenge. London, Routledge, 1996.

Yan Jiaqi, and Gao Gao. *Turbulent Decade: A History of the Cultural Revolution*. Honolulu, HI, University of Hawaii Press, 1996.

Yang, Dali L. *Calamity and Reform in China: State, Rural Society, and Institutional Change since the Great Leap Famine*. Stanford, CA, Stanford University Press, 1996.

Yang, Rae. *Spider Eaters*. Berkeley, CA, University of California Press, 1997.

Yee, Herbert, and Storey, Ian (Eds). *The China Threat: Perceptions, Myths, and Reality*. London, RoutledgeCurzon, 2002.

Yao Shujie and Liu Shaming (Eds). *Sustaining China's Economic Growth in the Twenty-first Century*. London, RoutledgeCurzon, 2003.

Yin, Jason Z., Lin Shuanglin, and Gates, David F. (Eds). *Social Security Reform; Options for China*. River Edge, NJ, World Scientific Publishing, 2000.

Zang Xiaowei. *Elite Dualism and Leadership Selection in China*. London, RoutledgeCurzon, 2003.

Zhang Mei. *China's Poor Regions*. London, RoutledgeCurzon, 2003.

Zhang Kaiyuan. *Eyewitness Accounts of the Nanjing Massacre*. Armonk, NY, M. E. Sharpe, 2000.

Zhang Xiao-guang. *China's Trade Patterns and International Comparative Advantage*. Basingstoke, Macmillan Press, 1999.

Zhang Zhongxiang. *The Economics of Energy Policy in China—Implications for Global Climate Change*. Cheltenham, Edward Elgar, 1998.

Zheng Shiping. *Party vs. State in Post-1949 China: The Institutional Dilemma*. Cambridge, Cambridge University Press, 1997.

Zhou, Kate Xiao. *How Farmers Changed China*. Boulder, CO, Westview Press, 1996.

Zweig, David. *Internationalizing China: Domestic Interests and Global Linkages*. Cornell University Press, 2002.

CHINESE SPECIAL ADMINISTRATIVE REGIONS

HONG KONG

Physical and Social Geography

MICHAEL FREEBERNE

With additions by the editorial staff

The population of the Special Administrative Region (SAR) of Hong Kong occupies a total land area of only 1,098 sq km (423.9 sq miles). The territory, which reverted to Chinese sovereignty in July 1997, is situated off the south-east coast of Guangdong Province of the People's Republic of China, to the east of the mouth of the Zhujiang (Pearl River), between latitudes 22° 9′ and 22° 37′ N and longitudes 113° 52′ and 114° 30′ E. The SAR comprises the island of Hong Kong, ceded to the United Kingdom by China in 1842, the Kowloon peninsula, ceded in 1860, and the New Territories, which are part of the mainland and were leased to the United Kingdom between 1898 and 1997, together with Deep Bay and Mirs Bay and some 236 outlying islands and islets. The fine anchorages between the capital of Victoria, on the northern shore of Hong Kong Island, and Kowloon provided an ideal situation for the growth of one of the world's leading entrepôt ports.

PHYSICAL FEATURES

Hong Kong Island is approximately 17 km long and between 3 km and 8 km wide. An irregular range of hills rises abruptly from the sea; several peaks are over 300 m in height, and Victoria Peak reaches 554 m. Granites, basalt and other volcanic rocks account for the main geological formations. These rocks are most common, too, on Lantau and Lamma islands and in the Kowloon peninsula and New Territories, which are mostly hilly, rising to 957 m in Tai Mo Shan, and have rugged, deeply indented coastlines. The territory is poor in minerals and, apart from hillside areas of dense scrub, was largely stripped of natural vegetation by indiscriminate tree-felling during the Japanese occupation of 1941–45. The resultant erosion has been extensively repaired under a vigorous programme of reafforestation. Flat land and agricultural land are scarce everywhere. Reclamation of land from the sea for building purposes is very important, and since 1945 much additional land has been made available for housing and commercial development, as well as for projects like the international airport at Chek Lap Kok, off Lantau Island, which was opened in 1998. In 1996 major reclamation work at West Kowloon, which included the extension of Stonecutters Island, was substantially completed. Reclamation also continued to progress on Hong Kong Island and in the New Territories.

CLIMATE

The climate of Hong Kong is subtropical and governed by monsoons. Winter lasts from October to April, when the winds are from the north or north-east, while during the summer months from May to September south or south-westerly winds predominate. Average daily temperatures are highest in July with 29°C and lowest in January with 16°C. The wet summer is very humid. Annual rainfall averages 2,214 mm, some 80% of which falls between May and September. Devastating typhoons occasionally strike in summer.

Despite the high rainfall, it has proved increasingly difficult to supply sufficient domestic and industrial water, and most supplies are piped from the neighbouring Guangdong Province of China. The Plover Cove reservoir, inaugurated in 1969, trebled Hong Kong's reservoir capacity, and further reservoirs subsequently came into operation, including the world's first seabed reservoir. Nevertheless, in 1997 Hong Kong was dependent upon China for 70% of its water supply, compared with 45% in 1984. By the year 2000 the annual supply from China was scheduled to increase to 840m. cu m, from 690m. cu m in 1995, and by the year 2010 to 1,100m. cu m. Additional purchases may be made in years of low rainfall.

POPULATION

The population of Hong Kong in mid-2003 was officially estimated at 6,803,100, giving an average density of 6,167.8 persons per sq km. However, average density in the New Territories in mid-2000 was 3,520 per sq km, whereas for Hong Kong Island and Kowloon it was 17,200 and 44,210 respectively. In the Mong Kok district of Kowloon in the 1980s the population density exceeded 200,000. These figures represent some of the highest population densities in the world. In 1979 the population was greatly swelled by the entry of many thousands of illegal immigrants from China and 'boat people' from Viet Nam. The influx continued throughout the 1980s.

Hong Kong has experienced an extraordinary growth in population. Between 1841, when only about 5,000 people lived on the island, and 1941, the colony received successive waves of migrants; then the population was estimated at about 1.5m. There was a drastic reduction during the Japanese occupation (1941–45), but by 1949 the population had increased to 1,857,000. After the establishment of the People's Republic of China in 1949, large numbers of refugees arrived in Hong Kong, where the rate of natural increase was already high. Hong Kong's crude birth rate in 2003 was estimated at 6.8 per 1,000 (compared with 18.3 in 1975). At an estimated 2.7 per 1,000 registered live births in 2001, the infant mortality rate (of under-fives) was among the lowest in Asia. The crude death rate in 2003 was estimated at 5.4 per 1,000 (compared with 4.9 in 1975). The proportion of the population under 15 years of age declined from 23% in 1986 to 17.2% in 2000, while those aged 65 years and over rose from 8% to 11.2%.

About 95% of the territory's population are of predominantly Chinese descent. The Cantonese form the largest community. About 60% of the population in 1991 were born in Hong Kong. The prospect of the territory's transfer to Chinese sovereignty in mid-1997 led to an exodus of skilled personnel, the total number of emigrants reaching a record 66,000 in 1992. It was estimated, however, that at least 12% of those who had emigrated in the 10 years to 1994 had subsequently returned to Hong Kong, many having secured residency rights in countries such as Australia and Canada.

History

N. J. MINERS

Revised by JAMES TANG

INTRODUCTION

On 1 July 1997 Hong Kong became a Special Administrative Region (SAR) of the People's Republic of China, thus ending more than 150 years of British colonial rule. The agreement for the reversion of Hong Kong's sovereignty to China was reached in the Sino-British Joint Declaration of 1984. The arrangement, known as 'one country, two systems', promised capitalist Hong

Kong 'a high degree of autonomy', with the exception of foreign relations and defence, while the territory's political and economic systems, as well as its existing way of life, were to remain unchanged for at least 50 years, until 2047. The territory of the Hong Kong SAR (see below) is almost exactly the same as that acquired by the British in the 19th century.

EARLY DEVELOPMENT TO 1945

The colonial territory of Hong Kong was acquired by the United Kingdom (UK) in three stages. The First Opium War of 1840–42 began after the Chinese commissioner in Guangzhou (Canton) had seized and destroyed large stocks of opium held by the British traders there, who then left the city. The British Government demanded compensation and a commercial treaty, and an expedition was dispatched to enforce these demands. During the hostilities a naval force occupied the island of Hong Kong, which was ceded to the UK 'in perpetuity' by the Treaty of Nanjing (Nanking) of 1842. As soon as this was ratified, a colony was formally proclaimed in June 1843. Continuing disputes between the UK and China over trade and shipping led to renewed warfare in 1856. This was ended by the Convention of Beijing (Peking) of 1860, by which the peninsula of Kowloon on the mainland opposite the island was annexed.

Following China's defeat in the Sino-Japanese War of 1895, the Western powers seized the opportunity to extract further concessions. The UK demanded, and obtained in 1898, a 99-year lease on the mainland north of Kowloon, together with the adjoining islands. These New Territories increased the area of the colony from about 110 sq km to more than 1,000 sq km. The terms of the 1898 Convention of Beijing allowed the existing Chinese magistrates to remain in the old walled city of Kowloon, but in 1899 they were unilaterally expelled on the pretext that they had encouraged resistance to the British occupation. The Chinese Government protested at the time and reasserted a claim to jurisdiction over this small area in 1933, 1948 and 1962, although this was rejected by the Hong Kong courts. In 1994, with the agreement of the People's Republic of China, the area was cleared, levelled and converted into a public park.

The main reason for the British occupation of Hong Kong in 1841 was its magnificent harbour. Attracted by its free port status, the entrepôt trade between the West and China grew steadily for the next 100 years. The great trading companies set up their headquarters under the British flag; banks, insurance companies and other commercial enterprises were established to serve the China traders as well as shipbuilding, ship-repairing and other industries dependent on the port. At the same time the population grew from about 5,000 in 1841 to more than 500,000 in 1916 and over 1m. by 1939, of which fewer than 20,000 were non-Chinese. Chinese were allowed free access and the flow of migrants increased whenever China was disturbed by wars or rebellions, the process being reversed when peaceful conditions had been restored on the mainland. Apart from the settled farming population of the New Territories, relatively few Chinese regarded Hong Kong as their permanent home until after the Second World War. Most came to trade or seek employment and then returned to their home towns. Europeans were similarly transient, whether they were government officials or in private employment.

The colony's administration followed the usual crown colony pattern, with power concentrated in the hands of a Governor advised by nominated executive and legislative councils, on which government officials had an overall majority over the unofficial members. The first unofficial members were appointed to the Legislative Council in 1850, and the first Chinese in 1880; the first 'unofficials' in the Executive Council were appointed in 1896, and the first Chinese in 1926. In 1894, 1916 and 1922 the British residents pressed for an unofficial majority in the Legislative Council and the election of some or all of the 'unofficials' on a franchise confined to British subjects, citing the constitutional progress made in other colonies; but on all occasions the British Government was unwilling to allow the Chinese majority to be politically subjected to a small European minority. A sanitary board was set up in 1883 and this was made partly elective in 1887. In 1936 it was renamed the Urban Council, although its powers were not significantly increased.

Little of note happened in Hong Kong throughout this period, apart from commercial expansion, land reclamation and the building of reservoirs. There were a number of large-scale strikes in the early 1920s, but otherwise anti-foreigner agitation in China had little effect on Hong Kong's prosperity.

The growing threat of war in the late 1930s led to an increase in defence expenditure, which forced the imposition of income and profits taxes for the first time; this wartime expedient was made permanent in 1947. Japanese forces occupied most of the Chinese Province of Guangdong, north of the colony, in 1938, and in December 1941 overran Hong Kong. In August 1945 the Japanese authorities handed power back to the surviving colonial officials who had been interned with the rest of the British community throughout the occupation. A British naval force arrived in late August to install an interim military administration, thus forestalling pressures from the US Government for Hong Kong to be returned to China.

POST-WAR ISSUES

After the Japanese capture of Hong Kong in December 1941, a planning unit was formed in London to prepare for the post-war rehabilitation of the colony. Its members later staffed the interim military administration, which restored public services on a minimum basis. Civil government, on the traditional colonial pattern, was re-established in May 1946. Meanwhile, China was disrupted by civil war between the nationalist and communist armies, which ended with the communist victory in 1949. The UK recognized the new communist Government of China in 1950, having heavily reinforced the garrison in Hong Kong in 1949 to deter any possible Chinese attack. The only serious violation of the frontier occurred in 1967. The strength of the garrison was reduced at successive defence reviews. In 1992 special units of the Hong Kong Police Force assumed responsibility for the security of the border. All Gurkha troops left the territory in November 1996; one British battalion provided security until mid-1997, when it was replaced by soldiers of the Chinese People's Liberation Army.

During the Second World War the colony's population had declined to about 600,000, as a result of privation and mass deportations by the Japanese. The population quickly regained pre-war levels and rose to about 2m. in 1950, owing to a massive influx of refugees from the civil war in China. The pressures resulting from this inflow forced the colony to abandon its policy of free access, and the frontier was closed in 1950. Movement over the border was subsequently tightly controlled by the Chinese authorities, with the exception of a period during May 1962, when the frontier was unexpectedly opened and 120,000 refugees were allowed to leave. Individual escapees also attempted to enter clandestinely. From 1974 to 1980 any illegal immigrants who were apprehended in the frontier region were transferred to the Chinese authorities, but those who succeeded in reaching the urban areas were allowed to remain. This concession continued to encourage escape attempts. After a massive surge in illegal immigration in 1979–80 (when it was estimated that more than 200,000 people succeeded in settling in the colony in spite of the fact that 170,000 were captured and repatriated), it was announced in October 1980 that, in future, all illegal immigrants who were discovered anywhere in the colony would be repatriated. This announcement caused a sharp decline in illegal attempts to enter, but such immigration continued, and in 1997 an average of some 49 illegal immigrants were arrested daily and forcibly repatriated to China. The Chinese authorities permitted 150 people a day to cross the border, and more were entitled to settle in Hong Kong after China's resumption of sovereignty. From 1990 a total of 25,000 workers were allowed to enter Hong Kong each year on fixed-term contracts to relieve the labour shortage.

Meanwhile, the problem of feeding the refugees and providing employment was made worse by the outbreak of the Korean War in 1950, which led to the imposition of an embargo on the export of strategic goods to China and gravely damaged Hong Kong's entrepôt trade. However, the refugees provided a pool of compliant, hard-working labour. Local businessmen and industrialists who had fled from Shanghai took advantage of this and, by making use of the colony's existing financial infrastructure and world-wide trading connections, they reorientated the economy towards manufacturing for export.

The refugees put an immense strain on all public services, and the newcomers were left to build themselves shanty towns which spread over the hillsides. A devastating fire at one of these shanty towns in 1953 prompted the Government to initiate a resettlement programme; huge estates were built, with rooms allocated on the scale of 2.3 sq m for each adult. The early designs provided few amenities, as the main consideration was speed of construction. The housing programme continued steadily, with additional expansion from 1972, when a 10-year programme to house a further 1.5m. people, mainly in new towns in the New Territories, was announced. At the same time, various government housing agencies were amalgamated into a new housing authority to assume responsibility for the planning, construction and management of all public housing in Hong Kong. By 1991 more than 50% of the population were living in government-provided housing.

From 1976 many refugees from Viet Nam attempted to reach Hong Kong by sea. Initially, most were accepted for resettlement in the USA, Canada, Australia and Europe. However, in 1982, after the resettlement countries had reduced their immigration quotas, Hong Kong adopted a policy of confining all newly arrived Vietnamese in closed camps, in order to deter others from landing in the territory. Nevertheless, the numbers continued to increase. A particularly large influx in 1988 led the Hong Kong Government to abandon its policy of automatically granting refugee status to newly arrived Vietnamese. All arrivals were subjected to a test, and those considered to be 'economic migrants' were classified as illegal immigrants and confined in detention centres, to await eventual repatriation to Viet Nam. This new policy, however, failed to act as a deterrent. Vietnamese continued to arrive in increasing numbers, in the hope that they could pass the screening procedure and secure passages to the USA, since the Government of Viet Nam refused to accept the forced repatriation of those who had been classified as economic migrants. At the end of 1991 Viet Nam finally agreed to accept a limited programme of mandatory repatriation. This significantly reduced the number of Vietnamese arriving in Hong Kong. About 12,000 of those facing mandatory repatriation volunteered to return to Viet Nam, and China demanded that all Vietnamese should be removed before July 1997. Many of those remaining in Hong Kong were determined to resist forcible repatriation by any available means, and there were intermittent riots and demonstrations in the camps. From early 1998 it was announced that Hong Kong would no longer conduct a 'port of first asylum' policy, whereby refugees were permitted to apply for asylum upon arriving in the territory. Some 1,200 refugees, 659 migrants and 743 illegal immigrants from Viet Nam remained in Hong Kong in March and negotiations were under way between the Hong Kong and British Governments to determine responsibility for their repatriation. Hong Kong's last remaining camp for Vietnamese refugees was closed on 31 May 2000. Inmates were granted residency in the SAR. Several of the refugees, however, tried to remain inside the camp on the grounds that they could not afford the price of housing in the territory and were without jobs.

From 1945 until 1982 the people of Hong Kong showed a noticeable apathy towards any form of political activity or agitation for democratic self-government. This political calm was disturbed only three times: in 1956 there were faction fights between communist and nationalist supporters; in 1966 there were three nights of rioting, provoked by an increase in fares on the cross-harbour ferry; and for several months in 1967 there were disturbances and bomb attacks, led by communist sympathizers who were inspired by the example of the 'Cultural Revolution' in China. These had ended by late 1967, with the restoration of order in China itself.

The next 15 years were a period of political calm and rapidly growing economic prosperity, until 1982, when the start of the negotiations on Hong Kong's future caused an upsurge of political activity. Anxiety at the prospect of rule by China after 1997 led to an increase in emigration and investment overseas by those residents able to do so. The collapse of the property market and the consequent depressed sales of leases of crown land resulted in budget deficits in 1983, 1984 and 1985, which were financed by increases in taxation and government borrowing. Confidence in Hong Kong's future recovered, and the property market revived following the signing of the Sino-British agreement of 1984 (see below). The economy continued to expand, in spite of periodic crises in the relationship with China. There were no budget deficits between 1985 and 1995, and taxes were steadily reduced.

In 1986 plans by China to construct a nuclear power plant at Daya Bay, about 50 km from the centre of Hong Kong, aroused great public anxiety. A petition against the project attracted more than 1m. signatures. Despite this, the Chinese authorities concluded the major contracts in September 1986, thereby demonstrating their lack of concern for Hong Kong opinion. The plant began operations in 1994. Construction of a second nuclear plant at Daya Bay began in 1995.

The massacre of students in Tiananmen Square, in Beijing, in June 1989 and the forcible suppression of the pro-democracy movement throughout China had a devastating impact on local confidence in the future. Demonstrations involving up to 1m. people were held to protest against the slaughter. The property market and the stock exchange suffered very sharp falls. The number of emigrants, mostly to Canada, Australia and the USA, reached a record 66,200 in 1992. The outflow fluctuated thereafter, and the number of departures declined to 30,900 in 1997. Many sought to obtain passports of foreign countries to enable them to leave Hong Kong before China's resumption of sovereignty in mid-1997. In an attempt to restore confidence, the British Government agreed in 1990 to grant full British passports with the right of abode in the UK to 50,000 business executives, administrators and professional people, together with their immediate family members, making a total of 225,000 individuals. It was hoped that this would encourage key personnel to remain in Hong Kong, since they would have the assurance that they could leave for the UK at any time. However, China denounced this move as a plot to entrench British influence in Hong Kong, and insisted that it would not recognize these passports after 1997.

In other moves to boost confidence, in October 1989 the Governor announced plans to build a new international airport at Chek Lap Kok, near Lantau Island, a railway to link the airport to the city, and large new port facilities and container terminals in the west of the harbour. China objected strongly to the proposals, and without Beijing's agreement it would not have been possible to raise the private finance required. In July 1991 the UK and China signed a memorandum of understanding on the arrangements for building the new airport, but China continued to raise objections to the costs of the project, and tried to use its power to withhold consent in order to extract concessions on political issues. In particular, China opposed the proposals made by the new Governor, Chris Patten, to extend the franchise for the 1995 Legislative Council elections. In 1995 China finally agreed to the financial arrangements and to the composition of the Airport Authority. The opening of the new airport, built at a cost of some US $20,000m., was initially scheduled for April 1998, but was postponed until July, owing to delays in the construction of the express rail link. The inauguration of the airport was overshadowed by problems in the cargo handling systems, which resulted in the loss of thousands of consignments of food and other goods. The authorities were forced to divert cargo to the old airport or to facilities on the mainland.

In June 1991 the Hong Kong Government enacted a Bill of Rights to give effect in local law to the relevant provisions of the UN International Covenant on Civil and Political Rights. These provisions remained in force following the territory's transfer to Chinese sovereignty in mid-1997.

POLITICAL AND ADMINISTRATIVE DEVELOPMENT

Following the restoration of civil government in May 1946, the returning Governor promised a greater measure of self-government and, after inviting suggestions from the public, proposed that an elected Municipal Council, with wide powers over local affairs in the urban area, should be established. These plans for major constitutional reform were deferred after the communist victory in China in 1949, and were finally abandoned in 1952. From then until 1985, there were no further moves towards democratic government, largely in deference to China's dislike of any such changes. Instead, the number of appointed unofficial members on the Legislative Council was successively increased from eight in 1951 to 30 in 1984.

Under the terms of the 1984 Joint Declaration (see below), the legislature of the Hong Kong SAR would be constituted by election. Therefore, in 1985 the composition of the Legislative Council was substantially changed, to include 24 indirectly elected members, 12 chosen by the district boards, Urban Council and Regional Council, and 12 by 'functional constituencies', composed of the representatives of the commercial and industrial sectors, trade unions and various professional bodies. These 24 elected members were outnumbered by the 22 appointed members and the 10 officials. In deference to China's wishes, only minor constitutional changes were made in 1988. In 1990 the British Foreign Secretary agreed with China that the 1991 Legislative Council would consist of 21 functional constituency members, 18 directly elected by universal franchise, 18 members appointed by the Governor, and three civil servants. Elections were held in September 1991. Of the 18 seats open to election by universal suffrage, 15 were won by the United Democrats (a liberal grouping led by Martin Lee—who subsequently became the Chairman of the Democratic Party of Hong Kong) and their allies. The new Legislative Council, inaugurated in October, had a majority of elected members for the first time in the territory's history, although most of the 21 members representing functional constituencies had been chosen by only a few hundred voters. Contrary to the normal constitutional practice in British colonies, none of the leaders of the United Democrats was invited to join the Executive Council.

In 1992 Sir David Wilson was replaced as Governor by Chris Patten, a Conservative politician who had lost his seat in the British general election. Three months after his arrival he announced detailed proposals for the conduct of the 1995 elections, without first consulting China. He proposed that the Legislative Council should conform to the model laid down in the Basic Law (see below), with 30 members elected by functional constituencies, 20 directly elected by geographical constituencies and 10 by an electoral college, but that the electorate for the nine new functional constituencies should be enlarged to 2.7m. voters and that the electorate for the 10 electoral college seats should consist of all the elected members of the District Boards. China denounced this plan as an attempt to increase the number of directly elected seats beyond that laid down in the Basic Law. Negotiations were held with China during 1993, but no compromise could be found. Patten's original proposals were enacted into law by the Legislative Council in June 1994.

At the elections to the Legislative Council held in September 1995, for the first time all 60 seats were determined by election. A total of 920,567 people (36% of registered electors) voted for the 20 seats open to direct election on the basis of geographical constituencies. The Democratic Party of Hong Kong won 19 seats (12 by direct election, five by functional constituencies and two chosen by electoral college). About one-half of the members of the Council usually supported the Democratic Party of Hong Kong, but the Government was able to avoid any significant defeats in the legislature by energetic lobbying for the support of smaller parties and independent members.

From 1896 to 1992 members of the Legislative Council who were not officials were appointed by the Governor to sit on the Executive Council. After 1946 the unofficial members outnumbered the officials. In 1992 Patten ended this practice, making the membership of the Executive Council entirely separate from the Legislative Council. The Governor was empowered to reject the advice given to him by the majority of the Council (of which he also was a member) but, in practice, this never occurred.

In 1996 the Urban Council remained responsible for public health and sanitation, recreation, amenities and cultural services in the urban area. From 1995 the Council consisted entirely of elected members, with nine indirectly chosen by District Boards and 32 elected by all those over 18 years of age who had been resident in Hong Kong for at least seven years. A similar Regional Council served the New Territories. The Chinese Government objected strongly to the abolition of the seats for appointed members, and threatened to reconstitute both councils after 1997. Elections were held for the two councils in March 1995, at which 26% of the registered voters cast their ballots. The Democratic Party of Hong Kong and its allies won 31 of the 59 directly elective seats.

Hong Kong is divided into 18 districts. In each of these there is a management committee of officials from various government departments working in the area, presided over by a senior administrative officer. This committee is assisted by an advisory district board. Between 1994 and mid-1997 all members of District Boards were elected by universal suffrage. However, from July 1997 the District Boards were replaced by Provisional District Boards, comprising members appointed by the Chief Executive of Hong Kong. The interests of the rural indigenous inhabitants of the New Territories are served by an elected advisory body, the Heung Yee Kuk. There are also advisory committees attached to most government departments.

SOVEREIGNTY NEGOTIATIONS

After the communist victory in the Chinese civil war in 1949, the People's Republic asserted that all the unequal treaties, forced upon China, were no longer recognized as binding; but the treaties of 1842, 1860 and 1898 were not formally abrogated. The Chinese Government was unwilling to clarify its intentions after 1997, when the lease on the New Territories expired, apart from giving intermittent assurances that the interests of investors would be protected.

In 1982 the UK decided to press China for a decision. Following the visit of the British Prime Minister, Margaret Thatcher, to China in September, negotiations commenced through diplomatic channels and continued for two years. In August 1984 agreement was reached on a Joint Declaration, which was subsequently approved by the British Parliament and ratified in May 1985. Under this agreement, the UK undertook to restore sovereignty over the whole of Hong Kong to China on 1 July 1997, upon the expiry of the lease on the New Territories. Until that date, the British Government was to continue to be responsible for the administration of the territory, but a Joint Liaison Group (JLG—see below), consisting of British and Chinese diplomatic representatives, was formed to consult on the implementation of the agreement, and to ensure a smooth transfer of sovereignty in 1997.

China, for its part, undertook that, after 1997, Hong Kong would be constituted as a Special Administrative Region (SAR, designated 'Hong Kong, China'), governed by its own inhabitants in accordance with its own legal code, except in matters of foreign affairs and defence, for a period of 50 years. It was to retain its status as a free port and separate customs territory, and the Hong Kong dollar was to remain a freely convertible currency. The region's social and economic systems were to remain unchanged, and freedom of speech, of the press, of association, of travel and of religion was to be guaranteed by law. Existing leases of land were to be recognized, and were to be extended to the year 2047.

From mid-1997 all existing Chinese residents of Hong Kong became citizens of the People's Republic of China. As such, they were forbidden by Chinese law to hold dual British nationality. Following approval of the Hong Kong agreement, the British Government announced a new form of nationality, to be effective from 1997, designated 'British National (Overseas)', which would entitle the holders to British consular protection when travelling outside China. This status conferred no right of abode in the UK and was not transferable to descendants. Non-Chinese residents of Hong Kong were not to be granted Chinese nationality after 1997. They were to be entitled to hold only the new British National (Overseas) passport, and there were fears that they might become, in effect, stateless. In September 1995 the Governor aroused much controversy when he urged the UK to give the right of abode to more than 3m. Hong Kong citizens.

A 59-member Basic Law Drafting Committee (BLDC) was formed in Beijing in June 1985, with the aim of drawing up a new Basic Law (Constitution) for Hong Kong, in accordance with Article 31 of the Chinese Constitution, which provides for special administrative regions within the People's Republic. The BLDC included 25 representatives from Hong Kong itself. In April 1988 the first draft of the Basic Law for Hong Kong was published, and public comments were invited. In the light of these criticisms, the draft was further revised and a second version was published in February 1989.

The final draft was adopted by the National People's Congress (NPC) of China in April 1990. The Basic Law gives China the right to declare a state of emergency in the territory and the right to station Chinese troops there. The SAR is also required to enact laws prohibiting any acts of subversion against the Government of China and forbidding any political groups in the SAR from establishing ties with any foreign organizations. Because of these provisions, the Legislative Council passed, by a large majority, a motion in April 1990 expressing its disapproval of the Basic Law.

A Joint Liaison Group (JLG), consisting of five representatives each from China and the UK, was established to hold consultations on the implementation of the Joint Declaration and to oversee arrangements for a smooth transition to Chinese sovereignty. Agreement was reached on a number of issues between 1985 and 1989, including the clearance of the Kowloon Walled City. Following the Tiananmen massacre, and particularly after the arrival of Chris Patten, progress was much slower. China asserted its right to veto any policy decisions that encompassed 1997 and withheld its approval of proposals to adapt legislation and revise international agreements to take account of China's resumption of sovereignty. In 1994, after seven years of negotiation, China and the UK finally reached agreement on the disposal of the land occupied by the British garrison. More than one-half of the barracks sites were to be handed over to the Hong Kong Government for redevelopment. The remainder was to be used by the Chinese army and navy. In 1995 agreement was reached on the procedure for the establishment of a Court of Final Appeal, which was to replace the Privy Council after 1997.

In September 1996 China and the UK agreed on the arrangements for the ceremony marking the transfer of sovereignty, which was to take place at midnight on 30 June 1997. In January 1997 the two Governments reached agreement on defining the borders of the territory of Hong Kong and formally signed a memorandum on the boundary in June.

PREPARATIONS FOR THE RESUMPTION OF CHINESE SOVEREIGNTY

Following the breakdown of the 1993 talks over Chris Patten's reform proposals, the Chinese Government declared that it would establish a 'second stove' in order to ensure a smooth political transition. This unilateral action to create new political institutions for the SAR, from the Chinese perspective, was to be consistent with the Joint Declaration, the Basic Law and the understanding reached in 1990 between the Chinese Minister of Foreign Affairs and the British Foreign Secretary. In July 1993 the Chinese Government established a Preliminary Working Committee (PWC) to advise on and formulate proposals for the transition. The PWC, chaired by the Vice-Premier and Minister of Foreign Affairs, Qian Qichen, comprised key Chinese officials in charge of Hong Kong affairs, as well as prominent pro-Beijing government figures. There were five sub-groups with responsibilities for economic, political, legal, cultural, and social and security matters. Since the Chinese Government refused to accept the Legislative Council elected in 1995, the PWC proposed the establishment of a provisional council in order to avoid a power vacuum in the territory.

In January 1996 a 150-member Preparatory Committee (PC) for the Hong Kong SAR was formally established, to succeed the PWC. The PC, also headed by Qian Qichen, included 94 members from Hong Kong and 56 members from the mainland. Most Hong Kong members were known to be sympathetic to the Beijing Government's position on Hong Kong. While a number of legislative councillors were appointed to the PC, the Democratic Party of Hong Kong, the largest political party in the Legislative Council, was excluded. In March the PC decided that a 60-member Provisional Legislative Council (PLC) was to be established, which would commence operation after the election of the first Chief Executive of the SAR. It also decided that the PLC should cease operation upon the formation of the first Legislative Council of the SAR, while the term of the PLC was not to extend beyond 30 June 1998. All PLC members had to be permanent residents of Hong Kong, with up to 12 members holding non-Chinese nationality or the right of abode in foreign countries. Members of the PLC were to be chosen by a 400-member Selection Committee, which would also elect the SAR's first Chief Executive. The responsibilities of the PLC included: to enact, amend or appeal laws to ensure the proper functioning of the Hong Kong SAR; to examine and approve budgets proposed by the administration; to approve taxation and public expenditure; to receive and debate the policy address of the Chief Executive of the SAR; to endorse the appointment of the judges of the Court of Final Appeal and the Chief Judge of the High Court; to deal with other necessary legislative matters before the formation of the first SAR Legislative Council. The President of the PLC also had to participate in the nomination of the six Hong Kong members of the Committee for the Basic Law of the Standing Committee of the National People's Congress in Beijing.

In November 1996 the PC elected the 400-member Selection Committee from among 5,789 candidates. On 11 December Tung Chee-hwa, having defeated two other candidates (the former Chief Justice, (Sir) Yang Ti-liang, and businessman Peter Woo) in the final round of the selection process, was elected the first Chief Executive of the SAR Government by the Selection Committee. On 21 December the PLC's 60 members were elected from among 134 candidates. Thirty-three served concurrently in the existing Legislative Council, but the Democratic Party of Hong Kong boycotted the election. On 14 March 1997 the NPC approved a PC report on the establishment of the PLC. Soon after its establishment, the PLC began operation in parallel to the Legislative Council, in the neighbouring town of Shenzhen, scrutinizing and passing bills. In June the PLC introduced legislation to restore colonial restrictions and to impose new conditions on public demonstrations and the establishment of political organizations in the SAR. Under the legislation, public demonstrations and political organizations could be banned in the interests of national security, defined as 'the safeguarding of the territorial integrity and the independence of the People's Republic of China'. Political organizations were, moreover, not permitted to have connections with foreign political organizations or with Taiwan.

THE SAR GOVERNMENT

On 1 July 1997, following the handover ceremony at midnight, the SAR Government was inaugurated at 1.30 a.m. The ceremonies were attended by 4,000 dignitaries, including ministers of foreign affairs from more than 40 countries and senior representatives of more than 40 international organizations. The newly elected British Prime Minister and his Foreign Secretary, as well as the US Secretary of State, however, did not attend the inauguration, in an expression of their disapproval of the swearing-in of PLC members during the ceremony. Legislators from the Democratic Party of Hong Kong and a number of independent Legislative Councillors protested against the abolition of the Legislative Council on the balcony of the Council building, shortly after midnight, vowing to return by means of elections in 1998.

More than 4,000 People's Liberation Army troops were deployed to the Hong Kong garrison. With the consent of the British, a small number of unarmed military personnel entered the territory in April and May 1997, and an advance party of some 500 troops crossed the border a few hours before the handover ceremony. Twenty-two of the 23 principal officials were retained by the SAR administration. The only new appointment was that of the Secretary for Justice, who replaced a retired expatriate. The incoming Executive Council had 14 members, with a former senior Executive Council member as convener, three ex-officio members, the Chief Secretary, the Financial Secretary and the Secretary for Justice, and 10 unofficial members. In addition to the Chief Secretary and the Financial Secretary, two outgoing members of the previous Executive Council were reappointed. The Chief Executive also appointed a special adviser with close connections to the Beijing Government. In March 1999 Anson Chan, regarded by the international media as representing Hong Kong's conscience, agreed to continue serving as Chief Secretary for Administration for two years beyond her normal retirement age, until 2002, when the term of office of the Chief Executive was to end. In June 1999 the Government appointed Dr E. K. Yeoh, hitherto Chief Executive of the Hospital Authority, as Secretary for Health and Welfare.

After the Secretary for Justice, Dr Yeoh was the second non-civil servant to be appointed to a senior civil service position by the SAR Government. The convener of the Executive Council, Dr Chung Sze-yuen, retired in late June. He was replaced by a fellow member of the Executive Council, Leung Chun-ying. In May 2000 Elsie Leung was reappointed as Secretary of Justice, for a further two years. In January 2001 Anson Chan unexpectedly announced that for personal reasons she would step down at the end of April, well before the expiry of Tung's first term. She was replaced by the hitherto Financial Secretary, Donald Tsang. Antony Leung, a former banker and member of the Executive Council, was appointed Financial Secretary. In 1999, meanwhile, the Government announced a civil service reform programme, including the introduction of performance-based pay, new entry and exit procedures and a lower starting salary. The initiative encountered resistance. In July 2002 thousands of civil servants took to the streets to demonstrate against the pay decreases, introduced by the Government in an effort to reduce civil service costs (as a measure to counter the SAR's budget deficit). While the public appeared to be supportive of the pay decreases, there were concerns that the administration had not handled the issue well and that as a result morale in the civil service had suffered.

The SAR Government replaced the existing local administration with the Provisional Urban Council, a Provisional Regional Council and 18 Provisional District Boards. While their responsibilities remained the same and all existing councillors and board members continued to serve, membership of these elected bodies was expanded with appointed members, many of whom had been unsuccessful in earlier elections. In April 1999 the SAR administration decided to restructure local government by abolishing the municipal and regional councils. The Government intended to assume responsibility for the public services offered by the councils. Existing district boards were to be replaced by district councils. While the public was dissatisfied with the performance of the two municipal councils, critics of the Government maintained that the restructuring was a retrograde step in terms of local democracy. In November the Democratic Party of Hong Kong threatened legal action if the plan were not abandoned, as it contravened the Basic Law. The Hong Kong SAR's first district elections took place on 28 November. The Democratic Party of Hong Kong won the largest number of elected seats (86), but the pro-Beijing Democratic Alliance for the Betterment of Hong Kong (DAB) substantially increased its representation, from 37 seats to 83.

One of the first acts of the SAR administration was the introduction of legislation to prevent an influx of mainland-born children of Hong Kong residents. These children were granted the right of abode in the territory by the Basic Law, but the Government insisted that they could be admitted only upon verification of their identities and in an orderly fashion, according to a quota system controlled by the mainland authorities.

Elections for the first Legislative Council of the SAR were held on 24 May 1998 and were conducted under new electoral arrangements. Of the 60 seats, 30 were elected by narrowly defined functional constituencies, 20 were elected by proportional representation in geographically based constituencies and 10 were chosen by an 800-member Election Committee. Pro-democracy parties criticized the reduction of the franchise under these new arrangements. Despite heavy rain, the people of Hong Kong turned out to vote in large numbers. With almost 1.5m. registered voters casting their ballots, the participation rate of 53.3% was the highest since the introduction of direct elections in Hong Kong. Most former Legislative Councillors who had boycotted the Provisional Legislative Council were elected to the new legislature.

The Democratic Party of Hong Kong, led by Martin Lee, returned to the Legislative Council with 13 seats, of which nine were obtained in the geographical constituencies. Although the Democratic Party of Hong Kong and other pro-democracy parties received solid support from the electorate and won most seats in the geographical constituencies (14 of the 20 seats) their overall political strength in the legislature was reduced. Together with the Frontier and the Citizens' Party, and liberal independents, the pro-democracy camp secured a total of 19 seats. Another pro-democracy group, the Association for Democracy and People's Livelihood, which had participated in the PLC, lost all of its seats on the Council.

Pro-Beijing supporters dominated the functional constituencies and the electoral committee ballot and also secured some seats in the geographical constituencies. The DAB, which won 10 seats, with five from geographical constituencies, benefited most from the new electoral system. The leader of the pro-business Liberal Party, Allen Lee, failed to win a seat in the geographical elections, but his party obtained 10 seats through the functional and election committee constituencies, while the Hong Kong Progressive Alliance managed to obtain five seats.

The powers of the new legislature were curbed by the Basic Law. Henceforth Legislative Councillors were not permitted to introduce bills related to public expenditure, the political structure or the operation of the government. The passage of private members' bills or motions also required the majority of votes of both groups of councillors—those elected through geographical constituencies, and those returned through functional constituencies and the election committee.

Although political forces in the first legislature were fragmented, the Legislative Council was more assertive than the PLC. Even before the new legislature convened, the seven major political parties, together with a number of like-minded independents, formed a temporary alliance in June 1998, demanding that the administration adopt new measures to address the economic downturn in Hong Kong. Tung Chee-hwa responded by announcing a series of government initiatives to improve the economic situation, including the abolition of savings tax for commercial corporations, the postponement of the sale of land by auction and tender until 31 March 1999, an increase in loans to home-buyers, and allocation of additional government funding to assist small and medium enterprises. The coalition was not maintained over constitutional and political issues. Senior civil servants were often required to respond to questions presented by the Legislative Council. In June 2000 the legislature overwhelmingly approved a vote of 'no confidence' in two senior officials, following a series of public housing scandals. One of the officials involved, the Secretary for Housing, Rosanna Wong, resigned shortly before the vote took place. The term of the Legislative Council ended on 30 June 2000. All the major political parties presented candidates for the elections to the second post-1997 legislature, which were held on 10 September 2000. The number of directly elective seats was increased from 20 to 24, while the number of Council members selected by the Election Committee was reduced from 10 to six. (The ballot for Election Committee members, who were responsible for choosing those who were to occupy the Election Committee seats, had taken place in July.) Of the 60 seats on the Legislative Council the Democratic Party of Hong Kong secured a total of 12 (including nine by direct election), the DAB won 11 seats (including eight by direct election), and the Liberal Party won eight seats. The level of voter participation was 43.6% of the electorate. Following the resignation from the incoming Legislative Council on 19 September of a newly elected member, Gary Cheng, the DAB's Vice-Chairman (owing to his admission of a conflict of interest between his role as a legislator and previously undeclared business assets), a by-election took place on 10 December. The seat was won by Audrey Eu, an independent candidate and former chairperson of the Hong Kong Bar Association.

In March 2001 the Hong Kong Government introduced the Chief Executive Election Bill, interpreting the powers of the central Government in Beijing to remove the Chief Executive and providing detailed arrangements for the next election for the latter. Although the Government amended the Bill's provisions relating to the removal of the Chief Executive (after it was criticized by some legislators for giving unnecessary powers to the Chinese Government), following which the Bill was duly passed by the Legislative Council by 36 votes to 18 in July 2001, the approval of the new legislation was widely regarded as a setback for Hong Kong's democratic progress.

An economic recession followed the financial turmoil that affected the Asian region from mid-1997, leading to the collapse of the Hong Kong property market, rising unemployment and negative economic growth. In August 1998 the Government launched a massive intervention in the financial markets, spending more than US $15,000m. in an attempt to maintain the stability of the currency and stock markets, in the face of manipulation by foreign speculators. Although the intervention was supported by the local business community, it undermined Hong Kong's reputation as a free economy, and at the time was criticized by the international media. However, many observers later accepted as necessary the authorities' involvement in the financial markets, owing to the exceptional situation. While the economic situation stabilized, the Government warned that economic difficulties were structural in nature. At the beginning of 2002 the Government forecast a modest real growth of 1% for the year, and in June the unemployment rate reached a record 7.4% (see Economy).

The issue of the right of abode raised questions as to the foundation of the rule of law and Hong Kong's autonomous status. On 29 January 1999 the Court of Final Appeal (CFA) ruled against the SAR Government's position that, according to Article 24 of the Basic Law, mainland children born of a Hong Kong permanent resident should be granted the right of abode in the territory only if at the time of their birth their parents had already become permanent residents. In delivering their verdict, the judges also maintained that the CFA had the authority to interpret the Basic Law. The ruling was questioned by mainland legal experts who had been involved with the drafting of the Basic Law and by other senior Chinese officials. In February, in response to a motion filed by the Department of Justice, the CFA declared that its ruling did not question the authority of the Standing Committee of the NPC. Human-rights groups argued that the Government's action, in seeking clarification from the Court, undermined the rule of law and judicial independence in Hong Kong. Claiming that the decision would lead to the influx of more than 1.6m. people from the mainland, the SAR authorities suggested that the CFA had misinterpreted the Basic Law. Supported by opinion polls, in May the SAR Government requested the State Council (Cabinet) of China to ask the Ninth NPC to interpret the relevant articles of the Basic Law. Pro-democracy legislators walked out of the Legislative Council Chamber in protest at the Government's decision. In late June the Standing Committee of the Ninth NPC ruled that the CFA's interpretation of the Basic Law was wrong and provided a new interpretation, which reduced the number of mainland children with the right of abode in Hong Kong to about 200,000.

In April 2000 a deputy director of the Liaison Office of the Central People's Government (CPG), formerly part of the Hong Kong branch of the Xinhua News Agency, suggested that the Hong Kong media should not disseminate the views of those who advocated the independence of Taiwan. In June another official from the Liaison Office warned the local business community not to deal with Taiwanese companies that support independence for Taiwan. Following both incidents, the Hong Kong Government reaffirmed that the local media remained free to report and comment on public issues, and that businesses in the SAR were at liberty to choose their business partners.

In January 2001 a CPG official stated that no organization or individual should be allowed to turn Hong Kong into the centre of activities for Falun Gong, a controversial religious group banned in the mainland since 1999. In June 2001 the Chief Executive declared the group to be an 'evil cult'. Nevertheless, the Government announced that it did not intend to introduce anti-cult legislation.

In July 2001 Li Shaomin, a US national who had been convicted in China of spying for Taiwan and who had spent five months in a mainland prison, was permitted by the Hong Kong Government to return to the SAR, where he worked as an academic. His case was widely seen as a test of the 'one country, two systems' arrangement with China. Li's employer, the City University of Hong Kong, decided to permit him to resume his teaching and research duties. He later returned to the USA when his leave application was unsuccessful.

In December 2001 Tung Chee-hwa announced that he would stand for a second five-year term as Chief Executive, the election having been scheduled for 24 March 2002. The central Government in Beijing endorsed his candidacy. On the closing date for nominations at the end of February 2002, however, Tung was the only candidate with overwhelming support, having received 714 nominations from the 794-strong Selection Committee. On 4 March, therefore, the central Government formally appointed Tung for a second term, beginning on 1 July 2002.

The most important political change that Tung introduced was a new scheme for the appointment of senior officials. Under this scheme, all principal officials, defined as the Chief Secretary for Administration, the Financial Secretary, the Secretary for Justice, along with the 11 secretaries in charge of policy bureaux, were to become political appointees, with their terms of office not exceeding that of the Chief Executive who had appointed them. These senior officials, widely regarded as 'ministers', were also to be members of the Executive Council, thus turning it into a cabinet-style body. The principal officials would report directly to the Chief Executive and would have to accept total responsibility for their respective portfolios. The civil service remained a permanent service based on meritocracy. Amid criticisms that the 'ministers' were accountable only to the Chief Executive and not to the Legislative Council or to the public, in addition to concerns that the arrangements might not be consistent with the Basic Law, the new scheme was approved by the Legislative Council on 29 May 2002, less than six weeks after its first formal introduction there. In mid-June the Council approved the funding, and a Government resolution transferring statutory powers to the principal officials was passed by a vote of 36 to 21.

The 14 principal officials appointed by Tung Chee-hwa included eight who had served under Tung in his first term. The three most senior officials, Donald Tsang (Chief Secretary), Antony Leung (Financial Secretary) and Elsie Leung (Secretary for Justice) remained unchanged. Tung appointed five new 'ministers' from business and the professions: the Secretary for Commerce, Industry and Technology, Henry Tang Ying-yen; the Secretary for Education and Manpower, Arthur Li Kwok-cheung; the Secretary for Home Affairs, Patrick Ho Chi-ping; the Secretary for the Environment, Transport and Works, Sarah Liao Sau-tung; and the Secretary for Financial Services and the Treasury, Frederick Ma Si-hang. In addition to the 14 principal officials, Tung appointed five non-official members to the Executive Council, including the leaders of two political parties—James Tien from the Liberal Party and Jasper Tsang Yok-sing from the DAB.

The Government tried to allay concerns about the political neutrality of the civil service under the new scheme by announcing that the Secretary for the Civil Service would be selected from within the civil service and would not lose civil servant status if he or she decided to return to the service upon the expiry of his or her term. The new scheme represented a significant political change. The accountability system was tested in mid-2002 when principal officials responsible for financial matters appeared to have underestimated the implications of a new proposal to eliminate stocks valued below HK $0.5. In September, after a review of the incident had been completed, the Secretary for Financial Services and the Treasury, Frederick Ma, apologized to the public. Following the introduction of the accountability system, the Chief Executive delivered his policy address in January 2003 instead of in October, as in the past.

In early 2003 Hong Kong was badly affected by an outbreak of Severe Acute Respiratory Syndrome (SARS), a little-known form of pneumonia, which had originated in neighbouring Guangdong. By March the spread of the virus had become a global health crisis. In Hong Kong more than 1,750 people were infected and 298 died. As a result, the World Health Organization (WHO) listed Hong Kong as an infected area and issued a travel advisory against the city, which was not lifted until June.

The Government's proposal to introduce a National Security Bill as stipulated by Article 23 of the Basic Law led to much disquiet. According to Article 23, the Hong Kong SAR 'shall enact laws on its own to prohibit any act of treason, secession, sedition, subversion against the Central People's Government (CPG) or theft of state secrets, to prohibit foreign political organisations or bodies from conducting political activities in the Region, and to prohibit political organisations or bodies of the region from establishing ties with foreign political organisations or bodies'. In a consultation document released in September 2002, the SAR Government suggested that the proposed legislation was necessary to protect sovereignty and safeguard territorial integrity, unity and national security. It would meet the requirements of Article 23 but would also respect the fundamental rights and freedoms of the people of Hong Kong.

While most people did not appear to question the constitutional responsibility of the Government to introduce a National Security Bill, concerns about specific provisions were expressed, and many demanded draft legislation in the form of a 'White Bill' to allow for more public discussion of the proposed legislation. On 15 December 2002 60,000 people demonstrated against the proposals. In January 2003 the SAR Government announced a series of amendments, but rejected the demand for a 'White Bill'. The draft National Security Bill was tabled on 14 February and had its first reading in the Legislative Council on the following day. After discussion in the Bills Committee, the Government announced further amendments to the Bill in June, but those who opposed the proposals still believed that the amendments did not address major issues concerning the protection of fundamental rights and freedoms. The annual gathering to commemorate the anniversary of the Tiananmen Square tragedy of 4 June 1989 turned into a demonstration against the Government's proposals. When the Government rejected demands to delay the introduction of the Bill for a second reading in the Legislative Council, on 1 July 2003 about 500,000 people took to the streets of Hong Kong in a demonstration against Article 23. This developed into an anti-Government protest, demonstrators demanding the resignation of the Chief Executive and of a number of principal officials. Among the officials targeted were Secretary for Security Regina Ip, who had been responsible for the legislative work of the draft National Security Bill, and Financial Secretary Anthony Leung, who had purchased a luxury car prior to introducing new taxes for such vehicles in his budget in March. The protesters also objected to the appointment of the Secretary for Health,Welfare and Food, Yeoh Eng-kiong, as chair of the independent inquiry on the SARS outbreak in Hong Kong. (Yeoh resigned in July 2004—see below.) The size of the demonstration and the intensity of the anger of the protesters surprised the Government and many observers. The demonstration created a political crisis in Hong Kong, and the popularity of the Chief Executive and his team of principal officials declined to a new low point, as shown in a public opinion poll conducted by the University of Hong Kong.

In early July 2003 Tung Chee-hwa continued to refuse to delay the introduction of the Bill to the Legislative Council but agreed to address specific concerns by deleting provisions regarding the possible proscription of Hong Kong organizations subordinate to those proscribed in the mainland on national security grounds, as well as those regarding additional emergency investigation powers for the police and the introduction of a 'public interest' defence for disclosure of certain official information. The Chief Executive reversed his position when James Tien, Chairman of the Liberal Party (part of the 'ruling coalition') resigned from the Executive Council, thus depriving the Government of the number of votes necessary in the Legislative Council for the approval of the Bill. The political crisis took another dramatic turn on the evening of 16 July when Tung Chee-hwa announced first that Secretary for Security Regina Ip had resigned, citing personal reasons, and two hours later that Financial Secretary Antony Leung had also resigned. The Chief Executive pledged that he would reopen consultation on Article 23, improve dialogue with the public, remove Secretary Yeoh from the chair of the investigation panel on the outbreak of SARS and improve Hong Kong's economic performance. Later that month Tung Chee-hwa met senior Chinese leaders, including President Hu Jintao and Premier Wen Jiabiao, during a visit to Beijing. (The new Chinese leaders had assumed power in March 2003.) Premier Wen Jiabao had recently returned from a successful visit to Hong Kong, where he had reaffirmed the principle of 'one country, two systems', leaving just hours prior to the demonstrations of 1 July. The Chinese leaders confirmed their support for Tung Chee-hwa, following his earlier indication that he had no intention of resigning. In early September, however, it was announced that the National Security Bill had been withdrawn, pending further public consultations.

In response to public pressure for the introduction of universal suffrage at the election for the Chief Executive in 2007 and for the members of the Legislative Council in 2008, the Government suggested that a timetable for public consultations would be announced by the end of 2003. In November pro-democracy candidates gained new seats in Hong Kong's second post-1997 District Council election at the expense of candidates from the pro-Beijing DAB. The election attracted a record turn-out of 1m. voters. The Chief Executive, however, subsequently exercised his power to appoint an additional 102 District Councillors.

During an official visit to Beijing in December 2003, Chief Executive Tung met President Hu Jintao and Premier Wen Jiabao. President Hu declared that Hong Kong's political system should be developed gradually, in accordance with the Basic Law and the 'actual situation' in Hong Kong. Xinhua, the official Chinese news agency, also reported on the views of four mainland legal experts, who suggested that changes to the electoral arrangements in Hong Kong required approval by the NPC and should not affect China's national interests. The legal experts also stated that Hong Kong's social stability and economic development should be maintained. The US Government issued a statement supporting the demand for direct elections in Hong Kong. The statement was criticized by the Chinese Government as interference in China's internal affairs.

The political atmosphere remained tense in 2004. The year began with a demonstration of 100,000 people demanding political reform. On 7 January, in his Policy Address, the Chief Executive announced the establishment of a three-member Task Force on Constitutional Development headed by the Chief Secretary, with the support of the Secretary for Constitutional Affairs and the Secretary for Justice. The aims of the Task Force were to examine the 'relevant principles and legislative process in the Basic Law' relating to constitutional development, to consult the central Government and to take account of the views of Hong Kong people. On the same day the Hong Kong and Macao Affairs Office in Beijing issued a statement expressing serious concerns over constitutional developments in Hong Kong and requesting that the Beijing Government be consulted.

In addition to conducting public consultations on constitutional reform in Hong Kong, the Task Force visited Beijing in February 2004 to confer with the central Government. After the departure of the Task Force, the Xinhua news agency published a series of commentaries on the principles governing the Basic Law and suggested that only 'patriots' should comprise the main body of the Government. The commentaries generated intense debates in Hong Kong when some pro-democracy figures were accused of being 'unpatrotic'. Chinese officials were particularly critical of the visit by former Democratic Party chairman Martin Lee to the USA, where he testified at a Congressional hearing on Hong Kong and met the US Secretary of State.

In March 2004 the central Government announced that the Standing Committee of the NPC would provide interpretations of Annexes I and II of the Basic Law on the methods for selecting the Chief Executive and Legislative Councillors. In April the Standing Committee of the NPC blocked movement towards faster democratization in Hong Kong with the ruling that there should not be universal suffrage for the election of the Chief Executive nor for all 60 members of the Legislative Council. It also ruled that the number of functional seats for the Legislative Council should remain unchanged at 30 and that the procedure for voting on the motions and bills in the Council should not be altered.

The general political mood in Hong Kong continued to be turbulent. In May 2004 the Government had to defend in public its commitment to freedom of speech in Hong Kong when two popular radio talk-show hosts, Albert Cheng and Wong Yuk-man, resigned, claiming that they were being placed under political pressure. Cheng's replacement, Allan Lee, a Hong Kong deputy to the NPC in Beijing, who also resigned, subsequently revealed to the Legislative Council that mainland officials had contacted him about the programme. The departure of the three high-profile radio presenters raised concerns that freedom of speech in Hong Kong was being threatened.

Although both Beijing and the pan-democratic camp seemed interested in adopting a more conciliatory stance, a large number of people demonstrated again on 1 July 2004, demanding more democracy for Hong Kong. The organizers put the number of demonstrators at 530,000, whereas the police suggested 200,000. The demonstration confirmed the people of Hong Kong's aspiration for faster democratization and a willingness to express that aspiration in public.

In July 2004 the Tung administration lost another member of the Executive Council, the third minister to resign since the introduction of the 'ministerial accountability system'. Following the release of a Legislative Council Select Committee report on the Government's management of the SARS outbreak, which criticized a number of senior government officials, Dr Yeoh Eng-kiong, the Secretary for Health, Welfare and Food, tendered his resignation on 7 July. While the Chief Executive insisted that there was no evidence that any officials were derelict in their duties, he accepted Dr Yeoh's decision to take political responsibility and 'bring closure' to the SARS episode in Hong Kong.

The forthcoming Legislative Council elections were expected to be the most critical for the citizens' political aspiration for greater democracy. Many observers anticipated that the pan-democratic camp would gain more seats, partly because the six electoral committee seats were to be replaced by directly elective seats, and partly because the Hong Kong population seemed to have become far more politicized and willing to fight for its democratic rights. The electoral arrangements, however, incorporating a system of proportional representation and the 30 functional seats, were likely to make the achievement of a simple majority very difficult for the pan-democracy camp. The democrats hoped to progress towards winning almost one-half of the Legislative Council's 60 seats, whereupon they would be able to exert greater influence over public policy and make a more significant impact on Hong Kong's political development.

In the event, at the elections of 12 September 2004, a record number of more than 1.78m. voters participated in the poll, resulting in the highest ever turn-out, of 55.63%. While popular support for the pan-democracy camp remained high, at 60%, the grouping won only 25 of the 60 seats on the Legislative Council. The pro-Government camp secured a total of 33 seats. The pro-Beijing/government parties, namely the DAB with 12 seats and the Liberal Party with 10 seats, became the largest party and second largest party respectively in the legislature. The Liberal Party, known also for its pro-business stance, succeeded in winning two directly elected seats in the geographical constituencies. However, the Hong Kong Progressive Alliance, a pro-Government party of which the members were elected by the Election Committee, failed to secure any seats following the abolition of the Election Committee Constituency. The pro-democracy camp secured a modest 25 seats, gaining ground not only in the geographical seats but also in two functional constituencies, representing medical and accountancy interests. However, within the pro-democratic camp, the Democratic Party of Hong Kong failed to maintain its position as the largest party in the chamber. The Concern for Article 45 group (consisting of lawyers who had come to the fore during the anti-Article 23 campaign) emerged as a significant political force. Several outspoken anti-Government critics, including Leung Kwok-hung, an anti-Beijing radical from April 4 Action (widely known as Long Hair because of his hair style), and the radio talk-show host Albert Cheng also won seats. The political balance remained tilted towards the Government, which would nevertheless be obliged to deal with a Legislative Council that incorporated a wider spectrum of moderate and radical pro-democracy members. Furthermore, some of the political parties and groups might possibly form ad hoc alliances against the Government on critical issues. In an effort to attract support, the central Government invited a number of newly elected members of the Legislative Council, including moderates from the pan-democracy camp, to Beijing to attend National Day celebratory activities.

Economy

FRANÇOIS GIPOULOUX

ECONOMIC DEVELOPMENT

After recovering quickly from the financial crisis that affected Asia from mid-1997 to 1998, the problems facing the Special Administrative Region (SAR) of Hong Kong from mid-2003 were largely related to the downturn in the world economy and to the outbreak of Severe Acute Respiratory Syndrome (SARS) in that year. The average annual growth rate of the territory's gross domestic product (GDP) between 1996 and 2001 was 2.5%, half the rate achieved between 1990 and 1996. Taking into account the price deflation from 1999, the value of Hong Kong's GDP at the end of 2002 was below the level of 1997. Hong Kong remains, however, a major international commercial centre, being one of the largest trading entities in goods and services and among the most vibrant financial centres in the world. Hong Kong was the second largest source of outward foreign direct investment (FDI) in Asia and the 10th in the world in 2002. Hong Kong continued to be classified as the world's freest economy by the US Heritage Foundation 2004 Index of Economic Freedom. Hong Kong also enjoyed the second highest per caput holding of foreign currency. Compared with the previous year, GDP was officially forecast to increase by 6% in 2004, following a 3.3% increase in 2003. According to preliminary figures, per caput GDP was estimated at HK $181,527 in 2003 and projected at HK $184,500 for 2004. Hong Kong is thus placed among the highest group of world economic rankings.

Under the 'one country, two systems' concept, Hong Kong has so far charted its own course, with the exception of foreign affairs and defence. The SAR's 'mini-constitution' (the Basic Law) guarantees that from 1997 the capitalist system and way of life in Hong Kong will remain unchanged for 50 years. Hong Kong was promised a high degree of autonomy and continues, among other responsibilities, to manage its own economic policies and finances. It issues its own currency, enjoys a low and simple tax regime, maintains its own laws and common-law legal system, employs its own civil servants and remains a separate customs territory. In 2004, however, seven years after Hong Kong's return to Chinese sovereignty, concerns persisted in relation to China's intervention in the functioning of the SAR's independent judiciary, increasing self-censorship of the press, and interference in public opinion research that was critical of the Government.

The Asian financial crisis of 1997/98 provoked a downturn in real estate prices. Office and residential rental prices declined by 65% and 55% respectively from the peak reached in the third quarter of 1997. The regional financial turmoil brought about a decline in the Hong Kong stock and property markets, and a sharp rise in unemployment. The depressed economic conditions in most countries of the region in the late 1990s affected intra-regional trade. The subdued sentiment, risk sensitivity and financial stringency, which hampered consumption, investment and normal business operations, compounded the uncertainties. Although the SARS outbreak had a negative impact on Hong Kong's development in 2003, the economic consequences were mainly restricted to the aviation, tourism and hotel sectors. The retail industry suffered to a lesser degree from the repercussions of the SARS epidemic. Exports were affected, visits to overseas markets by Hong Kong business representatives became less convenient, and overseas buyers cancelled or postponed their trips to Hong Kong and the neighbouring region.

Hong Kong's overall economy can be characterized as 'strong externally, weak internally'. Continued economic growth in China also contributed to Hong Kong's external dynamism. China's entry into the World Trade Organization (WTO) in December 2001 brought an additional stimulus to trade and foreign investment in China. Moreover, Hong Kong took advantage of growing exchanges between China and the other countries of East Asia, and in 2003 the SAR recorded growth of 14% in its total exports of goods (including re-exports and domestic exports), a further increase from the already notable growth of 8.6% in 2002. Exports of services maintained an appreciable progression in 2003 (rising by 5.5% throughout the period of recession). On the other hand, various structural factors restrained Hong Kong's economic growth in 2003: rising unemployment, declining income, downward pressures on property prices and deeper deflation. The internal weakness was a consequence of depressed internal demand. Private consumption decreased by 1.2% in 2002 and showed virtually no change in 2003, while investment declined by 4.4% in 2002 and by 0.1% in 2003. Inflation, which in 1997 had stood at an average annual rate of 5.8% and was threatening to become a major problem, was replaced by deflation in the late 1990s. Compared with its peak in 1998, by 2003 the consumer price index had fallen by 14%. It continues to decline, by 1.7%, in the first four months of

2004. Among the factors responsible, the more significant were the decrease in import prices in Hong Kong terms, a collapse of 60% in property prices (accounting for half of the overall decline in the index), output levels running below potential, and a downward pressure from across the border with the mainland (as a result of the increasing mobility of people and goods). The rate of petitions for personal bankruptcy decreased by 43% between the first quarter of 2003 and first quarter of 2004, but remained high, at 7.1%, compared with 2.5% before the SARS crisis. However, the economy as a whole displayed the profound flexibility that was required for it to withstand the ensuing difficulties. Furthermore, the enormous growth potential of the mainland was expected to instil confidence in the longer-term outlook for Hong Kong.

From the early 1980s the Hong Kong economy underwent considerable structural change. The contribution of the local manufacturing sector has declined significantly as a percentage of GDP, from almost 24% in 1980 to 19% in 1989 and 4.6% in 2002. There has been a corresponding growth in the services sector, from 68% of GDP in 1980 to 73% in 1989 and to 87.4% in 2002. Within the latter sector are the categories of finance, insurance, real estate and business services, which are indicative of the sophistication of a metropolitan economy. The contribution to GDP of this cluster of services more than doubled in value between 1985 and 1997, while the number of jobs provided by the services sector tripled. Hong Kong's predominance as a service-based economy is reflected in its communications network, one of the densest in the world. In turn, this shift to services requires measures to maintain and increase Hong Kong's international competitiveness in the sector, including training, language competence and the availability of reasonably-priced commercial office space. However, these figures can be misleading, and do not mean that Hong Kong does not depend significantly on manufacturing activities. For example, thousands of companies have switched some or all of their production to the mainland, and have been officially reclassified as trading rather than manufacturing entities, even though they may be local producers and their major activity remains the organization of production.

The four pillars of Hong Kong's economy (trade and logistics, financial services, business services—including computer, legal, accounting, auditing, architecture, design and security services—and tourism) accounted for 53.3% of GDP in 2002, compared with a contribution of 51.7% in 1995. They represent the most dynamic part of the economy, particularly since the 1997 Asian crisis. Labour productivity in those sectors is superior to the rest of the economy. Consequently, their share in employment is much smaller (44% in 2002) than their contribution to GDP. Hong Kong is a co-ordinating and service centre for an offshore manufacturing sector employing more than 12m. people all over the world. There is a broad consensus regarding the factors underlying Hong Kong's economic performance. For more than 40 years the colonial Government maintained reasonable stability of political institutions and supported a market regime that allowed for great flexibility of resource use and provided strong incentives to wealth acquisition, in the form of low income taxes and no taxes on capital gains. In addition, the world economy provided a favourable environment for growth through trade. The manufacturing base in Hong Kong was developed following the communists' assumption of power in China in 1949. Having been expelled from the mainland, the capitalists of Shanghai transferred their expertise in labour-intensive manufacturing, their commercial networks and some of their capital resources to Hong Kong. Their skills, combined with the availability of cheap labour, the entrepreneurial character of these refugees and the *laissez-faire* environment of the territory, produced impressive industrial results, propelling Hong Kong to a leading position in the labour-intensive manufacture of products such as textiles, and subsequently plastic flowers, rattan furniture, watches and clocks, electronics and precision machinery. By the early 1980s the manufacturing labour force that supported Hong Kong's production and export performance totalled almost 1m. workers.

During the post-War era from 1945, the Hong Kong Government adopted a policy of minimal interference, allowing market forces to prevail. The tax system is simple, and rates of taxation are low. Owing to a system of generous allowances, about 60% of the work-force pay no tax at all on their salaries. Furthermore, there are no taxes on capital gains, dividends or interest. The Government's policy of low taxation and prudent fiscal management enabled it, generally, to achieve surpluses on its consolidated account during the 1990s. Nevertheless, the Government plays a significant role in infrastructural development, healthcare services, public housing and land sales. Policy intervention was prompted mainly by strong popular pressure. For example, the post-War population inflow, and the huge shanty towns that it created, ultimately led to the establishment of a public housing programme. By 1982 40% of the population had been housed under the programme. Similarly, riots in 1966 led to an expansion of social services, such as a public assistance scheme covering the elderly, and the passage of employment legislation restricting child labour and mandating four rest days per month. From the 1990s the Government also became more pro-active in its support of high-technology development. It encourages applied research and development, and has plans to establish a science park.

Hong Kong is now a major regional centre for business in Asia. The total number of regional headquarters rose from 948 in 2002 to 966 in 2003. The USA headed the list of countries with the largest number of regional headquarters in Hong Kong, with a total of 233 companies. This was followed by Japan with 159 companies, and China with 96 companies. Among the most often quoted reasons for a foreign installation in Hong Kong are the tax system, the freedom of information, political stability, low level of corruption and rule of law. However, recent surveys also show that more than half of foreign establishments consider that the business environment has deteriorated in Hong Kong. Although office and residential rental prices have declined, Causeway Bay ranks third in the world for rental cost, after New York and Paris.

POPULATION AND WORK-FORCE

The quality of Hong Kong's human resources has been an important factor in economic growth. In mid-2003, the population of the SAR totalled 6.8m., representing an increase of 0.3m., compared with mid-1997. Ageing of the population—the average age was 32.2 years in 1992 and 38.1 in 2003—and decline in fertility has continued. In recent projections, Hong Kong's population was expected to increase at an even lower average annual rate, of 0.9%, to reach 8.72m. in mid-2031. According to the Census and Statistics Department, migration would be the main source of growth in population, accounting for 93% during the 2002–31 period. Natural increase of the population would account for 7% during the same period. With a land area of less than 1,100 sq km, Hong Kong is one of the most densely populated places in the world, the population density reaching 6,124.5 people per sq km in mid-2001, rising to 6,167.8 in mid-2003. The highest density is found in Kowloon (43,201 inhabitants per sq km in 2001), while Hong Kong Island's density reached 16,635 persons per sq km and that of the New Territories and outlying islands 3,443. The proportion of Hong Kong's population living in Kowloon decreased from 32% in 1996 to 30.2% in 2001, while the proportion living in the New Territories increased from 46.8% to almost 50%, a reflection of the Government's policy of developing new towns in the New Territories. The proportion of the population living on Hong Kong Island decreased from 21.1% to 19.9% during the same period. The annual population growth decreased from 3.3% in the 1960s, to 2.2% in the 1970s and 2.0% in the 1980s. The fall in population growth is attributed to a decline in the birth rate and to net migration. During the period 1990–2000 annual population growth averaged 1.8%. In 1986 the crude birth rate was 13.0 per 1,000 people; the rate declined to an estimated 7.1 per 1,000 in 2002. The population will continue to follow a continuous ageing trend. In 2003 15.7% of the population was aged under 15. This rate was expected to decrease to 12% by 2031. The proportion of those aged 65 and over was projected to rise significantly, from 11% in 2001 to 24% in 2031. Population by ethnicity (in October 2001) was predominantly Chinese (94.9%), with 2.1% Filipinos, 0.8% Indonesians, 0.3% Indians, 0.3% British, 0.2% Japanese, 0.2% Thais, 0.2% Nepalese, 0.2% Pakistani and 0.9% from other nationalities.

The population of Hong Kong is one of the most well-educated in the region. Moreover, the general educational level of the population improved markedly during 1990–2000. In 2003 the total labour force was 3.5m. As the sectoral composition of the economy and the skill level of the population have changed, so has the occupational distribution of employment. The decline of the manufacturing sector has been dramatic: in 1990 it accounted for 28% of total employment, but by 2003 a mere 8.5% of those employed were engaged in the sector. The expansion of the services sector, meanwhile, has been impressive. There is no legal minimum wage in Hong Kong. In March 2003 the average wage rate (including allowances and bonuses) for a computer systems analyst/programmer was HK $20,667 per month (compared with HK $21,798 in March 2001). An accounting supervisor was earning an average of HK $18,128 per month, while a receptionist or a sales clerk was earning HK $9,755 and HK $8,671 respectively. Hong Kong has always had one of the most fluid labour markets in the world. This is particularly true of the highly skilled work-force, which has formed a large proportion of the immigrants to and emigrants from Hong Kong. The right to emigrate is guaranteed in the Basic Law. From 1992 (when there were 66,200 emigrants) to 2001 (10,600 emigrants), there was a downward trend in emigration.

By international standards, Hong Kong's unemployment rate has traditionally been low, but the situation deteriorated in the late 1990s. As unemployment was unable to match the growth in labour supply amidst more widespread corporate downsizing and lay-offs, the unemployment rate rose sharply. The unemployment rate was at a historic high of 8.3% in May–July 2003. The average unemployment rate in 2003 as a whole stood at 7.9%, appreciably above the rate of 7.3% in 2002.This increase has been mainly caused by a conjunctural unemployment provoked by a strong deceleration of growth. Moreover, structural unemployment has been the result of the difficulties experienced by unqualified young people when trying to enter the labour market. In the 15–19-years-old age group, unemployment is very high, at 30.2%, compared with 8.8% for the 20–29-years age group. Finally, at 10.9%, unemployment is much higher for persons without a high educational level. Structural unemployment seems therefore partly related to the shortcomings of the educational system. The underemployment rate also rose appreciably, soaring to a peak of 4.3% in the second quarter of 2003. This was the result of a significant proportion of employees having been temporarily suspended from work during the SARS period. Unemployment and underemployment are found mainly among the semi-skilled, such as production-related workers. Also noteworthy was the relative decline of Hong Kong's entrepôt function and its growing importance as a major financial and business services hub. The wholesale and import-export trade lost 36,000 jobs between 1996 and 2002, while the finance, insurance, real estate and business services sectors created almost 33,000 jobs. Finally, the household service sector has been the main job provider for private employment: 146,000 jobs were created over five years, mainly in education, health and social services.

INDUSTRY

Hong Kong has long had a reputation for being a producer and exporter of manufactured goods. Although declining, manufacturing continues to be an important sector of Hong Kong's economy. Mechanization, automation, and the relocation of labour-intensive and lower value-added manufacturing processes to mainland China have contributed to the decline in manufacturing employment. This has facilitated Hong Kong's development of more knowledge-based and higher value-added manufacturing. Thus, manufacturing productivity (gross output per employee) significantly increased (by more than 400%) between 1983 and 1999, with a further 2% increase in 2000.

The opening of mainland China, combined with low labour and land costs there, significantly altered the industrial landscape in Hong Kong. Local manufacturers relocated most of their productive capacities across the border through subcontract processing arrangements in mainland China. Therefore, the number of establishments and persons engaged in the local manufacturing sector decreased significantly between 1990 and 2002, from 565,000 to 185,000, while the average number of employees per establishment also declined. This decline has been offset, however, by a strong rise in labour productivity. Over the period 1990–99 the average annual growth of the labour productivity index for the whole manufacturing sector was 7.1%. Among selected industry groups like electrical and electronic products, this ratio reached 16%. It is worth noting, moreover, that most of the firms that relocated their manufacturing processes to mainland China have continued to conduct export operations in Hong Kong as import-export firms. In doing so, they facilitate the import of goods produced by their associate manufacturing firms in the mainland for subsequent re-export to overseas markets. While those firms are now registered in the services sector, they often provide manufacturing-related technical support services (e.g. product design, sample and mould working, production planning, quality control) to the production activities in the mainland. The frontiers between manufacturing and non-manufacturing activities are thus becoming less distinct. It may be appropriate to take into account the contribution of trading firms with manufacturing-related activities in assessing the share of manufacturing in the economy. A survey of 123,000 industrial and commercial establishments registered in Hong Kong, conducted by the Federation of Hong Kong Industries, found that 11m. mainland Chinese workers were employed by Hong Kong companies in 2001. Of these, 10m. were working in Guangdong Province, in southern China. While 700,000 jobs in industry had disappeared since 1980, Hong Kong companies had created 12m. jobs abroad (half directly and half through sub-contracting arrangements).

The textile industry, which has played a pivotal role in Hong Kong's development, faces challenges from continued global economic restructuring and the accession of China to the World Trade Organization (WTO) in December 2001. With all quotas being abolished in 2004, China's removal of quota restrictions will allow Hong Kong manufacturers to market their mainland-origin products freely. Therefore, the relocation of Hong Kong manufacturers to the mainland to take advantage of its lower cost base will continue. Re-exports of apparel, clothing, textile yarn and fabrics, accounting for more than 20% of total Hong Kong re-exports in 1992, decreased to 14% in 2002. Among the 10 leading destinations of textile exports, nine were within Asia, with China in first place.

The electronics industry exhibits a particular strength in the manufacture of consumer electronics (especially audio and video equipment, information technology and multi-media products). Although other Asian manufacturers pose an increasing threat to that sector, Hong Kong's electronics industry is the largest merchandise export earner, which accounted for 42% of Hong Kong's total exports in 2003. In 2001, Hong Kong was the largest world exporter of calculators, and the second largest exporter of radios and telephone sets in value terms. Hong Kong's advantage comes from efficient low-cost manufacturing, consumer product trend identification and aesthetic design capabilities. As electronic companies are moving into more technology-intensive products, the shift from low-profit-margin, mass-produced, labour-intensive products to more capital items and equipment products has sharpened the competition with other Asian suppliers. Owing to the rapid expansion in internet applications and sustained demand for telecommunication services, the market for related equipment is expected to be strong, and more intense competition from mainland suppliers will affect Hong Kong's exports to this market.

Taken together with re-exports, Hong Kong is the world's largest toy exporter. The SAR produces a wide range of these products, with a particular strength in plastic toys. Hong Kong is also a major exporter of other plastic products. Additionally, there are numerous local end-users, including plastic factories producing toys, packaging materials, shoes, housewares and casings of electrical and electronic consumer goods. Despite the challenges coming from mainland Chinese manufacturers, Hong Kong has remained one of the world's principal manufacturers of watches and clocks. In 2001 the SAR was the second largest exporter of complete clocks (after China) and the second largest exporter of complete watches (after Switzerland).

TRADE AND INVESTMENT

The strength of Hong Kong lies in its efficient management of international business. Strong global and intra-regional demand, as well as a weak US dollar, has underpinned the good performance of external trade. In 2003 Hong Kong was the world's 11th largest trading economy. While China's opening up and implementation of economic reforms might have been expected to lead to a multiplication of channels for import and export all over the mainland, at least in coastal areas, and to a subsequent decline of the relative position of Hong Kong in China's trade, in fact the reverse occurred. The proportion of China's trade sent to or through Hong Kong reached 25% in 2003, compared with 11% in 1978. In 2003 60% of re-exports originated in China and 44% were destined for the Chinese mainland. The reasons for this situation are linked to Hong Kong's remarkable efficiency in providing a range of highly sophisticated services and to the economy of scale and of scope inherent in an urban centre. In other words, China depends on Hong Kong for numerous services in the value-added chain, such as order-processing, financing, sourcing, production management, product design, quality control, marketing and shipping.

Hong Kong's total exports of goods (comprising both re-exports and domestic exports) increased by 14% in 2003 after an already remarkable growth of 8.6% in 2002. Among total exports of goods, re-exports continued to provide the principal impetus of export performance, recording a growth rate of 16.1% in 2003. Reflecting the ongoing shift towards re-exports and offshore trade, domestic exports continued to contract, declining by 7.4% in 2003, compared with the 11.3% decrease recorded in 2002. As a facilitator in global sourcing and export manufacturing, Hong Kong has played a major role in intra-Asian trade since the late 1990s. In 2003 Hong Kong's total exports to Asia recovered further to increase by 20.9%, compared with a 13.4% growth rate in 2002. Total exports to North America declined by 1.4% in 2003, having moderately recovered by 3.9% in 2002. Exports to the countries of the European Union (EU) surged by 14.1% in 2003, in contrast to the 0.4% decline in 2002. The SAR's re-export growth was partly sustained by the outward processing activities in Guangdong Province. Raw materials and semi-manufactures are exported to the mainland for processing, and the final products are subsequently returned to Hong Kong before being exported to overseas markets. In 2001 48% of Hong Kong's total exports to the mainland were destined for outward processing. On the other hand, 78% of Hong Kong imports from the mainland were related to outward processing.

The substantial contribution of re-exports reflects Hong Kong's role as an entrepôt. Re-exports used to be important in the pre-War and 1950s economy, but in the 1960s and 1970s their significance faded quite markedly. Revival began as links with China grew, and by the late 1980s re-exports drew level again with domestic exports. In the late 1990s, while domestic exports were stagnant or decreasing, re-exports were still increasing dynamically, providing the momentum of growth for the Hong Kong economy. The re-exports sector attained double-digit growth, of 10.9%, in 2002 followed by a distinct surge of 16.1% in 2003. Mainland China remained the largest market for Hong Kong's re-exports, accounting for 43.5% of its total re-exports in 2003.Other major markets included the USA (17.6%), Japan (5.6%), Germany (3.2%) and the United Kingdom (3.1%). Hong Kong's crucial role as a regional business and service hub was reflected by the predominant share of Asia in the territory's export of services in 2002. The value of total exports of services was severely affected by the spread of SARS in the second quarter. It rose by 5.3% in 2003, compared with the 12.7% surge recorded in 2002.

Foreign Investment

According to the UNCTAD World Investment Report, Hong Kong is the second largest recipient of inward FDI in Asia and the 15th largest in the world. However, FDI inflows to Hong Kong decreased from US $23,800m. in 2001 to US $13,700m. in 2002, while FDI outflows from Hong Kong grew from US $11,300m. in 2001 to US $17,800m. in 2002. Although Hong Kong remains the main investor in the mainland, its relative share is declining. Excluding tax haven economies (British Virgin Islands, Bermuda and Cayman Islands), mainland China is the principal investor in Hong Kong (22.6% of the total in 2002), followed by the Netherlands (7.8%), the USA (7.1%) and Japan (5.4%).

Hong Kong's total stock of inward investment was estimated at HK $3,269,000m. at the end of 2001, equivalent to 256% of GDP. Excluding 'tax haven' economies (British Virgin Islands, Bermuda and Cayman Islands, which accounted for 42.3% of the total stock), China appeared to be the main investor in Hong Kong (accounting for 29.3% of the total). In 2001–02 there were about 2,000 mainland-related enterprises operating in Hong Kong. Mainland-related enterprises are prominent in a wide range of activities, including property, car and life insurance, transport, finance and construction. In terms of manufacturing, mainland investment was largelyly concentrated in two industries—transport equipment and food and beverages.

Several major mainland-related companies have taken up equity stakes in some major infrastructure ventures in Hong Kong, including the Western Harbour Crossing, the Tate's Cairn Tunnel, Kwai Chung Container Terminals 4, 6, 7, and 8, the Northwest New Territories Landfill Project and the Tuen Mun River Trade Terminal. Moreover, Hong Kong is still the major source of overseas direct investment in mainland China, with 224,509 overseas projects funded (48.3% of the total) at the end of 2003. The stock of utilized capital inflow from Hong Kong (US $222.6m.) accounted for 44% of the national total.

TOURISM

Tourism is a major source of foreign-exchange earnings and employment in Hong Kong. By the early 21st century the tourism industry accounted for more than 20% of Hong Kong's exports of services. Hong Kong is an important transit destination in Asia. One factor in the long-term growth of the tourism industry is that the territory has been increasingly successful in attracting international conventions. The number of visitor arrivals in Hong Kong increased at an annual average rate of 7.1% between 1990 and 2000. With 15.5m. arrivals in 2003, the SAR registered a significant decrease (of 6.2%) in the number of visitors compared with 2002. However, tourist arrivals recovered by 5.8% in the second half of 2003 and by 14.7% in the first quarter of 2004.The Individual Visit Scheme, an arrangement to allow residents of Guangdong to make up to two visits to Hong Kong within a period of three months and to stay in Hong Kong each time for up to seven days, was launched on July 2003. The Scheme was extended to residents of Beijing and Shanghai in September 2003. In 2002 each tourist from mainland China spent an average of HK $5,000 per trip. However, tourism's contribution to GDP is marginal (3% in 2002).

INFRASTRUCTURE AND TRANSPORT

A modern and efficient infrastructure has supported Hong Kong's role as a trading entrepôt and regional financial and services centre. The SAR has the best natural deep-water port on the Chinese coast. For much of the 1990s and in the early 21st century Hong Kong was the busiest container port in the world, maintaining a narrow lead over Singapore in most years. In 2003 Hong Kong's eight privately operated container terminals and mid-stream operators handled 20.4m. 20-foot equivalent units (TEUs) of cargo. One vessel is arriving or departing every 1.2 minutes. The Port Development Board expects container throughput to reach 32.8m. TEUs by the year 2016. This will require additional container terminals. The current eight terminals have 19 berths. The ninth container terminal (CT9), under construction on Tsing Yi Island, was scheduled for completion in late 2004.

The container traffic at Hong Kong consists of three distinct types. The largest category is direct shipment of containers carrying imports and exports to and from China (mainly Guangdong Province). A second category is transhipment of containers for other countries, and the final category is river transport, which is, strictly speaking, inland transport, in the same category as road and rail transport. The growth of direct traffic averaged 4.0% annually in the period 1995–99. The importance of transhipment declined over that period, contracting by an annual average of 0.4%, while river trade experienced a strong growth, averaging 25% during the same period.

Hong Kong faces increasing competition from neighbouring ports, with growing volumes of traffic hitherto transhipped to Hong Kong being rerouted to Shenzhen and other ports on the East China coast. Moreover, the eventual implementation of direct transport links between mainland China and Taiwan will lead to a decrease in Hong Kong's share of this trade (1m. TEUs or 6% of Hong Kong's total container throughput). In terms of basic monetary cost comparisons, Hong Kong is adversely affected by higher terminal handling charges than the Shenzhen ports and by the relatively high haulage costs to and from mainland China. However, Hong Kong enjoys several advantages in terms of non-monetary cost competitiveness. It benefits from a high port productivity, and has a higher frequency of calling (more than 80 international shipping lines with 400 container line services per week to over 500 destinations, and over 310 daily feeder services linking the Pearl River Delta area). Furthermore, Hong Kong enjoys well-developed and efficient logistics services allied to straightforward and transparent customs. This is combined with world-class banking and financial institutions and experience in international trade practices, securing timely payments and document processing.

Despite start-up problems during the first few weeks of operations, particularly for cargo movements, Hong Kong's new international airport at Chek Lap Kok, which was financed by the private sector and opened in July 1998, subsequently settled into a routine of efficiency. By 2001 Chek Lap Kok was the busiest international air cargo terminal in the world. The second runway, which began to operate in May 1999, brought the flight capacity from 37 to 45 movements per hour. About 65 international airlines operate some 3,800 flights weekly to some 130 destinations. Chek Lap Kok airport serves 40 destinations in China through 600 flights a week. In 2004 the Hong Kong Airport Authority was considering both privatization and external growth for further development. In order to strengthen its position in relation to its Pearl River Delta competitors, Chek Lap Kok airport could either participate in Shenzhen airport (which handled 10.84m. passengers in 2003) or buy the Government's HK $6,000m. share. The marine cargo terminal in the airport came into operation in March 2001. With 24-hour operations, two all-weather runways, an ability to handle all types of commercial aircraft, high-speed transport links from the terminal to the city, and the largest cargo-handling facility in the world, the new airport appeared to be well positioned to fulfil Hong Kong's aviation needs in the early part of the 21st century.

The two principal air carriers of the SAR registered mitigated results for 2001. Cathay Pacific recorded a 5% decrease in its passenger activity and an 8.5% decline in cargo handling. Dragonair, however, achieved a 12% increase in passenger traffic and 30% in air cargo. Hong Kong Air Cargo Terminals (HACTL) was planning two new operation centres in southern China, at Huangpu (near Guangzhou) and Futian (near Shenzhen). About 80% of air cargo being dispatched from Hong Kong originates in southern China, in particular Guangdong Province, and these new centres were to handle the packaging and transport of cargo shipped to Hong Kong prior to re-export.

The privatization of the underground Mass Transit Railway (MTR) was approved in February 2000, when the network extended to 77 km, with five lines and 44 stations. Compared with the previous year, in 2002 the MTR recorded a 2.32% increase in the number of passengers transported (which reached 777.17m.). This rise was the consequence of the opening in August 2002 of the new Tseung Kwan O line, linking south-east Kowloon to Victoria island. The number of Airport Express passengers, however, decreased by 6.75% in 2002, to 8.45m., owing to the closure of downtown check-in counters for flights to the USA, following the terrorist attacks there in September 2001. Complementing the MTR are the modernized Kowloon–Canton (Guangzhou) Railway, which in 1999 handled more than 738,000 passenger journeys daily, and the Light Rail Transit system, which operates in the north-western New Territories and carried about 385,000 passengers each day. An ambitious transport infrastructure programme involves a total investment of more than HK $300,000m. on 12 railway projects and on road projects. The rail projects will add more than 60 km to the existing 143 km of railway lines in Hong Kong and open up the north-west and north-east New Territories for further development. Additional rail projects proposed in the Railway Development Strategy 2000 will further expand the railway network from 200 km to 250 km by 2016. The road projects will add about 50 km of new roads (including two bridges and six tunnels) to Hong Kong's transport network. These will help to alleviate congestion in urban areas and provide vital new links to the New Territories. Among the major infrastructure projects for 2004 are the building of a new bridge at the Sha Tan Kok frontier checkpoint. After formal approval by China's State Council in August 2003, the Hong Kong-Zhuhai Macao Bridge project continued with feasibility studies in 2004.

Hong Kong is the largest teleport in Asia, and the territory has ambitions to become a leading regional telecommunication centre and internet and broadcasting hub. Total international telephone traffic grew at an annualized rate of 6.9% between 1999 and 2003. Mobile phone suscription reached 7.3m.in number, more than the total population, while broadband internet traffic increased five-fold between 2002 and 2003. Satellite-based communications and television broadcasting services are provided via 36 satellite earth antennas. Hong Kong provides dedicated relay services for multinational companies, international press agencies and television channels to downlink or uplink their satellite signals over the Asia-Pacific region. Hong Kong is connected to nine submarine cable systems. Three overland systems have also been put into operation to service the growing traffic between Hong Kong and the mainland.

BANKING AND FINANCE

Hong Kong is one of the world's most important international banking centres. It has a particularly large representation of international banks. The Hong Kong Monetary Authority (HKMA) was established in 1993 by the merger of the Office of the Commissioner of Banking and the Office of the Exchange Fund. Hong Kong has a three-tier system of licensed banks, restricted licence banks and deposit-taking companies, collectively known as Authorized Institutions. Along with the increase in merger and acquisitions activities, the number of licensed banks has decreased slightly in recent years. At the end of 2002 there were 133 licensed banks, 46 restricted licensed banks and 45 deposit-taking companies in Hong Kong. A total of 94 foreign banks were represented. The high number of branches demonstrates the need for even the largest banks with world-wide operations to seek funds at the primary level; otherwise Hong Kong would surely be considered 'over-banked' at the retail level. This growth after 1978 reflected in part the Government's lifting of its 10-year moratorium on new banks in Hong Kong (there was one exception). In recent years the mainland has continued to increase its presence in Hong Kong's financial sector. Upon completion in 2001 of the merger of the Bank of China, (which started issuing Hong Kong dollar bank notes in May 1994) and 11 associate banks, the new entity was named the Bank of China (Hong Kong) and became the second largest banking group in the SAR after the Hongkong and Shanghai Banking Corporation. China's other three specialized banks, namely the People's Construction Bank of China, the Agricultural Bank of China, and the Industrial and Commercial Bank of China, obtained banking licences in 1995 and subsequently opened their first branch operations in Hong Kong. In June 1996 Shenzhen Development Bank, the only listed bank in the mainland, opened a representative office in Hong Kong. In 2003 there were 26 mainland-related authorized institutions operating in Hong Kong, along with seven representative offices. The Bank of China was, in 2003, the second largest bank in Hong Kong after the Hong Kong Bank Group.

In the early 21st century more than one-half of the assets and liabilities of the Hong Kong banks were foreign. This 'internationalization' of Hong Kong banking is an essential part of the strategy to underpin Hong Kong's political independence as an SAR with international economic ties. Hong Kong joined the Bank for International Settlements (BIS) in 1996. In mid-1997 the SAR retained its membership of the Asian Development Bank (ADB) and of Asia-Pacific Economic Co-operation (APEC). In July 1999 the Government announced a series of initiatives based on a commissioned consultancy to suggest a reform programme to improve competitiveness and enhance the security and soundness of Hong Kong's banking sector. Among the major initiatives are simplification of the three-tier system, relaxation of the one branch (to three) policy for foreign banks, deregulation of remaining interest rate rules, measures to enhance depositor protection and the adoption of more formalized risk-based supervision. A further easing of criteria for foreign banks to enter the Hong Kong market was approved in May 2002. The level of asset requirements was lowered from US $16,000m. to US $513m. The requirement to maintain a local representative office for a period of one–two years was also abolished. Furthermore, the full deregulation of interest rates was implemented in July 2001. Individual banks became free to determine their own interest rates on savings and current accounts. Finally, permission to conduct personal renminbi business has been granted to banks in Hong Kong. This covers deposit-taking, exchange, remittances and renminbi credit cards. This measure, the scope of which is expected to be broadened in the future, will enhance the position of Hong Kong as a key financial centre for mainland China.

The Stock Market and The Exchange Rate

After the 1973 share market speculative upsurge, more attention was paid to controlling the potential excesses of the financial sector. In August 1980 the Stock Exchange Unification Ordinance was enacted, which provided for the establishment of one exchange in place of the previous four. Under the ordinance, a unified Stock Exchange of Hong Kong began operations in April 1986, with the exclusive right to operate a market. All members of the existing exchanges were invited to apply for shares in the new exchange. This unification aimed to achieve a broader market, to increase the attractiveness of Hong Kong securities to overseas investors and to assist the better management of the market and more effective regulation of the stockbrokers. The overall supervisory body is the Securities and Futures Commission, which began work in 1989. During the regional stock-market turmoil of late 1997 the Government was obliged to intervene after the securities and futures markets in Hong Kong experienced the most severe volatility ever recorded. A 1,438-point fall (13.7%) in the Hang Seng Index was registered on 28 October and a 1,705-point rebound (18.8%) on the following day. In 2000 the Stock Exchange of Hong Kong, the Hong Kong Futures Exchange and the Hong Kong Securities Clearing Company merged to form Hong Kong Exchanges and Clearing Ltd. Hong Kong has also developed a thriving capital market. In terms of market capitalization, the Hong Kong stock market was the second largest in Asia, behind Japan, in 2003. As measured by the Hang Seng Index, the market underwent wide fluctuations during 2003, reflecting investor concerns about the US-led war in Iraq. The outbreak of SARS led to a further set-back. The Hang Seng Index plunged to a four-and a half-year low on 25 April, representing a 9.8% fall from its level at the end of 2002. At the end of February 2004 Hong Kong's stock-market capitalization was US $791,000m. with 1,043 companies listed. Over 100 mainland-backed enterprises are listed on the Hong Kong stock exchange and its growth enterprise market.

Linked to the pound sterling, then to the US dollar, the Hong Kong dollar floated freely between 1974 and 1983. During that period the Hong Kong dollar lost half of its value. In that context, therefore, the Hong Kong monetary authorities returned to the policy of a fixed exchange rate (pegging) to the US dollar. Since then, the parity has remained unchanged at HK $7.8 to US $1. The Hong Kong dollar is pegged to the US dollar through the linked exchange rate (LER) system. This pegging of the exchange rate came under severe attack during the financial turbulence of 1997, when the Hong Kong dollar was subjected to speculative pressure on several occasions. The overnight interbank interest rate surged briefly to 280% on 23 October 1997, under the automatic adjustment mechanism of the currency board system. The combination of the interest-rate rise and the asset price adjustments in both the securities and property sectors caused serious concern within the local community. In September 1999 the territory reaffirmed its commitment to the linked exchange rate mechanism, while other measures aimed to increase liquidity in the interbank money market. The currency board's system relies on three pillars: a fixed exchange rate; a monetary base covered at 100% at least by the currency reserves; and a money creation strictly subordinated to the variation of exchange reserves. Although the 'peg' has been denounced as a factor of rigidity by the Financial Secretary, and a *de facto* obstacle to economic recovery, the resistance of the 'peg' to speculative attacks during the Asian crisis gave the fixed exchange rate a strong credibility and a pivotal element in Hong Kong's autonomy within the framework of 'one country, two systems'.

Housing and the Property Market

About 741,000 residential flats were completed to meet the housing needs of the population between 1992 and 2002. From 1999 the completion of new flats per annum averaged a total of 80,000. The proportion of owner-occupiers increased sharply, from 42.6% in 1990 to 52.6% in 2002, while the proportion of tenants decreased from 51.9% to 42.7%. Among the 2.13m. domestic households in 2002, 30.9% resided in public rental housing and 50.9% in private permanent housing. The proportion of domestic households in subsidized sale flats increased from 7.8% in 1992 to 16.9% in 2002, while the proportion of domestic households in temporary housing declined from 4.2% to 1.4%. As the SARS impact waned, markets for residential property and shopping space improved significantly in the second half of 2003, especially following the signing of the Closer Economic Partnership Arrangement (CEPA) with mainland China and the implementation of the Individual Visit Scheme. Prices of flats on average decreased by 2% for 2003 as a whole, a much smaller decline than the 12% fall registered in 2002. For 2003 as a whole, private housing rentals were down by 9%, compared with the 14% fall of 2002. Planned developments of all types of property in the private sector increased sharply, by 23% in 2003, after a decline of 25% in 2001. By contrast, planned developments of commercial and of industrial property plummeted by 45% and 99% respectively in 2003.

OTHER RECENT DEVELOPMENTS

Fiscal Policy

In his maiden budget speech on 10 March 2004, Financial Secretary Henry Tang did not announce any drastic policy changes. The aim to eliminate the budget deficit by 2008/09 appeared reasonably optimistic. The Government's plan to reduce the deficit is based on the assumption that there will be a nominal GDP growth of 4.5% per annum, for the medium term, and that expenditure reductions will be achieved despite a possible return of inflation. Nevertheless, with the tax base remaining too narrow, Hong Kong still faces crucial challenges in reforming its existing fiscal structure. The key debate over what core fiscal solutions need to be implemented was only briefly addressed in the speech. The Financial Secretary proposed to study the possible introduction of a 3% goods-and-services tax (GST) as a means to broaden the tax base and to generate revenues. Henry Tang also announced that the Government would issue up to HK $20,000m. in government bonds in 2004/05, and cautioned that the funds so derived should be used exclusively for specified infrastructure projects.

Economic Relations with China

Paradoxically, six years after the return of Hong Kong to Chinese sovereignty, Beijing was giving the impression that it was rescuing the Hong Kong economy from serious difficulties. The Closer Economic Partnership Arrangement (CEPA) was signed by Hong Kong and China on 29 June 2003, and took effect in 2004. The agreement covers three main aspects of economic exchanges between Hong Kong and mainland China: lowering or abolishing of tariffs for 273 goods exported by Hong Kong, preferential opening of China's market for Hong Kong enterprises in the service sector (legal, accounting, financial, construction, information technology and business services), and facilitation of bilateral exchanges and investment. Among the causes of Hong Kong long-term economic problems, SARS has certainly taken its toll, reducing GDP growth by an estimated 1.4%, while there was a sharp decline in consumer spending and tourism. More generally, Hong Kong's international position is threatened by China's emerging trading centres, especially Shanghai. However, Hong Kong's unemployment does not seem to be affected by low-cost sectors in mainland China. Hong Kong's intermediary position remains strong, as exemplified by the continuing willingness of multinational companies to maintain a presence there. The future of Hong Kong lies in focusing on high-value services rather than trying to keep its wages in line with mainland Chinese levels. The key to Hong Kong's future success lies in developing its expertise in fields such as legal services, financial services, and trade infrastructures, as well as improving education and retraining programmes.

Statistical Survey

Source (unless otherwise stated): Census and Statistics Department, 19/F Wanchai Tower, 12 Harbour Rd, Hong Kong; tel. 25825073; fax 28271708; e-mail genenq@censtatd.gcn.gov.hk; internet www.info.gov.hk/censtatd/eng/hkstat/index2.html/.

Area and Population

AREA, POPULATION AND DENSITY

Land area (sq km)	1,103*
Population (census results)†	
15 March 1996	6,412,937
15 March 2001	
Males	3,285,344
Females	3,423,045
Total	6,708,389
Population (official estimates at mid-year)‡	
2001	6,724,900
2002	6,787,000
2003	6,803,100
Density (per sq km) at mid-2003	6,167.8

* 424 sq miles.

† All residents (including mobile residents) on the census date, including those who were temporarily absent from Hong Kong.

‡ Revised figures, referring to resident population, including mobile residents.

DISTRICTS
(2001 census)

	Area (sq km)	Population*	Density (per sq km)
Hong Kong Island	80.28	1,335,469	16,635
Kowloon	46.85	2,023,979	43,201
New Territories	970.91	3,343,046	3,443
Total	1,098.04	6,702,494	6,104

* Excluding marine population (5,895).

PRINCIPAL TOWNS
(population at 1996 census)

Kowloon*	1,988,515	Tai Po	271,661
Victoria (capital)	1,011,433	Tseun Wan	268,659
Tuen Mun	445,771	Sheung Shui	192,321
Sha Tin	445,383	Tsing Yu	185,495
Kwai Chung	285,231	Aberdeen	164,439

* Including New Kowloon.

BIRTHS, MARRIAGES AND DEATHS*

	Known live births		Registered marriages		Known deaths	
	Number	Rate (per '000)	Number	Rate (per '000)	Number	Rate (per '000)
1996	64,559†	10.2	37,045	5.9	32,049†	5.1
1997‡	60,379†	9.1	37,593	5.8	32,079†	4.9
1998‡	53,356†	7.9	31,673	4.7	32,680†	4.8
1999‡	50,513	7.5	31,287	4.6	33,387	4.8
2000‡	53,720	8.1	30,879	4.6	33,993	5.1
2001‡	49,144	7.2	32,825	4.9	33,305	5.0
2002‡	48,200	7.1	32,070	4.7	34,300	5.0
2003‡	46,200	6.8	35,400	5.2	36,500	5.4

* Excluding Vietnamese migrants.

† Figure calculated by year of registration.

‡ Provisional. Figures prior to 2000 have not been revised to take account of the results of the 2001 population census.

Expectation of life (years at birth, 2002, provisional): Males 78.7; Females 84.7.

ECONOMICALLY ACTIVE POPULATION
('000 persons aged 15 years and over, excl. armed forces)

	2001	2002	2003
Agriculture and fishing	7.2	9.9	7.5
Mining and quarrying	0.2	0.4	0.4
Manufacturing	326.4	289.5	272.4
Electricity, gas and water	15.7	15.9	16.2
Construction	291.4	286.7	266.1
Wholesale, retail and import/export trades, restaurants and hotels	981.1	983.7	993.2
Transport, storage and communications	353.4	346.0	346.4
Financing, insurance, real estate and business services	478.1	474.7	470.2
Community, social and personal services	798.9	825.6	850.8
Total employed	3,252.3	3,232.3	3,223.3
Unemployed	174.8	255.5	277.6
Total labour force	3,427.1	3,487.8	3,500.9
Males	1,965.2	1,964.6	1,962.3
Females	1,461.9	1,523.2	1,538.5

Source: ILO.

Health and Welfare

KEY INDICATORS

Total fertility rate (children per woman, 1995–2000)	1.2
Under-5 mortality rate (per 1,000 live births, provisional, 2001)	2.7
HIV/AIDS (% of persons aged 15–49, 2003)	0.1
Physicians (per 1,000 head, provisional, 2002)	1.6
Hospital beds (per 1,000 head, provisional, 2002)	5.2
Human Development Index (2002): ranking	23
Human Development Index (2002): value	0.903

For sources and definitions, see explanatory note on p. vi.

Agriculture

PRINCIPAL CROPS
(FAO estimates, '000 metric tons)

	2000	2001	2002
Lettuce	5	5	5
Spinach	11	11	11
Onions and shallots (green)	4	4	4
Other vegetables	24	14	14
Fruit	4	4	4

Source: FAO.

LIVESTOCK
(FAO estimates, '000 head, year ending September)

	2000	2001	2002
Cattle (head)	1,500	1,500	1,500
Pigs	100	100	100
Chickens	1,000	1,000	1,000
Ducks	250	250	250

Source: FAO.

LIVESTOCK PRODUCTS
('000 metric tons)

	2000	2001	2002
Beef and veal*	18	15	14
Pig meat*	161	165	163
Poultry meat	67	62	61
Game meat†	6	6	6
Cattle hides (fresh)†	3	2	2

* Unofficial figures.
† FAO estimates.

Source: FAO.

Fishing

(FAO estimates, '000 metric tons, live weight)

	2000	2001	2002
Capture	157.0	174.0	169.8
Lizardfishes	5.8	6.4	6.3
Threadfin breams	17.0	18.8	18.3
Shrimps and prawns	4.6	5.1	5.0
Squids	7.3	8.1	7.9
Aquaculture	5.0	5.6	4.3
Total catch	162.0	179.6	174.1

Source: FAO.

Industry

SELECTED PRODUCTS
('000 metric tons, unless otherwise indicated)

	1999	2000	2001
Crude groundnut oil	18	n.a.	n.a.
Uncooked macaroni and noodle products	127	83	62
Cotton yarn (pure and mixed)	n.a.	n.a.	90.9
Cotton woven fabrics (million sq m)	366	306	378
Knitted sweaters ('000)	117,738	103,519	151,965
Men's and boys' jackets ('000)	6,414	2,027	1,838
Men's and boys' trousers ('000)	32,937	20,457	14,139
Women's and girls' blouses ('000)	63,596	93,323	187,209
Women's and girls' dresses ('000)	5,668	1,375	2,242
Women's and girls' skirts, slacks and shorts ('000)	77,344	35,141	23,823
Men's and boys' shirts ('000)	41,040	149,824	105,254
Watches ('000)	31,926	28,729	16,774
Electric energy (million kWh)	29,496	31,329	n.a.

Source: UN, *Industrial Commodity Statistics Yearbook*.

Finance

CURRENCY AND EXCHANGE RATES

Monetary Units
100 cents = 1 Hong Kong dollar (HK $).

Sterling, US Dollar and Euro Equivalents (31 May 2004)
£1 sterling = HK $14.298;
US $1 = HK $7.793;
€1 = HK $9.543;
HK $1,000 = £69.94 = US $128.32 = €104.79.

Average Exchange Rate (HK $ per US $)
2001 7.7988
2002 7.7990
2003 7.7868

BUDGET
(HK $ million, year ending 31 March)

Revenue	2001/02	2002/03	2003/04*
Direct taxes:			
Earnings and profits tax	77,749	73,028	78,080
Indirect taxes:			
Estate duty	1,928	1,403	1,500
Duties on petroleum products, beverages, tobacco and cosmetics	6,981	6,620	6,539
General rates (property tax)	12,727	8,923	11,131
Motor vehicle taxes	2,676	2,510	2,712
Royalties and concessions	1,881	1,726	1,654
Others	21,077	19,464	22,595
Fines, forfeitures and penalties	926	843	822
Receipts from properties and investments	8,621	8,015	7,708
Reimbursements and contributions	4,154	4,405	3,212
Operating revenue from utilities:			
Water	2,471	1,458	2,093
Others	895	610	757
Fees and charges	10,916	9,687	10,369
Interest receipts (operating revenue)	225	2,766	5,886
Land Fund (investment income)	106	13,281	17,150
Capital Works Reserve Fund (land sales and interest)	10,683	12,190	6,025
Capital Investment Fund	2,816	2,432	2,423
Loan funds	5,382	4,464	17,551
Other capital revenue	3,345	3,664	5,630
Total government revenue	175,559	177,489	203,837

Expenditure†	2001/02	2002/03	2003/04*
Economic affairs and services	13,714	13,748	15,561
Internal security	21,679	21,085	21,151
Immigration	2,301	2,531	2,805
Other security services	3,574	3,452	3,500
Social welfare	30,059	32,282	33,997
Health services	34,213	33,199	34,485
Education	52,232	54,785	57,748
Environmental services	11,207	11,443	11,213
Recreation, culture and amenities	6,381	6,026	6,433
Other community and external affairs	1,844	2,051	2,093
Transport	6,399	7,732	9,920
Land and buildings	10,768	9,385	9,686
Water supply	7,711	7,473	6,757
Support	35,222	34,297	35,980
Housing	32,055	24,031	27,854
Total	269,359	263,520	279,183
Recurrent	210,445	211,728	214,460
Capital	58,914	51,792	64,723

* Revised estimates.
† Figures refer to consolidated expenditure by the public sector.

INTERNATIONAL RESERVES
(US $ million at 31 December)

	2001	2002	2003
Gold*	19	23	28
Foreign exchange†	111,155	111,898	118,640
Total	111,174	111,921	118,668

* National valuation.
† Including the foreign exchange reserves of the Hong Kong Special Administrative Region Government's Land Fund.

Source: IMF, *International Financial Statistics*.

MONEY SUPPLY
(HK $ '000 million at 31 December)

	2001	2002	2003
Currency outside banks	101.4	113.0	127.6
Demand deposits at banking institutions	109.5	127.4	196.5
Total money	210.9	240.4	324.1

Source: IMF, *International Financial Statistics*.

COST OF LIVING
(Consumer price index; base: 1990 = 100)

	2000	2001	2002
Food	157.5	156.2	152.9
Housing	202.3	196.1	184.8
Electricity, gas and other fuels	154.2	151.1	140.7
Clothing and footwear	126.6	120.8	121.7
All items (incl. others)	168.5	165.8	160.7

Source: ILO.

NATIONAL ACCOUNTS
(HK $ million at current prices)

Expenditure on the Gross Domestic Product

	2001	2002	2003
Government final consumption expenditure	128,846	131,279	130,067
Private final consumption expenditure	765,105	728,092	704,863
Change in stocks	−4,060	5,660	9,471
Gross domestic fixed capital formation	333,036	286,020	269,127
Total domestic expenditure	1,222,927	1,151,051	1,113,528
Net exports of goods and services	46,969	96,330	106,321
GDP in purchasers' values	1,269,896	1,247,381	1,219,849
GDP at constant 2000 prices	1,294,306	1,318,743	1,361,036

Gross Domestic Product by Economic Activity
(at factor price)

	2000	2001	2002
Agriculture and fishing	920	1,003	1,002
Mining and quarrying	241	174	136
Manufacturing	71,655	63,519	54,848
Electricity, gas and water	38,853	40,126	41,540
Construction	64,026	58,971	53,089
Wholesale, retail and import/export trades, restaurants and hotels	324,622	324,654	324,131
Transport, storage and communications	125,724	124,260	128,278
Financing, insurance, real estate and business services	291,062	274,030	267,537
Community, social and personal services	252,435	265,081	267,659
Ownership of premises*	59,358	63,537	67,258
Sub-total	1,228,897	1,215,354	1,205,479
Taxes on production and imports	57,908	53,917	43,325
Statistical discrepancy	1,533	625	−1,423
GDP in market prices	1,288,338	1,269,896	1,247,381

* Including adjustments for financial intermediation services indirectly measured.

Source: Asian Development Bank, *Key Indicators of Developing Asian and Pacific Countries*.

BALANCE OF PAYMENTS
(US $ million)

	2001	2002	2003
Exports of goods f.o.b.	190,926	200,300	224,656
Imports of goods f.o.b.	−199,257	−205,353	−230,435
Trade balance	−8,331	−5,053	−5,779
Exports of services	39,449	43,333	45,002
Imports of services	−24,677	−24,800	−24,304
Balance on goods and services	6,441	13,480	14,919
Other income received	49,315	43,223	42,198
Other income paid	−44,036	−41,083	−37,813
Balance on goods, services and income	11,721	15,621	19,303
Current transfers received	605	777	574
Current transfers paid	−2,385	−2,673	−2,463
Current balance	9,941	13,725	17,414
Capital account (net)	−1,174	−2,011	−1,016
Direct investment abroad	−11,345	−17,463	−3,747
Direct investment from abroad	23,776	9,682	13,538
Portfolio investment assets	−40,133	−37,702	−31,458
Portfolio investment liabilities	−1,161	−1,084	991
Financial derivatives assets	17,971	20,035	25,435
Financial derivatives liabilities	−12,888	−13,424	−15,224
Other investment assets	59,137	46,617	−22,181
Other investment liabilities	−41,985	−26,412	15,193
Net errors and omissions	2,543	5,660	2,049
Overall balance	4,684	−2,377	994

Source: IMF, *International Financial Statistics*.

External Trade

PRINCIPAL COMMODITIES
(US $ million)

Imports c.i.f.	2000	2001	2002
Food and live animals	7,328.5	6,949.3	6,869.9
Chemicals and related products	13,203.8	11,355.3	11,852.4
Basic manufactures	36,684.6	32,925.6	33,589.9
Textile yarn, fabrics, made-up articles, etc.	13,726.0	12,183.7	12,072.2
Machinery and transport equipment	90,774.2	88,151.0	95,670.8
Office machines and automatic data-processing equipment	18,341.5	18,938.5	20,100.9
Telecommunications and sound recording and reproducing apparatus and equipment	20,742.3	20,348.1	23,254.3
Electrical machinery, apparatus and appliances n.e.s., and electrical parts thereof	37,407.0	33,980.2	38,101.1
Miscellaneous manufactured articles	55,304.6	53,127.0	51,060.7
Clothing (excl. footwear)	16,028.4	16,121.1	15,718.4
Footwear	5,318.2	4,863.1	4,766.8
Photographic apparatus, equipment and supplies, optical goods, watches and clocks	7,166.3	6,998.3	6,955.2
Total (incl. others)	214,041.7	202,008.0	207,969.0

Exports f.o.b.	2000	2001	2002
Chemicals and related products	10,318.3	9,166.0	9,558.6
Basic manufactures	30,203.5	27,065.0	27,919.5
Textile yarn, fabrics, made-up articles, etc.	13,455.4	12,233.8	12,431.2
Machinery and transport equipment	77,699.1	75,981.0	87,605.2
Office machines and automatic data-processing equipment	16,402.5	17,746.5	20,398.7
Telecommunications and sound recording and reproducing apparatus and equipment	19,618.2	18,696.5	22,877.8
Electrical machinery, apparatus and appliances n.e.s., and electrical parts thereof	32,357.4	30,649.9	34,018.3
Miscellaneous manufactured articles	76,710.3	71,670.6	69,126.0
Clothing (excl. footwear)	24,239.6	23,472.3	22,453.6
Photographic apparatus, equipment and supplies, optical goods, watches and clocks	9,058.3	8,563.2	8,381.8
Total (incl. others)	202,683.2	191,066.2	201,927.8

Source: UN, *International Trade Statistics Yearbook*.

2003 (HK $ million): Total imports 1,805,800; Domestic exports 121,700; Re-exports 1,620,700; Total exports 1,742,400.

PRINCIPAL TRADING PARTNERS
(HK $ million, excl. gold)

Imports	2001	2002	2003*
China, People's Repub.	681,980	717,074	785,600
Germany	33,309	32,997	n.a.
Japan	176,599	182,569	214,000
Korea, Repub.	70,791	75,955	n.a.
Malaysia	39,200	39,729	n.a.
Singapore	72,898	75,740	90,600
Taiwan	107,929	115,906	125,200
Thailand	27,370	29,556	n.a.
United Kingdom	28,877	26,082	n.a.
USA	104,941	91,478	98,700
Total (incl. others)	1,568,194	1,619,419	1,805,800

Domestic exports	2001	2002	2003*
Canada	3,093	2,411	n.a.
China, People's Repub.	49,547	41,374	36,800
France	n.a.	n.a.	n.a.
Germany	5,818	4,273	4,900
Japan	4,060	2,969	n.a.
Netherlands	4,619	3,470	n.a.
Singapore	2,650	2,161	n.a.
Taiwan	5,346	4,388	3,700
United Kingdom	8,578	7,588	7,800
USA	47,589	41,908	39,100
Total (incl. others)	153,520	130,926	121,700

Re-exports	2001	2002	2003*
China, People's Repub.	496,574	571,870	705,800
France	21,516	n.a.	n.a.
Germany	45,774	44,567	51,400
Japan	83,551	80,743	91,200
Korea, Repub.	24,640	29,264	n.a.
Netherlands	20,693	22,775	n.a.
Singapore	26,929	29,424	n.a.
Taiwan	30,021	30,193	n.a.
United Kingdom	46,764	46,644	49,600
USA	282,189	291,043	285,100
Total (incl. others)	1,327,467	1,429,590	1,620,700

* Preliminary figures, rounded.

Transport

RAILWAYS
(traffic)

	2001	2002	2003
Passenger train journeys	3,001	n.a.	n.a.
Freight (metric tons):			
Loaded	97,139	102,000	76,000
Unloaded	273,051	283,000	253,000

ROAD TRAFFIC
(registered motor vehicles at 31 December)

	2001	2002*	2003*
Private cars	381,757	341,000	339,000
Buses (private and public)	13,297	13,000	13,000
Light buses (private and public)	6,448	6,000	6,000
Taxis	18,138	18,000	18,000
Goods vehicles	126,233	111,000	110,000
Motorcycles	36,191	28,000	30,000
Government vehicles (excl. military vehicles)	7,127	7,000	7,000
Total (incl. others)	589,808	526,000	524,000

* Preliminary figures, rounded.
Note: Figures do not include tramcars.

SHIPPING

Merchant Fleet
(registered at 31 December)

	2001	2002	2003
Number of vessels	646	766	901
Total displacement ('000 grt)	13,709.7	16,164.3	20,507.5

Source: Lloyd's Register-Fairplay, *World Fleet Statistics*.

Traffic
(2003)

	Ocean-going vessels	River vessels
Total capacity (million nrt)	591	181
Cargo landed ('000 metric tons)*	97,600	28,900
Cargo loaded ('000 metric tons)*	48,100	29,400

* Provisional figures.

Passenger traffic ('000, ocean-going vessels and river vessels, 2003): Passengers landed 8,879; Passengers embarked 9,765.

CIVIL AVIATION

	2001	2002	2003
Passengers:			
Arrivals	11,533,000	11,841,000	9,486,000
Departures	11,488,000	11,722,000	9,361,000
Freight (in metric tons):*			
Landed	894,000	1,004,000	1,035,000
Loaded	1,180,000	1,475,000	1,608,000

* Provisional figures.

Tourism

VISITOR ARRIVALS BY COUNTRY OF RESIDENCE

	2000	2001	2002
Australia	352,409	324,156	343,294
Canada	253,095	249,707	264,967
China, People's Repub.	3,785,845	4,448,583	6,825,199
Germany	193,837	173,359	172,654
Indonesia	236,275	212,260	223,590
Japan	1,382,417	1,336,538	1,395,020
Korea, Repub.	372,639	425,732	457,438
Macao	449,947	532,391	534,590
Malaysia	314,857	286,338	318,854
Philippines	278,460	293,105	329,604
Singapore	450,569	421,513	426,166
Taiwan	2,385,739	2,418,827	2,428,776
Thailand	228,774	241,480	259,336
United Kingdom	367,938	360,581	379,965
USA and Guam	966,008	935,717	1,000,844
Total (incl. others)	13,059,477	13,725,332	16,566,382

Receipts from tourism (US $ million): 7,886 in 2000; 8,282 in 2001; 10,117 in 2002.

Communications Media

	2001	2002	2003
Telephones ('000 in use)	3,897.6	3,831.8	3,801.3
Facsimile machines (number in use)	411,000	546,000	n.a.
Mobile cellular telephones ('000 subscribers)	5,776.4	6,395.7	7,241.4
Personal computers ('000 in use)	2,600	2,864	n.a.
Internet users ('000)	2,601.3	2,918.8	3,212.8

1997 ('000 in use): Radio receivers 4,450.

2000 ('000 in use): Television receivers 3,105.

2002: Daily newspapers 52; Periodicals 754.

Sources: partly UNESCO, *Statistical Yearbook;* UN, *Statistical Yearbook*; International Telecommunication Union.

Education

(2002/03)*

	Institutions	Full-time teachers§	Students‖
Kindergartens	777	9,159	143,700
Primary schools	803	22,845	483,200
Secondary schools	542	25,093	465,900
Special schools	74	1,671	9,900
Institute of Vocational Education†	1	1,033	56,100
Approved post-secondary college	2	117	4,000
Other post-secondary colleges	11	—	3,000
UGC-funded institutions‡	8	5,620	86,500
Open University Institute	1	105	25,100
Adult education institutions	1,774	—	199,500

* Provisional figures.

† Formed by merger of two technical colleges and seven technical institutes in 1999.

‡ Funded by the University Grants Committee.

§ As of 2000/01.

‖ Figures are rounded to the nearest 100.

Adult literacy rate (UNESCO estimates): 93.5% (males 96.9%; females 89.6%) in 2001 (Source: UN Development Programme, *Human Development Report*).

Directory

The Constitution

Under the terms of the Basic Law of the Hong Kong Special Administrative Region, the Government comprises the Chief Executive, the Executive Council and the Legislative Council. The Chief Executive must be a Chinese citizen of at least 40 years of age; he is appointed for a five-year term, with a limit of two consecutive terms; in 2002 he was chosen by an 800-member Election Committee; he is accountable to the State Council of the People's Republic of China, and has no military authority; he appoints the Executive Council, judges and the principal government officials; he makes laws with the advice and consent of the legislature; he has a veto over legislation, but can be overruled by a two-thirds' majority; he may dissolve the legislature once in a term, but must resign if the legislative impasse continues with the new body. The Legislative Council has 60 members; in September 2004 candidates for 30 seats were directly elected under a system of proportional representation and 30 seats were determined by elections within 'functional constituencies' (comprising professional and special interest groups). The Legislative Council is responsible for enacting, revising and abrogating laws, for approving the budget, taxation and public expenditure, for debating the policy address of the Chief Executive and for approving the appointment of the judges of the Court of Final Appeal and of the Chief Justice of the High Court.

The Government

Chief Executive: TUNG CHEE-HWA (assumed office 1 July 1997; re-elected unopposed 28 February 2002).

EXECUTIVE COUNCIL

(October 2004)

Chairman: The Chief Executive.

Ex-Officio Members (Principal Officials)

Chief Secretary for Administration: DONALD TSANG YAM-KUEN.

Financial Secretary: HENRY TANG YING-YEN.

Secretary for Justice: ELSIE LEUNG OI-SIE.

Secretary for Commerce, Industry and Technology: JOHN TSANG CHUN-WAH.

Secretary for Housing, Planning and Lands: MICHAEL SUEN MING-YEUNG.

Secretary for Education and Manpower: ARTHUR LI KWOK-CHEUNG.

Secretary for Health, Welfare and Food: DR YORK CHOW.

Secretary for the Civil Service: JOSEPH WONG WING-PING.

Secretary for Home Affairs: PATRICK HO CHI-PING.

Secretary for Security: AMBROSE LEE SIU-KWONG.

Secretary for Economic Development and Labour: STEPHEN IP SHU-KWAN.

Secretary for the Environment, Transport and Works: SARAH LIAO SAU-TUNG.

Secretary for Financial Services and the Treasury: FREDERICK MA SI-HANG.

Secretary for Constitutional Affairs: STEPHEN LAM SUI-LUNG.

Non-Official Members

LEUNG CHUN-YING, SELINA CHOW LIANG SHUK-YEE, JASPER TSANG YOK-SING, CHENG YIU-TONG, ANDREW LIAO CHEUNG-SING.

GOVERNMENT OFFICES

Executive Council: Central Government Offices, Lower Albert Rd, Central; tel. 28102545; fax 28450176.

Office of the Chief Executive: 5/F Main Wing, Central Government Offices, Lower Albert Rd, Central; tel. 28783300; fax 25090577.

Government Secretariat: Central Government Offices, Lower Albert Rd, Central; tel. 28102900; fax 28457895.

Government Information Services: Murray Bldg, Garden Rd, Central; tel. 28428777; fax 28459078; internet www.info.gov.hk.

Legislature

LEGISLATIVE COUNCIL

The third Legislative Council to follow Hong Kong's transfer to Chinese sovereignty was elected on 12 September 2004. The Legislative Council comprises 60 members: 30 chosen by functional constituencies and 30 (increased from 24 in the previous legislature) by direct election in five geographical constituencies. The term of office of the Legislative Council was to last for four years.

President: RITA FAN HSU LAI-TAI.

Election, 12 September 2004

Party	Directly-elective seats	Functional Constituency seats	Total seats
Democratic Alliance for the Betterment of Hong Kong	9	3	12
Liberal Party	2	8	10
Democratic Party of Hong Kong	8	1	9
Article 45 Concern Group	3	1	4
Hong Kong Confederation of Trade Unions	2	—	2
The Frontier	1	—	1
Association for Democracy and People's Livelihood	1	—	1
Neighbourhood and Worker's Service Centre	1	—	1
April Fifth Action	1	—	1
Hong Kong Federation of Trade Unions	—	1	1
Independents	2	16	18
Total	30	30	60

Political Organizations

April Fifth Action: internet act.to/45; socialist group, anti-Beijing; Spokesperson LEUNG KWOK-HUNG.

Article 45 Concern Group: internet www.article45.org; pro-democracy, supports election of Hong Kong's Chief Executive by universal suffrage.

Association for Democracy and People's Livelihood (ADPL): Sun Beam Commercial Bldg, Room 1104, 469–471 Nathan Rd, Kowloon; tel. 27822699; fax 27823137; e-mail info@adpl.org.hk; internet www.adpl.org.hk; advocates democracy; Chair. FREDERICK FUNG KIN-KEE; Gen. Sec. TAM KWOK-KIU.

Citizens' Party: GPOB 321, Central, Hong Kong; tel. 28930029; e-mail enquiry@citizensparty.org; internet www.citizensparty.org; f. 1997; urges mass participation in politics; established by Christine Loh, Chairwoman 1997–2000; Chair ALEX CHAN.

Democratic Alliance for the Betterment of Hong Kong (DAB): SUP Tower, 12/F, 83 King's Rd, North Point; tel. 25280136; fax 25284339; e-mail info@dab.org.hk; internet www.dab.org.hk; f. 1992; pro-Beijing; supported return of Hong Kong to the motherland and implementation of the Basic Law; Chair. MA LIK; Sec.-Gen. KAN CHI HO.

Democratic Party: Hanley House, 4/F, 776–778 Nathan Rd, Kowloon; tel. 23977033; fax 23978998; e-mail dphk@dphk.org; internet www.dphk.org; f. 1994 by merger of United Democrats of Hong Kong (UDHK—declared a formal political party in 1990) and Meeting Point; liberal grouping; advocates democracy; Chair. YEUNG SUM; Sec.-Gen. CHEUNG YIN-TUNG.

The Frontier: Flat B, 1/F, Kam Wah Bldg, 514-516 Nathan Rd, Yau Ma Tei, Kowloon; tel. 25249899; fax 25245310; e-mail frontier@frontier.org.hk; internet www.frontier.org.hk; f. 1996; pro-democracy movement, comprising teachers, students and trade unionists; Spokesperson EMILY LAU.

Hong Kong Democratic Foundation: Hong Kong House, Room 301, 17–19 Wellington St, Central; GPOB 12287; tel. 28696443; fax 28696318; advocates democracy; Chair. ALAN LUNG.

Hong Kong Progressive Alliance: c/o The Legislative Council, Hong Kong; tel. 25262316; fax 28450127; f. 1994; advocates close relationship with mainland China; 52-mem. organizing cttee drawn from business and professional community; Spokesman AMBROSE LAU.

Hong Kong Voice of Democracy: 7/F, 57 Peking Rd, Tsimshatsui; tel. 92676489; fax 27915801; e-mail editor@democracy.org.hk; internet www.democracy.org.hk; pro-democracy movement; Dir LAU SAN-CHING.

Liberal Democratic Foundation (LDF): Hong Kong; pro-Beijing.

Liberal Party: 4/F Henley Bldg, 5 Queen's Road, Central; tel. 28696833; fax 25334239; e-mail liberal@liberal.org.hk; internet www.liberal.org.hk; f. 1993 by mems of Co-operative Resources Centre (CRC); business-orientated; pro-Beijing; Leader ALLEN LEE PENG-FEI; Chair. JAMES TIEN.

Neighbourhood and Worker's Service Centre: e-mail nwsc@netvigator.com; internet www.nwsc.org.hk.

New Hong Kong Alliance: 4/F, 14–15 Wo On Lane, Central; fax 28691110; pro-China.

At the elections of September 2004, the Hong Kong Confederation of Trade Unions and the Hong Federation of Trade Unions also secured seats in the Legislative Council (see Trade and Industry section for details of these organizations).

The **Chinese Communist Party** (based in the People's Republic) and the **Kuomintang** (Nationalist Party of China, based in Taiwan) also maintain organizations.

Judicial System

The Court of Final Appeal was established on 1 July 1997 upon the commencement of the Hong Kong Court of Final Appeal Ordinance. It replaced the Privy Council in London as the highest appellate court in Hong Kong to safeguard the rule of law. The Court comprises five judges—the Chief Justice, three permanent judges and one non-permanent Hong Kong judge or one judge from another common-law jurisdiction.

The High Court consists of a Court of Appeal and a Court of First Instance. The Court of First Instance has unlimited jurisdiction in civil and criminal cases, while the District Court has limited jurisdiction. Appeals from these courts lie to the Court of Appeal, presided over by the Chief Judge or a Vice-President of the Court of Appeal with one or two Justices of Appeal. Appeals from Magistrates' Courts are heard by a Court of First Instance judge.

COURT OF FINAL APPEAL

1 Battery Path, Central; tel. 21230123; fax 21210300; internet www.judiciary.gov.hk.

Chief Justice of the Court of Final Appeal: ANDREW K. N. LI.

Permanent Judges of the Court of Final Appeal: R. A. V. RIBEIRO, PATRICK S. O. CHAN, K. BOKHARY.

HIGH COURT

38 Queensway; tel. 28690869; fax 28690640; internet www.judiciary.gov.hk.

Chief Judge of the High Court: GEOFFREY T.L. MA.

Justices of Appeal: K. H. WOO, M. STUART-MOORE, F. STOCK, Mrs D. LE PICHON, P. C. Y. CHEUNG, A. G. ROGERS, M. C. K. N. YUEN, W. C. K. YEUNG.

Judges of the Court of First Instance: G. J. LUGAR-MAWSON, A. O. T. CHUNG, T. M. GALL, D. Y. K. YAM, W. S. Y. WAUNG, M. P. BURRELL, Ms C. F. L. CHU, Mrs V. S. BOKHARY, K. K. PANG, W. D. STONE, Ms C. M. BEESON, P. V. T. NGUYEN, M. J. HARTMANN, A. R. SUFFIAD, A. H. SAKHRANI, L. P. S. TONG, Miss S. S. H. KWAN, M. A. MCMAHON, J. M. H. LAM, A. K. N. CHEUNG, M. V. LUNN, A. T. BARMA, A. F. T. REYES.

OTHER COURTS

District Courts: There are 34 District Judges.

Magistrates' Courts: There are 59 Magistrates and 11 Special Magistrates, sitting in 9 magistracies.

Religion

The Chinese population is predominantly Buddhist. In 1994 the number of active Buddhists was estimated at between 650,000 and 700,000. Confucianism and Daoism are widely practised. The three religions are frequently found in the same temple. In 1999 there were some 527,000 Christians, approximately 80,000 Muslims, 12,000 Hindus, 1,000 Jews and 1,200 Sikhs. The Bahá'í faith and Zoroastrianism are also represented.

BUDDHISM

Hong Kong Buddhist Association: 1/F, 338 Lockhart Rd; tel. 25749371; fax 28340789; internet www.hkbuddhist.org; Pres. Ven. KOK KWONG.

CHRISTIANITY

Hong Kong Christian Council: 9/F, 33 Granville Rd, Kowloon; tel. 23687123; fax 27242131; e-mail hkcc@hkcc.org.hk; internet www.hkcc.org.hk; f. 1954; 22 mem. orgs; Chair. Rev. LI PING-KWONG; Gen. Sec. Rt Rev. Dr SOO YEE-PO THOMAS.

The Anglican Communion

Primate of Hong Kong Sheng Kung Hui and Bishop of Hong Kong Island and Macao: Most Rev. PETER K. K. KWONG, Bishop's House, 1 Lower Albert Rd, Central; tel. 25265355; fax 25212199; e-mail office1@hkskh.org.

Bishop of Eastern Kowloon: Rt Rev. LOUIS TSUI, Holy Trinity Bradbury Centre, 4/F, 139 Ma Tau Chung Rd, Kowloon; tel. 27139983; fax 27111609; e-mail ekoffice@ekhkskh.org.hk.

Bishop of Western Kowloon: Rt Rev. THOMAS SOO, Ultra Grace Commercial Bldg, 15/F, 5 Jordan Rd, Kowloon; tel. 27830811; fax 27830799; e-mail hkskhdwk@netvigator.com.

The Lutheran Church

Evangelical Lutheran Church of Hong Kong: 50A Waterloo Rd, Kowloon; tel. 23885847; fax 23887539; e-mail info@elchk.org.hk; internet www.elchk.org.hk; 13,000 mems; Pres. Rev. TSO SHUI-WAN.

The Roman Catholic Church

For ecclesiastical purposes, Hong Kong forms a single diocese, nominally suffragan to the archdiocese of Canton (Guangzhou), China. According to Vatican sources, in December 2002 there were an estimated 367,402 adherents in the territory, representing more than 5% of the total population.

Bishop of Hong Kong: JOSEPH ZEN ZE-KIUN, Catholic Diocese Centre, 12/F, 16 Caine Rd; tel. 257652; fax 254707; e-mail bishopzen@catholic.org.uk.

The Press

Hong Kong has a thriving press. At the end of 2002, according to government figures, there were 52 daily newspapers, including 25 Chinese-language and eight English-language dailies, and 754 periodicals.

PRINCIPAL DAILY NEWSPAPERS

English Language

Asian Wall Street Journal: GPOB 9825; tel. 25737121; fax 28345291; f. 1976; business; Editor REGINALD CHUA; circ. 85,000.

China Daily: Hong Kong edition of China's official English-language newspaper; launched 1997; Editor LIU DIZHONG; circ. 11,000.

International Herald Tribune: 1201 K Wah Centre, 191 Java Rd, North Point; tel. 29221188; fax 29221190; internet www.iht.com; Correspondent KEVIN MURPHY.

South China Morning Post: Morning Post Centre, Dai Fat St, Tai Po Industrial Centre, Tai Po, New Territories; tel. 26808888; fax 26616984; internet www.scmp.com; f. 1903; CEO OWEN JONATHAN; Editor ROBERT KEATLEY; circ. 118,000.

The Standard: Sing Tao Bldg, 3/F, 1 Wang Kwong Rd, Kowloon Bay, Kowloon; tel. 27982798; fax 27953009; e-mail webmaster@thestandard.com.hk; internet www.thestandard.com.hk; f. 1949; Editor KARL WILSON; circ. 45,000.

Target Intelligent Report: Suite 2901, Bank of America Tower, 12 Harcourd Rd, Central; tel. 25730379; fax 28381597; e-mail info@targetnewspapers.com; internet www.targetnewspapers.com; f. 1972; financial news, commentary, politics, property, litigations, etc.

Chinese Language

Ching Pao: 3/F, 141 Queen's Rd East; tel. 25273836; f. 1956; Editor MOK KONG; circ. 120,000.

Hong Kong Commercial Daily: 1/F, 499 King's Rd, North Point; tel. 25905322; fax 25658947.

Hong Kong Daily News: All Flats, Hong Kong Industrial Bldg, 17/F, 444–452 Des Voeux Rd West; tel. 28555111; fax 28198717; internet www.hkdailynews.net; f. 1958; morning; CEO RODDY YU; Chief Editor K. K. YEUNG; circ. 120,000.

Hong Kong Economic Journal: North Point Industrial Bldg, 22/F, 499 King's Rd; tel. 28567567; fax 28111070; e-mail info@hkej.com; Editor-in-Chief H. C. CHIU; circ. 70,000.

Hong Kong Economic Times: Kodak House, Block 2, Room 808, 321 Java Rd, North Point; tel. 28802888; fax 28111926; f. 1988; Publr PERRY MAK; Chief Editor ERIC CHAN; circ. 64,565.

Hong Kong Sheung Po (Hong Kong Commercial Daily): 499 King's Rd, North Point; tel. 25640788; f. 1952; morning; Editor-in-Chief H. CHEUNG; circ. 110,000.

Hsin Wan Pao (New Evening Post): 342 Hennessy Rd, Wanchai; tel. 28911604; fax 28382307; f. 1950; Editor-in-Chief CHAO TSE-LUNG; circ. 90,000.

Ming Pao Daily News: Block A, Ming Pao Industrial Centre, 15/F, 18 Ka Yip St, Chai Wan; tel. 25953111; fax 28982534; e-mail mingpao@mingpao.com; internet www.mingpao.com; f. 1959; morning; Chief Editor PAUL CHEUNG; circ. 96,579.

Oriental Daily News: Oriental Press Centre, Wang Tai Rd, Kowloon Bay, Kowloon; tel. 27951111; fax 27955599; Chair. C. F. MA; Editor-in-Chief MA KAI LUN; circ. 650,000.

Ping Kuo Jih Pao (Apple Daily): Hong Kong; tel. 29908685; fax 23708908; f. 1995; Propr JIMMY LAI; Publr LOH CHAN; circ. 400,000.

Seng Weng Evening News: f. 1957; Editor WONG LONG-CHAU; circ. 60,000.

Sing Pao Daily News: Sing Pao Bldg, 101 King's Rd, North Point; tel. 25702201; fax 28870348; f. 1939; morning; Chief Editor HON CHUNG-SUEN; circ. 229,250.

Sing Tao Daily: Sing Tao Bldg, 3/F, 1 Wang Kwong Rd, Kowloon Bay, Kowloon; tel. 27982575; fax 27953022; f. 1938; morning; Editor-in-Chief LUK KAM WING; circ. 60,000.

Ta Kung Pao: 342 Hennessy Rd, Wanchai; tel. 25757181; fax 28345104; e-mail tkp@takungpao.com; internet www.takungpao.com; f. 1902; morning; supports People's Republic of China; Editor T. S. TSANG; circ. 150,000.

Tin Tin Yat Pao: Culturecom Centre, 10/F, 47 Hung To Rd, Kwun Tong, Kowloon; tel. 29507300; fax 23452285; f. 1960; Chief Editor IP KAI-WING; circ. 199,258.

Wen Wei Po: Hing Wai Centre, 2–4/F, 7 Tin Wan Praya Rd, Aberdeen; tel. 28738288; fax 28730657; internet www.wenweipo.com; f. 1948; morning; communist; Dir ZHANG GUO-LIANG; First Editor-in-Chief CHEUNG CHING-WAN; Editor-in-Chief WONG BAK YAO; circ. 200,000.

SELECTED PERIODICALS

English Language

Asian Business: c/o TPL Corporation (HK) Ltd, Block C, 10/F, Seaview Estate, 2–8 Watson Rd, North Point; tel. 25668381; fax 25080197; e-mail absales@asianbusiness.com.hk; internet www.asianbusinessnet.com; monthly; Publr and Executive Editor JAMES LEUNG; circ. 75,000.

Asian Medical News: Pacific Plaza, 8/F, 410 Des Voeux Rd West; tel. 25595888; fax 25596910; e-mail amn@medimedia.com.hk; internet www.amn.com; f. 1979; monthly; Man. Editor ROSS GARBETT; circ. 28,300.

Asian Profile: Asian Research Service, GPOB 2232; tel. 25707227; fax 25128050; f. 1973; 6 a year; multi-disciplinary study of Asian affairs.

Business Traveller Asia/Pacific: Unit 404, Printing Hse, 6 Duddell St, Central; tel. 25119317; fax 25196846; e-mail enquiry@businesstravellerasia.com; f. 1982; consumer business travel; 12 a year; Publr PEGGY TEO; Editor JONATHAN WALL; circ. 23,320.

Far Eastern Economic Review: Central Plaza, 25/F, 18 Harbour Rd, Wanchai, GPOB 160; tel. 25084338; fax 25031549; e-mail review@feer.com; internet www.feer.com; f. 1946; weekly; Editor DAVID PLOTT; circ. 95,570.

Hong Kong Electronics: Office Tower, Convention Plaza, 38/F, 1 Harbour Rd; tel. 25844333; fax 28240249; e-mail hktdc@tdc.org.hk; internet www.tdctrade.com; f. 1985; bi-monthly; publ. by the Hong Kong Trade Development Council; Editor GEOFF PICKER; circ. 90,000.

Hong Kong Enterprise: Office Tower, Convention Plaza, 38/F, 1 Harbour Rd; tel. 25844333; fax 28240249; e-mail hktdc@tdc.org.hk; internet www.tdctrade.com; f. 1967; monthly; publ. by the Hong Kong Trade Development Council; Editor GEOFF PICKER; circ. 150,000.

Hong Kong Government Gazette: Govt Printing Dept, Cornwall House, Taikoo Trading Estate, 28 Tong Chong St, Quarry Bay; tel. 25649500; weekly.

Hong Kong Household: Office Tower, Convention Plaza, 38/F, 1 Harbour Rd, Wanchai; tel. 25844333; fax 28240249; e-mail hktdc@tdc.org.hk; internet www.tdctrade.com; f. 1983; publ. by the Hong Kong Trade Development Council; household and hardware products; 2 a year; Editor GEOFF PICKER; circ. 90,000.

Hong Kong Industrialist: Federation of Hong Kong Industries, Hankow Centre, 4/F, 5–15 Hankow Rd, Tsimshatsui, Kowloon; tel. 27323188; fax 27213494; e-mail fhki@fhki.org.hk; monthly; publ. by the Federation of Hong Kong Industries; Editor JAMES MANNING; circ. 6,000.

Hong Kong Trader: Office Tower, Convention Plaza, 38/F, 1 Harbour Rd, Wanchai; tel. 25844333; fax 28243485; e-mail trader@tdc.org.hk; internet www.tdc.org.hk/hktrader; f. 1983; publ. by the Hong Kong Trade Development Council; trade, economics, financial and general business news; monthly; Man. Editor SOPHY FISHER; circ. 70,000.

Official Hong Kong Guide: Wilson House, 3/F, 19–27 Wyndham St, Central; tel. 25215392; fax 25218638; f. 1982; monthly; information on sightseeing, shopping, dining, etc. for overseas visitors; Editor-in-Chief DEREK DAVIES; circ. 9,300.

Orientations: 17/F, 200 Lockhart Rd; tel. 25111368; fax 25074620; e-mail omag@netvigator.com; internet www.orientations.com.hk; f. 1970; 8 a year; arts of East Asia, the Indian subcontinent and South-East Asia; Publr and Editorial Dir ELIZABETH KNIGHT.

Reader's Digest (Asia Edn): Reader's Digest Association Far East Ltd, 19/F Cyber Centre, 3 Tung Wong Rd, Shau Kei Wan; tel. 96906381; fax 96906389; e-mail friends@rdasia.com.hk; f. 1963; general topics; monthly; Editor PETER DOCKRILL; circ. 332,000.

Spike: 2606 Westley Sq., 48 Hoi Yuen Rd, Kwun Tong, Kowloon; tel. 31058921; fax 31058927; e-mail editorial@spikehk.com; internet www.spikehk.com; f. 2003; weekly magazine; Publr STEPHEN VINES; Editor JON MARSH.

Sunday Examiner: Catholic Diocese Centre, 11/F, 16 Caine Rd; tel. 25220487; fax 25369939; e-mail sundayex@catholic.org.hk; internet sundayex.catholic.org.hk; f. 1946; religious; weekly; Deputy Editor-in-Chief Fr JIM MULRONEY; circ. 6,500.

Textile Asia: c/o Business Press Ltd, California Tower, 11/F, 30–32 D'Aguilar St, GPOB 185, Central; tel. 25233744; fax 28106966; e-mail texasia@netvigator.com; f. 1970; monthly; textile and clothing industry; Publr and Editor-in-Chief KAYSER W. SUNG; circ. 17,000.

Tradefinance Asia: Hong Kong monthly; Editor RICHARD TOURRET.

Travel Business Analyst: GPO Box 12761; tel. 25072310; e-mail TBAoffice@aol.com; internet www.travelbusinessanalyst.com; f. 1982; travel trade; monthly; Editor MURRAY BAILEY.

Chinese Language

Affairs Weekly: Hong Kong; tel. 28950801; fax 25767842; f. 1980; general interest; Editor WONG WAI MAN; circ. 130,000.

Cheng Ming Monthly: Hennessy Rd, POB 20370; tel. 25740664; Chief Editor WAN FAI.

City Magazine: Hang Seng Bldg, 7/F, 200 Hennessy Rd, Wanchai; tel. 28931393; fax 28388761; f. 1976; monthly; fashion, wine, cars, society, etc.; Publr JOHN K. C. CHAN; Chief Editor PETER WONG; circ. 30,000.

Contemporary Monthly: Unit 705, Westlands Centre, 20 Westlands Rd, Quarry Bay; tel. 25638122; fax 25632984; f. 1989; monthly; current affairs; 'China-watch'; Editor-in-Chief CHING CHEONG; circ. 50,000.

Disc Jockey: Fuk Keung Ind. Bldg, B2, 14/F, 66–68 Tong Mei Rd, Taikoktsui, Kowloon; tel. 23905461; fax 27893869; e-mail vinpres@netvigator.com; f. 1990; monthly; music; Publr VINCENT LEUNG; Editor ALGE CHEUNG; circ. 32,000.

Elegance HK: Aik San Bldg, 14/F, 14 Westlands Rd, Quarry Bay; tel. 2963011; fax 25658217; f. 1977; monthly; for thinking women; Chief Editor WINNIE YUEN; circ. 75,000.

Kung Kao Po (Catholic Chinese Weekly): 16 Caine Rd; tel. 25220487; fax 25213095; e-mail kkp@catholic.org.hk; internet kkp.catholic.org.hk; f. 1928; religious; weekly; Editor-in-Chief Fr LOUIS HA.

Lisa's Kitchen Bi-Weekly: Fuk Keung Ind. Bldg, B2, 14/F, 66–68 Tong Mei Rd, Taikoktsui, Kowloon; tel. 23910668; fax 27893869; f. 1984; recipes; Publr VINCENT LEUNG; circ. 50,000.

Metropolitan Weekly: f. 1983; weekly; entertainment, social news; Chief Editor CHARLES YOU; circ. 130,000.

Ming Pao Monthly: Ming Pao Industrial Centre, 15/F, Block A, 18 Ka Yip St, Chai Wan; tel. 25155107; fax 28982566; Chief Editor KOO SIU-SUN.

Motor Magazine: Prospect Mansion, Flat D, 1/F, 66–72 Paterson St, Causeway Bay; tel. 28822230; fax 28823949; f. 1990; Publr and Editor-in-Chief KENNETH LI; circ. 32,000. (Publication suspended, 2001.).

Next Magazine: 8 Chun Ying St, T. K. O. Industrial Estate West, Tseung Kwan O, Hong Kong; tel. 27442733; fax 29907210; internet www.nextmedia.com; f. 1989; weekly; news, business, lifestyle, entertainment; Editor-in-Chief CHEUNG KIM HUNG; circ. 172,708.

Open Magazine: Causeway Bay, POB 31429; tel. 28939197; fax 28915591; e-mail open@open.com.hk; internet www.open.com.hk; f. 1990; monthly; Chief Editor JIN CHONG; circ. 15,000.

Oriental Sunday: Oriental Press Centre, Wang Tai Rd, Kowloon Bay, Kowloon; tel. 27951111; fax 27952299; f. 1991; weekly; leisure magazine; Chair. C. F. MA; circ. 120,000.

Reader's Digest (Chinese Edn): Reader's Digest Association Far East Ltd, 19/F Cyber Centre, 3 Tung Wong Rd, Shau Kei Wan; tel. 28845678; fax 25671479; e-mail chrd@netvigator.com; f. 1965; monthly; Editor-in-Chief JOEL POON; circ. 200,000.

Today's Living: Prospect Mansion, Flat D, 1/F, 66–72 Paterson St, Causeway Bay; tel. 28822230; fax 28823949; e-mail magazine@todayliving.com; f. 1987; monthly; interior design; Publr and Editor-in-Chief KENNETH LI; circ. 35,000.

TV Week: 1 Leighton Rd, Causeway Bay; tel. 28366147; fax 28346717; f. 1967; weekly; Publr PETER CHOW; circ. 59,082.

Yazhou Zhoukan: Block A, Ming Pao Industrial Centre, 15/F, 18 Ka Yip St, Chai Wan; tel. 25155358; fax 25059662; e-mail loppoon@mingpao.com; internet www.yzzk.com; f. 1987; international Chinese news weekly; Chief Editor YAU LOP-POON; circ. 110,000.

Young Girl Magazine: Fuk Keung Ind. Bldg, B2, 14/F, 66–68 Tong Mei Rd, Taikoktsui, Kowloon; tel. 23910668; fax 27893869; f. 1987; biweekly; Publr VINCENT LEUNG; circ. 65,000.

Yuk Long TV Weekly: Hong Kong; tel. 25657883; fax 25659958; f. 1977; entertainment, fashion, etc.; Publr TONY WONG; circ. 82,508.

NEWS AGENCIES

International News Service: 2E Cheong Shing Mansion, 33–39 Wing Hing St, Causeway Bay; tel. 25665668; Rep. AU KIT MING.

Xinhua (New China) News Agency, Hong Kong SAR Bureau: 387 Queen's Rd East, Wanchai; tel. 28314126; f. 2000; from fmr news dept of branch office of Xinhua (responsibility for other activities being assumed by Liaison Office of the Central People's Government in the Hong Kong SAR); Dir ZHANG GUOLIANG.

Foreign Bureaux

Agence France-Presse (AFP): Telecom House, Room 1840, 18/F, 3 Gloucester Rd, Wanchai, GPOB 5613; tel. 28020224; fax 28027292; Regional Dir YVAN CHEMLA.

Agencia EFE (Spain): 10A Benny View House, 63–65 Wong Nai Chung Rd, Happy Valley; tel. 28080199; fax 28823101; Correspondent MIREN GUTIÉRREZ.

Associated Press (AP) (USA): 1282 New Mercury House, Waterfront Rd; tel. 25274324; Bureau Chief ROBERT LIU.

Central News Agency (CNA) Inc (Taiwan): Hong Kong Bureau Chief CONRAD LU.

Jiji Tsushin-Sha (Japan): 3503 Far East Finance Centre, 16 Harcourt Rd; tel. 25237112; fax 28459013; Bureau Man. KATSUHIKO KABASAWA.

Kyodo News Service (Japan): Unit 1303, 13/F, 9 Queen's Rd, Central; tel. 25249750; fax 28105591; e-mail tyoko@po.iijnet.or.jp; Correspondent TSUKASA YOKOYAMA.

Reuters Asia Ltd (United Kingdom): Hong Kong; tel. 258436363; Bureau Man. GEOFF WEETMAN.

United Press International (UPI) (USA): 1287 Telecom House, 3 Gloucester Rd, POB 5692; tel. 28020221; fax 28024972; Vice-Pres. (Asia) ARNOLD ZEITLIN; Editor (Asia) PAUL H. ANDERSON.

PRESS ASSOCIATIONS

Chinese Language Press Institute: Tower A, Sing Tao Bldg, 1 Wang Kwong Rd, Kowloon Bay, Kowloon; tel. 27982501; fax 27953017; Pres. AW SIAN.

Hong Kong Chinese Press Association: Rm 2208, 22/F, 33 Queen's Rd Central; tel. 28613622; fax 28661933; 13 mems; Chair. HUE PUE-YING.

Hong Kong Journalists Association: GPOB 11726, Henfa Commercial Bldg, Flat 15A, 348–350 Lockhart Rd, Waichai; tel. 25910692; fax 25727329; e-mail hkja@hk.super.net; internet www.hkja.org.hk; f. 1968; 413 mems; Chair. CHEUNG PING LING.

Newspaper Society of Hong Kong: Rm 904, 75–83 King's Rd, North Point; tel. 25713102; fax 25712676; f. 1954; Chair. LEE CHO-JAT.

Publishers

Art House of Collectors HK Ltd: 37 Lyndhurst Terrace, Ground Floor, Central; tel. 28818026; fax 28904304; Dir. LI LAP FONG.

Asia 2000 Ltd: 15B The Parkside, 263 Hollywood Rd, Sheung Wan; tel. 25301409; fax 25261107; e-mail sales@asia2000.com.hk; internet www.asia2000.com.hk; Asian studies, politics, photography, fiction; Man. Dir MICHAEL MORROW.

Asian Research Service: GPOB 2232; tel. 25707227; fax 25128050; f. 1972; maps, atlases, monographs on Asian studies and journals; Dir NELSON LEUNG.

Chinese University Press: Chinese University of Hong Kong, Sha Tin, New Territories; tel. 26096508; fax 26036692; e-mail cup@cuhk.edu.hk; internet www.cuhk.edu.hk/cupress; f. 1977; studies on China and Hong Kong and other academic works; Dir Dr STEVEN K. LUK.

Commercial Press (Hong Kong) Ltd: Eastern Central Plaza, 8/F, 3 Yiu Hing Rd, Shau Kei Wan; tel. 25651371; fax 25645277; e-mail webmaster@commercialpress.com.hk; internet www.commercialpress.com.hk; f. 1897; trade books, dictionaries, textbooks, Chinese classics, art, etc.; Man. Dir and Chief Editor CHAN MAN HUNG.

Excerpta Medica Asia Ltd: 8/F, 67 Wyndham St; tel. 25243118; fax 28100687; f. 1980; sponsored medical publications, abstracts, journals etc.

Hoi Fung Publisher Co: 125 Lockhart Rd, 2/F, Wanchai; tel. 25286246; fax 25286249; Dir. K. K. TSE.

Hong Kong University Press: Hing Wai Centre, 14/F, 7 Tin Wan Praya Rd, Aberdeen; tel. 25502703; fax 28750734; e-mail hkupress@hkucc.hku.hk; internet www.hkupress.org; f. 1956; Publr COLIN DAY.

International Publishing Co: Rm 213–215, HK Industrial Technology Centre, 72 Tat Chee Ave, Kowloon Tong, Kowloon; tel. 23148882; fax 23192208; Admin. Man. KAREN CHOW.

Ismay Publications Ltd: C. C. Wu Bldg; tel. 25752270; Man. Dir MINNIE YEUNG.

Ling Kee Publishing Co Ltd: Zung Fu Industrial Bldg, 1067 King's Rd, Quarry Bay; tel. 25616151; fax 28111980; e-mail admin@lingkee.com; internet www.lingkee.com; f. 1956; educational and reference; Chair. B. L. AU; Man. Dir K. W. AU.

Oxford University Press (China) Ltd: Warwick House East, 18/F, 979 King's Rd, Taikoo Place, Quarry Bay; tel. 25163222; fax 25658491; e-mail oupchina@oupchina.com.hk; internet www.oupchina.com.hk; f. 1961; school textbooks, reference, academic and general works relating to Hong Kong, Taiwan and China; Regional Dir SIMON LI.

Taosheng Publishing House: Lutheran Bldg, 3/F, 50A Waterloo Rd, Yau Ma Tei, Kowloon; tel. 23887061; fax 27810413; e-mail taosheng@elchk.org.hk; Dir CHANG CHUN WA.

Textile Asia/Business Press Ltd: California Tower, 11/F, 30–32 D'Aguilar St, GPOB 185, Central; tel. 25233744; fax 28106966; e-mail texasia@netvigator.com; internet www.textileasia-businesspress.com; f. 1970; textile magazine; Man. Dir KAYSER W. SUNG.

The Woods Publishing Co: Li Yuen Building, 2/F, 7 Li Yuen St West, Central; tel. 25233002; fax 28453296; e-mail tybook@netvigator.com; Production Man. TONG SZE HONG.

Times Publishing (Hong Kong) Ltd: Seaview Estate, Block C, 10/F, 2–8 Watson Rd, North Point; tel. 25668381; fax 25080255; e-mail abeditor@asianbusiness.com.hk; internet www.asianbusinessnet.com; trade magazines and directories; CEO COLIN YAM; Executive Editor JAMES LEUNG.

Government Publishing House

Government Information Services: see Government Offices.

PUBLISHERS' ASSOCIATIONS

Hong Kong Publishers' and Distributors' Association: National Bldg, 4/F, 240–246 Nathan Rd, Kowloon; tel. 23674412; 45 mems; Chair. HO KAM-LING; Sec. HO NAI-CHI.

Society of Publishers in Asia: c/o Worldcom Hong Kong, 502–503 Admiralty Centre, Tower I, 18 Harcourt Rd, Admiralty; tel. 28654007; fax 28652559; e-mail worldcom@hkstar.com.

Broadcasting and Communications

TELECOMMUNICATIONS

Asia Satellite Telecommunications Co Ltd (AsiaSat): East Exchange Tower, 23/F, 38–40 Leighton Rd; tel. 28056666; fax 25043875; e-mail wpang@asiasat.com; internet www.asiasat.com; CEO PETER JACKSON.

Pacific Century CyberWorks (PCCW): 39/F PCCW Tower, Taikoo Place, 979 King's Rd, Quarry Bay; tel. 28882888; fax 28778877; e-mail general@pccw.com; internet www.pccw.com; fmrly Cable and Wireless HKT Ltd, acquired by PCCW August 2000; telecommunications/internet services provider; CEO JACK SO.

Regulatory Authority

Telecommunications Authority: statutory regulator, responsible for implementation of the Govt's pro-competition and pro-consumer policies; Dir.Gen. ANTHONY S. K. WONG.

Hutchison Telecom, New T and T Hong Kong Ltd, and New World Telecom also operate local services. In 2000 six companies were licensed to provide mobile telecommunications services, serving over 5.2m. customers.

BROADCASTING

Regulatory Authority

Broadcasting Authority: 39/F Revenue Tower, 5 Gloucester Rd, Wanchai; tel. 25945721; fax 25072219; e-mail ba@tela.gov.hk; internet www.hkba.org.hk; regulatory body; administers and issues broadcasting licences; Chair. DANIEL R. FUNG.

Radio

Hong Kong Commercial Broadcasting Co Ltd: 3 Broadcast Drive, KCPOB 73000; tel. 23365111; fax 23380021; e-mail comradio@crhk.com.hk; internet www.crhk.com.hk; f. 1959; broadcasts in English and Chinese on three radio frequencies; Chair. G. J. HO; Dir and CEO WINNIE YU.

Metro Broadcast Corpn Ltd (Metro Broadcast): Hong Kong; tel. 23649333; fax 23646577; e-mail tech@metroradio.com.hk; internet www.metroradio.com.hk; f. 1991; broadcasts on three channels in English, Cantonese and Mandarin; Gen. Man. CRAIG B. QUICK.

Radio Television Hong Kong: Broadcasting House, 30 Broadcast Drive, Kowloon; tel. 23396300; fax 23380279; e-mail admin@rthk.org.hk; internet www.rthk.org.hk; f. 1928; govt-funded; 24-hour service in English and Chinese on seven radio channels; service in Putonghua inaugurated in 1997; Dir CHU PUI-HING.

Star Radio: Hutchison House, 12/F, 10 Harcourt Rd, Central; f. 1995; satellite broadcasts in Mandarin and English; Gen. Man. MIKE MACKAY.

Television

Asia Television Ltd (ATV): Television House, 81 Broadcast Drive, Kowloon; tel. 29928888; fax 23380438; e-mail atv@hkatv.com; internet www.hkatv.com; f. 1973; operates two commercial television services (English and Chinese) and produces television programmes; Dir and CEO FENG XIAO PING.

STAR Group Ltd: One Harbourfront, 8/F, 18 Tak Fung St, Hunghom, Kowloon; tel. 26218888; fax 26213050; e-mail corp_aff@startv.com; internet www.startv.com; f. 1990; subsidiary of the News Corpn Ltd; broadcasts over 40 channels in English, Hindi, Tamil, Mandarin, Cantonese, Korean and Thai, including a range of sports programmes, music, movies, news, entertainment and documentaries. STAR reaches more than 300m. people in 53 countries across Asia, India and the Middle East, with a daily audience of about 100m. people. STAR also owns a contemporary Chinese film library with more than 600 titles. Additionally, STAR has interests in cable systems in India and Taiwan. Its services also extend to interactive cable TV, radio, wireless and digital media platforms.; Chair. and CEO MICHELLE GUTHRIE; Pres. STEVE ASKEW.

Radio Television Hong Kong: see Radio; produces drama, documentary and public affairs programmes; also operates an educational service for transmission by two local commercial stations; Dir CHU PUI-HING (acting).

Television Broadcasts Ltd (TVB): TVB City, 77 Chun Choi St, Tseung Kwan O Industrial Estate, Kowloon; tel. 23352288; fax 23581300; e-mail external.affairs@tvb.com.hk; internet www.tvb.com; f. 1967; operates Chinese and English language services; two colour networks; Exec. Chair. Sir RUN RUN SHAW; Man. Dir LOUIS PAGE.

Wharf Cable Ltd: Wharf Cable Tower, 4/F, 9 Hoi Shing Rd, Tsuen Wan; tel. 26115533; fax 24171511; f. 1993; 24-hour subscription service of news, sport and entertainment on 35 channels; carries BBC World Service Television; Chair. PETER WOO; Man. Dir STEPHEN NG.

Finance

(cap. = capital; res = reserves; dep. = deposits; m. = million; brs = branches; amounts in Hong Kong dollars unless otherwise stated)

BANKING

In December 2002 there were 133 licensed banks, of which 26 were locally incorporated, operating in Hong Kong. There were also 46 restricted licence banks (formerly known as licensed deposit-taking companies), 45 deposit-taking companies, and 94 foreign banks' representative offices.

Hong Kong Monetary Authority (HKMA): 55/F, Two International Finance Centre, 8 Finance St, Central, Hong Kong; tel. 28788196; fax 28788197; e-mail hkma@hkma.gov.hk; internet www.hkma.gov.hk; f. 1993 by merger of Office of the Commissioner of Banking and Office of the Exchange Fund; carries out central banking functions; maintains Hong Kong dollar stability within the framework of the linked exchange rate system; supervises licensed banks, restricted licence banks and deposit-taking cos, their overseas brs and representative offices; manages foreign currency reserves; Chief Exec. JOSEPH YAM; Deputy Chief Execs PETER PANG, WILLIAM RYBACK, NORMAN CHAN.

Banks of Issue

Bank of China (Hong Kong) Ltd (People's Repub. of China): Bank of China Tower, 1 Garden Rd, Central; tel. 28266888; fax 28105963; internet www.bochk.com; f. 1917; became third bank of issue in May 1994; merged in Oct. 2001 with the local branches of 11 mainland banks (incl. Kwangtung Provincial Bank, Sin Hua Bank Ltd, China and the South Sea Bank Ltd, Kincheng Banking Corpn, China State Bank, National Commercial Bank Ltd, Yien Yieh Commercial Bank Ltd, Hua Chiao Commercial Bank Ltd and Po Sang Bank Ltd, to form the Bank of China (Hong Kong); cap. 43,043m., res 11,314m., dep. 660,044m. (Dec. 2002); CEO HE GUANGBIE; 369 brs.

The Hongkong and Shanghai Banking Corporation Ltd: 1 Queen's Rd, Central; tel. 28221111; fax 28101112; internet www.asiapacific.hsbc.com; f. 1865; personal and commercial banking; cap. 51,603m., res 12,855m., dep. 1,823,109m. (Dec. 2003); Chair. DAVID ELDON; CEO A. MEHTA; more than 600 offices world-wide.

Standard Chartered Bank: Standard Chartered Bank Bldg, 4–4A Des Voeux Rd, Central; tel. 28203333; fax 28569129; internet www.standardchartered.com.hk; f. 1859; CEO PETER WONG.

Other Commercial Banks

Asia Commercial Bank Ltd: Asia Financial Centre, 120 Des Voeux Rd, Central; tel. 25419222; fax 25410009; internet www.asia-commercial.com; f. 1934; fmrly Commercial Bank of Hong Kong; cap. 810.0m., res 460.5m., dep. 11,369.8m. (Dec. 2002); Chair. and CEO ROBIN Y. H. CHAN; Gen. Man. and Exec. Dir STEPHEN TAN; 13 domestic, 1 overseas br.

Bank of East Asia Ltd: Bank of East Asia Bldg, 16/F, 10 Des Voeux Rd, Central; tel. 28423200; fax 28459333; internet www.hkbea.com; inc in Hong Kong in 1918, absorbed United Chinese Bank Ltd in Aug. 2001, and First Pacific Bank (FPB) in Apr. 2002; cap. 3,615.9m., res 13,302.86m., dep. etc. 157,694.1m. (Dec. 2002); Chair. and Chief Exec. DAVID K. P. LI; 99 brs in Hong Kong and 18 overseas brs.

Chiyu Banking Corpn Ltd: 78 Des Voeux Rd, Central; tel. 28430111; fax 25267420; f. 1947; cap. 300m., res 2,970.5m., dep. 24,244.3m. (Dec. 2002); Chair. TAN KONG PIAT; 15 brs.

CITIC Ka Wah Bank Ltd: 232 Des Voeux Rd, Central; tel. 25457131; fax 25417029; e-mail info@citickawahbank.com; internet www.citickawahbank.com; f. 1922; cap. 2,393.3m., res 2,202.4m., dep. 62,783m. (Dec. 2002); acquired Hong Kong Chinese Bank Ltd Jan. 2002; Chair. DAN KONG; CEO KENNETH KONG SIU CHEE; Dir JU WEIMIN; 26 domestic brs, 2 overseas brs.

Dah Sing Bank Ltd: Dah Sing Financial Centre, 36/F, 108 Gloucester Rd, Central; tel. 25078866; fax 25985052; e-mail ops@dahsing.com.hk; internet www.dahsing.com; f. 1947; cap. 800.0m., res 1,057.7m., dep. 47,387.4m. (Dec. 2003); Chair. DAVID S. Y. WONG; Man. Dir DEREK H. H. WONG; 39 domestic brs.

DBS Bank (Hong Kong) Ltd: 11/F, The Center, 99 Queen's Rd, Central; tel. 22188822; fax 21678222; e-mail hkcs@dbs.com; internet www.dbs.com.hk; f. 1938; inc 1954 as Kwong On Bank, name changed 2000; subsidiary of the Development Bank of Singapore; cap. 5,200m., res 10,656.2m., dep. 130,540.8m. (Dec. 2003); acquired Dao Heng Bank and Overseas Trust Bank July 2003; Chair. FRANK WONG; CEO RANDOLPH GORDON SULLIVAN; 32 brs.

Hang Seng Bank Ltd: 83 Des Voeux Rd, Central; tel. 21981111; fax 28684047; e-mail ccd@hangseng.com; internet www.hangseng.com; f. 1933; The Hongkong and Shanghai Banking Corporation Ltd, 62.14%; cap. 9,559m., res 34,005m., dep. 414,765m. (Dec. 2002); Chair. DAVID ELDON; Vice-Chair. and CEO VINCENT CHENG; 155 domestic brs, 7 overseas brs.

Industrial and Commercial Bank of China (Asia): ICBC Tower, 122–126 Queen's Rd, Central; tel. 25343333; fax 28051166; internet www.icbcasia.com; f. 1964; fmrly Union Bank of Hong Kong; cap. 2,259.8m., res 3,691.5m., dep. 52,514.2m. (Dec. 2002); Chair. JIANG JIANQING; 20 brs.

International Bank of Asia Ltd: International Bank of Asia Bldg, 38 Des Voeux Rd, Central; tel. 28426222; fax 28401190; e-mail iba-info@iba.com.hk; internet www.iba.com.hk; f. 1982 as Sun Hung Kai Bank Ltd, name changed 1986; subsidiary of Arab Banking Corpn; cap. 1,172.2m., res 1,124.5m., dep. 33,501.1m. (Dec. 2003); Man. Dir and CEO MIKE M. MURAD; 26 brs.

Jian Sing Bank Ltd: F/41, Tower 1, Lippo Centre, 89 Queensway, Hong Kong; tel. 25410088; fax 25447145; e-mail admin@jsb.com.hk; internet www.jsb.com.hk; f. 1964 as Hongkong Industrial and Commercial Bank Ltd, acquired by Dah Sing Bank Ltd in 1987, 40% interest acquired by China Construction Bank in 1994, resulting in name change, China Construction Bank acquired 100% stake in 2002; subsidiary of China Construction Bank; cap. 300.0m., res 62.5m., dep. 2,361.9m. (Dec. 2002); Chair. LUO ZHEFU; CEO and Dir PATRICK P. T. HO.

Liu Chong Hing Bank Ltd: POB 2535, G/F New World Tower, 16–18 Queen's Rd, Central; tel. 28417417; fax 28459134; e-mail info@lchbank.com; internet www.lchbank.com; f. 1948; cap. 217.5m., res 5,726.9m., dep. 33,627.0m. (Dec. 2003); Chair. and Man. Dir LIU LIT-MAN; 34 domestic brs, 3 overseas brs.

Nanyang Commercial Bank Ltd: Nanyang Commercial Bank Bldg, 151 Des Voeux Rd, Central; tel. 28520888; fax 28153333; e-mail webmaster@ncb.com.hk; internet www.ncb.com.hk; f. 1949; cap. p.u. 600m.(Dec. 2002), res 9,342.1m., dep. 73,041.4m. (June 2003); Chair. HE GUANGBEI; 41 brs, 6 mainland brs, 1 overseas br.

Shanghai Commercial Bank Ltd: 12 Queen's Rd, Central; tel. 28415415; fax 28104623; e-mail contact@shacombank.com.hk; internet www.shacombank.com.hk; f. 1950; cap. 2,000m., res 7,349.4m., dep. 62,143.6m. (Dec. 2002); CEO, Man. Dir and Gen. Man. JOHN KAM-PAK YAN; 40 domestic brs, 4 overseas brs.

Standard Bank Asia Ltd (Standard Jardine Fleming Bank): 36/F, Two Pacific Place, 88 Queensway; tel. 28227888; fax 28227999; e-mail ashbanking@standardbank.com.hk; internet www.standardbank.com; f. 1970 as Jardine Fleming & Company Ltd, renamed Jardine Fleming Bank Ltd in 1993, absorbed by Standard Bank Investment Corpn Ltd and name changed as present in July 2001; cap. 66m., res 52.3m., dep. 1,571.4m. (Dec. 2001); Chair. PIETER PRINSLOO.

Tai Yau Bank Ltd: 16/F Tak Shing House, 20 Des Voeux Rd, Central; tel. 25229002; fax 28685334; f. 1947; cap. 300.0m., res 129.2m., dep 1,495.2m. (Dec. 2003); Chair. KO FOOK KAU.

Wing Hang Bank Ltd: 161 Queen's Rd, Central; tel. 28525111; fax 25410036; e-mail whbpsd@whbhk.com; internet www.whbhk.com; f. 1937; cap. 293.4m., res 5,751.9m., dep. 48,597.2m. (Dec. 2001); acquired Chekiang First Bank Ltd, Aug. 2004; Chair. and Chief Exec. PATRICK Y. B. FUNG; 26 domestic brs, 12 overseas brs.

Wing Lung Bank Ltd: 45 Des Voeux Rd, Central; tel. 28268333; fax 28100592; e-mail wlb@winglungbank.com; internet www.winglungbank.com; f. 1933; cap. 1,161.0m., res 7,447.7m., dep. 55,200.6m. (Dec. 2003); Chair. MICHAEL PO-KO WU; Exec. Dir and Gen. Man. CHE-SHUM CHUNG; 33 domestic brs, 2 representative offices in China.

Principal Foreign Banks

ABN AMRO Bank NV (Netherlands): Edinburgh Tower, 3–4/F, Landmark, 15 Queen's Rd, Central; tel. 28429211; fax 28459049; CEO (China) SERGIO RIAL; 3 brs.

American Express Bank Ltd (USA): One Pacific Place, 36/F, 88 Queensway, Central; tel. 28440688; fax 28453637; Senior Country Exec. DOUGLAS H. SHORT III; 3 brs.

Australia and New Zealand Banking Group Ltd: 27/F, One Exchange Square, 8 Connaught Place, Central; tel. 28437111; fax 28680089; Gen. Man. PETER RICHARDSON.

Bangkok Bank Public Co Ltd (Thailand): Bangkok Bank Bldg, 28 Des Voeux Rd, Central; tel. 28016688; fax 28451805; Gen. Man. CHEN MAN YING; 2 brs.

Bank of America (Asia) Ltd: GPOB 799, 2/F, Bank of America Tower, 12 Harcourt Rd; tel. 28476666; fax 28100821; internet www.bankofamerica.com.hk; Chair. COLM MCCARTHY; Man. Dir and CEO FREDERICK CHIN.

Bank of Communications, Hong Kong Branch: 20 Pedder St, Central; tel. 28419611; fax 28106993; f. 1934; Gen. Man. FANG LIANKUI; 41 brs.

Bank of India: Ruttonjee House, 2/F, 11 Duddell St, Central; tel. 25240186; fax 28106149; e-mail boihk@netvigator.com; Chief Exec. O. P. GUPTA.

Bank of Scotland: Jardine House, 15/F, 1 Connaught Place, Central; tel. 25212155; fax 28459007; Regional Dir I. A. MCKINNEY; 1 br.

Barclays Capital Asia Ltd: Citibank Tower, 42/F, 3 Garden Rd, Central; tel. 29032000; fax 29032999; internet www.barclayscapital.com; f. 1972; Chair. and CEO ROBERT A. MORRICE.

BNP Paribas (France): Central Tower, 4–14/F, 28 Queen's Rd, Central; tel. 29098888; fax 25302707; e-mail didier.balme@bnpgroup.com; internet www.bnpparibas.com.hk; f. 1958; Man. DIDIER BALME; 2 brs.

Citibank, NA (USA): Citibank Tower, 39–40/F and 44–50/F, Citibank Plaza, 3 Garden Rd, Central; tel. 28688888; fax 23068111; 20 brs.

Commerzbank AG (Germany): Hong Kong Club Bldg, 21/F, 3A Chater Rd, Central; tel. 28429666; fax 28681414; 1 br.

Crédit Agricole Indosuez (France): One Exchange Square, 42–45/F, 8 Connaught Rd, Central; tel. 28489000; fax 28681406; Sr Country Officer Charles Reybet-Degat; 1 br.

Deutsche Bank AG (Germany): 51–56/F, Cheung Kong Center, 2 Queen's Rd, Central; tel. 22038888; fax 28459056; Gen. Mans Dr Michael Thomas, Reiner Rusch; 1 br.

Equitable PCI Bank (Philippines): 7/F, No. 1, Silver Fortune Plaza, Wellington St; tel. 28680323; fax 28100050; Vice-Pres. Paul Lang; 1 br.

Fortis Bank (Belgium): Fortis Bank Tower, 26/F, 77–79 Gloucester Rd, Wanchai; tel. 28230456; fax 25276851; e-mail info@fortisbank.com.hk; internet www.fortisbank.com.hk; Gen. Man. David Yu; 28 brs.

Indian Overseas Bank: POB 182, Ruttonjee House, 3/F, 11 Duddell St, Central; tel. 25227249; fax 28450159; 2 brs.

JP Morgan Chase Bank (USA): 39/F, One Exchange Square, Connaught Place, Central; tel. 28431234; fax 28414396.

Malayan Banking Berhad (Malaysia): Entertainment Bldg, 18–19/F, 30 Queen's Rd, Central; tel. 25227141; fax 28106013; trades in Hong Kong as Maybank; Man. Hwan Woon Han; 2 brs.

Mevas Bank: 36/F, Dah Sing Financial Centre, 108 Gloucester Rd; tel. 31013286; fax 31013298; e-mail contactus@mevas.com; internet www.mevas.com; Chair. David S. Y. Wong.

Mizuho Corporate Asia (HK) Ltd (Japan): 17/F, 2 Pacific Place, 88 Queensway, Admiralty; tel. 21033040; fax 28101326; Man. Dir and CEO Noboru Akatsuka; 1 br.

National Bank of Pakistan: 18/F, ING Tower, 308–320 Des Voeux Rd, Central; tel. 25217321; fax 28451703; e-mail nbphkkm@netvigator.com; CEO Ghulam Hussain Azhar; 2 brs.

Oversea-Chinese Banking Corpn Ltd (Singapore): 9/F, 9 Queen's Rd, Central; tel. 28682086; fax 28453439; Gen. Man. Benjamin Yeung; 3 brs.

Philippine National Bank: Regent Centre Bldg, 7/F, 88 Queen's Rd, Central; tel. 25253638; fax 25253107; e-mail itdept@pnbhk.com; Sr Vice-Pres. and Gen. Man. Articer O. Quebal; 1 br.

N. M. Rothschild and Sons (Hong Kong) Ltd: 16/F, Alexandra House, 16–20 Chater Rd, Central; tel. 25255333; fax 28681728; e-mail jackson.woo@rothschild.com.hk; internet www.rothschild.com.hk; Dir Jackson Woo.

Société Générale Asia Ltd (France): 42/F, Edinburgh Tower, The Landmark, 15 Queen's Rd; tel. 25838600; fax 28400738; internet www.sgcib.com; CEO Jackson Cheung.

Sumitomo Mitsui Banking Corpn (SMBC) (Japan): 7–8F/, One International Finance Centre, 1 Harbour View St, Central; tel. 22062000; fax 22062888; Gen. Man. Toshio Morikawa; 1 br.

Tokyo-Mitsubishi International (HK) Ltd (Japan): 14/F, Tower 1, Admiralty Centre, 18 Harcourt Rd; tel. 28627888; fax 25291550; Man. Dir and CEO Yoshiaki Watanabe.

UBAF (Hong Kong) Ltd (France): Far East Finance Centre, 18/F, 16 Harcourt Rd, Central; tel. 25201361; fax 25274256; e-mail info@ubafhk.com; internet www.ubafhk.com; Man. Dir G. Alejandro.

UFJ International Finance Asia Ltd (Japan): 6/F, Hong Kong Club Bldg, 3A Chater Rd, Central; tel. 25334300; fax 28453518; internet www.ufjifal.com; CEO and Man. Dir Akihiko Kobayashi.

United Overseas Bank Ltd (Singapore): United Overseas Bank Bldg, 54–58 Des Voeux Rd, Central; tel. 28425666; fax 28105773; Sr Vice-Pres. and CEO Robert Chan Tze Leung; 5 brs.

Banking Associations

The Chinese Banks' Association Ltd: South China Bldg, 5/F, 1–3 Wyndham St, Central; tel. 25224789; fax 28775102; 1,666 mems; Chair. Bank of East Asia.

The DTC Association (The Hong Kong Association of Restricted Licence Banks and Deposit-Taking Companies): Suite 3738, 37/F, Sun Hung Kai Centre, 30 Harbour Rd; tel. 25264079; fax 25230180.

The Hong Kong Association of Banks: Rm 525, Prince's Building, Central, Hong Kong; tel. 25211169; fax 28685035; e-mail info@hkab.org.hk; internet www.hkab.org.hk; f. 1981 to succeed The Exchange Banks' Assen of Hong Kong; all licensed banks in Hong Kong are required by law to be mems of this statutory body, the function of which is to represent and further the interests of the banking sector; 133 mems; Chair. Standard Chartered Bank (Hong Kong) Ltd; Sec. Rona Morgan.

STOCK EXCHANGE

Honk Kong Exchanges and Clearing Ltd: 1 International Finance Centre, 12/F, 1 Harbour View St, Central; tel. 25221122; fax 22953106; e-mail info@hkex.com.hk; internet www.hkex.com.hk; f. 2000 by unification of the Stock Exchange of Hong Kong, the Hong Kong Futures Exchange and the Hong Kong Securities Clearing Co; 572 mems; Chair. Lee Yeh Kwong; CEO Kwong Ki Chi.

In 1998 Exchange Fund Investment Ltd was established by the Government to manage its stock portfolio acquired during the intervention of August (Chair. Yang Ti-liang).

SUPERVISORY BODY

Securities and Futures Commission (SFC): Edinburgh Tower, 12/F, The Landmark, 15 Queen's Rd, Central; tel. 28409222; fax 28459553; e-mail enquiry@hksfc.org.hk; internet www.hksfc.org.hk; f. 1989 to supervise the stock and futures markets; Chair. Andrew Sheng; Dep. Chair. Laura Cha.

INSURANCE

In December 2002 there were 195 authorized insurance companies, including 99 overseas companies. The following are among the principal companies:

Asia Insurance Co Ltd: World-Wide House, 16/F, 19 Des Voeux Rd, Central; tel. 28677988; fax 28100218; e-mail kclau@asiainsurance.com.hk; internet www.asiainsurance.com.hk; Chair. Sebastian Ki Chit Lau.

CGU International Insurance plc: Cityplaza One, 9/F, Taikoo Shing; tel. 28940555; fax 28905741; e-mail cguasia.com; Gen. Man. Andrew Lo.

Hong Kong Export Credit Insurance Corpn: South Seas Centre, Tower I, 2/F, 75 Mody Rd, Tsim Sha Tsui East, Kowloon; fax 27226277; Commr D. K. Dowding.

Mercantile and General Reinsurance Co PLC: 13c On Hing Bldg, 1 On Hing Terrace, Central; tel. 28106160; fax 25217353; Man. T. W. Ho.

Ming An Insurance Co (HK) Ltd: Ming An Plaza, 19/F, 8 Sunning Rd, Causeway Bay; tel. 28151551; fax 25416567; e-mail mai@mingan.com.hk; internet www.mingan.com; Dir and Gen. Man. K. P. Cheng.

National Mutual Insurance Co (Bermuda) Ltd: 151 Gloucester Rd, Wanchai; tel. 25191111; fax 25987204; life and general insurance; Chair. Sir David Akers-Jones; CEO Terry Smith.

Prudential Assurance Co Ltd: Cityplaza 4, 10/F, 12 Taikoo Wan Rd, Taikoo Shing; tel. 29773888; fax 28776994; life and general; CEO James C. K. Wong.

Royal and Sun Alliance (Hong Kong) Ltd: Dorset House, 32/F, Taikoo Place, 979 King's Road, Quarry Bay; tel. 29683000; fax 29685111; Man. Dir Keith Land.

Summit Insurance (Asia) Ltd: Sunshine Plaza, 25/F, 253 Lockhart Rd, Wanchai; tel. 21059000; fax 25166992; e-mail psi@hcg.com.hk; internet www.hsinchong.com/summit; CEO Iu Po Sing.

Willis China (Hong Kong) Ltd: 3502, The Lee Gardens, 33 Hysan Ave, Causeway Bay; tel. 28270111; fax 28270966; internet www.willis.com; Man. Dir Kirk Austin.

Winterthur Swiss Insurance (Asia) Ltd: Dah Sing Financial Centre, 19/F, 108 Gloucester Rd, Wanchai; tel. 25986282; fax 25985838; Man. Dir Allan Yu.

Insurance Associations

Hong Kong Federation of Insurers (HKFI): First Pacific Bank Centre, Room 902, 9/F, 56 Gloucester Rd, Wanchai; tel. 25201868; fax 25201967; e-mail hkfi@hkfi.org.hk; internet www.hkfi.org.hk; f. 1988; 117 general insurance and 42 life insurance mems; Chair. Choy Chung Foo; Exec. Dir Louisa Fong.

Insurance Institute of Hong Kong: GPO Box 6747; tel. 25825601; fax 28276033; internet www.iihk.org.hk; f. 1967; Pres. Stephen Law.

Trade and Industry

Hong Kong Trade Development Council: Office Tower, 38/F, Convention Plaza, 1 Harbour Rd, Wanchai; tel. 1830668; fax 28240249; e-mail hktdc@tdc.org.hk; internet www.tdctrade.com; f. 1966; Chair. Peter Woo Kwong-ching; Exec. Dir Michael Sze.

Trade and Industry Department: Trade and Industry Department Tower, 700 Nathan Rd, Kowloon; tel. 23985333; fax 27892491; e-mail enquiry@tid.gov.hk; internet www.tid.gov.hk; Dir-Gen. Kevin Ho.

DEVELOPMENT ORGANIZATIONS

Hong Kong Housing Authority: 33 Fat Kwong St, Homantin, Kowloon; tel. 27615002; fax 27621110; f. 1973; plans, builds and manages public housing; Chair. Dominic Wong; Dir of Housing J. A. Miller.

Hong Kong Productivity Council: HKPC Bldg, 78 Tat Chee Ave, Yau Yat Chuen, Kowloon Tong, Kowloon; tel. 27885678; fax 27885900; e-mail bettylee@hkpc.org; internet www.hkpc.org; f. 1967 to promote increased productivity of industry and to encourage optimum utilization of resources; council of 23 mems appointed by the Government, representing management, labour, academic and professional interests, and govt depts associated with productivity matters; Chair. KENNETH FANG; Exec. Dir THOMAS TANG.

Kadoorie Agricultural Aid Loan Fund: c/o Director of Agriculture, Fisheries and Conservation, Cheung Sha Wan Govt Offices, 5/F, 303 Cheung Sha Wan Rd, Kowloon; tel. 21506666; fax 23113731; e-mail mailbox@afcd.gov.hk; f. 1954; provides low-interest loans to farmers; HK $10,395,000 was loaned in 2003/04.

J. E. Joseph Trust Fund: c/o Director of Agriculture, Fisheries and Conservation, Cheung Sha Wan Govt Offices, 5/F, 303 Cheung Sha Wan Rd, Kowloon; tel. 21506666; fax 23113731; e-mail mailbox@afcd.gov.hk; f. 1954; grants low-interest credit facilities to farmers and farmers' co-operative socs; HK $7,845,000 was loaned in 2003/04.

CHAMBERS OF COMMERCE

Chinese Chamber of Commerce, Kowloon: 2/F, 8–10 Nga Tsin Long Rd, Kowloon; tel. 23822309; f. 1936; 234 mems; Chair. and Exec. Dir YEUNG CHOR-HANG.

The Chinese General Chamber of Commerce: 4/F, 24–25 Connaught Rd, Central; tel. 25256385; fax 28452610; e-mail cgcc@cgcc.org.hk; internet www.cgcc.org.hk; f. 1900; 6,200 mems; Chair. Dr TSANG HIN-CHI.

Hong Kong General Chamber of Commerce: United Centre, 22/F, 95 Queensway, POB 852; tel. 25299229; fax 25279843; e-mail chamber@chamber.org.hk; internet www.chamber.org.hk; f. 1861; 4,000 mems; Chair. ANTHONY NIGHTINGALE; CEO EDEN WOON.

Kowloon Chamber of Commerce: KCC Bldg, 3/F, 2 Liberty Ave, Homantin, Kowloon; tel. 27600393; fax 27610166; e-mail kcc02@hkkcc.biz.com.hk; internet www.hkkcc.org.hk; f. 1938; 1,640 mems; Chair. TONG KWOK-WAH; Sec. of Gen. Affairs CHENG PO-WO.

FOREIGN TRADE ORGANIZATIONS

Hong Kong Chinese Importers' and Exporters' Association: Champion Bldg, 7–8/F, 287–291 Des Voeux Rd, Central; tel. 25448474; fax 25444677; e-mail info@hkciea.org.hk; internet www.hkciea.org.hk; f. 1954; 3,000 mems; Pres. WONG TING KWONG.

Hong Kong Exporters' Association: Room 824–825, Star House, 3 Salisbury Rd, Tsimshatsui, Kowloon; tel. 27309851; fax 27301869; e-mail exporter@exporters.org.hk; internet www.exporters.org.hk; f. 1955; 630 mems comprising leading merchants and manufacturing exporters; Pres. CLIFF K. SUN; Exec. Dir SHIRLEY SO.

INDUSTRIAL AND TRADE ASSOCIATIONS

Chinese Manufacturers' Association of Hong Kong: CMA Bldg, 64 Connaught Rd, Central; tel. 25456166; fax 25414541; e-mail info@cma.org.hk; internet www.cma.org.hk; f. 1934 to promote and protect industrial and trading interests; operates testing and certification laboratories; 3,700 mems; Pres. CHAN WING KEE; Exec. Dir FRANCIS T. M. LAU.

Federation of Hong Kong Garment Manufacturers: Cheung Lee Commercial Bldg, Room 401–3, 25 Kimberley Rd, Tsimshatsui, Kowloon; tel. 27211383; fax 23111062; e-mail fhkgmfrs@hkstar.com; internet www.garment.org.hk; f. 1964; 180 mems; Pres. NORMAN TAM; Sec.-Gen. MICHAEL LEUNG.

Federation of Hong Kong Industries (FKHI): Hankow Centre, 4/F, 5–15 Hankow Rd, Tsimshatsui, Kowloon; tel. 27323188; fax 27213494; e-mail fhki@fhki.org.hk; internet www.fhki.org.hk; f. 1960; 3,000 mems; Chair. ANDREW LEUNG.

Federation of Hong Kong Watch Trades and Industries Ltd: Peter Bldg, Room 604, 58–62 Queen's Rd, Central; tel. 25233232; fax 28684485; e-mail hkwatch@netvigator.com; internet www.hkwatch.org; f. 1947; 650 mems; Chair. LUTHER WONG.

Hong Kong Association for the Advancement of Science and Technology Ltd: 2A, Tak Lee Commercial Bldg, 113–17 Wanchai Rd, Wanchai; tel. 28913388; fax 28381823; e-mail info@hkaast.org.hk; internet www.hkaast.org.hk; Pres. Prof. LEE WING BUN.

Hong Kong Biotechnology Association Ltd: Rm 789, HITEC, 1 Trademart Drive, Kowloon Bay, Kowloon; tel. 26209955; fax 26201238; e-mail etang@hkbta.org.hk; internet www.hkbta.org.hk; Chair. LO YUK LAM.

Hong Kong Chinese Enterprises Association: Harbour Centre, Room 2104–6, 25 Harbour Rd, Wanchai; tel. 28272831; fax 28272606; e-mail info@hkcea.com; internet www.hkcea.com; f. 1991; 1,000 mems; Chair. LIU JINBAO; Gen. Sec. LIU GUOYUAN.

Hong Kong Chinese Textile Mills Association: 11/F, 38–40 Tai Po Rd, Sham Shiu Po, Kowloon; tel. 27778236; fax 27881836; f. 1931; 150 mems; Pres. LEE CHUNG-CHIU.

Hong Kong Construction Association Ltd: 3/F, 180–182 Hennessy Rd, Wanchai; tel. 25724414; fax 25727104; e-mail admin@hkca.com.hk; internet www.hkca.com.hk; f. 1920; 372 mems; Pres. BILLY WONG; Sec.-Gen. PATRICK CHAN.

Hong Kong Electronic Industries Association Ltd: Rm 1201, 12/F, Harbour Crystal Centre, 100 Granville Rd, Tsimshatsui, Kowloon; tel. 27788328; fax 27882200; e-mail hkeia@hkeia.org; internet www.hkeia.org; 350 mems; Chair. Dr K. B. CHAN; Exec. Dir CHARLES CHAPMAN.

Hong Kong Garment Manufacturers Association: 401–3, Cheung Lee Commercial Bldg, 25 Kimberley Rd, Tsimshatsui, Kowloon; tel. 23052893; fax 23052493; e-mail mleung@textilecouncil.com; f. 1987; 40 mems; Chair. PETER WANG.

Hong Kong Information Technology Federation Ltd: The Center, 21/F, 99 Queen's Rd, Central; tel. 22878017; fax 22878038; e-mail info@hkitf.com; internet www.hkitf.org.hk; 316 mems; f. 1980; Pres. CHARLES MOK.

Hong Kong Jewellery and Jade Manufacturers Association: Flat A, 12/F, Kaiser Estate Phase 1, 41 Man Yue St, Hunghom, Kowloon; tel. 25430543; fax 28150164; e-mail hkjja@hkstar.com; internet www.jewellery-hk.org; f. 1965; 227 mems; Pres. CHARLES CHAN; Chair. KING LI; Gen. Man. CATHERINE CHAN.

Hong Kong Jewelry Manufacturers' Association: Unit G, 2/F, Kaiser Estate Phase 2, 51 Man Yue St, Hunghom, Kowloon; tel. 27663002; fax 23623647; e-mail hkjma@jewelry.org.hk; internet www.jewelry.org.hk; f. 1988; 260 mems; Chair. AARON SHUM.

Hong Kong Knitwear Exporters and Manufacturers Association: Cheung Lee Commercial Bldg, Rm 401–03, Tsimshatsui, Kowloon; tel. 27552621; fax 27565672; f. 1966; 80 mems; Chair. WILLY LIN; Exec. Sec. SHIRLEY LIU.

Hong Kong and Kowloon Footwear Manufacturers' Association: Kam Fung Bldg, 3/F, Flat D, 8 Cleverly St, Sheung Wan; tel. and fax 25414499; 88 mems; Pres. LOK WAI-TO; Sec. LEE SUM-HUNG.

Hong Kong Optical Manufacturers' Association Ltd: 2/F, 11 Fa Yuen St, Mongkok, Kowloon; tel. 23326505; fax 27705786; e-mail hkoma@netvigator.com; internet www.hkoptical.org.hk; f. 1982; 112 mems; Pres. HUI LEUNG-WAH.

Hong Kong Plastics Manufacturers Association Ltd: Fu Yuen Bldg, 1/F, Flat B, 39–49 Wanchai Rd; tel. 25742230; fax 25742843; f. 1957; 200 mems; Chair. JEFFREY LAM; Pres. DENNIS H. S. TING.

Hong Kong Printers Association: 1/F, 48–50 Johnston Rd, Wanchai; tel. 25275050; fax 28610463; e-mail printers@hkprinters.org; internet www.hkprinters.org; f. 1939; 437 mems; Chair. HO KA-HUN.

Hong Kong Rubber and Footwear Manufacturers' Association: Kar Tseuk Bldg, Block A, 2/F, 185 Prince Edward Rd, Kowloon; tel. 23812297; fax 23976927; e-mail hkrfma@netvigator.com; f. 1948; 180 mems; Chair. CHEUNG KAM; Pres. BENJAMIN KO.

Hong Kong Sze Yap Commercial and Industrial Association: Cosco Tower, Unit 1205–6, 183 Queen's Rd, Central; tel. 25438095; fax 25449495; f. 1909; 1,082 mems; Chair. LOUIE CHICK-NAN; Sec. WONG KA CHUN.

Hong Kong Toys Council: Hankow Centre, 4/F, 5–15 Hankow Rd, Tsimshatsui, Kowloon; tel. 27323188; fax 27213494; e-mail fhki@fhki.org.hk; internet www.toyshk.org; f. 1986; 200 mems; Chair. SAMSON CHAN.

Hong Kong Watch Manufacturers' Association: Yu Wing Bldg, 3/F and 11/F, Unit A, 64–66 Wellington St, Central; tel. 25225238; fax 28106614; e-mail hkwma@netvigator.com; internet www.hkwma.org; 616 mems; Pres. KELVIN K. W. LAU; Sec.-Gen. RICKY W. K. LAW.

Information and Software Industry Association Ltd: Suite 2, 8/F, Tower 6, China Hong Kong City, 33 Canton Rd, Tsimshatsui; tel. 26222867; fax 26222731; e-mail info@isia.org.hk; internet www.isia.org.hk; Chair. SATTI WONG.

Internet and Telecom Association of Hong Kong: GPOB 13461; tel. 25042732; fax 25042752; e-mail info@itahk.org.hk; internet www.itahk.org.hk; 130 mems; Chair. TONY HAU.

New Territories Commercial and Industrial General Association Ltd: Cheong Hay Bldg, 2/F, 107 Hoi Pa St, Tsuen Wan; tel. 24145316; fax 24934130; f. 1973; 2,663 mems; Pres. LAU YUE SUN; Chair. HOK LIM WAN; Sec.-Gen. KAM CHUEN NGAN.

Real Estate Developers Association of Hong Kong: Worldwide House, Room 1403, 19 Des Voeux Rd, Central; tel. 28260111; fax 28452521; f. 1965; 779 mems; Pres. Dr STANLEY HO; Chair. KEITH KERR; Sec.-Gen. LOUIS LOONG.

Textile Council of Hong Kong Ltd: 401-3, Cheung Lee Commercial Bldg, 25 Kimberley Rd, Tsimshatsui , Kowloon; tel. 23052893; fax 23052493; e-mail mleung@textilecouncil.com; internet www.textilecouncil.com; f. 1989; 10 mems; Chair. ANDREW LEUNG; Exec. Dir MICHAEL LEUNG.

Toys Manufacturers' Association of Hong Kong Ltd: Room 1302, Metroplaza, Tower 2, 223 Hing Fong Rd, Kwai Chung, New Territories; tel. 24221209; fax 24221639; e-mail tm_hk@hotmail.com; internet www.tmhk.net; 250 mems; Pres. ARTHUR CHAN; Sec. Y. M. KO.

EMPLOYERS' ORGANIZATIONS

Employers' Federation of Hong Kong: Suite 2004, Sino Plaza, 255–257 Gloucester Rd, Causeway Bay; tel. 25280536; fax 28655285; e-mail efhk@efhk.org.hk; internet www.efhk.org.hk; f. 1947; 446 mems; Chair. VICTOR APPS; Exec. Dir JACKIE MA.

Hong Kong Factory Owners' Association Ltd: Wing Wong Bldg, 11/F, 557–559 Nathan Rd, Kowloon; tel. 23882372; fax 23857129; f. 1982; 1,179 mems; Pres. HWANG JEN; Sec. CHA KIT YEN.

UTILITIES

Electricity

CLP Power Ltd: 147 Argyle St, Kowloon; tel. 26788111; fax 27604448; internet www.clpgroup.com; f. 1918; fmrly China Light and Power Co Ltd; generation and supply of electricity to Kowloon and the New Territories; Chair. MICHAEL D. KADOORIE; Man. Dir ROSS SAYERS.

The Hongkong Electric Co Ltd: 44 Kennedy Rd; tel. 28433111; fax 28100506; e-mail mail@hec.com.hk; internet www.hec.com.hk; generation and supply of electricity to Hong Kong Island, and the islands of Ap Lei Chau and Lamma; Chair. GEORGE C. MAGNUS; Man. Dir K. S. TSO.

Gas

Gas Authority: all gas supply cos, gas installers and contractors are required to be registered with the Gas Authority. At the end of 2002 there were seven registered gas supply cos.

Hong Kong and China Gas Co Ltd: 23/F, 363 Java Rd, North Point; tel. 29633388; fax 25632233; internet www.towngas.com; production, distribution and marketing of town gas and gas appliances; operates two plants; Chair. LEE SHAU-KEE; Man. Dir ALFRED W. K. CHAN.

Water

Drainage Services Department: responsible for planning, designing, constructing, operating and maintaining the sewerage, sewage treatment and stormwater drainage infrastructures.

Water Supplies Department: tel. 28294709; fax 25881594; e-mail wsdinfo@wsd.gov.hk; internet www.info.gov.hk/wsd/; responsible for water supplies; approx. 2.4m. customers (2001).

MAJOR COMPANIES

The following are among Hong Kong's leading companies. Capital, reserves and sales are given in HK dollars unless otherwise stated.

Akai Holdings Ltd: Two Exchange Sq., 30/F, 8 Connaught Place, Central; tel. 25241043; fax 28453558; cap. and res US $405m.(1998/99), sales US $2,224m. (2000); fmrly Semi-Tech (Global) Co Ltd; investment holding co, its subsidiaries' activities incl. manufacturing, retailing and distribution of sewing machines, consumer durables, electronics and audio and video products; Chair. JAMES H. TING.

Amoy Properties Ltd: Standard Chartered Bank Bldg, 28/F, 4 Des Voeux Rd, Central; tel. 28790111; fax 28686086; e-mail amoy@hanglung.com; internet www.hanglung.com/amoy/home.html; f. 1949; cap. and res 27,477m. (1999/2000), sales 257m. (1997/98); property investment and management, investment holding and car-park management; Chair. RONNIE CHICHUNG CHAN; Man. Dir NELSON WAI LEUNG YUEN; 807 employees.

Brilliance China Automotive Holdings: jt ventures with several mainland vehicle manufactures and Toyota (Japan); see Major Companies of the People's Republic of China.

Cheung Kong (Holdings) Ltd: Cheung Kong Centre, 7/F, 2 Queen's Rd, Central; tel. 21288888; fax 28452940; internet www.ckh.com.hk; cap. and res 165,473m., sales 7,486m. (2001); investment holding, project management, property development; Chair. Dr LI KA-SHING; Man. Dir VICTOR LI; 5,460 employees.

China Resources Enterprise Ltd: Rm 3908, China Resources Bldg, 26 Harbour Road, Wanchai; tel. 28271028; fax 25988453; e-mail charlie@cre.com.hk; internet www.cre.com.hk; f. 1992; revenue 24,196m., cap. and res 11,978m. (2001); food and beverages, distribution and trading, retailing; Chair. NING GAONING; Man. Dir YAN BIAO.

Chow Sang Sang Holdings International Ltd: Chow Sang Sang Bldg, 4/F, 229 Nathan Rd, Kowloon; tel. 27300111; fax 27309683; internet www.chowsangsang.com; cap. 110m.(1999), sales 4,686.2m. (2001); manufacturing, retailing and trading of jewellery, gold and other precious metals; Chair. CHOW KWEN-LIM; 1,389 employees.

CITIC Pacific Ltd: CITIC Tower, 32/F, 1 Tim Mei Ave, Central; tel. 28202111; fax 28772771; e-mail contact@citicpacific.com; internet www.citicpacific.com; cap. and res 38,178m., sales 26,424m. (1999), revenue 17,251m. (2001); investment holding, power generation, construction of roads, bridges and tunnels; Chair. LARRY YUNG CHI KIN; Man. Dir HENRY FAN HUNG LING; 11,354 employees.

Crocodile Garments Ltd: Lai Sun Commercial Centre, 11/F, 680 Cheung Sha Wan Rd, Kowloon; tel. 27853898; fax 27860190; e-mail raymond@crocodile.com.hk; internet www.crocodile.com.hk; cap. and res 470.7m. (1998/9), sales 679.9m. (2001); manufacture and sale of garments; Chair. LIM POR YEN; 1,270 employees.

Dah Chong Hong Holdings Ltd: 8/F 20 Kai Cheung Rd, Kowloon Bay, Kowloon; tel. 27683388; fax 27968838; e-mail dch@dch.com.hk; internet www.dch.com.hk; f. 1946; sales 220.8m. (2001/02); manufacture and retail of foods; Chief Exec. JOSEPH PANG CHO HUNG; 600 employees.

Daido Concrete (H.K.) Ltd: Tai Po Industrial Estate, 3 Dai Shing St, New Territories; tel. 26673630; fax 26648125; internet www.daidohk.com; cap. and res 118.8m. (1999), sales 1,991m. (2001); manufacture and sale of concrete piles; trading in construction materials; Chair. PANG TAK CHUNG; 212 employees.

Dairy Farm International Holdings Ltd: Devon House, 7/F, Taikoo Place, 979 King's Rd, Quarry Bay; tel. 22991888; fax 22994888; internet www.dairyfarmgroup.com; sales US $4,924.5m. (2001); international food and drugstore retailing; Chair. SIMON KESWICK.

Dickson Concepts (International) Ltd: East Ocean Centre, 4/F, 98 Granville Rd, Tsimshatsui East, Kowloon; tel. 23113888; fax 23113323; internet www.irasia.com/listco/hk/dickson/index.htm; cap. and res 1,064m., sales 2,233m. (2001); investment holding; trading of luxury goods; Chair. DICKSON POON.

Gold Peak Industries (Holdings) Ltd: Gold Peak Bldg, 8/F, 30 Kwai Wing Rd, Kwai Chung, New Territories; tel. 24271133; fax 24891879; e-mail gp@goldpeak.com; internet www.goldpeak.com; cap. and res 1,025.5m., sales 1,752.8m. (2001); manufacture and sale of batteries, car audio equipment, other electrical and electronic products; Chair. VICTOR C. W. LO; 15,000 employees.

Guangdong Investment Ltd: Guangdong Investment Tower, 29–30/F, 148 Connaught Rd, Central; tel. 28604368; fax 25284386; internet www.gdi.com.hk; cap. and res 4,414.8m., sales 5,359m. (1999), revenue 7,271m. (2001); holding company, activities incl. travel, hotels, property, industrial investment and energy; Chair. WU JIESI; Man. Dir KANG DIAN.

Henderson Investment Ltd: World-Wide House, 6/F, 19 Des Voeux Rd, Central; tel. 28265222; fax 29088838; e-mail henderson@hld.com; internet www.hld.com; cap. and res 18,459.1m. (1999/2000), sales 1,051m. (2001); property development and investment, investment holding; Chair. and Man. Dir Dr LEE SHAU KEE.

Hong Kong Land Holdings Ltd: 8/F, 1 Exchange Sq., 8 Connaught Place, Central; tel. 28428428; fax 28459226; e-mail gpobox@hkland.com; internet www.hkland.com; sales US $387m. (2000), cap. and res US $4,957m.; property investment and development; part of Jardine Matheson group; CEO NICHOLAS SALLNOW-SMITH.

Hopewell Holdings Ltd: Hopewell Centre, 64/F, 183 Queen's Rd East; tel. 25284975; fax 28656276; e-mail ir@hopewellholdings.com; internet www.hopewellholdings.com; cap. and res 13,687.7m., sales 1,750.1m. (2001); property investment and management, road infrastructure and power station projects; Chair. Sir GORDON WU YING SHEUNG.

Hutchison Whampoa Ltd: Hutchison House, 22/F, 10 Harcourt Rd; tel. 21281188; fax 21281705; internet www.hutchison-whampoa.com; cap. and res 218,273m., sales 89,038m. (2001); investment holding and management company; Chair. LI KA-SHING; Man. Dir CANNING FOK; 39,860 employees.

Hysan Development Co Ltd: Manulife Plaza, 49/F, The Lee Gardens, 33 Hysan Ave; tel. 28955777; fax 25775219; e-mail hysan@hysan.com.hk; internet www.hysan.com.hk; cap. and res 19,086m. (2002), sales 1,233m. (2002); property investment, management and development; Chair. PETER T. C. LEE; Dir MICHAEL T.H. LEE.

Inchcape Pacific Ltd: Standard Chartered Bank Bldg, 17/F, 4 Des Voeux Rd, Central; tel. 28424666; fax 28100031; f. 1987; motors, consumer and industrial, shipping services, buying services, business machines, inspection and testing, insurance services; Chair. Dr RAYMOND CH'IEN.

Jardine Matheson Ltd: Jardine House, 48/F, GPOB 70; tel. 28438288; fax 28459005; e-mail jml@jardines.com; internet www.jardines.com; f. 1832; cap. and res US $3,013m., sales US $9,413m. (2001); property, hotels, consumer marketing, engineering and construction, insurance broking, supermarkets and motor trading, transportation; Chair. HENRY KESWICK; Man. Dir PERCY WEATHERALL; 130,000 employees (world-wide).

Jardine International Motor Holdings Ltd: 31/F The Lee Gardens, 33 Hysan Ave, GPO Box 209; tel. 28957218; fax 28949234; e-mail sec@jardines.com; internet www.jardines.com; cap. and res US $396m., sales US $2,807m. (1999); sales and servicing of motor vehicles; Chair. ANTHONY NIGHTINGALE; 10,000 employees.

Jardine Pacific: Devon House, 25/F, Taikoo Place, 979 King's Rd; tel. 25792888; fax 28569674; internet www.jardines.com; f. 1997; sales US $1,588m. (1998); marketing and distribution, engineering and construction, aviation and shipping, property and financial services; Chair. ANTHONY NIGHTINGALE; 53,000 employees.

Johnson Electric Holdings Ltd: Johnson Bldg, 6–22 Dai Shun St, Tai Po Industrial Estate, Tai Po, New Territories; tel. 26636688; fax 28972054; internet www.johnsonelectric.com; f. 1959; cap. and res 556.2m., sales 1,000m. (2003); design, manufacture and marketing of motors for automotive and commercial applications; Chair. and Chief Exec. PATRICK WANG SHUI CHUNG; 30,000 employees (world-wide).

Kader Holdings Co Ltd: 22 Kai Cheung Rd, Kowloon; tel. 27981688; fax 27961126; e-mail kader@kader.com.hk; internet www.kader.com.hk; f. 1989; cap. and res 494.2m., sales 430m. (2001); manufacture and sale of plastic and stuffed toys, electronic toys and model trains, property investment, investment holding and trading; Chair. DENNIS H. S. TING; Man. Dir KENNETH W. S. TING; 6,300 employees.

Lai Sun Garment (International) Ltd: Lai Sun Commercial Centre, 11/F, 680 Cheung Sha Wan Rd, Kowloon; tel. 27410391; fax 27852775; e-mail advpr@laisun.com.hk; internet www.laisun.com.hk; cap. and res 3,132.2m., sales 2,966.9m. (1999/2000); manufacture and sales of garments; Chair. and Man. Dir LIM POR YEN; 2000 employees.

Lam Soon (Hong Kong) Ltd: 21 Dai Fu St, Tai Po Industrial Estate, New Territories; tel. 26803388; fax 26804069; e-mail webmaster@lamsoon.com.hk; internet www.lamsoon.com.hk; sales 1,663m. (2001); investment holding co, its subsidiaries' activities incl. the processing and trading of edible oils, flour products, fruit juices, food products, detergents, electronic products and packaging; Chair. WHANG TAR CHOUNG; 1,700 employees.

Li & Fung Ltd: 11/F LiFung Tower, Cheung Sha Wan Rd, Kowloon; tel. 23002300; fax 23002000; internet www.lifung.com; f. 1906; sales 33,028.6m. (2001); manufacture and trade of consumer products; Chair. VICTOR K. FUNG; Man. Dir WILLIAM K. FUNG.

Luks Industrial Co Ltd: Cheong Wah Factory Bldg, 5/F, 39–41 Sheung Heung Rd, Kowloon; tel. 23620297; fax 27643067; cap. and res 1,276.6m. (1995), sales 175m. (2001); manufacture and sales of printed circuit boards, colour TVs; Chair. and Man. Dir LUK KING TIN.

New World Development Co Ltd: New World Tower, 30/F, 18 Queen's Rd, Central; tel. 25231056; fax 28104673; e-mail newworld@nwd.com.hk; internet www.nwd.com.hk; cap. and res 58,208m. (1999/2000), sales 24,382.4m. (2000/01); property investment, construction and hotels; Exec. Chair. Dato' Dr CHENG YU-TUNG; Man. Dir CHENG KAR-SHUN; 16,512 employees.

Shanghai Industrial Holdings Ltd: Harcourt House, 26/F, 39 Gloucester Rd, Wanchai; tel. 25295652; fax 25295067; internet www.sihl.com.hk; cap. and res 12,438.4m., revenue 3,199.4m. (2001); manufacture, distribution and sale of cigarettes, packaging materials and printed products; Chair. CAI LAI XING.

Shell Electric Mfg (Holding) Co Ltd: Shell Industrial Bldg, 12 Lee Chung St, Chai Wan Industrial District; tel. 25580181; fax 29750720; internet www.smc.com.hk; sales 2,212m. (2001); manufacturing and marketing of electric fans and other household appliances; Chair. Dr YUNG YAU; Man. Dir BILLY YUNG KWOK KEE; 5,000 employees.

Shougang Concord International Enterprises Co Ltd: First Pacific Bank Centre, 7/F, 51–57 Gloucester Rd, Wanchai; tel. 28612832; fax 28613972; revenue 1,942m. (2001); manufacture and sale of steel products; Chair. WANG QINGHAI; Man. Dir CAO ZHONG; 4,500 employees.

Shui On Holdings Ltd: Shui On Centre, 34/F, 6–8 Harbour Rd; tel. 28791888; fax 28024396; e-mail corpcomm@shuion.com.hk; internet www.shuion.com; f. 1965; sales 5,556m. (2000/01); leading construction co, specializes in public-sector projects; Chair. VINCENT H. S. LO; Man. Dir FRANKIE Y. L. WONG.

Sime Darby Hong Kong Ltd: East Wing, Hennessy Centre, 28/F, 500 Hennessy Rd, Causeway Bay; tel. 28950777; fax 28905896; e-mail simedarbyHK@simenet.com; internet www.simenet.com.hk; cap. and res 1,352m., sales 5,673m. (1999/2000); distribution of motor vehicles and heavy construction equipment, industrial, electrical and mechanical contracting, etc.; Chair. Tan Sri Dato' Seri AHMAD SARJI BIN ABDUL HAMID; Man. Dir JOHN HICKMAN BELL.

Sino Land Co Ltd: Tsim Sha Tsui Centre, 12/F, Salisbury Rd, Kowloon; tel. 27218388; fax 27235901; e-mail info@sino-land.com; internet www.sino-land.com; cap. and res 27,412.6m. (2000/01); investment holding, share investment, property development and investment; Chair. ROBERT NG CHEE-SIONG.

Sun Hung Kai Properties Ltd: Sun Hung Kai Centre, 45/F, 30 Harbour Rd, Wanchai; tel. 28278111; fax 28272862; e-mail shkp@shkp.com.hk; internet www.shkp.com.hk; cap. and res 126,007.0m., sales 17,701.0m. (2000/01); investment holding, property development and management; Chair. and Chief Exec. WALTER KWOK; Man. Dir THOMAS KWOK; 18,000 employees.

Swire Group (John Swire and Sons (HK) Ltd): Swire House, 4/F, 9 Connaught Road, Central; tel. 28408888; fax 2845544; internet www.swire.com; shipping managers and agents, airline operators, aviation services, marine and aviation engineering, trading, China trade development, property development, operators of offshore oil drilling support equipment, and mfrs of soft drinks and paints, packagers and distributors of sugar; waste management; Chair. J. W. J. HUGHES-HALLETT.

Swire Pacific Ltd: Two Pacific Place, 35/F, 88 Queensway; tel. 28408867; fax 28455445; internet www.swirepacific.com; cap. and res 68,076m., sales 17,568m. (2003); real estate; Chair. J. W. J. HUGHES-HALLETT.

Tem Fat Hing Fung Holdings: Cheung Fat Bldg, 16/F, 7–9 Hill Rd, Western District; tel. 28032888; fax 28581799; e-mail tfhf@tfhf.com.hk; internet www.tfhf.com.hk; sales 5,550.6m. (1999/2000); production and sale of gold bullion, gold ornaments and jewellery; Chair. RAYMOND CHAN FAT CHU; Man. Dir. ALEXANDER CHAN FAT LEUNG.

Tse Sui Luen Jewellery (International) Ltd: Summit Bldg, Ground Floor, Block B, 30 Man Yue St, Hunghom, Kowloon; tel. 23334221; fax 27640753; cap. and res 62.9m., sales 983.5m. (2001/02); investment holding co, its subsidiaries' activities incl. manufacuring and marketing of jewellery products and property investment; Chair. TOMMY TSE; 850 employees.

Tsim Sha Tsui Properties Ltd: Tsim Sha Tsui Centre, 12/F, Salisbury Rd, Kowloon; tel. 27218388; fax 27235901; e-mail info@sino-land.com; internet www.sino-land.com; cap. and res 10,073.7m. (2000/01), sales 2,039.7m. (1999/2000); investment holding, property development; Chair. ROBERT NG CHEE SIONG; 4,700 employees.

Unibros FE Ltd: Jardine House, Rm 2106, 1 Connaught Place, Central; tel. 25252072; fax 28453446; internet www.unibros.com; sale of steel products.

Vitasoy International Holdings Ltd: 1 Kin Wong St, Tuen Mun, New Territories; tel. 24660333; fax 24563441; internet www.vitasoy.com; sales 2,012.4m. (2000/01); manufacture and distribution of food and beverages; Chair. and Man. Dir WINSTON L. Y. LAI; 2,320 employees.

A S Watson and Co Ltd: Watson House, 1–5 Wo Lin Hang Rd, Fo Tan, Shatin, New Territories; tel. 26068833; fax 26958833; sales 10,800m. (1995, est.); mfr of mineral water, fruit juices and ice cream; CEO SIMON MURRAY; 16,500 employees.

Wharf (Holdings) Ltd (The): Ocean Centre, 16/F, Harbour City, Canton Rd, Kowloon; tel. 21188118; fax 21188018; internet www.wharfholdings.com; f. 1886; cap. and res 48,713m. (2002), sales 11,333m. (2002); property investment and development, hotels, transport, telecommunications and multimedia; Chair. and Chief Exec. PETER K. C. WOO.

Wheelock & Co Ltd: Wheelock House, 23/F, 20 Pedder St; tel. 21182118; fax 21182018; e-mail teresatsang@wharfholdings.com; internet www.wheelockcompany.com; f. 1857; cap. and res 26,544.2m. (2003/04), sales 7,115.9m. (2003/04); merchant house and property investment; Chair. PETER K. C. WOO.

Winsor Industrial Corpn Ltd: East Ocean Centre, 2/F, 98 Granville Rd, Tsimshatsui East, Kowloon; tel. 27311888; fax 28101199; e-mail winsor@winsorindustrial.com; internet www.winsorindustrial.com; cap. and res 914.2m., sales 905.9m. (2000/01); investment holding co, its subsidiaries' activities incl. manufacture of textiles, knitwear and other garments; Chair. and Man. Dir W. H. CHOU; 3,400 employees.

TRADE UNIONS

In December 2000 there were 638 trade unions in Hong Kong, comprising 594 employees' unions, 25 employers' associations and 19 mixed organizations.

Hong Kong and Kowloon Trades Union Council (TUC): Labour Bldg, 11 Chang Sha St, Kowloon; tel. 23845150; f. 1949; 66 affiliated unions, mostly covering the catering and building trades; 28,200 mems; supports Taiwan; affiliated to ICFTU; Officer-in-Charge WONG YIU KAM.

Hong Kong Confederation of Trade Unions: Wing Wong Commercial Bldg, 19/F, 557–559 Nathan Rd, Kowloon; tel. 27708668; fax 27707388; e-mail hkctu@hkctu.org.hk; internet www.hkctu.org.hk; registered Feb. 1990; 63 affiliated independent unions and federations; 165,000 mems; Chair. CHENG CHING-FAT.

Hong Kong Federation of Trade Unions (FTU): 7/F, 50 Ma Tau Chung Rd, Tokwawan, Kowloon; tel. 27120231; fax 27608477; e-mail external@ftu.org.hk; internet www.ftu.org.hk; f. 1948; 176 member unions, mostly concentrated in shipyards, public transport, textile mills, construction, department stores, printing and public utilities; supports the People's Republic of China; 310,000 mems; Chair. WONG KWOK-KIN; Gen. Sec. LAM SHUK-YEE.

Also active are the **Federation of Hong Kong and Kowloon Labour Unions** (31 affiliated unions with 21,700 mems) and the **Federation of Civil Service Unions** (29 affiliated unions with 12,000 mems).

Transport

Transport Department: Immigration Tower, 41/F, 7 Gloucester Rd, Wanchai; tel. 28042600; fax 28240433; internet www.info.gov.hk/td.

RAILWAYS

Kowloon–Canton Railway Corpn: KCRC House, 9 Lok King St, Fo Tan, Sha Tin, New Territories; tel. 26881333; fax 26880983; internet www.kcrc.com; operated by the Kowloon–Canton Railway Corpn, a public statutory body; f. 1982; operates both heavy and light rail systems; the 34-km East Rail runs from the terminus at Hung Hom to the frontier at Lo Wu; through passenger services to Guangzhou (Canton), suspended in 1949, were resumed in 1979; the electrification and double-tracking of the entire length and redevelopment of all stations has been completed, and full electric train service came into operation in 1983; in 1988 a light railway network serving the north-western New Territories was opened; passenger service extended to Foshan in 1993, Dongguan in 1994, Zhaoqing in 1995 and Shanghai in 1997; direct Kowloon–Beijing service commenced in May 1997; also freight services to several destinations in China; West Rail, a domestic passenger line, links Tuen Mun and Yuen Long with Kowloon; three East Rail extensions were due for completion in 2004–07; a Kowloon Southern Link connecting West Rail and East Rail was due for completion in 2008/09; a new 17-km railway linking Sha Tin and Central via a new cross-harbour tunnel was due to be completed in 2011, providing the first direct rail route from the Chinese border to Hong Kong Island; Chair. and CEO SAMUEL M. H. LAI.

MTR Corporation: MTR Tower, Telford Plaza, Kowloon Bay; tel. 29932111; fax 27988822; internet www.mtr.com.hk; f. 1975; privatized in 2000, shares commenced trading on Hong Kong Stock Exchange in Oct. 2000; network of 87.7 km of railway lines and 50 stations; the first section of the underground mass transit railway (MTR) system opened in 1979; a 15.6-km line from Kwun Tong to Central opened in 1980; a 10.5-km Tsuen Wan extension opened in 1982; the 12.5-km Island Line opened in 1985–86; in 1989 a second harbour crossing between Cha Kwo Ling and Quarry Bay, known as the Eastern Harbour Crossing, commenced operation, adding 4.6 km to the railway system; 34-km link to new airport at Chek Lap Kok and to Tung Chung New Town opened in mid-1998; an additional line, the Tseung Kwan O Extension, was completed in August 2002; additional lines were also planned for 2006; CEO C. K. CHOW; Chair. Dr RAYMOND K. F. CH'IEN.

TRAMWAYS

Hong Kong Tramways Ltd: Whitty Street Tram Depot, Connaught Rd West, Western District; tel. 21186338; fax 21186038; f. 1904; operates six routes and 161 double-deck trams between Kennedy Town and Shaukeiwan; Dir and Gen. Man. FRANKIE YICK.

ROADS

At the end of 2002 there were 1,924 km of roads and 1,044 highway structures. Almost all of them are concrete or asphalt surfaced. Owing to the hilly terrain, and the density of building development, the scope for substantial increase in the road network is limited. A new 29-km steel bridge linking Hong Kong's Lantau Island with Macao and Zhuhai City, in the Chinese province of Guangdong, was being planned in in the early 21st century, with studies being carried out to determine the project's financial and technological feasibility.

Highways Department: Ho Man Tin Government Offices, 5/F, 88 Chung Hau St, Ho Man Tin, Kowloon; tel. 27623304; fax 27145216; e-mail hyd@hyd.gov.hk; internet www.hyd.gov.hk; f. 1986; planning, design, construction and maintenance of the public road system; co-ordination of major highway and railway projects; Dir MAK CHAI-KWONG.

FERRIES

Conventional ferries, hoverferries and catamarans operate between Hong Kong, China and Macao. There is also an extensive network of ferry services to outlying districts.

Hongkong and Yaumati Ferry Co Ltd: 98 Tam Kon Shan Rd, Ngau Kok Wan, North Tsing Yi, New Territories; tel. 23944294; fax 27869001; e-mail hkferry@hkf.com; internet www.hkf.com; licensed routes on ferry services, incl. excursion, vehicular and dangerous goods; Gen. Man. DAVID C. S. HO.

Hongkong Macao Hydrofoil Co Ltd: Turbojet Ferry Services (Guangzhou) Ltd, 83 Hing Wah St West, Lai Chi Kok, Kowloon; operates services to Macao, Fu Yong (Shenzhen airport) and East River Guangzhou.

'Star' Ferry Co Ltd: Kowloon Point Pier, Tsimshatsui, Kowloon; tel. 21186223; fax 21186028; e-mail sf@starferry.com.hk; f. 1898; operates 13 passenger ferries between the Kowloon Peninsula and Central, the main business district of Hong Kong; between Central and Hung Hom; between Tsimshatsui and Wanchai; between Tsimshatsui and Central; and between Wanchai and Hung Hom; Man. JOHNNY LEUNG.

SHIPPING

Hong Kong is one of the world's largest shipping centres and is a major container port. Hong Kong was a British port of registry until the inauguration of a new and independent shipping register in December 1990. Following Hong Kong's reunification with the People's Republic of China, Hong Kong maintains full autonomy in its maritime policy. At the end of 2002 the register comprised a fleet of 766 vessels, totalling 16.2m. grt. The eight container terminals at Kwai Chung, which are privately-owned and operated, comprised 18 berths in 1998. The construction of a ninth terminal (CT9) commenced in 1998 and was expected to be operational by 2005. Lantau Island has been designated as the site for any future expansion.

Marine Department, Hong Kong Special Administrative Region Government: Harbour Bldg, 22/F, 38 Pier Rd, Central, GPOB 4155; tel. 28523001; fax 25449241; e-mail mdenquiry@mardep.gov.hk; internet www.mardep.gov.hk; Dir of Marine S. Y. TSUI.

Shipping Companies

Anglo-Eastern Ship Management Ltd: Universal Trade Centre, 14/F, 3 Arbuthnot Rd, Central, POB 11400; tel. 28636111; fax 28612419; e-mail allhx470@gncomtext.com; internet www.webhk.com/angloeastern/; Chair. PETER CREMERS; Man. Dir MARCEL LIEDTS.

Chung Gai Ship Management Co Ltd: Admiralty Centre Tower 1, 31/F, 18 Harcourt Rd; tel. 25295541; fax 28656206; Chair. S. KODA; Man. Dir K. ICHIHARA.

Fairmont Shipping (HK) Ltd: Fairmont House, 21/F, 8 Cotton Tree Drive; tel. 25218338; fax 28104560; Man. CHARLES LEUNG.

Far East Enterprising Co (HK) Ltd: China Resources Bldg, 18–19/F, 26 Harbour Rd, Wanchai; tel. 28283668; fax 28275584; f. 1949; shipping, chartering, brokering; Gen. Man. WEI KUAN.

Gulfeast Shipmanagement Ltd: Great Eagle Centre, 9/F, 23 Harbour Rd, Wanchai; tel. 28313344; Finance Dir A. T. MIRMOHAMMADI.

Hong Kong Borneo Shipping Co Ltd: 815 International Bldg, 141 Des Voeux Rd, Central; tel. 25413797; fax 28153473; Pres. Datuk LAI FOOK KIM.

Hong Kong Ming Wah Shipping Co: Unit 3701, China Merchants Tower, 37/F, Shun Tak Centre, 168–200 Connaught Rd, Central; tel. 25172128; fax 25473482; e-mail mwins@cmhk.com; Chair. CHEUNG KING WA; Man. Dir and Vice-Chair. Capt. MAO SHI JIAN.

Island Navigation Corpn International Ltd: Harbour Centre, 28–29/F, 25 Harbour Rd, Wanchai; tel. 28333222; fax 28270001; Man. Dir F. S. SHIH.

Jardine Ship Management Ltd: Jardine Engineering House, 11/F, 260 King's Rd, North Point; tel. 28074101; fax 28073351; e-mail jsmhk@ibm.net; Man. Dir Capt. PAUL UNDERHILL.

Oak Maritime (HK) Inc Ltd: 2301 China Resources Bldg, 26 Harbour Rd, Wanchai; tel. 25063866; fax 25063563; Chair. STEVE G. K. HSU; Pres. FRED C. P. TSAI.

Ocean Tramping Co Ltd: Hong Kong; tel. 25892645; fax 25461041; Chair. Z. M. GAO.

Orient Overseas Container Line Ltd: Harbour Centre, 31/F, 25 Harbour Rd, Wanchai; tel. 28333888; fax 25318122; internet www.oocl.com; member of the Grand Alliance of shipping cos (five partners); Chair. C. C. TUNG.

Teh-Hu Cargocean Management Co Ltd: Unit B, Fortis Bank Tower, 15/F, 77 79 Gloucester Rd, Wanchai; tel. 25988688; fax 28249339; e-mail tehhuhk@on-nets.com; f. 1974; Man. Dir KENNETH K. W. LO.

Wah Kwong Shipping Agency Co Ltd: Shanghai Industrial Investment Bldg, 26/F, 48–62 Hennessy Rd, POB 283; tel. 25279227; fax 28656544; e-mail wk@wahkwong.com.hk; Chair. GEORGE S. K. CHAO.

Wah Tung Shipping Agency Co Ltd: China Resources Bldg, Rooms 2101–5, 21/F, 26 Harbour Rd, Wanchai; tel. 28272818; fax 28275361; e-mail mgr@watunship.com.hk; f. 1981; Dir and Gen. Man. B. L. LIU.

Wallem Shipmanagement Ltd: Hopewell Centre, 46/F, 183 Queen's Rd East; tel. 28768200; fax 28761234; e-mail rgb@wallem.com; Man. Dir R. G. BUCHANAN.

Worldwide Shipping Agency Ltd: Wheelock House, 6–7/F, 20 Pedder St; tel. 28423888; fax 28100617; Man. J. WONG.

Associations

Hong Kong Cargo-Vessel Traders' Association: 21–23 Man Wai Bldg, 2/F, Ferry Point, Kowloon; tel. 23847102; fax 27820342; 978 mems; Chair. CHOW YAT-TAK; Sec. CHAN BAK.

Hong Kong Shipowners' Association: Queen's Centre, 12/F, 58–64 Queen's Rd East, Wanchai; tel. 25200206; fax 25298246; e-mail hksoa@hksoa.org.hk; internet www.hksoa.org.hk; 220 mems; Chair. K.H. KOO; Dir ARTHUR BOWRING.

Hong Kong Shippers' Council: Rm 2407, Hopewell Centre, 183 Queen's Rd East; tel. 28340010; fax 28919787; e-mail shippers@hkshippers.org.hk; internet www.hkshippers.org.hk; 63 mems; Chair. WILLY LIN; Exec. Dir SUNNY HO.

CIVIL AVIATION

By the end of 2002 Hong Kong was served by 73 scheduled airlines. A new international airport, on the island of Chek Lap Kok, near Lantau Island, to replace that at Kai Tak, opened in July 1998, following delays in the construction of a connecting high-speed rail-link. The airport has two runways, with the capacity to handle 35m. passengers and 3m. metric tons of cargo per year. The second runway commenced operations in May 1999. A helicopter link with Macao was established in 1990.

Airport Authority of Hong Kong: Cheong Yip Rd, Hong Kong International Airport, Lantau; tel. 21887111; fax 28240717; f. 1995; Chair. Dr VICTOR FUNG KWOK-KING; CEO Dr DAVID J. PANG.

Civil Aviation Department: Queensway Government Offices, 46/F, 66 Queensway; tel. 28674203; fax 28690093; e-mail enquiry@cad.gov.hk; internet www.info.gov.hk/cad/; Dir-Gen. NORMAN LO SHUNG-MAN.

AHK Air Hong Kong Ltd: Units 3601–8, 36/F, Tower 1, Millennium City, 388 Kwun Tong Rd, Kowloon; tel. 27618588; fax 27618586; e-mail ahk.hq@airhongkong.com.hk; f. 1986; international cargo carrier; Chief Operating Officer HUNTER CRAWFORD.

Cathay Pacific Airways Ltd: South Tower, 5/F, Cathay Pacific City, 8 Scenic Rd, Hong Kong International Airport, Lantau; tel. 27475000; fax 28106563; internet www.cathaypacific.com/hk; f. 1946; services to more than 40 major cities in the Far East, Middle East, North America, Europe, South Africa, Australia and New Zealand; Chair. and CEO DAVID TURNBULL.

Hong Kong Dragon Airlines Ltd (Dragonair): Dragonair House, 11 Tung Fai Rd, Hong Kong International Airport, Lantau; tel. 31933193; fax 31933194; internet www.dragonair.com; f. 1985; scheduled and charter flights to 25 destinations in Asia, 16 of which are in mainland China; scheduled regional services include Phuket (Thailand), Hiroshima and Sendai (Japan), Kaohsiung (Taiwan), Phnom-Penh (Cambodia), Dhaka (Bangladesh), Bandar Seri Begawan (Brunei), and Kota Kinabalu (Malaysia); five additional international routes, Seoul (Republic of Korea), Tokyo (Japan), Manila (Philippines), Bangkok (Thailand), and Sydney (Australia) to commence in 2003; 43% owned by China National Aviation Corpn; Dir and CEO STANLEY HUI.

Tourism

Tourism is a major source of foreign exchange, tourist receipts reaching US $10,117m. in 2002. Some 16.6m. people visited Hong Kong in 2002. In December 2002 there were some 98 hotels, and the number of rooms available totalled 38,949. In November 1999 it was agreed that a new Disneyland theme park would be constructed in Hong Kong, the first phase of which was scheduled to open in 2005. The Government expected the park to create a huge influx of tourists to the territory.

Hong Kong Tourism Board: Citicorp Centre, 9–11/F, 18 Whitfield Rd, North Point; tel. 28076543; fax 28076595; e-mail info@discoverhongkong.com; internet www.DiscoverHongKong.com; f. 1957; reconstituted as Hong Kong Tourism Board 1 April 2001; co-ordinates and promotes the tourist industry; has govt support and financial assistance; up to 20 mems of the Board represent the Govt, the private sector and the tourism industry; Chair. SELINA CHOW; Exec. Dir CLARA CHONG.

Defence

In July 1997 a garrison of 4,800 PLA troops was established in Hong Kong. The garrison can intervene in local matters only at the request of the Hong Kong Government, which remains responsible for internal security. In August 2003 a total of 7,000 Chinese troops were deployed in Hong Kong.

Defence Expenditure: Projected expenditure on internal security in 2003/04 totalled HK $21,151m.

Commander of the PLA Garrison in Hong Kong: Maj.-Gen. WANG JITANG.

Education

Full-time education is compulsory in Hong Kong between the ages of six and 15. Schools fall into three main categories: those wholly maintained by the Government; those administered by non-government organizations with government financial aid; and those administered independently by private organizations. There are also government-aided schools for children with special educational needs; in 2002/03 they provided education for 9,900 children. The adult literacy rate in 2001 was estimated at 93.5% (males 96.9%; females 89.6%). Budgetary expenditure on education was estimated at HK $54,893m. for 2002/03.

PRE-PRIMARY AND PRIMARY SCHOOLS

Kindergartens are administered by private bodies without direct government assistance for children between the ages of three and five. The Government provides indirect assistance through rent and rate rebates to non-profit-making kindergartens, fee assistance for needy parents, etc. In 2002/03 there were 777 such schools, with a combined enrolment of 143,700. The age of entry into primary school is six, and the schools provide a six-year course of basic primary education. Primary school pupils totalled 483,200 in 2002/03. Compulsory primary education was first introduced in 1971 when fees were abolished in most of the primary schools in the public sector. There are nine government-subsidized primary schools and a number of private international schools catering for the education of English-speaking children. At the end of six years, every primary school-leaver is allocated a free place in a secondary school for three years. The method of allocation is based on parental choice and schools' internal assessments, monitored by an Academic Aptitude Test under the Secondary School Places Allocation scheme.

SECONDARY SCHOOLS

Junior secondary education (Secondary 1–3), which became compulsory in September 1979, has been free since September 1978. A centralized system of selection and allocation of subsidized school places for senior secondary education (Secondary 4–5), known as the Junior Secondary Education Assessment System, was first introduced in 1981, and was enhanced in 1988 by the adoption of the Mean Eligibility Allocation Method, which relieved all students from taking any public scaling test. Both the performance of students in the school internal assessments and parental choice form the basis for selection and allocation of Secondary 3 students to subsidized Secondary 4 places. Students may also choose to continue their studies in post-Secondary 3 craft courses offered by technical institutes and industrial training centres.

In 2002/03 there were 465,900 secondary school pupils. There are three main types of secondary school in Hong Kong: grammar, technical and prevocational schools. The Hong Kong Certificate of Education Examination may be taken after a five-year course; a further course of two years leads to the Hong Kong Advanced Level Examination. The prevocational schools provide a five-year secondary course. Sixth-form classes were introduced in September 1992. Following the resumption of Chinese sovereignty in mid-1997, Cantonese was gradually to replace English as the official medium of instruction. Some 100 schools were to be allowed to retain English as the medium of instruction, provided that they fulfil certain criteria (eg. language capability of the students).

HIGHER EDUCATION

In late 1999 there were 10 institutions of higher education: City University of Hong Kong (CUHK), Hong Kong Baptist University (HKBU), Hong Kong Polytechnic University (HKPU), Lingnan University (LU), the Chinese University of Hong Kong (CUHK), the Hong Kong University of Science and Technology (HKUST); the University of Hong Kong (HKU), the Open University of Hong Kong, the Hong Kong Academy for the Performing Arts and the Hong Kong Institute of Education. In 2002/03 an estimated total of 199,500 full-time and part-time students were enrolled at the 10 institutions.

Technological training is also provided by the Vocational Training Council (VTC), which operates the Hong Kong Institute of Vocational Education (IVE) and 24 training centres. The IVE was founded in 1999 and incorporated two technical colleges and seven institutes. A total of 56,100 students were enrolled in 2002/03.

The four government-run colleges of education and the Institute of Languages in Education of merged to form the Hong Kong Institute of Education in 1994. The Institute provides training for teachers of kindergartens, primary and secondary schools, and offers full- and part-time courses of two and three years' duration. There were some 9,000 teacher-training students during 1998/99. The Government provides loans and grants for needy students.

Bibliography

GENERAL

Balke, G. *Hong Kong Voices.* Hong Kong, Longman, 1989.

Blyth, Sally, and Wotherspoon, Ian. *Hong Kong Remembers.* Oxford, Oxford University Press, 1997.

Bristow, Roger. *Hong Kong's New Towns: A Selective Review.* Hong Kong and New York, Oxford University Press, 1990.

Chan, Anthony. *Li Ka-shing.* Hong Kong, Oxford University Press, 1997.

Cheek-Milby, K., and Mushkat, Miron. *Hong Kong, The Challenge of Transformation.* Centre of Asian Studies, University of Hong Kong, 1989.

Cheng, Joseph Y. S., and Lo, Sonny S. H. (Eds). *From Colony to SAR: Hong Kong's Challenges Ahead.* Hong Kong, Chinese University Press, 1996.

Cheung, Fanny M. (Ed.). *Engendering Hong Kong Society: A Gender Perspective of Women's Status.* Hong Kong, Chinese University Press, 1997.

Cheung, Stephen Y. L., and Sze, Stephen M. H. (Eds). *The Other Hong Kong Report 1996.* Hong Kong, Chinese University Press, 1996.

Cottrell, Robert. *The End of Hong Kong.* London, John Murray, 1993.

Cradock, Percy. *Experiences of China.* London, John Murray, 1994.

Evans, Gareth, and Tam, Maria. *Hong Kong: The Anthropology of a Chinese Metropolis.* Honolulu, HI, University of Hawaii Press, 1997.

Gauld, Robin, and Gould, Derek. *The Hong Kong Health Sector: Development and Change.* Otago, Otago University Press, 2003.

Government Information Services Department. *Hong Kong 1998.* Hong Kong, Government Publications Centre, 1999.

Hughes, R. *Hong Kong: Borrowed Place, Borrowed Time.* London, André Deutsch, 1976.

Jao, Y. C., Leung Chi-keung, Wesley-Smith, P., and Wong Siu-lun (Eds). *Hong Kong and 1997: Strategies for the Future.* University of Hong Kong, 1985.

Jones, Catherine. *Promoting Prosperity: The Hong Kong Way of Social Policy.* Hong Kong, Chinese University Press, 1991.

Lau Siu-kai and Kuan Hsin-chi. *The Ethos of the Hong Kong Chinese.* Hong Kong, Chinese University Press, 1988.

Le Corre, Philippe. *Après Hong Kong.* Editions Autrement, 1997.

Lee, Eliza W. Y. *Gender and Change in Hong Kong: Globalization, Postcolonialism, and Chinese Patriarchy.* Vancouver, University of British Columbia Press, 2003.

Lee, James. *Housing, Home Ownership and Social Change in Hong Kong.* Burlington, VT, Ashgate Publishing Co, 1999.

Lee, Rance P. L. (Ed.). *Corruption and Its Control in Hong Kong.* Hong Kong, Chinese University Press, 1981.

Leung, Benjamin K. P. (Ed.). *Social Issues in Hong Kong.* Oxford University Press, 1990.

Ma, Eric Kit-wai. *Culture, Politics and Television in Hong Kong.* London, Routledge, 1999.

Mok, Joshua K. H., and Chan, David K. K. (Eds). *Globalization and Education: The Quest for Quality Education in Hong Kong.* Hong Kong University Press, 2002.

Osgood, Cornelius. *The Chinese: A Study of a Hong Kong Community.* 3 vols. Tucson, University of Arizona Press, 1975.

Rooney, Nuala. *At Home With Density.* Hong Kong University Press, 2002.

Segal, Gerald. *The Fate of Hong Kong.* London, Simon and Schuster, 1993.

Wacks, Raymond (Ed.). *Civil Liberties in Hong Kong.* Oxford University Press, 1988.

The New Legal Order in Hong Kong. Hong Kong, Hong Kong University Press, 2000.

Wesley-Smith, Peter. *An Introduction to the Hong Kong Legal System.* Hong Kong, Oxford University Press, 3rd Edn, 1999.

Wilson, Dick. *Hong Kong! Hong Kong!* London, Unwin Hyman, 1990.

HISTORY AND POLITICS

Ash, Robert. F., Ferdinand, Peter, Hook, Brian, and Porter, Robin (Eds). *Hong Kong in Transition: One Country, Two Systems.* Richmond, Surrey, Curzon Press, 2002.

Banham, Tony. *Not the Slightest Chance: The Defence of Hong Kong, 1941.* Hong Kong University Press, 2003.

Beatty, Bob. *Democracy, Asian Values, and Hong Kong: Evaluating Political Elite Beliefs.* Westport, CT, Praeger, 2003.

Bonavia, David. *Hong Kong 1997—The Final Settlement.* Bromley, Columbus Books, 1985.

Brown, J. M., and Foot, R. *Hong Kong's Transitions, 1842–1997.* London, Macmillan, 1997.

Buckley, Roger. *Hong Kong: The Road to 1997.* Cambridge, Cambridge University Press, 1997.

Butenhoff, Linda. *Social Movements and Political Reform in Hong Kong.* New York, Praeger, 1999.

Byrnes, Andrew, and Chan, Johannes (Eds). *Public Law and Human Rights: A Hong Kong Sourcebook.* Singapore, Butterworth, 1994.

Callick, Rowan. *Comrades and Capitalists, Hong Kong Since the Handover.* Sydney, University of New South Wales Press, 1998.

Cameron, Nigel. *The Illustrated History of Hong Kong.* Hong Kong, Oxford University Press, 1991.

Chan Lau Kit-ching. *China, Britain and Hong Kong: 1895–1945.* Hong Kong, Chinese University Press, 1991.

Chan, Ming K., Postiglione, Gerard A., and Vogel, Ezra F. (Eds). *The Hong Kong Reader: Passage to Chinese Sovereignty.* Armonk, NY, M. E. Sharpe, 1996.

Chan, Ming K., and Young, John D. (Eds). *Precarious Balance: Hong Kong Between China and Britain, 1842–1992.* Armonk, NY, M. E. Sharpe, 1994.

Chang, David. W., and Chuang, Richard. Y. *The Politics of Hong Kong's Reversion to China.* Basingstoke, Macmillan, 1997.

Cheek-Milby, Kathleen. *A Legislature Comes of Age: Hong Kong's Search for Influence and Identity.* Hong Kong, Oxford University Press, 1995.

Cheung, Anne S.Y. *Self-Censorship and the Struggle for Press Freedom in Hong Kong.* New York, NY, Kluwer Law International, 2003.

Chi Kuen Lau. *Hong Kong's Colonial Legacy.* Hong Kong, Chinese University Press, 1998.

Clarke, David. *Reclaimed Land: Hong Kong in Transition.* Hong Kong University Press, 2002.

Crowell, Todd. *Farewell, My Colony: Last Days in the Life of British Hong Kong.* Hong Kong, Asia 2000, 1998.

Dimbleby, Jonathan. *The Last Governor—Chris Patten and the Handover of Hong Kong.* London, Little, Brown and Co, 1997.

Endacott, G. B. A. *A History of Hong Kong.* London, Oxford University Press, 1958.

Government and People in Hong Kong, 1841–1962: A Constitutional History. Hong Kong University Press, 1964.

Endacott, G. B., and Birch, A. H. *Hong Kong Eclipse.* Oxford University Press, 1978.

Fairbrother, Gregory P. *Toward Critical Patriotism: Student Resistance to Political Education in Hong Kong and China.* Hong Kong University Press, 2003.

Fenby, Jonathan. *Dealing With the Dragon—A Year in the New Hong Kong.* London, Little, Brown and Co, 2000.

Flowerdew, John. *The Final Years of British Hong Kong—The Discourse of Colonial Withdrawal.* London, Macmillan, 1997.

Fok, K. C. *Lectures on Hong Kong History*. The Commercial Press (Hong Kong), 1990.

Ghai, Y. *Hong Kong's New Constitutional Order*. Hong Kong, Hong Kong University Press, 1997.

Hamilton, Gary G. (Ed.). *Cosmopolitan Capitalists*. Seattle, WA, University of Washington Press, 2000.

Harris, Peter. *Hong Kong: A Study in Bureaucratic Politics*. Heinemann Asia, 1978.

Hayes, James W. *The Hong Kong Region 1850–1911, Institutions and Leadership in Town and Countryside*. New Haven, CT, and London, William Dawson, 1977.

Hong Kong Journalists Association and Article 19. *False Security: Hong Kong's National Security Laws Pose a Grave Threat to Freedom of Expression*. Hong Kong, Hong Kong Journalists Association and Article 19, 2003.

Horlemann, Ralf. *Hong Kong's Transition to Chinese Rule: The Limits of Autonomy*. Richmond, Surrey, Curzon Press, 2002.

Hsiung, James C. *Hong Kong the Super Paradox: Life After Return to China*. New York, St Martin's Press, 2000.

Keay, John. *Last Post: The End of the Empire in the Far East*. London, John Murray, 1997.

Kuan Hsin-chi, Lau Siu-kai, Louie Kin-shuen and Wong Ka-ying. *The 1995 Legislative Council Elections in Hong Kong*. Hong Kong, Chinese University Press, 1997.

Power Transfer and Electoral Politics: The First Legislative Election in the Hong Kong Special Administrative Region. Hong Kong, Chinese University Press, 1999.

Kwok, Reginald Yin-Wang, and So, Alvin Y. (Eds). *The Hong Kong-Guangdong Link: Partnership in Flux*. Armonk, NY, M. E. Sharpe, 1995.

Kwok, Rowena Y. F., Leung, Joan Y. H., and Scott, Ian (Eds). *Votes without Power: The Hong Kong Legislative Council Elections 1991*. Hong Kong University Press, 1992.

Ku, Agnes, and Ngai Pun (Eds). *Remaking Citizenship in Hong Kong: Community, Nation and the Global City*. London, RoutledgeCurzon, 2004.

Lam, Jermain T. M. *The Political Dynamics of Hong Kong Under the Chinese Sovereignty*. Huntington, NY, Nova Science Publishers, 2000.

Lau Siu-kai and Louie Kin-shuen. *Hong Kong Tried Democracy: The 1991 Elections in Hong Kong*. Hong Kong, Chinese University Press, 1997.

The First Tung Chee-hwa Administration: The First Five Years of the Hong Kong Special Administrative Region. Hong Kong, Chinese University Press, 2002.

Lee Pui Tak. *Hong Kong Reintegrating with China: Political, Cultural and Social Dimensions*. Hong Kong University Press, 2001.

Lo Shiu-hing. *The Politics of Democratization in Hong Kong*. London, Macmillan, 1997.

Luard, E. *Britain and China*. London, Chatto and Windus, 1962.

McGurn, William. *Perfidious Albion: The Abandonment of Hong Kong 1997*. Washington, Ethics and Public Policy Center, 1992.

McMillen, Donald H., and DeGolyer, Michael E. (Eds). *One Culture, Many Systems: Politics in the Reunification of China*. Hong Kong, Chinese University Press, 1993.

Meyer, David R. *Hong Kong as a Global Metropolis*. Cambridge, Cambridge University Press, 2000.

Mills, A. *British Rule in Eastern Asia*. London, Oxford University Press, 1942.

Miners, N. J. *Hong Kong under Imperial Rule 1912–1941*. Oxford University Press, 1987.

The Government and Politics of Hong Kong (with post-handover update by James T. H. Tang). 5th Edn, Oxford University Press, 1998.

Sing Ming. *Hong Kong's Tortuous Democratization*. London, RoutledgeCurzon, 2003.

Morris, J. *Hong Kong: Epilogue to an Empire*. London, Penguin Books Ltd, 1988, revised 1997.

Ngo Tak-Wing. *Hong Kong's History: State and Society Under Colonial Rule*. London, Routledge, 1999.

Patten, Christopher. *East and West—The Last Governor of Hong Kong*. London, Macmillan, 1998.

Postiglione, G. A., and Tang, J. T. H. *Hong Kong's Reunion with China: The Global Dimensions*. Armonk, NY, M. E. Sharpe, 1997.

Pottinger, George. *Sir Henry Pottinger—First Governor of Hong Kong*. Far Thrupp, Sutton, 1997.

Rafferty, Kevin. *City on the Rocks: Hong Kong's Uncertain Future*. London and New York, Penguin-Viking, 1989.

Roberti, Mark. *The Fall of Hong Kong: China's Triumph and Britain's Betrayal*. Chichester, John Wiley, 1994, revised 1997.

Sayer, G. R. *Hong Kong: Birth, Adolescence and Coming of Age*. London, Oxford University Press, 1937.

Hong Kong, 1862–1919. Hong Kong University Press, 1975.

Scott, Ian. *Political Change and The Crisis of Legitimacy in Hong Kong*. Oxford University Press, 1989.

Scott, Ian (Ed.). *Institutional Change and the Political Transition in Hong Kong*. London, Palgrave, 1997.

Snow, Philip. *The Fall of Hong Kong: Britain, China and the Japanese Occupation*. London, Yale University Press, 2003.

So, Alvin Y. *Hong Kong's Embattled Democracy*. Baltimore, MD, The Johns Hopkins University Press. 1999.

Thomas, Nicholas. *Democracy Denied: Identity, Civil Society and Illiberal Democracy in Hong Kong*. Burlington, VT, Ashgate Publishing Co, 1999.

Tsai Jung-Fang. *Hong Kong in Chinese History: Community and Social Unrest in the British Colony, 1842–1913*. New York, Columbia University Press, 1995.

Tsang, Steve Y. S. *Democracy Shelved: Great Britain, China and Attempts at Constitutional Reform in Hong Kong 1945–1952*. Oxford University Press, 1988.

An Appointment with China. London, Tauris, 1997.

A Modern History of Hong Kong: 1841–1998. London, Tauris, 2002.

Vickers, Edward. *In Search of an Identity: The Politics of History Teaching in Hong Kong, 1960s-2000*. London, Routledge, 2003.

Vines, Stephen. *Hong Kong: China's New Colony*. London, Aurum Press, 1998.

Wang Enbao. *Hong Kong 1997: The Politics of Transition*. Boulder, CO, Lynne Rienner, 1995.

Wang Gungwu and Wong Siu-lun (Eds). *Hong Kong's Transition—A Decade after the Deal*. London, Oxford University Press, 1997.

Welsh, Frank. *A History of Hong Kong*. London, HarperCollins, 1993, revised 1997.

Wesley-Smith, P. *Unequal Treaty 1898–1997*. Oxford University Press, 1980.

Wesley-Smith, P., and Chen, Albert (Eds). *The Basic Law and Hong Kong's Future*. London, Butterworth, 1988.

Yahuda, Michael. *Hong Kong: China's Challenge*. London, Routledge, 1996.

ECONOMY

Cohen, Warren I., and Li, Zhao (Eds). *Hong Kong under Chinese Rule: The Economic and Political Implications of Reversion*. Cambridge University Press, 1998.

England, J., and Rear, J. *Industrial Relations and Law in Hong Kong*. Oxford University Press, 1981.

Enright, Michael, Scott, Edith, and Dodwell, David. *The Hong Kong Advantage*. Hong Kong, Oxford University Press, 1997.

Federation of Hong Kong Industries. *Hong Kong's Industrial Investment in the Pearl River Delta*. Hong Kong, 1992.

Fosh, Patricia, Chan, Andy W., and Chow, Wilson W. S. (Eds). *Hong Kong Management and Labour: Change and Continuity*. London, Routledge, 2000.

Freris, Andrew. *The Financial Markets of Hong Kong*. London, Routledge, 1991.

Goodhart, Charles, and Lu, Dai. *Intervention to Save Hong Kong: The Authorities' Counter-Speculation in Financial Markets*. Oxford University Press, 2003.

Ho, Simon S. M., Scott, Robert Haney, and Wong, Kie Ann (Eds). *The Hong Kong Financial System*. Hong Kong, Oxford University Press China, 2003.

Hong Kong Government. *The Economic Background*. 1992.

Estimates of Gross Domestic Product 1961–1995. Census and Statistics Department, 1996.

Hong Kong in Figures. Census and Statistics Department, annual.

Hsia, Ronald, and Chau, L. *Industrialisation, Employment and Income Distribution*. London, Croom Helm, 1978.

Jao, Y. C. *Banking and Currency in Hong Kong, a Study of Post-War Financial Development*. London, Macmillan, 1974.

The Asian Financial Crisis and the Ordeal of Hong Kong. Westport, CT, Quorum Books, 2001.

Lam, Pun-Lee, and Chan, Sylvia. *Competition in Hong Kong's Gas Industry*. Hong Kong, Chinese University Press, 2000.

Lethbridge, David (Ed.). *The Business Environment in Hong Kong*. Oxford University Press, 4th Edn, 2000.

Low, C. K. (Ed.). *Financial Markets in Hong Kong*. Singapore, Springer-Verlag Singapore, 2003.

Mann, Richard I. (Ed.). *Business in Hongkong: Signposts for the 90s; A Positive View*. Toronto, Gateway Books, 1992.

McGuinness, Paul. *A Guide to the Equity Markets of Hong Kong*. New York, Oxford University Press, 2000.

Peebles, Gavin. *The Economy of Hong Kong*. New York, Oxford University Press, 1988.

Rowley, Chris, and Fitzgerald, Robert (Eds). *Managed in Hong Kong*. Ilford, Frank Cass, 2000.

Schenk, Catherine R. *Hong Kong as an International Financial Centre: Emergence and Development, 1945–65*. London, Routledge, 2001.

Tan, Chwee Huat. *Financial Sourcebook for Southeast Asia and Hong Kong*. Singapore, Singapore University Press, 2000.

Yu, Tony Fu-Lai. *Entrepreneurship and Economic Development in Hong Kong*. London, Routledge, 1997.

MACAO

Physical and Social Geography

The Special Administrative Region (SAR) of Macao (or Macau as it was also known prior to its reversion from Portuguese to Chinese sovereignty in December 1999), is situated on the south-eastern coast of the People's Republic of China, at latitude 22°14′ N and longitude 113°35′ E. The territory comprises the narrow, hilly Macao peninsula of the Chinese district of Foshan, on which is situated the Cidade do Santo Nome de Deus de Macau, together with two small islands to the south, Taipa and Coloane. The highest peak, of 170.6m, is situated on the island of Coloane. The SAR covered an area of 27.3 sq km (10.54 sq miles) in 2003 (compared with 17.32 sq km in 1989). A major land-reclamation programme continued to progress in the early 21st century. Macao lies some 64 km west of Hong Kong (across the Zhujiang (Pearl River) estuary), and 145 km south of the city of Guangzhou (Canton), the capital of Guangdong Province. In early 1998 it was announced that Macao and the Pearl River Water Resources Committee (PRWRC) of China were to conduct a joint study on the realignment of local waters, the land-reclamation projects having had a negative impact on water flow. In late 1998 Macao and the Land Department of Guangdong conducted a joint aerial land survey. The Macao peninsula is linked by two bridges (the first spanning 2.6 km and the second 4.4 km) to the island of Taipa, which in turn is connected to Coloane by a 2.2-km causeway. A second link with the mainland, the 1.5-km Lotus Bridge connecting Macao to Zhuhai (Guangdong Province), opened in December 1999. Plans for a 29-km bridge linking Macao with Hong Kong's Lantau Island and Zhuhai were under consideration in 2001, and plans for a third bridge between Macao and Taipa Island were approved in 2002. The climate is subtropical, with temperatures averaging 15°C in January and 29°C in July. The average annual rainfall is between 100 cm and about 200 cm. The highest levels of humidity and precipitation occur between April and September.

The census of August 2001 enumerated the population at 435,235, of whom 414,200, or 95.2%, held Chinese nationality, while 8,793 (2.0%) were of Portuguese nationality. (About 60% of the population were between 15 and 50 years of age at the census of 1996.) By December 2003 the population was estimated to have increased to 448,495. In 1990 more than 90% of the population resided on the Macao peninsula. At 16,428 persons per sq km in December 2003, the territory's population density was one of the highest in the world. In 2003 the birth rate stood at 7.2 per 1,000, and a death rate of 3.3 per 1,000 was recorded. The official languages are Chinese (Cantonese being the principal dialect) and Portuguese. English is also widely spoken, speakers of English outnumbering Portuguese-speakers. The predominant religions are Chinese Buddhism, Daoism, Confucianism and Roman Catholicism. The capital, the city of Macao, is situated on the peninsula.

History

FRANK H. H. KING

Revised since 1980 by the editorial staff

The territory was established as one of several trading posts by the Portuguese as early as 1537, the first Portuguese sailors having anchored in the Zhujiang (Pearl River) in 1513. A permanent settlement was established in 1557. Motivated by trade and missionary zeal, the Portuguese developed Macao as a base for their operations both in China and Japan, and penetration during Japan's 'Christian century' from 1543 involved close relations with Macao. The first Portuguese Governor was appointed in 1680. However, sovereignty remained vested in China; the Chinese residents were subject to a Chinese official, and Macao's Portuguese administration, virtually autonomous for the first 200 years, concerned itself with the governance of the Portuguese and, until the establishment of Hong Kong, with the growing presence of other European trading nations. The Portuguese paid an annual rent to China.

Macao was an uncertain base from which to expand trade with China. The Roman Catholic administration was unfriendly to Protestants, the Chinese authority was too close and restrictive, and the opium question and growing restlessness of the 'private' merchants undermined a system that had developed during the years of controlled and relatively limited 'company' trade.

The ceding of Hong Kong to Britain in 1842 revealed China's weakness, and in 1845 Portugal declared Macao a free port. In the consequent disputes Portugal drove out the Chinese officials, and the settlement was proclaimed Portuguese territory. This unilateral declaration was recognized by China in 1887 in return for provisions intended to facilitate the enforcement of its customs laws, particularly in regard to opium, by the Imperial Maritime Customs.

However, Macao's establishment as a colony did not restore prosperity. With the silting of its harbour, the diversion of its trade to Hong Kong, and the opening of the treaty ports as bases of trade and missionary work, Macao was left to handle the local distributive trade, while developing a reputation as a base for smuggling, gambling and other unsavoury activities. With the closing of the Hong Kong–China border in 1938, Macao's trade boomed, but this prosperity declined after 1942, when the colony became isolated as the only European settlement on the China coast not occupied by the Japanese during the Second World War.

In 1951 Macao was declared an overseas province of Portugal and elected a representative to the Portuguese legislature in Lisbon. Macao's economy depended largely on the gold trade, at that time illegal in Hong Kong, on gambling and tourism, and on an entrepôt business with China.

Macao's tranquillity was disrupted by communist riots in 1966–67, inspired by the 'Cultural Revolution' in mainland China. These were contained only after the Macao Government signed an agreement with Macao's Chinese Chamber of Commerce outlawing the activities of Chinese loyal to the Taiwan regime, paying compensation to the families of Chinese killed in the rioting, and refusing entry to refugees from China. However, China wished the Portuguese administration to continue, and on admission to the UN the Beijing Government affirmed that it regarded the future of Macao as an internal matter.

SUBSEQUENT POLITICAL DEVELOPMENTS

After the military coup in Portugal in April 1974, the only Governor of an overseas territory to be retained in office was that of Macao. China refused to discuss the future of Macao with Portugal, and the revolutionary leaders became convinced that there was no demand for an independence which China would not, in any case, have tolerated. Nevertheless, the revolution caused considerable political activity in Macao. The Centro Democrático de Macau (CDM) was established to press for radical political reform and the removal of those connected with the former regime in Portugal. The conservative Associação para a Defesa dos Interesses de Macau (ADIM) was established, and in April 1975 defeated the CDM's candidate for Macao's representative to the Lisbon assembly. The Macao electorate reaffirmed its conservative bias when over 65% voted for right-wing parties in the Lisbon constituency.

Col Garcia Leandro, who succeeded Nobre de Carvalho at the end of his much-extended term as Governor in late 1974, correctly assessed China's position while being sufficiently flexible to recognize that Macao needed capitalism and was not ready for socialism. He initiated policies which, in effect, insulated Macao from the direct influence of a series of Portuguese governments which contained elements sympathetic to the USSR and which might have disrupted the delicate balance necessary for the continuation of Macao's existence. Thus, Leandro, who had at first dealt with the CDM, expelled many of its supporters, effectively assuring the success of the ADIM in elections to the Legislative Assembly in July 1976. Leandro also supported the enactment of legislation in Portugal (the Organic Law) which, in February 1976, appeared to give Macao virtual political autonomy.

Macao had thus become a special territory of Portugal with a Governor of ministerial rank appointed by the President of the Portuguese Republic, to whom he was responsible. The Governor was the executive authority and could issue decrees, a subject of continued controversy, and he remained independent of the Legislative Assembly with the right to veto legislation not passed by a two-thirds' majority. The composition of the Assembly reflected Macao's social and political structure, which was still divided into 'Chinese' and 'Portuguese', the latter self-determined on the basis of culture, language and religion—not on place of origin. Of the 17 members at that time, six were elected on a proportional basis, six were elected indirectly by designated organizations, and five were appointed by the Governor.

The 1976 elections confirmed Macao's conservative preference, with the ADIM receiving 55% of the 2,700 votes cast and winning four of the six seats. A group of young independents stood as the Grupo de Estudos para o Desenvolvimento Comunitário de Macau (GEDEC), winning 17% of the vote, while the radical and once dominant CDM was less successful, but also elected a member. The Governor's appointees and those indirectly elected created an Assembly which was still recognizably Portuguese, confirming the Portuguese nature of the Government and avoiding the anomalous position of a Chinese population voting for non-communist parties on China's very borders.

Macao's international relations continued to be subject to the *de jure* approval of the President of Portugal and the *de facto* tolerance of China. In April 1978 the Chinese authorities acted on a long-standing request of Governor Leandro and invited him to visit several Chinese cities (excluding Beijing). It was the first visit by a Governor of Macao since the communist revolution but, while interpreted as confirmation of China's acceptance of Macao's status at that time, the visit was seen in the context of China's overall policy rather than as an endorsement of Leandro himself. Meanwhile, leaders of the Chinese community in Macao continued to participate in political events in China, including regular attendance at sessions of the National People's Congress (NPC) in Beijing. Portugal and China established diplomatic relations in February 1979.

Governor Leandro, who correctly appraised Macao's needs in 1974, was frustrated in his desire to obtain greater local Chinese political participation, and his efforts to have the Macao population involved in the nomination of his successor were ineffective. His four-year administration was, however, marked by a more realistic planning orientation, marred in execution by administrative problems. Nevertheless, he changed patterns of thinking, set in motion basic economic reforms, and secured the passage of legislation that resulted in a more efficient taxation system and improvements to the local civil service.

The next Governor, Gen. Nuno de Melo Egídio, arrived in February 1979. Elections for the Legislative Assembly were due in late 1979, but in June the Assembly extended the term of the seven elected representatives for another year. This temporarily averted the constitutional crisis confronting the Governor in 1980, when the four-year-old Organic Law was due for review. A delegation of the representatives had visited Lisbon to discuss proposals to revise the 1976 Organic Law with politicians in the Portuguese legislature. Their proposals, which were to give the local population a greater influence in decisions concerning the administration of Macao, included plans to enlarge the Legislative Assembly from 17 to 21 members. All of these would, they proposed, be elected, thus reducing the Governor to a merely titular status. In March 1980, when Gen. Melo Egídio visited Beijing on the first 'official' visit by a Governor of Macao since its establishment as a Portuguese colony in 1557, the Chinese leader Deng Xiaoping, while expressing his approval of the stability of Macao, also made it clear that the Chinese Government opposed any change in the Organic Law. His views were reiterated by sections of the local population in Macao; 97% of the population did not speak Portuguese, they took very little part in political life and were largely unaffected by the Portuguese administration.

The appointment in 1981 of Cdre (later Rear-Adm.) Vasco Almeida e Costa as Governor was evidence of an unspoken agreement between Beijing and Lisbon not to alter the legal status of Macao. Governor Almeida e Costa was determined to extend voting rights in Macao, to produce a more representative Assembly. Following a constitutional dispute in March 1984, after he had used his controversial authority to issue two administrative decrees without the approval of the Legislative Assembly (and had successfully introduced electoral reforms despite vigorous opposition), he requested President Eanes of Portugal to dissolve the Assembly. In August 1984 elections for a new Assembly were held, in which the Chinese majority were allowed to vote for the first time, regardless of their length of residence in the territory. Two of the six directly elected seats were won by Chinese candidates, while the six indirectly elected members, all Chinese, were returned unopposed. The Governor appointed four government officials and a Chinese businessman to complete the Assembly, which was thus, for the first time, dominated by ethnic Chinese deputies.

Governor Almeida e Costa resigned in January 1986. He was replaced as Governor by Joaquim Pinto Machado, who had hitherto been a professor of medicine and was little-known as a political figure. His appointment marked a break in the tradition of military governors for Macao, but his political inexperience placed him at a disadvantage. Within weeks of Pinto Machado's arrival, the likelihood of his departure was widely rumoured, and in May 1987, one year after his appointment, he resigned for reasons of 'institutional dignity'. He was replaced in August by Carlos Melancia, a former socialist deputy in the Assembly of the Republic, who had held ministerial posts in several Portuguese Governments led by Dr Mário Lopes Soares, Prime Minister and subsequently the President of Portugal.

PREPARATIONS FOR THE RESUMPTION OF CHINESE SOVEREIGNTY

In May 1985 President Eanes visited Beijing and Macao, and it was announced that the Portuguese and Chinese Governments had agreed to hold formal talks about the territory's future. The first session of negotiations took place in June 1986, in Beijing, and further talks were held in September and October, when it was reported that 'broad agreement' had been reached. Portugal's acceptance of Chinese sovereignty greatly simplified subsequent negotiations, and on 13 April 1987, following the conclusion of the fourth round of talks, a joint declaration was formally signed in Beijing by the Portuguese and Chinese Governments, during an official visit to China by the Prime Minister of Portugal. According to the agreement, which was formally ratified in January 1988, Macao was to become a 'Special Administrative Region' (SAR) of the People's Republic, to be known as Macao, China, on 20 December 1999. Macao was thus to have the same status as that agreed (with effect from mid-1997) for Hong Kong, and was to enjoy autonomy in most matters except defence and foreign policy. A Sino-Portuguese Joint Liaison Group (JLG), established to oversee the transfer of power, held its inaugural meeting in Lisbon in April 1988.

Under the detailed arrangements for the transfer, a chief executive for Macao was to be appointed in 1999 by the Chinese Government, following 'elections or consultations to be held in Macao', and the territory's legislature was to contain 'a majority of elected members'. The inhabitants of Macao were to become Chinese citizens; the Chinese Government refused to allow the possibility of dual Sino-Portuguese citizenship, although Macao residents in possession of Portuguese passports were apparently to be permitted to retain them for travel purposes. The agreement provided for a 50-year period during which Macao would be permitted to retain its free-enterprise capitalist economy, and to be financially independent of China.

In August 1988 the establishment of a Macao Basic Law Drafting Committee was announced by the Chinese Government. Comprising 30 Chinese members and 19 representatives from Macao, the Committee was to draft a law determining the territory's future constitutional status within the People's Republic of China.

Triennial elections to the six directly elective seats in Macao's 17-seat Legislative Assembly were held in October 1988. In a low turn-out (representing fewer than 30% of the 67,492 registered voters), an informally constituted 'liberal' grouping increased its representation from one seat in the previous Assembly to three seats. These gains were achieved at the expense of the long-dominant 'grand alliance' of pro-Beijing and Macanese business interests. The new members hoped to influence the administration's policies on housing, education and workers' welfare.

In January 1989 it was announced that Portuguese passports were to be issued to about 100,000 ethnic Chinese inhabitants, born in Macao before October 1981, and it was anticipated that as many as a further 100,000 would be granted before 1999. Unlike their counterparts in the neighbouring British dependent territory of Hong Kong, therefore, these Macao residents (but not all) were to be granted the full rights of a citizen of the European Community (EC, now European Union—EU). In February 1989 President Mário Soares of Portugal visited Macao, in order to discuss the transfer of the territory's administration to China.

Following the violent suppression of the pro-democracy movement in China in June 1989, as many as 100,000 residents of Macao participated in demonstrations in the enclave to protest against the Chinese Government's action. The events in the People's Republic gave rise to much concern in Macao, and it was feared that many residents would wish to leave the territory prior to 1999. In August 1989, however, China assured Portugal that it would honour the agreement to maintain the capitalist system of the territory after 1999.

In March 1990 the implementation of a programme to grant permanent registration to parents of 4,200 Chinese residents, the latter having already secured the right of abode in Macao, developed into chaos when other illegal immigrants demanded a similar concession. The authorities decided to declare a general amnesty, but were unprepared for the numbers of illegal residents who rushed to take advantage of the scheme, thereby revealing the true extent of previous immigration from China. In the ensuing stampede by 50,000 illegal immigrants, desperate to obtain residency rights, about 200 persons were injured and 1,500 arrested, as the police attempted to control the situation. Border security was increased, in an effort to prevent any further illegal immigration from China.

In late March 1990 the Legislative Assembly approved the final draft of the territory's revised Organic Law. The Law was approved by the Portuguese Assembly of the Republic in mid-April, and granted Macao greater administrative, economic, financial and legislative autonomy, in advance of 1999. The powers of the Governor and of the Legislative Assembly, where six additional seats were to be created, were therefore increased. The post of military commander of the security forces was abolished, responsibility for security being assumed by a civilian Under-Secretary.

In June 1990 the Under-Secretary for Justice, Dr Manuel Magalhães e Silva, resigned, owing to differences of opinion on the issues of Macao's political structure and Sino-Portuguese relations. In the same month, while on a visit to Lisbon for consultations with the President and Prime Minister, Carlos Melancia rebuked the Chinese authorities for attempting to interfere in the internal affairs of Macao. This unprecedented reproach followed criticism of the Governor's compromising attitude towards the People's Republic of China.

Meanwhile, in February 1990, Carlos Melancia had been implicated in a financial scandal. It was alleged that the Governor had accepted 50m. escudos from a Federal German company which hoped to be awarded a consultancy contract for the construction of the new airport in Macao. In September Melancia was served with a summons in connection with the alleged bribery. Although he denied any involvement in the affair, the Governor resigned, and was replaced on an acting basis by the Under-Secretary for Economic Affairs, Dr Francisco Murteira Nabo. In September 1991 it was announced that Melancia and five others were to stand trial on charges of corruption. The trial opened in Lisbon in April 1993. At its conclusion in August the former Governor was acquitted on the grounds of insufficient evidence. In February 1994, however, it was announced that Melancia was to be retried, owing to irregularities in his defence case.

The ability of Portugal to maintain a stable administration in the territory had once again been called into question. Many observers believed that the enclave was being adversely affected by the political situation in Lisbon, as differences between the socialist President and centre-right Prime Minister were being reflected in rivalries between officials in Macao. In an attempt to restore confidence, therefore, President Soares visited the territory in November 1990. In January 1991, upon his re-election as Head of State, the President appointed Gen. Vasco Rocha Vieira (who had served as the territory's Chief of Staff in 1973/74 and as Under-Secretary for Public Works and Transport in 1974/75) to be the new Governor of Macao. In March 1991 the Legislative Assembly was expanded from 17 to 23 members. All seven Under-Secretaries were replaced in May.

Following his arrival in Macao, Gen. Rocha Vieira announced that China would be consulted on all future developments in the territory. The 10th meeting of the Sino-Portuguese JLG took place in Beijing in April 1991. Topics under regular discussion included the participation of Macao in international organizations, progress towards an increase in the number of local officials employed in the civil service (hitherto dominated by Portuguese and Macanese personnel) and the status of the Chinese language. The progress of the working group on the translation of local laws from Portuguese into Chinese was also examined, a particular problem being the lack of suitably qualified bilingual legal personnel. (The training of civil servants was duly improved: the University of Macao opened new courses in administration, law and translation; and hundreds of civil servants were dispatched to Beijing or Lisbon for training.) It was agreed that Portuguese was to remain an official language after 1999. The two sides also reached agreement on the exchange of identity cards for those Macao residents who would require them in 1999. Regular meetings of the JLG continued.

In July 1991 the Macao Draft Basic Law was published by the authorities of the People's Republic of China. Confidence in the territory's future was enhanced by China's apparent flexibility on a number of issues. Unlike the Hong Kong Basic Law, that of Macao did not impose restrictions on holders of foreign passports assuming senior posts in the territory's administration after 1999, the only exception being the future Chief Executive. Furthermore, the draft contained no provision for the stationing of troops from China in Macao after the territory's return to Chinese administration.

In November 1991 the Governor of Macao visited the People's Republic of China, where it was confirmed that the 'one country, two systems' policy would operate in Macao from 1999. Following a visit to Portugal by the Chinese Premier in February 1992, the Governor of Macao stated that the territory was to retain 'great autonomy' after 1999. In March 1993 the final draft of the Basic Law of the Macao SAR was ratified by the NPC in Beijing, which also approved the design of the future SAR's flag. The adoption of the legislation was welcomed by the Governor of Macao, who reiterated his desire for a smooth transfer of power in 1999. The Chief Executive of the SAR was to be selected by local representatives. The SAR's first Legislative Council was to comprise 23 members, of whom eight would be directly elected. Its term of office would expire in October 2001, when it would be expanded to 27 members, of whom 10 would be directly elected.

Meanwhile, elections to the Legislative Assembly were held in September 1992. The level of participation was higher than on previous occasions, with 59% of the registered electorate (albeit only 13.5% of the population) attending the polls. Fifty candidates contested the eight directly elective seats, four of which were won by members of the main pro-Beijing parties, the União Promotora para o Progresso and the União para o Desenvolvimento.

Relations between Portugal and China remained cordial. In June 1993 the two countries reached agreement on all outstanding issues regarding the construction of the territory's airport and the future use of Chinese air space. Furthermore, Macao was to be permitted to negotiate air traffic agreements with other countries. In October, upon the conclusion of a three-day visit to Macao, President Soares expressed optimism regarding the territory's smooth transition to Chinese administration. In November President Jiang Zemin of China was warmly received in Lisbon, where he had discussions with both the Portuguese President and Prime Minister. In February 1994 the Chinese Minister of Communications visited Macao to discuss with the Governor the progress of the airport project.

In April 1994, during a visit to China, the Portuguese Prime Minister received an assurance that Chinese nationality would not be imposed on Macanese people of Portuguese descent, who would be able to retain their Portuguese passports. Speaking in Macao itself, the Prime Minister expressed confidence in the territory's future. Regarding the issue as increasingly one of foreign policy, he stated his desire to transfer jurisdiction over Macao from the Presidency of the Republic to the Government, despite the necessity for a constitutional amendment.

In July 1994 a group of local journalists dispatched a letter, alleging intimidation and persecution in Macao, to President Soares, urging him to intervene to defend the territory's press freedom. The journalists' appeal followed an incident involving the director of the daily *Gazeta Macaense*, who had been obliged to pay 300,000 escudos for reproducing an article from *Semanário*, a Lisbon weekly newspaper, and now faced trial. The territory's press had been critical of the Macao Supreme Court's decision to extradite ethnic Chinese to the mainland (despite the absence of any extradition treaty) to face criminal charges and the possibility of a death sentence.

Gen. Rocha Vieira embarked upon a second visit to China in August 1994. The Governor of Macao had discussions with the Chinese Minister of Foreign Affairs, who declared Sino-Portuguese relations to be sound but, as a result of a gaffe relating to the delegation's distribution to the press of a biography of Premier Li Peng containing uncomplimentary remarks, stressed the need for vigilance.

The draft of the new penal code for Macao did not incorporate the death penalty. In January 1995, during a visit to Portugal, Vice-Premier Zhu Rongji of China confirmed that the People's Republic would not impose the death penalty in Macao after 1999, regarding the question as a matter for the authorities of the future SAR. The new penal code, prohibiting capital punishment, took effect in January 1996.

On another visit to the territory in April 1995, President Soares emphasized the need for Macao to assert its identity, and stressed the importance of three issues: the modification of the territory's legislation; the rights of the individual; and the preservation of the Portuguese language. Travelling on to Beijing, accompanied by Gen. Rocha Vieira, the Portuguese President had successful discussions with his Chinese counterpart on various matters relating to the transition.

In May 1995, during a four-day visit to the territory, Lu Ping, the director of the mainland Hong Kong and Macao Affairs Office, proposed the swift establishment of a preparatory working committee (PWC) to facilitate the transfer of sovereignty. He urged that faster progress be made on the issues of the localization of civil servants and of the law, and on the use of Chinese as the official language. Lu Ping also expressed his desire that the reorganized legislative and municipal bodies to be elected in 1996–97 conform with the Basic Law.

In November 1995, following the change of government in Lisbon, the incoming Portuguese Minister of Foreign Affairs, Jaime Gama, urged that the rights and aspirations of the people of Macao be protected. In December, while attending the celebrations to mark the inauguration of the territory's new airport, President Soares had discussions with the Chinese Vice-President, Rong Yiren. During a four-day visit to Beijing in February 1996, Jaime Gama met President Jiang Zemin and other senior officials, describing the discussions as positive. While acknowledging the sound progress of recent years, Gama and the Chinese Minister of Foreign Affairs agreed on an acceleration in the pace of work of the Sino-Portuguese JLG. In March 1996 Gen. Rocha Vieira was reappointed Governor of Macao by the newly elected President of Portugal, Jorge Sampaio. António Guterres, the new Portuguese Prime Minister, confirmed his desire for constitutional consensus regarding the transition of Macao. The JLG's 26th meeting took place in June 1996 in Macao.

At elections to the Legislative Assembly in September 1996, a total of 62 candidates from 12 electoral groupings contested the eight directly elective seats. The pro-Beijing União Promotora para o Progresso received 15.2% of the votes and won two seats, while the União para o Desenvolvimento won 14.5% and retained one of its two seats. The business-orientated groups were more successful: the Associação Promotora para a Economia de Macau took 16.6% of the votes and secured two seats; the Convergência para o Desenvolvimento and the União Geral para o Desenvolvimento de Macau each won one seat. The pro-democracy Associação de Novo Macau Democrático also won one seat. The level of voter participation was 64%. The 23-member legislature was to remain in place beyond the transfer of sovereignty in 1999.

In October 1996 Portugal and China announced the establishment of a mechanism for regular consultation on matters pertaining to international relations. In the same month citizens of Macao joined a flotilla of small boats carrying activists from Taiwan and Hong Kong to protest against a right-wing Japanese group's construction of a lighthouse on the disputed Daioyu (or Senkaku) Islands, situated in the East China Sea (see the chapter on Taiwan for further details). Having successfully evaded Japanese patrol vessels, the protesters raised the flags of China and Taiwan on the disputed islands. In November activists from around the world attended a three-day conference in Macao, in order to discuss their strategy for the protection of the islands.

During 1996 the rising level of violent criminal activity became a cause of increasing concern. Between January and December there were 14 bomb attacks, in addition to numerous brutal assaults. In November a Portuguese gambling inspector narrowly survived an attempt on his life by an unidentified gunman and, as attacks on local casino staff continued, three people were killed and three wounded in six separate incidents. Criminal violence continued to gather momentum in 1997, giving rise to fears for the future of the territory's vital tourism industry. Many attributed the alarming increase in organized crime to the opening of the airport in Macao, which was believed to have facilitated the entry of rival gangsters from mainland China, Taiwan and Hong Kong. In May, following the murder of three men believed to have associations with one such group of gangsters, the Chinese Government expressed its concern at the deterioration of public order in Macao and urged Portugal to observe its responsibility, as undertaken in the Sino-Portuguese joint declaration of 1987, to maintain the enclave's social stability during the transitional period, whilst pledging the enhanced co-operation of the Chinese security forces in the effort to curb organized crime in Macao.

The freedom of Macao's press was jeopardized in June 1997, when several Chinese-language newspapers, along with a television station, received threats instructing them to cease reporting on the activities of the notorious 14K triad, a 10,000-member secret society to which much of the violence had been attributed. In July, during a night of arson and shooting, an explosive device was detonated in the grounds of the Governor's palace, although it caused no serious damage. In the following month China deployed 500 armed police-officers to reinforce the border with Macao in order to intensify its efforts to combat illegal immigration, contraband and the smuggling of arms into the enclave. Despite the approval in July of a law further to restrict activities such as extortion and 'protection rackets', organized crime continued unabated. In early October the police forces of Macao and China initiated a joint campaign against illegal immigration. In late October Leong Kwok-hon, an alleged leader of the 14K triad, was shot dead.

Meanwhile, the slow progress of the 'three localizations' (civil service, laws and the implementation of Chinese as an official language) continued to concern the Government of China. In mid-1996 almost 50% of senior government posts were still held by Portuguese expatriates. In January 1997 the Governor pledged to accelerate the process with regard to local legislation, the priority being the training of the requisite personnel. In the same month, during a visit to Portugal, the Chinese Minister of Foreign Affairs reiterated his confidence in the future of Macao. In February President Sampaio travelled to both Macao and China, where he urged respect for Macao's identity and for the Luso-Chinese declaration regarding the transfer of sovereignty. In December 1997 details of the establishment in Macao of the office of the Chinese Ministry of Foreign Affairs, which was to commence operations in December 1999, were announced. In January 1998 the Macao Government declared that 76.5% of 'leading and directing' posts in the civil service were now held by local officials.

In March 1998 the murder of a Portuguese gambling official, followed by the killing of a marine police-officer, prompted the Chinese authorities to reiterate their concern at the deteriorating situation in Macao. In the following month the driver of the territory's Under-Secretary for Public Security was shot dead. In April, by which month none of the 34 triad-related murders committed since January 1997 had been solved, the Portuguese and Chinese Governments agreed to co-operate in the exchange of information about organized criminal activities. Also in April 1998 the trial, on charges of breaching the gaming laws, of the head of the 14K triad, Wan Kuok-koi ('Broken Tooth'), was adjourned for two months, owing to the apparent reluctance of witnesses to appear in court. Following an attempted car-bomb attack on Macao's chief of police, António Marques Baptista, in early May Wan Kuok-koi was rearrested. The charge of the attempted murder of Marques Baptista, however, was dismissed by a judge three days later on the grounds of insufficient evidence. Wan Kuok-koi remained in prison, charged with other serious offences. (In April 1999 he was acquitted of charges of the coercion of croupiers, but was to stand trial again on charges of triad membership, illegal gambling activities and 'money laundering'.) The renewed detention in May 1998 of Wan Kuok-koi led to a spate of arson attacks. The Portuguese Government was reported to have dispatched intelligence officers to the enclave to reinforce the local security forces. In June Marques Baptista travelled to Beijing and Guangzhou for discussions on the problems of cross-border criminal activity and drugs-trafficking.

In April 1998 the Portuguese Prime Minister, accompanied by his Minister of Foreign Affairs and a business delegation, paid an official visit to Macao, where he expressed confidence that after 1999 China would respect the civil rights and liberties of the territory. The delegation travelled on to China, where the Prime Minister had cordial discussions with both President Jiang Zemin and Premier Zhu Rongji.

The Preparatory Committee for the Establishment of the Macao SAR, which was to oversee the territory's transfer to Chinese sovereignty and was to comprise representatives from both the People's Republic and Macao, was inaugurated in Beijing in May 1998. Four subordinate working groups (supervising administrative, legal, economic, and social and cultural affairs) were subsequently established. The second plenary session of the Preparatory Committee was convened in July 1998, discussions encompassing issues such as the 'localization' of civil servants, public security and the drafting of the territory's fiscal budget for the year 2000. In July 1998, during a meeting with the Chinese Premier, the Governor of Macao requested an increase in the mainland's investment in the territory prior to the 1999 transfer of sovereignty.

In July 1998, as abductions continued and as it was revealed, furthermore, that the victims of kidnapping and ransom had included two serving members of the Legislative Assembly, President Jiang Zemin of China urged the triads of Macao to cease their campaign of intimidation. The police forces of Macao, Hong Kong and Guangdong Province launched 'S Plan', an operation aiming to curb the activities of rival criminal gangs. In August, in an apparent attempt to intimidate the judiciary, the territory's Attorney-General and his pregnant wife were shot and slightly wounded. In the following month five police-officers and 10 journalists, who were investigating a bomb attack, were injured when a second bomb exploded.

In August 1998 representatives of the JLG agreed to intensify Luso-Chinese consultations on matters relating to the transitional period. In September, in response to the increasing security problems, China unexpectedly announced that, upon the transfer of sovereignty, it was to station troops in the territory. This abandonment of a previous assurance to the contrary caused much disquiet in Portugal, where the proposed deployment was deemed unnecessary. Although the Basic Law made no specific provision for the stationing of a mainland garrison, China asserted that it was to be ultimately responsible for the enclave's defence. By October, furthermore, about 4,000 soldiers of the People's Liberation Army (PLA) were on duty at various Chinese border posts adjacent to Macao. During a one-week visit to Beijing, the territory's Under-Secretary for Public Security had discussions with senior officials, including the Chinese Minister of Public Security. In mid-October the detention without bail of four alleged members of the 14K triad in connection with the May car-bombing and other incidents led, later in the day, to an outburst of automatic gunfire outside the courthouse.

In November 1998 procedures for the election of the 200 members of the Selection Committee were established by the Preparatory Committee. Responsible for the appointment of the members of Macao's post-1999 Government, the delegates of the Selection Committee were required to be permanent residents of the territory: 60 members were to be drawn from the business and financial communities, 50 from cultural, educational and professional spheres, 50 from labour, social service and religious circles and the remaining 40 were to be former political personages.

About 70 people were arrested in November 1998, when the authorities conducted raids on casinos believed to be engaged in illegal activities. In December an off-duty Portuguese prison warder was shot dead and a colleague wounded by a gunman, the pair having formed part of a contingent recently dispatched from Lisbon to improve security at the prison where Wan Kuok-koi was being held. At the end of December it was confirmed that Macao residents of wholly Chinese origin would be entitled to full mainland citizenship, while those of mixed Chinese and Portuguese descent would be obliged to decide between the two nationalities. In January 1999 protesters clashed with police during demonstrations to draw attention to the plight of numerous immigrant children, who had been brought illegally from China to Macao to join their legitimately resident parents. Several arrests were made. The problem had first emerged in 1996 when, owing to inadequate conditions, the authorities had closed down an unofficial school attended by 200 children, who because of their irregular status were not entitled to the territory's education, health and social services.

In January 1999 a grenade attack killed one person, and the proprietor of a casino and suspected member of 14K was shot dead. In that month details of the composition of the future PLA garrison were disclosed. The troops were to comprise solely ground forces, totalling fewer than 1,000 soldiers and directly responsible to the Commander of the Guangzhou Military Unit. They would be permitted to intervene to maintain social order in the enclave only if the local police were unable to control major triad-related violence or if street demonstrations posed a threat of serious unrest. In March, during a trip to Macao (where he had discussions with the visiting Portuguese President), Qian Qichen, a Chinese Vice-Premier, indicated that an advance contingent of PLA soldiers would be deployed in Macao prior to the transfer of sovereignty. Other sources of contention between China and Portugal remained the unresolved question of the post-1999 status of those Macao residents who had been granted Portuguese nationality and also the issue of the court of final appeal.

In April 1999 an alleged member of the 14 Carats triad was shot dead by a gunman on a motor cycle. Also in April, at the first plenary meeting of the Selection Committee, candidates for the post of the SAR's Chief Executive were elected. Edmund Ho received 125 of the 200 votes, while Stanley Au garnered 65 votes. Three other candidates failed to secure the requisite minimum of 20 votes. Edmund Ho and Stanley Au, both bankers and regarded as moderate pro-business candidates, thus proceeded to the second round of voting by secret ballot, held in May. The successful contender, Edmund Ho, received 163 of the 199 votes cast, and confirmed his intention to address the problems of law and order, security and the economy. The Chief Executive-designate also fully endorsed China's decision to deploy troops in Macao.

During 1999, in co-operation with the Macao authorities, the police forces of Guangdong Province, and of Zhuhai in particular, initiated a new offensive against the criminal activities of the triads. China's desire to deploy an advance contingent of troops prior to December 1999, however, reportedly continued to be obstructed by Portugal. Furthermore, the announcement that, subject to certain conditions, the future garrison was to be granted law-enforcement powers raised various constitutional issues. Some observers feared the imposition of martial law, if organized crime were to continue unabated. Many Macao residents, however, appeared to welcome the mainland's decision to station troops in the enclave. In a further effort to address the deteriorating security situation, from December 1999 Macao's 5,800-member police force was to be restructured.

In July 1999 the penultimate meeting of the JLG took place in Lisbon. In August, in accordance with the nominations of the Chief Executive-designate, the composition of the Government of the future SAR was announced by the State Council in Beijing. Appointments included that of Florinda da Rosa Silva Chan as Secretary for Administration and Justice. Also in August an outspoken pro-Chinese member of the Legislative Assembly was attacked and injured by a group of unidentified assailants. This apparently random assault on a serving politician again focused attention on the decline in law and order in the enclave. In September the Governor urged improved co-operation with the authorities of Guangdong Province in order to combat organized crime, revealing that more than one-half of the inmates of Macao's prisons were not residents of the territory. In the same month it was reported that 90 former Gurkhas of the British army were being drafted in as prison warders, following the intimidation of local officers. In September the Chief Executive-designate announced the appointment of seven new members of the Legislative Council, which was to succeed the Legislative Assembly in December 1999. While the seven nominees of the Governor in the existing Legislative Assembly were thus to be replaced, 15 of the 16 elected members (one having resigned) were to remain in office as members of the successor Legislative Council. The composition of the 10-member Executive Council was also announced.

In October 1999 President Jiang Zemin paid a two-day visit to Portugal, following which it was declared that the outstanding question of the deployment of an advance contingent of Chinese troops in Macao had been resolved. The advance party was to be restricted to a technical mission, which entered the territory in early December. In November the 37th and last session of the JLG took place in Beijing, where in the same month the Governor of Macao held final discussions with President Jiang Zemin.

Meanwhile, in April 1999 Wan Kuok-koi had been acquitted of charges of coercing croupiers. In November the trial of Wan Kuok-koi on other serious charges concluded: he was found guilty of criminal association and other illegal gambling-related activities and sentenced to 15 years' imprisonment. Eight co-defendants, including Wan Kuok-koi's brother, received lesser sentences. In a separate trial Artur Chiang Calderon, a former police-officer alleged to be Wan Kuok-koi's military adviser, received a prison sentence of 10 years and six months for involvement in organized crime. While two other defendants were also imprisoned, 19 were released on the grounds of insufficient evidence. As the transfer of the territory's sovereignty approached, by mid-December almost 40 people had been murdered in triad-related violence on the streets of Macao since January 1999.

THE SAR GOVERNMENT

In late November 1999 representatives of the JLG reached agreement on details regarding the deployment of Chinese troops in Macao and on the retention of Portuguese as an official language. At midnight on 19 December 1999, therefore, in a ceremony attended by the Presidents and heads of government of Portugal and China, the sovereignty of Macao was duly transferred; 12 hours later (only after the departure from the newly inaugurated SAR of the Portuguese delegation), 500 soldiers of the 1,000-strong force of the PLA, in a convoy of armoured vehicles, crossed the border into Macao, where they were installed in a makeshift barracks in a vacant apartment building. Prior to the ceremony, however, it was reported that the authorities of Guangdong Province had detained almost 3,000 persons, including 15 residents of Macao, suspected of association with criminal gangs. The celebrations in Macao were also marred by the authorities' handling of demonstrations by members of Falun Gong, a religious movement recently outlawed in China. The expulsion from Macao of several members of the sect in the days preceding the territory's transfer and the arrest of 30 adherents on the final day of Portuguese sovereignty prompted strong criticism from President Jorge Sampaio of Portugal. Nevertheless, in an effort to consolidate relations with the EU, in May 2000 the first official overseas visit of the SAR's Chief Executive was to Europe, his itinerary including Portugal.

Meanwhile, a spate of arson attacks on vehicles in February 2000 was followed by the fatal shooting, in a residential district of Macao, of a Hong Kong citizen believed to have triad connections. In March, in an important change to the immigration rules, it was announced that children of Chinese nationality whose parents were permanent residents of Macao would shortly be allowed to apply for residency permits. A monthly quota of 420 successful applicants was established, while the youngest children were to receive priority.

In May 2000 hundreds of demonstrators took part in a march to protest against Macao's high level of unemployment. This shortage of jobs was attributed to the territory's use of immigrant workers, mainly from mainland China and South-East Asia, who were estimated to total 28,000. During the ensuing clashes several police-officers and one demonstrator were reportedly injured. Trade unions continued to organize protests, and in July (for the first time since the unrest arising from the Chinese Cultural Revolution of 1966) tear gas and water cannon were used to disperse about 200 demonstrators who were demanding that the immigration of foreign workers be halted by the Government. In the same month it was announced that, in early 2001, an office of the Macao SAR was to be established in Beijing, in order to promote links between the two Governments. In Guangzhou in August 2000, as cross-border crime continued to increase, senior officials of Macao's criminal investigation unit met their counterparts from China and Hong Kong for discussions on methods of improving co-operation. It was agreed that further meetings were henceforth to be held twice a year, alternately in Beijing and Macao.

Celebrations to mark the first anniversary in December 2000 of Macao's reversion to Chinese sovereignty were attended by President Jiang Zemin, who made a speech praising the local administration, but warning strongly against those seeking to use either of the SARs as a base for subversion. A number of Falun Gong adherents from Hong Kong who had attempted to enter Macao for the celebrations were expelled. The same fate befell two Hong Kong human rights activists who had hoped to petition Jiang Zemin during his stay in Macao about the human rights situation in the People's Republic. A group of Falun Gong members in Macao held a protest the day before the Chinese

President's arrival. They were detained in custody and subsequently alleged that they had suffered police brutality.

In January 2001 China urged the USA to cease interfering in its internal affairs, following the signature by President Bill Clinton of the US Macao Policy Act, which related to the control of Macao's exports and the monitoring of its autonomy. In the same month voter registration began in Macao, in preparation for the expiry of the first Legislative Council's term of office in October and the election of a new assembly. In his Chinese Lunar New Year address on 23 January, Edmund Ho called for new efforts to revitalize the economy and achieve social progress.

The Governor of Guangdong Province, Lu Ruihua, made an official visit to Macao in early February to improve links between the two regions. At the same time, the Legislative Council announced plans to strengthen ties with legislative bodies in the mainland, and the Chairwoman of the Legislative Council, Susana Chou, visited Beijing and held talks with Vice-Premier Qian Qichen. Also in February, a Macao resident was charged with publishing online articles about the Falun Gong.

Edmund Ho visited Beijing in early March 2001 to attend the fourth session of the Ninth NPC, and held talks with President Jiang Zemin, who praised the former's achievements since the reversion of Macao to Chinese rule. On returning to Macao, Ho received the President of Estonia, Lennart Meri, who was touring the mainland and who thus became the first head of state to visit the SAR since its return to China. The two leaders discussed co-operation in the fields of tourism, trade, information technology (IT), and telecommunications, with Ho apparently seeking to learn from Estonia's experience in opening the telecommunications market. The EU announced in mid-March that SAR passport holders would, from May 2001, no longer require visas to enter EU countries. In the same month Jorge Neto Valente, a prominent lawyer and reputedly the wealthiest Portuguese person in Macao, was kidnapped by a gang, but freed in a dramatic police operation. The incident was the highest-profile kidnapping case in Macao since the return to Chinese rule.

The Macao, Hong Kong, and mainland police forces established a working group in mid-March 2001 to combat cross-border crime, with a special emphasis on narcotics, and in late March the Macao, Hong Kong, and Guangdong police forces conducted a joint anti-drugs operation, 'Spring Thunder', resulting in the arrest of 1,243 suspected traffickers and producers, and the seizure of large quantities of heroin, ecstasy, and marijuana. As part of the growing campaign against crime, a Shanghai court sentenced to death a Macao-based gangster, Zeng Jijun, on charges of running a debt-recovering group, members of which had committed murder. Three of Zeng's associates were given long prison sentences.

In May 2001 Macao and Portugal signed an agreement to strengthen co-operation in the fields of economy, culture, public security, and justice during the visit of the Portuguese Minister of Foreign Affairs, Jaime Gama, the highest-ranking Portuguese official to visit Macao since its reversion to Chinese rule.

In early June 2001 Macao's Secretary for Security, Cheong Kuoc Va, visited Beijing and signed new crime-fighting accords aimed at reducing the trafficking of drugs, guns, and people. In mid-June Chief Executive Edmund Ho made his first official visit to the headquarters of the EU in Brussels, where he sought to promote contacts and exchanges between the SAR and the EU.

At the beginning of July 2001 China's most senior representative in Macao, Wang Qiren, died of cancer. Later in the month, another major campaign against illegal activities related to the triads was conducted by the Macao, Hong Kong, and Guangdong police forces, and was part of ongoing attempts to eradicate organized crime. At the end of the month, the Secretary for Security, Cheong Kuoc Va, reported that cases of violent crime had declined by 37.3% year-on-year in the first half of 2001, and murders, robberies, arson, drugs-trafficking, and kidnapping had all decreased significantly over the same period. In a further sign of co-operation between Macao and the mainland against crime, the two sides signed an agreement on mutual judicial co-operation and assistance in late August, the first of its kind.

In September 2001 José Proença Branco and Choi Lai Hang were appointed police commander and customs chief respectively. Following the terrorist attacks in New York and Washington, DC, on 11 September, several Pakistanis were detained in Macao; however, it was quickly announced that the detentions were not connected to the world-wide terrorist searches that had been initiated. In the mean time, the Macao, Hong Kong, and Guangdong police departments began examining measures and directing activities aimed at fighting terrorism.

Elections to the Legislative Assembly (Council) were held on 24 September 2001, the first since Macao's reversion to Chinese rule. The number of seats was increased from 23 to 27: seven members were appointed by the Chief Executive, 10 elected directly and 10 indirectly. Of the 10 directly elective seats, two seats each were won by the business-orientated CODEM, the pro-Beijing factions UPP and UPD and the pro-democracy ANMD. Two other factions won one seat each. Of the 10 indirectly elective seats, four were won by the OMKC (a group representing business interests), and two seats each were one by the DCAR (a group representing welfare, cultural, educational, and sports interests), the CCCAE (a group representing labour), and the OMCY (a group representing professionals).

In mid-October 2001 China appointed Bai Zhijian as director of its liaison office in Macao, and later in the month Cui Shiping was selected as Macao's representative in the NPC, replacing the late Wang Qiren. At the same time, Edmund Ho attended the summit meeting of Asia-Pacific Economic Co-operation (APEC) in Shanghai, and the EU-Macao Joint Committee held a meeting in the SAR, aimed at improving trade, tourism and legal co-operation between the two entities. During late 2001, meanwhile, Macao increased co-operation with Hong Kong and the mainland in fighting crime and combating terrorism, amid reports that Russian mafias were becoming increasingly active in the SARs, and in mid-November the three police departments held an anti-drugs forum in Hong Kong. Later in the month, Edmund Ho announced that personal income tax would be waived and industrial and commercial taxes reduced for 2002, in order to alleviate the impact of the economic downturn. Ho also pledged to create 6,000 new jobs and invest more in infrastructure, and urged employers to avoid staff reductions.

In December 2001 the Government moved finally to break the 40-year monopoly on casinos and gambling held by Stanley Ho and his long-established company, the Sociedade de Turismo e Diversões de Macau (STDM). Under the new arrangements, some 21 companies, none of which was Chinese-owned, were to be permitted to bid for three new operating licences for casinos in the SAR. The intention was to improve the image of the gambling industry, ridding the territory of its reputation for vice and making it more business- and family-orientated. Meanwhile, Stanley Ho's daughter Pansy was playing an increasingly prominent role in managing the family businesses (which included the shipping, property and hotel conglomerate, Shun Tak holdings); in December the group opened a new convention and entertainment centre.

Also in December 2001 Edmund Ho paid a visit to Beijing, where he and President Jiang Zemin discussed the situation in Macao. In early January 2002 Ho visited the mainland city of Chongqing, seeking to reinforce economic ties between the two places, and stating that Macao would play a more active role in developing the region. Also in January, the Government granted permission to the Taipei Trade and Cultural Office (TTCO) to issue visas for Taiwan-bound visitors from Macao and the mainland. In February Li Peng, Chairman of the Standing Committee of the NPC, paid an official visit to Macao, where he held discussions with the Chief Executive of the SAR. During Li's visit, a leading Macao political activist, along with several activists from the Hong Kong-based 'April 5th Action Group', were arrested for planning to stage protests against Li for his role in the Tiananmen Square suppression of 1989 and in favour of the release of mainland political dissidents. The Hong Kong activists were immediately deported. At the same time the Hong Kong media reported that a Hong Kong-based cameraman had been beaten and had his camera destroyed by a Macao policeman when he attempted to film the interception of the activists. Other journalists also claimed to have been treated aggressively, their allegations being disputed by the Macao police.

In early March 2002 a new representative office of the Macao SAR was established in Beijing, with the aim of enhancing ties between the SAR and the central Government and mainland. Wu Beiming was named as its director. At its inaugural ceremony, Edmund Ho and Chinese Vice-Premier Qian Qichen praised the 'one country, two systems' model, and the director of the central government liaison office in Macao, Bai Zhijian, suggested that Macao might become a model for Taiwan's eventual reunification with the mainland.

On 1 April 2002 Stanley Ho's STDM formally relinquished its 40-year monopoly on casinos. However, Ho retained influence in the gambling sector after his Macao Gaming Holding Company (SJM) won an 18-year licence to operate casinos (see Economy, below). Also in early April, Edmund Ho attended the first annual conference of the Bo'ao Forum for Asia (BFA—a non-profit NGO), held on Hainan Island, China, where he met Hong Kong Chief Executive Tung Chee-hwa and Chinese Premier Zhu Rongji, as well as business leaders from both places.

The US Government in early April 2002 issued its second annual 'United States-Macao Policy Act Report', which stated that the SAR continued to develop in a positive direction, citing its support for the USA's anti-terrorism campaign, the opening of the economy, the reorganization of its customs services, efforts to counter organized crime, and the preservation of its own identity, including maintaining basic civil and human rights. As a result, Macao would continue to be accorded a special status distinct from mainland China under US law and policy. In the middle of the month the United Kingdom announced that it was granting visa-free access to holders of Macao SAR passports, and in late May the visiting Portuguese Minister of Foreign Affairs, António Martins da Cruz, also expressed confidence in Macao's future. In early June Macao hosted the Euro-China Business meeting, aimed at promoting small- and medium-sized enterprises in China to European investors. Also at this time, the Taiwanese Government eased restrictions on residents of Hong Kong and Macao applying for landing visas, essentially allowing those persons to obtain such visas on their first visit to Taiwan. However, in late July a Macao official criticized Taiwanese President Chen Shui-bian and accused him of seeking independence from the mainland.

In mid-June 2002 the Procurator-General, Ho Chio Meng, visited Portugal to promote judicial co-operation between the two territories, the first such visit by a Macao delegation since its return to Chinese rule. Later in the month José Chu was appointed director of the Public Administration and Civil Services Bureau of the SAR.

Meanwhile, the Macao police force continued to maintain co-operation with Hong Kong and the mainland. In late June it was announced that almost 1,000 people had been arrested in Hong Kong during a one-month operation with the Macao police aimed at reducing cross-border and organized crime. At the end of July the Secretary for Security, Cheong Kuoc Va, stated that although overall crime had increased by 1.8% during the first half of the year, serious crimes had registered significant decreases.

In early July 2002 Beijing appointed Wan Yongxiang as the special commissioner of the Office of Special Commissioner of the Chinese Ministry of Foreign Affairs in the Macao SAR, succeeding Yuan Tao, who had held that post since 1999, when the office was established. At the end of July the Government announced the introduction of new identity cards, to be introduced in December 2002. The cards were expected to function additionally as driving licences, border and medical access passes, and electronic payment methods.

The Russian Minister of Foreign Affairs, Igor Ivanov, visited Macao in late July 2002, mainly seeking to strengthen bilateral economic and trading links, and encouraging Macao businesses to invest in Russia. In early August 2002 Edmund Ho visited the Chinese Autonomous Region of Nei Mongol (Inner Mongolia) in order to examine the possibility of developing ties with the Sino-Russian border region, and later in the month he visited Guangzhou, in southern China, to discuss further economic co-operation and the joint development of Hengqin island, which is under the jurisdiction of Zhuhai City but located very close to Macao. In late September Ho visited Mozambique, where he and President Joaquim Alberto Chissano agreed to strengthen bilateral economic relations. At the same time, Secretary for Security Cheong Kuoc Va visited Portugal and signed security co-operation agreements aimed at combating transnational crime. In mid-October Edmund Ho visited the Republic of Korea (South Korea) and met President Kim Dae-Jung. While there, Ho sought increased investment and to promote tourism from South Korea. The two leaders also agreed to extend reciprocal visa-free visits from a maximum of 30 days to 90 days, effective from January 2003.

In early December 2002 Macao selected its 12 candidates for the 10th NPC, to be convened in Beijing in March 2003. At the same time, Edmund Ho paid a routine visit to Beijing, where he held discussions with outgoing President Jiang Zemin and Vice-President Hu Jintao. The Macao SAR and the EU signed a four-year legal co-operation programme in early December, aimed at consolidating Macao's legal system. In early January 2003 the Chairman of the National Committee of the Chinese People's Political Consultative Conference (CPPCC), Li Ruihuan, visited Macao and praised the way in which the territory had been administered. Observers noted that Beijing was more satisifed with the governance of the Macao SAR than that of Hong Kong. In March 2003 deputies from Macao duly attended the first plenary session of the 10th NPC in Beijing. In the same month Edmund Ho held a meeting with the new Chinese Premier, Wen Jiabao, at which the future development of Macao under the policy of 'one country, two systems' was discussed.

Following the outbreak of a hitherto-unknown virus, Severe Acute Respiratory Syndrome (SARS), on the Chinese mainland and in neighbouring Hong Kong in early 2003, strict measures were introduced to prevent the disease from spreading to Macao, including travel restrictions and health checks on passengers entering the SAR. The disease containment programme appeared to have been successful by June 2003, with Macao's only confirmed case of SARS having been discharged from hospital.

In October 2003 Chinese Vice-President Zeng Qinghong visited Macao to attend the signing ceremony of the Closer Economic Partnership Arrangement (CEPA—see below) between Macao and the Chinese mainland. Also in October, Chinese Vice-Premier Wu Yi attended a trade forum in Macao relating to China's trade relations with Portuguese-speaking countries (see below). Celebrations took place in December 2003 to mark the fourth anniversary of the establishment of the Macao SAR.

In a Policy Address in November 2003, Edmund Ho outlined plans for administrative reform of Macao's Government, alongside legal reform, to begin in 2004. Also in November 2003, Xiao Yang, President of China's Supreme People's Court, delivered a speech emphasizing the importance of judicial co-operation between Macao and the mainland. Draft legislation detailing the procedure for the election of Macao's next Chief Executive, who was to be chosen in 2004 by a 300-member Election Committee, was approved by the Government in February 2004. In June 2004 it was announced that the re-election of the Chief Executive would take place in late August. Edmund Ho began his election campaign in mid-August, and was duly re-elected on 29 August, securing 296 of the 300 votes of the members of the Election Committee.

Economy

Revised by the editorial staff

Macao's economy is based on tourism and textile manufacturing. Various development projects have been initiated in an attempt at diversification. Industries such as plastics, toy-making and electronics have been introduced, while efforts have been made to develop other areas of the services sector, especially finance, again with limited success. Export-orientated manufacturing has performed well, the value of merchandise exports increasing sharply in the 1980s and 1990s. However, exports decreased in 2001, reflecting a weakening demand in global markets. In 2002 there was a slight increase in exports, and this trend continued in 2003.

In 1990, according to official estimates, Macao's gross domestic product (GDP), measured at current prices, rose to 27,895.5m. patacas. By 1995 GDP had increased to 55,333m. patacas, but it declined to 55,294m. in 1996. In 1997 GDP reached 55,894m. patacas. In 1998, however, GDP decreased to 51,902m. patacas. GDP declined further in 1999, to stand at 49,021m. patacas, but increased to 49,742m. patacas in 2000 and to 49,862m. in 2001. In 2002 Macao's GDP rose to an estimated 54,295m., and in 2003 there was a further increase to 63,365m. patacas. In real terms, during 1990–95 the territory's GDP increased at an average annual rate of 6.0%. In 1996, however, real GDP contracted by 0.5% compared with the previous year. In 1997 and in 1998 negative rates, of 0.3% and of 4.6% respectively, were again recorded, and in 1999 GDP contracted by 2.9%. The lower growth rates (which followed the particularly strong expansion of 13.3% recorded in 1992) and subsequent contraction of GDP were attributed to weak domestic demand, the lack of capital and of new technology, the shortage of skilled labour and the higher cost of salaries. GDP increased by 4.6% in 2000, but expanded by only 2.2% in 2001, reflecting the global economic slowdown. GDP increased by 10.0% in 2002, and by 15.6% in 2003. Economic growth was initially forecast by the Government at about 7% in 2004. Between 1990 and 2001 the population increased at an average annual rate of 1.6%, while GDP per head increased, in real terms, at an average rate of 0.9% annually. According to the census of August 2001, the population totalled 435,235, rising to an estimated 448,495 in December 2003. The average annual rate of inflation (excluding rents) between 1990 and 1997 was 6.7%, declining to 3.5% in 1997 and to only 0.2% in 1998. Deflation rates of 3.2% in 1999, 1.6% in 2000, 2.0% in 2001, and 2.6% in 2002 were recorded. In 2003 the deflation rate was estimated at 1.5%. In 1998 the rate of unemployment rose to 4.6% of the labour force, many job losses being in the trade, tourism and construction sectors. In July 1998 the Government allocated 50m. patacas to various training initiatives, in an attempt to lessen the effects of unemployment. The unemployment rate was officially estimated at 6.4% in 1999, 6.8% in 2000, 6.4% in 2001 and 6.3% in 2002. Some sources, however, suggested that the level was much higher. The unemployment rate was estimated at 6.0% in 2003.

During the first half of 2002 Macao experienced a sharp increase in foreign investment, with inflows totalling around 1,000m. patacas. Macao continues its attempts to present attractive prospects to foreign investors. Tax and banking reform measures have been passed. The gambling industry is on a sound base, having been liberalized in 2002. The Macao Government Economic Committee was founded in February 1994. Its functions included the formulation of policies on industrial and commercial development, as well as on investment promotion. A law enacted in April 1995 aimed to attract overseas investment by offering the right of abode in Macao to entrepreneurs with substantial funds (at least US $250,000) at their disposal. The World Trade Center was established in 1995, and opened for business (on a site of newly reclaimed land) in early 1996. In addition to providing information services, the Center offers exhibition and conference facilities.

In February 2003, following a detailed assessment of Macao's economic, social and political situation, a leading US agency raised the credit rating of the Special Administrative Region (SAR) to the same level as that of the People's Republic of China and of Hong Kong, declaring that Macao's overall economic prospects had improved since the territory's reversion to Chinese sovereignty in late 1999. In October 2003 the Closer Economic Partnership Arrangement (CEPA), a free-trade agreement, was signed between Macao and mainland China.

TOURISM

Despite attempts to diversify, the economy of Macao remains heavily dependent on tourism and the related gambling industry, with clientele coming mainly from Hong Kong at weekends. The Government's receipts from direct taxes on gambling increased from 2,048m. patacas in 1990, when they accounted for 41.2% of total budgetary revenue, to 5,269m. patacas in 1995, thus accounting for 60.8% of government revenue in that year. In 1996 the Government's receipts from gambling decreased to 4,954m. patacas, before rising to 6,013m. in 1997. In 1998, however, direct taxes from gambling yielded only 5,100m. patacas, contributing nevertheless more than 50% to the Government's revenue. In 2000 taxes on gambling accounted for an estimated 60% of the Government's total recurrent revenue. In the same year the contribution of tourism and gambling to GDP was estimated at 38%, while the 10 licensed casinos employed 6% of the labour force. The Macau Financial Services Bureau estimated that taxes from the casinos had raised revenue of 6,100m. patacas in 2001, 7,600m. in 2002, and 10,100m. in 2003 (the 2003 figure being roughly equivalent to 70% of budgetary revenue). Unofficial estimates predicted a revenue of 12,000m. patacas from the gambling industry in 2004.

Visitor arrivals decreased from 8.15m. in 1996, to 7.0m. in 1997 and to 6.9m. in 1998, as a result of the Asian currency crisis and the upsurge of casino-related crime in the territory. The sharpest declines, of 43.6% in 1997 and of 74.0% in 1998, were in the numbers of Japanese visitors, the Government of Japan having been among those to warn citizens not to visit Macao. In 1999 total visitor arrivals rose to 7.4m., almost 57% of whom were residents of Hong Kong, and in 2000 total visitor arrivals had risen to 9,162,212, of whom 54% were residents of Hong Kong. In 2001 total visitor arrivals increased to an unprecedented 10,278,973, of whom 51% were residents of Hong Kong. In 2002 visitor arrivals increased further, to a total of 11,530,841, of whom 5,101,437 were from Hong Kong. In 2003 the total number of visitors was 11,887,876, of whom 4,623,162 were from Hong Kong. Although a helicopter link between Macao and Hong Kong was established in 1990, most visitors from Hong Kong continue to travel by sea (51% in 2001). From 1993 the number of visitors from the People's Republic of China increased rapidly, tourists from the mainland reaching a total of 816,816 in 1998 (excluding visitors using a double-entry visa between January and July of that year), more than doubling to 1,645,193 in 1999, and rising to 2,274,713 in 2000, 3,005,722 in 2001, 4,240,446 in 2002 and 5,742,036 in 2003. Average per caput spending (excluding mainland Chinese visitors) declined to US $171.63 in 1999, a decrease of 1.38% compared with 1998. The average visitor from China spent US $332.63 in 1999, a decline of 5.07% compared with the previous year. Following the territory's transfer of sovereignty in December 1999, passports remained necessary for mainland Chinese visitors to Macao, along with special passes issued by the Ministry of Public Security of the People's Republic. From August 2003 residents of Guangzhou, Shenzhen and Zhuhai were to be permitted to visit Macao individually instead of being required to join tour groups. The average per caput visitor spending in 2003 was 1,518 patacas, with average expenditure by tourists from mainland China reaching 2,847 patacas.

Efforts are being made to broaden the base of the tourist industry, to encourage longer stays (the average length of sojourn being only 1.35 nights in 2001) and to attract overseas visitors to the territory's unique heritage. Programmes to restore and develop Macao's historic and religious monuments have attempted to offer visitors opportunities for cultural insights perhaps lost in Hong Kong. The first preservation statutes were passed in 1976. Numerous sites have since been similarly protected. During the 1990s several new hotels were built, in which major US chains have interests. In 2000 and 2001, however, the hotel occupancy rate reached only 57.6% and 60.1% respectively, compared with 79% in 1991. At the end of 2002 there were 25 hotels rated two-stars or above, and the number of hotel rooms available was 9,185 in 2003. Macao's tourist industry was adversely affected by travel restrictions imposed following the outbreak of Severe Acute Respiratory Syndrome (SARS) on the Chinese mainland in early 2003. Tourist arrivals in Macao decreased by 70% at the start of Easter 2003, compared with the corresponding Friday of the previous year. However, in mid-2003 the Macao Government initiated a tourism promotion scheme, costing 30m. patacas, and by the end of the year recovery of the industry was evident, with a record number of tourist arrivals being recorded for the year as a whole.

By the mid-1990s a tourist activity centre was in operation, incorporating the Macao Grand Prix Museum, the Wine Museum, restaurants, conference halls, exhibition areas and other facilities. Other projects included a 360m.-patacas shopping centre (a joint venture with a major Japanese retailing company). In collaboration with a Taiwan-based enterprise, a 2,000m.-patacas complex, incorporating the territory's largest casino and office building, was constructed by the Sociedade de

Turismo e Diversões de Macau (STDM). A new sports stadium, covering an area of 44,000 sq m on the island of Taipa, was inaugurated in early 1997. In 1998 plans for the redevelopment of the greyhound racecourse (which had a turnover of US $44m. in 1997) were announced. The construction of a 16,000-sq m cultural centre, on a huge site of reclaimed land around the Outer Harbour, was completed in 1999. The complex comprises a museum block, housing the historical, archaeological and architectural collections of Macao, and a theatre block. A new 17-ha marine park, including a dolphin tank and aquarium, was planned for reclaimed land on Taipa Island. In July 2000 the SAR Government established a new body to prepare Macao for the Fourth East Asian Games in 2005, which had been awarded to Macao in 1996. In September 2001 construction began on a new 900m.-patacas (US $112m.) amusement park, Fisherman's Wharf, to be completed in 2004. The 140,000-sq-m park will incorporate a seafood section (Dynasty Wharf), and the main amusements and rides area will be on 40,000 sq m of reclaimed land; the third section will comprise a marina, restaurants, bars and a nightclub (Legend Wharf). The centrepiece will be an 'artificial volcano'. The tourist project, Macao's largest since its return to Chinese rule, was being implemented by David Chow Kam Fai, a wealthy entertainment magnate who was also a member of the Legislative Assembly. Chow was widely believed to be a rival of STDM president Stanley Ho, even though Ho owned a stake in the project. Meanwhile, in December 2001 STDM opened the 338m-high Macao Tower Convention and Entertainment Centre, as part of plans to make the SAR more family-orientated and to shed its unsavoury image.

The tourism sector has been dominated by STDM, a syndicate to which the territory's gambling monopoly was transferred in 1962 (although STDM's monopoly has since been withdrawn—see below). The company has interests in the jetfoil service between Hong Kong and Macao, the local airport and Air Macau, various infrastructural operations and property projects, the electricity company, a television network and a supermarket chain. In 1996 the company's assets were estimated at HK $21,500m. STDM's contribution to Macao's budget is substantial, taxes levied on the company accounting for a large percentage of total government revenues. In June 1997 the Sino-Portuguese Joint Liaison Group (JLG) reached agreement on the revision of Macao's gambling franchise. Under the revision, STDM was to pay tax of 31.8% on its revenues from casinos (increased from 30% and backdated to January 1996). In addition to an initial contribution of US $22.7m., STDM was also to contribute 1.6% of its annual gross takings to a new government controlled Macao development foundation—see below (responsible for the operation of the cultural centre). In 1998 this annual contribution reached US $116m. Furthermore, the contract obliged STDM to guarantee marine transport operations and maintenance, to meet the cost of various public works and to match any government spending on tourism promotion. In 1997, however, STDM's net profits declined by 19.9% compared with the previous year. Profits continued to decline sharply in 1998, by an estimated 50%, but were estimated to have increased from US $98m. in 1999 to $184m. in 2000. STDM's gross takings rose by 21% in 2000, compared with the previous year, to reach US $2m.

Meanwhile, the sharp increase in casino-related crime in the late 1990s led to speculation that STDM might lose its long-standing monopoly in 2001 upon the expiry of its gambling franchise. In July 2000 it was announced that a special committee, chaired by the Chief Executive of the SAR, was to be established to study the future development and management of the gambling industry. STDM's gambling franchise had been renewed once in 1996. However, in December 2001 the Government decided against renewing STDM's monopoly, and the agreement finally expired in April 2002. Under the new arrangements, the Government announced in February 2002 that three companies had successfully bid for casino licences—the Las Vegas-backed Wynn Resorts (Macao) Ltd, the Galaxy Casino Co Ltd (a Hong Kong-Macao joint venture), and the Macao Gaming Holding Co (SJM—Sociedade de Jogos de Macau, a subsidiary of STDM), managed by Stanley Ho. The Gaming Industrial Regime, approved by the Legislative Assembly in August 2001, stipulated that casinos should contribute to the development of tourism, and that the SJM was to pay 35% of its total income to the SAR Government annually—up from 31.8% previously. Despite the loss of his monopoly, Stanley Ho was expected to retain a strong influence in the gambling sector, as his competitors needed time to develop a presence in the SAR. (STDM employed more than 10,000 people in its 11 casinos in 2001, and contributed approximately 30% of Macao's GDP.) In late June 2002 the Government formally awarded Wynn Resorts (Macao) Ltd a 20-year contract to operate casinos. As a first step, Wynn was to invest 4,000m. patacas in Macao over seven years, and open its first casino and hotel complex in 2006. Also in June 2002, a similar contract was signed with Galaxy Casino Co Ltd. In May 2004 Las Vegas Sands Inc, operating under a sub-concession granted by Galaxy Casino, opened a new casino, The Sands, representing the effective end of Stanley Ho's monopoly on the gambling industry. Las Vegas Sands planned to open a second casino, The Venetian Macau, by 2006, and also planned to develop tourism projects on the Cotai strip (see below).

In June 1997 the JLG reached agreement on the establishment of a foundation, to be funded in part by a tax on gambling revenue (see above). The Macao Development and Co-operation Foundation, which replaced the Orient Foundation, became responsible for the promotion of academic research, science and technology, the arts, education and welfare, with particular emphasis on the preservation and development on the territory's cultural heritage. In August 2002 the Macao Foundation announced the construction of a new Macao Science Center, which would mainly serve educational purposes, but would also function as a convention centre and a tourist attraction. The Center, designed by renowned architect I. M. Pei, was to accommodate physics laboratories, an exhibition hall, a multi-media studio, a science ground for youngsters, an astronomical hall and a multi-purpose conference room.

In co-operation with the territory's large hotels, a tourism-training institute was established in the territory in 1995. Macao is a member of the Pacific Asia Travel Association (PATA) and of the World Tourism Organization.

INDUSTRY AND POWER

The manufacturing sector employed 18.3% of the economically active population in 2003. The production of clothing and of knitwear remain by far the most important manufacturing industries. Clothing output decreased from more than 191.8m. units in 1998 to 182.0m. units in 1999, but increased to 222.4m. units in 2000, and to 227.3m. units in 2001, before decreasing again to 223.27m. units in 2002. Output of knitwear rose from 27.7m. units in 1998 to 30.6m. units in 1999, 37.9m. units in 2000, 38.3m. units in 2001, and 39.46m. units in 2002. From 1998, however, many garment manufacturers reported a substantial decrease in export orders for Japan (Macao's leading non-quota market), owing to that country's economic recession. Other products include footwear (10.5m. pairs being manufactured in 2002, compared with 10.2m. pairs in 2001, 13.3m. pairs in 2000, 8.9m. pairs in 1999, and 5.4m. in 1998), toys, fireworks, plastics and electronics. The number of manufacturing establishments decreased from 1,381 in 1998 to an estimated 1,227 in 1999 and to 1,212 in 2000.

The Concordia Industrial Park was established in the early 1990s on reclaimed land near the new airport, with the aim of stimulating investment and enhancing industrial diversification, with an emphasis on small and medium enterprises, and new technology ventures being particularly encouraged. Investors benefited from numerous property, industrial, and corporate tax exemptions, as well as low land prices. By 2000 total investment had risen to 643m. patacas, and the various companies' sales were projected at 867m. patacas in 2000, and at 1,058m. in 2001. The industrial estate is promoted and managed by the Sociedade do Parque Industrial da Concórdia—SPIC (established in 1993), of which 60% is owned by the Macao Government. The SPIC has in recent years established ties with the Chinese Academy of Sciences and the China Association of Science and Technology. Future plans envision an expansion of the Park to meet future demands for such space, and using the Park to promote foreign direct investment in the mainland, which would extend into Macao.

In early 2003 the authorities of Macao and of the Chinese Special Economic Zone (SEZ) of Zhuhai planned jointly to develop a new industrial zone later that year. The Macao-Zhuhai industrial project was officially inaugurated in December 2003. It was hoped that the new zone, located between Macao and Zhuhai would bring benefits to both cities.

Regulations governing the entry of foreign workers were relaxed in 1988. In 2001 the total number of foreign workers was estimated at 26,000, mainly from the People's Republic of China and the Philippines, compared with 35,000 in 1995.

Macao possesses few natural resources. The People's Republic of China provides part of the SAR's water supply. In mid-1996 Guangdong Province announced the implementation of new measures to improve the quality of its water supplies to Macao. Four reservoirs in the Chinese city of Zhuhai supplied 150,000 cu m of water daily in 1999. In June 1984 Macao's electricity supply was linked to a China-based source. The People's Republic was initially to supply 10% of Macao's power requirement. In 1999 87% of the territory's electricity was domestically produced, Macao's two power stations having a total capacity of more than 400 MW. In April 2003 a new power plant with a capacity of 136 MW began operating on the island of Coloane. Total gross output of electricity rose from 1,477.7m. kWh in 2000 to 1,510.4m. kWh in 2001, 1,611.4m. kWh in 2002 and to 1,719.2m. kWh in 2003. Imports of fuels and lubricants accounted for 7.2% of total import costs in 2003 and 2002, compared with 7.9% in 2001.

FOREIGN TRADE

Macao's foreign trade expanded strongly in the 1980s, the trade surplus reaching a peak of 2,217m. patacas in 1987. The value of merchandise exports rose from 2,742m. patacas in 1980 to 13,638.2m. in 1990, 17,580.0m. in 1999 and to 20,380.4m. in 2000. In 2001, however, total exports decreased to 18,473.0m. patacas, owing to weaker demand in the USA and the European Union (EU). In 2002 there was a slight rise in exports to 18,925.4m. patacas. In 2003 the value of exports increased to 20,700.1 patacas. The trade surplus stood at 1,279.8m. patacas in 1999, when the cost of imports increased to 16,300.2m. patacas, and the surplus increased to 2,282.8m. patacas in 2000, the cost of imports in that year having risen to 18,097.6m. patacas. However, in 2001, when imports reached 19,170.4m. patacas, a trade deficit of 697.4m. patacas was recorded. In 2002 imports reached 20,323.4m. patacas, thus giving a trade deficit of 1,398.0m. patacas. In 2003 imports totalled 22,097.2m. patacas, implying a trade deficit of 1,397.1 patacas.

As a result of the policy of diversifying exports away from their heavy reliance on the textile sector, non-textile products accounted for almost 30% of overall shipments in 1989, compared with only 13% in 1980. During the 1990s, however, many non-textile operations were relocated to China and to South-East Asia, where labour costs were lower. From mid-1997 Macao's position was further weakened by the relative appreciation of the pataca against the currencies of its South-East Asian competitors. Exports of clothing decreased from 14,623.0m. patacas in 2000 to 13,202m. in 2001 and 13,158.1m. in 2002 (69.5% of the total). Exports of textile fabrics, however, increased from 1,245.1m. patacas in 2000 to 1,266.5m. in 2001 and to 1,615.9m. in 2002. In 2003 exports of clothing increased to 14,640.0 patacas (70.7% of total exports), whilst exports of textile fabrics decreased to 1,531.5 patacas. The toy-making sector showed strong growth in the late 1980s, providing almost 10.2% of all exports in 1989. In the 1990s, however, exports of toys declined sharply, reaching only 35.0m. patacas (less than 0.2% of total exports) in 1999. Sales of electronic goods showed strong expansion in the early 1990s, the value of exports of radio and television sets rising from 321.5m. patacas in 1991 to 565.6m. in 1994. However, exports of electronics subsequently decreased, earning only 15.0m. patacas in 1998. Exports of machinery and apparatus declined from 388.0m. patacas in 1998 to 337.0m. in 1999, but rose to 418.3m. in 2000 and to 555.6m. in 2001. In 2002 exports of machinery and apparatus decreased slightly, to 507.7m. patacas; however, the figure increased to 684.7 patacas in 2003.

The textile and garment sector therefore remains dominant, most exports having been regulated by the General Agreement on Tariffs and Trade (GATT, superseded by the World Trade Organization—WTO), an agency of which Macao became a member in January 1991. The quota limits established by the Multi-Fibre Arrangement (MFA) were due for renewal in 1991, but the existing agreement was extended until the end of 1992, pending a successful conclusion to the 'Uruguay Round' of GATT negotiations. Shortly before this conclusion in December 1993, the MFA was extended for a further year. It was envisaged that the provisions of the MFA would be phased out over a 10-year period, with final expiry of the arrangement scheduled for 2005. Meanwhile, an extension agreement between Macao and the USA (a major importer of the territory's textiles and garments) was signed in mid-1991. A two-year bilateral textile agreement was signed in January 1994. In September 1998, following the USA's expression of doubt that not all of Macao's textile exports were in fact manufactured in the enclave (in possible violation of country-of-origin regulations), the territory was warned that the situation was to be closely monitored.

Macao became an associate member of the Economic and Social Commission for Asia and the Pacific (ESCAP) in 1991. In June 1992 Macao and the European Community (EC, now European Union—EU) signed a five-year trade and economic co-operation agreement, granting mutual preferential treatment on tariffs and other commercial matters. The agreement was extended in December 1997. Macao remained a 'privileged partner' of the EU after December 1999. Wishing to enhance links with the EU, in May 2000 the new Chief Executive of the SAR, Edmund Ho, paid an official visit to Portugal, which in January had assumed the EU's rotating presidency for a six-month period. Ho made his first official visit to the EU's headquarters in Brussels in mid-June 2001, seeking to promote contacts and exchanges between the two entities. Macao also retained its membership of the WTO after December 1999. Upon its return to Chinese sovereignty in December 1999, Macao was allowed to retain its system of free enterprise for a period of 50 years. The territory's status as a free port was also retained. Macao remained a separate customs territory, and its trade with China continued to be classified as foreign trade. (It was estimated that between the late 1970s and 1995 mainland Chinese companies had invested a total of US $5,800m. in Macao, whilst in 1997 alone Macao companies invested a total of US $10,000m. in 6,333 projects in China.) Economic co-operation between Macao and the neighbouring Chinese SEZ of Zhuhai has continued to develop (see above). In October 2003 the Closer Economic Partnership Arrangement (CEPA) was signed between Macao and mainland China. In August 2004 the first CEPA certificates were issued to Macao retailers, enabling them to take advantage of preferential policies when investing in the mainland.

In 1997 the People's Republic of China had become Macao's principal supplier, providing 28.6% of Macao's imports, surpassing Hong Kong (25.2%), and with Japan ranked third (8.5%). In 1998 China supplied almost 32.2% of Macao's imports, while Hong Kong provided 23.7% and Japan 7.7%. In 1999 more than 35.6% of Macao's imports were purchased from China, with Hong Kong accounting for 18.1% and Japan for 6.7%. In 2000 some 41.1% of Macao's imports originated from China, with 15.2% coming from Hong Kong, while Taiwan ranked third (9.5%). In 2001 China supplied 42.6% of Macao's imports, followed by Hong Kong (13.9%) and Taiwan (6.7%). In 2002 China's exports to Macao totalled 8,477.2m. patacas, accounting for 41.7% of the SAR's total imports. Hong Kong provided 14.5% and Taiwan 6.7% of the total. In 2003 imports from China accounted for 42.9% of the value of Macao's total imports, whereas imports from Hong Kong and Taiwan accounted for 12.6% and 5.8% respectively. Imports from Japan amounted to 9.0% of Macao's imports in 2003. Exports to the USA increased rapidly from the late 1970s, reaching 537m. patacas in 1980. By 1989 this figure had risen to 4,946m. patacas, representing more than one-third of Macao's total export trade. After fluctuating in subsequent years, the territory's exports to the USA rose to 8,140.7m. patacas (47.7% of the total) in 1998, 8,249.1m. (46.9% of the total) in 1999, and a record 9,836.7m. (48.3% of the total) in 2000. In 2001, however, Macao's exports to the USA decreased to 8,907.1m. patacas (48.2% of the total). In 2002 exports to the USA increased to 9,151.5m. patacas (48.4% of the total), and the figure increased again to 10,320.2 patacas (49.9% of the total) in 2003. China purchased exports worth 2,155.0m. patacas (11.7% of the total) from Macao in 2001, compared with 2,079.8m. (10.2%) in 2000. In 2002 China purchased exports to the value of 2,948.1m. patacas from Macao (15.6% of the total). In 2003 the value of exports to China was 2,844.2 patacas (13.7% of the total). Other important trading partners were Germany and the United Kingdom. Macao actively seeks new markets through the promotion of official trade delegations.

FINANCE AND BANKING

In the early 1990s overall economic buoyancy was reflected in the increased levels of expenditure in successive budgets, total spending rising from 5,489.9m. patacas in 1990 to 11,251.3m. in 1994, when a budgetary surplus of 1,559.9m. was recorded. In 1995, however, expenditure was reduced to 10,314.9m. patacas, while revenue declined to 11,033.8m. (from 12,811.2m. in 1994) to give a surplus of 718.9m. By 1997 expenditure had been reduced to 14,134.1m. patacas, while revenue had also declined, to 14,134.1m. (compared with 14,681.3m. and 14,711.3m. respectively in 1996). In 1998 the budget was projected to balance at 14,831.1m. patacas. Revenue was lower than anticipated, however, owing to a shortfall in receipts from gambling and from land leases (95% of land in Macao being owned by the Government). A budgetary deficit, equivalent to an estimated 1.5% of GDP, thus resulted. In 1999 expenditure was projected at only 9,805.4m. patacas, while revenue was expected to total 10,111.8m. patacas. The provisional budget for 2000 envisaged a reduction in expenditure to 8,501.7m. patacas, although it aimed to increase spending on anti-corruption measures; expenditure on health and education was to remain at a level similar to that of the previous year. Revenue was provisionally estimated at 8,815.9m. patacas in 2000. The provisional budget for 2001 envisaged expenditure of 15,220.8m. patacas and revenue of 15,641.6m. patacas. For 2002 expenditure was estimated at 9,577.7m. patacas and revenue at 11,317.7m. patacas. For 2003 the equivalent figures were 11,621.5m. patacas and 14,279.1m. patacas respectively.

In April 1977 the local currency's link with the Portuguese escudo was ended and a new parity of 1.075 (subsequently 1.030) patacas = HK $1 was established, an appreciation of about 38%. Exporters of merchandise were required to surrender 50% (subsequently 40%) of their foreign-exchange earnings to the official Exchange Fund. The banking system therefore had sufficient resources to maintain the value of the pataca, which continues to float with the Hong Kong dollar. The latter currency also circulates widely in Macao. Upon the territory's transfer of sovereignty, the pataca was retained and remained freely convertible.

A state-owned institution responsible for currency issue, the Instituto Emissor de Macau (IEM), was founded in 1980, nominally taking over this function from the Banco Nacional Ultramarino, which remained the Government's banker for the territory. In 1989 the IEM was replaced by the Autoridade Monetária e Cambial de Macau (AMCM, Monetary and Foreign Exchange Authority of Macao), now the Autoridade Monetária de Macau (Monetary Authority of Macao). Banknotes will continue to be issued by the Banco Nacional Ultramarino, acting as the Government's agent until at least 2010. Since October 1995 notes have also been issued by the Bank of China's branch in Macao. In August 2004 representatives of the Monetary Authority of Macao and the People's Bank of China signed a memorandum allowing business in Macao to be conducted in renminbi.

Plans to revitalize the banking sector began in July 1982, when a new banking ordinance was passed, allowing for the establishment of development banks. The financial sector was declared open to competition, and six international and three Portuguese banks were granted full commercial licences. The aim was to establish an enlarged financial sector, providing a widening of financial services to support the growth of various business sectors. In 1987 legislation to permit offshore banking was introduced, allowing foreign companies to operate tax-free, apart from the payment of an annual fee. The level of offshore activity, however, has been disappointing. Also in 1987 Macao's largest bank, the Nan Tung Bank, was acquired by the Bank of China, the official foreign exchange bank of the People's Republic.

In 2003 a total of 23 registered banks were operating in Macao. In addition, a postal savings bank, established in 1935, is operated by the Government. The total assets of the commercial banking sector reached the equivalent of 295% of GDP in September 1998, compared with 207% in 1995.

Macao has no foreign-exchange controls or restrictions on capital flows. As a result, balance-of-payments data are not comprehensive. During 1990–98, however, the current account showed a consistent surplus. This reached US $2,700m. (equivalent to 37.4% of GDP) in 1996, before declining to US $2,400m. (38.8% of GDP) in 1998.

The Financial System Act, which aimed to improve the reputation of the territory's banks, took effect in September 1993. The legislation required banks to record the identity of those making unusually large transactions, in an attempt to curb the unauthorized acceptance of deposits. In June 1996, in an effort to combat organized crime, the first guide-lines on the prevention of 'money-laundering' operations were issued. These underwent revision in 1999. In May 2002 the Monetary Authority of Macao updated the Anti-Money Laundering Guideline for Credit Institutions, which was fully enforced in August of the same year. The revision required credit institutions to formulate clear policies and procedures, with customers being divided into different categories according to their risk levels. More stringent regulations to govern large cash transactions were also introduced.

From December 1999 Macao began administering its own finances, and was to be exempt from taxes imposed by central government. Consumption taxes are levied on only a limited number of items, such as cigarettes and alcohol. The rates of income tax are lower than in the neighbouring SAR of Hong Kong. In late 2002 the Government announced a series of tax reduction initiatives aimed at assisting the development of small and medium enterprises.

DEVELOPMENT AND INFRASTRUCTURE

In 1998 the Government allocated 1,900m. patacas (of a total budget of 10,700m.) to infrastructural projects, an increase of 300m. patacas compared with the previous year. Until the 1990s, however, Macao's development was hindered by the territory's lack of infrastructure, the limitations of the electricity and water supplies, the shortage of skilled labour and the dearth of suitable building land.

The shortage of land for development purposes has presented a serious problem. In August 1979 the administration announced that it had prepared a land-use plan. The master plan was based on a massive land-reclamation scheme which required soil from China. However, none of the feasibility studies was made in consultation with the Chinese authorities, thus undermining the plan's validity. Nevertheless, between 1989 and 2003 alone Macao's land area was increased from 17.32 sq km to 27.3 sq km.

In the early 21st century work was under way on a reclamation scheme on both sides of an existing causeway joining Taipa and Coloane Islands. A new town called Cotai was to be built on this site, which was to include a railway terminal serving a new planned passenger and freight route to Zhuhai and thence all the way to Guangzhou. In July 2003 Las Vegas Sands Inc announced plans for a large tourist development on the Cotai strip, for which investment was expected to amount to between US $5,000m. and US $10,000m. Another land-reclamation project, the Nam Van Lakes project, was scheduled to be completed in four stages and enlarge the peninsular area by 20%. Details of this land-reclamation scheme, which originally was to cost US $1,400m., were announced in July 1993 by a prominent entrepreneur, following the project's approval by the People's Republic of China. The development was to incorporate hotels, offices, apartments and commercial malls, with residential and business accommodation for 60,000 people along a six-lane waterfront highway. The Nam Van Lakes project was initially scheduled for completion in 2001, but was subsequently jeopardized by a sharp decline in the property market. The project was reported to be in its final stages in March 2004.

As a result of the establishment of diplomatic relations between Portugal and China in February 1979 and the subsequent growth of confidence in the future of the territory, there was a considerable rise in property prices in Macao. Land sold by the Portuguese administration a few years previously to encourage residential development had more than quadrupled in value by 1981. In March 2000 the transfer to the SAR's Government of the assets of the Sino-Portuguese Land Group, established in 1988 to handle land concession matters, was completed. Comprising income from land leases and attendant earnings from banking deposits and investments, the Land Fund totalled 10,185m. patacas.

The Government's receipts from land leases rose sharply in the early 1990s, reaching almost 40% of its total revenue (equivalent to 8.25% of GDP) in 1992. Having accounted for 25% of government revenue in 1993–94, receipts from land leases declined substantially during 1995–97 to make an average annual contribution to revenue of only 6% (or 1% of GDP). Mainland Chinese speculation in both residential and commercial property resulted in a surplus of buildings. By 1996 there were an estimated 30,000–50,000 vacant apartments, mainly on the island of Taipa. In August 1996, in order to address the problem of the property surplus, the Government began to offer financial incentives, in the form of mortgage subsidies, to first-time buyers. The decline in the property market, following a period of very strong growth in the construction sector, was also demonstrated by the sharp decrease in the number of buildings completed in 1997—only 1,150,000 sq m (of which 632,000 sq m represented residential property), compared with 1,908,000 sq m in 1996 (1,619,000 sq m being for residential purposes). Between January and June 1998 the number of real-estate transactions totalled 6,683, an increase of 1.6% compared with the corresponding period of 1997. The value of these transactions, however, declined by 56.9%, to stand at 4,300m. patacas. During the 1990s real interest rates increased markedly, reaching a level of 12.3% in September 1998. It was hoped that a series of reductions in interest rates in 1998/99 would stimulate the housing market. New gross floor area completed fell to 969,192 sq m in 1998, and to 668,778 sq m in 1999, reaching a new low of 370,315 sq m in 2000. However, in 2001 the figure increased to 404,325 sq m of new floor area. In 2002 the property market showed further signs of recovery, with transactions increasing by 56.1% on an annual basis during the first three quarters of that year.The upward trend seemed set to continue, with the volume of property transactions in the first quarter of 2004 being reported to have been 96% higher than the volume of transactions in the corresponding period of the previous year.

In March 1980, meanwhile, Chinese approval was secured for Macao's most ambitious project, an international airport. Following a feasibility study for its construction on reclaimed land and on piles off the island of Taipa, it was confirmed in 1987 that the project would proceed. By 1989 construction work at the new airport site had commenced. The cost of the airport scheme was originally estimated at some 3,500m. patacas, of which one-third was to be provided by the Government of the People's Republic of China. It was envisaged that the airport would not only enhance Macao's role as a point of entry to China and afford the enclave access to international transport networks, but also improve investment conditions and provide new opportunities for the development of industries such as tourism. In 1990, however, it was reported that the consortium of Chinese companies involved in the project had been forced to reduce its proposed equity from 500m. patacas to just 130m., owing to the shortage of foreign exchange in China. Furthermore, ballast required for the reclamation work had not been delivered on schedule by the consortium. Nevertheless, the future of the project was assured, when three Macao businessmen agreed to finance the shortfall. In June 1993 Portugal and China reached agreement on all outstanding issues regarding the building of the airport. A regional airline, Air Macau, was formally established in 1994, with investment from the Civil Aviation Administration of China (CAAC) and with Portuguese and Macao interests. The airport was officially inaugurated in December 1995, at a final cost of 8,900m. patacas. The terminal has an annual capacity of 6m. passengers. By 2000 Air Macau was operating flights to several cities in China, Taiwan, Thailand, Japan, the Republic of Korea and the Philippines. Of particular significance was the opening by Air Macau of a direct route between Taiwan and mainland China, passengers no longer being required to change aircraft. Flights to and from the Democratic People's Republic of Korea (North Korea) by Koryo Air resumed in April 2002, and a new Air Macau route to Singapore was inaugurated in August 2002. In 2002 a total of nine airlines operated 436 scheduled flights weekly to 21 destinations. The airport handled some 4.2m. passengers in 2002, compared with 3.8m. in 2001 and 3.2m. in 2000. In 2002 the airport processed 111,268 tons of cargo. In 2001 the airport's franchise was extended to 2039.

The regular catamaran, jetfoil and high-speed ferry services between Macao and Hong Kong are complemented by a helicopter service, which commenced in 1990. A scheduled helicopter service between Macao and Shenzhen began operations in January 2003. In conjunction with the construction of the airport, a new 4.4-km four-lane bridge providing an additional link between the island of Taipa and the peninsula of Macao opened in April 1994, having been completed at a cost of more than 600m. patacas. Other completed development plans include the construction of a deep-water harbour. The new port of Kao-ho, on the island of Coloane, began operations in 1990, handling both container and oil tanker traffic.

In mid-1997 a preliminary agreement was reached by the Sino-Portuguese Infrastructure Co-ordination Commission, which was established in April of that year, concerning the construction of a six-lane road bridge linking Macao with Zhuhai. The bridge, which was to cost US $12.29m. (financed equally by Macao and Zhuhai), was to extend from an area of reclaimed land between Taipa and Coloane to the island of Hengqin, situated off the coast of Zhuhai, and would provide a link with the planned Beijing–Zhuhai highway. Construction work began in June 1998, and the 1.5-km Lotus Bridge duly opened to traffic in December 1999. The Chinese Government was to develop the island of Hengqin as an international free trade zone. In July 2003 detailed plans for the connection between the Beijing–Zhuhai highway and Macao's highway system were announced, with the project expected to be completed by 2007. Following Macao's transfer of sovereignty in December 1999, the Chinese authorities planned greater integration of the SAR within the Zhujiang (Pearl River) Delta, particularly with regard to infrastructural development. Suggestions for the construction of a 37.9-km bridge between Macao and Hong Kong were made by the private sector in 1997. These plans developed into a proposal to build a new 29-km steel bridge linking Macao with Hong Kong's Lantau Island and Zhuhai, situated within Guangdong Province. Feasibility studies to determine the project's financial and technical viability were being conducted in 2003. In August 2002 the Chon Tit (Macao) Investment and Development Co was chosen for the construction of a third bridge connecting downtown Macao with Taipa Island. The new 1,720-m, six-lane bridge would have two levels and cost 560.2m. patacas to build, over a 28-month period. The project would create 1,000 jobs and relieve congestion on the existing bridges.

Macao's telecommunications network has been extensively modernized. In May 1984 a new satellite ground station was inaugurated, establishing direct communications with Portugal, Japan and the United Kingdom, and in 1986 plans were announced to install a 15-km fibre optic cable between Macao and the Zhuhai SEZ, thus improving telecommunications links between the two territories. A fully digital telephone network was established in 1991. In 1995 the Companhia de Telecomunicações de Macau (CTM) began to provide internet services. By early 1996 CTM had signed agreements for the provision of mobile telephone services with three of China's inland localities: Beijing, Guangdong Province and Shanghai. In 1994, meanwhile, CTM announced a four-year plan, costing HK $1,700m., to develop new facilities, many of which formed part of the telecommunications infrastructure of the territory's new airport. Following Macao's transfer of sovereignty in December 1999, Portugal Telecom retained its 28% share in CTM. The former's management concession was renewed until 2001, thus granting CTM exclusive rights to provide fixed-line telephone services, data communications and leased lines. In June 2000 a new body, the Office for the Development of Telecommunications and Information Technology (GDTTI), was established, with a view to building a legal framework for the telecommunications sector, ensuring that services met the needs of the market, developing the communications infrastructure, issuing licences, and maintaining quality control, price regulation and standardization of the network.

Statistical Survey

Source (unless otherwise indicated): Direcção dos Serviços de Estatística e Censos, Alameda Dr Carlos d'Assumpção 411–417, Dynasty Plaza, 17° andar, Macao; tel. 3995311; fax 307825; e-mail info@dsec.gov.mo; internet www.dsec.gov.mo.

AREA AND POPULATION

Area (2003): 27.3 sq km (10.54 sq miles).

Population: 435,235 (males 208,865, females 226,370) at census of 23 August 2001 (414,200 inhabitants were of Chinese nationality and 8,793 inhabitants were of Portuguese nationality); 448,495 (official estimate) at 31 December 2003.

Density (31 December 2003): 16,428 per sq km.

Births, Marriages and Deaths (2003): Registered live births 3,212 (birth rate 7.2 per 1,000); Registered marriages 1,309 (marriage rate 2.9 per 1,000); Registered deaths 1,474 (death rate 3.3 per 1,000).

Expectation of Life (years at birth, 1998–2001): 78.9 (males 77.2; females 81.5).

Economically Active Population (2003): Manufacturing 37,077; Production and distribution of electricity, gas and water 1,333; Construction 16,283; Wholesale and retail trade; repair of motor vehicles, motorcycles and personal and household goods 32,824; Hotels, restaurants and similar activities 22,114; Transport, storage and communications 14,215; Financial activities 6,211; Real estate, renting and services to companies 11,883; Public administration, defence and compulsory social security 17,812; Education 9,617; Health and social work 4,643; Other community, social and personal service activities 23,469; Private households with employed persons 4,278; Others 829; *Total employed* 202,588.

HEALTH AND WELFARE

Key Indicators

Under-5 Mortality Rate (per 1,000 live births, 2003): 1.6.

HIV/AIDS (% persons aged 15–49, 2003): 0.001.

Physicians (per 1,000 head, 2003): 2.13.

Hospital Beds (per 1,000 head, 2003): 2.24.

Human Development Index (2001): value 0.898.

For definitions, see explanatory note on p. vi.

AGRICULTURE, ETC.

Livestock ('000 head, 2002): Poultry 600 (FAO estimate). Source: FAO.

Livestock Products ('000 metric tons, 2002): Beef and veal 1.0*; Pig meat 9.4; Poultry meat 5.0*; Hen eggs 1.0*.

* FAO estimate.

Source: FAO.

Fishing (FAO estimates, metric tons, live weight, 2002): Marine fishes 1,020; Shrimps and prawns 230; Other marine crustaceans 210; Total catch (incl. others) 1,500. Source: FAO.

INDUSTRY

Production (2002, unless otherwise indicated): Wine 1,291,061 litres; Knitwear 39.46m. units; Footwear 10.47m. pairs; Clothing 223.27m. units; Furniture 4,431 units; Electric energy 1,719.2 million kWh (2003).

FINANCE

Currency and Exchange Rates: 100 avos = 1 pataca. *Sterling, Dollar and Euro Equivalents* (31 May 2004): £1 sterling = 14.728 patacas; US $1 = 8.028 patacas; €1 = 9.831 patacas; 1,000 patacas = £67.90 = $124.57= €101.72. *Average Exchange Rate* (patacas per US dollar): 8.034 in 2001; 8.033 in 2002; 8.021 in 2003. Note: The pataca has a fixed link with the value of the Hong Kong dollar (HK $1 = 1.030 patacas).

Budget (million patacas, 2003, provisional): *Total revenue*: 14,279.1 (direct taxes 11,342.8, indirect taxes 987.6, others 1,948.7). *Total expenditure*: 11,621.5.

International Reserves (US $ million at 31 December 2003): Foreign exchange 4,343.4; Total 4,343.4.

Money Supply (million patacas at 31 December 2003): Currency outside banks 2,361.7; Demand deposits at commercial banks 6,427.8; Total money 8,789.5. Source: IMF, *International Financial Statistics*.

Cost of Living (Consumer Price Index; base: Oct. 1999–Sept. 2000 = 100): All items 97.52 in 2001; 94.94 in 2002; 93.46 in 2003.

Gross Domestic Product (million patacas at current prices): 49,862 in 2001; 54,295 in 2002; 63,365 in 2003.

Expenditure on the Gross Domestic Product (provisional, million patacas at current prices, 2003): Government final consumption expenditure 6,600.2; Private consumption expenditure 21,900.2; Changes in inventories 262.0; Gross fixed capital formation 8,412.2; *Total domestic expenditure* 37,174.5; Exports of goods and services 62,469.8; *Less* Imports of goods and services 36,279.0; *GDP in purchasers' values* 63,365.4.

Gross Domestic Product by Economic Activity (provisional, million patacas at current prices, 2002): Mining and quarrying 7.1; Manufacturing 3,217.3; Electricity, gas and water supply 1,235.2; Construction 1,144.1; Trade, restaurants and hotels 5,398.0; Transport, storage and communications 3,003.5; Financial intermediation, real estate, renting and business activities 9,409.6; Public administration, other community, social and personal services (incl. gaming services) 23,310.4; *Sub-total* 46,725.2; *Less* Financial intermediation services indirectly measured 2,342.5; *GDP at basic prices* 44,382.6; Taxes on products (net) 9,453.0; *GDP in purchasers' values* 53,835.6.

Balance of Payments (US $ million): Exports of goods f.o.b. 2,357; Imports of goods f.o.b. –3,191; *Trade balance* –833; Exports of services 4,467; Imports of services –1,050; *Balance on goods and services* 2,584; Other income received 449; Other income paid –469; *Balance on goods, services and income* 2,564; Current transfers received 99; Current transfers paid –97; *Current balance* 2,565; Capital account (net) 128; Direct investment abroad –60; Direct investment from abroad 418; Portfolio investment assets –903; Portfolio investment liabilities 1; Financial derivatives assets 118; Other investment assets –866; Other investment liabilities 213; Net errors and omissions –1,320; *Overall balance* 293. Source: IMF, *International Financial Statistics*.

EXTERNAL TRADE

Principal Commodities (million patacas, 2003): *Imports c.i.f.*: (distribution by SITC): Food and live animals 1,422.2; Beverages and tobacco 1,414.4 (Beverages 930.9); Mineral fuels, lubricants, etc. 1,593.0 (Petroleum, petroleum products, etc. 1,299.8); Chemicals and related products 898.4; Basic manufactures 7,349.2 (Textile yarn, fabrics, etc. 6,175.0); Machinery and transport equipment 4,721.7 (Electrical machinery, apparatus, etc. 1,932.2; Transport equipment and parts 1,283.3); Miscellaneous manufactured articles 4,327.1 (Clothing and accessories 2,454.5); Total (incl.others) 22,097.2. *Exports f.o.b.*: Textile yarn and thread 733.4; Textile fabrics 1,531.5; Machinery and mechanical appliances 684.7; Clothing 14,640.0; Footwear 755.9; Total (incl. others) 20,700.1.

Principal Trading Partners (million patacas, 2003): *Imports c.i.f.*: Australia 307.9; China, People's Republic 9,489.9; France 751.4; Germany 603.3; Hong Kong 2,794.4; Italy 281.5; Japan 1,986.8; Korea, Republic 696.2; Singapore 777.1; Taiwan 1,282.0; United Kingdom 408.7; USA 871.9; Total (incl. others) 22,097.2. *Exports f.o.b.*: Canada 337.7; China, People's Republic 2,844.2; France 690.4; Germany 1,697.3; Hong Kong 1,361.7; Netherlands 433.4; United Kingdom 898.8; USA 10,320.2; Total (incl. others) 20,700.1.

TRANSPORT

Road Traffic (motor vehicles in use, December 2003): Light vehicles 59,556; Heavy vehicles 4,517; Motorcycles 66,399.

Shipping (international sea-borne freight traffic*, '000 metric tons, 2003): Goods loaded 159.2; Goods unloaded 74.5.

* Containerized cargo only.

Civil Aviation (2003): Passenger arrivals 780,522; Passenger departures 777,830; Goods landed (tons) 20,220; Goods loaded (tons) 68,449.

TOURISM

Visitor Arrivals by Country of Residence (2003): China, People's Republic 5,742,036; Hong Kong 4,623,162; Japan 85,613, Taiwan 1,022,830; Total (incl. others) 11,887,876.

Receipts from Tourism (US $ million): 3,205 in 2000; 3,745 in 2001; 4,415 in 2002.

COMMUNICATIONS MEDIA

Radio Receivers (1997): 160,000 in use.

Television Receivers (2000): 125,115 in use.

Daily Newspapers (2003): 11.

Telephones (Dec. 2003): 174,621 main lines in use.

Facsimile Machines (1999): 6,290 in use.

Mobile Cellular Telephones (2003): 364,031 subscribers.

Personal Computers (2001): 60,390 households.

Internet Users (2003): 59,401.
Sources: partly International Telecommunication Union and UNESCO, *Statistical Yearbook*.

EDUCATION

(2002/03)

Kindergarten: 62 schools; 465 teachers; 12,737 pupils.

Primary: 83 schools; 1,619 teachers; 41,535 pupils.

Secondary (incl. technical colleges): 56 schools; 1,995 teachers; 41,551 pupils.

Higher: 12 institutes; 1,112 teachers; 11,995 students.

Adult Literacy Rate: 91.3% at 2001 census.
Notes: Figures for schools and teachers refer to all those for which the category is applicable. Some schools and teachers provide education at more than one level. Institutions of higher education refer to those recognized by the Government of Macao Special Administrative Region.

Directory

The Constitution

Under the terms of the Basic Law of the Macao Special Administrative Region (SAR), which took effect on 20 December 1999, the Macao SAR is an inalienable part of the People's Republic of China. The Macao SAR, which comprises the Macao peninsula and the islands of Taipa and Coloane, exercises a high degree of autonomy and enjoys executive, legislative and independent judicial power, including that of final adjudication. The executive authorities and legislature are composed of permanent residents of Macao. The socialist system and policies shall not be practised in the Macao SAR, and the existing capitalist system and way of life shall not be changed for 50 years. In addition to the Chinese language, the Portuguese language may also be used by the executive, legislative and judicial organs.

The central people's Government is responsible for foreign affairs and for defence. The Government of Macao is responsible for maintaining social order in the SAR. The central people's Government appoints and dismisses the Chief Executive, principal executive officials and Procurator-General.

The Chief Executive of the Macao SAR is accountable to the central people's Government. The Chief Executive shall be a Chinese national of no less than 40 years of age, who is a permanent resident of the region and who has resided in Macao for a continuous period of 20 years. He or she is elected locally by a broadly-representative Selection Committee and appointed by the central people's Government.

The Basic Law provides for a 300-member Election Committee, which serves a five-year term. The Election Committee shall be composed of 300 members from the following sectors; 100 members from industrial, commercial and financial sectors; 80 from cultural, educational, and professional sectors; 80 from labour, social welfare and religious sectors; and 40 from the Legislative Council, municipal organs, Macao deputies to the National People's Congress (NPC), and representatives of Macao members of the National Committee of the Chinese People's Political Consultative Conference (NCCPPCC). The term of office of the Chief Executive of the Macao SAR is five years; he or she may serve two consecutive terms. The Chief Executive's functions include the appointment of a portion of the legislative councillors and the appointment or removal of members of the Executive Council.

With the exception of the first term (which expired on 15 October 2001) the term of office of members of the Legislative Council (commonly known as the Legislative Assembly) shall be four years. The second Legislative Council shall be composed of 27 members, of whom 10 shall be returned by direct election, 10 by indirect election and seven by appointment. The third and subsequent Legislative Councils shall comprise 29 members, of whom 12 shall be returned by direct election, 10 by indirect election and seven by appointment.

The Macao SAR shall maintain independent finances. The central people's Government shall not levy taxes in the SAR, which shall practise an independent taxation system. The Macao pataca will remain the legal currency. The Macao SAR shall retain its status as a free port and as a separate customs territory.

The Government

Chief Executive: Edmund Ho Hau Wah (assumed office 20 December 1999; re-elected 29 August 2004).

EXECUTIVE COUNCIL

(September 2004)

Secretary for Administration and Justice: Florinda da Rosa Silva Chan.

Secretary for Economy and Finance: Francis Tam Pak Yuen.

Secretary for Security: Cheong Kuoc Va.

Secretary for Social and Cultural Affairs: Fernando Chui Sai On.

Secretary for Transport and Public Works: Ao Man Long.

Other Members: Tong Chi Kin (Spokesman), Leong Heng Teng, Victor Ng, Liu Chak Wan, Ma Iao Lai.

GOVERNMENT OFFICES

Office of the Chief Executive: Headquarters of the Government of the Macao Special Administrative Region, Av. da Praia Grande; tel. 726886; fax 726665; internet www.macau.gov.mo.

Office of the Secretary for Administration and Justice: Headquarters of the Government of the Macao Special Administrative Region, Av. da Praia Grande; tel. 9895-179; internet www.macau.gov.mo.

Office of the Secretary for Economy and Finance: Headquarters of the Government of the Macao Special Administrative Region, Av. da Praia Grande; tel. 726886; fax 726665.

Office of the Secretary for Security: Calçada dos Quarteis, Quartel de S. Francisco; tel. 7997518; fax 580702.

Office of the Secretary for Social and Cultural Affairs: Headquarters of the Government of the Macao Special Administrative Region, Av. da Praia Grande; tel. 726886; fax 726168.

Office of the Secretary for Transport and Public Works: 1° andar, Edif. dos Secretários, Rua de S. Lourenço 28; tel. 726886; fax 727566.

Macao Government Information Bureau: Gabinete de Comunicação Social do Governo de Macau, Rua de S. Domingos 1, POB 706; tel. 332886; fax 355426; e-mail info@macau.gov.mo; internet www.macau.gov.mo; Dir Victor Chan Chi Ping.

Economic Services Bureau: Direcção dos Serviços de Economia, Rua Dr Pedro José Lobo 1–3, Edif. Luso Internacional, 25/F; tel. 386937; fax 590310; e-mail info@economia.gov.mo; internet www.economia.gov.mo.

Legislature

LEGISLATIVE COUNCIL (LEGISLATIVE ASSEMBLY)

Following the election of 23 September 2001, the Legislative Assembly comprised 27 members: seven appointed by the Chief Executive, 10 elected directly and 10 indirectly. Members serve for four years. The Assembly chooses its President from among its members, by secret vote. At the election of September 2001, the business-orientated Convergência para o Desenvolvimento de Macau (CODEM) won two of the 10 directly-elective seats. The Associação dos Empregados e Assalariados (AEA) and Associaçao Reforma Social-Economia de Macau (ARSEM) each won one of the 10 directly-elective seats. The pro-Beijing candidates of the União Promotora para o Progresso (UPP) and of the União para o Desenvolvimento (UPD) won two seats each. The pro-democracy Associação de Novo Macau Democrático (ANMD) also took two seats. Four of the 10 directly-elective seats were taken by the OMKC, a group representing business interests. Groups representing various other interests occupied the remaining six seats. The Legislative Assembly was superseded, under the terms of the Basic Law, by the Legislative Council. In practice, however, the legislature continues to be referred to as the Legislative Assembly.

Legislative Council (Legislative Assembly): Edif. da Assembléia Legislativa, Praça da Assembléia Legislativa, Aterros da Baía da Praia Grande; tel. 728377; fax 727857.

President: Susana Chou.

Political Organizations

There are no formal political parties, but a number of registered civic associations exist and may participate in elections for the Legislative Assembly by presenting a list of candidates. These include the União Promotora para o Progresso (UNIPRO), Associação Promotora para a Economia de Macau (APPEM), União para o Desenvolvimento (UPD), Associação de Novo Macau Democrático (ANMD), Convergência para o Desenvolvimento (CODEM), União Geral para o Desenvolvimento de Macau (UDM), Associação de Amizade (AMI), Aliança para o Desenvolvimento da Economia (ADE), Associação dos Empregados e Assalariados (AEA) and Associação pela Democracia e Bem-Estar Social de Macau (ADBSM). Civic associations that are considerably active in civic, educational, and charity activities and services are: Associação Geral dos Operários de Macau (General Workers' Association of Macao), União Geral das Associações dos Moradores de Macau (Union of Neighbourhood Associations), Instituto do Novo Macau (New Macao Institute), Santa Casa Misericórdia (Charity Organization), Tong Sin Tong (Charity Organization), and Associação dos Trabalhadores da Função Pública de Macau (Macao Civil Servants Association).

Judicial System

Formal autonomy was granted to the territory's judiciary in 1993. A new penal code took effect in January 1996. Macao operates its own five major codes, namely the Penal Code, the Code of Criminal Procedure, the Civil Code, the Code of Civil Procedure and the Commercial Code. In March 1999 the authority of final appeal was granted to the supreme court of Macao, effective from June. The judicial system operates independently of the mainland Chinese system.

Court of Final Appeal

Praçeta 25 de Abril, Edif. dos Tribunais de Segunda Instância e Ultima Instância; tel. 3984117; fax 326744; e-mail ptui@court.gov.mo; internet www.court.gov.mo.

Pres. SAM HOU FAI.

Procurator-General: HO CHIO MENG.

Religion

The majority of the Chinese residents profess Buddhism, and there are numerous Chinese places of worship, Daoism and Confucianism also being widely practised. The Protestant community numbers about 2,500. There are small Muslim and Hindu communities.

CHRISTIANITY

The Roman Catholic Church

Macao forms a single diocese, directly responsible to the Holy See. At 31 December 2002 there were 18,122 adherents in the territory.

Bishop of Macao: Rt Rev. JOSÉ LAI HUNG SENG, Paço Episcopal, Largo da Sé s/n, POB 324; tel. 309954; fax 309861; e-mail mdiocese@macau.ctm.net.

The Anglican Communion

Macao forms part of the Anglican diocese of Hong Kong (q.v.).

The Press

A new Press Law, prescribing journalists' rights and obligations, was enacted in August 1990.

PORTUGUESE LANGUAGE

Boletim Oficial: Rua da Imprensa Nacional, POB 33; tel. 573822; fax 596802; e-mail info@imprensa.macau.gov.mo; internet www.imprensa.macau.gov.mo; f. 1838; weekly govt gazette; Dir Dr ANTÓNIO GOMES MARTINS.

O Clarim: Rua Central 26-A; tel. 573860; fax 307867; e-mail clarim@macau.ctm.net; f. 1948; weekly; Editor ALBINO BENTO PAIS; circ. 1,500.

Hoje Macau: Av. Dr Rodrigues 600E, Edif. Centro Comercial First National, 14° andar, Sala 1408; tel. 752401; fax 752405; e-mail hoje@macau.ctm.net; internet www.hojemacau.com; daily; Dir CARLOS MORAIS JOSÉ; circ. 1,000.

Jornal Tribuna de Macau: Av. Almeida Ribeiro 99, Edif. Comercial Nam Wah, 6 andar, Salas 603–05; tel. 378057; fax 337305; internet www.jtm.com.mo; f. 1998; through merger of Jornal de Macau (f. 1982) and Tribuna de Macau (f. 1982); daily; Dir JOSÉ FIRMINO DA ROCHA DINIS; circ. 1,000.

Ponto Final: Rua Cidade do Porto, 315, Edif. Kam Yuen, Loja AL; tel. 339566; fax 339563; e-mail editor@pontofinalmacau.com ; internet www.pontofinalmacau.com; Dir RICARDO PINTO; circ. 1,500.

CHINESE LANGUAGE

Boletim Oficial: see above; Chinese edn.

Cheng Pou: Av. da Praia Grande, 63, Edif. Hang Cheong, E–F; tel. 965972; fax 965741; daily; Dir KUNG SU KAN; Editor-in-Chief LEONG CHI CHUN; circ. 5,000.

Jornal Informação: Rua de Francisco 22, 1° C, Edif. Mei Fun; tel. 561557; fax 566575; weekly; Dir CHAO CHONG PENG; circ. 8,000.

Jornal San Wa Ou: Av. Venseslau de Morais 221, Edif. Ind. Nam Fong, 2a Fase, 15°, Bloco E; tel. 717569; fax 717572; e-mail correiro@macau.ctm.net; daily; Dir LAM CHONG; circ. 1,500.

Jornal 'Si-Si': Rua de Brás da Rosa 58, 2/F; tel. 974354; weekly; Dir and Editor-in-Chief CHEANG VENG PENG; circ. 3,000.

Jornal Va Kio: Rua da Alfândega 7–9; tel. 345888; fax 580638; f. 1937; daily; Dir CHIANG SAO MENG; Editor-in-Chief LEONG CHI SANG; circ. 21,000.

O Pulso de Macau: Rua Oito do Bairro Iao Hon S/N; Edif. Hong Tai, Apt F058 R/C; tel. 400190; fax 400284; weekly; Dir HO SI VO.

Ou Mun Iat Pou (Macao Daily News): Rua Pedro Nolasco da Silva 37; tel. 371688; fax 331998; f. 1958; daily; Dir LEI SENG CHUN; Dir and Editor-in-Chief LEI PANG CHU; circ. 100,000.

Semanário Desportivo de Macau: Estrada D. Maria II, Edif. Kin Chit Garden, 2 G–H; tel. 718259; fax 718285; weekly; sport; Dir FONG NIM LAM; Editor-in-Chief FONG NIM SEONG; circ. 2,000.

Semenário Recreativo de Macau: Av. Sidónio Pais 31 D, 3/F A; tel. 553216; fax 516792; weekly; Dir IEONG CHEOK KONG; Editor-in-Chief TONG IOK WA.

Seng Pou (Star): Travessa da Caldeira 9; tel. 938387; fax 388192; f. 1963; daily; Dir and Editor-in-Chief KUOK KAM SENG; Deputy Editor-in-Chief TOU MAN KUM; circ. 6,000.

Si Man Pou (Jornal do Cidadão): Rua dos Pescadores, Edif. Ind. Ocean, Bl. 2, 2/F–B; tel. 722111; fax 722133; f. 1944; daily; Dir and Editor-in-Chief KUNG MAN; circ. 8,000.

Tai Chung Pou: Rua Dr Lourenço; P. Marques 7A, 2/F; tel. 939888; fax 934114; f. 1933; daily; Dir VONG U. KUONG; Editor-in-Chief CHAN TAI PAC; circ. 8,000.

Today Macau Journal: Pátio da Barca 20, R/C; tel. 215050; fax 210478; daily; Dir LAM VO I; Editor-in-Chief IU VENG ION; circ. 6,000.

NEWS AGENCIES

Associated Press (AP) (USA): POB 221; tel. 5994187; fax 519423; e-mail adamlee4@macau.ctm.net; Correspondent ADAM LEE.

China News Service: Av. Gov. Jaime Silveiro Marques, Edif. Zhu Kuan, 14/F, Y/Z; tel. 594585; fax 594585.

LUSA (Agência de Noticias de Portugal): Av. Conselheiro Ferreira de Almeida 95-A; tel. 967601; fax 967605; e-mail jcsantos@lusa.pt; internet www.lusa.pt; Dir JOSÉ COSTA SANTOS.

Reuters (United Kingdom): Rua de Penha, No. 22 Edif. Pearl Terrace, 7/F; tel. 345888; fax 930076; Correspondent HARALD BRUNING.

Xinhua (New China) News Agency Macao SAR Branch: Av. Gov. Jaime Silvério Marques, Edif. Zhu Kuan, 13 andar-V; tel. 727710; fax 700548; Dir PAN GUO JUN.

PRESS ASSOCIATIONS

Macao Chinese Media Workers Association: Travessa do Matadouro, Edif. 3, 3B; tel. 939486; e-mail mcju@macau.ctm.net; Pres. LEI PANG CHU.

Macao Journalists Association: Rua de Jorge Alvares, 7–7B, Viva Court, 17F, Flat A; tel. 569819; fax 569819; e-mail macauja@macau.ctm.net; internet home.macau.ctm.net/~macauja; f. 1999; Pres. CHEANG UT MENG.

Macao Journalists Club: Estrada do Repouso, Edif. Tak Fai 18B; tel. 921395; fax 921395; e-mail cjm@macau.ctm.net; Pres. DAVID CHAN CHI WA.

Macao Media Club: Rua de Santa Clara 5–7E, Edif. Ribeiro, 4B; tel. 330035; fax 330036; Pres. CHEONG CHI SENG.

Macao Sports Press Association: Av. Olímpica, Estádio de Macau 33; tel. 838206, ext. 151; fax 718285; e-mail macsport@macau.ctm.net; Pres. LAO LU KONG.

Publishers

Associação Beneficência Leitores Jornal Ou Mun: Nova-Guia 339; tel. 711631; fax 711630.

Fundação Macau: Av. República 6; tel. 966777; fax 968658; internet www.fmac.org.mo.

Instituto Cultural de Macau: see under Tourism; publishes literature, social sciences and history.

Livros do Oriente: Av. Amizade 876, Edif. Marina Gardens, 15 E; tel. 700320; fax 700423; e-mail rclilau@macau.ctm.net; internet www.loriente.com; f. 1990; publishes in Portuguese, English and Chinese on regional history, culture, etc.; Gen. Man. ROGÉRIO BELTRÃO COELHO; Exec. Man. CECÍLIA JORGE.

Universidade de Macau—Centro de Publicações: POB 3001; tel. 3974504; fax 3974506; e-mail PUB_GRP@umac.mo; internet www.umac.mo/pc; f. 1993; art, economics, education, political science, history, literature, management, social sciences, etc.; Head Dr RAYMOND WONG.

GOVERNMENT PUBLISHER

Imprensa Oficial: Rua da Imprensa Nacional s/n; tel. 573822; fax 596802; e-mail info@imprensa.macau.gov.mo.

Broadcasting and Communications

TELECOMMUNICATIONS

The Government initiated a liberalization of the mobile telecommunications market in 2001.

Companhia de Telecomunicações de Macau, SARL (CTM): Rua de Lagos, Edif. Telecentro, Taipa; tel. 833833; fax 8913031; e-mail comm@macau.ctm.net; internet www.ctm.net; holds local telecommunications monopoly; shareholders include Cable and Wireless (51%) Portugal Telecom (28%), and CITIC Pacific (20%); Chair. JAMES CHEESEWRIGHT; CEO DAVID KAY; 900 employees.

Hutchison Telephone (Macau) Co Ltd: Macao; e-mail cs_web@htmac.com; internet www.hutchisonmacau.com; mobile telecommunications operator; subsidiary of Hutchison Whampoa Group (Hong Kong); local capital participation.

SmarTone Mobile (Macau) Ltd: Macao; e-mail cnquiry@smartone.com.mo; internet www.smartone.com.mo; mobile telecommunications provider; subsidiary of SmarTone Mobile Communications Ltd (Hong Kong); local capital participation.

Regulatory Authority

Office for the Development of Telecommunications and Information Technology (GDTTI): Av. da Praia Grande 789, 3/F; tel. 3969161; fax 356328; e-mail ifx@gdtti.gov.mo; internet www.gdtti.gov.mo.

BROADCASTING

Radio

Rádio Vila Verde: Macao Jockey Club, Taipa; tel. 822163; private radio station; programmes in Chinese; Man. KOK HOI.

Television

Teledifusão de Macau, SARL (TDM): Rua Francisco Xavier Pereira 157-A, POB 446; tel. 335888 (Radio), 519188 (TV); fax 520208; privately owned; two radio channels: **Rádio Macau** (Av. Dr Rodrigo Rodrigues, Edif. Nam Kwong, 7/F; fax 343220) in Portuguese, broadcasting 24 hours per day in Portuguese on **TDM Canal 1** (incl. broadcasts from RTP International in Portugal) and 17 hours per day in Chinese on **TDM Channel 2**; Chair. STANLEY HO; Exec. Vice-Chair. Dr MANUEL GONÇALVES.

Macao Satellite Television: c/o Cosmos Televisão por Satélite, Av. Infante D. Henrique 29, Edif. Va Iong, 4/F A; commenced transmissions in 2000; operated by Cosmos Televisão por Satélite, SARL; domestic and international broadcasts in Chinese aimed at Chinese-speaking audiences world-wide.

Macao is within transmission range of the Hong Kong television stations.

Cosmos Televisão por Satélite, SARL: Av. Infante D. Henrique 29, Edif. Va Iong, 4/F A; tel. 785731; fax 788234; commenced trial satellite transmissions in 1999, initially for three hours per day; by the year 2003 the company planned to provide up to six channels; Chair. NG FOK.

Finance

(cap. = capital; res = reserves; dep. = deposits; m. = million; brs = branches; amounts in patacas unless otherwise indicated)

BANKING

Macao has no foreign-exchange controls, its external payments system being fully liberalized on current and capital transactions. The Financial System Act, aiming to improve the reputation of the territory's banks and to comply with international standards, took effect in September 1993.

Issuing Authority

Autoridade Monetária de Macau (AMCM) (Monetary Authority of Macao): Calçada do Gaio 24–26, POB 3017; tel. 568288; fax 325432; e-mail general@amcm.gov.mo; internet www.amcm.macau.gov.mo; f. 1989 as Autoridade Monetária e Cambial de Macau (AMCM), to replace the Instituto Emissor de Macau; govt-owned; Pres. ANSELMO L. S. TENG.

Banks of Issue

Banco Nacional Ultramarino (BNU), SA: Av. Almeida Ribeiro 22, POB 465; tel. 355111; fax 355653; e-mail markt@bnu.com.mo; internet www.bnu.com.mo; f. 1864; est. in Macao 1902; Head Office in Lisbon; agent of Macao Government; Gen. Man. Dr HERCULANO J. SOUSA; 10 brs.

Bank of China: Bank of China Bldg, Av. Dr Mário Soares; tel. 781828; fax 781833; e-mail bocmacau@macau.ctm.net; f. 1950 as Nan Tung Bank, name changed 1987; authorized to issue banknotes from Oct. 1995; Gen. Man. ZHANG HONGYI; 24 brs.

Other Commercial Banks

Banco da América (Macau), SA: Av. Almeida Ribeiro 70–76, POB 165; tel. 568821; fax 570386; f. 1937; fmrly Security Pacific Asian Bank (Banco de Cantão); cap. 100m., res 49.9m., dep. 1,061.3m. (Dec. 2001); Chair. SAMUEL NG TSIEN; Man. Dir KIN HONG CHEONG.

Banco Comercial de Macau, SA: Av. da Praia Grande 572, POB 545; tel. 7910000; fax 595817; e-mail bcmbank@bcm.com.mo; f. 1995; cap. 225m., res 316m., dep. 6,082m. (Dec. 2001); Chair. JORGE JARDIM GONÇALVES; CEO Dr MANUEL MARECOS DUARTE; 17 brs.

Banco Delta Asia, SARL: Av. Conselheiro Ferreira de Almeida 79; tel. 559898; fax 570068; e-mail contact@bdam.com; internet www.delta asia.com; f. 1935; fmrly Banco Hang Sang; cap. 210.0m., res 122.8m., dep. 3,291.2m. (Dec. 2003); Chair. STANLEY AU; Exec. Dir PHILIP NG; 10 brs.

Banco Seng Heng, SARL: Av. da Amizade 555, Macau Landmark, Torre Banco Seng Heng; tel. 555222; fax 338064; e-mail sengheng@macau.ctm.net; internet www.senghengbank.com; f. 1972; cap. 150.0m., res 755.8m., dep. 11.9m. (Dec. 2001); Chair. STANLEY HO; Gen. Man. ALEX LI; 6 brs.

Banco Tai Fung, SARL: Tai Fung Bank Bldg, Av. Alameda Dr Carlos d'Assumpção 418; tel. 322323; fax 570737; e-mail tfbsecr@taifungbank.com; internet www.taifungbank.com; f. 1971; cap. 1,000m., dep. 21,427m. (June 2003); Chair. FUNG KA YORK; Gen. Man. LONG RONGSHEN; 20 brs.

Banco Weng Hang, SA: Av. Almeida Ribeiro 241; tel. 335678; fax 576527; e-mail bwhhrd@whbmac.com; internet www.whbmac.com; f. 1973; subsidiary of Wing Hang Bank Ltd, Hong Kong; res 522m., dep. 8,766m. (Dec. 2003); Chair. PATRICK FUNG YUK-BUN; Gen. Man. and Dir LEE TAK LIM; 10 brs.

Guangdong Development Bank: Av. Dr Carlos d'Assumpção, 181–187, Centro Comercial do Grupo Brihantismo, 18° andar; tel. 750328; fax 750728; Gen. Man. GUO ZHI-HANG.

Luso International Banking Ltd: Av. Dr Mário Soares 47; tel. 378977; fax 711100; e-mail lusobank@lusobank.com.mo; internet www.lusobank.com.mo; f. 1974; cap. 151.5m., res 184m., dep. 6,245.8m. (Dec. 2000); Chair. WONG XI CHAO; Gen. Man. IP KAI MING; 10 brs.

Foreign Banks

Banco Comercial Português (Portugal): Av. da Praia Grande 594, BCM Bldg, 12/F; tel. 786769; fax 786772; Gen. Man. Dr ANTONIO MATOS.

Banco Espírito Santo do Oriente (Portugal): Av. Dr Mário Soares 323, Bank of China Bldg, 28/F, E–F; tel. 785222; fax 785228; e-mail besor@macau.ctm.net; f. 1996; subsidiary of Banco Espírito Santo, SA (Portugal); Exec. Dir JOÃO MANUEL AMBRÓSIO; Chair. and CEO JOSÉ MORGADO.

Bank of East Asia Ltd (Hong Kong): Av. da Praia Grande 697, Edif. Tai Wah R/C; tel. 335511; fax 337557; Gen. Man. LAI TZE HIM.

BNP Paribas (France): Av. Central Plaza, 10/F, Almeida Ribeiro 61; tel. 562777; fax 560626; f. 1979; Man. SANCO SZE.

Citibank NA (USA): Rua da Praia Grande 251–53; tel. 378188; fax 578451; Pres. DANIEL CHOW SHIU LUN.

DBS Bank (Hong Kong) Ltd: Rua de Santa Clara, 5-7E, Edif. Ribeiro, Lojas C&D; tel. 329338; fax 323711.

Hang Seng Bank: 20/F, Central Plaza, 61 Av. de Almeida Ribeiro; tel. 323321; fax 330612; internet main.hangseng.com/eng/abo/cu/macbro/index.html; Macao branch opened March 2004.

The Hongkong and Shanghai Banking Corporation Ltd (Hong Kong): Av. da Praia Grande 639; tel. 553669; fax 315421; e-mail hsbc@macau.ctm.net; f. 1972; CEO THOMAS YAM.

Industrial and Commercial Bank of China: Alameda Dr Carlos d'Assumpção 393–437, Edif. Dynasty Plaza, E, F, G e H; tel. 786338; fax 786328.

International Bank of Taipei (Taiwan): Av. Infante D. Henrique 52–58; tel. 715175; fax 715035; e-mail tppmonx@macau.ctm.net; f. 1996; fmrly Taipei Business Bank; Gen. Man. KEVIN CHIOU.

Liu Chong Hing Bank Ltd (Hong Kong): Av. da Praia Grande 693, Edif. Tai Wah, R/C; tel. 339982; fax 339990; Gen. Man. LAM MAN KING.

Macau Chinese Bank Ltd: Av. da Praia Grande 811; tel. 322678; fax 322680; fmrly Finibanco (Portugal).

Overseas Trust Bank Limited (Hong Kong): Rua de Santa Clara 5–7E, Edif. Ribeiro, Loja C e D; tel. 329338; fax 323711; e-mail otbmacau@macau.ctm.net; Senior Man. LAU CHI KEUNG.

Standard Chartered Bank (UK): 8/F Office Tower, Macao Landmark, Av. de Amizade; tel. 786111; fax 786222; f. 1982; Man. KIN YIP CHAN.

Banking Association

Associação de Bancos de Macau (ABM) (The Macao Association of Banks): Av. da Praia Grande 575, Edif. 'Finanças', 15/F; tel. 511921; fax 346049; Chair. ZHANG HONGYI.

INSURANCE

ACE Seguradora, SA: Rua Dr. Pedro José Lobo 1–3, Luso Bank Bldg, 17/F, Apt 1701–02; tel. 557191; fax 570188; Rep. ANDY AU.

American Home Assurance Co: Av. Almeida Ribeiro 61, Central Plaza, 15/F, 'G'; non-life insurance.

American International Assurance Co (Bermuda) Ltd: Av. Almeida Ribeiro 61, Central Plaza, 13/F; tel. 9881888; fax 315900; life insurance; Rep. ALEXANDRA FOO CHEUK LING.

Asia Insurance Co Ltd: Rua do Dr Pedro José Lobo 1–3, Luso International Bank Bldg, 11/F, Units 1103–04; tel. 570439; fax 570438; non-life insurance; Rep. S. T. CHAN.

AXA China Region Insurance Company: Rua de Xangai 175, Edif. da Associação Comercial de Macau, 17/F; tel. 781188; fax 780022; life insurance; Rep. KANE CHOW.

CGU International Insurance plc: Av. da Praia Grande 693, Edif. Tai Wah A & B, 13/F; tel. 923329; fax 923349; non-life insurance; Man. VICTOR WU.

China Insurance Co Ltd: Av. Dr. Rodrigo Rodgrigues, Edif. Seguros da China, 19/F; non-life insurance.

China Life Insurance Co Ltd: Av. Dr Rodrigo Rodrigues Quarteirão 11, Lote A, Zape, China Insurance Bldg, 15/F; tel. 558918; fax 787287; e-mail cic@macau.ctm.net; Rep. CHENG MINGJIN.

Companhia de Seguros Fidelidade: Av. Almeida Ribeiro 22–38 (BNU); tel. 374072; fax 511085; life and non-life insurance; Man. LEONEL ALBERTO RANGE RODRIGUES.

Companhia de Seguros Delta SA: Av. da Praia Grande 369–71, Edif. Keng Ou, 13/F, D; tel. 337036; fax 337037; Rep. JOHNNY CHENG.

Crown Life Insurance Co: Av. da Praia Grande 287, Nam Yuet Commercial Centre, Bl. B, 8/F; tel. 570828; fax 570844; Rep. STEVEN SIU.

Delta Asia Insurance Company Ltd: Av. da Praia Grande, 369–371, Edif. Keng Ou, 13/F.

HSBC Life Insurance (Asia) Ltd: Av. da Praia Grande 619, Edif. Comercial Si Toi, 1/F ; tel. 212323; fax 217162; non-life insurance; Rep. NORA CHIO.

Ing Life Insurance Co (Macao) Ltd: Av. Almeida Ribeiro 61, 11/F, Unit C and D; tel. 9886060; fax 9886100; internet www.ing.com.mo; Man. STEVEN CHIK YIU KAI.

Insurance Co of North America: Av. Almeida Ribeiro 32, Tai Fung Bank Bldg, Rm 806–7; tel. 557191; fax 570188; Rep. JOSEPH LO.

Luen Fung Hang Insurance Co Ltd: Rua de Pequim 202A–246, Macao Finance Centre, 6/F–A; tel. 700033; fax 700088; e-mail lfhins@macau.ctm.net; internet www.lfhins.com; non-life insurance; Rep. SI CHI HOK.

Macao Life Insurance Co (Macao Insurance Company): Av. da Praia Grande 574, Edif. BCM, 10-11F; tel. 555078; fax 551074; life and non-life insurance (Macao Insurance Company); Rep. MANUEL BALCÃO REIS.

Manulife (International) Ltd: Av. da Praia Grande 517, Edif. Comercial Nam Tung, 8/F, Unit B & C; tel. 3980388; fax 323312; internet www.manulife.com.hk; Rep. DANIEL TANG.

MassMutual Asia Ltd: Av. da Praia Grande 517, Edif. Nam Tung 16, 6/F; life insurance.

Min Xin Insurance Co Ltd: Rua do Dr Pedro José Lobo 1–3, Luso International Bank Bldg, 27/F, Rm 2704; tel. 305684; fax 305600; non-life insurance; Rep. PETER CHAN.

Mitsui Sumitomo Insurance Co Ltd: Rua Dr Pedro José Lobo 1–3, Edif. Banco Luso, 11/F, Apartment 1202; tel. 385917; fax 596667; non-life insurance; Rep. TAKAO YASUKOCHI.

QBE Insurance (International) Ltd: Av. da Praia Grande 369–71, Edif. Keng On 'B', 9/F; tel. 323909; fax 323911; non-life insurance; Rep. SALLY SIU.

The Wing On Fire & Marine Insurance Co Ltd: Av. Almeida Ribeiro 61, Central Plaza, 7/F, Block E; tel. 356688; fax 333710; non-life insurance; Rep. CHIANG AO LAI LAI.

Winterthur Swiss Insurance (Macao) Ltd: Av. da Praia Grande 369-371, Edif. Keng Ou, 13/F, C; tel. 356618; fax 356800; non-life insurance; Man. ALLAN YU KIN NAM.

Insurers' Association

Federation of Macao Professional Insurance Intermediaries: Rua de Pequim 244–46, Macao Finance Centre, 6/F, G; tel. 703268; fax 703266; Rep. DAVID KONG.

Macao Insurance Agents and Brokers Association: Av. da Praia Grande 309, Nam Yuet Commercial Centre, 8/F, D; tel. 378901; fax 570848; Rep. JACK LI KWOK TAI.

Macao Insurers' Association: Av. da Praia Grande 575, Edif. 'Finanças', 15/F; tel. 511923; fax 337531; e-mail minsa@macau.ctm.net; Pres. VICTOR WU.

Trade and Industry

CHAMBER OF COMMERCE

Associação Comercial de Macau (Macao Chamber of Commerce): Rua de Xangai 175, Edif. ACM, 5/F; tel. 576833; fax 594513; internet www.acm.org.mo; Pres. MA MAN KEI.

INDUSTRIAL AND TRADE ASSOCIATIONS

Associação dos Construtores Civis e Empresas de Fomento Predial de Macau (Macao Association of Building Contractors and Developers): Rua do Campo 9–11; tel. 323854; fax 345710; Pres. CHUI TAK KEI.

Associação dos Exportadores e Importadores de Macau (Macao Importers and Exporters Association): Av. Infante D. Henrique 60–62, Centro Comercial 'Central', 3/F; tel. 375859; fax 512174; e-mail aeim@macau.ctm.net; exporters' and importers' asscn; Chair. LIU CHAK WAN.

Associação dos Industriais de Tecelagem e Fiação de Lã de Macau (Macao Weaving and Spinning of Wool Manufacturers' Asscn): Av. da Amizade 271, Edif. Kam Wa Kok, 6/F–A; tel. 553378; fax 511105; Pres. WONG SHOO KEE.

Associação Industrial de Macau (Industrial Association of Macau): Rua Dr Pedro José Lobo 34–36, Edif. AIM, 17/F, POB 70; tel. 574125; fax 578305; e-mail aim@macau.ctm.net; internet www.madeinmacau.net; f. 1959; Pres. PETER PAN.

Centro de Produtividade e Transferência de Tecnologia de Macau (Macao Productivity and Technology Transfer Centre): Rua de Xangai 175, Edif. ACM, 6/F; tel. 781313; fax 788233; e-mail cpttm@cpttm.org.mo; internet www.cpttm.org.mo; vocational or professional training; Dir Dr ERIC YEUNG.

Euro-Info Centre Macao: Av. da Amizade, 918, Edif. World Trade Center, 2° andar; tel. 713338; fax 713339; e-mail eic@macau.ctm.net; internet www.ieem.org.mo/eic/eicmacau.html; promotes trade with EU; Man. YVONNE SUN; Pres. JOSÉ LUÍS DE SALES MARQUES.

Instituto de Promoção do Comércio e do Investimento de Macau (IPIM) (Macao Trade and Investment Promotion Institute): Av. da Amizade 918, World Trade Center Bldg, 2°–4° andares; tel. 710300; fax 590309; e-mail ipim@ipim.gov.mo; internet www.ipim.gov.mo; Pres. LEE PENG HONG.

SPIC (Concordia Industrial Park Ltd): Av. da Amizade 918, World Trade Center Bldg, 13/F A & B; tel. 786636; fax 785374; e-mail spic@macau.ctm.net; internet www.concordia-park.com; f. 1993; industrial park, promotion of investment and industrial diversification; Pres. of the Bd PAULINA Y. ALVES DOS SANTOS.

World Trade Center Macao, SARL: Av. da Amizade 918, Edif. World Trade Center, 16/F–19/F; tel. 727666; fax 727633; e-mail wtcmc@macau.ctm.net; internet www.wtc-macau.com; f. 1995; trade information and business services, office rentals, exhibition and conference facilities; Man. Dir Dr ANTÓNIO LEÇA DA VEIGA PAZ.

UTILITIES

Electricity

Companhia de Electricidade de Macau, SARL (CEM): Estrada D. Maria II 32–36, Edif. CEM; tel. 339933; fax 719760; f. 1972; sole distributor; 6% owned by China Power International Holding Ltd; Pres. Eng. Custódio Miguens.

Water

Sociedade de Abastecimento de Aguas de Macau, SARL (SAAM) (Macao Water): Av. do Conselheiro Borja 718; tel. 233332; fax 220150; e-mail mwater@macau.ctm.net; internet www.macaowater.com; f. 1985 as jt venture with Suez Lyonnaise des Eaux; Dir Philippe Wind.

TRADE UNIONS

Macao Federation of Trade Unions: Rua Ribeira do Patane 2; tel. 576231; fax 553110; Pres. Pun Iok Lan.

Transport

RAILWAYS

There are no railways in Macao. A plan to connect Macao with Zhuhai and Guangzhou (People's Republic of China) is under consideration. Construction of the Zhuhai–Guangzhou section was under way in the early 21st century.

ROADS

In 2000 the public road network extended to 341 km. The peninsula of Macao is linked to the islands of Taipa and Coloane by two bridges and by a 2.2-km causeway respectively. The first bridge (2.6 km) opened in 1974. In conjunction with the construction of an airport on Taipa (see below), a 4.4-km four-lane bridge to the Macao peninsula was opened in April 1994. A second connection to the mainland, the 1.5-km six-lane road bridge (the Lotus Bridge) linking Macao with Hengqin Island (in Zhuhai, Guangdong Province), opened to traffic in December 1999. A new 29-km steel bridge linking Macao with Hong Kong's Lantau Island and Zhuhai City, Guangdong Province, was being planned in 2003, with studies being carried out to determine the project's financial and technological feasibility. A third link between Macao and Taipa, a double-deck bridge, was under construction in 2003 and was expected to be completed by late 2004.

SHIPPING

There are representatives of shipping agencies for international lines in Macao. There are passenger and cargo services to the People's Republic of China. Regular services between Macao and Hong Kong are run by the Hong Kong-based **New World First Ferry** and **Shun Tak–China Travel Ship Management Ltd** companies. A new terminal opened in late 1993. The new port of Kao-ho (on the island of Coloane), which handles cargo and operates container services, entered into service in 1991.

CTS Parkview Holdings Ltd: Av. Amizade, Porto Exterior, Terminal Marítimo de Macau, Sala 2006B; tel. 726789; fax 727112; purchased by STDM in 1998.

STDM Shipping Dept: Av. da Amizade Terminal Marítimo do Porto Exterior; tel. 726111; fax 726234; affiliated to Sociedade de Turismo e Diversões de Macau; Gen. Man. Alan Ho; Exec. Man. Capt. Augusto Lizardo.

Association

Associação de Agências de Navegação e Congêneres de Macau: Av. Horta e Costa 7D–E, POB 6133; tel. 528207; fax 302667; e-mail macshpg@hotmail.com; Pres. Vong Kok Seng.

Port Authority

Capitania dos Portos de Macau: Rampa da Barra, Quartel dos Mouros, POB 47; tel. 559922; fax 511986; e-mail webmaster@marine.gov.mo; internet www.marine.gov.mo.

CIVIL AVIATION

In August 1987 plans were approved for the construction of an international airport, on reclaimed land near the island of Taipa, and work began in 1989. The final cost of the project was 8,900m. patacas. Macau International Airport was officially opened in December 1995. The terminal has the capacity to handle 6m. passengers a year. By 2002 a total of 9 airlines operated 436 scheduled flights weekly to 21 destinations, mostly in China, but also to the Democratic People's Republic of Korea, the Philippines, Singapore, Taiwan, and Thailand. Between January and December 2003 Macau International Airport handled a total of 2,905,811 passengers and 141,223 tons of cargo. A helicopter service between Hong Kong and Macao was put into operation in 1990 by East Asia Airlines, and a scheduled helicopter flight service between Macao and Shenzhen provided by the same company commenced in January 2003. East Asia Airlines transported a total of 99,985 helicopter passengers on its Macao–Hong Kong route and 2,691 helicopter passengers on their Macao–Shenzhen route in 2002. In late 2003 Air Macau launched direct flights between Macao and Chengdu, capital of Sichuan Province.

AACM (Civil Aviation Authority Macao): Rua Dr Pedro José Lobo 1–3, Luso International Bldg, 26/F; tel. 511213; fax 338089; e-mail aacm@macau.ctm.net; internet www.macau-airport.gov.mo; f. 1991; Pres. Rui Alfredo Balacó Moreira.

Administração de Aeroportos, Lda (ADA): Av. de João IV, Centro Comercial Iat Teng Hou, 5/F; tel. 711808; fax 711803; e-mail adamkt@macau.ctm.net; internet www.ada.com.mo; airport administration; Pres. Sun Bo; Dir Carlos Seruca Salgado.

CAM (Sociedade do Aeroporto Internacional de Macau, SARL): Av. Dr Mário Soares, Bank of China Bldg, 29/F; tel. 785448; fax 785465; e-mail cam@macau.ctm.net; internet www.macau-airport.gov.mo; f. 1989; airport owner, responsible for design, construction, development and international marketing of Macao International Airport; Chair. Deng Jun.

Air Macau: 398 Alameda Dr Carlos d'Assumpção, 12°-18° andar; tel. 3966888; fax 3966866; e-mail airmacau@airmacau.com.mo; internet www.airmacau.com.mo; f. 1994; controlled by China National Aviation Corporation (Group) Macao Co Ltd; services to several cities in the People's Republic of China, the Republic of Korea, the Philippines, Taiwan and Thailand; other destinations planned; Chair. Gu Tiefei; CEO David H.J. Fei.

Tourism

Tourism is now a major industry, a substantial portion of the Government's revenue being derived from the territory's casinos. The other attractions are the cultural heritage and museums, dog-racing, horse-racing, and annual events such as Chinese New Year (January/February), the Macao Arts Festival (February/March), Dragon Boat Festival (May/June), the Macao International Fireworks Festival (September/October), the International Music Festival, (October) the Macao Grand Prix for racing cars and motorcycles (November) and the Macao International Marathon (December). At the end of 2002 there were 25 hotels of two-stars and above. A total of 9,185 hotel rooms were available in December 2003. Average per caput visitor spending in 2002 was 1,454 patacas. Total visitor arrivals rose from 11.53m. in 2002 to 11.88m. in 2003. Of the former figure, 5.1m. were arrivals from Hong Kong and 4.2m. from the People's Republic of China. Receipts from tourism were US $4,415m. in 2002.

Macao Government Tourist Office (MGTO): Alameda Dr Carlos d'Assumpção, 335-341, Edif. Hot Line, 12° andar; tel. 315566; fax 510104; e-mail mgto@macautourism.gov.mo; internet www.macautourism.gov.mo; Dir Eng. João Manuel Costa Antunes.

Instituto Cultural de Macau: Praçeta de Miramar 87U, Edif. San On; tel. 700391; fax 700405; e-mail postoffice@icm.gov.mo; internet www.icm.gov.mo; f. 1982; organizes performances, concerts, exhibitions, festivals, etc.; library facilities; Pres. Heidi Ho.

Macao Hotels Association: Rua Luís Gonzaga Gomes s/n, Bl. IV, r/c, Centro de Actividades Turísticas, Cabinet A; tel. 703416; fax 703415.

Sociedade de Turismo e Diversões de Macau (STDM), SARL: Hotel Lisboa, 9F, Old Wing, POB 3036; tel. 566065; fax 371981; e-mail stdmmdof@macau.ctm.net; fmrly operated 10 casinos; since ending of STDM's monopoly franchise in 2002, operation of the casinos has been handled by STDM's subsidiary Macao Gaming Company Ltd; Man. Dir Dr Stanley Ho.

Defence

The budget for defence is allocated by the Chinese Government. Upon the territory's reversion to Chinese sovereignty in December 1999, troops of the People's Liberation Army (PLA) were stationed in Macao. The force comprises around 1,000 troops: a maximum of 500 soldiers are stationed in Macao, the remainder being positioned in Zhuhai, China, on the border with the SAR. The unit is directly responsible to the Commander of the Guangzhou Military Region and to the Central Military Commission. The Macao garrison is composed mainly of ground troops. Naval and air defence tasks are performed by the naval vessel unit of the PLA garrison in Hong Kong and by the air force unit in Huizhou. Subject to the request of the Macao SAR, the garrison may participate in law-enforcement and rescue operations in the SAR.

Commander of the PLA Garrison in Macao: Senior Col Liu Lianhua.

Education

The rate of literacy among the population aged 15 and over in Macao was 91.3% in 2001, illiteracy being confined mainly to elderly women. The education system in Macao is structured as follows: pre-school education (lasting two years); primary preparatory year (one year); primary education (six years); secondary education (five–six years, divided into junior secondary of three years and senior secondary of two–three years). Schooling normally lasts from the ages of three to 17. In 2002/03 schools enrolled a total of 98,271 pupils: kindergarten 12,737; primary 41,535; secondary 43,999 (including technical and vocational students). From 1995/96 free education was extended from government schools to private schools. Private schools provide education for more than 90% of children. Of these schools 77% have joined the free education system, and together with the government schools they form the public school system, in which all pupils from primary preparatory year up to the junior secondary level (10 years) receive free tuition. Based on the four years of free education, compulsory education was implemented from 1999/2000. By the end of 2001/02 the enrolment rate was as follows: pre-school education and primary preparatory year 91.2% (some families leave their children in China); primary education 105.4%; secondary education 78.5%; and higher education 27.7%. The government budget for education and training in 2003 was 1,886.6m. patacas (including administration costs), which was equal to 16.23% of the entire 2003 government budget.

In higher learning, there are 12 public and private universities, polytechnic institutes and research centres, namely: the University of Macao, Macao Polytechnic Institute, Institute for Tourism Studies, Macao Security Forces Academy, Inter-University Institute of Macao, Asia International Open University (Macao), Institute of European Studies of Macao, Kiang Wu Nursing College of Macao, United Nations University/International Institute for Software Technology, Macao University of Science and Technology, Macao Institute of Management, and Macao Millennium College. Some 22,571 students attended courses offered by those institutions in the academic year 2001/02, ranging from the bacharelato (three-year courses) to doctorate programmes.

Bibliography

Berlie, J. A. (Ed.). *Macao 2000.* Oxford, Oxford University Press, 2000.

Boxer, C. R. *The Portuguese Seaborne Empire*. London, Hutchinson and Co, 1969.

Braga, J. M. *O primeiro acordo Luso-Chinês.* Macao, 1939.

The Western Pioneers and their Discovery of Macau. Macao, 1949.

Brookshaw, D. *Visions of China: Stories from Macau*. Hong Kong, Hong Kong University Press, 2002.

Do Carmo, Maria Helena. *Os Interesses dos Portugueses em Macau, na Primeira Metade do Século XVIII.* Macao, University of Macau Publications Centre, 1999.

Chan, S. S. *The Macau Economy.* Macao, University of Macau Publications Centre, 2000.

Cheng, Christina Miu Bing. *Macau: A Cultural Janus.* Hong Kong, Hong Kong University Press, 1999.

Coates, Austin. *A Macao Narrative*. Hong Kong, Oxford University Press, 1998.

Macao and the British, 1637–1842: Prelude to Hong Kong. Hong Kong, Oxford University Press, 1989.

Gomes, L. G. *Bibliografia Macaense*. Macao, Imprensa Nacional, 1973.

Guillén Núñez, César, and Leong Kai Tai. *Macao Streets.* Hong Kong, Oxford University Press, 1999.

Gunn, Geoffrey C. *Encountering Macau: A Portuguese City-state on the Periphery of China, 1557–1999*. Boulder, CO, Westview Press, 1996.

Ieong Wan Chong and Ricardo Chi Sen Siu. *Macau: A Model of Mini-Economy.* Macao, University of Macau Publications Centre, 1997.

Jesus, C. A. Montalto de. *Historic Macau*. Macao, Salesean Printing Press, 1926.

Leong Ka Tai, and Davies, S. *Macau.* Singapore, Times Editions, 1986.

Ljungstedt, A. *An Historical Sketch of the Portuguese Settlements in China*. Boston, 1936.

Lui Kwok Man. *Macau in Transition.* Macao, University of Macau Publications Centre, 2000.

Macau Research Group. *A Strategic Assessment of Macau, 2000 Edition*. Icon Group International, 2000.

McGivering, Jill. *Macao Remembers.* Hong Kong, Oxford University Press, 1999.

Mo, Timothy. *An Insular Possession*. London, Chatto and Windus, 1998.

Pons, Philippe. *Macao.* Paris, Le Promeneur, 2000.

Porter, Jonathan. *Macau: The Imaginary City*. Boulder, CO, Westview Press, 2000.

Ramos, Rufino et al. (Eds). *Population and Development in Macau.* Macao, University of Macau Publications Centre, 1994.

Macau and its Neighbors in Transition. Macao, University of Macau Publications Centre, 1997.

Macau and its Neighbors Towards the 21st Century. Macao, University of Macau Publications Centre, 1998.

Shipp, Steve. *Macau, China: A Political History of the Portuguese Colony's Transition to Chinese Rule*. Jefferson, NC, McFarland & Co, 1997.

Da Silva Diaz de Seabra, Isabel Leonor. *Relações Entre Macau e O Sião (Séculos XVIII–XIX).* Macao, University of Macau Publications Centre, 1999.

Sit, Victor F. S., et al. *Entrepreneurs and Enterprises in Macao: A Study in Industrial Development*. Hong Kong, Hong Kong University Press, 1991.

USA International Business Publications. *Macau Country Study Guide.* International Business Publications, 2000.

Macao Government and Policy Guide. International Business Publications, 2000.

Wesley-Smith, P. 'Macao' in Albert P. Blaustein (Ed.), *Constitutions of Dependencies and Special Sovereignties, Vol. III*. New York, Oceana Press, 1977.

Wong Hon Keung. *Economic Interaction Between Guangxi and Macau.* Macao, University of Macau Publications Centre, 1998.

Yee, Herbert S. *Macau in Transition: From Colony to Autonomous Region.* New York, St. Martin's Press, 2001.

CHINA (TAIWAN)

Physical and Social Geography

The Republic of China has, since 1949, been confined mainly to the province of Taiwan (comprising one large island and several much smaller ones), which lies off the south-east coast of the Chinese mainland. The territory under the Republic's effective jurisdiction consists of the island of Taiwan (also known as Formosa) and nearby islands, including the P'enghu (Pescadores) group, together with a few other islands which lie just off the mainland and form part of the province of Fujian (Fukien), west of Taiwan. The largest of these is Kinmen (Jinmen), also known as Quemoy, which (with three smaller islands) is about 10 km from the port of Xiamen (Amoy), while five other islands under Taiwan's control, notably Matsu (Mazu), lie further north, near Fuzhou. The island of Taiwan itself is separated from the mainland by the Taiwan Strait, which is about 220 km wide at its broadest point and 130 km at the narrowest point. Taiwan is 36,188 sq km in area, measuring 394 km from north to south and, at its widest point, 144 km from east to west. The island straddles the Tropic of Cancer. The Central Range of mountains occupies almost 50% of the island, extending 270 km from north to south. At 3,952 m, Mount Jade is the island's highest point. Owing to the mountainous character of the relief, in 2002less than 24% of the land area was cultivated, while forests covered more than 58%. The climate is subtropical in the north and tropical in the south, being strongly modified by oceanic and relief factors. Apart from the mountainous core, winter temperatures average 15°C and summer temperatures about 26°C. Monsoon rains visit the north-east in winter (October to March) but come to the south in summer, and are abundant, the mean annual average rainfall being 2,580 mm. Typhoons are often serious, particularly between July and September, when windward mountain slopes may receive as much as 300 mm of rain within 24 hours.

The population was 22,604,550 at 31 December 2003, giving Taiwan a population density of 624.6 per sq km, one of the highest in the world. In 1964 the rate of natural increase fell below 3.0% for the first time, declining steadily thereafter to stand at 0.51% in 2002. The crude birth rate in 2003 was 10.1 per 1,000 (compared with 44.8 per 1,000 in 1956), and the crude death rate in 2003 was 5.8 per 1,000 (compared with 8.0 per 1,000 in 1956). The death rate is one of the lowest in Asia, as is the infant mortality rate (deaths under one year of age per 1,000 live births), which was estimated at 5.9 in 2000. In 2003 19.83% of Taiwan's population were estimated to be under 15 years of age; 70.94% were aged between 15 and 64; and 9.24% of the population were 65 and over (compared with 2.5% in 1962). With the expansion of industry, Taiwan's population has become increasingly urbanized. Between 1966 and 2002 the proportion living in towns of 100,000 or more inhabitants increased from 31.0% to 60.6%.

History

HISTORICAL BACKGROUND

The geographic location of the island of Taiwan has determined its history. Situated between the Malay archipelago, China and Japan, the island has had an eventful past. The original inhabitants were tribes of Malayan origin. China's relations with the island date from AD 607, but the first small Chinese settlements were not established there until the 14th century. During the 17th century Portuguese, Spanish and Dutch traders visited the island from time to time. In 1624 the Dutch settled the southern part. Two years later came the Spanish, who occupied the northern part. In 1642 the Dutch expelled the Spanish, and in 1661 the Dutch were, in turn, driven out by the Chinese Ming loyalist Zheng Zheng Gong (Coxinga) who ruled, with his sons, for 22 years. In 1663 the Qing emperor, Kang Xi, invaded and conquered the island, which became a part of his empire until it was ceded to the Japanese at the end of the Sino-Japanese war of 1895. During the period of independence and Qing rule, massive immigration from the mainland established the ethnic Chinese character of the island.

Following Japan's defeat in the Second World War, Taiwan became one of the provinces of the Republic of China, ruled by the Kuomintang (KMT, Nationalist Party). The leader of the KMT was Gen. Chiang Kai-shek, President of the Republic since 1928. Virtually all of Taiwan's exportable surpluses went to the Chinese mainland. In 1947 misgovernment by mainland officials led to a large-scale, but peaceful, political uprising, which was repressed with great brutality. In early 1949 the KMT regime, driven from the mainland by the Communists, moved to Taiwan's capital, Taipei, along with approximately 2m. soldiers, officials and their dependants. Thus, the island's population increased from 6.8m. to 7.5m. in 1950, excluding military personnel numbering 600,000. The KMT regime continued to assert that it was the rightful Chinese Government, in opposition to the People's Republic of China (proclaimed by the victorious Communists in 1949), and declared its intention to recover control of the mainland from the Communists. (For further details of events prior to 1949, see the chapter on the People's Republic of China.)

THE ESTABLISHMENT OF CHINESE NATIONALIST RULE AND SUBSEQUENT FOREIGN RELATIONS

Although its effective control was limited to Taiwan, the KMT regime continued to be dominated by politicians who had formerly been in power on the mainland. In support of the regime's claim to be the legitimate government of all China, Taiwan's legislative bodies were filled mainly by surviving mainland members, and the representatives of the island's native Taiwanese majority occupied only a minority of seats. Unable to replenish their mainland representation, the National Assembly (last elected fully in 1947) and other organs extended their terms of office indefinitely, although fewer than one-half of the original members were alive on Taiwan by the 1980s. While it promised eventually to reconquer the mainland, the KMT regime was largely preoccupied with ensuring its own survival, and promoting economic development, on Taiwan. Under the KMT Government, Taiwan achieved a remarkable record of economic growth, although the island's regeneration in the period after 1949 was assisted by massive US aid. The years 1951 and 1952 were notable for government reorganization. There followed a four-year (1953–56) period of adjustment and planning. In 1957 and 1958 the cumulative effect of domestic reform and US aid brought a great improvement in economic and other fields. The prominent developments of this period include the land reform programme, the rapid development of industry and the establishment of a system of nine-year free public education. The political domination of the island by immigrants from the mainland caused resentment among Taiwanese, and led to demands for increased democratization and for the recognition of Taiwan as a state independent of China. The KMT, however, consistently rejected demands for independence, constantly restating the party's long-standing policy of seeking political reunification, under KMT terms, with the mainland.

Membership of International Organizations and Diplomatic Recognition

The KMT Government continued to represent China at the UN (and as a permanent member of the UN Security Council) until October 1971, when the People's Republic of China was admitted in place of Taiwan. Nationalist China was subsequently expelled from several other international organizations. In November 1991, however, as 'Chinese Taipei', Taiwan joined the Asia-Pacific Economic Co-operation forum (APEC). As 'Taipei, China', Taiwan also remained a member of the Asian Development Bank (ADB). In September 1992, under the name of the 'Separate Customs Territory of Taiwan, P'enghu, Kinmen and Matsu', Taiwan was granted observer status at the General Agreement on Tariffs and Trade (GATT), and discussions on its application for full membership of the successor World Trade Organization (WTO) followed. Taiwan was finally approved for membership of the body in late September 2001, after 11 years of bidding, and on 1 January 2002 Taiwan formally became a member of the WTO. In June 1995, meanwhile, Taiwan offered to make a donation of US $1,000m., to be used for the establishment of an international development fund, if the island were permitted to rejoin the UN. In September 2003, for the 11th consecutive year, the General Committee of the UN General Assembly rejected a proposal urging Taiwan's participation in the UN. However, Taiwan's leaders pledged to continue the island's campaign to gain re-entry. In April 1997 the Minister of Foreign Affairs announced that Taiwan had applied to be an observer of the World Health Organization (WHO) and would be seeking similar status in other UN agencies. Its application was unsuccessful, however, as was a second attempt to gain observer status in May 1998 and subsequent annual attempts. In May 2003 the People's Republic of China strongly opposed a US proposal that Taiwan be permitted to participate in WHO.

After 1971 a number of countries broke off diplomatic relations with Taiwan and recognized the People's Republic. The British Government established full diplomatic relations with Beijing in March 1972. When Tokyo sought a *rapprochement* with Beijing in September 1972, Taiwan broke off diplomatic relations with Japan. In August 1992 the Republic of Korea accorded recognition to the People's Republic of China, but unofficial links were to be maintained. In July 2002 Taiwan broke off diplomatic relations with the Pacific nation of Nauru. In October 2003, following the withdrawal of Liberia's diplomatic recognition, the Government of Taiwan maintained formal diplomatic relations with only 26 countries. In November 2003 diplomatic recognition of Taiwan by Kiribati brought the number to 27; however, the figure decreased again to 26 following withdrawal of diplomatic recognition by Dominica in March 2004. Nevertheless, Taiwan's commercial relations with numerous countries continued to flourish.

Relations with the USA

In 1954 the USA, which refused to recognize the People's Republic of China, signed a mutual security treaty with the KMT Government, pledging to protect Taiwan and the Pescadores. In 1955 the islands of Kinmen (Quemoy) and Matsu, lying just off shore from the mainland, were included in the protected area. However, US documents declassified in 2002 revealed that in 1971 the USA had secretly pledged to China that it would not support Taiwan's independence, in exchange for Chinese assistance in ending the Viet Nam War. This had marked the beginning of a US shift away from Taiwan in favour of the People's Republic. In February 1973 the US Government announced that it would continue to maintain diplomatic relations with Taiwan but, at the same time, would set up an 'American mission' in Beijing and allow a 'Chinese liaison office' to open in Washington, DC. In December 1978, however, prior to a visit to Washington by Deng Xiaoping, the leader of the People's Republic, there was a dramatic change in US policy towards Taiwan. Diplomatic recognition was withdrawn from Taiwan as constituting the Republic of China, and the US embassy was closed. However, Taiwan, as an unrecognized but existing state, remained assured of continuing US protection and trade until some form of reconciliation between China and Taiwan could be brought about. China abandoned its denunciations of the Taiwan leadership, and, instead, offered suggestions for forms of autonomy for Taiwan within the People's Republic. These approaches were rejected by Taiwan, where public indignation was manifested against US policy. The USA also terminated the mutual security treaty with Taiwan. Commercial links were to be maintained. Taiwan's purchase of armaments from the USA remained a controversial issue. In August 1982 a joint Sino-US communiqué was published, in which the USA pledged to reduce gradually its sale of armaments to Taiwan. In April 1984 the US President, Ronald Reagan, gave an assurance that he would continue to support Taiwan, despite the improved relations between the USA and the People's Republic. In mid-1989 US relations with the People's Republic deteriorated sharply, following the Tiananmen Square massacre in Beijing. In September 1992 President George Bush announced the sale of up to 150 F-16 fighter aircraft to Taiwan. The announcement was condemned by the People's Republic. In December Carla Hills, the US trade representative, became the first senior US government official to visit the island since the severance of diplomatic relations in early 1979.

In September 1994 the USA announced a modification of its policy towards Taiwan, henceforth permitting senior-level bilateral meetings to be held in US government offices. In December the US Secretary of Transportation visited the Ministry of Foreign Affairs in Taipei, the first US official of cabinet rank to visit Taiwan for more than 15 years. In June 1995 President Lee Teng-hui was permitted to make a four-day unofficial visit to the USA, where he gave a speech at Cornell University, his alma mater, and met members of the US Congress. This highly significant visit by the Taiwanese Head of State provoked outrage in Beijing, and the Chinese ambassador to Washington was recalled. In January 1996 the Taiwanese Vice-President was granted a transit visa permitting him to disembark in the USA en route to Guatemala. In the following month he was accorded a similar privilege while travelling to Haiti and from El Salvador. In August the new Vice-President received a transit visa enabling him to spend two days in the USA, en route to the Dominican Republic, again arousing disapproval in Beijing. He was granted a similar concession in January 1997, on a journey to Nicaragua, and again in May 1998, en route to visit Taiwan's Central American allies.

In March 1996, as the mainland began a series of missile tests off the Taiwanese coast (see below), the USA stationed two naval convoys in waters east of the island, representing the largest US deployment in Asia since 1975. US President Clinton agreed to the sale of Stinger anti-aircraft missiles and other defensive weapons. In September 1996 President Lee and the US Deputy Treasury Secretary, Lawrence Summers, met in Taipei for discussions, the most senior-level contact between the two sides since 1994. In November 1996 Taiwan welcomed the US State Department's pledge to continue to sell defensive weapons to the island. Some controversy arose in late 1996, however, when irregularities in the financing of President Clinton's re-election campaign, involving Taiwanese donors, were reported. Furthermore, it was alleged that a senior KMT official had offered an illicit contribution of US $15m. to the Democratic Party in Washington. In early 1997 the first of the Patriot anti-missile air defence systems, purchased from the USA under an arrangement made in 1993, were reported to have been deployed on Taiwan. The first of the F-16s were delivered to the island in April 1997. The Taiwanese Government expressed satisfaction at the USA's continued commitment to Taiwan's security, confirmed following the visit to the USA of the President of the People's Republic, Jiang Zemin, in October, and again in June 1998, during President Clinton's visit to the People's Republic. However, a statement by Clinton affirming that the USA would not support Taiwan's membership of the UN was sharply criticized in Taiwan. In late October 1998 the Taiwanese Chief of Staff, Gen. Tang Fei, made a secret two-week visit to the USA. This was regarded as an extremely sensitive matter, in view of the fact that Koo Chen-fu (see below) had recently met with Jiang Zemin. In the following month the People's Republic complained to the USA following the US Energy Secretary's visit to Taiwan. In January 1999 the People's Republic was angered by Taiwan's proposed inclusion in the US-led Theater Missile Defence (TMD) anti-missile system. Tensions continued throughout the year, and in August the USA reaffirmed its commitment to defend Taiwan against Chinese military action. In the following month, however, the USA again refused to support Taiwan's application for UN membership. In October of that year the island welcomed the adoption, albeit in modified form, of the Taiwan Security Enhancement Act (TSEA), establishing direct military links, by the International Relations Committee of the US House of Representatives, despite opposition from the Clinton Administration. Reaction from the People's Republic was unfavourable, and its displeasure increased in early February 2000 when the House of Representatives overwhelmingly approved the Act. In April of that year the US Senate postponed consideration of the Act at the behest of the Taiwanese President-elect, Chen Shui-bian, in order not to antagonize Beijing during the sensitive period preceding his inauguration. In the same month the US Government announced that it had decided to defer the sale of four naval destroyers to Taiwan, although it was prepared to supply long-range radar and medium-range air-to-air missiles. In September the USA granted a transit visa to Taiwanese Vice-President Annette Lu to stay in New York en route to Central America. Lu subsequently declared that the stopover had marked a breakthrough in talks between the USA and Taiwan. In October China was angered by a resolution passed by the US Congress supporting Taiwan's participation in the UN and other international organizations.

The election of George W. Bush to the US Presidency in late 2000 was widely expected to boost US-Taiwan relations at the expense of US relations with the mainland, since Bush had used uncompromising rhetoric against the latter in his election campaign. This became more apparent after the crisis over the detention of a US spy plane and its crew following its collision with a Chinese fighter plane on 1 April 2001 (see the chapter on the People's Republic of China). Following that incident, in late April the USA agreed to sell Taiwan US $4,000m.-worth of arms consisting of Kidd-class navy destroyers, Orion anti-submarine aircraft, diesel submarines, amphibious assault vehicles, and surface-to-air missiles and torpedoes, all of which would bolster the island's defences. However, the USA stopped short of selling Taiwan the advanced Aegis combat-radar system, for fear of provoking Beijing. Plans by the USA to sell diesel submarines to Taiwan were blocked by Germany and the Netherlands in May 2001. The two countries, being the main owners of the design technology, were unwilling to raise tensions with Beijing by approving such a sale. At the same time President Bush said that the US would do whatever was necessary to defend Taiwan, in the event of an invasion by the mainland. Beijing was further angered by the visit of President Chen Shui-bian to the USA in late May 2001, when he met business leaders and members of the US Congress. Chen's visit was followed by that of his predecessor, Lee Teng-hui, in late June, as well as by a group of Taiwanese military and intelligence officials on an exchange programme, the first such exchange since 1979. Also in June 2001, Taiwan successfully tested its US-made 'Patriot' air defence missiles for the first time. In September an Australian firm emerged as the most likely supplier of diesel submarines, when it was rumoured that a US defence company would take a 40% stake in the former.

In late August 2001 two US Navy aircraft-carrier battle groups staged a one-day exercise in the South China Sea coinciding with Chinese military exercises in the Taiwan Straits, a further reminder that the USA was committed to protecting Taiwan. In October the Minister of National Defence, Wu Shih-wen, began finalizing the purchase of the naval destroyers offered earlier in the year, and the USA also offered Taiwan anti-tank missiles. In December the US House of Representatives approved the 2002 Defense Authorization Act, which included weapons sales to Taiwan and the promise of US help in acquiring submarines.

In January 2002 the Bush Administration rejected demands by a former US State Department official, Richard Holbrooke, that a fourth communiqué on US-Taiwan relations was needed, stating that the existing 1982 communiqué was satisfactory. At the same time, a delegation from a US 'think tank', consisting of retired generals and officials, visited the mainland, and subsequently Taiwan, where they met President Chen and other senior officials and discussed the island's security. Also in January, Vice-President Annette Lu made a brief stopover in New York, en route to South America, and former President Lee Teng-hui announced that he planned to visit the USA in May to raise Taiwan's profile. During his visit to Beijing in late February 2002, President Bush pledged to adhere to the 1979 Taiwan Relations Act—the first time a US President had stated this in China itself.

There were indications that the USA's long-standing 'strategic ambiguity' regarding Taiwan was gradually coming to an end in early 2002. In mid-March the Minister of National Defense, Gen. (retd) Tang Yao-ming visited the USA to attend a private three-day defence and security conference in Florida, where he met the US Deputy Secretary of Defense, Paul Wolfowitz, and Assistant Secretary of State James Kelly. During the conference, Wolfowitz reportedly stated that the USA would assist in training Taiwan's military in areas of command and doctrine. Several days later the Asia-Pacific Center for Security Studies, a 'think tank' operating under the US Navy's Pacific Command, issued invitations for Taiwanese military personnel to attend a 12-week course on security issues—another indication of the greater access being given to the Taiwanese military by the USA. Tang was the first incumbent to make a non-transit visit to the USA since 1979, and the event emphasized the increasingly important security links between the two sides. China condemned Tang's visit as interference in its affairs and responded by denying permission for a US warship to visit Hong Kong in April. Meanwhile, the scandal in late March concerning former President Lee Teng-hui's co-operation with the island's intelligence services in secretly establishing an unauthorized fund (see below) embarrassed Taiwan and several US lobbying groups, which had received sums of this money. Lee subsequently postponed his planned visit to the USA. In mid-April, however, the US Under-Secretary of Commerce for International Trade, Grant Aldonas, became the highest-ranking official of George W. Bush's Administration to visit Taiwan, where he met Chen Shui-bian. The latter proposed establishing a free-trade pact with the USA and Japan.

In June 2002 it became known that a US defence contractor would probably build eight new diesel-electric submarines for Taiwan's navy, having purchased a share of a German company that produced the relevant designs; a high-level US delegation visited Taiwan in late July to discuss these arrangements. Also in July, the USA considered accelerating the transfer of advanced air-to-air missiles purchased by Taiwan but undelivered, and a US Department of Defense report warned that preparations for a conflict with Taiwan was the main factor behind Beijing's military build-up. However, the USA refrained from commenting on Chen's mention in early August of a possible referendum to determine the island's future. Premier Yu Shyi-kun and the Chairwoman of the Mainland Affairs Council, Tsai Ing-wen, made a brief stopover in New York at the same time, their visits being denounced by Beijing.

In early September 2002 a Vice-Minister of National Defense, Kang Ning-hsiang, and the Vice-Commander of the Navy both visited Washington, DC, to discuss the planned acquisition of four new Kidd-class destroyers. However, the high costs of the vessels jeopardized their purchase, at a time when Taiwan's defence budget was being reduced. Later in the month Wu Shu-chen, the wife of President Chen, began a private 10-day visit to the USA, seeking to raise the island's profile.

Meanwhile, reflecting the occasional political aspects of commercial relations with the USA, the Taiwanese Government in September 2002 urged the state-owned China Airlines (CAL) to place a US $2,000m. order for US-manufactured Boeing aircraft for its fleet, rather than the rival European-manufactured Airbus. However, in October CAL awarded the contract to Airbus, while also purchasing 10 Boeing aircraft for cargo and long-haul routes.

In September 2002 the US Congress approved the Foreign Relations Authorization Act, Fiscal Year 2003, which for the first time allowed US State Department and other government agencies' staff to work at the American Institute in Taiwan, the unofficial US mission to the island. Previously, US officials were required to embark upon sabbatical leave before serving at the institute. US military officers would also be allowed to work there, albeit not in uniform. The bill also recognized Taiwan as a major non-NATO ally of the USA, and served to strengthen US-Taiwanese relations.

President Chen in early January 2003 noted that only nine members of the US Congress had visited Taiwan during 2002, the lowest number in 20 years, and far fewer than the number visiting mainland China. Also in January 2003, the USA was considering participating in Taiwan's annual military exercises. In mid-February the Vice-Minister of National Defense, Chen Chao-min (who had succeeded Kang), attended a defence industry meeting in San Antonio, Texas, which was also attended by several US defence officials. In March 2003 a US delegation, led by Pentagon official Mary Tighe, visited Taipei to assess Taiwan's defence capability in the event of a Chinese missile attack. In mid-May 2003 the first round of Taiwan's 19th Hankuang military exercises took place. The US delegation sent to observe these exercises was reported to have been the largest ever, and included senior officials from the US Department of Defense.

In August 2003 plans for Taiwan's purchase of four new Kidd-class destroyers from the USA, first discussed in September 2002, were finalized. Taiwan also made an advance payment towards the acquisition of eight new diesel submarines from the USA, which had been agreed in June 2002. Also in August, again reflecting the political influences on Taiwan's trade relations with the USA, CAL purchased 12 aircraft engines from US company General Electric, rejecting an alternative purchase from European company Rolls-Royce. At the end of August the USA expressed concern that China's increased military capability could pose a threat to Taiwan and reaffirmed US willingness to sell weapons to Taipei. At the same time, the Taiwanese Ministry of National Defense stated that its budget for 2005 would include special funds for the purchase of US-made Patriot PAC-3 anti-missile weaponry. Taiwan conducted further large-scale military exercises in early September, involving the use of US-made Harpoon missiles and F-16 fighter jets.

In November 2003 President Chen made a visit to New York, in the course of which he was permitted to make a public speech. However, during a visit by Chinese Premier Wen Jiabao to Washington, DC, in December the Bush Administration appeared to support China's opposition to referendum plans in Taiwan, reaffirming its 'one China' policy and opposing any unilateral decision by either China or Taiwan to change the status quo. In January 2004 President Chen attempted to placate US concerns over his proposed referendum on Chinese missiles by suggesting that the referendum might be presented as an opportunity to reduce tensions between Taiwan and mainland China (see below). In April Therese Shaheen, head of the American Institute in Taiwan (the unofficial US mission to Taiwan), resigned, after reportedly having misrepresented US policy on Taiwanese independence. Her departure prompted the resignation of Taiwan's Minister of Foreign Affairs, Eugene Y. H. Chien. In June 2004 the Executive Yuan approved a special defence budget for the purchase of weapons from the USA, which had been agreed in April 2001 (see above).

Relations with Japan

In February 1993, for the first time in two decades, the Taiwanese Minister of Foreign Affairs paid a visit to Japan, leading to a strong protest from the People's Republic. In September 1994 pressure from Beijing resulted in the withdrawal of President Lee's invitation to attend the forthcoming Asian Games in Hiroshima. Instead, however, the Taiwanese Vice-Premier was permitted to visit Japan. Similarly, in July 1995 Japan announced that the Taiwanese Vice-Premier would not be permitted to attend a meeting of APEC members to be held in Osaka in November. Instead, President Lee was represented by Koo Chen-fu, Chairman of the Straits Exchange Foundation. (The latter also attended the APEC meeting in the Philippines in November 1996.) At the APEC summit meeting in Kuala Lumpur in November 1998, President Lee was represented by a Minister without Portfolio, Chiang Ping-kun, who also attended the 1999 APEC conference, held in September in Auckland, New Zealand. Perng Fai-nan, the Governor of the Central Bank, represented President Chen Shui-bian at the APEC meeting in Brunei in November 2000.

In 1996 Taiwan's relations with Japan continued to be strained by the issue of adequate compensation for the thousands of Asian (mostly Korean) women used by Japanese troops for sexual purposes during the Second World War. In October Taiwan rejected a Japanese offer of nominal compensation for Taiwanese women. Relations deteriorated further in 1996 on account of a dispute relating to a group of uninhabited islets in the East China Sea: known as the Tiaoyutai (Diaoyu Dao) in Chinese, or Senkaku in Japanese, and situated about 200 km north-east of Taiwan and 300 km west of the Japanese island of Okinawa, the islands were claimed by Taiwan, China and Japan. In July, following the construction of a lighthouse on one of the islands by a Japanese right-wing group, the Taiwanese Ministry of Foreign Affairs lodged a strong protest over Japan's decision to incorporate the islands within its 200-mile (370-km) exclusive economic zone. Taiwan continued to urge that the dispute be settled by peaceful means. In early October further discussions with Japan on the question of Taiwanese fishing rights within the disputed waters ended without agreement. In the same month a flotilla of small boats, operated by activists from Taiwan, Hong Kong and Macao, succeeded in evading Japanese patrol vessels. Having reached the disputed islands, protesters raised the flags of Taiwan and of China. In May 1997 the Taiwanese Minister of Foreign Affairs expressed grave concern, following the landing and planting of their national flag on one of the disputed islands by a Japanese politician and three aides. A flotilla of 20 ships carrying about 200 protesters and journalists from Taiwan and Hong Kong set sail from the port of Shenao, ostensibly to participate in an international fishing contest. The boats were intercepted by Japanese coastguard vessels and failed to gain access to the islands. In September 1997 an attempted parachute landing by Taiwanese activists also ended in failure. Reports in October that Japanese patrol boats were forcibly intercepting Taiwanese fishing vessels were a further cause for concern for the Taiwanese authorities.

There was a significant development in relations between the two countries in November 1999, when the Governor of Tokyo paid an official visit to Taiwan. He was the most senior Japanese official to visit the island since the severing of diplomatic relations. The People's Republic condemned the visit, claiming that it undermined Sino-Japanese relations. Former President of Taiwan Lee Teng-hui made a private visit to Japan in late April 2001, ostensibly for medical treatment. In order to minimize tensions with Beijing, the Japanese authorities forbade Lee from making political statements while in Japan. In March 2002 it was revealed that Japanese politicians, including former Prime Minister Ryutaro Hashimoto, had received money from an unauthorized fund established by Lee in order to procure influence in Japan. However, the incident did not damage bilateral relations, and in late 2001–early 2002 officials from both sides were investigating the possibility of establishing a free-trade agreement. In November 2002 former President Lee Teng-hui was denied a visa to visit Japan to make a speech at Keio University, reportedly owing to Chinese pressure. In January 2003 Taiwan, along with China, condemned Japan's moves to assert sovereignty over the Tiaoyutai islands in the East China Sea by renting them out to private companies. In May Japan's Minister of Foreign Affairs allegedly told a Chinese official that Japan supported Taiwan's bid to join WHO with observer status. Discussions on a free-trade agreement between Japan and Taiwan remained unresolved in mid-2003; this was thought to have been due partly to China's opposition to such an agreement. In September 2003 the second Taiwan-Japan Forum opened in Tokyo, at which issues of mutual concern were discussed.

Other Foreign Relations

In March 1989, meanwhile, President Lee paid a state visit to Singapore, the first official visit overseas by a President of Taiwan for 12 years. In a further attempt to end diplomatic isolation, Lee said that he would visit any foreign country, even if it maintained diplomatic relations with Beijing. In January 1993 official confirmation of Taiwan's purchase of 60 Mirage fighter aircraft from France provoked strong protest from Beijing. In January 1994 Taiwan suffered a reverse when (following pressure from the People's Republic) France recognized Taiwan as an integral part of Chinese territory and agreed not to sell weapons to the island. In February President Lee embarked upon an eight-day tour of South-East Asia. His itinerary incorporated the Philippines, Indonesia and Thailand, all three of which maintained diplomatic relations with the People's Republic of China. Although the tour was described as informal, President Lee had meetings with the three heads of state, leading to protests from Beijing. In May the Taiwanese President visited Nicaragua, Costa Rica, South Africa (the island's only remaining major diplomatic ally) and Swaziland. In March 1995 it was reported that Taiwan was to purchase shoulder-fired anti-aircraft missiles from a French company, despite France's previous assurances to the People's Republic that it would not sell weapons to Taiwan (the purchase was confirmed in 1996). In April President Lee travelled to the United Arab Emirates and to Jordan. Although accompanied by senior members of the Executive Yuan, the visits were described as private (Taiwan having no diplomatic relations with these countries). In June the Taiwanese Premier visited Austria, Hungary and the Czech Republic, where he had private meetings with his Czech counterpart and with President Václav Havel. Again, the visits provoked strong protest from China.

In August 1996, as Beijing continued to urge Pretoria to sever its diplomatic links with Taipei, Vice-Premier Hsu Li-teh led a delegation of government and business representatives to South Africa. In September, however, the Taiwanese Minister of Foreign Affairs was obliged to curtail an ostensibly private visit to Jakarta (where he was reported to have had discussions with his Indonesian counterpart), following protests from the People's Republic of China. Also in 1996, the first delivery of Mirage aircraft to Taiwan from France (first announced in 1993) took place. At the same time, it was confirmed that France would sell six Lafayette-class frigates to Taiwan (delivery of these was completed in March 1998). In January 1997 the Vice-President was received by the Pope during a visit to the Holy See, the only European state that continued to recognize Taiwan. In the same month the Minister of Foreign Affairs embarked upon a tour of seven African nations, in order to consolidate relations. His itinerary included South Africa, despite that country's recent announcement of its intention to sever diplomatic relations with Taiwan (a major set-back to the island's campaign to gain wider international recognition). In March 1997 a six-day visit to Taiwan by the Dalai Lama was strongly condemned by the People's Republic of China, which denounced a meeting between President Lee and the exiled spiritual leader of Tibet as a 'collusion of splittists'. A second visit by the Dalai Lama, scheduled for July 1998, was postponed, following mainland China's criticism of the opening of a representative office of the Dalai Lama's religious foundation in Taiwan.

In July 1997, following the Bahamas' withdrawal of recognition from Taiwan and establishment of diplomatic relations with the People's Republic of China, the Taiwanese Minister of Foreign Affairs undertook an extensive tour of the countries of Central America and the Caribbean, in an effort to maintain their support. In September, during a tour of Central America (the six nations of the region having become the core of Taiwan's remaining diplomatic allies), President Lee attended an international conference on the development of the Panama Canal. The USA granted a transit visa to the Taiwanese President, enabling him to stop over in Hawaii en route to and from Central America. A visit to Europe in October by Vice-President Lien Chan was curtailed when pressure from Beijing forced the Spanish Government to withdraw an invitation. The Malaysian and Singaporean Prime Ministers met their Taiwanese counterpart in Taiwan in November, on their return from the APEC forum in Canada. The People's Republic expressed concern at the meetings.

In January 1998 the Taiwanese Premier, Vincent Siew, met senior officials during a visit to the Philippines and Singapore. It was believed that discussions had focused on the possibility of Taiwan extending economic assistance to those countries. Moreover, the Taiwanese Government expressed its intention to pursue the creation of a multilateral Asian fund, under the auspices of APEC, to support the ailing Asian economies. Relations with Taiwan's Asian neighbours were further strengthened in February, when a leading member of the KMT visited the Republic of Korea, again, it was understood, to discuss financial assistance in the wake of the Asian economic crisis. Negotiations between the Malaysian Deputy Prime Minister and Finance Minister and the Taiwanese Premier, held in Taiwan in February, also concentrated on economic and financial issues. Vincent Siew made a return visit to Malaysia in April. China was highly critical of these visits, accusing Taiwan of seeking to gain political advantage from the regional economic crisis. At the APEC meeting held in Kuala Lumpur, Malaysia, in November 1998 President Lee was represented by the Minister without Portfolio, Chiang Ping-kun, who also attended the 1999 APEC conference, held in September in Auckland, New Zealand.

In August 1999 the Taiwanese Government threatened to refuse entry to the island to Philippine labourers in retaliation for Manila's unilateral termination of its aviation agreement with Taipei. In October the Philippine authorities decided to close the country to Taiwanese aircraft. Access was, however, temporarily granted in November during talks on the renewal of the agreement. Negotiations faltered in December, but reached a successful conclusion in early 2000, thus permitting flights to the Philippines to resume. In March, however, the agreement collapsed, and flights were suspended again. In June the Taiwanese Government imposed a three-month ban on new work permits for Philippine nationals, citing interference by the Philippines' representative office in Taiwan concerning labour disputes. Any connection with the aviation dispute was denied. In September a new aviation agreement, enabling flights between Taipei and Manila to restart, was signed.

Following the withdrawal of diplomatic recognition by South Africa, in 1998 Taiwan made extensive efforts to maintain relations with the island's other allies. The Minister of Foreign Affairs visited eight African countries in February, and in April it was announced that Taiwan's overseas aid budget was to be substantially increased in an attempt to retain diplomatic support. In May the Vice-President, Lien Chan, visited Taiwan's Central American and Caribbean allies. China expressed serious concern in November at New Zealand's decision to grant Taiwanese officials in Wellington privileges accorded to accredited diplomats. Taiwan's strong economic position was a deciding factor in its diplomatic fortunes in late 1998 and in 1999. Following the loss of four allies during 1998, relations were established with the Marshall Islands in November of that year, and with the former Yugoslav republic of Macedonia in January 1999. The latter rapidly benefited from its contacts with Taiwan, receiving extensive aid following the Kosovo conflict. The Prime Minister of Papua New Guinea initiated diplomatic relations with Taiwan in early July 1999. Following his resignation later that month, recognition was withdrawn from Taiwan, allegedly because of a failure to adhere to correct procedures for the establishment of relations. Full diplomatic relations were established with Palau in December of that year, and Taiwan opened an embassy on the island in March 2000. In August 2000 the new Taiwanese President, Chen Shui-bian, embarked upon a tour of diplomatic allies in Central America and West Africa. In October Taiwan was concerned that it might lose a diplomatic ally when the Minister of Foreign Affairs of Solomon Islands unexpectedly cancelled a visit to Taipei and travelled instead to Beijing. The premier of Solomon Islands did not attend a regional forum, instead visiting Taipei to make amends and to reiterate his commitment to maintaining relations. There was speculation that this commitment arose from a need for substantial financial assistance from Taiwan.

In November 2000, after eight years' suspension, the 25th Joint Conference of Korea-Taiwan Business Councils took place in Seoul. It was agreed that henceforth conferences would be held annually, alternately in Taipei and Seoul, and Taiwan ultimately hoped for a resumption of ministerial-level discussions on the establishment of bilateral air links. In the same month the European Parliament passed a resolution on strengthening relations with Taiwan, and in December Taiwan and Egypt agreed to exchange representative offices. Also in December, France refused to grant a visa to the Taiwanese Minister of Justice.

Throughout 2001 Taiwan continued to seek a higher diplomatic profile. President Chen in late May began a tour of five Latin American nations—El Salvador, Guatemala, Panama, Paraguay, and Honduras—as part of Taiwan's 'dollar diplomacy'—a practice of giving aid and bringing investment in return for diplomatic recognition. While in El Salvador, Chen met eight regional leaders, who pledged support for Taiwan. Both Taiwan and the People's Republic of China were increasingly bidding for support in this region to boost their overall global standing. Taiwan was also in competition with China for support from Pacific island nations, five of which (Marshall Islands, Nauru, Palau, Solomon Islands, and Tuvalu) recognized Taiwan. Two of these nations, Solomon Islands and the Marshall Islands, were thought to be wavering in their support of Taiwan. The Dalai Lama visited Taiwan in March–April 2001 and held talks with President Chen, much to the anger of Beijing. In June the former Yugoslav republic of Macedonia announced that it would recognize the People's Republic of China, thus leading to a break in relations with Taiwan. In November Wu Shu-chen, wife of President Chen, travelled to Strasbourg, France, to accept the 'Prize for Freedom' awarded to her husband by Liberal International, a world grouping of liberal parties. Chen himself had been refused a visa by the European Union (EU).

In January 2002 it was reported that Taiwan had secretly been developing military and intelligence links with India, through mutual co-operation including bilateral visits of military personnel and the exchange of intelligence data. Also in January, Vice-President Lu visited Nicaragua and Paraguay, having visited The Gambia in December 2001. Vice-President Lu also visited Indonesia in mid-August 2002 and met several ministers, but not President Megawati Sukarnoputri, owing to pressure on the latter from Beijing. While in Indonesia, Lu discussed possible liquefied natural gas projects, investment and migrant labour. Taiwanese companies had in previous years invested US $17,000m. in Indonesia, the home of some 100,000 of Taiwan's migrant workers. President Chen completed a four-nation tour of Africa in early July 2002, having visited Senegal, São Tomé and Príncipe, Malawi and Swaziland. While en route to Swaziland, he was refused permission to land in South Africa, Pretoria being concerned not to offend Beijing. Diplomatic relations between Taiwan and Nauru were terminated in late July 2002 after the latter established relations with the People's Republic of China. In September 2002 Taiwan announced the opening of a trade and economic affairs representative office in Mongolia. In late 2002 Taiwan and the Democratic People's Republic of Korea (North Korea) were planning to open economic liaison offices in the respective capitals, the latter keen to attract new sources of investment. President Chen was forced to cancel a trip to Indonesia in December after the South-East Asian country came under heavy pressure from Beijing.

In March 2003 President Chen was again refused a visa by the EU, and was thus unable to address members of the European Parliament in Brussels. Also in March, the Taiwanese Government issued a warning that several Pakistani citizens with links to terrorist networks might be attempting to enter Taiwan. Taiwan expressed concern over the political situation in the West African state of São Tomé and Príncipe following a military coup there in in July 2003. There were also concerns in mid-2003 that the violent conflict in Liberia might affect that nation's diplomatic relations with Taiwan, and in October Liberia transferred its recognition to the People's Republic of China, in the hope of obtaining reconstruction aid from Beijing. In August, meanwhile, Taiwan signed a free-trade agreement with Panama, which was expected to take effect in January 2004. In November 2003 Kiribati granted diplomatic recognition to Taiwan, and a Taiwanese embassy opened on the Pacific island in January 2004. Also in January, there were tensions in Taiwan's relations with France after French President Jacques Chirac expressed his support for China's opposition to a proposed referendum in Taiwan (see above) during a visit by Chinese President Hu Jintao to Paris. President Chirac declared that the holding of a referendum in Taiwan on the issue of Chinese missiles would be a 'grave mistake'. Following President Chirac's remarks, two state visits by Taiwanese ministers to France were cancelled. In March 2004 Dominica withdrew its diplomatic recognition of Taiwan (see above).

South China Sea Dispute

The question of the sovereignty of the Spratly Islands, situated in the South China Sea and believed to possess petroleum and natural gas resources, to which Taiwan and five other countries laid claim, remained unresolved in mid-2003. A contingent of Taiwanese marines is maintained on Taiping Island, the largest of the disputed islands, located some 1,574 km south-west of Taiwan. In August 1993 Taiwan announced its intention to construct an airbase on Taiping Island, but in January 1996 the scheme was postponed. A satellite telecommunications link between Taiping and Kaohsiung was inaugurated in October 1995. In late December 1998 the Legislative Yuan approved the first legal definition of Taiwan's sea borders. The Spratly Islands were claimed, as were the disputed Tiaoyutai Islands, within the 12- and 24-nautical mile zones. In November 2002, in Phnom-Penh, Cambodia, members of the Association of South East Asian Nations (ASEAN) signed a landmark 'declaration on the conduct of parties in the South China Sea', which aimed to avoid conflict in the area. Under this agreement, which was similar to one drafted in late 1999 but not implemented, claimants would practise self-restraint in the event of potentially hostile action (such as inhabiting the islands), effect confidence-building measures and give advance notice of military exercises in the region. However, the agreement did not include the Paracel Islands, and disagreements continued among the signatories on what the accord should encompass. Additionally, China introduced a provision requiring consensus to resolve outstanding issues, thereby allowing for future indecision among ASEAN members. It was unclear how the declaration affected Taiwan, since it was not a member of ASEAN. In August 2003 Minister of the Interior Yu Cheng-hsien visited Taiping Island and reaffirmed Taiwanese sovereignty.

RELATIONS WITH THE PEOPLE'S REPUBLIC AND REUNIFICATION INITIATIVES

In October 1981 Taiwan rejected the latest in a series of proposals from the People's Republic of China for reunification, whereby Taiwan would become a 'special administrative region' of China and would enjoy a high degree of autonomy, including the retention of its own armed forces. In 1982 the Taiwan Government implied that eventual reunification could be made possible by narrowing the economic gap between the two sides over time, and on the basis of Sun Yat-sen's 'Three Principles of the People'. In 1983 Deng Xiaoping indicated the possibility that, following reunification, Taiwan would retain the right to purchase military equipment from abroad, would be free to export where necessary to sustain its economic growth, would make its own legal decisions, fly its own flag, and issue passports and visas, while the mainland authorities would not send civilian or military personnel to Taiwan. Limitations on this autonomy would be that the People's Republic would speak for China in international affairs, while Taiwan would be designated Chinese Taipei, as agreed for the summer Olympic Games in 1984 and 1988, or China-Taiwan.

In October 1984, following the agreement between the People's Republic of China and the United Kingdom that the former would regain sovereignty over Hong Kong after 1997, Chinese leaders urged Taiwan to accept similar proposals for reunification on the basis of 'one country, two systems'. The KMT Government, however, insisted that Taiwan would never negotiate with Beijing until the mainland regime renounced communism, thus reasserting the rigid and fundamental 'three no's' policy (of 'no compromise, no contact and no negotiation with the mainland') on which its relationship with the People's Republic was based. In May 1986 Taiwan was induced to adopt a more flexible policy after a Taiwanese cargo aircraft was diverted to the mainland by a pilot who wished to defect; representatives from the airlines of the two countries held negotiations in Hong Kong, leading to the return of the aircraft (and two other crew members) to Taiwan. These discussions represented the first-ever direct contact between the two countries (although Taipei insisted that the talks were for humanitarian reasons only and did not indicate a change of policy). In March 1987, however, in accordance with the 'three no's' policy, Taiwan declared the agreement that had been concluded between the People's Republic of China and Portugal, regarding the return of Macao to Chinese sovereignty, to be null and void.

Questions of Taiwan's political evolution and of its future relations with the People's Republic of China were the principal determinants of the pace at which the KMT proceeded with the programme of political reform. While the opposition parties and liberal elements within the KMT generally recognized that Taiwan should adopt a more pragmatic foreign policy if it were to end its diplomatic isolation, the KMT was, until its 13th National Congress in July 1988, dominated by 'conservative' members who had fled the mainland with Chiang Kai-shek in 1947, and who feared that reform would undermine the *raisons d'être* of the KMT: that it constituted 'the legitimate Government of all China and that Taiwan formed a province of this polity', not an independent state. They thus opposed the increasing dominance of the KMT by native Taiwanese.

In October 1987 the Government announced the repeal of the 38-year ban on visits to the mainland by Taiwanese citizens, with the exception of civil servants and military personnel. (The latter regulations were relaxed in November 1998.) In so doing, the Government tacitly recognized that Taiwanese had, for some years, been making illegal visits to the mainland via Hong Kong. Between November 1987 and October 1990, according to the mainland authorities, almost 1.8m. Taiwanese visited (mainly via Hong Kong) the People's Republic. In late 1988 permission was extended to include visits by mainland Chinese to Taiwan for humanitarian purposes. In August 1991 it was announced that restrictions on immigration by ethnic Chinese persons from Hong Kong and Macao were to be relaxed. In 1991 a total of 948,800 Taiwan residents visited the mainland, followed by 1.5m. in 1992. In 1995 almost 1.3m. Taiwanese citizens travelled to the mainland, while visitors from the People's Republic to Taiwan numbered 42,634, the latter figure representing an increase of more than 80% compared with the previous year. A total of 42,491 mainland residents were permitted to visit Taiwan in 1997.

In November 1987 the opposition Democratic Progressive Party (DPP) approved a resolution declaring that Taiwanese had the freedom to demand independence for Taiwan. It was disclaimed, however, that this constituted a pro-independence policy on the part of the DPP. In January 1988 two opposition activists were imprisoned, on charges of sedition, for advocating Taiwanese independence. While there was evidence of a more flexible policy towards the People's Republic at the 13th National Congress of the KMT in July (when the ruling party authorized an increase in indirect imports from the mainland and in direct Taiwanese investment in mainland projects), President Lee's restatement of the party's long-standing policy of seeking reunification under the KMT underlined the continuing necessity to satisfy 'conservative' sentiment within the party.

In April 1989 the Government announced that it was considering a 'one China, two governments' formula for its future relationship with the mainland, whereby China would be a single country under two administrations, one in Beijing and one in Taipei. In May a delegation led by the Minister of Finance attended a meeting of the ADB in Beijing, under the name of Taipei, China. Although the Government stressed that the delegation would not be allowed to hold talks with Chinese government officials, the visit, together with one made by a party of Taiwanese gymnasts a month earlier, represented a considerable relaxation in Taiwan's stance. Reconciliation initiatives were abruptly halted, however, by the violent suppression of the pro-democracy movement in Beijing in June 1989. The actions of the Chinese Government were strongly condemned by Taiwan. Nevertheless, in May 1990 President Lee suggested the opening of direct dialogue on a government-to-government basis with the People's Republic. The proposal, however, was rejected by Beijing, which continued to maintain that it would negotiate only on a party-to-party basis with the KMT.

In October 1990 the National Unification Council, chaired by President Lee, was formed. In the same month the Mainland Affairs Council, comprising heads of government departments and led by the Vice-Premier of the Executive Yuan, was founded. The DPP urged the Government to renounce its claim to sovereignty over mainland China and Mongolia. In November the Straits Exchange Foundation (SEF) was established for the purpose of handling civilian contacts with the People's Republic. In December President Lee announced that Taiwan would formally end the state of war with the mainland; the declaration of emergency was to be rescinded by May 1991, thus opening the way to improved relations with Beijing.

In February 1991 the National Unification Council put forward radical new proposals whereby Taiwan and the People's Republic of China might recognize each other as separate political entities. In March a national unification programme, which incorporated the demand that Taiwan be acknowledged as an independent and equal entity, was approved by the Central Standing Committee of the KMT. The programme also included a proposal for direct postal, commercial and shipping links between Taiwan and the mainland, this suggestion being well received in the People's Republic.

In April 1991 a delegation from the SEF travelled to Beijing for discussions, the first such delegation ever to visit the People's Republic. The talks were reported to have promoted understanding and consensus. In early May it was announced that the large financial rewards hitherto offered to members of the armed forces of the People's Republic who defected to Taiwan would no longer be available. In the following month the Premier of Taiwan reaffirmed that unification with the mainland would be pursued by peaceful and democratic means. Upon the second anniversary of the Tiananmen Square massacre, in June 1991, the Taiwanese authorities urged the Government of the People's Republic to cease its alleged persecution of pro-democracy activists.

In August 1991 a Beijing magazine published an informal 10-point plan for the eventual reunification of China, whereby Taiwan would become a special administrative region and retain its own legislative, administrative and judicial authority. Thus, for the first time, the 'one country, two systems' policy of the People's Republic was clearly stated. In the same month Deng Xiaoping and other senior leaders reportedly offered to travel from the mainland to Taiwan for the purpose of reunification talks.

Two senior envoys of the mainland Chinese Red Cross were allowed to enter Taiwan in August 1991 on a humanitarian mission, the first ever visit by official representatives of the People's Republic of China. As the Beijing Government continued to warn against independence for Taiwan, in September the island's President asserted that conditions were not appropriate for reunification wth the mainland and that Taiwan was a *de facto* sovereign and autonomous country. The President of the People's Republic indicated that force might be used to prevent the separation of Taiwan. In December the non-governmental Association for Relations across the Taiwan Straits (ARATS) was established in Beijing. In January 1992 the SEF protested to the People's Republic over the detention of a former pilot of the mainland air force who had defected to Taiwan in 1965 and, upon returning to his homeland for a family reunion in December 1991, had been arrested. He subsequently received a 15-year prison sentence. In May 1992 the National Unification Council's proposal for a non-aggression pact between Taiwan and the People's Republic was rejected.

In July 1992 the Taiwanese Government reiterated that it would not consider party-to-party talks with Beijing. In the same month President Lee urged the establishment of 'one country, one good system'. In mid-July statutes to permit the further expansion of economic and political links with the People's Republic were adopted by the Legislative Yuan. In August the vice-president of the mainland Red Cross travelled to the island, thus becoming the most senior representative of the People's Republic to visit Taiwan since 1949. Delegates from the SEF and ARATS met in Hong Kong in October 1992 for discussions. The Chairman of the Mainland Affairs Council, however, insisted upon the People's Republic's renunciation of the use of military force prior to any dialogue on the reunification question. At the end of the month the Government of Taiwan announced a further relaxation of restrictions on visits to the mainland by state employees. Upon taking office in February 1993, the new Premier of Taiwan confirmed the continuation of the 'One China' policy.

Historic talks between the Chairmen of the SEF and of the ARATS were held in Singapore in April 1993. Engaging in the highest level of contact since 1949, Taiwan and the People's Republic agreed on the establishment of a formal structure for future negotiations on economic and social issues. Agreements on the verification of official documents and on the registration of mail were also signed. One issue that remained unresolved, however, was that of adequate legal protection for Taiwanese investments in the People's Republic, the rapid increase in capital outflow to the mainland being of growing concern to the island's authorities.

In 1993 divisions between Taiwan's business sector and political groupings (the former advocating much closer links with the People's Republic, the latter urging greater caution) became evident. In January, and again later in the year, the Secretary-General of the SEF resigned, following disagreement with the Mainland Affairs Council.

In August 1993 the People's Republic issued a document entitled *The Taiwan Question and the Reunification of China*, reiterating its claim to sovereignty over the island. Relations were further strained by a series of aircraft hijackings to Taiwan from the mainland. A SEF-ARATS meeting, held in Taiwan in December 1993, attempted to address the issue of the repatriation of hijackers. Incidents of air piracy continued in 1994, prison sentences of up to 13 years being imposed on the hijackers by the Taiwanese authorities.

Further meetings between delegates of the SEF and ARATS were held in early 1994. Relations between Taiwan and the mainland deteriorated sharply in April, however, upon the disclosure of a tragedy in Zhejiang Province in the People's Republic: 24 Taiwanese tourists were among those robbed and killed on board a pleasure boat plying Qiandao Lake. Taiwanese outrage was compounded by the mainland's insensitive handling of the incident. Taiwan suspended all commercial and cultural exchanges with the People's Republic. In June three men were convicted of the murders and promptly executed. In February 1995 compensation totalling 1.2m. yuan was awarded to the victims' families by the People's Republic.

In July 1994 the Taiwanese Government released a White Paper on mainland affairs, urging that the division be acknowledged and the island accepted as a separate political entity. In August the SEF-ARATS talks were resumed when Tang Shubei, Vice-Chairman and Secretary-General of the ARATS, flew to Taipei for four days of discussions with his Taiwanese counterpart, Chiao Jen-ho. Tang thus became the most senior Communist Chinese official ever to visit the island. Although the visit was marred by opposition protesters, the two sides reached tentative agreement on several issues, including the repatriation of hijackers and illegal immigrants from Taiwan to the mainland. Procedures for the settlement of cross-Straits fishing disputes were also established. In mid-November relations were strained once again when, in an apparent accident during a training exercise, Taiwanese anti-aircraft shells landed on a mainland village, injuring several people. Nevertheless, in late November a further round of SEF-ARATS talks took place in Nanjing, at which agreement in principle on the procedure for the repatriation of hijackers and illegal immigrants was confirmed. Further progress was made at meetings in Beijing in January 1995, although no accord was signed.

It was announced in March 1995 that the functions of the SEF were to be enhanced. To improve co-ordination, the SEF board of directors would henceforth include government officials, while meetings of the Mainland Affairs Council would be attended by officials of the SEF. In the same month the Mainland Affairs Council approved a resolution providing for the relaxation of restrictions on visits by mainland officials and civilians.

President Jiang Zemin's Lunar New Year address, incorporating the mainland's 'eight-point' policy on Taiwan, was regarded as more conciliatory than hitherto. In April 1995, in response, President Lee proposed a 'six-point' programme for cross-Straits relations: unification according to the reality of separate rules; increased exchanges on the basis of Chinese culture; increased economic and trade relations; admission to international organizations on an equal footing; the renunciation of the use of force against each other; and joint participation in Hong Kong and Macao affairs. In late April, however, the eighth round of working-level SEF-ARATS discussions was postponed, owing to disagreement over the agenda.

In May 1995 the SEF Chairman, Koo Chen-fu, and his mainland counterpart, Wang Daohan, formally agreed to meet in Beijing in July. In June, however, this proposed second session of senior-level negotiations was postponed by the ARATS, in protest at President Lee's recent visit to the USA. Tension between the two sides increased in July, when the People's Republic unexpectedly announced that it was about to conduct an eight-day programme of guided missile and artillery-firing tests off the northern coast of Taiwan. A second series of exercises took place in August, again arousing much anxiety on the island. In mid-August President Jiang Zemin confirmed that the People's Republic would not renounce the use of force against Taiwan. Nevertheless, at the end of that month President Lee reaffirmed the KMT's commitment to reunification. In October President Jiang Zemin's offer to visit Taiwan in person was cautiously received on the island. President Lee confirmed his Government's anti-independence stance in November.

In January 1996 the Taiwanese Premier again urged the early resumption of cross-Straits dialogue. In February unconfirmed reports indicated that as many as 400,000 mainland troops had been mobilized around Fujian Province. In the same month, upon his appointment as Chairman of the Mainland Affairs Council, Chang King-yuh pledged to attempt to improve relations with the mainland. In early March, as Taiwan's first direct presidential election approached, the People's Republic began a new series of missile tests, including the firing of surface-to-surface missiles into an area off Taiwan's south-western coast, adjacent to the port of Kaohsiung, and into a zone off the north-eastern coast, near the port of Keelung. Live artillery exercises continued in the Taiwan Strait until after the election, arousing international concern. The USA deployed two naval task forces in the area (see above). In early April, as tension eased, the Mainland Affairs Council removed the ban on visits to Taiwan by officials of the People's Republic. At the end of April the SEF, which had lodged a strong protest with the ARATS during the missile tests of March, urged the resumption of bilateral discussions.

Upon his inauguration in May 1996, Taiwan's re-elected Head of State declared his readiness to visit the People's Republic for negotiations. In July President Lee reaffirmed his commitment to peaceful reunification, and urged the mainland to renounce the use of violence and to resume dialogue. In the same month visits to Taiwan by executives of the mainland's port authorities and of Air China (the flag carrier of the People's Republic) led to speculation that direct travel links between the two sides might be established. Other business delegations followed. The national oil corporations of Taiwan and of the People's Republic signed a joint exploration agreement. At the end of July, as bilateral relations continued to improve, the Mainland Affairs Council announced that Taiwanese governors and mayors were to be permitted to attend cultural activities and international functions in the People's Republic.

In October 1996 the Taiwanese Vice-Minister of Education, the most senior official to date, visited the People's Republic for discussions with his mainland counterparts. In November the Mainland Affairs Council announced that the permanent stationing of mainland media representatives in Taiwan was to be permitted. President Lee's renewed offer to travel to the People's Republic was rejected and, despite repeated SEF requests, Tang Shubei of the ARATS continued to assert that cross-Straits discussions could not resume owing to Taiwan's pursuit of its 'two Chinas' policy (a reference to President Lee's attempts to raise the diplomatic profile of the island). In January 1997, however, as the reversion of the entrepôt of Hong Kong to Chinese sovereignty approached, shipping representatives of Taiwan and of the People's Republic reached a preliminary consensus on the establishment of direct sea links. Under the terms of the agreement, which permitted mainland cargoes to be transhipped at Kaohsiung for onward passage to a third country but did not allow goods to enter the island's customs, five Taiwanese and six mainland shipping companies were granted permission to conduct cross-Straits cargo services. In April 1997, following the arrival in Kaohsiung of the first ship in 48 years to sail directly from the Chinese mainland to the island, the first Taiwanese-owned (but Panamanian-registered) vessel set sail for the port of Xiamen on a similarly historic voyage across the Taiwan Strait.

In March 1997 an unemployed journalist hijacked a Taiwanese airliner on an internal flight and, citing political repression on the island, forced the aircraft to fly to the mainland, where he requested asylum. The mainland authorities were commended for their handling of the incident, the aircraft returning to Taiwan later the same day. In May the SEF agreed to accept the

Taiwanese hijacker, who was to face criminal charges upon his return to the island.

In July 1997, upon the reversion to Chinese sovereignty of the British colony of Hong Kong, President Lee firmly rejected the concept of 'one country, two systems' and any parallel with Taiwan, and strenuously refuted a suggestion by President Jiang Zemin that Taiwan would eventually follow the example of Hong Kong. In August Liu Gangchi, Deputy Secretary-General of the ARATS, arrived in Taipei, at the head of a 32-member delegation, to attend a seminar on the subject of China's modernization. In September the Taiwanese Minister of Finance and the Governor of the Central Bank were obliged to cancel a visit to Hong Kong, where they had planned to have informal discussions with delegates to the forthcoming IMF/World Bank meeting, owing to Hong Kong's failure to issue them with visas. The affair compounded fears that Taiwan's business dealings with Hong Kong might be jeopardized.

A call for the opening of political negotiations, made by the Minister of Foreign Affairs of the People's Republic in September 1997, was welcomed by the Mainland Affairs Council. However, the Taiwanese authorities continued to insist that Beijing remove all preconditions before the opening of dialogue. In November the Secretary-General of the SEF was invited by the ARATS to attend a seminar on the mainland in the following month. The SEF proposed instead that its Chairman head a delegation to the People's Republic. An interview given by President Lee to a Western newspaper, in which he referred to Taiwan's independence, caused anger on the part of the Beijing authorities.

The declaration by an ARATS official, in January 1998, that Taiwan did not need to recognize the Government of the People's Republic as the central Government as a precondition for dialogue was regarded as a significant concession on the part of the mainland authorities. However, Taiwan continued to insist that China abandon its demand that talks be conducted under its 'One China' principle. In February the ARATS sent a letter to the SEF requesting the resumption of political and economic dialogue between the two sides, and inviting a senior SEF official to visit the mainland. The SEF responded positively to the invitation in March, and proposed that a delegation be sent to the People's Republic to discuss procedural details, prior to a visit to the mainland by the SEF Chairman.

In April 1998 a delegation chaired by the newly appointed Deputy Secretary-General of the SEF, Jan Jyh-horng, visited the People's Republic. Following negotiations with the ARATS, it was announced that the Chairman of the SEF would visit the People's Republic later in 1998 formally to resume the dialogue, which had been suspended since 1995. The visit was regarded as an important step towards restoring stability to Taiwan-China relations. The arrest on the mainland in May 1998 of four Taiwanese business executives on charges of espionage, and the visit to Malaysia by the Taiwanese Premier, in April, threatened to reverse the improvement in relations. However, in July the Chinese Minister of Science and Technology visited Taiwan, the first such visit by a mainland Minister since the civil war. This was followed later in the month by a formal visit to Taiwan by the Deputy Secretary-General of the ARATS. The kidnap and murder in August in the People's Republic of a Taiwanese local government official caused serious concern for the Taiwanese authorities. Furthermore, later in that month a Beijing court convicted the four Taiwanese businessmen on charges of espionage. In October Koo Chen-fu, the SEF Chairman, duly travelled to the People's Republic, where he met with President Jiang Zemin (the highest level of bilateral contact since 1949) and had discussions with his mainland counterpart and other senior officials. A four-point agreement was reached, allowing for increased communications between the two sides, but little was achieved in terms of a substantive breakthrough. However, the talks were considered to mark an important improvement in cross-Straits relations, and Wang Daohan accepted an invitation to visit Taiwan in March 1999. In January 1999 the ARATS invited the SEF Deputy Secretary-General to visit the People's Republic for talks in order to prepare for Wang Daohan's visit. The SEF made a counter-proposal that ARATS officials visit Taiwan to discuss preparations. In the following month, during a flight to the island of Kinmen, an SEF official was attacked by four Chinese convicted aircraft hijackers, who were part of a group being transferred to the nearby island prior to their repatriation to the mainland. During the following months Taiwan repatriated several hundred Chinese illegal immigrants. In March an ARATS delegation led by Deputy Secretary-General Lin Yafei visited Taiwan. It was agreed that Wang Daohan's visit would take place later in the year, but no date was set. In April President Lee reaffirmed that Beijing should recognize Taiwan as being of equal status. An SEF group went to Beijing in March, and preliminary agreement was reached that Wang Daohan would visit Taiwan in either mid-September or mid-October. In August, however, the ARATS suspended contacts with the SEF, following President Lee's insistence on the 'two-state theory' (see below), and it was confirmed in October that Wang Daohan would not visit Taiwan as long as it adhered to the theory.

Meanwhile, China was becoming increasingly demonstrative in its opposition to Taiwan's inclusion in the TMD system (see above). Ballistic missiles were deployed in mainland coastal regions facing Taiwan, and fears were heightened within the international community in July 1999, when the People's Republic announced that it had developed a neutron bomb, after declaring itself ready for war should Taiwan attempt to gain independence. This declaration was prompted by an interview given by President Lee to a German radio station, during which he asserted that relations with the People's Republic were 'state-to-state'. Chinese military exercises took place in the Taiwan Strait later that month, allegedly to intimidate Taiwan. Faced with this aggression and a lack of US support, Taiwan promised that it would not amend its Constitution to enshrine its claim to statehood in law. In August the USA reaffirmed its readiness to defend Taiwan against Chinese military action. Shortly afterwards the Taiwanese Government refused a request by Beijing that it retract the 'state-to-state' theory with regard to cross-Straits relations, and tension increased in late August when the KMT incorporated the 'two-state theory' into the party resolution, claiming that this would henceforth become the administrative guideline and priority of the Taiwanese authorities. Later that month the Mainland Affairs Council announced that former Taiwan government officials involved in affairs related to national intelligence or secrets were not to be permitted to travel to China within three years of leaving their posts. In September, however, following a severe earthquake in Taiwan that killed or injured several thousand people, China was among the many countries to offer emergency assistance to the island. Taiwan, however, accused the People's Republic of contravening humanitarian principles by trying to force other countries to seek its approval before offering help.

In February 2000 the People's Republic threatened to attack Taiwan if it indefinitely postponed reunification talks. The approval in the Legislative Yuan in March of a law providing for the first direct transport links between Taiwan's outlying islands and mainland China for 50 years did not substantially improve matters, and in April the Vice-President-elect, Annette Lu, was denounced by the Chinese media after she made 'separatist' remarks, televised in Hong Kong, declaring that Taiwan was only a 'remote relative and close neighbour' of China. A Taiwanese opposition group subsequently endorsed a motion to dismiss Lu for putting the island at risk by provoking China. In an attempt to improve relations with the People's Republic, President Chen offered to compromise and to reopen negotiations on the basis that each side was free to interpret the 'one China' formula as it saw fit. Chinese Premier Zhu Rongji rejected the suggestion, and questioned Taiwan's motives, effectively dispelling all hopes of restarting negotiations in the near future. In July 2000 the Chinese authorities responded angrily when the United Kingdom issued a visa to former President Lee Teng-hui, who continued to be perceived as a dissident by the People's Republic. Despite the fact that Lee's visa was for the purposes of a private visit by an individual, and specified as a condition that no public statements would be made during the trip, China cancelled ministerial-level contacts and official meetings with the United Kingdom, and threatened to take action against British trade interests.

In October 2000 Beijing published a policy document on its national defence, which confirmed that the People's Republic would use force to prevent Taiwanese secession, to stop occupation of the island, and also in the event of Taiwan indefinitely postponing reunification with the mainland. In November, however, there were signs of an improvement in relations when Wu Po-hsiung, the Vice-Chairman of the KMT, travelled to Beijing and met unofficially with Chinese Vice-Premier Qian Qichen. Wu was the most senior KMT official to visit mainland China for more than 50 years, and it was thought that Beijing was attempting to isolate Chen Shui-bian by consorting with his political rivals. During the meeting both sides agreed to hold important academic forums to discuss cross-Straits relations and to attempt to devise common positions. Qian stressed, however, the importance of Taiwan's recognition of the 'One China' principle before official negotiations could resume.

In November 2000 Taiwan announced that journalists from the People's Republic were to be granted permission to stay in Taiwan for periods of up to one month, during which time they would be invited to attend any press conferences called by the President's office and the Executive Yuan, in order to enhance cross-Straits exchanges and understanding. In the following month plans were announced for 'mini three links' with China, providing for direct trade, transport and postal links between Kinmen and Matsu islands and the mainland. The Taiwanese Government stressed, however, that any future direct links with Taiwan itself would be subject to rigorous security checks. In early January 2001 groups sailed from Kinmen and Matsu to Xiamen and Fuzhou, respectively, in the People's Republic. Beijing's response toward the initiative was guarded.

In March 2001 the press spokesman for the Fourth Session of the Ninth National People's Congress reiterated that China did not favour the confederal system of reunification, but rather the 'one country, two systems' model. Also in that month, exiled pro-democracy activist Wei Jingsheng visited Taiwan and had talks with Vice-President Lu. Cross-Straits relations had become strained by April following the visit of the Dalai Lama to Taiwan, and the forthcoming sale of weapons by the USA to Taiwan (see above).

In May 2001 President Chen announced that he aspired to become the first Taiwanese leader to visit the mainland since 1949 by attending the APEC forum to be held there in October; however, Beijing rejected Chen's offer. Although official cross-Straits relations were often hostile, ties continued to develop between Taiwan and the mainland, particularly in the business sphere. Many Taiwanese companies continued to invest in the mainland, and it was hoped that the growing economic interdependence between the two entities would reduce the risk of war. President Chen in late August endorsed a plan by a special advisory committee to expand economic and commercial links with the mainland, a reversal of the previous government's 'no haste, be patient' policy of limiting trade with the mainland for fear of becoming over-dependent on its main political enemy. The new policy of 'aggressive opening' included the removal of a US $50m. limit on individual investments in the mainland. Beijing responded coolly to Chen's initiative, stating that direct full transport links between the mainland and Taiwan would have to wait until Chen respected the 'One China' reunification formula, but a senior Chinese trade official stated that China would not block further Taiwanese investment. At the same time, state-owned oil companies from Taiwan and China announced plans to resume co-operative exploration of the Taiwan Straits. Beijing's reluctance to accept Chen's proposals reflected the fact that it had for years called for greater links with Taiwan which the latter had rejected, and was thus unwilling to embrace Taiwan too quickly. Chen's critics warned that his new policy would make Taiwan too economically dependent on the mainland. In September 2001 the Taiwanese Government approved a proposal allowing Chinese investment in Taiwan's land and property market, as part of the new opening to the mainland. The limit on individual investments in the mainland was formally removed on 7 November; restrictions on direct remits to and from the mainland via Taiwanese banks were also abolished.

In late October 2001, meanwhile, Taiwan boycotted the APEC summit meeting in Shanghai, following Beijing's refusal to allow Taiwan's chosen delegate, former Vice-President Li Yuan-tsu, to attend, on the grounds that he was not an 'economic' official. Despite this, the DPP deleted from its charter a vow to achieve the island's formal independence, since it was already a *de facto* separate entity. The change indicated a growing acceptance of the status quo by the DPP. It was hoped that the acceptance of China and Taiwan into the WTO in mid-November would improve cross-Straits relations, by enhancing communications in the field of trade.

The heavy defeat of the pro-mainland KMT by the pro-independence DPP at the legislative elections of December 2001 (see below) was initially seen as a reverse to cross-Straits relations; however, Beijing reacted with moderation to the event, in contrast to past bellicosity during elections, but insisted that Taiwan accept the 'One China' principle as a precondition for bilateral dialogue. Meanwhile, the Control Yuan reported that some 200 recently retired Taiwanese military and intelligence officials had visited Hong Kong and the mainland in violation of laws stipulating that they wait three years before doing so. There were fears that a number of these officials had divulged military secrets to mainland military officials, thereby jeopardizing the island's security.

In January 2002 the Government announced proposals for a new passport design incorporating the words 'issued in Taiwan' on the cover, ostensibly to differentiate clearly Taiwanese passports from mainland ones. The initiative was regarded disapprovingly by Beijing, as a sign of symbolic statehood. In mid-January the Government announced a list of more than 2,000 items that would thenceforth be legally importable from the mainland, mostly consumer but also agricultural goods, and at the same time facilitated direct transport links with the mainland. Later in the month, Chinese Vice-Premier Qian Qichen invited members of the DPP to visit the mainland, stating that most DPP members were not independence activists. President Chen welcomed Qian's remarks, and Premier Yu stated that he was planning to send a delegation to the mainland, but ruled out negotiations on the so-called '1992 consensus'. In February an unnamed senior Chinese official reportedly suggested, privately, that China was prepared to abandon its insistence that Taipei accept the 'One China' principle before commercial links could be realized.

In a sign of improving financial links, in early March 2002 Beijing announced that, for the first time, two Taiwanese banks would be allowed to open offices on the mainland. Taiwan would also allow mainland banks to establish offices on the island. At the end of the month, Taiwan eased restrictions on the island's companies investing in computer-chip manufacturing on the mainland; however, restrictions would remain on the number of plants established and type of chips produced. Underlying these restrictions was a fear that Taiwan's valuable electronics industry might become dependent on the mainland, and that the mainland might gain access to advanced semiconductor technology used to guide missiles. It was thought that Taiwan's efforts to open up to the mainland were being hampered by former President Lee Teng-hui's new political party, upon which the DPP depended to maintain a majority in the legislature and which generally favoured a slower approach to improving bilateral commercial relations.

At the beginning of May 2002 an official Chinese newspaper published Beijing's strongest criticism to date of President Chen, describing him as a 'troublemaker' who sought to damage bilateral relations, and criticizing his efforts to promote a separate 'Taiwanese' identity. The comments were thought to be a response to the strengthening of relations between Taiwan and the USA (see above). None the less, Taiwan at the same time reluctantly allowed China to ship more than 2,300 tons of water to its outlying islands in order to help relieve the worst drought in many years, an arrangement that would have been unthinkable a few years previously. Meanwhile, Chen in early May announced that he planned to send a DPP delegation to the mainland later in the year, in response to Qian Qichen's conciliatory speech in January. At the end of May the Mainland Affairs Council planned to introduce changes that would allow non-governmental organizations a greater role in promoting cross-Straits dialogue. Two state-owned oil companies from both sides at this time agreed upon a joint venture to explore petroleum and gas deposits in the straits. Taiwan's arrest in mid-June of one of its own military officers for passing military secrets to China failed to damage these improving commercial links, but offered a reminder that the intelligence 'war' between the two sides was far from over. Earlier, in mid-April, Taiwan had dispatched several navy vessels to monitor a Chinese research vessel operating just outside Taiwanese waters, and had tested an indigenously developed air-defence missile in May. Despite mutual suspicions, several Taiwanese legislators and retired generals secretly travelled to Beijing in June and discussed defence issues with their mainland counterparts.

In late June 2002 Beijing urged business groups to play the major role in establishing direct transport links between the two territories. However, in early July the mainland's Bank of China demanded that Taiwanese banks sign an acknowledgement of the 'One China' principle before cross-Straits banking services could begin. Taipei immediately rejected such demands, accusing Beijing of seeking to introduce a political element into the financial links between them, and questioning the sincerity of China's goodwill.

Bilateral relations again deteriorated in late July 2002, however, when the Pacific island nation of Nauru transferred its recognition from Taiwan to the People's Republic of China, thereby undermining Taipei diplomatically and prompting Chen to warn that Taiwan might have to chart its own future path. At the same time Taiwan's Ministry of Defense warned in a biannual report that China's military spending was accelerating rapidly and that it would possess 600 short-range missiles targeting the island by 2005. Chen further incensed Beijing in early August by supporting demands for a referendum to determine the island's future, and referring to China and Taiwan as two countries. Although China warned against any such moves, Chen's rhetoric was believed to have reflected his frustration at the lack of a political breakthrough in cross-Straits relations, and was probably aimed at raising the DPP's popularity ahead of forthcoming mayoral elections. However, DPP officials stated that there had been no change in policy, and Chen subsequently softened his rhetoric. Taiwan then cancelled planned military exercises as a gesture of good faith, but Chen's comments delayed the introduction of direct transport links. At the end of July the Government announced that Chinese products could be advertised on the island and that Chinese employees of Taiwanese or foreign companies would be allowed to work in Taiwan. A Taiwanese semiconductor manufacturer, the world's largest, also announced plans to build a new factory on the Chinese mainland, in Shanghai.

The political atmosphere between China and Taiwan was likely to remain volatile, however, and in early September 2002 Chen described the mainland's threats against the island as a form of 'terrorism'. Later that month Beijing accused Taiwan of allowing the banned Falun Gong sect to use the island as a base for disrupting mainland Chinese television and satellite broadcasts, although did not accuse Taipei of directly supporting the group. A Chinese official stated that the source of the sect's propaganda transmissions had been traced to the island, but a Taiwanese investigation revealed nothing, and the unofficial leader of Falun Gong in Taiwan, Chang Ching-hsi, emphasized that the sect had no desire to exacerbate the tensions in cross-Straits relations. At the same time, a retired Taiwanese air force officer, his wife and their son were charged with spying for China in the late 1980s. In mid-October 2002 a Taiwanese army officer fled to China, via Thailand, with his wife and family.

Attempts to restore direct transport links between China and Taiwan gained momentum in mid-October 2002 when Beijing abandoned the 'One China' principle as a condition for such links and agreed that these could be described as 'cross-Straits' rather than 'domestic'. However, two weeks later President Chen seemingly retreated from the proposal. A report by the Control Yuan in September 2002 had stated that only 2% of the money invested by Taiwanese firms in China was repatriated, and it was believed that Chen was reassessing the economic benefits of cross-Straits links. Elements in the DPP had long argued that these links with China were to Taiwan's economic disadvantage.

The 16th Congress of the Chinese Communist Party (CCP), held in Beijing in November 2002, stressed continuity in cross-Straits relations. However, the 'arms race' between China and Taiwan continued during 2002, with the two sides acquiring advanced weaponry from Russia and the USA respectively. In mid-October two Chinese ships and a submarine sailed around the east coast of Taiwan en route to the South China Sea, where the Chinese navy was staging military exercises. The voyage served to warn Taiwan that Beijing could encircle the island if necessary, prompting the Minister of National Defense, Tang Yao-ming, to urge a strengthening of Taiwan's navy. Analysts noted, however, that Taiwan's defence spending had been reduced in recent years, owing to the economic downturn.

Tension between Taipei and Beijing intensified in January 2003, however, when President Chen stated that Taiwan was a sovereign state, and would never accept a Hong Kong-style solution nor federate with the mainland. At the same time, Vice-President Lu described China as being of a 'terrorist nature', referring to its missile build-up across the Taiwan Straits and to its coercive diplomacy. In late January 2003, none the less, following discussions earlier in that month, the first Taiwanese airliner in more than 50 years flew to mainland China, via Hong Kong, landing in Shanghai. The flight was one of 16 charter operations organized by the Taiwanese carrier, CAL, to transport Taiwanese visitors in China home for Chinese New Year celebrations. The flights were a special arrangement, however, and the instigation of direct cross-Straits flights on a regular basis appeared unlikely to be realized in the near future. Also in late January a member of the DPP made a private visit to the mainland, ostensibly to promote exchanges of views.

In late February 2003 the Minister of National Defense stated that Taiwan would not reduce weapons purchases from the USA in return for the dismantling of Chinese missiles targeting the island. His remarks came in response to Chinese President Jiang Zemin's offer of such an arrangement, made to President Bush in October 2002.

Relations between the two sides were antagonized following the outbreak of a pneumonia-like virus, Severe Acute Respiratory Syndrome (SARS), on the mainland, which spread to Taiwan in early 2003. Taipei accused Beijing of concealing the seriousness of the outbreak, thereby endangering Taiwan and other neighbouring countries. In May Taiwan formally rejected an aid programme offered by China to assist in countering the SARS epidemic on the island. Also in May, China continued to stress its opposition to Taiwan's membership of WHO at a meeting of the World Health Assembly in Geneva. However, Beijing did agree to Taiwan's participation in a global conference on SARS in June 2003.

In mid-2003 it seemed likely that cross-Straits relations would continue to be dominated by political tensions, despite increased trade and educational connections between the two sides. In early August the island's authorities arrested two Taiwanese citizens and a Taiwanese American on charges of spying for China. Also in August, a Chinese surveillance vessel was seen to the north of Taiwan, in advance of large-scale military exercises in that area. Military exercises simulating an attack on Taiwan by China were conducted in early September 2003. Also in September, in a move likely to anger Beijing further, Taiwan announced that it was to issue new passports for its citizens, which for the first time would have the word 'Taiwan' printed on the front. Cross-Straits tensions in September were again heightened by mass protests in Taipei at the beginning of the month, demanding that the name 'Taiwan' be used officially by government agencies and private companies (see below). Political tensions notwithstanding, President Chen predicted in August 2003 that direct links with mainland China for purposes such as cargo transport would be in place by the end of 2004. In October 2003 the Taiwanese Legislative Yuan passed the so-called Act Governing Relations between Peoples of the Taiwan Area and the Mainland Area, which was to provide a framework for the regulation of travel and business links between Taiwan and the mainland.

In November 2003 tensions between Taiwan and the mainland were further exacerbated after Taiwan's legislature approved a bill allowing referendums to take place on the island (see below). Although the legislation in its final version prohibited a vote on Taiwan's sovereignty except in the event of an external attack, Beijing feared that the new law might lead to the holding of a vote on Taiwan's independence. An official at China's Taiwan Affairs Office warned that President Chen would be 'risking war' should he initiate any move towards Taiwanese independence. In December the issue of Taiwan dominated a visit by Chinese Premier Wen Jiabao to Washington, DC, with Wen seeking assurance from the USA that it would not support moves towards Taiwanese independence, following the introduction of the new referendum law. In the same month President Chen defended plans to hold a referendum in March 2004 on the subject of a proposed request to China that it remove hundreds of missiles aimed at the island. In January 2004 President Chen suggested that the referendum might be presented as an opportunity to reduce bilateral tensions. He stated that the referendum might include 'counter-proposals' such as direct transport links between Taiwan and China if the latter were to withdraw its missiles. In February, furthermore, President Chen suggested the establishment of a demilitarized zone between the mainland and Taiwan in order to facilitate joint discussions. Chen also stated that he would not declare Taiwanese independence, were he to win the presidential election in March, although his attitude to relations with the Chinese mainland, for example in rejecting the 'one country, two systems' model advocated by Beijing, was regarded by some as tantamount to stating that independence already existed. The Beijing Government did not attempt directly to influence the election when it was held in March, resulting in a narrow victory for Chen. However, following widespread controversy and protests over the election result, the Chinese Government stated that it would not tolerate social instability in Taiwan. In the event, President Chen's referendum, held concurrently with the presidential election, did not produce a valid result (see below). The Beijing Government remained suspicious of Chen. At the end of July 2004 a senior Chinese official warned of an attack on Taiwan by 2008, were Chen to proceed with constitutional reforms outlined in Chen's inaugural speech in May, despite pledges from Chen that the amendments would not address the question of Taiwan's sovereignty.

INTERNAL POLITICAL DEVELOPMENTS

President Chiang Kai-shek remained in office until his death in April 1975. He was succeeded by the former Vice-President, Dr Yen Chia-kan, as Head of State, and by his son, Gen. Chiang Ching-kuo (who had been Premier since May 1972), as Chairman of the KMT. Chiang Ching-kuo succeeded Dr Yen as President in May 1978, when Sun Yun-suan became Premier.

Meanwhile, legislative elections were held in December 1972, for the first time in 24 years, to fill 53 seats in the National Assembly. The new members, elected for a fixed term of six years, joined 1,376 surviving 'life-term' members of the Assembly. In the December 1983 elections for 71 local seats in the Legislative Yuan, whose original republican membership had been reduced by attrition from 760 to 274 (with an average age of 77), the ruling KMT won 62 seats. In addition, 27 supplementary vacancies for overseas Chinese were filled by presidential appointment. Younger, well-qualified Taiwan-orientated members thus entered the Legislative Yuan. In March 1984 President Chiang Ching-kuo was re-elected for a second six-year term by the National Assembly. Lee Teng-hui, hitherto Governor of Taiwan Province, was elected Vice-President. Yu Kuo-hwa, a former Governor of the Central Bank, was appointed Premier in May, in place of Sun Yun-suan, and a major reshuffle of the Executive Yuan (cabinet) brought several younger politicians into the Government. At local elections in November 1985 the KMT won 80% of the seats.

The Ending of Martial Law and Progress towards Democracy

In 1986 the KMT announced its readiness to discuss four controversial issues: the possible establishment of new political parties, the status of the martial law (in force since 1949), the structure of provincial government in Taiwan, and the problem of the ageing political leadership. In September 135 leading opposition politicians formed the Democratic Progressive Party (DPP), in defiance of the KMT, and in preparation for the legislative elections that were due to be held in December. In October the KMT announced its intention to suspend martial law. The formation of rival political groups was also permitted. During 1987 three such organizations emerged: the Chinese Freedom Party (CFP), favouring improved relations with mainland China; the Democratic Liberal Party (DLP); and the Kungtang, or Labour Party. The Chinese Republican Party (CRP) was formed in 1988.

Elections for 84 seats in the National Assembly and 73 seats in the Legislative Yuan were held in December 1986. The KMT won 68 seats in the National Assembly and 59 in the Legislative Yuan, but the DPP received about one-quarter of the total votes cast, winning 11 seats in the Assembly and 12 in the Legislative Yuan, thus more than doubling the non-KMT representation. Following the opening of a new session of the Legislative Yuan in February 1987, the KMT began to implement its programme of reform. The most significant change was the termination of martial law, and its replacement in July by a new national security law, whereby political parties other than the KMT were permitted, civilians were removed from the jurisdiction of military courts, and military personnel no longer had the right to determine the acceptability of persons entering and leaving Taiwan. Despite the fact that the removal of martial law had been the principal aim of the DPP, the party remained opposed to the conditions with which opposition groups had to comply in order to gain legal recognition. The new law stated that opposition parties should honour the Constitution, support the Government's anti-communist policy and oppose separatism, but many DPP leaders argued that acceptance of these conditions would effectively recognize the right of the ruling KMT to regulate its own opposition. The question of the DPP's legal status thus remained unresolved.

The KMT also attempted to rejuvenate Taiwan's ageing leadership. Within the party, younger members advocating reform were promoted to positions of influence, and in April 1987 secured seven major posts in a reshuffle of the Executive Yuan. Among the younger members of the KMT, a strong movement in favour of the increased democratization of Taiwan's parliamentary system developed. In February 1988 a plan to restructure the legislative bodies, which had been initiated by President Chiang Ching-kuo in 1986, was approved by the Central Standing Committee of the KMT. Under its provisions, 'life-term' members of the Legislative Yuan and the National Assembly would be progressively reduced in number through death and voluntary retirement. Seats in the National Assembly and the Legislative Yuan would no longer be reserved for representatives of mainland constituencies, and there would be a corresponding increase in members representing Taiwanese constituencies.

Chiang Ching-kuo died in January 1988, and was succeeded by the Vice-President, Lee Teng-hui, who was designated to serve the remaining two years of President Chiang's term of office. Before his death, President Chiang had stated that Taiwan's future leadership should be provided by constitutional means, thus signalling the end of the Chiang 'dynasty'. President Lee was the first native Taiwanese to serve as President and was thus, potentially, at variance with the 'old guard' of mainland-orientated members of the KMT and the Legislative Yuan. The new President sought to strengthen the KMT's commitment to the programme of political reform and to the rejuvenation and 'Taiwanization' of the country's leadership. However, the slow pace at which the Government proceeded attracted criticism, not only from the DPP, but also from liberal elements within the KMT. In January 1988 the DPP sought to obstruct the adoption of two draft laws regarding the freedom of assembly and demonstration, and the formation of new political parties, claiming that their provisions would regulate too strictly the activities of the opposition. In April liberal representatives of the KMT in the Legislative Yuan began to intensify the campaign for a restructuring of the party, demanding that all 150 members of its Central Committee, and at least some of the 31 members of its Central Standing Committee, be elected to office. In May an initially peaceful demonstration by farmers in Taipei resulted in the most serious riots in Taiwan's recent history.

At the 13th National Congress of the KMT, held in July 1988, Lee promised to accelerate reform and to fortify 'the substance and function of democracy'. The party congress confirmed him in the chairmanship of the party and went on to elect most of his 180 nominees for membership of an expanded Central Committee. Two-thirds of the members of the new Central Committee were chosen, for the first time in the history of the KMT, by free elections and had not previously been members. The proportion of members who were native Taiwanese also increased sharply, from one-fifth to almost two-thirds. President Lee's appointments to the Central Standing Committee followed the same trend, in that 12 of its 31 members were replaced by representatives of the younger, liberal faction of the KMT. For the first time, also, the number of members of the Central Standing Committee who were native Taiwanese was greater than that of those born on the mainland. A reshuffle of the Executive Yuan in late July reflected the changes that had been accomplished at the National Congress of the KMT: new ministerial appointments resulted in a government comprising younger members. At the same time, President Lee promoted three urgent legislative measures: a draft revision of regulations concerning the registration of civic organizations and political parties; a retirement plan (with generous pensions) for those members of the Legislative Yuan, the Control Yuan and the National Assembly who had been elected by mainland constituencies in 1947; and a new law aiming to give greater autonomy to the Taiwan Provincial Government and its assembly.

The three measures became law in January 1989. In the following month the KMT became the first political party to register under the new legislation. However, the new laws were severely criticized by the DPP, which protested at the size of the retirement pensions being offered and at the terms of the Civic Organizations Law, which required that, in order to register, political parties undertook to reject communism and any notion of official political independence for Taiwan. Despite these objections, the DPP applied for official registration in April 1989. In May Yu Kuo-hwa resigned as Premier of the Executive Yuan and was replaced by Lee Huan, the Secretary-General of the KMT.

Partial elections to the Legislative Yuan and the Taiwan Provincial Assembly were held on 2 December 1989. A total of 101 seats in the Legislative Yuan were contested by the KMT, the DPP and several independent candidates. The KMT obtained 72 seats and the DPP won 21. By virtue of achieving more than 20 seats, the DPP secured the prerogative to propose legislation in the Legislative Yuan.

In February 1990 the opening of the National Assembly's 35-day plenary session, convened every six years to elect the country's President, was disrupted by DPP members' violent action in a protest against the continuing domination of the Assembly by elderly KMT politicians, who had been elected on the Chinese mainland prior to 1949 and who had never been obliged to seek re-election. At the Legislative Yuan, demonstrators attempted to prevent senior KMT members from entering the building, and the election of a KMT veteran as President of the legislature had to be postponed, when opposition members deliberately delayed the procedure. More than 80 people were injured during the ensuing street clashes between riot police and demonstrators.

In March 1990 DPP members were barred from the National Assembly for refusing to swear allegiance to 'The Republic of China', attempting instead to substitute 'Taiwan' upon taking the oath. A number of amendments to the Temporary Provisions, which for more than 40 years had permitted the effective suspension of the Constitution, were approved by the National Assembly in mid-March. Revisions included measures to strengthen the position of the mainland-elected KMT members, who were granted new powers to initiate and veto legislation, and also an amendment to permit the National Assembly to meet annually. The revisions were opposed not only by the DPP but also by more moderate members of the KMT, and led to a large protest rally in Taipei, which attracted an estimated 10,000 demonstrators, who continued to demand the abolition of the National Assembly and the holding of direct presidential elections. Nevertheless, President Lee was duly re-elected, unopposed, by the National Assembly for a six-year term, two rival KMT candidates having withdrawn from the contest. In April President Lee and the Chairman of the DPP met for discussions.

There was renewed unrest in May 1990, however, following President Lee's unexpected appointment as Premier of Gen. (retd) Hau Pei-tsun, the former Chief of the General Staff and, since December 1989, the Minister of National Defense. Outraged opposition members prevented Hau from addressing the National Assembly, which was unable to approve his nomination until the session was reconvened a few days later, police being summoned to the Assembly to restore order. Angry demonstrators, fearing a reversal of the process of democratic reform, again clashed with riot police on the streets of Taipei. A new Executive Yuan was appointed at the end of the month, and included a civilian as Minister of National Defense. New Ministers of Foreign Affairs and of Finance were appointed, but the majority of ministers retained their previous portfolios.

The National Affairs Conference (NAC), convened in late June 1990, was attended by 150 delegates from various sections of society. At the historic meeting, proposals for reform were presented for discussion. A Constitutional Reform Planning Group was subsequently established. The NAC also reached consensus on the issue of direct presidential elections, which would permit the citizens of Taiwan, rather than the ageing members of the National Assembly, to select the Head of State. Conservative members of the KMT were strongly opposed to this proposal.

Meanwhile, the Council of Grand Justices had ruled that elderly members of the National Assembly and of the Legislative Yuan should step down by the end of 1991. Constitutional reform was to be implemented in several stages: in April 1991 the Temporary Provisions, adopted in 1948, were to be abolished; in late 1991 a new National Assembly was to be elected by popular vote, the number of members being reduced and all elderly mainland-elected delegates being obliged to relinquish their seats; elections to the new Legislative Yuan were to take place in 1992. Meanwhile, in early December 1990 Huang Hwa, the leader of a faction of the DPP and independence activist, had received a 10-year prison sentence upon being found guilty of 'preparing to commit sedition'.

In April 1991 the National Assembly was convened, the session again being marred by violent clashes between KMT and DPP members. The DPP subsequently boycotted the session, arguing that a completely new constitution should be introduced and that elderly KMT delegates, who did not represent Taiwan constituencies, should not have the right to make amendments to the existing Constitution. As many as 20,000 demonstrators attended a protest march organized by the DPP. Nevertheless, the National Assembly duly approved the constitutional amendments, and at midnight on 30 April the 'period of mobilization for the suppression of the Communist rebellion' and the Temporary Provisions were formally terminated. The existence, but not the legitimacy, of the Government of the People's Republic was officially acknowledged by President Lee. Furthermore, Taiwan remained committed to its 'One China' policy. Martial law remained in force until November 1992 on the islands of Kinmen (Quemoy) and Matsu where, owing to their proximity to the mainland, it had not been lifted in mid-1987 as elsewhere in Taiwan. In May 1991 widespread protests, following the arrest of four advocates of independence, led to the abolition of the Statute of Punishment for Sedition. The law had been adopted in 1949 and had been frequently employed by the KMT to suppress political dissent.

A senior UN official arrived on the island in August 1991, the first such visit since Taiwan's withdrawal from the organization in 1971. Large-scale rallies calling for a referendum to be held on the issue of Taiwan's readmission to the UN, resulting in clashes between demonstrators and the security forces, took place in September and October.

In August 1991 the opposition DPP officially announced its alternative draft constitution for 'Taiwan', rather than for 'the Republic of China', thus acknowledging the *de facto* position regarding sovereignty. In late September, only one day after being reinstated in the Legislative Yuan, Huang Hsin-chieh, the Chairman of the DPP, relinquished his seat in the legislature and urged other senior deputies to do likewise. Huang Hsin-chieh had been deprived of his seat and imprisoned in 1980, following his conviction on charges of sedition. At the party congress in October 1991, Huang Hsin-chieh was replaced as DPP Chairman by Hsu Hsin-liang. Risking prosecution by the authorities, the DPP congress adopted a resolution henceforth to advocate the establishment of 'the Republic of Taiwan', and urged the Government to declare the island's independence.

Elections to the new 405-member National Assembly, which was to be responsible for amending the Constitution, were held on 21 December 1991. The 225 seats open to direct election were contested by a total of 667 candidates, presented by 17 parties. The campaign was dominated by the issue of whether Taiwan should become independent or seek reunification with the mainland. The opposition's independence proposal was overwhelmingly rejected by the electorate, the DPP suffering a humiliating defeat. The KMT secured a total of 318 seats (179 of which were won by direct election), while the DPP won 75 seats (41 by direct election).

In late 1991, in a new campaign to curb illegal dissident activity, the authorities arrested 14 independence activists, including members of the banned, US-based World United Formosans for Independence (WUFI). Several detainees were indicted on charges of sedition. Furthermore, the four dissidents, whose arrest in May had provoked widespread unrest, were brought to trial and found guilty of sedition, receiving short prison sentences. In January 1992 four WUFI members were found guilty of plotting to overthrow the Government. In February 20,000 demonstrators took part in a march in Taichung. The protesters' demands included the abolition of the sedition laws and the holding of a referendum on the issue of independence for the island. In the same month the Government released a report on the 1947 massacre of 18,000–28,000 civilians. For the first time the KMT leadership admitted responsibility for the violent suppression of the alleged communist rebels.

In March 1992, at a plenary session of the KMT Central Committee, agreement was reached on several issues, including a reduction in the President's term of office from six to four years. The principal question of arrangements for future presidential elections, however, remained unresolved. Liberal members continued to advocate direct election, while conservatives favoured a complex proxy system. In April street demonstrations were organized by the DPP to support demands for direct presidential elections. In May the National Assembly adopted eight amendments to the Constitution, one of which empowered the President to appoint members of the Control Yuan.

Meanwhile, the radical dissident, (Stella) Chen Wan-chen, who had established the pro-independence Organization for Taiwan Nation-Building upon her return from the USA in 1991, was sentenced to 46 months' imprisonment in March 1992, having been found guilty of 'preparing to commit sedition'. In May, however, Taiwan's severe sedition law was amended, non-violent acts ceasing to be a criminal offence. As a result, several independence activists, including Chen Wan-chen and Huang Hwa, were released from prison. Other dissidents were able to return from overseas exile. Nevertheless, in June (George) Chang Tsang-hung, the chairman of WUFI, who had returned from exile in the USA in December 1991, received a prison sentence of five (commuted from 10) years upon conviction on charges of sedition and attempted murder, involving the dispatch of letter-bombs to government officials in 1976. Chang was released for medical treatment in October 1992, and in March 1993 he was acquitted of the sedition charges, on the grounds of insufficient evidence.

The First Full Elections and KMT Disunity

Taiwan's first full elections since the establishment of Nationalist rule in 1949 were held in December 1992. The KMT retained 102 of the 161 seats in the Legislative Yuan. The DPP, however, garnered 31% of the votes and more than doubled its representation in the legislature, winning 50 seats. Following this set-back, the Premier and the KMT Secretary-General resigned. In February 1993 President Lee nominated the Governor of Taiwan Province, Lien Chan, for the premiership. The Legislative Yuan duly approved the appointment of Lien Chan, who thus became the island's first Premier of Taiwanese descent. The incoming Executive Yuan incorporated numerous new ministers.

There were violent scenes in the National Assembly in April 1993, when deputies of the DPP (which in recent months had modified its aggressive pro-independence stance, placing greater emphasis on the issues of corruption and social welfare) accused members of the KMT of malpractice in relation to the election of the Assembly's officers.

In May 1993 the growing rift between conservative and liberal members of the ruling party was illustrated by the resignation from the KMT of about 30 conservative rebels, and their formation of the New Alliance Nationalist Party. Furthermore, in June the Government was defeated in the Legislative Yuan, when a group of KMT deputies voted with the opposition to approve legislation on financial disclosure requirements for elected and appointed public officials. The unity of the KMT was further undermined in August, when six dissident legislators belonging to the New Kuomintang Alliance, which had registered as a political group in March, announced their decision to leave the ruling party in order to establish the New Party. The rebels included Wang Chien-shien, the former Minister of Finance. Nevertheless, in the same month, at the 14th KMT Congress, Lee Teng-hui was re-elected Chairman of the party. A new 31-member Central Standing Committee and 210-member Central Committee, comprising mainly Lee's supporters, were selected. In a conciliatory gesture by the KMT Chairman, four vice-chairmanships were created, the new positions being filled by representatives of different factions of the party.

In September 1993, following a series of bribery scandals, the Executive Yuan approved measures to combat corruption. The administrative reform plan included stricter supervision of public officials and harsher penalties for those found guilty of misconduct. In the same month a KMT member of the Legislative Yuan was sentenced to 14 years' imprisonment for bribery of voters during the 1992 election campaign; similar convictions followed. At local government elections held in November 1993, although its share of the votes declined to 47.5%, the KMT fared better than anticipated, securing 15 of the 23 posts at stake. The DPP, which accused the KMT of malpractice, received 41.5% of the votes cast, but won only six posts; it retained control of Taipei County. The DPP Chairman, Hsu Hsin-liang, resigned, and was replaced by Shih Ming-teh. Following allegations of extensive bribery at further local polls in early 1994 (at which the DPP and independent candidates made strong gains), the Ministry of Justice intensified its campaign against corruption. Proposals for constitutional amendments to permit the direct election in 1996 of the Taiwanese President by popular vote (rather than by electoral college) and to limit the powers of the Premier were approved by the National Assembly in July 1994.

At gubernatorial and mayoral elections in December 1994 the DPP took control of the Taipei mayoralty, in the first such direct polls for 30 years, while the KMT succeeded in retaining the provincial governorship of Taiwan, in the first ever popular election for the post, and the mayoralty of Kaohsiung. The New Party established itself as a major political force, its candidate for the mayoralty of Taipei receiving more votes than the KMT incumbent. Almost 77% of those eligible voted in the elections. A government reorganization followed.

In March 1995, in response to continuing allegations of corruption (the number of indictments now having exceeded 2,000), President Lee announced the appointment of a committee to investigate the financial activities of the KMT. In the same month, following the President's formal apology at a ceremony of commemoration in February, the Legislative Yuan approved a law granting compensation to the relatives of the victims of a massacre by Nationalist troops in 1947 (the 'February 28 Incident'), in which an estimated 18,000 native Taiwanese had been killed. (In June 1997 the Executive Yuan approved draft legislation to grant an amnesty to those involved in the incident.)

Fewer than 68% of those eligible voted at the elections to the Legislative Yuan held on 2 December 1995. A major campaign issue was that of corruption. The KMT received only 46% of the votes cast, and its strength declined to 85 of the 164 seats. The ruling party fared particularly badly in Taipei. Although it performed less well than anticipated, the DPP increased its representation to 54 seats. The New Party, which favoured reconciliation with the mainland, secured 21 seats. At the Legislative Yuan's first session in February 1996, Liu Sung-pan of the KMT only narrowly defeated a strong challenge from Shih Ming-teh of the DPP to secure re-election as the chamber's President.

The Presidential Election of 1996 and Beyond

The first direct presidential election was scheduled for March 1996, to coincide with the National Assembly polls. President Lee had declared his intention to stand for re-election in August 1995. In January 1996 Lien Chan offered to resign as Premier in order to support President Lee and to concentrate on his own vice-presidential campaign. He remained in office in an interim capacity. Other contenders for the presidency included the independent candidate and former President of the Judicial Yuan, Lin Yang-kang, supported by former Premier Hau Pei-tsun (both conservative former KMT Vice-Chairmen having campaigned on behalf of New Party candidates at the December elections and therefore having had their KMT membership revoked); Peng Ming-min of the DPP; and Chen Li-an, former President of the Control Yuan and previously Minister of National Defense, an independent Buddhist candidate. Wang Chien-shien of the New Party withdrew his candidacy in favour of the Lin-Hau alliance. The campaign was dominated by the issue of reunification with the mainland. In mid-March a DPP demonstration on the streets of Taipei, in support of demands for Taiwan's independence, was attended by 50,000 protesters.

At the presidential election, held on 23 March 1996, the incumbent President Lee received 54.0% of the votes cast, thus securing his re-election. His nearest rival, Peng Ming-min of the DPP, took 21.1% of the votes. The independent candidates, Lin Yang-kang and Chen Li-an, received 14.9% and 10.0% of the votes respectively. At the concurrent elections for the National Assembly, the KMT garnered 55% of the votes and took 183 of the 334 seats. The DPP won 99 seats and the New Party 46 seats. The Chairman of the DPP, Shih Ming-teh, resigned and Hsu Hsin-liang subsequently returned to the post.

On 20 May 1996 President Lee was sworn in for a four-year term. In June, however, the President's announcement of the composition of the new Executive Yuan aroused much controversy. Although several members retained their previous portfolios, the President (apparently under pressure from within the KMT and disregarding public concern at the rising levels of corruption and organized crime) demoted the popular Ministers of Justice (Ma Ying-jeou) and of Transportation and Communications, who had exposed malpractice and initiated campaigns against corruption. Other changes included the replacement of the Minister of Foreign Affairs, Fredrick Chien (who became Speaker of the National Assembly), by John Chang, the grandson of Chiang Kai-shek. The most controversial nomination, however, was the reappointment as Premier of Lien Chan, despite his recent election as the island's Vice-President. As fears of a constitutional crisis grew, opposition members of the Legislative Yuan, along with a number of KMT delegates, demanded that the President submit the membership of the Executive Yuan to the legislature for approval, and threatened to boycott the chamber's business. In October the Constitutional Court opened its hearing regarding the question of the island's Vice-President serving concurrently as Premier.

In October 1996 the Taiwan Independence Party was established by dissident members of the DPP. In the same month the Legislative Yuan approved the restoration of funding for a controversial fourth nuclear power plant, construction of which had been suspended in 1986. Thousands of anti-nuclear protesters demonstrated at the legislature, clashing with police and preventing Lien Chan from entering the building.

In December 1996 the multi-party National Development Conference (NDC), established to review the island's political system, held its inaugural meeting. The convention approved KMT proposals to abolish the Legislative Yuan's right to confirm the President's choice of Premier, to permit the legislature to introduce motions of 'no confidence' in the Premier and to empower the President to dismiss the legislature. The Provincial Governor, (James) Soong Chu-yu, subsequently tendered his resignation in protest at the NDC's recommendations that elections for the provincial governorship and assembly be abolished, as the first stage of the dissolution of the provincial apparatus. An historical legacy duplicating many of the functions of central and local government, the Provincial Government was responsible for the entire island, with the exception of the cities of Taipei and Kaohsiung. In January 1997 President Lee refused to accept the Governor's resignation, but the affair drew attention to the uneasy relationship between the island's President and its Governor, and brought to the fore the question of reunification with the mainland. The Provincial Government was abolished in December 1998.

In May 1997 more than 50,000 demonstrators, protesting against the Government's apparent inability to address the problem of increasing crime, demanded the resignation of President Lee. Three members of the Executive Yuan resigned, including the popular Minister without Portfolio and former Minister of Justice, Ma Ying-jeou, who expressed his deep shame at recent events. The appointment of Yeh Chin-feng as Minister of the Interior (the first woman to oversee Taiwan's police force) did little to appease the public, which remained highly suspicious of the alleged connections between senior politicians and the perpetrators of organized crime. In mid-May thousands of protesters, despairing of the rapid deterioration in social order, again took to the streets of Taipei, renewing their challenge to President Lee's leadership and demanding the immediate resignation of Premier Lien Chan. In late June, prior to the return to Chinese sovereignty of Hong Kong, a 'Say No to China' rally attracted as many as 70,000 supporters.

In July 1997 the National Assembly approved various constitutional reforms, including the 'freezing' of the Provincial Government. Other revisions that received approval were to empower the President of Taiwan to appoint the Premier without the Legislative Yuan's confirmation; the legislature was to be permitted to hold a binding vote of 'no confidence' in the Executive Yuan, while the President gained the right to dissolve the Legislative Yuan. In August the Premier and his entire Government resigned in order to permit a reallocation of portfolios. Vincent Siew, former Chairman of the Council for Economic Planning and Development and also of the Mainland Affairs Council, replaced Lien Chan as Premier. (Lien Chan retained the post of Vice-President.) John Chang was appointed Vice-Premier, and the new Minister of Foreign Affairs was Jason Hu. Following the installation of the new Executive Yuan, the Premier pledged to improve social order, further develop the economy, raise the island's standard of living and improve links with the People's Republic of China. In the same month President Lee Teng-hui was re-elected unopposed as Chairman of the ruling KMT.

The KMT experienced a serious set-back in elections at mayoral and magistrate levels, held on 29 November 1997. The opposition DPP, which had campaigned on a platform of more open government, secured 43% of the total votes, winning 12 of the 23 constituency posts contested, while the KMT achieved only 42% (eight posts). Voter turn-out was 65.9%. The outcome of the elections meant that more than 70% of Taiwan's population would come under DPP administration. Following the KMT's poor performance in the ballot, the Secretary-General of the party resigned, and was replaced by John Chang. A major reorganization of the party followed. Liu Chao-shiuan was appointed Vice-Premier in place of Chang.

At local elections, held in January 1998, the KMT won an overwhelming majority of the seats contested, while the DPP, in a reversal of fortune, performed badly. A minor cabinet reshuffle was carried out in early February.

In late March the Minister of Transportation and Communications resigned, assuming responsibility for two aeroplane crashes in Taiwan in early 1998, in which more than 200 people had died. In the following month the Minister of Justice tendered, and then subsequently withdrew, his resignation, citing pressure from lawmakers with connections to organized crime. However, in July he was forced to resign, following his mishandling of an alleged scandal concerning the acting head of the Investigation Bureau.

In June 1998 the first-ever direct election for the leadership of the DPP was held. Lin Yi-hsiung won a convincing victory, assuming the chairmanship of the party in August. Meanwhile, in local elections in June, the KMT suffered a set-back, winning fewer than 50% of the seats contested. Independent candidates performed well. In August 17 new members were elected to the KMT Central Standing Committee, the 16 others being appointed by President Lee.

The Legislative Elections of December 1998 and Constitutional Issues

Elections to the newly expanded 225-member Legislative Yuan took place on 5 December 1998; 68.1% of the electorate participated in the poll. The KMT won 46.4% of the votes cast, securing 125 seats, the DPP received 29.6% of the votes and won 72 seats, while the pro-unification New Party secured only 7.1% of the votes and 11 seats. The New Nation Alliance, a breakaway group from the DPP (formed in September), won only one seat (with 1.6% of the votes cast). The KMT's victory was widely attributed to its management of the economy, in view of the Asian financial crisis, the developments in cross-Straits dialogue and a decline in factionalism within the party in 1998. In the election (held simultaneously) to select the mayor of Taipei, the KMT candidate, Ma Ying-jeou (a former Minister of Justice), defeated the DPP incumbent, Chen Shui-bian. However, the

DPP candidate for the office of mayor of Kaohsiung, Frank Hsieh, narrowly defeated the KMT incumbent. The KMT retained control of both city councils.

Owing to the DPP's poor performance at the elections, Lin Yi-hsiung offered his resignation as Chairman of the DPP, but withdrew it following overwhelming party support for his leadership. However, Chiou I-jen resigned as Secretary-General in December 1998 and was replaced by Yu Shyi-kun. Later that month Chao Shu-po, a Minister without Portfolio, was appointed Governor of Taiwan Province, replacing the elected incumbent, James Soong, as part of the plans to dismantle the Provincial Government, agreed in 1997. A minor reorganization of the Executive Yuan took place in late January 1999.

In March 1999 an unprecedented vote of 'no confidence' in the leadership of Premier Vincent Siew was defeated in the Legislative Yuan. The motion was presented by the opposition following Siew's reversal of his earlier position and his decision to reduce the tax on share transactions, apparently as a result of pressure from President Lee. The National Assembly passed a controversial constitutional amendment on 4 September, which, *inter alia*, extended the terms of the deputies from May 2000 to June 2002. Election to the assembly was henceforth to be on the basis of party proportional representation. Several politicians and critical citizens condemned the move as being 'against the public will'. Shortly afterwards, the KMT leadership expelled the Speaker of the National Assembly, Su Nan-cheng, from the party on the grounds that he had violated its policy on the tenure extension, thereby also removing him from his parliamentary seat and the post of speaker. There was widespread dissatisfaction regarding the National Assembly's action, and in March 2000 the Council of Grand Justices of the Judicial Yuan ruled it to be unconstitutional. Later that month the DPP and the KMT reached an agreement on the abolition of the body and the cancellation of elections scheduled for early May. In April the National Assembly convened, and approved a series of constitutional amendments, which effectively deprived the body of most of its powers, and reduced it to an ad hoc institution. The capacity to initiate constitutional amendments, to impeach the President or Vice-President, and to approve the appointment of senior officials, was transferred to the Legislative Yuan. The National Assembly was to retain the functions of ratifying constitutional amendments and impeachment proceedings, in which case 300 delegates, appointed by political parties according to a system of proportional representation, would convene for a session of a maximum duration of one month.

The Presidential Election of 2000 and Beyond

In November 1999 it was announced that a presidential election was to be held in March 2000. Five candidates registered: Lien Chan (with Vincent Siew as candidate for Vice-President) was the KMT nominee, while Chen Shui-bian, a former mayor of Taipei, was to stand for the DPP (with the feminist Annette Lu as vice-presidential candidate), and Li Ao was to represent the New Party; the former DPP Chairman, Hsu Hsin-liang, qualified as an independent candidate, as did James Soong, who was consequently expelled from the KMT, along with a number of his supporters. Jason Hu resigned as Minister of Foreign Affairs in November 1999 in order to direct the KMT's election campaign. He was replaced by Chang Che-shen, hitherto Director-General of the Government Information Office. It became evident in the following month, when Soong was publicly accused of embezzlement, that the election would be bitterly contested. Although Soong denied the charges, his popularity was affected. In January 2000 Lien Chan, in an attempt to regain the support of disillusioned voters, proposed that the KMT's extensive business holdings be placed in trust and that the party terminate its direct role in the management of the numerous companies in which it owned shares. The KMT adopted the proposal shortly afterwards. In a reflection of the tense political situation between Taiwan and China, Chen Shui-bian of the DPP modified the party's stance and pledged not to declare formal independence for the island unless Beijing attacked.

The presidential election, held on 18 March 2000, was won by Chen Shui-bian, who obtained 39.3% of the votes cast. James Soong, his closest rival, received 36.8% of the votes. (On the day after the election he founded the People First Party, in an attempt to take advantage of his popularity.) Lien Chan of the KMT secured only 23.1% of the votes. The remaining candidates obtained less than 1%. The poll attracted a high level of participation, 82.7% of the electorate taking part. (Upon his inauguration in May, Chen would thus become Taiwan's first non-KMT President since 1945.) Violence erupted as disappointed KMT supporters besieged the party's headquarters, attributing the KMT's defeat to the leadership's expulsion of James Soong and the resultant division of the party. Lee Teng-hui subsequently accepted responsibility for the defeat and resigned from the chairmanship of the party. Lien Chan assumed the leadership. As the KMT continued to dominate the Legislative Yuan, however, the party did not entirely relinquish its influence, and in early April gave permission for Tang Fei, a KMT member and hitherto Minister of National Defense, to serve as Premier (although he was to be suspended from party activities while in the post). Following protracted negotiations, the membership of the new Executive Yuan, which incorporated 11 DPP members and 13 KMT members, was approved in early May. The incoming Government largely lacked ministerial experience. Furthermore, the DPP's lack of a legislative majority impeded the passage of favourable legislation. The size of the budget deficit also made it difficult for the DPP to fulfil specific electoral pledges on health, housing and education. In July Frank Hsieh replaced Lin Yi-hsiung as Chairman of the DPP.

In July 2000 the Government was heavily criticized after a river accident in which four workers, stranded by a flash flood, drowned as a result of the authorities' failure to provide a rescue helicopter. While the various government departments deliberated over the allocation of responsibility, the victims' final hours were broadcast live on national television. The Vice-Premier and Chairman of the Consumer Protection Commission, Yu Shyi-kun, subsequently resigned, as did senior officials of the emergency services. The Premier's offer of resignation was refused by President Chen Shui-bian. In October 2000, however, Tang Fei resigned as Premier, ostensibly owing to ill health. It was suggested that his departure from the post was due to the Government's failure to agree upon the fate of Taiwan's fourth nuclear power plant, the DPP being opposed to the project. (Later that month the Executive Yuan announced that construction of the plant was to be halted.) Vice-Premier Chang Chun-hsiung was appointed Premier, and a minor government reorganization was effected. Changes included the appointment of Yen Ching-chang as Minister of Finance, his predecessor having resigned following a sharp decline in the stock market.

Political disputes over the construction of Taiwan's fourth nuclear power plant intensified later in October 2000. The Minister of Economic Affairs, Lin Hsin-yi, was expelled from the KMT for 'seriously opposing KMT policies and impairing the people's interests' after he had demonstrated his support for the cancellation of the project. At the end of the month Chang Chun-hsiung announced that the Executive Yuan had decided to halt construction of the plant for financial and economic reasons. Although environmentalists were pleased, citing Taiwan's inability to process nuclear waste and to cope with accidents, the KMT reacted furiously, rejecting the Government's right to cancel a project approved by the legislature, and, together with the New Party and the People First Party, immediately began collecting legislators' signatures for the recall (dismissal) of Chen Shui-bian. The opposition was not mollified by a subsequent apology from Chen, and shortly afterwards the Legislative Yuan passed revised legislation on the process for presidential impeachment. Owing to the controversy, KMT member Vincent Siew refused to act as the President's representative to the annual APEC forum in Brunei in November and was replaced by Perng Fai-nan, the Governor of the Central Bank. The dispute became so serious that the business community issued an unprecedented public statement that economic recovery should take priority over political differences, but in December some 10,000 protesters in Taipei demanded that Chen resign. In November the Government requested a constitutional interpretation on the issue from the Council of Grand Justices, which it agreed to be bound by. The Council ruled in mid-January 2001 that the Government should have sought the legislature's approval before halting construction of the plant, and in mid-February 2001 the Government decided immediately to resume construction of the plant. In March 2001 a minor government reorganization was effected. The most notable change was the appointment of Hu Ching-piao, hitherto a Minister without Portfolio, as Chairman of the Atomic Energy Council, replacing Hsia Der-yu. It was rumoured that Hsia had disagreed with Chang Chun-hsiung over the future of the nuclear plant.

Meanwhile, in September 2000 four retired naval officers and one still serving were arrested in connection with the suspected murder in 1993 of a naval captain, Yin Ching-feng, to prevent him from revealing a scandal surrounding Taiwan's 1991 purchase of French-built frigates. It was alleged that bribery had influenced the award of the contract, which had been abruptly withdrawn from a South Korean firm. In October the Control Yuan impeached three former naval admirals, including the former Commander-in-Chief of the Navy, Adm. (retd) Yeh Chang-tung, for their involvement in the affair. In December the Taiwanese authorities appealed to the French Government for information to assist their investigations. The report of a two-year investigation by the Control Yuan, released in March 2002, strongly criticized the Ministry of National Defense and the navy command for failing to investigate fully Capt. Yin's murder, and recommended the court-martial of Adm. (retd) Yeh and former Prime Minister and Minister of National Defense, Hau Pei-tsun. The report also revealed that France had divulged to China confidential information regarding the deal. Several retired admirals went on trial for corruption in late April 2002.

By March 2001 President Chen had lost popularity within his DPP for softening his stance on two of its key policies—independence for Taiwan, and commitment to a nuclear power-free island. Fears emerged that the DPP could lose seats in legislative elections due in December, though polls showed that some 60% of the public supported construction of the nuclear plant. However, polls also showed Chen's popularity to be in decline, with much of the optimism that had greeted his inauguration being replaced with concerns about his inexperience and lack of power, his handling of the economy, and the political gridlock in domestic affairs, as well as the impasse in cross-Straits relations.

Concerns about the economy were heightened in May 2001 when 20,000 demonstrators from 18 trade unions marched in Taipei to protest against the Government's inability to reduce the unemployment rate, which at nearly 4% was at a 16-year high; furthermore, economic growth had declined to its lowest rate in 26 years. Later in the month Chen announced that he was planning to form the island's first coalition government after the December elections, in order to end the political infighting. In late June Chen received a major political boost when former President Lee Teng-hui offered his public support and suggested that he would back Chen in the December elections. Lee had disagreed with KMT leader Lien Chan, and his possible defection to Chen's support base threatened significantly to weaken the KMT. The emerging alliance between the two pro-independence leaders signalled a potential realignment in Taiwanese politics into pro-mainland and pro-independence forces, and seemed to end Beijing's goals of developing a strong pro-mainland political bloc on the island. There were also concerns that the realignment could further polarize society in this regard.

Moves toward such a political environment gained momentum in early July 2001, when the KMT issued a policy paper arguing that Taiwan's best option in terms of its relations with China was to form a 'confederation' with the mainland—the furthest that any political party had moved in calling for a union with China. The architect of the new KMT policy, Su Chi, described this as being 'somewhere in the middle ground between independence and unification', though the party had adopted a noticeably more pro-reunification stance under Lien Chan. However, the KMT's Central Standing Committee in late July refrained from adopting the proposal, reflecting the party's uncertainty over mainland policy. On 12 August 2001 a new political party, the Taiwan Solidarity Union (TSU), was formally launched with the support of former President Lee, and consisting of breakaway members of the KMT and DPP, led by former Interior Minister Huang Chu-wen. The party was formed in order to help President Chen win a majority in December's elections, an important goal given that the KMT had used its majority to block many of Chen's reforms during the previous one-year period. However, there were fears that the new party would drain support from Chen's DPP.

In July 2001 President Chen called for major governmental reforms, including reorganization, streamlining measures, anti-corruption action, and improved inter-departmental co-operation. At the end of August Chen also accepted proposals by a special advisory committee on closer economic relations with the mainland (see above).

The Legislative Elections of December 2001 and Subsequent Events

The last months of 2001 were dominated by campaigning for legislative elections. As support for the KMT waned, politics increasingly became an ethnic issue, with the KMT and People First Party (PFP) drawing their support from those who had fled the mainland in 1949 and their descendants (approximately 15% of the population), while native Taiwanese (who comprised 65% of the population) supported the DPP and the TSU. Despite the economic recession, Chen's popularity rose in the period prior to the elections, amid rumours that the DPP would form a coalition to secure a majority. The DPP hoped that a coalition would enable reforms to the legislature, notably the abolition of the multi-member constituencies in favour of a single-seat, 'first-past-the-post' system.

At the elections held on 1 December 2001 the DPP emerged as the biggest single party in the new legislature, having won 36.6% of the votes cast and 87 seats, but failed to win a majority. The KMT won 31.3% of votes and 68 seats, thereby losing its dominance of the Legislative Yuan for the first time in its history. The PFP came third, winning 20.3% of the votes and 46 seats, while the newly formed TSU came fourth, with 8.5% of the votes and 13 seats. The New Party won only 2.9% of the votes and one seat, while independents took nine seats. The level of voter participation was registered as 66.2%.

In the immediate aftermath of the elections, political manoeuvring to establish a coalition began, as the DPP and KMT sought to have their candidate elected as Vice-President of the Legislative Yuan. On 21 January 2002 President Chen reorganized the Executive Yuan, appointing his hitherto Secretary-General, Yu Shyi-kun, as Premier. The move was seen as a consolidation of the President's power, aimed at improving his prospects for re-election in 2004; the new Premier was also thought to be a more efficient administrator than his predecessor. Other notable appointments included Chen Shih-meng, hitherto deputy governor of the Central Bank, as the Secretary-General to the President, replacing Yu. Lin Hsin-i, hitherto Minister of Economic Affairs, was appointed Vice-Premier and Chairman of the Council for Economic Planning and Development; Eugene Y. H. Chien, hitherto Deputy Secretary-General to the President, was appointed as Minister of Foreign Affairs; and Gen. Tang Yao-ming, hitherto Chief of the General Staff, became the new Minister of National Defense. Tang was the first native-born Taiwanese to hold the newly augmented defence post in a mainlander-dominated military, and he was to oversee military reforms in early 2002. Lee Yung-san, Chairman of the International Commercial Bank of China, was appointed Minister of Finance, while Lee Ying-yuan, hitherto deputy representative in Washington, DC, was appointed Secretary-General to the Executive Yuan; however, the ministers in charge of mainland and overseas Chinese affairs were retained, suggesting a desire for continuity in relations with China.

In late January 2002 elections were held for provincial city and township councillors. Although the KMT won the largest number of seats in these elections, it failed to expand its popularity on a national level, with the results reflecting the KMT's competent organizational mobilization methods. At the beginning of February an alliance of the KMT and PFP ('pan-blue camp') successfully blocked the DPP-TSU ('pan-green camp') candidate for the post of Vice-President of the Legislative Yuan, electing Chiang Ping-kun of the KMT to that position. At the same time, the incumbent President of the legislature, Wang Jin-pyng of the KMT, was re-elected to his post. Following their success, officials from the KMT and PFP stated that they would consider presenting joint candidates for the mayoralties of Taipei and Kaohsiung in late 2002, and possibly the presidency itself in 2004, with Wang as their candidate for the latter post.

In February 2002 the KMT-PFP alliance immediately challenged the new Government by attempting to force it to accept revisions to legislation concerning local budget allocations, which had been approved by the outgoing Legislative Yuan in December 2001. However, in the decisive vote in the new Legislative Yuan, the KMT and PFP failed to secure the majority necessary to accomplish this, thereby giving the Government a minor victory. The narrowness of this victory, however, indicated that the KMT continued to pose a formidable obstacle to the new Government. An early set-back for the Government came in late March when the Minister of Economic Affairs, Christine Tsung, resigned, citing a hostile political environment, particularly in the Legislative Yuan. She was replaced by her deputy, Lin Yi-fu.

A major political scandal erupted in late March 2002 when one daily and one weekly newspaper reported that the Government of former President Lee Teng-hui had, in co-operation with the island's intelligence service (National Security Bureau—NSB), clandestinely established an unauthorized fund worth US $100m. to finance covert operations on the mainland and to further Taiwanese interests among influential lobby groups abroad, including the USA. Prosecutors immediately raided the offices of the two newspapers and seized the offending copies, accusing the editors of leaking state secrets, amid fears that freedom of the press would come under threat. The scandal threatened to damage Lee's position, as well as jeopardize espionage missions on the mainland and Taiwan's reputation for maintaining secret intelligence—it was widely believed that the source of the 'leaks' was a former NSB colonel who had embezzled US $5.5m. and then fled the island. Following the revelations, President Chen reiterated his commitment to press freedom and proposed new oversights for the NSB.

In early April 2002 the Executive Yuan approved plans to abolish the posts of Speaker and Deputy Speaker of the National Assembly, and replace them with that of a chairman of the session. In early May the Government revealed plans to reform the electoral system, which would reduce the number of seats in the Legislative Yuan from 225 to 150 and extend the term of legislators from three to four years. Some 90 seats would be filled from single-seat constituencies (thereby eliminating the need for candidates from the same party to compete against each other, as in the existing multi-seat constituencies), with the remaining seats divided proportionally among parties that gained more than 5% of the total vote. It was hoped that such reforms would make government more efficient and less dependent upon the availability of finance. A disadvantage of the system of multi-seat constituencies was that it required the participation of larger numbers of candidates and thus greater funding, thereby encouraging corruption.

Also in early May 2002, KMT Chairman Lien Chan announced that his party would form an official alliance with the PFP in order to strengthen opposition to the ruling DPP. However, plans for a joint candidacy in the 2004 presidential elections were hampered by regulations stipulating that the presidential and vice-presidential candidates must come from the same party. Meanwhile, in mid-May thousands of people demonstrated in favour of changing Taiwan's official name from the 'Republic of China' to 'Taiwan'—an initiative that was supported by 70% of respondents in an opinion poll conducted by the Ministry of Foreign Affairs in 2001, but strongly opposed by China.

In mid-June 2002 Premier Yu Shyi-kun appointed Liu Shyh-fang as the first female Secretary-General of the Executive Yuan, replacing Lee Ying-yuan, who was standing as the DPP's candidate in elections for the mayoralty of Taipei, in December. In late July President Chen Shui-bian formally assumed the chairmanship of the DPP, in a move designed to bring party policy into line with the Government. However, critics suggested that the dual leadership style was reminiscent of the excessively strong executive that characterized the decades of KMT rule.

In mid-September 2002 the Executive Yuan approved drafts of the new Political Party Law, which would ban political parties from operating or investing in profit-making enterprises, and allow the Government to investigate and confiscate assets unlawfully obtained by political parties. Although ostensibly aimed at creating greater political fairness and financial openness, the draft legislation was viewed as being aimed at the KMT which, during the decades of its rule, had amassed a vast commercial fortune worth an estimated NT $53,750m. (US $1,600m.) in 2001. As a result, the KMT might be forced to sell many of its assets.

A new political dispute emerged in late September 2002 over planned reforms of the debt-ridden agricultural and fishermen's credit co-operatives. The Government sought to reduce the activities of these local financial bodies, which had traditionally been used as a source of funds and influence for local KMT politicians. However, opposition to the reforms from farmers, KMT politicians and also Lee Teng-hui and the TSU was so intense that by late November President Chen was forced to suspend the plans. At the same time, more than 120,000 farmers and fishermen marched through Taipei to protest against the reforms, in what was the largest demonstration on the island since Chen took office. The protesters also demanded the establishment of a new agricultural development fund to alleviate the difficulties caused by Taiwan's entry into the WTO. Supporters of the reforms warned that Chen's failure to deliver them raised doubts about his Government's commitment to broader financial reforms needed to promote growth. Premier Yu Shyi-kun offered to resign, but was retained by Chen, who instead accepted the resignations of the Minister of Finance, Lee Yung-san, and the Chairman of the Council of Agriculture, Fan Chen-tsung. They were replaced by Lin Chuan and Lee Chin-lung, respectively. The departure of the respected Lee Yung-san, and the appointment of the third finance minister in as many years, raised concerns about political stability.

Meanwhile, Taiwan's political forces were also increasingly focused on elections for the mayoralties of Taipei and Kaohsiung, which took place in early December 2002, and were widely regarded as a preparation for the 2004 presidential elections. In Taipei the incumbent mayor, Ma Ying-jeou of the KMT, defeated his DPP rival, Lee Ying-yuan, winning 64.1% of the votes cast. Thus, he immediately emerged as a potential presidential candidate. His popularity complicated efforts by the KMT and PFP ('pan-blue camp') to select one of respective party chairmen Lien Chan and James Soong as their joint presidential candidiate. In mid-December the two formally committed themselves to this goal. In Kaohsiung the incumbent mayor, Frank Hsieh of the DPP, narrowly beat his KMT rival Huang Chun-ying, by 50.0% to 46.8% of the votes cast.

In late December 2002–early January 2003 a political scandal in the Kaohsiung City Council embarrassed both the DPP and KMT, when it was alleged that Chu An-hsiung, a businessman and politician, had bribed councillors from both parties in order to secure his election as speaker of the city council. This was the second scandal in Kaohsiung in recent months. In late 2002 it was alleged that a local businesswoman, Su Hui-chen, had bribed officials in the DPP and KMT in order to win favourable treatment for her property company, Zanadau Development. The incidents led to renewed demands for the reform of campaign finance laws, to limit the use of irregular funding in Taiwan's politics.

In January 2003 the release from death row of three men accused of a double-murder in 1991 prompted fresh calls for a substantial reform of the judicial system. The men were released on the grounds of insufficient evidence, and human rights activists and lawyers hoped that the case would set a precedent for establishing a clearer burden of proof in future capital cases. New legislation was to take effect from September 2003, which would give a greater role for defence counsels in courts. Also on the agenda were reforms aimed at divesting political parties of their media interests, namely in radio and television stations. During the decades of KMT rule, the party had amassed control or ownership of various media outlets, which the DPP now sought to dismantle.

In early February 2003 President Chen appointed Chiou I-jen, hitherto Secretary-General of the National Security Council, as Secretary-General to the President, while Kang Ning-hsiang, hitherto Vice-Minister of National Defense, succeeded Chiou in his former position. Later that month the leaders of the KMT and PFP again pledged to field a joint candidate for the presidency of Taiwan in 2004, with Lien Chan ostensibly the presidential candidate, and Soong as his running mate. Although the rivalry between the two initially appeared to place the alliance in jeopardy, in April the two parties confirmed the joint campaign and announced that KMT Chairman Lien Chan would be the presidential candidate and PFP Chairman James Soong the vice-presidential candidate.

Following the outbreak of SARS in early 2003, in April the Executive Yuan issued a bill containing urgent measures to prevent the spread of the disease, including restrictions on travel between Taiwan and Hong Kong. In May the Taiwanese Minister of Health, Twu Shiing-jer, resigned over the SARS crisis, and the Legislative Yuan announced that US $3,100m. would be spent on combating the effects of the disease, which by June had killed more than 80 people on the island. The epidemic was believed to be under control by mid-2003.

The issue of Taiwan's very identity came to the fore once again in early September 2003, when as many as 150,000 independence activists, led by former President Lee Teng-hui, rallied in Taipei to demand that the island's name be formally changed from 'Republic of China' to 'Taiwan'. The continued use of the former had remained the source of much international confusion and diplomatic tensions, and it was believed that the adoption of 'Taiwan' would facilitate wider diplomatic recognition. On the following day a rival rally, attended mainly by descendants of Chiang Kai-shek and veterans of the civil war (who continued to favour eventual reunification with the mainland), attracted several thousand supporters. The two rallies highlighted divisions within Taiwan over the issue of independence. Also in September, President Chen made proposals for the drafting of a new Taiwanese constitution to replace the operative constitution that had been introduced by Chiang Kai-shek and the KMT, declaring Taiwan to be a part of China. President Chen stated that he wanted a new constitution to be drafted by 2006, then approved by referendum and enacted by 2008. Furthermore, in October 2003 around 1,000 people, mostly ex-servicemen, gathered to commemorate Soong Mei-ling (widow of Chiang Kai-shek), following her death in New York.

In November 2003 Taiwan's legislature passed a new law to allow referendums to be held on the island, prompting strong criticism from China (see above). The approval of the legislation came shortly after President Chen formally declared his intention to stand for re-election in the March 2004 presidential election. In December 2003 President Chen announced plans to hold a referendum on the issue of China's deployment of missiles against Taiwan (see above), to coincide with the forthcoming presidential poll. Also in December 2003, President Chen confirmed that Vice-President Annette Lu, a vocal supporter of Taiwanese independence, would stand alongside him in the presidential election. Opposition presidential candidate Lien Chan and vice-presidential candidate James Soong, of the KMT and of the allied PFP respectively, as well as the popular mayor of Taipei, Ma Ying-jeou of the KMT, voiced strong opposition to the proposed referendum. In February 2004 popular demonstrations involving more than 70,000 people took place in southern Taiwan in support of the referendum, amid demands for China's removal of the missiles targeted at the island. At the end of February, in Taiwan's largest-ever demonstration, an estimated 1.5m. people formed a 500-km 'human chain', linking hands across the island to protest against China's deployment of missiles. Meanwhile, the wording of the referendum question had been modified to ask voters whether, in the event of China's refusal to withdraw its missiles, they would favour a strengthening of Taiwan's defence system.

The Presidential Election of 2004

In a dramatic development on the day prior to the presidential election, scheduled for 20 March 2004, while campaigning in Tainan both President Chen and Vice-President Lu were wounded in a shooting incident. Neither was critically injured in the apparent assassination attempt, and the poll proceeded as planned. President Chen was re-elected by a narrow margin of 29,518 votes, equivalent to only 0.2% of valid votes, with 337,297 ballot papers being declared invalid. A turn-out of 80.28% of the electorate was recorded. In the concurrent referendum on the issue of Taiwan's response to the deployment of mainland missiles, however, the majority of citizens declined to vote on President Chen's proposals, thus indicating widespread division within Taiwan on the issue of the island's relationship with China. President Chen's re-election was immediately challenged by opposition candidate Lien Chan, who claimed that there had been irregularities in the electoral process, and demanded a recount of the votes. Supporters of Lien rioted in Taipei and other cities. There was also some speculation that the assassination attempt on President Chen had been staged in order to further his election prospects. Ballot papers were subsequently reported to have been seized by the judicial authorities. President Chen's re-election was none the less officially confirmed by Taiwan's Central Election Commission, amid further protests by supporters of Lien Chan.

In early April 2004 Minister of the Interior Yu Cheng-hsien resigned, citing the failure to prevent the shooting of Chen during the election campaign as the reason for his departure. In the same month, in a separate development, Minister of Foreign Affairs Eugene Y. H. Chien resigned in connection with the resignation of the head of the American Institute in Taiwan (see above). A recount of votes cast in the presidential election was completed on 19 May. It was estimated that 38,000 ballots were controversial, and that of these 23,000 were votes for Chen. This conclusion was not likely to reverse Chen's re-election, however; in addition to the controversial ballots there remained unanswered questions surrounding the apparent attempt to assassinate Chen on the day before his re-election. In July US forensic expert Henry Lee, who had been investigating the shooting, stated that the truth about the incident might never be established, although the likelihood that the attack had been staged was reportedly believed to be minimal. Meanwhile, President Chen was none the less inaugurated for his second term on 20 May, and on the same day announced appointments to his new Executive Yuan. Yu Shyi-kun was retained as Premier.

Economy

ROBERT F. ASH

Based on earlier contributions by PHILIP E. T. LEWIS and ATHAR HUSSAIN

ECONOMIC DEVELOPMENT

Since 1949 the economic development of Taiwan has been shaped by historical, geo-strategic and geo-economic factors. The main historical influences have been twofold. First, Taiwan's early post-1949 development capacity was enhanced by the legacy of important agricultural, industrial and infrastructural initiatives, undertaken by Japan during its colonization of Taiwan between 1895 and 1945. Second, the defeat during the Chinese Civil War of the forces of the Chinese Nationalist Party (Kuomintang—KMT) by those of the Chinese Communist Party (CCP), and the subsequent transfer, in 1949, of the Government of the Republic of China to Taiwan were cathartic events that prompted fundamental shifts in economic policy. These shifts, reinforced and underwritten by the USA, made possible a swift economic recovery and thereafter helped promote a pattern of rapid, sustained and equitable growth that contrasted markedly with the KMT Government's disappointing growth record during the 'Nanking Decade' in China (1928–37). Meanwhile, until the 1970s, China's alienation from most Western countries (especially the USA) further served Taiwan's economic interests. Although almost universal recognition of the People's Republic as the sole legitimate government of China subsequently left Taiwan's international diplomatic and political status dangerously exposed, its buoyant economic growth (including its role as one of the world's most important trading nations) continued unaffected. Since the late 1980s, however, the accelerated expansion of cross-Straits trade and investment has increasingly become a key determinant of Taiwan's development trajectory. Even more recently, political uncertainties associated with the incumbency of a President and Government headed by a political party other than the KMT and the associated decline in the influence of the KMT itself have also impinged on Taiwan's economic performance.

Natural and geographical factors have played an important part in shaping Taiwan's economic development. Its geographical location has given Taiwan a central, strategically important position in the Asia-Pacific trading region—especially along trade routes between Japan, Korea, China and South-East Asia. Also significant was the small size of Taiwan's economy ('smallness' being defined to embrace natural and human resources, as well as mere surface area). In contrast to the economic self-sufficiency to which mainland China has often aspired (and the fulfilment of which its continental scope has facilitated), inherent natural and market constraints forced Taiwan to look outwards in order to maintain its economic development momentum. Its external orientation has been one of the most important characteristic features, as well as a critical determinant, of Taiwan's post-1949 economic development.

Taiwan generally lacks mineral and energy resources—indeed, no important mineral is found in significant quantities on the island. Thus, its energy requirements, as well as most raw materials for domestic industry, have had to be imported. The domestic coal industry is tiny, while onshore and offshore sources of petroleum and natural gas are also of minor significance. Taiwan imports approximately 95% of its energy requirements.

ECONOMIC STRATEGY, GROWTH AND STRUCTURAL CHANGE

Following the recovery from wartime economic destruction and from the dislocation associated with the arrival in Taiwan of the exiled KMT Government, economic policy in the 1950s was dominated by efficiency-enhancing institutional and economic initiatives in agriculture (not least, the implementation of land reform) and import-substituting industrialization (ISI). One result of ISI was a significant rise in industry's share in gross domestic product (GDP), from 20% in 1952 to 27% in 1960. By the end of the 1950s, however, import substitution was exhausted, and the Government of Chiang Kai-shek embarked upon a new strategy of export-led growth. This outward orientation—one that was to drive Taiwan's subsequent techno-industrial transformation—was perhaps the single most important watershed in Taiwan's post-1949 economic development. It has lasted to the present day and has been accompanied by one of the fastest-growing performances of any economy in the world.

Between the early 1950s and the first half of the 1960s, Taiwan succeeded in laying the foundations of self-sustaining growth. Economic growth rates were among the highest in the world (GDP growth averaged 8.3% per annum during 1952–65), while inflationary pressures remained modest. Until the mid-1960s, however, a large part of the increase in output was absorbed by a rapid rise in population, as the rate of natural increase remained high (averaging over 3% a year). Whether viewed through estimates of per caput income or through more aggregate measures of national product, investment, exports and growth, data for the late 1960s and throughout the 1970s offer evidence of an even more buoyant economic performance. Between 1966 and 1980 GDP grew at an average annual rate of 9.8%, while per caput income growth accelerated even more sharply owing to a reduction in population expansion from 2.7% to 1.8%.

Economic structural change after the 1960s reflected the impact of two new policy impulses: first, in the 1970s, the process of industrial deepening; second, in the 1980s, that of industrial upgrading and diversification. The economic implications of these initiatives were most readily seen in terms of their impact on markets, ownership and industrial composition. In particular, as aid and policy guidance from the USA receded, so Taiwan's own economic technocrats identified net market opportunities, which they sought to realize by fostering indigenous industrial entrepreneurship.

Even allowing for a declining trend in Taiwan's economic growth from 1980, GDP expansion during the next two decades remained buoyant (averaging 7.9% annually during 1981–90; and 6.4% during 1991–2000). Unlike almost all its East and South-East Asian neighbours, Taiwan emerged relatively unscathed from the Asian financial crisis of the late 1990s, its real GDP growth falling from 6.4% to 4.6% between 1997 and 1998. This was a performance that compared very favourably with that of each of the other three Asian 'tiger economies' (Hong Kong and the Republic of Korea (South Korea) having both suffered serious contractions in GDP—down by 5.3% and 6.7% respectively in 1998, and Singapore's GDP having risen by a mere 0.1%). In 1999 and 2000 the Taiwanese economy maintained its buoyant growth momentum, with GDP increasing by 5.4% and 5.9% respectively. There was, however, a portent of future difficulties in the sharp decline in economic growth from 6.7% in the first three quarters of 2000 to a much more lacklustre performance in the final quarter. This contraction signalled the onset of a deep recession, Taiwan's first in a generation, and in 2001 GDP registered a decrease of 2.2%. This absolute decline (from NT $9,663,000m. to NT $9,507,000m.) was unprecedented in Taiwan's post-1949 experience, the previous low point having been reached in 1974 when, under the impact of the global oil crisis, GDP growth was just 1.2%. The seriousness of the situation was highlighted in large absolute declines—of 17.3% and 30.9% respectively—in exports and investment during the third quarter of 2001. Despite the persistence of economic problems (see below), positive growth resumed in 2002, with GDP rising by 3.6%—a figure that would have been considered respectable by the standards of most countries, but which (2001 apart) was still lower than in any year since 1974. Initial forecasts pointed to a marginal improvement on this performance during 2003, but the negative impact on spending resulting from the onset of Severe Acute Respiratory Syndrome (SARS) caused the economy to contract in the second quarter, and, despite recovery in the second half of the year, GDP growth in 2003 reached only 3.2% (lower than the corresponding global figure of 3.9%, but higher than in the USA, Japan and EU member states). The most important economic consequence of SARS was its impact on private consumption growth; by contrast, the effect on exports of information technology (IT) products proved to be less significant. In addition, however, SARS had a serious effect on tourism, albeit one that was relatively brief. Taiwan suffered more than 80 deaths as a result of SARS. Despite these problems, Taiwan's longer-term economic prospects are quite encouraging. In the mean time, renewed overseas demand for IT products, as well as further improvement in domestic aggregate demand conditions, seem likely to generate growth averaging around 5% in both 2004 and 2005. If such projections prove correct, Taiwan's GDP growth will significantly exceed average global economic expansion.

Both internal and external factors contributed to Taiwan's poor economic performance at the beginning of the 21st century. Taiwan's strong external economic orientation has made it particularly vulnerable to developments overseas, and the global economic downturn associated with the bursting of the 'high-tech bubble' in the USA—still, along with mainland China, one of Taiwan's two most important export markets—took a severe toll on the economy in 2001. By the same token, external economic recovery subsequently contributed significantly to renewed growth in 2002 and 2003. Domestic factors also played a part. Recessionary conditions in 2001 were underlined by contracting private-sector investment, slowing consumption growth, rising unemployment and deflation. Although inflationary pressures have remained weak, improved consumer confidence has since been translated into stronger domestic demand and export growth has encouraged private-sector investment. Having peaked at 5.2% in 2002, unemployment has also started to decline.

Recent difficulties notwithstanding, Taiwan's post-war record of economic growth is almost unparalleled globally. From a base of profound and widespread poverty at the end of the 1940s, it has become one of the most important trading nations—with a trade-GDP ratio of 92.5% in 2003. It also possesses one of the most competitive economies in the world: in the World Economic

Forum's (WEF's) Global Competitiveness Report for 2002/03, Taiwan was ranked third in terms of overall growth competitiveness, and second on the basis of the WEF's technology index ranking. Reflected in this outstanding performance is Taiwan's transformation into one of the major alternative centres of high-technology capability outside the USA, Western Europe and Japan—a position that the Chen Shui-bian Government was committed to strengthening through its plan to develop a knowledge-based economy. In 2003 Taiwan ranked fourth in the world—behind the USA, Japan and Germany—in terms of the number of US patents awarded to it (the US Patent and Trademark Office awarded 6,676 patents to Taiwanese enterprises in 2003). The island's economic vibrancy is demonstrated by the fact that, with less than 0.4% of the world's population, Taiwan contributes more than 1% of global GDP (in terms of purchasing-power parity—PPP) and accounts for more than 2% of the value of global exports. This transformation has been reflected in improvements in living standards for an increasing share of Taiwan's population. In constant domestic price terms, average per caput income increased five-and-a-half fold between 1952 and 1980. Thereafter, on the basis of market exchange rates but measured in nominal terms, it rose from US $2,344 to US $8,111 between 1980 and 1990 (by an average of 13.2% annually). On the same basis, it peaked in 2000 (at US $14,188), before recessionary conditions reduced it to US $12,876 in 2001. Thereafter, per caput GDP recovered to reach US $13,157 in 2003. PPP estimates, however, indicate that per caput GDP, having decreased slightly in 2001, subsequently recovered to reach a new peak of US $24,500 in 2003. Contained within these estimates are substantial improvements in welfare. Even allowing for the deterioration that has taken place since the 1980s by international standards among developed and developing countries, Taiwan still enjoys a high degree of equality in income distribution. By the end of the 1990s, furthermore, the island had less than 1% of its population living in poverty, enjoyed one of the highest literacy rates (with almost universal enrolment in primary and secondary schools) of any Asian country and provided amongst the highest standards of health and nutrition. In 2003 average life expectancy was 79.1 years for females and 73.4 for males. The infant mortality rate was 6.8 per 1,000.

Underlying Taiwan's rapid, long-term growth has been a consistently high rate of domestic capital formation. Between the mid-1950s and the 1970s domestic savings were insufficient to finance domestic investment, and Taiwan was forced to borrow from abroad. Between 1951 and 1965 US aid—cumulatively worth US $1,440m.—was also an important source of investment and defence spending. Since 1975, however, Taiwan has been a net overseas lender, and domestic savings have more than paid for domestic investment. The national gross savings rate peaked at 38.5% in 1986 and 1987, since when it has followed a downward trend to reach 26% in 2003. The rate of gross fixed investment in 2003 was 17.5%. In general, the high interest rate available to savers has reflected official policies, designed to yield a real return on their deposits. The gross investment rate in 2003 was 16.6% (the corresponding figure in 1995 was 24.9%), generating excess savings of NT $953,679m.—or 9.4% of gross national product (GNP).

Rapid economic growth from the 1950s was accompanied by major changes in the sectoral composition of the Taiwanese economy. Agriculture's share of total output decreased from almost 30% in 1960 to 1.82%in 2003. During the same period the contribution of industry, which had reached a high point of 47.1% in 1986, contracted to 30.38% (25.54% from manufacturing alone), as the service sector overtook the industrial sector, to account for more than two-thirds (67.79%) of GDP in 2003. The emerging pre-eminence of the service sector was largely due to the buoyant growth performance of Taiwan's financial, insurance and commercial sectors.

Growth-driven structural changes were also accompanied by inter-industry shifts. For example, between the 1960s and the 1980s food and food-processing, textiles, and wood and paper manufacturing all declined at the expense of increases in the output of machinery and electrical goods. The most noteworthy development, however, has been the more recent and spectacular enhancement of Taiwan's status as a manufacturer of electronic and IT hardware and equipment of all kinds. By 1995 the island had become the third largest producer of such goods in the world, after Japan and the USA. Alongside declining production of televisions, radios, telephones and calculators in the 1990s, integrated circuit production recorded impressive expansion. Despite a sharp reduction in demand during and after the Asian financial crisis of the late 1990s, output has averaged close to 15% growth annually since the early 1990s. Between 1995 and 2000 Taiwan's share of global semiconductor manufacturing rose from 21% to an astonishing 71%—this against the background of a doubling in the global revenue from this market. The pre-eminence of the IT industry, however, has made Taiwan vulnerable to the exigencies of changing global economic conditions. More recently, a new threat to Taiwan's dominant position in semiconductor manufacturing has emerged in mainland China, where new large-scale capacity has been created.

Economic Policies

In the years immediately after 1949 the Taiwanese economy depended heavily on the production and processing of food and agricultural products, using labour-intensive, low-technology manufacturing processes. The surplus output of such goods was exported mainly to the USA. Meanwhile, Japan and the USA provided more than two-thirds of much-needed foreign investment. The importance of these two countries, especially as a source of imports, has continued to the present day, although since the late 1980s their significance as a destination for Taiwanese exports has declined as the role of China (including Hong Kong) has become increasingly important. In its initial stage, Taiwan's manufacturing growth, particularly in the production of electronic and electrical goods, was also mainly financed by US and Japanese capital.

After its removal to Taiwan, the KMT initially maintained tight controls over the economy. Nevertheless, the need to restructure itself following the débâcle of the Civil War—an aspiration emphasized in advice from the USA—eventually led it to promote a more market-orientated, free-enterprise system. This shift in emphasis at a time when its Chinese Communist rivals were putting in place a Soviet-style central planning system was, however, accompanied by a continuing and significant degree of interventionism by the Government, characterized by the maintenance of state ownership and the implementation of trade restrictions to protect the nascent manufacturing sector against imports. Macroeconomic management was also highly interventionist by modern standards. Meanwhile, in the countryside, land reforms redistributed property rights, enabling former landowners to invest in urban businesses and encouraging farmers to increase productivity and output.

Beginning around the end of the 1950s, many earlier ISI policies were dismantled, as the Taiwanese economy was opened up and reorientated towards international markets. The immediate goal was to expand industrial exports in order to create employment and absorb surplus labour that was beginning to emerge in the agricultural sector; in the longer run, it was hoped that foreign-exchange earnings would become a major source of capital for financing further sustained growth. In the event, a combination of short-term export financing, tax incentives, import duty rebates and foreign direct investment (FDI) were the policy means whereby export industries were further expanded. In particular, the 1961 Statute for the Encouragement of Investment offered tax benefits to export firms, including those in government-sponsored 'pioneer' sectors, such as electronics. At the same time, in export-processing zones, firms enjoyed a duty-free environment that favoured both production and export activities, and encouraged further FDI inflows. Significant too was the Government's establishment of 'industrial estates' throughout Taiwan, designed to make use of the abundant labour force.

Numerous companies benefited from these initiatives, which facilitated the establishment of new firms and laid the foundation for what became one of the hallmarks of Taiwan's industrialization—namely, the dominance of small and medium-scale enterprises (SMEs), rather than the concentration of large firms, such as the South Korean *chaebol*. Over time, the SMEs became the mainstay of Taiwan's industrial economy. By the mid-1980s their share of manufacturing output had reached almost half, their share of non-farm employment was more than

60% and they contributed a similar proportion of export earnings. At the end of the 1990s SMEs accounted for 98% of all enterprises and accounted for 78% of total employment.

As a result of the Government's export promotion strategy, labour-intensive, export-orientated industries became the main engine of Taiwan's economic growth. In the 1970s, however, concern about the perceived 'shallowness' of the economy, as well as the threat posed by the declining supply of surplus labour—and consequent upward pressure on wages—and evidence of growing protectionism in some major markets (most notably, that of the USA), were reflected in proposals to embark on a programme of heavy and chemical industrialization (HCI). In the event, the impact of this initiative was much less marked than that of the much more extensive HCI in South Korea. There were two important consequences. First, Taiwan avoided massive foreign borrowing and reliance on big business in order to implement the policy of HCI, and thereby escaped the problems of high inflation, huge foreign debt and deteriorating income distribution that were later to confront South Korea. Second, Taiwan's mild HCI prevented the emergence of dominant, large-scale conglomerates and, instead of basic and petrochemical industries becoming the mainstay of the economy, Taiwan's industrial and export base continued to be dominated by labour-intensive items, such as textiles, garment and consumer electronics.

Against the background of calls for industrial upgrading, in the early 1980s the Government identified two 'strategic areas'—machinery and information industries—to play a key role in Taiwan's future industrial development. These were considered activities with good market potential, favourable linkage effects, high value-added, high-technology input, low energy use and a low pollution impact. Later, other high-technology sectors were added to the list. This was to prove a watershed, as was the Government's decision for the first time to reward spending on research and development.

The most notable success of the new strategy was reflected in the rapid expansion of the IT industry (especially, semiconductors and computers). Here, industrial policy embraced a comprehensive range of measures, including tax incentives, concessionary loans, research and development sponsorship, technology transfer, and infrastructural initiatives (exemplified in the establishment of Hsinchu Industrial Park). The rise of the semiconductor sector was attributable to a variety of factors, including the existence of buoyant global demand for components for the computer industry, access to a 'reverse brain drain' of Taiwanese-born, but US-trained experts (many of whom had experience of having worked in California's Silicon Valley), and the benefits of SMEs' flexibility and adaptability to the industry's short product cycle. The Government's continuing role in facilitating technology transfer and technology diffusion also did much to help generate rapid growth.

After 1987 previous restrictions on trade with, and subsequently investment in, mainland China were removed, precipitating a rapid increase in cross-Straits economic relations. The outcome was to make Taiwan's evolving relationship with the People's Republic of China a major economic, as well as political, issue—one that has remained a constant source of debate. China's insistence that Taipei must accept Beijing's interpretation of the 'one China' principle has long been a major obstacle to political *rapprochement* across the Taiwan Straits. Following the presidential election of 2000, fears that the new incumbent (Chen Shui-bian) might implement policies in support of Taiwanese independence associated with the new ruling party (the Democratic Progressive Party—DPP) added to the political difficulties. Chen's re-election in March 2004 merely served to increase political tensions between Taiwan and China. Such profound problems have not, however, halted the momentum of cross-Straits trade and investment. To what extent Taiwan's growing trade dependence on, and economic integration with, the mainland will facilitate Beijing's ability to control its economy, and thereby accelerate national reunification, remains uncertain. The Government in Taipei is inherently trapped in the dilemma of whether to strengthen investment and trade links with China, at the risk of excessive dependence on its political adversary; or to seek to limit Taiwanese companies' access to the world's fastest growing economy, thereby sacrificing higher rates of growth at home. The island's accession (in January 2002) to the World Trade Organization (WTO) has become an additional major issue impinging on Taiwan's economic security. On the one hand, membership was expected to help diversify Taiwan's overseas markets and reduce its economic dependence on the People's Republic, whilst providing a firmer legal basis for the implementation of market access agreements. On the other hand, it was also expected to force Taiwan to accept direct trade links with the mainland and thereby increase the island's dependence. Meanwhile, cross-Straits economic interactions have had major implications for domestic development through their impact on manufacturing industry and employment, manifested in widespread concern over a possible 'hollowing out' of Taiwan's economy.

In May 2002 the Executive Yuan formally approved a comprehensive six-year National Development Plan (2002–08), referred to as 'Challenge 2008' and designed to transform Taiwan into a 'green silicon island'. The Plan's broad goals include the creation of 700,000 new jobs, a reduction in the rate of unemployment to under 4%, an increase in research and development spending to 3% of GDP, and the attainment of annual GDP growth in excess of 5%. Total expenditure under the Plan is estimated to be NT $2,600,000m., of which 35% is expected to come from the private sector (central and local government contributing 48% and 4% respectively, with the balance coming from 'special funds'). 'Challenge 2008' also seeks to double the number of foreign tourists visiting Taiwan and to raise the number of broadband internet users to more than 6m. The principal areas of planned investment spending will be human resource development, research and development and innovation (including developing at least 15 products or technologies that will rank among the world's best), the promotion of global logistics channels, and improvements in living standards (not least through placing more emphasis on environmentally-sound and sustainable development).

POPULATION AND WORK-FORCE

Taiwan is one of the most heavily populated places in the world, the density of its population (624.6 persons per sq km in December 2003) exceeding that of any province in mainland China. Its total population at the end of 2003 was 22.6m. Since 1949 demographic pressures have derived largely from the influx of more than 1.5m. mainlanders during 1948–49 and a subsequent, albeit quite short-term, rise in the birth rate (in the early 1950s Taiwan's birth rate was among the highest in the world). At the end of 2003 Taiwan's vital rates were as follows: crude birth rate, 10.06 per 1,000; crude death rate, 5.80 per 1,000; and rate of natural increase, 4.27 per 1,000. The total fertility rate in 2003 was 1.57 live births per female. In and after the 1960s deliberate fertility control measures facilitated a sharp decline in population growth, enabling Taiwan rapidly to complete its demographic transition. Underlying these demographic changes there has been a major contraction in the share of population under the age of 15 (from 40% to 20.1% of total population between 1970 and 2003). At the same time, however, as life expectancy has increased, so the proportion of the over-65s has risen (from 3% to 9.3%—a relatively high figure by international standards). Taiwan's dependency ratio in 2002 was 0.42; this is unlikely to change much in the short term. The proportion of those under 15 years of age, however, is likely to decline but to be offset by a rise in the corresponding share of the over-65s (Taiwan's elderly population now numbers almost 2.1m.). Because more men than women fled the mainland after the KMT defeat in the Civil War, there is still an imbalance of males over females in Taiwan (1.04 males: 1.00 females in 2003). Official projections suggest that Taiwan's population will peak at around 24.5m. towards the end of the 2020s, after which it will decline to reach little more than 22m. by 2050.

As the process of economic modernization gathered pace, Taiwan's population became increasingly urbanized. In 1920 the urban share of total population was a mere 4%, a figure that changed little in succeeding decades. Not until the 1960s, under the impact of rising farm productivity and expanding industrialization, did rural–urban migration get under way on a significant scale. Such, however, was the impact of structural change within the economy that by the early 1970s Taiwan's population had already become about two-thirds urban. By 1980

78% of Taiwanese lived in cities with a population of 50,000 or more, a higher proportion than in both Japan and the USA.

The labour participation rate (labour force as a proportion of the population aged 15 and over) rose steadily from 55.2% in 1970 to 60% by the end of the 1980s, since when it has followed a declining trend (in May 2004 it was 57.6%). Between 1952 and 1975 the total labour force increased from 3.06m. to 5.66m., an average annual growth rate of 2.7%; since 1975 the rate of expansion has slowed considerably (rising by only 1.4% annually since 1993), and at the end of 2003 the total labour force numbered 10.076m. From 1965 until 2000 the unemployment rate never exceeded 3%, and as recently as 1995 it was as low as 1.8%. However, as severely recessionary conditions emerged, it rose from 3.0% to 4.6% between 2000 and 2001, and increased again, to 5.2%, in 2002. In 2003 the Government launched two employment creation schemes, intended to generate some 130,000 new jobs. In fact, employment increased by 119,000 in 2003 (more than 100,000 of these positions reportedly resulting from the government's job creation programme), as a result of which the rate of unemployment declined to 4.99%—and to 4.47% by May 2004. A more buoyant economy no doubt also helped employment recovery. However, recovery was also aided by improvement in external demand conditions: in November 2003, for example, Taiwan Semiconductor Manufacturing (TSMC) announced its intention to recruit some 3,500 new employees by the end of 2004. Even so, the likelihood of a swift and sustained reduction in unemployment in the foreseeable future faces major obstacles, including the impact of WTO accession, the continued relocation of manufacturing activities to the mainland and the increase in redundancies associated with privatization.

The Government has long sought to generate not only rapid, but also efficient, economic growth. Accordingly, it has placed a high premium on maintaining a productive and highly skilled work-force. Between the mid-1980s and mid-1990s labour productivity in manufacturing grew by well over 6% annually. During the next eight years (1995–2003), this momentum was just about maintained, the corresponding increase being 5.95% (but 7.4% during 2001–03). Average employee earnings in manufacturing have lagged behind productivity growth: between 1995 and 2003 wages in manufacturing rose, on average, by only 2.5% a year. Concealed in the estimates of manufacturing labour productivity growth are, however, significant variations: for example, since the mid-1990s labour productivity in apparel and textile production has declined by almost 2% annually, while in electronics and electrical equipment it has risen by well over 9% per year. Recent legislative priorities have focused on the rights of workers, including workers' welfare, labour-management relations, health and safety issues, and fixing appropriate quotas for the hiring of foreign workers.

The changing distribution of employment reflects changing sectoral contributions to GDP. The share of agriculture in total employment, which exceeded 50% in the 1950s, has since declined to stand at a mere 7.3% in 2003. The momentum of construction activity means that the absolute level of industrial employment has not yet peaked, although the industrial share of employment has been in decline since 1987 (decreasing from 42.8% to 34.8% in 2003). Manufacturing employment alone has also contracted, its share in total employment having declined from a peak of 35.2% in 1987 to 27.1% in 2003. These changes highlight both the transfer of manufacturing activity overseas (especially to eastern coastal regions of China) and improvements in productivity. The most important area of employment growth has meanwhile become the service sector, which in 2003 had a work-force of 5.54m., or 57.9% of the total. In the first half of 2004 non-agricultural employment accounted for about 95% of total employment.

In 1990 Taiwan liberalized its foreign labour policy in an attempt to remedy a serious labour shortage. At the end of 2000 310,000 foreign workers were employed on the island, 59% of whom were engaged in manufacturing and 13% in the construction industry. Most overseas workers came from Thailand, although the Philippines was also a major source of domestic helpers. Foreign workers also play an important role in the nursing, health care and catering sectors. Their positive economic contribution notwithstanding, foreign workers have at times been viewed as a threat to local employment, and at the end of 1998 the Government in Taipei signalled its intention to reduce the foreign labour quota for manufacturing industries by 10% in order to combat the rising level of unemployment.

Education has made a critical contribution to Taiwan's modern economic growth. Since 1968 education has been compulsory for all children for nine years, and an increasingly wide range of educational opportunities has been made available for those who have left school. In 2003 there were 8,252 registered schools, colleges and universities (including 3,306 kindergartens and 2,638 primary schools), staffed by 274,837 teachers and with a total enrolment of 5.38m. students. Of the total student population, 35.5% were enrolled at the primary level; and 25% at junior and senior high school level. The overall student/teacher ratio was 19.59, compared with 24.22 in 1991.

Enrolment rates for children of primary school (aged 6–11) and secondary school (12–17) ages are close to 100%. The number of students undertaking higher education (including those in junior colleges) per 1,000 members of total population rose from 24.8 to 56.2 between 1990 and 2003, as a result of which Taiwan enjoys among the highest university and college enrolment rates in the world. Education at primary and junior high school levels is overwhelmingly dominated by the public sector; most kindergarten, senior high school and tertiary educational institutions are also in the public sector, although the role of the private sector looms large in these areas. In 2003 a total of 208,659 students graduated from colleges and universities; an additional 110,208 graduated from junior colleges. In the same year, about 24% of Taiwan's total population were enrolled in some kind of educational institution. Despite criticisms of the higher education sector for its supposed inflexibility and failure to address Taiwan's economic and social needs, it is noteworthy that in 2002 the most popular discipline in higher education was engineering, followed (in decreasing order) by commerce and business administration, mathematics and computer science, and medical science. Together, these four branches accounted for two-thirds of all students undergoing higher education. Almost 40,000 Taiwanese students were also enrolled in overseas universities: the most popular destination was the USA, although in recent years a growing number of Taiwanese students have participated in courses in EU member states (especially the United Kingdom).

MANUFACTURING

At the beginning of the 1950s the manufacturing sector 'proper' accounted for a mere 13% of GDP, compared with 32% for agriculture, and for the remainder of the decade farm-processing remained the main industrial activity in Taiwan. However, following the switch from ISI to an export-led strategy, the 1960s and 1970s witnessed a rapid increase in the growth of manufacturing, as a result of which its share in GDP almost doubled, rising from 19.1% in 1960 to 36.0% in 1980. Its highest ever share, of 39.4%, was attained in 1986, since when it has consistently declined, year-on-year, reaching 25.5% in 2003.

The changing pattern of industrial growth in Taiwan followed a familiar trajectory. Initially, export-driven growth was centred on the production of labour-intensive goods, such as textiles, plastics, plywood and electronic products. Subsequently, the shift to higher value-added activities was reflected in the accelerated development of capital-intensive, heavy industrial goods, including synthetic fibres, steel, machinery, cars and ships—a process given further impetus by the HCI programme (see above). The most recent and, in terms of its economic impact, most significant, change was towards knowledge- and skill-intensive production, the leading branches being semiconductors and associated products. The impact of this final shift is well illustrated in the finding that in the 1990s the share of industrial exports characterized by low technology intensity decreased from 49% to 17%, while that of high intensity increased from 18% to 42% (in 2002 the corresponding figures were 13.3% and 46.6%). In 2003 the share of heavy and technology-intensive industrial products in total exports was 75.3%. With labour costs rising and capital costs declining, the underlying pattern of industrial growth implied in such figures has accorded with the principle of comparative advantage.

The role of research and development in supporting the expansion of knowledge- and skill-intensive activities is self-evident, and maintaining and increasing research and development expenditure is critically important to the fulfilment of Taiwan's aspiration to promote the accelerated growth of even higher value-added production. Between 1991 and 2001 research and development spending in Taiwan grew, on average, by 9.9% annually to reach NT $205,000m. (2.2% of GDP) in 2001—a figure that is higher than that of the United Kingdom (at 1.9%), but significantly lower than in the USA (2.7%) and Japan (3.0%). Since 1998 a particular focus of the Government's research and development strategy has been the promotion of industrial technology research and development, designed to stimulate the development of knowledge-intensive industries and upgrade Taiwan's research and development capacity. Between 1998 and 2002 expenditure on industrial technology research and development increased from NT $13,700m. to NT $15,860m., the highest shares being absorbed by communications and opto-electronics (23.6%), machinery and aerospace (19.3%), materials and chemical engineering (14.4%) and biotechnology and pharmaceuticals (8.9%). The Government's intention is that research and development spending should constitute 3% of GDP by 2008. Thanks to institutional and fiscal support measures, considerable success has been achieved in recent years in reducing the Government's share of research and development expenditure in Taiwan, but the private sector's contribution, at 63% in 2001, is still low when compared with that of major industrialized countries.

SMEs have made a vital contribution to industrial development in Taiwan. Not only have they contributed more to foreign trade growth than their larger-scale counterparts, but (unlike large firms) they have done so without having been major recipients of government protection and incentives. About 97% of manufacturing enterprises are SMEs; they employ almost 80% of the manufacturing labour force and generate around 30% of the sector's sales (and about one-quarter of the value of associated exports). SMEs' strong outward orientation is highlighted in the finding that up to two-thirds of their output is exported, compared with about 35% for large-scale enterprises.

The promotion of SMEs may have reflected the KMT Government's wish to avoid the emergence of 'predatory capitalism' by encouraging competition among many firms and minimizing the discretionary element in industrial policy. The SMEs also maximized their potential for creativity and flexibility in production, enabling them to adjust quickly to changes in market conditions. At the same time, however, they have sometimes faced serious financial constraints, and their ability to engage in research and development is quite severely limited, necessitating direct and indirect support from the Government. In their frequent role as sub-contractors for foreign firms, SMEs have also struggled to create brand names in international markets.

Textiles

By the end of its period of colonization, Taiwan had become up to 30% self-sufficient in textiles, and after 1949 indigenous textile mills were further supplemented by the removal to Taiwan of equipment from the mainland. Between 1950 and 1980 textiles, both cotton and wool production, were a dominant sector, constituting the largest share of industrial employment and generating more exports than any other single manufacturing activity. Initially, textiles benefited from abundant supplies of cheap labour, although rising wages subsequently shifted production towards synthetic and other 'speciality' products (at one time Taiwan ranked second to the USA in terms of its production of synthetic fibres). Textiles were also important because of the industry's linkages with the machinery and petrochemical sectors. At the beginning of the 1980s textile products (including finished garments) still accounted for about one-fifth of the total value of Taiwan's exports. Thereafter, however, the sector's export contribution declined steadily—to 15% in 1990 and just 9.3% in 2002. Even so, garments and textile products still rank third in terms of the absolute value of exports (earning US $12,150m. in 2002). Since the 1990s many of Taiwan's factories producing cheaper, lower-quality goods have been relocated to China (especially to the Pearl River Delta region of Guangdong) and to South-East Asia.

Machine Tools

Machine tools are one of the most important heavy industries in a developing economy. In Taiwan's case, not only have they generated important forward linkages with the rest of the economy, but they have also contributed much to its defence infrastructure. Machine tool production grew rapidly in the 1970s, and by the following decade had gained a significant share in the world market—even to the extent of invoking a US request that it introduce export restrictions. Although widely regarded as a large-scale, heavily capital-intensive industry, machine tool firms in Taiwan are predominantly small-scale, and their success has reflected their flexibility and ability to deliver high-quality products punctually. Between 1981 and 2000 machine tool exports increased, on average, by 12.3% annually; in 2003 the value of exports of machinery (including electrical machinery products) was US $15,877m.

Petrochemicals

As with machine tools, the 1970s witnessed the 'take-off' of petrochemical production in Taiwan, investment projects being sponsored by both government and the private sector. Following a period of rapid growth, by the end of the 1990s the industry's annual production value was in excess of US $12,000m. By the late 1990s Taiwan ranked among the leading 20 international producers of petrochemicals and was first in terms of polyvinyl chloride (PVC). This was an impressive performance, but in spite of domestic output having been directed overwhelmingly towards meeting local requirements, large-scale petrochemical imports have still been necessary in order to address what remains a large domestic demand deficit. Newly completed projects—not least those of the dominant Formosa Plastics Group—promise significantly to relieve such difficulties in the near future (in 2002 chemical purchases overseas cost Taiwan a total of US $11,340m. and ranked second among its imports). Environmental regulations have meanwhile become an increasingly important part of the background against which petrochemical output growth has taken place in recent years.

Motor Vehicles

The performance of the motor vehicle industry (mainly cars, but also some buses and trucks) has, by Taiwan's standards, been disappointing. Production began in the 1960s and subsequently expanded rapidly, peaking at over 436,000 vehicles in 1992 (an average annual rate of expansion of almost 20% over the previous 27 years, albeit from a minimal initial base). Since 1992, in response to the vagaries of changing domestic demand, output growth has been more erratic; in particular, in 2001 the onset of recession provoked a sharp decline, with production of motor vehicles decreasing by almost 28% (from 371,000 to 269,000). This precipitate decline was, however, brief, and output rebounded to 330,000 in 2002. Further expansion is expected to take place, as economic recovery and renewed growth continue. In general, rising incomes have made car ownership more popular: registered passenger cars numbered more than 5m. in 2003—almost all of them for private use—and in 2002 new car registrations numbered 250,000. A total of 10 car manufacturers were in operation in Taiwan in mid-2003, these being dominated by two local firms (Yulon Motors and China Motor Corporation).

For many years domestic automobile production benefited from protectionist policies, including the imposition of high imports tariffs and quantitative import bans and restrictions. The less favourable consequence of protection has, however, been low efficiency and poor competitiveness. Japan has already succeeded in significantly penetrating the island's car market, and Taiwan's accession to the WTO and the consequent need to reduce import tariffs is expected to intensify competition. Under the impact of the consequent rise in imports, smaller firms are likely to be pushed out of the market, leaving those that survive more dependent on overseas markets. Of these, China is likely to become increasingly important.

Information Technology

The development of high-technology industries can be dated from two initiatives that started in 1980: first, the construction of a manufacturing plant by United Microelectronics Corporation; second, the beginning of operations at Taiwan's first science-based industrial park at Hsinchu. By the end of the 1980s more than 70 research-based companies had established plants at Hsinchu, and 10 years later the corresponding figure was in excess of 200 (with sales worth US $11,600m.). Meanwhile, the 1990s witnessed the establishment of a second government-sponsored science park in Tainan, while private-sector sites, such as that opened near Taipei by the Formosa Plastics Corporation, also contributed to the development of high-technology manufacturing. Such facilities, assisted by the Ministry of Economic Affairs' Industrial Technology Research Institute (ITRI), are the principal means whereby technologies have been developed and transferred to domestic private enterprises.

By the end of the 1990s Taiwan had become the world's third largest manufacturer of IT products, behind the USA and Japan. With a total production value of US $39,900m. (domestic and overseas revenue combined), the IT industry had also become the single most important source of foreign exchange. Taiwan's 900 computer hardware manufacturers, meanwhile, provided jobs for some 100,000 employees. Laptop computers, monitors, desktop personal computers (PCs), and motherboards have come to account for about 80% of the production value of the IT industry. The rapid expansion of such activities is highlighted in the following cumulative rates of output growth between 1995 and 2002: PCs, 112%; mother boards, 191%; printed circuit boards, 113%; interface cards, 214%; and electronic condensers, 244%. The shift towards higher-end activities is also reflected in the fact that during the same period production of monitors and colour picture tubes declined by 69% and 72% respectively. Coinciding with the onset of SARS, IT output experienced absolute decline during the second quarter of 2003, although subsequent recovery—especially in the production of liquid crystal displays (LCDs) and integrated circuits (annual sales of each of which are planned to reach NT $1,000,000m. by 2006)—was swift, outstripping growth in other industries. The strength of Taiwan's IT industry is reflected in the high global market shares it has attained in key areas: in 2003 66% for notebook computers, 50% for personal digital assistants, and 6.7% for mobile telephones. That domestic electronics output should have maintained such a buoyant momentum since the mid-1990s, alongside large-scale Taiwanese FDI in electronics production in mainland China, suggests that the domestic IT industry was upgrading in 'classic' fashion, shifting from less sophisticated, lower-value to more sophisticated higher-value activities. Such adjustments indicate the need for caution in accepting the argument that 'hollowing out' has been taking place in Taiwan's industrial sector. Meanwhile, the administration of President Chen Shui-bian has been consistent and explicit in its determination further to develop Taiwan as a knowledge-based economy.

The spectacular foreign-exchange contribution of IT and related products is illustrated by the fact that associated exports increased almost 20-fold between 1986 and 2000, registering an average annual growth rate of 21.4%. The value of IT exports recorded in 2000 (US $19,562m.) has not subsequently been reattained. Indeed, in the wake of the bursting of the US technology 'bubble' and increasing competition both from existing IT companies and newly established firms (not least in China), recent years have seen a substantial decline in export earnings. In 2003 such earnings contracted by over 12% to US $14,061m.—the lowest level since 1998. These figures reflect the fact that the share of the IT sector in Taiwan's global export sales has fallen from over 13.0% in 2000 to 9.7% in 2003.

The structure of Taiwan's IT industry resembles a pyramid. At the top, a small number of companies pursue costly and time-consuming product-innovating research and development activities, the results of which are translated into production by SMEs located at the bottom of the pyramid (SMEs account for about 85% of total domestic IT output). As in other areas of manufacturing, inherent size constraints have prevented SMEs from themselves undertaking major research and development and marketing investment. In addition, heavy reliance on imports of key components and advanced technology from overseas has tied Taiwan's IT industry closely to the industries of the USA and Japan.

Since the late 1990s the benefits of cheap, but adequately skilled, labour and low rents have attracted Taiwanese entrepreneurs to relocate IT production facilities to China. Half or more of Taiwanese high-technology goods are now assembled on the mainland, including most digital cameras, visual display units (VDUs) and an increasing share of other computer hardware. In 2003 47.5% of Taiwan's total IT output was produced in China, compared with 37% in 2001—a figure which is expected to reach well over 60% by 2005. By contrast, only 36% of Taiwanese IT goods were actually produced on the island in 2003. Meanwhile, in 2003 the share of total computer (desktop and notebook) output by Taiwanese firms produced in mainland China had exceeded 60%—and in the case of some individual firms, was as high as 80%. Of course the large-scale relocation of IT manufacturing to the mainland has implications for China too: in 2002, for example, more than 60% of China's IT output was produced by Taiwanese firms. A controversial initiative in 2003 was the Government's decision to allow Taiwan Semiconductor Manufacturing Company (TSMC) to establish a semiconductor fabrication plant in China in order to produce wafer fabs. Important conditions, however, were attached to this decision: first, only the production of 200mm wafers—not the more advanced 300mm wafers—was allowed; second, relocation of production was permitted only on condition that the firm in question undertook sufficient investment in China to upgrade technological levels there. For Taiwanese semiconductor manufacturers, burgeoning demand on the mainland (already worth well over US $15,000m. a year, and projected to reach US $40,000m. by 2025) represents a way of maintaining revenue sales through the simultaneous penetration of the Chinese market, offsetting possible future falls in demand elsewhere in the world. Recent trends can also be interpreted as evidence of a shift in emphasis, as Taiwan focuses more on design and logistics, managing production and distribution for multinational companies from Taiwanese-owned enterprises on the mainland.

INFRASTRUCTURE

Taiwan benefited from important infrastructural initiatives undertaken by the Japanese during their colonization of the island. Between the mid-1950s and the end of the 1970s the Government directed large-scale investment funds towards extending Taiwan's physical economic infrastructure. Most important of all were the 'Ten Major Construction Projects' implemented between 1973 and 1980, which facilitated major improvements in roads, and air and sea transport. In the 1980s infrastructural initiatives were accorded much lower priority, as a result of which quite serious problems of congestion began to emerge. Accordingly, from 1986 economic plans once more gave a high priority to infrastructural construction, but with a new added emphasis on the need to accommodate environmental pressures. At the start of the 21st century, such initiatives have continued to be regarded as a vital factor in generating continuing economic growth. In particular, the Chen Shui-bian administration made clear its commitment to improving Taiwan's basic infrastructure through the expansion of airport passenger and cargo transport capacities, the extension of link roads between airports and harbours, and the construction of broadband networks. Under a six-year development plan ('Challenge 2008') announced in 2002, the Government is committed to spending NT $2,700,000m. on public projects, including a high-speed rail link between Taipei and Kaohsiung (described as the largest build-operate-transfer (BOT) project in the world and planned to begin operation in 2005), and an express rail service between central Taipei and Chiang Kai-shek International Airport. The importance of such initiatives was highlighted in June 2004 with the acknowledgement by the Council for Economic Planning and Development (CEPD) of Taiwan's Executive Yuan that inadequate infrastructure remained a major impediment to further improvements in Taiwan's economic competitiveness.

Since 1995 successive Taiwanese Governments have sought to promote the development of Taiwan into an Asia-Pacific Regional Operations Centre (APROC). This long-term proposal is intended to transform Taiwan into a business and investment hub in the region, embracing services and high technology, and providing multinational corporations with the manufacturing, air and sea transport, financial, telecommunications and media facilities essential for their efficient operation. The successful implementation of this project would greatly enhance Taiwan's regional economic status and, by making it a centre for transshipping, financial and communications activities, improve its strategic security and its ability to compete with other service and high-technology hubs, such as Hong Kong and Shanghai.

APROC's implementation is premised on further realizing Taiwan's core strengths: its strategic location within the Asia-Pacific region, its well-educated and highly skilled work-force, and its high level of scientific and technological expertise.

Transport and Communications

From the 1950s until the mid-1980s the growth of transport was rapid (there was an average annual rate of expansion of 9.8% during 1952–84). Subsequently, however, progress was much slower—and in many years entirely absent. Nevertheless, since 1975 the total length of roads in Taiwan has more than doubled, reaching more than 37,000 km (including almost 21,000 km of highways) in 2002. Even so, the pressure on major trunk routes is severe, making their further extension a high priority. Major highways, such as the Sun Yat-sen Freeway (the country's principal north–south route, connecting Taipei and Keelung), are often congested, and improvements are regarded as a major priority. The expansion of Taiwan's rail network has been much less impressive than that of its roads, although the opening of a high-speed rail link between Taipei and Kaohsiung (see above) will make available an important new transport artery. Total track length (including private railways used mainly for freight operations) in 2002 was 2,573 km—almost unchanged since 1980. As a result, whereas average road length per head of total population had increased by over two-thirds since 1980, the corresponding figure for railways showed a decline of almost 20%. These trends are reflected in figures for cargo and passenger transport. In 2002, for example, road and rail passengers totalled 1.229m., of whom over 85% were carried by roads; roads carried 96% of combined rail and road traffic in the same year.

Taiwan's six international ports (Keelung, Kaohsiung, Taichung, Hualien, Suao and Anping) are capable of handling marine traffic, and service ocean vessels, of all sizes. In 2003 the total volume of cargo handled (loaded and unloaded) by these ports approached 636m. tons to reach a record level.The single most important of these ports is Kaohsiung which, in 2002, accounted for 68% of all cargo handled by the six ports. Since the 1960s there has been a rapid expansion of Taiwan-registered ships, their gross registered tonnage having risen from a mere 1.13m. tons in 1970 to a peak of 6.6m. tons in 1993, although subsequently declining to 4.3m. tons in 2002.The peak level of cargo tonnage (130m. tons) was attained in 1995; in 2002 the corresponding figure was 100.5m. tons.

There are two international airports in Taiwan: one is Chiang Kai-shek International Airport in Taoyuan, which serves Taipei and northern regions of the island; the other is Kaohsiung International Airport, which serves the south. More than 50 airlines fly to and within Taiwan. Between the 1950s and the second half of the 1990s the numbers of domestic and international flights increased steadily. Since 1997, however, while the number of international flights has continued to rise, that of domestic flights has contracted. In 2002 some 549,000 flights were recorded, of which 156,000 (28%) were international. During the same year airports handled 44.2m. passengers—45% of them on international flights; freight traffic meanwhile totalled 1.51m. tons, of which 75% was international cargo. Although SARS took a heavy toll on the number of visitors to Taiwan in 2003 (the number of passengers handled decreased to 37.9m.), both Eva Air and China Airlines—the two local carriers—predicted that their profits would rise substantially in 2004.

Fixed telephone line penetration in Taiwan has reached about 60%, a level that is unlikely to change markedly in the foreseeable future. The most significant recent change in telecommunications has been the dramatic increase in the ownership of mobile phones, the penetration of which rose from a mere 6.7% in 1997 to more than 100% by 2003. The change in usage is highlighted in statistics that show that while the number of landline subscribers increased from 5.8m. to 13.4m. between 1989 and 2003, during the same period the corresponding rise for mobile telephone users was from close to zero to more than 25m. Further expansion in the mobile phone market will be mainly dependent on technological breakthroughs, such as the extended use of 3G technology (which permits users to receive and deliver internet data and video images on cellular handsets). Between 1995 and 2003 the number of internet service subscribers increased from 21,000 to more than 7.8m., making Taiwan's internet penetration among the highest in the world. Further expansion is likely to be quite slow, although the change to broadband use ('Challenge 2008' seeks to bring broadband facilities to more than 6m. households) will be a major element in such expansion. Foreign investment is already taking place and will continue to be encouraged in order to promote the development of broadband telecommunications services.

Energy

Taiwan has very limited natural resources, forcing it to import most of its raw materials and energy. Coal reserves, which are found in northern Taiwan, are negligible (102.65m. tons at end-2002); with reserves, respectively, of 190,000 Kl and 8,227m. cu m, petroleum and natural gas reserves are also scarce, being concentrated mainly in two northern counties of the island (Hsinchu and Miaoli). Total hydroelectric potential has been estimated at 5,047 MW; in 2002 installed hydro-capacity was 4.51m. kW.

In 1955 domestic sources provided almost three-quarters of Taiwan's total commercial energy supplies, 60% of which were derived from domestically produced coal. By 1970 well over one-half of energy needs were being met from imports, and the overseas share subsequently continued to rise steadily. At the same time, dependence on coal gave way to dependence on oil, overall responsibility for which lies with the state-owned Chinese Petroleum Corporation. Taiwan's energy imports account for almost 98% of total commercial energy supplies. In 2002 energy imports included 55,720m. kilolitres (oil equivalent) of crude petroleum, which cost US $6,752.5m., and 37,476m. kilolitres (oil equivalent) of coal. The fact that energy consumption growth has been marginally less than GDP growth since the mid-1970s points to improved conservation and/or efficiency in the use of energy during this period. The decline of heavy industry in Taiwan may lead to a slowing in the growth of energy demand in the coming years.

From 1990 the average annual increase in electricity consumption was 13.3%, reaching 141,317m. kWh in 2002 (6,260 kWh per household). Taiwan Power is the main domestic provider of electricity in Taiwan. Meanwhile, the likelihood that continuing growth of consumption will generate excess demand in the near future has led to the opening up of electricity generation to investment from overseas in order to enhance production.

From the late 1970s until the mid-1980s there was a sharp increase in the share of nuclear energy in total commercial energy supplies, from a mere 0.1% in 1977 to 18.1% in 1985. From the 1980s, however, plans to extend Taiwan's nuclear energy capacity became increasingly controversial and eventually, in late 2000, the Executive Yuan announced that a long-standing proposal to construct a fourth nuclear power plant would not, after all, be implemented. In 2002 Taiwan's commercial energy supplies totalled more than 112m. kilolitres: of this, the share of oil, coal, nuclear power and natural gas (including liquefied natural gas—LNG) were 49.6%, 33.4%, 8.7% and 6.9% respectively. In the same year, the industrial sector accounted for 52.7% of total energy consumption, compared with 23.9% for residential and commercial purposes, 15.6% for transport, 6.3% for the energy sector itself and a mere 1.5% for agriculture.

TRADE AND INVESTMENT

The outward orientation of Taiwan's development strategy and, in particular, the role of foreign trade (especially export promotion) have been important factors in the island's economic growth. Foreign trade expansion was a necessary condition of Taiwan's sustained economic growth, and from the 1960s the island's economy became increasingly trade-dependent. Indeed, along with the other first-échelon Asian newly-industrializing countries (NICs), Taiwan's experience has highlighted many traditional hallmarks of trade dependence. Foreign investment has also played an important role in Taiwan's economic transformation. Since the late 1980s cross-Straits interactions have added a new and crucial dimension to Taiwan's external economic relations, one that both complements and challenges its economic ties with other parts of the world.

The exhaustion of ISI by the late 1950s and the increasingly severe constraints imposed by limited domestic markets left export markets as the obvious alternative target for manufactured goods. Accordingly, at the end of the 1950s, the Government embarked on a more aggressive external-orientated strategy, as tariffs for domestic producers were lowered, protective measures for exporters introduced and other accommodating policies (such as the establishment of export-processing zones) implemented. Education and labour training were also accorded high priority in order to enhance the growth of manufacturing production for export.

Between 1960 and 1970 the value of Taiwan's merchandise trade rose six-and-a-half-fold, with annual growth averaging 20.6% (25% per annum for exports, and almost 18% for imports). For the time being, the balance of merchandise trade remained in deficit, although the trend was a declining one. Moreover, by 1970 industrial goods had already come to dominate exports, constituting almost 79% of total export value. Raw materials accounted for 63% of imports, compared with 32% for capital goods and 5% for consumption. While Japan generated 43% of Taiwan's imports, but absorbed only 9% of its exports, the corresponding figures for the USA were 24% and 38% respectively.

From the 1970s onwards, efforts were directed towards improving the quality and sophistication of exports, the most crucial feature of this process being the emergence of Taiwan as a major producer of electrical and electronics goods and, subsequently, of high-technology IT products (see above). The dual processes of industrialization and technological upgrading facilitated the accelerated growth of exports. In 1984 the total value of Taiwan's foreign trade reached US $52,400m., making it the 10th largest exporter and 15th largest trading nation in the world. By the 1980s Taiwan traded even more than Japan relative to the size of its economy: the combined value of imports and exports constituting 95% of GNP in 1980, compared with about 30% for Japan. Taiwan's export share alone was around 48%.

Until the late 1980s, the direction of Taiwan's foreign trade was overwhelmingly towards capitalist industrial countries and, above all, the USA and Japan. Fears that Taiwan might fall prey to the pessimistic predictions of dependency theory proved unfounded. Indeed, a major problem for Taiwan was its growing trade surplus vis-à-vis the USA, which peaked at more than US $16,000m. in 1987, compared with a mere US $1,200m. in 1976. The size of this bilateral surplus elicited protests from Washington, which were translated into the subsequent removal of Taiwan from preferential tariff treatment by the USA. Yet although the USA's deficit was later reduced, it is noteworthy that even in 2003 the Taiwanese surplus totalled US $9,121.5m. (for the period January–May 2004 it was US $2,043.5m.). These figures are taken from official Taiwanese government sources—it is instructive that foreign trade statistics published by the US Bureau of Census indicate a US trade deficit of US $14,151.5m.—a remarkable 55% higher than the figure cited above.

Taiwan's balance of trade with Japan, by contrast, has consistently deteriorated since the 1950s. By the mid-1980s its bilateral deficit averaged about US $3,000m.; by the end of that decade, it was more than US $6,000m.; and in 2003, US $20,722.9m. (US $5,429m.above the level of 2002, and only marginally below the record US $21,959m. deficit of 2000). Underlying these figures is the finding that successful efforts to increase Taiwanese exports to Japan were offset by soaring imports. Meanwhile, Japan's continuing economic problems make it an unlikely source of export growth for Taiwan in the immediate future.

Besides Japan and the USA, and with China not a protagonist of significance until the end of the decade, trade with other Asian countries accounted for around 15% of Taiwan's foreign trade in the 1980s. The remaining balance was roughly shared between Western Europe and the Middle East. During and after the 1980s, however, two new developments occurred: one (of minor importance) was Taiwan's willingness to enter into trade relations with Russia and Eastern Europe; the other, which rapidly became a key determinant, was the initiation for the first time since 1949 of indirect trade links with the Chinese mainland (see below).

In global terms, Taiwan has enjoyed a trade surplus in every year since 1976, although it is noteworthy that, were it not for the opening of trade links with China, a return to a global trade deficit would have taken place. The peak level of its global surplus was attained in 1987, when it reached US $18,700m. (and when foreign-exchange reserves totalled US $77,000m.—a remarkable rise on the US $9,000m. of just five years earlier, although well below the US $202,630m. of 2003). In the previous year Taiwan's export-to-GDP ratio had also reached its highest level (standing at 56.7%), and its import-to-GDP ratio its lowest level (–37.4%). This burgeoning surplus put great pressure on the NT dollar and, following the devaluation of the US dollar in September 1985, there was a sharp appreciation of the Taiwanese currency—by 42% against the US dollar by the end of 1987 (the biggest such appreciation among all Asia's major currencies). The fact that 1987 marked the beginning of a contraction in Taiwan's merchandise trade surplus is, of course, no coincidence. In 2003 Taiwan's global trade surplus was US $19,928m.

In general, until the onset of the Asian financial crisis in the late 1990s, Taiwan's foreign trade growth was consistently buoyant. During 1990–97, for example, the rate of expansion of Taiwan's global merchandise trade was almost 10% annually (8.9% for exports and 11.1% for imports). In 1998, however, overall trade decreased by 9%, and exports by 9.4%. The following two years witnessed recovery and further growth, which took the value of trade to US $288,300m. (exports reaching US $148,300m. and imports US $140,000m. in 2000)—levels that have not subsequently been reattained. In 2001 foreign trade once more contracted sharply (by over 20%), although, as a result of a more precipitate decline in imports than in exports, Taiwan's trade balance rose from US $8,310m. to US $15,629m. Recovery again took place, however, and in 2003 the total value of merchandise trade was US $271,419.8m., generating a surplus of US $16,928m. From this 2003 estimate it can be determined that the export-to-GDP ratio had reached the 51% targeted level set by the CEPD in the 1980s, even though the corresponding import-to-GNP ratio (44.8%) remained well below the CEPD's planned 52% level.

Exports

For many years Taiwan has relied heavily on industrial products for its exports. In 1965 agricultural goods (including processed farm products) accounted for 54% of all exports. Thereafter, however, the share of industrial goods rose rapidly to reach 90% by the end of the 1970s and 98.5% in 2003. The USA, Hong Kong, Japan and (more recently) mainland China have been the major purchasers of Taiwanese goods. Other important export destinations include Singapore, South Korea, the Netherlands, Germany and the United Kingdom.

The USA has consistently been the principal export destination for Taiwanese products, although indirect trade with China via Hong Kong increasingly rivalled—and has eventually overtaken—the position of the USA in recent years. Between 1995 and 2000 the value of Taiwanese exports to the USA rose by 32% from US $26,507m. to US $34,815m. Subsequently, however, exports declined sharply, and in 2003 the value of shipments to the USA was only US $25,942m.The most telling indicator of the reorientation of Taiwan's trade, however, is the fact that in 2002 and 2003 exports to Hong Kong (worth US $30,845m. and US $28,354m., respectively) outstripped those to the USA.

As early as 1990 Hong Kong replaced Japan as the second largest export destination for Taiwanese products—a position that it maintained until overtaking that of the USA in 2002. Two main factors underlie Hong Kong's increased importance to Taiwan. The first reflects the successful implementation of the Government's efforts to diversify its export markets. The second, and more important, stems from Taiwan's growing trade with China, which the Government insists must be carried out indirectly through a third party (usually Hong Kong). Although trade with Hong Kong has expanded, the annual rate of growth in recent years has fluctuated considerably, primarily because of the growing number of Taiwan-based industries that have invested in the mainland since the legalization of private exchanges in 1987.

Following 14 years of consistent increase, from 1996 export growth to Japan followed an erratic path, veering between expansion and contraction. In 1997 and 1998, for example, exports to Japan fell back cumulatively by 32%, before recovering to reach a record level of US $16,600m. in 2000—an improvement that was attributed to increasing exports of

Taiwanese IT products to Japan, as well as a rise in the output (mainly of electrical items and electronics) of Taiwanese-Japanese joint ventures based in Taiwan. These items were estimated to have accounted for almost half of all Taiwanese exports to Japan during these years. The 2000 surge, however, was not maintained: during the next three years exports once more contracted to a level (US $11,912m.) in 2003 that was almost identical with those of 1997 and 1999.

European markets have been targeted as part of Taiwan's export diversification policy. The Netherlands, Germany and the United Kingdom were the leading European destinations for Taiwanese goods in 2003, their purchases (worth US $11,217.7m.) accounting for well over 60% of the value of exports to all European Union (EU) member states in 2003. In order to overcome the trade restrictions created by the EU single market, several of Taiwan's largest companies have gained footholds in the EU by establishing their own factories or by merging with existing local companies. Such investment has mostly been directed to the production of electric appliances, electronics and IT products.

In North-East Asia, South Korea has become an important trading partner of Taiwan. Between 1995 and 2003 exports to South Korea grew, on average, by 5.9% annually—a figure that would have been higher but for the intervention of the Asian financial crisis, which caused a major export contraction in 1997 and 1998. In 2003 the value of exports to South Korea was US $4,573.5m., or 18% above the level of the previous year.

Countries in South-East Asia have also emerged as strong trading partners, and become an important destination for Taiwanese foreign investment. Abundant cheap labour, raw materials and lower land prices are the familiar factors that have attracted Taiwanese entrepreneurs. Between 1986 and 1996 the share of Taiwanese exports purchased by member countries of the Association of South East Asian Nations (ASEAN) rose from 5% to just over 12%. Having subsequently declined in the aftermath of the Asian financial crisis, only in 2003 did it almost reattain the previous peak (of 11.8%).

Imports

Most of Taiwan's imports comprise raw materials and semi-finished goods required for production, although there has recently been a small increase in imports of consumer goods. Capital goods imports have steadily risen, their share accounting for 20.5% in 2003. In 2000 the value of Taiwan's imports rose sharply to US $140,000m. (an annual rise of over 26% and 22% higher than the previous peak level of 1997). However, in 2001 total imports declined to less than US $107,238m., a 24.4% decrease on the previous year, and only in 2003 did they reattain and indeed exceed the 1997 figure (of US $127,248.5m.). In 2003 Asian countries accounted for about 57% of Taiwan's imports, followed by members of the North American Free Trade Agreement (NAFTA), ASEAN and the EU.

Cross-Straits Economic Ties

Trade relations with the mainland were historically extremely close, and in the early 1930s such exchanges probably accounted for about one-half of Taiwan's total trade, excluding Japan. Such relations were interrupted by the establishment in 1949 of the People's Republic of China and were only resumed in the 1980s, following an official announcement in July 1985 that while neither direct trade nor entrepreneurial contact would be permitted across the Taiwan Straits, entrepôt trade would no longer be prohibited. Since the end of the 1980s Taiwan has become increasingly trade-dependent on the mainland economy. A re-assessment of cross-Straits trade published by the Mainland Affairs Council (MAC) in Taipei indicates that in the first four months of 2004 Taiwan's total trade dependency ratio vis-à-vis China had risen to 17.5%—a figure that contains a relatively low import dependence (9.2%), but a much higher (25.3%) reliance on the mainland as an export market. Taiwan's degree of export dependence on China is uniquely high. By contrast, China's trade dependence on Taiwan is both lower and more stable, although it is instructive that by 2001 Taiwan had become the second largest source of China's imports, accounting for 11.2% of the total.

Government statistics from Taipei show cross-Straits trade having risen from US $8,100m. to $31,233m. between 1991 and 2000, implying an average annual growth rate of 16.7%. The recent pace of merchandise trade growth has been maintained. Thus, estimates made available by Taipei's Bureau of Foreign Trade indicate that during the first half of 2004 the value of two-way trade reached US $23,807.8m., or 84.7% more than in the corresponding period of 2003. Disaggregated data show that Taiwanese exports to the mainland more than doubled to reach US $16,187.3m., while imports also rose substantially (by 56%) to $7,620.6m. Implied in these figures is a bilateral trade surplus in the first half of 2004 of $8,566.7m.—174% above that of the period January–June 2003. It is notable, however, that if Taiwan's trade with Hong Kong and Macao is included, the surplus for the first half of 2004 falls to a mere US $129.5m.—that is, 13% below that of the first six months of 2003. Although these estimates no doubt accord with the reality of trade developments across the Taiwan Strait, their precision is subject to a significant margin of error. One certainty is that the rapid expansion in exports to China has been driven by increasing shipments of electrical machinery, equipment and parts (reported to have risen by 75% in 2002 to account for almost 32% of all Taiwan's exports to China).

In recent years huge amounts of FDI have also flowed to China from Taiwan, although quantifying these capital flows is extremely difficult. A major difficulty is that much investment by Taiwanese entrepreneurs flows through the Virgin and Cayman Islands, which, according to official mainland sources, together accounted for almost 14% of all inward FDI to China in 2002. Such considerations should be kept in mind when interpreting official data released by the Investment Commission of the Ministry of Economic Affairs in Taipei, which show that Taiwanese FDI to China totalled US $4,595m. in 2003 (19% more than in 2002), bringing the cumulative total for 1991–2003 to US $34,310m. What can be said with more certainty is that with around 40% of all Taiwanese FDI now being directed to the mainland, China has become an important link in the production chain of many Taiwanese firms. The Changjiang (Yangtze River) Delta (the regions of Jiangsu and Shanghai), Guangdong, Fujian and Zhejiang are, and will remain, the principal investment destinations within China. In 2002 these areas are thought to have absorbed about 90% of all Taiwanese FDI in China, although the geographical dispersion of Taiwanese FDI penetration in China is highlighted in the finding that Taiwan is also a major source of capital in Sichuan—the province at the heart of China's western regional development initiative. In 2002 plans were announced to construct two Taiwan-invested industrial parks in Shanghai. As for the functional distribution of Taiwanese FDI in China, industry remains dominant, although its share of FDI decreased from 99% to 81% between 1991 and 2002. The share of services, meanwhile, has increased from a negligible base to 9%. In the same year electronics and related activities accounted for almost 40% of FDI. With China's accession to WTO, the attraction to Taiwanese investors of tertiary sector activities was expected to grow in the coming years.

The extent of Taiwan's involvement in the mainland is also highlighted by the sheer number of Taiwanese business executives who live and work there. Several hundred thousand Taiwanese are permanently based in Shanghai, while there are said to be 30,000 resident Taiwanese managers, running 5,000 electronics factories that employ 2m. workers in Dongguan City in Guangdong's Pearl River Delta. Throughout China, in 2000–02 Taiwan directly employed at least 5m. Chinese workers—a figure that would be considerably higher if related employment was included.

The linked accession to the WTO of China and Taiwan in December 2001 and January 2002 respectively seems certain to accelerate the process of cross-Straits trade and investment. From Taiwan's perspective, a major implication of WTO membership will be the necessity, in accordance with WTO rules, to eliminate non-tariff barriers on Chinese goods. This process was already under way in January 2002. The Taiwanese Bureau of Foreign Trade issued a list of 2,126 items (including 901 agricultural products), which it stated would be added to the list of legal imports from China to Taiwan. The final list was expected to embrace almost 5,700 products, or 44% of all those included in Taiwan's tariff schedule. A further implication of the removal of such barriers will be to encourage more imports from China, which, in turn, is likely to reduce Taiwan's trade surplus vis-à-vis China.

Meanwhile, President Chen Shui-bian continued to face increasing pressure to relax investment restrictions. In October 2001 the DPP announced its decision to replace the supposedly outmoded 'no haste, be patient' policy on investment in China with a strategy of 'active opening and effective management'. Associated with this were the abandonment of previous restrictions on single investments of over US $50m. and a reduction in restricted investment categories, including PCs and semiconductors.

WTO membership is also likely to end Taiwan's long-standing bans on direct shipping and air links across the Straits. Although such links are not embraced by the 'Uruguay Round' of the negotiations of the predecessor General Agreement on Tariffs and Trade (GATT), their prevention would seem incompatible with the principle of most-favoured-nation treatment that lies at the heart of WTO rules. With direct links in place, freight and passenger transport costs will decrease, and trade can be expected to rise as a result of associated savings.

FINANCE

Taiwan's financial sector remained under notably tight central control until the end of the 1970s. By the following decade, however, changing circumstances—not least the oil crisis—dictated the first steps towards financial liberalization. Accordingly, price deregulation, including the abolition of interest rate controls, began to take place; and restrictions on capital movements were relaxed. Foreign exchange reforms were also introduced. Such changes served to facilitate more competitive financial behaviour and more efficient management. It was against this background that, in 1989, Taiwan's Banking Law was revised, the principal effect being to remove controls on deposit and lending interest rates.

Loosening control by the Central Bank of China (CBC, Taiwan's central bank) over the money supply was, however, confined to the mid-1980s (the rate of expansion of money supply in 1986–88 averaged 37.8%, compared with 13.3% for the previous three years). By the end of the 1980s quite tight control had been reimposed, and since then money supply expansion has not exceeded 16.9%.

A matter of growing concern in the 1990s was the deterioration in the financial position of Taiwan's banking sector. By the end of 1998 overdue bank loans had reached 4.5% of total lending, and while the major banks remained in profit, they felt compelled to set aside large reserves to meet lending losses. In 2000 some 10 Taiwanese banks were reported to have made losses, totalling NT $16,600m. By the end of that year non-performing loans (NPLs) were estimated to be NT $773,500m., equivalent to 5.34% of all domestic bank loans, and a year later the figure was estimated to have reached an unprecedented 8%. There did seem subsequently to have been some improvement in the situation: in 2001 there was reportedly a 60% rise in Taiwanese banks' cancellation of bad loans, and a further increase of 60% in 2002. By the end of 2002 official sources suggested that the NPL ratio had fallen to 6.1%—a figure that reportedly contracted further to about 5% by November 2003, following cancellation of debts amounting to NT $194,000m.

Since the beginning of 2001 the slow growth of money supply has been a source of anxiety, encouraging the CBC to ease its monetary policy (for example, the benchmark rediscount rate was lowered five times in 2001 to reach a record low). In the recessionary conditions of 2001, firms' demand for investment credit slackened, while the level of NPLs did little to encourage a pro active lending stance by banks. Having risen by 10.8% in 2000, in 2001 money supply growth declined by 0.9%, before recovering and again expanding, by 17.1%, in 2002. The upward momentum was maintained, and by November 2003, the rate of increase of money supply had reached 19%.

Membership of the WTO imposed new financial burdens on Taiwan, and from the late 1990s the Government introduced more financial deregulatory initiatives in anticipation of its accession. It has also shown itself to be in favour of the 'privatization' of banks—especially the three main commercial banks—most of which have traditionally been under state ownership or control (until 1991, 21 out of 24 Taiwanese banks were government-owned). Despite opposition from some quarters, legislative and institutional initiatives have helped to accelerate this process.

Meanwhile, in October 2000 an amendment to the Banking Law relaxed restrictions on the investment activities of Taiwanese banks, while increasing the ownership limit in other banks or financial companies from 15% to 25%. In December of the same year, in an attempt to encourage mergers involving foreign as well as domestic institutions, the 'Merger Act of Financial Institutions' was approved. In mid-2001 new financial reform laws were proposed, involving further liberalization and an acceleration of the consolidation of domestic banks. The 'Financial Holding Company Law' seeks to permit the establishment of integrated financial groups capable of offering a wide range of services, including banking, insurance and brokering. Insurance companies will also gain more freedom to formulate investment policies and develop new products. No less significant, in September 2001, control of 35 local financial institutions was transferred to 10 commercial banks. Financial reform will remain a priority for the Government, although progress is likely, as in the past, to be impeded by inter-party political conflict. Meanwhile, in the wake of recent money supply growth and the strengthening of domestic demand, increasing, albeit mild, inflationary pressures may be expected.

In the 1990s financial liberalization was accompanied by a parallel process in Taiwan's stock exchange, which was opened to direct investment by foreign institutions (in 1991) and to foreign individuals (in 1996). Many domestic households now own stocks, although only a quite small proportion of business firms in Taiwan seek to issue shares for public trading. Despite the earlier initiatives, in October 1997 little more than 3% of Taiwan's stocks were owned by foreigners. In the event, this no doubt protected Taiwan from damaging capital flight of the kind that was so injurious to other economies during the Asian crisis. Conditions have, however, since changed, not least, under the impact of further legislative initiatives designed to facilitate share ownership by foreigners, and by mid-2000 foreign investors accounted for the ownership of 12.5% of Taiwanese stocks. Since the early 1990s the number of activities not open to overseas investors has declined from more than 100 to 18 at mid-2003. Deregulation can be expected to continue in Taiwan's stock market.

The year 1997 was very eventful for the domestic stock market. Following a 50% rise over three months to take the index to a record high of 10,256 points, it then suffered as a result of panic selling associated with the onset of the Asian financial crisis. By the end of 1997 the index had fallen back to 8,228, and by the end of 1998 to 6,418 points. These movements should, however, be put into the perspective of a 52% rise in the stock price index between 1995 and 1997 (1966 as the base).

Government efforts to halt the stock market decline (for example, through the establishment of a US $9,000m. stabilization fund) were less than successful, and the 1997 peak has not subsequently been reattained. The huge scale of capital outflows since 1997, alongside ongoing securities liberalization, has led some members of Taiwan's business community to question the logic of greater exposure to international markets. As it turned out, under the impact of economic and political factors (such as, in the latter case, uncertainties associated with the emergence of the DPP as a major new political force), stock market volatility has continued, albeit around a downward trend. In 2000 the market lost more than 30% of its value, and although it recovered in 2001, it fell again in 2002 (when the market value was only 11% above the low point of two years earlier). In the first half of 2003 there were few signs of recovery, but a more buoyant momentum emerged in the second half of the year, which took the Taipei Stock Exchange Capitalisation Weighted Index (Taiex) to around 6,000 at the end of the year. The expectation is that as long as domestic and external economic recovery continues, share prices will continue to move upwards.

Statistical Survey

Source (unless otherwise stated): Bureau of Statistics, Directorate-General of Budget, Accounting and Statistics (DGBAS), Executive Yuan, 2 Kwang Chow St, Taipei 10729; tel. (2) 23710208; fax (2) 23319925; e-mail sicbs@emc.dgbas.gov.tw; internet www.dgbasey.gov.tw.

Area and Population

AREA, POPULATION AND DENSITY

Area (sq km)	36,188*
Population (census results)	
16 December 1990	20,393,628
16 December 2000	
Males	11,386,084
Females	10,914,845
Total	22,300,929
Population (official figures at 31 December)	
2001	22,405,568
2002	22,520,776
2003	22,604,550
Density (per sq km) at 31 December 2003	624.6

* 13,972 sq miles.

PRINCIPAL TOWNS
(population at 31 December 2003)

Taipei (capital)	2,627,138	Hsinchu	382,897
Kaohsiung	1,509,350	Taoyuan	357,647
Taichung	1,009,387	Jhongli	339,586
Tainan	749,628	Fongshan	328,878
Banciao	539,356	Sindian	280,661
Jhunghe	406,325	Chiayi	269,594
Keelung	392,242	Jhanghua	233,435
Sanchong	384,618	Yonghe	231,816
Sinjhuang	383,745	Pingdong	216,338

BIRTHS, MARRIAGES AND DEATHS
(registered)

	Live births		Marriages		Deaths	
	Number	Rate (per 1,000)	Number	Rate (per 1,000)	Number	Rate (per 1,000)
1996	325,545	15.18	169,424	7.90	122,489	5.71
1997	326,002	15.07	166,216	7.68	121,000	5.59
1998	271,450	12.43	145,976	6.69	123,180	5.64
1999	283,661	12.89	173,209	7.87	126,113	5.73
2000	305,312	13.76	181,642	8.19	125,958	5.68
2001	260,354	11.65	170,515	7.63	127,647	5.71
2002	247,530	11.02	172,655	7.69	128,636	5.73
2003	227,070	10.06	171,483	7.60	130,801	5.80

Expectation of life (years at birth, 2003): Males 73.4; Females 79.1.

ECONOMICALLY ACTIVE POPULATION
(annual averages, '000 persons aged 15 years and over)*

	2001	2002	2003
Agriculture, forestry and fishing	706	709	696
Mining and quarrying	10	9	8
Manufacturing	2,587	2,563	2,590
Construction	746	725	702
Electricity, gas and water	35	35	35
Trade, hotels and restaurants	2,207	2,268	2,283
Transport, storage and communications	487	477	484
Finance, insurance and real estate	432	438	442
Professional and technical services	267	285	285
Social, personal and related community services	918	950	988
Public administration	327	329	369
Total employed (incl. others)	9,383	9,454	9,573
Unemployed	450	515	503
Total labour force	9,832	9,970	10,076
Males	5,855	5,896	5,904
Females	3,977	4,074	4,172

* Excluding members of the armed forces and persons in institutional households.

Health and Welfare

KEY INDICATORS

Total fertility rate (children per woman, 2003)	1.24
Under-5 mortality rate (per 1,000 live births, 2002)	1.32
HIV/AIDS (% of persons aged 15–49, 2003)	0.03
Physicians (per 1,000 head, 2002)	1.58
Hospital beds (per 1,000 head, 2002)	5.92
Health expenditure (2000): US $ per head (PPP)	752
Health expenditure (2001): % of GDP	5.84
Health expenditure (2001): public (% of total)	64.9
Human Development Index (2002): value	0.891

For definitions, see explanatory note on p. vi.

Agriculture

PRINCIPAL CROPS
('000 metric tons)

	2001	2002	2003
Potatoes	32.1	38.3	44.3
Rice*	1,396.3	1,460.7	1,338.3
Sweet potatoes	188.7	191.4	199.8
Sorghum	21.7	17.9	17.5
Maize	166.0	188.9	167.9
Tea	19.8	20.3	20.7
Tobacco	9.2	7.6	5.2
Groundnuts	56.1	77.5	73.5
Sugar cane	2,180.3	1,973.1	1,695.7
Bananas	204.7	226.5	223.1
Pineapples	388.7	416.3	447.8
Citrus fruit	463.5	459.6	529.1
Vegetables	3,045.6	3,461.8	3,094.0

* Figures are in terms of brown rice. The equivalent in paddy rice (in '000 metric tons) was: 1,723.9 in 2001; 1,803.2 in 2002; 1,648.3 in 2003.

LIVESTOCK
('000 head at 31 December)

	2001	2002	2003
Cattle	146.0	144.1	144.0
Buffaloes	6.5	5.4	4.9
Pigs	7,164.6	6,793.9	6,778.8
Sheep and goats	184.7	161.9	241.0
Chickens	117,310	118,846	113,048
Ducks	10,104	10,124	10,112
Geese	2,613	2,542	2,757
Turkeys	235	191	186

LIVESTOCK PRODUCTS

	2001	2002	2003
Beef (metric tons)	5,057	5,303	5,523
Pig meat (metric tons)	1,165,998	1,125,721	1,082,224
Goat meat (metric tons)	7,219	6,789	6,591
Chickens ('000 head)*	376,196	377,522	371,420
Ducks ('000 head)*	32,142	31,012	31,040
Geese ('000 head)*	6,330	6,178	6,402
Turkeys ('000 head)*	458	387	364
Milk (metric tons)	346,079	357,804	354,421
Duck eggs ('000)	481,789	472,326	477,041
Hen eggs ('000)	7,325,125	7,069,644	7,018,647

* Figures refer to numbers slaughtered.

Forestry

ROUNDWOOD REMOVALS
('000 cubic metres)

	2001	2002	2003
Industrial wood	26.4	31.1	26.1
Fuel wood	6.0	4.0	15.0
Total	32.4	35.1	41.1

Fishing

('000 metric tons, live weight, incl. aquaculture)

	2001	2002	2003
Tilapias	82.9	85.1	85.4
Other freshwater fishes	19.2	18.8	16.3
Japanese eel	34.2	34.9	35.1
Milkfish	59.4	72.4	77.9
Pacific saury	39.8	51.3	91.5
Skipjack tuna	186.7	232.6	180.6
Albacore	64.1	62.8	54.1
Yellowfin tuna	109.6	97.2	96.8
Bigeye tuna	81.2	104.3	108.4
Chub mackerel	42.7	57.0	53.8
Sharks, rays, skates, etc.	44.2	45.3	67.5
Other fishes (incl. unspecified)	260.7	282.6	315.6
Total fish	1,024.7	1,144.3	1,183.0
Marine shrimps and prawns	29.0	25.2	42.0
Other crustaceans	9.8	9.9	17.4
Pacific cupped oyster	16.6	19.6	23.5
Common squids	13.7	14.1	17.2
Argentine shortfin squid	147.6	111.7	147.2
Flying squids	4.5	2.4	3.1
Other molluscs	0.4	0.5	0.8
Other aquatic animals	47.5	60.5	53.5
Total catch	1,301.2	1,388.2	1,487.7

Note: Figures exclude aquatic plants, totalling (in '000 metric tons) : 31.8 in 2001; 43.6 in 2002; 40.9 in 2003.

Mining

(metric tons, unless otherwise indicated)

	2001	2002	2003
Crude petroleum ('000 litres)	44,380	51,106	45,760
Natural gas ('000 cu m)	849,158	886,828	830,893
Salt	66,150	56,720	191
Sulphur	223,659	212,343	225,006
Marble (raw material)	20,475,479	23,736,157	21,040,986
Dolomite	70,698	54,913	53,815

Industry

SELECTED PRODUCTS
('000 metric tons, unless otherwise indicated)

	2001	2002	2003
Wheat flour	784.3	781.2	798.1
Granulated sugar	201.8	176.3	210.6
Carbonated beverages ('000 litres)	442,241	434,998	447,417
Alcoholic beverages—excl. beer ('000 hectolitres)	2,482.4	562.2	482.4
Cigarettes (million)	22,226	19,383	20,493
Cotton yarn	313.6	306.4	286.7
Paper	803.6	801.0	781.9
Paperboard	2,590.5	3,235.3	3,403.8
Sulphuric acid	895.5	917.4	1,067.3
Spun yarn	360.3	316.0	308.8
Cement	18,127.6	19,362.9	18,474.2
Steel ingots	16,399.0	17,423.2	17,638.3
Sewing machines ('000 units)	194.2	235.6	243.2
Electric fans ('000 units)	15,644.3	14,844.4	13,375.5
Personal computers ('000 units)	14,130.1	14,462.2	10,511.7
Monitors ('000 units)	6,752.7	5,104.9	4,214.9
Radio cassette recorders ('000 units)	4,023.4	3,955.7	4,358.4
Radio receivers ('000 units)	1,390.1	426.2	371.9
Television receivers ('000 units)	927.6	1,014.8	1,248.0
Picture tubes ('000 units)	8,444	3,784	233
Integrated circuits (million units)	4,924.2	6,160.0	6,532.2
Electronic condensers (million units)	121,670.8	149,959.0	199,089.3
Telephone sets ('000 units)	1,911.0	215.2	352.5
Passenger motor cars (units)	266,237	326,270	388,002
Trucks and buses (units)	2,654	3,432	4,106
Bicycles ('000 units)	4,746.5	4,674.6	4,616.0
Ships ('000 dwt)*	1,014.3	833.7	n.a.
Electric energy (million kWh)	178,358	187,910	n.a.
Liquefied petroleum gas	1,278.5	1,366.1	n.a.

* Excluding motor yachts.

Finance

CURRENCY AND EXCHANGE RATES

Monetary Units
100 cents = 1 New Taiwan dollar (NT $).

Sterling, US Dollar and Euro Equivalents (25 June 2004)
£1 sterling = NT $61.50;
US $1 = NT $33.68;
€1 = NT $40.99;
NT $1,000 = £16.26 = US $29.69 = €24.39.

Average Exchange Rate (NT $ per US $)
2001 33.800
2002 34.575
2003 34.418

BUDGET
(NT $ million, year ending 31 December)

Revenue	2002	2003	2004*
Taxes	820,051	828,551	903,088
Monopoly profits	346	—	—
Non-tax revenue from other sources	484,270	491,444	431,635
Total	1,304,667	1,319,995	1,334,723

* Forecasts.

Expenditure	2002	2003	2004*
General administration	162,255	167,738	168,258
National defence	225,243	227,742	250,949
Education, science and culture	267,008	300,196	306,446
Economic development	291,166	295,569	244,883
Social welfare	262,241	284,680	281,919
Community development and environmental protection	23,433	28,680	25,139
Pensions and survivors' benefits	124,288	125,414	123,757
Obligations	152,240	144,634	133,362
Subsidies to provincial and municipal governments	37,909	39,957	42,695
Other expenditure	6,159	4,303	14,162
Total	1,551,943	1,618,913	1,591,570

* Forecasts.

Note: Figures refer to central government accounts, including Taiwan Province.

INTERNATIONAL RESERVES

(US $ million at 31 December)

	2001	2002	2003
Gold*	4,361	4,390	4,508
Foreign exchange	122,211	161,656	206,632
Total	126,572	166,046	211,140

* National valuation.

MONEY SUPPLY

(NT $ million at 31 December)

	2001	2002	2003
Currency outside banks	525,659	527,278	608,205
Demand deposits at deposit money banks	4,500,201	4,964,311	5,944,627
Total money	5,025,860	5,491,589	6,552,832

COST OF LIVING

(Consumer Price Index; base: 2001 = 100)

	2000	2002	2003
Food	100.93	99.80	99.72
Clothing	101.70	100.58	101.96
Housing	100.34	98.88	97.81
Transport and communications	98.95	97.78	98.38
Medicines and medical care	98.68	101.29	104.65
Education and entertainment	97.92	100.11	98.81
All items (incl. others)	100.01	99.80	99.52

NATIONAL ACCOUNTS

(NT $ million in current prices)

National Income and Product

	2000	2001	2002
Compensation of employees	4,939,685	4,809,210	4,765,265
Operating surplus	3,163,763	3,141,257	3,321,091
Domestic factor incomes	8,103,448	7,950,467	8,086,356
Consumption of fixed capital	878,482	932,853	997,948
Gross domestic product (GDP) at factor cost	8,981,930	8,883,320	9,084,304
Indirect taxes	727,193	683,302	715,190
Less Subsidies	45,735	59,998	50,683
GDP in purchasers' values	9,663,388	9,506,624	9,748,811
Factor income from abroad	286,688	314,956	357,203
Less Factor income paid abroad	146,728	123,533	102,973
Gross national product (GNP)	9,803,348	9,698,047	10,003,041
Less Consumption of fixed capital	878,482	932,853	997,948
National income in market prices	8,924,866	8,765,194	9,005,093
Other current transfers from abroad	99,990	88,131	90,620
Less Other current transfers paid abroad	181,477	180,353	176,770
National disposable income	8,843,379	8,672,972	8,918,943

Expenditure on the Gross Domestic Product

	2001	2002	2003
Government final consumption expenditure	1,240,437	1,232,676	1,261,414
Private final consumption expenditure	6,042,628	6,149,507	6,186,554
Increase in stocks	-99,598	-84,111	-34,617
Gross fixed capital formation	1,781,752	1,728,279	1,724,373
Total domestic expenditure	8,965,219	9,026,351	9,137,724
Exports of goods and services	4,839,820	5,245,948	5,721,983
Less Imports of goods and services	4,298,415	4,523,488	5,012,152
GDP in purchasers' values	9,506,624	9,748,811	9,847,555
GDP at constant 1996 prices	9,349,923	9,685,551	9,999,787

Gross Domestic Product by Economic Activity

	2001	2002	2003
Agriculture, hunting, forestry and fishing	185,182	181,000	179,657
Mining and quarrying	37,986	41,986	36,597
Manufacturing	2,431,213	2,519,581	2,515,260
Construction	277,651	249,632	219,924
Electricity, gas and water	208,871	215,610	220,325
Transport, storage and communications	656,292	675,988	674,329
Trade, restaurants and hotels	1,833,533	1,895,476	1,954,911
Finance, insurance and real estate*	1,948,157	2,034,401	2,068,970
Business services	269,092	270,235	277,721
Community, social and personal services	962,939	1,009,031	1,040,492
Government services	1,011,122	1,024,869	1,064,656
Other services	114,981	121,283	119,692
Sub-total	9,937,019	10,239,092	10,372,534
Value-added tax	167,247	182,762	180,732
Import duties	119,100	126,370	128,710
Less Imputed bank service charge	716,742	799,413	834,421
GDP in purchasers' values	9,506,624	9,748,811	9,847,555

* Including imputed rents of owner-occupied dwellings.

BALANCE OF PAYMENTS
(US $ million)

	2001	2002	2003
Exports of goods f.o.b.	122,079	129,850	143,447
Imports of goods f.o.b.	−102,215	−105,657	−118,548
Trade balance	19,864	24,193	24,899
Exports of services	19,547	21,635	23,102
Imports of services	−24,465	−24,719	−25,635
Balance on goods and services	14,946	21,109	22,366
Other income received	9,327	10,334	12,991
Other income paid	−3,648	−3,321	−3,436
Balance on goods, services and income	20,625	28,122	31,921
Current transfers received	2,607	2,621	2,673
Current transfers paid	−5,341	−5,113	−5,392
Current balance	17,891	25,630	29,202
Capital account (net)	−163	−139	−87
Direct investment abroad	−5,480	−4,886	−5,679
Direct investment from abroad	4,109	1,445	453
Portfolio investment assets	−12,427	−15,711	−35,620
Portfolio investment liabilities	11,136	6,644	29,693
Other investment assets	−1,770	11,990	2,455
Other investment liabilities	4,048	9,268	15,438
Net errors and omissions	9	−577	1,237
Overall balance	17,353	33,664	37,092

External Trade

PRINCIPAL COMMODITIES
(US $ million)

Imports c.i.f.	2001	2002	2003
Mineral products	12,763.6	12,617.8	16,332.1
Crude petroleum	6,808.6	6,752.5	9,564.4
Products of chemical or allied industries	10,231.9	11,340.2	13,494.7
Organic chemicals	3,901.2	4,368.1	n.a.
Base metals and articles thereof	7,783.6	9,187.0	11,292.0
Iron and steel products	3,788.2	4,852.0	6,259.1
Metal products (excl. iron and steel)	3,995.4	4,335.0	5,032.9
Machinery and mechanical appliances; electrical equipment; sound and television apparatus	47,549.3	50,115.4	52,933.2
Electronic products	21,026.8	23,105.2	25,391.9
Machineries	10,489.0	9,764.7	11,468.6
Electrical machinery products	4,282.4	4,615.7	4,942.7
Information and communication products	8,119.5	8,246.4	6,093.9
Vehicles, aircraft, vessels and associated transport equipment	4,237.9	3,469.5	3,886.8
Optical, photographic, cinematographic, measuring, precision and medical apparatus; clocks and watches; musical instruments	6,213.5	6,614.4	8,627.4
Total (incl. others)	107,237.4	112,530.1	127,248.5

Exports f.o.b.	2001	2002	2003
Chemicals	4,137.9	4,667.0	5,661.4
Plastics, rubber and articles thereof	7,992.8	8,798.6	9,975.4
Textiles and textile articles	12,630.1	12,149.7	11,877.5
Fibre and yarn	8,981.7	8,654.1	8,373.5
Base metals and articles thereof	11,330.9	12,542.7	14,330.3
Iron and steel	6,888.0	7,825.7	9,351.0
Metal products (excl. iron and steel)	4,442.8	4,716.8	4,979.3
Machinery and mechanical appliances; electrical equipment; sound and television apparatus	66,851.5	70,633.3	75,352.3
Electronic products	23,601.1	25,838.0	31,158.4
Machineries	8,348.1	9,258.4	9,864.9
Electrical machinery products	4,665.1	5,898.7	6,012.3
Information and communication products	15,668.1	16,039.0	14,056.7
Vehicles, aircraft, vessels and associated transport equipment	4,441.6	4,830.5	5,672.0
Total (incl. others)	122,866.3	130,596.8	144,179.5

PRINCIPAL TRADING PARTNERS
(US $ million)

Imports c.i.f.	2001	2002	2003
Australia	3,084.9	2,832.6	2,726.6
Canada	996.1	945.0	1,078.8
China, People's Republic	5,901.9	7,947.7	10,960.5
France	2,130.5	1,551.3	1,628.3
Germany	4,246.0	4,421.5	4,964.5
Hong Kong	1,848.9	1,738.6	1,725.2
Indonesia	2,523.4	2,588.1	2,921.5
Italy	1,084.1	1,089.7	1,131.6
Japan	25,848.4	27,277.3	32,635.4
Korea, Republic	6,705.1	7,711.1	8,687.9
Malaysia	4,213.7	4,151.9	4,749.0
Netherlands	1,524.2	1,438.0	1,295.0
Philippines	3,250.5	3,651.6	3,081.0
Russia	603.5	927.1	1,299.3
Saudi Arabia	2,745.4	2,406.3	4,275.5
Singapore	3,367.2	3,543.6	3,860.9
Switzerland	865.8	807.0	1,074.9
Thailand	2,181.0	2,170.8	2,364.9
United Kingdom	1,442.8	1,356.8	1,416.1
USA	18,229.2	18,094.3	16,820.1
Total (incl. others)	107,237.4	112,530.1	127,248.5

Exports f.o.b.	2001	2002	2003
Australia	1,362.7	1,586.6	1,884.4
Canada	1,564.3	1,533.6	1,470.4
China, People's Republic	4,745.4	9,945.0	21,417.3
France	1,166.0	1,122.8	1,251.0
Germany	4,480.4	3,836.0	4,207.5
Hong Kong*	26,961.4	30,845.3	28,353.6
Indonesia	1,474.6	1,462.9	1,514.0
Italy	1,254.5	1,254.0	1,460.2
Japan	12,759.0	11,983.8	11,912.5
Korea, Republic	3,275.6	3,866.5	4,573.6
Malaysia	3,061.4	3,132.6	3,046.2
Netherlands	4,229.2	3,772.0	4,126.2
Philippines	2,148.7	1,971.5	2,300.4
Singapore	4,051.5	4,377.8	4,982.7
Thailand	2,125.7	2,292.7	2,565.3
United Kingdom	3,329.3	2,908.8	2,884.3
USA	27,654.5	26,763.7	25,941.5
Viet Nam	1,726.9	2,287.2	2,664.3
Total (incl. others)	122,866.3	130,596.8	144,179.5

* The majority of Taiwan's exports to Hong Kong are re-exported.

Transport

RAILWAYS
(traffic)

	2001	2002	2003
Passengers ('000)	476,214	500,464	478,228
Passenger-km ('000)	12,268,691	12,147,500	11,177,092
Freight ('000 metric tons)	19,287	18,217	16,735
Freight ton-km ('000)	1,009,863	940,603	863,912

ROAD TRAFFIC
(motor vehicles in use at 31 December)

	2001	2002	2003
Passenger cars	4,825,581	4,989,336	5,169,733
Buses and coaches	24,053	25,079	25,628
Goods vehicles	830,673	856,783	885,780
Motorcycles and scooters	11,733,202	11,983,757	12,366,864

SHIPPING

Merchant Fleet
(at 31 December)

	2001	2002	2003
Number of vessels	656	649	637
Total displacement ('000 grt)	4,617.9	4,289.0	3,477.1

Source: Lloyd's Register-Fairplay, *World Fleet Statistics*.

Sea-borne freight traffic
('000 metric tons)

	2001	2002	2003
Goods loaded	221,154	243,914	254,640
Goods unloaded	331,719	362,796	380,937

CIVIL AVIATION
(traffic on scheduled services)

	2001	2002	2003
Passengers carried ('000)	46,084.4	44,185.6	37,879.4
Freight carried ('000 metric tons)	1,310.2	1,513.9	1,622.7

Tourism

TOURIST ARRIVALS BY COUNTRY OF ORIGIN

	2001	2002	2003
Hong Kong	80,752	76,690	56,076
Indonesia	89,476	87,728	38,078
Japan	971,190	998,497	657,053
Korea, Republic	82,684	83,624	92,893
Malaysia	56,834	66,304	67,014
Philippines	69,118	79,261	80,026
Singapore	96,777	111,024	78,739
Thailand	116,420	110,650	98,390
USA	339,390	377,470	272,858
Overseas Chinese*	325,266	623,675	436,083
Total (incl. others)	2,617,137	2,977,692	2,248,117

* i.e. those bearing Taiwan passports.

Tourism receipts (US $ million): 4,152 in 2001; 4,584 in 2002; 2,976 in 2003.

Communications Media

	2001	2002	2003
Book production (titles)	36,546	38,953	39,138
Newspapers	454	474	602
Magazines	7,236	8,140	n.a.
Telephone subscribers ('000)	12,858	13,099	13,355
Mobile telephones ('000 in use)	21,633	23,905	25,090
Personal computers ('000 in use)*	5,000	n.a.	n.a.
Internet users ('000)*	6,232	7,459	7,828

Television receivers (2002): Colour television receivers per 100 households: 99.56; Cable television receivers per 100 households: 74.81.

Radio receivers (1994): more than 16 million in use.

* Source: International Telecommunication Union.

Education

(2003/04)

	Schools	Full-time teachers	Students
Pre-school	3,306	21,251	240,926
Primary	2,638	103,793	1,912,791
Secondary (incl. vocational)	1,192	97,738	1,679,970
Higher	158	47,472	1,270,194
Special	24	1,687	5,921
Supplementary	934	2,896	278,124
Total (incl. others)	8,252	274,837	5,384,926

Directory

The Constitution

On 1 January 1947 a new Constitution was promulgated for the Republic of China. When the Chinese Communist Party established the People's Republic of China on the Chinese mainland in 1949, the Government of the Republic of China, led by the Kuomintang (KMT), relocated to Taiwan, where it maintained jurisdiction over Taiwan, Penghu, Kinmen, Matsu and numerous other islets. The two sides of the Taiwan Strait have since been governed as separate territories. The form of government that was incorporated in the Constitution is based on a five-power system and has the major features of both cabinet and presidential government. A process of constitutional reform, initiated in 1991, continued in the early 21st century. The following is a summary of the Constitution, as subsequently amended:

PRESIDENT

The President shall be directly elected by popular vote for a term of four years. Both the President and Vice-President are eligible for re-election to a second term. The President represents the country at all state functions, including foreign relations; commands land, sea and air forces, promulgates laws, issues mandates, concludes treaties, declares war, makes peace, declares martial law, grants amnesties, appoints and removes civil and military officers, and confers honours and decorations. The President convenes the National Assembly and, subject to certain limitations, may issue emergency orders to deal with national calamities and ensure national security; may dissolve the Legislative Yuan; also nominates the Premier (who may be appointed without the Legislative Yuan's confirmation), and the officials of the Judicial Yuan, the Examination Yuan and the Control Yuan.

EXECUTIVE YUAN

The Executive Yuan is the highest administrative organ of the nation and is responsible to the Legislative Yuan; has three categories of subordinate organization:

Executive Yuan Council (policy-making organization);

Ministries and Commissions (executive organization);

Subordinate organization (19 bodies, including the Secretariat, Government Information Office, Directorate-General of Budget, Accounting and Statistics, Council for Economic Planning and Development, and Environmental Protection Administration).

LEGISLATIVE YUAN

The Legislative Yuan is the highest legislative organ of the State, empowered to hear administrative reports of the Executive Yuan, and to change government policy. It may hold a binding vote of 'no confidence' in the Executive Yuan. It comprises 225 members: 168 are chosen by direct election from two special municipalities and other cities and counties, eight members are elected from and by aborigines, and eight from and by overseas Chinese. The remaining 41 members are elected from a nation-wide constituency on the basis of proportional representation. Members serve for three years and are eligible for re-election.

JUDICIAL YUAN

The Judicial Yuan is the highest judicial organ of state and has charge of civil, criminal and administrative cases, and of cases concerning disciplinary measures against public functionaries (see Judicial System).

EXAMINATION YUAN

The Examination Yuan supervises examinations for entry into public offices, and deals with personnel matters of the civil service, implements training and protection measures for public functionaries, and supervises the public service pension fund.

CONTROL YUAN

The Control Yuan is the highest control organ of the State, exercising powers of impeachment, censure and audit, comprising 29 members serving a six-year term. (According to the Additional Articles of the Constitution of the Republic of China in April 2000, the members of the Control Yuan shall be nominated by the President of the State, and with the consent of the Legislative Yuan). The Control Yuan may impeach or censure a public functionary at central or local level, who is deemed guilty of violation of law or dereliction of duty, and shall refer the matter to the law courts for action in cases involving a criminal offence; the Control Yuan may propose corrective measures to the Executive Yuan or to its subordinate organs.

The Government

HEAD OF STATE

President: CHEN SHUI-BIAN (inaugurated 20 May 2000; re-elected 20 March 2004).

Vice-President: HSU-LIEN ANNETTE LU.

Secretary-General: SU TSENG-CHANG.

THE EXECUTIVE YUAN
(September 2004)

Premier: YU SHYI-KUN.

Vice-Premier and Minister of the Consumer Protection Commission: YEH CHU-LAN.

Secretary-General: ARTHUR IAP.

Ministers without Portfolio: LIN YI-FU, HU SHENG-CHENG (Minister of the Council for Economic Planning and Development), CHEN CHI-MAI, FU LI-YEH, LIN SHENG-FENG, KUO YAO-CHI (Minister of the Public Construction Commission), LIN FERNG-CHING.

Minister of the Interior: SU JIA-CHYUAN.

Minister of Foreign Affairs: CHEN TAN-SUN.

Minister of National Defense: LEE JYE.

Minister of Finance: LIN CHUAN.

Minister of Education: TU CHENG-SHENG.

Minister of Justice: CHEN DING-NAN.

Minister of Economic Affairs: HO MEI-YUEH.

Minister of Transportation and Communications: LIN LING-SAN.

Minister of the Mongolian and Tibetan Affairs Commission: HSU CHIH-HSIUNG.

Minister of the Overseas Chinese Affairs Commission: CHANG FU-MEI.

Minister of the Directorate-General of Budget, Accounting and Statistics: HSU JAN-YAU.

Minister of Central Personnel Administration: LEE YI-YANG.

Minister of the Government Information Office: LIN CHIA-LUNG.

Minister of Health: CHEN CHIEN-JEN.

Minister of the Environmental Protection Administration: CHANG JUU-EN.

Minister of the Coast Guard Administration: SHI HWEI-YOW.

Minister of the Mainland Affairs Council: JAUSHIEH JOSEPH WU.

Minister of the Veterans' Affairs Commission: KAO HUA-CHU.

Minister of the National Youth Commission: CHENG LI-CHIUN.

Minister of the Atomic Energy Council: OUYANG MIN-SHEN.

Minister of the National Science Council: WU MAW-KUEN.

Minister of the Research, Development and Evaluation Commission: YEH JIUNN-RONG.

Minister of the Council of Agriculture: LEE CHING-LUNG.

Minister of the Council for Cultural Affairs: CHEN CHI-NAN.

Minister of the Council of Labor Affairs: CHEN CHU.

Minister of the Fair Trade Commission: HWANG TZONG-LEH.

Minister of the Council of Indigenous Peoples: CHEN CHIEN-NIEN.

Minister of the National Council on Physical Fitness and Sports: CHEN CHUAN-SHOW.

Minister of the Council for Hakka Affairs: LUO WEN-JIA.

Governor of the Central Bank of China: PERNG FAI-NAN.

Chairperson of the Central Election Commission: MASA J. S. CHANG.

Chairperson of the Co-ordination Council for North American Affairs: LIN FANG-MEI.

Director of the National Palace Museum: SHIH SHOU-CHIEN.

MINISTRIES, COMMISSIONS, ETC.

Office of the President: Chiehshou Hall, 122 Chungking South Rd, Sec. 1, Taipei 100; tel. (2) 23718889; fax (2) 23611604; e-mail public@mail.oop.gov.tw; internet www.president.gov.tw.

Ministry of Economic Affairs: 15 Foo Chou St, Taipei; tel. (2) 23212200; fax (2) 23919398; e-mail service@moea.gov.tw; internet www.moea.gov.tw.

Ministry of Education: 5 Chung Shan South Rd, Taipei 10040; tel. (2) 23566051; fax (2) 23976978; internet www.moe.gov.tw.

Ministry of Finance: 2 Ai Kuo West Rd, Taipei; tel. (2) 23228000; fax (2) 23965829; e-mail root@www.mof.gov.tw; internet www.mof.gov.tw.

Ministry of Foreign Affairs: 2 Chiehshou Rd, Taipei 10016; tel. (2) 23119292; fax (2) 23144972; internet www.mofa.gov.tw.

Ministry of the Interior: 5–9/F, 5 Hsu Chou Rd, Taipei; tel. (2) 23565005; fax (2) 23566201; e-mail gethics@mail.moi.gov.tw; internet www.moi.gov.tw.

Ministry of Justice: 130 Chungking South Rd, Sec. 1, Taipei 100 10036; tel. (2) 23146871; fax (2) 23896759; internet www.moj.gov.tw.

Ministry of National Defense: 2/F, 164 Po Ai Rd, Taipei; tel. (2) 23116117; fax (2) 23144221; internet www.ndmc.edu.tw.

Ministry of Transportation and Communications: 2 Chang Sha St, Sec. 1, Taipei; tel. (2) 23492900; fax (2) 23118587; e-mail motceyes@motc.gov.tw; internet www.motc.gov.tw.

Mongolian and Tibetan Affairs Commission: 4/F, 5 Hsu Chou Rd, Sec. 1, Taipei; tel. (2) 23566166; fax (2) 23566432; internet www.mtac.gov.tw.

Overseas Chinese Affairs Commission: 4/F, 5 Hsu Chou Rd, Taipei; tel. (2) 23566166; fax (2) 23566323; e-mail ocacinfo@mail.ocac.gov.tw; internet www.ocac.gov.tw.

Directorate-General of Budget, Accounting and Statistics: 2 Kwang Chow St, Taipei 100; tel. (2) 23710208; fax (2) 23319925; e-mail sicbs@dgbas.gov.tw; internet www.dgbas.gov.tw.

Government Information Office: 2 Tientsin St, Taipei; tel. (2) 33568888; fax (2) 23568733; e-mail service@mail.gio.gov.tw; internet www.gio.gov.tw.

Council of Indigenous Peoples: 16–17/F, 4 Chung Hsiao West Rd, Sec. 1, Taipei; tel. (2) 23882122; fax (2) 23891967.

Council of Agriculture: see under Trade and Industry—Government Agencies.

Atomic Energy Council (AEC): 80 Cheng Kung Rd, Sec. 1, Taipei 234; tel. (2) 82317919; fax (2) 82317864; internet www.aec.gov.tw.

Central Personnel Administration: 109 Huai Ning St, Taipei; tel. (2) 23111720; fax (2) 23715252; internet www.cpa.gov.tw.

Consumer Protection Commission: 1 Chung Hsiao East Rd, Sec. 1, Taipei; tel. (2) 23566600; fax (2) 23214538; e-mail tcpc@ms1.hinet.net; internet www.cpc.gov.tw.

Council for Cultural Affairs: 102 Ai Kuo East Rd, Taipei; tel. (2) 25225300; fax (2) 25519011; e-mail wwwadm@ccpdunx.ccpd.gov.tw; internet expo96.org.tw/cca/welcome_c.html.

Council for Economic Planning and Development: 9/F, 87 Nanking East Rd, Sec. 2, Taipei; tel. (2) 25225300; fax (2) 25519011; internet www.cepd.gov.tw.

Environmental Protection Administration: 41 Chung Hua Rd, Sec. 1, Taipei; tel. (2) 23117722; fax (2) 23116071; e-mail www@sun.epa.gov.tw; internet www.epa.gov.tw.

Fair Trade Commission: 12–14/F, 2-2 Chi Nan Rd, Sec. 2, Taipei; tel. (2) 23517588; fax (2) 23974997; e-mail ftcse@ftc.gov.tw; internet www.ftc.gov.tw.

Department of Health: 12/F, 100 Ai Kuo East Rd, Taipei; tel. (2) 23210151; fax (2) 2392-9723; internet www.doh.gov.tw.

Council of Labor Affairs: 5–15/F, 132 Min Sheng East Rd, Sec. 3, Taipei; tel. (2) 27182512; fax (2) 25149240; internet www.cla.gov.tw.

Mainland Affairs Council: 5–13/F, 2-2 Chi Nan Rd, Sec. 1, Taipei; tel. (2) 23975589; fax (2) 23975700; e-mail macst@mac.gov.tw; internet www.mac.gov.tw.

National Science Council: 17–22/F, 106 Ho Ping East Rd, Sec. 2, Taipei; tel. (2) 27377501; fax (2) 27377668; e-mail nsc@nsc.gov.tw; internet www.nsc.gov.tw.

National Youth Commission: 14/F, 5 Hsu Chou Rd, Taipei; tel. (2) 23566271; fax (2) 23566290; internet www.nyc.gov.tw.

Research, Development and Evaluation Commission: 7/F, 2-2 Chi Nan Rd, Sec. 1, Taipei; tel. (2) 23419066; fax (2) 23928133; e-mail service@rdec.gov.tw; internet rdec.gov.tw.

Veterans' Affairs Commission: 222 Chung Hsiao East Rd, Sec. 5, Taipei; tel. (2) 27255700; fax (2) 27253578; e-mail hsc@www.vac.gov.tw; internet vac.gov.tw.

Co-ordination Council for North American Affairs: 133 Po-ai Rd, Taipei; tel. (2) 23119212; fax (2) 23822651; headquarters for Taipei Economic and Cultural Representative Office (TECRO) in the USA.

President and Legislature

PRESIDENT

Election, 20 March 2004 (provisional results)

Candidate	Votes	% of votes
Chen Shui-bian (Democratic Progressive Party—DPP)	6,471,970	50.11
Lien Chan (Kuomintang—KMT)	6,442,452	49.89
Total	12,914,422*	100.00

*Not including invalid or spoiled ballot papers, which numbered 337,297. The vice-presidential candidates were respectively Annette Hsiu-lien Lu of the DPP and James C. Y. Soong of the People First Party (PFP). Results were subject to revision, pending a recount conducted in May 2004 and subsequent review.

LI-FA YUAN
(Legislative Yuan)

The Legislative Yuan is the highest legislative organ of the State. It comprises 225 seats. The 168 directly elected members come from two special municipalities and other cities and counties. Eight members are elected from and by aborigines. Eight members are overseas Chinese. Overseas Chinese members and the remaining 41 members are elected from a nation-wide constituency on the basis of proportional representation. Members serve for three years and are eligible for re-election.

President: WANG JIN-PYNG.

General Election, 1 December 2001

Party	% of votes	Seats
Democratic Progressive Party (DPP)	36.6	87
Kuomintang (KMT)	31.3	68
People First Party (PFP)	20.3	46
Taiwan Solidarity Union (TSU)	8.5	13
New Party (NP)	2.9	1
Other	0.5	1
Independents	—	9
Total	100.0	225

Political Organizations

Legislation adopted in 1989 permitted political parties other than the KMT to function. By mid-2004 a total of 105 parties had registered with the Ministry of the Interior.

China Democratic Socialist Party (CDSP): 6/F, 7 Heping East Rd, Sec. 3, Taipei; tel. (2) 27072883; f. 1932 by merger of National Socialists and Democratic Constitutionalists; aims to promote democracy, to protect fundamental freedoms, and to improve public welfare and social security; Chair. I BUH-LUEN; Sec.-Gen. SUEN SHANN-HAUR.

China Young Party: 12/F, 2 Sinsheng South Rd, Sec. 3, Taipei; tel. (2) 23626715; f. 1923; aims to recover sovereignty over mainland China, to safeguard the Constitution and democracy, and to foster understanding between Taiwan and the non-communist world; Chair. JEAN JYI-YUAN.

Chinese Republican Party (CRP): 3/F, 26 Lane 90, Jongshun St, Sec. 2, Taipei; tel. (2) 29366572; f. 1988; advocates peaceful struggle for the salvation of China and the promotion of world peace; Chair. WANG YING-CHYUN.

Democratic Liberal Party (DLP): 4/F, 20 Lane 5, Cingtian, Taipei; tel. (2) 23121595; f. 1989; aims to promote political democracy and economic liberty for the people of Taiwan; Chair. HER WEI-KANG.

Democratic Progressive Party (DPP): 10/F, 30 Beiping East Rd, Taipei; tel. and fax (2) 23929989; e-mail foreign@dpp.org.tw; internet www.dpp.org.tw; f. 1986; advocates 'self-determination' for the people of Taiwan and UN membership; supports establishment of independent Taiwan following plebiscite; 140,000 mems; Chair. CHEN SHUI-BIAN; Sec.-Gen. CHANG CHUN-HSIUNG.

Democratic Union of Taiwan (DUT): 16/F, 15-1 Hangjhou South Rd, Sec. 1, Taipei; tel. (2) 23211531; f. 1998; Chair. HSU CHERNG-KUEN.

Green Party: 11/F-1, 273 Roosevelt Rd, Sec. 3, Taipei; tel. (2) 23621362; f. 1996 by breakaway faction of the DPP; Chair. CHEN GUANG-YEU.

Jiann Gwo Party (Taiwan Independence Party—TAIP): 9/F, 15–8 Nanjing East Rd, Sec. 5, Taipei; tel. (2) 22800879; internet www.taip.org/index.htm; f. 1996 by dissident mems of DPP; Chair. HUANG CHIEN-MING; Sec.-Gen. LI SHENG-HSIUNG.

Kungtang (KT) (Labour Party): 1 Lane 44, Jianguo South Rd, Sec. 1, Taipei; tel. (2) 27739044; fax (2) 27739050; f. 1987; aims to become the main political movement of Taiwan's industrial work-force; 10,000 mems; Chair. WU CHING-FU; Sec.-Gen. JENG JIN-YANG.

Kuomintang (KMT) (Nationalist Party of China): 11 Jongshan South Rd, Taipei 100; tel. (2) 23121472; fax (2) 23434561; internet www.kmt.org.tw; f. 1894; fmr ruling party; aims to supplant communist rule in mainland China; supports democratic, constitutional government, and advocates the unification of China under the 'Three Principles of the People'; aims to promote market economy and equitable distribution of wealth; plans for merger with PFP announced May 2004; 2,523,984 mems; Chair. LIEN CHAN; Sec.-Gen. LIN FONG-CHENG.

Nationwide Democratic Non-Partisan Union (NDNU): 13/F, 42-1 Huaining St, Taipei; Chair. YEH SHIANN-SHIOU.

New Nation Alliance: 14/F, 9 Songjiang Rd, Taipei; tel. (2) 23585643; f. 1998; promotes independence for Taiwan and the establishment of a 'new nation, new society and new culture'; Chair. PERNG BAE-SHEAN.

New Party (NP): 4/F, 65 Guangfu South Rd, Taipei; tel. (2) 27562222; fax (2) 27565750; e-mail webmaster@mail.np.org.tw; internet www.np.org.tw; f. 1993 by dissident KMT legislators (hitherto mems of New Kuomintang Alliance faction); merged with China

Social Democratic Party in late 1993; advocates co-operation with the KMT and DPP in negotiations with the People's Republic, the maintenance of security in the Taiwan Straits, the modernization of the island's defence systems, measures to combat government corruption, the support of small and medium businesses and the establishment of a universal social security system; 80,000 mems; Chair. YOK MU-MING.

People First Party (PFP): 1/F, 63 Chang-an East Rd, Sec. 2, Taipei; tel. (2) 25068555; internet www.pfp.org.tw; f. 2000; plans for merger with KMT announced May 2004; Chair. JAMES C. Y. SOONG; Sec.-Gen. TSAI CHUNG-HSIUNG.

Taiwan Solidarity Union (TSU): 9/F, 65 Guancian Rd, Taipei; tel. (2) 23706686; fax (2) 23706616; internet www.tsu.org.tw; f. 2001 by a breakaway faction of the Kuomintang (KMT); Chair. HUANG CHU-WEN; Sec.-Gen. LIN CHIH-JIA.

Workers' Party: 2/F, 181 Fusing South Rd, Sec. 2, Taipei; tel. (2) 27555868; f. 1989 by breakaway faction of the Kungtang; radical; Leader LOU MEIWEN.

Various pro-independence groups (some based overseas and, until 1992, banned in Taiwan) are in operation. These include the **World United Formosans for Independence** (WUFI—4,000 mems world-wide; Chair. GEORGE CHANG) and the **Organization for Taiwan Nation-Building.**

Diplomatic Representation

EMBASSIES IN THE REPUBLIC OF CHINA

Belize: 11/F, 9 Lane 62, Tien Mou West Rd, Taipei 111; tel. (2) 28760894; fax (2) 28760896; e-mail embelroc@ms41.hinet.net; internet www.embassyofbelize.org.tw; Ambassador WILLIAM QUINTO.

Burkina Faso: 6/F, 9-1 Lane 62, Tien Mou West Rd, Taipei 111; tel. (2) 28733096; fax (2) 28733071; e-mail abftap94@ms17.hinet.net; Ambassador JACQUES Y. SAWADOGO.

Chad: 8/F, 9 Lane 62, Tien Mou West Rd, Taipei; tel. (2) 28742943; fax (2) 28742971; e-mail amchadtp@ms23.hinet.net; Ambassador HISSEIN BRAHIM TAHA.

Costa Rica: 5/F, 9-1 Lane 62, Tien Mou West Rd, Taipei 111; tel. (2) 28752964; fax (2) 28753151; e-mail oscaralv@ttn.net; Ambassador DR OSCAR ALVAREZ.

Dominican Republic: 6/F, Lane 62, Tien Mou West Rd, Taipei 111; tel. (2) 28751357; fax (2) 28752661; Ambassador CARLOS J. GUZMÁN.

El Salvador: 2/F, 9 Lane 62, Tien Mou West Rd, Shih Lin, Taipei 111; tel. (2) 28763606; fax (2) 28763514; e-mail embasal.taipei@msa.hinet.net; Ambassador FRANCISCO RICARDO SANTANA BERRÍOS.

The Gambia: 9/F, 9-1 Lane 62, Tien Mou West Rd, Taipei 111; tel. (2) 28753911; fax (2) 28752775; e-mail gm.roc@msa.hinet.net; Ambassador JOHN-PAUL BOJANG.

Guatemala: 3/F, 9-1 Lane 62, Tien Mou West Rd, Taipei 111; tel. (2) 28756952; fax (2) 28740699; e-mail embchina@minex.gob.gt; Ambassador MANUEL ERNESTO GALVEZ CORONADO.

Haiti: 8/F, 9-1 Lane 62, Tien Mou West Rd, Taipei 111; tel. (2) 28766718; fax (2) 28766719; e-mail haiti@ms26.hinet.net; Ambassador MARIO CHOULOUTE (acting).

Holy See: 87 Ai Kuo East Rd, Taipei 106 (Apostolic Nunciature); tel. (2) 23216847; fax (2) 23911926; e-mail aposnunc@seed.net.tw; Chargé d'affaires a.i. Mgr AMBROSE MADTHA.

Honduras: 9/F, 9 Lane 62, Tien Mou West Rd, Taipei 111; tel. (2) 28755507; fax (2) 28755726; e-mail honduras@ms9.hinet.net; Ambassador MARLENE VILLELA DE TALBOTT.

Malawi: 2/F, 9-1 Lane 62, Tien Mou West Rd, Taipei 111; tel. (2) 28762284; fax (2) 28763545; Ambassador THENGO MALOYA.

Marshall Islands: 4/F, 9-1 Lane 62, Tien Mou West Rd, Taipei 111; tel. (2) 28734884; fax (2) 28734904; internet www.rmiembassy.org.tw; Ambassador ALEX CARTER BING.

Nicaragua: 3/F, Jyi Shyan Rd, Lu Chow, Taipei; tel. (2) 82814512; fax (2) 82814515; e-mail icaza@ms13.hinet.net; Ambassador SALVADOR STADTHAGEN.

Palau: 5/F, 9 Lane B2, Tien Mou West Rd, Taipei 111; tel. (2) 28765415; fax (2) 28760436; Ambassador JOHNSON TORIBIONG.

Panama: 6/F, 111 Sung Kiang Rd, Taipei 104; tel. (2) 25099189; fax (2) 25099801; Ambassador JOSÉ ANTONIO DOMÍNGUEZ.

Paraguay: 7/F, 9-1 Lane 62, Tien Mou West Rd, Taipei 111; tel. (2) 28736310; fax (2) 28736312; e-mail eptaipei@seed.net.tw; Ambassador JORGE CRAVID (acting).

São Tomé and Príncipe: 3/F, 18 Chi-lin Rd, Taipei 104; tel. (2) 25114111; fax (2) 25116255; e-mail stptw@ms69.hinet.net; Ambassador OVIDIO M. PEQUENO.

Senegal: 10/F, 9-1 Lane 62, Tien Mou West Rd, Taipei 111; tel. (2) 28766519; fax (2) 28734909; e-mail sngol.taipei@msa.hinet.net; Ambassador YOUSSOU DIAGNE.

Solomon Islands: 7/F, 9-1 Lane 62, Tien Mou West Rd, Taipei 111; tel. (2) 28731168; fax (2) 28735224; Ambassador BERAKI JINO.

Swaziland: 10/F, 9 Lane 62, Tien Mou West Rd, Taipei 111; tel. (2) 28725934; fax (2) 28726511; Chargé d'affaires a.i. MOSES LINDIWE MNDEBELE.

Judicial System

Under Articles 78 and 79 of the Constitution and Article 5 of the Constitutional Amendments, 15 Justices of the Constitutional Court* (including the President and Vice-President of the Judicial Yuan) have the authority to interpret the Constitution, and to render uniform interpretations of statutes and regulations whenever they are in conflict, and to declare the dissolution of any political party in violation of the Constitution. All of the Justices of the Constitutional Court are nominated and, with the consent of the Legislative Yuan (Congress), appointed by the President of the Republic of China. The Justices serve an eight-year term and may not be reappointed. However, the President and the Vice-President of the Judicial Yuan do not have the same term of protection. Due to a transition to a staggered system of appointment, of 15 Justices appointed in 2003, eight, including the President and Vice-President of the Judicial Yuan, serve a four-year term, while the others serve an eight-year term.

Judicial Yuan: 124 Chungking South Rd, Sec. 1, Taipei; tel. (2) 23618577; fax (2) 23898923; e-mail judicial@mail.judicial.gov.tw; internet www.judicial.gov.tw; Pres. WENG YUEH-SHENG; Vice-Pres. CHENG CHUNG-MO; Sec.-Gen. FAN KUANG-CHUN; the highest judicial organ, and the interpreter of the constitution and national laws and ordinances; supervises the following:

Supreme Court: 6 Chang Sha St, Sec. 1, Taipei; tel. (2) 23141160; fax (2) 23114246; e-mail tpsemail@mail.judicial.gov.tw; Court of third and final instance for civil and criminal cases; Pres. WU CHII-PIN.

High Courts: Courts of second instance for appeals of civil and criminal cases.

District Courts: Courts of first instance in civil, criminal and non-contentious cases.

Supreme Administrative Court: 1 Lane 126, Chungking South Rd, Sec. 1, Taipei; tel. (2) 23113691; fax (2) 23111791; e-mail jessie@judicial.gov.tw; Court of final resort in cases brought against govt agencies; Pres. CHANG DENG-KE.

High Administrative Courts: Courts of first instance in cases brought against govt agencies.

Commission on Disciplinary Sanctions Against Functionaries: 124 Chungking South Rd, 3/F, Sec. 1, Taipei 10036; tel. (2) 23111639; fax (2) 23826255; decides on disciplinary measures against public functionaries impeached by the Control Yuan; Chief Commissioner LIN KUO-HSIEN.

Religion

According to the Ministry of the Interior, in 2002 24% of the population were adherents of Buddhism, 20% of Daoism (Taoism), 3.7% of I-kuan Tao and 2.6% of Christianity.

BUDDHISM

Buddhist Association of Taiwan: Mahayana and Theravada schools; 1,613 group mems and more than 9.61m. adherents; Leader Ven. CHIN-HSIN.

CHRISTIANITY

The Roman Catholic Church

Taiwan comprises one archdiocese, six dioceses and one apostolic administrative area. In December 2002, according to official figures, there were 298,451 adherents.

Bishops' Conference

Chinese Regional Bishops' Conference, 34 Lane 32, Kuangfu South Rd, Taipei 10552; tel. (2) 25782355; fax (2) 25773874; e-mail bishconf@ms1.hinet.net; internet www.catholic.org.tw.

f. 1967; Pres. Cardinal PAUL SHAN KUO-HSI (Bishop of Kaohsiung).

Archbishop of Taipei: Most Rev. JOSEPH TI-KANG, Archbishop's House, 94 Loli Rd, Taipei 10668; tel. (2) 27371311; fax (2) 27373710.

The Anglican Communion

Anglicans in Taiwan are adherents of the Protestant Episcopal Church. In 1999 the Church had 2,000 members.

Bishop of Taiwan: Rt Rev. Dr JOHN CHIEN, 7 Lane 105, Hangchow South Rd, Sec. 1, Taipei 100; tel. (2) 23411265; fax (2) 23962014; e-mail skhtpe@ms12.hinet.net; internet www.dfms.org/taiwan.

Presbyterian Church

Tai-oan Ki-tok Tiu-Lo Kau-Hoe (Presbyterian Church in Taiwan): No. 3, Lane 269, Roosevelt Rd, Sec. 3, Taipei 106; tel. (2) 23625282; fax (2) 23628096; f. 1865; Gen. Sec. Rev. L. K. LO; 224,679 mems (2000).

DAOISM (TAOISM)

In 2002 there were about 4.54m. adherents. Temples numbered 8,604, and clergy totalled 33,850.

I-KUAN TAO

Introduced to Taiwan in the 1950s, this 'Religion of One Unity' is a modern, syncretic religion, drawn mainly from Confucian, Buddhist and Daoist principles and incorporating ancestor worship. In 1998 there were 108 temples and 18,000 family shrines. Adherents totalled 845,000.

ISLAM

Leader MOHAMMED NI GUO-AN; 53,000 adherents in 2002.

The Press

In 1999 the number of registered newspapers stood at 367. The majority of newspapers are privately owned.

PRINCIPAL DAILIES

Taipei

Central Daily News: 260 Pa Teh Rd, Sec. 2, Taipei; tel. (2) 27765368; fax (2) 27775835; internet www.cdn.com.tw; f. 1928; morning; Chinese; official Kuomintang organ; Publr and CEO SHAW YU-MING; circ. 600,000.

The China Post: 8 Fu Shun St, Taipei 104; tel. (2) 25969971; fax (2) 25957962; e-mail cpost@msl.hinet.net; internet www.chinapost.com.tw; f. 1952; morning; English; Publr and Editor JACK HUANG; readership 250,000.

China Times: 132 Da Li St, Taipei; tel. (2) 23087111; fax (2) 23063312; f. 1950; morning; Chinese; Chair. YU CHI-CHUNG; Publr YU ALBERT CHIEN-HSIN; circ. 1.2m.

China Times Express: 132 Da Li St, Taipei; tel. (2) 23087111; fax (2) 23082221; e-mail chinaexpress@mail.chinatimes.com.tw; f. 1988; evening; Chinese; Publr S. F. LIN; Editor C. L. HUANG; circ. 400,000.

Commercial Times: 132 Da Li St, Taipei; tel. (2) 23087111; fax (2) 23069456; e-mail commercialtimes@mail.chinatimes.com.tw; f. 1978; morning; Chinese; Publr PENG CHWEI-MING; Editor-in-Chief PHILLIP CHEN; circ. 300,000.

Economic Daily News: 555 Chung Hsiao East Rd, Sec. 4, Taipei; tel. (2) 27681234; fax (2) 27600129; f. 1967; morning; Chinese; Publr WANG PI-CHEN.

The Great News: 216 Chen Teh Rd, Sec. 3, Taipei; tel. (2) 25973111; f. 1988; morning; also *The Great News Daily-Entertainment* (circ. 460,000).

Liberty Times: 11/F, 137 Nanking East Rd, Sec. 2, Taipei; tel. (2) 25042828; fax (2) 25042212; f. 1988; Publr WU A-MING; Editor-in-Chief ROGER CHEN.

Mandarin Daily News: 2 Foo Chou St, Taipei; tel. (2) 23921133; fax (2) 23410203; f. 1948; morning; Publr LIN LIANG.

Min Sheng Daily: 555 Chung Hsiao East Rd, Sec. 4, Taipei; tel. (2) 27681234; fax (2) 27560955; f. 1978; sport and leisure; Publr WANG SHAW-LAN.

Taiwan Hsin Sheng Pao: 260 Pa Teh Rd, Sec. 2, Taipei; tel. (2) 87723058; fax (2) 87723026; f. 1945; morning; Chinese; Publr LIU CHZ-SHIEN.

Taiwan News: 7/F, 88 Hsin Yi Rd, Sec. 2, Taipei 106; tel. (2) 23517666; fax (2) 23518389; e-mail editor@etaiwannews.com; internet www.etaiwannews.com; f. 1949; morning; English; Chair. T. C. KAO; Publr LUIS KO.

United Daily News: 555 Chung Hsiao East Rd, Sec. 4, Taipei; tel. (2) 27681234; fax (2) 27632303; e-mail secretariat@udngroup.com.tw; f. 1951; morning; Publr WANG SHAW-LAN; Editor-in-Chief HUANG SHU-CHUAN; circ. 1.2m.

Provincial

China Daily News (Southern Edn): 57 Hsi Hwa St, Tainan; tel. (6) 2202691; fax (6) 2201804; f. 1946; morning; Publr LIU CHZ-SHIEN; circ. 670,000.

The Commons Daily: 180 Min Chuan 2 Rd, Kaohsiung; tel. (7) 3363131; fax (7) 3363604; f. 1950; fmrly Min Chung Daily News; morning; Executive-in-Chief WANG CHIN-HSIUNG; circ. 148,000.

Keng Sheng Daily News: 36 Wuchuan St, Hualien; tel. (38) 340131; fax (38) 329664; f. 1947; morning; Publr HSIEH YING-YIN; circ. 50,000.

Taiwan Daily News: 361 Wen Shin Rd, Sec. 3, Taichung; tel. (4) 22958511; fax (4) 2958950; f. 1964; morning; Publr ANTONIO CHIANG; Editor-in-Chief LIU CHIH TSUNG; circ. 250,000.

Taiwan Hsin Wen Daily News: 3 Woo Fu I Rd, Kaohsiung; tel. (7) 2226666; f. 1949; morning; Publr CHANG REI-TE.

Taiwan Times: 32 Kaonan Rd, Jen Wu Shan, Kaohsiung; tel. (7) 3428666; fax (7) 3102828; f. 1978; Publr WANG YUH-FA.

SELECTED PERIODICALS

Artist Magazine: 6/F, 147 Chung Ching South Rd, Sec. 1, Taipei; tel. (2) 23886715; fax (2) 23317096; e-mail artvenue@seed.net.tw; f. 1975; monthly; Publr HO CHENG KUANG; circ. 28,000.

Better Life Monthly: 11 Lane 199, Hsin-yih Rd, Sec. 4, Taipei; tel. (2) 27549588; fax (2) 27016068; e-mail bettlife@ms14.hinet.net; f. 1987; Publr JACK S. LIN.

Brain: 9/F, 47 Nanking East Rd, Sec. 4, Taipei; tel. (2) 27132644; fax (2) 27137318; f. 1977; monthly; Publr JOHNSON WU.

Business Weekly: 21/F-A, 333 Tun Hua South Rd, Sec. 2, Taipei; tel. (2) 25056789; fax (2) 27364620; f. 1987; Publr JIN WEI-TSUN.

Car Magazine: 1/F, 3 Lane 3, Tung-Shan St, Taipei; tel. (2) 23218128; fax (2) 23935614; e-mail carguide@ms13.hinet.net; f. 1982; monthly; Publr H. K. LIN; Editor-in-Chief TA-WEI LIN; circ. 85,000.

Central Monthly: 7/F, 11 Chung Shan South Rd, Taipei; tel. (2) 23433140; fax (2) 23435417; f. 1950; Publr HUANG HUI-TSEN.

China Times Weekly: 5/F, 25 Min Chuan East Rd, Sec. 6, Taipei; tel. (2) 27936000; fax (2) 27912238; f. 1978; weekly; Chinese; Editor CHANG KUO-LI; Publr CHUANG SHU-MING; circ. 180,000.

Commonwealth Monthly: 11/F, No. 139, Sec. 2, Nanking East Rd, Taipei 104; tel. (2) 85078629; fax (2) 25079011; f. 1981; monthly; business; Pres. CHARLES H. C. KAO; Publr and Editor DIANE YING; circ. 83,000.

Cosmopolitan: 5/F, 8 Lane 181, Jiou-Tzung Rd, Nei Hu Area, Taipei; tel. (2) 287978900; fax (2) 287978990; e-mail hwaker@ms13.hinet.net; f. 1992; monthly; Publr MINCHUN CHANG.

Country Road: 14 Wenchow St, Taipei; tel. (2) 23628148; fax (2) 23636724; e-mail h3628148@ms15.hinet.net; internet www.harvest.org.tw; f. 1975; monthly; Editor YU SHU-LIEN; Publr SUN MING-HSIEN.

Crown Magazine: 50 Alley 120, Tun Hua North Rd, Sec. 4, Taipei; tel. (2) 27168888; fax (2) 25148285; f. 1954; monthly; literature and arts; Publr PING HSIN TAO; Editor CHEN LIH-HWA; circ. 76,000.

Defense Technology Monthly: 6/F, 6 Nanking East Rd, Sec. 5, Taipei; tel. (2) 27669628; fax (2) 27666092; f. 1894; Publr J. D. BIH.

Earth Geographic Monthly: 4/F, 16 Lane 130, Min Chuan Rd, Hsin-Tien, Taipei; tel. (2) 22182218; fax (2) 22185418; f. 1988; Publr HSU CHUNG-JUNG.

Elle-Taipei: 9/F, 5 Lane 30, Sec. 3, Min Sheng East Rd, Taipei; tel. (2) 87706168; fax (2) 87706178; e-mail jdewitt@hft.com.tw; f. 1991; monthly; women's magazine; Publr JEAN DE WITT; Editors-in-Chief CINDY HU, DORIS LEE; circ. 50,000.

Evergreen Monthly: 11/F, 2 Pa Teh Rd, Sec. 3, Taipei; tel. (2) 25782321; fax (2) 25786838; f. 1983; health care knowledge; Publr LIANG GUANG-MING; circ. 50,000.

Excellence Magazine: 3/F, 15 Lane 2, Sec. 2, Chien Kuo North Rd, Taipei; tel. (2) 25093578; fax (2) 25173607; f. 1984; monthly; business; Man. LIN HSIN-JYH; Editor-in-Chief LIU JEN; circ. 70,000.

Families Monthly: 11/F, 2 Pa Teh Rd, Sec. 3, Taipei; tel. (2) 25785078; fax (2) 25786838; f. 1976; family life; Editor-in-Chief THELMA KU; circ. 155,000.

Foresight Investment Weekly: 7/F, 52 Nanking East Rd, Sec. 1, Taipei; tel. (2) 25512561; fax (2) 25119596; f. 1980; weekly; Dir and Publr SUN WUN HSIUNG; Editor-in-Chief WU WEN SHIN; circ. 55,000.

Global Views Monthly: 2/F, 1 Lane 93, Taipei; tel. (2) 25173688; fax (2) 25078644; f. 1986; Pres. CHARLES H. C. KAO; Publr and Editor-in-Chief WANG LI-HSING.

Gourmet World: 3/F, No. 53, Sec. 1, Jen-Ai Rd, Taipei; tel. (2) 23972215; fax (2) 23412184; f. 1992; Publr HSU TANG-JEN.

Harvest Farm Magazine: 14 Wenchow St, Taipei; tel. (2) 23628148; fax (2) 23636724; e-mail h3628148@ms15.hinet.net; internet www.harvest.org.tw; f. 1951; every 2 weeks; Publr YU SHU-LIEN; Editor SUN MING-HSIEN.

Information and Computer: 10/F, 116 Nang King East Rd, Taipei; tel. (2) 25422540; fax (2) 25310760; f. 1980; monthly; Chinese; Publr LIN FERNG-CHIN; Editor JENNIFER CHIU; circ. 28,000.

Issues and Studies: A Social Science Quarterly on China, Taiwan and East Asian Affairs: Institute of International Relations, National Chengchi University, 64 Wan Shou Rd, Wenshan, Taipei 116; tel. (2) 29386763; fax (2) 82377231; e-mail issues@nccu.edu.tw; internet http://iir.nccu.edu.tw/English/I&S.htm; f. 1965; quarterly; English; Chinese studies and East Asian affairs; Editor ANDREW D. MARBLE.

Jade Biweekly Magazine: 7/F, 222 Sung Chiang Rd, Taipei; tel. (2) 25811665; fax (2) 25210586; f. 1982; economics, social affairs, leisure; Publr HSU CHIA-CHUNG; circ. 98,000.

The Journalist: 16/F, 218 Tun Hua South Rd, Sec. 2, Taipei; tel. (2) 23779977; fax (2) 23775850; f. 1987; weekly; Publr WANG SHIN-CHING.

Ladies Magazine: 11/F, 3, 187 Shin Yi Rd, Sec. 4, Taipei; tel. (2) 27026908; fax (2) 27014090; f. 1978; monthly; Publr CHENG CHIN-SHAN; Editor-in-Chief THERESA LEE; circ. 60,000.

Living: 6/F, 100 Ai Kuo East Rd, Hsin Tien, Taipei; tel. (2) 23222266; fax (2) 33225050; f. 1997; monthly; Publr LISA WU.

Madame Figaro Taiwan: f. 2001; published by Stone Media.

Management Magazine: 5/F, 220 Ta Tung Rd, Sec. 3, Hsichih, Taipei; tel. (2) 86471828; fax (2) 86471466; e-mail frankhung@mail.chinamgt.com; internet www.harment.com; f. 1973; monthly; Chinese; Publr and Editor FRANK L. HUNG; Pres. KATHY T. KUO; circ. 65,000.

Money Monthly: 10/F, 289 Chung Hsiao East Rd, Taipei; tel. (2) 25149822; fax (2) 27154657; f. 1986; monthly; personal financial management; Publr PATRICK SUN; Man. Editor JENNIE SHUE; circ. 55,000.

Music and Audiophile: 2/F, 2 Kingshan South Rd, Sec. 1, Taipei; tel. (2) 25684607; fax (2) 23958654; f. 1973; Publr CHANG KUO-CHING; Editor-in-Chief CHARLES HUANG.

National Palace Museum Bulletin: 211 Chih-shan Rd, Wai Shuang Hsi, Sec. 2, Taipei 11102; tel. (2) 28812021; fax (2) 28821440; e-mail service01@npm.gov.tw; internet www.npm.gov.tw; f. 1965; every 3 months; Chinese art history research in English; Publr and Dir TU CHENG-SHENG; Editor-in-Chief SHIH SHOU-CHIEN; circ. 1,000.

National Palace Museum Monthly of Chinese Art: No. 221, Sec. 2, Jishan Rd, Shrlin Chiu, Taipei 11102; tel. (2) 28821230; fax (2) 28821440; f. 1983; monthly in Chinese; Publr TU CHENG-SHENG; circ. 10,000.

National Geographic/The Earth: 4/F, 319, Sec. 4, Bade Rd, Taipei 105; tel. (2) 27485988; fax (2) 27480188.

Nong Nong Magazine: 11/F, 141, Sec. 2, Minsheng East Rd, Taipei 104; tel. (2) 25058989; fax (2) 25001986; e-mail group@nongnong.com.tw; f. 1984; monthly; women's interest; Publr ANTHONY TSAI; Editor VIVIAN LIN; circ. 70,000.

PC Home: 4/F, 141, Sec. 2, Minsheng East Rd, Taipei 104; tel. (2) 25000888; fax (2) 25006609; f. 1996; monthly; Publr HUNG-TZE JANG.

PC Office: 11/F, 8 Tun Hua North Rd, Taipei; tel. (2) 25007779; fax (2) 25007903; f. 1997; monthly; Publr HUNG-TZE JANG.

Reader's Digest (Chinese Edn): 2/F, 2 Ming Sheng East Rd, Sec. 5, Taipei; tel. (2) 27607262; fax (2) 27461588; monthly; Editor-in-Chief VICTOR FUNG.

Sinorama: 5/F, 54 Chunghsiao East Rd, Sec. 1, Taipei 100; tel. (2) 23922256; fax (2) 23970655; f. 1976; monthly; cultural; bilingual magazine with edns in Chinese with Japanese, Spanish and English; Publr SU TZA-PING; Editor-in-Chief ANNA Y. WANG; circ. 110,000.

Studio Classroom: 10 Lane 62, Ta-Chih St, Taipei; tel. (2) 25338082; fax (2) 25331009; internet www.studioclassroom.com; f. 1962; monthly; Publr DORIS BROUGHAM.

Taipei Journal: 2 Tientsin St, Taipei 10041; tel. (2) 23970180; fax (2) 23568233; e-mail tj@mail.gio.gov.tw; internet taipeijournal.nat.gov.tw; f. 1964; fmrly Free China Journal; weekly; English; news review; Publr ARTHUR IAP; Exec. Editor-in-Chief MICHAEL CHEN; circ. 30,000.

Taipei Times: 5/F, 137 Nanking East Rd, Sec. 2, Taipei; tel. (2) 25182728; fax (2) 25189154; internet www.taipeitimes.com; f. 1999; Publr ANTONIO CHIANG.

Taiwan Review: 2 Tientsin St, Taipei 100; tel. (2) 23516419; e-mail tr@gio.gov.tw; f. 1951; fmrly *Taipei Review*, renamed as above March 2003; monthly; English; illustrated; Publr Dr LIN CHIA-LUNG; Editor-in-Chief ANDREW T. H. CHENG.

Time Express: 7/F, 2, 76 Tun Hua South Rd, Sec. 2, Taipei; tel. (2) 27084410; fax (2) 27084420; f. 1973; monthly; Publr RICHARD C. C. HUANG.

Unitas: 10/F, 180 Keelung Rd, Sec. 1, Taipei; tel. (2) 27666759; fax (2) 27567914; e-mail unitas@udngroup.com.tw; monthly; Chinese; literary journal; Publr CHANG PAO-CHING; Editor-in-Chief HSU HUI-CHIH.

Vi Vi Magazine: 7/F, 550 Chung Hsiao East Rd, Sec. 5, Taipei; tel. (2) 27275336; fax (2) 27592031; f. 1984; monthly; women's interest; Pres. TSENG CHING-TANG; circ. 60,000.

Vogue/GQ Conde Nast Interculture: 15/F, 51, Sec. 2, Keelung Rd, Sinyi District, Taipei 110; tel. (2) 27328899; fax (2) 27390504; f. 1996; monthly; Publr BENTHAM LIU.

Wealth Magazine: 7/F, 52 Nanking East Rd, Sec. 1, Taipei; tel. (2) 25816196; fax (2) 25119596; f. 1974; monthly; finance; Pres. TSHAI YEN-KUEN; Editor ANDY LIAN; circ. 75,000.

Win Win Weekly: 7/F, 52 Nanking East Rd, Taipei; tel. (2) 25816196; fax (2) 25119596; f. 1996; Publr GIN-HO HSHIE.

Youth Juvenile Monthly: 3/F, 66-1 Chung Cheng South Rd, Sec. 1, Taipei; tel. (2) 23112832; fax (2) 23612239; e-mail youth@ms2.hinet.net; internet www.youth.com.tw; f. 1965; Publr LEE CHUNG-GUAI.

NEWS AGENCIES

Central News Agency (CNA): 209 Sung Chiang Rd, Taipei; tel. (2) 25051180; fax (2) 25078839; e-mail cnamark@ms9.hinet.net; internet www.cna.com.tw; f. 1924; news service in Chinese, English and Spanish; feature and photographic services; 12 domestic and 30 overseas bureaux; Pres. HU YUAN-HUI.

Foreign Bureaux

Agence France-Presse (AFP): Rm 617, 6/F, 209 Sung Chiang Rd, Taipei; tel. (2) 25016395; fax (2) 25011881; e-mail AFPTPE@ms11.hitnet.tw; Bureau Chief YANG HSIN-HSIN.

Associated Press (AP) (USA): Rm 630, 6/F, 209 Sung Chiang Rd, Taipei; tel. (2) 25036651; fax (2) 25007133; Bureau Chief WILLIAM FOREMAN.

Reuters (UK): 10/F, 196 Chien Kuo North Rd, Sec. 2, Taipei; tel. (2) 25033034; fax (2) 25092624; Bureau Chief TIFFANY WU.

Publishers

There are 7,810 publishing houses. In 2001 a total of 40,235 titles were published.

Art Book Co: 1/F, 18 Lane 283, Roosevelt Rd, Sec. 3, Taipei; tel. (2) 23620578; fax (2) 23623594; e-mail artbook@ms43.hinet.net; Publr HO KUNG SHANG.

Cheng Wen Publishing Co: 3/F, 277 Roosevelt Rd, Sec. 3, Taipei; tel. (2) 23628032; fax (2) 23660806; e-mail ccicncwp@ms17.hinet.net; Publr LARRY C. HUANG.

Children Publication Co Ltd: 1F-1, 314 Nei-Hu Rd, Taipei; tel. (2) 87972799; fax (2) 87972700; e-mail cplemail@ms11.hinet.net; internet www.012book.com.tw; f. 1994.

China Times Publishing Co: 2/F, 240 Hoping West Rd, Sec. 3, Taipei; tel. (2) 23686222; fax (2) 23049302; e-mail ctpc@mse.hinet.net; internet www.readingtimes.com.tw; f. 1975; Pres. MO CHAO-PING.

Chinese Culture University Press: 55 Hua Kang Rd, Yangming-shan, Taipei; tel. (2) 28611861; fax (2) 28617164; e-mail ccup@staff.pccu.edu.tw; Publr LEE FU-CHEN.

Cite Publishing Ltd: 2/F, 141 Mingsheng East Rd, Sec. 2, Taipei; tel. (2) 25000888; fax (2) 25001941; e-mail sales@cite.com.tw; internet www.cite.com.tw; f. 1996.

The Commercial Press Ltd: 37 Chungking South Rd, Sec. 1, Taipei; tel. (2) 23115538; fax (2) 23716274; e-mail cptw@ms12.hinet.net; f. 1897; Editor NANCY YI-YING CHIANG.

Commonwealth Publishing Co: 2/F, 1 Lane 93, Sung Chiang Rd, Taipei; tel. (2) 25076735; fax (2) 25082941; e-mail jeffchen@cwgv.com.tw; internet www.bookzone.com.tw; f. 1982.

Crown Publishing Group: 50 Lane 120, Tun Hua North Rd, Taipei; tel. (2) 27168888; fax (2) 27133422; e-mail edit3@crown.com

.tw; internet www.crown.com.tw; f. 1954; Publr PHILIP PING; 90 employees.

The Eastern Publishing Co Ltd: 121 Chungking South Rd, Sec. 1, Taipei; tel. (2) 23114514; fax (2) 23814132; Publr CHENG LI-TSU.

Elite Publishing Co: 1/F, 33-1 Lane 113, Hsiamen St, Taipei 100; tel. (2) 23671021; fax (2) 23657047; e-mail elite113@ms12.hinet.net; f. 1975; Publr KO CHING-HWA.

Far East Book Co: 10/F, 66-1 Chungking South Rd, Sec. 1, Taipei; tel. (2) 23118740; fax (2) 23114184; e-mail service@mail.fareast.com.tw; internet www.fareast.com.tw; art, education, history, physics, mathematics, law, literature, dictionaries, textbooks, language tapes, Chinese-English dictionary; Publr GEORGE C. L. PU.

International Cultural Enterprises: Rm 612, 6/F, 25 Po Ai Rd, Taipei 100; tel. (2) 23318080; fax (2) 23318090; e-mail itsits@ms69.hinet.net; internet www.itsits.com.tw; Publr LAKE HU.

Kwang Hwa Publishing Co: 5/F, 54 Chung Hsiao East Rd, Sec. 1, Taipei; tel. (2) 23922256; fax (2) 23970655; e-mail service@sinorama.com.tw; internet www.sinorama.com.tw; Publr ARTHUR IAP.

Li-Ming Cultural Enterprise Co: 2/F, 49 Chungking South Rd, Sec. 1, Taipei 100; tel. (2) 23310537; fax (2) 23821244; e-mail liming03@ms57.hinet.net; internet www.limingco.com.tw; f. 1971; Pres. SHEN FANG-SHIN.

Linking Publishing Co Ltd: 555, Sec. 4, Jhongsiao East Rd, Taipei; tel. (2) 27681234; fax (2) 27493734; e-mail linkingp@ms9.hinet.net; internet www.udngroup.com.tw/linkingp; Publr LIU KUO-JUEI.

Locus Publishing Co: 11/F, 25 Nanking East Rd, Sec.4, Taipei; tel. (2) 87123898; fax (2) 87123897; e-mail locus@locuspublishing.com; internet www.locuspublishing.com; f. 1996.

San Min Book Co Ltd: 386 Fushing North Rd, Taipei; tel. (2) 25006600; fax (2) 25064000; e-mail editor@sanmin.com.tw; internet www.sanmin.com.tw; f. 1953; literature, history, philosophy, social sciences, dictionaries, art, politics, law; Publr LIU CHEN-CHIANG.

Sitak Publishing Group: 10/F, 15 Lane 174, Hsin Ming Rd, Neihu Dist., Taipei; tel. (2) 27911197; fax (2) 27918606; e-mail readers@sitak.com.tw; internet www.sitak.com.tw; Publr CHU PAO-LOUNG; Dir KELLY CHU.

Taiwan Kaiming Book Co: 77 Chung Shan North Rd, Sec. 1, Taipei; tel. (2) 25510820; fax (2) 25212894; Publr LUCY CHOH LIU.

Tung Hua Book Co Ltd: 105 Emei St, Taipei; tel. (2) 23114027; fax (2) 23116615; e-mail service@bookcake.com.tw; internet www.bookcake.com.tw; f. 1965; Publr CHARLES CHOH.

The World Book Co: 6/F, 99 Chungking South Rd, Sec. 1, Taipei; tel. (2) 23113834; fax (2) 23317963; e-mail wbc.ltd@msa.hinet.com; internet www.worldbook.com.tw; f. 1921; literature, textbooks; Chair. YEN FENG-CHANG; Publr YEN ANGELA CHU.

Youth Cultural Enterprise Co Ltd: 3/F, 66-1 Chungking South Rd, Sec. 1, Taipei; tel. (2) 23112832; fax (2) 23113309; e-mail howei@hotmail.com; internet www.youth.com.tw; f. 1958; Publr LEE CHUNG-KUEI.

Yuan Liou Publishing Co Ltd: 6/F, 81, Sec. 2, Nanchang Rd, Taipei; tel. (2) 23926899; fax (2) 33223707; e-mail ylib@ylib.com; internet www.ylib.com; f. 1975; fiction, non-fiction, children's; Publr WANG JUNG-WEN.

Broadcasting and Communications

TELECOMMUNICATIONS

Directorate-General of Telecommunications: Ministry of Transportation and Communications, 16 Chinan Rd, Sec. 2, Taipei; tel. (2) 23433969; internet www.dgt.gov.tw; regulatory authority.

Chunghwa Telecommunications Co Ltd: 21 Hsinyi Rd, Sec. 1, Taipei; tel. (2) 23445385; fax (2) 23919166; e-mail chtir@cht.com.tw; internet www.cht.com.tw; f. 1996; state-controlled company, privatization commenced 2000; sale of 2% of shares at domestic auction and 13.8% of shares to overseas investors scheduled for July 2003; Chair. HO-CHEN TAN.

Far EasTone Telecom: 334 Sze Chuan Rd, Sec. 1, Taipei; tel. (2) 29505478; internet www.fareastone.com.tw; mobile telephone services.

KG Telecom: 43 Kuan Chien Rd, Taipei; tel. (2) 23888800; e-mail kgtweb@kgt.com.tw; internet www.kgt.com.tw; mobile telephone services.

Taiwan Cellular Corpn: internet www.twngsm.com.tw; f. 1998 as Pacific Cellular Corpn; mobile telephone and internet services; Pres. JOSEPH FAN.

BROADCASTING

Broadcasting stations are mostly commercial. The Ministry of Transportation and Communications determines power and frequencies, and the Government Information Office supervises the operation of all stations, whether private or governmental.

Radio

In July 2004 there were 155 radio broadcasting corporations in operation.

Broadcasting Corpn of China (BCC): 7/F, Sung Chiang Rd, Taipei 104; tel. (2) 25019688; fax (2) 25018545; internet www.bcc.com.tw; f. 1928; domestic (6 networks and 1 channel) services; 9 local stations, 131 transmitters; Pres. LEE CHING-PING; Chair. CHAO SHOU-PO.

Central Broadcasting System (CBS): 55 Pei An Rd, Tachih, Taipei 104; tel. (2) 28856168; fax (2) 28852315; e-mail rtm@cbs.org.tw; internet www.cbs.org.tw; domestic and international service; Dir LIN FONG-CHEN.

Cheng Sheng Broadcasting Corpn Ltd: 7/F, 66-1 Chungking South Rd, Sec. 1, Taipei; tel. (2) 23617231; fax (2) 23715665; internet www.csbc.com.tw; f. 1950; 6 stations, 3 relay stations; Chair. WENG YEN-CHING; Pres. PANG WEI-NANG.

International Community Radio Taipei (ICRT): 2/F, 373 Sung Chiang Rd, Taipei; tel. (2) 25184899; fax (2) 25183666; internet www.icrt.com.tw; predominantly English-language broadcaster; Gen. Man. DOC CASEY.

Kiss Radio: 34/F, 6 Min Chuan 2 Rd, Kaohsiung; tel. (7) 3365888; fax (7) 3364931; internet www.kiss.com.tw; Pres. HELENA YUAN.

M-radio Broadcasting Corpn: 810/F-1, 1-18 Taichung Kang Rd, Sec. 2, Taichung City; tel. (4) 23235656; fax (4) 23231199; e-mail jason@mradio.com.tw; internet www.mradio.com.tw; Pres. CHEN WEI-LIANG; Gen. Man. JASON C. LIN.

UFO Broadcasting Co Ltd: 25/F, 102 Roosevelt Rd, Sec. 2, Taipei; tel. (2) 23636600; fax (2) 23673083; internet www.ufo.net.tw; Pres. JAW SHAU-KONG.

Voice of Taipei Broadcasting Co Ltd: 10/F, B Rm, 15-1 Han Chou South Rd, Sec. 1, Taipei; tel. (2) 23957255; fax (2) 23947855; internet www.vot.com.tw; Pres. NITA ING.

Television

Legislation to place cable broadcasting on a legal basis was adopted in mid-1993, and by June 2004 63 cable television companies were in operation. A non-commercial station, Public Television (PTV), went on air in July 1998. Legislation to place satellite broadcasting on a legal basis was adopted in February 1999, and by May 2004 132 satellite broadcasting channels (provided by 60 domestic and 19 international companies) and 5 domestic and 3 international Digital Broadcasting System (DBS) channels were in operation.

China Television Co (CTV): 120 Chung Yang Rd, Nan Kang District, Taipei; tel. (2) 27838308; fax (2) 27826007; e-mail pubr@mail.chinatv.com.tw; internet www.chinatv.com.tw; f. 1969; Pres. JIANG FENG-CHYI; Chair. CHENG SUMING.

Chinese Television System (CTS): 100 Kuang Fu South Rd, Taipei 10658; tel. (2) 27510321; fax (2) 27775414; e-mail public@mail.cts.com.tw; internet www.cts.com.tw; f. 1971; cultural and educational; Chair. JOU RUNG-SHENG; Pres. SHI LU.

Formosa Television Co (FTV): 14/F, 30 Pa Teh Rd, Sec. 3, Taipei; tel. (2) 25702570; fax (2) 25773170; internet www.ftv.com.tw; f. 1997; Chair. TSAI TUNG-RONG; Pres. CHEN KANG-HSING.

Public Television Service Foundation (PTS): 90 Lane 95, Sec. 9, Kang Ning Rd, Neihu, Taipei; tel. (2) 26329533; fax (2) 26338124; e-mail pts@mail.pts.org.tw; internet www.pts.org.tw; Chair. FRANK WU; Pres. YUNG-PE LEE.

Taiwan Television Enterprise (TTV): 10 Pa Teh Rd, Sec. 3, Taipei 10560; tel. (2) 25781515; fax (2) 25799626; internet www.ttv.com.tw; f. 1962; Chair. LAI KUO-CHOU; Pres. JENG IOU.

Finance

(cap. = capital; dep. = deposits; m. = million; brs = branches; amounts in New Taiwan dollars unless otherwise stated)

BANKING

In June 1991 the Ministry of Finance granted 15 new banking licences to private banks. A 16th bank was authorized in May 1992; further authorizations followed. Restrictions on the establishment of offshore banking units were relaxed in 1994. The banking sector was undergoing consolidation during 2002–03. At the end of May 2002 there were 52 banks in Taiwan. In September 2002 the Ministry of

Finance announced plans to privatize government banks by 2006, and sell its stake in commercial banks by 2010.

Central Bank

Central Bank of China: 2 Roosevelt Rd, Sec. 1, Taipei 100; tel. (2) 23936161; fax (2) 23571974; e-mail adminrol@mail.cbc.gov.tw; internet www.cbc.gov.tw; f. 1928; bank of issue; cap. 80,000m., dep. 6,090,191m. (Dec. 2003); Gov. PERNG FAI-NAN.

Domestic Banks

Bank of Taiwan: 120 Chungking South Rd, Sec. 1, Taipei 10036; tel. (2) 23493456; fax (2) 23315840; e-mail botservice@mail.bot.com.tw; internet www.bot.com.tw; f. 1899; cap. 48,000m., dep. 2,004,733m. (Dec. 2003); Chair. TZE-KAING YANG (acting); Pres. SHENG-YANN LII; 134 brs, incl. 7 overseas.

Chiao Tung Bank: 91 Heng Yang Rd, Taipei 100; tel. (2) 23613000; fax (2) 23310398; e-mail dp092@ctnbank.com.tw; internet www.ctnbank.com.tw; f. 1907; fmrly Bank of Communications; cap. 24,400m., dep. 360,787m. (Dec. 2000); Chair. SHEN-CHIH CHENG; Pres. KUO HSIUNG-CHUANG; 34 brs, incl. 2 overseas.

Export-Import Bank of the Republic of China (Eximbank): 8/F, 3 Nan Hai Rd, Taipei 100; tel. (2) 23210511; fax (2) 23940630; e-mail eximbank@eximbank.com.tw; internet www.eximbank.com.tw; f. 1979; cap. 10,000m., dep. 14,075m. (Dec. 2000); Chair. PAULINE FU; Pres. JOSEPH W. TSAI; 9 brs.

Farmers Bank of China: 85 Nanking East Rd, Sec. 2, Taipei 104; tel. (2) 21003456; fax (2) 25515425; internet www.farmerbank.com.tw; f. 1933; cap. 12,474m., dep. 448,723m. (Dec. 2000); Chair. HENRY K.C. CHEN; Pres. PONG-LONG LIN; 76 brs.

International Commercial Bank of China (ICBC): 100 Chi Lin Rd, Taipei 10424; tel. (2) 25633156; fax (2) 25611216; e-mail service@icbc.com.tw; internet www.icbc.com.tw; f. 1912; cap. 33,157m., dep. 756,910m. (Dec. 2000); Chair. TZONG-YEONG LIN; Pres. Y. T. (MCKINNEY) TSAI; 82 brs, incl. 18 overseas.

Land Bank of Taiwan: 46 Kuan Chien Rd, Taipei 10038; tel. (2) 23483456; fax (2) 23757023; e-mail lbot@imail.landbank.com.tw; internet www.landbank.com.tw; f. 1946; plans to merge with Bank of Taiwan and Central Trust of China in 2002 suspended; to be privatized by 2006; cap. 25,000m., res 55,151m., dep. 1,424,968m. (Dec. 2001); Chair. CHI-LIN WEA; 110 brs.

Taiwan Co-operative Bank: 77 Kuan Chien Rd, Taipei 10038; tel. (2) 23118811; fax (2) 23890704; e-mail tacbid01@14.hinet.net; internet www.tcb-bank.com.tw; f. 1946; acts as central bank for co-operatives, and as major agricultural credit institution; to be privatized by Dec. 2003; cap. 20,835m., dep. 1,705,354m. (Dec. 2001); Chair. PATRICK C. J. LIANG; Pres. WILLIAM MING-CHUNG TSENG; 144 brs.

Commercial Banks

Bank of Kaohsiung: 168 Po Ai 2nd Rd, Kaohsiung 813; tel. (7) 5570535; fax (7) 5580529; e-mail service@mail.bok.com.tw; internet www.bok.com.tw; f. 1982; cap. 4,487m., dep. 172,442m. (Dec. 2000); Chair. FLANDY SU; Pres. S. H. CHUANG; 31 brs.

Bank of Overseas Chinese: 8 Hsiang Yang Rd, Taipei 10014; tel. (2) 23715181; fax (2) 23814056; e-mail plan@mail.booc.com.tw; internet www.booc.com.tw; f. 1961; cap. 16,752m., dep. 238,008m. (June 2001); Chair. HERBERT S. S. CHUNG; Pres. WEN-LONG LIN; 56 brs.

Bank of Panhsin: 18 Cheng Tu St, Ban Chiau, 220, Taipei; tel. (2) 29629170; fax (2) 29572011; internet www.bop.com.tw; f. 1997; cap. 6,000m., dep. 77,602m. (Dec. 1999); Chair. PING-HUI LIU; Pres. ROGER CHUANG; 28 brs.

Bank SinoPac: 9-1 Chien Kuo North Rd, Sec. 2, Taipei; tel. (2) 25082288; fax (2) 25083456; internet www.banksinopac.com.tw; f. 1992; cap. 17,577m., dep. 204,688m. (Dec. 2000); Chair. PAUL C. LO; Pres. ANGUS CHEN; 35 brs.

Bowa Bank: B1, 11, 2/F, 123, Sec. 2, Chung Hsiao East Rd, Taipei 100; tel. (2) 23279998; fax (2) 33931565; e-mail pabkdbu@ms4.hinet.net; internet www.pab.com.tw; f. 1992; cap. 14,700m., dep. 153,431m. (Dec. 2000); Chair. WEI-CHI LIU; Pres. SHI-YUAN CHENG; 35 brs.

Cathay United Commercial Bank: 1/F, 7, Sung Jen Rd, Taipei, 110; tel. (2) 87226666; fax (2) 87898789; internet www.cathaybk.com.tw; Chair. GREGORY K.H. WANG; Pres. ROGER WU.

Central Trust of China: 49 Wu Chang St, Sec. 1, Taipei 10006; tel. (2) 23111511; fax (2) 23611544; e-mail ctc17001@ctc.com.tw; internet www.ctoc.com.tw; f. 1935; cap. 10,000m., dep. 204,808m. (Dec. 2000); Chair. JIA-DONG SHEA; Pres. RUEY-SONG HUANG; 19 brs; plans to merge with Bank of Taiwan and Land Bank of Taiwan in 2002 suspended.

Chang Hwa Commercial Bank Ltd: 57, Sec. 2, Chungshan North Rd, Taipei; tel. (2) 25362951; fax (2) 25716871; e-mail customem@ms1.chb.com.tw; internet www.chb.com.tw; f. 1905; cap. 35,356m., dep. 1,072,837m. (Dec. 2001); Chair. PO-SHIN CHANG; Pres. MIKE S. E. CHANG; 153 brs, 7 overseas.

China Development Industrial Bank: F/3, 125 Nan King East Rd, Sec. 5, Taipei 105; tel. (2) 27638800; fax (2) 27660047; internet www.cdibank.com.tw; Chair. ANGELO J.Y. KOO; Pres. JEFFREY SUEN.

Chinatrust Commercial Bank: 3 Sung Shou Rd, Taipei 110; tel. (2) 27222002; fax (2) 27251499; internet www.chinatrust.com.tw; f. 1966; cap. 73,925m., dep. 610,252m. (Dec. 2001); Chinatrust Financial Holding Co; Chair. JEFFREY L. S. KOO; Pres. ERIC CHEN; 57 brs, 11 overseas.

The Chinese Bank: 68, Sec. 3, Nan King East Rd, Taipei 104; tel. (2) 55586666; fax (2) 55588673; internet www.chinesebank.com.tw; f. 1992; cap. 15,171m., dep. 190,400m. (Dec. 2000); Chair. FAN-HSIUNG KAO; 30 brs.

Chinfon Commercial Bank: 1 Nanyang St, Taipei 100; tel. (2) 23114881; fax (2) 23141068; e-mail ibd@chinfonbank.com.tw; internet www.chinfonbank.com.tw; f. 1971; cap 11,128m., dep. 150,201m. (Dec. 2000); Chair. HUANG SHI-HUI; Pres GREGORY C. P. CHANG; 34 brs, 2 overseas.

Chung Shing Bank: 30, Su Wei 4th Rd, Kaohsiung 802; tel. (7) 3386033; internet www.csbank.com.tw; f 1992; cap. 15,076m., dep. 186,529m. (Dec. 1999); Chair. NAN-HWA WANG; Pres. CHIEN-CHING KUO; 25 brs.

Cosmos Bank: 5-10F., Tun Hua South Rd, Sec. 2, Taipei 106; tel. (2) 27011777; fax (2) 27849848; e-mail ibd@cosmosbank.com.tw; internet www.cosmosbank.com.tw; f. 1992; cap. 14,009m., dep. 174,374m. (Mar. 2002); Chair. HSUI SHENG-FA; Pres. C. C. HU; 35 brs.

Cota Commercial Bank: 59, Shih Fu Rd, Taichung 400; tel. (4) 22245161; fax (4) 22275237; internet www.cotabank.com.tw; f. 1995; cap. 3,184m., dep. 53,906m. (Dec. 2001); Chair. LIAO CHUN-TSE; Pres CHANG YING-CHE; 18 brs.

E. Sun Commercial Bank: 77 Wuchang St, Sec. 1, Taipei; tel. (2) 23891313; fax (2) 23125125; e-mail esbintl@email.esunbank.com.tw; internet www.esunbank.com.tw; f. 1992; cap. 16,933m., dep. 215,066m. (Dec. 2000); Chair. HUANG YUNG-JEN; Pres. HOU YUNG-HSUNG; 36 brs.

Enterprise Bank of Hualien: 1-7 Kung Yuan Rd, Hualien, 970; tel. (38) 351101; fax (38) 359162; e-mail hebsecd@ms27.hinet.nt; internet www.banklotus.com.tw; Chair. JASON KAO; Pres. HO PING-TUNG.

Entie Commercial Bank: 158 Min Sheng East Rd, Sec. 3, Taipei; tel. (2) 27189999; fax (2) 27187843; internet www.entiebank.com.tw; f. 1993; cap. 14,093m., dep. 174,572m. (Dec. 2001); Chair. PAUL C.H. CHIU; Pres. ANDREW C.Y. TSAI; 39 brs.

Far Eastern International Bank: 207 Tun Hua South Rd, Sec. 2, Taipei; tel. (2) 23786868; fax (2) 23779000; e-mail 800@mail.feib.com.tw; internet www.feib.com.tw; f. 1992; cap. 15,248m., dep. 156,861m. (June 2001); Chair. DOUGLAS T. HSU; Pres. ELI HONG; 30 brs.

First Commercial Bank: 30, Sec. 1, Chung Ching South Rd, Taipei 100; tel. (2) 23481111; fax (2) 23892967; e-mail fcb@mail.firstbank.com.tw; internet www.firstbank.com.tw; f. 1899; cap. 38,216m., dep. 1,179,017m. (Dec. 2001); Chair. STEVE S. F. SHIEH; Pres. LONG-I LIAO; 159 brs, 11 overseas.

Fubon Commercial Bank: 2–4/F, 169 Jen Ai Rd, Sec. 4, Taipei 106; tel. (2) 27716699; fax (2) 27780065; e-mail fubon@fubonbank.com.tw; internet www.fubonbank.com.tw; f. 1992; cap. 21,857m., res 2,613m., dep. 218,674m. (2002); Chair. CHEN S. YU; Pres. JAMES WU; 39 brs.

Fuhwa Commercial Bank: 4, Sec. 1, Chung Hsiao West Rd, Taipei 100; tel. (2) 23801888; fax (2) 23801700; e-mail service@fuhwabank.com.tw; internet www.fuhwabank.com.tw; Chair. KANG-SHIEN KAO; Pres. CHUNG-HSING CHEN.

Hsinchu International Bank: 106 Chung Yang Rd, Hsinchu 300; tel. (3) 5245131; fax (35) 250977; internet www.hibank.com.tw; f. 1948; cap. 12,665m., dep. 245,378m. (Dec. 1999); Chair. S. Y. CHAN; Pres C. W. WU; 73 brs.

Hua Nan Commercial Bank: 38 Chung Ching South Rd, Sec. 1, Taipei; tel. (2) 23713111; fax (2) 23316741; e-mail service@ms.hncb.com.tw; internet www.hncb.com.tw; f. 1919; cap. 35,198m., dep. 1,085,254m. (Dec. 2000); Chair. LIN MING-CHEN; Pres. HSU TEH-NAN; 138 brs, 5 overseas.

Hwa Tai Commercial Bank: 246 Chang An E. Rd, Sec. 2, Taipei; tel. (2) 27525252; fax (2) 27775213; internet www.hwataibank.com.tw; f. 1999; cap. 3,300m., dep. 50,111m. (Dec. 1999); Chair. M. H. LIN; Pres. CHI-RONG HUANG; 18 brs.

Industrial Bank of Taiwan: F/3, 97 Sung Jen Rd, Taipei 110; tel. (2) 23451101; fax (2) 87933568; internet www.ibt.com.tw; Chair. KENNETH C. M. LO; Pres. HENRY W. PANG.

International Bank of Taipei: 36 Nanking East Rd, Sec. 3, Taipei; tel. (2) 25063333; fax (2) 25063744; e-mail b630@ibtpe.com.tw; internet www.ibtpe.com.tw; f. 1948; cap. 18,038m., dep. 285,131m. (Dec. 2001); Chair. S. C. HO; Pres. K. C. YU; 88 brs.

Jih Sun International Bank: 68 Sungchiang Rd, Taipei; tel. (2) 25615888; fax (2) 25218878; internet www.jihsunbank.com.tw; f. 1992 as Baodao Commercial Bank, assumed present name in December 2001; cap. 10,815m., dep. 148,405m. (May 2000); Chair. EDWARD K. H. CHEN; Pres. DOLLY YANG; 24 brs.

Kao Shin Commercial Bank: 75 Lih Wen Rd, Kaohsiung 813; tel. (7) 5580711; fax (7) 5592980; f. 1997; cap. 2,300m., dep. 48,244m. (Dec. 1999); Chair. F. T. CHAO; Pres. CHARLES W. CHUNG; 27 brs.

Kaohsiung Business Bank: 87 Chung Cheng 4th Rd, Kaohsiung; tel. (7) 2613030; fax (7) 2913422; e-mail web@kbb.com.tw; internet www.kbb.com.tw; Chair. J. M. CHEN; Pres. CHEN LI-CHANG.

Lucky Bank: 35 Chung Hua Rd, Sec. 1, Taichung 403; tel. (4) 22259111; fax (4) 22258624; f. 1997; cap. 3,146m., dep. 74,753m. (Dec. 1999); Chair. C. C. CHANG; Pres. T. Y. SU; 27 brs.

Macoto Bank: 134 Hsi Chang St, Taipei 108; tel. (2) 23812160; fax (2) 23752538; e-mail master@macoto.com.tw; internet www.macotobank.com.tw; f. 1997; cap. 7,090m., dep. 157,472m. (Dec. 2001); Chair. C. I. LIN; Pres. SHERMAN CHUANG; 49 brs.

Shanghai Commercial and Savings Bank: 2 Min Chuan East Rd, Sec. 1, Taipei 104; tel. (2) 25817111; fax (2) 25318501; internet www.scsb.com.tw; f. 1915; cap. 13,260m., dep. 282,726m. (Dec. 2000); Chair. H. C. YUNG; Pres. Y. P. CHEN; 58 brs.

Sunny Bank: 88 Shih Pai Rd, Sec. 1, Taipei 112; tel. (2) 28208166; fax (2) 28233414; internet www.esunnybank.com.tw; f. 1997; cap. 3,800m., dep. 82,129m. (Dec. 1999); Chair. S. H. CHEN; Pres. CHUEH-YANG HU; 22 brs.

Ta Chong Commercial Bank: 58 Chungcheng 2nd Rd, Kaohsiung; tel. (2) 8786988; fax (2) 2509883; e-mail service@tcbank.com.tw; internet www.tcbank.com.tw; f. 1992; cap. 13,563m., dep. 144,953m. (Dec. 1999); Chair. TIAN-MAO CHEN; Pres. CHIN-TANG HUANG; 38 brs.

Taichung Commercial Bank: 87 Min Chuan Rd, Taichung 403; tel. (4) 22236021; fax (4) 22240748; e-mail webmaster@ms1.tcbbank.com.tw; internet www.tcbbank.com.tw; f. 1953; cap. 15,380m. (Dec. 2000), dep. 207,440m. (Dec. 2002); Chair. Y. F. TSAI; Pres. Y. C. TSAI; 80 brs.

Tainan Business Bank: 506 His Men Rd, Sec. 1, Tainan 700; tel. (6) 2139171; fax (6) 2136885; e-mail tnb@ms5.hinet.net; internet www.tnb.com.tw; Chair. CHEN PING-CHUN; Pres. CHEN FANG-LIEH.

Taipei Bank: 50 Chung Shan North Rd, Sec. 2, Taipei 10419; tel. (2) 25425656; fax (2) 25237896; e-mail br180@ms1.taipeibank.com.tw; internet www.taipeibank.com.tw; f. 1969; fmrly City Bank of Taipei; acquired by Fubon Financial Holding in August 2002; cap. 22,307m., res 20,283m., dep. 582,370m. (Dec. 2001); Chair. CHI YUAN LIN; Pres. JESSE Y. DING; 81 brs, 1 overseas.

Taishin International Bank: 44 Chung Shan North Rd, Sec. 2, Taipei 104; tel. (2) 25683988; fax (2) 25234551; e-mail pr@taishinbank.com.tw; internet www.taishinbank.com.tw; f. 1992; absorbed Dah An Commercial Bank in Feb. 2002; cap. 23,874m., dep. 247,254m. (Dec. 2000); Chair. THOMAS T. L. WU; Pres. DANIEL TSAI; 39 brs.

Taitung Business Bank: 354 Chung Hwa Rd, Sec. 1, Taitung 950; tel. (89) 331191; fax (89) 331194; e-mail secretpb@ttbb.com.tw; internet www.ttbb.com.tw; f. 1955; Chair. TEIN-KON SHAW; Pres. K. T. CHEN.

Taiwan Business Bank: 30 Tacheng St, Taipei 103; tel. (2) 25597171; fax (2) 25507942; e-mail tbb3688@hotmail.com; internet www.tbb.com.tw; f. 1915; reassumed present name 1994; cap. 35,878m., dep. 811,675m. (Dec. 2000); Chair. RONG-JOU WANG; Pres. JIN-FONG SOO.

Union Bank of Taiwan: 109 Ming Sheng East Rd, Sec. 3, Taipei 105; tel. (2) 27180001; fax (2) 27174093; e-mail 014_0199@email.ubot.com.tw; internet www.ubot.com.tw; f. 1992; cap. US $426m., res US $69.7m., dep. US $4,201m. (Sept. 2002); Chair. C. C. HUANG; Pres. S. C. LEE; 40 brs.

United-Credit Commercial Bank: 126 Chung Hwa Rd, Sec. 1, Taichung 400; tel. (4) 22203176; fax (4) 22232912; e-mail ucbank@mail.ucbank.com.tw; internet www.ucbank.com.tw; Chair. CHIU CHING-TE; Pres. LEE TSENG-CHANG.

There are also a number of Medium Business Banks throughout the country.

Community Financial System

The community financial institutions include both credit co-operatives and credit departments of farmers' and fishermen's associations. These local financial institutions focus upon providing savings and loan services for the community. At the end of 2003 there were 35 credit co-operatives, 253 credit departments of farmers' associations and 25 credit departments of fishermen's associations, with a combined total deposit balance of NT $1,932,900m., while outstanding loans amounted to NT $818,800m.

Foreign Banks

In December 2002 a total of 36 foreign banks were in operation in Taiwan.

STOCK EXCHANGE

In January 1991 the stock exchange was opened to direct investment by foreign institutions, and in March 1996 it was also opened to direct investment by foreign individuals. By the end of June 2003 702 foreign institutional investors had been approved to invest in the local securities market. Various liberalization measures have been introduced since 1994. In March 1999 the limits on both single and aggregate foreign investment in domestic shares were raised to 50% of the outstanding shares of a listed company. In November of that year the 'ceiling' of investment amount for each qualified foreign institutional investor in domestic securities markets was increased from US $600m. to US $1,200m. In July 2003 the Securities and Futures Commission relaxed relevant rules and regulations on foreign institutional investors.

Taiwan Stock Exchange Corpn: 13/F, 17 Po Ai Rd, Taipei 100; tel. (2) 23485678; fax (2) 23485324; f. 1962; Chair. C. Y. LEE.

Supervisory Body

Securities and Futures Commission: 85 Hsin Sheng South Rd, Sec. 1, Taipei; tel. (2) 87734202; fax (2) 8734134; Chair. CHU JAW-CHYUAN; Sec.-Gen. CHEN WEI-LUNG.

INSURANCE

In 1993 the Ministry of Finance issued eight new insurance licences, the first for more than 30 years. Two more were issued in 1994.

Aegon Life Insurance (Taiwan) Inc: 8/F, 39 Chung Hua Rd, Sec. 1, Taipei; tel. (2) 23707270; fax (2) 23707280; internet www.aegon.com.tw; f. 2001; Chair. JAMES H. C. LIU.

Allianz President General Insurance Co Ltd: 11/F, 69 Ming Sheng East Rd, Sec. 3, Taipei; tel. (2) 25157177; fax (2) 25077506; e-mail azpl@ms2.seeder.net; internet www.allianz.com.tw; f. 1995; Chair. NAN-TEN CHUNG; Gen. Man. NICHOLAS CHANG.

Cathay Life Insurance Co Ltd: 296 Jen Ai Rd, Sec. 4, Taipei 10650; tel. (2) 27551399; fax (2) 27551322; e-mail master@cathlife.com.tw; internet www.cathlife.com.tw; f. 1962; Chair. TSAI HONG-TU; Pres. LIU CHIU-TE; Gen. Man. LIU QIU-DE.

Central Insurance Co Ltd: 6 Chung Hsiao West Rd, Sec. 1, Taipei; tel. (2) 23819910; fax (2) 23116901; e-mail mngnt@cins.com.tw; internet www.cins.com.tw; f. 1962; Chair. HSEIN KUANG SHE; Gen. Man. CHING CHIANG HUANG.

Central Reinsurance Corpn: 12th Floor, 53 Nanking East Rd, Sec. 2, Taipei 104; tel. (2) 25115211; fax (2) 25235350; e-mail centralre@centralre.com; internet www.centralre.com; f. 1968; Chair. CHENG-TUI YANG; Pres. SOLOMON C. F. CHIU; 124 employees (2003).

Central Trust of China, Life Insurance Dept: 3–8/F, 69 Tun Hua South Rd, Sec. 2, Taipei; tel. (2) 27849151; fax (2) 27052214; e-mail sectrl@ctclife.com.tw; internet www.ctclife.com.tw; f. 1941; life insurance; Pres. EDWARD LO; Gen. Man. MAN-HSIUNG TSAI.

China Life Insurance Co Ltd: 122 Tun Hua North Rd, Taipei; tel. (2) 27196678; fax (2) 27125966; e-mail services@mail.chinalife.com.tw; internet www.chinalife.com.tw; f. 1963; Chair. C. F. KOO; Gen. Man. CHESTER C. Y. KOO.

China Mariners' Assurance Corpn Ltd: 11/F, 2 Kuan Chien Rd, Taipei; tel. (2) 23757676; fax (2) 23756363; internet www.cmac.com.tw; f. 1948; Chair. VINCENT M. S. FAN; Pres. W. H. HUNG.

Chung Kuo Insurance Co Ltd: 10–12/F, ICBC Bldg, 100 Chilin Rd, Taipei 10424; tel. (2) 25513345; fax (2) 25414046; f. 1931; fmrly China Insurance Co Ltd; Chair. S. Y. LIU; Pres. C. Y. LIU.

Chung Shing Life Insurance Co Ltd: 18/F, 200 Keelung Rd, Sec. 1, Taipei 110; tel. (2) 27583099; fax (2) 23451635; f. 1993; Chair. T. S. CHAO; Gen. Man. DAH-WEI CHEN.

The First Insurance Co Ltd: 54 Chung Hsiao East Rd, Sec. 1, Taipei; tel. (2) 23913271; fax (2) 23930685; internet www.firsin.com.tw; f. 1962; Chair. CHENG HANG LEE; Pres. JAMES LAI.

Fubon Insurance Co Ltd: 237 Chien Kuo South Rd, Sec. 1, Taipei; tel. (2) 27067890; fax (2) 27042915; internet www.fubon-ins.com.tw; f. 1961; Chair. TSAI MING-CHUNG; Gen. Man. T. M. SHIH.

Fubon Life Assurance Co Ltd: 14/F, 108 Tun Hua South Rd, Sec. 1, Taipei; tel. (2) 87716699; fax (2) 87715919; f. 1993; Chair. RICHARD M. TSAI; Gen. Man. PEN-YUAN CHENG.

Global Life Insurance Co Ltd: 18 Chung Yang South Rd, Sec. 2, Peitou, Taipei 11235; tel. (2) 28967899; fax (2) 28958312; e-mail jimbolin@globallife.com.tw; internet www.globallife.com.tw; f. 1993; Chair. JOHN TSENG; Pres. GIN-CHUNG LIN.

Hontai Life Insurance Co Ltd: 7/F, 70 Cheng Teh Rd, Sec. 1, Taipei; tel. (2) 25595151; fax (2) 25562840; internet www.hontai.com.tw; f. 1994; fmrly Hung Fu Life Insurance Co; Chair. TONY SHE; Gen. Man. YU-CHIEH YANG.

Kuo Hua Insurance Co Ltd: 166 Chang An East Rd, Sec. 2, Taipei; tel. (2) 27514225; fax (2) 27819388; e-mail kh11601@kuohua.com.tw; internet www.kuohua.com.tw; f. 1962; Chair. and Gen. Man. J. B. WANG.

Kuo Hua Life Insurance Co Ltd: 42 Chung Shan North Rd, Sec. 2, Taipei; tel. (2) 25621101; fax (2) 25423832; internet www.khl.com.tw; f. 1963; Chair. JASON CHANG; Pres. WEN-PO WANG.

Mercuries Life Insurance Co Ltd: 6/F, 2 Lane 150, Hsin-Yi North Rd, Sec. 5, Taipei; tel. (2) 23455511; fax (2) 23456616; internet www.mli.com.tw; f. 1993; Chair. HARVEY TANG; Gen. Man. CHUNG-SHIN LU.

Mingtai Fire and Marine Insurance Co Ltd: 1 Jen Ai Rd, Sec. 4, Taipei; tel. (2) 27725678; fax (2) 27729932; internet www.mingtai.com.tw; f. 1961; Chair. LARRY P. C. LIN; Gen. Man. H. T. CHEN.

Nan Shan Life Insurance Co Ltd: 144 Min Chuan East Rd, Sec. 2, Taipei 104; tel. (2) 25013333; fax (2) 25012555; internet www.nanshanlife.com.tw; f. 1963; Chair. EDMUND TSE; Pres. SUNNY LIN.

Newa Insurance Co Ltd: 6/F, 458 Hsin Yi Rd, Sec. 4, Taipei; tel. (2) 27205522; fax (2) 87891190; internet www.newa.com.tw; f. 1999; Chair. KENNETH K. T. YEN; Gen. Man. CHUNG-KENG CHEN.

Prudential Life Assurance Co Ltd: 12/F, 550 Chung Hsiao East Rd, Sec. 4, Taipei; tel. (2) 27582727; fax (2) 27086758; internet www.prudential-uk.com.tw; f. 1999; Chair. DOMINIC LEUNG KA KUI; CEO DAN L. TING.

Shin Fu Life Insurance Co Ltd: 8/F, 6 Chung Hsiao West Rd, Sec. 1, Taipei; tel. (2) 23817172; fax (2) 23817162; f. 1993; Chair. and Gen. Man. SONG CHI CHIENG.

Shin Kong Insurance Co Ltd: 15 Chien Kuo North Rd, Sec. 2, Taipei; tel. (2) 25075335; fax (2) 25074580; internet www.shinkong.com.tw; f. 1963; Chair. ANTHONY T. S. WU; Pres. YIH HSIUNG LEE.

Shin Kong Life Insurance Co Ltd: 66 Chung Hsiao West Rd, Sec. 1, Taipei; tel. (2) 23895858; fax (2) 23758688; internet www.skl.com.tw; f. 1963; Chair. EUGENE T. C. WU; Gen. Man. HONG-CHI CHENG.

Sinon Life Insurance Co Ltd: 11-2/F, 155 Tsu Chih St, Taichung; tel. (4) 3721653; fax (4) 3722008; e-mail sinonlife@mail.sinonlife.com.tw; internet www.sinonlife.com.tw; f. 1993; Chair. PO-YEN HORNG; Gen. Man. P. T. LAI.

South China Insurance Co Ltd: 5/F, 560 Chung Hsiao East Rd, Sec. 4, Taipei; tel. and fax (2) 27298022; internet www.south-china.com.tw; f. 1963; Chair. C. F. LIAO; Pres. ALLAN I. R. HUANG.

Tai Ping Insurance Co Ltd: 3–5/F, 550 Chung Hsiao East Rd, Sec. 4, Taipei; tel. (2) 27582700; fax (2) 27295681; f. 1929; Chair. C. C. HUANG; Gen. Man. JAMES SUN.

Taian Insurance Co Ltd: 59 Kwantsien Rd, Taipei; tel. (2) 23819678; fax (2) 23315332; e-mail taian@mail.taian.com.tw; f. 1961; Chair. C. H. CHEN; Gen. Man. PATRICK S. LEE.

Taiwan Fire and Marine Insurance Co Ltd: 8–9/F, 49 Kuan Chien Rd, Jungjeng Chiu, Taipei; tel. (2) 23821666; fax (2) 23882555; e-mail tfmi@mail.tfmi.com.tw; internet www.tfmi.com.tw; f. 1948; Chair. W. Y. LEE; Gen. Man. JOSEPH N. S. CHANG.

Taiwan Life Insurance Co Ltd: 16–19/F, 17 Hsu Chang St, Taipei; tel. (2) 23116411; fax (2) 23759714; e-mail service1@twlife.com.tw; internet www.twlife.com.tw; f. 1947; Chair. PING-YU CHU; Pres. CHENG-TAO LIN.

Union Insurance Co Ltd: 12/F, 219 Chung Hsiao East Rd, Sec. 4, Taipei; tel. (2) 27765567; fax (2) 27737199; internet www.unionins.com.tw; f. 1963; Chair. S. H. CHIN; Gen. Man. FRANK S. WANG.

Zurich Insurance Taiwan Ltd: 56 Tun Hua North Rd, Taipei; tel. (2) 27752888; fax (2) 27416004; internet www.zurich.com.tw; f. 1961; Chair. DEAN T. CHIANG; Gen. Man. YUNG H. CHEN.

Trade and Industry

GOVERNMENT AGENCIES

Bureau of Foreign Trade (Ministry of Economic Affairs): 1 Houkow St, Taipei; tel. (2) 23510271; fax (2) 23513603; e-mail boft@trade.gov.tw; internet www.trade.gov.tw; Dir-Gen. HUANG CHIH-PENG.

Council of Agriculture (COA): 37 Nan Hai Rd, Taipei 100; tel. (2) 23812991; fax (2) 23310341; e-mail webmaster@www.coa.gov.tw; f. 1984; govt agency directly under the Executive Yuan, with ministerial status; a policy-making body in charge of national agriculture, forestry, fisheries, the animal industry and food administration; promotes technology and provides external assistance; Chair. Dr LEE CHING-LUNG; Chief Sec. CHEN CHIH-CHING.

Industrial Development Bureau (Ministry of Economic Affairs): 41-3 Hsin Yi Rd, Sec. 3, Taipei; tel. (2) 27541255; fax (2) 27030160; e-mail service@moeaidb.gov.tw; internet www.moeaidb.gov.tw; Dir-Gen. CHEN CHAO-YIH.

Industrial Development and Investment Center (Ministry of Economic Affairs): 8/F, 71 Guancian Rd, Taipei 10047; tel. (2) 23892111; fax (2) 23820497; e-mail idic@mail.idic.gov.tw; internet www.idic.gov.tw; f. 1959 to assist investment and planning; Dir-Gen. ANGELA T. CHU.

CHAMBER OF COMMERCE

General Chamber of Commerce of the Republic of China: 6/F, 390 Fu Hsing South Rd, Sec. 1, Taipei; tel. (2) 27012671; fax (2) 27542107; f. 1946; 65 mems, incl. 40 nat. feds of trade asscns, 22 district export asscns and 3 district chambers of commerce; Chair. Dr GARY WANG; Sec.-Gen. CHIU JAW-SHIN.

INDUSTRIAL AND TRADE ASSOCIATIONS

China Productivity Center: 2/F, 79 Hsin Tai 5 Rd, Sec. 1, Hsichih, Taipei County; tel. (2) 26982989; fax (2) 26982976; internet www.cpc.org.tw; f. 1956; management, technology, training, etc.; Pres. CHEN MING-CHANG.

Chinese National Association of Industry and Commerce: 13/F, 390 Fu Hsing South Rd, Sec. 1, Taipei; tel. (2) 27070111; fax (2) 27017601; Chair. JEFFREY L. S. KOO.

Chinese National Federation of Industries (CNFI): 12/F, 390 Fu Hsing South Rd, Sec. 1, Taipei; tel. (2) 27033500; fax (2) 27033982; e-mail cnfi@mail.industry.net.tw; internet www.cnfi.org.tw; f. 1948; 142 mem. asscns; Chair. LIN KUNG-CHUNG; Sec.-Gen. Y. H. KUO.

Taiwan External Trade Development Council: 4/F, 333 Keelung Rd, Sec. 1, Taipei 11003; tel. (2) 27255200; fax (2) 27576653; e-mail taitra@taitra.org.tw; internet www.taiwantrade.org.tw; trade promotion body; Pres. CHAO YUEN-CHUAN.

Taiwan Handicraft Promotion Centre: 1 Hsu Chou Rd, Taipei; tel. (2) 23933655; fax (2) 23937330; f. 1956; Pres. Y. C. WANG.

Trading Department of Central Trust of China: 49 Wuchang St, Sec. 1, Taipei 10006; tel. (2) 23111511; fax (2) 23821047; f. 1935; export and import agent for private and govt-owned enterprises.

UTILITIES

Electricity

Taiwan Power Co (Taipower): 242 Roosevelt Rd, Sec. 3, Taipei 10016; tel. (2) 23651234; fax (2) 23678593; e-mail service@taipower.com.tw; internet www.taipower.com.tw; f. 1946; electricity generation; in process of privatization from 2001, to be completed in 2005; Chair. LIN NENG-PAI; Pres. LIN CHING-CHI.

Gas

The Great Taipei Gas Corpn: 5/F, 35 Kwang Fu North Rd, Taipei; tel. (2) 27684999; fax (2) 27630480; supply of gas and gas equipment.

Water

Taipei Water Dept: 131 Changxing St, Taipei; tel. (2) 7352141; fax (2) 7353185; f. 1907; responsible for water supply in Taipei and suburban areas; Commr LIN WEN-YUAN.

CO-OPERATIVES

In December 2003 there were 5,161 co-operatives, with a total membership of 5,138,145 and total capital of NT $23,524m. Of the specialized co-operatives the most important was the consumers' co-operative (4,023 co-ops).

The Co-operative League (f. 1940) is a national organization responsible for co-ordination, education and training and the movement's national and international interests (Chair. T. C. HWANG).

MAJOR COMPANIES

(cap. = capital; res = reserves; m. = million; amounts in New Taiwan dollars unless otherwise stated)

State Enterprises

China Shipbuilding Corpn: 3 Chung Kang Rd, Hsiao-kang, Kaohsiung; tel. (7) 8010111; fax (7) 8020805; e-mail 1588@kao.csbcnet.com.tw; internet www.csbcnet.com.tw; f. 1973; shipbuilding and repairing up to 1m. dwt; machinery mfrs; sales 14,430m. (2003); Pres. FAN KUANG-NAN; 2,725 employees.

Chinese Petroleum Corpn: 3 Sung Ren Rd, Shinyi District, Taipei 10010; tel. (2) 87898989; fax (2) 87899000; e-mail ir@cpc.com.tw; internet www.cpc.com.tw; f. 1946; natural gas, petroleum products, petrochemical feedstocks; refineries at Kaohsiung, Taoyuan and Talin; sales 448,821.6m. (2003); privatization plan postponed in May 2003; Chair. C. T. KUO; Pres. WENENT W. P. PAN; 15,580 employees.

Taiwan Fertilizer Co Ltd: 90 Nan King East Rd, Sec. 2, Chung Shan District, Taipei 10408; tel. (2) 25422231; fax (2) 25634597; e-mail tfc@taifer.com.tw; internet www.taifer.com.tw; mfrs of compound fertilizers, urea, ammonium sulphate, calcium super-phosphate, melamine, sulphamic acid, etc.; share cap. 9,800.0m., sales 7,831.2m. (2001); Chair. CHENG TSUNG WU; Pres. YAU KUO; 2,292 employees.

Taiwan Machinery Manufacturing Corpn: 3 Tai Chi Rd, Hsiaokang 29-87, Kaohsiung 81235; tel. (7) 8020111; fax (7) 8022129; f. 1946; machine mfg, shipbuilding and repairing, pre-fabricated steel frameworks, steel and iron casting, various steel products, and marine diesel engines; sales 2,541m. (1996/97); Chair. LIN I-HSIUNG; Pres. C. W. YUAN; 597 employees.

Taiwan Salt Industrial Corpn: 297 Chien Kan Rd, Sec. 1, Tainan 70203; tel. (6) 2610551; fax (6) 2649710; e-mail service@mail.towns.com.tw; internet www.taiwansalt.com.tw; sales 2,900m. (2002); Chair. CHENG PAO-CHING; Pres. CHIEU WEN-AN; 583 employees.

Taiwan Sugar Corpn: 266 Jianguo South Rd, Da-an District, Sec. 1, Taipei 10656; tel. (2) 23261300; fax (2) 27067038; e-mail tsc01@taisugar.com.tw; internet www.taisugar.com.tw; f. 1964; sugar, edible oils, pork, beverages, snacks, yeast, etc.; sales 29,955m. (2003); Pres. WEI WEI; 5,537 employees.

Selected Private Companies

Cement

Asia Cement Corpn: 30/F, Taipei Metro Tower, 207 Tun Hwa South Rd, Sec. 2, Taipei; tel. (2) 27338000; fax (2) 23785197; e-mail accacunt@metro.feg.com.tw; internet www.asiacement.com.tw; f. 1957; cement mfr and exporter; share cap. 20,383.9m., sales 9,958.3m. (2001); Chair. DOUGLAS TONG HSU; Pres. H. J. FENG; 1,332 employees.

Taiwan Cement Corpn: 113 Chung Shan North Rd, Sec. 2, Taipei; tel. (2) 25865101; fax (2) 25316650; e-mail finance@tcc.com.tw; internet www.tcc.com.tw; f. 1950; cement mfr and exporter; share cap. 26,073.9m., sales 20,244.7m. (2001); Chair. C. F. KOO; Pres. C. Y. KOO; Man. Dir L. S. KOO; 1,460 employees.

Chemicals

Chi Mei Corpn: 59–1 San Chia Tsun, Jenteh Hsiang, Tainan; tel. (6) 2663000; fax (6) 2665588; f. 1960; mfr of resins and other chemical products; sales 39,173m. (1995); Chair. W. L. CHI; Man. Dir C. S. LIAO; 1,500 employees.

China Petrochemical Development Corpn: 8–11/F, 12 Dong Hsing Rd, Taipei; tel. (2) 23969600; fax (2) 23517224; mfr of petroleum-related chemicals and their derivatives; cap. and res 20,049m., sales 10,778m. (1998); Pres. C. Y. HUANG.

Formosa Chemicals and Fibre Corpn: 2/F, 201 Tunhwa North Rd, Taipei; tel. (2) 27122211; fax (2) 27133229; e-mail management@fcfc.com.tw; internet www.fcfc.com.tw; f. 1965; mfrs of chemicals, pulp, rayon staple, yarns, cloth and nylon filament; share cap. 57,273m., sales 40,615m. (2001); Chair. WANG YONG-QING; Pres. WANG WANG; 7,700 employees.

Kaohsiung Ammonium-Sulphate Corpn Ltd: 100–2 Chung Shan 3rd Rd, POB 52, Kaohsiung 80614; tel. (7) 3819369; fax (7) 3352346; mfrs of ammonium sulphate, nitric acid, oleum and sulphuric acid; Chair. RICHARD M. CHEN; Gen. Man. CHEN HSIEN-HSIUNG; 890 employees.

Electrical and Computing

Acer Inc: 21/F, 88 Hsintaiwuh Rd, Sec. 1, Hsih Chih Cheng, Taipei; tel. (2) 26961234; fax (2) 25455308; e-mail stockaffairs@acer.com.tw; internet www.acer.com.tw; f. 1976; personal computers, multi-user systems, computer applications, laser printers, etc.; share cap. 43,261.1m., sales 113,189.6m. (2001); Chair. STAN SHIH; Pres. SIMON LIN; 4,000 employees.

Advanced Semiconductor Engineering Inc: Rm 1901, 19/F, 333 Keelung Rd, Sec. 1, Taipei; tel. (2) 87805489; fax (2) 27576121; e-mail ir@aseglobal.com; f. 1984; integrated circuit packaging and testing; sales US $2,240m. (2002); Chair. JASON CHANG CHIEN-SHENG; Pres. RICHARD CHANG HUNG-PENG; 24,000 employees.

Advanced Technology (Taiwan) Corpn: 1 Industry E, 6th Rd, SBIP, Hsinchu; tel. (35) 777300; fax (35) 776464; mfr of semiconductors, etc.

Chung Hwa Picture Tubes Ltd: 1127 Ho Ping Rd, Ta Nan Tsun, Pateh Hsiang, Taoyuan; tel. (3) 3675151; fax (3) 3667612; e-mail vpchien@cptt.com.tw; internet www.cptt.com.tw; f. 1971; mfr of electronic components; share cap. 43,734m., sales 69,000m. (1999); Pres. and Man. Dir C. Y. LIN; 12,000 employees.

CMC Magnetics Corpn: 104 Min Chuan West Rd, Taipei; tel. (2) 25536247; fax (2) 25535311; e-mail cmcnet@tptsl.seed.net.tw; internet www.cmcnet.com.tw; f. 1978; mfr of blank CDs for audio, video and CD-ROM uses; share cap. 21,732.5m., sales 18,614.4m. (Dec. 2001); Chair. BOB M. H. WONG; Gen. Man. WENG MING-KI; 2,000 employees.

Compal Electronics Inc: 7/F, 319 Ba De Rd, Sec. 4, Taipei; tel. (2) 27468446; fax (2) 27607903; internet www.compal.com; f. 1984; computers and accessories; share cap. 21,018.1m., sales 77,194.5m. (2001); Chair. ROCK SHENG-HSIUNG HSU; Pres. RAY JUICHENG CHEN; 4,225 employees.

Compeq Manufacturing Co Ltd: POB 9-22, 91 Lane 814, Ta Hsin Rd, Shin-chuang Vil., Lu Chu Hsiang, Taoyuan; tel. (3) 3231111; fax (3) 3235577; e-mail stock@compeq.com.tw; internet www.compeq.com.tw; f. 1973; mfr of computers and computer peripherals; share cap. 8,250.8m., sales 15,869.1m. (2001); Chair. H. W. CHEN; Pres. CHARLES CHIEN WU; 1,676 employees.

Delta Electronics Inc: 31–1 Sing Pang Rd, Kui Shan Siang, Shan Ting Chun, Tao Yuen; e-mail tse@delta.com.tw; internet www.delta.com.tw; f. 1971; electronic parts, colour monitors, etc.; share cap. 11,894.6m., sales 25,807.6m. (2001); Chair. BRUCE CHENG; Pres. MO HSIUNG CHEN; 3,500 employees.

Enlight Corpn: 11 Ting-Hu Rd, Taoyuan 333; tel. (3) 3977399; fax (3) 3973738; e-mail irene_cheng@enlightcorp.com.tw; internet www.enlightcorp.com.tw; f. 1973; mfr of computer parts and peripherals; share cap. 2,406.3m., sales 6,314.4m. (2001); Gen. Man. CHIH MING LIAO; 2,500 employees.

First International Computer Inc: 6/F, 201–04 Tun Hua North Rd, Sungshan District, Taipei; tel. (2) 27174500; fax (2) 27120231; e-mail sobin_chem@fic.com.tw; internet www.fic.com.tw; f. 1980; mfr of consumer electronics; share cap. 16,374.3m. sales 41,703.6m. (2001); Chair. MING JEN CHIEN; Pres. HSUEH LING WANG; 1,600 employees.

Fortronics International Co Ltd: 14/F, 110 Fu Hsing Rd, Taoyuan; tel. (3) 3353925; fax (3) 3328117; f. 1984; semiconductors, resistors, integrated circuits, etc.; Dir OLIVER YU.

Gold Circuit Electronics: 113 Shi Yuan Rd, Chung Li Industrial Park, Taoyuan; tel. (3) 3254591; fax (3) 4520673; e-mail 2c0@gce.com.tw; f. 1981; mfr of circuit boards; share cap. 3,908.7m., sales 6,006.7m. (2001); Chair. and Pres. HAROLD LIN; Gen. Man. MOLE LEE; 2,529 employees.

Goldentech Discrete Semiconductor Inc (TM): 4/F, 82 Pao Kao Rd, Hsintien, Taipei; tel. (2) 29178496; fax (2) 29149235; f. 1986; diodes, transistors, semiconductors, etc.; Dir JOHN LIN.

GVC Corpn: 14/F, 76 Tun Hua South Rd, Sec. 2, Ta-an District, Taipei; tel. (2) 27552888; fax (2) 27552413; e-mail jamesccho@gvc.com; internet www.gvc.com; f. 1980; personal computers, computer peripherals; share cap. 4,886.7m., sales 10,117.1m. (2001); Chair. MICHAEL CHIANG; 1000 employees.

Hon Hai Precision Industry Co (Foxconn Electronics Inc): 2 Zi You Rd, Tu Cheng City, Taipei; internet www.foxconn.com; f. 1974; computing equipment manufacturer; revenue US $2,800m. (2000); Chair. TERRY GOU.

Inventec Corpn: 66 Hou Kang St, Shih Lin District, Taipei; tel. (2) 28810721; fax (2) 28823605; e-mail iec@inventec.com; internet www.inventec.com.tw; f. 1975; computers and electronic products; share cap. 16,607.0m., sales 62,298.1m. (2001); Chair. KUO I. YEH; Pres. CHO TUNG HUA; 3,900 employees.

Lite-On Technology Corpn: 90 Chien I Rd, Chung Ho, Taipei; tel. (2) 22226181-8; fax (2) 25210660; e-mail optoservice@liteon.com.tw; internet www.liteon.com.tw; f. 1975; manufacturer of opto-electronics products; sales 15,627m. (2001); Chair. RAYMOND SOONG.

Macronix International Co Ltd: 16 Li-hsin Rd, Science-Based Industrial Park, Hsinchu; tel. (3) 25788888; fax (3) 25068616; e-mail mirandapeng@mxic.com.tw; internet www.macronix.com; f. 1989; mfr of semiconductors; share cap. 33,593.4m., sales 21,360.7m. (2001); Chair. DING-HUA HU; Pres. MING CHOI WU; 1,800 employees.

Matsushita Electric Co Ltd: 579 Yuan Shan Rd, Chung Ho City, Taipei; tel. (2) 22235121; fax (2) 22271197; f. 1962; audio equipment, cooking, heating and laundry appliances, air conditioners and refrigerators; sales 30,636m. (1995); Pres. CHI CHUNG TING; Gen. Man. FU SHAN TING; 4,000 employees.

Mitac International Corpn: 40, Wen Hua 2nd Rd, Kwei Shan Hsiang, Taoyuan; tel. (3) 3289000; fax (3) 3280928; design and manufacture of computers; cap. and res 6,614m., sales 61,546m. (1997); Chair. MATTHEW MIAU.

Philips Electronic Building Elements Industries (Taiwan): 23–30/F, 66 Chung Hsiao West Rd, Section 1, Taipei; internet www.philips.com.tw; f. 1967; mfr of integrated circuits; sales 37,996m. (1995); Chair. Y. C. LO; 2,806 employees.

Philips Electronic Industries (Taiwan) Ltd: Shih Kong Mitsukoshi Bldg, 22–24/F and 27–29/F, Chung Hsiao West Rd, Sec. 1, Taipei; tel. (2) 23887666; fax (2) 25155388; internet www.philips.com.tw; f. 1970; mfr of electronic components; sales 44,836m. (1995); Pres. YI CHIANG LO; 5,267 employees.

Quanta Computer Inc: 188 Wen Hwa 2nd Rd, Kuei Shan Hsiang, Tao Yuan Shien; tel. (3) 23280050; fax (3) 23271511; e-mail sheena.chien@quantatw.com; internet www.quantatw.com; f. 1988; manufacturer of portable personal computers; share cap. 20,825.0m., sales 112,313.4m. (2001); Chair. BARRY LAM; Pres. C. C. LEUNG; 800 employees.

Ritek Corpn: 42 Kuang Fu North Rd, Hsinchu Industrial Park, Hsinchu 30316; tel. (3) 5985696; fax (3) 5978684; e-mail personal@ritek.com.tw; internet www.ritek.com.tw; f. 1989; mfr of blank CDs for audio, video and CD-ROM uses; share cap. 16,618.4m., sales 23,285.1m. (2001); Chair. CHIN TAI YEH; Pres. TSUE GING YEH; 200 employees.

Taiwan Semiconductor Manufacturing Co Ltd (TSMC): 121 Park Ave 3, Hsinchu Science Industrial Park, Hsinchu; tel. (3) 5780221; fax (3) 5781546; e-mail invest@tsmc.com.tw; internet www.tsmc.com.tw; f. 1987; mfr of integrated circuits; share cap. 181,325.5m., sales 125,888.0m. (2001); Chair. and CEO MORRIS CHANG; Pres. DONALD W. BROOKS; 3,412 employees.

Tatung Co Ltd: 22 Chung Shan North Rd, Sec. 3, Taipei 104; tel. (2) 25925252; fax (2) 25915185; e-mail webmaster@tatung.com.tw; internet www.tatung.com.tw; f. 1918; household electric appliances, audio equipment, computers, telecommunications, wires and cables, heavy electrical apparatus, steel and machinery, material industry, construction and transport equipment; share cap. 42,100.0m., sales 71,444.5m. (2001); Chair. Dr LIN TING-SHEN; Pres. LIN WEI-SHAN; 3,970 employees.

TECO Electric & Machinery Co Ltd: 5/F, 19–9 San Chong Rd, Nan-Kang, Taipei; tel. (2) 25621111; fax (2) 25312796; e-mail andyliu@teco.com.tw; internet www.teco.com.tw; f. 1956; household appliances, commercial air conditioners, industrial motors and applications; share cap. 19,421.2m., sales 20,310.3m. (2001); Chair. THEODORE M. H. HUANG; Pres. T. S. HSIEH; 3,400 employees.

United Microelectronics Corpn Ltd (UMC): 3 Li-shin Rd 1, Science-Based Industrial Park, Hsinchu; tel. (3) 5782258; fax (3) 5781789; e-mail ibs@umc.com; internet www.umc.com; f. 1980; semiconductors, microcomputers, communications, etc.; share cap. 133,357.0m., sales 64,493.4m. (2001); Chair. ROBERT TSAO; Pres. HSUAN MING-CHIN; 9,370 employees.

Winbond Electronics Corpn: 4 Creation 3rd Rd, Science-Based Industrial Park, Hsinchu; tel. (3) 5770066; fax (3) 5789467; e-mail ckliu@winbond.com.tw; internet www.winbond.com.tw; f. 1987; design and production of very large-scale integrated circuits; share cap. 44,252.5m., sales 23,886.8m. (Dec. 2001); Pres. YANG DIN-YUAN; 1,903 employees.

Yageo Corpn: 3/F, 223-1 Pao Chiao Rd, Hsin Tien, Taipei; tel. (2) 29177555; fax (2) 29174285; e-mail fang.chang@yageo.com.tw; internet www.yageo.com.tw; f. 1987; resistors; share cap. 20,507.9m., sales 5,666.6m. (2001); Pres. CHEN TIE-MIN; 973 employees.

Engineering

Aerospace Industrial Development Corpn: 111 Fu-Hsing North Rd, Lane 68, Taichung 407; tel. (4) 2590001; fax (4) 2562265; f. 1969; aircraft, aircraft engines design and manufacturing; sales 23,506m. (1996/97); Chair. CHUEN HUEI-TSAI; Pres. CHIN HU; 4,243 employees.

China Motor Corpn: 11/F, 2 Tung Hua South Rd, Sec. 2, Ta-an District, Taipei; tel. (2) 23250000; fax (2) 27082913; e-mail spokesman@ms1.china-motor.com.tw; internet www.china-motor.com.tw; f. 1969; mfr of motor vehicles; share cap. 12,253.1m., sales 43,014.7m. (2001); Chair. VIVIAN W. YEN; Pres. H. Y. LIN; 2,471 employees.

Ford Lio Ho Motor Co: 705 Chung Hua Rd, Sec. 1, Chung Li City, Taoyuan; tel. (3) 4553131; fax (3) 4551474; f. 1972; motor vehicles; sales 47,100m. (1995); Chair. CHENG LI CHUNG; Pres. M. MCKELVIE; 2,700 employees.

Fortune Motors Co Ltd: 5/F, 270 Nanking East Rd, Sec. 3, Taipei; tel. (2) 27731111; fax (2) 27319436; f. 1975; wholesaler of motor vehicles; sales 34,889m. (1995); Pres. HSI JUI LIN; 3,000 employees.

Ho Tai Motor Co Ltd: 8–14/F, 121 Sung Chiang Rd, Chung Shan District, Taipei; tel. (2) 25062121; fax (2) 25041749; e-mail steven@mail.hotaimotor.com.tw; internet www.hotaimotor.com.tw; f. 1947; mfr of motor vehicles; cap. and res 11,809m., sales 60,300m. (2002); Pres. C. Y. CHANG; 588 employees.

Kuozui Motors Ltd: 11/F 121 Sung Chiang Rd, Taipei; tel. (3) 24529172; fax (3) 24519180; f. 1984; cars and trucks; sales 37,444m. (1995); Chair. YEN HUEI SU; 2,000 employees.

Kwang Yang Motor Co Ltd: 35 Wan Hsing St, Sanmin District, Kaohsiung; tel. (7) 3822526; fax (7) 3852583; f. 1963; mfr of motor cycles; sales 24,175m. (1995); Pres. S. C. WANG; 2,819 employees.

Nan Yang Industries Co Ltd: 46 Fu Hsing North Rd, Taipei; tel. (2) 27526571; fax (2) 27724465; f. 1965; motor vehicle components and motor cycles; sales 24,125m. (1995); Pres. CHI YUNG HSU; 2,710 employees.

Ret-Ser Engineering Agency: 207 Sung Chiang Rd, Taipei; tel. (2) 25032233; fax (2) 25031113; f. 1956; construction and design; sales 29,700m. (1995); Pres. YUAN YI TSENG; 9,451 employees.

San Yang Industry: 3 Chung Hua Rd, Hukou, Hsinchu; tel. (2) 27912161; fax (2) 27912160; e-mail omd@sym.com.tw; internet www.sym.com.tw; f. 1954; mfr of cars, motorcycles, etc.; share cap. 8,053.7m., sales 18,286.5m. (2001); Chair. S. H. HUANG; Pres. I-HSIUNG LIU; 3,550 employees.

Taiwan Aerospace Corpn (TAC): 17/F, 169 Jen-Ai Rd, Sec. 4, Taipei; tel. (2) 27716681; fax (2) 27716727; f. 1991; jet aircraft; Chair. JACK SUN; Pres. GEORGE K. LIU.

Yulon Motor Co Ltd: 39–1 Tsuen Po Kong Keng, West Lake San-yi Village, Miaoli County; tel. (3) 7871801; internet www.yulon-motor.com.tw; f. 1953; cars and pick-up trucks; share cap. 15,665.8m., sales 46,355.6m. (2000); Chair. VIVIAN W. SHUN-WEN; Pres. CHEN HWA LEE; Man. Dir and CEO KENNETH K. T. YEN; 3,130 employees.

Food and Drink

President Enterprises Corpn: 301 Chung Chen Rd, Yan Harng, Yeong Kang Shiang, Tainan Hsien; tel. (6) 2532121; fax (6) 2532661; f. 1967; noodles, processed foods, soft drinks, etc.; cap. and res 39,202m., sales 29,204m. (1998); Pres. JASON C. S. LIN; CEO KAO CHIN-YEN; 6,135 employees.

Ve Dan Enterprises Corpn: POB 9, 65 Hsin An Rd, Shalu, Taichung; tel. (4) 6622111; fax (4) 6627351; monosodium glutamate, instant noodles and canned foods; sales 5,880m. (1998); Pres. JENG YANG; 2,000 employees.

Ve Wong Corpn: 5/F, 79 Chung Shan North Rd, Sec. 2, Taipei; tel. (2) 25717271; fax (2) 25629689; e-mail tradep@vewong.com.tw; internet www.vewong.com; f. 1959; processed food and drinks; share cap. 2,092.5m., sales 2,443.3m. (2001); Chair. KUNG-PIN CHEN; Pres. W. C. HSU; 1,000 employees.

Weichuan Foods Corpn: 125 Sung Chiang Rd, Taipei; tel. (2) 25078221; fax (2) 25070623; e-mail viviankuo@weichuan.com.tw; internet www.weichuan.com.tw; f. 1953; milk products, monosodium glutamate, canned foods and soy sauce; share cap. 5,060.6m., sales 9,173.9m. (2001); Chair. KO MING HUANG; Pres. THOMAS NANTU HUANG; 3,390 employees.

Metals

China Steel Corpn: Lin Hai Industrial District, POB 47-29, 1 Chung Kang Rd, Hsiao Kang, Kaohsiung 81233; tel. (7) 8021111; fax (7) 8022511; e-mail f1000@mail.csc.com.tw; internet www.csc.com.tw; f. 1971; steel; state holding reduced to 40.6% in 1999; 16 subsidiaries; share cap. 91,089.4m., sales 85,101.3m. (2001); Chair. C. Y. WANG; Pres. J. Y. CHEN; 8,796 employees.

China Wire & Cable Co Ltd: 4/F, 54-6 Chung Shan North Rd, Sec. 3, Taipei; tel. (2) 25917111; fax (2) 25922765; aluminium doors and windows; sales 2,686m., cap. and res 4,134m. (1997); Chair. CHEN CHIN-CHUN; 862 employees.

First Copper and Iron Industry Co: 11/F, 210 Nanking East Rd, Sec. 3, Taipei; tel. (2) 27717611; fax (2) 27213467; e-mail he630@hegroup.com.tw; internet www.fcht.com.tw; manufacture and sale of metals and alloys for industrial purposes; share cap. 3,578.3m., sales 2,487.3m. (2001); Chair. HONG CHENG-TAI; Pres. WANG YUH-JEN.

Great China Metal Industry Co Ltd: 533 Mingchih Rd, Sec. 3, Taishan, Taipei; tel. (2) 29015153; fax (2) 29037168; e-mail gcm1@ms17.hinet.net; aluminium cans; share cap. 3,080.0m., sales 1,578.2m. (2001); Chair. CHIANG CHING-YI; Gen. Man. CHENG SHING CHIANG; 389 employees.

Tang Eng Iron Works Co Ltd: 458 Hsin Hsing Rd, Hu Kou Hsiang, Hsinchu Hsien; tel. (3) 5981721; fax (3) 5981646; f. 1940; stainless steel sheets and coils, steel bars, shapes etc., railway rolling stock, buses, general machinery, construction, bridge projects land development and transport business; sales 15,306m. (1997); Chair. YIN SHIU-HAU; Pres. YEN WEN-E; 2,259 employees.

Plastics and Glass

China General Plastics Corpn: 7/F, 37 Ji Hu Rd, Nei Hu district, Taipei 114; tel. (2) 25773661; fax (2) 26599553; e-mail cgpcstk@cgpc.com.tw; internet www.cgpc.com.tw; f. 1964; mfr of PVC products; share cap. 4,248,0m., sales 6,029.4m. (2001); Chair. CHANG CHIH-CHIEH; Pres. CHANG CHI; 1,470 employees.

Formosa Plastics Corpn: 39 Chong Shang Rd, Kaohsiung; tel. (7) 3331101; e-mail pjlau@fpc.com.tw; internet www.fpc.com.tw; f. 1958; PVC products, footwear, polyester fibre, etc.; several affiliates; share cap. 42,380.7m., sales 59,813.8m. (2001); Chair. YUNG-CHING WANG; Pres. C. T. LEE; 4,591 employees.

Nan Ya Plastics Corpn: 201 Tun Hwa North Rd, Taipei; tel. (2) 27122211; fax (2) 27178533; e-mail nanya@npc.com.tw; internet www.npc.com.tw; f. 1958; largest affiliate, mfr of plastic products; share cap. 58,079.2m., sales 95,400.6m. (2001); Chair. YUNG-CHING WANG; Pres. WU CHIN-JEN; 16,400 employees.

Taiwan Glass Industrial Corpn: 11/F, Taiwan Glass Bldg, 261 Nanking East Rd, Sec. 3, Sungshan District, Taipei 105; tel. (2) 27130333; fax (2) 27150333; e-mail stock@mail.taiwanglass.com.tw; internet www.taiwanglass.com.tw; f. 1964; share cap. 12,000.0m., sales 10,559.5m. (2001); Chair. LIN YU-CHIA; Pres. LIN PO-FENG; 3,670 employees.

Textiles and Garments

Chung Shing Textile Co Ltd: 10/F, 123 Chung Hsiao East Rd, Sec. 2, Chengchung District, Taipei; tel. (2) 23971188; fax (2) 23963346; e-mail chga@mail.chung-shing.com.tw; f. 1956; textiles and garments; share cap. 9,084.5m., sales 8,633.1m. (2001); Chair. I. S. CHOU; Pres. WILLIAM W. SHANG; 2,500 employees.

Far Eastern Textile Ltd: 38/F, Taipei Metro Tower, 207 Tun Hua Rd, Sec. 2, Taipei 106; tel. (2) 27338000; fax (2) 27369621; e-mail service@metro.feg.com.tw; internet www.feg.com.tw; f. 1951; polyester staple, polyester filament, texturized yarn, cotton yarn, blended yarn, cotton clothing, shirts, underwear, pyjamas, pants/suits, bedsheets, etc; share cap. 2,691.6m., sales 31,882.8m. (2001); Chair. DOUGLAS TONG HSU; Pres. JOHNNY SHI; 7,971 employees.

Hualon-Teijran: 9/F, 351 Chung Shan Rd, Sec. 2, Taipei; tel. (2) 22266801; fax (2) 22266851; e-mail archives@hualon.com.tw; internet www.hualon.com.tw; f. 1967; mfr of fabrics and yarns; share cap. 24,345.6m., sales 18,668.9m. (2001); Chair. Y. M. WANG; Pres. ZHUANG MING QI; Man. Dir LIANG CHING-HSIUNG; 6,663 employees.

Pou Chen Corpn: 2 Fukong Rd, Fu Hsing Industrial Zone, Chang Hwa Hsien; tel. (4) 7695147; fax (4) 7695150; e-mail finance@mail.pouchen.com.tw; internet www.pouchen.com.tw; mfr of footwear; share cap. 13,520.9m., sales 10,418.6m. (2001); Chair. CHI CHIEH TSAI; Pres. NAI FANG TSAI; 2,800 employees.

Shin Kong Synthetic Fibres Corpn: 8/F, 123 Nan King East Rd, Sec. 2, Chung Shan, Taipei; tel. (2) 25071251; fax (2) 25072264; e-mail jnku@shinkong.com.tw; internet www.shinkong.com.tw; f. 1967; mfr of fabrics, yarns and silk; share cap. 13,253.4m., sales 12,667.3m. (2001); Chair. TUNG CHIN WU; Pres. TUNG LIANG WU; 2,475 employees.

Tai Yuen Textile Co Ltd: 8/F, 2 Tun Hua South Rd, Sec. 2, Taipei 106; tel. (2) 27552222; fax (2) 27061277; e-mail tyt070@email.taiyuen.com; internet www.taiyuen.com; f. 1951; yarn, cloth, denim, knitting fabrics, garments and sewing thread; Chair. VIVIAN WU YEN; Pres. P. C. LEE; Man. Dir KENNETH K. T. YEN; 1,939 employees.

Tainan Spinning Co Ltd: 511 Yu Nung Rd, Tung District, Tainan 701; tel. (2) 27589888; fax (2) 27582804; e-mail general@mail.tainanspin.upeg.com.tw; internet www.tainanspin.com.tw; f. 1955; cotton, blended and synthetic yarns, etc; share cap. 15,013.4m., sales 9,446.6m. (2001); Chair. SHIU-CHI WU; Pres. PO-MIN HOU; 2,800 employees.

Miscellaneous

Cheng Loong Corpn: 1 Ming Sheng Rd, Sec. 1, Panchiao, Taipei Hsien; tel. (2) 22225131; fax (2) 22226110; e-mail stocks@mail.clc.com.tw; internet www.clc.com.tw; f. 1959; mfr of paper and paper products; share cap. 9,900m., sales 14,536.8m. (2001); Chair. CHENG-LOONG CHENG; Pres. and Man. Dir TONG-HO TSAI; 2,350 employees.

Cheng Shin Rubber Industry: 215 Meei-Kong Rd, Ta-Sun Hsiang, Chang-Hwa Hsien; tel. (4) 28525151; fax (4) 28526468; e-mail cst001@ms1.hinet.net; internet www.cst.com.tw; f. 1967; mfr of rubber goods; share cap. 8,360.1m., sales 7,761.9m. (2001); Chair. CHIEH LO; Pres. JUNG HUA CHEN; Gen. Man. CHENG RONG HWA; 2,663 employees.

Kunnan Enterprises Ltd: 33 Hsiang Ho Rd, Lee Lin Village, Tan Tzu Hsiang, Taichung; tel. (4) 5360183; fax (4) 9256491; f. 1969; sports equipment; sales 4,800m. (1991); Chair. LO KUN-NAN; 2,000 employees.

Yuen Foong Yu Paper Manufacturing Co Ltd: 4/F, 51 Chung Ching South Rd, Sec. 2, Taipei; tel. (2) 23961166; fax (2) 23966771; e-mail webmaster@yfy.com.tw; internet www.yfy.com.tw; f. 1950; mfr of paper products; share cap. 12,590.4m., sales 18,613.2m. (2001); Chair. S. S. HO; Pres. S. C. HO; 3,807 employees.

TRADE UNIONS

Chinese Federation of Labour: 7/F, 17 Jin Shan South Rd, Sec.1, Taipei 100; tel. (2) 33225111; fax (2) 33225121; e-mail cfllabor@ms10.hinet.net; internet www.cfl.org.tw; f. 1954; mems: 43 federations of unions representing 1,000,000 workers; Pres. LIN HUI-KUAN.

National Federations

Chunghwa Postal Workers' Union: 9/F, 45 Chungking South Rd, Sec. 2, Taipei 100; tel. (2) 23921380; fax (2) 23414510; e-mail cpwu8331@ms18.hinet.net; internet www.cpwu.org.tw; f. 1930; fmrly Chinese Federation of Postal Workers; restructuring completed July 2003; 25,000 mems; Pres. TSAI LIANG-CHUAN.

National Chinese Seamen's Union: 8/F, 25 Nanking East Rd, Sec. 3, Taipei; tel. (2) 25150265; fax (2) 25078211; f. 1913; 21,705 mems; Pres. FANG FU-LIANG.

Taiwan Railway Labor Union: Rm 6044, 6/F, 3 Peiping West Rd, Taipei; tel. (2) 23896115; fax (2) 23896134; f. 1947; 15,579 mems; Pres. CHEN HAN-CHIN.

Regional Federations

Taiwan Federation of Textile and Dyeing Industry Workers' Unions (TFTDWU): 2 Lane 64, Chung Hsiao East Rd, Sec. 2, Taipei; tel. (2) 23415627; f. 1958; 11,906 mems; Chair. CHANG MING-KEN.

Taiwan Provincial Federation of Labour: 92 Sungann Rd, Sec.1, Taichung; tel. (4) 22309009; fax (2) 22309012; f. 1948; 81 mem. unions and 1,571,826 mems; Pres. CHEN JEA; Sec.-Gen. HUANG YAO-TUNG.

Transport

RAILWAYS

Taiwan Railway Administration (TRA): 3 Peiping West Rd, Taipei 10026; tel. (2) 23815226; fax (2) 23831367; f. 1891; a public utility under the Ministry of Communications and Transportation; operates both the west line and east line systems, with a route length of 1,097.2 km, of which 685.0 km are electrified; the west line is the main trunk line from Keelung, in the north, to Fangliao, in the south, with several branches; electrification of the main trunk line was completed in 1979; the east line runs along the east coast, linking Hualien with Taitung; the north link line, with a length of 79.2 km from Suao Sing to Hualien, was opened in 1980; the south link line, with a length of 98.2 km from Taitung Shin to Fangliao, opened in late 1991, completing the round-the-island system; Dir-Gen. T. W. HSU.

Bureau of High Speed Rail (BOHSR): 9/F, 7 Sianmin Blvd, Sec. 2, Banchiao City, Taipei County; tel. (2) 80723333; high-speed rail link between Taipei and Kaohsiung under construction and scheduled for completion in 2005; Dir-Gen. N. H. HO.

Taipei Rapid Transit Corporation (TRTC): 7 Lane 48, Zhong Shan North Rd, Sec. 2, Taipei; tel. (2) 21812345; fax (2) 25115003; internet www.trtc.com.tw; f. 1994; 66 km (incl. 10.5 km light rail) open, with further lines under construction; Chair. RICHARD C. L. CHEN; Pres. DR HUEL-SHENG TSAY.

ROADS

There were 20,947 km of highways in 2002, most of them asphalt-paved. The Sun Yat-sen Freeway was completed in 1978. Construction of a 505-km Second Freeway, which was to extend to Pingtung, in southern Taiwan, began in July 1987 and was completed in late 2003. The Nantou branch, from the Wufeng system interchange of the Second Freeway to Puli in central Taiwan, with a length of 38 km, was scheduled for completion in 2008. Work on the Taipei–Ilan freeway began in 1991. The 31-km freeway and its 24.1-km extension, from Toucheng to Suao, were scheduled to be completed in 2005.

Taiwan Area National Expressway Engineering Bureau: 1 Lane 1, Hoping East Rd, Sec. 3, Taipei; tel. (2) 27078808; fax (2) 27017818; e-mail neebeyes@taneeb.gov.tw; internet www.taneeb.gov.tw; f. 1990; responsible for planning, design, construction and maintenance of provincial and county highways; Dir-Gen. CHENG WEN-LON.

Taiwan Area National Freeway Bureau: POB 75, Hsinchuang, Taipei 242; tel. (2) 29096141; fax (2) 29093218; e-mail tanfb1@freeway.gov.tw; internet www.freeway.gov.tw; f. 1970; Dir-Gen. LIANG YUEH.

Taiwan Motor Transport Co Ltd: 5/F, 17 Hsu Chang St, Taipei; tel. (2) 23715364; fax (2) 23820634; f. 1980; operates national bus service; Gen. Man. CHEN WU-SHIUNG.

SHIPPING

Taiwan has six international ports: Anping, Kaohsiung, Keelung, Taichung, Hualien and Suao. In 2002 the merchant fleet comprised 649 vessels, with a total displacement of 4,289,028 grt.

Evergreen International Storage & Transport Corpn: 899, Ching Kuo Rd, Taoyuan; tel. (3) 3252060; fax (3) 3252059.

Evergreen Marine Corpn: 166 Ming Sheng East Rd, Sec. 2, Taipei 104; tel. (2) 25057766; fax (2) 25055256; e-mail prd@evergreen-marine.com; internet www.evergreen-marine.com; f. 1968; world-wide container liner services; Chair. KUO SHIUAN-YU; Pres. ARNOLD WONG.

Taiwan Navigation Co Ltd: 29 Chi Nan Rd, Sec. 2, Taipei 104; tel. (2) 23941769; fax (2) 23919316; e-mail tnctpe@taiwanline.com.tw; internet www.taiwanline.com.tw; Chair. HSIU-PING HSIEH; Pres. I. Y. CHANG.

U-Ming Marine Transport Corpn: 29/F, Taipei Metro Tower, 207 Tun Hua South Rd, Sec. 2, Taipei; tel. (2) 27338000; fax (2) 27359900; e-mail uming@metro.feg.com.tw; internet www.feg.hinet.net/u-ming.html; world-wide transportation services; Chair. DOUGLAS HSU; Pres. C. K. ONG.

Wan Hai Lines Ltd: 10/F, 136 Sung Chiang Rd, Taipei; tel. (2) 25677961; fax (2) 25216000; e-mail serv@wanhai.com.tw; internet www.wanhai.com.tw; f. 1965; regional container liner services; Chair. CHEN CHAO HON; Pres. CHEN PO TING.

Yang Ming Marine Transport Corpn (Yang Ming Line): 271 Ming De 1st Rd, Chidu, Keelung 206; tel. (2) 24550357; fax (2) 24550781; e-mail richardyu@yml.com.tw; internet www.yml.com.tw; f. 1972; world-wide container liner services, bulk carrier and supertanker services; Chair. FRANK LU; Pres. HUANG WANG-HSIU.

CIVIL AVIATION

There are two international airports, Chiang Kai-shek at Taoyuan, near Taipei, which opened in 1979 (a second passenger terminal and expansion of freight facilities being completed in 2000), and Hsiaokang, in Kaohsiung (where an international terminal building was inaugurated in 1997). There are also 16 domestic airports. In March 2004 Air Macau opened a new passenger air route between Shenzhen Airport on the mainland with Taipei via Macao.

Civil Aeronautics Administration: 340 Tun Hua North Rd, Taipei; tel. (2) 23496000; fax (2) 23496277; e-mail gencaa@mail.caa.gov.tw; internet www.caa.gov.tw; Dir-Gen. BILLY K. C. CHANG.

China Airlines Ltd (CAL): 131 Nanking East Rd, Sec. 3, Taipei; tel. (2) 25062345; fax (2) 25145754; e-mail ju-reng_chen@email.china-airlines.com; internet www.china-airlines.com; f. 1959; international services to destinations in the Far East, Europe, the Middle East, the USA and Australia; Chair. Capt. Y. L. LEE; Pres. WEI HSING-HSIUNG.

EVA Airways (EVA): Eva Air Bldg, 376 Hsin-nan Rd, Sec. 1, Luchu, Taoyuan Hsien; tel. (3) 3515151; fax (3) 3510023; e-mail sammykao@evaair.com; internet www.evaair.com.tw; f. 1989; subsidiary of Evergreen Group; commenced flights in 1991; services to destinations in Asia (incl. Hong Kong and Macao), the Middle East, Europe, North America, Australia and New Zealand; Chair. CHANG KUO-CHENG; Pres. LIN BOU-SHIU.

Far Eastern Air Transport Corpn (FAT): 5 Alley 123, Lane 405, Tun Hua North Rd, Taipei 10592; tel. (2) 27121555; fax (2) 27122428; internet www.fat.com.tw; f. 1957; domestic services and regional international services; Chair. STEPHEN J. TSUEI; Pres. M. W. CHENG.

Mandarin Airlines (MDA): 13/F, 134 Ming Sheng East Rd, Sec. 3, Taipei; tel. (2) 27171188; fax (2) 27170716; e-mail mandarin@mandarin-airlines.com; internet www.mandarin-airlines.com; f. 1991; subsidiary of CAL; merged with Formosa Airlines 1999; domestic and regional international services; Chair. Y. L. LEE; Pres. MICHAEL LO.

TransAsia Airways (TNA): 9/F, 139 Chengchou Rd, Taipei; tel. (2) 25575767; fax (2) 25570643; internet www.tna.com.tw; f. 1951; fmrly Foshing Airlines; domestic flights and international services; Chair. FAN CHIEH-CHIANG; Pres. SUN HUANG-HSIANG.

UNI Airways Corpn (UIA): 7/F, 100 Chang An East Rd, Sec. 2, Taipei; tel. (2) 25135533; fax (2) 25133202; internet www.uniair.com.tw; f. 1989; fmrly Makung Airlines; merged with Great China Airlines and Taiwan Airlines 1998; domestic flights and international services (to Kota Kinabalu, Malaysia; Bali, Indonesia; Phuket, Thailand); Chair. JENG KUNG-YEUN; Pres. TONY SU.

Tourism

The principal tourist attractions are the cuisine, the cultural artefacts and the island scenery. In 2003 there were 2,248,117 visitor arrivals (including 436,083 overseas Chinese) in Taiwan. Receipts from tourism in 2003 totalled US $2,976m. To strengthen the development of the tourism industry, a 'Doubling Tourist Arrivals Plan', targeting 5m. arrivals in 2008, was inaugurated in September 2002. The plan also proclaimed 2004 as 'Visit Taiwan Year', in which travel incentives were to be offered to foreign visitors.

Tourism Bureau, Ministry of Transportation and Communications: 9/F, 290 Chung Hsiao East Rd, Sec. 4, Taipei 106; tel. (2) 23491635; fax (2) 27717036; e-mail tbroc@tbroc.gov.tw; internet taiwan.net.tw; f. 1972; Dir-Gen. SU CHERNG TYAN.

Taiwan Visitors Association: 5/F, 9 Min Chuan East Rd, Sec. 2, Taipei 104; tel. (2) 25943261; fax (2) 25943265; internet www.tva.org.tw; f. 1956; promotes domestic and international tourism; Chair. CHANG SHUO LAO.

Defence

In August 2003, according to Western sources, the armed forces totalled an estimated 290,000: army 200,000 (with deployments of 15,000–20,000 and 8,000–10,000, respectively, on the islands of Kinmen (Quemoy) and Matsu), navy 45,000 (including 15,000 marines), and air force 45,000. Paramilitary forces totalled 26,650. Reserves numbered 1,657,500. Military service is for 20 months (reduced from 24 months in January 2004.

Defence Expenditure: Budgeted at NT $230,000m. for 2002.

Chief of the General Staff: Adm. LI TIEN-YU.

Commander-in-Chief of the Army: Gen. JU KAI-SHENG.

Commander-in-Chief of the Navy: Adm. MIAO YUNG-CHING.

Commander-in-Chief of the Air Force: Gen. LIU KUEI-LI.

Education

Taiwan's educational policy stresses national morality, the Chinese cultural tradition, scientific knowledge and the ability to work and to contribute to the community. Hence government policies have been mainly directed towards improving the quality and availability of education, accelerating the in-service training of teaching staff and co-ordinating education to the economic and social needs of the country. The central budget for 2004 allocated NT $309,947m. to education, science and culture.

ELEMENTARY AND SECONDARY EDUCATION

Pre-school kindergarten education is optional, although in 2003/04 240,926 children attended kindergartens. In 1968 an educational development programme was begun; this extended compulsory education for children of school-age to nine years. Children above school-age and adults who have had no education whatsoever receive education in the form of supplementary courses of six months' to one year's and 18 months' to two years' duration, which are held in national elementary schools. In 2003/04 there were 2,638 primary schools, with a total of 103,793 teachers and 1,912,791 pupils. With the extension of compulsory education to junior high school, the junior high school entrance examinations have been abolished.

Secondary, including vocational, education has shown substantial growth in past years, with 1,192 schools, 97,738 teachers and 1,676,970 pupils in 2003/04. There are three types of school: junior high, senior high and senior vocational. Senior high schools admit junior high school graduates and prepare them for higher education. They offer a three-year programme. Vocational schools also offer a three-year programme and provide training in agriculture, fishery, commerce and industry, etc.

HIGHER AND ADULT EDUCATION

In 2003/04 there were 158 universities, junior colleges and independent colleges. Most of them offer postgraduate facilities. The great majority of courses are of four years' duration. Junior colleges provide courses of between two and five years' duration. Enrolment in higher education in 2003/04 was 1,270,194 students, with 47,472 teachers.

As part of the government policy to promote advanced education and academic standards, it has been encouraging existing universities to establish graduate schools with special budgets made available for the purpose. The Government also encourages the establishment of overseas Chinese institutes of higher learning for the study of Chinese as a means of promoting international understanding. During the 1980s rules were considerably relaxed to allow more students to go abroad for further education, mostly to the USA, Japan and Europe. Government lectureships and research professorships have been established to encourage Chinese scholars abroad to return to Taiwan.

In 1987 the nine teachers' junior colleges were upgraded to teachers' colleges. These admit senior secondary graduates for a four-year course. High school teachers are trained at normal universities.

In adult education, the main aim has been to raise the literacy rate and standard of general knowledge. In 2003/04 there were 934 supplementary schools, with a total enrolment of 278,124. Chinese language, general knowledge, arithmetic, music and vocational skills are taught. Radio and television are an important component in the expansion of education. The National Education Radio is supervised by the Ministry of Education to broadcast cultural and educational programmes.

Bibliography

See also the Bibliography of the People's Republic of China

GENERAL

Alagappa, Muthiah (Ed.). *Taiwan's Presidential Politics: Democratization and Cross-Strait Relations in the 21st Century*. Armonk, NY, M. E. Sharpe, 2001.

Aspalter, Christian. *Democratization and Welfare State Development in Taiwan*. Aldershot, Ashgate, 2002.

Aspalter, Christian, and Kepler, Johannes. *Understanding Modern Taiwan: Essays in Economics, Politics, and Social Policy*. Aldershot, Ashgate, 2001.

Bullard, Monte. *The Soldier and the Citizen—The Role of the Military in Taiwan's Development*. Armonk, NY, M. E. Sharpe, 1997.

Chao, Linda, and Myers, Ramon H. (Eds). *The First Chinese Democracy: Political Life in the Republic of China on Taiwan*. Baltimore, MD, Johns Hopkins University Press, 1999.

Cheng Tun-jen, and Haggard, Stephen (Eds). *Political Change in Taiwan*. Boulder, CO, Lynne Rienner Publishers, 1993.

Chin Ko-Lin. *Heijin: Organized Crime, Business, and Politics in Taiwan*. Armonk, NY, M. E. Sharpe, 2003.

Ching Cheong. *Will Taiwan Break Away: The Rise of Taiwanese Nationalism*. Singapore, World Scientific Publishing, 2001.

Ching, Leo T. S. *Becoming 'Japanese': Colonial Taiwan and the Politics of Identity Formation*. Berkeley, University of California Press, 2001.

Clough, Ralph N. *Island China*. Cambridge, MA, Harvard University Press, 1978.

Cohen, Marc J. *Taiwan at the Crossroads: Human Rights, Political Development and Social Change on the Beautiful Island*. Washington, DC, Asia Resource Centre, 1989.

Copper, John F. *China Diplomacy: The Washington-Taipei-Beijing Triangle*. Boulder, CO, Westview Press, 1993.

Words Across The Taiwan Straits. Lanham, MD, University Press of America, 1995.

Taiwan: Nation-State or Province? Boulder, CO, Westview Press, 1999.

As Taiwan Approaches the New Millennium. Lanham, MD, University Press of America, 2002.

Corcuff, Stephane (Ed.). *Memories of the Future: National Identity Issues and the Search for a New Taiwan*. Armonk, NY, M. E. Sharpe, 2002.

Crozier, Brian. *The Man Who Lost China: The First Full Biography of Chiang Kai-Shek*. London, Angus and Robertson, 1977.

Dell'Orto, Alessandro. *Place and Spirit in Taiwan*. Richmond, Surrey, Curzon Press, 2002.

Edmonds, Martin, and Tsai, Michael (Eds). *Defending Taiwan: The Future Vision of Taiwan's Defence Policy and Military Strategy*. London, RoutledgeCurzon, 2002.

Taiwan's Maritime Security. London, RoutledgeCurzon, 2002.

Taiwan's Security and Air Power. London, RoutledgeCurzon, 2003.

Edmonds, Richard L., and Goldstein, Steven M. (Eds). *Taiwan in the Twentieth Century: A Retrospective*. New York, Cambridge University Press, 2001.

Feldman, Harvey J. (Ed.). *Constitutional Reform and the Future of the Republic of China*. Armonk, NY, M. E. Sharpe, 1993.

Fenby, Jonathan. *Generalissimo: Chiang Kai-shek and the China he Lost*. New York, NY, Free Press, 2003.

Garver, John W. *The Sino-American Alliance—Nationalist China and American Cold War Strategy in Asia*. Armonk, NY, M. E. Sharpe, 1997.

Goddard, W. G. *Formosa: A Study in Chinese History*. London, Macmillan, 1966.

Herschensohn, Bruce. (Ed.). *Across the Taiwan Strait: Democracy: The Bridge Between Mainland China and Taiwan*. Lanham, MD, Lexington, 2002.

Hickey, Dennis Van Vranken. *Taiwan's Security in the Changing International System*. Boulder, CO, Lynne Rienner Publishers, 1997.

Hsiau A-Chin. *Contemporary Taiwanese Cultural Nationalism*. London, Routledge, 2000.

Hsu Long-Hsuen, and Chang Ming-Kai. *History of the Sino-Japanese War*. Chung Wu Publishing Co, 1971.

Kaplan, David E. *Fires of the Dragon: Politics, Murder and the Kuomintang*. New York, Atheneum, 1993.

Kiang, Clyde Y. *The Hakka Search for a Homeland*. Elgin, PA, Allegheny Press, 1991.

Klintworth, Gary. *New Taiwan, New China: Taiwan's Changing Role in the Asia-Pacific Region*. Melbourne, Longman, 1996.

Klintworth, Gary (Ed.). *Taiwan in the Asia-Pacific in the 1990s*. St Leonards, NSW, Allen and Unwin, 1994.

Lai Tse-han, Myers, Ramon H., and Wei Wou. *A Tragic Beginning: The Taiwan Uprising of February 28, 1947*. CA, Stanford University Press, 1992.

Lasater, Martin L. *The Changing of the Guard: President Clinton and the Security of Taiwan*. Boulder, CO, Westview Press, 1996.

The Taiwan Conundrum. Boulder, CO, Westview Press, 1999.

Lasater, Martin L., Yu, Peter Kien-Hong, Hsu, Kuang-Min, and Lym, Robyn (Eds). *Taiwan's Security in the Post-Deng Xiaoping Era*. London, Frank Cass, 2001.

Lee, David Tawei. *The Making of the Taiwan Relations Act: Twenty Years in Retrospect*. Oxford, Oxford University Press, 2000.

Leng Shao-chuan (Ed.). *Chiang Ching-Kuo's Leadership in the Development of the Republic of China on Taiwan*. Lanham, MD, University Press of America, 1994.

Long, Simon. *Taiwan: China's Last Frontier*. London, Macmillan, 1991.

Marsh, Robert M. *The Great Transformation—Social Change in Taipei, Taiwan Since the 1960s*. Armonk, NY, M. E. Sharpe, 1996.

Nadeau, Jules. *Twenty Million People—Made in Taiwan*. Montréal, Montreal Press, 1990.

Rawnsley, Gary D. *Taiwan's Informal Diplomacy and Propaganda*. New York, St Martin's Press, 2000.

Republic of China Yearbook. Annual. Taipei, Kwang Hwa Publishing Co.

Rigger, Shelley. *Politics in Taiwan*. London, Routledge, 1999.

From Opposition to Power: Taiwan's Democratic Progressive Party. Boulder, CO, Lynne Rienner Publishers, 2001.

Roy, Denny. *Taiwan: A Political History*. Ithaca, Cornell University Press, 2003.

Rubinstein, Murray A. (Ed.). *Taiwan: A New History*. Armonk, NY, M. E. Sharpe, 1999.

Shambaugh, David (Ed.). *Contemporary Taiwan*. Oxford, Clarendon Press, 1998.

Sheng, Lijun. *China's Dilemma: The Taiwan Issue*. London, I. B. Tauris, 2001.

Cross-Strait Relations Under Chen Shui-bian. Singapore, Institute of Southeast Asian Studies, 2002.

Skoggard, Ian A. *The Indigenous Dynamic in Taiwan's Postwar Development*. Armonk, NY, M. E. Sharpe, 1996.

Sutter, Robert G., and Johnson, William R. (Eds). *Taiwan in World Affairs*. Boulder, CO, Westview Press, 1996.

Swaine, Michael. *Taiwan: Foreign and Defense Policymaking 2001*. Santa Monica, CA, Rand, 2001.

Tan, Alexander C., Chan, Steven, and Jillson, Calvin (Eds). *Taiwan's National Security: Dilemmas and Opportunities*. Aldershot, Ashgate, 2001.

Taylor, Jay. *The Generalissimo's Son: Chiang Ching-Kuo and the Revolutions in China and Taiwan*. Cambridge, MA, Harvard University Press, 2000.

Tehpen, Tsai. *Elegy of Sweet Potatoes: Stories of Taiwan's White Terror*. Irvine, CA, Taiwan Publishing Company, 2003.

Tien Hung-mao. *The Great Transition: Political and Social Change in the Republic of China*. Stanford, CA, Hoover Institution Press, 1990.

Tien Hung-mao (Ed.). *Taiwan's Electoral Politics and Democratic Transition: Riding the Third Wave*. Armonk, NY, M. E. Sharpe, 1995.

Tien Hung-mao and Steve Tsang (Eds). *Democratization in Taiwan: Implications for China*. New York, St Martin's Press, 1999.

Tsang, Steve. *Peace and Security Across the Taiwan Strait*. Basingstoke, Palgrave Macmillan, 2004.

Tsang, Steve (Ed.). *Political Developments in Taiwan since 1949*. Honolulu, HI, University of Hawaii Press, 1993.

Wachman, Alan M. *Taiwan: National Identity and Democratization*. Armonk, NY, M. E. Sharpe, 1994.

Wakeman, Frederic E. *Spymaster: Dai Li and the Chinese Secret Service*. Berkeley, CA, University of California Press, 2003.

Wu, Jaushieh Joseph. *Taiwan's Democratization: Forces Behind the New Momentum*. Hong Kong, Oxford University Press, 1995.

Wu Zhuoliu. *The Fig Tree: Memoirs of a Chinese Patriot*. Bloomington, IN, 1stBooks Library, 2003.

Yang, Maysing H. *Taiwan's Expanding Role in the International Arena*. Armonk, NY, M. E. Sharpe, 1997.

Yu, Peter Kien-Hong. *The Crab and Frog Motion Paradigm Shift: Decoding and Deciphering Taipei and Beijing's Dialectical Politics*. Lanham, MD, University Press of America, 2002.

Zhan Jun. *Ending the Chinese Civil War*. New York, St Martin's Press, 1994.

Zhao Suisheng (Ed.). *Across the Taiwan Strait: Mainland China, Taiwan, and the 1995–96 Crisis*. London, Routledge, 1999.

ECONOMY

Aberbach, Joel D., Dollar, David, and Sokoloff, Kenneth L. (Eds). *The Role of the State in Taiwan's Development*. Armonk, NY, M. E. Sharpe, 1995.

Center for Quality of Life Studies, Ming Teh Foundation. *Economic Development in Taiwan: A Selected Bibliography*. Taipei, 1984.

Champion, Steven R. *The Great Taiwan Bubble: The Rise and Fall of an Emerging Stock Market*. Berkeley, CA, Pacific View Press, 1998.

Chang Chun-yen, Yu Po-lung, Zhang Junyan, and Yu P. L. (Eds). *Made By Taiwan: Booming in the Information Technology Era*. River Edge, NJ, World Scientific Publishing, 2001.

Chang Han-yu, and Myers, R. H. *Japanese Colonial Development Policy in Taiwan 1895–1906: A Case of Bureaucratic Entrepreneurship*. The Journal of Asian Studies, Vol. XXII, No. 4, 1963.

Chen Fen-Ling. *Working Women and State Policies in Taiwan: A Study in Political Economy*. Basingstoke, Palgrave, 2000.

Cho, Hui-wan, and Foreman, William (Foreword). *Taiwan's Application to GATT/WTO: Significance of Multilateralism for an Unrecognized State*. Westport, CO, Praeger, 2002.

Chow, Peter Y. C., and Bates, Gill (Eds). *Weathering the Storm: Taiwan, Its Neighbours, and the Asian Financial Crisis*. Washington, DC, Brookings Institute Press, 2000.

Chow, Peter Y. C., and Liao, Kuang-sheng (Eds). *Taiwan in the Global Economy: From an Agrarian Economy to an Exporter of High-Tech Products*. New York, Praeger, 2002.

Directorate-General of Budget, Accounting and Statistics, Executive Yuan. *Statistical Yearbook of the Republic of China*. Taipei.

Dwyer, Gerald P. Jnr, Lin Jin-long, Shea Jia-dong and Wu Chung-shu (Eds). *Monetary Policy and Taiwan's Economy*. Northampton, MA, Edward Elgar, 2002.

Galenson, Walter. *Economic Growth and Structural Change in Taiwan*. Ithaca, NY, Cornell University Press, 1979.

Gold, Thomas. *State and Society in the Taiwan Miracle*. Armonk, NY, M. E. Sharpe, 1986.

Ho, Samuel P. S. *Economic Development of Taiwan, 1860–1970*. New Haven, CT, Yale University Press, 1978.

Ho Yhi-Min. *Agricultural Development of Taiwan: 1903–1960*. Nashville, TN, Vanderbilt University Press, 1966.

Hsueh Li-Min, Hsu Chen-kuo, and Perkins, Dwight H. (Eds). *Industrialization and the State: The Changing Role of Government in Taiwan's Economy, 1945–1998*. Cambridge, MA, Harvard University Press, 2000.

International Business Publications. *Taiwan Business and Investment Opportunities Yearbook*. International Business Publications, 2002.

Jacoby, N. H. *An Evaluation of US Economic Aid to Free China 1951–1965*. New York, Praeger, 1967.

Kuo, Shirley W. Y. *The Taiwan Economy in Transition*. Boulder, CO, Westview Press, 1983.

Lee Teng-hui. *Intersectoral Capital Flows in the Economic Development of Taiwan 1895–1960*. Ithaca, NY, Cornell University Press, 1971.

Lin Ching-yuan. *Industrialization in Taiwan 1946–72*. New York, Praeger, 1973.

Mai, Chao-cheng, and Shih, Chien-sheng (Eds). *Taiwan's Economic Success Since 1980*. Cheltenham, Edward Elgar Publishing, 2001.

McBeath, Gerald. A. *Wealth and Freedom: Taiwan's New Political Economy*. Brookfield, VT, Ashgate, 1998.

Ranis, Gustav, Chu Yun-peng and Hu Sheng-cheng (Eds). *The Political Economy of Taiwan's Development Into the 21st Century*. Cheltenham, Edward Elgar Publishing, 1999.

Rubinstein, Murray A. (Ed.). *The Other Taiwan: 1945 to the Present*. Armonk, NY, M. E. Sharpe, 1995.

Shen Tsung-Han. *Agricultural Development on Taiwan since World War II*. Taipei, Mei Ya Publications Inc, 1971.

Wade, Robert. *Governing the Market: Economic Theory and the Role of Government in East Asian Industrialization*. Princeton, NJ, Princeton University Press, 1990.

INDONESIA

Physical and Social Geography

HARVEY DEMAINE

With revisions by ROBERT CRIBB

The Republic of Indonesia, which today comprises the same area as the former Netherlands East Indies, lies along the Equator between the south-eastern tip of the Asian mainland and Australia. Its western and southern coasts abut on the Indian Ocean; to the north it faces the Straits of Melaka (Malacca) and the South China Sea; and the remote northern shore of Papua (formerly Irian Jaya) province has frontage on to the Pacific Ocean. Indonesia's only land frontiers are with Papua New Guinea, to the east of (West) Papua, with Timor-Leste (known as East Timor prior to its accession to independence in May 2002), and with the Malaysian states of Sarawak and Sabah, which occupy northern Borneo; almost all of the remainder of Borneo comprises the Indonesian territory of Kalimantan.

Indonesia extends more than 4,800 km from east to west and 2,000 km from north to south. However, nearly four-fifths of the area between these outer extremities consists of sea, and the total land surface of Indonesia covers 1,922,570 sq km (742,308 sq miles).

PHYSICAL FEATURES

The country consists of about 18,108 islands of extremely varied size and character, of which some 6,000 are inhabited. The largest exclusively Indonesian island is Sumatra (Sumatera), covering 482,393 sq km, though this is exceeded by the Indonesian segment, comprising 547,891 sq km, or about two-thirds, of Borneo. These islands are followed in size by the 421,981 sq km of Papua, then by Sulawesi (Celebes), with 191,800 sq km, and by Java (Jawa), which, with the neighbouring island of Madura, totals 127,499 sq km. The remaining areas are much smaller islands, comprising Bali, the Nusa Tenggara group and the small scattered islands of the Maluku (Moluccas) group, which lie between Sulawesi and Papua.

Differences in size also reflect fundamental differences in geological structure. All the large islands except Sulawesi stand on one of two great continental shelves: the Sunda Shelf, representing a prolongation of the Asian mainland, covered by the shallow waters of the Straits of Melaka, the Java Sea and the southernmost part of the South China Sea; and the Sahul Shelf, which is covered by the shallow Arafura Sea and links New Guinea with Australia. In Sumatra, Java, north-eastern Borneo and western Papua there are pronounced mountain ranges facing the deep seas along the outer edges of the shelves, and extensive lowland tracts, facing the shallow inner seas whose coastlines reveal evidence of recent submergence. In contrast to the larger islands of western and eastern Indonesia, most of those lying between the two shelves, including Sulawesi as well as those of the Nusa Tenggara and Maluku groups, rise steeply from deep seas on all sides, with only extremely narrow coastal plains.

The recent mountain building in most parts of the archipelago is related to widespread vulcanicity, much of which is still in the active stage. Except in Borneo and western Papua, the culminating relief normally consists of volcanic cones, many of which exceed 3,000 m in altitude, though the loftiest peaks of all are the non-volcanic Punjak Jaya (5,000 m) and Idenburg-top (4,800 m) in the Snow mountains of western Papua. The archipelago is also subject to earthquakes and associated tsunami ('tidal waves'). In December 1992 approximately 2,500 people died in Flores as a consequence of tremors and inundation, and in 1993 and 1994 more than 400 died after earthquakes off the south-west coast of Sumatra and the south coast of Java.

Although the most extensive lowlands occur along the eastern coast of Sumatra and the southern coasts of Borneo and Papua, the larger part of all three lowland areas consists of tidal swamp which, until very recently, has been virtually ignored for cultivation purposes, and still constitutes a major obstacle to the opening-up of the better-drained areas further inland. Reclamation is under way in some of these lands, as part of the transmigration programme, but it still cannot match that of the narrower coastal lowlands of the smaller central island of Java.

NATURAL RESOURCES

The much greater fertility of the soils of the eastern two-thirds of Java and nearby Bali, by comparison with nearly all the rest of Indonesia, except a small part of interior and coastal north-eastern Sumatra, arises from the neutral-basic character (as opposed to the prevailingly acidic composition elsewhere) of the volcanic ejecta from which they are derived. In the remaining nine-tenths or more of Indonesia, the soils—whether volcanically derived or not—are altogether poorer in quality than they are popularly assumed to be; indeed, they are not noticeably better than in most other parts of the humid tropics.

There is much controversy about the scale of Indonesia's mineral wealth. Exploration is still continuing in many parts of the archipelago, and substantial deposits certainly exist, particularly of hydrocarbons. Despite the country's wealth of hydrocarbon deposits, growing domestic consumption of petroleum products is leading Indonesia to diversify its energy sources. Exploration and extraction of coal are being expanded, and sources of geothermal energy are being exploited on a small scale. Although overshadowed by the hydrocarbon sector, which makes a major contribution to the economy, there are significant reserves of tin, bauxite, copper, nickel, gold and silver.

CLIMATE AND VEGETATION

Climatically the greater part of Indonesia may be described as maritime equatorial, with consistently high temperatures (except at higher altitudes) and heavy rainfall in all seasons, though in many parts of western Indonesia there are distinct peak periods of exceptionally heavy rain when either the north-east or the south-west monsoon winds are blowing on shore. However, the eastern half of Java, Bali, southern Sulawesi and Nusa Tenggara, which lie further to the south and nearer to the Australian desert, experience a clearly marked dry season during the period of the south-east monsoon between June/July and September/October. The south-east monsoon subsequently changes direction, to become the south-west monsoon over western Indonesia. In Pontianak, situated almost exactly on the Equator on the west coast of Borneo, the monthly mean temperature varies only from 25.6°C in December to 26.7°C in July, and average monthly rainfall varies from 160 mm in July to 400 mm in December. The total annual rainfall is 3,200 mm. Surabaya, in eastern Java, shows even less variation in mean monthly temperature, which ranges between 26.1°C and 26.7°C throughout the year. It has four months (December–March) with over 240 mm of rain, and four others (July–October) with less than 50 mm, out of an annual total of 1,735 mm.

Nearly all of Indonesia, in its natural state, supports very dense vegetation, with significant variations, including tidal swamps, normal lowlands, lower slopes, and higher altitudes. Natural forests (which covered 109.8m. ha in 1995, or approximately 57% of Indonesia's land area) become progressively thinner as one goes eastwards from central Java to Timor, and over much of Nusa Tenggara the vegetation is better described as scrub.

POPULATION AND CULTURE

With a population of 206,264,595 according to the results of the June 2000 census, Indonesia ranks as the fourth most populous country in the world, after the People's Republic of China, India and the USA. Indonesia's population increased at an average rate of 1.4% per year over the period 1990–2002. So large a population, spread over so vast and fragmented a territory, presents wide variations, notably in ethnic type, religion and language. Java, Madura and Bali, which comprise about one-thirteenth of the total area of Indonesia, contain almost two-thirds of its population. This situation has persisted, despite extensive efforts to shift population out of Java and Bali from as early as 1905 under colonization schemes, latterly known as 'transmigration'.

Archaeological evidence suggests that the archipelago was sparsely populated from at least 50,000 years ago by people of Austromelanesian (Papuan) stock. The Austromelanesians achieved higher population densities only in New Guinea, where there is evidence of intensive cultivation of taro and other crops from about 7000 BC. In the western two-thirds of the archipelago, the Austromelanesians were later largely displaced by Austronesians, seafaring people from the island of Taiwan, who began to move southwards from about 4000 BC and who went on, in a series of great migrations, to colonize Polynesia and Madagascar. The considerable physical differences between ethnic groups in Indonesia is now thought to be related to different levels of mixing between the two peoples, with the older Austromelanesian elements progressively stronger in the eastern part of the archipelago.

Nearly 90% of the Indonesian population was Muslim in 1993. The character of Indonesian Islam, however, varies widely from region to region. The peoples of the coastal regions of Western Indonesia are predominantly Muslim, although their religious practice varies considerably according to the influence of local custom or *adat*. The strongly Muslim Minangkabau of West Sumatra, for instance, maintain a matrilineal system of inheritance directly at variance with normal Islamic practice, and local customs are important even in Aceh (far northern Sumatra), which is traditionally regarded as the staunchest Muslim region. In the interior of Java, earlier Hindu and animist influences are strong, producing a belief system known as *Kejawen* which is only broadly recognizable as Muslim. In the interior of the other larger islands, some of the formerly animist ethnic groups have been converted to Islam (such as the Gayo and Alas in northern Sumatra) or Christianity (such as the neighbouring Bataks and the Minahasans of North Sulawesi). Christian missionary influence is strongest in eastern Indonesia. Hinduism remains the religion of the island of Bali and of a few small enclaves in Java, but the traditional religion of the Dayaks of Borneo has also been given official recognition as a form of Hinduism, although it bears only a passing resemblance to the religion practised in Bali. Christianity, Buddhism and Confucianism are strong amongst the Chinese minority. Small animist communities remain in many regions, but official policy requires all Indonesians in time to accept one of the larger religions.

Over 270 Austronesian languages and 180 Papuan languages have been recognized in Indonesia, but only 13 of these have more than 1m. speakers. The national language is a development of Malay, the language of Srivijaya and of other early states on the Melaka Strait. In the pre-colonial era Malay had spread widely as a trading lingua franca, and as the language of Islam in the archipelago. The Dutch strengthened its position by using it as a major language of administration, law and education. The Indonesian nationalist movement formally adopted Malay as the national language in 1929, and called it 'Indonesian' (Bahasa Indonesia). Since independence the vocabulary of Indonesian has expanded enormously. Indonesian is now used exclusively at all but the initial levels of the education system and is the only language of public affairs.

Besides its indigenous population, Indonesia has one of the largest Chinese communities in South-East Asia. This community may have totalled nearly 3m. Although a substantial proportion of the Chinese were born and brought up in Indonesia, accepted citizenship and became 'Indonesianized', up to one-half remained without citizenship and a source of friction in the society, as demonstrated during the riots of 1998 (see History). The Chinese and smaller non-indigenous groups, such as Arabs and Eurasians, are largely concentrated in the urban areas. The cities have grown rapidly in recent years, with the population of the capital, Jakarta, rising from 4.6m. in 1971 to an estimated 12.3m. in 2003.

History

ROBERT CRIBB

HISTORICAL BACKGROUND

From about AD 100 the rise of Asia's great maritime trade route, linking India and China through South-East Asia, drew the small hunting, fishing and agricultural communities of the Indonesian archipelago into a broader world of civilization and commerce. Newly rich local rulers, trading in spices, resins and fragrant woods, justified their acquired wealth by adopting Indian political ideology, turning themselves into Hindu god-kings and providing a conduit through which Hindu ideas spread unevenly to the mass of the people. Srivijaya on the Sumatra coast (c. 700–1200) and Majapahit in the interior of Java (c. 1300–1450) emerged as the greatest early states, their influence covering much of the western archipelago. Islam arrived on the trade routes from India in about 1100, and during subsequent centuries largely displaced Hinduism and Buddhism as the formal religion of the courts and of society, although in practice it fused with Indian and local religions into distinctive Indonesian forms.

Colonial Rule

European traders and raiders were present in the archipelago from the early 16th century, but only gradually did the Dutch East Indies Company turn its scattered forts and trading posts into a colonial empire. The company focused initially on trade and soon became involved in plantations, but it was finally unable to cope with the complexities of colonial administration and the metropolitan Dutch Government took over the colony in 1799. After a brief period of British occupation under Sir Stamford Raffles (1811–16), the Dutch authorities launched an era of intensive colonial exploitation on Java known as the Cultivation System. Many authorities have argued that this exploitation brought about the long-term impoverishment of the island, though empirical evidence is equivocal. From the 1870s Western private enterprise was allowed to operate in the colony, leading to a spectacular expansion of plantation agriculture outside Java (Jawa), especially in East Sumatra (Sumatera). Petroleum extraction in Sumatra and Borneo (Kalimantan) also became important. During the four decades to 1910, the Dutch largely completed their military conquest of the archipelago. The Dutch preserved many of the traditional rulers of the archipelago as agents of colonial rule, and established a complicated legal system under which the traditional native law (*adat*) in each region was codified and applied to indigenous people. The result was a form of racial classification in which the population was divided into Europeans, Natives and 'Foreign Orientals' (principally Chinese), with differing rights and duties. Only Natives were permitted to own land, but they had fewer political and legal rights. The Chinese community, which included both recent immigrants and locally born families with a long history in the archipelago, came to occupy a middle position in society, dominating small- and medium-scale commerce to the exclusion of indigenous Indonesians.

The disruption of indigenous society by Western economic penetration and the emergence of a small educated élite, trained especially to serve the increasingly complex government and private bureaucracies, led to the rise of a nationalist movement in the early 20th century. Islamic and communist influences were initially strong in this movement, but by the late 1920s the movement was dominated by 'secular' nationalists, notably Sukarno, Hatta and Sjahrir. The movement extracted few concessions from the colonial authorities and remained politically weak until the Japanese occupation (1942–45), which removed the Dutch and raised hopes of rapid independence.

INDEPENDENCE

In the confusion following the Japanese surrender, Sukarno and Hatta declared independence on 17 August 1945, becoming President and Vice-President respectively of the new Republic of Indonesia. More than four years of fighting and negotiation with the Dutch ensued, however, before the formal transfer of sovereignty on 27 December 1949. By January 1949 the Dutch had reconquered most of the archipelago, but they were defeated by a combination of guerrilla resistance and foreign pressure; the USA, in particular, became convinced of the moderate credentials of the Indonesian nationalists, and wished to avoid a prolonged struggle which might encourage the growth of communism. The transfer of sovereignty left Indonesia with a federal system, devised by the Dutch to isolate radical forces on Java and Sumatra; the Republic of Indonesia declared in 1945 was thus a constituent state of the 14-member Republic of the United States of Indonesia. The other states were quickly dissolved, and in August 1950 Indonesia returned to a unitary structure. Against Indonesian wishes, however, the Dutch retained the territory of West New Guinea (later West Irian, renamed Irian Jaya and subsequently Papua) on the grounds that it was ethnically distinct from the rest of Indonesia.

Parliamentary Democracy, 1950–57

Under the 1950 Constitution, political power was vested in a legislature, with Prime Ministers and their cabinets responsible for executive government. Although the legislature was unelected (most members were present by virtue of their roles in one or other of the many deliberative bodies created by both sides during the war of independence), it was diverse and appears to have represented the main social forces and currents of political thought in the country. Because of this diversity, governments were invariably coalitions and were vulnerable to defections; few lasted longer than a year in office. Indonesia lacked the human and infrastructural resources to deal rapidly and effectively with the social and economic problems left by colonialism, and the performance of these governments was thus generally disappointing. Expectations had been raised by the independence struggle and could not easily be met. The political parties themselves undermined their own standing and performance by discord, corruption and partisan appointments to the civil service. The country also faced rebellions: in South Maluku (the Moluccas) a conservative, largely Christian movement attempted to secede, while in parts of Java, Sumatra and Celebes (Sulawesi) the Darul Islam movement attempted to impose an Islamic state. Many people also resented the failure of successive governments to dislodge the Dutch from West New Guinea, though Indonesia's international status was improved when it hosted the Afro-Asian Conference in Bandung in 1955.

At the first national elections in 1955 Java-based parties and those towards the left gained the greatest support, although no party came close to a majority. The continuing impasse in Jakarta, together with growing alarm in the islands outside Java, led in 1956 to local mutinies in Sumatra, an island that generated much of Indonesia's export revenue. The mutinies developed into full rebellions in 1957, although their aim was to recover central power, not to secede. In response to the crisis, President Sukarno declared martial law in 1957, and during the next two years gradually replaced the parliamentary system with authoritarian rule based on the original 1945 Constitution adopted at the beginning of the war for independence.

'Guided Democracy', 1959–65

Sukarno called his authoritarian system 'Guided Democracy', but for the most part it was a retreat from democracy. The elected legislature was replaced by an appointed one; cabinets were chosen by and responsible to the President. Political activity was restricted, and two parties were banned. The army, which had by then defeated the regional rebellions, became deeply involved in the administration under martial law, and also assumed a major role in the economy as managers of Dutch enterprises that had been nationalized in 1957. Although conservative in structure and social policy, 'Guided Democracy' was radical in rhetoric and ideology. Sukarno proclaimed a continuing revolution in the name of the poor and oppressed, and he increasingly incorporated Marxist elements into state ideology. Although he enjoyed unrivalled oratorical power over the Indonesian masses, he also made extensive use of the Partai Komunis Indonesia (PKI—Indonesian Communist Party) in mobilizing popular support. The communists had created by far the most effective party structure in the country and had won 16.4% of the vote in the 1955 elections, but probably enjoyed the support of between one-third and one-half of the population by 1965. 'Guided Democracy' delivered few tangible benefits to Indonesians. Sukarno showed little interest in day-to-day administration, and the fabric of the country gradually deteriorated; by 1965 Indonesia was one of the poorest countries in the world. The little radical legislation that he did sponsor (notably in the area of land reform) failed to be implemented, owing to bureaucratic inertia and the resistance of local powers. Sukarno's recovery of West New Guinea in 1963 by a mixture of bluff and shrewd international diplomacy won him enormous credit with Indonesians, but his confrontation with Malaysia (1963–66), which he regarded as an undemocratic and neo-colonial 'puppet' of the British, brought no such success.

Although the PKI held virtually none of the levers of power, there was widespread speculation that its broader ideological influence would deliver it control of Indonesia on the death of Sukarno, who by 1965 was visibly ageing and appeared to be losing political control. Such a prospect, however, provoked concern within the Muslim community and in the army, which remained the most powerful institution in the country. Deepening social and ideological conflict, exacerbated by declining economic conditions, led to a sense of impending crisis. On 30 September 1965 a group of left-wing junior army officers staged a limited coup against the more conservative High Command in order to forestall a rumoured army coup a few days later. Both Sukarno and the PKI probably had some knowledge of the plot, although there is no reliable evidence that either took part in planning it. Poor execution of the coup (most but not all of the targeted generals were killed), however, apparently led the plotters to expand it into a full-scale seizure of power, for which they were utterly unprepared. Their movement, therefore, was easily suppressed by the senior surviving general, Suharto.

Sukarno remained formally in office, but between October 1965 and March 1966 Suharto steadily eroded his power, preventing his exercise of authority in the Government and countering his still formidable oratorical skills by establishing what became known as the 'New Order' coalition of Muslims, students, economic managers and the armed forces, which demanded a radical change of policy direction towards economic recovery. During the same period the PKI was proscribed. Probably about 500,000 party members and sympathizers (from a claimed membership of 3m.) were killed, some by the armed forces, some by anti-communist vigilantes, and more than 1.5m. were detained for various periods. Suharto became acting President in March 1967 and full President a year later. Sukarno was held under house arrest until his death in 1970.

SUHARTO AND THE NEW ORDER

Suharto's new regime presented itself as a managerial government with the task of restoring stability to politics and growth to the economy. Suharto left much of the management of the Indonesian economy to a team of US-trained economists, who worked closely with the Inter-Governmental Group for Indonesia (IGGI), an international consortium of aid donors to Indonesia. Although they maintained the large state sector of the economy inherited from the period of 'Guided Democracy', they opened the Indonesian economy to foreign investment, first in mining and forestry, and later in industrial manufacture. Indonesia benefited especially from the rapid increase in petroleum prices, and was able to underwrite a massive programme of infrastructural investment as a base for growth (see Economy). The new Government also began programmes aimed at addressing the problems of rapid population growth. A major and effective family-planning campaign was launched. The long-standing policy of encouraging people from densely populated Java and Bali to move to less-populated regions in the other islands was expanded, with World Bank finance, to a massive population transfer that encompassed 1.5m. people between 1969 and 1982. New techniques for cultivating rice were also introduced, leading to a spectacular expansion of production, with self-sufficiency being achieved in 1982.

Suharto swiftly abandoned the left-wing rhetoric of 'Guided Democracy', but refrained from installing any ideology in its place. Instead, he elevated as a national symbol the *Pancasila*, five principles expounded by Sukarno just before the declaration of independence. These principles were: belief in God, national unity, humanitarianism, social justice and democracy. Sukarno had expressed them in deliberately general terms (belief in God, for instance, referred to belief in any supreme deity, not just belief in Allah or the Christian God) in order to transcend emerging ideological differences within the nationalist movement on the eve of independence. Suharto now used the *Pancasila* to deny the central symbolic space in Indonesian politics to any ideology, and he often portrayed his regime as representing a desirable middle course between the left (communism) and the right (fundamentalist Islam).

The armed forces played a crucial role in the new Government. They provided the main security and intelligence apparatus in the form of the Operational Command for the Restoration of Security and Order (Kopkamtib). Those who had been involved in the abortive 1965 coup were rigorously pursued, and involvement was interpreted so broadly that most leftist dissent was encompassed. The armed forces also increased their role in administration through the doctrine of *dwifungsi* or dual function, which held that the concern of the armed forces was not simply national defence but also the actual conduct of government in the interests of good administration. A powerful myth began to develop that the armed forces had played the principal role in securing independence in the 1940s, that it had done so in the face of obstruction and betrayal by the civilian authorities and that it therefore retained a special right and duty to supervise and take charge of the conduct of government. Under *dwifungsi*, serving and retired army officers took a wide variety of posts, from the level of cabinet minister down to village head.

In order to meet domestic and international expectations of a restoration of democracy, elections took place in July 1971, but under conditions designed to ensure a government victory. During the preceding months and years, the authorities comprehensively undermined the surviving parties, often intervening directly in their internal affairs to ensure that they were controlled by groups sympathetic to the armed forces. More important, the Government introduced the doctrine of 'monoloyalty', under which government officials were permitted to support only the Government at election time; other political parties thus lost the influential backing they had previously received from various sections of the bureaucracy, especially in the countryside. At the same time, the Government decreed the so-called 'floating mass policy', under which the mass of Indonesians were not to be exposed to political campaigning except during the brief, defined campaign period before elections; this was ostensibly to avoid distracting people from the tasks of national development. For the election campaign itself, contestants were screened by security officials before being permitted to stand, and public questioning of government policy was not permitted (although the implementation of policy might be challenged). In addition, inter-ethnic, inter-religious or inter-'group' (i.e. class) issues were not permitted to be discussed publicly at any time. In 1973, after the elections, the Government further weakened the parties by forcing them to merge into two fractious federations: the Partai Persatuan Pembangunan (PPP—United Development Party), comprising the four Muslim parties, and the Partai Demokrasi Indonesia (PDI—Indonesian Democratic Party), composed of the remaining parties.

Government candidates contested the 1971 elections as the Sekretariat Bersama Golongan Karya (Joint Secretariat of Functional Groups, also known as Sekber Golkar, or simply Golkar), an organization that had played a minor role under 'Guided Democracy' as an army-dominated co-ordinating body for anti-communist trade unions and other associations. Golkar, however, was not a political party; it had no individual membership and virtually no identity except during the election period, and was thus exempt from many of the restrictions on the parties, as well as enjoying the support of the bureaucracy and the armed forces. Golkar received 63% of the vote in the 1971 elections, and its share of the vote remained largely unchanged in the elections until the end of Suharto's 'New Order'. Serious discontent began to arise in Indonesia during the mid-1970s, however. Student groups, in particular, protested against the perceived lack of true democracy, the hardships being caused to ordinary Indonesians by the rapid modernization programme and the corruption of the ruling group. The President's wife, Siti Hartinah (Tien), was among those accused of appropriating funds from the Government and of abusing her position to obtain favours and to attract exorbitant commissions. Concern that the exceptionally high revenue derived from petroleum exports was being squandered was later vindicated when the state oil company, Pertamina, came close to bankruptcy in 1975–76. There was much resentment, too, of the business relations between 'New Order' figures and Chinese Indonesian business-owners (*cukong*). This dissent culminated in the Malari riots of 1974.

Towards a 'Pancasila State', 1975–88

Disturbed by this dissent, Suharto launched an ambitious programme to turn the *Pancasila* from a symbol into a national ideology. In a major programme of ideological construction, the *Pancasila*'s set of vague unifying principles became a comprehensive ideology, with the Government exclusively responsible for formulation and interpretation. The *Pancasila* was held to prescribe a society modelled on the traditional family, in which parental authority was respected and in which individual interest was subordinate to the well-being of the community. *Pancasila* labour relations, for instance, implied that there could be no conflict of interests between workers and management and that strikes were inherently anti-social. In emphasizing authority, the *Pancasila* also reinforced Suharto's presidential authority, and some observers began to liken his style to that of a traditional Javanese king.

Opposition to the new interpretation of the *Pancasila* came first from a group of older politicians associated with the parliamentary period and the early 'New Order'. In 1980 they signed the Petition of 50, objecting to Suharto's new doctrines and urging the legislature to review (i.e. reject) them. The strongest reaction to the new *Pancasila*, however, came from religious groups, when the Government insisted in the early 1980s that all social organizations adopt the *Pancasila* as their sole basic principle (*azas tunggal*), placing it even above religious principles. Deep disquiet over this requirement prompted violent protests. One of the largest factions of the Muslim PPP withdrew from the party in consequence of the new requirement.

The years 1975–88 were also a time of continued political repression. Muslim militants especially were often arrested and tried, but the Government maintained that the banned PKI remained a threat and periodically dismissed officials on the grounds of tenuous left-wing connections. In late 1982, in response to an escalation of crime in Indonesia's major cities, the authorities launched the *petrus*, or 'mysterious killings' campaign, in which known and suspected small-scale criminals were sought out and killed in vigilante-style operations.

None the less, these were also years in which the Suharto Government deftly refined economic and social policy. Pertamina was rehabilitated, and the regime reacted promptly and efficiently to the collapse of petroleum prices. As the impossibility of significantly diminishing Java's population by means of transmigration became apparent, the goals of the programme were shifted to economic development in the other islands, the resettlement of people displaced by development projects and natural disasters, and the establishment of politically reliable Javanese settler communities in potentially secessionist regions. Economic development proceeded and, by the late 1980s, the country was in a position to emulate the achievements of the East Asian newly-industrializing countries.

Suharto's Final Decade, 1988–98

From the late 1980s, the question of the succession increasingly overshadowed the political agenda in President Suharto's Indonesia. Born in 1921, Suharto had long passed normal retirement age; however, he gave no indication of selecting a successor. During the approach to each of the successive five-yearly Majelis Permusyawaratan Rakyat (MPR) sessions, at which the President was elected, Suharto hinted that he expected his next term to be his last; but once he was elected, these hints gradually gave way to expressions of willingness to stand for a further term, if that were the will of the MPR. Few observers inside or outside the country could foresee any circumstances that would force Suharto from office, and most believed that only death or serious illness would end his rule. The political order was commonly described as being in 'a state of waiting', and it was widely felt that policy-making was being seriously hampered by the efforts of major political figures to position themselves in preparation for Suharto's demise.

Within élite circles, much attention focused on the Vice-Presidency. This position is one of little power in itself, but is of significance as the Vice-President automatically succeeds to the presidency if the President dies or steps down. Suharto's early Vice-Presidents had all been men with weak power bases of their own, who were therefore unlikely to challenge his power. By the late 1980s, however, the military had become increasingly concerned that Suharto's sudden death might leave them with an unacceptable President. In 1993, therefore, the army engineered a series of public endorsements of its Commander, Gen. Try Sutrisno, in order to secure his election to the Vice-Presidency. As a former presidential adjutant without conspicuous presidential talents, Try was by no means unacceptable to Suharto, and his presence in office did little to allay the general feeling that the future of Indonesia after Suharto was still very uncertain.

Outside élite circles, there was a growing perception that Indonesia might be destined for a more liberal political order. During the 1990s the Government's emphasis on the *Pancasila* as the central principle for all social activity perceptibly diminished. Although the *azas tunggal* requirement remained in place, the Government seemed not to be attempting to impose ideological conformity with any great vigour. Instead, it appeared to accept that decision-making in a complex and diverse society must be the outcome of negotiation among various ideological, religious and social groups. In his independence day speech in August 1990 the President stated that there was no need to fear differences of opinion within society. His remarks were followed by measures both to relax the censorship of the press and to limit the sanctions against publications deemed to have offended political proprieties. In May 1996 the People's Consultative Assembly began to discuss legislation to end the Government's monopoly of news services. Although there was no more than a modest easing of political repression, these statements and actions were in sharp contrast to the official position in the late 1970s when the *Pancasila* state was being constructed.

These policy alterations reflected broader changes in Indonesian society under the 'New Order'. Growing prosperity and an increasingly complex economy and society had led to better education, improved living standards and greater international awareness for a substantial new middle class of bureaucrats, professionals and entrepreneurs. These groups were keen to obtain greater political influence and to see a more regularized economic and administrative order; without necessarily wanting a democracy that enfranchised the poor, they desired a political system in which power and influence were no longer reserved for a small civilian and military élite. These aspirations were generally described as a wish for 'openness' (*keterbukaan*, itself a direct translation of *glasnost*).

More or less independent political organizations became increasingly prominent. A new Muslim organization, the Ikatan Cendekiawan Muslim Indonesia (ICMI—Association of Muslim Intellectuals), which was founded in December 1990 and led by Suharto's protégé, the Minister of State for Research and Technology, Prof. Dr Ir Bucharuddin Jusuf (B. J.) Habibie, became a frequent contributor to public debate. In 1991 the Chairman of the Nahdlatul Ulama (NU—Council of Scholars, Indonesia's largest Muslim organization), Abdurrahman Wahid, founded the Democracy Forum, comprising 45 prominent Indonesian intellectuals, including a number of Christians. In 1993, moreover, the formerly ineffective PDI elected to its chair Megawati Sukarnoputri (a daughter of the late President Sukarno), whose dynamic leadership appeared to be winning greater public support for the party. There even appeared to be limited space for more radical organizations, with the Government failing to suppress the student-worker coalition, Infight, or the illegal Partai Rakyat Demokrasi (PRD—People's Democratic Party), while several nominally social organizations were founded with names close to or even identical with those of parties from the 1950s. In May 1996 Sri Bintang Pamungkas, a former PPP parliamentarian who had been expelled from the legislature in 1995 for allegedly challenging the *Pancasila*, founded the Partai Uni Demokrasi Indonesia (PUDI—United Democratic Party of Indonesia). In 1995 it was announced that parliamentary seats reserved for the armed forces would be reduced from 100 to 75 at the next election. In early 1996 there was increasing public discussion of the electoral system, and suggestions for a number of reforms that would remove some of Golkar's advantages were made. In March a group of political activists established an independent election monitoring committee.

In the early phase of this apparent political relaxation, Suharto appeared to be moving closer to Islam than earlier in his presidency. He adopted the given name Mohamed after his pilgrimage to Mecca in 1991; in 1993 the number of Christians in the Cabinet was sharply reduced; and in November of that year he halted a controversial state lottery, which had been strongly criticized by Muslims on religious and social grounds. Habibie, moreover, was often mentioned as Suharto's likely choice for Vice-President in 1998. Several incidents illustrated the Government's determination not to allow greater openness to move very far towards political liberalization. In June 1994 the Minister of Information revoked the publishing licences of three widely-read weekly newspapers, including the well-established and respected journal, *Tempo*. The ban was believed to stem from *Tempo*'s coverage of a controversial decision to purchase for the armed forces 39 warships from the former East German navy, in which Habibie had played a major role. The ban was overruled by a Jakarta court in May 1995, but in June 1996 the Supreme Court upheld a government appeal against this decision. In May 1996 the founder of the Partai Uni Demokrasi Indonesia, Sri Bintang Pamungkas, was sentenced to nearly three years' imprisonment on dubious charges of insulting the President. Leading military figures also continued to draw public attention to an allegedly dangerous clandestine opposition linked to the banned communist party. Accusations of membership of what were called *organisasi tanpa bentuk* or 'formless organizations' were used in an attempt to discredit critics of the Government. Government figures also publicly threatened several non-governmental organizations (NGOs) active in human rights and environmental issues, especially the Legal Aid Institute (LBH) and the environmental forum (WALHI). The security forces also targeted radical Islamic groups for arrests and interrogations. On several occasions in the 1990s members of the security forces were punished for their roles in killing demonstrators, but for the most part the penalties imposed were slight in comparison with those imposed on the demonstrators themselves.

With the *Pancasila*'s plausibility on the wane, and in the context of increased popular expectations of government propriety, these events provoked an increasing dissatisfaction with the 'New Order', despite its clear economic achievements. This dissatisfaction was compounded by a growing awareness of the extent to which Suharto's family members had profited from their position. Three of Suharto's six children, in particular, had built up substantial and diverse business interests, based apparently on privileged access to government contracts, licences and subsidies: the President's youngest son, Hutomo ('Tommy') Mandala Putra, for example, had obtained exclusive rights for the purchase of cloves from farmers, which he then exploited in order to make substantial profits from resale to industrial users. Even in the late 1980s, there was a consensus that curbing the privileges of Suharto's children would be the single most popular measure that any new President could take; it was also widely felt that Suharto was holding on to power to give his children time to consolidate their positions as legitimate business-owners, so that they would continue to prosper once he was no longer President.

During the 1990s, however, there were increasing indications that Suharto was not simply aiming to consolidate his family's economic position, but was planning in due course to transfer power to a small coterie of his family and friends. In October 1993 leadership positions in Golkar were assumed by two of Suharto's own children and seven children of former government figures close to Suharto. At the same time, the President also removed or transferred a number of senior military figures linked to the campaign that had installed Gen. Try Sutrisno as Vice-President, and he appeared to be promoting his son-in-law, Gen. Prabowo Subianto, the head of the army's special forces, as a military guarantor for his family's fortunes under the next President.

As the 1997 parliamentary elections approached, the new leader of the PDI, Megawati Sukarnoputri, appeared to be the main beneficiary of this discontent. Taking advantage of somewhat romanticized memories of her father as a President who cared for the underprivileged, she appeared likely to lead the PDI to a record vote, and there was even speculation that her party would nominate her as a presidential candidate in 1998. To avoid this possibility, government figures began to organize a series of manoeuvres to destroy Megawati's power. First, officials in East Java allowed internal party opponents of Megawati to establish a rival regional board, and prohibited her from visiting the province to seek a resolution. In June 1996 government supporters within the PDI engineered a special party congress of dubious legality, which removed Megawati as leader and installed her predecessor, Soerjadi. A sharp attack on Megawati by the Commander-in-Chief of the Armed Forces, Gen. Feisal Tanjung, an invited speaker at the congress, indicated the end of a period in which army figures saw her as a useful counter-balance to Habibie and the dominant civilian group in Golkar. As the congress met, however, Megawati mobilized her supporters in rallies in Jakarta and other cities, prepared court cases to reclaim the party leadership and asserted publicly that the PDI would win 80%–85% of the vote if free and fair elections were held. She also announced that her opponents had been expelled from the party, and refused to surrender the party headquarters to them. The NU leader, Abdurrahman Wahid, meanwhile signed a remarkable public petition warning that the Government's authoritarian nature, and its tolerance of social injustice, threatened to create a culture of violence in Indonesia.

On 27 July 1996 armed vigilantes from the Soerjadi faction stormed the PDI headquarters in Jakarta, with military support, evicting Megawati's supporters and provoking two days of rioting in the capital. Disturbances on a smaller scale continued in other cities during the following weeks. To defuse the growing conflict with the Megawati faction, the Government moved quickly to blame the disturbances on the underground PRD, which was portrayed as a successor to the Communist Party. The PRD leader, Budiman Sudjatmiko, and other prominent activists were arrested; in April 1997 they were convicted of subversion and given sentences ranging from three to 15 years. By contrast, most PDI members arrested at the same time were acquitted or given token sentences. Megawati herself was careful not to encourage violence, and tried to have her exclusion from the party overruled in the courts. None the less, in October 1996 the courts refused to hear Megawati's case; she was not permitted to recover any of her former influence in the party, and she and her supporters were excluded from the PDI candidate lists for the forthcoming general election. The authorities also prevented her supporters from opening a rival party headquarters.

The misfortunes of the PDI reinvigorated the PPP, which had suffered from internal division and lack of funds and which had been humiliated in previous months by the public defection to Golkar of several former leading supporters. The dour personality of the party Chairman, Ismael Hassan Metareum, once a liability, now enhanced the party's image as a serious contestant, but the PPP benefited most of all from growing signs that it was winning the support of discontented PDI supporters. As the election approached, a widespread, illicit and parallel campaign developed under the slogan 'Mega-Bintang', literally 'superstar', but in this context an encouragement to Megawati supporters to vote for the PPP, whose electoral symbol is a star. PPP rallies were massively attended, and it appeared that the PPP might possibly oust Golkar from first place in the Jakarta capital district, which it had not done since 1982.

Soerjadi's standing was further weakened in October 1996, when the National Commission on Human Rights published a report on the events of July which concluded that the raid on the PDI headquarters had not been carried out by Soerjadi supporters but by specially recruited thugs with military backing. This analysis was confirmed shortly before the 1997 election, when persons involved in the action complained publicly that they had still not been paid for their services. The report blamed the subsequent riots directly on government provocation, rather than on allegedly subversive groups such as the PRD, and concluded with a recommendation that those involved in the seizure should be prosecuted.

The months leading up to the election witnessed several outbreaks of religious and ethnic tension. In various towns in East and West Java Muslim mobs burned churches, temples and shops associated with local Chinese Indonesians. Yet more serious was a prolonged clash between indigenous Dayaks and Madurese immigrants in the province of West Kalimantan, which lasted from December 1996 until March 1997.

The Government formally permitted an election campaign period of 25 days before the poll on 29 May 1997, but Golkar (and, to a much lesser extent, the PPP) began campaigning earlier by holding supposedly closed 'meetings with cadres'. During the official campaign period itself the Government imposed a strict roster so that no two parties were campaigning in the same region on the same day, thus effectively allowing each party only eight or nine days campaigning in each province. As in previous elections, the authorities maintained a visibly heavy security presence, both to prevent disturbances and to create an atmosphere of threat, but there was extensive localized violence in Jakarta and some other centres. The campaign was notable for the extent to which both Golkar and the PPP campaigned with promises of specific actions, such as improving education or public transport (in past elections, the contestants were generally barred from making specific policy proposals and had instead resorted to more general appeals); Golkar figures also issued barely-veiled threats that areas failing to support Golkar would receive fewer development funds. The PDI campaign, by contrast, was lacklustre, and was further undermined by Megawati's announcement shortly before polling that she would abstain from voting.

In the event, the election delivered an unprecedented victory to Golkar, which won every province and received 74.51% of the votes nationally. The PPP also improved its performance, winning 22.43% (up from 15.97% in 1992), while the PDI vote plummeted to 3.07% from its previous figure of 14.9%. Soerjadi himself failed to win a seat. The result greatly disappointed PPP supporters, who had expected to win about 35% of the vote and who at once accused the Government of fraud and manipulation. According to official figures, 93.38% of the electorate took part in the election. However, it was widely rumoured that many government officials had voted twice—at home and at the office—and that counting procedures had been fraudulent. Party supporters in Madura rioted and destroyed ballot boxes, forcing a repoll in some districts.

The election victory appeared to leave President Suharto firmly in control of the political order. The dissident PRD was formally banned in September and the formal process leading to Suharto's nomination for a seventh term as President began in October. However, the social tensions that had preceded the election failed to dissipate; three days of anti-Chinese violence rocked Ujung Pandang and there was widespread anticipation of further unrest. The sense of foreboding was exacerbated by serious forest fires in Sumatra, Kalimantan, Irian Jaya and elsewhere, which covered large areas of the archipelago with a thick blanket of smoke, often for periods of several weeks. The unrestrained clearing of forest (much of it previously logged) and of swamp areas for agricultural projects was acknowledged as the principal cause of the fires. The smoke seriously hampered communications, forcing the cancellation of hundreds of commercial airline flights and paralysing tourism in many areas of the country. The cost in terms of damage to the health of 60m.–80m. people cannot be calculated. Although fires affected some areas in Java, the capital, Jakarta, was largely unaffected, and the authorities showed little intention of taking any action to address the problem, relying instead on the anticipated onset of the annual rains in December.

During the same period, Indonesia was being drawn steadily into the economic crisis that had begun in Thailand in mid-1997. Although the decline in the value of the rupiah was partly a consequence of 'contagion' from the rest of the region, it was also exacerbated by international concerns over the high levels of indebtedness amongst large Indonesian companies, many of which were associated with President Suharto's circle. The decline in the rupiah worsened the debt crisis, forcing Indonesia to request assistance from the IMF in October. The resulting IMF rescue programme, however, went much further than the Government had expected, demanding a substantial limitation of the privileges enjoyed by Suharto's family and friends and also a reduction in government expenditure in areas that were an important source of patronage, both within and outside élite circles.

The crisis had immediate and far-reaching implications for the economy. External credit for Indonesian companies dried up, and even those firms that were not seriously indebted were unable to obtain finance for routine operations. Factories and construction projects were closed down, forcing millions of workers out of employment. The price of imported goods (including staples such as wheat flour) rose dramatically, compounding the hardship. Unrest spread rapidly, precipitating riots in dozens of cities and towns across the archipelago. Shops and other businesses owned by members of the Chinese community, who were blamed for increasing prices and accused of sending money out of the country, were especially targeted. The crisis worsened as it became clear that the Suharto Government did not intend to implement the IMF programme rapidly or sincerely, and as international money markets forced the value of the rupiah still lower. By mid-January 1998 an extensive range of public figures had begun to urge Suharto to step down at the MPR session scheduled for March 1998. In the last week of January the markets drove the rupiah down to an unprecedented 15,000 to the US dollar, after Suharto indicated that he would choose as his Vice-President the Minister of State for Research and Technology, B. J. Habibie, whose reputation as a proponent of economic nationalism and expensive technological projects ran directly counter to the IMF's insistence on economic austerity and openness.

Both inside and outside Indonesia, at this point, there was hope that the President would respond to the obvious discontent with imaginative and responsible measures, but also fear of cataclysmic disorder if he were to step down. Two of the main opposition leaders, the ousted PDI leader, Megawati Sukarnoputri, and the leader of the Muhammadiyah, Amien Rais, began to collaborate and both gave support to a common opposition movement called Siaga. A third influential leader, Abdurrahman Wahid of the NU, was more hesitant, however, and his political career received a temporary set-back when he suffered a severe stroke in January 1998. As the presidential election approached, the growing sentiment in favour of a new President also fuelled a short-lived movement to secure the selection of Emil Salim, a respected economist and former Minister of the Environment who had spoken out publicly in favour of reform, as Vice-President; with Suharto firmly in control of the MPR, however, this movement made no progress. On 10 March Suharto was duly re-elected as President. Habibie was installed as Vice-President. As a concession to the President's increasing frailty, it was announced that Habibie would take charge of multilateral international relations; however, there was no indication that he would take any significant role in administration. In February, prior to the election, Suharto had consolidated his position by appointing his son-in-law, Gen. Prabowo, as head of the army's strategic reserve, KOSTRAD. Following his election victory, Suharto further strengthened his position by appointing a Cabinet that concentrated power in his immediate circle of friends and family but which left Habibie's followers rather poorly represented: Suharto's business associate, Mohamad 'Bob' Hasan, became Minister of Trade and Industry; his daughter, Tutut, became Minister of Social Affairs; and Tutut's close friend, Gen. (retd) Hartono, assumed the influential home affairs portfolio.

The composition of the new Cabinet, together with the fact that the main points of continuing dispute between the IMF and the Suharto Government concerned the economic privileges that continued to be enjoyed by Suharto's family members and associates, largely exhausted any hopes that Suharto himself might permit significant political and economic reforms in Indonesia. The IMF appeared to acknowledge this by conceding much ground in a new agreement with the Indonesian Government reached in April 1998. Within the country, however, a groundswell of discontent was emerging, expressed first in a growing number of student demonstrations in the cities and subsequently in riots in centres across the archipelago. Meanwhile, the political opposition to Suharto grew more confident, with Amien Rais taking a particularly prominent role and abandoning his former Islamist stridency. In a process reminiscent of the events of 1965, disparate opposition groups played down their differences and adopted a single slogan, *reformasi* (reform), as a code for the simple goal of removing Suharto from office. As the popular movement against Suharto gathered momentum, signs of disunity began to appear within the military: elements of the armed forces launched a campaign of kidnapping, torture and intimidation of student activists, but in May 1998 the armed forces chief of staff, Gen. Wiranto, disavowed the kidnappings and announced an inquiry into them. The Government appeared also to be giving ground: in late April the Minister of Justice, Muladi, promised unspecified reforms; his promises were echoed by Hartono. The timetable for and the extent of such reforms was kept vague, however.

On 5 May 1998, when the Government announced a 70% increase in fuel prices as part of its agreement with the IMF to cut state subsidies, the violence escalated abruptly. As in January, the main targets of the violence were members of Indonesia's ethnic Chinese minority. In April and early May tens of thousands of Chinese Indonesians fled to neighbouring Singapore and Malaysia in search of refuge from the violence. Within days of the riots of 5 May, members of the legislature, senior figures from within the regime, and even Vice-President Habibie's ICMI, began to demand immediate reform. Some military units were even seen enthusiastically greeting student protesters and protecting them from riot police. Suharto's departure from the country on 9 May for a meeting of the G-15 group (of leaders of developing countries) in Cairo, Egypt, aroused speculation that he might choose to stay in exile. As the violence continued, however, alternative speculation emerged that pro-Suharto elements in the military were deliberately exacerbating the unrest to provide the pretext for a military crack-down, out of which Gen. Prabowo would emerge to wield greater power. Street violence in Jakarta reached a peak on 14 May, when armed gangs took control of large parts of the city. An estimated 500 people died; 3,000 buildings were burnt and many thousands of vehicles were also destroyed. A further 700 people were reported to have been killed in other centres.

On 18 May 1998, three days after Suharto's return from Cairo, his former supporter, Harmoko, Golkar Chairman and Speaker of the Legislature, astounded observers by demanding the President's resignation and by then publicly threatening Suharto with impeachment if he failed to step down. Suharto endeavoured to recover the initiative by offering immediate political reforms, including the appointment of a new Cabinet, and the holding of an election in early 1999, followed by an MPR session to elect a new President; he also reversed the fuel price increase, the original announcement of which had precipitated the worst disturbances. Although these pledges won the apparent support of Gen. Wiranto and Abdurrahman Wahid, both of whom feared that instability would ensue if Suharto were to resign abruptly, they failed to appease the protest movement. Finally, following the resignation on the evening of 20 May of the 14 economic ministers in Suharto's Cabinet, Gen. Wiranto visited the President late at night to inform him that the military could no longer guarantee security in Jakarta if he remained in office. The following morning Suharto announced his resignation, stating 'I find it difficult to carry out my duty as the country's ruler'. Vice-President Habibie was sworn in immediately as his successor, and Gen. Wiranto promised to defend Suharto and his family.

PRESIDENT B. J. HABIBIE, 1998–99

Suharto's resignation certainly saved Indonesia from even greater turmoil; however, it also raised new uncertainties. President Habibie's relations with the armed forces were considered poor following the naval acquisitions issue of 1994, his capacity to manage the economy was widely doubted in view of his reputation for promoting prestigious but costly projects, and he was also widely perceived as a protégé of his predecessor and as possibly no more than a tool, subject to continuing manipulation by Suharto. In the event, Habibie managed his first months in office rather astutely, balancing the continuing demand for reform against the power of entrenched interests. He at once dubbed his Government the 'Reform Order', and supported Gen. Wiranto in his dismissal of Suharto's son-in-law, Gen. Prabowo, from his command position in KOSTRAD (Prabowo was dismissed from the armed forces in August 1998 and subsequently went into exile in Jordan). Habibie also released prominent political prisoners, including Muchtar Pakpahan and Sri Bintang Pamungkas, encouraged government departments to sever their business ties with enterprises owned by the Suharto family, and foreshadowed the implementation of electoral reform, with the announcement that new elections would take place in 1999. In June the Government announced a liberalization of the censorship laws and an investigation into the wealth of state officials, and it was widely rumoured that the Suharto family interests were also to be subject to investigation. However, both the extent and the timetable of most of these reforms fell short of the demands that students and others had made of the Suharto regime, and there was widespread impatience with the pace of change under Habibie.

In the early days of his presidency, Habibie announced that he would not serve out the remainder of Suharto's term (1998–2003), and called a special session of the MPR on 10–13 November 1998 to authorize the holding of new legislative elections. The MPR limited the President and Vice-President to a maximum of two five-year terms of office each, decreed that the number of seats in the Dewan Perwakilan Rakyat (DPR) held by nominated representatives of the military should be reduced, and provided for the establishment of a Commission on General Elections (Komisi Pemilihan Umum, KPU), within which all contesting parties would be represented, to oversee the poll. Greatest attention focused, however, on the revised election laws, which were drafted by a committee of academic experts known as the Team of Seven and approved with some significant amendments by the DPR on 28 January 1999. The new law relating to political parties provided that only parties with branches in at least nine of the country's 27 provinces would be permitted to contest the election and that they should have branches in at least half the districts in those provinces; this measure was intended to ensure that only parties with a national orientation would contest the election. In addition, the law provided that a party should win at least 2% of the seats in the MPR to be permitted to compete in the elections scheduled for 2004. Although candidates were required to be 'loyal to the *Pancasila*' and the Communist Party remained banned, the former requirement that all political organizations accept the *Pancasila* as their sole basic principle gave way to the milder provision that the parties should certify that their principles and programmes did not contradict the *Pancasila*. In contrast with previous elections, at which government employees had been impelled to support Golkar, the new laws prohibited Indonesia's 4.1m. civil servants on active duty from campaigning. In addition, the armed forces announced that they would maintain neutrality in the election.

Although the Team of Seven proposed that a predominantly district-based system of election should replace the proportional system of the Suharto era (in which each province effectively constituted a single electorate), the DPR retained the proportional system, with the complex proviso that each party should link each candidate to a district and that the successful candidates would be drawn from the districts where the party had performed best. The DPR also preserved a long-standing arrangement which gave extra weighting to provinces in the more sparsely populated outer islands over those in densely populated Java. Following intense debate and vehement student protests, the DPR agreed to reduce the number of seats in the legislature allocated to the military from 75 to 38; in addition, the military presence in provincial and district assemblies was reduced from 20% to 10%. It was also announced that the membership of the MPR was to be reduced from 1,000 to 700 (500 of whom would comprise the members of the DPR while, of the remainder, 135 were to be elected by the provincial assemblies and 65 appointed by the KPU to represent social organizations).

Soon after May 1998, in anticipation of the new laws, political parties began to form on a wide variety of ideological, ethnic and religious bases. By early 1999 more than 200 new parties had been created; however, only 48 met the criteria for competing in the forthcoming election. Most observers agreed that Megawati's wing of the PDI, reconstituted as the PDI—Perjuangan (PDI—P—Struggle PDI) would perform well, and that the Partai Kebangkitan Bangsa (PKB—National Awakening Party) of Abdurrahman Wahid had a secure base amongst the traditionalist rural Muslims of East Java; the potential of the other parties, including Golkar, however, remained extremely uncertain. Particularly unclear was whether Golkar's prestige as the party of government would help it to survive the loss of military and civil service support and the defection of former supporters to new parties. This uncertainty bedevilled attempts by the parties to anticipate the kind of partners they might seek to work with after the election, in the likely event that no party won a majority. A semi-formal alliance between party leaders Megawati, Amien Rais and Abdurrahman Wahid showed signs of serious strains as soon as it was announced. A further source of tension and uncertainty was the position of former President Suharto. Although Gen. Wiranto had promised to defend Suharto and his family following the transfer of the presidency to Habibie, the latter's Government faced enormous public pressure to investigate the Suharto family's accumulation of wealth during the era of the New Order. Suharto himself denied any wrongdoing, but the slow progress of investigations by the Attorney-General and the lack of any clear results gave rise to strong suspicions that the Habibie Government was not being as thorough in its investigation as perhaps it might, possibly out of fear that too close an examination might compromise other leading government figures. There was also suspicion that Suharto retained considerable informal influence over the military and in some civilian circles and that he was using this influence not only to protect himself but to maintain a sway over political developments.

Meanwhile, serious violence broke out in several parts of the country. In Jakarta in mid-November 1998, in the worst violence since the riots of May, at least 16 people were killed and more than 400 injured when students and civilians clashed with soldiers outside the building where a four-day special session of the MPR was being held. Further riots occurred in Jakarta in late November, in which at least another 14 people died during confrontations between Muslims and Christians. In East Java in late 1998 at least 182 Muslim clerics and alleged black magicians were killed over a period of several months in a spate of savage and apparently organized murders blamed on unidentified 'ninja' and verified by an official investigation in December 1998 (the investigation failed, however, to draw any conclusion as to the identity of the perpetrators). In Maluku violent clashes between Muslims and Christians broke out in early 1999, with several hundred deaths reported; a number of those who died were killed by the military, who were drafted into the region in an attempt to restore order. Meanwhile, in West Kalimantan, indigenous Dayaks resumed their hunting-down of Madurese settlers, this time with support from the majority Malay population; by late March, more than 165 Madurese were reported to have been killed, with numerous reports of the mutilation and cannibalization of the bodies of many of the victims. Many thousands of Madurese fled the province.

In the countryside, moreover, dozens of land disputes erupted, a legacy of the high-handed expropriation of landowners during the Suharto era and a disturbing reminder of the bitter land conflicts of the early 1960s. In some regions there were violent protests against corrupt or oppressive local officials, or against destructive or polluting factories. Gen. Wiranto's proposal to counter such violence by establishing a 40,000-strong civilian auxiliary militia was met with some incredulity. Also in early 1999 it emerged that the military's prestige and self-confidence had been damaged by its association with political repression during the Suharto years to a greater extent than anticipated, and the armed forces remained uncertain of how best to fulfil their role. Particularly significant in provoking a questioning of the role and conduct of the armed forces was the publication in November 1998 of a report containing the findings of a panel appointed by the Government to investigate the riots in May: while the report gave no single figure for the number of people killed or injured, the panel found that elements of the military had acted as provocateurs during the riots (with particular suspicion falling on a unit led by Gen. Prabowo) and concluded that the riots had been 'created as part of a political struggle at the level of the élite'. Military leaders announced that they would 'redefine, reposition and reactualize' their place in Indonesian society, implying that their role in politics and administration would be diminished, but the precise nature of any intended changes was far from clear. In an attempt to reduce the military's association with petty repression, the police force (part of the military since 1962) was formally separated from the armed forces on 1 April 1999; however, the police force remained under the control of the Ministry of Defence. The armed forces also resumed their revolutionary-era name, Tentara Nasional Indonesia (TNI—the Indonesian National Defence Forces), in place of the Suharto-era name, ABRI (the Armed Forces of the Republic of Indonesia).

Although the election campaign dominated politics during the first half of 1999, the Habibie Government continued an extensive programme of reform. In April the DPR passed a new law providing for the election of district heads (*bupati*) by the district assemblies, rather than through appointment by Jakarta, and transferring a wide range of administrative functions to the provinces and districts (*kabupaten* and *kotamadya*); another law provided that provinces would receive a greater share of revenue from natural resources such as oil, gas, fisheries and timber. In a presidential decree announced in May, Habibie removed a ban on the use and teaching of the Mandarin Chinese language and also outlawed discrimination on the grounds of ethnic origin. Given the level of tension prevalent throughout the archipelago, the election campaign itself proceeded with surprising calm. Although some violent clashes occurred, the various parties were aware that widespread violence might well result in the cancellation of the elections and consequently kept outdoor rallies to a minimum. The PDI—P's campaign emphasized the party's commitment to extensive reform and the righting of the injustices of the Suharto era, while the Golkar campaign focused on a promise of moderate reform and stability, and the administrative experience of the party. To varying degrees, the Muslim parties stressed the need to strengthen the Islamic character of the State, and accused the PDI—P of being dominated by Christians, Hindus and non-committed Muslims. These accusations, in turn, reinforced the PDI—P's standing as the party most likely to defend religious and ethnic minorities, although initially it remained vehemently opposed to any fragmentation of the country.

The day on which the polls were held, 7 June 1999, was declared a public holiday in order to eliminate any potential intimidation of voters by their employers. Approximately 118m. people voted (a turn-out of 91%). The final results of the poll were not announced until mid-July. Although foreign observers had pronounced the election itself fair, the subsequent delay in the announcement of the outcome gave rise to suspicions that results were being tampered with. None the less, most observers concluded that the slowness was a consequence of lack of infrastructure (the military communications facilities used in previous counts were not available) and of inexperience. The slow accumulation of voting tallies allowed the public to adjust to some unexpected results: although Megawati's PDI—P led the poll as expected, with 34% of the votes, its nearest rival proved to be Golkar, whose sluggish performance in much of Java, especially in the cities, was balanced by the achievement of impressive results in the outer islands, especially in eastern Indonesia. Although Golkar received 20% of the national vote, the weighting given to outer islands meant that the party won 120 seats (26% of the total) as compared with the PDI—P's 154. As expected, Abdurrahman Wahid's PKB did well in East Java, but the Partai Amanat Nasional (PAN—National Mandate Party) of Amien Rais performed relatively poorly, receiving only 7% of the vote. Formerly closely associated with the modernist Muslim Muhammadiyah movement and with arguments for making Indonesia more clearly Muslim, Rais had campaigned energetically and had sought to broaden his appeal to Christians, Chinese and more secular Muslims; however, this strategy appeared to cause the alienation of his own base while failing to allay the suspicions of those other constituencies. In third place in the poll was the PPP, the principal Muslim party of the Suharto era, with 11% of the vote and 59 seats. The party had campaigned on general Islamic principles without defining a clear ideology, but it appeared to have benefited from its reputation for defying Suharto during the New Order era and its consistent reformist position in the MPR following Habibie's accession to the Presidency, as well as from its well-developed national organization. In addition to these five parties, some 14 smaller parties won seats in the MPR (although their combined share of the vote was only about 7%). A number of small parties, some of which had failed to win a seat, withheld formal ratification of the election results, but they were subsequently overruled by President Habibie, who finally endorsed the results on 3 August.

During the election campaign, each of the parties nominated a preferred candidate for the presidency, which was originally scheduled to be decided by the MPR in November 1999. The Indonesian Constitution provided no clear guidance on whether the successful candidate would require a majority of votes in the MPR or merely a plurality; however, in mid-1999 there existed a general consensus that the next President would need a base of support broader than any single party and that extensive negotiations would be necessary to construct any potential coalition. With the PDI—P and Golkar leading the polls, Megawati and Habibie initially emerged as the main candidates for the presidency. Habibie's prospects, however, were seriously damaged by factionalism within Golkar and by allegations that he had helped to channel funds originally intended for the recapitalization of the struggling Bank Bali into his wing of Golkar. He was also widely perceived to have mishandled the issue of East Timor (now Timor-Leste): many Indonesians felt that the country had been humiliated by the overwhelming vote in favour of independence by the East Timorese in August 1999 and by the subsequent introduction of foreign peace-keeping troops into the territory. Megawati's prospects, meanwhile, were impaired by her apparent remoteness: she made few public speeches during the election campaign and showed no interest in lobbying other parties for support. Concealed threats from PDI—P members that there would be violence if she were not elected, and indications of 'money politics' in the PDI—P also raised doubts about Megawati's capacity to improve the standards of government if she were to come to power. Megawati's position was further weakened by the public statements of some Muslim leaders, including Abdurrahman Wahid, to the effect that a woman should not become President in a predominantly Muslim country.

Meanwhile, Amien Rais recovered from his electoral set-back by assembling a coalition of Muslim parties called Poros Tengah (Central Axis), which aimed to give its member parties a powerful collective voice in the proceedings, as a means of exerting pressure with a view to increasing future influence in government. As the date of the MPR vote approached, both Habibie and Megawati appeared increasingly unlikely to succeed; however, no clear alternatives had emerged. On 20 October 1999, only hours before the rescheduled vote was due to take place, Habibie withdrew his candidacy, following the rejection of his presidential record by the MPR in a secret ballot. Poros Tengah and much of Golkar then transferred their support to Abdurrahman Wahid (also known as Gus Dur), even though Wahid's party, the PKB, was not part of the Poros Tengah coalition. Wahid was subsequently elected President by the MPR, securing a total of 373 votes; Megawati received 313 votes. The announcement of Wahid's victory provoked outrage among Megawati's supporters, leading to violence in Jakarta and elsewhere. On 21 October, however, the MPR voted to appoint Megawati as Vice-President.

PRESIDENT ABDURRAHMAN WAHID, 1999–2001

President Wahid's new Cabinet, announced on 26 October 1999, was large (consisting of 36 members) and was drawn from many parties, reflecting both Wahid's desire to include in the new political order all the major forces in Indonesian society and to discharge the political debts he had incurred in winning the presidency. Poros Tengah, however, with more than one-third of the seats in the DPR, secured half the posts in the Cabinet, whereas Golkar, which held one-quarter of the elected seats in the DPR, was allocated only four positions in the Cabinet. Many observers questioned whether such a diverse and unwieldy Cabinet, led by a President who was blind and in ill health, would be capable of effectively addressing the country's pressing economic problems. However, President Wahid's tolerant, inclusive style of leadership initially appeared ideal for easing the political tensions that had developed during and after the Suharto era. Wahid's credentials as an opposition leader during the Suharto era were strong, but he made it clear that he had no interest in avenging the wrongs of the New Order; as a pious Muslim he was acceptable to Islamic modernists, but he also had a strong record of respect and tolerance for other religious traditions. Wahid continued President Habibie's policy of seeking to reintegrate the Chinese community into Indonesian political and cultural life, lifting restrictions on the public celebration of Chinese festivals and including a Chinese Indonesian, Kwik Kian Gie, in his Cabinet, in the senior post of Co-ordinating Minister for the Economy, Finance and Industry. He also invited Indonesians exiled by Suharto to return home. The President abolished the Ministries of Information and Social Affairs (both of which had been closely associated with the Suharto-era apparatus for the political control of society), and he regularly appeared before the legislature to answer questions on his administration. While the new President attempted a radical reform of the notoriously corrupt judiciary, he also appeared to have achieved some success in reducing the political power of the military. Although Gen. Wiranto became Co-ordinating Minister for Political Affairs and Security, and although six members of the new Cabinet had a military background, a civilian, Yuwono Sudarsono, held the post of Minister of Defence, and for the first time in recent history a naval officer, Adm. Widodo Adi Sutjipto, was appointed Commander-in-Chief of the Armed Forces. The number of military officers included in the presidential staff was also sharply reduced.

The President's commitment to regularizing the political order was undermined, however, by his own opaque political style. His public statements were often impulsive, contradictory and made without consulting his advisers or Cabinet. In November 1999, only a month after installing the Cabinet, Wahid dismissed the Co-ordinating Minister for People's Welfare and Poverty Alleviation, and PPP leader, Hamzah Haz, on unspecified charges of corruption, which were never pursued. The President appeared to veer between detachment from the process of government and undue intervention in administrative detail. Wahid was also rumoured to be manipulated by 'whisperers' who took advantage of his blindness to further their own political advantage, and doubts over his state of health were raised repeatedly. Meanwhile, hopes that Vice-President Megawati would take a more active role than her predecessors were disappointed, as she continued her inactive approach to politics. On the other hand, with a divided Cabinet, a passive Vice-President, and a vast civilian and military bureaucracy whose capacity and loyalty was in doubt, the President had little choice but to embark upon a complex series of manoeuvres, amid shifting alliances, to ensure that at least some of his policy goals were achieved.

In early 2000 Wahid removed Gen. Wiranto from his powerful cabinet position after Indonesia's National Human Rights Commission on 31 January found him responsible for the violence that followed East Timor's referendum on independence in August 1999. Travelling in Europe and Asia at the time, the President announced that he had asked Wiranto to resign. Wiranto resisted, however, and amid rumours of the possibility of a military coup, Wahid agreed on 13 February 2000 to allow him to retain his post until the Attorney-General had investigated the issue. Nevertheless, on the following day the Cabinet Secretary announced that Wiranto had been suspended from his position. He was replaced by the Minister of Home Affairs, Gen. (retd) Suryadi Sudirja. Wiranto subsequently resigned in mid-May.

Rivalry between ministers and lack of co-ordination between government departments became a serious problem. Even before the end of 1999, moreover, allegations had begun to emerge of corruption and the abuse of power within the new Government. Many cabinet ministers appeared to be turning their departments into party fiefdoms, as had happened in the 1950s. In April 2000 President Wahid dismissed the Minister of State for Investment and Development of State Enterprises, Laksamana Sukardi of the PDI—P, and the Minister of Trade and Industry, Yusuf Kalla of the so-called 'black' (pro-Habibie) wing of Golkar, later suggesting that both Ministers were guilty of corruption. Such accusations had often been heard previously against Kalla. Sukardi, however, was a respected and capable Minister; his dismissal appeared to reflect the fact that he had pressed too hard for rapid reform and had thus lost support within his own party. Wahid shocked the legislature in July when he uncharacteristically refused to explain his reasons for dismissing the Ministers. Misgivings also grew over the increasing business activities of Wahid's own organization, the NU, and of the President's brother, Hasyim Wahid, who, despite a lack of economic experience, had been secretly appointed adviser to the Indonesian Bank Restructuring Agency (IBRA). Still more disturbing was an affair in early 2000 in which the President's masseur, Suwondo, was alleged to have solicited US $4.1m. from the deputy chief of the state logistical agency, BULOG. According to official accounts, the funds were a 'private' bribe paid to Suwondo by the deputy chief of the agency, Sapuan, who was subsequently formally dismissed from his post. The Sultan of Brunei was also reported to have illegally given $2m. to Wahid for unspecified purposes. Wahid's appointment of the NU chairperson, Rozy Munir, as Minister for State Industries also encouraged rumours that he intended to exploit those industries—traditionally a major source of graft—for the NU's use in future election campaigns. In addition, the President appeared to be seeking ways to protect some of the most heavily indebted Suharto-era business conglomerates from the full effects of economic restructuring. Observers suspected that he anticipated substantial donations to the PKB in return.

By mid-2000 President Wahid's political 'honeymoon' was clearly over, and there was growing speculation over whether the forces unhappy with his policies and style of leadership might combine to remove him from office. Wahid was summoned to give an account of his actions at the annual MPR session on 7–18 August, but escaped formal censure, as party leaders feared that action against him might precipitate mass unrest, and because they obtained what they believed was a promise from Wahid to relinquish major policy-making powers to Megawati. With the MPR session safely over, however, he claimed that only 'administrative' duties, and not decision-making powers, were to be transferred to the Vice-President, and on 23 August 2000 he announced the formation of a new 26-member Cabinet. The incoming Cabinet included considerably fewer representatives of Megawati's PDI—P, reportedly resulting in serious tensions between the President and the Vice-President. Two of the most influential cabinet posts were allocated to Wahid loyalists: Susilo Bambang Yudhoyono was appointed Co-ordinating Minister for Political Affairs, Security and Social Welfare, while Rizal Ramli was designated Co-ordinating Minister for the Economy, Finance and Industry, replacing Kwik Kian Gie, who had previously resigned from the post. The lack of a clear alternative candidate for the presidency, however, remained one of Wahid's strengths.

The investigation of alleged wrongdoing by former President Suharto, which had stalled under President Habibie, resumed under President Wahid. Suharto suffered a stroke in 1999 and twice failed to appear for questioning by the Attorney-General's Office, his lawyers arguing that he was too ill. In April 2000 the former President was banned from leaving Jakarta. The Attorney-General announced in July that Suharto would go on trial in August. President Wahid had previously stated, however, that the formal establishment of Suharto's guilt was more important than any punishment and had declared that he would pardon the former President if he were found guilty.

The most serious problem facing the Wahid Government, however, was the rise of communal violence in Maluku, where Christians and Muslims were in approximately equal numbers. Hostility between the two communities was based partly on traditional rivalries, exacerbated by a rise of religious orthodoxy on both sides. Migration into the province by Muslim traders during recent decades and the growing administrative dominance of Muslim officials had also served to accentuate the Christians' increasing sense of being beleaguered. However, the violence was precipitated principally by rivalry between Christian and Muslim gangs in Jakarta and in the provincial capital, Ambon, and then exacerbated by rumours of the massacre, rape and torture of members of each religious community and the desecration of places of worship. Sporadic violence following the fall of Suharto developed into a local civil war, with Christian and Muslim militias launching attacks on each other's villages and places of worship. By the end of 2001 more than 6,000 people were reported to have been killed, some 500,000 had been displaced, downtown Ambon had been destroyed, and the province was becoming increasingly partitioned into separate Christian and Muslim cantons. The formal separation of the predominantly Muslim northern districts of the region as the new province of North Maluku in late 1999 did nothing to resolve the tensions. Given the army's record of aggravating tensions in Aceh and Irian Jaya (now Papua), the President was reluctant to yield to military calls for the introduction of martial law in Maluku. In July 2000, however, Wahid also rejected suggestions that UN troops be sent to the province to quell the violence. Wahid placed Vice-President Megawati in charge of seeking a solution to the violence, but her limited activities had no apparent effect on the tensions, and her demonstrable lack of commitment to the project was widely criticized. In May 2000 local Muslims were joined by a so-called Laskar Jihad (Holy War Militia) from Java, which eventually numbered an estimated 3,000 fighters. There was much speculation that the violence in the province was being encouraged by external forces, perhaps with a view to destabilizing the Wahid Government, or perhaps simply for local commercial reasons. By mid-2000 clear evidence had emerged that local army units were supporting the Muslims whilst the local police backed the Christians.

Meanwhile, levels of violence throughout the archipelago increased considerably. Bombs exploded in Jakarta on several occasions. In January 2000 Muslim demonstrators launched arson attacks on churches and Christian businesses in Lombok, and there were many reports of new militia groups training in Java. Members of the Cabinet also regularly received death threats. In May the offices of the Surabaya-based newspaper, the *Jawa Pos*, which had published accusations of corruption against President Wahid and his relatives, were attacked by a paramilitary force associated with Wahid's supporters. A series of apparently co-ordinated bomb blasts at Jakarta churches on Christmas Eve 2000 claimed 16 lives. Communal violence reportedly resulted in the deaths of 200 people in Sulawesi in June. Furthermore, clashes between indigenous Dayaks and immigrant Madurese in West Kalimantan claimed an estimated 500 lives in February 2001, and about 50,000 Madurese were displaced. Many of the bombings showed a degree of technical sophistication that aroused suspicions that sections of the military might have been involved. The violence and general insecurity across the archipelago was estimated to have displaced perhaps 1m. people in 18 provinces.

During his first year in office, Wahid attempted to establish greater control over the military by supporting the promotion of reforming officers, but these attempts generated a hostile reaction in military circles. The August 2000 session of the MPR was also perceived to have made a number of concessions to the military, including the introduction of a constitutional amendment that excluded military personnel from prosecution for crimes committed prior to the enactment of the legislation used to prosecute them; this particular amendment was feared by many international observers seriously to threaten the possibility of the prosecution of members of the Indonesian military believed responsible for recent human rights violations in the former province of East Timor. Doubts concerning the commitment of the Indonesian Government to the trial of military personnel suspected of involvement in gross human rights violations in East Timor were further consolidated in early September when Gen. Wiranto's name was not included in a list announced by the Attorney-General's office of 19 suspects named in connection with such violations. Legislation was also passed to extend military representation in the MPR until 2009, four years after the date at which the military had been scheduled to lose its remaining 38 seats in the House.

Potentially even more important than reforms to the military, however, was the process of administrative decentralization implemented under Wahid's presidency. Planning for a greater delegation of authority to district and municipality governments (the administrative level below the provinces) had begun under Suharto, and enabling legislation was passed during Habibie's term. A specifically appointed State Minister for Regional Autonomy, Ryaas Rasyid, oversaw early preparations for decentralization. In one of Wahid's arbitrary changes of direction, however, Rasyid was shifted to an unrelated portfolio in August 2000. The decision to devolve authority to Indonesia's approximately 350 districts and municipalities, rather than to provinces, was taken to diminish the risk that the provinces might develop a desire for self-rule and attempt to secede, but it raised serious questions over the administrative capacity of such small units and over the likely social consequences for resource-poor districts. Popular aspirations, moreover, continued to focus on the provincial level, with public pressure leading to the establishment of three new provinces (Gorontalo, Banten and Bangka-Belitung) in late 2000, with strong campaigns under way for the creation of several more.

In early August 2000 former President Suharto was formally charged with corruption arising from his 30 years in power. However, in late September all corruption charges against Suharto were dismissed after he was declared mentally and physically unfit to stand trial. In early November, however, the Jakarta High Court ruled that the trial was to resume. In mid-September, meanwhile, President Wahid announced that he had ordered the arrest of Suharto's youngest son, Tommy. Supporters of the former President and his family were suspected of involvement in a series of bomb threats and explosions in Jakarta in August and September. In one attack in mid-September at least 15 people were killed when a bomb exploded in the basement of the Jakarta Stock Exchange. The police declined to arrest Tommy without evidence, whereupon President Wahid sought to dismiss the police chief. Tommy was later sentenced to 18 months' imprisonment on separate corruption charges, but disappeared before he could be detained. In July 2001 Syafiuddin Kartasasmita, the judge who had sentenced Tommy, was assassinated by apparently professional gunmen in Jakarta; the gunmen later admitted in custody that they had been paid by Tommy.

In February 2001 the DPR formally censured Wahid over the BULOG and Brunei corruption allegations, thus taking the first step towards formal impeachment. The impeachment process was accompanied by demonstrations by both opponents and supporters of Wahid. In May the DPR issued a second formal memorandum claiming unsatisfactory performance on the part of Wahid (just two days previously he had been cleared of corruption charges in the Brunei and BULOG scandals by the Attorney-General, Marzuki Darusman), and on this basis requested the MPR to call a special session to impeach the President. Since the MPR was in practice a somewhat augmented DPR, the working body of the assembly agreed, calling the session for 1 August. Wahid responded by describing the DPR's action as unconstitutional and threatening that his supporters would resort to violence. He reshuffled his Cabinet on 3 June and again on 12 June; the first reshuffle was a major reorganization that removed Susilo Bambang Yudhoyono and the second led to the replacement of his Minister for Finance, Prijadi Praptosuharjo. The President also attempted to dismiss the head of the national police, Gen. Surojo Bimantoro, although Wahid made no attempt to obtain the requisite approval of the DPR and Bimantoro subsequently refused to accept his dismissal.

The President announced on 9 July 2001 that he would declare a state of emergency on 20 July if a compromise were not reached. The declaration was not made on the latter date, however, and Wahid merely delayed the deadline for reconciliation to 31 July, the day before the impeachment hearing of the MPR. A special session convened on 21 July, but was boycotted by Wahid's PKB and the small Christian party, Partai Demokrasi Kasih Bangsa (PDKB). The assembly speaker, Amien Rais, requested the President to deliver an account of his actions, to which Wahid replied that the session was illegal and that he would not attend. He separately met military MPR representatives and threatened them with dismissal if they continued to support the session. In the early hours of 23 July Wahid declared a state of civil emergency, suspended the MPR, the DPR, and Golkar, and announced that new elections would be held within a year. Later that morning, however, the Supreme Court issued an advisory opinion that the President's declaration was illegal, while the chief of the Jakarta police force announced that he would accept orders only from the Vice-President and would protect the security of the MPR session. The MPR itself then declared that the President had no constitutional authority to attempt to suspend it and that he had violated his oath of office in trying to do so; Amien Rais demanded an immediate impeachment hearing. That afternoon the assembly formally voted unanimously to dismiss Abdurrahman Wahid and to elevate Megawati Sukarnoputri to the country's presidency. Three days later the PPP leader, Hamzah Haz, defeated four other candidates, including the Golkar leader, Akbar Tandjung, and the former Co-ordinating Minister for Political Affairs, Security and Social Welfare, Susilo Bambang Yudhoyono, for election as Vice-President. Although Hamzah Haz had been among those rejecting a female president on Islamic principles in 1999, he appeared to be Megawati's preferred candidate and received strong PDI—P support in the ballot.

PRESIDENT MEGAWATI SUKARNOPUTRI, 2001–04

Delays in the announcement of the composition of the new Cabinet led to widespread fears that Megawati was already weakened by the bargaining among the various power groups. However, her Cabinet, announced on 9 August 2001, offered an impressive blend of technocratic skill and political connections. Susilo Bambang Yudhoyono returned as Co-ordinating Minister for Political Affairs, Security and Social Welfare, with Dorodjatun Kuntjoro-Jakti as Co-ordinating Minister for the Economy, Finance and Industry and Yusuf Kalla as Co-ordinating Minister for People's Welfare. Many observers were disappointed, however, with her selection as Attorney-General of M. A. Rachman, generally described as unremarkable. Vigorous reform of Indonesia's notoriously corrupt legal system was widely considered essential for the restoration of domestic and international confidence in the country. International risk surveys consistently identified Indonesia as having one of the highest perceived levels of corruption in Asia and the world, and addressing this problem through the courts was considered a high priority.

None the less, Megawati's election brought with it a powerful sense of relief and hope for stability and reconciliation. Her placid personality appeared to satisfy a widespread desire that the President should play a calming and moderating role, rather than spearheading changes to the Constitution, the legal system, the administrative structure or Indonesian political culture in general. However, the Government remained seriously constrained by a lack of revenue, which meant that the implementation of many desirable measures simply remained beyond its capacity. The far-reaching decentralization of 1 January 2001 had also succeeded in placing many important areas of policy beyond the reach of the central Government. In November 2001 the MPR agreed to amend the Constitution to provide for the direct election of the President and Vice-President, for a constitutional court to review legislation and for a bicameral legislature, with a new upper house representing the regions. In August 2002 the MPR finally approved more detailed plans providing for the direct election of the President—provision was made for the establishment of an electoral process by which, if no candidate polled more than 50% of the votes, a second round of voting would be held. It was also decided that the 38 seats reserved for the military within the legislature would be abolished by 2004, five years ahead of the previously established schedule. The new electoral procedure removed the system of preferential weighting of seats in sparsely populated provinces outside Java; the representation of the province of Papua in the MPR would thus decline from 13 seats to five. However, requests for the introduction of Islamic law were rejected.

In August 2001 Megawati's decision to revoke Wahid's dismissal of the corrupt Bimantoro as Chief of Police raised fears that efforts to combat official corruption would end. In January 2002 the Government announced that the payment period granted to major shareholders of failed banks—all of them associates of Suharto—would be extended to 10 years, meaning that these debts were unlikely ever to be repaid (this decision was reversed in March). The Ministry of Information, which had sought to manage public access to news under Suharto and which had been abolished by Habibie, was to some extent re-established in the form of the State Ministry of Communications and Information (an office under the control of the Co-ordinating Minister for Political Affairs, Security and Social Welfare) in August 2001. Shortly after Megawati came to power, moreover, the authorities arrested a number of activists under laws against 'sowing hatred' from the colonial era, which had not been used since the fall of Suharto. During 2003 several people were charged with the colonial-era offence of insulting the head of state, which Suharto had often invoked to protect his dignity. Megawati surprised observers in July 2001 by establishing *ad hoc* human rights tribunals to try TNI members for excesses in Tanjung Priok in 1984 and in East Timor between April and August 1999. Over time, however, her Government seemed to become less sympathetic to the ideals of an open society. A new Broadcasting Law, passed in November 2002, restored government censorship of films and advertisements, restricted foreign ownership and content and required the fragmentation of private television companies into several local stations. A new armed forces bill, presented to the legislature in February 2003, gave the Commander-in-Chief of the Armed Forces the authority to deploy troops anywhere in the country to meet an emergency without having previous approval from the President. In June the Government announced that it would screen all 4.1m. civil servants in order to test their loyalty to the State.

In March 2002 the trial began of 18 military officers, civilian officials and militia members accused of complicity in an attack on a church in Suai, East Timor, in which 26 people were killed. The defendants, none of whom were in custody during the trial, included: Abílio Soares, the last Indonesian Governor of East Timor; Maj.-Gen. Adam Damiri, the regional military commander; and Eurico Guterres, the most prominent militia leader. In August 2002 the Indonesian judiciary was subjected to widespread international condemnation when the special tribunal returned its first verdicts relating to the violence in East Timor. While Abílio Soares was convicted and given a three-year prison term for gross human rights violations, a sentence criticized for its leniency, the former Chief of Police in East Timor, Timbul Silaen, and five other officers were acquitted of all the charges against them. A further four Indonesian security officials were acquitted of similar charges in November, whereas the East Timorese militia leader, Eurico Guterres, was found guilty of crimes against humanity and sentenced to 10 years in prison. The verdicts prompted criticism that only East Timorese were being found guilty and demands from human rights organizations for UN intervention in the process. In February 2003 the UN indicted 32 people, including 16 soldiers, for crimes against humanity in East Timor. Meanwhile, the authorities in Timor-Leste also indicted several people, including the former chief of staff of the armed forces, Gen. Wiranto. and the former head of military intelligence in East Timor, Gen. Zacky Anwar Makarim. Indonesia, however, made it clear that it would not extradite its citizens to face trial abroad. The conviction of Brig.-Gen. Noer Muis in March was seen in some quarters as an attempt to allay foreign misgivings about the Indonesian trial process. In August 2003 Maj.-Gen. Adam Damiri, the most senior military officer to be tried in connection with the violence in East Timor, was found guilty of crimes against humanity and sentenced to a three-year prison term. The sentence was the last to be handed down by the court, which was widely condemned for returning only six guilty verdicts.

Indonesia's dysfunctional legal system remained a major problem. Tommy Suharto remained at large for 12 months following his conviction on corruption charges. While a fugitive, he was declared to be the main suspect in the murder of Justice Syafiuddin Kartasasmita, one of the Supreme Court judges who had found him guilty. In October 2001, however, the Supreme

Court overruled Tommy's conviction and, in the following month, Tommy was finally arrested. In March 2002 he stood trial, charged with possession of weapons and with ordering the murder of Kartasasmita. During the trial, which was broadcast on both television and radio, Tommy boasted that the police had helped him to visit his family regularly during his flight. One of Tommy's lawyers was herself charged with trying to bribe prosecution witnesses, and the Jakarta police officer who had arrested him was investigated for smuggling luxury cars into Indonesia. Tommy was found guilty in July and sentenced to 15 years in prison; it emerged later that during his trial he had been living in considerable comfort in a well-appointed cell in Jakarta's Cipinang prison, protected by his own bodyguard, visited freely by friends and family, and permitted to take regular leave. In early August Tommy announced that he did not intend to appeal against his conviction. He promised, however, to spend his time in prison 'deepening his religious knowledge'. Meanwhile, Suharto's half-brother, Probosutejo, was found guilty of embezzlement and fraud, but remained free pending an appeal. Suharto's own prosecution, suspended in 2001, was delayed again when his doctors reported that he was unable to communicate properly. In mid-2002 Datuk Param Cumaraswamy, UN Special Rapporteur on the Independence of Judges and Lawyers, commented after visiting Indonesia that its legal system was the worst that he had seen. The Indonesian Audit Commission on State Officials' Wealth reported that most of the judges whom it had examined could not account for the sources of their wealth.

In March 2002 Akbar Tandjung, Speaker of the DPR and Chairman of Golkar, was arrested and put on trial on charges of sequestering funds for his party's election campaign from the state logistical agency, BULOG. During the course of the trial, it emerged that most of the parties had received a share of these illegal BULOG funds, said to amount to US $350m. In September 2002 Tandjung was found guilty of misusing state funds and sentenced to three years in prison. He remained at liberty pending an appeal. In July 2003 the country's political parties agreed that convicted criminals were eligible to stand for the presidency in the 2004 elections, clearing the way for Akbar Tandjung's expected candidature, but in February 2004 the Supreme Court overturned his conviction. Meanwhile, the Governor of the Central Bank, Syahril Sabirin, was sentenced to a three-year prison term in March for his role in the Bank Bali scandal, and Ginanjar Kartasasmita, a senior economic minister in the Suharto era, was questioned over decisions he made in the 1990s. Many of these trials and investigations, however, were criticized: in some cases, the defendants seemed to have been selected for political reasons; in others the charges appeared to have been chosen in order to increase the possibility of an acquittal or to minimize the severity of the eventual sentence. The criticism was to some extent borne out in August 2002 when the Court of Appeal overruled Sabirin's conviction for corruption. Sabirin had retained his position as Governor throughout the appeals process.

In June 2003 reports emerged of irregularities in Indonesia's decision to purchase *Sukhoi* jet fighters from Russia to replace its current, ageing, British aircraft. It seemed that President Megawati had approved the purchase on a recent state visit to Russia; some figures alleged that illegal bribes had played a part in influencing the decision. Complaints about the extent of high-level corruption in Megawati's Government continued to increase, in particular concerning a decision taken in August 2003 by the Attorney-General, M. A. Rachman, to halt investigations into corruption allegations against high-profile 'New Order' figures, including Suharto's daughter Tutut. The National Audit Board claimed that the Attorney-General and the police had investigated only 8% of the 6,162 cases of suspected corruption referred to them since 2001. Furthermore, Indonesia was placed 12th from last in a list of 133 countries ranked by the global organization, Transparency International, according to their perceived levels of corruption. In December 2003 Megawati installed a new Corruption Eradication Commission (KPK), chaired by Taufiequrrachman Ruki. In its first six months the KPK received more than 500 complaints and began a number of high-profile investigations, including one into allegations against the Governor of Aceh, Abdullah Puteh. Trials conducted on the basis of these investigations became possible only after Megawati installed an *ad hoc* corruption court in July 2004. The most noteworthy action against corruption, however, was in May 2004, when a court in West Sumatra sentenced 43 of the 55 members of the provincial parliament to prison terms, having convicted them of embezzlement from the funds of the 2002 provincial budget.

Excluding such prominent cases, the legal system seemed incapable of addressing the high levels of violent crime in society, political violence or the deep-seated corruption of the judiciary. Suspected petty criminals were routinely beaten to death by crowds in many centres. Piracy was a serious problem in the Straits of Melaka (Malacca). The police also failed to prevent attacks on the offices of the human rights organization Kontras in March 2002 and May 2003. Despite the widespread perception that justice was 'for sale' in the courts, no judges were charged successfully with any offence. In December 2001 the DPR approved legislation placing the police under the direct command of the country's President, but this measure seemed unlikely to improve the probity or effectiveness of police operations. The new legislation gave the police exclusive responsibility for internal security, but the military authorities made it clear that they considered this area to remain within their remit.

It also became apparent that Indonesia was, in some respects, beginning to emulate Singapore, where those in or close to the government took advantage of strict libel laws to punish criticism of their actions. In several cases, Indonesian police and security forces used criminal libel laws against journalists who had accused them of human rights abuses, while prominent Chinese-Indonesian businessman Tomy Winata sued the well-known journalist Goenawan Mohamad and the newspaper *Koran Tempo* for the equivalent of nearly US $25m. after an article was published alleging that Winata had used violence to intimidate his enemies. In January 2004 *Koran Tempo* was ordered to pay damages totalling approximately $1m. to Winata and was also found guilty of libelling the businessman Marimutu Sinivasan. In early June 2004 the head of the National Intelligence Agency (BIN), A. M. Hendropriyono, ordered the expulsion from Indonesia, on unspecified grounds, of Sidney Jones, a respected and highly knowledgeable analyst with the International Crisis Group (ICG) in Jakarta.

Decentralization also seriously undermined the unity of the armed forces and police by opening up new opportunities for local and regional commanders to develop lucrative commercial arrangements with regional authorities. The Government planned a review of the decentralization process with the aim of restoring key powers to the centre, but that review was unlikely to be completed before the 2004 general election.

Progress was, however, made in resolving two of Indonesia's worst communal conflicts. In central Sulawesi, where violence had broken out in late 1999 and had cost an estimated 2,000 lives over a two-year period, representatives from the two sides agreed to meet in the southern town of Malino, where they reached a peace settlement. The success of this agreement encouraged the Government to schedule a similar meeting in Malino in February 2002, bringing together the warring parties in Maluku. This meeting led to an agreement, known as 'Malino II', which included a provision that the Laskar Jihad forces should leave Maluku. The Laskar Jihad itself rejected this provision and promised to stay in the region to undertake 'humanitarian' work. In April an unidentified gang attacked the Christian village of Soya, killing 12 people and setting fire to a 450-year-old church. Vice-President Hamzah Haz supported the Laskar Jihad decision to remain in Maluku, commenting that the province was not yet safe for Muslims, but he visited Soya and made a donation to the cost of rebuilding the church. Bombs exploded in Christian areas of Ambon in July and September 2002, and in Poso there were fresh attacks on Christian communities, in which six people were killed and two churches burnt. In both cases suspicions were voiced that the security forces were responsible for the violence. The state of civil emergency was lifted in North Maluku in May 2003, but remained in force in Maluku. In January 2003 two Christian separatist leaders from Maluku were sentenced to three-year prison terms for treason. They had not been present at their trial, which was held in Jakarta, while they remained in Maluku.

The efforts of politicians to position themselves for the forthcoming elections became increasingly apparent during 2002. In May Hamzah Haz visited Laskar Jihad commander Ja'far Umar Thalib, who had recently been arrested for agitation and for slandering the President. Later in the month the Vice-President also visited an Islamic boarding school run by Abu Bakar Bashir, accused by the Government of Singapore of having links with the international terrorist organization al-Qa'ida (Base).

On 12 October 2002 a devastating car bomb exploded outside a night-club on the island of Bali, killing 202 people, including 88 Australian tourists and 38 Indonesians. Suspicion immediately fell on the regional Islamist terrorist organization Jemaah Islamiah, reputed to be the South-East Asian affiliate of al-Qa'ida, although it was unclear whether Jemaah Islamiah was a genuine organization or simply a loose network of radicals. The bombing appeared to have been motivated by a general level of hostility towards 'Western' decadence as it manifested itself in the tourist resorts of Bali, as well as by specific resentment of US policies and of Australia for its military role in East Timor and Afghanistan and its support of the USA. The police were remarkably swift in detaining the first suspects, whose trials began in mid-May 2003. The first defendant, Amrozi, was found guilty and sentenced to death in August. Two other defendants, Imam Samudra and Mukhlas, were sentenced to death in September and October, while a fourth, Ali Imron, was sentenced to life imprisonment after expressing remorse over his actions.

Meanwhile, in response to international pressure, an emergency anti-terrorism decree was passed. In March 2003 the DPR approved two bills specifically designating terrorism as a crime (the legislation was also made retrospective in order to cover the Bali bombing) and providing for detention without trial and for the use of intelligence reports in court cases. Abu Bakar Bashir was formally detained on suspicion of involvement in the bombings of several churches in December 2000. His trial on treason charges began in April 2003, becoming akin to a public forum in which he defended the right of Muslims to take up arms in order to defend their religion. In September Bashir was found guilty of subversion and sentenced to a four-year prison term.

During 2003 several bombs exploded in Jakarta—at the national police headquarters in February, at Jakarta airport in April and at the parliament building in July. A bomb outside the Marriott Hotel in early August killed 12 people. Although the bombings were generally assumed to be the work of Islamist militants (the Marriott bomb exploded shortly before the sentencing of Amrozi for his involvement in the Bali bombing), suspicion lingered that sections of Indonesian military intelligence with close links to Islamist radicals might have been involved. The sudden withdrawal of Laskar Jihad from Maluku and its abrupt disbandment in October 2002 added to speculation that the organization was acting on instructions from above. The Front Pembela Islam (FPI—Islamic Defenders' Front), widely known for its attacks on bars and other places of entertainment, also disbanded in early November.

Despite some high-profile convictions, Indonesia faced difficulty in meeting conflicting expectations of its judicial system. On the one hand, it was under pressure to reform the harsh, often arbitrary and generally corrupt, legal practices of the 'New Order', but, when a greater sensitivity to procedure and to the rights of the accused led to acquittals and short sentences, the authorities were criticized for not taking their security role seriously. Upon closer examination, it often seemed that the law enforcement authorities had been careless in their collection of evidence or in their preparation of indictments, and there were sometimes indications of collusion between the authorities and the defendants. In August 2003 the Human Rights Court in Jakarta initiated hearings into the 1984 shooting of Muslim demonstrators in Tanjung Priok. A year later the Court acquitted two generals on charges of ordering troops to open fire on demonstrators at Tanjung Priok, although a third was convicted. In July 2004 the Constitutional Court, which had been installed in the previous August as a consequence of a constitutional change, declared unconstitutional the law passed in 2003 that had been used to convict three of the Bali bombers. A spokesperson for the Department of Justice stated that the ruling did not overturn the convictions, but rather made it impossible to continue using the law in future cases. None the less, the status of the convictions remained uncertain. The four-year sentence of Abu Bakar Bashir was eventually reduced to 18 months. In August 2004 an Indonesian appeals court overturned the convictions of three army officers and one policeman for crimes against humanity in East Timor in 1999, as well as reducing from 10 to five years the prison sentence of the militia leader Eurico Guterres.

Religious tensions were inflamed further by a new education bill, proposed in early 2003, under which all schools were obliged to provide religious instruction according to the faith of each student. The measure was widely perceived to be an attempt to prevent the proselytization of Muslim students attending Christian schools, which had a better reputation for both quality and discipline than government-run schools. Christians, however, saw the move both as a burden on their schools and as an attempt to impose common Islamic orthodoxy, which was considered likely to be hostile to Christianity and other minority religions. The bill prompted demonstrations by Christians and even threats of secession by predominantly Christian provinces. The legislation was passed in June during a session boycotted by the PDI—P, which was the only party to oppose it. However, it required regulations for its implementation before it could come into force, and these had not been drafted by August 2004.

The campaign against radical Islam remained politically sensitive, because many Indonesians were unconvinced of the existence of Islamist terrorist organizations and were inclined to suspect that reports of Islamist involvement in the Bali and Marriott Hotel bombings, and in several smaller bombing incidents, were ominous propaganda disseminated to serve Western interests. In September 2003 Vice-President Hamzah Haz described the USA as a 'terrorist king' in a speech that he gave in Central Java. Although public opinion polls suggested widespread nostalgia for the administrative efficiency of the 'New Order', there appeared to be little enthusiasm for bolstering the military's security powers. In particular, there was considerable public opposition to a new bill on military affairs, which would confirm the army's territorial role and authorize the Commander-in-Chief of the Armed Forces to take emergency action for up to three days before seeking the approval of the President. There were also signs of a new police sensitivity to public opinion, at least outside Aceh and Papua, when the National Police Chief, Gen. Da'i Bachtiar, dismissed four senior police officers after a brutal police attack at the Indonesia Muslim University (UMI) in Makassar in May 2004, in which 65 students were injured. The attack took place after students had taken one policeman hostage and had assaulted another in protest at the rearrest of Abu Bakar Bashir in Jakarta on terrorism charges.

The violence between Muslims and Christians that had wracked the provinces of Maluku, North Maluku and Central Sulawesi during the Wahid presidency largely abated as a result of peace agreements negotiated by national politicians, including Yusuf Kalla, and the state of civil emergency in Maluku was finally removed in September 2003. Mutual suspicion between Christians and Muslims, however, remained at a high level, fuelled by rampant conspiracy theories and by a draft law on religious tolerance, which proposed to recognize only five religions (Islam, Protestantism, Catholicism, Hinduism and Buddhism), to ban conversion of, and marriage between, their followers, and to prohibit people from attending religious ceremonies of a different religion from their own. A further clash took place in Poso in October 2003, and Ambon was once again affected by riots in April 2004, after a radical Christian group staged a rally marking the 54th anniversary of the separatist Republic of the South Moluccas. Thirty-seven people died and more than 100 homes and churches were destroyed in the riots.

A year before the 2004 presidential election was due to take place, President Megawati's authority seemed relatively secure, bolstered by the fact that her inaction did not directly alienate any significant groups, by her hard line in refusing to make concessions to the regional separatist movements in Aceh and Papua, and by the absence of a convincing alternative candidate for the presidency. The two figures considered most likely to secure the Golkar nomination for the presidential election, Akbar Tandjung and retired Gen. Wiranto, were both tarnished, Akbar by his conviction on corruption charges and Wiranto by his links with the 'New Order' and by the fact that he had been indicted for war crimes by a UN-supported human rights tribunal in Timor-Leste. Akbar was able to secure nomination, in fact, only because his conviction was overturned on technical grounds. Golkar postponed choosing its presidential candidate until after the April 2004 legislative election, in order to encourage party discipline amongst the seven nominated challengers.

During the run-up to the legislative election there was considerable doubt over the ability of the KPU to manage the operation effectively. However, in the event, there were no serious problems at the polls, which took place on 5 April 2004, and there was no indication of systematic fraud or manipulation. The island of Java, with 59% of the 147m. eligible voters, elected just over one-half of the seats in the new legislature. A 'National Movement for Not Electing Rotten Politicians' drew much attention early in the campaign, but there was little evidence that its activities affected the outcome of the election. One of the first decisions taken by the new Constitutional Court, moreover, was to overturn a regulation prohibiting those accused of 'direct or indirect involvement' in the 1965 coup from standing as candidates. This regulation, which encompassed the family members of people with connections to the former PKI and its associated organizations, was said to affect between 10m. and 20m. people. While the ruling came after nominations for the election had closed, it put an end to the official investigation of four candidates whose eligibility might thereby have been removed. The authorities announced that they had no intention of rescinding the 30 or so other regulations that discriminated against former communists and their families, and the teaching of communism remained illegal.

The result of the legislative election significantly altered President Megawati's position. Her PDI—P received only 18.5% of the votes, down from 34% in 1999, and came second after Golkar, which secured 21.6%, a similar result to that which it had achieved in the previous election. The result both confirmed the public's loss of enthusiasm for Megawati and suggested that voters would turn instead to new parties, rather than revert to Golkar as a nostalgic representative of the stability and prosperity of the 'New Order'. Two new parties performed particularly well in the election. These were the Partai Demokrat (DP) of the former Co-ordinating Minister for Political Affairs, Security and Social Welfare, Gen. (retd) Susilo Bambang Yudhoyono, which obtained 7.5% of the votes, and the Partai Keadilan Sejahtera (PKS—Prosperous Justice Party), an Islamist party whose members took a strong stand against corruption, which received 7.4%. The PKS headed the poll in Jakarta, where the DP also did well. Golkar's strength remained in the outer islands, while the PDI—P performed best in Central and East Java and Bali. Other parties that had achieved a significant share of the vote in 1999—Abdurrahman Wahid's PKB, the PPP of Hamzah Haz and the PAN of Amien Rais—all experienced a slight decline in votes.

The election produced a national legislature that no longer included either any appointed military members or any members from East Timor (now Timor-Leste), and which had acquired greater powers than any legislature since the 1950s. Moreover, the number of parties represented decreased from 32 to 25. The result was overshadowed, however, by the forthcoming presidential election campaign. Five teams of candidates for the posts of President and Vice-President competed in this election. (A sixth team, with former President Abdurrahman Wahid as a candidate, was ruled invalid by the KPU under a rule that required all candidates to pass a medical test.) Susilo Bambang Yudhoyono recruited as his running mate Yusuf Kalla, who had been one of the seven candidates for the Golkar nomination, while Megawati chose the NU leader, Ahmad Hasyim, in place of Vice-President Hamzah Haz, who stood in his own right with Minister of Transportation and Telecommunications, Lt-Gen. (retd) Agum Gumelar. As predicted, Amien Rais was the candidate of the PAN, with Siswono Yudohusodo as his deputy. Gen. Wiranto, with Solahuddin Wahid (younger brother of Abdurrahman Wahid and former deputy chairman of the National Human Rights Commission), was Golkar's candidate.

The strong performance of the DP in the legislative election especially strengthened the presidential prospects of Susilo Bambang Yudhoyono. He had not been directly associated with either human rights abuses or corruption in the Suharto era and increasingly projected an image of competence and commitment to democracy. His candidacy had been confirmed when he resigned from his cabinet position as Co-ordinating Minister for Political Affairs, Security and Social Welfare in March 2004, claiming that President Megawati had failed to include him in her decision-making. Most polls soon showed Yudhoyono to be the leading contender, although his campaign was hampered by the lack of a strong party organization. In the first round of the presidential election, held on 5 July 2004, he received 33.5% of the vote, with Megawati in second place on 26.6%. Wiranto received 22.2% of the votes cast. Some 32m. eligible voters failed to vote, a significant increase on the 23.5m. who had declined to vote in the legislative election of April. With no candidate securing more than the requisite 50% of the vote, the result necessitated a second round of voting, between Yudhoyono and Megawati, which was held on 20 September 2004. In early October Yudhoyono was declared the winner of the country's first direct presidential election, having received 61% of the votes cast. The new President's inauguration took place on 20 October. The composition of the new Cabinet was announced on the following day.

SEPARATIST MOVEMENTS

Despite the archipelago's ethnic diversity, Indonesia experienced surprisingly few separatist movements during the first 50 years of independence. Regional movements generally sought only greater autonomy or aimed primarily to change the nature of the central Government rather than to achieve independence. During the 1970s, however, important separatist movements emerged in the provinces of Aceh and Irian Jaya (now Papua), and from 1975 Indonesia faced persistent nationalist resistance in the occupied territory of East Timor. After the fall of President Suharto in 1998, all these movements gathered momentum, and in 1999 a referendum on independence was held in East Timor, leading to Indonesia's withdrawal from the territory in October of that year (see the chapter on Timor-Leste). Demands for independence were voiced for the first time in other provinces, notably in the resource-rich provinces of Riau and East Kalimantan, but also in Bali and South Sulawesi. It seemed likely, however, that these new calls were intended principally to stake a claim for greater autonomy within Indonesia, rather than being envisaged as a serious contribution to the disintegration of the Republic.

Aceh

Aceh, the northernmost region of Sumatra, was an independent sultanate until the late 19th century, when it was conquered by the Dutch in a ferocious campaign. With a reputation as Indonesia's most staunchly Muslim region, Aceh was never fully subdued by the colonial power, and was one of the first areas where Indonesians took effective control from the Japanese after the declaration of independence in 1945. The Dutch never attempted to reoccupy the region, and Aceh ended the war of independence as a full province of the Indonesian Republic. After 1950 the Acehnese quickly became disillusioned with the Republic's leadership, which was generally perceived as corrupt, neglectful and 'un-Islamic'. The removal of Aceh's provincial status was also resented. A rebellion erupted in September 1953 under Daud Beureu'eh, and the Aceh revolt formally joined the broader Darul Islam movement for an Islamic Indonesia. Conciliatory policies by the central Government and a willingness to compromise on the part of the Acehnese, however, ended the revolt in the late 1950s. In 1959 Aceh became a special territory (Daerah Istimewa), with considerable autonomy in religious and educational affairs.

Dissent re-emerged in the mid-1970s, provoked by the exploitation by the central Government of natural gas and coal fields in Aceh; many Acehnese felt that Aceh was receiving none of the benefits of these operations. Migration and transmigration of other Indonesians into the province, together with the growing power of the central Government, led to a feeling that the region's autonomy was being eroded. Hasan di Tiro formed the Gerakan Aceh Merdeka (GAM—Free Aceh Movement) in 1976 and declared independence in 1977. This small rebellion was quickly suppressed by the army, although Tiro later established a government-in-exile in Sweden.

Opposition to the central Government arose again in 1989, led by the National Liberation Front Acheh Sumatra. In 1990 Aceh was made a 'military operations zone', thus allowing the armed forces far greater freedom to counter the rebellion. The armed forces were accused of using excessive and indiscriminate force in their subsequent operations, and it was estimated that by mid-1991, when the rebellion had been largely suppressed, about 1,000 Acehnese had been killed. This figure continued to rise over subsequent years, and there were persistent reports of torture, kidnapping and sexual assault by the security forces. In July 1993 the human rights group Amnesty International produced a report accusing the Government of protecting those members of the armed forces responsible for atrocities in Aceh, thus enabling them to act with impunity. Following the downfall of Suharto in May 1998, Aceh's status as a military operations zone was revoked in June, and in August 1,000 troops were withdrawn from the province. Following rioting in September, however, the withdrawal of troops was suspended. Although President Habibie's decentralization measures were intended to defuse some of the resentment underlying the Aceh revolt by giving the province a greater share of oil and gas revenues, public opinion in Aceh became increasingly sympathetic towards the notion of independence. Violence continued to escalate in the province over the months following the accession of President Habibie. In January 1999 27 soldiers were court-martialled over the deaths in custody of four Acehnese earlier the same month, but military violence in the countryside continued. Tension remained high in the province, and was exacerbated by the discovery of several mass graves of people killed by the army during security operations. Voter turn-out in the legislative elections of 7 June was low in many districts, and violence continued to escalate following the poll as GAM guerrillas intensified their campaign for independence for the province. By late August 200 Acehnese were reported to have been killed by the Indonesian armed forces since the instigation in June of a renewed military campaign in an attempt by the central Government to suppress the uprising.

On coming to power in October 1999, President Wahid foreshadowed a referendum on independence in the territory. In response, Acehnese nationalists belonging to the Aceh Referendum Information Centre (SIRA) organized a pro-referendum rally in the provincial capital, Banda Aceh, on 7–8 November, reportedly attended by 1m. people. The prospect of a referendum, however, was rejected by the army. Whereas East Timor was, in some respects, accepted as having been a special case, independence for Aceh seemed likely to precipitate the disintegration of Indonesia. Sections of the armed forces were also widely believed to have interests in Aceh's lucrative but illegal marijuana industry. Furthermore, the Acehnese independence movement was seriously divided and lacked the international support that had helped the East Timorese movement to achieve its aim. Although there were signs of some material support from Acehnese living in Malaysia, Malaysia itself emphatically declined to support independence for the province. By early 2000 President Wahid appeared to have withdrawn from the idea of a referendum on independence for Aceh, suggesting instead greater autonomy for the province, together with a larger share of revenue from natural resources and the limited introduction of Islamic law. This offer failed to satisfy the nationalists, however, and violence continued in Aceh. By mid-2000 more than 300,000 people out of the province's total population of 4m. were reported to have been displaced by the violence, and GAM was said to control about half the villages in the province. In May 2000, in Geneva, Switzerland, the Indonesian Government and Acehnese rebel representatives agreed upon a three-month cease-fire ('humanitarian pause'), which took effect on 2 June. Also in May 24 junior soldiers and one civilian were sentenced to various terms of imprisonment for the murder in July 1999 of 58 Acehnese in Beutong Ateuh; the trial was criticized by both Acehnese community groups, who had demanded the death penalty for the accused, and human rights lawyers, who called for a wider trial of crimes against humanity in the province. The cease-fire was later extended to January 2001, but appeared to have little influence on the level of violence. Pressure on the Government was increased by continued attacks on the major LNG plant in Aceh (see Economy). In April 2001 President Wahid signed an instruction for the police to assist the military in Aceh, effectively signalling a return to repressive strategies. An estimated 1,500 people were killed in 2001, mostly by the military, but some by GAM, which increasingly targeted Javanese settlers.

Aside from maintaining military pressure on GAM, the Government's main strategy for combating the insurgency was to prepare an enhanced autonomy plan for Aceh, giving the provincial government a greater share of gas revenue and imposing Islamic law within the province. The necessary legislation was drafted under Wahid's presidency but was signed into law by Megawati as one of her first acts in office. Although apparently generous in the autonomy it gives to Aceh, the legislation was widely criticized by the Acehnese for failing to deal with the military presence. Many Acehnese, moreover, regarded the implementation of Islamic law as a cynical attempt to alienate modern-minded Acehnese from the traditional religious leaders whose power was thereby enhanced. Although there was a common perception in Jakarta that the main orientation of the Acehnese movement was Islamic, GAM had never proposed an Islamic state, and its relations with radical Muslim groups in Java were poor. GAM, in fact, condemned the terrorist attacks on the USA of 11 September 2001 (see below) more strongly than the Indonesian Government. The autonomy measures came into force on 1 January 2002, when the province was renamed Nanggroë Aceh Darussalam (NAD—Islamic State of Aceh). The new regional government was permitted to retain 70% of provincial revenues and began in March to implement its version of Islamic law by imposing a dress code. Aceh opened its first Shari'a (Islamic law) court in March 2003, promising to implement Islamic law in a 'moderate' way. It was not clear, however, whether implementation would be restricted to narrow, family-law issues (divorce and inheritance) or whether it would be extended to matters such as adultery and theft. There continued to be no movement on the issue of human rights. Although President Megawati apologized to the Acehnese for past oppression in August 2001, she insisted that the province would not be permitted to leave Indonesia and that armed separatist movements would not be tolerated. She also eventually obtained the release of five GAM negotiators who had been arrested by police in July 2001 during negotiations with the Government, but the time taken to achieve this release indicated the depth of resistance within the armed forces to any compromise with the rebels. From early 2002 military leaders adopted an increasingly stringent public policy towards separatism, and the military presence in Aceh rose to about 17,000. In January 2002 army units trapped and killed the GAM military leader, Abdullah Syafei, and in the same month the Iskandar Muda military command (Kodam) covering Aceh was restored. None the less, discussions between government and GAM representatives on the issue of restoring security took place in Geneva in May 2002.

Indonesia came under some pressure from the US envoy for the Middle East, Gen. Anthony Zinni, to maintain the dialogue with GAM. For a time, the Government discussed splitting Aceh into two, with a new province, Leuser Antara, being created to incorporate regions relatively unaffected by the rebellion. At the same time, the restoration of martial law was threatened if GAM did not co-operate; the Speaker of the MPR, Amien Rais, commented, 'if necessary, cut off the hands of those troublemakers'. During 2002 the Government's tone hardened, and it insisted that GAM accept the autonomy proposals by 9 December or face a resumption of military action. Meanwhile, a Scottish researcher, Lesley McCulloch, was arrested in Aceh in September while investigating the separatist movement and charged with violating the provisions of her tourist visa. On 9 December 2002 the Indonesian Government and GAM signed a Cessation of Hostilities Agreement in Geneva, under the auspices of the Henri Dunant Centre and with the support of Japan, the USA and the European Union (EU). The agreement provided for a Joint Security Committee to monitor the cease-fire and for further discussions aimed at moving towards a peaceful solution to the conflict. Levels of violence dropped dramatically after the agreement, but each side accused the other of using the cease-fire to consolidate its position, and the fundamental disagreement over Aceh's future remained unresolved. Gradually, the level of violence on both sides increased, and in April the army revealed that it planned a military operation to crush the rebels. The Government issued a deadline of 17 May for GAM to disarm and renounce its aim of independence. Peace talks resumed on that date in Tokyo, Japan, but neither side changed its position and on 19 May the Government declared martial law in Aceh. Widespread fighting broke out, some 200 schools were burnt down and hundreds of deaths were reported, although reliable information on the conflict was difficult to obtain owing to Indonesia's stringent control of external media. By early July 2003 the army claimed to be in full control of the province, but reports of fighting continued. In June a military court sentenced three soldiers to short prison terms for committing violent acts against Acehnese civilians.

Indonesia presented the operation in Aceh as an attack on Islamist terrorism and blamed GAM for several of the bomb attacks that had taken place in other parts of Indonesia in preceding months, but there was no clear evidence that GAM had undertaken any operations outside the province. Although the army claimed by the end of the year to have killed 800 rebels and captured another 1,700, reports from the province suggested that the GAM infrastructure had dispersed rather than disappeared. The military commander in Aceh, Maj.-Gen. Bambang Darmono, responded to reports of violence against civilians and of the plundering of property by troops by commenting that the beating of suspects was acceptable as long as it did not cause serious physical harm. In November 2003, when martial law was extended for a further six months, the TNI announced that it would use smaller-scale operations to attack the remnants of GAM and would co-operate more closely with civilians. The 30,000-strong Aceh expeditionary force was a major drain on army resources. Although martial law was lifted in May 2004, the province remained under civilian emergency rule.

In June 2004 Indonesia welcomed Sweden's decision to arrest two GAM leaders who had been in exile in that country since 1979 and who had played a major role in marshalling international support for independence for Aceh. Indonesia had long argued that these figures directly supported terrorist activities in Indonesia, and in February Sweden had begun investigations into allegations that the men, who were Swedish citizens, had been involved in kidnapping, arson and bombings in Indonesia.

Papua

The western half of the island of New Guinea had been a part of the Netherlands East Indies before the Second World War, but it was scarcely integrated with the remainder of the colony and was inhabited by Melanesians who were ethnically different from the Malay peoples who dominated the rest of Indonesia. Partly on these grounds and partly to assuage the humiliation of defeat in the Indonesian war of independence, the Netherlands decided in 1949 to retain control of the region, envisaging its eventual self-determination. This decision, bitterly resented by Indonesians, led to more than a decade of diplomatic activity, which culminated in the transfer of the colony to Indonesia under UN auspices in 1962–63. Indonesia named the region Irian Barat (West Irian), but in 1972 changed it to Irian Jaya (Victorious Irian).

Under the UN agreement, the opinion of the indigenous population was to be heard in an 'Act of Free Choice' five years after integration, but this act was not internationally guaranteed or supervised and Indonesia carried it out in 1968 in a way that made acceptance of Indonesian rule mandatory. Resentment against Indonesian rule soon emerged. Indonesian authorities had little respect for traditional Papuan dress and custom, and attempted to impose Indonesian culture. The region became a major destination for internal migration and transmigration, and the less-educated local people often found themselves unable to compete with new inhabitants for bureaucratic, professional and other skilled jobs. Logging and copper-mining in the region, moreover, became major sources of export revenue for the Indonesian Government, while few of these funds seemed to reach the province.

A rebellion erupted in 1965, led by the Organisasi Papua Merdeka (OPM—Free Papua Movement) and unrest has continued in the province ever since. Seth Rumkorem declared a Republic of West Papua in 1971 and a major uprising took place in 1977. Despite grandiose claims, the movement controlled no significant territory by the early 1990s, although its forces were able to range widely in the difficult terrain and take sanctuary across the border in Papua New Guinea. In 1984 about 10,000 refugees crossed the border into Papua New Guinea (see also Foreign Relations, below).

Human rights organizations continue to report allegations of torture, killing, intimidation and cultural suppression in the province. Most recently, attention has focused on the huge Freeport mine near the province's southern coast, which is Indonesia's largest source of gold and copper. Since mid-1994 the mine authorities have been accused of complicity with Indonesian military forces in the deaths of more than 60 local people who protested at environmental degradation and what they saw as the privileged position of outsiders in the mining industry. In January 1996 the OPM launched a bid for greater international attention by kidnapping 26 people, including seven foreigners, in the south-east of the province. The OPM later released 15 of them, but the remainder were freed in an operation by Indonesian special forces in May, during which the OPM killed two Indonesian hostages. In August 1996 Indonesian forces freed several Indonesian forestry workers taken hostage by another OPM group.

In the months immediately after the fall of President Suharto, Irian Jaya was relatively quiet. In early October 1998 the Indonesian Government revoked the status of Irian Jaya as a 'military operations zone' following the conclusion of a cease-fire agreement with the OPM in late September; however, this action was not followed by any withdrawal of Indonesian troops from the province. The movement towards independence for East Timor (see below), however, encouraged the Irianese to begin pressing the case for independence for the region. In late February 1999 100 tribal leaders raised the issue of independence at a meeting with President Habibie, and pro-independence banners began to appear in the larger centres. However, Irian Jaya's immediate prospects for independence were weaker than those of East Timor, because 40% of the province's 2m. people are of non-Irianese descent, because its copper and gold mines are important to the national economy, and because it was formerly a part of the Netherlands East Indies.

In September 1999 President Habibie announced that the province of Irian Jaya would be divided into three. Although ostensibly intended to bring administration closer to the people, the move was widely perceived as a device to split the region's independence movement. On coming to power in October, President Wahid revoked the partition, and in December announced that the province's name would be changed from Irian Jaya to Papua. (The central authorities, however, subsequently declined to ratify the change; nationalists, meanwhile, preferred the name of West Papua, Papua alone being widely used within the province itself but also being the traditional name for the southern part of Papua New Guinea.) The President, nevertheless, resisted any suggestion of independence for the territory, and the police and army continued to arrest Papuans for raising the nationalist flag; the flag was formally banned again on 1 December 2000. In February 2000 an unofficial Congress of the Papuan People met in Jayapura, formally repudiated the 1968 'Act of Free Choice' and began planning strategies for achieving independence for the province. Although, according to various reports, Indonesian military and intelligence operations had largely defeated or compromised the OPM as a resistance force, the example of East Timor had demonstrated to independence movements in other parts of the archipelago the critical importance of an international campaign and of a visible protest movement in the cities. The congress held in Jayapura was therefore dominated by urban NGOs rather than guerrillas. In response to the activities of the independence movement, the Indonesian military authorities apparently began to develop and support local pro-Indonesia militia groups similar to those used in East Timor. In 2001 these militia began receiving help from the Java-based radical Muslim group Laskar Jihad. In June a political congress held in the province, attended by about 3,000 Papuans, declared that the territory had never been legally integrated into Indonesia. The claim was immediately rebuffed by the Indonesian Government. In July, however, members of the Presidium Dewan Papua (PDP—Papua Presidium Council) held talks with President Wahid in Jakarta and renewed their demand for independence for the territory.

In October 2000 the Minister of Defence announced that a tougher line on separatism would be taken in the province. In November Theys Eluay, a moderate leader of the PDP, and four others were charged with sedition; shortly afterwards Eluay was assassinated after attending a dinner with the local army commander. An official investigation later concluded that members of the military were responsible for the murder. In April 2003 a court in Surabaya convicted four officers and three soldiers of responsibility for the death; they received sentences of between two and three-and-a-half years. Legislation was also drafted, however, for special autonomy for the province, along lines broadly similar to those in Aceh, though without the implementation of Islamic law in the mainly Christian province. The progress of this law through the legislature was delayed by the long struggle in Jakarta to oust President Wahid, and by disagreement over the percentage of resource income that was to be retained by the provincial government. When the special autonomy bill was finally approved on 22 October 2001, it provided that 80% of revenue from forestry and fisheries and 70% from oil and gas would be retained by Papua. The bill confirmed the province's name as Papua and legalized the use of the Papuan flag. The legislation took effect on 1 January 2002.

In February 2003, however, the Government announced that it intended to implement the law, passed in 1999 but never enforced, to divide the province of Papua into three. The move was presented by the Government as one that would enable the closer involvement of the people in the province's administration, but it was implemented without public consultation and against the wishes of virtually all Papuan leaders. The move was also in conflict with the 2001 special autonomy law, which specifically provided that no new provinces should be created without the approval of the Papuan consultative council, which awaited formation. Acting governors for the new provinces were appointed in May, in a measure generally seen as being intended to undermine Papuan separatism by promoting disunity.

In late August 2002 two US teachers and an Indonesian teacher were killed and 11 others wounded in an attack on a vehicle convoy near the Freeport mine. The military blamed the attack on OPM separatists, but OPM groups denied responsibility. Suspicion was raised by the police chief for Papua, I Made Pastika, that the military had engineered the attack in order to incriminate the OPM. (Gun cartridges of a type used by the military were reported to have been found at the scene of the attack.) The Commander-in-Chief of the Armed Forces, Gen. Endriartono Sutarto, later threatened to sue the US newspaper, the *Washington Post,* for US $1,000m. for reporting the allegation as a fact. A later investigation by the US Federal Bureau of Investigation (FBI) suggested that the attack had been carried out by soldiers angered by a reduction in the payments they received from the mining company, and in February 2003 the *Washington Post* published a retraction, stating that there was 'no substantiation' in its original claim. The Indonesian army subsequently announced that it would no longer provide special security for Freeport against payment, on the grounds that it was not a mercenary force.

The creation of three provinces out of the former province of Papua appeared to have proceeded in August 2003, prompting a three-day confrontation between supporters and opponents of the policy in Timika, the designated capital of Central Papua. The Co-ordinating Minister for Political Affairs, Security and Social Welfare, Susilo Bambang Yudhoyono, then declared in late August that the division had been reversed and promised a review of the contradictory laws concerning the region, although this review had not taken place by the time of the 2004 legislative election. The creation of a province of West Irian Jaya, based on Manokwari, however, appeared to go ahead, until the State Administrative Court overturned a law appointing its governor. Alarm over the Government's intentions in Papua was also raised by an announcement from the East Timorese militia leader, Eurico Guterres, that he planned to establish a pro-Indonesia militia in Papua, and by the appointment of Brig.-Gen. Timbul Silaen as the head of police in Papua in December 2003. Silaen had been chief of police in East Timor at the time of the 1999 violence, but was acquitted of charges of human rights abuses relating to that time shortly after his appointment to the Papuan position.

FOREIGN RELATIONS

Since independence Indonesia has prided itself on maintaining an 'active and independent' foreign policy. During the 1950s Indonesia was a founder-member of the Non-aligned Movement (NAM), and assumed the chairmanship of the organization in September 1992. However, Indonesia's dependent position in the international economy has never permitted it the freedom of action that it professes. During 'Guided Democracy', Sukarno turned to the USSR and the People's Republic of China for support, and from 1965–66 'New Order' Indonesia solicited funds from Western donors.

The centre-piece of Indonesian foreign policy is its membership of the Association of South East Asian Nations (ASEAN), which it founded with Malaysia, the Philippines, Singapore and Thailand in 1967. Indonesia also played a major role in establishing the Asia-Pacific Economic Co-operation forum (APEC), and hosted the November 1994 APEC summit meeting in Bogor. ASEAN's original aims emphasized regional economic co-operation as a means of diminishing threats to internal security, but it has worked most effectively in defusing conflicts between its members and in creating a united diplomatic front on broader international issues. ASEAN support, for instance, helped to limit the damage to Indonesia's international reputation caused by its invasion and protracted pacification of East Timor and by its repeated disregard for UN resolutions on the colony.

Indonesian suspicion of China has been a major feature of foreign policy since 1965–66, when the military accused China of supporting the PKI and the attempted coup. Indonesia severed diplomatic relations with China in 1967. Poor relations were compounded by long-term doubts about the loyalty of Indonesia's ethnic Chinese minority and a belief that China might seek to take advantage of their support. To prevent such intervention, Indonesia banned the import of material written in Chinese characters. In May 1985, however, the two countries began to discuss the resumption of direct trade links, and in April 1988 Indonesia announced that it was willing to establish full diplomatic relations, provided China gave an assurance that it would not seek to interfere in Indonesia's internal affairs. Diplomatic relations were finally restored in August 1990, after Indonesia undertook to repay debts to China incurred during the period of 'Guided Democracy'. Relations between Indonesia and China suffered a set-back in 1998 when, some six weeks after the perpetration of terrible violence against ethnic Chinese Indonesians around the time of the fall of Suharto in May, China issued a sharp diplomatic protest against the violence and gave prominent coverage in the state-controlled media to reports of the gruesome rapes, arson and murders. China's response was perhaps prompted by the indignation and anger at the violence expressed by overseas Chinese in general; however, for those who wished to emphasize the primary loyalty of ethnic Chinese Indonesians to Indonesia, rather than to China, this display of Chinese interest was less than welcome. Following his accession to the presidency, President Habibie publicly expressed his sympathy for the plight of the ethnic Chinese victims of violence. Subsequently, in May 1999, as part of an ongoing programme of general reform, Habibie lifted a ban that had existed on the use and teaching of the Mandarin Chinese language within Indonesia.

Indonesia aspires to be recognized as the major power in South-East Asia, and welcomed the withdrawal of the former USSR from its base in Cam Ranh Bay in Viet Nam and of the USA from Subic Bay in the Philippines. Indonesia has also been involved in diplomatic efforts to resolve long-standing regional conflicts. From June 1991 it hosted several informal meetings attended by representatives of Brunei, China, Malaysia, the Philippines, Taiwan and Viet Nam to discuss these countries' conflicting claims to all or parts of the Spratly Islands in the South China Sea, which are strategically important and show considerable potential for petroleum exploitation. The meetings resulted in a joint statement agreeing to resolve the dispute peacefully. Paradoxically, Indonesia's bilateral relations with Malaysia became strained in 1993 by a dispute over two islands, Sipadan and Ligitan, off the coast of Sabah. (The dispute was submitted to the International Court of Justice (ICJ) in the Hague, the Netherlands, in June 2002, and in December the ICJ ruled in favour of Malaysia's claim.) In April 1995 the Indonesian armed forces increased patrols in the South China Sea after China re-emphasized its claim to seas near the Natuna archipelago, which Indonesia has traditionally claimed. Indonesia also assumed an active diplomatic role in the long-running conflicts in Indo-China. From July 1988 Indonesia hosted a series of informal meetings between the contending Cambodian factions and, with France, Indonesia jointly chaired the Paris International Conference on Cambodia from August 1989. Indonesia contributed troops to the UN Transitional Authority in Cambodia prior to the elections of 1993. In July 1996 Indonesia hosted a series of meetings between ASEAN ministers of foreign affairs and inter-nation 'dialogue partners' (Australia, Canada, China, the EU, India, Japan, the Republic of Korea, New Zealand, Russia and the USA) to discuss regional security and international trade. Smoke from forest fires in Indonesia seriously disrupted commerce and communications and also damaged public health in Singapore and Malaysia during the second half of 1997. Although both countries played down suggestions of tension with Indonesia, they were clearly irritated by Indonesia's failure to address the problem more promptly. President Suharto twice apologized publicly to the neighbouring countries for the haze. Indonesia's relations with Malaysia were also strained by the forced repatriation in early 1998 of thousands of Indonesian workers from Malaysia as a result of the impact of the regional economic crisis; many of those repatriated reported brutal treatment in detention centres prior to their removal. Under Malaysian regulations introduced in 2002, illegal immigrants faced up to five years in prison and six strokes of the cane. In mid-2003 some 22,000 illegal workers who had fled Malaysia were still living in refugee camps at Nunukan in East Kalimantan.

Indonesia's relations with the West have been periodically disrupted by disagreements over human rights and over East Timor (now Timor-Leste), whose annexation by Indonesia remained widely unrecognized internationally in the late 1990s, prior to the referendum held in August 1999. In March 1992, after the Dutch Government linked the continuation of aid to an improvement in Indonesia's human rights performance, Indonesia angrily rejected all Dutch aid and dissolved IGGI, which had been chaired by the Netherlands. A new international aid consortium, the Consultative Group on Indonesia (CGI—chaired by the World Bank and including 18 countries and 13 multilateral aid agencies, but excluding the Netherlands), met in Paris, France, in July 1992 to assume the functions of IGGI. In early 1994 the USA threatened to remove Indonesia's trade privileges unless it took steps to conform to International Labour Organization (ILO) standards for labour conditions. The events surrounding Megawati's expulsion from the PDI leadership led the USA to postpone the sale to Indonesia of nine F-16 fighter aircraft in September 1996; the US position was weak, however, as it was known to have had difficulty in finding a buyer for the aircraft. In June 1997 Indonesia itself cancelled the sale, on the grounds of unacceptable US criticism over the alleged abuse of human rights in the country. The British Government pledged to review the sale of 16 *Hawk* fighter aircraft to Indonesia, but allowed the sale to proceed after the Indonesian Government made it clear that it would simply buy from elsewhere. The continuing resistance in East Timor strengthened the international campaign to maintain pressure on Indonesia over its occupation of the territory.

Within the EU, Portugal was perhaps the most vociferous in condemning Indonesia and lobbying for a UN-supervised referendum in East Timor; in July 1992 Portugal blocked an economic co-operation treaty between ASEAN and the European Community (EC) on these grounds. Portugal also began proceedings against Australia in the ICJ, seeking a ruling against the so-called Timor Gap Treaty, concluded between Australia and Indonesia in 1991. The Treaty provided a legal framework for petroleum and gas exploration in the maritime zone between Australia and East Timor, which had not been covered by earlier Indonesian-Australian treaties. Portugal claimed that the agreement infringed both Portuguese sovereignty and the East Timorese right to self-determination. (Only Australia was named because Indonesia does not come under the court's jurisdiction.) In a judgment brought in June 1995, however, the Court ruled that it could not exercise jurisdiction because the central issue was the legality of actions by Indonesia, which had refused to present a case. Formal contacts between the Indonesian and Portuguese Governments, especially over the status of Portuguese culture in East Timor, took place in 1995 and 1996 but were, for the most part, inconclusive. In January 1996 Portugal began direct satellite television broadcasts to East Timor. Indonesia's relations with Australia, which had been increasingly close in the early 1990s, soured when widespread public protests in Australia forced Indonesia to withdraw the nomination of Lt-Gen. Herman Mantiri as ambassador to Canberra. Mantiri was targeted because of remarks he made in 1991, apparently defending the Dili massacre. The rift with Australia appeared to have been healed when the two countries signed a security agreement in December 1995. Although not amounting to a formal alliance, the agreement binds the two sides to consult on security matters and to consider possible joint measures. In March 1997 the two countries signed a treaty concerning sea-bed and 'economic zone' boundaries. Following the downfall of President Suharto in May 1998, relations between Indonesia and Australia were significantly affected by the issue of East Timor; in January 1999 the Indonesian Government expressed its 'deep regret' at Australia's announcement earlier the same month that it was to change its policy on the territory and actively promote 'self-determination' in East Timor. Following the announcement in March 1999 that the Indonesian Government was to allow the East Timorese people to vote on the issue of independence, Australia continued to attempt to establish a role for itself in the process of the definition of the future of the province. In early May Indonesia and Portugal signed a UN-sponsored accord on East Timor which allowed for total independence for the territory if the East Timorese people voted to reject autonomy proposals offered by Indonesia in the referendum scheduled to be held in August. On 28 December, following Indonesia's withdrawal from East Timor in October after the territory's population voted in favour of independence, diplomatic relations between Indonesia and Portugal (which had been severed in 1975) were restored. Meanwhile, however, Indonesia's relations with Australia deteriorated, following the latter's involvement in peace-keeping operations in East Timor after the referendum. Furthermore, many members of the Indonesian élite deeply resented Australia's wider role in the detachment of East Timor from Indonesia, as well as the public claim made by the Australian Prime Minister, John Howard, that he had 'stood up' to Indonesia over the issue. There was also widespread suspicion in Indonesia that official Australian support for the territorial integrity of Indonesia might not endure in response to increasing public pressure in favour of independence for the Indonesian province of Irian Jaya (now Papua—see above). Relations improved in June 2001, when Wahid finally made an official visit to Australia, the first by an Indonesian President for 26 years.

Relations between the new Megawati Government and Australia were strained almost immediately after the President's accession in mid-2001 by the issue of refugees, asylum-seekers and illegal immigrants, mainly from South Asia and the Middle East, who passed through Indonesia while attempting to reach Australia. The issue was one of great controversy in Australia but was of little direct importance to Indonesia, which objected to Australian claims that it should be doing more to prevent unauthorized migrants from passing through its territory.

On 31 January 2000 the UN Commission on Human Rights recommended that an international tribunal be established to try Indonesian military personnel and other individuals suspected of involvement in the violence in East Timor in August and September 1999. In late February 2000, however, UN Secretary-General Kofi Annan stated that the Indonesian legal process should be allowed to run its course before the establishment of any such tribunal. In January the EU lifted an embargo on military assistance to Indonesia, which it had imposed in an attempt to force Indonesia to repatriate displaced East Timorese from West Timor and to bring to justice those responsible for the violence in the territory. A similar embargo imposed by the USA remained in place, however. During a visit to Indonesia in April, the US commander in the Pacific, Adm. Dennis Blair, publicly criticized the human rights record of the Indonesian military and predicted that military co-operation between the USA and Indonesia would not be resumed in the foreseeable future.

Indonesia's relations with Papua New Guinea were strained by the disputed common border issue and by the question of the

OPM. The Papua New Guinea Government scrupulously refrained from assisting the separatists: in May 1990 the OPM leader, Melkianus Salossa, was arrested in Papua New Guinea and deported to Indonesia, where he received a life sentence, and since October 1990 a new border agreement has enabled the exchange of military intelligence and the conduct of joint border patrols. However, the insecure border and widespread sympathy for the OPM in Papua New Guinea meant that the rebels periodically obtained support and sanctuary across the border. Papua New Guinea was angered by regular incursions of Indonesian troops and aircraft into its territory and by Indonesia's crude and corrupt attempts to buy influence in Port Moresby. In March 2003 Papua New Guinea announced that it would return to Indonesia about 300 asylum-seekers who had crossed the border in 2000, claiming refugee status.

The economic crisis of 1997 and the fall of President Suharto in 1998 turned Indonesian attention inward and distracted attention and energy from international politics. ASEAN, in particular, was notably less effective during this period without the driving force of Indonesia behind it. Although Western countries generally welcomed President Habibie's reform programme, they expressed some doubts about the depth of Indonesia's commitment to fundamental change. The early months of President Wahid's leadership, in late 1999 and early 2000, were characterized by a hectic programme of foreign travel intended to strengthen his Government's credentials, to win international support for Wahid's own political position from within Indonesia itself, and to restore business confidence in the country's future. In February 2000 President Wahid visited East Timor for the first time since the Timorese people's vote for independence, in an attempt to begin the process of Indonesia's reconciliation with the territory. Relations with Singapore were strained by occasional dismissive remarks from the Indonesian President, while relations with the USA deteriorated over repeated outspoken criticism of Indonesian policy made by the US ambassador to Jakarta, Robert Gelbard. President Wahid shared former President Suharto's long-term aim of establishing Indonesia's credentials as a leader in both East Asia and the Islamic world, and he publicly commented on the need of both groupings to establish a higher profile and a greater degree of independence from external powers, especially the USA.

On 19 September 2001 a scheduled visit to Washington, DC, by President Megawati proceeded despite the terrorist attacks on New York and Washington only eight days previously. In her discussions with President George W. Bush, Megawati offered support for the USA's campaign against terrorism, receiving in exchange promises of financial aid and improved access to defence equipment. The US response to the attacks, however, added greatly to Megawati's difficulties. The US attack on Afghanistan was widely condemned in Muslim circles. There were protests against the US bombing in several Indonesian cities and demands for a boycott of US goods and firms. Radical Muslim groups began 'sweeping' tourist areas in Jakarta, Central Java and Makassar to tell US citizens to leave the country. The authorities later estimated that 1.3m. tourists had cancelled visits to Indonesia as a result. None the less, with a widespread perception in Indonesia that the USA had overreacted to the attacks, the Government was reluctant to be seen to be acting against Islam at the behest of the USA. Whereas individuals with al-Qa'ida connections, including at least three Indonesians, were arrested in neighbouring South-East Asian countries, Megawati's Government sought to downplay the issue, arguing that there was no proven Indonesian involvement with Islamic terrorism. In October Megawati stated publicly that she did not wish to see any country use violence against another, even in retaliation. Several moderate Muslim leaders argued that a harsh reaction would strengthen Islamic radicalism, rather than controlling it, but some reports indicated that the authorities were quietly keeping Muslim radicals under tighter control.

In early 2002 the USA and Indonesia agreed that the former would provide US $10m. to train Indonesian police in counter-terrorism techniques and to train bank officials in the tracing of terrorist financial transactions. Australia also increased co-operation with Indonesia in police training and encouraged closer co-operation in police matters between Indonesia and its ASEAN neighbours. In July the US Senate Appropriations Committee voted to abandon conditions on providing US military training to the Indonesian armed forces, thus enabling the Indonesian military to participate in the US Department of Defense's International Military Education and Training programme (IMET). The Pentagon had argued that it needed Indonesia's co-operation in the global war on terrorism and that this war should take precedence over human rights considerations. President Bush, however, undertook to consult the US Congress before resuming formal co-operation, and the Congress authorization was withdrawn in 2003 when strong suspicions arose that the Indonesian military had been responsible for the murder of two US citizens at the Freeport mine in Papua. These suspicions were allayed only in July 2004, when an alleged OPM leader, Anthonius Wamang, was charged over the incident.

The terrorist attacks on Bali in October 2002 were generally perceived to constitute a further reason for the strengthening of Indonesia's military links with the USA. Although Indonesia was not believed to be as important a base for terrorist groups as the neighbouring southern Philippines, the country's relatively porous borders and the political difficulty of monitoring Islamic organizations made it an important element in terrorist networks in the view of the USA and Australia. Australia was also concerned to prevent Indonesia from being used as a conduit for illegal immigrants from the Middle East and sought the help of the Indonesian authorities in preventing immigrant boats from leaving Indonesia. Australia's return of a boat-load of Kurdish refugees to Indonesia in November 2003, however, led to Indonesian complaints that it was being used by Australia as a 'dumping ground' for unwanted migrants.

Human rights issues remained a source of contention between Indonesia and the West. The EU, Japan and the USA issued a statement of regret at the extension of martial law in Aceh in November 2003. In January 2004 the US State Department placed presidential candidate Gen. Wiranto on a 'watch list' of indicted war criminals, thus barring him from entering the USA. In March 2003 Indonesia revoked visa-free entry for most Western tourists and reduced the length of the maximum stay for tourists from 60 to 30 days.

Economy

SARWAR O. H. HOBOHM

Revised for this edition by the editorial staff

Indonesia is the fourth most populous country in the world, with its population officially having exceeded 200m. people on 4 February 1997. According to the results of the census of June 2000, the total population had risen to almost 206.3m., and by mid-2003 it was estimated to have reached more than 219m. With substantial deposits of petroleum, natural gas and other minerals, as well as the capacity to produce a wide range of agricultural commodities, Indonesia is also one of the world's most richly endowed countries in terms of natural resources. Yet the country remains relatively poor, and achieved the status of a 'middle-income' country, by World Bank criteria, only in 1981, when its national income per caput first exceeded the threshold of US $500. After some brief set-backs in the mid-1980s the Indonesian economy recorded almost a decade of rapid economic growth, and by 1996 the country's gross domestic product (GDP) had risen to $1,155 per head. However, the Asian financial crisis, which began in the latter half of 1997 and intensified during 1998, prompted a sharp regression in this figure, both because of a decline in the volume of output and a fall in its value in terms of US dollars, owing to a severe devaluation of the national currency, the rupiah. Official national accounts data showed that GDP per caput decreased to $1,080 in 1997, and declined sharply to some $470 in 1998, before increasing to $680 in 1999 and $715 in 2000, as a result of both a recovery in the value of the rupiah and a real growth in output. This trend towards recovery suffered a set-back in 2001, when a slowdown in economic growth and a sharp depreciation of the rupiah resulted in a decline in GDP per caput to $690, but was restored in 2002, when a steady rate of real GDP growth and a significant appreciation of the rupiah resulted in an increase in GDP per head to $807. These favourable trends continued into 2003, resulting in an increase in GDP per caput to an estimated $963.

The Indonesian economy is also characterized by a variety of distributional imbalances. The results of the 2000 population census showed that 60.4% of Indonesia's population lived on the three islands of Java (Jawa), Madura and Bali, which together comprised only 7.2% of the country's total land area. These islands also represent the most fertile regions of Indonesia, and are the most heavily urbanized and most developed areas, both industrially and infrastructurally. These advantages are, to a considerable extent, offset by the three islands' high population density, and they also contain areas of extreme poverty. This variation in income levels is emphasized by periodic surveys of consumption expenditure conducted by the Central Bureau of Statistics: these indicated that in 2002, for example, the poorest 15% of the population accounted for only 12.8% of total consumption expenditure, while the wealthiest 15% of the population accounted for 59.6% of the total.

ECONOMIC DECLINE AND RECOVERY, 1949–69

At independence, Indonesia inherited an economy that had already been seriously disrupted by three years of Japanese wartime occupation and a four-year armed struggle for independence. The following two decades were characterized by dislocation and decline, as weak international commodity markets combined with domestic political upheavals and economic mismanagement to prevent the effective rehabilitation of the economy. This culminated in the virtual collapse of the economy by the mid-1960s.

Following the political crisis of 1965–66, President Suharto assumed office and adopted economic stability and development as the principal objectives of his 'New Order' Government. A comprehensive programme for the restoration of economic stability and growth was prepared with the assistance of the IMF, and significant measures were adopted to restore the economy's external balance. The rupiah was devalued, a relatively liberal Foreign Investment Law was introduced in 1967, and a debt-rescheduling agreement was reached in the same year with Indonesia's foreign creditors. The Western creditors, who had combined to form the Inter-Governmental Group for Indonesia (IGGI) also agreed to resume aid payments to Indonesia, which facilitated the rapid stabilization of the economy in the late 1960s. By 1969 it had become possible for Indonesia's economic policy-makers to move towards the more ambitious goal of generating economic growth and development.

ECONOMIC PLANNING AND GROWTH, 1969–97

This goal was pursued through a series of five-year Development Plans, known by the acronym Repelita, the first of which was initiated on 1 April 1969 at the beginning of the 1969/70 fiscal year. These Plans did not constitute comprehensive proposals for thoroughgoing state-controlled change with fixed sectoral targets; rather, they represented guidelines for public-sector investment projects, and also sought to encourage and guide private investment. Public expenditure targets under these Plans were generally quite ambitious, and their implementation required considerable quantities of external financial resources, although this dependence on foreign aid declined gradually. Five such plans had been implemented in full by the end of the 1993/94 fiscal year on 31 March 1994. The sixth plan was disrupted by the onset of the Asian financial crisis of 1997–98, and the ensuing economic and political upheavals, which resulted in a breakdown of the planning system.

Repelita I

The formulation of Repelita I (1969/70–73/74) by the National Planning Development Board (Bappenas) established the pattern for all subsequent development planning in Indonesia. The Plan involved expenditure of US $2,000m. and aimed to achieve an average real annual GDP growth rate of 4.7%. Its main emphasis was on agricultural and infrastructural development. With only a few exceptions, such as rice production, most of the Plan targets were achieved or surpassed, with a real annual GDP growth rate of 8.6% being recorded.

Repelita II

Repelita II (1974/75–78/79) granted a high priority to the generation of employment and the even distribution of economic development. Some 28% of the Government's domestically funded capital expenditure was thus committed to regional and social development. Agriculture, and especially the production of rice and other food crops, continued to receive priority, as did infrastructural development. Supported by earnings from petroleum and liquefied natural gas (LNG), which was produced and exported on a large scale from 1978 onwards, an impressive average annual real GDP growth rate of 7.7% was achieved during this period.

Repelita III

Repelita III (1979/80–83/84) retained many of the goals of the two previous five-year Plans, including the achievement of rice self-sufficiency, a more equal distribution of the fruits of development, and increased private-sector participation in industrial development. The Plan also envisaged an increase in the production of non-oil and non-gas traded goods, such as agricultural commodities, forestry products, non-petroleum minerals and, above all, labour-intensive manufactures. The implementation of the Plan was disrupted by persistent volatility in world petroleum prices, which prompted the Government to devalue the rupiah sharply in November 1978, shortly before the Plan came into effect, and in March 1983. The latter devaluation was accompanied by a rephasing of 47 large capital- and import-intensive development projects, involving a projected investment of US $21,000m. Despite these difficulties, real GDP growth over the Plan period averaged 5.7% per year.

Repelita IV

Repelita IV (1984/85–88/89) had three major objectives: the absorption of 9.3m. new workers; the growth of high-technology industries; and the promotion of non-oil and non-gas exports, to help to achieve a real annual average GDP growth rate of 5%. In order to promote private investment, an extensive liberalization of the economy was effected. As a result of these measures, the value of non-oil and non-gas exports exceeded the combined value of oil and natural gas exports for the first time in more than a decade in 1987, and continued to expand further in 1988 and 1989. In overall terms, the economy achieved a real average annual growth rate of 5.2% over the Plan period.

Repelita V

Repelita V (1989/90–93/94) established an annual GDP growth target of 5% in real terms, in order to enable the Indonesian economy to absorb the anticipated annual increase of 3% in the labour force between 1988 and 1993. A particularly high target of 10% average annual growth was set for the non-oil and non-gas manufacturing sector, which was expected to become the main source of employment and export growth during the Plan period. About 55% of the total investment of Rp. 239,000,000m. projected by the Plan was to be provided by the private sector, with public funding being aimed primarily at the development of Indonesia's infrastructure and human resources. Many of the Plan's targets were achieved or exceeded and the economy recorded a real average growth rate of about 8.3% per year. The manufacturing sector grew by an average 10.2% per year, with its non-hydrocarbon component expanding by 11.6% per year. This resulted in manufacturing superseding agriculture for the first time in 1991, and becoming the most important sector in the Indonesian economy.

Repelita VI

The introduction of Repelita VI (1994/95–98/99) on 1 April 1994 coincided with the inauguration of the projected 25-year Second Long-Term Development Period, known by its Indonesian acronym as PJPT II, of which Repelita VI represented the first phase. PJPT II provided for a real annual average GDP growth rate of more than 7%, which was to be driven by a rapid growth in manufacturing industry, and was intended to lead to Indonesia's emergence as a modern industrialized economy. This was projected to result in a fourfold increase in real per caput incomes and an increase in the share of manufacturing industry in GDP to more than 32.5% by the end of PJPT II.

The targets for Repelita VI itself were more modest, and called for an annual average GDP growth rate of 6.2% in real terms. The most rapid growth, of 9.4% per year, was expected to be recorded by the manufacturing sector, with its non-oil and non-gas component projected to expand by 10.3% per year. The achievement of these targets was expected to require a total investment of Rp. 660,000,000m., of which the private sector was expected to account for 73.4%. Some 95% of the total planned investment was to be generated within Indonesia itself. As in Repelita V, investment in the productive sectors was left mainly to private entrepreneurs, allowing the Government to concentrate its development expenditure in such fields as infrastructure and human resources.

Official statistics indicated that Indonesia's economic performance significantly exceeded the Government's targets during the first three years of Repelita VI. GDP was officially estimated to have grown, in real terms, by an annual average of almost 8% in 1994–96, with manufacturing industry expanding by an average of some 11.6% per annum. The utilities, construction and financial services industries also recorded double-digit growth in these years. With the population estimated to be increasing at less than 2% per year, this implied a substantial annual increase in per caput income levels. A reversal of this favourable trend began in mid-1997, however, as Indonesia succumbed to the Asian economic crisis. From 8.5% in the first quarter of the year, the quarterly rate of real GDP growth declined steadily to a mere 1.3% in the fourth quarter, causing the annual rate to average only 4.7% in 1997. This deterioration in Indonesia's economic performance continued into 1998, when GDP contracted by 13.1%.

INDONESIA AND THE ASIAN FINANCIAL CRISIS, 1997–98

Indonesia was affected particularly seriously by the Asian financial crisis, which began in mid-1997. In response to mounting speculative attacks on the rupiah, the Indonesian authorities were forced to float the currency on 14 August, causing it to depreciate from the previously-established trading range of Rp. 2,378–2,682 per US dollar to Rp. 2,830 by the end of the day. This effective devaluation had a serious impact on Indonesia's corporate and banking sector, which carried substantial volumes of unsecured short-term offshore loans that became much more difficult to service from their largely rupiah-based earnings.

During the weeks that followed the Government introduced a wide-ranging set of policy measures to restore stability and boost market confidence. However, the threat to the international solvency of Indonesia's banks and businesses caused by the depreciation of the rupiah continued to undermine market sentiment. Following a renewed assault on the rupiah in late September by both the international markets and domestic companies seeking dollars to pay for or secure their foreign debts, the Government turned to the IMF for assistance in October. A US $23,000m. rescue programme was subsequently announced by the IMF, the World Bank and the Asian Development Bank (ADB), which included a $5,000m. commitment of Indonesia's own external assets and was supported by a second tier of bilateral aid commitments of $15,000m. from Japan, Singapore, the USA, Australia and Malaysia.

The programme required the Indonesian Government to implement a wide range of reforms in the areas of fiscal and monetary policy, the restructuring of the financial sector, and the deregulation of the economy. The enforcement of these conditions proved extremely difficult, however, as they frequently conflicted with the interests of the politically influential conglomerates owned by President Suharto's relatives and associates, which dominated the Indonesian business sector. Increasingly frequent cases of the Government reneging on its reform commitments provoked a further loss of confidence in international financial markets, which were already under pressure as the continued weakness of the rupiah threatened to push many of Indonesia's corporations to the brink of insolvency. As the financial crisis began to jeopardize the real economy, causing inflation and unemployment to increase, political and social tensions also began to rise, most often manifesting themselves in riots and attacks on the economically powerful ethnic Chinese business community, which served further to erode public confidence.

This confidence was damaged further over the following months by the Government's adoption of a controversial proposal to peg the rupiah to the US dollar by means of a currency board mechanism. It was jolted anew by the outcome of the presidential election of March 1998, which resulted in the appointment of the long-standing Minister of State for Research and Technology, B. J. Habibie, who was known to have championed a number of controversial high-cost development programmes, as Vice-President, and the inclusion of several close business associates of President Suharto, including his daughter Siti Hardiyanti Rukmana, a controversial businesswoman, in the new Cabinet. These events also undermined the Government's relations with Indonesia's major donors, culminating in the suspension of the IMF support programme on 5 March. Although the conclusion of a new agreement with the IMF in April 1998 and modest progress on the increasingly pressing issues of banking reform and debt rescheduling appeared to offer some hope of an end to the crisis, the markets remained wary and reports soon began to emerge of renewed government regression in the reform process. In May, partly to assuage these concerns, the Government announced a sharp cut in subsidies on fuel, energy and public transport, which resulted in a dramatic increase in the prices of these goods and services, shortly before the governing body of the IMF was scheduled to decide on the disbursement of the next tranche of its financial support programme. These price increases precipitated widespread rioting and an intensification of public protest by the country's students, which eventually led to the abdication of President Suharto on 21 May, after 32 years in power, and his succession by Vice-President Habibie. The change in leadership prompted a renewed moratorium on official aid payments, as the IMF and other multilateral and bilateral donors sought to assess the implications of Habibie's accession to the presidency; this provided a further impetus to business confidence and economic activity.

In view of this severe economic downturn and the attendant social and humanitarian threats, both the IMF and the World Bank reversed their earlier hard-line stance in the latter half of 1998 and began to encourage the Indonesian Government to pursue more expansionary fiscal and economic policies. Following a further revision of the IMF rescue programme in June 1998, a new letter of intent was signed on 20 October between the IMF and the Government, which provided, *inter alia*, for a strengthening of the latter's employment-creation and food-distribution programmes. These developments, combined with intensified efforts to restructure the domestic banking sector and to resolve the corporate offshore debt problem, helped to arrest the economic downturn in the third quarter of 1998.

Notwithstanding the incipient improvement in the economic situation by the end of 1998, the damage caused to the economy as a cumulative effect of the events after mid-1997 proved to be extremely serious. The principal economic indicators for 1998 showed that GDP had declined by 13.2% and inflation had surged to 58.4%, while unemployment had risen to almost 15% of the total work-force of some 90m. At the same time, Indonesia's international trade had stalled owing to lack of access to finance, causing a shift in the balance of the current account from a traditional deficit of some US $4,900m. in 1997 to a surplus of $4,100m., owing to a decline in the value of imports, from $46,200m. to $31,900m.

FALTERING RECOVERY, 1999–2002

The incipient recovery of the latter half of 1998 continued to strengthen during 1999. It was supported by the comparatively peaceful nature of the legislative election held in June (with the benchmark Jakarta stock index rising by 12% to a 23-month high on the day after the election) and the election of Abdurrahman Wahid to the presidency in October, both of which events generated hopes of a comprehensive break with the past and a greater commitment to economic reform. Official data thus indicated a modest expansion of GDP by 0.8% in real terms in 1999, with inflation easing dramatically to 20.5%, interest rates declining to about 20%–25%, and both the rupiah and the stock market gaining ground.

However, the sustainability of this recovery remained fraught with a number of problems. The banking sector continued to be burdened with high levels of bad debt, estimated in some cases to account for as much as 75%–85% of total loans, and was further undermined in mid-1999 by revelations of a major political scandal implicating one of the country's leading banks, Bank Bali. Meanwhile, the corporate sector continued to be constrained by its high level of offshore debt, officially estimated at US $65,600m. in mid-1999. Unemployment remained high, and at mid-1999 some 25% of the population were reported to be living in absolute poverty. Furthermore, although the current account of the balance of payments remained in surplus, this principally reflected a high degree of import compression, which itself threatened to restrain the short-term growth prospects of the economy. In addition, shifts in the relative production costs and prices of a wide range of products caused by the regional economic crisis significantly altered the competitiveness structures of various Indonesian industries, and gave rise to the need for extensive inter- and intra-sectoral restructuring.

Despite these lingering problems, the economy recorded a relatively promising performance in 2000, with GDP expanding by 4.9% in real terms, on the basis of a strong growth in exports and a recovery in domestic demand. The rate of inflation also slowed to a mere 3.7%, allowing the monetary authorities to ease interest rates to an average of some 13.5% in an effort to support a continued acceleration of GDP growth. However, these favourable trends were increasingly undermined by a steady deterioration in the political situation, which manifested itself in numerous scandals within the Government, infighting within the political élite in Jakarta, increasing separatist and sectarian violence in several of Indonesia's outlying provinces, and rising crime levels throughout the country. These political difficulties also hampered the Government's ability to meet the deadlines set by the IMF for its economic reform programme, which resulted in a renewed moratorium on the disbursement of IMF funds in December 2000.

During the first half of 2001 the Indonesian economy's capacity to sustain its improved performance of the previous year had become increasingly weak, as political stability was damaged by parliamentary efforts to impeach President Abdurrahman Wahid for alleged corruption and incompetence, while Indonesia's access to foreign financial support continued to be blocked by the IMF's refusal to release payments. The resulting erosion of business confidence led to a sharp downturn in the capital and currency markets, with the exchange rate of the rupiah falling from an average of some Rp. 8,400 to the US dollar in 2000 to more than Rp. 11,000 from April 2001 onwards. By mid-2001 this dramatic deterioration in Indonesia's financial markets had begun to jeopardize the country's prospects for continued economic recovery, as the depreciation of the rupiah significantly raised the costs to both the Government and the corporate sector of servicing their substantial external debts.

The inauguration of Megawati Sukarnoputri as the country's President in August 2001 resulted in the restoration of a degree of political stability. The new Government submitted a revised letter of intent to the IMF, which enabled the release of previously unavailable loans and led to hopes of a revival of the faltering economic recovery. The rehabilitation of the economy was interrupted, however, by the global economic uncertainty precipitated by the terrorist attacks on the USA on 11 September 2001. The real rate of GDP growth thus declined to 3.5% in 2001, with inflation rising to 11.5% and causing a renewed, if modest, increase in Bank Indonesia's discount rate to 14.5% by the end of the year. The year-end exchange rate subsequently stood at Rp. 10,400 to the US dollar.

The first three quarters of 2002 witnessed an improvement in Indonesia's economic performance, largely in response to the increased political stability and the resulting growth in market confidence as the Government continued to implement its agreement with the IMF approximately on schedule. This improvement was marked by a strengthening of the rupiah, a significant rise in the composite index of the Jakarta Stock Exchange, the easing of inflationary pressures, and an acceleration in the rate of economic growth. However, these favourable trends were arrested by the terrorist attacks on the island of Bali in October 2002, which led to a sharp decline not only in tourist arrivals but in business confidence as a whole. The full-year data for 2002 consequently showed only a modest increase in the rate of GDP growth, to 3.7%, and a slight rise in inflation, to 11.9%, although the rupiah appreciated significantly, to a rate of Rp. 8,940 to the US dollar by the end of the year.

CONSOLIDATION OF RECOVERY, 2003–04

Economic developments in the first half of 2003 were overshadowed by the negative impact of the Bali bombings of October 2002, the outbreak of the epidemic of Severe Acute Respiratory Syndrome (SARS) in East and South-East Asia, and the renewed violence in Aceh, which combined to weaken business confidence. These concerns were heightened by fears of further terrorist attacks in response to the US-led invasion of Iraq and the legal proceedings being pursued in Indonesia against the alleged perpetrators of the Bali bombings. This resulted in a sharp slowdown in economic activity, with GDP expanding by a mere 3.4% in the first quarter of 2003, compared with growth rates of approximately 4% in each of the preceding three quarters, and the rate of inflation slowing to only 0.2% in May 2003, compared with 6.9% in May 2002. As the year proceeded, however, the economic environment experienced a gradual improvement, driven by the persistence of broad political stability—which was only disrupted by a bomb attack at the prestigious Marriott Hotel in Jakarta in July—good macroeconomic management, and a number of favourable external developments. The initial slowdown in consumer demand, and the resulting decrease in inflation, permitted a reduction in interest rates, which was reinforced by substantial inflows of capital, comprising both a repatriation of capital that had been transferred off shore during the crisis and, to a lesser extent, new investments. This resulted in the average lending rate falling to below 15.5% in the final quarter of the year, from approximately 18.5% in the corresponding quarter of 2002. Inflationary pressures were eased further by a continued appreciation of the rupiah, to a rate of Rp. 8,577 to the US dollar by the end of the year, which reduced import costs. This development, combined with a significant strengthening in the world price of most of Indonesia's leading export commodities, also helped to encourage economic activity, with GDP growing by 4.1% in real terms.

In order to signal that it had finally recovered from the crisis of 1997–98 and was resuming full control of the economic management of the country, the Indonesian Government allowed the IMF-supported reform programme that had guided economic policy for most of the past six years to expire at the end of 2003. In order to reassure investors and donors about its continued commitment to the economic reform process, however, in September 2003 the Government issued a White Paper on Economic Policy containing a list of 57 reform objectives, and also agreed to sign a Post-Programme Monitoring (PPM) arrangement with the IMF, under which the Fund would continue to offer policy advice but would no longer be able to set binding conditions. Despite some initial fears to the contrary, this assertion of economic independence has not prompted a significant decline in business confidence, with most observers continuing to award the Indonesian Government high praise for its economic policy measures in the first half of 2004. The resulting strength in market sentiment was underlined in January by a surge in the Jakarta Stock Exchange index to above the pre-crisis high point. Despite concerns that the political uncertainties associated with the legislative and presidential elections scheduled for the middle of that year might undermine these positive trends, or that the rapid pace of economic recovery and continued increase in world commodity prices might precipitate increased inflation, macroeconomic stability was largely maintained and business confidence remained strong during the first half of 2004.

AGRICULTURE

Until it was superseded by manufacturing in 1991, agriculture was the predominant sector of the Indonesian economy in terms of output, although it still accounted for almost 20% of GDP in that year. It remains the most important sector in terms of employment, however, engaging 44.3% of the working population in 2002. The agricultural sector is divided into five subsectors—food crops, cash crops, animal husbandry, fishing, and forestry—of which the food-crop subdivision is the most important in terms of income and employment generation. The other subdivisions play a significant role as earners of non-oil and non-gas export revenues, and were estimated to account for almost 25% of such earnings in 2003, in both unprocessed and processed forms.

Food Crops

The predominant food crops are rice and other 'secondary' (*palawija*) food staples such as maize, cassava, sweet potatoes and soybeans. Food crops form the most important component of the agricultural sector, accounting for approximately 53.0% of annual agricultural GDP in 1998–2003. These crops are usually cultivated on relatively small land-holdings (seldom exceeding 1 ha in size) and under highly complex traditional tenurial relationships, which have historically ensured an adequate level of subsistence for all members of the rural community.

Rice is by far the most important food crop, and is also the preferred staple of the vast majority of Indonesians. Successive governments in the first four decades after independence therefore viewed the achievement and maintenance of rice self-sufficiency as a principal policy objective, and launched several programmes aimed at encouraging farmers to apply productivity-enhancing cultivation techniques and the increased use of purchased inputs such as higher-yielding seeds, chemical fertilizers and pesticides. The impact of these programmes was dramatic, and was enhanced by substantial investment in irrigation infrastructure during the 1970s, which permitted an expansion of cultivated area and multiple cropping. Between 1969 and 1984 gross annual production of milled rice increased at an average annual rate of 5.1%, from 12.3m. metric tons to 25.9m. tons, while the harvested area increased by almost 22%, from 8m. ha to approximately 9.8m. ha. As a result of these developments, the long-cherished goal of self-sufficiency in rice was reached by 1984/85.

After 1985 the growth of rice production became much more erratic, with sharp increases of output in 1992 and 1995 being interspersed with periods of stagnation or contraction. This was due in part to unfavourable weather conditions and pest infestations, and in part to shifts in government policy, which aimed to match output growth with the growth in demand. Official production data suggest that the steady recovery in production of milled rice, from 26.3m. metric tons in 1994 to 28.8m. tons in 1996, was reversed in 1997, when production fell to 27.7m. tons, owing to the protracted droughts caused by the climatic phenomenon known as El Niño. A further small decline was recorded in 1998, largely as a result of continued adverse weather conditions and the inability of farmers to pay the sharply increased prices of imported fertilizers and other production inputs. Improved climatic and economic conditions permitted a recovery in rice production to 28.4m. tons in 1999 and 29.0m. tons in 2000, temporarily allaying fears about Indonesia's food security, which had been heightened in the aftermath of the economic crisis of 1997–98 as a consequence of serious concerns regarding the country's ability to cover the gap in domestic production through imports. These fears subsequently re-emerged, however, following a 4.5% decline in production to 27.7m. tons in 2001 and only a modest recovery, to 28.2m. tons, in 2002. In response, in 2002 the Government proposed a number of measures intended to provide domestic producers with incentives to raise their production levels and to enhance their capacity to compete with foreign suppliers. These measures included: a 15% increase in the minimum price guaranteed to local farmers; an increase in the fertilizer subsidy designed to reduce the price paid by farmers for urea by 15%–20%; and, more controversially, an increase in the tariff on imported rice from Rp. 430 per kg to Rp. 753 per kg. These attempts to support domestic producers were reinforced in 2003 by the introduction of a new policy banning rice imports in the months of the main rice harvest, thereby supporting domestic prices during this period of glut. Partly as a result of these measures, output was estimated to have risen to 28.7m. tons in 2003.

In contrast to the emphasis placed on the intensification of rice production, relatively little effort was made to expand production of the secondary food staples. Consequently, output of these crops remained virtually stagnant during the 1970s, and the production increases that did occur were due to an expansion in planted area rather than to improved yields, which remained low by international standards. Considerable production increases were, however, achieved in the 1980s and early 1990s, with output of maize rising from 3.6m. metric tons in 1980 to 8.0m. tons in 1992, while annual output of cassava increased from 13.8m. tons to 16.5m. tons, and that of soybeans from 680,000 tons to 1.9m. tons, during the same period. Meanwhile, production of groundnuts increased from 424,000 tons to 739,000 tons. The output of most of these products, except cassava, declined sharply in 1993, owing to adverse weather conditions, but recovered, albeit erratically, in 1994–96. Official data for 1996 showed that the output of maize had risen to the unprecedented level of 9.3m. tons, while that of cassava had reached a record high of 17.0m. tons. The output of sweet potatoes, groundnuts and soybeans remained close to the average of the previous five years, however, at 2.0m. tons, 738,000 tons and 1.5m. tons, respectively. The output of all of these crops declined significantly in 1997 as a result of the droughts induced by El Niño, with only a modest and partial recovery being recorded in 1998–2000, followed by further declines in 2001 as climatic and economic conditions deteriorated again. There was, however, a steady recovery in the output of maize, groundnuts and sweet potatoes in 2002–03, with provisional data for the latter year indicating production levels of 10.9m. tons, 785,000 tons and almost 2.0m. tons, respectively. Meanwhile, production of soybeans declined to an estimated 672,000 tons in 2003 from almost 827,000 tons in 2001, while the output of cassava decreased from 17.1m. tons in 2001 to 16.9m. tons in 2002, before recovering to an estimated 18.5m. tons in 2003.

Cash Crops

Indonesia is an important producer of a wide range of cash crops, including rubber, palm oil, copra, coffee, tea, cocoa, sugar and tobacco. Most of these commodities are grown on commercial plantations as well as smallholdings, and are destined primarily for export. The growing importance of petroleum export earnings in the 1970s caused a relative decline in the contribution to the economy of agricultural exports: their share of total export value fell from 70% in 1969 to less than 9% in 1983. Increasing anxiety about an impending decline in world petroleum prices at the end of the 1970s prompted a reversal of the neglect suffered by the cash crops sector during the early 1970s. Considerable efforts were therefore made during the 1980s and 1990s to promote the expansion of cash-crop production.

To increase the productivity of existing plantings, 'intensification' programmes were introduced for annual crops grown principally by smallholders, such as sugar and tobacco. These programmes, involving the provision of extension services and subsidized credit and inputs, were accompanied by similar programmes covering smallholder producers of major perennial cash crops, who were encouraged to replant their fields with higher-yielding varieties, and to employ yield-increasing cultivation techniques. The development of new acreage was also encouraged, *inter alia*, through a 'nucleus estate and smallholder' programme, providing for the establishment of smallholdings supported by commercial plantations.

Production of most major agricultural commodities expanded considerably during the 1980s and 1990s. National income statistics show that the combined real value (in constant prices) of 'farm non-food crops' and 'estate crops' more than doubled between the late 1970s and the mid-1990s, and continued to rise in the latter half of the decade. There has been considerable variation between crops, however, largely in response to market conditions and to the degree of government support granted to particular commodities. These factors have significantly affected the performance of the principal agricultural export commodities.

Owing to previous neglect of the industry, rubber production increased only sporadically during the 1970s, while efforts were made to rehabilitate the industry. These efforts resulted in rapid increases in production from the late 1970s onwards. This growth trend was cut short, however, by a sharp decline in international rubber prices in 1982, which reduced tappers' incentives; output declined to only 900,000 metric tons in 1982, compared with 1.05m. tons in the previous year. By 1985 production had recovered to approximately 1981 levels (despite continuing weak prices), as more of the area planted and replanted during the late 1970s became productive. Production subsequently increased steadily to an officially estimated figure of 1.57m. tons in 1996, before declining marginally to around 1.55m. tons in 1997 as a result of the regional economic crisis and adverse weather conditions. Thereafter, Indonesia's production of rubber rose steadily to an estimated 1.75m. tons in 2000, a level that was exceeded only by that of Thailand. Since then, however, Indonesia's rubber production has declined significantly, to approximately 1.6m. tons per annum in 2001–03. This decline was, in part, a consequence of a regional agreement signed in 2001 by Thailand, Indonesia and Malaysia, the three largest producers of natural rubber in the world, to limit rubber production and exports from 2002 onwards in an effort to boost international rubber prices. This was followed in October 2003 by the establishment by the three countries of a trading cartel known as the International Rubber Consortium Ltd (IRCo), in order to reinforce this effort.

The production of palm oil received considerable encouragement from the Government from 1970, and annual output increased from 216,500 metric tons in that year to almost 8.2m. tons in 2002. Similarly, efforts were also made to promote the growth of coconut oil production, particularly from the mid-1980s. Although palm oil was intended primarily for export, shipments have occasionally been restricted in times of poor harvest in order to ensure adequate supplies of vegetable oil in the domestic market. An attempt to restrain the rising domestic price of cooking oil was made in January 1998, when the Government introduced an indefinite ban on exports of palm oil. This measure contradicted the open-markets policies pursued by the IMF, however, and an agreement subsequently reached in April of that year specifically provided for the ban to be replaced by a 40% export tax later in the same month, which would itself be gradually phased out as circumstances permitted. As a result of these measures, exports of palm oil declined to less than 1.5m. tons in 1998 from almost 2.9m. tons in 1997, before increasing steadily to an estimated 6.4m. tons in 2003.

Indonesia is one of the world's largest producers of coffee, and output of its mainly *robusta* varieties increased steadily, from 186,300 metric tons in 1970 to more than 459,000 tons in 1996. Although output declined to 428,000 tons in 1997, the subsequent years to 2002 witnessed a steady recovery to almost 569,000 tons. Until 1989 Indonesia's exports were subject to stringent quotas imposed by the International Coffee Organization (ICO), which often amounted to less than one-half of total output. The suspension of the ICO's export quotas in July 1989 permitted Indonesia to increase the volume of its coffee exports by almost 18%, from 352,300 tons in 1989 to 414,900 tons in 1990, which enabled it to recover some of the losses incurred as a result of the ensuing fall in prices. Over the following years Indonesia's exports of coffee fluctuated sharply as a result of changing conditions in domestic and international markets. Between 1993 and 2003 the volume of exports ranged from a high of 363,000 tons in 1996 to a low of 254,800 tons in 2001.

The production of cocoa, which was first planted on a large scale in the mid-1970s, has gained dramatically in importance in Indonesia, which has now become one of the world's largest producers. Output was estimated at 207,150 metric tons in 1992 and increased steadily to 374,000 tons in 1996, before declining to some 330,000 tons in 1997 as a result of El Niño and the economic crisis. Output growth was resumed in 1998, however, and output was estimated to have reached 432,000 tons by 2002. The crop has also become an increasingly important export commodity, with the volume of cocoa-bean exports having risen from a mere 6,800 tons in 1981 to 367,500 tons in 2002, before decreasing to an estimated 266,300 tons in 2003 owing to declining global prices, increased competition from Viet Nam and a rise in domestic processing.

Animal Husbandry and Fishing

Serious efforts have been made to promote the growth of the animal husbandry sector in recent decades, *inter alia* through the import of large numbers of breeding animals, including pigs, dairy and beef cattle, and poultry, in the late 1980s and early 1990s. All the main products of this sector have consequently recorded impressive growth since the early 1970s, with annual meat production increasing from 314,000 metric tons to 1.58m. tons, egg production from 59,000 tons to almost 909,000 tons, and milk from 29m. litres to approximately 521m. litres during the period 1970–2002. Despite this growth, animal husbandry remains relatively underdeveloped in Indonesia, contributing only about 2.0% of GDP in 2003. The sector was also seriously affected by the economic crisis in 1997–98, and in particular by the depreciation in the value of the rupiah, which raised the cost of imported animal feed and affected the profitability of the commercial animal husbandry enterprises. This resulted in the culling of substantial numbers of dairy animals and poultry during the crisis and the immediate post-crisis period. Extensive additional culling of poultry was also necessitated in the last quarter of 2003 and the first quarter of 2004 by the epidemic of avian influenza that spread through the region during that period.

In 1982 international recognition of the concept of the archipelagic state permitted Indonesia to declare the waters separating its many islands to be an exclusive economic zone (EEZ), giving the country undisputed control over the vast marine fisheries resources of this sea area. These resources began to be developed in 1987, both through the issuing of licences to foreign fishing fleets, and the encouragement of private (including foreign) investment in the fishing industry; particular emphasis was given to shrimp and tuna fisheries. The sea-fishing industry remains beset by the problem of widespread illegal fishing, however, owing to the Government's inability to control the activities of domestic and foreign fishing vessels in the country's extensive waters. Meanwhile, efforts have also been made to enhance the potential of Indonesia's freshwater fisheries, which provide the bulk of animal protein consumed in Indonesia, through the promotion of freshwater shrimp hatcheries and a nucleus fishpond scheme, similar to the expansion programme for cash crops. The shrimp hatcheries, which operated predominantly on a commercial basis, suffered serious set-backs as a result of the economic crisis.

Forestry

With the latest available data indicating that 110m. ha, or approximately 57% of its land area, was covered by forest in 1995, Indonesia has some of the most extensive concentrations of tropical hardwoods in the world. These have been exploited at a rapid rate since the late 1960s, resulting in an extensive degree of deforestation over the subsequent three decades. Although no firm statistics are available, it has been estimated that some 2.1m. ha of forest cover is lost in Indonesia every year. Having attracted a high level of foreign investment into its forestry industry during the 1970s, Indonesia's annual output of industrial logs increased from only 4m. cu m in 1963 to an average of 22.2m. cu m between 1974 and 1980, remaining at about this level for much of the following two decades. As the bulk of this production was destined for export, timber became the third most important export commodity, after petroleum and natural gas, during this period.

Following a government decision in 1980 to increase the content of domestic value added in its forestry exports, the export of raw logs was gradually reduced, and in 1985 it was prohibited altogether until June 1992, when the ban was replaced by a high export tax. The promotion of local processing resulted in a rapid increase in plywood production, which rose from about 1m. cu m in 1980 to a peak of 9.9m. cu m in the year to March 1994. After remaining in excess of 9m. cu m during the following two years, however, output declined sharply as the economic crisis drove many producers out of business, and by 2002 production had fallen to approximately 1.2m. cu m. The production of sawn timber also expanded rapidly, reaching 10.9m. cu m in 1989/90, and thereafter fluctuated within the 10m.–12m. cu m range annually until 1997–98, before declining to less than 415,000 cu m in 2002. The remainder of the domestic wood-processing industry is devoted to the manufacture of veneers, and smaller quantities of furniture and wooden handicrafts.

Despite the continued exploitation of Indonesia's forestry resources, the Government reacted with concern to the environmental implications of indiscriminate logging practices, which characterized the industry during its early years of growth. From the early 1980s the Government increasingly required logging companies to introduce selective cutting policies, and in 1985 the practice of total tree felling was banned. Emphasis was also placed on reafforestation, and logging companies were given the choice of either providing reafforestation deposits, or of carrying out compulsory reafforestation themselves. In 2000 the Government reimposed a total ban on the export of raw logs. These efforts to protect the country's forestry resources have been only partially successful, however, as witnessed by widespread forest fires in Sumatra and Kalimantan that have become an almost regular feature since the mid-1990s; these frequently generate huge clouds of smoke, which are carried over much of South-East Asia. Many of these fires are set deliberately, both by subsistence farmers practising traditional 'slash-and-burn' methods of cultivation and by large-scale plantation owners seeking to clear the forest for new cash-crop estates. In addition, illegal logging and the smuggling of timber remain serious problems. In an effort to overcome these issues, the Government has supplemented its domestic enforcement efforts by signing international agreements with the European Union (EU), Malaysia, Japan and the People's Republic of China to ban the import of illegal logs from Indonesia.

In addition to tropical rain forests, Indonesia also has sizeable teak plantations. Other significant forestry products are rattan, resins and copal. In order to promote the domestic rattan-processing industry, the Government banned all exports of raw rattan (with effect from January 1987) and of semi-processed rattan products (from July 1988). As with logs, however, this ban was replaced by an export tax in June 1992.

MINING

Petroleum and Natural Gas

The petroleum and natural gas industries have played a crucial role in providing Indonesia with resources to fund its developmental programmes. Owing to the heavy dependence of the economy on such industries for both its foreign exchange and fiscal revenues, the country's prosperity has been closely linked to fluctuations in international petroleum markets. Even after several years of efforts to promote alternative sectors of the economy and the collapse in petroleum and gas revenues, following the international decline in petroleum prices in 1986, the petroleum and natural gas industries have continued to make a disproportionately large contribution to government resources.

According to the Indonesian Constitution, sovereign ownership of the country's mineral resources is vested collectively in the Indonesian people, on whose behalf the State (currently through the Ministry of Energy and Mineral Resources) administers all aspects of their exploitation. In the case of petroleum and natural gas, this function was historically delegated to state-owned oil companies, of which the latest, Pertamina, was formed in 1968 by the merger of two existing companies. This constitutional provision resulted in a revocation, in 1960, of the concessions that had earlier been granted to foreign oil companies, but, as a result of the technical and financial constraints facing state-owned corporations, foreign oil companies were permitted to continue operating in Indonesia as contractors of state companies under several forms of agreement, the most common of which being the production-sharing contract. Pertamina's control over the hydrocarbons industry was significantly weakened following the promulgation of a new oil and gas law in October 2001. This law provided for a separation of the upstream and downstream activities formerly carried out by Pertamina, with the former (i.e. the regulatory functions and the management of the production-sharing contracts) being assigned to a new government agency known as BP Migas and the latter (i.e. the distribution and marketing of oil and fuel products) to another new agency, BPH Migas, both of which were established in mid-2002. In addition, Pertamina's monopoly over the downstream sector was ended, allowing other companies to compete in the retail market. These changes implied a dramatic transformation of the oil and gas sector, although not all of the regulations required to implement them had been put in place by mid-2004.

Foreign investment in the Indonesian petroleum industry has been actively encouraged since 1966, and 346 production-sharing and other contracts had been signed by the end of 2003, of which about 161 have subsequently lapsed. Heavy investment, encouraged by the oil price rises of the 1970s, resulted in a rapid increase in output, from 189m. barrels in 1967 to a record 615m. barrels in 1977. Thereafter, fluctuating international supply and demand caused production levels gradually to be reduced to an average of 585m. barrels per year during the period 1978–81. The Organization of the Petroleum Exporting Countries (OPEC) subsequently attempted to maintain prices through the imposition of output restraints, which led to a further decline in Indonesia's production, with annual output levels fluctuating between 484m. barrels and 588m. barrels between 1982 and 2001. A sharp decline to 458.5m. barrels was recorded in 2002, owing largely to the maturing of existing oilfields and a decline in exploration and investment during the crisis years. All available indicators suggested that this decline continued in 2003–04, with the volume of imports of oil exceeding that of exports in March 2004 for the first time in living memory.

With proven and probable reserves of some 9,700m. barrels of oil, Indonesia is well placed to recover its position as a net exporter. The Government has recognized that this will require a considerable volume of new investment, however, and in 2004 was therefore planning to open 50 new concessions over the next five years. Fifteen new production-sharing contracts were signed in 2003, which were expected to result in a 55% increase in expenditure on exploration and production, to US $7,000m.

In addition to its petroleum resources, Indonesia had known reserves of natural gas amounting to 176,600,000m. cu ft (equivalent to approximately 31,000m. barrels of petroleum) at the end of 2003. More than one-half of these reserves are located offshore, especially near the Natuna Islands in the South China Sea and near the Bird's Head area of Papua. The commercial exploitation of these reserves began in 1977/78, when two major natural gas fields at Arun, in the province of Aceh in northern Sumatra, and Badak, in East Kalimantan, were brought into production, and facilities were established to permit the production of LNG.

The initial development of the LNG industry was linked with the signing of several long-term export contracts to Japan in 1973 and 1981. These contracts were extended until 2011 in mid-1995, and have been supplemented by a number of other sales agreements with Japanese, South Korean and Taiwanese utilities in the intervening period. This growing international demand for LNG, reinforced by the increasing domestic use of natural gas to generate power and to fuel large manufacturing plants, has resulted in a steady expansion of the natural gas production and liquefaction industries. By 1997 Indonesia's production of natural gas and LNG had risen to 3,166,750m. cu ft (corresponding to about 542m. barrels) and 1,402,621,000m. British thermal units (20.6m. tons), respectively, although output of both natural gas and LNG declined modestly in 1998, as a result of the economic crisis. After a recovery in 1999, the production of both natural gas and LNG came under severe threat after 2000 owing to mounting political unrest in Aceh, which caused the operator of the Arun field to shut down production in March 2001. Although this stoppage was only temporary and a phased resumption of production began in July, Indonesia's total output amounted to only 2,803,034m. cu ft of natural gas and 1,257,446m. British thermal units of LNG in 2001. Although output rose to 3,031,028m. cu ft of natural gas and 1,352,878m. British thermal units in 2002, widespread concerns over security issues continued to threaten production in the following years.

With the Arun and Badak fields being gradually depleted, the need to develop new fields had become increasingly urgent in the early 1990s if Indonesia was to be able to meet these commitments. Negotiations on the development of the Natuna field were begun in 1991 between the Indonesian authorities and the US corporation, Exxon, which led to the signing of a formal agreement between the two parties in January 1995. The development of the Natuna field, involving an expected investment of US $35,000m.–$40,000m. over a period of 30 years, proceeded rapidly, and commercial production began in January 2001 with the inauguration of a 650-km undersea pipeline to Singapore with a capacity of 325m. cu ft/day under a 22-year contract. This was followed in August 2002 with the opening of a 96-km pipeline to Malaysia under a 20-year contract, through which Indonesia's sales were expected to rise from 100m. cu ft per day in 2002 to 250m. cu ft/day by 2007. Meanwhile, a second, 477-km pipeline to Singapore from South Sumatra was opened in August 2003, in connection with a contract to supply Singapore with up to 350m. cu ft/day of natural gas over a 20-year period. In addition, considerable efforts were also being made to develop a number of new offshore fields in Papua. In September 2002 a 25-year contract was signed, under which Indonesia would supply 2.6m. tons of LNG annually from the Tangguh field off Papua to Fujian Province in China from 2007 onwards.

Indonesia is also seeking to increase production and exports of liquefied petroleum gas (LPG), which until 1987 was produced almost entirely as a by-product of the refining of crude petroleum. A 1986 agreement to supply 2m. metric tons of LPG per year to Japan from 1988 led to the construction of specifically designed LPG facilities at the Arun and Badak natural-gas-processing plants. In response to an earlier contract signed with Japan in 1983, for the delivery of some 450,000 tons of butane and propane per year, a fractionation plant was also established in 1986 on Bintan island, near Singapore. Between 1981 and 1995 the production of LPG increased from 560,000 tons to 3.9m. tons, before decreasing gradually to less than 1.8m. tons in 2002, in response to declining demand and increased external competition in export markets.

Other Minerals

In addition to petroleum and natural gas, Indonesia has significant (but, in many cases, unquantified) reserves of a variety of other minerals, including coal, tin, bauxite, copper, nickel, iron sands, gold and silver. Most of these are localized, and are frequently situated in remote areas, so that their exploitation often involves high costs. The sector grew rapidly between the early 1970s and late 1990s, but was seriously affected by the financial crisis of 1997–98 and by the subsequent security threats and legal uncertainties. In most cases this has led to declines in investment and output in the industry, which appear unlikely to be reversed in the foreseeable future unless the underlying problems can be resolved.

Indonesia has substantial reserves of coal, estimated at 36,300m. metric tons at the end of 1995. In order to restrict the growth of domestic petroleum consumption (especially in the production of electricity), a major effort has been made to expand the use of coal as a source of fuel, thereby increasing coal production from 304,000 tons in 1980 to 114.6m. tons in 2003. This effort has involved the expansion of the main existing coalfields at Bukit Asam, in South Sumatra, and Ombilin, in West Sumatra, the promotion of increased exploration, and the development of new fields. It has been accompanied by a sharp increase in coal exports, from 2.5m. tons in 1989 to almost 85.7m. tons in 2003, making Indonesia the world's third largest exporter of coal after Australia and South Africa. The country appears likely to consolidate this position in the coming years as a result of China's decision in 2003 to halt coal exports, together with continuing strong demand for coal in the world market.

Indonesia is one of the world's leading producers of tin. This is mined primarily by a state-owned company, PT Tambang Timah, which was partially privatized in October 1995, on the islands of Bangka and Belitung, off the eastern coast of Sumatra. The devaluation of the rupiah by 30.7% in relation to the US dollar in September 1986, combined with stringent cost-cutting, enabled PT Tambang Timah to remain competitive in the latter half of the 1980s, despite the collapse in world prices following the 1985–86 tin trading crisis; Indonesian tin production rose steadily to a peak of 31,300 metric tons in 1989. Output fluctuated between 28,200 tons and 30,700 tons in the following years to 1994, largely in response to the imposition of supply controls by the Association of Tin Producing Countries (ATPC) from 1 March 1987. Having been restored to profitability, a proportion of the company's shares were floated in London, United Kingdom, and Jakarta in October 1995. A firm market recovery in the second half of 1995 and a decision by the ATPC countries to abandon their export restraints with effect from 1 July 1996 resulted in a significant improvement in the industry's outlook. This was underlined by a steady increase in production, from 30,610 tons in 1994 to almost 55,175 tons in 1997. Output declined to less than 54,000 tons in 1998, however, largely as a result of the impact of the regional economic crisis, and continued to contract, to 47,753 tons, in 1999, before recovering steadily to 88,142 tons in 2002 and declining to 71,695 tons in 2003.

West Borneo and the Riau archipelago are the main sources of Indonesia's bauxite, which is mined on the islands of Tembeling, Kelong, Dendang and Bintan by another state-owned company, PT Aneka Tambang. Other major hard minerals include nickel, which is mined at several locations in Celebes (Sulawesi), principally by the Canadian-owned Inco at Soroako and PT Aneka Tambang at Pomalaa, and copper, which is mined solely by a US company, PT Freeport Indonesia Inc, in a controversial large-scale mining operation in the province of Papua. Copper deposits have also been found in Sumatra, Borneo, Java and Celebes. A recovery in international prices for most of these metals after mid-1987 prompted significant increases in mining output and mineral exports. By 1992 production of nickel ore had increased to 2.50m. metric tons (gross weight) and, although it declined to 1.98m. tons in 1993, it had risen again to almost 3.43m. tons by 1996, before falling back to approximately 2.83m. tons in 1997. This was followed by a steady recovery, which resulted in an output of just under 4.4m. tons in 2003. Meanwhile, output of copper concentrate increased from approximately 190,000 tons (gross weight) in 1984 to 3.79m. tons in 2002, before declining to 3.24m. tons in 2003.

In the mid-1980s Indonesia initiated a major programme to expand its output of precious metals, and of gold in particular. Accordingly, more than 100 contracts were awarded to private companies to explore for gold during 1985–87. It was hoped that as many as 600 further contracts would be signed in the following years, and that the programme would increase Indonesia's production of gold to 150 metric tons per year by the late 1990s, from 3.3 tons in 1986. Although the implementation of this programme was hindered by a number of infrastructural constraints and the growing problem of illegal mining on concessions offered to private mining companies, output increased substantially to 129.0 tons in 1999 and, after declining to 117.6 tons in 2000, rose sharply to reach 166.1 tons in 2001. More than one-half of this total amount was produced by PT Freeport Indonesia at its copper mine in Papua. The gold sector has not been spared the general malaise that has affected the mining industry in more recent years, however, with output falling to 142.2 tons in 2002 and 141.0 tons in 2003.

MANUFACTURING

During the first two decades after independence, few efforts were made to develop Indonesia's embryonic manufacturing sector. Indeed, many of the relatively unsophisticated industries that did exist fell into decline, as a result not only of the general economic difficulties but also of specific measures such as the expropriation of many foreign-owned enterprises and the replacement of expatriate managements with poorly-trained and ineffectual bureaucracies. By the mid-1960s, therefore, Indonesia's manufacturing sector, which consisted almost entirely of handicrafts and a small textile industry, accounted for less than 10% of GDP.

Following the accession to power of the 'New Order' Government of President Suharto in 1966, however, high priority began to be given to industrial development. While the Government played an active part in this process, through the establishment of numerous state-owned enterprises and through equity participation in large numbers of private enterprises, it also sought to encourage private investors, both domestic and foreign, to contribute to Indonesia's industrialization. This resulted in the rapid expansion of manufacturing industry, so that by the mid-1990s Indonesia had the capacity to produce a wide variety of goods ranging from handicrafts to high-technology aerospace products. In 1991 the manufacturing sector superseded agriculture as the largest contributor to GDP, and by 1997 it had acquired a share of almost 27% of GDP.

The industrialization strategy initially adopted by the Indonesian Government was characterized by the establishment of import-substitution enterprises at all levels of manufacturing, including several in which the country has only a questionable comparative advantage. These enterprises therefore required considerable protection, through tariff and non-tariff barriers, which tended to perpetuate inefficiency. As industrialization became increasingly 'upstream', these inefficiencies were compounded by the obligation of 'downstream' producers to procure inputs from domestic producers of capital and intermediate goods, or from officially licensed importers, who had little cause to reduce their selling prices much below those charged by domestic suppliers. This gave rise to the much-debated 'high-cost economy' and the lack of international competitiveness suffered by much of Indonesia's manufacturing industry, and also stimulated a series of liberalizing reforms aimed at improving the export potential of Indonesia's manufactured products in response to the 1986 depression in world petroleum prices.

These measures proved very successful and a strong growth in manufactured exports took place after 1987. Particularly encouraging was the fact that the increase in traditional exports of semi-processed raw materials was accompanied by a corresponding rise in exports of manufactured end-use products, such as cement, iron and steel products, and motor vehicles.

Of Indonesia's main products, most consumer goods, such as processed foods and beverages, tobacco products, textiles and garments, motor vehicle components and assemblies, and electrical appliances, are produced by the private manufacturing sector, often in joint ventures with foreign companies. In many of the capital and intermediate goods industries, such as chemicals, cement, glass, fertilizers, ceramics, machinery and basic metal products, the Government plays an important role, with the private sector excluded from some industries, such as those producing fertilizers. In addition to more 'formal' industries, the Government has also promoted the establishment of small-scale enterprises, often involving the production of handicrafts and organized on a co-operative basis, as a means of generating employment in rural areas.

One of the major objectives of Indonesia's industrialization policy has been to add value to domestically produced raw materials. Consequently, a variety of processing industries has been set up, ranging from relatively small-scale units for the processing of agricultural commodities and forestry products (including food-processing enterprises, crumb rubber factories, processing plants for crude palm oil, plywood mills and sawmills and establishments for the manufacture of rattan products) to several large-scale petrochemical plants located in various parts of Indonesia, as well as processing units for Indonesia's hard minerals, of which the aluminium smelter on the Asahan river in North Sumatra is the most significant.

Another important goal has been the installation of a domestic manufacturing capacity for industries defined as having strategic importance. This has resulted in the establishment of a number of (usually state-owned) high-technology industries in such fields as telecommunications and shipbuilding, as well as an aerospace industry, based in Bandung. However, under the terms of the financing agreements reached with the IMF in 1997–98, this approach to industrialization has been reconsidered.

The manufacturing sector was particularly badly affected by the regional financial crisis from 1997. Having drawn heavily on offshore loans in earlier years, the sector was largely insolvent by mid-1998, and many jobs had been lost. Declining demand in the domestic market and the inability to raise trade credits to finance imports of components or exports of finished products exacerbated the situation. Consequently, manufacturing value added decreased by 11.4% in 1998, with its share in GDP declining from almost 27% in 1997 to approximately 25% in 1998. With the sector recording growth of 3.8% in 1999, 5.5% in 2000 and 4.3% in 2001, its share of GDP had recovered to about 30% by 2003. In the aftermath of the financial crisis, moreover, the Government began to review the conglomerate-based industrialization strategy adopted in the past, and planned to focus more on the promotion of small and medium-sized enterprises in the future.

In 2002 the need for such a shift in strategy was underlined by a sharp increase of 22% in bankruptcies amongst large and medium-sized companies—a result of continuing political instability, rising costs, labour unrest and growing competition from other South-East Asian countries and China. These difficulties were also beginning to affect foreign companies, including suppliers to the international garment and sports-shoe companies. In a particularly important related development, in November 2002 Japan's Sony Corporation announced that it would be shutting down its manufacturing operations in Indonesia with effect from March 2003. These closures reportedly resulted in a vast loss of jobs, with the Sony closure alone affecting 1,000 employees. In overall terms, 100,000 manufacturing jobs were reported to have been lost in 2001, even before the acceleration in the shutdown of manufacturing enterprises in 2002–03.

INFRASTRUCTURE

Transport

At the time of the 'New Order' Government's accession to power, Indonesia's transport infrastructure was limited, and in an extremely poor state of repair. In the following three decades, however, great emphasis was placed on the rehabilitation and expansion of transport facilities, with public investment in the transportation and tourism sectors averaging 18% of total government capital spending during Repelitas I and II, almost 13% during Repelita III, 15% during Repelita IV and 20% during Repelita V. Priority continued to be given to infrastructure development in Repelita VI (1994/95–98/99), under which the transport sector was scheduled to receive 21.5% of total projected capital expenditure. This target was exceeded in the first four years of the Plan period, when the share of public spending on transport and communications averaged almost 22% in each year. Despite significant improvements brought about by these investments, many transport facilities remain inadequate and continue to inhibit economic development. Their effectiveness has suffered further in the aftermath of the 1997/98 crisis, as the resulting financial constraints have reduced repair and maintenance activities and contributed to a deterioration of the existing infrastructure.

In 2001 Indonesia had a total road length of 361,782 km, almost 60% of which was located in Sumatra, Java and Bali. On these islands most cities are now connected by highways or secondary roads, and several major motorway projects have been completed. Indonesia's railway services are limited to the islands of Java (including Madura), where most major cities are connected, and Sumatra, where three separate networks, with a total track length of some 6,500 km, operate in the northern, western and southern parts of the island. A major programme to rehabilitate the country's rail services, initiated in 1981, had resulted, by 2000, in the renovation or replacement of about 3,000 passenger carriages and some 27,000 freight wagons. Proposals to establish a 150-km network to facilitate commuter travel between Jakarta and its suburbs, as well as an underground mass rapid transit system in central Jakarta, have also been developed. However, the implementation of such projects has been delayed by the effects of the economic crisis.

Considerable efforts have been made to expand and upgrade ports and shipping since the mid-1980s. An ambitious six-year port modernization programme, financed in part by the World Bank and the ADB, was initiated in 1985. In 1984, in an attempt to modernize the country's ageing merchant fleet, the authorities introduced a policy whereby all cargo vessels of more than 25 years of age were to be replaced with locally designed and locally built freighters. This policy had to be suspended in 1988, owing to the inability of the local shipbuilding industry to provide adequate numbers of replacement vessels. Inter-island passenger shipping was also being improved through the introduction of new liners, funded partially with German aid. Major investments are currently being made in the Jakarta international container terminal to make it into one of the world's largest full-service international seaports.

Air transport in Indonesia is relatively well developed, with 61 recognized main airports in mid-2002, most of which were capable of handling aircraft of at least Fokker F-28 size. The largest airline is the state-owned Garuda Indonesia, which operated an all-jet fleet of 49 aircraft in mid-2002. A number of other airlines, mostly privately owned, operate domestic and regional services. The country's civil aviation sector was negatively affected by the regional economic crisis and political unrest in late 1997 and early 1998, with a number of the smaller, privately owned carriers being particularly severely hit as a result of rising costs and declining revenues.

Electric Power

Indonesia's power-generating capacity expanded rapidly between 1966 and 2001, from approximately 650 MW to 22,347 MW, most of which was operated by the state-owned power generating company, Perusahaan Listrik Negara (PLN). As late as 1980, some 95% of Indonesia's electricity was generated by oil- or gas-fuelled plants, but by 2000, as a result of a deliberate diversification policy, oil- and gas-fuelled plants (including combined-cycle plants) accounted for less than 50% of Indonesia's power-generating capacity, with the remainder being generated by coal-fired steam, hydropower and geothermal power plants. Concerns over impending power shortages in the Java-Bali grid prompted the signing of several agreements in 2002–03 between PLN and private contractors regarding the expansion of local power plants. In order to provide a price incentive for investors, in 2002 the Government agreed to permit a gradual increase in electricity tariffs to the equivalent of 7 US cents per kilowatt hour (kWh). In mid-2004 a new pricing policy was introduced, under which electricity prices would be allowed to vary, within a broad band, according to changes in fuel prices and variations in the inflation and exchange rates.

TRADE

For more than a decade from the mid-1970s, Indonesia's export trade was dominated by crude petroleum and natural gas, which, between 1980 and 1985, accounted for an average of 76% of total export revenue. Consequently, Indonesia's export earnings became heavily dependent on the vagaries of the international energy markets during this period. After 1986, however, the Indonesian Government adopted a wide range of measures aimed at promoting the expansion of non-hydrocarbon exports, which resulted in a significant shift in the composition of Indonesia's exports. In 1987 the value of non-oil and non-gas exports exceeded the value of petroleum and gas exports for the first time in almost 15 years. This trend was reinforced in the following years, and the share of petroleum and gas exports declined steadily, with only a brief interruption in 1990 caused by the sharp increase in international energy prices as a result of the Iraqi invasion of Kuwait; the share had fallen to approximately 24% by 1996. Another temporary increase in this share, to 29%, occurred in 1997, owing to the high energy prices prevailing in the early part of that year, following which the sector's share declined to a mere 16% of total export revenue in 1998. In 1999–2001, however, this share increased markedly, to approximately 23%, owing initially to the severe contraction of non-oil and non-gas exports caused by the continuing effects of the economic crisis, and subsequently to the sharp increase in world oil prices in 2000 and their continued stability in 2001. This share declined only modestly in 2002, to 21%, before rising to 22% in 2003 as international oil prices strengthened again.

Indonesia's merchandise trade has consistently recorded a surplus, although this surplus narrowed through most of the early 1990s, from US $6,664m. in 1989 to $5,129m. in 1996, as a result of rising imports, associated with a rapid acceleration in the growth of domestic consumption and investment. A sharp reversal in this trend occurred in 1997 as a result of the Asian financial crisis, which prompted an increase in the surplus to $25,040m. by 2000, largely as a result of the strong compression of imports precipitated by the devaluation of the rupiah and the general economic downturn during this period. However, the surplus on the merchandise account declined again to $22,695m. in 2001, reflecting a significant decline in exports owing to the deceleration of economic growth in that year, and expanded only modestly, to $23,513m. in 2002 and $23,702m. in 2003 as export earnings and import spending increased broadly in line with each other.

The surplus on the merchandise account has usually been more than offset by a consistent deficit on invisible items, resulting from interest payments on Indonesia's external debt and the repatriation of profits by foreign investors, and also from Indonesia's heavy dependence on foreign transport services. Indonesia's current account was therefore in deficit for most of the 1980s and 1990s. This deficit increased particularly sharply from US $2,790m. in 1994 to $7,663m. in 1996, before narrowing to $4,889m. in 1997 as the regional economic crisis prompted a reduction in trade and a slowdown in debt repayments and profit repatriation. These developments resulted in the recording of a highly unusual surplus of $4,096m. on Indonesia's current account in 1998, which increased steadily to $7,985m. by 2000, before easing to $6,899m. in 2001. A renewed increase, to $7,822m., was recorded in 2002, followed by a slight reduction, to $7,424m., in 2003.

Apart from petroleum and gas, which, despite their diminished importance, remain the country's principal export products, the bulk of Indonesia's exports consists of primary and semi-processed agricultural and mineral commodities. The most important of these have traditionally been rubber, coffee, tin, shrimps and palm oil. Exports of logs, which accounted for 7.6% of total exports in 1980, declined into virtual insignificance in the following years and were replaced by exports of processed wood, which in 1994 contributed almost 13% of total export value. This share declined gradually to approximately 9% in 1997 and 1998, and further, to some 7% in 2000 and 2001, before fading into virtual obscurity by 2003, owing to marketing problems and increased foreign competition in Indonesia's main export markets. Another group of manufactured exports that expanded rapidly during the 1980s but suffered a similar contraction in the mid-1990s was textiles and garments. Its share of total exports increased from 0.4% in 1980 to 17.8% in 1992, but subsequently declined to 9.7% by 1997, before recovering sharply to 14.8% in 1998 and 17.8% in 1999 as exports of other products contracted more rapidly. This trend was reversed in 2000, however, when the increase in petroleum prices raised the value of the oil and gas component of Indonesia's external trade. In 2002–03 this share amounted to approximately 11.7%. Following efforts to promote non-oil and non-gas exports, a wide range of other manufactured goods also began to assume an increasingly important share of the country's exports in the early and mid-1990s, although this trend was arrested by the impact on the manufacturing sector of the regional economic crisis which began in 1997. A major exception to this general trend was electronic products, the export value of which rose steadily from US $2.97m. in 1997 to $6.30m. in 2003. Indonesia's imports, meanwhile, consisted largely of capital and intermediate goods between the mid-1980s and 2003, as the domestic expansion of rice production and import-substituting manufacturing caused imports of consumer goods to be sharply reduced.

Indonesia's trade was traditionally biased towards Japan, the USA and Singapore, which together accounted for approximately 77% of exports and 52% of imports during the 1980s. This pattern changed only modestly in the following decade. Japan remained the most important trading partner in 2003, purchasing 22.3% of Indonesia's exports and supplying 13.0% of imports, while the USA accounted for 12.1% of Indonesia's exports and 8.3% of its imports in the same year. Singapore remained the third most important trading partner in 2003, accounting for 8.8% and 12.8% of the country's total exports and imports, respectively. Since the early 1990s the Republic of Korea and China have also emerged as increasingly important trading partners. The opening of a natural gas pipeline linking the Natuna gasfield with Malaysia in 2002 has also boosted Malaysia's role as a destination for Indonesian exports.

TOURISM

The tourism industry in Indonesia developed rapidly from the 1980s, as the country emerged as an increasingly important long-haul destination for tourists from Europe, the Americas and the Asia-Pacific region. This development was actively encouraged by the Government, which took a conscious decision in the early 1980s to promote the industry as a means of creating employment, generating foreign exchange and stimulating the economic and social development of some of the country's more remote, but touristically attractive, regions. In the following years a number of measures were taken to support the growth of the industry, including an easing of the previously highly restrictive visa requirements for foreign visitors in 1983, the encouragement of private-sector investment in tourism-related facilities, the establishment of an increasing number of 'gateway' airports and seaports and the development of the associated infrastructure, and the progressive opening of regional airports to foreign carriers. Considerable efforts were also made to publicize Indonesia's tourism potential in the major overseas markets.

These measures had a dramatic impact, with the number of tourist arrivals exceeding 1m. in 1987, 2m. in 1990, 3m. in 1992, 4m. in 1994, and 5m. in 1996 and 1997 (in which years the value of receipts from tourism was recorded as US $6,308m. and $5,321m., respectively). This growth in tourist arrivals and spending was accompanied by a corresponding increase in tourist facilities, as exemplified by a rise in the number of star-rated hotel rooms from 34,300 in 1980 to almost 70,000 in 1996, and an increase in the number of travel agencies operating in Indonesia from 330 to 2,225 during the same period. However, the tourism industry suffered a serious set-back as a result of the political unrest and economic crisis of 1997/98, which led to a sharp decline in the number of tourist arrivals to approximately 4.6m. in 1998, in which year the value of receipts from tourism was recorded as $3,459m. In 1999 the number of tourist arrivals rose marginally, to 4.7m., before increasing further in 2000 and 2001, to 5.1m. and 5.2m., respectively. In 2002 sectoral growth was undermined by the terrorist bomb attack in the tourist resort of Kuta on the island of Bali in October, which resulted in the deaths of 202 people, including a large number of foreign (mainly Australian) tourists, and precipitated a sharp reduction in the number of tourist arrivals in the following months, so that the figure for the year as a whole amounted to just 5.0m. The reintroduction of visa requirements for nationals of most Western countries in April 2003 affected the industry further, with the number of tourist arrivals (through 13 main entry points) declining to 3.7m. in that year. However, in the first half of 2004, despite the introduction of further visa restrictions, the industry experienced a significant recovery, with arrivals increasing by more than 30%, to 2.13m., compared with 1.39m. in the corresponding period of 2003.

FOREIGN AID AND INVESTMENT

The ambitious development programmes introduced in Indonesia after the late 1960s necessitated enormous investments of capital. Since these capital needs were considerably in excess of Indonesia's own resources, a significant proportion of the country's investment requirements had to be met from abroad, in the form of commercial and concessional loans, as well as direct and portfolio foreign investments.

Aid

Indonesia has substantial foreign borrowings, which peaked at some US $150,000m. in 1998 and were estimated at approximately $135,000m. at the end of 2003. About 60% of this debt was held by the Government and state-owned enterprises, with the remainder being held by the private sector. The majority of Indonesia's public-sector debt derives from loans that were disbursed through the IGGI, which co-ordinated the official financial assistance that Western donors provided to Indonesia until 1991. In March 1992, angered by the human rights conditions attached to its aid by the Dutch Government (which traditionally held the chairmanship of IGGI), the Indonesian Government rejected all further Dutch aid and requested the dissolution of IGGI. IGGI was replaced by the Consultative Group for Indonesia (CGI), chaired by the World Bank, which held its first meeting in Paris, France, in July 1992, when it pledged an annual total of $4,940m. Over the following years the annual commitment was progressively increased to $5,360m. by 1995, and then retained at approximately this level until 1997, before being increased to approximately $5,900m. in 1999. At its pledging session for 2001, held in October 2000, the CGI approved budget support loans of $4,800m. for the Government, as well as $530m. in loans for technical co-operation projects implemented through non-governmental organizations (NGOs). Much of the budgetary support loan remained undisbursed as late as mid-2001, however, owing to the lenders' concerns over the slow implementation of the IMF-sponsored reform programme and the unstable political situation in Indonesia. Although the disbursement rate began to improve following the accession of Megawati Sukarnoputri to the presidency in August 2001, the level of CGI commitments subsequently declined steadily, to $3,100m. for 2002 and $2,700m. for 2003. A modestly increased sum of $2,800m. was allocated for 2004, of which $1,000m. was contingent upon progress with policy reform, with another $500m. being allocated for expenditure by NGOs and provincial governments.

The liberalization of regulations governing foreign borrowing by Indonesian banks and firms also prompted a rapid rise in the private sector's foreign indebtedness. The volume of this debt was estimated to have increased by almost US $6,000m. in 1990 alone. This caused considerable concern in the Government, which sought to exercise increasingly forceful forms of moral suasion to limit overseas borrowing by private enterprises, and in September 1991 formed a special inter-ministerial team to examine the commercial borrowing plans of all investment projects involving official participation. However, these measures failed to restrain the growth of corporate offshore debt incurred by the Indonesian banking sector and many of the country's business conglomerates, and this debt played a significant role in precipitating and exacerbating the financial and economic crisis of 1997–98.

Forecasts prepared by the World Bank suggested that Indonesia's debt repayment schedule would have peaked in the first half of 2002. In order to ease the resulting pressures, the Government sought to negotiate debt-rescheduling agreements with both its official and private creditors. The first such agreement was concluded in April 2002 with the 'Paris Club' of official creditors, which provided for a rescheduling of US $5,400m. from the total debt-service obligations of $7,500m. falling due between 1 April 2002 and 31 December 2003. This was followed by an announcement in June 2002 that the 'London Club' of commercial creditors had similarly agreed, in principle, to extend the servicing of a number of loans taken out by the Indonesian Government in 1995–97. The rehabilitation of the Indonesian Government in international capital markets culminated in March 2004 with its first global sovereign bond issue since the 1997–98 crisis, which was eight times oversubscribed. The outlook for the repayment of Indonesia's large volume of private-sector debt appeared more complex. While most Indonesian companies sought to reach some agreement with their creditors involving a combination of debt restructuring, debt forgiveness and deferred payments, there have been cases where local debtors have simply stopped paying, secure in the knowledge that their creditors would have little recourse to redress, given the inadequacies of the Indonesian legal system.

Investment

During the early years of President Suharto's 'New Order' Government, high priority was given to attracting foreign investment, and the enactment of the Foreign Investment Law of 1967 granted investors a high degree of protection and a wide range of privileges. Between 1967 and the end of 2003 foreign investment projects worth about US $273,000m. had been approved by a national investment co-ordinating board, the Badan Koordinasi Penanaman Modal (BKPM), although this total excluded investments in the petroleum, natural gas and financial sectors, which were beyond the BKPM's jurisdiction. A particularly strong growth in foreign investment interest occurred between 1987 and 1997, following the introduction of a deregulation process by the Government aimed at relaxing the remaining restrictions on foreign investment. As a result, the value of projects approved by the BKPM increased steadily from 1987 onwards, with only a slight downturn in 1993, and peaked at almost $40,000m. in 1995. This upward trend was arrested in 1996, when the value of approved projects declined to approximately $30,000m., and although a modest recovery to a figure of $33,833m. was recorded in 1997, the deepening financial and economic crisis resulted in a sharp reduction in the value of approved projects to $13,563m. in 1998 and $10,891m. in 1999. A subsequent recovery to approximately $15,200m. in 2000 and 2001 proved temporary, however, and the volume of approved investment fell again, to about $9,700m., in 2002 before recovering to $13,200m. in 2003. Furthermore, the available data suggested that only a relatively small proportion (estimated at about 33%) of approved investment projects had actually been implemented by the end of 2003.

The sectoral distribution of these investments has been heavily biased towards manufacturing, which had received about 70% of approved domestic and 65% of approved foreign capital investment by the end of 2003. Apart from the manufacturing sector, such investment has been channelled mainly into agriculture, tourism, infrastructure and financial services. Interest in infrastructural investment, including power generation and transport services, rose particularly rapidly after mid-1994, when new deregulation measures opened these sectors to foreign participation.

FINANCE

The economic turmoil of the early 1960s left Indonesia with a thoroughly disrupted financial system. Rampant inflation, caused by persistently large budgetary deficits, an absence of monetary restraint, and the isolation of the state-run banking system from market forces, had caused major disruptions to the country's financial structure. Between 1966 and the mid-1990s considerable efforts were made to stabilize the financial sector, but these efforts were undermined by the liberalization policies introduced after the mid-1980s and their abuse by an increasingly corrupt and politically well-connected business élite. The financial sector was consequently ill-prepared to withstand the financial crisis of 1997/98, and its poor handling by the Government and multinational agencies such as the IMF.

Fiscal Policy

In response to the economic crisis of the mid-1960s, President Suharto's 'New Order' Government committed itself to the principle of a balanced budget—implying the absence of budget deficits financed from domestic sources—shortly after assuming office in 1966. The somewhat unorthodox definition of a balanced budget employed by the Suharto Government permitted receipts of foreign aid to be classified as revenues, rather than as means through which to finance a deficit, however, and the Government retained the option of using surplus revenues to increase its deposits at the central bank. These deposits could then be drawn upon for extrabudgetary funding of government spending in times of reduced revenues. Consequently, the nominal budgetary balance often concealed quite substantial deficits. This pattern persisted until 1997, when an agreement was reached with the IMF introducing more standard budgeting practices; it allowed a (domestically funded) budget deficit equivalent to 3.5% of GDP in the 1998/99 fiscal year to counteract the contraction caused by the ongoing economic crisis. Such deficits, albeit with a gradually declining magnitude, have been permitted in subsequent years; the deficit for 2002 was fixed at 2.4% of GDP and that for 2003 was initially set at 1.3% of GDP, but later raised to 1.8% of GDP, in order to allow the Government greater scope to stimulate the economy in the wake of the Bali bombings. The budget for 2004, announced in August 2003, set the deficit at 1.2% of projected GDP, while the draft budget for 2005, presented to the Dewan Perwakilan Rakyat (DPR—House of Representatives) in May 2004, envisaged an even smaller deficit, in the range of 0.7%–0.9% of GDP. Meanwhile, provisional figures suggested that the actual deficit amounted to only 1.6% of GDP in 2002, but exceeded the target for 2003, reaching 1.9%.

A number of other major fiscal policy reforms have also been introduced in the post-Suharto era. After a transitional nine-month budget period from 1 April to 31 December 2000, the fiscal year was adjusted to coincide with the calendar year from 2001. At the same time a wide-ranging decentralization of the budgetary process was introduced, with the generation of public revenues and expenditure being devolved to the provincial and district level to a large extent. This resulted in a dramatic reduction in the central Government's revenues, and threatened to increase the prosperity gap between the densely populated and resource-poor provinces of Java and the less heavily populated, resource-rich provinces in the outer islands. The devolution of many fiscal powers to the district level caused particularly serious problems, as it created considerable differences in local taxation levels and generated widespread uncertainty among personal and corporate taxpayers. More recently the Government has been attempting to remove some of these excesses by shifting many of the powers held at district level to the provincial level.

Government spending is divided into the usual categories of recurrent and capital expenditure, and has historically been an important element in the generation of employment and national income. The State employs some 17% of the labour force, and public contracts have been a major source of income for a number of private sector suppliers. Changes in Indonesia's economic fortunes in the mid-1980s necessitated a deceleration in the rate of growth of public expenditure and caused greater emphasis to be shifted from capital expenditure to the less capital- and import-intensive recurrent expenditure, in an effort to maintain domestic demand and the economic growth rate. Even within the capital budget, increasing emphasis was given to agricultural and infrastructural development, which provided the greatest domestic benefits. In the late 1980s the growth of capital expenditure began to accelerate again, as structural adjustments and rising inflows of foreign funds made more resources available, and as mounting population pressure emphasized the urgent need for an acceleration of economic growth rates. The volume of capital expenditure in 1999/2000 was thus more than five times as high in nominal rupiah terms as the recorded capital expenditure in 1989/90. In 2001–03 the central Government's development expenditures amounted to some 15% of total government expenditures, and the budget for 2004 foresaw a small rise in this share, to 17.6%.

Money and Banking

Indonesia's financial sector is headed by Bank Indonesia, the central bank. For almost two decades, until the late 1980s, commercial banking was dominated by five large state-owned banks. In addition, there were 11 foreign branch banks (the activities of which were, however, confined to the Jakarta area) and some 70 private banks, of which 10 were licensed to deal in foreign exchange. The banking industry also comprised a state-owned national development bank, as well as provincial development banks in each of Indonesia's 27 provinces, and a state-owned savings bank. Fourteen 'non-bank financial institutions', performing a variety of merchant-banking functions, had also been established. By the end of 1988 there were also more than 80 leasing companies, 90 insurance companies, and a stock exchange, which had traded since 1977.

Faced with considerable inefficiencies in the mobilization and allocation of resources, the Government initiated a significant reform of the banking sector from the mid-1980s onwards. Most of the previously-imposed credit and interest-rate restrictions were abolished by 1987, and replaced by a variety of indirect means of regulating liquidity, including money-market securities and discount window facilities to generate liquidity and assist banks facing temporary shortages, as well as special forms of discount instruments, known as Bank Indonesia Certificates, to reduce surplus liquidity.

The reform process continued into the late 1980s and early 1990s, and culminated in the introduction of new banking legislation in 1992. The reforms were aimed at simplifying the structure of the banking system and facilitating an enhanced mobilization of resources available for investment, while simultaneously ensuring the efficient use of these resources in productive investments. The categories of recognized banking institutions were reduced to two—commercial banks and small-scale credit banks—and many of the remaining restraints on banking were withdrawn: foreign branch banks operating in Jakarta were given permission to open sub-branches in six regional capitals; domestic banks were allowed to form joint-venture banks with international institutions which maintained representative offices in Jakarta; the upper limit on offshore borrowing by Indonesian commercial banks was removed; privately owned banks were granted permission to compete with state-owned banks in attracting deposits from the parastatal enterprises; and licensing requirements for dealing in foreign exchange were relaxed.

Efforts were also made to invigorate the Indonesian Stock Exchange by partially opening it to foreign institutional investors. This policy was reinforced by the provision of a wide range of incentives to encourage the flotation on the Jakarta Stock Exchange of domestically incorporated firms and equalizing the taxes applicable to interest earned from time and savings deposits and dividends received on equity holdings. A special 'over-the-counter' stock exchange was established in Jakarta, and private entrepreneurs were given permission to create privately owned regional stock markets in several provincial capitals, with one stock exchange being established in Surabaya. Measures were also taken to encourage the entry of new firms, including foreign joint ventures, into the leasing, factoring, venture capital, securities trading, credit card and consumer finance industries.

The new reforms stimulated a dramatic expansion of the Indonesian financial sector. By the end of 1996 the number of commercial banks had risen to 239, while the number of secondary banks had risen to 9,276. In the capital markets the number of companies listed on the Jakarta Stock Exchange, which began computerized trading in mid-1995, increased from 24 in 1988 to almost 290 by the end of 1996. Its growth was accompanied by the emergence of a strong overseas interest in the Indonesian capital markets, with a number of major international brokerages establishing a presence in Indonesia.

This unrestrained growth of the financial sector inevitably gave rise to fears about its viability, which were reinforced by a number of bank failures and several well-publicized scandals involving banks and securities firms. Although judicious policy measures enabled the Government to maintain a comparatively high degree of monetary stability until mid-1997, it proved unable to withstand the turmoil precipitated by the Thai financial crisis which then erupted. This caused the Indonesian currency and capital markets to come under severe speculative pressure in the second half of the year, and a corresponding collapse of the rupiah's exchange rate and the prices of shares in companies listed on the Indonesian stock markets.

The rationalization of the financial sector, involving in particular a restructuring of the banking system, emerged as a major priority of the reform measures the Indonesian Government agreed to undertake as part of the rescue programme negotiated with the IMF in 1997–98. The first step in this direction was taken in November 1997, when the Government announced plans to close 16 banks known to be burdened by a high level of bad debt. This was followed in January 1998 by the establishment of the Indonesian Bank Restructuring Agency (IBRA), which was assigned the task of rehabilitating the highly indebted financial and corporate sectors, and acquired control of almost US $70,000m. in the form of both equity and debt from the banks that had failed during the financial crisis. IBRA was intended to strengthen the industry by consolidating the large number of existing, and often troubled, banks into a few solid entities through a combination of closures, mergers, recapitalization, privatizations and other divestments of the assets it had acquired by the end of 2003. This rescue plan, which also provided for the surviving banks to raise their capital adequacy ratios (CARs) to 8% by 2001, was to be funded through the issue of government bonds worth a total of Rp. 157,600,000m.

Although the financial sector rehabilitation programme generally lagged behind the somewhat ambitious targets set for it, IBRA had, for the most part, achieved its ultimate objectives by the time it was dissolved, on schedule, in February 2004. The major achievements of IBRA included the fact that it had met, or exceeded, the revenue collection targets set for it and that it had succeeded in restructuring and privatizing many of the banks and other commercial enterprises that it had acquired. The remaining assets, worth some Rp. 40,000,000m., were to be transferred to a new, government-owned, holding company.

ECONOMIC PROSPECTS

The Asian financial crisis of the late 1990s took a heavy toll on Indonesia. Although by 2000 the economy had stabilized to some degree—with inflation having been reduced almost to a single-digit figure, interest rates having been lowered and a surplus having been achieved on the current account of the balance of payments—the situation deteriorated again in 2001, largely owing to the ongoing political uncertainty prevailing in the country. This continued to erode both domestic and international confidence in Indonesia's future, and played a major role in precipitating a serious weakening of the capital and currency markets. By mid-2001 this deterioration in the financial markets had begun to place in jeopardy Indonesia's prospects for continued economic recovery.

Megawati Sukarnoputri's appointment as President in August 2001 helped to restore a degree of investor confidence, and her Government's management of the economy received widespread praise both within Indonesia and beyond. The country's economic performance improved markedly in 2002 compared with 2001, despite the impact of, for example, the Bali bombings, and it has remained robust in the following years despite the challenges posed by terrorism, the SARS epidemic, the continuing violence in Aceh, and the uncertainties generated by the legislative and presidential elections that took place in 2004. While it is also widely acknowledged that much still needs to be done to enhance the Indonesian economy's competitiveness, through reforms in such areas as labour policy, the legal system, investment and trade regulations, and the financial sector, there is also growing confidence both domestically and internationally concerning the sustainability of the economic recovery that had taken place during 2002–03. This confidence was underlined in October 2003 by the upgrading of Indonesia's rating by two leading international credit agencies, and by the continued strength of the index of the Jakarta Stock Exchange. An important guarantor of this confidence was Indonesia's decision to enter into a PPM agreement with the IMF following the expiry of its support programme at the end of 2003, and the broadly positive tenor of the IMF report issued in July 2004 following the first set of PPM discussions, which had been held in May. With Indonesian policy-makers themselves very conscious of the need to continue the process of economic, institutional and legal reform, and with pressure from investors and donors remaining strong, the prospects of an early recovery of the Indonesian economy to its pre-crisis levels appeared to be improving progressively.

Statistical Survey

Source (unless otherwise stated): Badan Pusat Statistik (Central Bureau of Statistics/Statistics Indonesia), Jalan Dr Sutomo 6–8, Jakarta 10710; tel. (21) 3507057; fax (21) 3857046; e-mail bpshq@bps.go.id; internet www.bps.go.id.

Note: Unless otherwise stated, figures for East Timor (now Timor-Leste, occupied by Indonesia between July 1976 and October 1999) are not included in the tables.

Area and Population

AREA, POPULATION AND DENSITY

Area (sq km)	1,922,570*
Population (census results)	
31 October 1990	178,631,196
30 June 2000	
Males	103,417,180
Females	102,847,415
Total	206,264,595
Population (UN estimates at mid-year)†	
2001	214,356,000
2002	217,131,000
2003	219,883,000
Density (per sq km) at mid-2003	114.4

* 742,308 sq miles.

† Source: UN, *World Population Prospects: The 2002 Revision*.

ISLANDS

(population at 2000 census)*

	Area (sq km)	Population	Density (per sq km)
Jawa (Java) and Madura	127,499	121,352,608	951.8
Sumatera (Sumatra)	482,393	43,309,707	89.8
Sulawesi (Celebes)	191,800	14,946,488	77.9
Kalimantan	547,891	11,331,558	20.7
Nusa Tenggara†	67,502	7,961,540	117.9
Bali	5,633	3,151,162	559.4
Maluku (Moluccas)	77,871	1,990,598	25.6
Papua‡	421,981	2,220,934	5.3
Total	1,922,570	206,264,595	107.3

* Figures refer to provincial divisions, each based on a large island or group of islands but also including adjacent small islands.

† Comprising most of the Lesser Sunda Islands, principally Flores, Lombok, Sumba, Sumbawa and part of Timor.

‡ Formerly Irian Jaya (West Papua).

PRINCIPAL TOWNS

(estimated population at 31 December 1996)

Jakarta (capital)	9,341,400	Malang	775,900
Surabaya	2,743,400	Padang	739,500
Bandung	2,429,000	Banjarmasin	544,700
Medan	1,942,000	Surakarta	518,600
Palembang	1,394,300	Pontianak	459,100
Semarang	1,366,500	Yogyakarta (Jogjakarta)	421,000
Ujung Pandang (Makassar)	1,121,300		

Mid-2003 (UN estimates, '000 persons, incl. suburbs): Jakarta 12,296; Bandung 3,765; Surabaya 2,616; Medan 2,010; Palembang 1,569; Ujung Pandang 1,140 (Source: UN, *World Urbanization Prospects: The 2003 Revision*).

BIRTHS AND DEATHS

(UN estimates, annual averages)

	1985–90	1990–95	1995–2000
Birth rate (per 1,000)	28.0	24.9	22.5
Death rate (per 1,000)	9.3	8.3	7.6

Source: UN, *World Population Prospects: The 2002 Revision*.

Expectation of life (WHO estimates, years at birth): 66.4 (males 64.9; females 67.9) in 2002 (Source: WHO, *World Health Report*).

Birth rate (per 1,000): 22.9 in 1997; 22.8 in 1998; 22.4 in 1999 (Source: UN, *Statistical Yearbook for Asia and the Pacific*).

Death rate (per 1,000): 7.5 in 1997; 7.7 in 1998; 7.5 in 1999 (Source: UN, *Statistical Yearbook for Asia and the Pacific*).

ECONOMICALLY ACTIVE POPULATION

(persons aged 15 years and over)

	2000*	2001	2002‡
Agriculture, hunting, forestry and fishing	40,676,713	39,743,908	40,633
Mining and quarrying	n.a.	n.a.	632
Manufacturing	11,641,756	12,086,122	12,110
Electricity, gas and water	n.a.	n.a.	178
Construction	3,497,232	3,837,554	4,274
Trade, restaurants and hotels	18,489,005	17,469,129	17,795
Transport, storage and communications	4,553,855	4,448,279	4,673
Financing, insurance, real estate and business services	882,600	1,127,823	992
Public services	9,574,009	11,003,482	10,360
Activities not adequately defined	522,560†	1,091,120†	—
Total employed	89,837,730	90,807,417	91,647
Unemployed	5,813,231	8,005,031	9,132
Total labour force	95,650,961	98,812,448	100,779

* Excluding Maluku province.

† Includes mining and quarrying and electricity, gas and water.

‡ '000 persons. Source: ILO.

Health and Welfare

KEY INDICATORS

Total fertility rate (children per woman, 2002)	2.4
Under-5 mortality rate (per 1,000 live births, 2002)	45
HIV/AIDS (% of persons aged 15–49, 2003)	0.10
Physicians (per 1,000 head, 1994)	0.16
Hospital beds (per 1,000 head, 1994)	0.66
Health expenditure (2001): US $ per head (PPP)	77
Health expenditure (2001): % of GDP	2.4
Health expenditure (2001): public (% of total)	25.1
Access to water (% of persons, 2000)	76
Access to sanitation (% of persons, 2000)	66
Human Development Index (2002): ranking	111
Human Development Index (2002): value	0.692

For sources and definitions, see explanatory note on p. vi.

Agriculture

PRINCIPAL CROPS

('000 metric tons, incl. East Timor)

	2000	2001	2002
Rice (paddy)	51,898	50,461	51,579
Maize	9,677	9,347	9,527
Potatoes	977	831	826
Sweet potatoes	1,828	1,749	1,772
Cassava (Manioc)	16,089	17,055	16,913
Other roots and tubers*	350	350	350
Sugar cane†	23,900	25,185	25,530
Dry beans	290	288	304
Cashew nuts	90	80	80
Soybeans	1,018	827	653
Groundnuts (in shell)†	974	710	722
Coconuts†	15,120	15,164	16,086
Oil palm fruit*	36,380	38,300	44,000
Cabbages	1,634	1,399	1,438
Tomatoes	593	484	452
Pumpkins, squash and gourds	159	135	172
Cucumbers and gherkins	400*	432	406
Aubergines (Eggplants)	271	244	273
Chillies and green peppers	728	861	772
Dry onions	773	861	772
Garlic	59	50	49
Green beans	716	644	635
Carrots	327	301	300
Other vegetables*	1,128	1,014	1,073
Oranges	644	691	664
Avocados	146	142	151
Mangoes	876	923	892
Pineapples	393	495	463
Bananas	3,747	4,300	3,683
Papayas	429	501	491
Other fruits and berries*	2,040	2,140	2,240
Coffee (green)	625	622	623
Cocoa beans	466	381	450†
Tea (made)	163	165	165
Tobacco (leaves)	136	134	143
Natural rubber	1,610	1,607	1,630

* FAO estimate(s).
† Unofficial figure(s).

Source: FAO.

LIVESTOCK

('000 head, incl. East Timor, year ending September)

	2000	2001	2002
Cattle	11,008	10,275	10,435
Sheep	7,427	7,401	7,661
Goats	12,566	12,464	13,045
Pigs	5,357	5,867	5,927
Horses	412	422	419
Buffaloes	2,405	2,333	2,436
Chickens	859,497	960,164	1,071,948
Ducks	28,076	32,068	46,000

Source: FAO.

LIVESTOCK PRODUCTS

('000 metric tons, incl. East Timor)

	2000	2001	2002
Beef and veal	340	339	324
Buffalo meat	46	44	44
Mutton and lamb	36	33	37
Goat meat	45	49	51
Pig meat*	413	463	471
Poultry meat	831	844	843
Cows' milk	498	505	521
Sheep's milk*	89	88	92
Goats' milk*	200	200	200
Hen eggs	642	693	742
Other poultry eggs	141	158	170
Wool: greasy*	22	22	23
Cattle and buffalo hides*	62	60	58
Sheepskins*	8	8	9
Goatskins*	10	11	12

Note: Figures for meat refer to inspected production only, i.e. from animals slaughtered under government supervision.
* FAO estimates.

Source: FAO.

Forestry

ROUNDWOOD REMOVALS

('000 cubic metres, excl. bark)

	2000	2001	2002
Sawlogs, veneer logs and logs for sleepers	27,000	20,000	26,500
Pulpwood*	3,248	3,248	3,248
Other industrial wood*	3,249	3,249	3,249
Fuel wood*	88,981	85,712	82,556
Total	122,478	112,209	115,553

* FAO estimates.

Source: FAO.

SAWNWOOD PRODUCTION

('000 cubic metres, incl. railway sleepers)

	2000	2001	2002
Total (all broadleaved)	6,500	6,750	6,500

Source: FAO.

Fishing

('000 metric tons, live weight)

	2000	2001	2002
Capture	4,120.1	4,273.7	4,505.5
Scads	255.4	258.4	267.1
Goldstripe sardinella	172.2	185.9	191.3
'Stolephorus' anchovies	173.9	190.2	197.8
Skipjack tuna	247.5	228.6	238.4
Indian mackerels	207.0	214.4	220.4
Aquaculture	788.5	864.3	914.1
Common carp	180.2	194.9	199.6
Milkfish	217.2	209.5	222.3
Total catch	4,908.6	5,137.9	5,419.5

Note: Figures exclude aquatic plants ('000 metric tons): 247.9 (capture 42.7, aquaculture 205.2) in 2000; 247.0 (capture 34.5, aquaculture 212.5) in 2001; 258.9 (capture 35.8, aquaculture 223.1) in 2002. Also excluded are crocodiles, recorded by number rather than by weight. The number of crocodiles caught was: 10,387 in 2000; 13,402 in 2001; n.a. in 2002.

Source: FAO.

Mining

('000 metric tons, unless otherwise indicated)

	2000	2001	2002
Crude petroleum (million barrels)‡	516.1	489.5	432.0
Natural gas (million cubic metres)	82,334	79,470	90,200
Bauxite	1,151	1,237	1,283
Coal	77,015	92,500	103,329
Nickel*	98.2	102.0	123.0
Copper*	1,012.1	1,048.7	1,171.7
Tin ore (metric tons)*	55,624	61,863	88,142
Gold (kg)†	123,994	162,605	142,238
Silver (kg)†	255,578	269,825	293,520

* Figures refer to the metal content of ores and concentrates.
† Including gold and silver in copper concentrate.
‡ Including condensate.

Source: US Geological Survey.

Industry

SELECTED PRODUCTS

('000 metric tons, unless otherwise indicated)

	1998	1999	2000
Refined sugar	1,793	1,657	1,745
Cigarettes (million)	271,177	254,276	n.a.
Veneer sheets ('000 cubic metres)	1,110	50	69
Plywood ('000 cubic metres)	7,800	7,500	8,200
Newsprint	478	532	477
Other printing and writing paper	1,855	2,733	2,818
Other paper and paperboard*	3,154	3,713	3,682†
Nitrogenous fertilizers*†‡	2,899	2,842	2,853
Jet fuel	1,150	721	2,746
Motor spirit (petrol)	7,302	8,125	7,997
Naphthas	522	794	1,344
Kerosene	6,956	7,635	7,494
Gas-diesel oil	13,423	13,984	14,401
Residual fuel oils	11,560	11,420	11,421
Lubricating oils	211	332	364
Liquefied petroleum gas	1,937	1,979	1,762
Rubber tyres ('000)§	25,701	33,089	21,000†
Cement	22,344†	22,806	22,789
Aluminium (unwrought)‖¶	133.4	111.7	190.5
Tin (unwrought, metric tons)†‖¶	54,000	48,300	46,400
Radio receivers ('000)	80	4,937	n.a.
Passenger motor cars ('000)[1]	6	39	n.a.
Electric energy (million kWh)	90,027	95,944	99,511
Gas from gasworks (terajoules)	26,715	27,421	33,380†

2001: Refined sugar ('000 metric tons) 1,863; Veneer sheets ('000 cubic metres) 94; Plywood ('000 cubic metres) 7,300; Newsprint ('000 metric tons) 511; Other printing and writing paper ('000 metric tons) 2,697; Other paper and paperboard ('000 metric tons) 3,787*†; Nitrogenous fertilizers ('000 metric tons) 2,396*†‡; Rubber tyres ('000) 20,500§; Aluminium (unwrought, '000 metric tons) 208.8‖¶; Tin (unwrought, metric tons) 44,600‖¶.

* Data from FAO.

† Provisional or estimated production.

‡ Production in terms of nitrogen.

§ For road motor vehicles, excluding bicycles and motorcycles.

‖ Primary metal production only.

¶ Data from *World Metal Statistics*.

[1] Vehicles assembled from imported parts.

Source: mainly UN, *Industrial Commodity Statistics Yearbook*.

Palm oil ('000 metric tons): 5,385 in 1997; 5,902 in 1998; 6,250 (unofficial figure) in 1999; 7,276 in 2000; 8,015 (unofficial figure) in 2001; 9,750 (unofficial figure) in 2002 (Source: FAO).

Raw sugar (centrifugal, '000 metric tons): 2,160 in 1996; 2,187 in 1997; 1,846 in 1998; 1,494 in 1999; 1,896 in 2000; 2,025 in 2001; 1,963 (unofficial figure) in 2002 (Source: FAO).

Finance

CURRENCY AND EXCHANGE RATES

Monetary Units

100 sen = 1 rupiah (Rp.).

Sterling, Dollar and Euro Equivalents (31 May 2004)

£1 sterling = 16,897.6 rupiah;
US $1 = 9,210.0 rupiah;
€1 = 11,278.6 rupiah;
100,000 rupiah = £5.918 = $10.858 = €8.866.

Average Exchange Rate (rupiah per US $)

2001 10,260.9
2002 9,311.2
2003 8,577.1

BUDGET

('000 million rupiah, year ending 31 March)

Revenue	1998/99	1999/2000	2000/01*
Tax revenue	147,600	183,281	196,720
Taxes on income, profits, etc.†	97,313	118,164	94,462
Individual	52,995	15,571	19,146
Corporate‡	41,368	99,638	70,463
Domestic taxes on goods and services	38,750	55,453	78,254
General sales, turnover or VAT	27,803	33,087	55,863
Excises	7,733	10,381	17,394
Taxes on international trade	6,936	5,035	9,568
Import duties	2,306	4,177	9,026
Export duties	4,630	858	542
Non-tax revenue	9,781	15,330	111,121
Entrepreneurial and property income	6,281	10,292	15,834
Other	1,967	3,347	92,787
Total	157,381	198,611	307, 841
Capital revenue	31	62	35

Expenditure	1998/99	1999/2000	2000/01*
General public services	16,148	11,425	16,607
Defence	8,955	8,576	10,673
Public order and safety	3,080	4,453	7,400
Education	11,918	14,349	13,433
Health	3,889	5,186	4,542
Social security and welfare	9,220	12,006	30,766
Housing and community amenities	23,435	33,787	4,726
Recreational, cultural and religious affairs	2,992	3,347	2,257
Economic affairs and services	22,164	23,250	12,747
Agriculture, hunting, forestry and fishing	11,511	8,610	4,652
Transportation and communication	6,395	5,580	3,709
Other economic affairs and services	2,481	5,509	858
Other	72,297	109,494	255,886
Interest payments	31,264	42,910	—
Total	174,097	225,874	359,038
Current expenditure	114,412	170,684	n.a.
Capital expenditure	59,686	55,190	n.a.

* Figures are provisional.

† Prior to 1999, tax on income, profits, etc. could not be broken into its components, and almost all income tax was classified as taxes paid by individuals.

‡ Prior to 2000, revenues from natural resources and oil and gas revenue were classified as corporate income tax.

Source: IMF, *Government Finance Statistics Yearbook*.

INTERNATIONAL RESERVES

(US $ million at 31 December)

	2001	2002	2003
Gold*	772	1,077	1,291
IMF special drawing rights	16	19	4
Reserve position in IMF	183	198	216
Foreign exchange	27,048	30,753	34,742
Total	28,019	32,047	36,253

* Valued at market-related prices.

Source: IMF, *International Financial Statistics*.

MONEY SUPPLY

('000 million rupiah at 31 December)

	2001	2002	2003
Currency outside banks	76,342	80,659	94,539
Demand deposits at deposit money banks	97,746	106,558	125,329
Total money (incl. others)	175,110	188,008	220,552

Source: IMF, *International Financial Statistics*.

COST OF LIVING
(Consumer Price Index; base: 1996 = 100)

	2001	2002	2003
Food (incl. beverages)	267.1	296.0	310.0
Non-food	213.4	251.7	270.6
All items	249.2	274.1	279.6

Source: Asian Development Bank, *Key Indicators of Developing Asian and Pacific Countries.*

NATIONAL ACCOUNTS
('000 million rupiah at current prices)

National Income and Product

	1999	2000	2001
Domestic factor incomes*	1,044,745	1,217,917	1,416,425
Consumption of fixed capital	54,987	64,101	74,549
Gross domestic product (GDP) at factor cost	1,099,732	1,282,018	1,490,974
Indirect taxes, *less* subsidies	—	—	—
GDP in purchasers' values	1,099,732	1,282,018	1,490,974
Net factor income from abroad	–83,765	–92,162	–58,079
Gross national product (GNP)	1,015,967	1,189,856	1,432,895
Less Consumption of fixed capital	54,987	64,101	74,549
National income in market prices	960,980	1,125,755	1,358,346

* Compensation of employees and the operating surplus of enterprises. The amount is obtained as a residual.

Source: UN, *National Accounts Statistics.*

Expenditure on the Gross Domestic Product

	2001	2002	2003
Government final consumption expenditure	113,416	132,219	163,701
Private final consumption expenditure	972,938	1,120,164	1,238,892
Increase in stocks*	–53,624	–73,876	–67,258
Gross fixed capital formation	314,066	326,165	352,361
Total domestic expenditure	1,346,796	1,504,672	1,687,696
Exports of goods and services	624,341	577,082	558,091
Less Imports of goods and services	503,482	471,188	459,097
GDP in purchasers' values	1,467,655	1,610,565	1,786,691
GDP at constant 1993 prices	411,752	426,943	444,453

* Figures obtained as a residual.

Source: Asian Development Bank, *Key Indicators of Developing Asian and Pacific Countries.*

Gross Domestic Product by Economic Activity

	2001	2002	2003
Agriculture, forestry and fishing	244,722	275,271	296,238
Mining and quarrying	193,541	178,197	191,177
Manufacturing	372,916	409,666	440,452
Electricity, gas and water	22,170	30,492	39,665
Construction	85,602	93,966	107,119
Trade, hotels and restaurants	235,738	265,535	291,590
Transport, storage and communications	74,247	92,797	111,728
Finance, insurance, real estate and business services	94,819	110,158	123,001
Public administration	81,851	83,294	101,606
Other services	62,049	71,189	84,117
Total	1,467,655	1,610,565	1,786,691

Source: Asian Development Bank, *Key Indicators of Developing Asian and Pacific Countries.*

BALANCE OF PAYMENTS
(US $ million)

	2000	2001	2002
Exports of goods f.o.b.	65,406	57,364	59,165
Imports of goods f.o.b.	–40,366	–34,669	–35,652
Trade balance	25,040	22,695	23,513
Exports of services	5,213	5,500	6,661
Imports of services	–15,011	–15,880	–17,054
Balance on goods and services	15,242	12,315	13,120
Other income received	2,456	2,004	1,318
Other income paid	–11,529	–8,940	–8,366
Balance on goods, services and income	6,169	5,379	6,072
Current transfers received	1,816	1,520	2,255
Current balance	7,985	6,899	7,823
Direct investment abroad	–150	–125	–500
Direct investment from abroad	–4,550	–3,278	145
Portfolio investment liabilities	–1,909	–243	1,222
Other investment liabilities	–1,287	–3,968	–2,040
Net errors and omissions	3,637	700	–1,692
Overall balance	3,726	–15	4,958

Source: IMF, *International Financial Statistics.*

External Trade

PRINCIPAL COMMODITIES
(distribution by SITC, US $ million)

Imports c.i.f.	2000	2001	2002
Food and live animals	2,781.3	2,496.6	2,851.7
Cereals and cereal preparations	1,100.0	746.0	1,214.6
Crude materials (inedible) except fuels	3,296.6	3,171.3	2,658.9
Textile fibres and waste	1,005.8	1,332.1	914.8
Mineral fuels, lubricants, etc.	6,179.0	5,644.1	6,687.9
Crude petroleum oils, etc.	2,524.9	2,887.5	3,216.9
Refined petroleum products	3,562.5	2,652.9	3,409.3
Gas oils	1,701.3	1,355.2	1,728.4
Chemicals and related products	5,753.6	5,253.0	5,120.4
Organic chemicals	2,476.3	2,188.8	2,005.5
Artificial resins and plastic materials, etc	1,074.5	984.4	961.4
Basic manufactures	5,152.1	4,345.6	4,333.3
Textile yarn, fabrics, etc.	1,259.0	1,096.1	887.9
Iron and steel	1,684.5	1,300.3	1,458.3
Machinery and transport equipment	9,194.5	9,030.3	8,581.2
Machinery specialized for particular industries	1,610.4	1,546.9	1,340.7
General industrial machinery, equipment and parts	1,950.8	1,941.4	1,926.0
Road vehicles and parts*	1,851.8	1,838.9	1,639.0
Parts and accessories for cars, buses, lorries, etc.*	1,120.9	936.4	829.7
Total (incl. others)	33,514.8	30,962.1	31,288.8

Exports f.o.b.	2000	2001	2002
Food and live animals	3,499.9	3,252.3	3,604.5
Fish, crustaceans and molluscs	1,583.3	1,532.2	1,487.0
Crude materials (inedible) except fuels	4,317.1	4,187.7	4,522.0
Ores and concentrates of non-ferrous base metals	1,943.2	1,987.2	1,889.3
Mineral fuels, lubricants, etc.	15,683.5	14,274.0	13,910.3
Coal, coke and briquettes	1,296.4	1,625.2	1,770.8
Coal, lignite and peat	1,276.1	1,617.7	1,762.4
Petroleum, petroleum products, etc.	7,762.2	6,916.6	6,561.9
Crude petroleum oils, etc.	6,090.1	5,714.7	5,227.6
Petroleum gases, etc., in the liquefied state	6,624.9	5,732.2	5,577.6
Animal and vegetable oils, fats and waxes	1,772.1	1,446.2	2,656.3
Fixed vegetable oils, fluid or solid, crude, refined or purified	1,648.9	1,341.1	2,511.3
Palm oil	1,087.3	1,080.9	2,092.4
Chemicals and related products	3,114.8	2,803.5	2,936.9
Basic manufactures	12,452.5	11,272.4	10,998.7
Wood and cork manufactures (excl. furniture)	3,226.6	2,891.9	2,811.0
Veneers, plywood, etc.	2,286.6	2,073.4	2,023.4
Plywood of wood sheets	1,988.9	1,837.9	1,748.3
Paper, paperboard, etc.	2,300.5	2,048.4	2,118.2
Textile yarn, fabrics, etc.	3,517.3	3,214.4	2,911.2
Machinery and transport equipment	10,740.8	9,079.3	9,766.9
Telecommunications and sound equipment	3,500.1	3,353.7	3,359.4
Miscellaneous manufactured articles	9,913.0	9,271.1	8,171.6
Clothing and accessories (excl. footwear)	4,806.7	4,597.6	4,000.9
Total (incl. others)	62,124.0	56,316.9	57,158.8

* Excluding tyres, engines and electrical parts.

Source: UN, *International Trade Statistics Yearbook*.

PRINCIPAL TRADING PARTNERS
(US $ million)*

Imports c.i.f.	2001	2002	2003
Australia	1,814.1	1,587.2	1,648.4
Canada	356.5	411.9	321.8
China, People's Republic	1,842.7	2,427.4	2,957.5
France	396.9	406.3	453.2
Germany	1,300.5	1,224.3	1,181.2
Hong Kong	257.3	240.7	222.2
Italy	407.5	401.7	323.7
Japan	4,689.5	4,409.3	4,228.3
Korea, Republic	2,209.3	1,646.8	1,527.9
Malaysia	1,005.5	1,037.4	1,138.2
Netherlands	343.8	352.2	369.6
Singapore	3,147.1	4,099.6	4,155.1
Taiwan	1,071.0	1,010.4	877.1
Thailand	986.0	1,190.7	1,701.7
United Kingdom	643.0	656.2	463.7
USA	3,207.5	2,639.9	2,694.8
Total (incl. others)	30,962.1	31,288.9	32,550.7

Exports f.o.b.	2000	2001	2002
Australia	1,519.4	1,844.9	1,924.4
Belgium/Luxembourg	840.6	772.1	793.8
China, People's Republic	2,767.7	2,200.7	2,902.9
France	718.3	662.7	648.9
Germany	1,443.1	1,297.0	1,269.9
Hong Kong	1,554.1	1,290.3	1,242.3
Italy	757.8	621.8	719.8
Japan	14,415.2	13,010.2	12,045.1
Korea, Republic	4,317.9	3,772.5	4,107.2
Malaysia	1,971.8	1,778.6	2,029.9
Netherlands	1,837.4	1,498.2	1,618.4
Philippines	819.5	814.8	778.2
Singapore	6,562.4	5,363.9	5,349.1
Spain	932.2	903.6	996.4
Taiwan	2,378.3	2,188.0	2,067.5
Thailand	1,026.5	1,063.6	1,227.4
United Kingdom	1,507.9	1,383.1	1,252.4
USA	8,475.4	7,748.7	7,558.6
Total (incl. others)	62,124.0	56,320.9	57,158.8

* Imports by country of production; exports by country of consumption. Figures include trade in gold.

Transport

RAILWAYS
(traffic)

	2000	2001	2002
Passengers embarked (million)	192	187	176
Freight loaded ('000 tons)	19,545	18,702	17,099

Source: Indonesian State Railways.

ROAD TRAFFIC
(motor vehicles registered at 31 December)

	1998*	1999	2000†
Passenger cars	2,769,375	2,897,803	3,038,913
Lorries and trucks	1,586,721	1,628,531	1,707,134
Buses and coaches	626,680	644,667	666,280
Motorcycles	12,628,991	13,053,148	13,563,017

* Including East Timor (now Timor-Leste).
† Preliminary figures.

Source: State Police of Indonesia.

SHIPPING

Merchant Fleet
(registered at 31 December)

	2001	2002	2003
Number of vessels	2,528	2,628	2,700
Displacement ('000 grt)	3,613.1	3,723.1	3,840.4

Source: Lloyd's Register-Fairplay, *World Fleet Statistics*.

Sea-borne Freight Traffic
('000 metric tons)

	1999	2000	2001*
International:			
Goods loaded	139,340	141,528	143,750
Goods unloaded	43,477	45,040	46,659
Domestic:			
Goods loaded	113,633	127,740	163,685
Goods unloaded	122,368	137,512	138,667

* Preliminary figures.

CIVIL AVIATION
(traffic on scheduled services)

	1997	1998	1999
Kilometres flown (million)	199	155	122
Passengers carried ('000)	12,937	9,603	8,047
Passenger-km (million)	23,718	15,974	14,544
Total ton-km (million)	2,797	1,826	1,560

Source: UN, *Statistical Yearbook*.

Tourism

FOREIGN TOURIST ARRIVALS

Country of Residence	2000	2001	2002
Australia	459,994	397,982	346,245
Germany	151,897	159,881	142,649
Japan	643,794	611,314	620,722
Korea, Republic	213,762	212,233	210,581
Malaysia	475,845	484,692	475,163
Netherlands	105,109	114,656	110,631
Singapore	1,427,886	1,477,132	1,447,315
Taiwan	356,436	391,696	400,334
United Kingdom	161,662	189,027	160,077
USA	176,379	177,869	160,982
Total (incl. others)	5,064,217	5,153,620	5,033,400

Receipts from tourism (US $ million): 4,710 in 1999; 5,749 in 2000; 5,411 in 2001 (Source: World Bank).

Communications Media

	2000	2001	2002
Telephones ('000 main lines in use)	6,662.6	7,218.9	7,750.0
Mobile cellular telephones ('000 subscribers)	3,669.3	6,520.9	11,700.0
Personal computers ('000 in use)	2,100	2,300	2,519
Internet users ('000)	2,000	4,000	8,000

Facsimile machines (estimated number in use, 1997): 185,000.

Radio receivers ('000 in use, 1997): 31,500 (Source: UN, *Statistical Yearbook*).

Television receivers ('000 in use, 2000): 31,700.

Daily newspapers (1998): Number: 172; Average circulation: 4,713,000 (Source: UN, *Statistical Yearbook*).

Non-daily newspapers (1998): Number: 433; Average circulation: 7,838,000 (Source: UN, *Statistical Yearbook*).

Source (unless otherwise indicated): International Telecommunication Union.

Education

(2001/02)

	Institutions	Teachers	Pupils and Students
Primary schools	148,516	1,164,808	25,850,849
General junior secondary schools	20,842	476,827	7,466,458
General senior secondary schools	7,785	224,149	3,024,176
Vocational senior secondary schools	4,522	139,359	2,027,464
Universities*	1,634	194,828	3,126,307

* Figures from 1999/2000.

Source: Ministry of National Education.

Adult literacy rate (UNESCO estimates): 87.9% (males 92.5%; females 83.4%) in 2002 (Source: UN Development Programme, *Human Development Report*).

Directory

The Constitution

Indonesia had three provisional Constitutions: in August 1945, February 1950 and August 1950. In July 1959 the Constitution of 1945 was re-enacted by presidential decree. The General Elections Law of 1969 supplemented the 1945 Constitution, which has been adopted permanently by the Majelis Permusyawaratan Rakyat (MPR—People's Consultative Assembly). Amendments made to the Constitution in 2001 and 2002 took effect in 2004, when Indonesia held its next general election. The following is a summary of the Constitution's main provisions, with subsequent amendments:

GENERAL PRINCIPLES

The 1945 Constitution consists of 37 articles, four transitional clauses and two additional provisions, and is preceded by a preamble. The preamble contains an indictment of all forms of colonialism, an account of Indonesia's struggle for independence, the declaration of that independence and a statement of fundamental aims and principles. Indonesia's National Independence, according to the text of the preamble, has the state form of a Republic, with sovereignty residing in the People, and is based upon five fundamental principles, the *pancasila*:

1. Belief in the One Supreme God.
2. Just and Civilized Humanity.
3. The Unity of Indonesia.
4. Democracy led by the wisdom of deliberations (*musyawarah*) and consensus among representatives.
5. Social Justice for all the people of Indonesia.

STATE ORGANS

Majelis Permusyawaratan Rakyat—MPR (People's Consultative Assembly)

Sovereignty is in the hands of the People and is exercised in full by the MPR as the embodiment of the whole Indonesian People. The MPR is the highest authority of the State, and is to be distinguished from the legislative body proper (Dewan Perwakilan Rakyat—DPR, see below), which is incorporated within the MPR. The bicameral MPR, with a total of 678 members (reduced from 700 in 2004), is composed of the 550 members of the DPR and the members of the Dewan Perwakilan Daerah—DPD (see below). Elections to the MPR are held every five years. The MPR sits at least once every five years, and its primary competence is to determine the Constitution and the broad lines of the policy of the State and the Government. It also inaugurates the President and Vice-President, who are responsible for implementing that policy. All decisions are taken unanimously in keeping with the traditions of *musyawarah*.

The President

The highest executive of the Government, the President, holds office for a term of five years and may be re-elected once. As Mandatory of the MPR he/she must execute the policy of the State according to the Decrees determined by the MPR during its Fourth General and Special Sessions. In conducting the administration of the State, authority and responsibility are concentrated in the President. The Ministers of the State are his/her assistants and are responsible only to him/her. The President and Vice-President are to be directly elected on a single ticket (until November 2001 the MPR had exercised the power to elect them). If no candidate succeeds in obtaining more than one-half of the votes cast in a presidential election, a second round of voting shall be held. The President and Vice-President may be dismissed by the MPR on the proposal of the Dewan Perwakilan Rakyat if it is proven that he/she has either violated the law or no longer meets the requirements of his/her office. The President may not freeze or dissolve the Dewan.

Dewan Perwakilan Rakyat—DPR (House of Representatives)

The legislative branch of the State, the Dewan Perwakilan Rakyat, sits at least once a year. Its members are all directly elected. Every statute requires the approval of the Dewan. Members of the Dewan have the right to submit draft bills which require ratification by the President, who has the right of veto. In times of emergency the President may enact ordinances which have the force of law, but such Ordinances must be ratified by the Dewan during the following session or be revoked.

Dewan Perwakilan Daerah—DPD (House of Representatives of the Regions)

The Dewan Perwakilan Daerah is the second chamber of the MPR. Its members are directly elected from every province. Each province has an equal number of members and total membership of the DPD is no more than one-third of the total membership of the DPR. The DPD sits at least once a year. It may propose to the DPR bills relating to regional autonomy, the relationship between central and local government, the formation, expansion and merger of regions, the management of natural and other economic resources, and the financial balance between the centre and the localities. It may also participate in the discussion of such bills and oversee the implementation of regional laws, as well as the state budget, taxation, education and religion.

Dewan Pertimbangan Agung—DPA (Supreme Advisory Council)

The DPA is an advisory body assisting the President, who chooses its members from political parties, functional groups and groups of prominent persons.

Mahkamah Agung (Supreme Court)

The judicial branch of the State, the Supreme Court and the other courts of law (public courts, religious courts, military tribunals, administrative courts and a Constitutional Court) are independent of the Executive in exercising their judicial powers. There is an independent Judicial Commission which is authorized to propose candidates for appointment as justices of the Supreme Court and to ensure the good behaviour of judges. Its members are appointed and dismissed by the President with the approval of the DPR.

Badan Pemeriksa Keuangan (Supreme Audit Board)

Controls the accountability of public finance, enjoys investigatory powers and is independent of the Executive. Its findings are presented to the Dewan.

The Government

HEAD OF STATE

President: Susilo Bambang Yudhoyono (inaugurated 20 October 2004).

Vice-President: Yusuf Kalla.

CABINET
(October 2004)

Co-ordinating Minister for Political Affairs, Security and Social Welfare: Adm. (retd) Widodo Adi Sutjipto.

Co-ordinating Minister for the Economy, Finance and Industry: Aburizal Bakrie.

Co-ordinating Minister for People's Welfare: Alwi Shihab.

Minister of Home Affairs and Regional Autonomy: Lt-Gen. (retd) Muhammad Ma'aruf.

Minister of Foreign Affairs: Hassan Wirayuda.

Minister of Defence: Matori Juwono Sudarsono.

Minister of Justice and Human Rights Affairs: Hamid Awaluddin.

Minister of Finance and State Enterprises Development: Jusuf Anwar.

Minister of Energy and Mineral Resources: Dr Ir Purnomo Yusgiantoro.

Minister of Industry: Adung Nitimiharja.

Minister of Trade: Mari E. Pangestu.

Minister of Agriculture: Anton Apriyantono.

Minister of Forestry: M. S. Ka'ban.

Minister of Transportation and Telecommunication: Hatta Radjasa.

Minister of Maritime Affairs and Fisheries: Freddy Numberi.

Minister of Manpower and Transmigration: Fahmi Idris.

Minister of Public Works: Joko Kirmanto.

Minister of Health: Siti Fadila Supari.

Minister of National Education: Bambang Soedibyo.

Minister of Social Affairs: Bachtiar Chamsyah.

Minister of Religious Affairs: M. Maftuh Basyuni.

Minister of Culture and Tourism: Jero Watjik.

Minister of State for Research and Technology: Kusmayanto Kadiman.

Minister of State for Co-operatives and Small and Medium-Sized Businesses: Suryadarma Ali.

Minister of State for the Environment: Rachmat Witoelar.

Minister of State for Women's Empowerment: Meutia Hatta.

Minister of State for Administrative Reform: Taufik Effendi.

Minister of State for State Enterprises: Soegiarto.

Minister of State for Communications and Information: Sofyan A. Djalil.

Minister of State for Acceleration of Development in Backwards Regions: Syaifulah Yusuf.

Minister of State for National Development Planning: Sri Mulyani Indrawati.

Minister of State for Public Housing: M. Jusuf Anshari.

Minister of State for Youth and Sports Affairs: Adyaksa Dault.

Officials with the rank of Minister of State:

Attorney-General: Muhammad Abdurrahman Saleh.

State Secretary: Yusril Ihza Mahendra.

Cabinet Secretary: Sudi Silalahi.

MINISTRIES

Office of the President: Instant Merdeka, Jakarta; tel. (21) 3840946.

Office of the Vice-President: Jalan Merdeka Selatan 6, Jakarta; tel. (21) 363539.

Office of the Attorney-General: Jalan Sultan Hasanuddin 1, Kebayoran Baru, Jakarta; tel. (21) 7221377; fax (21) 7392576; e-mail kejagung@kejaksaan.go.id; internet www.kejaksaan.go.id.

Office of the Cabinet Secretary: Jalan Veteran 18, Jakarta Pusat; tel. (21) 3810973.

Office of the Co-ordinating Minister for the Economy, Finance and Industry: Jalan Taman Suropati 2, Jakarta 10310; tel. (21) 3849063; fax (21) 334779.

Office of the Co-ordinating Minister for Political Affairs, Security and Social Welfare: Jalan Medan Merdeka Barat 15, Jakarta 10110; tel. (21) 3849453; fax (21) 3450918.

Office of the State Secretary: Jalan Veteran 17, Jakarta 10110; tel. (21) 3849043; fax (21) 3452685; internet www.ri.go.id.

Ministry of Agriculture and Forestry: Jalan Harsono R. M. 3, Gedung D-Lantai 4, Ragunan, Pasar Minggu, Jakarta Selatan 12550; tel. (21) 7822638; fax (21) 7816385; e-mail eko@deptan.go.id; internet www.deptan.go.id.

Ministry of Defence: Jalan Medan Merdeka Barat 13–14, Jakarta Pusat; tel. (21) 3456184; fax (21) 3440023; e-mail postmaster@dephan.go.id; internet www.dephan.go.id.

Ministry of Energy and Mineral Resources: Jalan Merdeka Selatan 18, Jakarta 10110; tel. (21) 3804242; fax (21) 3847461; e-mail pulahta@setjen.dpe.go.id; internet www.dpe.go.id.

Ministry of Finance and State Enterprises Development: Jalan Lapangan Banteng Timur 2–4, Jakarta 10710; tel. (21) 3814324; fax (21) 353710; internet www.depkeu.go.id.

Ministry of Foreign Affairs: Jalan Taman Pejambon 6, Jakarta 10410; tel. (21) 3441508; fax (21) 3805511; e-mail guestbook@dfa-deplu.go.id; internet www.deplu.go.id.

Ministry of Health and Social Welfare: Jalan H. R. Rasuna Said, Block X5, Kav. 4–9, Jakarta 12950; tel. (21) 5201587; fax (21) 5203874; e-mail webadmin@depkes.go.id; internet www.depkes.go.id.

Ministry of Home Affairs and Regional Autonomy: Jalan Merdeka Utara 7–8, Gedung Utama Lt. 4, Jakarta Pusat 10110; tel. (21) 3842222; fax (21) 372812; e-mail dpod@indosat.net.id; internet www.depdagri.go.id.

Ministry of Industry and Trade: Jalan Jenderal Gatot Subroto, Kav. 52–53, 2nd Floor, Jakarta Selatan; tel. (21) 5256458; fax (21) 5229592; e-mail men-indag@dprin.go.id; internet www.dprin.go.id.

Ministry of Justice and Human Rights Affairs: Jalan H. R. Rasuna Said, Kav. 4–5, Kuningan, Jakarta Pusat; tel. (21) 5253004; fax (21) 5253095; internet www.depkehham.go.id.

Ministry of Manpower and Transmigration: Jalan Jenderal Gatot Subroto, Kav. 51, Jakarta Selatan 12950; tel. (21) 5255683; fax (21) 515669; internet www.nakertrans.go.id.

Ministry of National Education: Jalan Jenderal Sudirman, Senayan, Jakarta Pusat; tel. (21) 5731618; fax (21) 5736870; internet www.depdiknas.go.id.

Ministry of Religious Affairs: Jalan Lapangan Banteng Barat 3–4, Jakarta Pusat; tel. (21) 3811436; fax (21) 380836; internet www.depag.go.id.

Ministry of Transportation and Telecommunication: Jalan Merdeka Barat 8, Jakarta 10110; tel. (21) 3811308; fax (21) 3451657; e-mail pusdatin@rad.net.id; internet www.dephub.go.id.

Office of the Minister of State for Co-operatives and Small and Medium-Sized Businesses: Jalan H. R. Rasuna Said, Kav. 3–5, POB 177, Jakarta Selatan 12940; tel. (21) 5204366; fax (21) 5204383; internet www.depkop.go.id.

Office of the Minister of State for the Environment: Jalan D. I. Panjaitan, Kebon Nanas Lt. II, Jakarta 134110; tel. (21) 8580103; fax (21) 8580101; internet www.bapedal.go.id.

Office of the Minister of State for Research and Technology: BPP Teknologi II Bldg, Jalan M. H. Thamrin 8, Jakarta Pusat 10340; tel. (21) 3169166; fax (21) 3101952; e-mail webmstr@ristek.go.id; internet www.ristek.go.id.

Office of the Minister of State for Women's Empowerment: Jalan Medan Merdeka Barat 15, Jakarta 10110; tel. (21) 3805563; fax (21) 3805562; e-mail biroren@cbn.net.id.

OTHER GOVERNMENT BODIES

Dewan Pertimbangan Agung (DPA) (Supreme Advisory Council): Jalan Merdeka Utara 15, Jakarta; tel. (21) 362369; internet www.dpa.go.id; Chair. Gen. (retd) ACHMAD TIRTOSUDIRO; Sec.-Gen. SUTOYO.

Badan Pemeriksa Keuangan (BPK) (Supreme Audit Board): Jalan Gatot Subroto 31, Jakarta; tel. (21) 584081; internet www.bpk.go.id; Chair. Prof. Dr SATRIO BUDIHARDJO JUDONO; Vice-Chair. Drs BAMBANG TRIADJI.

President and Legislature

PRESIDENT

Presidential Election, First Ballot, 5 July 2004

Candidate	Votes	% of votes
Susilo Bambang Yudhoyono (DP)	39,838,184	33.57
Megawati Sukarnoputri (PDI—P)	31,569,104	26.61
Gen. Wiranto (Golkar)	26,286,788	22.15
Amien Rais (PAN)	17,392,931	14.66
Hamzah Haz (PPP)	3,569,861	3.01
Total	118,656,868	100.00

Presidential Election, Second Ballot, 20 September 2004

Candidate	Votes	% of votes
Susilo Bambang Yudhoyono (DP)	69,266,350	60.62
Megawati Sukarnoputri (PDI—P)	44,990,704	39.38
Total	114,257,054	100.00

LEGISLATURE

Majelis Permusyawaratan Rakyat (MPR) (People's Consultative Assembly)

Jalan Jendral Gatot Subroto 6, Jakarta 10270; tel. (21) 5715268; fax (21) 5715611; e-mail kotaksurat@mpr.go.id; internet www.mpr.go.id.

In late 2002 the Constitution was amended to provide for the direct election of all members of the Majelis Permusyawaratan Rakyat (MPR—People's Consultative Assembly) at the next general election, held in 2004. The MPR thus became a bicameral institution comprising the Dewan Perwakilan Daerah (DPD—House of Representatives of the Regions) and the Dewan Perwakilan Rakyat (DPR—House of Representatives). The MPR consists of the 550 members of the DPR and 128 regional delegates (subject to confirmation).

Speaker: HIDAYAT NUR WAHID.

	Seats
Members of the Dewan Perwakilan Rakyat	550
Regional representatives	128
Total	678

Dewan Perwakilan Rakyat (DPR) (House of Representatives)

Jalan Gatot Subroto 16, Jakarta; tel. (21) 586833; e-mail humas-dpr@dpr.go.id; internet www.dpr.go.id.

Speaker: AGUNG LAKSONO.

General Election, 5 April 2004

	Seats
Partai Golongan Karya (Golkar)	128
Partai Demokrasi Indonesia Perjuangan (PDI—P)	109
Partai Persatuan Pembangunan (PPP)	58
Partai Demokrat	57
Partai Amanat Nasional (PAN)	52
Partai Kebangkitan Bangsa (PKB)	52
Partai Keadilan Sejahtera (PKS)	45
Partai Bintang Reformasi (PBR)	13
Partai Damai Sejahtera (PDS)	12
Partai Bulan Bintang (PBB)	11
Partai Persatuan Demokrasi Kebangsaan (PPDK)	5
Partai Karya Peduli Bangsa (PKPB)	2
Partai Keadilan dan Persatuan Indonesia (PKPI)	1
Others	5
Total	550

Political Organizations

Prior to 1998, electoral legislation permitted only three organizations (Golkar, the PDI and PPP) to contest elections. Following the replacement of President Suharto in May 1998, political restrictions were relaxed and new parties were allowed to form (with the only condition being that all parties must adhere to the *pancasila* and reject communism). A total of 24 parties registered to contest the legislative elections of April 2004.

Barisan Nasional (National Front): Jakarta; f. 1998; committed to ensuring that Indonesia remains a secular state; Sec.-Gen. RACHMAT WITOELAR.

Indonesian National Unity: Jakarta; f. 1995 by fmr mems of Sukarno's National Party; seeks full implementation of 1945 Constitution; Chair. SUPENI.

Indonesian Reform Party (PPI): Jakarta; f. 1998; Gen. Chair. CHANDRA KUWATLI; Sec.-Gen. Dr H. ACE MULYADI.

Islamic Indonesian Party (PII): Jakarta; f. 1998; Pres. SUUD BAJEBER; Sec.-Gen. SYAIFUL MUNIR.

National Brotherhood Foundation: Jakarta; f. 1995; Chair. KHARIS SUHUD.

New Indonesian National Party: Jakarta; f. 1998.

Partai Amanat Nasional (PAN) (National Mandate Party): Jalan H. Nawi 15, Jakarta Selatan 12420; tel. (21) 72794535; fax (21) 7268695; internet www.geocities.com/CapitolHill/Congress/6678; f. 1998; aims to achieve democracy, progress and social justice, to limit the length of the presidential term of office and to increase autonomy in the provinces; Gen. Chair. Dr AMIEN RAIS; Sec.-Gen. HATTA RADJASA.

Partai Bhinneka Tunggal Ika (PBI): c/o Dewan Perwakilan Rakyat, Jalan Gatot Subroto 16, Jakarta.

Partai Bintang Reformasi (PBR) (Reform Star Party): Jalan Radio IV 5, Kramat Pela, Kebayoran Baru, Jakarta; tel. (21) 7211132; fax (21) 7209734; f. 2002 by fmr members of the Partai Persatuan Pembangunan (PPP); Islamic party; Chair. ZAENUDDIN; Sec.-Gen. DJA'FAR BADJEBER.

Partai Bulan Bintang (PBB) (Crescent Moon and Star Party): Jalan Raya Pasar Minggu 1B, Km 18, Jakarta Selatan; tel. (21) 3106739; f. 1998; Leader Dr YUSRIL IHZA MAHENDRA.

Partai Buruh Sosial Demokrat (PBSD) (Socialist Democratic Labour Party): Jalan Kramat Raya 91A, Jakarta Pusat; tel. (21)

3154092; fax (21) 3909834; f. 2002; Chair. MUCHTAR PAKPAHAN; Sec.-Gen. DIAH INDRIASTUTI.

Partai Damai Sejahtera (PDS) (Prosperous Peace Party): Jalan Rukan Artha Gading Niaga 10, Blok B, Kelapa Gading, Jakarta Utara; tel. (21) 45850517; fax (21) 45850518; e-mail berita@partaidamaisejahtera.com; internet www.partaidamaisejahtera.com; f. 2001; Chair. RUYANDI MUSTIKA HUTASOIT; Sec.-Gen. MAGIT LES DENNY TEWU.

Partai Demokrasi Indonesia (PDI) (Indonesian Democratic Party): Jalan Diponegoro 58, Jakarta 10310; tel. (21) 336331; fax (21) 5201630; f. 1973 by merger of five nationalist and Christian parties; Chair. SOERJADI (installed to replace Megawati Sukarnoputri as leader of the party in a government-orchestrated coup in 1996).

Partai Demokrasi Indonesia Perjuangan (PDI—P) (Indonesian Democratic Struggle Party): Jalan Raya Pasar Minggu, Lenteng Agung, Jakarta; established by Megawati Sukarnoputri, fmr leader of the Partai Demokrasi Indonesia (PDI—see above), following her removal from the leadership of the PDI by the Government in 1996; Chair. MEGAWATI SUKARNOPUTRI; Sec.-Gen. SUTJIPTO.

Partai Demokrasi Kasih Bangsa Indonesia (PDKB) (The Nation Compassion Democratic Party of Indonesia): Kompleks Widuri Indah Blok A-4, Jalan Palmerah Barat 353, Jakarta Selatan 12210; tel. and fax (21) 5330973; e-mail info@kasihbangsa.net; internet www.kasihbangsa.net.

Partai Demokrat (DP): Jalan Pemuda 712A, Jakarta; tel. (21) 4755254; fax (21) 4754959; f. 2001; Chair. Dr S. BUDHISANTOSO; Sec.-Gen. Dr UMAR SAID.

Partai Golongan Karya (Golkar) (Functional Group): Jalan Anggrek Nellimurni, Jakarta 11480; tel. (21) 5302222; fax (21) 5303380; f. 1964; reorg. 1971; 23m. mems (1999); Co-ordinator of Advisers Haji HARMOKO; Pres. and Chair. Ir AKBAR TANDJUNG; Sec.-Gen. BUDI HARSONO.

Partai Karya Peduli Bangsa (PKPB) (Concern for the Nation Functional Party): Jalan Cimandiri 30, Raden Saleh Cikini, Jakarta Pusat 13033; tel. (21) 31927421; fax (21) 31937417; f. 2002; Chair. R. HARTONO; Sec.-Gen. H. ARY MARDJONO.

Partai Keadilan (PK) (Justice Party): c/o Dewan Perwakilan Rakyat, Jalan Gatot Subroto 16, Jakarta; e-mail partai@keadilan.or.id; internet www.keadilan.or.id; f. 1998; Islamic; Pres. Dr NUR MAHMUDI ISMA'IL; Sec.-Gen. LUTHFI HASAN ISHAAQ.

Partai Keadilan dan Persatuan Indonesia (PKPI) (Justice and Unity Party): Jalan Cilandak Raya KKO 32, Jakarta 12560; f. 2002; Chair. EDI SUDRADJAT; Sec.-Gen. SEMUEL SAMSON.

Partai Keadilan Sejahtera (PKS) (Prosperous Justice Party): Jalan Mampang Prapatan Raya 98 D–E–F, Jakarta Selatan; tel. (21) 7995425; fax (21) 7995433; e-mail partai@pk-sejahtera.org; internet www.pk-sejahtera.org; f. 2002; Islamic party; Chair. TIFATUL SEMBIRING (acting); Sec.-Gen. ANIS MATTA.

Partai Kebangkitan Bangsa (PKB) (National Awakening Party): Jalan Kalibata Timur 12, Jakarta Selatan 12519; tel. (21) 7974353; fax (21) 7974263; e-mail fahmi201@yahoo.com; internet www.kebangkitanbangsa.org; Islamic; f. 1998; Chair. ALWI ABDURRAHMAN SHIHAB; Sec.-Gen. SAIFULLAH YUSUF.

Partai Kebangkitan Umat (PKU) (Islamic Awakening Party): c/o Dewan Perwakilan Rakyat, Jalan Gatot Subroto 16, Jakarta; f. 1998 by clerics and members of the Nahdlatul Ulama (NU), with the aim of promoting the adoption of Islamic law in Indonesia.

Partai Merdeka (Freedom Party): Jalan Majapahit, Kav. 26, Jakarta Pusat; tel. (21) 3861464; fax (21) 3861465; e-mail info@partaimerdeka.or.id; internet www.partaimerdeka.or.id; f. 2002; Chair. ADI SASONO; Sec.-Gen. DHARMA SETIAWAN.

Partai Nahdlatul Ummah Indonesia (PNUI) (Indonesian Nahdlatul Community Party): Jalan Cipinang Cempedak 4, 1 Jatinegara, Jakarta Timur 13340; tel. and fax (21) 8571736; f. 2003; Islamic party; Chair. SYUKRON MA'MUN; Sec.-Gen. ACHMAD SJATARI.

Partai Nasional Banteng Kemerdekaan (PNBK) (Freedom Bull National Party): Jalan Penjernihan I 50, Jakarta Utara; tel. (21) 5739550; fax (21) 5739519; f. 2002; Chair. EROS DJAROT; Sec.-Gen. SOEHARDI SOEDIRO.

Partai Nasional Indonesia Marhaenisme (PNI Marhaenisme): Jalan Cikoko 15, Pancoran, Jakarta 12770; tel. (21) 8971241; fax (21) 7900489; f. 2002; Chair. SUKMAWATI SUKARNOPUTRI; Sec.-Gen. ACHMAD MARHAEN.

Partai Patriot Pancasila (Pancasila Patriot Party): Gedung Tri Tangguh, 3rd Floor, Jalan Haji Samali 31, Kalibata, Jakarta Selatan; tel. (21) 79198510; fax (21) 79198520; f. 2001; Chair. YAPTO SULISTIO SOERJOSOEMARNO; Sec.-Gen. SOPHAR MARU.

Partai Pelopor (Pioneer Party): Jalan K. H. Syafei A22, Gudang Peluru, Tebet, Jakarta Selatan; tel. (21) 8299112; fax (21) 8301569; internet www.rachmawati.com; f. 2002; Chair. RACHMAWATI SUKARNOPUTRI; Sec.-Gen. EKO SURYO SANTJOJO.

Partai Pembauran (Assimilation Party): Jakarta; f. 1998; Chinese.

Partai Penegak Demokrasi Indonesia (PPDI): Jalan R. E. Martadinata, Kompleks Rukan Permata, Blok E-1, Ancol, Jakarta Utara; tel. (21) 6456215; fax (21) 6456216; f. 2003; Chair. DIMMY HARYANTO; Sec.-Gen. JOSEPH WILLIEM LEA.

Partai Perhimpunan Indonesia Baru (PIB) (New Indonesia Alliance Party): Jalan Teuku Cik Ditiro 31, Jakarta; tel. (21) 3108057; f. 2002; Chair. Dr SYAHRIR; Sec.-Gen. AMIR KARAMOY.

Partai Persatuan Daerah (Regional United Party): Jalan Dr Satrio C-4 18, Jakarta Selatan 12940; tel. (21) 5205764; fax (21) 5273249; f. 2002; Chair. OESMAN SAPTA; Sec.-Gen. H. RONGGO SOENARSO.

Partai Persatuan Demokrasi Kebangsaan (PPDK) (National Democratic Unity Party): Jalan Ampera Raya 99, Jakarta Selatan 12560; tel. (21) 7807432; fax (21) 7817341; f. 2002; Chair. Dr RYAAS RASYID; Sec. Gen. RIVAI PULUNGAN.

Partai Persatuan Pembangunan (PPP) (United Development Party): Jalan Diponegoro 60, Jakarta 10310; tel. (21) 336338; fax (21) 3908070; e-mail dpp@ppp.or.id; internet www.ppp.or.id; f. 1973 by the merger of four Islamic parties; Leader HAMZAH HAZ; Sec.-Gen. YUNUS YOSFIAH.

Partai Rakyat Demokratik (PRD) (People's Democratic Party): Jalan Tebet Barat Dalam VIII Nomor 4, Jakarta Selatan; tel. (21) 8296467; e-mail prd@centrin.net.id; internet www.prd.4-all.org; f. 1996; Chair. BUDIMAN SUJATMIKO.

Partai Reformasi Tionghoa Indonesia (Chinese Indonesian Reform Party): Jakarta; e-mail parti_id@usa.net; f. 1998; Chinese.

Partai Sarikat Indonesia (PSI) (Indonesia Unity Party): Jalan Ampera Raya 65, Cilandak, Jakarta Selatan; tel. (21) 78847138; fax (21) 7800106; e-mail dpppsi@indosat.net.id; internet www.psi.online.or.id; f. 2002; Chair. RAHARDJO TJAKRANINGRAT; Sec.-Gen. MUHAMMAD JUMHUR HIDAYAT.

Partai Uni Demokrasi Indonesia (PUDI) (Democratic Union Party of Indonesia): Jalan Raya Tanjung Barat 81, Jakarta 12530; tel. (21) 7817565; fax (21) 7814765; e-mail pudi@pudi.or.id; internet www.pudi.or.id; f. 1996; Chair. Sri BINTANG PAMUNGKAS; Sec.-Gen. ESA HARUMAN.

Other groups with political influence include:

Ikatan Cendekiawan Muslim Indonesia (ICMI) (Association of Indonesian Muslim Intellectuals): Gedung BPPT, Jalan M. H. Thamrin 8, Jakarta; tel. (21) 3410382; e-mail nama_anda@icmi.or.id; internet www.icmi.or.id; f. 1990 with government support; Chair. ACHMAD TIRTOSUDIRO; Sec.-Gen. ADI SASONO.

Masyumi Baru: Jalan Pangkalan Asem 12, Cempaka Putih Ba, Jakarta Pusat; tel. (21) 4225774; fax (21) 7353077; Sec.-Gen. RIDWAN SAIDI.

Muhammadiyah: Jalan Menteng Raya 62, Jakarta Pusat; tel. (21) 3903021; fax (21) 3903024; e-mail ppmuh@indosat.net.id; internet www.muhammadiyah.or.id; second largest Muslim organization; f. 1912; 28m. mems; Chair. Dr AHMAD SYAFII MAARIF.

Nahdlatul Ulama (NU) (Council of Scholars): Jalan H. Agus Salim 112, Jakarta Pusat; tel. (21) 336250; largest Muslim organization; 30m. mems; Chair. AHMAD HASYIM.

Syarikat Islam: Jalan Taman Amir Hamzah Nomor 2, Jakarta; tel. (21) 31906037.

The following groups remain in conflict with the Government:

Gerakan Aceh Merdeka (GAM) (Free Aceh Movement): based in Aceh; f. 1976; seeks independence from Indonesia; Leader HASAN DI TIRO; Military Commdr MUZZAKIR MANAF.

National Liberation Front Acheh Sumatra: based in Aceh; f. 1989; seeks independence from Indonesia.

Organisasi Papua Merdeka (OPM) (Free Papua Movement): based in Papua; f. 1963; seeks unification with Papua New Guinea; Chair. MOZES WEROR; Leader KELLY KWALIK.

Presidium Dewan Papua (PDP) (Papua Presidium Council): based in Papua; internet www.westpapua.net; seeks independence from Indonesia; Chair. (vacant); Sec.-Gen. TOM BEANAL.

Diplomatic Representation

EMBASSIES IN INDONESIA

Afghanistan: Jalan Dr Kusuma Atmaja 15, Jakarta Pusat 10310; tel. (21) 3143169; fax (21) 335390; e-mail afghanembassy_jkk@yahoo.com; Ambassador ABDUL RAHEEM SAYED JAN.

Algeria: Jalan H. R. Rasuna Said, Kav. 10-1, Kuningan, Jakarta 12950; tel. (21) 5254719; fax (21) 5254654; e-mail ambalyak@rad.net.id; internet www.algeria-id.org; Ambassador SOUFIANE MIMOUNI.

Argentina: Menara Mulia, Suite 1901, Jalan Jenderal Gatot Subroto, Kav. 9–11, Jakarta 12930; tel. (21) 5265661; fax (21) 5265664; e-mail embargen@cbn.net.id; Ambassador JOSÉ LUIS MIGNINI.

Australia: Jalan H. R. Rasuna Said, Kav. C15–16, Kuningan, Jakarta 12940; tel. (21) 25505555; fax (21) 25505467; e-mail public-affairs-jakt@dfat.gov.au; internet www.austembjak.or.id; Ambassador DAVID JAMES RITCHIE.

Austria: Jalan Diponegoro 44, Menteng, Jakarta 10310; tel. (21) 31938101; fax (21) 3904927; e-mail jakarta-ob@bmaa.gv.at; internet www.austrian-embassy.or.id; Ambassador Dr BERNHARD ZIMBURG.

Bangladesh: Jalan Denpasar Raya 3, Block A-13, Kav. 10, Kuningan, Jakarta 12950; tel. (21) 5221574; fax (21) 5261807; e-mail bdootjak@dnet.net.id; internet www.bangladeshembassyjakarta.or.id; Ambassador NASIM FIRDAUS.

Belgium: Deutsche Bank Bldg, 16th Floor, Jalan Imam Bonjol 80, Jakarta 10310; tel. (21) 3162030; fax (21) 3162035; e-mail jakarta@diplobel.org; Ambassador HANS-CHRISTIAN KINT.

Bosnia and Herzegovina: Menara Imperium, 11th Floor, Suite D-2, Metropolitan Kuningan Super Blok, Kav. 1, Jalan H. R. Rasuna Said, Jakarta 12980; tel. (21) 83703022; fax (21) 83703029; Ambassador ZDRAVKO RAJIĆ.

Brazil: Menara Mulia, Suite 1602, Jalan Jenderal Gatot Subroto, Kav. 9, Jakarta 12390; tel. (21) 5265656; fax (21) 5265659; e-mail brasemb@rad.net.id; Ambassador CARLOS EDUARDO SETTE.

Brunei: Jalan Tanjung Karang 7, Jakarta Pusat 10230; tel. (21) 31906080; fax (21) 31905070; e-mail kbjindo@cbn.net.id; Ambassador Dato' Seri Setia Dr Haji MOHAMMAD AMIN BIN PEHI.

Bulgaria: Jalan Imam Bonjol 34–36, Jakarta 10310; tel. (21) 3904048; fax (21) 3904049; e-mail bgemb.jkt@centrin.net.id; Chargé d'affaires a.i. Dr PANTELEY M. SPASOV.

Cambodia: 33 Jalan Kintamani Raya C-15, Jakarta 12950; tel. (21) 9192895; fax (21) 5202673; e-mail recjkt@cabi.net.id; Ambassador KHEM BUNNEANG.

Canada: World Trade Centre, 6th Floor, Jalan Jenderal Sudirman, Kav. 29–31, POB 8324/JKS, Jakarta 12920; tel. (21) 25507800; fax (21) 25507811; e-mail canadianembassy.jkrta@dfait-maeci.gc.ca; internet www.dfait-maeci.gc.ca/jakarta/; Ambassador RANDOLPH MANK.

Chile: Bina Mulia Bldg, 7th Floor, Jalan H. R. Rasuna Said, Kav. 10, Kuningan, Jakarta 12950; tel. (21) 5201131; fax (21) 5201955; e-mail emchijak@indosat.net.id; Ambassador SINCLAIR MANLEY JAMES.

China, People's Republic: Jalan Mega Kuningan 2, Karet Kuningan, Jakarta 12950; tel. (21) 5761038; fax (21) 5761034; e-mail enbsychn@cbn.net.id; internet www.chinaembassy-indonesia.or.id; Ambassador LU SHUMIN.

Croatia: Menara Mulia, Suite 2101, Jalan Gatot Subroto, Kav. 9–11, Jakarta 12930; tel. (21) 5257822; fax (21) 5204073; e-mail croemb@rad.net.id; internet www.croatemb.or.id; Ambassador ALEKSANDAR BROZ.

Cuba: Taman Puri, Jalan Opal, Blok K-1, Permata Hijau, Jakarta 12210; tel. (21) 5304293; fax (21) 53676906; e-mail cubaindo@cbn.net.id; Ambassador MIGUEL ANGEL RAMÍREZ RAMOS.

Czech Republic: Jalan Gunwarman 2, Kebayoran Baru, Jakarta; tel. (21) 7231226; fax (21) 7231436; e-mail jakarta@embassy.mzv.cz; internet www.mfa.cz/jakarta; Ambassador JAROSLAV VESELÝ.

Denmark: Menara Rajawali, 25th Floor, Jalan Mega Kuningan, Lot 5.1, Jakarta 12950; tel. (21) 5761478; fax (21) 5761535; e-mail jktamb@um.dk; internet www.emb-denmark.or.id; Ambassador GEERT AAGAARD ANDERSEN.

Egypt: Jalan Denpasar Raya, Blok A12, No 1, Kuningan Timur, Setiabudi, Jakarta Selatan 12950; tel. (21) 5204359; fax (21) 5204792; e-mail egypt@indosat.net.id; Ambassador EZZAT SAAD.

Finland: Menara Rajawali, 9th Floor, Jalan Mega Kuningan, Kawasan Mega Kuningan, Jakarta 12950; tel. (21) 5761650; fax (21) 5761631; e-mail sanomat.jak@formin.fi; internet www.finembjak.com; Ambassador MARKKU NIINIOJA.

France: Jalan M. H. Thamrin 20, Jakarta 10350; tel. (21) 3142807; fax (21) 3143338; e-mail ambassade@ambafrance-id.org; internet www.ambafrance-id.org; Ambassador RENAUD VIGNAL.

Germany: Jalan M. H. Thamrin 1, Jakarta 10310; tel. (21) 3901750; fax (21) 3901757; e-mail germany@rad.net.id; internet www.deutschebotschaft-jakarta.or.id; Ambassador (vacant).

Greece: Plaza 89, 12th Floor, Suite 1203, Jalan H. R. Rasuna Said, Kav. X-7 No. 6, Kuningan, Jakarta 12540; tel. (21) 5207776; fax (21) 5207753; e-mail grembas@cbn.net.id; internet www.greekembassy.or.id; Ambassador ALEXIOS G. CHRISTOPOULOS.

Holy See: Jalan Merdeka Timur 18, POB 4227, Jakarta Pusat (Apostolic Nunciature); tel. (21) 3841142; fax (21) 3841143; e-mail vatjak@cbn.net.id; Apostolic Nuncio Most Rev. PATABENDIGE DON ALBERT MALCOLM RANJITH (Titular Archbishop of Umbriatico).

Hungary: 36 Jalan H. R. Rasuna Said, Kav. X/3, Kuningan, Jakarta 12950; tel. (21) 5203459; fax (21) 5203461; e-mail huembjkt@rad.net.id; internet www.huembjkt.or.id; Ambassador GYÖRGY BUSZTIN.

India: Jalan H. R. Rasuna Said, Kav. S/1, Kuningan, Jakarta 12950; tel. (21) 5204150; fax (21) 5204160; e-mail eoiisi@indo.net.id; internet www.eoijakarta.or.id; Ambassador HEMANT KRISHAN SINGH.

Iran: Jalan Hos Cokroaminoto 110, Menteng, Jakarta Pusat 10310; tel. (21) 331391; fax (21) 3107860; e-mail irembjkt@indo.net.id; internet www.iranembassy.or.id; Ambassador SHABAN SHAHIDI MOADAB.

Iraq: Jalan Teuku Umar 38, Jakarta 10350; tel. (21) 3904067; fax (21) 3904066; e-mail iraqembi@rad.net.id.

Italy: Jalan Diponegoro 45, Jakarta 10310; tel. (21) 337445; fax (21) 337422; e-mail italemba@italambjkt.or.id; internet www.italambjkt.or.id; Ambassador FRANCESCO MARIA GRECO.

Japan: Menara Thamrin, 7th–10th Floors, Jalan M. H. Thamrin, Kav. 3, Jakarta 10350; tel. (21) 324308; fax (21) 325460; internet www.id.emb-japan.go.jp; Ambassador YUTAKA IIMURA.

Jordan: Jalan Denpasar Raya, Blok A XIII, Kav. 1–2, Jakarta 12950; tel. (21) 5204400; fax (21) 5202447; e-mail jordanem@cbn.net.id; internet www.jordanembassy.or.id; Ambassador MOHAMED ALI DAHER.

Korea, Democratic People's Republic: Jalan Teluk Betung 1–2, Jakarta Pusat 12050; tel. (21) 31908425; fax (21) 31908427; Ambassador JANG CHANG-CHON.

Korea, Republic: Jalan Jenderal Gatot Subroto 57, Jakarta Selatan; tel. (21) 5201915; fax (21) 5254159; e-mail koremb_in@mofat.go.kr; internet www.mofat.go.kr/indonesia; Ambassador YUN HAI-JUNG.

Kuwait: Jalan Teuku Umar 51, Menteng, Jakarta 10310; tel. (21) 3919916; fax (21) 3912285; e-mail ami@Kuwait-toplist.com; Ambassador MOHAMMED FADEL KHALAF.

Laos: Jalan Patra Kuningan XIV 1-A, Kuningan, Jakarta 12950; tel. (21) 5229602; fax (21) 5229601; e-mail laoemjktof@hotmail.com; Ambassador LEUANE SOMBOUNKHAN.

Lebanon: Jalan YBR V 82, Kuningan, Jakarta 12950; tel. (21) 5253074; fax (21) 5207121; e-mail lebanon_embassy_jkt@yahoo.com; Ambassador TAUFIK JABER.

Libya: Jalan Pekalongan 24, Jakarta Pusat 10310; tel. (21) 31935308; fax (21) 31935726; e-mail gsplaj@cbn.net.id; Chargé d'affaires a.i. ALI MAMBRUK ASH SHERIQY.

Malaysia: Jalan H. R. Rasuna Said, Kav. X/6, 1–3 Kuningan, Jakarta 12950; tel. (21) 5224947; fax (21) 5224974; e-mail mwjkarta@indosat.net.id; Ambassador Dato' HAMIDON ALI.

Mali: Jalan Mendawai III 18, Kebayoran Baru, Jakarta 12130; tel. (21) 7208472; fax (21) 7229589; e-mail ambamali@indosat.net.id; Ambassador AMADOU N'DIAYE.

Marshall Islands: Jalan Pangeran Jayakarta 115, Blok A-11, Jakarta Pusat 10730; tel. (21) 7248565; fax (21) 7248566; e-mail marshall@idola.net.id; Ambassador CARL L. HEINE.

Mexico: Menara Mulia, Suite 2306, Jalan Gatot Subroto, Kav. 9–11, Jakarta 12930; tel. (21) 5203980; fax (21) 5203978; e-mail embmexic@rad.net.id; Ambassador PEDRO GONZÁLES-RUBIO SÁNCHEZ Sr.

Morocco: Suite 512, 5th Floor, South Tower, Kuningan Plaza, Jalan H. R. Rasuna Said C-11–14, Jakarta 12940; tel. (21) 5200773; fax (21) 5200586; e-mail sifamajakar@cbn.net.id; internet www.morocco-embassy.or.id; Ambassador M. ABDERAHMAN DRISSI ALAMI.

Mozambique: Wisma GKBI, 37th Floor, Suite 3709, Jalan Jenderal Sudirman 28, Jakarta 10210; tel. (21) 5740901; fax (21) 5740907; e-mail embamoc@cbn.net.id; Ambassador GERALDO ANTONIO CHIRINZA.

Myanmar: Jalan Haji Agus Salim 109, Jakarta Selatan; tel. (21) 327684; fax (21) 327204; e-mail myanmar@cbn.net.id; Ambassador U KYAW MYINT.

Netherlands: Jalan H. R. Rasuna Said, Kav. S/3, Kuningan, Jakarta 12950; tel. (21) 5251515; fax (21) 5700734; e-mail jak-cdp@minbuza.nl; internet www.netherlandsembassy.or.id; Ambassador RUDOLF JAN TREFFERS.

New Zealand: BRI II Bldg, 23rd Floor, Jalan Jenderal Sudirman, Kav. 44–46, Jakarta; tel. (21) 5709460; fax (21) 5709457; e-mail nzembjak@cbn.net.id; Ambassador CHRIS ELDER.

Nigeria: Jalan Tamam Patra XIV/11–11A, Kuningan Timur, POB 3649, Jakarta Selatan 12950; tel. (21) 5260922; fax (21) 5260924; e-mail embnig@centrin.net.id; Ambassador MUHAMMED BUBA AHMED.

Norway: Menara Rajawali, 25th Floor, Kawasan Mega Kuningan, Jakarta 12950; tel. (21) 5761523; fax (21) 5761537; e-mail emb.jakarta@mfa.no; internet www.norwayemb-indonesia.org; Ambassador BJØRN BLOKHUS.

Pakistan: Jalan Lembang 10, Menteng, Jakarta; tel. (21) 3144008; fax (21) 3103945; e-mail parepjkt@link.net.id; Ambassador Maj.-Gen. (retd) SYED MUSTAFA ANWER HUSAIN.

Panama: World Trade Centre, 8th Floor, Jalan Jenderal Sudirman, Kav. 29–31, Jakarta 12920; tel. (21) 5711867; fax (21) 5711933; e-mail panacon@pacific.net.id; Ambassador VIRGINIA WEDEN DE ACOSTA.

Papua New Guinea: Panin Bank Centre, 6th Floor, Jalan Jenderal Sudirman 1, Jakarta 10270; tel. (21) 7251218; fax (21) 7201012; e-mail kdujkt@cbn.net.id; Ambassador CHRISTOPHER MERO.

Peru: Menara Rajawali, 12th Floor, Jalan Mega Kuningan, Lot 5.1, Kawasan Mega Kuningan, Jakarta 12950; tel. (21) 5761820; fax (21) 5761825; e-mail embaperu@cbn.net.id; Ambassador NILO FIGUEROA CORTAVARRIA.

Philippines: Jalan Imam Bonjol 6–8, Jakarta Pusat 10310; tel. (21) 3100334; fax (21) 3151167; e-mail phjkt@indo.net.id; Ambassador SHULAN O. PRIMAVERA.

Poland: Jalan H. R. Rasuna Said, Kav. X, Blok IV/3, Jakarta Selatan 12950; tel. (21) 2525948; fax (21) 2525958; e-mail plembjkt@net.id; internet www.polandembjak.org; Ambassador KRZYSZTOF SZUMSKI.

Portugal: Bina Mulia Bldg I, 7th Floor, Jalan H. R. Rasuna Said, Kav. X, Kuningan, Jakarta 12950; tel. (21) 5265103; fax (21) 5271981; e-mail porembjak@cbn.net.id; Ambassador JOSÉ MANUEL SANTOS BRAGA.

Qatar: Jalan Taman Ubud I, No. 5, Kuningan Timur, Jakarta 12920; tel. (21) 5277751; fax (21) 5277754; e-mail qataremj@indosat.net.id; Ambassador ABDULLA MOHAMMED TALIB AL-MARRI.

Romania: Jalan Mas Putih 29, Kompleks D, Permata Hijau, Jakarta Selatan; tel. (21) 5305073; fax (21) 5305074; e-mail romind@cbn.net.id; Ambassador GHEORGHE SAVUICA.

Russia: Jalan H. R. Rasuna Said, Kav. X-6, Jakarta; tel. (21) 5222912; e-mail rusembjkt@dnet.net.id; Chargé d'affaires a.i. SERGEI PETLYAKV.

Saudi Arabia: Jalan M. T. Haryono, Kav. 27, Cawang Atas, Jakarta Timur; tel. (21) 8011533; fax (21) 3905864; e-mail idemb@mofa.gov.sa; Ambassador ABDULLAH BIN ABDULRAHMAN A'ALIM.

Serbia and Montenegro: Jalan Hos Cokroaminoto 109, Jakarta 10310; tel. (21) 3143560; fax (21) 3143613; e-mail ambajaka@rad.net.id; Chargé d'affaires a.i. DRAGAN PETROVIĆ.

Singapore: Jalan H. R. Rasuna Said, Blok X/4, Kav. 2, Kuningan, Jakarta 12950; tel. (21) 5201489; fax (21) 5201486; e-mail denpasar@pacific.net.id; internet www.mfa.gov.sg/jkt; Ambassador EDWARD LEE.

Slovakia: Jalan Prof. Mohammed Yamin 29, POB 1368, Jakarta Pusat; tel. (21) 3101068; fax (21) 3101180; e-mail slovemby@indo.net.id; Ambassador PETER HOLASEK.

South Africa: Suite 705, Wisma GKBI, Jalan Jenderal Sudirman 28, Jakarta 10210; tel. (21) 5740660; fax (21) 5740661; e-mail saembhom@mweb.co.id; internet www.saembassy-jakarta.or.id; Ambassador NORMAN M. MASHABANE.

Spain: Jalan H. Agus Salim 61, Jakarta 10350; tel. (21) 335937; fax (21) 325996; e-mail embespid@mail.mae.es; Ambassador DÁMASO DE LARIO RAMÍREZ.

Sri Lanka: Jalan Diponegoro 70, Jakarta 10320; tel. (21) 3161886; fax (21) 3107962; e-mail lankaemb@rad.net.id; Ambassador GRACE NESAMALAR NILIAH.

Sudan: Wisma Bank Dharmala, 7th Floor, Suite 01, Jalan Jenderal Sudirman, Kav. 28, Jakarta 12920; tel. (21) 5212099; fax (21) 5212077; e-mail leen@cbn.net.id; Ambassador SIDIQ YOUSIF ABU-AGLA.

Suriname: Gedung Plaza Central, 16th Floor, Jalan Jenderal Sudirman, Kav. 47, Jakarta 12930; tel. (21) 5742878; fax (21) 5740015; e-mail suramjkt@cbn.net.id; Ambassador SAHIDI RASAM.

Sweden: POB 2824, Jakarta 10001; tel. (21) 55535900; fax (21) 5762691; e-mail ambassaden.jakarta@foreign.ministry.se; internet www.swedenabroad.com/jakarta; Ambassador LENNART LINNÉR.

Switzerland: Jalan H. R. Rasuna Said, Kav. X-3/2, Kuningan, Jakarta Selatan 12950; tel. (21) 5256061; fax (21) 5202289; e-mail vertretung@jak.rep.admin.ch; internet www.eda.admin.ch/jakarta; Ambassador GEORGES MARTIN.

Syria: Jalan Karang Asem I/8, Jakarta 12950; tel. (21) 5255991; fax (21) 5202511; e-mail syrianemb@cbn.net.id; Ambassador Dr BASHAR JAAFARI.

Thailand: Jalan Imam Bonjol 74, Jakarta 10310; tel. (21) 3904052; fax (21) 3107469; e-mail thaijkt@indo.net.id; Ambassador ATCHARA SERIPUTRA.

Timor-Leste: Gedung Surya, 11th Floor, Jalan M. H. Thamrin, Kav. 9, Jakarta; tel. (21) 3902678; fax (21) 3902660; Ambassador Rev. ARLINDO MARÇAL.

Tunisia: Wisma Dharmala Sakti, 11th Floor, Jalan Jenderal Sudirman 32, Jakarta 10220; tel. (21) 5703432; fax (21) 5700016; e-mail atjkt@uninet.net.id; Ambassador MOHAMED MOULDI KEFI.

Turkey: Jalan H. R. Rasuna Said, Kav. 1, Kuningan, Jakarta 12950; tel. (21) 5256250; fax (21) 5226056; e-mail cakartabe@telkom.net; Ambassador FERYAL ÇOTUR.

Ukraine: WTC Bldg, 8th Floor, Jalan Jenderal Sudirman, Kav. 29–31, Jakarta 12084; tel. (21) 5211700; fax (21) 5211710; e-mail uaembas@indo.net.id; Chargé d'affaires a.i. VALERIY KRAVCENKO.

United Arab Emirates: Jalan Prof. Dr Satrio, Kav. 16–17, Jakarta 12950; tel. (21) 5206518; fax (21) 5206526; e-mail uaeemb@rad.net.id; Ambassador YOUSIF RASHID ALSHRAM.

United Kingdom: Jalan M. H. Thamrin 75, Jakarta 10310; tel. (21) 3156264; fax (21) 3926263; e-mail britem2@ibm.net.id; internet www.britain-in-indonesia.or.id; Ambassador CHARLES HUMFREY.

USA: Jalan Merdeka Selatan 4–5, Jakarta 10110; tel. (21) 34359000; fax (21) 3857189; e-mail jakconsul@state.gov; internet jakarta.usembassy.gov; Ambassador B. LYNN PASCOE (designate).

Uzbekistan: Menara Mulia, Suite 2401, 24th Floor, Jalan Jenderal Gatot Subroto, Kav. 9–11, Jakarta 12930; tel. (21) 5222581; fax (21) 5222582; e-mail registan@indo.net.id; Ambassador KHAMIS DJABBAROV.

Venezuela: Menara Mulia, 20th Floor, Suite 2005, Jalan Jenderal Gatot Subroto, Kav. 9–11, Jakarta Selatan 12930; tel. (21) 5227547; fax (21) 5227549; e-mail evenjakt@indo.net.id; Ambassador LUIS EDUARDO SOTO.

Viet Nam: Jalan Teuku Umar 25, Jakarta; tel. (21) 3100358; fax (21) 3100359; e-mail embvnam@uninet.net.id; Ambassador NGUYEN DANG QUANG.

Yemen: Jalan Yusuf Adiwinata 29, Menteng, Jakarta 10350; tel. (21) 3108029; fax (21) 3904946; e-mail yemenemb@rad.net.id; internet www.yemenembassyindonesia.com; Ambassador Dr AHMAD SALIM AL-WAHISHI.

Judicial System

There is one codified criminal law for the whole of Indonesia. In December 1989 the Islamic Judicature Bill, giving wider powers to Shari'a courts, was approved by the Dewan Perwakilan Rakyat (House of Representatives). The new law gave Muslim courts authority over civil matters, such as marriage. Muslims may still choose to appear before a secular court. Europeans are subject to the Code of Civil Law published in the State Gazette in 1847. Alien orientals (i.e. Arabs, Indians, etc.) and Chinese are subject to certain parts of the Code of Civil Law and the Code of Commerce. The work of codifying this law has started, but, in view of the great complexity and diversity of customary law, it may be expected to take a considerable time to achieve.

Supreme Court
(Mahkamah Agung)

Jalan Merdeka Utara 9–13, Jakarta 10110; tel. (21) 3843348; fax (21) 3811057; e-mail pansekjen@mari.go.id; internet www.mari.go.id.

The final court of appeal.

Chief Justice: Prof. BAGIR MANAN.

Deputy Chief Justice: H. TAUFIK.

High Courts in Jakarta, Surabaya, Medan, Makassar, Banda Aceh, Padang, Palembang, Bandung, Semarang, Banjarmasin, Menado, Denpasar, Ambon and Jayapura deal with appeals from the District Courts. **District Courts** deal with marriage, divorce and reconciliation.

Religion

All citizens are required to state their religion. According to a survey in 1985, 86.9% of the population were Muslims, while 9.6% were Christians, 1.9% were Hindus, 1.0% were Buddhists and 0.6% professed adherence to tribal religions.

Five national religious councils—representing the Islamic, Catholic, Protestant, Hindu and Buddhist religious traditions—were established to serve as liaison bodies between religious adherents and the Government and to advise the Government on the application of religious principles to various elements of national life.

ISLAM

In 1993 nearly 90% of Indonesians were Muslims. Indonesia has the world's largest Muslim population.

Majelis Ulama Indonesia (MUI) (Indonesian Ulama Council): Komp. Masjid Istiqlal, Jalan Taman Wijaya Kesuma, Jakarta 10710; tel. (21) 3455471; fax (21) 3855412; internet www.mui.or.id; central Muslim organization; Chair. SAHAL MAHFUDZ; Sec.-Gen. DIEN SYAMSUDDIN.

CHRISTIANITY

Persekutuan Gereja-Gereja di Indonesia (Communion of Churches in Indonesia): Jalan Salemba Raya 10, Jakarta 10430; tel. (21) 3908119; fax (21) 3150457; e-mail pgi@bit.net.id; internet www.pgi.or.id; f. 1950; 70 mem. churches; Chair. Rev. Dr SULARSO SOPATER; Gen. Sec. Rev. Dr JOSEPH M. PATTIASINA.

The Roman Catholic Church

Indonesia comprises 10 archdioceses and 26 dioceses. At 31 December 2002 there were an estimated 6,375,984 adherents in Indonesia, representing 3.3% of the population.

Bishops' Conference

Konferensi Waligereja Indonesia (KWI), Jalan Cut Meutia 10, POB 3044, Jakarta 10002; tel. (21) 336422; fax (21) 3918527; e-mail kwi@parokinet.org.

f. 1973; Pres. Cardinal JULIUS RIYADI DARMAATMADJA (Archbishop of Jakarta).

Archbishop of Ende: Most Rev. ABDON LONGINUS DA CUNHA, Keuskupan Agung, POB 210, Jalan Katedral 5, Ndona-Ende 86312, Flores; tel. (381) 21176; fax (381) 21606; e-mail uskup@ende.parokinet.org.

Archbishop of Jakarta: Cardinal JULIUS RIYADI DARMAATMADJA, Keuskupan Agung, Jalan Katedral 7, Jakarta 10710; tel. (21) 3813345; fax (21) 3855681.

Archbishop of Kupang: Most Rev. PETER TURANG, Keuskupan Agung Kupang, Jalan Thamrin, Oepoi, Kupang 85111, Timor NTT; tel. (380) 826199; fax (380) 833331.

Archbishop of Makassar: Most Rev. JOHANNES LIKU ADA', Keuskupan Agung, Jalan Thamrin 5–7, Makassar 90111, Sulawesi Selatan; tel. (411) 315744; fax (411) 326674; e-mail pseupg@indosat.net.id.

Archbishop of Medan: Most Rev. ALFRED GONTI PIUS DATUBARA, Jalan Imam Bonjol 39, POB 1191, Medan 20152, Sumatra Utara; tel. (61) 4519768; fax (61) 4145745; e-mail mar39@indosat.net.id.

Archbishop of Merauke: Most Rev. NICOLAUS ADI SEPTURA, Keuskupan Agung, Jalan Mandala 30, Merauke 99602, Papua; tel. (971) 321011; fax (971) 321311.

Archbishop of Palembang: Most Rev. ALOYSIUS SUDARSO, Keuskupan Agung, Jalan Tasik 18, Palembang 30135; tel. (711) 350417; fax (711) 314776.

Archbishop of Pontianak: Most Rev. HIERONYMUS HERCULANUS BUMBUN, Keuskupan Agung, Jalan A. R. Hakin 92A, POB 1119, Pontianak 78011, Kalimantan Barat; tel. (561) 732382; fax (561) 738785; e-mail kap@pontianak.wasantara.net.id.

Archbishop of Samarinda: Most Rev. FLORENTINUS SULUI HAJANG HAU, Keuskupan Agung, POB 1062, Jalan Gunung Merbabu 41, Samarinda 75010; tel. (541) 741193; fax (541) 203120.

Archbishop of Semarang: Most Rev. IGNATIUS SUHARYO HARDJOATMODJO, Keuskupan Agung, Jalan Pandanaran 13, Semarang 50231; tel. (24) 312276; fax (24) 414741; e-mail uskup@semarang.parokinet.org.

Other Christian Churches

Protestant Church in Indonesia (Gereja Protestan di Indonesia): Jalan Medan Merdeka Timur 10, Jakarta 10110; tel. (21) 3519003; fax (21) 34830224; consists of 10 churches of Calvinistic tradition; 2,789,155 mems, 3,841 congregations, 1,965 pastors (1998); Chair. Rev. Dr D. J. LUMENTA.

Numerous other Protestant communities exist throughout Indonesia, mainly organized on a local basis. The largest of these (1985 memberships) are: the Batak Protestant Christian Church (1,875,143); the Christian Church in Central Sulawesi (100,000); the Christian Evangelical Church in Minahasa (730,000); the Christian Protestant Church in Indonesia (210,924); the East Java Christian Church (123,850); the Evangelical Christian Church in Irian Jaya (Papua—360,000); the Evangelical Christian Church of Sangir-Talaud (190,000); the Indonesian Christian Church/Huria Kristen Indonesia (316,525); the Javanese Christian Churches (121,500); the Kalimantan Evangelical Church (182,217); the Karo Batak Protestant Church (164,288); the Nias Protestant Christian Church (250,000); the Protestant Church in the Moluccas (575,000); the Simalungun Protestant Christian Church (155,000); and the Toraja Church (250,000).

BUDDHISM

All-Indonesia Buddhist Association: Jakarta.

Indonesian Buddhist Council: Jakarta.

HINDUISM

Hindu Dharma Council: Jakarta.

The Press

In August 1990 the Government announced that censorship of both the local and foreign press was to be relaxed and that the authorities would refrain from revoking the licences of newspapers that violated legislation governing the press. In practice, however, there was little change in the Government's policy towards the press. Following the resignation of President Suharto in May 1998, the new Government undertook to allow freedom of expression.

PRINCIPAL DAILIES

Bali

Harian Pagi Umum (Bali Post): Jalan Kepudang 67A, Denpasar 80232; e-mail balipost@indo.net.id; internet www.balipost.co.id; f. 1948; daily (Indonesian edn), weekly (English edn); Editor K. NADHA; circ. 25,000.

Java

Angkatan Bersenjata: Jalan Kramat Raya 94, Jakarta Pusat; tel. (21) 46071; fax (21) 366870.

Bandung Post: Jalan Lodaya 38A, Bandung 40264; tel. (22) 305124; fax (22) 302882; internet www.bandung-post.com; f. 1979; Chief Editor AHMAD SAELAN; Dir AHMAD JUSACC.

Berita Buana: Jalan Tahah Abang Dua 33–35, Jakarta 10110; tel. (21) 5487175; fax (21) 5491555; internet www.beritabuana.net; f. 1970; relaunched 1990; Indonesian; circ. 150,000.

Berita Yudha: Jalan Letjenderal Haryono MT22, Jakarta; tel. (21) 8298331; f. 1971; Indonesian; Editor SUNARDI; circ. 50,000.

Bisnis Indonesia: Wisma Bisnis Indonesia, Jalan Letjenderal S. Parman, Kav. 12, Slipi, Jakarta 11480; tel. (21) 5305869; fax (21) 5305868; e-mail iklana@bisnis.co.id; internet www.bisnis.com; f. 1985; available online; Indonesian; Editor SUKAMDANI S. GITOSARDJONO; circ. 60,000.

Harian Indonesia (Indonesia Rze Pao): Jalan Toko Tiga Seberang 21, POB 4755, Jakarta 11120; tel. (21) 6295948; fax (21) 6297830; e-mail info@harian-indonesia.com; internet www.harian-indonesia.com; f. 1966; Chinese; Editor W. D. SUKISMAN; Dir HADI WIBOWO; circ. 42,000.

Harian Terbit: Jalan Pulogadung 15, Kawasan Industri Pulogadung, Jakarta 13920; tel. (21) 4602953; fax (21) 4602950; f. 1972; Indonesian; Editor H. R. S. HADIKAMAJAYA; circ. 125,000.

Harian Umum AB: CTC Bldg, 2nd Floor, Kramat Raya 94, Jakarta Pusat; f. 1965; official armed forces journal; Dir GOENARSO; Editor-in-Chief N. SOEPANGAT; circ. 80,000.

The Indonesia Times: Jalan Pulo Lentut 12, Jakarta Timur; tel. (21) 4611280; fax (21) 375012; e-mail info@webpacific.com; internet www.indonesiatimes.com; f. 1974; English; Editor TRIBUANA SAID; circ. 35,000.

Indonesian Observer: Wisma Indovision, 11th Floor, Jalan Raya Panjang Blok Z/III, Green Garden, Jakarta 11520; tel. (21) 5818855; fax (21) 58302414; internet www.indonesian-observer.com; f. 1955; English; independent; Editor TAUFIK DARUSMAN; circ. 25,000.

Jakarta Post: Jalan Palmerah Selatan 15, Jakarta 10270; tel. (21) 5300476; fax (21) 5492685; e-mail editorial@thejakartapost.com; internet www.thejakartapost.com; f. 1983; English; Chief Editor RAYMOND TORUAN; circ. 50,000.

Jawa Pos: Graha Pena Bldg, 4th and 5th Floors, Achmad Yani 88, Surabaya 60234; tel. (31) 8283333; fax (31) 8285555; internet www .jawapos.co.id; f. 1949; Indonesian; CEO DAHLAN ISKAN; circ. 120,000.

Jepara Pos: Jepara; internet www.jeparapos.com; Indonesian.

Kedaulatan Rakyat: Jalan P. Mangkubumi 40–42, Yogyakarta; tel. (274) 65685; fax (274) 63125; internet www.kr.co.id; f. 1945; Indonesian; independent; Editor IMAN SUTRISNO; circ. 50,000.

Kompas: Jalan Palmerah Selatan 26–28, Jakarta; tel. (21) 5483008; fax (21) 5305868; internet www.kompas.com; f. 1965; Indonesian; Editor Drs JAKOB OETAMA; circ. 523,453.

Koran Tempo: Gedung Tempo, Jalan H. R. Rasuna Said, Kav. C-17, Kuningan, Jakarta 10270; tel. (21) 5201022; fax (21) 5200092; internet www.korantempo.com; f. 2001; Indonesian.

Media Indonesia Daily: Jalan Pilar Mas Raya, Kav. A–D, Kedoya Selatan, Kebon Jeruk, Jakarta 11520; tel. (21) 5812088; fax (21) 5812105; e-mail redaksi@mediaindonesia.co.id; internet www .mediaindo.co.id; f. 1989; fmrly Prioritas; Indonesian; Publr SURYA PALOH; Editor DJAFAR H. ASSEGAFF; circ. 2,000.

Merdeka: Jalan Raya Kebayoran Lama 17, Jakarta Selatan 12210; tel. (21) 5556059; fax (21) 5556063; f. 1945; Indonesian; independent; Dir and Chief Editor B. M. DIAH; circ. 130,000.

Neraca: Jalan Jambrut 2–4, Jakarta; tel. (21) 323969; fax (21) 3101873.

Pelita (Torch): Jalan Jenderal Sudirman 65, Jakarta; f. 1974; Indonesian; Muslim; Editor AKBAR TANDJUNG; circ. 80,000.

Pewarta Surabaya: Jalan Karet 23, POB 85, Surabaya; f. 1905; Indonesian; Editor RADEN DJAROT SOEBIANTORO; circ. 10,000.

Pikiran Rakyat: Jalan Asia-Afrika 77, Bandung 40111; tel. (22) 51216; internet www.pikiran-rakyat.com; f. 1950; Indonesian; independent; Editor BRAM M. DARMAPRAWIRA; circ. 150,000.

Pos Kota: Yayasan Antar Kota, Jalan Gajah Mada 100, Jakarta 10130; tel. (21) 6290874; e-mail iklankilat@poskota.net; internet www.poskota.co.id; f. 1970; Indonesian; Editor H. SOFYAN LUBIS; circ. 500,000.

Republika: Jalan Warung Buncit Raya 37, Jakarta 12510; tel. (21) 7803747; fax (21) 7800420; e-mail kepemimpinan@republika.co.id; internet www.republika.co.id; f. 1993; organ of ICMI; Chief Editor PARNI HADI.

Sinar Pagi: Jalan Letjenderal Haryono MT22, Jakarta Selatan.

Suara Karya: Jalan Bangka II/2, Kebayoran Baru, Jakarta Selatan 12720; tel. (21) 7192656; fax (21) 71790784; internet www .suarakarya-online.com; f. 1971; Indonesian; Editor SYAMSUL BASRI; circ. 100,000.

Suara Merdeka: Jalan Pandanaran 30, Semarang 50241; tel. (24) 412660; fax (24) 411116; internet www.suaramerdeka.com; f. 1950; Indonesian; Publr Ir BUDI SANTOSO; Editor SUWARNO; circ. 200,000.

Suara Pembaruan: Jalan Dewi Sartika 136/D, Cawang, Jakarta 13630; tel. (21) 8013208; fax (21) 8007262; e-mail koransp@suarapembaruan.com; internet www.suarapembaruan.com; f. 1987; licence revoked in 1986 as Sinar Harapan (Ray of Hope); Publr Dr ALBERT HASIBUAN.

Surabaya Post: Jalan Taman Ade Irma Nasution 1, Surayaba; tel. (31) 45394; fax (31) 519585; internet www.surabayapost.co.id; f. 1953; independent; Publr TUTY AZIS; Editor IMAM PUJONO; circ. 115,000.

Kalimantan

Banjarmasin Post: Jalan Haryono MT 54–143, Banjarmasin 70111; tel. (511) 54370; fax (511) 66123; e-mail banjarmasin_post@persda.co.id; internet www.indomedia.com/bpost/; f. 1971; Indonesian; Editor-in-Chief H. PRAMONO; circ. 50,000.

Gawi Manuntung: Jalan Pangeran Samudra 97B, Banjarmasin; f. 1972; Indonesian; Editor M. ALI SRI INDRADJAYA; circ. 5,000.

Harian Umum Akcaya: Pontianak Post Group, Jalan Gajah Mada 2–4, Pontianak; tel. (561) 735071; e-mail redaksi@pontianakpost .com; internet www.pontianakpost.com; Editor B. SALMAN.

Lampung Post: Jalan Pangkal Pinang, Lampung.

Manuntung: Jalan Jenderal Sudirman RT XVI 82, Balikpapan 76144; tel. (542) 35359; internet www.manuntung.co.id; largest newspaper in East Borneo.

Maluku

Ambon Ekspres: Ambon.

Pos Maluku: Jalan Raya Pattimura 19, Ambon; tel. (911) 44614.

Suara Maluku: Komplex Perdagangan Mardikas, Blok D3/11A, Ternate; tel. (911) 44590.

Nusa Tenggara

Pos Kupang: Jalan Kenari 1, Kupang; tel. (380) 833820; fax (380) 831801; e-mail poskupang@persda.co.id; internet www.indomedia .com/poskup.

Papua

Berita Karya: Jayapura.

Cendrawasih Post: Jayapura; Editor RUSTAM MADUBUN.

Teropong: Jalan Halmahera, Jayapura.

Riau

Batam Pos: Jalan Raja Ali Haji, Kompleks Orchid Point, Blok B, No. 9, Batam 29432; tel. (778) 424543; fax (778) 452002; e-mail redaksi@harianbatampos.com; internet www.harianbatampos.com; Editor-in-Chief AMZAR.

Riau Pos: Pekanbaru, Riau; internet www.riaupos.com; circ. 40,000.

Sulawesi

Bulletin Sulut: Jalan Korengkeng 38, Lt II Manado, 95114, Sulawesi Utara.

Cahaya Siang: Jalan Kembang II 2, Manado, 95114, Sulawesi Utara; tel. (431) 61054; fax (431) 63393.

Fajar (Dawn): Jalan Racing Centre 101, Makassar; tel. (411) 441441; fax (411) 441224; e-mail fajar@fajar.co.id; internet www .fajar.co.id; circ. 35,000.

Manado Post: Manado Post Centre, Jalan Babe Palar 54, Manado; tel. (431) 855558; fax (431) 860398; internet www.mdopost.net.

Pedoman Rakyat: Jalan H. A. Mappanyukki 28, Makassar; f. 1947; independent; Editor M. BASIR; circ. 30,000.

Suluh Merdeka: Jalan R. W. Mongsidi 4/96, POB 1105, Manado, 95110; tel. and fax (431) 866150.

Tegas: Jalan Mappanyukki 28, Makassar; tel. (411) 3960.

Sumatra

Harian Analisa: Jalan Jenderal A. Yani 37–43, Medan; tel. (61) 326655; fax (61) 514031; internet www.analisadaily.com; f. 1972; Indonesian; Editor SOFFYAN; circ. 75,000.

Harian Berita Sore: Jalan Letjen Suprapto 1, Medan; tel. (61) 4158787; fax (61) 4150383; e-mail redaksi@beritasore.com; internet www.beritasore.com; Indonesian; Editor-in-Chief H. TERUNA JASA SAID.

Harian Haluan: Jalan Damar 59 C/F, Padang; f. 1948; Editor-in-Chief RIVAI MARLAUT; circ. 40,000.

Harian Umum Nasional Waspada: Jalan Brigjenderal 1 Katamso, Medan 20151; tel. (61) 4150858; fax (61) 4510025; e-mail waspada@indosat.net.id; internet www.waspada.co.id; f. 1947; Indonesian; Editor-in-Chief H. PRABUDI SAID.

Mimbar Umum: Merah, Medan; tel. (61) 517807; f. 1947; Indonesian; independent; Editor MOHD LUD LUBIS; circ. 55,000.

Serambi Indonesia: Jalan T. Nyak Arief 159, Lampriek, Banda Aceh; e-mail serambi@indomedia.com; internet www.indomedia .com/serambi/.

Sinar Indonesia Baru: Jalan Brigjenderal Katamso 66, Medan 20151; tel. (61) 4512530; fax (61) 438150; e-mail redaksi@hariansib .com; internet www.hariansib.com; f. 1970; Indonesian; Chief Editor G. M. PANGGABEAN; circ. 150,000.

Sriwijaya Post: Jalan Jenderal Basuki Rahmat 1608 B-C-D, Palembang; tel. (711) 310088; fax (711) 312888; e-mail sripo@mdp .net.id; internet www.indomedia.com/sripo; f. 2002.

Suara Rakyat Semesta: Jalan K. H. Ashari 52, Palembang; Indonesian; Editor DJADIL ABDULLAH; circ. 10,000.

Waspada: Jalan Letjen Suprapto, cnr Jalan Brigjen Katamso 1, Medan 20151; tel. (61) 4150868; fax (61) 4510025; e-mail waspada@waspada.co.id; internet www.waspada.co.id; f. 1947; Indonesian; Chief Editors ANI IDRUS, MOHAMMAD SAID; circ. 60,000 (daily), 55,000 (Sunday).

PRINCIPAL PERIODICALS

Amanah: Jalan Garuda 69, Kemayoran, Jakarta; tel. (21) 410254; fortnightly; Muslim current affairs; Indonesian; Man. Dir MASKUN ISKANDAR; circ. 180,000.

Ayahbunda: Jalan H. R. Rasuna Said, Blok B, Kav. 32–33, Jakarta 12910; tel. (21) 5209370; fax (21) 5209366; e-mail info@femina-online.com; internet www.ayahbunda-online.com; fortnightly; family magazine.

Berita Negara: Jalan Pertjetakan Negara 21, Kotakpos 2111, Jakarta; tel. (21) 4207251; fax (21) 4207251; f. 1951; 2 a week; official gazette.

Bobo: PT Gramedia, Jalan Kebahagiaan 4–14, Jakarta 11140; tel. (21) 6297809; fax (21) 6390080; f. 1973; weekly; children's magazine; Editor Tineke Latumeten; circ. 240,000.

Bola: Tunas Bola, Jalan Palmerah Barat 33–37, Jakarta 10270; tel. (21) 53677835; fax (21) 5301952; e-mail redaksi@bolanews.com; internet www.bolanews.com; 2 a week; Tue. and Fri.; sports magazine; Indonesian; Chief Editor Ian Situmorang; circ. 715,000.

Buana Minggu: Jalan Tanah Abang Dua 33, Jakarta Pusat 10110; tel. (21) 364190; weekly; Sunday; Indonesian; Editor Winoto Parartho; circ. 193,450.

Business News: Jalan H. Abdul Muis 70, Jakarta 10160; tel. (21) 3848207; fax (21) 3454280; f. 1956; 3 a week (Indonesian edn), 2 a week (English edn); Chief Editor Sanjoto Sastromihardjo; circ. 15,000.

Cita Cinta: Jalan H. R. Rasuna Said, Blok B, Kav. 32–33, Jakarta 12910; tel. (21) 5209370; fax (21) 5209366; internet www.citacinta.com; f. 2001; teenage lifestyle magazine.

Citra: Gramedia Bldg, Unit 11, 5th Floor, Jalan Palmerah Selatan 24–26, Jakarta 10270; tel. (21) 5483008; fax (21) 5494035; e-mail citra@gramedia-majalah.com; internet www.tabloid-citra.com; f. 1990; weekly; TV and film programmes, music trends and celebrity news; Chief Editor H. Maman Suherman; circ. 239,000.

Depthnews Indonesia: Jalan Jatinegara Barat III/6, Jakarta 13310; tel. (21) 8194994; fax (21) 8195501; f. 1972; weekly; publ. by Press Foundation of Indonesia; Editor Sumono Mustoffa.

Dunia Wanita: Jalan Brigjenderal, Katamso 1, Medan; tel. (61) 4150858; fax (61) 4510025; e-mail waspada@indosat.net.id; f. 1949; fortnightly; Indonesian; women's tabloid; Chief Editor Dr Rayati Syafrin; circ. 10,000.

Economic Review: Bank BNI, Strategic Planning Division, Gedung Bank BNI, Jalan Jenderal Sudirman, Kav. 1, POB 2955, Jakarta 10220; tel. (21) 5728692; fax (21) 5728456; e-mail renkek01@bni.co.id; internet www.bni.co.id; f. 1946; 3 a year; English; economic and business research and analysis; Editor-in-Chief Darwin Suzandi.

Ekonomi Indonesia: Jalan Merdeka, Timur 11–12, Jakarta; tel. (21) 494458; monthly; English; economic journal; Editor Z. Achmad; circ. 20,000.

Eksekutif: Jalan R. S. Fatmawati 20, Jakarta 12430; tel. (21) 7659218; fax (21) 7504018; e-mail eksek@pacific.net.id; internet www.pacific.net.id/eksekutif/.

Femina: Jalan H. R. Rasuna Said, Blok B, Kav. 32–33, Jakarta Selatan 12910; tel. (21) 5209370; fax (21) 5209366; e-mail svida.alisjahbana@feminagroup.com; internet www.femina-online.com; f. 1972; weekly; women's magazine; Publr Svida Alisjahbana; Editor Petty S. Fatimah; circ. 130,000.

Forum: Kebayoran Centre, 12a–14, Jalan Kebayoran Baru, Welbak, Jakarta 12240; tel. (21) 7255625; fax (21) 7255645; internet www.forum.co.id.

Gadis: Jalan H. R. Rasuna Said, Blok B, Kav. 32–33, Jakarta 12910; tel. (21) 5253816; fax (21) 5262131; e-mail gadis@indosat.net.id; internet www.gadis-online.com; f. 1973; every 10 days; Indonesian; teenage lifestyle magazine; Editor-in-Chief Petty S. F.; circ. 100,000.

Gamma: Jakarta; internet www.gamma.co.id; f. 1999 by fmr employees of *Tempo* and *Gatra*.

Gatra: Gedung Gatra, Jalan Kalibata Timur IV/15, Jakarta 12740; tel. (21) 7973535; fax (21) 79196923; e-mail gatra@gatra.com; internet www.gatra.com; f. 1994 by fmr employees of *Tempo* (banned 1994–98); Gen. Man. Beny Suharsono; Editor-in-Chief Iwan Qodar Himawan.

Gugat (Accuse): Surabaya; politics, law and crime; weekly; circ. 250,000.

Hai: Gramedia, Jalan Palmerah Selatan 22, Jakarta 10270; tel. (21) 5483008; fax (21) 6390080; f. 1973; weekly; youth magazine; Editor Arswendo Atmowiloto; circ. 70,000.

Indonesia Business News: Wisma Bisnis Indonesia, Jalan Letjenderal S. Parman, Kav. 12, Slipi, Jakarta 11410; tel. (21) 5304016; fax (21) 5305868; English.

Indonesia Business Weekly: Jalan Letjenderal S. Parman, Kav. 12, Slipi, Jakarta 11410; tel. (21) 5304016; fax (21) 5305868; English; Editor Taufik Darusman.

Indonesia Magazine: 20 Jalan Merdeka Barat, Jakarta; tel. (21) 352015; f. 1969; monthly; English; Chair. G. Dwipayana; Editor-in-Chief Hadely Hasibuan; circ. 15,000.

Intisari (Digest): Jalan Palmerah Selatan 24–26, Gedung Unit II, 5th Floor, Jakarta 10270; tel. (21) 5483008; fax (21) 53696525; e-mail intisari@gramedia-majalah.com; internet www.intisari-online.com; f. 1963; monthly; Indonesian; popular science, health, technology, crime and general interest; Editors Al. Heru Kustara, Irawati; circ. 141,000.

Jakarta Jakarta: Gramedia Bldg, Unit II, 5th Floor, Jalan Palmerah Selatan No. 24–26, Jakarta 10270; tel. (21) 5483008; fax (21) 5494035; f. 1985; weekly; food, fun, fashion and celebrity news; circ. 70,000.

Jurnal Indonesia: Jalan Hos Cokroaminoto 49a, Jakarta 10350; tel. (21) 31901774; fax (21) 3916471; e-mail jurnal@cbn.net.id; internet www.jurnalindonesia.com; monthly; political, economic and business analysis.

Keluarga: Jalan Sangaji 11, Jakarta; fortnightly; women's and family magazine; Editor S. Dahono.

Kontan: Jalan Kebayoran Lama 1119, Jakarta 12210; tel. (21) 5357636; fax (21) 5357633; e-mail red@kontan-online.com; internet www.kontan-online.com; weekly; Indonesian; business newspaper.

Majalah Ekonomis: POB 4195, Jakarta; monthly; English; business; Chief Editor S. Arifin Hutabarat; circ. 20,000.

Majalah Kedokteran Indonesia (Journal of the Indonesian Medical Assen): Jalan Kesehatan 111/29, Jakarta 11/16; f. 1951; monthly; Indonesian, English.

Manglé: Jalan Lodaya 19–21, 40262 Bandung; tel. (22) 411438; f. 1957; weekly; Sundanese; Chief Editor Drs Oejang Darajatoen; circ. 74,000.

Matra: Grafity Pers, Kompleks Buncit Raya Permai, Kav. 1, Jalan Warung, POB 3476, Jakarta; tel. (21) 515952; f. 1986; monthly; men's magazine; general interest and current affairs; Editor in Chief (vacant); circ. 100,000.

Mimbar Kabinet Pembangunan: Jalan Merdeka Barat 7, Jakarta; f. 1966; monthly; Indonesian; publ. by Dept of Information.

Mutiara: Jalan Dewi Sartika 136d, Cawang, Jakarta Timur; general interest; Publr H. G. Rorimpandey.

Nova: PT Gramedia, Gedung Unit II, Lantai V, Jalan Palmerah Selatan No. 24–26, Jakarta 10270; tel. (21) 5483008; fax (21) 5483146; e-mail nova@gramedia-majalah.com; internet www.tabloidnova.com; weekly; Wed.; women's interest; Indonesian; Editor Koes Sabandiyah; circ. 618,267.

Oposisi: Jakarta; weekly; politics; circ. 400,000.

Otomotif: Gramedia Bldg, Unit II, 5th Floor, Jalan Palmerah Selatan 24–26, Jakarta 10270; tel. (21) 5490666; fax (21) 5494035; e-mail iklanmjl@ub.net.id; f. 1990; weekly; automotive specialist tabloid; circ. 215,763.

PC Magazine Indonesia: Jalan H. R. Rasuna Said, Blok B, Kav. 32–33, Jakarta 12910; tel. (21) 5209370; fax (21) 5209366; internet www.pcmag.co.id; computers; Editor-in-Chief Svida Alisjahbana.

Peraba: Bintaran Kidul 5, Yogyakarta; weekly; Indonesian and Javanese; Roman Catholic; Editor W. Kartosoeharsono.

Pertani PT: Jalan Pasar Minggu, Kalibata, POB 247/KBY, Jakarta Selatan; tel. (21) 793108; f. 1974; monthly; Indonesian; agricultural; Pres. Dir Ir Rusli Yahya.

Petisi: Surabaya; weekly; Editor Choirul Anam.

Rajawali: Jakarta; monthly; Indonesian; civil aviation and tourism; Dir R. A. J. Lumenta; Man. Editor Karyono Adhy.

Selecta: Kebon Kacang 29/4, Jakarta; fortnightly; illustrated; Editor Samsudin Lubis; circ. 80,000.

Simponi: Jakarta; f. 1994 by fmr employees of *DeTik* (banned 1994–98).

Sinar Jaya: Jakarta Selatan; fortnightly; agriculture; Chief Editor Ir Suryono Projopranoto.

Swasembada: Gedung Chandra Lt 2, Jalan M. H. Thamrin 20, Jakarta 10310; tel. (21) 3103316.

Tempo: Gedung Tempo, 8th Floor, Jalan H. R. Rasuna Said, Kav. C-17, Kuningan, Jakarta 12940; tel. (21) 5201022; fax (21) 5200092; internet www.tempo.co.id; f. 1971; weekly; Editor-in-Chief Bambang Harymurti.

Tiara: Gramedia Bldg, Unit 11, 5th Floor, Jalan Palmerah Selatan 24–26, Jakarta 10270; tel. (21) 5483008; fax (21) 5494035; f. 1990; fortnightly; lifestyles, features and celebrity news; circ. 47,000.

Ummat: Jakarta; Islamic; sponsored by ICMI.

Wenang Post: Jalan R.W. Mongsidi 4/96, POB 1105, Manado 95115; tel. and fax (431) 866150; weekly.

NEWS AGENCIES

Antara (Indonesian National News Agency): Wisma Antara, 3rd, 19th and 20th Floors, 17 Jalan Merdeka Selatan, POB 1257, Jakarta 10110; tel. (21) 3802383; fax (21) 3840970; e-mail antara@antara.co.id; internet www.antara.co.id; f. 1937; 20 radio, seven television, 96 newspaper, eight foreign newspaper, seven tabloid, seven magazine, two news agency, nine embassy and seven dotcom subscribers in 2001; 26 brs in Indonesia, six overseas brs/correspondents; four bulletins in Indonesian and one in English; monitoring service of stock exchanges world-wide; photo service; Exec. Editor HERU PURWANTO; Man. Dir MOHAMAD SOBARY.

Kantorberita Nasional Indonesia (KNI News Service): Jalan Jatinegara Barat III/6, Jakarta Timur 13310; tel. (21) 811003; fax (21) 8195501; f. 1966; independent national news agency; foreign and domestic news in Indonesian; Dir and Editor-in-Chief Drs SUMONO MUSTOFFA; Exec. Editor HARIM NURROCHADI.

Foreign Bureaux

Agence France-Presse (AFP): Jalan Indramayu 18, Jakarta Pusat 10310; tel. (21) 3336082; fax (21) 3809186; Chief Correspondent PASCAL MALLET.

Agenzia Nazionale Stampa Associata (ANSA) (Italy): Jalan Petogogan 1 Go-2 No, 13 Kompleks RRI, Kebayoran Baru, Jakarta Selatan; tel. (21) 7391996; fax (21) 7392247; Correspondent HERYTNO PUJOWIDAGDO.

Associated Press (AP) (USA): Deutsche Bank Bldg, 14th Floor, No. 1403–1404, Jalan Imam Bonjol 80, Jakarta 10310; tel. (21) 39831269; fax (21) 39831270; e-mail gspencer@ap.org; Chief of Bureau GEOFF SPENCER.

Central News Agency Inc (CNA) (Taiwan): Jalan Gelong Baru Timur 1–13, Jakarta Barat; tel. and fax (21) 5600266; Bureau Chief WU PIN-CHIANG.

Informatsionnoye Telegrafnoye Agentstvo Rossii—Telegrafnoye Agentstvo Suverennykh Stran (ITAR—TASS) (Russia): Jalan Surabaya 7 Menteng, Jakarta Pusat 10310; tel. and fax (21) 3155283; e-mail ab1952@indosat.net.id; Correspondent ANDREY ALEKSANDROVICH BYTCHKOV.

Inter Press Service (IPS) (Italy): Gedung Dewan Pers, 4th Floor, Jalan Kebon Sirih 34, Jakarta 10110; tel. (21) 3453131; fax (21) 3453175; Chief Correspondent ABDUL RAZAK.

Jiji Tsushin (Japan): Jalan Raya Bogor 109B, Jakarta; tel. (21) 8090509; Correspondent MARGA RAHARJA.

Kyodo Tsushin (Japan): Skyline Bldg, 11th Floor, Jalan M. H. Thamrin 9, Jakarta 10310; tel. (21) 345012; Correspondent MASAYUKI KITAMURA.

Reuters (United Kingdom): Wisma Antara, 6th Floor, Jalan Medan Merdeka Selatan 17, Jakarta 10110; tel. (21) 3846364; fax (21) 3448404; e-mail jerry.norton@reuters.com; internet www.reuters.com; Bureau Chief JERRY NORTON.

United Press International (UPI) (USA): Wisma Antara, 14th Floor, Jalan Medan Merdeka Selatan 17, Jakarta; tel. (21) 341056; Bureau Chief JOHN HAIL.

Xinhua (New China) News Agency (People's Republic of China): Jakarta.

PRESS ASSOCIATIONS

Aliansi Jurnalis Indpenden (AJI) (Alliance of Independent Journalists): Jakarta; internet www.aji.or.id; f. 1994; unofficial; aims to promote freedom of the press; Sec.-Gen. AHMAD TAUFIK.

Persatuan Wartawan Indonesia (Indonesian Journalists' Asscn): Gedung Dewan Pers, 4th Floor, Jalan Kebon Sirih 34, Jakarta 10110; tel. (21) 353131; fax (21) 353175; f. 1946; government-controlled; 5,041 mems (April 1991); Chair. TARMAN AGAM; Gen. Sec. H. SOFJAN LUBIS.

Serikat Penerbit Suratkabar (SPS) (Indonesian Newspaper Publishers' Asscn): Gedung Dewan Pers, 6th Floor, Jalan Kebon Sirih 34, Jakarta 10110; tel. (21) 3459671; fax (21) 3862373; e-mail sps-pst@dnet.net.id; internet www.dnet.net.id/sps; f. 1946; Exec. Chair. Drs S. L. BATUBARA; Sec.-Gen. Drs AMIR E. SIREGAR.

Yayasan Pembina Pers Indonesia (Press Foundation of Indonesia): Jalan Jatinegara Barat III/6, Jakarta 13310; tel. (21) 8194994; f. 1967; Chair. SUGIARSO SUROYO, MOCHTAR LUBIS.

Publishers

JAKARTA

Aries Lima/New Aqua Press PT: Jalan Rawagelan II/4, Jakarta Timur; tel. (21) 4897566; general and children's; Pres. TUTI SUNDARI AZMI.

Aya Media Pustaka PT: Wijaya Grand Centre C/2, Jalan Dharmawangsa III, Jakarta 12160; tel. (21) 7206903; fax (21) 7201401; f. 1985; children's; Dir Drs ARIANTO TUGIYO.

PT Balai Pustaka Peraga: Jalan Gunung Sahari Raya 4, Gedung Balai Pustaka, 7th Floor, Jakarta 10710; tel. (21) 3447003; fax (21) 3446555; e-mail bp1917@hotmail.com; internet www.balaiperaga.com; f. 1917; children's, school textbooks, literary, scientific publs and periodicals; Dir R. SISWADI.

Bhratara Niaga Media PT: Jalan Cipinang Bali 17, Jakarta Timur 13420; tel. (21) 8520319; fax (21) 8191858; f. 1986; fmrly Bhratara Karya Aksara; university and educational textbooks; Man. Dir ROBINSON RUSDI.

Bina Rena Pariwara PT: Jalan Pejaten Raya 5-E, Pasar Minggu, Jakarta 12510; tel. (21) 7901931; fax (21) 7901939; e-mail hasanbas@softhome.net; f. 1988; financial, social science, economic, Islamic, children's; Dir Drs HASAN BASRI.

Bulan Bintang PT: Jalan Kramat Kwitang 1/8, Jakarta 10420; tel. (21) 3901651; fax (21) 3107027; f. 1954; Islamic, social science, natural and applied sciences, art; Pres. AMRAN ZAMZAMI; Man. Dir FAUZI AMELZ.

Bumi Aksara PT: Jalan Sawo Raya 18, Rawamanguu, Jakarta 13220; tel. (21) 4717049; fax (21) 4700989; f. 1990; university textbooks; Dir H. AMIR HAMZAH.

Cakrawala Cinta PT: Jalan Minyak I/12B, Duren Tiga, Jakarta 12760; tel. (21) 7990725; fax (21) 7974076; f. 1984; science; Dir Drs M. TORSINA.

Centre for Strategic and International Studies (CSIS): Jalan Tanah Abang III/23–27, Jakarta 10160; tel. (21) 3865532; fax (21) 3847517; e-mail csis@pacific.net.id; internet www.csis.or.id; f. 1971; political and social sciences; Dir Dr DAOED JOESOEF.

Cipta Adi Pustaka: Graha Compaka Mas Blok C 22, Jalan Cempaka Putih Raya, Jakarta Pusat; tel. (21) 4213821; fax (21) 4269315; f. 1986; encyclopedias; Dir BUDI SANTOSO.

Dian Rakyat PT: Jalan Rawagelas I/4, Kaw. Industri Pulo Gadung, Jakarta; tel. (21) 4604444; fax (21) 4609115; f. 1966; general; Dir MARIO ALISJAHBANA.

Djambatan PT: Jalan Wijaya I/39, Jakarta 12170; tel. (21) 7203199; fax (21) 7227989; f. 1954; children's, textbooks, social sciences, fiction; Dir SJARIFUDIN SJAMSUDIN.

Dunia Pustaka Jaya: Jalan Rawa Bambu Rt. 003/07 38, Komp. BATAN, Pasar Minggu, Jakarta Selatan 12520; tel. (21) 7891875; fax (21) 3909320; f. 1971; fiction, religion, essays, poetry, drama, criticism, art, philosophy and children's; Man. A. RIVAI.

EGC Medical Publications: Jalan Agung Timur 4, 39 Blok 0–1, Jakarta 14350; tel. (21) 65306283; fax (21) 6510480; e-mail contact@egc-arcan.com; f. 1978; medical and public health, nursing, dentistry; Dir IMELDA DHARMA.

Elex Media Komputindo: Jalan Palmerah Selatan 22, Kompas–Gramedia Bldg, 6th Floor, Jakarta 10270; tel. (21) 53699059; fax (21) 5326219; e-mail elex@elexmedia.co.id; internet www.elexmedia.co.id; f. 1985; management, computing, software, children's, parenting, self-development and fiction; Dir AL. ADHI MARDHIYONO.

Erlangga PT: Kami Melayani II, Pengetahuan, Jalan H. Baping 100, Ciracas, Jakarta 13740; tel. (21) 8717006; fax (21) 8717011; e-mail erlprom@rad.net.id; internet www.erlangga.com; f. 1952; secondary school and university textbooks; Man. Dir GUNAWAN HUTAURUK.

Gaya Favorit Press: Jalan Rawagelam II/4, Kawasan Industri Pulogadung, Jakarta 13930; tel. (21) 46821321; fax (21) 46821419; f. 1971; fiction, popular science, lifestyle and children's; Vice-Pres. MIRTA KARTOHADIPRODJO; Man. Dir WIDARTI GUNAWAN.

Gema Insani Press: Jalan Kalibata Utara II/84, Jakarta 12740; tel. (21) 7984391; fax (21) 7984388; e-mail gipnet@indosat.net.id; internet www.gemainsani.co.id; f. 1986; Islamic; Dir UMAR BASYARAHIL.

Ghalia Indonesia: Jalan Pramuka Raya 4, Jakarta 13140; tel. (21) 8584330; fax (21) 8502334; f. 1972; children's and general science, textbooks; Man. Dir LUKMAN SAAD.

Gramedia Widyasarana Indonesia: Jalan Palmerah Selatan 22, Lantai IV, POB 615, Jakarta 10270; tel. (21) 5483008; fax (21) 5300545; f. 1973; university textbooks, general non-fiction, children's and magazines; Gen. Man. ALFONS TARYADI.

Gunung Mulia PT: Jalan Kwitang 22–23, Jakarta 10420; tel. (21) 3901208; fax (21) 3901633; e-mail corp.off@bpkgm.com; internet www.bpkgm.com; f. 1951; general, children's, Christian; Pres. Dir STEPHEN Z. SATYAHADI; Dir V. N. LEIMENA.

Hidakarya Agung PT: Jalan Kebon Kosong F/74, Kemayoran, Jakarta Pusat; tel. (21) 4241074; Dir MAHDIARTI MACHMUD.

Ichtiar: Jalan Majapahit 6, Jakarta Pusat; tel. (21) 3841226; f. 1957; textbooks, law, social sciences, economics; Dir JOHN SEMERU.

Indira PT: Jalan Borobudur 20, Jakarta 10320; tel. (21) 882754; f. 1953; general science and children's; Man. Dir BAMBANG P. WAHYUDI.

Kinta CV: Jalan Kemanggisan Ilir V/110, Pal Merah, Jakarta Barat; tel. (21) 5494751; f. 1950; textbooks, social science, general; Man. Drs MOHAMAD SALEH.

Midas Surya Grafindo PT: Jalan Kesehatan 54, Cijantung, Jakarta 13760; tel. (21) 8400414; fax (21) 8400270; f. 1984; children's; Dir Drs FRANS HENDRAWAN.

Mutiara Sumber Widya PT: Jalan Kramat II 55, Jakarta; tel. (21) 3926043; fax (21) 3160313; f. 1951; textbooks, Islamic, social sciences, general and children's; Pres. FADJRAA OEMAR.

Penebar Swadya PT: Jalan Gunung Sahari III/7, Jakarta Pusat; tel. (21) 4204402; fax (21) 4214821; agriculture, animal husbandry, fisheries; Dir Drs ANTHONIUS RIYANTO.

Penerbit Universitas Indonesia: Jalan Salemba Raya 4, Jakarta; tel. (21) 335373; f. 1969; science; Man. S. E. LEGOWO.

Pradnya Paramita PT: Jalan Bunga 8–8A, Matraman, Jakarta 13140; tel. (21) 8504944; e-mail pradnya@centrin.net.id; f. 1973; children's, general, educational, technical and social science; Pres. Dir KONDAR SINAGA.

Pustaka Antara PT: Jalan Taman Kebon Sirih III/13, Jakarta Pusat 10250; tel. (21) 3156994; fax (21) 322745; e-mail nacelod@cbn.net.id; f. 1952; textbooks, political, Islamic, children's and general; Man. Dir AIDA JOESOEF AHMAD.

Pustaka Binaman Pressindo: Bina Manajemen Bldg, Jalan Menteng Raya 9–15, Jakarta 10340; tel. (21) 2300313; fax (21) 2302047; e-mail pustaka@bit.net.id; f. 1981; management; Dir Ir MAKFUDIN WIRYA ATMAJA.

Pustaka LP3ES Indonesia: Jalan Letjen. S. Parman 81, Jakarta 11420; tel. (21) 5663527; fax (21) 56964691; e-mail puslp3es@indo.net.id; f. 1971; general; Dir M. D. MARUTO.

Pustaka Sinar Harapan PT: Jalan Dewi Sartika 136D, Jakarta 13630; tel. (21) 8093208; fax (21) 8091652; f. 1981; general science, fiction, comics, children's; Dir W. M. NAIDEN.

Pustaka Utma Grafiti PT: Jalan Wahid Hasyim Nomor 166A, Jakarta 10250; tel. (21) 31902906; fax (21) 31902435; f. 1981; social sciences, humanities and children's books; Dir ZULKIFLY LUBIS.

Rajagrafindo Persada PT: Jalan Pelepah Hijau IV TN-1 14–15, Kelapa Gading Permai, Jakarta 14240; tel. (21) 4529409; fax (21) 4520951; f. 1980; general science and religion; Dir Drs ZUBAIDI.

Rineka Cipta PT: Blok B/5, Jalan Jenderal Sudirman, Kav. 36A, Bendungan Hilir, Jakarta 10210; tel. (21) 5737646; fax (21) 5711985; f. 1990 by merger of Aksara Baru (f. 1972) and Bina Aksara; general science and university texts; Dir Dr H. SUARDI.

Rosda Jayaputra PT: Jalan Kembang 4, Jakarta 10420; tel. (21) 3904984; fax (21) 3901703; f. 1981; general science; Dir H. ROZALI USMAN.

Sastra Hudaya: Jalan Kalasan 1, Jakarta Pusat; tel. (21) 882321; f. 1967; religious, textbooks, children's and general; Man. ADAM SALEH.

Tintamas Indonesia: Jalan Kramat Raya 60, Jakarta 10420; tel. and fax (21) 3911459; f. 1947; history, modern science and culture, especially Islamic; Man. MARHAMAH DJAMBEK.

Tira Pustaka: Jalan Cemara Raya 1, Kav. 10D, Jaka Permai, Jaka Sampurna, Bekasi 17145; tel. (21) 8841277; fax (21) 8842736; e-mail Tirapus@cbn.net.id; f. 1977; translations, children's; Dir ROBERT B. WIDJAJA.

Toko Buku Walisongo PT: Gedung Idayu, Jalan Kwitang 13, Jakarta 10420; tel. (21) 3154890; fax (21) 3154889; e-mail tokowalisongo@plasa.com; f. 1986; fmrly Masagung Group; general, Muslim religious, textbooks, science; Pres. H. KETUT ABDURRAHMAN MASAGUNG.

Widjaya: Jalan Pecenongan 48C, Jakarta Pusat; tel. (21) 3813446; f. 1950; textbooks, children's, religious and general; Man. DIDI LUTHAN.

Yasaguna: Jalan Minangkabau 44, POB 422, Jakarta Selatan; tel. (21) 8290422; f. 1964; agricultural, children's, handicrafts; Dir HILMAN MADEWA.

Bandung

Alma'arif: Jalan Tamblong 48–50, Bandung; tel. (22) 4264454; fax (22) 4239194; f. 1949; textbooks, religious and general; Man. H. M. BAHARTHAH.

Alumni PT: Jalan Bukit Pakar Timur II/109, Bandung 40197; tel. (22) 2501251; fax (22) 2503044; f. 1968; university and school textbooks; Dir EDDY DAMIAN.

Angkasa: Jalan Merdeka 6, POB 1353 BD, Bandung 40111; tel. (22) 4204795; fax (22) 439183; Dir H. FACHRI SAID.

Armico: Jalan Madurasa Utara 10, Cigereleng, Bandung 40253; tel. (22) 5202234; fax (22) 5201972; f. 1980; school textbooks; Dir Ir ARSIL TANJUNG.

Citra Aditya Bakti PT: Jalan Geusanulun 17, Bandung 40115; tel. (22) 438251; fax (22) 438635; f. 1985; general science; Dir Ir IWAN TANUATMADJA.

Diponegoro Publishing House: Jalan Mohammad Toha 44–46, Bandung 40252; tel. and fax (22) 5201215; e-mail dpnegoro@indosat.net.id; f. 1963; Islamic, textbooks, fiction, non-fiction, general; Dir HADIDJAH DAHLAN.

Epsilon Group: Jalan Marga Asri 3, Margacinta, Bandung 40287; tel. (22) 7567826; f. 1985; school textbooks; Dir Drs BAHRUDIN.

Eresco PT: Jalan Megger Girang 98, Bandung 40254; tel. (22) 5205985; fax (22) 5205984; f. 1957; scientific and general; Man. Drs ARFAN ROZALI.

Ganeca Exact Bandung: Jalan Kiaracondong 167, Pagauban, Bandung 40283; tel. (22) 701519; fax (22) 775329; f. 1982; school textbooks; Dir Ir KETUT SUARDHARA LINGGIH.

Mizan Pustaka PT: Jalan Yodkali 16, Bandung 40124; tel. (22) 7200931; fax (22) 7207038; e-mail info@mizan.com; internet www.mizan.com; f. 1983; Islamic and general books; Pres. Dir HAIDAR BAGIR; Man. Dir PUTUT WIDJANARKO.

Penerbit ITB: Jalan Ganesa 10, Bandung 40132; tel. and fax (22) 2504257; e-mail itbpress@bdg.centrin.net.id; f. 1971; academic books; Dir EMMY SUPARKA; Chief Editor SOFIA MANSOOR-NIKSOLIHIN.

Putra A. Bardin: Jalan Ganesa 4, Bandung; tel. (22) 2504319; f. 1998; textbooks, scientific and general; Dir NAI A. BARDIN.

Remaja Rosdakarya PT: Jalan Ciateul 34–36, POB 284, Bandung 40252; tel. (22) 5200287; fax (22) 5202529; textbooks and children's fiction; Pres. ROZALI USMAN.

Sarana Panca Karyam PT: Jalan Kopo 633 KM 13/4, Bandung 40014; f. 1986; general; Dir WIMPY S. IBRAHIM.

Tarsito PT: Jalan Guntur 20, Bandung 40262; tel. (22) 304915; fax (22) 314630; academic; Dir T. SITORUS.

Flores

Nusa Indah: Jalan El Tari, Ende 86318, Flores; tel. (0381) 21502; fax (0381) 23974; e-mail namkahu@yahoo.com; f. 1970; religious and general; Dir LUKAS BATMOMOLIN.

Kudus

Menara Kudus: Jalan Menara 4, Kudus 59315; tel. (291) 371143; fax (291) 36474; f. 1958; Islamic; Man. CHILMAN NAJIB.

Medan

Hasmar: Jalan Letjenderal Haryono M. T. 1, POB 446, Medan 20231; tel. (61) 24181; f. 1962; primary school textbooks; Dir FAUZI LUBIS; Man. AMRAN SAID RANGKUTI.

Impola: Jalan H. M. Joni 46, Medan 20217; tel. (61) 711415; f. 1984; school textbooks; Dir PAMILANG M. SITUMORANG.

Madju Medan Cipta PT: Jalan Amaliun 37, Medan 20215; tel. (61) 7361990; fax (61) 7367753; e-mail koboi@indosat.net; f. 1950; textbooks, children's and general; Pres. H. MOHAMED ARBIE; Man. Dir Drs DINO IRSAN ARBIE.

Masco: Jalan Sisingamangaraja 191, Medan 20218; tel. (61) 713375; f. 1992; school textbooks; Dir P. M. SITUMORANG.

Monora: Jalan Letjenderal Jamin Ginting 583, Medan 20156; tel. (61) 8212667; fax (61) 8212669; e-mail monora_cv@plasa.com; f. 1962; school textbooks; Dir CHAIRIL ANWAR.

Semarang

Aneka Ilmu: Jalan Raya Semarang Demak Km 8.5, Sayung, Demak; tel. (24) 6580335; fax (24) 6582903; e-mail aneka@semarang.wasantara.net.id; f. 1983; general and school textbooks; Dir H. SUWANTO.

Effhar COY PT: Jalan Dorang 7, Semarang 50173; tel. (24) 3511172; fax (24) 3551540; e-mail effhar_dahara@yahoo.com; f. 1976; general books; Dir H. DARADJAT HARAHAP.

Intan Pariwara: Jalan Beringin, Klaten Utara, Kotak Pos III, Kotif Klaten, Jawa-Tengah; tel. (272) 22441; fax (272) 22021; school textbooks; Pres. Soetikno.

Mandira PT: Jalan Letjenderal M. T. Haryono 501, Semarang 50241; tel. (24) 316150; fax (24) 415092; f. 1962; Dir Ir A. Hariyanto.

Mandira Jaya Abadi PT: Jalan Letjenderal M. T. Haryono 501, Semarang 50241; tel. (24) 519547; fax (24) 542189; e-mail mjabadi@indosat.net.id; f. 1981; Dir Ir A. Hariyanto.

Solo

Pabelan PT: Jalan Raya Pajang, Kertasura KM 8, Solo 57162; tel. (271) 743975; fax (271) 714775; f. 1983; school textbooks; Dir Agung Sasongko.

Tiga Serangkai PT: Jalan Dr Supomo 23, Solo; tel. (271) 714344; fax (271) 713607; f. 1977; school textbooks; Dir Abdullah.

Surabaya

Airlangga University Press: Kampus C, Jalan Mulyorejo, Surabaya; tel. (31) 5992246; fax (31) 5992248; e-mail aupsby@rad.net.id; academic; Dir Dr Arifaan Razak.

Bina Ilmu PT: Jalan Tunjungan 53E, Surabaya 60275; tel. (31) 5323214; fax (31) 5315421; f. 1973; school textbooks, Islamic; Pres. Ariefin Noor.

Bintang: Jalan Potroagung III/41C, Surabaya; tel. (31) 3770687; fax (31) 3715941; school textbooks; Dir Agus Winarno.

Grip PT: Jalan Rungkut Permai II/C–11, Surabaya; tel. (31) 22564; f. 1958; textbooks and general; Man. Suripto.

Jaya Baya: Jalan Embong Malang 69H, POB 250, Surabaya 60001; tel. (31) 41169; f. 1945; religion, philosophy and ethics; Man. Tadjib Ermadi.

Sinar Wijaya: Jalan Raya Sawo VII/58, Bringin-Lakarsantri, Surabaya; tel. (31) 706615; general; Dir Dulradjak.

Yogyakarta

Andi Publishers: Jalan Beo 38–40, Yogyakarta 55281; tel. (274) 561881; fax (274) 588282; e-mail andi_pub@indo.net.id; f. 1980; Christian, computing, business, management and technical; Dir J. H. Gondowijoyo.

BPFE PT: Jalan Gambiran 37, Yogyakarta 55161; tel. (274) 373760; fax (274) 380819; f. 1984; university textbooks; Dir Drs Indriyo Gitosudarmo.

Centhini Yayasan: Gg. Bekisar UH V/716 E–1, Yogyakarta 55161; tel. (274) 383148; f. 1984; Javanese culture; Chair. H. Karkono Kamajaya.

Gadjah Mada University Press: Jalan Grafika 1, Campus UGM, Bulaksumur, Yogyakarta 55231; tel. and fax (274) 561037; e-mail gmupress@ugm.ac.id; f. 1971; university textbooks; Dir Ana Nadhya Abrar.

Indonesia UP: Gg. Bekisar UH V/716 E–1, Yogyakarta 55161; tel. (274) 383148; f. 1950; general science; Dir H. Karkono Kamajaya.

Kanisius Printing and Publishing: Jalan Cempaka 9, Deresan, POB 1125, Yogyakarta 55281; tel. (274) 588783; fax (274) 563349; e-mail office@kanisiusmedia.com; internet www.kanisiusmedia.com; f. 1922; children's, textbooks, Christian and general; Pres. Dir S. J. Sarwanto.

Kedaulatan Rakyat PT: Jalan P. Mangkubumi 40–42, Yogyakarta; tel. (274) 2163; Dir Drono Hardjusuwongso.

Penerbit Tiara Wacana Yogya: Jalan Kaliurang KM 7, 8 Kopen 16, Banteng, Yogyakarta 55581; tel. and fax (274) 880683; f. 1986; university textbooks and general science; Dir Sitoresmi Prabuningrat.

Government Publishing House

Balai Pustaka PT (Persero) (State Publishing and Printing House): Jalan Gunung Sahari Raya 4, Gedung Balai Pustaka, 7th Floor, Jakarta Pusat 10710; tel. (21) 3447003; fax (21) 3446555; e-mail mail@balaiperaga.com; internet www.balaiperaga.com; history, anthropology, politics, philosophy, medical, arts and literature; Pres. Dir H. R. Siswadi Idris.

PUBLISHERS' ASSOCIATION

Ikatan Penerbit Indonesia (IKAPI) (Assen of Indonesian Book Publishers): Jalan Kalipasir 32, Jakarta 10330; tel. (21) 3141907; fax (21) 3146050; e-mail sekretariat@ikapi.or.id; internet www.ikapi.or.id; f. 1950; 496 mems (Aug. 2002); Pres. Makfudin Wirya Atmaja; Sec.-Gen. Robinson Rusdi.

Broadcasting and Communications

TELECOMMUNICATIONS

Directorate-General of Posts and Telecommunications (Postel): Gedung Sapta Pesona, Jalan Medan Merdeka Barat 17, Jakarta 10110; tel. (21) 3835912; fax (21) 3860754; e-mail admin@postel.go.id; internet www.postel.go.id; Dir-Gen. Djamhari Sirat.

PT Indosat (Persero) Tbk: Jalan Medan Merdeka Barat 21, POB 2905, Jakarta 10110; tel. (21) 30003001; fax (21) 3804045; e-mail sant@indosat.com; internet www.indosat.com; f. 1967; telecommunications; partially privatized in 1994; 41.94% stake sold to Singapore Technologies Telemedia in 2002; Pres. Dir Widya Purnama.

PT Satelit Palapa Indonesia (SATELINDO): Jalan Daan Mogot Km 11, Jakarta 11710; tel. (21) 5451745; fax (21) 5451748; e-mail webmaster@satelindo.co.id; internet www.satelindo.co.id; f. 1993; telecommunications and satellite services; Pres. Dir Johny Swandi Sjam.

PT Telekomunikasi Indonesia Tbk (TELKOM): Corporate Office, Jalan Japati No. 1, Bandung 40133; tel. (22) 4521510; fax (22) 440313; internet www.telkom.co.id; domestic telecommunications; 24.2% of share capital was transferred to the private sector in 1995; Pres. Dir and CEO Kristiono.

BROADCASTING

Regulatory Authority

Directorate-General of Radio, Television and Film: Jalan Merdeka Barat 4–5, Jakarta 10110; tel. (21) 3849091; fax (21) 3457132; Dir-Gen. Suryanta Saleh.

Radio

Radio Republik Indonesia (RRI): Jalan Medan Merdeka Barat 4–5, Jakarta 10110; tel. (21) 3846817; fax (21) 3457134; e-mail rri@rrionline.com; internet www.rrionline.com; f. 1945; 49 stations; Dir Suryanta Saleh; Dep. Dirs Fachruddin Soekarno (Overseas Service), Abdul Rochim (Programming), Sukri (Programme Development), Sazli Rais (Administration), Chaerul Zen (News).

Voice of Indonesia: Jalan Medan Merdeka Barat 4–5, POB 1157, Jakarta; tel. (21) 3456811; international service provided by Radio Republik Indonesia; daily broadcasts in Arabic, English, French, German, Bahasa Indonesia, Japanese, Bahasa Malaysia, Mandarin, Spanish and Thai.

Television

In March 1989 Indonesia's first private commercial television station began broadcasting to the Jakarta area. In 1996 there were five privately-owned television stations in operation.

PT Cakrawala Andalas Televisi (ANTEVE): Gedung Sentra Mulia, 18th Floor, Jalan H. R. Rasuna Said, Kav. X-6 No. 8, Jakarta 12940; tel. (21) 5222086; fax (21) 5222087; e-mail humas@anteve.co.id; internet www.anteve.co.id; f. 1993; private channel; broadcasting to 10 cities; Pres. Dir H. R. Agung Laksono; Gen. Man. Nenny Soemawinata.

PT CIPTA TPI: Jalan Pintu II—Taman Mini Indonesia Indah, Pondok Gede, Jakarta Timur 13810; tel. (21) 8412473; fax (21) 8412470; e-mail info@tpi.co.id; internet www.tpi.co.id; f. 1991; private channel funded by commercial advertising; Pres. Dir Siti Hardijanti Rukmana.

PT Rajawali Citra Televisi Indonesia (RCTI): Jalan Raya Pejuangan 3, Kebon Jeruk, Jakarta 11000; tel. (21) 5303540; fax (21) 5493852; e-mail webmaster@rcti.co.id; internet www.rcti.co.id; f. 1989; first private channel; 20-year licence; Pres. Dir M. S. Ralie Siregar; Vice-Pres. Alex Kumara.

PT Surya Citra Televisi (SCTV): GRHA SCTV, 2nd Floor, Jalan Gatot Subroto, Kav. 21, Jakarta 12930; tel. (21) 5225555; fax (21) 5224777; e-mail pr@sctv.co.id; internet www.sctv.co.id; f. 1990; private channel broadcasting nationally; Pres. Dir Dr Ir Agus Mulyanto.

Televisi Republik Indonesia (TVRI): TVRI Senayan, Jalan Gerbang Pemuda, Senayan, Jakarta; tel. (21) 5733135; fax (21) 5732408; e-mail info@tvrisby.com; internet www.tvrisby.com; f. 1962; state-controlled; Pres. Dir Nasirwan Uyun.

Finance

(cap. = capital; auth. = authorized; p.u. = paid up; res = reserves; dep. = deposits; m. = million; brs = branches; amounts in rupiah)

BANKING

In December 1996 there were 9,276 banks, with 14,956 branches, in operation in Indonesia, including seven state commercial banks, 27 regional government banks, 164 private national banks and 41 foreign and joint-venture banks. At the end of March 1995 total bank deposits stood at 167,123,000m. rupiah. A programme of extensive reform of the banking sector was ongoing from 1998 onwards, under the auspices of the Indonesian Bank Restructuring Agency (IBRA). Having satisfied its mandate, the IBRA was dissolved in February 2004. However, continued reform of the sector remained a priority for the Government.

Central Bank

Bank Indonesia: Jalan M. H. Thamrin 2, Jakarta Pusat 10002; tel. (21) 2310408; fax (21) 2311058; e-mail humasbi@bi.go.id; internet www.bi.go.id; f. 1828; nationalized as central bank in 1953; cap. 2,948,029m., res 130,820,882m., dep. 341,944,951m. (Dec. 2002); Gov. BURHANUDDIN ABDULLAH; 42 brs.

State Banks

In late December 1997 the Government announced that four of the state-owned banks—Bank Bumi Daya, Bank Dagang Negara, Bank Ekspor Impor Indonesia and Bank Pembangunan Indonesia—were to merge; the four banks subsequently merged into a single new institution, PT Bank Mandiri, which was established in July 1999.

PT Bank Ekspor Indonesia (Persero): Gedung Brs Efek, Menara II, Lt. 8, Jalan Sudirman, Kav. 52–53, Jakarta; tel. (21) 5154638; fax (21) 5154639; e-mail prbei@bexi.co.id; internet www.bexi.co.id; cap. 3,000,000m., res 221,634m., dep. 370,679m. (Dec. 2002); Chair. MANSJURDIN NURDIN; Pres. Dir BAMBANG HENDRAJATIN.

PT Bank Mandiri (Persero): Plaza Mandiri, Jalan Jenderal Gatot Subroto, Kav. 36–38, Jakarta 12190; tel. (21) 5265045; fax (21) 5268246; internet www.bankmandiri.co.id; f. 1999 as a result of the merger of four state-owned banks—PT Bank Bumi Daya, PT Bank Dagang Negara, PT Bank Ekspor Impor Indonesia and PT Bank Pembangunan Indonesia; cap. 10,000,000m., res 7,166,651m., dep. 194,916,764m. (Dec. 2003); Pres. Commr BINHADI; Pres. Dir E. C. W. NELOE.

PT Bank Negara Indonesia (Persero) Tbk: Jalan Jenderal Sudirman, Kav. 1, Jakarta 10220; tel. (21) 2511946; fax (21) 2511221; e-mail hin@bni.co.id; internet www.bni.co.id; f. 1946; commercial bank; specializes in credits to the industrial sector; cap. 7,042,194m., res 2,554,253m., dep. 111,025,247m. (Dec. 2003); Pres. Dir SAIFUDDIEN HASAN; 585 local brs, 5 overseas brs.

PT Bank Rakyat Indonesia (Persero): Jalan Jenderal Sudirman, Kav. 44–46, POB 94, Jakarta 10210; tel. (21) 2510244; fax (21) 2500077; internet www.bri.co.id; f. 1895; present name since 1946; commercial and foreign exchange bank; specializes in agricultural smallholdings and rural development; state-owned; cap. 1,728,000m., res 29,168,432m., dep. 73,435,981m. (Dec. 2002); Pres. Dir RUDJITO; 325 brs.

PT Bank Tabungan Negara (Persero): 10th Floor, Bank BTN Tower, Jalan Gajah Mada 1, Jakarta 10130; tel. (21) 6336789; fax (21) 6336704; e-mail webadmin@btn.co.id; internet www.btn.co.id; f. 1964; commercial bank; state-owned; cap. 1,250,000m., res 13,843,540m., dep. 20,694,532m. (Dec. 2002); Chair. DARMIN NASUTION; 44 brs.

PT BPD Jawa Timur (Bank Jatim): Jalan Basuki Rachmad 98–104, Surabaya; tel. (31) 5310090; fax (31) 5312226; internet www.bankjatim.co.id; f. 1961; cap. and res 196,413m., dep. 4,915,594m. (Dec. 2001); Chair. NANIK KUSMINI.

Commercial Banks

PT ANZ Panin Bank: Panin Bank Centre, Jalan Jenderal Sudirman (Senayan), Jakarta 10270; tel. (21) 5750300; fax (21) 5727447; internet www.anz.com/indonesia; f. 1990 as Westpac Panin Bank; 85%-owned by the Australia and New Zealand Banking Group Ltd; cap. 50,000m., res. 10,000m., dep. 1,112,743m. (Dec. 2002); Pres. Dir SCOTT ARMSTRONG.

PT Bank Artha Graha: Bank Artha Graha Tower, 5th Floor, Jalan Jenderal Sudirman, Kav. 52–53, Jakarta 12920; tel. (21) 5152168; fax (21) 5152162; e-mail agraha@rad.net.id; f. 1967; merged with PT Bank Arta Pratama in 1999; cap. 75,000m., dep. 340,344m. (Dec. 1993); Pres. and Dir ANTON B. S. HUDYANA; Chair. LETJEN; 64 brs.

PT Bank Buana Indonesia Tbk: Jalan Asemka 32–36, Jakarta 11110; tel. (21) 2601015; fax (21) 2601014; e-mail bbitdiv@bankbuanaina.co.id; internet www.bankbuana.com; f. 1956; cap. 744,494m., res 264,530m., dep. 11,802,231m. (Dec. 2002); Chair. R. RACHMAD; Pres. Dir JIMMY KURNJAWAN LAIHAD; 31 brs.

PT Bank Bumiputera Indonesia Tbk: Wisma Bumiputra, 14th Floor, Jalan Jenderal Sudirman, Kav. 75, Jakarta 12910; tel. (21) 5701626; fax (21) 5255244; e-mail bank@bumiputera.co.id; internet www.bumiputera.co.id; f. 1989; cap. 200,000m., res 5,649m., dep. 2,070,454m. (Dec. 2002); Pres. and CEO WINNY ERWINDIA HASSAN.

PT Bank Central Asia Tbk (BCA): Wisma BCA, 5th Floor, Jalan Jenderal Sudirman, Kav. 22–23, Jakarta 12920; tel. (21) 5711250; fax (21) 5701865; e-mail humas@bca.co.id; internet www.klikbca.com; f. 1957; placed under the supervision of the IBRA in May 1998; 51% share sold to Farallon Capital Management (USA) in March 2002; cap. 1,532,784m., res 5,331,095m., dep. 119,308,970m. (Dec. 2003); Pres. Dir D. E. SETIJOSO; 780 local brs, 2 overseas brs.

PT Bank Chinatrust Indonesia: 16th and 17th Floors, Wisma Tamara, Jalan Jenderal Sudirman, Kav. 24, Jakarta 12920; tel. (21) 5207878; fax (21) 5206767; e-mail ctcbjak@rad.net.id; f. 1995; cap. 150,000m., dep. 1,179,015m. (Dec. 2002); Pres. Dir RAY KUO; Chair. JEFFREY L. S. KOO.

PT Bank Danamon Indonesia Tbk: Menara Danamon, Jalan Prof. Dr Satrio 6, Kav. E/4, Mega Kuningan, Jakarta 12930; tel. (21) 57991142; fax (21) 57991008; internet www.danamon.co.id; f. 1956; placed under the supervision of the IBRA in April 1998; merged with PT Bank Tiara Asia, PT Tamara Bank, PT Bank Duta and PT Bank Nusa Nasional in 2000; 51% share sold to consortium led by Singapore's Temasek Holdings in May 2003; cap. 3,562,261m., res 28,524m., dep. 37,017,617m. (Dec. 2002); Pres. Commr SIM KEE BOON; Pres. Dir FRANCIS ANDREW ROZARIO; 737 brs.

PT Bank Internasional Indonesia Tbk (BII): Plaza BII, Jalan M. H. Thamrin 51, Kav. 22, Jakarta 10350; tel. (21) 2300888; fax (21) 2301494; e-mail bii-info@idola.net.id; internet www.bii.co.id; cap. 17,867,731m., res 1,339,320m., dep. 29,743,726m. (Dec. 2003); Chair. SOEDARJONO; Pres. Dir SIGIT PRAMONO; 249 brs; 3 overseas brs.

PT Bank Mayapada Internasional Tbk: Gedung Arthaloka, Ground and 1st Floors, Jalan Jenderal Sudirman, Kav. 2, Jakarta 10220; tel. (21) 2511588; fax (21) 2511539; e-mail mayapada@bankmayapada.com; internet www.bankmayapada.com; f. 1989; cap. 164,145m., dep. 925,394m. (Dec. 2000); Chair. TAHIR; Pres. Dir HARYONO TJAHJARIJADI; 5 brs, 7 sub-brs.

PT Bank Mizuho Indonesia: Plaza B11, 24th Floor, Menara 2, Jalan M. H. Thamrin 51, Jakarta 10350; tel. (21) 3925222; fax (21) 3926354; fmrly PT Bank Fuji International Indonesia; f. 1989; cap. 396,250m., res 11,121m., dep. 4,176,926m. (Dec. 2002); Pres. Dir KEIICHI KOGURE.

PT Bank Muamalat Indonesia (BMI): Arthaloka Bldg, Jalan Jenderal Sudirman 2, Jakarta 10220; tel. (21) 2511414; fax (21) 2511453; internet www.muamalatbank.com; Indonesia's first Islamic bank; cap. 165,593m. (July 2000); Pres. Dir A. RIAWAN AMIN; Chair. ABBAS ADHAR.

PT Bank Niaga Tbk: Graha Niaga, Jalan Jenderal Sudirman, Kav. 58, Jakarta 12190; tel. (21) 2505252; fax (21) 2505205; e-mail caniaga@attglobal.net; internet www.bankniaga.com; f. 1955; cap. 746,907m., res 994,801m., dep. 20,214,100m. (Dec. 2003); Pres. Commr SUKANTO REKSOHADIPRODJO; Pres. Dir PETER B. STOK; 181 brs.

PT Bank NISP Tbk: Jalan Taman Cibeunying Selatan 31, Jakarta 40114; tel. (22) 7234123; fax (22) 7100466; e-mail nisp@banknisp.com; internet www.banknisp.com; f. 1941; cap. 516,747m., res 158,925m., dep. 12,594,401m. (Dec. 2003); Pres. Dir PRAMUKTI SURJAUDAJA; Chair. KARMAKA SURJANDAJA; 135 brs.

PT Bank Permata Tbk: PermataBank Tower I, Jalan Jenderal Sudirman, Kav. 27, Jakarta 12920; tel. (21) 5237888; fax (21) 2500702; internet www.permatabank.com; f. 1954 as Bank Persatuan Dagang Indonesia; became PT Bank Bali in 1971 and PT Bank Bali Tbk in 1990; name changed as above Sept. 2002 following merger with PT Bank Prima Express, PT Bank Universal Tbk, PT Arthamedia Bank and PT Bank Patriot; cap. 1,300,534m., res 867,217m., dep. 24,879,795m. (Dec. 2003); Chair. and Pres. Dir AGUS D. W. MARTOWARDOJO; 302 brs.

PT Bank Rabobank International Indonesia: Plaza 89, 9th Floor, Jalan H. R. Rasuna Said, Kav. X-7, Jakarta 12940; tel. (21) 2520876; fax (21) 2520875; internet www.rabobank.com; cap. 350,000m., res 2,741m., dep. 1,010,012m. (Dec. 1999); Pres. and Dir ANTONIO COSTA.

PT Bank Sumitomo Mitsui Indonesia: Summitmas II, 10th Floor, Jalan Jenderal Sudirman, Kav. 61–62, Jakarta 12069; tel. (21) 5227011; fax (21) 5227022; internet www.smbc-jkt.co.id; f. 1989; fmrly PT Bank Sumitomo Indonesia, merged with PT Bank Sakura Swadharma in April 2001; cap. 1,502,441m. (Dec. 2003), res –517,404m., dep. 3,773,680m. (Dec. 2002); Pres. Dir KIYOZUMI NAKAMURA; 1 br.

PT Bank UFJ Indonesia: PermataBank Tower I, 4th–5th Floors, Jalan Jenderal Sudirman, Kav. 27, Jakarta 12920; tel. (21) 2500401; fax (21) 2500410; f. 1989 as PT Sanwa Indonesia Bank; fmrly PT Bank Sanwa Indonesia; name changed as above Oct. 2001 following merger with PT Tokai Lippo Bank; cap. 817,449m., res 11,043m., dep. 3,475,818m. (Dec. 2002); Pres. Commr TAKESHI OGASAWARA; Pres. Dir SEIJI OSAKI; 1 br., 6 sub-brs.

PT Bank UOB Indonesia: Menara BCD, 1st–3rd Floors, Jalan Jenderal Sudirman, Kav. 26, Jakarta 12920; tel. (21) 2506330; fax (21) 2506428; e-mail uob.jakarta@uobgroup.com; internet www.uobgroup.com; 99%-owned by United Overseas Bank Ltd, Singapore; cap. 50,000m., res 10,000m., dep. 1,757,128m. (Dec. 2001); Chair. WEE CHO YAW; Pres. Dir CHUA KIM HAY.

PT Hagabank: Jalan Abdul Muis 28, Jakarta 10160; tel. (21) 2312888; fax (21) 2312250; e-mail info@hagabank.com; internet www.hagabank.com; f. 1989; cap. 65,000m., res –170m., dep. 2,277,305m. (Dec. 2002); Pres. DANNY HARTONO; Chair. TIMOTY E. MARNANDUS.

PT Korea Exchange Bank Danamon: Suite 1201, 12th Floor, Wisma GKBI, Jalan Jenderal Sudirman, Kav. 28, Selatan, Jakarta; tel. (21) 5741030; fax (21) 5741032; e-mail kebd@idola.net.id; cap. 150,000m., dep. 1,126,764m. (Dec. 2002); Pres. KANG IN KOO; Chair. KIM KWANG SEOB.

PT Lippo Bank Tbk: Gedung Menara Asia, 20th Floor, Jalan Raya Diponegoro 101, Lippo Karawaci, Tangerang 15810; tel. (21) 5460555; fax (21) 5460816; e-mail info_crc@lippobank.co.id; internet www.lippobank.co.id; f. 1948; cap. 811,494m., res 10,700,381m., dep. 22,080,993m. (Dec. 2002); Pres. Dir Dr JOS LUHUKAY; 379 brs.

PT Pan Indonesia Tbk (Panin Bank): Panin Bank Centre, 11th Floor, Jalan Jenderal Sudirman, Senayan, Jakarta 10270; tel. (21) 2700545; fax (21) 2700340; e-mail jasman@panin.co.id; internet www.panin.co.id; f. 1971; cap. 1,488,934m., res 1,852,792m., dep. 11,356,973m. (Dec. 2002); Chair. ENRIQUE VALDEZ BERNARDO; Pres. H. ROSTIAN SJAMSUDIN; 138 brs, 2 overseas brs.

PT Woori Bank Indonesia: Jakarta Stock Exchange Bldg, 16th Floor, Jalan Jenderal Sudirman, Kav. 52–53, Jakarta 12190; tel. (21) 5151919; fax (21) 5151477; e-mail wooridealingroom@yahoo.com; fmrly PT Hanvit Bank Indonesia; cap. 170,000m., res 34,000m., dep. 765,073m. (Dec. 2002); Pres. IN CHUL PARK.

Foreign Banks

ABN AMRO Bank NV (Netherlands): Jalan Ir H. Juanda 23–24, POB 2950, Jakarta 10029; tel. (21) 2312777; fax (21) 2313222; internet www.abnamro.co.id; Man. C. J. DE KONING; 15 brs.

Bangkok Bank Public Company Ltd (Thailand): POB 4165, Jakarta 11041; tel. (21) 2311008; fax (21) 3853881; f. 1968; Gen. Man. PRASARN TUNTASOOD.

Bank of America NA (USA): Jakarta Stock Exchange Bldg Tower 1, 22nd Floor, Jalan Jenderal Sudirman, Kav. 52–53, Jakarta 12190; tel. (21) 5158000; fax (21) 5158088; e-mail rahul.goswamy@bankofamerica.com; f. 1968; Man. Dir and Country Man. RAHUL GOSWAMY.

Bank of China (China): Jakarta.

Bank of Tokyo-Mitsubishi Ltd (Japan): Midplaza Bldg, 1st–3rd Floors, Jalan Jenderal Sudirman, Kav. 10–11, POB 2711, Jakarta 10227; tel. (21) 5706185; fax (21) 5731927; e-mail pip@botm.co.id; internet www.btmjkt.com; Gen. Man. HIDEYUKI ABE.

Citibank NA (USA): Citibank Tower, Jalan Jenderal Sudirman, Kav. 54–55, Jakarta 12910; tel. (21) 52962277; fax (21) 52969303; internet www.citibank.co.id; f. 1912; Vice-Pres JAMES F. HUNT, EDWIN GERUNGAN, ROBERT THORNTON.

Deutsche Bank, AG (Germany): POB 1135, Jakarta 10011; tel. (21) 3904792; fax (21) 335252; Gen. Man. HEINZ POEHLSEN.

Hongkong and Shanghai Banking Corpn Ltd (Hong Kong): POB 2307, Jakarta 10023; tel. (21) 5246222; fax (21) 5211103; internet www.hsbc.co.id; CEO P. C. L. HOLBERTON; 6 brs.

JP Morgan Chase Bank, NA (USA): Chase Plaza, 5th Floor, Jalan Jenderal Sudirman, Kav. 21, POB 311/JKT, Jakarta 12920; tel. (21) 5712213; fax (21) 5703690; Vice-Pres. and Sr Officer PETER NICE.

Standard Chartered Bank (United Kingdom): Wisma Standard Chartered Bank, Jalan Jenderal Sudirman, Kav. 33-A, Jakarta 10220; tel. (21) 2513333; fax (21) 5721234; internet www.standardchartered.com/id/index.html; Chief Exec. DAVID HAWKINS; 5 brs.

Banking Association

The Association of Indonesian National Private Commercial Banks (Perhimpunan Bank-Bank Umum Nasional Swasta—PERBANAS): Jalan Perbanas, Karet Kuningan, Setiabudi, Jakarta 12940; tel. (21) 5223038; fax (21) 5223037; e-mail secretariat@perbanas.web.id; internet www.perbanas.web.id; f. 1952; 94 mems; Chair. GUNARNI SOEWORO; Sec.-Gen. TIMOTY E. MARNANDUS.

STOCK EXCHANGES

At the end of January 2000 278 companies were listed on the Jakarta Stock Exchange, and market capitalization was 410,520,769m. rupiah. At the end of August 2002 there were 287 companies listed on the Jakarta Stock Exchange.

Bursa Paralel: PT Bursa Paralel Indonesia, Gedung Bursa, Jalan Medan Merdeka Selatan 14, Jakarta; tel. (21) 3810963; fax (21) 3810989; f. 1987.

Jakarta Stock Exchange (JSX): PT Bursa Efek Jakarta, Jakarta Stock Exchange Bldg, 4th Floor, Jalan Jenderal Sudirman, Kav. 52–53, Jakarta 12190; tel. (21) 5150515; fax (21) 5150330; e-mail webmaster@jsx.co.id; internet www.jsx.co.id; PT Bursa Efek Jakarta, the managing firm of the JSX, was transferred to the private sector in April 1992; 197 securities houses constitute the members and the shareholders of the exchange, each company owning one share; Pres. Dir ERRY FIRMANSYAH.

Surabaya Stock Exchange: 5th Floor, Gedung Medan Pemuda, Jalan Pemuda 27–31, Surabaya 60271; tel. (31) 5340888; fax (31) 5342888; e-mail helpdesk@bes.co.id; internet www.bes.co.id; f. 1989; Chair. NATAKOESOEMAH.

Regulatory Authority

Badan Pengawas Pasar Modal (BAPEPAM) (Capital Market Supervisory Agency): Jalan Medan Merdeka Selatan 14, Jakarta 10110; tel. (21) 365509; fax (21) 361460; e-mail info@bapepam.go.id; internet www.bapepam.go.id; Chair. HERWIDAYATMO; Exec. Sec. PANDE PUTU RAKA.

INSURANCE

In August 1996 there were 163 insurance companies, comprising 98 non-life companies, 56 life companies, four reinsurance companies and five social insurance companies.

Insurance Supervisory Authority of Indonesia: Directorate of Financial Institutions, Ministry of Finance, Jalan Dr Wahidin, Jakarta 10710; tel. (21) 3451210; fax (21) 3849504; Dir H. FIRDAUS DJAELANI.

Selected Life Insurance Companies

PT Asuransi AIA Indonesia: Gedung Bank Panin Senayan, 7th and 8th Floors, Jalan Jenderal Sudirman, Senayan, Jakarta 10270; tel. (21) 5721388; fax (21) 5721389; e-mail hrd.aiai@aig.com; f. 1975; Pres. Dir HARRY HARMAIN DIAH.

PT Asuransi Allianz Life Indonesia: Gedung Summitmas II, 20th Floor, Jalan Jenderal Sudirman, Kav. 61–62, Jakarta 12190; tel. (21) 52998888; fax (21) 2526953; e-mail general@allianz.co.id; internet www.allianz.co.id; f. 1996; Pres. Dir Dr JENS REISCH.

Asuransi Jiwa Bersama Bumiputera 1912: Wisma Bumiputera, 17th–21st Floors, Jalan Jenderal Sudirman, Kav. 75, Jakarta 12910; tel. (21) 5703812; fax (21) 5712837; e-mail spw@bumiputera.com; internet www.bumiputera.com; Pres. Dir Dr H. BASRI HASANUDIN.

PT Asuransi Jiwa Central Asia Raya: Wisma Asia, 10th–11th Floors, Jalan S. Parman, Kav. 79, Slipi, Jakarta 11420; tel. (21) 5637901; fax (21) 5637902; e-mail service@car.co.id; internet www.car.co.id; Man. Dir DJONNY WIGUNA.

PT Asuransi Jiwa 'Panin Putra': Jalan Pintu Besar Selatan 52A, Jakarta 11110; tel. (21) 672586; fax (21) 676354; f. 1974; Pres. Dir SUJONO SOEPENO; Chair. NUGROHO TJOKROWIRONO.

PT Asuransi Jiwasraya (Persero): Jalan H. Juanda 34, Jakarta 10120; tel. (21) 3845031; fax (21) 3862344; e-mail asuransi@jiwasraya.co.id; internet www.jiwasraya.co.id; f. 1959; Pres. and CEO HERRIS B. SIMANDJUNTAK.

PT Asuransi Lippo Life: Menara Matahari Lippo Life, Lt. 7, Jalan Bulevar Palem Raya 7, Lippo Karawaci, Tangerang 15811; tel. (21) 5475433; fax (21) 5475401; f. 1983; life insurance, pensions, health-care.

PT Asuransi Panin Life: Panin Bank Plaza, 5th Floor, Jalan Palmerah Utara 52, Jakarta 11480; tel. (21) 5484870; fax (21) 5484570; Pres. NUGROHO TJOKROWIRONO; Dir FADJAR GUNAWAN.

Bumi Asih Jaya Life Insurance Co Ltd: Jalan Matraman Raya 165–167, Jakarta 13140; tel. (21) 2800700; fax (21) 8509669; e-mail baj@bajlife.com; internet www.bajlife.co.id; f. 1967; Chair. P. SITOMPUL; Pres. VIRGO HUTAGALUNG.

Koperasi Asuransi Indonesia: Jalan Iskandarsyah I/26, Jakarta; tel. (21) 7207879; fax (21) 7207451; Dir H. J. V. SUGIMAN.

Selected Non-Life Insurance Companies

PT Asuransi Bina Arta Tbk: Wisma Dharmala Sakti, 8th Floor, Jalan Jenderal Sudirman, Kav. 32, Jakarta 10220; tel. (21) 5708157; fax (21) 5708166; e-mail dharins@uninet.net.id; Pres. SUHANDA WIRAATMADJA; Vice-Pres. M. MULYATNO.

PT Asuransi Bintang Tbk: Jalan R. S. Fatmawati 32, Jakarta 12430; tel. (21) 7504872; fax (21) 7506197; e-mail bintang@asuransibintang.com; internet www.asuransibintang.com; f. 1955; general insurance; Pres. Dir ARIYANTI SULIYANTO.

PT Asuransi Buana Independen: Jalan Pintu Besar Selatan 78, Jakarta 11110; tel. (21) 6904331; fax (21) 6263005; e-mail abipst99@indosat.net.id; Exec. Vice-Pres. SUSANTY PURNAMA.

PT Asuransi Central Asia: Wisma Asia, 12th–15th Floors, Jalan Letjen S. Parman, Kav. 79, Slipi, Jakarta 11420; tel. (21) 5637933; fax (21) 5638029; e-mail cust-aca@aca.co.id; internet www.aca.co.id; Pres. ANTHONY SALIM; Dir TEDDY HAILAMSAH.

PT Asuransi Danamon: Gedung Danamon Asuransi, Jalan H. R. Rasuna Said, Kav. C10, Jakarta 12920; tel. (21) 516512; fax (21) 516832; Chair. USMAN ADMADJAJA; Pres. Dir OTIS WUISAN.

PT Asuransi Dayin Mitra: Jalan Raden Saleh Raya, Kav. 1B–1D, Jakarta 10430; tel. (21) 3153577; fax (21) 39129; e-mail nuning@dayinmitra.co.id; internet www.dayinmitra.co.id; f. 1982; general insurance; Man. Dir LARSOEN HAKER.

PT Asuransi Indrapura: Jakarta; tel. (21) 5703729; fax (21) 5705000; f. 1954; Pres. Dir ROBERT TEGUH.

PT Asuransi Jasa Indonesia: Jalan Letjenderal M. T. Haryono, Kav. 61, Jakarta 12780; tel. (21) 7994508; e-mail jasindo@jasindo.co.id; internet www.jasindo.co.id; Pres. Dr Ir BAMBANG SUBIANTO; Dir AMIR IMAM POERO.

PT Asuransi Parolamas: Komplek Golden Plaza, Blok G 39–42, Jalan R. S. Farmawati 15, Jakarta 12420; tel. (21) 7508983; fax (21) 7506339; internet www.parolamas.co.id; Pres. TJUT RUKMA; Dir Drs SYARIFUDDIN HARAHAP.

PT Asuransi Ramayana: Jalan Kebon Sirih 49, Jakarta 10340; tel. (21) 31937148; fax (21) 31934825; f. 1965; Chair. R. G. DOERIAT; Pres. Dir Dr A. WINOTO DOERIAT.

PT Asuransi Tri Pakarta: Jalan Paletehan I/18, Jakarta 12160; tel. (21) 711850; fax (21) 7394748; Pres. Drs M. MAINGGOLAN; Dir HUSNI RUSTAM.

PT Asuransi Wahana Tata: Jalan H. R. Rasuna Said, Kav. C-4, Jakarta 12920; tel. (21) 5203145; fax (21) 5203149; e-mail aswata@aswata.co.id; internet www.aswata.co.id; Chair. A. R. RUMLY; Pres. RUDY WANANDI.

Berdikari Insurance Company: Jalan Merdeka Barat 1, Jakarta 10002; tel. (21) 3841339; fax (21) 3440586; e-mail ho@berdikari-insurance.com; internet www.berdikari-insurance.com; Pres. HOTBONAR SINAGA.

PT Lloyd Indonesia: Jalan Tiang Bendera 34-1, Jakarta 11230; tel. (21) 677195; Dir JOHNY BASUKI.

PT Maskapai Asuransi Indonesia (MAI): Jalan Sultan Hasanuddin 53–54, Kebayoran Baru, Jakarta Selatan 12160; tel. (21) 7204250; fax (21) 7256980; e-mail ptmai@cbn.net.id; Pres. Dir J. TRI WAHONO.

PT Maskapai Asuransi Jasa Tania: Gedung Agro Bank, Lt. 4, Jalan Teuku Cik Ditiro 14, Jakarta 10350; tel. (21) 3101912; fax (21) 323089; Pres. H. R. SUTEDJIO; Dir Drs H. ABELLAH.

PT Maskapai Asuransi Timur Jauh: Jalan Medan Merdeka Barat 1, Jakarta Pusat; tel. (21) 370266; f. 1954; Pres. Dir BUSTANIL ARIFIN; Dirs V. H. KOLONDAM, SOEBAKTI HARSONO.

PT Pan Union Insurance: Panin Bank Plaza, Lt. 6, Jalan Palmerah Utara 52, Jakarta 11480; tel. (21) 5480669; fax (21) 5484047; e-mail paninins@cbn.net.id; Chair. CHANDRA R. GUNAWAN; Pres. NIZARWAN HARAHAP.

PT Perusahaan Maskapai Asuransi Murni: Jalan Roa Malaka Selatan 21–23, Jakarta Barat; tel. (21) 679968; f. 1953; Dirs HASAN DAY, HOED IBRAHIM, R. SOEGIATNA PROBOPINILIH.

PT Pool Asuransi Indonesia: Blok A–IV Utara, Jalan Muara Karang Raya 293, Jakarta 14450; tel. (21) 6621946; fax (21) 6678021; f. 1958; Pres. BAMBANG GUNAWAN TANUJAYA; Dir TANDJUNG SUSANTO.

PT Tugu Pratama Indonesia: Wisma Tugu I, Jalan H. R. Rasuna Said, Kav. C8–9, Jakarta 12940; tel. (21) 52961777; fax (21) 52961555; e-mail tpi@tugu.com; internet www.tugu.com; f. 1981; general insurance; Chair. AINUN NA'IM; Pres. BAHDER MUNIR SJAMSOEDDIN.

Joint Ventures

PT Asuransi AIG Lippo Life: Matahari AIG Lippo Cyber Tower, 5th-7th Floors, Jalan Bulevar Palem Raya 7, Lippo Karawaci 1200, Tangerang 15811; tel. (21) 54218888; fax (21) 5475415; e-mail contact@aig-lippo.com; internet www.aig-lippo.com; jt venture between American International Group, Inc, and PT Asuransi Lippo Life; life insurance; Dep. Pres. Dir S. BUDISUHARTO.

PT Asuransi AIU Indonesia: Gedung Bank Panin Senayan, 3rd Floor, Jalan Jenderal Sudirman, Jakarta 10270; tel. (21) 5720888; fax (21) 5703759; Pres. Dir PETER MEYER; Vice-Pres. Dir SWANDI KENDY; Dir GUNAWAN TJIU.

PT Asuransi Allianz Utama Indonesia: Gedung Summitmas II, 9th Floor, Jalan Jenderal Sudirman, Kav. 61–62, Jakarta Selatan 12190; tel. (21) 2522470; fax (21) 2523246; e-mail general@allianz.co.id; internet www.allianz.co.id; f. 1989; non-life insurance; Pres. Dir KLAUS-DIETER VOESTE.

PT Asuransi Jayasraya: Jalan M. H. Thamrin 9, Jakarta; tel. (21) 324207; Dirs SUPARTONO, SADAO SUZUKI.

PT Asuransi Jiwa EKA Life: Wisma EKA Jiwa, 9th Floor, Jalan Mangga Dua Raya, Jakarta 10730; tel. (21) 6257808; fax (21) 6257837; e-mail cs@ekalife.co.id; internet www.ekalife.co.id; Pres. Dir HENRY C. SURYANAGA.

PT Asuransi Jiwa John Hancock Indonesia: Plaza DM, 7th Floor, Jalan Jenderal Sudirman, Kav. 25, Jakarta 12920; tel. (21) 5228857; fax (21) 5229730; e-mail info@jhancock.co.id; internet www.jhancock.co.id; f. 1987; life insurance and pension schemes; CEO DAVID W. COTTRELL.

PT Asuransi Jiwa Manulife Indonesia: Jalan Pegangsaan Timur 1A, Jakarta 10320; tel. (21) 2303224; fax (21) 2303225; e-mail communication_id@manulife.com; internet www.manulife-indonesia.com; f. 1985; life insurance; Pres. Dir JOHN HARRISON.

PT Asuransi Mitsui Marine Indonesia: Menara Thamrin, 14th Floor, Jalan M. H. Thamrin, Kav. 3, Jakarta 10340; tel. (21) 2303432; fax (21) 2302930; internet www.mitsuimarine.co.id; Pres. Dir S. AOSHIMA; Vice-Pres. PUTU WIDNYANA.

PT Asuransi Royal Indrapura: Jakarta Stock Exchange Bldg, 29th Floor, Jalan Jenderal Sudirman, Kav. 52–53, Jakarta 12190; tel. (21) 5151222; fax (21) 5151771; Pres. Dir Ir MINTARTO HALIM; Man. Dir MORAY B. MARTIN.

Insurance Association

Dewan Asuransi Indonesia (Insurance Council of Indonesia): Jalan Majapahit 34, Blok V/29, Jakarta 10160; tel. (21) 363264; fax (21) 354307; e-mail dai@dai.or.id; internet www.dai.or.id; f. 1957; Chair. MUNIR SIAMSOEDDIN; Gen. Sec. SOEDJIWO.

Trade and Industry

GOVERNMENT AGENCIES

Badan Pelaksana Kegiatan Usaha Hulu Minyak dan Gas Bumi (BP Migas): Gedung Patra Jasa, Lantai 1,2,13,14,16,21,22, Jalan Gatot Subroto, Kav. 32–34, Jakarta Selatan; tel. (21) 52900245; fax (21) 52901323; e-mail helpdesk@bpmigas.com; internet www.bpmigas.com; f. 2002; regulates upstream petroleum and natural gas industry; Chair. Ir RACHMAT SUDIBJO.

Badan Pengatur Hilir Minyak dan Gas Bumi: Jakarta; f. 2002; regulates downstream petroleum and gas industry; Chair. TUBAGUS HARYONO.

Badan Pengembangan Industri Strategis (BPIS) (Agency for Strategic Industries): Gedung Arthaloka, 3rd Floor, Jalan Jenderal Sudirman 2, Jakarta 10220; tel. (21) 5705335; fax (21) 3292516; f. 1989; co-ordinates production of capital goods.

Badan Pengkajian dan Penerapan Teknologi (BPPT) (Agency for the Assessment and Application of Technology): Jalan M. H. Thamrin 8, Jakarta 10340; tel. (21) 3162222; e-mail pusyantis@bppt.go.id; internet www.bppt.go.id; Chair. M. HATTA RAJASA.

Badan Tenaga Nuklir Nasional (BATAN) (National Nuclear Energy Agency): Jalan Kuningan Barat, Mampang Prapatan, Jakarta 12710; tel. (21) 5251109; fax (21) 5251110; e-mail humas@batan.go.id; internet www.batan.go.id; Chair. Dr SOEDYARTOMO.

Badan Urusan Logistik (BULOG) (National Logistics Agency): Jalan Jenderal Gatot Subroto, Kav. 49, Jakarta 12950; tel. (21) 5252209; fax (21) 5266482; e-mail rotu03@bulog.go.id; internet www.bulog.go.id; Chair. WIDJANARKO PUSPOYO.

National Agency for Export Development (NAFED): Jalan Gajah Mada 8, Jakarta 10310; tel. (21) 3841082; fax (21) 6338360; e-mail nafed@nafed.go.id; internet www.nafed.go.id.

National Economic Council: Jakarta; f. 1999; 13-member council formed to advise the President on economic policy; Chair. EMIL SALIM; Sec.-Gen. Sri MULYANI INDRAWATI.

DEVELOPMENT ORGANIZATIONS

Badan Koordinasi Penanaman Modal (BKPM) (Investment Co-ordinating Board): Jalan Jenderal Gatot Subroto 44, POB 3186, Jakarta 12190; tel. (21) 5252008; fax (21) 5254945; e-mail sysadm@bkpm.go.id; internet www.bkpm.go.id; f. 1976; Chair. THEO F. TOEMION.

Badan Perencanaan Pembangunan Nasional (Bappenas) (National Planning Development Board): Jalan Taman Suropati 2, Jakarta 10310; tel. (21) 336207; fax (21) 3145374; e-mail admin@bappenas.go.id; internet www.bappenas.go.id; formulates Indonesia's national economic development plans; Chair. Sri MULYANI INDRAWATI.

Commercial Advisory Foundation in Indonesia (CAFI): Jalan Probolinggo 5, Jakarta 10350; tel. (21) 3156013; fax (21) 3156014; f. 1958; information, economic regulations bulletin, consultancy and translation services; Chair. JOYCE SOSROHADIKOESOEMO; Man. Dir LEILA RIDWAN SOSROHADIKOESOEMO.

CHAMBER OF COMMERCE

Kamar Dagang dan Industri Indonesia (KADIN) (Indonesian Chamber of Commerce and Industry): Chandra Bldg, 3rd–5th Floors, Jalan M. H. Thamrin 20, Jakarta 10350; tel. (21) 324000; fax (21) 3150241; internet www.kadinnet.com; f. 1969; 27 regional offices throughout Indonesia; Chair. Ir ABURIZAL BAKRIE; Sec.-Gen. Ir IMAN SUCIPTO UMAR.

INDUSTRIAL AND TRADE ASSOCIATIONS

Association of Indonesian Automotive Industries (GAIKINDO): Jalan H. O. S. Cokroaminoto 6, Jakarta Pusat 10350; tel. (21) 3157178; fax (21) 3142100; e-mail gaikindo@cbn.net.id.

Association of Indonesian Beverage Industries (ASRIM): Jalan M. Ikhwan Mais 8, Jakarta Pusat; tel. (21) 3841222; fax (21) 3842294.

Association of Indonesian Coal Industries (ABBI): Perum Batu Bara Bldg, Jalan Supomo 10, Jakarta Selatan; tel. (21) 8295608.

Association of Indonesian Coffee Exporters (AICE): Jalan R. P. Soeroso 20, Jakarta 10330; tel. (21) 3106765; fax (21) 3144115; Chair. OESMAN SOEDARGO.

Association of Indonesian Heavy Equipment Industries (HINABI): c/o PT Traktor Nusantara, Jalan Pulogadung 32, Jakarta 13930; tel. (21) 4703932; fax (21) 4713940.

Association of State-Owned Companies: CTC Bldg, Jalan Kramat Raya 94–96, Jakarta; tel. (21) 346071; co-ordinates the activities of state-owned enterprises; Pres. ODANG.

Electric and Electronic Appliance Manufacturers' Association: Jalan Pangeran, Blok 20/A-1D, Jakarta; tel. (21) 6480059.

GINSI (Importers' Asscn of Indonesia): Oil Center Bldg, 1st Floor, Jalan M. H. Thamrin 55, Jakarta 10350; tel. (21) 39837395; fax (21) 39837394; f. 1956; 2,921 mems (1996); Chair. AMIRUDIN SAUD; Sec.-Gen. MUSTAFA KEMAL.

Indonesian Cement Association: Graha Purnayudha, Jalan Jenderal Sudirman, Jakarta; tel. (21) 5207603; fax (21) 5207188.

Indonesian Cocoa Association (AKI): Jalan Brawijaya VII/5, Kebayoran Baru, Jakarta 12160; tel. (21) 771721; fax (21) 7203487.

Indonesian Exporters' Federation: Menara Sudirman, 8th Floor, Jalan Jenderal Sudirman, Kav. 60, Jakarta 12190; tel. (21) 5226522; fax (21) 5203303; Chair. HAMID IBRAHIM GANIE.

Indonesian Footwear Association (APRISINDO): Gedung Adis Dimension Footwear, Jalan Tanah Abang 111/18, Jakarta 10610; tel. (21) 3447575; fax (21) 3447572; e-mail aprisindo@vision.net.id; internet www.aprisindo.com; Pres. HARIJANTO; Sec.-Gen. YUDHI KOMARUDIN.

Indonesian Furniture Industry and Handicraft Association (ASMINDO): Jalan Pegambiran 5A, 3rd Floor, Rawamangun Jakarta 13220; tel. (21) 47864028; fax (21) 47864031; e-mail asmindo@indo.net.id; internet www.furniture-indonesia.com/asmindo; Chair. DJALAL KAMAL.

Indonesian Nutmeg Exporters' Association: c/o PT Berdirari (Persero) Trading Division, Jalan Yos Sudarso 1, Jakarta; tel. (21) 4301625; e-mail bnuina@indosat.net.id.

Indonesian Palm Oil Producers' Association: Jalan Pulo Mas IIID/1, Jakarta; tel. (21) 4892635; Chair. NUKMAN NASUTION.

Indonesian Precious Metals Association: Jalan Wahid Hasyim 45, Jakarta; tel. (21) 3841771.

Indonesian Pulp and Paper Association: Jalan Cimandiri 6, Jakarta 10330; tel. (21) 326084; fax (21) 3140168.

Indonesian Tea Association: Wisma 77, 20th Floor, Jalan S. Parman, Kav. 77, Jakarta 11410; tel. (21) 53660366; fax (21) 53660365; e-mail insyaf@hotmail.com; internet www.indotea.org; Chair. INSYAF MALIK.

Indonesian Textile Association (API): Panin Bank Centre, 3rd Floor, Jalan Jenderal Sudirman 1, Jakarta Pusat 10270; tel. (21) 7396094; fax (21) 7396341; f. 1974; Sec.-Gen. DANANG D. JOEDONAGORO.

Indonesian Tobacco Association: Jalan H. Agus Salim 85, Jakarta 10350; tel. (21) 3140627; fax (21) 325181; Pres. H. A. ISMAIL.

Masyarakat Perhutanan Indonesia (MPI) (Indonesian Forestry Community): Gedung Manggala Wanabakti, 9th Floor, Wing B/Blok IV, Jalan Jenderal Gatot Subroto, Jakarta Pusat 10270; tel. (21) 5733010; fax (21) 5732564; f. 1974; nine mems; Pres. M. HASAN.

National Board of Arbitration (BANI): Wahana Graha, 2nd Floor, Jalan Mampang Prapatan 2, Jakarta 12760; tel. (21) 7940542; fax (21) 7940543; e-mail bani-arb@indo.net.id; internet www.bani-arb.org; f. 1977; resolves company disputes; Chair. Prof. H. PRIYATNA ABDURRASYID.

Rubber Association of Indonesia (Gapkindo): Jalan Cideng Barat 62A, Jakarta; tel. (21) 3846813; fax (21) 3846811; e-mail karetind@indosat.net.id; Exec. Dir SUHARTO HONGGOKUSUMO.

Shippers' Council of Indonesia: Jalan Kramat Raya 4–6, Jakarta; Pres. R. S. PARTOKUSUMO.

UTILITIES

Electricity

PT Perusahaan Listrik Negara (Persero) (PLN): Jalan Trunojoyo, Blok M1/135, Kebayoran Baru, Jakarta Selatan 12160; tel. (21) 7261875; fax (21) 7204929; e-mail webmaster@pln.co.id; internet www.pln.co.id; state-owned electricity co; Pres. EDDIE WIDIONO.

Gas

PT Perusahaan Pertambangan Minyak dan Gas Bumi Negara (PERTAMINA): Jalan Medan Merdeka Timur 1A, Jakarta 10110; tel. (21) 3815111; fax (21) 3846865; e-mail mthakky@pertamina.com; internet www.pertamina.com; f. 1968; state-owned petroleum and natural gas mining enterprise; Pres. Dir and CEO WIDYA PURNAMA.

Perusahaan Gas Negara (PGN) (Public Gas Corporation): Jakarta; e-mail webmaster@pgn.co.id; internet www.pgn.co.id; monopoly of domestic gas distribution; Pres. Dir W. M. P. SIMANDJUNTAK.

Water

PDAM DKI Jakarta (PAM JAYA): Jalan Penjernihan 11, Pejompongan, Jakarta 10210; tel. (21) 5704250; fax (21) 5711796; f. 1977; responsible for the water supply systems of Jakarta; Pres. Dir Ir H. MUZAHIEM MOKHTAR.

PDAM Kodya Dati Il Bandung: Jalan Badaksinga 10, Bandung 40132; tel. (22) 2509030; fax (22) 2508063; e-mail pdambdg@elga.net.id; f. 1974; responsible for the water supply and sewerage systems of Bandung; Pres. Dir Ir SOENITIYOSO HADI PRATIKTO.

PDAM Tirtanadi Medan: Jalan Sisingamangaraja 1, Medan 20212; tel. (61) 571666; fax (61) 572771; f. 1979; manages the water supply of Medan and nearby towns and cities; Man. Dir Ir KUMALA SIREGAR.

CO-OPERATIVES

In 1996 there were 48,391 primary and secondary co-operatives in Indonesia; membership of primary co-operatives was 27,006,000 in the same year.

Dewan Koperasi Indonesia (DEKOPIN) (Indonesian Co-operative Council): Jakarta; internet www.dekopin.coop; Pres. Sri EDY SWASONO.

MAJOR COMPANIES

(cap. = capital; res = reserves; m. = million)

State Enterprises

PT ASEAN Aceh Fertilizer: Jalan Medan Banda Aceh, Krueng Geukueh, POB 09, Lhokseumawe, North Aceh; tel. (645) 56933; fax (645) 56660; e-mail aaf@aaf.co.id; internet www.aaf.co.id; f. 1979; company owned by ASEAN governments (Indonesia 60%, Malaysia 13%, Thailand 13%, Philippines 13%, Singapore 1%); produces ammonia and urea; sales Rp. 341,500m. (1996); Pres. Dir Ir RAUF PURNAMA; 850 employees.

PT Barata Indonesia: Jalan Ngagel 109, Surabaya 60246; tel. (31) 5019060; fax (31) 5019077; e-mail info@barata.co.id; internet www.barata.co.id; f. 1971; manufacture of heavy construction and industrial equipment; Pres. Dir IMAM KARTONO; 3,500 employees.

PT Boma Bisma Indra (BBI): Jalan K. H. Mas Mansyur 229, Surabaya 60162; tel. (31) 3555798; fax (31) 3535884; e-mail pr@ptbbi.co.id; internet www.ptbbi.co.id; f. 1971; engineering services and the manufacture of industrial plant equipment; sales Rp. 210,000m. (1998); Pres. Dir TJOKRO SUPRIJADI; 2,100 employees.

PT Cipta Niaga: Jalan Malaka 7–9, Jakarta 11230; tel. (21) 6912823; fax (21) 6911162; e-mail ciptajkt@indosat.net.id; internet www.ciptaniaga.co.id; f. 1964; import and distribution of basic goods, bulk articles, sundries, provisions and drinks, and export of Indonesian produce; Pres. Dir EDDIE M. GUNADI.

PT Dahana: Jalan Letkol Basir Surya, POB 117, Tasikmalaya 46196, Jawa Barat; tel. (265) 331853; fax (265) 332425; e-mail dahana@ibm.net.id; internet www.dahana.co.id; f. 1966; manufacture of dynamite and other industrial explosives and accessories; Pres. Dir Ir RUSBANDI.

PT Dharma Niaga Ltd: Jalan Kalibesar Barat 11, Jakarta 11230; tel. (21) 6903430; fax (21) 6906533; f. 1970; import, export, distribution, installation, after sales service; sales Rp. 450,000m. (1996); Pres. Drs BENARTO; 1,200 employees.

PT Dirgantara Indonesia: Gedung GPM, Jalan Pajajaran 154, Bandung 40174; tel. (22) 632043; fax (22) 631696; e-mail santos@telkom.go.id; internet www.dprin.go.id/links/pt_iptn/main.htm; f. 1976; fmrly PT Industri Pesawat Terbang Nusantara; name changed as above 2001; aircraft manufacture; Chair. ASHADI TJAHJADI; 16,000 employees.

INKA: Jalan Yos Sudarso 71, Madiun 63122; tel. (351) 452271; fax (351) 452275; e-mail sekretariat@inka.go.id; internet www.inka.co.id; f. 1981; manufacture of railway rolling stock; Exec. Dir Ir ISTANTORO.

INTI: Jalan Moch Toha 77, Bandung 40253; tel. (22) 502784; f. 1974; manufacture of telecommunications equipment.

PT Jasa Marga: Plaza Tol Taman Mini, Jalan Tol Jagorawi, Jakarta 13550; tel. (21) 8413630; fax (21) 8401533; e-mail jasmar@jasamarga.com; internet www.jasamarga.com; f. 1978; construction, maintenance and operation of toll roads and bridges; cap. and res Rp. 1,492,000m., sales Rp. 928,492m. (2001); Pres. Dir SYARIFUDDIN ALAMBAI; 6,364 employees.

PT Krakatau Steel: Wisma Baja, 4th–8th Floors, 54 Jalan Gatot Subroto, Jakarta Selatan 12960; tel. (21) 510266; fax (21) 5204208; f. 1971; mfr of metals and metal products; sales Rp. 2,600,000m. (1996); Chair. T. ARIWIBOWO; Pres. Dir DAENUL HAY; 7,000 employees.

LEN Industri: Jalan Sukarno Hatta 442, Bandung 40254; tel. (22) 5202682; fax (22) 5202695; e-mail marketing@len.co.id; internet www.len.co.id; f. 1965 as the National Electrotechnics Research Institute, became a state-owned company in 1991; manufactures electronic products and components; Pres. Dir Ir DODI HIDAYAT RIVAI.

PT Mega Eltra: Jalan Menteng Raya 27, Jakarta 10340; tel. (21) 3909018; fax (21) 3102937; e-mail sekretariat@megaeltra.co.id; internet www.megaeltra.co.id; f. 1960; building co and trader in cement, pharmaceuticals, chemicals and machinery; sales Rp. 280,000m. (1996); Pres. Commr Ir ERAWAN ASIKIN; Pres. Dir CODDY SOERONO; 600 employees.

PT PAL Indonesia (Persero): POB 1134, Ujung, Surabaya 60155; tel. (31) 3292275; fax (31) 3292530; e-mail palsub@pal.co.id; internet www.pal.co.id; f. 1980; shipbuilding and general engineering; Pres. Dir ADWIN H. SURYOHADIPROJO; 3,369 employees.

PT Perhutani (Persero) (State Forest Corporation): Gedung Manggala Wanabakti, Blok IV/Lantai 4, Jalan Gatot Subroto, Senayan, Jakarta Pusat 10270; tel. (21) 587090; fax (21) 583616; internet www.perhutani.co.id; f. 1973; Pres. Dir Ir MARSANTO.

PT Petrokimina Gresik: Jalan Jenderal Akhmad Yani, Gresik 61119, East Java; tel. (31) 3981811; fax (31) 3981722; e-mail pkg@petrokimia-gresik.com; internet www.petrokimia-gresik.com; f. 1972; produces fertilizers, ammonia and other related products; cap. 994,059m., sales Rp. 2,875,000m. (Dec. 2003); Pres. Dir ARIFIN TASRIF; 3,709 employees.

PT Pindad (Persero): Jalan Jenderal Gatot Subroto 517, Bandung 40284; tel. (22) 7312073; fax (22) 7304095; e-mail info@pindad.com; internet www.pindad.com; manufacture of products for commercial and military markets; Pres. Dir BUDI SANTOSO.

PT Pos Indonesia (Persero): Jalan Banda 30, Bandung 40115; tel. (21) 4213640; fax (22) 4219012; e-mail piol-3@posindonesia.co.id; internet www.posindonesia.co.id; f. 1927; provides communication, logistic and financial services; sales Rp. 559,000m. (1996); Pres. Dir ALINAFIAH; 27,000 employees.

PT Pupuk Iskandar Muda: Jalan Medan-Banda Aceh, Lhok Seumawe, North Aceh; tel. (645) 56222; fax (645) 56095; e-mail info@pim.co.id; internet www.pim.co.id; f. 1982; produces ammonia and urea; sales Rp. 185,000m. (1996); Pres. Dir Ir HIDAYAT NYAKMAN; 1,200 employees.

PT Pupuk Sriwwijaya (PT Pusri): Jalan Mayor Zen, Sungai Selayur, Palembang 30118; tel. (711) 712111; fax (711) 712100; e-mail info@pusri.co.id; internet www.pusri.co.id; f. 1959; produces ammonia and urea; sales Rp. 2,495,000m. (1996); Pres. Dir ZAENAL SOEDJAIS; 6,000 employees.

PT Semen Gresik (Persero) Tbk: Gedung Utama Semen Gresik, Jalan Veteran, Gresik 61122, ; tel. (31) 3981732; fax (31) 3983209; e-mail ptsg@sg.sggrp.com; internet www.sggrp.com; f. 1969; cement plant; cap. and res Rp. 8,763,075m., sales Rp. 4,659,203m. (2001); Pres. Dir SATRIYO.

PT Tambang Batubara Bukit Asam (PTBA): Plaza Setiabudi II, 5th Floor, Jalan H. R. Rasuna Said, Kuningan, Jakarta 12940; tel. (21) 5254014; fax (21) 5253992; e-mail milawarma@bukitasam.co.id; internet www.ptbukitasam.com; f. 1981; merged with Perum Tambang Batubara in 1990; coal-mining; Pres. Dir ISMET HARMAINI.

PT Timah Tbk: Jalan Jenderal Sudirman 51, Pangkalpinang 33121, Bangka; tel. (717) 431335; fax (717) 432323; e-mail timah@pt.timah.co.id; internet www.pttimah.com; tin-mining; partially privatized in 1995; Pres. Dir ERRY RIYANA HARDJAPAMEKAS; 5,400 employees.

Management Board

General Management Board of the State Trading Corporations (BPU-PNN): Jakarta; f. 1961; Pres. Col SUHARDIMAN.

Private Companies

Agribusiness

PT Central Proteinaprima: Gedung SHS, Jalan Ancol Barat, Blok A/5E 10 Ancol, Jakarta Utara; tel. (21) 5152858; fax (21) 6905640; e-mail cpprima@indoexchange.com; f. 1988; mfr of animal feeds; cap. and res Rp. 96,815m., sales Rp. 5,567,615m. (2001); Pres. Dir FRANCISCUS AFFANDY; 4,100 employees.

PT Charoen Pokphand: Jalan Ancol VIII/I, Ancol Barat, Jakarta 14430; tel. (21) 5152858; fax (21) 6905640; e-mail charoen@indoexchange.com; f. 1972; mfr of livestock feeds, poultry farming; cap. and res Rp. 778,205m., sales Rp. 3,513,123m. (2001); Pres. Dir SUMET JIARAVANON; 4,400 employees.

PT Dharmala Agrifood: Wisma Dharmala Sakti, 20th Floor, Jalan Jenderal Sudirman, Kav. 32, Jakarta; tel. (21) 5704434; fax (21) 5708166; e-mail dharagri@indoexchange.com; f. 1970; mfr of animal feeds, poultry farming, trading of rice and maize; cap. and res Rp. 177,573m., sales Rp. 270,603m. (1996); Pres. Dir THOMAS CHANDRA; 1,000 employees.

PT Japfa Comfeed Indonesia: Graha Praba Samanta, Jalan Daan Mogot 9, Km 12, Jakarta 11730; tel. (21) 5448710; fax (21) 5448709; e-mail info@japfacomfeed.co.id; internet www.japfacomfeed.co.id; f. 1971; mfr of animal feeds, poultry and aquaculture farming; total assets Rp. 3,116,219m., sales Rp. 4,407,906m. (2003); Pres. Dir HANDOJO SANTOSA; Pres. Commr SYAMSIR SIREGAR; 10,500 employees.

PT Sinar Mas Multiartha: BII Plaza, Tower III, 7th Floor, Suite 702, Jalan M. H. Thamrin 51, Jakarta 10350; tel. (21) 3925660; fax (21) 3925788; e-mail peter.c@sinarmas.com; internet www.sinarmas.com; cap. and res Rp. 9,233,307m., sales Rp. 8,404,602 (1998); Chair. EDWARD H. HADIDJAJA.

Food Processing

PT Indofood Sukses Makmur Tbk: Gedung Ariobimo Central, 12th Floor, Jalan H. R. Rasuna Said, X-2 Kav. 5, Kuningan, Jakarta 12950; tel. (21) 5228822; fax (21) 5226014; e-mail ism@indofood.com; internet www.indofood.co.id; f. 1971; mfr of food, incl. noodles; total assets Rp. 15,308,900m., sales Rp. 17,871,400m. (2003); Pres. Dir and CEO ANTHONI SALIM; Pres. Commr MANUEL V. PANGILINAN; 23,000 employees.

PT Mayora Indah: Gedung Mayora, Jalan Tomang Raya 21–23, Jakarta 11440; tel. (21) 5655311; fax (21) 5655323; e-mail mayora@indoexchange.com; f. 1977; mfr of processed food products; cap. and res Rp. 627,522m., sales Rp. 833,977m. (2001); Chair. JOGI HENDRA ATMADJA; 7,000 employees.

PT Medan Tropical Canning and Frozen Industries: Jalan K. L. Yos Sudarso, Km 10.5, Kawasan Industri Medan, Medan 20242, North Sumatra; tel. (61) 6850038; fax (61) 6851330; e-mail sales@indonesiaseafood.com; internet www.indonesiaseafood.com; f. 1984; processing and packaging of seafood; Man. Dir BINTARNA TARDY; 1,500 employees.

PT Multi Bintang Indonesia: Gedung Ratu Plaza, 21st Floor, Jalan Jenderal Sudirman 9, Jakarta 10270; tel. (21) 7207511; fax (21) 7207864; e-mail info_mbi@multibintang.co.id; internet www.multibintang.co.id; producer and distributor of alcoholic and non-alcoholic beverages; cap. and res Rp. 268,297m., sales Rp. 562,852m. (2003); Pres. Dir HERMAN P. P. M. HOFHUIS; 850 employees.

PT SMART Corporation (Sinar Mas Agro Resources and Technology): Gedung JITC, 10th Floor, Jalan Mangga Dua Raya, Jakarta 14430; tel. (21) 2601088; fax (21) 2601059; e-mail investor@smart-corp.com; internet www.smart-corp.com; f. 1962; food and soft drinks mfr and plantation owner (tea, banana, coconut, rubber); cap. and res Rp. -599,753m., sales Rp. 2,294,284m. (2001); Pres. MUKTAR WIDJAJA; 10,000 employees.

Heavy Equipment

PT Bakrie and Brothers Tbk: Wisma Bakrie, Lot 6, Jalan H. R. Rasuna Said, Kav. B-1, Jakarta 12920; tel. (21) 5250192; fax (21) 5200864; e-mail bakri@indoexchange.com; internet www.bakrie-brothers.com; f. 1942; steel and pipe mfr, automotives construction, telecommunications, infrastructure support, etc.; cap. and res Rp. 2,433,286m., sales Rp. 1,362,885m. (2001); Pres. Dir BOBBY GAFUR S. UMAR; 15,000 employees.

PT Komatsu Indonesia: Jalan Raya Cakung, Cilincing Km 4, Jakarta 14140; tel. (21) 4400611; fax (21) 4400615; e-mail corpsec@komi.co.id; internet www.komi.co.id; f. 1982; mfr of construction equipment; cap. and res Rp. 635,117m., sales Rp. 789,753m. (2003/04); Pres. Dir BUDIARDJO SOSROSUKARTO; 840 employees.

PT United Tractors: Jalan Raya Bekasi, Km 22, Cakung, Jakarta 13910; tel. (21) 4605949; fax (21) 4600657; e-mail ir@unitedtractors.com; internet www.unitedtractors.com; f. 1972; distributor of heavy equipment; cap. and res Rp. 814,974m., sales Rp. 7,058,396m. (2001); Pres. HAGIANTO KUMALA; 7,200 employees.

Mining

PT Adaro Indonesia: Suite 704, World Trade Centre, Jalan Jenderal Sudirman, Kav. 29–31, Jakarta 12920; tel. (21) 5211255; fax (21) 5211266; e-mail marketing@ptadaro.com; internet www.ptadaro.com; f. 1982; coal mining; sales US $400m. (2001); Pres. Dir GRAEME L. ROBERTSON; Gen. Man. A. H. CHIA; 5,000 employees.

PT Aneka Tambang Tbk: Gedung Aneka Tambang, Jalan Letjen T. B. Simatupang 1, Lingkar Solatan, Tanjung Barat, Jakarta 12530; tel. (21) 7805119; fax (21) 7812822; e-mail corsec@antam.com; internet www.antam.com; f. 1968; exploration, mining, processing and marketing of nickel ore, ferro-nickel, bauxite, iron, sand, gold and silver; cap. and res Rp. 1,919,725m.; sales Rp. 1,735,224m. (2001); Pres. ADITYA SUMANAGARA; Chair. FIRMAN TAMBOEN; 3,593 employees.

PT Freeport Indonesia Co: Plaza 89, 5th Floor, Jalan H. R. Rasuna Said, Kav. X-7/6, Jakarta 12940; tel. (21) 2520727; fax (21) 5225976; f. 1991; copper, gold and silver mining; Pres. Dir ADRIANTO MAHRIBIE; 16,000 employees.

PT International Nickel Indonesia Tbk: Plaza Bapindo-Citibank Tower, 22nd Floor, Jalan Jenderal Sudirman, Kav. 54–55, Jakarta; tel. (21) 5249000; fax (21) 5249030; e-mail inco@indoexchange.com; f. 1968; nickel mining and processing; cap. and res Rp. 7,608,085m., sales Rp. 3,066,195m. (2001); Pres. and CEO RUMENGAN MUSU; 2,360 employees.

Motor Vehicle Assembly

ADR Group: Wisma ADR, Jalan Pluit Raya I, Jakarta 14440; tel. (21) 6690244; fax (21) 6605071; e-mail adr@adr-group.com; internet www.adr-group.com; f. 1973; mfr of motor vehicle parts; Pres. Dir EDDY HARTONO.

PT Astra International Tbk: Jalan Ir Juanda 22, Jakarta 10120; tel. (21) 3750008; fax (21) 3848102; internet www.astra.co.id; f. 1957; car mfr, heavy industry, financial and non-financial services; cap. and res Rp. 1,146,233m., sales Rp. 1,417,491m. (2001); Pres. Dir BUDI SETIADHARMA; 91,000 employees.

PT Federal Motor: 1 Jalan Yos Sudarso Sunter, Jakarta 14350; tel. (21) 6518080; fax (21) 6521889; e-mail yozardi@federal.co.id; internet www.federal.co.id; f. 1971; mfr of motor cycles, parts and accessories; sales US $250m. (1997); Pres. Dir BUDI SETIADHARMA; 1,650 employees.

Paper

PT Indah Kiat Pulp and Paper Corpn Tbk: Bll Plaza, Menara II Lantai 7, Jalan M. H Thamrin 52, Jakarta 10350; tel. (21) 5380001; fax (21) 5380200; e-mail indahkiat@indoexchange.com; f. 1976; cap. and res Rp. 7,847,512m. (1997), sales Rp. 14,829,330m. (2000); Pres. Dir TEGUH GANDA WIDJAYA; 16,000 employees.

PT Pabrik Kertas Tjiwi Kimia: Jalan Raya Surabaya, Mojokerto Km 44, Sidoardjo, Jawa Timur; tel. (321) 21574; fax (321) 21615; internet www.tjiwi.co.id; f. 1972; mfr of writing and printing paper, stationery products and general office products; sales Rp. 6,678,877m. (1999); Chair. YUDI SETIAWAN LIN; 10,500 employees.

PT Surabaya Agung Industri Pulp and Kertas: Jalan Kedungdoro 60, 8th–10th Floors, Surabaya 60251; tel. (31) 5482003; fax (31) 5482039; e-mail bsd@suryakertas.com; internet www.suryakertas.com; f. 1973; producer of paper and packaging materials; cap. and res Rp. –1,364,365m., sales Rp. 634,852m. (2001); Pres. Dir TIRTOMULYADI SULISTYO; 1,800 employees.

Petrochemicals

PT Continental Carbon Indonesia: Jalan Kebon Sirih 63, Jakarta; tel. (21) 3907939; fax (21) 3907929; production of carbon black; Pres. Dir LILI SOEMANTRI.

PT Eastern Polymer: Jalan Cilincing Raya, Tanjung Priok, North Jakarta; tel. (21) 4301167; fax (21) 496083; jt venture with Mitsubishi Corpn (50%) and PT Anugrah Daya Laksana (50%); production of PVC resin; Pres. Dir MOTONOBU TOKUDA.

PT Indochlor Prakarsa: Wisma Bimoli, 3rd Floor, Jalan Jembatab Tiga, Blocks F and G, Jakarta 14440; tel. (21) 6603601; fax (21) 6603610; produces caustic soda, hydrochloric acid and liquid chlorine; Pres. Dir GAUTAMA SETIAWAN.

PT Petrokimia Nusantara Interindo: Plaza 89, Suite 705, Jalan H. R. Rasuna Said, Kav. X-7 No. 6, Jakarta 12490; tel. (21) 5222722; fax (21) 5209360; production of polyethylene; Pres. Dir Dr JAMES WHITE.

PT Standard Toyo Polymer: Wisma Permata Plaza, 9th Floor, Jalan M. H. Thamrin 57, Jakarta 10310; tel. (21) 3903132; jt venture between Tosoh Corpn and Mitsui Corpn (50%) and local companies (50%); production of PVC resin; Pres. Dir NORIO MARSUYAMA.

PT Tri Polyta Indonesia Tbk: Menara Kebon Sirih, 12th Floor, Jalan Kebon Sirih 17–19, Jakarta 10340; tel. (21) 3929828; fax (21) 3929818; internet www.tripolyta.com; production of polypropylene, acrylic acid and acrylic ester; sales Rp. 1,134,115m. (1999); Pres. Dir IMAM SUCIPTO UMAR.

PT Unggul Indah Cahaya Tbk: Wisma UIC, 2nd Floor, Jalan Jenderal Gatot Subroto, Kav. 6–7, Jakarta 12930; tel. (21) 57905100; fax (21) 57905111; e-mail corp_sect@uic.co.id; internet www.uic.co.id; f. 1983; chemicals mfr and processor; cap. and res Rp. 495,646m. (1999), sales US $250.9m. (2003); Pres. HARTONO GUNAWAN; 320 employees.

Pharmaceuticals

PT Dankos Laboratories Tbk: Jalan Rawa Gatel Blok III-S, Kav. 37–38, Kawasan Industri Pulogadung, Jakarta 13930; tel. (21) 4600158; fax (21) 4611301; e-mail dankos@dankoslabs.com; internet www.dankoslabs.com; f. 1974; mfr and marketing of human health pharmaceutical products; cap. and res 826,778m., sales Rp. 1,191,273m. (2003); Pres. Dir HERMAN WIDJAJA; 2,882 employees.

PT Darya-Varia Laboratoria Tbk: Graha Darya-Varia, 2nd–3rd Floors, Jalan Melawai Raya 93, Kebayoran Baru, Jakarta Selatan 12130; tel. (21) 7258010; fax (21) 7258011; e-mail info@darya-varia.com; internet www.darya-varia.com; f. 1976; mfr of pharmaceuticals and consumer healthcare products; cap. and res Rp. 280,000m. (2001), sales Rp. 390,635m. (2003); Chair. MANUEL P. ENGWA; 1,165 employees.

PT Enseval Putera Megatrading Tbk (EPM): Gedung Enseval, Jalan Letjen, Suprapto Kav. 4, Jakarta 10510; tel. (21) 4243908; fax (21) 4244812; e-mail enseval@indoexchange.com; internet www.enseval.com; f. 1988; distribution of pharmaceuticals, cosmetics, medical products, consumer products and raw materials; cap. and res Rp. 128,682m., sales Rp. 2,063,696m. (2001); Pres. Dir BUDI DHARMA WREKSOATMODJO; 2,000 employees.

PT Kalbe Farma: Jalan Jend A. Yani, Pulo Mas, Jakarta 13210; tel. (21) 4892808; fax (21) 4893549; e-mail kalbeibd@pacific.net.id; internet www.kalbefarma.com; f. 1966; mfr of healthcare products, pharmaceuticals and veterinary products; cap. and res. Rp. 220,774m., sales Rp. 2,046,499m. (2001); Pres. Dir JOHANNES SETIJONO; 1,550 employees.

Merck Indonesia Tbk: Jalan Raya Gedong 8, Pasar Rebo, Jakarta 13760; tel. (21) 8400081; fax (21) 8400492; e-mail merck@indoexchange.com; f. 1970; fmrly PT Merck Indonesia; mfr of pharmaceuticals and medicines; cap. and res Rp. 86,229m., sales Rp. 224,073m. (2001); Pres. Dir H. U. WOLF; 515 employees.

PT Nellco Indopharma: Jalan Kebon Jeruk 18/6, Jakarta 11160; tel. (21) 6297562; fax (21) 6297753; f. 1963; production of pharmaceutical goods; Pres. Dir RUDY CHANDRA.

PT Surya Hidup Satwa: Jalan Angol Barat Blok A-5E/10, Jakarta 14430; tel. (21) 6909958; fax (21) 6909957; e-mail suryahidup@indoexchange.com; f. 1976; mfr and distribution of veterinary products and animal husbandry equipment; cap. and res Rp. 454,667m., sales Rp. 3,840,857m. (1998); Pres. Dir FREDDIE HADIWIBOWO; 150 employees.

PT Tempo Scan Pacific Tbk: Gedung Bina Mulia II, Jalan H. R. Rasuna Said, Kav. 11, Jakarta 12950; tel. (21) 5201858; fax (21) 5201857; e-mail tempo@indoexchange.com; f. 1970; mfr of pharmaceutical, personal care and cosmetic products; cap. and res Rp. 1,270,581m., sales Rp. 1,785,230m. (2001); Chair. DIAN PARAMITA TAMZIL; Pres. Dir HANDOJO SELAMET; 3,200 employees.

PT Unilever Indonesia: Graha Unilever, Jalan Gatot Subroto, Kav. 15, Jakarta 12930; tel. (21) 5262112; fax (21) 5262044; e-mail unilever@indoexchange.com; f. 1933; mfr of personal care and home care products; cap. and res Rp. 1,728,199m., sales Rp. 6,012,611m. (2001); Pres. Dir NIHAL KAVIRATNE.

Plastics

PT Industri Dinar Makmur: Jalan Palmerah Utara 69–71, Jakarta 10270; tel. (21) 5481205; fax (21) 5483412; specializes in styrofoam and air bubble film; Dir ICHWAN HARTONO.

PT Pioneer Plastic Ltd: Jalan Bandengan Utara 43, Jakarta Utara 14440; tel. (21) 6690908; fax (21) 6694431; f. 1954; mfr of plastic hardware products; Pres. PANDJI WISAKSANA; 200 employees.

PT Sinar Panah Industry Co: Jalan Padamulya IV, Gg. Karung 39, Jakarta 11330; tel. (21) 6317947; fax (21) 6318168; f. 1963; Man. HENDRA WIDJAJA.

Rubber

PT Bakrie Sumatera Plantations Tbk: Kisaran 21202, Kabupaten Asahan, North Sumatera; tel. (623) 41434; fax (623) 41066; e-mail bspkis@kisaran.wasantara.net.id; internet www.bakrie-brothers.com; f. 1911; owner and operator of rubber and oil palm plantations; cap. and res Rp. –57,001m. (2001), sales Rp. 357,756m. (2002); Pres. Dir AMBONO JANURIANTO; 4,816 employees.

PT Gajah Tunggal: Wisma Hayam Wuruk, 10th Floor, Jalan Hayam Wuruk, Kav. 8, Jakarta 10120; tel. (21) 3805916; fax (21) 3804908; e-mail gajah@indoexchange.com; internet www.gt-tires.com; f. 1951; produces automotive tyres; cap. and res Rp. –589,346m., sales Rp. 5,078,432m. (2000); Pres. Dir R. KASENDA; 13,600 employees.

PT Goodyear Indonesia: POB 5, Jalan Pemuda 27, Bogor 16161; tel. (251) 326593; fax (251) 328088; e-mail goodyear@indoexchange.com; internet www.goodyear-indonesia.com; f. 1935; mfr of tyres, inner tubes and other related rubber products; cap. and res Rp. 258,725m., sales Rp. 593,045m. (2001); Pres. Dir GOTTFRIED HESS; 850 employees.

PT Hevea Latex and Rubber Works: Jalan Dr Setiabudhi 276A, Bandung 40143; tel. (22) 211149; fax (22) 212840; f. 1949; mfrs and exporters of sports and other shoes, rubber articles; Dir Ir SUGIRI.

PT Sepatu Bata: Jalan Taman Pahlawan Kalibata, Jakarta 12750; tel. (21) 7992008; fax (21) 7995679; e-mail bata@indoexchange.com; internet www.bata.com; f. 1931; mfr of shoes and other footwear products; cap. and res Rp. 141,738m., sales Rp. 407,887m. (2001); Chair. G. L. ZANACCO; 2,300 employees.

Textiles

PT Apac Citra Centertex Tbk: Gedung Graha BIP, 10th Floor, Jalan Jenderal Gatot Subroto, Kav. 23, Jakarta 12930; tel. (21) 5258180; fax (21) 5258400; e-mail kusno_erudi@apacinti.com; internet www.apacinti-online.com; f. 1987; mfr of textiles and textile products; sales Rp. 2,164,637m. (2001); Pres. Dir BENNY SOETRISNO; 15,732 employees.

PT Argo Pantes Tbk: Wisma Argo Manunggal, 16th Floor, Jalan Gatot Subroto 95, Kav. 22, Jakarta 12930; tel. (21) 2520065; fax (21) 2520028; e-mail argo@indoexchange.com; internet www.argo.co.id; f. 1977; textiles mfr; sales Rp. 1,089,820m. (2000); Pres. Dir A. MOEIS; 6,500 employees.

PT Dan Liris: Kelurahan Banaran, Kecarnatan Grogol, Kabupaten Sukoharjo 57193, Central Java; tel. (271) 714400; fax (271) 717178; e-mail marketing-div@danliris-id.com; internet www.danliris.com; f. 1974; mfr of garments and integrated textile factory; Pres. Dir HANDIANTO TJOKROSAPUTRO; 9,000 employees.

PT Panasia Indosyntec: Jalan Garuda 153/74, Bandung 40184; tel. (22) 6034123; fax (22) 6036434; e-mail panafil@panasiagroup.co.id; internet www.panasiagroup.co.id; f. 1973; producer of integrated textiles and selected clothing; cap. and res Rp. 167,241m., sales Rp. 1,309,066m. (2001); Pres. Dir AWONG HIDJALA; 8,000 employees.

PT Indorama Synthetics: Graha Irama, 17th Floor, Jalan H. R. Rasuna Said, Blok X-1, Kav. 1–2, Jakarta 12950; tel. (21) 5261555; fax (21) 5261501; e-mail corporate@indorama.com; internet www.indorama.com; f. 1974; mfr of polyester products and spun yarns; cap. and res US $223m., sales US $319m. (2001); Man. Dir S. P. LOHIA; 8,000 employees.

PT Polysindo Eka Perkasa Tbk: Kiara Payung Village, Klari District, Karawang, West Java; tel. (267) 431971; fax (267) 431975; e-mail polysindo@indoexchange.com; mfr of raw materials for the textile industry; cap. and res Rp. –8,126,974m., sales Rp. 4,012,064m. (2001); Pres. MARIMUTU SINIVASAN; 2,650 employees.

PT Textile Manufacturing Co Jaya Tbk: Kiara Payung Village, Klari District, West Java; tel. (267) 432400; fax (267) 432307; mfr of textiles; cap. and res Rp. 96,796m. (1997), sales Rp. 2,352,933m. (1999); Pres. Dir MARIMUTU SINIVASAN; 2,650 employees.

Timber

PT Barito Pacific Timber: Wisma Barito Pacific, Tower B, 9th Floor, Jalan Jenderal S. Parman, Kav. 62–63, Jakarta 11410; tel. (21) 5306711; fax (21) 5306680; e-mail barito@indoexchange.com; internet www.barito.co.id; f. 1979; producer and processor of timber and wood; cap. and res Rp. 387,378m., sales Rp. 1,410,630m. (2000); Pres. Dir YOHANNES HARDIAN WIDJANARKO; 28,500 employees.

PT Daya Sakti Unggul Corporation Tbk: Gedung BSG, 9th Floor, Jalan Abdul Muis 40, Jakarta 10160; tel. (21) 3505380; fax (21) 3505381; e-mail dayasakti@indoexchange.com; f. 1982; mfr of timber; cap. and res Rp. 81,981m., sales Rp. 600,323m. (2001); Pres. Dir NJOTO SUHARDJOJO; 4,000 employees.

PT Inti Indorayon Utama: 20th Floor, Wisma BNI, Jalan Jenderal Sudirman, Kav. 1, Jakarta 10220; tel. (21) 5706047; fax (21) 5702606; e-mail indorayon@indoexchange.com; f. 1983; mfr of processed wood products; cap. and res Rp. 207,699m. (1997), sales Rp. 768,292m. (1999); Pres. Dir H. DARLIN; 5,500 employees.

PT Surya Dumai Industri Tbk: Wisma GKBI, Lantai 31, Jalan Jenderal Sudirman 28, Jakarta 10210; tel. (21) 5740888; fax (21) 5740383; e-mail surdumind@indoexchange.com; producer of wood-based products and operator of forest and timber estate concessions; cap. and res Rp. –532,726m., sales Rp. 495,362m. (2001); Chair. CITRA GUNAWAN.

Tobacco

PT Gudang Garam (Perusahaan Rokok Tjap): Jalan Semampir II/1, Kediri 64121, East Java; tel. (354) 682091; fax (354) 681555; e-mail gg@indoexchange.com; f. 1973; producer of clove cigarettes; cap. and res Rp. 6,111,108m., sales Rp. 14,964,674m. (2000); Pres. Dir DJAJUSMAN SUROWIJONO; 43,000 employees.

PT Hanjaya Mandala Sampoerna: Jalan Rungkut Industri Raya, Kav. 14–18, Surabaya 60293; tel. (31) 8431699; fax (31) 8430986; e-mail sampoerna@indoexchange.com; internet www.indoexchange.com/sampoerna; f. 1964; producer of clove cigarettes; cap. and res Rp. 3,821,862m., sales Rp. 10,029,401m. (2000); Pres. Dir PUTERA SAMPOERNA; 19,000 employees.

PT Perusahaan Rokok Tjap Bentoel: Jalan Raya Karanglo, Banjar Arum, Singosari, Malang 65153, East Java; tel. (341) 49000; fax (341) 54710; f. 1930; mfr of clove cigarettes; sales Rp. 610,000m. (1994); Pres. BUDHIWIDJAYA KUSUMANEGARA; 8,000 employees.

Miscellaneous

PT Aneka Kimia Raya: Wisma AKR, 8th Floor, Jalan Panjang 5, Kabon Geruk, Jakarta 11530; tel. (21) 5311110; fax (21) 5311388; e-mail akr@indoexchange.com; f. 1977; mfr of industrial chemicals; cap. and res Rp. 383,612m., sales Rp. 1,455,125m. (2001); Pres. Dir H. ADIKOSOEMO; 428 employees.

PT Asahimas Flat Glass Co Ltd: Jalan Ancol IX/5, Ancol Barat, Jakarta 14430; tel. (21) 6904041; fax (21) 6904705; e-mail corporate-secretary@amfg.co.id; internet www.amfg.co.id; f. 1971; mfr of plate glass and other glass products; cap. and res Rp. 539,413m., sales Rp. 1,226,821m. (2001); Pres. Dir MITSURU JIBIKI; 2,617 employees.

PT Bayu Buana: Jalan Ir H. Juanda III 2A, Jakarta 10120; tel. (21) 3801705; fax (21) 3861955; e-mail office@bayubuanatravel.com; internet www.bayubuanatravel.com; f. 1972; operator of travel agency, retail and leisure services; cap. and res Rp. 61,959m., sales Rp. 636,544m. (2001); Pres. Dir TYRONE KASKAM AWAN; 403 employees.

PT Bimantara Citra Tbk: Menara Kebon Sirih, Jalan Kebon Sirih 17–19, Jakarta 10340; tel. (21) 3909211; fax (21) 3909207; e-mail alex@bimantara.co.id; internet www.bimantara.co.id; f. 1981; holding company with primary interests in advertising and mass media; cap. and res Rp. 1,369,658m., sales Rp. 1,660,939m. (2001); Pres. Dir JOSEPH DHARMABRATA; 9,750 employees.

PT Budi Acid Jaya: Wisma Budi, 8th–9th Floors, Jalan H. R. Rasuna Said, Kav. C-6, Jakarta 12940; tel. (21) 5213383; fax (21) 5213392; e-mail budiacid@centrin.net.id; internet www.budiacidjaya.co.id; f. 1979; mfr and marketer of tapioca starch, sulphuric acid, citric acid and polypropylene woven bags; cap. and res Rp. 133,229m., sales Rp. 823,660m. (2001); Pres. Dir SANTOSO WINATA; 4,200 employees.

PT Ciputra Development Tbk: Jalan Prof. Dr Satrio, Kav. 6, Karet Kuningan, Jakarta 12940; tel. (21) 5225858; fax (21) 5205262; e-mail investor@ciputra.com; internet www.ciputra.com; f. 1981; real estate and property development; cap. and res Rp. 806,250m., sales Rp. 329,418m. (2001); Pres. Dir Ir CIPUTRA; 3,865 employees.

PT Duta Pertiwi Tbk: ITC Mangga Dua, 7th and 8th Floors, Jalan Mangga Dua Raya, Jakarta 14430; tel. (21) 6019788; fax (21) 6017039; e-mail duti@simasred.com; internet www.simasred.com; f. 1972; property and real estate development and investment; cap. and res Rp. 1,615,708m., sales Rp. 1,209,416m. (2003); Pres. Dir MUKTAR WIDJAJA; 3,415 employees.

PT Indocement Tunggal Prakarsa Tbk: Wisma Indocement, 8th Floor, Jalan Jenderal Sudirman, Kav. 70–71, Jakarta 12910; tel. (21) 2512121; fax (21) 2510066; e-mail corpsec@ibm.net; internet www.indocement.co.id; f. 1985; cement producer; cap. and res Rp. 2,763,087m., sales Rp. 3,453,411m. (2001); Pres. Dir SUDWIKATMONO; 7,000 employees.

PT Matahari Putra Prima Tbk: Menara Matahari, 20th Floor, Jalan Bulevar Palem Raya 7, Lippo Karawari 12000, Tangerang 15811; tel. (21) 5469333; fax (21) 5475673; e-mail dannykjg@matahari.co.id; internet www.matahari.co.id; f. 1986; owner and operator of Indonesia's largest department store chain (81 stores); cap. and res Rp. 1,709,743m., sales Rp. 5,430,465m. (2001); Pres. Dir BENJAMIN J. MAILOOL; 18,022 employees.

PT Modern Photo Film Tbk: Jalan Matraman Raya 12, Jakarta 13150; tel. (21) 2801000; fax (21) 8581620; e-mail modernpho@indoexchange.com; f. 1971; mfr of photography products; cap. and res Rp. 198,831m., sales Rp. 1,912,996m. (2001); Pres. Dir SUNGKONO HONORIS; 3,400 employees.

PT Mulialand Tbk: Plaza Kuningan, Menara Utara, 10th Floor, Jalan H. R. Rasuna Said, Kav. C11–14, Jakarta 12940; tel. (21) 5207729; fax (21) 5200795; e-mail mulialand@indoexchange.com; f. 1987; property development and real estate investment, leasing of office blocks and residential apartments; cap. and res Rp. –173,198m., sales Rp. 614,928m. (2001); Pres. Dir JOKO SOERGIARTO TJANDRA; 140 employees.

PT Ramayana Lestari Sentosa Tbk: Jalan K. H. Wahid Hasyim, Kav. 220A–B, Jakarta 10250; tel. (21) 3920480; fax (21) 3920484; e-mail ramayana@indoexchange.com; f. 1983; operator of department stores and supermarkets; cap. and res Rp. 1,175,302m., sales Rp. 2,878,059m. (2001); Chair. AGUS MAKMUR; 16,500 employees.

PT Supreme Cable Manufacturing Corporation Tbk: Jalan Kabon Sirih 71, Jakarta 10340; tel. (21) 3100525; fax (21) 6192628; e-mail sucaco@indoexchange.com; f. 1970; mfr of cable, metal goods and alloys and plastic products and materials; cap. and res Rp. –30,922m. (1997), sales Rp. 461,666m. (2000); Pres. Dir ELLY SOEPONO; 800 employees.

PT Tigaraksa Satria Tbk: Wisma Tira, Jalan H. R. Rasuna Said, Kav. B3, Jakarta 12920; tel. (21) 5254208; fax (21) 5222413; e-mail tigaraksa@indoexchange.com; f. 1986; distributor of food products, personal care products, household products and garments; cap. and res Rp. 341,590m., sales Rp. 1,085,848m. (2000); Pres. Dir GIN SUGIANTO; 700 employees.

PT Warna Agung: Gedung Kompleks Delta, Blok C 3–6, Jalan Suryopranoto, Kav. 1–9, Jakarta 10160; tel. (21) 3808711; fax (21) 3809721; f. 1969; mfr of paints, emulsions and other related products; sales Rp. 24,000m. (1994); Chair. L. MOELJONO; Dir F. SUTADJI; 400 employees.

PT Wicaksana Overseas International: Jalan Ancol Barat VII, Blok A5D 2, Jakarta 14430; tel. (21) 6927293; fax (21) 6909436; e-mail wicaksana@indoexchange.com; internet www.wicaksana.co.id; f. 1964; distributor of consumer goods; cap. and res Rp. –315,877m., sales Rp. 2,242,495m. (2001); Pres. BACHTIAR YUSUF; 2,800 employees.

TRADE UNIONS

Serikat Buruh Sejahtera Indonesia (SBSI) (Indonesian Prosperity Trade Union Central Board): Jakarta; f. 1998; application for official registration rejected in May 1998; 1,228,875 mems in 168 branches in 27 provinces throughout Indonesia; Gen. Chair. MUCHTAR PAKPAHAN; Gen. Sec. SUNARTY.

Federasi Serikat Pekerja Seluruh Indonesia (FSPSI) (Federation of All Indonesian Trades Unions): Jalan Raya Pasar Minggu Km 17 No. 9, Jakarta 12740; tel. (21) 7974359; fax (21) 7974361; f. 1973; renamed 2001; sole officially recognized National Industrial Union; 5.273m. mems in June 2004; Gen. Chair. JACOB NUWA WEA; Gen. Sec. SYUKUR SARTO.

Transport

Directorate General of Land Transport and Inland Waterways: Ministry of Transportation and Telecommunication, Jalan Medan Merdeka Barat 8, Jakarta 10110; tel. (21) 3502971; fax (21) 3503013; e-mail hubdat@hubdat.go.id; Dir-Gen. SOEJONO.

RAILWAYS

There are railways on Java, Madura and Sumatra, totalling 6,458 km in 1996, of which 125 km were electrified.

In 1995 a memorandum of understanding was signed by a consortium of European, Japanese and Indonesian companies for the construction of a 15-km Mass Rapid Transport (MRT) rail system in Jakarta, part of which would be underground. However, in 2004 construction had not yet begun, owing to the Government's inability to fund the project, which was expected to cost US $476m. to implement.

PT Kereta Api (Persero) (PERUMKA): Jalan Perintis Kermedekaan 1, Bandung 40117; tel. (22) 4230031; fax (22) 4241370; e-mail info@kereta-api.com; internet www.kereta-api.com; six regional offices; transferred to the private sector in 1991; Chief Dir Drs ANWAR SUPRIADI.

ROADS

There is an adequate road network on Java, Sumatra, Sulawesi, Kalimantan, Bali and Madura, but on most of the other islands traffic is by jungle track or river boat. In 1997 the road network in Indonesia totalled 342,700 km; 27,357 km were main roads and 40,490 km were secondary roads. About 158,670 km of the network were paved.

In July 2003 the Government announced that it finally intended to implement a series of infrastructure projects postponed since the Asian financial crisis of 1997. These included the planned construction of a series of toll roads around the city of Surabaya, at a cost of Rp. 4,900,000m.

Directorate General of Highways: Ministry of Public Works, Jalan Pattimura 20, Kebayoran Baru, Jakarta Selatan 12110; tel. (21) 7262805; fax (21) 7260769; Dir-Gen. Ir SURYATIN SASTROMIJOYO.

SHIPPING

The Ministry of Communications controls 349 ports and harbours, of which the four main ports of Tanjung Priok (near Jakarta), Tanjung Perak (near Surabaya), Belawan (near Medan) and Makassar (formerly Ujung Pandang, in South Sulawesi) have been designated gateway ports for nearly all international shipping to deal with Indonesia's exports and are supported by 15 collector ports. Of the ports and harbours, 127 are classified as capable of handling ocean-going shipping.

Directorate General of Sea Communications: Ministry of Communications, Jalan Medan Merdeka Barat 8, Jakarta 10110; tel. (21) 3456332; Dir-Gen. SOENTORO.

Indonesian National Ship Owners' Association (INSA): Jalan Gunung Sahari 79, Jakarta Pusat; tel. (21) 414908; fax (21) 416388; Pres. H. HARTOTO HADIKUSUMO.

Shipping Companies

Indonesian Oriental Lines, PT Perusahaan Pelayaran Nusantara: Jalan Raya Pelabuhan Nusantara, POB 2062, Jakarta 10001; tel. (21) 494344; Pres. Dir A. J. SINGH.

PT Jakarta Lloyd: Jalan Agus Salim 28, Jakarta Pusat 10340; tel. (21) 331301; fax (21) 333514; f. 1950; services to USA, Europe, Japan, Australia and the Middle East; Pres. Dir Drs M. MUNTAQA.

PT Karana Line: Jalan Kali Besar Timur 30, POB 1081, Jakarta 11110; tel. (21) 6907381; fax (21) 6908365; Pres. Dir BAMBANG EDIYANTO.

PT Pelayaran Bahtera Adhiguna (Persero): Jalan Kalibesar Timur 10–12, POB 4313, Jakarta 11043; tel. (21) 6912613; fax (21) 6901450; f. 1971; Pres. H. DJAJASUDHARMA.

PT Pelayaran Nasional Indonesia (PELNI): Jalan Gajah Mada 14, Jakarta; tel. (21) 6334342; fax (21) 63854130; internet www.pelni.co.id; state-owned; national shipping co; Pres. Dir ISNOOR HARYANTO.

PT Pelayaran Samudera Admiral Lines: POB 1476, Jakarta 10014; tel. (21) 4247908; fax (21) 4206267; e-mail admiral@uninet.net.id; Pres. Dir DJOKO SOETOPO.

PT (Persero) Pann Multi Finance: Pann Bldg, Jalan Cikini IV 11, POB 3377, Jakarta 10330; tel. (21) 322003; fax (21) 322980; state-controlled; Pres. Dir W. NAYOAN; Dir HAMID HADIJAYA.

PT Pertamina (Persero): Downstream Directorate for Shipping, Jalan Yos Sudarso 32–34, POB 14020, Tanjung Priok, Jakarta; tel. (21) 4301086; fax (21) 4301492; e-mail marketing@pertaminashipping.com; internet www.pertaminashipping.com; f.

1959; state-owned; maritime business services; Pres. and Chair. Dr IBNU SUTOWO; Senior Vice-Pres. DEDENG WAHYU EDI.

PT Perusahaan Pelayaran Gesuri Lloyd: Gesuri Lloyd Bldg, Jalan Tiang Bendera IV 45, Jakarta 11230; tel. (21) 6904000; fax (21) 6904190; e-mail gesuri@indosat.net.id; internet www.gesuri.co.id; f. 1963; Pres. FRANKIE NURIMBA.

PT Perusahaan Pelayaran 'Nusa Tenggara': Kantor Pusat, Jalan Raya Pelabuhan Benoa, POB 3069, Denpasar 80222, Bali; tel. (361) 723608; fax (361) 722059; e-mail ntship@indo.net.id; Man. Dir KETUT DERESTHA.

PT Perusahaan Pelayaran Samudera 'Samudera Indonesia': Jalan Kali Besar Barat 43, POB 1244, Jakarta; tel. (21) 671093; fax (21) 674242; Chair. and Dir SOEDARPO SASTROSATOMO; Exec. Dir RANDY EFFENDI.

PT Perusahaan Pelayaran Samudera Trikora Lloyd: POB 4076, Jakarta 11001; tel. (21) 6907751; fax (21) 6907757; f. 1964; Pres. Dir GANESHA SOEGIHARTO; Man. Dir P. R. S. VAN HEEREN.

CIVIL AVIATION

The first stage of a new international airport, the Sukarno-Hatta Airport, at Cengkareng, near Jakarta, was opened in 1985, to complement Halim Perdanakusuma Airport, which was to handle charter and general flights only. A new terminal was opened at Sukarno-Hatta in 1991, vastly enlarging airport capacity. Construction of an international passenger terminal at the Frans Kaisepo Airport, in Papua (then Irian Jaya), was completed in 1988. Other international airports include Ngurah Rai Airport at Denpasar (Bali), Polonia Airport in Medan (North Sumatra), Juanda Airport, near Surabaya (East Java), Sam Ratulangi Airport in Manado (North Sulawesi) and Hasanuddin Airport, near Makassar (formerly Ujung Pandang, South Sulawesi). There are a total of 179 commercial airports, 61 of which are capable of accommodating wide-bodied aircraft. Domestic air services link the major cities, and international services are provided by the state airline, PT Garuda Indonesia, by its subsidiary, PT Merpati Nusantara Airlines, and by numerous foreign airlines. In December 1990 it was announced that private airlines equipped with jet-engined aircraft would be allowed to serve international routes.

In 2000 the Government announced a policy of liberalization for the airline industry. This led to a dramatic increase in the number of airlines operating in Indonesia; by January 2004 Indonesia had 27 domestic airlines, including four that were scheduled to begin operations in 2004, compared with five in 2000.

Directorate-General of Civil Aviation: Jalan Medan Merdeka Barat 8, Jakarta 10110; tel. (21) 3505133; fax (21) 3505139; Dir-Gen. CUCUK SURYO SUPROJO.

Air Paradise International: Jalan Tangkuban Perahu 66, Kerobokan, Kuta Bali 80361; tel. (361) 756666; fax (361) 766100; e-mail flywith@airparadise.co.id; internet www.airparadise.co.id; f. 2002; international services; Man. Dir I. MADE WIRANATHA SITA.

Awair International: Graha Aktiva Bldg, Jalan H. R. Rasuna Said, Jakarta; tel. and fax (21) 5203598; e-mail contact@awairlines.com; internet www.awairlines.com; f. 2000; scheduled domestic passenger services from Jakarta; Pres. YASSIR ISMAIL.

Bayu Indonesia Air (BYU): Jalan Bikatamsu 29E, Jakarta; tel. (21) 4515588; fax (21) 4515777; f. 1975; international services; Man. Dir PERMADI WIRATANUNINGRAT.

PT Bouraq Indonesia Airlines (BOU): Jalan Angkasa 1–3, POB 2965, Kemayoran, Jakarta 10720; tel. (21) 6288815; fax (21) 6008729; e-mail info@bouraq.com; internet www.bouraq.com; f. 1970; private company; scheduled regional and domestic passenger and cargo services linking Jakarta with points in Java, Borneo, Sulawesi, Bali, Timor and Tawau (Malaysia); Pres. Dir DANNY SUMENDAP.

Carstensz Papua Airlines (CPA): Papua; f. 2002; domestic services.

Citilink: Jakarta; internet www.ga-citilink.com; f. 2001; subsidiary of PT Garuda Indonesia; provides shuttle services between domestic destinations.

Deraya Air Taxi (DRY): Terminal Bldg, 1st Floor, Rm 150/HT, Halim Perdanakusuma Airport, Jakarta 13610; tel. (21) 8093627; fax (21) 8095770; e-mail drytech@centrin.net.id; internet www.boedihardjogroup.com; scheduled and charter passenger and cargo services to domestic and regional destinations; Pres. Dir SITI RAHAYU SUMADI.

Dirgantara Air Service (DAS): POB 6154, Terminal Bldg, Halim Perdanakusuma Airport, Rm 231, Jakarta 13610; tel. (21) 8093372; fax (21) 8094348; charter services from Jakarta, Barjarmas and Pontianak to destinations in West Kalimantan; Pres. MAKKI PERDANAKUSUMA.

PT Garuda Indonesia: Garuda Indonesia Bldg, Jalan Merdeka Selatan 13, Jakarta 10110; tel. (21) 2311801; fax (21) 2311962; internet www.garuda.co.id; f. 1949; state airline; operates scheduled domestic, regional and international services to destinations in Europe, the USA, the Middle East, Australasia and the Far East; Pres. and CEO INDRA SETIAWAN.

Indonesia Air Transport (IAT) (IDA): Pondok Cabe Aerodrome, POB 2485, Jakarta; tel. (21) 7490213; fax (21) 7491287; charter passenger and cargo services to domestic, regional and international destinations; f. 1968; Man. Dir AZMAR MUALIM.

Indonesian Airlines: Jakarta; f. 1999; domestic services; Pres. RUDY SETYOPURNOMO.

Lion Mentari Airlines: Gedung Jaya, 7th Floor, Jalan M. H. Thamrin 12, Jakarta 10340; tel. (21) 331838; fax (21) 327808; internet www.lionairlines.com; f. 1999; domestic and international services; Pres. Dir RUSDI KILANA.

PT Mandala Airlines: Jalan Tomang Raya, Kav. 33–37, Jakarta 11440; tel. (21) 5665434; fax (21) 5663788; e-mail widya@mandalaair.com; internet www.mandalaair.com; f. 1969; privately-owned; scheduled regional and domestic passenger and cargo services; Pres. and Dir SUHADI.

PT Merpati Nusantara Airlines: Jalan Angkasa, Blok 15, Kav. 2–3, Jakarta 10720; tel. (21) 6546789; fax (21) 6540620; e-mail lt13@indosat.net.id; internet www.merpati.co.id; f. 1962; subsidiary of PT Garuda Indonesia; scheduled for privatization in 2004; domestic and regional services to Australia and Malaysia; Pres. Dir HOTASI NABABAN.

Pelita Air Service: Jalan Abdul Muis 52–56A, Jakarta 10160; tel. (21) 2312030; fax (21) 2312216; internet www.pelita-air.co.id; f. 1970; domestic services; Pres. Dir SOERATMAN.

Sriwijaya Air: Jakarta; f. 2003; domestic services; Dir CHANDRA LIE.

PT Star Air: Jalan Gunung Sahari Raya 57 A-B, POB 4724, Jakarta Pusat 10610; tel. (21) 4222622; fax (21) 4249538; e-mail starair@cbn.net.id; internet www.starair-online.com; f. 2000; scheduled domestic passenger services; CEO ALE SUGIARTO.

Sumut Airlines: North Sumatra; f. 2002; domestic services.

Tourism

Indonesia's tourist industry is based mainly on the islands of Java, famous for its volcanic scenery and religious temples, and Bali, renowned for its scenery and Hindu-Buddhist temples and religious festivals. Lombok, Sumatra and Sulawesi are also increasingly popular. Domestic tourism within Indonesia has also increased significantly. Following the terrorist attack on Bali in October 2002 (see History) it was feared that the tourist industry would be significantly affected; in 2002 tourist arrivals declined to 5.0m., compared with 5.2m. in 2001. Tourist receipts totalled US $5,411m. in 2001. In 2003 tourist arrivals increased steadily, prompting hopes that the industry might have begun to recover by the end of the year. However, it was feared that the bombing of the Marriott Hotel in Jakarta in August would impact adversely upon this recovery. In early 2004 the Government introduced stringent visa requirements for many foreign nationals wishing to visit Indonesia; it was feared that the regulations would further discourage tourists from visiting the country.

Ministry of Culture and Tourism: Jalan Medan Merdeka Barat 17, Jakarta Pusat 10110; tel. (21) 3838000; fax (21) 3848245; e-mail pusdatin@budpar.go.id; internet www.budpar.go.id; f. 1957; Chair. I. GEDE ARDIKA.

Indonesia Tourism Promotion Board: Wisma Nugra Santana, 9th Floor, Jalan Jenderal Sudirman 7–8, Jakarta 10220; tel. (21) 5704879; fax (21) 5704855; e-mail itpb@cbn.net.id; internet www.goindo.com; private body; promotes national and international tourism; Chair. PONTJO SUTOWO; CEO GATOT SOEMARTONO.

Defence

In August 2003 the total strength of the armed forces was an estimated 302,000: army 230,000, navy 45,000, and air force 27,000; paramilitary forces comprised some 195,000 police, 12,000 marine police and an estimated 40,000 trainees of KAMRA (People's Security). Military service, which is selective, lasts for two years.

Defence Expenditure: Budgeted at an estimated 16,000,000m. rupiah in 2003.

Commander-in-Chief of the Armed Forces: Gen. RYAMIZARD RYACUDU (acting).

Chief of Staff of the Army: Gen. Ryamizard Ryacudu.

Chief of Staff of the Navy: Adm. Bernard Kent Sondakh.

Chief of Staff of the Air Force: Air Chief Marshal Chappy Hakim.

Education

Education is controlled mainly by the Ministry of Education and Culture, but the Ministry of Religious Affairs also operates Islamic religious schools (*madrasahs*) at the primary level.

Primary education, beginning at seven years of age and lasting for six years, was made compulsory in 1987. In 1993 it was announced that compulsory education was to be expanded to nine years. Secondary education begins at 13 years of age and lasts for a further six years, comprising three years of junior secondary education and a further three years of senior secondary education. A further three years of academic level or five years of higher education may follow.

As a proportion of children in the relevant age-group, enrolment at primary level in 1996 was equivalent to 113% (males 115%; females 110%); enrolment at secondary level in the same year was equivalent to 56% of children in the relevant age-group.

Technical and vocational education is the least developed aspect of the educational system, but vocational subjects have been introduced in the secondary schools. In 2001/02 there were 2,027,464 pupils at 4,522 vocational senior secondary schools.

In 1999/2000 there were 1,634 universities, with a total enrolment of 3.1m. In 1996 enrolment at tertiary level was equivalent to 11% of the relevant population.

In 2000/01 the Government spent an estimated Rp. 13,433,000m., representing approximately 3.7% of total estimated budgetary expenditure, on education.

Bibliography

GENERAL

Cribb, Robert. *Historical Dictionary of Indonesia.* Metuchen, NJ, Scarecrow Press, 1992.

Historical Atlas of Indonesia. Richmond, Surrey, Curzon Press, 2000.

Hardjono, Joan (Ed.). *Indonesia: Resources, Environment, Ecology.* Kuala Lumpur, Oxford University Press, 1991.

Wilhelm, Donald. *Emerging Indonesia.* London, Quiller Press, 1985.

HISTORY AND POLITICS

Abaza, Mona. *Changing Images of Three Generations of Azharites in Indonesia.* Singapore, Institute of Southeast Asian Studies, 1993.

Alatas, Ali. *A Voice for a Just Peace.* Singapore, Institute of Southeast Asian Studies, 2001.

Ananta, Aris. *The Indonesian Crisis: A Human Development Perspective.* Singapore, Institute of Southeast Asian Studies, 2002.

Ananta, Aris, Arifin, Evi Nurvidya, and Suryadinata, Leo. *Indonesian Electoral Behaviour: A Statistical Perspective.* Singapore, Institute of Southeast Asian Studies, 2004.

Andaya, Leonard Y. *The World of Maluku: Eastern Indonesia in the Early Modern Period.* Honolulu, University of Hawaii Press, 1993.

Anderson, Benedict. *Language and Power: Exploring Political Cultures in Indonesia.* Ithaca, NY, and London, Cornell University Press, 1991.

Antlöv, Hans, and Cederroth, Sven. *Elections in Indonesia: The New Order and Beyond.* London and New York, RoutledgeCurzon, 2004.

Anwar, Dewi Fortuna. *Indonesia in ASEAN: Foreign Policy and Regionalism.* Singapore, Institute of Southeast Asian Studies, 1994.

Aragon, Lorraine V. *Fields of the Lord: Animism, Christian Minorities, and State Development in Indonesia.* Richmond, Surrey, Curzon Press, 2000.

Aspinall, Edward, Feith, Herb, and van Klinken, Gerry (Eds). *The Last Days of President Suharto.* Clayton, Vic, Monash Asia Institute, 1999.

Aspinall, Edward, and Fealy, Greg (Eds). *Local Power and Politics in Indonesia: Decentralization and Democratization.* Singapore, Institute of Southeast Asian Studies, 2003.

Baker, R. W., Soesastro, M. H., Kristiadi, J., and Ramage, D. E. (Eds). *Indonesia—The Challenge of Change.* Singapore, Institute of Southeast Asian Studies, 1999.

Barber, Charles Victor, and Schweithelm, James. *Trial by Fire: Forest Fires and Forestry Policy in Indonesia's Era of Crisis and Reform.* Washington, DC, World Resources Institute, 2000.

Barton, Greg. *Abdurrahman Wahid: Muslim Democrat, Indonesian President.* Sydney, University of New South Wales Press, 2002.

Bellwood, Peter. *Prehistory of the Indo-Malaysian Archipelago.* Sydney, Academic Press, 1985.

Bertrand, Jacques. *Nationalism and Ethnic Conflict in Indonesia.* New York, NY, Cambridge University Press, 2004.

Boland, B. J. *The Struggle of Islam in Modern Indonesia.* Leiden, Koninklijk Instituut voor Taal-, Land- en Volkenkunde, 1971.

Bourchier, David, and Legge, John (Eds). *Democracy in Indonesia: 1950s and 1990s.* Clayton, Vic, Monash University Centre of Southeast Asian Studies, 1994.

Bourchier, David, and Hadiz, Vedi (Eds). *Indonesian Politics and Society.* London, Routledge, 2001.

Bresnan, John. *Managing Indonesia: the Modern Political Economy.* New York, NY, Columbia University Press, 1993.

Buchholt, Helmut, and Mai, Ulrich (Eds). *Continuity, Change and Aspirations: Social and Cultural Life in Minahasa, Indonesia.* Singapore, Institute of Southeast Asian Studies, 1994.

Challis, Roland. *Shadow of a Revolution: Indonesia and the Generals.* Stroud, Glos., Sutton Publishing, 2001.

Chauvel, Richard. *Essays on West Papua.* Clayton, Vic, Monash Asia Institute, 2003.

Cleary, Mark, and Eaton, Peter. *Borneo: Change and Development.* Kuala Lumpur, Oxford University Press, 1995.

Colombijn, Freek, and Lindblad, Thomas J. (Eds). *Roots of Violence in Indonesia.* Singapore, Institute of Southeast Asian Studies, 2002.

Coppel, Charles A. *Indonesian Chinese in Crisis.* Kuala Lumpur, Oxford University Press, 1983.

Coppel, Charles (Ed.). *Violent Conflicts in Indonesia: Analysis, Representation, Resolution.* Richmond, Surrey, Curzon Press, 2001.

Cribb, Robert (Ed.). *The Indonesian Killings of 1965–1966: Studies from Java and Bali.* Clayton, Vic, Monash University, 1991.

Cribb, Robert, and Brown, Colin. *Modern Indonesia: A History Since 1945.* London, Longman, 1995.

Crouch, Harold. *The Army and Politics in Indonesia.* Ithaca, NY, Cornell University Press, 1978.

Effendy, Bahtiar. *Islam and the State: The Transformation of Islamic Political Ideas and Practices in Indonesia.* Singapore, Institute of Southeast Asian Studies, 2001.

Eklöf, Stefan. *Indonesian Politics in Crisis: The Long Fall of Suharto 1996–98.* Copenhagen, Nordic Institute of Asian Studies, 1999.

Power and Political Culture in Suharto's Indonesia: The Indonesian Democratic Party and Decline of the New Order (1986–98). Copenhagen, Nordic Institute of Asian Studies, 2003.

Eliraz, Giora. *Islam in Indonesia: Modernism, Radicalism and the Middle East Dimension.* Brighton, Sussex Academic Press, 2004.

Elson, R. E. *Suharto: A Political Biography.* Cambridge, Cambridge University Press, 2001.

Emmerson, Donald K. (Ed.). *Indonesia beyond Suharto: Polity, Economy, Society, Transition.* New York, NY, M. E. Sharpe, 1999.

Feith, Herbert. *The Decline of Constitutional Democracy in Indonesia.* Ithaca, NY, Cornell University Press, 1970.

Forrester, Geoff (Ed.). *Post-Soeharto Indonesia: Renewal or Chaos?* Singapore, Institute of Southeast Asian Studies, 1999.

Frederick, William H. *Visions and Heat: The Making of the Indonesian Revolution.* Athens, OH, Ohio University Press, 1988.

Friend, Theodore. *The Blue-eyed Enemy: Japan against the West in Java and Luzon, 1942–1945.* Princeton, NJ, Princeton University Press, 1988.

Indonesian Destinies. Cambridge, MA, Harvard University Press, 2003.

Gardner, Paul F. *Shared Hopes, Separate Fears: Fifty Years of US-Indonesian Relations.* Boulder, CO, Westview Press, 1997.

Gooszen, Hans. *A Demographic History of the Indonesian Archipelago, 1880–1942.* Singapore, Institute of Southeast Asian Studies, 2000.

Hadiwinata, Bob S. *The Politics of NGOs in Indonesia: Developing Democracy and Managing a Movement.* London, RoutledgeCurzon, 2002.

Hadiz, Vedi. *Workers and the State in New Order Indonesia*. London, Routledge, 1997.

Halldorsson, Jon O. *Authoritarian Imperatives: The Political Economy of State and Democratization in Indonesia*. Richmond, Surrey, Curzon Press, 2001.

Hefner, Robert W. *Civil Islam: Muslims and Democratization in Indonesia*. Princeton, NJ, Princeton University Press, 2000.

Hill, Hal (Ed.). *Indonesia's New Order: The Dynamics of Socio-economic Transformation*. St Leonards, NSW, Allen and Unwin, 1994.

Honna, Jun. *Military and Democracy in Indonesia*. London, RoutledgeCurzon, 2002.

Hooker, Barry. *Indonesian Islam: Social Change Through Contemporary Fatawa*. Honolulu, University of Hawaii Press, 2003.

Houseman, Gerald L. *Researching Indonesia: A Guide to Political Analysis*. New York, NY, Edwin Mellen Press, 2004.

Hughes, John. *The End of Sukarno: A Coup that Misfired: A Purge that Ran Wild*. Archipelago Press, 2003.

Jackson, Karl D., and Pye, Lucian W. (Eds). *Political Power and Communication in Indonesia*. Berkeley, University of California Press, 1978.

Jenkins, David. *Suharto and his Generals: Indonesian Military Politics 1975–1983*. Ithaca, NY, Cornell Modern Indonesia Project, 1984.

Jones, Gavin W., and Hull, Terence H. *Indonesia Assessment—Population and Human Resources*. Singapore, Institute of Southeast Asian Studies, 1997.

Kahin, Audrey R. (Ed.). *Regional Dynamics of the Indonesian Revolution: Unity from Diversity*. Honolulu, University of Hawaii Press, 1985.

Kahin, Audrey R., and Kahin, George McTurnan. *Subversion as Foreign Policy: The Secret Eisenhower and Dulles Debacle in Indonesia*. New York, NY, New Press, 1995.

Kahin, George McTurnan. *Nationalism and Revolution in Indonesia*. Ithaca, NY, Cornell University Press, 1952.

Kell, Tim. *The Roots of the Acehnese Rebellion 1989–1992*. Ithaca, NY, Cornell University Southeast Asia Program, 1995.

King, Dwight Y. *Half-Hearted Reform: Electoral Institutions and the Struggle for Democracy in Indonesia*. Westport, CT, Greenwood Press, 2003.

Kingsbury, Damien (Ed.). *The Presidency of Abdurrahman Wahid: An Assessment After the First Year*. Clayton, Vic, Monash Asia Institute, 2001.

Power Politics and the Indonesian Military. London, RoutledgeCurzon, 2003.

Kingsbury, Damien, and Aveling, Harry (Eds). *Autonomy and Disintegration in Indonesia*. London, RoutledgeCurzon, 2002.

Kivimaki, Timo. *US-Indonesian Hegemonic Bargaining: Strength or Weakness?* Aldershot, Ashgate, 2003.

Laffan, Michael F. *Islamic Nationhood and Colonial Indonesia: The Umma Below the Winds*. London, RoutledgeCurzon, 2003.

Legge, J. D. *Sukarno: A Political Biography*. Harmondsworth, Penguin, 1973.

Leifer, Michael. *Indonesia's Foreign Policy*. London, Allen and Unwin, 1983.

Leithe, Denise. *The Politics of Power: Freeport in Suharto's Indonesia*. Honolulu, University of Hawaii Press, 2002.

Lloyd, Grayson J., and Smith, Shannon L. (Eds). *Indonesia Today: Challenges of History*. Singapore, Institute of Southeast Asian Studies, 2001.

Lowry, Bob. *Indonesian Defence Policy and the Indonesian Armed Forces*. Canberra, Australian National University, 1993.

Lubis, Todung Mulya. *In Search of Human Rights: Legal-Political Dilemmas of Indonesia's New Order, 1966–1990*. Jakarta, Gramedia Pustaka Utama, 1994.

MacFarling, Ian. *The Dual Function of the Indonesian Armed Forces: Military Politics in Indonesia*. Canberra, Australian Defence Studies Centre, 1996.

Mackie, J. A. C. *Konfrontasi: The Indonesia-Malaysia Dispute, 1963–1966*. Kuala Lumpur, Oxford University Press, 1974.

Manning, Chris, and Van Diermen, Peter (Eds). *Indonesia in Transition: Social Aspects of Reformasi and Crisis*. Singapore, Institute of Southeast Asian Studies, 2000.

May, Brian. *The Indonesian Tragedy*. London, Routledge and Kegan Paul, 1978.

McDonald, Hamish. *Suharto's Indonesia*. Melbourne, Fontana, 1980.

Moertono, Soemarsaid. *State and Statecraft in Old Java: a Study of the Later Mataram Period, 16th to 19th Century*. Ithaca, NY, Cornell Modern Indonesia Project, 1974.

Mortimer, Rex (Ed.). *Showcase State: The Illusion of Indonesia's 'Accelerated Modernisation'*. Sydney, Angus and Robertson, 1973.

Mortimer, Rex. *Indonesian Communism under Sukarno: Ideology and Politics, 1959–1965*. Ithaca, NY, Cornell University Press, 1974.

Nasution, Adnan Buyung. *The Aspiration for Constitutional Government in Indonesia: A Socio-legal Study of the Indonesian Konstituante 1956–1959*. Jakarta, Pustaka Sinar Harapan, 1993.

Nguyen, Thang D., and Richter, Frank-Jurgen. *Indonesia Matters: Diversity, Unity and Stability in Fragile Times*. Singapore, Times Editions, 2004.

Nishihara, Masashi. *Golkar and the Indonesian Elections of 1971*. Ithaca, NY, Cornell Modern Indonesia Project, 1972.

Ocy-Gardiner, Mayling, and Bianpoen, Carla (Eds). *Indonesian Women: The Journey Continues*. Canberra, Australian National University, Research School of Pacific and Asian Studies, 2000.

O'Rourke, Kevin. *Reformasi: the Struggle for Power in post-Soeharto Indonesia*. Crow's Nest, NSW, Allen and Unwin, 2002.

Osborne, Robin. *Indonesia's Secret War: The Guerrilla Struggle in Irian Jaya*. Sydney, Allen and Unwin, 1985.

Otten, Mariël. *Transmigrasi, Myths and Realities: Indonesian Resettlement Policy, 1965–1985*. Copenhagen, International Workshop for Indigenous Affairs, 1986.

Pangaribuan, Robinson. *The Indonesian State Secretariat 1945–1993*. Perth, Murdoch University, 1995.

Pemberton, John. *On the Subject of 'Java'*. Ithaca, NY, Cornell University Press, 1994.

Penders, C. L. M. *Indonesia 1945–1962: Dutch Decolonisation and the West New Guinea Debacle*. Richmond, Surrey, Curzon Press, 2001.

Philpott, Simon. *Rethinking Indonesia: Postcolonial Theory, Authoritarianism and Identity*. New York, St Martin's Press, 2000.

Porter, Donald. *Managing Politics and Islam in Indonesia*. Richmond, Surrey, Curzon Press, 2002.

Ramage, Douglas. *Politics in Indonesia: Democracy, Islam and the Ideology of Tolerance*. London, Routledge, 1995.

Ramstedt, Martin (Ed.). *'Hinduism' in Modern Indonesia: Hindu Dharma Indonesia between Local, National and Global Interest*. Richmond, Surrey, Curzon Press, 2001.

Reeve, David. *Golkar of Indonesia: An Alternative to the Party System*. Singapore, Oxford University Press, 1985.

Reid, Anthony J. S. *The Indonesian National Revolution 1945–1950*. Hawthorn, Vic, Longmans, 1974.

Southeast Asia and the Age of Commerce, 1450–1680, Volume one: The Land Below the Winds. New Haven, CT, Yale University Press, 1988.

Southeast Asia and the Age of Commerce, 1450–1680, Volume two: Expansion and Crisis. New Haven, CT, Yale University Press, 1993.

Reuter, Thomas (Ed.). *Inequality, Crisis and Social Change in Indonesia: The Muted Worlds of Bali*. London, RoutledgeCurzon, 2002.

Ricklefs, M. C. *A History of Modern Indonesia Since c. 1200*. Basingstoke, Hampshire, Palgrave Macmillan, 2001.

Robinson, Geoffrey. *The Dark Side of Paradise: Political Violence in Bali*. Ithaca, NY, Cornell University Press, 1995.

Robinson, Kathryn, and Bessell, Sharon (Eds). *Women in Indonesia: Gender, Equity and Development*. Singapore, Institute of Southeast Asian Studies, 2002.

Robison, Richard, and Hadiz, Vedi (Eds). *Reorganising Power in Indonesia: The Politics of Oligarchy in an Age of Markets*. London, RoutledgeCurzon, 2004.

Roeder, O. G. *The Smiling General: President Soeharto of Indonesia*. Jakarta, Gunung Agung, 1969.

Romano, Angela. *Politics and the Press in Indonesia: Understanding an Evolving Political Culture*. Richmond, Surrey, Curzon Press, 2002.

Rosser, Andrew. *The Politics of Liberalization in Indonesia: State, Market and Power*. Richmond, Surrey, Curzon Press, 2001.

Sajoo, Amyn B. *Pluralism in 'Old Societies and New States': Emerging ASEAN Contexts*. Singapore, Institute of Southeast Asian Studies, 1994.

Salim, Arskal, and Azra, Azyumardi (Eds). *Shari'a and Politics in Modern Indonesia*. Singapore, Institute of Southeast Asian Studies, 2002.

Saltford, John. *The United Nations and the Indonesian Takeover of West Papua, 1962–1969: The Anatomy of a Betrayal.* Richmond, Surrey, Curzon Press, 2002.

Schwarz, Adam. *A Nation in Waiting: Indonesia in the 1990s.* St Leonards, NSW, Allen and Unwin, 1994.

Indonesia: The 2004 Election and Beyond. Singapore, Institute of Southeast Asian Studies, 2004.

Sen, Krishna, and Hill, David T. *Media, Culture and Politics in Indonesia.* Melbourne, Oxford University Press, 2000.

Internet and Democracy in Indonesia. London, RoutledgeCurzon, 2003.

Singh, Bilveer. *ABRI and the Security of Southeast Asia: The Role and Thinking of General L. Benny Murdani.* Singapore Institute of International Affairs, 1994.

Sjamsuddin, Nazaruddin. *The Republican Revolt: a Study of the Acehnese Rebellion.* Singapore, Institute of Southeast Asian Studies, 1985.

Smith, Anthony L. *Strategic Centrality: Indonesia's Changing Role in ASEAN.* Singapore, Institute of Southeast Asian Studies, 2000.

Soesastro, Hadi, Smith, Anthony L., and Ling, Han Mui (Eds). *Governance in Indonesia: Challenges Facing the Megawati Presidency.* Singapore, Institute of Southeast Asian Studies, 2002.

Southwood, Julie, and Flanagan, Patrick. *Indonesia: Law, Propaganda and Terror.* London, Zed Books, 1983.

Sukma, Rizal. *Indonesia and China: The Politics of a Troubled Relationship.* London, Routledge, 1999.

Islam in Indonesian Foreign Policy: Domestic Weakness and the Dilemma of Dual Identity. London, Routledge, 2001.

Sundhaussen, Ulf. *The Road to Power: Indonesian Military Politics 1945–1967.* Kuala Lumpur, Oxford University Press, 1982.

Suryadinata, Leo. *Political Parties and the 1982 General Election in Indonesia.* Singapore, Institute of Southeast Asian Studies, 1982.

Elections and Politics in Indonesia. Singapore, Institute of Southeast Asian Studies, 2001.

Suryadinata, Leo, et al. (Eds). *Indonesia's Population: Ethnicity and Religion in a Changing Political Landscape.* Singapore, Institute of Southeast Asian Studies, 2003.

Sutherland, Heather. *The Making of a Bureaucratic Elite: The Colonial Transformation of the Javanese Priyayi.* Singapore, Heinemann, 1979.

Tan, T. K. (Ed.). *Sukarno's Guided Indonesia.* Brisbane, Jacaranda Press, 1967.

Taylor, Jean Gelman. *The Social World of Batavia: European and Eurasian in Dutch Asia.* Madison, WI, University of Wisconsin Press, 1983.

Indonesia: Peoples and Histories. New Haven, CT, Yale University Press, 2003.

Thoolen, Hans (Ed.). *Indonesia and the Rule of Law: Twenty Years of 'New Order' Government.* London, Pinter, 1987.

Uhlin, Anders. *Indonesia and the 'Third Wave of Democratization': The Indonesian Pro-Democracy Movement in a Changing World.* Richmond, Curzon Press, 1998.

van Dijk, C. *Rebellion under the Banner of Islam: The Darul Islam in Indonesia.* The Hague, Martinus Nijhoff, 1981.

A Country in Despair: Indonesia between 1997 and 2000. Leiden, KITLV Press, 2001.

Vatikiotis, Michael R. J. *Indonesian Politics Under Suharto: Order, Development and Pressure for Change.* London and New York, Routledge, 1993.

Vickers, Adrian. *Bali: a Paradise Created.* Ringwood, Vic, Penguin, 1989.

Way, Wendy (Ed.). *Australia and the Indonesian Incorporation of Portuguese Timor, 1974-1976.* Carlton, Vic, Melbourne University Press, 2000.

Weatherbee, Donald E. *Ideology in Indonesia: Sukarno's Indonesian Revolution.* New Haven, CT, Yale University Southeast Asian Studies, 1966.

Weinstein, Franklin B. *Indonesian Foreign Policy and the Dilemma of Dependence: from Sukarno to Suharto.* Ithaca, NY, Cornell University Press, 1974.

Wessel, Ingrid, and Wimhöfer, Georgia (Eds). *Violence in Indonesia.* Hamburg, Abera, 2001.

Wiener, Margaret J. *Visible and Invisible Realms: Power, Magic and the Colonial Conquest of Bali.* Chicago, University of Chicago Press, 1995.

ECONOMY

Alexander, P., Boomgaard, P., and White, B. *In the Shadow of Agriculture: Non-farm Activities in the Javanese Economy, Past and Present.* Amsterdam, Royal Tropical Institute, 1991.

Anmar, M. A., and Omura, K. (Eds). *Local Development in Indonesia.* Tokyo, Institute of Developing Economies, 1994.

Arndt, H. W. *The Indonesian Economy: Collected Papers.* Singapore, Chopmen Publishers, 1983.

Arndt, H. W., and Hill, Hal. *Southeast Asia's Economic Crisis—Origins, Lessons and the Way Forward.* Singapore, Institute of Southeast Asian Studies, 1999.

Australian National University. *Bulletin of Indonesian Economic Studies.* Canberra, 3 a year.

Bank Indonesia. *Report for the Financial Year.* Jakarta, annually.

Barlow, Colin, and Hardjono, Joan (Eds). *Indonesia Assessment 1995: Development in Eastern Indonesia.* Singapore, Institute of Southeast Asian Studies, 1996.

Basri, M. Chatib, and Van Der Eng, Pierre (Eds). *Business in Indonesia: New Challenges, Old Problems.* Singapore, Institute of Southeast Asian Studies, 2004.

Binhadi. *Financial Sector Deregulation, Banking Development and Monetary Policy: The Indonesian Experience (1983–1993).* Jakarta, Indonesian Bankers' Institute, 1995.

Booth, A. *Agricultural Development in Indonesia.* Sydney, Wellington and London, Allen and Unwin, 1988.

Booth, A. (Ed.). *The Oil Boom and After: Indonesian Economic Policy and Performance in the Soeharto Era.* Singapore, Oxford University Press, 1992.

Booth, A., and McCawley, P. (Eds). *The Indonesian Economy during the Soeharto Era.* Kuala Lumpur, Oxford University Press, 1981.

Booth, A., O'Malley, W. J., and Weidemann, A. (Eds). *Indonesian Economic History of the Dutch Colonial Era.* New Haven, CT, Yale University Press, 1988.

Breman, Jan, and Wiradi, Gunawan (Eds). *Good Times and Bad Times in Rural Java.* Singapore, Institute of Southeast Asian Studies, 2002.

Bresnan, John. *Managing Indonesia—The Modern Political Economy.* New York, Columbia University Press, 1993.

Chalmers, Ian, and Hadiz, Vedi (Eds). *The Politics of Economic Development in Indonesia: Contending Perspectives.* London, Routledge, 1997.

Cole, David C., and Slade, Betty F. *Building a Modern Financial System—The Indonesian Experience.* Cambridge, Cambridge University Press, 1996.

Cribb, Robert (Ed.). *The Late Colonial State in Indonesia: Political and Economic Foundations of the Netherlands Indies, 1880–1942.* Leiden, KITLV Press, 1994.

Dick, H. *The Indonesian Interisland Shipping Industry. An Analysis of Competition and Regulation.* Singapore, Institute of Southeast Asian Studies, 1987.

Dick, H., Fox, J. J., and Mackie, J. A. C. *Balanced Development: East Java in the New Order.* Singapore, Oxford University Press, 1993.

Dick, H., et al. *The Emergence of a National Economy: An Economic History of Indonesia, 1800–2000.* Honolulu, University of Hawaii Press, 2002.

Dickie, R. B., and Layman, T. A. *Foreign Investment and Government Policy in the Third World: Forging Common Interests in Indonesia and Beyond.* London, Macmillan, 1988.

Dirkse, J. P., et al. (Eds). *Development and Social Welfare: Indonesia's Experiences Under the New Order.* Leiden, KITLV Press, 1993.

Drake, C. *National Integration in Indonesia: Patterns and Policies.* Honolulu, University of Hawaii Press, 1989.

Economist Intelligence Unit. *Country Forecast: Indonesia.* London, quarterly.

Country Profile: Indonesia. London, annually.

Country Report: Indonesia. London, quarterly.

Quarterly Economic Review: Indonesia. London, quarterly.

Faulkner, George. *Business Indonesia: A Practical Insight into Doing Business in Indonesia.* Sydney, Business & Professional Publishing, 1995.

Fenton, Robert. *The Indonesian Plywood Industry: A Study of the Statistical Base, the Value-added Effects and the Forest Impact.* Singapore, Institute of Southeast Asian Studies, 1996.

Financial Times Business Information Ltd. *Banking in the Far East 1993: Structures and Sources of Finance.* London, 1993.

Gérard, Françoise, and Ruf, François (Eds). *Agriculture in Crisis: People, Commodities and Natural Resources in Indonesia 1996–2001.* Richmond, Surrey, Curzon Press, 2001.

Glover, David, and Jessup, Timothy (Eds). *Indonesia's Fires and Haze—The Cost of Catastrophe.* Singapore, Institute of Southeast Asian Studies, 1999.

Goeltom, Miranda S. *Indonesia's Financial Liberalization: An Analysis of 1981–88 Panel Data.* Singapore, Institute of Southeast Asian Studies, 1995.

Hardjono, J. M. *Transmigration in Indonesia.* Kuala Lumpur, Oxford University Press, 1977.

Hardjono, J. M. (Ed.). *Indonesia: Resources, Ecology and Environment.* Singapore, Oxford University Press, 1991.

Hayami, Yujiro, and Kawagoe, Toshihito. *The Agrarian Origins of Commerce and Industry: A Study of Peasant Marketing in Indonesia.* New York, St Martins Press, 1993.

Heij, Gitte. *Tax Administration and Compliance in Indonesia.* Murdoch University, Western Australia, Asia Research Centre on Social, Political and Economic Change, 1994.

Hicks, G. L., and McNicoll, G. *The Indonesian Economy, 1950–67: A Bibliography.* New Haven, CT, Yale University Press, 1968.

Hill, Hal. *Foreign Investment and Industrialization in Indonesia.* Singapore, Oxford University Press, 1988.

'Indonesia's Textile and Garment Industries: Developments in an Asian Perspective'. *Occasional Paper No. 87.* Singapore, Institute of Southeast Asian Studies, 1992.

The Indonesian Economy since 1966: Southeast Asia's Emerging Giant. Melbourne, Cambridge University Press, 1996.

Indonesia's Industrial Transformation. Singapore, Institute of Southeast Asian Studies, 1997.

The Indonesian Economy in Crisis—Causes, Consequences and Lessons. Singapore, Institute of Southeast Asian Studies, 1999.

The Indonesian Economy. Cambridge, Cambridge University Press, 2000.

Hill, Hal (Ed.). *Unity and Diversity: Regional Economic Development in Indonesia since 1970.* Singapore, Oxford University Press, 1989.

Hill, Hal, and Thee Kian Wie. *Indonesia Assessment—Indonesia's Technological Challenge.* Singapore, Institute of Southeast Asian Studies, 1998.

Hoadley, Mason C. *Towards a Feudal Mode of Production in West Java, 1680–1800.* Singapore, Institute of Southeast Asian Studies, 1994.

Hobohm, S. O. H. *Indonesia to 1991: Can Momentum be Regained?* London, Economist Intelligence Unit, 1987.

Indonesia to 1993: Breakthrough in the Balance. London, Economist Intelligence Unit, 1989.

Iqbal, Farrukh, and James, William E. *Deregulation and Development in Indonesia.* Westport, CT, Praeger, 2002.

Kenward, Lloyd. *From the Trenches: The First Year of Indonesia's Crisis of 1997/98 as Seen From the World Bank's Office in Jakarta.* Jakarta, Centre for Strategic and International Studies, 2002.

Khatkhate, Deena. 'The Regulatory Impediments to the Private Industrial Sector Development in Asia—A Comparative Study'. *World Bank Discussion Paper 177.* Washington, DC, World Bank, 1993.

Leinbach, Thomas (Ed.). *Indonesia's Rural Economy: Mobility, Work and Enterprise.* Singapore, Institute of Southeast Asian Studies, 2003.

Li, Tania (Ed.). *Transforming the Indonesian Uplands: Marginality, Power and Production.* Singapore, Institute of Southeast Asian Studies, 1999.

Lindblad, J. Th. *Historical Foundations of a National Economy in Indonesia, 1890s–1990s.* Amsterdam, North Holland and the Royal Netherlands Academy of Arts and Sciences, 1996.

MacIntyre, Andrew. *Business and Politics in Indonesia.* Sydney, Allen and Unwin, in association with the Asian Studies Association of Australia, 1991.

Mackie, J. A. C. *The Indonesian Economy: 1950–1963.* In *Studien zur Entwicklung in Süd- und Ostasien, Neue Folge, Teil 3, Indonesien.* Frankfurt am Main, Berlin, Alfred Metzner Verlag, 1964.

Maddison, A., and Prince, G. (Eds). *Economic Growth in Indonesia 1820–1940.* Dordrecht, Foris Publications, 1989.

Mann, Richard. *Economic Crisis in Indonesia—The Full Story.* Jakarta, Gateway Books, 1998.

Plots & Schemes that brought down Soeharto. Jakarta, Gateway Books, 1998.

Business in Indonesia—Changes, Challenges, Opportunities. Jakarta, Gateway Books, 1999.

Manning, Chris. *Indonesian Labour in Transition: An East Asian Success Story?* Cambridge, Cambridge University Press, 1998.

Manning, C., and Hardjono, J. (Eds). *Indonesia Assessment 1993—Labour: Sharing in the Benefits of Growth?* Canberra, Australian National University, 1993.

Manning, Chris, and Van Diermen (Eds). *Indonesia in Transition: Social Aspects of Reformasi and Crisis.* Singapore, Institute of Southeast Asian Studies, 2000.

Martokoesoemo, S. B. *Beyond the Frontiers of Indonesian Banking and Finance: Financial Intermediation to Mobilize the Potential of Small Entrepreneurs.* Rotterdam, Labyrint Publication, 1993.

McLeod, Ross H. *Indonesia's Crisis and Future Prospects.* In Karl D. Jackson (Ed.). *Asian Contagion—The Causes and Consequences of a Financial Crisis.* Singapore, Institute of Southeast Asian Studies, 1999.

Indonesia. In Ross H. McLeod and Ross Garnaut (Eds). *East Asia in Crisis—From Being a Miracle to Needing One.* London, Routledge, 1998.

McLeod, Ross H. (Ed.). *Indonesia Assessment 1994: Finance as a Key Sector in Indonesia's Development.* Canberra, Australian National University, and Singapore, Institute of Southeast Asian Studies, 1994.

Mears, L. A. *Rice Marketing in the Republic of Indonesia.* Jakarta, PT Pembangunan, 1961.

The New Rice Economy of Indonesia. Yogyakarta, Gadjah Mada University Press, 1981.

Montes, Manuel F., and Abdusalamov, Muhammad Ali. *Indonesia: Reaping the Market.* In K. S. Jomo (Ed.). *Tigers in Trouble—Financial Governance, Liberalisation and Crises in East Asia.* London, Zed Books, 1998.

Montes, Manuel. *The Currency Crisis in Southeast Asia.* Singapore, Institute of Southeast Asian Studies, 1999.

Mortimer, R. (Ed.). *Showcase State: The Illusion of Indonesia's 'Accelerated Modernization'.* Sydney, Angus and Robertson, 1973.

Mukherjee, N., Hardjono, J., and Carriere, E. *People, Poverty and Livelihoods: Links for Sustainable Poverty Reduction in Indonesia.* Jakarta, World Bank and Department for International Development, 2002.

Palmer, Ingrid. *The Indonesian Economy since 1965: A Case Study of Political Economy.* London, Frank Cass, 1977.

Pangestu, Mari, and Sato, Yuri. *Waves of Change in Indonesia's Manufacturing Industry.* Tokyo, Institute of Developing Economies, 1997.

Patten, R. H., and Rosengard, J. K. *Progress with Profits: The Development of Rural Banking in Indonesia.* San Francisco, International Centre for Economic Growth and Harvard Institute for International Development, 1991.

Pearson, Scott, et al. *Rice Policy in Indonesia.* Ithaca, NY, and London, Cornell University Press, 1991.

Pierce Colfer, Carol J., and Resosudarmo, Ida Pradnja (Eds). *Which Way Forward? Forests, Policy and People in Indonesia.* Singapore, Institute of Southeast Asian Studies, 2002.

Piggott, R. R., et al. *Food Price Policy in Indonesia.* Canberra, Australian Centre for International Agricultural Research, 1993.

Poot, H., Kuyvenhoven, A., and Jansen, J. *Industrialization and Trade in Indonesia.* Yogyakarta, Gadjah Mada University Press, 1990.

Prawiro, Radius. *Indonesia's Struggle for Economic Development—Pragmatism in Action.* Kuala Lumpur, Oxford University Press, 1998.

Robison, R. *Indonesia: The Rise of Capital.* Sydney, Allen and Unwin, 1986.

Power and Economy in Suharto's Indonesia. Manila and Wollongong, Journal of Contemporary Asia Publishers, 1990.

Singh, Bilveer. *The Indonesian Military Business Complex: Origins, Course and Future.* Canberra, Strategic and Defence Studies Centre, Australian National University, 2001.

Smith, Anthony L. (Ed.). *Gus Dur and the Indonesian Economy.* Singapore, Institute of Southeast Asian Studies, 2001.

Smith, Shannon L. D. *Indonesian Political Economy: Developing Batam.* Singapore, Institute of Southeast Asian Studies, 2001.

Soemardjan, S. *Indonesia: A Socio-Economic Profile.* New Delhi, Sterling Publishers, 1988.

Strauss, John, et al. *Indonesian Living Standards: Before and After the Financial Crisis.* Singapore, Institute of Southeast Asian Studies, 2004.

Thee Kian Wie (Ed.). *Recollections: The Indonesian Economy, 1950s–1990s.* Singapore, Institute of Southeast Asian Studies, 2003.

Thee Kian Wie and Yoshihara, K. 'Foreign and Domestic Capital in Indonesian Industrialization' in *Southeast Asian Studies* Vol 24, No. 4, March 1987.

Thorbecke, Erik, et al. *Adjustment and Equity in Indonesia*. Paris, OECD Development Centre Studies, 1992.

Turner, Sarah. *Indonesia's Small Entrepreneurs: Trading on the Margins*. London, RoutledgeCurzon, 2002.

United Nations Industrial Development Organization (UNIDO). *Indonesia: Industrial Development Review*. London, Economist Intelligence Unit, 1993.

Van Dierman, Peter. *Small Business in Indonesia*. Aldershot, Ashgate, 1997.

Winters, Jeffrey A. *Power in Motion: Capital Mobility and the Indonesian State*. Ithaca, NY, Cornell University Press, 1996.

Woo, W. T., Glassburner, B., and Nasution, A. *Macroeconomic Policies, Crises, and Long-Term Growth in Indonesia, 1965–90*. Washington, DC, World Bank, 1994.

JAPAN

Physical and Social Geography

JOHN SARGENT

The archipelago comprising Japan, or Nihon Koku (Land of the Rising Sun), lies to the east of the Asian mainland in an arc stretching from latitude 45° N to latitude 24° N, covering a land area of 377,864 sq km (145,894 sq miles). The Tsushima Strait, which separates Japan from Korea, is about 190 km wide, while 800 km of open sea lie between Japan and the nearest point on the coast of the Chinese mainland. Four large and closely grouped islands—Hokkaido, Honshu, Shikoku, and Kyushu—constitute 98% of the territory of Japan, the remainder being made up by numerous smaller islands.

PHYSICAL FEATURES

The Japanese islands belong to a belt of recent mountain-building which extends around the rim of the Pacific Ocean, and which is characterized by frequent volcanic activity and crustal movement. Around the fringes of the western Pacific, this belt takes the form of a complex series of island arcs, stretching southwards from the Aleutians and including Japan. In the Japanese islands, the Kurile, Kamchatka, Bonin, Ryukyu and Korean arcs converge. Where two or more of these major arcs meet, as in Hokkaido and in central Honshu, conspicuous knots of highland occur. In the latter area, the Japan Alps, which rise to more than 3,000 m above sea-level, form the highest terrain in the country, although the highest single peak, Mt Fuji (3,776 m), is an extinct volcano unrelated to the fold mountains of the Alps.

Three major zones of active volcanoes and hot springs occur: in Hokkaido, in northern and central Honshu, and in southern Kyushu. Further evidence of crustal instability is provided by the occurrence, each year, of over a thousand earth tremors. Earthquakes strong enough to cause damage to buildings are, however, less frequent, and occur, on average, once every five years.

While the major arcs determine the basic alignment of the main mountain ranges, complex folding and faulting has resulted in an intricate mosaic of landform types, in which rugged, forested mountains alternate with small pockets of intensively cultivated lowland.

In the mountains, short, fast-flowing torrents, fed by melt-water in the spring and by heavy rains in the summer, have carved a landscape which is characterized by steep and sharply angled slopes. Narrow, severely eroded ridges predominate and rounded surfaces are rare. Although the mountain torrents provide many opportunities for the generation of hydroelectric power, marked seasonal changes in precipitation cause wide fluctuations in the rate of flow, and consequently hinder the efficient operation of hydroelectric plant throughout the year.

The extreme scarcity of level land is one of the salient features of the geography of Japan. In a country where the population was the eighth largest in the world in mid-1998, only 15% of the total land area is cultivable. Thus, the small areas of lowland, which contain not only most of the cultivated land but also all the major concentrations of population and industry, are of vital importance.

Most Japanese lowlands consist of small coastal plains which have been formed through the regular deposition of river-borne alluvium. On encountering the low-lying land of the coastal plain, the typical torrent becomes a sluggish river which meanders across the gently sloping surface of the plain, to terminate in a shallow estuary. The river bed is usually raised above the surface of the surrounding plain, and the braided channel is contained by levees, both man-made and natural. Most alluvial plains are bounded inland by rugged upland, and many are flanked by discontinuous benches of old and poorly consolidated alluvial material. None of the alluvial plains of Japan is extensive: the Kanto, which is the largest, has an area of only 12,800 sq km. Many plains are merely small pockets of nearly level land, closely hemmed in by the sea and the steeply sloping mountains.

The coastline of Japan is long and intricate. On the Pacific coast, where major faults cut across the prevailing grain of the land, large bays, flanked by relatively extensive alluvial plains, are conspicuous features. Three of these bay-head plains—the Kanto, the Nobi and the Kansai—contain more than one-third of the population of the country, and more than one-half of its industrial output. Further west along the Pacific coast, two narrow channels lead into the sheltered waters of the Inland Sea, which occupies a zone of subsidence between Shikoku and western Honshu. By contrast with the Pacific coasts, the Japan Sea coastline is fairly smooth. The overall insularity of Japan may be indicated by reference to the fact that very few parts of the country are more than 100 km from the sea.

CLIMATE

While relief conditions in Japan often impose severe limits upon economic activity, climatic conditions are, on the whole, more favourable. Japanese summers are of sufficient warmth and humidity to allow the widespread cultivation of paddy rice; yet cold, dry winters clearly differentiate Japan from those countries of subtropical and tropical Asia, where constant heat prohibits prolonged human effort.

The climate of Japan, like the climates of the rest of Monsoon Asia, is characterized by a marked seasonal alternation in the direction of the prevailing winds. In winter, in association with the establishment of a centre of high atmospheric pressure over Siberia, cold, dry air masses flow outwards from the continent. During their passage over the Sea of Japan (or the East Sea as it is known on the Korean Peninsula), these air masses are warmed in their lower layers, and pick up moisture, which, when the air masses rise on contact with the Japanese coast, is precipitated in the form of snow. Thus, winter weather along the Sea of Japan coastlands is dull, cloudy and characterized by heavy falls of snow. By contrast, the Pacific side of the country experiences cold, dry weather, with low amounts of cloud. Near the Pacific coast, winter temperatures are ameliorated by the influence of the warm Kuro Shio sea current.

Besides this contrast between the two sides of the country, a latitudinal variation in temperature, similar to that of the Atlantic seaboard of the USA, is also apparent. Thus, north of latitude 38° N, average January temperatures fall below 0°C, and reach −10°C in Hokkaido. In this northern zone, winter weather conditions prohibit the double cropping that is elsewhere characteristic of Japanese agriculture. South of latitude 38° N, January temperatures gradually rise, reaching 4°C at Tokyo, and 6°C at Kagoshima in southern Kyushu.

After mid-March, the winter pattern of atmospheric circulation begins to change, with high pressure developing over the Aleutians, and low pressure over Siberia. In association with these unstable conditions, the first of the two annual rainfall maxima occurs, with the onset of the Bai-u rains in June. By July, however, the high pressure centre to the east of Japan has fully developed, and, with low pressure prevailing over the continent, a south-easterly flow of warm, moist air covers the entire country. On the Pacific coast, August temperatures rise to over 26°C, and the weather becomes unpleasantly hot and humid. To the north, however, August temperatures are lower, reaching only 18°C in Hokkaido.

In late August and early September the Pacific high pressure centre begins to weaken, and the second rainfall maximum occurs, with the arrival of typhoons, or tropical cyclones, which travel northwards to Japan from the equatorial regions of the Pacific Ocean. These severe storms, which frequently coincide with the rice harvest, cause widespread damage. By October, high pressure has again developed over Siberia, and the north-westerly winter monsoon is consequently re-established.

Annual precipitation in Japan varies from 850 mm in eastern Hokkaido to more than 3,000 mm in the mountains of central Honshu, and in those parts of the Pacific coast which are fully exposed to the force of the late summer typhoons.

RESOURCES

Although about 67% of the total area of Japan is forested, not all of the forest cover is commercially valuable, and large areas of woodland must be preserved to prevent soil erosion. Because many houses are still built of wood, the demand for timber is high, and the domestic output is supplemented by imports.

In terms of value, and also of volume, the Japanese fish catch is one of the world's largest. Seafood provides a large proportion of the protein content of the average Japanese diet, and demand is therefore high. Rich fishing grounds occur in both the Sea of Japan and the Pacific Ocean to the east of Japan.

Japan has few mineral resources, and industry is heavily dependent upon imported raw materials and fuels. Japan's coal is of poor to medium quality, and seams are thin and badly faulted. The two main coalfields are located towards the extremities of the country, in Hokkaido and in northern Kyushu. Japanese coal deposits are particularly weak in coking coal, much of which is imported, mainly from the USA and Australia. The small Japanese oilfields, which are located in north-east Honshu, supply a minimal percentage of domestic fuel demand. Japan is heavily dependent on foreign iron ores, imported mainly from Australia, Brazil and India. Many other minerals are mined, but none exists in large quantities. Japan is self-sufficient only in limestone and sulphur.

POPULATION

In 1867, on the eve of modernization, the population of Japan was already approximately 30m., a level at which it had remained, with little fluctuation, for the preceding 150 years. With industrialization, the population increased rapidly and by 1930 had reached 65m. After the Second World War, the population policy initiated by the Japanese Government succeeded in drastically lowering the rate of population increase, and during the second half of the 20th century the growth rate closely corresponded to the rates prevailing in Western Europe. In the early 21st century, however, Japan's fertility rate continued to decline steadily, prompting the authorities to express concern.

The population of Japan, according to census results, was 126,925,843 on 1 October 2000. By mid-2003 the population had risen to an estimated 127,619,000, representing an average density of 337.7 persons per sq km. Only 15% of Japan's land area is cultivable lowland, and the population density in these areas is among the highest in the world. Three conspicuous urban-industrial concentrations are centred upon Tokyo, Osaka and Nagoya. Japan is by far the most urbanized country in Asia. Tokyo, the capital of Japan and one of the largest cities in the world, had an estimated population of 8,130,408 on 1 October 2000.

Apart from the very small number of Ainu (a people who exhibit certain Caucasian characteristics), the Japanese population has been, since early times, ethnically and linguistically uniform. The racial origins of the Japanese remain conjectural, but both Mongol and southern Pacific strains are apparent in the present-day population.

History up to 1952

RICHARD STORRY

ANTIQUITY AND THE MIDDLE AGES

It is generally agreed that the ancestors of the Japanese must have been immigrants from the mainland of Asia. It is also claimed that there was probably some migration to Japan from the islands of South-East Asia, but the whole subject is still one of pure conjecture. What does seem undeniable is that the forebears of the small and dwindling Ainu communities of Hokkaido once occupied the whole country and were in fact the original inhabitants. Be that as it may, an elaborate mythology surrounds the origins of Japan and the Japanese. This declares, for example, that the country itself was created by the gods, and that the first Emperor, Jimmu (c. 660 BC), was a direct descendant, in the fifth generation, of the sun goddess.

Yamato Period

At all events, it seems probable that the invading immigrants from Asia, who no doubt crossed over from Korea, gradually forced their way eastward from Kyushu along the shores of the Inland Sea, until, around the beginning of the Christian era, they found themselves in the fertile Kansai plain (the modern Kyoto-Osaka region). Here, in the Yamato district, they established an ordered society under chieftains who became priest-kings, dedicated to the cult of the Sun.

This early Japanese society was profoundly influenced by the civilizations of Korea and China. The Chinese ideographic script is only one important and very striking example of many cultural importations from or through Korea. Of even greater significance was the introduction of Buddhism in the sixth century AD. It was at this stage that the existing body of religious practices, associated with sun worship and animism, became known as Shinto, or 'The Way of the Gods'. Neither the theology of Buddhism, nor the ethics of Confucianism (another import from the continent) made Shinto superfluous. Old beliefs existed side by side with the new; and in course of time, as one would expect, Chinese ideas of religion, of morality, of artistic excellence, of good government, of sound agriculture, were adapted to Japanese conditions and thus underwent a degree of change in the process.

Nara and Heian Periods

At the beginning of the eighth century Nara became the capital, being built on the contemporary Chinese model. This was the heyday of the early Buddhist sects in Japan; and the splendid temples surviving at Nara have a particular interest today, since they are the best remaining examples anywhere of Chinese architecture of the Tang period. Nara was intended to be a permanent capital. This was, in fact, Heian-kyo, later to be known as Kyoto, founded in 794 and constructed, like Nara, on the model of the Chinese capital. It was to be the home of the Japanese imperial family until 1868. The establishment of this city marks the opening of the Heian age (794–1185), a period remarkable for the artistic sophistication of the court and metropolitan aristocracy.

By the middle of the 12th century effective power in Kyoto was in the hands of a warrior household, the Taira. Their great rivals were another family, the Minamoto. At first the Taira carried all before them; and Kiyomori, the head of the family, ruled Japan in the Emperor's name for a generation. After his death in 1181, however, the tables were turned; and in a final battle, in 1185, the Minamoto annihilated their enemies. Thereafter the leader of the Minamoto, Yoritomo, set up a new system of government, known as the *Bakufu* (literally 'camp office'), at Kamakura in the east of the country, far from the imperial capital. The Emperor gave Yoritomo the title *Sei-i Tai Shogun*, or 'Barbarian-subduing Generalissimo'—usually abbreviated, in Western use, as 'Shogun'.

Kamakura Period

The original purpose of Yoritomo's *Bakufu* was the control and administration of the Japanese warrior class, which was now a distinct entity, and one that was rapidly becoming all-powerful in society. The Japanese fighting man was already a member of an élite class by the 12th century. The true rulers of the country from that time forward, until the late 19th century, would tend nearly always to belong to the warrior class. Not for a moment did this class seek to overthrow the imperial dynasty. The idea was indeed unthinkable, since the Emperor's line was descended from the sun goddess. So ceremonious respect was always paid to the Kyoto court; but it was exceptional, and usually a sign of uncharacteristic weakness, for any warrior administration to allow the reality of power to slip back into the hands of the imperial household. Every Shogun governed in the Emperor's name and received his appointment from the Emperor.

The Kamakura *Bakufu* lasted until well into the 14th century. Yoritomo was a man of exceptional energy, organizing ability, and ruthlessness, who did not hesitate to pursue a vendetta against his own younger half-brother Yoshitsune, who as a military commander had been chiefly responsible for the ultimate defeat of the Taira. Yoritomo died in 1199. His successors in the office of Shogun were leaders of inferior calibre, and the *Bakufu* was run by the house of Hojo, related to the Minamoto by marriage. It was the Hojo who rallied the country in resistance to the Mongol invasions of 1274 and 1281. Japanese martial courage was a vital element in the discomfiture of the invaders; but the decisive factor, both in 1274 and 1281, seems to have been the storms which wrecked the Mongol ships lying off the coast. With some justice the Japanese described the great typhoon of 1281 as a *kami-kaze*, or 'divine wind'.

Some 50 years later both the Hojo family and the Kamakura *Bakufu* were overthrown in the course of a civil war. The climax occurred in 1331 when, with their enemies overrunning Kamakura, the Hojo and their supporters—more than 800 in all—committed *seppuku*, the formal term for the act of *hara-kiri*, the warrior's suicide by self-disembowelling.

Muromachi Period

Over the succeeding 250 years and more there was great disorder, including much bitter fighting in and near Kyoto. A new *Bakufu* was established, this time in the Muromachi district of the capital, with members of the Ashikaga house (of the Minamoto line) holding office as Shogun. From the fall of Kamakura to the latter half of the 16th century political events, so often shaped by domestic warfare, were extremely complicated. This period was marked not only by civil war but also by economic growth and artistic achievement. The breakdown of central government gave at least some provincial lords the freedom and incentive to embark on foreign trade on their own account, especially with China. One consequence of this commerce was a substantial importation in the 15th century of copper cash from China, which promoted the growth of money instead of rice as a medium for exchange. At the same time painting, classical drama, architecture, landscape gardening, ceramics, the tea ceremony, flower arrangement—a great deal of what is recognized today as Japan's magnificent cultural heritage—blossomed in these stormy years. Here Zen Buddhism, in all its manifestations, played a central part. Japan presented a paradoxical scene of savagery and civilization, of barbarism and beauty, intertwined.

TOKUGAWA RULE

Effective central government and internal peace were not finally secured until the early years of the 17th century, after Ieyasu founded the Tokugawa *Bakufu* in Yedo (the modern Tokyo), giving the whole country a domestic order that would endure until the coming of US and European men-of-war in the 1850s. Tokugawa Ieyasu built on the work already performed by two notable captains, Oda Nobunaga (1534–82) and Toyotomi Hideyoshi (1536–98). The former contrived, before his death, to unify about one-half of the provinces of Japan. Hideyoshi, the son of a foot-soldier, was one of Nobunaga's commanders. Within 10 years of Nobunaga's death he made himself master of the whole country, with the help of a wise and cautious ally, Tokugawa Ieyasu.

The 'Closed Country'

After Hideyoshi's death Ieyasu lost little time in making his own position supreme. He defeated his most formidable rivals in battle in 1600, and three years later he was appointed Shogun by the Emperor. The history of Japan in Ashikaga days taught him, no doubt, the lesson that the Shogun's government was best conducted, like Yoritomo's regime, well away from Kyoto. At any rate, he made Yedo Castle the headquarters of his administration.

Ieyasu and his immediate descendants adopted a number of important measures to buttress the dominant position of the Tokugawa house (a branch of the seemingly indestructible Minamoto line). Careful watch at all times was kept on those lords considered to be unreliable. Yet a more effective way of controlling all feudatories was the rule, strictly enforced, that they spend part of every year in the Shogun's capital at Yedo. It was also decreed that when a lord returned to his own province he must leave his wife and family behind him, in Yedo.

Moreover, the Tokugawa *Bakufu* adopted a policy of severe national isolation. From 1628 only the Chinese and Dutch were allowed in, as traders, and their commerce was confined to the port of Nagasaki, where the handful of Dutch merchants was restricted to the tiny island of Deshima. No other foreigners were granted access. No Japanese was permitted to go abroad. Vessels above a certain tonnage could not be built. The modest foreign trade at Nagasaki was a Tokugawa monopoly, controlled by officials appointed by Yedo.

This situation, known as *sakoku* or the 'closed country', was not broken until 1853 and 1854, when Cdre Matthew Perry's squadron of US warships visited Yedo (now Tokyo) Bay. On his second visit Perry secured *Bakufu* consent to the opening of two ports and the acceptance, at a future date, of a resident US consul. The door having been forced ajar, it was soon widened. Other powers lost no time in following the example of the USA, and a decade after Perry's expedition a community of foreign diplomats and traders had settled on Japanese soil.

While none of Japan's leaders really welcomed this intrusion by the West, some were implacable in their hostility, insisting that the 'barbarians' be expelled. Others perceived the weakness of their country and argued that it must come to terms with the situation, learning the techniques of modern Western civilization. Only then would Japan attain the necessary power to hold its own. At the cost of much humiliation—a great deal of pride had to be swallowed—the second, more realistic, course was adopted as the national policy.

THE RISE OF IMPERIAL JAPAN

Modernization followed the domestic transformation known as the Meiji Restoration. The Tokugawa Shogunate had lost face from the moment the first concessions were made to Perry and other intruders. Eventually, in 1868, the much weakened *Bakufu* was overthrown by provincial lords from the south-west, acting in concert and impelled by a coalition of their own most vigorous, far-sighted, warrior retainers. The Emperor, still in his teens, was persuaded to leave Kyoto for Yedo, which was renamed Tokyo and became the new capital. Nominally, full governing powers were 'restored' to the ancient monarchy, but the young Emperor, Meiji, reigned rather than governed. Real power was exercised by an oligarchy composed almost entirely of the provincial warriors (all of them young or still in the prime of life) who had engineered the downfall of the Shogunate. These men, the Meiji modernizers, dominated Japanese politics, actively or from their retirement, for the best part of 50 years. The pace of modernization, with the abolition of so many cherished customs and privileges, inevitably gave acute offence to many conservatives. There was more than one unsuccessful armed rising against the Government in the decade following the Restoration of 1868.

The heritage of Confucian ethics, with their strong emphasis on loyalty to seniors and superiors, fortified the traditions of Shinto, with its veneration of the imperial house, in sustaining a spirit of harmony and hard work, deeply influencing the great majority of the people. Educational indoctrination played a significant part here. The Meiji Government founded an impressive structure of schools, colleges and universities. In 1890 the Emperor issued his famous *Rescript on Education*, an exhortation commending the nation's fundamental ethical code to all young people. The Rescript, stressing the patriotic virtues of obedience and self-sacrifice, was read aloud in all schools on days of national festival and commemoration.

A constitution promulgated in 1889, setting up a bicameral legislature, represented a concession by the oligarchy to the growing demand for some form of national legislative assembly. However, the powers of the Diet (as the new legislature was known) were modest. Nevertheless, the party leaders in the Diet soon became a serious irritant to the Government.

Wars with China and Russia

Domestic political squabbles, however, were put aside in the face of a crisis with China over Korea, in 1894, which led to a war in which Japan won spectacular victories on land and at sea. By the Treaty of Shimonoseki (1895) China surrendered to Japan the island of Taiwan (Formosa) and the Liaodong peninsula in South Manzhou (Manchuria), including Lushun (Port Arthur). Within a few days Japan was forced by Russia, Germany and France to waive its claim to Manzhou. A few years later Russia established itself in control of Lushun and its hinterland. Revenge came in the Russo-Japanese War of 1904–05, which was a much more costly affair for Japan in terms of men and material resources than the Sino-Japanese War had been, 10 years earlier. However, Japan's victories, including the destruction of the Tsar's Baltic fleet in the Tsushima Strait in May 1905, were dramatic, and Asia in particular was deeply moved by what happened.

THE TAISHO ERA

The death of Emperor Meiji in 1912 was decidedly a landmark, the end of a not inglorious chapter. For the Meiji era was Japan's Victorian age, when despite set-backs and disappointments everything seemed to move forward. The new Emperor proved to be mentally unstable, and in 1921 his eldest son, Crown Prince Hirohito, became regent, succeeding to the throne at the end of 1926.

The period 1912–26 is known as the Taisho era, after the title chosen for the reign of Meiji's successor. It is noteworthy for three important developments and one shocking disaster. In the first place, thanks to the World War of 1914–18, the nation's economic power began to swell in dynamic fashion, as Japanese shipyards, factories and foundries were overwhelmed with orders from the Allied countries. Secondly, as Britain's ally, Japan invaded and occupied Shandong (Shantung), Germany's leased territory in China, bringing Japan firmly into China's affairs. The temptation to dictate to China could not be resisted, with the result that Chinese dislike and distrust of Japan increased dramatically, setting the tone of relations between the two countries for years to come. Thirdly, Lenin's triumph in Russia gave some impetus to protest movements created by the contrast in the standards of living between those who had been enriched by the war and the poorer sections of the urban working class. Left-wing groups began to obtain a measure of representation in the Diet; democracy appeared to be coming into vogue. Then, in September 1923, more than one-half of the city of Tokyo and the whole of Yokohama were destroyed in a series of earth tremors and subsequent fires. In recorded history there have been few comparable natural disasters so calamitous in loss of life and destruction of property.

THE PRE-WAR SHOWA PERIOD

After Emperor Taisho's death in 1926 his successor chose as the title for the new reign two Chinese characters, Sho Wa, which can be translated as 'Bright Harmony'. The years that followed, however, belied the promise implicit in these words. A Prime Minister, Hamaguchi, was shot and wounded by a nationalist fanatic in 1930, and died some months later from his injuries. In 1932 another Prime Minister, Inukai, was assassinated; and in 1936 two former premiers, Saito Makoto and Takahashi Korekiyo, were shot down in their homes by parties of mutinous troops. These and other instances of civil bloodshed and violence were among the more lurid symptoms of a wave of irrational nationalist hysteria prompted partly by events on the continent of Asia and in part by the economic consequences for Japan of the world depression, which hit the country hard at the beginning of the 1930s. Unrest and dissatisfaction exploded in anger against Diet politicians, wealthy capitalists, and liberal-minded men at the palace and in other influential positions. Public opinion came to regard such figures as weak, corrupt and incompetent. The man in uniform, on the other hand, was back in favour, and in power. For in the early autumn of 1931 Japanese forces in South Manzhou carried out a coup against the Chinese in Shenyang (Mukden), and this soon developed into the forcible seizure of all Manzhou.

Military Expansion in Asia

Condemned by the League of Nations for aggression in Manzhou, Japan left that body in 1933. Manzhou became a vassal state (Manchukuo), an apanage of the Empire, dominated by the army and only in name more independent than Taiwan, or Korea, which had been annexed in 1910. Domination of Manzhou led to involvement in northern China, and out of this came undeclared war in mid-1937. By the end of that year Japan and China were locked in a combat that did not end until 1945. As the war continued, Japan's relations with other powers underwent a change. It drew closer to Nazi Germany and Fascist Italy, eventually joining them in full alliance in September 1940. Increasingly, both the United Kingdom and the USA, powers that supported the Chinese Government in Chongqing, were seen as potential enemies.

In July 1941 Vichy France agreed to the Japanese occupation of bases in the Saigon area—bases in northern French Indo-China had been occupied by Japan in the previous year. The move southward seemed a clear threat to both Malaya and the Netherlands East Indies. It indicated, too, that for the time being, at least, Japan was not going to join its ally Germany in the assault on the USSR. There was an immediate response by the British Commonwealth, the USA and the Netherlands, in the form of a virtual embargo on all trade with Japan. This was serious, for it meant that petroleum imports into that country had to cease. US-Japanese talks were inconclusive. Inexorably, Japan drifted towards armed confrontation with the Western powers.

The Sino-Japanese War of 1894–95 and the Russo-Japanese War of 1904–05 had started with surprise attacks by the Japanese navy. On 7 December 1941 Japan followed the same strategy, attacking the US fleet at Pearl Harbor in Hawaii. Later in the same month, Japanese forces invaded Hong Kong, Malaya, Singapore and the Netherlands East Indies.

THE PACIFIC WAR

For the first six months the Japanese advance was virtually unhindered. Hong Kong, Malaya, Singapore, Java and the Indies, the Philippines, Burma and the Andaman Islands, New Britain and the Solomons all fell to Japanese arms. There had, however, been a grave miscalculation of the spirit and resources of the nation's principal enemies. Allied submarines, US island-hopping strategy and superior fire-power led to a reversal of Japan's position. From mid-1944 the tide turned against Japan. By mid-1945 military collapse was imminent. US air raids had inflicted fearful punishment. The merchant fleet, like the battle fleet, had practically ceased to exist. Germany was out of the war. The USSR was an unknown but menacing factor, returning no answer to pleas that it should act as a mediator.

The Potsdam proclamation at the end of July 1945 seemed to leave the Government unmoved, although in reality the Premier, the aged Baron Suzuki, was seeking ways and means of ending the war short of abject capitulation. On 6 August the first atomic bomb laid waste Hiroshima. On 9 August the second descended on the suburbs of Nagasaki. Between those dates the Soviet army overran Manchukuo. In this supreme crisis the nation's leaders were divided between those who favoured surrender (with the proviso that the monarchy be maintained) and those who were ready to fight on in spite of everything. It was the Emperor, invited to give an unprecedented decision of his own, who tipped the balance by declaring that the Potsdam terms must be accepted.

THE OCCUPATION

US Gen. Douglas MacArthur represented all the Allies in Japan, but the occupation was nevertheless an almost exclusively US undertaking and to a very great extent MacArthur took his own decisions, without direct reference to Washington. He rejected the view that the Japanese would be better off without the age-old institution of the monarchy. He felt that the Emperor was a stabilizing factor in a society shaken to its roots by the capitulation. Popular regard for the Emperor, however, no longer rested on the belief that he partook of divinity because of his descent from the sun goddess. When, at the beginning of 1946, the Emperor formally renounced his 'divinity', it created little interest among most Japanese.

In his administration of Japan, MacArthur acted through the Japanese Government, a procedure that worked smoothly in nearly every instance. Between conquerors and conquered there was indeed a harmony that nobody could have foreseen during the years of warfare. The Japanese, however, can be intensely pragmatic. The events of 1945 seemed to demonstrate that their own way of conducting affairs was inefficient and harmful to themselves. So when the US occupying forces arrived, and once it was clear that their general behaviour was by no means vengeful and oppressive, the Japanese were ready to be their pupils in all manner of activities.

Political, Economic and Social Reforms

The guiding theme of the occupation, in the early days especially, was disarmament and democratization. A new Constitution, promulgated in 1946, reflected both these aims. One clause stated that the Japanese people renounced war; and it went on to say that 'land, sea, and air forces, as well as other war potential, will never be maintained'. A further clause laid down that the Prime Minister and his Cabinet colleagues must be civilians. The other articles of the Constitution reflected the authentic spirit of North American democracy, with full emphasis on the rights of the individual. Sovereignty of the people was declared. The Emperor was made 'the symbol of the State and of the unity of the people', and it was affirmed that he derived his position 'from the will of the people'.

Although undeniably US-inspired, the post-war Constitution captured the imagination of the Japanese. To this day its defenders are sufficiently numerous to make it unlikely that the Constitution will be so radically amended as to change its basic character. Amendment requires the assent of two-thirds of the members of both houses of the Diet, confirmed by a referendum of the people as a whole.

Another measure of profound social and political importance, instigated by MacArthur's headquarters, was the land reform programme. Thousands of tenant farmers were able to obtain ownership of the land they cultivated. Up to the war a depressed class, the farmers of Japan, thanks to the land reform, became firm, if not always satisfied, upholders of the political status quo. Left-wing parties found the farming vote difficult to entice. The average farmer, freed from the burden of rent and assured of sales for his crop at guaranteed prices, was not impressed by advocates of collectivization and other projects of agrarian socialism.

The educational system was comprehensively reformed. In terms of organization and syllabus it was reworked to a pattern resembling that of the USA. The famous *Rescript on Education*, needless to say, was discarded, and there was a thorough revision of school-books concerned with history, political science and ethics.

These manifold and generally liberating changes were not far short of revolutionary in character. Political freedom gave the parties of the Left an opportunity to exploit these changes, and to make them even more far-reaching. However, except for the period between May 1947 and March 1948, when a coalition Government under Tetsu Katayama of the Socialist Party was in office, electoral success always attended the conservative parties. Until the end of 1954 the political scene was dominated by Shigeru Yoshida; and, even after his retirement, the old man was influential, as adviser to successive Governments, until his death in 1967.

Consolidation of Relations with the West

As the international situation hardened into the Cold War, the attitude of MacArthur's headquarters underwent a subtle but definite change. The emphasis shifted from reform to rehabilitation. In particular, as the armies of Mao Zedong began to gain ground in China, and as it became clear that US influence on the Chinese mainland might soon be eliminated, the importance of Japan's future role in the non-communist world was perceived with growing clarity. After the Korean War broke out in June 1950 it seemed all the more desirable, and in fact urgent, to nourish the revival of Japan. In other words Japan was now regarded not as a recent enemy but rather as a new friend and junior ally. In these circumstances the disarmament clause of the Constitution appeared as an embarrassment. In practice, however, it was to be blandly ignored.

The US military occupation lasted until 1952. This was much longer than had been planned. Soon after his basic reforms had been introduced, MacArthur had decided that the situation called for a treaty of peace. When he was dismissed by President Truman in 1951 the Japanese feared that progress towards a peace treaty would be checked. However, on 8 September the treaty was concluded at San Francisco between Japan and 48 nations (but not the USSR). It was a magnanimous settlement, free from punitive clauses, although Japan's territorial losses were confirmed. On the same day a bilateral security pact was signed by Japan and the USA. In this, Japan asked the USA to retain its forces in and around the Japanese islands as a defence against outside attack. When all the signatories of the San Francisco Treaty had ratified the document, it came into force; on 28 April 1952 Japan became, once again, formally an independent state.

Recent History

AKIRA YAMAZAKI

Revised by LESLEY CONNORS

THE POST-WAR POLITICAL ORDER

The reforms introduced during the US military occupation created the institutional framework within which Japan was to conduct its social and political life. The conservative political forces, under the guidance of Shigeru Yoshida (Prime Minister in 1946–47 and 1948–54) and bolstered by the return to the Kokkai (Diet) of pre-war political leaders who had been purged, were divided between two main parties, the Liberals and the Democrats, but were united in their desire to change the institutional framework and to reverse many of the democratic reforms. The conservatives were confronted by a vigorous opposition, dominated by the Japan Socialist Party (JSP), which regarded any attempt to tamper with the reforms and, above all, with the Constitution (particularly Article 9, the 'peace clause', whereby Japan renounces the use of war) as a potential threat to return Japan to the militarism of the 1930s and 1940s.

The contest between these groups was resolved in 1955, at which time the foundations of the post-war political order were established. At a general election for the House of Representatives (the lower house of the Diet) in February 1955, the JSP received 29.2% of the total votes and won 156 of the 467 seats in the chamber. The JSP thus controlled the minimum number of seats (one-third of the total) necessary to obstruct any proposed revision of the post-war Constitution. However, the radical policies (and even more radical rhetoric) of the JSP alarmed business interests, who were organized in the Japan Federation of Economic Organizations (KEIDANREN). Business leaders urged the Liberals and the Democrats (who had together won 63.2% of the votes and 297 seats at the election) to merge into a single party, which they did in November 1955. The new party, the Liberal-Democratic Party (LDP), governed Japan from 1955 until 1993 (returning to office in 1994, initially as part of a coalition Government), leading the country during a period of remarkable economic expansion.

The advantages that the LDP enjoyed were overwhelming: as the incumbent governing party, it obtained the political benefits of the 'economic miracle', while Japan's electoral arrangements provided an additional asset for the LDP. At the national level, the opposition parties were unable to achieve as much support, even temporarily, as they had locally. A high degree of ideological fragmentation precluded any serious possibility that the opposition could replace the LDP in government. Some of the opposition parties, notably the Komeito ('Clean Government Party'), the Democratic Socialist Party (DSP) and the New Liberal Club (NLC), were more inclined to co-operate with the LDP than with the increasingly isolated Japanese Communist Party (JCP).

Furthermore, the opposition parties were hampered by the uneven distribution of parliamentary constituencies. For many years successive LDP Governments steadfastly refused to allow any revision of electoral boundaries, to take account of population movements, with the result that the rural areas, where support for the LDP was strong, were substantially over-represented in the Diet. In April 1976 the Supreme Court ruled that the allocation of seats in the House of Representatives was unconstitutional, owing to 'mal-apportionment', which denied equal rights to urban voters (as guaranteed by the Japanese Constitution). In 1990 the LDP leadership announced its commitment to the implementation of comprehensive changes to the procedure for the allocation of seats to the lower house (see below).

Finance was another factor behind the success of the LDP. Politics in Japan, and particularly elections, are spectacularly expensive: the officially reported income of the political parties in 1990 (not an unusually expensive year) was US $700m., which was probably little more than one-quarter of their actual income. To win elections, a political party must be able to raise enormous funds. The LDP traditionally enjoyed a close relationship with the business community, thus the party was well placed to obtain large-scale financial support. One estimate suggested that Japanese business interests transferred more than 45,000m. yen per year to the LDP.

LIBERAL-DEMOCRATIC GOVERNMENTS

Nobusuke Kishi became Prime Minister in February 1957, and held office until July 1960, when he was succeeded by Hayato Ikeda. In November 1964 Ikeda resigned, owing to ill health, and was replaced by Eisaku Sato. Japan enjoyed strong economic growth under Ikeda and Sato. The latter remained in office until July 1972, when he was succeeded by Kakuei Tanaka, hitherto the Minister of International Trade and Industry.

Meanwhile, many of Japan's outer islands, surrendered to the USA at the time of the 1945 armistice, were restored to Japanese sovereignty. The Tokara Archipelago and the Amami Islands (parts of the Ryukyu group) had been restored to Japan in December 1951 and December 1953, respectively. The Bonin Islands and the remainder of the Ryukyu Islands (including Okinawa) reverted to Japan in June 1968 and May 1972, respectively.

1972–82

Tanaka's period of tenure as Prime Minister was characterized primarily by scandals, illustrating the problem of widespread corporate involvement in Japanese politics, although initially more noteworthy was Japan's recognition of the People's Republic of China in September 1972. During his premiership, Tanaka allegedly accepted bribes totalling 500m. yen from the Marubeni Corporation, a representative in Japan of the Lockheed Aircraft Corporation (a leading US aerospace company). Following a severe reduction in the LDP's majority in the House of Councillors (upper house of the Diet) at elections held in July 1974, Tanaka's hold on the premiership became tenuous. In December he resigned in favour of Takeo Miki, a former Deputy Prime Minister. Tanaka was subsequently arrested, in July 1976, on charges of accepting bribes, and resigned from the LDP. Largely as a result of voters' disapproval of the LDP's alleged involvement in corruption, the party lost its majority in the House of Representatives in December. Miki was forced to resign, and was succeeded by Takeo Fukuda, who had resigned in November as Deputy Prime Minister.

At the elections held in July 1977 for the House of Councillors Fukuda was unable to reverse the LDP's decline. He was defeated in the LDP presidential election by Masayoshi Ohira, the party's Secretary-General, who became Prime Minister in December 1978. Although Ohira managed to retain his position as Prime Minister following the election of October 1979 to the House of Representatives (when the LDP once again failed to obtain a majority and there were significant gains by the JCP), the Government was defeated on a motion of 'no confidence' in the House of Representatives in May 1980. The lower house was thus dissolved, and at elections held in June the LDP received 47.9% of the total votes and won 284 of the 511 seats. A compromise candidate, Zenko Suzuki, was elected President of the LDP in July, and was subsequently appointed Prime Minister. Suzuki's Government was beset by serious economic problems and growing factionalism within the LDP. As the economic crisis worsened, Suzuki was forced to resign as Prime Minister and President of the LDP in October 1982.

The Nakasone Administration

Suzuki's successor was Yasuhiro Nakasone, who was supported by the Suzuki and Tanaka factions of the LDP. At elections in June 1983 the LDP increased its strength in the upper house from 134 to 137 members in the 252-seat chamber. This result was seen as an endorsement of Nakasone's policies of increased spending on defence, closer ties with the USA and greater Japanese involvement in international affairs.

In October 1983, after judicial proceedings lasting seven years, a Tokyo court found Tanaka, the former Prime Minister, guilty of accepting bribes. He immediately began appeal proceedings against the conviction and the sentence, and refused to relinquish his legislative seat, which led to a boycott of the Diet by the opposition. This forced Nakasone to dissolve the House of Representatives in preparation for a premature general election in December 1983. The election campaign was dominated by the issues of political ethics and Nakasone's forthright style of leadership. The LDP suffered its worst reverse to date, losing 36 seats (and its majority) in the lower house. Nakasone was placed second (behind Takeo Fukuda) in his district, whereas Tanaka was returned with an overwhelming majority. The Komeito, the DSP and the JSP gained seats, while the JCP and the NLC lost seats. A coalition was formed between the LDP, the NLC (which had split from the LDP over the Tanaka affair in 1976) and several independents, and Nakasone remained as President of the LDP, promising to reduce Tanaka's influence. Six members of Tanaka's faction, however, held posts in Nakasone's new Cabinet.

Nakasone's domestic policy was based on the 'three reforms': administrative reforms, particularly of government-controlled enterprises such as telecommunications and railways; fiscal reforms, to enable the Government to balance its budget after many years of persistent deficit; and educational reforms, to liberalize the rigid examination-dominated system.

In November 1984 Nakasone was re-elected as President of the LDP, guaranteeing him a further two years in office as Prime Minister, the first to serve a second term since Eisaku Sato (1964–72). Nakasone was committed to raising Japan's international status by fostering friendly relations with other world leaders. He made successful tours to the USA, Australia and South-East Asia in 1984, and to Europe in 1985. However, there was continued concern in the European Community (EC, now European Union—EU) over trade protectionism in Japan, and in the USA over the imbalance of bilateral trade.

In May 1986 an agreement was reached on a redistribution of seats in the House of Representatives between urban and rural constituencies, reducing the maximum ratio of discrepancy in constituency size to less than 3:1, the limit that the Supreme Court had stipulated as permissible. Following this agreement, the lower house was dissolved in June, enabling the holding of a general election to coincide with the triennial election for one-half of the seats in the House of Councillors. Polling for both houses of the Diet took place in July and produced decisive victories for the LDP. In the election for the House of Representatives, the LDP received 49.4% of the votes, its highest level of electoral support since 1963, and won a record 304 of the 512 seats. The LDP, therefore, was able to dispense with its coalition partner, the NLC (which disbanded in August and rejoined the LDP).

Nakasone's second term as President of the LDP (a position that had to be held in order to be Prime Minister) ended in October 1986, and party rules prohibited a third term. However, the leaders of the five main LDP factions agreed to change the party's rules to permit a one-year extension of a President's term, with the approval of two-thirds of the LDP members of the Diet. Nakasone was confirmed as party President in September.

In September 1983, meanwhile, Masashi Ishibashi became Chairman of the JSP and initiated a shift from the party's traditional left-wing policies towards a position closer to the centre of the political spectrum. The JSP's moderation resulted in a slight increase in support at the 1983 general election, and a policy of further moderation was implemented, which brought about the end of the domination of the Marxist-orientated Shakaishugi Kyokai (Socialist Association) within the party. At the 1986 election, however, the JSP's less extreme image failed to attract voters, and the party lost 26 of the 112 seats won in 1983. This disastrous result brought about Ishibashi's resignation and the appointment, in September 1986, of Takako Doi, the first woman to lead a major Japanese political party.

Following the 1986 elections, Nakasone's priorities were to oversee an untroubled transfer to private-sector control of the Japanese National Railways (JNR), to alleviate economic tensions with the USA, while maintaining a high level of Japanese exports in spite of the sharp rise in the value of Japan's currency, and to reform both the education and tax systems. The JNR was successfully reorganized in April 1987, but Nakasone failed to make significant progress in the three remaining areas. Educational reforms remained under discussion by a specially established council, while proposed reforms to the tax system were withdrawn in April 1987, after the LDP suffered a serious defeat in the unified local elections. In spite of Nakasone's efforts, economic tensions with the USA continued, as the current-account surplus on Japan's balance of payments maintained its growth.

Meanwhile, Tanaka's illness led to the disintegration of his faction. In July 1987 Noboru Takeshita (who had been appointed Secretary-General of the LDP after the 1986 election) announced the formation of a major new grouping within the LDP, the Takeshita faction. He gained the support of 113 other members of the Tanaka faction, while around 20 Tanaka faction members remained with Susumu Nikaido, a former Vice-President of the LDP and the second most powerful man in the Tanaka faction. In October Nakasone nominated Takeshita as his successor.

The Takeshita Administration

On 6 November 1987 the Diet was convened and Takeshita was formally elected as Prime Minister. In the new Cabinet, Takeshita carefully maintained a balance among the five major factions of the LDP. He claimed that he would work to continue Nakasone's domestic and foreign policies (seeking the former Prime Minister's advice on foreign affairs), with particular emphasis on correcting the external trade imbalance, further liberalizing the financial market, reforming the tax system, land policy and the education system.

In contrast to its failure to stem the rapid rise in land prices, the Takeshita administration initially achieved steady progress in easing friction with overseas trading partners. The restrictions on imports in the agricultural sector were abolished, while US companies were permitted greater access to the Japanese domestic construction market. Trade tensions worsened, however, during Takeshita's tenure of office. The primary cause of concern was the continuing trade imbalance between Japan and the USA (the US trade deficit with Japan still accounted for about one-third of the USA's total world trade deficit). In May 1989 the situation deteriorated when the USA named Japan, together with India and Brazil, as unfair trading partners.

The implementation of a programme of tax reform, which Nakasone had failed to achieve, was one of the most important issues confronting Takeshita's Government. In June 1988 the LDP's tax deliberation council proposed the introduction of a new indirect tax (a general consumption tax or a form of value-added tax), which was to be levied at a 3% rate. This proposal encountered widespread disapproval, however, both from the general public and from the opposition parties.

Takeshita and the LDP suffered a serious political reversal in June 1988, when several leading figures in the party, including Nakasone, Shintaro Abe, Kiichi Miyazawa and Takeshita himself, were alleged to have been indirectly involved in share-trading irregularities with the Recruit Cosmos Company. Although these politicians strenuously denied any knowledge of, or involvement in, such transactions, the Prime Minister expressed his concern that these allegations would alienate public opinion from the proposals for tax reform and hinder their progress towards approval by the Diet. As the situation regarding the Recruit affair became increasingly serious, the opposition demanded the resignation of the alleged participants and the commissioning of a full parliamentary investigation into the alleged share transactions. In November, in exchange for the establishment of a committee to investigate the affair, the House of Representatives approved the tax reform measures (which constituted the most wide-ranging revision of the tax system for 40 years); they were approved by the House of Councillors in the following month.

In late 1988 and early 1989 three ministers, including the Deputy Prime Minister and Finance Minister, Kiichi Miyazawa, were forced to resign from their posts, owing to their alleged involvement in the Recruit affair. Hiromasa Ezoe, who founded the Recruit group, was said to have given large amounts of shares and money, totalling some 1,300m. yen, to many leading politicians and bureaucrats, in an attempt to buy influence and to help to expand his business empire.

In late April 1989 Takeshita suddenly resigned from his post. There were several factors leading to his decision: Takeshita, personally, was found to have received political contributions worth more than 150m. yen from the Recruit group in the form of pre-listed shares and money; the Takeshita Cabinet, whose position was severely damaged by the shares scandal and public outrage over the introduction of the 3% consumption tax in April, saw its public approval rating fall to less than 10%; and, finally, there was a growing consensus among LDP officials that Takeshita's continued leadership would adversely affect the party's prospects in the elections to the House of Councillors at the end of July. Hard-pressed to find a candidate who was not only willing to accept the position but also suitable (three of the leading contenders, Michio Watanabe, Kiichi Miyazawa and Shintaro Abe, were, temporarily at least, out of the question, since they had all received pre-listed Recruit shares), Takeshita finally nominated Sosuke Uno, the incumbent Minister of Foreign Affairs. Uno was elected Prime Minister at a Diet session on 2 June, becoming the first Japanese Prime Minister, since the LDP was founded in 1955, not to command his own political faction.

At the end of May 1989, following an eight-month investigation, public prosecutors indicted 13 people (eight on charges of offering bribes, and five for allegedly accepting them). Two of those indicted were politicians: Takao Fujinami, an LDP member belonging to the Nakasone faction, and Katsuya Ikeda, a former Deputy Secretary-General of the Komeito. At the same time, under heavy criticism from members of the opposition as well as from his own party, Nakasone resigned from his faction and from the LDP, assuming complete moral responsibility for the Recruit affair, since it had occurred during his administration. However, on the same day as his resignation, Nakasone announced that he would continue to undertake his political activities and that he would not resign from his seat in the Diet.

The Showa era came to an end when, after a long illness, Emperor Hirohito, who had reigned since 1926 (and who was, thus, the longest reigning monarch in Japan's history), died in January 1989. He was succeeded by his son, Akihito, and the Government announced that the new era was to be known as the Heisei ('achievement of universal peace') era.

The Uno Administration

Within days of Uno's appointment in June 1989, the LDP was confronted with further scandal, when a Japanese magazine published allegations that Uno had paid a geisha girl for a five-month sexual affair in 1985/86. In response to these allegations (on which Uno refused to comment), there were demands for the immediate resignation of the new Prime Minister from outraged women's groups and from various members of the opposition. The LDP's fears of waning support appeared to be confirmed in July 1989, when the ruling party lost its majority in the House of Councillors for the first time in history. At elections for one-half of the seats in the upper house, the LDP obtained only 27% of the total votes, while the JSP received 35%. The leader of the JSP, Takako Doi, attracted widespread support during the election campaign. Women voters, expressing disgust at the corruption in male-dominated politics, were attracted by the image of an intellectual female political leader. Consequently, Uno offered to resign as soon as the LDP had decided on a suitable successor, assuming total responsibility for his party's defeat. On 8 August the LDP chose the relatively unknown Toshiki Kaifu, a former Minister of Education and a member of the small faction led by Toshio Komoto, to replace Uno as the party's President and as the new Prime Minister. Although the House of Councillors' ballot rejected Kaifu as the new Prime Minister in favour of Takako Doi, the decision of the lower house was adopted (in accordance with constitutional procedures). This was the first time in 41 years that the two houses of the Diet had disagreed over the choice of Prime Minister.

The Kaifu Administration

At the end of August 1989 the LDP suffered another reversal when the Chief Cabinet Secretary was forced to resign, owing to a sex scandal. Nevertheless, Kaifu swiftly gained the approval of the electorate, owing to his untainted political record and his promise to revise the consumption tax. In October Kaifu was re-elected unopposed as President of the LDP for a two-year term.

Meanwhile, in August 1989 the JSP, in an apparent move to broaden its base of support, unanimously approved a change of policy that would commit it to retaining the Japanese-US bilateral security treaty (see below) and a free-market economy in the event of the establishment of a coalition government.

At the election for the House of Representatives held in February 1990 the LDP was returned to power with a large measure of support. The LDP received 46.1% of the votes cast and secured 275 of the 512 seats in the lower house. The JSP made substantial gains, winning 136 seats. The election results were significant in demonstrating not only the willingness of the electorate to forgive past indiscretions (many major politicians who had been implicated in recent scandals were returned to the lower house, including former Prime Ministers Nakasone and Uno) but also the possibility of a future polarization of voters, with the role of the smaller parties becoming increasingly insignificant.

In May 1990 Prime Minister Kaifu announced his commitment to the implementation of the electoral reforms that had been proposed in April. The proposals, for the House of Representatives, included a plan to replace the present multi-seat constituencies with a combination of single seats and proportional representation. Although the proposals were presented as an attempt to counter electoral corruption, LDP members expressed fears that the changes would invest more power in party committees responsible for nominating candidates, and would therefore increase the scope for bribery.

Kaifu's domestic and international standing altered significantly following Iraq's invasion of Kuwait in August 1990, and the subsequent outbreak of hostilities in the Persian (Arabian) Gulf region. In September Japan announced a US $4,000m.-contribution to the international effort to force an unconditional Iraqi withdrawal from Kuwait. Controversial legislation, which provided for the dispatch to the Gulf area of some 2,000 non-combatant personnel, encountered severe opposition and provoked widespread discussion on the constitutional legitimacy of the deployment of Japanese personnel (in any capacity) in such a conflict. The proposals were withdrawn in November. Nevertheless, Kaifu was reported to be considering resignation until, later that month, the LDP won a by-election to the upper house. In January 1991, following repeated demands by the USA for a greater financial commitment (and a swifter disbursement of funds already pledged), the Kaifu Government announced plans to increase its contribution by $9,000m. and to provide aircraft for the transport of refugees in the Gulf region. Opposition to the proposal within Japan was again vociferous.

Meanwhile, a developing power struggle within the LDP had also weakened Kaifu's political authority. Former Prime Minister Takeshita took advantage of diplomatic and political errors committed by Shin Kanemaru, the leader of the LDP faction that supported Kaifu, to ease his way back into a position of prominence in the LDP. In order to bolster his preparations for a return to the political forum, Takeshita also promoted Nakasone's return to the party, together with that of Kiichi Miyazawa. In June 1991 Kaifu broached the controversial issue of political reform by requesting that the LDP endorse a series of electoral reform bills. This initiative, following a year's deliberation of the issue within the party, was viewed as an attempt not only to regain public confidence but also to obstruct the efforts of tainted party leaders to return to office when Kaifu's LDP presidential term expired in October. Later that month, however, the Diet rejected Kaifu's proposals, leaving the way open for Kiichi Miyazawa.

Having successfully prevented the Government from dispatching Self Defence Force (SDF) personnel overseas during the Gulf War, the JSP was unable to sustain popular support for its strict pacifist stance. Although the party, in a bid to attract wider support, had changed the English rendering of its name to the Social Democratic Party of Japan (SDPJ) in February 1991, it suffered defeat on an unprecedented scale in the unified local elections in April. The set-back forced the party to discard the abolition of the consumption tax, and to approve the LDP's proposal to revise the tax in the following month. In July the SDPJ leader, Takako Doi, resigned and the party selected a new leader, Makoto Tanabe, from its conservative wing.

The Miyazawa Administration

At the election for the presidency of the LDP in October 1991 Miyazawa (with the support of Shin Kanemaru, a former Deputy Prime Minister) defeated Michio Watanabe and Hiroshi Mitsuzuka. Miyazawa attempted to win control of the ruling party by strengthening his long-established links with Takeshita, Watanabe and Mitsuzuka on the one hand, and by establishing new ties with Kanemaru on the other. Watanabe was appointed to the posts of Deputy Prime Minister and Minister of Foreign Affairs. However, a bitter feud developed between Kanemaru and Takeshita over the allocation of posts within the party executive and the Cabinet. The new Cabinet comprised the same proportion of the LDP's four major factions as its predecessor, but with the Takeshita faction obtaining more of the senior portfolios than it had previously held.

The new Prime Minister attempted to enact controversial legislation to allow SDF troops to serve abroad on UN peace-keeping missions. Realizing the need to gain Kanemaru's full support, Miyazawa repeatedly requested that he assume the post of LDP Vice-President. In January 1992 Kanemaru at last accepted the offer, together with Miyazawa's pledge of loyalty. Subsequently, the peace-keeping legislation was approved, with the support of the Komeito and the DSP, on the condition that SDF personnel join only non-military operations. The legislation was enacted in June 1992, in spite of opposition from the SDPJ.

In July 1992 the LDP recovered from a period of apparent unpopularity to achieve unexpected gains in elections to the upper house of the Diet. Although the LDP failed to recapture the majority that it had lost in 1989, the results provided the Miyazawa Government with significant encouragement. The dominant Takeshita faction of the LDP, however, was beset by crisis when, following further investigations into the activities of the Sagawa Kyubin transport company, it emerged that the latter's former president, Hiroyasu Watanabe, had made an unreported 500m.-yen donation to Shin Kanemaru in February 1990. In late August 1992 Kanemaru was forced to resign as LDP Vice-President, offering also to resign as head of the Takeshita faction, since he faced the charge of violating the Political Funds Control Law. However, he avoided trial by issuing a written statement to the authorities, in which he admitted the charge. The Tokyo Summary Court responded with a fine of 200,000 yen, prompting a public outcry over the failure to prosecute this case more resolutely. Kanemaru returned to his duties as head of the Takeshita faction for two weeks, only to yield to mounting pressure and resign from the Diet in mid-October. Two days prior to the opening of an extraordinary Diet session in late October, Ichiro Ozawa, a protégé of Kanemaru, announced the split of his group of 36 lower house members from the Takeshita faction, following his failure to obtain the chairmanship. Further changes in the political map occurred with the formation in early November of the Sirius Group of 27 reform-minded Diet members belonging to the SDPJ, Social Democratic Federation and RENGO, the trade union organization. Other such party and cross-party groups increased in prominence and number, reflecting the growing demand for political and electoral reform. As the trials connected to the Sagawa Kyubin scandal continued, seven leading LDP politicians, including Kanemaru, were accused of dealings with an extreme rightist group and with an organized crime gang in 1987, during Takeshita's ascent to the leadership of the party.

In a bid to restore confidence in the embattled Government, Miyazawa reshuffled the Cabinet in December 1992. Changes included the appointment of Hajime Funada as Director-General of the Economic Planning Agency, at 39 the youngest-ever member of a post-war cabinet. Seiroku Kajiyama, Ozawa's rival, became the LDP's Secretary-General. Political and electoral reforms were postponed until after the completion of the budget in April 1993.

With the arrest of Kanemaru and his secretary, Masahisa Haibara, in March 1993, on suspicion of evading the payment of 1,040m. yen in tax, groups seeking political reform within the Diet gained fresh momentum. Each of the established parties, apart from the Communists, assembled proposals for changes to the electoral system. While the LDP's preference was for a single-member constituency system, the opposition parties produced a variety of suggestions, which incorporated elements of proportional representation. Within the ruling party itself, differences between younger Diet members and their seniors emerged, as the former, led by Ozawa and his ally, Tsutomu Hata, encouraged an agreement with the opposition parties. A cross-party consensus was needed for any legislation to pass through the upper house, where the LDP was in the minority.

In June 1993 the LDP confirmed that it would not compromise its proposal to meet the demands of the opposition, thus effectively abandoning the reforms. The lower house adopted a no-confidence motion against the Miyazawa Government by 255 to 220 votes; 39 LDP members, including 34 members of the Hata-Ozawa group, voted against their party, while 16 other LDP politicians did not vote. The Hata-Ozawa group, consisting of 44 members of the LDP, immediately formed a new party called the Japan Renewal Party (JRP, Shinseito), in order to contest the forthcoming general election, called for July. Another group of 10 LDP Diet members also broke away to form the New Party (Shinto/Sakigake).

The election for the House of Representatives, held on 18 July 1993, marked the end of uninterrupted LDP rule. Apart from the record low turn-out of 67%, voting patterns changed surprisingly little. The LDP won 223 of the 511 seats, slightly more than it had held immediately prior to the election, but still well short of an overall majority. The SDPJ fared particularly badly, its number of seats being almost halved to 70. Of the parties formed by ex-LDP members, the JRP and New Party Sakigake won 55 and 13 seats respectively, most of which had been occupied by LDP members. The performance of the other recently established party, the Japan New Party (JNP) led by Morihiro Hosokawa, which fielded no incumbent candidates, was impressive, its 35 seats being secured mainly in urban constituencies and in the same marginal seats where new political parties had enjoyed success in the past. The remaining parties managed to maintain their strength in the lower house: the Komeito took 51 seats, the JCP 15, the DSP 15 and the Social Democratic Federation/United Social Democratic Party (USDP) four, while candidates with no party affiliations won 30 seats. Nevertheless, the LDP remained a potent electoral force.

THE END OF UNINTERRUPTED LDP RULE AND ESTABLISHMENT OF A COALITION GOVERNMENT

Following the election of July 1993, therefore, Prime Minister Miyazawa resigned, since there was no prospect of his party regaining its majority in the lower house. Amidst widespread expectations that the LDP would remain in power, either as a minority government, or as part of a coalition government, the other non-communist parties formed an alliance to oust the LDP. Thus, Morihiro Hosokawa, possessor of an illustrious aristocratic lineage and leader of the JNP, became Japan's first non-LDP Prime Minister for 38 years, defeating the new LDP President, Yohei Kono, by 262 votes to 224. Hosokawa formed a coalition Government, which included representatives of all the coalition partners. However, there was some dissatisfaction within the SDPJ concerning the distribution of key government posts. In spite of being the largest coalition partner, the SDPJ secured relatively few important positions, although its Chairman, Sadao Yamahana, was appointed Minister Responsible for Political Reform; the party was also allocated the Home Affairs, Construction and Transport portfolios. Tsutomu Hata, one of the JRP's leaders, became Deputy Prime Minister and Minister of Foreign Affairs, while his party also took responsibility for the crucial Ministries of Finance, and of International Trade and Industry. The sense of irritation within the SDPJ was exacerbated by its failure to obtain the position of Chief Cabinet Secretary, or even one of the two Deputy Chief Cabinet Secretary posts, leaving the party with no influence over the direction and co-ordination of policy. The former SDPJ Chairman, Takako Doi, who might have articulated such disaffection, was silenced by her reluctant acceptance of the post of Speaker of the lower house. Yamahana was replaced as Chairman of the SDPJ by Tomiichi Murayama in September 1993.

One of the Hosokawa Government's outstanding achievements was the approval of legislation on political reform, which was surrounded by much controversy until its final passage through the Diet in late January 1994. Under the new electoral system, Japan's 511 medium-sized, multi-member constituencies in the House of Representatives were to be replaced by 500 seats, 300 of which were to be filled from single-member constituencies in a 'first-past-the-post' contest, and the remaining 200 in 11 regional blocks by proportional representation. New funding rules allowed politicians to receive a maximum of 500,000 yen per year from each company wishing to donate, for the next five years, after which time payments to individual politicians from businesses would be illegal. Under the reforms, a 40% increase in state funding for political parties was also promised. The legislation was approved despite strong opposition from within the coalition, owing to a late agreement between Hosokawa and the LDP President. Their compromise thus met LDP demands for an increased number of single-member constituencies, for 11 regional rather than one national proportional representation list, and for no immediate termination of corporate donations to individual politicians.

Encouraged by record public approval ratings, Hosokawa promised further, wide-ranging reform. Economic strategy was revised by the Economic Reform Research Council, under the chairmanship of Gaishi Hiraiwa. The co-operation of Hiraiwa, who was Chairman of Japan's leading business association, KEIDANREN, reinforced the impression of change in the relationship between business and politics. This had already been suggested by KEIDANREN's decision in mid-1993 to abandon its role as a conduit for funds to the LDP, although its members remained free to contribute at their own discretion. These developments reflected trends in trade unions, which were disengaging themselves from exclusive relationships with their former political partners.

Although the Socialists were recognized as a potential source of instability within the coalition, the breakdown of the relationship between Hosokawa and his close ally, Masayoshi Takemura, leader of New Party Sakigake, was unexpected. The apparent cause of the rift was the sudden announcement by the Prime Minister in February 1994 of his intention to introduce a 6,000,000m.-yen reduction in income and residential taxes, while planning to establish a national welfare tax of 7% in 1997. This initiative was taken without wide consultation of the coalition partners. The rise in indirect taxation, from its current level of 3%, was strongly opposed not only by the SDPJ, which threatened to leave the coalition, but also by Chief Cabinet Secretary Takemura, who made public his view that the Prime Minister should reconsider this policy. Hosokawa, therefore, was forced to withdraw the proposed tax increases.

Perceptions of relationships within the coalition altered. After Hosokawa tried to reorganize the Cabinet in late February 1994, he appeared to be drawing closer to Ichiro Ozawa and the JRP-

Komeito grouping within the coalition, and away from Takemura's New Party Sakigake group, with which Hosokawa had previously suggested he wanted the JNP to merge. Hosokawa was understood to be seeking to remove Takemura as Chief Cabinet Secretary. In early March, however, Hosokawa was forced to abandon his planned reshuffle amidst mounting opposition within the coalition.

The 1994/95 budget was agreed by the Cabinet in mid-February 1994. The opposition, however, prevented its being approved by directing political discussion to details of a 100m.-yen loan received by Hosokawa, during his time as Governor of Kumamoto, from the scandal-tainted Sagawa Kyubin distribution company. This, together with speculation surrounding the Prime Minister's former share dealings, fuelled intense discussion in the Diet. In June Hosokawa was forced to give sworn testimony in the Diet concerning his financial activities.

With the coalition becoming increasingly fragile, Hosokawa resigned on 8 April 1994 amid controversy over his financial dealings. During the subsequent negotiations between the coalition partners regarding the succession to the premiership, the differences between the SDPJ and the JRP over tax policy and the North Korean nuclear issue (see below) became apparent. Tsutomu Hata of the JRP and Michio Watanabe were both potential candidates for the premiership. Watanabe, however, eventually decided against leaving the LDP along with some of his supporters, and on 25 April Tsutomu Hata was chosen as Prime Minister. The SDPJ was offended by the creation in the lower house of a parliamentary organization, Kaishin (Reform), comprising the JRP, JNP, DSP and two smaller political groupings (subsequently joined by the Komeito), which had facilitated Hata's appointment. Hata was therefore obliged to form a minority Government, without the SDPJ and New Party Sakigake, which pledged support only until the passage of all budget legislation through the Diet in mid-June.

This political realignment encouraged Satsuki Eda in May 1994 to dissolve the USDP and merge the party with the JNP. Within weeks, however, the JNP itself faced an uncertain future. Keigo Ouchi was replaced as leader of the DSP by Takashi Yonezawa in early June. Largely paralysed by the Government's minority status, Hata resigned as Prime Minister on 25 June; a general election was not called. In a surprise development, the SDPJ united with the LDP and New Party Sakigake to secure the election of Tomiichi Murayama, the SDPJ Chairman, as Prime Minister on 29 June, the first Socialist Prime Minister for 47 years. In the House of Representatives' ballot Murayama defeated Toshiki Kaifu, the former LDP Prime Minister, who had left the LDP to join Ozawa, by 261 to 214 votes. Many LDP members opposed Murayama's election, and even within the SDPJ there was disapproval, especially following Murayama's statement that the unarmed neutrality policy of the SDPJ was outdated and that he no longer considered the SDF to be unconstitutional.

In August 1994 leaders of the major Japanese opposition parties, at the instigation of former Prime Ministers Hata, Hosokawa and Kaifu, agreed to establish a consultative body, as a first step towards founding a joint party to counter the LDP-New Party Sakigake-SDPJ coalition Government. In the same month Murayama appointed Sohei Miyashita, the former Director-General of the Defence Agency, as the Director-General of the Environment Agency in place of Shin Sakurai, who had angered China and the Republic of Korea (South Korea) by denying Japan's 'war of aggression' in the Pacific during the Second World War. Sakurai was the second Cabinet member to resign within three months over controversial remarks about Japan's war record; Shigeto Nagano, the Minister of Justice, had been forced to resign in May after describing the 1937 Nanjing massacre (in which more than 300,000 Chinese citizens were killed by Japanese soldiers) as a 'fabrication'.

The longer the SDPJ remained in government, the more difficult it became for it to appeal to traditional supporters. In late September 1994 Murayama announced an increase in the consumption tax, from 3% to 5%, to take effect in April 1997. Although, in mitigation, a two-tiered reduction in income and residential taxes of some 5,500,000m. yen was to be introduced, it seemed ironic that a Socialist Prime Minister should announce this increase in tax, the introduction of which his party had so long opposed. The SDPJ was widely considered the party most likely to threaten the coalition Government's stability, but rather than its pacifist wing, it was those on the right of the SDPJ who began to endanger the party. In October 1994 Murayama was forced to question another of his party's former policy commitments, when he suggested that the construction of further nuclear power plants might be unavoidable. The following day, to Murayama's apparent consternation, the Secretary-General of the SDPJ, Wataru Kubo, echoed those proposing the dissolution of the party and the formation of a new party of social democrats and liberals. Kubo planned to co-operate with Sadao Yamahana, who was poised to lead his own faction out of the SDPJ.

Meanwhile, among the opposition parties, although the Shinseito, the Komeito, DSP, JNP and LDP splinter groups had largely agreed by September 1994 that they should merge, misgivings remained among some of the smaller partners that the Shinseito and the Komeito would dominate a new opposition party. Difficulties were also experienced in negotiating the inclusion of the Komeito's party machinery in these plans, the party being the only one of the prospective partners to have a national organization. In late October significant differences on foreign policy emerged between the Komeito and the Shinseito. Ozawa urged Japan to become an 'ordinary country', whereas the Komeito reaffirmed its position that Japan should not become involved in peace-keeping operations or any collective security system that required the use of force.

Driven by electoral realities, however, the Komeito opted to divide into two groups. The first comprised all 52 Komeito lower-house members and a majority (24) of its upper-house members. This group was to participate in the anticipated merger of the opposition parties. The second group, which included those not standing for election to the House of Councillors in 1995 and all Komeito local politicians (approximately 3,000), retained the party's machinery, including its newspaper and its headquarters. This arrangement was approved at the party's extraordinary congress in early December 1994. A few days later the New Frontier Party (Shinshinto), comprising all the major opposition parties with the exception of the JCP, held its inaugural congress, during which the former LDP Prime Minister, Toshiki Kaifu, was elected leader, defeating Hata and Yonezawa. Ozawa was elected unopposed as the new party's Secretary-General.

In January 1995 Yamahana began negotiations in preparation for the creation of a new party, claiming the support of as many as 30 SDPJ Diet members. Unable to ignore the momentum that these developments were creating, the SDPJ leadership sought a delay and promised to introduce measures to form a new party at a party congress scheduled to be held in February. Yamahana, however, made it clear that a split in the party was imminent.

On 18 January 1995 the country suffered its worst disaster since the Second World War when a massive earthquake struck the Kobe region, killing more than 6,000 people. The scale of this calamity effectively postponed the dissolution of the SDPJ; politics was thrown into hiatus for several weeks, as public and media attention focused on the efforts to bring relief to the disaster area. In the aftermath of the earthquake, the Government was severely criticized (and subsequently acknowledged responsibility) for the poor co-ordination of the relief operation.

In mid-February 1995 it emerged that two credit unions had been involved in questionable financial dealings with two senior members of the New Frontier Party (Shinshinto) and a number of senior officials of the Ministry of Finance. At the same time, the coalition became involved in a dispute with the Ministry of Finance over attempts by the Cabinet to streamline special public corporations. At issue was a proposal, opposed by the Ministry of Finance, to merge the Japan Development Bank with the Export-Import Bank. In the following month the Minister of Finance, Masayoshi Takemura, was forced to retract the proposal, promising instead to merge the Export-Import Bank with the Overseas Co-operation Fund.

In March 1995 12 people died and more than 5,000 were injured, when a poisonous gas, sarin, was released into the Tokyo underground railway system. The religious sect, Aum Shinrikyo, which was believed to be responsible for a similar incident in Matsumoto in 1994, was widely suspected of launching the attack, although the sect's leader, Shoko Asahara, initially denied that Aum Shinrikyo had been involved. Following a further gas attack in Yokohama in April 1995, a number of sect members were detained by the authorities, and in June Asahara was indicted on a charge of murder. The sect was declared bankrupt in March 1996 and the trial of Asahara opened in the following month. In September Asahara and two other members of the sect were instructed to pay some US $7.3m. in compensation to victims of the Tokyo incident. Attempts by the Ministry of Justice to outlaw the sect, on the grounds that it had engaged in subversive activities, were unsuccessful; however, the sect was denied legal status as a religious organization.

The gas attack in Tokyo, and the sensation generated in the media by the sporadic acts of terrorism that followed, kept party politics at a low ebb. This was reflected in the unified local elections held in April 1995. Despite the election of independent candidates, Yukio Aoshima and Nokku Yokoyama (both former comedians), in the Tokyo and Osaka gubernatorial elections, it was hard to detect anything but apathy in the record low turn-out registered in the vote for members of Japan's local assemblies and local chief executives. In accordance with his campaign pledges, Aoshima took a controversial decision in May to halt the Tokyo Exposition, despite the 98,200m.-yen loss that was expected to ensue. In the same month the SDPJ established a working group to study policy in preparation for the creation of a new party. Yamahana had resigned from the SDPJ earlier in the month.

Following considerable disagreement in the Diet, during which New Party Sakigake threatened to withdraw from the coalition if an apology for Japanese actions in the war were not adopted, a resolution was passed in June 1995 to commemorate the 50th anniversary of the end of the Second World War. The New Frontier Party (Shinshinto) boycotted the vote, while the resolution was openly criticized by a group of 160 LDP Diet members, led by Seisuke Okuno, who objected to the labelling of Japan as an aggressor, preferring to characterize the war as one of liberation for the peoples of Asia. The resolution was also widely criticized as insufficiently explicit by nations whose citizens had been prisoners of the Japanese army during the Second World War.

In late June 1995 a total of some 29,900m. yen was allocated to political organizations in the country's first distribution of subsidies to political parties. In elections to the House of Councillors, held in July, the coalition parties suffered as the electorate registered a post-war record low turn-out of 44.5%. With one-half of the 252 seats being contested, the LDP won 49 seats, the SDPJ 16 and New Party Sakigake three. The New Frontier Party (Shinshinto), benefiting from the strong organizational support of the Soka Gakkai religious organization, was aided by the low turn-out and was able to win some 40 seats. The defeat prompted some LDP Diet members to question Kono's leadership. In August Murayama undertook a major reorganization of the Cabinet. Isamu Miyazaki, an independent civil servant, was appointed Director-General of the Economic Planning Agency. Yohei Kono announced that he would not seek re-election in September to the presidency of the LDP, and was succeeded by Ryutaro Hashimoto, the Minister of International Trade and Industry.

In October 1995 the Minister of Justice was obliged to resign, following allegations that he had accepted an unreported loan of 200m. yen from a Buddhist group. The Director-General of the Management and Co-ordination Agency resigned in November, owing to controversy arising from his suggestion that Japanese colonial rule over Korea had been of some benefit.

Murayama announced his resignation as Prime Minister on 5 January 1996. The three-party coalition continued to govern under the premiership of the LDP President, Ryutaro Hashimoto, whose experience as Minister of Finance and of International Trade and Industry was well respected by Japanese business leaders and US officials. Hashimoto's first task as Prime Minister was to gain Diet approval for the 1996 draft budget, which included expenditure of 685,000m. yen for the liquidation of seven insolvent housing loan companies (*jusen*). The use of public funds for the settlement of the *jusen* issue aroused considerable opposition. In March the New Frontier Party (Shinshinto) organized a parliamentary 'sit-in' to obstruct the Budget Committee meetings of the House of Representatives. Budget deliberations resumed when the Government agreed to the party's demand for an inquiry regarding the alleged acceptance by the LDP Secretary-General, Koichi Kato, of illegal political donations. No action was taken against Kato as a result of this inquiry, and the budget proposals were approved eventually in May with little revision; however, the liquidation of the *jusen* was postponed, pending the introduction of a tighter financial regulatory system.

In early 1996 the Hashimoto administration conducted an investigation into the Ministry of Health and Welfare's poor management of HIV-tainted blood products, which had infected about 1,800 haemophiliac patients with the virus in the 1980s. Led by the Minister of Health and Welfare, Naoto Kan (a Sakigake member), the investigation revealed the irresponsible conduct of the Ministry's officials and of the medical experts involved in delaying the decision to recall tainted products. In August charges of professional negligence were brought against the former head of a government advisory body on AIDS, and in the following month several senior officials from the pharmaceutical company involved in the sale of the contaminated blood products were also arrested. Their trial opened in March 1997.

The opposition New Frontier Party (Shinshinto) elected Ichiro Ozawa as President in December 1995. The contest for the party leadership, however, aggravated the internal division between the Ozawa group and the Hata-Hosokawa group, inviting speculation about a possible dissolution of the party. While this did not occur, individual members expressed opposition to Ozawa's policies. Other parties also experienced internal problems. Some local organizations opposed the transformation of the SDPJ, under Murayama's chairmanship, into a moderate liberal party, and an independent New Socialist Party was formed by left-wing members of the SDPJ. In September 1996, as the end of her term as Speaker of the House of Representatives approached, Takako Doi agreed to resume the leadership of the SDPJ, replacing Murayama. In August Masayoshi Takemura resigned as leader of the New Party Sakigake, the smallest coalition partner, and in the same month the Secretary-General of the party, Yukio Hatoyama, also resigned. With Naoto Kan and several members of the SDPJ, he formed a new party in September, the Democratic Party of Japan (DPJ). Hatoyama felt that New Party Sakigake's electoral prospects had been damaged by the poor performance of its leader, Takemura, during the latter's tenure as Minister of Finance. In late August Shoichi Ide and Hiroyuki Sonoda became leader and Secretary-General, respectively, of New Party Sakigake.

THE ELECTION OF OCTOBER 1996

Such developments indicated that the parties and individual politicians were preparing for a general election, which Hashimoto duly declared would take place on 20 October 1996. The election was to be the first for the lower house to be held under the new electoral system of 300 single-seat constituencies and 200 proportional-representation seats. In an effort to strengthen the electoral bases in new single-member districts, the parties promoted co-operation even with formerly rival organizations. The LDP, for example, approached labour unions, and the New Frontier Party (Shinshinto) received some support from local labour union organizations that refused to support the new SDPJ. At the election the LDP won 239 of the 500 seats in the House of Representatives. The New Frontier Party (Shinshinto) secured 156 seats, the recently formed DPJ 52 seats, the JCP 26 seats and the SDPJ 15 seats. New Party Sakigake won only two seats. The low turn-out (59%) was regarded as an indication of widespread electoral disillusionment with all the political parties.

At an extraordinary session of the Diet, convened on 7 November 1996, Ryutaro Hashimoto was re-elected Prime Minister, winning a majority in the first ballot in both houses, with the support not only of his own LDP, but also members of the SDPJ and New Party Sakigake. Both the SDPJ, which attributed its loss of seats to its political relationship with the LDP, and New Party Sakigake entered into a policy agreement with the LDP, but decided to remain outside the Government. Thus, the new Cabinet, inaugurated later that day, consisted entirely of LDP members—the first single-party LDP administration since August 1993. The LDP also secured the co-operation of both the New Frontier Party (Shinshinto) and the DPJ, while maintaining the basic framework of the LDP-SDPJ-New Party Sakigake policy accord. Despite criticism from party members seeking yet closer co-operation with the New Frontier Party (Shinshinto), the LDP leadership emphasized the importance of maintaining the unofficial coalition agreement, which highlighted administrative reform as a priority.

The incoming Government was almost immediately beset by corruption scandals, when it was alleged that some 10 LDP members, among them the Ministers of Finance and of Health and Welfare, had received political donations from Junichi Izui, the owner of an oil company, who had been detained on charges of tax evasion. Although not illegal in themselves, the payments should have been disclosed, following the introduction of new legislation concerning political funds. In a further corruption scandal, a Deputy Minister of Health and Welfare was forced to resign after being accused of accepting bribes from the developer of a nursing home. He was subsequently arrested and, at the opening of his trial in March 1997, pleaded guilty to charges of bribery.

THE INTRODUCTION OF ADMINISTRATIVE AND ECONOMIC REFORM

A programme of comprehensive administrative and economic reforms was inaugurated by the new administration, with the establishment of special commissions, chaired by the Prime Minister, to examine a reorganization of ministries and to investigate ways of reducing government expenditure. These commissions proposed a reduction in the number of central ministries, an increase in the powers of the Prime Minister, the establishment of an effective crisis management system to be headed by the Prime Minister, and a review of the structure of public investment. In February 1997 the Government released details of a series of financial deregulation measures: among other reforms, government control over the financial sector was to be reduced and a new financial supervisory agency was to be established, to assume responsibility for some of the Ministry of Finance's regulatory duties. Social welfare reforms were also introduced, with revisions to the national health insurance system and the introduction of a nursing assistance insurance scheme. The Government's management of the nuclear programme was comprehensively reviewed, following two accidents, in December 1995 and March 1997, at plants managed by the Power Reactor and Nuclear Fuel Development Corporation, a public organization supervised by the Science and Technology Agency. Allegations that the corporation had failed to report a further 11 radiation leaks over the previous three years served to heighten public disquiet over Japan's nuclear research and development programme.

In elections to the Tokyo Metropolitan Assembly, held in June 1997, the LDP performed well; moreover, with the defection to the party of several New Frontier Party (Shinshinto) members, the LDP increased its number of seats in the House of Representatives to 247. Meanwhile, the New Frontier Party (Shinshinto) experienced serious internal difficulties as junior members began openly to criticize Ozawa, following the party's dismal performance in the 1996 general election. In January 1997, together with several supporters, Tsutomu Hata, a co-founder of the party, left the New Frontier Party (Shinshinto) to form a new party, the Sun Party (Taiyoto), and in June Hosokawa also relinquished his membership. The New Frontier Party (Shinshinto) sustained a defeat in the elections to the Tokyo Metropolitan Assembly, failing to win a single seat. In July, faced with the apparent decline in the party's influence, several New Frontier Party (Shinshinto) members formed a parliamentary faction with members of the DPJ and the Sun Party. The JCP continued to attract voters who were frustrated with the major parties, and became the second largest party in the Tokyo Metropolitan Assembly.

In September 1997 Hashimoto was re-elected unopposed to the presidency of the LDP. A wide-ranging cabinet reorganization was effected on 11 September, but the appointment of Koko Sato, who had been convicted on charges of bribery, as Director-General of the Management and Co-ordination Agency caused widespread anger, and he was forced to resign shortly afterwards. Meanwhile, the LDP regained its majority in the House of Representatives, following a series of defections by members of the New Frontier Party (Shinshinto). Following further defections from the party, the New Frontier Party (Shinshinto) was dissolved in December. Several new opposition parties were established by former members of the New Frontier Party (Shinshinto), including the Liberal Party (LP), led by Ichiro Ozawa. A subsequent realignment of six political parties, including the DPJ, in January 1998, resulted in the formation of an opposition grouping, Minyuren, that was to constitute the largest opposition force in the Diet. In March the members of Minyuren announced their integration into the DPJ, to form a single party, led by Naoto Kan. The new DPJ was formally established in late April, and became the second largest party in the House of Representatives; Naoto Kan and Tsutomu Hata were elected as President of the party and Secretary-General, respectively.

Meanwhile, in late 1997 a series of corruption scandals, involving substantial payments to corporate racketeers by leading financial institutions, had a severe impact on the Japanese economy. The crisis was exacerbated by an increase in the rate of the unpopular consumption tax, in April, from 3% to 5%, and a decrease in public expenditure (as part of the Government's fiscal reforms), which resulted in a significant weakening in consumer demand. The collapse of several prominent financial institutions in November, and the threat of further bankruptcies, deepened the economic crisis. The Government announced a series of measures designed to encourage economic growth, including a reduction in taxes and, in a major reversal of policy, proposed the use of public money to support the banking system. The credibility of the Ministry of Finance was, however, weakened in late January 1998, when two senior officials were arrested on suspicion of accepting bribes from banks. The Minister of Finance, Hiroshi Mitsuzuka, resigned, accepting full moral responsibility for the affair. He was replaced by Hikaru Matsunaga. The repercussions of the bribery scandal widened in early 1998, as other banks were implicated in the affair. The Bank of Japan, the central bank, subsequently began an internal investigation into its own practices, and in March the Governor resigned, as a result of further bribery allegations made against a senior bank official. He was replaced by Masaru Hayami. Trials of those implicated in the financial scandals continued in 1998–2000.

Attempts to stimulate the Japanese economy continued in early 1998. Several important financial deregulation measures, including those concerning the liberalization of the telecommunications market, retail outlets and foreign-exchange transactions, became effective on 1 April, designated as the Japanese equivalent of the 'Big Bang' liberalization process that had already been undertaken by the London and New York financial markets. Legislation for the reorganization of government ministries was approved by the Diet in June, whereby, from the year 2001, the number of central ministries and agencies was to be reduced from 22 to 13. Hashimoto, however, was not able to enforce effective measures to revive the Japanese economy. Restricted by his commitment to achieve a balanced budget, and confronted with the problem of non-performing loans, totalling some 77,000,000m. yen (subsequently revised upwards, to 87,500,000m. yen), that had been accumulated by Japanese financial institutions, Hashimoto was unable to introduce significant tax cuts, the measure that many analysts considered necessary for economic revival. The economy continued to stagnate during 1998, with declines recorded in consumer spending, the construction industry and manufacturing. Hashimoto was increasingly criticized for his lack of decisive action, and in June the SDPJ and New Party Sakigake left the governing coalition.

In preparation for the election for one-half of the seats in the House of Councillors, the Government issued proposals in late May 1998 for public spending totalling some 16,000,000m. yen. The plan, however, failed to revive Hashimoto's popularity and boost public confidence in his administration of the economy, and in the election to the upper house, held on 12 July, the LDP performed poorly, losing 17 of its 61 seats contested. The DPJ, by contrast, won 27 seats, bringing its total to 47, and the JCP more than doubled its representation, taking 15 seats. Electoral turnout was low, at some 58%. Hashimoto resigned as President of the LDP, assuming responsibility for the party's failure at the election, and was replaced by Keizo Obuchi, formerly Minister of Foreign Affairs, and leader of the largest faction in the LDP.

Despite the party's preference for a consensus candidate, the election for LDP President was unusually open, and two other candidates, Seiroku Kajiyama, whose economic policies were favoured by business leaders, and Junichiro Koizumi, who was popular with the electorate, also contested the ballot of party members in the Diet. All three candidates were unanimous, however, in emphasizing the need for permanent tax cuts and reform of the tax system. Obuchi was subsequently elected Prime Minister, on 30 July, at an extraordinary session of the Diet, despite the election by the House of Councillors of an opposition candidate, Naoto Kan, as their choice for Prime Minister. The decision of the lower house, nevertheless, prevailed.

Obuchi's Government, appointed on his election, and designated an 'economic reform' cabinet, comprised a large number of hereditary politicians. The appointment of Kiichi Miyazawa, a former Prime Minister, as Minister of Finance, was, however, welcomed by some observers, owing to his financial expertise. Doubts nevertheless remained about the new Government's commitment to economic reform, and Obuchi's administration failed to attract the support of the electorate, achieving extremely low approval ratings. In his inaugural policy address to the legislature, Obuchi announced that attempts to achieve a balanced budget were to be postponed, and proposed additional tax cuts, to the value of 7,000,000m. yen. His reluctance to commit the Government to the closure of failing banks and to a fundamental restructuring of the banking sector, however, led to a further weakening of confidence in the Japanese economy. Following weeks of negotiations, in October 1998 the Diet approved banking legislation which included provisions for the nationalization of failing banks, as demanded by the opposition. In November the Government presented a 24,000,000m.-yen programme aimed at revitalizing the country's economy, but ruled out a reduction in the consumption tax.

In November 1998 Komei merged with Shinto Heiwa (New Peace Party, founded in 1997) to form New Komeito, which thus became the second-largest opposition party. In the same month Fukushiro Nukaga, the Director-General of the Defence Agency, resigned from the Government to assume responsibility for a procurement scandal involving his agency. In mid-November the LDP and the LP reached a basic accord on the formation of a coalition, which would still remain short of a majority in the upper house. In early January 1999, following intense discussions, agreement was reached on coalition policies. Ozawa appeared to have won concessions on a number of proposals that the LP wanted to be submitted for consideration by the Diet in forthcoming sessions, including a reduction in the number of seats determined by proportional representation in the House of Representatives from 200 to 150 and provision for an expansion of Japan's participation in UN peace-keeping operations. The Cabinet was reshuffled to include the LP, with the number of ministers reduced from 20 to 18 (excluding the Prime Minister). Takeshi Noda of the LP, the only new member of the Cabinet, was appointed as Minister of Home Affairs. At the end of January the Government adopted an administrative reform plan, which aimed to reduce further the number of cabinet ministers and public servants and to establish an economic and fiscal policy committee. Draft legislation on the implementation of the plan was introduced to the Diet in April. In March Shozaburo Nakamura resigned as Minister of Justice following allegations of repeated abuse of power.

Local elections in April 1999 were largely unremarkable; 11 of the 12 governorships contested were won by the incumbents, all standing as independents. The 19-candidate gubernatorial election for Tokyo created by far the most interest. The convincing victory of Shintaro Ishihara, a nationalist writer and a former Minister of Transport under the LDP (although now unaffiliated), was regarded as an embarrassment for the ruling party, which had supported Yasushi Akashi, a former senior UN official. Ishihara immediately provoked controversy, making a series of inflammatory comments about the 1937 Nanjing massacre and criticizing the Chinese Government, which responded angrily, prompting the Japanese Government to distance itself publicly from the new Governor's remarks.

In June 1999 the Government voted to grant official legal status to the *de facto* national flag (*Hinomaru*) and anthem (*Kimigayo*), despite considerable opposition owing to their association with Japan's militaristic past. The necessary legislation was subsequently approved by the Diet, however, and became effective in mid-August. Meanwhile, in July New Komeito agreed to join the ruling LDP-LP coalition, giving the Government a new majority in the upper house and expanding its control in the lower house to more than 70% of the seats. Negotiations on policy initiatives proved difficult, however, and were still continuing in September, owing to differences over a number of contentious issues such as constitutional revision and New Komeito's opposition to a reduction in the number of seats in the lower house, as favoured by the LP. Obuchi was re-elected to the presidency of the LDP in September, defeating Koichi Kato, a former Secretary-General of the party, and Taku Yamasaki, a former policy chief. Naoto Kan, however, failed to retain the presidency of the DPJ, and was replaced by Yukio Hatoyama.

A new Cabinet was appointed in October 1999. The Minister of Finance, Kiichi Miyazawa, and the Director-General of the Economic Planning Agency retained their portfolios, while Michio Ochi was appointed Chairman of the Financial Reconstruction Commission. The LP and New Komeito each received one cabinet post. A basic accord on coalition policy included an agreement to seek a reduction in the number of seats in the House of Representatives, initially by 20 and subsequently by a further 30.

At the end of September 1999 a serious accident at a uranium-processing plant at Tokaimura, which raised levels of radiation to 15,000 times the normal level, severely undermined public confidence in the safety of Japan's nuclear industry. Furthermore, in November it was revealed that in recent inspections by government officials, 15 of 17 nuclear facilities had failed to meet the required health and safety standards. In December legislation was enacted that aimed to prevent accidents and to improve procedures for the management of any future incidents at nuclear power facilities. In February 2000 the Government announced that a total of 439 people had been exposed to radiation in the Tokaimura accident (compared with initial estimates of 69), and by April two workers from the plant had died from radiation exposure.

Trials continued in 1997–99 of members of Aum Shinrikyo, the cult believed to be responsible for the sarin gas attack on the Tokyo underground railway system in 1995. In September 1999 Masato Yokoyama, a leading member of the cult, became the first of those accused to receive the death sentence; at least five other cult members had been sentenced to death by mid-2000. In late 1999 the Diet enacted legislation aimed at curbing the activities of Aum Shinrikyo and, in an attempt to prevent any such restriction, the cult's leaders announced a suspension of all external activities, and acknowledged culpability for a number of crimes, including the gas attack. In January 2000 the cult announced that its name was to change to Aleph, and that it no longer considered Shoko Asahara, on trial for his part in the gas attack, to be its leader. In February, following a police raid on Aum Shinrikyo premises, it was revealed that a number of major companies and government agencies had placed orders for computer software with a firm believed to be a major source of revenue for Aum Shinrikyo; the Defence Agency subsequently announced that it was to abandon software purchased from the company for use by the SDF.

At the end of 1999 media reflections on the past 12 months reported that the popular mood was still pessimistic, despite the Economic Planning Agency's October statement that the economy was 'continuing a moderate improvement'. In fact, the country was back in recession, following a second consecutive quarter of negative growth from October to December. Consumer spending remained weak, overcapacity was still a problem, and unemployment continued to rise. The very fabric of Japan seemed to be fragmenting, as juvenile involvement in serious crimes rose, various police scandals were uncovered, rocket launches by Japan's Space Agency failed, railway tunnels collapsed and nuclear-power workers ladled radioactive material with buckets. There was persistent scepticism regarding the fate of the administrative reforms that had been adopted in July and about the Government's commitment to fiscal discipline. The new coalition, which was regularly threatened with withdrawal by Ozawa and the LP, did not inspire confidence in rational decision-making, and the promise made to Ozawa that priority would be given during the forthcoming Diet session to the bill to reduce the number of seats in the House of Representatives indicated difficulties to come.

Unpopular as the three-party alliance was, its significance became apparent as Obuchi was forced, in the interests of keeping his coalition together, to honour his promise to Ozawa of electoral reform. The coalition between the LP and the LDP had been underpinned by Obuchi's agreement to reduce the number of seats determined by proportional representation by 50. The New Komeito, a party that was heavily dependent on these seats and passionately opposed to the proposed legislation, was, in turn, promised a lesser reduction in proportional-representation seats, of only 20, to be followed by a reduction of 30 single-member seats at a later date. The nature of the divisions between the alliance parties meant that no compromise was possible with the principal opposition party, the DPJ, which, as the potential second party in a two-party system, was strongly in favour of a reduction in the number of proportional-representation seats from 200 to 150. However, the size of the coalition rendered a compromise unnecessary. In February 2000, without committee debate, and despite a boycott of the Diet by opposition parties, legislation was passed to reduce the number of proportional-representation seats in the House of Representatives from 500 to 480. The action strongly recalled an earlier era of 'snap votes'. The reinforced majority of the expanded coalition, and the Government's success in enacting the controversial legislation on security, the flag and the anthem, aroused old fears, and the opposition parties responded by reverting to the old tactics of boycott. Between 20 January and 9 February the Government ruled unopposed.

Meanwhile, in January 2000 multi-party commissions had been established in both houses to review the Constitution over a five-year period. The commissions were discussion bodies only and were not permitted to submit legislation to the Diet. However, their range of discussion was broad and included both the structure of the bicameral legislature itself and Article 9. The inauguration of the commissions was a significant development, which marked a new openness and a wide agreement on the need for debate. A poll taken as the commissions began their work suggested that 50.6% of people supported constitutional amendment, compared with 44.9% in 1997, while only 24.1% opposed any amendment, compared with 31.0% three years earlier.

The first ordinary session of the Diet in 2000 was different, owing not only to the boycott and the 'alternative Diet' established by the opposition, but also to the implementation of changes to Diet procedure, which ended the system of responses being given by bureaucrats on behalf of ministers. The end of the 100-year-old practice was expected to enliven Diet debate, and was regarded as a means of educating a new style of politician and facilitating the shift to a more politician-led pattern of policy-making. To that end, the cabinet reshuffle in October 1999 had brought the appointment of parliamentary vice-ministers with particular expertise in the areas covered by their ministries. Vice-ministerial appointments thus included senior people who might previously have been appointed to a cabinet position. Among these was Nishimura Shingo, of the LP, who was forced to resign from the Defence Agency after arguing that the Diet should debate the possibility of Japan acquiring nuclear weapons. The goal of politician-led governance was also behind the inauguration of a Prime Minister's question time. However, the new Diet standing committee, which became the forum for the 40-minute weekly debate between the Prime Minister and the leaders of the opposition parties, failed to win critical acclaim.

In addition to the strong parliamentary opposition, the Government was also undermined by a number of scandals during 2000. In February the Chairman of the Financial Reconstruction Commission, Michio Ochi, was forced to resign after he remarked that he would endeavour to ensure that bank inspections were lenient, and in that month a personal aide to the Prime Minister was accused of misappropriating shares, which he later traded. In March a number of incidents of police misconduct at a senior level resulted in disciplinary action being taken against the chief of the National Police Agency.

In February 2000, meanwhile, Japan's first female governor took office in Osaka, having won a by-election necessitated by the resignation of the previous incumbent after his indictment in December 1999 for sexually harassing a female member of his staff during his election campaign earlier in that year. In April 2000 a second female governor was voted into office in southern Japan. Japan's governors, and local government, played an increasingly prominent role, drawing attention to their desire for greater autonomy. Growing opposition to a number of government projects became evident in 2000; a project to construct a nuclear-power plant in Mie Prefecture was stopped, and in a referendum held in Tokushima, on Shikoku island, in January 90% of the electorate voted against a proposed dam across the Yoshino River (although senior ministers announced that the project would proceed, as local referendums are not legally binding). In February Governor Ishihara of Tokyo proposed that a tax be levied on major banks based on the size of their business rather than their profits; he was criticized by the central Government, which had previously been the only body able to introduce taxes. Ishihara attracted criticism from the foreign residents of Tokyo, and from neighbouring countries, in April, when, in an address to members of the SDF, he blamed foreign residents for a number of serious crimes and referred to them as *sangokujin*, a derogatory wartime term for people from Taiwan and Korea. He also angered the People's Republic of China by visiting Taiwan in November 1999, following an earthquake, and again in May 2000 for the inauguration of the latter's new President, Chen Shui-bian, despite having refused an invitation to visit China as part of celebrations to mark the anniversary of the establishment of relations between Beijing and Tokyo.

Friction within the ruling coalition continued throughout the early months of 2000, as the LP pushed for the consideration of elements of its policy programme, such as the upgrading of the Defence Agency to a ministry and the revision of education legislation, policies that were opposed by New Komeito. As elections to the House of Representatives drew inexorably closer, tension mounted over which coalition candidates would be endorsed. Finally, on 1 April Ozawa told Obuchi that he intended to withdraw the LP from the coalition. On the following day Obuchi suffered a stroke and went into a coma from which he never regained consciousness.

The handling of Obuchi's collapse, and the manner of the subsequent appointment of Secretary-General Yoshiro Mori, head of the third-largest faction in the LDP, as Prime Minister, led many to believe that nothing had changed in the conception of the nature of leadership since the end of the 1955 system. There was criticism of the secrecy surrounding Obuchi's hospitalization, which was kept from the public for more than 20 hours. There was criticism, too, that Mori was selected to succeed Obuchi as leader of the LDP, and consequently as Prime Minister, by a few politicians in the traditional smoke-filled rooms. The events in the first days following Obuchi's stroke exposed the lack of adequate legislation to prevent a political vacuum in such circumstances. (The Government subsequently clarified the law, designating a ranking of five ministers to assume the premiership, starting with the Chief Cabinet Secretary.)

Mori immediately announced his commitment to the economic and political reform initiatives of his predecessor, and in early April formed a coalition with New Komeito and a new party, Hoshuto (New Conservative Party, formed by 26 members of the LP on 3 April 2000). All ministers from the Obuchi administration were reappointed to their posts.

Obuchi's death was just one element of a dramatic change in the cast of leading political characters. During April 2000 both Seiroku Kajiyama, a former Chief Cabinet Secretary, and former Prime Minister Noboru Takeshita, hospitalized since April 1999 but still involved in behind-the-scenes manoeuvring, announced their retirement from politics and from the LDP; both men died shortly afterwards. Former Prime Minister Ryutaro Hashimoto was appointed head of the Takeshita faction of the LDP, which had been led by Obuchi prior to his stroke. The physical decline of the old guard, combined with party realignments and a need for expertise that the older politicians lacked, at a time when the demand for greater political input into, and accountability for, policy-making was growing, resulted in greater prominence for the younger generation of politicians. The generational rifts became embarrassingly public in 1993. By 2000, however, the 'policy-making new generation', *seisaku shinjinrui*, was appearing in live television debates, drafting private members' bills, demanding structural change within the parties and rejecting factional discipline.

Following his appointment as Prime Minister, Mori made a number of controversial public statements, expressing imperialist views. He was heavily criticized by the opposition and the media and was forced to issue apologies, although he did not retract his remarks. Shortly prior to the elections, newspaper surveys showed that some 50% of the electorate remained undecided on how to cast their vote; Mori expressed the view that these voters should 'sleep in' on election day, later claiming that his intent was to urge unaligned voters to abstain rather than vote against the LDP, although some concern was expressed as Japan tried to combat voter apathy.

The election, which was held on 25 June 2000, was the first that the LDP had contested as part of a coalition and the first in which it did not present candidates for every seat. The LDP won the most seats (233), although its representation was reduced (from 239 at the 1996 election), and many of its key political figures, including current and former cabinet ministers, lost their seats, particularly in metropolitan areas. The DPJ increased its representation to 127 seats. New Komeito won 31 seats, the LP 22 seats, Hoshuto 20 seats and the SDPJ 19 seats. Only the JCP among the opposition parties suffered a loss of seats. Down six seats, from 26 to 20, the JCP was left without the minimum number of members required to propose legislation to the lower house. The party subsequently demonstrated a greater inclination to co-operate with the other opposition parties. The LDP benefited from the electoral system, which gives rural areas (where the LDP maintained strong support) disproportionately high representation. Tradition and family connections remained important, with two candidates (the brother of Noboru Takeshita and the daughter of Keizo Obuchi) winning seats despite their lack of political credentials and experience. Only 62.5% of those eligible participated in the election.

Following the June 2000 election, questions were immediately raised about how long the new Mori administration could survive in the light of the difficulties it faced: further bribery scandals, a reduced mandate, Mori's own poor ratings, LDP factionalism and rivalries, and the need to balance the coalition both in the Cabinet and in terms of policy-making. One week after the appointment of his second Cabinet, in July, only 11% of the public wanted Mori to remain as Prime Minister, while 38.2% wanted him to go and 46.9% were unhappy but resigned (*shikata ga nai*). Meanwhile, within the LDP itself, there were growing demands for Mori's resignation from a cross-factional group of younger Diet members, who were fearful of the coalition losing its majority in the next elections to the House of Councillors.

A summit of the G-8 industrialized nations was held in Japan in July 2000. Although riots had been predicted, and Obuchi had been widely condemned for selecting Okinawa as the host site, the summit was uneventful and produced little of substance. At the end of August the Atomic Energy Commission released its long-term programme for 'research, development and utilization of nuclear energy'. In a clear response to the accident at Tokaimura in the previous year, and the debate that it had provoked, numerical targets were removed from the draft plan, with important implications for the future of the nuclear energy programme.

The Diet reconvened in September 2000. In the new session electoral reform again became a source of conflict between the Government and opposition parties. In October an amendment to the existing electoral system for the upper house, whereby voters cast their ballots for parties, to one in which electors could choose to vote either for a party or for an individual candidate, was proposed by the governing coalition. The issue was again the subject of dispute within the coalition, and compromise between the Government and the opposition, which argued that the proposed changes would make campaigning in elections more costly, was elusive. In a repetition of the events of January, the opposition parties commenced a boycott of Diet proceedings in protest against the Government forcing the bill through the upper house. The boycott, which lasted 18 days, ended following an agreement between the governing coalition and the opposition to debate the bill in the House of Representatives. The legislation, which was enacted in late October, also reduced the number of seats in the upper house, from 252 to 247.

Throughout September and October 2000 there was considerable public demand for the suspension of outdated, meaningless public-works projects. In November the coalition parties approved the cancellation of 255 such projects, at an estimated saving of 2,500,000m. yen. Popular alienation from the major parties continued, and trust was undermined by ongoing revelations of misconduct. A leading LDP politician, the Chief Cabinet Secretary, Hidenao Nakagawa, resigned in October after it was alleged, *inter alia*, that he had links to a right-wing activist and had conducted an extramarital affair. Once again, the scandal provoked demands for Mori's resignation and criticism of the Prime Minister by senior officials within his own party.

In early November 2000 the leaders of two LDP factions, Koichi Kato and Taku Yamasaki, joined the campaign to force Mori to resign and threatened to abstain if the opposition parties were to propose a vote of no confidence. With a total of 72 members in the House of Representatives, abstention by the Kato and Yamasaki factions was not sufficient to result in the approval of such an opposition motion. Opposition from Miyazawa (Kato's former faction leader and political mentor), a lack of public support for the move, together with splits within the factions over the wisdom of leaving the relative safety of the LDP prior to elections to the upper house, which were scheduled for July 2001, brought the rebellion to a muted end and left Kato seriously, if not permanently, damaged.

Mori survived the November no-confidence vote and two weeks later, in December 2000, reshuffled his Cabinet to include former Prime Minister Hashimoto in the post of Minister of State for the Development of Okinawa and the Settlement of the Northern Territories, areas of particular involvement for Hashimoto during his own premiership. The appointment was well received in the business community, but provoked suggestions that Hashimoto was positioning himself for a return to power.

The new Diet session opened at the end of January 2001, with apologies for the most recent scandal (the misuse of funds by officials of the Ministry of Foreign Affairs) and a plea for the speedy passage of the budget. The new Diet was declared by Mori to be a 'Reform Diet', which would bring about the rebirth of the country. Mori particularly emphasized the role of information technology and education reform in this process. He was also positive about the prospects for structural reforms and for Japan's economic recovery. When the Diet reconvened, the restructuring of central government ministries and agencies was largely complete. A new Cabinet Office had been created, under the control of the Prime Minister, which was ranked higher than other ministries and charged with inter-ministry policy co-ordination. Within the new Cabinet Office, a new Council on Economic and Fiscal Policy, led by the Prime Minister, had been established to address budgetary, financial and economic issues. The Cabinet Secretariat had been strengthened and expanded, and 22 other ministries had been consolidated down to 12. New political posts were established in each of the ministries and agencies to increase political input into the policy-making process. The administrative reforms had a number of objectives, including increased transparency and streamlining, but the major aim was to enable greater political leadership to be assumed by the Prime Minister and the Cabinet.

Despite the enhanced powers afforded to the Prime Minister by the administrative reforms (the ability to propose policy, majority voting in the Cabinet, greater control over appointments, etc.), Mori continued to be a 'lame-duck' Prime Minister throughout the early months of the year, suffering from one scandal after another. The coalition parties were made even less sanguine about their prospects in the election when the DPJ, the LP and the SDPJ agreed, at the end of March 2001, to put forward joint opposition candidates in 13 of 27 single-seat constituencies. In the mean time fear was mounting, both inside and outside Japan, that continuing economic problems could lead to a political and social crisis, and also that, without a transformation of the nature of political leadership, the economic problems could not be addressed.

At the beginning of April 2001 another reform came into effect, requiring the disclosure of administrative documents on request, although there were suggestions that many documents had been disposed of in the January restructuring. These institutional changes were important, but so too were the people who were to implement them. In April 2001 four candidates contested the presidency of the LDP, a first since the election of Yasuhiro Nakasone as President in 1982. The four were Ryutaro Hashimoto, Junichiro Koizumi, Taro Aso and Shizuka Kamei. The favourite to win was former Prime Minister Hashimoto, the leader of the largest faction of the LDP, with more than 100 members. Koizumi, who had just resigned as Chairman of the Mori faction, had high public ratings, but had alienated significant sections of the LDP with his plans for the privatization of the postal services. At a time when other traditional sources of LDP support, such as the construction industry and agricultural unions, were losing their ability to gather votes, this was regarded as important and worried Koizumi's supporters, who included Koichi Kato and Taku Yamazaki. Kamei and Aso were outsiders. The system used by the LDP to elect its President has varied over time. On this occasion, the election took place in two stages: a prefectural membership vote, which accounted for 30% of the total, followed by a vote by LDP Diet members. The nature of the election process proved crucial. Although there was much political manoeuvring between the LDP factions, and within the factions between the generations, the overwhelming support of the local branches of the party for Koizumi, who ran on a platform of 'Change the LDP, Change Japan', was decisive, and the LDP Diet members also backed him. Overall, Koizumi won 298 of the 478 presidency votes, while Hashimoto won 155. With the support of the coalition, Koizumi was confirmed as Prime Minister by the Diet.

The Koizumi Administration

Junichiro Koizumi, perceived as eccentric, a reformist and a nationalist, rapidly became a political phenomenon, with unparalleled popularity ratings. His appointment was widely regarded in Japan as signalling a 'seismic shift' within the LDP and, by extension, within the political world as a whole. Nevertheless, Koizumi's ostensibly non-factional appointments to leadership positions within the LDP and to the Cabinet demonstrated the need to keep the party and the coalition, as well as the nation, behind him. Although seven ministers were retained from the previous Mori administration, the new Cabinet contained an unprecedented five women, including Makiko Tanaka as Minister of Foreign Affairs. It also included three non-Diet members, most notably Heizo Takenaka, a pro-reformist economics professor at Keio University and adviser to two previous Cabinets, as Minister of State in charge of Economic and Fiscal Policy and Information Technology Policy, and Nobuteru Ishihara, the son of the controversial Governor of Tokyo, and one of the LDP 'Young Turks', as Minister of State in charge of Administrative Reform and Regulatory Reform. The perception that Koizumi's inaugural speech lacked policy detail, however, did nothing to undermine his cult status, although it may have contributed to his decision to create teams of personal advisers to draft reform proposals on a range of subjects.

Koizumi's depiction in the foreign press as a nationalist was further encouraged by his announcement that he would visit the Yasukuni Shrine on 15 August 2001, the 56th anniversary of the end of the Second World War, and by his vocal support for constitutional revision, beginning with the public election of the Prime Minister, but also embracing changes that would allow Japan to exercise the right of collective security. On this, as on many other issues, Koizumi's political philosophy and policies were close to those of former Prime Minister Nakasone, who strongly supported Koizumi, comparing the new Prime Minister's reforms with his own 'final settlement of post-war politics'. Early indications that Koizumi was prepared to exercise political leadership came in May 2001, with his agreement, against the advice of the Ministry of Justice and the Ministry of Health, Labour and Welfare, to the payment of compensation to leprosy sufferers detained in sanitation centres. More significantly, in June, Koizumi moved to impose his own guide-lines for the 2002/03 budget.

In June 2001 an initiative drastically to reform the legal system, which had been overshadowed and delayed by the drama of the LDP presidential election and the appointment of the new Prime Minister, resulted in the Final Report of the Judicial Reform Council and government support for its recommendations. A bill to increase the number of lawyers (from 20,000 to 50,000 by 2018); to open postgraduate law schools in 2004; to increase the pass rate in law studies from 30% to 70%–80%; to introduce juries or lay judges in criminal trials; and to shorten trials and introduce trial deadlines was to be considered by the Diet later in the year. These legal reforms were likely to be of great social and economic significance as Japan endeavoured to become a law-based society, competitive internationally and able to attract investment. The Koizumi reform plan, drafted by the Council on Economic and Fiscal Policy, was also published in June. The overall short-term objectives of the plan were to eradicate bad loans and to implement reforms in seven areas, including privatization, deregulation, the encouragement of entrepreneurial activity and fiscal reform. The longer-term aim was for private-sector demand to become the driving force of the economy. The proposals reiterated the promise to reduce the issuance of new government bonds in the 2002/03 budget, to 30,000,000m. yen, and to that end, to decrease public works, reform special public corporations and reduce central government grants to local administrations.

Elections to the Tokyo Metropolitan Assembly in early July 2001, in which the LDP secured five additional seats, were regarded as an expression of 'Koizumi fever' and an indication of what the LDP might expect in the elections to the House of Councillors scheduled for later in the month. The new electoral system for the upper house, which allowed votes for individuals as well as parties, led, predictably, to a large number of celebrity candidates. The results of the elections, which were held on 29 July, were even better than anticipated for the LDP, which regained Tokyo and Osaka (lost in 1998), and took 25 of the 27 contested single-seat constituencies. Overall, the coalition won 79 of the 121 contested seats, giving the Government a useful majority of 140 seats in the 247-member chamber. Although Koizumi hailed the results as a mandate for reform, the stock market fell on the following day to a 16-year low, driven by two conflicting fears: firstly, that Koizumi would fail to carry out reform owing to LDP opposition, and secondly, that his reforms would lead rapidly to short-term bankruptcies and loss of corporate earnings. Two consecutive quarters of decline in industrial production since the beginning of the year also compounded these concerns. There were questions in some quarters about the prospects for the Prime Minister's structural reforms. However, Koizumi's public support, the absence of a viable challenger within the LDP and the lack of concerted opposition from outside placed him in a strong position.

At the beginning of September 2001 the Ministry of Finance, together with the Council on Economic and Fiscal Policy, under the Prime Minister, began to consider ministry budget requests for 2002/03. Early indications suggested that the initial requests, which had increased by 3.6% from the previous year, despite the Council's guide-lines urging a 10% reduction, demonstrated heavy involvement by the bureaucracy and LDP policy committees (*zoku*), and little response to Koizumi's request for a new balance to be achieved in budget allocations through targeted, rather than across-the-board, cuts. Koizumi's attempts to give budgetary priority to seven specific areas (environment, the ageing society, local revitalization, urban redevelopment, science, education and information technology) were undermined by equivocation between ministries and by a failure to persuade the new so-called 'super ministries' to make any changes to the budgetary share of their constituent ministries and agencies. This was particularly evident in the requests from the newly created Ministry of Land, Infrastructure and Transport. In the first few days of September, the main stock market index declined to its lowest point since 1989, unemployment rose to the highest levels on record and gross domestic product (GDP) for the second quarter of 2001 showed 0.8% negative growth.

The crisis precipitated by the terrorist attacks on the USA on 11 September 2001 gave the Prime Minister the opportunity to exercise the sort of decisive leadership that the structural reforms of the last several years had been designed to enhance. However, the impact of the terrorist attacks on the Japanese stock market was instantaneous. On 12 September the Nikkei index plummeted again, this time to below 10,000 points (its lowest level in 17 years), thus threatening the already beleaguered banking sector and prompting an announcement from

Koizumi that the attacks on the USA would not be allowed to slow down the pace of Japan's reforms, including the disposal of non-performing loans. The events of September prolonged Koizumi's political 'honeymoon' for a short while and suppressed growing opposition within the LDP to his political and economic reforms, but the pace of reform was slow. However, Koizumi's non-consensual and decisive leadership style and his reliance on private advisory bodies created a groundswell of opposition within the LDP policy committees and within the LDP leadership itself. His attacks on the faction system in general, and the *Keiseikai* (a faction of the LDP) in particular, left him isolated within the party and caused the pace of reform to decelerate.

Koizumi's weakness within the party was in some ways his strength. His relative lack of dependency on the factional leadership, his distance from the vested interests that maintain themselves through public works and his independent sources of policy expertise gave him public popularity that, in turn, protected him and his reform programme from attack by the LDP. For Koizumi, therefore, maintaining his popularity was vital. Polls conducted in November 2001 showed that his approval ratings had risen even higher for the first three months of the administration before suffering a downturn during mid-2001.

Although Koizumi retained his exceptionally high popularity ratings throughout late 2001, the media began to focus on his failure to deliver reform. The restructuring of the public corporations, one of Koizumi's central aims and one of his more popular reforms, showed signs of failing under concerted opposition both from the ministries, which provide more than half the corporations' directors, and the *zoku,* to which the corporations provide funding in return for political favours. Koizumi's efforts to dissolve or privatize public corporations, the details for which were to be submitted to the Diet in January 2002 had, by September, produced plans for the dissolution of only five of the 163 corporations. The reforms suffered a further set-back with the report by the Government Secretariat for the Promotion of Administrative Reform in early October that recommended the abolition or privatization of only 34 public corporations. In mid-November, Koizumi attempted to take the initiative with an announcement, firstly, that the Housing Loan Corporation should be abolished, rather than privatized, and, secondly, that the four major road corporations should suspend construction of 2,383 km of highway and expect an end to government subsidies in the fiscal year 2003.

An outbreak of bovine spongiform encephalopathy (BSE, or so-called mad cow disease), the first outside Europe, was confirmed in September and brought another ministry into the glare of adverse publicity. The Ministry of Agriculture, Forestry and Fisheries (MAFF) stood accused of failing to heed WHO warnings on the use of meat and bone meal feed, and of failing to provide adequate crisis management. The failures were attributed to the lack of cohesion within the ministry, despite previous efforts to restructure it. The Ministry's structural problems were greatly exacerbated by the strength of LDP Diet member links with agricultural interests and their influence on civil servants. One-half of the annual MAFF budget of around 3,000,000m. yen was spent on public works, a sum equal to 20% of total spending.

The outbreak of BSE, along with the decline in IT production and the impact of the terrorist attacks in the USA, forced the Cabinet Office to lower its projected economic growth rate from 1.7% to −1% and led the Ministry of Finance to warn of the possibility of a contraction of GDP for the fiscal year 2001 and of a fourth consecutive year of negative economic growth—the longest experienced since the beginning of the Meiji era. The ongoing recession was blamed for a rise of nearly 18% in serious crimes between 2000 and 2001, mainly the result of a massive increase in armed robbery and arson. The unemployment rate for September rose to 5.3%, with the construction sector and IT industries continuing to suffer the most severe job losses. In addition to the redundancies, many larger companies began to implement wage reductions. The response of the Minister for Health, Labour and Welfare to the September figures was to declare that the employment situation had entered a state of emergency. The situation continued to worsen especially for males in the transport, communication, manufacturing and construction sectors. The response of RENGO (Japanese Trade Union Confederation) was to abandon the post-war *shunto* system of annual unified wage increase demands and to fight wage cuts and job insecurity. The shift towards a merit-based system is a function of restructuring and unemployment but it also reflects a growing desire for a more competition-orientated society and a decline in the number of people who want an egalitarian society with a minimal gap between the rich and the poor.

Demographic issues, in particular declining population growth and an ageing society, continued to attract attention. Official figures showed population growth at its slowest since the Second World War, the first ever decline in the 'productive age group' (aged between 15 and 64 years), and those in the 'old' category (aged over 65 years) exceeding the 'young' (children under 15 years) for the first time since records began in 1920. Related to this, the average age of marriage continued to rise, although unmarried cohabitation and the number of children born out of wedlock remained low. Figures released in February 2002 suggested that the situation was graver than previously predicted and that an ongoing decline in the fertility rate would result in Japan's population peaking in 2006, one year earlier than expected. By 2050 the population will have declined to 100.59m. people, 35% of whom will be over 65, double the figure at the beginning of the 21st century. Of the total population in 2050, it is anticipated that only 54% will be of working age.

The implications of a decreasing population for the social welfare system are stark. One immediate effect of the downward revision of predicted fertility rates to 1.39 from 1.61 is on the percentage of salary that will need to be paid in pension contributions, which will increase as the declining numbers of those in work have to support a growing population of retirees. The percentage cost of healthcare to GDP is expected to more than double, to 18%, by 2025; public expenditure will grow by 7.5% of GDP over the same period, in order to pay for the ageing population. Despite the urgent need for a sustainable social security system, plans for medical insurance reform and pension reform put forward by the Government met strong opposition from LDP *zoku*.

Since the mid-1970s Japan has also witnessed a net outflow of skilled workers through migration. One possible way of addressing the demographic stagnation that is such an obstacle to economic growth would be changes to the immigration laws to encourage inward migration. Minor amendments in the early 1990s brought slight increases in the numbers of certain types of immigrant. However, following the attacks on the USA in September 2001, the Government's seven-point plan strengthened international co-operation and information exchange on immigration controls, and further opening of national borders was not anticipated in the near future. Koizumi instructed the Ministry of Health, Labour and Welfare to produce plans by September 2002 to encourage higher birth rates. These are expected to go well beyond the provision of nursery care and to include financial incentives, paternity leave and low-interest loans for higher education.

The birth of Princess Toshi (Aiko), the first child of the Crown Prince and Princess, took place in December 2001. In accordance with the Imperial Household Law, the younger brother of the Crown Prince remained second in line to the throne, but the birth drew further debate on the need for revision of the law to allow female succession. The proposal attracted widespread support from the public and also among members of the Diet who, under the post-war Constitution, would be required to enact the changes.

Media evaluations of the workings of the new government structure, one year after its inception, reported rigidity in bureaucratic structures, and continued bureaucratic dominance. There appeared to be little transfer of actual power from the bureaucracy, despite increased formal powers for the Prime Minister's Office, the creation of the Financial Services Agency and the Council on Economic and Fiscal Policy (CEFP), and the introduction of senior vice-ministers and parliamentary secretaries. The media viewed the reshuffle of a number of these vice-ministers and parliamentary secretaries in January 2002 as a retreat by Koizumi on faction-based appointments and a victory for Hashimoto. In July 2002 a review of the functions of the CEFP was ordered amid fears that it was not fulfilling its original purpose of transferring responsibility for economic and fiscal policy from the bureaucracy to the Prime Minister and

Cabinet. Reforms of the Diet, such as the introduction of Prime Minister's 'question time' and the abolition of the practice whereby bureaucrats answered on behalf of ministers, did not attract greater public attention to the Diet nor render its work any more meaningful.

Koizumi's dismissal of Makiko Tanaka on 29 January 2002 after weeks of pressure from the LDP, despite his pledge not to reshuffle his Cabinet during his term of office, appeared to precipitate a sharp decline in the Prime Minister's popularity. Predictions of a return to factional politics and an end to the prospects for reform caused falls in the stock market, in government bonds and in the value of the yen. Tanaka claimed that the Prime Minister had broken his pledges on reform and was allowing his Cabinet to be influenced by former Prime Minister Mori. The DPJ began to conduct itself more in the manner of an opposition party and criticized Koizumi's handling of the economy, rising unemployment and the growing number of scandals involving senior members of the LDP. For the first time Koizumi was no longer able to use the promise of an approach to the DPJ as a credible threat to control the LDP.

In early March 2002 one of Koizumi's two closest allies, Kato Koichi, resigned following the arrest of the former head of his Tokyo office in response to charges of tax evasion and accepting bribes for facilitating the allocation of public works. Kato's resignation from the LDP followed that of the former Director-General of the Hokkaido and Okinawa Development Agency, Suzuki Muneo, on charges of malpractice. The scandals focused attention on the urgent need to address the issues of the relationship of LDP Diet members and the bureaucracy with regard to policy-making. Koizumi's rising unpopularity, along with a decline in support for the LDP, prompted coalition fears of adverse voter reaction at the forthcoming elections.

Almost simultaneously with the resignations, a draft report was produced by the LDP National Vision Project Headquarters under Koizumi, and submitted to the Prime Minister without prior consideration by the General Council. The report, which proposed a study of measures to create a cabinet-led system and deny influence to politicians with vested interests, drew an angry response from the LDP, both for its content and the manner of its introduction. The proposals included cabinet drafting of all bills and policies, restrictions on contact between back-bench politicians and bureaucrats, and an end to approval of bills by party committees before submission to the Diet. The report, and its rejection by the LDP, was best viewed in the context of Koizumi's attempt to open up the postal delivery service to private competition, which, under the LDP operating methods, might be diverted by one dissenting LDP vote. In an attempt to re-establish his reformist credentials, the Prime Minister also proposed restricting political donations from public works contractors and reviewing the fund-raising role of local branches of political parties.

Restructuring of the postal services remained the basis of Koizumi's reform programme because of the implications for controlling vested-interest politics. It was opposed by many in the LDP for the same reasons. In an unusual move, the four postal bills were submitted to the Diet without the support of the LDP, and the passage of the legislation was secured only after some modification of the content and the promise of a cabinet reshuffle. The compromises resulted in requirements for entering the mail delivery sector being set at a level that would exclude most private companies. The extended Diet session ended on 31 July 2002.

In September 2002, as a direct result of the Prime Minister's North Korean initiative (see below), his popularity rose dramatically. The recovery strengthened Koizumi's position in preparation for a cabinet reorganization in late September. Changes included the removal of Hakuo Yanagisawa, Minister of State responsible for the Financial Services Agency and widely regarded as an opponent of reform, and his replacement by Heizo Takenaka. Takenaka, an unelected official with few allies within the LDP, also continued as Minister of State for Economic and Fiscal Policy. The reallocation of portfolios, Koizumi's first reorganization following a pledge to maintain the same cabinet membership, was limited. The six new cabinet appointments and a new list of senior vice-ministers were generally perceived as having been carried out without consultation or regard for factional considerations and as a reaffirmation of Koizumi's commitment to reform. Koizumi's adherence to Takenaka's firm policy on the disposal of bad loans was given a further fillip by the five LDP by-election victories at the end of October, which did not, however, put an end to criticism of Koizumi's reform policies by senior LDP Diet members. By November 2002 senior party officials were reminding Koizumi that failure to co-operate might jeopardize his prospects at the LDP presidential election scheduled for the following September. The threats provoked speculation that Koizumi might pre-empt this by dissolving the Diet and precipitating a general election.

The closing months of 2002 were even less satisfactory for the DPJ and its leader. The party re-elected Yukio Hatoyama as leader in a close contest with Naoto Kan. The electorate included both DPJ Diet members and ordinary supporters who had paid 1,000 yen. The four candidates in the election all favoured electoral co-operation with the LP and SDPJ but, like a microcosm of the DPJ itself, differed on economic policy and constitutional revision. Barely three months later losses in the by-elections, compounded by a failed attempt to merge with the LP to form a new opposition party, led to Hatoyama's resignation and the election, this time by an electorate consisting solely of Diet members, of Naoto Kan. Kan commanded greater support among the public, many of whom regarded him as a potential replacement prime minister, but support for the DPJ remained static. The disarray in the DPJ also contributed to criticism of Koizumi among the senior ranks of the LDP. In late December four defectors from the DPJ and nine members of the New Conservative Party (NCP), formed a new NCP, which aligned itself with the governing coalition. This minor change represented only the latest development in 10 years of unfulfilled hopes for proper party realignment that would deliver an alternating system of government.

Figures released in the latter part of 2002 showed that in 2001 donations to all political parties had declined for the fifth consecutive year. The purchase of tickets by individual companies to party fund-raising events, however, had increased to exceed all other forms of corporate donations combined, provoking fears of a decline in transparency. State subsidies to political parties, as revised in the Hosokawa reforms of 1994 and having increased to account for more than 28% of total funds raised, were a factor in the timing and nature of party realignment. Koizumi's proposals in March 2002 further to restrict donations from companies involved in public works had made no progress in the face of LDP antagonism and mixed sentiments among the opposition parties. Political funding once again became a major issue in early 2003, as the LDP struggled to deal with several difficulties: first, the arrest of a former secretary-general of the LDP Nagasaki Prefectural chapter for accepting illegal donations; then the arrest of a sitting Diet member belonging to Koizumi's own faction for concealing political donations; and finally the resignation of the Minister of Agriculture, Forestry and Fisheries over the misuse of funds by a former aide. Senior members of the LDP, including Koizumi, therefore welcomed a decision in May 2003 by the Japan Business Federation (JBF—the former Keidanren, the influence of which on government policy-making had weakened since its suspension of fund-raising for political parties in 1993), to return to brokering political contributions in 2004. The avowed aim of the JBF was to use donations to encourage tax and social security reforms. To this end, LDP members drafted guide-lines for evaluating party policy. The anticipated shift from party ticket sales, which depended on the strength of individuals and factions, to corporate grants direct to the parties, was expected to bolster the power of central party officials. The decline in the ability of factions to provide either funding or position was matched by a loss of factional loyalty, especially among younger party members, and even within the biggest LDP faction led by former Prime Minister Hashimoto. The creation of a tax system study group by more than 70 younger members of the party was evidence of this.

The debate on constitutional revision was revived in November 2002 by the release, against the wishes of the JCP and SDPJ, of an interim report by the Research Commission on the Constitution, which had been established in the House of Representatives in January 2000. The report set out the arguments for and against amending Article 9 of the Constitution, the popular election of the Prime Minister, changes to the bicameral legislature and the introduction into the Constitution of new human rights. The report devoted more than 100 of its 706 pages to issues related to Article 9 but did not reach any conclusions. The relevance of the debate was brought into focus in April by the conflict in Iraq and developments in North Korea. Former Prime Ministers Nakasone and Miyazawa found themselves, unusually, agreeing that under certain circumstances, collective self-defence was allowed under the present Constitution.

The extraordinary Diet session convened in mid-October and ended in mid-December 2002, having failed for a second time to approve government bills on defence and privacy. Both items of proposed legislation were rescheduled for the middle of the ordinary Diet session, which was reconvened in late January 2003. The media criticized the poor level of debate, the lack of opposition and the resurgence of bureaucratic dominance of policy-making. By February the cabinet approval rating had declined to 46%. It fell again, but only slightly, in the first polls taken after Koizumi's decision to support the US-led campaign in Iraq. Koizumi's own approval ratings were also little changed by the decision, not least because of a perceived absence of other suitable candidates.

A combination of the international situation, divisions in the DPJ and concessions by the Government resulted in the privacy bill being enacted into law in late May 2003 and the defence bills in early June. The ordinary legislative session was extended by the governing coalition by 40 days to 28 July to enable the passage of an Iraq reconstruction bill that would permit the dispatch of SDF personnel, but only after a week of opposition boycotting of the Diet. The last few days of this extended session were enlivened by the announcement of a planned merger of the DPJ and Ichiro Ozawa's LP (which, upon its conclusion in September, brought the total representation of the two parties to 136 in the lower and 66 in the upper house). The DPJ was eager to persuade the SDPJ, which was beset by a funding scandal, to join them. The opposition presented non-binding censure motions in the House of Councillors and a no-confidence motion in the House of Representatives, finally resorting to physical obstruction of the committee vote stage in an unsuccessful attempt to block the Iraq bill. The Government's announcement of plans for an extraordinary Diet session from mid-September 2003 to amend the Anti-terrorism Law before its expiry on 1 November 2003, and a planned visit by US President George W. Bush provoked rumours of a forthcoming dissolution of the Diet and demands for detailed manifestos from the various parties. Koizumi promised that if he were re-elected as President of the LDP, his policy pledges would become the party platform. In late September Koizumi was re-elected leader of the LDP, winning 399 of the 657 votes cast—260 more than Shizuka Kamei, the nearest of his three rivals. Koizumi then effected a major reorganization of the Cabinet without consulting the LDP factions on the appointments The new appointments brought a more youthful cast to the Cabinet, with three members in their forties and three in their fifties. This promotion of younger, reform-minded politicians was part of a broader attempt to produce a generation-shift, which included the appointment of 49-year-old Shinzo Abe as the Secretary-General of the party and the introduction of a mandatory retirement age of 73 for LDP Diet members elected from constituencies determined by means of proportional representation (PR). Former Prime Minister Miyazawa (aged 84) accepted the proposal, but former Prime Minister Nakasone (85), who in 1996 had been promised a position at the head of the PR list for the rest of his life and whose long-standing ambition was to see reform of the Constitution during his political career, condemned his enforced retirement as 'a kind of political terrorism'.

At the beginning of October 2003 the legislation to extend the Anti-terrorism Law by two years was approved by both Houses; on 10 October the Prime Minister obtained a Rescript from the Emperor and announced the dissolution of the House of Representatives. A general election, the 20th since the new Constitution came into effect in 1947, the first since Koizumi became Prime Minister and the third under the predominantly single-seat constituency system, was scheduled for 9 November. Campaigning began by 1,159 candidates for 300 seats from single-seat constituencies and for 180 seats from the 11 PR nation-wide blocs.

One of the last acts of the Diet before its dissolution was to amend the Public Offices Election Law to permit the distribution by candidates of their party manifesto during the election campaign from designated places, although not from within newspapers. The first official use of party manifestos in a Japanese election was seen as enhancing the prospects for a two-party alternating system of government, although only 34% of respondents to a poll conducted by a leading publication *Nikkei* thought manifestos were important, compared with 45% who cited candidates' personality and policies. In addition to clarifying the differences between the parties and countering public apathy, the introduction of manifestos was expected to limit the influence of the LDP *zoku* and the bureaucracy over policy-making. The election itself was seen as a potential watershed in post-war Japanese politics. This was partly because the DPJ, which had recently merged with Ozawa's Liberal Party, was presenting 277 candidates (267 in single-seat constituencies, of whom 246 were in direct conflict with the LDP), and partly because Koizumi had promised not to remain in power if the ruling coalition did not maintain a majority.

Despite the availability of manifestos and the attendant publicity, the turn-out at the election on 9 November 2003 decreased to just under 60% of the electorate. With the LDP benefiting in a significant number of single-seat constituencies from the backing of its coalition partner, New Komeito, and its supporting group, the Soka Gakkai, the distribution of seats among the main political parties in the House of Representatives after the election was: LDP 237 seats (compared with 247 before the election); DPJ 177 (previously 137); New Komeito 34 (31); JCP nine (20); SDPJ six (18); and NCP four (10). The NCP, one of the governing coalition parties in partnership with New Komeito and the LDP, disbanded and merged with the LDP after Hiroshi Kumagai and other party leaders lost their seats. The merger thus gave the LDP a simple majority in the lower house where it now held a total of 241 of the 480 seats. In addition, four independent members of the House of Representatives joined the LDP after the election. With 279 seats, together the new LDP-New Komeito coalition still had an absolutely stable majority, giving it control of all committee chairmanships and a majority in all standing committees. Major individual losses in the election included LDP Vice-President Taku Yamazaki and SDPJ leader Takako Doi, who nevertheless was returned to a PR seat in the legislature. The big gains made by the DPJ and the fact that it even outperformed the LDP in PR voting, suggested that a proper two-party system might indeed be emerging. In the event, 84% of the seats in the election were won by the two major parties.

Policy debate during the election, sharpened somewhat by the manifestos though still lacking in specific proposals, focused on the issues of retirement pensions and tax reform, the privatization of the postal system and of road-building corporations, and the dispatch of the SDF to Iraq. These questions continued to dominate Japanese domestic politics in 2004. On the subject of the revision of the Constitution, the manifestos of both the LDP and the DPJ favoured an expansion of the debate. This was facilitated by the substantial losses of seats suffered by the SDPJ and JCP. The issue was placed firmly on the agenda in January 2004 when Koizumi promised that the LDP would draft a revised constitution by 2005, the 50th anniversary of the founding of the party. The DPJ announced its own intention to produce a draft constitution by 2006. Revision would require the approval of a two-thirds' majority in both houses, followed by a referendum. Legislation for establishing the legal procedures for constitutional amendment would be the first stage. Some of the central issues were the right to collective defence, the introduction of a constitutional court, changes to the requirements for constitutional revision and the shift to a unicameral legislature. In early 2004 a public opinion survey showed 55% to be in favour of constitutional revision.

Following his re-election as party President and the return of the LDP to government at the lower house election of November 2003, Koizumi was generally regarded as having a brief opportunity for reform, which would last until the upper house election in July 2004; beyond that date there would be no further need for his rivals in the party to support him. Even during this period, however, the contest for power within the party remained intense. During his three years in office Koizumi had undermined the LDP factions by disregarding their recommendations for cabinet and party appointments and had made particular efforts to weaken the largest grouping, the Hashimoto faction. The elections for LDP President in September 2003 divided the Hashimoto faction into those led by Mikio Aoki, who supported Koizumi, and those led by Hiromu Nonaka, who did not favour the incumbent leader. Those who had supported Koizumi were rewarded with cabinet positions. The November election further directed the factional balance in the lower house away from the Hashimoto faction, which lost eight seats to leave it with 51 in the new legislature, and towards Koizumi's former faction, the Mori faction, which gained 12 seats to give it a total of 52. When the Hashimoto faction lost 11 members in the upper house elections in July 2004, Hashimoto announced his resignation as faction leader and his intention not to seek re-election from a single-seat constituency in the next election. Hashimoto, whose faction was directly descended from the scandal-ridden Tanaka and Takeshita factions, was at the time under investigation for allegedly receiving undisclosed political contributions of 100m. yen from the Japan Dental Association.

Koizumi's 2001 promise to reform the LDP had been an undertaking to change the conduct of politics by destroying the interest-dominated, 'pork-barrel' politics embodied by the Hashimoto faction. To this end, in addition to successfully targeting that faction specifically, Koizumi also undermined the policy-making capacity of the LDP committees and other top-level party bodies by limiting their involvement at the pre-legislative stage and promoting the activities of cabinet-led groups such as the Council on Economic and Fiscal Policy (CEFP). In mid-2004 it remained to be seen whether or not he had been successful. Plans for the privatization of the postal services, which was viewed as the clearest measure of the success of the reforms, were drawn up by the CEFP in August 2004; however, the reforms were widely judged to have been diluted to the point of evisceration.

Other domestic reforms also seemed to lose momentum as Koizumi's declining political assets were increasingly spent on dealing with the problem of Iraq. Bills to privatize the four public highway corporations, which cleared a lower house plenary session in April 2004, were condemned by the *Nikkei* weekly as 'an affront to political decency' which failed to conform to the model drawn up in late 2002. Pension reforms enabling an increase in contributions and the reduction of benefits were belatedly pushed through the House of Councillors as scandals raged about the non-payment of national pension premiums by Diet members of all hues, including cabinet ministers and party leaders. The 150-day ordinary Diet session, which was punctuated by DPJ boycotts over issues such as pension reform and political scandals, enacted 135 laws, including 120 out of 127 government bills. In a trend towards an increased number of Diet member bills that had begun in the early 1990s, the laws enacted also incorporated a range of private member's legislation on various matters including foreign policy and intellectual property rights. The session ended in mid-June, the day after an opposition motion of 'no confidence' had failed, and campaigning began immediately for the House of Councillors election due on 11 July 2004. Special sessions of the Diet were scheduled for late July and late September to elect officials after the election and to clear a backlog of legislation.

The DPJ entered the upper house election campaign under the leadership of Katsuya Okada, after first Naoto Kan and then the leader-in-waiting, Ichiro Ozawa, stepped down because of their failure to pay pension contributions. Speculation was rife that Ozawa's fluctuating influence in politics would increase under Okada's 'uncertain' leadership. The election campaign focused on only two main issues—pension reform and SDF participation in Iraq; a discussion of postal reform was eschewed by both major parties. A revision of the Public Offices Election Law reduced the number of upper house seats by five, to 242, leaving 121 seats to be contested (one-half of the members being elected every three years for a six-year term, with the chamber not being subject to dissolution).

The outcome of the election, in which the DPJ for the second time won more votes than the LDP, was regarded as another milestone on the road to a two-party system. The DPJ secured 50 seats and the LDP, which had lost its majority in the upper house in 1989, failed not only to win the 56 it had needed to regain it, but even to achieve the 'safe' 51 seat target it had set for itself. (The ruling coalition as a whole, however, including New Komeito, did continue to hold a majority in the House of Councillors.) The results were seen as a serious set-back to Koizumi's leadership, not least within his own party, and represented a major threat to his reform programme, particularly to his plans for privatization of public financial services supplied by post offices. Encouraged by these results, the DPJ planned to ask the SDPJ to refrain from presenting candidates in seats it had no chance of winning in order to maximize DPJ chances of achieving a change of government in the future. The extraordinary Diet session that followed the election saw the appointment of the first female speaker of the upper house, LDP member Chikage Ogi, and the defeat of an opposition bill to abolish the new pension legislation that had been forced through at the end of the previous session.

In September 2004 the Prime Minister effected a major reorganization of the Cabinet. Nobutaka Machimura was appointed Minister of Foreign Affairs. Sadakazu Tanigaki remained as Minister of Finance, but other changes included the establishment of a new portfolio to handle the privatization of the postal services, the position being assumed by Heizo Takenaka, who retained responsibility for economic and fiscal policy. A new portfolio with responsibility for addressing the country's declining birth rate was also created.

FOREIGN RELATIONS

In the 20 years following the Second World War, successive Japanese Governments sought to shelter the country behind its alliance with the USA, and to avoid independent commitments in foreign policy. When Japan signed the San Francisco Treaty in 1951, it also signed a bilateral security treaty with the USA, whereby the Japanese Government granted the use of military bases in Japan exclusively to the USA in return for a US commitment to provide military support to Japan in the event of external aggression. Japan has since then functioned as a major base for US forces in eastern Asia, its sole military obligation being to defend its own territory. Thus, Japan will not go to the defence of its ally if the USA is attacked elsewhere. The insistence of the USA that Japan rearm was a significant factor in the reawakening of Japanese foreign policy.

Prime Minister Hosokawa expended every effort to promote Japan's international status, embarking in September 1993 upon a visit to New York, where he met President Clinton and also addressed the General Assembly of the United Nations (UN). Although a long-standing US demand to liberalize the Japanese rice market was met in December, it was significant that the agricultural trade accord was drafted by the Secretariat of the General Agreement on Tariffs and Trade (GATT), and accepted by Japan in the context of the successful attempt to complete the Uruguay Round of negotiations. The agreement gave foreign rice-producers access to Japanese domestic markets, beginning with 4% of the market and rising to 8% over six years, after which rice would be subject to a tariff system.

In its relations with the USA, Japan was unable to demonstrate any shift in policy, resisting efforts by the Clinton administration to introduce numerical targets as 'objective criteria' in the trade negotiations at the Washington Summit in July 1994. The failure of these talks, with Hosokawa's rejection of US demands, was heralded in Japan as an indication of a new maturity in the bilateral relationship. Japan's policy towards the USA subsequently began to follow two different directions. On the one hand, in negotiations on the automobile trade, an uncompromising position was maintained, despite US demands for concessions. On the other hand, however, Japan was prepared to meet US expectations in areas such as Japanese involvement in UN activities, and security ties with the USA (for the latter see below).

After 10 months of trade negotiations at the sub-cabinet level, the USA expressed strong dissatisfaction with the proportion of Japanese government contracts awarded to non-Japanese firms. A 60-day consultation period was established, following the expiry of which the USA promised that retaliatory sanctions would be imposed against Japan. While agreement was reached by early October 1994 in three of the four main areas under discussion, the two sides had yet to resolve their dispute over the automobile trade. Following the failure of the trade negotiations in early May 1995, the USA threatened to impose severe sanctions on a number of luxury car models. Japan responded by lodging a complaint with the World Trade Organization (WTO), on the grounds that the USA's actions violated the WTO Agreement and other international accords. The USA retaliated by filing a complaint of its own. Prime Minister Murayama found some support for Japan's position on a visit to EU leaders in Paris in June, but the USA added the threat of further sanctions on Japanese air cargo after access to flights was not granted to a US carrier. In late June, immediately prior to the introduction of US sanctions, an agreement was reached, whereby Japan promised to allow an increase in the proportion of US-manufactured parts bought by Japanese car-makers in North America, and to guarantee that Japanese manufacturers would not contravene their 1994 purchase plans. An agreement on the air cargo dispute was also reached in July 1995, although discussions continued on perceived obstacles to foreign competition in the Japanese photographic film market.

Japanese-US trade negotiations in 1996 continued to focus on the opening of the Japanese market. Specific areas of contention included photographic film and insurance, and discussions began on the renegotiation of two previous agreements on semiconductors and airline routes, which were due to expire in that year. Agreement on the issue of semiconductors was reached in August, whereby two new bodies were to be established to regulate the market. However, in the wake of the events in Okinawa (see below), Prime Minister Hashimoto and President Clinton were more concerned with reinforcing the bilateral security relationship than trade disputes in late 1996 and early 1997. At a meeting in April 1997 they discussed co-operation in regional affairs, focusing on the issue of stability in the Korean peninsula. As Japan's trade surplus with the USA increased in early 1997, with notably, a growth in the export of automobiles, it was feared that the friction over trade issues might intensify. The USA continued to put pressure on Japan to open its market further and to transform its export-orientated economy to one based on the domestic market. In September the USA imposed large fines on three Japanese shipping companies, following complaints about restrictive harbour practices in Japan; however, an agreement to reform Japanese port operations was concluded shortly thereafter. Negotiations on increased access to airline routes for Japanese and US carriers were also successfully concluded in January 1998.

During 1998 the USA became increasingly concerned about the deceleration of the Japanese economy and its impact on other Asian economies, already severely depressed by the regional currency crisis. Discussion of Japan's economic and financial problems was the focus of the G-7 summit meeting in April. During a two-day visit to Japan in November, President Clinton urged the Government to implement measures rapidly to encourage domestic demand, reform the banking sector and liberalize the country's markets, reinforcing earlier warnings that it risked provoking protectionist measures. The USA was also critical of Japan's refusal to lower tariffs on rice and forestry and fisheries products, or to curb low-priced steel exports to the USA. Japan's trade surplus with the USA continued to increase, growing by some 33% in 1998, to reach its highest level since 1987. In May 1999, during a six-day visit by Prime Minister Obuchi to the USA (the first such official state visit in 12 years), Clinton praised Obuchi's efforts to stimulate economic recovery and welcomed Japanese plans for further deregulation in several sectors, including telecommunications, energy, housing and financial services. Meanwhile, the dispute over Japanese steel exports had escalated. A ruling, in April, by the US Department of Commerce that Japan had 'dumped' hot-rolled steel into the US market was endorsed, in June, by the US International Trade Commission, and punitive duties were subsequently imposed. In November Japan brought a complaint before the WTO against the US ruling.

Stability in the Asia-Pacific region is a vital consideration in Japanese foreign policy, since Japan depends on regional markets for a substantial proportion of its foreign trade, as well as for its imports of vital raw materials. In January 1993, on a tour of member countries of the Association of South East Asian Nations (ASEAN), Prime Minister Miyazawa outlined his policy for the area, which included regional co-operation, a commitment to economic openness in the Asia-Pacific area, and a fuller political and security dialogue among ASEAN countries. In August the incoming Prime Minister, Hosokawa, promised to initiate a new era in Japan's relations with its neighbours by making a full apology for Japan's war record in his first policy speech. However, political pressure forced him to moderate his language and to state that Japan would not pay compensation to victims of Japanese aggression. On a tour of the Philippines, Viet Nam, Malaysia and Singapore in late August 1994, Prime Minister Murayama emphasized Japan's responsibility and remorse for its actions in the Second World War, and its desire for reconciliation. Shortly afterwards, and on the same theme, Murayama announced the 'Peace, Friendship and Exchange Initiative', a 100,000m.-yen programme to promote historical studies and exchanges among Asian nations. Murayama chaired the Asia-Pacific Economic Co-operation (APEC) meeting in Osaka in November 1995, as part of Japan's commitment to regional co-operation. Although Japan's close security co-operation with the USA sometimes caused concern among its Asia-Pacific neighbours, Japan's relations with these countries remained stable in the late 1990s. Hashimoto visited ASEAN countries in January 1997, and the Japanese Minister of Foreign Affairs participated in the meeting of ASEAN and other foreign ministers in July. However, the Asian economic crisis that began in late 1997 threatened to disrupt Japan's trading relations with its neighbours. At the annual IMF-World Bank conference, held in late 1997 in Hong Kong, Japan proposed the establishment of an Asian Monetary Fund, a regional organization in which only Asian countries would participate. Strong objections were voiced by the USA and European countries, which advocated a more international response to the crisis. Japan responded by co-operating with the USA and international financial institutions in providing aid for, among other countries, Thailand and Indonesia. However, Asian Governments remained dissatisfied with Japan's response to the crisis, fearing that the weakness of the Japanese currency would prevent regional economic recovery by inhibiting a growth in exports. In response to increasing international pressure, in October 1998 the Japanese Government announced a US $30,000m.-aid 'package' for Asian countries, and in November the USA and Japan presented a joint initiative for growth and economic recovery in the region. In addition, at an ASEAN summit meeting, held in Hanoi, Viet Nam, in December, Japan pledged further assistance, in the form of loans worth some $5,000m., to be disbursed over a three-year period.

In May 1990 Japan's relations with the Republic of Korea, which had been strained since the Second World War, were greatly improved following a visit by President Roh Tae-Woo, during which Prime Minister Kaifu offered an unequivocal apology for Japanese colonial aggressions on the Korean peninsula in the past, and promised to improve legislation protecting the basic rights of those Koreans and their descendants resident in Japan, by 1993. In February 1995 Prime Minister Murayama publicly acknowledged that Japan had been responsible, in part, for the post-war division of the Korean peninsula. He was forced to retract the statement, however, following bitter controversy in the Diet. In June the Diet issued a resolution expressing deep regret for the atrocities committed by Japanese troops during the Second World War. The resolution, which was timed to coincide with the 50th anniversary of the end of the war, was widely criticized, however, by former prisoners of war of the Japanese, as insufficiently explicit and for being a personal statement by the Prime Minister, rather than a representation of the views of the Government as a whole. In August Murayama formally reiterated the statement of remorse, expressing 'a heartfelt apology' and admitting that Japanese national policy during the Second World War had been 'mistaken'. He discounted, however, the possibility of individual compensation payments by the Government to Asian (mostly Korean) women, used by Japanese troops for sexual purposes during the war ('comfort women'), preferring to advocate the creation of a private fund to collect donations from the Japanese people.

In February 1996 a report issued by the UN criticized Japan's treatment of the 'comfort women' and urged it to accept full responsibility for its actions. The first payments from the private fund, created to provide compensation to the victims, were disbursed in August, together with a letter of apology from Hashimoto, to four Philippine women. Further payments were made in January 1997 to several South Korean victims, but the majority of groups representing the women refused to accept payment from the fund, demanding that compensation be forthcoming from official, rather than private, sources. In April 1998, however, in the first such ruling, a Japanese district court ordered the Government to compensate three former 'comfort women' from South Korea. Meanwhile, in June 1996 Hashimoto met the South Korean President, Kim Young-Sam, to discuss bilateral co-operation in economic and security affairs. Relations with the Republic of Korea were strained in late 1996 over a territorial dispute concerning a group of islands, to which both countries laid claim.

Relations with the Republic of Korea deteriorated in early 1998, when Japan unilaterally terminated a bilateral fisheries agreement, following the failure of negotiations concerning the renewal of the accord. Discussions were held at intervals during 1998 to attempt to renegotiate the terms of the agreement. The Japanese Government contributed financial aid to the Republic of Korea as part of the international effort to stimulate its economic recovery. Japan's relations with the Republic of Korea improved considerably in October, during a four-day visit to Tokyo by President Kim Dae-Jung. A joint declaration was signed by the South Korean President and Prime Minister Obuchi, in which Japan apologized for its conduct towards Korea during the period of Japanese colonial rule. Emperor Akihito also expressed deep sorrow for the suffering inflicted on the Korean people. In addition, the Republic of Korea agreed to revoke a ban on the import of various Japanese goods, while Japan promised US $3,000m. in aid to the Republic of Korea in support of its efforts to stimulate economic recovery. In November the two countries concluded negotiations on the renewal of their bilateral fisheries agreement, which came into effect in January 1999. An agreement to modify some of the terms of the accord was reached in March, following a series of differences over its implementation. Increased co-operation was emphasized during a visit by Obuchi to the Republic of Korea later that month, when both countries agreed to strengthen bilateral economic relations, and Japan pledged further aid to the Republic of Korea. At the end of May 2000 Prime Minister Mori visited Seoul and held a meeting with President Kim Dae-Jung, in which he affirmed Japan's support for the forthcoming inter-Korean summit and advocated close co-operation between Japan and the Republic of Korea with a view to establishing peace and stability on the Korean peninsula.

Japan's attempts to establish full diplomatic relations with the Democratic People's Republic of Korea (DPRK—North Korea) in early 1991 were hindered by the latter's insistence that Japan make financial reparations for losses sustained during and following Japan's colonial rule of Korea in 1910–45. The refusal of the DPRK to allow inspection of its nuclear facilities featured prominently on Japan's national and international agenda. Japan stressed the need to find a diplomatic solution and any substantial support for military intervention was ruled out. In March 1995 a delegation comprising members of all three parties in the Japanese Government visited North Korea to prepare for negotiations on the establishment of normal relations between the two countries. Japan provided emergency aid to the DPRK in 1995/96, when a serious food shortage appeared to threaten the stability of the Korean peninsula. The Japanese Government supported a US initiative of dialogue involving the DPRK, China and the USA, aimed at negotiating a formal peace treaty between the two Korean states. Concerns that the DPRK had developed a missile capable of reaching Japanese territory resulted in the suspension of food aid in early 1997, but relations improved slightly later in that year, when agreement was reached concerning the issue of visits to relatives in Japan by Japanese nationals resident in the DPRK. The first such visits took place in November. It was announced in August that the two countries were to conduct negotiations aimed at restoring full diplomatic relations. It was subsequently reported that the Japanese Government had pledged some US $27m. in food aid to the DPRK. However, food aid and negotiations on the resumption of full diplomatic relations, which had commenced following the visit to the DPRK by an LDP delegation, were suspended in mid-1998, following the testing by North Korea of a suspected missile in the sea near Japan. Tensions were exacerbated in March 1999, when two suspected North Korean spy ships, which had infiltrated Japanese waters, were pursued and fired on by Japanese naval forces, in the first such operation since the establishment of the Japanese SDF. In September the Japanese Government welcomed North Korea's agreement with the USA to suspend its reported plans to test a new long-range missile, but remained cautious regarding any easing of sanctions against Pyongyang. In October unofficial talks were held in Singapore between Japanese and North Korean government officials, and Japan subsequently lifted a ban on charter flights to North Korea. Following a visit to North Korea by a multi-party delegation of Diet members in December, the Japanese Government announced that it would resume the provision of food aid. Later that month intergovernmental preparatory talks on re-establishing diplomatic relations were held in China, following an agreement between Japanese and North Korean Red Cross officials on humanitarian issues, most notably the commitment by the Red Cross organization of North Korea to urge its Government to co-operate in an investigation into the fate of some 10 missing Japanese nationals, believed by Japan to have been abducted by North Korean agents in the 1970s and 1980s. Official talks on the normalization of diplomatic relations were held in April and August 2000. A number of issues were discussed, including compensation for the Japanese colonial rule of Korea, although no substantive progress was made; a third round of talks was held in October.

In 1978 a treaty of peace and friendship was signed with the People's Republic of China. During an official visit to China in August 1988, Prime Minister Takeshita announced that Japan would advance 810,000m. yen in loans to China between 1990 and 1995. These loans were withheld following the massacre by Chinese troops in Tiananmen Square, Beijing, in June 1989, but were released in mid-1990 following the Chinese Government's declaration, in January, that a state of martial law no longer existed. There was further dissatisfaction with the Chinese Government in early 1992, when Chinese sovereignty was declared over the Ryukyu island group, which was claimed by Japan. Nevertheless, good relations between the two countries were bolstered by visits to China by Emperor Akihito in October 1992 and by Prime Minister Hosokawa in March 1994. Having protested strongly against the resumption of French nuclear testing in the Pacific, in August Japan announced that it would suspend economic aid to China, following renewed nuclear testing by the Chinese Government. The provision of economic aid was resumed in early 1997, following a moratorium on Chinese nuclear testing.

In mid-1996 Japan's relations with both China and Taiwan were strained when a group of nationalists, the Japan Youth Federation, built a lighthouse and war memorial on the Senkaku Islands (or Diaoyu Islands in Chinese), a group of uninhabited islets situated in the East China Sea, to which all three countries laid claim. The situation was further aggravated in September by the drowning of a Hong Kong citizen during a protest against Japan's claim to the islands. In October a flotilla of small boats, operated by 300 activists from Taiwan, Hong Kong and Macao evaded Japanese patrol vessels and raised the flags of China and Taiwan on the disputed islands. However, Japan sought to defuse tensions with China and Taiwan by withholding official recognition of the lighthouse.

In September 1997 Prime Minister Hashimoto visited China to commemorate the 25th anniversary of the normalization of relations between the two countries. China expressed concern at the revised US-Japanese security arrangements, following a statement by a senior Japanese minister that the area around Taiwan might be covered under the new guide-lines. Procedures for the removal of chemical weapons, deployed in China by Japanese forces during the Second World War, were also discussed. Japan's economic policy was criticized by the Chinese Government in 1998, which feared that the weakening yen would force a devaluation of the Chinese currency. In November, during a six-day state visit by the Chinese head of state, Prime Minister Obuchi and President Jiang Zemin issued (but declined to sign) a joint declaration on friendship and co-operation, in which Japan expressed deep remorse for past aggression against China. China was reported to be displeased by the lack of a written apology, however, and remained concerned by the implications of US-Japanese defence arrangements regarding Taiwan. A subsequent US-Japanese agreement to initiate joint technical research on the development of a theatre missile defence system, followed by the Japanese Diet's approval, in May 1999, of legislation on the implementation of the revised US-Japanese defence guide-lines provoked severe criticism from China, despite Japan's insistence that military co-operation with the USA was purely defensive. In July a meeting in Beijing between Obuchi and the Chinese Premier, Zhu Rongji, resulted in the formalization of a bilateral agreement on China's entry to the WTO, following several months of intense negotiations on the liberalization of trade in services.

Japan and the USSR were, historically and geopolitically, rivals for supremacy in north-eastern Asia, over which they disputed control in large- and small-scale wars. Mutual mistrust was maintained from the end of the Second World War until the mid-1980s. Japan's treaty of peace and friendship, signed with the People's Republic of China in 1978, signalled an end to Japan's policy of maintaining equal distance from both China and the USSR. Japan's relations with the USSR were further strained by the Soviet military intervention in Afghanistan, in December 1979, and by the concomitant reinforcement of Soviet territory to the north of Japan. There was, however, a noticeable improvement in relations between Japan and the USSR in 1986. In January the Soviet Minister of Foreign Affairs, Eduard Shevardnadze, visited Japan (the first such visit by a Soviet Minister of Foreign Affairs for 10 years). The two countries agreed to improve economic and trade relations and to resume regular ministerial consultations.

The major obstacle to any substantial improvement in Soviet-Japanese relations was the seemingly intractable dispute over the Northern Territories. Japan has a strong claim to two islands, Habomai and Shikotan, which were formerly administered as part of Hokkaido. In addition, Japan claims Etorofu (Iturup) and Kunashiri (Kunashir), which, together with the rest of the Chishima (Kurile) chain and the southern part of Karafuto (Sakhalin), were captured by Soviet forces during the closing stages of the Second World War. Following a visit to Tokyo by the Soviet Minister of Foreign Affairs in December 1988, Japan and the USSR agreed to establish a high-level joint working group to negotiate the future of the disputed territory and the conclusion of a peace treaty. After the dissolution of the Soviet Union at the end of 1991, the President of the Russian Federation, Boris Yeltsin, reconfirmed the existence of this territorial dilemma.

The G-7 Summit of industrialized nations, held in Tokyo in July 1993, provided the opportunity for Yeltsin to visit Japan, two proposed trips having been cancelled in the previous 12 months. However, despite his presence, no solution was found to the dispute over the Northern Territories. The US $500m. in bilateral aid promised to Yeltsin at the Summit fell far short of the $4,000m. that President Clinton had initially suggested as a privatization fund. Despite US pressure, Japan's continuing reservations towards the provision of aid for Russia reflected both its frustration that the resolution of its territorial claims was still distant, as well as the development of more assertive diplomacy. Yeltsin also visited Tokyo in October, when the Northern Territories were discussed, although to little effect.

Bilateral negotiations over the status of the disputed territory opened in March 1995. Relations between the two countries steadily improved, and a commitment was made to resume negotiations on the signing of a peace treaty to bring a formal conclusion to the Second World War. In November 1996 Japan indicated that it was prepared to resume the disbursement of a US $500m. aid 'package', withheld since 1991, and in May 1997 the Japanese Government abandoned its opposition to Russia's proposed membership of the G-7 group. Russian plans for joint development of the mineral and fishing resources of the disputed territory were followed, in July, by an outline agreement on the jurisdiction of the islands. A meeting between Hashimoto and Yeltsin later in that month resulted in the forging of a new diplomatic policy, based on 'trust, mutual benefit and long-term prospects'. At an informal summit meeting, held between Yeltsin and Hashimoto in Krasnoyarsk, Russia, in November, the two parties agreed to work towards the conclusion of a formal peace treaty by the year 2000. A series of measures aimed at encouraging Japanese assistance in the revival of the Russian economy were also discussed. Bilateral negotiations resulted in the conclusion of a framework fisheries agreement in December 1997. Yeltsin visited Japan in April 1998, and the Japanese Government's commitment to an improvement in bilateral relations was confirmed by its offer of financial aid, in the form of loans, and an expansion in economic co-operation. No further progress was made on the status of the disputed islands. The Japanese Government indicated its support for Russia's application for membership of the WTO and APEC. Further discussions between Yeltsin and Obuchi, held during the G-8 summit meeting in Cologne, Germany, in June 1999, were to be followed by a visit to Tokyo by the Russian President later that year. At the beginning of September Japan agreed to resume lending to Russia, which had been suspended since the Russian Government had effectively devalued the rouble and defaulted on some of its debts in mid-1998. At the same time an accord was concluded on improved access to the disputed islands for former Japanese inhabitants.

Negotiations on the territorial dispute achieved little progress during 2000, a major obstacle being the issue of how many of the islands should be returned. Despite Russian President Vladimir Putin's repudiation of Japan's claim to any of the islands during his first official visit to Tokyo in September, Russia subsequently offered to abide by a 1956 declaration that it would relinquish two of the islands after the signature of a peace treaty, but Japan initially rejected this partial solution. Talks held in March 2001 on the interpretation of the 1956 declaration proved inconclusive, although Prime Minister Mori and President Putin reaffirmed that the declaration would ultimately form the basis for a peace treaty. Further discussions were to be held in October 2001, at which the Japanese Government was expected to insist on negotiating the return of all four disputed islands, rather than only two. Meanwhile, in November 2000 a former Japanese naval officer on trial in Tokyo admitted spying for Russia. There was outrage in Japan at the fact that the crime carried possible punishments that were perceived as being extremely lenient in relation to the severity of the offence.

The sharp increase in world petroleum prices in 1973–74 illustrated the vulnerability of the Japanese economy to developments in the international arena. Japan's increasingly pragmatic approach in foreign affairs was demonstrated by its support for the Arab countries in its pronouncements on Middle East issues, thus ensuring a continuing supply of petroleum. Moreover, its growing involvement in world affairs was signalled by the financial aid granted to the international effort to force the Iraqi withdrawal from Kuwait in 1991. In September 1995 Prime Minister Murayama visited the Middle East. During meetings with the Syrian President, Hafiz al-Assad, the Israeli Prime Minister, Itzhak Rabin, and the Palestinian leader, Yasser Arafat, Murayama pledged economic aid to the Middle East, the promotion of trade with the region and Japan's involvement in the UN Disengagement Observer Force (UNDOF) operation in the Golan Heights.

In September 1994 the possibility that Japan was soon to be given a permanent seat on the UN Security Council receded, following a UN report that called for further discussion on the issue. Earlier, Japan's Minister of Foreign Affairs had anticipated that it would take time for Japan to be accorded a permanent seat, and he emphasized that this could only be on condition that its contributions to UN activities did not involve the use of force. During the late 1990s the Japanese Government campaigned for a greater proportion of senior-level positions within the UN to be allocated to Japanese personnel, as a reflection of its contribution to the UN budget. From January 1997 until January 1999 Japan held a non-permanent seat on the UN Security Council, while continuing to seek permanent membership. Prime Minister Mori addressed the UN General Assembly on this issue in October 2000 and also advocated Japan's membership to a gathering of Asian and African nations. Japan had the strong backing of the United Kingdom and lukewarm support from Russia and China, but controversy over the whole idea of permanent membership meant that no new developments were expected in the near future, despite the fact that Japan was the second largest contributor to the UN in 2000, being responsible for more than 20% of the total budget.

Japan's foreign relations generally failed to flourish under the leadership of the Mori Cabinet, which seemed to lack a coherent national vision or strategy. Mori's propensity to make inappropriate remarks and his mishandling of various issues drew a constant barrage of criticism in the domestic media, which was taken up by the foreign press, thereby undermining respect for him abroad and any capacity for international leadership. The Okinawa summit of G-8 nations, which contributed little of value to Japan's international standing, saw Mori become the butt of Western jokes. Unable to overcome its economic problems, Japan suffered a loss of confidence domestically and also found its only diplomatic means seriously curtailed. Even Japan's official development assistance (ODA), which since the early 1990s had assumed an increasing significance in the country's efforts to define a more proactive international role for itself, came under pressure from 1997 and was reduced in the 2002/03 budget by 10%. Revelations of scandals involving misuse of funds within the Ministry of Foreign Affairs and the subsequent dismissal of top officials, including the ambassadors to the USA and the United Kingdom, left the ministry weakened. Disputes between senior officials and the new Minister of Foreign Affairs in the Koizumi Cabinet, Makiko Tanaka, also made them less effective.

Economic issues were also a source of friction between Japan and its allies in 2001, with Japan subjected to conflicting demands for economic reform and deregulation and for disposal of its bad debts and adoption of import targets. Measures taken by Japan in April against the import of three agricultural products—leeks, shiitake mushrooms and *tatami* reeds—from China led to broad tariff increases by the Chinese against a number of Japanese industrial products, including automobiles. Trade difficulties with China are symptomatic of attempts to deal with the shifting power balance in the region, as Japan formulates its strategy to position itself advantageously as China's economic and military strength grows. The Ministry of Economy, Trade and Industry's policy document on international trade, which was published in May, warned of heightening competition with China for economic leadership within the region, as China experienced outstanding levels of growth, and predicted that Japanese industry would relocate to China. The Japanese Government welcomed the new Administration of President George W. Bush in the USA and its shift from a policy of 'strategic partnership' with China to one of 'engagement'. A number of regional conferences in 2001 took as their theme the need for a new regional framework and for efforts to balance the emergent strength of China. The Malaysian Prime Minister, Mahathir bin Mohamad, proposed an international currency for use in international trade. Sub-regional economic co-operation and bilateral free trade, as well as the creation of an East Asia free-trade area, were all under discussion. Diplomatic and security questions were identified as possible areas of difficulty for Prime Minister Koizumi. Disputes over the contents of Japanese textbooks continued to strain relations with both China and the Republic of Korea under the Koizumi Cabinet, which, in early June, refused to make any of a number of requested revisions to a textbook produced by the Japanese Society for History Textbook Reform, which seeks to teach historical pride in the nation. Tensions were exacerbated by the perception of Koizumi as a dangerous nationalist by those countries.

The repercussions of the terrorist attacks on the USA in September 2001, three days after the 50th anniversary of the signing of the San Francisco Peace Treaty and the US-Japan Security Treaty, dominated many of Japan's subsequent international relations. The US-Japan relationship improved to a level not seen since the days of the 'Ron-Yasu' relationship enjoyed by US President Ronald Reagan and Yasuhiro Nakasone in the 1980s. Koizumi's own aggressive image meant that his normalization of Japan's foreign policy necessitated skilful diplomacy if it were not to be perceived as a threat to the stability of the region. As a consequence, at the same time as Japan was drawn into a more active global role, for example in its contributions to peace-keeping operations and in its initiatives for a UN-sponsored World Summit on the reconstruction of Afghanistan, it also become more closely involved in regional activity on both a bilateral and multilateral level. Koizumi's regional offensive encompassed a meeting with Jiang Zemin in October 2001, visits to five ASEAN nations in January 2002, the signing of a New-Age Economic Partnership with Singapore and participation in the first annual Boao Economic Forum for Asia in the Chinese province of Hainan in April when he argued in favour of an Asian free-trade zone. These initiatives were also driven by concern over China's rapid economic and military growth and by that country's proposals for a China-ASEAN free-trade agreement; demands were made in Japan for drastic reductions in ODA to China. In May 2002 tensions were exacerbated when the Chinese seized North Korean nationals seeking refuge in the Japanese consulate in Shenyang.

An unexpected official visit to the Yasukuni Shrine by Koizumi in April 2002 resulted in the cancellation of a visit to Japan by the South Korean Minister of National Defence and led 800 South Koreans to join a pending legal action against the Prime Minister for unconstitutional behaviour. The joint hosting of the 2002 football World Cup in June brought an increase in co-operation and cultural exchange between Japan and South Korea. This, along with the initiation of a joint history study group, contributed to a steady improvement in relations between the two countries. Economic relations were strengthened by an investment alliance and by progress in talks for a broader free-trade agreement. Although Japan did not agree with the Bush Administration's description of North Korea, Iraq and Iran as an 'axis of evil', North Korea continued to be a source of growing concern, with a worsening situation on the Korean Peninsula and incursions into Japanese waters by North Korean spy ships. The relationship took a dramatic turn in mid-September when Koizumi returned from an unprecedented visit to Pyongyang with an apology for the abduction of 12 Japanese nationals by North Korea during the 1970s and 1980s, and the news that eight of the abductees had died. Apparent progress was made in the other major area of concern, ballistic missiles and nuclear weapons, with the signing of the Pyongyang Declaration whereby the North Korean Government promised to continue its moratorium on missile test launches, to respect international agreements on nuclear weapons inspections and to halt operations on spy ships in Japanese waters. For its part, Japan expressed remorse and apologies for its colonial past, and a readiness to pursue negotiations on the restoration of normal relations and resumption of economic assistance.

The continuing impact of the 'war on terror' provoked by the events of 11 September 2001 made the distinction between international relations and national security debatable. Nowhere was this truer than in the case of Japan where, as the extensive coverage in the Japanese print media attested, North Korea continued to dominate Japan's foreign relations/national security debate in subsequent months., The situation in Iraq was also much debated in Japan. The two issues were clearly related and the underlying common concern, the potential for damage to Japan's relationship with the USA (the 'cornerstone of Japanese diplomacy'), was critical in determining Japan's responses to relations with both countries. Only one month after the signing of the Pyongyang Declaration came the US announcement that North Korea had admitted to having an ongoing nuclear-weapons programme. This was followed in short order by North Korea's expulsion of inspectors of the International Atomic Energy Agency (IAEA), and an announcement that it would withdraw from the Non-Proliferation Treaty and resume operations at nuclear facilities. The sharp increase in public fear and anger was reflected at official levels by a heightened state of alert against the threat of North Korean missiles. In April 2003 North Korea's admission of its possession of nuclear weapons and its offer, during three-way talks with the USA and China, to end nuclear weapons development and the export of missiles after the establishment of normal relations with the USA and Japan, further complicated the issue for Japanese officials, who were caught between an aversion to imposing sanctions on North Korea and a deep reluctance to normalize links without a final settlement of the issue of Japanese abductees. In late June Japan unexpectedly took the initiative with proposals for wider negotiations encompassing also Japan and South Korea and suggestions that Prime Minister Koizumi might make a second visit to the North in the near future. This was taken one stage further in July with a proposal for a resumption of bilateral negotiations alongside the multilateral talks. The USA, in the mean time, took the harder line of presenting UN Security Council members with a draft statement denouncing North Korea's nuclear ambitions.

The issue was made more difficult for Japan by its need to remain on favourable terms with South Korea and China. In early June 2003 discussions were held in Tokyo with South Korean President Roh Moo-Hyun. The talks overcame controversial claims by the LDP policy chief, Taro Aso, that Koreans had chosen to take Japanese names during the colonial period, and stressed a 'forward-looking' approach to bilateral relations and shared concerns over North Korea. A visit to Japan was scheduled for August for the Chinese Minister of Foreign Affairs, Li Zhaoxing, to discuss the options on North Korea and an exchange of visits by Koizumi and Chinese Premier Wen Jiabao was proposed for later in the year. Wen's previous planned visit had been cancelled following disputes over the leasing by Japan of the Senkaku/Diaoyu Islands and a visit by Koizumi to the Yasukuni Shrine earlier in the year. The 25th anniversary of the signing of the Peace and Friendship Treaty with China appeared likely to be marked by a more positive note, as negotiations on the waiving of visas for Japanese visitors approached completion. In 2002 a total of 2.98m. Japanese tourists had travelled to China, an increase of 25% compared with the previous year (although tourism in China was badly affected in 2003 by the outbreak of Severe Acute Respiratory Syndrome (SARS) and the obfuscation that followed the discovery of the virus). As was acknowledged by a prime ministerial task force on foreign policy in November 2002, China thus remained 'the most important theme' for Japan's diplomacy in the 21st century.

Despite this, Iraq and North Korea undoubtedly remained the main foreign policy/national security preoccupations in the last quarter of 2003 and during 2004. Tension in relations with North Korea increased in late October 2003 with the firing of two *Silkworm* missiles into the Sea of Japan (or the East Sea as it is known on the Korean peninsula), but bilateral efforts to resolve the remaining issues related to the Japanese nationals abducted by North Korea continued, with Japanese diplomats making visits to Pyongyang in January and February 2004 prior to the six-nation talks on North Korea's nuclear programme held in Beijing at the end of February. Throughout the six-nation talks and the associated bilateral discussions, progress on the abduction issue and the dismantling of North Korea's nuclear arms and missile programme remained obstacles to Japanese participation in any provision of aid or agreement to normalization of relations, and the talks ended without any decisions being reached. Diplomatic discussions continued, alongside a hardening of the Japanese position that included the introduction of legislation to allow the imposition of economic sanctions on North Korea, as well as, subsequently, legislation to ban port visits from foreign ships deemed a threat to public security. In May, to a mixed domestic reaction, Koizumi made a second trip to Pyongyang and returned to Tokyo with five children of abductees who had themselves been repatriated in 2002; but Koizumi returned without the daughters or the US husband, Charles Jenkins, of one abductee. (In July, however, Charles Jenkins, who was facing charges of desertion in the USA, arrived with his daughters in Tokyo, where he began a programme of medical treatment.) There was no further clarification of the fate of the remaining 10 missing Japanese. None the less, Japan promised food and financial aid in return for the release of the abductees' children. In June Koizumi stated his goal as the normalization of diplomatic relations within one to two years, and in August his Cabinet gave authorization for dispatch, via UN agencies, of the first batch of the 250,000 tons of food and US $10m.-worth of medical supplies he had promised. Diplomatic negotiation continued, and a fourth round of six-nation talks was scheduled for the end of September (a third round having taken place in June, without reaching any significant conclusion).

Relations with China suffered several upsets from October 2003, although there were some positive developments. The Japanese Government undertook to send more experts to China to facilitate the ongoing removal of chemical weapons left behind during World War II; the yearly tripartite summit meeting with South Korea begun under Prime Minister Obuchi committed itself to greater closeness on security and economic issues, and visa restrictions for Chinese visitors were loosened. Nevertheless, Japan continued to be concerned about the potential for trouble between China and Taiwan following the re-election of President Chen Shui-bian in March 2004. There were also heightened tensions over the disputed territories of the Senkaku/Diayu islands. Japan arrested and then immediately deported without prosecution seven Chinese activists who had landed on one of the islands. The US Government, possibly in gratitude for Japan's support in Iraq, made clear that the islands were included in the scope of the US-Japan Security Treaty. In response, China cancelled or postponed various ceremonies, visits and conferences including one on incursions by its ocean survey ships into Japanese exclusive economic zone waters (EEZ). There was also concern within Japan that China was accessing natural gas in Japan's EEZ from installations close by in the South China Sea. A dispute over the boundary between the EEZs claimed by the two countries seemed possible in July when Japanese ships surveying in the area were prevented from carrying out their activities by the Chinese navy. There was a new development on the continuing issue of official visits to the Yasukuni Shrine when in April 2004 a Japanese District Court ruled Prime Minister Koizumi's August 2001 visit to have been unconstitutional. Both China and Korea welcomed the ruling that the visit violated the separation of state and religion.

NATIONAL SECURITY, DEFENCE AND REARMAMENT

Serious consideration of rearmament, in order to make Japan both an economic and a military power, would have been welcomed by the USA in the 1970s, but there were many fundamental obstacles. Constitutional revision was one prerequisite, since Article 9 of Japan's post-war Constitution is usually interpreted to mean that Japan's armed forces are for defence only, must remain relatively small in number and cannot be equipped with inter-continental ballistic missiles. However, any attempt to revise the Constitution would cause a major domestic political crisis. Moreover, rearmament would be bitterly contested in South-East Asia, where anti-Japanese sentiment remains very strong. Attention, therefore, was turned to alternative solutions, which included, in 1979, Prime Minister Ohira's plan for 'comprehensive security'. Briefly, this proposal envisaged a commitment by Japan to promote the integrity of the geopolitical structures within which it conducts its economic relations. For example, Egypt was identified as the key to regional stability in the Middle East, and so became an important beneficiary of Japan's foreign aid, as did Indonesia, a major supplier of petroleum to Japan after 1973. Prime Minister Nakasone favoured the creation of a strong defence force and a broader role for Japan in regional affairs. In matters of defence, Nakasone sought to make concessions to the USA, so as to reduce friction in (or at least divert attention from) the less tractable problems of trade.

In Japan, as elsewhere, global economic recession led to fiscal austerity, which, in turn, compromised the defence programme for 1983–87. Despite an average increase of 6.5% per year in military expenditure, the programme's aims were not fully realized. At the same time, Nakasone's decision to permit the transfer to the USA of new technologies with military applications indicated a change of policy, which was welcomed in the USA. In January 1987 the Japanese Government announced that it was to abandon its self-imposed limit on defence expenditure of 1% of the gross national product (GNP), which had operated since 1976. Defence spending equivalent to 1.004% of the forecast GNP was proposed for 1987/88; it was also announced that defence expenditure would be maintained at about this level until 1991. This decision was welcomed by the USA but harshly criticized by the USSR and the People's Republic of China. Nakasone, however, stressed that the Government did not intend to re-establish Japan as a major military power.

In March 1991 the Government reduced defence expenditure from the budget proposals for the financial year ending 31 March 1992, partly to persuade the Komeito to approve Japan's pledged additional US $9,000m.-contribution to the Gulf War effort, but also to reflect domestic concern that previous levels of defence expenditure were no longer necessary. Compared with the previous year, proposed defence spending was to rise by less than 4% in 1992/93.

After failing to win the approval of the Diet for proposals to send SDF personnel overseas in November 1990, the LDP agreed with the Komeito and the DSP that the Government would establish a new body to participate in UN peace-keeping operations. Although the Komeito initially insisted that SDF personnel be excluded from the new body, the proposal was finally approved on the condition that SDF personnel be confined to non-combat duties. In July 1991 the three political parties dispatched a mission to Cambodia and other Asian countries to assess the possibility of Japanese involvement in a UN-sponsored peace plan for Cambodia. In September 1992 683 Japanese personnel, including some civilian police officers, were sent to Cambodia to participate in the UN Transitional Authority in Cambodia (UNTAC) mission. The death of two Japanese personnel in Cambodia, in early 1993, heightened domestic opposition to Japan's involvement in UN peace-keeping operations. In September 1995 it was announced that SDF personnel would participate in a UN observer force in Israel from February 1996. In November 1999 the ruling parties agreed to postpone the consideration of a proposal to expand Japan's participation in UN peace-keeping operations.

The report of the Advisory Group on Defence Issues, commissioned by Hosokawa's Government, was published in August 1994. The report urged a reduction in SDF personnel from some 274,000 to 240,000, recognizing the difficulty that the SDF ground forces had faced in achieving full recruitment. The proposals were largely adopted in early January 1995 in a draft document on new defence policy guide-lines. The report also called for reductions in aircraft and ships, the modernization of Japan's defence capability, and improvements in command, communications and intelligence. A case was also made for co-operation with the USA on the development of weapons systems such as anti-ballistic missile defences.

In a bid to counter the difficulties being experienced in their economic relationship, Japan and the USA sought to strengthen their security relationship. In early 1995 Japan's Defence Agency undertook a review of the bilateral security treaty to expand the financial support accorded by Japan to US operations overseas.

In September 1995 a 12-year-old girl was raped by three US servicemen in Okinawa. The incident led to nation-wide demonstrations against the three men (who were found guilty and sentenced to prison terms in March 1996) and against the Government's policy on Japanese-US security arrangements. Public support for the 35-year-old mutual security treaty declined sharply. In view of such public discontent, Governor Masahide Ota of Okinawa refused to approve the continued use of land by US military forces when the Defence Agency requested an extension of the leases demanding a review of the Japanese-US security arrangements in general, and of the Status of Forces Agreement in particular. A referendum, held in Okinawa in September 1996, revealed that the majority of residents favoured a reduction in the number of US bases. Nevertheless, later in that month Ota reversed his decison and agreed to sign the documents renewing the leases. In December it was announced that the USA was to return some 20% of the land used for military bases and to build a floating offshore helicopter base; there was, however, to be no significant reduction in the number of troops stationed in Okinawa. In order to reduce the financial and other obligations borne by Okinawa, some US military exercises were relocated to other prefectures, and the Government also began to negotiate special measures to promote economic development in the region. In December 1997, in a non-binding referendum held in Nago, Okinawa, to assess public opinion concerning the construction of the offshore helicopter base, the majority of voters rejected the proposal. The Mayor of Nago, who advocated the construction of the base in return for measures to stimulate the region's economy, tendered his resignation. Governor Ota stated his opposition to the proposed base. The new Mayor, elected in February 1998, initially approved of the helicopter base, but subsequently announced that he would support Ota in opposing the construction. In November 1998 Keiichi Inamine defeated Ota in the Okinawa gubernatorial election. Inamine, who had been supported by the LDP, presented an alternative solution, in an attempt to gain government support for the local economy, proposing that a military-commercial airport be built in northern Okinawa and leased to the USA for a period of 15 years. In December a US military site was officially returned to the Japanese Government, the first of the 11 bases to be returned under the 1996 agreement. In December 1999 Inamine's proposal for the relocation of the US air base to northern Okinawa was approved by both the local authorities and the Japanese Government, with the Henoko district of Nago chosen as the site for the new airport; at the same time funding was allocated for a 10-year development plan for the area. Negotiations with the USA, which opposed any time limit on its use of the airport, subsequently took place; although it had been hoped that negotiations would be concluded by July 2000, when the summit of G-8 nations was held in Nago, no agreement was reached by that time. In March, during a visit by the US Secretary of Defense, William Cohen, some agreements were concluded, giving Japan control of the US Kadena air base and resolving an air pollution problem in Kanagawa prefecture, but the central issues of a time limit on US use of the new airport and of the level of Japan's payments to the USA as a host nation to its military forces remained unresolved.

Following the signing of a US-Japan Joint Declaration on Security in April 1996 by President Clinton and Prime Minister Hashimoto, the review of the Guidelines for Japan-US Defense Co-operation (compiled in November 1978) continued in 1997, and culminated in the issuing of a joint statement detailing the new Guidelines in mid-September. China expressed concern at the provisions of the new agreement, which envisaged enhanced co-operation between the USA and Japan, not only on Japanese territory, but also in situations in unspecified areas around Japan. Despite opposition from its coalition partners, in April 1998 the LDP Government approved legislation to define the operations of the SDF under the revised Guidelines. The legislation was enacted in May 1999, prompting criticism from China and Russia. Its approval was ensured by an agreement between the LDP, its new coalition partner, the LP, and New Komeito to exclude a clause that would have allowed the inspection of unidentified foreign ships by the SDF, with the aim of enforcing economic sanctions; separate legislation on this issue was expected to be proposed later in the year. In August 1999 the Japanese Government formally approved a memorandum of understanding with the USA stipulating details of joint technical research on the development of a theatre missile defence system, which aims to detect and shoot down incoming ballistic missiles within a 3,000-km radius. The Defence Agency estimated that Japan would have to allocate up to 30,000m. yen to the controversial research project over a period of five years.

Instability on the Korean peninsula was a cause of concern for the Japanese Government in 1998, particularly following the testing of a suspected missile by the DPRK (see above). Officials from the ministries of foreign affairs and defence of the Republic of Korea and Japan agreed to convene regular bilateral security meetings. The incursion into Japanese waters of two suspected North Korean spy ships in March 1999 prompted the first-ever invocation of legislation allowing naval forces to engage in maritime policing operations (see above). The incident also provided the first opportunity for the South Korean and Japanese armed forces to operate a new 'emergency liaison system', which had been established as a result of a recent bilateral security agreement. The Japanese Government criticized India and Pakistan for conducting nuclear tests in mid-1998, and suspended grants of non-humanitarian aid and loans to India in response to the tests. A series of missile tests carried out by India and Pakistan in April 1999 again provoked criticism from Japan. Relations had improved by the end of 1999; in late 1999 and early 2000 the Indian Ministers of External Affairs and of Defence paid official visits to Tokyo, and in February 2000 former Prime Minister Ryutaro Hashimoto visited India as a special envoy. During his visit he urged India to sign the UN's Comprehensive Test Ban Treaty, although he also indicated that Japan had no desire for bilateral relations to be determined by India's response to the Treaty. In late August Yoshiro Mori visited India as part of a tour of South Asia, in his first overseas trip as Prime Minster.

As part of efforts to combat increasing piracy in South-East Asia, Japan advocated the creation of a regional patrol for these waters. Although constitutionally unable to provide members of its navy for this task, Japan proposed the involvement of its coastguard, a non-military organization, in a regional force. In May 2000 Japan hosted a conference of coastguard officials from 15 Asian nations, at which the problem of piracy, and possible solutions, were discussed.

A new five-year mid-term defence programme was adopted in December 2000, in response to the changing nature of security threats in the post-Cold War world and to demands for a greater international contribution to peace by Japan. The new defence programme was to develop readiness to deal with nuclear, biological and chemical threats and terrorist attacks and was intended to bring Japan's defence capability to the levels prescribed in the national defence programme outline.

Under the Mori Cabinet, the old issues that bedevilled the US-Japanese relationship continued. Military bases in Okinawa contributed to anti-American feeling in the wake of rising crime in the area. US demands for a greater contribution to the maintenance of regional security from Japan grew under the new Bush Administration, and the question of what role Japan should play under the new Guidelines remained unresolved. The sinking of the fisheries training ship, the *Ehime-Maru*, by a US nuclear submarine in February 2001 exacerbated this growing anti-US nationalism.

In 2001 Japan maintained its official position of 'understanding' the missile defence initiative and its support for joint US-Japanese technical studies, while steadfastly avoiding any decision on whether it would enter into joint development of the Theater Missile Defense system. However, criticisms of the initiative by Minister of Foreign Affairs Makiko Tanaka and a hardline stance on the part of the Bush Administration, which refused to distinguish between support for short-range theatre missile defence and long-range national missile defence, suggested that the issue would prove difficult to resolve.

Unusually, security and defence issues were the main focus of Junichiro Koizumi's first press conference after his appointment as Prime Minister in April 2001. Koizumi emphasized the need to avoid the sort of international isolation that had led to the last war, the importance of the US-Japan Security Alliance, and the value of changing the Constitution over time to recognize the right to existence of the SDF and to take part in collective self-defence.

The terrorist attacks on the USA in September 2001 and President Bush's insistence on co-operation from the rest of the world in the so-called 'war on terror' revived unpleasant memories in Japan of the tensions created by the character and timing of Japan's contributions to the Gulf War in the early 1990s. Such memories clearly contributed to the speed and strength of the Government's response. Despite the USA's later omission of Japan from the list of countries contributing to the 'war on terror', the Prime Minister's response was rapid and positive, going further than ever before towards normalizing Japan's international role, although without leading to any change in the Government's interpretation of its ability to exercise the right to collective defence. The Government's seven-point response plan, which included legislation to allow deployment of the SDF as theatre support, the dispatch of information-gathering warships as far as the Indian Ocean and the resumption of aid to India and Pakistan, received a mixed response domestically. Overall, however, the terrorist attacks contributed to the continuing erosion of domestic hostility towards a broader international role for Japan.

Anti-terrorism legislation was implemented on 2 November 2001, effective for two years but with the possibility of subsequent renewal. The legislation allowed for the dispatch of the SDF to non-combat zones in areas of conflict, subject to consent by the host government and to approval by the Diet within 20 days of troops being dispatched, or at the start of a new session. Such troops were given expanded powers to use weapons to protect refugees and injured foreign servicemen. In mid-November two destroyers and a supply ship sailed for the Indian Ocean in the first of several SDF contributions.

With various restraints being eroded by the events of September 2001 and the subsequent 'war on terror', the other major legislation introduced in the 2002 ordinary Diet session consisted of three defence bills, one setting out the response to a direct foreign military attack or expected attack, and the others revising the Self Defence Forces Law and the Law on the Establishment of the Security Council. Efforts to create a new framework to deal with national emergencies had begun in the mid-1970s, one reason perhaps why the proposed legislation did not include responses to terrorism or incursions into Japan's territorial waters, despite the urgings of Koizumi. The new legislation provided for an emergency headquarters in the Cabinet Office, headed by the Prime Minister, established that government would make plans to deal with military emergencies on the basis of UN Security Council recommendations, granted the Prime Minister extended powers to direct the operations of local government and reduced restrictions on the domestic activities of the SDF. Flaws in the drafting, a scandal involving the Defence Agency and LDP internal disputes meant that debate on the bills was initially postponed until the October 2002 session of the Diet, but the proposed legislation was still pending in early 2003 at the height of the conflict in Iraq and the confrontation with North Korea. A bipartisan agreement was finally reached in mid-May, following the inclusion of opposition provisions on protection of human rights, and the bills became law with the support of more than 80% of the Diet in early June. The passage of the military emergency bills was indicative of the dramatic changes in both party politics and public opinion over the previous 10 years. Such, indeed, were the changes that Prime Minister Koizumi felt able to suggest that Japan could attack foreign missile bases if it perceived there to be an imminent threat and to argue that the Constitution should be revised to acknowledge the SDF to be a military force. Following its launch of the first two of four spy satellites in March 2003, Japan's capacity to monitor any threatening activity on the part of North Korea was greatly enhanced.

The establishment of guide-lines for responding to foreign attack was not the only difficult issue to be tackled. The controversial decision in December 2002 to dispatch to the Indian Ocean a destroyer capable of sharing military data with the US fleet was perceived by many opponents as an erosion of the constitutional prohibition against collective self-defence. Similar criticisms were made of the Government's decision in June 2003 to move to the development and deployment stage of the missile defence initiative with the USA. The decision, which raised a number of legal and constitutional issues, was facilitated by evidence of nuclear and missile development on the part of North Korea. Joint exercises with the USA on missile interception were scheduled to begin in fiscal 2004, and a decision on moving to the final stage was to be made immediately afterwards. The Defence Agency announced that the Constitution would allow Japan to intercept missiles aimed at a third country but overflying Japan. The deployment of a missile defence system was likely to be part of a US 'military transformation', which would include a significant reduction in the 37,000 US forces stationed in South Korea but not in the number of troops deployed in Japan itself.

There was little support for Koizumi's strong, early backing of the US decision to oust the regime of Saddam Hussain in Iraq, either among the Japanese public, 49% of whom were against the operation, or in the Diet, where 68% opposed any action without a UN Security Council resolution. In defence of his position, Koizumi cited both the importance of the US-Japan Alliance and the North Korean nuclear and missile crisis. The same combination of factors stood behind Japan's contributions to post-war reconstruction work in Iraq. The controversial legislation needed to authorize the sending of SDF personnel to help in the rebuilding of Iraq raised issues about the use of weapons, conditions for dispatch and parliamentary approval.

The Prime Minister continued to jeopardize his position as a consequence of his efforts to convince the USA that Japan was a 'trustworthy ally'. The overwhelming importance of Japan's security relationship with the USA was evident in the Government's downplaying of the need for international co-operation through the UN and in the high political risks taken by Koizumi in his decision to send troops to Iraq. Deployment of the Air Self-Defence Force (ASDF) to provide reconstruction and humanitarian assistance to Iraq was delayed by domestic opposition that grew as the situation in Iraq deteriorated, with the killing in November 2003 of two Japanese diplomats, the first

Japanese casualties since the start of the conflict. Parliamentary debate on the deployment began during the legislature's recess at the end of 2003. In accordance with the special law passed in the previous July, the deployment of all SDF forces was given *ex post facto* approval in a forced vote in the House of Representatives in January 2004. This led to a short opposition boycott of the House of Councillors. The dispatch of the SDF was seen as a further step in the 'normalization' of Japan's foreign and security policies, a step that Koizumi insisted did not violate the constitution. Ground Self-Defence Force (GSDF) units totalling 550 men arrived in southern Iraq in February and March. The taking of Japanese hostages in April, widely seen as a crisis that could bring down the Government, ended without incident and without the withdrawal of Japanese forces. At a meeting with President Bush prior to a UN Security Council resolution in June, Koizumi announced that the SDF would, for the first time, participate in a multinational force following the return of sovereignty to Iraq. The conditions set out by Koizumi were that there would be no use of force and no activities outside non-combat zones, and that the Japanese forces would act within the existing special law and remain under Japanese command. Japan was expected to host a donors' conference on Iraq in October 2004.

In December 2003, meanwhile, commitment to the missile defence initiative was taken a stage further when the Government made a formal decision to introduce the system and authorized the signing of contracts. The Defence Agency made known its intention to propose legislation that would give the Prime Minister the power to mobilize the missile defence system without the authorization of the Cabinet or the Security Council of Japan in case of emergency. A new defence plan proposed by the Defence Agency provided for further shifts away from the defence posture of the 1995 outline towards missile defence and anti-terrorism measures more fitting to the international situation following the events of 11 September 2001 and to the heightened threat from a nuclear-capable North Korea. The Government's new defence guide-lines were to be finalized in the latter part of 2004 and adopted in December. It was also announced that restrictions on the export of weapons were to be revised and that the Acquisition and Cross-Servicing Agreement (ACSA), regulating relations between the SDF and US forces, was to be extended. Special anti-terrorist units were created by the Defence Agency, and the 2004 budget made extra funds available to the police to expand internal security measures. The war contingency law approved in June 2003 was expanded with further legislation in June 2004 that set down detailed rules on responses by US forces in Japan and by the Marine Self-Defence Force (MSDF) in the event of a foreign attack on Japan. The legislation was supported by both the ruling coalition and the DPJ. A defence 'white paper' published in July set out the strategic background to all these changes. In addition to these substantial developments in Japan's defence posture, there were various proposals on issues including the upgrading of the Defence Agency to a ministry, SDF units dedicated to international operations, a comprehensive, permanent law on the dispatch abroad of the SDF and the revision of Article 9 of the Constitution to allow collective defence. Japan announced its bid for a permanent seat on the UN Security Council at a meeting of the UN General Assembly in September 2004, but the country's pacifist Constitution was seen as a potential obstacle to achieving this aim.

Economy

BRUCE HENRY LAMBERT

INTRODUCTION AND OVERVIEW

Japan's economic condition is now developing positively on many fronts, with resurgent growth of GDP in real terms, increased new investment by businesses, rising exports and gradual structural reform. There has been a progressive opening-up to outside investment, and sound progress in reducing the long-term chronic problems of non-performing loans (NPL) by Japanese banks has been made. Critics complain about the slow pace of such changes, and the Japanese press regularly specifies the weakening of major reforms stemming from bureaucrats and politicians linked to vested interests. Yet major change is afoot, if only in that much of the population is eager for positive systematic modification and fervently seeks to end nearly 15 years of national economic stagnation.

Since the early 1990s Japan has been attempting to adjust to demands for deregulation and for a scaling-down of government, while conversely responding to calls for visionary leadership and the effective priming of a weakened economy. The Japanese populace has had to contend with economic stagnation, a gradual exposure of corporate financial mismanagement and continuing revelations of deficiencies in regulation. Associated problems have included rising unemployment, political instability, a steady dismantling of familiar institutional arrangements and a seeming loss of confidence in ethical values. It has become increasingly difficult for people to remember, let alone to reconcile, the exhilarating days of the late 1980s, when the economy was booming and Japanese corporate expansion was seen as a major threat to other businesses around the world. Nevertheless, Japan remains a wealthy nation. In the early 21st century Japan accounted for about 70% of total East Asian gross domestic product (GDP) and for about 60% of total combined GDP for East, South-East and South Asia. According to data from the Cabinet Office, Japan's overall real GDP for 2003 was 553,633,900m. yen (at 1995 prices), a rise of 3.2% compared with the previous year. The IMF projected that Japan's real GDP growth rate for 2004 would be 3.4%. In 2002 Japan's per caput GDP was equivalent to US $31,300, thus ranking it sixth among the world's 30 wealthiest nations belonging to the Organisation for Economic Co-operation and Development—OECD. Japan's adjusted GDP per caput when considered in terms of purchasing-power parity (PPP—comparing costs for a common basket of goods and services) was $27,000, ranking it 15th of the 30 OECD nations. Even this figure was still 4% higher than the average for the 15 members of the European Union (EU), although only 75% of the $36,100 average for the USA. (Luxembourg had the OECD's highest PPP-adjusted GDP at $50,600.)

Japan has comparatively low crime rates. The people receive good basic health care and enjoy the longest average life expectancy of any nation in the world. Tax and interest rates are low, and income is quite equitably distributed; the people are well-educated, city streets are clean, and government is small and contracting. The national infrastructure is up-to-date and well-developed (though notably exposed to potential disasters such as earthquakes). Therefore, Japan is in a relatively good position. The people have high standards of living, the Government is actively seeking to 'achieve a vibrant society', and history has given the nation reason to expect to persevere through various challenges. Prime Minister Junichiro Koizumi came to power in April 2001, promising major changes and 'structural reform with nothing off-limits'. During his early months in office, while doing little of substance, his popularity rating exceeded 80%; but by June 2004 his approval rating in opinion polls had decreased to 45%, with a 42% disapproval rating. The causes of this decline included a perceived inability to implement substantive reforms, delays in achieving privatization of costly public corporations, the dismissal of popular Minister of Foreign Affairs Makiko Tanaka in early 2002 and a series of government scandals that increased the public's distrust of both civil servants and politicians. The most recent major scandal arose in the midst of pension system reform, when it was revealed in 2004 that many major politicians, including the Prime Minister, had for many years managed to avoid paying required national pension premiums. The Japanese people want to believe that there are better days ahead, but a major problem at present is that Japan is now near the technological frontier. It is more difficult to find proven and unambiguous models to emulate, there is growing distrust of government, the national homogeneity is increasingly a weakness, and the path ahead is unclear.

After the following historical background, this essay provides an overview of the contemporary Japanese economy along with an explanation for many of the more novel features that affect the economy, including national policy, Japanese management systems, the influence of government, and relationships with the wider outside world. A crucial point might be borne in mind (from the 1989 work of one analyst, Tessa Morris-Suzuki, *A History of Japanese Economic Thought*): the term 'economics' in Japan (*keizai*) has its roots in *keikoku saimin/keisei saimin*: 'administering the nation and relieving the suffering of the people'. Economics in Japan is thus not simply a mechanistic allocation analysis of scarce resources, but the Confucian ideal of holding together the social fabric of the nation.

HISTORICAL BACKGROUND

Planning and public administration in Japan have a long history, and at times systemic controls have been extremely comprehensive. At the start of the Edo period in the 17th century, public administration systems put a number of stringent demands on society. One policy imposed was national seclusion (whereby leaving or entering Japan became punishable by death). Another policy required detailed reporting of all vital statistics for each area and family (and many of these records survive today). The system of *sankin kotai* that existed between 1635 and 1862 imposed limits on ambitious regional lords, who were required to spend alternate years in Edo (now Tokyo) and to leave their families behind as hostages when they returned to their domains. Other controls limited regional financial power: local lords were responsible for collecting and paying taxes imposed by the central Government, and their personal funds were deliberately kept in check through the expenditures involved in visiting the capital, maintaining two households, offering periodic bountiful gifts to the Shogunate, and maintaining roads and infrastructure to specifications and standards set by the central Government.

In the mid-19th century, Japan was forced by foreign powers to end more than 200 years of seclusion. This led to the disintegration of many of the above systems and relaxed a wide variety of rules. Feudal institutions drastically broke down: common people were allowed to take surnames and to travel freely, samurai were released from service and had to find new work, and boisterous anti-order dances swept the country (where, for example, people entered the homes of those more prosperous and helped themselves to food and drink, chanting 'eejanaika'—or 'what's the problem?'). The change from a feudal system to a more market-orientated economy brought a large measure of confusion and even chaos to the world of commerce. Numerous severe market shocks developed from the large-scale resumption of foreign trade and from the introduction of new foreign products. A revolutionary change to Western-style garments brought sharp alterations in demand for many items, with concurrent effects on employment. Social and technological rigidities made it difficult to adapt quickly. Many industries were suddenly and directly undercut by the import of competing items, while other industries (such as sword-making) suffered massive decreases in demand, based on new regulations and changing tastes and style. Such huge social displacement and change brought with it not only distribution channel breakdown and widespread malpractice that continued for many years, but also more serious problems: selective assassination and violent civil uprising (for the first 20 years after Japan's reopening, foreigners and those who supported them were often under threat). Market order was reimposed by co-ordinated self-regulation among businesses, supported by state authority; and these early corporatist restraining mechanisms were a precursor to some that can be found today.

Early Manufacturing and Trade

From 1854, when the so-called 'Black Ships' of the US navy threatened to bombard Tokyo and forced the Japanese to trade with the outside world, until well into the 1880s, most of Japan's actual import and export transactions were being conducted by foreign traders. This system was slow to change, as these foreign traders were not keen to share their knowledge and expertise with Japanese who might supplant them. The Meiji Government also found it difficult to give potential domestic traders substantial assistance, owing to Japan's international treaties of commerce. The Government itself was at first quite active in promoting industrialization through its programme to 'increase production and promote industry'. Modern technology was introduced from overseas, and several thousand foreign experts from numerous fields were hired by the Government to teach, consult, and assist Japan in modernizing. The central Government had inherited various operations such as shipyards and mines from the Tokugawa Shogunate, and to these were added railways and a number of model factories, the most notable in the textile industry, including silk-reeling, cotton-spinning, and woollen mills.

During the 1870s the authorities rushed to modernize Japan by way of an increasingly wide range of government projects. In the early 1880s, however, it became clear that state resources were seriously overextended. The Government decided to divest itself of direct administration of many factories and production facilities, and offered them to private investors. At first, only loss-making factories were scheduled to be sold, but by 1884 the Government had begun to sell off profitable facilities such as mines, often at nominal prices. Strategic munitions and communications facilities remained under government control, however, later to be joined by the huge Yawata Iron and Steel Works. Most companies were not sold on the open market. The key to procuring government-owned operations was often an existing special relationship between business insiders and government (a *seisho* relationship). Such divestiture to protégés would now not be tolerated, and even then it was a problem: insider dealings precipitated substantial public outrage in the Hokkaido Colonization Office Scandal of 1881. The Government did not develop a more equitable system or maximize immediate government income; certain buyers were favoured because the Meiji leaders wanted the businesses to succeed. To that purpose they minimized uncertainty and used a shortlist of preferred contractors: businesses and individuals with proven entrepreneurial and managerial skills, whose efforts were then supported. It must be remembered that in Japan at that time few people had substantive relevant experience of managing these new large-scale industries. The country had been closed off and fragmented in many ways up to the time of the Meiji Restoration in 1868; the clan-related strife of the Ansei Purge of 1858 and the Boshin War of the late 1860s had a negative effect on many organizations, and others were damaged by the loss of feudal patronage with the fall of the Shogunate. Two of the three major merchant houses (Ono and Shimada) collapsed under these changes, and Mitsui barely survived the transition; smaller merchant houses greatly suffered or became insolvent. The Meiji Government of the 1880s chose a risk-avoiding strategy favouring preferred firms, and such privileges were based upon an expectation of future service to the state.

The Government followed up its divestiture sales with another important move: becoming the major customer to many of these newly-privatized firms. This may have been particularly important with novel products with which people were unfamiliar, such as new electrical, glass or chemical products. Government purchases in these areas allowed such industries to generate important economies-of-scale with production and made many businesses viable. At the same time that the Government was selling various business operations to the private sector, it was becoming more active in both supporting and regulating private enterprise. Government adjusted its oversight of private-sector institutions in the direction of more interference. Businesses learned to exist under trying circumstances, or they perished. The main complaints by Japanese businessmen in the latter half of the 19th century were with what the business world saw as onerous taxation by government and their need to compete with imports. Yet a revised tariff convention signed in June 1866 between the Shogunate and the Five Powers reduced tariffs to a flat 5% of declared value, and this remained in force throughout the period of Japan's modernization. Full tariff autonomy was not achieved until the conclusion of a new set of treaties in 1911. After waiting so long to regain control of such sovereign protectionist measures, it is not surprising that subsequent Japanese governments were loath to lay them aside.

Relationships between Business and Government

Business in Japan, wary of charges of profiteering and selfishness, has needed carefully to state its case that investment risk and hard work justify a reasonable return. Mobilizing resources for production and serving the nation is far more attractive than the pursuit of profit. The motives of businessmen were harshly questioned in Japan, while politicians and bureaucrats were honoured for their service to others. Merchants in feudal times were deliberately and explicitly ranked below warriors, farmers and artisans on the social scale (*shi-no-ko-sho*). Japanese society in the modern period has not been widely recognized as inhospitable to mercantile pursuits, though from the Meiji Restoration in 1868 through to the end of the Second World War in 1945, Japanese business needed at times to struggle creatively in order to maintain a system of private enterprise. Private capital and independent direction and management of enterprise were severely and repeatedly threatened by the State. Such efforts at government hegemony into management were carefully resisted, a process that gave rise to protective mechanisms among businesses.

In the early 20th century Japanese business owners developed increasingly astute ways of dealing with government and with their own employees. Industrial disturbances and the organization of labour occurred to a minor extent, but an ideal workplace format was expected to operate as a family, under paternalistic ownership with similarities to feudal times. The Government repeatedly sought to guide and even control business strategy, but was typically hesitant to interfere logistically in such matters as friction with labour. To some degree, a mutually beneficial collaboration between major Japanese business interests and government was strengthened by the divestiture programme of the 1880s. Congenial relationships between the highest ranks of business and of government subsequently proved difficult to modify, and some businesses came to expect treatment with largesse. Many relationships were further consolidated through intermarriage and adult adoption. The resulting family networks often remain vigorous and powerful today. Small and medium-sized firms, on the other hand, tend to receive little assistance.

In the aftermath of the First World War in 1918, Japan's economy was in particular need of adjustment. Exports to Europe had quickly declined with the advent there of peace and reconstruction. This foreign market contraction raised pressures in Japan for adjustment in manufacturing and marketing techniques and pricing, for co-ordination between firms, and for government to take a more active role in encouraging trade. A major early dispute (in what was to be a long struggle) between business and the national government bureaucracy came in 1918, when the Government proposed the Munitions Industries Mobilization Law. The law sought to impose government administration of wartime factory production in the event of future hostilities. The business world, supported by the newly formed Industry Club of Japan, strongly resisted what it saw as a bureaucratic venture into its domain, but the Government prevailed. In 1929 (in the midst of peace-time) the Resources Investigation Law was passed, requiring firms to report on their resource and production potential. Later, as warfare on the Asian continent intensified, in September 1937 the Diet approved the Law for Temporary Measures for Imports and Exports which allowed the Commerce Ministry to regulate trade without consultation. Change pervaded Japanese society at all levels after 1939. The war in Manchuria had been waged since 1931, and over the course of the decade the people of Japan had become accustomed to not only escalating commodity deprivations but also the concepts of domestic propaganda and 'spiritual mobilization'. Exhortations to behave as Japanese and not to mimic the West went so far as to include a 1939 ban on permanent waves for the hair and a condemnation of tennis. The war footing required popular co-operation and self-censorship, and external overt controls continued to expand.

Control Associations

The Important Industries Association Ordinance of 30 August 1941 established a format of control associations (*toseikai*) for all vital industries. At first each control association was under the leadership of the chief executive of the industry's top firm (which in effect transferred control to the major *zaibatsu* groups such as Mitsui, Mitsubishi, Yasuda and Sumitomo). This was not the leadership format desired by government economic planners but a compromise forged with business leaders with the aim of eliciting their support and participation. This format soon changed however: each control association was required to have a full-time president, and each such president became a quasi-government official whose directives had force of law.

So-called reform bureaucrats repeatedly tried to extend their command over business, hoping to take control from the 'selfish' interests of the capitalist owners. Direct administration was partly stymied by the political influence of big business, and partly by the military, who demanded their own direct influence over munitions-related industries. More radical change was successfully resisted owing to the unacceptability of expected costs: direct bureaucratic administration would involve great uncertainty and a huge short-term set-back as administrators would have to learn production system basics.

While industry was shaken by these various developments, not all pressures were generated from outside the business world. The *zaibatsu* and the largest firms were each actively seeking to increase their group domains, often at the expense of the small and medium-sized producers. Many huge newer companies (called *shin-zaibatsu*) had grown quickly in the 1930s with the expansion of opportunities in Japan's colonial markets. Yet as war escalated and continued, large shares of many businesses were lost to the old established *zaibatsu*. The control association system also furthered the decline of the newer ventures because the associations were generally under the direction of the largest firms, usually from the largest *zaibatsu* groups. The control system actively encouraged the consolidation or voluntary domination of small firms. Control association rules changed quickly, were often vague, and were difficult to resist without special assistance.

One question that has yet to be adequately charted in Japanese economic history is the extent to which the wartime control organization left a legacy to the post-war world. It has been common in Japan to belittle links with repudiated wartime institutions, yet institutional innovations of that period influence trade associations and economic federations even to the present. During wartime those who had a stake in private enterprise were forced to combine their energies and work together both for the good of the nation and for their own self-preservation. The struggles of internal factionalism were fought privately; publicity could threaten the survival of the overall system of private administration of capital and property. The control associations served as high-level forums (throughout and past the end of the war) where the large industrial capitalist cliques thrashed out differences. Control association leadership demanded excellent managerial skills such as resourcefulness, agility and creativity, the ability to keep informed, and an advanced aptitude for mediation. Such talents are always of use, and it is thus not surprising that many of these leaders continued to direct Japan's business world in the post-war years.

The Early Post-War Years

From 1945 the post-war occupation authorities ordered the purging of certain remaining wartime Japanese political and economic leaders. Further changes in the business world emanated from the dissolution of the *zaibatsu* and economic deconcentration programmes, and new anti-trust legislation. At the end of the war, wealth and power were also targeted. The peerage was dissolved. Living expenses were limited by statute, and 56 key people from 10 *zaibatsu* families were required to submit full monthly details of their household accounts. The occupation authorities chose to use the Japanese central government bureaucracy to administer the country and implement their directives, and the balance of power in Japanese society shifted in a number of ways. The bureaucracy quickly became stronger than ever through the restriction or elimination of competing administrative elements: politicians were purged, *zaibatsu* were dissolved, and the military was disbanded. Senior-level wartime bureaucrats were purged to some extent, but personnel in the economic ministries were barely affected. The bureaucracy was instead suddenly in a very favourable position to receive timely intelligence. It had a measure of power in being able selectively to screen information it collected for the occupation authorities, and bureaucrats could choose the level of vigour with which directives were applied and implemented.

The *zaibatsu* dissolution programme of the post-war period yielded mixed results. One of the major changes was an influx of new talent into the senior managerial ranks of the major corporations, and a concurrent decline in the influence of the founding families. The widespread advent of professional managers took away some of the onus of corporations as vehicles merely for the private profit of big capital, and increasingly open public ownership gave corporations a more social character. Undesired bureaucratic manipulation continued, but companies now could claim more convincingly to be caretakers of the savings of many from throughout society. Of course, the post-war *keiretsu* business groups rose from the remains of the *zaibatsu*, and to this day exert substantial influence on the economy.

These early post-war years were extremely difficult owing to lack of capital, raw materials and markets. The occupation forces themselves generated an increasing amount of business, and this improved considerably with the advent of special procurement orders for the Korean War of the early 1950s and for reconstruction work afterwards. Overall, Japan's business world rather quickly recovered influence, and was heavily involved with determining details of post-war development, including framing much of the detail of what became the San Francisco Peace Treaty of September 1951. The Korean War provided an economic boost for Japan, and the steadily-increasing involvement of the USA in Viet Nam and Indo-China propelled the Japanese economy even further. This era of high-speed growth was marked by protected domestic markets, export promotion and subsidies, high levels of investment and low corporate interest rates. Japan's hard-working populace was rewarded for its diligence by recognizable improvements in living standards. From the foundation of the Liberal Democratic Party (LDP) in 1955, a pact had been agreed by business, politicians and the bureaucracy, each supporting the other. The People's Political Association (*Kokumin Seiji Kyokai*) was the funding conduit from business to the political world, closely linked with political fund-raising by the Keidanren business federation. The People's Political Association underwent a number of organizational changes after its beginning as the Economic Reconstruction Council in January 1955. It played an important role in the so-called '1955 system' by which the LDP came to be formed and through which it held uninterrupted power for 38 years. This Keidanren political funding programme was substantial: in 1991 the system transferred 13,000m. yen (US $125.0m.) into politics and in 1989 reportedly contributed 30,000m. yen ($241.5m.).

Business helped the bureaucracy in offering lucrative *amakudari* (or 'descent from heaven') appointments to retired bureaucrats. The politicians, each with only a small personal staff, were beholden to the bureaucrats (and to some extent the party and faction leadership) for research and background support in their efforts to draft laws and effectively implement policy. Throughout all of this, the US Government was a background ally, with tens of thousands of troops in Japan and a vested interest in Japan remaining a bastion of market-driven capitalism and a staging area for US anti-Communist activities in the Asian region. Although relations between Japan and the USA were quite solid, the so-called 'Nixon shocks' of departure from the gold standard and of ending the Bretton Woods agreement (which in 1944 had established a system of fixed exchange rates), and *rapprochement* with the People's Republic of China, caught many in Japan unawares during the 1970s. These factors, combined with the effects of the oil shocks and an economic slowdown, forced many in Japan to recognize that the strategic interests of Japan and the USA might considerably differ.

The focus of economic progress in the post-war years has shifted from heavy industry to lighter manufacturing to value-added high technology and services. There is hope to move even more to knowledge-intensive industries such as software development or cutting-edge biotechnology. One of the weak dimensions of the Japanese infrastructure, however, is that it has not been very accepting of diversity, or supportive of exploration. The rule of orthodoxy and consensus has taken strong root in Japan, to the detriment of that which is different, creative and unorthodox. Opportunity for the inward migration of non-Japanese talent is also limited, and the local environment is not well prepared for approaches in other than the Japanese language. Singapore or the San Francisco Bay area of the USA, for example, seem far more fertile ground than Japan for future-orientated global knowledge-intensive industries.

AGRICULTURE AND FOOD SUPPLY

Although Japanese agriculture is declining in relative significance, that which remains is highly cherished and considered by many to be vital to national well-being and security. Japan is now a modern society, and a large proportion of the citizenry have a complex diet and cosmopolitan tastes. An enduring key indicator, sure to make news and generate comment in Japan, is the annual announcement of rice consumption per caput, a figure steadily decreasing. In 1960 annual consumption per head was 114.9 kg. By 1980 this figure had declined to 78.9 kg, by 2000 annual consumption averaged only 64.6 kg, and for the fiscal year 2003 (to 31 March 2004) was 59.5 kg, or a decrease of 48% since 1960.

Rice is often called the staple food of the Japanese people, but judging from consumption patterns it appears that rice is now merely one of many important foods. The importance of rice-derived calories and rice-derived protein has dwindled. Japanese rice prices are four to seven times higher than world prices, owing to import controls. Wheat-based products, potential substitute foods, remain under partial control and are also as yet highly priced in Japan. Rice has without doubt been quite important in the past, both in terms of diet and also in terms of supplying a livelihood for a large number of rural people who were often otherwise untrained. Yet rice (along with tea-leaves and silkworms) is no longer a cornerstone for the tax system or the overall economy, as was the case in the past. Protection for rice farming is becoming progressively more difficult to maintain. Under the WTO Agreement on Agriculture, Japan retains 'special treatment' privileges toward applying barriers to rice imports. The compromise has been that both tariffication and market access guarantees exist in tandem; the latter was set at 7.2% of domestic consumption for 2004 (at 682,000 metric tons).

On a calorific basis, domestic food production supplies only 40% of total consumption. Notwithstanding government pledges to raise the figure, it continues to decline, owing to a widespread demand for low consumer prices. Japan is officially resistant to accepting its dependence on external, non-Japanese suppliers (there is similar sentiment in the labour market limiting the use of foreign specialists and labourers). The issue of food security involves questions of the nutritive value of different crops, and the efficient use of land area. Various forms of fish-farming, soybean production, multiple cropping, etc., are alternative farming-type uses of land that might be more efficient and better suit Japan.

Since the early 1990s agriculture has contracted in relative importance in all the OECD nations when compared with industry and services. In terms of its contribution to the overall economy, in 2001 the Japanese agricultural sector contributed 1.4% of gross value added (compared with 2.6% in 1989); industry contributed 31.0% (down from 40.9% in 1989) and services 67.7% (up from 56.5% in 1989). Agriculture in the USA had only slightly more relative weight than in Japan, of 1.6% in 2001 (down from 2.0% in 1989). Yet the costs involved with agriculture in these two nations differ enormously. *OECD Agricultural Policies 2004* estimated total agricultural producer supports in Japan in 2003 at 58% of the value of production (US producer supports were 18% of production value; the figure for the 15 EU members was 37%).

It is important to consider the scarcity of land in Japan and the cost of occupation of such land by any form of marginally-productive farming. Japan is crowded. National average population density was almost 338 people per sq km at October 2003, but much of Japan is mountainous; 65% of the population live on just 3.3% of the overall land. In these densely inhabited districts (mostly along the coasts), population density averages 6,650 people per sq km.

INDUSTRY

Owing to Japan's relatively poor economic environment from the early 1990s, many firms reduced their capital investment, and much corporate capital stock became obsolete. Between 1995 and 2000 Japan's total manufacturing output grew by 5.2%, a moderate improvement that did not compare well with the average for the EU countries (which recorded an increase of 16.6%), the OECD overall (23.0%) or such growth in the USA (30.9%). Overseas production was 13.4% of Japan's total manufacturing output in 2000. Conditions further deteriorated for Japanese manufacturing in 2001, with seasonally-adjusted manufacturing production for the third quarter of that year being equivalent to only 94.4% of 1995 levels. Japan's index of industrial production, set at 100 in 2000, was 94.9 for the whole of 2003 and was 97.4 at the end of February 2004. (By way of contrast, South Korea's index developed from 100 in 2000 to 122.6 in January 2004). About 10% of manufacturing output in Japan relates to motor vehicles.

Japan is a nation with few natural resources, but the country has been remarkably slow to embrace the global trading regime. Protectionist import barriers have been erected and dismantled repeatedly over Japan's history, but at present the number of such barriers is low and declining, and there is new enthusiasm for trade. Some raw materials are still regulated, but markets for finished goods are open and at least officially in conformity with World Trade Organization (WTO) rules. Japan's government and its exporting manufacturers seem to have belatedly realized that highly-priced raw materials lead to highly-priced finished products that limit markets. Where market controls exist, Japanese firms find themselves losing market share to low-cost imports by foreign-based firms using raw materials purchased at standard world prices. In other cases (especially with foods), Japanese firms import semi-finished goods rather than basic raw materials because of a gap between domestic and world prices. If domestic products are expensive to process or manufacture, there is little possibility of developing export markets for such products. Import controls speed up the 'hollowing out' of industry as factories move abroad in an attempt to remain competitive.

In Western Europe and also in the USA, waning or non-competitive industries have had their eclipse prolonged through domestic policies involving subsidies or protectionist tariffs. Political pressures are an unavoidable but not insurmountable part of the policy process, involving considerations of employment, regional development, and economic parity problems, along with the concerns of vested interests. In many cases it is only possible to delay, not completely to avoid, the temporary problem of displacement and loss of jobs. Governments often act to ease such a period of industrial reorganization through information, education, or capital-transfer endowments. Efforts to ignore underlying structural weakness, however, do nothing to stimulate improvement. Instead, the structural weakness becomes a continuous and often increasing drain on the nation's economic vitality.

For manufacturers, Japan's high relative costs of intermediate inputs and labour do not compare well with those available in nearby China or the countries of the Association of South East Asian Nations (ASEAN). A major development of the 1980s was the improved economic and investment climate in these neighbouring countries, leading to a major movement of manufacturers and their suppliers to the lower-cost venues. Some in Japan have resented such moves, as foreign overseas workers undercut Japanese labour; firms can use the threat of relocation as a way to exact concessions over wages and benefits. For many firms, however, there has been little choice but to move abroad: there is no other way to produce quality manufactured goods at the low cost required to compete in the increasingly global markets both at home and abroad. A persisting aspect of this 'hollowing out', and a cause for complaint by some overseas nations, has been that most of the more highly-skilled jobs have remained in Japan, as have many of the major research and development functions.

Japan's future industrial competitiveness will be affected by the consequences of an ageing population: in coming decades a diminishing proportion of working-age people will be supporting a larger proportion of citizens who have retired. Female workers are often unable to continue in their jobs after childbirth; parental leave is meagre, and subsidized day-care facilities have been reported to have 100,000 families on the waiting lists. On the positive side, Japan has been largely successful with its general education system and preparing the next generation of productive adults; OECD figures give Japan an upper secondary graduation rate of 94.9%, the highest of any member state (the rate in the USA is 78.2%).

TRADE AND BALANCE OF PAYMENTS

Japan has a very high reliance on trade, and a strong vested interest in the continuation of the international trading system. Modern Japan is highly dependent on imported foodstuffs, and would not easily or stoically support itself on rice and roots if its borders were closed. Ploughing, cultivation and harvesting become quite complicated problems without imported tractor fuel and fertilizers. It therefore seems an inescapable fact that Japan must trade. Productive capacity is geared to trade (domestic and international), and the population has various raw material requirements that can only be satisfied by it. Like it or not, Japan's prosperity seems to be contingent on the continuity of peaceful international trade, and this would seem to deserve an expenditure of resources toward maintaining the trading system.

Japan continues to be in a good position in terms of foreign-exchange reserves, with official reserve assets valued at US $819,203m. at the end of July 2004. The balance of payments remains strongly positive, and Japan's current-account surplus at the end of March 2004 was 17,297,200m. yen, or 29.2% more than in the previous financial year. The positive balance in fiscal 2003 from trade in goods was 13,299,200m. yen (up 14.7% from the previous year, following a 28.9% rise the year before). Trade with China contributed greatly to these figures, and it is becoming increasingly clear that China is a key trading partner for Japan both as a market and as a supplier. Trade between Japan and China grew by 31% in 2003 and is now worth some US $134,000m. (exports from China to Japan reaching US $74,150m., and from Japan to China US $59,430m., according to Chinese customs data). According to the Japanese Ministry of Finance, the nation's world-wide exports increased in fiscal 2003 by 6.3%, to total 56,060,900m. yen in value, and imports rose by 4.1% to reach 44,833,000m. yen (both figures on customs clearance basis). Since 1980 Japan's imports of manufactured goods have risen from the equivalent of 6% of GDP to almost 9%. (The USA's imports in 1980 were also equivalent to 6% of GDP and had increased to 16% by the early 2000s.) Japan is also, of course, an important market for many Asian producers. About 40% of Japan's imports come from Asia, and it had been hoped that an improvement in Japan's vitality might help neighbouring nations more fully to recover from the regional economic crisis of 1997–98. In July 2001 *The Oriental Economist Report* claimed that 75% of the increase in Japan's imports emanated from 'captive imports' from overseas affiliates of Japanese firms (a further claim was that 80% of export growth was a result of shipments to overseas affiliates). As regards services, Japan is still a net importer; the negative balance of 3,904,300m. yen in fiscal 2003 represented a 26% improvement from the previous year. The positive balance (of 12,259,600m. yen) from trade in goods to the year ending March 2004 was 3.1 times Japan's net negative balance for services.

Japan's international investment position also remains strongly positive, with net assets at the end of 2003 totalling 172,818,000m. yen, down by 1.4% from the previous year. Private-sector funds represented 58.3% of this total, of which banks accounted for 21.3%, or 12.4% of the overall total; the overseas net assets of Japanese banks contracted by 14.6% compared with the previous year, but both the US and euro-area economies were relatively weak during this period.

Bilateral and International Negotiations

Trade specialists from both Japan's public and private sectors are regularly involved in bilateral trade discussions. Probably the most important are those between Japan and the USA (where a 'Framework for a New Economic Partnership'—agreed in 1993 and renewed in 1995—promises an 'Enhanced Initiative on Deregulation'), and between Japan and the EU (which holds annual summit meetings and a joint Regulatory Reform Dialogue, and established a 10-year 'Action Plan' from the year

2000). There are differing tones to these discussions, with the USA typically being much more confrontational and the EU seeking mutual understanding. None the less, in both cases the Governments usually exchange suggestions, with hundreds of proposals for reforms.

Japan has been widely involved in studies and discussions of free trade agreements (FTA), but as of the third quarter of 2004 only two agreements had been finalized, with Singapore and Mexico. In the latter case, Japan's agricultural protections were the most difficult area of negotiation. The FTA with Singapore came into force in November 2002, whereas the agreement with Mexico was expected to enter into effect in 2005.

Among major achievements with the EU was the signing of Japan's first Mutual Recognition Agreement (MRA) on 4 April 2001. The MRA streamlines export procedures for telecommunications equipment, electrical products, chemicals and pharmaceuticals by allowing approved overseas facilities to assess a product's conformity to legal standards and technical regulations. Quality re-examination is no longer required for customs entry, and the European Commission expects savings to be substantial: up to €400m. on €21,000m. of such annual trade. Japan has participated energetically in ASEAN + 3, bringing together the 10 ASEAN member states with Japan, China and South Korea. Their Chiang Mai Initiative, signed in May 2000, has led Japan to become partner to various regional bilateral swap arrangements, whereby ASEAN + 3 Governments will support the financial markets of partner states in times of crisis. Japan is active in other major international organizations and multilateral forums such as the UN and OECD, and is an active participant in the WTO. By the end of July 2003, Japan had been respondent (the country complained against) in 13 of 313 developed-country disputes filed with the WTO, and had brought 11 cases as complainant (with no new respondent or complainant cases since March 2002). One highly-publicized case followed the decision by the US Government in March 2002 to impose tariffs of up to 30% on imports of steel; Japan stepped back from its complaint in August of that year after the USA declared certain steel items exempt. Japan thus appears to make pragmatic use of the trading regime, in preference to adhering inflexibly to an ideology of free trade. (The WTO dispute over US steel tariffs continued with other complainants and was largely upheld to be inconsistent with trade rules, a ruling that ultimately will benefit Japan.)

FINANCE

The 'Bubble' Economy and its Aftermath

In the late 1980s rampant real estate speculation combined with loose lending policies to elevate both the Japanese stock market and the overall economy. Real estate steadily and spectacularly rose in value from 1985 to late 1990, in many cases tripling or quadrupling in value. Interest rates were low, and regulatory oversight was often remarkably lax during these 'bubble years'. In general expectation of continuing growth, it became possible to borrow 120%–150% of the value of land. Businesses found they could borrow easily for expansion or for speculation, although many companies shifted their sourcing for funds from bank loans to equity financing, issuing convertible or warrant bonds. The stock market was also booming. In 1987 capital gains from securities and real estate transactions exceeded nominal GDP by 40%, and it was estimated that total Japanese land valuation was four times the value of the entire USA. After the Plaza Accord of September 1985, as the value of the yen greatly increased (from 243 yen to 120 yen per US dollar by the end of 1987), imports became relatively less expensive and personal consumption of luxury goods grew rapidly. One negative aspect of the real estate boom was that those renting apartments or homes in formerly less-affluent urban areas were sometimes forced to move out by developers' intimidatory tactics. Working people seeking to buy a home or apartment did not find the situation much easier: many gave up, others took out novel multi-generation 75-year mortgages with their adult offspring as co-signers. Seeking to curb speculation, the Bank of Japan on 31 May 1989 raised the official discount rate (ODR) to 3.25% from a then-historic low of 2.5%. Four further rate increases had soon raised the ODR to 6% by 30 August 1990. This compounded concerns among Japanese banks about implementation of the 1988 Basle Capital Accord, which required banks doing international business to show capital adequacy by maintaining a reserve (or 'BIS ratio', as determined by the Bank for International Settlements) of 8% on risk-adjusted assets.

Introduction of a 3% national consumption tax in April 1989 (along with promised further rate increases) restrained consumer spending (Japan's consumption tax stood at 5% in 2004, and a figure of 10% was being recommended). Meanwhile, regulations were introduced in 1990 setting lending limits on real estate projects, and a new land tax was announced in December 1990 (which took effect in 1992). These all contributed to price reversals. On the stock market, the Nikkei 225 Average suffered a 20-month collapse, the index declining from 38,915.87 points to 14,309.41 between December 1989 and August 1992; and as the economy experienced massive asset deflation, non-payment of loans began to become a problem for banks. The yen/dollar exchange rate also remained volatile; the historic high for the yen of 79.75 reached on 19 April 1995 adversely affected those repatriating foreign investments. The full extent of the banking problem was largely concealed, which increased uncertainty and led to many Japanese becoming vastly more prudent with their money. Banks reportedly became much stricter with new borrowers, even while they steadily supported and cancelled existing NPLs. Efforts to revive consumer spending were largely unsuccessful, although the government tried 'pump-priming' with public works and other deficit-funded projects, as the ODR was repeatedly reduced. By February 1993 it had returned to 2.5%. Following further successive decreases, in August 2004 the ODR had remained at an astounding 0.10% for three years. In that month the short-term uncollateralized overnight call rate stood at 0.001%.

Bad Loans and Bad Lenders

The bursting of this 'bubble' left many banks exposed and in trouble, owing to bad debts. A marked lack of economic hopefulness and relative stagnation continued. From the early 1990s it was widely held that Japanese banks were vastly under-reporting their NPLs, and this was belatedly confirmed; the cumulative bad debts led eventually to several bank failures. A 'Japan Premium' surcharge (which rose to 100 basis points in late 1997) came to be imposed by international markets on Japanese borrowing because of uncertainty over bad debt accounting. The banking crisis came to a head in 1998 (in which year real economic growth contracted by 2.5%). The Financial Function Early Strengthening Law allowed the use of public funds to recapitalize financial institutions seeking to dispose of bad loans. Institutions applying for a capital injection were required to construct and submit a 'plan for restoring sound management', with the Prime Minister and Cabinet Office to monitor implementation. This substantial inflow of public funds provided at least a temporary respite, and finally led to changes in government oversight structure. Dozens of Japanese financial institutions went into receivership under the Financial Reconstruction Law, and 15 banks received public funds injections, being under the loose supervision of the Financial Services Agency. Liabilities from firms declared bankrupt peaked in 2000, declining to about half in value terms for the fiscal year ending March 2004 (when the number of bankrupt firms was also in decline, but still stood at 16,255). While the Japanese people had witnessed gradual revelations of severely negative information for more than 10 years, ordinary bank depositors had not yet lost money. Japan has a comprehensive system of deposit insurance. This has increasingly become a source of friction: such insurance has been unlimited, and ordinary (current or demand) deposits are still thus protected (until 31 March 2005, and beyond if they bear no interest). From April 2002 government guarantees for time deposits became limited to 10m. yen per account. This 'pay-off' limit attracted a huge amount of attention, as many believed such withdrawal of government support was a prelude to turmoil. The insurance limit was due to be similarly introduced for ordinary savings accounts in 2003, but the Government retreated for fear that depositors might withdraw funds from the most weakly performing banks and so cause them to collapse.

The Financial Services Agency reported that NPLs at Japanese banks had decreased from 42,200,000m. yen in 2001/02 to 26,600,000m. yen (US $252,870m.) for the fiscal year ending in

March 2004. Combined net losses for the 134 Japanese banks in fiscal year 2003 were 4,851,500m. yen (an almost identical sum to that of the year before). It was indicative of the extent of the bad loan problem, and of the problems in terms of forcing liquidation, that the fiscal year witnessed only a 1.6% recovery rate (700,000m. of the 42,200,000m. yen outstanding at the start of the year). Banks removed (wrote off) 15,100,000m. yen of bad loans from their balance sheets, which was partially offset by 6,700,000m. yen in newly defined bad loans (deemed 'doubtful and bankrupt/*de facto* bankrupt'). It was clear that the problem continued to be extensive, with few confidently knowing its true scale. In 2001 the IMF estimated Japan's bad debt at 95,000,000m. yen (US $766,000m.). The likelihood of loan recovery was graded according to two different four-category systems, but only 30% of firms that went bankrupt in 2000 had been rated as 'at risk of failure' (according to *The Oriental Economist Report*, July 2001). Special inspections by the Financial Services Agency from late 2001 of 149 major debtors led to 71 of them (and 58% of the total loans) being reclassified downwards. Clearly, therefore, much is yet doubtful. The Resolution and Collection Corporation has been criticized for failing to act decisively with foreclosures; criticism has arisen that they have overpaid for purchases of non-performing assets, and permission has even been sought to allow them deliberately to pay higher-than-market values. This provides public subsidies to bad managers, and adds layers of politicization to any hopes of recovery. The final accounting is still unclear, but the long delay in dealing with these problems effectively has allowed substantial transfers of both debts and assets between firms, the securitization of debt and a diffusion of responsibility for the uncollectable debt. This is a highly remarkable point: from 1992 to March 2004 Japan's financial institutions have officially cancelled the astounding sum of 93,572,400m. yen in bad loans (some US $886,000m. if converted at the exchange rate of 31 March 2004), and the end is yet to come. However, very little attention has been directed at personal accountability or mismanagement among lenders, and barely more among borrowers. The Bank of Japan has undermined the market by seeking to save nearly-bankrupt banks by purchasing shares from them. The Industrial Revitalization Corporation of Japan (IRCJ) became responsible for identifying ailing firms capable of being rehabilitated, buying their loans from the banks, and formulating salvage plans. The IRCJ was expected to be able to assist about 100 firms during its five-year mandate, but was slow to identify suitable rescue targets. There has been much discussion of selection standards, one important question being why public agencies and public funds should support bad performers at all, when rescues and liquidations could possibly be better handled by the private sector. Government rescues may maintain continuity in employment, for example, for some period, but ultimately a firm must be able successfully to perform in its market. In some sense, the Government is training skilled domestic 'turn-around' specialists, as many in Japan have learned a reflexive fear and resistance to further inroads by foreign capital and non-Japanese experts. Prior infusions of Japanese public funds have been cited as justification for limiting foreign buy-outs, but by limiting buyers, rescue becomes less likely. During the period that Japan has steadily delayed and dissembled, Sweden and South Korea, by comparison, have instituted substantive banking system rescue operations and reforms. Foreclosure or buy-out arrangements may also prove necessary in Japan.

From 'Big Bang' to Protracted Decline

Japan's 'Big Bang' was a diverse range of measures for financial-sector deregulation that was announced in 1996 and gradually implemented from April 1998. One major change came with foreign-exchange liberalization, and another gave the central bank (the Bank of Japan) greater autonomy. On the level of commercial finance reform, a further change was the introduction of mark-to-market accounting, whereby financial assets such as land and securities began to be measured at market price rather than as formerly by book (purchase) value. Furthermore, it was previously the case that banks, trust banks, life and casualty insurance companies, and securities brokers all had clearly defined domains. This is no longer true, and the previously compartmentalized financial services system has now been widely deregulated. New competition has struck at the core of the formerly tight relationships among such firms. All main financial firms are part of extended industry groups, *keiretsu*, which have become increasingly engaged in direct internal competition. The lowering of barriers has created some anomalies. For example, banks have posted their employees to affiliated securities firms to learn about the business (in order to develop the ability to offer such functions themselves). The securities firms, partly owned by and beholden to the banks, were thus in the difficult position of having to train their future competition. An ongoing major realignment has also been taking place among banks, with numerous mergers creating by 2003 just five major banking groups from the 23 major banks that were operating in 1990, with further reorganization under way in 2004. Public confidence in these new 'mega-banks' has been undermined by the highly-publicized logistical problems of merging computer systems and transaction security arrangements, as well as by the revelation of various scandals and details of bad business decisions.

Savings and Pensions

A special institution in the Japanese financial world is Japan Post, which includes the Postal Savings System. This Postal Savings System exists alongside private banks, and has long been a point of convenience in that any post office was an access point to personal savings. However, with national mutual online access now widely available amongst banks via automatic teller machines (ATMs), the system is no longer as remarkable for providing access to funds. What remains remarkable is that Japan Post is the world's largest financial organization, with total assets as of 1 April 2003 of 415,525,300m. yen (US $3,522,595m.). Many claim that there is now no reason for government to be in the banking business, and that privatization will clear away stifling regulations that have unreasonably limited normal bank opening hours, use of ATMs and account promotional activities. The Government has long utilized postal savings as a source of funds outside the regular budget, but public corporations are now increasingly issuing their own bonds (*zaito*) in order to raise funds. The Government has also made greater use of postal funds to purchase securities, both domestic and overseas. While only a small proportion of such capital is thus used, some argue that this should be more interventionist and expanded by a factor of 10. Prime Minister Koizumi has sought to split Japan Post into component parts (mail, savings and life insurance) with privatization. While a future break-up is as yet uncertain, privatization in stages that would run from April 2007 to 2017 has been agreed.

Japan's rate of gross national saving as a percentage of nominal GDP is high: of the 22 OECD nations reporting for 2001, Japan (at 26.4%) was ranked fourth, behind Norway (35.1%), South Korea (29.8%) and Finland (27.9%). On an annual basis, the household savings rate as a percentage of disposable household income had declined sharply in Japan, to 5.8%, or 15th of the 21 OECD nations reporting (Italy was highest at 16.0%). In Japan there is still a perceived need to save in order to provide for an uncertain future, and many efforts have been made by government to boost consumer spending. Some believed that a new-denomination 2,000-yen banknote would help encourage spending (the notes proved unpopular, however); tax reductions similarly had little impact; neither did the widespread issuance of 20,000-yen shopping vouchers (to families with children aged 15 and under, and to elderly people).

The demographic trend of Japan's ageing society was one of the concerns of Japan's fourth Economic Structural Reform Plan (running until 2005). The total fertility rate as of April 2004 was a historically low 1.29%, far below replacement rate, and suggesting a future where there will be insufficient productive workers to support retirees. (The birth rate for Tokyo residents is 0.99, which might be indicative of a wider future trend.) Such observations, combined with the banking and finance difficulties, leads many to worry about the possibility that their pensions are under threat. Their fears are not misplaced. Returns on investment tend to be low. Mismanagement threatens both public and private pension funds. On the implementation level, there is further concern about the many who fail to contribute to the National Pension Programme (national politicians have set very poor examples). There is also uncertainty about pensions keeping pace with an unpredictable future cost of living.

Prices, Wages and Unemployment

From a 1995 base, the consumer price index in Japan had risen by an average annual rate of only 0.1% over the six years to December 2001, having begun to decrease in 1999. Compared with 2000, when a decrease of 0.6% was recorded, consumer prices declined by 1.0% in 2001 and by 0.6% in 2002. Prices declined by a further 0.2% in the 12 months to 31 March 2004, thus representing five consecutive years of deflation. Consumer prices continued to decrease throughout mid-2004, and average land prices were declining for the 13th consecutive year. Total employee compensation decreased by an estimated 4.1% over the two years to March 2004. According to the IMF, the level of unemployment rose from only 2.1% of the labour force in 1990 to 3.2% in 1995, and continued to increase steadily during the late 1990s. By December 2001, however, Japan's rate of unemployment had exceeded 5.0%, rising to about 5.5% in mid-2003, and slightly down to 5.2% in mid-2004. While the Japanese consider this level very high, the US rate had improved from 6.3% in June 2003 to 5.6% in June 2004; while the euro-area average was 9.0% (with 10.5% being recorded in Germany). Re-employment of middle-aged workers is a problem in Japan, with many large firms' rates of pay based on seniority, and promotion possible on an internal basis only. The inflexibility of the system operates against the recruitment of employees mid-way through their career who try to join a new company, and helps explain the fervour with which many workers cling to their jobs.

Government Funding

Fortunately for the public, the Japanese Government takes a relatively small percentage of income in tax: total government tax receipts as a proportion of GDP placed Japan 28th of the 30 OECD nations in 2001 (above Mexico and South Korea). In September 2002 the Government announced plans to reduce taxation by around 1,500,000m. yen (US $12,000m.) in an effort to stimulate the stagnant economy. For 2001, Japan was ninth among the OECD nations for lowest percentage of GDP expended on general government (final government consumption of 17.4%; their reported figure for 1999 was the OECD's lowest at 10.2%). Japan's proportion of government employment to total employment is about 6%, which compares favourably to figures for the USA, Germany and the United Kingdom (all at 15%), France (25%) and Sweden (32%). Furthermore, the Japanese figure has not varied by a full percentage point since 1970. In the smaller subset of public administration and defence employment, Japan and Korea are lowest in the OECD at 3.3% of total employment (2001); 14 OECD nations have more than twice that proportion (Belgium is highest at 10.9%).

One problem, however, is that Japan also maintains many state-supported public corporations and semi-governmental agencies. For decades they have drawn substantial off-budget funds from the Fiscal Investment and Loan Program (FILP); funds come largely from the Postal Savings System, with distributions handled by the Trust Fund Bureau of the Ministry of Finance. Originally, these special public corporations were mostly for post-war infrastructural development projects, when private capital was unavailable, but now there are alternative funding sources and often competition from the private sector for such business. The organizations have also been criticized for providing lucrative posts for retired bureaucrats. Their outstanding loans at the end of June 2004 totalled 143,335,700m. yen (of the total 377,391,700m. yen of FILP 'assets'). This growing amount is equal to approximately 26% of overall national GDP; it is unlikely to be repaid, as many of these organizations reportedly have huge sums in other unreconciled bad debts as well, and have been able to evade inquiry into their long-term strategic viability. The process of rationalization initiated by the Government in December 2001 provoked much discussion about the extent to which government subsidies and public services should operate in a market economy. Nationwide universal service by the postal system, for example, is highly desirable but often inefficient; there are fears that privatization will lead to the closing of rural services and accelerate migration away from the countryside. Attention has largely been focused, however, on the intransigence of vested interests within these organizations, on their ties to the ruling LDP, and on links to each associated ministry, where retiring bureaucrats still expect offers of employment. Future fund-raising and lending rates by FILP are to follow market principles, with disclosure of information on expected taxpayer burdens. There is also debate on reform focusing on the format and role of state-run universities, which are of varying quality, operate completely differently from private universities, and will need to respond to a severe demographic change in the coming years.

Foreign Aid

Japan was the world's principal aid donor in terms of total flow of funds from 1992 until 2000, but budget reductions and a depreciation of the yen in 2001 led to the USA regaining the primary position, as Japan's net official development assistance (ODA) declined by 18.1% in that year and by another 1.8% in 2002. The budgeted amount in yen decreased by a further 5.8% for the fiscal year 2003 and was projected to decline by a further 4.8% in the year to March 2005. Both the USA and Japan are often criticized for not giving more: in 2002 the combined EU national donations of ODA totalled US $29,093m., the USA gave US $12,900m., while Japan gave US $9,220m. According to 2002 data, Japan ranked equal 17th of the 22 member nations within OECD's Development Assistance Committee (DAC) in the proportion of gross national income (GNI) allocated to ODA (0.23%), while the USA was last among DAC nations, with 0.12%; Denmark gave the highest proportion of GNI at 0.96% (four times the ratio of Japan and eight times that of the USA). Controversy continues to beset Japan for its use of 'soft' loans, and for blunt statements such as that in 2001 by an official of the Fisheries Agency that Japan was using ODA as a reward to poorer, smaller nations that supported its policy goals in international forums such as the International Whaling Commission. Within Japan, attention has further focused on China, a quickly growing economic partner and also a rival. Japan is substantially benefiting from Chinese development, but China is an anomaly in that it receives substantial aid while also disbursing aid to other nations. Some are critical: it is as if Japanese money were flowing to third countries but in China's name, and Japanese finance may also be said to be freeing up funds that China then uses for military development.

Defence Spending

Expenditures on Japan's national defence are substantial in gross terms, but a relatively low 0.98% of GDP (a 1976 cabinet decision seeks to keep the figure below 1%). Such spending levels are made possible by the alliance with the USA under the Japan-US Defense Guidelines. Defence expenditure in fiscal 2004 accounted for 6.0% of Japan's national finance expenses, falling slightly from the year before.

The Quest for Fiscal Responsibility

The Bank of Japan highlights the fact that in 2002/03 discounts and interest on Japanese government bonds consumed 22.8% of the national budget, in contrast to 13.5% in the USA, 16.5% in Germany, and 6.3% in France. Japan's bond dependency for the year ending March 2004 was budgeted to be 44.6% (deficit as percentage of expenditures). One continuing cause for concern has been Japan's rising long-term public debt, which was scheduled at the end of fiscal year 2004 to reach 144% of GDP (about 719,000,000m. yen; the OECD estimated this gross debt as 161.2% of Japan's GDP). This ratio is markedly higher than that of other major nations (US government gross debt is about 66% of GDP), and foreign credit-rating agencies have cited such anomalies as reasons for downgrading Japan's credit rating, a further part being that it is still uncertain to what extent the Government will guarantee the growing amount of *zaito* bonds being issued by semi-public organizations. There has been some fiscal tightening, but regardless of government rhetoric, the figure shows every sign of continuing to grow. One danger in this is that increased interest rates could quickly lead to severe problems with servicing the debt, and to instability throughout the economy.

Public works projects are the traditional outlet for special large-scale government expenditures, and these continue, although there is increasing public awareness of overspending and occasional repercussions concerning environmental impact. Prime Minister Koizumi promised a number of far-reaching reforms, but one that was uncommonly specific was a budgetary pledge to keep government bond issues below 30,000,000m. yen for fiscal 2002 (the revised amount reached 34,968,000m. yen). The initial budget for the year ending 31 March 2004 envisaged the issue of government bonds of 36,445,000m. yen. There is new interest in resource allocation and investment efficiency, as economic impact and demand multipliers can vary greatly among projects. A further consideration is to develop an expanded use of private finance initiatives (PFIs), where private capital is allowed to develop, to construct and/or to manage major projects of the type now administered by government. The United Kingdom reportedly uses the PFI method for 40% of national projects, and many such public-private systems have been successfully developed in other parts of Asia. Within Japan, a municipal government facility being constructed via PFI in the city of Chiba in the early 2000s was expected to cost an estimated 50% less than if managed by the public sector. It would be beneficial to the economy if the national Government could realize better efficiencies through such methods, but as with open public bidding being undermined by collusion, oversight would remain a problem.

The private sector is also actively proposing novel ways to revitalize the economy. A bid to raise standard household electricity voltage from 100 to 230 volts has been promoted since late 2000 by the Federation of Electric Power Companies and the Japan Electrical Manufacturers' Association. The upgrading and change-over would require 15 years, improving power transmission efficiency and saving 7,000m. kWh of electricity annually, or 0.8% of total electrical output. The cost for new infrastructure would be substantial, however, requiring around 80,000m. yen annually for new heavy electrical equipment, and similar expenditures for household appliances (which would almost all need to be replaced). While producers are keen on this demand-boosting measure, having proposed that tax concessions would offset expected consumer resistance, the Ministry of Economy, Trade and Industry (METI) has thus far taken a slow and cautious approach.

Private demand strengthened at the end of 2003, and went up by a further 1.3% in the first quarter of 2004. Public demand, however, was depressed at –0.1% (the eighth straight zero or negative quarter); deflation had reduced disposable income in workers' households for five consecutive years, with continued contraction through the second quarter. The value of public works contracts was similarly decreasing for five years running, with a continued decline through the first half of 2004.

The Stock Market

Although since the 1990s there has been general pessimism about the economy and the market, some industries and shares have performed well. Booms and reversals often occur in parallel over world-wide markets and, following the trend in the US market, there was a general withdrawal from high-technology and information technology shares in mid-2001. Although there have been substantive fluctuations, full recovery has been elusive. The benchmark Nikkei 225 Stock Average year-end close has fluctuated as follows: 18,934.34 points (1999), 13,785.69 (2000), 10,542.62 (2001), 8,578.95 (2002) and 10,676.64 (2003). On 28 April 2003 the index closed at 7,607.88, its lowest level since 9 November 1982 and a very long way from the historic high set on 29 December 1989 of 38,915.87. By August 2004 the market had strengthened, with the index in the region of 11,000, and analysts continued to be widely at variance with market predictions. The TOPIX index (all First Section firms on the Tokyo Stock Exchange) had declined by 18% at the end of 2002 compared with the end of the previous year, and ended the fiscal year down a further 7% (standing at 788 points on 31 March 2003). Although continuing to be volatile, by mid-August 2004 the TOPIX had greatly recovered, to reach the level of 1,115.

Exchange Rate

Japan is the world's second largest national economy, and much attention and influence emanate from its relationship to the largest: the USA. The yen–dollar exchange rate has varied substantially over past years. In 1998 the annual daily average per US dollar was 130.9 yen; this went up by 13% to 113.9 yen in 1999; up by 5% to 107.8 in 2000; down by 13% to 121.5 in 2001; down by 3% to 125.3 in 2002; and up by 8% to 115.9 in 2003. The yen–dollar rate stood at 110.5 in mid-August 2004. Yen-selling intervention by the Japanese Government in 2003 took place on an unprecedented scale. However, a reported temptation to make the yen even weaker (to stimulate exports and restrict imports) would lower capital adequacy (BIS-ratios) among the Japanese banks with exposure to dollar-denominated loans on their books, and possibly force foreclosure on domestic loans to sustain capital.

Foreign Direct Investment

Japan's officially reported inward foreign direct investment (FDI) flow for 2002 was equivalent to 0.22% of GDP. (This was equal in percentage terms to Mexico, as the lowest of the 28 OECD nations reporting; the US figure plummeted in the wake of the terrorist attacks of 11 September 2001 to 0.29%, down nearly 89% in gross terms from the year 2000; the combined figure for Belgium-Luxembourg was highest at 54.0% of GDP.) Japan's outward flow of FDI was 0.78% of GDP (in gross terms, one-third less than in 2000). The stock of foreign investment in Japan (US $50,320m. at the end of 2001) was one-thirtieth the sum of foreign investment in the USA. One explanation for Japan's distance from the norm is that Japanese language and society are perceived as barriers to outsiders, who fail to recognize opportunity; even if they do so, it is none the less difficult to establish operations and do business in Japan. Until recently there were onerous restrictions on foreign businesses in Japan, although the worst of such legal barriers have now been removed. The improved business climate is due in part to pressures being put on the Japanese Government by foreign governments, at the urging of Japan-based foreign businesses. The American Chamber of Commerce in Japan is a good example of an activist, proactive organization systematically working for improvement; it collects anecdotes and data, and has steadily built up a network of local and home contacts to promote identified problems it seeks to change. Yet there was still both tacit and more organized resistance to foreign capital in 2002. Shinsei Bank (purchased in 2000 by the Ripplewood investment fund of the USA) encountered problems with regulators and resorted to complaining publicly about being excluded from repayment schedules by borrower Daiei, while Japanese creditors were being repaid. When Nestlé was being considered as a possible rescuer for Snow Brand Milk Products Co (after the Japanese firm and its subsidiary experienced food poisoning and mislabelling scandals), a Ministry of Agriculture, Forestry and Fisheries official was quoted as saying that partnership with a foreign firm would be unwise. Japan's Minister of State responsible for the Financial Services Agency voiced similar sentiments in opposing the sale of a large interest in Aozora Bank by Softbank to US investment fund Cerberus, stating that the sale would call into question Softbank's sense of responsibility.

DEREGULATION AND REORGANIZATION

The business world is increasingly pluralistic, yet Japan remains highly adept at exchange of information, co-ordination of efforts, and co-operation. Much of this success is due to the astute use of trade associations and the major business federation, the Keidanren. Business forums, such as the Keizai Doyukai and the Japan Chamber of Commerce and Industry, along with various government advisory committees, are also important to bring together and brief key leaders from various parts of the economy.

For many hundreds of years Japan has had a highly-regulated society, with government exercising active control through laws and licensing, as well as through the offering of extra-judicial 'guidance'. Deregulation has now occurred in many realms, but such facts take time to be fully recognized by those who might grasp new opportunities. Vagueness and diffused responsibility are well-known attributes of the Japanese political economy. Many believe that a measure of ambiguity allows for smoother, more harmonious relations. The inter-relationships of the Japanese ministries, and between ministries and those they regulate, can still often be vague and ill-defined. One example is in the practice of 'administrative guidance', where a regulatory office within the government bureaucracy strongly suggests that a firm should operate in a certain way. Such guidance does not have the rule of law. In some cases it will indicate forthcoming legislation, in other cases it is simply a warning for self-restraint so as not to cause 'excessive competition' or 'market instability'. Most firms seem to find it in their better interest to keep the regulators happy through compliance. The professional bureaucracy of the Japanese Government has been damaged and in decline since the early 1990s. Staff numbers have been reduced, and the highly popular process of deregulation has shifted the initiative away from bureaucratic officialdom. The vast majority of Japanese government officials are hired directly after graduating from non-technical undergraduate university programmes, and then trained at work; mid-career appointments from outside government are rare. These are people who have been considered the brightest and the best in Japan, with a high sense of professionalism and *esprit de corps*. During the 1990s, however, scandal affected many ministries. Deregulation has also disrupted the post-retirement system of *amakudari*, where early-retiring officials are systematically placed into high-level private-sector posts. Many individual officials have thus been led to feel that their financial future may be in question. These career problems have the potential to give rise to even more misconduct, scandal, and a further downward spiral for Japanese officialdom.

In early 2001 the Japanese government structure was substantially reorganized. The restructuring led to the separating of some services as independent offices and to a decrease in the number of ministries and agencies from 23 to 13. There was to be a gradual rationalization and realization of efficiencies and an eventual 10% reduction in government staff totals. The immediate effect was a reorganization of supervisory responsibilities. The newly formed Cabinet Office combined the functions of the former Prime Minister's Office, the Economic Planning Agency, the Okinawa Development Agency and the Financial Services Agency (FSA). The Cabinet Office also included the Council on Economic and Fiscal Policy, which was henceforth charged with drafting the national budget (a responsibility formerly in the realm of the Ministry of Finance—MOF), and the newly created Financial Crisis Management Meeting, responsible for policies to forestall any threatened systemic financial crisis. Another change involved postal operations: mail, postal savings and postal insurance were moved from being part of the Postal Services Agency (an external organ of the Ministry of Public Management, Home Affairs, Posts and Telecommunications) to become Japan Post, a public corporation with increasingly privatized components. In July 2002 legislation allowing private firms to enter the state-run delivery market was approved. Various other major organizational changes have occurred; but other than eventual cost savings, their main substantive impact has been to question the work-life complacency and institutional loyalties of the government employees affected.

A further area of change has been the exertion of pressures by the central Government in Tokyo to change its fiscal relationship with local governments. This is being accomplished through reducing state subsidies and seeking new efficiencies, such as promoting mergers between cities, towns and villages; the number of such municipalities was projected to decrease from 3,200 to 1,000. Various vested interests and political considerations continued to impede Prime Minister Koizumi's efforts with deregulation. One evolving area of promise has been the creation of Special Regulatory Reform Zones; these have allowed experimental deregulation on a limited scale. As results are compared with mainstream conditions, where areas outside the Zones become less competitive, they can be expected to press for regulatory changes, broadening the deregulatory coalition.

The Telecommunications Sector

Japanese telecommunication services are said to be among the most expensive in the world, costing several times more than similar services in, for example, the USA and the United Kingdom. Although competition has been permitted since 1985, and fees have declined sharply, many still complain about the high interconnection fees charged by Nippon Telegraph and Telephone (NTT), the former national carrier. The NTT group is gradually facing competition from local networks and markets. NTT's rates must be approved by the Japanese Government (which by law must hold at least one-third of NTT shares, and owned 46% in 2003). Following pressure from the US Government, which claimed that the high rates stifled competition and cross-subsidized other services, NTT interconnection charges were scheduled for further reductions, but complaints continue. It has been argued that high access costs limit some households' access to internet services and thus curtail the development of a computer-literate population, and that, as a result, Japan's international competitiveness with regard to information technology suffers directly. The NTT firms must provide universal service (as designated 'qualified telecommunications carriers') for which they can receive subsidies.

One of Japan's most-watched companies recently has been NTT DoCoMo, now the nation's largest firm in terms of market capitalization. Newspapers and magazines regularly feature details of its products and prospects, especially new developments in its 3-G mobile services for the i-mode system, which allow convenient internet, e-mail and 'i-shot picture mail' functions in fashionable handsets. In August 2004 there were nearly 42m. i-mode subscribers in NTT DoCoMo's 46.6m. subscriber base (among 88.1m. wireless subscribers in all Japan). The technology at the consumer end is appealing; its development has benefited from synergies with Japan's electronic manufacturers and highly advanced production engineering, as well as a technically literate populace. The industry is becoming less regulated, but is still linked to officialdom; in mid-2004 NTT DoCoMo was 62%-owned by NTT, which (in turn) was 46%-owned by the Japanese Government: not a condition of market virtue. When government maintains such a position in the market, smaller-scale businesses and the effective promotion of entrepreneurship are hampered.

MANAGEMENT SYSTEMS AND SHAREHOLDER RELATIONS

The role of individual shareholders in Japan has been minimal, in part because of logistic limitations where most major corporate shareholders' meetings are staged simultaneously in late June. On 29 June 2004 a total of 1,120 firms, accounting for 64% of the Tokyo Stock Exchange, held meetings at the same time (throughout the 1990s more than 90% of such meetings were held on the same day). The typical explanation for this is that professional racketeers, *sokaiya*, sometimes disrupt shareholder meetings; their attendance is limited by holding about 2,000 shareholder meetings all at once. Yet the opinions of legitimate shareholders are also stifled. This serves the interests of corporate managers rather than those of shareholders, who miss the chance to question and guide management strategy. A related artifice is that these firms release their financial results on a common day one month before the meeting; owing to the sudden volume of information, poor results often escape careful media scrutiny. This approach to shareholder relations is very different from elsewhere in the world, with important ramifications, including a firm's ability to retain earnings and thus to lower its cost of capital, which has an impact on international competitiveness. Such corporate governance results in unique costs and benefits across Japan's mixed group of corporate stakeholders; most directly it is a problem for information flow in that the annual shareholders' meetings may be deliberately orchestrated so as not to operate as a check on management. The high reliance of Japanese corporate finance on retained earnings, derived from the relative freedom of managers from shareholder governance, also allows firms to avoid the discipline that might be imposed by other sources of external finance.

Shareholders are beginning to complain, and there is a growing trend toward activism. This includes pressure from foreign shareholders. The California Public Employees' Retirement System (CalPERS), with many thousands of millions of US dollars of investments in Japan, has published a list of guidelines for better Japanese corporate governance. They have led unsuccessful proxy efforts to keep certain managers who have admitted corruption from being readmitted to corporate boards. Institutional Shareholder Services have made similar efforts, thus far without success. The Japanese domestic group Shareholders Ombudsman has also been working for change, and domestic pension fund managers have been driven to increased activism by legal requirements that they vote to maximize the value of their portfolio components.

In December 1997 Japan's Ministry of Justice requested that corporations avoid the practice of holding annual shareholders' meetings on the same day, and also noted problems with efforts at keeping meetings as short as possible. Japan's National

Police Agency, the Business Sector Advisory Group on Corporate Governance of OECD, and the Tokyo Stock Exchange have also raised the call for reform, focusing on the use of *sokaiya* but also including criticism of meeting timing, the independence of auditors, the low returns on equity and low corporate dividend rate in Japan, etc. The system is increasingly difficult to justify. Yet change would alter many fundamental Japanese business formulas and relationships. It is interesting that the Japanese government bureaucracy and the Cabinet have only recently begun to take an active role in condemning the practice of holding simultaneous annual shareholders' meetings. A cynical argument is that the belated interest in part might be retaliation for the fact that business has been critical of bureaucratic corruption; business has also been agitating for limits on public official staffing numbers under the banner of deregulation.

Most large Japanese firms have a core group of institutional shareholders (major banks, insurance companies, institutional customers and suppliers) who hold stable blocks of stock and participate in cross-shareholding relationships. These shareholders are most interested in managing their own firms—they are not 'activist' shareholders. Their investment interest has often been to develop an existing business relationship rather than to seek dividends or capital gains. With cross-shareholding, teams of managers from different firms do not publicly criticize the others because to do so might result in the same thing happening to them. They instead provide their proxy votes in support of management. Cross-shareholding between firms thus contributes to a mutually-forgiving approach to excesses. Such an approach is convenient for a management team, but short-sighted when a firm loses the benefit of checks by expert shareholders that might improve its operations. In most cases, however, the senior management teams logistically are unable to attend each other's shareholder meetings because they are conducting their own at the same time.

Some reform will eventually come about, but because modifying the system threatens the stability of existing management, hitherto there has been little support for change among business leaders. Reform of the basic structure of shareholder relationships in Japan strikes at the heart of corporate stability. There have been efforts in the past few years among a few dozen firms to stagger their meetings away from the common day, but quiet efforts have yet to solve the problem; it has not disappeared, and a growing stridency in calls for change by shareholders will probably lead to stronger legal measures towards instituting reform and an improvement in accurate information flow.

Overall, in most Japanese firms, foreigners have yet to levy much direct influence in terms of corporate governance, but the foreign-owned component of listed firms is steadily increasing. According to *The Oriental Economist*, foreign investors owned 22% of all shares in 2004, compared with 8% in 1994. Pressure is steadily building to bring Japan's low corporate dividend payments more into line with those paid by firms elsewhere in the world. The Government has also scheduled introduction (from 2007) of the ability for foreign firms to exchange domestic office equity with Japanese firms, a move expected substantially to boost mergers and acquisitions.

There are several other noteworthy Japanese management practices which have a substantial impact on corporate performance. These include leadership personnel policies, extensive use of ancillary or affiliate organizations by government ministries and agencies, and the already-mentioned practice of adopting retired government officials into private firms as *amakudari* corporate executives. Notable also is the growing number of recent shareholder lawsuits in Japan, where corporate managers and directors have been ordered to pay thousands of millions of yen in compensation for losses resulting from mismanagement or breach of trust. The Keidanren business federation was seeking legal reform to restrict potential individual liability, but huge corporate collapses elsewhere (such as in the USA, notably at Enron and WorldCom) have provided salutary lessons. Domestic scandals continue to emerge, such as *tobashi* or 'window-dressing' of losses, the use of deferred tax credits and of other impaired assets to bolster capital reserves by Resona and other banks and the reckless hidden book-keeping of UFJ Bank. These examples show that Japan's business leaders might better focus their efforts on encouraging ever more diligent corporate oversight.

EMERGING ISSUES

Research and Development Spending

Japan continues to invest heavily in research and development, with total spending in 2002 at 3.09% of GDP, in third place within OECD behind Sweden (4.27%) and Finland (3.40%). The Council for Science and Technology, a government advisory committee, designed the basis of the current five-year research and development plan with specific focus on four strategic areas for coming decades: nanotechnology, life sciences, telecommunications, and the environment. There was to be a major increase in the award of competitively based research grants. Overall, government is expected to contribute an average of 4,800,000m. yen per year, compared with 3,500,000m. yen in 1999. In Japan's private sector, many dozens of firms are among the best in the world, investing heavily in both technical research and human capital. Japan's private sector financed 73.0% of research and development spending in 2002 (government financed 18.5%). This was the highest proportion of such private spending in OECD. Japan is highly active with patenting new technology, and has a high positive technology balance of payments (US $5,703m), third in the OECD behind the USA and United Kingdom.

Neighbourhood Retailers

Japan has many small-scale elderly retailers whose post-retirement businesses have been termed a 'tolerated inefficiency', and whose work is supported by government as a means both to maintain employment and to nurture social welfare. Neighbourhood shops are often the most convenient for elderly people who go shopping on foot, and the shopkeepers contribute to maintaining a sense of local community. Yet uncompetitive businesses are increasingly criticized as contributing to congestion. They take up important space; their supplier deliveries slow down traffic. Until the recent past they were able to obstruct the introduction of larger-scale retailers into their neighbourhoods (although this is now more difficult). The fact that there is both friction and discussion over acceptable means of doing business would seem to be a good sign for Japan's economy, although resolution seems unlikely without a great amount of distress.

Increased Transparency

Increased transparency and improved administrative responsiveness can significantly reduce corporate overheads and stimulate competition, thus leading to lower costs for consumers. Japanese governmental efforts at regulatory reform and transparency have led to the development of a new and important responsiveness. The Economic Structural Reform Plan, adopted on 1 December 2000 and scheduled for implementation by 2005, specifies some 260 reforms. One of the more far-reaching is that the Japanese Government will adopt a 'no-action letter' system in which firms can request clarification of regulations, and ministries and agencies will be required within a fixed period to respond in writing. The FSA has already begun such a service, and in the early 2000s the METI was considering the possibility of posting its responses on the internet. This type of transparency will further undermine *amakudari*, as there will be less need to employ retired bureaucrats to help navigate through opaque regulations.

Since April 2001 banks have been required to revise their accounting systems and adopt mark-to-market rules (allowing for adjustment of contract prices to market prices). At issue is compliance with the need to maintain minimum capital adequacy reserves (BIS ratios of capital/risk assets) of 8%, as required to conduct international business. The further effect is that banks are opening up their books with contemporary valuations, revealing previously hidden pitfalls and non-performing investments.

The format of developing the Special Regulatory Reform Zones has also been important: proposals for regulatory reform are posted on the website of their Cabinet Office headquarters and forwarded to the relevant Ministry, which has 30 days in which to respond. Their response is then posted and followed up so that the public can see the logic (or lack thereof) to these regulations. The fact that deregulation would be applied to a limited area, at their request, overcomes some of the conservatism of the ministries, which had previously tended to delay change in consideration of safety or unknown outcome. Now, locales can request experimentation upon themselves, and numerous unparalleled projects and novel experiments are in development.

In today's Japan, most sectors and industries have become accustomed to the fact that favoured insider relationships are increasingly difficult to develop and expensive to maintain. Structural changes that require strategic adjustment are occurring more and more regularly. Unlike in the past, areas of imperfect competition are more quickly identified and seized upon by either domestic or international firms—in many cases as a matter of survival. Since the 1990s substantial deregulation in retailing, telecommunications, and the air transport industries (among others) has taken place; competition has markedly increased, more choices are available and prices are lower.

In summary, one point is clear: Japan does not have a history of seeking market-based solutions. Governments of the past have been prone to intervention, and the people are accustomed to such a role for government. Vested interests in Japan are strong and astute and will continue to resist any threats to their advantageous position. Few people ask why public agencies and public funds should be necessary to sustain bad performers. Something is amiss if politicized bureaucrats can do a better job than can private-sector 'turn-around' specialists. Bad managers have learned to politicize, and thus delay, foreclosure or hostile buy-out. Such choices make Japan's economy markedly sub-optimal, and likely to remain economically enfeebled for many years to come, especially in comparison with dynamic neighbouring economies such as South Korea and China. Great effort is being expended on redesigning economic systems to be more robustly competitive, but a necessary infusion of confidence is elusive. For future prosperity, the Japanese people may yet wish to pay more attention to employing the resources of non-traditional outsiders, attending to transparency, and giving greater heed to compliance and accountability.

Statistical Survey

Source (unless otherwise stated): Statistics Bureau and Statistics Center, Ministry of Public Management, Home Affairs, Posts and Telecommunications, 2-1-2, Kasumigaseki, Chiyoda-ku, Tokyo 100-8926; tel. (3) 5253-5111; fax (3) 3504-0265; e-mail webmaster@stat.go.jp; internet www.stat.go.jp.

Area and Population

AREA, POPULATION AND DENSITY

Area (sq km)	377,864*
Population (census results)†	
1 October 1995	125,570,246
1 October 2000	
Males	62,110,764
Females	64,815,079
Total	126,925,843
Population (official estimates at 1 October)	
2001	127,291,000
2002	127,435,000
2003	127,619,000
Density (per sq km) at 1 October 2003	337.7

* 145,894 sq miles.

† Excluding foreign military and diplomatic personnel and their dependants.

PRINCIPAL CITIES
(population at census of 1 October 2000)*

Tokyo (capital)†	8,130,408	Utsunomiya	443,787
Yokohama	3,426,506	Nishinomiya	438,129
Osaka	2,598,589	Oita	436,490
Nagoya	2,171,378	Kurashiki	430,239
Sapporo	1,822,300	Yokosuka	428,836
Kobe	1,493,595	Nagasaki	423,163
Kyoto	1,467,705	Gifu	402,748
Fukuoka	1,341,489	Hirakata	402,586
Kawasaki	1,249,851	Toyonaka	391,732
Hiroshima	1,126,282	Wakayama	386,501
Kitakyushu‡	1,011,491	Fujisawa	379,151
Sendai	1,008,024	Fukuyama	378,793
Chiba	887,163	Machida	377,546
Sakai	792,034	Nara	366,196
Kumamoto	662,123	Toyohashi	364,868
Okayama	626,534	Iwaki	360,143
Sagamihara	605,555	Nagano	360,117
Hamamatsu	582,120	Asahikawa	359,526
Kagoshima	552,099	Takatsuki	357,440
Funabashi	550,079	Toyota	351,068
Hachioji	536,000	Suita	347,938
Higashiosaka	515,055	Okazaki	336,570
Niigata	501,378	Koriyama	334,845
Urawa	484,834	Takamatsu	332,866
Himeji	478,312	Kawagoe	330,737
Matsuyama	473,397	Kochi	330,613
Shizuoka	469,679	Tokorozawa	330,152
Amagasaki	466,161	Kashiwa	327,868
Matsudo	464,836	Toyama	325,693
Kawaguchi	459,952	Akita	317,563
Kanazawa	456,434	Koshigaya	308,277
Omiya	456,164	Miyazaki	305,777
Ichikawa	448,553	Naha	301,107

* Except for Tokyo, the data for each city refer to an urban county (*shi*), an administrative division which may include some scattered or rural population as well as an urban centre.

† The figure refers to the 23 wards (*ku*) of the old city. The population of Tokyo-to (Tokyo Prefecture) was 12,059,237.

‡ Including Kokura, Moji, Tobata, Wakamatsu and Yahata (Yawata).

Source: UN, *Demographic Yearbook*.

BIRTHS, MARRIAGES AND DEATHS*

	Registered live births		Registered marriages†		Registered deaths	
	Number	Rate (per 1,000)	Number	Rate (per 1,000)	Number	Rate (per 1,000)
1996	1,206,555	9.7	795,080	6.4	896,211	7.2
1997	1,191,665	9.5	775,651	6.2	913,402	7.3
1998	1,203,147	9.6	784,595	6.3	936,484	7.5
1999	1,177,669	9.4	762,011	6.1	982,031	7.8
2000	1,190,547	9.5	798,138	6.4	961,653	7.7
2001	1,170,662	9.3	799,999	6.4	970,331	7.7
2002	1,153,900	9.2	757,300	6.0	982,400	7.8
2003	1,123,800	8.9	740,220	5.9	1,015,000	8.0

* Figures relate only to Japanese nationals in Japan.
† Data are tabulated by year of registration rather than by year of occurrence.

Expectation of life (WHO estimates, years at birth): 81.9 (males 78.4; females 85.3) in 2002 (Source: WHO, *World Health Report*).

ECONOMICALLY ACTIVE POPULATION*
(annual averages, '000 persons aged 15 years and over)

	2001	2002	2003
Agriculture and forestry	2,860	2,680	2,660
Fishing and aquaculture	270	280	270
Mining and quarrying	50	50	50
Manufacturing	12,840	12,220	11,780
Electricity, gas and water	340	340	320
Construction	6,320	6,180	6,040
Wholesale and retail trade, restaurants and hotels	14,730	14,380	14,830
Transport, storage and communications	4,070	4,010	4,960
Financing, insurance, real estate and business services	2,400	2,410	2,320
Community, social and personal services	17,680	18,040	17,050
Government (not elsewhere classified)	2,110	2,170	2,270
Activities not adequately defined	450	540	610
Total employed	64,120	63,300	63,160
Unemployed	3,400	3,590	3,500
Total labour force	67,520	66,890	66,660
Males	39,920	39,560	39,340
Females	27,600	27,330	27,320

* Figures are rounded to the nearest 10,000 persons.

Health and Welfare

KEY INDICATORS

Total fertility rate (children per woman, 2002)	1.3
Under-5 mortality rate (per 1,000 live births, 2002)	5
HIV/AIDS (% of persons aged 15–49, 2003)	<0.10
Physicians (per 1,000 head, 2000)	1.9
Hospital beds (per 1,000 head, 2000)	16.5
Health expenditure (2001): US $ per head (PPP)	2,131
Health expenditure (2001): % of GDP	8.0
Health expenditure (2001): public (% of total)	77.9
Human Development Index (2002): ranking	9
Human Development Index (2002): value	0.938

For sources and definitions, see explanatory note on p. vi.

Agriculture

PRINCIPAL CROPS
('000 metric tons)

	2000	2001	2002
Wheat	688.2	699.9	827.8
Rice (paddy)	11,863.0	11,320.0	11,111.0
Barley	214.3	206.4	217.2
Potatoes	2,898.0	2,959.0	3,069.0
Sweet potatoes	1,073.4	1,063.0	1,030.0
Taro (Coco yam)	230.5	218.1	209.2
Yams	201.2	182.4	181.7
Other roots and tubers	75.5	75.3	75.3
Sugar cane	1,395.0	1,497.0	1,328.0
Sugar beets	3,673.0	3,796.0	4,098.0
Dry beans	103.5	94.6	99.9
Soybeans (Soya beans)	235.0	290.6	270.2
Cabbages	2,485.0	2,472.0	2,392.0
Lettuce	537.2	553.8	561.6
Spinach	316.4	319.3	311.6
Tomatoes	806.3	797.6	784.5
Cauliflowers	114.7	120.9	124.6
Pumpkins, squash and gourds	253.6	227.5	219.5
Cucumbers and gherkins	766.7	735.0	728.9
Aubergines (Eggplants)	476.9	448.0	432.4
Chillies and green peppers	171.4	159.3	161.0
Green onions and shallots	537.0	526.9	518.7
Dry onions	1,247.0	1,259.0	1,274.0
Green beans	63.9	62.1	58.8
Carrots	681.7	690.3	644.5
Green corn	289.0	290.0*	290.0*
Mushrooms	67.2	67.2	67.0*
Other vegetables†	2,900.0	2,800.0	2,800.0
Watermelons	580.6	573.3	526.9
Cantaloupes and other melons	317.5	307.4	286.7
Grapes	237.5	225.4	231.7
Apples	799.6	930.0	925.8
Pears	392.9	368.2	406.7
Peaches and nectarines	174.6	175.8	175.1
Plums	121.2	123.7	112.7
Oranges†	104.0	102.0	105.0
Tangerines, mandarins, clementines and satsumas	1,143.0	1,281.0	1,131.0
Other citrus fruit	256.0	246.0	202.0
Persimmons	278.8	281.8	269.3
Strawberries	205.3	208.6	210.5
Other fruits and berries*	80.7	86.8	89.9
Tea (made)	85.0	85.0	84.0
Tobacco (leaves)	60.8	61.0	58.0

* FAO estimate(s).
† Unofficial figure.

Source: FAO.

LIVESTOCK
('000 head at 30 September)

	2000	2001	2002
Horses*	25	21	20
Cattle	4,588	4,531	4,564
Pigs	9,806	9,788	9,612
Sheep*	10	10	11
Goats*	35	35	35
Poultry	295,792	292,437	287,404

* FAO estimates.

Source: FAO.

LIVESTOCK PRODUCTS
('000 metric tons)

	2000	2001	2002
Beef and veal	530.4	458.0	535.1
Pig meat	1,255.8	1,231.6	1,244.1
Poultry meat	1,194.5	1,216.4	1,229.1
Cows' milk	8,497.0	8,301.0	8,385.3
Butter	87.6	79.5	82.7
Cheese	126.2	123.4	122.8
Hen eggs	2,535.4	2,514.2	2,513.7
Cattle hides (fresh)*	33.0	28.5	33.2

* FAO estimates.

Source: FAO.

Forestry

ROUNDWOOD REMOVALS
('000 cubic metres, excl. bark)

	2000	2001	2002
Sawlogs, veneer logs and logs for sleepers	12,936	11,948	11,421
Pulpwood	4,717	3,826	3,671
Other industrial wood	334	0	0
Fuel wood*	134	129	124
Total	18,121	15,903	15,216

* FAO estimates.

Source: FAO.

SAWNWOOD PRODUCTION
('000 cubic metres, incl. railway sleepers)

	2000	2001	2002
Coniferous (softwood)	16,479	14,974	13,970
Broadleaved (hardwood)	615	511	432
Total	17,094	15,485	14,402

Source: FAO.

Fishing

('000 metric tons, live weight)

	2000	2001	2002
Capture	4,984.8	4,713.0	4,443.0
Chum salmon (Keta or Dog salmon)	163.5	217.4	211.7
Alaska (Walleye) pollock	300.0	241.9	213.3
Atka mackerel	165.1	161.2	154.7
Pacific saury (Skipper)	216.5	269.8	205.3
Japanese jack mackerel	246.0	214.4	196.0
Japanese pilchard (sardine)	149.6	178.4	50.3
Japanese anchovy	381.0	301.2	443.2
Skipjack tuna (Oceanic skipjack)	341.4	276.7	276.1
Chub mackerel	346.2	375.3	279.0
Yesso scallop	304.3	291.0	306.7
Japanese flying squid	337.3	298.2	273.6
Aquaculture	762.8	801.9	828.4
Pacific cupped oyster	221.3	231.5	221.4
Yesso scallop	210.7	235.6	272.0
Total catch	5,747.6	5,515.0	5,271.4

Note: Figures exclude aquatic plants ('000 metric tons): 647.9 (capture 119.0, aquaculture 528.9) in 2000; 633.9 (capture 122.1, aquaculture 511.8) in 2001; 686.1 (capture 127.9, aquaculture 558.2) in 2002. Also excluded are aquatic mammals (generally recorded by number rather than by weight), pearls, corals and sponges. The number of whales caught was: 923 in 2000; 1,131 in 2001; 796 in 2002. The catch of other aquatic mammals (in '000 metric tons) was: 1.8 in 2000; 1.9 in 2001; 2.0 in 2002. For the remaining categories, catches (in kilograms) were: Pearls 29.9 in 2000; 34.5 in 2001; 31.9 in 2002; Corals (including FAO estimates) 8.9 in 2000; 3.5 in 2001; 4.2 in 2002; Sponges (FAO estimate) 4.0 in 2000; n.a. in 2001 and 2002.

Source: FAO.

Mining

('000 metric tons, unless otherwise indicated)

	2000	2001	2002*
Hard coal	3,126	3,198	1,367
Zinc ore†	64	45	43
Iron ore‡	1.5	0.8	0.7
Silica stone	15,578	14,213	13,571
Limestone	185,569	182,255	171,948
Copper ore (metric tons)†	1,211	744	n.a.
Lead ore (metric tons)†	8,835	4,997	5,723
Gold ore (kg)†	8,400	7,815	8,615
Crude petroleum ('000 barrels)	4,656	4,782	4,548
Natural gas (million cu m)	2,453	2,521	2,571

* Provisional figures.

† Figures refer to the metal content of ores.

‡ Figures refer to gross weight. The estimated iron content is 54%.

Source: US Geological Survey.

Industry

SELECTED PRODUCTS
('000 metric tons, unless otherwise indicated)

	1999	2000	2001
Refined sugar*	2,242	2,281	2,257
Cotton yarn—pure	157.9[1]	144.5[1]	125.8
Cotton yarn—mixed	13.1[1]	14.3[1]	13.7
Woven cotton fabrics—pure and mixed (million sq m)	774	664	603
Flax, ramie and hemp yarn	1.0	0.9	0.5
Linen fabrics (million sq m)	4.0	3.9	3.2
Woven silk fabrics—pure and mixed ('000 sq m)	34,564	33,444	30,837
Wool yarn—pure and mixed	42.0	33.7	29.6
Woven woollen fabrics—pure and mixed (million sq m)	199.1[2]	98.2[2]	94.9
Rayon continuous filaments	34	30	18
Acetate continuous filaments	24	25	23
Rayon discontinuous fibres	77	72	66
Acetate discontinuous fibres[3]	81	84	89
Woven rayon fabrics—pure and mixed (million sq m)[2]	325	241	213
Woven acetate fabrics—pure and mixed (million sq m)[2]	27	32	28
Non-cellulosic continuous filaments (metric tons)	664	669	640
Non-cellulosic discontinuous fibres	754	764	727
Woven synthetic fabrics (million sq m)[2, 4]	1,581	1,573	1,484
Leather footwear ('000 pairs)	37,546	35,961	34,667
Mechanical wood pulp	1,474	1,405	1,394
Chemical wood pulp[5]	9,497	9,968	9,398
Newsprint	3,295	3,419	3,464
Other printing and writing paper	11,330	11,740	11,163
Other paper	3,769	3,877	3,758
Paperboard	12,238	21,791	12,332
Synthetic rubber	1,577	1,590	1,465
Road motor vehicle tyres ('000)	163,705	166,709	165,063
Rubber footwear ('000 pairs)	15,480	12,421	10,999
Ethylene—Ethene	7,087	7,614	7,361
Propylene—Propene	5,520	5,453	5,342
Benzene—Benzol	4,459	4,426	4,261
Toluene—Toluol	1,488	1,489	1,423
Xylenes—Xylol	5,520	4,681	4,798
Ethyl alcohol—95% (kilolitres)	263,633	n.a.	n.a.
Sulphuric acid—100%	6,493	7,059	6,727
Caustic soda—Sodium hydroxide	4,215	4,337	4,162
Soda ash—Sodium carbonate	722	669	680
Ammonia	1,685	1,715	1,604
Nitrogenous fertilizers	751	756	683
Phosphate fertilizers	399	369	334
Liquefied petroleum gas	4,885	4,946	n.a.
Naphtha	13,251	13,234	n.a.
Motor spirit—gasoline[6]	41,448	41,776	n.a.
Kerosene	21,709	22,699	n.a.
Jet fuel	8,187	8,323	n.a.
Gas–diesel oil	60,904	60,032	n.a.
Residual fuel oil	37,432	35,668	n.a.
Lubricating oil	2,399	2,363	n.a.
Petroleum bitumen—Asphalt	5,597	5,504	n.a.

— *continued*	1999	2000	2001
Coke-oven coke	36,473	39,954	n.a.
Cement	80,120	81,097	76,550
Pig-iron	74,520	81,071	78,836
Ferro-alloys[7]	984	1,049	1,056
Crude steel	94,192	106,444	102,866
Aluminium—unwrought	1,158	1,214	1,171
Refined copper—unwrought	1,342	1,437	1,426
Refined lead—unwrought	227.1	239.4	236.0
Electrolytic, distilled and rectified zinc—unwrought	633.4	654.4	644.4
Calculating machines ('000)	2,402	1,841	1,301
Still cameras ('000)	10,326	9,743	7,907
DVD players ('000)	4,524	4,517	2,832
Television receivers ('000)	4,386	3,382	2,862
Cellular telephones ('000)	43,350	55,272	53,652
Personal computers (incl. servers, '000)	9,254	12,149	11,350
Passenger motor cars ('000)	8,100	8,363	8,118
Lorries and trucks ('000)	1,742	1,720	1,596
Motorcycles, scooters and mopeds ('000)	2,252	2,415	2,328
Watches	531,289	554,130	515,302
Construction: new dwellings started ('000)	1,215	1,230	1,174
Electric energy (million kWh)*	1,066,130	1,091,499	1,075,890

* Twelve months beginning 1 April of the year stated.
[1] Including condenser cotton yarn.
[2] Including finished fabrics.
[3] Including cigarette filtration tow.
[4] Including blankets made of synthetic fibres.
[5] Including pulp prepared by semi-chemical processes.
[6] Not including aviation gasoline.
[7] Including silico-chromium.

Source: UN, *Industrial Commodity Statistics Yearbook*.

Finance

CURRENCY AND EXCHANGE RATES

Monetary Units

100 sen = 1 yen.

Sterling, Dollar and Euro Equivalents (31 May 2004)

£1 sterling = 202.73 yen;
US $1 = 110.50 yen;
€1 = 135.32 yen;
1,000 yen = £4.93 = $9.05 = €7.39.

Average Exchange Rate (yen per US $)

2001 121.53
2002 125.39
2003 115.93

BUDGET

('000 million yen, year ending 31 March)*

Revenue	2002/03†	2003/04‡	2004/05‡
Tax and stamp revenues	44,276	41,786	41,747
Individual income tax	14,708	13,810	13,778
Corporation tax	9,990	9,114	9,407
Consumption tax	9,592	9,489	9,563
Liquor tax	1,735	1,733	1,588
Stamp revenue	1,444	1,129	1,148
Government bond issues	34,968	36,445	36,590
Total (incl. others)	83,689	81,789	82,111

Expenditure‡	2002/03	2003/04	2004/05
Defence	4,956	4,953	4,903
Social security	18,280	18,991	19,797
Public works	8,424	8,097	7,816
Servicing of national debt§	16,671	16,798	17,569
Transfer of local allocation tax to local governments	17,012	17,399	16,494
Total (incl. others)	81,230	81,789	82,111

* Figures refer only to the operations of the General Account budget. Data exclude transactions of other accounts controlled by the central Government: two mutual aid associations and four special accounts (including other social security funds).
† Revised forecasts.
‡ Initial forecasts.
§ Including the repayment of debt principal and administrative costs.

Source: Ministry of Finance, Tokyo.

INTERNATIONAL RESERVES

(US $ million at 31 December)

	2001	2002	2003
Gold*	1,082	1,171	1,280
IMF special drawing rights	2,377	2,524	2,766
Reserve position in IMF	5,051	7,203	7,733
Foreign exchange	387,727	451,458	652,790
Total	396,237	462,357	664,569

* Valued at SDR 35 per troy ounce.

Source: IMF, *International Financial Statistics*.

MONEY SUPPLY

('000 million yen at 31 December)

	2001	2002	2003
Currency outside banks	66,676	71,328	72,455
Demand deposits at deposit money banks	215,109	276,651	291,144
Total money	281,785	347,979	363,599

Source: IMF, *International Financial Statistics*.

COST OF LIVING

(Consumer Price Index; average of monthly figures; base: 2000 = 100)

	2001	2002	2003
Food (incl. beverages)	99.4	98.6	98.4
Housing	100.2	100.1	100.0
Rent	100.4	100.4	100.4
Fuel, light and water charges	100.6	99.4	98.9
Clothing and footwear	97.8	95.6	93.8
Miscellaneous	99.8	100.0	100.9
All items	99.3	98.4	98.1

NATIONAL ACCOUNTS

('000 million yen at current prices, year ending 31 December)

National Income and Product

	2000	2001	2002
Compensation of employees	275,048.2	273,372.8	265,368.4
Operating surplus and mixed income	94,912.0	87,045.4	91,461.4
Domestic primary incomes	369,960.2	360,418.2	356,829.8
Consumption of fixed capital	97,995.1	99,380.4	98,568.8
Statistical discrepancy	5,089.6	7,214.0	5,021.3
Gross domestic product (GDP) at factor cost	473,044.9	467,012.6	460,419.9
Indirect taxes	43,136.1	42,911.7	41,464.1
Less Subsidies	4,718.6	4,076.8	3,781.9
GDP in purchasers' values	511,462.4	505,847.4	498,102.0
Primary incomes received from abroad	11,574.8	13,783.4	12,791.9
Less Primary incomes paid abroad	5,153.5	5,462.7	4,598.5
Gross national income (GNI)	517,883.7	514,168.1	506,295.4

Expenditure on the Gross Domestic Product

	2001	2002*	2003†
Government final consumption expenditure	86,418.5	87,972.9	87,191.3
Private final consumption expenditure	285,965.5	284,796.6	283,214.7
Changes in stocks	−21.8	−1,335.5	390.8
Gross fixed capital formation	130,310.9	120,429.7	119,048.6
Total domestic expenditure	502,672.8	491,863.7	489,845.4
Exports of goods and services	52,567.0	55,829.1	58,882.4
Less Imports of goods and services	49,392.8	49,417.2	50,906.9
GDP in purchasers' values	505,847.4	498,275.6	497,821.0
GDP at constant 1995 prices	534,851.5	533,041.6	546,177.0

* Revised figures.
† Preliminary figures.

Gross Domestic Product by Economic Activity

	2000	2001	2002
Agriculture, hunting, forestry and fishing	7,109.9	6,780.9	6,613.0
Mining and quarrying	661.8	683.5	622.8
Manufacturing	112,114.0	105,220.9	102,299.0
Electricity, gas and water	14,218.2	14,505.7	14,134.8
Construction	37,936.3	36,268.0	34,318.2
Wholesale and retail trade	70,070.2	69,275.2	68,482.0
Transport, storage and communications	32,619.5	32,651.1	31,546.4
Finance and insurance	31,119.0	33,210.0	33,944.1
Real estate*	66,342.2	67,383.8	68,398.8
Public administration	26,268.3	26,879.2	27,349.1
Other government services	18,271.1	18,466.3	18,690.7
Other business, community, social and personal services	103,751.9	103,840.8	103,536.8
Private non-profit services to households	9,342.8	9,433.2	9,825.7
Sub-total	529,825.2	524,598.6	519,761.4
Import duties	3,165.0	3,242.9	3,157.3
Less Imputed bank service charge	23,204.1	25,777.8	26,815.1
Less Consumption taxes for gross capital formation	3,413.3	3,430.3	3,022.9
Statistical discrepancy	5,089.6	7,214.0	5,021.3
GDP in purchasers' values	511,462.4	505,847.4	498,102.0

* Including imputed rents of owner-occupied dwellings.
Source: Economic and Social Research Institute, Tokyo.

BALANCE OF PAYMENTS
(US $ million)*

	2001	2002	2003
Exports of goods f.o.b.	383,600	395,580	449,120
Imports of goods f.o.b.	−313,380	−301,750	−342,720
Trade balance	70,210	93,830	10,640
Exports of services	64,520	65,710	77,620
Imports of services	−108,250	−107,940	−111,530
Balance on goods and services	26,480	51,600	72,490
Other income received	103,090	91,480	95,210
Other income paid	−33,870	−25,710	−23,970
Balance on goods, services and income	95,700	117,370	143,730
Current transfers received	6,150	10,040	6,510
Current transfers paid	−14,060	−14,960	−14,020
Current balance	87,790	112,450	136,220
Capital account (net)	−2,870	−3,320	−4,000
Direct investment abroad	−38,500	−32,020	−28,770
Direct investment from abroad	6,190	9,090	6,240
Portfolio investment assets	−106,790	−85,930	−176,290
Portfolio investment liabilities	60,500	−20,040	81,180
Financial derivatives assets	102,790	77,250	64,960
Financial derivatives liabilities	−101,400	−74,770	−59,380
Other investment assets	46,590	36,410	149,890
Other investment liabilities	−17,550	26,630	34,100
Net errors and omissions	3,720	390	−16,990
Overall balance	40,490	46,130	187,150

* Figures are rounded to the nearest US $10m.

Source: IMF, *International Financial Statistics*.

JAPANESE DEVELOPMENT ASSISTANCE
(net disbursement basis, US $ million)

	2000	2001	2002
Official:			
Bilateral assistance:			
Grants	5,813	4,849	4,473
Grant assistance	2,109	1,907	1,718
Technical assistance	3,705	2,943	2,754
Loans	3,827	2,603	2,253
Total	9,640	7,452	6,726
Contributions to multilateral institutions	3,779	2,448	2,633
Total	13,419	9,900	9,359
Other official flows:			
Export credits	−1,552	−495	−469
Equities and other bilateral assets, etc.	−3,052	−1,328	−2,360
Transfers to multilateral institutions	−252	−875	−2,512
Total	−4,855	−2,698	−5,341
Total official	8,564	7,202	4,018
Private flows:			
Export credits	−358	593	−1,078
Direct investment and others	6,191	12,127	12,108
Bilateral investment in securities, etc.	478	−4,133	−3,413
Transfers to multilateral institutions	−52	−355	−2,804
Grants from private voluntary agencies	231	235	157
Total private	6,490	8,467	4,970
Grand total	15,053	15,669	8,989

External Trade

PRINCIPAL COMMODITIES
(million yen)

Imports c.i.f.	2001	2002	2003
Food and live animals	5,250,600	5,282,300	5,104,600
Fish and fish preparations*	1,626,474	1,658,025	1,475,463
Crude materials (inedible) except fuels	2,586,100	2,522,000	n.a.
Mineral fuels, lubricants, etc.	8,523,700	8,173,900	9,350,000
Crude and partly refined petroleum	4,718,360	4,573,000	5,328,000
Liquefied natural gas	1,594,000	1,492,000	1,695,000
Chemicals	3,101,100	3,239,000	n.a.
Machinery and transport equipment	13,215,900	13,434,000	n.a.
Office machines	2,764,027	2,698,000	2,745,000
Thermionic valves, tubes, etc.	1,909,535	1,910,000	2,015,000
Textiles	2,890,000	2,752,000	n.a.
Clothing (excl. footwear)	2,318,293	2,189,000	2,224,100
Total (incl. others)†	42,415,533	42,227,500	443,620

* Including crustacea and molluscs.
† Including re-imports not classified according to kind.

Exports f.o.b.	2001	2002	2003
Chemicals	3,738,800	4,173,700	4,525,000
Metals and metal products	2,889,100	3,227,400	3,388,500
Iron and steel	1,649,543	1,939,592	2,066,009
Machinery and transport equipment	32,895,700	35,523,000	n.a.
Non-electric machinery	10,229,500	10,599,000	n.a.
Office machines	2,820,710	3,005,000	2,619,000
Electrical machinery, apparatus, etc.	11,533,300	11,924,000	n.a.
Thermionic valves, tubes, etc.	3,647,382	3,867,000	4,074,000
Transport equipment	11,132,900	13,000,000	n.a.
Road motor vehicles	7,210,812	8,775,000	n.a.
Road motor vehicle parts	1,880,380	2,117,000	2,300,000
Precision instruments	2,629,101	2,019,000	n.a.
Scientific instruments and photographic equipment	2,504,480	1,897,000	n.a.
Total (incl. others)*	48,979,244	52,109,000	545,484

* Including re-exports not classified according to kind.

PRINCIPAL TRADING PARTNERS
(million yen)*

Imports c.i.f.	2000	2001	2002
Australia	1,595,908	1,755,871	1,753,000
Canada	938,485	941,469	895,000
China, People's Republic	5,941,358	7,026,677	7,727,793
France	691,297	750,358	817,000
Germany	1,371,925	1,505,798	1,553,000
Indonesia	1,766,187	1,805,632	1,774,000
Iran	577,787	609,819	593,000
Ireland	399,108	441,373	401,000
Italy	572,761	654,978	679,000
Korea, Republic	2,204,703	2,088,356	1,936,787
Kuwait	538,281	538,096	526,000
Malaysia	1,562,726	1,561,324	1,401,350
Philippines	776,247	778,879	818,000
Qatar	632,000	731,801	657,000
Russia	493,791	468,419	410,000
Saudi Arabia	1,531,277	1,496,299	1,455,000
Singapore	693,625	653,684	626,767
Taiwan	1,930,161	1,722,643	1,698,926
Thailand	1,142,346	1,260,472	1,314,594
United Arab Emirates	1,599,649	1,559,855	1,450,000
United Kingdom	709,180	729,016	677,000
USA	7,778,861	7,671,481	7,237,000
Total (incl. others)	40,938,423	42,415,533	42,228,000

Exports f.o.b.	2000	2001	2002
Australia	923,830	933,178	1,039,000
Belgium	564,615	555,798	575,000
Canada	805,939	797,113	918,000
China, People's Republic	3,274,448	3,763,723	4,979,796
France	803,801	758,786	768,000
Germany	2,155,178	1,896,740	1,766,000
Hong Kong	2,929,696	2,826,044	3,176,000
Indonesia	817,745	777,704	780,000
Italy	624,309	584,615	562,000
Korea, Republic	3,308,751	3,071,871	3,572,439
Malaysia	1,496,627	1,337,217	1,377,609
Mexico	561,557	496,995	472,000
Netherlands	1,356,814	1,393,132	1,323,000
Panama	695,408	586,630	579,000
Philippines	1,105,654	995,303	1,058,000
Singapore	2,243,914	1,786,059	1,775,000
Taiwan	3,874,042	2,942,227	3,281,188
Thailand	1,469,397	1,442,488	1,648,577
United Kingdom	1,598,434	1,474,989	1,498,000
USA	15,355,867	14,711,055	14,873,000
Total (incl. others)	51,654,198	48,979,244	52,109,000

* Imports by country of production; exports by country of last consignment.

2003 (selected trading partners, million yen): *Imports c.i.f.:* People's Republic of China 8,731,139; Republic of Korea 2,071,182; Malaysia 1,458,086; Singapore 628,794; Taiwan 1,655,700; Thailand 1,375,905; USA 6,825,000; Total (incl. others) 44,362,000. *Exports f.o.b.:* People's Republic of China 6,635,482; Republic of Korea 4,022,469; Malaysia 1,301,741; Taiwan 3,609,890; Thailand 1,853,752; Total (incl. others) 54,548,000.

Transport

RAILWAYS
(traffic, year ending 31 March)

	2000	2001	2002
National railways:			
Passengers (million)	8,670	8,650	8,590
Passenger-km (million)	240,659	241,133	239,240
Freight ('000 tons)	39,620	39,026	38,200
Freight ton-km (million)	21,855	21,907	21,860
Private railways:			
Passengers (million)	12,980	13,070	12,980
Passenger-km (million)	384,442	144,288	142,990
Freight ('000 tons)	19,654	19,642	18,400
Freight ton-km (million)	280	286	270
Total:			
Passengers (million)	21,646	21,720	21,560
Passenger-km (million)	384,441	385,421	382,240
Freight ('000 tons)	59,274	58,668	56,600
Freight ton-km (million)	22,135	22,193	22,130

ROAD TRAFFIC
('000 motor vehicles owned, year ending 31 March)

	2000/01	2001/02	2002/03
Passenger cars	42,365	42,528	42,655
Buses and coaches	236	234	233
Trucks, incl. trailers	8,106	7,907	7,666
Special use vehicles	1,754	1,754	1,720
Heavy use vehicles	323	325	n.a.
Light two-wheeled vehicles	1,308	1,334	1,352
Light motor vehicles	21,755	22,513	23,266
Total	75,525	76,271	76,893

SHIPPING

Merchant Fleet
(registered at 31 December)

	2001	2002	2003
Number of vessels	7,924	7,458	7,151
Total displacement ('000 grt)	16,653	13,918	13,562

Source: Lloyd's Register-Fairplay, *World Fleet Statistics*.

International Sea-borne Traffic
('000 metric tons)

	2000	2001	2002
Goods loaded	101,727	106,986	119,385
Goods unloaded	787,987	772,996	762,329

CIVIL AVIATION
(traffic on scheduled services)

	2000	2001	2002
Kilometres flown (million)	942	929	n.a.
Passengers carried ('000)	112,177	111,690	114,590
Passenger-km (million)	176,628	165,618	156,770
Total ton-km (million)	8,728	7,630	8,150

Tourism

FOREIGN VISITOR ARRIVALS
(excl. Japanese nationals resident abroad)

Country of nationality	2001	2002	2003
Australia	149,621	164,896	172,134
Canada	125,570	131,542	126,065
China, People's Republic	391,384	452,520	448,782
Germany	87,740	93,936	93,571
Hong Kong	262,229	290,624	260,214
Korea, Republic	1,133,971	1,271,835	1,459,333
Philippines	124,072	129,914	137,584
Taiwan	807,202	877,709	785,379
United Kingdom	197,965	219,271	200,543
USA	692,192	731,900	655,821
Total (incl. others)	4,771,555	5,238,963	5,211,725

Source: Japan National Tourist Organization.

Receipts from tourism (US $ million): 3,374 in 2000; 3,301 in 2001; 3,499 in 2002.

Communications Media

	1999	2000	2001
Television receivers ('000 in use)	91,000	92,000	n.a.
Telephones ('000 main lines in use)	70,530.0	74,343.6	76,000.0
Mobile telephones ('000 subscribers)	47,308.0	56,845.6	66,784.4
Personal computers ('000 in use)	36,300	40,000	44,400
Internet users ('000)	27,060	47,080	57,900
Book production:			
titles	65,026	65,065	71,073
copies (million)	1,368	1,420	1,390
Daily newspapers:			
number	121	122	124
circulation ('000 copies)	72,218	n.a.	53,681

Radio receivers ('000 in use): 120,500 in 1997.

Facsimile machines ('000 in use): 16,000 in 1997.

2002: Mobile telephones ('000 subscribers) 74,819; Personal computers ('000 in use) 48,700; Internet users ('000) 57,200.

2003: Mobile telephones ('000 subscribers) 81,118; Internet users ('000) 77,300.

Sources: Foreign Press Center, *Facts and Figures of Japan*; UNESCO, *Statistical Yearbook*; UN, *Statistical Yearbook*; International Telecommunication Union.

Education

(2002)

	Institutions	Teachers	Students
Elementary schools	23,808	411,000	7,239,000
Lower secondary schools	11,159	254,000	3,863,000
Upper secondary schools	5,472	262,000	3,930,000
Colleges of technology	62	4,000	57,000
Junior colleges	541	14,000	267,000
Graduate schools and universities	686	155,000	2,786,000

Directory

The Constitution

The Constitution of Japan was promulgated on 3 November 1946 and came into force on 3 May 1947. The following is a summary of its major provisions, with subsequent amendments:

THE EMPEROR

Articles 1–8. The Emperor derives his position from the will of the people. In the performance of any state act as defined in the Constitution, he must seek the advice and approval of the Cabinet, though he may delegate the exercise of his functions, which include: (i) the appointment of the Prime Minister and the Chief Justice of the Supreme Court; (ii) promulgation of laws, cabinet orders, treaties and constitutional amendments; (iii) the convocation of the Diet, dissolution of the House of Representatives and proclamation of elections to the Diet; (iv) the appointment and dismissal of Ministers of State, the granting of amnesties, reprieves and pardons, and the ratification of treaties, conventions or protocols; (v) the awarding of honours and performance of ceremonial functions.

RENUNCIATION OF WAR

Article 9. Japan renounces for ever the use of war as a means of settling international disputes.

Articles 10–40 refer to the legal and human rights of individuals guaranteed by the Constitution.

THE DIET

Articles 41–64. The Diet is convened once a year, is the highest organ of state power and has exclusive legislative authority. It comprises the House of Representatives (480 seats—300 single-seat constituencies and 180 determined by proportional representation) and the House of Councillors (247 seats). The members of the former are elected for four years whilst those of the latter are elected for six years and election for approximately one-half of the members takes place every three years. If the House of Representatives is dissolved, a general election must take place within 40 days and the Diet must be convoked within 30 days of the date of the election. Extraordinary sessions of the Diet may be convened by the Cabinet when one-quarter or more of the members of either House request it. Emergency sessions of the House of Councillors may also be held. A quorum of at least one-third of the Diet members is needed to carry out parliamentary business. Any decision arising therefrom must be passed by a majority vote of those present. A bill becomes law having passed both Houses, except as provided by the Constitution. If the House of Councillors either vetoes or fails to take action within 60 days upon a bill already passed by the House of Representatives, the bill becomes law when passed a second time by the House of Representatives, by at least a two-thirds' majority of those members present.

The Budget must first be submitted to the House of Representatives. If, when it is approved by the House of Representatives, the House of Councillors votes against it or fails to take action on it within 30 days, or failing agreement being reached by a joint committee of both Houses, a decision of the House of Representatives shall be the decision of the Diet. The above procedure also applies in respect of the conclusion of treaties.

THE EXECUTIVE

Articles 65–75. Executive power is vested in the Cabinet, consisting of a Prime Minister and such other Ministers as may be appointed. The Cabinet is collectively responsible to the Diet. The Prime Minister is designated from among members of the Diet by a resolution thereof.

If the House of Representatives and the House of Councillors disagree on the designation of the Prime Minister, and if no agreement can be reached even through a joint committee of both Houses, provided for by law, or if the House of Councillors fails to make designation within 10 days, exclusive of the period of recess, after the House of Representatives has made designation, the decision of the House of Representatives shall be the decision of the Diet.

The Prime Minister appoints and may remove other Ministers, a majority of whom must be from the Diet. If the House of Representatives passes a no-confidence motion or rejects a confidence motion, the whole Cabinet resigns, unless the House of Representatives is dissolved within 10 days. When there is a vacancy in the post of Prime Minister, or upon the first convocation of the Diet after a general election of members of the House of Representatives, the whole Cabinet resigns.

The Prime Minister submits bills, reports on national affairs and foreign relations to the Diet. He exercises control and supervision over various administrative branches of the Government. The Cabinet's primary functions (in addition to administrative ones) are to: (a) administer the law faithfully; (b) conduct State affairs; (c) conclude treaties subject to prior (or subsequent) Diet approval; (d) administer the civil service in accordance with law; (e) prepare and present the budget to the Diet; (f) enact Cabinet orders in order to make effective legal and constitutional provisions; (g) decide on amnesties, reprieves or pardons. All laws and Cabinet orders are signed by the competent Minister of State and countersigned by the Prime Minister. The Ministers of State, during their tenure of office, are not subject to legal action without the consent of the Prime Minister. However, the right to take that action is not impaired.

Articles 76–95. Relate to the Judiciary, Finance and Local Government.

AMENDMENTS

Article 96. Amendments to the Constitution are initiated by the Diet, through a concurring vote of two-thirds or more of all the members of each House and are submitted to the people for ratification, which requires the affirmative vote of a majority of all votes cast at a special referendum or at such election as the Diet may specify.

Amendments when so ratified must immediately be promulgated by the Emperor in the name of the people, as an integral part of the Constitution.

Articles 97–99 outline the Supreme Law, while Articles 100–103 consist of Supplementary Provisions.

The Government

HEAD OF STATE

His Imperial Majesty AKIHITO, Emperor of Japan (succeeded to the throne 7 January 1989).

THE CABINET
(September 2004)

A coalition of the Liberal-Democratic Party (LDP) and New Komeito; the New Conservative Party, previously a coalition partner, was absorbed into the LDP in November 2003.

Prime Minister: JUNICHIRO KOIZUMI.

Minister of Public Management, Home Affairs, Posts and Telecommunications: TARO ASO.

Minister of Justice and Minister of State for Youth Affairs and Measures for Declining Birthrate: CHIEKO NOHNO.

Minister of Foreign Affairs: NOBUTAKA MACHIMURA.

Minister of Finance: SADAKAZU TANIGAKI.

Minister of Education, Culture, Sports, Science and Technology: NARIAKI NAKAYAMA.

Minister of Health, Labour and Welfare: HIDEHISA OTSUJI.

Minister of Agriculture, Forestry and Fisheries: YOSHINOBU SHIMAMURA.

Minister of Economy, Trade and Industry: SHOICHI NAKAGAWA.

Minister of Land, Infrastructure and Transport: KAZUO KITAGAWA.

Minister of the Environment and Minister of State for Okinawa and Northern Territories Affairs: YURIKO KOIKE.

Chairman of the National Public Safety Commission and Minister of State for Disaster Management and for National Emergency Legislation: YOSHITAKA MURATA.

Chief Cabinet Secretary and Minister of State for Gender Equality: HIROYUKI HOSODA.

Minister of State for Defence: YOSHINORI OHNO.

Minister of State for Financial Services: TATSUYA ITO.

Minister of State for Economic and Fiscal Policy and for Privatization of the Postal Services: HEIZO TAKENAKA.

Minister of State (Administrative Reform, Regulatory Reform): SEIICHIRO MURAKAMI.

Minister of State (Science and Technology Policy, Food Safety, Information Technology): YASUFUMI TANAHASHI.

MINISTRIES

Imperial Household Agency: 1-1, Chiyoda, Chiyoda-ku, Tokyo 100-8111; tel. (3) 3213-1111; fax (3) 3282-1407; e-mail information@kunaicho.go.jp; internet www.kunaicho.go.jp.

Prime Minister's Office: 1-6-1, Nagata-cho, Chiyoda-ku, Tokyo 100-8968; tel. (3) 3581-2361; fax (3) 3581-1910; internet www.kantei.go.jp.

Cabinet Office: 1-6-1, Nagata-cho, Chiyoda-ku, Tokyo 100-8914; tel. (3) 5253-2111; internet www.cao.go.jp.

Ministry of Agriculture, Forestry and Fisheries: 1-2-1, Kasumigaseki, Chiyoda-ku, Tokyo 100-8950; tel. (3) 3502-8111; fax (3) 3592-7697; e-mail white56@sc.maff.go.jp; internet www.maff.go.jp.

Ministry of Economy, Trade and Industry: 1-3-1, Kasumigaseki, Chiyoda-ku, Tokyo 100-8901; tel. (3) 3501-1511; fax (3) 3501-6942; e-mail webmail@meti.go.jp; internet www.meti.go.jp.

Ministry of Education, Culture, Sports, Science and Technology: 3-2-2, Kasumigaseki, Chiyoda-ku, Tokyo 100-8959; tel. (3) 5253-4111; fax (3) 3595-2017; internet www.mext.go.jp.

Ministry of the Environment: 1-2-2, Kasumigaseki, Chiyoda-ku, Tokyo 100-8975; tel. (3) 3581-3351; fax (3) 3502-0308; e-mail moe@eanet.go.jp; internet www.env.go.jp.

Ministry of Finance: 3-1-1, Kasumigaseki, Chiyoda-ku, Tokyo 100-8940; tel. (3) 3581-4111; fax (3) 5251-2667; e-mail info@mof.go.jp; internet www.mof.go.jp.

Ministry of Foreign Affairs: 2-11-1, Shiba-Koen, Minato-ku, Tokyo 105-8519; tel. (3) 3580-3311; fax (3) 3581-2667; e-mail webmaster@mofa.go.jp; internet www.mofa.go.jp.

Ministry of Health, Labour and Welfare: 1-2-2, Kasumigaseki, Chiyoda-ku, Tokyo 100-8916; tel. (3) 5253-1111; fax (3) 3501-2532; internet www.mhlw.go.jp.

Ministry of Justice: 1-1-1, Kasumigaseki, Chiyoda-ku, Tokyo 100-8977; tel. (3) 3580-4111; fax (3) 3592-7011; e-mail webmaster@moj.go.jp; internet www.moj.go.jp.

Ministry of Land, Infrastructure and Transport: 2-1-3, Kasumigaseki, Chiyoda-ku, Tokyo 100-8918; tel. (3) 5253-8111; fax (3) 3580-7982; e-mail webmaster@mlit.go.jp; internet www.mlit.go.jp.

Ministry of Public Management, Home Affairs, Posts and Telecommunications: 2-1-2, Kasumigaseki, Chiyoda-ku, Tokyo 100-8926; tel. (3) 5253-5111; fax (3) 3504-0265; internet www.soumu.go.jp.

Defence Agency: 5-1 Ichigaya, Honmura-cho, Shinjuku-ku, Tokyo 162-8801; tel. (3) 3268-3111; e-mail info@jda.go.jp; internet www.jda.go.jp.

Financial Services Agency: 3-1-1 Kasumigaseki, Chiyoda-ku, Tokyo 100-8967; tel. (3) 3506-6000; internet www.fsa.go.jp.

National Public Safety Commission: 2-1-2, Kasumigaseki, Chiyoda-ku, Tokyo 100-8974; tel. (3) 3581-0141; internet www.npsc.go.jp.

Legislature

KOKKAI
(Diet)

The Diet consists of two Chambers: the House of Councillors (upper house) and the House of Representatives (lower house). The members of the House of Representatives are elected for a period of four years (subject to dissolution). Following the enactment of reform legislation in December 1994, the number of members in the House of Representatives was reduced to 500 (from 511) at the general election of October 1996. Further legislation was enacted in February 2000, reducing the number of members in the House of Representatives to 480, comprising 300 single-seat constituencies and 180 seats determined by proportional representation. For the House of Councillors, which has 242 members (reduced from 247 at the 2004 election), the term of office is six years, with approximately one-half of the members elected every three years.

House of Councillors

Speaker: CHIKAGE OGI.

Party	Seats after elections*	
	29 July 2001	11 July 2004
Liberal-Democratic Party	110	115
Democratic Party of Japan	60	82
New Komeito	23	24
Japanese Communist Party	20	9
Social Democratic Party of Japan	8	5
Liberal Party†	8	—
New Conservative Party‡	5	—
Independents	8	7
Other parties	5	—
Total	247	242

* Approximately one-half of the seats are renewable every three years. At the 2004 election 48 of the 121 seats were allocated on the basis of proportional representation.

† Absorbed by the Democratic Party of Japan in September 2003.

‡ Absorbed by the Liberal-Democratic Party in November 2003.

House of Representatives

Speaker: YOHEI KONO.

General Election, 9 November 2003

Party	Seats
Liberal-Democratic Party	241*
Democratic Party of Japan	177
New Komeito	34
Japanese Communist Party	9
Social Democratic Party of Japan	6
New Conservative Party †	4
Independents and others	9
Total	480

* Including four independents who joined the party after the election.

† The New Conservative Party was subsequently absorbed by the Liberal-Democratic Party.

Political Organizations

The Political Funds Regulation Law provides that any organization wishing to support a candidate for an elective public office must be registered as a political party. There are more than 10,000 registered parties in the country, mostly of local or regional significance.

Dai-Niin Club: Rm 531, Sangiin Kaikan, 2-1-1, Nagata-cho, Chiyoda-ku, Tokyo 100-0014; tel. (3) 3508-8531; e-mail info@niinkuraba.gr.jp; successor to the Green Wind Club (Ryukufukai), which originated in the House of Councillors in 1946–47.

Democratic Party of Japan (DPJ): 1-11-1, Nagata-cho, Chiyoda-ku, Tokyo 100-0014; tel. (3) 3595-9960; fax (3) 3595-7318; e-mail dpjenews@dpj.or.jp; internet www.dpj.or.jp; f. 1998 by the integration into the original DPJ (f. 1996) of the Democratic Reform League, Minseito and Shinto Yuai; advocates a cabinet formed and controlled by the people; absorbed Party Sakigake in March 2001; absorbed Liberal Party in Sept. 2003; Pres. KATSUYA OKADA; Sec.-Gen. HIROHISA FUJII.

Japanese Communist Party (JCP): 4-26-7, Sendagaya, Shibuya-ku, Tokyo 151-8586; tel. (3) 3403-6111; fax (3) 3746-0767; e-mail intl@jcp.jp; internet www.jcp.or.jp; f. 1922; 400,000 mems (2004); Chair. of Cen. Cttee TETSUZO FUWA; Chair. of Exec. Cttee KAZUO SHII.

Liberal-Democratic Party—LDP (Jiyu-Minshuto): 1-11-23, Nagata-cho, Chiyoda-ku, Tokyo 100-8910; tel. (3) 3581-6211; e-mail koho@ldp.jimin.or.jp; internet www.jimin.jp; f. 1955; advocates the establishment of a welfare state, the promotion of industrial development, the improvement of educational and cultural facilities and constitutional reform as needed; absorbed New Conservative Party in Nov. 2003; 2,369,252 mems (2001); Pres. JUNICHIRO KOIZUMI; Sec.-Gen. TSUTOMU TAKEBE; Acting Sec.-Gen. SHINZO ABE; Chair. of Gen. Council FUMIO KYUMA.

New Komeito: 17, Minami-Motomachi, Shinjuku-ku, Tokyo 160-0012; tel. (3) 3353-0111; internet www.komei.or.jp/; f. 1964 as Komeito, renamed as Komei 1994 following defection of a number of mems to the New Frontier Party (Shinshinto, dissolved Dec. 1997); absorbed Reimei Club Jan. 1998; renamed as above Nov. 1998 following merger of Komei and Shinto Heiwa; advocates political moderation, humanism and globalism, and policies respecting 'dignity of human life'; 400,000 mems (2003); Chief Representative TAKENORI KANZAKI; Sec.-Gen. TETSUZO FUYUSHIBA.

New Socialist Party: Sanken Bldg, 6th Floor, 4-3-7, Hatchobori, Chuo-ku, Tokyo 104-0032; tel. (3) 3551-3980; e-mail honbu@sinsyakai.or.jp; internet www.sinsyakai.or.jp; f. 1996 by left-wing defectors from SDPJ; opposed to US military bases on Okinawa and to introduction in 1996 of new electoral system; seeks to establish an ecological socio-economic system; Chair. TATSUKUNI KOMORI; Sec.-Gen. KEN-ICHI UENO.

Social Democratic Party of Japan—SDPJ (Shakai Minshuto): 1-8-1, Nagata-cho, Chiyoda-ku, Tokyo 100-0014; tel. (3) 3580-1171; fax (3) 3580-0691; e-mail sdpjmail@omnics.co.jp; internet www.sdp.or.jp; f. 1945 as the Japan Socialist Party (JSP); adopted present name in 1996; seeks the establishment of collective non-aggression and a mutual security system, including Japan, the USA, the CIS and the People's Republic of China; 115,000 mems (1994); Chair. MIZUHO FUKUSHIMA; Sec.-Gen. SADAO FUCHIGAMI.

Diplomatic Representation

EMBASSIES IN JAPAN

Afghanistan: Matsumoto International House (MIH), 37-8 Nishihara 3-chome, Shibuya-ku, Tokyo 151-0066; tel. (3) 5465-1219; fax (3) 5465-1229; e-mail akbary6373@hotmail.com; Ambassador HAROUN AMIN.

Algeria: 2-10-67, Mita, Meguro-ku, Tokyo 153-0062; tel. (3) 3711-2661; fax (3) 3710-6534; e-mail ambalgto@twics.com; Ambassador BOUDJEMAA DELMI.

Angola: 2-10-24 Daizawa, Setagaya-ku, Tokyo 155-0032; tel. (3) 5430-7879; fax (3) 5712-7481; e-mail embassy@angola.or.jp; internet www.angola.or.jp; Ambassador VICTOR MANUEL RITA DA FONSECA LIMA.

Argentina: 2-14-14, Moto Azabu, Minato-ku, Tokyo 106-0046; tel. (3) 5420-7101; fax (3) 5420-7109; internet www.embargentina.or.jp; Ambassador ALFREDO VICENTE CHIARADIA.

Australia: 2-1-14, Mita, Minato-ku, Tokyo 108-8361; tel. (3) 5232-4111; fax (3) 5232-4149; internet www.australia.or.jp; Ambassador MURRAY MCLEAN; (designate).

Austria: 1-1-20, Moto Azabu, Minato-ku, Tokyo 106-0046; tel. (3) 3451-8281; fax (3) 3451-8283; e-mail austria@gol.com; internet www.austria.or.jp; Ambassador HANS DIETMAR SCHWEISGUT.

Bangladesh: 4-15-15, Meguro, Meguro-ku, Tokyo 153-0063; tel. (3) 5704-0216; fax (3) 5704-1696; Ambassador S. M. RASHED AHMED.

Belarus: 4-14-12, Shirogane K House, Shirogane, Minato-ku, Tokyo 108-0072; tel. (3) 3448-1623; fax (3) 3448-1624; e-mail belarus@japan.co.jp; Ambassador PETR K. KRAVCHANKA.

Belgium: 5, Niban-cho, Chiyoda-ku, Tokyo 102-0084; tel. (3) 3262-0191; fax (3) 3262-0651; e-mail tokyo@diplobel.org; Ambassador JEAN-FRANÇOIS BRANDERS.

Benin: Sougo Nagatacho Bldg 2F N-3, 1-11-28 Nagata-cho, Chiyoda-ku, Tokyo 100-0014; tel. (3) 3591-6565; Ambassador BANTOLÉ YABA.

Bolivia: Kowa Bldg, No. 38, Room 804, 4-12-24, Nishi Azabu, Minato-ku, Tokyo 106-0031; tel. (3) 3499-5441; fax (3) 3499-5443; e-mail emboltk@interlink.or.jp; Ambassador EUDORO GALINDO ANZE.

Bosnia and Herzegovina: 3-4 Rokuban-cho, Chiyoda-ku, Tokyo 102-0085; tel. (3) 3556-4151; Ambassador VLADIMIR RASPUDIĆ.

Brazil: 2-11-12, Kita Aoyama, Minato-ku, Tokyo 107-0061; tel. (3) 3404-5211; fax (3) 3405-5846; internet www.brasemb.or.jp; Ambassador FERNANDO GUIMARÃES REIS.

Brunei: 6-5-2, Kita Shinagawa, Shinagawa-ku, Tokyo 141-0001; tel. (3) 3447-7997; fax (3) 3447-9260; Ambassador P. S. N. YUSUF.

Bulgaria: 5-36-3, Yoyogi, Shibuya-ku, Tokyo 151-0053; tel. (3) 3465-1021; fax (3) 3465-1031; e-mail bulemb@gol.com; Ambassador Prof. BLAGOVEST SENDOV.

Burkina Faso: Apt 301, Hiroo Glisten Hills, 3-1-17, Hiroo, Shibuya-ku, Tokyo 150-0012; tel. (3) 3400-7919; fax (3) 3400-6945; Chargé d'affaires a.i. PATRICE KAFANDO.

Burundi: 3-27-16, Nozawa, Setagaya-ku 154, Kita-Shinagawa, Shinagawa-ku, Tokyo 141; tel. (3) 3443-7321; fax (3) 3443-7720; Ambassador GABRIEL NDIHOKUBWAYO.

Cambodia: 8-6-9, Akasaka, Minato-ku, Tokyo 107-0052; tel. (3) 5412-8521; fax (3) 5412-8526; e-mail aap33850@hkg.odn.ne.jp; Ambassador Ing Kieth.

Cameroon: 3-27-16, Nozawa, Setagaya-ku, Tokyo 154-0003; tel. (3) 5430-4381; fax (3) 5430-6489; e-mail ambacamtokyo@gol.com; Chargé d'affaires a.i. Mbella Mbella Lejeune.

Canada: 7-3-38, Akasaka, Minato-ku, Tokyo 107-8503; tel. (3) 5412-6200; fax (3) 5412-6249; internet www.canadanet.or.jp; Ambassador Robert G. Wright.

Chile: Nihon Seimei Akabanebashi Bldg, 8th Floor, 3-1-14, Shiba, Minato-ku, Tokyo 105-0014; tel. (3) 3452-7561; fax (3) 3452-4457; e-mail embajada@chile.or.jp; internet www2.tky.3web.ne.jp/~oficomtc/main.htm; Ambassador Demetrio Infante.

China, People's Republic: 3-4-33, Moto Azabu, Minato-ku, Tokyo 106-0046; tel. (3) 3403-3380; fax (3) 3403-3345; internet www.china-embassy.or.jp; Ambassador Wang Yi.

Colombia: 3-10-53, Kami Osaki, Shinagawa-ku, Tokyo 141-0021; tel. (3) 3440-6451; fax (3) 3440-6724; e-mail embajada@emcoltokyo.or.jp; internet www.colombianembassy.org; Ambassador Francisco J. Sierra.

Congo, Democratic Republic: Harajuku Green Heights, Room 701, 3-53-17, Sendagaya, Shibuya-ku, Tokyo 151-0051; tel. (3) 3423-3981; fax (3) 3423-3984; Chargé d'affaires Ngambani Zi-Mizele.

Costa Rica: Kowa Bldg, No. 38, Room 901, 4-12-24, Nishi Azabu, Minato-ku, Tokyo 106-0031; tel. (3) 3486-1812; fax (3) 3486-1813; Chargé d'affaires a.i. Ana Lucía Nassar Soto.

Côte d'Ivoire: 2-19-12, Uehara, Shibuya-ku, Tokyo 151-0064; tel. (3) 5454-1401; fax (3) 5454-1405; e-mail ambacijp@gol.com; Chargé d'affaires a.i. M. Kouadio Fry.

Croatia: 3-3-100, Hiroo, Shibuya-ku, Tokyo 150-0012; tel. (3) 5469-3014; fax (3) 5469-3015; e-mail veltok@hpo.net; Ambassador Drago Bovač.

Cuba: 1-28-4 Higashi-Azabu, Minato-ku, Tokyo 106-0044; tel. (3) 5570-3182; internet www.cyborg.ne.jp/~embcubaj; Ambassador Ernesto Meléndez Bachs.

Czech Republic: 2-16-14, Hiroo, Shibuya-ku, Tokyo 150-0012; tel. (3) 3400-8122; fax (3) 3400-8124; e-mail tokyo@embassy.mzv.cz; internet www.mzv.cz/tokyo; Ambassador Karel Žebrakovský.

Denmark: 29-6, Sarugaku-cho, Shibuya-ku, Tokyo 150-0033; tel. (3) 3496-3001; fax (3) 3496-3440; e-mail embassy.tokyo@denmark.or.jp; internet www.denmark.or.jp; Ambassador Poul Hoiness.

Djibouti: 5-18-10, Shimo Meguro, Meguro-ku, Tokyo 153-0064; tel. (3) 5704-0682; fax (3) 5725-8305; Ambassador (vacant); Chargé d'Affaires a.i. Yacin Houssein Douale.

Dominican Republic: Kowa Bldg, No. 38, Room 904, 4-12-24, Nishi Azabu, Minato-ku, Tokyo 106-0031; tel. (3) 3499-6020; fax (3) 3499-2627.

Ecuador: Kowa Bldg, No. 38, Room 806, 4-12-24, Nishi Azabu, Minato-ku, Tokyo 106-0031; tel. (3) 3499-2800; fax (3) 3499-4400; internet www.embassy-avenue.or.jp/ecuador/index-j.htm; Ambassador Marcelo Avila.

Egypt: 1-5-4, Aobadai, Meguro-ku, Tokyo 153-0042; tel. (3) 3770-8022; fax (3) 3770-8021; internet embassy.kcom.ne.jp/egypt/index.html; Ambassador Dr Mahmoud Karem.

El Salvador: Kowa Bldg, No. 38, 8th Floor, 4-12-24, Nishi Azabu, Minato-ku, Tokyo 106-0031; tel. (3) 3499-4461; fax (3) 3486-7022; e-mail embesal@gol.com; Ambassador Ricardo Paredes-Osorio.

Estonia: 2-6-15, Jingu-mae, Shibuya-ku 150-0001; tel. (3) 5412-7281; fax (3) 5412-7282; e-mail embassy.tokyo@mfa.ee; internet www.estemb.or.jp; Chargé d'affaires a.i. Argo Kangro.

Ethiopia: 3-4-1, Takanawa, Minato-ku, Tokyo 108-0074; tel. (3) 5420-6860; fax (3) 5420-6866; e-mail ethioemb@gol.com; Ambassador Dr Koang Tutam Dung.

Fiji: Noa Bldg, 14th Floor, 2-3-5, Azabudai, Minato-ku, Tokyo 106-0041; tel. (3) 3587-2038; fax (3) 3587-2563; e-mail fijiemb@hotmail.com; Ambassador Ratu Tevita Momoedonu.

Finland: 3-5-39, Minami Azabu, Minato-ku, Tokyo 106-8561; tel. (3) 5447-6000; fax (3) 5447-6042; e-mail info@finland.or.jp; internet www.finland.or.jp; Ambassador Eero Kalevi Salovaara.

France: 4-11-44, Minami Azabu, Minato-ku, Tokyo 106-8514; tel. (3) 5420-8800; fax (3) 5420-8917; e-mail ambafrance.tokyo@diplomatie.fr; internet www.ambafrance-jp.org; Ambassador Bernard de Montferrand.

Gabon: 1-34-11, Higashigaoka, Meguro-ku, Tokyo 152-0021; tel. (3) 5430-9171; fax (3) 5430-9175; e-mail info@gabonembassy-tokyo.org; Ambassador Jean-Christian Obame.

Germany: 4-5-10, Minami Azabu, Minato-ku, Tokyo 106-0047; tel. (3) 5791-7700; fax (3) 3473-4243; e-mail germtoky@ma.rosenet.ne.jp; internet www.germanembassy-japan.org; Ambassador Henrik Schmiegelow.

Ghana: 1-5-21, Nishi Azabu, Minato-ku, Tokyo 106-0031; tel. (3) 5410-8631; fax (3) 5410-8635; e-mail mission@ghanaembassy.or.jp; internet www.ghanaembassy.or.jp; Ambassador Dr Barfuor Adjei-Barwuah.

Greece: 3-16-30, Nishi Azabu, Minato-ku, Tokyo 106-0031; tel. (3) 3403-0871; fax (3) 3402-4642; e-mail greekemb@gol.com; Ambassador Kyriakos Rodoussakis.

Guatemala: Kowa Bldg, No. 38, Room 905, 4-12-24, Nishi Azabu, Minato-ku, Tokyo 106-0031; tel. (3) 3400-1830; fax (3) 3400-1820; e-mail embguate@twics.com; Ambassador Antonio Roberto Castellanos López.

Guinea: 12-9, Hachiyama-cho, Shibuya-ku, Tokyo 150-0035; tel. (3) 3770-4640; fax (3) 3770-4643; e-mail ambagui-tokyo@gol.com; Ambassador Ousmane Tolo Thiam.

Haiti: Kowa Bldg, No. 38, Room 906, 4-12-24, Nishi Azabu, Minato-ku, Tokyo 106; tel. (3) 3486-7096; fax (3) 3486-7070; Ambassador Marcel Duret.

Holy See: Apostolic Nunciature, 9-2, Sanban-cho, Chiyoda-ku, Tokyo 102-0075; tel. (3) 3263-6851; fax (3) 3263-6060; Apostolic Nuncio Most Rev. Ambrose B. de Paoli (Titular Archbishop of Lares).

Honduras: Kowa Bldg, No. 38, Room 802, 8th Floor, 4-12-24, Nishi Azabu, Minato-ku, Tokyo 106-0031; tel. (3) 3409-1150; fax (3) 3409-0305; e-mail honduras@interlink.or.jp; Ambassador Carlos Manuel Zerón.

Hungary: 2-17-14, Mita, Minato-ku, Tokyo 108-0073; tel. (3) 3798-8801; fax (3) 3798-8812; e-mail huembtio@gol.com; internet www.hungary.or.jp; Ambassador Dr Gyula Dabrónaki.

India: 2-2-11, Kudan Minami, Chiyoda-ku, Tokyo 102-0074; tel. (3) 3262-2391; fax (3) 3234-4866; internet embassy.kcom.ne.jp/embnet/india.html; Ambassador Siddharth Singh.

Indonesia: 5-2-9, Higashi Gotanda, Shinagawa-ku, Tokyo 141-0022; tel. (3) 3441-4201; fax (3) 3447-1697; Ambassador Soemadi D. M. Brotodiningrat.

Iran: 3-10-32, Minami Azabu, Minato-ku, Tokyo 106-0047; tel. (3) 3446-8011; fax (3) 3446-9002; internet www2.gol.com/users/sjei/indexjapanese.html; Ambassador Ali Majedi.

Iraq: 8-4-7, Akasaka, Minato-ku, Tokyo 107-0052; tel. (3) 3423-1727; fax (3) 3402-8636.

Ireland: Ireland House, 2-10-7, Kojimachi, Chiyoda-ku, Tokyo 102-0083; tel. (3) 3263-0695; fax (3) 3265-2275; internet www.embassy-avenue.jp/ireland/; Ambassador Pádraig Murphy.

Israel: 3, Niban-cho, Chiyoda-ku, Tokyo 102-0084; tel. (3) 3264-0911; fax (3) 3264-0791; e-mail consular@tky.mfa.gov.il; internet tokyo.mfa.gov.il; Ambassador Eli Cohen.

Italy: 2-5-4, Mita, Minato-ku, Tokyo 108-8302; tel. (3) 3453-5291; fax (3) 3456-2319; internet www.embitaly.jp; Ambassador Mario Bova.

Jamaica: Toranomon Yatsuka Bldg, 2nd Floor, 1-1-11, Atago, Minato-ku, Tokyo 105-0002; tel. (3) 3435-1861; fax (3) 3435-1864; e-mail secrat@jamaicaemb.or.jp; Ambassador Paul Anthony Robotham.

Jordan: Chiyoda House, 4th Floor, 2-17-8, Nagata-cho, Chiyoda-ku, Tokyo 100-0014; tel. (3) 3580-5856; fax (3) 3593-9385; internet www2.giganet.net/private/users/emb-jord; Ambassador Samir Naouri.

Kazakhstan: 5-9-8 Himonya, Meguro-ku, Tokyo 152-0023; tel. (3) 3791-5273; fax (3) 3791-5279; e-mail embkazjp@gol.com; internet www.embkazjp.org; Ambassador Bolat Nurgaliyev.

Kenya: 3-24-3, Yakumo, Meguro-ku, Tokyo 152-0023; tel. (3) 3723-4006; fax (3) 3723-4488; e-mail kenrepj@ma.kcom.ne.jp; internet embassy.kcom.ne.jp/kenya; Ambassador Dennis N.O. Awori.

Korea, Republic: 1-2-5, Minami Azabu, Minato-ku, Tokyo 106-0047; tel. (3) 3452-7611; fax (3) 5232-6911; internet www.mofat.go.kr/japan; Ambassador Ra Jong-Yil.

Kuwait: 4-13-12, Mita, Minato-ku, Tokyo 108-0073; tel. (3) 3455-0361; fax (3) 3456-6290; internet kuwait-embassy.or.jp; Ambassador Sheikh Azzam Mubarak Sabah al-Sabah.

Kyrgyzstan: Tokyo; Ambassador Askar Kutanov.

Laos: 3-3-22, Nishi Azabu, Minato-ku, Tokyo 106-0031; tel. (3) 5411-2291; fax (3) 5411-2293; Ambassador Soukthavone Keola.

Lebanon: Chiyoda House, 5th Floor, 2-17-8, Nagata-cho, Chiyoda-ku, Tokyo 100-0014; tel. (3) 3580-1227; fax (3) 3580-2281; e-mail ambaliba@cronos.ocn.ne.jp; Ambassador Jaafar Moawi.

Liberia: Sugi Terrace 201, 3-13-11, Okusawa, Setagaya-ku, Tokyo 158; tel. (3) 3726-5711; fax (3) 3726-5712; Chargé d'affaires a.i. HARRY TAH FREEMAN.

Libya: 10-14, Daikanyama-cho, Shibuya-ku, Tokyo 150-0034; tel. (3) 3477-0701; fax (3) 3464-0420; Secretary of the People's Bureau SULAIMAN ABU BAKER BADI (acting).

Lithuania: 2-11-25 Oyamadai, Setagaya-ku, Tokyo 158-0086; tel. (3) 3703-6000; fax (3) 5758-8281; e-mail lithemb@gol.com; internet www2.gol.com/users/lithemb; Ambassador Dr ALGIRDAS KUDZYS.

Luxembourg: 1/F Luxembourg House, 1st Floor, 8–9, Yonban-cho, Chiyoda-ku, 102-0081; tel. (3) 3265-9621; fax (3) 3265-9624; internet www.luxembourg.or.jp; Ambassador MICHÈLE PRANCHÈRE-TOMASSINI.

Madagascar: 2-3-23, Moto Azabu, Minato-ku, Tokyo 106; tel. (3) 3446-7252; fax (3) 3446-7078; Ambassador CYRILLE FIDA.

Malawi: Takanawa-Kaisei Bldg, 7th Floor, 3-4-1, Takanawa, Minato-ku, Tokyo 108-0074; tel. (3) 3449-3010; fax (3) 3449-3220; e-mail malawi@mx1.ttcn.ne.jp; internet embassy.kcom.ne.jp/malawi; Ambassador BRIGHT S. M. MANGULAMA.

Malaysia: 20-16, Nanpeidai-cho, Shibuya-ku, Tokyo 150-0036; tel. (3) 3476-3840; fax (3) 3476-4971; e-mail maltokyo@kln.gov.my; Ambassador Dato' MARZUKI MOHAMMAD NOOR.

Mali: 15-15 Sikazawa, 5-chome, Sitagaya-ku, Tokyo; tel. (3) 3705-3433; fax (3) 3705-3489.

Marshall Islands: Meiji Park Heights 101, 9-9, Minamimoto-machi, Shinjuku-ku, Tokyo 106; tel. (3) 5379-1701; fax (3) 5379-1810; Ambassador AMATLAIN ELIZABETH KABUA.

Mauritania: 5-17-5, Kita Shinagawa, Shinagawa-ku, Tokyo 141-0001; tel. (3) 3449-3810; fax (3) 3449-3822; Ambassador (vacant).

Mexico: 2-15-1, Nagata-cho, Chiyoda-ku, Tokyo 100-0014; tel. (3) 3581-1131; fax (3) 3581-4058; e-mail embamex@mexicoembassy.jp; internet www.sre.gob.mx/japon; Chargé d'affaires MERCEDES FELICITAS RUÍZ ZAPATA.

Micronesia: Reinanzaka Bldg, 2nd Floor, 1-14-2, Akasaka, Minato-ku, Tokyo 107-0052; tel. (3) 3585-5456; fax (3) 3585-5348; e-mail fsmemb@fsmemb.or.jp; Ambassador KASIO MIDA (designate).

Mongolia: Pine Crest Mansion, 21-4, Kamiyama-cho, Shibuya-ku, Tokyo 150-0047; tel. (3) 3469-2088; fax (3) 3469-2216; e-mail embmong@gol.com; internet www.embassy.avenue.jp/mongolia/index-j.htm; Ambassador JAMBYU BATJARGAL.

Morocco: Silva Kingdom Bldg, 5th–6th Floors, 3-16-3, Sendagaya, Shibuya-ku, Tokyo 151-0051; tel. (3) 3478-3271; fax (3) 3402-0898; Ambassador SAAD EDDIN TAIB.

Mozambique: Shiba Amerex Bldg, 6th Floor, 3-12-17 Mita, Minato-ku, Tokyo 108-0073; tel. (3) 5419-0973; fax (3) 5442-0556; internet home.att.ne.jp/kiwi/mozambique; Ambassador ANTONIO FERNANDO MATERULA.

Myanmar: 4-8-26, Kita Shinagawa, Shinagawa-ku, Tokyo 140-0001; tel. (3) 3441-9291; fax (3) 3447-7394; Ambassador U SAW LLA MIN.

Nepal: 7-14-9, Todoroki, Setagaya-ku, Tokyo 158-0082; tel. (3) 3705-5558; fax (3) 3705-8264; e-mail nepembjp@big.or.jp; internet www.nepal.co.jp/embassy.html; Ambassador KEDAR BHAKTA MATHEMA.

Netherlands: 3-6-3, Shiba Koen, Minato-ku, Tokyo 105-0011; tel. (3) 5401-0411; fax (3) 5401-0420; e-mail nlgovtok@oranda.or.jp; internet www.oranda.or.jp; Ambassador EGBERT F. JACOBS.

New Zealand: 20-40, Kamiyama-cho, Shibuya-ku, Tokyo 150-0047; tel. (3) 3467-2271; fax (3) 3467-2278; e-mail nzemb.tky@mail.com; internet www.nzembassy.com./japan; Ambassador PHILIP GIBSON.

Nicaragua: Kowa Bldg, No. 38, Room 903, 9th Floor, 4-12-24, Nishi Azabu, Minato-ku, Tokyo 106; tel. (3) 3499-0400; fax (3) 3499-3800; Ambassador Dr HARRY BODÁN-SHIELDS.

Nigeria: 5-11-17, Shimo-Meguro, Meguro-ku, Tokyo 153-0064; tel. (3) 5721-5391; fax (3) 5721-5342; internet www.crisscross.com/users/nigeriaemb/home.htm; Ambassador ADAMU ALIYU.

Norway: 5-12-2, Minami Azabu, Minato-ku, Tokyo 106-0047; tel. (3) 3440-2611; fax (3) 3440-2620; e-mail emb.tokyo@mfa.no; internet www.norway.or.jp; Ambassador ÅGE BERNHARD GRUTLE.

Oman: 2-28-11, Sendagaya, Shibuya-ku, Tokyo 151-0051; tel. (3) 3402-0877; fax (3) 3404-1334; e-mail omanemb@gol.com; Ambassador MOHAMMED ALI AL-KHUSAIBY.

Pakistan: 2-14-9, Moto Azabu, Minato-ku, Tokyo 106-0046; tel. (3) 3454-4861; fax (3) 3457-0341; e-mail pakemb@gol.com; Ambassador TOUQIR HUSSAIN.

Palau: Rm 201, 1-1, Katamachi, Shinjuku-ku, Tokyo 160-0001; tel. (3) 3354-5500; Ambassador SANTOS OLIKONG.

Panama: Kowa Bldg, No. 38, Room 902, 4-12-24, Nishi Azabu, Minato-ku, Tokyo 106-0031; tel. (3) 3499-3741; fax (3) 5485-3548; e-mail panaemb@gol.com; internet www.embassy-avenue.jp/panama/index-j.html; Ambassador JOSÉ A. SOSA.

Papua New Guinea: Mita Kokusai Bldg, Room 313, 3rd Floor, 1-4-28, Mita, Minato-ku, Tokyo 108; tel. (3) 3454-7801; fax (3) 3454-7275; Ambassador AIWA OLMI.

Paraguay: 3-12-9, Kami-Osaki, Shinagawa-ku, Tokyo 141-0021; tel. (3) 5485-3101; fax (3) 5485-3103; e-mail embapar@gol.com; internet www.embassy-avenue.jp/paraguay/index-j.htm; Ambassador ISAO TAOKA.

Peru: 4-4-27, Higashi, Shibuya-ku, Tokyo 150-0011; tel. (3) 3406-4243; fax (3) 3409-7589; e-mail 1-tokio@ma.kcom.ne.jp; Ambassador LUIS MACHIAVELLO.

Philippines: 5-15-5, Roppongi, Minato-ku, Tokyo 106-8537; tel. (3) 5562-1600; e-mail phpjp@gol.com; internet www.rptokyo.org; Ambassador ROMEO ABELARDO ARGUELLES.

Poland: 2-13-5, Mita, Meguro-ku, Tokyo 153-0062; internet www.poland.or.jp; tel. (3) 5794-7020; Ambassador MARCIN RYBICKI.

Portugal: Kamiura-Kojimachi Bldg, 5th Floor, 3-10-3, Kojimachi, Chiyoda-ku, Tokyo 102-0083; tel. (3) 5212-7322; fax (3) 5226-0616; e-mail emportj@zb.so-net.ne.jp; internet www.pnsnet.co.jp/users/cltembpt; Ambassador MANUEL GERVÁSIO DE ALMEIDA LEITE.

Qatar: 2-3-28, Moto-Azabu, Minato-ku, Tokyo 106-0046; tel. (3) 5475-0611; Ambassador REYAD ALI AL-ANSARI.

Romania: 3-16-19, Nishi Azabu, Minato-ku, Tokyo 106-0031; tel. (3) 3479-0311; fax (3) 3479-0312; e-mail romembjp@gol.com; internet www2.gol.com/users/romembjp/; Ambassador ION PASCU.

Russia: 2-1-1, Azabu-dai, Minato-ku, Tokyo 106-0041; tel. (3) 3583-4224; fax (3) 3505-0593; internet www.embassy-avenue.jp/russia/index-j.html; e-mail rosconsl@ma.kcom.ne.jp; Ambassador ALEKSANDR LOSYUKOV.

Rwanda: Kowa Bldg, No. 38, 4-12-24, Nishi Azabu, Minato-ku, Tokyo 106; tel. (3) 3486-7801; fax (3) 3409-2434; Ambassador MATANGUHA ZEPHYR.

Saudi Arabia: 1-8-4, Roppongi, Minato-ku, Tokyo 106-0032; tel. (3) 3589-5241; fax (3) 3589-5200; Ambassador MOHAMED BASHIR KURDI.

Senegal: 1-3-4, Aobadai, Meguro-ku, Tokyo 153-0042; tel. (3) 3464-8451; fax (3) 3464-8452; e-mail senegal@senegal.jp; Ambassador GABRIEL ALEXANDRE SAR.

Serbia and Montenegro: 4-7-24, Kita-Shinagawa, Shinagawa-ku, Tokyo 140-0001; tel. (3) 3447-3571; fax (3) 3447-3573; e-mail embassy@embassy-serbia-montenegro.jp; internet www.embassy-serbia-montenegro.jp; Ambassador Dr PREDRAG FILIPOV.

Singapore: 5-12-3, Roppongi, Minato-ku, Tokyo 106-0032; tel. (3) 3586-9111; fax (3) 3582-1085; Ambassador LIM CHIN BENG.

Slovakia: POB 35, 2-16-14, Hiroo, Shibuya-ku, Tokyo 150-8691; tel. (3) 3400-8122; fax (3) 3406-6215; e-mail zutokio@twics.com; internet www.embassy-avenue.jp/slovakia/index-j.html; Ambassador JÚLIUS HAUSER.

Slovenia: 7-5-15, Akasaka, Minato-ku, Tokyo 107-0052; tel. (3) 5570-6275; fax (3) 5570-6075; e-mail vto@mzz-dkp.gov.si; Ambassador ROBERT BASEJ.

South Africa: 414 Zenkyoren Bldg, 4th Floor, 2-7-9, Hirakawa-cho, Chiyoda-ku, Tokyo 102-0093; tel. (3) 3265-3366; fax (3) 3265-1108; e-mail sajapan@rsatk.com; internet www.rsatk.com; Ambassador KARAMCHUND MACKERDHUJ.

Spain: 1-3-29, Roppongi, Minato-ku, Tokyo 106-0032; tel. (3) 3583-8531; fax (3) 3582-8627; e-mail embspjp@mail.mae.es; Ambassador FRANCISCO JAVIER CONDE DE SARO.

Sri Lanka: 2-1-54, Takanawa, Minato-ku, Tokyo 108-0074; tel. (3) 3440-6911; fax (3) 3440-6914; e-mail lankaemb@sphere.ne.jp; internet www.embassy-avenue.jp/srilanka/index.html; Ambassador KARUNATILAKA AMUNUGAMA.

Sudan: 2-7-11, Shirogane, Minato-ku, Tokyo 108-0072; tel. (3) 3280-3161; Ambassador Dr AWAD MURSI TAHA.

Sweden: 1-10-3-100, Roppongi, Minato-ku, Tokyo 106-0032; tel. (3) 5562-5050; fax (3) 5562-9095; e-mail ambassaden.tokyo@foreign.ministry.se; internet www.sweden.or.jp; Ambassador KRISTER KUMLIN.

Switzerland: 5-9-12, Minami Azabu, Minato-ku, Tokyo 106-8589; tel. (3) 3473-0121; fax (3) 3473-6090; e-mail vertretung@tok.rep.admin.ch; internet www.eda.admin.ch/Tokyo; Ambassador JACQUES REVERDIN.

Syria: Homat Jade, 6-19-45, Akasaka, Minato-ku, Tokyo 107-0052; tel. (3) 3586-8977; fax (3) 3586-8979; Chargé d'affaires a.i. HAMZAH HAMZAH.

Tanzania: 4-21-9, Kami Yoga, Setagaya-ku, Tokyo 158-0098; tel. (3) 3425-4531; fax (3) 3425-7844; e-mail tzrepjp@gol.com; Ambassador ELLY E. E. MTANGO.

Thailand: 3-14-6, Kami Osaki, Shinagawa-ku, Tokyo 141-0021; tel. (3) 3447-2247; fax (3) 3442-6750; e-mail thaitke@crisscross.com; Ambassador CHAWAT ARTHAYUKTI.

Tunisia: 3-6-6, Kudan-Minami, Chiyoda-ku, Tokyo 102-0074; tel. (3) 3511-6622; fax (3) 3511-6600; Ambassador SALAH HANNACHI.

Turkey: 2-33-6, Jingumae, Shibuya-ku, Tokyo 150-0001; tel. (3) 3470-5131; fax (3) 3470-5136; e-mail embassy@turkey.jp; internet www.turkey.jp; Ambassador YAMAN BAŞKUT.

Uganda: 4-15-3, Shimomeguro, Meguro-ku, Tokyo 153-0064; tel. (3) 5773-0481; fax (3) 5725-3720; e-mail ugabassy@hpo.net; internet www.ugandaembassy.jp; Ambassador JAMES B. BABA.

Ukraine: 3-15-6, Nishi Azabu, Minato-ku, Tokyo 106-0046; tel. (3) 5474-9770; fax (3) 5474-9772; e-mail ukremb@rose.ocn.ne.jp; internet ukremb-japan.gov.ua; Ambassador YURIY KOSTENKO.

United Arab Emirates: 9-10, Nanpeidai-cho, Shibuya-ku, Tokyo 150-0036; tel. (3) 5489-0804; fax (3) 5489-0813; e-mail info@uaeembassy.jp; internet www.uaeembassy.jp; Ambassador AHMED ALI HAMAD ALMUALLA.

United Kingdom: 1, Ichiban-cho, Chiyoda-ku, Tokyo 102-8381; tel. (3) 5211-1100; fax (3) 5275-3164; e-mail embassy.tokyo@fco.gov.uk; internet www.uknow.or.jp; Ambassador GRAHAM FRY.

USA: 1-10-5, Akasaka, Minato-ku, Tokyo 107-8420; tel. (3) 3224-5000; internet tokyo.usembassy.gov; Ambassador HOWARD H. BAKER.

Uruguay: Kowa Bldg, No. 38, Room 908, 4-12-24, Nishi Azabu, Minato-ku, Tokyo 106-0031; tel. (3) 3486-1888; fax (3) 3486-9872; e-mail urujap@luck.ocn.ne.jp; Ambassador CARLOS CLULOW.

Uzbekistan: 5-11-8, Shimo-Meguro, Meguro-ku, Tokyo 153-0064; tel. (3) 3760-5625.

Venezuela: Kowa Bldg, No. 38, Room 703, 4-12-24, Nishi Azabu, Minato-ku, Tokyo 106-0031; tel. (3) 3409-1501; fax (3) 3409-1505; e-mail embavene@interlink.or.jp; internet sunsite.sut.ac.jp/embassy/venemb/embvenez.html; Ambassador Dr CARLOS ENRIQUE NONES.

Viet Nam: 50-11, Moto Yoyogi-cho, Shibuya-ku, Tokyo 151-0062; tel. (3) 3466-3313; fax (3) 3466-3391; Ambassador NGUYEN TAM CHIEN.

Yemen: Kowa Bldg, No. 38, Room 807, 4-12-24, Nishi Azabu, Minato-ku, Tokyo 106-0031; tel. (3) 3499-7151; fax (3) 3499-4577; Chargé d'affaires a.i. ABDULRAHMAN M. AL-HOTHI.

Zambia: 1-10-2, Ebara, Shinagawa-ku, Tokyo 142-0063; tel. (3) 3491-0121; fax (3) 3491-0123; e-mail emb@zambia.or.jp; Ambassador GODFREY S. SIMASIKU.

Zimbabwe: 5-9-10, Shiroganedai, Minato-ku, Tokyo 108; tel. (3) 3280-0331; fax (3) 3280-0466; e-mail zimtokyo@chive.ocn.ne.jp; Ambassador STUART H. COMBERBACH.

Judicial System

The basic principles of the legal system are set forth in the Constitution, which lays down that judicial power is vested in the Supreme Court and in such inferior courts as are established by law, and enunciates the principle that no organ or agency of the Executive shall be given final judicial power. Judges are to be independent in the exercise of their conscience, and may not be removed except by public impeachment, unless judicially declared mentally or physically incompetent to perform official duties. The justices of the Supreme Court are appointed by the Cabinet, the sole exception being the Chief Justice, who is appointed by the Emperor after designation by the Cabinet.

The Court Organization Law, which came into force on 3 May 1947, decreed the constitution of the Supreme Court and the establishment of four types of lower court—High, District, Family (established 1 January 1949) and Summary Courts. The constitution and functions of the courts are as follows:

SUPREME COURT

4-2, Hayabusa-cho, Chiyoda-ku, Tokyo 102-8651; tel. (3) 3264-8111; fax (3) 3221-8975; internet www.courts.go.jp.

This court is the highest legal authority in the land, and consists of a Chief Justice and 14 associate justices. It has jurisdiction over Jokoku (Jokoku appeals) and Kokoku (Kokoku appeals), prescribed in codes of procedure. It conducts its hearings and renders decisions through a Grand Bench or three Petty Benches. Both are collegiate bodies, the former consisting of all justices of the Court, and the latter of five justices. A Supreme Court Rule prescribes which cases are to be handled by the respective Benches. It is, however, laid down by law that the Petty Bench cannot make decisions as to the constitutionality of a statute, ordinance, regulation, or disposition, or as to cases in which an opinion concerning the interpretation and application of the Constitution, or of any laws or ordinances, is at variance with a previous decision of the Supreme Court.

Chief Justice: AKIRA MACHIDA.

Secretary-General: HIRONOBU TAKESAKI.

LOWER COURTS

High Court

A High Court conducts its hearings and renders decisions through a collegiate body, consisting of three judges, though for cases of insurrection the number of judges must be five. The Court has jurisdiction over the following matters:

Koso appeals from judgments in the first instance rendered by District Courts, from judgments rendered by Family Courts, and from judgments concerning criminal cases rendered by Summary Courts.

Kokoku appeals against rulings and orders rendered by District Courts and Family Courts, and against rulings and orders concerning criminal cases rendered by Summary Courts, except those coming within the jurisdiction of the Supreme Court.

Jokoku appeals from judgments in the second instance rendered by District Courts and from judgments rendered by Summary Courts, except those concerning criminal cases.

Actions in the first instance relating to cases of insurrection.

Presidents: TOKUJI IZUMI (Tokyo), YOSHIO OKADA (Osaka), REISUKE SHIMADA (Nagoya), TOYOZO UEDA (Hiroshima), TOSHIMARO KOJO (Fukuoka), FUMIYA SATO (Sendai), KAZUO KATO (Sapporo), FUMIO ARAI (Takamatsu).

District Court

A District Court conducts hearings and renders decisions through a single judge or, for certain types of cases, through a collegiate body of three judges. It has jurisdiction over the following matters:

Actions in the first instance, except offences relating to insurrection, claims where the subject matter of the action does not exceed 900,000 yen, and offences liable to a fine or lesser penalty.

Koso appeals from judgments rendered by Summary Courts, except those concerning criminal cases.

Kokoku appeals against rulings and orders rendered by Summary Courts, except those coming within the jurisdiction of the Supreme Court and High Courts.

Family Court

A Family Court handles cases through a single judge in case of rendering judgments or decisions. However, in accordance with the provisions of other statutes, it conducts its hearings and renders decisions through a collegiate body of three judges. A conciliation is effected through a collegiate body consisting of a judge and two or more members of the conciliation committee selected from among citizens.

It has jurisdiction over the following matters:

Judgment and conciliation with regard to cases relating to family as provided for by the Law for Adjudgment of Domestic Relations.

Judgment with regard to the matters of protection of juveniles as provided for by the Juvenile Law.

Actions in the first instance relating to adult criminal cases of violation of the Labour Standard Law, the Law for Prohibiting Liquors to Minors, or other laws especially enacted for protection of juveniles.

Summary Court

A Summary Court handles cases through a single judge, and has jurisdiction in the first instance over the following matters:

Claims where the value of the subject matter does not exceed 900,000 yen (excluding claims for cancellation or change of administrative dispositions).

Actions which relate to offences liable to a fine or lesser penalty, offences liable to a fine as an optional penalty, and certain specified offences such as habitual gambling and larceny.

A Summary Court cannot impose imprisonment or a graver penalty. When it deems proper the imposition of a sentence of imprisonment or a graver penalty, it must transfer such cases to a District Court, but it can impose imprisonment with labour not exceeding three years for certain specified offences.

Religion

The traditional religions of Japan are Shintoism and Buddhism. Neither is exclusive, and many Japanese subscribe at least nominally to both. Since 1945 a number of new religions (Shinko Shukyo) have evolved, based on a fusion of Shinto, Buddhist, Daoist, Confucian and Christian beliefs. In 1995 there were some 184,000 religious organizations registered in Japan, according to the Ministry of Education.

SHINTOISM

Shintoism is an indigenous religious system embracing the worship of ancestors and of nature. It is divided into two cults: national Shintoism, which is represented by the shrines; and sectarian Shintoism, which developed during the second half of the 19th century. In 1868 Shinto was designated a national religion and all Shinto shrines acquired the privileged status of a national institution. Complete freedom of religion was introduced in 1947, and state support of Shinto was prohibited. In the mid-1990s there were 81,307 shrines, 90,309 priests and 106.6m. adherents.

BUDDHISM

World Buddhist Fellowship: Hozenji Buddhist Temple, 3-24-2, Akabane-dai, Kita-ku, Tokyo; Rev. Fuji Nakayama.

CHRISTIANITY

In 1993 the Christian population was estimated at 1,050,938.

National Christian Council in Japan: Japan Christian Centre, 2-3-18-24, Nishi Waseda, Shinjuku-ku, Tokyo 169-0051; tel. (3) 3203-0372; fax (3) 3204-9495; e-mail ncc-j@jca.apc.org; internet www.jca.apc.org/ncc-j; f. 1923; 14 mems (churches and other bodies), 19 assoc. mems; Chair. Reiko Suzuki; Gen. Sec. Rev. Toshimasa Yamamoto.

The Anglican Communion

Anglican Church in Japan (Nippon Sei Ko Kai): 65, Yarai-cho, Shinjuku-ku, Tokyo 162-0805; tel. (3) 5228-3171; fax (3) 5228-3175; e-mail general-sec.po@nskk.org; internet www.nskk.org; f. 1887; 11 dioceses; Primate of Japan Rt Rev. James Toru Uno (Bishop of Osaka); Gen. Sec. Laurence Y. Minabe; 57,878 mems (2001).

The Orthodox Church

Japanese Orthodox Church (Nippon Haristosu Seikyoukai): Holy Resurrection Cathedral (Nicolai-Do), 4-1-3, Kanda Surugadai, Chiyoda-ku, Tokyo 101; tel. (3) 3291-1885; fax (3) 3291-1886; e-mail ocj@gol.com; three dioceses; Archbishop of Tokyo, Primate and Metropolitan of All Japan Most Rev. Daniel; 24,821 mems.

Protestant Church

United Church of Christ in Japan (Nihon Kirisuto Kyodan): Japan Christian Center, Room 31, 2-3-18, Nishi Waseda, Shinjuku-ku, Tokyo 169-0051; tel. (3) 3202-0541; fax (3) 3207-3918; e-mail ecumeni-c@uccj.org; f. 1941; union of 34 Congregational, Methodist, Presbyterian, Reformed and other Protestant denominations; Moderator Rev. Nobuhisha Yamakita; Gen. Sec. Rev. Noboru Takemae; 196,044 mems (2002).

The Roman Catholic Church

Japan comprises three archdioceses and 13 dioceses, and the Apostolic Prefecture of Karafuto. There were an estimated 516,176 adherents at 31 December 2002.

Catholic Bishops' Conference of Japan (Chuo Kyogikai)

2-10-10, Shiomi, Koto-ku, Tokyo 135-8585; tel. (3) 5632-4411; fax (3) 5632-4457; e-mail info@cbcj.catholic.jp; internet www.cbcj.catholic.jp.

Pres. Most Rev. Augustine Jun-ichi Nomura (Bishop of Nagoya).

Archbishop of Nagasaki: Joseph Mitsuaki Takami, Catholic Center, 10–34 Uenomachi, Nagasaki-shi 852-8113; tel. (95) 846-4246; fax (95) 848-8310.

Archbishop of Osaka: Most Rev. Leo Jun Ikenaga, Archbishop's House, 2-24-22, Tamatsukuri, Chuo-ku, Osaka 540-0004; tel. (6) 6941-9700; fax (6) 6946-1345.

Archbishop of Tokyo: Most Rev. Peter Takeo Okada, Archbishop's House, 3-16-15, Sekiguchi, Bunkyo-ku, Tokyo 112-0014; tel. (3) 3943-2301; fax (3) 3944-8511; e-mail peter2000@nifty.com.

Other Christian Churches

Japan Baptist Convention: 1-2-4, Minami Urawa, Minami-ku-Saitama-shi, Saitama 336-0017; tel. (48) 883-1091; fax (48) 883-1092; f. 1947; Gen. Sec. Rev. Makoto Kato; 33,734 mems (March 2003).

Japan Baptist Union: 2-3-18, Nishi Waseda, Shinjuku-ku, Tokyo 169-0051; tel. (3) 3202-0053; fax (3) 3202-0054; e-mail generalsecretary@jbu.or.jp; f. 1958; Moderator Yoshihisa Sawano; Gen. Sec. Kazuo Oya; 4,600 mems.

Japan Evangelical Lutheran Church: 1-1, Sadohara-cho, Ichigaya-shi, Shinjuku-ku, Tokyo 162-0842; tel. (3) 3260-8631; fax (3) 3268-3589; e-mail s-matsuoka@jelc.or.jp; internet www.jelc.or.jp; f. 1893; Moderator Rev. Masatoshi Yamanouchi; Gen. Sec. Rev. Shunichiro Matsuoka; 21,967 mems (2000).

Korean Christian Church in Japan: Room 52, Japan Christian Center, 2-3-18, Nishi Waseda, Shinjuku-ku, Tokyo 169-0051; tel. (3) 3202-5398; fax (3) 3202-4977; e-mail kccj@kb3.so-net.ne.jp; f. 1909; Moderator Choi Jung-kang; Gen. Sec. Park Soo-kil; 7,119 mems (2002).

Among other denominations active in Japan are the Christian Catholic Church, the German Evangelical Church and the Tokyo Union Church.

OTHER COMMUNITIES

Bahá'í Faith

The National Spiritual Assembly of the Bahá'ís of Japan: 7-2-13, Shinjuku, Shinjuku-ku, Tokyo 160-0022; tel. (3) 3209-7521; fax (3) 3204-0773; e-mail nsajpn@tka.att.ne.jp; internet www.bahaijp.org.

Judaism

Jewish Community of Japan: 3-8-8 Hiro-o, Shibuya-ku, Tokyo 150-0012; tel. (3) 3400-2559; fax (3) 3400-1827; e-mail office@jccjapan.or.jp; internet www.jccjapan.or.jp; Man. Lior Jacobi; Leader Rabbi Henri Noach.

Islam

Islam has been active in Japan since the late 19th century. There is a small Muslim community, maintaining several mosques, including those at Kobe, Nagoya, Chiba and Isesaki, the Arabic Islamic Institute and the Islamic Center in Tokyo. The construction of Tokyo Central mosque was ongoing in 1999.

Islamic Center, Japan: 1-16-11, Ohara, Setagaya-ku, Tokyo 156-0041; tel. (3) 3460-6169; fax (3) 3460-6105; e-mail islamcpj@islamcenter.or.jp; internet www.islamcenter.or.jp; f. 1965; Chair. Dr Salih Samarrai.

The New Religions

Many new cults have emerged in Japan since the end of the Second World War. Collectively these are known as the New Religions (Shinko Shukyo), among the most important of which are Tenrikyo, Omotokyo, Soka Gakkai, Rissho Kosei-kai, Kofuku-no-Kagaku, Agonshu and Aum Shinrikyo. (Following the indictment on charges of murder of several members of Aum Shinrikyo, including its leader, Shoko Asahara, the cult lost its legal status as a religious organization in 1996. In January 2000 the cult announced its intention to change its name to Aleph. At that time it named a new leader, Tatsuko Muraoka.)

Kofuku-no-Kagaku (Institute for Research in Human Happiness): Tokyo; f. 1986; believes its founder to be reincarnation of Buddha; 8.25m. mems; Leader Ryuho Okawa.

Rissho Kosei-kai: 2-11-1, Wada Suginami-ku, Tokyo 166-8537; tel. (3) 3380-5185; fax (3) 3381-9792; internet www.kosei-kai.or.jp; f. 1938; Buddhist lay organization based on the teaching of the Lotus Sutra, active inter-faith co-operation towards peace; Pres. Rev. Dr Nichiko Niwano; 6.3m. mems with 245 brs world-wide (2000).

Soka Gakkai: 32, Shinano-machi, Shinjuku-ku, Tokyo 160-8583; tel. (3) 5360-9830; fax (3) 5360-9885; e-mail webmaster@sokagakkai.info; internet sokagakkai.info; f. 1930; society of lay practitioners of the Buddhism of Nichiren; membership of 8.21m. households (2000); group promotes activities in education, international cultural exchange and consensus-building towards peace, based on the humanist world view of Buddhism; Hon. Pres. Daisaku Ikeda; Pres. Einosuke Akiya.

The Press

In December 2000 there were 122 daily newspapers in Japan. Their average circulation was the highest in the world, and the circulation per head of population was also among the highest, at 573 copies per 1,000 inhabitants in 1999. The large number of weekly news journals is a notable feature of the Japanese press. At December 1998 a total of 2,763 periodicals were produced, 85 of which were weekly publications. Technically the Japanese press is highly advanced, and the major newspapers are issued in simultaneous editions in the main centres.

The two newspapers with the largest circulations are the *Yomiuri Shimbun* and *Asahi Shimbun*. Other influential papers include *Mainichi Shimbun, Nihon Keizai Shimbun, Chunichi Shimbun* and *Sankei Shimbun*.

NATIONAL DAILIES

Asahi Shimbun: 5-3-2, Tsukiji, Chuo-ku, Tokyo 104-8011; tel. (3) 3545-0131; fax (3) 3545-0358; internet www.asahi.com; f. 1879; also published by Osaka, Seibu and Nagoya head offices and Hokkaido branch office; Pres. SHINICHI HAKOSHIMA; Dir and Exec. Editor MASAO KIMIWADA; circ. morning 8.3m., evening 4.1m.

Mainichi Shimbun: 1-1-1, Hitotsubashi, Chiyoda-ku, Tokyo 100-8051; tel. (3) 3212-0321; fax (3) 3211-3598; internet www.mainichi.co.jp; f. 1882; also published by Osaka, Seibu and Chubu head offices, and Hokkaido branch office; Pres. MASATOU KITAMURA; Man. Dir and Editor-in-Chief TATSUAKI HASHIMOTO; circ. morning 4.0m., evening 1.7m.

Nihon Keizai Shimbun: 1-9-5, Otemachi, Chiyoda-ku, Tokyo 100-8066; tel. (3) 3270-0251; fax (3) 5255-2661; internet www.nikkei.co.jp; f. 1876; also published by Osaka head office and Sapporo, Nagoya and Seibu branch offices; Pres. TAKUHIKO TSURUTA; Dir and Man. Editor YASUO HIRATA; circ. morning 3.0m., evening 1.7m.

Sankei Shimbun: 1-7-2, Otemachi, Chiyoda-ku, Tokyo 100-8077; tel. (3) 3231-7111; internet www.sankei.co.jp; f. 1933; also published by Osaka head office; Man. Dir and Editor NAGAYOSHI SUMIDA; circ. morning 2.0m., evening 905,771.

Yomiuri Shimbun: 1-7-1, Otemachi, Chiyoda-ku, Tokyo 100-8055; tel. (3) 3242-1111; e-mail webmaster@yomiuri.co.jp; internet www.yomiuri.co.jp; f. 1874; also published by Osaka, Seibu and Chubu head offices, and Hokkaido and Hokuriku branch offices; Pres. and Editor-in-Chief TSUNEO WATANABE; circ. morning 10.2m., evening 4.3m.

PRINCIPAL LOCAL DAILIES

Tokyo

Daily Sports: 1-20-3, Osaki, Shinagawa-ku, Tokyo 141-8585; tel. (3) 5434-1752; f. 1948; morning; Man. Dir HIROHISA KARUO; circ. 400,254.

The Daily Yomiuri: 1-7-1, Otemachi, Chiyoda-ku, Tokyo 100-8055; tel. (3) 3242-1111; f. 1955; morning; Man. Editor TSUTOMU YAMAGUCHI; circ. 51,421.

Dempa Shimbun: 1-11-15, Higashi Gotanda, Shinagawa-ku, Tokyo 141-8790; tel. (3) 3445-6111; fax (3) 3444-7515; f. 1950; morning; Pres. TETSUO HIRAYAMA; Man. Editor TOSHIO KASUYA; circ. 298,000.

Hochi Shimbun: 4-6-49, Kohnan, Minato-ku, Tokyo 108-8485; tel. (3) 5479-1111; internet www.yomiuri.co.jp/hochi/home.htm; f. 1872; morning; Pres. MASARU FUSHIMI; Man. Editor TATSUE AOKI; circ. 755,670.

The Japan Times: 4-5-4, Shibaura, Minato-ku, Tokyo 108-8071; tel. (3) 3453-5312; internet www.japantimes.co.jp; f. 1897; morning; English; Chair. and Pres. TOSHIAKI OGASAWARA; Dir and Editor-in-Chief YUTAKA MATAEBARA; circ. 61,929.

The Mainichi Daily News: 1-1-1, Hitotsubashi, Chiyoda-ku, Tokyo 100-8051; tel. (3) 3212-0321; f. 1922; morning; English; also publ. from Osaka; Man. Editor TETSUO TOKIZAWA; combined circ. 49,200.

Naigai Times: 1-1-15, Ariake, Koto-ku, Tokyo 135-0063; tel. (3) 5564-7021; fax (3) 5564-1022; e-mail info@naigai-times.co.jp; f. 1949; evening; Pres. MITSUGU ONDA; Vice-Pres. and Editor-in-Chief KENICHIRO KURIHARA; circ. 410,000.

Nihon Kaiji Shimbun (Japan Maritime Daily): 5-19-2, Shimbashi, Minato-ku, Tokyo 105-0004; tel. (3) 3436-3221; internet www.jmd.co.jp; f. 1942; morning; Man. Editor OSAMI ENDO; circ. 55,000.

Nihon Kogyo Shimbun: 1-7-2, Otemachi, Chiyoda-ku, Tokyo 100-8125; tel. (3) 3231-7111; internet www.jij.co.jp; f. 1933; morning; industrial, business and financial; Man. Editor YOSHIMI KURA; circ. 408,444.

Nihon Nogyo Shimbun (Agriculture): 2-3, Akihabara, Taito-ku, Tokyo 110-8722; tel. (3) 5295-7411; fax (3) 3253-0980; f. 1928; morning; Man. Editor YASUNORI INOUE; circ. 423,840.

Nihon Sen-i Shimbun (Textile and Fashion): 13-10, Nihombashi-kobunacho, Chuo-ku, Tokyo 103-0024; tel. (3) 5649-8711; f. 1943; morning; Man. Editor KIYOSHIGE SEIRYU; circ. 116,000.

Nikkan Kogyo Shimbun (Industrial Daily News): 1-8-10, Kudankita, Chiyoda-ku, Tokyo 102-8181; tel. (3) 3222-7111; fax (3) 3262-6031; internet www.nikkan.co.jp; f. 1915; morning; Man. Editor HIDEO WATANABE; circ. 533,145.

Nikkan Sports News: 3-5-10, Tsukiji, Chuo-ku, Tokyo 104-8055; tel. (3) 5550-8888; fax (3) 5550-8901; internet www.nikkansports.com; f. 1946; morning; Man. Editor YUKIHIRO MORI; circ. 993,240.

Sankei Sports: 1-7-2, Otemachi, Chiyoda-ku, Tokyo 100-8077; tel. (3) 3231-7111; internet www.xusxus.com; f. 1963; morning; Man. Editor YUKIO INADA; circ. 809,245.

Shipping and Trade News: Tokyo News Service Ltd, Tsukiji Hamarikyu Bldg, 5-3-3, Tsukiji, Chuo-ku, Tokyo 104-8004; tel. (3) 3542-6511; fax (3) 3542-5086; internet www.tvguide.or.jp; f. 1949; English; Man. Editor TAKASHI INOUE; circ. 15,000.

Sports Nippon: 2-1-30, Etchujima, Koto-ku, Tokyo 135-8735; tel. (3) 3820-0700; internet www.mainichi.co.jp/suponichi; f. 1949; morning; Man. Editor SUSUMU KOMURO; circ. 929,421.

Suisan Keizai Shimbun (Fisheries): 6-8-19, Roppongi, Minato-ku, Tokyo 106-0032; tel. (3) 3404-6531; fax (3) 3404-0863; f. 1948; morning; Man. Editor KOSHI TORINOUMI; circ. 61,000.

Tokyo Chunichi Sports: 2-3-13, Kohnan, Minato-ku, Tokyo 108-8010; tel. (3) 3471-2211; f. 1956; evening; Head Officer TETSUO TANAKA; circ. 330,431.

Tokyo Shimbun: 2-3-13, Kohnan, Minato-ku, Tokyo 108-8010; tel. (3) 3471-2211; fax (3) 3471-1851; internet www.tokyo-np.co.jp; f. 1942; Man. Editor KATSUHIKO SAKAI; circ. morning 655,970, evening 354,191.

Tokyo Sports: 2-1-30, Etchujima, Koto-ku, Tokyo 135-8721; tel. (3) 3820-0801; f. 1959; evening; Man. Editor YASUO SAKURAI; circ. 1,321,250.

Yukan Fuji: 1-7-2, Otemachi, Chiyoda-ku, Tokyo 100-8077; tel. (3) 3231-7111; fax (3) 3246-0377; internet www.zakzak.co.jp; f. 1969; evening; Man. Editor MASAMI KATO; circ. 268,984.

Osaka District

Daily Sports: 1-18-11, Edobori, Nishi-ku, Osaka 550-0002; tel. (6) 6443-0421; f. 1948; morning; Man. Editor TOSHIAKI MITANI; circ. 562,715.

The Mainichi Daily News: 3-4-5, Umeda, Kita-ku, Osaka 530-8251; tel. (6) 6345-1551; f. 1922; morning; English; Man. Editor KATSUYA FUKUNAGA.

Nikkan Sports: 5-92-1, Hattori-kotobuki-cho, Toyonaka 561-8585; tel. (6) 6867-2811; internet www.nikkansports.com/osaka; f. 1950; morning; Man. Editor KATSUO FURUKAWA; circ. 513,498.

Osaka Shimbun: 2-4-9, Umeda, Kita-ku, Osaka 530-8279; tel. (6) 6343-1221; internet www.osakanews.com; f. 1922; evening; Man. Editor KAORU YURA; circ. 88,887.

Osaka Sports: Osaka Ekimae Daiichi Bldg, 4th Floor, 1-3-1-400, Umeda, Kita-ku, Osaka 530-0001; tel. (6) 6345-7657; f. 1968; evening; Head Officer KAZUOMI TANAKA; circ. 470,660.

Sankei Sports: 2-4-9, Umeda, Kita-ku, Osaka 530-8277; tel. (6) 6343-1221; f. 1955; morning; Man. Editor MASAKI YOSHIDA; circ. 552,519.

Sports Nippon: 3-4-5, Umeda, Kita-ku, Osaka 530-8278; tel. (6) 6346-8500; f. 1949; morning; Man. Editor HIDETOSHI ISHIHARA; circ. 477,300.

Kanto District

Chiba Nippo (Chiba Daily News): 4-14-10, Chuo, Chuo-ku, Chiba 260-0013; tel. (43) 222-9211; internet www.chibanippo.co.jp; f. 1957; morning; Man. Editor NOBORU HAYASHI; circ. 190,187.

Ibaraki Shimbun: 2-15, Kitami-cho, Mito 310-8686; tel. (292) 21-3121; internet www.ibaraki-np.co.jp; f. 1891; morning; Pres. and Editor-in-Chief TADANORI TOMOSUE; circ. 117,240.

Jomo Shimbun: 1-50-21, Furuichi-machi, Maebashi 371-8666; tel. (272) 54-9911; internet www.jomo-news.co.jp; f. 1887; morning; Man. Editor MUTSUO ODAGIRI; circ. 296,111.

Joyo Shimbun: 2-7-6, Manabe, Tsuchiura 300-0051; tel. (298) 21-1780; internet www.tsukuba.com; f. 1948; morning; Pres. MINEO IWANAMI; Man. Editor AKIRA SAITO; circ. 88,700.

Kanagawa Shimbun: 6-145, Hanasaki-cho, Nishi-ku, Yokohama 220-8588; tel. (45) 411-2222; internet www.kanagawa-np.co.jp; f. 1890; morning; Man. Editor NOBUYUKI CHIBA; circ. 238,203.

Saitama Shimbun: 6-12-11, Kishi-cho, Urawa 336-8686; tel. (48) 862-3371; internet www.saitama-np.co.jp; f. 1944; morning; Man. Editor YOTARO NUMATA; circ. 162,071.

Shimotsuke Shimbun: 1-8-11, Showa, Utsunomiya 320-8686; tel. (286) 25-1111; internet www.shimotsuke.co.jp; f. 1884; morning; Man. Dir and Editor-in-Chief EISUKE TODA; circ. 306,072.

Tohoku District
(North-east Honshu)

Akita Sakigake Shimpo: 1-1, San-no-rinkai-machi, Akita 010-8601; tel. (18) 888-1800; fax (188) 23-1780; internet www.sakigake.co.jp; f. 1874; Man. Editor SHIGEAKI MAEKAWA; circ. 263,246.

Daily Tohoku: 1-3-12, Shiroshita, Hachinohe 031-8601; tel. (178) 44-5111; f. 1945; morning; Man. Editor TOKOJU YOSHIDA; circ. 104,935.

Fukushima Mimpo: 13-17, Ota-machi, Fukushima 960-8602; tel. (245) 31-4111; internet www.fukushima-minpo.co.jp; f. 1892; Pres. and Editor-in-Chief TSUTOMU HANADA; circ. morning 308,353, evening 9,489.

Fukushima Minyu: 4-29, Yanagi-machi, Fukushima 960-8648; tel. (245) 23-1191; internet www.minyu; f. 1895; Man. Editor KENJI KANNO; circ. morning 201,414, evening 6,066.

Hokuu Shimpo: 3-2, Nishi-dori-machi, Noshiro 016-0891. (185) 54-3150; f. 1895; morning; Chair. KOICHI YAMAKI; circ. 31,490.

Ishinomaki Shimbun: 2-1-28, Sumiyoshi-machi, Ishinomaki 986; tel. (225) 22-3201; f. 1946; evening; Man. Editor MASATOSHI SATO; circ. 13,050.

Iwate Nichi-nichi Shimbun: 60, Minamishin-machi, Ichinoseki 021-8686; tel. (191) 26-5114; internet www.isop.ne.jp/iwanichi; f. 1923; morning; Pres. TAKESHI YAMAGISHI; Man. Editor SEIICHI WATANABE; circ. 59,850.

Iwate Nippo: 3-7, Uchimaru, Morioka 020-8622; tel. (196) 53-4111; internet www.iwate-np.co.jp; f. 1876; Man. Editor TOKUO MIYAZAWA; circ. morning 230,073, evening 229,815.

Kahoku Shimpo: 1-2-28, Itsutsubashi, Aoba-ku, Sendai 980-8660; tel. (22) 211-1111; fax (22) 224-7947; internet www.kahoku.co.jp; f. 1897; Exec. Dir and Man. Editor MASAHIKO ICHIRIKI; circ. morning 503,318, evening 133,855.

Mutsu Shimpo: 2-1, Shimo-shirogane-cho, Hirosaki 036-8356; tel. (172) 34-3111; f. 1946; morning; Man. Editor YUJI SATO; circ. 53,500.

Shonai Nippo: 8-29, Baba-cho, Tsuruoka 997-8691; tel. (235) 22-1480; f. 1946; morning; Pres. TAKAO SATO; Man. Editor MASAYUKI HASHIMOTO; circ. 19,100.

To-o Nippo: 78, Kanbayashi, Yatsuyaku, Aomori 030-0180; tel. (177) 39-1111; internet www.toonippo.co.jp; f. 1888; Exec. Dir YOSHIO WAJIMA; Man. Editor TAKAO SHIOKOSHI; circ. morning 262,532, evening 258,590.

Yamagata Shimbun: 2-5-12, Hatagomachi, Yamagata 990-8550; tel. (236) 22-5271; internet www.yamagata-np.co.jp; f. 1876; Man. Editor TOSHINOBU SHIONO; circ. morning 213,057, evening 213,008.

Yonezawa Shimbun: 3-3-7, Monto-cho, Yonezawa 992-0039; tel. (238) 22-4411; f. 1879; morning; Man. Dir and Editor-in-Chief MAKOTO SATO; circ. 13,750.

Chubu District
(Central Honshu)

Chubu Keizai Shimbun: 4-4-12, Meieki, Nakamura-ku, Nagoya 450-8561; tel. (52) 561-5215; f. 1946; morning; Man. Editor NORIMITSU INAGAKI; circ. 97,000.

Chukyo Sports: Chunichi Kosoku Offset Insatsu Bldg, 4-3-9, Kinjo, Naka-ku, Nagoya 460-0847; tel. (52) 982-1911; f. 1968; evening; circ. 289,430; Head Officer OSAMU SUETSUGU.

Chunichi Shimbun: 1-6-1, San-no-maru, Naka-ku, Nagoya 460-8511; tel. (52) 201-8811; internet www.chunichi.ne.jp; f. 1942; Man. Editor NOBUAKI KOIDE; circ. morning 2.7m., evening 748,635.

Chunichi Sports: 1-6-1, San-no-maru, Naka-ku, Nagoya 460-8511; tel. (52) 201-8811; f. 1954; evening; Head Officer YASUHIKO AIBA; circ. 631,429.

Gifu Shimbun: 10, Imakomachi, Gifu 500-8577; tel. (582) 64-1151; internet www.jic-gifu.or.jp/np; f. 1881; Exec. Dir and Man. Editor TADASHI TANAKA; circ. morning 170,176, evening 31,775.

Higashi-Aichi Shimbun: 62, Torinawate, Shinsakae-machi, Toyohashi 441-8666; tel. (532) 32-3111; f. 1957; morning; Man. Editor YOSHIYUKI SUZUKI; circ. 52,300.

Nagano Nippo: 3-1323-1, Takashima, Suwa 392-8611; tel. (266) 52-2000; f. 1901; morning; Man. Editor ETSUO KOIZUMI; circ. 73,000.

Nagoya Times: 1-3-10, Marunouchi, Naka-ku, Nagoya 460-8530; tel. (52) 231-1331; f. 1946; evening; Man. Editor NAOKI KITO; circ. 146,137.

Shinano Mainichi Shimbun: 657, Minamiagata-machi, Nagano 380-8546; tel. (26) 236-3000; fax (26) 236-3197; internet www.shinmai.co.jp; f. 1873; Man. Editor SEIICHI INOMATA; circ. morning 469,801, evening 55,625.

Shizuoka Shimbun: 3-1-1, Toro, Shizuoka 422-8033; tel. (54) 284-8900; internet www.sbs-np.co.jp; f. 1941; Man. Editor HISAO ISHIHARA; circ. morning 730,746, evening 730,782.

Yamanashi Nichi-Nichi Shimbun: 2-6-10, Kitaguchi, Kofu 400-8515; tel. (552) 31-3000; internet www.sannichi.co.jp; f. 1872; morning; Man. Editor KATSUHITO NISHIKAWA; circ. 205,758.

Hokuriku District
(North Coastal Honshu)

Fukui Shimbun: 1-1-14, Haruyama, Fukui 910-8552; tel. (776) 23-5111; internet www.fukuishimbun.co.jp; f. 1899; morning; Man. Editor KAZUO UCHIDA; circ. 202,280.

Hokkoku Shimbun: 2-5-1, Korinbo, Kanazawa 920-8588; tel. (762) 63-2111; internet www.hokkoku.co.jp; f. 1893; Man. Editor WATARU INAGAKI; circ. morning 328,532, evening 97,051.

Hokuriku Chunichi Shimbun: 2-7-15, Korinbo, Kanazawa 920-8573; tel. (762) 61-3111; internet www.hokuriku.chunichi.co.jp; f. 1960; Man. Editor KANJI KOMIYA; circ. morning 116,719, evening 12,820.

Kitanippon Shimbun: 2-14, Azumi-cho, Toyama 930-8680; tel. (764) 45-3300; internet www.kitanippon.co.jp; f. 1884; Dir and Man. Editor MINORU KAWATA; circ. morning 223,033, evening 29,959.

Niigata Nippo: 772-2, Zenku, Niigata 950-1189; tel. (25) 378-9111; internet www.niigata-nippo.co.jp; f. 1942; Dir and Man. Editor MICHIEI TAKAHASHI; circ. morning 496,567, evening 66,836.

Toyama Shimbun: 5-1, Ote-machi, Toyama 930-8520; tel. (764) 91-8111; internet www.toyama.hokkoku.co.jp; f. 1923; morning; Man. Editor SACHIO MIYAMOTO; circ. 42,988.

Kinki District
(West Central Honshu)

Daily Sports: 1-5-7, Higashikawasaki-cho, Chuo-ku, Kobe 650-0044; tel. (78) 362-7100; morning; Man. Editor TAKASHI HIRAI.

Ise Shimbun: 34-6, Honmachi, Tsu 514-0831; tel. (592) 24-0003; internet www.isenp.co.jp; f. 1878; morning; Man. Editor FUJIO YAMAMOTO; circ. 100,550.

Kii Minpo: 100, Akizucho, Tanabe 646-8660; tel. (739) 22-7171; internet www.agara.co.jp; f. 1911; evening; Man. Editor KAZUSADA TANIGAMI; circ. 38,165.

Kobe Shimbun: 1-5-7, Higashikawasaki-cho, Chuo-ku, Kobe 650-8571; tel. (78) 362-7100; internet www.kobe-np.co.jp; f. 1898; Man. Editor MASAO MAEKAWA; circ. morning 545,854, evening 268,787.

Kyoto Shimbun: 239, Shoshoi-machi, Ebisugawa-agaru, Karasuma-dori, Nakagyo-ku, Kyoto 604-8577; tel. (75) 241-5430; internet www.kyoto-np.co.jp; f. 1879; Man. Editor OSAMU SAITO; circ. morning 505,723, evening 319,313.

Nara Shimbun: 606, Sanjo-machi, Nara 630-8686; tel. (742) 26-1331; internet www.nara-shimbun.com; f. 1946; morning; Dir and Man. Editor HISAMI SAKAMOTO; circ. 118,064.

Chugoku District
(Western Honshu)

Chugoku Shimbun: 7-1, Dobashi-cho, Naka-ku, Hiroshima 730-8677; tel. (82) 236-2111; fax (82) 236-2321; e-mail denshi@hiroshima-cdas.or.jp; internet www.chugoku-np.co.jp; f. 1892; Man. Editor NOBUYUKI AOKI; circ. morning 734,589, evening 85,089.

Nihonkai Shimbun: 2-137, Tomiyasu, Tottori 680-8678; tel. (857) 21-2888; internet www.nnn.co.jp; f. 1976; morning; Man. Editor KOTARO TAMURA; circ. 101,768.

Okayama Nichi-Nichi Shimbun: 6-30, Hon-cho, Okayama 700-8678; tel. (86) 231-4211; f. 1946; evening; Man. Dir and Man. Editor TAKASHI ANDO; circ. 45,000.

San-In Chuo Shimpo: 383, Tono-machi, Matsue 690-8668; tel. (852) 32-3440; f. 1882; morning; Man. Editor MASAMI MOCHIDA; circ. 172,605.

Sanyo Shimbun: 2-1-23, Yanagi-machi, Okayama 700-8634; tel. (86) 231-2210; internet www.sanyo.oni.co.jp; f. 1879; Man. Dir and Man. Editor TAKAMASA KOSHIMUNE; circ. morning 454,263, evening 71,200.

Ube Jiho: 3-6-1, Kotobuki-cho, Ube 755-8557; tel. (836) 31-1511; f. 1912; evening; Exec. Dir and Man. Editor KAZUYA WAKI; circ. 42,550.

Yamaguchi Shimbun: 1-1-7, Higashi-Yamato-cho, Shimonoseki 750-8506; tel. (832) 66-3211; internet www.minato-yamaguchi.co.jp; f. 1946; morning; Man. Editor SHOICHI SASAKI; circ. 84,000.

Shikoku Island

Ehime Shimbun: 1-12-1, Otemachi, Matsuyama 790-8511; tel. (899) 35-2111; internet www.ehime-np.co.jp; f. 1876; morning; Man. Editor RYOJI YANO; circ. 319,522.

Kochi Shimbun: 3-2-15, Honmachi, Kochi 780-8572; tel. (888) 22-2111; internet www.kochinews.co.jp; f. 1904; Dir and Man. Editor KENGO FUJITO; circ. morning 233,319, evening 146,276.

Shikoku Shimbun: 15-1, Nakano-cho, Takamatsu 760-8572; tel. (878) 33-1111; internet www.shikoku-np.co.jp; f. 1889; morning; Man. Editor JUNJI YAMASHITA; circ. 208,816.

Tokushima Shimbun: 2-5-2, Naka-Tokushima-cho, Tokushima 770-8572; tel. (886) 55-7373; fax (866) 54-0165; internet www.topics.or.jp; f. 1944; Dir and Man. Editor HIROSHI MATSUMURA; circ. morning 253,184, evening 52,203.

Hokkaido Island

Doshin Sports: 3-6, Odori-nishi, Chuo-ku, Sapporo 060-8711; tel. (11) 241-1230; internet douspo.aurora-net.or.jp; f. 1982; morning; Pres. KOSUKE SAKAI; circ. 139,178.

Hokkai Times: 10-6, Nishi, Minami-Ichijo, Chuo-ku, Sapporo 060; tel. (11) 231-0131; f. 1946; Man. Editor KOKI ITO; circ. morning 120,736.

Hokkaido Shimbun: 3-6, Odori-nishi, Chuo-ku, Sapporo 060-8711; tel. (11) 221-2111; internet www.aurora-net.or.jp; f. 1942; Man. Editor RYOZO ODAGIRI; circ. morning 1.2m., evening 740,264.

Kushiro Shimbun: 7-3, Kurogane-cho, Kushiro 085-8650; tel. (154) 22-1111; internet www.hokkai.or.jp/senshin/index.html; f. 1946; morning; Man. Editor YUTAKA ITO; circ. 55,686.

Muroran Mimpo: 1-3-16, Hon-cho, Muroran 051-8550; tel. (143) 22-5121; internet www.muromin.mnw.jp; f. 1945; Man. Editor TSUTOMO KUDO; circ. morning 60,300, evening 52,500.

Nikkan Sports: 3-1-30, Higashi, Kita-3 jo, Chuo-ku, Sapporo 060-0033; tel. (11) 242-3900; fax (11) 231-5470; internet www.kita-nikkan.co.jp; f. 1962; morning; Pres. SATOSHI KATO; circ. 160,355.

Tokachi Mainichi Shimbun: 8-2, Minami, Higashi-Ichijo, Obihiro 080-8688; tel. (155) 22-2121; fax (155) 25-2700; internet www.tokachi.co.jp; f. 1919; evening; Dir and Man. Editor TOSHIAKI NAKAHASHI; circ. 89,264.

Tomakomai Mimpo: 3-1-8, Wakakusa-cho, Tomakomai 053-8611; tel. (144) 32-5311; internet www.tomamin.co.jp; f. 1950; evening; Dir and Man. Editor RYUICHI KUDO; circ. 60,676.

Yomiuri Shimbun: 4-1, Nishi, Kita-4 jo, Chuo-ku, Sapporo 060-8656; tel. (11) 242-3111; f. 1959; Head Officer TSUTOMO IKEDA; circ. morning 261,747, evening 81,283.

Kyushu Island

Kagoshima Shimpo: 7-28, Jonan-cho, Kagoshima 892-8551; tel. (99) 226-2100; internet www.kagoshimashimpo.com; f. 1959; morning; Dir and Man. Editor JUNSUKE KINOSHITA; circ. 39,330.

Kumamoto Nichi-Nichi Shimbun: 172, Yoyasu-machi, Kumamoto 860-8506; tel. (96) 361-3111; internet www.kumanichi.co.jp; f. 1942; Man. Editor HIROSHI KAWARABATA; circ. morning 389,528, evening 101,795.

Kyushu Sports: Fukuoka Tenjin Center Bldg, 2-14-8, Tenjin-cho, Chuo-ku, Fukuoka 810-0001; tel. (92) 781-7401; f. 1966; morning; Head Officer HIROSHI MITOMI; circ. 449,850.

Minami Nippon Shimbun: 1-9-33, Yojirou, Kagoshima 890-8603; tel. (99) 813-5001; fax (99) 813-5016; e-mail tuusin@po.minc.ne.jp; internet www.minaminippon.co.jp; f. 1881; Man. Editor YASUSHI MOMIKI; circ. morning 401,938, evening 27,959.

Miyazaki Nichi-Nichi Shimbun: 1-1-33, Takachihodori, Miyazaki 880-8570; tel. (985) 26-9315; internet www.the-miyanichi.co.jp; f. 1940; morning; Man. Editor MASAAKI MINAMIMURA; circ. 236,083.

Nagasaki Shimbun: 3-1, Mori-machi, Nagasaki 852-8601; tel. (958) 44-2111; internet www.nagasaki-np.co.jp; f. 1889; Dir and Man. Editor SADAKATSU HONDA; circ. morning 200,128.

Nankai Nichi-Nichi Shimbun: 10-3, Nagahama-cho, Naze 894-8601; tel. (997) 53-2121; internet www.amami.or.jp/nankai; f. 1946; morning; Man. Editor TERUMI MATSUI; circ. 24,038.

Nishi Nippon Shimbun: 1-4-1, Tenjin, Chuo-ku, Fukuoka 810-8721; tel. (92) 711-5555; internet www.nishinippon.co.jp; f. 1877; Exec. Dir and Man. Editor TAKAMICHI TAMAGAWA; circ. morning 834,800, evening 188,444.

Nishi Nippon Sports: 1-4-1, Tenjin, Chuo-ku, Fukuoka 810; tel. (92) 711-5555; f. 1954; Man. Editor KENJI ISHIZAKI; circ. 184,119.

Oita Godo Shimbun: 3-9-15, Funai-machi, Oita 870-8605; tel. (975) 36-2121; internet www.oita-press.co.jp; f. 1886; Dir and Man. Editor MASAKATSU TANABE; circ. morning 245,257, evening 245,227.

Okinawa Times: 2-2-2, Kumoji, Naha 900-8678; tel. (98) 860-3000; internet www.okinawatimes.co.jp; f. 1948; Dir and Man. Editor MASAO KISHIMOTO; circ. morning 204,420, evening 204,420.

Ryukyu Shimpo: 1-10-3, Izumizaki, Naha 900-8525; tel. (98) 865-5111; internet www.ryukyushimpo.co.jp; f. 1893; Man. Editor TOMOKAZU TAKAMINE; circ. 200,936.

Saga Shimbun: 3-2-23, Tenjin, Saga 840-8585; tel. (952) 28-2111; fax (952) 29-4829; internet www.saga-s.co.jp; f. 1884; morning; Man. Editor TERUHIKO WASHIZAKI; circ. 138,079.

Yaeyama Mainichi Shimbun: 614, Tonoshiro, Ishigaki 907-0004; tel. (9808) 2-2121; internet www.cosmos.ne.jp/~mainichi; f. 1950; morning; Exec. Dir and Man. Editor YOSHIO UECHI; circ. 14,500.

WEEKLIES

An-An: Magazine House, 3-13-10, Ginza, Chuo-ku, Tokyo 104-03; tel. (3) 3545-7050; fax (3) 3546-0034; f. 1970; fashion; Editor MIYOKO YODOGAWA; circ. 650,000.

Asahi Graphic: Asahi Shimbun Publishing Dept, 5-3-2, Tsukiji, Chuo-ku, Tokyo 104-11; tel. (3) 3545-0131; f. 1923; pictorial review; Editor KIYOKAZU TANNO; circ. 120,000.

Diamond Weekly: Diamond Inc, 1-4-2, Kasumigaseki, Chiyoda-ku, Tokyo 100; tel. (3) 3504-6250; f. 1913; economics; Editor YUTAKA IWASA; circ. 78,000.

Focus: Shincho-Sha, 71, Yaraicho, Shinjuku-ku, Tokyo 162; tel. (3) 3266-5271; fax (3) 3266-5390; politics, economics, sport; Editor KAZUMASA TAJIMA; circ. 850,000.

Friday: Kodan-Sha Co Ltd, 2-12-21, Otowa, Bunkyo-ku, Tokyo 112; tel. (3) 5395-3440; fax (3) 3943-8582; current affairs; Editor-in-Chief TETSU SUZUKI; circ. 1m.

Hanako: Magazine House, 3-13-10, Ginza, Chuo-ku, Tokyo 104-03; tel. (3) 3545-7070; fax (3) 3546-0994; f. 1988; consumer guide; Editor KOJI TOMONO; circ. 350,000.

Nikkei Business: Nikkei Business Publications Inc, 2-7-6, Hirakawa-cho, Chiyoda-ku, Tokyo 102-8622; tel. (3) 5210-8101; fax (3) 5210-8520; internet www.nikkeibp.co.jp; f. 1969; Editor-in-Chief HIROTOMO NOMURA; circ. 350,000.

Shukan Asahi: Asahi Shimbun Publishing Dept, 5-3-2, Tsukiji, Chuo-ku, Tokyo 104-8011; tel. (3) 3545-0131; f. 1922; general interest; Editor-in-Chief AKIRA KATO; circ. 482,000.

Shukan Bunshun: Bungei-Shunju Ltd, 3-23, Kioicho, Chiyoda-ku, Tokyo 102; tel. (3) 3265-1211; fax (3) 3234-3964; general interest; Editor SEIGO KIMATA; circ. 800,000.

Shukan Gendai: Kodan-Sha Co Ltd, 2-12-21, Otowa, Bunkyo-ku, Tokyo 112; tel. (3) 5395-3438; fax (3) 3943-7815; f. 1959; general; Editor-in-Chief TETSU SUZUKI; circ. 930,000.

Shukan Josei: Shufu-To-Seikatsu Sha Ltd, 3-5-7, Kyobashi, Chuo-ku, Tokyo 104; tel. (3) 3563-5130; fax (3) 3563-2073; f. 1957; women's interest; Editor HIDEO KIKUCHI; circ. 638,000.

Shukan Post: Shogakukan Publishing Co Ltd, 2-3-1, Hitotsubashi, Chiyoda-ku, Tokyo 101-01; tel. (3) 3230-5951; f. 1969; general; Editor NORIMICHI OKANARI; circ. 696,000.

Shukan Shincho: Shincho-Sha, 71, Yarai-cho, Shinjuku-ku, Tokyo 162-8711; tel. (3) 3266-5311; fax (3) 3266-5622; f. 1956; general interest; Editor HIROSHI MATSUDA; circ. 521,000.

Shukan SPA: Fuso-Sha Co, 1-15-1, Kaigan, Minato-ku, Tokyo 105; tel. (3) 5403-8875; f. 1952; general interest; Editor-in-Chief TOSHIHIKO SATO; circ. 400,000.

Shukan ST: Japan Times Ltd, 4-5-4, Shibaura, Minato-ku, Tokyo 108-0023; tel. (3) 3452-4077; fax (3) 3452-3303; e-mail shukanst@japantimes.co.jp; f. 1951; English and Japanese; Editor MITSURU TANAKA; circ. 150,000.

Shukan Yomiuri: Yomiuri Shimbun Publication Dept, 1-2-1, Kiyosumi, Koto-ku, Tokyo 135; tel. (3) 5245-7001; f. 1938; general interest; Editor SHINI KAGEYAMA; circ. 453,000.

Sunday Mainichi: Mainichi Newspapers Publishing Dept, 1-1-1, Hitotsubashi, Chiyoda-ku, Tokyo 100-51; tel. (3) 3212-0321; fax (3) 3212-0769; f. 1922; general interest; Editor KENJI MIKI; circ. 237,000.

Tenji Mainichi: Mainichi Newspapers Publishing Dept, 3-4-5, Umeda, Osaka; tel. (6) 6346-8386; fax (6) 6346-8385; f. 1922; in Japanese braille; Editor TADAMITSU MORIOKA; circ. 12,000.

Weekly Economist: Mainichi Newspapers Publishing Dept, 1-1-1, Hitotsubashi, Chiyoda-ku, Tokyo 100-51; tel. (3) 3212-0321; f. 1923; Editorial Chief NOBUHIRO SHUDO; circ. 120,000.

Weekly Toyo Keizai: Toyo Keizai Inc, 1-2-1, Hongoku-cho, Nihombashi, Chuo-ku, Tokyo 103-8345; tel. (3) 3246-5655; fax (3) 3270-0159; e-mail sub@toyokeizai.co.jp; internet www.toyokeizai.co.jp; f. 1895; business, economics, finance, and corporate information; Editor TOSHIKI OTA; circ. 62,000.

PERIODICALS

All Yomimono: Bungei-Shunju Ltd, 3-23, Kioicho, Chiyoda-ku, Tokyo 102; tel. (3) 3265-1211; fax (3) 3239-5481; f. 1930; monthly; popular fiction; Editor KOICHI SASAMOTO; circ. 95,796.

Any: 1-3-14, Hirakawa-cho, Chiyoda-ku, Tokyo 102; tel. (3) 5276-2200; fax (3) 5276-2209; f. 1989; every 2 weeks; women's interest; Editor YUKIO MIWA; circ. 380,000.

Asahi Camera: Asahi Shimbun Publishing Dept, 5-3-2, Tsukiji, Chuo-ku, Tokyo 104-8011; tel. (3) 3545-0131; fax (3) 5565-3286; f. 1926; monthly; photography; Editor HIROSHI HIROSE; circ. 90,000.

Balloon: Shufunotomo Co Ltd, 2-9, Kanda Surugadai, Chiyoda-ku, Tokyo 101; tel. (3) 3294-1132; fax (3) 3291-5093; f. 1986; monthly; expectant mothers; Dir MARIKO HOSODA; circ. 250,000.

Brutus: Magazine House, 3-13-10, Ginza, Chuo-ku, Tokyo 104-03; tel. (3) 3545-7000; fax (3) 3546-0034; f. 1980; every 2 weeks; men's interest; Editor KOICHI TETSUKA; circ. 250,000.

Bungei-Shunju: Bungei-Shunju Ltd, 3-23, Kioicho, Chiyoda-ku, Tokyo 102-8008; tel. (3) 3265-1211; fax (3) 3221-6623; internet bunshun.topica.ne.jp; f. 1923; monthly; general; Pres. MASARU SHIRAISHI; Editor KIYONDO MATSUI; circ. 656,000.

Business Tokyo: Keizaikai Bldg, 2-13-18, Minami-Aoyama, Minato-ku, Tokyo 105; tel. (3) 3423-8500; fax (3) 3423-8505; f. 1987; monthly; Dir TAKUO IDA; Editor ANTHONY PAUL; circ. 125,000.

Chuokoron: Chuokoron-Shinsha Inc, 2-8-7, Kyobashi, Chuo-ku, Tokyo 104–8320; tel. (3) 3563-2751; fax (3) 3561-5929; internet www.chuko.co.jp; f. 1887; monthly; general interest; Chief Editor JUN MAYIMA; circ. 90,000.

Croissant: Magazine House, 3-13-10, Ginza, Chuo-ku, Tokyo 104-03; tel. (3) 3545-7111; fax (3) 3546-0034; f. 1977; every 2 weeks; home; Editor MASAAKI TAKEUCHI; circ. 600,000.

Fujinkoron: Chuokoron-Sha Inc, 2-8-7, Kyobashi, Chuo-ku, Tokyo 104; tel. (3) 3563-1866; fax (3) 3561-5920; f. 1916; women's literary monthly; Editor YUKIKO YUKAWA; circ. 185,341.

Geijutsu Shincho: Shincho-Sha, 71, Yarai-cho, Shinjuku-ku, Tokyo 162-8711; tel. (3) 3266-5381; fax (3) 3266-5387; e-mail geishin@shinchosha.co.jp; f. 1950; monthly; fine arts, music, architecture, films, drama and design; Editor-in-Chief KAZUHIRO NAGAI; circ. 50,000.

Gendai: Kodan-Sha Ltd, 2-12-21, Otowa, Bunkyo-ku, Tokyo 112; tel. (3) 5395-3517; fax (3) 3945-9128; f. 1966; monthly; cultural and political; Editor SHUNKICHI YABUKI; circ. 250,000.

Ginza: Magazine House, 3-13-10, Ginza, Chuo-ku, Tokyo 104-8003; tel. (3) 3545-7080; fax (3) 3542-6375; internet webmaster.magazine.co.jp; f. 1997; monthly; women's interest; Editor MIYOKO YODOGAWA; circ. 250,000.

Hot-Dog Press: Kodan-Sha Ltd, 2-12-21, Otowa, Bunkyo-ku, Tokyo 112-01; tel. (3) 5395-3473; fax (3) 3945-9128; every 2 weeks; men's interest; Editor ATSUHIDE KOKUBO; circ. 650,000.

Ie-no-Hikari (Light of Home): Ie-no-Hikari Asscn, 11, Ichigaya Funagawaramachi, Shinjuku-ku, Tokyo 162-8448; tel. (3) 3266-9013; fax (3) 3266-9052; e-mail hikari@mxd.meshnet.or.jp; internet www.mediagalaxy.co.jp/ienohikarinet; f. 1925; monthly; rural and general interest; Pres. SHUZO SUZUKI; Editor KAZUO NAKANO; circ. 928,000.

Japan Company Handbook: Toyo Keizai Inc, 1-2-1, Nihombashi Hongoku-cho, Chuo-ku, Tokyo 103-8345; tel. (3) 3246-5621; fax (3) 3246-5473; e-mail sub@toyokeizai.co.jp; internet www.toyokeizai.co.jp; f. 1974; quarterly; English; Editor MASAKI HARA; total circ. 100,000.

Jitsugyo No Nihon: Jitsugyo No Nihon-Sha Ltd, 1-3-9, Ginza, Chuo-ku, Tokyo 104; tel. (3) 3562-1967; fax (3) 2564-2382; f. 1897; monthly; economics and business; Editor TOSHIO KAWAJIRI; circ. 60,000.

Junon: Shufu-To-Seikatsu Sha Ltd, 3-5-7, Kyobashi, Chuo-ku, Tokyo 104; tel. (3) 3563-5132; fax (3) 5250-7081; f. 1973; monthly; television and entertainment; circ. 560,000.

Kagaku (Science): Iwanami Shoten Publishers, 2-5-5, Hitotsubashi, Chiyoda-ku, Tokyo 102; tel. (3) 5210-4070; fax (3) 5210-4073; f. 1931; Editor NOBUAKI MIYABE; circ. 29,000.

Kagaku Asahi: Asahi Shimbun Publishing Dept, 5-3-2, Tsukiji, Chuo-ku, Tokyo 104-8011; tel. (3) 5540-7810; fax (3) 3546-2404; f. 1941; monthly; scientific; Editor TOSHIHIRO SASAKI; circ. 105,000.

Keizaijin: Kansai Economic Federation, Nakanoshima Center Bldg, 6-2-27, Nakanoshima, Kita-ku, Osaka 530-6691; tel. (6) 6441-0101; fax (6) 6443-5347; internet www.kankeiren.or.jp; f. 1947; monthly; economics; Editor M. YASUTAKE; circ. 2,600.

Lettuce Club: SS Communications, 11-2, Ban-cho, Chiyoda-ku, Tokyo 102; tel. (3) 5276-2151; fax (3) 5276-2229; f. 1987; every 2 weeks; cookery; Editor MITSURU NAKAYA; circ. 800,000.

Money Japan: SS Communications, 11-2, Ban-cho, Chiyoda-ku, Tokyo 102; tel. (3) 5276-2220; fax (3) 5276-2229; internet www.sscom.co.jp/money; f. 1985; monthly; finance; Editor TOSHIO KOBAYASHI; circ. 500,000.

Popeye: Magazine House, 3-13-10, Ginza, Chuo-ku, Tokyo 104-8003; tel. (3) 3545-7160; fax (3) 3545-9026; f. 1976; every 2 weeks; fashion, youth interest; Editor KATSUMI NAMAIZAWA; circ. 320,000.

President: President Inc, Bridgestone Hirakawacho Bldg, 2-13-12, Hirakawa-cho, Chiyoda-ku, Tokyo 102; tel. (3) 3237-3737; fax (3) 3237-3748; internet www.president.co.jp; f. 1963; monthly; business; Editor KAYOKO ABE; circ. 263,308.

Ray: Shufunotomo Co Ltd, 2-9, Kanda Surugadai, Chiyoda-ku, Tokyo 101; tel. (3) 3294-1163; fax (3) 3291-5093; f. 1988; monthly; women's interest; Editor TATSURO NAKANISHI; circ. 450,000.

Ryoko Yomiuri: Ryoko Yomiuri Publications Inc, 2-2-15, Ginza, Chuo-ku, Tokyo 104; tel. (3) 3561-8911; fax (3) 3561-8950; f. 1966; monthly; travel; Editor TETSUO KINUGAWA; circ. 470,000.

Sekai: Iwanami Shoten Publishers, 2-5-5, Hitotsubashi, Chiyoda-ku, Tokyo 101; tel. (3) 5210-4141; fax (3) 5210-4144; internet www.iwanami.co.jp/sekai; f. 1946; monthly; review of world and domestic affairs; Editor ATSUSHI OKAMOTO; circ. 120,000.

Shinkenchiku: Shinkenchiku-Sha Co Ltd, 2-31-2, Yushima, Bunkyo-ku, Tokyo 113-8501; tel. (3) 3811-7101; fax (3) 3812-8229; e-mail ja-business@japan-architect.co.jp; internet www.japan-architect.co.jp; f. 1925; monthly; architecture; Editor AKIHIKO OMORI; circ. 87,000.

Shiso (Thought): Iwanami Shoten Publishers, 2-5-5, Hitotsubashi, Chiyoda-ku, Tokyo 101-8002; tel. (3) 5210-4055; fax (3) 5210-4037; e-mail shiso@iwanami.co.jp; internet www.iwanami.co.jp/shiso; f. 1921; monthly; philosophy, social sciences and humanities; Editor KIYOSHI KOJIMA; circ. 20,000.

Shosetsu Shincho: Shincho-Sha, 71, Yarai-cho, Shinjuku-ku, Tokyo 162-8711; tel. (3) 3266-5241; fax (3) 3266-5412; internet www.shincho.net/magazines/shosetsushincho; f. 1947; monthly; literature; Editor-in-Chief TSUYOSHI MENJO; circ. 80,000.

Shufunotomo: Shufunotomo Co Ltd, 2-9, Kanda Surugadai, Chiyoda-ku, Tokyo 101; tel. (3) 5280-7531; fax (3) 5280-7431; f. 1917; monthly; home and lifestyle; Editor KYOKO FURUTO; circ. 450,000.

So-en: Bunka Publishing Bureau, 4-12-7, Hon-cho, Shibuya-ku, Tokyo 151; tel. (3) 3299-2531; fax (3) 3370-3712; f. 1936; fashion monthly; Editor KEIKO SASAKI; circ. 270,000.

NEWS AGENCIES

Jiji Tsushin (Jiji Press Ltd): Shisei-Kaikan, 1-3, Hibiya Park, Chiyoda-ku, Tokyo 100-8568; tel. (3) 3591-1111; e-mail info@jiji.co.jp; internet www.jiji.com; f. 1945; Pres. MASATOSHI MURAKAMI; Man. Dir and Man. Editor MASAKI SUGIURA.

Kyodo Tsushin (Kyodo News): 2-2-5, Toranomon, Minato-ku, Tokyo 105-8474; tel. (3) 5573-8081; fax (3) 5573-2268; e-mail kokusai@kyodonews.jp; internet home.kyodo.co.jp; f. 1945; Pres. TOYOHIKO YAMANOUCHI; Man. Editor TOSHIEI KOKUBU.

Radiopress Inc: R-Bldg Shinjuku, 5F, 33-8, Wakamatsu-cho, Shinjuku-ku, Tokyo 162-0056; tel. (3) 5273-2171; fax (3) 5273-2180; e-mail rptokyo@oak.ocn.ne.jp; f. 1945; provides news from China, the former USSR, Democratic People's Repub. of Korea, Viet Nam and elsewhere to the press and govt offices; Pres. AKIO IJUIN.

Sun Telephoto: Palaceside Bldg, 1-1-1, Hitotsubashi, Chiyoda-ku, Tokyo 100-0003; tel. (3) 3213-6771; e-mail photo@suntelephoto.com; internet www.suntelephoto.com; f. 1952; Pres. KOZO TAKINO; Man. Editor ASAMI SAKURAI.

Foreign Bureaux

Agence France-Presse (AFP): Asahi Shimbun Bldg, 11th Floor, 5-3-2, Tsukiji, Chuo-ku, Tokyo 104-0045; tel. (3) 3545-3061; fax (3) 3546-2594; Bureau Chief PHILIPPE RIES.

Agencia EFE (Spain): Kyodo Tsushin Bldg, 9th Floor, 2-2-5, Toranomon, Minato-ku, Tokyo 105-0001; tel. (3) 3585-8940; fax (3) 3585-8948; Bureau Chief CARLOS DOMÍNGUEZ.

Agenzia Nazionale Stampa Associata (ANSA) (Italy): Kyodo Tsushin Bldg, 9th Floor, 2-2-5, Toranomon, Minato-ku, Tokyo 105-0001; tel. (3) 3584-6667; fax (3) 3584-5114; Bureau Chief ALBERTO ZANCONATO.

Antara (Indonesia): Kyodo Tsushin Bldg, 9th Floor, 2-2-5, Toranomon, Minato-ku, Tokyo 105-0001; tel. (3) 3584-4234; fax (3) 3584-4591; Correspondent MARIA ANDRIANA.

Associated Press (AP) (USA): Shiodome Media Tower, 7/F, 1-7-1 Higashi-Shimbashi, Minato-ku, Tokyo 105-7207; tel. (3) 6215-8930; fax (3) 6215-8949; internet www.ap.org; Gen. Man. KAZUO ABIKO.

Central News Agency (Taiwan): 3-7-3-302, Shimo-meguro, Meguro-ku, Tokyo 153-0064; tel. (3) 3495-2046; fax (3) 3495-2066; Bureau Chief CHANG FANG MIN.

Deutsche Presse-Agentur (dpa) (Germany): Nippon Press Center, 3rd Floor, 2-2-1, Uchisaiwai-cho, Chiyoda-ku, Tokyo 100-0011; tel. (3) 3580-6629; fax (3) 3593-7888; Bureau Chief LARS NICOLAYSEN.

Informatsionnoye Telegrafnoye Agentstvo Rossii—Telegrafnoye Agentstvo Suverennykh Stran (ITAR—TASS) (Russia): 1-5-1, Hon-cho, Shibuya-ku, Tokyo 151-0071; tel. (3) 3377-0380; fax (3) 3378-0606; Bureau Chief Vasilii Golovnin.

Inter Press Service (IPS) (Italy): 1-15-19, Ishikawa-machi, Ota-ku, Tokyo 145-0061; tel. (3) 3726-7944; fax (3) 3726-7896; Correspondent Suvendrini Kakuchi.

Magyar Távirati Iroda (MTI) (Hungary): 1-3-4-306, Okamoto, Setagaya-ku, Tokyo 157-0076; tel. (3) 3708-3093; fax (3) 3708-2703; Bureau Chief János Marton.

Reuters (UK): Shuwa Kamiya-cho Bldg, 5th Floor, 4-3-13, Toranomon, Minato-ku, Tokyo 105-0001; tel. (3) 3432-4141; fax (3) 3433-2921; Editor William Sposato.

Rossiiskoye Informatsionnoye Agentstvo—Novosti (RIA—Novosti) (Russia): 3-9-13 Higashi-gotanda, Shinagawa-ku, Tokyo 141-0022; tel. (3) 3441-9241; fax (3) 3447-8443; e-mail riatokyo@ma.kcom.ne.jp; Bureau Chief Vyacheslav Bantin.

United Press International (UPI) (USA): Ferrare Bldg, 4th Floor, 1-24-15, Ebisu, Shibuya-ku, Tokyo 150-0013; tel. (3) 5421-1333; fax (3) 5421-1339; Bureau Chief Ruth Youngblood.

Xinhua (New China) News Agency (People's Republic of China): 3-35-23, Ebisu, Shibuya-ku, Tokyo 150-0013; tel. (3) 3441-3766; fax (3) 3446-3995; Bureau Chief Wang Dajun.

Yonhap (United) News Agency (Republic of Korea): Kyodo Tsushin Bldg, 2-2-5, Toranomon, Minato-ku, Tokyo 105-0001; tel. (3) 3584-4681; fax (3) 3584-4021; f. 1945; Bureau Chief Moon Young Shik.

PRESS ASSOCIATIONS

Foreign Correspondents' Club of Japan: 20th Floor, 1-7-1, Yuraku-cho, Chiyoda-ku, Tokyo 100-0006; tel. (3) 3211-3161; fax (3) 3211-3168; e-mail yoda@fccj.or.jp; internet www.fccj.or.jp; f. 1945; 193 companies; Pres. Anthony Rowley; Man. Seishi Yoda.

Foreign Press Center: Nippon Press Center Bldg, 6th Floor, 2-2-1, Uchisaiwai-cho, Chiyoda-ku, Tokyo 100-0011; tel. (3) 3501-3401; fax (3) 3501-3622; e-mail nsuzuki@fpcjpn.or.jp; internet www.fpcj.jp; f. 1976; est. by the Japan Newspaper Publrs' and Editors' Asscn and the Japan Fed. of Economic Orgs; provides services to the foreign press; Pres. Terusuke Terada.

Nihon Shinbun Kyokai (The Japan Newspaper Publishers and Editors Asscn): Nippon Press Center Bldg, 2-2-1, Uchisaiwai-cho, Chiyoda-ku, Tokyo 100-8543; tel. (3) 3591-3462; fax (3) 3591-6149; e-mail s_intl@pressnet.or.jp; internet www.pressnet.or.jp; f. 1946; mems include 151 companies (111 daily newspapers, 4 news agencies and 36 radio and TV companies); Chair. Shin-ichi Hakoshima; Man. Dir Shigemi Murakami.

Nihon Zasshi Kyokai (Japan Magazine Publishers Asscn): 1-7, Kanda Surugadai, Chiyoda-ku, Tokyo 101-0062; tel. (3) 3291-0775; fax (3) 3293-6239; f. 1956; 85 mems; Pres. Haruhiko Ishikawa; Sec. Genya Inui.

Publishers

Akane Shobo Co Ltd: 3-2-1, Nishikanda, Chiyoda-ku, Tokyo 101-0065; tel. (3) 3263-0641; fax (3) 3263-5440; f. 1949; juvenile; Pres. Masaharu Okamoto.

Akita Publishing Co Ltd: 2-10-8, Iidabashi, Chiyoda-ku, Tokyo 102-8101; tel. (3) 3264-7011; fax (3) 3265-5906; f. 1948; social sciences, history, juvenile; Chair. Sadao Akita; Pres. Sadami Akita.

ALC Press Inc: 2-54-12, Eifuku, Suginami-ku, Tokyo 168-0064; tel. (3) 3323-1101; fax (3) 3327-1022; e-mail menet@alc.co.jp; internet www.alc.co.jp; f. 1969; linguistics, educational materials, dictionary, juvenile; Pres. Terumaro Hiramoto.

Asahi Shimbun Publications Division: 5-3-2, Tsukiji, Chuo-ku, Tokyo 104-8011; tel. (3) 3545-0131; fax (3) 5540-7682; f. 1879; general; Pres. Muneyuki Matsushita; Dir of Publications Hisao Kuwashima.

Asakura Publishing Co Ltd: 6-29, Shin Ogawa-machi, Shinjuku-ku, Tokyo 162-8707; tel. (3) 3260-0141; fax (3) 3260-0180; e-mail edit@asakura.co.jp; internet www.asakura.co.jp; f. 1929; natural science, medicine, social sciences; Pres. Kunizo Asakura.

Baifukan Co Ltd: 4-3-12, Kudan Minami, Chiyoda-ku, Tokyo 102-8260; tel. (3) 3262-5256; fax (3) 3262-5276; f. 1924; engineering, natural and social sciences, psychology; Pres. Itaru Yamamoto.

Baseball Magazine-Sha: 3-10-10, Misaki-cho, Chiyoda-ku, Tokyo 101-8381; tel. (3) 3238-0081; fax (3) 3238-0106; internet www.bbm-japan.com; f. 1946; sports, physical education, recreation, travel; Chair. Tsuneo Ikeda; Pres. Tetsuo Ikeda.

Bijutsu Shuppan-Sha Ltd: Inaoka Kudan Bldg, 6th Floor, 2-36, Kanda Jimbo-cho, Chiyoda-ku, Tokyo 101-8417; tel. (3) 3234-2151; fax (3) 3234-9451; f. 1905; fine arts, graphic design; Pres. Atsushi Oshita.

Bonjinsha Co Ltd: 1-3-13, Hirakawa-cho, Chiyoda-ku, Tokyo 102-0093; tel. (3) 3263-3959; fax (3) 3263-3116; f. 1973; Japanese language teaching materials; Pres. Hisamitsu Tanaka.

Bungeishunju Ltd: 3-23, Kioi-cho, Chiyoda-ku, Tokyo 102-8008; tel. (3) 3265-1211; fax (3) 3239-5482; internet www.bunshun.co.jp; f. 1923; fiction, general literature, recreation, economics, sociology; Dir Masaru Shiraishi.

Chikuma Shobo: Chikumashobo Bldg, 2-5-3, Kuramae, Taito-ku, Tokyo 111-8755; tel. (3) 5687-2671; fax (3) 5687-1585; e-mail webinfo@chikumashobo.co.jp; internet www.chikumashobo.co.jp; f. 1940; general literature, fiction, history, juvenile, fine arts; Pres. Akio Kikuchi.

Child-Honsha Co Ltd: 5-24-21, Koishikawa, Bunkyo-ku, Tokyo 112-8512; tel. (3) 3813-3785; fax (3) 3813-3765; e-mail ehon@childbook.co.jp; internet www.childbook.co.jp; f. 1930; juvenile; Pres. Yoshiaki Shimazaki.

Chuokoron-Shinsha Inc: 2-8-7, Kyobashi, Chuo-ku, Tokyo 104-8320; tel. (3) 3563-1261; fax (3) 3561-5920; internet www.chuko.co.jp; f. 1886; philosophy, history, sociology, general literature; Pres. Junichi Hayakawa.

Corona Publishing Co Ltd: 4-46-10, Sengoku, Bunkyo-ku, Tokyo 112-0011; tel. (3) 3941-3131; fax (3) 3941-3137; e-mail info@coronasha.co.jp; internet www.coronosha.co.jp; f. 1927; electronics business publs; Pres. Tatsumi Gorai.

Dempa Publications Inc: 1-11-15, Higashi Gotanda, Shinagawa-ku, Tokyo 141-8755; tel. (3) 3445-6111; fax (3) 3447-4666; f. 1950; electronics, personal computer software, juvenile, trade newspapers, English and Japanese language publications; Pres. Tetsuo Hirayama.

Diamond Inc: 6-12-17, Jingumae, Shibuya-ku, Tokyo 150-8409; tel. (3) 5778-7203; fax (3) 5778-6612; e-mail mitachi@diamond.co.jp; internet www.diamond.co.jp; f. 1913; business, management, economics, financial; Pres. Yutaka Iwasa.

Dohosha Ltd: TAS Bldg, 2-5-2, Nishikanda, Chiyoda-ku, Tokyo 101-0065; tel. (3) 5276-0831; fax (3) 5276-0840; e-mail intl@doho-sha.co.jp; internet www.doho-sha.co.jp; f. 1997; general works, architecture, art, Buddhism, business, children's education, cooking, flower arranging, gardening, medicine.

Froebel-Kan Co Ltd: 6-14-9, Honkomagome, Bunkyo-ku, Tokyo 113-8611; tel. (3) 5395-6614; fax (3) 5395-6639; e-mail info-e@froebel-kan.co.jp; internet www.froebel-kan.co.jp; f. 1907; juvenile, educational; Pres. Mamoru Kitabayashi; Dir Mitsuhiro Tada.

Fukuinkan Shoten Publishers Inc: 6-6-3, Honkomagome, Bunkyo-ku, Tokyo 113-8686; tel. (3) 3942-2151; fax (3) 3942-1401; f. 1952; juvenile; Pres. Shiro Tokita; Chair. Katsumi Sato.

Gakken Co Ltd: 4-40-5, Kamiikedai, Ota-ku, Tokyo 145-8502; tel. (3) 3726-8111; fax (3) 3493-3338; f. 1946; juvenile, educational, art, encyclopaedias, dictionaries; Pres. Kazuhiko Sawada.

Graphic-sha Publishing Co Ltd: 1-9-12, Kudan Kita, Chiyoda-ku, Tokyo 102-0073; tel. (3) 3263-4318; fax (3) 3263-5297; e-mail info@graphicsha.co.jp; internet www.graphicsha.co.jp; f. 1963; art, design, architecture, manga techniques, hobbies; Pres. Seiichi Sugaya.

Gyosei Corpn: 4-30-16, Ogikubo, Suginami-ku, Tokyo 167-8088; tel. (3) 5349-6666; fax (3) 5349-6677; e-mail business@gyosei.co.jp; internet www.gyosei.co.jp; f. 1893; law, education, science, politics, business, art, language, literature, juvenile; Pres. Motoo Fujisawa.

Hakusui-Sha Co Ltd: 3-24, Kanda Ogawa-machi, Chiyoda-ku, Tokyo 101-0052; tel. (3) 3291-7821; fax (3) 3291-7810; f. 1915; general literature, science and languages; Pres. Kazuaki Fujiwara.

Hayakawa Publishing Inc: 2-2, Kanda-Tacho, Chiyoda-ku, Tokyo 101-0046; tel. (3) 3252-3111; fax (3) 3254-1550; f. 1945; science fiction, mystery, autobiography, literature, fantasy; Pres. Hiroshi Hayakawa.

Heibonsha Ltd: 2-29-4 Hakusan, Bunkyo-ku, Tokyo 112-0001; tel. (3) 3818-0641; fax (3) 3818-0754; internet www.heibonsha.co.jp; f. 1914; encyclopaedias, art, history, geography, literature, science; Pres. Naoto Shimonaka.

Hirokawa Publishing Co: 3-27-14, Hongo, Bunkyo-ku, Tokyo 113-0033; tel. (3) 3815-3651; fax (3) 5684-7030; f. 1925; natural sciences, medicine, pharmacy, nursing, chemistry; Pres. Setsuo Hirokawa.

Hoikusha Publishing Co: 1-6-12, Kawamata, Higashi, Osaka 577-0063; tel. (6) 6788-4470; fax (6) 6788-4970; internet www.hoikusha.co.jp; f. 1947; natural science, juvenile, fine arts, geography; Pres. Yuki Imai.

Hokuryukan Co Ltd: 3-8-14, Takanawa, Minato-ku, Tokyo 108-0074; tel. (3) 5449-4591; fax (3) 5449-4950; e-mail hk-ns@mk1.macnet.or.jp; internet www.macnet.or.jp/co/hk-ns; f. 1891; natural science, medical science, juvenile, dictionaries; Pres. HISAKO FUKUDA.

The Hokuseido Press: 3-32-4, Honkomagome, Bunkyo-ku, Tokyo 113-0021; tel. (3) 3827-0511; fax (3) 3827-0567; e-mail info@hokuseido.com; f. 1914; regional non-fiction, dictionaries, textbooks; Pres. MASAZO YAMAMOTO.

Ie-No-Hikari Association: 11, Funagawara-cho, Ichigaya, Shinjuku-ku, Tokyo 162-8448; tel. (3) 3266-9000; fax (3) 3266-9048; e-mail hikari@mxd.mesh.ne.jp; internet www.ienohikari.or.jp; f. 1925; social science, agriculture, cooking; Exec. Man. MASAYUKI YAMAMOTO; Pres. HARUTO SATO.

Igaku-Shoin Ltd: 5-24-3, Hongo, Bunkyo-ku, Tokyo 113-8719; tel. (3) 3817-5610; fax (3) 3815-4114; e-mail info@igaku-shoin.co; internet www.igaku-shoin.co.jp; f. 1944; medicine, nursing; Pres. YU KANEHARA.

Institute for Financial Affairs Inc (KINZAI): 19, Minami-Motomachi, Shinjuku-ku, Tokyo 160-8519; tel. (3) 3358-1161; fax (3) 3359-7947; e-mail JDI04072@nifty.ne.jp; internet www.kinzai.or.jp; f. 1950; finance and economics, banking laws and regulations, accounting; Pres. MASATERU YOSHIDA.

Ishiyaku Publishers Inc: 1-7-10, Honkomagome, Bunkyo-ku, Tokyo 113-8612; tel. (3) 5395-7600; fax (3) 5395-7606; internet www.ishiyaku.co.jp; f. 1921; medicine, dentistry, rehabilitation, nursing, nutrition and pharmaceutics; Pres. KATSUJI FUJITA.

Iwanami Shoten, Publishers: 2-5-5, Hitotsubashi, Chiyoda-ku, Tokyo 101-8002; tel. (3) 5210-4000; fax (3) 5210-4039; e-mail rights@iwanami.co.jp; internet www.iwanami.co.jp; f. 1913; natural and social sciences, humanities, literature, fine arts, juvenile, dictionaries; Pres. NOBUKAZU OTSUKA.

Japan Broadcast Publishing Co Ltd: 41-1, Udagawa-cho, Shibuya-ku, Tokyo 150-8081; tel. (3) 3464-7311; fax (3) 3780-3353; e-mail webmaster@npb.nhk-grp.co.jp; internet www.nhk-grp.co.jp/npb; f. 1931; foreign language textbooks, gardening, home economics, sociology, education, art, juvenile; Pres. TAKESHI MATSUO.

Japan External Trade Organization (JETRO): 2-2-5, Toranomon, Minato-ku, Tokyo 105-8466; tel. (3) 3582-5511; fax (3) 3587-2485; internet www.jetro.go.jp; f. 1958; trade, economics, investment.

Japan Publications Trading Co Ltd: 1-2-1, Sarugaku-cho, Chiyoda-ku, Tokyo 101-0064; tel. (3) 3292-3751; fax (3) 3292-0410; e-mail jpt@po.iijnet.or.jp; internet www.jptco.co.jp; f. 1942; general works, art, health, sports; Pres. SATOMI NAKABAYASHI.

The Japan Times Ltd: 4-5-4, Shibaura, Minato-ku, Tokyo 108-0023; tel. (3) 3453-2013; fax (3) 3453-8023; e-mail jt-books@kt.rim.or.jp; internet bookclub.japantimes.co.jp; f. 1897; linguistics, culture, business; Pres. TOSHIAKI OGASAWARA.

Japan Travel Bureau Inc: Shibuya Nomura Bldg, 1-10-8, Dogenzaka, Shibuya-ku, Tokyo 150-8558; tel. (3) 3477-9521; fax (3) 3477-9538; internet www.jtb.co.jp; f. 1912; travel, geography, history, fine arts, languages; Vice-Pres. MITSUMASA IWATA.

Jimbun Shoin: 9, Nishiuchihata-cho, Takeda, Fushimi-ku, Kyoto 612-8447; tel. (75) 603-1344; fax (75) 603-1814; e-mail edjimbun@mbox.kyoto-inet.or.jp; internet www.jimbunshoin.co.jp; f. 1922; general literature, philosophy, fiction, social science, religion, fine arts; Pres. MUTSUHISA WATANABE.

Kadokawa Shoten Publishing Co Ltd: 2-13-3, Fujimi, Chiyoda-ku, Tokyo 102-0071; tel. (3) 3238-8611; fax (3) 3238-8612; f. 1945; literature, history, dictionaries, religion, fine arts, books on tape, compact discs, CD-ROM, comics, animation, video cassettes, computer games; Pres. TSUGUHIKO KADOKAWA.

Kaibundo Publishing Co Ltd: 2-5-4, Suido, Bunkyo-ku, Tokyo 112-0005; tel. (3) 5684-6289; fax (3) 3815-3953; e-mail LED04737@nifty.ne.jp; f. 1914; marine affairs, natural science, engineering, industry; Pres. YOSHIHIRO OKADA.

Kaiseisha Publishing Co Ltd: 3-5, Ichigaya Sadohara-cho, Shinjuku-ku, Tokyo 162-8450; tel. (3) 3260-3229; fax (3) 3260-3540; e-mail foreign@kaiseisha.co.jp; internet www.kaiseisha.co.jp; f. 1936; juvenile; Pres. MASAKI IMAMURA.

Kanehara & Co Ltd: 2-31-14, Yushima, Bunkyo-ku, Tokyo 113-8687; tel. (3) 3811-7185; fax (3) 3813-0288; f. 1875; medical, agricultural, engineering and scientific; Pres. SABURO KOMURO.

Kenkyusha Ltd: 2-11-3, Fujimi, Chiyoda-ku, Tokyo 102-8152; tel. (3) 3288-7711; fax (3) 3288-7821; e-mail hanbai@kenkyusha.co.jp; internet www.kenkyusha.co.jp; f. 1907; bilingual dictionaries, books on languages; Pres. KUNIKATSU ARAKI.

Kinokuniya Co Ltd: 5-38-1, Sakuragaoka, Setagaya-ku, Tokyo 156-8691; tel. (3) 3439-0172; fax (3) 3439-0173; e-mail publish@kinokuniya.co.jp; internet www.kinokuniya.co.jp; f. 1927; humanities, social science, natural science; Pres. OSAMU MATSUBARA.

Kodansha International Ltd: 1-17-14, Otowa, Bunkyo-ku, Tokyo 112-8652; tel. (3) 3944-6492; fax (3) 3944-6323; e-mail sales@kodansha-intl.co.jp; f. 1963; art, business, cookery, crafts, gardening, language, literature, martial arts; Pres. SAWAKO NOMA.

Kodansha Ltd: 2-12-21, Otowa, Bunkyo-ku, Tokyo 112-8001; tel. (3) 5395-3574; fax (3) 3944-9915; f. 1909; fine arts, fiction, literature, juvenile, comics, dictionaries; Pres. SAWAKO NOMA.

Kosei Publishing Co Ltd: 2-7-1, Wada, Suginami-ku, Tokyo 166-8535; tel. (3) 5385-2319; fax (3) 5385-2331; e-mail dharmaworld@kosei-shuppan.co.jp; internet www.kosei-shuppan.co.jp; f. 1966; general works, philosophy, religion, history, pedagogy, social science, art, juvenile; Pres. TEIZO KURIYAMA.

Kyoritsu Shuppan Co Ltd: 4-6-19, Kohinata, Bunkyo-ku, Tokyo 112-8700; tel. (3) 3947-2511; fax (3) 3947-2539; e-mail kyoritsu@po.iijnet.or.jp; internet www.kyoritsu-pub.co.jp; f. 1926; scientific and technical; Pres. MITSUAKI NANJO.

Maruzen Co Ltd: 3-9-2, Nihombashi, Chuo-ku, Tokyo 103-8244; tel. (3) 3272-0521; fax (3) 3272-0693; internet www.maruzen.co.jp; f. 1869; general works; Pres. SEISHIRO MURATA.

Medical Friend Co Ltd: 3-2-4, Kudan Kita, Chiyoda-ku, Tokyo 102-0073; tel. (3) 3264-6611; fax (3) 3261-6602; f. 1947; medical and allied science, nursing; Pres. KAZUHARU OGURA.

Minerva Shobo: 1, Tsutsumi dani-cho, Hinooka, Yamashina-ku, Kyoto 607-8494; tel. (75) 581-5191; fax (75) 581-0589; e-mail info@minervashobo.co.jp; internet www.minervashobo.co.jp; f. 1948; general non-fiction and reference; Pres. KEIZO SUGITA.

Misuzu Shobo Ltd: 5-32-21, Hongo, Bunkyo-ku, Tokyo 113-0033; tel. (3) 3815-9181; fax (3) 3818-8497; e-mail info@msz.co.jp; internet www.msz.co.jp; f. 1947; general, philosophy, history, psychiatry, literature, science, art; Pres. TAKASHI ARAI.

Morikita Shuppan Co Ltd: 1-4-11, Fujimi, Chiyoda-ku, Tokyo 102-0071; tel. (3) 3265-8341; fax (3) 3264-8709; e-mail info@morikita.co.jp; internet www.morikita.co.jp; f. 1950; natural science, engineering; Pres. HAJIME MORIKITA.

Nakayama-Shoten Co Ltd: 1-25-14, Hakusan, Bunkyo-ku, Tokyo 113-8666; tel. (3) 3813-1100; fax (3) 3816-1015; e-mail eigyo@nakayamashoten.co.jp; internet www.nakayamashoten.co.jp; f. 1948; medicine, biology, zoology; Pres. TADASHI HIRATA.

Nanzando Co Ltd: 4-1-11, Yushima, Bunkyo-ku, Tokyo; tel. (3) 5689-7868; fax (3) 5689-7869; e-mail info@nanzando.com; internet www.nanzando.com; medical reference, paperbacks; Pres. HAJIME SUZUKI.

Nigensha Publishing Co Ltd: 2-2, Kanda Jimbo-cho, Chiyoda-ku, Tokyo 101-8419; tel. (3) 5210-4733; fax (3) 5210-4723; e-mail sales@nigensha.co.jp; internet www.nigensha.co.jp; f. 1953; calligraphy, fine arts, art reproductions, cars, watches; Pres. TAKAO WATANABE.

Nihon Keizai Shimbun Inc, Publications Bureau: 1-9-5, Otemachi, Chiyoda-ku, Tokyo 100-0004; tel. (3) 3270-0251; fax (3) 5255-2864; f. 1876; economics, business, politics, fine arts, video cassettes, CD-ROM; Pres. TOYOHIKO KOBAYASHI.

Nihon Vogue Co Ltd: 3-23, Ichigaya Honmura-cho, Shinjuku-ku, Tokyo 162-8705; tel. (3) 5261-5139; fax (3) 3269-8726; e-mail asai@tezukuritown.com; internet www.tezukuritown.com; f. 1954; quilt, needlecraft, handicraft, knitting, decorative painting, pressed flowers; Pres. NOBUAKI SETO.

Nippon Jitsugyo Publishing Co Ltd: 3-2-12, Hongo, Bunkyo-ku, Tokyo 113-0033; tel. (3) 3814-5651; fax (3) 3818-2723; e-mail int@njg.co.jp; internet www.njg.co.jp; f. 1950; business, management, finance and accounting, sales and marketing; Chair. and CEO YOICHIRO NAKAMURA.

Obunsha Co Ltd: 78, Yarai-cho, Shinjuku-ku, Tokyo 162-0805; tel. (3) 3266-6000; fax (3) 3266-6291; internet www.obunsha.co.jp; f. 1931; textbooks, reference, general science and fiction, magazines, encyclopaedias, dictionaries; software; audio-visual aids; CEO FUMIO AKAO.

Ohmsha Ltd: 3-1, Kanda Nishiki-cho, Chiyoda-ku, Tokyo 101-8460; tel. (3) 3233-0641; fax (3) 3233-2426; e-mail kaigaika@ohmsha.co.jp; internet http://www.ohmsha.co.jp/index_e.htm; f. 1914; engineering, technical and scientific; Pres. SEIJI SATO; Dirs M. MORI, O. TAKEO.

Ondorisha Publishers Ltd: 11-11, Nishigoken-cho, Shinjuku-ku, Tokyo 162-8708; tel. (3) 3268-3101; fax (3) 3235-3530; f. 1945; knitting, embroidery, patchwork, handicraft books; Pres. HIDEAKI TAKEUCHI.

Ongaku No Tomo Sha Corpn (ONT): 6-30, Kagurazaka, Shinjuku-ku, Tokyo 162-0825; tel. (3) 3235-2111; fax (3) 3235-2119; internet www.ongakunotomo.co.jp; f. 1941; compact discs, videograms, music magazines, music books, music data, music textbooks; Pres. JUN MEGURO.

PHP Institute Inc: 11, Kitanouchi-cho, Nishikujo, Minami-ku, Kyoto 601-8411; tel. (75) 681-4431; fax (75) 681-9921; internet www.php.co.jp; f. 1946; social science; Pres. MASAHARU MATSUSHITA.

Poplar Publishing Co Ltd: 5, Suga-cho, Shinjuku-ku, Tokyo 160-8565; tel. (3) 3357-2216; fax (3) 3351-0736; e-mail henshu@poplar.co.jp; internet www.poplar.co.jp; f. 1947; children's; Pres. HARUO TANAKA.

Sanseido Co Ltd: 2-22-14, Misaki-cho, Chiyoda-ku, Tokyo 101-8371; tel. (3) 3230-9411; fax (3) 3230-9547; f. 1881; dictionaries, educational, languages, social and natural science; Chair. HISANORI UENO; Pres. TOSHIO GOMI.

Sanshusha Publishing Co Ltd: 1-5-34, Shitaya, Taito-ku, Tokyo 110-0004; tel. (3) 3842-1711; fax (3) 3845-3965; e-mail maeda_k@sanshusha.or.jp; internet www.sanshusha.co.jp; f. 1938; languages, dictionaries, philosophy, sociology, electronic publishing (CD-ROM); Pres. KANJI MAEDA.

Seibundo-Shinkosha Co Ltd: 3-3-1, Hongo, Bunkyo-ku, Tokyo 113-0033; tel. (3) 5800-5775; fax (3) 5800-5773; f. 1912; technical, scientific, design, general non-fiction; Pres. MINORU TAKITA.

Sekai Bunka Publishing Inc: 4-2-29, Kudan-Kita, Chiyoda-ku, Tokyo 102-8187; tel. (3) 3262-5111; fax (3) 3221-6843; internet www.sekaibunka.com; f. 1946; history, natural science, geography, education, art, literature, juvenile; Pres. TSUTOMU SUZUKI.

Shincho-Sha Co Ltd: 71, Yarai-cho, Shinjuku-ku, Tokyo 162-8711; tel. (3) 3266-5411; fax (3) 3266-5534; e-mail shuppans@shinchosha.co.jp; internet www.shinchosha.co.jp; f. 1896; general literature, fiction, non-fiction, fine arts, philosophy; Pres. TAKANOBU SATO.

Shinkenchiku-Sha Co Ltd: 2-31-2, Yushima, Bunkyo-ku, Tokyo 113-8501; tel. (3) 3811-7101; fax (3) 3812-8229; e-mail ja-business@japan-architect.co.jp; internet www.japan-architect.co.jp; f. 1925; architecture; Pres. AKIHIKO OMORI.

Shogakukan Inc: 2-3-1, Hitotsubashi, Chiyoda-ku, Tokyo 101-8001; tel. (3) 3230-5526; fax (3) 3288-9653; internet www.shogakukan.co.jp; f. 1922; juvenile, education, geography, history, encyclopaedias, dictionaries; Pres. MASAHIRO OHGA.

Shokabo Publishing Co Ltd: 8-1, Yomban-cho, Chiyoda-ku, Tokyo 102-0081; tel. (3) 3262-9166; fax (3) 3262-7257; e-mail info@shokabo.co.jp; internet www.shokabo.co.jp; f. 1895; natural science, engineering; Pres. TATSUJI YOSHINO.

Shokokusha Publishing Co Ltd: 25, Saka-machi, Shinjuku-ku, Tokyo 160-0002; tel. (3) 3359-3231; fax (3) 3357-3961; e-mail eigyo@shokokusha.co.jp; f. 1932; architectural, technical and fine arts; Pres. TAKESHI GOTO.

Shucisha Inc: 2-5-10, Hitotsubashi, Chiyoda-ku, Tokyo 101-8050; tel. (3) 3230-6320; fax (3) 3262-1309; f. 1925; literature, fine arts, language, juvenile, comics; Pres. and CEO TAMIO KOJIMA.

Shufunotomo Co Ltd: 2-9, Kanda Surugadai, Chiyoda-ku, Tokyo 101-8911; tel. (3) 5280-7567; fax (3) 5280-7568; e-mail international@shufunotomo.co.jp; internet www.shufunotomo.co.jp; f. 1916; domestic science, fine arts, gardening, handicraft, cookery and magazines; Pres. KUNIHIKO MURAMATSU.

Shunju-Sha: 2-18-6, Soto-Kanda, Chiyoda-ku, Tokyo 101-0021; tel. (3) 3255-9614; fax (3) 3255-9370; f. 1918; philosophy, religion, literary, economics, music; Pres. AKIRA KANDA; Man. RYUTARO SUZUKI.

Taishukan Publishing Co Ltd: 3-24, Kanda-Nishiki-cho, Chiyoda-ku, Tokyo 101-8466; tel. (3) 3294-2221; fax (3) 3295-4107; internet www.taishukan.co.jp; f. 1918; reference, Japanese and foreign languages, sports, dictionaries, audio-visual aids; Pres. KAZUYUKI SUZUKI.

Tankosha Publishing Co Ltd: 19-1, Miyanishi-cho Murasakino, Kita-ku, Kyoto 603-8691; tel. (75) 432-5151; fax (75) 432-0273; e-mail tankosha@magical.egg.or.jp; internet tankosha.topica.ne.jp; f. 1949; tea ceremony, fine arts, history; Pres. YOSHITO NAYA.

Teikoku-Shoin Co Ltd: 3-29, Kanda Jimbo-cho, Chiyoda-ku, Tokyo 101-0051; tel. (3) 3262-0834; fax (3) 3262-7770; e-mail kenkyu@teikokushoin.co.jp; f. 1926; geography, atlases, maps, textbooks, history, civil studies; Pres. MISAO MORIYA.

Tokai University Press: 2-28-4, Tomigaya, Shibuya-ku, Tokyo 151-8677; tel. (3) 5478-0891; fax (3) 5478-0870; f. 1962; social science, cultural science, natural science, engineering, art; Pres. TATSURO MATSUMAE.

Tokuma Shoten Publishing Co Ltd: 1-1-16, Higashi Shimbashi, Minato-ku, Tokyo 105-8055; tel. (3) 3573-0111; fax (3) 3573-8788; e-mail info@tokuma.com; internet www.tokuma.com; f. 1954; Japanese classics, history, fiction, juvenile; Pres. YASUYOSHI TOKUMA.

Tokyo News Service Ltd: Tsukiji Hamarikyu Bldg, 5-3-3, Tsukiji, Chuo-ku, Tokyo 104; tel. (3) 3542-6511; fax (3) 3545-3628; f. 1947; shipping, trade and television guides; Pres. T. OKUYAMA.

Tokyo Shoseki Co Ltd: 2-17-1, Horifune, Kita-ku, Tokyo 114-8524; tel. (3) 5390-7513; fax (3) 5390-7409; internet www.tokyo-shoseki.co.jp; f. 1909; textbooks, reference books, cultural and educational books; Pres. YOSHIKATSU KAWAUCHI.

Tokyo Sogen-Sha Co Ltd: 1-5, Shin-Ogawa-machi, Shinjuku-ku, Tokyo 162-0814; tel. (3) 3268-8201; fax (3) 3268-8230; f. 1954; mystery and detective stories, science fiction, literature; Pres. YASUNOBU TOGAWA.

Tuttle Publishing Co Inc: Yaekari Bldg, 3rd Floor, 5-4-12 Osaki, Shinagawa-ku, Tokyo 141-0032; tel. (3) 5437-0171; fax (44) 5437-0755; e-mail tuttle@gol.com; internet www.tuttlepublishing.com; f. 1948; books on Japanese and Asian religion, history, social science, arts, languages, literature, juvenile, cookery; Pres. ERIC OEY.

United Nations University Press: 5-53-70, Jingumae, Shibuya-ku, Tokyo 150-8925; tel. (3) 3499-2811; fax (3) 3499-2828; e-mail sales@hq.unu.edu; internet www.unu.edu/unupress; f. 1975; social sciences, humanities, pure and applied natural sciences; Rector HANS J. H. VAN GINKEL.

University of Tokyo Press: 7-3-1, Hongo, Bunkyo-ku, Tokyo 113-8654; tel. (3) 3811-0964; fax (3) 3815-1426; e-mail info@utp.or.jp; f. 1951; natural and social sciences, humanities; Japanese and English; Chair. MASARU NISHIO; Man. Dir TADASHI YAMASHITA.

Weekly Toyo Keizai: 1-2-1, Nihombashi, Hongoku-cho, Chuo-ku, Tokyo 103-8345; tel. (3) 3246-5655; fax (3) 3231-0906; e-mail sub@toyokeizai.co.jp; internet www.toyokeizai.co.jp; f. 1895; economics, business, finance and corporate, information; Pres. HIROSHI TAKAHASHI.

Yama-Kei Publishers Co Ltd: 1-1-33, Shiba-Daimon, Minato-ku, Tokyo 105-0012; tel. (3) 3436-4021; fax (3) 3438-1949; f. 1930; natural science, geography, mountaineering; Pres. YOSHIMITSU KAWASAKI.

Yohan: 3-14-9, Okubo, Shinjuku-ku, Tokyo 169-0072; tel. (3) 3208-0181; fax (3) 3209-0288; internet www.yohan.co.jp; f. 1963; social science, language, art, juvenile, dictionary; Pres. MASANORI WATANABE.

Yuhikaku Publishing Co Ltd: 2-17, Kanda Jimbo-cho, Chiyoda-ku, Tokyo 101-0051; tel. (3) 3264-1312; fax (3) 3264-5030; f. 1877; social sciences, law, economics; Pres. TADATAKA EGUSA.

Yuzankaku Shuppan: 2-6-9, Fujimi, Chiyoda-ku, Tokyo 102; tel. (3) 3262-3231; fax (3) 3262-6938; e-mail yuzan@cf.mbn.or.jp; internet www.nepto.co.jp/yuzankaku; f. 1916; history, fine arts, religion, archaeology; Pres. KEIKO NAGASAKA.

Zoshindo Juken Kenkyusha Co Ltd: 2-19-15, Shinmachi, Nishi-ku, Osaka 550-0013; tel. (6) 6532-1581; fax (6) 6532-1588; e-mail zoshindo@mbox.inet-osaka.or.jp; internet www.zoshindo.co.jp; f. 1890; educational, juvenile; Pres. AKITAKA OKAMATO.

Government Publishing House

Government Publications' Service Centre: 1-2-1, Kasumigaseki, Chiyoda-ku, Tokyo 100-0013; tel. (3) 3504-3885; fax (3) 3504-3889.

PUBLISHERS' ASSOCIATIONS

Japan Book Publishers Association: 6, Fukuro-machi, Shinjuku-ku, Tokyo 162-0828; tel. (3) 3268-1301; fax (3) 3268-1196; internet www.jbpa.or.jp; f. 1957; 499 mems; Pres. TAKAO WATANABE; Exec. Dir TOSHIKAZU GOMI.

Publishers' Association for Cultural Exchange, Japan: 1-2-1, Sarugaku-cho, Chiyoda-ku, Tokyo 101-0064; tel. (3) 3291-5685; fax (3) 3233-3645; e-mail office@pace.or.jp; internet www.pace.or.jp; f. 1953; 135 mems; Pres. Dr TATSURO MATSUMAE; Man. Dir YASUKO KORENAGA.

Broadcasting and Communications

TELECOMMUNICATIONS

International Digital Communications: 5-20-8, Asakusabashi, Taito-ku, Tokyo 111; tel. (3) 5820-5080; fax (3) 5820-5363; f. 1985; 53% owned by Cable and Wireless Communications (UK); Pres. SIMON CUNNINGHAM.

Japan Telecom Co Ltd: 4-7-1, Hatchobori, Chuo-ku, Tokyo 104-8508; tel. (3) 5540-8417; fax (3) 5540-8485; internet www.japan-telecom.co.jp; fixed-line business acquired by Ripplewood Holdings Aug. 2003; acquired by Softbank Corpn in 2004; Chair. HARUO MURAKAMI; Pres. HIDEKI KURASHIGE.

KDDI Corpn: KDDI Bldg, 2-3-2, Nishi Shinjuku, Shinjuku-ku, Tokyo 163-03; tel. (3) 3347-7111; fax (3) 3347-6470; internet www.kddi.com; f. 2000 by merger of DDI Corpn, Kokusai Denshin Denwa Corpn (KDD) and Nippon Idou Tsushin Corpn (IDO); major international telecommunications carrier; Pres. TADASHI ONODERA.

Nippon Telegraph and Telephone Corpn (NTT): 2-3-1, Otemachi, Chiyoda-ku, Tokyo 100-0004; tel. (3) 5359-2122; internet www.ntt.co.jp; operates local, long-distance and international services; largest telecommunications co in Japan; Pres.and CEO NORIO WADA.

NTT DoCoMo: 2-11-1 Nagatacho, Chiyoda-ku, Tokyo 100-6150; tel. (3) 5156-1111; fax (3) 5156-0271; internet www.nttdocomo.com; f. 1991; operates mobile phone network; Pres. MASAO NAKAMURA.

Tokyo Telecommunication Network Co Inc: 4-9-25, Shibaura, Minato-ku, Tokyo 108; tel. (3) 5476-0091; fax (3) 5476-7625.

KDDI, Digital Phone and Digital TU-KA also operate mobile telecommunication services in Japan.

BROADCASTING

NHK (Japan Broadcasting Corporation): 2-2-1, Jinnan, Shibuya, Tokyo 150-8001; tel. (3) 3465-1111; fax (3) 3469-8110; e-mail webmaster@www.nhk.or.jp; internet www.nhk.or.jp; f. 1925; fmrly Nippon Hose Kyokai, NHK (Japan Broadcasting Corpn); Japan's sole public broadcaster; operates five TV channels (incl. two terrestrial services—general TV and educational TV, two digital satellite services—BS-1 and BS-2 and a digital Hi-Vision service—HDTV), three radio channels, Radio 1, Radio 2, and FM Radio, and three worldwide services, NHK World TV, NHK World Premium and NHK World Radio Japan; headquarters in Tokyo, regional headquarters in Osaka, Nagoya, Hiroshima, Fukuoka, Sendai, Sapporo and Matsuyama; Pres. KATSUJI EBISAWA.

National Association of Commercial Broadcasters in Japan (NAB-J): 3-23, Kioi-cho, Chiyoda-ku, Tokyo 102-8577; tel. (3) 5213-7727; fax (3) 5213-7730; internet www.nab.or.jp; f. 1951; asscn of 201 companies (133 TV cos, 110 radio cos). Among these companies, 42 operate both radio and TV, with 664 radio stations and 8,315 TV stations (incl. relay stations); Pres. SEIICHIRO UJIIE; Exec. Dir AKIRA SAKAI.

In June 2000 there were a total of 99 commercial radio broadcasting companies and 127 commercial television companies operating in Japan.

Some of the most important companies are:

Asahi Hoso—Asahi Broadcasting Corpn: 2-2-48, Ohyodominami, Kita-ku, Osaka 531-8501; tel. (6) 6458-5321; fax (6) 6458-3672; internet www.asahi.co.jp; Pres. TOSHIHARU SHIBATA.

Asahi National Broadcasting Co Ltd—TV Asahi: 6-9-1, Roppongi, Minato-ku, Tokyo 106-8001; tel. (3) 6406-1275; fax (3) 3405-3714; internet www.tv-asahi.co.jp; f. 1957; Pres. MICHISADA HIROSE.

Bunka Hoso—Nippon Cultural Broadcasting, Inc: 1-5, Wakaba, Shinjuku-ku, Tokyo 160-8002; tel. (3) 3357-1111; fax (3) 3357-1140; internet www.joqr.co.jp; f. 1952; Pres. SHIGEKI SATO.

Chubu-Nippon Broadcasting Co Ltd: 1-2-8, Shinsakae, Naka-ku, Nagoya 460-8405; tel. (052) 241-8111; internet hicbc.com.co.jp; Pres. KEN-ICHI YOKOYAMA.

Fuji Television Network, Inc: 2-4-8, Daiba, Minato-ku, Tokyo 137-8088; tel. (3) 5500-8888; fax (3) 5500-8027; internet www.fujitv.co.jp; f. 1959; Pres. HISASHI HIEDA.

Kansai TV Hoso (KTV)—Kansai: 2-1-7, Ogimachi, Kita-ku, Osaka 530-8408; tel. (6) 6314-8888; internet www.ktv.co.jp; Pres. MICHIO IZUMA.

Mainichi Hoso (MBS)—Mainichi Broadcasting System, Inc: 17-1, Chayamachi, Kita-ku, Osaka 530-8304; tel. (6) 6359-1123; fax (6) 6359-3503; internet mbs.co.jp; Pres. MASAHIRO YAMAMOTO.

Nippon Hoso—Nippon Broadcasting System, Inc: 2-4-8, Daiba, Minato-ku, Tokyo 137-8686; tel. (3) 5500-1234; internet www.1242.com; f. 1954; Pres. MICHIYASU KAWAUCHI.

Nippon TV Hoso-MO (NTV)—Nippon Television Network Corpn: 1-6-1 Higashi Shimbashi, Minato-ku, Tokyo 105-7444; tel. (3) 6215-3156; fax (3) 6215-3157; internet www.ntv.co.jp; f. 1953; Pres. SEIICHIRO UJIIE.

Okinawa TV Hoso (OTV)—Okinawa Television Broadcasting Co Ltd: 1-2-20, Kumoji, Naha 900-8588; tel. (988) 63-2111; fax (988) 61-0193; internet www.otv.co.jp; f. 1959; Pres. BUNKI TOMA.

Radio Tampa—Nihon Short-Wave Broadcasting Co: 1-9-15, Akasaka, Minato-ku, Tokyo 107-8373; tel. (3) 3583-8151; fax (3) 3583-7441; internet www.tampa.co.jp; f. 1954; Pres. TAMIO IKEDA.

Ryukyu Hoso (RBC)—Ryukyu Broadcasting Co: 2-3-1, Kumoji, Naha 900-8711; tel. (98) 867-2151; fax (98) 864-5732; internet www.rbc-ryukyu.co.jp; f. 1954; Pres. YOSHIO ISHIGAKE.

TV Osaka (TVO)—Television Osaka, Inc: 1-2-18, Otemae, Chuo-ku, Osaka 540-8519; tel. (6) 6947-0019; fax (6) 6946-9796; internet www.tv-osaka.co.jp; f. 1982; Pres. MAKOTO FUKAGAWA.

TV Tokyo (TX)—Television Tokyo Channel 12 Ltd: 4-3-12, Toranomon, Minato-ku, Tokyo 105-8012; tel. (3) 3432-1212; fax (3) 5473-3447; internet www.tv-tokyo.co.jp; f. 1964; Pres. YUTAKA ICHIKI.

Tokyo -Hoso (TBS)—Tokyo Broadcasting System, Inc: 5-3-6, Akasaka, Minato-ku, Tokyo 107-8006; tel. (3) 3746-1111; fax (3) 3588-6378; internet www.tbs.co.jp/index.html; f. 1951; Chair. HIROSHI SHIHO; Pres. YUKIO SUNAHARA.

Yomiuri TV Hoso (YTV)—Yomiuri Telecasting Corporation: 2-2-33, Shiromi, Chuo-ku, Osaka 540-8510; tel. (6) 6947-2111; internet www.ytv.co.jp; f. 1958; 20 hrs colour broadcasting daily; Pres. TOMONARI DOI.

Satellite, Cable and Digital Television

In addition to the two broadcast satellite services that NHK introduced in 1989, a number of commercial satellite stations are in operation. Cable television is available in many urban areas, and in 1996/97 there were some 12.6m. subscribers to cable services in Japan. Satellite digital television services, which first became available in 1996, are provided by Japan Digital Broadcasting Services (f. 1998 by the merger of PerfecTV and JSkyB—now Sky Perfect Communications Inc) and DirecTV. Terrestrial digital broadcasting was launched in Japan in December 2003.

Finance

(cap. = capital; p.u. = paid up; res = reserves; dep. = deposits; m. = million; brs = branches; amounts in yen)

BANKING

Japan's central bank and bank of issue is the Bank of Japan. More than one-half of the credit business of the country is handled by 131 private commercial banks, 32 trust banks and three long-term credit banks, collectively designated 'All Banks'.

Of the private commercial banks, the most important are the eight city banks, some of which have a long and distinguished history, originating in the time of the *zaibatsu*, the private entrepreneurial organizations on which Japan's capital wealth was built before the Second World War. Although the *zaibatsu* were abolished as integral industrial and commercial enterprises during the Allied Occupation, the several businesses and industries that bear the former *zaibatsu* names, such as Mitsubishi, Mitsui and Sumitomo, continue to operate and to give each other mutual assistance through their respective banks and trust corporations.

Among the commercial banks, the Bank of Tokyo-Mitsubishi specializes in foreign-exchange business, while the Industrial Bank of Japan (now part of Mizuho Bank Ltd) finances capital investment by industry. Shinsei Bank and Nippon Credit Bank (now Aozora Bank) also specialize in industrial finance; the work of these three privately-owned banks is supplemented by the government-controlled Development Bank of Japan.

The Government has established a number of other specialized institutions to provide services that are not offered by the private banks. Thus the Japan Bank for International Cooperation advances credit for the export of heavy industrial products and the import of raw materials in bulk. A Housing Loan Corporation assists firms in building housing for their employees, while the Agriculture, Forestry and Fisheries Finance Corporation provides loans to the named industries for equipment purchases. Similar services are provided for small enterprises by the Japanese Finance Corporation for Small Business (now the Japanese Finance Corporation for Small and Medium Enterprise).

An important financial role is played by co-operatives and by the many small enterprise institutions. Each prefecture has its own federation of co-operatives, with the Central Co-operative Bank of Agriculture and Forestry as the common central financial institution. This bank also acts as an agent for the government-controlled Agriculture, Forestry and Fisheries Finance Corporation.

There are also two types of private financial institutions for small business. There were 277 Credit Co-operatives, with total assets of 22,000,000m. yen, and 369 Shinkin Banks (credit associations), which lend only to members. The latter also receive deposits.

The most common form of savings is through the government-operated Postal Savings System, which collects small savings from the public by means of the post office network. Total deposits amounted to 248,000,000m. yen in November 1998. The funds thus made available are used as loan funds by government financial institutions, through the Ministry of Finance's Trust Fund Bureau. In 2004 plans were announced for privatization of Japan Post, which was expected to be split into four companies by April 2007.

In June 1998 the Financial Supervisory Agency (now the Financial Services Agency) was established to regulate Japan's financial institutions.

Central Bank

Nippon Ginko (Bank of Japan): 2-1-1, Hongoku-cho, Nihombashi, Chuo-ku, Tokyo 100-8630; tel. (3) 3279-1111; fax (3) 5200-2256; e-mail prd@info.boj.or.jp; internet www.boj.or.jp; f. 1882; cap. and res 2,404,840m., dep. 64,278,090m. (March 2003); Gov. Toshihiko Fukui; Dep. Govs Toshiro Muto, Kazumasa Iwata; 32 brs.

Principal Commercial Banks

Ashikaga Bank Ltd: 4-1-25, Sakura, Utsunomiya, Tochigi 320-8610; tel. (286) 22-0111; e-mail ashigin@ssctnet.or.jp; internet www.ashikagabank.co.jp; f. 1895; nationalized Nov. 2003 owing to insolvency; cap. 147,429m., res 1,859m., dep. 5,055,042m. (March 2003); Pres. Norito Ikeda; 140 brs.

Bank of Fukuoka Ltd: 2-13-1, Tenjin, Chuo-ku, Fukuoka 810-8727; tel. (92) 723-2591; fax (92) 711-1746; f. 1945; cap. 58,658m., res 245,835m., dep. 6,411,752m. (March 2003); Chair. Kiyoshi Teramoto; Pres. Ryoji Tsukuda.

Bank of Tokyo-Mitsubishi Ltd: 2-7-1, Marunouchi, Chiyoda-ku, Tokyo 100-8388; tel. (3) 93240-1111; fax (3) 93240-4197; internet www.btm.co.jp; subsidiary of Mitsubishi Tokyo Financial Group; f. 1996 as a result of merger between Bank of Tokyo Ltd (f. 1946) and Mitsubishi Bank Ltd (f. 1880); specializes in international banking and financial business; agreement signed in Aug. 2004 for merger of Mitsubishi Tokyo Financial Group with UFJ Holdings by Oct. 2005; cap. 871,973m., res 819,583m., dep. 64,347,975m. (March 2003); Chair. Satoru Kishi; Pres. Shigemitsu Miki; 805 brs.

Bank of Yokohama Ltd: 3-1-1, Minatomirai, Nishi-ku, Yokohama, Kanagawa 220-8611; tel. (45) 225-1111; fax (45) 225-1160; e-mail iroffice@hamagin.co.jp; internet www.boy.co.jp; f. 1920; cap. 184,803m., res 193,371m., dep. 9,663,689m. (March 2003); Pres. and CEO Sadaaki Hirasawa; 184 brs.

Chiba Bank Ltd: 1-2, Chiba-minato, Chuo-ku, Chiba 260-8720; tel. (43) 245-1111; e-mail int@chibabank.co.jp; internet www.chibabank.co.jp; f. 1943; cap. 121,019m., res 189,524m., dep. 7,463,569m. (March 2003); Dir Kenji Yasui; Pres. Tsuneo Hayakawa; 163 brs.

Hachijuni Bank: 178-8 Okada, Nagano-shi, Nagano 380-8682; tel. (26) 227-1182; fax (26) 226-5077; internet www.82bank.co.jp; f. 1931; cap. 52,243m., res 78,334m., dep. 5,225,771m. (March 2003); Pres. Kazuyuki Narusawa.

Hokuriku Bank Ltd: 1-2-26, Tsutsumichodori, Toyama 930-8637; tel. (764) 237-111; fax (764) 915-908; e-mail kokusaibu@hokugin.co.jp; internet www.hokugin.co.jp; f. 1877; cap. 140,409m., res 14,667m., dep. 5,245,753m. (March 2003); Pres. Shigeo Takagi; 191 brs.

Japan Net Bank: internet www.japannetbank.co.jp; f. 2000; Japan's first internet-only bank.

Joyo Bank Ltd: 2-5-5, Minami-machi, Mito-shi, Ibaraki 310-0021; tel. (29) 231-2151; fax (29) 255-6522; e-mail joyointl@po.net-ibaraki.ne.jp; internet www.joyobank.co.jp; f. 1935; cap. 85,113m., res 263,511m., dep. 6,203,315m. (March 2003); Chair. Toranosuke Nishino; Pres. Isao Shibuya; 175 brs.

Mizuho Bank Ltd: 1-5 Uchisaiwai-cho, 1-chrome, Chiyodu-ku, Tokyo 100-0011; tel. (3) 3596-1111; fax (3) 3596-2179; internet www.mizuhobank.co.jp; f. 1971 as Dai-Ichi Kangyo Bank; merged with Fuji Bank and Industrial Bank of Japan to form above in 2002; cap. 650,000m., res 1,598,666m., dep. 63,418,353m. (March 2003); Pres. and CEO Katsuyuki Sugita; 334 brs.

North Pacific Bank (Hokuyo Bank): 3-11 Odori Nishi, Chuo-ku, Sapporo 060-8661; tel. (11) 261-1416; fax (11) 232-6921; internet www.hokuyobank.co.jp; f. 1917 as Hokuyo Sogo Bank Ltd; assumed present name in 1989; cap. 54,993m., res 64,662m., dep. 5,404,646m. (March 2003); Chair. Masanao Takei; Pres. Iwao Takamuki.

Resona Bank Ltd: 2-2-1, Bingo-machi, Chuo-ku, Osaka 540-8610; tel. (6) 6271-1221; internet www.resona-gr.co.jp; f. 1918; merged with Asahi Bank in 2002 and changed name as above; cap. 443,158m., res 325,013m., dep. 28,903,844m. (March 2003); Chair. Eiji Hosoya; Pres. Masaaki Nomura; 367 brs.

Shizuoka Bank Ltd: 1-10, Gofuku-cho, Shizuoka 420-8760; tel. (54) 261-3131; fax (54) 344-0090; internet www.shizuokabank.co.jp; f. 1943; cap. 90,845m., res 69,975m., dep. 7,259,476m. (March 2003); Chair. Soichiro Kamiya; Pres. Yasuo Matsuura; 199 brs.

Sumitomo Mitsui Banking Corpn: 1-2 Yuraku-cho, Chiyoda-ku, Tokyo 100-0006; tel. (3) 3230-8811; fax (3) 3239-4170; internet www.smbc.co.jp; f. 1895; merged with Sakura Bank Ltd in April 2001 and assumed present name; cap. 559,985m., res 1,323,869m., dep. 88,691,977m. (March 2003); Chair. Akishige Okada; Pres. Yoshifumi Nishikawa; 351 brs.

Tokyo Star Bank: 1-6-16 Akasaka, Minato-ku, Tokyo; tel. (3) 3586-3111; fax (3) 3582-8949; internet www.tokyostarbank.co.jp; f. 1950 as Tokyo Sogo Bank, name changed as above in 2001; cap. 21,000m., res 19,080m., dep. 1,093,419m. (March 2003); Chair. Shoichi Osada; Pres. Todd Budge; 100 brs.

UFJ Bank Ltd: 3-5-6, Fushimi-machi, Chuo-ku, Osaka-shi, Osaka 541-8530; tel. (6) 6206-8111; fax (6) 6299-1066; internet www.ufj.co.jp; subsidiary of UFJ Holdings, Inc.; f. April 2001 following merger of Sanwa Bank, Tokai Bank, and Toyo Trust and Banking; agreement signed in Aug. 2004 for merger of UFJ Holdings with Mitsubishi Tokyo Financial Group by Oct. 2005; cap. 843,582m., res 1,469,757m., dep. 61,028,460m. (March 2003); Pres.and CEO Takamune Okihara.

Principal Trust Banks

Chuo Mitsui Trust and Banking Co Ltd: 3-33-1, Shiba, Minato-ku, Chuo-ku, Tokyo 105-8574; tel. (3) 5232-3331; fax (3) 5232-8879; internet www.chuomitsui.co.jp; f. 1962 as Chuo Trust and Banking Co Ltd, name changed as above in 2000, following merger with Mitsui Trust and Banking Co Ltd; cap. 349,894m., res 53,419m., dep. 9,650,394m. (March 2003); Pres. Kazuo Tanabe; 169 brs.

Mitsubishi Trust and Banking Corporation: 1-4-5 Marunouchi, Chiyoda-ku, Tokyo 100-8212; tel. (3) 3212-1211; fax (3) 3514-6660; internet www.mitsubishi-trust.co.jp; subsidiary of Mitsubishi Tokyo Financial Group; f. 1927; absorbed Nippon Trust Bank Ltd and Tokyo Trust Bank Ltd in Oct. 2001; agreement signed in Aug. 2004 for merger of Mitsubishi Tokyo Financial Group with UFJ Holdings by Oct. 2005; cap. 324,279m., res 264,428m., dep. 16,848,004m. (March 2003); Chair. Toyoshi Nakano; Pres. Akio Utsumi; 54 brs.

Mizuho Trust and Banking Co Ltd: 1-2-1, Yaesu, Chuo-ku, Tokyo 103-8670; tel. (3) 3278-8111; fax (3) 3281-6947; internet www.mizuho-tb.co.jp; fmrly Yasuda Trust and Banking Co Ltd; f. 1925; cap. 247,231m., res 118,487m., dep. 4,200,412m. (March 2003); Chair. Takahiko Kiminami; Pres. and CEO Hiroaki Etoh; 50 brs.

Sumitomo Trust and Banking Co Ltd: 4-5-33, Kitahama, Chuo-ku, Osaka 540-8639; tel. (6) 6220-2121; fax (6) 6220-2043; e-mail ipda@sumitomotrust.co.jp; internet www.sumitomotrust.co.jp; f. 1925; cap. 287,015m., res 272,021m., dep. 12,587,229m. (March 2003); Chair. Hitoshi Murakami; Pres. Atsushi Takahashi; 57 brs.

UFJ Trust Bank: 1-4-3, Marunouchi, Chiyoda-ku, Tokyo 100-0005; tel. (3) 3287-2211; fax (3) 3201-1448; internet www.ufjtrustbank.co.jp; subsidiary of UFJ Holdings, Inc; f. 1959 as Toyo Trust and Banking; merged with Sanwa Bank Ltd and Tokai Bank Ltd in April 2001 (see above); agreement signed in Aug. 2004 for merger of UFJ Holdings with Mitsubishi Tokyo Financial Group by Oct. 2005; cap. 280,536m., res 19,966m., dep. 4,270,193m. (March 2003); Pres. Shintaro Yasuda; 56 brs.

Long-Term Credit Banks

Aozora Bank: 3-1, Kudan-minami, Chiyoda-ku, Tokyo 102-8660; tel. (3) 3263-1111; fax (3) 3265-7024; e-mail sora@aozora.co.jp; internet www.aozorabank.co.jp; f. 1957; nationalized Dec. 1998, sold to consortium led by Softbank Corpn in Aug. 2000; fmrly The Nippon Credit Bank, name changed as above 2001; 62% owned by Cerberus Group; cap. 419,781m., res 34,439m., dep. 4,954,245m. (March 2003); Pres. and CEO Hiroshi Maruyama; 17 brs.

Mizuho Corporate Bank Ltd (The Industrial Bank of Japan Ltd): 3-3, Marunouchi, Chiyoda-ku, Tokyo 100-8210; tel. (3) 3214-1111; fax (3) 3201-7643; internet www.mizuhocbk.co.jp; f. 1902; renamed as above in 2002 following merger of the Dai-Ichi Kangyo Bank, the Fuji Bank and the Industrial Bank of Japan; medium- and long-term financing; cap. 1,070,965m., res 1,883,632m., dep. 51,879,870m. (March 2003); Pres. and CEO Hiroshi Saito; 23 domestic brs, 20 overseas brs.

Shinsei Bank Ltd: 2-1-8, Uchisaiwai-cho, Chiyoda-ku, Tokyo 100-8501; tel. (3) 5511-5111; fax (3) 5511-5505; internet www.shinseibank.co.jp; f. 1952 as The Long-Term Credit Bank of Japan; nationalized Oct. 1998, sold to Ripplewood Holdings (USA), renamed as above June 2000; cap. 180,853m., res 304,956m., dep. 5,171,864m. (March 2003); Chair. Masamoto Yashiro; 23 brs.

Co-operative Bank

Shinkin Central Bank: 3-8-1, Kyobashi, Chuo-ku, Tokyo 104-0031; tel. (3) 3563-4111; fax (3) 3563-7553; internet www.shinkin.co.jp; f. 1950; cap. 290,998m., res 494,202m. (March 2001), dep. 25,526,993m. (March 2003); Chair. Yukihiko Nagano; Pres. Yasutaka Miyamoto; 17 brs.

Principal Government Credit Institutions

Agriculture, Forestry and Fisheries Finance Corporation: Koko Bldg, 1-9-3, Otemachi, Chiyoda-ku, Tokyo 100-0004; tel. (3) 3270-2261; e-mail intl@afc.go.jp; internet www.afc.go.jp; f. 1953;

finances mainly plant and equipment investment; Gov. TOSHIHIKO TSURUOKA; Dep. Gov. SHIGEO OHARA; 22 brs.

Development Bank of Japan: 1-9-1, Otemachi, Chiyoda-ku, Tokyo 100-0004; tel. (3) 3244-1770; fax (3) 3245-1938; e-mail safukas@dbj.go.jp; internet www.dbj.go.jp; f. 1951 as the Japan Development Bank; renamed Oct. 1999 following consolidation with the Hokkaido and Tohoku Development Finance Public Corpn; provides long-term loans; subscribes for corporate bonds; guarantees corporate obligations; invests in specific projects; borrows funds from Govt and abroad; issues external bonds and notes; provides market information and consulting services for prospective entrants to Japanese market; cap. 1,182,286m., res 1,000,908m., dep. 1,596,630m. (March 2003); Gov. TAKESHI KOMURA; Dep. Govs NORITADA TERASAWA, KIMIO YAMAGUCHI; 10 domestic brs, 6 overseas brs.

Housing Loan Corporation: 1-4-10, Koraku, Bunkyo-ku, Tokyo 112-8570; tel. (3) 3812-1111; fax (3) 5800-8257; internet www.jyukou.go.jp; f. 1950 to provide long-term capital for the construction of housing at low interest rates; cap. 97,200m. (1994); plans announced by Japanese Government to abolish the Housing Loan Corporation in 2006; Pres. SUSUMU TAKAHASHI; Vice-Pres. HIROYUKI ITOU; 12 brs.

Japan Bank for International Cooperation (JBIC): 1-4-1, Otemachi, Chiyoda-ku, Tokyo 100-8144; tel. (3) 5218-3101; fax (3) 5218-3955; internet www.jbic.go.jp; f. 1999 by merger of The Export-Import Bank of Japan (f. 1950) and The Overseas Economic Co-operation Fund (f. 1961); governmental financial institution, responsible for Japan's external economic policy and co-operation activities; cap. 7,489,844m., res 889,055m., dep. 1,589,084m. (March 2003); Gov. KYOSUKE SHINOZAWA.

Japan Finance Corporation for Small and Medium Enterprise: 9-3, Otemachi 1-chome, Chiyoda-ku, Tokyo 100-0004; tel. (3) 3270-0505; fax (3) 3279-5910; internet www.jasme.go.jp; f. 1953 to promote long-term growth and development of small businesses by providing the necessary funds and information on their use in accordance with national policy; cap. 433,715m. (Jan. 2002); wholly subscribed by Govt; Gov. TOMIO TSUTSUMI; Vice-Gov. SOHEI HIDAKA; 58 brs.

National Life Finance Corporation: Koko Bldg, 1-9-3, Otemachi, Chiyoda-ku, Tokyo 100-0004; tel. (3) 3270-1361; internet www.kokukin.go.jp; f. 1999 following consolidation of The People's Finance Corpn (f. 1949 to provide business funds, particularly to small enterprises unable to obtain loans from banks and other private financial institutions) and the Environmental Sanitation Business Finance Corpn (f. 1967 to improve sanitary facilities); cap. 347,971m. (July 2003); Gov. NOBUAKI USUI; 152 brs.

Norinchukin Bank (Central Co-operative Bank for Agriculture, Forestry and Fisheries): 1-13-2, Yuraku-cho, Chiyoda-ku, Tokyo 100; tel. (3) 3279-0111; fax (3) 3218-5177; internet www.nochubank.or.jp; f. 1923; main banker to agricultural, forestry and fisheries co-operatives; receives deposits from individual co-operatives, federations and agricultural enterprises; extends loans to these and to local govt authorities and public corpns; adjusts excess and shortage of funds within co-operative system; issues debentures, invests funds and engages in other regular banking business; cap. 1,224,999m., res 419,329m., dep. 55,072,310m. (March 2003); Pres. HIROFUMI UENO; 39 brs.

Shoko Chukin Bank (Central Co-operative Bank for Commerce and Industry): 2-10-17, Yaesu, Chuo-ku, Tokyo 104-0028; tel. (3) 3272-6111; fax (3) 3272-6169; e-mail JDK06560@nifty.ne.jp; internet www.shokochukin.go.jp; f. 1936 to provide general banking services to facilitate finance for smaller enterprise co-operatives and other organizations formed mainly by small- and medium-sized enterprises; issues debentures; cap. 511,265m., res 118,869m., dep. 11,797,809m. (March 2003); Pres. TADASHI EZAKI; Dep. Pres. SHIGENORI SHIODA; 99 brs.

Other government financial institutions include the Japan Finance Corpn for Municipal Enterprises, the Small Business Credit Insurance Corpn and the Okinawa Development Finance Corpn.

Principal Foreign Banks

In 2004 there were 84 foreign banks operating in Japan.

ABN AMRO Bank NV (Netherlands): Atago Green Hills MORI Tower, 32nd Floor, 2-5-1, Atago, Minato-ku, Tokyo 105-6231; tel. (3) 5405-6500; fax (3) 5405-6900; Br. Man. ATSUSHI WATANABE.

Bangkok Bank Public Co Ltd (Thailand): Bangkok Bank Bldg, 2-8-10, Nishi Shinbashi, Minato-ku, Tokyo 105-0003; tel. (3) 3503-3333; fax (3) 3502-6420; Senior Vice-Pres. and Gen. Man. (Japan) THAWEE PHUANGKETKEOW; br. in Osaka.

Bank of America NA: Sanno Park Tower, 15th Floor, 2-11-1, Nagatacho, Chiyoda-ku, Tokyo 100-6115; tel. (3) 3508-5800; fax (3) 3508-5811.

Bank of India: Mitsubishi Denki Bldg, 2-2-3, Marunouchi, Chiyoda-ku, Tokyo 100-0005; tel. (3) 3212-0911; fax (3) 3214-8667; e-mail boitok@gol.com; CEO (Japan) P. SIVARAMAN; br. in Osaka.

Bank Negara Indonesia (Persero): Rm 117-18, Kokusai Bldg, 3-1-1, Marunouchi, Chiyoda-ku, Tokyo 100-0005; tel. (3) 3214-5621; fax (3) 3201-2633; e-mail tky-br@ptbni.co.jp; Gen. Man. SURYO DANISWORO.

Bank One NA (USA): Hibiya Central Bldg, 7th Floor, 1-2-9, Nishi Shimbashi, Minato-ku, Tokyo 105; tel. (3) 3596-8700; fax (3) 3596-8744; Sr Vice-Pres. and Gen. Man. YOSHIO KITAZAWA.

Barclays Bank PLC (UK): Urbannet Otemachi Bldg, 15th Floor, 2-2-2, Otemachi, Chiyoda-ku, Tokyo 100-0004; tel. (3) 3276-5100; fax (3) 3276-5085; CEO ANDY SIMMONDS.

Bayerische Hypo- und Vereinsbank AG (Germany): Otemachi 1st Sq. East Tower, 17th Floor, 1-5-1, Otemachi, Chiyoda-ku, Tokyo 100-0004; tel. (3) 3284-1341; fax (3) 3284-1370; Exec. Dirs Prof. PETER BARON, KENJI AKAGI.

BNP Paribas (France): Tokyo Sankei Bldg, 22rd Floor, 1-7-2, Otemachi, Chiyoda-ku, Tokyo 100-0004; tel. (3) 5290-1000; fax (3) 5290-1111; internet www.bnpparibas.co.jp; CEO (Japan) ERIC MARTIN; Representative in Japan HIROAKI INOUE; br. in Osaka.

Citibank NA (USA): Pan Japan Bldg, 1st Floor, 3-8-17, Akasaka Minato-ku, Tokyo 107; tel. (3) 3584-6321; fax (3) 3584-2924; Country Corporate Officer MASAMOTO YASHIRO; 20 brs.

Commerzbank AG (Germany): Nippon Press Center Bldg, 2nd Floor, 2-2-1, Uchisaiwai-cho, Chiyoda-ku, Tokyo 100-0011; tel. (3) 3502-4371; fax (3) 3508-7545; Gen. Man. NORIO YATOMI.

Crédit Agricole Indosuez (France): Indosuez Bldg, 3-29-1, Kanda Jimbo-cho, Chiyoda-ku, Tokyo 101; tel. (3) 3261-3001; fax (3) 3261-0426; Sr Country Exec. FRANÇOIS BEYER.

Deutsche Bank AG (Germany): Sanno Park Tower, 2-11-1 Nagatacho, Chiyoda-ku, Tokyo 100-6170; tel. (3) 5156-4000; fax (3) 5156-6070; CEO and Chief Country Officer JOHN MACFARLANE; brs in Osaka and Nagoya.

The Hongkong and Shanghai Banking Corpn Ltd (Hong Kong): HSBC Bldg, 3-11-1, Nihombashi, Chuo-ku, Tokyo 103-0027; tel. (3) 5203-3000; fax (3) 5203-3108; CEO NORMAN A. WILSON; br. in Osaka.

International Commercial Bank of China (Taiwan): Togin Bldg, 1-4-2, Marunouchi, Chiyoda-ku, Tokyo 100; tel. (3) 3211-2501; fax (3) 3216-5686; Sr Vice-Pres. and Gen. Man. SHIOW-SHYONG LAI; br. in Osaka.

JP Morgan Chase Bank (USA): Akasaka Park Bldg, 11th–13th Floors, 5-2-20, Akasaka, Minato-ku, Tokyo 107; tel. (3) 5570-7500; fax (3) 5570-7960; Man. Dir and Gen. Man. NORMAN J. T. SCOTT; br. in Osaka.

Korea Exchange Bank (Republic of Korea): Shin Kokusai Bldg, 3-4-1, Marunouchi, Chiyoda-ku, Tokyo 100; tel. (3) 3216-3561; fax (3) 3214-4491; f. 1967; Acting Gen. Man. CHO YOUNG-HYO; brs in Osaka and Fukuoka.

Lloyds TSB Bank PLC (UK): Akasaka Twin Tower, New Tower 5th Floor, 11-7, Akasaka 2-chome, Minato-ku, Tokyo 107-0052; tel. (3) 3589-7700; fax (3) 3589-7722; e-mail tokyo@lloydstsb.co.jp; internet www.lloydstsb.co.jp; Principal Man. (Japan) KAH HIN LIM.

Morgan Guaranty Trust Co of New York (USA): Akasaka Park Bldg, 5-2-20, Akasaka, Minato-ku, Tokyo 107-6151; tel. (3) 5573-1100; Man. Dir TAKESHI FUJIMAKI.

National Bank of Pakistan: S. K. Bldg, 3rd Floor, 2-7-4, Nishi Shimbashi, Minato-ku, Tokyo 105; tel. (3) 3502-0331; fax (3) 3502-0359; f. 1949; Gen. Man. ZIAULLAH KHAN.

Oversea-Chinese Banking Corpn Ltd (Singapore): Akasaka Twin Tower, 15th Floor, 2-17-22, Akasaka, Minato-ku, Tokyo 107-0052; tel. (3) 5570-3421; fax (3) 5570-3426; Gen. Man. ONG SING YIK.

Société Générale (France): Ark Mori Bldg, 1-12-32, Akasaka, Minato-ku, Tokyo 107-6014; tel. (3) 5549-5800; fax (3) 5549-5809; Chief Operating Officer SHOZO NURISHI; br. in Osaka.

Standard Chartered Bank (UK): 21st Floor, Sanno Park Tower, 2-11-1, Nagata-cho, Chiyoda-ku, Tokyo 100-6155; tel. (3) 5511-1200; fax (3) 5511-9333; Chief Exec. (Japan) JULIAN WYNTER.

State Bank of India: 352 South Tower, Yuraku-cho Denki Bldg, 1-7-1, Yuraku-cho, Chiyoda-ku, Tokyo 100-0006; tel. (3) 3284-0085; fax (3) 3201-5750; e-mail sbitok@gol.com; internet www.sbijapan.com; CEO J. K. SINHA; br. in Osaka.

UBS AG: Urbannet Otemachi Bldg, 2-2-2, Otemachi, Chiyoda-ku, Tokyo 100-0004; tel. (3) 5201-8585; fax (3) 5201-8099; Man. MITSURU TSUNEMI.

Union de Banques Arabes et Françaises (UBAF) (France): Sumitomo Jimbocho Bldg, 8th Floor, 3-25, Kanda Jimbocho, Chiyoda-ku, Tokyo 101-0051; tel. (3) 3263-8821; fax (3) 3263-8820;

e-mail antoine.homsy@ubaf.fr; Gen. Man. (Japan) ANTOINE R. HOMSY; br. in Osaka.

WestLB AG (Germany): Fukoku Seimei Bldg, 2-2-2, Uchisaiwaicho, Chiyoda-ku, Tokyo 100-0011; tel. (3) 5510-6200; fax (3) 5510-6299; internet www.westlb.co.jp; Man. Dir and Gen. Man. MICHAEL KRAMER.

Bankers' Associations

Japanese Bankers Association: 1-3-1, Marunouchi, Chiyoda-ku, Tokyo 100-8216; tel. (3) 3216-3761; fax (3) 3201-5608; internet www.zenginkyo.or.jp; f. 1945; fmrly Federation of Bankers Associations of Japan; 133 full mems, 49 associate mems, 65 special mems; Chair. YOSHIFUMI NISHIKAWA.

Tokyo Bankers Association, Inc: 1-3-1, Marunouchi, Chiyoda-ku, Tokyo 100-8216; tel. (3) 3216-3761; fax (3) 3201-5608; f. 1945; 109 mem. banks; conducts the above Association's administrative business; Chair. YOSHIFUMI NISHIKAWA.

National Association of Labour Banks: 2-5-15, Kanda Surugadai, Chiyoda-ku, Tokyo 101-0062; tel. (3) 3295-6721; fax (3) 3295-6752; Pres. TETSUEI TOKUGAWA.

Regional Banks Association of Japan: 3-1-2, Uchikanda, Chiyoda-ku, Tokyo 101-0047; tel. (3) 3252-5171; fax (3) 3254-8664; f. 1936; 64 mem. banks; Chair. SADAAKI HIROSAWA.

Second Association of Regional Banks: 5, Sanban-cho, Chiyoda-ku, Tokyo 102-0075; tel. (3) 3262-2181; fax (3) 3262-2339; f. 1989; fmrly National Asscn of Sogo Banks; 65 commercial banks; Chair. MASANAO TAKEI.

STOCK EXCHANGES

Fukuoka Securities Exchange: 2-14-2, Tenjin, Chuo-ku, Fukuoka 810-0001; tel. (92) 741-8231; internet www.fse.or.jp; Pres. FUBITO SHIMOMURA.

Hiroshima Stock Exchange: 14-18, Kanayama-cho, Naka-ku, Hiroshima 730-0022; tel. (82) 541-1121; f. 1949; 20 mems; Pres. MASARU NANKO.

Jasdaq Market: 1-14-8 Nihombashi-Ningyocho, Chuo-ku, Tokyo 103-0013; tel. (3) 5641-1818; internet www.jasdaq.co.jp.

Kyoto Securities Exchange: 66, Tachiuri Nishimachi, Shijodori, Higashitoin Higashi-iru, Shimogyo-ku, Kyoto 600-8007; tel. (75) 221-1171; Pres. IICHI NAKAMURA.

Nagoya Stock Exchange: 3-3-17, Sakae, Naka-ku, Nagoya 460-0008; tel. (52) 262-3172; fax (52) 241-1527; e-mail kikaku@nse.or.jp; internet www.nse.or.jp; f. 1949; Pres. HIROSHI FUJITA; Sr Exec. Dir KAZUNORI ISHIMOTO.

Nasdaq Japan Market: 23rd Floor, Akasaka, Minato-ku, Tokyo 107-6023; tel. (3) 5563-8210; internet www.nasdaq-japan.com.

Niigata Securities Exchange: 1245, Hachibancho, Kami-Okawamaedori, Niigata 951-8068; tel. (252) 222-4181; Pres. KYUUZOU NAKATA.

Osaka Securities Exchange: 1-6-10, Kitahama, Chuo-ku, Osaka 541-0041; tel. (6) 4706-0875; fax (6) 6231-2639; internet www.ose.or.jp; f. 1949; 103 regular mems, 5 special participants; Chair. HIROTARO HIGUCHI; Pres. GORO TATSUMI.

Sapporo Securities Exchange: 5-14-1, Nishi, Minami Ichijo, Chuo-ku, Sapporo 060-0061; tel. (11) 241-6171; Pres. YOSHIRO ITOH.

Tokyo Stock Exchange, Inc: 2-1, Nihombashi-kabuto-cho, Chuo-ku, Tokyo 103-8220; tel. (3) 3666-0141; fax (3) 3662-0547; internet www.tse.or.jp; f. 1949; 107 general trading participants, 79 bond futures trading participants, 2 stock index futures trading participants (Nov. 2003); Pres. and CEO MASAAKI TSUCHIDA; Sr Man. Dir and CFO SADAO YOSHINO.

Supervisory Body

The Securities and Exchange Surveillance Commission: 3-1-1, Kasumigaseki, Chiyoda-ku, Tokyo 100; tel. (3) 3581-7868; fax (3) 5251-2136; f. 1992 for the surveillance of securities and financial futures transactions; Chair. TOSHIHIRO MIZUHARA.

INSURANCE

Principal Life Companies

Aetna Heiwa Life Insurance Co Ltd: 3-2-16, Ginza, Chuo-ku, Tokyo 104-8119; tel. (3) 3563-8111; fax (3) 3374-7114; f. 1907; Pres. BARRY S. HALPERN.

American Family Life Assurance Co of Columbus AFLAC Japan: Shinjuku Mitsui Bldg, 12th Floor, 2-1-1, Nishishinjuku, Shinjuku-ku, Tokyo 163-0456; tel. (3) 3344-2701; fax (3) 0424-41-3001; f. 1974; Chair. YOSHIKI OTAKE; Pres. HIDEFUMI MATSUI.

American Life Insurance Co (Japan): 1-1-3, Marunouchi, Chiyoda-ku, Tokyo 100-0005; tel. (3) 3284-4111; fax (3) 3284-3874; f. 1972; Pres. TOMIO MIYAMOTO.

Aoba Life Insurance Co Ltd: 3-6-30, Aobadai, Meguro-ku, Tokyo 153-8523; tel. (3) 3462-0007; fax (3) 3780-8169; Pres. TAKASHI KASAGAMI.

Asahi Mutual Life Insurance Co: 1-7-3, Nishishinjuku, Shinjuku-ku, Tokyo 163-8611; tel. (3) 3342-3111; fax (3) 3346-9397; internet www.asahi-life.co.jp; f. 1888; Pres. YUZURU FUJITA.

AXA Japan Holding Co Ltd: 1-2-19 Higashi, Shibuya-ku, Tokyo 150-8020; tel. (3) 3407-6210; fax (3) 5466-7131; internet www.axa.co.jp; Pres. and CEO PHILIPPE DONNET.

Cardif Assurance Vie: 3-25-2, Toranomon, Minato-ku, Tokyo 105-0001; tel. (3) 5776-6230; fax (3) 5776-6236; f. 2000; Pres. ATSUSHI SAKAUCHI.

Chiyoda Mutual Life Insurance Co: 2-19-18, Kamimeguro, Meguro-ku, Tokyo 153-8611; tel. (3) 5704-5111; fax (3) 3719-6605; internet www.chiyoda-life.co.jp; f. 1904; declared bankrupt Oct. 2000; Pres. REIJI YONEYAMA.

Chiyodakasai EBISU Life Insurance Co Ltd: Ebisu MF Bldg, 6th Floor, 4-6-1, Ebisu Shibuya-ku, Tokyo 150-0013; tel. (3) 5420-8282; fax (3) 5420-8273; f. 1996; Pres. SHIGEJI MINOSHIMA.

Daido Life Insurance Co: 1-2-1, Edobori Nishi-ku, Osaka City, Osaka 550-0002; tel. (6) 6447-6111; fax (6) 6447-6315; f. 1902; Pres. NAOTERU MIYATO.

Daihyaku Mutual Life Insurance Co: 3-1-4, Shibuya, Shibuya-ku, Tokyo 150-8670; tel. (3) 3498-2294; fax (3) 3400-9313; e-mail kikaku@daihyaku-life.co.jp; internet www.daihyaku-life.co.jp; f. 1914; declared bankrupt June 2000; Pres. MITSUMASA AKIYAMI.

Dai-ichi Mutual Life Insurance Co: 1-13-1, Yuraku-cho, Chiyoda-ku, Tokyo 100-8411; tel. (3) 3216-1211; fax (3) 5221-8139; f. 1902; Chair. TAKAHIDE SAKURAI; Pres. TOMIJIRO MORITA.

Dai-Tokyo Happy Life Insurance Co Ltd: Shinjuku Square Tower, 17th Floor, 6-22-1, Nishishinjuku, Shinjuku-ku, Tokyo 163-1131; tel. (3) 5323-6411; fax (3) 5323-6419; f. 1996; Pres. HITOSHI HASUNUMA.

DIY Life Insurance Co Ltd: 5-68-2, Nakano, Nakano-ku, Tokyo 164-0001; tel. (3) 5345-7603; fax (3) 5345-7608; f. 1999; Pres. HITOSHI KASE.

Fuji Life Insurance Co Ltd: 1-18-17, Minamisenba, Chuo-ku, Osaka-shi 542-0081; tel. (6) 6261-0284; fax (6) 6261-0113; f. 1996; Pres. YOSHIAKI YONEMURA.

Fukoku Mutual Life Insurance Co: 2-2-2, Uchisaiwai-cho, Chiyoda-ku, Tokyo 100-0011; tel. (3) 3508-1101; fax (3) 3597-0383; f. 1923; Chair. TAKASHI KOBAYASHI; Pres. TOMOFUMI AKIYAMA.

GE Edison Life Insurance Co: Shibuya Markcity, 1-12-1, Dogenzaka, Shibuya-ku, Tokyo 150-8674; tel. (3) 5457-8100; fax (3) 5457-8017; acquired by American International Group (AIG) August 2003; Chair. MICHAEL D. FRAIZER; Pres. and CEO K. RONE BALDWIN.

Gibraltar Life Insurance Co Ltd: 4-4-1, Nihombashi, Hongoku-cho, Chuo-ku, Tokyo 103-0021; tel. (3) 3270-8511; fax (3) 3231-8276; internet www.gib-life.co.jp; f. 1947; fmrly Kyoei Life Insurance Co Ltd, declared bankrupt Oct. 2000; Pres. KAZUO MAEDA.

ING Life Insurance Co Ltd: 26th Floor, New Otani Garden Court, 4-1, Kioi-cho, Chiyoda-ku, Tokyo 102-0094; tel. (3) 5210-0300; fax (3) 5210-0430; f. 1985; Pres. MAKOTO CHIBA.

Koa Life Insurance Co Ltd: 3-7-3, Kasumigaseki, Chiyoda-ku, Tokyo 100-0013; tel. (3) 3593-3111; fax (3) 5512-6651; internet www.koa.co.jp; f. 1996; Pres. AKIO OKADA.

Kyoei Kasai Shinrai Life Insurance Co Ltd: J. City Bldg, 5-8-20, Takamatsu, Nerima-ku, Tokyo 179-0075; tel. (3) 5372-2100; fax (3) 5372-7701; f. 1996; Pres. YOSHIHIRO TOKUMITSU.

Manulife Life Insurance Co: 4-34-1, Kokuryo-cho, Chofu-shi, Tokyo 182-8621; tel. (3) 2442-7120; fax (3) 2442-7977; e-mail geoff_crickmay@manulife.com; internet www.manulife.co.jp; f. 1999; fmrly Manulife Century Life Insurance Co; Pres. and CEO GEOFF CRICKMAY.

Meiji Life Insurance Co: 2-1-1, Marunouchi, Chiyoda-ku, Tokyo 100-0005; tel. (3) 3283-8111; fax (3) 3215-5219; internet www.meiji-life.co.jp; f. 1881; merged with Yasuda Mutual Life Insurance Co in 2004; Chair. KENJIRO HATA; Pres. RYOTARO KANEKO.

Mitsui Mirai Life Insurance Co Ltd: Mitsui Kaijyo Nihombashi Bldg, 1-3-16, Nihombashi, Chuo-ku, Tokyo 103-0027; tel. (3) 5202-2811; fax (3) 5202-2997; f. 1996; Pres. KATSUYA WATANABE.

Mitsui Mutual Life Insurance Co: 1-2-3, Otemachi, Chiyoda-ku, Tokyo 100-8123; tel. (3) 3211-6111; fax (3) 5252-7265; internet www.mitsui-seimei.co.jp; f. 1927; Chair. KOSHIRO SAKATA; Pres. AKIRA MIYAKE.

NICOS Life Insurance Co Ltd: Hongo MK Bldg, 1-28-34, Hongo, Bunkyo-ku, Tokyo 113-8414; tel. (3) 5803-3111; fax (3) 5803-3199; internet www.nicos-life.co.jp; f. 1986; Pres. RENE MULLER.

Nippon Fire Partner Life Insurance Co Ltd: 3-4-2, Tsukiji, Chuo-ku, Tokyo 104-8407; tel. (3) 5565-8080; fax (3) 5565-8365; f. 1996; Pres. HIRONOBU HARA.

Nippon Life Insurance Co (Nissay): 3-5-12, Imabashi, Chuo-ku, Osaka 541-8501; tel. (6) 6209-4500; e-mail hosokawa15560@nissay.co.jp; internet www.nissay.co.jp; f. 1889; Chair. JOSEI ITOH; Pres. IKUO UNO.

Orico Life Insurance Co Ltd: Sunshine 60, 26th Floor, 3-1-1, Higashi Ikebukuro, Toshima-ku, Tokyo 170-6026; tel. (3) 5391-3051; fax (3) 5391-3060; f. 1990; Chair. HIROSHI ARAI; Pres. TAKASHI SATO.

ORIX Life Insurance Corpn: Shinjuku Chuo Bldg, 5-17-5, Shinjuku, Shinjuku-ku, Tokyo 160-0022; tel. (3) 5272-2700; fax (3) 5272-2720; f. 1991; Chair. SHOGO KAJINISHI; Pres. SHINOBU SHIRAISHI.

Prudential Life Insurance Co Ltd: 1-7, Kojimachi, Chiyoda-ku, Tokyo 102-0083; tel. (3) 3221-0961; fax (3) 3221-2305; f. 1987; Chair. KIYOFUMI SAKAGUCHI; Pres. ICHIRO KONO.

Saison Life Insurance Co Ltd: Sunshine Sixty Bldg, 39th Floor, 3-1-1, Higashi Ikebukuro, Toshima-ku, Tokyo 170-6067; tel. (3) 3983-6666; fax (3) 2980-0598; internet www.saison-life.co.jp; f. 1975; Chair. and Pres. TOSHIO TAKEUCHI.

Skandia Life Insurance Co (Japan) Ltd: 5-6-6, Hiroo, Shibuya-ku, Tokyo 150-0012; tel. (3) 5488-1500; fax (3) 5488-1501; f. 1996; Pres. and CEO IAIN MESSENGER.

Sony Life Insurance Co Ltd: 1-1-1, Minami-Aoyama, Minato-ku, Tokyo 107-8585; tel. (3) 3475-8811; fax (3) 3475-8914; Chair. TSUNAO HASHIMOTO; Pres. KEN IWAKI.

Sumitomo Life Insurance Co: 7-18-24, Tsukiji, Chuo-ku, Tokyo 104-8430; tel. (3) 5550-1100; fax (3) 5550-1160; f. 1907; Chair. TOSHIOMI URAGAMI; Pres. KOICHI YOSHIDA.

Sumitomo Marine Yu-Yu Life Insurance Co Ltd: 2-27-1, Shinkawa, Chuo-ku, Tokyo 104-0033; tel. (3) 5541-3111; fax (3) 5541-3976; f. 1996; Pres. KATSUHIRO ISHII.

T & D Financial Life Insurance Co: 1-5-2, Uchisaiwai-cho, Chiyoda-ku, Tokyo 100-8555; tel. (3) 3504-2211; fax (3) 3593-0785; f. 1895; fmrly Tokyo Mutual Life Insurance Co; Pres. OSAMU MIZUYAMA.

Taiyo Mutual Life Insurance Co: 2-11-2, Nihombashi, Chuo-ku, Tokyo 103-0027; tel. (3) 3272-6211; fax (3) 3272-1460; Pres. MASAHIRO YOSHIIKE.

Tokio Marine & Nichido Life Insurance Co Ltd: 5-3-16, Ginza, Chuo-ku, Tokyo 106-0041; tel. (3) 5223-2111; fax (3) 5223-2165; internet www.tmn-anshin.co.jp; Pres. SUKEAKI OHTA.

Yamato Mutual Life Insurance Co: 1-1-7, Uchisaiwai-cho, Chiyoda-ku, Tokyo 100-0011; tel. (3) 3508-3111; fax (3) 3508-3118; f. 1911; Pres. KEIJI NONOMIYA.

Yasuda Kasai Himawari Life Insurance Co Ltd: 2-1-1, Nishi-Shinjuku, Shinjuku-ku, Tokyo 163-0434; tel. (3) 3348-7011; fax (3) 3346-9415; f. 1981; fmrly INA Himawari Life Insurance Co Ltd; Chair. (vacant); Pres. MAKOTO YOSHIDA.

Yasuda Mutual Life Insurance Co: 1-9-1, Nishi-Shinjuku, Shinjuku-ku, Tokyo 169-8701; tel. (3) 3342-7111; fax (3) 3349-8104; f. 1880; merged with Meiji Life Insurance Co in 2004; Chair. YUJI OSHIMA; Pres. MIKIHIKO MIYAMOTO.

Zurich Life Insurance Co Ltd: Shinanomachi Rengakan, 35, Shinanomachi, Shinjuku-ku, Tokyo 160-0016; tel. (3) 5361-2700; fax (3) 5361-2728; f. 1996; Pres. KENICHI NOGAMI.

Principal Non-Life Companies

ACE Insurance: Arco Tower, 1-8-1, Shimomeguro, Meguro-ku, Tokyo 153-0064; tel. (3) 5740-0600; fax (3) 5740-0608; internet www.ace-insurance.co.jp; f. 1999; Chair. FUMIO TOKUHIRA; Pres. TAKASHI IMAI.

Allianz Fire and Marine Insurance Japan Ltd: MITA N. N. Bldg, 4th Floor, 4-1-23, Shiba, Minato-ku, Tokyo 108-0014; tel. (3) 5442-6500; fax (3) 5442-6509; e-mail admin@allianz.co.jp; internet www.allianz.co.jp; f. 1990; Chair. HEINZ DOLLBERG; Pres. ALEXANDER ANKEL.

The Asahi Fire and Marine Insurance Co Ltd: 2-6-2, Kaji-cho, Chiyoda-ku, Tokyo 101-8655; tel. (3) 3254-2211; fax (3) 3254-2296; e-mail asahifmi@blue.ocn.ne.jp; f. 1951; Pres. MORIYA NOGUCHI.

AXA Non-Life Insurance Co Ltd: Ariake Frontier Bldg, Tower A, 3-1-25, Ariake Koto-ku, Tokyo 135-0063; tel. (3) 3570-8900; fax (3) 3570-8911; f. 1998; Pres. GUY MARCILLAT.

The Chiyoda Fire and Marine Insurance Co Ltd: 1-28-1, Ebisu, Shibuya-ku, Tokyo 150-8488; tel. (3) 5424-9288; fax (3) 5424-9382; bought by Dai-Tokyo Fire and Marine Insurance Co in 2000, to combine in April 2001; f. 1897; Pres. KOJI FUKUDA.

The Daido Fire and Marine Insurance Co Ltd: 1-12-1, Kumoji, Naha-shi, Okinawa 900-8586; tel. (98) 867-1161; fax (98) 862-8362; f. 1971; Pres. MUNEMASA URA.

The Dai-ichi Property and Casualty Insurance Co Ltd: 1-2-10, Hirakawa-cho, Chiyoda-ku, Tokyo 102-0093; tel. (3) 5213-3124; fax (3) 5213-3306; f. 1996; Pres. TSUYOSHI SHINOHARA.

The Dai-Tokyo Fire and Marine Insurance Co Ltd: 3-25-3, Yoyogi, Shibuya-ku, Tokyo 151-8530; tel. (3) 5371-6122; fax (3) 5371-6248; internet www.daitokyo.index.or.jp; f. 1918; bought Chiyoda Fire and Marine Insurance Co in 2000; Chair. HAJIME OZAWA; Pres. AKIRA SESHIMO.

The Dowa Fire and Marine Insurance Co Ltd: St Luke's Tower, 8-1, Akashi-cho, Chuo-ku, Tokyo 104-8556; tel. (3) 5550-0254; fax (3) 5550-0318; internet www.dowafire.co.jp; f. 1944; Chair. MASAO OKAZAKI; Pres. SHUICHIRO SUDO.

The Fuji Fire and Marine Insurance Co Ltd: 1-18-11, Minamisenba, Chuo-ku, Osaka 542-8567; tel. (6) 6271-2741; fax (6) 6266-7115; internet www.fujikasai.co.jp; f. 1918; Pres. YASUO ODA.

The Japan Earthquake Reinsurance Co Ltd: Kobuna-cho, Fuji Plaza, 4th Floor, 8-1, Nihombashi, Kobuna-cho, Chuo-ku, Tokyo 103-0024; tel. (3) 3664-6107; fax (3) 3664-6169; e-mail kanri@nihonjishin.co.jp; f. 1966; Pres. KAZUMOTO ADACHI.

JI Accident & Fire Insurance Co Ltd: A1 Bldg, 20-5, Ichiban-cho, Chiyoda-ku, Tokyo 102-0082; tel. (3) 3237-2045; fax (3) 3237-2250; internet www.jihoken.co.jp; f. 1989; Pres. TSUKASA IMURA.

The Kyoei Mutual Fire and Marine Insurance Co: 1-18-6, Shimbashi, Minato-ku, Tokyo 105-8604; tel. (3) 3504-2335; fax (3) 3508-7680; e-mail reins.intl@kyoeikasai.co.jp; internet www.kyoeikasai.co.jp; f. 1942; Pres. WATARU OZAWA.

Meiji General Insurance Co Ltd: 2-11-1, Kanda-tsukasa-cho, Chiyoda-ku, Tokyo 101-0048; tel. (3) 3257-3141; fax (3) 3257-3295; e-mail uwredept@meiji-sonpo.co.jp; internet meiji-general.aaapc.co.jp; f. 1996; Pres. SEISUKE ADACHI.

Mitsui Marine and Fire Insurance Co Ltd: 3-9, Kanda Surugadai, Chiyoda-ku, Tokyo 101-8011; tel. (3) 3259-3111; fax (3) 3291-5467; internet www.mitsuimarine.co.jp; f. 1918; Pres. TAKEO INOKUCHI.

Mitsui Seimei General Insurance Co Ltd: 2-1-1, Toranomon, Minato-ku, Tokyo 105-0001; tel. (3) 3224-2830; fax (3) 3224-2677; f. 1996; Pres. KIYOSHI MATSUOKA.

The Nichido Fire and Marine Insurance Co Ltd: 5-3-16, Ginza, Chuo-ku, Tokyo 104-0061; tel. (3) 3289-1066; fax (3) 3574-0646; e-mail nichido@mu2.so-net.ne.jp; internet www.mediagalaxy.co.jp/nichido; f. 1914; Chair. IKUO EGASHIRA; Pres. TAKASHI AIHARA.

The Nipponkoa Insurance Co Ltd: 3-7-3, Kasumigaseki, Chiyoda-ku, Tokyo 100-8965; tel. (3) 3593-3111; fax (3) 3593-5388; internet www.nipponkoa.co.jp; f. 1892; fmrly The Nippon Fire and Marine Insurance Co Ltd before merging with The Koa Fire and Marine Insurance Co Ltd; Pres. and CEO KEN MATSUZAWA.

Nissay General Insurance Co Ltd: Shinjuku NS Bldg, 25th Floor, 2-4-1, Nishi-Shinjuku, Shinjuku-ku, Tokyo 163-0888; tel. (3) 5325-7932; fax (3) 5325-8149; f. 1996; Pres. TADAO NISHIOKA.

The Nisshin Fire and Marine Insurance Co Ltd: 2-3 Kanda Surugadai, Chiyoda-ku, Tokyo 100-8329; tel. (3) 5282-5534; fax (3) 5282-5582; e-mail nisshin@mb.infoweb.ne.jp; internet www.nisshinfire.co.jp; f. 1908; Pres. MICHIO NODA.

Saison Automobile and Fire Insurance Co Ltd: Sunshine 60 Bldg, 3-1-1, Higashi Ikebukuro, Toshima-ku, Tokyo 170-6068; tel. (3) 3988-2572; fax (3) 3980-7367; internet www.ins-saison.co.jp; f. 1982; Pres. TOMONORI KANAI.

Secom General Insurance Co Ltd: 2-6-2, Hirakawa-cho, Chiyoda-ku, Tokyo 103-8645; tel. (3) 5216-6129; fax (3) 5216-6149; internet www.secom-sonpo.co.jp; Pres. SEIJI YAMANAKA.

Sompo Japan Insurance Inc: 26-1, Nishi-Shinjuku 1-chome, Shinjuku-ku, Tokyo 160-8338; tel. (3) 3349-3111; fax (3) 3349-4697; internet www.sompo-japan.co.jp; Pres. HIROSHI HIRANO; formed by merger of Yasuda Fire and Marine Insurance and Nissan Fire and Marine Insurance 2002.

Sony Assurance Inc.: Aromia Square 11/F, 5-37-1, Kamata, Ota-ku, Tokyo 144-8721; tel. (3) 5744-0300; fax (3) 5744-0480; internet www.sonysonpo.co.jp; f. 1999; Pres. SHINIEH YAMAMOTO.

The Sumi-Sei General Insurance Co Ltd: Sumitomo Life Yotsuya Bldg, 8-2, Honshio-cho, Shinjuku-ku, Tokyo 160-0003; tel. (3) 5360-6229; fax (3) 5360-6991; f. 1996; Chair. HIDEO NISHIMOTO; Pres. HIDEKI ISHII.

The Sumitomo Marine and Fire Insurance Co Ltd: 2-27-2, Shinkawa, Chuo-ku, Tokyo 104-8252; tel. (3) 3297-6663; fax (3)

3297-6882; internet www.sumitomomarine.co.jp; f. 1944; Chair. TAKASHI ONODA; Pres. HIROYUKI UEMURA.

Taiyo Fire and Marine Insurance Co Ltd: 7-7, Niban-cho, Chiyoda-ku, Tokyo 102-0084; tel. (3) 5226-3117; fax (3) 5226-3133; f. 1951; Chair. YUJI YAMASHITA; Pres. TSUNAIE KANIE.

The Toa Reinsurance Co Ltd: 3-6, Kanda Surugadai, Chiyoda-ku, Tokyo 101-8703; tel. (3) 3253-3177; fax (3) 3253-5298; f. 1940; Dir TAKAYA IMASHIMIZU.

The Tokio Marine and Fire Insurance Co Ltd (Tokio Kaijo): 1-2-1, Marunouchi, Chiyoda-ku, Tokyo 100-8050; tel. (3) 3285-1900; fax (3) 5223-3040; internet www.tokiomarine.co.jp; f. 1879; Chair. SHUNJI KONO; Pres. KOUKEI HIGUCHI.

The Yasuda General Insurance Co Ltd: Shinjuku MAYNDS Tower, 29th Floor, 2-1-1, Yoyogi, Shibuya-ku, Tokyo 151-0053; tel. (3) 5352-8129; fax (3) 5352-8213; e-mail uwdept@mx7.mesh.ne.jp; f. 1996; Chair. SHIGEO FUJINO; Pres. IEJI YOSHIOKA.

The Post Office also operates life insurance and annuity plans.

Insurance Associations

The General Insurance Association of Japan Inc (Nihon Songai Hoken Kyokai): Non-Life Insurance Bldg, 2-9, Kanda Awaji-cho, Chiyoda-ku, Tokyo 101-8335; tel. (3) 3255-1437; fax (3) 3255-1234; e-mail kokusai@sonpo.or.jp; internet www.sonpo.or.jp; f. 1946; 23 mems (Nov. 2003); Chair. KEN MATSUZAWA; Exec. Dir EIJI NISHIURA.

Japan Trade and Investment Insurance Organization (Boeki Hoken Kiko): 6th Floor, 2-8-6, Nishi-Shinjuku, Minato-ku, Tokyo 105-0003; tel. (3) 3580-0321; internet www.jtio.or.jp; Pres. YUKIO OTSU.

The Life Insurance Association of Japan (Seimei Hoken Kyokai): New Kokusai Bldg, 3-4-1, Marunouchi, Chiyoda-ku, Tokyo 100-0005; tel. (3) 3286-2652; fax (3) 3286-2630; internet www.seiho.or.jp; f. 1908; 42 mem. cos; Chair. SHINICHI YOKOYAMA; Senior Man. Dir SHIGERU SUWA.

Non-Life Insurance Rating Organization of Japan: 1-9, Kanda-nishikicho, Chiyoda-ku, Tokyo 101-0054; tel. (3) 3233-4762; fax (3) 3295-9301; e-mail service@grp.nliro.or.jp; internet www.nliro.or.jp; f. 1964; 39 mems (July 2004); Chair. AKIO MORISHIMA; Senior Exec. Dir MASAHIRO ISHII.

Trade and Industry

CHAMBERS OF COMMERCE AND INDUSTRY

The Japan Chamber of Commerce and Industry (Nippon Shoko Kaigi-sho): 3-2-2, Marunouchi, Chiyoda-ku, Tokyo 100-0005; tel. (3) 3283-7851; fax (3) 3216-6497; e-mail info@jcci.or.jp; internet www.jcci.or.jp/home-e; f. 1922; the cen. org. of all chambers of commerce and industry in Japan; mems 521 local chambers of commerce and industry; Chair. KOSAKU INABA; Pres. SHOICHI TANIMURA.

Principal chambers include:

Kobe Chamber of Commerce and Industry: 6-1, Minatojima-nakamachi, Chuo-ku, Kobe 650-8543; tel. (78) 303-5806; fax (78) 306-2348; e-mail info@kobe-cci.or.jp; internet www.kobe-cci.or.jp; f. 1878; 12,700 mems; Chair.(acting) TOSHIRO OTA; Pres. HIROSHI MIYAMICHI.

Kyoto Chamber of Commerce and Industry: 240, Shoshoi-cho, Ebisugawa-agaru, Karasumadori, Nakakyo-ku, Kyoto 604-0862; tel. (75) 212-6450; fax (75) 251-0743; e-mail kyoto@kyo.or.jp; f. 1882; 13,008 mems; Chair. KAZUO INAMORI; Pres. OSAMU KOBORI.

Nagoya Chamber of Commerce and Industry: 2-10-19, Sakae, Naka-ku, Nagoya, Aichi 460-8422; tel. (52) 223-5722; fax (52) 232-5751; f. 1881; 20,622 mems; Chair. SEITARO TANIGUCHI; Pres. YOSHIKI KOBAYASHI.

Naha Chamber of Commerce and Industry: 2-2-10, Kume Naha, Okinawa; tel. (98) 868-3758; fax (98) 866-9834; e-mail cci-naha@cosmos.ne.jp; f. 1927; 4,874 mems; Chair. AKIRA SAKIMA; Pres. KOSEI YONEMURA.

Osaka Chamber of Commerce and Industry: 2-8, Hommachi-bashi, Chuo-ku, Osaka 540-0029; tel. (6) 6944-6400; fax (6) 6944-6248; e-mail intl@osaka.cci.or.jp; internet www.osaka.cci.or.jp; f. 1878; 35,069 mems; Chair. WA TASHIRO; Pres. TAKAO OHNO.

Tokyo Chamber of Commerce and Industry: 3-2-2, Marunouchi, Chiyoda-ku, Tokyo 100-0005; tel. (3) 3283-7756; fax (3) 3216-6497; e-mail webmaster@tokyo-cci.or.jp; f. 1878; 118,642 mems; Chair. KOSAKU INABA; Pres. SHOICHI TANIMURA.

Yokohama Chamber of Commerce and Industry: Sangyo Boueki Center Bldg, 8th Floor, Yamashita-cho, Naka-ku, Yokohama 231-8524; tel. (45) 671-7400; fax (45) 671-7410; e-mail info@yokohama-cci.or.jp; f. 1880; 14,965 mems; Chair. MASAYOSHI TAKANASHI; Pres. NAMIO OBA.

INDUSTRIAL AND TRADE ASSOCIATIONS

General

The Association for the Promotion of International Trade, Japan (JAPIT): 1-26-5, Toranomon, Minato-ku, Tokyo; tel. (3) 3506-8261; fax (3) 3506-8260; f. 1954 to promote trade with the People's Repub. of China; 700 mems; Chair. YOSHIO NAKATA; Pres. YOSHIO SAKURAUCHI.

Industry Club of Japan: 1-4-6, Marunouchi, Chiyoda-ku, Tokyo; tel. (3) 3281-1711; f. 1917 to develop closer relations between industrialists at home and abroad and promote expansion of Japanese business activities; c. 1,600 mems; Pres. GAISHI HIRAIWA; Exec. Dir KOUICHIROU SHINNO.

Japan Commercial Arbitration Association: Taishoseimei Hibiya Bldg, 1-9-1, Yurakucho, Chiyoda-ku, Tokyo 100-1006; tel. (3) 3287-3061; fax (3) 3287-3064; e-mail info@jcaa.or.jp; internet www.jcaa.or.jp; f. 1950; 1,012 mems; provides facilities for mediation, conciliation and arbitration in international trade disputes; Pres. NOBUO YAMAGUCHI.

Japan External Trade Organization (JETRO): 2-2-5, Toranomon, Minato-ku, Tokyo 105-8466; tel. (3) 3582-5511; fax (3) 3582-5662; e-mail seh@jetro.go.jp; internet www.jetro.go.jp; f. 1958; information for international trade, investment, import promotion, exhibitions of foreign products; Chair. and CEO OSAMU WATANABE; Pres. HIROSHI TSUKAMOTO.

Japan Federation of Smaller Enterprise Organizations (JFSEO) (Nippon Chusokigyo Dantai Renmei): 2-8-4, Nihombashi, Kayaba-cho, Chuo-ku, Tokyo 103-0025; tel. (3) 3669-6862; f. 1948; 18 mems and c. 1,000 co-operative socs; Pres. MASATAKA TOYODA; Chair. of Int. Affairs SEIICHI ONO.

Japan General Merchandise Exporters' Association: 2-4-1, Hamamatsu-cho, Minato-ku, Tokyo; tel. (3) 3435-3471; fax (3) 3434-6739; f. 1953; 40 mems; Pres. TADAYOSHI NAKAZAWA.

Japan Productivity Center for Socio-Economic Development (JPC-SED) (Shakai Keizai Seisansei Honbu): 3-1-1, Shibuya, Shibuya-ku, Tokyo 150-8307; tel. (3) 3409-1112; fax (3) 3409-1986; f. 1994 following merger between Japan Productivity Center and Social Economic Congress of Japan; 10,000 mems; concerned with management problems and research into productivity; Chair. SUGIICHIRO WATARI; (acting); Pres. YASUO SAWAMA.

Keizai Doyukai (Japan Association of Corporate Executives): 1-4-6, Marunouchi, Chiyoda-ku, Tokyo 100-0005; tel. (3) 3211-1271; fax (3) 3212-3774; e-mail contact@doyukai.or.jp; internet www.doyukai.or.jp; f. 1946; mems: c. 1,400; corporate executives concerned with national and international economic and social policies; Chair. KAKUTARO KITASHIRO.

Nihon Boeki-Kai (Japan Foreign Trade Council, Inc): World Trade Center Bldg, 6th Floor, 2-4-1, Hamamatsu-cho, Minato-ku, Tokyo 105-6106; tel. (3) 3435-5952; fax (3) 3435-5969; e-mail mail@jftc.or.jp; internet www.jftc.or.jp; f. 1947; 192 mems; Chair. KENJI MIYAHARA; Exec. Man. Dir KEISUKE TAKANASHI; Man. Dir HISAO IKEGAMI.

Chemicals

Federation of Pharmaceutical Manufacturers' Associations of Japan: Tokyo Yakugyo Bldg, 2-1-5, Nihombashi Honcho, Chuo-ku, Tokyo 103-0023; tel. (3) 3270-0581; fax (3) 3241-2090; Pres. TADASHI SUZUKI.

Japan Chemical Industry Association: Sumitomo Fudosan Rokko Bldg, 4-1 Shinkawa 1-chome, Chuo-ku, Tokyo 104-0033; tel. (3) 3297-2576; fax (3) 3297-2606; e-mail chemical@jcia-net.or.jp; internet www.nikkakyo.org; f. 1948; 266 mems; Pres. Dr MITSUO OHASHI.

Japan Cosmetic Industry Association: Hatsumei Bldg, 2-9-14, Toranomon, Minato-ku, Tokyo 105-0001; tel. (3) 3502-0576; fax (3) 3502-0829; f. 1959; 687 mem. cos; Chair. REIJIRO KOBAYASHI.

Japan Gas Association: 1-15-12, Toranomon, Minato-ku, Tokyo 105-0001; tel. (3) 3502-0116; fax (3) 3502-3676; f. 1947; Chair. SHINICHIRO RYOKI; Vice-Chair. and Sr Man. Dir KOSHIRO GODA.

Japan Perfumery and Flavouring Association: Saeki No. 3 Bldg, 3rd Floor, 37 Kandakonya-cho, Chiyoda-ku, Tokyo 101-0035; tel. and fax (3) 3526-7855; f. 1947; Chair. YONEJIRO KORAYASHI.

Photo-Sensitized Materials Manufacturers' Association: JCII Bldg, 25, Ichiban-cho, Chiyoda-ku, Tokyo 102-0082; tel. (3) 5276-3561; fax (3) 5276-3563; f. 1948; Pres. MASAYUKI MUNEYUKI.

Fishing and Pearl Cultivation

Japan Fisheries Association (Dainippon Suisankai): Sankaido Bldg, 1-9-13, Akasaka, Minato-ku, Tokyo 107-0052; tel. (3) 3585-6683; fax (3) 3582-2337; internet www.suisankai.or.jp; Pres. HIROYA SANO.

Japan Pearl Export and Processing Co-operative Association: 3-7, Kyobashi, Chuo-ko, Tokyo; f. 1951; 130 mems.

Japan Pearl Exporters' Association: 122, Higashi-machi, Chuo-ku, Kobe; tel. (78) 331-4031; fax (78) 331-4345; e-mail jpeakobe@lime.ocn.ne.jp; internet www.japan-pearl.com; f. 1954; 56 mems; Pres. HIDEO KANAI.

Paper and Printing

Japan Federation of Printing Industries: 1-16-8, Shintomi, Chuo-ku, Tokyo 104; tel. (3) 3553-6051; fax (3) 3553-6079; Pres. HIROMICHI FUJITA.

Japan Paper Association: Kami Parupu Bldg, 3-9-11, Ginza, Chuo-ku, Tokyo 104-8139; tel. (3) 3248-4801; fax (3) 3248-4826; internet www.jpa.gr.jp; f. 1946; 54 mems; Chair. MASAO KOBAYASHI; Pres. KIYOSHI SAKAI.

Japan Paper Exporters' Association: Kami Parupu Bldg, 3-9-11, Ginza, Chuo-ku, Tokyo 104-8139; tel. (3) 3248-4831; fax (3) 3248-4834; e-mail japex@green.an.egg.or.jp; f. 1952; 36 mems; Chair. IWAO NAKAJIMA.

Japan Paper Importers' Association: Kami Parupu Bldg, 3-9-11, Ginza, Chuo-ku, Tokyo 104-8139; tel. (3) 3248-4832; fax (3) 3248-4834; e-mail japim@yacht.ocn.ne.jp; f. 1981; 27 mems; Chair. KYOUICHI HIRATO.

Japan Paper Products Manufacturers' Association: 4-2-6, Kotobuki, Taito-ku, Tokyo; tel. (3) 3543-2411; f. 1949; Exec. Dir KIYOSHI SATOH.

Mining and Petroleum

Asbestos Cement Products Association: Takahashi Bldg, 7-10-8, Ginza, Chuo-ku, Tokyo; tel. (3) 3571-1359; f. 1937; Chair. KOSHIRO SHIMIZU.

Japan Cement Association: Hattori Bldg, 1-10-3, Kyobashi, Chuo-ku, Tokyo 104-0031; tel. (3) 3561-8632; fax (3) 3567-8570; f. 1948; 20 mem. cos; Chair. KAZUTSUGU HIRAGA; Exec. Man. Dir HIROFUMI YAMASHITA.

Japan Coal Association: Hibiya Park Bldg, 1-8-1, Yuraku-cho, Chiyoda-ku, Tokyo 100; tel. (3) 3271-3481; fax (3) 3214-0585; Chair. TADASHI HARADA.

Japan Mining Industry Association: Shuwa Toranomon Bldg, No. 3, 1-21-8 Toranomon, Minato-ku, Tokyo 105-0001; tel. (3) 3502-7451; fax (3) 3591-9841; f. 1948; 53 mem. cos; Chair. HIROSHI MAKIHARA; Pres. HIROE TAKAHARA; Dir-Gen. HIROE TAKAHARA.

Japan Petrochemical Industry Association: 2nd Floor, 2-1-1, Uchisaiwai-cho, Chiyoda-ku, Tokyo 100-0011; tel. (3) 3501-2151; internet www.jpca.or.jp; Chair. MITSUO OHASHI.

Japan Petroleum Development Association: Keidanren Bldg, 1-9-4, Otemachi, Chiyoda-ku, Tokyo 100; tel. (3) 3279-5841; fax (3) 3279-5844; f. 1961; Chair. TAMOTSU SHOYA.

Metals

Japan Aluminium Association (JAA): Tsukamoto-Sozan Bldg, 4-2-15, Ginza, Chuo-ku, Tokyo 104-0061; tel. (3) 3538-0221; fax (3) 3538-0233; f. 1999 by merger of Japan Aluminium Federation and Japan Light Metal Association; Chair. SHINJI YANO.

Japan Brass Makers' Association: 1-12-22, Tsukiji, Chuo-ku, Tokyo 104-0045; tel. (3) 3542-6551; fax (3) 3542-6556; e-mail jbmajwcc@copper-brass.gr.jp; internet www.copper-brass.gr.jp; f. 1948; 62 mems; Pres. S. SATO; Man. Dir J. HATANO.

The Japan Iron and Steel Federation: Tekko Kaikan Bldg, 3-2-10, Nihombashi Kayaba-cho, Chuo-ku, Tokyo 103-0025; tel. (3) 3669-4818; fax (3) 3661-0798; internet www.jisf.or.jp; f. 1953; mems 62 mfrs, 65 dealers; Chair. AKIO MIMURA.

Japan Stainless Steel Association: Tekko Bldg, 3-2-10, Nihombashi Kayaba-cho, Chuo-ku, Tokyo 103; tel. (3) 3669-4431; fax (3) 3669-4431; e-mail yabe@jssa.gr.jp; internet www.jssa.gr.jp; Pres. MIKIO KATOH; Exec. Dir TAKEO YABE.

The Kozai Club: Tekko Bldg, 3-2-10, Nihombashi Kayaba-cho, Chuo-ku, Tokyo 103-0025; tel. (3) 3669-4815; fax (3) 3667-0245; f. 1947; mems 39 mfrs, 69 dealers; Chair. AKIRA CHIHAYA.

Steel Castings and Forgings Association of Japan (JSCFA): Shikoku Bldg Bekkan, 1-14-4, Uchikannda, Chiyoda-ku, Tokyo 101-0047; tel. (3) 5283-1611; fax (3) 5283-1613; e-mail cf@jscfa.gr.jp; f. 1972; mems 42 cos, 44 plants; Exec. Dir TOMOO TAKENOUCHI.

Machinery and Precision Equipment

Electronic Industries Association of Japan: 3-2-2, Marunouchi, Chiyoda-ku, Tokyo 100-0005; tel. (3) 3213-5861; fax (3) 3213-5863; e-mail pao@eiaj.or.jp; internet www.eiaj.or.jp; f. 1948; 540 mems; Chair. FUMIO SATO.

Japan Camera Industry Association: JCII Bldg, 25, Ichibancho, Chiyoda-ku, Tokyo 102-0082; tel. (3) 5276-3891; fax (3) 5276-3893; internet www.photo-jcia.gr.jp; f. 1954; Pres. MASATOSHI KISHIMOTO.

Japan Clock and Watch Association: Kudan Sky Bldg, 1-12-11, Kudan-kita, Chiyoda-ku, Tokyo 102-0073; tel. (3) 5276-3411; fax (3) 5276-3414; internet www.jcwa.or.jp; Chair. HIROSHI HARUTA.

Japan Electric Association: 1-7-1, Yuraku-cho, Chiyoda-ku, Tokyo 100-0006; tel. (3) 3216-0551; fax (3) 3214-6005; f. 1921; 4,610 mems; Pres. TATSUO KAWAI.

Japan Electric Measuring Instruments Manufacturers' Association (JEMIMA): 1-9-10, Toranomon, Minato-ku, Tokyo 105-0001; tel. (3) 3502-0601; fax (3) 3502-0600; e-mail o-mitani@jemima.or.jp; internet www.jemima.or.jp; 117 mems; Gen. Man. OSAMU MITANI.

Japan Electrical Manufacturers' Association: 2-4-15, Nagata-cho, Chiyoda-ku, Tokyo 100-0014; tel. (3) 3581-4841; fax (3) 3593 3198; internet www.jema-net.or.jp; f. 1948; 245 mems; Chair. TAIZO NISHIMURO.

Japan Energy Association: Houwa Mita Tsunasaka Bldg, 2-7-7, Mita, Minato-ku, Tokyo 108-0073; tel. (3) 3451-1651; fax (3) 3451-1360; e-mail common@jea-wec.or.jp; internet www.jea-wec.or.jp; f. 1950; 142 mems; Chair. SHIGE-ETSU MIYAHARA; Exec. Dir HAJIME MURATA.

Japan Machine Tool Builders' Association: Kikai Shinko Bldg, 3-5-8, Shiba Koen, Minato-ku, Tokyo 105-0011; tel. (3) 3434-3961; fax (3) 3434-3763; f. 1951; 112 mems; Chair. TOYO KATO; Exec. Dir S. ABE.

Japan Machinery Center for Trade and Investment (JMC): Kikai Shinko Bldg, 3-5-8, Shiba Koen, Minato-ku, Tokyo 105-0011; tel. (3) 3431-9507; fax (3) 3436-6455; Pres. ISAO YONEKURA.

The Japan Machinery Federation: Kikai Shinko Bldg, 3-5-8, Shiba Koen, Minato-ku, Tokyo 105-0011; tel. (3) 3434-5381; fax (3) 3434-2666; f. 1952; Pres. SHOICHI SADA; Exec. Vice-Pres. SHINICHI NAKANISHI.

Japan Machinery Importers' Association: Koyo Bldg, 8th Floor, 1-2-11, Toranomon, Minato-ku, Tokyo 105-0001; tel. (3) 3503-9736; fax (3) 3503-9779; f. 1957; 94 mems; Pres. ISAO YONEKURA.

Japan Microscope Manufacturers' Association: c/o Olympus Optical Co Ltd, 2-43-2, Hatagaya, Shibuya-ku, Tokyo 151-0072; tel. (3) 3377-2139; fax (3) 3377-2139; e-mail jmma@olympus.co.jp; f. 1954; 31 mems; Chair. T. SHIMOYAMA.

Japan Motion Picture Equipment Industrial Association: Kikai Shinko Bldg, 3-5-8, Shiba Koen, Minato-ku, Tokyo 105; tel. (3) 3434-3911; fax (3) 3434-3912; Pres. MASAO SHIKATA; Gen. Sec. TERUHIRO KATO.

Japan Optical Industry Association: Kikai Shinko Bldg, 3-5-8, Shiba Koen, Minato-ku, Tokyo 105-0011; tel. (3) 3431-7073; f. 1946; 7 mems; Chair. SHOICHIRO YOSHIDA; Exec. Sec. SHIRO IWAHASHI.

The Japan Society of Industrial Machinery Manufacturers: Kikai Shinko Bldg, 3-5-8, Shiba Koen, Minato-ku, Tokyo 105-0011; tel. (3) 3434-6821; fax (3) 3434-4767; e-mail obd@jsim.or.jp; internet www.jsim.or.jp; f. 1948; 213 mems; Exec. Man. Dir KOJI FUJISAKI; Pres. KENTARO AIKAWA.

Japan Textile Machinery Association: Kikai Shinko Bldg, Room 310, 3-5-8, Shiba Koen, Minato-ku, Tokyo 105; tel. (3) 3434-3821; fax (3) 3434-3043; f. 1951; Pres. JUNICHI MURATA.

Textiles

Central Raw Silk Association of Japan: 1-9-4, Yuraku-cho, Chiyoda-ku, Tokyo; tel. (3) 3214-5777; fax (3) 3214-5778.

Japan Chemical Fibers Association: Seni Kaikan, 3-1-11, Nihombashi-Honcho, Chuo-ku, Tokyo 103-0023; tel. (3) 3241-2311; fax (3) 3246-0823; internet www.fcc.co.jp/JCFA; f. 1948; 41 mems, 9 assoc. mems; Pres. KATSUHIKO HIRAI; Dir-Gen. KUNIO YAGI.

Japan Cotton and Staple Fibre Weavers' Association: 1-8-7, Nishi-Azabu, Minato-ku, Tokyo; tel. (3) 3403-9671.

Japan Silk Spinners' Association: f. 1948; 95 mem. firms; Chair. ICHIJI OHTANI.

Japan Spinners' Association: Mengyo Kaikan Bldg, 2-5-8, Bingo-machi, Chuo-ku, Osaka 541-0051; tel. (6) 6231-8431; fax (6) 6229-1590; e-mail spinas@cotton.or.jp; internet www.jsa-jp.org; f. 1948; Exec. Dir HARUTA MUTO.

Transport Machinery

Japan Association of Rolling Stock Industries: Awajicho Suny Bldg, 1-2, Kanda-Sudacho, Chiyoda-ku, Tokyo 101-0041; tel. (3) 3257-1901.

Japan Auto Parts Industries Association: 1-16-15, Takanawa, Minato-ku, Tokyo 108-0074; tel. (3) 3445-4211; fax (3) 3447-5372; e-mail japiaint@green.am.egg.or.jp; f. 1948; 530 mem. firms; Chair. Tsuneo Ishimaru; Exec. Dir K. Shibasaki.

Japan Automobile Manufacturers Association, Inc (JAMA): Otemachi Bldg, 1-6-1, Otemachi, Chiyoda-ku, Tokyo 100-0004; tel. (3) 5219-6660; fax (3) 3287-2073; e-mail kaigai_tky@mta.jama.or.jp; internet www.jama.or.jp; f. 1967; 14 mem. firms; Chair. Yoshihide Munekuni; Pres. Takao Suzuki.

Japan Bicycle Manufacturers' Association: 1-9-3, Akasaka, Minato-ku, Tokyo 107; tel. (3) 3583-3123; fax (3) 3589-3125; f. 1955.

Japan Ship Exporters' Association: Nippon-Zaidan Bldg, 1-15-16, Toranomon, Minato-ku, Tokyo 105-0001; tel. (3) 3502-2094; fax (3) 3508-2058; e-mail postmaster@jsea.or.jp; 38 mems; Exec. Man. Dir Yuichi Watanabe.

Japanese Marine Equipment Association: Kaiyo Senpaku Bldg, 15-16, Toranomon, Minato-ku, Tokyo 105-0001; tel. (3) 3502-2041; fax (3) 3591-2206; e-mail info@jsmea.or.jp; internet www.jsmea.or.jp; f. 1956; 240 mems; Pres. Tadao Yamaoka.

Japanese Shipowners' Association: Kaiun Bldg, 2-6-4, Hirakawa-cho, Chiyoda-ku, Tokyo 102-0093; tel. (3) 3264-7171; fax (3) 3262-4760; Pres. Kentaro Kawamura.

Shipbuilders' Association of Japan: 1-15-16, Toranomon, Minato-ku, Tokyo 105-0001; tel. (3) 3502-2010; fax (3) 3502-2816; internet www.sajn.or.jp; f. 1947; 21 mems; Chair. Mototsugu Ito.

Society of Japanese Aerospace Companies Inc (SJAC): Toshin-Tameike Bldg, 2nd Floor, 1-1-14, Akasaka, Minato-ku, Tokyo 107-0052; tel. (3) 3585-0511; fax (3) 3585-0541; e-mail miwa-shuichi@sjac.or.jp; internet www.sjac.or.jp; f. 1952; reorg. 1974; 117 mems, 41 assoc. mems; Chair. Toshifumi Takei; Pres. Masamoto Tazaki.

Miscellaneous

Communications Industry Association of Japan (CIA-J): Sankei Bldg, 1-7-2, Otemachi, Chiyoda-ku, Tokyo 100-0004; tel. (3) 3231-3005; fax (3) 3231-3110; e-mail admin@ciaj.or.jp; internet www.ciaj.or.jp; f. 1948; non-profit org. of telecommunications equipment mfrs; 236 mems; Chair. Tadashi Sekizawa; Pres. Yutaka Hayashi.

Japan Canners' Association: Yurakucho Denki Bldg, 1-7-1, Yuraku-cho, Chiyoda-ku, Tokyo 100-0006; tel. (3) 3213-4751; fax (3) 3211-1430; Pres. Keinosuke Hisai.

Japan Hardwood Exporters' Association: Matsuda Bldg, 1-9-1, Ironai, Otaru, Hokkaido 047; tel. (134) 23-8411; fax (134) 22-7150; 7 mems.

Japan Lumber Importers' Association: Yushi Kogyo Bldg, 3-13-11, Nihombashi, Chuo-ku, Tokyo 103; tel. (3) 3271-0926; fax (3) 3271-0928; f. 1950; 130 mems; Pres. Shoichi Tanaka.

Japan Plastics Industry Federation: Kaseihin-Kaikan, 5-8-17, Roppongi, Minato-ku, Tokyo 106-0032; tel. (3) 3586-9761; fax (3) 3586-9760; internet www.jpif.gr.jp; Chair. Akio Sato.

Japan Plywood Manufacturers' Association: Meisan Bldg, 1-18-17, Nishi-Shimbashi, Minato-ku, Tokyo 105; tel. (3) 3591-9246; fax (3) 3591-9240; f. 1965; 92 mems; Pres. Hiroshi Inoue.

Japan Pottery Manufacturers' Federation: Toto Bldg, 1-1-28, Toranomon, Minato-ku, Tokyo; tel. (3) 3503-6761.

The Japan Rubber Manufacturers' Association: Tobu Bldg, 1-5-26, Moto Akasaka, Minato-ku, Tokyo 107-0051; tel. (3) 3408-7101; fax (3) 3408-7106; f. 1950; 125 mems; Pres. Yasuo Tominaga.

Japan Spirits and Liquors Makers' Association: Koura Dai-ichi Bldg, 7th Floor, 1-1-6, Nihombashi-Kayaba-cho, Chuo-ku, Tokyo 103; tel. (3) 3668-4621.

Japan Sugar Import and Export Council: Oshima Bldg 1-3, Nihonbashi Koamicho, Chuo-ku, Tokyo; tel. (3) 3571-2362; fax (3) 3571-2363; 16 mems.

Japan Sugar Refiners' Association: 5-7, Sanban-cho, Chiyoda-ku, Tokyo 102; tel. (3) 3288-1151; fax (3) 3288-3399; f. 1949; 17 mems; Sr Man. Dir Katsuyuki Suzuki.

Japan Tea Exporters' Association: 17, Kitaban-cho, Shizuoka, Shizuoka Prefecture 420-0005; tel. (54) 271-3428; fax (54) 271-2177; e-mail japantea@sound.jp; 44 mems.

Japan Toy Association: 4-22-4, Higashi-Komagata, Sumida-ku, Tokyo 130; tel. (3) 3829-2513; fax (3) 3829-2549; Chair. Makoto Yamashina.

Motion Picture Producers' Association of Japan, Inc: Tokyu Ginza Bldg, 2-15-2, Ginza, Chuo-ku, Tokyo 104-0061; tel. (3) 3547-1800; fax (3) 3547-0909; e-mail eiren@mc.neweb.ne.jp; internet www.eiren.org; Pres. Isao Matsuoka.

EMPLOYERS' ORGANIZATION

Japan Business Federation (JBF) (Keidanren Kaikan): 1-9-4, Otemachi, Chiyoda-ku, Tokyo 100-8188; tel. (3) 5204-1920; fax (3) 5204-1943; e-mail intlab@keidanren.or.jp; internet www.keidanren.or.jp; f. 2002 by merger of Keidanren (f. 1946) and Nikkeiren (f. 1948); 1,540 mem. asscns; Chair. Hiroshi Okuda; Dir-Gen. Ryukoh Wada.

UTILITIES

Electricity

Chubu Electric Power Co Inc: 1, Higashi-Shincho, Higashi-ku, Nagoya 461-8680; tel. (52) 951-8211; fax (52) 962-4624; internet www.chuden.co.jp; Chair. Kohei Abe; Pres. Hiroji Ota.

Chugoku Electric Power Co Inc: 4-33, Komachi, Naka-ku, Hiroshima 730-8701; tel. (82) 241-0211; fax (82) 523-6185; e-mail angel@inet.energia.co.jp; internet www.energia.co.jp; f. 1951; Chair. Shitomi Takasu; Pres. Shigeo Shirakura.

Hokkaido Electric Power Co Inc: internet www.hepco.co.jp; Chair. Kazuo Toda; Pres. Seiji Izumi.

Hokuriku Electric Power Co Inc: internet www.rikuden.co.jp.

Kansai Electric Power Co Inc: 3-3-22, Nakanoshima, Kita-ku, Osaka 530-8270; tel. (6) 6441-8821; fax (6) 6441-8598; e-mail postmaster@kepco.co.jp; internet www.kepco.co.jp; Chair. Yoshihisa Akiyama; Pres. Y. Fuji.

Kyushu Electric Power Co Inc: 2-1-82, Watanabe-dori, Chuo-ku, Fukuoka 810-8726; tel. (92) 726-1649; fax (92) 731-8719; internet www.kyuden.co.jp; Chair. Michisada Kamata.

Shikoku Electric Power Co Inc: 2-5, Marunouchi, Takamatsu 760-8573; tel. (878) 21-5061; fax (878) 26-1250; e-mail postmaster@yonden.co.jp; internet www.yonden.co.jp; Chair. Hiroshi Yamamoto; Pres. Kozo Kondo.

Tohoku Electric Power Co Inc: 3-7-1, Ichiban-cho, Aoba-ku, Sendai 980; tel. (22) 225-2111; fax (22) 222-2881; e-mail webmaster@tohoku-epco.co.jp; internet www.tohoku-epco.co.jp; Chair. Teruyuki Akema; Pres. Toshiaki Yashima.

Tokyo Electric Power Co Inc: 1-1-3, Uchisaiwai-cho, Chiyoda-ku, Tokyo 100; tel. (3) 3501-8111; fax (3) 3592-1795; internet www.tepco.co.jp; Chair. Shoh Nasu; Pres. N. Minami.

Gas

Osaka Gas Co Ltd: e-mail intlstaff@osakagas.co.jp; internet www.osakagas.co.jp.

Toho Gas Co Ltd: 19-18, Sakurada-cho, Atsuta-ko, Nagoya 456; tel. (52) 871-3511; internet www.tohogas.co.jp; f. 1922; Chair. Susumu Ogawa; Pres. Sadahiko Simizu.

Tokyo Gas Co Inc: 1-5-20, Kaigan, Minato-ku, Tokyo 105; tel. (3) 3433-2111; fax (3) 5472-5385; internet www.tokyo-gas.co.jp; f. 1885; Chair. Hiroshi Watanabe; Pres. H. Uehara.

CO-OPERATIVE ORGANIZATION

Nikkenkyo (Council of Japan Construction Industry Employees' Unions): Moriyama Bldg, 1-31-16, Takadanobaba, Shinjuku-ku, Tokyo 169; tel. (3) 5285-3870; fax (3) 5285-3879; Pres. Noboru Sekiguchi.

MAJOR COMPANIES

(cap. = capital; res = reserves; m. = million; amounts in yen, unless otherwise indicated)

Ajinomoto Co Inc: 1-15-1, Kyobashi, Chuo-ku, Tokyo 104-8315; tel. (3) 5250-8111; fax (3) 5250-8293; internet www.ajinomoto.com; f. 1909; cap. and res 381,017m., sales 943,540m. (2001/02); mfrs and distributors of seasonings, edible oils, processed foods, beverages, dairy products, pharmaceuticals, amino acids, speciality chemicals; Pres. Kunio Egashira; 24,326 employees (group).

Asahi Breweries Ltd: 3-7-1 Kyobashi, Chuo-Ku, Tokyo 104–8323; internet www.asahibeer.co.jp; f. 1949; sales 182,530m. (2001); alcoholic beverages; Pres. and CEO Kouichi Ikeda; 3,799 employees.

Asahi Glass Co Ltd: 1-12-1, Yurakucho, Chiyoda-ku, Tokyo 100-8405; tel. (3) 3218-5555; fax (3) 3201-5390; internet www.agc.co.jp; f. 1907; cap. and res 585,975m., sales 1,263,196m. (2001/02); manufacture and sale of flat glass, TV bulbs, alkali and other chemicals, refractories and electronics; associated companies and subsidiaries in Belgium, India, Indonesia, Singapore, Thailand and the USA; Chair. Hiromichi Seya; Pres. Shinya Ishizu; 7,453 employees.

Asahi Kasei Corpn: Hibiya-Mitsui Bldg, 1-1-2, Yuraku-cho, Chiyoda-ku, Tokyo 100-8440; tel. (3) 3507-2060; fax (3) 3507-2495; e-mail asahi@om.asahi-kasei.co.jp; internet www.asahi-kasei.co.jp; f. 1931; cap. and res 496,825m., sales 1,195,393m. (2002/03); manufacture and sale of chemicals and plastics, housing and construction materials, fibres and textiles, electronics, membranes and systems, biotechnology and medical products, engineering, and others; Chair. Nobuo Yamaguchi; Pres. Shiro Hiruta; 26,227 employees.

Bridgestone Corpn: 1-10-1, Kyobashi, Chuo-ku, Tokyo 104-8340; tel. (3) 3567-0111; fax (3) 3567-4615; internet www.bridgestone.co.jp; f. 1931; cap. and res 835,143m., sales 2,133,825m. (2001); mfrs of rubber tyres and tubes, shock absorbers, conveyor belts, hoses, foam rubber, polyurethane foam, golf balls; Chair. and Pres. Shigeo Watanabe; 104,700 employees.

Canon Inc: 3-30-2, Shimomaruko, Ohta-ku, Tokyo 146-8501; tel. (3) 3758-2111; fax (3) 5482-5130; internet www.canon.com; cap. and res 1,298,914m. (2000), sales 2,907,573m. (2001); mfrs of cameras, business machines, etc.; Pres. Fujio Mitarai; 93,620 employees.

Casio Computer Co Ltd: 1-6-2, Hon-machi, Shibuya-ku, Tokyo 151-8543; tel. (3) 5334-4111; fax (3) 5334-4669; e-mail webmaster@casio.co.jp; internet www.casio.co.jp; f. 1957; cap. and res 134,317m., sales 382,154m. (2001/02); manufacture and sale of electronic calculators, digital watches, electronic musical instruments, liquid crystal televisions, Japanese language word processors; Chair. Toshio Kashio; Pres. Kazuo Kashio; 19,325 employees.

Citizen Watch Co Ltd: 6-1-12, Tanashi-cho, Nishi-Tokyo City, Tokyo 188-8511; tel. (4) 2466-1231; fax (3) 2466-1280; e-mail info@citizen.co.jp; internet www.citizen.co.jp; f. 1930; cap. and res 204,589m., sales 327,555m. (2001/02); manufacture and sale of wristwatches and parts, machine tools and tools, jewellery and eyeglasses, information and electronic equipment, precision machine and precision measuring instruments; Pres. Hiroshi Haruta; 17,459 employees.

Cosmo Oil Co Ltd: 1-1-1, Shibaura, Minato-ku, Tokyo 105-8528; tel. (3) 3798-3211; fax (3) 3798-3411; internet www.cosmo-oil.co.jp; f. 1986; cap. and res 194,302m., sales 1,813,838m. (2001/02); importing of petroleum, refining, sales and distribution of petroleum products and related activities; Chair. and CEO Keiichiro Okabe; 1,970 employees.

Dai Nippon Printing Co Ltd: 1-1-1, Ichigaya Kaga-cho, Shinjuku-ku, Tokyo 162-8001; tel. (3) 3266-2111; fax (3) 5225-8239; e-mail info@mail.dnp.co.jp; internet www.dnp.co.jp; f. 1876; cap. and res 946,998m., sales 1,311,934m. (2001/02); printing, packaging, paper products, plastics, precision electronic products; Chair. and Pres. Yoshitoshi Kitajima; 35,868 employees.

Daido Steel Co Ltd: 1-11-18, Nishiki, Naka-ku, Nagoya, 460-8581; tel. (52) 201-5112; fax (52) 221-9268; internet www.daido.co.jp; f. 1950; cap. and res 37,172m. (2000/01), sales 203,449m. (2001/02); metal refining, steel, etc.; Chair. Kanji Tomita; Pres. Tsuyoshi Takayama; 4,662 employees.

Daihatsu Motor Co Ltd: 1-1, Daihatsu-cho, Ikeda, Osaka 563-8651; tel. (727) 51-8811; fax (727) 53-6880; internet www.daihatsu.co.jp; f. 1907; subsidiary of Toyota Motor Corpn; cap. and res 184,265m., sales 943,938m. (2001/02); Chair. Iichi Shingu; Pres. Takaya Yamada; 25,804 employees.

Dainippon Ink & Chemicals Inc: DIC Bldg, 3-7-20, Nihonbashi, Chuo-ku-ku, Tokyo 103-8233; tel. (3) 3272-4511; fax (3) 3273-7586; e-mail webmaster@dic.co.jp; internet www.dic.co.jp; f. 1937; cap. and res 175,162m., sales 979,779m. (2001/02); manufacture and sale of printing inks, printing supplies, machinery, chemicals, imaging and reprographic products, synthetic resins, resin-related products, petrochemicals, packaging materials, plastic compounds, colourants, plastic moulded products, building materials, pressure-sensitive adhesive materials and biochemicals; Pres. Kozo Okumura; 30,972 employees.

Denso Corpn: 1-1, Showa-cho, Kariya-shi, Aichi 448-8661; tel. (566) 25-5519; fax (566) 25-4537; e-mail admin@web.denso.co.jp; internet www.denso.co.jp; f. 1949; cap. and res 1,421,212m., sales 2,401,098m. (2001/02); car electrical equipment, air conditioners, automobile parts; Chair. Akira Takahashi; Pres. and CEO Koichi Fukaya; 80,795 employees.

Fuji Electric Co Ltd: Gate City Ohsaki, East Tower, 1-11-2, Ohsaki, Shinagawa-ku, Tokyo 141-0032; tel. (3) 5435-7111; fax (3) 5435-7493; e-mail info@fujielectric.co.jp; internet www.fujielectric.co.jp; cap. and res 248,061m., sales 839,135m. (2001/02); manufacture of electrical machinery; Chair. Takeo Kato; Pres. and CEO Kunihiko Sawa; 11,060 employees.

Fuji Heavy Industries Co Ltd: 1-7-2, Nishishinjuku, Shinjuku-ku, Tokyo 160-8316; tel. (3) 3347-2111; fax (3) 3347-2295; internet www.fhi.co.jp; cap. and res 396,112m., sales 1,362,493m. (2001/02); motor vehicles and industrial products; Pres. Kyoji Takenaka; Chair. Takeshi Tanaka; 13,600 employees.

Fuji Photo Film Co Ltd: 2-26-30, Nishi Azabu, Minato-ku, Tokyo 106-8620; tel. (3) 3406-2111; fax (3) 3406-2193; internet www.fujifilm.co.jp, http://home.fujifilm.com; f. 1934; cap. and res 1,698,063m., sales 2,401,144m. (2001/02); films and photographic materials, magnetic tapes, carbonless copying paper; Chair. and CEO Minoru Ohnishi; Pres. Shigetaka Komori; 37,151 employees.

Fujitsu Ltd: Marunouchi Center Bldg, 1-6-1, Marunouchi, Chiyoda-ku, Tokyo 100-8211; tel. (3) 3216-3211; fax (3) 3216-9365; e-mail pr@fujitsu.com; internet www.fujitsu.com; f. 1935; cap. and res 853,756m., sales 5,006,977m. (2001/02); manufacture and sale of electronic computers and data processing equipment, telephone equipment, etc.; Chair. Naoyuki Akikusa; Pres. Hiroaki Kurokawa; 170,111 employees.

Furukawa Electric Co Ltd: 2-6-1, Marunouchi, Chiyoda-ku, Tokyo 100-8322; tel. (3) 3286-3001; fax (3) 3286-3694; e-mail pub@ho.furukawa.co.jp; internet www.furukawa.co.jp; f. 1896; cap. and res 494,777m., sales 771,411m. (2001/02); manufacture and sale of electric, telephone and optic-fibre wires, cables and non-ferrous metal products; Pres. Hiroshi Ishihara; Chair. Junnosuke Furukawa; 23,323 employees.

Hino Motors Ltd: 3-1-1, Hinodai, Hino-shi, Tokyo 191-8660; tel. (42) 586-5011; fax (3) 5419-9363; internet www.hino.co.jp; f. 1942; cap. and res 180,267m., sales 758,640m. (2001/02)); diesel trucks and buses; Chair. Iwao Okijima; Pres. Tadaaki Jagawa; 23,687 employees.

Hitachi Ltd: 4-6, Kanda Suragadai, Chiyoda-ku, Tokyo 101-8010; tel. (3) 3258-1111; fax (3) 3258-5480; e-mail webmaster@hitachi.co.jp; internet www.hitachi.co.jp; f. 1910; cap. and res 2,404,224m., sales 7,993,784m. (2001/02); manufacture and sale of power systems, information and communication systems, electronic devices, industrial machinery, metals, chemicals, wire, cable and other products; Chair. Tsutomu Kanai; Pres. Etsuhiko Shoyama; 340,939 employees.

Hitachi Zosen Corpn: 1-7-89, Nanko-kita, Suminoe-ku, Osaka 559-8559; tel. (6) 6569-0001; fax (6) 6569-0002; internet www.hitachizosen.co.jp; f. 1881; cap. and res 50,294m., sales 439,108m. (2001/02); ship-building, ship repairing, conversion, manufacture of diesel engines, offshore equipment, marine auxiliary machinery and fittings; mfrs of industrial machinery and plant for chemicals, paper, petroleum, sugar, cement and iron, steel bridges and steel structures, environmental equipment; Chair. Junichiro Kojima; Pres. Takenao Shigefuji; 10,867 employees.

Honda Motor Co Ltd: 2-1-1, Minami-Aoyama, Minato-ku, Tokyo 107-8556; tel. (3) 3423-1111; fax (3) 5412-1515; internet www.honda.co.jp/english; f. 1948; cap. and res 2,573,941m., sales 7,362,438m. (2001/02); mfrs of automobiles, motorcycles, power tillers, general purpose engines, outboard motors, lawn mowers and portable generators; 24 foreign subsidiaries; Chair. Yoshihide Munekuni; Pres.and CEO Takeo Fukui; 101,100 employees.

Hoya Corpn: 2-7-5, Naka-Ochiai, Shinjuku-ku, Tokyo 161-8525; tel. (3) 3952-1151; fax (3) 3952-1314; internet www.hoya.co.jp; cap. and res 219,180m., sales 235,265m. (2001/02); mfrs of medical and ophthalmic equipment; Pres. and CEO Hiroshi Suzuki; 2,381 employees.

Idemitsu Kosan Co Ltd: 3-1-1, Marunouchi, Chiyoda-ku, Tokyo 100; tel. (3) 3213-3115; internet www.idemitsu.co.jp/e; f. 1911; cap. and res 80,111m., sales 2,186,726m. (1999/2000); manufacture and sale of petroleum products and petrochemicals, and related enterprises; Pres. Akira Idemitsu; 4,592 employees.

Ishikawajima-Harima Heavy Industries Co Ltd: 2-2-1, Otemachi, Chiyoda-ku, Tokyo 100-8182; tel. (3) 3244-5111; fax (3) 3244-5131; internet www.ihi.co.jp; f. 1853; sales 1,082,402m. (2001/02); rocket and satellite propulsion systems, jet engines, gas turbine power generation systems, storage systems, process plants, solid waste treatment systems, container cranes, unloaders, physical distribution systems, bridges, industrial machinery, compressors, semiconductor and LCD panel equipment, parking systems, ozone-based deodorizing and disinfecting equipment, shipbuilding and ship repair service, manufactures, aircraft gas turbines, nuclear power equipment, material handling equipment, iron and steel manufacturing plant, mining and civil engineering machinery, hydro- and thermal electric generating equipment, pneumatic and hydraulic machinery, chemical plant, steel structures, power plants, aero-engines, space utilities, turbochargers, construction machinery; Pres. Mototsugu Ito; 10,966 employees.

Isuzu Motors Ltd: 6-26-1, Minami-Oi, Shinagawa-ku, Tokyo 140-8722; tel. (3) 5471-1111; fax (3) 5471-1042; e-mail pr@notes.isuzu.co.jp; internet www.isuzu.co.jp; f. 1937; cap. and res 61,084m., sales 1,597,701m. (2001/02); manufacture and sale of trucks, buses, sports utility vehicles, components and engines; Chair. TAKESHI INOH; Pres. YOSHINORI IDA; 30,232 employees.

Japan Energy Corpn: 2-10-1, Toranomon, Minato-ku, Tokyo 105-8407; tel. (3) 5573-6100; fax (3) 5573-6784; e-mail ask@j-energy.co.jp; internet www.j-energy.co.jp; f. 1905; present name 1993, fmrly Nikko Kyodo Co (following merger of Nippon Mining Co Ltd and Kyodo Oil Co Ltd in 1992); cap. and res 132,482m., sales 1,592,759m. (2003/04); petroleum resource exploration and development; refining and marketing of petroleum products; pharmaceuticals and biotechnologies; manufacture and marketing of electronic materials, optoelectronics and electronics components; Chair. AKIHIKO NOMIYAMA; Pres. MITSUNORI TAKAHAGI; 2,700 employees.

Japan Tobacco Inc (JT): 2-2-1, Toranomon, Minato-ku, Tokyo 105-8422; tel. (3) 3582-3111; fax (3) 5572-1441; internet www.jti.co.jp; f. 1985; cap. and res 1,613,104m., sales 4,544,174m. (2001/02); tobacco, pharmaceuticals, food, agribusiness, real estate, engineering; Chair. TADASHI OGAWA; Pres. and CEO KATSUHIKO HONDA; 15,588 employees.

JFE Steel Corpn: 2-3-3, Uchisaiwaicho, Chiyoda-ku, Tokyo; internet www.jfe-steel.co.jp; f. 2003; cap. 239,600m. (2003); member of JFE Group; formed by consolidation of NKK and Kawasaki Steel; manufacture and sale of pig iron, steel ingots, tubes, plates, sheets, bars, special steels and ferro-alloys, coal-derived chemicals, refractories and slag wool; engineering and construction of pipelines, steel plants, steel structures, water treatment plants, waste incineration plants, ships; Chair. MASAYUKI HANMMYO; Pres. and CEO FUMIO SUDO.

Kanebo Ltd: 3-20-20, Kaigan, Minato-ku, Tokyo 108-8080; tel. (3) 5446-3002; fax (3) 5446-3027; e-mail webmaster@kanebo.co.jp; internet www.kanebo.co.jp; f. 1887; cap. and res 926m., sales 528,816m. (2001/02); manufacture, bleaching, dyeing, processing and sale of cotton yarns, cloth and thread, worsted and woollen yarns, woollen fabrics, nylon and polyester yarns and fabrics, carpets, spun silk yarns, silk thread spun from waste, silkworm eggs, silk fabrics, rayon staple, spun rayon yarns and fabrics, synthetic resins; cosmetics, pharmaceuticals and industrial materials; Pres. AKIYOSHI NAKAJIMA; 2,860 employees.

Kao Corpn: 1-14-10, Nihonbashi Kayabacho, Chuo-ku, Tokyo 103-8210; tel. (3) 3660-7111; fax (3) 3660-7103; internet www.kao.co.jp; cap. and res 459,731m., sales 839,026m. (2001/02); health and household; Pres. TAKUYA GOTO; 6,086 employees.

Kawasaki Heavy Industries Ltd: Kobe Crystal Tower 1-1-3, Higashi-Kawasakicho, Chuo-ku, Kobe 650-8680; tel. (78) 371-9530; fax (3) 3432-4759; e-mail webadmin@khi.co.jp; internet www.khi.co.jp; f. 1896; cap. and res 167,730m., sales 1,144,534m. (2001/02); manufacture and sale of ships, rolling stock, aircraft, machinery, engines and motorcycles, plant engineering; Chair. TOSHIO KAMEI; Pres. MASAMOTO TAZAKI; 29,772 employees.

Kawasaki Microelectronics Inc: 1-3, Nakase, Mihama-ku, Chiba 261–8501; internet www.k-micro.co.jp; f. 2001; cap. 5,000m. (2001); member of JFE Group; silicon wafers, gases, opto-electronic products, consumer durables and chemical products, sale of super-microcomputers, provision of construction, information, computer software, data communications and engineering services; Pres. SUSUMU HIRANO; 480 employees.

Kirin Brewery Co Ltd: 2-10-1, Shinkawa, Chuo-ku, Tokyo 104-8288; tel. (3) 5540-3411; fax (3) 5540-3547; internet www.kirin.co.jp; f. 1907; cap. and res 782,902m., sales 1,561,879m. (2001); production and sale of beer, soft drinks, dairy foods, pharmaceuticals, engineering and information systems; Chair. YASUSHIRO SATO; Pres. KOICHIRO ARAMAKI; 6,502 employees.

Kobe Steel Ltd: 5-9-12, Kita-Shinagawa, Shinagawa-ku, Tokyo 141-8688; tel. (3) 5739-6000; fax (3) 5739-6903; e-mail www-admin@kobelco.co.jp; internet www.kobelco.co.jp; f. 1905; cap. 218,163m., sales 1,219,179m. (2003/04); manufacture and sale of iron and steel products, aluminium and copper products, industrial machinery, construction machinery; real estate; Chair. KOSHI MIZUKOSHI; Pres. and CEO YASUO INUBUSHI; 26,179 employees.

Komatsu Ltd: 2-3-6, Akasaka, Minato-ku, Tokyo 107-8414; tel. (3) 5561-2616; fax (3) 3505-9662; e-mail info@komatsu.co.jp; internet www.komatsu.co.jp; f. 1921; cap. and res 395,143m., sales 1,035,891m. (2001/02); mfrs of construction equipment and industrial machinery including bulldozers, motor graders, wheel loaders, dump trucks, hydraulic excavators, presses, machine tools, arc welding robots and diesel engines; Chair. TOSHITAKA HAGIWARA; Pres. MASAHIRO SAKANE; 28,522 employees.

Konica Corpn: Shinjuku Nomura Bldg, 1-26-2, Nishi-Shinjuku, Shinjuku-ku, Tokyo 163-0512; tel. (3) 3349-5251; fax (3) 3349-5290; internet www.konica.co.jp; cap. and res 171,226m., sales 539,571m. (2001/02); Chair. TOMIJI UEMATSU; Pres. and CEO FUMIO IWAI; 4,180 employees.

Kubota Corpn: 1-2-47, Shikitsuhigashi, Naniwa-ku, Osaka 556-8601; tel. (6) 6648-2111; fax (6) 6648-3862; internet www.kubota.co.jp; f. 1890; cap. and res 434,979m. (2000/01), sales 976,097m. (2001/02); manufacture and sale of ductile iron pipes, pumps, valves, spiral-welded steel pipes, polyvinyl chloride pipes, tractors, combines, engines, mini-excavators, general farming equipment, cement roofing materials, fire-resistant sidings, sale and installation of environmental control plant and other steel structures, building materials; Pres. DAISUKE HATAKAKE; 14,594 employees.

Kyocera Corpn: 6, Takeda Tobadono-cho, Fushimi-ku, Kyoto 612-8501; tel. (75) 604-3500; fax (75) 604-3501; e-mail webmaster@kyocera.co.jp; internet www.kyocera.co.jp; cap. and res 1,039,478m., sales 1,034,574m. (2001/02); manufacture of fine ceramic parts, semiconductor parts, electronic components and equipment, optical instruments and consumer-related products; Chair. KENSUKE ITOH; Pres. YASUO NISHIGUCHI; 53,000 employees.

Maruha Corpn (Maruha k.k.): 1-1-2, Otemachi, Chiyoda-ku, Tokyo 100-8608; tel. (3) 3216-0821; fax (3) 3216-2082; internet www.maruha.co.jp; f. 1880; name changed from Taiyo Fishery Co Ltd Sept. 1993; cap. and res 31,783m. (2001/02), sales 804,174m. (2002/03); fishing, processing and sale of agricultural marine and meat products; canned and frozen salmon, crab, etc.; food processing, marine transport, export and import; refrigeration, ice production and cold storage; manufacture and sale of pharmaceuticals, organic fertilizers and sugar; Pres. YUJI IGARASHI; 1,084 employees.

Matsushita Electric Industrial Co Ltd: 1006 Oaza Kadoma, Kadoma-shi, Osaka 571-8501; tel. (6) 6908-1121; fax (6) 6908-2351; internet www.matsushita.co.jp, www.panasonic.co.jp/global; f. 1918; cap. and res 3,243,084m., sales 6,876,688m. (2001/02); manufacture of electrical and electronic home appliances; 11 major subsidiaries in Japan; manufacturing and sales companies in 47 countries; Chair. YOICHI MORISHITA; Pres. KUNIO NAKAMURA; 290,448 employees.

Matsushita Electric Works Ltd: 1048 Kadoma, Kadoma-shi, Osaka 571-8686; tel. (6) 6908-1131; fax (6) 6909-6244; e-mail webmaster@mew.co.jp; internet www.mew.co.jp; f. 1918; cap. and res 559,090m., sales 1,199,371m. (2000/01); lighting equipment, housing and building materials, electrical construction materials, electric appliances, plastic and electronic materials and automation components; Chair. KIYOSUKE IMAI; Pres. KAZUSHIGE NISHIDA; 41,234 employees.

Mazda Motor Corpn: 3-1, Shinchi, Fuchu-cho, Aki-gun, Hiroshima 730-8670; tel. (82) 282-1111; fax (82) 287-5190; internet www.mazda.co.jp; f. 1920; fmrly Toyo Kogyo Co Ltd; 33%-owned by Ford Motor Co (USA); cap. and res 172,837m., sales 2,094,914m. (2001/02); manufacture and sale of 'Mazda' passenger cars and commercial vehicles; subsidiaries in Japan, Australia, Belgium, the USA, Canada, Colombia, Indonesia, New Zealand, Italy, Portugal, Spain and Germany; Chair. KAZUHIDE WATANABE; Pres. and CEO HISAKAZU IMAKI; 43,818 employees.

Mitsubishi Chemical Corpn: 2-5-2, Marunouchi, Chiyoda-ku, Tokyo 100-0005; tel. (3) 3283-6254; fax (3) 3283-6287; e-mail mccpr@cc.m-kagaku.co.jp; internet www.m-kagaku.co.jp; f. 1994 by merger; cap. and res 343,749m., sales 1,780,346m. (2001/02); Japan's largest integrated chemical co; manufacture and sale of coke and coal-tar derivatives, dyestuffs and intermediates, caustic soda, organic solvents and chemicals, reagents, ammonia derivatives, inorganic chemicals, pesticides and herbicides, fertilizers, food additives and pharmaceutical intermediates; Pres. and CEO RYUICHI TOMIZAWA; Chair. KANJI SHONO; 10,430 employees.

Mitsubishi Electric Corpn: Mitsubishi Denki Bldg, 2-2-3, Marunouchi, Chiyoda-ku, Tokyo 100; tel. (3) 3218-2111; fax (3) 3218-2431; e-mail prd.prdesk@hq.melco.co.jp; internet www.mitsubishielectric.com; f. 1921; cap. and res 541,710m., sales 3,648,986m. (2001/02); manufacture and sale of electrical machinery and equipment (for power plant, mining, ships, locomotives and other rolling stock, aircraft), electronic products and systems, domestic electric appliances, radio communication equipment, radio and television sets, meters and relaying equipment, fluorescent lamps, lighting, fixtures, refrigerators, lifts, electric tools, sewing machines; Chair. ICHIRO TANIGUCHI; Pres. TAMOTSU NOMAKUCHI; 116,588 employees.

Mitsubishi Heavy Industries Ltd: 16-5 Konan 2-chome, Minato-ku, Tokyo 108-8215; tel. (3) 6716-3111; fax (3) 6716-5800; internet www.mhi.co.jp; f. 1870; cap. and res 1,282,727m., sales 2,863,985m. (2001/02); shipbuilding, ship repairing, power systems, chemical plant and machinery, industrial machinery, heavy machinery, rolling stock, precision machinery, steel structures, construction machinery, refrigerating and air-conditioning machinery, engines, aircraft, special purpose vehicles, space systems; major subsidiaries in Japan, Brazil and other countries; Pres. KAZUO TSUKUDA; 35,426 employees.

Mitsubishi Materials Corpn: 1-5-1, Otemachi, Chiyoda-ku, Tokyo 100-8117; tel. (3) 5252-5201; fax (3) 5252-5272; e-mail www.adm@mmc.co.jp; internet www.mmc.co.jp; cap. and res 206,412m., sales 1,046,807m. (2001/02); metal and metal forming; Chair. YUMI AKIMOTO; Pres. AKIRA NISHIKAWA; 6,556 employees.

Mitsubishi Motors Corpn: 5-33-8, Shiba, Minato-ku, Tokyo 108-8410; tel. (3) 3456-1111; fax (3) 5232-7747; internet www.mitsubishi-motors.co.jp; cap. and res 270,663m., sales 3,200,699m. (2001/02); manufacture of motor vehicles; Pres. YOICHIRO OKAZAKI; 25,846 employees.

Mitsui Chemicals, Inc.: Kasumigaseki Bldg, 3-2-5, Kasumigaseki, Chiyoda-ku, Tokyo 100-6070; tel. (3) 3592-4105; fax (3) 3592-4213; internet www.mitsui-chem.co.jp; f. 1997 by merger of Mitsui Petrochemical Industries Ltd and Mitsui Toatsu Chemicals; cap. and res 366,988m., sales 952,680m. (2001/02); industrial chemicals, fertilizers, dyestuffs, fine chemicals, agricultural and pharmaceuticals, adhesives, electric materials and resins, etc.; Chair. SHIGENORI KODA; Pres. HIROYUKI NAKANISHI; 5,792 employees.

Mitsui Engineering & Shipbuilding Co Ltd: 5-6-4, Tsukiji, Chuo-ku, Tokyo 104-8439; tel. (3) 3544-3147; fax (3) 3544-3050; e-mail prdept@mes.co.jp; internet www.mes.co.jp; f. 1917; cap. and res 105,315m., sales 457,352m. (2001/02); shipbuilding and industrial machinery; Pres. TAKAO MOTOYAMA; 3,931 employees.

NEC Corpn: 5-7-1, Shiba, Minato-ku, Tokyo 108-8001; tel. (3) 3454-1111; fax (3) 3798-1510; internet www.nec-global.com; f. 1899; cap. and res 564,915m., sales 5,101,022m. (2001/02); integrating computers and communications, manufacture and sale of telephone switching systems, carrier transmission and terminals, digital radio and satellite communications, broadcasting electronic data processing and industrial electronic systems, electronic devices and consumer electronic products; Chair. HAJIME SASAKI; Pres. KOJI NISHIGAKI; 152,450 employees.

Nintendo Co Ltd: 60 Fukuine, Kamitoba Hokotate-cho, Minami-ku, Kyoto 601-8501; tel. (75) 662-9600; fax (75) 662-9615; internet www.nintendo.com; cap. and res 935,075m., sales 554,886m. (2001/02); manufacture of electronic video games systems; Pres. SATORU IWATA; 3,073 employees.

Nippon Meat Packers Inc: 3-6-14, Minami-honmachi, Chuo-ku, Osaka 541-0054; tel. (6) 6282-3031; fax (6) 6282-1056; internet www.nipponham.co.jp; cap. and res 257,776m., sales 945,099m. (2001/02); Pres. YOSHIKIYO FUJII; 3,441 employees.

Nippon Oil Corpn (fmrly Nippon Mitsubishi Oil Corpn): 1-3-12, Nishi Shimbashi, Minato-ku, Tokyo 105-8412; tel. (3) 3502-1184; fax (3) 3502-9862; internet www.eneos.co.jp; f. 1999 by merger of Mitsubishi Oil Co Ltd and Nippon Oil Co Ltd; cap. and res 898,083m. (2000/01), sales US $34,939.6m. (2002/03); refining and marketing of petroleum products; Chair. FUMIAKI WATARI; 13,882 employees.

Nippon Paper Industries Co Ltd: 1-12-1, Yuraku-cho, Chiyoda-ku, Tokyo 100-0006; tel. (3) 3218-8000; fax (3) 3214-5226; internet www.npaper.co.jp; f. 1993 by merger between Jujo Paper and Sanyo-Kokusaku Pulp Co Ltd; paper, pulp, chemical, wood products; cap. 104,873 (2002/03), sales 906,041m. (1999/2000); Pres. TAKAHIKO MIYOSHI; 6,009 employees.

Nippon Steel Corpn: Shin Nittetsu Bldg, 2-6-3, Otemachi, Chiyoda-ku, Tokyo 100-8071; tel. (3) 3242-4111; fax (3) 3275-5641; e-mail www-info@www.nsc.co.jp; internet www.nsc.co.jp; f. 1950; cap. and res 907,150m., sales 2,581,399m. (2001/02); Chair. TAKASHI IMAI; Pres. AKIO MIMURA; 19,816 employees.

Nippon Suisan Kaisha Ltd: 2-6-2, Otemachi, Chiyoda-ku, Tokyo 100-8686; tel. (3) 3244-7000; fax (3) 3244-7085; e-mail home@nissui.co.jp; internet www.nissui.co.jp; f. 1911; cap. and res 68,518m., sales 482,953m. (2001/02); marine fisheries and fish products; food processing; cargo and tanker services; Pres.and CEO NAOYA KAKIZOE; 1,790 employees.

Nissan Motor Co Ltd: 2, Takara-cho, Kanagawa-ku, Yokohama; tel. (5) 5565-2147; internet www.nissan.co.jp; f. 1933; cap. and res 1,620,822m., sales 6,196,241m. (2001/02); manufacture and sale of automobiles, rockets, textile machinery, other machines and appliances and parts; Chair. YOSHIKAZU HANAWA; Pres. and CEO CARLOS GHOSN; 136,397 employees.

Nissan Shatai Co Ltd: 10-1, Amanuma, Hiratsuka-shi, Kanagawa-ken 254-8610; tel. (463) 21-8001; fax (463) 21-8155; internet www.nissan-shatai.co.jp; f. 1949; cap. and res 49,767m., sales 462,975m. (2001/02); auto-bodies for passenger cars and small trucks; Pres. KAZUTAKA KOBATAKE; 4,836 employees.

Nisshin Steel Co Ltd: Shinkokusai Bldg, 3-4-1, Marunouchi, Chiyoda-ku, Tokyo 100-8366; tel. (3) 3216-5511; fax (3) 3214-1895; internet www.nisshin-steel.co.jp; f. 1928; cap. and res 233,500m., sales 394,494m. (2001/02); mfrs of coated steel, stainless steel, special steel and various secondary products; Chair. KAZUO HOSHINO; Pres. and CEO TOSHIHIKO ONO; 5,040 employees.

Oji Paper Co Ltd: 4-7-5, Ginza, Chuo-ku, Tokyo 104-0061; tel. (3) 3563-1111; fax (3) 3563-1135; e-mail info@ojipaper.co.jp; internet www.ojipaper.co.jp; f. 1873; name changed 1996; cap. and res 424,256m., sales 1,203,797m. (2001/02); newsprint, packing paper and printing paper; Chair. MASAHIKO OHKUNI; Pres.and CEO SHOICHIRO SUZUKI; 14,044 employees.

Oki Electric Industry Co Ltd: 1-7-12, Toranomon, Minato-ku, Tokyo 105-8460; tel. (3) 3501-3111; fax (3) 3581-5522; e-mail www-admin@www.oki.co.jp; internet www.oki.co.jp; f. 1949; cap. and res 109,066m., sales 604,572m. (2001/02); Pres. and CEO KATSUMASA SHINOZUKA; 8,760 employees.

Omron Corpn: Karasuma Nanajo, Shimogyo-ku, Kyoto 600-8530; tel. (75) 344-7000; fax (75) 344-7001; internet www.omron.co.jp; cap. and res 298,234m., sales 533,964m. (2001/02); mfr of advanced computer, communications and control technologies; Chair. YOSHIO TATEISI; Pres. and CEO HISAO SAKUTA; 23,742 employees.

Pioneer Corpn: 1-4-1, Meguro, Meguro-ku, Tokyo 153-8654; tel. (3) 3494-1111; fax (3) 3495-4431; e-mail pioneer_ir@post.pioneer.co.jp; internet www.pioneer.co.jp; f. 1938; cap. and res 347,003m., sales 651,311m. (2001/02); electronics; Chair. KANYA MATSUMOTO; Pres. KANEO ITO; 27,414 employees.

Ricoh Co Ltd: 1-15-5, Minami-Aoyama, Minato-ku, Tokyo 107-8544; tel. (3) 3479-3111; fax (3) 3403-1578; internet www.ricoh.co.jp; cap. and res 633,020m., sales 1,672,340m. (2001/02); electronics; Chair. HIROSHI HAMADA; Pres. MASAMITSU SAKURAI; 12,392 employees.

Sankyo Co Ltd: 3-5-1, Nihonbashi Honcho, Chuo-ku, Tokyo 103-8426; tel. (3) 5255-7111; fax (3) 5255-7035; internet www.sankyo.co.jp; f. 1899; cap. and res 652,220m., sales 548,893m. (2001/02); health and household; Chair. TETSUO TAKATO; Pres. TAKASHI SHODA; 10,760 employees.

Sanyo Electric Co Ltd: 2-5-5, Keihan Hondori, Moriguchi City, Osaka 570-8677; tel. (6) 6991-1181; fax (6) 6991-5411; internet www.sanyo.co.jp; f. 1947; cap. and res 602,175m., sales 2,112,127m. (2001/02); manufacture and sale of electrical and electronic machinery and appliances—refrigerators, washing machines, electric fans, television and radio sets, bicycle dynamos, personal computers, commercial air conditioning systems, etc.; Chair. SATOSHI IUE; Pres. YUKINORI KUWANO KONDO; 76,176 employees.

Seiko Epson Corpn: Shinjuku NS Bldg, 2-4-1 Nishi-Shinjuku, Tokyo 163-0811; tel. (3) 3348-8531; internet www.epson.co.jp; f. 1942; sales 1,322,453m. (2002/03); development, manufacturing, sales, marketing and servicing of information-related equipment; Pres. SABURO KUSAMA; Chair. HIDEAKI YASUKAWA; 13,084 employees.

Sekisui Chemical Co Ltd: 2-4-4, Nishi-Tenma, Kita-ku, Osaka 5308565; tel. (6) 6365-4122; fax (6) 6365-4370; internet www.sekisui.co.jp; cap. and res 271,287m., sales 845,496m. (2001/02); chemicals, building materials, etc.; Chair. KAORU HIROTA; Pres. NAOTAKE OHKUBO; 5,176 employees.

Sharp Corpn: 22-22, Nagaike-cho, Abeno-ku, Osaka 545-8522; tel. (6) 6621-1221; fax (6) 6628-1653; internet www.sharp.co.jp; f. 1912; cap. and res 926,856m., sales 1,803,798m. (2001/02); manufacture and sale of consumer electronic products, information systems and electronic components; Pres. KATSUHIKO MACHIDA; 49,748 employees.

Shin-Etsu Chemical Co Ltd: 2-6-1, Otemachi, Chiyoda-ku, Tokyo 100-0004; tel. (3) 3246-5011; fax (3) 3246-5350; e-mail sec-pr@shinetsu.co.jp; internet www.shinetsu.co.jp; cap. and res 812,068m., sales 775,096m. (2001/02); Pres. CHIHIRO KANAGAWA; 19,398 employees.

Shiseido Co Ltd: 7-5-5, Ginza, Chuo-ku, Tokyo 104-8010; tel. (3) 3572-5111; fax (3) 3572-6973; internet www.shiseido.co.jp/e; f. 1872; cap. and res 345,667m., sales 589,962m. (2001/02); manufacture and export of cosmetics and toiletries; Chair. AKIRA GEMMA; Pres. and CEO MORIO IKEDA; 23,688 employees.

Showa Denko KK: 1-13-9, Shiba Daimon, Minato-ku, Tokyo 105-8518; tel. (3) 5470-3111; fax (3) 3436-2625; e-mail pr_office@hq.sdk.co.jp; internet www.sdk.co.jp; f. 1939; cap. and res 139,457m., sales 708,900m. (2001); manufacture and sale of bulk and speciality chemicals, plastics, ferro-alloys, electronics materials, electrodes and abrasives; Pres. MITSUO OHASHI; 12,475 employees.

Showa Shell Sekiyu KK: 2-3-2, Daiba, Minato-ku, Tokyo 135-8074; tel. (3) 5531-5601; fax (3) 5531-5609; f. 1942; cap. and res 212,168m., sales 1,664,954m. (2001); petroleum; Chair. HARUYUKI NIIMI; Pres. JOHN S. MILLS; 1,130 employees.

Snow Brand Milk Products Co Ltd: 13, Honshio-cho, Shinjuku-ku, Tokyo 160-8575; tel. (3) 3226-2111; fax (3) 3226-2150; internet www.snowbrand.co.jp; f. 1950; cap. and res 30,371m., sales 1,164,715m. (2001/02); mfrs of liquid milk, condensed and powdered milk, butter, cheese, ice-cream, infant foods, instant foods, margarine, fruit juices, frozen foods; also imported wine distribution; Chair. KATSUYA SHONO; Pres. TADAAKI KONOSE; 15,380 employees.

Sony Corpn: 6-7-35, Kitashinagawa, Shinagawa-ku, Tokyo 141-0001; tel. (3) 5448-2111; fax (3) 5448-2244; internet www.world.sony.com; f. 1946; cap. and res 2,370,410m., sales 7,578,258m. (2001/02); manufacture and sale of electronic appliances, including professional and consumer audio and video equipment; production and distribution of music, motion pictures and television programmes;

Pres. and Chief Operating Officer KUNITAKE ANDO; Chair. and CEO NOBUYUKI IDEI; 181,800 employees.

Sumitomo Chemical Co Ltd: 2-27-1, Shinkawa, Chuo-ku, Tokyo 104-8260; tel. (3) 5543-5102; fax (3) 5543-5901; internet www.sumitomo-chem.co.jp; f. 1913; cap. and res 444,579m., sales 1,018,352m. (2001/02); manufacture and sale of chemical fertilizers, dyestuffs, agricultural chemicals, intermediates, organic and inorganic industrial chemicals, synthetic resins, finishing resins, synthetic rubber and rubber chemicals; many subsidiaries; Chair. AKIO KOSAI; Pres. HIROMASA YONEKURA; 5,410 employees.

Sumitomo Electric Industries Ltd: 4-5-33, Kitahama, Chuo-ku, Osaka 541-0041; tel. (6) 6220-4141; fax (6) 6222-3380; e-mail www@prs.sei.co.jp; internet www.sei.co.jp; f. 1911; cap. and res 690,562m., sales 1,485,021m. (2001/02); mfrs of electric wires and optical-fibre cables, high carbon steel wires; sintered alloy products; rubber and plastic products; disc brakes; radio-frequency products; Chair. and CEO NORITAKA KURAUCHI; Pres. NORIO OKAYAMA; 70,936 employees.

Sumitomo Heavy Industries Ltd: 5-9-11, Kitashinagawa, Shinagawa-ku, Tokyo 141-8686; tel. (3) 5488-8335; fax (3) 5488-8056; e-mail webadmin@shi.co.jp; internet www.shi.co.jp; f. 1934; cap. and res 87,493m., sales 517,137m. (2001/02); industrial machinery and shipbuilding; Chair. MITOSHI OZAWA; Pres. and CEO YOSHIO HINO; 13,794 employees.

Sumitomo Metal Industries Ltd: 4-5-33, Kitahama, Chuo-ku, Osaka 541-0041; tel. (6) 6220-5111; fax (6) 6223-0305; internet www.sumitomometals.co.jp; f. 1897; cap. and res 274,432m., sales 1,349,528m. (2001/02); manufacture and sale of pig iron, steel ingots, steel bars, shapes, wire rods, tubes, pipes, castings, forgings, rolling stock parts, engineering; 100 subsidiaries in Japan; 8 offices abroad; Chair. REIJIRO MORI; Pres. HIROSHI SHIMOZUMA; 11,655 employees.

Sumitomo Metal Mining Co Ltd: 5-11-3, Shimbashi, Minato-ku, Tokyo 105-8716; tel. (3) 3436-7701; fax (3) 3434-2215; internet www.smm.co.jp; cap. and res 236,313m., sales 330,194m. (2001/02); Pres. KOICHI FUKUSHIMA; 2,670 employees.

Suzuki Motor Corpn: 300 Takatsuka, Hamamatsu, Shizuoka 432-8611; tel. (53) 440-2030; fax (53) 440-2776; internet www.suzuki.co.jp; f. 1920; cap. and res 620,004m., sales 1,668,251m. (2001/02); motor vehicles, outboard motors, power products, prefabricated houses; Pres. and CEO MASAO TODA; Chair. and CEO OSAMU SUZUKI; 14,620 employees.

Taiheiyo Cement Corpn: 3-8-1, Nishi Kanda, Chiyoda-ku, Tokyo 101-8357; tel. (3) 5214-1520; fax (3) 5214-1707; e-mail webmaster@taiheiyo-cement.co.jp; internet www.taiheiyo-cement.co.jp; fmrly Chichibu Onoda Cement Corpn; cap. and res 212,666m., sales 979,574m. (2001/02); building materials; Pres. FUMIO SAMESHIMA; Chair. MICHIO KIMURA; 2,600 employees.

Taisei Corpn: 1-25-1, Nishi-Shinjuku, Shinjuku-ku, Tokyo 163-0606; tel. (3) 3348-1111; fax (3) 3345-1386; internet www.taisei.co.jp; f. 1873; cap. and res 177,930m. (1999/2000), sales 1,673,834m. (2001/02); engineering, construction; Chair. OSAMU HIRASHIMA; Pres. HAYAMA KANJI; 10,190 employees.

Takeda Chemical Industries Ltd: 4-1-1, Doshomachi, Chuo-ku, Osaka 540-8645; tel. (6) 6204-2111; fax (6) 6204-2880; internet www.takeda.co.jp; f. 1925; cap. and res 1,420,081m., sales 1,005,060m. (2001/02); mfrs and distributors of pharmaceuticals, industrial chemicals, OTC drugs, food additives; enriched foods and drinks, agricultural chemicals, fertilizers; Pres. and CEO YASUCHIKA HASEGAWA; Chair. and CEO KUNIO TAKEDA; 16,254 employees.

TDK Corpn: 1-13-1, Nihonbashi, Chuo-ku, Tokyo 103-8272; tel. (3) 3278-5111; fax (3) 5201-7110; internet www.tdk.co.jp; f. 1935; cap. and res 583,927m., sales 575,029m. (2001/02); mfrs of recording media and electronic materials and components; Chair. YUTAKA OTOSHI; Pres. and CEO HAJIME SAWABE; 29,747 employees.

Teijin Ltd: 1-6-7, Minami-Honmachi, Chuo-ku, Osaka 541-8587. (6) 6268-2132; fax (6) 6268-3205; internet www.teijin.co.jp; f. 1918; cap. and res 311,468m., sales 923,446m. (2001/02); mfrs of fibres, yarns and fabrics from polyester, fibres (Teijin Tetoron), nylon, polyvinyl chloride fibre (Teijin Teviron), acetate, acrylic fibre (Teijin Beslon), polycarbonate resin (Panlite), acetate resin (Tenex), petrochemicals, pharmaceuticals; 75 subsidiaries; Chair. SHODAKU YASUI; Pres.and CEO TORU NAGASHIMA; 5,220 employees.

Tomen Corpn: 3-2-18, Nakanoshimai, Kita-ku, Osaka 530-8622; tel. (6) 6447-9333; fax (6) 5208-9062; internet www.tomen.co.jp; f. 1920; cap. and res 8,278m., sales 2,516,523m. (2000/01); distributors of natural resources and manufactured goods; Chair. SHIGERU SHIMAZAKI; Pres. MAHITO KAGEYAMA; 1,475 employees.

Toppan Printing Co Ltd: 1, Kanda Izumi-cho, Chiyoda-ku, Tokyo 101-0024; tel. (3) 3835-5741; fax (3) 3835-0674; e-mail kouhou@toppan.co.jp; internet www.toppan.co.jp; f. 1900; cap. and res 707,489m., sales 1,296,195m. (2001/02); Chair. H. FUJITA; Pres. NAOKI ADACHI; 1,610 employees.

Toray Industries Inc: Toray Bldg, 2-2-1, Nihonbashi-Muromachi, Chuo-ku, Tokyo 103-8666; tel. (3) 3245-5111; fax (3) 3245-5459; internet www.toray.co.jp; f. 1926; cap. and res 413,140m., sales 1,015,713m. (2001/02); mfrs of nylon, Toray Tetoron (polyester fibre), Toraylon (acrylic fibre), Torayca (carbon fibre), pharmaceuticals and medical equipment, plastics and chemicals; Chair. KATSUNOSUKE MAEDA; Pres. and CEO SADAYUKI SAKAKIBARA; 8,790 employees.

Toshiba Corpn: 1-1-1, Shibaura, Minato-ku, Tokyo 105-8001; tel. (3) 3457-2096; fax (3) 5444-9202; internet www.toshiba.co.jp; f. 1875; cap. and res 705,314m., sales 5,394,033m. (2001/02); manufacture, sale and export of electric appliances, apparatus and instruments; heavy electric machinery; overseas offices in 24 countries; Chair. TAIZO NISHIMURO; Pres. and CEO TADASHI OKAMURA; 52,265 employees.

Tostem Inax Holding Corpn: 2-1-1, Ojima, Koto-ku, Tokyo 136-8535; tel. (3) 3638-8115; fax (3) 3638-8343; internet www.ithd.co.jp; cap. and res 548,909m., sales 833,522m. (2001/02); housing and building sashes, housing materials, fabricated home products; Chair. and CEO KENJIRO USHIODA; Pres. and Chief Operating Officer CHIKAHISA MIZUTANI; 13,536 employees.

Toyo Seikan Kaisha Ltd: 1-3-1, Uchisaiwai-cho, Chiyoda-ku, Tokyo 100-8522; tel. (3) 3508-2113; fax (3) 3592-9471; internet www.toyo-seikan.co.jp; cap. and res 557,597m., sales 696,395m. (2001/02); metal products; Chair. YOSHIRO TAKASAKI; Pres. HIROFUMI MIKI; 5,825 employees.

Toyobo Co Ltd: 2-2-8, Dojima Hama, Kita-ku, Osaka 530-8230; tel. (6) 6348-3137; fax (6) 6348-3149; internet www.toyobo.co.jp; f. 1882; cap. and res 96,603m., sales 383,078m. (2001/02); manufacture; Chair. MINORU SHIBATA; Pres. JUNJI TSUMURA; 4,080 employees.

Toyota Auto Body Co Ltd: 100, Kanayama, Ichiriyama-cho, Kariya, Aichi 448-8666; tel. (566) 36-2121; fax (566) 36-9113; cap. and res 117,701m., sales 788,755m. (2001/02); Chair. AKIRA IIJIMA; Pres. RISUKE KUBOCHI; 7,889 employees.

Toyota Industries Corpn: 2-1, Toyodacho Kariya-shi, Aichi 448-8671; tel. (566) 22-2511; fax (566) 27-5650; internet www.toyota-industries.com; f. 1926; cap. and res 878,812m., sales 980,163m. (2001/02); transport manufacture, industrial equipment, textile machinery; Chair. AKIRA YOKOI; Pres. TADASHI ISHIKAWA; 9,580 employees.

Toyota Motor Corpn: 1, Toyota-cho, Toyota, Aichi 471-8571; tel. (565) 28-2121; fax (565) 23-5800; internet www.toyota.co.jp; f. 1937; cap. and res 7,325,072m., sales 15,106,297m. (2001/02); manufacture and sale of passenger cars, trucks, forklifts and parts; Chair. HIROSHI OKUDA; Pres. FUJIO CHO; 66,000 employees.

UBE Industries Ltd: 1978–96, Kogushi, Ube City, Yamaguchi, 775-8633; tel. (8) 3631-1111; fax (8) 5419-6230; internet www.ube.co.jp; f. 1897; cap. and res 96,947m., sales 537,548m. (2001/02); production, processing and sale of coal, limestone, chemical fertilizers, sulphuric acid, nitric acid, oxalic acid, ammonium nitrate, ammonia, pharmaceuticals, cement, caprolactam, high pressure polyethylene, industrial machinery and equipment, cast steel products, synthetic rubbers; Pres. KAZUMASA TSUNEMI; 3,630 employees.

Yamaha Corpn: 10-1, Nakazawa-cho, Hamamatsu, Shizuoka 430-8650; tel. (3) 5488-6602; fax (3) 5488-5060; e-mail ngc-y@post.yamaha.co.jp; internet www.yamaha.co.jp; cap. 28,533m., sales 524,763m. (2002/03); musical instruments, electronic devices; Pres. SHUJI ITO; 23,563 employees.

Yamaha Motor Co Ltd: 2500, Shingai Iwata-shi, Shizuokaken 438-8501; tel. (538) 32-1117; fax (538) 32-1131; internet www.yamaha-motor.co.jp; f. 1955; cap. and res 163,591m., sales 946,817m. (2001/02); mfrs of motorcycles, outboard motors, boats, snowmobiles; Pres. TORU HASEGAWA; 26,464 employees.

TRADE UNIONS

A feature of Japan's trade union movement is that the unions are usually based on single enterprises, embracing workers of different occupations in that enterprise. In June 1994 there were 32,581 unions; union membership stood at 12.5m. workers in 1996. In November 1989 the two largest confederations, SOHYO and RENGO, merged to form the Japan Trade Union Confederation (JTUC—RENGO).

Japanese Trade Union Confederation (JTUC–RENGO): 3-2-11, Kanda Surugadai, Chiyoda-ku, Tokyo 101-0062; tel. (3) 5295-05226; fax (3) 5295-0548; e-mail jtuc-kokusai@sv.rengo-net.or.jp; internet www.jtuc-rengo.org; f. 1989; 6.8m. mems; Pres. KIYOSHI SASAMORI.

Principal Affiliated Unions

Ceramics Rengo (All-Japan Federation of Ceramics Industry Workers): 3-11, Heigocho, Mizuho-ku, Nagoya-shi, Aichi 467; tel. (52) 882-4562; fax (52) 882-9960; 30,083 mems; Pres. TSUNEYOSHI HAYAKAWA.

Chain Rokyo (Chain-store Labour Unions' Council): 3rd Floor, 2-29-8, Higashi-ikebukuro, Toshima-ku, Tokyo 170; tel. (3) 5951-1031; fax (3) 5951-1051; 40,015 mems; Pres. TOSHIFUMI HIRANO.

Denki Rengo (Japanese Electrical, Electronic & Information Union): Denkirengo Bldg, 1-10-3, Mita, Minato-ku, Tokyo 108-8326; tel. (3) 3455-6911; fax (3) 3452-5406; internet www.jeiu.or.jp; f. 1953; 688,436 mems; Pres. NOBUAKI KOGA.

Denryoku Soren (Federation of Electric Power Related Industry Workers' Unions of Japan): TDS Mita 7-13, 3rd Floor, Mita 2-Chome, Minato-ku, Tokyo 108-0073; tel. (3) 3454-0231; fax (3) 3798-1470; e-mail info@denryokusoren.or.jp; internet www.denryokusoren.or.jp; 255,278 mems; Pres. NORIO TSUMAKI.

Dokiro (Hokkaido Seasonal Workers' Union): Hokuro Bldg, Kita 4, Nishi 12, Chuo-ku, Sapporo, Hokkaido 060; tel. (11) 261-5775; fax (11) 272-2255; 19,063 mems; Pres. YOSHIZO ODAWARA.

Gomu Rengo (Japanese Rubber Workers' Union Confederation): 2-3-3, Mejiro, Toshima-ku, Tokyo 171; tel. (3) 3984-3343; fax (3) 3984-5862; 60,070 mems; Pres. YASUO FURUKAWA.

Hitetsu Rengo (Japanese Metal Mine Workers' Union): Gotanda Metalion Bldg, 5-21-15, Higashi-gotanda, Shinagawa-ku, Tokyo 141; tel. (3) 5420-1881; fax (3) 5420-1880; 23,500 mems; Pres. SHOUZOU HIMENO.

Insatsu Roren (Federation of Printing Information Media Workers' Unions): Yuai-kaikan, 7th Floor, 2-20-12, Shiba, Minato-ku, Tokyo 105-0014; tel. (3) 5442-0191; fax (3) 5442-0219; 22,303 mems; Pres. HIROFUMI NAKABAYASHI.

JA Rengo (All-Japan Agriculture Co-operative Staff Members' Union): 964-1, Toyotomicho-mikage, Himeji-shi, Hyogo 679-21; tel. and fax (792) 64-3618; 2,772 mems; Pres. YUTAKA OKADA.

Japan Federation of Service and Distributive Workers' Unions: New State Manor Bldg, 3rd Floor, 2-23-1, Yoyogi, Shibuya-ku, Tokyo 151-0053; tel. (3) 3370-4121; fax (3) 3370-1640; e-mail honda@jsd-union.org; internet www.jsd-union.org; 170,000 mems; Pres. TAKAAKI SAKURADA.

Jichi Roren (National Federation of Prefectural and Municipal Workers' Unions): 1-15-22, Oji-honcho, Kita-ku, Tokyo 114; tel. and fax (3) 3907-1584; 5,728 mems; Pres. NOBUO UENO.

Jichiro (All-Japan Prefectural and Municipal Workers' Union): Jichiro Bldg, 1, Rokubancho, Chiyoda-ku, Tokyo 102-0085; tel. (3) 3263-0263; fax (3) 5210-7422; internet www.jichiro.gr.jp; f. 1951; 1,004,000 mems; Pres. MORISHIGE GOTO.

Jidosha Soren (Confederation of Japan Automobile Workers' Unions): U-Life Center, 1-4-26, Kaigan, Minato-ku, Tokyo 105-8523; tel. (3) 3434-7641; fax (3) 3434-7428; internet www.jaw.or.jp; f. 1972; 728,000 mems; Pres. YUJI KATO.

Jiunro (Japan Automobile Drivers' Union): 2-3-12, Nakameguro, Meguro-ku, Tokyo 153; tel. (3) 3711-9387; fax (3) 3719-2624; 1,958 mems; Pres. SADAO KANEZUKA.

JR-Rengo (Japan Railway Trade Unions Confederation): TOKO Bldg, 9th Floor, 1-8-10, Nihonbashi-muromachi, Chuo-ku, Tokyo 103; tel. (3) 3270-4590; fax (3) 3270-4429; 78,418 mems; Pres. KAZUAKI KUZUNO.

JR Soren (Japan Confederation of Railway Workers' Unions): Meguro-satsuki Bldg, 3-2-13, Nishi-gotanda, Shinagawa-ku, Tokyo 141-0031; tel. (3) 3491-7191; fax (3) 3491-7192; internet www.jr-souren.com; 87,000 mems; Pres. YUJI ODA.

Jyoho Roren (Japan Federation of Telecommunications, Electronic Information and Allied Workers): Zendentsu-rodo Bldg, 3-6, Kanda Surugadai, Chiyoda-ku, Tokyo 101-0062; tel. (3) 3219-2231; fax (3) 3253-3268; 265,132 mems; Pres. KAZUO SASAMORI.

Kagaku League 21 (Japanese Federation of Chemistry Workers' Unions): Senbai Bldg, 5-26-30, Shiba, Minato-ku, Tokyo 108-8389; tel. (3) 3452-5591; fax (3) 3454-7464; internet www.jec-u.com; formed by merger of Goka Roren and Zenkoku Kagaku; 104,000 mems; Pres. KATUTOSHI KATO.

Kagaku Soren (Japanese Federation of Chemical Workers' Unions): Kyodo Bldg, 7th Floor, 2-4-10, Higashi-shimbashi, Minato-ku, Tokyo 105; tel. (3) 5401-2268; fax (3) 5401-2263; Pres. HIROKAZU IWASAKI.

Kaiin Kumiai (All-Japan Seamen's Union): 7-15-26, Roppongi, Minato-ku, Tokyo 106-0032; tel. (3) 5410-8330; fax (3) 5410-8336; internet www.jsu.or.jp; 35,000 mems; Pres. SAKAE IDEMOTO.

Kamipa Rengo (Japanese Federation of Pulp and Paper Workers' Unions): 2-12-4, Kita Aoyama, Minato-ku, Tokyo 107-0061; tel. (3) 3402-7656; fax (3) 3402-7659; 50,858 mems; Pres. TUNEO MUKAI.

Kensetsu Rengo (Japan Construction Trade Union Confederation): Yuai Bldg, 7th Floor, 2-20-12, Shiba, Minato-ku, Tokyo 105; tel. (3) 3454-0951; fax (3) 3453-0582; 13,199 mems; Pres. MASAYASU TERASAWA.

Kinzoku Kikai (National Metal and Machinery Workers' Unions of Japan): 6-2, Sakuraokacho, Shibuya-ku, Tokyo 150-0031; tel. (3) 3463-4231; fax (3) 3463-7391; f. 1989; 205,082 mems; Pres. MASAOKI KITAURA.

Kokko Soren (Japan General Federation of National Public Service Employees' Unions): 1-2-1, Kasumigaseki, Chiyoda-ku, Tokyo 100; tel. (3) 3508-4990; fax (3) 5512-7555; 40,370 mems; Pres. MARUYAMA KENZO.

Koku Domei (Japanese Confederation of Aviation Labour): Nikko-kiso Bldg, 2nd Floor, 1-6-3, Haneda-kuko, Ota-ku, Tokyo 144; tel. (3) 3747-7642; fax (3) 3747-7647; 16,310 mems; Pres. KATSUMI UTAGAWA.

Kokuzei Roso (Japanese Confederation of National Tax Unions): R154, Okurasho Bldg, 3-1-1, Kasumigaseki, Chiyoda-ku, Tokyo 100; tel. (3) 3581-2573; fax (3) 3581-3843; 40,128 mems; Pres. TATSUO SASAKI.

Kotsu Roren (Japan Federation of Transport Workers' Unions): Yuai Bldg, 3rd Floor, 2-20-12, Shiba, Minato-ku 105-0014; tel. (3) 3451-7243; fax (3) 3454-7393; 97,239 mems; Pres. SHIGEO MAKI.

Koun-Domei (Japanese Confederation of Port and Transport Workers' Unions): 5-10-2, Kamata, Ota-ku, Tokyo 144-0052; tel. (3) 3733-5285; fax (3) 3733-5280; f. 1987; 1,638 mems; Pres. SAKAE IDEMOTO.

Leisure Service Rengo (Japan Federation of Leisure Service Industries Workers' Unions): Zosen Bldg, 4th Floor, 3-5-6, Misaki-cho, Chiyoda-ku, Tokyo 101-0061; tel. (3) 3230-1724; fax (3) 3239-1553; 47,601 mems; Pres. HIROSHI SAWADA.

NHK Roren (Federation of All-NHK Labour Unions): NHK, 2-2-1, Jinnan, Shibuya-ku, Tokyo 150; tel. (3) 3485-6007; fax (3) 3469-9271; 12,526 mems; Pres. YASUZO SUDO.

Nichirinro (National Forest Workers' Union of Japan): 1-2-1, Kasumigaseki, Chiyoda-ku, Tokyo 100; tel. (3) 3580-8891; fax (3) 3580-1596; Pres. KOH IKEGAMI.

Nikkyoso (Japan Teachers' Union): Japan Education Hall, 2-6-2, Hitotsubashi, Chiyoda-ku, Tokyo 101-0003; tel. (3) 3265-2171; fax (3) 3230-0172; internet www.jtu-net.or.jp; f. 1947; 400,000 mems; Pres. NAGAKAZU SAKAKIBARA.

Rosai Roren (National Federation of Zenrosai Workers' Unions): 2-12-10, Yoyogi, Shibuya-ku, Tokyo 151; tel. (3) 3299-0161; fax (3) 3299-0126; 2,091 mems; Pres. TADASHI TAKACHI.

Seiho Roren (National Federation of Life Insurance Workers' Unions): Tanaka Bldg, 3-19-5, Yushima, Bunkyo-ku, Tokyo 113-0034; tel. (3) 3837-2031; fax (3) 3837-2037; 414,021 mems; Pres. YOHTARU KOHNO.

Seiroren (Labour Federation of Government-Related Organizations): Hasaka Bldg, 4th-6th Floors, 1-10-3, Kanda-ogawacho, Chiyoda-ku, Tokyo 101; tel. (3) 5295-6360; fax (3) 5295-6362; Chair. MITSURU WATANABE.

Sekiyu Roren (Japan Confederation of Petroleum Industry Workers' Union): NKK Bldg, 7th Floor, 2-18-2, Nishi-shinmbashi, Minato-ku, Tokyo 105; tel. (3) 3578-1315; fax (3) 3578-3455; 28,807 mems; Pres. HIROSHI MOCHIMARU.

Sen'i Seikatsu Roren (Japan Federation of Textile Clothing Workers' Unions of Japan): Katakura Bldg, 3-1-2, Kyobashi, Chuo-ku, Tokyo 104; tel. (3) 3281-4806; fax (3) 3274-3165; 4,598 mems; Pres. KATSUYOSHI SAKAI.

Shigen Roren (Federation of Japanese Metal Resources Workers' Unions): Roppongi Azeria Bldg, 1-3-8, Nishi-azabu, Minato-ku, Tokyo 106; tel. (3) 3402-6666; fax (3) 3402-6667; Pres. MINORU TAKAHASHI.

Shin Unten (F10-Drivers' Craft Union): 4th Floor, 3-25-6, Negishi, Taito-ku, Tokyo 110; tel. (3) 5603-1015; fax (3) 5603-5351; 4,435 mems; Pres. SHOHEI SHINOZAKI.

Shinkagaku (National Organization of All Chemical Workers): MF Bldg, 2nd Floor, 2-3-3, Fujimi, Chiyoda-ku, Tokyo 102; tel. (3) 3239-2933; fax (3) 3239-2932; 8,400 mems; Pres. HISASHI YASUI.

Shinrin Roren (Japanese Federation of Forest and Wood Workers' Unions): 3-28-7, Otsuka, Bunkyo-ku, Tokyo 112; tel. (3) 3945-6385; fax (3) 3945-6477; 13,928 mems; Pres. ISAO SASAKI.

Shitetsu Soren (General Federation of Private Railway Workers' Unions): 4-3-5, Takanawa, Minato-ku, Tokyo 108-0074; tel. (3) 3473-0166; fax (3) 3447-3927; f. 1947; 160,000 mems; Pres. RYOICHI IKEMURA.

Shokuhin Rengo (Japan Federation of Foods and Tobacco Workers' Unions): Hiroo Office Bldg, 8th Floor, 1-3-18, Hiroo, Shibuya-ku, Tokyo 150; tel. (3) 3446-2082; fax (3) 3446-6779; f. 1991; 116,370 mems; Pres. SHIGERU MASUDA.

Shokuhin Rokyo (Food Industry Workers' Union Council—FIWUC): ST Bldg, 6th Floor, 4-9-4, Hatchobori, Chuo-ku, Tokyo 104; tel. (3) 3555-7671; fax (3) 3555-7760; Pres. TAROU FUJIE.

Sonpo Roren (Federation of Non-Life Insurance Workers' Unions of Japan): Kanda MS Bldg, 4th Floor, 27, Kanda-higashimatsushitacho, Chiyoda-ku, Tokyo 101; tel. (3) 5295-0071; fax (3) 5295-0073; Pres. KUNIO MATSUMOTO.

Tanro (Japan Coal Miners' Union): Hokkaido Rodosha Bldg, 2nd Floor, Kita-11, Nishi-4, Kita-ku, Sapporo-shi, Hokkaido 001; tel. (11) 717-0291; fax (11) 717-0295; 1,353 mems; Pres. KAZUO SAKUMA.

Tanshokukyo (Association of Japan Coal Mining Staff Unions): 2-30, Nishiminatomachi, Omuta-shi, Fukuoka 836; tel. (944) 52-3883; fax (944) 52-3853; Pres. KEIZO UMEKI.

Tekko Roren (Japan Federation of Steel Workers' Unions): I&S Riverside Bldg, 4th Floor, 1-23-4, Shinkawa, Chuo-ku, Tokyo 104-0033; tel. (3) 3555-0401; fax (3) 3555-0407; internet www.tekko-roren.or.jp; 135,000 mems; Pres. TAKESHI OGINO.

Tokei Roso (Statistics Labour Union Management and Co-ordination Agency): 19-1, Somucho, Wakamatsucho, Shinjuku-ku, Tokyo 162; tel. (3) 3202-1111; fax (3) 3205-3850; Pres. TOSHIAKI MAGARA.

Toshiko (The All-Japan Municipal Transport Workers' Union): 3-1-35, Shibaura, Minato-ku, Tokyo 108; tel. (3) 3451-5221; fax (3) 3452-2977; 43,612 mems; Pres. SHUNICHI SUZUKI.

Ui Zensen (Japanese Federation of Textile, Chemical, Food, Commercial, Service and General Workers' Unions): 4-8-16, Kudanminami, Chiyoda-ku, Tokyo 102-0074; tel. (3) 3288-3723; fax (3) 3288-3728; e-mail kokusai@uizensen.or.jp; internet www.uizensen.or.jp; f. 2002 by merger of CSG Rengo and Zensen Domei; 1,986 affiliates; 790,289 mems (Jan. 2003); Pres. TSUYOSHI TAKAGI.

Unyu Roren (All-Japan Federation of Transport Workers' Union): Zennittsu Kasumigaseki Bldg, 5th Floor, 3-3-3, Kasumigaseki, Chiyoda-ku, Tokyo 100-0013; tel. (3) 3503-2171; fax (3) 3503-2176; f. 1968; 143,084 mems; Pres. KAZUMARO SUZUKI.

Zeikan Roren (Federation of Japanese Customs Personnel Labour Unions): 3-1-1, Kasumigaseki, Chiyoda-ku, Tokyo 100; tel. and fax (3) 3593-1788; Pres. RIKIO SUDO.

Zen Insatsu (All-Printing Agency Workers' Union): 3-59-12, Nishigahara, Kita-ku, Tokyo 114; tel. (3) 3910-7131; fax (3) 3910-7155; 5,431 mems; Chair. TOSHIO KATAKURA.

Zen Yusei (All-Japan Postal Labour Union): 1-20-6, Sendagaya, Shibuya-ku, Tokyo 151; tel. (3) 3478-7101; fax (3) 5474-7085; 77,573 mems; Pres. NOBUAKI IZAWA.

Zenchuro (All-Japan Garrison Forces Labour Union): 3-41-8, Shiba, Minato-ku, Tokyo 105; tel. (3) 3455-5971; fax (3) 3455-5973; Pres. EIBUN MEDORUMA.

Zendensen (All- Japan Electric Wire Labour Union): 1-11-6, Hatanodai, Shinagawa-ku, Tokyo 142; tel. (3) 3785-2991; fax (3) 3785-2995; Pres. NAOKI TOKUNAGA.

Zen-eien (National Cinema and Theatre Workers' Union): Hibiya Park Bldg, 1-8-1, Yurakucho, Chiyoda-ku, Tokyo 100; tel. (3) 3201-4476; fax (3) 3214-0597; Pres. SADAHIRO MATSUURA.

Zengin Rengo (All-Japan Federative Council of Bank Labour Unions): R904, Kyodo Bldg, 16-8, Nihonbashi-Kodenmacho, Chuo-ku, Tokyo 103; tel. and fax (3) 3661-4886; 32,104 mems; Pres. KIKUO HATTORI.

Zenjiko Roren (National Federation of Automobile Transport Workers' Unions): 3-7-9, Sendagaya, Shibuya-ku, Tokyo 151; tel. (3) 3408-0875; fax (3) 3497-0107; Pres. OSAMU MIMASHI.

Zenkairen (All-Japan Shipping Labour Union): Shinbashi Ekimae Bldg, No. 1, 8th Floor, 2-20-15, Shimbashi, Minato-ku, Tokyo 105; tel. (3) 3573-2401; fax (3) 3573-2404; Chair. MASAHIKO SATO.

Zenkin Rengo (Japanese Federation of Metal Industry Unions): Yuai Bldg, 5th Floor, 2-20-12, Shiba, Minato-ku, Tokyo 105-0014; tel. (3) 3451-2141; fax (3) 3452-0239; f. 1989; 310,818 mems; Pres. MITSURO HATTORI.

Zenkoku Gas (Federation of Gas Workers' Unions of Japan): 5-11-1, Omori-nishi, Ota-ku, Tokyo 143; tel. (3) 5493-8381; fax (3) 5493-8216; 31,499 mems; Pres. AKIO HAMAUZU.

Zenkoku Keiba Rengo (National Federation of Horse-racing Workers): 2500, Mikoma, Miho-mura, Inashiki-gun, Ibaragi 300-04; tel. (298) 85-0402; fax (298) 85-0416; Pres. TOYOHIKO OKUMURA.

Zenkoku Nodanro (National Federation of Agricultural, Forestry and Fishery Corporations' Workers' Unions): 1-5-8, Hamamatsucho, Minato-ku, Tokyo 105; tel. (3) 3437-0931; fax (3) 3437-0681; 26,010 mems; Pres. SHIN-ICHIRO OKADA.

Zenkoku Semento (National Federation of Cement Workers' Unions of Japan): 5-29-2, Shimbashi, Minato-ku, Tokyo 105; tel. (3) 3436-3666; fax (3) 3436-3668; Pres. KIYONORI URAKAWA.

Zenkoku-Ippan (National Council of General Amalgamated Workers' Unions): Zosen Bldg, 5th Floor, 3-5-6, Misakicho, Chiyoda-ku, Tokyo 101-0061; tel. (3) 3230-4071; fax (3) 3230-4360; 54,708 mems; Pres. YASUHIKO MATSUI.

Zenkyoro (National Race Workers' Union): Nihon Kyoiku Kaikan, 7th Floor, 2-6-2, Hitotsubashi, Chiyoda-ku, Tokyo 101; tel. (3) 5210-5156; fax (3) 5210-5157; 24,720 mems; Pres. SHIMAKO YOSHIDA.

Zennitto (Japan Painting Workers' Union): Shin-osaka Communication Plaza, 1st Floor, 1-6-36, Nishi-miyahara, Yodogawa-ku, Osaka-shi, Osaka 532; tel. (6) 6393-8677; fax (6) 6393-8533; Pres. SEIICHI UOZA.

Zenrokin (Federation of Labour Bank Workers' Unions of Japan): Nakano Bldg, 3rd Floor, 1-11, Kanda-Awajicho, Chiyoda-ku, Tokyo 101; tel. (3) 3256-1015; fax (3) 3256-1045; Pres. EIICHI KAKU.

Zensuido (All-Japan Water Supply Workers' Union): 1-4-1, Hongo, Bunkyo-ku, Tokyo 113; tel. (3) 3816-4132; fax (3) 3818-1430; 33,522 mems; Pres. KAZUMASA KATO.

Zentanko (National Union of Coal Mine Workers): Yuai Bldg, 6th Floor, 2-20-12, Shiba, Minato-ku, Tokyo 105; tel. (3) 3453-4721; fax (3) 3453-6457; Pres. AKIRA YASUNAGA.

Zentei (Japan Postal Workers' Union): 1-2-7, Koraku, Bunkyo-ku, Tokyo 112-0004; tel. (3) 3812-4260; fax (3) 5684-7201; internet www.zentei.or.jp; 156,784 mems; Pres. MASAYUKI ISHIKAWA.

Zenzohei (All-Mint Labour Union): 1-1-79, Temma, Kita-ku, Osaka-shi, Osaka 530; tel. and fax (6) 6354-2389; Pres. CHIKASHI HIGUCHI.

Zenzosen-kikai (All-Japan Shipbuilding and Engineering Union): Zosen Bldg, 6th Floor, 3-5-6, Misakicho, Chiyoda-ku, Tokyo 101; tel. (3) 3265-1921; fax (3) 3265-1870; Pres. YOSHIMI FUNATSU.

Zosen Juki Roren (Japan Confederation of Shipbuilding and Engineering Workers' Unions): Yuai Kaikan Bldg, 4th Floor, 2-20-12, Shiba, Minato-ku, Tokyo 105-0014; tel. (3) 3451-6783; fax (3) 3451-6935; e-mail zosenjuki@mth.biglobe.ne.jp; 111,405 mems; Pres. MASAYUKI YOSHII.

Transport

RAILWAYS

Japan Railways (JR) Group: 1-6-5, Marunouchi, Chiyoda-ku, Tokyo 100-0005; tel. (3) 3215-9649; fax (3) 3213-5291; internet www.japanrail.com; fmrly the state-controlled Japanese National Railways (JNR); reorg. and transferred to private-sector control in 1987; the high-speed Shinkansen rail network consists of the Tokaido line (Tokyo to Shin-Osaka, 552.6 km), the Sanyo line (Shin-Osaka to Hakata, 623.3 km), the Tohoku line (Tokyo to Morioka, 535.3 km) and the Joetsu line (Omiya to Niigata, 303.6 km). The 4-km link between Ueno and Tokyo stations was opened in June 1991. The Yamagata Shinkansen (Fukushima to Yamagata, 87 km) was converted in 1992 from a conventional railway line. It is operated as a branch of the Tohoku Shinkansen with through trains from Tokyo, though not at full Shinkansen speeds. In 1997 the total railway route length was about 36,634 km. In 2003 magnetically levitated (Maglev) trains were being tested for use in commercial service.

Central Japan Railway Co: Yaesu Center Bldg, 1-6-6, Yaesu, Chuo-ku, Tokyo 103-8288; tel. (3) 3274-9727; fax (3) 5255-6780; internet www.jr-central.co.jp; f. 1987; also operates travel agency services, etc.; Chair. HIROSHI SUDA; Pres. YOSHIYUKI KASAI.

East Japan Railway Co: 2-2-2, Yoyogi, Shibuya-ku, Tokyo 151-8578; tel. (3) 5334-1151; fax (3) 5334-1110; internet www.jreast.co.jp; privatized in 1987; Pres. MUTSUTAKE OTSUKA.

Hokkaido Railway Co: West 15-chome, Kita 11-jo, Chuo-ku, Sapporo 060-8644; tel. (11) 700-5717; fax (11) 700-5719; e-mail keieki@jrhokkaido.co.jp; internet www.jrhokkaido.co.jp; Chair. YOSHIHIRO OHMORI; Pres. SHINICHI SAKAMOTO.

Japan Freight Railway Co: 2-3-19, Koraku, Bunkyo-ku, Tokyo 112-0004; tel. (3) 3816-9722; internet www.jrfreight.co.jp; Chair. MASASHI HASHIMOTO; Pres. YASUSHI TANAHASHI.

Kyushu Railway Co: 3-25-21, Hakataekimae, Hakata-ku, Fukuoka 812-8566; tel. (92) 474-2501; fax (92) 474-9745; internet www.jrkyushu.co.jp; Chair. K. TANAKA; Pres. S. ISHIHARA.

Shikoku Railway Co: 8-33, Hamano-cho, Takamatsu, Kagawa 760-8580; tel. (87) 825-1622; fax (87) 825-1623; internet www.jr-shikoku.co.jp; Chair. HIROATSU ITO; Pres. TOSHIYUKI UMEHARA.

West Japan Railway Co: 2-4-24, Shibata, Kita-ku, Osaka 530-8341; tel. (6) 6375-8981; fax (6) 6375-8919; e-mail wjr01020@mxy.meshnet.or.jp; internet www.westjr.co.jp; scheduled for privatization in 2001; Chair. MASATAKA IDE; Pres. SHOJIRO NANYA.

Other Principal Private Companies

Hankyu Corpn: 1-16-1, Shibata, Kita-ku, Osaka 530-8389; tel. (6) 6373-5092; fax (6) 6373-5670; e-mail koho@hankyu.co.jp; internet www.hankyu.co.jp; f. 1907; links Osaka, Kyoto, Kobe and Takarazuka; Chair. Kohei Kobayashi; Pres. T. Ohashi.

Hanshin Electric Railway Co Ltd: 1-1-24, Ebie, Fukushima-ku, Osaka 553; tel. (6) 6457-2123; f. 1899; Chair. S. Kuma; Pres. M. Tezuka.

Keihan Electric Railway Co Ltd: 1-2-27, Shiromi, Chuo-ku, Osaka 540; tel. (6) 6944-2521; fax (6) 6944-2501; internet www.keihan.co.jp; f. 1906; Chair. Minoru Miyashita; Pres. A. Kimba.

Keihin Express Electric Railway Co Ltd: 2-20-20, Takanawa, Minato-ku, Tokyo 108-8625; tel. (3) 3280-9120; fax (3) 3280-9199; internet www.keikyu.co.jp; f. 1899; Chair. Ichiro Hiramatsu; Pres. M. Kotani.

Keio Electric Railway Co Ltd: 1-9-1, Sekido, Tama City, Tokyo 206-8052; tel. (42) 337-3106; fax (42) 374-9322; internet www.keio.co.jp; f. 1913; Chair. K. Kuwayama; Pres. H. Nishiyama.

Keisei Electric Railway Co Ltd: 1-10-3, Oshiage, Sumida-ku, Tokyo 131; tel. (3) 3621-2242; fax (3) 3621-2233; internet www.keisei.co.jp; f. 1909; Chair. (vacant); Pres. M. Sato.

Kinki Nippon Railway Co Ltd: 6-1-55, Uehommachi, Tennoji-ku, Osaka 543-8585; tel. (6) 6775-3444; fax (6) 6775-3468; internet www.kintetsu.co.jp; f. 1910; Chair. Wa Tashiro; Pres. Akio Tsujii.

Nagoya Railroad Co Ltd: 1-2-4, Meieki, Nakamura-ku, Nagoya-shi 450; tel. (52) 571-2111; fax (52) 581-6060; e-mail info@meitetsu.co.jp; internet www.meitetsu.co.jp; Chair. S. Taniguchi; Pres. S. Minoura.

Nankai Electric Railway Co Ltd: 5-1-60, Namba, Chuo-ku, Osaka 542; tel. (6) 6644-7121; internet www.nankai.co.jp; Pres. Shigeru Yoshimura; Vice-Pres. K. Okamoto.

Nishi-Nippon Railroad Co Ltd: 1-11-17, Tenjin-cho, Chuo-ku, Fukuoka 810; tel. (92) 761-6631; fax (92) 722-1405; internet www.nnr.co.jp; serves northern Kyushu; Chair. H. Yoshimoto; Pres. G. Kimoto.

Odakyu Electric Railway Co Ltd: 1-8-3, Nishi Shinjuku, Shinjuku-ku, Tokyo 160; tel. (3) 3349-2151; fax (3) 3346-1899; internet www.odakyu-group.co.jp; f. 1948; Chair. Tatsuzo Toshimitsu; Pres. M. Kitanaka.

Sanyo Electric Railway Co Ltd: 3-1-1, Oyashiki-dori, Nagata-ku, Kobe 653; tel. (78) 611-2211; Pres. T. Watanabe.

Seibu Railway Co Ltd: 1-11-1, Kasunokidai, Tokorozawa-shi, Saitama 359; tel. (429) 26-2035; fax (429) 26-2237; internet www.seibu-group.co.jp/railways; f. 1894; Pres. Yoshiaki Tsutsumi.

Tobu Railway Co Ltd: 1-1-2, Oshiage, Sumida-ku, Tokyo 131-8522; tel. (3) 3621-5057; internet www.tobu.co.jp; f. 1897; Chair. Kaichiro Nezu; Pres. Takashige Uchida.

Tokyo Express Electric Railway (Tokyu) Co Ltd : 5-6, Nanpeidai-cho, Shibuya-ku, Tokyo 150; tel. (3) 3477-6111; fax (3) 3496-2965; e-mail public@tokyu.co.jp; internet www.tokyu.co.jp; f. 1922; Pres. S. Shimuzu.

Principal Subways, Monorails and Tunnels

Subway services operate in Tokyo, Osaka, Kobe, Nagoya, Sapporo, Yokohama, Kyoto, Sendai and Fukuoka. A subway was being planned for Kawasaki by 2010. Most subway lines operate reciprocal through-services with existing private railway lines which connect the cities with suburban areas.

The first commercial monorail system was introduced in 1964 with straddle-type cars between central Tokyo and Tokyo International Airport, a distance of 13 km. Monorails also operate in other cities, including Chiba, Hiroshima, and Kitakyushu.

In 1985 the 54-km Seikan Tunnel (the world's longest undersea tunnel), linking the islands of Honshu and Hokkaido, was completed. Electric rail services through the tunnel began operating in March 1988.

Fukuoka City Subway: Fukuoka Municipal Transportation Bureau, 2-5-31, Daimyo, Chuo-ku, Fukuoka 810-0041; tel. (92) 732-4107; fax (92) 721-0754; internet subway.city.fukuoka.jp; 2 lines of 17.8 km open; Dir Kennichirou Nishi.

Kobe Rapid Transit: 6-5-1, Kanocho, Chuo-ku, Kobe 650; tel. (78) 331-8181; 22.7 km open; Dir Yasuo Maeno.

Kyoto Rapid Transit: 48, Bojocho Mibu, Nakakyo-ku, Kyoto 604; tel. (75) 822-9115; fax (75) 822-9240; 26.4 km open; Chair. T. Tanabe.

Nagoya Subway: Transportation Bureau City of Nagoya, Nagoya City Hall, 3-1-1, Sannomaru, Naka-ku, Nagoya 460-8508; tel. (52) 972-3824; fax (52) 972-3849; internet www.kotsu.city.nagoya.jp; 78.2 km open (2001); Dir-Gen. Takayasu Tsukamoto.

Osaka Monorail: 5-1-1, Higashi-machi, Shin-Senri, Toyonakashi, Osaka 565; tel. (6) 871-8280; fax (6) 871-8284; 113.5 km open; Gen. Man. S. Oka.

Osaka Underground Railway: Osaka Municipal Transportation Bureau, 1-11-53, Kujominami, Nishi-ku, Osaka 550; tel. (6) 6582-1101; fax (6) 6582-7997; f. 1933; 120 km open in 1998; the 6.6 km computer-controlled 'New Tram' service began between Suminoekoen and Nakafuto in 1981; a seventh line between Kyobashi and Tsurumi-ryokuchi was opened in 1990; Gen. Man. Harumi Sakai.

Sapporo Transportation Bureau: Higashi, 2-4-1, Oyachi, Atsubetsu-ku, Sapporo 004; tel. (11) 896-2708; fax (11) 896-2790; f. 1971; 3 lines of 48 km open in 1993/94; Dir T. Ikegami.

Sendai City Subway: Sendai City Transportation Bureau, 1-4-15, Kimachidori, Aoba-ku, Sendai-shi, Miyagi-ken 980-0801; tel. (22) 224-5502; fax (22) 224-6839; internet www.comminet.or.jp/~kotsu-s; 15.4 km open; Dir T. Iwama.

Tokyo Metropolitan Government (TOEI) Underground Railway: Bureau of Transportation, Tokyo Metropolitan Government, 2-8-1 Nishi-Shinjuku, Tokyo 163-8001; tel. (3) 5320-6026; internet www.kotsu.metro.tokyo.jp; operates four underground lines, totalling 105 km.

Tokyo Underground Railway (TEITO): Teito Rapid Transit Authority (TRTA), 3-19-6, Higashi Ueno, Taito-ku, Tokyo 110-0015; tel. (3) 3837-7046; fax (3) 3837-7048; internet www.tokyometro.go.jp; f. 1941; operates nine lines; 183.2 km open (March 2003); Pres. Yasutoshi Tsuchisaka.

Yokohama Rapid Transit: Municipal Transportation Bureau, 1-1, Minato-cho, Naka-ku, Yokohama 231-80; tel. (45) 671-3201; fax (45) 664-3266; 40.4 km open; Dir-Gen. Michinori Kishida.

ROADS

In December 2000 Japan's road network extended to an estimated 1,166,340 km, including 6,617 km of motorways and 53,777 km of highways. In May 1999 work was completed on a 29-year project to construct three routes, consisting of a total of 19 bridges, between the islands of Honshu and Shikoku across the Seto inland sea, at a cost of some US $25,000m.

There is a national omnibus service, 60 publicly-operated services and 298 privately-operated services.

Japan Highway Public Corpn: 3-3-2, Kasumigaseki, Chiyoda-ku, Tokyo 100-8979; tel. (3) 3506-0111; privatization plans announced in Dec. 2001.

SHIPPING

Shipping in Japan is subject to the supervision of the Ministry of Transport. At 31 December 2003 the Japanese merchant fleet (7,151 vessels) had a total displacement of 13,561,521 grt. The main ports are Tokyo, Yokohama, Nagoya and Osaka.

Principal Companies

Daiichi Chuo Kisen Kaisha: Dowa Bldg, 3-7-13, Toyoi, Koto-ku, Tokyo 103-8271; tel. (3) 5634-2276; fax (3) 5634-2262; f. 1960; liner and tramp services; Pres. Mahiko Saotome.

Iino Kaiun Kaisha Ltd: Iino Bldg, 2-1-1, Uchisaiwai-cho, Chiyoda-ku, Tokyo 100; tel. (3) 3506-3037; fax (3) 3508-4121; f. 1918; cargo and tanker services; Chair. T. Chiba; Pres. A. Karino.

Kansai Kisen KK: Osaka Bldg, 3-6-32, Nakanoshima, Kita-ku, Osaka 552; tel. (6) 6574-9131; fax (6) 6574-9149; f. 1942; domestic passenger services; Pres. Toshikazu Eguchi.

Kawasaki Kisen Kaisha Ltd (K Line): 1-2-9, Nishi Shimbashi, Minato-ku, Tokyo 105-8421; tel. (3) 3595-5082; fax (3) 3595-5001; e-mail otaki@email.kline.co.jp; internet www.kline.co.jp; f. 1919; containers, cars, LNG, LPG and oil tankers, bulk carriers; Chair. of Bd I. Shintani; Exec. Vice-Pres. Z. Wakabayashi.

Nippon Yusen Kaisha (NYK) Line: 2-3-2, Marunouchi, Chiyoda-ku, Tokyo 100-0005; tel. (3) 3284-5151; fax (3) 3284-6361; e-mail prteam@jp.nykline.com; internet www.nykline.com; f. 1885; merged with Showa Line Ltd in 1998; world-wide container, cargo, pure car and truck carriers, tanker and bulk carrying services; Chair. Jiro Nemoto; Pres. Kentaro Kawamura.

Nissho Shipping Co Ltd: 33, Mori Bldg, 7th Floor, 3-8-21, Toranomon, Minato-ku, Tokyo 105; tel. (3) 3438-3511; fax (3) 3438-3566; f. 1943; Pres. Minoru Ikeda.

OSK Mitsui Ltd: Shosen Mitsui Bldg, 2-1-1, Toranomon, Minato-ku, Tokyo 105-91; tel. (3) 3587-7092; fax (3) 3587-7734; f. 1942; merged with Navix Line Ltd in 1999; world-wide container, liner, tramp, and specialized carrier and tanker services; Chair. Susumu Temporin; Pres. Masahuru Ikuta.

Ryukyu Kaiun KK: 1-24-11, Nishi-machi, Naha, Okinawa 900; tel. (98) 868-8161; fax (98) 868-8561; cargo and passenger services on domestic routes; Pres. M. AZAMA.

Taiheiyo Kaiun Co Ltd: Mitakokusai Bldg, 23rd Floor, 1-4-28, Minato-ku, Tokyo 100; tel. (3) 5445-5805; fax (3) 5445-5806; f. 1951; cargo and tanker services; Pres. SANROKURO YAMAJI.

CIVIL AVIATION

There are international airports at Tokyo (Haneda and Narita), Osaka, Nagoya, and Fukuoka. In 1991 the Government approved a plan to build five new airports, and to expand 17 existing ones. In September 1994 the world's first offshore international airport (Kansai International Airport) was opened in Osaka Bay, and a second runway was due for completion in 2007. In April 2002 a second runway was opened at Narita. In December 2001 plans were approved for a fourth runway at Haneda. Nagoya, Kansai, and Narita airports were scheduled to be privatized in 2004. Japan's second international offshore airport was planned to open at Chubu in 2005. An international airport at Shizuoka was scheduled to open in 2006. At March 1999 a total of 85 airports were in operation.

Air Do: 6, Nishi, Kita 5, Chuo-ku, Sapporo; tel. (11) 252-5533; fax (11) 252-5580; e-mail postbear@airdo.co.jp; internet www.airdo.co.jp; f. 1996; domestic service between Tokyo and Sapporo; Pres. AKIRA NAKAMURA.

Air Nippon: 3-5-10, Haneda Airport, Ota-ku, Tokyo 144-0041; tel. (3) 5756-4710; fax (3) 5756-4788; internet www.air-nippon.co.jp; f. 1974; fmrly Nihon Kinkyori Airways; international and domestic passenger services; Pres. and CEO YUZURU MASUMOTO.

All Nippon Airways (ANA): Shiodome City Center, 1-5-2 Higashi-Shimbashi, Minato-ku, Tokyo 105-7133; tel. (3) 6735-1111; fax (3) 6735-1115; internet www.ana.co.jp; f. 1952; operates domestic passenger and freight services; scheduled international services to the Far East the USA and Europe; charter services world-wide; Pres. and CEO YOJI OHASHI.

Hokkaido Air System: New Chitose Airport, Bibi Chitose City, Hokkaido 066-0055; tel. (123) 46-5533; fax (123) 46-5534; internet www.hac-air.co.jp; f. 1997; domestic services on Hokkaido; Pres. YASAYUKI BABA.

JALways Co Ltd: Sphere Tower Tennoz 23 F. 2-8, Higashi-Shinagawa 2-chome, Shinagawa-ku, Tokyo, Japan 140-0002; tel. (3) 5460-6830; fax (3) 5460-8660; e-mail jazgz.jaz@jal.com; internet www.jalways.co.jp; f. 1990; subsidiary of JAL; domestic and international scheduled and charter services; Chair. KAZUNARI YASHIRO; Pres. KATSUMI CHIYO.

Japan Air Commuter: 8-2-2, Fumoto, Mizobe-cho, Aira-gun, Kagoshima 899-64; tel. (995) 582151; fax (995) 582673; e-mail info@jac.co.jp; internet www.jac.co.jp; f. 1983; subsidiary of Japan Air System; domestic services; Chair. YOSHITOMI ONO.

Japan Air System: JAS M1 Bldg, 3-5-1, Haneda Airport, Ota-ku, Tokyo 144-0041; tel. (3) 5756-4022; fax (3) 5473-4109; internet www.jas.co.jp; f. 1971; domestic and international services; plans to coordinate services with Northwest Airlines (USA) announced in 1995; merging with JAL in late 2002 to form Japan Airlines System; Chair. TAKESHI MASHIMA; Pres. HIROMI FUNABIKI.

Japan Airlines Co Ltd (JAL) (Nihon Koku Kabushiki Kaisha): JAL Bldg, 2-4-11, Higashi-Shinagawa, Shinagawa-ku, Tokyo 140; tel. (3) 5460-3121; fax (3) 5460-3936; internet www.jal.co.jp; f. 1951; fully transferred to private-sector control in 1987; domestic and international services to Australasia, the Far East, North America, South America and Europe; merging with Japan Air System in late 2002 to form Japan Airlines System; Pres. ISAO KANEKO.

Japan Asia Airways Co: JAL Bldg, 19th Floor, 2-4-11. Higashi-shinagawa, Shinagawa-ku, Tokyo 140-0002; tel. (3) 5460-7285; fax (3) 5460-7286; e-mail jaabz@jaa.jalgroup.or.jp; internet www.japanasia.co.jp; f. 1975; subsidiary of JAL; international services from Tokyo, Osaka, Nagoya and Okinawa to Hong Kong and Taiwan; Chair. TEIICHI KURIBAYASHI; Pres. OSAMU IGARASHI.

Japan Transocean Air: 3-24, Yamashita-cho, Naha-shi, Okinawa 9000027; tel. (98) 857-2112; fax (98) 857-9396; internet www.jal.co.jp/jta; f. 1967; present name since 1993; subsidiary of JAL; domestic passenger services; 8 brs; Chair. KEIICHI INAMINE; Pres. TAKESHI IOHINOSAWA.

Nakanihon Airlines (NAL): Nagoya Airport, Toyoyama-cho, Nishikasugai-gun, Aichi 480-0202; tel. (568) 285405; fax (568) 285417; internet www.nals.co.jp; f. 1988; regional and domestic services; Pres. AKIRA HIRABAYASHI.

Skymark Airlines: World Trade Center, Bldg 3F, 2-4-1, Hamamatsucho, Minato-ku, Tokyo 105-6103; tel. (3) 5402-6767; fax (3) 5402-6770; e-mail info@skymark.co.jp; internet www.skymark.co.jp; f. 1997; domestic services; Chair. HIDEO SAWADA; Pres MASAYUKI INOUE.

Tourism

The ancient capital of Kyoto, pagodas and temples, forests and mountains, traditional festivals and the classical Kabuki theatre are some of the many tourist attractions of Japan. The number of foreign visitors to Japan declined from 5,238,963 in 2002 to 5,211,725 in 2003. Receipts from tourism in 2002 totalled US $3,499m.

Department of Tourism: 2-1-3, Kasumigaseki, Chiyoda-ku, Tokyo 100; tel. (3) 3580-4488; fax (3) 3580-7901; f. 1946; a dept of the Ministry of Transport; Dir-Gen. KIMITAKA FUJINO.

Japan National Tourist Organization: Tokyo Kotsu Kaikan Bldg, 2-10-1, Yuraku-cho, Chiyoda-ku, Tokyo 100-0006; tel. (3) 3216-1901; fax (3) 3216-1846; internet www.jnto.go.jp; Pres. MINORU NAKAMURA.

Japan Travel Bureau Inc: 1-6-4, Marunouchi, Chiyoda-ku, Tokyo 100-0005; tel. (3) 3284-7028; f. 1912; 10,297 mems; Chair. I. MATSUHASHI; Pres. R. FUNAYAMA.

Defence

The total strength of the Japanese Self-Defence Forces at 1 August 2003 was some 239,900: army 148,200, navy 44,400, air force 45,600 and central staff 1,700. Military service is voluntary. US forces stationed in Japan totalled 40,680 at 1 August 2003.

Defence Expenditure: Budgeted for 2004/05: 4,903,000m. yen.

Chairman of the Joint Staff Council: Adm. HAJIME MASSAKI.

Chief of Staff of Ground Self-Defence Force: Gen. TSUTOMU MORI.

Chief of Staff of Maritime Self-Defence Force: Adm. KOICHI FURUSHO.

Chief of Staff of Air Self-Defence Force: Gen. YOSHIMITSU TSUMAGARI.

Education

Immediately after the Second World War, with the introduction of democratic ideas into Japan, the educational system underwent extensive reform. General standards of education are very high, especially in mathematics and foreign languages. The standard of literacy among the Japanese has been almost 100% since before 1900. The Ministry of Education administers education at all levels and provides guidance, advice and financial assistance to local authorities. In each of the 47 prefectures and 3,255 municipalities, boards of education are responsible for upper secondary and special schools, while municipal boards maintain public elementary and lower secondary schools. Central government expenditure on education and science was expected to amount to 6,473,100m. yen for the 1999/2000 financial year. Each level of government provides for education with funds derived from its own revenue including taxes. The central Government may also grant subsidies where appropriate. The Government offers a scholarship system to able students with financial difficulties, who are expected to return the money within 20 years of graduation. Steadily increasing numbers of young people from Asian countries are coming to Japan for technical training at scientific and technological institutes and at factories.

PRE-SCHOOL EDUCATION

In 2000 there were 14,451 kindergartens (yochien) for children between three and five years of age, in which 1.8m. children were enrolled. Most kindergartens are privately-controlled.

ELEMENTARY AND LOWER SCHOOL EDUCATION

All children between six and 15 are required to attend six-year elementary schools (Shogakko) and three-year lower secondary schools (Chugakko). All children are provided with textbooks free of charge, while children of needy families are assisted in paying for school lunches and educational excursions by the Government and the local bodies concerned. Enrolment is almost 100%; there were 7.2m. pupils in 23,808 elementary schools and 3.7m. pupils in 11,159 lower secondary schools in 2002.

SECONDARY EDUCATION

There are three types of course available: full-time (which last for three years), part-time and correspondence (both of which last for four years). In 2002 there were 5,472 upper secondary schools (or high schools) in Japan, with an enrolment of 3.9m.

HIGHER EDUCATION

There were 686 universities and graduate schools, and 603 junior and technical colleges in 2002. The universities offer courses extending from three to four years and, in most cases, postgraduate courses for a master's degree in two years and a doctorate in three years. Junior colleges offer two- or three-year courses, credits for which can count towards a first degree. The technical colleges admit lower secondary school students for five years. The number of students in graduate schools and universities in 2002 was 2.8m., and in junior and technical colleges 267,000. Teacher training is offered in both universities and junior colleges.

Bibliography

GENERAL

Barrett, Brendan F. D., and Therivel, Riki. *Environmental Policy and Impact Assessment in Japan*. London and New York, Routledge, 1992.

Beasley, W. G. (Ed.). *Modern Japan: Aspects of History, Literature and Society*. London, Allen and Unwin, 1975.

Befu, Harumi, and Guichard-Anguis, Sylvie (Eds). *Globalizing Japan—Ethnography of the Japanese Presence in Asia, Europe and America*. London, Routledge, 2001.

Bestor, Theodore C. *Neighbourhood Tokyo*. Stanford, CA, Stanford University Press, 1989.

Bowring, Richard, and Kornicki, Peter (Eds). *The Cambridge Encyclopedia of Japan*. Cambridge, Cambridge University Press, 1993.

Buck, David N. *Responding to Chaos—Tradition, Technology, Society and Order in Japanese Design*. London, Routledge, 2000.

Buckley, Roger. *Japan Today*. Cambridge, Cambridge University Press, revised edn, 1999.

Buckley, Sandra. *Encyclopedia of Contemporary Japanese Culture*. London, Routledge, 2001.

Bunce, W. K. *Religions in Japan*. Rutland, VT, Charles E. Tuttle.

Clamoner, John. *Difference and Modernity: Social Theory and Contemporary Japanese Society*. Tokyo, Sophia University, 1996.

Craig, Timothy. J. *Inside the World of Japanese Popular Culture*. Armonk, NY, M. E. Sharpe, 2000.

Dore, Ronald. *Taking Japan Seriously: A Confucian Perspective on Leading Social Issues*. London, The Athlone Press, 1988.

Dore, R., and Sinha, R. (Eds). *Japan and the World Depression: Then and Now*. New York, St Martin's Press, 1987.

Downer, Lesley. *On the Narrow Road to the Deep North: Journey into a Lost Japan*. London, Jonathan Cape, 1989.

Fingleton, E. *Blindside: Why Japan is Still on Track to Overtake the U.S. by the Year 2000*. New York, Houghton Mifflin, 1995.

Hall, Ivan P. *Cartels of the Mind: Japan's Intellectual Closed Shop*. New York, W. W. Norton, 1998.

Hendry, Joy (Ed.). *Interpreting Japanese Society—Anthropological Approaches*. London, Routledge, 1998.

Understanding Japanese Society. London, RoutledgeCurzon, 2003.

Hendry, Joy, and Raveri, Massimo (Eds). *Japan at Play*. London, Routledge, 2001.

Hood, Christopher, P. *Japanese Education Reform—Nakasone's Legacy*. London, Routledge, 2001.

Hosokawa, Morihiro. *The Time to Act is Now: Thoughts for a New Japan*. Tokyo, NTT Mediascope Inc, 1994.

Ishida, R. *Geography of Japan*. Tokyo, Kokusai Bunka Shinkokai, 1961.

Johnson, Sheila K. *The Japanese through American Eyes*. Stanford, CA, Stanford University Press, 1988.

Kaplan, David E., and Dubro, Alec. *Yakuza: Japan's Criminal Underworld*. Berkeley, CA, University of California Press, 2003.

Kaplan, David E., and Marshal, A. *The Cult at the End of the World*. London, Arrow Books, 1996.

Kato, Shuichi. *Japan: Spirit and Form*. Tokyo, Charles E. Tuttle, 1995.

Katzenstein, Peter J., and Shiraishi, Takashi. *Network Power: Japan and Asia*. Ithaca, NY, Cornell University Press, 1997.

Kerr, Alex. (trans. Fishman, B.). *Lost Japan*. Hawthorn, Vic, Lonely Planet, 1996.

Dogs and Demons: Tales From the Dark Side of Japan. New York, Hill and Wang, 2001.

Kingston, Jeff. *Japan's Quiet Transformation: Politics, Economics and Society*. London, RoutledgeCurzon, 2004.

Kodansha. *Japan: An Illustrated Encyclopedia*. London, Kodansha Europe, 1993.

Japan: Profile of a Nation. London, Kodansha Europe, 1995.

Lie, John. *Multi-Ethnic Japan*. Cambridge, MA, Harvard University Press, 2001.

Lockwood, William M. *The Economic Development of Japan*.- Princeton, NJ, Princeton University Press, 1954.

Lockwood, William W. (Ed.). *The State and Economic Enterprise in Postwar Japan*. Princeton, NJ, Princeton University Press, 1965.

Maher, John C., and MacDonald, Gaynor (Eds). *Diversity in Japanese Culture and Language*. London, Kegan Paul International, 1996.

Maswood, Javed S. *Japan and East Asian Regionalism*. London, Routledge, 2000.

McGregor, Richard. *Japan Swings: Politics, Culture and Sex in the New Japan*. St Leonards, NSW, Allen and Unwin, 1996.

McVeigh, Brian J. *The Nature of the Japanese State—Rationality and Rituality*. London, Routledge, 1998.

Minichiello, Sharon A. (Ed.). *Japan's Competing Modernities—Issues in Culture and Democracy, 1900–1930*. Honolulu, HI, University of Hawaii Press, 1998.

Mulgan, Aurelia George. *The Politics of Agriculture in Japan*. London, Routledge, 2000.

Naff, C. *About Face: How I Stumbled onto Japan's Social Revolution*. New York, Kodansha International, 1995.

Nakane, C. *Japanese Society*. London, Weidenfeld and Nicolson, 1970.

Nakayama, Shigeru. *Science, Technology and Society in Postwar Japan*. London, Kegan Paul International, 1992.

Nathan, John. *Japan Unbound: A Volatile Nation's Quest for Pride and Purpose*. Boston, MA, Houghton Mifflin, 2004.

Oe, Kenzaburo (trans. Yamanouchi, Hisaki). *Japan, the Ambiguous and Myself*. Tokyo, Kodansha International, 1996.

Ozawa, Ichiro. *Blueprint for a New Japan*. New York, Kodansha International, 1994.

Pezeu-Massabuau, J. (trans. Blum, P. C.). *The Japanese Islands: A Physical and Social Geography*. Rutland, VT, Charles E. Tuttle.

Pharr, Susan J., and Kraus, Ellis S. *Media and Politics in Japan*. Honolulu, HI, University of Hawaii Press, 1996.

Reischauer, Edwin O. *The Japanese*. Cambridge, MA, Harvard University Press, 1977.

Richie, Donald. *Different People: Pictures of some Japanese*. Tokyo, Kodansha, 1988.

Sale, Murray. *A Day in the Life of Japan*. London, William Collins and Sons Ltd, 1986.

Sargent, John. *Perspectives on Japan—Towards the Twenty-First Century*. Richmond, Surrey, Curzon Press, 1999.

Schreiber, Mark. *The Dark Side: Infamous Japanese Crimes and Criminals*. Tokyo, Kodansha International, 2002.

Starr, Don. Japan: *A Historical and Cultural Dictionary*. Richmond, Surrey, Curzon Press, 2001.

Takeda, Kiyoko. *The Dual Image of the Japanese Emperor*. London, Macmillan Education, 1989.

Thomsen, H. *The New Religions of Japan*. Rutland, VT, Charles E. Tuttle, 1963.

Tipton, Elise. *Modern Japan—A Social and Political History*. London, Routledge, 2002.

Tsunoda de Bary, Keene. *Sources of the Japanese Tradition*. New York, Columbia University Press, 1958.

Upham, Frank K. *Law and Social Change in Postwar Japan*. Cambridge, MA, Harvard University Press, 1988.

Weiner, Michael. *Japan's Minorities—The Illusion of Homogeneity*. London, Routledge, 1996.

Wilson, W. *The Sun at Noon: An Anatomy of Modern Japan*. London, Hamish Hamilton, 1986.

Williams, Dominic. *A Dictionary of Japanese Financial Terms*. Richmond, Surrey, Curzon Press, 1995.

Wong, Anny. *The Roots of Japan's Environmental Policies*. London, Routledge, 2001.

HISTORY

Allen, Louis. *The End of the War in Asia*. London, Hart-Davis, MacGibbon, 1976.

Allinson, Gary D. *Japan's Postwar History*. Ithaca, NY, Cornell University Press, 1997.

Beasley, William G. *The Meiji Restoration*. London, Oxford University Press, 1973.

Behr, Edward. *Hirohito: Behind the Myth*. London, Hamish Hamilton, 1989.

Bix, Herbert P. *Peasant Protest in Japan 1590–1884*. New Haven, CT, Yale University Press, 1986.

Hirohito and the Making of Modern Japan. New York, NY, Harper Collins, 2000.

Blacker, Carmen. *The Japanese Enlightenment*. Cambridge, Cambridge University Press, 1964.

Buruma, Ian. *Inventing Japan: From Empire to Economic Miracle 1853–1964*. London, Weidenfeld & Nicolson, 2003.

Coaldrake, William H. (Ed.). *Japan from War to Peace: The Coaldrake Records 1939–1956*. London, RoutledgeCurzon, 2003.

Dore, Ronald P. (Ed.). *Aspects of Social Change in Modern Japan*. Princeton, NJ, Princeton University Press, 1967.

Dower, John W. *Origins of the Modern Japanese State*. New York, Pantheon Books, 1975.

War Without Mercy: Race and Power in the Pacific War. New York, Pantheon Books, 1986.

Embracing Defeat: Japan in the Wake of World War II. New York, W. W. Norton, 1999.

Drifte, Reinhard. *Japan's Foreign Policy in the 1990s: From Economic Superpower to What Power?* Basingstoke, Macmillan, 1996.

Duus, Peter. *The Rise of Modern Japan*. Boston, MA, Houghton Mifflin, 1976.

Duus, Peter (Ed.). *The Cambridge History of Japan, Volume 6: The Twentieth Century*. New York, Cambridge University Press, 1989.

Duus, Peter, Myers, Ramon H., and Peattie, Mark R. (Eds). *The Japanese Wartime Empire 1931–1945*. Ewing, NJ, Princeton University Press, 1996.

Giffard, Sydney. *Japan among the Powers: 1890–1990*. New Haven, CT, Yale University Press, 1994.

Gordon, Andrew (Ed.). *Postwar Japan as History*. Berkeley, University of California Press, 1993.

Hall, John W. *Japan: From Prehistory to Modern Times*. London, Weidenfeld and Nicolson, 1970.

Harries, Meirion and Susie. *Soldiers of the Sun*. London, Heinemann, 1992.

Harvey, Robert. *The Undefeated: The Rise, Fall and Rise of Greater Japan*. London, Macmillan, 1995.

Havens, Thomas R. H. *Valley of Darkness: The Japanese People and World War Two*. New York, W. W. Norton, 1978.

Fire Across the Sea: the Vietnam War and Japan 1965–75. Princeton, NJ, Princeton University Press, 1988.

Hirschmeier, Johannes, and Yui, Tsunehiko. *The Development of Japanese Business, 1600–1973*. London, Allen and Unwin, 1975.

Honjo, Yukiko Alison. *Japan's Early Experience of Contract Management in the Treaty Ports*. London, RoutledgeCurzon, 2003.

Hoyt, Edwin P. *Japan's War: The Great Pacific Conflict*. London, Hutchinson, 1986.

Inkster, Ian. *Japanese Industrialisation—Historical and Cultural Perspectives*. London, Routledge, 2001.

Jansen, Marius B. *The Making of Modern Japan*. Harvard, MA, Harvard University Press, 2001.

Junju Banno. *Democracy in Pre-War Japan—Concepts of Government, 1871–1937: Collected Essays*. London, Routledge, 2000.

Kamija, Morinosuke. *The Emergence of Japan as a World Power: 1895–1925*. Rutland, VT, Charles E. Tuttle, 1968.

Modern Japan's Foreign Policy. Rutland, VT, Charles E. Tuttle, 1969.

Korhonen, Pekka. *Japan and Asia-Pacific Integration: Pacific Romances 1968–1996*. London, Routledge, 1998.

Kornicki, Peter (Ed.). *Meiji Japan—Political, Economic and Social History 1868–1912*. London, Routledge, 1998.

LaFeber, Walter. *The Clash: A History of US-Japan Relations*. New York, W. W. Norton, 1997.

Large, Stephen S. (Ed.). *Showa Japan—Political, Economic and Social History 1926–1989*. London, Routledge, 1998.

Lehmann, Jean-Pierre. *The Image of Japan 1850–1905: From Feudal Isolation to World Power*. London, Allen and Unwin, 1978.

Livingston, Jon, Moore, Joe, and Oldfather, Felicia (Eds). *The Japan Reader 1: Imperial Japan: 1800–1945*. London, Penguin, 1976.

The Japan Reader 2: Postwar Japan 1945 to the Present. London, Penguin, 1976.

Lu, David J. *Japan—A Documentary History (Two Vols)*. Armonk, NY, M. E. Sharpe, 1996.

Maddox, Robert J. *Weapons for Victory: The Hiroshima Decision Fifty Years Later*. Columbia, MO, University of Missouri Press, 1995.

Martin, Peter. *The Chrysanthemum Throne—A History of the Emperors of Japan*. Honolulu, HI, University of Hawaii Press, 1998.

Morley, James W. *Dilemmas of Growth in Prewar Japan*. Princeton, NJ, Princeton University Press, 1971.

Murakami, Haruki. *Underground: The Tokyo Gas Attack and the Japanese Psyche*. London, Harvill, 2001.

Murdoch, James. *A History of Japan (Three Vols)*. London, Routledge, 1999.

Nakamura, James I. *Agricultural Production and the Economic Development of Japan 1873–1922*. Princeton, NJ, Princeton University Press, 1966.

Nish, Ian. *The Story of Japan*. London, Faber, 1968.

Japanese Foreign Policy, 1869–1942: Kasumigaseki to Miyakezaka. London, Routledge and Kegan Paul, 1977.

Packard, Jerrold M. *Sons of Heaven: A Portrait of the Japanese Monarchy*. London, MacDonald Queen Anne Press, 1988.

Rose, Caroline. *Interpreting History in Sino-Japanese Relations—A Case Study in Political Decision Making*. London, Routledge, 1998.

Ruoff, Kenneth J. *The People's Emperor: Democracy and the Japanese Monarchy, 1945–1995*. Harvard, MA, Harvard University Press, 2002.

Sansom, George B. *The Western World and Japan*. London, Cresset Press, 1950.

A History of Japan. London, Cresset Press, 1958–64.

Shibusawa, Masahide. *Japan and the Asian Pacific Region: Profile of Change*. London, Croom Helm, 1984.

Shillony, Ben-Ami. *Politics and Culture in Wartime Japan*. Oxford, Clarendon Press, 1988.

Smith, Dennis B. *Japan since 1945*. Basingstoke, Macmillan, 1995.

Steele, M. William. *Alternative Narratives in Modern Japanese History*. London, RoutledgeCurzon, 2003.

Storry, Richard. *A History of Modern Japan*. London, Penguin.

Japan and the Decline of the West in Asia 1894–1943. London, Macmillan Press, 1979.

Tsouras, Peter G. *Rising Sun Victorious*. London, Greenhill Books, 2001.

Weintraub, Stanley. *The Last Great Victory: The End of World War II July–August 1945*. New York, E.P. Dutton, 1995.

Wetzler, Peter. *Hirohito and War—Imperial Tradition and Military Decision Making in Pre-war Japan*. Honolulu, HI, University of Hawaii Press, 1998.

Williams, David. *Defending Japan's Pacific War*. London, RoutledgeCurzon, 2004.

Wilson, Sandra. *The Manchurian Crisis and Japanese Society, 1931–33*. London, Routledge, 2001.

Yahara, Col Hiromichi. *The Battle for Okinawa*. New York, Wiley and Sons, 1995.

Yoshiaki Yoshimi. *Comfort Women: Sexual Slavery in the Japanese Military During World War II*. New York, NY, Columbia University Press, 2001.

POLITICS

Abe, Hitoshi, Shindo, Muneyuki, and Kawato, Sadafumi. *The Government and Politics of Japan*. Tokyo, Tokyo University Press, 1994.

Allinson, Gary D., and Sone, Yasuhiro (Eds). *Political Dynamics in Contemporary Japan*. Princeton, NJ, Princeton University Press, 1993.

Christensen, Ray. *Ending the LDP Hegemony: Party Cooperation in Japan*. Honolulu, University of Hawaii Press, 2000.

Curtis, Gerald L. *The Logic of Japanese Politics: Leaders, Institutions, and the Limits of Change*. New York, Columbia University Press, 1999.

Curtis, Gerald L. (Ed.). *Japan's Foreign Policy After the Cold War: Coping With Change*. Armonk, NY, M. E. Sharpe, 1993.

Dobson, Hugo. *Japan and UN Peacekeeping: Pressures and New Responses.* London, RoutledgeCurzon, 2003.

Dore, Ronald. *Japan, Internationalism and the UN.* London, Routledge, 1997.

Funabashi, Yoichi (Ed.). *Japan's International Agenda.* New York, New York University Press, 1994.

Garby, Craig C., and Brown Bullock, Mary (Eds). *Japan: A New Kind of Superpower?* Washington, DC, The Woodrow Wilson Center Press, 1994.

Harries, M., and Harries, S. *Sheathing the Sword: The Demilitarization of Japan.* London, Hamish Hamilton, 1987.

Hayes, Louis D. *Introduction to Japanese Politics.* Armonk, NY, M. E. Sharpe, 2000.

Hook, Glenn D., et al. *Japan's International Relations—Politics, Economics and Security.* London, Routledge, 2000.

Hook, Glenn D., and McCormack, Gavan (Eds). *The Japanese Constitution—Documents and Analysis.* London, Routledge, 2000.

Horiuchi, Yusaku. *Institutions, Incentives and Electoral Behaviour in Japan.* London, RoutledgeCurzon, 2004.

Hrebenar, Ronald J. *Japan's New Party System.* Boulder, CO, Westview Press, 2000.

Hughes, Christopher W. *Japan's Economic Power and Security—Japan and North Korea.* London, Routledge, 1999.

Japan's Security Policy and Ballistic Missile Defence System. London, RoutledgeCurzon, 2004.

Hunsberger, Warren S. *Japan's Quest—The Search for International Role, Recognition and Respect.* Armonk, NY, M. E. Sharpe, 1996.

Inoguchi, Takashi, and Jain, Purnendra (Eds). *Japanese Foreign Policy Today.* New York, St Martin's Press, 2000.

Ishihara, Shintaro. *The Japan That Can Say No.* London, Simon and Schuster, 1991.

Jain, Purnendra, and Inoguchi, Takashi. *Japanese Politics Today.* Basingstoke, Macmillan, 1997.

Johnson, Chalmers. *Japan: Who Governs? The Rise of the Developmental State.* New York, W.W. Norton, 1995.

Johnsonn, Stephen. *Opposition Politics in Japan—Strategies under a One-Party Dominant Regime.* London, Routledge, 2000.

Jones, Christopher. *The Political Philosophy of Japan.* London, RoutledgeCurzon, 2004.

Kataoka, Tetsuya (Ed.). *Creating Single-Party Democracy: Japan's Postwar Political System.* Stanford, CA, Hoover Institution Press, 1993.

Kersten, Rikki, and Williams, David. *The Left in Japanese Politics.* London, RoutledgeCurzon, 2004.

Kishima, Takako. *Political Life in Japan: Democracy in a Reversible World.* Princeton University Press, 1992.

Kohno, Masaru. *Japan's Postwar Party Politics.* Princeton, NJ, Princeton University Press, 1997.

Kyogoku, Jun-ichi. *The Political Dynamics of Modern Japan.* Tokyo, University of Tokyo Press, 1987.

Lee, Chong-Sik. *Japan and Korea: The Political Dimension.* Stanford, CA, Hoover Institution Press, 1986.

Masumi, Junnosuke. *Contemporary Politics in Japan.* Berkeley, University of California Press, 1995.

McCormack, Gavan, and Sugimoto, Yoshio (Eds). *Democracy in Contemporary Japan.* Armonk, NY, M. E. Sharpe, 1986.

Mendl, Wolf. *Japan's Asia Policy—Regional Security and Global Interests.* London, Routledge, 1997.

Millard, Mike. *Leaving Japan—Observations on the Dysfunctional U.S.-Japan Relationship.* Armonk, NY, M. E. Sharpe, 2000.

Mitchell, Richard H. *Political Bribery in Japan.* Honolulu, HI, University of Hawaii Press, 1996.

Nakano, Minoru. *The Policy-Making Process in Contemporary Japan.* Basingstoke, Macmillan, 1996.

Nakasone, Yasuhiro. *Japan—A State Strategy for the Twenty-First Century.* London, RoutledgeCurzon, 2002.

Pempel, T. J. (Ed.). *Policymaking in Contemporary Japan.* Ithaca, NY, Cornell University Press, 1977.

Reed, Steven (Ed.). *Japanese Electoral Politics: Creating a New Party System.* London, RoutledgeCurzon, 2003.

Reischauer, E. O. *The United States and Japan.* Cambridge, MA, Harvard University Press.

Richardson, Bradley M. *Japanese Democracy: Power, Coordination and Performance.* New Haven, CT, Yale University Press, 1997.

Rose, Caroline. *Sino-Japanese Relations: Towards a Future-Oriented Diplomacy.* London, RoutledgeCurzon, 2004.

Schwartz, Frank J. *Advice and Consent—The Politics of Consultation in Japan.* Cambridge, Cambridge University Press, 1999.

Stegewerns, Dick. *Nationalism and Internationalism in Imperial Japan.* London, RoutledgeCurzon, 2002.

Stephan, John J. *The Kuril Islands: Russo-Japanese Frontier in the Pacific.* Oxford, Clarendon Press, 1975.

Stockwin, J. A. A. *Governing Japan: Divided Politics in a Major Economy.* Oxford, Blackwell, 1999.

Dictionary of the Modern Politics of Japan. London, Routledge, 2003.

Stockwin, J. A. A., et al. *Dynamic and Immobilist Politics in Japan.* Honolulu, University of Hawaii Press, 1988.

Upham, Frank K. *Law and Social Change in Postwar Japan.* Cambridge, MA, Harvard University Press, 1987.

Van Wolferen, Karel. *The Enigma of Japanese Power: People and Politics in a Stateless Nation.* London, Macmillan, 1989.

Yamamoto, Mari. *Pacifism and Revolt in Post-War Japan.* London, RoutledgeCurzon, 2004.

Yanaga, Chitoshi. *Japanese People and Politics.* New York, John Wiley, 1956.

Yoshida, Shigeru. *Japan's Decisive Century.* New York, Praeger, 1967.

ECONOMY

Abegglen, James C., and Stalk, George, Jr. *Kaisha—The Japanese Corporation—How Marketing, and Manpower Strategy, Not Management Style, Make the Japanese World Pace-setters.* New York, Basic Books, 1985.

Allen, George C. *An Economic History of Modern Japan.* London, Macmillan, 1981.

Japan's Economic Expansion. London, Oxford University Press, 1965.

Amyx, Jennifer, and Drysdale, Peter (Eds). *Japanese Governance: Beyond Japan Inc.* London, RoutledgeCurzon, 2002.

Aoki, M. (Ed.). *The Economic Analysis of the Japanese Firm.* Amsterdam, Elsevier Science Publishers, 1984.

Aoki, Masahiko. *Information, Incentives, and Bargaining in the Japanese Economy.* New York, Cambridge University Press, 1988.

Argy, Victor. *The Japanese Economy.* Basingstoke, Macmillan, 1996.

Balassa, Bela, and Noland, Marcus. *Japan in the World Economy.* Washington, Institute for International Economics, 1990.

Basu, Dipak R., and Miroshnik, Victoria. *Japanese Foreign Investments, 1970–1998: Perspectives and Analyses.* Armonk, NY, M. E. Sharpe, 2000.

Beechler, Schon, and Stucker, Kristin (Eds). *Japanese Business.* London, Routledge, 1997.

Bergsten, Fred, and Cline, William R. *The United States–Japan Economic Problem.* Washington, Institute for International Economics, 1985.

Boltho, Andrea. Japan: *An Economic Survey 1953–1973.* Oxford University Press, 1976.

Boyer, Robert, and Yamanda, Toshio (Eds). *Japanese Capitalism in Crisis—A 'Regulationist' Interpretation.* London, Routledge, 2000.

Calder, Kent E. *Strategic Capitalism—Private Business and Public Purpose in Japanese Industrial Finance.* Princeton, NJ, Princeton University Press, 1993.

Carlile, Lonny E., and Tilton, Mark (Eds). *Is Japan Really Changing its Ways? Regulatory Reform and the Japanese Economy.* Washington, DC, Brookings Institution Press, 1998.

Choate, Pat. *Agents of Influence: How Japan's Lobbyists in the United States Manipulate America's Political and Economic System.* New York, Alfred A. Knopf, 1990.

Crump, John. *Nikkeiren and Japanese Capitalism.* London, RoutledgeCurzon, 2003.

Dattel, Eugene. *The Sun that Never Rose—The Inside Story of Japan's Failed Attempt at Global Financial Dominance.* Chicago, Probus, 1994.

Drysdale, Peter. *International Economic Pluralism: Economic Policy in Asia and the Pacific.* Sydney, Columbia University Press and Allen and Unwin, 1988.

Drysdale, Peter, and Gower, Luke (Eds). *The Japanese Economy.* London, Routledge, 1998.

Drysdale, Peter, Viviani, Nancy, Akio, Watanabe, and Ippei, Yamazawa. *The Australia-Japan Relationship: Towards the Year 2000.* Canberra, Australia-Japan Research Centre/Japan Centre for Economic Research, 1989.

Economic Planning Agency (Japanese Government). *Economic Survey of Japan.* Tokyo, Okurasho Insatsu Kyoku, annual.

Ezrati, Milton. *Kawari.* Reading, MA, Perseus Books, 1999.

Fallows, James. *Looking at the Sun—The Rise of the New East Asian Economic and Political System*. New York, Pantheon Books, 1994.

Flath, David. *The Japanese Economy*. Oxford, Oxford University Press, 2000.

Francks, Penelope. *Japanese Economic Development: Theory and Practice*. 2nd Edn, London, Routledge, 1999.

Freedman, Craig (Ed.). *Why Did Japan Stumble? Causes and Cures*. Cheltenham, Edward Elgar, 1999.

Economic Reform in Japan—Can the Japanese Change? Cheltenham, Edward Elgar Publishing, 2001.

Funabishi, Yoichi (Ed.). *Japan's International Agenda*. New York, New York University Press, 1994.

Gerlach, Michael. *Alliance Capitalism—The Social Organization of Japanese Business*. Berkeley, University of California Press, 1992.

Hamada, Koichi, and Kasuya, Munehisa. *Financial Crises in Japan: Past, Present and Future*. Cheltenham, Edward Elgar Publishing, 2003.

Hartcher, Peter. *The Ministry*. Boston, MA, Harvard Business School Press, 1998.

Hasegawa, Harukiyo, and Hook, Glenn D. (Eds). *Japanese Business Management—Restructuring for Low Growth and Globalisation*. London, Routledge, 1997.

Hayes, Declan. *Japan's Big Bang: The Deregulation and Revitalisation of the Japanese Economy*. Rutland, VT, Charles E. Tuttle, 2000.

Hester, William R. *Japanese Industrial Targeting*. London, Macmillan, 1992.

Higashi, Chikara, and Lauter, Peter. *The Internationalisation of the Japanese Economy*. Boston, Kluwer Academic Publishers, 1990.

Holstein, William J. *The Japanese Power Game: What It Means For America*. Maxwell-Macmillan International, 1991.

Hook, Glenn, D. (Ed.). *Political Economy of Japanese Globalization*. London, Routledge, 2001.

Horne, James. *Japan's Financial Markets—Conflict and Consensus in Policymaking*. Sydney, Allen and Unwin, 1985.

Hsu, Robert C. *The MIT Encyclopedia of the Japanese Economy*. Cambridge, MA, MIT Press, 1995.

Hughes, Christopher W. *Japan's Economic Power and Security*. London, Routledge, 1999.

Hunter, Janet (Ed.). *Japanese Economic History, 1930–1960*. London, Routledge, 2000.

Inoguchi, Takashi, and Okimoto, Daniel I. (Eds). *The Political Economy of Japan: The Changing International Context, Vol. 2*. Stanford, CA, Stanford University Press, 1988.

Ito, Takatoshi. *The Japanese Economy*. Cambridge, Massachusetts Institute of Technology, 1992.

Jackson, Keith, and Tomika, Mikuki. *The Changing Face of Japanese Management*. London, Routledge, 2003.

Johnson, Chalmers. *MITI and the Japanese Miracle: The Growth of Industrial Policy, 1925–1975*. Stanford, CA, Stanford University Press, 1982.

Katz, Richard. *Japan: The System that Soured—The Rise and Fall of the Japanese Economic Miracle*. Armonk, NY, M. E. Sharpe, 1998.

Japanese Phoenix: The Long Road to Economic Recovery. Armonk, NY, M. E. Sharpe, 2003.

Komiya, Ryutaro, Masahiro, Okuno, and Kotaro, Suzumura (Eds). *Industrial Policy in Japan*. Tokyo, Academic Press, 1988.

Koppel, Bruce M., and Orr, Jr, Robert M. (Eds). *Japan's Foreign Aid: Power and Policy in a New Era*. Boulder, CO, Westview Press, 1993.

Kosai, Yutaka, and Ogino, Yoshitaro. *The Contemporary Japanese Economy*. Armonk, NY, M. E. Sharpe, 1985.

Kumon, Shumpei, and Rosovsky, Henry (Eds). *The Political Economy of Japan: Cultural and Social Dynamics, Vol. 3*. Stanford, CA, Stanford University Press, 1992.

Lincoln, Edward J. *Japan Facing Economic Maturity*. Washington, DC, Brookings Institution, 1988.

Makiko, Yamada. *Japan's Top Management from the Inside (Studies in the Modern Japanese Economy)*. Hampshire, Palgrave, 1998.

Makoto, Itoh. *The Japanese Economy Reconsidered*. New York, NY, St Martins Press, 2001.

Masafumi, Matsuba. *The Contemporary Japanese Economy—Between Civil Society and Corporation-Centered Society*. Singapore, Institute of Southeast Asian Studies, 2001.

Masasuke, Ide. *Japanese Corporate Finance and International Competition: Japanese Capitalism Versus American Capitalism(Studies in The Modern Japanese Economy)*. Hampshire, Palgrave, 1998.

Matanle, Peter. *Japanese Capitalism and Modernity in a Global Era*. London, RoutledgeCurzon, 2003.

Matsumoto, Koji. *The Rise of the Japanese Corporate System*. Kegan Paul International, 1992.

Matsushita, Mitsuo. *International Trade and Competition Law in Japan*. Oxford, Oxford University Press, 1993.

McKenzie, C., and Stutchbury, M. (Eds). *The Yen and Japanese Financial Markets*. Sydney, Allen and Unwin, 1992.

Mendl, Wolf. *Japan and Southeast Asia: International Relations*. London, Routledge, 2001.

Mikanagi, Yumiko. *Japan's Trade Policy—Action or Reaction?* London, Routledge, 1996.

Minami, Ryoshin. *The Economic Development of Japan*. London, Macmillan, 1986.

Mirza, Hafiz. *Japan's Economic Empire—Foreign Investment by Japanese Companies Before the Pacific War*. Richmond, Surrey, Curzon Press, 2000.

Miyashita, Kenichi, and Russell, David. Keiretsu: *Inside the Hidden Japanese Conglomerates*. New York, McGraw-Hill, 1994.

Morris-Suzuki, Tessa. *The Technological Transformation of Japan*. Cambridge, Cambridge University Press, 1995.

Nakamura, Takafusa. *The Postwar Japanese Economy: Its Development and Structure*. Tokyo, University of Tokyo Press, 1982.

Nathan, John. Sony: *The Private Life*. Boston, MA, and New York, Houghton Mifflin, 1999.

Nihon Keizai Shimbunsha (Ed.). *Industrial Review of Japan*. Tokyo, Nihon Keizai Shimbunsha, annual.

Ohkawa, Kazushi, and Rosovsky, Henry. *Japanese Economic Growth in Trend Acceleration in the Twentieth Century*. Stanford, CA, Stanford University Press, 1973.

Ohno, Kenichi, and Ohno, Izumi (Eds). *Japanese Views on Economic Development—Diverse Paths to the Market*. London, Routledge, 1998.

Okabe, Mitsuaki. *Cross Shareholdings in Japan—A New Unified Perspective of the Economic System*. Cheltenham, Edward Elgar Publishing, 2002.

Okimoto, Daniel I. *Between MITI and the Market: Japanese Industrial Policy for High Technology*. Stanford, CA, Stanford University Press, 1990.

Orr, Robert M. *The Emergence of Japan's Foreign Aid Power*. New York, Columbia University Press, 1991.

Patrick, Hugh, and Meissner, Larry (Eds). *Japanese Industrialization and its Social Consequences*. Berkeley and London, University of California Press, for the Social Sciences Research Council, 1977.

Pempel, T. J. *Regime Shift: Comparative Dynamics of the Japanese Political Economy, Cornell Studies in Political Economy*. Ithaca, NY, Cornell University Press, 1998.

Prestowitz, Jr, Clyde V. *Trading Places—How We are Giving our Future to Japan and How to Reclaim it*. New York, Basic Books, 1989.

Rafferty, Kevin. *Inside Japan's Power Houses: The Culture, Mystique and Future of Japan's Greatest Corporations*. London, Weidenfeld and Nicolson, 1995.

Reszat, Beate. *The Japanese Foreign Exchange Market*. London, Routledge, 1997.

Samuels, Richard J. *Rich Nation, Strong Army—National Security and the Technological Transformation of Japan*. Ithaca, NY, Cornell University Press, 1994.

Sato, Kazuo. *The Transformation of the Japanese Economy*. Armonk, NY, M. E. Sharpe, 1999.

The Japanese Economy—A Primer. Armonk, NY, M. E. Sharpe, 1999.

Sato, Ryuzo. *The Chrysanthemum and the Eagle: The Future of US-Japan Relations*. New York, New York University Press, 1994.

Schmieglow, Michele and Henrik. *Strategic Pragmatism: Japanese Lessons in the Use of Economic Theory*. New York, Praeger, 1990.

Schodt, Frederik L. *America and the Four Japans: Friend, Foe, Model, Mirror*. Berkeley, CA, Stone Bridge Press, 1994.

Sheard, Paul (Ed.). *International Adjustment and the Japanese Firm*. Sydney, Allen and Unwin, 1992.

Shibata, Tokue (Ed.). *Japan's Public Sector—How the Government is Financed*. Tokyo, University of Tokyo Press, 1993.

Stern, Robert M. (Ed.). *Japan's Economic Recovery: Commercial Policy, Monetary Policy and Corporate Governance*. Cheltenham, Edward Elgar Publishing, 2003.

Suzuki, Yoshio. *Money and Banking in Contemporary Japan: The Theoretical Setting and its Applications*. New Haven, CT, Yale University Press, 1980.

Money, Finance and Macroeconomic Performance in Japan. New Haven, CT, Yale University Press, 1986.

Suzuki, Yoshio, (Ed.). *The Japanese Financial System*, Tokyo, Bank of Japan, 1995.

Tachibanaki, Toshiaki. *Public Policies and the Japanese Economy: Savings, Investments, Unemployment, Inequality*. Basingstoke, Macmillan, 1996.

Taggart Murphy, R. *The Real Price of Japanese Money*. London, Weidenfeld & Nicolson, 1996.

Tasker, Peter. *Inside Japan: Wealth, Work and Power in the New Japanese Empire*. London, Sidgwick and Jackson, 1988.

The Weight of the Yen. New York, Norton, 1996.

Tett, Gillian. *Saving the Sun: A Wall Street Gamble to Rescue Japan from its Trillion-Dollar Meltdown*. New York, NY, HarperBusiness, 2003.

Thurow, Lester. *Head to Head: The Coming Economic Battle Among Japan, Europe and America*. New York, William Morrow, 1992.

Tolliday, Steven. *The Economic Development of Modern Japan, 1868–1945—From the Meiji Restoration to the Second World War*. Cheltenham, Edward Elgar Publishing, 2001.

The Economic Development of Modern Japan, 1945–1995. Cheltenham, Edward Elgar Publishing, 2001.

Tsutsui, William M. (Ed.). *Banking in Japan*. London, Routledge, 1999.

van Wolferen, Karel. *The Enigma of Japanese Power—People and Politics in a Stateless Nation*. London, Macmillan, 1989.

Vestel, James. *Planning for Change: Industrial Policy and Japanese Economic Development 1945–1990*. Oxford, Oxford University Press, 1994.

Williams, David. *Japan: Beyond the End of History*. London, Routledge, 1994.

Wood, Christopher. *The Bubble Economy: The Japanese Economic Collapse*. London, Sidgwick and Jackson, 1992.

The End of Japan Inc.: How the New Japan Will Look. New York, Simon & Schuster, 1995.

Woronoff, Jon. *Japan's Commercial Empire*. Armonk, NY, M. E. Sharpe, 1985.

The Japanese Economic Crisis. Basingstoke, Macmillan, 1996.

Yamamura, Kozo (Ed.). *Policy and Trade Issues of the Japanese Economy—American and Japanese Perspectives*. Seattle, University of Washington Press, 1982.

Yamamura, Kozo, and Yasukichi, Yasuba (Eds). *The Political Economy of Japan: Vol. 1, The Domestic Transformation*. Stanford, CA, Stanford University Press, 1985.

Yamashita, Shoichi (Ed.). *Transfer of Japanese Technology and Management to Asean Countries*. Tokyo, University of Tokyo Press, 1991.

Yoshihara, Kunio. *Japanese Economic Development*. Tokyo, Oxford University Press, 1986.

KOREA

Physical and Social Geography

JOHN SARGENT

The total area of Korea is 223,337 sq km (86,231 sq miles), comprising the Democratic People's Republic of Korea (North Korea), the Republic of Korea (South Korea) and the demilitarized zone (DMZ) between them. North Korea has an area of 122,762 sq km (47,399 sq miles) and South Korea an area of 99,313 sq km (38,345 sq miles). The DMZ covers 1,262 sq km (487 sq miles). The Korean peninsula is bordered to the north by the People's Republic of China, and has a very short frontier with Russia in the north-east.

PHYSICAL FEATURES

Korea is predominantly an area of ancient folding, although in the south-east, where a relatively small zone of recent rocks occurs, a close geological similarity with Japan may be detected. Unlike Japan, the peninsula contains no active volcanoes and earthquakes are rare.

Although, outside the extreme north, few mountains rise to more than 1,650 m, rugged upland, typically blanketed in either pine forest or scrub, predominates throughout the peninsula. Cultivated lowland forms only 20% of the combined area of North and South Korea.

Two broad masses of highland determine the basic relief pattern of the peninsula. In the north the Changpai Shan and Tumen ranges form an extensive area of mountain terrain, aligned from south-west to north-east, and separating the peninsula proper from the uplands of eastern Manzhou (Manchuria) in the People's Republic of China. A second mountain chain runs for almost the entire length of the peninsula, close to, and parallel with, the eastern coast. Thus, in the peninsula proper, the main lowland areas, which are also the areas of maximum population density, are found in the west and south.

The rivers of Korea, which are short and fast-flowing, drain mainly westwards into the Huang Hai (Yellow Sea). Of the two countries, North Korea, with its many mountain torrents, is especially well endowed with opportunities for hydroelectric generation. Wide seasonal variations in the rate of flow, however, tend to hamper the efficient operation of hydroelectric plants.

In contrast with the east coast of the peninsula, which is smooth and precipitous, the intricate western and southern coasts are well endowed with good natural harbours, an asset which, however, is partly offset by an unusually wide tidal range.

CLIMATE

In its main elements, the climate of Korea is more continental than marine, and is thus characterized by a wide seasonal range in temperature. In winter, with the establishment of a high pressure centre over Siberia and Mongolia, winds are predominantly from the north and north-west. North Korea in winter is extremely cold, with January temperatures falling, in the mountains, to below −13°C. Owing to the warming influence of the surrounding seas, winter temperatures gradually rise towards the south of the peninsula, but only in the extreme southern coastlands do January temperatures rise above freezing point. Winter precipitation is light, and falls mainly in the form of snow, which, in the north, lies for long periods.

In the southern and western lowlands summers are hot and humid, with July temperatures rising to 26°C. In mid-summer violent cloudbursts occur, often causing severe soil erosion and landslides. In the extreme north-east summers are cooler, and July temperatures rarely rise above 17°C.

Annual precipitation, of which more than one-half falls in the summer months, varies from about 600 mm in the north-east to more than 1,500 mm in the south.

NATURAL RESOURCES

Although 70% of the total area of Korea is forested, high-quality timber is virtually limited to the mountains of North Korea, where extensive areas of larch, pine and fir provide a valuable resource. Elsewhere, excessive felling has caused the forest cover to degenerate into poor scrub.

Korea is fairly rich in mineral resources, but most deposits are concentrated in the north, where large-scale mining operations were begun by the Japanese before the Second World War. In North Korea the main iron-mining areas are located south of Pyongyang, and in the vicinity of Chongjin in the extreme north-east.

Throughout Korea many other minerals, including copper, sulphur, lead, zinc, tungsten, gold, silver and magnesite, are mined.

POPULATION

At mid-2003 the estimated population of North Korea was 22,522,000, while that of South Korea was 47,925,318, giving a combined total of some 70m. The population of the Korean peninsula has thus more than doubled since 1954, when the combined total was 30m.

Population density is higher in South Korea (482.6 per sq km in 2003) than in North Korea (an estimated 183.5 per sq km in 2003), but mean density figures conceal the crowding of population on the limited area of agricultural land, which is a salient characteristic of the geography of South Korea.

In 1970 about 40% of the population of South Korea was concentrated in cities with populations of 100,000 and over. According to the final results of the census of 1 November 2000, Seoul, the capital of South Korea, had a population of 9,853,972, while Busan, with 3,655,437 inhabitants, was the second largest city, followed by Daegu (2,473,990) and Incheon (2,466,338). In 1985 the urbanization rate was 65%, with 24% of the population concentrated in Seoul.

According to the 1993 census results, the population of Pyongyang, the capital of North Korea, was 2,741,260. Apart from Pyongyang's port of Nampo (population 731,448 in 1993), the other two principal cities are Hamhung (709,730) and Chongjin, the leading port of the north-east coast (582,480).

History up to the Korean War

ANDREW C. NAHM

HISTORICAL BACKGROUND

Political History

Tribal units of the Puyo people (later known as Koreans) emerged in c. 3000 BC, when the Tungusic people migrated into south Manzhou (Manchuria) and the Korean peninsula, bringing with them their Ural-Altaic tongue, shamanistic religion and a palaeolithic culture. A mythological figure named Dan'gun is said to have consolidated tribal units into a 'kingdom' named Joseon in the northern part of Korea in 2333 BC. Ancient Joseon of the Dan'gun, the Kija and the Wiman dynasties lasted some 2,225 years, but it was overthrown by the Chinese in 108 BC. The Chinese colonies in the north-western region of Korea lasted until the fourth century AD.

A new kingdom of Goguryeo, which emerged in 37 BC in the southern region of Manzhou, along the Yalu River, and later extended into the Korean peninsula, ended Chinese domination in Korea and successfully defended its territory against Chinese aggression in the late sixth and early seventh centuries. Meanwhile, tribal federations which existed in the central and southern regions of the peninsula were consolidated into the kingdoms of Baekje and Silla in 18 BC and 57 BC respectively, ushering in the 'Three Kingdom' period in Korean history.

Silla destroyed Baekje in AD 663 and Goguryeo in 668, in collaboration with China, so unifying Korea. However, the kingdom of Goryeo, which rose in the central region in 918, brought about the demise of Silla in 935. Korea (Corea) is the Western version of Goryeo. During the Goryeo period the political system became similar to that of China, and in the 13th century Korea became a vassal to China, then ruled by the Yuan dynasty (established by the Mongol conquest).

In 1392 Gen. Yi Song-Gye overthrew the Goryeo dynasty, which had suffered invasions by the Khitans in the 10th century and by the Mongols in the 13th century. He established the Yi dynasty and renamed the kingdom Joseon, with Seoul as the new capital. The Yi dynasty brought the entire Korean peninsula and the island of Jeju (Cheju) under its rule, and it governed the kingdom with a Confucian bureaucracy manned by an élite class of scholar-officials. Korea became increasingly Confucianized as a vassal to China, then under the Ming dynasty.

While the power struggle among scholar-officials and between the monarchy and bureaucracy weakened the foundation of the nation, Korea suffered much from Japanese invasions in the late 16th century and Manzhou invasions in the early 17th century. Following the opening of Korea to the West in 1882, an international power struggle developed in Korea, initially between China and Japan, and then between Russia and Japan. The Japanese victories in the Sino-Japanese War of 1894–95 and the Russo-Japanese War of 1904–5 virtually sealed the fate of Korea, although nationalistic reformists made gallant efforts to save the sinking nation.

Despite the repeated invasions of foreign aggressors, Korea maintained its independence and preserved its national territory. However, in 1905 Korea became a Japanese protectorate, and in 1910 was annexed by Japan, ending the rule of the Yi dynasty as well as Korea's independence. The Japanese colonial rule in Korea was highly repressive and exploitative. Freedom of speech and press was non-existent, human rights were completely disregarded, farm lands were confiscated under various pretexts, economic and educational opportunities were extremely limited, and Korean workers and peasants alike were exploited under the repressive rule of the Japanese.

The Koreans retaliated in various ways. On 1 March 1919 some 2m. Koreans demonstrated peacefully, expressing their desire to be free from Japanese colonialism and to restore national independence under the principle of self-determination and the concept of 'one people, one nation'. A provisional government of Korea in exile (established in Shanghai in April 1919), and various non-violent as well as militant organizations of overseas Koreans, kept alive hopes for the eventual restoration of the Korean nation.

Economic and Social Development

An agricultural life developed during the bronze and iron ages (c. 2000 BC–AD 200), and a fully-fledged agricultural economy grew during the 'Three Kingdom' period, when land was monopolized by the aristocrats and cultivated by peasants who constituted the majority of the population. Domestic commerce did not develop until after the 10th century, but foreign trade with China and Japan had flourished during the 'Three Kingdom' and the unified periods.

The anti-commercialism of Confucianism did not encourage commercial economy to grow. However, as cities and towns expanded during the Yi period, government-approved commercial enterprises, as well as rural markets and fairs, increased in number, and cottage industries developed rapidly. Land continued to be owned by the gentry class (*yangban*), and cultivated by peasants and slaves.

Social evolution brought about the stratification of the people into the landed gentry and the toiling masses. The toiling masses were classified into 'good people' or 'common people', who were engaged in agriculture, and 'the low-born', who were engaged in trade, manufacturing of goods and other lesser occupations. The *yangban* formed the educated and land-owning class of the Yi period, and provided all high-ranking government officials. The 'middle people' class provided the middle- and lower-ranking officials. Only those who passed the civil service examinations were qualified to be government officials during the Yi period.

Modern commerce and industry developed during the Japanese colonial period. Food production was accelerated to feed the ever-growing number of Japanese as an increasing amount of Korean rice was exported to Japan. Rapid industrial growth came after the Japanese invasions of Manzhou and China proper in the 1930s. With this, the number of industrial workers increased rapidly as the influence of the gentry class diminished.

Cultural History

During and after the period of Chinese domination of the north-western region of Korea, the sinification of Korean culture occurred. Buddhism, which migrated from China to Korea during the third century AD, reached its zenith following the unification of Korea by Silla. Many historic and renowned Buddhist temples, pagodas, statues of Buddha and Buddhist writings were produced by the Koreans during the 'Three Kingdom' and the unified Silla periods.

Buddhism became the state religion of Goryeo. With the growing influence of Buddhism, book-printing techniques became advanced and sophisticated. *Tripitaka Koreana*, a Buddhist text of over 81,000 pages, and other Buddhist works were printed in Korea with movable type (wooden and metal blocks) in the 13th and 14th centuries.

Confucianism spread slowly but, after the establishment of the Tang dynasty in China in the early seventh century, Confucian influence grew strong in Korea. The increasing influence of Chinese culture had led to the development of native songs, called *hyangga*, as well as the creation of a system of writing Korean words in Chinese, called *idu*, during the unified Silla period. During the Goryeo and Yi periods, scholarship grew rapidly as more books on early Korean history, geography and other subjects were published by Confucian scholars. All scholarly books were written in the Chinese language.

The adoption of neo-Confucianism as the state creed and rapid Confucianization of political and social patterns and institutions during the Yi period brought about a sudden decline of Buddhism. At the same time, the growing number of public and private schools of Confucian and Chinese learning, and the introduction of the Chinese civil service examination system, resulted in the rise of an educated élite.

The adoption and promulgation of a new Korean script, commonly called *han'geul* ('Korean Letters'), by King Sejong in 1446 constituted an important milestone in the cultural history of Korea. With this, the Korean form of poetry, called *sijo*, flourished, as it enabled more Koreans to become literate.

The *Sirhak* ('Practical Learning') school of reformist Confucian scholars not only stimulated the development of a new interpretation of Confucianism, but also Korean studies, including historical and geographical studies, during and after the 17th century. Genre and folk painters, together with folk musicians, dancers and players, contributed greatly, not only to the preservation, but also to the growth, of a distinctive Korean folk culture during the Yi period, as novels and travelogues, written in *han'geul*, appeared.

The arrival and growth of Roman Catholicism and the 'Western Learning' after the 18th century, and the establishment of contacts with the West after 1882, led to the modernization of Korea. The creation of a modern educational system and the introduction of Western culture during the late Yi and Japanese colonial periods brought about a rapid increase of an educated population, and the growth of modern culture and the number of Christians.

The Japanese endeavoured to impose their culture on Korea by forcing Korean people to adopt Shintoism and to change their names to read like Japanese names. Efforts made by the Japanese to destroy the language and the racial and cultural identity of the Koreans were to no avail.

Liberation and Partition

On 15 August 1945 the Japanese surrendered to the Allies. The Cairo Declaration of December 1943, issued by the British and US leaders and Chiang Kai-shek of China, had stated that 'in due course Korea shall become free and independent'. The USSR accepted the Cairo agreement, but proposals made by the USA in 1945 led to the division of Korea into two military zones: the area south of the 38th parallel line (latitude 38°N) under US occupation and the northern area under Soviet control.

The Japanese Governor-General in Korea had persuaded Yo Un-Hyong, a prominent left-wing nationalist (socialist), to form a political body to maintain law and order at the end of the Japanese colonial rule. The Committee for the Preparation of the National Construction of Korea was thus organized. After Japan's surrender, Korean political prisoners were freed and the committee began to function as a government. Provincial, district and local committees were organized to maintain law and order. On 6 September 1945, two days before the arrival of US occupation forces, the committee called a 'National Assembly' and established a 'People's Republic of Korea', claiming jurisdiction over the whole country. Meanwhile, Soviet troops, which had entered Korea in early August, quickly moved southward as they crushed Japanese resistance, and within a month the entire northern half of Korea had come under Soviet occupation.

The US occupation authority accepted the surrender of Korea from the Japanese Governor-General, but, unlike the Soviet authorities in the North, refused to recognize the legitimacy of either the 'People's Republic' or the provisional Government of Korea, based in China. The United States Army Military Government in Korea (USAMGIK) was established and operated until the proclamation of South Korea's independence in August 1948.

Exiled political leaders returned to Korea toward the end of 1945—Dr Syngman Rhee from the USA, Kim Ku and Dr Kim Kyu-Sik from China, and Kim Il Sung and other communists from the USSR and China. Pak Hon Yong, a communist who had been released from Japanese imprisonment, quickly formed the communist South Korean Workers' Party in the US-occupied zone. Freedom of political activity permitted by USAMGIK resulted in a proliferation of political parties and social organizations of all political orientations, each vying for prominence. USAMGIK attempted in vain to bring about a coalition of moderate nationalists and the non-communist left wing.

In December 1945 representatives of the United Kingdom, the USA and the USSR entered into the Moscow agreement, providing for a five-year trusteeship for Korea under a four-power regime (China was the fourth power), with a view to establishing an independent and united nation of Korea. Despite violent anti-trusteeship demonstrations, the Allied occupation authorities resolved to implement the Moscow plan. Then, abruptly, the communists throughout Korea changed their attitude in favour of the Moscow plan, splitting the Korean people into two opposing camps. In the US-occupied zone left-wing organizations created serious political and economic problems. Communist-directed labour strikes became widespread, and terrorism of both right- and left-wing organizations became rampant.

A joint US-Soviet commission was formed to establish a national government of Korea in consultation with Korean political and social organizations. The first session of the joint commission was held in Seoul, the capital of the South, in March–May 1946. The Soviet delegate insisted that only 'democratic' organizations should participate and that only organizations which supported the Moscow agreement were 'democratic'. It became clear that the USSR sought to establish a national government of Korea dominated by the communists. In May 1947 the second session of the joint commission, held in the northern capital, Pyongyang, similarly failed to achieve any agreement and in June the commission's business was suspended indefinitely.

Realizing that the establishment of Korean unity and of a national government was a remote possibility, USAMGIK adopted new plans for South Korea. The Soviet occupation authority likewise proceeded to establish a client regime under Kim Il Sung. All anti-Soviet and anti-communist organizations were either dissolved or placed under communist leadership. A centralized, communist state began to emerge in the North, as Kim Il Sung formed his own party in defiance of Pak Hon Yong, head of the South Korean Workers' Party, whose headquarters were in Seoul.

The USA established a South Korean interim Legislative Assembly in late 1946, and in May 1947 an interim Government was created, both under moderate nationalists. These actions were bitterly criticized by right-wing leaders such as Dr Rhee and Kim Ku. The relationship between the USA and the right-wing nationalists worsened, while terrorist activities created an extremely uneasy situation. Several prominent politicians were assassinated. Neither the interim Legislative Assembly nor the interim Government was effective, for both were regarded by the conservative nationalists as US protégés attempting to prolong the US military occupation of Korea.

In September 1947 the US Government discarded the Moscow plan and placed the Korean question before the UN. The UN General Assembly formed the UN Temporary Commission on Korea in November and authorized it to conduct a national election in Korea to create a national government for the whole country.

The UN decision was welcomed by the USA and by most people in South Korea. The Soviet occupation authority and the Korean communists in the North, however, rejected the UN plan, and did not allow the UN Temporary Commission to visit North Korea. It soon became apparent that the UN plan would not work in the whole of Korea, and the Commission adopted an alternative plan to hold elections in those areas where it was possible, namely in South Korea only. It was assumed by the Commission that UN-sponsored and supervised elections would be held in the North in the near future, that a National Assembly created by the first democratic elections in Korea would represent the entire country, that the government to be established would be that of all Korea, and that the people in the North would elect their representatives to the National Assembly at a later date.

Whereas the right-wing nationalists welcomed such an alternative plan, the moderate and progressive nationalists, such as Kim Kyu-Sik, the head of the Democratic Independence Party, as well as Kim Ku, an extreme right-wing nationalist, vehemently opposed it, fearing that it would turn the temporary division of Korea into a permanent political partition. They visited North Korea and talked with Kim Il Sung and other communists, but failed to achieve their objective.

The Soviet authorities in the North had already begun to transfer power to the Supreme People's Assembly and the Central People's Committee, both established in early 1947. Dr Rhee's organization, the National Society for the Acceleration of Korean Independence, advocated the immediate independence of South Korea. The UN-sponsored elections held in the South in May 1948 created a National Assembly heavily dominated by the right wing. About 7.5m. people, or 75% of the electorate, elected 198 of 210 representatives from the South, while 100 unfilled seats in the 310-member Assembly were reserved for North Korean representatives. The National Assembly drew up a democratic Constitution for the Republic of Korea. Dr Rhee was elected the first President of the Republic of Korea, whose legitimacy was immediately recognized by the UN. On 15 August 1948 the Republic of Korea was inaugurated, and the US occupation came to an end.

In August 1948 the communists in the North held an election and established the new 527-member Supreme People's Assembly of the Democratic People's Republic of Korea (DPRK), which was proclaimed on 9 September.

THE KOREAN WAR, 1950–53

(This section was contributed by the editorial staff)

US forces were withdrawn from South Korea in June 1949, leaving only a small military mission. South Korea's own forces were weaker than those of the North, which had been built up with Soviet help. Increased tension between the North and the South culminated in the Korean War, beginning on 25 June 1950, when a North Korean force of over 60,000 troops, supported by Soviet-built tanks, crossed the 38th parallel and invaded the South. Four days later the North Koreans captured Seoul; US forces, whose assistance was requested by the Seoul Government, arrived on 30 June.

In response to North Korea's attack, the UN mounted a collective defence action in support of South Korea. Armed forces from 16 UN member states, attached to a unified command under the USA, were sent to help repel the invasion. Meanwhile, the North Koreans continued their drive southwards, advancing so rapidly that they soon occupied most of South Korea, leaving UN troops confined to the south-east corner of the peninsula. Following sea-borne landings by UN forces at Incheon, near Seoul, in September 1950, the invaders were driven back and UN troops advanced into North Korea, capturing Pyongyang in October and reaching the Chinese frontier on the Yalu River in November.

In October 1950 the People's Republic of China sent troops to assist North Korea: 200,000 Chinese crossed the Yalu River into Korea, forcing the evacuation of South Korean and UN troops. The Chinese advanced into South Korea but were driven back by a UN counter-attack in April 1951.

Peace negotiations began in July 1951, but hostilities continued until an armistice agreement was concluded on 27 July 1953. The war caused more than 800,000 casualties in South Korea and enormous damage to property. The 1953 cease-fire line, roughly along the 38th parallel, remains the boundary between North and South Korea, with a narrow demilitarized zone (DMZ) separating the two frontiers.

THE DEMOCRATIC PEOPLE'S REPUBLIC OF KOREA

History

AIDAN FOSTER-CARTER

Based on an earlier article by ANDREW C. NAHM

INTRODUCTION

Strong nationalist leadership, with potentially large popular support, was available in North Korea when the Second World War ended. It consisted chiefly of democratically inclined, Western missionary-educated individuals, of whom the most outstanding leader was Cho Man Shik. In August 1945 the Japanese Governor in Pyongyang relinquished control to Cho and a newly formed Provincial People's Committee. In the same month Soviet troops reached Pyongyang, accepted the legitimacy of the committee, and approved Cho as Chairman of the Five Provinces Administrative Bureau, formed to act as the indigenous government organ for North Korea.

In September 1945 Kim Il Sung, a young communist, who had led a guerrilla group of Korean communists in south-eastern Manzhou (Manchuria), returned to Korea with Soviet troops. Kim, however, had to cope with the 'domestic' communists who challenged his 'Kapsan' or 'partisan' faction. Two further groups of communists returned to North Korea following its liberation from the Japanese: one associated with the Soviet Army known as the 'Soviet faction', the other from Yanan, China, under the leadership of Kim Tu Bong of the Korean Independence League. In the early power struggle among the communists, Hyon Chun Hyok, the leader of the 'domestic' faction, was assassinated in September 1945. In the following month Kim Il Sung formed the North Korean Central Bureau of the Korean Communist Party (KCP) in order to consolidate his political position. In this he received covert support from the USSR.

Cho Man Shik organized the Korean Democratic Party (KDP), which received the support of the majority of the people, but his uncompromising stand against the Moscow plan for a five-year trusteeship of the Allied powers led to his downfall in January 1946. He was promptly placed under house arrest, and many members of the KDP fled to South Korea.

After the departure of the nationalists, a North Korean Provisional People's Committee was established in February 1946, with Kim Il Sung as Chairman and Kim Tu Bong as Vice-Chairman. The USSR accorded government status to the Committee. Kim Tu Bong formed the New People's Party (NPP) in March, to expand his power base, and managed to increase his party's membership. In July the North Korean Central Bureau of the KCP and the NPP merged to form the North Korean Workers' Party, with Kim Tu Bong as its Chairman and Kim Il Sung as Vice-Chairman. Real power, however, was in the hands of the latter.

In early 1947 the Supreme People's Assembly (SPA) was established as the highest legislative body in North Korea, and the Assembly, in turn, established a Central People's Committee to exercise executive authority. The Committee's first major act was to direct land reforms. No real attempts were made to establish collective farms, and the land which was distributed to landless peasants became the private property of the cultivators. It was not until the end of the Korean War, in 1953, that the agricultural 'co-operativization' programme was inaugurated. Land reform was followed by the nationalization of industry, transport, communications and financial institutions.

In early 1948 Pak Hon Yong, with other leaders of the communist South Korean Workers' Party, fled from the South when the party was outlawed by the US occupation authority. Pak, who enjoyed strong support from the 'domestic' faction, felt that he, instead of Kim Il Sung, should lead the movement in Korea. However, he was unable to achieve his objectives, and he grudgingly accepted a position subordinate to that of Kim.

THE ESTABLISHMENT OF THE DEMOCRATIC PEOPLE'S REPUBLIC

After refusing to allow the United Nations Temporary Commission on Korea to visit North Korea and to conduct elections there, Kim Il Sung established a separate, pro-USSR state. In August 1948 elections were held in the North for a new SPA. The newly created Assembly drafted a Constitution, ratified it on 8 September and proclaimed the Democratic People's Republic of Korea (DPRK) on the following day. Kim Il Sung was named Premier, while Pak Hon Yong was made Vice-Premier and Minister for Foreign Affairs. In June 1949 the merger of the North Korean Workers' Party and the South Korean Workers' Party brought about a unified communist party, the Korean Workers' Party (KWP), with Kim as its Chairman and Pak as its Vice-Chairman. The establishment of the DPRK entrenched the temporary military division of Korea as a permanent political partition. The USSR announced the withdrawal of its troops from North Korea, completing the process in December 1949. However, a large number of Soviet advisers in various fields remained.

During 1950 North Korea substantially increased the size and strength of its armed forces with Soviet supplies. In June the North Korean invasion of the South precipitated the Korean War (see p. 485), inflicting great damage on both sides. During and after the unsuccessful attempt to conquer South Korea, Kim Il Sung purged many of his enemies, including Pak Hon Yong. Conflict among the surviving communist leaders, however, did not end, and Kim Tu Bong remained a formidable figure. Kim Il Sung's economic reconstruction programme, which emphasized the development of heavy industry, met strong opposition. The debate lasted until 1956, when Kim Tu Bong fell from power. Meanwhile, the 'Yanan' faction attacked the growing personality cult of Kim Il Sung; he counter-attacked, forcing some 'Yanan' communists to flee to China. The USSR and China effected a temporary reconciliation, but leaders of the 'Yanan' faction were systematically relegated to less important posts or eased entirely out of power. By 1958 it had ceased to pose any further threat, and Kim Il Sung continued to consolidate his position of unassailability during the following decade.

DEVELOPMENTS DURING THE 1970s

Following the announcement of the 1972 joint North-South agreement to open dialogue for the peaceful unification of the peninsula, the KWP proposed amendments to the Constitution. General elections to the fifth SPA were held in December. The newly elected representatives adopted a socialist Constitution, and elected Kim Il Sung and Kim Il as President and Premier, respectively. For the first time, the North Korean Constitution stated the capital to be Pyongyang, not Seoul. It also elevated the Central People's Committee, headed by the President, to become the highest organ of state, while an Administration Council, headed by the Premier, was established as the DPRK's cabinet.

In 1973 North Korea gained observer status at the UN. This status nullified both the branding of North Korea as an aggressor in 1950 and the view that the Government of South Korea was the only lawful government in Korea. Both Koreas were invited to the UN General Assembly in November 1973 for a debate on the Korean question. North Korea was also given membership of the UN Conference on Trade and Development in May 1973.

In February 1974 the Central Committee of the KWP launched the 'Three Great Revolutions': ideological, technical and cultural. It emphasized the promotion of a self-orientated, self-reliant and independent ideology, or *juche* thought. The KWP also reorganized the structure of the Administration Council and reshuffled its membership twice in 1974. Kim Yong Ju, younger brother of Kim Il Sung, who had been regarded as heir apparent, was demoted in the party hierarchy, while Kim Il Sung's son, Kim Jong Il, rose in rank as a possible successor to his father. Significantly, a military leader, Gen. O Jin U, also rose in rank within the KWP.

One of North Korea's major objectives in the mid-1970s was the intensification of diplomatic activity to strengthen ties between non-aligned nations. The ministerial conference of the Non-aligned nations in Lima, Peru, in August 1975 voted to accept North Korea's application for participation.

Economic problems increasingly troubled the Pyongyang regime. Critical shortages of food and commodities were reported in the mid-1970s. In 1977 North Korea signed a new trade pact with the People's Republic of China, and an economic and technical co-operation agreement with the USSR. It was reported that the USSR had made large shipments of military goods to North Korea and had sent technical advisers in 1977.

During 1978 an important ideological-political campaign was undertaken to strengthen *juche* thought. A renewed drive to promote the 'Three Revolutions' was reportedly led by Kim Jong Il. In April a new socialist labour law was promulgated, which called for a change in the way of life for workers—eight hours of work, eight hours of rest, and eight hours of study of Kim Il Sung's *juche* thought.

The visits to Pyongyang by China's then Premier, Hua Guofeng, and his Vice-Premier, Deng Xiaoping, in 1978, and the trade agreement between Pyongyang and Beijing, seemed to improve Sino-North Korean ties. China reportedly promised to supply more petroleum and greater economic assistance. Pyongyang dealt cautiously with the new Sino-US relationship and, while criticizing the Vietnamese invasion of Kampuchea (now Cambodia), eschewed any comments on China's punitive war against Viet Nam.

The new trade agreement between Moscow and Pyongyang, Vice-President Pak Song Chol's visit to the Soviet capital in January 1979, and increasing contacts between Soviet and North Korean military leaders appeared to indicate a growing solidarity between the two countries. North Korea provided special privileges to the USSR in the port of Rajin on the north-eastern coast, making it a Soviet 'leased' territory and a Soviet naval base.

THE EMERGENCE OF KIM JONG IL

After conducting intense campaigns to select reliable and loyal supporters of Kim Il Sung's son, Kim Jong Il, as delegates, the Sixth Congress of the KWP met in early October 1980, the first such congress since 1970. Many significant structural and personnel changes in the party hierarchy were made. Although Kim Jong Il was not officially designated as successor to his father, as anticipated, he became a key member of several crucial committees in the party. A new five-member Standing Committee of the Political Committee (Politburo) of the Central Committee of the KWP was established, and Kim Il Sung became its Chairman, thus strengthening the concentration of power in the hands of a few. Many new members of the Central Committee and its subcommittees were supporters of Kim Jong Il, and it was reported that those who opposed his succession to power were removed from other key positions in the party, Government and military.

In April 1981 elections were held for members of provincial, city and county people's assemblies in the usual manner: one candidate named by the KWP for each position and a 100% turn-out of voters. In February 1982 elections were held to the SPA, and the Seventh Assembly emerged in early April. It approved the reappointment of the President, three Vice-Presidents, the Premier and 13 Vice-Premiers, but, contrary to expectations, it failed to name Kim Jong Il as a Vice-President.

Shortly before the convening of the SPA, the Central Committee of the KWP met, followed by a joint conference in mid-April 1982 of the Central Committee and the SPA. Both meetings failed to resolve the question of succession. Meanwhile, the power struggle intensified between the respective supporters of Kim Jong Il and his half-brother, Kim Pyong Il, son of Kim Il Sung's second wife. While the political turmoil surrounding the issue of succession increased instability in North Korea, several armed clashes between the military and workers occurred in Chongjin, in the north-east, in September 1981, followed in June 1982 by civil disturbances in Nampo, near Pyongyang, involving Koreans who had come to North Korea from Japan. Some 500 workers were reported to have been killed in the clashes in the north-eastern regions, while many fled into Soviet territory. For undisclosed reasons, an emergency meeting of the Central Committee of the KWP was convened in April 1982, and it was reported that some 12 generals and a large number of party leaders were purged in July. Furthermore, Choe Hyon, an experienced politician and a key member of the Central Committee of the ruling KWP, died in April, amid rumours that he had been murdered. However, Gen. O Jin U, a staunch supporter of Kim Jong Il, was retained as Minister of the People's Armed Forces.

It became known in April 1982 that there were several concentration camps for political dissidents and 'undesirables' in north-east Korea. More than 100,000 persons, including 23,000 Koreans who emigrated from Japan to North Korea, were reported to have been among the internees. (A report issued in January 1992 by the US Department of State estimated there to be a total of 12 concentration camps in North Korea, in which between 105,000 and 150,000 political prisoners were being detained.)

THE 1980s: SURVIVAL OF THE STALINIST REGIME

Elections were held in March 1983 to choose members for the local people's committees. The second session of the Seventh SPA met in April and elected Rim Chun Chu as Vice-President, succeeding Kang Ryang Uk. At the same time, it elected Yang Hyong Sop as Chairman of the SPA. Both were believed to be trusted supporters of Kim Jong Il.

The defection of a North Korean airman, in his fighter aircraft, to South Korea in February 1983, a labour uprising in Yanggang (North Hamgyong Province) and mass riots among workers in Wonsan (Kangwon Province) in April created serious domestic problems. It was reported that some 500 air force officers were purged on charges of disloyalty. As a result of the Rangoon (Yangon) bombing incident of October 1983 (see below), which was allegedly planned by North Korean agents, the intra-party struggle between moderates and radicals intensified.

In January 1984 Ri Jong Ok was dismissed from the premiership, a post he had held since 1977, and replaced by Kang Song San. Ri became a Vice-President. In early March 1984 Vice-President Kim Il, who had been critical of the junior Kim, died after a long illness. His death effectively marked the conclusion of the period of dominance by the 'old guard' of political leaders who had been associates of Kim Il Sung before he came to power.

The Government took a significant economic step in September 1984, announcing a new joint-venture law. It sought capital investment in North Korea on the part of foreign nations, particularly those of the West and Japan. This action was regarded as an admission of the failure of North Korea's 'self-reliant', closed economic policy. According to a Chinese source, the North Korean Government allowed some farmers to have private plots and fishermen to have private shops on a limited basis.

In August 1984 North Korea's official radio station confirmed, for the first time in a public broadcast, that Kim Jong Il would succeed his father as President, claiming that the transfer of power had been 'internationally acknowledged'. When the Central Committee met in early December, Kim Il Sung resumed direct control over economic and international affairs, including policy toward South Korea, which had been under the direction of Kim Jong Il. It was reported that there was conflict between the hard-line, pro-Soviet faction of the junior Kim, and the moderates who had been pro-China.

Frequent changes of personnel in the KWP Politburo and the Administration Council, effected during 1985 and 1986, were interpreted as an indication of North Korea's complex economic problems. In November 1986 unfounded reports that Kim Il Sung had been assassinated focused international attention on Pyongyang, and led to speculation that there had been an attempted *coup d'état*. However, stability was apparently restored promptly, and, after elections to the Eighth SPA, held in the same month (when its 655 members were elected unopposed), Kim Il Sung was re-elected President, and a new Administration Council was formed. Ri Kun Mo, a member of the Central People's Committee, became Premier, replacing Kang Song San.

In February 1988 Gen. Choe Kwang replaced Gen. O Kuk Ryol as Chief of General Staff of the Korean People's Army (KPA). Gen. O was regarded as a close ally of Kim Jong Il. His replacement by Gen. Choe, a veteran associate of Kim Il Sung, led to speculation among foreign observers that the President had strengthened his position to the detriment of his son, owing possibly to a series of economic failures for which the latter was allegedly responsible.

In December 1988 Ri Kun Mo resigned as Premier, reportedly because of ill health, and was replaced by Yon Hyong Muk, a former Vice-Premier and a member of the KWP Politburo. In late 1988 suspicions increased that Kim Il Sung was reclaiming much of Kim Jong Il's power in favour of Kim Pyong Il. However, in April 1989, when the then General Secretary of the Chinese Communist Party, Zhao Ziyang, visited Pyongyang, Kim Jong Il played a conspicuous diplomatic role.

In July 1989 North Korea hosted the 13th World Festival of Youth and Students, both in an attempt to enhance the country's international image, and, it was surmised, as a rival event to the Olympic Games held in Seoul in 1988. More than 15,000 delegates from 165 countries (including one student from South Korea) participated in the festival, which was the largest international event ever staged in North Korea.

The announcement in February 1990 that elections to the SPA would be held on 22 April, six months ahead of schedule, led to renewed speculation that Kim Il Sung was preparing to transfer presidential power to Kim Jong Il. In the event, however, Kim Il Sung was re-elected President, although the junior Kim did acquire his first state (as distinct from party) post: in late May he was elected First Vice-Chairman of the National Defence Commission, a body responsible to the Central People's Committee.

Following the elections to the SPA, the number of seats in the Assembly was increased from 655 to 687. In a concession to statistical impossibility, the electoral turn-out was put at only 99.78%, rather than the usual 100%, excluding those abroad or at sea. All 100% of those who did vote, however, were claimed to have supported the single approved list of candidates.

CRACKS IN THE MONOLITH

Compared with the momentous developments in inter-Korean and foreign relations in the early 1990s (see below), North Korean domestic politics showed few overt signs of change during the same period. Recurrent speculation that Kim Jong Il would formally take over from Kim Il Sung reached a peak in early 1992, when father and son celebrated, respectively, their 80th and 50th birthdays, but once again proved premature. The younger Kim's role did, however, become more emphasized, and there was a noticeable intensification of 'loyalty campaigns' for the promotion of his personality cult. In December 1991 Kim Jong Il was appointed Supreme Commander of the KPA, a post hitherto constitutionally reserved for the President. One month later, a major policy statement issued in the junior Kim's name declared North Korea's unwavering allegiance to socialism (notwithstanding its demise elsewhere), including an explicit rebuttal of any market-orientated economic reforms.

Such public continuity and defiance, however, scarcely concealed the pressures for change in a country increasingly isolated and impoverished. Defectors (including the first diplomat ever to do so) revealed that even major enterprises were often inoperative owing to shortages of power and raw materials, and that senior officials were now openly critical of party economic policy. Rumours of unrest also continued. Japanese press reports that young Soviet-trained army officers had, in February 1991, attempted a coup against Kim Jong Il were predictably denied in Pyongyang. Better attested were demonstrations in the north-western border town of Sinuiju in August by some 7,000 people, protesting against food shortages and working conditions.

During 1992 several measures, seemingly designed to quell discontent, were adopted by the Government. In February, immediately before Kim Jong Il's customarily lavish birthday celebrations, it was announced that wages were to be increased by an average of 43.4%. Two months later, the SPA approved the budget for 1992, which included an increase of 11.6% in expenditure on social welfare (compared with an increase of only 3.5% in the 1991 budget). Efforts to placate the armed forces were also apparent. In April 1992 the celebrations of Armed Forces Day were given an unusually high profile, and included the first military parade in seven years and promotions for several hundred senior officers. In the same month Kim Il Sung was given the title of Grand Marshal, while his former rank of Marshal was conferred on Kim Jong Il and on O Jin U, Minister of the People's Armed Forces and hitherto Vice-Marshal.

There was a minor reshuffle of the Administration Council in December 1992, which included the replacement of Yon Hyong Muk as Premier by Kang Song San, an economist who had previously held the post in 1984–86. His appointment was interpreted by foreign observers as an attempt to provide fresh stimulus for economic reform. However, the fact that Kang was neither seen in public nor mentioned for several months subsequently led to speculation that his reputed reformist tendencies had fallen foul of close confidants of Kim Jong Il.

Two further promotions at the same time raised hopes of a liberalization in Pyongyang. Both Vice-Premier Kim Tal Hyon and Kim Yong Sun attained candidate (alternate) membership of the Politburo. Moreover, the former, who had impressed his hosts during a business-orientated visit to South Korea in July 1992, relinquished responsibility for external economic affairs to become Chairman of the State Planning Commission. In April 1993 Kim Yong Sun, who (as the KWP's international secretary) had also impressed foreign opinion, replaced Choe Tae Bok as the party's secretary for reunification. Initial hopes for a new openness proved unfounded, however, since from late 1992 North Korea reverted to more hard-line positions.

Legislative changes showed a similar ambivalence. In October 1992 the SPA adopted three new laws on foreign investment and joint ventures. However, the overall commitment to centrally planned socialism remained unchanged, and indeed was constantly reaffirmed. Meanwhile, certain constitutional amendments had been made in April 1992, although they were not published (except several months later in South Korea). Principal among the changes were the deletion of the last remaining references to Marxism-Leninism, and the upgrading of the National Defence Commission (now chaired by Kim Jong Il, and the highest military organ of state power) to become the most senior executive body below the President.

The first party youth congress in 12 years was held in February 1993, amid fulsome pledges of loyalty to both the 'Great Leader' (Kim Il Sung) and the 'Dear Leader' (Kim Jong Il). No full KWP congress had been convened since 1980, possibly because North Korea's crises were by then so deep that such a gathering could not be successfully staged. Furthermore, the country's growing economic hardships were reflected in the budget for 1993, which envisaged substantially reduced increases in expenditure as compared with 1992. Both Kim Il Sung's birthday and that of his son were much less lavishly celebrated in 1993 than in the previous year.

Kim Jong Il was not seen in public between late April and late July 1993, when he re-emerged for the 40th anniversary of what North Korea proclaims as its 'victory' in the Korean War. There were reports that he had been treated for a heart condition (related to his alleged unhealthy lifestyle). An intriguing alternative version was that he had suffered a nervous breakdown after an upbraiding by his father on account of various policy errors, above all North Korea's threatened withdrawal from the Treaty on the Non-Proliferation of Nuclear Weapons (see below).

Important economic and political changes took place at the end of 1993. In what may have been the first admission ever of failure by Pyongyang, the KWP Central Committee announced, in early December, that the Seven-Year Plan (1987–93) had not been fulfilled. It was to be followed by a three-year 'adjustment period', giving priority to agriculture, light industry and foreign trade. As scapegoat for these economic failures, Kim Tal Hyon was removed from his post of Chairman of the State Planning Commission and reportedly allocated the even more onerous task of directing the Sunchon Vinalon Works—a perennially underperforming favourite project of Kim Il Sung. His replacement was the previously unknown Hong Sok Hyong, who formerly managed the Kimchaek Iron and Steel Complex (the largest in the country).

What attracted most attention was the return to political life of Kim Il Sung's younger brother, Kim Yong Ju, after a 17-year absence, as both a full member of the Politburo and one of four state Vice-Presidents. Kim Yong Ju's reappearance now, however, was interpreted less as the emergence of a rival to Kim Jong Il than as evidence of a continuing need to bolster the 'Dear Leader' as successor, with the overt backing of such a senior figure. Also appointed as Vice-President was Kim Pyong Shik, a returnee from Japan, who earlier in 1993 had assumed the chairmanship of the 'puppet' Korean Social Democratic Party.

The hardships of ordinary North Koreans, meanwhile, intensified, and in 1994 the trickle of defectors increased both in quantity and 'quality', the latter including two sons-in-law of government ministers. All painted a grim picture of deteriorating economic conditions and tight political control. The former was confirmed by the 1994 budget, which for the second successive year anticipated only modest increases in expenditure. It was revealed, however, that actual spending on social services in 1993 had needed to be greater than that originally budgeted for.

THE DEATH OF KIM IL SUNG AND BEYOND

Questions as to how long the world's last remaining Stalinist regime could endure unchanged acquired a sharp new focus in mid-1994. On 8 July, after almost 46 years in power, Kim Il Sung died, reportedly of a heart attack. Amid extraordinary scenes of mass mourning, Kim Jong Il was named as head of the funeral committee and, as in the past, he was generally referred to as the inheritor of his father's work. North Korea's first year without its founding 'Great Leader' presented a mixture of continuity and ambiguity. Continuity was evident in the style of the regime, where internally the cult of Kim Il Sung continued unabated, culminating in his embalmed body being placed on display in his former palace, now referred to as the 'holy land of *juche*'. There was continuity, too, in personnel, with few major new appointments among the ruling élite. The second most powerful figure in the regime, Marshal O Jin U, died in February 1995; his successor as Minister of the People's Armed Forces, Marshal Choe Kwang, was announced only in October. North Korea's stance toward the wider world also displayed elements of continuity, with characteristic militancy of rhetoric shown toward South Korea, Japan and the USA.

However, there was also growing ambiguity in North Korean political life. More than a year after the death of Kim Il Sung, his son and heir-presumptive, Kim Jong Il, had still not been officially appointed to any of the three top posts: General Secretary of the KWP, state President, and Chairman of the party's Central Military Commission. This was not the only failure of due process. General elections to the SPA (due to have been held by April 1995) did not take place, and the Assembly (still operational, albeit technically unconstitutional) failed to hold its annual meeting to consider the state budget. Amid increasingly unconvincing excuses of the need to observe a period of national mourning, the official media continued to treat Kim Jong Il as *de facto* leader, particularly in 1995, referring to him as the 'Great Leader'. Kim, however, remained as reclusive as ever, invisible to public view for long periods at a time. His appearances were confined mainly to army units and military occasions, which suggested that Kim's acceptance by the armed forces was not yet complete. The rise to prominence in early 1995 of several Vice-Marshals of the KPA, who were believed to be sympathetic to Kim, was interpreted as a further attempt by Kim to enhance his prestige among the military. Meanwhile, speculation continued that Kim Jong Il was in poor health. As in the past, there were rumours of coup attempts and popular unrest.

At the same time, however, there were signs of attempts to change. These included an international sports festival, held in Pyongyang in April 1995, which was attended by several thousand foreign visitors. More substantially, the nuclear agreement with the USA, signed in October 1994 (see below), gradually began to be implemented in the ensuing months, despite many difficulties. North Korea also expanded efforts to attract foreign (including South Korean) firms to invest in its only free economic and trade zone, at Rajin-Sonbong, in the north-east of the country. However, hopes for a more general turn toward economic reform, as in China and Viet Nam, remained unfulfilled. Rather, works by Kim Jong Il continued to inveigh against private ownership, pluralism and any effort to 'pollute' pure socialism.

The impression that North Korea was lacking in any form of effective leadership continued into 1996. Despite several occasions when Kim Jong Il might have been officially inaugurated, no ceremony took place. Kim Jong Il continued to be treated by the media as *de facto* leader, and to appear periodically, both at major state occasions and local 'guidance' visits. As before, these visits were confined mainly to army units.

Analysts' opinions were divided as to whether the 'Dear Leader' was quietly consolidating his power, was the hesitant arbiter of an ongoing struggle between hard-liners and reformers, or merely a figurehead for senior officers in the KPA. There were several indications that the military were in the ascendant. In October 1995 the 50th anniversary of the KWP was celebrated more as an army than a party affair. At official events, senior Vice-Marshals were given higher precedence than party and state representatives. One of these, Kim Kwang Jin, made two extremely bellicose speeches in March and July 1996, warning that it was only a matter of time before an inter-Korean war broke out, and threatening that North Korea would be the first to attack. In other respects, however, such as the nuclear issue (see below), North Korea was pragmatic and co-operative.

There was little overt sign of the problems attendant upon any political succession. Two senior figures, Vice-President Kim Yong Ju and Premier Kang Song San, hardly appeared in public in the first half of 1996; yet in July both were cited as still in office, with their membership of the Politburo intact. Kim Jong Il was said to favour some Vice-Ministers, such as the First Deputy Minister of Foreign Affairs, Kang Sok Ju, who seemed to be acting as Minister—for example, with regard to nuclear negotiations—while his nominal superior, Kim Yong Nam, appeared on ceremonial occasions.

The uncertainties over the situation in North Korea were not confined to the higher echelons of society. Ordinary citizens saw their already spartan living standards further eroded as the economy, still unreformed, contracted throughout the 1990s. Furthermore, the main farming areas in the west of the country were badly damaged in August 1995 by the worst floods of the century, forcing the 'hermit kingdom' to appeal for help from the international community. Foreign aid workers were given unprecedented access to the country, and were impressed by the degree of organization of the Government and the stoicism of the people, in the face of ever-worsening living conditions. Further flooding in mid-1996 threatened to exacerbate an already critical situation.

It was also clear, however, that, apart from the natural disaster, North Korea's major problems—both in agriculture, and on a more general level—were structural, and thus demanded bold new policies of reform. Without such steps, even in this most controlled of societies, there could be no guarantee that an ever more impoverished citizenry would remain quiescent indefinitely. The growing numbers of defectors, albeit perhaps surprisingly few, were an indication of the worsening situation.

This curious combination of political uncertainty and economic decline persisted well into 1997. Despite earlier indications, the third anniversary of Kim Il Sung's death passed in July without the formal inauguration of Kim Jong Il. However, the mourning period was officially declared to be over, prompting renewed speculation that the junior Kim would soon succeed his father officially, although some analysts regarded it as too risky to essay the pomp and circumstance of a 'coronation' while the country suffered serious famine. Nevertheless, in April 1997 Kim Jong Il presided over a typical parade in Pyongyang, taking the salute (in front of Western television cameras, for the first time) from goose-stepping soldiers who did not appear to be noticeably underfed.

None the less, questions persisted as to who, if anyone, was really leading North Korea. A 'leaked' speech by Kim Jong Il, published in Seoul in April 1997, revealed him to be well aware of his country's plight (not least, the food shortages) and its vulnerability. Yet he disavowed any responsibility for the economy, declaring that his job was to guide the party and army. The latter aspect was clearly more in evidence: most of Kim's appearances were still on military occasions and the role of the KPA (both symbolic and real) continued to increase, including the drafting of soldiers to assist with farming.

The long-delayed reshuffle seemed to be under way in February 1997, when the defection of the senior leader, Hwang Jang Yop (see North-South Relations), appeared to precipitate a spate of morbidity. Both the Minister of the People's Armed Forces, Marshal Choe Kwang, and his deputy, Kim Kwang Jin, died within a week of one another. The latter was replaced by the head of the navy, Vice-Marshal Kim Il Chol, who by late 1998 had also assumed the responsibilities of Choe Kwang. The most powerful man in the military was evidently Vice-Marshal Jo Myong Rok, head of the KPA Political Bureau and Kim Jong Il's constant companion.

Meanwhile, North Korea's condolences on the death of the Chinese leader, Deng Xiaoping, in February 1997 were sent by Hong Song Nam as acting Premier, implying that Kang Song San had been dismissed. The impression of a purge was strengthened by Choe Kwang's funeral committee, from which several senior, pro-reform figures were absent. Besides Kang, these included the former Premier, Yon Hyong Muk, and the Minister of the Metal Industry, Choe Yong Rim. Perplexingly, however, all these absentees subsequently reappeared (if briefly) at major ceremonies in April and July. To mark the third anniversary of the passing of the 'Great Leader', North Korea introduced a new 'Juche' calendar, starting from 1912, his year of birth. If Pyongyang's aim was to keep the world guessing as to its true intentions, it certainly succeeded.

From late 1997 there were some signs of moves towards political normalization, at least by North Korea's own standards. The official rationale was that, following the end of the three years of official mourning for Kim Il Sung, normal life could now resume. Thus, on 8 October, Kim Jong Il at last assumed one of his late father's two vacant posts, becoming General Secretary of the KWP. The manner of his elevation was, however, unorthodox, being by acclamation at a series of provincial party conferences, rather than through election by the Central Committee, as laid down in the KWP's rules. There was still no sign that the Central Committee had actually met since Kim Il Sung's death, and there was no immediate prospect of a full party Congress, none having been convened since 1980. The legislature, however, was revived. The elections for the SPA, which should have been held in 1995, finally took place on 26 July 1998. As ever, the electorate was presented with a single list of candidates, which it reportedly endorsed unanimously: some 99.85% of electors voted (that is, all North Koreans except those who were abroad or at sea), and fully 100% of voters supported the candidates.

The 10th SPA was duly convened in September 1998, the 50th anniversary of the foundation of the DPRK; the expected and long-awaited appointment of Kim Jong Il as President did not, however, occur. In an unexpected move, the Constitution was amended to elevate Kim Il Sung posthumously to the rank of 'Eternal President' and thus perpetual Head of State. The chairmanship of the National Defence Commission (the highest military office, to which Kim Jong Il had been re-elected, having stood for election in a military constituency) was defined as the most senior position in the state hierarchy. Kim Jong Il consequently assumed the role of *de facto* Head of State, while remaining in the shadow of his father. The 10th SPA, like the Ninth, had 687 members. Usually the number increases at each election, so there was speculation that famine had taken its toll on population growth. Of the 687 members, as many as 443 (or 64%) were new, indicating that Kim Jong Il was at last able to promote his own generation and supporters. Military predominance was striking, with some 50 younger generals included among the new deputies. The composition of the assembly was also a guide to the ongoing power struggles in Pyongyang. Absentees included Kim Song Ae, Kim Il Sung's widow (and Kim Jong Il's stepmother); Kang Song San, at last officially replaced as Premier by Hong Song Nam (as part of the extensive government reorganization of September 1998); and several officials who in the past had handled relations with South Korea, including Kim Tal Hyon (who had been regarded as a possible candidate for the premiership). On the other hand, Vice-President Kim Yong Ju, Kim Jong Il's uncle and once his rival for the succession, retained an SPA seat despite being all but invisible since the death of his brother, Kim Il Sung.

Politics in North Korea also proceeded behind the scenes and by harsher means. Several reports claimed that So Kwan Hi, the long-serving party secretary for agriculture, had been executed during late 1997, presumably as a scapegoat for the ongoing food crisis. The head of the youth league, Choe Ryong Hae (a friend of Kim Jong Il since childhood), was purged in January 1998 after reports that several youth league officials had been executed for spying for South Korea. It was reported in October that Kim Jong U, formerly the reformist Chairman of the DPRK Committee for the Promotion of External Economic Co-operation, had been executed.

In April 1999 the SPA resumed the annual sessions to consider the budget, which had lapsed since the death of Kim Il Sung in 1994. Unusually, some 50 of the newly elected members were absent, prompting speculation of continuing purges. The budget itself gave only broad magnitudes, but these revealed that both revenue and expenditure had declined by one-half in the five years since figures were last published: a fact that passed almost without comment. The SPA approved a new economic planning law during the same session. Several other economic laws were also announced during 1999, covering specific areas such as agriculture, forestry, and even fish-breeding. Most of this legislation appeared to codify rather than alter existing arrangements, and gave the impression of a 'rearguard action' by the centre, to try to exert control over an economy which *de facto* had become increasingly anarchic, as the old planned economy had broken down. The revised Constitution of September 1998 gave slightly more scope to private enterprise and market forces, but the amendments failed to keep pace with the actual situation, even though in theory the DPRK remained bonded to communism and hostile to any explicit market reforms. In June 1999 a major policy statement proclaimed a 'military-first' policy, giving defence absolute priority.

Kim Jong Il himself remained elusive, although his 'on the spot guidance' broadened from visits to military bases to include more economic sites (many of them run by the military). He continued to delegate the task of meeting foreigners to Kim Yong Nam, the former Minister of Foreign Affairs, in his new capacity as President of the SPA Presidium. From mid-2000, however, Kim Jong Il adopted a less reclusive attitude and a startling change of image, if only for the purposes of external display. Within two months he met the heads of state of China, South Korea and Russia, while his polite but affable manner at the North-South summit meeting, which took place in mid-June, impressed television viewers in South Korea. His behaviour was also revealing when entertaining South Korean media executives, whom he had invited to Pyongyang in August. Accounts of this lengthy luncheon showed the North Korean leader as effusive, yet still enigmatic and somewhat eccentric.

How, when, or even whether North Korea's apparent new openness to the outside world would be translated into domestic political and/or economic change remained ambiguous. Hitherto Kim Jong Il had explicitly and frequently inveighed against reform as a betrayal of socialism. He hinted to his Southern guests, however, that a long-overdue KWP Congress would be held in late 2000, at which the party's statutes might be changed to excise the goal of communizing South Korea, despite the fact that this would necessitate the purging of officials loyal to his late father, Kim Il Sung. However, the Congress was not held or further mentioned. While the KWP formally remained the ruling party, its real power in relation to either the military or Kim Jong Il's circle was not evident. Nor was it clear who served on the Politburo, or even whether the Central Committee met regularly.

The ambiguity continued in 2001, which began promisingly with a clearly business-orientated visit by Kim Jong Il to Shanghai, China (see below), and the publication by him of maxims emphasizing the need to adapt to new times. However, by late 2001 none of this had been implemented. In fact, the year was largely devoid of overt domestic political activity of any kind. In April 2001 the SPA met for just one day, instead of the usual three. As well as passing a budget, the few figures of which suggested that the economy might have stabilized, the SPA ratified laws on copyright and the processing trade, although no details of these were given. This failure to change suggested fierce debate behind the scenes. Seoul press sources reported that Kim Yong Sun, the party secretary in charge of dialogue with South Korea, spent a week under arrest in March 2001 before Kim Jong Il ordered his release. The implication was that the initial hostility of the new US Administration under President George W. Bush undermined North Korea's more reform-minded figures and strengthened the position of its hard-liners: this interpretation was supported by the suspension of talks with South Korea (see below). Meanwhile, the world had its first glimpse of yet another Kim, the 'Dear Leader's' son and reputed heir, Kim Jong Nam, in an incident that did little to improve North Korea's reputation for bizarre behaviour. In May 2001 the young Kim and his family were detained on entering Japan at Narita airport, having been found to be travelling under false names and on fake Dominican Republic passports. They admitted their identity, stated that they hoped to visit the Disneyland theme park, and were swiftly deported to Beijing. This embarrassment seemed a set-back to possible plans to appoint Kim Jong Nam as his father's official successor.

In March 2002 the Kim Il Sung Socialist Youth League met for the first time in six years. Amid much mention of the need to inherit revolutionary traditions, no details were announced. In April it became known that Shin Il Nam, hitherto a Vice-Minister of People's Security, had been appointed Vice-Premier and Chairman of the Commission for Capital Construction. In August Japanese sources reported that a different son, Kim Hyon (also known as Kim Hyon Nam), had been appointed head of the KWP's propaganda and agitation department. If confirmed, this was the same route by which the young Kim Jong Il had begun his ascent to power, and might herald Kim Hyon's position as heir-apparent.

The year 2002 also witnessed two major anniversaries: Kim Jong Il's 60th birthday—in Korea a key event, called *hwan'gap*—in February, and the 90th anniversary of his late father's birth, in April. Both were celebrated with the usual lavish displays. This ceremonial mode continued with North Korea's largest ever mass arts and gymnastics festival, *Arirang*, held from May to August. Despite denials, this appeared as if it was intended to counter the football World Cup in South Korea; yet poor marketing meant that few foreign visitors attended. Substantive politics was harder to discern. As in 2001, the SPA convened for just a single day, in March. Besides approving the budget, it discussed an (unrevealed) 'organizational matter', and approved a law on land management, thought to be aimed at curbing illicit private use of land and resources, a practice that had spread since the famine of the late 1990s.

The latter part of 2002, however, brought radical changes in economic management, possibly auguring a long-awaited definitive turn to reform. No formal announcements had been made; officials, when pressed, spoke of 'perfecting socialism'. Thus, the full scope of the innovations was unclear. Their core consisted of drastic increases in prices, broadly to match those in the 'black market', and concomitant—but lesser, and uneven—wage increases. Some subsidies remained in place, but in general people and firms alike were thenceforth required to pay, and firms allowed to charge, the real cost of goods and services.

Sceptics suggested that the aim was to curb, not enhance, the free market: for instance, to get household savings, often held at home in US dollars or Chinese yuan, back into won and into banks. Yet other reports implied deeper change, including rumours of experimental private farm projects (the form in which market reform began in China). Also hopeful was the designation in the latter part of 2002 of three new special economic zones. One, at Sinuiju in the north-west on the Chinese border, was swiftly aborted by China's arrest of its first head (see Foreign Relations, below). The other two were the Hyundai group's established tourist concession at Mount Kumgang in the south-east, and a planned industrial export zone at Kaesong in the south-west, close to Seoul (see North-South Relations). Remarkably, in the domestic economy downsizing of unproductive labour was said to extend even to the Party, at least at the 'grass roots'; with drastic reductions in the hitherto ubiquitous and intrusive apparatus of political and guidance secretaries, agitprop teams, and the like in enterprises and even in the KPA. Such measures were highly likely to have political repercussions. It was believed that, if successful and well received, the measures might be extended, and in due time proclaimed as a new turn; perhaps at the long-awaited KWP Congress. If the measures were to prove unsuccessful, on the other hand, some analysts feared hyperinflation and that the result might destabilize the regime.

Instability was not, however, in evidence in 2003: a year when domestic politics was overshadowed by a growing nuclear crisis (see Foreign Relations). The year began in militant style, with a joint editorial of the three main daily newspapers hailing the 'Great Banner of Army-Based Policy' and warning soldiers to 'combat illusion about the enemy and peace'. State television reported on 14 January that of Kim Jong Il's 207 'on-the-spot guidance' visits in 2002, more than half (120) were to military units. In brief mention of the economy, the new year joint editorial called for seeking 'the largest profitability while firmly adhering to socialist principles': a seeming effort to square the circle. As nuclear tensions grew, Kim Jong Il—perhaps fearing the fate of Saddam Hussain in Iraq—disappeared from public view for seven weeks between 12 February and 3 April: his second longest absence after his three months' sequestration in 1994 following the death of his father Kim Il Sung. He thus missed the celebrations for his 61st birthday on 16 February, and—unusually—the annual SPA spring session on 26 March. As had become the norm, the SPA met for just one day, which sufficed to approve both that year's budget and the report by the Minister of Finance on the previous year's budget. Never generous with statistics, this time Mun Il Bong did not provide a single useful figure, but only percentages—presumably because of problems in reconciling data before and after the massive devaluation of July 2002. For 2003, a planned rise in spending of 14.4% was to be financed by North Korea's first bond issue in half a century. This was unlikely to be voluntary, suggesting yet another effort by the regime to gain access to private household savings. The SPA also adopted—unanimously, as always—five bills earlier approved by its Presidium (which meets when the SPA is not in session): on military service, city planning, accounting, rivers, and 'structures'. No further details, much less full texts, were given. Conditions of military service, hitherto lengthy and harsh, were said to have been eased, but with action against draft evaders.

In February South Korean sources claimed that a campaign was under way to promote another of Kim Jong Il's sons, Kim Jong Chul, born in 1981, as his heir. The evidence was typically indirect and esoteric: a KPA study document in August 2002 praised an unnamed 'respected mother' who 'assists the comrade supreme commander closest to his body'. This was interpreted as a reference to Kim Jong Il's current partner, Ko Yong Hui, a returnee from Japan and former dancer, who is Kim Jong Chul's mother. Little was known of Kim Jong Chul, except that he studied in Switzerland and was said to hold—despite his youth, but like his father before him—a key post in the KWP's agitation and propaganda department. Kim Jong Il's marital history appeared complex, with reports of at least two other sons by different partners.

Normality, by North Korean standards, was seen in elections held on 3 August for the SPA as well as provincial and local councils. Unlike the last SPA election in 1998, which was three years late, this reverted to the five-yearly intervals stipulated in the Constitution. Typically, a 99.9% turnout was proclaimed—only those abroad or 'working on the far-off seas' could not participate—and 100% voted for the single list of candidates. Kim Jong Il cast his ballot at Kim Il Sung Military University, whose candidate (unsurprisingly) was a KPA officer. The 'Dear Leader' himself was elected by another military constituency; he apologized to voters elsewhere that the rules forbade him to accept the nominations which all 687 constituencies had unanimously bestowed upon him. As in 1998, at least half the deputies were reported to be new members. Separately, in July the elderly Minister of People's Security, Paek Hak Rim, had been replaced by Choe Ryong Su, whose background was unknown. Other posts rotated in 2003 included the Ministry of Education, where Kim Yong Jin replaced Pyon Yong Rip, and the heads of the Sinuiju and Kaesong special economic zones and of Nampo, the port for Pyongyang. It was believed that this last trio of appointments might lead to more economic opening. In an important development, in June, the role of markets had for the first time been officially acknowledged; the official Korean Central News Agency (KCNA) carried a photograph of a large covered market under construction in Pyongyang.

When the 11th SPA convened a month after its 'election', on 3 September 2003, it endorsed a partial cabinet reorganization, mainly of economic portfolios. Pak Pong Ju, hitherto Minister of Chemical Industry, replaced Hong Song Nam as Premier. In his former post Pak had visited South Korea in 2002 in an economic delegation, impressing his hosts. Also on that trip were Pak Nam Gi, who moved from managing the State Planning Commission (SPC) to chairing the SPA budget committee; and Kim Kwang Rin, promoted from SPC Vice-Chairman to replace him. Two new Vice-Premiers were appointed: Ro Tu Chol, whose background was unknown, and Jon Sung Hun, who was previously Minister of Metal and Machine-Building Industries; he was replaced by his Vice-Minister, Kim Sung Hyon. There were also new Ministers of Agriculture, Ri Kyong Sik; Power and Coal Industry, Ju Tong Il; and Culture, Choe Ik Gyu. Not for the first time, the composition of the new Cabinet raised hopes that pragmatic technocrats might be gaining in power over ideologues.

There were changes, too, on the National Defence Commission (which outranks the Cabinet). Three aged generals left the body. They were replaced by the new Minister of People's Security, Choe Ryong Su, and by a hitherto unknown official, Paek Se Bong. Yon Hyong Muk, a former Premier and rare civilian on the National Defence Commission, became a Vice-Chairman, replacing the Minister of the People's Armed Forces, Vice-Marshal Kim Il Chol, who remained an ordinary member of the Commission.

As the full SPA met for only a few days each year, most legislative business continued to be conducted by its Presidium. The latter's officials were unchanged, but four of the 11 ordinary members were new. Most notable was a return by Hong Sok Hyong, a former chief planner and alternate Politburo member, appointed party secretary in North Hamgyong in the north-east, the province worst affected by famine and the source of most migrants to China. Other new members were Kim Kyong Ho, secretary of the youth league, which underwent a major purge; Pak Sun Hui, who chaired the women's union; and Pyon Yong Rip, the former Minister of Education who had moved earlier in the year to head the Academy of Sciences. Of the four Presidium members ousted, only one was known to have transferred to a new post: Ri Kil Song, formerly party secretary in South Pyongan province, was appointed as Procurator-General. While it is hard to identify any political trend in this turnover, there was satisfaction in Seoul that at least five figures prominent in inter-Korean dialogue had also became members of the SPA.

Soon after the SPA meeting, Kim Jong Il again went to ground. After the DPRK 55th anniversary ceremonies on 9 September 2004, he was not seen again until 21 October. He thus missed, unusually, the other main autumn festival: the KWP's 58th anniversary, on 21 October. Once more rumours abounded: that his consort Ko Yong Hui was ill with cancer, and that a power struggle was taking place for the succession between Kim Jong Chul and another hitherto unknown son, Kim Jong Un, aged only 20. According to a book by Kim Jong Il's former sushi chef, Kenji Fujimoto, the 'Dear Leader' favoured the latter, regarding the former as effeminate.

Questions were also raised by the death of one of Kim Jong Il's closest aides, Kim Yong Sun, on 26 October. A party secretary and long-serving senior diplomat, Kim had latterly been in charge of North-South links. He had been in a coma since a road accident in June; there was speculation that this followed one of Kim Jong Il's parties. The urbane and able Kim was likely to be much missed.

Hopes of a technocratic turn were not borne out by the 1 January joint editorial of three main daily newspapers, which set a political course for the New Year. As summarized by KCNA, this called for 'giving top priority to increasing the military muscle'. Other goals included strengthening the defence industry, and mounting 'a strong counter-attack on the imperialists in the ideological and cultural fields'. Three 'fronts' of struggle were highlighted: politics and ideology, anti-imperialism and military affairs, and economy and science—in that order. Reflecting these priorities, a Japanese source claimed that among Kim Jong Il's 86 'on-the-spot guidance' trips in 2003—said to be a reduction from 117 in 2002; the official figure was 207 (see above)—the number that were military-related increased from 24 to 56, or 65% of the total.

In April 2004 Japanese reports claimed that Kim Jong Il's brother-in-law Jang Song Taek, hitherto a key confidant, had been demoted following a disagreement with the Premier. Jang had reportedly criticized recent reforms as inviting 'unhealthy ideas from outside'. Jang was not part of Kim Jong Il's entourage when he visited Beijing in April. Ambivalence over reform was evident at the SPA's usual spring meeting to review the budget, in March. As in 2003, not a single meaningful figure was given, but only a few percentages. Premier Pak Pong Ju called for profitability to be ensured—alongside socialism and 'the fighting trait and working method created by the People's Army', with priority going to the defence industry. There was thus no change in the avowed military-first policy; nor was any reference made to partial market reforms introduced since mid-2002.

On 22 April 2004 a serious railway explosion occurred at Ryongchon, near the Chinese border, just hours after Kim Jong Il had passed through on his way back from Beijing. Initial reports claimed 3,000 casualties, but the eventual official toll was 161 dead and more than 1,300 injured. This might have been lower, had the Government accepted international aid immediately. Nevertheless, for a state where news of accidents was routinely suppressed, the fact that North Korea allowed access to foreign aid workers two days after the accident was some progress. The official account stated that this had been a shunting accident: a short circuit had ignited flammable cargoes of fuel and fertilizer. Speculation of foul play, however, was rife. A ban in May on mobile telephones, permitted only 18 months earlier, was interpreted by some as reflecting suspicions that such a device had been used to trigger the explosion. A less conspiratorial view was that Kim Jong Il's special train had disrupted the usual timetables, thus contributing to the accident. Given the dire condition of North Korea's rail network and other infrastructure, the catastrophe was no surprise. Yet the dismissal on 10 July of the Minister of People's Security, Choe Ryong Su, after just a year in office suggested that concerns remained in this area. His successor was an army general, Ju Sang Song.

On 8 July 2004 a nation-wide three minutes' silence marked the 10th anniversary of the death of Kim Il Sung. In 1994 few had expected the DPRK to endure without its founding leader. Yet a decade later, Kim Jong Il appeared firmly in control of a state which, while belatedly embarking on partial reform, was still desperately poor and defiant of the international community. North Korea's longer-term prospects thus remained as uncertain as ever.

NORTH-SOUTH RELATIONS

Relations (or the lack thereof) between North and South Korea must be understood against a very particular and, indeed, unique background. Both the Democratic People's Republic of Korea and the Republic of Korea still claim, as they have done ever since their founding in 1948, to be the sole legitimate government on the peninsula. Constitutionally and legally, each still defines the other as an enemy. Indeed, they remain technically at war, inasmuch as the Korean War (1950–53) concluded only with an armistice, not a peace treaty (and South Korea did not even sign the armistice).

Against that background, for most of the subsequent four decades both regimes not only eschewed all mutual contacts but also forced their citizens, on pain of draconian penalties, to do likewise. Quite unlike former East and West Germans, North and South Koreans have never been able to write to or telephone one another, let alone visit. Several million people have thus spent more than half a century utterly cut off from close relatives, not knowing if they are alive or dead. Each regime has also suppressed all but negative information about the other. As a result, over the years the two Koreas (unlike the two former Germanys) have become strangers as well as enemies, a fact that has added to the already huge incubus of mutual mistrust.

Only in the early 1990s did real signs of change begin to emerge. Previously there had been several attempts at dialogue, all abortive. The first began with secret visits in 1971, at which time both Korean states were alarmed by the recent US-Chinese *rapprochement*. This led to a joint statement on 4 July 1972, and the establishment of a South-North Co-ordinating Committee (SNCC). Red Cross talks also began, with the aim of arranging family reunions. Although SNCC meetings continued until 1975 and Red Cross talks until 1978, neither produced any result. The same was true of a further brief round of dialogue in 1979–80, during the democratic interlude in South Korea between the assassination of President Park Chung-Hee and Chun Doo-Hwan's coup.

Meanwhile, North Korea did nothing to enhance its trustworthiness, committing regular acts of aggression against the South. In August 1974 a North Korean resident of Japan shot at President Park, killing his wife. In the late 1970s South Korea discovered the existence of several tunnels, dug under the demilitarized zone (DMZ) separating the two countries, large enough for an invading force from the North. In October 1983, while on a visit to Rangoon, Burma (now Yangon, Myanmar), President Chun Doo-Hwan narrowly escaped an assassination attempt, which, however, killed 17 members of his entourage, including four cabinet ministers. The attack was believed to have been perpetrated by North Korean agents.

Perhaps because of the opprobrium incurred by the Rangoon bombing incident, North Korea subsequently adopted a different approach. What may have been intended only as a propagandist offer of 'aid' to South Korea, after severe floods occurred there in late 1984, was shrewdly accepted by Seoul. This led to a year of three-tiered dialogue in 1985, comprising economic, parliamentary and Red Cross talks. Only the last session yielded results, in the first reunion of separated families, which took place in September.

North Korea suspended all dialogue in early 1986 in protest against the annual 'Team Spirit' US-South Korean military exercises, and resumed its duplicitous policy: negotiating (unsuccessfully) with the International Olympic Committee to host part of the 1988 Seoul Olympic Games, only to bomb a South Korean civilian airliner in November 1987, causing the loss of 115 lives. None the less, South Korea's growing economic and diplomatic strength (symbolized by the full participation in the 1988 Olympics by China, the USSR and the countries of Eastern Europe, even though, at that point, none recognized the legitimacy of Seoul) brought the North back into the negotiating process, albeit sporadically. Red Cross talks resumed in 1988, although planned family reunions in late 1989 failed to take place. There was, however, some progress in the sporting arena. The first ever inter-Korean football matches were held in Pyongyang and Seoul in late 1990, and in the following year joint Korean teams participated in two international sporting events. However, North and South Korea sent separate teams to Spain to participate in the 1992 Barcelona Olympics.

Meanwhile, although there was no resumption of economic talks, in 1988 South Korea initiated indirect trade with the North. Direct inter-Korean trade, which began in late 1990, increased rapidly in 1991, and was equivalent to some US $210m. annually by 1992, when the South became the North's fourth largest trading partner (after China, Russia and Japan).

In the political arena, the first ever talks between the countries' heads of government finally took place in September 1990. The three such meetings in that year were largely symbolic, and the premiers did not meet again until October 1991. However, progress was made in mid-1991, when North Korea withdrew its long-standing objection to both Korean states joining the UN as separate entities (if only because it became clear that neither the USSR nor China would any longer veto South Korea's unilateral application). Both Koreas were thus admitted to the UN in September 1991. (They were already members of most of its specialized agencies, including FAO, WHO and UNESCO.)

At the resumption of the premier-level talks in October 1991, both parties agreed on the title, and envisaged provisions, of an accord governing future inter-Korean relations. Under the 'Agreement on Reconciliation, Non-aggression and Exchanges and Co-operation', which was signed in December and ratified in February 1992, North and South pledged, *inter alia*, to desist from mutual slander and sabotage, and to promote economic and other co-operation, as well as the reunion of separated family members. The accord was widely hailed as a milestone, and subsequent premiers' meetings, in May and September 1992, resulted in agreements to fulfil its provisions, as well as to establish several joint commissions.

Thereafter, however, relations worsened and the agreement remained largely unimplemented. In late 1992 South Korea was angered by the discovery on its territory of a large-scale Northern espionage operation, while North Korea criticized the South's decision to resume in 1993 the 'Team Spirit' exercises (which had been suspended in 1992). The projected ninth meeting of premiers in December 1992 was cancelled, and such meetings were not subsequently resumed.

The main impediment to a real improvement in inter-Korean relations was the question of suspected nuclear ambitions (see below). Although the accord of December 1991 omitted any reference to nuclear issues, a separate agreement was signed later that month to create a bilateral Joint Nuclear Control Committee (JNCC). This made no progress, however, with North Korea opposing the South's demand for unannounced inspections, while Seoul resisted Pyongyang's demand to open all the US bases stationed on its territory to Northern scrutiny.

The visit to South Korea of Kim Tal Hyon, the North Korean Vice-Premier in charge of trade and investment, who toured a range of factories and met prominent business leaders as well as the then President, Roh Tae-Woo, in July 1992, brought hopes of a breakthrough. While North Korea desperately needed Southern aid and investment, it still seemed reluctant or unable to convince the South (and indeed the world) that it had unequivocally abandoned any nuclear ambitions; indeed, prior to October 2002, it had always denied ever having had a military-nuclear programme. Such suspicions were only enhanced by Pyongyang's announced intention, in March 1993, of withdrawing from the Treaty on the Non-Proliferation of Nuclear Weapons, which it had signed in 1985 (see below). The nuclear issue continued to blight inter-Korean relations throughout the remainder of 1993, not least because of Pyongyang's insistence on negotiating with Washington rather than Seoul. There were contacts in the 'peace village' of Panmunjom (in the DMZ) in October 1993 and March 1994, but both proved abortive; at the latter, the chief Northern delegate threatened to reduce Seoul to 'a sea of fire'. This was on a par with other North Korean rhetoric, which included regular denunciations of the Southern President, not to mention a call by the Chief of General Staff of the KPA, Vice-Marshal Choe Kwang, in a speech to soldiers in late 1993, for 'reunification with guns'.

Despite tense relations between the two Governments, however, their companies continued to do business. Inter-Korean trade in 1993 totalled some US $195m., slightly less than in 1992. With over 90% comprising Southern purchases, this represented an important source of revenue for Pyongyang. Although Seoul still banned its companies from strengthening these ties by investing in the North, the fact that businessmen from both Koreas were now in regular contact (mainly in China) provided some hope—as with Taiwan and the People's Republic of China—of a slow improvement in their political relationship, too.

Just such an improvement became dramatically apparent in June 1994, when the former US President, Jimmy Carter, returned from Pyongyang (see below) with an offer from Kim Il Sung for a summit meeting with his South Korean counterpart. Kim Young-Sam accepted with alacrity, and two highly successful planning meetings were held at Panmunjom. Following the death of Kim Il Sung on 8 July, however, the summit (which had been arranged for 25–27 July) was postponed. Not only did the summit not take place in the ensuing months (technically North Korea remained without a Head of State), but North-South relations became markedly worse. Pyongyang professed outrage when Seoul failed to issue any condolence on the death of Kim Il Sung and acted harshly against the few Southern radicals who did so. Northern denunciations of Kim Young-Sam, former dissident though he was, were, if anything, more virulent than they ever had been against the generals who preceded him.

The new bilateral relationship with Washington, engendered by the Geneva nuclear agreement of October 1994 (see below), had the advantage, from Pyongyang's viewpoint, of excluding Seoul—at least formally. Although this remained the North's official position, in practice by mid-1995 North Korea had accepted South Korean light-water reactors (LWRs) and engineers to build them. The North also continued business links with South Korean companies (if not with their Government), and in July 1995 13 technicians of the Daewoo conglomerate became the first Southerners since 1953 to settle in the North with both Governments' approval. The technicians were to supervise Daewoo's new export factory at the port of Nampo, a pioneer venture which other Southern firms were thought likely to emulate.

Yet the North's policy of ignoring or bypassing the South Korean Government was hardly sustainable as a long-term strategy, particularly in the light of the agreements signed in 1991–92 (albeit only partially implemented). In June 1995 North Korea appeared to have changed its tactics, accepting an offer of rice aid from the South, following a request for similar aid from Japan in the previous month (see below). However, the first ship was forced to fly the Northern flag on its arrival in the port of Chongjin (for which Pyongyang later apologized), while in August another Southern vessel was detained on spying charges. South Korea was aggrieved that its generosity did not elicit a similar spirit on the part of the North.

Such incidents embittered both the public and official moods in Seoul. When the floods of 1995 (see above) led to fresh appeals by the UN and the Red Cross for food aid to North Korea, South Korea was reluctant to oblige. The Government gave US $3m. in June 1996 and allowed small shipments from the Red Cross, but tried to prevent church groups and other private organizations from sending food aid to the North independently.

In so far as the rice aid of 1995 failed to herald the desired improvement in inter-Korean relations, both states shared the responsibility, albeit not equally. North Korea, which was willing to hold discussions and even sign agreements with the South in the early 1990s, had no convincing reason for its subsequent refusal of such dialogue. Its professed insistence on signing a peace treaty with the USA to the exclusion of South Korea, was patently unrealistic, while its continuous denigration of former President Kim Young-Sam was unacceptable practice by international standards. Yet the South Korean President was widely seen as having no clear or consistent strategy towards Pyongyang, oscillating between an uncompromising attitude and a more relaxed stance.

As ever, spies and refugees enlivened inter-Korean relations in the mid-1990s, while doing nothing to improve them. The report of Kim Dong Sik, a Northern agent captured after a gun battle in October 1995, was worthy of any spy novel; he landed by midget submarine after a decade of training, which had included the use of full-scale models of areas of Seoul. In July 1996 another alleged spy was arrested after living for more than a decade in South Korea disguised as a Lebanese-Filipino history professor. Two dramatic incidents affected inter-Korean relations during 1996–97. In mid-September 1996 a North Korean submarine ran aground off South Korea's east coast. In the ensuing manhunt, all but two of its crew of 26 died (some by their own hands, apparently; one was captured alive and one escaped). Seoul's fury was only assuaged in December when, pressed by the USA, Pyongyang perfunctorily apologized for what, it still claimed, was an accident caused by engine trouble. Then, in mid-February 1997, Hwang Jang Yop, one of North Korea's most senior leaders, sought asylum in the South Korean embassy in Beijing, while returning from a visit to Japan. Ranked 25th in the hierarchy, Hwang was the main theorist of North Korea's official ideology of *juche*, and was currently serving as party secretaryand Chairman of the SPA foreign affairs committee. This defection was awkward for China, but the situation was eventually resolved by sending Hwang first to a third country, the Philippines, before allowing him to enter South Korea. Once in Seoul, Hwang warned that his former comrades were serious in threatening to attack the South. Even before Hwang's arrival, an earlier high-ranking defector had been assassinated near Seoul by unknown gunmen.

While both these events were set-backs for inter-Korean relations, there were also more positive signs. In particular, nuclear co-operation through the Korean Peninsula Energy Development Organization (KEDO) reached the stage where ground-breaking for construction of the new LWRs began in mid-August 1997. Several dozen South Korean engineers were already on site at Shinpo, Southern ships had delivered machinery and materials, and a telephone link with South Korea was in service. KEDO's office at Shinpo included the first South Korean diplomats ever to be based in North Korea.

In 1997 South Korea also appeared to be easing its restrictions on Southern businesses wishing to invest in North Korea. In May five more companies received permission to explore joint ventures. Though the parlous state of the Northern economy dampened optimism somewhat, in April North-South trade reached US $37.4m., its highest-ever monthly total. While infinitesimal to South Korea, annual trade of some $200m.–$300m. in the late 1990s sufficed to make it North Korea's third largest trading partner, after China and Japan. South Korea also announced its support for the North's bid to join the Asian Development Bank, which, however, was obstructed by Japanese opposition. Meanwhile, negotiations concerning the opening of North Korean airspace—only maritime hitherto—to flights to and from Seoul finally yielded results in April 1998, so that the tiny numbers of North and South Koreans in regular contact with one another began to include air traffic controllers.

In general, North Korea continued its venomous diatribes against the South and, in particular, Kim Young-Sam during 1997, while South Korea, for its part, appeared ambivalent, not to say inconsistent, over how to approach the North. This was particularly evident over the issue of North Korea's food shortage. Although South Korea did give some relatively small amounts of aid, both in response to UN appeals and bilaterally through the Red Cross (the latter including actual visits by Southern officials to the North, from China, to deliver 50,000 metric tons of grain), it often seemed more concerned either to play down the severity of the North's suffering, or to suggest that Pyongyang should help itself by spending less on the military and by privatizing the country's farms. This uncompromising approach placed Seoul somewhat at odds with the USA, where the Administration of President Bill Clinton remained committed to seeking engagement with Pyongyang and, in particular, to persuading it to enter four-way talks (see below).

The prospects for inter-Korean relations improved markedly with the election of Kim Dae-Jung as South Korea's President in December 1997. Kim Dae-Jung had long preached, and once in office immediately began to implement, a so-called 'sunshine' policy towards Pyongyang, involving consistent openness towards the North (while maintaining a strong security posture), in the belief that this would eventually elicit a positive response. The policy entailed a distinction between governmental and private (including business) contacts, and a much more relaxed attitude towards the latter whatever the vicissitudes of the former. The acknowledged model here is relations between China and Taiwan. The first official North-South talks since the death of Kim Il Sung (there had been others which were quasi-governmental) took place in April 1998. Held in Beijing to discuss a Northern request for fertilizer, the talks failed when South Korea linked this issue to its own demands for progress on family reunions. Yet, in a break from the past, Seoul made no effort to prevent the transfer of private Southern aid to the North. The pace of civilian and business contacts thus increased in 1998, even though the South's economic crisis took its toll on inter-Korean trade, which declined in the first half of the year by almost 50%, to US $77m.

During mid-1998 inter-Korean relations were dominated by dramatic, if contradictory, developments. In mid-June Chung Ju-Yung, the founder of the Hyundai group, South Korea's largest conglomerate, crossed the normally impenetrable DMZ at Panmunjom, bringing 500 cattle as a gift for his home town near Wonsan. During his week-long visit, Chung Ju-Yung also discussed a wide range of potential joint ventures between Hyundai and North Korean interests. The most dramatic involved a plan to run daily tour boats to Mount Kumgang, just north of the DMZ; this commenced in November 1998 and represented the first opportunity for South Korean tourists to set foot in the North. This breakthrough seemed jeopardized, however, when, during Chung Ju-Yung's visit to North Korea, a Southern fishing boat caught a small Northern spy submarine in its nets. When the submarine was eventually towed to port, its crew of nine were found to have killed themselves (or each other). Then in July, a dead North Korean frogman was found on a Southern beach. In the past such provocations would certainly have led Seoul to forbid Hyundai to continue with its plans. It was indicative of Kim Dae-Jung's imagination and courage that on this occasion no such linkage was made, and, after a short delay, Hyundai was allowed to proceed.

During 1999 Hyundai's tourism project proved a major advancement for the 'sunshine' policy, with more than 80,000 South Korean tourists making the journey north in the first eight months alone. There was, however, criticism that Hyundai's payments to Pyongyang—which were to total almost US $1,000m. over six years—might be funding the North's military. The growing scepticism in Seoul reflected the failure of 'sunshine' to generate wider warmth. In June 1999 fresh talks in Beijing on fertilizer and family reunions broke down, even though South Korea softened its stance and sent fertilizer without preconditions. Meanwhile, on 15 June the two Koreas' navies fought a brief gun battle for the first time since the Korean War. North Korean boats were fishing for crab in the Yellow Sea south of the Northern Limit Line (NLL), which Pyongyang did not accept and, unusually, held their positions when challenged by Southern patrol boats. After several days of confrontation the South resorted to ramming, and the North opened fire; however, one of its boats was sunk (with a reported 80 dead) and three others were badly damaged. Remarkably, both the Beijing talks and fertilizer deliveries—by sea, close to the combat area—continued throughout, and despite, this contretemps. The gun battle may have reassured Kim Dae-Jung's domestic critics that 'sunshine' did not mean appeasement, and the policy remained in place. In August a workers' team from the militant Korean Confederation of Trade Unions was allowed to travel north to play soccer with Northern counterparts.

The 'sunshine' policy finally achieved results in 2000. In March Kim Dae-Jung's offer in his 'Berlin Declaration' of Southern aid to rebuild infrastructure in North Korea led to secret talks in China, and the announcement in April of the first ever North-South summit meeting. This momentous event duly took place on 13–15 June, after a last-minute 24-hour delay. Kim Dae-Jung made the first ever official direct flight between Seoul and Pyongyang. Kim Jong Il met him at the airport with a full honour guard of the KPA's three services. South Korean television viewers—but not their Northern counterparts, who saw only the formalities—marvelled at the friendly persona of a man hitherto viewed as a completely evil figure. After two days of public affability and private tough negotiations, the two leaders signed a brief declaration pledging further progress. The document's only substantive stipulation was the reunion of separated families, duly held on 15–18 August when two sets of 100 elderly Koreans flew in each direction to meet relatives whom they had not seen for 50 years. Two further reunions took place in December 2000 and February 2001 (but a third, scheduled for October, was cancelled). In the following month the first exchange of personal mail in over 50 years—involving some 300 letters from each side—was permitted between North and South Korean families.

The bilateral summit meeting and family reunions ushered in a wider inter-Korean peace process. Ministerial meetings to follow the summit began in Seoul at the end of July 2000, with a second round held in Pyongyang at the end of August and two further sessions thereafter. The first session agreed to reconnect railway lines across the DMZ: a goal endorsed by Kim Jong Il soon afterwards, with work inaugurated by Kim Dae-Jung in September. In a related development, Hyundai won Kim Jong Il's approval to build a vast industrial estate in, and run tour buses to, Kaesong, the ancient Korean capital just north of the DMZ. The project was envisaged as being comparable to the relationship between Shenzhen Special Economic Zone, in the People's Republic of China, and Hong Kong: serving both to link the two economies and as a basis for growth in the North. There were also positive security implications if the DMZ were to become a thoroughfare, instead of an all but impassable barrier. Economic talks were also held, during which a basic framework for business co-operation was agreed. North Korea's Minister of the People's Armed Forces, Vice-Marshal Kim Il Chol, visited Seoul in September, but would only discuss railways (the relinking of which was being carried out by the army on both sides); no return visit was arranged.

These promising beginnings came to a halt in 2001. In January economic talks broke down when the South refused a technically unfeasible Northern demand for the immediate supply of electricity. In the following month North Korea agreed, but failed to ratify, a protocol on joint railway building within the DMZ; construction work on its side had barely begun. In March the North withdrew from the fifth ministerial talks only hours before they were due to start. Thereafter it refused all official contact with South Korea for six months, seemingly as a corollary of its annoyance with the new US Administration, except for sending a condolence delegation on the death of the Hyundai patriarch, Chung Ju-Yung. A further Southern donation of fertilizer in May did not soften this stance. Indeed, June saw a reversion to provocation, when several Northern merchant ships took short cuts through Southern waters. The restrained response of the South Korean navy angered hardliners in the South; as did a later incident, when a few members of a South Korean unification activists' delegation, allowed to visit Pyongyang for Liberation Day celebrations on 15 August, appeared to support North Korean positions. The controversy that this generated brought down South Korea's ruling coalition (see below).

All this gravely weakened the position of Kim Dae-Jung. By late 2001 most South Koreans endorsed the opposition's criticism of the 'sunshine' policy as appeasement. North Korea's erratic behaviour did not help matters. It accepted a fifth round of ministerial talks in September, only to cancel family reunions, at short notice in October, on the pretext of South Korea's heightened state of alert after the 11 September terrorist attacks on the USA. A sixth round of ministerial talks in November, held (at the North's insistence) at Mount Kumgang, ended with no agreement to meet further; the South's unification minister was dismissed soon after. Official relations thus remained suspended; although, in a major change from past policy, business and private contacts continued. So did links through KEDO: an unpublicized Northern team inspected South Korean nuclear facilities in December, and other delegations followed in 2002. In December 2001 also, Hyundai sharply reduced its Mount Kumgang tours because of falling demand.

In February 2002 North Korea cancelled a joint celebration of the Lunar New Year, for which the Southern civic delegates had already arrived at Mount Kumgang. President George W. Bush's designation in January of North Korea as part of an 'axis of evil' alarmed Kim Dae-Jung; in April he sent his adviser Lim Dong-Won, the 'sunshine' policy's *éminence grise*, to try to persuade Kim Jong Il to agree to talks with the USA. This visit also revived North-South dialogue: April witnessed a fourth round of family reunions, this time at Mount Kumgang (on the North's insistence, again) rather than in the two capitals. In May, however, North Korea withdrew from economic discussions at a day's notice. It also disregarded all entreaties to share in the football World Cup co-hosted by South Korea and Japan, but did broadcast highlights of some matches held in the South.

Yet private and semi-official contacts burgeoned. Official subsidies revived Mount Kumgang tourism. South Korean firms established several business and educational joint ventures in information technology (IT), including a college and Pyongyang's first internet café. In June 2002 South Korea's assistant Minister of Information and Communication led a delegation from major companies to Pyongyang; but North Korea later denied reports that agreement had been reached to install mobile telephone services in Pyongyang and Nampo, and no further detail emerged from this. Provincial links continued: Jeju sent oranges, and Gangwon (divided by the DMZ, and later to be badly hit by typhoon Rusa in late August) jointly sprayed against pine pests. In June 320 members of a South Korean Christian aid non-governmental organization (NGO), when denied a promised church service in Pyongyang, held their own impromptu worship in a hotel; they were not impeded. In May, in an encounter that would have startled their parents, Kim Jong Il hosted a dinner for Park Chung-Hee's daughter, Park Geun-Hye, herself a possible presidential candidate. Kim also agreed to a friendly football match in September.

On 29 June 2002 relations sharply deteriorated when North Korean warships without warning fired on and sank a Southern patrol boat in the Yellow Sea, killing six crew members. As in the 1999 incident, for which this may have been revenge, this occurred in disputed maritime border seas during the crab-fishing season. Yet it was wholly unexpected, prompting further criticism of the 'sunshine' policy and of the South Korean navy's lack of preparedness; the Southern Minister of National Defence was dismissed a fortnight later. Within a month, however, North Korea expressed 'regret', and this sufficed for dialogue to resume. The seventh ministerial talks, held in Seoul in August, arranged a full roster of further meetings in specific areas. Economic discussions later that month set a timetable to open two cross-border road and rail links: not only the previously agreed route near Seoul, work on which had stalled, but a second corridor near the east coast. On 18 September 2002 ceremonies were held on both sides of the DMZ to mark the beginning of reconstruction of the two rail links between the Koreas. North Korea also agreed to take part in the Asian Games to be staged in Busan in October—the first time it had ever participated in an international event held in South Korea—in contrast to its eschewal of the football World Cup. Nearly 700 North Korean athletes and supporters went to Busan; but speculation that Kim Jong Il might attend the opening ceremony proved unfounded.

From October 2002 the growing nuclear crisis cast a pall over inter-Korean relations, but did not derail them. North Korea firmly refused to discuss nuclear matters with the South, and on 12 May 2003 even declared null and void a North-South agreement on a nuclear-free peninsula signed in 1991. Yet most of the now established post-summit channels continued to operate, with some interruptions. Thus, regular ministerial talks reached their 11th round in July 2003, with repeated pledges to expedite substantive co-operation. The other major change in 2003 was the appointment of a new Government in Seoul. President Roh Moo-Hyun, who took office in February (when North Korea upstaged his inauguration by testing a 'cruise missile') pledged to continue Kim Dae-Jung's 'sunshine' policy; but first indications suggested that in practice he might take a tougher line, or a more erratic one. Tallying the results of 'sunshine', the South's Ministry of Unification found 2002 to be the most intensive year for North-South interaction since regular contacts began in 1989. Of a cumulative 39,433 South Koreans who had travelled to the North since 1989, nearly a third did so in 2002. (This excluded tourists to Mount Kumgang, whose total since tours began in 1998 had exceeded 500,000.) By category, the largest group (31%) was involved in KEDO's light-water reactors, followed by non-Kumgang tourism (24%). Aid workers made up 11%, business people 9%, and family reunions just 5%. Travel in the opposite direction was less brisk; but again, of 2,568 North Koreans to visit the South since 1989, 40% went in 2002 alone. A total of 34 sets of North-South talks were held in 2002, or 9% of the cumulative total of 400. Unlike in the past, when inter-Korean contacts were often outside the peninsula, almost all were now taking place in Korea: either in Seoul or Pyongyang, or at Mount Kumgang. For many in Seoul, however, the pace was too slow, and the results disappointing. Thus, family reunions stalled in the last quarter of 2002, as North Korea took umbrage at the South raising the issue of South Koreans abducted to the North (who might number as many as 80,000, mainly taken during the Korean War). Reunions for a fortunate few resumed in 2003, with a sixth and seventh round (since the 2000 summit) in February and June. Yet these isolated encounters had no follow-up; and with a mere 100 from each side meeting each time, most of those affected would die without ever being reunited with their long-lost kin. Plans to build a permanent reunion centre at Mount Kumgang raised hopes of the pace quickening.

A significant development in 2003 was the partial opening to civilians of the long-sealed North-South border. Here again, progress was fitful, as North Korea delayed matters by denying the jurisdiction of the UN Command (UNC) in the two new road and rail corridors through the DMZ. In February a temporary road to Mount Kumgang in the eastern (Donghae) corridor was used for the sixth round of family reunions and by a few tourists, before North Korea abruptly closed it for unspecified repairs, and then suspended tours altogether as part of stringent quarantine measures against Severe Acute Respiratory Syndrome (SARS). Work continued on linking railways in both corridors, and on 14 June both were ceremonially joined in the DMZ: a merely symbolic gesture, thus far, as gaps remained in both lines on the Northern side. With genuine completion expected in the western (Kyongui) corridor by 2004, the prospect arose, politics permitting, of trains running for the first time in half-a-century from Seoul to Pyongyang, and on into China or Russia. For regular service to be possible, however, North Korea's decrepit internal rail network would need massive and costly repairs and upgrading. The Kyongui corridor was also crucial for the viability of the industrial zone at Kaesong. After many delays, ground-breaking took place on 30 June: South Korean attendees crossed for the day from Seoul. Such cross-border commuting, unthinkable previously, had thus become normal for railway work and some other meetings, such as economic talks held in Kaesong in July. Yet, at these, North Korea baulked at making such access—essential if Southern business was to invest in Kaesong as planned—routine. It did, however, agree finally to implement four basic business accords arranged in December 2000, but never subsequently ratified; only to fail to turn up as promised at Panmunjom on 18 August to exchange the documents (they were eventually exchanged on 20 August). This seemed part of renewed displeasure with South Korea; a day earlier, the North withdrew from the Daegu Universiade (world student games) in protest at the burning of its flag—not a rare event—at a conservative rally in Seoul. It relented after an apology from President Roh Moo-hyun, which in turn drew right-wing wrath at home, and in the event the Northern team, complete with cheerleaders, did in fact take part in the games. These were marred, however, by an assault by North Korean 'journalists' on peaceful human rights protesters, with the police slow to intervene. The nadir of gamesmanship was reached in late October in Jeju, an island province which had given generous aid to the North. The Northern team arrived late, minus the promised cheerleaders, the main attraction; at the end they refused to leave unless paid more.

In a more serious instance of cash diplomacy, the 'sunshine' policy was now tarnished for many South Koreans by allegations that the June 2000 North-South summit had, in effect, been paid for by South Korea. In late June 2003 a special counsel, appointed by the opposition-controlled legislature, confirmed that the previous Kim Dae-Jung administration had secretly sent US $100m. to North Korea just before the summit via the Hyundai group, which itself remitted a further $400m., supposedly as a business fee. Kim Dae-Jung was not questioned, but those charged included the former Minister of Unification, Lim Dong-Won, and the Chairman of Hyundai, Chung Mong-Hun. In early August Chung jumped to his death from Hyundai's headquarters. This tragedy seemed sadly symbolic of the disillusionment and compromises (or worse) in the current state of inter-Korean ties. In September Lim and four others were convicted, but received suspended sentences as they had acted in the national interest. However Park Jie-Won, a former presidential chief of staff, was later sentenced to 12 years in prison, since his case involved corruption.

In the event, despite these set-backs and the continuing nuclear crisis, the Roh Moo-hyun Government persisted with a 'sunshine' approach, now renamed as 'policy for peace and prosperity'; the more so after the pro-Roh Uri Party won a majority in legislative elections on 15 April 2004. In late 2003 it was estimated that on any given day up to 1,000 South Koreans were visiting the North: tourists, divided families, business people, NGOs and aid workers, civic organizations, educators and other professionals, journalists, cultural figures, government officials, railway inspectors and technicians, nuclear engineers and others. What was once newsworthy and exceptional—direct flights between Seoul and Pyongyang or land travel across the DMZ—was becoming regular, even mundane. A gradual *de facto* normalization of North-South links was gathering pace, even though the broader political and security context remained anything but normal. Most South Koreans, especially younger people, increasingly saw the North more as a junior neighbour in need of succour than a mortal enemy to be guarded against.

Thus, by late 2003 and into 2004, inter-Korean dialogue seemed institutionalized. Its highest normal level was that of so-called ministerial meetings (although the North Korean delegates were described as 'cabinet counsellors'), held quarterly in each capital alternately. The 12th, 13th, and 14th sessions since the June 2000 summit meeting took place in October 2003 and February and May 2004, respectively. On each occasion South Korea sought to raise the nuclear issue but was rebuffed.

The other regular high-level meeting was of the Economic Co-operation Promotion Committee (ECPC). This, too, met quarterly, alternating between Pyongyang and Seoul. Its sixth session in August 2003 had a typically wide agenda: pledging to expedite cross-border transport, the Kaesong industrial zone, Mount Kumgang tourism, direct trade and processing on commission, Imjin river flood control, mutual economic visits, food aid and inspections, and more. Crucially, unlike in the past, these were not mere aspirations, but projects actually under way, if not always very rapidly. Most were the subject of lower-level working meetings, which were increasingly held in towns near the border (Kaesong or Mount Kumgang in the North; Munsan, Paju or Sokcho in the South) with the visiting side commuting daily across the DMZ, a development that had previously been unimaginable.

Family reunions were another regular event: held at Mount Kumgang in September of 2003 and in March–April and July of 2004. In each case, 100 elderly Koreans from each side met briefly and publicly with relatives from across the DMZ. At this rate, the target group would become extinct before most saw their wish fulfilled. By late 2003, of 122,000 Southerners who had originally applied for reunions, more than 20,000 had already died. The spring meeting was curtailed after a South Korean official made a joke about Kim Jong Il, but this did not prevent the programme resuming with a 10th round in July.

Besides regular meetings, one-off events added to a sense of progress. In October 2003 a 1,100-strong bus convoy drove from Seoul to Pyongyang for the dedication of a new US $50m. gymnasium, built by Hyundai. In September 114 Southern tourists flew from Seoul to Pyongyang for a five-day tour on the first ever direct flight. Presented as the first trip of many, this programme was suspended after just one month; as of mid-August 2004 it had yet to resume. In October 2003 the first private-sector truck delivery was made across the DMZ when Korea Express, South Korea's largest logistics operator, took 100,000 roof tiles to Kaesong: the first batch of 400,000 was donated by South Korean Buddhists, to help restore a temple in this ancient capital city located just across the border.

The Kaesong industrial zone made slow progress. In December 2003 North Korea published detailed regulations for the zone. Some were restrictive, and South Korea continued to demand an investment pact, plus free passage for Southern businesspersons and equipment. Practically, this would also entail completion of road and rail links, which were ready on the Southern side and due for finishing and test runs by late 2004. A first pilot mini-zone with 15 South Korean firms was due to open in November. The USA pressed South Korea to ensure no breach of the Wassenaar Arrangement (to which both nations were signatories), restricting exports of sensitive technology to communist regimes. A 50-year lease for the first full phase of development was signed in April. It was planned that a business complex covering 3.3 sq km would open in 2007, accommodating 250 South Korean firms. The complex was to be developed by South Korean companies Korea Land and Hyundai.

South Korea was encouraged in August 2003 by the fact that North Korea's newly elected SPA included at least five figures prominent in inter-Korean dealings. These were: Kim Ryong Song, the North's chief delegate to inter-Korean ministerial talks; Song Ho Kyong, who as Vice-Chairman of the Asia-Pacific Peace Committee was in charge of projects with Hyundai; Pak Chang Ryon, head of the Northern side in the ECPC; Jung Un Up, chief of another North-South cooperation body; and Choi Seung Chol, in charge of inter-Korean Red Cross contacts.

In an important step forward, in May 2004 North Korea accepted a long-standing Southern demand for bilateral military talks. After just two meetings, a week apart, the two sides agreed in June that their Yellow Sea fleets would keep daily radio contact, to avoid any risk of fatal firefights like those in 1999 and 2002. In return, South Korea accepted a Northern demand to dismantle all propaganda at the DMZ. Loudspeakers fell silent after half a century on 15 June, and all structures were initially due to come down by Liberation Day (15 August). However, dismantling was suspended by North Korea in July following an incident in which a South Korean ship fired warning shots at a North Korean vessel after the latter crossed a maritime demarcation between North and South Korea. (It was subsequently revealed that information regarding radio communication with the North Korean ship had been intentionally withheld by the South Korean navy; the incident led to the resignation of South Korean Minister of National Defence, Cho Young-Kil.) North Korea resumed removal of its propaganda facilities in August.

A rail explosion on 22 April at Ryongchon, near the North Korean border with China, led through tragedy to some progress in North Korea's opening, including southwards. The South at once offered an aid convoy, but the North rejected this, insisting that relief goods come by sea; the first boat reached Nampo only on 29 April. However, a day later a Southern cargo plane flew to the North with 70 tons of aid, in the first ever inter-Korean humanitarian direct flight. Not until 7 May was a land convoy permitted; it had to unload just across the border in Kaesong. By then, the cargo was equipment and materials to rebuild schools and other facilities. Useful as these were, the North's initial delay meant that the chance to save more burns victims was lost.

A more divisive development involved Song Du-Yul, a long-time South Korean dissident, who in September 2003 came home after 37 years exile in Germany, where he taught philosophy and held citizenship. Accused of doubling under a pseudonym as a senior KWP official (that he had often visited Pyongyang was no secret), he was arrested, charged, and on 30 March sentenced to seven years in prison. His case divided South Koreans: some thought him a traitor, others a reunification hero, yet others misguided. Freed on appeal in July, he returned to Germany. It was unclear why South Korea had not simply deported him in the first place.

North Korean human rights were a perennially difficult issue, which South Korea, as ever, was reluctant to raise. In April 2003 it absented itself from a vote (which was carried) on this at the UN Commission on Human Rights in Geneva; a year later it was present, but abstained. Yet, as one instance, the plight of refugees from North Korea continued to draw attention. In October 2003 South Korea twice temporarily closed its consulate in Beijing to clear a backlog of about 100 Northern refugees camping there. After a spate of incidents involving North Koreans seeking sanctuary, China had tightened security around embassies in Beijing; but any who successfully ran this gauntlet eventually gained safe passage to Seoul. Many more, however, were forced to pass through China to seek refuge elsewhere. On 26 and 27 July 2004, no fewer than 468 North Korean refugees were flown to Seoul from Viet Nam (not officially named, for reasons of protocol), which had reportedly threatened to repatriate them. This was by far the largest group of defectors ever to reach South Korea, which some saw as a portent of things to come.

Despite the general progress, North Korea on occasion reverted to previous tactics. In March it missed a meeting in the South, later claiming (by radio) that Roh Moo-Hyun's impeachment meant South Korea was in anarchy. In August the North failed to organize the scheduled 15th ministerial meeting, seemingly for two reasons: anger at the Viet Nam refugee airlift; and displeasure at the South Korean refusal to let pro-North activists attend joint Liberation Day celebrations (which it cancelled) due to be held in Pyongyang on 15 August. None the less, in August the two Koreas' athletes marched together, as agreed, at the opening of the Olympic Games in Athens, as they had in Sydney in 2000; but once again competed separately.

South Korea also could be negative. In January 2004 it revoked the licence of a Southern firm, Hoonnet, which had installed Pyongyang's first internet café and public server, with online gaming sites (illegal in the South) and an informal North-South internet chat room. Despite the 'sunshine' policy, telephone, fax and e-mail contact with the North remained technically illegal in the South under the still-unrepealed National Security Law.

The fourth anniversary of 2000's first ever North-South summit meeting in Pyongyang, was celebrated on both sides of the DMZ in June 2004. To mark the anniversary, a senior North Korean emissary made a rare visit Seoul. Ri Jong Hyok's bland title—Vice-Chairman of the Asia-Pacific Peace Committee—belied his role as a key confidant of Kim Jong Il. He met President Roh Moo-hyun, and gave him a message from Kim Jong Il, prompting speculation that the 'Dear Leader' might at last fulfil his promised but long-delayed return visit southwards. This seemed unlikely while the nuclear crisis persisted.

South Korea's Ministry of Unification used this anniversary to sum up and quantify progress. If including the summit meeting, in the previous four years there had been 111 official inter-Korean meetings: 47 economic, 27 military (but usually about cross-border road and rail links), 19 political and 18 'humanitarian and athletic'. Visitors from South to North, excluding Hyundai's tours to Mount Kumgang, which had brought some 680,000 visitors since 1998, more than doubled, from 7,280 in 2000 to 15,280 in 2003. Traffic in the other direction was smaller, predictably, and more erratic, decreasing sharply from 706 in 2000 to 191 in 2001, rising to 1,052 in 2002, then back to 1,023 in 2003. (None of this was remotely on the scale of interchange between, for example, China and Taiwan.)

In mid-2004, with inter-Korean talks suspended and North Korean propaganda attacks intensifying, the South's National Intelligence Service (NIS) warned of a risk of North Korean terrorist attacks. It seemed that, even after four years, the light of the 'sunshine' policy remained at risk of being extinguished in an instant.

FOREIGN RELATIONS

Despite (or, perhaps, because of) North Korea's roots in Soviet military government, it has been the regime's consistent goal to emphasize and maximize what its own slogans call *chaju, chalip, chawi*: independence in politics, economics and defence. In practice, this has largely meant a refusal to be beholden to—let alone a satellite of—either of its giant neighbours and erstwhile sponsors, the USSR and China, while simultaneously exhibiting unremitting hostility towards the USA, Japan and, of course, South Korea. Although the Sino-Soviet dispute enabled Kim Il Sung for many years to play off Moscow against Beijing and receive aid from both, the end of the Cold War and the collapse of the USSR exposed Pyongyang's vaunted self-reliance as ultimately self-defeating isolation and friendlessness.

North Korean foreign policy has undergone several phases over the years. During the 1950s Pyongyang emphasized its adherence to the communist bloc, receiving both military aid during the Korean War (1950–53) and assistance for reconstruction thereafter from the USSR, China and Eastern Europe. Yet already there were disputes with Moscow over how best to use Soviet aid, with North Korea preferring to develop its own heavy industry rather than join the Council for Mutual Economic Assistance. Meanwhile, although Japan had relations with neither Korean government at this stage, more than 75,000 pro-communist Koreans in Japan (mainly of southern origin) emigrated to the new socialist fatherland in the late 1950s.

In the wake of the public Sino-Soviet split in the early 1960s, Pyongyang demonstrated broad sympathy with Beijing's more revolutionary position, which led to the temporary suspension of Soviet aid. North Korea's own bellicosity peaked in 1968, with its seizure of the *USS Pueblo* and its dispatch of a commando unit to attack the presidential mansion in Seoul.

By contrast, the 1970s were an era of broadening contacts. Suspicious of China's amenability to US 'ping-pong diplomacy', North Korea not only repaired relations with Moscow but sought new allies, particularly in the developing countries and within the Non-aligned Movement (which it joined in 1975). Other initiatives were less successful. Pyongyang's breakthrough in establishing diplomatic relations with the four Nordic countries (its first such ties in the West) was marred shortly thereafter when, in 1976, all four of its ambassadors were expelled, their staff accused of systematic smuggling (a practice in which North Korean diplomats have allegedly been engaged ever since). Similarly, what seemed a useful development of economic ties with Japan and Western European countries came to an abrupt halt when it became clear that Pyongyang had no overt intention of paying for several hundred million dollars' worth of capital equipment imported in the early 1970s.

By the early 1980s North Korea's relations with many of its communist allies showed signs of deterioration. Neither Moscow nor Beijing approved at first of the official designation of Kim Jong Il as his father's successor, although China at length relented and, in 1983, invited the 'Dear Leader' on his first known trip there.

This did not, however, prevent a distinct inclination towards the USSR in the mid-1980s, inspired perhaps by suspicion of Deng Xiaoping's reforms, yet continuing into the era of Mikhail Gorbachev. In 1984 Kim Il Sung visited Moscow for the first time in 23 years, and also spent several weeks touring Eastern Europe. He returned to meet Gorbachev in 1986, in which year joint Soviet-North Korean naval exercises were undertaken. In addition, the North Korean Government granted port facilities and overflying rights to the Soviet fleet and air force, reportedly in return for the supply of Soviet MiG-23 fighters and surface-to-air missiles. Soviet-North Korean trade grew rapidly, with North Korean imports more than quadrupling in the four years between 1984 and 1988.

However, within a period of less than five years, a series of setbacks comprehensively undermined North Korea's foreign policy orientations of the previous four decades. It was inevitable that the lure of South Korea's far greater economic prospects (coupled with the skilful diplomacy of Seoul's 'nordpolitik') would eventually lead pragmatists such as Deng and Gorbachev to qualify their inherited Cold War loyalties towards a Pyongyang that they increasingly considered to be a political and economic liability. Although China moved first to begin trading with Seoul, it was the USSR under Gorbachev's leadership that dealt both the diplomatic and financial *coup de grâce*: first by establishing full diplomatic relations with South Korea in September 1990, and then by stipulating that, from January 1991, its trade with North Korea would be conducted in convertible currencies at world market prices. This caused Pyongyang's total trade volume to decline by more than US $1,100m. (almost one-quarter) in 1991. Although the demise of the USSR itself afforded a certain grim satisfaction, Russia's President Boris Yeltsin had no vestige of comradeship with the Pyongyang regime.

The deterioration in relations with Moscow left China as North Korea's only major ally, although even this relationship was qualified by increasing impatience in Beijing, as much over Kim Il Sung's failure to embrace economic reform as for his suspected nuclear ambitions (see below). With reformers once again dominant in the Chinese leadership, economic ties with Seoul rapidly increased and, in August 1992, full diplomatic relations were established between China and South Korea, much to the consternation of Pyongyang. Sino-North Korean relations subsequently deteriorated significantly. Although China opposed UN action against North Korea over the nuclear issue, its support was tenuous at best.

In the light of these shifts of allegiance, North Korea had no option but to try to repair relations with its traditional enemies. In late 1990 a breakthrough with Japan seemed likely, after a highly successful visit to Pyongyang by Shin Kanemaru, the senior mediator of the ruling Liberal-Democratic Party. Yet, eight rounds of talks, held in 1991–92, on the possible normalization of diplomatic relations between North Korea and Japan made no progress, owing to intransigent demands from both sides. None the less, regular charter flights between the two countries led to an increase in unofficial contacts and visits, although not yet to the aid and investment which North Korea desperately needs. Relations with Japan were severely strained in May 1993, following North Korea's successful testing of the *Rodong-1* medium-range missile in the Sea of Japan. According to US intelligence reports, the missile would be capable of reaching most of Japan's major cities (and possibly of carrying either a conventional or a nuclear warhead).

Contacts with the USA also increased considerably in the early 1990s, although in the early 2000s diplomatic relations had yet to be established. North Korea on four occasions returned the remains of US soldiers who went missing in action during the Korean War, and several high-level US delegations, including retired senior political and military figures (and even the evangelist Rev. Billy Graham), visited Pyongyang. In January 1992 the KWP's international secretary, Kim Yong Sun, visited New York for discussions with the US Under-Secretary of State (although one year later he was refused a visa to attend ceremonies marking the inauguration of President Clinton). US-North Korean discussions did, however, resume in mid-1993 over the nuclear issue.

For both Washington and Tokyo, a major obstacle to better relations with Pyongyang was and remained their suspicion that North Korea was seeking to develop nuclear weapons. In July 1991, after several years of prevarication, North Korea finally agreed a draft Nuclear Safeguards Agreement (NSA) with the International Atomic Energy Agency (IAEA), permitting the outside inspection of North Korean nuclear facilities. Following the announcement by President George Bush, Sr of proposals to withdraw all US tactical nuclear weapons worldwide, and President Roh Tae-Woo's confirmation that none remained in South Korea, North Korea signed the NSA in January 1992. Moreover, Pyongyang subsequently submitted an unexpectedly detailed report on its nuclear facilities, almost one month ahead of schedule. The IAEA Director-General visited North Korea in May 1992 and formal IAEA inspections began later that month.

Yet, despite this unprecedented progress, suspicions were not allayed. Indeed, one large building at the Yongbyon installation, north of Pyongyang, was believed by some outside observers to be a nuclear-reprocessing plant. Likewise, North Korea's apparent attempts to obstruct the separate inter-Korean mutual nuclear inspections (see below) aroused widespread mistrust. Finally, though not part of the nuclear issue as such, the fact that North Korea sold improved *Scud* missiles to Iran and Syria in exchange for petroleum, did nothing to enhance relations with the West.

All these issues came to a head in 1993. In January North Korea refused to allow special inspections (as demanded by the IAEA) of two sites at Yongbyon, which were thought likely to reveal that more plutonium had been extracted than Pyongyang had admitted. Then, in an unprecedented move, on 12 March North Korea announced that it was to withdraw from the Treaty on the Non-Proliferation of Nuclear Weapons (the Non-Proliferation Treaty—NPT), which it had signed with the IAEA in 1985. This led to protracted diplomatic activity between Washington, Seoul and Tokyo, as well as muted criticism by the UN Security Council (in part because China would not support decisive action at this stage). However, the main channel for defusing the crisis was the holding of two rounds of direct talks between North Korea and the USA in mid-1993, which resulted in North Korea suspending implementation of its withdrawal from the NPT.

Despite this hopeful sign, the nuclear crisis continued for a further year. In May 1993 international concern about North Korea's weapons programme was heightened following the successful testing of an intermediate-range missile, the *Rodong-1* (see above). Negotiations between North Korea and the USA, and separate inter-Korean talks, continued during the latter part of the year, but with no tangible results. Although some IAEA monitoring and intermittent inspection activities at Yongbyon were permitted, it remained uncertain whether North Korea had already succeeded in producing a nuclear weapon. Alarm was aroused again in May 1994, when North Korea began replacing spent fuel rods without effective supervision, prompting the IAEA to suggest the imposition of international sanctions. The situation deteriorated further on 13 June, when North Korea retaliated by announcing its complete withdrawal from the IAEA (and not merely the NPT), although in fact inspectors subsequently remained at Yongbyon. Meanwhile, the USA began to lobby the UN Security Council to impose sanctions on North Korea, despite objections by China and Russia and an unenthusiastic response from Japan.

The rising sense of crisis was defused in mid-June 1994, when former US President Jimmy Carter visited Pyongyang. After 10 hours of talks with Kim Il Sung, he returned with two offers: a summit meeting with South Korea (see above), and a pledge to suspend North Korea's nuclear programme. While the latter was vague, it sufficed for Washington to resume high-level talks. When the third round commenced in Geneva, Switzerland, on 8 July, however, Kim Il Sung was already dead—although the announcement of his death came only on the following day. The talks were duly postponed. None the less, following several subsequent rounds, a 'framework agreement' between North Korea and the USA was signed in Geneva in October. In essence, Pyongyang agreed to close down its nuclear site at Yongbyon in exchange for substantial compensation, principally in the form of new LWRs worth some US $4,500m., as well as up to 500,000 metric tons of heavy fuel oil annually during the estimated 10 years' construction period of the LWRs.

Although not written into the agreement, it was understood that South Korea would play a key part, both as supplier and main financier of the LWRs, and as a core member, with the USA and Japan, of KEDO, the consortium that was to supervise the entire project. North Korea initially protested bitterly against Seoul's involvement, and for several months it seemed as if the Geneva accord might collapse. By August 1995, however, Pyongyang tacitly abandoned its opposition to the South's *de facto* participation.

The Geneva nuclear accord perhaps set the pattern for a new development in North Korean foreign policy, which might be termed 'militant mendicancy'. Thus, in May 1995, a North Korean delegation visiting Japan made an unprecedented request for rice as aid. Not only was this granted, but South Korea insisted that the North should also, and first, accept free rice from fellow Koreans (see above). This produced the remarkable spectacle of the North Korean regime being in effect sustained by its three oldest and bitterest foes—which it continued to denounce—without any proviso requiring Pyongyang to reform or mend its ways.

North Korea continued to undermine the existing armistice agreements. In March 1995 it expelled the Polish observers of the Neutral Nations Supervisory Commission at Panmunjom, following which it closed its side of the Joint Security Area to all comers from the South. The North's professed aim was a bilateral peace treaty with the USA, excluding the South. Pyongyang continued to pursue this quixotic quest in 1996, declaring on 4 April that it would no longer observe protocol in the DMZ, and raising tension with a few symbolic incursions into the Southern half of the zone. This prompted the USA and South Korea to propose four-way talks with North Korea and China.

In any case, Pyongyang had *de facto* achieved the direct line of communication to Washington that it had long sought. While there were no formal ties, and talks on an exchange of liaison offices stalled, there were now regular contacts between the US State Department and the North Korean UN mission in New York. Discussions on missile control were held in Berlin, Germany, in April 1996, albeit without result. Progress was also made on 'missing in action' (MIA) issues: the USA paid US $2m. (in cash, at Panmunjom) for remains that had already been returned, and US investigators were, for the first time, allowed to search directly inside North Korea. In early 1996 there were also many visits to the USA, including separate delegations led by two leading reformists, Kim Jong U and Ri Jong Hyok.

However, all this was regarded as appeasement in some quarters, both in Seoul and in the Republican Party in the USA. In June 1996 Congress granted barely one-half of the modest US $25m. sought by the White House as the US contribution to KEDO's oil shipments. This was regrettable, since KEDO had achieved the remarkable feat of turning what had been the peninsula's worst risk into its best hope. Not only had North Korea abandoned its initial hostility to the South's leading role in the project, but for much of 1995–96 its delegates were in New York taking constructive part in negotiations over the text of the agreement. South Korean engineers were now travelling routinely to the LWR site at Shinpo, albeit via Beijing, and in August 1997 construction work began at the site.

Pyongyang also made overtures to Tokyo about resuming negotiations towards establishing diplomatic relations. Cautious as ever, Japan insisted that North Korea must first accept the four-way talks with South Korea and the USA. None the less, regular contacts continued. In July 1996 Kim Jong U toured several Japanese cities to try to encourage investment in the Rajin-Sonbong free zone, but had little success. (Kim Jong U subsequently fell from favour, and in 1998 was reported to have been executed; other reports denied this.)

With its old (and former) allies, too, North Korea experienced mixed fortunes. China's avowed support for inter-Korean dialogue annoyed the North, but China, for its part, was equally irritated at Kim Jong Il's refusal of economic reform, not least because Beijing had been continually obliged to offer the country vital assistance. In July 1996 the 35th anniversary of the Sino-North Korean friendship treaty was marked by the exchange of middle-ranking delegations, as well as by a rare visit to Nampo by a Chinese naval flotilla.

Meanwhile, Moscow, which created North Korea and sustained it until the beginning of the collapse of the USSR in 1990, occupied a much less significant position in North Korea's foreign relations. In April 1996 a Deputy Chairman of the Russian Government, Vitalii Ignatenko, led the first major delegation from the new Russia to Pyongyang. There was talk of resuming economic co-operation: most major North Korean industrial installations were originally Soviet aid projects; but since both countries were in economic difficulties, not to mention Pyongyang's huge debts to Moscow, prospects for co-operation were not bright. They did not improve in 1997, when North Korea criticized Russia for supplying ultra-modern armaments (including tanks) to South Korea, in payment of debts.

North Korea's foreign relations neither changed nor advanced greatly during 1996–97. With no Head of State yet inaugurated, there were few high-ranking visitors either to or from Pyongyang, which appeared more isolated than ever. Only with China were there regular exchanges of delegations, and most of these were low-level. Still, as seen in the defection of Hwang Jang Yop (see above), Beijing remained concerned for North Korea's sensitivities, even as its economic and other links with South Korea continued to progress smoothly. China was also the biggest provider of food aid to North Korea. This was so, despite China's displeasure at the somewhat unlikely warming of relations between North Korea and Taiwan: worlds apart ideologically, yet perhaps united in their pariah status. In January 1997 an agreement was signed, whereby North Korea would dispose of low-grade nuclear waste from Taiwan, in a contract believed to be worth more than US $100m. South Korea protested vociferously, and this project seemed to have been abandoned.

Elsewhere, in the first half of 1997 Japan began to adopt a more rigid policy towards North Korea than it had previously. The seizure in April of amphetamines worth US $90m. on a North Korean ship in a Japanese port did not improve relations. Unlike the USA, or even South Korea, Japan declined to respond to the UN's increasingly desperate appeals for food aid for North Korea. (It had supplied 500,000 metric tons of rice in 1995, and received scant gratitude.) Japan insisted that improved relations, including aid, would depend on Pyongyang making concessions, including the disclosure of details of the alleged kidnappings of Japanese citizens in the 1970s. In late August, however, agreement was reached between the two countries, whereby some now-elderly Japanese wives of Koreans who settled in North Korea in the 1950s and 1960s were to be allowed to visit their native land for the first time. Only with the USA, ironically, did North Korea enjoy improving relations. Washington was swift to respond to UN appeals for food aid, and bilateral contacts continued in areas ranging from further joint excavation for MIA remains to talks about Pyongyang's missile development and sales. Several senior US politicians visited North Korea, usually by military aircraft, which would have been unthinkable in the past. Washington's main aim was to persuade North Korea to agree to attend the four-way talks first proposed in April 1996. After much prevarication by Pyongyang, preliminary discussions were held in New York on 5 August 1997. China, the fourth party (which was also initially hesitant), became much more positive about the proposal during 1997.

Full four-way talks finally commenced in Geneva in December 1997, and were followed by more substantial discussions in March 1998. These were, however, unsuccessful, owing to wide differences over the agenda, with North Korea demanding that this should include the withdrawal of US troops from the peninsula. Further four-way talks took place in October; agreement was reached on the establishment of two subcommittees, with a view to instituting a permanent peace mechanism. Otherwise, North Korea's foreign relations remained fairly constant during 1998. Of the four major powers, Russia continued to count the least, in stark contrast to its predominance in the Soviet era. Efforts to revise the 1961 friendship treaty, not least in order to strike out its commitment to mutual military assistance, stalled for several years, until 1999 (see below). China, by contrast, continued to shore up Pyongyang with aid, even as its ties with Seoul become ever closer. Senior-level dialogue with the North went into abeyance for a decade, whereas Chinese and South Korean leaders exchanged regular bilateral visits as well as meeting in multilateral forums such as Asia-Pacific Economic Co-operation (APEC) and Asia-Europe Meeting (ASEM). Beijing's long-term aim was to displace the USA as the broker of choice between the two Koreas.

Relations with Japan continued to follow an uneven course. The first ever home visits of two groups of elderly Japanese-born wives of North Koreans were finally realized in late 1997 and early 1998. Thereafter, however, the kidnap issue—in particular, anger in Japan at what was seen as a perfunctory 'investigation' of the matter by Pyongyang—once again cast a shadow, preventing the resumption of talks towards restoring diplomatic ties. At the end of August 1998 relations deteriorated, when Japan accused Pyongyang of test-firing an unarmed *Taepo Dong* medium-range missile over its territory. In a strange twist, the North Korean regime subsequently claimed that the object launched had, in fact, been a satellite intended to broadcast patriotic music. That rocket launch, whatever its purpose, had a lasting effect on North Korea's relations with its main foes. Japanese opinion and policy have hardened, both towards Pyongyang and on defence issues more generally; all the more so after a further provocation in March 1999, when the Japanese navy pursued and fired on two intruding boats, which were later traced to the North Korean port of Chongjin. Unlike South Korea and the USA, Japan still gave no food aid to the DPRK, and considered further sanctions such as banning remittances sent by pro-North Koreans living in Japan. The possibility, widely canvassed in mid-1999, that Pyongyang might test another missile carried a real risk that Japan would withdraw from KEDO; which in turn threatened to reopen the North Korean nuclear issue and reactivate the tensions of mid-1994.

The USA shared Japan's alarm at North Korea's missile activities, albeit more with regard to proliferation: purchasers included Libya, Syria, Iran and Pakistan. In the past Pyongyang had hinted it might be 'bought off', as with its nuclear programme; but this seemed less likely after the NATO bombing of Yugoslavia in early 1999. Missiles were only one cause of US concern during 1998–99; the other being a large construction site at Kumchang-ri, near the disused nuclear site at Yongbyon, which was feared to be a covert continuation of nuclear activity. Kumchang-ri dominated US-North Korean relations for many months, until in May 1999 a US inspection team pronounced it 'clean'. While denying any link, Washington simultaneously announced a further 400,000 metric tons of grain in aid. The USA continued to be the mainstay of the UN World Food Programme's aid to North Korea, this operation being its largest ever, prior to Afghanistan in 2001–02. Meanwhile, under pressure from Republican critics in the US Congress, in November 1998 the Clinton Administration appointed William Perry, a former Secretary of Defense, to carry out a full review of US policy towards North Korea. Perry visited Pyongyang in May 1999, and his report was published in September of that year. It offered substantial incentives, in exchange for a definitive and verifiable end to North Korea's nuclear and missile ambitions. During talks in Berlin in mid-September the USA lifted sanctions on trade and travel in return for a promise by North Korea that it would refrain from testing long-range missiles until 2003. As for allies (past or present), in March 1999 a new treaty with Russia was at last agreed, to replace that of 1961 with the former USSR. This was assumed to exclude the military support provided for in the old version. The treaty was signed during a visit to Pyongyang by the Russian Minister of Foreign Affairs, Igor Ivanov, in February 2000; it was ratified by North Korea in April of that year. With China effectively remaining North Korea's sole ally and source of finance, a high-level delegation visited Beijing in June 1999 for the first time in eight years. Led by Kim Yong Nam, the delegates included the Premier and the ministers responsible for defence and foreign affairs—but no economic cadres, to China's reported annoyance. Multilaterally, the quadripartite talks involving the two Koreas, China and the

USA continued to be held in Geneva, the sixth round taking place in August 1999, but with no obvious progress.

From the latter part of 1999 onwards, North Korea's diplomacy took a striking new turn, with conscious efforts made both to restore old relationships and foster new alliances. The first sign was a wide range of meetings held at the UN in September 1999 by the North Korean Minister of Foreign Affairs, Paek Nam Sun. In January 2000 Italy became the first of the Group of Seven (G-7) industrialized nations to establish full diplomatic relations with the DPRK. Later in the year Kuwait, the Philippines and Australia followed suit—the last resuming ties first forged in 1975 but abruptly severed by Pyongyang soon after. In October 2000 the United Kingdom announced that it planned to normalize diplomatic relations with North Korea, and duly did so in December. Meanwhile, in July, Paek travelled to Bangkok, Thailand, for North Korea's admission as the 23rd member of the Regional Forum (ARF) of the Association of South East Asian Nations (ASEAN), at which Canada and New Zealand also announced plans to establish relations. Paek also held unprecedented meetings with his South Korean, US and Japanese counterparts.

Progress continued in the development of relations with the USA. In October 2000 Vice-Marshal Jo Myong Rok, North Korea's most powerful military figure, visited Washington as a special envoy of Kim Jong Il. This led in the same month to a visit to Pyongyang by the US Secretary of State, Madeleine Albright. President Clinton was ready to follow to sign an agreement on missiles, but this foundered on verification difficulties. Agreement also eluded high-level bilateral talks with Japan, in abeyance since 1992, of which three rounds were held in 2000. Despite a show of cordiality, the two sides' agendas remained far apart. North Korea demanded compensation of up to US $10,000m. for Japanese colonial rule prior to 1945, whereas Japan continued to prioritize its missile and abduction concerns. There was no obvious way to overcome this impasse. However, in 2000 Japan gave 100,000 metric tons of food aid to North Korea for the first time in several years, and was permitted to monitor its distribution. A third visit home by Japanese wives of North Koreans—the first since 1998—took place in September.

The most important development was a new effort to improve relations with the DPRK's original major allies. In May 2000 Kim Jong Il made an unofficial, and initially secret, visit to China, his first overseas trip since 1983. As well as meeting President Jiang Zemin and other leaders, he toured a computer factory and was quoted—for the first time—explicitly praising China's reform programme. This visit prepared Kim Jong Il for hosting Kim Dae-Jung a fortnight afterwards. One month later the 'Dear Leader' welcomed the President of Russia, Vladimir Putin, as the first ever Russian or Soviet leader to visit the DPRK. By this simple gesture, in a trip lasting less than 24 hours, Putin reversed a decade of hostility which had begun when the then Soviet leader, Mikhail Gorbachev, hastily forged ties with Seoul in 1990; further, the Russian President reconfirmed Moscow's importance with regard to the possible reunification of Korea, later presenting the G-8 (the G-7 plus Russia) with an offer by Kim Jong Il to abandon his missile programme in return for access to satellite-launching facilities. The 'Dear Leader' later commented that this was said in jest, which left Moscow unamused, and was a reminder that rebuilding ties might not be straightforward. Furthermore, North Korea owed Russia some US $3,000m.

The year 2001 was also an active one for Northern diplomacy, but with more mixed results. Ties with the USA suffered a setback when the new Republican administration under President George W. Bush expressed mistrust towards North Korea, prompting Pyongyang to suspend inter-Korean dialogue as well. No further talks were held with Japan, which Pyongyang denounced (as did Seoul) for a new schools' history textbook that glossed over pre-1945 atrocities, and for Prime Minister Junichiro Koizumi's visit to the Yasukuni shrine which commemorates (among others) convicted war criminals. With other Western countries, however, there was more progress. By June 2001 13 of the 15 member states of the European Union (EU), and the organization as such, had restored full diplomatic relations with the DPRK; the exceptions were France, which cites human rights concerns, and Ireland. In May a high-level EU delegation led by the Swedish Prime Minister, Göran Persson, had visited Pyongyang and met Kim Jong Il, in what was interpreted as a prompt to President Bush to resume dialogue. It was reported that the DPRK's leader had pledged to maintain his country's moratorium on missile-testing until 2003. Other states establishing relations with North Korea during 2001 included Brazil and Turkey.

As in 2000, however, the main trend was restoring good relations with old allies. In January 2001 Kim Jong Il paid his second visit to China in nine months. Again nominally secret, this was mainly to Shanghai and clearly business-related: the week-long itinerary took in several joint-venture factories and even the Stock Exchange, raising hopes that North Korea might at last adopt market reforms. In September President Jiang Zemin reciprocated, with the first visit to Pyongyang by a senior Chinese leader in over a decade. Behind the formal warmth, Jiang would have pressed the case for reform (not least to save Beijing the cost of supporting the DPRK's economy, and coping with an outflow of hungry North Korean refugees); and openly called on his hosts to resume dialogue with South Korea, which they promptly did.

Relations with Russia also deepened. An expected visit by Kim Jong Il in April 2001 did not materialize, but his defence minister returned from Moscow with reported pledges of unspecified new military co-operation. The 'Dear Leader' finally made the journey to Russia in August: by special train, taking over three weeks and causing many complaints as stations along the route were cleared of normal traffic as he passed. A joint statement with President Putin referred to Russian 'understanding' of the DPRK's demand for US troops to leave South Korea. Reports also suggested that a deal had been concluded to repay the DPRK's Soviet-era debts, mainly by sending contract labour (whose working conditions the *Moscow Times* likened to serfdom). As this implied, not all Russians endorsed their Government's efforts for renewed friendship with this relic of Stalinism.

The events of 11 September 2001 cast an especially long shadow for North Korea. The country remained on the US Department of State's list of nations alleged to sponsor terrorism, if mainly for sheltering a number of Japanese hijackers since 1970, rather than for any recent transgressions. No links to the Islamist militant group al-Qa'ida were seriously alleged, and North Korea swiftly condemned the suicide attacks on the mainland USA—but went on to criticize the US war in Afghanistan. For its part, Washington for the first time cited North Korea's suspected biological weapons programme as a major threat, in addition to its nuclear and missile concerns. Besides such weapons of mass destruction, the Bush Administration also gave notice that North Korea's conventional force posture was unacceptably threatening. In January 2002 President Bush notoriously grouped North Korea with Iraq and Iran as forming what he termed an 'axis of evil'. None the less, MIA co-operation and US food aid continued. In December 2001 the pursuit and sinking by the Japanese navy of a suspected North Korean spy ship, later salvaged and put on public display, did nothing for either Pyongyang's relations with Tokyo or its wider reputation.

The DPRK's image also suffered from a growing diplomatic problem in which it was, curiously, both cause and bystander. North Korean fugitives in China—numbering up to 300,000, none of whom Beijing acknowledged as refugees—became more militant in 2002; aided by foreign NGOs, several sought asylum in foreign embassies. Chinese police intrusions to seize them caused disputes with both South Korea and Japan; yet in the end all, including those arrested, were allowed to travel to Seoul via third countries. The price of freedom for the few was a campaign of suppression in the border region, with many more migrants arrested and deported to an uncertain fate. It was doubtful if signs of reform in North Korea, or repression there and in China, could stem this tide from swelling over time, to destabilizing effect. The number of North Korean defectors reaching South Korea is still tiny compared to most refugee situations elsewhere, but is nearly doubling each year. From 148 in 1999, the total rose to 312 in 2000, 583 in 2001 and 1,141 in 2002.

In early 2002 DPRK diplomacy temporarily turned away from the major powers, and seemed mainly to be motivated by economic needs. The Minister of Foreign Trade toured Western

Europe, while Kim Yong Nam, President of the SPA Presidium, went to Thailand and Malaysia, securing rice and palm oil on generous terms. At home Kim Jong Il hosted Indonesia's President, Megawati Sukarnoputri (their late fathers had been close associates), who then travelled on to Seoul. The ever less reclusive 'Dear Leader' also visited the Russian embassy in Pyongyang three times in as many months. In July Russia's Minister of Foreign Affairs, after visiting both Koreas, declared Kim Jong Il ready for dialogue 'without preconditions'.

This duly ushered in a new bout of North Korean goodwill diplomacy. In July Paek Nam Sun returned to the ARF, in Brunei, having been absent in 2001. An informal meeting with the US Secretary of State, Colin Powell, was the first high-level contact with the Bush Administration. A month later a less senior envoy, Jack Pritchard, represented the USA at KEDO's LWR ground-breaking ceremony in Shinpo; he urged North Korea to submit to full IAEA inspections.

With Tokyo there was faster movement. In Brunei Paek signed a joint statement with his Japanese counterpart, Yoriko Kawaguchi. Within a month, bilateral Red Cross and diplomatic talks were held in Pyongyang, cordially, but with no visible progress. However, Junichiro Koizumi made an historic visit to Pyongyang on 17 September 2002, becoming the first Japanese Prime Minister ever to do so. This was a considerable risk, in view of the outstanding disputes between the two countries. The most significant outcome of Koizumi's summit meeting with Kim Jong Il was the latter's admission that North Korean agents had indeed kidnapped 11 or 12 Japanese nationals in the 1970s and 1980s, and of these, only five were still alive. Pyongyang's long denial of this had been an obstacle to the normalization of diplomatic relations, and the revelation shocked the Japanese public. Kim apologized for the kidnappings, blaming rogue military elements, and agreed to return the surviving Japanese; Koizumi apologized for Japan's colonization of the Korean peninsula. Kim also told Koizumi that he would allow international inspections of North Korea's nuclear facilities, and would maintain the moratorium on missile-testing beyond 2003. The new bonhomie did not last long. Amid the emerging nuclear crisis, Japanese public anger at North Korea's failure to divulge the facts of the deaths of most of the abductees, and its refusal to let the survivors' children join them in Japan, returned relations to their prior state of hostility.

Koizumi's sudden enthusiasm for summit meetings might have been inspired by Vladimir Putin, who in Vladivostok, Russia, in August 2002 met Kim Jong Il for the third time in as many years. The Russian President pressed for inter-Korean rail links, which could create an 'iron silk road'—a freight route from South Korea to Europe via Siberia, which Putin hoped would encourage economic development in Russia's depressed Far East. For his part, Kim Jong Il reportedly sought modern weapons, and visited factories in the region as well as a Russian Orthodox church. (Although not officially acknowledged, Kim was born in Khabarovsk in 1942, where his father was in exile.)

From October 2002, however, any optimism was reversed by a growing nuclear crisis which dominated North Korea's foreign relations for the next 10 months, with no sign of any early resolution. In mid-October the USA revealed that, two weeks earlier, the Bush Administration's first senior envoy to Pyongyang—the Assistant Secretary of State for East Asian and Pacific Affairs, James Kelly—had confronted his hosts with evidence that they were pursuing a second covert nuclear weapons programme, this time based on highly enriched uranium, in violation of the 1994 Agreed Framework (AF); and that North Korea had, against expectations, admitted this. In November KEDO's executive board, at US urging, voted to suspend heavy fuel oil shipments due under the AF until North Korea returned to full compliance. In December Spanish and US forces detained a North Korean vessel carrying *Scud* missiles off the Horn of Africa; but later let it proceed to its destination, Yemen, since no laws had been broken (neither buyer nor seller being a signatory to the Missile Technology Control Regime). Thereafter the crisis escalated rapidly. North Korea stated that it would reopen its nuclear site at Yongbyon, suspended under the AF, and asked the IAEA to remove seals and monitoring cameras. Later that month it expelled the IAEA's inspectors, and in January 2003 announced its immediate withdrawal from the NPT: the first signatory state ever to do so. In February and March North Korea test-fired two short-range missiles, and one of its jets briefly entered South Korean airspace. More seriously, in early March four North Korean jet fighters harassed a US spy plane west of the peninsula, in an apparent effort to force it to land in Pyongyang. Such provocations dissipated with the start of hostilities in Iraq.

These developments aroused widespread concern, but initially elicited no unified policy response. The Bush Administration, preoccupied with Iraq, said it sought a peaceful solution, but seemed in no hurry to negotiate. A further problem was North Korea's demand for bilateral talks with the USA, which insisted on a wider multilateral forum. Not until April 2003 did talks take place, in Beijing, involving the USA, North Korea and China, which under its new leader, Hu Jintao, had become an active mediator. China of late had become vexed by its old ally's maverick behaviour, and not only on the nuclear issue. In September 2002, meanwhile, North Korea declared the establishment of a free economic zone at Sinuiju on its north-western border with China, and appointed a wealthy Sino-Dutch orchid dealer, Yang Bin, to run it. Furious at not being consulted about either the zone or the appointment of its head, China promptly arrested Yang; in July 2003 he was imprisoned for 18 years on fraud and corruption charges. On the nuclear issue, North Korea's statements remained ambiguous. Before the Beijing meeting, it seemed to claim to have finished reprocessing spent fuel rods from Yongbyon to produce plutonium. At the talks, its chief delegate unofficially told the USA that North Korea had nuclear weapons and reserved the right to sell them. In the absence of hard intelligence data (but with the release of much politically motivated information) on how far along either of Pyongyang's two nuclear programmes might be, it was unclear the extent to which such brinkmanship was bluff. In its public statements North Korea asserted a right to nuclear defence, abandoning its earlier claim that Yongbyon's purpose was to generate electricity.

From May 2003 US policy, insofar as there was one, shifted more towards interdicting North Korean shipments regarded as suspicious, with a new Proliferation Security Initiative (PSI). The 11 nations supporting the PSI included Australia, which in April seized a North Korean freighter suspected of landing heroin worth US $125m., and Japan, which tightened port inspections of North Korean ships, hitherto laxly supervised despite suspicions of smuggling methamphetamines to Japan and missile parts and other strategic materials into North Korea. South Korea did not join the PSI, but fears of a breach with the USA were allayed after its new President, Roh Moo-Hyun, adopted a more aggressive stance on his first visit to Washington in May. Yet with younger South Koreans inclined to blame the USA for exacerbating tensions on the peninsula, acute awareness of the closeness of greater Seoul's 20m. people to the front line—Seoul is within range of KPA heavy artillery, the arsenal of which is believed to include chemical shells—led to much anxiety. Despite this, inter-Korean contacts continued as normal throughout the crisis (see above).

After much diplomatic activity, it was agreed that six-party talks—involving the two Koreas, China, the USA, Japan and Russia—would be held in Beijing in August 2003. While, by any measure, this was a full complement of concerned parties, and despite relief at this prospect, the chances for substantive progress were unclear. Kim Jong Il faced a difficult decision: whether to exchange nuclear programmes—or other activities, such as missile exports—for aid and security guarantees, or keep them as a deterrent against sharing the fate of the regime of Saddam Hussain in Iraq. For his interlocutors, the many dilemmas included: how to prioritize an agenda, with so many concerns; how any nuclear agreement could realistically be verified; and what combination of incentives and punishments to apply. Six-party talks were held in August as scheduled, but ended without resolution. The USA appeared isolated in insisting that North Korea not be compensated for nuclear disarmament. Divisions in the Bush Administration, between those willing to engage Kim Jong Il and 'hard-liners' who sought regime change, seemed likely to come to a head. On 25 August, on the eve of the six-party talks, Charles 'Jack' Pritchard —who had since the Clinton era been considered the US 'point man' on North Korea, meeting regularly in New York with North Korean diplomats at the UN—resigned. Taking up a post at the Brookings Institution in Washington, DC, he confirmed speculation that he had quit in dismay by urging the USA to recommit to serious and sustained engagement with North Korea.

Once launched, the six-party process continued, albeit at a desultory pace and with scant substantive progress. Despite much shuttle diplomacy (especially by China, which as host had invested heavily in this initiative), hopes of holding a second meeting before the end of 2003 were not fulfilled. In November KEDO's executive board voted to suspend the consortium's LWR project for a year. While South Korea hoped this would resume in due course, the USA—which had pushed for formal suspension—made clear that it saw no future for the LWRs. In January 2004 a senior private US delegation visited Pyongyang and the Yongbyon nuclear site, which was once more operational. One US scientist was allowed to handle a lump of what appeared to be plutonium: a signal that North Korea's nuclear intentions were in earnest. A second round of six-party talks was eventually held in Beijing in February, but again made little headway. This time, China and even South Korea were dismayed that the USA merely repeated its demand of CVID—complete, verified, irreversible dismantling of nuclear facilities—and refused to entertain North Korea's step-by step offer, starting with a 'freeze'. However, it was agreed to meet again by the end of June, with prior working groups to discuss details. The latter met briefly in April, and again just before the third round of full talks in June.

At the third round talks the USA modified its position to offer a phased process, including interim incentives. Details were not published, but this seemed to be a hopeful sign. By now, however, the US electoral cycle had become a factor. Even if a fourth round was held as scheduled by the end of September, there was little expectation of North Korea conceding before November, since Kim Jong Il was thought to expect a better deal if the Democratic candidate, Senator John Kerry, were to defeat George W. Bush at the presidential election. Other US incentives, such as a donation of 50,000 metric tons of grain in July, failed to mollify North Korea; which instead reacted angrily to US urgings to emulate Libya and definitively surrender all of its weapons of mass destruction. In August North Korea said it saw no purpose at this time in working groups, but it did not rule out attending a fourth round of six-party talks.

Meanwhile, in April 2004 Kim Jong Il visited Beijing for the third time in four years. As ever, he travelled by train, his journey being shrouded in secrecy: his brief trip (of just two nights) was confirmed only after it was over. All was publicly cordial, as usual. This was Kim's first encounter with China's new leadership under Hu Jintao; while the main aim of the People's Republic was to ensure North Korea's commitment to the six-party talks, and its amenability more generally. This all came at a price: China admitted offering aid, but details were not given. By contrast to the China visit, in neither 2003 nor 2004, unlike the two previous years, did Kim's train roll north for summits with Vladimir Putin. However, Russia continued to cultivate North Korea, and to take a soft line within the six-party talks. In July 2004 Sergei Lavrov, Russia's Minister of Foreign Affairs, returned home by way of Seoul and Pyongyang from the ARF held in the Indonesian capital of Jakarta.

Japan, meanwhile, displayed its own combination of persuasion and incentives. The Koizumi Government passed laws restricting transfers to hostile nations (though North Korea was not named, nor was the legislation implemented), and tightened inspections of North Korean vessels in Japanese ports. Yet this did not stop Koizumi making a second day-trip to Pyongyang in May 2004. This time he brought back five children of two abductee couples returned in 2002; the apparent price being 250,000 metric tons of rice. In a more complex case, the US husband of another ex-abductee, a former sergeant who had apparently deserted across the DMZ in 1965, and their two daughters were brought to Jakarta for the family to be reunited. They all then travelled to Japan, despite the possibility of the husband, Charles Jenkins, facing a US court-martial. Japanese public opinion, quick to anger over the fate of dead abductees, in this case desired a happy ending. There was a risk of this affair disrupting US-Japanese relations, should the US Department of Defense take a hard line. In August Japan and North Korea resumed bilateral talks; but despite Koizumi airily predicting diplomatic relations as feasible within a year or two, the abduction, nuclear and missile issues were all expected to slow any progress.

In December 2003 North Korea and Ireland established full diplomatic relations, leaving France as the sole EU member still resisting this. France's ground remained the issue of human rights abuses: also a major concern for the EU majority which had recognized the Democratic People's Republic. (None the less, there was some suggestion in June 2004 that France was likely to establish diplomatic relations with the DPRK.) In 2003 and again in 2004, European nations successfully introduced resolutions condemning such abuses at the UN Commission on Human Rights in Geneva. China opposed on both occasions, while South Korea absented itself in 2003 and abstained in 2004. The latter resolution provided for the appointment of a special rapporteur, ensuring that the issue would remain on the agenda. In similar vein, in 2004 two bills on North Korean human rights were before the US Congress. One passed the House of Representatives unanimously in July; yet it was not clear if either would become law during the current session, owing to pressure of time. As of late 2004, not for the first time, North Korea appeared beleaguered on all sides and on many issues. Unless the country altered its behaviour, that would remain the case.

Economy

ROBERT F. ASH

Based on an earlier article by JOSEPH S. CHUNG

Following the introduction of a highly centralized planning system based on the former Soviet model, and radical economic initiatives introduced during both the interim post-war Soviet occupation of North Korea (1945–48) and the formal establishment of the Democratic People's Republic of Korea (DPRK) on 9 September 1948, the economy has been characterized by a strong orientation towards the socialization of production and productive relationships. By the end of 1958, the full socialist transformation of North Korea had been completed, laying the foundations of what was one of the world's most highly centralized and planned economic systems, and what has become, since the implementation of market-orientated reform in China, the former USSR and other previously socialist countries, the most monocratic and autarchic economic regime left in the world.

There is a consensus that, even allowing for the impressive rates of economic growth that were sometimes achieved, central planning systems have generated severe problems of waste and inefficiency. In the case of North Korea, there is no doubt that the Government's rigid adherence to central planning has inhibited economic growth and constrained its long-term economic performance. Until the early 1960s, it is true that economic expansion was rapid, but this performance owed more to recovery from a war-ravaged economy (following Japan's surrender and withdrawal in 1945, and, subsequently, the Korean War truce agreement in 1953), than to net growth. Even so, the rate of economic expansion in North Korea (averaging some 12% annually) exceeded that of the South until the late 1960s, and not until around the middle of the following decade did South Korea's aggregate gross national product (GNP) surpass that of North Korea. In per caput terms, the emerging pre-eminence of the South occurred even later.

Economic growth in all communist regimes has been driven by the physical accumulation of resources (especially capital), rather than by qualitative improvements. In this respect, North Korea is no exception, a capital accumulation rate—and its corollary, minimal consumption improvements—rather than productivity gains having facilitated economic expansion. Over time, infrastructural shortcomings and associated difficulties have grown increasingly serious, highlighting the need to enhance efficiency through the more effective use of resources and the adoption of more advanced technology in order to maintain and increase the growth momentum.

Throughout its existence, North Korea's development strategy has been guided by the ideology of *juche*, or 'self-reliance' (a notion whose application has extended also to the realm of politics). *Juche* has quasi-mythical overtones, although the idea of '*juche* for economic development' was first articulated by Kim Il Sung in December 1956. It dominated the formulation and implementation of economic development until well after Kim's death in 1994, and, under the pretext of making working people the masters of agricultural and industrial undertakings, was reflected in a continuing emphasis on the need to generate an independent, self-reliant economy. Indeed, despite a willingness to embrace modest market-orientated policies in recent years, the notion of *juche* has remained a dominant economic organizational principle in North Korea.

Following the death of his father, Kim Il Sung, North Korea's new paramount leader, Kim Jong Il, continued to defend socialist ownership and the consolidation and development of a socialist economic system embodying the *juche* ideal of self-reliance, while emphatically rejecting the introduction of capitalist methods. These principles have been echoed in various official pronouncements, and underlined in the advocacy of reliance on exhortations, mass social mobilization and production campaigns designed to increase output.

The flexibility with which *juche* can be interpreted does not, however, necessarily make it incompatible with economic liberalization. From this perspective, it is significant that a constitutional amendment, introduced in 1998, made reference to privatization, material incentives and profitability, while at the same time apparently endorsing North Korea's pursuit of a planned, socialist and self-reliant economy. In the same year Pyongyang sought advice from the World Bank on the establishment of a market economy. Even so, it remains premature to speculate about the outcome of such initiatives: on the one hand, the North Korean Government had shown itself eager to expand foreign trade, encourage foreign investment and embrace modern technology; on the other hand, it had also continued to show considerable determination not to allow market-orientated reforms to jeopardize central control over key national assets. In newspaper editorials that appeared in January 2004, no mention was made of market reforms; instead, the emphasis shifted back towards the need for enhanced central controls in order to facilitate supposedly scientific and technological initiatives. Heavy industrial development was also given a higher priority than that of light, consumer goods industries. Such policy priorities contrast curiously with Kim Jong Il's apparent decision, as reported in November 2003 by Swedish visitors to North Korea, to embark on a three-year experimental programme of opening up the economy to the outside world. Seemingly significant too were Kim Jong Il's reportedly enthusiastic remarks, made during a visit to a machine tool plant in June 2004, about the role of profits, even if his subsequent praise for the 'socialist method of industrial management' (made in the course of the same visit) raises questions about their real import. Contradictions such as these, as well as the absence of published statistics, are an integral part of the mixture of rhetoric and obfuscation that make it so difficult to assess current economic developments, let alone predict the country's future trajectory. The ongoing international dispute over North Korea's nuclear weapons programme is another factor that impinges on the country's economic prospects. Put simply, a strategy that accords overwhelming priority to military construction (an estimated 23% of GDP was thought to have been directed towards military expenditure in 2003) cannot fail seriously to constrain the potential for welfare-enhancing economic growth.

Estimates by the Bank of Korea (BOK) in Seoul indicated a real GDP growth rate of 1.8% during 2003, compared with 1.2% in 2002. In contrast to 2002, when there was thought to have been contraction in mining, manufacturing and power, all sectors shared in this positive growth performance. Thus, BOK data suggested the following rates of expansion in 2003: agriculture (incl. forestry and fisheries), 1.7%; mining, 3.2%; manufacturing, 2.6%; construction, 2.1%; and services, 0.7%. Accelerated growth in mining, manufacturing, power and services was, however, partly offset by sharp declines in agricultural and construction growth. In addition, while heavy industrial expansion accelerated sharply (from –4.2% to 2.7%), that of light industry slowed down (from 2.7% to 2.3%).

Despite the encouragement suggested by improved economic growth, North Korea continued to face severe economic problems throughout 2003 and into 2004. These were largely attributable to the inward-looking policies of the past, which, as in other former socialist countries, placed undue emphasis on heavy industry at the expense of welfare-enhancing consumer goods and services, as well as downgrading the role of foreign trade and inhibiting technology transfer. The outcome was to offer protection to domestic industries behind artificial barriers, and thereby prevent the emergence of internationally competitive industries driven by the principle of comparative advantage. This, in turn, not only led to a serious shortage of foreign exchange, but also exacerbated North Korea's long-standing and growing foreign indebtedness. In the first half of the 1970s North Korea's outstanding foreign debt doubled to well over US $500m., resulting in its becoming the first communist country to default on its debt. More than three-quarters of this debt was owed to non-communist countries and an important consequence was that, from 1975, North Korea was for many years cut off from access to advanced Western technology. Notwithstanding the critical need to maintain imports of foreign technology and capital in the interests of economic modernization, North Korea's isolation only served to underline the emphasis on self-reliance.

North Korea has also faced persistent shortages in skilled labour, modern equipment and technology. Capacity under-utilization, reflecting congestion in the energy, transport and mining sectors, and inadequate infrastructural investment have further impeded development. In addition, excessive bureaucracy, a shortage of fertile arable land and the absence of modern farming methods and equipment have resulted in a disappointing agricultural performance, giving rise to continued food shortages, widespread malnutrition and starvation. After repeated reports of food supply problems and rationing for several years, North Korea officially acknowledged that, as a result of damage caused in 1995 and 1996 by the heaviest floods of the century, the country faced serious food shortages. The consensus view is that such conditions gave rise to widespread famine during 1996–98. Food shortages have persisted, although by early 2004, their worst effects appeared to have been overcome (see also below).

Many would argue that North Korea's economic salvation lies in the direction of gradual, if not radical economic reform, embracing Chinese-style market-orientated policies and opening the economy to the outside world. As a long-term goal, this is a process that may be facilitated by Korean unification, although the short-term economic and social costs of integration are likely to be considerable (even supposing that unification can be achieved). Meanwhile, important initiatives have been taken to encourage foreign investment in North Korea. For example, as early as 1991, the Government in Pyongyang urged the creation of special economic zones and designated the 621-sq km Rajin-Sonbong Strip on the north-eastern coast as a 'Free Economic and Trade Zone', in imitation of the Chinese model. Whether, however, the North Korean Government will prove itself capable of using reform to turn around its economy, remains impossible to determine and the possibility of complete collapse still cannot be ruled out. Thus, in November 2003 the chairman of Standard and Poor's sovereign ratings committee suggested that economic collapse in North Korea was inevitable, although whether this would occur imminently or in the longer term was much more difficult to predict.

PHYSICAL RESOURCE BASE, ECONOMIC GROWTH AND STRUCTURAL CHANGE

North Korea occupies about 55% of the total area of the Korean peninsula. Of its total surface area, only about 14% is arable. A generally harsh climate restricts the output of arable farming to one crop per year, although a high irrigation ratio helps offset the high summer concentration of rainfall, as well as providing relief in the face of frequent spring droughts. Natural soil fertility is less favourable than in South Korea and the average 'natural' farm size is small (probably less than 2 ha—and more likely 0.55 ha—per member of the agricultural labour force). Yet the emergence of serious food shortages in recent years cannot simply be attributed to a lack of fertile land, nor to the tiny scale of farming (as the experiences of other Asian countries demonstrate, small farm size is not necessarily inimical to the achievement of high yields and sustained agricultural growth).

North Korea is well endowed with mineral resources, compared with the South, although its mineral industry remains underdeveloped and primitive because of the obsolescence and physical shortage of modern equipment and technology. Telecommunications, too, have remained seriously underdeveloped. The lack of modern equipment and vehicles appears to be the main problem facing both land and marine transport. In the absence of known deposits of crude petroleum, North Korea's electricity production (about 30,010m. kWh in 2001, according to one estimate) derived from fossil fuel (29%) and hydropower sources (71%). The implied energy constraint—especially the problem of stagnating coal production—has inhibited industrial expansion. The development of the mining, power and metal industries, as well as rail transport, was given the highest economic priority by the Government in the late 1980s and early 1990s. However, in the mid-1990s emphasis was switched to agriculture, light industry and foreign trade: a somewhat surprising development and an implicit admission that past economic policy had failed.

Throughout North Korea's existence, the pursuit of an unbalanced industrial strategy, geared towards maximizing heavy industrial growth, has characterized its economic development. This, in turn, has been reflected in significant structural change within its economy. At the time of partition, agriculture (including fishing and forestry) generated almost two-thirds of gross domestic product (GDP), compared with an industrial contribution of under 20%. By 2003, according to BOK estimates, the agricultural share had fallen to 27.2%, although mining added a further 8.3% to GDP, so that the primary sector's contribution overall was just over 35%. By contrast, despite the industrial thrust of North Korea's development strategy, the share of secondary sector activities was still quite low, at 31.7%; this figure comprised manufacturing (only 18.5%), construction (8.7%) and electricity, gas and water (4.5%). Services, meanwhile, accounted for an estimated 32.8% of GDP, of which 22.9% reflected government activities. Even allowing for a significant margin of error (see below), such figures reveal that the agricultural contribution to overall economic activity is significantly higher than that of low-income countries taken as a whole, while the shares of industry and services (especially non-government services) are lower.

Beneath the positive, albeit modest, industrial growth performance, qualitative indicators suggest a much less buoyant situation. For example, industrial plants, most of which were under strict state ownership and control, reportedly operated at an average 30% below capacity for years prior to Kim Il Sung's death, and, by the late 1990s, were believed to be functioning at about 20% of capacity. Nor was industrial growth in North Korea translated into significant improvements in living standards, owing to the disproportionately high allocation of resources to heavy industry and military expenditure.

Judged by its record of published economic data, North Korea is one of the most secretive countries in the world. The questionable reliability and ambiguities of limited official data pose additional problems in assessing the country's economic performance. Despite such problems and other, more familiar, difficulties associated with estimating the national output of a communist country on a US dollar basis, quantitative insights are available into North Korea's recent economic performance. The general picture that emerges is one of long-term positive growth during the early decades of the regime (for example, average national income growth was probably around 9% per annum during 1960–89), followed by a serious contraction during most of the 1990s, after economic links with the USSR and Eastern bloc countries had collapsed. Contraction during this period was exacerbated by problems of energy shortages, the poor maintenance of existing industrial facilities, technological backwardness and inadequate investment. (Some suggest that by the second half of the 1990s, investment had fallen below the replacement level, leading to a contraction of capital stock.) During 1990–98 the average rate of contraction of GDP was 3.8% annually, the sharpest declines occurring in 1992 (of 4.2%) and 1997 (6.3%), as a result of which aggregate GDP contracted by just over one-third. This trend was eventually reversed in 1999, owing mainly to higher farm production and, more significantly, to much-needed inflows of foreign aid (estimated at US $650m.). Thus, GDP growth in 1999 rose by an estimated 6.2%, and maintained a positive momentum during the next four years (recording an average annual increase of around 2% between 2000 and 2003). The improved economic performance in recent years reflects efforts by the Government in Pyongyang to stabilize production in the traditional core heavy industries, while seeking to expand the output of consumer goods and promote the development of information and communications activities. A major initiative in 2002 was the Government's announcement of a policy 'package', designed to strengthen a still-nascent orientation towards limited marketization (see below). Gross national income (GNI) in 2003 was an estimated 21,947,000m. South Korean won, implying an average per caput income of 974,000 won (2.1% more than in 2002). The US Central Intelligence Agency (CIA) suggested that in terms of purchasing-power parity (PPP), North Korea's GDP in 2003 was US $22,850m., while average GDP per head was around US $1,000.

With its goal of achieving self-sufficiency and its pursuit of an inward-looking developmental policy that has rejected integration into the international economic order, North Korea's strategy has most closely resembled that of China under the leadership of Mao Zedong. North Korea is also one of the world's most highly defence-constrained economies. Official estimates suggest that the share of spending on defence has fallen steadily, from a peak of 32% in 1968 to under 15% in the early 21st century. However, defence expenditure is notoriously difficult to assess, and some Western sources suggest that as much as 25%–30% of GDP may still be allocated to the military sector. CIA estimates suggested that in 2003 the ratio of military expenditure to GDP was 22.9%.

In 1984, in order to stimulate the economy, North Korea made an unprecedented change to its previously rigid insistence on self-sufficiency, by announcing its willingness to implement joint ventures with foreign companies (including those from capitalist countries). This development was a logical progression from the new foreign economic policy that had been contained in the January 1984 decision of the Supreme People's Assembly (SPA), in which reference was made to the need to expand economic co-operation and technical exchange, as well as trade, with Western countries. Some viewed this readiness to accept inflows of foreign capital as the clearest indication of North Korea's slow but steady progression towards a modernization of its economy. It remains to be seen, however, to what extent recent diplomatic initiatives and co-operative commercial projects reflect Pyongyang's real readiness to accept the limits to national self-sufficiency.

POPULATION, LABOUR AND WAGES

Serious demographic losses (estimated at around 1.5m. persons) during the Korean War (1950–53) and migration to the South, as well as a relatively low population density, have exacerbated labour scarcities in North Korea. The average rate of natural increase of total population was close to 2% per year in the 1990s, although there is also evidence of a declining trend rate. As of July 2004 a CIA estimate put North Korea's total population at 22,697,553—less than half that of South Korea—and growing at an average annual rate of 0.98% (birth rate 16.77 per 1,000; death rate 6.99 per 1,000). North Korea's rate of natural increase is below the world average (1.14%), although the low rate of population expansion of recent years may reflect the impact of food shortages more than that of family control policies. The CIA estimates indicated a dependency ratio of 45.2% in 2004. This figure conceals a remarkably low age-dependency ratio, with a mere 7.6% of the population aged 65 years and above (almost twice as many women as men fell into this category in 2004), alongside a more 'normal' young age dependency (24.6% of population being under the age of 15). In 2004 the infant mortality rate was estimated by the CIA to be 24.84 per 1,000 live births, compared with 22.8 per 1,000 in 2002; average life expectancy at birth was 71.08 years.

The constraint on manpower resulting from the maintenance of large armed forces (with over 1m. members in 2002, North Korea had the fifth largest army in the world) has been offset by the ability of a strong central state to mobilize military personnel for civilian economic purposes. The demographic estimates cited above suggest that in 2004 67.8% of the population, some 15.388m. men and women, were of working age (15–64). However, only about 62% of these (9.6m.) were reckoned to constitute the labour force, of whom 36% were engaged in farming. Estimates of unemployment are not available, although, as in pre-reform China, the likelihood is that the organizational framework within which the economy functions has concealed large numbers of people who, using orthodox criteria, would be considered to be unemployed.

Information on wages is scarce. Reports make clear, however, that in 2003 there were dramatic wage increases, especially among some privileged groups such as officials of the Korean Workers' Party (KWP), military personnel, scientists, farmers and miners, all of whom received rises that were well above average. There is evidence, for example, that miners and soldiers were awarded a wage rise of 1,500%, while the corresponding figure for agricultural workers was 900%. One source has speculated that such major adjustments reflected the Government's wish to enhance the role of material incentives in the allocation of labour. In assessing the impact of these wage rises on living standards, it should be kept in mind that sharp price increases took place simultaneously (see below).

PLANNING

Even now, the legacy of central planning is very apparent in North Korea, where some 90% of economic activity remains under state control in the guise of collectivized agriculture and state-controlled industrial and service activities. As early as the interim post-war Soviet occupation of North Korea (1945–48), the North Korean Provisional People's Committee introduced a series of radical economic reforms (including land reform and the nationalization of major industries) which signalled a strong orientation towards the socialization of productive relationships and the introduction of a centralized planning system on the Soviet model. Following the formal establishment of the DPRK on 9 September 1948, the new Government launched its first Two-Year Plan (1949–50), whereby it sought to consolidate the foundations of a self-reliant national economy. This process was, however, interrupted by the Korean War.

From the end of the Korean War in 1953 until 1984 North Korea's economic policy was conducted within the framework of five development plans. As a result of revisions, early fulfilment and delays, the dates of these plans are somewhat confusing and discontinuous: a Three-Year (Post-war Reconstruction) Plan (1954–56); a Five-Year Plan (actually 1957–60); the first Seven-Year Plan (actually 1961–70); a Six-Year Plan (1971–76); and, with the designation of 1977 as a year of readjustment, the second Seven-Year Plan (1978–84). Following a two-year adjustment period, the third Seven-Year Plan (1987–93), was announced by President Kim Il Sung in 1986. However, the process of orderly planning was seriously undermined as a result of the severe economic problems of the 1990s and beyond. The dominance of planning has also been dented by recent market-orientated initiatives.

The division of Korea after Japan's defeat in 1945 left North Korea in possession of about two-thirds of the peninsula's heavy industrial facilities and infrastructure. With post-war economic rehabilitation completed during the 1954–56 Plan, the country made substantial progress under the 1957–60 Plan towards establishing a firm foundation for industrialization, as envisaged in Plan targets. During this period the 'socialization' process was completed, and the *chollima* movement (which took its name from a legendary flying horse, symbolizing rapid progress) was introduced. Not until the first Seven-Year Plan, however, did the process of industrialization begin in earnest. This was the period in which economic and infrastructural shortcomings first emerged, forcing the Government to extend the Plan and to adjust its policy in order to accommodate economic and military needs. As a result, some of the planned targets for 1961–70 remained unfulfilled. In the subsequent 1971–76 Plan, the basic objectives were unchanged, although greater emphasis was placed on technological upgrading, the attainment of self-sufficiency in industrial raw materials, the development of energy industries, raising the quality of products and restoring sectional balances.

The Six-Year Plan (1971–76) was notable for its attempt to revise North Korea's policy of self-reliance in favour of seeking greater access to foreign capital and technology. The success of this strategy was reflected in the fulfilment of major economic targets ahead of plan. At the same time, however, increased capital and technology imports generated increasing foreign debts, which, in turn, led to problems of debt repayment. As a result, North Korea became the first communist country to default on its debt, preventing further purchases of advanced Western technology and forcing it to return to the previous strategy of self-reliance.

Against this background, the major thrust of the 1978–84 Plan was the achievement of self-reliance—an associated aim being the lessening of dependence on Soviet economic aid—modernization, and 'scientization', as well as the promotion of the export of manufactured goods. Modernization meant increasing mechanization and automation, while 'scientization' was the North Korean term for introducing more modern production and management techniques. Available evidence suggests that the results of the 1978–84 Plan were disappointing—all the more so, given that major targets were set at levels no higher than those of the previous Plan. The seriousness of continuing economic problems was reflected in the paucity of public official statements or proclamations on the 1978–84 Plan. Even North Korean officials recognized that the targeted growth of net material product (NMP) and several major commodities had not been fulfilled, and their suggestion that NMP growth had averaged 8.8% per year has also been widely questioned. Nor is it likely that total industrial output increased at an average rate of 12.2% per annum, as claimed, thereby virtually fulfilling the planned rate of expansion of 12.1%. The fact that three years elapsed before the formulation of the next economic plan is another indicator of the failure of the 1978–84 Plan and of the severity of the economic problems that already confronted North Korea in the mid-1980s.

The targets of the third Seven-Year Plan (1987–93), announced in April 1987, were generally less ambitious than those of previous Plans, reflecting persistent economic shortcomings and a more realistic approach by economic planners. Modernization and 'scientization' were again at the heart of the Plan, which called for average annual industrial growth of 10%. The country's NMP was targeted to increase by 70% during the Plan period, compared with the reported 80% under the previous Plan. Annual production of crude steel was projected to reach 10m. metric tons—significantly lower than the previous target of 15m. tons per year announced in 1980. By contrast, the projected annual production of non-ferrous metals and aquatic products, two of North Korea's most important export commodities, was increased to 1.7m. tons and 11m. tons, respectively, compared with earlier goals of 1.5m. tons and 5m. tons.

At the conclusion of the Plan, in December 1993, it was officially conceded that the Plan's targets had not been fulfilled, as a result of the difficulties associated with the demise of the USSR and the former socialist bloc. Gross industrial output was said to have expanded by only 50% over the Plan period (an average annual growth rate of 5.6%), although even this disappointing figure was questioned by Western sources. The only other production figures released by the North Korean authorities for 1987–93 were: power generation (an increase of 30%); coal (40%); non-ferrous metal ores (60%); steel (30%); and chemical fertilizers (50%). Significantly, all five were below the planned growth rates.

The enormity of North Korea's economic problems—the result of years of planning deficiencies, and now exacerbated by developments taking place elsewhere in the communist world—were reflected in the Government's inability to formulate a new long-term economic programme. After 1994 formal economic planning was replaced by greater emphasis on rhetoric, urging the work-force to 'rally under the Red Banner, continue on their arduous march and demonstrate their revolutionary zeal'. Simultaneously, market initiatives began to play a minor, but still significant, role supplementary to that of planning. Meanwhile, the years 1994–97 were designated a 'period of adjustment in socialist economic construction'. Few quantitative targets were announced for this transitional period, other than planned increases in textile production (20%), fruit and meat output (both by 30%), chemical fibre and synthetic resin production (10%) and power output (30%). In recognition of North Korea's need to produce more and better consumer goods and to solve its increasingly severe food problem, greater emphasis was directed to agriculture, light industry and foreign trade, while efforts were urged to maintain the development momentum of the coal and power industries, and to expand rail transport.

This transitional programme must be judged as another failure because, rather than consolidation taking place, the mid-1990s witnessed continuing economic decline. Nevertheless, significant developments were also in evidence. They included the introduction of constitutional amendments (September 1998), providing for greater private enterprise and market reform, and, early the following year, the Government's adoption of a new economic planning law, which codified the principles and system of the centralized economy. As the experience of China shows, efforts to integrate reform into a socialist system pose huge difficulties. From this perspective, North Korea's simultaneous commitment to planning—much greater than that of post-1978 Chinese Governments—and a degree of economic liberalization lie uneasily with one another. The unambiguous commitment of Pyongyang to radical reform has yet to be demonstrated, and in the mean time there has been evidence of the Government using diplomacy (not least, taking advantage of regional and global aspirations to bring peace and stability to the Korean peninsula) in order to obtain secure foreign exchange and maximize additional resources under the pretext of humanitarian assistance. Thus, it seems more than coincidental that North Korea should have become the main recipient of US aid in Asia.

ECONOMIC REFORM AND KOREAN REUNIFICATION

From the beginning, Pyongyang's verbal commitment to economic reform—a clear prerequisite of inter-Korean economic integration—was greeted by widespread scepticism. Some, for example, have argued that the true motive behind North Korean diplomatic activity in recent years has been to enhance its economic advantage by maximizing foreign exchange and foreign aid. The difficulties of interpreting developments in North Korea were exemplified by the different reactions to reports in July 2002 to the effect that tax measures abolished in 1974 were to be revived. Suffice to say that statements by authoritative sources claiming that such moves were evidence of Pyongyang's determination to introduce market elements into its economy were matched by those of others who remained convinced that the North Korean Government's true intent was to reinforce state control over an expanding 'underground' economy.

Also effective from July 2002, prices and wages were simultaneously increased in an effort to enhance economic management and thereby improve living standards. Alongside major rises in prices of basic goods, including staple foods (rice up from 0.08 to 44 won per kg), transport (bus fares up from 0.1 to 2 won), accommodation and energy (electricity rates up from 0.035 to 1.8 won per kWh), workers' basic wages were also raised from 110 won to 2,000 won. The underlying rationale of the price increases was to eliminate the serious drain on fiscal resources resulting from the use of state subsidies and to allow prices to reflect the true cost of production. It is notable that the price increases implied in the figures cited above—ranging from 1,800% to an astonishing 55,000%—are far greater than those that were sanctioned by the Chinese Government during the early stages of its economic reforms. In 1979, for example, China agreed to rises of 20%–50% in the price of grain; by way of comparison, the increase in the price of North Korean rice cited above for 2002 was of the order of 55,000%—incidentally, much more than the 1,800% rise in workers' average wage.

The extent of market-orientated reforms is suggested in statistics that show that, as of mid-2004, there existed 300 profit-maximizing private markets in North Korea. Agricultural reforms are evidenced in the recent extension of farm households' permitted private landholdings from a maximum of 160 sq m to almost 1,300 sq m. Some 300 North Korean officials are reported to have gone overseas in 2003 in order to receive Western-style training in economics. Yet even accepting the North Korean Government's genuine commitment to a programme of market-orientated reforms, China's experience since the early 1980s is a warning of the high risks of such a strategy—not least, in terms of widening income differentials and the emergence of attendant economic and social strains. Recent measures adopted in North Korea do seem to have improved incentives and led to output growth (especially in light industry) and enhanced commercial transactions. If the momentum of price reform is maintained, however, widening income gaps may generate new economic, as well as social, problems. In any case, the seriousness of shortages of essential raw materials and energy is such that it cannot simply be taken for granted that even a genuine reformist commitment by Kim Jong Il will automatically be translated into the successful implementation of economic reform. China may appear to be, and probably is, the most obvious model for North Korea to follow (in January 2001 a delegation from Pyongyang, led by Kim Jong Il himself, visited Beijing and Shanghai in order to familiarize itself with the Chinese reformist development strategy). However, the political obstacles to market-orientated reform faced by North Korea—not least, the existence of a fundamental ideological challenge from Seoul—are much greater than those that have confronted Beijing. Nor are North Korea's base economic conditions as favourable as those enjoyed by China at the end of the 1980s. In short, it is not impossible that, far from facilitating sustained growth, reform could face insurmountable barriers and lead to economic collapse in North Korea.

The economic implications for North Korea of reunification with the South have been the subject of much speculation. So opaque is the screen behind which North Korea operates that it is extremely difficult to interpret its motives with regard to recent diplomatic initiatives—the most important of which include normalization of relations with various countries, the June 2000 Pyongyang Summit meeting between the two Korean leaders (Kim Dae-Jung and Kim Jong Il) and subsequent ministerial-level talks.

In assessing the scale of the economic challenge implied in reunification, the relative sizes of the two economies should be borne in mind. In 2003 North Korea's GDP was a mere 3% of that of South Korea, while the ratio of the two countries' average per caput incomes was about 15.5:1. The infrastructural gap is highlighted in transport and energy statistics: for example, in 2003 there were 24,879 km. of roads in the North, compared with 97,252 km. in the South (the corresponding figures for vehicle production were 4,800 and 3,178,000); South Korean power generation was 322.4 kWh. from a capacity of 56,050 MW, while for North Korea it was just 19.6 kWh. from 7,770 MW. Most dramatic of all were the contrasting levels of foreign trade of the two countries: US $372,600m. for South Korea; less than $ 2,400m. for North Korea – a 155-fold gap.

After the first contacts between North and South Korea, the establishment of discussions by working-level groups facilitated discussion of a number of issues such as the settlement of payments, mutual investment protection, the avoidance of double taxation and the arbitration of disputes. It would, however, be premature to predict the outcome of such talks, and in 2004, more than four years after the Pyongyang Summit, there was no guarantee of deepening bilateral economic co-operation through trade liberalization, let alone through the establishment of a free-trade area or some other form of economic union. The agreement of the two Governments to resume contacts, following the visit to Pyongyang, in April 2002, by the South Korean presidential envoy, Lim Dong-Won, was no doubt a promising development, although the election in February 2003 of a new President, Roh Moo-Hyun, and the appointment of a new administration in South Korea were viewed with suspicion in the North. The cancellation of economic talks, scheduled to take place in Pyongyang in March 2003, was a less-than-encouraging start to the further development of inter-Korean economic relations. However, the meetings of the North-South Committee for the Promotion of Economic Cooperation (NSCPEC) have continued—the ninth such having taken place in Pyongyang in June 2004. In all, almost 50 meetings on economic affairs were held between North and South Korean representatives during the four years that followed the first summit meeting in Pyongyang.

Expressions of co-operation between North Korea and the outside world have been few. In 1991 the Rajin-Sonbong Special Economic Zone (SEZ) was established, and in 2002 the Sinuiju Special Administrative Zone (SAZ) was inaugurated. However, as of mid-2004, neither the SEZ nor the SAZ had had any significant economic impact: in the case of the former, mainly because of infrastructural and regulatory problems; in that of the latter (see below), because of its peculiar administrative status.

More noteworthy than these so far abortive efforts was the 1998 agreement with Hyundai to develop the tourist potential of Mount Kumgang and, potentially far more important, to develop a new industrial park in North Korea (at Kaesong). If the target of attracting 1.5m. visitors to Mount Kumgang by 2005 is fulfilled, associated revenue could reach US $450m. per year—sufficient, if not misappropriated for party purposes, to pay for imports of essential goods, such as food. Progress towards fulfilling the goals of these co-operative ventures has, however, been halting. The opening of a cross-border road at last enabled Hyundai to organize three tour-group visits to Mount Kumgang in the first half of 2003, only for the road subsequently and without warning to be closed by the North Korean authorities in order to undertake unspecified repairs. At the end of 2003 work on the construction of the Kaesong Industrial Zone (KIZ) had yet to begin, although in June 2004 representatives attending the ninth meeting of NSCPEC agreed that the pilot phase would get under way by the end of 2004 and a KIZ management office be established. They also announced plans to open two roads across the Demilitarized Zone in October 2004, and to institute a full North–South rail service in 2005. It remains to be seen whether these targets will be fulfilled. Meanwhile, the difficulties that the implementation of such 'local' co-operative projects have encountered merely highlight the far greater challenges inherent in other more grandiose schemes, such as building a gas pipeline from Russia (Siberia) through North Korea to its southern neighbour, let alone that of constructing a freight road corridor from Europe to South Korea via Russia and North Korea.

Attempts have been made to estimate the cost of rehabilitation programmes in the North, designed to facilitate economic integration between the two Koreas. For example, rural energy rehabilitation has been costed at US $2,000–$3,000m. over five years, while the corresponding figure for a more comprehensive programme aimed at stimulating energy improvements throughout the economy has been estimated at $20,000m.–$50,000m. over a 20-year period. A South Korean source has suggested that some $6,000m. would be required in order to upgrade North Korea's economic infrastructure even to South Korea's 1990 level. Elsewhere, it has been suggested that unification might require the expenditure of more than 8% of South Korea's national product in order to bring per caput income in the North to about the South's level.

The massive economic, developmental and welfare gaps that exist between North and South Korea seem certain to make the process of economic integration as difficult, if not more so, than that either of the reunification of Germany, or of the process of integration within the European Union (EU). The income disparities between North and South Korea are, for example, much greater than those between the former West and East Germany, and the costs of reunification are likely to be much higher. At the beginning of the 1990s, the Korea Development Institute (KDI) estimated that total reunification costs would be US $240,000m. In view of South Korea's subsequent growth and the North's economic stagnation, that figure doubtless needs to be revised considerably upwards. The clear message contained in all these estimates is that the economic challenge facing Pyongyang and Seoul should not be underestimated.

AGRICULTURE, FORESTRY AND FISHING

About 80% of North Korea's total surface area consists of mountains and uplands, which extend close to the east coast. Plains are mainly concentrated in the west of the country and constitute the major agricultural base. Arable land constitutes about 18% of the total surface area. Since the mid-1990s the agricultural growth rate has shown sharp annual fluctuations, varying from a contraction of 3.8% in 1997 to expansion of 9.2% in 1999. The average rate of farm growth was probably around 2% annually during 1996–2002.

In the far north, summer lasts for only about two months, shortening the growing season considerably, but further south a typical growing season lasts for a minimum of four months. North Korea's moist climate is conducive to forest growth, but high-quality timber is limited to the northern interior, where extensive forests of larch, spruce, fir and pine provide the basis of the timber industry. Domestic timber shortages are offset by substantial imports from the eastern territories of Russia. Years of clearing forest lands for farming reduced the total forest area from 9.9m. ha in the 1970s to 9.2m. ha in 2000. In an attempt to meet the increasing demand for timber, as well as to conserve forest resources, a new forestry law was enacted in December 1992, since when the Government has consistently implemented mass mobilization afforestation campaigns. Even so, it would be premature to claim any significant improvement in reforestation efforts.

The Land Reform Act of March 1946 sought to abolish tenancy and redistribute land to the farmers. The outcome was to improve the distribution of land ownership, but at the expense of temporarily further reducing the average size of already small farms. Private farming persisted until the end of the Korean War, but thereafter, as collectivization intensified, it fell into decline until, by August 1958, it had disappeared. Thereafter, except on the small private plots allocated to households (the average size of which was later increased from a maximum 160 to 1,300 sq m), farming was conducted within the framework of co-operative farms (collectives) or state farms. Alongside agricultural collectivization, quantitative planning for production and state marketing and distribution of grain was also instituted. In the 1980s it was estimated that North Korea had about 3,800 co-operative and 180 state farms, the former managing more than 90% of the total cultivated land. One impact of collectivization was to increase average farm size. In the mid-1990s, however, the Government began to advocate the gradual transfer of collective farms to state ownership, which was considered to be more ideologically sound. Estimates by the UN's Food and Agriculture Organization (FAO) suggest that North Korea's total arable area is about 1.9m. ha. Of this, 0.4m. ha comprise high-quality, permanently-irrigated land; 0.7m. ha comprise medium-quality, semi-irrigated land capable of supporting rice production; and 0.8m. ha is low-quality land suitable for cultivating other cereal crops.

In 1976, in an effort to extend the arable area, North Korea launched a so-called 'nature remaking programme'. As its name implies, the aim was to reconfigure the countryside by establishing fields of 'regular shapes like a chess-board'. Its major economic objectives were to complete the irrigation of non-paddy lands, to reclaim 100,000 ha of new land, to build 150,000 ha–200,000 ha of terraced fields, to reclaim tidal land, and to conduct work on afforestation and water conservation projects. Subsequently, the third Seven-Year Plan (1987–93) set out the goal of reclaiming some 300,000 ha of tidal land, most of it to be used for growing rice. In fact, reclamation work proved to be slow and, according to South Korean estimates, only 28,400 ha had been reclaimed by March 1995.

Central to North Korea's agricultural strategy has been the extension of irrigation and the application of modern farm inputs (in the 1960s, agriculture in North Korea was among the most input-intensive systems in the world). Following success in providing permanent irrigation for rice farmers, the focus of irrigation projects shifted to non-paddy fields, although despite a 40% extension of the irrigated area, irrigation provision has remained less well developed on such land. In early 1994 there were reportedly 40,000 km of irrigation waterways, which, together with 1,770 reservoirs and 26,000 pumping stations, supplied water to about 70% of the country's agricultural land. In 1989 a project was initiated to build a 400 km-long canal, by diverting the flow of the Taedong river along the west coast of North Korea, thereby providing water to rural areas and newly reclaimed tideland in South Hwanghae and South Pyongan provinces. By the end of the 1990s, Korea had constructed 80,000 artificial lakes, 1,700 reservoirs, 25,210 pumping stations, 124,000 groundwater facilities and 40,000 km of flumes. Subsequently, however, the combined impact of flooding and inadequate investment resulted in quite significant deterioration of the irrigation infrastructure. For example, FAO data suggest that in mid-2001 North Korea's largest reservoir contained less than one-sixth of its total water capacity. The effectiveness of the country's irrigation system has also been affected by shortages of fuel and repair facilities.

If improved irrigation provision has been central to North Korea's farm strategy, the Government has also sought to raise farm yields through an expansion of mechanization (rice planting was already basically—95%—mechanized in the 1980s) and electrification, deep ploughing, close planting and the intensive use of fertilizers. In the face of recent declines in food production, there have also been attempts to introduce double cropping, although the success of this initiative has been constrained by associated higher resource requirements, in terms of labour and working capital. The number of tractors per 100 ha of arable land rose from 11 to 37 between 1970 and 1997, although concealed in this figure is a selective approach towards farm mechanization. In any case, in more recent years, shortages of fuel and spare parts have resulted in a decline in the use of farm machinery. Chemical fertilizer applications rose quite strongly during the 1970s and 1980s, and it was reported that in 1989 the average quantity of fertilizer applied to cereal crops was about one metric ton per ha. Under the impact of North Korea's increasing economic isolation and growing domestic economic problems, in the 1990s the situation deteriorated, so that by 1994 the corresponding figure had fallen to 750 kg, and to 500 in 1995. Estimates made available by FAO suggest that the downward trend in modern fertilizer use has continued. They show that in 2001 chemical fertilizer applications averaged 123 kg/ha for paddy, and 91 kg/ha for maize, although such supplies were supplemented by organic fertilizers. Such levels of application, though high by the standards of many developing countries, are nevertheless well below the average rate for cereals (215 kg/ha) specified by the North Korean Government. The further implication is that, in the absence of increased supplies of chemical fertilizers, it will be difficult to generate sufficient growth to return grain yields to their previous peak levels, attained in the 1980s. Other agricultural chemicals also remain in short supply. One positive aspect of the decline in chemical usage is that it must have reduced the problems of land degradation and water pollution that were said to be corollaries of excessive chemical fertilizer applications at the end of the 1980s.

Agriculture in North Korea is dominated by grain farming. Reliable production data are scarce, but what is beyond doubt is that from the mid-1990s, under the impact of policy failures and natural disasters, falling output resulted in widespread hunger and starvation. The severity of the situation is highlighted in Western estimates, showing that annual grain production averaged a mere 4m. metric tons in 1993–94, far below the planned level of 15m. tons. This figure had decreased to just 3.7m. tons by 1996–98, and although North Korean sources claimed a degree of recovery in 1999, production declined again to 3.6m. tons in 2000. Such figures may be compared with domestic requirements at this time of over 6m. tons. Estimates point to annual grain harvests of 3.9m. and 4.1m. tons in 2001 and 2002; and according to a FAO report, in 2003 domestic cereal availability was 4.2m. tons.

In terms of both production and consumption, paddy rice and maize are the two most important crops in North Korea, followed by potatoes. Of much less importance as a source of energy is the production of wheat, barley, millet, sorghum, oats and rye. Potatoes have high calorific yields and, against the background of developing food shortages, it is significant that recent years have witnessed an emphasis on the expansion of the sown area under this crop (although potato production (in cereal equivalent) accounted for only 9% of total cereal output in 2001, compared with 53% for rice and 33% for maize). Reports in May 2004 suggested that a 'potato farming revolution' was under way in North Korea, with subsidies being extended to potato farmers. Typically, the calorific content of 5 kg of potatoes equals that of 1 kg of rice. North Korean plans to increase potato output to 8m. tons by 2006 would therefore generate energy equal to that of 1.6m. tons of rice, and thereby go a long way towards overcoming current food shortages.

Rice was an important export item until the mid-1980s, although from the late 1990s North Korea began to import rice (from Thailand, South Korea and other countries), as well as some wheat. FAO estimates that, between 1997 and 1998, rice imports rose from US $104m. to $112m. (the value of wheat imports, meanwhile, having fallen sharply from $80.5m. to $48m.). In 2000 North Korea commercially imported an estimated 200,000 metric tons of grain, although a further 500,000 tons were made available on a concessionary basis by China and South Korea. In recent years financial constraints have limited commercial imports, and humanitarian assistance has become an increasingly important source of food supplies to North Korea.

The initial response of the North Korean Government to food shortages caused by deteriorating grain production in the 1990s was to urge its people to follow a 'two-meals-a-day' campaign; but subsequently—and unprecedentedly—it sought assistance from various UN and other international organizations, as well as from developed countries. As early as 1992–93, grain imports are estimated already to have accounted for almost one-third of food requirements. As the food crisis deepened, North Korean policy-makers accorded higher priority to agriculture, allocating additional resources to affected farms and making available extra fertilizer, farm machinery and manpower (including members of the armed forces). The relief provided by such measures was minor and temporary relief, and the late 1990s saw a worsening of famine conditions. During May and August 2000, insufficient rainfall and unseasonably hot weather caused the worst drought for 50 years, and in September strong tropical storms resulted in further crop destruction (an official source later stating that 1.4m. metric tons of grain had been lost from the annual harvest).

In October 2003 FAO estimated North Korea's total cereal demand for all uses (human food consumption, animal feed and seeds) during 2003/04 to be 5.1 metric tons. Given a projected domestic availability of 4.16m. tons, the expected deficit was estimated at 944,000 tons. After deducting expected commercial imports (0.1m. tons), concessional imports (0.3m. tons) and food aid (0.14m. tons), an outstanding deficit of 0.404m. tons would remain.

Such is the background against which, since the mid-1990s, severe food shortages have persisted in North Korea, resulting in malnutrition, starvation and famine-related diseases. According to Hwang Jang Yop, a senior diplomat who defected from North Korea in 1997, some 2.5m. people died as a result of famine in 1995–98. This figure is close to other estimates, such as that given in 2003 by the US Agency for International Development (USAID), indicating that famine had resulted in the deaths of 2.5m. people (i.e. about 10% of the North Korean population as it had been before the 1990s food crisis). Admittedly, others have argued that such figures are likely to have significantly exaggerated reality and that the actual loss of lives was between 600,000 and 1m. lives (3%–5% of the pre-crisis population). Whatever the truth, it is beyond dispute that, food aid notwithstanding, malnutrition rates in North Korea have been among the highest in the world since the early 1990s. Food shortages also gave rise to large-scale emigration, and by 1998 between 100,000 and 400,000 were estimated to have tried, mainly unsuccessfully, to flee across the Tumen river border into China.

Although climatic factors have played their part in causing food deficits in recent years, systemic policy defects have been a more important contributory factor. Until such problems are successfully addressed, food aid will be needed for many years to come. Meanwhile, the continuing severity of food shortages was confirmed by data provided by FAO and the World Food Programme (WFP), which indicated that daily rations for the 70% of total population—some 23m. people—living in urban areas were likely to decrease from 319 grams per caput to a maximum of 300 grams. This figure is sufficient to provide about half of calorific requirements. Nor, ironically, has the potential positive effect of recent market reforms on food supplies been forthcoming. Rather, because food prices have risen faster than wages, consumers have found it impossible to buy marketed rice and maize, let alone more nutritious non-staples, as well as paying for other non-food necessities. Hence, the finding of a FAO-WFP report that malnutrition rates have remained 'alarmingly high', forcing households to supplement rations with edible grasses, acorns, tree bark and sea algae. Although child malnutrition has declined, a survey in 2002 found that 40% of North Korean children were stunted, 20% underweight and 8% wasted.

As the above analysis indicates, in addition to increased grain imports, food aid has played a critical role in preventing famine from becoming even more dire. Implicit in statements by FAO, WFP and other authoritative sources is the reality that North Korea is still dependent on international food donations to feed its population. WFP has been working in North Korea since 1995 and now has access to 161 out of 203 counties. By 2004 it had supplied more than 3.5m. metric tons of food, valued at US $1,500m.—mainly to small children, pregnant and nursing mothers, and the elderly. WFP has also extended support through the implementation of food-for-work projects. Other international bodies associated with the activities of WFP in North Korea include UNICEF, the UN Development Programme (UNDP), the Adventist and Relief Agency (ADRA), International Fund for Agricultural Development (IFAD) and others. Bilateral assistance has also been important (the most important donors have been the USA, South Korea and Japan), although political tensions have sometimes affected food supplies. For example, US food aid fell sharply following the election of George W. Bush as US President. A more stable source of food has been the South Korean Government, which has made available 300,000–400,000 tons of rice a year.

With a coastline of some 17,000 km, a mixture of warm and cold ocean currents, and many rivers and streams, North Korea has considerable potential for fishery development. The principal fishing grounds are in the coastal areas of the Sea of Japan (or the East Sea as it is known on the Korean peninsula) to the east, and the Yellow Sea to the west. The main fishery ports are Sinpo, Kimchaek and the deep-sea fishery bases of Yanghwa and Hongwon. Besides the fishery stations, smaller fishery co-operatives are located along both coasts in traditional fishing centres. It is estimated that some 129,000 people were engaged in fishing and aquaculture in 2000. Total aquaculture production in that year was 467,700 metric tons, according to one estimate, and average per caput consumption of fish and fishery products was 8 kg per annum (half of the world average), providing a mere 3% of total protein supplies. The number of docked fishery vessels in North Korea was estimated to be 1,553. Despite emerging energy shortages (see below), the value of fish exports doubled between 1980 and 2000, to reach US $87m. (but $74m. in net terms).

MINING

With 80%–90% of the important mineral deposits of the peninsula concentrated within its borders, North Korea is relatively well endowed in natural resources. Most important among the minerals are coal, iron ore, lead, magnesite, graphite, zinc, tungsten, mica, fluorite and precious metals. Between 1990 and 1998, the share of mining in GDP fell from 9.0% to 6.6%, although by 2003 it had recovered to 8.3%. BOK estimates highlight the generally disappointing performance of the mining sector, at least until the late 1990s. During 1990–98 the average annual rate of decline was 3%, and although recovery subsequently took place (average annual growth averaging 9.6% from the late 1990s), in 2003 mining GDP was, in nominal terms, only 23% higher than in 1990. Problems in mining are largely attributable to energy shortages and the increasing obsolescence of mining equipment. The transformation from negative (–3.8%) to positive (3.2%) growth between 2002 and 2003 no doubt reflected an improvement in labour morale resulting from a major wage increase.

Iron ore, with estimated reserves of about 3,000m. metric tons, is very important to domestic industry and has also been a major source of foreign exchange. According to Western estimates, output increased from 8m. tons in 1985 to 10m. tons in 1988, reflecting the opening of new mines at Tokson and Sohaeri, both of which benefited from large-scale state investment. Thereafter, however, despite the commissioning in 1991 of the Chongpyong mine in South Hamgyong Province, production stagnated; indeed, in the mid-1990s it declined significantly, from 4.6m. tons in 1994 to a mere 2.9m. tons in 1997. Recovery subsequently took place, although output remained far below previous peak levels. In 2001 estimated iron ore production was 4.21m. tons.

North Korea is especially rich in magnesite, with reserves—mainly concentrated in the Tanchon District in the north-east—estimated at 6.5m. metric tons. After China, it is the second largest producer of magnesia products in the world. Mining of magnesite and production of its derivative, clinker, are important both for supporting the domestic refractory industry and for exports. The completion of expansion projects at the Tanchon magnesia plant, and the construction of the Unsong crushing and screening plant in the Tanchon District in 1987, raised the production capacity of clinker to 2m. tons per year. The 1994 agreement by North Korea and the USA to lift restrictions on bilateral trade offered an important new market for clinker, facilitating a large-scale expansion of North Korea's magnesite production capacity. In June 1995 the first North Korean trade mission to visit the USA concluded an agreement to export 100,000 tons of magnesite to a US company. In August 2000 a US mining company announced that it had formed a joint venture with the Korea Magnesia Clinker Industry group to mine, process and export magnesia products from North Korea. The first phase of the project envisaged the export of 200,000 tons of magnesia products to the USA and Asia.

Other minerals produced in large quantities by North Korea are lead, zinc, tungsten, graphite, mercury, phosphates, nickel, gold, silver, fluorspar and sulphur. Both zinc and lead ore are smelted domestically (at Tanchon, Nampo, Komdok, Haeju and Mungyong), and zinc and lead ingots are major exports. In 1994 North Korea produced an estimated 200,000 metric tons of high-grade electrolytic zinc and 80,000 tons of lead. A joint-venture project to redevelop the Unsan gold mine began operation in April 1987, and the first shipment of gold to Japan (totalling some 100 kg) was reported three months later. Unsan is potentially one of the world's major gold mines, with deposits estimated at more than 1,000 tons. The 1990 target was 2 tons, with an eventual annual production target of 10 tons. In 2001 total non-ferrous metals production was estimated to be 92,000 tons.

MANUFACTURING

Under the impact of its Soviet-style planning system, North Korea has consistently prioritized the development of heavy industry, in the ostensible belief that this would ultimately benefit light and domestic consumer industries. This said, the share of heavy industry in total industrial output has fallen significantly in recent years, from 80% in 1990 to 62% in 2003. Yet, as a comprehensive development strategy, the emphasis on heavy industry has been a failure, and even when rapid heavy industrial growth has taken place, quantitative expansion has often been at the expense of major qualitative deficiencies. In particular, excessive emphasis on machine-building (the basis for the munitions industry) generated a distorted industrial structure, and the orthodoxy that rapid heavy industrial development would enhance living standards through its ultimate promotion of agricultural and consumer goods industries has proven wholly unfounded. In general, the inwardly orientated nature of North Korea's industrial strategy for many years contributed to inefficiency and low productivity. Hence, recent reform initiatives, whereby the Government has indicated a willingness to embrace a greater role for prices, markets and other 'orthodox' economic criteria, such as profitability, in order to enhance manufacturing production efficiency.

Notwithstanding North Korean claims to have fulfilled the second Seven-Year Plan (1978–84) target of increasing industrial gross value output by 120%, it is unlikely that the implied average annual growth of 12.2% is a reflection of reality, given that most major industrial commodities expanded by much less than this during the same period (for example, output of electric power increased by 78%, coal by 50%, steel by 85%, cement by 78%, tractors by 50%, textiles by 45% and chemical fibres by 80%). In the third Seven-Year Plan (1987–93), official claims that industrial growth rose by 5.6% per annum were also contested by Western sources, which suggested that industrial output had actually declined. The years from 1994 to 1997 were designated a period of 'adjustment in socialist economic construction', with emphasis shifting from mining, power and metallurgical industries towards agriculture, light industry and trade. The readjustment of economic priorities, which incidentally was similar to those undertaken by China in the wake of its own 'great famine' (1959–61), no doubt reflected the severity of North Korea's agricultural crisis and the urgent need to restore food supplies. That difficulties facing farmers had severe cumulative effects on industry can hardly be doubted, and even official estimates acknowledged that industrial production had fallen sharply between 1992 and 1996. Construction is likely to have been similarly affected, some sources suggesting that investment had fallen below replacement level, thereby causing a contraction in the capital stock. According to BOK statistics, manufacturing output decreased cumulatively by almost 36% between 1990 and 1998 (by an annual average rate of 5.4%). The same source suggested that construction experienced an even sharper decline, of –37%. Subsequently, there was rapid recovery of construction, which reattained its 1990 level in 2001 and in 2002 and 2003 expanded by almost 16% annually. By contrast, the growth performance of the manufacturing sector was less impressive: despite positive growth since 1997 (2000 excluded), manufacturing GDP in 2003 remained 22% below the 1990 level (measured in nominal terms).

Machine-Building and Metallurgy

Through its provision of machinery for domestic industry and agriculture, the machine-building industry is central to any strategy designed to achieve industrial self-sufficiency. In North Korea, it has also made available an extensive range of military equipment, including rifles, mortars, machine-guns, multiple rocket launchers, artillery, anti-aircraft weapons, tanks, personnel carriers, patrol craft and frigates, missile-equipped fast attack craft and amphibious vessels, submarines, medium-range *Scud*-type surface-to-surface missiles, and long-range *Taepo Dong* missiles. The industry accounted for only 5.1% of total industrial output in 1946, but its contribution had expanded to about one-third of the total by the early 1980s. In more recent years declines in production have reflected those affecting other branches of the heavy industrial sector.

The machine-building industry is capable of producing 5,000-metre boring machines, 300-hp bulldozers, 10,000-ton power presses, 50,000-kilovolt-ampere generators, 200,000-kilovolt-ampere transformers for power production, 3,000-hp locomotives, 20,000-ton ships, 7,000-hp electric locomotives, 100-ton freight cars and 10-cu m excavators for construction and mining. Such impressive quantitative measures are, however, belied by the poor quality of North Korean machinery, the technological level of which is well below international standards. From the mid-1970s North Korea began to import advanced machinery and equipment from developed Western countries. The 1987–93 Plan sought to modernize the machine-building industry by introducing high-speed, precision machines and equipment, and in early 1990 it was claimed that the Huichon machine-building complex had initiated a flexible manufacturing process by adding robots to its numerically controlled machine tools—an example expected to be followed by other plants. From the present perspective, accelerated economic reform and, in particular, the extension of joint ventures would appear to offer North Korea's best opportunity to pursue much-needed technological upgrading in its machine-building industry.

With an annual production capacity of 10m. metric tons of steel and a work-force of 20,000–30,000, the Kimchaek Iron and Steel Works replaced the Hwanghae Iron Works(in Songrim) as the largest North Korean centre for the production of iron and steel. Kimchaek is located in Chongjin (North Hamgyong Province), near the Musan iron ore mine, and has estimated deposits of 100m. tons. It currently operates well below capacity, its annual production (an estimated 6m. tons) accounting for around 60% of national steel output. Other steel mills are found elsewhere in Chongjin, as well as at Taean. The disappointing performance of the metallurgical industry is highlighted in estimates for 1995, made by the US Bureau of Mines, which showed total output of pig iron, crude steel and rolled steel to have been, respectively, 6.6m., 8.1m. and 4.0m. tons—figures that can be compared with the unfulfilled targets for 1984 of 6.4m.–7.0m. tons (pig-iron and granulated iron), 7.4m.–9.0m. tons (crude steel) and 5.6m.–6.0m. tons (rolled steel). Outdated technology, shortages of coking coal and the low purity of domestic iron ore have posed serious constraints on the growth of the iron and steel industry. In the past, in addition to the construction of new and/or extension of existing facilities, mass mobilization campaigns to collect scrap iron have also been used in order to alleviate the steel shortage. At best, however, such efforts have been only partially successful, and in 2001 Kim Jong Il himself spoke of the need for technological modernization of the steel industry. It is significant too that in March 2001 Yonhap (the South Korean news agency) quoted North Korean sources to the effect that South Korean steel firms should be encouraged to set up steel mills in the North. Meanwhile, domestic steel production decreased by 24,000 tons (2.3%) between 2001 and 2002, from 1.062m. tons to 1.038m. tons in the latter year. It is instructive, too, that in 2000 steel imports from China rose from 9,600 tons to 24,000 tons (a rise of 150%—but an increase of 70% in value terms).

Chemicals

Important integrated chemical plants include the Chongyun Works (located in the Anju District, north of Pyongyang), the construction of which began, with French assistance, during the Six-Year Plan (1971–76). It was the country's first petrochemical complex, designed to produce ethylene, polyethylene, acrylonitrite and urea, using crude petroleum supplied by the nearby refinery at Sonbong (formerly Unggi). The output capacity of ammonium nitrate and nitric acid is thought to be 200,000 metric tons, three-quarters of which is produced at the Hungnam Fertilizer Complex. Most chemical products are polymer-based (e.g. polyethylene, acrylonitrile, and vinalon). In 1995 there were still no facilities capable of producing nylon and polyester in significant quantities, although North Korea may now possess a small-scale production capability for refined chemicals (especially organophosphates). In general, energy shortages and capital obsolescence have made chemical production an under-performing sector.

Because of its importance to agriculture, the chemical fertilizer industry has been the beneficiary of large-scale investment. Most fertilizers are produced by the giant Hungnam fertilizer plant, which has an annual capacity of 1m. metric tons, generating 770,000 tons in 1997. Official sources in Pyongyang suggest that domestic production of chemical fertilizers rose from 3.0m.tons to 4.7m. tons between 1976 and 1984 (an average annual increase of 5.9%), and increased further to 5.6m. tons by 1989 (3.6% per annum). Such figures are viewed sceptically in the West, where North Korea's annual production capacity was estimated to be only 3.5m. tons in 1995. It is clear too that industrial collapse in the 1990s had serious implications for fertilizer supplies: in particular, oil shortages have impacted seriously on the production of urea and ammonium sulphate (the main fertilizers used in North Korea). Imports have helped to a limited extent in offsetting domestic shortages and in 2000, in a gesture intended to demonstrate its support for economic rehabilitation in the North, South Korea sent 300,000 tons of fertilizer, valued at about US $90m. Other international donations of fertilizers, designed to enhance food production, have also assisted in increasing supplies of this vital input. Indeed, FAO data show that in 1999–2000 almost 77% of nutrient availability derived from humanitarian assistance, compared with less than a quarter from domestic production. Between 2001 and 2002 aggregate domestic fertilizer production decreased from 546,000 tons to 503,000 tons (down by almost 8%).

Textiles

At the heart of North Korea's textile industry is the Pyongyang integrated textile mill, built with Soviet assistance in the late 1950s. Its production derives mainly from locally produced synthetics, such as vinalon and petrochemically based fibres, as well as cotton and silk. Although Pyongyang remains the centre of the national textile industry, plants in Sinuiju and Sariwon have become increasingly important in recent years. North Korea's output of textile fabrics increased by 78% during the 1978–84 Plan (an average annual growth rate of 8.6%), thereby reportedly fulfilling the 1984 target of 800m. m of cloth. Meanwhile, a combination of machine imports from Japan and upgrading of existing equipment facilitated an expansion of knitwear products, derived from domestically produced orlon. The intention was to increase the annual output of textile fabrics to 1,500m. m by 1993. This target is, however, unlikely to have been fulfilled, and overseas estimates indicate that textile production at the beginning of the 1990s was no more than 680m. m. A salutary reminder of the severity of economic conditions in more recent years is the suggestion, made by South Korean sources, that North Korea's textile production had fallen to a mere 120m. m by the late 1990s.

Despite the difficulties faced by the industry, textiles have remained one of the most important sources of foreign exchange. In 1995 textiles replaced steel and non-ferrous metals as the single most important export category, accounting for 31.8% of export earnings (US $233.7m. out of $736m.). Thereafter, however, in absolute and relative terms, the value of textile exports declined: to $137.4m. (24.7%) in 2000, and $123.1m. (16.7%) in 2002. The main reasons for this disappointing performance were stagnating trade with Japan and difficulties in maintaining North Korea's competitiveness in textiles. It deserves stating, however, that in 2003, although exports to Japan continued to decline, those to China and India rose significantly. Processing on commission between North and South Korea has also increased sharply in recent years.

SERVICES

Until 1984 all outlets for the provision of services and the distribution of goods, including retail shops, were either state-owned or run as co-operatives. Following the introduction in 1984 of the so-called '3 August Consumer Goods Movement', local governments were permitted to establish, within their districts, direct sales outlets for the distribution of consumer goods produced locally with locally available resources. By the early 1990s the total number of shops, service establishments and 'food-processing and storage bases' was estimated to be 130,000. However, outside the special economic zones, markets (in which farmers were allowed to sell surplus farm products, products raised on private plots and a range of non-farm products at free-market prices reflecting scarcity values) have been the only exception to monopoly powers exercised by the State. As economic reforms have continued, the number of such outlets has increased. Perhaps the most recent significant initiative in this regard was the statement by a senior official of North Korea's State Planning Commission, in March 2003, that farmers' markets should simply be called 'markets', since the scope of their transactions transcended farming activities. The same source was also explicit in pointing out that such markets were no longer subject to controls and had become part of the 'socialist distribution system'. Between 1990 and 1998 tertiary activities grew rapidly—by 9.8% annually—as a result of which their share in GDP doubled (from 18.0% to 35.6%). Thereafter, however, annual growth slowed to 2.8% in 1998–2003, and the service sector's contribution to GDP decreased to less than 32.8%. In assessing the extent to which distribution in North Korea is still state-controlled, it is salutary to discover that the Government's share in the output of services increased from 61% in 1990 to 69.9% in 2003.

INFRASTRUCTURE

Transport

By the 1970s shortcomings in the transport sector were being blamed for the failure to fulfil major targets under the Six-Year Plan (1971–76). In particular, difficulties in delivering raw materials (especially coal) and semi-finished goods limited the expansion of the mining and manufacturing sectors, as well as adversely affecting energy supplies and impeding foreign trade. The existence of such problems was reflected in renewed emphasis on modernizing and extending the freight capacity of railway, road and marine transport. The expansion and renovation of port facilities also benefited from large-scale investment funding in order to alleviate problems in the handling of cargo at North Korea's ports. Even so, transport problems persisted during the 1990s, the development of rail services being singled out for special attention.

With a total route length of over 5,200 km., the railway system has remained North Korea's principal means of transport, handling about 90% of the country's freight, and 70% of its passenger traffic. Most of the rail system comprises only single track, and much of it would be widely regarded as obsolescent. Nevertheless, recent statistics indicate an overall electrification ratio of almost 70%, but higher for standard gauge railways, which account for some 80% of the total. This is important for a mountainous country like North Korea, since electrification enhances the traction capability of railways. Meanwhile, efforts continue to improve rail access to remote areas, and to areas near the Chinese border.

Following the 15 June Declaration of 2000, construction work began to reconnect the 486-km inter-Korean Kyongui railway line, which, prior to the division of the Korean peninsula, ran from Sinuiju in the North through Pyongyang to Seoul in the South. Associated with this initiative was a parallel project to build a highway that would run alongside the railway. To these ends, the South Korean Government allocated US $143m. to the two projects. Once opened, the railway was expected to facilitate not only inter-Korean travel, but also to enhance rail links with Russia, China and Japan. (During a visit to Pyongyang in March 2001 the Russian President, Vladimir Putin, spoke of his Government's interest in assisting in a project that would reconnect the Trans-Korean and Trans-Siberian railways, and subsequently the two countries signed an agreement on railway transport co-operation.) Such agreements and commitments notwithstanding, and despite the official opening of the Kyongui line in June 2003, the introduction of a full service has been delayed, owing mainly to the vagaries of North Korea's foreign relations.

Fuel constraints and the near-absence of private automobiles have relegated road transport to a secondary role. Depending on the source of information, estimates suggest that in 2002 the length of North Korea's road network was between 24,449 km and 31,200 km, although only a tiny proportion of such roads was properly surfaced and paved. There also exists a network of approaching 700 km of multi-lane highways, the construction of which began in the 1970s. This includes an expressway, connecting Pyongyang and Kaesong (a distance of 170 km), which was completed in April 1992. Other expressways link Pyongyang and Nampo (53 km), Pyongyang and Wonsan on the east coast (172 km), Pyongyang and Sunan (15 km), and Wonsan and Mount Kumgang (114 km). A 135-km highway, known as the Tourist Expressway, connecting Pyongyang and Hyangsan, via Anju, was also completed in October 1995. Local transport between villages is provided by rural bus services, while bus and tram services operate in towns and cities. In 1973 an extravagantly furbished underground railway system was completed in Pyongyang.

Water transport plays a minor but growing role in freight and passenger traffic. The total length of inland waterways is 2,253 km, most of which is navigable only by small craft. In November 1995 the creation of a marine transport system in the port of Rajin-Sonbong Free Economic and Trade Zone was announced, as part of the Tumen Delta development project. This initiative, jointly implemented by North Korea and China's Yanbian Sea Transport Corporation, may be seen as part of North Korea's economic reform efforts, designed to connect Rajin with Yanji City (China) and Busan (South Korea) and thereby save shipping time and transport costs in the interests of enhancing inter-Korean trade. No less important as a sign of the modest opening of the North Korean economy—and perhaps a portent for the future—has been the opening of container operations between Incheon (South Korea) and Nampo in the North. As of 2001 North Korea's merchant fleet comprised 176 ships (1,000 grt or over), including more than 90 cargo vessels, four bulk carriers and one oil tanker. Its shipping tonnage totalled 870,000 tons, and its harbour loading and unloading capacity was 35.5m. tons.

In 2003 there were 83 airports, of which only 35 possessed paved runways, and North Korea's international air connections remain largely undeveloped. Regular flights (once or twice a week) connect Pyongyang to Moscow and Khabarovsk (Russia), Beijing, Macao and Shenyang (China), Nagoya (Japan), Bangkok (Thailand), Berlin (Germany) and Sofia (Bulgaria). In addition, there are irregular flights between Pyongyang and Eastern European, Middle Eastern and African destinations. Internal flights are very restricted, serving mainly to connect Pyongyang with the port cities of Hamhung and Chongjin. Two significant developments were the signing in Bangkok (in October 1997) of an agreement with South Korea, allowing foreign commercial flights through North Korean airspace; and subsequently the opening of a direct air route between Seoul and Pyongyang, albeit one that has remained little used, in order to facilitate the North-South summit meeting of June 2000.In March 2001 it was also revealed that DHL Korea was offering delivery services for parcels, but not letters or cash, to selected destinations including Pyongyang, Rajin-Sonbong, Hamhung and Nampo. Only corporations authorized to do business with North Korea are, however, able to take advantage of such facilities.

Energy

North Korea's importance as a producer of fuel and energy emerged in the pre-partition period, when the North, through the Japanese-built Supung hydroelectric plant, supplied more than 90% of electricity in the peninsula. Following partition, several large hydroelectric plants, as well as smaller localized power plants (both thermal and hydroelectric) were constructed. From the 1970s, North Korea sought increasingly to use coal as an energy source, and by the beginning of the 21st century coal accounted for an estimated 86% of primary energy consumption—the rest coming from oil (8%), hydropower (4%) and other sources (2%). This major shift in energy policy was a response to the perceived disadvantages of hydroelectric plants, such as high initial costs, the long construction period and instability engendered by prolonged drought. However, against the advantage of being able to site coal-based power plants close to industrial and heavily populated areas are disadvantages associated with the poor quality and inaccessibility of coal deposits.

North Korea's recoverable reserves of coal are estimated to be 661m. short tons, most of which is anthracite, bituminous coal requirements being imported, mainly from China. The country's coal mines are located mainly in South Pyongan Province. Since the 1978–84 Plan investment has focused on the Anju District Coal Mining Complex, constructed in the 1970s with Soviet aid. In early 1991 a second mining facility, with a potential of 300,000 tons per year, was added to this complex, as part of a plan to increase the annual capacity to 650,000 tons. Following a steady increase in national coal production in the 1980s, output stagnated. In 2002, for example, total output of coal (37m. short tons) was still 3m. tons below the 1993 level. Contained in these figures is the finding that annual output decreased by an average of 7.6% during 1993–98, but in the next four years (1998–2002) there was growth of 8.2% annually. In general, the expansion of coal production has lagged behind the growing energy needs of the industrial sector, creating a persistent energy shortage and thereby constraining economic growth. In the 1990s declining imports, caused by shortages of foreign exchange, exacerbated the situation. Ageing mining equipment, inefficiencies associated with the need to mine deeper seams, the low level of mechanization and the lack of advanced equipment have contributed to the poor performance of the domestic coal industry. According to International Energy Agency (IEA) estimates, in 2000 North Korea's average per caput electricity consumption was 1,288 kWh; the corresponding figure per US dollar of GDP was 3.45 kWh. It is, however, a telling reminder of the continuing importance of coal that the 2003 government budget designated a 30% increase in expenditure on the coal industry, representing the single biggest rise for any economic sector.

Problems in the coal industry have impinged seriously on the provision of electricity. In the 1980s widespread construction of small and medium-scale hydroelectric plants got under way, especially in the mountainous provinces of Hamgyong, Jagang and Ryanggang. Some 1,300 of such plants were reported to be operational by the end of 1994. Such construction has continued, a North Korean report in 2000 revealing that some 250 small and medium-scale power plants, with a generating capacity of 50,000 kWh, had been constructed in the previous few years, reportedly facilitating the attainment of electricity self-sufficiency in Jagang Province. In 2000 North Korea's electricity production was estimated by one source to be 33,400m. kWh.

In 2001 hydroelectric power plants generated about 69% of North Korea's electricity, with the remaining 31% met by thermal plants. Thermal generating capacity is, however, underutilized because of fuel shortages, and in 2001 total electricity consumption was only 58% of the 1991 level. Indeed, between 1992 and 2001 in only one year did electricity consumption register an increase—by a mere 1% in 1999. The outcome has been a persistent problem of black-outs lasting for extended periods of time. Meanwhile, North Korea's antiquated transmission grid has also been the source of serious losses of power. In 2001 per caput electricity consumption was 760 kWh, equivalent to 33% of the global average but 38% above the average for all Asia. The possibility of South Korean facilities being used to help resolve North Korea's electricity shortages, perhaps by linking the two countries' electricity grids, was discussed at the June 2000 Summit meeting. By mid-2004, however, this had not materialized.

Oil accounts for less than 10% of North Korea's total primary energy consumption. Petroleum exploration in North Korea began as long ago as 1965 and, despite high drilling costs, continues to the present day. The most likely locations of crude petroleum deposits are the West Sea Bay and Anju Basin. In recent times, Swedish, British and Australian oil companies have owned petroleum exploration concessions off the west and east coasts of North Korea, but following disappointing results these have lapsed. Joint explorations with South Korea's Korean National Oil Company (KNOC) may also be expected to expand in the foreseeable future.

So far, however, exploration has generated zero output, necessitating reliance on imports. In 1990 North Korea and the USSR concluded an agreement, at the latter's insistence, under which, from 1 January 1991, bilateral trade was to be conducted in convertible currencies and payments for Soviet oil were to be made at world market prices. As a result of the agreement and in the face of deteriorating relations between the two countries, between 1990 and 1992 North Korean imports of crude petroleum from Russia declined from 500,000 metric tons to 30,000 tons. The outcome was to make China the single most important source of oil imports, although when, in late 1992, Beijing imposed similar trading regulations, North Korea was forced to turn to the Middle East for crude petroleum supplies. Such temporizing manoeuvres were not, however, wholly successful and, in March 1996, worsening energy shortages prompted Kim Jong Il to instruct all factories and enterprises to reduce their use of coal and other energy sources, emphasizing that the deficit in raw materials and energy supply had become the most urgent problem facing the country.

Crude petroleum imports in 2001 and 2002 were 580,000 metric tons and 597,000 tons respectively—most of it coming from China (in 2002 and 2003 China supplied North Korea with about 70% of its oil consumption requirements). Such heavy dependence on a single source carries obvious risks, as the temporary suspension of oil supplies from China in March 2003—allegedly as a political warning—highlighted. A significant recent development was the resumption, in 2003, of Russian oil supplies. Not least, such imports helped fill the gap caused by the suspension of shipments from a US government-led consortium—the Korean Peninsula Energy Development Organization (KEDO).

A Soviet-designed nuclear research reactor, with a generating capacity of about 3 MW, was reportedly installed in Pyongyang as early as 1959, although a North Korean-Soviet agreement signed in 1985, whereby the USSR would construct a 1,760-MW nuclear power plant in North Korea, is thought never to have been fulfilled (for more details of North Korea's nuclear installations, see History). In October 1994 North Korea and the USA signed a 'framework agreement' whereby Pyongyang agreed to suspend the development of nuclear weapons in exchange for the construction of two light-water reactors (LWRs) worth US $4,500m., in addition to receiving as much as 500,000 metric tons of oil annually for the duration of the construction period (forecast at 10 years). The construction project is managed by KEDO, and South Korea's state-run Korea Electric Power Corporation is its prime contractor. Initial construction work commenced in August 1997, with some 5,000 South Korean engineers expected to work on the project. Progress towards its completion has, however, all too predictably been bedevilled by political disagreements, causing the completion date to be postponed from 2003 to 2008.

Some have questioned whether the 1994 agreement was really sufficient to meet North Korea's economic needs, rather than simply accommodating international concerns about its nuclear weapons programme. It has, for example, been argued that a better approach would have been to upgrade the existing electrical grid and to construct new, efficient electricity-generating systems that ultimately might even facilitate exports to the South. In any case, the 1994 agreement was placed in serious doubt following Pyongyang's admission, in October 2002, that it had maintained a secret nuclear programme. Tensions associated with North Korea's nuclear programme were intensified when, in April 2003, Pyongyang formally withdrew from the Nuclear Non-Proliferation Treaty (NPT). This was the background against which, starting in December 2003, the reactor project was suspended for one year.

Such problems highlight the peculiar obstacles that impede progress towards enhancing energy supplies in North Korea. As a result, even allowing for the success of ongoing efforts—continuing domestic construction of small- and medium-scale facilities, increased investment in the power industry (for example, investment under the 2003 budget was intended to rise by 14.4%) and higher imports—improvements in the provision of energy are likely to be slow. Large-scale infrastructural construction will also be necessary. (North Korea possesses a mere 37 km and 180 km of pipelines to carry, respectively, crude petroleum and its products.) In short, it is safe to suppose that energy constraints will continue to impede North Korean economic growth into the foreseeable future.

Telecommunications

Telecommunications remain seriously underdeveloped in North Korea. In 1997 only 1.1m. main telephone lines were estimated to be in use. The development of mobile phones has yet to get under way on a significant scale (a very limited service began in late 2002), although in March 2003 an unconfirmed Russian report quoted a senior North Korean official to the effect that a nation-wide mobile telephone service would be put in place by 2006. Meanwhile, international telephone connections are mainly made through Beijing and Moscow, although since the second half of the 1990s a direct telephone line has been established with South Korea. In March 2001 a Chinese source cited another North Korean official to the effect that North Korea possessed a national e-mail network, known as 'Kuang Myong'. The report revealed, however, that the network only allowed users to log on from within North Korea and that its content was limited to providing science and technology information. It was subsequently reported that the restricted-use SiliBank international e-mail service had begun to relay e-mails 24 hours a day rather than once every hour. It was announced in April 2003 that North Korea was about to launch hand-held computers.

Recent estimates suggest that there were about 3.36m. radios and 1.2m. television sets available to the North Korean population in 1997—an average of 15.5 and 5 per 100 persons, respectively. Only about 10% of radios and 30% of television sets are in private households, although most urban households have access to both. Domestic media censorship is strictly enforced, and neither radios nor TV sets can be tuned to anything other than official programming. Listening to foreign media broadcasts, except by senior party cadres, is forbidden; nor are any foreign media permitted free access to North Korea.

FINANCE

In the absence of a private sector, public (or government) finance in a planned economy plays the principal role in seeking to mobilize domestic savings for investment purposes and ensure financial equilibrium in the economy. There exist in North Korea a central government budget and local government budget—the two being consolidated into the national (state) budget. The central government budget is largely financed from the central Government's net income, transfers from local governments and other revenue sources. Revenue for the local government budget derives from local industrial and other sources, as well as from central government subsidies. Traditionally, the principal sources of revenue under the state budget have been turnover (sales) taxes and state enterprise profits.

In March 2004 the North Korean Minister of Finance, Mun Il Bong, told the Second Session of the 11th SPA that the planned level of state budgetary revenue in 2003 had been overfulfilled by 0.9%, while the expenditure target had been underfulfilled by 1.8%. He revealed that 15.7% of total spending had been allocated to defence, 23.3% to economic construction, and 40.5% to 'various popular policies for the promotion of the people's welfare'. In 2004 state revenue was expected to increase by 5.7% (compared to 13.6% planned for 2003), and state expenditure by 8.6% (compared with 14.4% in 2003).The principal beneficiaries of increased spending in 2004 would be power, coal and metal industries, as well as railways; but forestry, mining, machine-building, chemical and building materials industries would also receive increased allocations. Funds for scientific and technological development would rise by 60%, and disbursements for education and public health by 9.5% and 5.9%.

Interpreting the 2004 budgetary estimates is extremely difficult, and not all the figures can be taken at face value. In particular, there is a strong presumption that the projected rise in defence spending is understated, some sources suggesting that North Korea allocates US $1,400m.—around 25% of its GDP—to the military (the highest in the world). Peaceful reunification—or even simply political accommodation between North and South Korea—would permit a reduction in military expenditure and thereby allow a significant reallocation of resources to more productive economic ends. Apart from fiscal benefits, demobilization would also, by making more labour available, boost the expansion of labour-intensive light industrial manufactures, perhaps along the lines of China's township and village enterprises.

Like other socialist countries, North Korea has traditionally pursued a conservative budgetary policy, through the maintenance of a balanced budget or generation of a budget surplus. Data released in April 2001 suggested that a small surplus (of some 61m. won) was in fact achieved in 2000. One interpretation of budgetary plans for 2004 pointed to the generation of a small budget surplus, although difficulties in assessing and analysing the data demand a large margin of error in making such an inference.

Another way of improving North Korea's financial viability is to seek integration in international financial institutions. In 2000 North Korea submitted an application to join the Asian Development Bank (ADB). Hostility to such membership from the USA is one factor that has impeded efforts to improve North Korea's international financial profile.

FOREIGN TRADE AND FOREIGN INVESTMENT

A corollary of North Korea's emphasis on self-reliance has, at least until recently, been a downgrading of the role of foreign trade. As in other communist countries, the Government has pursued an essentially conservative and passive trade policy that emphasizes import-substitution in support of its programme of heavy industrialization. In other words, using its foreign relations merely to provide the goods, trade has been primarily orientated towards importing the machinery, equipment and raw materials necessary to fulfil the output targets of successive economic plans, and to relieve unplanned shortages. Exports have been used mainly to pay for such imports, as well as to dispose of unplanned surpluses. This passive strategy has been reflected in a level of foreign trade that is much smaller than might have been expected of a country of North Korea's size and structure. The trade share, having peaked at 29.4% of GNP in 1975, thereafter fell sharply to 20% in 1985, and to just 14.4% in 2001. By way of comparison, at the end of the 1990s the corresponding figure for South Korea was 50%–55%.

Even allowing for an expansion in trade with capitalist countries, these figures also reflect the declining role of foreign trade resulting from the onset of economic difficulties in the 1990s. Thus, having grown on average by 9.4% annually during 1960–90 to reach US $4,800m. in 1990, the combined value of exports and imports subsequently declined sharply. According to the Korea Trade-Investment Promotion Agency (KOTRA) in Seoul, between 1990 and 1998 the value of North Korea's merchandise trade decreased from US $4,170m. to $559m.—a cumulative fall of 86.6% (or of 22.2% annually). Contained within these figures are annual contractions of 38%, 21% and 34% in 1991, 1994 and 1998 respectively. A degree of recovery subsequently took place, as a result of which the value of merchandise trade reached $2,391m. in 2003,66% above the low point of 1998 but still only 54.2% of the 1990 level. Imports decreased by an annual average of 11.9% in 1990–98, and increased by 12.8% in 1998–2003. The lowest point for exports was in 1999, when their value had declined to a mere 29.7% of the corresponding 1990 level (implying average annual decline of 12.6%); since 1999 exports have consistently risen year-on-year (averaging 10.8% annually) to reach US $777m. in 2003. Implied in these figures has been a consistent merchandise trade deficit, which peaked at $970m. in 2001.

In support of its economic modernization programme, in the early 1970s North Korea sought to use massive imports of advanced machinery and equipment from Western Europe and Japan. During the first half of the 1970s its trade balance steadily worsened and, unable to repay loans incurred as a result of such imports, in January 1976 it attracted world-wide notoriety when it became the first communist country to default on its foreign debt. Declining prices for North Korea's principal exports, rising import costs associated with the 1973 oil price rise, global recession and domestic transport difficulties—all contributed to the growing trade deficit and debt problem. In any case, the default resulted in a severe curtailment of Western credits to North Korea. Thereafter, its debt position continued to worsen, rising to US $5,200m. in 1988, and $11,900m. in 1997. After the decision, in early 1999, by the Dutch investment bank, ING Barings, to cease operations in North Korea, the Government was forced to conduct international banking transactions through a limited number of banks in Japan and Macao. The current level of North Korea's external debt is not known.

Ideological and economic imperatives ensured that until the 1990s North Korea's principal trading partners were other communist countries. Between 1955 and 1988, for example, trade with such countries rose, on average, by 11% per annum to reach US $3,290m. (this compared with a trade turnover of $600m. in 1995). Most important of all was the former Soviet Union, which in 1990 still accounted for 57% of North Korea's total foreign trade. The subsequent collapse of the USSR was a major watershed and in a single year (1991) the former USSR's share fell to a mere 17.3%. Nor has this decline been halted: by 1995 Russia was the source of 5% of North Korea's imports, and the destination for a mere 1% of its exports. Some recovery has since taken place, and in 2003 trade with Russia accounted for 4.9% of the total. However, this overall expansion conceals a 23.3% decline in North Korean exports to Russia, alongside a 50% rise in imports (especially of oil), the further implication being a North Korean deficit of $112.8m.

Well before the collapse of DPRK-USSR trade, another watershed occurred in the 1970s, when North Korea began to purchase much-needed capital goods from capitalist countries (especially Japan). Between 1965 and 1975 bilateral trade with Japan rose by 23% annually (North Korean imports increasing by 27% annually), and during the next decade by a further 6% per year. Indeed, for some years, Japan was North Korea's second largest trading partner (but see below). Data for 2003 showing it still to be the second most important destination for North Korean exports are misleading to the extent that they conceal the major contraction that has occurred in bilateral trade in recent years. In 2003, for example, exports to Japan decreased by 25.8% (to US $173.8m.), while imports declined by 32.3% ($91.5m.). KOTRA estimates show that purchases of Japanese textiles, machinery and non-ferrous metals all shared in the 2003 contraction.

For the 20 years after 1965 bilateral trade between North Korea and China rose by over 5% per annum, until it reached US $2,100m., accounting for almost 15% of North Korea's foreign trade. Thereafter, China's importance as a trading partner declined: by 1990 the value of bilateral trade had fallen to $545m. (about 10%), and by 1999 it was just $370m. Such, however, was the impact of the dramatic contraction in trade with the former USSR that China swiftly emerged as North Korea's largest trading partner, contributing about 30% of its total trade in 1999. China has since remained North Korea's principal trade partner. In 2003 goods worth US $395.3m. (46% more than in 2002) were shipped to China, while North Korean imports rose by 34% to reach $627.6m. This was the first year in which bilateral trade exceeded $1,000m. The outcome was a bilateral deficit with China of $232.3m. Other important trading partners, the combined share of which in North Korea's foreign trade has risen sharply in recent years, are Thailand and India. In 2003 these two countries accounted for 17.3% of all North Korean merchandise trade. Trade with Thailand rose by an astonishing 62% in 2002, and again by 17% in 2003.

KOTRA deliberately excludes South Korea from its ranking of North Korea's trading partners on the grounds that inter-Korean trade is internal. To this extent, the foregoing analysis is misleading, since South Korea has in fact become the North's second largest partner. In November 1988, for the first time since the Korean War, limited indirect trade with South Korea was resumed, and just two years later direct inter-Korean trade got under way, its value rising to US $210m. by 1992. By the end of the decade—and notwithstanding the temporary negative impact of the Asian financial crisis—inter-Korean trade had reached a record level of $333.4m. (including some $200m. in tour fees associated with the Mount Kumgang project), making South Korea the North's third largest trading partner. In 2000 bilateral trade increased further, to $425m., pushing South Korea into second place after China, a place it has since retained. From a base of $425.2m. in 2000, bilateral trade fell to $403m. the following year, but thereafter rose sharply—to $641.8m. in 2002 and $724.3m. in 2003. In 2003 the value of South Korean exports was $435.0m., compared with $289.3m. for imports.The result was a South Korea trade surplus of $145.7m. In interpreting these figures, however, it is important to bear in mind that around two-thirds of exports from South Korea are more in the nature of aid than of true trade.

The fall of communist regimes in Eastern Europe and the dissolution of the USSR, as well as the insistence by both Russia and China on abandoning barter trading systems in favour of charging international prices for oil and other products, along with the rapid expansion of inter-Korean trade, all suggest that a significant realignment of North Korea's economic relations is under way. At the same time, inefficiencies and infrastructural constraints (for example, low product quality and lack of variety, poor packaging, failure to meet delivery dates, and limited transport and harbour facilities) have continued to impede North Korean efforts to expand exports and increase much-needed foreign-exchange earnings.

Major increases in wages and prices took place in the second half of 2002 (see above). These adjustments have had major implications for exchange rates. Thus, in February 2003 a dramatic devaluation was confirmed, as a result of which the won's exchange rate in relation to the US dollar is estimated by the BOK to have increased to 2.2 won: 1 US $ in 2002 and to 145 won: 1 US $ in 2003.

North Korea's principal exports include non-ferrous metals (mostly zinc, lead, barytes and gold), iron and steel, textile yarn and fabrics, military equipment, cement, vegetables and fishery products. The main imports are advanced machinery, transport equipment, high-grade iron and steel products, crude petroleum, coking coal, grain, chemicals and some consumer goods.

Throughout its existence, North Korea has almost consistently suffered an unfavourable trade balance (see above for estimates of its merchandise trade deficit). Suffice it to say that this deficit has been reflected in its current-account balance. In recent years North Korea's current-account deficit and foreign debt difficulties have been alleviated by inflows of convertible currency from pro-Pyongyang Koreans living in Japan. Such contributions and remittances reportedly ranged from US $600m. to $1,000m. per year, the latter figure approximating to total annual North Korean exports in the early 1990s.

The introduction, in 1984, of a joint-venture law (revised in 1994) signalled North Korea's willingness to use radical means in order to acquire Western capital and technology. Initially, such efforts achieved only limited success. Indeed, the Ministry of Joint-Venture Industry, established in 1988, was closed down in 1990 and its responsibilities transferred to the General Bureau of Joint Venture, under the External Economic Affairs Commission. Of an estimated 140 joint ventures under way by 1994, about three-quarters involved companies owned by pro-Pyongyang Korean residents of Japan. Most of them were small, with an average capitalization of less than US $1m.

It was expected that government plans to establish special economic zones would stimulate further joint-venture activities, as well as encourage involvement by South Korean companies. Attempts to attract foreign capital into the Rajin-Sonbong Free Economic and Trade Zone, which was established in the Tumen Delta in 1991, were initially largely unsuccessful. By the end of 1997, for example, it was reported that although 111 contracts (worth some US $751m.) had been agreed, only 77 of them were under way, with associated utilized investment of $57m. The Tumen river project was sponsored by UNDP, and the first specifically inter-Korean joint-venture agreement—with the South Korean Daewoo conglomerate—was concluded in January 1992, when the North Korean Government began to enact new laws designed to offer even more preferential treatment to foreign investors than that available under Chinese laws. In practice, foreign-exchange shortages, infrastructural deficiencies, unco-operative bureaucrats and low levels of skilled labour have prevented North Korea from realizing any significant benefits from the Rajin-Sonbong project, leading some to judge it to be a failure. Its future, according to some outside observers, may lie more in the direction of acting as a regional trans-shipment base for North Korean exports to China.

The construction, in the late 1980s and early 1990s, of an international telecommunications centre, a new international airport and international hotels is a clear expression of North Korea's willingness to open the country to foreign tourism in order to improve its invisible trade balance. The most significant initiative in this context was the commencement, in November 1998, of a joint venture involving the South Korean conglomerate, Hyundai, to send daily tour boats to Mount Kumgang, representing the first ever opportunity for South Korean tourists to visit the North. By May 2000 North Korea had earned some US $180m. from the project, and in July of that year the Pyongyang authorities put forward the suggestion that Hyundai should develop a high-technology zone in Kaesong city (just north of the demilitarized zone). As of mid-2004, neither of these projects had yielded any significant benefit to either side. Indeed, work to develop the site at Kaesong had yet to begin, and Hyundai had meanwhile suffered major losses on the Mount Kumgang resort project.

Such difficulties notwithstanding, it seems likely that North Korea will maintain its efforts to encourage greater foreign involvement in its economic development. The most recent high-profile joint venture, the Pyongso J-V Company Ltd, was opened in June 2004, with financial backing from Peter Zuellig, the Chief Executive of Interpacific of Switzerland. Meanwhile, in 2001 a South Korean source reported that China and North Korea had reached agreement on the creation of a 'second Shenzhen' through the establishment of an economic zone based on Sinuiju and Dandong, which would act as a distribution centre for light industrial goods. (In 1999 Sinuiju had already been proposed to Hyundai as a special economic zone and industrial complex.) In late September 2002 Yang Bin, a wealthy Chinese-born businessman with Dutch citizenship and Chairman of the Euro-Asia Group, was appointed Chief Executive of the Sinuiju Special Administrative Region, seemingly confirming Pyongyang's commitment to Chinese-style economic reforms (although Yang Bin was subsequently imprisoned by the Chinese authorities). In addition, since 2000 Pyongyang has made extensive efforts to establish diplomatic relations with wealthier nations such as the United Kingdom, Italy, Australia, the Philippines and other countries—initiatives clearly not unconnected with its desire to accelerate economic growth and development. In July 2000 North Korea also became a member of the ASEAN Regional Forum (ARF), established by the Association of South East Asian Nations.

Statistical Survey

Area and Population

AREA, POPULATION AND DENSITY*

Area (sq km)	122,762†
Population (census results)	
31 December 1993	
Males	10,329,699
Females	10,883,679
Total	21,213,378
Population (unofficial estimates)‡	
2001	22,253,000
2002	22,369,000
2003	22,522,000
Density (per sq km) in 2003	183.5

* Excluding the demilitarized zone between North and South Korea, with an area of 1,262 sq km (487 sq miles).

† 47,399 sq miles.

‡ Source: Bank of Korea (Republic of Korea).

PRINCIPAL TOWNS

(population at 1993 census)

Pyongyang (capital)	2,741,260	Wonsan	300,148
Nampo	731,448	Pyongsong	272,934
Hamhung	709,730	Sariwon	254,146
Chongjin	582,480	Haeju	229,172
Kaesong	334,433	Kanggye	223,410
Sinuiju	326,011	Hyesan	178,020

Source: UN, *Demographic Yearbook*.

BIRTHS AND DEATHS

(UN estimates, annual averages)

	1985–90	1990–95	1995–2000
Birth rate (per 1,000)	20.6	20.8	18.6
Death rate (per 1,000)	5.8	7.0	10.4

Expectation of life (years at birth): 65.8 (males 64.4; females 67.1) in 2002 (Source: WHO, *World Health Report*).

ECONOMICALLY ACTIVE POPULATION

(ILO estimates, '000 persons at mid-1990)

	Males	Females	Total
Agriculture, etc.	2,027	1,877	3,904
Industry	2,206	1,043	3,249
Services	1,577	1,549	3,126
Total labour force	5,810	4,469	10,279

Source: ILO, *Economically Active Population Estimates and Projections, 1950–2010*.

Mid-2002 (estimates in '000): Agriculture, etc. 3,323; Total labour force 11,595 (Source: FAO).

Health and Welfare

KEY INDICATORS

Total fertility rate (children per woman, 2002)	2.0
Under-5 mortality rate (per 1,000 live births, 2002)	55
HIV/AIDS (% of persons aged 15–49, 1994)	<0.01
Health expenditure (2001): US $ per head (PPP)	44
Health expenditure (2001): % of GDP	2.5
Health expenditure (2001): public (% of total)	73.4
Access to water (% of persons, 2000)	100
Access to sanitation (% of persons, 2000)	99

For sources and definitions, see explanatory note on p. vi.

Agriculture

PRINCIPAL CROPS
('000 metric tons)

	2000	2001	2002
Wheat	50*	124	130
Rice (paddy)	1,690	2,060	2,186
Barley	29*	70	69
Maize	1,041	1,483	1,651
Rye	60	68	60
Oats†	20	11	15
Millet	45†	50*	65*
Sorghum	10†	14*	20*
Potatoes	1,870	2,268	1,884
Sweet potatoes	290*	320†	340†
Pulses†	290	300	300
Soybeans (Soya beans)*	350	350	360
Cottonseed†	23	23	24
Cabbages†	630	650	680
Tomatoes†	62	70	70
Pumpkins, squash and gourds†	85	88	88
Cucumbers and gherkins†	64	65	65
Aubergines (Eggplants)†	43	45	45
Chillies and green peppers†	55	57	57
Green onions and shallots†	90	95	95
Dry onions†	82	84	84
Garlic†	80	85	85
Other vegetables†	2,406	2,406	2,431.2
Apples†	650	660	660
Pears†	130	130	134.5
Peaches and nectarines†	110	115	122
Watermelons†	104	105	108
Cantaloupes and other melons†	110	112	115
Other fruits and berries†	460	480	480
Tobacco (leaves)†	63	63	64
Hemp fibre†	13	13	13
Cotton (lint)†	12	13	12

* Unofficial figure(s).
† FAO estimate(s).

Source: FAO.

LIVESTOCK
('000 head)

	2000	2001	2002
Horses*	46	47	48
Cattle	579	570	575
Pigs	3,120	3,137	3,152
Sheep	185	189	170
Goats	2,276	2,566	2,693
Chickens	15,733	16,894	18,506
Ducks	2,078	3,158	4,189
Rabbits	11,475	19,455	19,482

* FAO estimates.

Source: FAO.

LIVESTOCK PRODUCTS
(FAO estimates, '000 metric tons)

	2000	2001	2002
Beef and veal	20.0	21.5	21.8
Goat meat	10.2	10.5	11.0
Pig meat	140.0	145.0	145.7
Poultry meat	26.8	30.8	33.7
Cows' milk	90.0	92.0	92.0
Poultry eggs	110.0	120.0	130.0
Cattle hides (fresh)	2.8	3.0	3.0

Source: FAO.

Forestry

ROUNDWOOD REMOVALS
(FAO estimates, '000 cubic metres, excl. bark)

	2000	2001	2002
Sawlogs, veneer logs and logs for sleepers	1,000	1,000	1,000
Other industrial wood	500	500	500
Fuel wood	5,503	5,561	5,620
Total	7,003	7,061	7,120

Sawnwood production ('000 cubic metres, incl. railway sleepers): 280 (coniferous 185, broadleaved 95) per year in 1970–2002 (FAO estimates).

Source: FAO.

Fishing

(FAO estimates, '000 metric tons, live weight)

	2000	2001	2002
Capture	212.9	206.5	205.0
Freshwater fishes	8.0	4.9	5.0
Alaska pollock	60.0	60.0	60.0
Other marine fishes	112.2	109.0	107.6
Marine crustaceans	15.6	16.2	16.0
Squids	9.5	9.5	9.5
Aquaculture	66.7	63.7	63.7
Molluscs	63.0	60.0	60.0
Total catch	279.6	270.2	268.7

Note: Figures exclude aquatic plants (FAO estimates, '000 metric tons, aquaculture only): 401.0 in 2000; 391.0 in 2001; 444.3 in 2002.
Source: FAO.

Mining

(estimates, '000 metric tons, unless otherwise indicated)

	2000	2001	2002
Hard coal	22,500	23,000	24,000
Brown coal and lignite	6,500	7,000	7,000
Iron ore: gross weight	3,800	4,200	4,100
Iron ore: metal content	1,100	1,200	1,150
Copper ore*	13	13	13
Lead ore*	60	60	60
Zinc ore*	100	100	100
Tungsten concentrates (metric tons)*	500	500	600
Silver (metric tons)*	40	40	40
Gold (kg)*	2,000	2,000	2,000
Magnesite (crude)	1,000	1,000	1,000
Phosphate rock†	350	350	300
Fluorspar‡	25	25	25
Barite (Barytes)	70	70	70
Salt (unrefined)	500	500	500
Graphite (natural)	30	25	25
Talc, soapstone and pyrophyllite	120	120	110

Note: No recent data are available for the production of molybdenum ore and asbestos.

* Figures refer to the metal content of ores and concentrates.

† Figures refer to gross weight. The phosphoric acid content (estimates, '000 metric tons) was: 105 in 2000; n.a. in 2001; n.a. in 2002.

‡ Metallurgical grade.

Source: US Geological Survey.

Industry

SELECTED PRODUCTS

('000 metric tons, unless otherwise indicated)

	1999	2000	2001
Nitrogenous fertilizers*	72	72	n.a.
Motor spirit (petrol)	848†	902	n.a.
Kerosene	168†	185	n.a.
Gas-diesel (distillate fuel) oils	925†	994	n.a.
Residual fuel oils	539†	573	n.a.
Coke-oven coke (excl. breeze)	3,098†	3,098†	n.a.
Cement‡	6,000	6,000	5,160
Pig-iron†‡	250	250	250
Crude steel†‡	1,000	1,000	1,000
Refined copper (unwrought)†‡	16	14	14
Lead (primary metal)†‡	75	75	75
Zinc (primary metal)†‡	180	200	180

* Output is measured in terms of nitrogen.

† Provisional or estimated figure(s).

‡ Data from the US Geological Survey.

Electric energy (million kWh): 31,450 in 1999; 32,815 in 2000.

Source: UN, *Industrial Commodity Statistics Yearbook*.

Finance

CURRENCY AND EXCHANGE RATES

Monetary Units

100 chon (jun) = 1 won.

Sterling, Dollar and Euro Equivalents (30 April 2004)

£1 sterling = 241.864 won;
US $1 = 136.400 won;
€1 = 162.957 won;
1,000 won = £4.13 = $7.33 = €6.14.

Note: In August 2002 it was reported that a currency reform had been introduced, whereby the exchange rate was adjusted from US $1 = 2.15 won to $1 = 150 won: a devaluation of 98.6%.

BUDGET

(projected, million won)

	1992	1993	1994
Revenue	39,500.9	40,449.9	41,525.2
Expenditure	39,500.9	40,449.9	41,525.2
Economic development	26,675.1	27,423.8	28,164.0
Socio-cultural sector	7,730.6	7,751.5	8,218.3
Defence	4,582.1	4,692.2	4,816.9
Administration and management	513.1	582.4	326.0

1998 (estimates, million won): Total revenue 19,790.8; Total expenditure 20,015.2.

1999 (estimates, million won): Total revenue 19,801.0; Total expenditure 20,018.2.

2000 (estimates, million won): Total revenue 20,955.0; Total expenditure 20,903.0.

2001 (projected, million won): Total revenue 21,571.0; Total expenditure 21,571.0.

2002 (projected, million won): Total revenue 22,174.0; Total expenditure 22,174.0.

NATIONAL ACCOUNTS

Gross Domestic Product by Economic Activity

(unofficial estimates, '000 million won)*

	2001	2002	2003
Agriculture, forestry and fishing	6,139	6,429	5,961
Mining	1,617	1,652	1,820
Manufacturing	3,651	3,825	4,043
Electricity, gas and water	968	939	995
Construction	1,410	1,699	1,896
Government services	4,483	4,678	5,017
Other services	1,953	2,054	2,156
Total	20,223	21,277	21,887

* Totals may not be equal to sum of component parts, owing to rounding.

Source: Bank of Korea (Republic of Korea).

External Trade

PRINCIPAL COMMODITIES

(US $ million)*

Imports	2000	2001	2002
Live animals and animal products	20.3	73.9	103.4
Vegetable products	159.0	221.0	118.4
Animal or vegetable fats and oils; prepared edible fats; animal or vegetable waxes; Prepared foodstuffs; beverages, spirits and vinegar; tobacco and manufactured substitutes	89.1	89.9	72.3
Mineral products	171.2	231.1	235.9
Products of chemical or allied industries	108.4	123.4	122.1
Plastics, rubber and articles thereof	67.5	66.0	66.0
Textiles and textile articles	171.9	203.9	158.5
Base metals and articles thereof	85.2	100.4	88.2
Machinery and mechanical appliances; electrical equipment; sound and television apparatus	205.1	243.8	234.7
Vehicles, aircraft, vessels and associated transport equipment	146.2	88.4	76.1
Total (incl. others)	1,406.5	1,620.3	1,525.4

Exports	2000	2001	2002
Live animals and animal products	97.9	158.4	261.1
Vegetable products	30.3	42.0	27.5
Mineral products	43.2	50.5	69.8
Products of chemical or allied industries; Plastics, rubber and articles thereof	44.9	44.6	42.4
Wood, cork and articles thereof; wood charcoal; manufactures of straw, esparto, etc.	10.9	5.6	10.2
Textiles and textile articles	140.0	140.5	123.1
Natural or cultured pearls, precious or semi-precious stones, precious metals and articles thereof; imitation jewellery; coin	9.8	14.1	14.6
Base metals and articles thereof	43.9	60.2	57.4
Machinery and mechanical appliances; electrical equipment, sound and television apparatus	105.2	97.9	85.6
Total (incl. others)	565.8	650.2	735.0

* Excluding trade with the Republic of Korea (US $ million): Imports 272.8 in 2000, 226.8 in 2001, 370.2 in 2002; Exports 152.4 in 2000, 176.2 in 2001, 271.6 in 2002.

Source: Korea Trade-Investment Promotion Agency (KOTRA), Republic of Korea.

PRINCIPAL TRADING PARTNERS
(US $ million)*

Imports	2001	2002	2003
China, People's Republic	570.7	467.3	627.6
Germany	82.1	140.4	n.a.
Hong Kong	42.6	29.2	n.a.
India	154.8	186.6	157.9
Japan	249.1	135.2	91.5
Netherlands	9.1	27.6	n.a.
Russia	63.8	77.0	115.6
Singapore	112.3	83.0	n.a.
Spain	31.6	n.a.	n.a.
Thailand	106.0	172.0	203.6
United Kingdom	40.7	n.a.	n.a.
Total (incl. others)	1,620.3	1,524.1	1,614.4

Exports	2001	2002	2003
Bangladesh	38.0	32.3	n.a.
China, People's Republic	166.8	270.7	395.3
Germany	22.8	27.8	n.a.
Hong Kong	38.0	21.9	n.a.
India	3.1	4.8	1.6
Japan	225.6	234.4	173.8
Netherlands	10.4	6.4	n.a.
Russia	4.5	3.6	2.8
Spain	12.6	n.a.	n.a.
Thailand	24.9	44.6	50.7
Total (incl. others)	650.2	736.2	777.0

* Excluding trade with the Republic of Korea (US $ million): *Imports:* 226.8 in 2001; 370.2 in 2002; 435.0 in 2003. *Exports:* 176.2 in 2001; 271.6 in 2002; 289.3 in 2003.

Source: Korea Trade-Investment Promotion Agency (KOTRA), Republic of Korea.

Transport

SHIPPING

Merchant Fleet
(registered at 31 December)

	2001	2002	2003
Number of vessels	176	225	292
Total displacement ('000 grt)	697.8	870.5	959.0

Source: Lloyd's Register-Fairplay, *World Fleet Statistics*.

International Sea-borne Freight Traffic
(estimates, '000 metric tons)

	1988	1989	1990
Goods loaded	630	640	635
Goods unloaded	5,386	5,500	5,520

Source: UN, *Monthly Bulletin of Statistics*.

CIVIL AVIATION
(traffic on scheduled services)

	1995	1996	1997
Kilometres flown (million)	3	3	5
Passengers carried ('000)	254	254	280
Passenger-km (million)	207	207	286
Total ton-km (million)	22	22	30

Source: UN, *Statistical Yearbook*.

Tourism

	1996	1997	1998
Tourist arrivals ('000)	127	128	130

Source: World Tourism Organization, mainly *Yearbook of Tourism Statistics*.

Communications Media

	1994	1995	1996
Radio receivers ('000 in use)	2,950	3,000	3,300
Television receivers ('000 in use)	1,000	1,050	1,090
Telephones ('000 main lines in use)*	1,100	1,100	1,100
Telefax stations (number in use)	3,000*	n.a.	n.a.
Daily newspapers:			
number	11	11*	3
average circulation ('000 copies)*	5,000	5,000	4,500

* Estimate(s).

1997 ('000 in use): Radio receivers 3,360; Television receivers 1,200.

Sources: UNESCO, *Statistical Yearbook*; UN, *Statistical Yearbook*.

Education

(2000)

	Institutions	Students
Kindergartens	14,167	748,416
Primary	4,886	1,609,865
Senior middle schools	4,772	2,181,524

Source: mainly Government of the Democratic People's Republic of Korea, *UNESCO Education for All Assessment Report 2000*.

Universities and Colleges: The *UNESCO Education for All Assessment Report 2000* identified more than 300 universities and colleges with 1.89m. students and academics.

Teachers (1987/88, UNESCO, *Statistical Yearbook*): Pre-primary 35,000, Primary 59,000, Secondary 111,000, Universities and colleges 23,000, Other tertiary 4,000.

Directory

The Constitution

A new Constitution was adopted on 27 December 1972. According to South Korean sources, several amendments were made in April 1992, including the deletion of references to Marxism-Leninism, the extension of the term of the Supreme People's Assembly from four to five years, and the promotion of limited 'economic openness'. Extensive amendments to the Constitution were approved on 5 September 1998. The main provisions of the revised Constitution are summarized below:

The Democratic People's Republic of Korea is an independent socialist state; the revolutionary traditions of the State are stressed (its ideological basis being the *juche* idea of the Korean Workers' Party), as is the desire to achieve national reunification by peaceful means on the basis of national independence. The Late President Kim Il Sung is the Eternal President of the Republic.

National sovereignty rests with the working people, who exercise power through the Supreme People's Assembly and Local People's Assemblies at lower levels, which are elected by universal, equal and direct suffrage by secret ballot.

The foundation of an independent national economy, based on socialist and *juche* principles, is stressed. The means of production are owned solely by the State and socialist co-operative organizations.

Culture and education provide the working people with knowledge to advance a socialist way of life. Education is free, universal and compulsory for 11 years.

Defence is emphasized, as well as the rights of overseas nationals, the principles of friendly relations between nations based on equality, mutual respect and non-interference, proletarian internationalism, support for national liberation struggles and due observance of law.

The basic rights and duties of citizens are laid down and guaranteed. These include the right to vote and to be elected (for citizens who are more than 17 years of age), to work (the working day being eight hours), to free medical care and material assistance for the old, infirm or disabled, and to political asylum. National defence is the supreme duty of citizens.

THE STRUCTURE OF STATE

The Supreme People's Assembly

The Supreme People's Assembly is the highest organ of state power, exercises legislative power and is elected by direct, equal, universal and secret ballot for a term of five years. Its chief functions are: (i) to adopt, amend or supplement legal or constitutional enactments; (ii) to determine state policy; (iii) to elect the Chairman of the National Defence Commission; (iv) to elect the Vice-Chairmen and other members of the National Defence Commission (on the recommendation of the Chairman of the National Defence Commission); (v) to elect the President and other members of the Presidium of the Supreme People's Assembly, the Premier of the Cabinet, the President of the Central Court and other legal officials; (vi) to appoint the Vice-Premiers and other members of the Cabinet (on the recommendation of the Premier of the Cabinet); (vii) to approve the State Plan and Budget; (viii) to receive a report on the work of the Cabinet and adopt measures, if necessary; (ix) to decide on the ratification or abrogation of treaties. It holds regular and extraordinary sessions, the former being once or twice a year, the latter as necessary at the request of at least one-third of the deputies. Legislative enactments are adopted when approved by more than one-half of those deputies present. The Constitution is amended and supplemented when approved by more than two-thirds of the total number of deputies.

The National Defence Commission

The National Defence Commission, which consists of a Chairman, first Vice-Chairman, other Vice-Chairmen and members, is the highest military organ of state power, and is accountable to the Supreme People's Assembly. The National Defence Commission directs and commands the armed forces and guides defence affairs. The Chairman of the National Defence Commission serves a five-year term of office and has the most senior post in the state hierarchy.

The Presidium of the Supreme People's Assembly

The Presidium of the Supreme People's Assembly, which consists of a President, Vice-Presidents, secretaries and members, is the highest organ of power in the intervals between sessions of the Supreme People's Assembly, to which it is accountable. It exercises the following chief functions: (i) to convene sessions of the Supreme People's Assembly; (ii) to examine and approve new legislation, the State Plan and the State Budget, when the Supreme People's Assembly is in recess; (iii) to interpret the Constitution and legislative enactments; (iv) to supervise the observance of laws of State organs; (v) to organize elections to the Supreme People's Assembly and Local People's Assemblies; (vi) to form or abolish ministries or commissions of the Cabinet; (vii) to appoint or remove Vice-Premiers and other cabinet or ministry members, on the recommendation of the Premier, when the Supreme People's Assembly is not in session; (viii) to elect or transfer judges of the Central Court; (ix) to ratify or abrogate treaties concluded with other countries; (x) to appoint or recall diplomatic envoys; (xi) to confer decorations, medals, honorary titles and diplomatic ranks; (xii) to grant general amnesties or special pardon. The President of the Presidium represents the State and receives credentials and letters of recall of diplomatic representatives accredited by a foreign state.

The Cabinet

The Cabinet is the administrative and executive body of the Supreme People's Assembly and a general state management organ. It serves a five-year term and comprises the Premier, Vice-Premiers, Chairmen of Commissions and other necessary members. Its major functions are the following: (i) to adopt measures to execute state policy; (ii) to guide the work of ministries and other organs responsible to it; (iii) to establish and remove direct organs of the Cabinet and main administrative economic organizations; (iv) to draft the State Plan and adopt measures to make it effective; (v) to compile the State Budget and to implement its provisions; (vi) to organize and execute the work of all sectors of the economy, as well as education, science, culture, health and environmental protection; (vii) to adopt measures to strengthen the monetary and banking system; (viii) to adopt measures to maintain social order, protect State interests and guarantee citizens' rights; (ix) to conclude treaties; (x) to abolish decisions and directives of economic administrative organs which run counter to those of the Cabinet. The Cabinet is accountable to the Supreme People's Assembly.

Local People's Assemblies

The Local People's Assemblies and Committees of the province (or municipality directly under central authority), city (or district) and county are local organs of power. The Local People's Assemblies consist of deputies elected by direct, equal, universal and secret ballot. The Local People's Committees consist of a Chairman, Vice-Chairmen, secretaries and members. The Local People's Assemblies and Committees serve a four-year term and exercise local budgetary functions, elect local administrative and judicial personnel and carry out the decisions at local level of higher executive and administrative organs.

THE JUDICIARY

Justice is administered by the Central Court (the highest judicial organ of the State), local courts and the Special Court. Judges and other legal officials are elected by the Supreme People's Assembly. The Central Court protects state property and constitutional rights, guarantees that all state bodies and citizens observe state laws, and executes judgments. Justice is administered by the court comprising one judge and two people's assessors. The court is independent and judicially impartial. Judicial affairs are conducted by the Central Procurator's Office, which exposes and institutes criminal proceedings against accused persons. The Office of the Central Procurator is responsible to the Chairman of the National Defence Commission, the Supreme People's Assembly and the Central People's Committee.

The Government

HEAD OF STATE

President: President Kim Il Sung died on 8 July 1994 and was declared 'Eternal President' in September 1998.

Chairman of the National Defence Commission: Marshal Kim Jong Il.

First Vice-Chairman: Vice-Marshal Jo Myong Rok.

Vice-Chairmen: Yon Hyong Muk, Vice-Marshal Ri Yong Mu.

Other members: Vice-Marshal Kim Yong Chun, Kim Il Chol, Choe Ryong Su, Paek Se Bong, Jon Byong Ho.

CABINET
(September 2004)

Premier: PAK PONG JU.

Vice-Premiers: RO TU CHOL, KWAK POM GI, JON SUNG HUN.

Minister of Foreign Affairs: PAEK NAM SUN.

Minister of People's Security: JU SANG SONG.

Minister of the People's Armed Forces: Vice-Marshal KIM IL CHOL.

Chairman of the State Planning Commission: KIM KWANG RIM.

Minister of Power and Coal Industry: JU TONG IL.

Minister of Extractive Industries: RI KWANG NAM.

Minister of Metal and Machine-Building Industries: KIM SUNG HYON.

Minister of Construction and Building Materials Industries: JO YUN HUI.

Minister of the Electronics Industry: O SU YONG.

Minister of Railways: KIM YONG SAM.

Minister of Land and Marine Transport: KIM YONG IL.

Minister of Agriculture: RI KYONG SIK.

Minister of Chemical Industry: RI MU YONG.

Minister of Light Industry: RI JU O.

Minister of Foreign Trade: RI KYONG MAN.

Minister of Forestry: RI SANG MU.

Minister of Fisheries: RI SONG UNG.

Minister of City Management: CHOE JONG GON.

Minister of Land and Environmental Protection: JANG IL SON.

Minister of State Construction Control: PAE TAL JUN.

Minister of Commerce: RI YONG SON.

Minister of Procurement and Food Administration: CHOE NAM GYUN.

Minister of Education: KIM YONG JIN.

Minister of Post and Telecommunications: RI KUM BOM.

Minister of Culture: CHOE IK GYU.

Minister of Finance: MUN IL BONG.

Minister of Labour: RI WON IL.

Minister of Public Health: KIM SU HAK.

Minister of State Inspection: KIM UI SUN.

Chairman of the Physical Culture and Sports Guidance Committee: PAK MYONG CHOL.

President of the National Academy of Sciences: PYON YONG RIP.

President of the Central Bank: KIM WAN SU.

Director of the Central Statistics Bureau: KIM CHANG SU.

Chief Secretary of the Cabinet: JONG MUN SAN.

MINISTRIES

All Ministries and Commissions are in Pyongyang.

Legislature

CHOE KO IN MIN HOE UI
(Supreme People's Assembly)

The 687 members of the 11th Supreme People's Assembly (SPA) were elected unopposed for a five-year term on 3 August 2003. The SPA's permanent body is the Presidium.

Chairman: CHOE TAE BOK.

President of the Presidium: KIM YONG NAM.

Vice-Presidents of the Presidium: YANG HYONG SOP, KIM YONG DAE.

Political Organizations

Democratic Front for the Reunification of the Fatherland: Pyongyang; f. 1946; a vanguard organization comprising political parties and mass working people's organizations seeking the unification of North and South Korea; Mems of Presidium PAK SONG CHOL, RYOM TAE JUN, YANG HYONG SOP, JONG TU HWAN, RI YONG SU, JONG SHIN HYOK, KIM PONG JU, PYON CHANG BOK, RYU MI YONG, RYO WON GU, KANG RYON HAK.

The component parties are:

Chondoist Chongu Party: Pyongyang; tel. (2) 334241; f. 1946; supports policies of Korean Workers' Party; follows the guiding principle of *Innaechon* (the realization of 'heaven on earth'); Chair. RYU MI YONG.

Korean Social Democratic Party (KSDP) (Joson Sahoeminju-dang): Pyongyang; tel. (2) 5211981; fax (2) 3814410; f. 1945; advocates national independence and a democratic socialist society; supports policies of Korean Workers' Party; Chair. KIM YONG DAE; First Vice-Chair. KANG PYONG HAK.

Korean Workers' Party (KWP): Pyongyang; f. 1945; merged with the South Korean Workers' Party in 1949; the guiding principle is the *juche* idea, based on the concept that man is the master and arbiter of all things; 3m. mems; Gen. Sec. Marshal KIM JONG IL.

Sixth Central Committee

General Secretary: Marshal KIM JONG IL.

Politburo

Presidium: Marshal KIM JONG IL.

Full Members: KIM YONG NAM, PAK SONG CHOL, KIM YONG JU, KYE UNG TAE, JON BYONG HO, HAN SONG RYONG.

Alternate Members: HONG SONG NAM, YON HYONG MUK, YANG HYONG SOP, CHOE TAE BOK, KIM CHOL MAN, CHOE YONG RIM, RI SON SHIL.

Secretariat: Marshal KIM JONG IL, KYE UNG TAE, JON BYONG HO, HAN SONG RYONG, CHOE TAE BOK, KIM KI NAM, KIM KUK TAE, KIM JUNG RIN, JONG HA CHOL.

The component mass working people's organizations (see under Trade Unions) are:

General Federation of Trade Unions of Korea (GFTUK).

Kim Il Sung Socialist Youth League.

Korean Democratic Women's Union (KDWU).

Union of Agricultural Working People of Korea.

There is one opposition organization in exile, with branches in Tokyo (Japan), Moscow (Russia) and Beijing (People's Republic of China):

Salvation Front for the Democratic Unification of Chosun: f. early 1990s; seeks the overthrow of the Kim dynasty, the establishment of democracy in the DPRK and Korean reunification; Chair. PAK KAP DONG.

Diplomatic Representation

EMBASSIES IN THE DEMOCRATIC PEOPLE'S REPUBLIC OF KOREA

Algeria: Munsudong, Taedongkang District, Pyongyang; tel. (2) 90372; Ambassador MOKHTAR REGUIEG.

Benin: Pyongyang; Ambassador A. OGIST.

Bulgaria: Munsudong, Taedongkang District, Pyongyang; tel. (2) 3827343; fax (2) 3817342; Ambassador YORDAN MUTAFCHIYEV.

Cambodia: Munsudong, Taedongkang District, Pyongyang; tel. (2) 3817283; fax (2) 3817625.

China, People's Republic: Kinmauldong, Moranbong District, Pyongyang; tel. (2) 3823316; fax (2) 3813425; Ambassador WU DONGHE.

Cuba: Munsudong, Taedongkang District, Pyongyang; tel. (2) 3827380; fax (2) 3817703; Ambassador ESTEBAN LOBAINA ROMERO.

Czech Republic: Munsudong, Taedongkang District, Pyongyang.

Egypt: Pyongyang; tel. (2) 3817414; fax (2) 3817611; Ambassador AHMED RAMY AWAD ALHOSAINY.

Ethiopia: POB 55, Munsudong, Taedongkang District, Pyongyang; tel. (2) 3827554; fax (2) 3827550; Chargé d'affaires FEKADE S. G. MESKEL.

Germany: Munsudong, Taedongkang District, Pyongyang; tel. (2) 3817385; fax (2) 3817397; e-mail zreg@pjoe.auswaertiges-amt.de; Ambassador DORIS HERTRAMPF.

India: Block 53, Munsudong, Taehak St, Taedongkang District, Pyongyang; tel. (2) 3817274; fax (2) 3817619; e-mail indemhoc@di.chesin.com; Ambassador N. T. KHANKHUP.

Indonesia: 5 Foreigners' Bldg, Munsudong, Taedongkang District, Pyongyang; tel. (2) 3827439; fax (2) 3817620; e-mail kbripyg@public.east.cn.net; Ambassador HENDRATI SUKENDAR MUNTHE.

Iran: Munhungdong, Monsu St, Taedongkang District, Pyongyang; tel. (2) 3817492; fax (2) 3817612; Ambassador MUHAMMAD GANJI-DOOST.

Laos: Munhungdong, Taedongkang District, Pyongyang; tel. (2) 3827363; fax (2) 3817722; Ambassador KHAMKENG SAYAKEO.

Libya: Munsudong, Taedongkang District, Pyongyang; tel. (2) 3827544; fax (2) 3817267; Secretary of People's Bureau AHMED AMER AL-MUAKKAF.

Malaysia: Rm 1-17-05, Pyongyang Koryo Hotel, Tonghong-dong, Chaoyang District, Pyongyang; tel. (2) 3814397; fax (2) 3814422; e-mail malpygyang@kln.gov.my; Ambassador MD YUSOFF BIN MD ZAIN.

Mali: Pyongyang; Ambassador NAKOUNTE DIAKITÉ.

Mongolia: Munsudong, Taedongkang District, Pyongyang; tel. (2) 3827322; fax (2) 3817321; Ambassador J. LOMBO.

Nigeria: Munsudong, Taedongkang District, POB 535, Pyongyang; tel. (2) 3827558; fax (2) 3817293; Ambassador SULE BUBA.

Pakistan: Munsudong, Taedongkang District, Pyongyang; tel. (2) 3827478; fax 3817622; Ambassador SULTAN HABIB.

Poland: Munsudong, Taedongkang District, Pyongyang; tel. (2) 3817327; fax (2) 3817634; Ambassador WOJCIECH KALUZA.

Romania: Munhungdong, Taedongkang District, Pyongyang; tel. (2) 3827336; fax (2) 3817336; e-mail ambrophe@di.chesin.com; Chargé d'affaires a.i. EUGEN POPA.

Russia: Sinyangdong, Central District, Pyongyang; tel. (2) 3823102; fax (2) 3813427; e-mail rusembdprk@yahoo.com; Ambassador ANDREI KARLOV.

Sweden: Munsudung, Taedongkang District, Pyongyang; tel. (2) 3817908; fax (2) 3817258; e-mail ambassaden.pyongyang@foreign.ministry.se; Ambassador PAUL BEIJER.

Syria: Munsudong, Taedongkang District, Pyongyang; tel. (2) 3827473; fax (2) 3817635; Ambassador YASSER AL-FARRA.

Thailand: Pyongyang; Ambassador NIKHOM TANTEMSAPYA.

United Kingdom: Munsudong, Taedongkang District, Pyongyang; tel. (2) 3817980; fax (2) 3817985; Ambassador DAVID SLINN.

Viet Nam: Munsudong, Taedongkang District, Pyongyang; tel. (2) 3817353; fax (2) 3817632; Ambassador LE XUAN VINH.

Judicial System

The judicial organs include the Central Court, the Court of the Province (or city under central authority) and the People's Court. Each court is composed of judges and people's assessors.

Procurators supervise the ordinances and regulations of all ministries and the decisions and directives of local organs of state power to ensure that they conform to the Constitution, laws and decrees, as well as to the decisions and other measures of the Cabinet. Procurators bring suits against criminals in the name of the State, and participate in civil cases to protect the interests of the State and citizens.

Central Court

Pyongyang.

The highest judicial organ; supervises the work of all courts.

President: KIM BYONG RYUL.

First Vice-President: YUN MYONG GUK.

Central Procurator's Office

Supervises work of procurator's offices in provinces, cities and counties.

Procurator-General: RI KIL SONG.

Religion

The religions that are officially reported to be practised in the DPRK are Buddhism, Christianity and Chundo Kyo, a religion peculiar to Korea combining elements of Buddhism and Christianity. Religious co-ordinating bodies are believed to be under strict state control.

Korean Religious Believers Council: Pyongyang; f. 1989; brings together members of religious organizations in North Korea; Chair. JANG JAE ON.

BUDDHISM

In 1995, according to North Korean sources, there were some 60 Buddhist temples and an estimated 300 monks in the DPRK; the number of believers was about 10,000.

Korean Buddhists Federation: POB 77, Pyongyang; tel. (2) 43698; fax (2) 3812100; f. 1945; Chair. Cen. Cttee PAK TAE HWA; Sec. SHIM SANG RYON.

CHRISTIANITY

In 1995, according to North Korean sources, there were approximately 13,000 Christians (including 3,000 Roman Catholics) in the country, many of whom worshipped in house churches (of which there were about 500).

Korean Christians Federation: Pyongyang; f. 1946; Chair. Cen. Cttee KANG YONG SOP; Sec. O KYONG U.

The Roman Catholic Church

For ecclesiastical purposes, North and South Korea are nominally under a unified jurisdiction. North Korea contains two dioceses (Hamhung and Pyongyang), both suffragan to the archdiocese of Seoul (in South Korea), and the territorial abbacy of Tokwon (Tokugen), directly responsible to the Holy See.

Korean Roman Catholics Association: Changchung 1-dong, Songyo District, Pyongyang; tel. (2) 23492; f. 1988; Chair. Cen. Cttee JANG JAE ON; Vice-Chair. MUN CHANG HAK.

Diocese of Hamhung: Catholic Mission, Hamhung; 134-1 Waekwan-dong Kwan Eub, Chil kok kun, Gyeongbuk 718-800, Republic of Korea; tel. (545) 970-2000; Bishop (vacant); Apostolic Administrator of Hamhung and of the Abbacy of Tokwon Fr PLACIDUS DONG-HO RI.

Diocese of Pyongyang: Catholic Mission, Pyongyang; Bishop Rt Rev. FRANCIS HONG YONG HO (absent); Apostolic Administrator Most Rev. NICHOLAS CHEONG JIN-SUK (Archbishop of Seoul).

CHUNDO KYO

Korean Chundoists Association: Pyongyang; tel. (2) 334241; f. 1946; Chair. of Central Guidance Cttee RYU MI YONG.

The Press

PRINCIPAL NEWSPAPERS

Choldo Sinmun: Pyongyang; f. 1947; every two days.

Joson Inmingun (Korean People's Army Daily): Pyongyang; f. 1948; daily; Editor-in-Chief RI TAE BONG.

Kyowon Sinmun: Pyongyang; f. 1948; publ. by the Education Commission; weekly.

Minju Choson (Democratic Korea): Pyongyang; f. 1946; govt organ; 6 a week; Editor-in-Chief KIM JONG SUK; circ. 200,000.

Nongup Kunroja: Pyongyang; publ. of Cen. Cttee of the Union of Agricultural Working People of Korea.

Pyongyang Sinmun: Pyongyang; f. 1957; general news; 6 a week; Editor-in-Chief SONG RAK GYUN.

Rodong Chongnyon (Working Youth): Pyongyang; f. 1946; organ of the Cen. Cttee of the Kim Il Sung Socialist Youth League; 6 a week; Editor-in-Chief RI JONG GI.

Rodong Sinmun (Labour Daily): Pyongyang; f. 1946; organ of the Cen. Cttee of the Korean Workers' Party; daily; Editor-in-Chief CHOE CHIL NAM; circ. 1.5m.

Rodongja Sinmun (Workers' Newspaper): Pyongyang; f. 1945; organ of the Gen. Fed. of Trade Unions of Korea; Editor-in-Chief RI SONG JU.

Saenal (New Day): Pyongyang; f. 1971; publ. by the Kim Il Sung Socialist Youth League; 2 a week; Deputy Editor CHOE SANG IN.

Sonyon Sinmun: Pyongyang; f. 1946; publ. by the Kim Il Sung Socialist Youth League; 2 a week; circ. 120,000.

Tongil Sinbo: Kangan 1-dong, Youth Ave, Songyo District, Pyongyang; f. 1972; non-affiliated; weekly; Chief Editor JO HYON YONG; circ. 300,000.

PRINCIPAL PERIODICALS

Chollima: Pyongyang; popular magazine; monthly.

Choson (Korea): Pyongyang; social, economic, political and cultural; bi-monthly.

Choson Minju Juuiinmin Gonghwaguk Palmyonggongbo (Official Report of Inventions in the DPRK): Pyongyang; 6 a year.

Choson Munhak (Korean Literature): Pyongyang; organ of the Cen. Cttee of the Korean Writers' Union; monthly.

Choson Yesul (Korean Arts): Pyongyang; organ of the Cen. Cttee of the Gen. Fed. of Unions of Literature and Arts of Korea; monthly.

Economics: POB 73, Pyongyang; fax (2) 3814410; quarterly.

History: POB 73, Pyongyang; fax (2) 3814410; quarterly.

Hwahakgwa Hwahakgoneop: Pyongyang; organ of the Hamhung br. of the Korean Acad. of Sciences; chemistry and chemical engineering; 6 a year.

Jokook Tongil: Kangan 1-dong, Youth Ave, Songyo District, Pyongyang; organ of the Cttee for the Peaceful Unification of Korea; f. 1961; monthly; Chief Editor LI MYONG GYU; circ. 70,000.

Korean Medicine: POB 73, Pyongyang; fax (2) 3814410; quarterly.

Kunroja (Workers): 1 Munshindong, Tongdaewon, Pyongyang; f. 1946; organ of the Cen. Cttee of the Korean Workers' Party; monthly; Editor-in-Chief RYANG KYONG BOK; circ. 300,000.

Kwahakwon Tongbo (Bulletins of the Academy of Science): POB 73, Pyongyang; fax (2) 3814410; organ of the Standing Cttee of the Korean Acad. of Sciences; 6 a year.

Mulri (Physics): POB 73, Pyongyang; fax (2) 3814410; quarterly.

Munhwao Haksup (Study of Korean Language): POB 73, Pyongyang; fax (2) 3814410; publ. by the Publishing House of the Acad. of Social Sciences; quarterly.

Philosophy: PO Box 73, Pyongyang; fax (2) 3814410; quarterly.

Punsok Hwahak (Analysis): POB 73, Pyongyang; fax (2) 3814410; organ of the Cen. Analytical Inst. of the Korean Acad. of Sciences; quarterly.

Ryoksagwahak (Historical Science): Pyongyang; publ. by the Acad. of Social Sciences; quarterly.

Saengmulhak (Biology): POB, Pyongyang; fax (2) 3814410; publ. by the Korea Science and Encyclopedia Publishing House; quarterly.

Sahoekwahak (Social Science): Pyongyang; publ. by the Acad. of Social Sciences; 6 a year.

Suhakkwa Mulli: Pyongyang; organ of the Physics and Mathematics Cttee of the Korean Acad. of Sciences; quarterly.

FOREIGN LANGUAGE PUBLICATIONS

The Democratic People's Republic of Korea: Korea Pictorial, Pyongyang; f. 1956; illustrated news; Korean, Russian, Chinese, English, French, Arabic and Spanish edns; monthly; Editor-in-Chief HAN POM CHIK.

Foreign Trade of the DPRK: Foreign Trade Publishing House, Potonggang District, Pyongyang; economic developments and export promotion; English, French, Japanese, Russian and Spanish edns; monthly.

Korea: Pyongyang; f. 1956; illustrated; Korean, Arabic, Chinese, English, French, Spanish and Russian edns; monthly.

Korea Today: Foreign Languages Publishing House, Pyongyang; current affairs; Chinese, English, French, Russian and Spanish edns; monthly; Vice-Dir and Editor-in-Chief HAN PONG CHAN.

Korean Women: Pyongyang; English and French edns; quarterly.

Korean Youth and Students: Pyongyang; English and French edns; monthly.

The Pyongyang Times: Sochondong, Sosong District, Pyongyang; tel. (2) 51951; English, Spanish and French edns; weekly.

NEWS AGENCIES

Korean Central News Agency (KCNA): Potonggangdong 1, Potonggang District, Pyongyang; internet www.kcna.co.jp; f. 1946; sole distributing agency for news in the DPRK; publs daily bulletins in English, Russian, French and Spanish; Dir-Gen. KIM KI RYONG.

Foreign Bureaux

Informatsionnoye Telegrafnoye Agentstvo Rossii—Telegrafnoye Agentstvo Suverennykh Stran (ITAR—TASS) (Russia): Munsudong, Bldg 4, Flat 30, Taedongkang District, Pyongyang; tel. (2) 3817318; Correspondent ALEKSANDR VALIYEV.

The Xinhua (New China) News Agency (People's Republic of China) is also represented in the DPRK.

Press Association

Korean Journalists Union: Pyongyang; tel. (2) 36897; f. 1946; assists in the ideological work of the Korean Workers' Party; Chair. Cen. Cttee KIM SONG GUK.

Publishers

Academy of Sciences Publishing House: Nammundong, Central District, Pyongyang; tel. (2) 51956; f. 1953.

Academy of Social Sciences Publishing House: Pyongyang; Dir CHOE KWAN SHIK.

Agricultural Press: Pyongyang; labour, industrial relations; Pres. HO KYONG PIL.

Central Science and Technology Information Agency: Pyongyang; f. 1963; Dir JU SONG RYONG.

Education Publishing House: Pyongyang; f. 1945; Pres. KIM CHANG SON.

Foreign Language Press Group: Sochondong, Sosong District, Pyongyang; tel. (2) 841342; fax (2) 812100; f. 1949; Dir CHOE KYONG GUK.

Foreign Language Publishing House: Oesong District, Pyongyang; Dir SONG KI HYON.

Higher Educational Books Publishing House: Pyongyang; f. 1960; Pres. JU IL JUNG.

Industrial Publishing House: Pyongyang; f. 1948; technical and economic; Dir KIM TONG SU.

Kim Il Sung University Publishing House: Pyongyang; f. 1965.

Korea Science and Encyclopedia Publishing House: POB 73, Pyongyang; tel. (2) 18111; fax (2) 3814410; publishes numerous periodicals and monographs; f. 1952; Dir Gen. KIM YONG IL; Dir of International Co-operation JEAN BAHNG.

Korean People's Army Publishing House: Pyongyang; Pres. YUN MYONG DO.

Korean Social Democratic Party Publishing House: Pyongyang; tel. (2) 3818038; fax (2) 3814410; f. 1946; publishes quarterly journal *Joson Sahoemingjudang* (in Korean) and *KSDP Says* (in English); Dir KIM SOK JUN.

Korean Workers' Party Publishing House: Pyongyang; f. 1945; fiction, politics; Dir RYANG KYONG BOK.

Kumsong Youth Publishing House: Pyongyang; f. 1946; Dir HAN JONG SOP.

Literature and Art Publishing House: Pyongyang; f. by merger of Mass Culture Publishing House and Publishing House of the Gen. Fed. of Literary and Art Unions; Dir Gen. YUN KYONG NAM.

Transportation Publishing House: Namgyodong, Hyongjaesan District, Pyongyang; f. 1952; travel; Editor PAEK JONG HAN.

Working People's Organizations Publishing House: Pyongyang; f. 1946; fiction, government, political science; Dir PAK SE HYOK.

WRITERS' UNION

Korean Writers' Union: Pyongyang; Chair. Cen. Cttee KIM PYONG HUN.

Broadcasting and Communications

North Korea established a satellite communications station in Pyongyang, through an agreement with France in 1986, which enabled North Korea to communicate by satellite with Western countries. In 1990 an agreement was reached on satellite communications for the operation of telephone, telex and telegram services between North Korea and Japan. In October 2001 North Korea launched its first e-mail service provider in co-operation with China-based company Silibank.com, which was used for business and trade purposes. However, access to the internet remained severely limited, with information flow within North Korea still being conducted mainly via a closed intranet system (Kwangmyong).

TELECOMMUNICATIONS

Korea Post and Telecommunications Co: Pyongyang; Dir KIM HYON JONG.

BROADCASTING

Regulatory Authorities

DPRK Radio and Television Broadcasting Committee: see Radio, below.

Pyongyang Municipal Broadcasting Committee: Pyongyang; Chair. KANG CHUN SHIK.

Radio

DPRK Radio and Television Broadcasting Committee: Jonsungdong, Moranbong District, Pyongyang; tel. (2) 3816035; fax (2) 3812100; programmes relayed nationally with local programmes supplied by local radio cttees; loudspeakers are installed in factories and in open spaces in all towns; home broadcasting 22 hours daily;

foreign broadcasts in Russian, Chinese, English, French, German, Japanese, Spanish and Arabic; Chair. CHA SUNG SU.

Television

General Bureau of Television: Gen. Dir CHA SUNG SU.

DPRK Radio and Television Broadcasting Committee: see Radio.

Kaesong Television: Kaesong; broadcasts five hours on weekdays, 11 hours at weekends.

Korean Central Television Station: Ministry of Post and Telecommunications, Pyongyang; broadcasts five hours daily; satellite broadcasts commenced Oct. 1999.

Mansudae Television Station: Mansudae, Pyongyang; f. 1983; broadcasts nine hours of cultural programmes, music and dance, foreign films and news reports at weekends.

Finance

(cap. = capital; res = reserves; dep. = deposits; m. = million; brs = branches)

BANKING

During 1946–47 all banking institutions in North Korea, apart from the Central Bank and the Farmers Bank, were abolished. The Farmers Bank was merged with the Central Bank in 1959. The Foreign Trade Bank (f. 1959) conducts the international business of the Central Bank. Other banks, established in the late 1970s, are responsible for the foreign-exchange and external payment business of North Korean foreign trade enterprises.

The entry into force of the Joint-Venture Act in 1984 permitted the establishment of joint-venture banks, designed to attract investment into North Korea by Koreans resident overseas. The Foreign Investment Banking Act was approved in 1993.

Central Bank

Central Bank of the DPRK: Munsudong, Seungri St 58-1, Central District, Pyongyang; tel. (2) 3338196; fax (2) 3814624; f. 1946; bank of issue; supervisory and control bank; Pres. KIM WAN SU; 227 brs.

State Banks

Changgwang Credit Bank: Saemaeul 1-dong, Pyongchon District, Pyongyang; tel. (2) 18111; fax (2) 3814793; f. 1983; commercial, joint-stock and state bank; cap. 601.0m. won, res 1,194.2m. won, dep. 10,765.8m. won (Dec. 1997); Chair. KIM CHOL HWAN; Pres. KYE CHANG HO; 172 brs.

Credit Bank of Korea: Chongryu 1-dong, Munsu St, Otandong, Central District, Pyongyang; tel. (2) 3818285; fax (2) 3817806; f. 1986 as International Credit Bank, name changed 1989; Pres. LI SUN BOK; Vice-Pres. SON YONG SUN.

Foreign Trade Bank of the DPRK: FTB Bldg, Jungsongdong, Seungri St, Central District, Pyongyang; tel. (2) 3815270; fax (2) 3814467; f. 1959; deals in international settlements and all banking business; Pres. KIM JUN CHOL; 11 brs.

International Industrial Development Bank: Jongpyong-dong, Pyongchon District, Pyongyang; tel. (2) 3818610; fax (2) 3814427; f. 2001; Pres. SHIN DOK SONG.

Korea Daesong Bank: Segoridong, Gyongheung St, Potonggang District, Pyongyang; tel. (2) 3818221; fax (2) 3814576; f. 1978; Pres. RI HONG.

Koryo Bank: Ponghwadong, Potonggang District, Pyongyang; tel. (2) 3818168; fax (2) 3814033; f. 1989 as Koryo Finance Joint Venture Co, name changed 1994; co-operative, development, regional, savings and universal bank; Pres. RI CHANG HWAN; 10 brs.

Kumgang Bank: Jungsongdong, Central District, Pyongyang; tel. (2) 3818532; fax (2) 3814467; f. 1979; Chair. KIM JANG HO.

Joint-Venture Banks

Korea Commercial Bank: f. 1988; joint venture with Koreans resident in the USA.

Korea Joint Financial Co: f. 1988; joint venture with Koreans resident in the USA.

Korea Joint Venture Bank: Ryugyongdong, Potonggang District, Pyongyang; tel. (2) 3818151; fax (2) 3814410; f. 1989; with co-operation of the Federation of Korean Traders and Industrialists in Japan; cap. US $1,932.5m. (1994); Chair PAK IL RAK; Vice-Pres. KIM SONG HWAN; 6 brs.

Korea Nagwon Joint Financial Co: f. 1987 by Nagwon Trade Co and a Japanese co.

Korea Rakwon Joint Banking Co: Pyongyang; Man. Dir HO POK DOK.

Korea United Development Bank: Central District, Pyongyang; tel. (2) 3814165; fax (2) 3814497; f. 1991; 51% owned by Zhongce Investment Corpn (Hong Kong), 49% owned by Osandok General Bureau; cap. US $60m.; Pres. KIM SE HO.

Koryo Joint Finance Co: Pyongyang; Dir KIM YONG GU.

Foreign-Investment Banks

Daesong Credit Development Bank: Potonggang Hotel, 301 Ansan-dong, Pyongchon District, Pyongyang; tel. (2) 3814866; fax (2) 3814723; f. 1996 as Peregrine-Daesong Development Bank; jt venture between Oriental Commercial Holdings Ltd (Hong Kong) and Korea Daesong Bank; Man. NIGEL COWIE.

Golden Triangle Bank: Rajin-Sonbong Free Economic and Trade Zone; f. 1995.

INSURANCE

State Insurance Bureau: Central District, Pyongyang; tel. (2) 38196; handles all life, fire, accident, marine, hull insurance and reinsurance.

Korea Foreign Insurance Co (Chosunbohom): Central District, Pyongyang; tel. (2) 3818024; fax (2) 3814464; f. 1974; conducts marine, motor, aviation and fire insurance, reinsurance of all classes, and all foreign insurance; brs in Chongjin, Hungnam and Nampo, and agencies in foreign ports; overseas representative offices in Chile, France, Germany, Pakistan, Singapore; Pres. RI JANG SU.

Korea International Insurance Co: Pyongyang; Dir (vacant).

Korea Mannyon Insurance Co: Pyongyang; Pres. PAK IL HYONG.

Trade and Industry

GOVERNMENT AGENCIES

DPRK Committee for the Promotion of External Economic Co-operation: Jungsongdong, Central District, Pyongyang; tel. (2) 333974; fax (2) 3814498; Chair. KIM YONG SUL.

DPRK Committee for the Promotion of International Trade: Central District, Pyongyang; Pres. RI SONG ROK; Chair. KIM JONG GI.

Economic Co-operation Management Bureau: Ministry of Foreign Trade, Pyongyang; f. 1998; Dir KIM YONG SUL.

Korea International Joint Venture Promotion Committee: Pyongyang; Chair. CHAE HUI JONG.

Korean Association for the Promotion of Asian Trade: Pyongyang; Pres. RI SONG ROK.

Korean International General Joint Venture Co: Pyongyang; f. 1986; promotes joint economic ventures with foreign countries; Man. Dir RI KWANG GUN.

Korean General Merchandise Export and Import Corpn: Pyongyang.

INDUSTRIAL AND TRADE ASSOCIATIONS

Korea Building Materials Trading Co: Tongdaewon District, Pyongyang; tel. (2) 18111–3818085; fax (2) 38145555; chemical building materials, woods, timbers, cement, sheet glass, etc; Dir SHIN TONG BOM.

Korea Cement Export Corpn: Central District, Pyongyang; f. 1982; cement and building materials.

Korea Cereals Export and Import Corpn: Central District, Pyongyang; high-quality vegetable starches, etc.

Korea Chemicals Export and Import Corpn: Central District, Pyongyang; petroleum and petroleum products, raw materials for the chemical industry, rubber and rubber products, fertilizers, etc.

Korea Daesong Jei Trading Corpn: Pulgungori 1–dong, Potonggang District, Pyongyang; tel. (2) 18111-3818213; fax (2) 3814431; machinery and equipment, chemical products, textiles, agricultural products, etc.

Korea Daesong Jesam Trading Corpn: Pulgungori 1-dong, Potonggang District, Pyongyang; tel. (2) 18111–3818562; fax (2) 3814431; remedies for diabetes, tonics, etc.

Korea Ferrous Metals Export and Import Corpn: Potonggang 2–dong, Potonggang District, Pyongyang; tel. (2) 18111-3818078; fax (2) 381-4581; steel products.

Korea Film Export and Import Corpn: Daedongmundong, Central District, POB 113, Pyongyang; tel. (2) 180008034; fax (2)

3814410; f. 1956; feature films, cartoons, scientific and documentary films; Dir-Gen. CHOE HYOK U.

Korea First Equipment Export and Import Co: Central District, Pyongyang; tel. (2) 334825; f. 1960; export and import of ferrous and non-ferrous metallurgical plants, geological exploration and mining equipment, communication equipment, machine-building plant, etc.; construction of public facilities such as airports, hotels, tourist facilities, etc.; joint-venture business in similar projects; Pres. CHAE WON CHOL.

Korea Foodstuffs Export and Import Corpn: Kangan 2–dong, Songyo District, Pyongyang; tel. (2) 18111-3818289; fax (2) 3814417; cereals, wines, meat, canned foods, fruits, cigarettes, etc.

Korea Fruit and Vegetables Export Corpn: Central District, Pyongyang; tel. (2) 35117; vegetables, fruit and their products.

Korea General Corpn for External Construction (GENCO): Sungri St 25, Jungsong-dong, Central District, Pyongyang; tel. (2) 18111-3818090 ; fax (2) 3814611; e-mail gen122@co.chesin.com; f. 1961; construction of dwelling houses, public establishments, factories, hydroelectric and thermal power stations, irrigation systems, ports, bridges, and transport services, technical services; Gen. Dir CHOE BONG SU.

Korea General Export and Import Corpn: Central District, Pyongyang; plate glass, tiles, granite, locks, medicinal herbs, foodstuffs and light industrial products.

Korea General Machine Co: Tongsin 3-dong, Tongdaewon, Pyongyang; tel. (2) 18555-3818102; fax (2) 381-4495; Dir RA IN GYUN.

Korea Hyopdong Trading Corpn: Othan-dong, Kangan St, Central District, Pyongyang; tel. (2) 18111-3818011; fax (2) 3814454; fabrics, glass products, ceramics, chemical goods, building materials, foodstuffs, machinery, etc.

Korea Industrial Technology Co: Junsongdong, Central District, Pyongyang; tel. (2) 18111-3818025; fax (2) 3814537; Pres. KWON YONG SON.

Korea International Chemical Joint Venture Co: Pyongyang; Chair. RYO SONG GUN.

Korea International Joint Venture Co: Pyongyang; Man. Dir HONG SONG NAM.

Korea Jangsu Trading Co: Kyogudong, Central District, Pyongyang; tel. (2) 18111-3818834; fax (2) 3814410; medicinal products and clinical equipment.

Korea Jeil Equipment Export and Import Corpn: Jungsong-dong, Central District, Pyongyang; tel. (2) 334825; f. 1960; ferrous and non-ferrous metallurgical plant, geological exploration and mining equipment, power plant, communications and broadcasting equipment, machine-building equipment, railway equipment, construction of public facilities; Pres. CHO JANG DOK.

Korea Jesam Equipment Export and Import Corpn: Central District, Pyongyang; chemical, textile, pharmaceutical and light industry plant.

Korea Koryo Trading Corpn: Jongpyongdong, Pyongchon District, Pyongyang; tel. (2) 18111-3818104; fax (2) 3814646; Dir KIM HUI DUK.

Korea Kwangmyong Trading Corpn: Jungsongdong, Central District, Pyongyang; tel. (2) 18111-3818111; fax (2) 3814410; dried herbs, dried and pickled vegetables; Dir CHOE JONG HUN.

Korea Light Industry Import-Export Co: Juchetab St, Tongdaewon District, Pyongyang; tel. (2) 37661; exports silk, cigarettes, canned goods, drinking glasses, ceramics, handbags, pens, plastic flowers, musical instruments, etc.; imports chemicals, dyestuffs, machinery, etc.; Dir CHOE PYONG HYON.

Korea Machine Tool Trading Corpn: Tongdaewon District, Pyongyang; tel. (2) 18555-381810; fax (2) 3814495; Dir KIM KWANG RYOP.

Korea Machinery and Equipment Export and Import Corpn: Potonggang District, Pyongyang; tel. (2) 333449; f. 1948; metallurgical machinery and equipment, electric machines, building machinery, farm machinery, diesel engines, etc.

Korea Maibong Trading Corpn: Central District, Pyongyang; non-ferrous metal ingots and allied products, non-metallic minerals, agricultural and marine products.

Korea Manpung Trading Corpn: Central District, Pyongyang; chemical and agricultural products, machinery and equipment.

Korea Mansu Trading Corpn: Chollima St, Central District, POB 250, Pyongyang; tel. (2) 43075; fax (2) 812100; f. 1974; antibiotics, pharmaceuticals, vitamin compounds, drugs, medicinal herbs; Dir KIM JANG HUN.

Korea Marine Products Export and Import Corpn: Central District, Pyongyang; canned, frozen, dried, salted and smoked fish, fishing equipment and supplies.

Korea Minerals Export and Import Corpn: Central District, Pyongyang; minerals, solid fuel, graphite, precious stones, etc.

Korea Namheung Trading Co: Sinri-dong, Tongdaewon District, Pyongyang; tel. (2) 18111-3818974; fax (2) 3814623; high-purity reagents, synthetic resins, vinyl films, essential oils, menthol and peppermint oil.

Korea Non-ferrous Metals Export and Import Corpn: Potonggang 2-dong, Potonggang District, Pyongyang; tel. (2) 18111-3818247; fax (2) 3814569.

Korea Okyru Trading Corpn: Kansongdong, Pyongchon District, Pyongyang; tel. (2) 18111-3818110; fax (2) 3814618; agricultural and marine products, household goods, clothing, chemical and light industrial products.

Korea Ponghwa Contractual Joint Venture Co: Pyongyang; Dir RIM TONG CHON.

Korea Ponghwa General Trading Corpn: Jungsong-dong, Central District, Pyongyang; tel. (2) 18111-3818023; fax (2) 3814444; machinery, metal products, minerals and chemicals.

Korea Publications Export and Import Corpn: Yokjondong, Yonggwang St, Central District, Pyongyang; tel. (2) 3818536; fax (2) 3814404; f. 1948; export of books, periodicals, postcards, paintings, cassettes, videos, CDs, CD-ROMs, postage stamps and records; import of books; Pres. RI YONG.

Korea Pyongyang Trading Co Ltd: Central District, POB 550, Pyongyang; pig iron, steel, magnesia clinker, textiles, etc.

Korea Rungra Co: Sinwondong, Potonggang District, Pyongyang; tel. (2) 18111-3818112; fax (2) 3814608; Dir CHOE HENG UNG.

Korea Rungrado Trading Corpn: Segori-dong, Potonggang District, Pyongyang; tel. (2) 18111-3818022; fax (2) 3814507; food and animal products; Gen. Dir PAK KYU HONG.

Korea Ryongaksan General Trading Corpn: Pyongyang; Gen. Dir HAN YU RO.

Korea Samcholli General Corpn: Pyongyang; Dir JONG UN OP.

Korea Senbong Trading Corpn: Central District, Pyongyang; ferrous and non-ferrous metals, rolled steels, mineral ores, chemicals, etc.

Korea Somyu Hyopdong Trading Corpn: Oesong District, Pyongyang; clothing and textiles.

Korea Songhwa Trading Corpn: Oesong District, Pyongyang; ceramics, glass, hardware, leaf tobaccos, fruit and wines.

Korea Technology Corpn: Jungsongdong, Central District, Pyongyang; tel. (2) 18111-3818090; fax (2) 3814410; scientific and technical co-operation.

Korea Unha Trading Corpn: Rungra 1-dong, Taedonggang District, Pyongyang; tel. (2) 18111-3818236; fax (2) 3814506; clothing and fibres.

Korea Yonghung Trading Co: Tongan-dong, Central District, Pyongyang; tel. (2) 18111-3818223; fax (2) 3814527; e-mail greenlam@co.chesin.com; f. 1979; export of freight cars, vehicle parts, marine products, electronic goods, import of steel, chemical products; Pres. CHOE YONG DOK.

TRADE UNIONS

General Federation of Trade Unions of Korea (GFTUK): POB 333, Dongmun-dong, Daedonggang District, Pyongyang; fax (2) 3814427; f. 1945; 1.6m. mems (2003); seven affiliated unions (2003); Pres. RYOM SUN GIL.

Trade Union of Metal and Engineering Industries of Korea: Pyongyang; f. 1945; 332,800 mems (2003); Pres. CHOE GWANG HYON.

Trade Union of Mining and Power Industries of Korea: Pyongyang; f. 1945; 221,000 mems (2003); Pres. SON YONG JUN.

Trade Union of Light and Chemical Industries of Korea: Pyongyang; f. 1945; 372,500 mems (2003); Pres. RI JIN HAK.

Trade Union of Public Employees and Service Workers of Korea: Pyongyang; f. 1945; 305,900 mems (2003); Pres. KIM GANG HO.

Trade Union of Construction and Forestry Workers of Korea: Pyongyang; f. 1945; 160,000 mems (2003); Pres. WON HYONG GUK.

Trade Union of Educational and Cultural Workers: Dongmun-dong, Daedonggang District, Pyongyang, POB 333; fax (2) 3814427; f. 1946; 89,800 mems (2003); Pres. KIM YONG DO.

Trade Union of Transport and Fisheries Workers of Korea: Pyongyang; f. 1945; 119,800 mems (2003); Pres. CHOE RYONG SU.

General Federation of Agricultural and Forestry Technique of Korea: Chung Kuyuck Nammundong, Pyongyang; f. 1946; 523,000 mems.

General Federation of Unions of Literature and Arts of Korea: Pyongyang; f. 1946; seven br. unions; Chair. Cen. Cttee CHANG CHOL.

Kim Il Sung Socialist Youth League: Pyongyang; fmrly League of Socialist Working Youth of Korea; First Sec. KIM GYONG HO.

Korean Architects' Union: Pyongyang; f. 1954; 500 mems; Chair. Cen. Cttee PAE TAL JUN.

Korean Democratic Lawyers' Association: Ryonhwa 1, Central District, Pyongyang; fax (2) 3814644; f. 1954; Chair. HAM HAK SONG.

Korean Democratic Scientists' Association: Pyongyang; f. 1956.

Korean Democratic Women's Union: Jungsongdong, Central District, Pyongyang; fax (2) 3814416; f. 1945; Chief Officer PAK SUN HUI.

Korean General Federation of Science and Technology: Jungsongdong, Seungri St, Central District, Pyongyang; tel. (2) 3224389; fax (2) 3814410; f. 1946; 550,000 mems; Chair. Cen. Cttee CHOE HUI JONG.

Korean Medical Association: Pyongyang; f. 1970; Chair. CHOE CHANG SHIK.

Union of Agricultural Working People of Korea: Pyongyang; f. 1965 to replace fmr Korean Peasants' Union; 2.4m. mems; Chair. Cen. Cttee SUNG SANG SOP.

Transport

RAILWAYS

Railways were responsible for some 62% of passenger journeys in 1991 and for some 74% of the volume of freight transported in 1997. In 2002 the total length of track was estimated at 5,235 km, of which some 70% was electrified. There are international train services to Moscow (Russia) and Beijing (People's Republic of China). Construction work on the reconnection of the Kyongui (West coast, Sinuiju–Seoul) and East Coast Line (Wonsan–Seoul) began in September 2002. The two lines were officially opened in June 2003, but were not yet open to traffic, as construction work on the Northern side remained to be completed. Reconnection work was subject to disruption by the changing political situation. Eventually the two would be linked to the Trans-China and Trans-Siberian railways respectively, greatly enhancing the region's transportation links.

There is an underground railway system in Pyongyang, with two public lines. Unspecified plans to expand the system were announced in February 2002.

ROADS

In 2000, according to South Korean estimates, the road network totalled 23,407 km (of which only about 8% was paved), including 682 km of multi-lane highways.

INLAND WATERWAYS

The Yalu (Amnok-gang) and Taedong, Tumen and Ryesong are the most important commercial rivers. Regular passenger and freight services: Nampo–Chosan–Supung; Chungsu–Sinuiju–Dasado; Nampo–Jeudo; Pyongyang–Nampo.

SHIPPING

The principal ports are Nampo, Wonsan, Chongjin, Rajin, Hungnam, Songnim and Haeju. In 1997 North Korean ports had a combined capacity for handling 35m. tons of cargo. At 31 December 2002 North Korea's merchant fleet comprised 225 vessels, with a combined displacement of 870,458 grt.

Korea Chartering Corpn: Central District, Pyongyang; arranges cargo transportation and chartering.

Korea Daehung Shipping Co: Ansan 1–dong, Pyongchon District, Pyongyang; tel. (2) 18111 ext 8695; fax (2) 3814508; f. 1994; owns 6 reefers, 3 oil tankers, 1 cargo ship.

Korea East Sea Shipping Co: Pyongyang; Dir RI TUK HYON.

Korea Foreign Transportation Corpn: Central District, Pyongyang; arranges transportation of export and import cargoes (transit goods and charters).

Korean-Polish Shipping Co Ltd: Moranbong District, Pyongyang; tel. (2) 3814384; fax (2) 3814607; f. 1967; maritime trade mainly with Polish, Far East and DPRK ports.

Korea Tonghae Shipping Co: Changgwang St, Central District, POB 120, Pyongyang; tel. (2) 345805; fax (2) 3814583; arranges transportation by Korean vessels.

Ocean Maritime Management Co Ltd: Tonghungdong, Central District, Pyongyang.

Ocean Shipping Agency of the DPRK: Moranbong District, POB 21, Pyongyang; tel. (2) 3818100; fax (2) 3814531; Pres. O JONG HO.

CIVIL AVIATION

The international airport is at Sunan, 24 km from Pyongyang. In September 2003 the first tourist flight from Seoul to Pyongyang was completed by an Air Koryo plane, representing the first commercial flight between North and South Korea in more than 50 years. In January 2004 plans were announced for an aviation agreement between North and South Korea, which would allow regular inter-Korean flight routes to be opened.

Chosonminhang/General Civil Aviation Bureau of the DPRK: Sunan Airport, Sunan District, Pyongyang; tel. (2) 37917; fax (2) 3814625; f. 1954; internal services and external flights by Air Koryo to Beijing and Shenyang (People's Republic of China), Bangkok (Thailand), Macao, Nagoya (Japan), Moscow, Khabarovsk and Vladivostok (Russia), Sofia (Bulgaria) and Berlin (Germany); charter services are operated to Asia, Africa and Europe; Pres. KIM YO UNG.

Tourism

The DPRK was formally admitted to the World Tourism Organization in 1987. Tourism is permitted only in officially accompanied parties. In 1999 there were more than 60 international hotels (including nine in Pyongyang) with 7,500 beds. Tourist arrivals totalled 130,000 in 1998. A feasibility study was undertaken in 1992 regarding the development of Mount Kumgang as a tourist attraction. The study proposed the construction of an international airport at Kumnan and of a number of hotels and leisure facilities in the Wonsan area. Local ports were also to be upgraded. It was hoped that the development, scheduled to cost some US $20,000m. and to be completed by 2004, would attract 3m. tourists to the area each year. In November 1998 some 800 South Korean tourists visited Mount Kumgang, as part of a joint venture mounted by the North Korean authorities and Hyundai, the South Korean conglomerate. By November 2000 only 350,000 South Korean tourists had visited the attraction. In November 2002, in an effort to increase profitability, Mount Kumgang was designated a special tax-free economic zone. In 1996 it was reported that proposals had been made to create a tourist resort in the Rajin-Sonbong Free Economic and Trade Zone, in the north-east of the country. It was announced that hotels to accommodate some 5,000 people were to be constructed, as well as an airport to service the area. There were reports in 1998 that a heliport had been opened in the Zone, and in 1999 the resort was completed. Mount Chilbo, Mount Kuwol, Mount Jongbang and the Ryongmum Cave were transformed into new tourist destinations in that year. In August 2000 plans were announced for the development, jointly with China, of the western part of Mount Paektu, Korea's highest mountain, as a tourist resort. In September 2003 South Korean tourists were able to visit Pyongyang for the first time. In 2003 it was also estimated that around 1,500 Western tourists visited North Korea annually.

Korea International Tourist Bureau: Pyongyang; Pres. HAN PYONG UN.

Korean International Youth Tourist Co: Mankyongdae District, Pyongyang; tel. (2) 73406; f. 1985; Dir HWANG CHUN YONG.

Kumgangsan International Tourist Co: Central District, Pyongyang; tel. (2) 31562; fax (2) 3812100; f. 1988.

National Tourism Administration of the DPRK: Central District, Pyongyang; tel. (2) 3818901; fax (2) 3814547; e-mail nta@silibank.com; f. 1953; state-run tourism promotion organization; Dir RYO SUNG CHOL.

Ryohaengsa (Korea International Travel Company): Central District, Pyongyang; tel. (2) 3817201; fax (2) 3817607; f. 1953; has relations with more than 200 tourist companies throughout the world; Pres. CHO SONG HUN.

State General Bureau of Tourism: Pyongyang; Pres. RYO SUNG CHOL.

Defence

The estimated total strength of the armed forces in August 2003 was 1,082,000: army 950,000, air force 86,000, and navy 46,000. Security and border troops numbered 189,000, and there was a workers' and peasants' militia ('Red Guards') numbering about 3.5m. Military service is selective: army for five to eight years; navy for five to 10 years; and air force for three to four years.

Defence Expenditure: Budgeted at 15.5% of total spending for fiscal 2004.

Supreme Commander of the Korean People's Army and Chairman of the National Defence Commission: Marshal KIM JONG IL.

Chief of General Staff of the Korean People's Army: Vice-Marshal KIM YONG CHUN.

Commander of the Air Force: Col-Gen. O KUM CHOL.

Commander of the Navy: Gen. KIM YUN SHIM.

Education

Universal, compulsory primary and secondary education were introduced in 1956 and 1958, respectively, and are provided at state expense. Free and compulsory 11-year education in state schools was introduced in 1975. Children enter kindergarten at five years of age, and primary school at the age of six. After four years, they advance to senior middle school for six years. In 1987/88 there were 1.5m. primary school students and 2.5m. secondary school students. In that year some 325,000 students were enrolled in university-level institutions (of which there were 519 in 1986). In 1988 the Government announced the creation of new educational establishments, including one university, eight colleges, three factory colleges, two farmers' colleges and five special schools. English is compulsory as a second language from the age of 14. The adult literacy rate was estimated at 99% in 1984. In March 2001 the Ministry of Education announced plans for the establishment of a university of information science and technology in Pyongyang, in co-operation with a South Korean education foundation. The new university was due to offer a postgraduate degree course from September 2002 and undergraduate courses from 2003.

Bibliography

For Bibliography of the Democratic People's Republic of Korea and the Republic of Korea see p. 579

THE REPUBLIC OF KOREA

History*

AIDAN FOSTER-CARTER

Based on an earlier article by ANDREW C. NAHM

THE FIRST REPUBLIC, 1948–60

The foundation of the Republic was hardly settled when a communist-inspired military rebellion broke out in October 1948. The rebellion was crushed, but it demoralized the nation and increased the repressive character of the Government. The democratic aspirations and trends of the pre-Korean War period diminished as the Government became more autocratic during and after the war. Political and social conditions became chaotic as economic hardships multiplied.

Faced by a series of crises, President Syngman Rhee and his Liberal Party (LP, established in 1952) acted high-handedly towards their opponents, and various constitutional amendments were forced through the National Assembly. In July 1952 the National Assembly adopted one such amendment to elect the President by popular vote, and the election, conducted under martial law, was won by Dr Rhee. In 1954 the National Assembly adopted another series of amendments, including the exemption of the incumbent President from the two-term constitutional limitation in office, and the abolition of the post of Prime Minister.

In the 1956 presidential election, a new opposition Democratic Party (DP), founded in 1955, nominated candidates for the offices of President and Vice-President. The sudden death of the presidential candidate of the opposition party assured victory for the 81-year-old Dr Rhee, but the DP candidate, Chang Myon, defeated the LP candidate for the vice-presidency.

As corruption among government officials and LP members, as well as repression by the police, increased, a widespread desire for change developed, particularly among urban voters. At elections to the National Assembly in 1958, the DP substantially increased its number of seats. Aware of the danger of losing absolute control, the LP-dominated National Assembly repealed local autonomy laws, and passed a new national security law.

The death of the DP candidate, Dr Cho Pyong-Ok, some weeks before the fourth presidential elections, contributed to the re-election of Dr Rhee in March 1960, following a campaign characterized by violence and intimidation of opposition candidates and supporters. Popular reaction against the corrupt and fraudulent practices of the administration increased, and fierce student riots erupted throughout the country. The student uprising of 19 April forced President Rhee and his Government to resign one week subsequently. A caretaker Government was established under Ho Chong, and in mid-June the National Assembly adopted a constitutional amendment instituting a strong parliamentary system, reducing the presidency to a figurehead office, and resurrecting the office of Prime Minister. In August the National Assembly elected Yun Po-Son as President and Chang Myon as Prime Minister; thus, the Second Republic emerged.

With the exception of the Land Reform Law of 1949, the First Republic achieved no positive success in the economic field. In the post-Korean War period a degree of economic recovery was achieved with aid from UN agencies and the USA, but South Korea remained economically backward, suffering shortages of power, fuel, food and consumer goods.

THE SECOND REPUBLIC, 1960–61

The Second Republic was hampered from the start: it had no mandate from the people, and both President Yun and Prime Minister Chang lacked fortitude and practical ability. The Chang administration was indecisive in dealing with former leaders of the Rhee regime and proved unable to deal effectively with ideological and social differences between political and sectional groups, while gaining no new support nor the loyalty of the people. Divisions emerged within the DP, and no solutions to economic and social problems appeared imminent. With the exception of the (totally ineffective) Five-Year Plan, the Chang administration failed to adopt measures for solving the country's serious economic problems. Meanwhile, there were renewed demonstrations, as communist influence spread among students. Agitation by students for direct negotiations with their North Korean counterparts, aimed at reunification of the country, compounded by shortages of food and jobs, increased the perceived threat to national security.

MILITARY RULE, 1961–63

On 16 May 1961 a military junta, led by a small group of young army officers headed by Maj.-Gen. Park Chung-Hee, overthrew the Chang administration. The junta dissolved the National Assembly, banned all political activity, and declared martial law, prohibiting student demonstrations and censoring the press. Lt-Gen. Chang Do-Yong, the army chief of staff, became Chairman of a Supreme Council for National Reconstruction. President Yun remained in office, but the Government was in the hands of the military. Pledges were issued by the Supreme Council, upholding anti-communism and adherence to the UN, envisaging a strengthening of links with the USA and the Western bloc, and promising a wide-ranging programme of economic and political reform, as well as the eventual restoration of civilian rule.

The Supreme Council acted as a legislative body, and a 'national reconstruction extraordinary measures law' replaced the Constitution. In July 1961, when Gen. Chang was arrested for alleged anti-revolutionary conspiracy, Gen. Park assumed the chairmanship of the Supreme Council. In August Gen. Park announced that political activity would be permitted in early 1963, as a prelude to the restoration of a civilian government. A constitutional amendment was passed by national referendum in December 1962, restoring a strong presidential system while limiting presidential office to two four-year terms. When President Yun resigned in March 1962, Gen. Park was appointed acting President.

In January 1963 the revolutionaries formed the Democratic Republican Party (DRP), which nominated Gen. Park as its presidential candidate. In mid-March a plot to overthrow the military Government was allegedly uncovered and the acting President announced that a plebiscite would be held on a four-year extension of military rule. The reaction was strongly negative, and in July civilian government was promised within a year. In August Gen. Park retired from the army and became an active presidential candidate of the DRP. Freedom of political activity was restored for those not charged with past political crimes. The opposition forces were afflicted by divisions; Yun Po-Son eventually emerged as the candidate of the Civil Rule Party. The election in October resulted in victory by a narrow margin for Gen. Park, and at National Assembly elections, held in November, the DRP won an overwhelming majority of the votes. Civilian constitutional rule was restored on 17 December 1963, with the inauguration of President Park and the convening of the Assembly.

*Details of the Korean War and earlier history are given in the article History up to the Korean War (see p. 483).

THE THIRD REPUBLIC, 1963–72

Despite the establishment of a civilian government, all important positions in the administration were occupied by ex-military men, and the National Assembly was fully controlled by the DRP, headed by President Park. Although considerable economic development was achieved under the two Five-Year Plans (1962–66 and 1967–71), the Third Republic faced many domestic difficulties. In March 1964 large-scale student demonstrations broke out in Seoul, in protest at negotiations being conducted with Japan to normalize relations between the two countries. Despite demonstrations in opposition, the Government dispatched troops to South Viet Nam in co-operation with the USA, declared martial law in June 1965 in the Seoul area, and concluded the treaty normalizing relations with Japan.

In order to promote a parliamentary democracy, if not to weaken the power of the ruling party, minor parties formed a coalition grouping, the New Democratic Party (NDP), in January 1967. However, in the May 1967 presidential election, the incumbent President defeated Yun Po-Son, nominee of the NDP, again by a large margin, and the ruling party won a substantial majority of seats in the National Assembly. Following the disclosure of electoral irregularities involving the ruling party, the NDP demanded the nullification of the results and called for a fresh election.

Prompted by the growing popularity of the NDP in urban areas, the increase in threats from North Korea, and the realization that President Park's aims of 'national regeneration' were not forthcoming, the ruling party proposed a constitutional amendment in order to allow the incumbent President to serve a third term of office. This was adopted in September 1969 at a session of the National Assembly (boycotted by the NDP). A national referendum, held in October, approved the amendment. In the seventh presidential election, held in April 1971, President Park defeated Kim Dae-Jung, nominee of the NDP, by a narrow margin.

On 4 July 1972 Seoul and Pyongyang simultaneously issued a statement which announced the opening of dialogue between North and South to achieve national unification by peaceful means without outside intervention. A North-South Co-ordinating Committee was duly established for the purpose.

THE FOURTH REPUBLIC, 1972–79

The two Five-Year Plans, spanning the period 1962–71, had established a sound foundation, and the economic future of the nation seemed brighter. The sudden changes in the international situation, due to the Sino-US *détente* and new developments in North-South relations since 1972, provided the ruling party with convenient pretexts to perpetuate President Park's rule. As a result, the Government proclaimed martial law in October 1972, dissolved the National Assembly, and suspended the 1962 Constitution in order to pave the way for Park's continued rule. A new Constitution was proposed by the Extraordinary State Council and approved in a referendum in November.

The new Constitution, known as the *Yusin* ('Revitalizing Reform') Constitution, gave the President greatly expanded powers, authorizing him to issue emergency decrees and establish the National Conference for Unification (NCU) as an electoral college. In December 1972 the NCU, with 2,359 members, was established, and it elected Park to serve a new six-year term. Thus, the Fourth Republic emerged.

At the elections of February 1973, the DRP won 71 of the 146 directly elective seats of the National Assembly. Meanwhile, a new political movement, named *Yujonghoe* ('Political Fraternity for the Revitalizing Reform'), was established as a companion organization to the DRP, and 73 of its members were elected by the NCU, on the President's recommendation, to serve a three-year term in the National Assembly. Thus, President Park was assured an absolute majority.

South Korea witnessed tremendous economic growth during the period of the third Five-Year Plan (1972–76) and the fourth Five-Year Plan (which began in 1977), accompanied by rapid industrialization and an increase in per caput income. This, in turn, brought about remarkable educational and cultural development. However, the increasing autocracy and bureaucratism of the administration, coupled with corruption, caused the democratic movement to suffer, as freedom of speech and the press, and other civil rights, were suppressed or violated, and the number of political dissidents increased.

The kidnapping in 1973 of Kim Dae-Jung (who had been campaigning against Park in the USA and Japan) from Tokyo to South Korea by agents of the Korean Central Intelligence Agency (KCIA) created serious problems for Seoul with the US and Japanese Governments. On the domestic scene, anti-Government agitation and demands for the abolition of the 1972 *Yusin* Constitution continued to cause political instability in 1974 and after. To address the unrest, the Government banned all anti-Government activities and agitation for constitutional reform, rendering the political situation more unstable. Against this background of tension, President Park's wife was killed in August in an assassination attempt against Park by a North Korean agent. In the following two months, the ban was lifted, but the opposition NDP and others relentlessly pressed for constitutional reform and the release of political prisoners.

The Presidential Emergency Measure for Safeguarding National Security, which was proclaimed in May 1975 (ostensibly to strengthen national security against a mounting threat of aggression from North Korea, following the fall of South Viet Nam), only antagonized the dissidents further. The new measure imposed further prohibition on opponents of the 1972 Constitution and banned student demonstrations (with limited success). In March 1978 the three most prominent dissident leaders issued a joint statement, demanding the abolition of the 1972 Constitution and the complete restoration of human rights. The re-election in May 1978 of President Park to serve a further six-year term exacerbated the situation, as student unrest, supported by the opposition party, caused greater political turmoil.

At elections to the National Assembly in December 1978, the DRP received only 31.7% of the votes cast, while the NDP won 32.9%; however, the election of 22 independent candidates was a clear display of the voters' displeasure with both parties. President Park carried out a major ministerial reshuffle in that month, and released 1,004 prisoners, including Kim Dae-Jung.

From June 1979 until the complete military take-over of the Government in May 1980, South Korea encountered daunting political, social and economic problems. In July 1979 the NDP elected Kim Young-Sam as its new President. However, Kim's anti-Government speeches and press interviews led to the suspension of his presidency, and then to his expulsion from the National Assembly. In October all the NDP legislators tendered their resignations in protest. A power struggle within the NDP ensued, although the resignation notices were returned. Some conciliatory measures taken by President Park, such as the release of more political and 'model' prisoners in mid-1979, did not satisfy the dissidents and students. The resulting protests led to a serious uprising in Busan and other southern cities in October 1979, and students in Seoul prepared for a large-scale uprising towards the end of that month. In the midst of the crisis, Kim Chae-Kyu, director of the KCIA, shot and killed President Park on 26 October. The Prime Minister, Choi Kyu-Ha, was named acting President, as martial law was proclaimed. Kim Jong-Pil assumed the presidency of the DRP. The cancellation of Emergency Decree No. 9 was announced in December, and the termination of the *Yusin* rule was effected. A further 1,640 prisoners were pardoned in December.

THE INTERIM PERIOD, 1979–81

The NCU elected Choi Kyu-Ha as the new President of the Republic on 6 December 1979, and a new State Council (cabinet), headed by Shin Hyun-Hwack, emerged. Park's assassin, Kim Chae-Kyu, and his accomplices were executed in May 1980. Meanwhile, a power struggle within the DRP, as well as within the military leadership, developed. In December 1979 Lt-Gen. Chun Doo-Hwan, Commander of the Defence Security Command, led a coup within the armed forces, removing the martial law commander and making himself a new 'strong man' in the country.

In April 1980 there was more violent anti-Government agitation by students and the NDP (the presidency of which had been resumed by Kim Young-Sam). The appointment of Gen. Chun Doo-Hwan as acting director of the KCIA in April only inflamed the situation further. More campus rallies followed in May, demanding the immediate end of martial law, the adoption of a new constitution without delay, and the resignation of Gen. Chun. Troops were mobilized, and in mid-May martial law was extended throughout the country. Some 30 political leaders, including Kim Jong-Pil and Kim Dae-Jung, were arrested for interrogation. Kim Young-Sam was placed under house arrest and the National Assembly was closed, as were colleges, while all political activities, assemblages and public demonstrations were banned. In spite of these restrictions, students and dissidents took over the city of Gwangju on 19 May, after several days of bloody clashes with paratroopers and police. This uprising, which became known as the 'Gwangju Incident', was violently suppressed by the army, with the loss of nearly 200 lives.

On 20 May 1980 all members of the State Council tendered their resignation, and a new State Council (headed by the acting Prime Minister, Park Choong-Hoon) emerged. Meanwhile, as riots spread to other cities, the martial law command brought charges against Kim Dae-Jung for alleged seditious activities, including a plot to overthrow the Government by force, and for instigating student uprisings and the Gwangju rebellion. A Special Committee for National Security Measures (SCNSM) was formed on 31 May. President Choi became its chairman but real power rested with Gen. Chun and 15 other army generals, appointed by him to the SCNSM. With the establishment of the SCNSM, Gen. Chun resigned as acting director of the KCIA; however, as Chairman of the Standing Committee of the SCNSM, he still exercised absolute power.

President Choi unexpectedly stepped down in August 1980. On 27 August the electoral college chose Chun to be the next President, and on 2 September he was inaugurated. On that date an all-civilian State Council, headed by Nam Duck-Woo, took office. President Chun made it known that he intended to offer himself as a candidate for the presidency under the new Constitution. Kim Dae-Jung was sentenced to death, although this sentence was subsequently suspended. The National Assembly, which had been in recess since May, dissolved itself in late September. In October a national referendum was held to approve a new Constitution. Meanwhile, the Legislative Council for National Security (LCNS) was created to replace the SCNSM. All members of the LCNS were appointed by President Chun. The Government carried out intensive investigations and purged some 835 politicians in November. Some key political leaders of both parties, such as Kim Jong-Pil and Kim Young-Sam, were not only deprived of rights to participate in the political process, but were imprisoned during the investigation period. Although later released, they were placed under house arrest.

THE FIFTH REPUBLIC, 1981–88

The partial lifting of martial law was announced in January 1981. With this, new political parties were organized and a new electoral college of 5,278 members was created by popular election in February. On 25 February the new electoral college elected the incumbent President as the 12th President of the Republic, to serve a single seven-year term of office under the new Constitution, which banned re-election. On 3 March President Chun was inaugurated, and the Fifth Republic emerged. A new State Council was formed, with Nam Duck-Woo remaining as Prime Minister. Later in the month elections were held to the new 276-member National Assembly. The Democratic Justice Party (DJP), headed by President Chun, won 151 seats and became the majority party, while the newly formed Democratic Korea Party (DKP) secured 81 seats. With the establishment of the new National Assembly, the LCNS was dissolved. The KCIA was renamed the Agency for National Security Planning (ANSP) in April.

In January 1982 Yoo Chang-Soon, a former politician, replaced Nam Duck-Woo as Prime Minister and four other ministers were replaced. Some concessions to the wishes of the students and others were made by the Government, and Chun pledged that he would retire at the end of his term in 1988, thus becoming South Korea's first head of state to transfer power constitutionally. In January 1982 the midnight curfew, in force since September 1945, was lifted, except in the area near the demilitarized zone (DMZ) and along the coasts. In March 1982 some 2,860 prisoners were granted amnesty, which included the reduction of the life sentence for Kim Dae-Jung to a 20-year term.

A financial scandal, involving relatives of the wife of President Chun, precipitated another crisis in May 1982. As a result, there was a large-scale reorganization of the State Council, in which 11 ministers were replaced. Meanwhile, Kim Young-Sam, former leader of the now defunct NDP, was put under house arrest. In late June Chun appointed Kim Sang-Hyop, a respected educational leader, as the new Prime Minister. In December Kim Dae-Jung was released from prison and allowed to visit the USA for medical treatment.

South Korea's political climate was relatively calm for most of 1983, despite some campus disturbances in the latter part of the year. There was widespread shock and dismay in September at the shooting-down by the USSR of a Korean Air Lines passenger jet (which had apparently strayed into Soviet airspace), with the loss of 269 lives. In October President Chun embarked upon an overseas tour of several Asian and Australasian countries. However, his trip was cut short by an assassination attempt against him in Rangoon, Burma (now Yangon, Myanmar), allegedly perpetrated by North Korean agents. A bomb exploded at a mausoleum only minutes before the arrival of President Chun, killing 17 South Korean officials, among whom were the Deputy Prime Minister, three other ministers, three vice-ministers and two key members of the President's personal staff. Following the Rangoon incident, the surviving ministers tendered their resignations *en bloc*. Chin Iee-Chong, hitherto Chairman of the DJP, was appointed Prime Minister.

In February 1984 President Chun restored the political rights of 202 of the politicians who had been purged in 1981. However, 99 remained on the political blacklist, including Kim Jong-Pil, Kim Dae-Jung and Kim Young-Sam. In November Chun restored political rights to a further 84 persons. The drive launched by a group of former politicians who had regained their political rights brought about the formation of the New Korea Democratic Party (NKDP) in January 1985. Shortly after this, and just before the February general election, Kim Dae-Jung returned to Seoul from the USA, ending his self-imposed exile.

In the election to the National Assembly, held in mid-February 1985, the ruling DJP retained its majority, with 148 seats in the 276-member Assembly. Significantly, the NKDP (67 seats) won the majority of urban votes. President Chun reshuffled the State Council, appointing Lho Shin-Yong as Prime Minister and 12 other new ministers.

In April 1985 the Government restored political rights to the remaining 14 persons, including Kim Jong-Pil, Kim Dae-Jung and Kim Young-Sam, who had been on the political blacklist since 1980. Mass defections from the DKP and some defections from the Korea National Party increased the NKDP representation in the Assembly to 102 seats. Kim Young-Sam officially joined the NKDP in March 1986, and became adviser to the party President, Lee Min-Woo.

In March and April 1986 mass rallies were held, demanding constitutional reform and the resignation of President Chun. In June Chun finally agreed to the formation of a special parliamentary committee to discuss constitutional reform, which was to include members of the opposition parties; in the same month a special session of the National Assembly was convened to consider the findings of the committee, and negotiations continued for the remainder of the year. The NKDP proposed a new system of government, based on direct presidential elections. The DJP, however, favoured a system centred on a powerful Prime Minister, elected by the National Assembly, with greater responsibility to be accorded to the State Council, while the role of the President would be mainly ceremonial. The negotiations made little progress, despite a major concession by the NKDP when Kim Dae-Jung announced in November that he would not stand as a presidential candidate if the DJP accepted the NKDP's proposals.

In January 1987 the death of a university student, following torture in police custody, led to a new wave of anti-Government rallies and to the dismissal by President Chun of the Minister of Home Affairs and of the Chief of Police. Meanwhile, internal divisions were developing within the NKDP. In December 1986 the NKDP President, Lee Min-Woo, indicated his willingness to consider the Government's reform programme. While Lee's conditional endorsement of DJP proposals was supported by some members of the NKDP leadership, Kim Dae-Jung and Kim Young-Sam and 74 of the party's 90 National Assembly members left the NKDP and formed the Reunification Democratic Party (RDP).

In an unexpected move, in April 1987 President Chun announced the suspension of the reform process until the conclusion of the Seoul Olympic Games in 1988. While he reaffirmed his commitment to relinquish the presidency in February 1988, he indicated that the election of his successor would take place within the framework of the existing electoral college system. This precipitated an angry popular reaction against the Government, resulting in further violent clashes with riot police.

At its inaugural meeting in May 1987, the RDP elected Kim Young-Sam to the chairmanship and issued a strong denunciation of Chun's suspension of the reform process. The DJP responded by refusing to recognize the RDP as the main opposition party; in mid-May it reallocated committee chairmanships in the National Assembly, electing new chairmen without the participation of RDP members. In late May, following new disclosures about the circumstances of the death in January of the student under detention by the Seoul police, new riots erupted. In an attempt to stem the continued unrest, a reorganization of the State Council was effected, which included the appointment of a new Prime Minister, Lee Han-Key. The nomination, in early June, of Roh Tae-Woo, the Chairman of the DJP, as the ruling party's presidential candidate exacerbated anti-Government sentiment still further.

The RDP organized mass rallies in support of its demands for immediate constitutional reform, and violent confrontations between demonstrators and riot police became a daily occurrence. The US Government sent a diplomatic mission to advise the South Korean Government against the introduction of martial law. In late June 1987, after having conferred with former presidents Yun and Choi, as well as with Cardinal Stephen Sou-Hwan Kim (the Roman Catholic Archbishop of Seoul) and other religious leaders, President Chun met Kim Young-Sam in an unsuccessful attempt to seek solutions to the country's political crisis. Chun refused, however, to offer any concessions with regard to the opposition's principal demands. The RDP responded by mobilizing mass support for a 'great peace march', with tens of thousands taking to the streets of Seoul. Kim Young-Sam and other opposition leaders were arrested, and Kim Dae-Jung was returned to house arrest. This restriction on Kim Dae-Jung's freedom had been imposed more than 50 times since his return from exile in 1985.

Such was the extent of the national crisis that, in late June 1987, Roh Tae-Woo informed President Chun that he would relinquish both the DJP chairmanship and his presidential candidature if the main demands of the RDP for electoral reform were not met. Under pressure from the DJP leadership, from international (and particularly US) opinion and from the continuing public disorder, Chun acceded, and negotiations for a new constitutional framework were announced.

In a conciliatory move, in July 1987, the Government granted amnesty and the restoration of their civil rights to some 2,335 political prisoners, including Kim Dae-Jung. President Chun relinquished the presidency of the DJP (to which Roh Tae-Woo was elected in early August), and reorganized the State Council, appointing Kim Chung-Yul (a former air force chief of staff and Minister of Defence) as Prime Minister, and reorganizing eight other portfolios. In late August the DJP and the RDP agreed on the basic outline of a new Constitution; it was announced that a public referendum on a draft Constitution would be held in October, and that a direct presidential election would be conducted in December.

Following these announcements, industrial unrest increased, while students continued to hold anti-Government demonstrations. More than 500 industrial disputes broke out, mainly in the motor vehicle, mining and shipbuilding industries. By mid-October 1987, however, nearly all the disputes had been settled, the Government having conceded a hurried revision of labour laws, guaranteeing workers' rights to form trade unions and to conduct collective bargaining.

Negotiations between Kim Dae-Jung and Kim Young-Sam failed to achieve agreement on a single RDP presidential candidate. In mid-October 1987 Kim Young-Sam declared his candidacy, and in early November Kim Dae-Jung, together with 27 of the RDP's National Assembly members, formed the Peace and Democracy Party (PDP), which selected Kim Dae-Jung as its presidential candidate. The formation of the PDP resulted in the virtual dissolution of the Korea National Party and the NKDP. Meanwhile, Kim Jong-Pil revived the DRP, renaming it the New Democratic Republican Party (NDRP), and was chosen as its presidential candidate.

In October 1987 the National Assembly approved a constitutional amendment providing for direct presidential elections, and the new Constitution (to take effect in February 1988) was submitted to a national referendum. Some 20m., or 78.2%, of the eligible voters cast their ballots, 93.3% of which were in favour of the new Constitution. The first direct presidential election for 16 years took place on 16 December 1987. Some 23m. voters, representing 89.2% of the eligible electorate, cast their ballots; Roh Tae-Woo was elected President for a non-renewable five-year term of office, receiving 36.6% of the total votes cast. Kim Dae-Jung and Kim Young-Sam each received about 27% of the votes cast. While they both alleged electoral fraud, and although many irregularities were reported, it appeared that the principal cause of Roh Tae-Woo's victory was the opposition's failure to unite in support of a single candidate.

THE SIXTH REPUBLIC, 1988–

On 25 February 1988 Roh Tae-Woo was inaugurated as President. In his inaugural address, he proclaimed that the era of 'ordinary people' had arrived, and that 'the day when freedom and human rights could be relegated in the name of economic growth and national security has ended.' Shortly before his inauguration, Roh had appointed a new State Council, with Lee Hyun-Jae (a former President of Seoul University) as Prime Minister.

At the general election, which took place in late April 1988 under the newly adopted electoral law, four major parties (the DJP, the PDP, the RDP and the NDRP) competed for 299 seats. Of 26m. eligible voters, 75.8% turned out to elect 224 district representatives. The DJP secured the most seats but failed to win a majority in the National Assembly while the PDP, led by Kim Dae-Jung, became the main opposition party.

The Sixth Republic granted an increased measure of autonomy to national and private universities, and permitted the organization of student associations, thus expanding the initiatives taken during the period of the Fifth Republic. It also liberalized the press law, revoking the ban on the works of artists and writers who had defected to the North and allowing the circulation of certain North Korean publications. Restrictions on foreign travel were eased considerably. A campaign to bring to justice those who had been involved in political corruption resulted in the indictment of Chun Kyung-Hwan, a brother of former President Chun, and two of his brothers-in-law, who, as leading officials of the New Community Movement, were alleged to have embezzled US $9.7m. In April 1988, in response to this scandal, former President Chun resigned from all of the public offices that he held.

In late May and June 1988 thousands of students in Seoul and Gwangju took part in anti-Government and anti-US demonstrations, in commemoration of the uprising of May 1980. Demonstrations continued throughout June, July and August 1988. In many instances these led to violent confrontations between students and riot police, giving rise to fears that civil unrest would disrupt the Olympic Games in Seoul in September–October. In the event, the common perception of the Games as a matter affecting national prestige, shared by both the Government and the majority of the population, prevailed, and the Games were concluded successfully. A panel of the National Assembly began public hearings on alleged official corruption and violations of human rights during the Fifth Republic. As the opposition parties increased their pressure for the punishment of ex-President Chun and his aides, the anti-Government National Council of Student Representatives (Chondaehyop) intensified its activities, holding mass rallies and staging campus riots.

In November 1988 Chun apologized to the nation for the misdeeds of the Fifth Republic in a nation-wide televised address; he subsequently returned his property to the state and retreated with his wife to a Buddhist monastery in Gangwon Province. Meanwhile, the Government arrested 47 former advisers and officials of the Chun administration and put them on trial. However, the three opposition parties and the electorate were not satisfied with the measures taken. In order to alleviate tension, Roh reorganized the State Council in early December, replacing 21 of its 25 members and appointing Dr Kang Young-Hoon (a former ambassador to the United Kingdom) as Prime Minister.

In January 1989 some 200 anti-Government groups formed the Pan-National Coalition of the Democratic Movement (Chonminyon) and, in conjunction with Chondaehyop, intensified protest and strike activities. As these events, together with the clandestine visit in March to North Korea by a Presbyterian minister (the Rev. Moon Ik-Hwan) and three others, created a new political crisis, the Minister of Government Administration, Kim Yong-Kap, resigned, warning against the growing threat of 'leftist tendencies'. Confronted by this new crisis, Roh announced, in March 1989, the indefinite postponement of a referendum to provide an interim appraisal of his first year in office, causing a new wave of protests. The citizens of Gwangju held a week-long rally, to commemorate the ninth anniversary of the events of 1980, without resorting to violence. However, further demonstrations resulted in injuries to a large number of students and policemen.

The political situation altered dramatically in January 1990, when it was announced that the RDP and the NDRP would merge with the ruling DJP to form a new party, the Democratic Liberal Party. While this move secured for the Democratic Liberal Party control of more than two-thirds of the seats in the National Assembly (the DJP having lacked a majority), the broader aim of creating Japanese-style consensus politics was not attained in the months following the merger. Outside the new ruling bloc, the PDP, which was effectively isolated as the sole opposition party in the National Assembly, complained of a virtual coup, while the public responded by rejecting one Democratic Liberal Party candidate and nearly ousting another at by-elections for the National Assembly in April, in what should have been safe Democratic Liberal Party seats. A public opinion poll put the Democratic Liberal Party's popularity as low as 14%, and a new opposition party, the Democratic Party (DP), was formed, largely comprising members of the RDP opposed to the merger.

In March 1990 President Roh announced a major reshuffle of the State Council, in which 15 of its 27 members, including all the economic ministers, were replaced. In late April industrial unrest flared up again, followed by student demonstrations. In late July some 200,000 people participated in a rally in Seoul to protest at the approval by the National Assembly of several items of controversial legislation, which included proposals to restructure the military leadership and to reorganize the broadcasting media. Shortly afterwards, all the opposition members of the National Assembly tendered their resignation in protest at the contentious legislation. They also demanded the dissolution of the National Assembly and the holding of a general election two years before that scheduled for 1992. However, the Democratic Liberal Party claimed that the resignations were illegal and would not be accepted. The PDP deputies returned to the National Assembly only in mid-November, following an agreement with the Democratic Liberal Party that local council elections would be held, as demanded by the PDP, in the first half of 1991, to be followed by gubernatorial and mayoral elections in 1992. The Democratic Liberal Party also agreed to abandon plans for constitutional amendments, whereby executive power, currently vested in the President, would be transferred to the State Council. In late December there was an extensive government reshuffle, in which Kang Young-Hoon was replaced as Prime Minister by Ro Jai-Bong, hitherto chief presidential secretary.

The revelation of two new scandals dominated domestic political affairs in early 1991. The first incident involved the acceptance of bribes by high-ranking officials and prompted Roh to effect a minor government reshuffle in February. The second scandal was the beating to death by police of a student protester in an anti-Government rally in April, which precipitated weeks of widespread demonstrations, and also inspired the suicides of several students and others in protest. In response, Roh appointed a new Minister of Home Affairs in late April, and in the following month legislation was introduced to tighten control over the police and to relax the National Security Law. These concessions were, however, undermined by the hasty manner in which both measures were passed through the National Assembly, to the outrage of many opposition members who wished to debate more comprehensive reforms.

During May 1991 public unrest escalated to a level unprecedented during President Roh's tenure of office, as demonstrations by students and workers occurred throughout the country. The 11th anniversary of the 'Gwangju Incident' again occasioned widespread unrest, and in Gwangju itself more than 100,000 people were estimated to have participated in anti-Government activity. In late May the second government reshuffle of the year took place, and included the replacement of Ro Jai-Bong as Prime Minister by Chung Won-Shik, a former Minister of Education. Following the reorganization, an amnesty for more than 250 political detainees was announced.

Meanwhile, the Government drew some comfort from its results in the first local elections to be held in South Korea for 30 years. The Democratic Liberal Party won 65% of the seats in the elections to provincial and large city councils in June, securing control of 11 out of 15 assemblies (in fact, all except the opposition's south-western strongholds of the two Jeolla provinces and the city of Gwangju, and Jeju island, where independents gained a narrow majority). However, opposition parties secured almost as many votes but were disadvantaged by the 'first past the post' electoral system. The smaller opposition DP suffered particularly: its 14% of the votes cast secured it only 2.4% of the seats. However, its seats were at least distributed across the country, whereas the 19% of the seats and 22% of the votes cast obtained by the newly established New Democratic Party (NDP, created in April by a merger of the PDP with the smaller, dissident Party for New Democratic Alliance) remained overwhelmingly confined to the south-west, excluding some successes in Seoul. In September 1991 Kim Dae-Jung and Lee Ki-Taek agreed to a merger of their respective parties, the NDP and the DP, to form a stronger opposition front. The new party retained the latter's name: the Democratic Party (DP).

The main political development in the latter part of 1991 was a serious altercation between the Government and the Hyundai *chaebol* (conglomerate), in particular its founder and honorary chairman, Chung Ju-Yung. What was widely regarded as a politically motivated investigation into Hyundai share dealings resulted in claims for 136,000m. won (almost US $170m.) in unpaid taxes being brought against Chung and members of his family. In January 1992 Chung severed formal ties with Hyundai and formed a new political party, the Unification National Party (UNP), which attracted a mixed membership, including former dissidents, malcontents and media personalities. The UNP performed well, as did the DP, in the elections to the 14th National Assembly, held in late March. The Democratic Liberal Party suffered a humiliating set-back, securing only 38.5% of the votes cast, as opposed to the 73% which its then separate pre-merger component parties had totalled in the 1988 general elections. The Democratic Liberal Party thus emerged with 149 seats, one short of an absolute majority in the 299-member Assembly (although enough independents had been won over by the time the Assembly opened in July to ensure a working majority). The opposition DP obtained 97 seats (having won 29.2% of the votes cast), including 25 of the 44 seats in Seoul, as well as an expected clear majority of the seats in Kim Dae-Jung's heartland in the south-west; however, it gained only a handful of seats elsewhere. By contrast, the UNP's 31 seats and 17.4% of the votes cast were more evenly distributed nation-wide.

The emergence of the UNP (which subsequently changed its name to the United People's Party, UPP) added a new dimension to the presidential elections, which were due to be held in December 1992. In May the Democratic Liberal Party chose the former opposition leader, Kim Young-Sam, as its candidate, by a majority of two to one over his rival, Lee Jong-Chan. Initially there were fears that Lee, who had the support of Kim Young-Sam's many enemies among the ex-DJP old guard core of the Democratic Liberal Party, might split the party by leading a 'walk-out' of anti-Kim elements. This prospect receded, however, owing to divisions among these elements, pressure from Roh Tae-Woo, and above all the realization that such a split might allow the opposition to win. In late August Kim replaced Roh as Democratic Liberal Party President.

In other respects, the domestic political scene during 1992 appeared relatively stable. The Government succeeded in postponing a third round of elections (mayoral and gubernatorial), on the grounds that three elections in one year would be prohibitively expensive. Both student and labour activism were more muted than in previous years, except for a brief outbreak of pro-North Korean demonstrations in some universities in early 1992. In June and October President Roh effected the second and third partial government reshuffles of the year. The latter was presented as the formation of a politically neutral State Council to guarantee a fair presidential election, and included the odd spectacle of the entire Government, as well as the President, resigning from the ruling party.

The Presidency of Kim Young-Sam

The presidential election, on 18 December 1992, gave a convincing victory to Kim Young-Sam, who received some 42% of the votes cast. Kim thus became the first South Korean President since 1960 not to have a military background. Of the six other candidates, his nearest rivals were Kim Dae-Jung, with 34% of the votes cast, and Chung Ju-Yung, with 16%. Kim Dae-Jung, after his third presidential defeat, announced his retirement from politics. Chung Ju-Yung resigned as President of the UPP in early 1993, following allegations that he had embezzled Hyundai finances to fund his election campaign. Subsequent defections from the UPP caused the party to lose its status as a parliamentary negotiating group, and in 1994 the UPP merged with a smaller opposition party.

The opposition was further weakened by the new President's unexpected emergence as a radical reformer. In a campaign against corruption, which won him widespread approval, Kim publicly declared his own assets and forced the entire political élite to follow his example. This was an astute political move, since it exposed many of the ex-DJP old guard in the Democratic Liberal Party as possessing wealth that they were hard put to explain, and thus weakened their position *vis-à-vis* the President's own faction. At first it seemed as if this campaign might go awry, when three newly appointed ministers and several key presidential aides were also caught in the net and forced to resign on various charges of corruption. However, the main casualties were enemies of the President.

By mid-1993 the net appeared to be widening. The military were also targeted, and a number of senior officers were removed, charged either with corruption or association with the military coup of December 1979. The President's official redefinition of this event as a 'coup-like incident' raised the possibility that he might even bring to book his own two immediate predecessors as its instigators. However, Kim's real aim was more probably to impose his authority firmly on the ruling party without going so far as to risk splitting it. The popularity of this anti-corruption drive accounted for the Democratic Liberal Party winning five out of six by-elections held in April and June 1993, including seats in Seoul that the opposition had been expected to take.

Other areas of political life in the first half of 1993 showed continuity, even conservatism. The usual May student riots were firmly quelled, as was unrest in July at the Hyundai motor works in Ulsan—itself an exception to a generally quiescent labour situation. Elsewhere, a number of dissident figures expressed support for President Kim, whose overall position appeared strong. The same mixture of radicalism and conservatism continued during the latter part of 1993. The President's reform drive reached its peak in August, when a ban was announced on bank accounts held under false names. This was an issue that previous administrations had not dared to address, and in the event the severity of the initial decree was mitigated by various concessions.

Another bold step was Kim's announcement in December 1993 that his Government would ratify the recently concluded Uruguay Round of the General Agreement on Tariffs and Trade (GATT), even though this contravened his campaign pledge never to permit the opening of South Korea's rice market to foreign competition. The violent public demonstrations that followed this policy change prompted Kim to effect a major government reorganization in mid-December. The lack-lustre Hwang In-Sung was replaced as Prime Minister by Lee Hoi-Chang, who, as Chairman of the Board of Audit and Inspection, had played a major role in the President's crusade against corruption. The new Prime Minister's tenure lasted barely four months, however. He resigned on 22 April 1994, after a dispute over his exclusion from a new committee established to co-ordinate policy towards North Korea. Lee Yung-Duk succeeded him as Prime Minister, while the veteran Lee Hong-Koo resumed the unification portfolio (in which capacity he had earlier served under Roh Tae-Woo).

Lee Hoi-Chang's departure caused a degree of disappointment, as did two scandals that emerged in early 1994, which implicated Kim Young-Sam in financial malpractices. Although the President managed to evade prosecution, during 1994 his stance became more conservative. In June strikes staged by railway workers ended with mass arrests. One month subsequently radical students mourning the death of Kim Il Sung, the North Korean leader, received the same treatment—prompting some opposition legislators to accuse the Government of 'McCarthyism' and over-reaction.

In late 1994 and the first half of 1995 the Government's popularity declined markedly. One reason for this was a series of man-made disasters, of which three were especially perturbing. In October 1994 the Songsu road bridge across the Han river in Seoul collapsed, killing 32 people. In April 1995 a gas explosion on a subway construction site in Daegu caused more than 100 deaths. However, these disasters were overshadowed by the collapse, in late June, of the luxury Sampoong department store in Seoul (which had been built as recently as 1989), with the loss of 458 lives. In the resultant public outcry, the Government was blamed for inadequate safety regulations to prevent disasters, and was accused of ill-co-ordinated responses when they occurred. Moreover, officials were alleged to have accepted bribes to overlook shoddy work and malpractice in the construction industry.

Accidents aside, there was also a sense of instability within the administration. A major restructuring of the State Council in December 1994, only one year after the last reshuffle, prompted criticism that when in opposition Kim Young-Sam had condemned such frequent turnovers. Lee Hong-Koo was promoted to the post of Prime Minister, while other appointments were regarded as conciliatory gestures to the increasingly restive ex-DJP old guard in the ruling party.

Further ministerial changes were effected in 1995, including the dismissal of Lee Hyung-Koo, the Minister of Labour, who was accused of corruption. His dismissal was unfortunately timed, occurring amidst harsh government action against worker protest. In early June, in an act unprecedented under past military regimes or even Japanese colonial rule, riot police stormed the Catholic Myeongdong Cathedral and a leading Buddhist temple in Seoul to seize 13 trade union leaders from the state telecommunications agency, Korea Telecom, who had sought sanctuary there.

In August 1995 one of the President's closest confidants, Seo Seok-Jai, the Minister of Government Administration, provoked an outcry by an unguarded comment to journalists that a former President was in possession of a huge political 'slush fund', which had allegedly been deposited under false and borrowed names in various accounts. Seo's prompt resignation and partial retraction, however, did not quell rumours. For a while there had been no great public support for opposition attempts to have Chun Doo-Hwan and Roh Tae-Woo prosecuted for their role in the 1979 coup and the Gwangju massacre of 1980. (In October 1994 a tribunal investigating the 1979 coup had found that Chun and Roh had participated in a 'premeditated military rebellion' but it had decided not to prosecute the former Presidents; likewise, in July 1995, during an official investigation into the Gwangju events, Chun and Roh were cleared of having committed 'homicide aimed at achieving insurrection'.) The scandal deepened in late October 1995, when Roh Tae-Woo, in an emotional televised address, admitted to having amassed 500,000m. won (some US $650m.) in illicit political funds during his term of office. In early November Roh appeared before the Chief Justice's office for cross-examination. After a second interrogation in mid-November, Roh was arrested on charges of corruption.

Even before this scandal, and only days before the Sampoong disaster, the electorate delivered a rebuke to the Government in South Korea's first full local elections for 34 years, held on 27 June 1995. The Democratic Liberal Party won only five of the 15 major gubernatorial and mayoral posts, followed by the DP, which took four (including the mayorship of Seoul). Four posts were also won by the United Liberal Democrats (ULD), a grouping established in March by defectors from the Democratic Liberal Party; the new party's President was Kim Jong-Pil, who had resigned (or had been forced out) as Democratic Liberal Party Chairman in January.

The DP also performed well in major city and provincial council elections, winning 355 of the total 875 seats, followed by the Democratic Liberal Party (286), independents (151) and the ULD (83). However, the DP's satisfaction was short-lived: Kim Dae-Jung, the party's former leader and continuing *éminence grise*, announced in mid-July 1995 that he was returning to political life and would found his own party. This, the National Congress for New Politics (NCNP), was formally constituted in early September, severely undermining the DP, as 54 of its 96 deputies defected to the new party.

For the remainder of 1995 and into early 1996, however, party politics and local government were eclipsed by the public disgrace of those formerly in power. The charges against Roh Tae-Woo were widened to include the coup of December 1979 and the Gwangju massacre of May 1980; in December 1995 another ex-President, Chun Doo-Hwan, was also arrested on these charges and his trial began in February 1996. He was subsequently arraigned for accumulating illicit funds even greater than those of Roh. The prosecutors requested that Chun receive the death penalty and Roh life imprisonment, and in late August they were sentenced accordingly; however, these terms were subsequently commuted to life and 17 years, respectively.

This astonishing turn of events reflected a deliberate political decision by Kim Young-Sam, once the scandal of Roh's illicit funds had become public, to bring down, once and for all, those who had previously dominated the ruling party. Although a popular course of action to take—most of the nation was glad to see these former influential figures, in particular Chun, brought to justice—it was potentially risky, since it depended on the President being able to distance himself entirely from his predecessors' corrupt activities. Kim Dae-Jung, himself damaged by an admission that even he had taken US $2m. from Roh, made every attempt to draw Kim Young-Sam's name into the scandal. Yet despite the many uncertainties surrounding the source of funds for Kim Young-Sam's 1992 election campaign, and the arrest in March 1996 of his close aide, Chang Hak-Ro, on corruption charges, the reputation of the President, an adroit politician, remained untarnished.

In December 1995 Kim renamed the ruling party the New Korea Party (NKP) and Lee Soo-Sung, the president of Seoul National University, replaced Lee Hong-Koo as Prime Minister. Further government changes took place in an attempt to reassure the business community. In elections for the National Assembly, held on 11 April 1996, the NKP obtained 139 of the 299 seats, including most of the seats in Seoul (the first time that a ruling party had achieved this). The NCNP took 79, which was fewer than it had hoped, mostly in the south-west. Kim Dae-Jung failed to win a seat, having placed himself too far down the list of appointed candidates in a display of over-confidence. The ULD increased their tally of members from 32 to 50, while the DP, already poorly represented, was reduced to 15 seats.

Although the ruling party was thus 11 votes short of a working majority, it had acquired the necessary majority by the time the Assembly convened in June 1996, having persuaded 12 DP and independent members to join. This provoked protests from the NCNP and the ULD, who worked together to delay the Assembly's normal business until July. This unlikely co-operation—Kim Jong-Pil founded the fearsome KCIA, which had tried to assassinate Kim Dae-Jung in 1973—proved surprisingly durable.

August 1996 was notable not only for the sentencing of former Presidents Chun and Roh, but also for the severity with which the annual pro-unification rallies by radical students were suppressed. The students were besieged for several days in Yonsei University, Seoul, before an assault was launched by police, which destroyed the building. One riot policeman was killed and more than 5,000 students were taken into custody, the largest number ever arrested. The latter half of 1996 also witnessed a remarkable rate of attrition among ministers, with many dismissals and resignations taking place in the months prior to Kim Young-Sam's customary cabinet reorganization in mid-December, when a further nine ministers were relieved of their posts. In late December 1996, having spent much of that year failing to persuade both sides of industry to agree on labour law reform, the Government approved legislation that gave employers enhanced powers of engagement and dismissal, but which failed to legalize the powerful Korean Confederation of Trade Unions (KCTU). To add insult to injury, the legislation was passed at a swift dawn session of the National Assembly, of which the opposition was not informed. A predictable reaction followed, with the KCTU attracting wide support for strikes, which cost more than US $3,000m. in lost production. However, legislation incorporating a compromise was approved in March, which recognized the KCTU, while postponing the introduction of greater flexibility for employers.

By then, however, the nation's attention was elsewhere. In mid-January 1997 the Hanbo group, South Korea's 14th largest *chaebol* (conglomerate), was declared bankrupt with debts of some US $6,000m., largely incurred through the failure of a steel mill project, rapidly revealed to be both ill-advised and corrupt. For a President whose clean image had already been tarnished by accusations that his election campaign might have drawn on the illicit funds of his predecessor, the procession of senior aides, ministers, bank chairmen and others who were implicated in 'Hanbogate', as the scandal soon became known, was a severe set-back. Yet more serious were separate charges of influence-peddling against his own son, Kim Hyun-Chul; in October 1997 Kim junior was found guilty of receiving bribes and of tax evasion.

These humiliations severely weakened the President's position as his term in office drew to a close. Attempts to lessen the implications of the scandals included yet more cabinet reorganizations. In early March 1997 Goh Kun, a known incorruptible, became Kim Young-Sam's sixth Prime Minister in four years. The entire economics team was replaced, and an experienced former Minister of Finance, Kang Kyung-Shik, was appointed Deputy Prime Minister in charge of the economy, while Lim Chang-Yul became Minister of Trade. Efforts by the new team to promote financial reform, the urgency of which had been illustrated by the near or total collapse of several further conglomerates after Hanbo, none the less faced obstacles, not only through the reluctance of politicians to address unpopular issues in an election year, but also from bureaucratic wrangling between the Bank of Korea and the Ministry of Finance and the Economy over spheres of influence and regulatory responsibilities.

Kim Young-Sam's weakness benefited the democratic process, however, by ensuring that his successor (the Constitution allows only a single, five-year term) could not be personally chosen. In mid-July 1997 an NKP convention approved Lee Hoi-Chang, the former Prime Minister, as its candidate for the presidential election. In September the increasingly unpopular Kim resigned as President of the NKP and was replaced by Lee. Allegations that Lee's sons had evaded military service, however, proved damaging. Factional disunity in the NKP was highlighted by the decision of Rhee In-Je (who had challenged Lee for the party's nomination) to resign from the party and announce his intention to stand in the election. In the event Rhee's act was decisive, as it split the ruling bloc.

The Presidency of Kim Dae-Jung

The outcome of the presidential election, held on 18 December 1997, was a triumph at last for Kim Dae-Jung, some 26 years after his first bid for the presidency, and represented a new milestone for South Korean democracy, with the first ever transfer of power to the opposition. On a turn-out of 80% of the electorate, Kim won 40.3% of the votes cast, only just ahead of Lee (38.7%), while Rhee trailed in third place (19.2%). The contest had narrowed to these three main candidates, with Kim Jong-Pil of the ULD withdrawing his candidacy in favour of Kim Dae-Jung (representing the NCNP) in exchange for promises of the premiership and of constitutional change, to be effected by 2000, with a view to giving the Prime Minister more power. Similarly, Lee had been boosted by the decision of Cho Soon, a former governor of the Bank of Korea (and one-time political protégé of Kim Dae-Jung) who was representing the ailing DP, to merge that party with the NKP, thus forming the Grand National Party (GNP) and presenting a more effective challenge to the new NCNP-ULD alliance. Regional loyalties were once again much in evidence. As ever, the Jeolla provinces in the south-west voted *en masse* for their favourite politician, Kim Dae-Jung, and Kim Jong-Pil's heartland, the usually conservative Chungcheong provinces in the centre-west, also duly delivered a majority to Kim Dae-Jung. Conversely, in the south-eastern Gyeongsang provinces, which had furnished all previous presidents since 1961, the votes were split between Lee and Rhee. The outcome was thus decided in the greater Seoul region, comprising the capital, the surrounding Gyeonggi province and the port of Incheon, where 40% of voters lived.

The unprecedented opposition victory also reflected popular anger at the Kim Young-Sam administration's handling of the economy. This had reached crisis point in November 1997, when South Korea was forced (having hitherto denied any such intention) to seek emergency rescue loans from the IMF. At a total of over US $57,000m., this was the largest-ever such rescue 'package' released by the IMF and was perceived as a national humiliation by many Koreans. The 'IMF era' at once transformed the context for politics and policy, and thus presented significant difficulties for the new President. Kim Dae-Jung initially announced that he would renegotiate with the IMF. Once elected, however, he rapidly took a leading role in calming markets, and emerged as a champion of deregulation and liberalization, rather than as the populist that his background might have suggested. Such intervention, while not strictly constitutional (Kim did not take office formally until 25 February 1998), averted disaster during what would otherwise have been a dangerous power vacuum at a crucial time. This pattern continued after his inauguration, with the introduction of a series of reforms that were radical by past standards. New legislation to allow labour flexibility, which had provoked such repercussions under Kim Young-Sam, was passed, and a tripartite commission of management, labour and government was set up. By August the unemployment rate had tripled to over 7% without provoking more than sporadic strikes. The exception was a confrontation at Hyundai Motor, the first large *chaebol* to declare compulsory redundancies. Even here there were, by late August, signs that the new regime's preference for consensus over confrontation—itself a novelty in Korean politics—might avert a forcible solution.

Yet Kim Dae-Jung also faced great obstacles. In the formal political arena, the GNP held a majority in the National Assembly. It refused to confirm Kim Jong-Pil as Prime Minister until August 1998; prior to that he was designated as 'acting' premier. As such he was not permitted to form a cabinet, although, fortunately, the outgoing Prime Minister, Goh Kun, agreed to do this, appointing a State Council in March (see below). Goh later joined the NCNP, for whom, in June, he was elected Mayor of Seoul in local elections which boosted support for the new ruling coalition. Parliamentary infighting rendered the National Assembly largely inactive from May to August, provoking widespread popular disgust. The NCNP-ULD strategy was to break the GNP's hold on the legislative body by encouraging defections to the ruling camp, and this looked likely eventually to produce a working majority for the Government. Even so, society at large remained apprehensive of, if not hostile to, the new paeans to market capitalism. Importantly, suspicion was not confined to workers and civil servants, who feared the potential effects of downsizing, smaller government, and privatization. Equally hostile were the *chaebol*, whose reckless over-expansion was, by common consent, to blame for the nation's financial plight, but which showed little inclination to follow government advice and rectify their affairs by means of the rationalization of activities, sale of assets, debt reduction and more transparent accounting. Given the size and power of, in particular, the largest conglomerates, it was not obvious how an administration avowedly committed to market forces could intervene to force the *chaebol* to change.

On an organizational level, the new Government instituted several changes in early 1998. The two posts of Deputy Prime Minister, which formerly accompanied the economy and unification portfolios, were abolished. Responsibility for foreign trade and foreign affairs was combined under the new Ministry of Foreign Affairs and Trade, while the Ministry of Commerce, Industry and Economy replaced the former Ministry of Trade, Industry and Economy. The Ministries of Home Affairs and Government were merged. Two new bodies were created: the Financial Supervisory Commission, which replaced three old regulatory bodies for banks, insurance and the stock market, and the Planning and Budget Commission, within the Office of the President; those institutions that lost powers were resentful, notably the Ministry of Finance and the Economy and the Bank of Korea, resulting in rivalry and arguments over policy. Compromise was evident in Kim Dae-Jung's first Cabinet, announced in March 1998, which allocated seven posts to the NCNP and five to the ULD. Non-party appointees included the Minister of Finance and the Economy, Lee Kyu-Song, who had held the same post a decade previously and was regarded as a scion of the old 'Korea Inc.' Fears that he would clash with the Presidential Secretary for Economics, Kim Tae-Dong, a professor famously critical of the *chaebol*, were eased when the latter was quickly transferred to another post. (In general, Kim Dae-Jung pledged not to continue the habit of frequent changes of personnel.) Other interesting appointments included that of the unification Minister, Kang In-Duk, whose hardline reputation and KCIA background did not prevent his implementation of Kim Dae-Jung's 'sunshine policy' with regard to relations with North Korea. Another ex-KCIA figure, Lee Jong-Chan, returned to head the Agency for National Security Planning (NSP); it was subsequently further retitled, as the National Intelligence Service, in a bid to distance itself from an insalubrious past, which included efforts at every election to tarnish Kim Dae-Jung as a pro-communist, culminating in 1997 in alleged co-operation with its equivalent in Pyongyang to forge the required documents.

More generally, Kim Dae-Jung's desire for reconciliation was seen in the release from prison, four days after his election victory, of ex-Presidents Chun Doo-Hwan and Roh Tae-Woo. In a break with the tradition of persecuting one's presidential predecessors, he resisted demands for Kim Young-Sam to be called to account for the 1997 financial crisis, although hearings on this were promised. He also offered amnesty to political prisoners who promised to obey the law, a softening of the previous insistence that they renounce their leftist beliefs. Many, however, refused this offer, leaving South Korea still holding some of the world's longest-serving prisoners of conscience. Most have subsequently been released in any case.

The Government's drive for restructuring continued undiminished in 1999, winning praise for South Korea as the leading force of economic reform in Asia. A much faster economic recovery than expected also helped to ease the difficulties of transition. Kim Dae-Jung's insistence that 'capital has no nationality' won growing acceptance, and foreign investment was increasingly welcomed rather than feared. This change of attitude was facilitated by widespread recognition that the culprits of the 1997 crisis were the *chaebol*, and their stubborn refusal to reform. This applied especially to the 'big five'—Hyundai, Samsung, Daewoo, LG and SK—which deemed themselves too big to fail. In 1999 Daewoo in effect went bankrupt, with debts eventually revealed to total some US $80,000m. The Government intervened to prevent a second financial crisis, dismantling the group and attempting to sell its viable parts. The proposed sale of Daewoo Motor to the Ford Motor Co collapsed in September 2000, however, following Ford's investigation of the company's finances. Amidst criticism that reform of the *chaebol* was not proceeding fast enough, during 2000 Hyundai's financial position became a serious cause for concern. The controlling Chung family fought for supremacy within the group but then pledged in June to withdraw from the *chaebol's* management.

Further reorganization of government was effected in 1999, mainly in connection with Kim Dae-Jung's first major reshuffle in late May. The Planning and Budget Commission became a full ministry, with the reformer Jin Nyum retaining the portfolio. With another reform figure, Kang Bong-Kyun, appointed at its head, the normally conservative Ministry of Finance and the Economy was expected to vie with the new Financial Supervisory Commission for control of restructuring. Other cabinet changes affected a further 10 portfolios, including defence, unification, commerce and industry. Two months earlier the ministers responsible for science and maritime affairs had also been replaced. Overall, the changes were regarded as strengthening the positions of Kim Dae-Jung, the NCNP and those favouring reform. In mid-1999 a series of scandals ended Kim Dae-Jung's initial period of political harmony and lost the ruling coalition two by-elections in early June (although, at the end of March it had gained two seats from the GNP). The wife of the new Minister of Justice, Kim Tae-Joung, was accused of accepting fur coats from the wife of a tycoon being tried for corruption. Nothing was proved, but the Minister was later dismissed over another matter: during his tenure of the post of Prosecutor-General, a drunken underling had boasted of fomenting a strike at the national mint as a pretext to take action against trade-union militants. The new Minister of the Environment, Son Sook, an actress, resigned after only one month for accepting gifts of US $20,000 from businessmen during a performance in Moscow when already appointed to the post. All this, however, besides being rather bizarre, was of little significance compared with the large-scale corruption under previous regimes.

Parliamentary and party politics remained mostly unedifying in 1999, with the National Assembly idle for long periods owing to infighting between government and opposition. The ruling coalition too became strained, with the ULD frustrated at their leader Kim Jong-Pil's agreement not to press for constitutional changes which would have given him, as Prime Minister, more power. In January 2000 Kim Jong-Pil resigned as premier. He was replaced by another ULD figure, Park Tae-Joon, the founder of the Pohang Iron and Steel Co (POSCO), the world's largest steel-maker, still partly state-owned. A wider reorganization of the administration followed, with Lee Hun-Jai, the dynamic reformer heading the Financial Supervisory Commission, taking over as Minister of Finance and the Economy. The following portfolios were also reallocated: foreign affairs and trade; commerce, industry and energy; government administration and home affairs; education; construction and transportation; and maritime affairs and fisheries. Earlier, a new Minister of Unification and head of national intelligence had been appointed.

With legislative elections approaching, the ULD ended their coalition with Kim Dae-Jung's party, itself relaunched as the Millennium Democratic Party (MDP) in a bid to widen its appeal. The MDP had hoped to absorb the ULD, but Kim Jong-Pil objected to the MDP's support for the posting on the internet (use of which had spread rapidly in South Korea) by civic groups of blacklists of politicians deemed unfit for office, whether as corrupt, opportunists, idle, or involved with past military regimes. Kim Jong-Pil himself was so named, as were many other ULD representatives. Confusingly, Park Tae-Joon and other ULD ministers remained in government positions, while the ULD declared themselves to be an opposition party and put up their own candidates against the MDP. The result was the decimation of the ULD in elections for the 16th National Assembly on 13 April 2000, held under controversially revised rules that reduced the number of seats from 299 to 273: 227 were directly elective on a 'first-past-the-post' basis, and 46 were chosen from party lists based on the share of the overall vote. The ULD representation declined from 50 to 17 seats, with the party losing ground even in its regional heartland of Chungchong, south of Seoul. Three days prior to the election it was announced that a summit meeting was to take place in June between Kim Dae-Jung and the North Korean leader, Kim Jong Il. (Kim Dae-Jung was awarded the Nobel Peace Prize in October in recognition of his progress towards reconciliation with North Korea.) Although the opposition condemned the timing of the announcement as electioneering, it did not appear to benefit the MDP significantly as the MDP's tally only rose from 98 to 115, well short of the majority it had hoped for. The GNP remained the largest party, gaining 11 seats to reach 133, four short of overall control. A small new party, the Democratic People's Party (DPP), took two seats, and there were six independents (most pro-MDP). Regional loyalties persisted in some areas; the GNP took 64 out of 65 seats in the Gyeongsang provinces in the south-east, while the MDP won 25 out of 29 in Jeolla in the south-west.

This result meant that political manoeuvring would continue. After briefly approaching the GNP, Kim Dae-Jung chose to rebuild his alliance with the rump of the ULD, who furnished a third Prime Minister, Lee Han-Dong, in May 2000 when Park Tae-Joon had to resign over a tax scandal. The fact that Lee had lately been a senior figure in the GNP did nothing to improve relations between the Government and the opposition, and the new Assembly, like its predecessor, remained largely paralysed by boycotts and infighting. A cabinet reshuffle followed in August 2000. The security team—comprising foreign affairs, defence, unification, and national intelligence—was retained, to stress continuity and to reward success in North Korean relations after June's breakthrough summit meeting. Economics ministers, however, were removed, owing to a perceived slackening and drift in the progress towards reform. After only seven months, Lee Hun-Jai was replaced at the Ministry of Finance and the Economy by Jin Nyum, hitherto Minister of Planning and Budget, who was succeeded by Jeon Yun-Churl, previously head of the Fair Trade Commission, the main anti-trust body. The Ministers of Education, Labour, Health and Welfare, Agriculture, and Commerce, Industry and Energy were also replaced as well as the heads of several cabinet-level agencies. With both the Ministry of Finance and the Economy and the Ministry of Commerce, Industry and Energy now allocated to their fourth minister in 30 months, Kim Dae-Jung's pledge to avoid the over-frequent cabinet changes of previous administrations looked somewhat meaningless. The turnover of personnel continued when the new Minister of Education resigned after just three weeks, on suspicion of involvement in a scandal; this was closely followed by the resignation of the Minister of Culture and Tourism, Park Jie-Won, a key figure in the secret talks which led to the North-South summit meeting.

Despite the summit's success, in domestic affairs the Government appeared to lose some control. An ill-conceived health-care reform brought doctors and pharmacists out on strike in mid-2000, angered the public, and cost some US $3,000m. The parties continued to disagree, delaying the next session of the National Assembly for a month. Even the awarding in October 2000 to Kim Dae-Jung of the Nobel Peace Prize, the first ever of any Nobel prize to a Korean, did not improve the national mood.

The year 2001 witnessed a further weakening of the President's administration. In January the MDP lent four parliamentary deputies to the ULD, to give the latter the numbers needed to become a recognized floor group. In the same month two posts of Deputy Prime Minister were reinstated: one, as was the case previously, was allocated to the Minister of Finance and the Economy; the other was a new post of Education and Human Resources Development—with yet another new education minister. At the same time, the Presidential Commission on Women's Affairs, which was created under Kim Dae-Jung, was upgraded to become the Ministry of Gender Equality.

In March 2001 there was yet another reshuffle, as the MDP formalized its coalition with the ULD and the tiny DPP to gain a narrow majority in the National Assembly. Nine ministers, three other officials of cabinet rank and two senior presidential secretaries were replaced, with a similar removal of vice-ministers soon after. The DPP's Han Seung-Soo, a former ambassador to the USA, became the Minister of Foreign Affairs and Trade, bespeaking a need for better relations with the new US Administration under President George W. Bush (see below). Others replaced were the Ministers of Construction and Transportation, National Defence, Government Administration and Home Affairs, Health and Welfare, Information and Communication, Science and Technology, and Unification. The Government's economic team was retained, despite criticisms of falling growth and a retreat from reform, the latter including a series of subventions to ailing companies in the Hyundai group. Other commentators condemned the President for appointing nine professional politicians, including three ULD placemen, and six from his home region of Jeolla in the south-west of the country. The turnover continued: in May an incoming Minister of Justice set a new record, serving for just 43 hours after the 'leak' of a letter in which he pledged fealty to Kim Dae-Jung in feudal tones.

Politics outside the legislature also became more fractious, with a resurgence of violent strikes and demonstrations, including action at Daewoo Motor, where some workers (but not a majority) opposed its proposed sale to General Motors. A tax audit of leading newspapers, most critical of the Government, led to substantial fines totalling around US $390m. and accusations, including some from abroad, of an attack on press freedom. The *chaebol*, hitherto on the defensive, successfully demanded an easing of deadlines for restructuring. Dissension grew over the 'sunshine' policy, as Pyongyang broke off talks, provoked the South with naval incursions and manipulated a visiting southern delegation. As a result, in September 2001 the conservative ULD joined the GNP to pass a vote dismissing the Minister of Unification, Lim Dong-Won. Although this had no binding force, Lim resigned (he was promptly made a special adviser to the President), as did the entire State Council. ULD ministers were dismissed, while the ULD in turn expelled Lee Han-Dong for agreeing to stay on as Prime Minister. For separate reasons, the rapid turnover of ministers reached a new level of absurdity when South Korea had four different transport ministers in the space of six weeks during August and September 2001. In October the MDP lost three by-elections to the GNP. Soon after, Kim Dae-Jung symbolically resigned from the party presidency.

If 2001 was a difficult year for Kim Dae-Jung, the year 2002 proved to be even worse. In a curious repetition of Kim Young-Sam's final year and despite the major difference of a strong economy (for which he received scant credit) a President who had begun as a reformer left office regarded as a liability, beset by scandals. Furthermore, like his predecessor, Kim Dae-Jung had to witness the imprisonment of two sons for alleged corruption and influence-peddling, as part of a series of scandals which, if small by past standards, showed the limits of the reform process—involving as it did close associates, as well as kin, of a leader who had claimed moral superiority. Institutions too, notably the tax and intelligence services and the prosecution office, were shown as being involved in past or present corrupt practices. This remained as a challenge for the next President.

What South Korea surely did not need was endless cabinet reshuffles. Yet on 29 January 2002 there was yet another reallocation of portfolios, drawing general criticism except for the appointment of a woman, Park Sun-Sook, as presidential spokesperson. Lee Sang-Joo became the administration's seventh Minister of Education, while the Minister of Unification, Hong Soon-Young, was removed as a result of the failure of inter-Korean talks. The vaunted aim was to create a politically neutral cabinet, and four MDP-affiliated ministers lost their posts; however, two of the new ministers were also MDP members, with another ULD supporter. An unfortunate loss was the experienced Minister of Foreign Affairs and Trade, Han Seung-Soo, again ostensibly on the grounds of removing party political figures (although the DPP, as such, was of no consequence), just as he was arranging a delicate visit by US President George W. Bush. The new ministers responsible for unification and foreign affairs, Jeong Se-Hyun and Choi Sung-Hong, were both formerly vice-ministers for their respective portfolios.

As the presidential election approached, the MDP won some plaudits for introducing Korea's first-ever primary elections, which in March 2002 produced an unexpected momentum (dubbed the 'Roh wind') for an unlikely candidate: Roh Moo-Hyun, a populist lawyer and outsider from a poor farming background, without a university degree and with a left-wing image. This worried Washington, DC, as Roh had never visited the USA and had once demanded the withdrawal of US troops. His popularity then seemed to decline, however; by September, trailing in the opinion polls, he looked set to be withdrawn by leading power-brokers in the MDP as the usual pre-poll realignments to assemble a winning coalition gathered pace. An early bid by Park Geun-Hye, daughter of the late Park Chung-Hee, who in February left the GNP to form her own Korean Coalition for the Future, made little impact. Instead, the leading contender appeared to be Chung Mong-Joon, an independent legislator, scion of the Hyundai business conglomerate and, above all, organizer of South Korea's co-hosting of the association football World Cup 2002. Taking advantage of the national team's unexpected success (it reached the semi-finals) and the euphoria that this generated, Chung declared his candidacy in mid-September. It was at first unclear if he would seek to replace Roh as MDP candidate or would form a new party.

Two tests of public opinion were the local elections held on 13 June and by-elections on 8 August 2002. The GNP achieved a decisive victory at both, the former by 3.9m. votes, the largest margin in Korean electoral history. Winning 11 of the 13 by-elections brought control of the National Assembly, which the party then used to veto two successive presidential nominations for the post of Prime Minister after yet another government reorganization on 11 July. Kim Dae-Jung's initial choice was Chang Sang, the female President of Ewha Woman's University. The transgressions for which the legislature voted against first her and then the unrelated Chang Dae-Whan, a (male) newspaper proprietor, hardly seemed grave enough to warrant risking a power vacuum, should the President be incapacitated. Kim Dae-Jung, a weary 78, had twice been hospitalized in 2002, most recently with pneumonia. In early October Kim Suk-Soo, a career judge and former Head of the National Election Commission (formerly Central Election Management Committee), was appointed Prime Minister, his nomination having been confirmed by the National Assembly.

Kim's final ministerial reshuffle brought no respite. The premiership apart, the outgoing Minister of Justice indicated that he had been removed for failing to protect the President's sons, while the dismissed Minister of Health and Welfare blamed foreign pharmaceutical companies. Also replaced were the Ministers of Information and Communication, and National Defence, the latter as a consequence of an incident in which North Korean warships had fired on and sunk a southern patrol boat in the Yellow Sea (see the chapter on North Korea). Lee Jun, the new minister, was, as was customary, a former general—and also a previous head of Korea Telecom, as was the new Minister of Information and Communication, Lee Sang-Chul.

The close of 2002 brought fresh twists to the election campaign. With Lee Hoi-Chang's lead seemingly unassailable, in November Roh Moo-Hyun and Chung Mong-Joon agreed to join forces. They debated on television, leaving it to opinion polls to judge who performed the better. That proved to be Roh, who hitherto had been faring so badly that the MDP had begun to split, with up to 40 of its deputies abandoning the party and being expected to support Chung. The 'Roh wind' then rose again, fuelled by public anger at the acquittal in November by courts martial of the US Forces of Korea (USFK) of two soldiers whose vehicle in June had run over and killed two teenage girls. These protests, which continued into early 2003, reflected a widespread sense, especially among the young, of the USA's cavalier attitude to its Korean ally, both on this matter and in George W. Bush's uncompromising policy on North Korea. Roh Moo-Hyun was quoted as saying that he saw no need to visit the USA to 'kowtow', while Lee Hoi-Chang had been feted there (wishfully) as a prospective president.

In the event, against all predictions, Roh narrowly won the presidential election held on 19 December 2002, even though on the eve of the poll Chung withdrew his support over an alleged anti-US remark. Out of almost 35m. eligible voters, ballots were cast by fewer than 25m., or 70%, a record low turn-out (in 1997 and 1992 the level of participation had been more than 80%). Having received 12,014,277 votes, 48.9% of the total, compared with Lee Hoi-Chang's 11,443,297 (46.6%), Roh's margin was wider than that of Kim Dae-Jung in 1997. Kwon Young-Gil of the leftist Democratic Labour Party (DLP) more than doubled his 1997 result, but came a distant third, with 3.9% and fewer than 1m. votes. Three other candidates—an ex-premier, a Buddhist, and a socialist—received just 0.6% of the votes among them. Youth and technology were credited with this upset. South Korea is one of the world's most 'wired' societies. After early exit polls showed Lee in the lead, his supporters used the internet and mobile telephones to urge normally apolitical young electors to vote. This political generation gap thus added to the usual regional divisions, which were also apparent in the election result. Jeolla in the south-west voted overwhelmingly for Roh, but he also garnered 20%–30% of the vote in his native south-east, the GNP heartland. Chungcheong in the centre-west, whither Roh had promised to transfer the capital, favoured him more narrowly; as did greater Seoul, which contains 40% of the electorate.

The Presidency of Roh Moo-Hyun

In the context of the emerging North Korean nuclear crisis, the outcome of the presidential election caused unease. That was not allayed by the composition of Roh's 25-strong transition team, dominated by left-leaning provincial academics and including no politicians or officials. His inauguration on 25 February 2003 was overshadowed by a North Korean missile test, and also scaled down in response to a recent subway disaster in Daegu, South Korea's third city, in which some 200 people lost their lives. For two days he had no cabinet, as the opposition-controlled National Assembly refused to endorse his chosen Prime Minister, Goh Kun, a former premier, until a bill was passed to appoint a special counsel to investigate the 'cash for peace' scandal arising from revelations that the previous administration of Kim Dae-Jung had in 2000 secretly sent some US $100m. to North Korea via the Hyundai group, which itself had remitted a further $400m. (see the chapter on North Korea). Roh's first cabinet, which he pledged to retain (barring scandal) for two years, was less radical than his transition team. Kim Jin-Pyo, Deputy Premier for Finance and the Economy, was an experienced career bureaucrat, as were the Ministers of Commerce, Industry and Energy, Yoon Jin-Shik, and of Planning and Budget, Park Bong-Heum. The sole radical was Lee Joung-Woo, the head of policy planning in the Office of the President. The Minister of Foreign Affairs and Trade, Yoon Young-Kwan, was an academic with no diplomatic experience, who on a visit to Washington caused alarm when quoted—wrongly, he claimed—as preferring a nuclear North Korea to the collapse of the country. More reassuring, if surprising, was the choice of Han Sung-Joo (who had served as Minister of Foreign Affairs a decade earlier) as ambassador to the USA. Ra Jong-Yil, lately ambassador to the United Kingdom, took the new ministerial-level post of presidential national security adviser. In a signal of policy continuity, the Minister of Unification, Jeong Se-Hyun, was the sole incumbent to be retained from the Kim Dae-Jung administration. Otherwise, despite the formal continuity of the same ruling party, this was very much a new team in charge. Apart from the President, only two cabinet ministers were members of the MDP. In a challenge to Confucian hierarchies of age and sex, only two ministers were aged over 60, while four (twice as many as ever before) were female. The Minister of Justice, Kang Gum-Sil, was a radical lawyer whose appointment provoked an adverse reaction within her ministry. Other controversial choices were the Minister of Agriculture and Forestry, Kim Young-Jin, who had once shaved his head to protest at the opening of rice markets, and who resigned in short order; and Lee Chang-dong, a novelist and film-maker, as Minister of Culture and Tourism.

The freshness of what the Seoul press dubbed an 'experimental' cabinet was balanced, unfortunately, by a tendency to indiscretions. Roh Moo-Hyun himself, beset by challenges, within two months declared himself overwhelmed by the job: a view widely shared. On the recurrent issue of corporate reform, highlighted anew when a US $1,200m. accounting fraud was revealed at SK, the country's third largest *chaebol*, he seemed to retreat from his earlier radicalism, and it was hard to discern a consistent policy. Retreat was also evident in industrial relations. Although in June 2003 riot police broke up an illegal strike against privatization by railway workers, other strikes—truckers who in May blockaded the main ports, and bank workers who shut down computers to resist a merger—saw most of their demands granted. This encouraged militancy elsewhere, including a seven-week strike at Hyundai Motor, a major exporter, which won a five-day working week with no loss of pay or holidays and other concessions. Such unrest was unlikely to aid competitiveness, or attract foreign investment, to fulfil the official goal of making Seoul the business hub of East Asia. Furthermore, the new administration was soon beset by scandal. In August Roh Moo-Hyun sued four leading Seoul dailies over allegations of real estate deals involving him and his brother. His long-standing hostility to the press, which he accused of persecuting him, was widely seen as unseemly, if not alarming, in a democratic head of state. Separately, the MDP chairman, Chyung Dai-Chul, admitted taking money from a businessman charged with embezzlement, and embarrassingly claimed that the MDP's election campaign had received 20,000m. won (US $16.9m.) from business interests, much of it apparently undeclared. Also in August, Hyundai chairman Chung Mong-Hun, who had been charged in June in connection with the 'cash for peace' scandal, committed suicide. Compounding the air of volatility, tensions in both main parties made one of Seoul's regular political realignments ever more likely. On 7 July five legislators left the GNP after it elected a new leader—Choe Byung-Yul, a former editor, mayor of Seoul and cabinet minister—of conservative bent (his nickname being Choetler). (Lee Hoi-Chang had resigned after losing the 2002 presidential election, but in mid-2003 it was rumoured that he might essay a return.) The nominally ruling MDP was by now barely functioning, split between supporters of Roh Moo-Hyun, who openly planned to found a new party, and an old guard loyal to Kim Dae-Jung. Hopes that young radicals from both camps would unite in a new party, in order to win Roh a majority in National Assembly elections due in April 2004, seemed sanguine, to say the least. Thus, in the latter part of 2003, the prospect of more than four years of continuing weak leadership, amid formidable challenges, did not inspire confidence.

Subsequent events bore out this pessimism. On 19 September 2003 the MDP formally split, when 36 of its parliamentary members broke away and joined five from the GNP to form a new party, eventually named Uri ('We'). Ten days later Roh declared that he would leave the MDP, which accused him of betrayal, and joined the GNP in opposition. However, Roh did not at once join Uri, professing to remain above the fray. Earlier, the GNP had used its majority to force the removal of the Minister of Government Administration and Home Affairs, Kim Doo-Kwan, a key Roh ally, for not preventing an incursion by radical students into a US military base near Seoul in August. He was succeeded by the Minister of Maritime Affairs and Fisheries, Huh Sung-Kwan, whose deputy, Choi Lark-Jung, replaced him. The latter's ministerial career lasted less than a fortnight: he was dismissed after a series of gaffes egregious even for this administration, including referring to schoolteachers as 'bastards'. His successor was Chang Seung-Woo, a former Minister of Planning and Budget.

In October 2003 Roh Moo-hyun caused consternation by declaring that he would seek a public verdict on his stewardship and resign if this went against him. There was talk of a referendum in December, or if this were ruled unconstitutional, some less formal poll. In the event nothing came of this, but it cast a pall of anxiety over the last quarter of 2003. Roh's main motive was a sense of being beleaguered over corruption allegations against his close aides. Such charges escalated early in 2004, when investigations revealed that both main parties had accepted illicit funds from most major *chaebol* during the 2002 presidential election campaigns. However, with the GNP having received the larger share, Roh was able to engineer a deflection of public opprobrium onto the opposition and onto the tycoons and firms that had given the money. Fighting back, the GNP used its control of the legislature to create a special counsel to probe finances in the President's camp, thus implying that the public prosecution service could not be trusted to be politically even-handed. On 25 November Roh Moo-Hyun vetoed this, but the National Assembly overturned his veto in early December: the first time that this had happened since 1961.

Although Roh Moo-hyun had pledged to end the practice of frequent ministerial changes, by the end of 2003 eight of his original cabinet had been replaced within 10 months. In December Yoon Jin-Shik resigned as Minister of Commerce, Industry and Energy, taking responsibility for an ongoing dispute over a proposed nuclear waste dump which had led to violent protests. His replacement was Lee Hee-Beom, aged 54, president of Seoul National University of Technology and a former Vice-Minister. Five days later the Deputy Prime Minister for Education and Human Resources Development, Yoon Deok-Hong, resigned over the issue of management of personal information on students. Yoon's resignation was by common consent a disaster in a sector beset by problems. His pro-forma apologies were belied by his avowed ambition to stand in his native city of Daegu as a parliamentary candidate at the forthcoming legislative election. He was succeeded by Ahn Byung-Young, a professor aged 62 who had served as Minister of Education in 1995–97 before the post was upgraded to Deputy Prime Minister. This post had had 20 occupants in 23 years, so a degree of stability seemed badly needed. In late December Kim

Byung-Il was named Minister of Planning and Budget, in place of Park Bong-Heum, who was appointed chief policy planner at the Blue House (Office of the President). Kang Dong-Suk became Minister of Construction and Transportation, replacing Choi Jong-Chan who resigned to become a candidate for the National Assembly. Oh Myung replaced Park Ho-Koon as Minister of Science and Technology. There was some relief that all the new appointees were experienced administrators.

Reshuffling continued in 2004. On 15 January Roh Moo-Hyun dismissed the Minister of Foreign Affairs and Trade, Yoon Young-Kwan, followed by his defence and national security advisers, after a dispute where a number of foreign ministry officials, alarmed at growing anti-US tendencies, had allegedly criticized the President and dubbed some radicals in his entourage as 'Taliban' (a reference to the fundamentalists of Afghanistan). The new Minister of Foreign Affairs and Trade, Ban Ki-moon, was a seasoned diplomat who it was hoped would assuage any concerns in Washington. In February Lee Hun-Jai, who had spearheaded Kim Dae-Jung's reform drive as founding head of the Financial Supervisory Commission (FSC) before serving briefly as Deputy Prime Minister for Finance and the Economy, returned to the latter post after Kim Jin-Pyo resigned to stand as a parliamentary candidate. For the same reason (in South Korea cabinet ministers are not normally members of the National Assembly) Kim Dae-Hwan, an academic, replaced Kwon Ki-Hong as Minister of Labour. A week later the Minister of the Environment, Han Myun-Sook, likewise stepped down, to be replaced by her Vice-Minister, Kwak Kyul-Ho.

There followed, in March 2004, the most turbulent episode of Roh Moo-Hyun's career to date. Answering a question at a press conference, Roh Moo-Hyun expressed support for Uri. Although his views were hardly a secret, it was illegal under strict laws, drafted after years of abuse by dictators, for a government official, even a President elected on a party political platform, to be thus partisan. The National Election Commission reprimanded Roh but did not take any further action, seeing this as a minor infraction. However, the MDP seized the opportunity and with the GNP drew up a motion of impeachment, adding charges of corruption and economic mismanagement. An apology from Roh would have deflected the crisis; but whether from obstinacy or cunning, none was forthcoming until it was too late. Uri parliamentary members slept around the Speaker's podium to try to prevent a vote, but on 12 March they were ousted by the GNP's and MDP's larger numbers, in fierce scrimmages shown live on television. The impeachment motion—the first ever in the Republic's history—was then carried by 193 votes to two (Uri having walked out); easily surpassing the requisite two-thirds of the (then) 273-member assembly.

South Korean politics thus entered an unprecedented situation. Fortunately, all concerned duly followed the rules. Roh Moo-Hyun took a quasi-sabbatical period of leave, remaining in the Blue House, on full pay and with no duties, in uncharacteristic but prudent silence. The Prime Minister, Goh Kun, who had served the past six presidents as a 'safe pair of hands', took over as acting President while the Constitutional Court deliberated whether to endorse or reject Roh's impeachment. All other ministers remained in post; Lee Hun-Jai sent more than 1,000 e-mails to reassure foreign investors and international institutions that the economy was not at risk. After the initial shock, markets largely dismissed the matter.

What swiftly became clear was that the opposition had scored a massive 'own goal'. Protests and polls showed that 70% of South Koreans, whatever their view on Roh, regarded impeachment as an excessive response to a mere gaffe and deemed the other grounds dubious too. Support for Uri rose to 50%, while the GNP's declined to only 20% and the MDP's all but vanished. To stem the tide of discontent, the GNP picked a new leader: Park Geun-Hye, daughter of Park Chung-Hee, the dictator who in 1961–79 had established the basis of South Korea's industrial development. Hitherto marginal in the GNP, Park Geun-Hye was popular in her native south-east, the GNP heartland, and as a woman projected a new image for a conservative party. Not to be outdone, the MDP too, amid much internecine turmoil, appointed a female leader, Chu Mi-Ae.

All this was of no avail. On 15 April 2004, at the four-yearly scheduled elections for the National Assembly, now restored in size to 299 seats, the new Uri Party won 152 seats, more than tripling its representation, to wrest control from the GNP, which won 121 seats compared with its previous strength of 137. The MDP was decimated, from its 59 seats being reduced to nine; while the ULD, who under Kim Dae-Jung had held the balance of power, looked moribund, their representation declining from 10 to four seats. Meanwhile, the socialist DLP won 10 seats, its first ever, receiving 13% support nationwide. Turn-out, at just under 60% of the 35.6m electors, was 2% higher than in 2000, which had been a record low. Region remained a major divide: the GNP won 60 of 68 seats in the south-east (Gyeongsang), but none in Jeolla and just one in Chungcheong. Age too was thought to be a factor: Uri may have suffered from a gaffe by its chairman, Chung Dong-Young, who had suggested that the elderly—seen as pro-GNP—should stay home and rest on polling day. Of the new members of the National Assembly, 39 were women, twice as many as previously. Only 95 former legislators retained their seats, and 135 of the 204 newcomers were wholly new to party politics.

This clear popular verdict made Roh Moo-Hyun's restoration all but certain. In May 2004 the Constitutional Court overruled his impeachment, dismissing charges of economic mismanagement and corruption. It found that he had broken election law, if inadvertently, by expressing support for the Uri Party; and had also violated the Constitution with his abortive call in 2003 for a referendum;. however, these transgressions were not considered grave enough to warrant his dismissal.

Despite apologizing belatedly for this episode, once restored to office Roh Moo-Hyun continued much as before. He at once argued with the Prime Minister, Goh Kun, who refused to endorse a minor reshuffle and resigned on 25 May 2004. For a month South Korea had no Prime Minister, as Roh was determined to appoint Kim Hyuck-Kyu, a former provincial governor who had left the GNP, which therefore vowed to oppose him. Roh only relented after local by-elections on 5 June rebuffed Uri as firmly as April's general election had endorsed it, albeit on a low turn-out of 28.5%. The GNP took three of four mayoralties and governorships at stake, in Busan, Jeju and South Gyeongsang; while the MDP won in South Jeolla, its former heartland.

On 8 June 2004 Roh Moo-hyun chose as Prime Minister Lee Hae-Chan, a former education minister and a close ally. His appointment was confirmed by the National Assembly at the end of June by 200 votes to 84; the GNP allowed a free vote. This cleared the way for Roh to bring three Uri grandees into the State Council the next day. Chung Dong-Young, the former Uri Chairman who had managed the party's election campaign, became Minister of Unification; while Kim Geun-Tae, Uri's former floor leader, took the health and welfare portfolio. Chung Dong-Chae, an ex-journalist who was Roh Moo-Hyun's chief of staff before his election, replaced Lee Chang-Dong as Minister of Culture and Tourism. Apart from Lee, those evicted were not regarded as underperforming. Rather, Chung Dong-Young and Kim Geun-Tae were seen—already—as rivals to succeed President Roh in 2008, so the aim was to give them some direct experience of government.

In mid-2004 controversies erupted on several fronts. Roh Moo-Hyun renewed his election pledge to transfer the administrative capital, and almost all institutions of government, from Seoul to a site in Chungcheong. He resisted calls for more studies or a referendum on so major a step, which was expected to cost as much as US $100,000m., and accused opponents of trying to undermine him. This threatened to continue as a contentious issue for the remainder of his term. Inter-party relations worsened as Roh and Uri called for a re-evaluation of history, including former military regimes and collaboration with Japanese colonial rule before 1945. This clearly targeted Park Geun-Hye, whose father Park Chung-Hee was guilty on both counts.

Disinterring a difficult past seemed a distraction from present discontents. The usual summer of strikes in mid-2004 had a new focus: the introduction of a five-day working week, for which militant unions demanded no loss of pay. Areas affected included the familiar: car manufacturers, hospitals, subways—and some new spheres, including a 19-day strike (the longest ever in the financial sector) at KorAm Bank against its new owners, Citigroup. A walk-out at LG-Caltex, which refines 30% of the nation's petroleum, raised concern. The two latter were illegal actions, and in August the Government arrested leaders of both. A more harmonious framework for industrial relations therefore looked as elusive as ever.

In late July 2004 the Minister of National Defence, Cho Young-Kil, resigned after a complex episode where the South Korean navy had initially denied, but then allowed it to be known, that it had had prior radio contact with the North before firing warning shots at an intruding North Korean patrol boat. Roh Moo-Hyun publicly criticized the insubordination of those in the military who still found the 'sunshine' policy hard to accept. While as Commander-in-Chief he had to exert his authority, critics felt that Roh could have done so more tactfully. The President's defence adviser Yoon Kwang-Ung, a retired admiral, took over as Minister of National Defence and promptly pledged to bring in more civilians as senior officials.

Separately, in an unexpected development, Kim Seung-Kyu, a lawyer, became Minister of Justice, replacing the controversial Kang Kum-Sil, who resigned, reportedly as a result of fatigue, after her efforts to reform the prosecution service were continually blocked. The future of the Minister of Foreign Affairs and Trade, Ban Ki-Moon, was said to be in doubt after public outrage at the beheading in Iraq on 22 June 2004 of a South Korean hostage, Kim Sun-Il. However, the Government held fast to its plan to dispatch additional South Korean forces to Iraq. With just two of Roh Moo-Hyun's original cabinet appointees still in post after 17 months, amid challenges on many sides, there seemed little prospect that the remaining three-and-a-half years of Roh's term of office would prove tranquil.

RELATIONS WITH NORTH KOREA

A full account of North-South relations is given in the chapter on the Democratic People's Republic of Korea (see p. 486)

FOREIGN RELATIONS

South Korean foreign policy, ever since the state's proclamation in 1948, has been shaped by the circumstances of the Cold War partition of Korea. As the successor to three years of US military government, the Republic of Korea has consistently cleaved to the USA to an extent unique in Asia, symbolized by the continued presence of 37,000 US forces personnel (with nuclear arms until late 1991). Nor has this been a one-way relationship, since South Korea was the only other country to commit substantial troop levels to the US war effort in Viet Nam in the late 1960s. This relationship has been strengthened by regular senior-level visits over the years, including a trip to Seoul by President George Bush, Sr in January 1992 and a similarly positive visit by his successor, President Bill Clinton, in July 1993. It has had its costs, notably North Korea's propaganda victory in the 1970s, when it achieved its aim of excluding Seoul from the Non-aligned Movement as an alleged 'lackey' of Washington. Such allegations, however, underestimate the considerable freedom of manoeuvre afforded to successive leaders in Seoul, owing to a shrewd perception and manipulation of their small country's large strategic importance to the USA. South Korea's growth as an economic power has led to some friction with the USA over trade issues since the late 1980s, when Seoul benefited from an enormous trade surplus; by the late 1990s a large deficit had replaced the surplus, which was also a source of tension. There was also anxiety in mid-1992, when the targeting of ethnic Korean businesses during rioting in Los Angeles led not only to understandable concern in Seoul but also to what some in the USA perceived as unwarranted interference by South Korean political figures.

Much more significant in the longer term are the effects of successful 'nordpolitik' (see below) and the ending of the Cold War, both of which reflect a reassertion of geopolitics over ideology and have led to better relations with South Korea's close neighbours, Russia and the People's Republic of China. Yet the linchpin has remained Seoul's link to Washington, as testified by their close co-operation in the early 1990s in pursuing the issue of North Korea's suspected nuclear programme.

By contrast, South Korea's other major strategic relationship, that with Japan, has always been more problematic. Korean resentment against Japan's harsh colonial rule in the first half of the 20th century runs deep, and has been regularly reinforced by allegations, for example, of continued discrimination against the Korean minority in Japan or of attempts by the Japanese Government to 'whitewash' the imperialist period. One such issue concerned the so-called 'comfort women': as many as 200,000 mostly Korean young women and girls who were forcibly recruited in the late 1930s and early 1940s for the sexual use of Japanese troops. The problem lay in the Japanese Government's persistent efforts to deny the overwhelming evidence of official complicity, to rule out any question of compensation, and to avoid a full and frank apology. These issues plagued the visit by the Japanese Prime Minister, Kiichi Miyazawa, to Seoul in January 1992. However, in August 1993 the Japanese Government admitted for the first time that Korean and other Asian women had been forced to serve in Japanese military brothels, and offered full official apologies. None the less, such animosities did not prevent Japan from becoming South Korea's second largest trading partner and principal source of imports, especially of technology (although Seoul's large deficit on this trade had become yet another issue of contention). Despite the ties of geography, history, economics and culture that link South Korea and Japan, the political relationship would surely continue to be delicate.

A notable feature of South Korea's foreign policy has been its so-called 'nordpolitik', namely the replacement of unqualified anti-communism by a more subtle pursuit of improved relations with China and the former USSR. This has been highly successful in both its direct and indirect aims: to forge better ties with those powerful neighbours, and thereby to pressurize their ally, North Korea, into a more accommodating attitude. The key turning-point of 'nordpolitik' was the 1988 Seoul Olympiad. Of North Korea's communist allies, only Cuba, Ethiopia and Albania heeded its call for a boycott, while China, the USSR and the remaining Eastern European countries all participated. In the following year Hungary, Poland and Yugoslavia all established diplomatic relations with South Korea (prompting fierce denunciations by Pyongyang), a process which became more general with the collapse of communist rule in Eastern Europe.

The decisive step was the USSR's full recognition of South Korea in September 1990, accompanied by close personal ties between Presidents Gorbachev and Roh (who met three times within a 10-month period during 1990/91). With the subsequent demise of the USSR, these relations were continued not only with Russia but also with other republics of the CIS. President Yeltsin visited Seoul in November 1992, and gave assurances that Moscow no longer supported the North Korean regime. The Presidents of Kazakhstan and Uzbekistan (both of whose populations include Korean ethnic minorities, deported by Stalin to Central Asia from the Soviet Far East during the 1930s) had already visited South Korea.

With China, the *rapprochement* was more protracted. Trade began fitfully in the early 1980s, and a large Chinese team attended the Asian Games, held in Seoul in 1986. From the late 1980s Sino-South Korean trade increased rapidly, and trade offices were opened in Seoul and Beijing in 1991. In the following year there was a further strengthening of ties, culminating in the establishment of full diplomatic relations in August (and the consequent severance of ties by Taiwan). Still stronger links were subsequently forged between China and South Korea, including high-level visits in both directions and accelerating trade and investment.

Aside from these fundamental relationships with the abovementioned four powers—inevitable given Korea's geopolitical position as (in the words of a Korean proverb) a 'shrimp among whales'—South Korean foreign policy has mainly been dictated by two factors: growing economic success (until the economic decline of the late 1990s) and rivalry with North Korea. Sometimes these have gone in harness, for instance in ensuring good relations with Western European nations (although there had been some friction on trade issues with the European Union—EU). Elsewhere, they have diverged, as in Africa, where the contest for influence led both Koreas to open embassies wherever they could, at considerable expense. As Pyongyang began to draw back, Seoul extended its network, as former supporters of the North, such as Algeria, Angola and Tanzania, finally granted recognition to South Korea as well.

Closer to home, in December 1992 South Korea restored relations with Viet Nam, severed since 1975. As with China, trade links preceded diplomatic ties; and Hanoi seemingly bore no grudge for the South Korean involvement on the Saigon side during the Viet Nam War. The other Indo-Chinese states appeared less susceptible to South Korean advances, especially Cambodia: Prince Sihanouk was an old friend of Kim Il Sung, who had a palace built for him in Pyongyang.

While President Kim Young-Sam's initial priorities were domestic—like those of Bill Clinton, who also took office in early 1993—he subsequently visited all four major powers involved in Korea. Starting with the USA in November 1993, he then travelled to both Japan and China in March 1994. He completed his tour of the quartet with a visit to Russia (and also Uzbekistan) in June.

In general, this 'quadrangular diplomacy' (as it was officially called) inevitably revolved around the North Korean nuclear issue in 1993–95. Seoul pursued close policy co-ordination with Washington and Tokyo on the matter, while with Moscow and Beijing it became a question of strengthening what were still very new ties. The relationship with China, in particular, expanded rapidly, underpinned by increasing trade, and amounted to a *de facto* shift away from South Korea's traditional dependence on the USA.

As regards Japan, President Kim Young-Sam set himself the task of overcoming the legacy of bitterness dating back to Japan's colonial rule in Korea, and there was an easing of restrictions on Japanese cultural items, such as films and music. Kim developed very cordial relations with Morihiro Hosokawa, whose resignation as Japan's Prime Minister, in April 1994, was regretted in Seoul. The visit to South Korea in July of the new Japanese Prime Minister, Tomiichi Murayama, was significant in that his Social Democratic Party had traditionally maintained friendly ties with Pyongyang, while not recognizing Seoul. In August 1995 South Korea acknowledged a statement made by Murayama on the occasion of the 50th anniversary of the end of the Second World War, in which the Japanese Prime Minister expressed 'deep reflection and sincere apologies' for Japanese colonial aggression.

There were no major foreign policy developments during 1994–95. The core relationship with Washington was consolidated—as it perhaps needed to be, given the strains caused by the new US relationship with North Korea (see above)—by two visits to the USA by Kim Young-Sam in 1995: first in July for the dedication of the (long overdue) Korean War memorial in Washington, and then in October for the UN's 50th anniversary celebrations. Kim also made his first official visit to Europe in March, travelling to six countries: the United Kingdom, France, Germany, Belgium, Denmark and the Czech Republic. The meetings of the Asia-Pacific Economic Co-operation (APEC) forum in November 1994 provided an occasion for Kim to visit Australia and the Philippines, as well as the host nation, Indonesia.

The most important official visits to South Korea during this period were from former foes. In November 1994 China's Premier, Li Peng, became the most senior Beijing leader yet to come to Seoul; a year later this new friendship was sealed with a state visit by President Jiang Zemin. Scarcely less significant was the arrival in May 1995 of Gen. Pavel Grachev, the Russian Minister of Defence, accompanied by many of Moscow's élite. Various agreements were concluded, including one concerning the exchange of military intelligence, which was sure to anger Pyongyang (and possibly displease Beijing).

Relations with Japan were tested in 1996. In addition to dissatisfaction at Japan's failure to be properly contrite for its past aggression, two more immediate issues dominated. A dispute arose over a group of islets, called Dokdo (in Korean) or Takeshima (in Japanese), long claimed by both countries but newly salient with their adoption of 200-nautical-mile exclusive economic zones; there was also rivalry over staging the football World Cup in 2002, which became so hostile that in June 1996 FIFA, the international football federation, took the unprecedented step of offering it to them both to co-host. Despite these problems, practical co-operation on North Korea and other issues was not affected.

To a lesser extent, ties with the USA were tested too. In May 1996 students demonstrated against the presence of US forces in the country, demanding that the USA accept some responsibility for the Gwangju massacre of 1980. These were followed by further demonstrations in August 1996 at Yonsei University, Seoul, in which students were barricaded in the university, demanding reunification with North Korea and the withdrawal of US troops. Following a nine-day siege, riot police stormed the building and 5,715 students were detained. Seoul remained wary of the Clinton Administration's overtures to Pyongyang. As far as the security of the country was concerned, South Korea chafed alike at US procrastination in revising the Status of Forces Agreement (SOFA) so as to give Korean courts more jurisdiction over errant GIs, and at restrictions on its own right to develop missiles to counter the threat from North Korea. There was resentment at pressure from Washington for easier access to the Korean market for US goods, the more so since South Korea had accumulated a large trade deficit with the USA. Yet underlying relations remained sound, and were strengthened by Clinton's visit to Seoul in April 1996.

Seoul also extended its foreign policy interests beyond the peninsula and the four major powers, a reflection of its position as the world's 11th largest economy. Kim Young-Sam attended the first Asia-Europe meeting (ASEM) in February 1996 in Bangkok, Thailand, and took the opportunity to visit India and Singapore, where, as in the whole Asian region, South Korea has substantial and expanding business interests. Such interests also extend to the Middle East; in January 1996 South Korea became a non-permanent member of the UN Security Council (for a two-year term), and sanctions or other measures against Iran, Iraq or Libya could be problematic for Seoul, since it has or has had strong commercial ties with all three. South Korea is a dutiful supplier of personnel and funds to UN peace-keeping operations around the world.

The year 1996 was long ago set as the target for South Korea to become a full member of the Organisation for Economic Co-operation and Development (OECD). In the course of negotiations, the realization that this symbol of Seoul's attainment of full developed country status also carries responsibilities (in the form of faster market-opening and deregulation than might have been wished) prompted occasional doubts. South Korea was formally invited to join the OECD in October 1996 and became the 29th member of that organization in December.

The once extensive, but now dwindling, list of countries refusing to recognize South Korea decreased further in 1996. The involvement of the North Korean embassy in a forged currency scandal gave the Cambodian Government its chance to overrule King Sihanouk's objections to the recognition of South Korea. Hun Sen, one of Cambodia's then Co-Prime Ministers, visited Seoul in July, and it was agreed to exchange missions. South Korea already had flourishing ties with Viet Nam, despite having fought for the former South Viet Nam, and with Laos.

In September 1996 South Korea's foreign policy moved in a new direction when Kim Young-Sam visited Central and South America. During his time in Guatemala he held meetings with leaders of the five other Central American countries, who requested increased South Korean investment in their economies, and he then visited Chile, Argentina, Brazil and Peru, with all of which South Korea maintains good political relations and rapidly expanding commercial links. He omitted to visit Mexico, possibly because Roh Tae-Woo had been there in 1991, but paid a separate visit in June 1997. The Colombian President had, earlier, visited Seoul. Kim had also planned to visit several countries in Europe in March, but had to cancel his travels due to the internal political problems in South Korea.

For the most part, however, South Korea's foreign links centred on the quartet of major powers interested in maintaining stability on the peninsula. Relations with the USA remained broadly good, despite differences over how to deal with the North Korean leadership, and the occasional trade dispute. Several high-ranking US officials visited Seoul in early 1997: Vice-President Al Gore and the Speaker of the House of Representatives, Newt Gingrich, as well as the newly appointed Secretaries of State and of Defense, Madeleine Albright and William Cohen. In June Kim Young-Sam briefly met President Clinton, while attending the UN summit on the environment in New York. Relations with Japan were also mostly positive, and included close co-operation over policy towards North Korea. In June and July 1997, however, there was anger in Seoul when Japan detained five South Korean boats for allegedly fishing in Japanese waters.

Among South Korea's newer allies, relations with China were undamaged after the defection of Hwang Jang Yop (see the chapter on North Korea). Sino-South Korean amity is now regularly reinforced when the Presidents meet at APEC sessions, or when their Ministers of Foreign Affairs conduct meetings at the Association of South East Asian Nations (ASEAN) Regional Forum, whereas North Korea no longer has such ready access to the Chinese leadership. As far as Russia is concerned, the Minister of Foreign Affairs, Yevgenii Primakov, was the most senior of several Russian visitors to South Korea in mid-1997. He announced that a direct line of communication was to be set up between the respective leaderships; needless to say, the post-communist Russia has no such channel to the North Korean administration. Russia's hopes of selling an anti-missile defence system and jet fighter planes to South Korea, however, appeared likely to be vetoed by the USA, which expected its ally to continue to buy US-made military equipment.

The election of President Kim Dae-Jung in December 1997 signalled no major changes in South Korean external policy, with the important exception of North Korea. The fact that Kim had lived as a political exile in the USA and Japan, and so was well connected in both Washington and Tokyo, was expected to promote warmer relations with these two major allies. Thus, in June 1998, he made a cordial trip to the USA, where he was fêted as a rare Asian leader wholly in favour of both democracy and free markets. Unlike his predecessor, Kim Dae-Jung also shared the Clinton Administration's preference for engagement with North Korea. At another level, the US rescue of South Korea from the brink of default at the end of 1997 increased Seoul's debt to Washington, both literally and metaphorically, a situation which over time might need to be handled with care, given Korean pride and sensitivity to charges of 'flunkeyism'.

Kim Dae-Jung inherited a slightly more complex relationship with Tokyo, owing to long-running animosity over such issues as 'comfort women', fishing rights, and the Dokdo/Takeshima islets. In addition, South Koreans shared the rest of Asia's concern at Japan's economic stagnation, with the added twist that the Asian financial crisis pitted South Korean exports against Japanese. On the other hand, Kim Dae-Jung's connections with Japan, as well as the fact that Prime Minister Kim Jong-Pil was the very man who, in 1965, negotiated the first post-war formal relations between the two countries, were positive factors. In any case, a visit by Kim Dae-Jung to Tokyo in early October 1998 proved a huge success, resulting in greatly improved relations between the two countries and mutual pledges to co-operate fully henceforth in strengthening bilateral economic, security and cultural ties. The Japanese Prime Minister, Keizo Obuchi, publicly apologized (Tokyo's first such official expression of regret ever) for Japan's conduct towards South Korea during its occupation of the peninsula between 1910 and 1945. In the economic arena South Korea agreed to remove a number of restrictions on Japanese imports, while Tokyo announced that it would commit US $3,000m. in addition to existing financial assistance to its neighbour. A year of vigorous and successful presidential diplomacy ended busily, with a visit to China in November, followed by two regional summits—APEC in Kuala Lumpur, Malaysia, and ASEAN in Hanoi, Viet Nam—and a visit to Seoul by US President Clinton. Kim Dae-Jung's stay in China was as cordial as that in Japan.

In March 1999 Kim Dae-Jung welcomed the Japanese Prime Minister, Keizo Obuchi, to Seoul. His own first journey of the year was in May, to Russia and Mongolia. Relations with Moscow had been damaged in mid-1998 by the expulsion of a South Korean diplomat for spying, an affair which eventually led to the resignation of Seoul's Minister of Foreign Affairs. A longer-running issue was Russia's debt of US $1,700m. dating from loans extended by Roh Tae-Woo in 1990 as a reward for diplomatic relations. Seoul was reluctant to accept Moscow's offer to repay its debt in military equipment, preferably submarines. More widely, Russia resented being excluded from multilateral fora on the peninsula, be it the four-party talks or the Korean Peninsula Energy Development Organization (see the chapter on North Korea). In July 1999 Kim Dae-Jung made a second visit as President to the USA (and a first to Canada). This was not quite the success of the year before; his hosts were disconcerted when Kim requested that South Korea be allowed to develop missiles with a range of 500 km, rather than the 300 km recently agreed in principle. The steady strengthening of links with Beijing deepened further in August 1999, when for the first time a South Korean Minister of Defence visited China. Any military co-operation between these former Cold War adversaries was expected to proceed carefully, for fear of upsetting Pyongyang, and perhaps also Washington. The Russian Minister of Defence visited Seoul in September 1999. Relations with the USA remained good, and were reaffirmed in September by a three-way summit with Keizo Obuchi in Auckland, New Zealand, to show a common front against any new missile launch by North Korea. This took place just before the APEC meeting, which Kim Dae-Jung combined with bilateral visits to Australia and New Zealand. In November he visited Manila for the 'ASEAN + 3' meetings, where he met the Chinese and Japanese Prime Ministers: the first summit for that particular troika, and one of several signs of Kim's Asianist—but not anti-Western—proclivities. All this consolidated his reputation as South Korea's most internationally minded leader to date.

During 2000 Kim extended his focus to Europe, with visits in March to France, Germany, Italy, and the Vatican (as a devout Catholic). He had visited the United Kingdom in 1998 for the ASEM, which Seoul hosted in October 2000. In a speech in Berlin, he offered aid to rebuild North Korean infrastructure, an offer that led to inter-Korean summit talks being announced a month later. Relations with Japan continued to improve despite the death of Keizo Obuchi, whose funeral in June gave Kim Dae-Jung the chance to meet his successor, Yoshiro Mori, as well as Bill Clinton.

Relations with the USA faced new challenges in 2000. Allegations surfaced that US troops had massacred civilian refugees early in the Korean War at a village called Nogun-ri; both Governments set up official inquiries into this. It subsequently emerged that two US soldiers who had testified to being present at the massacre were in fact located elsewhere at the time of the killings. Anti-US sentiment was also heightened by an accident at a bombing range at Maehyang-ri, south-west of Seoul, which provoked a protest campaign; as well as by revelations of a toxic leak into the Han river from a US base in Seoul. The north-south summit implicitly raised the question of an eventual withdrawal of the 37,000 US troops stationed in South Korea; though Kim Dae-Jung insisted that Kim Jong Il had accepted his argument that they should stay to perform a regional peace-keeping role. This backdrop made it no easier to revise the Status of Forces Agreement (SOFA), which governs US troops in Korea. A fresh agreement was finally concluded in December, after five years of negotiation.

As for Seoul's newer links, relations with the People's Republic of China remained good, despite protests in January 2000 when Beijing repatriated seven young North Korean refugees (earlier expelled from Russia). There was criticism of the Government for being too meek towards China on the refugee issue and other matters, such as its refusal to grant a visa to the Dalai Lama (the spiritual leader of Tibet), or to congratulate the new President of Taiwan, Chen Shui-ban, whose triumph as a former opposition leader elected as President paralleled Kim Dae-Jung's own trajectory. Kim Dae-Jung publicly welcomed Vladimir Putin's election as Russian President, though the latter's prompt visit to Pyongyang discomforted some in Seoul. Multilaterally, the third ASEM, which South Korea hosted in October, was the largest event ever held in South Korea's diplomatic history, with over 20 heads of state or government attending. The United Kingdom and Germany used the occasion to announce their intention to open ties with North Korea, as a boost for the 'sunshine' policy. In November Kim Dae-Jung took a prominent role as usual in two regional summits, the APEC in Brunei and the 'ASEAN + 3' in Singapore, with state visits there and to Indonesia. The President rounded off a remarkable year with visits to Sweden and Norway, where he received his Nobel Peace Prize.

The year 2001 proved to be much less successful, however. Fears that the new US Administration would harm détente with North Korea proved correct in March, when Kim Dae-Jung became the first Asian leader to visit President George W. Bush, who publicly voiced mistrust of the DPRK. By contrast his Secretary of State, Colin Powell, expressed readiness to continue engagement, a view that subsequently prevailed, but not before damage had already been done. It did not help that before

Kim's visit a joint statement with Russia's President Putin, who visited Seoul in February, strongly supported the 1972 Anti-Ballistic Missile Treaty (ABM), which Bush's missile defence proposals were set to breach. Relations improved as a result of the massive terrorist attacks perpetrated against US targets in September 2001, which brought strong sympathy from South Korea and a reaffirmation of security ties with the USA.

Meanwhile, Kim Dae-Jung's efforts to forge better ties with Japan met a set-back. A new history textbook to be used in Japanese schools, which concealed that country's pre-1945 aggression, caused strong repercussions in both Koreas and in China: these were compounded by Japanese Prime Minister Junichiro Koizumi's visit to the Yasukuni Shrine, a controversial war memorial, in August 2001. By contrast, relations with China remained good despite the latter's persecution of North Korean refugees, and a growing perception of China as an economic competitor. Senior visitors to Seoul included Li Peng. Later in 2001 Kim Dae-Jung made his customary forays to the APEC and 'ASEAN + 3' meetings. Ever at ease in Europe, Kim's politics being essentially 'Christian Democrat' in nature, in December 2001 he visited the United Kingdom, Norway and Hungary; he became the first Asian leader to address the European Parliament in Strasbourg. A planned visit to Latin America, which would have been his first, was postponed.

In 2002 foreign affairs provided no solace from domestic problems which, along with ill health, prevented the President from travelling as much as he had hoped. In President George W. Bush's identification, in January, of North Korea, together with Iran and Iraq, as forming an 'axis of evil' was difficult to reconcile with the 'sunshine' policy. In the circumstances, Bush's visit to Seoul in February went off better than many had feared. There was dismay, however, at a decision to buy South Korea's next-generation combat aircraft from the USA. The choice of the latter's F-15K *Eagle* fighter was widely seen as a political decision, considering the technical superiority of France's Dassault *Rafale*. Other trade disputes were less fraught. A nationalist rebuff which prevented a US firm, Micron, from buying the chip-maker Hynix was balanced by General Motors' agreement to take over Daewoo Motor.

Relations with Japan were improved in 2002 by the two nations' co-hosting, albeit with largely separate organization, of the football World Cup. This afforded opportunities for mutual visits by Koizumi for the opening and Kim for the closing matches, as well as the first visit to post-war Korea by a member of the Japanese royal family (but not yet the Emperor or Crown Prince). Kim was also glad of Koizumi's unexpected decision to visit North Korea, as an indication of support for the 'sunshine' approach as against that of the 'axis of evil'. It was surprising, therefore, that Seoul persisted in a seemingly trivial campaign to have the sea between the two countries renamed from the Sea of Japan to the East Sea.

Relations with Russia were not especially close in 2002, despite agreement to settle a long-standing Moscow debt in part by South Korean weapons purchases. The Russian Minister of Foreign Affairs, Igor Ivanov, visited both Koreas in July. Kim Dae-Jung endorsed President Putin's decision to prioritize the improvement of relations with Pyongyang. One shared interest was in an 'iron silk road', a proposed rail freight route linking South Korea to Europe via Siberia. Yet this depended not only on Kim Jong Il's consent, but on the investment of US $3,000m. to upgrade North Korea's decrepit network. Moscow hoped that Seoul would finance this.

Also in 2002 some overdue realism entered South Korea's hitherto rather ingenuous view of China. Trade disputes enhanced the image of a tough economic competitor, while the sight of Chinese police beating South Korean diplomats in Beijing, who were protecting North Korean refugees, was salutary also. Even so, not only was the Dalai Lama again refused a visa, but Asiana Airlines declined even to carry him from India to Mongolia via Seoul. South Korea's priority with China remained a perceived need to maintain Beijing's engagement, especially in putting pressure on North Korea to make peace and introduce reforms.

From late 2002 four developments threatened South Korea's core alliance with the USA. The acquittal in November by USFK courts martial of two US soldiers, whose vehicle had in June killed two Korean girls in a road accident, provoked major demonstrations, especially by younger Koreans. Although these protests were mostly peaceful, the lack of equivalent outrage at the far greater and more deliberate threat posed by North Korea's nuclear defiance perplexed older Koreans and foreign observers. The unexpected election of Roh Moo-Hyun as President in February 2003, taking advantage of resentment, added to the unease. All this gave the US Department of Defense an opportunity to press for a redeployment of US forces, back from their 'tripwire' position near the DMZ to a line south of Seoul. This led to renewed concern in South Korea, not least over fears that such a redeployment would make it easier for the USA to contemplate an air strike on North Korea if its own troops were no longer in the firing line. In the event, the relationship looked set to survive these strains. In office Roh Moo-Hyun moved to improve relations with the USA, including an unpopular decision to send non-combat forces to Iraq in April 2003. His first (ever) visit to the USA in May, which was carefully prepared, went well; although here again Roh alienated supporters by the issue of a joint statement with President George W. Bush, warning North Korea of 'further steps' if its nuclear defiance persisted. On his return, radical students blocked his motorcade at a ceremony for victims of the 1980 Gwangju massacre. While in the USA, for a week in total, he also visited Silicon Valley and the New York stock exchange, and lobbied for investment; Samsung took the opportunity to announce a US $500m. addition to its chip plant in Austin, Texas. South Korea reluctantly accepted the USFK redeployment, to be phased over several years. Despite this closer convergence, as six-party talks with North Korea drew near, it remained possible that a robust US stance would clash with a softer South Korean approach.

The little-travelled Roh Moo-Hyun, whose provincial background contrasted with that of his predecessor Kim Dae-Jung's international experience and renown, followed his US trip—inevitably first—with visits to South Korea's neighbouring powers: Japan in June 2003, and then China in July. Both were successful, although with poor timing the Japanese Diet on the day of his arrival ratified three new wartime contingency laws, which had caused the usual friction over alleged Japanese militarism in some quarters in Seoul. Roh was more forward-looking, citing the EU as a case of neighbours transcending old enmities to forge a new partnership. Unusually, he also took questions from ordinary Japanese on live television at a 'town hall' meeting. North Korea was, naturally, a major agenda item in both Tokyo and Beijing. Roh may, tacitly, have felt closer to China's push for dialogue than Japan's rapidly hardening line. Trade ties with China were more dynamic, although South Korea's bilateral surplus was as much a concern in Beijing as Japan's persistent surplus was to Koreans. The China visit, which included a tour of a Hyundai car plant, had more of a business focus. Despite fears of competition in almost every field of industry, South Korea had a level of ease with China that was lacking with Japan. As trading links continued to grow, the question of how far politics and security might follow suit was current. If China continued to present itself as a peace-maker on the peninsula, while the Bush Administration was widely perceived as raising tensions, the long-term consequences for the strategic orientations of a more assertive South Korea (or of an eventual united Korea) could be profound. Old Korea had, after all, been under loose Chinese hegemony for many centuries, for the most part contentedly.

Six-party talks on North Korea's nuclear weapons programme, held in Beijing in late August, brought together the two Koreas with the four powers—the USA, China, Japan and Russia—for the first time in half a century, bound by geography and/or history. Now that the Cold War 'two triangles' were long gone, many permutations and nuances of alliance were possible. A degree of unity amongst the five seeking to curb North Korea's nuclear and other threats belied both tactical differences over how to accomplish this and longer-term divergences of strategy and goals. The Beijing talks ended without resolution, and efforts to reconvene them before the end of 2003 failed. An eventual second session in Beijing on 25–28 February 2004 again achieved little except agreeing to meet again by June, and to establish working groups to that end. South Korea reportedly shared the frustration of China and Russia at the US refusal to meet North Korea halfway. It was thus relieved when at the third round of six-party talks, held on 23–26 June, the USA for the first time offered a detailed proposal—said to be based on a South Korean draft—for phased nuclear disarmament, including incentives of immediate energy aid. North Korea did not reject this outright; it was due to be further discussed at a hoped-for fourth round in September, albeit with scant expectation that Kim Jong Il would yield much before the outcome of November's US presidential election was known.

Bringing together the five states most central to South Korean foreign policy, the six-party process, despite its lack of results, allowed for a wide range of consultations on the sidelines. Elsewhere, in October 2003 Roh Moo-hyun attended the ASEAN + 3 meetings on the Indonesian island of Bali and the APEC summit meeting in Bangkok. In early 2004 his impeachment curtailed diplomacy: a planned visit to Russia was now expected later in the year. In December Roh was due to pay the first ever state visit by a South Korean President to the United Kingdom, and would probably also visit other European countries.

However, the two main foci of South Korean foreign policy—North Korea apart—were, as ever, the USA and China. If on the surface relations with Washington remained better than many had feared, underlying tensions persisted. Although South Korea welcomed the Bush Administration's belated softer approach towards Pyongyang, it was anxious at the USA's plans to move its forces back from the DMZ to south of Seoul. In addition, in May 2004 came news that 3,600 troops would soon be redeployed from US Forces Korea (USFK) to Iraq, followed by word at the beginning of June that one-third of the total USFK strength of 37,000—12,500, including the 3,600—would leave permanently by the end of 2005. If in one sense, ironically, this forced the more independent defence posture that Roh Moo-Hyun had long advocated, Seoul felt a lack of consultation; while also worrying that the new US doctrine of mobile forces could see USFK used in the Taiwan Straits, to the detriment of South Korea's relations with China. As ever, too, the details of change involved much haggling. At the end of July it was announced that 13 US bases would revert to South Korea by 2006, far sooner than expected; while the main Yongsan garrison in Seoul, long a focus of complaint, would close by 2008. Many in Seoul believed that some of these issues, and the question of links with the USA in general, might become less difficult if, in November 2004, John Kerry were to be elected as the next US President.

In 2003 China displaced the USA as South Korea's main export market, and looked set soon to become its largest trade partner overall. This economic factor added to a diplomatic convergence over North Korea, where South Korea and China's wish to engage was at odds with the harder line (until recently) of the USA and Japan. Seoul remained deferential to Beijing's sensitivities, not querying its refusal to regard any North Koreans in China as refugees. (With similar caution, it did not name Viet Nam as the source of 468 North Korean refugees airlifted to Seoul in July: the largest single group since the Korean War. All had had to travel the length of China in secret, risking arrest and repatriation.) If current issues did not spoil Sino-South Korean amity, ancient history was—as ever in Korea—a more serious matter. By August, a major dispute was brewing over China's revisionist claim that the Goguryeo (Koguryo) kingdom (37 BC–668 AD), covering much of modern North Korea and Manchuria, was Chinese rather than Korean. This seemed to be part of a wider Chinese effort to secure its frontiers by claiming all its present borderlands as primordially Chinese in all aspects, thereby pre-empting any conceivable irredentist claims. Beijing might not have realized just how deep passions on such matters ran in Korea.

With Japan, with which Koreans have long had issues over history, there were no major developments during late 2003 or early 2004. There were hopes that a long-mooted free trade area (FTA) might be implemented in 2005, but some opposition to this remained. (South Korea's first ever FTA, with Chile, took more than a year to be ratified, owing to the hostility of farmers.) On 22 July 2004 Roh Moo-Hyun and Junichiro Koizumi held an informal one-day summit meeting on the South Korean resort island of Jeju. Roh was criticized at home for saying that he would not raise contentious issues of the past, such as Koizumi's visits to the Yasukuni shrine, revisionist Japanese school textbooks and so-called 'comfort women'.

Overall, unlike in the past, when a weak Korea had fallen victim to imperialism in 1905 and partition by the superpowers in 1945, both Korean states were now equal arbiters, along with the great powers, of the peninsula's future. While disarming North Korea was a tough enough immediate task, it was not too soon to start pondering how an eventually unified Korea might conceive its national interest and foreign alignments.

Economy

ROBERT F. ASH

Based on an earlier article by JOSEPH S. CHUNG

INTRODUCTION

Until the end of the 1980s, South Korea operated a mixed economic system. On the one hand, the country adhered to the basic tenets of private enterprise and a market economy; on the other hand, it followed a highly visible policy of government intervention. Through planning, direct or indirect ownership and control of enterprises and financial institutions, regulation of foreign exchange, and the implementation of appropriate monetary and fiscal policies, the Government played a crucial role in making market adjustments and maximizing incentives in pursuit of the fulfilment of its desired economic, social, political and cultural objectives. However, economic success, the increasing complexity of the economy, the emergence of a more democratic and pluralistic society (including greater participation in decision-making by different interest groups), and increasing international competitiveness were major factors in bringing about, from the late 1980s, a decline in the Government's role in the South Korean economy. Thus, the 1990s witnessed significant progress towards privatization and away from 'command capitalism'. The reformist thrust of government economic policy was underlined by renewed emphasis on the need for greater efficiency, improved labour productivity and enhanced competitiveness in order to meet the demands of globalization. The admission of South Korea, in December 1996, to membership of the Organisation for Economic Co-operation and Development (OECD) was another watershed, which heightened the importance of economic reform, as was, from a negative aspect, the impact in the late 1990s of the Asian financial crisis, which not only temporarily halted South Korea's growth momentum, but also, by highlighting long-standing structural weaknesses in the economy, threatened to undermine its previous economic achievements. Thanks to the subsequent implementation of an IMF programme and the successful negotiation of foreign debt restructuring with creditor banks, a significant degree of recovery followed and growth was relatively quickly resumed. South Korea's continuing international economic importance is reflected in its status as the 12th largest economy in the world by 2003 and a global player in shipbuilding, electronics, car and semiconductor manufacturing. Projections at the beginning of the 21st century suggest that its international position is likely to be maintained in coming decades, although it may well remain outside the G-10 group of countries. Meanwhile, in the more immediate term, the Roh Moo-Hyun administration faced a number of serious economic challenges.

In common with other first-rank Asian newly-industrializing countries (NICs), South Korea's modern economic transformation was driven by GDP growth sustained over a long period. From the beginning of the 1960s until the early 1990s, apart from one year of negative growth in 1980 caused by sharply rising oil prices, the pace of economic growth remained consistently buoyant—for much of that period, according to data provided by the Bank of Korea (BOK), averaging in excess of 9% per annum. From this perspective, a major recent watershed was the onset of the Asian financial crisis, which abruptly halted the rate of economic expansion and caused GDP to contract by 6.6% in 1998, compared with growth of 5.0% in 1997. Renewed GDP growth, averaging around 10% in 1999 and 2000, apparently signalled rapid recovery from this crisis, although the true significance of the events of 1997–98 was their manifestation of serious systemic and structural problems facing industry and the financial sector. Such difficulties, as well as others associated with the vagaries of South Korea's domestic politics, were reflected in an average rate of GDP growth of only 4.1% during 2001–03 (2.7% in 2003 alone). Authoritative sources pointed to only a slight acceleration in growth during 2004–05.

THE PHYSICAL AND HUMAN RESOURCE BASE: WAGES AND PRICES

South Korea contains quite rich concentrations of iron ore in the north-east of the country, although no more than 20% of these deposits are thought to be high-grade. It also has significant reserves of limestone, the basis of the cement industry and so critically important for construction purposes, as well as minor deposits of gold, silver and copper. Its energy resources—mainly located in the north-east—are limited to anthracite coal, firewood and hydroelectric power. Total coal reserves have been estimated at about 1,500m. metric tons, of which about one-third are recoverable. Around 20% of South Korea's primary energy supplies currently derive from coal.

South Korea occupies only about 45% of the total area of the Korean peninsula, although its total population (estimated at 48.2m. at mid-2004) is more than double that of North Korea. Its total area is some 99,313 sq km, but little more than 20% of this is cultivable land (the remainder being shared between forest, which represents 65%–70% of total surface area, and land used to meet urban and transport needs). Such statistics emphasize the high density of population—about 490 persons per sq km—a figure that is exceeded throughout the world, small islands and city states apart, only by Bangladesh and Taiwan.

As in the North, the main demographic impact of the Korean War was on mortality, which rose sharply between 1950 and 1953. Total military and civilian casualties attributable to the war were almost 2m., of which deaths totalled some 600,000. Thereafter, however, a remarkable transformation occurred. Between 1955 and 1975 the death rate declined sharply (from 33 per 1,000 to 15 per 1,000). A rapidly decelerating birth rate resulted in the annual rate of natural increase falling below 2%. The deceleration in population growth continued: the average rate of natural increase during 1995–2003 was little more than 76 per 1,000, and in 2003 it was a mere 60 per 1,000. Projections made by the National Statistical Office (NSO) indicate that South Korea's total population will peak at around 50.7m. in 2023, thereafter falling back to less than the 2004 level by mid-century.

At 0.4, South Korea's dependency ratio is remarkably low. In 2003 71.4% of the total population were between the ages of 15 and 64, compared with 63% in 1980. Children below the age of 15 constituted 20.3% of the population in 2003, while the share of the elderly (those aged 65 and above) was 8.3%. The ageing population is, however, growing rapidly, and OECD estimates suggest that in little more than two decades—marginally faster than in Japan, and much faster than in Germany, the United Kingdom, the USA and France—South Korea will, on the basis of UN criteria, have become an 'aged' society, with 14% of its population at or above the age of 65. NSO projections suggest that by 2030, there will be 11.6m. people—23.1% of the population—of 65 or more years of age, implying an aged dependency ratio of 35.7%.

Rapid urbanization has been a notable characteristic of demographic change in South Korea. In 1955 the urban share of total population was less than 25%. By 1975 the corresponding figure was already over 50%, and by the end of the 1980s it had reached 70%. At the beginning of the 21st century, the rate of urbanization was in excess of 80%, a figure that was expected to rise to 88% by 2015. The most important urban centres have long been Seoul and Busan, which, by 1990, already contained well over one-half of the urban population, and 33% of the total population. Against the background of almost half of the total population living in 'Greater Seoul', in 2003 the South Korean legislature gave its approval for a preliminary plan to relocate the national capital to a greenfield site in the middle of the country. Infrastructural pressures and security concerns are the main considerations behind the proposed move, and it seems unlikely that Seoul will lose its status as the industrial and financial centre of the country. From this perspective, even if the plan is fulfilled, hopes that a more balanced regional distribution of economic development can be attained are unlikely to be realized.

At the end of June 2004 the economically active population numbered 37.71m., the participation rate being 62.5% (but 75.3% for men)—of whom 22.8m. (61.3%) constituted the labour force. Of the labour force, 22.82m. were employed, implying an overall rate of unemployment of 3.2% (763,000 persons). By way of comparison, the corresponding figure in 1995 was a mere 2%. In mid-2004 the unemployment rate among men, at 3.5%, was higher than that for women, which was 2.9%. Less than 10% of the employed labour force are now engaged in agriculture (crop farming, forestry and fishing), while industry accounts for some 21% of total employment, most of it in manufacturing. It is a mark of the advanced level of development that South Korea has reached that around 70% of all employment now takes place in the service sector, especially in wholesale and retail business.

As in Taiwan, one factor that has exacerbated the employment situation in South Korea is the pressure to relocate manufacturing activities to cheaper foreign locations, especially the People's Republic of China. Certainly, major job creation efforts will be required if the targeted reduction in unemployment to around 2.5% by 2007 is to be fulfilled. For the time being, however, the most contentious employment issues are those affecting young people seeking jobs. In the second quarter of 2004, for example, alongside a national unemployment rate of 3.3%, the corresponding figures for those in the age categories of 15–19 and 20–29 were 13.6% and 17.4%.

In the second half of the 1980s the rate of increase in the urban consumer price index (CPI) averaged 4.1% per year, a figure that accelerated to 6% annually between 1990 and 1995. Rising prices during this period reflected the award of large wage rises, associated with emerging shortages of labour, growing unionization and a higher incidence of labour disputes. Thus, by 1995, the average monthly manufacturing wage had reached 1,123,895 won (some US $1,500).

The effect of the Asian financial crisis was quickly to exacerbate inflationary pressures. Thus, from an annual CPI increase of 4.5% in 1997, by February 1998 the index had risen to 9.5%; although the rate of increase subsequently slowed, for the whole of 1998 CPI rose by 7.5%. In 1999 consumer price rises fell back sharply (to 0.8%), since when inflationary pressures have remained stable and quite modest. In 2003 consumer price inflation averaged 3.5%, compared with 2.8% during the previous year.In the second quarter of 2004 the CPI stood at 114.2 (2000 = 100), although rising oil prices threatened to cause an accelerated increase in the index in the second half of the year.

Increasing labour unionization and the award of higher 'fringe' benefits have combined to undermine South Korea's export competitiveness in relation to a number of low-wage industrializing countries. In 2000 there was also a resurgence of strike activity, reflecting growing confidence by the trade unions in their ability to press claims for wage increases and enhanced job security on behalf of their members, as the economy achieved rapid recovery from the regional financial crisis of 1997. However, despite President Roh Moo-hyun's declared determination to address outstanding labour issues, serious tensions between employers and trade unions have persisted. Not least, as temporary job opportunities in the informal sector have come increasingly to dominate employment markets, the existence of wide earning differentials between those working on long-term contracts and those on temporary contracts has also become a major issue. Thus, notwithstanding generous wage settlements in

2003 (during which monthly wages rose by 9.4%), strikes and other forms of union agitation in support of greater job security continued into 2004. An important employment initiative taken in 2003 was the approval by the National Assembly of legislation reducing the six-day working week to five days.

ECONOMIC GROWTH AND STRUCTURAL CHANGE

The social and economic legacy of the Korean War of 1950–53 was devastating. In addition to casualties, some 1.5m. refugees from the North had to be accommodated in the South. The cost of physical damage to property is estimated to have been the equivalent of South Korea's entire gross national product (GNP) for 1953, or more than 10 times the then annual rate of fixed capital investment. In 1953, levels of production in all sectors were well below the previous peak levels of the early 1940s, and output was heavily skewed towards agriculture, with manufacturing production accounting for less than 9% of GNP. Such conditions ensured that following the war, rehabilitation, reconstruction and stabilization assumed the highest economic priority. In part owing to large-scale US aid, these goals were fulfilled fairly quickly, and by 1957 war damage had been repaired and prices stabilized. Thereafter, however, until the military coup of 16 May 1961 brought Park Chung-Hee to power, the pace of South Korea's economic expansion slowed considerably.

The launch of the First Five-Year Plan (1962–66) was an important watershed in South Korea's post-war development, not only because it marked a break between different magnitudes of growth, but also because it embodied a distinctly new policy approach, symbolized by a fundamental shift from inward-looking import-substitution to an outward-looking growth strategy of export promotion. The speed of transformation after the 1962 'take-off' was remarkable. By 1970 South Korea had already acquired the status of a newly-industrializing economy; by 1986 it had reached the stage of self-sustaining growth, with domestic savings more than able to finance investment and with the balance of payments having shifted from chronic deficit into surplus (thereby obviating the need for aid or overseas borrowing). Although the momentum of growth slowed in and after the 1990s, South Korea's development record remains almost unmatched in post-World War II experience. In 2003 average GDP per head in South Korea was US $10,885 (at market exchange rates); but in terms of purchasing power parity (PPP), the corresponding figure was about US $17,700, ranking it 49th in the world.

In the aftermath of the Korean War, massive inflows of aid facilitated rapid economic reconstruction, although recovery was accompanied by severe inflationary pressures. By the end of the 1950s, South Korea's productive facilities had been restored. Thereafter, between the First (1962–66) and Sixth (1987–91) Five-Year Plans and, driven by a rapid and sustained expansion of exports, South Korea's GNP grew, on average, by more than 9% per year. Although this average estimate conceals significant year-to-year fluctuations, the rapid and mutually-reinforcing expansion of output, income and exports during these years was accompanied by increasing shares of savings, investment and foreign trade in national income. Much slower economic expansion in the 1990s (GNP increased, on average, by 5.7% annually during 1990–99) reflected decelerating growth in the early 1990s and, in 1997–98, the impact of the Asian financial crisis. By 2000 South Korea's aggregate GDP had reached US $457,219m., a figure exceeded by only 11 other countries throughout the world, although that of a 12th country, India, was almost identical. The fact that aggregate GDP in 2000 was 44% higher than in 1998 was a sign of how far and how fast the Korean economy had recovered, in purely quantitative terms, from the worst point of the Asian financial crisis.

The momentum of rapid and sustained growth over almost three decades from the early 1960s reflected the deliberate choice of a strategy designed to maximize growth through the pursuit of outward-orientated policies. During this period, the imperative of export-led growth dominated South Korea's development plans and replaced the previous emphasis on indigenous import-substitution. Other complementary measures were adopted, and the mutually-reinforcing impact of a variety of factors—the existence of an abundant supply of highly skilled, highly educated, disciplined and, at least initially, cheap labour, a readiness to use advanced foreign capital and technology, the growth of an indigenous managerial class and a growing research and development capacity—contributed to the success of the new strategy. From a demand perspective, rising real wages that benefited a wide cross-section of the population, especially in the urban sector, also facilitated the emergence of a sizeable middle class that possessed considerable purchasing power.

Under the impact of rapid and sustained growth over several decades, the structure of the South Korean economy has undergone dramatic change. Predictably, the role of agriculture as a source of GDP has given way to that of industry (construction, mining and manufacturing) and, latterly, services. In the mid-1960s the farm sector still accounted for about 40% of GDP, compared with some 16% from industry and 44% from services. In 2003 the corresponding figures were 3.9%, 41.3%(almost 30% coming from manufacturing alone) and 54.8%. Projections suggest that the relative contributions of the three major sectors to GDP are unlikely to change significantly in the foreseeable future. Contained within the changing structure is the finding that, like its neighbouring 'dragon' economies, Hong Kong and Taiwan, South Korea has since the 1960s not only industrialized, but also subsequently transformed itself into a mature, services-dominated economy.

A characteristic institutional feature of South Korea's economy has been the dominant role played by *chaebol*, or conglomerates—a Korean version of the Japanese *zaibatsu*. The domination of these large-scale entities—in 1995, the top 30 *chaebol* accounted for 16% of GDP, 41% of manufacturing value-added and 50% of export value—may well have disadvantaged the growth of small and medium-scale enterprises (SMEs) of the kind that played such a vital role in Taiwan's post-War economic growth. Following almost 14,000 SME bankruptcies in South Korea in 1995, a Small and Medium Business Administration, intended to promote more sustained SME growth, was established in 1996. Its potential role was, however, overtaken by the onset of the financial crisis, and by the end of 1998 the failure of 22,828 SMEs had been recorded. Much more serious in terms of its economic consequences, however, was a series of corporate bankruptcies, affecting even major *chaebol* (for example, Hanbo, Kia and Yuwon). Precipitated by the 1997 financial crisis, the collapse of such enterprises had severe repercussions for the country's financial system, dramatically increasing the scale of outstanding non-performing loans (NPLs) and underlining the need for radical corporate and financial restructuring throughout the economy. Significant progress towards improving the state of the *chaebol* sector was achieved under the Kim Dae-Jung administration, and Kim's successor as President, Roh Moo-Hyun, maintained this policy thrust in his promise to make the conglomerates more open and accountable. However, against the background of decades during which the relationship between government and *chaebol* was such a close one, the challenge of corporate reform remains urgent. In 2003 accounting frauds were uncovered within the SK Corporation—the fourth largest business group in South Korea—which led to a three-year prison sentence for SK's chairman, Chey Tae-Won. The ramifications of this and other scandals, such as those associated with large-scale, illegal political donations made by major *chaebol* (including Samsung, Hyundai Motor and SK itself) continued throughout 2004. The evidence of such malpractice highlights the long-term nature of the challenge of corporate sector restructuring in South Korea.

At the beginning of the 21st century, even allowing for significant recovery since 1999, South Korea still faces major challenges, which must be met if the country is to return to high and sustained growth. In addition to the need for intensified corporate and financial restructuring (including bank privatization), the challenges include the implementation of measures, such as the further deregulation of services, in order to accelerate employment creation; and the formulation of effective ways of accommodating the growth of protective sentiments in the USA, Japan and European Union (EU) member states, precisely the countries that, in the past, have absorbed a high share of Korean exports. The nature of future political and economic development in China and North Korea will also be critically important determinants of South Korea's own development trajectory.

PLANNING

From 1962 the South Korean Government used formal economic planning as a framework within which to exercise influence on the behaviour of the private sector and to guide economic development. Although the Government's use of 'indicative' planning did not directly compel enterprises to adhere strictly to specific targets, it did bring indirect pressure to bear through the market mechanism. Thus, enterprises were expected to conform to the basic objectives of plans, while the Government resorted to fiscal, monetary and other measures in order to fulfil planned targets. The existence of a private sector and the absence of compulsion highlights the peculiarly 'managerial approach' inherent in the Korean economy—a major contrast with the planned system of the former Soviet bloc, which sought to integrate all major economic inputs and outputs in accordance with preconceived objectives and priorities through a vertical hierarchy in which orders are passed down from higher to lower administrative levels. In short, the Korean experience was one of planning conducted in the context of a market-based economic system. Although vestiges of this approach remain, structural reforms in the financial, corporate, public and labour sectors under way since the 1997 economic crisis highlight the continuing retreat by the Government from its previous interventionist role in economic activities and its commitment to a more 'orthodox' free market economy.

Except for the Fourth Five-Year Plan (1977–81), during which South Korea faced a serious recession, GDP growth has consistently over-fulfilled targeted rates. A common theme in all the national plans has been a strong emphasis on the outward orientation of national economic development—in particular, the need for export expansion. In other particulars, however, policy thrusts of successive plans differed. For example, under the First Five-Year Plan (1962–66), a major priority was the expansion of infrastructural capital in electric power, railways, ports and communications, in order to obviate resource and other physical obstacles to development. In the Second Plan (1967–71), special attention was directed to the growth of electronic and petrochemical industries, as well as to raising farm incomes by maintaining high prices for rice (the staple crop). Priority development of heavy and chemical industries—especially steel, petrochemicals and ship-building—was strongly in evidence in the Third Plan (1972–76) and defined a major new policy thrust in South Korea's development strategy. In the Fourth Plan initiatives focused on industrial development based on the intensive use of technology and skilled labour. The same Plan, for the first time, also stressed the importance of social development, based on higher government welfare spending as a means of promoting a more equal distribution of income.

Under the Fifth (1982–86) and Sixth (1987–91) Plans, the strategy of export-led growth was strongly reaffirmed alongside unprecedented efforts to liberalize domestic markets, intended to dismantle regulations that had previously constrained the South Korean economy in its attempt to adjust to changing internal and external environments. In the simultaneous search for welfare and efficiency improvements, the Plans also articulated the need for more balanced sectoral and regional development.

In 1993 a revised Five-Year Plan was introduced. Within the framework of a targeted growth rate of 6.9%, the new Plan sought to elevate South Korea into the ranks of advanced economies and to lay the economic foundation for eventual reunification with the North. In line with the requirements of OECD membership, measures were introduced in an effort to combat official corruption and to initiate structural reforms within the economy. To these ends, three separate sub-plans were formulated, embracing deregulation, financial liberalization, and the management of foreign exchange and capital flows.

THE FINANCIAL CRISIS OF 1997 AND BEYOND

With the benefit of hindsight, it is clear that despite its record over several decades of high growth and rising living standards, as of the mid-1990s the South Korean economy faced deeply rooted structural problems. In particular, it had become highly vulnerable to external shocks thanks to two structural features. The first was the highly leveraged nature of the corporate financial structure; the second was the disproportionate share, alongside an external debt-to-GDP ratio of around 25%, of short-term debt (which had reached almost 60% of the total by the end of 1996). Such problems were the major contributory factors which precipitated the severe financial dislocation and economic collapse that affected South Korea, and other Asian countries, during 1997–98. Even before the onset of the Asian crisis, bankruptcy proceedings against a major *chaebol*—the Hanbo group—had served notice of the scale of difficulties facing the Government in Seoul. In July 1997 another *chaebol*, the Kia Group, defaulted on loans amounting to 2,770m. won. Efforts to rescue the situation were unsuccessful and the crisis deepened, leading to severe pressure on the national currency and stock market. The banking sector was particularly affected by the conglomerate bankruptcies, and the incidence of non-performing loans (NPLs) increased sharply in the second half of 1997. Against this background, fears began to grow that the Government might find itself with insufficient foreign-currency reserves to service its foreign debt.

Central to understanding this rapid economic and financial deterioration are the close links that had evolved over many years between the Government, banks and *chaebol*. These links concealed underlying serious economic weaknesses, whilst creating an 'economic bubble', characterized by excessive investment in productive capacity and the overvaluation of the domestic currency. In the absence of adequate regulatory and supervisory controls, the financial system lacked transparency. With tacit acceptance by the Government, industrial projects were duplicated, generating excess capacity. Meanwhile, many *chaebol* had invested aggressively, and frequently in disregard of normal risk analysis, on the basis of profligate loans from domestic financial institutions, while also funding long-term investments by recourse to short-term foreign capital markets.

By the end of 1997 capital flight had become a serious problem, as foreign investors withdrew capital and many Koreans moved their savings overseas. Local firms sought to avoid hard-currency exposure, and export revenue declined. Such developments brought further pressure to bear on stock prices and on land values, which both fell sharply. Despite intervention by the central bank, the won continued to slide, and by the end of the year it had lost more than half of its value in relation to the US dollar.

After protracted negotiations, on 3 December 1997 IMF officials approved a US $57,000m. loan agreement ($10,000m. to be made available as 'accelerated aid'), the largest rescue programme ever undertaken by the Fund. In return, the South Korean Government committed itself to the introduction of more rigorous fiscal and monetary policies (including the maintenance of high interest rates), the strengthening of regulatory control mechanisms, the institution of financial and corporate restructuring, consolidation of labour market reform, and the enhancement of financial transparency in both public and private sectors. Central to the success of these reforms was the implementation of a policy package embracing economic and institutional measures. These sought to fulfil the following goals: to lower the level of corporate debt and institute a system of corporate governance that would improve competitiveness by facilitating greater transparency in management; to reduce and eliminate NPLs through financial restructuring (including bank purchases, mergers and acquisitions); to institute budgetary reforms in order to improve public sector efficiency; and to liberalize employment mechanisms in order to make the labour market more flexible.

In the wake of the financial crisis, severe economic and associated social strains were in evidence throughout 1998. Annual GDP contracted by 6.7%, while the unemployment rate rose from 2.6% to 7.0%; labour unrest and social demoralization were exacerbated by a rapid salary depreciation, mounting personal bankruptcies, and sharp increases in the prices of foreign consumer goods and energy imports. In July deepening recessionary conditions led the IMF to relax the fiscal and monetary conditions of its loan 'package'. Meanwhile, however, with the won successfully stabilized, in the second half of 1998 the Government, which was now headed by President Kim Dae-

Jung, introduced measures designed to stimulate domestic demand. Interest rates were lowered, spending on unemployment relief was increased, and efforts to promote financial and corporate restructuring were intensified.

As indicated above, a major issue in *chaebol* restructuring plans was the reduction of their extremely high level of indebtedness, which was thought to be one of the principal reasons for the severity of South Korea's financial crisis. For manufacturing industry as a whole, between 1997 and 2002 the debt ratio fell from 396.3% to 130.0%—well below the targeted 200% level set by the Government in 1999. However, such apparent success should be set against the reality that a significant part of this decline was attributable to equity issues, asset revaluations and selective debt exclusion, rather than to genuine debt reduction. In addition, the record of the largest *chaebol* was more disappointing: in 2002, for example, the debt-equity ratio of the 19 largest *chaebol* averaged 171.7%, and in the case of Hyundai was still an extraordinary 977.6%.

Initially, burdened with NPLs of 160,000,000m. won—far in excess of the 64,000,000m. won allocated by the Government to buy back bad loans and recapitalize the banking system—and preoccupied with mere survival, the banks themselves were in a weak position to assist in corporate restructuring. In 2000, despite the closure of five large commercial banks and 16 smaller merchant banks, as well as the provision of further large-scale government funding (some 102,000,000m. won were made available by the government in rescue programmes) it was clear that further reform and consolidation, including the disposal of government holdings in banking institutions, were needed. Rationalization of the banking sector continued, however, and in the first quarter of 2004 South Korea's 19 banks reported a collective net income of 1,680,000m. won.

South Korea's economic performance improved markedly in 1999, the contraction in GDP of the previous year (by 6.7%) being transformed into positive growth of 10.9%. An easing of fiscal and monetary policy kept interest rates low and helped stabilize the currency (the won reaching a two-year high). Meanwhile, private consumption spending and business investment rose sharply (by 10.1% and 46.8%, respectively, compared with 1998). As annual manufacturing growth accelerated to 21%, the rate of unemployment began to decline (from 6.8% in 1998 to 6.3% in 1999). Overseas demand for South Korean goods meanwhile strengthened, and by the end of 1999 the current-account surplus was US $24,500m., compared with a deficit of $8,200m. two years previously. Foreign-currency reserves continued their recovery from the low point of 1997, when they were just $19,710m., to reach almost $74,000m. Compared with foreign direct investment (FDI) inflows of $5,400m. in 1998, in the following year FDI inflows were restored to $9,300m. (from 1.7% to 2.3% of GDP). It was against this background that President Kim Dae-Jung claimed that his Government had 'completely overcome' the financial crisis; and had done so without full recourse to the IMF rescue 'package'.

Owing to a sharp slowdown in private consumption in the final quarter, GDP growth in 2000 fell to 9.3%—slightly lower than the 9.5% growth that had earlier been predicted. Indeed, by the end of 2000, it was clear that rapid expansion during 1999 and in the first half of 2000 had not subsequently been maintained. Stagnation in traditional manufacturing industries (for example, textiles and car production) was partly to blame, but the major contributory factor was declining domestic demand. A similar but less pronounced pattern was apparent in the external sector, where export growth of 23.5% during January–September slowed to around 18% in the final quarter. By the end of 2000 the current-account surplus had been halved from US $24,500m. (at the end of 1999) to $12,250m., subsequently declining further to $8,000m. and $5,400m. (2001 and 2002), before recovering sharply to reach $12,300m. in 2003. Stated differently, between 1998 and 2002 South Korea's current-account surplus decreased from 12.7% to a mere 1.1% of GDP, before rising again to 2.4% in 2003. Meanwhile, boosted by a strong performance on capital account, in 1999 national foreign-exchange reserves reached a record level of $96,100m. (30% above the level of the previous end-year and 85% above that of 1998), and thereafter increased further, until, in 2003, they totalled $155,281m.

Between 1997 and 1999, net inflows of investment, mainly FDI, rose sharply from US $3,323m. to $12,261m. Subsequently, however, inflows declined dramatically and in 2003 South Korea recorded a net outflow of investment of about US$1,000m. Under the impact of the financial crisis, the rate of unemployment rose sharply. Having peaked at 6.8% in 1998, it fell back slightly to 6.3% the following year before declining much more sharply to 4.1% in 2000. The downward trend was maintained until 2002, when the rate stood at 3.1%, although in 2003 it rose again to 3.4%.

If South Korea's recovery from the economic crisis of 1997–98 was striking, doubts about the long-run sustainability of the high rates of GDP growth attained in 1999 and 2000 have been reinforced by the volatility of domestic politics and perceptions of government policy confusion. The future growth trajectory will be determined by a combination of familiar internal and external forces. The former include the pattern of domestic consumer and investment demand. One effect of depressed conditions in 1998 was to reduce private consumption by 13.5%. Although the following year saw private consumption demand rebound (up by 11.5%), thereafter growth both slowed and fluctuated considerably—indeed, in 2003 it actually contracted by 1.2%. The growth of public sector demand was more consistent: although it declined in 1998, it was still positive (2.3%) and during 1999-2003 it averaged about 3.8% p.a. Interestingly, the rate of gross fixed investment (GFI) was already contracting in 1997 (by 2.3%), although the full impact of the crisis was not evident until 1998, when GFI decreased by 22.9%, with the demand for machinery and equipment declining by a remarkable 42.3%. Although construction demand continued to contract, during 1999 and 2000 the overall annual rate of GFI expansion averaged over 35%. A negative GFI trend in 2001 gave way to renewed expansion, albeit at a declining rate, of 4.9% and 2.1% in 2002 and 2003. Projections by the Economist Intelligence Unit suggested zero GFI growth in 2004, before recovering to 2.5% in 2005.

South Korea's strong outward economic orientation means that external forces have a major bearing on its overall economic performance. In the aftermath of the financial crisis, the value of exports rose, in terms of Korea's domestic currency, by 12.7% (15.9% for merchandise exports) in 1998, while the value of imports declined by 21.8% (minus 24.8% for foreign merchandise purchases). The effect of these changes was to transform a merchandise trade deficit of 30,000m. won in 1997 into a surplus of 25,300m. won in 1998. However, because of the sharp fall in the value of the domestic currency, in US dollar terms the effect was to change a deficit of US $3,179m. to a surplus of $41,665m.—a level that had not previously and has not subsequently been (re)attained. Meanwhile, South Korea's current-account balance moved from a negative $8,167m. in 1997 to a positive $40,371m. in 1998. Between 1999 and 2002 the trade performance, in terms of both current and merchandise trade accounts, deteriorated, although the balance of both showed a marked improvement in 2003 (with surpluses of $12,321m. and $22,161m.).

One of the effects of the 1997 Asian crisis was to bring about the demise of what used to be called 'Korea Inc'—the nexus of state-business relations, whereby industrial activities were subject to serious political interference and characterized by widespread corruption. Nevertheless, the legacy of such practices remains, and central to South Korea's future economic trajectory is the South Korean Government's ability to address the need for further corporate restructuring by enhancing competition in order to regulate the power of the *chaebol*, the activities of which will, after all, continue to dominate industry and exports. The election, in December 2002, of President Roh Moo-Hyun was expected to give a further impetus to corporate reform, since Roh's Millennium Democratic Party (MDR) was more committed to the need for such restructuring than the opposition Grand National Party. In that same year further salutary reminders of the need for greater self-regulation by the corporate sector were provided by several developments: the bankruptcy of Daewoo and the purchase of Daewoo Motor by General Motors, the separation of Hyundai into smaller constituent elements, and official endorsement of the purchase of South Korea's third largest life insurance company (Korea Life Insurance) by a *chaebol* (Hanhwa) with a reputation for financial misconduct. In the event, however, political preoccupations culminating in President Roh's ultimately unsuccessful impeachment in March 2004 by the National Assembly constrained government efforts to improve corporate governance. As a result, as of mid-2004, concerns remained about the ability to maintain an appropriate momentum of corporate restructuring. In the absence of the strict implementation of rigorous control mechanisms, the re-emergence of costly mismanagement and misconduct within the corporate business sector remained a real threat, especially as privatization offered new opportunities for corporate expansion.

The moral of such developments is clear: if South Korea is to retain, let alone improve, its place in the global economy, it is incumbent on the Government, as privatization leads it to downsize its interventionist economic role, to strengthen structural reforms (including banking reforms) and to put in place an effective legal and regulatory framework that will facilitate increased competitiveness by autonomous decision-making enterprises. At the same time, research and development and educational investment must be increased in order to make the South Korean economy a truly knowledge-based economy that can compete effectively in the global environment.

AGRICULTURE

South Korea is a mountainous country. Only 20% of the surface area (1.9m. ha) is arable, 65% being designated as forest land. The Taebaek range, which runs from north to south along the east coast, is the watershed of the Korean peninsula and the source of the country's principal rivers (the Han, Nakdong and Kum). South Korea's farming is concentrated mainly in these river basins and in the surrounding plains in the west and south. Owing to its more favourable climate and longer growing seasons (between 170 and 226 days), South Korea is more suited to farming than the North—especially in the cultivation of rice. This is still the main crop, with paddy fields accounting for 60% of the total cultivated area. Other food crops include barley, beans, potatoes and wheat, among which barley production has been promoted in order to conserve rice.

In the 1950s Korea remained a typical pre-industrial country, in which almost half of GDP derived from agriculture and an even larger proportion of its workforce was engaged in farming. Thereafter, however, the vigorous industrialization and export drive transformed the agrarian character of the economy: between 1960 and the early 1980s, the share of agriculture in GDP decreased to about 15%, while the countryside's share of total population declined from 58% to less than 25%. In 2003 agriculture (crop farming, forestry and fisheries) accounted for a mere 3.9% of GDP and 8.9% of total employment. Such figures highlight the relatively low productivity of the farming sector, although viewed from an international perspective South Korea's long-run record of agricultural growth since the early 1960s has exceeded the world average, as well as that of many Asian countries. Nevertheless, there was a contraction in the agricultural sector in both 2002 and 2003.

The Land Reform Acts of 1947 and 1948 abolished tenancy and defined the legal framework in which small owner-operated farms emerged. The rural population, which remained almost static at slightly more than 15m. until 1969, thereafter steadily declined. However, the rate of arable land contraction failed to match the decline in the size of the farm population, as a result of which the average size of farm actually increased slightly. Thus, in 2002 average availability of arable land per head of farm population was 0.52 ha., compared with 0.41 ha. in 1995. The tiny scale of farming in South Korea is highlighted in the finding that in 2002 one-third of all farmers had plots that were smaller than 0.5 ha. in size, while only 14% of farmers had plots that were greater than 2 ha.

Despite the use of price subsidies, government farm purchases and import restrictions, the economic role of agriculture has gradually weakened. In particular, deficiencies associated with the small-scale nature and labour-intensive operation of farms, an ageing farm population and inefficiencies in the farm marketing system have left improvements in agricultural productivity lagging substantially behind those in the industrial sector. One consequence—exacerbated by growing international pressure to open its domestic markets for rice and other agricultural products—was a rise in food imports to the extent that by the early 1990s the value of such sales was in excess of US $7,150m., making South Korea the sixth largest importer of farm products in the world. Between 1990 and 1995 the cost of all agricultural imports rose by 80%. In 2003 the deficit on the agricultural trade account was $6,715m., of which net imports of food and live animals accounted for almost 92%.

Food Production

Korea has used protectionist and farm-support mechanisms in search of food self-sufficiency as a basis on which to secure food security. Membership of the World Trade Organization (WTO) compelled the Government to adjust its farm policies, although it succeeded in maintaining import restrictions under special domestic regulations and negotiated a deal whereby its rice imports were restricted to a minimum of 4% of consumption up to 2004. As recently as 1998–2000 net imports constituted only 19% of domestic rice consumption and 21% of barley consumption (although the corresponding figures for soyabeans, wheat and maize were respectively 66%, 97% and 100%). Between 1998 and 2001 rice output rose steadily, but in 2002 a reduction in the sown area was mainly responsible for a 10.7% decline in production to 4.93m. metric tons.In 2002 average calorific intake was over 3,000 per day, most of which derived from vegetable products (although meat consumption is comparable with that of Japan, it remains well below that of the USA and other Western countries). The decline in average per caput grain consumption from 195.2 kg in 1980 to about 145 kg in 2002—and from 132 kg to 87.4 kg for rice alone in those years—highlights improved access to a more varied diet by the South Korean population, afforded by the impact of rises in income.

In recent years, net imports of live animals and meat have played an increasingly important role in meeting domestic consumption requirements. Indeed, the domestic output of most important animal categories has been in decline, most notably in the rearing of beef cattle, numbers of which more than halved between 1997 and 2002. The value of net imports of food and live animals rose from US $3,270m. in 1995 to $6,166m. in 2003.

Fisheries

South Korea has a long coastline, and fishing remains important for its contribution to national diet, livelihood and exports. Although accounting for only 0.5% of the labour force, South Korea's aquatic activities are profitable and make a significant contribution to export earnings. Until the second half of the 1990s, aquatic production exceeded 3m. metric tons annually, although under the dual impact of the imposition of exclusive fishing zones by countries such as Russia and the USA and the introduction of stricter fishery regulations for resource management, in more recent years it has been in decline. In 2002 the fishing catch was 2.0m. tons.

Forest Products

Although two-thirds of South Korea's surface area is forest land, and the country's moist climate is conducive to forest development, indiscriminate felling before 1945 depleted most of the original tree cover. Nation-wide afforestation and soil conservation campaigns have, however, successfully reversed the trend: since the early 1990s, for example, the forested area has remained quite stable (in 2002 it was 6.412m. ha). During the same period the growing stock of trees rose from 283.8m. cu m to 448.5m. cu m, and timber production from 0.84m. to 1.24m. cu m.Major tree species include red pine, Korean white pine, larch and oak. Lumbering, mainly of coniferous trees, is limited to the mountains of Gangwon and Gyeongsang Provinces, but contributes only a fraction of domestic timber needs.

MINING

In the absence of any known reserves of petroleum and possessing only 10%–20% of the Korean peninsula's mineral deposits, South Korea is poorly endowed with natural resources. Nevertheless, more than 50 different minerals have been found, of which the most important are graphite, anthracite coal, fluorite, salt, limestone, gold, silver, tungsten and some iron ore. In absolute terms, the value of mining production reached its peak level in 1997, although, since the early 1980s, the mining sector's share in GDP has never exceeded 1.5%. In 2003 mining production was still 15% below the pre-crisis level of 1997. As a result of ever-increasing domestic industrial requirements, the value of South Korea's imports of minerals (especially iron, zinc, copper and aluminium ores) has far exceeded that of exports (talc, agalmatolite, tungsten and graphite). In an attempt to enhance the provision of minerals, South Korean enterprises have in the past used joint ventures in order to participate in mineral extraction projects in resource-rich countries, such as Australia, Canada, Indonesia and the USA. In 2004, Korea Resources Corporation announced that the South Korean Government planned to invest US $340m. by 2010 in mines overseas (including the Philippines, Indonesia, Mongolia and Peru) in order to meet rising domestic demand for coal, copper and other raw materials. In 2003 South Korean companies had stakes in 16 foreign mines, a number that was expected to more than double by 2010.

Coal

Coal, mostly anthracite, is one of South Korea's leading mineral resources, with estimated deposits of 276m. tons. The production of anthracite peaked at 24.3m. tons in 1988, after which the closure of small-scale, uneconomic mines caused a sharp decline—to 5m. tons in 1995 and to a mere 3.3m tons in 2003. Rising domestic demand and declining production have necessitated imports of anthracite, mainly for domestic heating and cooking, as well as much greater quantities of bituminous coal for use in power plants, industrial boilers, and the iron and steel industry. The average annual cost of coal imports doubled between 1986 and 1990, and continued to rise thereafter (especially in the second half of the 1990s and beyond). In 2002 South Korea was, after Japan, the second largest coal importer in the world, its purchases of 70m. tons of coal, mainly from Australia and China, accounting for almost 11% of global imports. Coal currently contributes 21% of Korea's total energy needs.

Metallic Minerals

In general, South Korea lacks sufficient deposits of non-fuel minerals to service its industrial needs, the average import ratio of such minerals being about 70%. Around 98% of metallic mineral requirements must be supplied from abroad, and only 10 out of 45 Korean metallic minerals satisfy more than 90% of domestic demand. Notwithstanding estimated deposits of 120m. metric tons (mostly magnetite), the peak output level of iron ore, of 677,100 tons in 1989, has not subsequently been reattained, and since 1997 there has been a dramatic output contraction. South Korea is a major importer of iron ore, accounting for 8% of global imports in 2002. Between 1992 and 2002 iron ore imports rose from 35.6m. tons to 41m. tons. Ore reserves of lead, zinc, copper and tungsten, in terms of metal content, have been estimated at 492m., 738m., 105m. and 100m. tons, respectively. Production of lead and especially zinc ore has declined markedly since the early 1980s. Other ores mined in South Korea include copper, gold, silver, molybdenum and tungsten—the tungsten mine at Sangdong being the second largest in the world.

Non-metallic Minerals

With estimated deposits of 30m. metric tons of amorphous graphite and 2.6m. tons of crystalline graphite, South Korea is an important global source of natural graphite. In 2002 output of all types of graphite totalled 94,000 tons, compared with 65,000 tons the previous year.Production of fluorite has contracted dramatically since the 1980s. Limestone is abundant in South Korea, with reserves estimated at 1,500m. tons, and output rose steadily from the early 1980s to reach 88m. tons (at 50% purity) by the mid-1990s. Following the 1997 Asian financial crisis, production fell to 69.9m. tons in 1998, since when it has once more risen (to 83.8m tonsin 2002). Limestone deposits are the basis of the cement industry, and, in turn, of construction activities, which from the 1960s underwent rapid expansion in support of national economic growth momentum. By the end of the 1980s South Korea was the fourth largest cement producer after China, the US and Japan. Cement exports have been an important source of foreign-exchange earnings. In 1997, at the time of the onset of the financial crisis, cement production totalled 60.3m. tons, but it decreased sharply to 46.8m. tons in the following year. Recovery subsequently took place, although in 2002 total output (55.5m. tons) had still not reattained its previous peak. Kaolin production is also substantial, with a total output of 2.8m. tons in 2002. In addition, there are significant sources of uranium ore, with estimated reserves of 56m. tons (at 0.3%–0.4% uranium content).

MANUFACTURING

In the early stages of its industrialization and economic modernization, South Korea's labour-intensive consumer goods industry succeeded quite well in meeting the demands of growing domestic and foreign markets. In the early 1970s, however, the previous bias towards light manufacturing industry shifted towards more capital-intensive activities, such as machine-building, engineering, ship-building, whole plant construction, and the production of electronic goods, transport equipment and petrochemicals. Defence-related heavy industries were, as they have remained, also a high priority. From the 1980s, another watershed was passed, as increasing emphasis was given to the development of high-technology and knowledge-intensive industries, including telecommunications and information technology activities, which have become core export-orientated industries in South Korea. Implicit in strategic structural shifts has been a recognition of South Korea's changing comparative advantage, based on a scarcity of natural resources and ready access to a skilled and increasingly highly educated labour force.

As a stimulus to rapid and sustained national industrial and economic growth, the expansion of exports has played a critically important role. The penetration of global markets has required South Korean manufacturing enterprises to become efficient and internationally competitive. This was achieved through the provision of incentives that facilitated a reduction in costs, the attainment of optimal scales of production and the introduction of productivity-enhancing innovations. In addition, the forces of international competition gave impetus to specialized production in those areas in which the country maximized its comparative advantage. By such means, South Korea has been able to create world-class industries in car manufacturing, shipbuilding, nuclear energy, and the production of semiconductors, telecommunications and information technology equipment.

Between 1965 and 1980 the average annual rate of growth of manufacturing output was a remarkable 16.4%, a figure that decreased to a still-impressive 12% during 1980–90. In the 1990s manufacturing growth continued to decline, although by international standards it remained impressive, averaging 8.6% annually between 1990 and 1995. After 1995 growth decelerated sharply, especially in the wake of the financial crisis, contracting by 6.6% in 1998. Subsequent recovery was driven by shipbuilding, semiconductors and car production, and in 1999 and 2000 manufacturing output expanded by 25.0% and 17.1%, respectively. Such growth was, however, not wholly translated into sales, as revealed by an almost six-fold increase in industrial inventories in 2000. A deteriorating external environment almost halted the momentum of recovery in 2001, when manufacturing output increased by a mere 0.9%, but in the following year there was renewed expansion of 7.3%. In the first half of 2003 output growth again contracted slightly, but thanks to a surge in overseas demand for semiconductors and motor vehicles, thereafter it began to accelerate.

Textiles

South Korea is one of the world's largest producers of textiles, and during the early phase of modernization the textile industry made a major contribution to domestic employment, exports and economic growth. However, between 1980 and 1997 the textile sector's share of total manufacturing employment declined from 24.7% to 8.7%, and by the late 1980s textiles had been replaced by electronics as South Korea's largest single export item. In 1997, on the eve of the Asian financial crisis, the value of textile exports was US $18,500m. and accounted for 13.6% of total export value (compared with 28.6% in 1980). Under the impact of the financial crisis, in 1998 the value of textile exports decreased to $16,700m. (12.6% of total export value) and this trend has since continued. Growing protectionism as a result of increasing competition from other low-wage economies has, in recent years, encouraged major textile enterprises to diversify into other fields, embodying high technology and biotechnology. As a result, by 2001 textiles no longer ranked in the top five export categories. In 2003, mainly as a result of competition from China, the export value of textiles (clothing, woven and textile fabrics) was $9,734.5m., compared with $13,547.1m. in 2000.

Shipbuilding and Car Industries

Construction of new plants during the late 1960s laid the basis for a shift of emphasis in the 1970s, from production of light machinery products and simple metal-working machinery to the manufacture of transport and communications equipment, industrial machinery, precision machinery, textile machinery, and electric and electronic appliances.

South Korea is, after Japan, the second largest shipbuilding country in the world, and on occasions it has even overtaken Japan to become the global leader, as in 1993 when it received 38% of all shipbuilding orders. In 2002 a 19% increase in orders meant that South Korean companies took 32% of all new orders placed, compared with 43% for Japan, and a mere 5.5% for the EU. In terms of contract completions, however, South Korea was pre-eminent in 2002, accounting for 40% of the global total, compared with Japan's 37% and the EU's 13%. In the same year overseas sales of ships accounted for 6.7% of total exports. In 2003 South Korean shipbuilding companies won record orders of 16.75m. gross tons—sufficient to sustain them for at least three years. In addition to the positive impact of post-1997/98 crisis reforms, economic expansion in neighbouring China has been an important element in the recent buoyant performance of the South Korean shipbuilding industry.

The shipbuilding industry was badly affected by the Asian crisis of 1997, in the immediate aftermath of which there was a sharp reduction in South Korea's export sales. Yet recovery was swift, and by 2000 the value of shipping exports reached a record US $8,230m. Not least, the financial crisis served to highlight the weak financial position of many shipbuilding companies, which had for many years benefited from subsidies that took no account of commercial realities. For example, the competitiveness of Daewoo Shipbuilding and Marine Engineering (DSME), since 1995 the largest shipbuilding company in the world, was enhanced following the crisis by a far-reaching programme of restructuring, which enabled it to return to profit in 2002. Nevertheless, subsidies have remained an issue. In 2003, for example, the European Commission delivered a formal protest on behalf of European shipbuilders, who alleged that official subsidies were enabling their South Korean competitors to produce at between 15% and 40% below cost price, thereby undercutting competition from European shipyards. In 2004 rising steel prices also threatened to erode the profitability of South Korean shipbuilding firms, as revealed in DSME's announcement of a 21% decline in profits during the second quarter.

The origins of the South Korean automobile industry lie in developments that took place in the 1960s (Hyundai Motor was founded in 1967). The sustained growth of the motor industry reflects the successful exploration of both domestic as well as overseas markets. Until the 1980s, the main orientation of the industry was towards exports; however, during that recession-affected decade and into the 1990s greater emphasis was placed on meeting the needs of an increasingly affluent domestic market. Despite this, car ownership remains significantly below levels in the USA, Japan and many Western European countries. During 1988–91 South Korean car output grew, on average, by 11.4% annually; in 1993 total production exceeded 2m. units for the first time; and in 1995 South Korea became the world's fifth largest car producer (after the USA, Japan, Germany and France). Although its global ranking fell back under the impact of the financial crisis, and despite having been overtaken in 2002 by China, in 2003 it regained fifth place, after the USA, Japan, Germany and China. Since 1995 car output has risen year-on-year, with the sole exception of 1998; in 2000 annual production was around 3m. units, and in 2003 total sales reached 5.2m. Contained within this last figure was an exceptionally buoyant market for Korean cars in the US and EU, but a depressed domestic market. The single largest car producer in South Korea is Hyundai, which together with its affiliate (Kia Motors) set a sales target of 3.4m. units (worth $39,800m.) for 2004.

South Korea began to export motor vehicles to Canada in 1983, and had begun to penetrate the European market by 1985. The country's automobile producers passed another watershed when, in the mid-1980s, they entered the US market (in 1986 Hyundai's 'Excel' model proved to be the most successful new foreign car ever to enter the US automobile market). The value of exported passenger cars rose from US $87.4m. in 1983 to a record $3,336.1m. in 1988. There followed a short period of declining exports, but by 1993 recovery and renewed growth had taken exports to a new peak of $3,892.3m. The momentum of export growth slowed only marginally in 1998, and by 2002 the value of passenger car exports had reached $11,938m. (8.1% of all exports), making this industry the second largest source of export earnings after electronic products. In 2003 the export value of cars rose by an astonishing 29.3% to reach $15,436m.

Electronics

The origins of the electronics industry lie in the assembly, in the early 1960s, of vacuum tube radios from imported parts. Subsequently, South Korea's electronics industry expanded rapidly, and in 1969 it received full-scale government support through the enactment of the Electronics Industry Promotion Law, which designated electronics as a priority industry and recognized its strategic export potential. The implementation of a comprehensive eight-year development programme was accompanied by major inflows of foreign capital and technology, which played a crucial role in generating 'foreign-led' electronics exports. In the early 1970s the manufacture of electronic components (such as transistors, diodes, integrated circuits, radio receivers and parts for monochrome television receivers) dominated. After 1974 colour television receivers were produced, fuelling a rapid expansion in consumer electronics. By the end of the 1970s South Korean-made products included electronic calculators and watches, and in 1979 the country had become one of the most important global producers of video-cassette recorders (VCRs).

The production and export of computers and peripheral equipment also expanded rapidly, as computer manufacturers began to pursue a policy of import substitution. With a total output and export value of US $23,531m. and $15,200m., respectively, by 1988, the industry already accounted for 13.4% of GNP and 25.0% of total exports. In that year South Korea was the sixth largest exporter of electronic goods in the world and electronics had surpassed textiles as the country's principal export item. In 1992 South Korea became one of the five largest electronics-producing countries in the world, with overseas sales accounting for 25.9% of total exports—a figure that rose to 30.6% in the first half of 1999. In 2002 export earnings from electronic products totalled $51,200m. and accounted for 34.7% of the value of all exports. The corresponding figures for 2003 were $61,425m. (a year-on-year rise of 20%) and 35.2%. Disaggregated data show that overseas sales of semiconductors, information technology (IT) and associated products were $17,549m., $31,199m. and $8,158m.

The growth of semiconductor production is considered essential if a country is to secure an internationally competitive position in high-technology areas, especially in the production of automatic data-processing equipment. Thus, as in Taiwan, semiconductors have played a major role in the development of South Korea's electronics industry. Indeed, by the mid-1990s, it had become one of the foremost international producers of semiconductors, South Korean companies having established global brand name recognition for their products. Helped by strong growth in the global market for semiconductor chips, Samsung Electronics garnered 10.8% of the world's memory chip market in 1993, thereby becoming the seventh largest semiconductor chip producer in the world. In the same year South Korea's share in the global semiconductor market was 17.9%—and 23.6% in the dynamic random access memory (DRAM) chip market. Manufacturers have, however, remained dependent on Japan for the supply of many of the components that they use, although in 1993 a number of Japanese electronics and machinery manufacturers began themselves to purchase electronic parts and components from South Korean companies. The experience of the USA and other countries highlighted the critical need to allocate increasing investment to research and development in order to enhance international competitiveness, and it is instructive that in the early 1990s the three leading electronics manufacturers—Samsung, Daewoo and LG (formerly Lucky Goldstar)—substantially increased their research and development expenditure. Meanwhile, by the mid-1990s South Korea ranked second in the world in terms of production capacity of VCRs, microwave ovens, fax machines and videocassette tapes, and third for colour television receivers and telephones. In March 1996 Samsung initiated a DRAM plant in Texas—its first semiconductor facility to be located in the USA, attesting to the globalization of the country's electronic industry. Since the 1990s Samsung has also undertaken large-scale investment in China, and by 2005 it plans to have extended production facilities on the mainland to manufacture semiconductors, digital appliances and IT products, a sign of South Korean companies' expanding horizons in China. Electronics have also been a major destination for FDI inflows into Korea.

Chemicals

The construction, in 1959, of the Chungju Fertilizer Plant marked the beginning of South Korea's development of an indigenous chemical industry. Prior to this, all chemical fertilizer requirements were imported. By 1968, however, chemical fertilizer self-sufficiency had been secured, and in 1976 South Korea became a net exporter of fertilizers. The peak level of output (1.8m. metric tons in terms of primary nutrients) was reached in 1994, since when the production trend has been downward, reaching 1.2m. tons in 2002.Complex fertilizers now account for almost 70% of total consumption.

Diversification of the chemical industry has been under way since the 1960s. Major products include sulphuric acid, ammonia, compressed oxygen, dye, insecticides, polyethylene, polypropylene, polyvinyl chloride, polyester fibres and acrylic fibres. In 2003 chemicals and chemical products ranked fourth as a source of export earnings, the value of overseas sales being US $13,521m. (25.5% more than in 2002), equivalent to 7.8% of the value of all South Korean exports. The chemical industry has also been a significant recipient of FDI.

Metallurgy

The construction, in 1973, of the Pohang Iron and Steel Co (POSCO)—South Korea's first integrated steel plant—was an important watershed in the development of a national steel industry. POSCO subsequently became the second largest steel complex in the world, enjoying sales of $14,924m. in 2003. By 1996, with an annual production capacity of some 40m. metric tons, South Korea had become the world's fifth largest producer of steel after China, Japan, the USA and Russia—a position it retains to this day. Nevertheless, its performance was seriously affected by the 1997 crisis, as a result of which domestic demand for steel fell back sharply, and three of the country's largest producers were declared bankrupt. Recovery from an unprecedented output decline in 1998 was, however, swift, and thereafter crude production of steel increased steadily (averaging 3.2% per year during 1998–2001) to reach 45.4m. tons in 2002. A noteworthy feature of the steel industry's performance in 1999–2000 was that it was achieved against the background of a fall in the price of steel and consequent pressure on exports. Perhaps the clearest evidence of the metallurgical industry's continuing viability was the fact that in 2003 associated products were exported to the value of US $11,843m., 26.8% above the level of the previous year and accounting for 6.8% of the value of total exports.

INFRASTRUCTURE

Transport

Until quite recently, the principal means of freight and passenger traffic in South Korea was railway transport. With the rapid expansion of the motorway network, improvements in its quality and extended motor vehicle ownership, roads have now assumed the position previously occupied by railways. In 2002, for example, 584.6m. metric tons of domestic freight (three-quarters of the total) were transported by road—some 43% more than in 1995 and 9% more than in 2001. By contrast, the railway network carried less than 6% of all freight. Recent trends in passenger traffic display a different pattern, the share of passengers carried by road having declined, while those travelling by rail and air have increased. The contraction in road passengers is most dramatically highlighted in the finding that, whereas in 1973 roads accounted for 96% of all passenger traffic, by 2002 the figure had decreased to 76%. The increasing popularity of railways is partly attributable to the expansion of subway systems that are linked to the national overground rail network. Mass transit rail systems in the largest cities have been accorded special priority in recent years, as exemplified by the construction of underground railway systems in Seoul, Busan, Daegu, Daejeon, Gwangju and Incheon. In 2002 subways carried 2m. passengers and accounted for almost 16% of total passenger traffic.

Air transport capacity, both internal and external, has expanded markedly. Until the end of 1988 Korean Air was the only airline company to operate in and from South Korea. In December of that year, however, Asiana Airlines began operations on domestic routes, and in 1995 these were extended to include international destinations (in Japan, China, South-East Asia and the USA). A major development was the opening of a new international airport at Incheon in 2001, by which time the number of air passengers using South Korean airports was approaching 30m. In 2002 both outward-bound air passenger and air freight traffic increased by over 11%, generating profitability for both airlines.

The capacity of South Korea's ports, including newly constructed facilities at Bukbyong on the east coast, has also expanded rapidly, reaching 295m. metric tons per annum in 1996, compared with almost 100m. tons in 1986. In line with this expansion, the national merchant fleet has greatly increased: in 2002 its 2,532 vessels had a total displacement of 7.0m. grt. In 2001 sea vessels carried 140.54m. tons of freight, 19.4% of the total. Meanwhile, the southern city of Busan had now become the third largest container port in the world.

The vagaries of inter-Korean relations are reflected in the halting progress that has so far been made towards restoring and developing transport links between North and South Korea. Nevertheless, limited progress has been made, as in the agreement of both governments to reconnect North–South road and rail links. Subject to further *rapprochement* taking place, two roads (a western and an eastern corridor) will be permanently opened to facilitate passenger and freight traffic across the border. Other goals include the rebuilding of the Seoul–Sinuiju (North Korea) Railway Line and the construction of a four-lane highway linking Seoul and the North Korean capital, Pyongyang. Much more ambitious—and from the present perspective, remote—are plans to build a gas pipeline from Siberia to both Koreas. The idea has even been mooted of constructing a freight route, sometimes referred to as the 'iron silk road', from Europe to South Korea via Siberia and North Korea.

Telecommunications

Owing to the vigorous government-sponsored promotion of the industry, South Korea has emerged as a global leader in telecommunications. Investment in the industry in 2002 exceeded US $1,000m. and with 45.4m. sets and 48 main lines per head of total population, fixed-line telephone penetration had reached a very high level. The most dramatic development of recent years has been the extraordinarily rapid take-up of mobile cellular telephone subscribers. The number of such subscribers exceeded the number of wire telephone subscribers for the first time in September 1999. By the end of 2002 the number of mobile telephone subscribers totalled 32.5m. (a penetration rate of 68% and the highest in the Asia-Pacific region after Singapore, Taiwan and Hong Kong). It is expected that penetration will have exceeded 80% by 2008. The mobile telecommunications industry in South Korea is dominated by three companies—SK Telecom, KT Freetel and LG Telecom.

Personal computer (PC) ownership was over 17m. in 2002, and rose by around 15% annually in 2001 and 2002. Data provided by the International Telecommunication Union (ITU) indicate that some 26.5m. PCs were in use in 2002. Associated with the increasing popularity of PC ownership has been extraordinarily rapid growth in access to the internet, so that by 2002 the internet penetration rate had reached as much as 65% of the total population, equating to 31.0m. internet users (ITU put the number of users at 26.3m.). There has also been a rapid take-up of broadband usage, with some 9.2m. subscribers to high-speed internet access services in 2002. Consolidation is likely to be one consequence of the highly competitive environment of the internet industry. In January 2003 Hanaro Telecom, South Korea's second largest broadband provider, acquired a controlling share in its smaller rival, Thrunet. The deal gave Hanaro a 42% market share, bringing it close to the 47% share of its rival and Korea's largest company in the industry, KT (Korea Telecom) Corporation.

Energy

Total consumption of primary energy increased almost fourfold between 1964 and 1980, from 11.5m. tons (oil equivalent) to 43.9m. tons. Thereafter, down to 2003, consumption increased, on average, by 7.2% annually, to reach 215.2m. tons, although the strong upward trend was interrupted by a 8.1% decline in 1998 in the aftermath of the financial crisis. Such rapid long-term growth reflected the demands of a generally buoyant economy and increasingly prosperous society (not least in terms of car ownership), but also concealed considerable inefficiency in energy use. Thus, International Energy Agency data indicate that consumption of energy per unit of GDP in South Korea is almost twice as high as in most OECD countries in Europe.

Since 1990 the average annual rate of increase of domestic energy production has been less than half that of consumption, so placing an increasing burden on imports, which have grown (in gross terms) by 7.5% per annum. In 2003 domestic output (38.9m. metric tons of oil equivalent) was sufficient to meet just 18% of total consumption (215.2m. tons), necessitating imports of 214.9m. tons, but 185.1m. in net terms. In the same year South Korea's energy dependency ratio was 96.9%. The heavy burden of such imports is revealed in the finding that in 2002 crude petroleum purchases from overseas cost South Korea US $19,200m., and in 2003 $23,082m. Declining dependence on oil as a source of primary energy in recent years, from 62.5% in 1995 to 47.6% in 2003, reflects the Government's efforts to reduce its vulnerability to petroleum price rises and interruptions in supply, as well as public demands for cleaner, high-grade energy. The reduction has been offset by rises in the share of energy supplied by coal, which increased from 18.7% in 1995 to 21% in 2003, liquefied natural gas (South Korea being a major importer of LNG), and nuclear power (from 6.1% to over 10%).

South Korea lacks domestic petroleum reserves and relies entirely on imports. It is the one of the largest oil consumers in the world, with oil accounting for about 55% of total primary energy supplies. Despite the contraction in oil as a share of total primary energy supplies, crude petroleum imports grew by 11.2% annually between 1990 and 2000, although contracting by 11.5% during the next two years. In 2002 crude petroleum imports totalled 125.8m. kilolitres, of which 44% came from Saudi Arabia and the United Arab Emirates. In order to obviate shortages resulting from the interruption of imports, the Government has developed a strategic oil reserve, managed by the state-owned Korea National Oil Corporation (KNOC), which seeks to guarantee stocks sufficient to meet needs over a 90-day period. In addition, in an effort to minimize energy dependence, South Korea is developing its own petroleum and coal fields in some 13 countries through the implementation of joint-venture schemes. KNOC seeks to provide for 10% of national oil requirements by 2010.

LNG production has been promoted as part of a strategy to diversify South Korea's sources of energy for both domestic and industrial consumption. The first LNG terminal, in Pyeongtaek, was completed in April 1987 and has an annual processing capacity of 1m. metric tons. The Korea Gas Corpn (KGC), which operates the terminal, signed a 20-year agreement with Indonesia, providing for the import of 2m. tons of LNG annually from 1986 until 2006. A second terminal has since been established at Incheon, and construction of a third facility was due to be completed in 2005 at Kwangyang. In 1987 KGC began to supply LNG to the Seoul metropolitan area, making South Korea the seventh nation in the world to use LNG. It subsequently became the second largest importer of LNG in the world. In 2003 LNG supplied 11.2% of primary energy consumption—24.2m. tons (oil equivalent)—compared with 6.1% in 1995. Gross LNG imports in 2003 totalled 17.47m. tons, most of it coming from Indonesia, Malaysia, Qatar and Oman.

The importance of nuclear energy is revealed in the finding that South Korea's 16 nuclear power plants, the first of which was built in 1977, contribute some 40% of electricity supply. By 2014 it is expected that South Korea will have constructed a further eight nuclear power stations. In 2003 total electricity power generation (from thermal and hydroelectric, as well as nuclear sources) was in excess of 56 gigawatts.

INTERNATIONAL TRADE

Although South Korea lacks natural resources and is constrained by its small domestic market, it enjoys the benefits of a committed and highly educated labour force, which has helped facilitate the expansion of exports that has sustained the high economic growth rate over several decades. In every year between 1953 and 1985 imports exceeded exports, although accelerating export growth facilitated a sharp contraction in the trade deficit as a share of exports—from 772.2% in 1953 to 137.5% in 1970, and to a mere 2.8% in 1985. In 1986 South Korea recorded a trade surplus of US $4,210m.—the first to be achieved since the end of the Korean War—and by 1988 the corresponding figure had reached $11,440m. This surplus was virtually eliminated in 1989, after which, until 1998, the trade balance moved back into deficit. Among the factors responsible for the rapid rise in import costs during this period were the enormous increase in oil prices, following the Iraqi invasion of Kuwait in August 1990, the rising value of the Japanese yen, and the expansion of domestic demand in South Korea itself. Between 1996 and 1997 South Korea's merchandise trade deficit narrowed from $20,620m. to $8,460m. In the following year, as South Korea's demand for imports declined sharply during the recession, a sizeable surplus of $39,030m. emerged. Trade remained in surplus in subsequent years—although the absolute level declined from $23,940m. in 1999 to $9,340m. in 2001. This sharp contraction reflected a combination of factors, including adverse global economic conditions, reduced international demand for telecommunications and information technology products, and the effects of the 11 September terrorist attacks in the USA. South Korea's trade balance in 2001 would have been more adversely affected, had not imports also decreased (by 12%) as a result of slowing global economic growth and declining international crude petroleum prices. In the next two years the merchandise trade surplus expanded once again, to reach $14,991m. in 2003. Economic recovery in the USA and Japan, as well as the continuing growth of trade with China, were significant factors in South Korea's buoyant trade performance during these years.

Exports

Rapid export growth commenced in the 1960s, following a deliberate decision to pursue vigorous export promotion in pursuit of rapid and sustained economic growth. Between 1965 and 1980 exports grew, on average, by an astonishing 27.2% per year—a record unmatched anywhere in the world. In the 1980s the export sector encountered more difficulties and was subject to wide annual fluctuations—compare a nominal rate of increase of 36.2% in 1987 with one of just 2.8% in 1989—whilst also recording a sharp downturn in annual growth (to 15.3% per year).

The 1990s too were a decade of varying fortunes. Between 1990 and 2000 average annual export growth was a very creditable 10.5%. However, concealed within this figure were large annual fluctuations, ranging from 30.3% in 1994 to –2.8% in 1998: the former reflecting widening international demand for Korean electronic goods (especially semiconductors), cars and ships; the latter, not just a familiar reflection of the impact of the financial crisis, but also of falling international prices for Korean goods and of the weakness of the Japanese yen. Following the contraction of 1998, renewed expansion took place (exports rising, on average, by 7.9% between 1998 and 2003). This buoyant picture was, however, interrupted in 2001, when exports declined by 12.7% (see above). In 2003 the value of exports was US $193,817m., 19.3% more than in the previous year.

In recent years, South Korea has faced formidable challenges from a variety of sources. One is fierce competition from low-wage economies, such as the People's Republic of China and countries in South-East Asia. Another external pressure has been the impact of protectionist measures introduced by advanced countries. At home, labour unrest and upward pressure on wages have also taken their toll. Nevertheless, South Korean products have competed favourably with goods produced in developed countries, and in some cases (for example, cars, microwave ovens, television receivers, VCRs and personal computers) their brand names have gained ever-wider international customer recognition.

As recently as 1995 light industrial goods generated almost 25% of the total value of exports, with 7.4% coming from textiles and apparel; by 2003, under the impact of the forces of contraction, the corresponding figures were 14% and 5.8%. Competition from China has been a major factor behind the loss of foreign markets for Korean light industrial goods. In the same period average annual export growth for major heavy industries was as follows: iron and steel products 3.5%; machinery and precision equipment 7.9%; electrical goods and electronics 8.7%; automobiles 13.1%; and ships 9.1%. Concealed in these figures is, however, a distortion associated with the downturn in overseas demand for electronics and information technology goods in 2001. Thus, if growth is calculated between 1995 and 2000, electrical goods and electronics emerge as an export leader, alongside passenger cars, having grown by 12.1% annually. In particular, China has been a strong source of overseas demand for such goods (as it has of South Korean steel). In value terms, the main sources of export earnings in 2003 were: electronics goods(which contributed US $68,189.1m., equivalent to 44.15% of all exports); automobiles ($17,479.8m., or 11.3%); machinery and equipment ($16,007.8m., or 10.3%); chemicals and associated products ($14,781.6m., or 9.6%); and metallurgical goods ($13,089.8m., or 7.8%). In 2003 demand for Korean cars in the US and EU member states grew strongly. It is noteworthy too that in 2003 South Korean shipbuilders for the first time overtook Japan, their new orders taking 44% of the global market—more than those of Japan and the EU combined.

Export destinations are dominated by Asia, followed by North America and Europe. In terms of individual countries, in 2003, for the first time ever, the People's Republic of China overtook the USA as South Korea's most important export destination, these two being followed by Japan, Hong Kong, and Taiwan. In January-November 2003, exports of metal goods and ships to China increased by 27% and 11%, respectively. Meanwhile, within Europe, export trade is dominated by EU member states.

Imports

Annual import growth between 1965 and 1980 was 15.2%, falling to 10.8% per year during the 1980s. Between 1990 and 1996 the momentum of rapid growth was maintained, with imports expanding, on average, by almost 12% annually. Thereafter the collapse of domestic demand in the wake of the Asian financial crisis had a dramatic effect, and imports contracted by 3.8% in 1997, and—much more dramatically—by 35.5% in 1998. Recovery from the downturn was, however, swift. As demand for industrial raw materials, machinery and high-price petroleum recovered, so imports rose again—by 28.4% in 1999 and by a further 34% in 2000 (by which time the value of imports was almost 7% above the previous 1996 peak level). Another decline—by 12%—in 2001 was again followed by recovery, as a result of which the value of imports in 2003 reached a new peak of $178,826.7m.

The three most important categories of imports are electrical goods and electronic machinery (accounting for $42,528.5m., or 23.8% of all imports in 2003), fuel ($38,155.5m., or 21.3%), and machinery and equipment ($21,704.2m., or 12.1%). Other important import categories include chemicals and chemical compounds (worth $14,443.1m., or 8.1% of the total) and iron and steel products ($8,204.8m., or 4.6%).

Japan and the USA are the two most important sources of South Korean imports, and import dependence on Japan for machinery and electronic goods has become increasingly strong in recent years. In 2003 these two countries accounted for 34% of all imports. However, just as China has grown in significance as a destination for its exports, so South Korea has also become an increasingly important customer for Chinese products. Thus, between 1991 and 2003 the value of Chinese imports rose from US $3,400m. to $21,992.4m., registering an average increase of almost 17% per year. In 2003 China accounted for 12.3% of all South Korea's imports, compared with 11.4% in 2002. Projections suggest that by 2006 the value of all merchandise trade between China and South Korea could reach US$100,000m.

In response to the mounting trade deficit with South Korea in the late 1980s, the US Government began to exert strong pressure for greater US access to South Korean markets. Anxious to avoid the economic sanctions which the USA had already imposed on certain Japanese electronic products, South Korea introduced measures designed to liberalize its import trade, whilst also encouraging voluntary export restraints on selected products. In order to appease US protectionist sentiment, and to counter the effects of the appreciation of the Japanese yen, it began to import from the USA some 100 products which had previously come from Japan. The South Korean Government also sought to placate its US critics by allowing the exchange value of the won to appreciate, thereby enhancing the competitiveness of US imports. Between 1998 and 2003 the value of the won in relation to the US dollar declined from 1,401.4 to 1,191.6 (in July 2004 it was 1,171.3).

Except for 1986–89, from 1966 to 1997 South Korea suffered a deficit on its merchandise trade account. The effect of the Asian financial crisis was to shift the trade balance into surplus (US$39,031m.) in 1998—a position it subsequently retained (a surplus of $14,991m. being recorded in 2003). Throughout the 1980s and 1990s South Korea consistently ran a trade deficit with Japan (rising from US $3,000m. to $15,400m. between 1985 and 1996). By contrast, until 1990 it enjoyed a large surplus with the USA, although this was subsequently eliminated, facilitating the attainment of greater balance in bilateral trade, and even, during 1991–92, 1994 and 1996–97, the emergence of a deficit. Liberalization of South Korea's agricultural market and acceptance, in 1993, of rice imports were important contributory factor towards reducing the US deficit. Since 1998, however, the trade balance has moved back into surplus.

Like other countries in the region—not least China—South Korea has shown an active interest in pursuing bilateral trade agreements. In February 2003, in the face of considerable opposition (especially in farming and fisheries), South Korea signed its first free-trade agreement (FTA) with Chile. At the end of 2003 discussions with Japanese officials got under way, intended to lead to the establishment of a bilateral FTA. If successful, its establishment would signal the creation of one of the biggest trading zones in the world. There are, however, significant difficulties to be overcome, not least the persistence of the South Korean trade deficit with Japan.

Invisible Trade

With the single exception of 1975, during the period from the end of the Korean War to the end of the 1970s, South Korea consistently enjoyed a surplus in its invisible trade, at times sufficiently large to offset its merchandise trade deficit and move its current account into surplus. However, between 1980 and 1985 this position was reversed, as construction contracts with Middle Eastern countries declined and interest payments on outstanding foreign debt increased. Since 1986 the current account has moved between periods of surplus and deficit. In 1996 a record deficit of US $23,000m. was recorded, although in 1997 this figure contracted sharply and from 1998 it moved back into surplus. Since 1998 the current account has remained in surplus (averaging $18,120m.), although it declined from $40,371.2m. in 1998 to just $5,393.9m.in 2002, before recovering to $12,320.7m. in 2003.

Inflows of foreign investment capital into South Korea began in the 1970s, but only reached significant levels towards the end of the 1980s. In particular, the second half of the 1990s saw a high point in terms of investment capital inflows: during 1996–2000, the cumulative value of contractual inward investment was US $49,786m., exceeding the corresponding figure for the previous 30 years. However, having peaked at more than $15,000m. per annum in 1999 and 2000, contractual inflows have since declined sharply and totalled only $2,660m. in 2003. In any case, expressed as a share of GDP, the stock of FDI has risen from 6.3% to 9.5% (1998–2002). The surge in inward FDI in the late 1990s no doubt reflected the exigencies of the financial crisis and urgent need for financial support for ailing industries. More positively, however, it also reflected the Government's recognition of the need to accommodate the demands of economic globalization. South Korea is itself an investor overseas. Outflows first exceeded US $1,000m. in 1991, since when cumulative outflows have totalled over $42,000m. (the peak level—$5,043.7m.—having been reached in 2001). In 2003 58% of all outward investment went to Asia, compared with 28%

to North America and a mere 5.7% to Europe. Within Asia, China has become the single most important destination, and South Korea has become the fourth largest overseas investor in China. Preliminary statistics indicate that in the first quarter of 2004, South Korea accounted for 16% of all FDI inflows into China, exceeding the contributions of Japan, the USA and Taiwan.

FINANCE

Fiscal Policy

The South Korean Government has used tax reforms in order to influence resource allocation and guide the direction and pattern of national economic development. Early (pre-1960) measures were primarily designed to limit consumption and promote capital accumulation. By contrast, more comprehensive reforms introduced in the 1960s sought to put in place a more indigenously financed growth strategy, by replacing economic aid with increased domestic savings. Thus, between 1965 and 1975 the savings rate rose sharply (from 7.4% to 19.1% of GDP). By 1986 it had reached 30%, for the first time making it possible to finance investment wholly from domestic sources. Since that time, South Korea has been a net exporter of capital; for example, in 2003 the gross domestic savings ratio was 32.5%, compared with a gross fixed investment ratio of 19.4%. The difference between the two yielded a savings surplus of about US $5,500m.

South Korea's tax burden (national tax receipts as a share of GDP) is relatively low, although between 1995 and 2004 it rose slightly from 19.1% to 22.7%. From a longer-term perspective, the tax burden has risen steadily since 1973, when the ratio was just 12.2%. Fiscal policy was relaxed in mid-1998, as South Korea's economic recession deepened, in order to facilitate increased expenditure on social security benefits, job creation and financial restructuring. The central government budget, having been in surplus in 1993, during 1997–99 reverted to deficit (peaking at 18,800m. won in 1998, or 4.3% of GNP). In 2000, however, as a result of more stringent fiscal and monetary policies, it moved back into surplus. In 2003 the budget surplus was 57,069,000m. won.

Foreign Debt

Between 1985 and 1990 South Korea's total outstanding debt was reduced from US $47,100m. to $35,000m. Subsequently, however, financial liberalization encouraged increasingly large inflows of foreign capital. As a result, total debt rose sharply until, by 1997, it had reached US $139,100m. (an average annual growth of 18.8% since 1990). Meanwhile, many enterprises continued to finance their industrial expansion by borrowing large amounts of capital from abroad. The seriousness of the situation was highlighted in the finding that by 1996—on the eve of the financial crisis—some 57% of outstanding debt was in the form of short-term debt. This was a dangerously high figure: in 1996, for example, interest on short-term debt was $3,900m., compared with only $2,800m. on long-term; and by 1998 the cost of debt-servicing equated with 12.9% of revenue from exports of goods and services. The deterioration in its external debt situation was the background against which, during 1997–99, South Korea had recourse to IMF credit, cumulatively totalling over $34,000m.

Recovery from the effects of the 1997 crisis was, however, swift. Between 1998 and 2002 external debt (excluding government debt) was reduced from US $141,300m. (equivalent to 44.5% of GDP) to $122,200m. (25.7%). In 2003 the level of debt once more increased to $140,100m., although in relative terms it remained largely unchanged at 26.9% of GDP. Expressed differently, South Korea's debt-service ratio declined from over 25% in 1998 to 6.4% in 2003.

At the end of 2003 South Korea's foreign-exchange reserves totalled US $154,508.8m. This figure compared with $19,700m. in 1997, when reserves decreased by 39% compared with the previous year. By the second quarter of 2004 reserves had reached a record $167,029.7m.

Monetary Policy and Banking

Through the operations of the central bank and by means of direct ownership—or control through equity participation—of most financial institutions, until the 1980s the Government maintained very tight control over interest rate determination, the underwriting of private loans from abroad, and the allocation of financial resources to the private sector and other enterprises. In 1964 a government-sponsored foreign exchange rate initiative involving a major currency devaluation (of about 50%) and the introduction of a unified floating rate system greatly facilitated export expansion. In the wake of subsequent economic growth, the case in favour of financial liberalization became increasingly pressing, and in the 1980s significant monetary reforms took place. Amongst these were denationalization of the commercial banking sector, the first moves towards the freeing of interest rates from government control, and the abolition of credit 'ceilings' and quotas in order to control bank lending. By such means, banks were given unprecedented discretion in managing loanable funds in pursuit of profit maximization. More recent evidence of increasing reliance on indirect control is afforded by the BOK's involvement in limited open-market operations.

By 1991 interest rate deregulation had embraced most money market instruments, large certificates of deposits and repurchase agreements. By the end of 1993 lending rates at banks and non-banks had also been freed, while interest rates on policy loans and special credit facilities were scheduled for decontrol during 1994–97. As interest rate liberalization took place, the BOK began to shift from using direct monetary controls, involving the imposition of 'ceiling rates' on bank loans, to reliance on indirect policy instruments and open market operations. Meanwhile, starting in October 1993, the range of permissible daily interbank foreign exchange rate fluctuations was widened. December 1994 witnessed the introduction of the Foreign Exchange Reform Plan, designed to be implemented in three stages between 1995 and 1999, focusing, in turn, on economic globalization-related issues, the liberalization of cross-border transactions, and improvements in the foreign-exchange system.

A major reform, introduced in August 1993, was the implementation of a 'real-name' system of financial transactions, which replaced the previous system that had allowed financial accounts and property to be recorded under false, assumed or borrowed names. The desired effect was to curb corruption and enhance incentives, as well as to facilitate the emergence of a more equitable tax regime. Following the publication of the findings of a presidential commission on financial reform in mid-1997, the Government also undertook to seek to curb the influence of the powerful Ministry of Finance and the Economy through the creation of a Financial Supervisory Commission. The principal purpose of the new body, which would be directly responsible to the Prime Minister, was to strengthen supervision over the financial sector. Measures to increase the independence of the BOK were also proposed.

During 1986–89 a by-product of an appreciating won was a significant narrowing of the gap between official and black-market rates for foreign currencies. From 1990 until 1994, the value of the won registered a declining trend, although after mid-1994 its value began to rise again. This trend, combined with the depreciation of the Japanese yen in mid-1995, did not bode well for South Korean exports. Following the rapid depreciation of the won in late 1997, the Government adopted tight monetary policies in order to stabilize the currency. This having been achieved, from mid-1998 monetary policy was eased, with interest rates lowered substantially in a bid to stimulate domestic demand. More recently, concerns about flat private consumption growth and the low level of US interest rates have ruled out upward adjustments of domestic interest rates. The average deposit rate fell from 5% to 4.1% between the first quarter of 2002 and the third quarter of 2003; during the same period, the average lending rate fell from 6.8% to 6.1%. Given the uncertainty of domestic and international conditions, it seemed likely that the Government would seek to use a prudent monetary policy in 2004 in order to provide support to ongoing efforts to improve the macroeconomic environment, not least that of the financial sector itself.

The stability of the banking system was tested in 1997, upon the collapse of the Hanbo Group. Following revelations that, notwithstanding the lack of adequate collateral, several of the country's largest banks had lent money to Hanbo under government pressure, an investigation was launched in an attempt to discover how such loose credit creation could have taken place. Emergency funds were released into the banking system by the Ministry of Finance and the Economy in order to prevent further corporate bankruptcies.

In 1998 financial institutions and industrial corporations were radically restructured in accordance with the IMF programme. By mid-2000 the Government had closed 440 failing financial institutions, including five large commercial banks, as a result of which more than one-third of workers in the financial sector lost their jobs. In addition, the Government recapitalized 15 of the 17 remaining commercial banks to internationally required standards, and introduced regulatory reforms in an effort to improve the transparency of banking operations. Such measures notwithstanding, in 1999 the banking sector recorded losses of 5,000,000m. won, principally as a result of huge liabilities resulting from the collapse of the Daewoo Group. In September 1999 agreement was reached to sell Korea First Bank—one of five banks (the others were Hanvit Bank, Cho Hung Bank, Korea Exchange Bank and Seoulbank) that had been nationalized in 1998—to Newbridge Capital, a US investment fund.

Between late 1997 and early 2000 the Korean Government spent some 102,000,000m. won in an attempt to prevent the collapse of the banking sector. In May 2000, in the face of continuing severe difficulties, plans to inject a further 30,000,000m. won were announced. A consensus view was that significant progress has been made towards improving the state of South Korea's banking sector, even though the position of some individual institutions remains far from good. It is an essential condition of future sustained economic growth that the benefits of banking reforms in recent years should not be jeopardized. Measures are needed to address the high level of non-performing loans in the non-bank financial sector (especially investment trust companies). Most important of all, however, is the need to intensify the programme of bank privatization in order to enhance efficiency and transparency within the industry. The most serious recent problem to confront the financial market has been growing delinquency rates associated with the excessive issue of credit cards and made worse by slowing economic growth. Between the end of 2002 and November 2003 the average delinquency rate among eight credit card firms was reported to have more than doubled (from 6.6% to 13.5%). Indeed, in November 2003 lack of funds forced the largest credit card company—LG Card—to suspend cash advances, and bankruptcy was only averted by massive cash injections.

The opening, in 1967, of a branch of the Chase Manhattan Bank was a signal for other foreign banks to establish branches in South Korea. By the end of 1993 some 74 foreign banks had established offices throughout the country. Arising from the programme of recovery from the 1997 financial crisis has also been a willingness to open up domestic banks to foreign involvement, a process that has been strengthened by government recognition of the need to embrace globalization in a more active manner. The extent to which such involvement has taken place is revealed in BOK statistics, showing that in 2003 foreign ownership of South Korea's banks was 38.6%—much higher than in Japan (7%), the Philippines (15%) and Malaysia (19%). Against this background, the BOK urged the Government to restrain a further expansion of foreign ownership.

In the 1990s measures were also introduced to provide foreign access to the domestic securities market, and from January 1992 direct investments by foreigners in South Korea's stock market were permitted, albeit initially with an upper limit of 10% of total shares. In May 1996 it was announced that most restrictions on the entry of foreign firms into the stock market would be removed by the year 2000. Meanwhile, in July 1993 the limit for foreign equity investment had already been eliminated for companies with a 50% or greater foreign ownership, but in December 1997, the IMF insisted, as a condition of its rescue programme, that the limit for foreign equity investment should be abandoned entirely; this was in the hope that foreign investors would purchase stakes in the ailing *chaebol*.

The stock market, the main index of which, under the impact of the regional financial crisis, had declined dramatically at the end of 1997, subsequently rose sharply. By the end of 1999 the stock price index was 173% higher than it had been two years previously. In 2000, however, it lost 50% of its value—not least because of struggling investment trust companies' divestment of their portfolios in order to finance their debts. Substantial, but limited, recovery took place in 2001 (during which the index rose by more than 37% to outperform the stock market of every country in the world except Russia) and during January–September 2002. Thereafter, however, there was a substantial decline until, in the second half of 2003, renewed and accelerating recovery took place. Even so, in the last quarter of 2003 the Korea Composite Stock Price Index (KOSPI) was still almost 10% below the level prevailing at the beginning of 2002.

Statistical Survey

Source (unless otherwise stated): National Statistical Office, Bldg III, Government Complex-Daejeon 920, Dunsan-dong, Seo-gu, Daejeon 302-701; tel. (42) 481-4114; fax (42) 481-2460; internet www.nso.go.kr.

Area and Population

AREA, POPULATION AND DENSITY*

Area (sq km)	99,313†
Population (census results)‡	
1 November 1995	44,608,726
1 November 2000	
Males	23,158,582
Females	22,977,519
Total	46,136,101
Population (official estimates at mid-year)	
2001	47,342,828
2002	47,639,618
2003	47,925,318
Density (per sq km) at mid-2003	482.6

* Excluding the demilitarized zone between North and South Korea, with an area of 1,262 sq km (487 sq miles).

† 38,345 sq miles. The figure indicates territory under the jurisdiction of the Republic of Korea, surveyed on the basis of land register.

‡ Excluding adjustment for underenumeration, estimated at 1.4% in 1995.

PRINCIPAL TOWNS
(population at 1995 census)

Seoul (capital)	10,231,217	Jeonju (Chonju)	563,153
Busan (Pusan)	3,814,325	Jeongju (Chongju)	531,376
Daegu (Taegu)	2,449,420	Masan	441,242
Incheon (Inchon)	2,308,188	Jinju (Chinju)	329,886
Daejeon (Taejon)	1,272,121	Kunsan	266,559
Gwangju (Kwangju)	1,257,636	Jeju (Cheju)	258,511
Ulsan	967,429	Mokpo	247,452
Seongnam (Songnam)	869,094	Chuncheon (Chunchon)	234,528
Suwon	755,550		

2000 census: Seoul 9,853,972; Busan 3,655,437; Daegu 2,473,990; Incheon 2,466,338; Daejeon 1,365,961; Gwangju 1,350,948; Ulsan 1,012,110.

BIRTHS, MARRIAGES AND DEATHS*

	Registered live births		Registered marriages		Registered deaths	
	Number	Rate (per 1,000)	Number	Rate (per 1,000)	Number	Rate (per 1,000)
1995	721,074	16.0	398,484	8.7	248,089	5.4
1996	695,825	15.3	434,911	9.4	245,588	5.3
1997	678,402	14.8	388,591	8.4	247,938	5.3
1998	642,972	13.8	375,616	8.0	248,443	5.3
1999	616,322	13.2	362,673	7.7	246,539	5.2
2000	636,780	13.4	334,030	7.0	247,346	5.2
2001	557,228	11.6	320,063	6.7	242,730	5.1
2002	494,625	10.4	n.a.	n.a.	246,515	5.2

* Owing to late registration, figures are subject to continuous revision.

Expectation of life (WHO estimates, years at birth): 75.5 (males 71.8; females 79.4) in 2002 (Source: WHO, *World Health Report*).

ECONOMICALLY ACTIVE POPULATION*
(annual averages, '000 persons aged 15 years and over)

	1998	1999	2000
Agriculture, forestry and fishing	2,480	2,349	2,288
Mining and quarrying	21	20	18
Manufacturing	3,898	4,006	4,244
Electricity, gas and water	61	61	63
Construction	1,578	1,476	1,583
Trade, restaurants and hotels	5,571	5,724	5,943
Transport, storage and communications	1,169	1,202	1,260
Financing, insurance, real estate and business services	1,856	1,925	2,089
Community, social and personal services	3,339	3,499	3,551
Total employed (incl. others)	19,994	20,281	21,061
Unemployed	1,461	1,353	889
Total labour force	21,456	21,634	21,950
Males	12,893	12,889	12,950
Females	8,562	8,745	9,000

2001 ('000 persons aged 15 years and over): Total employed 21,572; Unemployed 845; Total labour force 22,417 (Males 13,142; Females 9,275).

2002 ('000 persons aged 15 years and over): Total employed 22,169; Unemployed 708; Total labour force 22,877 (Males 13,411; Females 9,466).

2003 ('000 persons aged 15 years and over): Total employed 22,139; Unemployed 777; Total labour force 22,916 (Males 13,518; Females 9,397).

* Excluding armed forces.

Health and Welfare

KEY INDICATORS

Total fertility rate (children per woman, 2002)	1.4
Under-5 mortality rate (per 1,000 live births, 2002)	5
HIV/AIDS (% of persons aged 15–49, 2003)	<0.10
Physicians (per 1,000 head, 2000)	1.3
Hospital beds (per 1,000 head, 2000)	6.1
Health expenditure (2001): US $ per head (PPP)	948
Health expenditure (2001): % of GDP	6.0
Health expenditure (2001): public (% of total)	44.4
Access to water (% of persons, 2000)	92
Access to sanitation (% of persons, 2000)	63
Human Development Index (2002): ranking	28
Human Development Index (2002): value	0.888

For sources and definitions, see explanatory note on p. vi.

Agriculture

PRINCIPAL CROPS
('000 metric tons)

	2000	2001	2002
Rice (paddy)	7,196.6	7,406.5	6,687.2
Barley	226.6	382.8	304.6
Maize	64.2	57.0	73.2
Potatoes	704.6	603.6	666.2
Sweet potatoes	344.9	273.1	316.7
Dry beans	21.0	21.9	19.8
Chestnuts	92.8	94.1	72.4
Soybeans (Soya beans)	113.2	117.7	115.0
Sesame seed	31.7	31.0	23.8
Other oilseeds	23.0	22.5	20.9
Cabbages	3,423.4	3,385.6	2,575.8
Lettuce	203.5	182.5	179.0
Spinach	120.8	126.7	112.7
Tomatoes	276.7	205.8	226.6
Pumpkins, squash and gourds	240.5	295.6	276.5
Cucumbers and gherkins	453.5	451.5	463.7
Chillies and green peppers	391.3	411.8	381.2
Green onions and shallots	657.9	635.7	566.8
Dry onions	877.5	1,073.7	933.1
Garlic	474.4	406.4	391.2
Carrots	155.1	153.4	136.1
Mushrooms	21.8	18.1	22.7
Other vegetables	3,689.0	3,760.0	3,800.0
Tangerines, mandarins, clementines and satsumas	563.5	644.7	642.5
Apples	489.0	403.6	433.2
Pears	324.2	417.2	386.3
Peaches and nectarines	170.0	166.3	187.5
Plums	51.7	57.9	75.6
Strawberries	180.5	203.0	210.0
Grapes	475.6	454.0	422.0
Watermelons	922.7	949.0	839.6
Cantaloupes and other melons	332.8	270.3	247.2
Persimmons	287.8	270.3	281.1
Other fruits	82.1	89.2	86.3
Tobacco (leaves)	68.2	55.6	47.5

Source: FAO.

LIVESTOCK
('000 head)

	2000	2001	2002
Cattle	2,134	1,954	1,954
Pigs	8,214	8,720	8,974
Goats	445	440	444
Rabbits	436	416	362
Chickens	102,547	102,393	101,693
Ducks	5,134	6,716	7,823

Source: FAO.

LIVESTOCK PRODUCTS
('000 metric tons)

	2000	2001	2002
Beef and veal	305.9	232.6	210.8
Pig meat	915.9	927.7	1,005.2
Chicken meat	373.6	377.1	381.4
Duck meat*	44.7	45.0	56.0
Other meat†	8.9	9.5	8.3
Cows' milk	2,252.8	2,338.9	2,537.0
Goats' milk†	4.6	4.4	4.4
Butter†	54.9	57.1	61.3
Hen eggs	478.8	529.2	536.6
Other poultry eggs	21.3	23.5†	25.0†
Honey	17.7	22.0	25.5
Cattle hides (fresh)†	50.9	38.3	34.8

* Unofficial figures.
† FAO estimate(s).

Source: FAO.

Forestry

ROUNDWOOD REMOVALS
(FAO estimates, '000 cubic metres, excl. bark)

	2000	2001	2002
Sawlogs, veneer logs and logs for sleepers	420	575	646
Pulpwood	552	376	373
Other industrial wood	620	582	586
Fuel wood	2,449	2,454	2,458
Total	4,041	3,987	4,063

Source: FAO.

SAWNWOOD PRODUCTION
('000 cubic metres, incl. sleepers)

	2000	2001	2002
Coniferous (softwood)	4,044	4,330	5,045
Broadleaved (hardwood)	500	90	149
Total	4,544	4,420	5,194

Source: FAO.

Fishing

('000 metric tons, live weight)

	2000	2001	2002
Capture	1,825.0	1,990.7	1,669.0
Alaska (walleye) pollock	86.1	197.4	24.8
Croakers and drums	67.6	50.8	48.1
Japanese anchovy	201.2	273.9	236.3
Skipjack tuna	137.0	137.6	173.7
Chub mackerel	145.9	203.7	142.1
Largehead hairtail	81.1	79.9	60.2
Argentine shortfin squid	150.1	142.6	98.6
Japanese flying squid	226.3	225.6	226.7
Aquaculture	293.4	294.5	296.8*
Pacific cupped oyster	177.1	174.1	170.3
Total catch	2,118.4	2,285.2	1,965.8

* FAO estimate.

Note: Figures exclude aquatic plants ('000 metric tons): 387.7 (capture 13.0, aquaculture 374.6) in 2000; 388.5 (capture 14.9, aquaculture 373.5) in 2001; 512.4 (capture 14.8, aquaculture 497.6) in 2002 (FAO estimates). Also excluded are aquatic mammals, recorded by number rather than by weight. The number of whales caught was: 174 in 2000; 377 in 2001; 281 in 2002.

Source: FAO.

Mining

('000 metric tons, unless otherwise indicated)

	2000	2001	2002
Hard coal (Anthracite)	4,174	3,817	3,318
Iron ore: gross weight	336	195	157
Iron ore: metal content	188	109	88
Lead ore (metric tons)*	2,724	988	28
Zinc ore (metric tons)*	11,474	5,129	99
Kaolin	2,098.5	2,384.0	2,831
Feldspar	330.4	389.4	415.6
Salt (unrefined)†	800	800	800
Mica (metric tons)	65,249	109,339	29,870
Talc (metric tons)	11,344	47,712	37,863
Pyrophyllite	918.0	1,101.8	890.0

* Figures refer to the metal content of ores.
† Estimated production.

Source: US Geological Survey.

Industry

SELECTED PRODUCTS
('000 metric tons, unless otherwise indicated)

	1999	2000	2001
Wheat flour	1,834	1,871	1,843
Refined sugar	1,182	1,257	1,264
Beer (million litres)	1,487	1,654	1,777
Cigarettes (million)	95,995	94,531	94,116
Cotton yarn—pure and mixed	282.6	294.1	303.5
Plywood ('000 cu m)	774	817	801
Newsprint	1,718	1,770	1,585
Rubber tyres ('000)*	67,120	71,348	68,728
Caustic soda (metric tons)	1,163	1,203	1,309
Liquefied petroleum gas	2,595	2,997	n.a.
Naphtha	18,815	19,109	n.a.
Kerosene	11,648	11,299	n.a.
Distillate fuel oil	30,245	31,535	n.a.
Bunker C oil (million litres)	36,509	n.a.	n.a.
Residual fuel oil	36,341	34,813	n.a.
Cement	48,579	51,417	53,062
Pig-iron	23,328	24,943	26,182
Crude steel	41,502	43,423	44,199
Television receivers ('000)	15,556	10,054	9,321
Passenger cars—produced ('000 units)	2,158	2,626	2,477
Lorries and trucks—produced (number)	264,212	265,448	254,233
Electric energy (million kWh)	266,818	295,156	n.a.

* Tyres for passenger cars and commercial vehicles.

Shipbuilding (merchant ships launched, '000 grt): 8,977 in 1999; 11,211 in 2000; 8,385 in 2001 (Source: UN, *Industrial Commodity Statistics Yearbook*).

Finance

CURRENCY AND EXCHANGE RATES

Monetary Units

100 chun (jeon) = 10 hwan = 1 won.

Sterling, Dollar and Euro Equivalents (31 May 2004)

£1 sterling = 2,128.44 won;
US $1 = 1,160.10 won;
€1 = 1,420.66 won;
10,000 won = £4.70 = $8.62 = €7.04.

Average Exchange Rate (won per US $)

2001 1,290.99
2002 1,251.09
2003 1,191.61

BUDGET
('000 million won)*

Revenue	1996	1997	1998
Taxation	72,385	78,434	78,310
Taxes on income, profits and capital gains	24,137	24,292	27,975
Income tax	14,767	14,868	17,194
Corporation tax	9,356	9,425	10,776
Social security contributions	7,425	8,506	10,512
Employees	2,804	3,433	n.a.
Employers	4,261	4,864	n.a.
Taxes on property	1,473	1,590	1,379
Domestic taxes on goods and services	27,478	30,650	27,159
Value-added tax	16,790	19,488	15,707
Excises	10,027	10,373	10,530
Import duties	5,309	5,798	3,836
Other current revenue	11,791	13,639	17,480
Entrepreneurial and property income	4,600	5,634	9,854
Administrative fees and charges, non-industrial and incidental sales	1,518	1,394	1,530
Fines and forfeits	3,502	5,032	4,646
Capital revenue	1,352	1,295	883
Total	85,528	93,368	96,673

Expenditure	1996	1997	1998
General public services	7,847	9,039	10,841
Defence	12,553	13,159	13,621
Education	14,435	16,249	17,779
Health	682	777	957
Social security and welfare	7,884	9,632	12,252
Housing and community amenities	7,077	6,677	7,336
Recreational, cultural and religious affairs and services	534	679	788
Economic affairs and services	21,965	24,334	28,453
Interest expenditures	2,241	2,258	3,399
Other	9,211	18,084	20,263
Total	84,429	100,888	115,689
Current	n.a.	62,812	70,631
Capital	n.a.	18,791	20,359
Net lending	n.a.	19,285	23,375
Other expenditures	n.a.	0	1,324

Source: IMF, *Republic of Korea: Statistical Appendix* (February 2000).

1999 ('000 million won)* Total revenue 107,923 (current 106,523; capital 1,386); Total expenditure (excl. net lending and other expenditures) 101,236 (current 76,798; capital 24,438) (Source: Bank of Korea, Ministry of Planning and Budget).

2000 ('000 million won)* Total revenue 133,584 (current 132,366; capital 1,218); Total expenditure (excl. net lending and other expenditures) 108,259 (current 82,667; capital 25,592) (Source: Bank of Korea, Ministry of Planning and Budget).

2001 ('000 million won)* Total revenue 139,890 (current 138,203; capital 1,697); Total expenditure (excl. net lending and other expenditures) 122,275 (current 99,755; capital 22,520) (Source: Bank of Korea, Ministry of Planning and Budget).

2002 ('000 million won)* Total revenue 151,873 (current 150,239; capital 1,633); Total expenditure (excl. net lending and other expenditures) 131,993 (current 109,123; capital 22,870) (Source: Bank of Korea, Ministry of Planning and Budget).

2003 ('000 million won): Total revenue 169,563 (current 168,159; capital 1,404); Total expenditure (excl. net lending and other expenditures) 144,785 (current 120,741; capital 24,044).

2004 ('000 million won): Total revenue 183,077 (current 181,724; capital 1,353); Total expenditure (excl. net lending and other expenditures) 168,272 (current 143,625; capital 24,647).

* Figures refer to the consolidated operations of the central Government, including extrabudgetary accounts, but excluding enterprise special accounts. Figures refer to the fiscal year.

INTERNATIONAL RESERVES

(US $ million at 31 December)

	2001	2002	2003
Gold*	68.3	69.2	70.9
IMF special drawing rights	3.3	11.8	21.1
Reserve position in IMF	262.4	522.0	754.4
Foreign exchange	102,487.5	120,811.4	154,508.8
Total	102,821.6	121,414.4	155,355.1

* National valuation.

Source: IMF, *International Financial Statistics*.

MONEY SUPPLY

('000 million won at 31 December)

	2001	2002	2003
Currency outside banks	18,702	19,863	20,111
Demand deposits at deposit money banks	34,918	43,265	45,226
Total money (incl. others)	53,506	63,151	65,481

Source: IMF, *International Financial Statistics*.

COST OF LIVING

(Consumer Price Index; base: 2000 = 100)

	2001	2002	2003
Food	103.5	107.7	112.4
Housing	103.9	109.2	113.2
Fuel, light and water	111.1	107.1	113.1
Furniture and utensils	102.4	104.0	106.7
Clothing and footwear	103.1	106.3	110.1
Medical treatment	112.3	111.4	114.1
Education	104.4	110.3	116.8
Culture and recreation	99.7	100.0	100.0
Transport and communications	102.0	101.4	102.7
All items (incl. others)	104.1	106.9	110.7

NATIONAL ACCOUNTS

('000 million won at current prices)

National Income and Product

	2000	2001	2002
Compensation of employees	248,167.3	270,469.8	294,480.9
Operating surplus	176,653.3	185,661.9	210,635.9
Domestic factor incomes	424,820.6	456,131.7	505,116.8
Consumption of fixed capital	83,416.1	88,112.6	91,113.4
Gross domestic product (GDP) at factor cost	508,236.7	544,244.3	596,230.2
Indirect taxes, *less* subsidies	70,427.9	77,878.3	88,033.3
GDP in purchasers' values	578,664.6	622,122.6	684,263.5
Net factor income from abroad	–2,504.6	–1,094.8	805.6
Gross national product	576,160.0	621,027.8	685,069.1
Less Consumption of fixed capital	83,416.1	88,112.6	91,113.4
National income in market prices	492,743.9	532,915.2	593,955.7
Other current transfers from abroad (net)	644.1	–496.2	–1,976.5
National disposable income	493,388.0	532,419.0	591,979.1

Expenditure on the Gross Domestic Product

	2001	2002	2003
Government final consumption expenditure	80,298	88,512	96,180
Private final consumption expenditure	343,417	381,063	388,417
Increase in stocks	–1,315	–41	–1,868
Gross fixed capital formation	183,792	199,047	213,844
Statistical discrepancy	1,657	6,238	6,575
Total domestic expenditure	607,849	674,819	703,148
Exports of goods and services	235,187	241,209	275,316
Less Imports of goods and services	220,914	231,765	257,118
GDP in purchasers' values	622,123	684,263	721,346
GDP at constant 2000 prices	600,866	642,748	662,474

Gross Domestic Product by Economic Activity

	2001	2002	2003
Agriculture, forestry and fishing	24,806.2	24,654.9	22,833.3
Mining and quarrying	2,020.7	2,051.4	2,136.9
Manufacturing	151,766.0	161,952.0	169,113.8
Electricity, gas and water	14,648.6	15,929.4	17,338.1
Construction	47,181.9	51,541.7	61,021.3
Trade, restaurants and hotels	59,212.3	62,656.7	62,071.4
Transport, storage and communications	41,190.5	45,133.8	47,467.7
Finance, insurance, real estate and business services	112,472.6	131,666.8	139,240.9
Public administration and defence, compulsory social security	32,207.4	35,557.2	38,704.8
Education	28,803.6	32,296.7	35,713.3
Health and social work	16,771.1	17,432.4	18,864.1
Other service activities	18,927.2	21,219.0	22,026.2
Sub-total	550,008.1	602,091.9	636,531.8
Taxes, less subsidies, on products	72,114.5	82,171.6	84,814.2
GDP in purchasers' values	622,122.6	684,263.5	721,346.0

BALANCE OF PAYMENTS
(US $ million)

	2000	2001	2002
Exports of goods f.o.b.	175,948	151,262	162,554
Imports of goods f.o.b.	–159,076	–137,770	–148,374
Trade balance	16,872	13,492	14,180
Exports of services	30,534	29,055	28,143
Imports of services	–33,423	–32,883	–35,603
Balance on goods and services	13,982	9,664	6,719
Other income received	6,375	6,650	6,807
Other income paid	–8,797	–7,848	–6,356
Balance on goods, services and income	11,561	8,466	7,171
Current transfers received	6,500	6,687	7,293
Current transfers paid	–5,820	–6,914	–8,372
Current balance	12,241	8,239	6,092
Capital account (net)	–615	–731	–1,091
Direct investment abroad	–4,999	–2,420	–2,674
Direct investment from abroad	9,283	3,528	1,972
Portfolio investment assets	–520	–5,521	–5,036
Portfolio investment liabilities	12,697	12,227	4,940
Financial derivatives assets	532	463	1,308
Financial derivatives liabilities	–711	–586	–1,029
Other investment assets	–2,289	7,099	–2,404
Other investment liabilities	–1,268	–11,650	5,538
Net errors and omissions	–561	2,629	4,155
Overall balance	23,790	13,278	11,770

Source: IMF, *International Financial Statistics*.

External Trade

PRINCIPAL COMMODITIES
(distribution by SITC, US $ million)*

Imports c.i.f.	2001	2002	2003
Food and live animals	6,789.3	7,620.7	8,330.9
Crude materials (inedible) except fuels	9,052.3	9,178.9	10,146.5
Mineral fuels, lubricants, etc.	33,790.2	32,128.8	38,155.5
Petroleum, petroleum products, etc.	26,485.6	23,183.1	28,192.4
Crude petroleum oils, etc.	21,367.8	19,200.3	23,081.6
Refined petroleum products	4,650.3	n.a.	n.a.
Gas (natural and manufactured)	5,236.6	5,394.3	6,468.7
Chemicals and related products	12,941.6	14,133.3	16,459.0
Organic chemicals	4,329.6	4,604.8	5,408.0
Basic manufactures	16,683.8	19,192.3	22,312.1
Iron and steel	4,420.0	5,533.3	7,355.0
Machinery and transport equipment	47,911.0	53,314.2	62,655.2
Machinery specialized for particular industries	3,713.1	4,110.5	5,080.6
General industrial machinery, equipment and parts	4,594.2	5,232.9	6,100.4
Office machines and automatic data-processing machines	5,640.8	5,486.7	5,433.6
Telecommunications and sound equipment	4,821.5	4,910.2	5,486.1
Other electrical machinery, apparatus, etc.	22,616.5	25,581.6	31,296.8
Thermionic valves and tubes, microprocessors, transistors, etc.	15,865.2	n.a.	n.a.
Electronic integrated circuits and microassemblies	13,356.8	n.a.	n.a.
Miscellaneous manufactured articles	11,166.6	13,358.9	16,234.0
Total (incl. others)	141,097.8	152,126.2	178,826.7

Exports f.o.b.	2001	2002	2003
Mineral fuels, lubricants, etc.	9,999.5	8,498.1	9,048.4
Petroleum, petroleum products, etc.	7,972.8	6,500.1	6,788.5
Refined petroleum products	7,736.3	n.a.	n.a.
Chemicals and related products	12,523.8	13,756.9	16,928.6
Plastics in primary forms	4,633.2	5,080.6	6,447.0
Basic manufactures	26,789.5	26,986.3	30,125.9
Textile yarn, fabrics, etc.	10,940.8	10,940.5	10,776.2
Iron and steel	5,825.8	5,704.1	7,782.6
Machinery and transport equipment	86,694.8	99,597.8	121,142.2
Office machines and automatic data-processing machines	13,498.7	16,445.1	18,069.3
Automatic data-processing machines and units, etc.	7,484.9	n.a.	n.a.
Parts and accessories for automatic data-processing equipment	5,640.2	n.a.	n.a.
Telecommunications and sound equipment	15,943.6	20,150.1	26,634.3
Transmission apparatus for radio or television	7,483.6	n.a.	n.a.
Other electrical machinery, apparatus, etc.	21,694.0	23,694.4	28,604.0
Thermionic valves and tubes, microprocessors, transistors, etc.	14,741.9	n.a.	n.a.
Electronic integrated circuits and microassemblies	11,142.3	n.a.	n.a.
Road vehicles and parts†	15,363.1	17,198.0	22,900.5
Passenger motor cars (excl. buses)	11,450.8	13,322.3	17,479.8
Other transport equipment and parts†	10,229.9	11,041.9	11,645.3
Ships, boats and floating structures	9,699.2	10,673.2	11,103.9
Miscellaneous manufactured articles	11,247.0	10,466.0	12,063.4
Total (incl. others)	150,439.1	162,470.5	193,817.4

* Figures exclude trade with the Democratic People's Republic of Korea (US $ million): *Total imports:* 176.2 in 2001; 271.6 in 2002; 289.3 in 2003. *Total exports:* 226.8 in 2001; 370.2 in 2002; 435.0 in 2003.

† Data on parts exclude tyres, engines and electrical parts.

Source: mainly Korea Trade Information Services.

PRINCIPAL TRADING PARTNERS
(US $ million)*

Imports c.i.f.	2001	2002	2003
Australia	5,534.1	5,973.4	5,915.7
Canada	1,821.3	1,845.5	1,860.2
China, People's Republic	13,302.7	17,399.8	21,909.1
France	2,092.3	2,116.2	2,220.3
Germany	4,473.4	5,472.4	6,821.7
Hong Kong	1,227.6	1,695.0	2,735.4
Indonesia	4,473.5	4,723.4	5,212.3
Iran	2,099.3	1,335.4	1,844.7
Italy	1,787.5	2,274.2	2,382.2
Japan	26,633.4	29,856.2	36,313.1
Kuwait	2,250.7	2,230.4	3,191.1
Malaysia	4,125.0	4,041.4	4,249.1
Oman	2,310.9	1,895.9	2,322.8
Philippines	1,819.0	1,867.4	1,964.0
Qatar	2,572.1	2,173.0	3,139.8
Russia	1,929.5	2,217.6	2,521.8
Saudi Arabia	8,058.0	7,550.8	9,267.8
Singapore	3,011.5	3,430.1	4,089.8
Taiwan	4,301.4	4,832.0	5,879.6
Thailand	1,589.2	1,702.5	1,897.7
United Arab Emirates	4,633.0	4,210.2	5,756.5
United Kingdom	2,353.5	2,437.4	2,703.3
USA	22,376.2	23,008.6	24,814.1
Total (incl. others)	141,097.8	152,126.2	178,826.7

Exports f.o.b.	2001	2002	2003
Australia	2,173.2	2,339.6	3,272.1
Brazil	1,611.2	1,247.2	1,137.4
Canada	2,035.7	2,340.6	2,682.1
China, People's Republic	18,190.2	23,753.6	35,109.7
France	1,541.2	1,629.0	1,755.4
Germany	4,321.8	4,287.2	5,603.3
Greece	1,222.4	1,653.6	1,765.0
Hong Kong	9,451.7	10,145.5	14,653.7
India	1,407.7	1,384.1	2,853.0
Indonesia	3,279.8	3,144.8	3,377.6
Italy	2,063.3	2,217.3	2,560.6
Japan	16,505.8	15,143.2	17,276.1
Malaysia	2,628.0	3,218.3	3,851.8
Mexico	2,148.9	2,230.8	2,455.0
Netherlands	2,532.1	2,567.2	2,535.0
Panama	1,719.0	1,184.4	1,252.3
Philippines	2,535.4	2,950.0	2,975.0
Singapore	4,079.6	4,221.6	4,636.0
Spain	1,518.2	1,552.5	2,015.8
Taiwan	5,835.3	6,631.6	7,044.6
Thailand	1,848.2	2,335.4	2,523.8
United Arab Emirates	2,169.1	2,268.8	2,207.6
United Kingdom	3,490.0	4,255.5	4,094.3
USA	31,210.8	32,780.2	34,219.4
Viet Nam	1,731.7	2,240.2	2,561.2
Total (incl. others)	150,439.1	162,470.5	193,817.4

* Excluding trade with the Democratic People's Republic of Korea.

Source: Korea Trade Information Services.

Transport

RAILWAYS
(traffic)

	2000	2001	2002
Passengers carried ('000)	814,472	912,149	983,266
Passenger-km (million)	27,787	29,172	28,743
Freight ('000 metric tons)	45,240	45,122	45,733
Freight ton-km (million)	10,803	10,492	10,784

ROAD TRAFFIC
(motor vehicles in use at 31 December)

	2001	2002	2003
Passenger cars	8,889,327	9,737,428	10,278,923
Goods vehicles	2,728,405	2,894,412	3,016,407
Buses and coaches	1,257,008	1,275,319	1,246,629
Motorcycles and mopeds	1,700,600	1,708,457	1,730,193

SHIPPING

Merchant Fleet
(registered at 31 December)

	2001	2002	2003
Number of vessels	2,426	2,532	2,604
Total displacement ('000 grt)	6,395.0	7,049.7	6,757.4

Source: Lloyd's Register-Fairplay, *World Fleet Statistics*.

Sea-borne Freight Traffic
('000 metric tons)*

	2000	2001	2002
Goods loaded	282,768	315,297	290,951
Goods unloaded	550,811	571,076	575,209

* Including coastwise traffic loaded and unloaded.

CIVIL AVIATION

	2000	2001	2002
Passengers ('000)	41,967	42,162	43,965
Passenger-km (million)	83,955	84,544	92,175
Freight ('000 metric tons)	2,383	2,295	2,510
Freight ton-km (million)	12,430	11,327	12,606

Tourism

FOREIGN VISITOR ARRIVALS*†

Country of nationality	2000	2001	2002
China, People's Republic	442,794	482,227	539,466
Hong Kong	200,874	204,959	179,299
Japan	2,472,054	2,377,321	2,320,837
Philippines	248,737	210,975	215,848
Russia	155,392	134,727	165,341
Taiwan	127,120	129,410	136,921
USA	458,617	426,817	459,362
Total (incl. others)	5,321,792	5,147,204	5,347,468

* Including same-day visitors (excursionists) and crew members from ships.

† Including Korean nationals resident abroad.

Source: Korean National Tourism Organization.

Receipts from tourism (US $ million): 6,811.3 in 2000; 6,373.2 in 2001; 5,276.9 in 2002 (Source: Bank of Korea).

Communications Media

	2001	2002	2003
Telephones ('000 main lines in use)	22,724.7	23,257.0	22,877.0
Mobile cellular telephones ('000 subscribers)	29,045.6	32,342.0	33,591.8
Personal computers ('000 in use)	22,495	23,500	26,700
Internet users ('000)	24,380	26,270	29,220
Book production:			
titles	25,146	27,113	n.a.
copies ('000)	74,914	81,513	n.a.

1996: Facsimile machines (estimate, '000 in use): 400; Daily newspapers: Number 60, Circulation ('000 copies) 17,700 (estimate).

1997: Radio receivers ('000 in use) 47,500.

2000: Television receivers ('000 in use) 17,229.

Sources: mainly UNESCO, *Statistical Yearbook*; UN, *Statistical Yearbook*; International Telecommunication Union.

Education

(2003)

	Institutions	Teachers	Pupils
Kindergarten	8,292	30,290	546,531
Primary schools	5,463	154,075	4,175,626
Middle schools	2,850	99,717	1,854,641
High schools	2,031	115,829	1,766,529
Junior vocational colleges	158	11,974	925,963
Teachers' colleges	11	740	23,552
Universities and colleges	169	45,272	1,808,539
Graduate schools	1,010	n.a.	272,331

Adult literacy rate (UNESCO estimates): 97.9% (males 99.2%; females 96.6%) in 2001 (Source: UN Development Programme, *Human Development Report*).

Directory

Note: from 2001 the romanization of place-names in South Korea was in the process of change, to be completed by 2005. Transliteration of names of people and corporations was to remain unchanged for the time being.

The Constitution

The Constitution of the Sixth Republic (Ninth Amendment) was approved by national referendum on 29 October 1987. It came into effect on 25 February 1988. The main provisions are summarized below:

THE EXECUTIVE

The President

The President shall be elected by universal, equal, direct and secret ballot of the people for one term of five years. Re-election of the President is prohibited. In times of national emergency and under certain conditions the President may issue emergency orders and take emergency action with regard to budgetary and economic matters. The President shall notify the National Assembly of these measures and obtain its concurrence, or they shall lose effect. He may, in times of war, armed conflict or similar national emergency, declare martial law in accordance with the provisions of law. He shall lift the emergency measures and martial law when the National Assembly so requests with the concurrence of a majority of the members. The President may not dissolve the National Assembly. He is authorized to take directly to the people important issues through national referendums. The President shall appoint the Prime Minister (with the consent of the National Assembly) and other public officials.

The State Council

The State Council shall be composed of the President, the Prime Minister and no more than 30 and no fewer than 15 others appointed by the President (on the recommendation of the Prime Minister), and shall deliberate on policies that fall within the power of the executive. No member of the armed forces shall be a member of the Council, unless retired from active duty.

The Board of Audit and Inspection

The Board of Audit and Inspection shall be established under the President to inspect the closing of accounts of revenue and expenditures, the accounts of the State and other organizations as prescribed by law, and to inspect the administrative functions of the executive agencies and public officials. It shall be composed of no fewer than five and no more than 11 members, including the Chairman. The Chairman shall be appointed by the President with the consent of the National Assembly, and the members by the President on the recommendation of the Chairman. Appointments shall be for four years and members may be reappointed only once.

THE NATIONAL ASSEMBLY

Legislative power shall be vested in the National Assembly. The Assembly shall be composed of not fewer than 200 members, a number determined by law, elected for four years by universal, equal, direct and secret ballot. The constituencies of members of the Assembly, proportional representation and other matters pertaining to the Assembly elections shall be determined by law. A regular session shall be held once a year and extraordinary sessions shall be convened upon requests of the President or one-quarter of the Assembly's members. The period of regular sessions shall not exceed 100 days and of extraordinary sessions 30 days. The Assembly has the power to recommend to the President the removal of the Prime Minister or any other Minister. The Assembly shall have the authority to pass a motion for the impeachment of the President or any other public official, and may inspect or investigate state affairs, under procedures to be established by law.

THE CONSTITUTIONAL COURT

The Constitutional Court shall be composed of nine members appointed by the President, three of whom shall be appointed from persons selected by the National Assembly and three from persons nominated by the Chief Justice. The term of office shall be six years. It shall pass judgment upon the constitutionality of laws upon the request of the courts, matters of impeachment and the dissolution of political parties. In these judgments the concurrence of six members or more shall be required.

THE JUDICIARY

The courts shall be composed of the Supreme Court, which is the highest court of the State, and other courts at specified levels (for further details, see section on Judicial System). The Chief Justice and justices of the Supreme Court are appointed by the President, subject to the consent of the National Assembly. When the constitutionality of a law is a prerequisite to a trial, the Court shall request a decision of the Constitutional Court. The Supreme Court shall have the power to pass judgment upon the constitutionality or legality of administrative decrees, and shall have final appellate jurisdiction over military tribunals. No judge shall be removed from office except following impeachment or a sentence of imprisonment.

ELECTION MANAGEMENT

Election Commissions shall be established for the purpose of fair management of elections and national referendums. The National Election Commission shall be composed of three members appointed by the President, three appointed by the National Assembly and three appointed by the Chief Justice of the Supreme Court. Their term of office is six years, and they may not be expelled from office except following impeachment or a sentence of imprisonment.

POLITICAL PARTIES

The establishment of political parties shall be free and the plural party system guaranteed. However, a political party whose aims or activities are contrary to the basic democratic order may be dissolved by the Constitutional Court.

AMENDMENTS

A motion to amend the Constitution shall be proposed by the President or by a majority of the total number of members of the National Assembly. Amendments extending the President's term of office or permitting the re-election of the President shall not be effective for the President in office at the time of the proposal. Proposed amendments to the Constitution shall be put before the public by the President for 20 days or more. Within 60 days of the public announcement, the National Assembly shall decide upon the proposed amendments, which require a two-thirds' majority of the National Assembly. They shall then be submitted to a national referendum not later than 30 days after passage by the National Assembly and shall be determined by more than one-half of votes cast by more than one-half of voters eligible to vote in elections for members of the National Assembly. If these conditions are fulfilled, the proposed amendments shall be finalized and the President shall promulgate them without delay.

FUNDAMENTAL RIGHTS

Under the Constitution all citizens are equal before the law. The right of habeas corpus is guaranteed. Freedom of speech, press, assembly and association are guaranteed, as are freedom of choice of residence and occupation. No state religion is to be recognized and freedom of conscience and religion is guaranteed. Citizens are protected against retrospective legislation, and may not be punished without due process of law.

Rights and freedoms may be restricted by law when this is deemed necessary for the maintenance of national security, order or public welfare. When such restrictions are imposed, no essential aspect of the right or freedom in question may be violated.

GENERAL PROVISIONS

Peaceful unification of the Korean peninsula, on the principles of liberal democracy, is the prime national aspiration. The Constitution mandates the State to establish and implement a policy of unification. The Constitution expressly stipulates that the armed forces must maintain political neutrality at all times.

The Government

HEAD OF STATE

President: ROH MOO-HYUN (took office 25 February 2003).

STATE COUNCIL
(September 2004)

Prime Minister: LEE HAE-CHAN.

Deputy Prime Minister for Finance and the Economy: LEE HUN-JAI.

Deputy Prime Minister for Education and Human Resources Development: AHN BYUNG-YOUNG.

Minister of Unification: CHUNG DONG-YOUNG.

Minister of Foreign Affairs and Trade: BAN KI-MOON.

Minister of Justice: KIM SEONG-KYU.

Minister of National Defence: YOON KWANG-WOONG.

Minister of Government Administration and Home Affairs: HUH SUNG-KWAN.

Minister of Science and Technology: OH MYUNG.

Minister of Culture and Tourism: CHUNG DONG-CHAE.

Minister of Agriculture and Forestry: HUH SANG-MAN.

Minister of Commerce, Industry and Energy: LEE HEE-BEOM.

Minister of Information and Communication: CHIN DAE-JAE.

Minister of Health and Welfare: KIM GEUN-TAE.

Minister of the Environment: KWAK KYUL-HO.

Minister of Labour: KIM DAE-HWAN.

Minister of Gender Equality: JI EUN-HEE.

Minister of Construction and Transportation: KANG DONG-SUK.

Minister of Maritime Affairs and Fisheries: CHANG SEUNG-WOO.

Minister of Planning and Budget: KIM BYUNG-IL.

Chairman of the Civil Service Commission: DR CHO CHANG-HYUN.

Chairman of the Financial Supervisory Commission: YOON JEUNG-HYUN.

Minister of Government Policy Co-ordination: HAN DUCK-SOO.

Minister of the Government Information Agency: JOUNG SOON-KYUN.

Chairman of the Korea Independent Commission Against Corruption: CHUNG SOUNG-JIN.

MINISTRIES

Office of the President: Chong Wa Dae (The Blue House), 1, Sejong-no, Jongno-gu, Seoul; tel. (2) 770-0055; fax (2) 770-0344; e-mail president@cwd.go.kr; internet www.bluehouse.go.kr.

Office of the Prime Minister: 77, Sejong-no, Jongno-gu, Seoul; tel. (2) 737-0094; fax (2) 739-5830; internet www.opm.go.kr.

Ministry of Agriculture and Forestry: 1, Jungang-dong, Gwacheon City, Gyeonggi Prov.; tel. (2) 503-7200; fax (2) 503-7238; internet www.maf.go.kr.

Ministry of Commerce, Industry and Energy: 1, Jungang-dong, Gwacheon City, Gyeonggi Prov.; tel. (2) 503-7171; fax (2) 503-3142; internet www.mocie.go.kr.

Ministry of Construction and Transportation: 1, Jungang-dong, Gwacheon City, Gyeonggi Prov. 427–712; tel. (2) 503-9405; fax (2) 503-9408; e-mail webmaster@moct.go.kr; internet www.moct.go.kr.

Ministry of Culture and Tourism: 82-1, Sejong-no, Jongno-gu, Seoul 110-050; tel. (2) 7704-9114; fax (2) 3704-9119; internet www.mct.go.kr.

Ministry of Education and Human Resources Development: 77, 1-ga, Sejong-no, Jongno-gu, Seoul 110-760; tel. (2) 720-3404; fax (2) 720-1501; internet www.moe.go.kr.

Ministry of Environment: 1, Jungang-dong, Gwacheon City, Gyeonggi Prov.; tel. (2) 2110-6576; fax (2) 504-9277; internet www.moenv.go.kr.

Ministry of Finance and the Economy: 1, Jungang-dong, Gwacheon City, Gyeonggi Prov.; tel. (2) 503-9032; fax (2) 503-9033; internet www.mofe.go.kr.

Ministry of Foreign Affairs and Trade: 77, 1-ga, Sejong-no, Jongno-gu, Seoul; tel. (2) 3703-2555; fax (2) 720-2686; internet www.mofat.go.kr.

Ministry of Gender Equality: 520-3, Banpo-dong, Seocho-gu, Seoul 137-756; tel. (2) 2106-5000; fax (2) 2106-5145; internet www.moge.go.kr.

Ministry of Government Administration and Home Affairs: 77-6, Sejong-no, Jongno-gu, Seoul; tel. (2) 3703-4110; fax (2) 3703-5501; internet www.mogaha.go.kr.

Ministry of Health and Welfare: 1, Jungang-dong, Gwacheon City, Gyeonggi Prov. 427-760; tel. (2) 503-7505; fax (2) 503-7568; internet www.mohw.go.kr.

Ministry of Information and Communication: 100, Sejong-no, Jongno-gu, Seoul 110-777; tel. (2) 750-2000; fax (2) 750-2915; internet www.mic.go.kr.

Ministry of Justice: 1, Jungang-dong, Gwacheon City, Gyeonggi Prov.; tel. (2) 503-7012; fax (2) 504-3337; internet www.moj.go.kr.

Ministry of Labour: 1, Jungang-dong, Gwacheon City, Gyeonggi Prov.; tel. (2) 503-9713; fax (2) 503-8862; internet www.molab.go.kr.

Ministry of Maritime Affairs and Fisheries: 139, Chungjeong-no 3, Seodaemun-gu, Seoul 120-715; tel. (2) 3148-6040; fax (2) 3148-6044; internet www.momaf.go.kr.

Ministry of National Defence: 1, 3-ga, Yonsan-dong, Yeongsan-gu, Seoul; tel. (2) 795-0071; fax (2) 796-0369; internet www.mnd.go.kr.

Ministry of Planning and Budget: 520-3, Banpo-dong, Seocho-gu, Seoul 137-756; tel. (2) 3480-7716; fax (2) 3480-7600; internet www.mpb.go.kr.

Ministry of Science and Technology: 1, Jungang-dong, Gwacheon City, Gyeonggi Prov.; tel. (2) 503-7619; fax (2) 503-7673; internet www.most.go.kr.

Ministry of Unification: 77-6, Sejong-no, Jongno-gu, Seoul 110-760; tel. (2) 720-2424; fax (2) 720-2149; internet www.unikorea.go.kr.

Civil Service Commission: Kolon Bldg, 35-34, Dongui-dong, Jongno-gu, Seoul 110-040; tel. (2) 3703-3633; fax (2) 3771-5027; internet www.csc.go.kr.

Financial Supervisory Commission: 27, Yeouido-dong, Yeongdeungpo-gu, Seoul 150-743; tel. (2) 3771-5000; fax (2) 3771-5027; e-mail webmaster@fsc.go.kr; internet www.fsc.go.kr.

Korea Independent Commission Against Corruption: Seoul City Tower, 581, 5-ga, Namdaemun-no, Jung-gu, Seoul 100-095; tel. (2) 2126-0114; fax (2) 2126-0310; internet www.kicac.go.kr.

President and Legislature

PRESIDENT

Election, 19 December 2002

Candidate	Votes	% of total
Roh Moo-Hyun	12,014,277	48.9
Lee Hoi-Chang	11,443,297	46.6
Kwon Young-Gil	957,148	3.9
Lee Han-Dong	74,027	0.3
Kim Gil-Su	51,104	0.2
Kim Yeong-Kyu	22,063	0.1
Total	24,561,916	100.0

LEGISLATURE

KUK HOE

(National Assembly)

1 Yeouido-dong, Yeongdeungpo-gu, Seoul; tel. (2) 788-2786; fax (2) 788-3375; internet www.assembly.go.kr.

Speaker: KIM ONE-KI.

General Election, 15 April 2004

Party	Elected representatives	Proportional representatives	Total seats
Uri Party	129	23	152
Grand National Party	100	21	121
Democratic Labour Party	2	8	10
Millennium Democratic Party	5	4	9
United Liberal Democrats	4	—	4
Others	3	—	3
Total	243	56	299

Political Organizations

Democratic Labour Party (DLP): Hanyang Bldg, 14-31, Yeouido-dong, Yeongdeungpo-gu, Seoul 150-748; tel. (2) 761-1333; fax (2) 761-4115; e-mail inter@kdlp.org; internet www.kdlp.org; f. 2000; Pres. KWON YOUNG-GIL.

Grand National Party (GNP) (Hannara Party): 17-7, Yeouido-dong, Yeongdeungpo-gu, Seoul 150-010; tel. (2) 3786-3373; fax (2) 3786-3610; internet www.hannara.or.kr; f. 1997 by merger of Democratic Party and New Korea Party; Chair. PARK GEUN-HYE.

Millennium Democratic Party (MDP): 15, Gisan Bldg, Yeongdeungpo-gu, Seoul; tel. (2) 784-7007; fax (2) 784-6070; internet www.minjoo.or.kr; f. 2000 following dissolution of National Congress for New Politics (f. 1995); Chair. CHOUGH SOON-HYUNG; Rep. CHOO MI-AE.

United Liberal Democrats (ULD): Insan Bldg, 103-4, Shinsu-dong, Mapo-gu, Seoul 121-110; tel. (2) 701-3355; fax (2) 707-1637; internet www.jamin.or.kr; f. 1995 by fmr mems of the Democratic Liberal Party; Pres. KIM HAK-WON.

Uri Party: c/o National Assembly, 1 Yeouido-dong, Yeongdeungpo-gu, Seoul; Chair. LEE BU-YOUNG; party founded Nov. 2003 by defectors from the MDP.

Civic groups play an increasingly significant role in South Korean politics. These include: the People's Solidarity for Participatory Democracy (Dir Jang Hasung); the Citizens' Coalition for Economic Justice; and the Citizens' Alliance for Political Reform (Leader Kim Sok-Su).

Diplomatic Representation

EMBASSIES IN THE REPUBLIC OF KOREA

Algeria: 2-6, Itaewon 2-dong, Yeongsan-gu, Seoul 140-202; tel. (2) 794-5034; fax (2) 792-7845; e-mail sifdja01@kornet.net; internet www.algerianemb.or.kr; Ambassador AHMED BOUTACHE.

Argentina: Chun Woo Bldg, 5th Floor, 534 Itaewon-dong, Yeongsan-gu, Seoul 140-861; tel. (2) 793-4062; fax (2) 792-5820; Ambassador RODOLFO IGNACIO RODRÍGUEZ.

Australia: Kyobo Bldg, 11th Floor, 1, 1-ga, Jongno-gu, Seoul 110-714; tel. (2) 2003-0100; fax (2) 722-9264; internet www.australia.or.kr; Ambassador COLIN STUART HESELTINE.

Austria: Kyobo Bldg, Rm 1913, 1-1, 1-ga, Jongno, Jongno-gu, Seoul 110-714; tel. (2) 732-9071; fax (2) 732-9486; e-mail seoul-ob@bmaa.gv.at; internet www.austria.or.kr; Ambassador Dr HELMUT BOECK.

Bangladesh: 7-18, Woo Sung Bldg, Dongbinggo-dong, Yeongsan-gu, Seoul; tel. (2) 796-4056; fax (2) 790-5313; e-mail dootrok@soback.kornet21.net; Ambassador HUMAYUN A. KAMAL.

Belarus: 432-1636 Sindang 2-dong, Jung-gu, Seoul; tel. (2) 2237-8171; fax (2) 2237-8174; e-mail consul_korea@belembassy.org; Ambassador ALYAKSANDR VIKTOROVICH SEMESHKO.

Belgium: 1-94, Dongbinggo-dong, Yeongsan-gu, Seoul 140-230; tel. (2) 749-0381; fax (2) 797-1688; e-mail seoul@diplobel.org; Ambassador KOENRAAD ROUVROY.

Brazil: Ihn Gallery Bldg, 4th and 5th Floors, 141 Palpan-dong, Jongno-gu, Seoul; tel. (2) 738-4970; fax (2) 738-4974; e-mail braseul@soback.kornet21.net; Ambassador PEDRO PAULO PINTO ASSUMPÇÃO.

Brunei: Gwanghwamun Bldg, 7th Floor, 211, Sejong-no, Jongnogu, Seoul 110-050; tel. (2) 399-3707; fax (2) 399-3709; e-mail kbnbd_seoul@yahoo.com; Ambassador Dato' ABD. RAHMAN HAMID.

Bulgaria: 723-42, Hannam 2-dong, Yeongsan-gu, Seoul 140-894; tel. (2) 794-8626; fax (2) 794-8627; e-mail ebdy1990@unitel.co.kr; Chargé d'affaires a.i. VALERY ARZHENTINSKI.

Cambodia: 657-162, Hannam-dong, Yeongsan-gu, Seoul 140–910; tel. (2) 3785-1041; fax (2) 3785-1040; e-mail camboemb@korea.com; Ambassador CHHEANG VUN.

Canada: Kolon Bldg, 10-11th Floors, 45, Mugyo-dong, Jung-gu, Seoul 100-662; tel. (2) 3455-6000; fax (2) 3455-6123; e-mail canada@cec.or.kr; internet www.korea.gc.ca; Ambassador DENIS COMEAU.

Chile: Heungkuk Life Insurance Bldg, 14th Floor, 226 Sinmun-no 1-ga, Jongno-gu, Seoul; tel. (2) 2122-2600; fax (2) 2122-2601; e-mail echilekr@unitel.co.kr; internet www.echilecor.or.kr; Ambassador FERNANDO SCHMIDT.

China, People's Republic: 54, Hyoja-dong, Jongno-gu, Seoul; tel. (2) 738-1038; fax (2) 738-1077; Ambassador LI BIN.

Colombia: Kyobo Bldg, 13th Floor, 1-ga, Jongno, Jongno-gu, Seoul; tel. (2) 720-1369; fax (2) 725-6959; Ambassador Dr MIGUEL DURÁN ORDÓÑEZ.

Congo, Democratic Republic: 702, Daewoo Complex Bldg, 167 Naesu-dong, Jongno-gu, Seoul; tel. (2) 6722-7958; fax (2) 6722-7998; e-mail congokrembassy@yahoo.com; Ambassador N. CHRISTOPHE NGWEY.

Côte d'Ivoire: Chungam Bldg, 2nd Floor, 794-4, Hannam-dong, Yeongsan-gu, Seoul; tel. (2) 3785-0561; fax (2) 3785-0564; e-mail abenikof@hotmail.com; Ambassador HONORAT ABENI KOFFI.

Czech Republic: 1-121, 2-ga, Shinmun-ro, Jongno-gu, Seoul 110-062; tel. (2) 725-6765; fax (2) 734-6452; e-mail seoul@embassy.mzv.cz; internet www.mzv.cz/seoul; Ambassador TOMAS SMETANKA.

Denmark: Namsong Bldg, 5th Floor, 260-199, Itaewon-dong, Yeongsan-gu, Seoul 140-200; tel. (2) 795-4187; fax (2) 796-0986; e-mail selamb@um.dk; Ambassador LEIF DONDE.

Dominican Republic: Taepyeong-no Bldg, 19th Floor, 2-ga, 310 Taepyeong-no, Jung-gu, Seoul; tel. (2) 756-3513; fax (2) 756-3514; Ambassador JOSÉ M. NUNEZ.

Ecuador: Korea First Bldg, 19th Floor, 100, Gongpyeong-dong, Jongno-gu, Seoul; tel. (2) 739-2401; fax (2) 739-2355; e-mail mecuadorcor1@kornet.net; Ambassador FRANKLIN ESPINOSA.

Egypt: 46-1, Hannam-dong, Yeongsan-gu, Seoul; tel. (2) 749-0787; fax (2) 795-2588; internet www.mfg.gov.eg; Ambassador AMR HELMY.

El Salvador: Samsung Life Insurance Bldg, 20th Floor, Taepyeong-no 2-ga, Jung-gu, Seoul 100-716; tel. (2) 753-3432; fax (2) 753-3456; e-mail koembsal@hananet.net; Ambassador ALFREDO FRANCISCO UNGO.

Finland: Kyobo Bldg, Suite 1602, 1-1, 1-ga, Jongno, Jongno-gu, Seoul 110-714; tel. (2) 732-6737; fax (2) 723-4969; e-mail sanomat.seo@formin.fi; internet www.finlandembassy.or.kr; Ambassador KIM LUOTONEN.

France: 30, Hap-dong, Seodaemun-gu, Seoul 120-030; tel. (2) 3149-4300; fax (2) 3149-4328; e-mail ambfraco@elim.net; internet www.ambafrance-kr.org; Ambassador FRANÇOIS DESCOUEYTE.

Gabon: Yoosung Bldg, 4th Floor, 738-20, Hannam-dong, Yeongsan-gu, Seoul; tel. (2) 793-9575; fax (2) 793-9574; e-mail amgabsel@unitel.co.kr; Ambassador EMMANUEL ISSOZE-NGONDET.

Germany: 308-5, Dongbinggo-dong, Yeongsan-gu, Seoul 140-816; tel. (2) 748-4114; fax (2) 748-4161; e-mail dboseoul@kornet.net; internet www.gembassy.or.kr; Ambassador MICHAEL GEIER.

Ghana: 5-4, Hannam-dong, Yeongsan-gu, Seoul (CPOB 3887); tel. (2) 3785-1427; fax (2) 3785-1428; e-mail ghana3@kornet.net; Ambassador EDWARD OBENG KUFUOR.

Greece: Hanwha Bldg, 27th Floor, 1, Janggyo-dong, Jung-gu, Seoul 100-797; tel. (2) 729-1401; fax (2) 729-1402; Ambassador CONSTANTIN DRAKAKIS.

Guatemala: 614, Lotte Hotel, 1, Sogong-dong, Jung-gu, Seoul 100-635; tel. (2) 771-7582; fax (2) 771-7584; e-mail embcorea@minex.gob.gt; Ambassador EMILIO R. MALDONADO.

Holy See: 2, Gungjeong-dong, Jongno-gu, Seoul (Apostolic Nunciature); tel. (2) 736-5725; fax (2) 739-2310; e-mail nunseoul@kornet.net; Apostolic Nuncio Most Rev. GIOVANNI BATTISTA MORANDINI (Titular Archbishop of Numida).

Honduras: Jongno Tower Bldg, 2nd Floor, 6, Jongno 2-ga, Jongno-gu, Seoul 110-160; tel. (2) 738-8402; fax (2) 738-8403; e-mail hondseul@kornet.net; Ambassador RENE FRANCISCO UMANA CHINCHILLA.

Hungary: 1-103, Dongbinggo-dong, Yeongsan-gu, Seoul 140-230; tel. (2) 792-2105; fax (2) 792-2109; e-mail huembsel@kornet.net; Ambassador Dr ISTVÁN TORZSA.

India: 37-3, Hannam-dong, Yeongsan-gu, CPOB 3466, Seoul 140-210; tel. (2) 798-4257; fax (2) 796-9534; e-mail eoiseoul@soback.kornet.nm.kr; Chargé d'affaires MOHINDER SINGH GROVER.

Indonesia: 55, Yeouido-dong, Yeongdeungpo-gu, Seoul 150-010; tel. (2) 783-5675; fax (2) 780-4280; e-mail bidpen@soback.kornet21.net; Ambassador ABDUL GHANI.

Iran: 726-126, Hannam-dong, Yeongsan-gu, Seoul; tel. (2) 793-7751; fax (2) 792-7052; e-mail iranssy@chollian.net; internet www.mfa.gov.ir; Ambassador MOHSEN TALAE'I.

Ireland: Daehan Fire and Marine Insurance Bldg, 15th Floor, 51-1, Namchang-dong, Jung-gu, Seoul; tel. (2) 774-6455; fax (2) 774-6458; e-mail hibernia@bora.dacom.co.kr; Ambassador CONOR MURPHY.

Israel: 18th Fl., Kabool Bldg, 149 Seorin-dong, Jongro-gu, Seoul 110-726; tel. (2) 739-8666 ; fax (2) 739-8667; e-mail seoul@israel.org; internet www.israelemb.or.kr; Ambassador UZI MANOR.

Italy: 1-398, Hannam-dong, Yeongsan-gu, Seoul 140-210; tel. (2) 796-0491; fax (2) 797-5560; e-mail ambseoul@italyemb.or.kr; Ambassador FRANCESCO RAUSI.

Japan: 18-11, Junghak-dong, Jongno-gu, Seoul; tel. (2) 2170-5200; fax (2) 734-4528; Ambassador TOSHIYUKI TAKANO.

Kazakhstan: 13-10, Seongbuk-dong, Seongbuk-gu, Seoul; tel. (2) 744-9714; fax (2) 744-9760; e-mail kazkor@chollian.net; Ambassador BORLAT K. NURGALIEV.

Kuwait: 309-15, Dongbinggo-dong, Yeongsan-gu, Seoul; tel. (2) 749-3688; fax (2) 749-3687; Ambassador FAWZI AL-JASEM.

Laos: 657-93, Hannam-dong, Yeongsan-gu, Seoul; tel. (2) 796-1713; fax (2) 796-1771; e-mail laoseoul@korea.com; Ambassador THONGSAVATH PRASEUTH.

Lebanon: 310-49, Dongbinggo-dong, Yongsan-ku, Seoul 140-230; tel. (2) 794-6482; fax (2) 794-6485; e-mail emleb@lebanonembassy.net; Ambassador HUSSEIN RAMMAL.

Libya: 4-5, Hannam-dong, Yeongsan-gu, Seoul; tel. (2) 797-6001; fax (2) 797-6007; e-mail libyaemb@kornet.net; Sec. of People's Bureau AHMED MOHAMED TABULI.

Malaysia: 4-1, Hannam-dong, Yeongsan-gu, Seoul 140-210; tel. (2) 795-9203; fax (2) 794-5480; e-mail mwseoul@kornet.net; internet www.malaysia.or.kr; Ambassador Dato' M. SANTHANANABAN.

Mexico: 33-6, Hannam 1-dong, Yeongsan-gu, Seoul 140-885; tel. (2) 798-1694; fax (2) 790-0939; Ambassador ROGELIO GRANGUILLHOME.

Mongolia: 33-5, Hannam-dong, Yeongsan-gu, Seoul; tel. (2) 794-1350; fax (2) 794-7605; e-mail monemb@uriel.net; internet www.mongoliaemb.or.kr; Ambassador URJINLHUNDEV PERENLEYN.

Morocco: S-15, UN Village, 270-3, Hannam-dong, Yeongsan-gu, Seoul; tel. (2) 793-6249; fax (2) 792-8178; e-mail sifamase@bora.dacom.co.kr; internet www.moroccoemb.or.kr; Ambassador JAAFAR ALJ HAKIM.

Myanmar: 724-1, Hannam-dong, Yeongsan-gu, Seoul 140-210; tel. (2) 792-3341; fax (2) 796-5570; e-mail myanmare@ppp.kornet.net; Ambassador U NYO WIN.

Netherlands: Kyobo Bldg, 14th Floor, 1-ga, Jongno, Jongno-gu, Seoul 110-714; tel. (2) 737-9514; fax (2) 735-1321; e-mail seo@minbuza.nl; Ambassador RADINCK J. VAN VOLLENHOVEN.

New Zealand: Kyobo Bldg, 18th Floor, 1, 1-ga, Jongno, Jongno-gu, CPOB 1059, Seoul 100-610; tel. (2) 730-7794; fax (2) 737-4861; e-mail nzembsel@kornet.net; internet www.nzembassy.com/korea; Ambassador DAVID TAYLOR.

Nigeria: 310-19, Dongbinggo-dong, Yeongsan-gu, Seoul; tel. (2) 797-2370; fax (2) 796-1848; e-mail chancery@nigeriaembassy.or.kr; Ambassador ABBA A. JIJANI.

Norway: 258-8, Itaewon-dong Yongsan-gu, Seoul 140-200; tel. (2) 795-6850; fax (2) 798-6072; e-mail emb.seoul@mfa.no; internet www.norway.or.kr; Ambassador ARILD BRAASTAD.

Oman: 309-3, Dongbinggo-dong, Yeongsan-gu, Seoul; tel. (2) 790-2431; fax (2) 790-2430; e-mail omanembs@ppp.kornet.nm.kr; Ambassador MOOSA HAMDAN AL-TAEE.

Pakistan: 124-13, Itaewon-dong, Yeongsan-gu, Seoul 140-200; tel. (2) 796-8252; fax (2) 796-0313; e-mail heamb@pakistan-korea-trade.org; internet www.pakistan-korea-trade.org; Ambassador SYED PERVEZ HUSSAIN.

Panama: Northgate Bldg, 6th Floor, 66, Jeokseon-dong, Jongno-gu, Seoul; tel. (2) 734-8610; fax (2) 734-8613; e-mail panaemba@kornet.net; Ambassador FÉLIX PÉREZ ESPINOSA.

Papua New Guinea: 36-1, Hannam 1-dong, Yeongsan-gu, Seoul; tel. (2) 798-9854; fax (2) 798-9856; e-mail pngembsl@ppp.kornet.nm.kr; Ambassador DAVID ANGGO.

Paraguay: SK Bldg, 2nd Floor, 99 Seorin-dong, Jongno-gu, Seoul 110-728; tel. (2) 730-8335; fax (2) 730-8336; e-mail pyemc2@hananet.net; Ambassador LUIS FERNANDO AVALOS GIMÉNEZ.

Peru: Namhan Bldg, 6th Floor, 76-42, Hannam-dong, Yeongsan-gu, Seoul 140-210; tel. (2) 793-5810; fax (2) 797-3736; e-mail ipruseul@uriel.net; Chargé d'affaires GUSTAVO LEMBCKE.

Philippines: Diplomatic Center, 9th Floor, 1376-1, Seocho 2-dong, Seocho-gu, Seoul; tel. (2) 577-6147; fax (2) 574-4286; e-mail phsk@soback.kornet.net; Ambassador JUANITO P. JARASA.

Poland: 70, Sagan-dong, Jongno-gu, Seoul; tel. (2) 723-9681; fax (2) 723-9680; e-mail embassy@polandseoul.org; internet www.polandseoul.org; Ambassador TADEUSZ CHOMICKI.

Portugal: Wonseo Bldg, 2nd Floor, 171, Wonseo-dong, Jongno-gu, Seoul; tel. (2) 3675-2251; fax (2) 3675-2250; e-mail ambport@chollian.net; Ambassador FERNANDO MACHADO.

Qatar: 309-5, Dongbinggo-dong, Yeongsan-gu, Seoul; tel. (2) 790-1308; fax (2) 790-1027; Ambassador ABDUL RAZZAK AL-ABDULKGHANI.

Romania: 1-42, UN Village, Hannam-dong, Yeongsan-gu, Seoul 140-210; tel. (2) 797-4924; fax (2) 794-3114; e-mail romemb@uriel.net; internet www.uriel.net/~romemb; Ambassador VALERIU ARTENI.

Russia: 34-16, Jeong-dong, Jung-gu, Seoul 100–120; tel. (2) 318-2116; fax (2) 754-0417; e-mail rusemb@uriel.net; internet www.russian-embassy.org; Ambassador TEYMURAZ O. RAMISHVILI.

Saudi Arabia: 1-112, 2-ga, Sinmun-no, Jongno-gu, Seoul; tel. (2) 739-0631; fax (2) 732-3110; Ambassador SALEH BIN MANSOUR AL-RAJHY.

Singapore: Seoul Finance Bldg, 28th Floor, 84, 1-ga, Taepyeong-no, Jung-gu, Seoul 100-102; tel. (2) 774-2464; fax (2) 773-2465; e-mail singemb@unitel.co.kr; Ambassador CALVIN EU MUN HOO.

Slovakia: 802, Hyundai Liberty House, 258, Hannam-dong, Yeongsan-gu, Seoul 140-210; tel. (2) 794-3981; fax (2) 794-3982; e-mail slovakemb@yahoo.com; Chargé d'affaires a.i. JÁN CHLÁDEK.

South Africa: 1-37, Hannam-dong, Yeongsan-gu, Seoul 140-210; tel. (2) 792-4855; fax (2) 792-4856; e-mail sae@saembasy.dacom.net; internet saembassy.dacom.net; Ambassador SYDNEY BAFANA KUBHEKA.

Spain: 726-52, Hannam-dong, Yeongsan-gu, Seoul; tel. (2) 794-3581; fax (2) 796-8207; Ambassador ENRIQUE PANES CALPE.

Sri Lanka: Kyobo Bldg, Rm 2002, 1-1, 1-ga, Jongno, Jongno-gu, Seoul 110-714; tel. (2) 735-2966; fax (2) 737-9577; e-mail lankaemb@chollian.net; Ambassador K. C. LOGESWARAN.

Sudan: 653-24, Hannam-dong, Yeongsan-gu, Seoul; tel. (2) 793-8692; fax (2) 793-8693; Ambassador BABIKER A. KHALIFA.

Sweden: Seoul Central Bldg, 12th Floor, 136, Seorin-dong, Jongno-gu, KPO Box 1154, Seoul 110-611; tel. (2) 738-0846; fax (2) 733-1317; e-mail embassy@swedemb.or.kr; internet www.swedenabroad.com/seoul; Ambassador HARALD SANDBERG.

Switzerland: 32-10, Songwol-dong, Jongno-gu, Seoul 110-101; tel. (2) 739-9511; fax (2) 737-9392; e-mail swissemb@elim.net; internet www.elim.net/~swissemb/; Ambassador CHRISTIAN MUEHLETHALER.

Thailand: 653-7, Hannam-dong, Yeongsan-gu, Seoul; tel. (2) 795-3098; fax (2) 798-3448; e-mail rteseoul@elim.net; Ambassador SOMBOON SANGIAMBUT.

Tunisia: 1-17, Dongbinggo-dong, Yeongsan-gu, Seoul 140-809; tel. (2) 790-4334; fax (2) 790-4333; e-mail ambtnkor@kornet.net; Ambassador OTHMAN JERANDI.

Turkey: Vivien Corpn Bldg, 4th Floor, 4-52, Seobinggo-dong, Yeongsan-gu, Seoul; tel. (2) 794-0255; fax (2) 797-8546; e-mail tcseulbe@kornet.net; Ambassador SELIM KUNERALP.

Ukraine: 1-97, Dongbinggo-dong, Yeongsan-gu, Seoul; tel. (2) 790-5696; fax (2) 790-5697; e-mail secretary@ukrembrk.com; internet www.ukrembrk.com; Ambassador VOLODYMYR V. FURKALO.

United Arab Emirates: 5-5, Hannam-dong, Yeongsan-gu, Seoul; tel. (2) 790-3235; fax (2) 790-3238; Ambassador ABDULLA MOHAMED ALI AL-SHURAFA AL-HAMMADY.

United Kingdom: Taepyeongno 40, 4, Jeong-dong, Jung-gu, Seoul 100-120; tel. (2) 3210-5500; fax (2) 725-1738; e-mail bembassy@britain.or.kr; internet www.britain.or.kr; Ambassador WARWICK MORRIS.

USA: 32, Sejong-no, Jongno-gu, Seoul 110-710; tel. (2) 397-4114; fax (2) 735-3903; internet usembassy.state.gov/seoul/; Ambassador CHRISTOPHER HILL.

Uruguay: Daewoo Bldg, 1025, 541, 5-ga, Namdaemun, Jung-gu, Seoul; tel. (2) 753-7893; fax (2) 777-4129; e-mail uruseul@kornet.net; Ambassador JULIO GIAMBRUNO.

Uzbekistan: Diplomatic Center, Rm. 701, 1376-1, Seocho 2-dong, Seocho-gu, Seoul; tel. (2) 574-6554; fax (2) 578-0576; Ambassador VITALI V. FEN.

Venezuela: 16th Floor, Korea First Bank Bldg, 100 Gongpyeong-dong, Jongno-gu, 110-702 Seoul; tel. (2) 732-1546; fax (2) 732-1548; e-mail emvesel@soback.kornet.net; internet www.venezuelaemb.or.kr; Ambassador GUILLERMO QUINTERO.

Viet Nam: 28-58, Samcheong-dong, Jongno-gu, Seoul 140-210; tel. (2) 738-2318; fax (2) 739-2064; e-mail vietnam@elim.net; Ambassador DUONG CHINH THUC.

Yemen: 11-444, Hannam-dong, Yeongsan-gu, Seoul 140-210; tel. (2) 792-9883; fax (2) 792-9885; internet www.gpc.org.ye; Chargé d'affaires YAHYA AHMED AL-WAZIR.

Judicial System

SUPREME COURT

The Supreme Court is the highest court, consisting of 14 Justices, including the Chief Justice. The Chief Justice is appointed by the President, with the consent of the National Assembly, for a term of six years. Other Justices of the Supreme Court are appointed for six years by the President on the recommendation of the Chief Justice. The appointment of the Justices of the Supreme Court, however, requires the consent of the National Assembly. The Chief Justice may not be reappointed. The court is empowered to receive and decide on appeals against decisions of the High Courts, the Patent Court, and the appellate panels of the District Courts or the Family Court in civil, criminal, administrative, patent and domestic relations cases. It is also authorized to act as the final tribunal to review decisions of courts-martial and to consider cases arising from presidential and parliamentary elections.

Chief Justice: KIM YOUNG-RAN, 967, Seocho-dong, Seocho-gu, Seoul; tel. (2) 3480-1002; fax (2) 533-1911; internet www.scourt.go.kr.

Justices: KOH HYUN-CHUL, KIM YONG DAM, CHO MOO-JEH, BYUN JAE-SEUNG, YOO JI-DAM, YOON JAE-SIK, LEE YONG-WOO, BAE KI-WON, KANG SHIN-WOOK, LEE KYU-HONG, LEE KANG-KOOK, SON JI-YOL, PARK JAE-YOON.

CONSTITUTIONAL COURT

The Constitutional Court is composed of nine adjudicators appointed by the President, of whom three are chosen from among persons selected by the National Assembly and three from persons nominated by the Chief Justice. The Court adjudicates the following matters: constitutionality of a law (when requested by the other courts); impeachment; dissolution of a political party; disputes between state agencies, or between state agencies and local governments; and petitions relating to the Constitution.

President: YUN YOUNG-CHUL, 83 Jae-dong, Jongno-gu, Seoul 110-250; tel. (2) 708-3456; fax (2) 708-3566; internet www.ccourt.go.kr.

HIGH COURTS

There are five courts, situated in Seoul, Daegu, Busan, Gwangju and Daejeon, with five chief, 78 presiding and 145 other judges. The courts have appellate jurisdiction in civil and criminal cases and can also pass judgment on administrative litigation against government decisions.

PATENT COURT

The Patent Court opened in Daejeon in March 1998, to deal with cases in which the decisions of the Intellectual Property Tribunal are challenged. The examination of the case is conducted by a judge, with the assistance of technical examiners.

DISTRICT COURTS

District Courts are established in 13 major cities; there are 13 chief, 241 presiding and 966 other judges. They exercise jurisdiction over all civil and criminal cases in the first instance.

MUNICIPAL COURTS

There are 103 Municipal Courts within the District Court system, dealing with small claims, minor criminal offences, and settlement cases.

FAMILY COURT

There is one Family Court, in Seoul, with a chief judge, four presiding judges and 16 other judges. The court has jurisdiction in domestic matters and juvenile delinquency.

ADMINISTRATIVE COURT

An Administrative Court opened in Seoul in March 1998, to deal with cases that are specified in the Administrative Litigation Act. The Court has jurisdiction over cities and counties adjacent to Seoul, and deals with administrative matters, including taxes, expropriations of land, labour and other general administrative matters. District Courts deal with administrative matters within their districts until the establishment of regional administrative courts is complete.

COURTS-MARTIAL

These exercise jurisdiction over all offences committed by armed forces personnel and civilian employees. They are also authorized to try civilians accused of military espionage or interference with the execution of military duties.

Religion

The traditional religions are Mahayana Buddhism, Confucianism and Chundo Kyo, a religion peculiar to Korea.

BUDDHISM

Korean Mahayana Buddhism has about 80 denominations. The Chogye-jong is the largest Buddhist order in Korea, having been introduced from China in AD 372. The Chogye Order accounts for almost two-thirds of all Korean Buddhists. In 1995 it had 2,426 out of 19,059 Buddhist temples and there were 12,470 monks.

Korean United Buddhist Association (KUBA): 46-19, Soosong-dong, Jongno-gu, Seoul 110-140; tel. (2) 732-4885; 28 mem. Buddhist orders; Pres. SONG WOL-JOO.

Won Buddhism

Won Buddhism combines elements of Buddhism and Confucianism. In 1995 there were 404 temples, 9,815 priests, and 86,823 believers.

CHRISTIANITY

National Council of Churches in Korea: Christian Bldg, Rm 706, 136-46, Yeonchi-dong, Jongno-gu, Seoul 110-736; tel. (2) 763-8427; fax (2) 744-6189; e-mail kncc@kncc.or.kr; internet www.kncc.or.kr; f. 1924 as National Christian Council; present name adopted 1946; eight mem. churches; Gen. Sec. Rev. PAIK DO-WOONG.

The Anglican Communion

South Korea has three Anglican dioceses, collectively forming the Anglican Church of Korea (founded as a separate province in April 1993), under its own Primate, the Bishop of Seoul.

Bishop of Pusan (Busan): Rt Rev. JOSEPH DAE-YONG LEE, 455-2, Oncheon-1-dong, Dongnae-gu, Busan 607-061; tel. (51) 554-5742; fax (51) 553-9643; e-mail bpjoseph@hanmail.net.

Bishop of Seoul: Most Rev. MATTHEW CHUNG CHUL-BUM, 3, Jeong-dong, Jung-gu, Seoul 100-120; tel. (2) 738-6597; fax (2) 723-2640; e-mail bishop100@hosanna.net.

Bishop of Taejon (Daejeon): Rt Rev. PAUL YOON HWAN, 88-1, Sonhwa 2-dong, POB 22, Daejeon 300-600; tel. (42) 256-9987; fax (42) 255-8918.

The Roman Catholic Church

For ecclesiastical purposes, North and South Korea are nominally under a unified jurisdiction. South Korea comprises three archdioceses, 11 dioceses, and one military ordinate. At 31 December 2002 some 4,262,263 people were adherents of the Roman Catholic Church.

Bishops' Conference

Catholic Bishops' Conference of Korea, 643-1, Junggok-dong, Gwangjin-gu, Seoul 143-912; tel. (2) 460-7500; fax (2) 460-7505; e-mail cbck@cbck.or.kr; internet www.cbck.or.kr.

f. 1857; Pres. Most Rev. ANDREAS CHOI CHANG-MOU (Archbishop of Gwangju).

Archbishop of Kwangju (Gwangju): Most Rev. ANDREAS CHOI CHANG-MOU, Archdiocesan Office, 5-32, Im-dong, Buk-gu, Gwangju 500-868; tel. (62) 510-2838; fax (62) 525-6873; e-mail biseo@kjcatholic.or.kr.

Archbishop of Seoul: Most Rev. NICHOLAS CHEONG JIN-SUK, Archdiocesan Office, 1, 2-ga, Myeong-dong, Jung-gu, Seoul 100-022; tel. (2) 727-2114; fax (2) 773-1947; e-mail ao@seoul.catholic.or.kr.

Archbishop of Taegu (Daegu): Most Rev. PAUL RI MOON-HI, Archdiocesan Office, 225-1, Namsan 3-dong, Jung-gu, Daegu 700-804; tel. (53) 253-7011; fax (53) 253-9441; e-mail taegu@tgcatholic.or.kr.

Protestant Churches

Korean Methodist Church: 64-8, 1-ga, Taepyeong-no, Jung-gu, Seoul 100-101; KPO Box 285, Seoul 110-602; tel. (2) 399-4300; fax (2) 399-4307; e-mail bishop@kmcweb.or.kr; internet www.kmcweb.or.kr; f. 1885; 1,417,213 mems (2003); Bishop KIM JIN HO.

Presbyterian Church in the Republic of Korea (PROK): 1501, Ecumenical Bldg, 136-56, Yeonchi-dong, Jongno-gu, Seoul 110-470; tel. (2) 708-4021; fax (2) 708-4027; e-mail prok3000@chollian.net; internet www.prok.org; f. 1953; 332,915 mems (2001); Gen. Sec. Rev. Dr KIM JONG-MOO.

Presbyterian Church of Korea (PCK): Korean Church Centennial Memorial Bldg; 135, Yeochi-dong, Jongno-gu, Seoul 110-470; tel. (2) 741-4350; fax (2) 766-2427; e-mail thepck@pck.or.kr; internet www.pck.or.kr; 2,328,413 mems (2001); Moderator Rev. CHOI BYUNG-GON; Gen. Sec. Rev. Dr. KIM SANG-HAK.

There are some 160 other Protestant denominations in the country, including the Korea Baptist Convention and the Korea Evangelical Church.

CHUNDO KYO

A religion indigenous and unique to Korea, Chundo Kyo combines elements of Shaman, Buddhist, and Christian doctrines. In 1995 there were 274 temples, 5,597 priests, and 28,184 believers.

CONFUCIANISM

In 1995 there were 730 temples, 31,833 priests, and 210,927 believers.

TAEJONG GYO

Taejong Gyo is Korea's oldest religion, dating back 4,000 years, and comprising beliefs in the national foundation myth, and the triune god, Hanul. By the 15th century the religion had largely disappeared, but a revival began in the late 19th century. In 1995 there were 103 temples, 346 priests, and 7,603 believers.

The Press

NATIONAL DAILIES

(In Korean, unless otherwise indicated)

Chosun Ilbo: 61, 1-ga, Taepyeong-no, Jung-gu, Seoul 100-756; tel. (2) 724-5114; fax (2) 724-5059; internet www.chosun.com; f. 1920; morning, weekly and children's edns; independent; Pres. BANG SANG-HOON; Editor-in-Chief KIM DAE-JUNG; circ. 2,470,000.

Daily Sports Seoul: 25, 1-ga, Taepyeong-no, Jung-gu, Seoul; tel. (2) 721-5114; fax (2) 721-5396; internet www.seoul.co.kr; f. 1985; morning; sports and leisure; Pres. LEE HAN-SOO; Man. Editor SON CHU-WHAN.

Dong-A Ilbo: 139-1, 3-ga, Sejong-no, Jongno-gu, Seoul 100-715; tel. (2) 2020-0114; fax (2) 2020-1239; e-mail newsroom@donga.com; internet www.donga.com; f. 1920; morning; independent; Pres. KIM HAK-JOON; Editor-in-Chief LEE HYUN-NAK; circ. 2,150,000.

Han-Joong Daily News: 91-1, 2-ga, Myeong-dong, Jung-gu, Seoul; tel. (2) 776-2801; fax (2) 778-2803; Chinese.

Hankook Ilbo: 14, Junghak-dong, Jongno-gu, Seoul; tel. (2) 724-2114; fax (2) 724-2244; internet www.hankooki.com; f. 1954; morning; independent; Pres. CHANG CHAE-KEUN; Editor-in-Chief YOON KOOK-BYUNG; circ. 2,000,000.

Hankyoreh Shinmun (One Nation): 116-25, Gongdeok-dong, Mapo-gu, Seoul 121-020; tel. (2) 710-0114; fax (2) 710-0210; internet www.hani.co.kr; f. 1988; centre-left; Chair. KIM DOO-SHIK; Editor-in-Chief SUNG HAN-PYO; circ. 500,000.

Ilgan Sports (The Daily Sports): 14, Junghak-dong, Jongno-gu, Seoul 110-792; tel. (2) 724-2114; fax (2) 724-2299; internet www.dailysports.co.kr; morning; f. 1969; Pres. CHANG CHAE-KEUN; Editor KIM JIN-DONG; circ. 600,000.

Jeil Economic Daily: 24-5 Yeouido-dong, Yeongdeungpo-gu, Seoul; tel. (2) 792-1131; fax (2) 792-1130; f. 1988; morning; Pres. HWANG MYUNG-SOON; Editor LEE SOO-SAM.

JoongAng Ilbo (JoongAng Daily News): 7, Soonhwa-dong, Jung-gu, Seoul; tel. (2) 751-5114; fax (2) 751-9709; internet www.joins.com; f. 1965; morning; Chair. and CEO HONG SEOK-HYUN; circ. 2,300,000.

Kookmin Ilbo: 12, Yeouido-dong, Yeongdeungpo-gu, Seoul; tel. (2) 781-9114; fax (2) 781-9781; internet www.kukminilbo.co.kr; Pres. CHA IL-SUK.

Korea Daily News: 25, 1-ga, Taepyeong-no, Jung-gu, Seoul; tel. (2) 2000-9000; fax (2) 2000-9659; internet www.kdaily.com; f. 1945; morning; independent; Publr and Pres. SON CHU-HWAN; Man. Editor LEE DONG-HWA; circ. 700,000.

Korea Economic Daily: 441, Junglim-dong, Jung-gu, Seoul 100-791; tel. (2) 360-4114; fax (2) 779-4447; internet www.ked.co.kr; f. 1964; morning; Pres. PARK YONG-JUNG; Man. Dir and Editor-in-Chief CHOI KYU-YOUNG.

The Korea Herald: 1-12, 3-ga, Hoehyeon-dong, Jung-gu, Seoul; tel. (2) 727-0114; fax (2) 727-0670; internet www.koreaherald.co.kr; f. 1953; morning; English; independent; Pres. KIM CHIN-OUK; Man. Editor MIN BYUNG-IL; circ. 150,000.

The Korea Times: 14, Junghak-dong, Jongno-gu, Seoul 110-792; tel. (2) 724-2114; fax (2) 732-4125; e-mail kt@koreatimes.co.kr; internet www.koreatimes.co.kr; f. 1950; morning; English; independent; Pres. YOON KOOK-BYUNG; Man. Ed. LEE SANG-SEOK; circ. 100,000.

Kyung-hyang Shinmun: 22, Jeong-dong, Jung-gu, Seoul; tel. (2) 3701-1114; fax (2) 737-6362; internet www.khan.co.kr; f. 1946; evening; independent; Pres. HONG SUNG-MAN; Editor KIM HI-JUNG; circ. 733,000.

Maeil Business Newspaper: 51-9, 1-ga, Bil-dong, Jung-gu, Seoul 100-728; tel. (2) 2000-2114; fax (2) 2269-6200; internet www.mk.co.kr; f. 1966; evening; economics and business; Pres. CHANG DAE-WHAN; Editor JANG BYUNG-CHANG; circ. 235,000.

Munhwa Ilbo: 68, 1-ga, Chungjeong-no, Jung-gu, Seoul 110-170; tel. (2) 3701-5114; fax (2) 722-8328; internet www.munhwa.co.kr; f. 1991; evening; Pres. NAM SI-UK; Editor-in-Chief KANG SIN-KU.

Naeway Economic Daily: 1-12, 3-ga, Hoehyon-dong, Jung-gu, Seoul 100; tel. (2) 727-0114; fax (2) 727-0661; internet www.naeway.co.kr; f. 1973; morning; Pres. KIM CHIN-OUK; Man. Editor HAN DONG-HEE; circ. 300,000.

Segye Times: 63-1, 3-ga, Hangang-no, Yeongsan-gu, Seoul; tel. (2) 799-4114; fax (2) 799-4520; internet www.segyetimes.co.kr; f. 1989; morning; Pres. HWANG HWAN-CHAI; Editor MOK JUNG-GYUM.

Seoul Kyungje Shinmun: 19, Junghak-dong, Jongno-gu, Seoul 100; tel. (2) 724-2114; fax (2) 732-2140; internet www.sed.co.kr; f. 1960; morning; Pres. KIM YOUNG-LOUL; Man. Editor KIM SEO-WOONG; circ. 500,000.

Sports Chosun: 61, 1-ga, Taepyeong-no, Jung-gu, Seoul; tel. (2) 724-6114; fax (2) 724-6979; internet www.sportschosun.com; f. 1964; Publr BANG SANG-HOON; circ. 400,000.

LOCAL DAILIES

Cheju Daily News: 2324-6, Yeon-dong, Jeju; tel. (64) 740-6114; fax (64) 740-6500; internet www.chejunews.co.kr; f. 1945; evening; Pres. KIM DAE-SUNG; Man. Editor KANG BYUNG-HEE.

Chonbuk Domin Ilbo: 207-10, 2-ga, Deokjin-dong, Deokjin-gu, Jeonju, N Jeolla Prov.; tel. (63) 251-7114; fax (63) 251-7127; internet www.domin.co.kr; f. 1988; morning; Pres. LIM BYOUNG-CHAN; Man. Editor YANG CHAE-SUK.

Chonju Ilbo: 568-132, Sonosong-dong, Deokjin-gu, Jeonju, N. Jeolla Prov.; tel. (63) 285-0114; fax (63) 285-2060; f. 1991; morning; Chair. KANG DAE-SOON; Man. Editor SO CHAE-CHOL.

Chonnam Ilbo: 700-5, Jungheung-dong, Buk-gu, Gwangju, 500-758; tel. (62) 527-0015; fax (62) 510-0436; internet www.chonnamilbo.co.kr; f. 1989; morning; Pres. LIM WON-SIK; Editor-in-Chief KIM YONG-OK.

Chunbuk Ilbo: 710-5, Kumam-dong, Deokjin-gu, Jeonju, N. Jeolla Prov.; tel. (63) 250-5500; fax (63) 250-5550; f. 1950; evening; Pres. SUH JUNG-SANG; Man. Editor LEE KON-WOONG.

Chungchong Daily News: 304, Sachang-dong, Hungduk-gu, Cheongju, N. Chungcheong Prov.; tel. (43) 279-5114; fax (43) 262-2000; internet www.ccnews.co.kr; f. 1946; morning; Pres. SEO JEONG-OK; Editor IM BAIK-SOO.

Halla Ilbo: 568-1, Samdo 1-dong, Jeju; tel. (64) 750-2114; fax (64) 750-2520; internet www.hallailbo.com; f. 1989; evening; Chair. KANG YONG-SOK; Man. Editor HONG SONG-MOK.

Incheon Ilbo: 18-1, 4-ga, Hang-dong, Jung-gu, Incheon; tel. (32) 763-8811; fax (32) 763-7711; internet www.inchonnews.co.kr; f. 1988; evening; Chair. MUN PYONG-HA; Man. Editor LEE JAE-HO.

Jungdo Daily Newspaper: 274-7, Galma-dong, Seo-gu, Daejeon; tel. (42) 530-4114; fax (42) 535-5334; internet www.joongdo.com; f. 1951; morning; Chair. KI-CHANG; Man. Editor SONG HYOUNG-SOP.

Kangwon Ilbo: 53, 1-ga, Jungang-no, Chuncheon, Gangwon Prov.; tel. (33) 252-7228; fax (33) 252-5884; internet www.kwnews.co.kr; f. 1945; evening; Pres. CHO NAM-JIN; Man. Editor KIM KEUN-TAE.

Kookje Daily News: 76-2, Goje-dong, Yeonje-gu, Busan 611-702; tel. (51) 500-5114; fax (51) 500-4274; e-mail jahwang@ms.kookje.co.kr; internet www.kookje.co.kr; f. 1947; morning; Pres. LEE JONG-DEOK; Editor-in-Chief JEONG WON-YOUNG.

Kwangju Ilbo: 1, 1-ga, Geumnam-no, Dong-gu, Gwangju; tel. (62) 222-8111; fax (62) 227-9500; internet www.kwangju.co.kr; f. 1952; evening; Chair. KIM CHONG-TAE; Man. Editor CHO DONG-SU.

Kyeonggi Ilbo: 452-1, Songjuk-dong, Changan-gu, Suwon, Gyeonggi Prov.; tel. (31) 247-3333; fax (31) 247-3349; internet www.kgib.co.kr; f. 1988; evening; Chair. SHIN SON-CHOL; Man. Editor LEE CHIN-YONG.

Kyeongin Ilbo: 1121-11, Ingye-dong, Paldal-gu, Suwon, Gyeonggi Prov.; tel. (31) 231-5114; fax (31) 232-1231; internet www.kyeongin.com; f. 1960; evening; Chair. SUNG BAEK-EUNG; Man. Editor KIM HWA-YANG.

Kyungnam Shinmun: 100-5, Sinwol-dong, Changwon, S. Gyeongsang Prov.; tel. (55) 283-2211; fax (55) 283-2227; internet www.knnews.co.kr; f. 1946; evening; Pres. KIM DONG-KYU; Editor PARK SUNG-KWAN.

Maeil Shinmun: 71, 2-ga, Gyesan-dong, Jung-gu, Daegu; tel. (53) 255-5001; fax (53) 255-8902; internet www.m2000.co.kr; f. 1946; evening; Chair. KIM BOO-KI; Editor LEE YONG-KEUN; circ. 300,000.

Pusan Daily News: 1-10, Sujeong-dong, Dong-gu, Busan 601-738; tel. (51) 461-4114; fax (51) 463-8880; internet www.pusanilbo.co.kr; f. 1946; Pres. JEONG HAN-SANG; Man. Editor AHN KI-HO; circ. 427,000.

Taegu Ilbo: 81-2, Sincheon 3-dong, Dong-gu, Daegu; tel. (53) 757-4500; fax (53) 751-8086; internet www.tgnews.go.kr; f. 1953; morning; Pres. PARK GWON-HEUM; Editor KIM KYUNG-PAL.

Taejon Ilbo: 1-135, Munhwa 1-dong, Jung-gu, Daejeon; tel. (42) 251-3311; fax (42) 253-3320; internet www.taejontimes.co.kr; f. 1950; evening; Chair. SUH CHOON-WON; Editor KWAK DAE-YEON.

Yeongnam Ilbo: 111, Sincheon-dong, Dong-gu, Daegu; tel. (53) 757-5114; fax (53) 756-9009; internet www.yeongnam.co.kr; f. 1945; morning; Chair. PARK CHANG-HO; Man. Editor KIM SANG-TAE.

SELECTED PERIODICALS

Academy News: 50, Unjung-dong, Bundang-gu, Seongnam, Gyeonggi Prov. 463-791; tel. (31) 709-8111; fax (31) 709-9945; organ of the Acad. of Korean Studies; Pres. HAN SANG-JIN.

Business Korea: 26-3, Yeouido-dong, Yeongdeungpo-gu, Seoul 150-010; tel. (2) 784-4010; fax (2) 784-1915; f. 1983; monthly; Pres. KIM KYUNG-HAE; circ. 35,000.

Eumak Dong-A: 139, Sejong-no, Jongno-gu, Seoul 110-715; tel. (2) 781-0640; fax (2) 705-4547; f. 1984; monthly; music; Publr KIM BYUNG-KWAN; Editor KWON O-KIE; circ. 85,000.

Han Kuk No Chong (FKTU News): Federation of Korean Trade Unions, FKTU Bldg, 35, Yeouido-dong, Yeongdeungpo-gu, Seoul; tel. (2) 786-3970; fax (2) 786-2864; e-mail fktuintl@nownuri.net; internet www.fktu.or.kr; f. 1961; labour news; circ. 20,000.

Hyundae Munhak: Mokjung Bldg, 1st Floor, 1361-5, Seocho-dong, Seocho-gu, Seoul; tel. (2) 3472-8151; fax (2) 563-9319; f. 1955; literature; Publr KIM SUNG-SIK; circ. 200,000.

Korea Business World: Yeouido, POB 720, Seoul 150-607; tel. (2) 532-1364; fax (2) 594-7663; f. 1985; monthly; English; Publr and Pres. LEE KIE-HONG; circ. 40,200.

Korea Buyers Guide: Rm 2301, Korea World Trade Center, 159, Samseong-dong, Gangnam-gu, Seoul; tel. (2) 551-2376; fax (2) 551-2377; e-mail info@buyersguide.co.kr; internet www.buykorea21.com; f. 1973; monthly, consumer goods; quarterly, hardware; Pres. YOU YOUNG-PYO; circ. 30,000.

Korea Journal: CPOB 54, Seoul 100-022; tel. (2) 776-2804; organ of the UNESCO Korean Commission; Gen. Dir CHUNG HEE-CHAE.

Korea Newsreview: 1-12, 3-ga, Hoehyeon-dong, Jung-gu, Seoul 100-771; tel. (2) 756-7711; weekly; English; Publr and Editor PARK CHUNG-WOONG.

Korea and World Affairs: Rm 1723, Daewoo Center Bldg, 5-541, Namdaemun-no, Jung-gu, Seoul 100-714; tel. (2) 777-2628; fax (2) 319-9591; organ of the Research Center for Peace and Unification of Korea; Pres. CHANG DONG-HOON.

Korean Business Review: FKI Bldg, 28-1, Yeouido-dong, Yeongdeungpo-gu, Seoul 150-756; tel. (2) 3771-0114; fax (2) 3771-0138; monthly; publ. by Fed. of Korean Industries; Publr KIM KAK-CHOONG; Editor SOHN BYUNG-DOO.

Literature and Thought: Seoul; tel. (2) 738-0542; fax (2) 738-2997; f. 1972; monthly; Pres. LIM HONG-BIN; circ. 10,000.

Monthly Travel: Cross Bldg, 2nd Floor, 46-6, 2-ga, Namsan-dong, Jung-gu, Seoul 100-042; tel. (2) 757-6161; fax (2) 757-6089; e-mail kotfa@unitel.co.kr; Pres. SHIN JOONG-MOK; circ. 50,000.

News Maker: 22, Jung-dong, Jung-gu, Seoul 110-702; tel. (2) 3701-1114; fax (2) 739-6190; e-mail hudy@kyunghyang.com; internet www.kyunghyang.com/newsmaker; f. 1992; Pres. JANG JUN-BONG; Editor PARK MYUNG-HUN.

Reader's Digest: 295-15, Deoksan 1-dong, Geumcheon-gu, Seoul 153-011; tel. (2) 866-8800; fax (2) 839-4545; f. 1978; monthly; general; Pres. YANG SUNG-MO; Editor PARK SOON-HWANG; circ. 115,000.

Shin Dong-A (New East Asia): 139, Chungjeong-no, Seodaemun-gu, Seoul 120–715; tel. (2) 361-0974; fax (2) 361-0988; f. 1931; monthly; general; Publr KIM HAK-JUN; Editor YOU YOUNG-EUL; circ. 170,000.

Taekwondo: Sinmun-no Bldg, 5th Floor, 238, Sinmun-no, 1-ga, Jongno-gu, Seoul 110-061; tel. (2) 566-2505; fax (2) 553-4728; e-mail wtf@unitel.co.kr; internet www.wtf.org; f. 1973; organ of the World Taekwondo Fed.; Pres. Dr KIM UN-YONG.

Vantage Point: 85-1, Susong-dong, Jongno-gu, Seoul, 110-140; tel. (2) 398-3519; fax (2) 398-3539; e-mail kseungji@yna.co.kr; internet www.yna.co.kr; f. 1978; monthly; developments in North Korea; Editor KWAK SEUNG-JI.

Weekly Chosun: 61, Taepyeong-no 1, Jung-gu, Seoul; tel. (2) 724-5114; fax (2) 724-6199; weekly; Publr BANG SANG-HOON; Editor CHOI JOON-MYONG; circ. 350,000.

The Weekly Hankook: 14, Junghak-dong, Jongno-gu, Seoul; tel. (2) 732-4151; fax (2) 724-2444; f. 1964; Publr CHANG CHAE-KUK; circ. 400,000.

Wolgan Mot: 139, Sejong-no, Jongno-gu, Seoul 110; tel. (2) 733-5221; f. 1984; monthly; fashion; Publr KIM SEUNG-YUL; Editor KWON O-KIE; circ. 120,000.

Women's Weekly: 14, Junghak-dong, Jongno-gu, Seoul; tel. (2) 735-9216; fax (2) 732-4125.

Yosong Dong-A (Women's Far East): 139, Sejong-no, Jongno-gu, Seoul 110-715; tel. (2) 721-7621; fax (2) 721-7676; f. 1933; monthly; women's magazine; Publr KIM BYUNG-KWAN; Editor KWON O-KIE; circ. 237,000.

NEWS AGENCIES

Yonhap News Agency: 85-1, Susong-dong, Jongno-gu, Seoul; tel. (2) 398-3114; fax (2) 398-3257; internet www.yonhapnews.co.kr; f. 1980; Pres. KIM KUN.

Foreign Bureaux

Agence France-Presse (AFP): Yonhap News Agency Bldg, 3rd Floor, 85-1, Susong-dong, Jongno-gu, Seoul; tel. (2) 737-7353; fax (2) 737-6598; e-mail seoul@afp.com; Bureau Chief TIM WITCHER.

Associated Press (AP) (USA): Yonhap News Agency Bldg, 85-1, Susong-dong, Jongno-gu, Seoul; tel. (2) 739-0692; fax (2) 737-0650; Bureau Chief REID MILLER.

Central News Agency (Taiwan): 33-1, 2-ga, Myeong-dong, Jung-gu, Seoul; tel. (2) 753-0195; fax (2) 753-0197; Bureau Chief CHIANG YUAN-CHEN.

Deutsche Presse-Agentur (Germany): 148, Anguk-dong, Jongno-gu, Seoul; tel. (2) 738-3808; fax (2) 738-6040; Correspondent NIKOLAUS PREDE.

Informatsionnoye Telegrafnoye Agentstvo Rossii—Telegrafnoye Agentstvo Suverennykh Stran (ITAR—TASS) (Russia): 1-302, Chonghwa, 22-2, Itaewon-dong, Yeongsan-gu, Seoul; tel. (2) 796-9193; fax (2) 796-9194.

Jiji Tsushin (Jiji Press) (Japan): Joong-ang Ilbo Bldg, 7, Soonhwa-dong, Jung-gu, Seoul; tel. (2) 753-4525; fax (2) 753-8067; Chief Correspondent KENJIRO TSUJITA.

Kyodo News Service (Japan): Yonhap News Agency Bldg, 85-1, Susong-dong, Jongno-gu, Seoul; tel. (2) 739-2791; fax (2) 737-1776; Bureau Chief HISASHI HIRAI.

Reuters (UK): Byuck San Bldg, 7th Floor, 12-5, Dongja-dong, Yeongsan-gu, Seoul 140-170; tel. (2) 727-5151; fax (2) 727-5666; Bureau Chief ANDREW BROWNE.

Rossiiskoye Informatsionnoye Agentstvo—Novosti (RIA—Novosti) (Russia): 14, Junghak-dong, Jongno-gu, Seoul; tel. (2) 737-2829; fax (2) 798-0010; Correspondent SERGEI KUDASOV.

United Press International (UPI) (USA): Yonhap News Agency Bldg, Rm 603, 85-1, Susong-dong, Jongno-gu, Seoul; tel. (2) 737-9054; fax (2) 738-8206; Correspondent JASON NEELY.

Xinhua News Agency (People's Republic of China): B-1, Hillside Villa, 726-111, Hannam-dong, Yeongsan-gu, Seoul; tel. (2) 795-8258; fax (2) 796-7459.

PRESS ASSOCIATIONS

Korean Newspaper Editors' Association: Korea Press Center, 13th Floor, 25, 1-ga, Taepyeong-no, Jung-gu, Seoul; tel. (2) 732-1726; fax (2) 739-1985; f. 1957; 416 mems; Pres. SEONG BYONG-WUK.

Korean Newspapers Association: Korea Press Center, 13th Floor, 25, 1-ga, Taepyeong-no, Jung-gu, Seoul 100-745; tel. (2) 733-2251; fax (2) 720-3291; e-mail ccy73_2000@yahoo.co.kr; f. 1962; 49 mems; Pres. HONG SEOK-HYUN.

Seoul Foreign Correspondents' Club: Korea Press Center, 18th Floor, 25, 1-ga, Taepyeong-no, Jung-gu, Seoul; tel. (2) 734-3272; fax (2) 734-7712; f. 1956; Pres. PARK HAN-CHUN.

Publishers

Ahn Graphics Ltd: 260-88, Seongbuk 2-dong, Seongbuk-gu, Seoul 136-012; tel. (2) 763-2320; fax (2) 743-3352; e-mail lbr@ag.co.kr; f. 1985; computer graphics; Pres. KIM OK-CHUL.

Bak-Young Publishing Co: 13-31, Pyeong-dong, Jongno-gu, Seoul; tel. (2) 733-6771; fax (2) 736-4818; f. 1952; sociology, philosophy, literature, linguistics, social science; Pres. AHN JONG-MAN.

BIR Publishing Co Ltd: 506, Sinsa-dong, Gangnam-gu, Seoul 135-120; tel. (2) 515-2000; fax (2) 514-3249.

Bobmun Sa Publishing Co: Hanchung Bldg, 4th Floor, 161-7, Yomni-dong, Mapo-gu, Seoul 121-090; tel. (2) 703-6541; fax (2) 703-6594; f. 1954; law, politics, philosophy, history; Pres. BAE HYO-SEON.

Bumwoo Publishing Co: 21-1, Kusu-dong, Mapo-gu, Seoul 121-130; tel. (2) 717-2121; fax (2) 717-0429; f. 1966; philosophy, religion, social science, technology, art, literature, history; Pres. YOON HYUNG-DOO.

Cheong Moon Gak Publishing Co Ltd: 486-9, Kirum 3-dong, Seongbuk-gu, Seoul 136-800; tel. (2) 985-1451; fax (2) 988-1456; e-mail cmgbook@cmgbook.co.kr; internet www.cmgbook.co.kr; f. 1974; science, technology, business; subsidiaries HanSeung Publishers, Lux Media; Pres. KIM HONG-SEOK; Man. Dir HANS KIM.

Design House Publishing Co: Paradise Bldg, 186-210, Jangchung-dong, 2-ga, Jung-gu, Seoul 100-392; tel. (2) 2275-6151; fax (2)

2275-7884; f. 1987; social science, art, literature, languages, children's periodicals; Pres. LEE YOUNG-HEE.

Dong-Hwa Publishing Co: 130-4, 1-ga, Wonhyoro, Yeongsan-gu, Seoul 140-111; tel. (2) 713-5411; fax (2) 701-7041; f. 1968; language, literature, fine arts, history, religion, philosophy; Pres. LIM IN-KYU.

Doosan Co-operation Publishing BG: 18-12, Ulchi-ro, 6-ga, Jeong-gu, Seoul 100-196; tel. (2) 3398-880; fax (2) 3398-2670; f. 1951; general works, school reference, social science, periodicals; Pres. CHOI TAE-KYUNG.

Eulyoo Publishing Co Ltd: 46-1, Susong-dong, Jongno-gu, Seoul 110-603; tel. (2) 733-8151; fax (2) 732-9154; e-mail eulyoo@chollian.net; internet www.eulyoo.co.kr; f. 1945; linguistics, literature, social science, history, philosophy; Pres. CHUNG CHIN-SOOK.

Hainaim Publishing Co Ltd: Minjin Bldg, 5th Floor, 464-41, Seokyo-dong, Mapo-gu, Seoul 121-210; tel. (2) 326-1600; fax (2) 333-7543; e-mail hainaim@chollian.net; f. 1983; philosophy, literature, children's; Pres. SONG YOUNG-SUK.

Hakwon Publishing Co Ltd: Seocho Plaza, 4th Floor, 1573-1, Seocho-dong, Seocho-gu, Seoul; tel. (2) 587-2396; fax (2) 584-9306; f. 1945; general, languages, literature, periodicals; Pres. KIM YOUNG-SU.

Hangil Publishing Co: 506, Sinsa-dong, Gangnam-gu, Seoul 135-120; tel. (2) 515-4811; fax (2) 515-4816; f. 1976; social science, history, literature; Pres. KIM EOUN-HO.

Hanul Publishing Company: 201, Hyuam Bldg, 503-24, Changcheon-dong, Seodaemun-gu, Seoul 120-180; tel. (2) 336-6183; fax (2) 333-7543; e-mail newhanul@nuri.net; f. 1980; general, philosophy, university books, periodicals; Pres. KIM CHONG-SU.

Hollym Corporation: 13-13, Gwancheol-dong, Jongno-gu, Seoul 110-111; tel. (2) 735-7551-4; fax (2) 730-5149; e-mail hollym@chollian.net; internet www.hollym.co.kr; f. 1963; academic and general books on Korea in English; Pres. HAM KI-MAN.

Hyang Mun Sa Publishing Co: 645-20, Yeoksam-dong, Gangnam-gu, Seoul 135-081; tel. (2) 538-5672; fax (2) 538-5673; f. 1950; science, agriculture, history, engineering, home economics; Pres. NAH JOONG-RYOL.

Hyonam Publishing Co Ltd: 627-5, Ahyun 3-dong, Mapo-gu, Seoul 121-013; tel. (2) 365-5056; fax (2) 365-5251; e-mail lawhyun@chollian.net; f. 1951; general, children's, literature, periodicals; Pres. CHO KEUN-TAE.

Il Ji Sa Publishing Co: 46-1, Junghak-dong, Jongno-gu, Seoul 110-150; tel. (2) 732-3980; fax (2) 722-2807; f. 1956; literature, social sciences, juvenile, fine arts, philosophy, linguistics, history; Pres. KIM SUNG-JAE.

Ilchokak Publishing Co Ltd: 9, Gongpyeong-dong, Jongno-gu, Seoul 110-160; tel. (2) 733-5430; fax (2) 738-5857; f. 1953; history, literature, sociology, linguistics, medicine, law, engineering; Pres. HAN MAN-NYUN.

Jigyungsa Publishers Ltd: 790-14, Yeoksam-dong, Gangnam-gu, Seoul 135-080; tel. (2) 557-6351; fax (2) 557-6352; e-mail jigyung@uriel.net; internet www.jigyung.co.kr; f. 1979; children's, periodicals; Pres. KIM BYUNG-JOON.

Jihak Publishing Co Ltd: 180-20, Dongkyo-dong, Mapo-gu, Seoul 121-200; tel. (2) 330-5220; fax (2) 325-5835; f. 1965; philosophy, language, literature; Pres. KWON BYONG-IL.

Jipmoondang: 95, Waryon-dong, Jongno-gu, Seoul 110-360; tel. (2) 743-3098; fax (2) 743-3192; philosophy, social science, Korean studies, history, Korean folklore; Pres. LIM KYOUNG-HWAN.

Jisik Sanup Publications Co Ltd: 35-18, Dongui-dong, Jongno-gu, Seoul 110-040; tel. (2) 738-1978; fax (2) 720-7900; f. 1969; religion, social science, art, literature, history, children's; Pres. KIM KYUNG-HEE.

Jung-Ang Publishing Co Ltd: 172-11, Yomni-dong, Mapo-gu, Seoul 121-090; tel. (2) 717-2111; fax (2) 716-1369; f. 1972; study books, children's; Pres. KIM DUCK-KI.

Kemongsa Publishing Co Ltd: 772, Yeoksam-dong, Gangnam-gu, Seoul 135-080; tel. (2) 531-5335; fax (2) 531-5520; f. 1946; picture books, juvenile, encyclopaedias, history, fiction; Pres. RHU SEUNG-HEE.

Ki Moon Dang: 286-20, Haengdang-dong, Seongdong-gu, Seoul 133-070; tel. (2) 2295-6171; fax (2) 2296-8188; f. 1976; engineering, fine arts, dictionaries; Pres. KANG HAE-JAK.

Korea Britannica Corpn: 117, 1-ga, Jungchung-dong, Seoul 100-391; tel. (2) 272-2151; fax (2) 278-9983; f. 1968; encyclopaedias, dictionaries; Pres. JANG HO-SANG, SUJAN ELEN TAPANI.

Korea University Press: 5-1, Anam-dong, 5-ga, Seongbuk-gu, Seoul 136-701; tel. (2) 3290-4231; fax (2) 923-6311; e-mail kupress@korea.ac.uk; internet www.korea.ac.kr/~kupress; f. 1956; philosophy, history, language, literature, Korean studies, education, psychology, social science, natural science, engineering, agriculture, medicine; Pres. EUH YOON-DAE.

Kum Sung Publishing Co: 242-63, Gongdeok-dong, Mapo-gu, Seoul 121-022; tel. (2) 713-9651; fax (2) 718-4362; f. 1965; literature, juvenile, social sciences, history, fine arts; Pres. KIM NAK-JOON.

Kyohak-sa Publishing Co Ltd: 105-67, Gongdeok-dong, Mapo-gu, Seoul 121-020; tel. (2) 717-4561; fax (2) 718-3976; f. 1952; dictionaries, educational, children's; Pres. YANG CHEOL-WOO.

Kyung Hee University Press: 1, Hoeki-dong, Dongdaemun-gu, Seoul 130-701; tel. (2) 961-0106; fax (2) 962-8840; f. 1960; general, social science, technology, language, literature; Pres. CHOE YOUNG-SEEK.

Kyungnam University Press: 28-42, Samchung-dong, Jongno-gu, Seoul 110-230; tel. (2) 370-0700; fax (2) 735-4359; Pres. PARK JAE-KYU.

Minumsa Publishing Co Ltd: 5/F Kangnam Publishing Culture Centre, 506, Sinsa-dong, Gangnam-gu, Seoul 135-120; tel. (2) 515-2000; fax (2) 515-2007; e-mail michell@bora.dacom.co.kr; f. 1966; literature, philosophy, linguistics, pure science; Pres. PARK MAENG-HO.

Munhakdongne Publishing Co Ltd: 6/F Dongsomun B/D 260, Dongsomundong 4-ga, Seongbuk-gu, Seoul 136-034; tel. (2) 927-6790; fax (2) 927-6793; e-mail etepluie@hotmail.com; internet www.munhak.com; f. 1993; art, literature, science, philosophy, non-fiction, children's, periodicals; Pres. KANG BYUNG-SUN.

Panmun Book Co Ltd: 923-11, Mok 1-dong, Yangcheon-gu, Seoul 158-051; tel. (2) 653-5131; fax (2) 653-2454; e-mail skliu@panmun.co.kr; internet www.medicalplus.co.kr; f. 1955; social science, pure science, technology, medicine, linguistics; Pres. LIU SUNG-KWON.

Sakyejul Publishing Ltd: 1-181, Sinmun-no-2-ga, Jongno-gu, Seoul 110-062; tel. (2) 736-9380; fax (2) 737-8595; e-mail sakyejul@soback.kornet.nm.kr; f. 1982; social sciences, art, literature, history, children's; Pres. KANG MAR-XILL.

Sam Joong Dang Publishing Co: 261-23, Soke-dong, Yeongsan-gu, Seoul 140-140; tel. (2) 704-6816; fax (2) 704-6819; f. 1931; literature, history, philosophy, social sciences, dictionaries; Pres. LEE MIN-CHUL.

Sam Seong Dang Publishing Co: 101-14, Non Hyun-dong, Gangnam-gu, Seoul 135-010; tel. (2) 3442-6767; fax (2) 3442-6768; e-mail kyk@ssdp.co.kr; f. 1968; literature, fine arts, history, philosophy; Pres. KANG MYUNG-CHAE.

Sam Seong Publishing Co Ltd: 1516-2, Seocho-dong, Seocho-gu, Seoul 137-070; tel. (2) 3470-6900; fax (2) 597-1507; internet www.howpc.com; f. 1951; literature, history, juvenile, philosophy, arts, religion, science, encyclopaedias; Pres. KIM JIN-YONG.

Samsung Publishing Co Ltd: Seocho-dong, Seocho-gu, Seoul 137-871; tel. (2) 3470-6900; fax (2) 521-8534; e-mail lisababy@ssbooks.com; internet www.ssbooks.com; www.samsungbooks.com; children's books, comics, cooking, parenting, health, travel; f. 1951; Chief Editor BOSUNG KONG.

Segyesa Publishing Co Ltd: Dasan Bldg 102, 494-85, Yeongkan-dong, Mapo-gu, Seoul 121-070; tel. (2) 715 1542; fax (2) 715-1544; f. 1988; general, philosophy, literature, periodicals; Pres. CHOI SUN-HO.

Se-Kwang Music Publishing Co: 232-32, Seogye-dong, Yeongsan-gu, Seoul 140-140; tel. (2) 719-2652; fax (2) 719-2656; f. 1953; music, art; Pres. PARK SEI-WON; Chair. PARK SHIN-JOON.

Seong An Dang Publishing Co: 4579, Singil-6-dong, Yeongdeungpo-gu, Seoul 150-056; tel. (2) 3142-4151; fax (2) 323-5324; f. 1972; technology, text books, university books, periodicals; Pres. LEE JONG-CHOON.

Seoul National University Press: 56-1, Sinrim-dong, Gwanak-gu, Seoul 151-742; tel. (2) 889-0434; fax (2) 888-4148; e-mail snubook@chollian.net; f. 1961; philosophy, engineering, social science, art, literature; Pres. LEE KI-JUN.

Si-sa-young-o-sa, Inc: 55-1, 2-ga, Jongno, Jongno-gu, Seoul 110-122; tel. (2) 274-0509; fax (2) 271-3980; internet www.ybmsisa.co.kr; f. 1959; language, literature; Pres. CHUNG YOUNG-SAM.

Sogang University Press: 1, Sinsu-dong, Mapo-gu, Seoul 121-742; tel. (2) 705-8212; fax (2) 705-8612; f. 1978; philosophy, religion, science, art, history; Pres. LEE HAN-TAEK.

Sookmyung Women's University Press: 53-12, 2-ga, Jongpa-dong, Yeongsan-gu, Seoul 140-742; tel. (2) 710-9162; fax (2) 710-9090; f. 1968; general; Pres. LEE KYUNG-SOOK.

Tam Gu Dang Publishing Co: 158, 1-ga, Hanggangno, Yeongsan-gu, Seoul 140-011; tel. (2) 3785-2271; fax (2) 3785-2272; f. 1950; linguistics, literature, social sciences, history, fine arts; Pres. HONG SUK-WOO.

Tong Moon Gwan: 147, Gwanhoon-dong, Jongno-gu, Seoul 110-300; tel. (2) 732-4355; f. 1954; literature, art, philosophy, religion, history; Pres. LEE KYUM-NO.

Woongjin.com Co. Ltd: Woongjin Bldg, 112-2, Inui-dong, Jongno-gu, Seoul; tel. (2) 3670-1832; fax (2) 766-2722; e-mail lois.kim@email.woongjin.com; internet www.woongjin.com; children's; Pres. YOON SUCK-KEUM.

Yearimdang Publishing Co Ltd: Yearim Bldg, 153-3, Samseong-dong, Gangnam-gu, Seoul 135-090; tel. (2) 566-1004; fax (2) 567-9610; e-mail yearim@yearim.co.kr; internet www.yearim.co.kr; f. 1973; children's; Pres. NA CHOON-HO.

Yonsei University Press: 134, Sincheon-dong, Seodaemun-gu, Seoul 120-749; tel. (2) 361-3380; fax (2) 393-1421; e-mail ysup@yonsei.ac.kr; f. 1955; philosophy, religion, literature, history, art, social science, pure science; Pres. KIM BYUNG-SOO.

Youl Hwa Dang: 506, Sinsa-dong, Gangnam-gu, Seoul 135-120; tel. (2) 515-3141; fax (2) 515-3144; e-mail horang2@unitel.co.kr; f. 1971; art; Pres. YI KI-UNG.

PUBLISHERS' ASSOCIATION

Korean Publishers' Association: 105-2, Sagan-dong, Jongno-gu, Seoul 110-190; tel. (2) 735-2702; fax (2) 738-5414; e-mail kpa@kpa21.or.kr; internet www.kpa21.or.kr; f. 1947; Pres. LEE JUNG-IL; Sec.-Gen. JUNG JONG-JIN.

Broadcasting and Communications

TELECOMMUNICATIONS

Dacom Corpn: Dacom Bldg, 706-1, Yeoksam-dong, Gangnam-gu, Seoul 135-610; tel. (2) 6220-0220; fax (2) 6220-0702; internet www.dacom.net; f. 1982; domestic and international long-distance telecommunications services and broadband internet services; CEO PARK UN-SUH.

Daewoo Telecom Co Ltd: 14-34, Yeouido-dong, Yeongdeungpo-gu, Seoul; tel. (2) 3779-7114; fax (2) 3779-7500; internet www.dwt.co.kr; Pres. (vacant).

Hanaro Telecom: Kukje Electronics Center Bldg, 24th Floor, 1445-3, Seocho-dong, Seocho-gu, Seoul 137-728; tel. (2) 6266-4114; fax (2) 6266-4379; internet www.hanaro.com; local telecommunications and broadband internet services; Pres. and CEO YOON CHANG-BUN.

Korea Mobile Telecommunications Corpn: 267, 5-ga, Namdaemun-no, Jung-gu, Seoul; tel. (2) 3709-1114; fax (2) 3709-0499; f. 1984; Pres. SEO JUNG-UK.

Korea Telecom: 206 Jungja-dong, Bundang-gu, Seongnam-si, Gyeonggi Prov. 463-711; tel. (2) 727-0114; fax (2) 750-3994; internet www.kt.co.kr; domestic and international telecommunications services and broadband internet services; privatized in June 2002; Pres. LEE YONG-KYUNG.

Korea Telecom (KT) Freetel: Seoul; internet www.ktf.co.kr; subsidiary of Korea Telecom; 10m. subscribers (2002); CEO JOONG SOO-NAM.

LG Telecom: LG Gangnam Tower, 19th Floor, 679 Yeoksam-dong, Gangnam-gu, Seoul 135-985; tel. (2) 2005-7114; fax (2) 2005-7505; internet www.lg019.co.kr; mobile telecommunications and wireless internet services; commenced commercial CDMA2000 1x service in May 2001; 4m. subscribers (2002); CEO NAM YONG.

Onse Telecom: 192-2, Gumi-dong, Bundang-gu, Seongnam-si, Gyeonggi Prov. 463-500; tel. and fax (31) 738-6000; internet www.onse.net; domestic and international telecommunications services; Pres. and CEO HWANG KEE-YEON.

SK Telecom Co Ltd: 99, Seorin-dong, Jongno-gu, Seoul 110-110; tel. (2) 2121-2114; fax (2) 2121-3999; internet www.sktelecom.com; cellular mobile telecommunications and wireless internet services; merged with Shinsegi Telecom in Jan. 2002; 16m. subscribers (2002); Chair. and CEO (vacant).

BROADCASTING

Regulatory Authority

Korean Broadcasting Commission: KBS Bldg, 923-5, Mok-dong, Yangcheon-gu, Seoul 158-715; tel. (2) 3219-5117; fax (2) 3219-5371; Chair. KANG DAE-IN.

Radio

Korean Broadcasting System (KBS): 18, Yeouido-dong, Yeongdeungpo-gu, Seoul 150-010; tel. (2) 781-1000; fax (2) 781-4179; internet www.kbs.co.kr; f. 1926; publicly-owned corpn with 26 local broadcasting and 855 relay stations; overseas service in Korean, English, German, Indonesian, Chinese, Japanese, French, Spanish, Russian and Arabic; Pres. JUNG YUN-JOO.

Buddhist Broadcasting System (BBS): 140, Mapo-dong, Mapo-gu, Seoul 121-050; tel. (2) 705-5114; fax (2) 705-5229; internet www.bbsfm.ko.kr; f. 1990; Pres. CHO HAE-HYONG.

Christian Broadcasting System (CBS): 917-1, Mok-dong, Yangcheon-gu, Seoul 158-701; tel. (2) 650-7000; fax (2) 654-2456; internet www.cbs.co.kr; f. 1954; independent religious network with seven network stations in Seoul, Daegu, Busan, Gwangju, Chonbuk, Jeonju and Chuncheon; programmes in Korean; Pres. Rev. KWON HO-KYUNG.

Educational Broadcasting System (EBS): 92-6, Umyeon-dong, Seocho-gu, Seoul 137-791; tel. (2) 526-2000; fax (2) 526-2179; internet www.ebs.co.kr; f. 1990; Pres. Dr PARK HEUNG-SOO.

Far East Broadcasting Co (FEBC): 89, Sangsu-dong, Mapo-gu, MPO Box 88, Seoul 121-707; tel. (2) 320-0114; fax (2) 320-0129; e-mail febcadm@febc.net; internet www.febc.net; Dir Dr BILLY KIM.

Radio Station HLAZ: MPO Box 88, Seoul 121-707; tel. (2) 320-0114; fax (2) 320-0129; e-mail febcadm@febc.net; internet www.febc.net; f. 1973; religious, educational service operated by Far East Broadcasting Co; programmes in Korean, Chinese, Russian and Japanese; Dir Dr BILLY KIM.

Radio Station HLKX: MPO Box 88, Seoul 121-707; tel. (2) 320-0114; fax (2) 320-0129; e-mail febcadm@febc.net; internet www.febc.net; f. 1956; religious, educational service operated by Far East Broadcasting Co; programmes in Korean, Chinese and English; Pres. Dr BILLY KIM.

Munhwa Broadcasting Corpn (MBC): 31, Yeouido-dong, Yeongdeungpo-gu, Seoul 150-728; tel. (2) 784-2000; fax (2) 784-0880; e-mail mbcir@imbc.com; internet www.imbc.com; f. 1961; public; Pres. KIM JOONG-BAE.

Pyong Hwa Broadcasting Corpn (PBC): 2-3, 1-ga, Jeo-dong, Jung-gu, Seoul 100-031; tel. (2) 270-2114; fax (2) 270-2210; internet www.pbc.co.kr; f. 1990; religious and educational programmes; Pres. Rev. PARK SHIN-EON.

Seoul Broadcasting System (SBS): 10-2, Yeouido-dong, Yeongdeungpo-gu, Seoul 150-010; tel. (2) 786-0792; fax (2) 780-2530; internet www.sbs.co.kr; f. 1991; Pres. SONG DO-KYUN.

US Forces Network Korea (AFN Korea): Seoul; tel. (2) 7914-6495; fax (2) 7914-5870; e-mail info@afnkorea.com; internet afnkorea.com; f. 1950; six originating stations and 19 relay stations; 24 hours a day.

Television

In late 1997 almost 40 domestic television channels were in operation.

Educational Broadcasting System (EBS): see Radio.

Inchon Television Ltd (ITV): 587-46, Hakik-dong, Nam-gu, Incheon; tel. (32) 830-1000; fax (32) 865-6300; internet www.itv.co.kr; f. 1997.

Jeonju Television Corpn (JTV): 656-3, Sonosong-dong, Deokjin-gu, Jeonju, N. Jeolla Prov.; tel. (63) 250-5231; fax (63) 250-5249; e-mail jtv@jtv.co.kr; f. 1997.

Korean Broadcasting System (KBS): 18, Yeouido-dong, Yeongdeungpo-gu, Seoul 150-790; tel. (2) 781-1000; fax (2) 781-4179; f. 1961; publicly-owned corpn with 25 local broadcasting and 770 relay stations; Pres. JUNG YUN-JOO.

Munhwa Broadcasting Corpn (MBC-R/TV): 31, Yeouido-dong, Yeongdeungpo-gu, Seoul 150-728; tel. (2) 789-2851; fax (2) 782-3094; e-mail song@mbc.co.kr; internet www.imbc.com; f. 1961; public; 19 TV networks; Pres. LEE KEUNG-HEE.

Seoul Broadcasting System (SBS): see Radio.

US Forces Network Korea (AFN Korea): Seoul; tel. (2) 7914-2711; fax (2) 7914-5870; f. 1950; main transmitting station in Seoul; 19 rebroadcast transmitters and translators; 168 hours weekly.

Finance

(cap. = capital; res = reserves; dep. = deposits; m. = million; brs = branches; amounts in won, unless otherwise indicated)

BANKING

The modern financial system in South Korea was established in 1950 with the foundation of the central bank, the Bank of Korea. Under financial liberalization legislation, adopted in the late 1980s, banks were accorded greater freedom to engage in securities or insurance operations. In 2003 there were 59 commercial banks in South Korea, comprising eight nation-wide banks, six regional commercial banks, five specialized banks and 40 branches of foreign banks. The Financial Supervisory Service oversees the operations of commercial banks and the financial services sector.

Specialized banks were created in the 1960s to provide funds for sectors of the economy not covered by commercial banks. There are also two development banks: the Korea Development Bank and the Export-Import Bank of Korea.

In late 1997 many merchant banks were forced to cease operations, after incurring heavy losses through corporate bankruptcies. In June 1998 five commercial banks were also required to cease operations.

Regulatory Authority

Financial Supervisory Service: 27, Yeouido-dong, Yeongdeungpo-gu, Seoul 150-743; tel. (2) 3771-5000; fax (2) 785-3475; internet www.fss.or.kr; Gov. YOON JEUNG-HYUN.

Central Bank

Bank of Korea: 110, 3-ga, Namdaemun-no, Jung-gu, Seoul 100-794; tel. (2) 759-4114; fax (2) 759-4139; e-mail bokdiri@bok.or.kr; internet www.bok.or.kr; f. 1950; bank of issue; res 5,456,500m., dep. 139,210,300m. (Dec. 2002); Gov. PARK SEUNG; Dep. Gov. LEE SEONG-TAE; 16 domestic brs, 7 overseas offices.

Commercial Banks

Chohung Bank: 14, 1-ga, Namdaemun-no, Jung-gu, Seoul 100-757; tel. (2) 733-2000; fax (2) 723-6473; internet www.chb.co.kr; f. 1897; merged with Chungbuk Bank in May 1999 and Kangwon Bank in Sept. 1999; 80.4% owned by Shinhan Financial Group; cap. 3,395,592m., dep. 45,125,839m. (Dec. 2002); Chair. SUNG BOK-WEE; Dir HONG CHIL-SUN; 446 domestic brs, 11 overseas brs.

Hana Bank: 101-1, 1-ga, Ulchi-no, Jung-gu, Seoul 100-191; tel. (2) 2002-1111; fax (2) 775-7472; e-mail webmaster@hanabank.com; internet www.hanabank.co.kr; f. 1991; merged with Boram Bank in Jan. 1999; merged with Seoulbank in Dec. 2002; cap. 987,161m., res 877,633m., dep. 60,814,076m. (Dec. 2002); Chair. and CEO KIM SEUNG-YU; 303 brs.

Kookmin Bank: 9-1, 2-ga, Namdaemun-no, Jung-gu, CPOB 815, Seoul 100-703; tel. (2) 317-2891; fax (2) 317-2885; e-mail corres@kookminbank.com; internet www.kookminbank.com; f. 1963 as Citizen's National Bank, renamed 1995; re-established Jan. 1999, following merger with Korea Long Term Credit Bank; merged with H&CB in November 2001; cap. 1,641,293m., res 6,251,573m., dep. 150,341.3m. (Dec. 2002); Chair. KIM SANG-HOON; Pres. and CEO (vacant); 1,122 domestic brs, 6 overseas brs.

KorAm Bank: 39, Da-dong, Jung-gu, Seoul 100-180; tel. (2) 3455-2114; fax (2) 3455-2966; e-mail shk@goodbank.com; internet www.goodbank.com; f. 1983; acquired by CitiGroup in 2004; cap. 1,093,334m., res 227,185m., dep. 25,760,831m. (Dec. 2002); CEO and Chair. HA YUNG-KU; 227 domestic brs, 4 overseas brs.

Korea Exchange Bank: 181, 2-ga, Ulchi-no, Jung-gu, Seoul 100-793; tel. (2) 729-0114; fax (2) 775-2565; internet www.keb.co.kr; f. 1967; merged with Korea International Merchant Bank in Jan. 1999; cap. 1,850,875m., res 92,637m., dep. 42,545,293m. (Dec. 2002); Chair. CHUNG MOON-SOO; Pres. LEE KANG-WON; 269 domestic brs.

Korea First Bank: 100, Gongpyeong-dong, Jongno-gu, Seoul 110-702; tel. (2) 3702-3114; fax (2) 3702-4934; e-mail master@kfb.co.kr; internet www.kfb.co.kr; f. 1929; 49% owned by Newbridge Capital Ltd (USA), 51% govt-owned; cap. 980,584m., res 149,838m., dep. 25,427,255m. (Dec. 2002); Pres. and CEO ROBERT COHEN; 389 domestic brs, 2 overseas brs.

Shinhan Bank: 120, 2-ga, Taepyeong-no, Jung-gu, Seoul 100-102; tel. (2) 756-0505; fax (2) 774-7013; e-mail corres@shinhan.com; internet www.shinhan.com; f. 1982; cap. 1,027,305m., res 858,703,694m., dep. 42,893,694m. (Dec. 2003); Pres. and CEO SHIN SANG-HOON; 356 domestic brs, 9 overseas brs.

Woori Bank: 203, 1-ga, Hoehyeon-dong, Jung-gu, Seoul; tel. (2) 2002-3000; fax (2) 2002-5687; internet www.wooribank.com; f. 2002; following the merger of Hanvit Bank and Peace Bank of Korea; 100% government-owned; cap. 2,764,400m., res 1,378,457m., dep. 68,126,751m. (Dec. 2002); Pres. LEE DUK-HOON; 668 domestic brs.

Development Banks

Export-Import Bank of Korea: 16-1, Yeouido-dong, Yeongdeungpo-gu, Seoul 150-873; tel. (2) 3779-6114; fax (2) 3779-6732; e-mail kexim@koreaexim.go.kr; internet www.koreaexim.go.kr; f. 1976; cap. 2,725,755m., res 83,884m. (Dec. 2002); Chair. and Pres. SHIN DONG-KYU; 8 brs.

Korea Development Bank: 16-3, Yeouido-dong, Yeongdeungpo-gu, Seoul 150-793; tel. (2) 787-4000; fax (2) 787-6191; internet www.kdb.co.kr; f. 1954; cap. 7,161,861m., dep. 42,345,324m. (Dec. 2002); Gov. JUNG KEUN-YONG; 32 domestic brs, 5 overseas brs.

Specialized Banks

Industrial Bank of Korea: 50, 2-ga, Ulchi-no, Jung-gu, Seoul 100-758; tel. (2) 729-6114; fax (2) 729-6402; e-mail ifd@ibk.co.kr; internet www.ibk.co.kr; f. 1961 as the Small and Medium Industry Bank; 85.5% govt-owned; cap. 2,291,385m., res 1,090,736m., dep. 42,737,135m. (Dec. 2002); Chair. and Pres. KIM JONG-CHANG; 387 domestic brs, 6 overseas brs.

Korean-French Banking Corpn (SogeKo): Marine Center, 118, 2-ga, Namdaemun-no, Jung-gu, CPOB 8572, Seoul 100-092; tel. (2) 777-7711; fax (2) 756-0464; f. 1977; cap. 130,000m., res 6,631m., dep. 255,359m. (Dec. 2002); Pres. KIM DOO-BAE.

National Agricultural Co-operative Federation (NACF): 75, 1-ga, Chungjeong-no, Jung-gu, Seoul 100-707; tel. (2) 397-5114; fax (2) 397-5140; e-mail nacfico@nuri.net; internet www.nonghyup.com; f. 1961; merged with National Livestock Co-operatives Federation in July 2000; cap. 2,564,400m., res 1,395,800m., dep. 64,063,500m. (2002); Chair. and Pres. CHUNG DAE-KUN; 2,025 brs and member co-operatives.

National Federation of Fisheries Co-operatives: 11-6, Sincheon-dong, Songpa-gu, Seoul 138-730; tel. (2) 2240-2114; fax (2) 2240-3049; internet www.suhyup.co.kr; f. 1962; cap. 1,158,100m., res 301,900m., dep. 4,371,000m. (2002); Chair. and Pres. CHANG BYUNG-KOO; 120 brs.

Provincial Banks

Cheju Bank: 1349, Ido-1-dong, Jeju 690-021, Jeju Prov.; tel. (64) 734-1711; fax (64) 720-0183; f. 1969; cap. 55,500m., res. 30,700m., dep. 1,044,100m. (2002); merged with Central Banking Co in 2000; Chair. and Pres. KANG JOON-HONG; 29 brs.

Daegu Bank Ltd: 118, 2-ga, Susong-dong, Susong-gu, Daegu 706-712; tel. (53) 756-2001; fax (53) 740-6902; internet www.daegubank.co.kr; f. 1967; cap. 660,625m., res 7,482m., dep. 12,064,956m. (Dec. 2002); Chair. and Pres. KIM KUK-NYON; 183 brs.

Jeonbuk Bank Ltd: 669-2, Geumam-dong, Deokjin-gu, Jeonju 561-711, N Jeolla Prov.; tel. (63) 250-7114; fax (63) 250-7078; internet www.jbbank.co.kr; f. 1969; cap. 165,300m., res 30,200m., dep. 2,784,100m. (2002); Chair. and Pres. HONG SUNG-JOO; 68 brs.

Kwangju Bank Ltd: 7-12, Daein-dong, Dong-gu, Gwangju 501-719; tel. (62) 239-5000; fax (62) 239-5199; e-mail kbjint1@nuri.net; internet www.kjbank.com; f. 1968; cap. 170,403m., res 598m., dep. 6,277,745m. (Dec. 2002); Chair. and Pres. UM JONG-DAE; 135 brs.

Kyongnam Bank: 246-1, Sokjeon-dong, Hoewon-gu, Masan 630-010, Gyeongsang Prov.; tel. (551) 290-8000; fax (551) 294-9426; internet www.knbank.co.kr; f. 1970 as Gyeongnam Bank Ltd, name changed 1987; cap. 259,000m., res 22,851m., dep. 7,847,571m. (Dec. 2002); Chair. and Pres. KANG SHIN-CHUL; 110 brs.

Foreign Banks

ABN-AMRO Bank NV (Netherlands): Seoul City Tower Bldg, 11–12th Floors, 581, 5-ga, Namdaemun-no, Jung-gu, Seoul; tel. (2) 2131-6000; fax (2) 399-6554; f. 1979; Gen. Man. CHUNG DUCK-MO.

American Express Bank Ltd (USA): Gwanghwamun Bldg, 15th Floor, 64-8, 1-ga, Taepyeong-no, Jung-gu, CPOB 1390, Seoul 100-101; tel. (2) 399-2929; fax (2) 399-2966; f. 1977; Gen. Man. CHOE JAE-ICK.

Arab Bank PLC (Jordan): Daewoo Center Bldg, 22nd Floor, 541, 5-ga, Namdaemun-no, Jung-gu, CPOB 1331, Seoul 100-714; tel. (2) 317-9000; fax (2) 757-0124; Gen. Man. JO SEUNG-SHIK.

Australia and New Zealand Banking Group Ltd (Australia): Kyobo Bldg, 18th Floor, 1, 1-ga, Jongno, Jongno-gu, CPOB 1065, Seoul 110-714; tel. (2) 730-3151; fax (2) 737-6325; f. 1987; Gen. Man. PHIL MICHELL.

Bank Mellat (Iran): Bon Sol Bldg, 14th Floor, 144-27, Samseong-dong, Gangnam-gu, Seoul; tel. (2) 558-4448; fax (2) 557-4448; e-mail info@bankmellat.co.kr; internet www.bankmellat.co.kr; f. 2001; Gen. Man. ALI AFZALI.

Bank of America (USA): Hanwha Bldg, 9th Floor, 1, Janggyo-dong, Jung-gu, Seoul 100-797; tel. (2) 729-4500; fax (2) 729-4400; Gen. Man. BANG CHOON-HO.

Bank of Hawaii (USA): Daeyonkak Bldg, 14th Floor, 25-5, 1-ga, Jungmu-no, Jung-gu, Seoul 100-011; tel. (2) 757-0831; fax (2) 757-3516; Man. PARK YONG-SOO.

Bank of Nova Scotia (Canada): KCCI Bldg, 9th Floor, 45, 4-ga, Namdaemun-no, Jung-gu, Seoul 100-094; tel. (2) 757-7171; fax (2) 752-7189; e-mail bns.seoul@scotiabank.com; Gen. Man. HENRY YONG.

Bank of Tokyo-Mitsubishi Ltd (Japan): Young Poong Bldg, 4th Floor, 33, Seorin-dong, Jongno-gu, Seoul; tel. (2) 399-6474; fax (2) 735-4897; f. 1967; Gen. Man. KAZUMASA KOGA.

Bankers Trust Co (USA): Center Bldg, 10th Floor, 111-5, Sokong-dong, Jung-gu, Seoul; tel. (2) 3788-6000; fax (2) 756-2648; f. 1978; Man. Dir LEE KEUN-SAM.

BNP Paribas (France): Dong Yang Chemical Bldg, 8th Floor, 50, Sogong-dong, Jung-gu, Seoul 100-070; tel. (2) 317-1700; fax (2) 757-2530; e-mail bnppseoul@asia.bnparibas.com; f. 1976; Gen. Man. ALAIN PÉNICAUT.

Citibank NA (USA): Citicorp Center Bldg, 89-29, 2-ga, Sinmun-no, Jongno-gu, CPOB 749, Seoul 110-062; tel. (2) 2004-1114; fax (2) 722-3644; f. 1967; Gen. Man. SAJJAD RAZVI.

Crédit Agricole Indosuez (France): Kyobo Bldg, 19th Floor, 1, 1-ga, Jongno, Jongno-gu, CPOB 158, Seoul 110-714; tel. (2) 3700-9500; fax (2) 738-0325; f. 1974; Gen. Man. PATRICE COUVEGNES.

Crédit Lyonnais SA (France): You One Bldg, 8th–10th Floors, 75-95, Seosomun-dong, Jung-gu, Seoul 100-110; tel. (2) 772-8000; fax (2) 755-5379; f. 1978; Gen. Man. GEOFFROY DE LASSUS.

DBS Bank Ltd (Development Bank of Singapore Ltd): Gwanghwamun Bldg, 20th Floor, 64-8, 1-ga, Taepyeong-no, Jung-gu, CPOB 9896, Seoul; tel. (2) 399-2660; fax (2) 732-7953; e-mail jeefun@dbs.com; f. 1981; Gen. Man. LOW JEE FUN.

Deutsche Bank AG (Germany): Sei An Bldg, 20th–22nd Floor, 116, 1-ga, Sinmun-no, Jongno-gu, Seoul 110-700; tel. (2) 724-4500; fax (2) 724-4645; f. 1978; Gen. Man. KIM JIN-IL.

First National Bank of Chicago (USA): Oriental Chemical Bldg, 15th Floor, 50, Sokong-dong, Jung-gu, Seoul 100-070; tel. (2) 316-9700; fax (2) 753-7917; f. 1976; Vice-Pres. and Gen. Man. MICHAEL S. BROWN.

Fuji Bank Ltd (Japan): Doosan Bldg, 15th Floor, 101-1, 1-ga, Ulchi-no, Jung-gu, Seoul 100-191; tel. (2) 311-2000; fax (2) 754-8177; f. 1972; Gen. Man. IKUO YAMAMOTO.

Hongkong and Shanghai Banking Corpn Ltd (Hong Kong): HSBC Bldg, 1-ga, Bongrae-dong, Jung-gu, CPOB 6910, Seoul 110-161; tel. (2) 2004-0000; fax (2) 381-9100; Gen. Man. G. P. S. CALVERT.

Indian Overseas Bank: Daeyungak Bldg, 3rd Floor, 25-5, 1-ga, Jungmu-no, Jung-gu, CPOB 3332, Seoul 100-011; tel. (2) 753-0741; fax (2) 756-0279; e-mail iobseoul@chollian.net; f. 1977; Gen. Man. K. P. MUNIRATHMAN.

Industrial and Commercial Bank of China (China): Taepyeong Bldg, 17th Floor, 310, 2-ga, Taepyeong-no, Seoul; tel. (2) 755-5688; fax (2) 779-2750; f. 1997; Gen. Man. ZHANG KEXIN.

ING Bank NV (Netherlands): Hungkuk Life Insurance Bldg, 15th Floor, 226, 1-ga, Sinmun-no, Jongno-gu, Seoul 110-061; tel. (2) 317-1800; fax (2) 317-1883; Man. YIM SANG-KYUN.

JP Morgan Chase Bank (USA): Chase Plaza, 34-35, Jeong-dong, Jung-gu, Seoul 100-120; tel. (2) 758-5114; fax (2) 758-5420; f. 1978; Gen. Man. KIM MYUNG-HAN.

Mizuho Bank Ltd (Japan): Nae Wei Bldg, 14th Floor, 6, 2-ga, Ulchi-no, Jung-gu, Seoul 100-192; tel. (2) 756-8181; fax (2) 754-6844; f. 1972; Gen. Man. TSUNEO KIKUCHI.

National Australia Bank Ltd: KDIC Bldg, 16th Floor, 33, Da-dong, Jung-gu, Seoul; tel. (2) 3705-4600; fax (2) 3705-4602; Gen. Man. MARK EDMONDS.

National Bank of Canada: Leema Bldg, 6th Floor, 146-1, Susong-dong, Jongno-gu, Seoul 110-140; tel. (2) 733-5012; fax (2) 736-1508; Vice-Pres. and Country Man. C. N. KIM.

National Bank of Pakistan: Kyobo Bldg, 12th Floor, 1, 1-ga, Jongno, Jongno-gu, CPOB 1633, Seoul 110-121; tel. (2) 732-0277; fax (2) 734-5817; f. 1987; Gen. Man. ABDUL GHAFOOR.

Overseas Union Bank Ltd (Singapore): Kyobo Bldg, 8th Floor, Suite 806, 1, 1-ga, Jongno, Jongno-gu, Seoul 110-714; tel. (2) 739-3441; fax (2) 732-9004; Vice-Pres. and Gen. Man. OOI KOOI KEAT.

Royal Bank of Canada: Kyobo Bldg, 22nd Floor, 1, 1-ga, Jongno, Jongno-gu, Seoul 110-714; tel. (2) 730-7791; fax (2) 736-2995; f. 1982; Gen. Man. THOMAS P. FEHLNER, Jr.

Société Générale (France): Sean Bldg, 10th Floor, 1-ga, Sinmun-no, Jongno-gu, Seoul 110-700; tel. (2) 2195-7777; fax (2) 2195-7700; f. 1984; CEO ERIC BERTHÉLEMY.

Standard Chartered Bank (UK): Seoul Finance Center, 22nd Floor, 84, 1-ga, Taepyeong-no, Jung-gu, Seoul; tel. (2) 750-6114; fax (2) 757-7444; Gen. Man. WILLIAM GEMMEL.

UBS AG (Switzerland): Young Poong Bldg, 10th Floor, 33, Seorin-dong, Jongno-gu, Seoul 110-752; tel. (2) 3702-8888; fax (2) 3708-8714; f. 1999; Gen. Man. LEE JAE-HONG.

UFJ Bank Ltd (Japan): Lotte Bldg, 22nd Floor, 1, 1-ga, Sogong-dong, Jung-gu, Seoul; tel. (2) 752-7321; fax (2) 754-3870; Gen. Man. HIDEKI YAMAUCHI.

Union Bank of California NA (USA): Kyobo Bldg, 12th Floor, 1, 1-ga, Jongno, Jongno-gu, CPOB 329, Seoul 110; tel. (2) 721-1700; fax (2) 732-9526; Gen. Man. KIM TAEK-JOONG.

Union de Banques Arabes et Françaises (France): ACE Tower, 3rd Floor, 1-170, Sunhwa-dong, Jung-gu, CPOB 1224, Seoul 100-742; tel. (2) 3455-5300; fax (2) 3455-5354; f. 1979; Gen. Man. PATRICK OBERREINER.

United Overseas Bank Ltd (Singapore): Kyobo Bldg, 20th Floor, 1, 1-ga, Jongno, Jongno-gu, Seoul 110-714; tel. (2) 739-3916; fax (2) 730-9570; Gen. Man. LIEW CHAN HARN.

Banking Association

Korea Federation of Banks: 4-1, 1-ga, Myeong-dong, Jung-gu, Seoul 100-021; tel. (2) 3705-5000; fax (2) 3705-5337; internet www.kfb.or.kr; f. 1928; Chair. SHIN DONG-HYUCK; Vice-Chair. KIM KONG-JIN.

STOCK EXCHANGE

Korea Stock Exchange: 33, Yeouido-dong, Yeongdeungpo-gu, Seoul 150-977; tel. (2) 3774-9000; fax (2) 786-0263; e-mail world@kse.or.kr; internet www.kse.or.kr; f. 1956; Chair. and CEO KANG YUNG-JOO.

Kosdaq Stock Market, Inc: 45-2, Yeouido-dong, Yeongdeungpo-gu, Seoul 150-974; tel. (2) 2001-5700; fax (2) 784-4505; e-mail webmaster@kosdaq.or.kr; internet www.kosdaq.or.kr; f. 1996; stock market for knowledge-based venture cos; 828 listed cos (Aug. 2002) with a market capitalization of US $39,945m.; Pres. and CEO SHIN HO-JOO.

INSURANCE

Principal Life Companies

Allianz Jeil Life Insurance Co Ltd: 1303-35, Seocho 4-dong, Seocho-gu, Seoul 137-074; tel. (2) 3481-3111; fax (2) 3481-0960; f. 1954; Pres. LEE TAE-SIK.

Choson Life Insurance Co Ltd: 111, Sincheon-dong, Dong-gu, Daegu 701-620; tel. (53) 743-3600; fax (53) 742-9263; f. 1988; cap. 12,000m.; Pres. LEE YOUNG-TAEK.

Daishin Life Insurance Co Ltd: 395-68, Sindaebang-dong, Dongjak-gu, Seoul 156-010; tel. (2) 3284-7000; fax (2) 3284-7451; internet www.dslife.co.kr; f. 1989; cap. 144,200m. (2002); Pres. PARK BYUNG-MYUNG.

Dong-Ah Life Insurance Co Ltd: Dong-Ah Life Insurance Bldg, 33, Da-dong, Jung-gu, Seoul; tel. (2) 317-5114; fax (2) 771-7561; f. 1973; cap. 10,000m.; Pres. KIM CHANG-LAK; 900 brs.

Dongbu Life Insurance Co Ltd: Dongbu Bldg, 7th Floor, 891-10, Daechi-dong, Gangnam-gu, Seoul 135-820; tel. (2) 1588-3131; fax (2) 3011-4100; internet www.dongbulife.co.kr; f. 1989; cap. 85,200m. (2002); Pres. CHANG KI-JE.

Dongyang Life Insurance Co Ltd: 185, Ulchi-no 2-ga, Jung-gu, Seoul 100-192; tel. (2) 728-9114; fax (2) 771-1347; internet www.myangel.co.kr; f. 1989; cap. 340,325m. (2002); Pres. KU JA-HONG.

Doowon Life Insurance Co Ltd: 259-6, Sokjon-dong, Hoewon-gu, Masan 630-500; tel. (55) 52-3100; fax (55) 52-3119; f. 1990; cap. 10,000m.; Pres. CHOI IN-YONG.

Han Deuk Life Insurance Co Ltd: 878-1, Bumchyun 1-dong, Busanjin-gu, Busan 641-021; tel. (51) 631-8700; fax (51) 631-8809; f. 1989; cap. 10,000m.; Pres. SUH WOO-SHICK.

Hanil Life Insurance Co Ltd: 118, 2-ga, Namdaemun-no, Jung-gu, Seoul 100-770; tel. (2) 2126-7777; fax (2) 2126-7631; internet www.hanillife.co.kr; f. 1993; cap. 115,000m. (2002); Pres. LEE MYUNG-HYUN.

Hankuk Life Insurance Co Ltd: Daehan Fire Bldg, 51-1, Namchang-dong, Jung-gu, Seoul 100-060; tel. (2) 773-3355; fax (2) 773-1778; f. 1989; cap. 10,000m.; Pres. PARK HYUN-KOOK.

Hansung Life Insurance Co Ltd: 3, Sujung-dong, Dong-gu, Busan 601-030; tel. (51) 461-7700; fax (51) 465-0581; f. 1988; Pres. CHO YONG-KEUN.

Hungkuk Life Insurance Co Ltd: 226, Sinmun-no 1-ga, Jongno-gu, Seoul 100-061; tel. (2) 2002-7000; fax (2) 2002-7804; internet www.hungkuk.co.kr; f. 1958; cap. 12,221m. (2002); Pres. RYU SEOK-KEE.

ING Life Insurance Co Korea Ltd: Sean Bldg, 116, Sinmun-no, Jongno-gu, Seoul 110-700; tel. (2) 3703-9500; fax (2) 734-3309; f. 1991; cap. 64,820m. (2002); Pres. JOOST KENEMANS.

Korea Life Insurance Co Ltd: 60, Yeouido-dong, Yeongdeungpo-gu, Seoul 150-603; tel. (2) 789-5114; fax (2) 789-8173; internet www.korealife.com; f. 1946; cap. 3,550,000m. (2002); Pres. LEE KANG-HWAN.

Korean Reinsurance Company: 80, Susong-dong, Jongno-gu, Seoul 110-733; tel. (2) 3702-6000; fax (2) 739-3754; internet www.koreanre.co.kr; f. 1963; Pres. PARK JONG-WON.

Kumho Life Insurance Co Ltd: 57, 1-ga, Sinmun-no, Jongno-gu, Seoul 110-061; tel. (2) 6303-5000; fax (2) 771-7561; internet www.kumholife.co.kr; f. 1988; cap. 211,249m. (2002); Pres. SONG KEY-HYUCK.

Kyobo Life Insurance Co Ltd: 1, 1-ga, Jongno, Jongno-gu, Seoul 110-714; tel. (2) 721-2121; fax (2) 737-9970; internet www.kyobo.co.kr; f. 1958; cap. 92,500m.; Pres. and CEO CHANG HYUNG-DUK; 84 main brs.

Lucky Life Insurance Co Ltd: 3, Sujung-dong, Dong-gu, Busan 601-716; tel. (51) 461-7700; fax (51) 465-0581; internet www.luckylife.co.kr; f. 1988; cap. 139,054m. (2002); Pres. CHANG NAM-SIK.

MetLife Insurance Co of Korea Ltd: Sungwon Bldg, 8th Floor, 141, Samseong-dong, Gangnam-gu, Seoul 135-716; tel. (2) 3469-9600; fax (2) 3469-9700; internet www.metlifekorea.co.kr; f. 1989; cap. 97,700m. (2002); Pres. STUART B. SOLOMON.

Pacific Life Insurance Co Ltd: 705-9, Yeoksam-dong, Gangnam-gu, Seoul 135-080; tel. (2) 3458-0114; fax (2) 3458-0392; internet www.pli.co.kr; f. 1989; cap. 10,000m.; Pres. KIM SUNG-MOO.

PCA Life Insurance Co Ltd: 142, Nonhyun-dong, Gangnam-gu, Seoul 135-749; tel. (2) 515-5300; fax (2) 514-3844; f. 1990; cap. 52,100m. (2002); Pres. MIKE BISHOP.

Prudential Life Insurance Co of Korea Ltd: Prudential Bldg, Yeoksam-dong, Gangnam-gu, Seoul; tel. (2) 2144-2000; fax (2) 2144-2100; internet www.prudential.or.kr; f. 1989; cap. 26,400m.; Pres. JAMES C. SPACKMAN.

Samshin All State Life Insurance Co Ltd: Samwhan Bldg, 5th Floor, 98-5, Unni-dong, Jongno-gu, Seoul 110-742; tel. (2) 3670-5000; fax (2) 742-8197; Pres. KIM KYUNG-YOP.

Samsung Life Insurance Co Ltd: 150, 2-ga, Taepyeong-no, Jung-gu, Seoul 100-716; tel. (2) 751-8000; fax (2) 751-8100; internet www.samsunglife.com; f. 1957; cap. 100,000m. (2002); Pres. BAE JUNG-CHOONG; 1,300 brs.

Shinhan Life Insurance Co Ltd: 120, 2-ga, Taepyeong-no, Jung-gu, Seoul 100-102; tel. (2) 3455-4000; fax (2) 753-9351; internet www.shinhanlife.co.kr; f. 1990; Pres. HAN DONG-WOO.

SK Life Insurance Co Ltd: 168, Gongduk-dong, Mapo-gu, Seoul 121-705; tel. (2) 3271-4114; fax (2) 3271-4400; internet www.sklife.com; f. 1988; cap. 246,275m. (2002); Pres. KANG HONG-SIN.

Non-Life Companies

Daehan Fire and Marine Insurance Co Ltd: 51-1, Namchang-dong, Jung-gu, Seoul 100-778; tel. (2) 3455-3114; fax (2) 756-9194; e-mail dhplane@daeins.co.kr; internet www.daeins.co.kr; f. 1946; cap. 19,500m.; Pres. LEE YOUNG-DONG.

Dongbu Insurance Co Ltd: Dongbu Financial Center, 891-10, Daechi-dong, Gangnam-gu, Seoul 135-840; tel. (2) 2262-3450; fax (2) 2273-6785; e-mail dongbu@dongbuinsurance.co.kr; internet www.idongbu.com; f. 1962; cap. 30,000m.; Pres. LEE SU-KWANG.

First Fire and Marine Insurance Co Ltd: 12-1, Seosomun-dong, Jung-gu, CPOB 530, Seoul 100-110; tel. (2) 316-8114; fax (2) 771-7319; internet www.insumall.co.kr; f. 1949; cap. 17,200m.; Pres. KIM WOO-HOANG.

Green Fire and Marine Insurance Co Ltd: Seoul City Tower, 581, 5-ga, Namdaemun-no, Jung-gu, Seoul 100-803; tel. (2) 1588-5959; fax (2) 773-1214; internet www.greenfire.co.kr; Pres. KIM JONG-CHEN.

Haedong Insurance Co Ltd: 1424-2, Seocho-dong, Seocho-gu, Seoul; tel. (2) 520-2114; e-mail webmaster@haedong.co.kr; internet www.haedong.co.kr; f. 1953; cap. p.u. 12,000m.; Chair. KIM DONG-MAN; CEO NAH BOO-WHAN.

Hankuk Fidelity and Surety Co Ltd: 51-1, Namchang-dong, Jung-gu, Seoul; tel. (2) 773-3355; fax (2) 773-1778; e-mail hfs025@unitel.co.kr; f. 1989; cap. 103,000m.; Pres. CHO AM-DAE.

Hyundai Marine and Fire Insurance Co Ltd: 8th Floor, 140-2, Kye-dong, Jongno gu, Seoul 110-793; tel. (2) 3701-8000; fax (2) 732-5687; e-mail webpd@hdinsurance.co.kr; internet www.hi.co.kr; f. 1955; cap. 30,000m.; Pres. KIM HO-IL.

Korean Reinsurance Co: 80, Susong-dong, Jongno-gu, Seoul 100-733; tel. (2) 3702-6000; fax (2) 739-3754; e-mail service@koreanre.co.kr; internet www.koreanre.co.kr; f. 1963; cap. 34,030m.; Pres. PARK JONG-WON.

Kukje Hwajae Insurance Co Ltd: 120, 5-ga, Namdaemun-no, Jung-gu, Seoul 100-704; tel. (2) 753-1101; fax (2) 773-1214; internet www.directins.co.kr; f. 1947; cap. 10,784m.; Chair. LEE BONG-SUH.

Kyobo Auto Insurance Co Ltd: 76-4, Jamwon-dong, Seocho-gu, Seoul 137-909; tel. (2) 3479-4900; fax (2) 3479-4800; internet www.kyobodirect.com; Pres. SHIN YONG-KIL.

LG Insurance Co Ltd: LG Da-dong Bldg, 85, Da-dong, Jung-gu, Seoul 100-180; tel. (2) 310-2391; fax (2) 753-1002; e-mail webmaster@lginsure.com; internet www.lginsure.com; f. 1959; Pres. KOO CHA-HOON.

Oriental Fire and Marine Insurance Co Ltd: 25-1, Yeouido-dong, Yeongdeungpo-gu, Seoul 150-010; tel. (2) 3786-1910; fax (2) 3886-1940; e-mail webmaster@ofmi.co.kr; internet www.insuworld.co.kr; f. 1922; cap. 42,900m.; Pres. CHUNG KUN-SUB.

Samsung Fire and Marine Insurance Co Ltd: Samsung Insurance Bldg, 87, 1-ga, Ulchi-no, Jung-gu, Seoul 100-191; tel. (2) 758-7948; fax (2) 758-7831; internet www.samsungfire.com; f. 1952; cap. 6,566m.; Pres. LEE SOO-CHANG.

Seoul Guarantee Insurance Co: 136-74, Yeonchi-dong, Jongno-gu, Seoul 110-470; tel. (2) 3671-7459; fax (2) 3671-7480; internet www.sgic.co.kr; Pres. PARK HAE-CHOON.

Shindongah Fire and Marine Insurance Co Ltd: 43, 2-ga, Taepyeong-no, Jung-gu, Seoul; tel. (2) 6366-7000; fax (2) 755-8006; internet www.sdafire.com; f. 1946; cap. 60,220m.; Pres. JEON HWA-SOO.

Ssangyong Fire and Marine Insurance Co Ltd: 60, Doryeom-dong, Jongno-gu, Seoul 110-716; tel. (2) 724-9000; fax (2) 730-1628; e-mail sfmi@ssy.insurance.co.kr; internet www.insurance.co.kr; f. 1948; cap. 27,400m.; Pres. LEE JIN-MYUNG.

Insurance Associations

Korea Life Insurance Association: Kukdong Bldg, 16th Floor, 60-1, 3-ga, Jungmu-no, Jung-gu, Seoul 100-705; tel. (2) 2262-6600; fax (2) 2262-6580; internet www.klia.or.kr; f. 1950; Chair. BAE CHAN-BYUNG.

Korea Non-Life Insurance Association: KRIC Bldg, 6th Floor, 80, Susong-dong, Jongno-gu, Seoul; tel. (2) 3702-8539; fax (2) 3702-8549; internet www.knia.or.kr; f. 1946; 13 corporate mems; Chair. PARK JONG-IK.

Trade and Industry

GOVERNMENT AGENCIES

Fair Trade Commission: 1, Jungang-dong, Gwacheon-si, Gyeonggi Prov. 427-760; internet www.ftc.go.kr; Chair. LEE NAM-KEE.

Federation of Korean Industries: FKI Bldg, 2nd Floor, 28-1, Yeouido-dong, Yeongdeungpo-gu, Seoul 150-756; tel. (2) 3771-0114; fax (2) 3771-0110; e-mail webmaster@fki.or.kr; internet www.fki.or.kr; f. 1961; conducts research and survey work on domestic and overseas economic conditions and trends; advises the Govt and other interested parties on economic matters; exchanges economic and trade missions with other countries; sponsors business conferences; 380 corporate mems and 65 business asscns; Chair. KANG SHIN-HO.

Korea Appraisal Board: 171-2, Samseong-dong, Gangnam-gu, Seoul; tel. (2) 555-1174; Chair. KANG KIL-BOO.

Korea Asset Management Corpn (KAMCO): 814, Yeoksam-dong, Gangnam-gu, Seoul; tel. (2) 3420-5049; fax (2) 3420-5100; internet www.kamco.co.kr; f. 1963; collection and foreclosure agency; appointed following Asian financial crisis as sole institution to manage and dispose of non-performing loans for financial institutions; Pres. CHUNG JAE-RYONG.

Korea Export Industrial Corpn: 33, Seorin-dong, Jongno-gu, Seoul; tel. (2) 853-5573; f. 1964; encourages industrial exports, provides assistance and operating capital, conducts market surveys; Pres. KIM KI-BAE.

Korea Export Insurance Corpn: 136, Seorin-dong, Jongno-gu, Seoul 110-729; tel. (2) 399-6800; fax (2) 399-6679; internet www.keic.or.kr; f. 1992; official export credit agency of Korea; Pres. LIM TAE-JIN.

Korea Institute for Industrial Economics and Trade (KIET): 206-9, Cheongnyangni-dong, Dongdaemun-gu, Seoul; tel. (2) 3299-3114; fax (2) 963-8540; internet www.kiet.re.kr; f. 1976; economic and industrial research; Pres. PAI KWANG-SUN.

Korean Intellectual Property Office: Government Complex-Daejeon, Dunsan-dong, Seo-gu, Daejeon; tel. (42) 481-5027; fax (42) 481-3455; internet www.kipo.go.kr; Commissioner KIM GWANG-LIM.

Korea Industrial Research Institutes: FKI Bldg, 28-1, Yeouido-dong, Yeongdeungpo-gu, Seoul; tel. (2) 780-7601; fax (2) 785-5771; f. 1979; analyses industrial and technological information from abroad; Pres. KIM CHAE-KYUM.

Korea National Oil Corpn (KNOC): tel. 380-2114 ; fax 387-9321; e-mail webmaster@knoc.co.kr ; internet www.knoc.co.kr; Pres. YI OK-SU.

Korea Trade-Investment Promotion Agency (KOTRA): 300-9, Yeomgok-dong, Seocho-gu, Seoul; tel. (2) 3460-7114; fax (2) 3460-7777; e-mail net-mgr@kotra.or.kr; internet www.kotra.or.kr; f. 1962; various trade promotion activities, market research, cross-border investment promotion, etc.; 98 overseas brs; Pres. OH YOUNG-KYO.

CHAMBER OF COMMERCE

Korea Chamber of Commerce and Industry: 45, 4-ga, Namdaemun-no, Jung-gu, Seoul 100-743; tel. (2) 316-3114; fax (2) 757-9475; internet www.korcham.net; f. 1884; over 80,000 mems; 63 local chambers; promotes development of the economy and of international economic co-operation; Pres. PARK YONG-SUNG.

INDUSTRIAL AND TRADE ASSOCIATIONS

Agricultural and Fishery Marketing Corpn: 191, 2-ga, Hangang-no, Yeongsan-gu, CPOB 3212, Seoul 140; tel. (2) 790-8010; fax (2) 798-7513; internet www.afmc.co.kr; f. 1967; integrated development for secondary processing and marketing distribution for agricultural products and fisheries products; Pres. AHN KYO-DUCK; Exec. Vice-Pres. KIM JIN-KYU.

Construction Association of Korea: Construction Bldg, 8th Floor, 71-2, Nonhyon-dong, Gangnam-gu, Seoul 135-701; tel. (2) 547-6101; fax (2) 542-6264; f. 1947; national licensed contractors' assen; 2,700 mem. firms (1995); Pres. CHOI WON-SUK; Vice-Pres. PARK KU-YEOL.

Electronic Industries Association of Korea: 648, Yeoksam-dong, Gangnam-gu, CPOB 5650, Seoul 135-080; tel. (2) 553-0941; fax (2) 555-6195; e-mail eiak@soback.kornet.nm.kr; internet www.eiak.org; f. 1976; 328 mems; Chair. JOHN KOO.

Korea Automobile Manufacturers Association: 658-4. Deungchon-dong, Gangseo-gu, Seoul; tel. (2) 3660-1800; fax (2) 3660-1900; e-mail cwkim@kama.or.kr; internet www.kama.or.kr; f. 1988; Chair. KIM NOI-MYUNG.

Korea Coal Association: 80-6, Susong-dong, Jongno-gu, Seoul; tel. (2) 734-8891; fax (2) 734-7959; f. 1949; 49 corporate mems; Chair. JANG BYEONG-DUCK.

Korea Consumer Goods Exporters Association: KWTC Bldg, Rm 1802, 159, Samseong-dong, Gangnam-gu, Seoul; tel. (2) 551-1865; fax (2) 551-1870; f. 1986; 230 corporate mems; Pres. YONG WOONG-SHIN.

Korea Federation of Textile Industries: 944-31, Daechi-dong, Gangnam-gu, Seoul; tel. (2) 528-4001; fax (2) 528-4069; e-mail kofoti@kofoti.or.kr; internet www.kofoti.or.kr; f. 1980; 50 corporate mems; Pres. PARK SANG-CHUL.

Korea Foods Industry Association: 1002-6, Bangbae-dong, Seocho-gu, Seoul; tel. (2) 585-5052; fax (2) 586-4906; internet www.kfia.or.kr; f. 1969; 104 corporate mems; Pres. CHUN MYUNG-KE.

Korea Importers Association (KOIMA): 218, Hangang-no, 2-ga, Yeongsan-gu, Seoul 140-875; tel. (2) 792-1581; fax (2) 785-4373; e-mail info@aftak.com; internet www.koima.or.kr; f. 1970; 11,903 mems; Chair. CHIN CHUL-PYUNG.

Korea International Trade Association: 159-1, Samseong-dong, Gangnam-gu, Seoul; tel. (2) 6000-5114; fax (2) 6000-5115; internet www.kita.org; f. 1946; private, non-profitmaking business org. representing all licensed traders in South Korea; provides foreign businessmen with information, contacts and advice; 80,000 corporate mems; Pres. KIM JAE-CHUL.

Korea Iron and Steel Association: 824, Yeoksam-dong, Gangnam-gu, Seoul; tel. (2) 559-3500; fax (2) 559-3508; internet www.kosa.or.kr; f. 1975; 39 corporate mems; Chair. YOO SANG-BOO.

Korea Oil Association: 28-1, Yeouido-dong, Yeongdeungpo-gu, Seoul; tel. (2) 3775-0520; fax (2) 761-9573; f. 1980; Pres. CHOI DOO-HWAN.

Korea Productivity Center: 122-1, Jeokseon-dong, Jongno-gu, Seoul 110-052; tel. (2) 724-1114; fax (2) 736-0322; internet www.kpc.or.kr; f. 1957; services to increase productivity of the industries, consulting services, education and training of specialized personnel; Chair. and CEO LEE HEE-BEOM.

Korea Sericultural Association: 17-9, Yeouido-dong, Yeongdeungpo-gu, Seoul; tel. (2) 783-6072; fax (2) 780-0706; f. 1946; improvement and promotion of silk production; 50,227 corporate mems; Pres. CHOI YON-HONG.

Korea Shipbuilders' Association: 65-1, Unni-dong, Jongno-gu, Seoul; tel. (2) 766-4631; fax (2) 766-4307; internet www.koshipa.or.kr; f. 1977; 9 mems; Chair. KIM HYUNG-BYUK.

Korea Textiles Trade Association: Textile Center, 16th Floor, 944-31, Daechi-dong, Gangnam-gu, Seoul; tel. (2) 528-5158; fax (2) 528-5188; f. 1981; 947 corporate mems; Pres. KANG TAE-SEUNG.

Korean Apparel Industry Association: KWTC Bldg, Rm 801, 159, Samseong-dong, Gangnam-gu, Seoul 135-729; tel. (2) 551-1454; fax (2) 551-1467; f. 1993; 741 corporate mems; Pres. PARK SEI-YOUNG.

Korean Development Associates: Seoul; tel. (2) 392-3854; fax (2) 312-3856; f. 1965; economic research; 25 corporate mems; Pres. KIM DONG-KYU.

Mining Association of Korea: 35-24, Dongui-dong, Jongno-gu, Seoul 110; tel. (2) 737-7748; fax (2) 720-5592; f. 1918; 128 corporate mems; Pres. KIM SANG-BONG.

Spinners and Weavers Association of Korea: 43-8, Gwancheol-dong, Jongno-gu, Seoul 110; tel. (2) 735-5741; fax (2) 735-5749; internet www.swak.org; f. 1947; 20 corporate mems; Pres. SUH MIN-SOK.

EMPLOYERS' ORGANIZATION

Korea Employers' Federation: KEF Bldg, 276-1 Daeheung-dong, Mapo-gu, Seoul 121-726; tel. (2) 3270-7310; fax (2) 706-1059; e-mail delee@kef.or.kr; internet www.kef.or.kr; f. 1970; advocates employers' interests with regard to labour and social affairs; 13 regional employers' asscns, 20 economic and trade asscns, and 4,000 major enterprises; Chair. LEE SOO-YOUNG.

UTILITIES

Electricity

Korea Electric Power Corpn (KEPCO): 167, Samseong-dong, Gangnam-gu, Seoul; tel. (2) 3456-3630; fax (2) 3456-3699; internet www.kepco.co.kr; f. 1961; transmission and distribution of electric power, and development of electric power sources; privatization pending; Pres. KANG DONG-SUK.

Gas

Korea Gas Corpn: 215, Jeongja-dong, Bundang-gu, Seongnam, Gyeonggi Prov.; tel. (31) 710-0114; fax (31) 710-0117; internet www.kogas.or.kr; state-owned; proposed transfer to private-sector ownership announced in July 1998; Pres. KIM MYUNG-KYU.

Samchully Co Ltd: 35-6, Yeouido-dong, Yeongdeungpo-gu, Seoul; tel. (2) 368-3300; fax (2) 783-1206; internet www.samchully.co.kr; f. 1966; gas supply co for Seoul metropolitan area and Gyeonggi Prov.; Chair. JIN JU-HWA.

Water

Korea Water Resources Corpn: 6-2, Yeonchuk-dong, Daedeok-gu, Daejeon; tel. (42) 629-3114; fax (42) 623-0963; internet www.kowaco.or.kr.

Office of Waterworks, Seoul Metropolitan Govt: 27-1 Hap-dong, Seodaemun-gu, Seoul; tel. (2) 390-7332; fax (2) 362-3653; f. 1908; responsible for water supply in Seoul; Head SON JANG-HO.

Ulsan City Water and Sewerage Board: 646-4, Sin-Jung 1-dong, Nam-gu, Ulsan; tel. (52) 743-020; fax (52) 746-928; f. 1979; responsible for water supply and sewerage in Ulsan; Dir HO KUN-SONG.

CO-OPERATIVES

Korea Computers Co-operative: Seoul; tel. (2) 780-0511; fax (2) 780-7509; f. 1981; Pres. MIN KYUNG-HYUN.

Korea Federation of Knitting Industry Co-operatives: 586-1, Sinsa-dong, Gangnam-gu, Seoul; tel. (2) 548-2131; fax (2) 3444-9929; internet www.knit.or.kr; f. 1962; Chair. JOUNG MAN-SUB.

Korea Federation of Non-ferrous Metal Industry Co-operatives: Backsang Bldg, Rm 715, 35-2, Yeouido-dong, Yeongdeungpo-gu, Seoul; tel. (2) 780-8551; fax (2) 784-9473; f. 1962; Chair. PARK WON-SIK.

Korea Federation of Small and Medium Business (KFSB): 16-2, Yeouido-dong, Yeongdeungpo-gu, Seoul 150-010; tel. (2) 2124-3114; fax (2) 782-0247; f. 1962; Chair. KIM YOUNG-SOO.

Korea Mining Industry Co-operative: 35-24, Dongui-dong, Jongno-gu, Seoul; tel. (2) 735-3490; fax (2) 735-4658; f. 1966; Chair. JEON HYANG-SIK.

Korea Steel Industry Co-operative: 915-14, Bangbae-dong, Seocho-gu, Seoul; tel. (2) 587-3121; fax (2) 588-3671; internet www.kosic.or.kr; f. 1962; Pres. KIM DUK-NAM.

Korea Woollen Spinners and Weavers Co-operatives: Rm 503, Seawha Bldg, 36, 6-ga, Jongno-gu, Seoul; tel. (2) 747-3871; fax (2) 747-3874; e-mail woollen@woolspd.or.kr; internet www.woolspd.or.kr; f. 1964; Pres. KIM YOUNG-SIK.

National Agricultural Co-operative Federation (NACF): 1, 1-ga, Chungjeong-no, Jung-gu, Seoul; tel. (2) 397-5114; fax (2) 397-5380; internet www.nacf.co.kr; f. 1961; international banking, marketing, co-operative trade, utilization and processing, supply, co-operative insurance, banking and credit services, education and research; Pres. WON CHUL-HEE.

National Federation of Fisheries Co-operatives: 11-6, Sincheon-dong, Songpa-gu, Seoul; tel. (2) 2240-3114; fax (2) 2240-3024; internet www.suhyup.co.kr; f. 1962; Pres. HONG JONG-MOON.

MAJOR COMPANIES

The following are some of South Korea's major industrial groups and companies, arranged by sector (cap. = capital; res = reserves; m. = million; amounts in won, unless otherwise indicated):

Major Industrial Groups (Chaebol)

Daelim Group: 23-9 Yeouido-dong, Yeongdeungpo-gu, Seoul; tel. (2) 368-7114; fax (2) 368-7700; internet www.dic.co.kr; mfrs of construction materials, light industrial goods; Chair. LEE YONG-KU.

Daewoo International Corpn: 541, 5-ga, Namdaemun-no, Jung-gu, Seoul; tel. (2) 759-2114; fax (2) 753-9489; internet www.daewoo.com; f. 1967; construction, machinery, shipbuilding, automobiles, electronics, financing, chemicals, light industry, etc.; collapsed with debts of US $50,000m. in 1999; Pres. and CEO LEE TAE-YONG.

Dongbu Group: 21-9, Jeo-dong, Jung-gu, Seoul; tel. (2) 2279-9426; fax (2) 278-3615; internet www.dongbu.co.kr; f. 1969; mfrs of chemicals, semiconductors, steel and steel products; civil engineering and construction; Pres. KIM JOON-KY.

Doosan Group: 18-12, 6-ga, Ulchi-no, Jung-gu, Seoul; tel. (2) 3398-1081; fax (2) 3398-1135; internet www.doosan.co.kr; industrial machinery, construction, electro-materials, glass; Chair. and CEO PARK YONG-OH.

Haitai Group: 131, Namyong-dong, Yongsan-gu, Seoul; tel. (2) 709-7766; fax (2) 790-8123; f. 1945; food, retailing, electronics; Chair. YANG JONG-SOK.

Hanjin Group: 51, Sogong-dong, Jung-gu, Seoul; tel. (2) 756-7739; fax (2) 757-7478; internet www.hanjin.net; f. 1945; transport, shipping, heavy industries; Chair. CHO CHOONG-HOON.

Hanwha Group: 1, Janggyo-dong, Jung-gu, Seoul; tel. (2) 729-2700; fax (2) 729-3000; e-mail webmaster@hanwha.co.kr; internet www.hanwha.co.kr; f. 1952; chemicals; Chair. KIM SEUNG-YOUN; Pres. LEE SOON-JONG.

Hyosung Group: Gongdeok Bldg, 450 Gongdeok-dong, Mapo-gu, Seoul 121-020; tel. (8222) 707-6228; fax (822) 707-6226; internet www.hyosung.co.kr; f. 1957; steel and metals, electronics, industrial equipment, chemicals, fabrics, leather goods; Chair. CHO SOOK-RAE; Pres. LEE DON-YOUNG; 25,000 employees.

Hyundai Group: 226, 1-ga, Sinmunno, Jongnogu, Seoul; tel. (2) 390-1114; internet www.hyundaicorp.com; f. 1953; electronics, construction, heavy industry, petrochemicals, automobile manufacture, finance and securities, etc.; Pres.and CEO MARK JUHN; Chair. HYUN JEONG-EUN; 180,000 employees.

Kolon Group: Kolon Bldg, 45, Mukyo-dong, Jung-gu, Seoul; tel. (2) 311-8114; fax (2) 754-5314; internet www.kolon.co.kr; f. 1954; chemicals, construction, electric machinery; Chair. LEE WOONG-YEUL.

Kumho Group: 10-1, 2-ga, Hoehyeon-dong, Jung-gu, Seoul; tel. (2) 758-1114; fax (2) 758-1515; internet www.kumho.net; construction, engineering, chemicals, textiles; Chair. and CEO PARK SAM-KOO.

LG Group: 20, Yeouido-dong, Yeongdeungpo-gu, Seoul 100; tel. (2) 787-1114; fax (2) 785-7762; internet www.lg.co.kr; f. 1947; fmrly Lucky-Goldstar Group; chemicals and energy, electronics and telecommunications, financial services, etc.; Chair. KOO BON MOO.

Lotte Group: 23, 4-ga, Yangpyeong-dong, Yeongdeungpo-gu, Seoul; tel. (2) 670-6114; fax (2) 6672-6600; internet www.lotte.co.kr; f. 1967; foods and beverages, distribution, tourism and leisure, chemicals, construction and machinery; Chair. SHIN KYUK-HO.

Samsung Group: Taepyeong-no Bldg, 310, 2-ga, Taepyeong-no, Jung-gu, Seoul; tel. (2) 751-3355; fax (2) 3706-1212; internet www.samsungcorp.com; f. 1945; electronics, service industries, financial services, etc.; CEO PAE CHONG-YEUL; 174,000 employees.

SK Group: 26-4, Yeouido-dong, Yeongdeungpo-gu, Seoul; tel. (2) 758-5114; fax (2) 788-7001; e-mail info@sk.com; internet www.sk.co.kr; f. 1956; engineering, electronics, petroleum and gas, industry; Pres. and CEO JUNG MAN-WON; 25,000 employees.

Ssangyong Group: 24-1, 2-ga, Jeo-dong, Jung-gu, Seoul; tel. (2) 270-8155; fax (2) 273-0981; e-mail webadm@www.ssy.co.kr; internet www.ssangyong.co.kr; f. 1954; cement, construction materials, iron and steel, electronic goods, machinery, chemicals, automobiles, garments, textiles, etc.; Chair. KIM SEOK-WON; Pres. SON MYOUNG-WON.

Cement

Asia Cement Manufacturing Co Ltd: 726, Yoksam-dong, Gangnam-gu, CPOB 5278, Seoul; tel. (2) 527-6400; fax (2) 563-5839; e-mail webmaster@asiacement.co.kr; internet www.asiacement.co.kr; f. 1957; manufactures and exports Portland cement, sulphate resistant cement, concrete; cap. and res 418,076m., sales 249,237m. (2001); Chair. LEE BYUNG-MOO; Pres. KIM DONG-RYUL; 480 employees.

Hanil Cement Co Ltd: 832-2, Yeoksam-dong, Gangnam-gu, Seoul; tel. (2) 531-7000; fax (2) 531-7115; internet www.hanilcement.co.kr; f. 1961; cap. and res 555,454m., sales 484,913m. (2001); Chair. HU JUNG-SUP; Pres. JEONG HWAN-JIN; 766 employees.

Hyundai Cement Co Ltd: 1424-2, Seocho-dong, Seocho-gu, Seoul; tel. (2) 520-2114; fax (2) 520-2118; internet www.hdcement.co.kr; f. 1970; cap. and res 295,105m., sales 367,822m. (2001); mfrs of Portland cement and various building materials; Pres. KIM KWANG-YONG; 918 employees.

Ssangyong Cement Industrial Co Ltd: 24-1, 2-ga, Jeo-dong, Jung-gu, Seoul 100-748; tel. (2) 2270-5114; fax (2) 2270-5577; internet www.ssangyongcement.co.kr; f. 1962; cap. and res 1,092,683m., sales 1,273,500m.m. (2003); cement mfrs; mine excavating, exporting and importing, civil engineering; Chair. IMAMURA KAZUSUKE; Pres. and CEO MYUNG HO-KEUN; 1,321 employees.

Tong Yang Cement Corpn: TYSEC Bldg, 23-8, Yeouido-dong, Yeongdeungpo-gu, Seoul; tel. (2) 3770-3000; fax (2) 3770-3305; e-mail pr@tycement.co.kr; internet www.tycement.co.kr; f. 1957; cap. and res 261,627m., sales 1,440,781m. (2001); mfrs of Portland cement and ready-mixed concrete; Pres. ROH YOUNG-IN; Chair. HYUN JAE-HYUN; 1,040 employees.

Chemicals

DC Chemical Co Ltd: 50, Sogong-dong, Jung-gu, Seoul 100-718; tel. (2) 727-9500; fax (2) 777-0615; internet www.dcchem.co.kr; f. 1959 as Oriental Chemical Industries; assumed present name in April 2001; absorbed Korea Steel Chemical in March 2000; production of basic chemicals, agrochemicals and fine chemicals; cap. 940,000m., sales 1,639,000m. (2000); Pres. LEE BOK-YOUNG; Chair. LEE SOO-YOUNG; 1,301 employees.

Hanwha Chemical Corpn: Hanwha Bldg, 1, Changgyo-dong, Jung-gu, Seoul 100-797; tel. (2) 729-2700; fax (2) 729-2997; internet www.hanwha.co.kr; f. 1974; fmrly Hanyang Chemical Corpn; cap. and res 1,356,437m., sales 3,410,096m. (2001); mfrs of dynamite and other industrial explosives, safety fuses, electric detonators, ammunition, precision machinery and chemicals; Chair. and CEO PARK WON-BAE; 1,869 employees.

Korea Kumho Petrochemical Co Ltd: 15th–16th Floors, Kumho Bldg, 57, 1-ga, Shinmun-no, Jongno-gu, Seoul; tel. (2) 399-7560; fax (2) 399-9248; internet www.kkpc.co.kr; f. 1976; cap. and res 252,356m. (1999), sales 787,142m. (2001); Chair. PARK CHAN-KOO; 750 employees.

LG Chemical Ltd: 20, Yeouido-dong, Yeongdeungpo-gu, Seoul; tel. (2) 3773-7223; fax (2) 3773-7899; e-mail chparkb@mail.lgchem.lg.co.kr; internet www.lgchem.co.kr; f. 1947; cap. 365,400m., sales 4,744,500m. (2001); petrochemical products; Pres. and CEO NO KI-HO; 8,183 employees.

Construction

Daewoo Engineering and Construction Co Ltd: 541, 5-ga, Namdaemun-no, Jung-gu, Seoul; tel. (2) 759-2114; fax (2) 753-9489; e-mail webmaster@mail.dwconst.co.kr; internet www.dwconst.co.kr; f. 1973; cap. and res –18,727,099m., sales 22,134,018m. (1999); construction projects; Chair. LEE JUNG-KOO; Pres. and CEO NAM SANG-KOOK; 3,050 employees.

Dong Ah Construction Industrial Co Ltd: 120-23, Sosomun-dong, Jung-gu, Seoul; tel. (2) 3709-2114; fax (2) 3709-3000; e-mail webmaster@dongah.co.kr; internet www.dongah.co.kr; cap. and res 250,294m., sales 2,205,741m. (1999); contracting, construction; Chair. CHOI DONG-SUP; 4,956 employees.

Hanjin Heavy Industries and Construction Co Ltd: 546-1, Guui-dong, Seongdong-gu, Seoul; tel. (2) 450-8114; fax (2) 450-8101; e-mail webmaster@hanjinsc.com; internet www.hanjinsc.com; f. 1967; sales 1,834,100m. (2001); construction, electrical work, mining, gas and petroleum transport; Pres. PARK JAE-YOUNG; 5,295 employees.

Hyundai Engineering & Construction Co Ltd: 140-2, Kye-dong, Jongno-gu, Seoul; tel. (2) 746-1114; fax (2) 743-8963; internet www.hec.co.kr; f. 1947; merged with LG Semicon in 1999; cap. and res –885,177m., sales 392,246m. (2000), revenue 6,297,300m. (2001); engineering, manufacture and supply of civil, architectural and industrial plants and electrical works; CEO BANG JUNG-SUP; 4,500 employees.

LG Engineering and Construction Co Ltd: 537, 5-ga, Namdaemun-no, Jung-gu, Seoul; tel. (2) 3777-1114; fax (2) 774-6610; internet www.lgenc.co.kr; f. 1969; cap. and res –2,639,800m., sales 3,153,100m. (2001); contracting, construction; Chair. HUH CHANG-SOO; Pres. and CEOs KIM KAB-RYUL, HAROLD J. SHIN; 3,584 employees.

Samsung Engineering and Construction Corpn: Samsung Plaza Bldg, 263 Seohyon-dong, Bundang-gu, Seongnam-shi, Gyeonggi Prov., 463-721; tel. (2) 2145-6338; fax (2) 2145-6343; e-mail encmaster@samsung.com; internet www.secc.co.kr; f. 1938; cap. and res 3,073,370m., sales 36,289,400m. (2001); electronics, chemicals; Pres. and CEO LEE SANG-DAE; 5,600 employees.

Electrical and Electronics

Anam Electronics Co Ltd: 280-8, 2-ga, Songsu-dong, Seongdong-gu, Seoul; tel. (2) 460-5114; fax (2) 460-5393; internet www.aname.co.kr; f. 1956; cap. and res 1,156,676m. (2000), sales US $166m. (2001); electronics; Chair. NAM KWI-HYEN; 9,000 employees.

Daewoo Electronics Co Ltd: 541, Daewoo Center Bldg, Namdaemun-no 5-ga, Jung-gu, Seoul; tel. (2) 360-7114; fax (2) 360-7979; internet www.dwe.co.kr; f. 1972; cap. and res 3,088,000m., sales 3,031,000m. (2001); mfrs of computers, TV, hi-fi, microwave ovens, refrigerators and other consumer electronic components; Pres. and CEO KIM CHOONG-HOON; 8,700 employees.

Hynix Semiconductor Inc: San 136-1, Ami-ri, Pubal-up, Ichon-shi, Gyeonggi; tel. (336) 630-4114; fax (336) 630-4101; internet www.hynix.co.kr; f. 1949; fmrly Hyundai Electronics Industrial Co Ltd, merged with LG Semiconductor in 1999; cap. and res. 5,242,434m, sales 3,983,461m. (2001); mfr of electronic equipment for industries; Chair. and CEO WOO EUI-JEI; 14,000 employees.

LG Electronics Inc: LG Twin Towers, 20, Yeouido-dong, Yeongdeungpo-gu, Seoul 150-721; tel. (2) 3777-1114; fax (2) 3777-5304; internet www.lge.co.kr; f. 1958; cap. and res 4,265,236m., sales 16,600,971m. (2001); mfrs and exporters of electric and electronic products, incl. computers and communications equipment; merged with LG Information and Communications in 2000; Chair. and CEO S. S. KIM; 26,789 employees.

Orion Electric Co Ltd: 165, Gongdan-dong, Gumi-shi, Gyeongsanbuk-do; tel. (54) 469-5000; fax (54) 461-8779; internet www.orion.co.kr; f. 1965; cap. and res 507,827m. (1998), sales 617,048m. (2001); electrical; Pres. KIM YOUNG-NAM; 5,074 employees.

Samsung Electro-Mechanics Co Ltd: 314, Maetan 3-dong, Paldal-gu, Suwon-shi, Gyeonggi; tel. (331) 210-5114; fax (331) 210-5992; internet www.sem.samsung.co.kr; f. 1973; cap. and res 1,587,040m. (2000), sales 3,112,000m. (2001); mfr of electronic components; Pres. and CEO KANG HO-MOON; 8,500 employees.

Samsung Electronics Co Ltd: 11/F Samsung Main Bldg, 250, Taepyeong-no 2-ga, Jung-gu, Seoul; tel. (2) 727-7114; fax (2) 727-7159; internet www.samsungelectronics.com; f. 1969; cap. and res 19,474,000m., sales 46,444,000m. (2001); world's largest producer of dynamic random access memory (DRAM) semiconductor chips; mfrs of wide range of electronic goods, incl. TV, washing machines, hi-fi, refrigerators and industrial electronic equipment; CEO YUN JONG-YONG; 55,000 employees.

Samsung SDI Co Ltd: 575 Shin-dong, Paldal-gu, Jung-gu, Suwon-shi, Gyeonggi-do; tel. (331) 210-1114; fax (331) 210-7146; internet www.samsungsdi.co.kr; mfr of colour television tubes; cap. and res 5,631,000m., sales 4,030,000m. (2001); Chair. LEE KUN-HEE; Pres. and CEO KIM SOON-TAEK; 8,207 employees.

Samsung Techwin Co Ltd: 647-9, Yeoksam 1-dong, Gangnam-gu, Seoul; tel. (2) 3467-7114; fax (2) 3467-7050; internet www.samsung-smt.com; fmrly Samsung Aerospace Industries; mfrs of aerospace parts, semiconductors, industrial robots, optical and defence-related products; cap. and res 435,697m., sales 1,430,899m. (2000); Chair. LEE JOONG-KOO.

Engineering

Daewoo Heavy Industries and Machinery Co Ltd: 541, Namdaemun-no 5-ga, Chung-gu, Seoul; tel. (2) 726-3114; fax (2) 726-3307; internet www.dhiltd.co.kr; f. 1937; cap. and res 839,787m., sales 1,540,261m. (2001); mfrs of diesel engines, industrial vehicles, railroad carriages, industrial automated machinery, aircraft components; Pres. YANG JAE-SHIN; 4,377 employees.

Hyundai Heavy Industries Co Ltd: 1, Jeonha-dong, Dong-gu, Ulsan-shi, Gyeonsang nam-do; tel. (52) 230-3899; fax (52) 230-3450; e-mail ir@hhi.co.kr; internet www.hhi.co.kr; f. 1973; cap. and res 2,832,400m., sales 7,404,200m. (2001); industrial and offshore construction and engineering, shipbuilding; Pres. and CEOs CHOI KIL-SEON, MIN KEH-SIK; 26,507 employees.

Hyundai Mobis Co Ltd: 140-2, Kye-dong, Jongno-gu, Seoul 110-793; tel. (2) 746-1114; fax (2) 741-4244; internet www.mobis.co.kr; f. 1977; cap. and res 486,775m. (1998), sales 4,134,697.7m. (2002); largest container producer in the world; also manufactures heavy machinery, rolling stock, machine tools; Chair. CHUNG MONG-KOO; Pres. PARK JUNG-IN; 7,200 employees.

Samsung Heavy Industries Co Ltd: Samsung Yeoksam, 17th Floor, Bldg 649-7, Yeoksam-dong, Gangnam-gu, Seoul 135-080; tel. (2) 3458-6000; fax (2) 3458-6501; e-mail isnam@samsung.co.kr; internet www.shi.samsung.co.kr; f. 1974; cap. and res 1,715,288m., sales 4,110,559m. (2001), shipbuilding, industrial and construction equipment; Pres. and CEO KIM JING-WAN; 8,533 employees.

Iron and Steel

Dongkuk Steel Mill Co Ltd: 50, Suha-dong, Jung-gu, Seoul; tel. (2) 317-1149; fax (2) 317-1391; internet www.dongkuk.co.kr; f. 1954; cap. and res 440,060m., sales 1,785,169m. (2001); iron and steel mfrs; Pres. and CEO CHANG SAE-JOO; 1,599 employees.

INI Steel Co Ltd: 1, Songhyun-dong, Dong-gu, Incheon; tel. (32) 760-2114; fax (32) 760-2814; internet www.inisteel.co.kr; f. 1953; fmrly Inchon Iron and Steel Co Ltd, assumed present name in Aug. 2001; cap. and res 646,622m., sales 2,608,218m. (2001); mfrs of iron and steel products; Chair. and CEO RYU IN-GYUN; 4,499 employees.

Pohang Iron & Steel Co (POSCO) Ltd: 1, Goedong-dong, Pohang-shi, Gyeongbuk; tel. (562) 220-0114; fax (562) 220-6000; internet www.posco.co.kr; f. 1968; state-owned; proposed transfer to private-sector ownership announced in July 1998; cap. and res 1,301,722m., sales 11,086,119m. (2001); mfr of steel and steel products; CEO YOO SANG-BOO; Pres. LEE KU-TAEK; 19,012 employees.

Sammi Steel Co Ltd: 1004, Daechi-dong, Gangnam-gu, Seoul; tel. (2) 2222-4115; fax (2) 538-3806; e-mail webmaster@sammi.co.kr; internet www.sammi.co.kr; f. 1966; cap. and res 131,311m., sales 312,077m. (1997); mfr of steel products; owned by INI Steel; Pres. KIM DONG-SIK; 3,321 employees.

Motor Vehicles

GM Daewoo Motor Co Ltd: 199, Chongchon-dong, Bupyeong-gu, Incheon; tel. (32) 520-2001; fax (32) 520-4606; e-mail webmaster@dm.co.kr; internet www.dm.co.kr; f. 1972; absorbed by General Motors (USA) in April 2002; cap. and res. 4,072,950m., sales 5,119,125m. (1998); mfr of buses, passenger cars, heavy-duty trucks; Pres. and CEO DAVID NICK REILLY; 13,555 employees.

Hyundai Motor Co Ltd: 231, Yangjae-dong, Seocho-gu, Seoul; tel. (2) 3464-1114; fax (2) 3464-3453; internet www.hyundai-motor.com; f. 1967; cap. and res 9,097,811m., sales 22,505,100m. (2001); mfrs and assemblers of passenger cars, trucks, buses, etc.; Chair. CHUNG MONG-KOO; Pres. KIM DONG-JIN; 48,831 employees.

Renault Samsung Motors: 25, 1-ga, Bongrae-dong, Jung-gu, Seoul; tel. (2) 3707-5000; fax (2) 757-4577; internet www.renaultsamsungm.com; f. 2000; mfr of passenger cars; cap. and res 440,000m., sales 836,300m. (2001); Chair. JEROME STOLL; 3,970 employees.

Ssangyong Motor Co Ltd: 150-3, Chilgoe-dong, Pyeongtaek-shi, Gyeonggi-do; tel. (31) 610-1114; fax (31) 610-3700; internet www.smotor.com; f. 1954; mfrs of passenger cars, minibuses and special-purpose vehicles; cap. and res 252,927m., sales 2,326,694m. (2001); Chair. SO JIN-KWAN; Man. Dir JIN CHANG-KI; 6,126 employees.

Textiles, Silk and Synthetic Fibres

Cheil Industries Inc: 290, Kandan-dong, Gumi-shi, Gyeongbuk; tel. (546) 468-2114; fax (546) 468-2229; internet www.cii.samsung.com; f. 1954; owned by Samsung Corp; mfrs of clothing, textiles and petrochemical products; cap. and res 250,000m., sales 1,736,000m. (2001); Pres. and CEO AHH BOK-HYUN; 1,800 employees.

Hanil Synthetic Fiber Co Ltd: 222, Yangdeok-dong, Hoewon-gu, Masan-shi, Gyeonsangnam-do; tel. (551) 90-3114; fax (551) 90-3114; internet www.hanilsf.com; f. 1964; cap. and res 307,270m., sales 345,027m. (2001); mfrs and exporters of synthetic fibre; Pres. SON BYUNG-SUK; 1,000 employees.

Ilshin Spinning Co Ltd: 15-15, Yeouido-dong, Yeongdeungpo-gu, Seoul; tel. (2) 3774-0114; fax (2) 786-5893; internet www.ilshin.co.kr; f. 1951; cap. and res 311,200m., sales 326,100m. (2001); cotton spinning and production of yarn and fabrics; one subsidiary with dyeing and finishing factories; import and export; Chair. KIM YOUNG-HO; 1,120 employees.

Taekwang Industries: 162-1, Jangchung-dong, Jung-gu, Seoul; tel. (2) 3406-0300; fax (2) 2273-9160; internet www.taekwang.co.kr; f. 1954; miscellaneous household goods and textiles; cap. and res 5,567m., sales 1,079,814m. (2001); Pres. LEE HO-JIN; 6,499 employees.

Miscellaneous

Daelim Industrial Co Ltd: 146-12 Susong-Dong, Jongro-Gu, Seoul; tel. (2) 2011-7114; fax (2) 2011-8000; internet www.daelim.co.kr; f. 1939; cap. and res 1,590,732m., sales 2,516,401m. (2001);

general contractor for all construction fields, engineering and petrochemical producer; Chair. LEE YONG-KU; 2,500 employees.

Hankuk Glass Industry Co Ltd: 64-5 Jungmu-no, 2-ga, Jung-gu, Seoul; tel. (2) 3706-9114; fax (2) 771-5340; internet www.hanglas.co.kr; f. 1957; cap. and res 634,889m., sales 333,987m. (2001); mfrs of flat glass, figured glass and tube glass; one subsidiary; Pres. KIM SUNG-MAN; 828 employees.

Korea Coal Corpn: 33, Yeouido-dong, Yeongdeungpo-gu, Seoul; tel. (2) 767-6600; fax (2) 782-4010; internet www.kocoal.or.kr; f. 1950; cap. and res 181,162m., sales 166,259m. (2001); operation and development of coal mines, and related research, sales, import, export, etc.; Chair. YOO SUNG-KYU; 3,204 employees.

Korea Land Corpn: 217, Jeongja-dong, Seongnam-shi, Gyeonggi Prov.; tel. (31) 738-7114; fax (31) 717-5431; internet www.koland.co.kr; f. 1975; land development; Chair. KIM JIN-HO.

Korea National Housing Corpn: 175, Gumi-dong, Bundang-gu, Seongnam-shi, Gyeonggi Prov.; tel. (31) 738-3114; internet www.knhc.co.kr; f. 1962; housing business; Chair. KWON HAE-OK.

Korea Tobacco and Ginseng Corpn: 100, Pyeongchon-dong, Daedeok-gu, Daejeon 306-130; tel. (42) 939-5122; fax (42) 933-5128; internet www.ktg.or.kr; cap. and res 2,615,947m., sales 4,713,340m. (2001); mfr of cigarettes, tobacco, and ginseng products; Pres. and CEO KWAK JOO-YOUNG; 4,430 employees.

Shinsegae Co Ltd: 25-5, 1-ga, Jungmu-no, Jung-gu, Seoul; tel. (2) 727-1449; fax (2) 727-1192; internet www.shinsegae.com; retailing and department stores; cap. and res 901,022m., sales 3,722,092m. (2000); Pres. and CEO KOO HAK-SUH.

SK Corpn: 99 Seorin-dong, Jongno-gu, Seoul; tel. (2) 2121-5114; fax (2) 2121-7001; internet www.skcorp.com; f. 1962; fmrly Yukong Ltd; production and marketing of petroleum, petrochemical and lubricating oil products; cap. and res 5,648,240m., sales 14,114,861m. (2001); Chair. (vacant); 4,541 employees.

SK Networks Co Ltd: 226, 1-ga, Shinmun-no, Jongno-gu, Seoul; tel. (2) 758-2114; fax (2) 754-9414; e-mail webmaster@skglobal.com; internet www.skglobal.com, www.sknetworks.co.kr; f. 1953; fmrly Sunkyong Ltd; fmrly SK Global; cap. and res 2,220,817m., sales 18,036,333m. (2001); provision of services to various sectors, incl. energy, chemicals, telecommunications, engineering and construction; Pres. MAN WON-CHUNG; 2,485 employees.

S-Oil Corpn: Yeouido POB 758, 60, Yeouido-dong, Yeongdeungpo-gu, Seoul 150-607; tel. (2) 3772-5151; fax (2) 786-4031; internet www.s-oil.com; oil refining and production of petrochemicals and lubricants; cap. and res 1,379,768m., sales 7,623,771m. (2001); Chair. and CEO KIM SUN-DONG; Pres. YOO HO-KI.

Whashin Industrial Co Ltd: 360-1, Magok-dong, Gangseo-gu, Seoul 157-210; tel. (2) 3661-5343; fax (2) 3661-5347; e-mail export@whashin.com; internet www.whashin.com; f. 1962; exporters, importers, domestic sales of stationery, textiles, electrical consumer products, commercial air-conditioning equipment and other merchandise; 8 subsidiaries; Pres. PARK JUNG-KYU; 2,000 employees.

TRADE UNIONS

Federation of Korean Trade Unions (FKTU): FKTU Bldg, 168-24, Chungam-dong, Yongsan-ku 141-050, Seoul; tel. (2) 715-3954; fax (2) 715-7790; e-mail fktuintl@fktu.or.kr; internet www.fktu.org; f. 1941; Pres. LEE YONG-DEUK; affiliated to ICFTU; 29 union federations are affiliated with a membership of some 960,000.

Federation of Foreign Organization Employees' Unions: 5-1, 3-ga, Dangsan-dong, Yeongdeungpo-gu, Seoul; tel. (2) 2068-1645; fax (2) 2068-1644; f. 1961; Pres. KANG IN-SIK; 22,450 mems.

Federation of Korean Apartment Workers' Unions: 922-1, Bangbae-dong, Seocho-gu, Seoul; tel. (2) 522-6860; fax (2) 522-4624; f. 1997; Pres. LEE DAE-HYUNG; 3,670 mems.

Federation of Korean Chemical Workers' Unions: Sukchun Bldg, 2nd Floor, 32-100, 4-ga, Dangsan-dong, Yeongdeungpo-gu, Seoul; tel. (2) 761-8251; fax (2) 761-8255; e-mail fkcu@chollian.net; internet www.fkcu.or.kr; f. 1961; Pres. PARK HUN-SOO; 116,286 mems.

Federation of Korean Metalworkers' Unions: 1570-2, Sinrim-dong, Gwanak-gu, Seoul; tel. (2) 864-2901; fax (2) 864-0457; e-mail fkmtu@chollian.net; internet www.metall.or.kr; f. 1961; Pres. LEE BYUNG-KYUN; 130,000 mems.

Federation of Korean Mine Workers' Unions: Guangno Bldg, 2nd Floor, 10-4, Karak-dong, Songpa-gu, Seoul; tel. (2) 403-0973; fax (2) 400-1877; f. 1961; Pres. KIM DONG-CHUL; 6,930 mems.

Federation of Korean Printing Workers' Unions: 201, 792-155, 3-ga, Guro-dong, Guro-gu, Seoul; tel. (2) 780-7969; fax (2) 780-6097; f. 1961; Pres. LEE KWANG-JOO; 5,609 mems.

Federation of Korean Public Construction Unions: 293-1, Kumdo-dong, Sujong-gu, Seongnam-si, Gyeonggi-do; tel. (2) 2304-7016; fax (2) 230-4602; f. 1998; Pres. HONG SANG-KI (acting); 9,185 mems.

Federation of Korean Public Service Unions: Sukchun Bldg, 3rd Floor, 32-100, 4-ga, Dangsan-dong, Yeongdeungpo-gu, Seoul; tel. (2) 769-1330; fax (2) 769-1332; internet www.fkpu.or.kr; f. 1997; Pres. LEE KWAN-BOO; 15,641 mems.

Federation of Korean Rubber Workers' Unions: 830-240, 2-ga, Bumil-dong, Dong-gu, Busan; tel. (51) 637-2101; fax (51) 637-2103; f. 1988; Pres. CHO YUNG-SOO; 6,600 mems.

Federation of Korean Seafarers' Unions: 544, Donhwa-dong, Mapo-gu, Seoul; tel. (2) 716-2764; fax (2) 702-2271; e-mail fksu@chollian.net; internet www.fksu.or.kr; f. 1961; Pres. KIM PIL-JAE; 60,037 mems.

Federation of Korean State-invested Corporation Unions: Sunwoo Bldg, 501, 350-8, Yangjae-dong, Seocho-gu, Seoul; tel. (2) 529-2268; fax (2) 529-2270; internet www.publicunion.or.kr; f. 1998; Pres. JANG DAE-IK; 19,375 mems.

Federation of Korean Taxi & Transport Workers' Unions: 415-7, Janan 1-dong, Dongdaemun-gu, Seoul; tel. (2) 2210-8500; fax (2) 2247-7890; internet www.ktaxi.or.kr; f. 1988; Pres. KWAN OH-MAN; 105,118 mems.

Federation of Korean Textile Workers' Unions: 274-8, Yeomchang-dong, Gangseo-gu, Seoul; tel. (2) 3665-3117; fax (2) 3662-4373; f. 1954; Pres. OH YOUNG-BONG; 6,930 mems.

Federation of Korean United Workers' Unions: Sukchun Bldg, 32-100, 4-ga, Dangsan-dong, Yeongdeungpo-gu, Seoul; internet www.fkuwu.or.kr; f. 1961; 51,802 mems.

Federation of Korean Urban Railway Unions: Urban Railway Station, 3-ga, Yeouido-dong, Yeongdeungpo-gu, Seoul; tel. (2) 786-5163; fax (2) 786-5165; f. 1996; Pres. HA WON-JOON; 9,628 mems.

Korea Automobile & Transport Workers' Federation: 678-27, Yeoksam-dong, Gangnam-gu, Seoul; tel. (2) 554-0890; fax (2) 554-1558; f. 1963; Pres. KANG SUNG-CHUN; 84,343 mems.

Korea Federation of Bank & Financial Workers' Unions: 88, Da-dong, Jung-gu, Seoul; tel. (2) 756-2389; fax (2) 754-4893; internet www.kfiu.org; f. 1961; 113,994 mems.

Korea Federation of Communication Trade Unions: 10th Floor, 106-6, Guro 5-dong, Guro-gu, Seoul; tel. (2) 864-0055; fax (2) 864-5519; internet www.ictu.co.kr; f. 1961; Pres. OH DONG-IN; 18,810 mems.

Korea Federation of Food Industry Workers' Unions: 106-2, 1-ga, Yanpyeong-dong, Yeongdeungpo-gu, Seoul; tel. (2) 679-6441; fax (2) 679-6444; f. 2000; Pres. BAEK YOUNG-GIL; 19,146 mems.

Korea Federation of Port & Transport Workers' Unions: Bauksan Bldg, 19th Floor, 12-5, Dongja-dong, Yeongsan-gu, Seoul; tel. (2) 727-4741; fax (2) 727-4749; f. 1980; Pres. CHOI BONG-HONG; 33,347 mems.

Korea National Electrical Workers' Union: 167, Samseong-dong, Gangnam-gu, Seoul; tel. (2) 3456-6017; fax (2) 3456-6004; internet www.knewu.or.kr; f. 1961; Pres. KIM JU-YOUNG; 16,741 mems.

Korea Professional Artist Federation: Hanil Bldg, 43-4, Donui-dong, Jongno-gu, Seoul; tel. (2) 764-5310; fax (2) 3675-5314; f. 1999; Pres. PARK IL-NAM; 2,395 mems.

Korea Tobacco & Ginseng Workers' Unions: 100, Pyeongchon-dong, Daedeok-gu, Daejeon; tel. (42) 932-7118; fax (42) 931-1812; f. 1960; Pres. KANG TAE-HEUNG; 6,008 mems.

Korea Unions of Teaching and Educational Workers: Dongin Bldg, 7th Floor, 65-33, Singil 1-dong, Yeongdeungpo-gu, Seoul; tel. (2) 849-1281; fax (2) 835-0556; internet www.kute.or.kr; f. 1999; Pres. SON KYUNG-SOON; 18,337 mems.

Korean Postal Workers' Union: 154-1, Seorin-dong, Jongno-gu, Seoul 110-110; tel. (2) 2195-1773; fax (2) 2195-1761; e-mail cheshin@chol.com; internet www.kpwu.or.kr; f. 1958; Pres. JUNG HYUN-YOUNG; 23,500 mems.

Korean Railway Workers' Union: 40, 3-ga, Hangang-no, Yeongsan-gu, Seoul; tel. (2) 795-6174; f. 1947; Pres. KIM JONG-WOOK; 31,041 mems.

Korean Tourist Industry Workers' Federation: 749, 5-ga, Namdaemun-no, Jung-gu, Seoul 100-095; tel. (2) 779-1297; fax (2) 779-1298; f. 1970; Pres. JEONG YOUNG-KI; 27,273 mems.

National Medical Industry Workers' Federation of Korea: 134, Sincheon-dong, Seodaemun-gu, Seoul; tel. (2) 313-3900; fax (2) 393-6877; f. 1999; Pres. LEE YONG-MOO; 5,610 mems.

Korean Confederation of Trade Unions: 139, 2-ga, Yeouido-dong, Yeongdeungpo-gu, Seoul; tel. (2) 635-1133; fax (2) 635-1134; internet www.kctu.org; f. 1995; legalized 1999; Chair. LEE SOO-HO; c. 600,000 mems.

Transport

RAILWAYS

At the end of 2001 there were 6,819 km (including freight routes) of railways in operation. Construction of a new high-speed rail system connecting Seoul to Busan (412 km) via Cheonan, Daejeon, Daegu, and Gyungju, was under way in 2003. The first phase, Seoul–Daejeon, was scheduled for completion in 2004. The second phase, Daejeon–Busan, was scheduled for completion in 2010. Construction work on the reconnection of the Kyongui (West coast, Sinuiju (North Korea)–Seoul) and East Coast Line (Wonsan (North Korea)–Seoul) began in September 2002. The two lines were officially opened in June 2003, but were not yet open to traffic, as construction work on the Northern side remained to be completed. Reconnection work was subject to disruption by the changing political situation. Eventually the two would be linked to the Trans-China and Trans-Siberian railways respectively, greatly enhancing the region's transportation links.

Korean National Railroad: 920, Dunsan-dong, Seo-gu, Daejeon 302-701; tel. (42) 1544-7788; fax (42) 481-373; internet www.korail.go.kr; f. 1963; operates all railways under the supervision of the Ministry of Construction and Transportation; total track length of 6,819 km (2001); Admin. SON HAK-LAE.

City Underground Railways

Busan Subway: Busan Urban Transit Authority, 861-1, Bumchun-dong, Busan 614-021; tel. (51) 633-8783; e-mail ipsubway@buta.or.kr; internet subway.busan.kr; f. 1988; length of 71.6 km (2 lines, with a further 3rd line under construction); Pres. LEE HYANG-YEUL.

Daegu Metropolitan Subway Corpn: 1500 Sangin 1-dong, Dalseo-gu, Daegu; tel. (53) 640-2114; fax (53) 640-2229; e-mail webmaster@daegusubway.co.kr; internet www.daegusubway.co.kr; length of 28.3 km (1 line, with a further five routes totalling 125.4 km planned or under construction); Pres. YOON JIN-TAE.

Incheon Rapid Transit Corpn: 67-2, Gansok-dong, Namdong-gu, Incheon 405-233; tel. (32) 451-2114; fax (32) 451-2160; internet www.irtc.co.kr; length of 24.6 km (22 stations, 1 line), with two further lines planned; Pres. CHOUNG IN-SOUNG.

Seoul Metropolitan Rapid Transit Corporation: Seoul; internet www.smrt.co.kr; operates lines 5-8.

Seoul Metropolitan Subway Corpn: 447-7, Bangbae-dong, Seocho-gu, Seoul; tel. (2) 520-5020; fax (2) 520-5039; internet www.seoulsubway.co.kr; f. 1981; length of 134.9 km (115 stations, lines 1-4); Pres. KIM JUNG-GOOK.

Underground railways were also under construction in Daejeon and Gwangju.

ROADS

At the end of 2001 there were 91,396 km of roads, of which 76.7% were paved. A network of motorways (2,637 km) links all the principal towns, the most important being the 428-km Seoul–Busan motorway. Improvements in relations with North Korea resulted in the commencement of work on a four-lane highway to link Seoul and the North Korean capital, Pyongyang, in September 2000. In February 2003 a road link between the two countries was reportedly opened.

Korea Highway Corpn: 293-1, Kumto-dong, Sujong-gu, Seongnam, Gyeonggi Prov.; tel. (822) 2230-4114; fax (822) 2230-4308; internet www.freeway.co.kr; f. 1969; responsible for construction, maintenance and management of toll roads; Pres. OH JUM-LOCK.

SHIPPING

In December 2003 South Korea's merchant fleet (2,604 vessels) had a total displacement of 6,757,400 grt. Major ports include Busan, Incheon, Donghae, Masan, Yeosu, Gunsan, Mokpo, Pohang, Ulsan, Jeju and Gwangyang.

Korea Maritime and Port Authority: 112-2, Inui-dong, Jongno-gu, Seoul 110; tel. (2) 3466-2214; f. 1976; operates under the Ministry of Maritime Affairs and Fisheries; supervises all aspects of shipping and port-related affairs; Admin. AHN KONG-HYUK.

Korea Shipowners' Association: Sejong Bldg, 10th Floor, 100, Dangju-dong, Jongno-gu, Seoul 110-071; tel. (2) 739-1551; fax (2) 739-1565; e-mail korea@shipowners.or.kr; internet www.shipowners.co.kr; f. 1960; 40 shipping co mems; Chair. HYUN YUNG-WON.

Korea Shipping Association: 66010, Dungchon 3-dong, Gangseo-gu, Seoul 157-033; tel. (2) 6096-2024; fax (2) 6096-2029; e-mail kimny@haewoon.co.kr; internet www.haewoon.co.kr; f. 1962; management consulting and investigation, mutual insurance; 1,189 mems; Chair. PARK HONG-JIN.

Principal Companies

Cho Yang Shipping Co Ltd: Chongam Bldg, 85-3, Seosomun-dong, Jung-gu, CPOB 1163, Seoul 100; tel. (2) 3708-6000; fax (2) 3708-6926; internet www.choyang.co.kr; f. 1961; Korea–Japan, Korea–China, Korea–Australia–Japan, Asia–Mediterranean–America liner services and world-wide tramping; Chair. N. K. PARK; Pres. J. W. PARK.

DooYang Line Co Ltd: 166-4, Samseong-dong, Gangnam-gu, Seoul 135-091; tel. (2) 550-1700; fax (2) 550-1777; internet www.dooyang.co.kr; f. 1984; world-wide tramping and conventional liner trade; Pres. CHO DONG-HYUN.

Hanjin Shipping Ltd: 25-11, Yeouido-dong, Yeongdeungpo-gu, Seoul; tel. (2) 3770-6114; fax (2) 3770-6740; internet www.hanjin.com; f. 1977; marine transportation, harbour service, warehousing, shipping and repair, vessel sales, harbour department and cargo service; Pres. CHOI WON-PYO.

Hyundai Merchant Marine Co Ltd: 66, Jeokseon-dong, Jongno-gu, Seoul 110-052; tel. (2) 3706-5114; fax (2) 723-2193; internet www.hmm.co.kr; f. 1976; Chair. HYUN YUNG-WON.

Korea Line Corpn: Dae Il Bldg, 43, Insa-dong, Jongno-gu, Seoul 110-290; tel. (2) 3701-0114; fax (2) 733-1610; f. 1968; world-wide transportation service and shipping agency service in Korea; Pres. JANG HAK-SE.

Pan Ocean Shipping Co Ltd: 51-1, Namchang-dong, Jung-gu, CPOB 3051, Seoul 100-060; tel. (2) 316-5114; fax (2) 316-5296; f. 1966; transportation of passenger cars and trucks, chemical and petroleum products, dry bulk cargo; Pres. CHIANG JIN-WON.

CIVIL AVIATION

There are seven international airports in Korea; at Incheon (Seoul), Gimpo (Seoul), Busan, Cheongju, Daegu, Gwangju, Jeju and Yangyang. The main gateway into Seoul is Incheon International Airport, which opened for service in March 2001. It is used by 30m. passengers annually, and has a capacity for 240,000 aircraft movements annually. The second phase began construction in 2002, with completion due by 2008. When complete, the airport will handle 44m. passengers and 4.5m. tons of cargo annually. The airport is located 52 km from Seoul. A new airport, Yangyang International Airport, opened in Gangwon province in April 2002.

Asiana Airlines Inc: 47, Osae-dong, Gangseo-gu, Seoul; tel. (2) 758-8114; fax (2) 758-8008; e-mail asianacr@asiana.co.kr; internet www.asiana.co.kr; f. 1988; serves 14 domestic cities and 36 destinations in 16 countries; CEO PARK SAM-KOO.

Korean Air: 1370, Gonghang-dong, Gangseo-gu, Seoul; tel. (2) 656-7092; fax (2) 656-7289; internet www.koreanair.com; f. 1962 by the Govt, privately owned since 1969; fmrly Korean Air Lines (KAL); operates domestic and regional services and routes to the Americas, Europe, the Far East and the Middle East, serving 73 cities in 26 countries; Pres. and CEO SHIM YI-TAEK.

Seoul Air International: CPOB 10352, Seoul 100-699; tel. (2) 699-0991; fax (2) 699-0954; operates domestic flights and routes throughout Asia.

Tourism

South Korea's mountain scenery and historic sites are the principal attractions for tourists. Jeju Island, located some 100 km off the southern coast, is a popular resort. In 2002 there were 5,347,468 visitors to South Korea, of whom more than 43% came from Japan. Receipts from tourism in 2002 amounted to US $5,277m.

Korea National Tourism Organization: KNTO Bldg, 10, Dadong, Jung-gu, CPOB 903, Seoul 100; tel. (2) 729-9600; fax (2) 757-5997; internet www.knto.or.kr; f. 1962 as Korea Tourist Service; Pres. LEE DEUK-RYUL.

Korea Tourism Association: Saman Bldg, 11th Floor, 945, Daechi-dong, Gangnam-gu, Seoul; tel. (2) 556-2356; fax (2) 556-3818; f. 1963; Pres. CHO HANG-KYU, KIM JAE-GI.

Defence

In August 2003 the strength of the active armed forces was 686,000 (including an estimated 159,000 conscripts): army 560,000, navy 63,000, air force 63,000. Paramilitary forces included a 3.5m.-strong civilian defence corps. Military service is compulsory and lasts for 26 months in the army, for 30 months in the navy and in the air force. In August 2003 38,500 US troops were stationed in South Korea.

Defence Expenditure: Budgeted at 17,900,000m. won for 2003.

Chairman of the Joint Chiefs of Staff: Gen. KIM JONG-HWAN.

Chief of Staff (Army): Gen. NAM JAE-JOON.

Chief of Staff (Air Force): LEE HAN-HO.

Chief of Naval Operations: Adm. MOON JUNG-IL.

Education

Education, available free of charge, is compulsory for nine years between the ages of six and 15 years. Primary education begins at six years of age and lasts for six years. In 2001 enrolment at primary schools included 98.2% of children in the appropriate age-group. Secondary education begins at 12 years of age and lasts for up to six years, comprising two cycles of three years each. Enrolment at secondary schools in 2001 included 96.7% of children (middle schools 98.0%, high schools 95.3%). A five-day school week was to be gradually introduced during 2003–05. In 2003 there were 169 colleges and universities with a student enrolment of 1.8m. There were 1,010 graduate schools in 2003. In 2001, according to UNESCO estimates, the rate of adult literacy averaged 97.9% (males 99.2%, females 96.6%). Expenditure on education by the central Government in 2002 was projected at 22,278,358m. won, representing 16.3% of total spending.

Bibliography of the Democratic People's Republic of Korea and the Republic of Korea

GENERAL

Akaha, Tsuneo (Ed.). *The Future of North Korea*. London, Routledge, 2002.

Armstrong, Charles. K. *Korean Society: Civil Society, Democracy, and the State*. London, Routledge, 2002.

Australian National Korean Studies Centre. *Korea to the Year 2000*. Canberra, East Asia Analytical Unit, Department of Foreign Affairs and Trade, 1992.

Becker, Jasper. *Rogue State: Understanding North Korea and its Continuing Threat*. New York, NY, Oxford University Press, 2004.

Bedeski, Robert E. *The Transformation of South Korea*. London, Routledge, 1994.

Belke, Thomas J. *Juche: A Christian Study of North Korea's State Religion*. Bartlesville OK, Living Sacrifice Book Co, 1999.

Bermudez, Joseph S., Jr. *The Armed Forces of North Korea*. London and New York, I. B. Tauris, 2001.

Breen, Michael. *The Koreans: Who They Are, What They Want, Where Their Future Lies*. London, Orion, 1999.

Kim Jong-Il: North Korea's Dear Leader. John Wiley & Sons, 2004.

Chamberlain, Paul F. and Kim Kihwan. *Korea 2010: The Challenges of the New Millennium*. Washington, DC, Center for Strategic and International Studies, 2001.

Cha, Victor and Kang, David. *Nuclear North Korea—A Debate on Engagement Strategies*. Irvington, NY, Columbia University Press, 2003.

Clark, Donald (Ed.). *Korea Briefing 1993*. Boulder, CO, Westview Press for the Asia Society, 1993. (Also other years.).

Connor, Mary. E. *The Koreas: A Global Studies Handbook*. Santa Barbara, CA, ABC-Clio, 2002.

Cornell, Erik. *North Korea Under Communism: Report of an Envoy to Paradise*. London, RoutledgeCurzon, 2002.

Cotton, James (Ed.). *Politics and Policy in the New Korean State: from Roh Tae-woo to Kim Young-sam*. Melbourne, Longman, and New York, St Martin's Press, 1995.

Covell, Jon C. *Korea's Cultural Roots*. Seoul, Moth House/Hollym, 1981.

Cumings, Bruce. *North Korea: Another Country*. New York, NY, New Press, 2003.

De Mente, Boye. *NTC's Dictionary of Korea's Business and Cultural Code Words*. Chicago, NTC, 1998.

Dong Wonmo (Ed.). *The Two Koreas and the United States—Issues of Peace, Security, and Economic Cooperation*. Armonk, NY, M. E. Sharpe, 2000.

Eberstadt, Nicholas and Ellings, Richard J. (Eds). *Korea's Future and the Great Powers*. Seattle and London, University of Washington Press, 2001.

Eder, Norman R. *Poisoned Prosperity—Development, Modernization, and the Environment in South Korea*. Armonk, NY, M. E. Sharpe, 1995.

Eder, Norman R., and Hong, Wuk-hee (Eds). *Re-inventing Han—Continuity and Change in Modern South Korea*. Armonk, NY, M. E. Sharpe, 1999.

European Union Chamber of Commerce in Korea. *A Practical Business Guide on the Democratic People's Republic of Korea*. Seoul, EUCCK, 1998.

Foster-Carter, Aidan. *Korea's Coming Reunification: Another East Asian Superpower?* London, Economist Intelligence Unit, 1992.

Grinker, Roy Richard. *Korea and Its Futures: Unification and the Unfinished War*. New York, St Martin's Press, 1998.

Harrison, Selig S. *Korean Endgame: A Strategy for Reunification and US Disengagement*. Princeton, NJ, Princeton University Press, 2002.

Henriksen, Thomas H., and Lho, Kyongsoo (Eds). *One Korea? Challenges and Prospects for Reunification*. Stanford, CA, Hoover Institution Press, 1994.

Hoare, James E., and Pares, Susan. *Conflict in Korea: An Encyclopedia*. Santa Barbara, CA, ABC-Clio, 1999.

Kang Chol Hwan and Rigoulot, Pierre. *The Aquariums of Pyongyang: Ten Years in the North Korean Gulag*. New York, Basic Books, 2001.

Kihl Young Whan. *Politics and Policies in Divided Korea: Regimes in Contrast*. Boulder, CO, and London, Westview Press, 1984.

(Ed.). *Korea and the World: Beyond the Cold War*. Boulder, CO, Westview Press, 1994.

Kim Byung-Lo Philo. *Two Koreas in Development*. NJ, Transaction Books, 1992.

Kim Dae Hwan and Kong Tat Yan (Eds). *The Korean Peninsula in Transition*. London, Macmillan, 1997.

Kim, Samuel S. (Ed.). *The North Korean System in the Post Cold War Era*. New York, Palgrave Macmillan, 2001.

Koo, Hagen (Ed.). *State and Society in Contemporary Korea*. Ithaca, NY, Cornell University Press, 1993.

Lee Hyangjin. *Contemporary Korean Cinema: Identity, Culture, and Politics*. Manchester, Manchester University Press, 2001.

Lewis, James B., and Sesay, Amadu (Eds). *Korea and Globalization: Politics, Economics, and Culture*. London, RoutledgeCurzon, 2002.

Macdonald, Donald S. (Clark, Donald, Ed.). *The Koreans: Contemporary Politics and Society*. Boulder, CO, Westview Press (3rd edn), 1996.

Mack, Andrew (Ed.). *Asian Flashpoint: Security and the Korean Peninsula*. St Leonards, NSW, Allen and Unwin, 1993.

McCune, Shannon. *Korea's Heritage: A Regional and Social Geography*. Tokyo and Rutland, VT, Charles E. Tuttle, 1966.

Moltz, James Clay, and Mansourov, Alexandre Y. (Eds). *The North Korean Nuclear Program: Security, Strategy, and New Perspectives from Russia*. New York and London, Routledge, 2000.

Moon Chung-In (Ed.). *Understanding Regime Dynamics in North Korea*. Seoul, Yonsei University Press, 1998.

Nahm, Andrew C. (Ed.). *Studies in the Developmental Aspects of Korea*. Kalamazoo, The Center of Korean Studies, Western Michigan University, 1969.

North Korea: Her Past, Reality, and Impression. Kalamazoo, The Center for Korean Studies, Western Michigan University, 1978.

Natsios, Andrew S. *The Great North Korean Famine*. Washington, DC, United States Institute of Peace, 2001.

Noland, Marcus. *Korea After Kim Jong Il*. Washington, DC, Institute for International Economics, 2003.

Oh, Kongdan, and Hassig, Ralph C. (Eds). *North Korea Through the Looking Glass*. Washington, DC, Brookings Institution, 2000.

Korea Briefing 2000–01: First Steps Toward Reconciliation and Reunification. Armonk, NY, M. E. Sharpe, 2002.

Park, Han S. *North Korea: The Politics of Unconventional Wisdom*. Boulder, CO, Lynne Rienner, 2002.

Park Kyung-Ae (Ed.). *Korean Security Dynamics in Transition*. New York, Palgrave Macmillan, 2001.

Pollack, Jonathan D., and Lee, Chung Min. *Preparing for Korean Unification: Scenarios and Implications*. Santa Monica, CA, RAND, 1999.

Potrzeba Lett, Denise. *In Pursuit of Status: The Making of South Korea's 'New' Urban Middle Class*. Cambridge, MA, Harvard University Press, 2002.

Ryang, Sonia. *North Koreans in Japan: Language, Ideology and Identity*. Boulder, CO, Westview Press, 1997.

Savada, Andrea Matles (Ed.). *North Korea: A Country Study*. Washington, DC, Library of Congress for Department of the Army, 1994.

Scalapino, Robert, and Kim, Jun-Yop (Eds). *North Korea Today: Strategic and Domestic Issues*. Los Angeles, University of California Press, 1984.

Shin Dong-Myeon. *Social and Economic Policies in Korea*. London, RoutledgeCurzon, 2003.

Sigal, Leon V. *Disarming Strangers: Nuclear Diplomacy with North Korea*. Princeton, NJ, Princeton University Press, 1998.

Smith, Hazel, et al (Eds). *North Korea in the New World Order*. London, Macmillan, 1996.

Snyder, Scott. *Negotiating on the Edge: North Korean Negotiating Behavior*. US Institute of Peace, 1999.

Soh, Chunghee Sarah. *Women in Korean Politics*. Boulder, CO, Westview Press, 1993.

Steinberg, David. *The Republic of Korea: Economic Transformation and Social Change*. Boulder, CO, Westview Press, 1991.

Won Dal Yang. *Korean Ways, Korean Mind*. Seoul, Tamgu Dang Book Centre, 1983.

Yang Sung-Chul. *The North and South Korean Political Systems: A Comparative Analysis*. Elizabeth, NJ, Hollym, 2001 (revised edition).

Yoon Chang-Ho and Lau, Lawrence J. *North Korea in Transition*. Cheltenham, Edward Elgar, 2001.

Yu Chai-Shin. *The Founding of Catholic Tradition in Korea*. Berkeley, CA, Asian Humanities Press, 2002.

HISTORY

Bandow, Doug, and Galen Carpenter, Ted (Eds). *The US-South Korean Alliance*. New Brunswick, NJ, and London, Transaction Publishers, 1993.

Bateman, Robert. *No Gun Ri: A Military History of the Korean War Incident*. Mechanicsburg, PA, Stackpole, 2002.

Breuer, William B. *Shadow Warriors*. New York, John Wiley and Sons, 1996.

Bridges, Brian. *Korea and the West*. London, Routledge and Kegan Paul, 1986.

Buzo, Adrian. *The Guerrilla Dynasty: Politics and Leadership in North Korea*. Boulder, CO, Westview Press, 1999.

The Making of Modern Korea: A History. London, Routledge, 2002.

Catchpole, Brian. *The Korean War 1950–1953*. London, Robinson, 2001.

Cho Soo-Sung. *Korea in World Politics 1940–1950*. Berkeley, CA, University of California Press, 1967.

Choy Bong-Youn. *Korea: A History*. Rutland, VT, Charles E. Tuttle.

Chung, Chin O. *Pyongyang Between Peking and Moscow: North Korea's Involvement in the Sino-Soviet Dispute 1958–75*. University of Alabama Press, 1978.

Clay, Blair (Jr). *The Forgotten War: America in Korea 1950–53*. Annapolis, MD, United States Naval Institute, 2003.

Conroy, F. H. *The Japanese Seizure of Korea, 1868–1910*. Philadelphia, University of Pennsylvania Press, 1960.

Cumings, Bruce. *Korea's Place in the Sun: A Modern History*. New York, W. W. Norton and Co, 1997.

The Origins of the Korean War. Princeton, NJ, Princeton University Press, Vol. 1, 1981, Vol. 2, 1990.

(Ed.). *Child of Conflict: The Korean-American Relationship 1943–53*. Seattle and London, University of Washington Press, 1983.

Deuchler, Martina. *Confucian Gentlemen and Barbarian Envoys: The Opening of Korea 1875–1885*. Seattle, University of Washington Press, 1978.

Diamond, Larry, and Kim, Byung-Kook (Eds). *Consolidating Democracy in South Korea*. Boulder, CO, Lynne Rienner, 1999.

Downs, Chuck. *Over the Line: North Korea's Negotiating Strategy*. Washington, DC, American Enterprise Institute, 1999.

Duus, Peter. *The Abacus and the Sword: The Japanese Penetration of Korea, 1895–1910*. Berkeley, CA, University of California Press, 1995.

Eberstadt, Nicholas. *The End of North Korea*. Washington, DC, American Enterprise Institute, 1999.

Eckert, Carter, et al. *Korea Old and New: a History*. Cambridge, MA, Harvard University Press, 1990.

Gibney, Frank. *Korea's Quiet Revolution: From Garrison State to Democracy*. New York, Walker and Co, 1994.

Gills, Barry. *Korea versus Korea—A Case of Contested Legitimacy*. London, Routledge, 1998.

Gragent, Edwin H. *Land Ownership under Colonial Rule—Korea's Japanese Experience, 1900–1935*. Honolulu, HI, University of Hawaii Press, 1994.

Grayson, James H. *Korea: A Religious History*. London, Routledge-Curzon, 2002.

Halliday, Jon, and Cumings, Bruce. *Korea: The Unknown War*. London, Viking, 1988.

Hamm, Taik-Young. *Arming the Two Koreas—State, Capital and Military Power*. London, Routledge, 1999.

Han Sungjoo. *The Failure of Democracy in South Korea*. Berkeley, CA, University of California Press, 1974.

Hanley, Charles J., Choe Sang-Hun, and Mendoza, Martha. *The Bridge at No Gun Ri*. New York, Henry Holt and Co, 2001.

Helgesen, Geir. *Democracy and Authority in Korea: The Cultural Dimension in Korean Politics*. Richmond, Surrey, Curzon Press, 1998.

Henderson, Gregory. *Korea: The Politics of the Vortex*. Cambridge, MA, Harvard University Press, 1968.

Henthorn, William E. *A History of Korea*. New York, NY, The Free Press, 1971.

Hickey, Michael. *The Korean War: The West Confronts Communism 1950–1953*. London, John Murray, 1999.

Hicks, George. *The Comfort Women: Japan's Brutal Regime of Enforced Prostitution in the Second World War*. New York, W. W. Norton, 1995.

Howard, Keith (Ed.). *True Stories of the Korean Comfort Women*. London, Cassell, 1995.

Hulbert, Homer B. *History of Korea*. London, Routledge, 1998 (reissue).

Hunter, Helen-Louise. *Kim Il-song's North Korea*. Westport, CT, Praeger, 1999.

Kim, C. I. Eugene, and Ki Han-Kyo. *Korea and the Politics of Imperialism, 1876–1910*. Berkeley, CA, University of California Press, 1970.

Kim, C. I. Eugene, and Mortimore, D. E. (Eds). *Korea's Response to Japan: The Colonial Period 1910–45*. Kalamazoo, MI, The Center for Korean Studies, Western Michigan University, 1977.

Kim, Ilpyong J. (Ed.). *Two Koreas in Transition*. Rockville, MD, Paragon House, 1998.

Kim Joong-Seop. *The Korean Paekjong under Japanese Rule: The Quest for Equality and Human Rights*. London, RoutledgeCurzon, 2003.

Kim, Samuel S. (Ed.). *North Korean Foreign Relations in the Post-Cold War Era*. Oxford, Oxford University Press, 1999.

Kim Se Jin. *The Politics of Military Revolution in Korea*. Chapel Hill, NC, University of North Carolina Press, 1971.

Kwak Tae-Hwan. *US-Korean Relations 1882–1982*. Seoul, Kyungnam University Press.

Lee Chang-Soo (Ed.). *Modernization of Korea and the Impact of the West*. Los Angeles, CA, University of Southern California Press, 1981.

Lee Chong-Sik. *The Politics of Korean Nationalism*. Berkeley, CA, University of California Press, 1963.

Lee Jong-Sup and Heo Uk. *The US-South Korean Alliance, 1961–88: Free-riding or Bargaining?* New York, Edwin Mellen Press, 2002.

Lee Ki-Baik. *A New History of Korea*. (trans. by Edward W. Wagner and Edward J. Shultz). Cambridge and London, Harvard University Press, 1984.

Lee, Peter H. (Ed.). *Sourcebook of Korean Civilization: From Early Times to the Sixteenth Century* (Vol. I). New York, Columbia University Press, 1993.

Sourcebook of Korean Civilization: From the Seventeenth Century to the Modern Period (Vol. II). New York, Columbia University Press, 1996.

Lewis, James B. *Frontier Contact between Choson Korea and Tokugawa Japan*. London, RoutledgeCurzon, 2003.

Lewis, Linda Sue. *Laying Claim to the Memory of May: A Look Back at the 1980 Kwangju Uprising*. Honolulu, University of Hawaii Press, 2002.

Lim, Un. *The Founding of a Dynasty in North Korea*. Tokyo, Jiyusha, 1982.

Lone, Stewart, and McCormack, Gavan. *Korea Since 1850*. Melbourne, Longman Cheshire Pty, 1993.

Lowe, Peter. *The Origins of the Korean War*. Harlow, Longman, 1986.

Malkasian, Carter. *The Korean War 1950–53*. London, Fitzroy Dearborn, 2001.

McCann, David R. (Ed.) *Korea Briefing—Toward Reunification*. Armonk, NY, M. E. Sharpe, 1996.

Myers, Ramon H., and Peattie, Mark R. (Eds). *The Japanese Colonial Empire, 1895–1914*. Princeton, NJ, Princeton University Press, 1984.

Nahm, Andrew C. (Ed.). *Korea Under Japanese Colonial Rule*. Kalamazoo, MI, The Center for Korean Studies, Western Michigan University, 1973.

The United States and Korea—American-Korean Relations 1866–1976. Kalamazoo, MI, The Center for Korean Studies, Western Michigan University.

A Panorama of 5,000 Years: Korean History. Seoul and Elizabeth, NJ, Hollym International Corpn, 1983.

Oberdorfer, Don. *The Two Koreas: A Contemporary History*. New York, Basic Books, 2002 (2nd edition).

Oh, Bonnie B. C. *Korea Under the American Military Government, 1945–48*. Westport, CT, Praeger, 2002.

Oh, John K. C. *Korea: Democracy on Trial*. Ithaca, NY, Cornell University Press, 1968.

Park, Tong Whan (Ed.). *The U.S. and the Two Koreas: A New Triangle*. Boulder, CO, Lynne Rienner, 1998.

Robinson, M. *Cultural Nationalism in Colonial Korea, 1920–25*. Seattle, University of Washington Press, 1989.

Rutt, Richard. *James Scarth Gale and his History of the Korean People*. Seoul, Royal Asiatic Society, Korea Branch, Seoul, 1972.

Scalapino, Robert A., and Lee, Chong-Sik. *Communism in Korea*. Berkeley, CA, University of California Press, 1972.

Shin, Doh C. *Mass Politics and Culture in Democratizing Korea*. Cambridge, Cambridge University Press, 1999.

Springer, Chris. *Pyongyang: The Hidden History of the North Korean Capital*. Budapest, Entente Bt., 2003.

Stueck, William. *The Korean War: An International History*. Princeton, NJ, Princeton University Press, 1996.

Suh Dae-Sook. *The Korean Communist Movement 1918–1948*. Princeton, NJ, Princeton University Press, 1967.

Kim Il Sung: The North Korean Leader. New York, Columbia University Press, 1989.

Suh Dae-Sook and Lee Chae-Jin (Eds). *North Korea after Kim Il-Sung*. Boulder, CO, Lynne Rienner, 1998.

Summers, Harry. G. *Korean War Almanac*. Bridgewater, NJ, Replica Books, 2001.

Toland, John. *In Mortal Combat: Korea, 1950–53*. New York, William Morrow, 1991.

Uden, Martin. *Times Past in Korea: An Illustrated Collection of Encounters, Customs and Daily Life Recorded by Foreign Visitors*. London, RoutledgeCurzon, 2003.

Wells, Kenneth M. (Ed.) *South Korea's Minjung Movement—The Culture and Politics of Dissidence*. Honolulu, HI, University of Hawaii Press, 1996.

Wickham, Gen. (retd) John A. *Korea on the Brink: A Memoir of Political Intrigue and Military Crisis*. Dulles, VA, Brassey's, 2001.

Wright, Edward R. (Ed.). *Korean Politics in Transition*. Seattle, University of Washington Press, 1975.

Yang Sung-Chul. *Korea and Two Regimes*. Cambridge, MA, Schenkman Publishing Co, 1981.

Young, Whan Kihl, and Hayes, Peter (Eds). *Peace and Security in Northeast Asia—The Nuclear Issue and the Korean Peninsula*. Armonk, NY, M. E. Sharpe, 1996.

ECONOMY

Amsden, Alice H. *Asia's Next Giant: South Korea and Late Industrialization*. New York, Oxford University Press, 1990.

Cho Lee-Jay and Kim Yoon-Hyung. *Economic Development in the Republic of Korea: A Policy Perspective*. Honolulu, HI, East-West Center, University of Hawaii Press, 1991.

(Eds). *Economic Systems in North and South Korea: The Agenda for Economic Integration*. Seoul, Korea Development Institute, 1995.

Choi, E. Kwan, Merrill, Yesook, and Kim, E. Han (Eds). *North Korea in the World Economy*. London, RoutledgeCurzon, 2003.

Choi Young-Back and Merrill, Yesook (Eds). *Perspectives on Korean Unification and Economic Integration*. Cheltenham, Edward Elgar, 2001.

Chung Jae-Yong and Kirkby, Richard J. *The Political Economy of Development and Environment in Korea*. London, Routledge, 2000.

Chung, Joseph Sang-Hoon. *The North Korean Economy: Structure and Development*. Stanford, CA, Hoover Institution Press, 1974.

Clifford, Mark L. *Troubled Tiger: The Unauthorised Biography of Korea, Inc.* Singapore, Butterworth-Heinemann Asia, 1994, reissued 1998.

Cyhn, Jin W. *Technology Transfer and International Production: The Development of the Electronics Industry in Korea*. Cheltenham, Edward Elgar, 2002.

Eberstadt, Nicholas. *Korea Approaches Reunification*. Armonk, NY, M. E. Sharpe, for National Bureau of Asian Research, 1995.

Eberstadt, Nicholas, and Banister, Judith. *The Population of North Korea*. Berkeley, CA, Institute of East Asian Studies, University of California, 1992.

Frank, Charles R., Kim, Kwang Suk, and Westphal, Larry. *Foreign Trade Régimes and Economic Development: South Korea*. New York, Columbia University Press (for the National Bureau of Economic Research), 1977.

Hillebrand, Wolfgang. *Shaping Competitive Advantages: Conceptual Framework and the Korean Approach*. London, Frank Cass, 1996.

Hwang Eui-Gak. *The Korean Economies: A Comparison of North and South*. London, Clarendon Press, 1993.

International Business Publications. *North Korea Business and Investment Opportunities Yearbook*. USA, International Business Publications, 2002.

International Business Publications. *North Korea–US Political and Economic Co-operation Handbook*. USA, International Business Publications, 2002.

Janelli, Roger, with Yim, Dawnhee. *Making Capitalism: The Social and Cultural Construction of a South Korean Conglomerate*. Stanford, CA, Stanford University Press, 1995.

Jwa Sung-Hee. *The Evolution of Large Corporations in Korea*. Cheltenham, Edward Elgar, 2002.

Kim Dong Ki and Kim Linsu. *Management Behind Industrialization: Readings in Korean Business*. Seoul, Korea University Press, 1990.

Kirk, Donald. *Korean Dynasty: Hyundai and Chung Ju Yung*. Hong Kong, Asia 2000, 1995.

Korean Crisis, Unraveling of the Miracle in the IMF Era. New York, NY, St Martin's Press, 2000.

Kong Tat Yan. *The Politics of Economic Reform in South Korea—A Fragile Miracle*. London, Routledge, 2000.

Kuznets, Paul W. *Economic Growth and Structure in the Republic of Korea*. New Haven, CT, Yale University Press, 1977.

Kwon Seung-Ho, and O'Donnell, Michael. *The Chaebol and Labour in Korea*. London, Routledge, 2001.

Lee, Chung H., and Yamazawa, Ippei (Eds). *The Economic Development of Japan and Korea: A Parallel with Lessons*. New York, Praeger Publishers, 1990.

Lee Yeon-Ho. *The State, Society and Big Business in South Korea*. London, Routledge, 1997.

Lee, You Il. *The Political Economy of Korean Crisis: A Turning Point or the End of the Miracle?* London, Ashgate Publishing, 2001.

Lie, John. *Han Unbound—The Political Economy of South Korea*. Cambridge, Cambridge University Press, 1998.

McNamara, Dennis L. (Ed.). *Corporatism and Korean Capitalism*. London, Routledge, 1999.

Market and Society in Korea: Interest, Institution, and the Textile Industry. London, RoutledgeCurzon, 2002.

Michell, T. *From a Developing to a Newly Industrialised Country: The Republic of Korea, 1961–82*. Geneva, International Labour Office, 1989.

Mo, Jongryn, and Moon, Chung Il (Eds). *Democracy and the Korean Economy*. Stanford, CA, Hoover Institute Press, 1999.

Moonjoong Tcha and Suh Chung-Sok. *The Korean Economy at the Crossroads: Triumphs, Difficulties and Triumphs Again*. London, RoutledgeCurzon, 2003.

Noland, Marcus. *Avoiding the Apocalypse: The Future of the Two Koreas*. Washington, DC, Institute for International Economics, 2000.

Noland, Marcus (Ed.) *Economic Integration of the Korean Peninsula*. Washington, DC, Institute for International Economics, 1998.

O Yul-Kwon (Ed.). *Korea's Economic Prospects: From Financial Crisis to Prosperity*. Cheltenham, Edward Elgar, 2001.

Ogle, George E. *South Korea: Dissent Within the Economic Miracle*. London, Zed Books, 1991.

Sakong, Il. *Korea in the World Economy*. Washington, DC, Institute for International Economics, 1994.

Sakong, Il, and Kim, Kwang Suk (Eds). *Policy Priorities for the Unified Korean Economy*. Seoul, Institute for Global Economics, 1998.

Shin Jang-Sup and Chang Ha-Joon. *Restructuring 'Korea Inc'*. London, RoutledgeCurzon, 2003.

Song Byung-Nak. *The Rise of the Korean Economy*. Oxford, Oxford University Press, 1997 (2nd edition).

Steers, Richard M., et al. *The Chaebol: Korea's New Industrial Might*. New York, Harper and Row, 1990.

Steers, Richard M. *Made in Korea—Chung Ju Yung and the Rise of Hyundai*. London, Routledge, 1999.

Suh Sang-Chul. *Growth and Structural Changes in the Korean Economy, 1910–40*. Cambridge, MA, Harvard University Press, 1978.

Woo Jung-En. *Race to the Swift: State and Finance in Korean Industrialization*. New York, Columbia University Press, 1993.

LAOS

Physical and Social Geography

HARVEY DEMAINE

The Lao People's Democratic Republic is a land-locked state in South-East Asia, bordered by the People's Republic of China to the north, by Viet Nam to the east, by Cambodia to the south, by Thailand to the west and by Myanmar (formerly Burma) to the north-west. Covering an area of 236,800 sq km (91,400 sq miles), Laos consists almost entirely of rugged upland, except for the narrow floors of the river valleys. Of these rivers, by far the most important is the Mekong, which forms the western frontier of the country for much of its length.

In the northern half of Laos, the deeply-dissected plateau surface is more than 1,500 m above sea-level over wide areas. The average altitude of the Annamite chain, which occupies most of the southern half of the country, is somewhat lower, but its rugged and more densely-forested surface makes it equally inhospitable. Temperatures on the plateau and in the Annamite chain are mitigated by altitude, but the more habitable lowlands experience tropical conditions throughout the year, and receive a total annual rainfall of about 1,250 mm, most of which falls between May and September. In Vientiane, the capital, temperatures range between 23°C and 38°C in April, the hottest month, and from 14°C to 28°C in January, the coolest month.

The natural resources of Laos have not been fully surveyed. In 2000 it was estimated by the World Bank that some 54% of the country was still covered with forest, but there was continuing concern about deforestation. Laos has considerable mineral resources: tin and gypsum are the principal minerals that are exploited and exported. Other mineral deposits include lead, zinc, coal, potash, iron ore and small quantities of gold, silver and precious stones. The Mekong River offers substantial potential for fisheries, irrigation and hydroelectricity. The Nam Ngum dam and power complex, built on the Mekong 80 km north of Vientiane, began operations in 1971. In 1991 a second major hydroelectric project, the Xeset dam in southern Laos, was completed.

Laos had an enumerated population of 3,584,803 at the census of 1 March 1985, although the migration of refugees, as a result of problems of internal security and government policies, may well have rendered this figure an overestimate. The population was enumerated at 4,581,258 at the census of 1 March 1995. According to UN estimates, the population of Laos reached 5,657,000 at mid-2003, with an average density of 23.9 per sq km. About 60% of the population are ethnic Lao, residing mainly in the western valleys. A further 35% belong to various hill tribes, although the important Hmong group has been affected by fighting. The remainder are either Vietnamese or Chinese. The urban population is an estimated 20% of the total; Vientiane, the capital, is the only large town. Its population was 176,637 in 1973, but had increased to 528,109 at the census of 1995. According to UN estimates, the population of Vientiane was 716,380 at mid-2003.

History

MARTIN STUART-FOX

EARLY HISTORY: THE KINGDOM OF LAN XANG

More than 60 different ethnic groups inhabit present-day Laos. Officially, they are divided into three broad categories on the basis of settlement patterns, culture and language. The earliest inhabitants were those minorities now known as 'Lao of the mountain slopes' (Lao Thoeng), who spoke Austro-Asiatic languages akin to Cambodian and farmed using slash-and-burn methods. From perhaps the 10th century they began to be displaced by speakers of Tai languages who entered Laos from the north-east. These included the now politically dominant lowland (or ethnic) Lao, along with other Tai groups, which together comprised the 'Lao of the plains' (Lao Lum). All practised wet-rice cultivation and most became Theravada Buddhists. The 'Lao of the mountain tops' (Lao Sung) were the last to arrive, from southern China, in the 19th century. They spoke either Hmong-Mien or Tibeto-Burman languages, followed their own various religions, and often cultivated opium as a cash crop.

Very early human remains have been discovered in northern Laos, but the first culture to have been archaeologically investigated was centred on the Plain of Jars, and was an iron-age megalithic culture which built both underground burial chambers and massive stone mortuary urns to hold the ashes of the dead. Some of these 'jars' (after which the plain was named) were more than 2 m high and weighed some 10 tons. Who these people were, however, and what became of them remains a mystery.

By the eighth or ninth century, north-eastern Thailand and Laos had begun to be influenced by the early Indianized kingdoms in mainland South-East Asia. Laos was situated at an important crossroads of trade routes along which travelled not only merchants, but also Buddhist monks. In the areas of Thakhaek in central Laos, Vientiane (Viang Chan) and Luang Prabang small principalities were formed, which at first owed allegiance to more powerful states. By the late 12th century the Cambodian Angkorian empire had begun to expand north to include not only southern Laos, but most of the Mekong basin perhaps as far north as Luang Prabang. Khmer power was short-lived, however, and as the Khmer empire contracted, Tai-speaking peoples established their own kingdoms in northern Laos and northern and central Thailand. By the late 13th century it is known that there was an established Lao principality in the region of Luang Prabang, and others in the central Mekong basin. By the mid-14th century an enterprising ruler, King Fa Ngoum—subsequently revered by the Lao—had incorporated the Lao principalities into the powerful kingdom of Lan Xang Hom Khao ('A Million Elephants and the White Parasol'), a title indicating both military might and royal kingship. Many of the symbols of the modern Laotian state originate in the Kingdom of Lan Xang. For two centuries Luang Prabang (then known as Xieng Dong Xieng Thong) served as the capital of the kingdom. Lan Xang was loosely organized as a tributary state comprising several *meuang* (semi-feudal principalities of variable extent), each of which not only paid an annual tribute of gold or other valuable products, but was also expected to raise an army in time of conflict.

The kings of Lan Xang were ardent Buddhists and, as in the other Tai states and Burma (now Myanmar) and Cambodia, Theravada Buddhism served to legitimize their rule. In return for his patronage of the *Sangha* (the monastic order), monks taught that the King ruled because he possessed superior *karma* (moral merit accumulated during previous lifetimes). The image of Buddha known as the *Phra Bang* became the much revered palladium of the kingdom. The kingdom of Lan Xang was the equal of other states in mainland South-East Asia at the time, and defeated and expelled invaders from Viet Nam and Burma. Lao settlers advanced south down the Mekong valley as far as Champasak and beyond, and on to the Korat plateau (now north-eastern Thailand). As they did so, the centre of Lao population shifted south. When Burmese armies threatened Luang Prabang in the mid-16th century, King Xetthathirat decided to re-establish the capital at the more strategically situated and less vulnerable site of Vientiane.

Lan Xang was at its apogee in the 17th century during the long reign of King Surinyavongsa (1637–94). It was then that the first European merchants and missionaries arrived in Vientiane. Their descriptions of the wealth of the city and the brilliance of its court were the first independent accounts, and portray a centre of art, culture and religion attracting scholars and artists from throughout the region. Surinyavongsa's death was followed by one of the periodic successional disputes that served to weaken the kingdom. Within two decades the kingdom had split into three parts, centred on Luang Prabang in the north, Vientiane in the centre with control over much of the Korat plateau, and Champasak in the south. Xiangkhouang, on the Plain of Jars, reluctantly acknowledged the suzerainty of Vientiane, but took every opportunity to assert its autonomy. The weakened Lao kingdoms could no longer match their powerful neighbours. Even before the disintegration of Lan Xang, the maritime states of mainland South-East Asia had been much better placed than the inland states to benefit from increased seaborne trade, particularly with the recently arrived Europeans. Their greater wealth enabled them to purchase new weapons and to equip larger armies; thus, well before all three Lao kingdoms were reduced to tributaries of Siam in the late 18th century, the balance of power had begun shifting against the Lao.

In 1827 Anuvong, the last King of Vientiane, made a desperate bid to free himself and his kingdom from Siamese hegemony. The war that followed was a disaster: Vientiane was captured and sacked, its people were forcibly resettled, and the king was removed to Bangkok where he subsequently died. Appeals by local Lao rulers to the Vietnamese were unsuccessful, except in Xiangkhouang where Viet Nam had an historic interest. In the 1870s the Laotian territories were ravaged by marauding bands of Chinese and, despite appeals for assistance, the Siamese—themselves hard-pressed by encroaching European powers—were slow to respond. The court in Bangkok was nevertheless determined to assert Siamese suzerainty even over the remote Tai highlands in what is now north-western Viet Nam. It was there in the late 1880s that they encountered the French.

LAOS UNDER FRENCH RULE

French interest in Laos derived initially from a belief that the Mekong River would provide a 'river road' to southern China. When the Mekong expedition of 1867–68 proved that the river was unnavigable, interest waned. However, French commercial interests in Saigon still hoped to tap the wealth of the Mekong basin and, with the annexation of central and northern Viet Nam as a French protectorate in 1884, French interest in Laos was revived. French control of the Mekong was deemed essential to the strategic defence and economic prosperity of the new possession; at a minimum, the Laotian territories east of the Mekong had to be annexed to 'round out' French Indo-China.

Attempts to assert Vietnamese claims to these Laotian lands were hardly convincing, and were contested by Siam. The creation of a French presence through exploration and commerce, however, was more successful. Eventually the determined efforts of Auguste Pavie, the French consul appointed to Luang Prabang, and gunboat diplomacy in Bangkok, forced the Siamese in 1893 to cede all territories east of the Mekong to France. Subsequent treaties in 1904 and 1907 established the present borders that both Cambodia and Laos share with Thailand.

For half a century Laos was ruled as a French colony, although the former kingdom of Luang Prabang nominally enjoyed the status of a protectorate. The French rebuilt Vientiane as the administrative capital of French Laos, and extended a minimal presence throughout the rest of the country. Their first priority in Laos was to cover the cost of administration. However, the depleted population that had escaped Siamese resettlement provided an insufficient tax base. Although taxes were high and all had to perform unpaid *corvée* labour, Laos was always dependent on a subsidy from the federal budget for Indo-China. The French devised various schemes designed to 'open up' Laos to economic exploitation, which required construction of a railway from the Vietnamese coast to the Mekong, and mass immigration of Vietnamese to provide an adequate work-force. From the beginning, the middle ranks of the civil service in Laos were largely filled by Vietnamese; the Great Depression in the 1930s, however, prevented construction of the railway, and relatively few Vietnamese migrated. Had the Second World War not intervened, however, the French might well have succeeded in their intention to reduce the Lao to a minority in their own country.

In 1939 Siam was renamed Thailand as part of a policy designed to appeal to all Tai-speaking peoples, including notably the Lao. To counter this appeal, for the first time the French administration began to encourage a weak Laotian nationalism among the small and educated élite. Thai seizure of Laotian territories west of the Mekong provoked Lao anger against both the Thai and the French for failing to protect the borders.

Under an agreement between the Japanese Government and Vichy France, administration of French Indo-China remained the responsibility of the French, while the Japanese had the right to move and station troops throughout the region. Not until 9 March 1945 did the Japanese forces stage a lightning pre-emptive attack to neutralize and intern all French military and civilian personnel. Only in Laos did a few French officers and men escape the Japanese net, their survival in jungle hideouts being due to the support they received from loyal Laotians.

As in other parts of South-East Asia, the élite in Laos was divided between opposition to the Japanese and opposition to colonialism. While some fought for the Allies, others collaborated. In Laos King Sisavang Vong, though pro-French, was forced on 8 April 1945 to declare the independence of his country, while, less reluctantly, Prince Phetsarath served as Prime Minister under the Japanese. The surrender of Japan in August left a political vacuum throughout Indo-China, which nationalists in all three countries moved to exploit: in Viet Nam the communist Viet Minh seized power; in Laos the nationalist Lao Itsara, or Free Laos movement. On 1 September Prince Phetsarath in Vientiane reaffirmed the independence of Laos, even though the King in Luang Prabang had welcomed the return of the French. When the King repudiated Phetsarath's actions, the Prince declared the reunification of the protectorate of Luang Prabang with the rest of the country. In October the King dismissed Phetsarath as Prime Minister but, in response, a newly constituted Lao Itsara Government deposed the King shortly after, thereby initiating a struggle for power.

Under the terms of the Potsdam Agreement between the Allies, the Japanese surrender in Indo-China was accepted by the Chinese Nationalists north of the 16th parallel and by the British to the south. While the British facilitated the return of the French to southern Laos, the Chinese favoured the Lao Itsara. Not until the Chongqing Agreement had been concluded between France and China did the Chinese withdraw, thus allowing French forces to reoccupy central and northern Laos. In March 1946 French forces and their Laotian allies moved north. Lao Itsara volunteers supported by local Vietnamese countered the incursion at Thakhaek, but the French successfully reoccupied Vientiane on 24 April and Luang Prabang on 13 May. Some 2,000 Lao Itsara supporters and thousands of Vietnamese crossed the Mekong to Thailand, where Phetsarath established a government-in-exile in Bangkok. The French, meanwhile, set about reasserting their control. Prince Bunum of Champasak, who had aided the French, renounced his claim to a separate southern kingdom, and Laos was unified under the royal house of Luang Prabang. The nominally independent Kingdom of Laos was incorporated into the French Union, but was still closely tied (in terms of, *inter alia*, defence, currency and customs regulations) to Viet Nam and Cambodia.

By early 1947 the new Kingdom of Laos had begun to take shape. The territories west of the Mekong seized by Thailand in 1940 had been returned, elections for a Constituent Assembly had been held, and in May a new Constitution was proclaimed.

The Government, which was both conservative and pro-French, was acutely aware of its nationalist opponents in Bangkok. Its members, drawn from the powerful families of the élite that had benefited most from the French presence, reflected a judicious balance from northern, central and southern Laos. While some attempt was made, with French support, to improve woefully neglected services such as education, health and agricultural extension, much of the politicking was between influential leaders who sought to obtain regional or family advantage. Perhaps the most serious shortcoming of the new Government, however, was its failure to generate any real sense of national unity or purpose. The country was still deeply divided, with regional and family loyalties counting for more than national allegiance. The only unifying symbol might have been the monarchy, as in Thailand or Cambodia, but the King remained in Luang Prabang, and the political opportunity to unify the royal and administrative capitals and overcome regionalism by moving to Vientiane was lost.

The outbreak of the First Indo-China War in December 1946 between the Viet Minh and their allies and the French, and the activities of the Lao Itsara, both served to complicate the establishment of governmental authority. The radical anti-French nationalism of the Viet Minh provided an example for the like-minded elements of the Lao Itsara, led by Phetsarath's younger half-brother, Prince Souphanouvong. Lao Itsara guerrilla raids targeting the French in Laos were co-ordinated by and carried out with the Viet Minh. The proximity of the alliance between Souphanouvong and the Viet Minh, however, caused divisions within the Lao Itsara. Criticism of Souphanouvong eventually led him to resign from the organization in March 1949, and two months later the moderate majority responded by formally relieving him of all his ministerial functions.

In July 1949 a Franco-Lao Convention was signed giving the Laotian Government much greater powers and sufficient independence to meet the demands of the moderates in the Lao Itsara. In October, led by Phetsarath's younger brother, Souvanna Phouma, the Lao Itsara was dissolved, and its members returned to Laos. Only Phetsarath himself remained in Bangkok. Souphanouvong went to Viet Nam where he made contact with the Viet Minh. While the moderate Lao Itsara joined the new Royal Lao Government (RLG) in August 1950, Souphanouvong established his own government of the 'Land of Laos' (Pathet Lao, the name by which the communist movement in Laos became known). Over the next four years, as communist Chinese military support flowed to the Viet Minh and the USA supported France, the First Indo-China War engulfed the region. Twice Laos was subjected to major Viet Minh invasions, during which large areas of the country were seized and turned over to the Pathet Lao. The French responded by granting Laos formal independence in October 1953. The defeat of French forces at Dien Bien Phu, located in a mountain valley close to the border with Laos, represented the end of French military involvement in Viet Nam. Meetings were already under way in Geneva to seek an end to the conflict, and with respect to Laos, the Geneva Conference agreed in July 1954 to an armistice and a regrouping of opposing forces, leading to elections within two years. The Pathet Lao were allotted the two north-eastern provinces of Phongsali and Xam Neua, while the RLG administered the rest of the country. By the end of 1954 Lao independence was reinforced by the abrogation of all agreements linking Laos with the other states of Indo-China.

THE QUEST FOR UNITY AND NEUTRALITY

For the Royal Lao Government in Vientiane, the first priority was to regain administrative control of the two Pathet Lao provinces so that elections could be held in accordance with the Geneva Agreement. As a *quid pro quo*, however, the Pathet Lao insisted on changes to the electoral law and on freedom for its front organization—the Lao Patriotic Front (behind which stood the secret Lao People's Party)—to operate as a political party. Negotiations continued throughout 1955, but failed to bring results. In December 1955 the Government proceeded with elections in the provinces that it controlled. These resulted in the formation of a new Government under the leadership, once again, of Souvanna Phouma, who immediately entered into new negotiations with his half-brother, Souphanouvong.

By early 1956 the USA had replaced France as the dominant Western power in Indo-China. As of early 1955 a US Operations Mission (USOM) was functioning in Vientiane and the USA was financing not only much of the Laotian budget, but also meeting the entire cost of the Lao National Army (LNA). Under the terms of the Geneva Agreement, a French Military Mission continued to train the LNA, but growing numbers of US military personnel in civilian guise were attached to the US embassy. Both the US Information Service and the Central Intelligence Agency (CIA) were also active. From 1955 to 1962 US aid to Laos, the overwhelming majority of it military, was, in per caput terms, the highest for any country in South-East Asia, including South Viet Nam. In the context of the Cold War the USA aimed to prevent Laos from becoming communist and thus was strongly opposed to the inclusion of communists in the Government. Souvanna Phouma's agreement with the Pathet Lao to include two of their members in a coalition government as the price for reintegration of the two provinces under their control met strenuous opposition from the USA. Souvanna Phouma pushed ahead, however, in the belief that national unity and a policy of neutrality were essential for the preservation of the Laotian state. An agreement was signed with the Pathet Lao in November 1957. Supplementary elections held the following May gave leftist candidates 21 seats in the National Assembly, much to the alarm of the USA.

With the active support of the US embassy and the CIA, right-wing politicians and army officers formed the Committee for the Defence of National Interests (CDNI). When US aid was withheld in July 1958, Souvanna Phouma's first coalition collapsed, and was replaced by a right-wing Government led by Phuy Sananikon. In January 1959, on the basis of fabricated reports of a North Vietnamese invasion, Phuy received emergency powers to govern for a year without reference to the National Assembly. Gen. Phoumi Nosavan, a powerful military figure, was appointed Vice-Minister of Defence in the new Government. The Pathet Lao watched these events with mounting concern. Negotiations to integrate two Pathet Lao battalions into the LNA stalled, and in May one battalion absconded to Xam Neua. As insurgent activity resumed, the Government responded by imprisoning Pathet Lao members of the National Assembly in Vientiane. However, the conservative Government was internally divided by personal and political differences, and in December 1959 Gen. Phoumi, with the support of the CDNI, mounted a coup. Although Gen. Phoumi failed to gain the premiership, the coup resulted in the formation of a new Government dominated by the military. New elections were held in April and, owing to malpractice, all left-wing candidates lost their seats, much to US satisfaction. As the country moved towards civil war, however, even elements of the military became concerned at the turn of events. In August 1960 a young army captain, Kong Lae, unexpectedly seized control of Vientiane under the banner of neutrality and an end to 'Lao killing Lao'. Souvanna Phouma was again invited to form a Government, which Gen. Phoumi refused to join. With the support of the CIA, but in defiance of the US embassy, Phoumi built up his forces in central Laos. In mid-December, after his efforts to prevent a civil war had failed, Souvanna flew to Phnom-Penh and Gen. Phoumi seized control of Vientiane. Kong Lae's neutralist force staged an orderly withdrawal to the Plain of Jars, where they were welcomed by the Pathet Lao. As Gen. Phoumi failed to follow up his victory, large areas of the country fell to the combined neutralist and Pathet Lao forces.

Following the undermining and collapse of Souvanna's Government in 1957, the communists made impressive gains. US policy, having clearly failed in this respect, took a new course. In March 1961, as a renewed neutralist and Pathet Lao offensive got under way, President Kennedy declared US support for a political settlement involving the neutralization of Laos. The communist response was conciliatory and a cease-fire came into effect in early May, followed by the convening of a second conference of interested powers at Geneva. It took more than a year, however, for the feuding Laotian factions to reach agreement on the formation of a second coalition Government of National Union, and then only after renewed fighting had resulted in a decisive defeat for the LNA at the battle of Nam Tha in northern Laos. The new Government, including two full ministers and two vice-ministers from the Pathet Lao, eventually signed the final document in Geneva, proclaiming the neutralization of Laos on 21 July 1962.

The second coalition, however, failed to achieve either of its two objectives: to reunify the country and, through neutralization, to insulate Laos from the gathering war in Viet Nam. Of the three factions comprising the coalition, the neutralists were the weakest and under the greatest political pressure. While the Pathet Lao and the right-wing elements enjoyed military and economic support from their principal supporters, North Viet Nam and the USA respectively, the neutralists enjoyed only diplomatic backing, principally from France. Neutralist armed forces were dependent for supplies on either the Pathet Lao or the LNA. As both the left and right wings attempted to absorb the neutralist centre, divisions developed between the neutralists themselves. Only a façade of coalition government and neutrality remained, as Laos was drawn increasingly into the Second Indo-China war. The Government lasted less than a year and, after the assassination in April 1963 of the neutralist Minister of Foreign Affairs, Kinim Phonsena, Pathet Lao ministers left Vientiane, fearing their own assassination or arrest. Thereafter, both the USA and North Viet Nam subverted Laotian neutrality for their own purposes in pursuing the war in Viet Nam. The North Vietnamese had two strategic aims in Laos: to prevent northern Laos, particularly the Plain of Jars, from being used by the USA to threaten North Viet Nam, and to control the Ho Chi Minh trail in the mountainous east of Laos down which combatants and weapons flowed to communist forces in South Viet Nam. US policy sought to challenge communist control of both areas: on the Plain of Jars by building up a clandestine army recruited mainly from the Hmong ethnic minority and commanded by the Hmong General, Vang Pao; and on the Ho Chi Minh trail through the (initially secret) massive use of air power. For a decade, until the cease-fire of February 1973 brought fighting to an end, the country was caught up in the Viet Nam War. The war in Laos was of much greater strategic significance to both sides than the war in Cambodia, and in the north the Hmong suffered considerable casualties against North Vietnamese regular forces and had to be reinforced by Thai 'volunteers'. A great number of North Vietnamese and their Pathet Lao allies were killed in the US bombing in both northern and southern Laos (200,000 killed and more than 400,000 wounded is a conservative estimate); by 1973 2m. tons of bombs had been dropped, making Laos the most heavily bombed country per caput in the history of warfare. A quarter of the entire population of the country was displaced by the war.

The Lao people themselves were unable to prevent the use of their territory by opposing forces. While the North Vietnamese were nominally 'aiding' the Pathet Lao, the US embassy was directing bombing 'in defence of Laotian neutrality'. Souvanna Phouma attempted to preserve a minimum freedom of action, but only retained office in the face of attempted right-wing coups with the support of foreign ambassadors. Not until peace talks got under way did the Laotians regain some influence over their own affairs. Negotiations lasting over a year eventually led to the formation in April 1974 of a third coalition Government, the composition of which reflected the gains made by the Pathet Lao during a decade of war: half the ministerial portfolios went to communists or left-leaning 'neutralists' and the remainder to right-wing figures. The only real neutralist left was Souvanna Phouma, who once again presided as Prime Minister. Souphanouvong headed the National Political Consultative Council (NPCC), a policy-making body which met in the royal capital of Luang Prabang.

The NPCC agreed upon a relatively moderate 18-point programme, and it seemed initially that Laos might gradually work towards its own form of mild socialism, preserving both the monarchy and traditional Buddhist values. However, events in other parts of Indo-China had repercussions for Laos. In April 1975 first Phnom-Penh, then Saigon, fell to Cambodian and Vietnamese communist forces. In Laos, Pathet Lao forces advanced towards Vientiane. As towns in southern Laos were progressively 'liberated', demonstrations were mounted in Vientiane against both the continuing US presence (in the form of the US Agency for International Development—USAID) and right-wing political and military figures. Five ministers and several generals fled the country, but even as power shifted to the Pathet Lao the facade of the third coalition was retained. On 23 August Vientiane was symbolically 'liberated' by the arrival of a contingent of 50 women soldiers of the Pathet Lao. Fearing Pathet Lao reprisals, some families fled to Thailand at this time, but many more elected to stay and to serve the new regime. Thousands of civil servants and military officers went willingly to re-education 'seminars' in the remote north-east of the country, which they were assured would last only a few weeks. The Pathet Lao took advantage of the absence of their opponents, however, to press ahead with the final stage of their takeover of power. On 2 December 1975 a National Congress of People's Representatives accepted the abdication of King Savang Vatthana and proclaimed the Lao People's Democratic Republic (LPDR). While Souphanouvong was named President, real power lay with the little-known Secretary-General of the (renamed) Lao People's Revolutionary Party (LPRP) and Chairman of the Council of Ministers (Prime Minister), Kaysone Phomvihane.

THE LAO PEOPLE'S DEMOCRATIC REPUBLIC

The leaders of communist Laos saw their task as building and defending socialism by means of 'bypassing' the capitalist stage of development through three simultaneous revolutions: in production, through nationalizing industry and collectivizing agriculture; in science and technology, through their application to the economy; and in ideology and culture, in order to create new Lao 'socialist men and women' devoted to the regime. Laos was seen by its new leaders as the vanguard of socialism in South-East Asia, and as such they believed the country was a prime target for destabilization by its principal opponents, the USA and Thailand. Support for both building and defending socialism in Laos came from the communist bloc countries, notably the USSR, Viet Nam and the People's Republic of China, but also from East European states and even Mongolia and Cuba, all of which established embassies in Vientiane.

The steady progress towards modernization expected by the new regime almost immediately encountered difficulties, however, largely owing to unimaginative and dogmatic policies. The end of US aid and Thailand's closure of its border with Laos, which curtailed the supply of consumer goods, both contributed to the collapse of the urban economy; new government restrictions also limited internal trade. Shops closed as their owners fled to Thailand and factories, deprived of imported inputs, had to cease production. An unpopular tax on agricultural production, introduced when much of the country was suffering from drought and poor harvests, further reduced the availability of food supplies as farmers withheld produce from markets. The new Government's policies, which sought to reduce religious expenditure and placed restrictions on freedom of movement because of security fears, further lost it support. However, the principal concern of the educated middle class, which had been prepared to co-operate with the new Government, was the latter's refusal to allow the return of those undergoing political re-education. Terms of imprisonment (in 're-education' camps), frequently extending into years, deprived many families of their principal income-earner and forced them to sell their possessions in order to survive. Many crossed the Mekong River into Thailand. Most of those eventually released also fled the country at the first opportunity. In a decade Laos lost 10% of its population, comprising an estimated 90% of its educated class, a loss that probably set the country back a generation in its development.

Increasing domestic disillusionment and the growing numbers of refugees in camps in Thailand created ideal conditions for the recruitment of opponents of the regime; a 'Lao National Revolutionary Front' was consequently established in Thailand to send anti-Government propaganda and sabotage teams into Laos. When in March 1977 rebel remnants of the Hmong 'secret army' in northern Laos briefly captured a village not far from Luang Prabang, the regime feared that the former King might instigate a revolt. The royal family was arrested and banished to a remote region of Xam Neua, where the elderly King and Queen subsequently both died, as did the crown prince, reportedly of malaria. Vietnamese forces assisted in putting down the Hmong revolt, and in July Laos and Viet Nam signed a 25-year Treaty of Friendship and Co-operation to formalize Vietnamese political, economic and military assistance, including the stationing of 30,000–40,000 Vietnamese troops in Laos over the next decade. Relations with Viet Nam and the USSR were close over these years, but those with China and Cambodia became increasingly difficult. As hundreds of Soviet technicians and Vietnamese advisers drew Laos firmly into the Soviet sphere of influence, Chinese aid was limited to projects in the north of the country, in an area where Chinese influence had always been strong. Though the Government tried hard to maintain good relations with both sides in the Sino-Soviet dispute, the Vietnamese invasion of Cambodia in January 1979 and China's response in invading northern Viet Nam eventually forced Laos to take sides. China was asked politely to terminate its aid projects and reduce the size of its embassy staff. Subsequent Chinese warnings of the risk that the Laotian regime was taking, and the acceptance of several thousand Lao refugees from Thailand for settlement in China, led to charges by Laos of a Chinese-supported insurgency. Not until the mid-1980s did Laotian-Chinese relations begin to improve.

As relations with China were becoming more strained, those with Thailand improved. Both sides agreed to suppress insurgent activity aimed at the other and to encourage bilateral trade. Relations with Thailand were variable, however, and it was not long before the Government in Vientiane was accusing Bangkok of collusion with Beijing in attempts to destabilize the regime. Security concerns were an important reason for the decision in July 1979 to suspend further co-operativization of agriculture. This unpopular programme had been pursued over two years, leading to the theoretical formation by highly motivated cadres of some 2,500 co-operatives. In fact, most were inadequately organized and the benefits promised did not materialize, not least because the Government had so few economic resources. Moreover, peasants objected to pooling their land and working for 'work points' rather than wages. Some peasants, mostly in the south, slaughtered their livestock and left for Thailand, whilst others joined the anti-Government resistance. By 1979 both the Soviet and Vietnamese Governments were advising that the programme be halted.

The external and internal threats by then confronting the regime forced a reappraisal of government policy. By the end of 1979 the decision had been taken to reform the currency (a new National Bank kip replaced the former 'liberation kip'), and to adopt a less rigid form of socialism. The change in direction was justified by reference to Lenin's 'new economic policy' of the early 1920s, but the regime could barely disguise the fact that earlier policies had been a failure. The economy of Laos was recognized as comprising a mix of modes of production, including subsistence farming and a capitalist sector, all of which were to be officially encouraged. Profit was introduced as a criterion of efficiency, economic decision-making was decentralized, and controls on the circulation and marketing of goods and produce were eliminated, preparing for the successful return of a market economy. On these new economic foundations the LPDR launched its first five-year economic development plan (1981–85), to coincide with those of other Soviet bloc countries. The plan's targets were over-ambitious, however, and partly owing to a shortage of trained personnel and to structural inadequacies, virtually none was met. Both party and governmental organization was weak, and cadres lacked training and commitment. At the Third Party Congress held in April 1982, the first since the formation of the LPDR, there was considerable criticism of the reforms. While the same seven-member Politburo was re-elected, the Central Committee was more than doubled in size to include new members from the provinces, the military and the Government. Provinces were given the right to trade directly with foreign countries and to determine their own development strategies in conformity with central government planning, a measure which considerably strengthened the political power of the Party's provincial secretaries. Party reform extended to the expulsion of members on the basis of ideological shortcomings or for corruption. In the early 1980s solidarity with Viet Nam was the measure of ideological orthodoxy and anyone questioning the proximity of relations with Viet Nam could expect to be purged for being 'pro-Chinese'. Meanwhile, the easing of economic restrictions, together with the patronage exercised by powerful party leaders, encouraged corruption.

Early in 1984 relations with Thailand deteriorated following the outbreak of fighting in the vicinity of three villages on the border of Sayabouri Province claimed by both Laos and Thailand. In June a full-scale attack by 1,500 Thai troops failed to take the disputed area. A cease-fire was agreed upon, but subsequent negotiations failed to resolve the problem, and border demarcation remained a matter of dispute, which four years later again erupted in border fighting. Relations with both China and the USA, however, improved. Border trade resumed with China, and an agreement in February 1985 with the USA led to the first excavation at the crash site of a US aircraft to determine the fate of US servicemen missing-in-action during the Viet Nam War.

In March 1985 the LPRP celebrated 30 years of 'correct and creative' leadership of the revolution, and in December the regime celebrated both its first decade in power and the completion of the first Five-Year Plan. However, relatively little had been accomplished. The country remained desperately poor, and dependent as ever on foreign aid. The per caput income of the country's 3.5m. people was estimated at just over US $100 a year, making Laos one of the poorest countries in the world. However, the country was more or less self-sufficient in rice—given favourable weather conditions—although only a tiny fraction of paddy land was irrigated. Industrial production stood at little more than 5% of gross domestic product (GDP), a figure low even for the least developed countries. Most state-owned enterprises continued to make losses. Claims of 100% literacy were manifestly false, and both educational and health standards had actually declined through loss of trained personnel. Moreover, according to the human rights organization, Amnesty International, some 6,000 political prisoners remained in detention in remote internment camps. The feeling grew within the Party that a change of direction was necessary—especially given the possible reduction in aid from the USSR and Viet Nam.

For some time the LPRP Secretary-General, Kaysone Phomvihane, had been warning of a 'two-line struggle' being waged, although between which factions and in pursuit of which policies remained unclear at the time. Only subsequently was it revealed that the struggle was between those who wanted to introduce more radical economic reforms, led by Kaysone himself, and those who resisted them in the name of socialist orthodoxy, led by Kaysone's deputy in charge of the economy, Nouhak Phoumsavanh. Kaysone eventually succeeded in securing his policy orientation, but only after intense debate within the Party had led to the postponement of the Fourth Party Congress and of the introduction of the second Five-Year Plan until November 1986. The Congress eventually endorsed what was called the 'new economic mechanism', which allowed the operation of a market economy. State-owned enterprises were granted greater autonomy and expected to make a profit, food subsidies were progressively eliminated and the civil service was reduced in size.

Economic difficulties in 1987, brought on by drought and a decrease in electricity production, most of which was sold to Thailand, served to strengthen the position of the reformers. As the currency continued to depreciate and revenue consistently failed to cover expenditure, so dependency on foreign aid increased. Pressure for further reform came from the IMF, from foreign aid donors, and even from the USSR and Viet Nam, which were then undertaking their own programmes of reform. Measures were subsequently taken to dismantle the remaining elements of the centralized socialist economy. Over the next two years, as communism began to collapse in Eastern Europe and the USSR, and as Viet Nam withdrew its forces from both Laos and Cambodia, the LPDR strengthened ties with Thailand (following border fighting early in 1988) and with capitalist states, notably Japan, Sweden and Australia. At the same time, the Government passed a liberal foreign investment law, and began to lay the necessary legal foundations to attract foreign capital.

In April 1988 the first local elections since 1975 were held, followed in November by district, municipal and provincial elections. In March 1989 national elections took place to replace the Supreme People's Assembly appointed in 1975. The new 79-member Assembly met in June and elected Nouhak Phoumsavanh as Chairman, an indication that it would no longer be a powerless institution. Its first task was to appoint a Constitution Drafting Committee to draw up the long-delayed Constitution. Even though all candidates were carefully screened by the Party, these elections did provide some popular legitimacy for the regime at a time when, in response to the collapse of communism in Eastern Europe and events in Cambodia, anti-communist opposition to the regime had increased. Bolstered by the possibility of international support, both Hmong and right-wing Lao guerrillas operating from Thailand intensified their activities, though never sufficiently to threaten the Party's hold on power. In May 1988 Laos and China re-established full diplomatic relations. The following October Kaysone led a high-ranking Laotian delegation to Beijing to reinforce the much-improved relations with China, a visit reciprocated in December 1990 when the Chinese Premier, Li Peng, arrived in Vientiane. Relations with Thailand also rapidly improved after the Thai Prime Minister, Gen. Chatichai Choonhavan, exchanged visits with Kaysone in November 1988 and February 1989, and Thai businessmen led foreign investment in Laos. At the same time the Laotian Government continued to proclaim its special, if somewhat weakened, relationship with Viet Nam. By the early 1990s, therefore, Laos had succeeded in developing good relations with all its neighbours, as the country reverted to its more traditional role as a buffer state between contending powers.

Although not all problems with neighbouring states have been solved, in 1990 the diplomatic efforts of the Government led to an agreed delineation of the Laotian–Vietnamese frontier. Border agreements were also signed with China and Myanmar. Only the Laotian–Thai border remained a matter of contention. In June 1991 an agreement was concluded with Thailand in conjunction with the UN High Commissioner for Refugees (UNHCR) for the repatriation or resettlement in third countries of the 60,000 Laotian refugees, the majority of them Hmong, remaining in Thai camps. Although the programme moved more slowly than envisaged, owing mainly to Hmong resistance, the closure of several camps proceeded with a corresponding reduction in support for anti-Government insurgents operating in Laos. Laotian and Thai Governors of provinces along the common border began to meet regularly to discuss border issues. Co-operation with the USA on the control of drugs and the search for missing US servicemen resulted in 1992 in the re-establishment of full ambassadorial relations. The resumption of US aid to Laos followed three years later, when an exception was made to the restrictions imposed by the US administration on granting aid to communist states. In mid-1998 US officials agreed to extend financial and logistical support for the ongoing programme to clear unexploded ordnance from the Viet Nam War (supported by the USA since 1996).

The broadening of diplomatic and economic relations that followed the Government's 'open-door' approach reflected both increased confidence on the part of the regime and a series of important political initiatives. By 1991 the LPRP had sufficiently recovered from the impact of events in the former USSR to hold its own Fifth Congress. During the previous year the Party had orchestrated debate on the new Constitution, which subsequently underwent three separate drafts. It had also indicated that it would permit no challenge to its monopoly hold on political power when it arrested and subsequently imprisoned three critics whom it accused of 'activities aimed at overthrowing the regime'. All three were members of a group of intellectuals who advocated greater democracy. The Fifth Congress not only endorsed the draft Constitution but also introduced major changes in Party structure and in relations between the Party and government. The Party Secretariat was abolished, and Kaysone was named Party President. Three members of the Politburo retired, and changes were also made to the membership of the Central Committee, with a number of younger members gaining election, several of whom were closely related to senior members of the Party. Equally important, the Party endorsed the direction of economic reform in accordance with free-market principles, symbolically replacing the communist red star in the national crest by an outline of the That Luang stupa and simultaneously eliminating the word 'socialism' from the national motto.

On 14 August 1991, more than 15 years after the formation of the LPDR, the Supreme People's Assembly finally adopted a Constitution that referred to the 'Party' only once (as the 'leading nucleus' of the political system), and formally established Laos as a 'people's democracy' in the Marxist sense. The Constitution also guaranteed basic freedoms and the right to private ownership of property. The Supreme People's Assembly was renamed the National Assembly, the title used by the former Royal Lao Government, and its powers were enunciated. The executive powers of the President of the State were greatly enhanced, and provision was made for the appointment of a Vice-President, though none was named. Kaysone was appointed President of the LPDR, while the former Minister of Defence, Khamtay Siphandone, took Kaysone's place as Prime Minister, so entitled in preference to Chairman of the Council of Ministers.

The internal cohesion of the Party and the strength of the new institutions were tested in November 1992 by the death of Kaysone. He had led the Lao communist movement since the formation of the Lao People's Party in 1955, and his power as President of both the State and Party had been unchallenged. On his death, power was divided between the Chairman of the National Assembly, Nouhak Phoumsavanh, who became State President, and Khamtay Siphandone, who added the presidency of the Party to his post as Prime Minister. The transition had clearly been worked out in advance and was remarkably smooth, thereby ensuring continued political stability. Within a month scheduled elections for the National Assembly went ahead as planned. The 85 seats were contested by 154 candidates, all endorsed by the Party-controlled Lao Front for National Construction (LFNC). At its initial sitting in February 1993, the new Assembly confirmed the appointments of Nouhak and Khamtay, elected Saman Vignaket as its President in place of Nouhak, and endorsed a new ministry based on an extensively reorganized structure of government.

The new Government undertook to continue the economic reform programme, to extend the rule of law, to limit the worst excesses resulting from earlier decentralization of power, and to curb corruption. Reforms to the tax system sought to increase revenue collection, and further to liberalize conditions for foreign investment. Thailand continued to be the largest source of investment (inflows having reached US $1,940m. by the end of 1995), followed by the USA, with Australia, France and China also being prominent. Inflation and the chronic underlying trade deficit continued to cause concern, although it was hoped that the future completion of major hydroelectricity and mining projects would improve the balance-of-payments situation.

In March 1996 the LPRP's Sixth Congress revealed the concern felt by senior members of the party over the pace and implications of economic change. Secret political discussions prior to the Congress led to an outcome that was unexpected by most observers: the ruling Politburo was reduced from 11 to nine members, six of whom were military generals; Khamtay Siphandone remained President of the Party, but Deputy Prime Minister Khamphoui Keoboualapha, widely identified as a leading proponent of economic reform, was excluded from both the Politburo and the Central Committee (itself reduced from 55 to 49 members), while Sisavat Keobounphan (a close associate of Khamtay) not only regained his membership of the Politburo, but was also named as the country's first Vice-President. As expected, Nouhak Phoumsavanh stepped down from the Politburo, leaving the Party firmly under the control of the army.

In July 1997 Laos formally joined the Association of South East Asian Nations (ASEAN); membership of ASEAN entailed costs as well as benefits for Laos, not the least of which was the financial burden of participating in the organization's many regional forums. Relations with neighbouring states continued to be friendly in the 1990s; Laos developed amicable ties with Myanmar and Cambodia, both of which shared a degree of suspicion of Thai ambitions in the region. Relations between Vientiane and Beijing remained cordial: delegations were frequently exchanged, and China provided not only economic and technical assistance to Laos, but also some military aid. Vientiane was careful, however, to balance its foreign relations. Historically close political ties with Viet Nam meant that Hanoi still wielded considerable influence: links between Vietnamese and Laotian parties and military establishments were strong, and several memorials have been dedicated to Vietnamese troops who died in Laos during the Viet Nam War. Japan, meanwhile, remained the largest foreign aid donor, followed by Germany, Sweden, France and Australia. Cordial relations with the USA were maintained through continued co-operation over the issues of drugs and of US servicemen missing-in-action since the Viet Nam War. Moves to restore Normal Trading Relations (NTR) encountered resistance, however, and the USA remained critical of Laos's human rights record. Full diplomatic relations had been established with the Republic of Korea and several Central Asian republics.

In December 1997 elections were held for the 99-seat National Assembly. Seventeen of the candidates elected were army officers, while only one was not a member of the ruling Party. At the first session of the Assembly in February 1998, Nouhak relinquished the presidency to Khamtay Siphandone, while Vice-President Sisavat replaced Khamtay as Prime Minister. Political developments since the Sixth Party Congress confirmed both the dominance of the army and its desire to exercise closer control over the economy, although political appointments still reflected family and clan influence, regionalism and the country's ethnic diversity. The military was determined to gain a substantial share of the benefits to be derived from resource exploration and development projects. Three military-controlled companies divided the country among them to develop commercial agriculture, logging, mining and tourism; contracts for construction projects constituted another profitable source of income. Along with other state-owned companies, however, these military enterprises failed to repay substantial bank loans. Another form of corruption was the use of political influence to gain lucrative appointments for family members or clients, a development that the Prime Minister's Anti-Corruption Commission seemed unable to prevent.

The strategic position of Laos makes it a key state in the economic integration of mainland South-East Asia. An Australian-financed bridge across the Mekong River, some 30 km downstream from Vientiane, was opened in April 1994, and was the first in a series of infrastructure projects, including roads and bridges, destined eventually to link Thailand, Viet Nam and southern China via Laos. The construction of a second, Japanese-financed bridge at Paksé in southern Laos was completed in 2000, and an agreement has been reached to build a third bridge at Savannakhet, to be completed by 2005. The railhead at Nong Khai, on the Thai side, will be extended as far as Vientiane, although a feasibility study for a proposed rail link through northern Laos into southern China has indicated that the project is not, at present, commercially viable. Meanwhile, airports have been upgraded to cope with increasing tourist arrivals, and a new air traffic control system has been installed.

Particular attention is being given to the development of the road network. The last stretch of the southbound Route 13, the highway following the Mekong River from Vientiane to the Cambodian border, has been surfaced, while several roads linking Laos with Viet Nam are being upgraded. These include Route 7, linking Houaphanh to the Laotian–Vietnamese border in the north, and an extension of Route 15 from Sekong across the southern highlands of Viet Nam to Danang, to be constructed at a cost of US $33m. by Malaysian timber interests. The upgrading of Route 9, which connected with the bridge at Savannakhet, was completed in early 2004. Meanwhile, Laos, Thailand and China have agreed to contribute equally to a $45m. road project, to be constructed mainly in Laos, to link the three countries. The project will be co-ordinated by the Asian Development Bank (ADB) and was to be completed by 2006.

Some Lao, however, are becoming concerned over the threat that rapid regional integration poses to their society and culture, as well as over the effects that such integration may have on the environment. There exists a growing awareness of the importance of environmental issues such as the impact of dam construction, the rate of timber extraction and threats to the country's natural biodiversity. Dam construction slowed in the late 1990s as a result of the Asian economic crisis: in 1999 it appeared unlikely that the series of hydroelectric projects planned to follow the completion of the major Nam Theun 2 project would go ahead, while the Nam Theun 2 project itself was delayed. It was not until February 2002 that protracted negotiations between the Laotian and Thai authorities and the consortium of companies building the massive dam resulted in the signing of a contract committing Thailand to the purchase of sufficient Laotian electricity to make the project viable. Construction was expected to take six years, however, so hydro-power was unlikely to provide a major source of government income for some time. Substantial levels of foreign aid are thus likely to continue to be crucial to Laos for the foreseeable future. In May 2000 the Government imposed a ban on all timber exports, in an attempt to reduce the endemic over-exploitation and smuggling of the country's dwindling timber reserves. Whether the measure would have any effect, given the voracious demand of foreign timber interests and the level of official corruption, remained to be seen.

Related government attempts to reduce swidden ('slash-and-burn') farming and encourage reafforestation have so far proved only partially successful. (In 1993, in an initiative to preserve unique Laotian flora and fauna, the Government set aside 17 'national biodiversity conservation areas', which together constituted just over 10% of Laos's total land area. However, these areas are not national parks; they are inhabited, and some timber extraction is permitted.) Illicit trade in exotic Lao wildlife, mainly to China, is a serious problem. Tourism is also likely to have an increasing impact on the Laotian environment (although environmental and ethnological tourism are being encouraged): for example, it is expected that the fragile beauty of the city of Luang Prabang, approved by UNESCO in February 1998 as a World Heritage site, will come under serious threat in the 21st century, as a result of the sheer numbers of tourists visiting the city (expected to attract more than one-third of the annual total of 0.6m. tourists travelling to the country). World Heritage site listing for Wat Phu in southern Laos may have a similar effect. Tourism, however, is viewed by the Government as an expandable source of national income, particularly in a time of economic stringency.

Another area of concern is that of the potential social impact of consumerism, drugs, HIV/AIDS and prostitution, as the country's boundaries become more porous. Despite the imposition of relatively strict controls by the Government, prostitution has reappeared. Laos has not been unaffected by the growing HIV/AIDS epidemic in the region, and HIV/AIDS awareness campaigns are now part of the national public health programme. Of equal concern in the spread of the disease is intravenous drug use. Despite the fact that the cultivation of opium was outlawed in Laos in 1997 as a result of pressure from foreign governments, and that the Laotian authorities intend to eradicate opium production by 2008, the drug is still readily available. Meanwhile, heroin and amphetamines also flood in across the border from neighbouring Myanmar, attracting Western 'drug tourists' and increasing numbers of young Lao. (An estimated 15% of Lao students aged between 15 and 20 are reported to be addicted.) In 2001 the Government introduced the death penalty for drugs-trafficking and launched an intensive anti-drugs campaign aimed at the country's youth. As an indication of the Government's mounting concern, a Central Committee for Drug Control has been established, chaired by the Prime Minister, and Laos has entered into drug-control agreements with neighbouring states.

In August 1999 the Asian economic crisis resulted in the removal from their posts of two leading finance officials, the Governor of the Central Bank of Laos, Cheuang Sombounkhan, and the Deputy Prime Minister and Minister of Finance, Khamphoui Keoboualapha, for alleged mismanagement of fiscal and banking policy. Economic conditions were subsequently slow to improve, however, and popular dissatisfaction continued to increase. In October students staged a rare anti-Government protest in Vientiane. A number of the demonstrators were promptly arrested, whilst others suspected of involvement in the protest fled across the border to Thailand. The Government refused even to acknowledge that either the incident or the arrests had ever taken place but, according to Amnesty International, some of the leaders of the demonstration remained in detention in 2004.

In June 1999 Hmong insurgent groups in north-eastern Laos renewed their anti-Government activities. It was speculated that the unrest had been fomented from abroad by Hmong activists linked to former Gen. Vang Pao. In March 2000 a bomb exploded in a restaurant in Vientiane, injuring a number of people, several of them foreigners. This attack was followed by a series of further explosions over subsequent months. Although no organization claimed responsibility, a government spokesperson attributed the explosions to Hmong insurgents and other 'bad elements'. The Hmong denied responsibility, however, and the incidents appeared instead to have been organized by Lao dissidents operating from abroad, with the aim of discrediting the Government by showing it to be incapable of maintaining security. A more serious attack took place in July 2000, when about 50 suspected royalist insurgents seized customs and immigration offices on the Laotian–Thai border, near the southern town of Paksé. At least six of the dissidents were subsequently killed by government troops, some were captured, while a further 28 were arrested by Thai authorities. The Lao Government immediately called for the extradition of 17 of those captured, who were Lao nationals. A long legal dispute followed, in the course of which Laos and Thailand signed their first extradition treaty. Finally, in July 2004 the remaining 16 Lao (one had died in the mean time) were handed over to the Laotian authorities, despite a ruling by the Thai courts rejecting their extradition on the grounds that the enabling legislation could not be applied retrospectively. Some opposition continued. Meanwhile, a small explosion occurred in January 2001 at the Laotian end of the Friendship Bridge over the Mekong River, although security was soon restored.

The principal response of the Laotian Government, both to the risk posed to national security by the bomb attacks and to the ongoing economic crisis, was to seek to consolidate relations with its communist neighbours. Both Viet Nam and China offered to increase economic assistance to Laos. Cross-border trade was stimulated in order to reduce Laotian dependency on Thai imports and the country's substantial negative trade balance with Thailand, while Chinese and Vietnamese investment both increased. China agreed to provide Laos with technical assistance and emergency loans, partly for the purchase of military equipment. In July 2000 President Khamtay made an official visit to China, and in November President Jiang Zemin became the first Chinese Head of State to visit Laos. Senior Laotian and Vietnamese officials also exchanged visits. Relations with Thailand, meanwhile, remained friendly. Demarcation of the Laotian–Thai border continued slowly, while work also commenced on the Laotian border with Cambodia.

Laos entered the new millennium with some trepidation. Short-term economic prospects were unfavourable, with little likelihood of any reduction of either the budgetary or trade deficits. Despite this, the value of the currency had stabilized by the end of 2000, and some growth in the economy was evident. Security was still a concern, however, and both the Government and the Party feared continuing popular dissatisfaction, precipitated by growing social and regional disparities and endemic corruption. Nevertheless, the Seventh Congress of the LPRP was held in March 2001. The outcome was as expected, with the military retaining control of the Politburo. All eight surviving members of the previous Politburo were re-elected, together with three new members. Of these 11 members, eight were either retired or serving military officers. Of the other three, two were senior Party officials in charge of mass mobilization and Party organization, while the other was the new first Deputy Prime Minister, Thongloun Sisolit. The only unexpected omission from the new Politburo was that of the second Deputy Prime Minister and Minister of Foreign Affairs, Somsavat Lengsavat. Twelve new members were appointed to the Central Committee, increasing total membership to 53, including all 16 provincial governors. Party membership was reported to have increased by 28%, to 100,000.

Barring unforeseen circumstances, this LPRP leadership, which was perceived to be both cautious and conservative with respect to political liberalization and economic reform, was to remain in power until 2006. The Party Congress reiterated its opposition to multi-party democracy, and proclaimed ambitious targets for economic growth (see Economy). The Congress endorsed a reduction in tariffs in accordance with its commitment to the ASEAN Free Trade Area (AFTA), and restated its aim to join the World Trade Organization (WTO). A projected improvement in living standards was likely to be assisted by ongoing funds promised to Laos under the IMF Poverty Reduction and Growth Facility.

Following the Congress in March 2001, the National Assembly convened to endorse a change of government. Bounghang Volachit, hitherto Deputy Prime Minister and Minister of Finance, replaced the elderly Sisavat Keobounphan as Prime Minister, with Thongloun Sisolit as the new first Deputy Prime Minister and President of the State Planning Committee. Lt-Gen. Choummali Saignason, hitherto the Minister of Defence, was elected the country's Vice-President (following the death of Oudom Khattigna in December 1999). Nevertheless, most government ministers retained their portfolios, thereby reinforcing the conservative continuity demonstrated by the LPRP Congress. If the Party's intention was to perpetuate existing policies, however, it still faced significant popular discontent, mainly over poor economic performance and increasing levels of corruption and crime (especially in Vientiane).

In an apparent attempt to reinforce its legitimacy, the Government advanced the date of elections to the National Assembly to 24 February 2002, several months ahead of schedule. The Assembly was enlarged from 99 to 109 members, elected from provincial lists numbering 166 candidates, all of whom were pre-endorsed by the LFNC. Of those elected, only one was not a member of the LPRP. A total of 25 were women, while 39 were members of minority ethnic groups (including six from upland Tai minorities). The remaining 70 were ethnic Lao. Overall, the members of the new Assembly were younger (with an average age of 51) and better educated (63% had received some tertiary education) than their predecessors.

In the conduct of its foreign relations Laos maintained an even balance between Viet Nam and China. Prime Minister Boungnang Volachit visited first Hanoi and then Beijing shortly after the elections of early 2002, as did the new Minister of Defence, Gen. Douangchai Phichit. On each occasion new bilateral agreements were signed, including one concerning security ties with Viet Nam. The month of July 2002 marked both the 40th anniversary of the establishment of diplomatic relations between Laos and Viet Nam, and the 25th anniversary (and expiration) of the Treaty of Friendship and Co-operation signed by the two countries in 1977. President Khamtay paid an official four-day visit to Hanoi in May, during which both sides agreed to hold joint celebrations and to protect and develop their 'traditional relations'. No mention was made of an extension of the Treaty, which seemed unnecessary as both states had now acceded to ASEAN membership.

Relations with Cambodia were relaxed as border demarcation progressed. Plans were announced for a new development zone in the border area shared by Laos, Cambodia and Viet Nam. Though demarcation of the border between Laos and Thailand also progressed, relations with Thailand continued to be uneasy, even after the visit of Thai Prime Minister Thaksin Shinawatra to Vientiane in June 2001. Tensions centred upon river works being undertaken by both countries, Thai naval patrols, drugs- and people-trafficking, the activities of Lao anti-Government insurgents in Thailand, and even a Thai film that the Lao considered insulting to the 19th-century King of Viang Chan, Chao Anou. While production of the film was halted, other matters proved more intractable.

By the end of 2002 the Government could congratulate itself on its political achievements, demonstrated by the Seventh Congress and the National Assembly elections. However, although the LPRP faced no overt political opposition, with the collapse of its Marxist ideology its legitimacy was more than ever dependent on its ability to manage the economy and to improve the performance of its own cadres. This perhaps explained an apparent increase in official support for Buddhist ceremonies and celebrations in the new millennium, and for a new emphasis on Lao nationalism. In January 2003 an elaborate pageant took place, during which a statue of King Fa Ngoum was erected, to commemorate the 650th anniversary of the founding of the Lao kingdom of Lan Xang in 1353. In response to the changing requirements of the times, in 2002 the Ministry of the Interior was renamed the Ministry of Security and placed under the direction of Gen. Soudchai Thammasith. Its mandate was soon challenged, however, by two fatal attacks on buses on the highway linking Vientiane with Luang Prabang in February and April 2003. In total, 23 people were killed, including two Swiss cyclists caught up in the first attack, with dozens more injured. The Government blamed Hmong 'bandits' for both attacks, but downplayed any suggestion that the Hmong insurgency that had been ongoing since 1975 had entered a new phase. A further bus attack took place in August 2003 in northern Laos, killing five people, while two other attacks were reported in the south of the country. These incidents were enough, at a time of global concern about terrorism, to attract the attention of the world press. For the first time, two foreign journalists managed to contact Hmong insurgents in the Xaisomboun Special Military Region, revealing an insurgency at the point of exhaustion and desperate for assistance. This was confirmed after two more journalists and their US-Hmong interpreter also made contact with Hmong insurgents, only to be arrested by the Laotian authorities, tried and sentenced to 15 years' imprisonment. Following widespread international condemnation of the severity of the sentences, all three were released and expelled from the country. Such international attention was not welcomed by the Laotian Government, which feared the effect that it might have on foreign investment and tourism. In fact, however, independent reports on the desperate state of the insurgents served to confirm the Government's claim that the Hmong insurgency had been reduced to only a few hundred scattered guerrillas. Claims by Hmong resident in the USA that fighting had become widespread were discounted by observers, while accusations of Vietnamese involvement in anti-insurgency measures were denied by the Government.

In February and March 2004 some 1,000 Hmong, mostly women, children and the elderly, surrendered to the Laotian authorities in Xiangkhouang and Luang Prabang Provinces. While this did not necessarily signify the end of the Hmong insurgency, it did indicate that, after years of skirmishing, government forces had gained the advantage in the conflict. Security in the rest of Laos did not appear to be a cause for major concern, despite a new spate of five small, home-made bombs that exploded in Vientiane between October 2003 and June 2004. These resulted in the death of one bomber, owing to the premature explosion of his device, but no other casualties. Two other minor bomb attacks were reported to have taken place in southern Laos, leaving one person dead and six injured. A covert organization that called itself the Free Democratic People's Government of Laos (FDPGL) claimed responsibility for all the explosions that had occurred since 2000, and threatened to carry out further attacks unless the Government reduced its dependence on Viet Nam and instituted democratic reforms. However, relations with Viet Nam continued to be close, with a bilateral agreement having been concluded to promote co-operation between the general staffs of the Laotian and Vietnamese armed forces. Laos also enjoyed friendly relations with significant donor countries, although some frustration was caused by delays in the granting of NTR status by the USA. These delays were largely attributable to lobbying by Lao-Americans and concerns over lack of democracy and religious freedom in Laos. Meanwhile, the European Union (EU) opened a representative office in Vientiane and increased its aid to the country.

By 2004 political interest had already begun to focus on the forthcoming Eighth Party Congress, to be held in 2006, as this would involve an important generational change in the Lao leadership. Most of the ageing generals who had dominated the Politburo were expected to retire, thus reducing military influence within the LPRP. Manoeuvring was already taking place between powerful figures and their supporters to ensure the protection of their interests in the new Politburo. Bousone Boupavanh, one of the younger members appointed to the Politburo in 2002, became fourth Deputy Prime Minister in October, while Venthong Luangvilay was placed in charge of the State Audit Authority in the Prime Minister's Office, in order, it was assumed, that he could make efforts to reduce the level of governmental corruption. In the only other government reorganization to have taken place in 2003, Chansy Phosikham became Minister of Finance in January. He was replaced as Governor of the Central Bank by the former Minister of Commerce and Tourism, Phoumi Thipphavone. The most serious effect of the political uncertainty associated with the run-up to the Eighth Congress was a slowing in the pace of economic reform, much to the dismay of international donors.

Economy

NICK FREEMAN

The establishment of the Lao People's Democratic Republic (Lao PDR) in December 1975 marked the end of a lengthy civil war in Laos. Caught within the vortex of Cold War rivalry and the more intensive conflict being waged in neighbouring Viet Nam, Laos's civil war was primarily fought between pro-communist Pathet Lao forces and anti-communist Royal Lao Government (RLG) forces. The chaos and division caused by the civil war had excluded the possibility of any significant economic development occurring in Laos following the end of French colonial rule in 1953. The preceding 70 years of French administration had resulted in very little economic development in the country. Having been regarded as little more than a colonial backwater within the French empire, Laos had been governed by France on a low budget. Compared with the relatively significant investment enacted in Viet Nam, French entrepreneurs identified few commercial projects in Laos, other than a limited number of isolated agriculture-related and mining operations. Rather, land-locked Laos's primary role was deemed to be strategic, with the country acting as a buffer between France's more lucrative occupation of Viet Nam and British interests in colonial Burma (now Myanmar) and the independent Kingdom of Siam (now Thailand).

Following the Pathet Lao victory in 1975, senior members of the political leadership, the Lao People's Revolutionary Party (LPRP), descended from their headquarters in the remote north-east of the country and took power in the capital, Vientiane. In addition to the social, political and security challenges confronting the new Government, the LPRP rapidly had to contend with an economy that was suddenly having to function without the massive levels of US assistance that had effectively financed the RLG regime, and grossly distorted Laos's economy for much of the previous decade. The immense damage inflicted on the country during the Viet Nam War, most notably as a result of US 'carpet bombing' of areas along Laos's eastern border with Viet Nam, needed to be addressed, in order that refugees could be resettled and agricultural activity could recommence. Furthermore, areas of upland Laos that had previously been under LPRP control now had to be integrated with the newly liberated lowland parts of the country. With Soviet-inspired central planning techniques, and under the guidance of fraternal sponsors, the LPRP leadership set about addressing these challenges with a degree of zeal (and insensitivity) perhaps to be expected of those still flush with victory. The Government's aim was to create a centrally planned economy in Laos, based on the notion of collective ownership. Most private companies in the nascent industry and service sectors were nationalized, as was the entire banking sector; the latter was also consolidated into a single mono-bank structure. It is estimated that up to 80% of Laos's small industrial sector was placed under direct state ownership during this period, and most companies found that market forces were replaced as the leading influence on business by output targets set by the State Planning Committee.

A state marketing board to which peasants were obligated to sell their surplus agricultural goods was established in 1976, setting prices for staple goods, including rice. The marketing board issued coupons in payment for agricultural products, which peasants could then use to buy goods in state-run stores. Adverse weather conditions in the period 1976–78, compounded by ill-conceived policies, resulted in a marked decline in agricultural output. The mass collectivization of agriculture was also attempted by the LPRP, albeit relatively briefly. Initially at least, the number of agricultural collectives was low, but by 1978 the total had exceeded 1,000. Although mandatory directives on collectivization were abandoned as early as 1979, the subsequent introduction of various inducements—such as privileged access to cheap credit, and lower tax rates—resulted in an increase in the number of collectives to almost 4,000 by 1986. Some of these collectives, however, were little more than quasi-formal groups of peasants providing mutual assistance, content to be classified as collectives in order to receive the privileges granted by the Government. Collectivization was particularly strong in areas that had previously been under the control of the Pathet Lao prior to 1975, or where the provincial Government's socialist vigour was most ardent, and weak in areas where swidden (shifting) cultivation was practised. For example, over 80% of the cultivated area of Champasak Province was under collectivization in 1984, compared with 8.5% in Vientiane and less than 6% in Luang Prabang. State-owned trading companies enjoyed a monopoly on both external and inter-provincial trading activity. A broad policy of import-substitution was introduced by Laos's policy-makers, despite the very obvious limitations of the domestic industrial sector.

The scale of the challenge confronting the relatively inexperienced leadership, together with the inappropriate use of economic planning methods and the lack of sensitivity shown to those who had previously lived in areas controlled by the RLG, cumulatively served to create an economic crisis. The exodus (primarily across the Mekong River to Thailand) of those Lao who had been directly associated with the RLG Government, its regular armed forces or irregular ethnic minority forces, was followed by the departure of those entrepreneurs and professionals who did not relish the prospect of living under a communist regime, where property rights were no longer respected and where business assets were arbitrarily seized. Within a few years, Laos thus lost a substantial proportion of its already small educated middle and entrepreneurial classes, and the resulting paucity of human resources has been a significant factor in the slow pace of the country's economic development ever since.

Prior to 1975, Laos had been heavily dependent upon trade with neighbouring Thailand for a wide range of goods. However, tension between Vientiane and Bangkok (particularly over the fate of the Laotian royal family) led to a disruption of border trade in the late 1970s. High inflation soon developed, exacerbated by an expansive monetary and fiscal policy implemented by the new leadership. The scale and vehemence of popular resistance to the leadership's attempt at the collectivization of lowland agriculture, as well as opposition to its efforts to halt swidden upland agriculture, along with various other elements of the country's 'socialist transformation', appear to have surprised the LPRP. Far from showing signs of the development of a vigorous new economy, and deriving the anticipated benefits of the peace dividend and central planning methods, the first years of the Lao PDR were characterized by declining output, lower living standards for the urban Lao (no longer supported by US aid), and stubborn rural resistance to the policies of the new leadership. Quite clearly, attempts to create a centrally planned economy, in accordance with the model advocated by the Soviet Union and other fraternal socialist allies, were not generating the required results in Laos. Consequently, in 1979 the LPRP was obliged to signal a less zealous approach to the socialist transformation of the Laotian economy, following a similar decision taken in neighbouring Viet Nam in that same year. The concept of obligatory collectivization was abandoned, and the limited role of the private sector was formally recognized. State enterprises were slowly granted a greater degree of independence over day-to-day decision-making. The local currency, the kip, was effectively devalued in late 1979, with the 'liberation kip' being replaced by the 'national bank kip'. The Third Congress of the LPRP, held in 1982, witnessed the affirmation by the leadership that Laos's transition to a socialist system could not be conducted according to the rapid pace envisaged after the party's military victory in 1975. Instead, it was decided that a more gradual and incremental approach would be pursued, although the ultimate goal of the socialist transformation of the economy was still very clearly envisaged by the LPRP. While these partial reforms helped withstand the economic crisis facing Laos in the late 1970s, they were insufficient in bringing about a sustained economic revival, and by the mid-1980s it had become clear that a more substantive programme of economic liberalization measures would be needed.

A more radical and comprehensive programme of reforms, referred to as the New Economic Mechanism (NEM), was endorsed by the Fourth Party Congress in November 1986. The NEM involved a transition of the economy from a broadly centrally planned economic system to a more market-orientated system, albeit with many socialist elements remaining in place, and from an economy striving towards collective ownership to one based on private ownership. The NEM principally entailed the introduction of a Constitution and a legal regime, the recognition of property rights, the privatization of most state enterprises, price reform, the creation of factor markets and the end of state distribution systems, financial and macro-economic reforms, and the opening of the economy to foreign trade and investment. As the period of central planning in Laos had endured for less than a decade, it was possible to dismantle much of the socialist economic system within the first five years of the reform programme. The collapse of the socialist bloc in the late 1980s also influenced Laos's economic reform programme, as aid from fraternal allies contracted sharply. The country was thus obliged to seek alternative sources of external assistance: in order to attract funding from the international donor community, the Laotian leadership had little choice but to make changes to its economic management methods. In 1987 the state marketing board was closed, allowing the prices and distribution of items like rice to be dictated by the market. The policy of import-substitution was also abandoned, in favour of greater external trade, and state-owned trading enterprises slowly began to lose their monopoly on import and export activities. One year later, the tax and credit privileges granted to agricultural collectives were halted. The local currency was again devalued in 1987, and the fixed exchange rate system was abandoned in the following year. Subsequently, the official exchange rate was adjusted to correspond to the unofficial market rate, ultimately allowing for a managed 'float' of the kip to be formally adopted in 1995. Today, the margin between the official bank and unofficial parallel exchange rate rarely exceeds 2%.

The first, and long overdue, Constitution of the Lao PDR was promulgated in August 1991, 16 years after the LPRP took power, and heralded the leadership's belated recognition that genuine economic reform and sustained development could not proceed without the rule of law and some respect for property rights. Foreign investors, in particular, who were perceived as an important source of both financial and non-financial contributions to the economic development of the country, had been unwilling to commit substantial sums of investment capital into Laos without some recognition for property rights and a legal process, however vague and untested. A property law (albeit based more on user rights than private ownership as such), contract law and inheritance law were all passed in 1990. A number of other reforms were enacted in 1990–91, including the reintroduction of central government control over regional bank branches and state enterprises (responsibility for these agencies had previously been with the relevant provincial authorities). The provincial authorities also lost control over the collection of revenues and the allocation of state expenditure. A minimum wage was introduced in 1991. A second foreign investment law was approved in 1994, along with a business law and customs law, and the remaining restrictions on internal movement were removed. A domestic investment law and tax law were passed one year later, with the latter undergoing slight revision in 1998.

The average per caput gross national income (GNI) in Laos remains low, at around US $300, albeit measuring more than double the figure recorded at the commencement of the economic reform programme. Some upland provinces, however, have average per caput incomes of less than $60. Roughly 30% of the population lives below the poverty line, with income of less than $1.50 per day; statistics more typical of sub-Saharan Africa than South-East Asia. According to the IMF, in 2003 the gross domestic product (GDP) of Laos expanded by 4.8%, compared with growth of 5.7% in each of the previous two years. The Asian Development Bank (ADB), however, estimated Laos's GDP growth to be 5.9% in 2003. The UN Development Programme (UNDP) estimated that in 2002 average life expectancy at birth in Laos was just 55.1 years (compared with around 69 years in neighbouring Viet Nam and Thailand); the country had an adult illiteracy rate of 49%, just 40% of the population had completed primary school, and only 30% had enrolled at secondary school level. Of a total population of around 5.7m. in 2003, only about 20% lived in urban areas. The annual rate of population growth was estimated at approximately 2.2%. Laos has a youthful demographic profile. Approximately 43% of the population is under the age of 15, and less than 3.5% is over 65 years of age. The distribution of the population in Laos is far from uniform, with 45% of the country's population living in just four of the 18 provinces (Vientiane, Savannakhet, Champasak and Luang Prabang) in 1995; 18% of the population resides within Vientiane Province alone. Population density in Laos as a whole is very low, at 23.9 people per sq km in 2003. The total work-force in Laos is around 2.2m., of whom about 10% are salaried. Approximately 130,000 people work in some element of government administration.

AGRICULTURE AND FORESTRY

Agriculture remains the primary source of income and employment in Laos, although much agricultural activity continues to be conducted on a subsistence basis, and productivity levels are relatively low. While the economy as a whole achieved relatively high average annual rates of growth during much of the 1990s, the agricultural sector generally recorded more modest growth rates over the decade. According to the ADB, agricultural GDP grew by 4.0% in 2002 and by 8.3% in 2003, owing largely to the onset of rains following a dry period in 2002. Despite the relatively rapid growth of the industrial and service sectors under the economic reform programme, the agricultural sector continues to dominate the Laotian economy, accounting for almost half of the country's total GDP, and around 80% of all employment. Rice production accounts for roughly 75% of total land under cultivation, although the Government has been encouraging farmers to diversify their crop production and, also, to pursue animal husbandry. Weather conditions have the potential to cause a marked impact on the country's overall economic performance, as witnessed during the drought of 1987–88 and the floods of 1995–96, 2000 and 2002. This is particularly true with reference to the upland areas of Laos, where subsistence farming dominates, most households are poor, there is a lack of capital for the upgrading of technology, environmental degradation appears to be accelerating, and transport to the limited number of markets remains wholly inadequate. The lowland areas of Laos began to show signs of greater vitality in the 1990s, however, with households able to sell a proportion of their increased output, as the revival of market forces began to generate benefits. Lowland farmers are now able to invest in new technology and to buy pesticides and fertilizer and have better access to markets to sell surplus output, thereby benefiting from a conducive spiral of improved income and living standards. The change is particularly evident in the lowland 'Mekong corridor', where the creation of all-weather roads and the development of increased border trade are identified as the main catalysts behind the achievement of improved productivity and income levels. This divergence between the two agro-economic zones—the uplands and the lowlands—warrants some concern, as all indicators suggest that the disparity between the two areas continued to widen throughout the 1990s.

In 2000 crop production in Laos accounted for 59% of total agricultural output, followed by livestock and fishery activities (35%), and forestry (6%). The dominant crop by far is rice, with approximately 2.4m. metric tons of paddy harvested in 2002 (compared with 1.7m. tons in 1998). Vientiane municipality has tended to be the most productive province for growing paddy, followed by Bokeo, Xiangkhouang, Saravan and Champasak Provinces. Sekong, Phongsali and Luang Prabang Provinces, by contrast, have tended to be the least productive. Factors influencing the level of productivity in each province include the quality of transport available, the level of foreign assistance provided, the presence of unexploded ordnance, and the extent to which swidden cultivation methods are still being used. Laos has tended to have only one main rice harvest per year, during the wet season, with a much smaller dry season harvest. Although the Lao population continued to increase during the 1990s, rice output has remained broadly static, resulting in a fairly substantial deficit that had to be offset through imports. However, in recent years the Government has sought to increase the size of the dry season rice crop substantially in a bid to make the country self-sufficient in rice. In 1999, for example, it was reported that the dry season rice crop totalled 300,000 tons, compared with less than 50,000 tons in 1995. The area being harvested during the 1999 dry season was reported to total 92,000 ha, more than double that harvested in the previous year, and a dramatic increase from only 15,500 ha in 1994. In 1997–98 the Government took a risk when it spent a considerable proportion of its foreign-exchange reserves on the purchase of a substantial number of diesel-powered water pumps to help farmers irrigate the dry season paddy fields in lowland areas. This purchase was followed up with training courses on dry season cultivation techniques for both paddy and other cash crops. The initiative constituted part of a wider 'strategic vision' for the agricultural sector, announced by the Ministry of Agriculture and Forestry in late 1999. This vision was reported to include the extension of policies and initiatives that had previously worked well in the lowland areas (including transport upgrading, access to micro-financing for farmers, and guidance on diversifying crop production) to the upland areas. Aid donors have noted the omission of some important aspects from the 'strategic vision'; little attention is focused on, for example, the fishery sub-sector (despite fish being an important source of protein in Laos), the use of livestock (one of the few consistent growth areas in upland agriculture), and forestry.

Other crops grown in relatively substantial quantities in Laos include corn, sweet potatoes, cassava, peanuts, soybeans, coffee and tobacco, as well as some tea and sugar cane. Although total coffee production remains quite small (24,000 metric tons were harvested in 2000, up from 10,000 tons in 1996), Laos has significant potential in this area, and it is likely that coffee will become a more prominent source of foreign-exchange earnings in the future, when the international price of coffee stabilizes. The development of a dry season harvest of cash crops should also result in an increase in the output of other agricultural products, providing a useful addition to farmers' incomes and increasing the foreign-exchange earnings of the country as a whole.

Swidden cultivation remains common in some upland provinces, despite the Government's attempts to halt this form of activity. (In 1998 it was estimated that more than 155,000 households continued this practice, compared with 280,000 in the mid-1990s.) Many practitioners of swidden cultivation tend to be from ethnic minorities that resist what they perceive as lowland Lao interference. However, under the Government's new 'strategic vision' for the agricultural sector, a renewed attempt to convert swidden farmers to more sedentary forms of cultivation was to be attempted through the introduction of feeder roads to allow improved market access, better rural savings mobilization and credit techniques, the development of land entitlement and land use zoning practices, support for crop diversification, and so on.

With assistance from UNDP, the Laotian Government has sought to clear substantial areas of southern and eastern Laos of unexploded ordnance, most of which was dropped by the USA during the Viet Nam War. Approximately one-half of Laos remains contaminated by over 0.5m. tons of unexploded ordnance, resulting in about 120 casualties per year, and effectively rendering parts of 15 provinces unsafe for habitation or cultivation. (The USA dropped a massive quantity of ordnance on Laos, partly in a covert bid to support the RLG Government, but also in an attempt to destroy the 'Ho Chi Minh trail', a supply route between North and South Viet Nam, which ran through large parts of eastern Laos.) This unexploded ordnance—and small anti-personnel devices known as 'bombies', in particular—has also hindered the expansion of agriculture in certain areas. (Despite efforts at clearance, just 6% of Laos's total land area was being cultivated in 1993.)

The Laotian Government has also sought to eradicate the cultivation of opium in remote upland areas of the country. In 2003 the UN Office on Drugs and Crime (formerly the UN Office of Drug Control and Crime Prevention) estimated that Laos was the third largest opium producer in the world, after Afghanistan and neighbouring Myanmar. Total opium production in that year was estimated to be 120 metric tons (reduced from 210 tons in 2001 and from 180 tons in 2002), cultivated across 11 northern provinces of Laos, particularly Phongsali and Houaphanh Provinces. (The Government's own statistics suggested that opium production was 78.5 tons in 2002, a reduction from 112 tons in the previous year.) The potential harvest area for opium declined from 28,150 ha in 1997 to 12,000 ha in 2002. Most opium cultivated in Laos is produced by small-scale subsistence farmers, both for use as a medicine and as a cash crop, and the country is believed to have the world's second highest opium addiction rate, after Iran. Most opium addicts live in remote upland areas, making it difficult for detoxification programmes to reach the estimated 60,000 addicts. The Government has boldly pledged to eradicate opium production by 2005, and all other drugs by 2015. It thus sought US $80m. from foreign donors in early 2002 to fund a three-year programme to rid Laos of opium.

About 42% of Laos (or 10m. ha) is covered in natural forestland, a marked reduction from about 50% in 1985 and 70% in 1940. Forest-related activities account for around 10% of GDP and 25% of foreign-exchange earnings, as well as a substantial proportion of government revenues from payments of forest royalties. In 2000, according to official figures, the production of logs totalled 378,000 cu m, compared with 819,000 cu m in 1995. A small number of companies connected to the Laotian military dominate the logging industry in Laos, and have been particularly active in those areas identified for eventual flooding under proposed hydropower projects. An Environment Protection Law, which is retroactive in effect, was passed in 1999, although it remains to be seen whether Laos's legal institutions have the capacity to enforce this law properly, particularly with regard to powerful vested interests. The Government aims to have achieved forest cover of 70% by 2020, although this target seems unlikely to be achieved.

INDUSTRY

The industrial sector (comprising all manufacturing, mining, construction, and electricity-generating activity) has recorded a faster rate of growth since the enactment of the NEM reforms than either the agriculture or service sectors. Between 1993 and 1996 the rate of growth of the industrial sector was in double figures, although this subsequently declined to 8% in 1999, before recovering to 14.6% in 2003, driven in part by robust growth in the construction and garment sectors. Foreign investment participation in the industrial sector—most notably in the energy, mining and garment sub-sectors—played a critical role in the sector's relatively impressive performance during the 1990s. Yet, despite commendable growth during that decade, in 2003 industry accounted for only 22% of GDP in Laos, a large proportion of which is light industry and handicrafts.

The manufacturing sector accounts for around 75% of industrial activity, and includes a relatively substantial export-orientated garment sector, which has enjoyed improved performance following the reinstatement by the European Union (EU) of its Generalized System of Preferences (GSP) for Laos in late 1997. The principal products of the manufacturing sector include beer and soft drinks, wood products, plastic products, tobacco and cigarettes. Heavy industry is virtually nonexistent in Laos, with the exception of cement production. The industrial sector is unevenly distributed within Laos: roughly 63% of all enterprises are located in Vientiane, Savannakhet, Champasak, and Luang Prabang Provinces, and two-thirds of all Laos's large-scale manufacturing companies are to be found within Vientiane Province alone. Similarly, the vast majority of Laos's garment, footwear and textiles companies are located within municipal Vientiane. The construction sector failed to prosper in the late 1990s, owing in large part to the suspension of most of the country's planned hydropower projects and a steep decline in foreign investment inflows, although it appeared to have experienced a slight recovery in 2002.

Despite having been granted partial managerial autonomy in the late 1970s, state enterprises continued to perform poorly during much of the 1980s. The Government responded by introducing a programme of privatization for all non-strategic state firms in 1989. The privatization programme has entailed a number of different forms of divestment, including leasing arrangements, management buy-outs, transfer to provincial authorities, outright sales to local or foreign enterprises, partial equity sales into joint-venture companies with foreign investors, and several liquidations. However, the programme has also faced a number of difficulties and criticisms relating to the methods used to value state assets, the opacity of sales procedures, resistance to the initiative from civil servants, state enterprise employees and powerful vested interests, and the rather inconclusive terms of fixed-term leasing arrangements. Nevertheless, by the late 1990s most state enterprises (including the state telecommunications company) had been partially or fully disposed of, with the exception of those strategic state organizations that the Government does not wish to see divested. By 2003 only 158 state-owned companies remained from among the 800 or so that had existed a decade earlier, prompting the IMF to describe the country's privatization programme as 'one of the most successful parts of Laos's structural reforms thus far'. Of the companies that remain, approximately 40 are strategic enterprises that will remain state-owned, whilst the remainder are non-strategic companies that have proven impossible to divest and will ultimately face liquidation. Since 1997 the strategic state enterprises—such as Electricité du Laos, Postes du Laos, the National Tourism Authority of Lao PDR, Nam Papa Lao (Lao Water Supply Authority), and the Banque pour le Commerce Extérieur Lao—have been undergoing a process of 'commercialization', entailing financial and managerial restructuring, in order that they might be better able to compete with their private-sector equivalents.

ENERGY

In 2000 Laos's total electricity generation was slightly over 3,000m. kWh (up from 1,085m. kWh in 1995), and electricity accounted for around 10% of the value of the country's total industrial output. However, this increased in 2001–02, with the commissioning of a number of new power stations. The vast majority of Laos's energy supplies are derived from hydroelectric generators, with a relatively small amount of electricity also being sourced from diesel-powered installations. About 60% of the electricity generated by Laos is exported to neighbouring Thailand, providing much-needed foreign-exchange earnings. Despite the presence of a growing number of major power installations in Laos, the country does not yet have a national power grid, and only 19% of the Lao population has access to electricity, with firewood remaining the dominant source of energy for those living in non-urban areas. Domestic electricity prices have been subsidized in the past, but the Government has gradually increased tariffs to cost-recovery levels since 2002, and should have fully achieved this aim by 2005. Large parts of the country remain completely without access to electricity supplies; indeed, some border provinces import small quantities of electricity from neighbouring provinces in Viet Nam and Thailand. Three regional transmission sub-systems serve Vientiane and Luang Prabang, Savannakhet and Thakhaek, and Champasak and Saravan. Domestic consumers of electricity in Laos enjoy low electricity prices, cross-subsidized from power export earnings.

Laos's potential power-generating capabilities are considerable, with the country's hydropower potential alone being estimated at approximately 18,000 MW. The Government aspires for Laos to become the 'battery of Asia'. Although there have been plans for the construction of a substantial number of hydropower plants on rivers feeding into the Mekong (a total of 24 projects were at, or beyond, the advanced planning stage in 1997), only seven of these had received investment licences, and just two had become fully operational, by 2003. Finance appears to have been the principal obstacle for a number of the projects, including the 1,088-MW Nam Theun 2 project, located 250 km east of Vientiane, which was expected to be the single biggest investment project ever undertaken in Laos, at a cost of US $1,100m. It had initially been intended that construction of Nam Theun 2 would commence in 1996 and be completed by 2000, but construction was subsequently expected to begin in earnest in mid-2005, with the commissioning of the project envisaged for around 2010, if sufficient financing could be found before the mid-2005 deadline. Much still depended on whether the World Bank would agree to provide partial risk cover for the build-own-operate-transfer project, which would be essential if the developers were to raise sufficient financing for the project. A final power-purchasing agreement between Vientiane and

Bangkok was approved by the Thai Government in July 2003. Thailand was committed to buy electricity generated by the power plant, at an agreed price, for 25 years. With Thailand expected to take 95% of the power generated by Nam Theun 2, the project was expected to raise more than $200m. in revenue for Laos per year. A reservoir extending about 450 sq km was to be created, necessitating the relocation of 16 villages. Since its commissioning in 1971, the ageing Nam Ngum hydropower plant, located 80 km north of Vientiane, has been the primary source of electricity in Laos, with an installed capacity of 150 MW. (Japan agreed to overhaul the power plant's turbines by 2005, at a cost of $9.5m.) A 45-MW plant located at Xeset was completed in 1991, and generates power solely for export purposes. In April 1998 the 210-MW Thuen Hinboun hydropower project became operational, raising export earnings through the sale of the electricity generated to Thailand. The 126-MW power project at Huay Ho and the 60-MW Nam Leuk power plant in north-western Laos were both commissioned in 2000.

Although Laos and Thailand had earlier signed a non-binding power-purchasing agreement, which envisaged the purchase by Thailand of 3,000 MW of Laotian electricity per year by 2006, the sharp economic downturn in Thailand following the onset of the regional economic crisis in 1997 prompted the Electricity Generating Authority of Thailand (EGAT) to revise downwards its future power demand forecasts. In 2001 EGAT intended to purchase just 1,880 MW of power per year from Laos by 2006, and 3,200 MW by 2008, with a new pricing formula that was no greater than the cost of purchasing power from independent power producers in Thailand (4.2 US cents per kWh). The revised pricing has had a significant impact on the viability of Laos's proposed power projects, as their funding is calculated in US dollars. In addition, both Cambodia and Myanmar are also keen to export power to Thailand; indeed, Myanmar is already piping natural gas from its offshore Yadana field to a 3,645-MW power station at Ratchaburi. China is also seeking to supply electricity to Thailand, using a power transmission line that will run right across Laos.

MINING

Although Laos is rich in mineral resources, the mining sector accounted for just 2.4% of industrial sector output in 2000 (compared with 10% for the construction sector). Substantial deposits of such natural resources as lignite, tin, gypsum, potash, iron ore, coal, gold, copper, silver, manganese, zinc and lead are believed to be present, along with some precious stones. However, the lack of supporting physical and legal infrastructure, together with the remote location of many of the deposits, has meant that mining has represented a challenging proposition for foreign investors. During the period of French colonial rule, tin mining was the largest single area of industry in Laos, accounting for roughly 40% of total export earnings in the years prior to the Second World War. In 1986 Laos's tin reserves were estimated at 70,000 metric tons, although some reports suggest that this may be a conservative figure. In 2003 Laos produced an estimated 360 tons of tin concentrates, down from a high of 736 tons in 1996. Gypsum production is centred on mines in Savannakhet Province. Potentially viable coal seams have been identified in Saravan Province and north of Vientiane, whilst high-quality iron-ore reserves are located on the Plain of Jars, 170 km north-east of the capital. The Newmont Mining Corporation of the USA signed a joint-venture agreement in 1993 to mine for gold in Vientiane Province. Rio Tinto, meanwhile, began exploration in 1992 at a wholly owned copper and gold mine at Sepon, east of Savannakhet, under a 36-year exploration and production agreement with the Government. The mine is believed to contain 1m. tons of high grade copper, more than 3.5m. oz of gold, and some silver, all of which can be extracted using open-pit mining techniques. In 2000 Rio Tinto sold an 80% stake in this mine, for US $22m., to Oxiana Resources of Australia, which conducted exploratory work and feasibility studies during 2001, prior to investing $150m. in the mine. Construction of the mine began in late 2001, and gold production commenced in 2003. In its first year of operation the mine produced 165,000 oz of gold. Copper extraction at the same mine was to begin in 2005. Other ongoing mining projects in Laos include: a granite and limestone mine in Bolikhamsai Province (awarded to a Lao-Taiwanese joint venture); a tin mine in Khammouane Province (awarded to a Lao-South Korean joint venture); a gypsum mine, also in Khammouane Province (awarded to a Thai company); and several copper and gold mines at various locations across the country. In late 2001 two Chinese companies jointly signed a $1m. agreement to mine for zinc and copper in Oudomxay Province.

Three production-sharing contracts for the exploration and production of petroleum have been signed between the Laotian Government and a number of foreign energy companies, relating to substantial concessions in the southern panhandle of the country and Vientiane Province. The first of these contracts was signed in 1989 between the Laotian Government and a consortium led by Enterprise Oil of the United Kingdom, and related to a 20,000-sq km area in the Savannakhet basin. However, the contract was halted after initial surveying work was completed. Hunt Oil of the USA had a 28,000-sq km concession around Paksé, and undertook exploratory drilling work in the mid-1990s, whilst Monument Oil of the United Kingdom took a controlling stake in a 37,000-sq km concession in Vientiane Province, originally issued to Shlapak Development. Survey work in the southern provinces was hampered, however, by the considerable quantities of unexploded ordnance that remain scattered over large parts of the country, necessitating intensive land clearance in advance of any exploration activity. Despite fairly extensive surveys and limited exploratory drilling, no commercial reserves of petroleum or gas have yet been identified in Laos.

TRANSPORT, COMMUNICATIONS AND TOURISM

In 2002 transport, storage and communications accounted for 24.5% of the GDP of the service sector, or 6.1% of total GDP. In the period 1995–97 about 40% of the Government's capital expenditure budget was allocated to transport and communications, compared with just 9% for education and 5% for public health. Despite this, the road and communications networks remain inadequate. Laos's national highway system is composed of about 22,300 km of roads, although a large proportion of these are in poor condition, and some can be used only during the dry season. It is estimated that about one-third of all villages in Laos, and 22% of the population, are located in areas that are not accessible by vehicles. About 4,500 km of roads have a metalled surface, while a further 7,600 km have a gravel surface. Around 13,200 km of roads are simply made of earth. The country's river network constitutes an important element of the transport system, with a total of about 4,600 km of navigable waterways available along stretches of the Mekong and its tributaries. A number of dock facilities were constructed or upgraded during the 1980s and 1990s, and improvements were also made to some river routes through the dynamiting of rapids.

Only one of Laos's 18 provinces does not have a border with a neighbouring country, and for some of the more outlying towns, transport and trading links with the neighbouring country are often better than those with Vientiane. Laos's poor transport infrastructure continues to keep most transaction costs high, to prevent the establishment of a more integrated national economy, and to hinder regions in their attempts at the development of specialized inputs. In 1994 the Friendship Bridge (also referred to as the Mitraphab Bridge)—the first bridge to span the Mekong River—was completed. This road bridge connects Nong Khai in Thailand with Tha Naleng (a relatively short distance east of Vientiane) in Laos, and was funded by the Australian Government. A second Mekong road bridge (a 1.4-km suspension bridge, costing US $50m. and funded largely by Japan), linking Paksé with Thailand's Phongthong district, was opened in August 2000. A third road bridge across the Mekong, extending 2 km and linking Savannakhet with Mukdahan in Thailand (costing $45m. and largely funded by the Japan Bank for International Cooperation—JBIC) is scheduled for completion in 2006, and there is the prospect of a fourth bridge being erected between Huay Xai in Paksé and Chiang Khong in Thailand's Chiang Rai province. The ADB envisages the third road bridge forming part of an east–west arterial road, often referred to as Route 9, which will run from the coastal port of Da Nang in Viet Nam right across Laos's southern panhandle and the Annamite cordillera into north-eastern Thailand. The main purpose of the road is to enable companies in both north-eastern Thailand and land-locked Laos to use the port at Da Nang as a conduit for exports, with the aid of favourable customs agreements at the various border crossing points. Indeed, an east–west economic corridor is being considered for the territory on either side of Route 9, in a bid to create a conducive business environment for new investment in these relatively neglected provinces of all three participating countries. A feasibility study of the east–west corridor proposal was completed by the ADB in 2001. The Governments of Laos, Thailand and China are also seeking to construct, by 2006, a transnational highway linking Kunming, in Yunnan Province, China, with Bangkok.

In May 2000 Laos signed a pact on the navigation of the Mekong River with China, Myanmar and Thailand. After more than five years of protracted negotiations, this pact, which took effect in 2001, finally allowed for the cross-border commercial navigation of an 886-km stretch of the upper Mekong, from Simao in China's Yunnan Province to Luang Prabang in Laos. Of the 14 river ports designated under the pact, six are in Laos. Authorized vessels using this stretch of the river are no longer obliged to pay transit fees, and it is therefore anticipated that the use of the river for both goods and passenger traffic will increase substantially. It is possible that attempts will be made to widen the course of this stretch of the Mekong River, which at present can take vessels of only 50 metric tons during the wet season (and vessels of just 15 tons during the dry season).

In 1997 the Laotian Government announced plans to develop a comprehensive railway network in the country. However, whilst a contract was awarded to a Thai company, no timetable for the implementation of the scheme was announced. The only railway line as yet completed in Laos was a short, narrow-gauge line near the Khone waterfalls, built and operated during the colonial period, as a means to transport goods around an unnavigable part of the Mekong River. More recently, the Friendship Bridge was designed to take a single railway track, in addition to a two-lane road. Although a railway track from the Thai rail terminus at Nong Khai to the centre of the Friendship Bridge was laid in 1996, the planned continuation of the track a further 14 km into Vientiane has not yet occurred. However, in February 2004 the Thai and French Governments agreed to provide US $5m. to fund a 3.2-km railway line across the Friendship Bridge to a loading terminal on the Lao side of the Mekong River. It is conceivable that the line could be extended as far as Vientiane at some point in the future, if additional funding can be found. A far more ambitious plan to extend the railway line 1,000 km further north, via Luang Prabang and across the border into China (as part of a scheme ultimately to link Singapore with Europe by rail), is unlikely to be implemented in the foreseeable future. On a much more modest level, the commencement of a bus service linking Vientiane with Nong Khai and Udon Thani Provinces in Thailand was agreed to by the Laotian and Thai Governments in May 2002.

Given Laos's challenging terrain and poor road system, there is a significant role for a domestic airline service. With the assistance of various bilateral donors, the country's airports were steadily upgraded during the 1990s. Domestic flights provided by Lao Airlines (formerly Lao Aviation) currently link Vientiane with Luang Prabang, Xiangkhouang, Luang Namtha, Oudomxay, Houeisay, Sam Neua, Sayabouri, Phongsali, Xaysomboune, Savannakhet and Paksé. International flights provided by Lao Airlines link Vientiane with Bangkok, Chiang Mai, Kunming, Hanoi and Phnom-Penh. The airline carried about 51,000 passengers on international flights in 2000. A small number of foreign airlines also fly to Vientiane, including Viet Nam Airlines, Yunnan Airlines of China and Thai Airways International. (Malaysia Airlines and SilkAir of Singapore both suspended their flights to Vientiane in 1998, as a result of low passenger volumes.) Previous attempts by Lao Airlines to expand have not proved very successful, having been hampered by various financial and managerial constraints, including subsidized prices for domestic flights. In 2002 the airline reportedly flew 200,000 passengers on both domestic and international routes, at well below its full capacity. In 2003 it leased an Airbus A-320. In the same year the Government embarked on a project to sell a strategic stake in Lao Airlines to a foreign airline that would be able to improve its performance, although no investor has yet been found. The upgrading of Luang Prabang airport, including an extension of the runway, in order to permit international flights, was largely funded by Thailand and was completed in 1998. Meanwhile, the upgrading of Vientiane's Wattai airport in the late 1990s, incorporating improvements to the runway and the construction of a new terminal building, received funding from Japan. With some 70 foreign airlines transiting Lao airspace, there are about 150 flights over Laotian territory each day, generating additional revenue for the Government in the form of overflight fees.

Improvements to both Luang Prabang and Vientiane airports were completed in anticipation of 'Visit Laos Year', which actually spanned both 1999 and 2000. In support of this tourism campaign, visas were for the first time made available on entry at both airports and at the Friendship Bridge crossing. In 2001 Laos attracted around 673,823 tourists, exceeding the Government's target of 600,000. Aggregate earnings from tourism in 2002 were estimated at US $113m. In 2003, however, tourism figures were negatively affected by the outbreak of Severe Acute Respiratory Syndrome (SARS) in the region. Having previously been very cautious in the development of its tourism industry, Laos began to place greater emphasis on tourism receipts in the late 1990s. However, its attempt to attract more tourists has not been assisted by occasional bomb attacks carried out in Vientiane since 2000, nor by ongoing concerns about the safety record of Lao Airlines. (In 2000 two of its aircraft crashed, in Sam Neua and Xieng Khuang.) Tourists are still cautioned against travelling on roads traversing areas known to be occupied by anti-Government elements and bandits, including parts of Xiangkhouang, Xaysomboune and Bolikhamsai Provinces, and Route 13, which links Vientiane and Luang Prabang. In July 2003 two foreign tourists were killed on this road, near the town of Vang Vieng. Tourist numbers in Laos are also constrained by a shortage of hotel accommodation, with only about 7,000 hotel rooms available in the country.

In 1994 the Laotian Government urged the involvement of foreign investors in the country's relatively primitive telephone network. Shortly after this, Vientiane licensed a joint-venture agreement between Enterprises des Postes et Télécommunications de Laos (a state-owned enterprise) and the Shinawatra Group of Thailand, the joint venture being known as Lao Télécommunications. The company has a 25-year concession to operate all telecommunications services in Laos, including pagers, mobile phones and internet services. In a partial privatization of the telecommunications system, Lao Télécommunications has been upgrading both the fixed-line and mobile telephone networks. Although the system nominally encompasses the whole country, only 58 out of 142 districts in Laos had fixed-line services. In 2001 there were just 15 fixed-line telephones per 1,000 people. The number of mobile telephone users reached about 55,200 in 2002, up from about 12,000 in 2000. In early 2002 it was reported that Millicom International had been awarded a licence to provide a national GSM wireless network in Laos, in collaboration with the Government. As with electricity tariffs, the Government has been gradually increasing domestic telephone charges in recent years, in order to bring pricing more into line with underlying operational costs. Conversely, the prices of international calls—which are amongst the highest in Asia—are being reduced, in order to comply with international standards. Laos's three internet service providers—GlobeNet, LaoTel (which dominates) and PlaNet Online—reportedly had a combined client base of no more than 3,000 subscribers by late 2001, and a further 6,000 internet users (Laos has just three computers for every 1,000 people). An ambitious international joint venture to launch a satellite—'Lao Star'—did not survive the impact of the Asian financial crisis, and in late 1999 the Bangkok-based United Communications Industry (Ucom) announced that it was cancelling investment in the project.

BANKING

Laos's small banking sector has experienced a number of transformations since economic reforms commenced in the late 1980s, yet it remains undeveloped. Prior to 1988, the two principal banks in Laos were the Banque d'Etat de la RDP Lao and its subsidiary, the Banque pour le Commerce Extérieur Lao

(BCEL). The former was responsible for all domestic banking, including local currency issuance and public-sector accounts, and subsidized lending to state enterprises. The latter's remit extended to all foreign-trade financing, foreign-exchange reserves, foreign loans and debt. The standard format for banking in command economies was revised in 1988, in order to conform with reforms being implemented in other areas of the economy. The Banque d'Etat de la RDP Lao lost its monopoly status and became a more conventional central bank, subsequently renamed the Bank of the Lao PDR. The BCEL became one of several state-owned commercial banks, losing its monopoly over all international currency transactions. However, in 2000 BCEL continued to be the principal participant in the area of trade financing, and remained the largest of Laos's new commercial banks. Between 1988 and 1991 a further six state-run commercial banks came into operation (two in Vientiane, and one each in Luang Prabang, Savannakhet, Champasak and Xiangkhouang), formed from various branches of the former central bank. These banks were joined by an eighth state-run commercial bank in 1993, the Agricultural Promotion Bank, which is designed to provide low-cost loans to farmers and to administer micro-financing credit schemes (funded by external donors) through its substantial branch network. In 1999 six of the eight state-run commercial banks were merged back into just two entities. The three banks in the northern half of the country were merged to form Lane Xang Bank, while their three counterparts in the south were merged to form Lao May Bank. In 2001 it was further decided that these two banks should also be merged into a single entity. As a result, the Bank of the Lao PDR governed three state-owned commercial banks (which accounted for about 70% of total bank assets in 1999), three foreign joint-venture banks (9% of total bank assets) and seven foreign bank branches (with 21% of total bank assets). One joint-venture bank (Joint Development Bank) has Thai investor participation, having commenced operations as early as 1989. Another joint-venture bank, Vientiane Commercial Bank Ltd, was established in 1993 and includes Taiwanese, Thai and Australian equity partners. The newest bank in Laos is a US $10m. joint venture between BCEL and the Bank for Investment and Development of Vietnam—one of Viet Nam's four state-run commercial banks—which commenced operations in 2000 as Lao-Viet Bank. Based in Vientiane, this bank's primary aim is to help finance burgeoning trade flows between Laos and Viet Nam, and it has opened branch offices in Champasak, Ho Chi Minh City and Hanoi. Both BCEL and Lao-Viet Bank joined the Worldwide Interbank Financial Telecommunications (SWIFT) payment system in late 2001, marking the entry of SWIFT into Laos for the first time. Six of the seven foreign bank branch licences issued in Laos during the early 1990s were held by Thailand's major commercial banks, with the Public Bank of Malaysia holding the remaining one. In 2002, however, Thai Farmers' Bank Public Co Ltd announced the closure of its banking operations in Laos, reducing the number of foreign banks active in the country to six. Standard Chartered Bank has operated a representative office in Vientiane since 1996. Total banking assets are estimated to be around $400m., or the equivalent of roughly 25% of GDP.

In the view of the IMF, Laos's banking sector remains largely insolvent, despite the fairly considerable efforts of both the ADB and the World Bank to recapitalize the country's state-run commercial banks in 1994. According to the IMF, these commercial banks remained deeply insolvent, with reported aggregate non-performing loan levels of 52% (the real figure was probably markedly higher). A four-year recapitalization programme for these banks was to be implemented between 2002 and 2005, funded by the Government, in a bid to raise the capital adequacy levels of these banks to 12%. It was estimated that around US $50m. would be required for this purpose, equivalent to 3% of GDP. Most Laotian citizens do not have bank accounts, preferring instead to keep their savings in gold, foreign currency, or even livestock in rural areas. The local currency is also unwieldy for transactions, with the highest denomination note (5,000 kip) worth about 50 US cents, although the central bank announced its intention to issue new 10,000 and 20,000 kip notes during 2002. These factors explain in part why the Laotian economy is highly 'dollarized', with over 75% of broad money in the form of foreign currency deposits (mostly US dollars and Thai baht), much of which is held outside the formal banking sector. Urban Lao remain generally reluctant to use bank savings accounts, owing in part to recent depreciations in the value of the local currency, the fragility of the banks themselves, and the fact that interest rates have not kept pace with high inflation in recent years. Laos has yet to develop a capital market, although it has experimented with bond issues in recent years in a bid to improve the mobilization of domestic savings and to absorb excess liquidity. In 1998 Laos issued its first retail-orientated bonds, which, in addition to providing a fixed rate of interest, also comprised a lottery component.

FOREIGN TRADE AND INVESTMENT

Laos has consistently recorded both a trade deficit and a current-account deficit, the latter equivalent to 5.6% of GDP in 2002. In 2003 Laos's merchandise exports rose by 23%, to approximately US $380m. (equivalent to about one-quarter of GDP). Major export-earners include timber and wood products (23% of total exports), coffee and other agricultural products (7%), garments (29%), and electricity (33%). With the mining sector burgeoning, gold and copper were also likely to become major sources of foreign-exchange earnings in the near future. Most data pertaining to Laos's external trade tend to be rather inaccurate, however, given the relatively substantial amount of smuggling and informal trading that is conducted across the country's five porous land borders.

Laos commenced garment exports in the late 1980s, and this sector of export activity developed into a major source of foreign-exchange earnings—as well as a major source of urban employment—in the 1990s, offsetting in part the decline in the export of wood-related items. Under the GSP, a large proportion of garment exports are directed to EU countries. Laos's GSP status was briefly revoked in the mid-1990s, after an EU delegation found that the country was not complying with the minimum 60% local content required for goods exported under the GSP programme; however, Laos subsequently regained its GSP privileges. Thailand has traditionally been the destination for about one-quarter of Laos's total exports, ahead of all other countries in the Asia-Pacific region and beyond. However, since the mid-1990s Viet Nam has become the leading recipient of Laos's exports, reflecting the development of closer relations between the two countries since the onset of the Asian economic crisis: in 2002 exports to Viet Nam accounted for more than one-third of Laos' total exports. In 2002 Laos's imports were valued at US $524m., compared with $689.6m. in 1996. Major import items include consumption goods (which accounted for about 49% of total imports in 2001), fuels (13%), materials for the garment industry (12%), construction and electrical equipment (9%), and motorcycle parts (7%). Thailand tends to provide over half of Laos's total imports (58% in 2002), followed by Viet Nam, China (including Hong Kong SAR), Singapore, Japan and France.

Having participated as an observer at meetings of the Association of South East Asian Nations (ASEAN) since 1992, Laos applied for full membership of the Association in 1993 and was inducted into the grouping in mid-1997. In July 2004 Laos assumed the rotating chairmanship of ASEAN. As part of its commitments to ASEAN, Laos must comply with a wide range of tariff reductions under the ASEAN Free Trade Area (AFTA) agreement; these include reducing tariff rates for most products imported from other ASEAN member countries to below 5% by 2008. The reduction of tariff rates will have some impact on government revenues, as around 20% of Laos's total budgetary revenues are derived from import duties. However, Laos's tariff rates have tended not to be particularly high (with the exception of cars and luxury items), and the country should be able to meet its deadline for complying with the agreement. Having signed a second bilateral trade accord with the USA in September 2003 (the first having been signed, but not ratified, in 1997), Laos continues to seek normal trading relations (NTR) status—formerly known as 'most favoured nation' (MFN) status—with the USA, although increasing concerns in the USA regarding Laos's record on human rights, particularly towards the ethnic Hmong, as well as insufficient religious and political freedoms, are proving to be the main obstacle in this regard. Laos also aspires to gain entry to the World Trade Organization (WTO).

Since the onset of the regional economic crisis, Laos's heavy reliance on Thailand as a source of investment and trade has been diminished by a reorientation towards the country's other immediate neighbours, particularly Viet Nam and China. While Vientiane's close ties with Hanoi and Beijing are most apparent in the political realm, they are also evident in the economic sphere, as exemplified by burgeoning cross-border investment and trading activity, and various other business-related initiatives. This development has been motivated in part by a shortage of foreign currency, obliging Laotian companies to enact barter and countertrade deals with other countries. Trade and investment relations between Laos and Yunnan Province in China also developed rapidly in the late 1990s, as exemplified by the construction of a number of new cement plants by Yunnan companies, in Vientiane and Saravan Provinces. The Laotian and Thai Governments have been working on an economic co-operation ageement (similar to those Thailand already has with Cambodia and Myanmar), focusing on the areas of electricity, agriculture, telecommunications and transport, and investment, since 1997. However, trade relations between Laos and Thailand have not always been straightforward, partly as a result of various differences of opinion between the two Governments on non-trade issues (such as border demarcation), but also because of a perception in Vientiane that Thailand has taken unfair advantage of Laos's land-locked state to exact onerous transhipment costs on the country's exports. Such concerns should have been allayed by a recent agreement to liberalize the transhipment business, thereby ending the oligopoly that six Thai transport companies had previously held.

Foreign direct investment (FDI) activity in Laos is regulated under the Law on the Promotion and Management of Foreign Investment, promulgated in June 1994; this law replaced a 1989 foreign investment law that heralded the opening up of the Laotian economy to foreign capital. The 1994 law outlines a relatively liberal foreign investment regime, at least in theory, with wholly foreign-owned projects permitted. Foreign investors also enjoy a corporate income tax rate of 20%, which is markedly below the 35% set for local companies, and expatriates face an attractive 10% flat rate for personal income tax. However, the foreign investment law lacks some of the necessary supporting implementing regulations, and elements of the law are not wholly compatible with various other pieces of legislation, including the country's mining and domestic investment laws. The Committee for Investment and Cooperation (CIC) is the government agency that approves, monitors and promotes all FDI activity in Laos and reports directly to the Prime Minister's office. By late 2001 Laos had approved 860 foreign investment projects, with an aggregate registered capital of more than US $7,000m., sourced from over 35 countries. Of these, one-half were wholly foreign-owned, and two-thirds were small-scale projects with registered capital of less than $1m. However, only around 25%–30% of approved FDI has actually been invested in Laos, and it is likely that many projects will never be implemented. None the less, it is estimated that the country's aggregate stock of FDI increased from virtually nil in the late 1980s to around $500m. in 1998. In terms of FDI inflows, in 1996 a peak inflow of around $1,300m. was recorded, compared with just $122m. in 1998. However, in 2002 and 2003 the cumulative value of new foreign investments was just $5m and $19.5m., respectively. Such anaemic inflows prompted the Government to decentralize and streamline the investment approval process in 2003, for both foreign and domestic investors alike. Under the new system, the national CIC office was permitted independently to approve foreign investment projects valued at $10m. or less, and domestic investments of 100,000m. kips or less. Projects valued at levels greater than this first had to gain national government approval. Provincial CIC offices could independently approve foreign investment projects valued at $1m. or less, and domestic investments of 10,000m. kips or less ($2m. and 20,000m. kips, respectively, for the CIC offices in Vientiane municipality, Savannakhet, Champasak and Luang Prabang Provinces). Not surprisingly perhaps, the largest single investor in Laos by far has been neighbouring Thailand (accounting for 268 projects and 74% of total FDI inflows), followed by the USA (47 projects), the Republic of Korea (42 projects), and Malaysia. (A proportion of Laos's US-sourced investment is likely to be relatively small-scale FDI projects enacted by overseas Lao now residing in the USA.) Laos's heavy dependence on Thailand for FDI inflows became apparent in 1997, when new investment pledges declined to almost nil and actual flows contracted by more than 40%, as Thailand's corporate sector was overwhelmed by the country's economic crisis.

In terms of sectoral distribution, the energy sector has witnessed the largest proportion of total FDI inflows by far, as a spectrum of foreign companies have sought to develop power generation (primarily hydropower) projects in Laos. While Laos's energy needs are small, anticipated demand from Thailand for additional electricity supplies has stimulated business interest in generating power for export from Laos. As at late 1998, about 65% of total approved foreign investment in Laos pertained to the hydropower sector (in connection with just seven projects, only two of which had been completed by 2000); the telecommunications and transport sector accounted for 9% of total approved foreign investment at that date, followed by the hotel and tourism sector (9%), and industry and handicrafts (7%). As measured by the number of foreign investment project licences issued, the industry and handicrafts sector has received 160 projects, while the services sector has secured 153 projects (with 124 projects in trade, 89 in agriculture, 87 in garments, 76 in processing and construction, 45 in hotels and tourism, 39 in consulting, 31 in petroleum and mining, and 17 in communications). As a percentage of gross fixed capital formation, FDI inflows in Laos averaged 18.7% between 1995 and 1999. As a member of ASEAN, Laos was expected to comply with the ASEAN Investment Area (AIA) initiative, which was launched in 1998. Under this agreement, Laos must provide all ASEAN investors with both 'national treatment' and access to most of the country's business sectors by 2010. Laos is also committed to an unrestricted flow of foreign investment activity, including national treatment and open access to all business sectors for foreign investors beyond ASEAN member countries, by 2020.

INTERNATIONAL AID AND DEBT

The economy is heavily dependent on external assistance and aid, both from multilateral agencies and bilateral donors. The largest bilateral donors in Laos include Japan, Sweden, Australia and France, while multilateral agencies active in Laos include the ADB, World Bank and IMF. International donations account for around 17%–18% of the country's total GDP. In particular, external grants fund both Laos's current-account deficit and its budget deficit, with budget revenues rarely exceeding more than 70% of government expenditure. In 2003 Laos's budget deficit was estimated to be the equivalent of 7.8% of GDP, a slight improvement on the 8.3% that had been recorded in 2002. In recent years the current-account deficit and the budget deficit have each been roughly equivalent to 8%–10% of total GDP. In the case of the latter, of the total tax revenues collected by the Government in 2000, about 10% was from taxes on trade (import duties), 20% from turnover tax, 22% from profit and income tax, 23% from excise duties, 8% from timber royalties, and 7% from overflight fees (collected from airlines using Laotian airspace). Thus, overseas grants tend to amount to more than the aggregate fiscal revenues emanating from trade, excise duties, timber royalties and overflight fees combined. Indeed, tax revenues have generally been decreasing since 1996, owing in part to a weakening tax administration. The ADB has described revenue mobilization as a 'major and worsening problem' for the Lao Government, and fiscal weakness as the biggest single challenge that it faces. With the encouragement of the IMF, the Government plans to introduce a value-added tax (VAT), or sales tax, in the near future. It has also increased taxes on petroleum, alcohol and tobacco products. However, these measures may not be sufficient to offset the loss of customs revenues that would result from Laos's compliance with AFTA by 2008. By this date most tariffs on imports from other ASEAN member countries must have been reduced to below 5%. At the end of 2003 Laos's aggregate external debt obligations (including Russian debt) amounted to US $1,328m. This compared with estimated foreign-exchange reserves (excluding gold) of $216m. in 2003 (equivalent to about four months of merchandise import cover). Laos's debt-servicing ratio, as a ratio of total exports of goods and services, was estimated to be a manageable 6.1% in 2003. Whilst much of this debt is on concessional terms to various multilateral agencies, it also includes some commercial debt (for example, $160m. incurred during the construction of the Theun Hinboun power plant). External debt statistics produced by the World Bank, the IMF, the Organisation for Economic Co-operation and Development (OECD) and the Bank for International Settlements indicated that Laos's total bank loans stood at $79m. at the end of 1999, and that multilateral claims amounted to $1,020m. In 2001 Laos's total external debt-service burden was around $44m. In June 2003 the Russian Government agreed to cancel 70% of Laos's $1,300m. debt and to permit Laos to clear the remaining $378m. over the following 33 years at preferential interest rates. Of the remaining external debt, 95% consists of long-term concessional loans with various bilateral and multilateral donor agencies, and just 5% is commercial debt.

Given the importance of the agricultural sector to the economy as a whole, it is not surprising that a significant proportion of total external assistance is focused on this sector. In 1999 there were reported to be around 80 agriculture-orientated aid projects active in Laos, with aggregate funds of US $140m. The ADB's Greater Mekong Subregion (GMS) programme has also contributed to Laos's economic development plans since its commencement in 1993. The GMS programme—which also encompasses Cambodia, Myanmar, Thailand, Viet Nam, and Yunnan Province in China—has sought to bring private-sector participants together with external assistance agencies to conduct a series of economic and business initiatives in the participant countries. These initiatives have ranged from broader tourism, trade and investment and human resource development projects to very specific power generation, road and rail projects. Meanwhile, in addition to supporting the unexploded ordnance clearing programme in Laos, the USA has also contributed to attempts to eradicate the production and trafficking of opium in the country. The USA continues to assist in the establishment of counter-narcotics enforcement units in each of the relevant provinces, as well as crop control and development projects in Houaphanh and Phongsali Provinces.

In recent years both multilateral agencies and bilateral donors have expressed their disappointment at the apparent deceleration in the pace of economic reform in Laos. A report by the ADB cited 'inadequate government capacity and commitment to pushing forward economic reforms', as well as 'insufficient transparency and accountability', as factors contributing to the undeveloped state of the economy of Laos. Furthermore, a recent UNDP survey of donor perceptions noted that the 'burdensome government decision-making process and the lack of transparency' are becoming 'matters of serious concern'. Since the external assistance agencies and donors play such an important role in supporting the Laotian economy, providing official development assistance that is equivalent to between 15% and 25% of GDP, their increasing disenchantment with the Government's lack of progress in instituting new reforms could, if not arrested, pose significant problems for Vientiane. In the same context, mounting concern over Laos's human rights record and the Government's treatment of political protesters could also indirectly prompt a further contraction in external assistance to the country. Notwithstanding the stance adopted by multilateral lending agencies and donors, the IMF announced in April 2001 that it was to implement a new three-year US $40.2m. Poverty Reduction and Growth Facility for Laos, which was generally designed to strengthen macro-economic stability and reduce poverty 'through growth with equity'. More specifically, the programme attached to the loan provided for 'continued fiscal and monetary restraint and is centred on the implementation of revenue enhancement, restraining credit of the central bank, restructuring state-owned commercial banks, commercializing large enterprises, and developing the enterprise sector'.

PROBLEMS AND PROSPECTS

There are clear indications that the pace of economic reform in Laos decelerated considerably after 1996–97, although there have been some signs of reform momentum since 2001. This slowdown was attributable in part to the impact on the country of the regional economic crisis, which has led to apparent indecision within the Laotian leadership over whether further economic liberalization represents the correct way to proceed. It may also have stemmed from recent changes in the profile of the leadership and the development of much greater military representation in the senior ranks of the LPRP: by 2000, all but one of the eight members of the Politburo were serving or retired military men.

The sharp economic deterioration in neighbouring Thailand in 1997 undoubtedly had an adverse impact on Laos, most notably through a very sharp downturn in the value of the kip against both the US dollar and the Thai baht. Relative to the US dollar, the kip lost 80% of its value in the two years following the onset of the regional economic crisis in July 1997. Not all of the depreciation in the value of the kip can be attributed directly to contagion from Thailand, however: a fairly large spending programme was implemented by Vientiane at this time in a bid substantially to increase the amount of dry season paddy through a mass acquisition of water pumps. Under this programme, funding from the central bank was directed both to state enterprises to enable them to import the pumps, and as credit to farmers to permit them to buy the imported pumps. Thus, at a time when fiscal restraint might have been most appropriate in order to restore confidence in the Laotian currency and economy, the Government embarked upon an expansionary monetary programme to fund the (off-budget) dry season irrigation programme. This, together with the Government's issuing of higher-denomination currency notes and a vigorous campaign against unlicensed foreign-exchange dealers (which prompted some Lao to fear an impending revaluation of the currency), served further to erode popular confidence in the local currency. The resulting panic exacerbated the precipitous decline in the kip, as Lao citizens and companies alike sought to convert into foreign currency or gold. Even though Laos did not suffer the massive political upheaval witnessed in Indonesia following the onset of the regional economic crisis, the damage done to the value of the kip in 1998–99 was even greater than that inflicted on the rupiah, and the Laotian currency depreciated from around 1,500 to more than 7,500 to the US dollar within just a few years. This decline in the value of the currency resulted in triple-digit inflation in 1998–99, as the price of imported goods rose sharply. Consequently, the spending power of the urban populace decreased considerably, forcing the Government to reduce the length of the working week for civil servants in order that they could devote more time to the pursuit of secondary incomes. By the end of 2003 the kip was trading at around 10,500 to the US dollar. The rate of inflation in Laos declined from 139.7% in 1999 to 20.5% in 2000 and 10.7% in 2002, before increasing slightly to 15.4% in 2003 as a result of higher fuel, electricity, water and food prices. Such a commendable contraction in the inflation rate allowed the Government to embark on upward price adjustments for heavily subsidized water (by 100%), electricity and domestic aviation tariffs, as well as a 25% increase in salaries for state employees (partially to offset the decline in their real income during 1998–99) at the beginning of 2002.

Since the mid-1990s Laos's post-socialist economic development has been focused less on the transition of the economy from one guided by central planning to one led by market forces, and more on the straightforward challenge of general economic progress for one of Asia's least developed countries. While the leadership of the Lao PDR remains the LPRP, and it articulates its policies in the vernacular of an avowedly socialist state, in the economic sphere little evidence remains of the initial post-1975 socialist zeal. The economic challenges currently facing Laos are those commonly experienced by numerous less-developed countries in Asia and beyond, namely: the identification of policy prescriptions to address the problem of economic underachievement; the coherent and competent enactment of those policies; the tackling of bureaucracy and corruption; and the radical improvement of the human development level of the country. Widening disparities in income between rural and urban Lao—and even between those living in rural upland and lowland areas—need to be addressed, particularly where they are congruent with ethnic divisions. The better integration of the activities of subsistence farmers into the national economy, the diversification of crops, the creation of specialized markets and the improvement of agricultural output levels in general are also required if the living standards of the Lao populace as a whole are to be improved. To date, the benefits of the economic reform process in Laos have mostly been witnessed in the urban areas of the country, largely through advances made in the service and industry sectors. There is, however, a need to ensure that the effects of future development initiatives are extended to the rural areas, both upland and lowland, through parallel advances in the agricultural sector. The challenges facing Laos in the construction of a more effective domestic transport system and communications infrastructure—and thereby the reduction of the high transaction costs faced by the agricultural sector—across a sparsely populated and topographically demanding country do not make this an easy task. Furthermore, economic development will also require a strengthening of the capacities of numerous state institutions, a move towards good governance, improved macro-economic management, and the enforcement of a relatively new legislative regime. The need to improve the education and skills levels of the country's populace remains a particularly critical issue in relation to the sustained economic development of Laos (which produces around 3,000 higher education graduates per year). The Laotian Government appears to recognize this pressing need, as evidenced by recent public-sector spending increases on health and education: of 11.4% in 2001 and 19.1% in 2002. (Fortunately for Laos, there were no reported cases of SARS during the regional epidemic in early 2003.) None the less, education-related spending has only accounted for 10% of total budget spending, or around 2% of GDP, in recent financial years. Health-related spending comprised 5% of total budget spending, or 1% of GDP. For the foreseeable future the Laotian economy will continue to depend in large part on external assistance.

At the LPRP's Seventh Party Congress, held in March 2001, the Government announced its five-year plan and its socio-economic development strategy to 2020. By that year Laos aimed to progress from its current status as a less developed country. Under the five-year plan, the Government aimed to record GDP growth of 7.0%–7.5% per year, to keep inflation in single figures, to restrain the budget deficit to below 5% of GDP, and to increase average per caput income to US $500–$550. Targets of 4%–5% growth per year for the agricultural sector, 10%–11% for the industrial and handicrafts sector, and 8%–9% for the services sector were presented under the plan. Whether these commendable targets can be achieved, however, remains to be seen. In 2001 the Government also embarked on a decentralization initiative, granting more autonomy to the provinces, and delegating greater budgetary planning responsibilities to the district level. A similar decentralization scheme had been attempted in 1986, but was subsequently reversed in 1991 after a sharp contraction in tax revenues. It will be interesting to see whether this second attempt at decentralization proves more successful, or results in another dramatic downturn in tax receipts. Initial evidence from 2003 suggests the latter, with some provinces reportedly tardy in remitting tax revenues to the national Government. Looking ahead, the pace of economic reform in Laos will continue to be gradual and relatively cautious, as the Government remains mindful of the real and potential social and political consequences.

Statistical Survey

Source (unless otherwise stated): Service National de la Statistique, Vientiane.

Area and Population

AREA, POPULATION AND DENSITY

Area (sq km)	236,800*
Population (census results)	
1 March 1985	3,584,803
1 March 1995	
Males	2,265,867
Females	2,315,391
Total	4,581,258
Population (UN estimates at mid-year)†	
2001	5,403,000
2002	5,529,000
2003	5,657,000
Density (per sq km) at mid-2003	23.9

* 91,400 sq miles.

† Source: UN, *World Population Prospects: The 2002 Revision*.

PROVINCES
(official estimates, mid-1995)

	Area (sq km)	Population	Density (per sq km)
Vientiane (municipality)	3,920	531,800	135.6
Phongsali	16,270	153,400	9.4
Luang Namtha	9,325	115,200	12.4
Oudomxay	15,370	211,300	13.7
Bokeo	6,196	114,900	18.5
Luang Prabang	16,875	367,200	21.8
Houaphanh	16,500	247,300	15.0
Sayabouri	16,389	293,300	17.9
Xiangkhouang	15,880	201,200	12.7
Vientiane	15,927	286,800	18.0
Bolikhamsai	14,863	164,900	11.1
Khammouane	16,315	275,400	16.9
Savannakhet	21,774	674,900	31.0
Saravan	10,691	258,300	24.2
Sekong	7,665	64,200	8.4
Champasak	15,415	503,300	32.7
Attopu	10,320	87,700	8.5
Special region	7,105	54,200	7.6
Total	236,800	4,605,300	19.4

PRINCIPAL TOWNS
(population at 1995 census)

Viangchan (Vientiane—capital)	160,000	Xam Nua (Sam Neua)	33,500
Savannakhet (Khanthaboury)	58,500	Luang Prabang	25,500
Pakxe (Paksé)	47,000	Thakek (Khammouan)	22,500

Source: Stefan Helders, *World Gazetteer* (internet www.world-gazetteer.com).

Mid-2003 (UN estimate, incl. suburbs): Vientiane 716,380 (Source: UN, *World Urbanization Prospects: The 2003 Revision*).

BIRTHS AND DEATHS
(UN estimates, annual averages)

	1985–90	1990–95	1995–2000
Birth rate (per 1,000)	44.6	41.3	38.2
Death rate (per 1,000)	18.2	15.8	14.1

Source: UN, *World Population Prospects: The 2002 Revision.*

Birth rate (2001): 35.7 per 1,000.

Death rate (2001): 12.6 per 1,000.

Source: UN, *Statistical Yearbook for Asia and the Pacific.*

Expectation of life (WHO estimates, years at birth): 55.1 (males 54.1; females 56.2) in 2002 (Source: WHO, *World Health Report*).

ECONOMICALLY ACTIVE POPULATION
(ILO estimates, '000 persons at mid-1980)

	Males	Females	Total
Agriculture, etc.	717	675	1,393
Industry	79	51	130
Services	193	123	316
Total labour force	990	849	1,839

Source: ILO, *Economically Active Population Estimates and Projections, 1950–2025.*

Mid-2002 (estimates in '000): Agriculture, etc. 2,113; Total labour force 2,776 (Source: FAO).

Health and Welfare

KEY INDICATORS

Total fertility rate (children per woman, 2002)	4.8
Under-5 mortality rate (per 1,000 live births, 2002)	100
HIV/AIDS (% of persons aged 15–49, 2003)	0.10
Physicians (per 1,000 head, 1996)	0.24
Hospital beds (per 1,000 head, 1990)	2.57
Health expenditure (2001): US $ per head (PPP)	51
Health expenditure (2001): % of GDP	3.1
Health expenditure (2001): public (% of total)	55.5
Access to adequate water (% of persons, 2000)	90
Access to adequate sanitation (% of persons, 2000)	46
Human Development Index (2002): ranking	135
Human Development Index (2002): value	0.534

For sources and definitions, see explanatory note on p. vi.

Agriculture

PRINCIPAL CROPS
('000 metric tons)

	2000	2001	2002
Rice (paddy)	2,202	2,335	2,417
Maize	117	112	124
Potatoes	33†	35†	35*
Sweet potatoes	118	101	194
Cassava (Manioc)	100	7	83
Sugar cane	297	209	222
Pulses*	14	14	14
Soybeans	5	3	3
Groundnuts (in shell)	13	17	16
Sesame seed	5*	3	4
Vegetables*	671	664	796
Watermelons	n.a.	4	83
Bananas*	37	46	53
Oranges*	29	28	29
Pineapples*	35	35	36
Other fruit (excl. melons)*	93	93	95
Coffee (green)	24	26	32
Tobacco (leaves)	40	30	27

* FAO estimate(s).
† Unofficial figure.

Source: FAO.

LIVESTOCK
('000 head, year ending September)

	2000	2001	2002
Horses*	29	29	30
Cattle	1,100	1,217	1,208
Buffaloes	1,028	1,051	1,089
Pigs	1,425	1,426	1,416
Goats	121	124	128
Chickens	13,095	14,063	15,274
Ducks*	1,630	1,700	1,900

* FAO estimates.

Source: FAO.

LIVESTOCK PRODUCTS
('000 metric tons)

	2000	2001	2002
Beef and veal	16	17	20
Buffalo meat	17	17	17
Pig meat	28	32	32
Poultry meat	12	13	13
Cows' milk*	6	6	6
Hen eggs	10	12	13
Cattle and buffalo hides (fresh)*	4	4	4

* FAO estimates.

Source: FAO.

Forestry

ROUNDWOOD REMOVALS
('000 cubic metres, excl. bark)

	2000	2001	2002
Sawlogs, veneer logs and logs for sleepers	435	438	260
Other industrial wood*	132	132	132
Fuel wood*	5,872	5,885	5,899
Total	6,439	6,455	6,291

* FAO estimates.

Source: FAO.

SAWNWOOD PRODUCTION
('000 cubic metres, incl. railway sleepers)

	2000	2001	2002
Total (all broadleaved)	208	227	182

Source: FAO.

Fishing

('000 metric tons, live weight)

	2000*	2001*	2002
Capture	29.3	31.0	33.4*
Cyprinids	4.4	4.7	5.0*
Other freshwater fishes	24.9	26.4	28.4*
Aquaculture	42.1	50.0	59.7*
Common carp	10.5	12.5	14.9
Roho labeo	1.8	2.1	2.5*
Mrigal carp	1.8	2.1	2.5*
Bighead carp	2.5	2.9	3.5*
Silver carp	2.5	2.9	3.5*
Nile tilapia	18.9	22.5	26.9*
Total catch	71.3	81.0	93.2*

* FAO estimate(s).

Source: FAO.

Mining

('000 metric tons, unless otherwise indicated)

	2001	2002	2003*
Coal (all grades)	122.9	233.8	230.0
Gemstones ('000 carats)	—	200.0*	—
Gypsum	121.2	110.3	120.0
Salt	2.6	5.4	5.0
Tin (metric tons)†	490	366	360

* Estimated production.
† Figures refer to metal content.

Source: US Geological Survey.

Industry

SELECTED PRODUCTS

	1999	2000	2001*
Beer ('000 hectolitres)	480	508	577
Soft drinks ('000 hectolitres)	123	143	142
Cigarettes (million packs)	38	41	41
Garments ('000 pieces)	21	24	32
Wood furniture (million kips)	12,725	12,700	15,240
Plastic products (metric tons)	3,900	3,850	4,350
Detergent (metric tons)	879	900	700
Agricultural tools ('000)	4	4	4
Nails (metric tons)	691	650	740
Bricks (million)	65	66	87
Electric energy (million kWh)	2,436	3,678	3,590
Tobacco (metric tons)	757	1,277	358
Plywood ('000 sheets)	2,086	2,150	2,200

* Estimates.

Source: IMF, *Lao People's Democratic Republic: Selected Issues and Statistical Appendix* (September 2002).

Finance

CURRENCY AND EXCHANGE RATES

Monetary Units

100 at (cents) = 1 new kip.

Sterling, Dollar and Euro Equivalents (31 March 2004)

£1 sterling = 19,189.7 new kips;
US $1 = 10,461.0 new kips;
€1 = 12,787.5 new kips;
100,000 new kips = £5.21 = $9.56 = €7.82.

Average Exchange Rate (new kips per US $)

2000 7,887.64
2001 8,954.58
2002 10,056.33

Note: In September 1995 a policy of 'floating' exchange rates was adopted, with commercial banks permitted to set their rates.

GENERAL BUDGET

('000 million new kips, year ending 30 September)*

Revenue†	1999/2000	2000/01‡	2001/02§
Tax revenue	1,367	1,629	2,043
Profits tax	187	205	362
Income tax	117	153	190
Turnover tax	290	318	452
Taxes on foreign trade	176	236	291
Import duties	135	179	229
Export duties	41	57	62
Excise tax	226	371	362
Timber royalties	273	182	165
Hydro royalties	22	51	55
Other revenue	324	372	438
Payment for depreciation or dividend transfers	42	67	89
Leasing income	15	39	57
Administrative fees	17	41	62
Overflight	123	114	153
Interest or amortization	79	76	42
Total	1,691	2,000	2,481

Expenditure	1999/2000‡	2000/01‡	2001/02§
Current expenditure	1,050	1,229	1,449
Wages and salaries	335	410	517
Transfers	130	243	292
Materials and supplies	174	330	370
Interest	103	134	145
Timber royalty-financed expenditure	242	—	—
Capital expenditure and net lending	1,704	1,911	2,165
Domestically-financed	481	872	1,017
Foreign-financed	1,302	1,200	1,256
Loan-funded projects	827	724	826
Grant-funded projects	475	476	430
Onlending (net)	−78	−160	−108
Total (incl. others)	2,754	3,140	3,614

* Since 1992 there has been a unified budget covering the operations of the central Government, provincial administrations and state enterprises.
† Excluding grants received ('000 million new kips): 475 in 1999/2000; 476 in 2000/01‡; 549 in 2001/02§.
‡ Estimates.
§ Budget forecasts.

Source: IMF, *Lao People's Democratic Republic: Selected Issues and Statistical Appendix* (September 2002).

INTERNATIONAL RESERVES

(US $ million at 31 December)

	2001	2002	2003
Gold*	2.53	2.53	4.10
IMF special drawing rights	3.42	6.07	19.13
Foreign exchange	127.51	185.51	189.46
Total	133.46	194.14	212.69

* National valuation.

Source: IMF, *International Financial Statistics.*

MONEY SUPPLY

(million new kips at 31 December)

	2001	2002	2003
Currency outside banks	113,080	228,810	399,100
Demand deposits at commercial banks	256,880	358,150	437,290
Total (incl. others)	371,840	587,000	836,540

Source: IMF, *International Financial Statistics.*

COST OF LIVING

(Consumer Price Index for Vientiane; base: 2000 = 100)

	2001	2002	2003
All items	107.8	119.3	137.7

Source: IMF, *International Financial Statistics.*

NATIONAL ACCOUNTS

Expenditure on the Gross Domestic Product

(million new kips at current prices)

	1989	1990	1991
Government final consumption expenditure	34,929	61,754	69,499
Private final consumption expenditure	414,639	558,437	647,826
Increase in stocks } Gross fixed capital formation }	55,560	75,572	91,435
Total domestic expenditure	505,128	695,763	808,760
Exports of goods and services	49,421	69,411	73,359
Less Imports of goods and services	128,613	150,154	156,550
GDP in purchasers' values	425,936	615,020	725,569
GDP at constant 1987 prices	213,769	228,105	237,098

Source: World Bank, *Historically Planned Economies: A Guide to the Data.*

Gross Domestic Product by Economic Activity
(million new kips at current prices)

	2001	2002	2003
Agriculture, hunting, forestry and fishing	7,974,629	9,173,517	10,828,834
Mining and quarrying	73,150	89,114	378,238
Manufacturing	2,786,838	3,483,192	4,276,550
Electricity, gas and water	450,414	536,315	619,398
Construction	376,985	389,893	508,363
Wholesale and retail trade, restaurants and hotels	1,506,869	1,792,015	2,291,722
Transport, storage and communications	929,724	1,114,964	1,408,139
Finance, insurance, real estate and business services	127,836	75,979	99,487
Public administration	517,137	633,063	804,925
Other services	820,391	930,824	1,080,569
GDP at factor cost	15,563,971	18,218,874	22,296,225
Indirect taxes *less* subsidies	140,899	171,501	239,882
GDP in purchasers' values	15,704,870	18,390,375	22,536,107

Source: Asian Development Bank, *Key Indicators of Developing Asian and Pacific Countries*.

BALANCE OF PAYMENTS
(US $ million)

	1999	2000	2001
Exports of goods f.o.b.	338.2	330.3	311.1
Imports of goods f.o.b.	–527.7	–535.3	–527.9
Trade balance	–189.5	–205.0	–216.8
Exports of services	130.0	175.7	166.1
Imports of services	–51.8	–43.1	–31.6
Balance on goods and services	–111.3	–72.4	–82.3
Other income received	10.5	7.3	5.8
Other income paid	–49.9	–59.7	–39.6
Balance on goods, services and income	–150.7	–124.7	–116.1
Current transfers received	80.2	116.3	33.7
Current transfers paid	–50.6	—	—
Current balance	–121.1	–8.5	–82.4
Other investment assets	–43.2	18.8	25.2
Other investment liabilities	–3.7	73.3	86.6
Net errors and omissions	–165.1	–74.2	–57.2
Overall balance	–333.1	43.4	–3.9

Source: IMF, *International Financial Statistics*.

External Trade

PRINCIPAL COMMODITIES
(US $ million)

Imports c.i.f.	1999	2000	2001*
Investment goods	184.0	161.8	166.4
Machinery and equipment	21.0	16.2	—
Vehicles†	35.8	23.3	—
Fuel†	36.7	79.1	—
Construction/electrical equipment	90.5	43.2	—
Consumption goods	252.7	288.0	280.1
Materials for garments industry	66.5	60.4	67.2
Motorcycle parts for assembly	38.4	22.6	30.0
Unrecorded imports	—	27.0‡	9.4
Total	554.3	569.4	567.4

* Estimates.
† Estimates based on the assumption that 50% of total are consumption goods.
‡ Estimate included due to weaknesses in customs data.

Exports f.o.b.	1999	2000	2001*
Wood products	84.9	87.1	81.0
Logs	20.0	26.0	n.a.
Timber	26.9	37.7	n.a.
Other and unrecorded	38.0	23.4	n.a.
Coffee	15.2	12.1	7.8
Other agricultural products	8.3	15.4	12.6
Manufactures†	27.9	9.6	9.7
Garments	72.0	91.6	100.0
Motorcycles	38.4	22.1	23.4
Electricity	90.5	112.2	114.2
Total (incl. others)	342.1	351.0	349.8

* Estimates.
† Excluding garments and wood products.

Source: IMF, *Lao People's Democratic Republic: Selected Issues and Statistical Appendix* (September 2002).

PRINCIPAL TRADING PARTNERS
(US $ million)

Imports	2001	2002	2003
Australia	8.3	12.6	9.5
China, People's Republic	59.9	59.7	108.1
France	8.5	8.9	11.8
Germany	7.4	4.1	5.9
Hong Kong	10.1	6.1	8.2
Japan	13.0	19.6	16.7
Korea, Republic	6.9	5.0	5.6
Singapore	28.9	29.1	22.4
Thailand*	451.7	444.0	501.8
United Kingdom	3.7	2.6	n.a.
Viet Nam	70.8	76.8	87.6
Total (incl. others)	719.4	730.9	844.9

Exports	2001	2002	2003
Belgium	10.4	13.6	17.6
China, People's Republic	6.8	8.8	10.2
France	33.7	33.8	33.5
Germany	25.5	22.0	25.2
Italy	10.9	10.1	10.3
Japan	6.3	6.1	7.3
Netherlands	9.7	10.6	10.2
Thailand*	81.0	85.0	94.4
United Kingdom	9.3	13.4	14.1
USA	3.6	2.6	n.a.
Viet Nam	61.9	67.1	76.6
Total (incl. others)	375.4	396.0	441.6

* Trade with Thailand may be overestimated, as it may include goods in transit to and from other countries.

Source: Asian Development Bank, *Key Indicators of Developing Asian and Pacific Countries*.

Transport

ROAD TRAFFIC
(motor vehicles in use at 31 December, estimates)

	1994	1995	1996
Passenger cars	18,240	17,280	16,320
Buses and coaches	440	n.a.	n.a.
Lorries and vans	7,920	6,020	4,200
Motorcycles and mopeds	169,000	200,000	231,000

Source: International Road Federation, *World Road Statistics*.

SHIPPING

Inland Waterways
(traffic)

	1993	1994	1995
Freight ('000 metric tons)	290	876	898
Freight ton-kilometres (million)	18.7	40.8	98.8
Passengers ('000)	703	898	652
Passenger-kilometres (million)	110.2	60.6	24.3

Source: Ministry of Communications, Transport, Post and Construction.

Merchant Fleet
(registered at 31 December)

	2001	2002	2003
Number of vessels	1	1	1
Displacement ('000 grt)	2.4	2.4	2.4

Source: Lloyd's Register-Fairplay, *World Fleet Statistics*.

CIVIL AVIATION
(traffic on scheduled services)

	1997	1998	1999
Kilometres flown (million)	1	1	1
Passengers carried ('000)	125	124	197
Passenger-kilometres (million)	48	48	78
Total ton-kilometres (million)	5	5	8

Source: UN, *Statistical Yearbook*.

Tourism

FOREIGN VISITOR ARRIVALS
(incl. excursionists)

Country of Nationality	1999	2000	2001
China, People's Republic	20,269	28,215	40,644
France	19,960	24,534	21,662
Japan	14,860	20,687	15,547
Thailand	356,105	442,564	376,685
United Kingdom	12,298	15,204	15,722
USA	24,672	32,869	25,779
Viet Nam	71,748	68,751	82,411
Total (incl. others)	614,278	737,208	673,823

Tourism receipts (US $ million): 114 in 2000; 104 in 2001; 113 in 2002.

Source: World Tourism Organization.

Communications Media

	2000	2001	2002
Television receivers ('000 in use)	52	n.a.	n.a.
Telephones ('000 main lines in use)	40.9	52.6	61.9
Mobile cellular telephones ('000 subscribers)	12.7	29.5	55.2
Personal computers ('000 in use)	14	16	18
Internet users ('000)	6.0	10.0	15.0

Radio receivers ('000 in use): 730 in 1997.

Facsimile machines (estimated number in use): 500 in 1994 (Source: UN, *Statistical Yearbook*).

Book production (1995): Titles 88; copies ('000) 995.

Daily newspapers (1996): 3 (average circulation 18,000).

Non-daily newspapers (1988, estimates): 4 (average circulation 20,000).

Sources (unless otherwise specified): International Telecommunication Union; UNESCO, *Statistical Yearbook*.

Education

(1996/97)

	Institutions	Teachers	Students: Males	Students: Females	Students: Total
Pre-primary	695	2,173	18,502	19,349	37,851
Primary	7,896	25,831	438,241	348,094	786,335
Secondary:					
general	n.a.	10,717	108,996	71,164	180,160
vocational*	n.a.	808	3,731	1,928	5,659
teacher training	n.a.	197	960	780	1,740
University level	n.a.	456	3,509	1,764	5,273
Other higher	n.a.	913	5,378	2,081	7,459

* Data for 1995/96.

Source: UNESCO, *Statistical Yearbook*.

Adult literacy rate (UNESCO estimates): 66.4% (males 77.4%; females 55.5%) in 2002 (Source: UN Development Programme, *Human Development Report*).

Directory

The Constitution

The new Constitution was unanimously endorsed by the Supreme People's Assembly on 14 August 1991. Its main provisions are summarized below:

POLITICAL SYSTEM

The Lao People's Democratic Republic (Lao PDR) is an independent, sovereign and united country and is indivisible.

The Lao PDR is a people's democratic state. The people's rights are exercised and ensured through the functioning of the political system, with the Lao People's Revolutionary Party as its leading organ. The people exercise power through the National Assembly, which functions in accordance with the principle of democratic centralism.

The State respects and protects all lawful activities of Buddhism and the followers of other religious faiths.

The Lao PDR pursues a foreign policy of peace, independence, friendship and co-operation. It adheres to the principles of peaceful co-existence with other countries, based on mutual respect for independence, sovereignty and territorial integrity.

SOCIO-ECONOMIC SYSTEM

The economy is market-orientated, with intervention by the State. The State encourages all economic sectors to compete and co-operate in the expansion of production and trade.

Private ownership of property and rights of inheritance are protected by the State.

The State authorizes the operation of private schools and medical services, while promoting the expansion of public education and health services.

FUNDAMENTAL RIGHTS AND OBLIGATIONS OF CITIZENS

Lao citizens, irrespective of their sex, social status, education, faith and ethnic group, are equal before the law.

Lao citizens aged 18 years and above have the right to vote, and those over 21 years to be candidates, in elections.

Lao citizens have freedom of religion, speech, press and assembly, and freedom to establish associations and to participate in demonstrations which do not contradict the law.

THE NATIONAL ASSEMBLY

The National Assembly is the legislative organ, which also oversees the activities of the administration and the judiciary. Members of the National Assembly are elected for a period of five years by universal adult suffrage. The National Assembly elects its own Standing Committee, which consists of the Chairman and Vice-Chairman of the National Assembly (and thus also of the National Assembly Standing Committee) and a number of other members. The National Assembly convenes its ordinary session twice annually. The National Assembly Standing Committee may convene an extraordinary session of the National Assembly if it deems this necessary. The National Assembly is empowered to amend the Constitution; to endorse, amend or abrogate laws; to elect or remove the President of State and Vice-Presidents of State, as proposed by the Standing Committee of the National Assembly; to adopt motions expressing 'no confidence' in the Government; to elect or remove the President of the People's Supreme Court, on the recommendation of the National Assembly Standing Committee.

THE PRESIDENT OF STATE

The President of State, who is also Head of the Armed Forces, is elected by the National Assembly for a five-year tenure. Laws adopted by the National Assembly must be promulgated by the President of State not later than 30 days after their enactment. The President is empowered to appoint or dismiss the Prime Minister and members of the Government, with the approval of the National Assembly; to appoint government officials at provincial and municipal levels; and to promote military personnel, on the recommendation of the Prime Minister.

THE GOVERNMENT

The Government is the administrative organ of the State. It is composed of the Prime Minister, Deputy Prime Ministers and Ministers or Chairmen of Committees (which are equivalent to Ministries), who are appointed by the President, with the approval of the National Assembly, for a term of five years. The Government implements the Constitution, laws and resolutions adopted by the National Assembly and state decrees and acts of the President of State. The Prime Minister is empowered to appoint Deputy Ministers and Vice-Chairmen of Committees, and lower-level government officials.

LOCAL ADMINISTRATION

The Lao PDR is divided into provinces, municipalities, districts and villages. Provincial governors and mayors of municipalities are appointed by the President of State. Deputy provincial governors, deputy mayors and district chiefs are appointed by the Prime Minister. Administration at village level is conducted by village heads.

THE JUDICIARY

The people's courts comprise the People's Supreme Court, the people's provincial and municipal courts, the people's district courts and military courts. The President of the People's Supreme Court and the Public Prosecutor-General are elected by the National Assembly, on the recommendation of the National Assembly Standing Committee. The Vice-President of the People's Supreme Court and the judges of the people's courts at all levels are appointed by the National Assembly Standing Committee.

The Government

HEAD OF STATE

President of State: Gen. KHAMTAY SIPHANDONE (took office February 1998).

Vice-President: Lt-Gen. CHOUMMALI SAIGNASON.

COUNCIL OF MINISTERS
(September 2004)

Prime Minister: BOUNGNANG VOLACHIT.

Deputy Prime Minister and President of the State Planning Committee: THONGLOUN SISOLIT.

Deputy Prime Minister and Minister of Foreign Affairs: SOMSAVAT LENGSAVAT.

Deputy Prime Ministers: Maj.-Gen. ASANG LAOLI, BOUSONE BOUPAVANH.

Minister of National Defence: Maj.-Gen. DOUANGCHAI PHICHIT.

Minister of Finance: CHANSY PHOSIKHAM.

Minister of Security: Maj.-Gen. SOUDCHAI THAMMASITH.

Minister of Justice: KHAMOUANE BOUPHA.

Minister of Agriculture and Forestry: SIANE SAPHANTHONG.

Minister of Communications, Transport, Post and Construction: BOUATHONG VONGLOKHAM.

Minister of Industry and Handicrafts: ONNEUA PHOMMACHANH.

Minister of Commerce and Tourism: SOULIVONG DARAVONG.

Minister of Information and Culture: PHANDOUANGCHIT VONGSA.

Minister of Labour and Social Welfare: SOMPHAN PHENGKHAMMI.

Minister of Education: PHIMMASONE LEUANGKHAMMA.

Minister of Public Health: Dr PONEMEKH DARALOY.

Minister to the Office of the President: SOUBANH SRITHIRATH.

Ministers to the Prime Minister's Office: BOUNTIEM PHITSAMAI, SOULI NANTHAVONG, SAISENGLI TENGBIACHU, SOMPHUNG MONGKHUNVILAI, VENTHONG LUANGVILAY.

MINISTRIES

Office of the President: rue Lane Xang, Vientiane; tel. (21) 214200; fax (21) 214208.

Office of the Prime Minister: Ban Sisavat, Vientiane; tel. (21) 213653; fax (21) 213560.

Ministry of Agriculture and Forestry: Ban Phonxay, Vientiane; tel. (21) 412359; fax (21) 412344.

Ministry of Commerce and Tourism: Ban Phonxay, Muang Saysettha, Vientiane; tel. (21) 412436; fax (21) 412434; internet www.moc.gov.la.

Ministry of Communications, Transport, Post and Construction: ave Lane Xang, Vientiane; tel. (21) 412251; fax (21) 414123.

Ministry of Education: BP 67, Vientiane; tel. (21) 216004; fax (21) 212108.

Ministry of Finance: rue That Luang, Ban Phonxay, Vientiane; tel. (21) 412401; fax (21) 412415.

Ministry of Foreign Affairs: rue That Luang 01004, Ban Phonxay, Vientiane; tel. (21) 413148; fax (21) 414009; e-mail sphimmas@laonet.net; internet www.mfa.laogov.net.

Ministry of Industry and Handicrafts: rue Nongbone, Ban Phai, BP 4708, Vientiane; tel. (21) 416718; fax (21) 413005; e-mail mihplan@laotel.com.

Ministry of Information and Culture: rue Sethathirath, Ban Xiengnheun, Vientiane; tel. (21) 210409; fax (21) 212408.

Ministry of Justice: Ban Phonxay, Vientiane; tel. (21) 414105.

Ministry of Labour and Social Welfare: rue Pangkham, Ban Sisaket, Vientiane; tel. (21) 213003.

Ministry of National Defence: rue Phone Kheng, Ban Phone Kheng, Vientiane; tel. (21) 412803.

Ministry of Public Health: Ban Simeuang, Vientiane; tel. (21) 214002; fax (21) 214001; e-mail cabinet.fr@moh.gov.la.

Ministry of Security: rue Nongbone, Ban Hatsady, Vientiane; tel. (21) 212500.

Legislature

At the election held on 24 February 2002 166 candidates, approved by the Lao Front for National Construction, contested the 109 seats in the National Assembly.

President of the National Assembly: Lt-Gen. SAMAN VIGNAKET.

Vice-President: PANY YATHOTU.

Political Organizations

Lao Front for National Construction (LFNC): BP 1828, Vientiane; f. 1979 to replace the Lao Patriotic Front; comprises representatives of various political and social groups, of which the LPRP (see below) is the dominant force; fosters national solidarity; Pres. Gen. SISAVAT KEOBOUNPHAN; Vice-Chair. SIHO BANNAVONG, KHAMPHOUI CHANTHASOUK, TONG YEUTHOR.

Phak Pasason Pativat Lao (Lao People's Revolutionary Party—LPRP): Vientiane; f. 1955 as the People's Party of Laos; reorg. under present name in 1972; Cen. Cttee of 53 full mems elected March 2001; Pres. Gen. KHAMTAY SIPHANDONE.

Political Bureau (Politburo)

Full members: Gen. KHAMTAY SIPHANDONE, Lt-Gen. SAMAN VIGNAKET, Lt-Gen. CHOUMMALI SAIGNASON, THONGSIN THAMMAVONG, BOUNGNANG VOLACHIT, Gen. SISAVAT KEOBOUNPHAN, Maj.-Gen. ASANG LAOLI,

THOUNGLONG SISOLIT, Maj.-Gen. DOUANGCHAI PHICHIT, BOUSONE BOUPAVANH.

Numerous factions are in armed opposition to the Government. The principal groups are:

Democratic Chao Fa Party of Laos: led by Pa Kao Her until his death in Oct. 2002; Pres. SOUA HER; Vice-Pres. TENG TANG.

Free Democratic Lao National Salvation Force: based in Thailand.

United Front for the Liberation of Laos: Leader PHOUNGPHET PHANARETH.

United Front for the National Liberation of the Lao People: f. 1980; led by Gen. PHOUMI NOSAVAN until his death in 1985.

United Lao National Liberation Front: Sayabouri Province; comprises an estimated 8,000 members, mostly Hmong (Meo) tribesmen; Sec.-Gen. VANG SHUR.

Diplomatic Representation

EMBASSIES IN LAOS

Australia: rue Pandit J. Nehru, quartier Phonxay, BP 292, Vientiane; tel. (21) 413600; fax (21) 413601; internet www.laos.embassy.gov.au; Ambassador ALISTAIR CHARLES MACLEAN.

Brunei: Unit 12, Ban Thoungkang, rue Lao-Thai Friendship, Muang Sisattanak, Xaysettha District, Vientiane; tel. (21) 352294; fax (21) 352291; e-mail embdlaos@laonet.com; Ambassador Pengiran Haji HAMDAN BIN Haji ISMAIL.

Cambodia: rue Thadeua, Km 2, BP 34, Vientiane; tel. (21) 314952; fax (21) 314951; e-mail recamlao@laotel.com; Ambassador HUOT PHAL.

China, People's Republic: rue Wat Nak, Muang Sisattanak, BP 898, Vientiane; tel. (21) 315100; fax (21) 315104; e-mail embassyprc@laonet.net; Ambassador LIU YONGXING.

Cuba: Ban Saphanthong Neua 128, BP 1017, Vientiane; tel. (21) 314902; fax (21) 314901; e-mail embacuba@laonet.net; Ambassador EDUARDO VALIDO GARCÍA.

France: rue Sethathirath, BP 06, Vientiane; tel. (21) 215253; fax (21) 215250; e-mail contact@ambafrance-laos.org; internet www.ambafrance-laos.org; Ambassador BERNARD POTTIER.

Germany: rue Sok Paluang 26, BP 314, Vientiane; tel. (21) 312110; fax (21) 351152; e-mail zreg@vien.diplo.de; Ambassador (vacant).

India: 2 Ban Wat Nak, rue Thadeua, Km 3, Sisattanak District, Vientiane; tel. (21) 352301; fax (21) 352300; e-mail indiaemb@laotel.com; internet www.indianembassylao.com; Ambassador TSEWANG TOPDEN.

Indonesia: ave Phone Keng, BP 277, Vientiane; tel. (21) 413909; fax (21) 214828; e-mail kbrivte@laotel.com; Ambassador ZAINUDDIN NASUTION.

Japan: rue Sisangvone, Vientiane; tel. (21) 414401; fax (21) 414406; Ambassador MAKOTO KATSURA.

Korea, Democratic People's Republic: quartier Wat Nak, Vientiane; tel. (21) 315261; fax (21) 315260; Ambassador KIM THAE JONG.

Korea, Republic: rue Lao-Thai Friendship, Ban Watnak, Sisattanak District, BP 7567, Vientiane; tel. (21) 415833; fax (21) 415831; e-mail koramb@laotel.com; Ambassador CHANG CHUL-KYOON.

Malaysia: rue That Luang, quartier Pholxay, BP 789, Vientiane; tel. (21) 414205; fax (21) 414201; e-mail mwvntian@laopdr.com; Ambassador AHMAD RASIDI HAZIZI.

Mongolia: rue Wat Nak, Km 3, BP 370, Vientiane; tel. (21) 315220; fax (21) 315221; e-mail embmong@laotel.com; Ambassador N. ALIASUREN.

Myanmar: Ban Thong Kang, rue Sok Palaung, BP 11, Vientiane; tel. (21) 314910; fax (21) 314913; e-mail mev@loxinfo.co.th; Ambassador U TIN OO.

Philippines: Ban Phonsinuane, Sisattanak, BP 2415, Vientiane; tel. (21) 452490; fax (21) 452493; e-mail pelaopdr@laotel.com; Ambassador ANTONIO CABANGON CHUA.

Poland: 263 Ban Thadeua, Km 3, quartier Wat Nak, BP 1106, Vientiane; tel. (21) 312940; fax (21) 312085; e-mail vieampol@laotel.com; Chargé d'affaires Dr TOMASZ GERLACH.

Russia: Ban Thadeua, quartier Thaphalanxay, BP 490, Vientiane; tel. (21) 312222; fax (21) 312210; e-mail rusemb@laotel.com; Ambassador YURI RAIKOV.

Singapore: Unit 12, Ban Naxay, rue Nong Bong, Muang Sat Settha, Vientiane; tel. (21) 416860; fax (21) 416855; e-mail sinemvte@laotel.com; internet www.mfa.gov.sg/vientiane; Ambassador KAREN TAN.

Thailand: ave Phone Keng, Vientiane; tel. (21) 214581; fax (21) 214580; e-mail thaivtn@mfa.go.th; Ambassador RATHAKIT MANATHAT.

USA: 19 rue Bartholonie, BP 114, Vientiane; tel. (21) 212581; fax (21) 212584; e-mail khammanpx@state.gov; internet usembassy.state.gov/laos; Ambassador PATRICIA HASLACH.

Viet Nam: 85 rue That Luang, Vientiane; tel. (21) 413409; fax (21) 413379; e-mail dsqvn@laotel.com; Ambassador HUYNH ANH DUNG.

Judicial System

President of the People's Supreme Court: KHAMMY SAYAVONG.

Vice-President: DAVON VANGVICHIT.

People's Supreme Court Judges: NOUANTHONG VONGSA, NHOTSENG LITTHIDETH, PHOUKHONG CHANTHALATH, SENGSOUVANH CHANTHALOUNNAVONG, KESON PHANLACK, KONGCHI YANGCHY, KHAMPON PHASAIGNAVONG.

Public Prosecutor-General: KHAMPANE PHILAVONG.

Religion

The 1991 Constitution guarantees freedom of religious belief. The principal religion of Laos is Buddhism.

BUDDHISM

Lao Unified Buddhists' Association: Maha Kudy, Wat That Luang, Vientiane; f. 1964; Pres. (vacant); Sec.-Gen. Rev. SIHO SIHAVONG.

CHRISTIANITY

The Roman Catholic Church

For ecclesiastical purposes, Laos comprises four Apostolic Vicariates. At 31 December 2002 an estimated 0.7% of the population were adherents.

Episcopal Conference of Laos and Cambodia
c/o Mgr Pierre Bach, Paris Foreign Missions, 254 Silom Rd, Bangkok 10500, Thailand.

f. 1971; Pres. Mgr JEAN KHAMSÉ VITHAVONG (Titular Bishop of Moglaena, Vicar Apostolic of Vientiane).

Vicar Apostolic of Luang Prabang: (vacant), Evêché, BP 74, Luang Prabang.

Vicar Apostolic of Paksé: Mgr LOUIS-MARIE LING MANGKHANEKHOUN (Titular Bishop of Proconsulari), Centre Catholique, BP 77, Paksé, Champasak; tel. (31) 212879.

Vicar Apostolic of Savannakhet: Mgr JEAN SOMMENG VORACHAK (Titular Bishop of Muzuca in Proconsulari), Centre Catholique, BP 12, Thakhek, Khammouane; tel. (51) 212184; fax (51) 213070.

Vicar Apostolic of Vientiane: Mgr JEAN KHAMSÉ VITHAVONG (Titular Bishop of Moglaena), Centre Catholique, BP 113, Vientiane; tel. (21) 216219; fax (21) 215085.

The Anglican Communion

Laos is within the jurisdiction of the Anglican Bishop of Singapore.

The Protestant Church

Lao Evangelical Church: BP 4200, Vientiane; tel. (21) 169136; Exec. Pres. Rev. KHAMPHONE KOUTHAPANYA.

BAHÁ'Í FAITH

National Spiritual Assembly: BP 189, Vientiane; tel. and fax (21) 216996; e-mail usme@laotel.com; f. 1956; Sec. SUSADA SENCHANTHISAY.

The Press

Aloun Mai (New Dawn): Vientiane; f. 1985; quarterly; theoretical and political organ of the LPRP.

Finance: rue That Luang, Ban Phonxay, Vientiane; tel. (21) 412401; fax (21) 412415; organ of Ministry of Finance.

Heng Ngan: 87 ave Lane Xang, BP 780, Vientiane; tel. (21) 212750; fortnightly; organ of the Federation of Lao Trade Unions; Editor BOUAPHENG BOUNSOULINH.

Lao Dong (Labour): 87 ave Lane Xang, Vientiane; f. 1986; fortnightly; organ of the Federation of Lao Trade Unions; circ. 46,000.

Laos: 80 rue Setthathirath, BP 3770, Vientiane; tel. (21) 21447; fax (21) 21445; internet www.laolink.com; quarterly; published in Lao and English; illustrated; Editor V. PHOMCHANHEUANG; English Editor O. PHRAKHAMSAY.

Meying Lao: rue Manthatoarath, BP 59, Vientiane; e-mail chansoda@hotmail.com; f. 1980; monthly; women's magazine; organ of the Lao Women's Union; Editor-in-Chief VATSADY KHUTNGOTHA; Editor CHANSODA PHONETHIP; circ. 7,000.

Noum Lao (Lao Youth): Vientiane; f. 1979; fortnightly; organ of the Lao People's Revolutionary Youth Union; Editor DOUANGDY INTHAVONG; circ. 6,000.

Pasason (The People): 80 rue Setthathirath, BP 110, Vientiane; f. 1940; daily; Lao; organ of the Cen. Cttee of the LPRP; Editor BOUABAN VOLAKHOUN; circ. 28,000.

Pasason Van Athit: Vientiane; weekly; circ. 2,000.

Pathet Lao: 80 rue Setthathirath, Vientiane; tel. (21) 215402; fax (21) 212446; f. 1979; monthly; Lao and English; organ of Khao San Pathet Lao (KPL); Dep. Dir SOUNTHONE KHANTHAVONG.

Sciences and Technics: Science, Technology and the Environment Agency (STEA), BP 2279, Vientiane; f. 1991 as Technical Science Magazine; quarterly; organ of the Dept of Science and Technology; scientific research and development.

Siang Khong Gnaovason Song Thanva (Voice of the 2nd December Youths): Vientiane; monthly; youth journal.

Sieng Khene Lao: Vientiane; monthly; organ of the Lao Writers' Association.

Suksa Mai: Vientiane; monthly; organ of the Ministry of Education.

Valasan Khosana (Propaganda Journal): Vientiane; f. 1987; organ of the Cen. Cttee of the LPRP.

Vannasinh: Vientiane; monthly; literature magazine.

Vientiane Mai (New Vientiane): rue Setthathirath, BP 989, Vientiane; tel. (21) 2623; fax (21) 5989; f. 1975; morning daily; organ of the LPRP Cttee of Vientiane province and city; Editor SICHANE (acting); circ. 2,500.

Vientiane Times: BP 5723, Vientiane; tel. (21) 216364; fax (21) 216365; internet www.vientianetimes.com; f. 1994; 2 a week; English; emphasis on investment opportunities; Editor SOMSANOUK MIXAY; circ. 3,000.

Vientiane Tulakit (Vientiane Business-Social): rue Setthathirath, Vientiane; tel. (21) 2623; fax (21) 6365; weekly; circ. 2,000.

There is also a newspaper published by the Lao People's Army, and several provinces have their own newsletters.

NEWS AGENCIES

Khao San Pathet Lao (Lao News Agency—KPL): 80 rue Setthathirath, BP 3770, Vientiane; tel. (21) 215090; fax (21) 212446; e-mail kplnews@yahoo.com; internet www.kplnet.net; f. 1968; organ of the Cttee of Information, Press, Radio and Television Broadcasting; news service; daily bulletins in Lao, English and French; teletype transmission in English; Gen. Dir KHAMSENE PHONGSA; English Editor BOUNLERT LOUANEDOUANGCHANH.

Foreign Bureaux

Rossiiskoye Informatsionnoye Agentstvo—Novosti (RIA—Novosti) (Russia): Vientiane; tel. (21) 213510; f. 1963.

Viet Nam News Agency (VNA): Vientiane; Chief DO VAN PHUONG.

Reuters (UK) is also represented in Laos.

PRESS ASSOCIATION

The Journalists' Association of the Lao PDR: BP 122, Vientiane; tel. (21) 212420; fax (21) 212408; Pres. BOUABANE VORAKHOUNE; Sec.-Gen. KHAM KHONG KONGVONGSA.

Publishers

Khoualuang Kanphim: 2–6 Khoualuang Market, Vientiane.

Lao-phanit: Ministry of Education, Bureau des Manuels Scolaires, rue Lane Xang, Ban Sisavat, Vientiane; educational, cookery, art, music, fiction.

Pakpassak Kanphin: 9–11 quai Fa-Hguun, Vientiane.

State Printing Enterprise: 314/C rue Samsemthai, BP 2160, Vientiane; tel. (21) 213273; fax (21) 215901; Dir NOUPHAY KOUNLAVONG.

Broadcasting and Communications

TELECOMMUNICATIONS

Entreprises des Postes et Télécommunications de Laos: ave Lane Xang, 01000 Vientiane; tel. (21) 215767; fax (21) 212779; e-mail laoposts@laotel.com; state enterprise, responsible for the postal service and telecommunications; Dir-Gen. KIENG KHAMKETH.

Lao Télécommunications Co Ltd: ave Lane Xang, BP 5607, 0100 Vientiane; tel. (21) 216465; fax (21) 216156; e-mail marketin@laotel.com; internet www.laotel.com; f. 1996; a jt venture between a subsidiary of the Shinawatra Group of Thailand and Entreprises des Postes et Télécommunications de Laos; awarded a 25-year contract by the Government in 1996 to undertake all telecommunications projects in the country; Dir-Gen. HOUMPHANH INTHARATH.

BROADCASTING

Radio

In addition to the national radio service, there are several local stations.

Lao National Radio: rue Phangkham, Km 6, BP 310, Vientiane; tel. (21) 212428; fax (21) 212432; e-mail natradio@laonet.net; internet www.lnr.org.la; f. 1960; state-owned; programmes in Lao, French, English, Thai, Khmer and Vietnamese; domestic and international services; Dir-Gen. BOUNTHANH INTHAXAY.

In 1990 resistance forces in Laos established an illegal radio station, broadcasting anti-Government propaganda: **Satthani Vithayou Kachai Siang Latthaban Potpoi Sat Lao** (Radio Station of the Government for the Liberation of the Lao Nation): programmes in Lao and Hmong languages; broadcasts four hours daily.

Television

A domestic television service began in December 1983. In May 1988 a second national television station commenced transmissions from Savannakhet. In December 1993 the Ministry of Information and Culture signed a 15-year joint-venture contract with a Thai firm on the development of broadcasting services in Laos. Under the resultant International Broadcasting Corporation Lao Co Ltd, IBC Channel 3 was inaugurated in 1994 (see below).

Lao National Television (TVNL): rue Chommany Neua, Km 6, BP 5635, Vientiane; tel. (21) 412183; fax (21) 412182; f. 1983; colour television service; Dir-Gen. BOUASONE PHONGPHAVANH.

Laos Television 3: BP 860, Vientiane; tel. (21) 315449; fax (21) 215628; operated by the International Broadcasting Corpn Lao Co Ltd; f. 1994 as IBC Channel 3; 30% govt-owned, 70% owned by the International Broadcasting Corpn Co Ltd of Thailand; programmes in Lao.

Finance

(cap. = capital; dep. = deposits; br.(s) = branch(es); m. = million)

BANKING

The banking system was reorganized in 1988–89, ending the state monopoly of banking. Some commercial banking functions were transferred from the central bank and the state commercial bank to a new network of autonomous banks. The establishment of joint ventures with foreign financial institutions was permitted. Foreign banks have been permitted to open branches in Laos since 1992. In 1998 there were nine private commercial banks in Laos, most of them Thai. In March 1999 the Government consolidated six state-owned banks into two new institutions—Lane Xang Bank Ltd and Lao May Bank Ltd; these merged in 2001.

Central Bank

Banque de la RDP Lao: rue Yonnet, BP 19, Vientiane; tel. (21) 213109; fax (21) 213108; e-mail bol@pan-laos.net.la; internet www.bankoflao.com; f. 1959 as the bank of issue, became Banque Pathetlao 1968, took over the operations of Banque Nationale du Laos 1975; known as Banque d'Etat de la RDP Lao from 1982 until adoption of present name; dep. 394,017m. kips, total assets US 361m. (Dec. 2002); Gov. PHOUMI THIPPHAVONE.

Commercial Banks

Agriculture Promotion Bank: 58 rue Hengboun, Ban Haysok, BP 5456, Vientiane; tel. (21) 212024; fax (21) 213957; e-mail apblaopdr@laonet.net; Man. Dir BOUNSONG SOMMALAVONG.

Banque pour le Commerce Extérieur Lao (BCEL): 1 rue Pangkham, BP 2925, Vientiane; tel. (21) 213200; fax (21) 213202; e-mail bcelhovt@hotmail.com; f. 1975; 100% state-owned; Chair. AKSONE BOUPHAKONEKHAM; Gen. Dir SONOXAY SITHPHAXAY.

Joint Development Bank: 82 ave Lane Xang, BP 3187, Vientiane; e-mail jdb@jdbbank.com; internet www.jdbbank.com; f. 1989; the first joint-venture bank between Laos and a foreign partner; 30% owned by Banque de la RDP Lao, 70% owned by Thai company, Phrom Suwan Silo and Drying Co Ltd; cap. US $4m.

Lao May Bank Ltd: 39 rue Pangkam, BP 2700, Vientiane; tel. (21) 213300; fax (21) 213304; f. 1999 as a result of the consolidation by the Government of ParkTai Bank, Lao May Bank and NakornLuang Bank; merged with Lane Xang Bank Ltd in 2001.

Lao-Viet Bank (LVB): 5 ave Lane Xang, Vientiane; tel. (21) 216316; fax (21) 212197; e-mail lvbho@laotel.com; internet www.laovietbank.com; f. 1999; joint venture between BCEL and the Bank for Investment and Development of Vietnam.

Vientiane Commercial Bank Ltd: 33 ave Lane Xang, Ban Hatsady, Chanthaboury, Vientiane; tel. (21) 222700; fax (21) 213513; f. 1993; privately-owned joint venture by Laotian, Thai, Taiwanese and Australian investors; Man. Dir SOP SISOMPHOU.

Foreign Banks

Bangkok Bank Public Co Ltd (Thailand): 38/13–15 rue Hatsady, BP 5400, Vientiane; tel. (21) 213560; fax (21) 213561; e-mail bblvte@laotel.com; f. 1993; Man. THEWAKUN CHANAKUN.

Bank of Ayudhya Public Co Ltd (Thailand): 79/6 Unit 17, ave Lane Xang, BP 5072, Vientiane; tel. (21) 213521; fax (21) 213520; e-mail baylaos@laotel.com; internet www.bay.co.th; f. 1994; Man. SUWAT TANTIPATANASAKUL.

Krung Thai Bank Public Co Ltd (Thailand): Unit 21, 80 ave Lane Xang, Ban Xiengngeuanthong, Chanthaboury, Vientiane; tel. (21) 213480; fax (21) 222762; e-mail ktblao@laotel.com; internet www.ktb.co.th; f. 1993; Gen. Man. SOMCHAI KANOKPETCH.

Public Bank Berhad (Malaysia): 100/1–4 rue Talat Sao, BP 6614, Vientiane; tel. (21) 216614; fax (21) 222743; e-mail pbbvte@laotel.com; Gen. Man. SIA KYUN MIN.

Siam Commercial Bank Public Co Ltd (Thailand): 117 ave Lane Xang-Samsenethai, BP 4809, Ban Sisaket Mouang, Chanthaboury, Vientiane; tel. (21) 213500; fax (21) 213502; Gen. Man. CHARANYA DISSAMARN.

Thai Military Bank Public Co Ltd: 69 rue Khoun Boulom, Chanthaboury, BP 2423, Vientiane; tel. (21) 217174; fax (21) 216486; the first foreign bank to be represented in Laos; Man. AMNAT KOSKTPON.

INSURANCE

Assurances Générales du Laos (AGL): Vientiane Commercial Bank Bldg, ave Lane Xang, BP 4223, Vientiane; tel. (21) 215903; fax (21) 215904; e-mail agl@agl-allianz.com; internet www.agl-allianz.com; Man. Dir PHILIPPE ROBINEAU.

Trade and Industry

GOVERNMENT AGENCY

National Economic Research Institute: rue Luang Prabang, Sithanneua, Vientiane; tel. (21) 216653; fax (21) 216660; e-mail neri@pan-laos.net; govt policy development unit; Dir SOUPHAN KEOMISAY.

DEVELOPMENT ORGANIZATIONS

Department of Livestock and Fisheries: Ministry of Agriculture and Forestry, Ban Phonxay, BP 811, Vientiane; tel. (21) 416932; fax (21) 415674; e-mail eulaodlf@laotel.com; public enterprise; imports and markets agricultural commodities; produces and distributes feed and animals; Dir-Gen. SINGKHAM PHONVISAY.

State Committee for State Planning: Office of the Prime Minister, Ban Sisavat, Vientiane; tel. (21) 213653; fax (21) 213560; Pres. THOUNGLONG SISOLIT.

CHAMBER OF COMMERCE

Lao National Chamber of Commerce and Industry: rue Sihom, Ban Haisok, BP 4596, Vientiane; tel. and fax (21) 219223; e-mail ccilcciv@laotel.com; internet www.lncci.laotel.com; f. 1989; 470 mems; Pres. KISSANA VONGSAY; Sec. Gen. KHAMPANH SENGTHONGKHAM.

INDUSTRIAL AND TRADE ASSOCIATION

Société Lao Import-Export (SOLIMPEX): 43–47 ave Lane Xang, BP 278, Vientiane; tel. (21) 213818; fax (21) 217054; Dir KANHKEO SAYCOCIE; Dep. Dir PHONGSAMOUTH VONGKOT.

UTILITIES

Electricity

Electricité du Laos: rue Nongbone, BP 309, Vientiane; tel. (21) 451519; fax (21) 416381; e-mail edlgmo@laotel.com; responsible for production and distribution of electricity; Gen. Man. VIRAPHONH VIRAVONG.

Lao National Grid Co: Vientiane; responsible for Mekong hydroelectricity exports.

Water

In 1998 the Government adopted a policy of decentralization with regard to water supply and sanitation in Laos. As a result, the national water supply authority, Nam Papa Lao, was divided into Nam Papa Vientiane, with jurisdiction over the capital, and a number of provincial authorities. Activities within the sector were to be co-ordinated by a newly established body, the Water Supply Authority.

Nam Papa Vientiane (Lao Water Supply Authority): rue Phone Kheng, Thatluang Neua Village, Sat Settha District, Vientiane; tel. (21) 412880; fax (21) 414378; f. 1962; fmrly Nam Papa Lao; authority responsible for the water supply of Vientiane; Gen. Man. Dr SOMPHON DETHOUDON; Dep. Man. KHAMPINH VORACHAKDAOVY.

Water Supply Authority (WASA): Dept of Housing and Urban Planning, Ministry of Communications, Transport, Post and Construction, ave Lane Xang, Vientiane; tel. (21) 415764; fax (21) 451826; e-mail mctpcwwa@laotel.com; f. 1998; Dir NOUPHEUAK VIRABOUTH.

STATE ENTERPRISES

Agricultural Forestry Development Import-Export and General Service Co: trading co of the armed forces.

Bolisat Phatthana Khet Phoudoi Import-Export Co: rue Khoun Boulom, Vientiane; tel. (21) 216234; fax (21) 215046; f. 1984; trading co of the armed forces.

Dao-Heuang Import-Export Co: Ban Thaluang, Champasak Province; tel. (31) 212250; fax (31) 212438; e-mail info@dao-heuang.com; internet www.dao-heuang.com; f. 1990; imports and distributes whisky, beer, mineral water and foodstuffs.

Lao Houng Heuang Export-Import Co: rue Nongbone, Vientiane; tel. (21) 217344; fax (21) 212107.

Luen Fat Hong Lao Plywood Industry Co: BP 83, Vientiane; tel. (21) 314990; fax (21) 314992; e-mail lfhsdsj@laotel.com; internet www.luenfathongyada.laopdr.com; development and management of forests, logging and timber production.

CO-OPERATIVES

Central Leading Committee to Guide Agricultural Co-operatives: Vientiane; f. 1978 to help organize and plan regulations and policies for co-operatives; by the end of 1986 there were some 4,000 co-operatives, employing about 74% of the agricultural labour force; Chair. (vacant).

TRADE UNION ORGANIZATION

Federation of Lao Trade Unions: 87 ave Lane Xang, BP 780, Vientiane; tel. (21) 313682; e-mail kammabanlao@pan-laos.net.la; f. 1956; 21-mem. Cen. Cttee and five-mem. Control Cttee; Pres. BOSAIKHAM VONGDALA (acting); 70,000 mems.

Transport

RAILWAYS

The construction of a 30-km rail link between Vientiane and the Thai border town of Nong Khai began in January 1996 but was indefinitely postponed in February 1998 as an indirect consequence of a severe downturn in the Thai economy. In 1997 the Government announced plans to develop a comprehensive railway network, and awarded a contract to a Thai company, although no timetable for the implementation of the scheme was announced. In 2003 the Thai Government agreed to finance a 3.5-km rail link from Tha Naleng (near Vientiane) to Nong Khai.

ROADS

The road network provides the country's main method of transport, accounting for about 90% of freight traffic and 95% of passenger traffic in 1993 (according to the Asian Development Bank—ADB). In 1999 there were an estimated 21,716 km of roads, of which 9,664 km were paved. The main routes link Vientiane and Luang Prabang with Ho Chi Minh City in southern Viet Nam and with northern Viet Nam and the Cambodian border, Vientiane with Savannakhet, Phongsali to the Chinese border, Vientiane with Luang Prabang and the port of Ha Tinh (northern Viet Nam), and Savannakhet with the port of Da Nang (Viet Nam). In 2002 Laos, Thailand and China agreed to a US $45m. road project intended to link the three countries; most of the road construction would be carried out in Laos. The project was to be completed by 2006. In February 2004 construction of a 245-km national road (Route 9) was completed, linking Laos with Thailand and Viet Nam.

The Friendship Bridge across the Mekong River, linking Laos and Thailand between Tha Naleng and Nong Khai, was opened in April 1994. In early 1998 construction work began in Paksé on another bridge across the Mekong River. The project was granted substantial funding from the Japanese Government, and was completed in August 2000. Construction commenced in December 2003 on a second Friendship Bridge, linking Savannakhet and Mukdahan. The bridge was due to be completed in 2005.

INLAND WATERWAYS

The Mekong River, which forms the western frontier of Laos for much of its length, is the country's greatest transport artery. However, the size of river vessels is limited by rapids, and traffic is seasonal. In April 1995 Laos, Cambodia, Thailand and Viet Nam signed an agreement regarding the joint development of the lower Mekong, and established a Mekong River Commission. There are about 4,600 km of navigable waterways.

CIVIL AVIATION

Wattai airport, Vientiane, is the principal airport. Following the signing of an agreement in 1995, the airport was to be upgraded by Japan; renovation work commenced in 1997, and a new passenger terminal was opened in 1998. The development of Luang Prabang airport by Thailand, at a cost of 50m. baht, began in May 1994 and the first phase of the development programme was completed in 1996; the second phase was completed in 1998. In April 1998 Luang Prabang airport gained formal approval for international flights. The airports at Paksé and Savannakhet were also scheduled to be upgraded to enable them to accommodate wide-bodied civilian aircraft; renovation work on the airport at Savannakhet was completed in April 2000. Construction of a new airport in Oudomxay Province was completed in the late 1990s.

In mid-2004 the Lao Government announced that a Memorandum of Understanding was to be signed with Thailand by the end of 2005 providing for the development of Savannakhet Airport into an international landing strip, to be used jointly by Japan, which would partially fund the development. The plan constituted part of the east–west economic corridor project, a proposed transport network linking Laos with Myanmar, Thailand and Viet Nam.

Lao Civil Aviation Department: BP 119, Vientiane; tel. and fax (21) 512163; e-mail laodca@laotel.com; Dir-Gen. YAKUA LOPANGKAO.

Lao Airlines: National Air Transport Co, 2 rue Pangkham, BP 6441, Vientiane; tel. (21) 212057; fax (21) 212056; e-mail laoairlines@laoairlines.com; internet www.laoairlines.com; f. 1975; state airline, fmrly Lao Aviation; operates internal and international passenger and cargo transport services within South-East Asia; Gen. Man. Dir POTHONG NGONPHACHANH.

Tourism

Laos boasts spectacular scenery, ancient pagodas and abundant wildlife. However, the development of tourism remains constrained by the poor infrastructure in much of the country. Western tourists were first permitted to enter Laos in 1989. In 1994, in order to stimulate the tourist industry, Vientiane ended restrictions on the movement of foreigners in Laos. Also in 1994 Laos, Viet Nam and Thailand agreed measures for the joint development of tourism. Luang Prabang was approved by UNESCO as a World Heritage site in February 1998. The years 1999–2000 were designated as Visit Laos Years. The number of visitors reached 737,208 in 2000, when receipts from tourism totalled an estimated US $114m. Visitor arrivals decreased to 673,823 in 2001, however, and receipts declined to $104m. in that year.

National Tourism Authority of Lao PDR: ave Lane Xang, BP 3556, Hadsady, Chanthaboury, Vientiane; tel. (21) 212251; fax (21) 212769; e-mail mtsc@mekongcenter.com; internet www .mekongcenter.com; 17 provincial offices; Dir CHENG SAYAVONG.

Defence

In August 2003 the total strength of the armed forces was estimated at 29,100: army 25,600; navy an estimated 600; air force 3,500. Conscription lasts a minimum of 18 months. Paramilitary forces comprise militia self-defence forces numbering about 100,000 men. Defence expenditure for 2003 was an estimated 115,000m. kips.

Supreme Commander of the Lao People's Army (Commander-in-Chief): Lt-Gen. CHOUMMALI SAIGNASON.

Chief of the General Staff: Maj.-Gen. KHENEKHAM SENGLATHONE.

Education

Education was greatly disrupted by the civil war, causing a high illiteracy rate, but educational facilities have since improved significantly. Lao is the medium of instruction. A comprehensive educational system is in force.

Education is officially compulsory for nine years, between six and 15 years of age. Primary education begins at six years of age and lasts for five years. Secondary education, beginning at the age of 11, lasts for six years, comprising two cycles of three years each (the second being a senior high school course). In 1996 enrolment at primary schools included 72% of the primary school-age population (males 76%; females 68%). Total enrolment at secondary schools in the same year was equivalent to 28% of the relevant age-group (males 34%; females 23%). The total enrolment at primary and secondary schools was equivalent to 72% of the school-age population (males 80%; females 63%) in 1996, compared with 68% in 1987. There are several regional technical colleges. The National University of Laos was founded in 1995. Enrolment in tertiary education in 1996 was equivalent to 3% of the relevant age-group (males 4%; females 2%). Government expenditure on education in 1997/98 was budgeted at 37,400m. kips, representing 6.9% of total budgetary expenditure.

In 1990 it was reported that the Government had permitted the establishment of five private primary schools.

Bibliography

See also Cambodia and Viet Nam

Anderson, Kym. *Lao Economic Reform and WTO Accession: Implications for Agriculture and Rural Development.* Singapore, Institute of Southeast Asian Studies, 1999.

Asian Development Bank. *Lao PDR and the Greater Mekong Subregion: Securing Benefits from Economic Cooperation.* Manila, 1996.

Reforming the Financial Sector in the Lao PDR. Manila, 1996.

Bourdet, Yves. *The Economics of Transition in Laos: From Socialism to ASEAN Integration.* Cheltenham, Edward Elgar, 2000.

Brahm, Laurence, and Macpherson, Neill. *Investment in the Lao People's Democratic Republic.* Hong Kong, Longman, 1992.

Brown, M., and Zasloff, J. *Apprentice Revolutionaries: The Communist Movement in Laos, 1930–1985.* Stanford, CA, Hoover Institution Press, 1986.

Castle, Timothy N. *A War in the Shadow of Vietnam.* New York, Columbia University Press, 1995.

Chazee, Laurent (Ed.). *The People of Laos.* Bangkok, White Lotus, 2001.

Conroy, Paul. *10 Months in Laos: A Vast Web of Intrigue, Missing Millions and Murder.* Melbourne, Vic, Crown Content, 2002.

Cooper, R., Tapp, N., Yia Lee, G., and Schwoer-Kohl, G. *The Hmong.* Bangkok, Artasia Press, 1992.

Deuve, J. *La Royaume de Laos, 1949–1965.* Paris, 1984.

Dommen, A. J. *Conflict in Laos: The Politics of Neutralization.* London, Pall Mall Press, 2nd Edn, 1971.

Evans, G. *Lao Peasants Under Socialism.* New Haven, CT, Yale University Press, 1991.

The Politics of Ritual and Remembrance: Laos since 1975. Chiang Mai, Silkworm Books, 1998.

A Short History of Laos: The Land in Between. St Leonards, NSW, Allen and Unwin, 2003.

Evans, Grant (Ed.). *Laos: Culture and Society.* Chiang Mai, Silkworm Books, 1999.

Evans, G., and Rowley, K. *Red Brotherhood at War.* London, Verso, Revised Edn, 1990.

Freeman, Nick J. 'Laos: No Safe Haven from the Regional Tumult' in *Southeast Asian Affairs, 1998.* Singapore, Institute of Southeast Asian Studies, 1998.

Gunn, G. C. *Political Struggles in Laos (1930–1954).* Bangkok, Editions Duang Kamol, 1988.

Rebellion in Laos. Boulder, CO, Westview Press, 1990.

Hamilton-Merritt, Jane. *Tragic Mountains: The Hmong, the Americans, and the Secret Wars for Laos, 1942–1992.* Bloomington, IN, Indiana University Press, 1993.

Hannah, Norman. *The Key to Failure.* New York, Madison Books, 1987.

Ireson-Doolittle, Carol, and Moreno-Black, Geraldine. *The Lao: Gender, Power and Livelihood.* Boulder, CO, Westview Press, 2003.

Kremmer, Christopher. *Stalking the Elephant Kings.* St Leonards, NSW, Allen and Unwin, 1998.

Lancaster, D. *The Emancipation of French Indo-China.* London, Oxford University Press, 1961.

Langer, P. F., and Zasloff, J. J. *North Vietnam and the Pathet Lao.* Cambridge, MA, Harvard University Press, 1970.

McCoy, A. W. *The Politics of Heroin in Southeast Asia.* New York, Harper and Row, 1972.

Mansfield, Stephen. *Lao Hill Tribes: Traditions and Patterns of Existence.* Kuala Lumpur, Oxford University Press, 2001.

Matelas, S. *Laos: A Country Report.* Washington, DC, US Government Printing Office, 1995.

Menon, Jayant. *Laos in the ASEAN Free Trade Area: Trade, Revenue and Investment Implications.* Pacific Economic Papers, No. 276, Canberra, Australia-Japan Research Centre, 1998.

Murphy, Dervla. *One Foot In Laos.* London, John Murray, 2000.

Ngaosrivathana, Mayoury. *Lao Women Yesterday and Today.* Vientiane, State Publishing Enterprise, 1993.

Ngaosrivathana, Mayoury and Pheuiphanh. *Kith and Kin Politics: The Relationship Between Laos and Thailand.* Manila, Journal of Contemporary Asia Publishers, 1994.

Ngaosrivathana, M., and Breazeale, K. (Eds). *Breaking New Ground in Lao History.* Seattle, WA, University of Washington Press, 2002.

Pham, Chi Do (Ed.). *Economic Development in Lao P.D.R.: Horizon 2000.* Vientiane, 1994.

Program for South-East Asian Studies. *New Laos, New Challenges.* Tempe, Arizona State University Press, 1998.

Quincy, Keith. *Harvesting Pa Chay's Wheat: The Hmong and America's Secret War in Laos.* Washington, DC, University of Washington Press, 2000.

Sagar, D. J. *Major Political Events in Indochina 1945–1990.* Oxford, Facts on File, 1991.

Simms, Peter and Sanda. *The Kingdoms of Laos: Six Hundred Years of History.* Richmond, Surrey, Curzon Press, 1998.

Sisouphanthong, Bounthavy, and Taillard, Christian. *Atlas of Laos.* Copenhagen, Nordic Institute of Asian Studies, 2000.

Stuart-Fox, Martin. *Laos: Politics, Economics and Society.* London, Francis Pinter, 1986.

'Laos: Towards Subregional Integration' in *Southeast Asian Affairs, 1995.* Singapore, Institute of Southeast Asian Studies, 1995.

Buddhist Kingdom, Marxist State: The Making of Modern Laos. Bangkok, White Lotus, 1996.

A History of Laos. Cambridge, Cambridge University Press, 1997.

The Lao Kingdom of Lan Xang: Rise and Decline. Bangkok, White Lotus, 1998.

Historical Dictionary of Laos. Metuchen, NJ, Scarecrow Press, 2nd Edn, 2000.

Stuart-Fox, Martin (Ed.). *Contemporary Laos.* University of Queensland Press, 1982.

Taillard, Christian. *Le Laos, stratégie d'un Etat-tampon.* Montpellier, Reclus, 1989.

Than, Mya, and Tan, Joseph L. H. (Eds). *Laos' Dilemmas and Options: The Challenge of Economic Transition in the 1990s.* Singapore, Institute of Southeast Asian Studies, 1996.

Toye, H. *Laos—Buffer State or Background.* London, Oxford University Press.

United Nations Development Programme. *Micro-finance in Rural Lao PDR: A National Profile.* Vientiane, UNDP, 1997.

Warner, Roger. *Back Fire: The CIA's Secret War in Laos.* New York, Simon & Schuster, 1995.

Zasloff, J. J. *The Pathet Lao: Leadership and Organisation.* Lexington, MA, Heath, 1973.

Zasloff, J. J., and Unger, L. (Eds). *Laos: Beyond the Revolution.* Basingstoke, Macmillan, 1991.

MALAYSIA

Physical and Social Geography

HARVEY DEMAINE

Malaysia covers a total area of 329,847 sq km (127,355 sq miles), comprising the 11 states of Peninsular Malaysia, with an area of 131,686 sq km (50,845 sq miles), together with the two states of Sarawak and Sabah (with the Federal Territory of Labuan), in northern Borneo, with areas of, respectively, 124,450 sq km (48,051 sq miles) and 73,711 sq km (28,460 sq miles). Peninsular Malaysia includes a number of islands, the largest being Langkawi and Pulau Pinang (Penang).

While Peninsular Malaysia, Sabah and Sarawak lie in almost identical latitudes between 1° N and 7° N of the Equator, and have characteristic equatorial climates with uniformly high temperatures and rain in all seasons, there is nevertheless a fundamental difference in their geographical position. Peninsular Malaysia forms the southern tip of the Asian mainland, bordered by Thailand to the north and by the island of Singapore at its southernmost point. On its western side, facing the sheltered and calm waters of the Straits of Melaka (Malacca), Peninsular Malaysia flanks one of the oldest and most frequented maritime highways of the world, whereas Sabah and Sarawak lie off the main shipping routes, along the northern fringe of the remote island of Borneo, bordered by Indonesia and, in north-eastern Sarawak, by Brunei.

PHYSICAL FEATURES

Structurally, both parts of Malaysia form part of the old stable massif of Sundaland, though whereas the dominant folding in the Malay peninsula is of Mesozoic age, that along the northern edge of Borneo dates from Tertiary times. In Peninsular Malaysia the mountain ranges, whose summit levels reach 1,200 m–2,100 m, run roughly north to south and their granitic cores have been widely exposed by erosion. The most continuous is the Main Range, which, over most of the peninsula, marks the divide between the relatively narrow western coastal plain draining to the Straits of Melaka, and the much larger area of mountainous interior and coastal lowland which drains to the South China Sea.

Because of the much greater accessibility of the western lowlands to the main sea-routes, and also of the existence of extensive areas of alluvial tin in the gravels deposited at the break of slope in the western foothills of the Main Range, the strip of country lying between the latter and the western coast of Peninsular Malaysia has been much more intensively developed than the remaining four-fifths of the country. The planting of rubber became concentrated in the vicinity of roads, railways and other facilities originally developed in connection with the tin industry. In contrast to the placid waters of the west coast, the east coast is open to the full force of the north-east monsoon during the period from October to March.

In many respects Sabah and Sarawak display similar basic geographical characteristics to eastern Peninsular Malaysia, but in a more extreme form. Thus, the lowlands are mostly wider, the rivers longer and even more liable to severe flooding, the coastline exposed to the north-east monsoon and avoided by shipping, and the equatorial forest cover even denser and more continuous than that of the peninsula. Moreover, while in general the mountains of Sabah and Sarawak are of comparable height to those in Peninsular Malaysia, there is one striking exception in Mt Kinabalu, a single isolated horst, which towers above the Croker Range of Sabah to an altitude of 4,101 m.

Throughout Malaysia, average daily temperatures range from about 21°C to 32°C, although in higher areas temperatures are lower and vary more widely. Rainfall averages about 2,540 mm throughout the year, although this is subject to regional variation.

NATURAL RESOURCES

Malaysia is endowed with an extremely rich natural resource base. The country has extensive tin deposits, and geophysical surveys in the eastern part of the peninsula's Main Range have suggested the presence of substantial deposits of other important minerals such as copper and uranium. Significant deposits have also been identified in East Malaysia, with copper mining established in Sabah, and bauxite and coal exploited in Sarawak.

East Malaysia's main wealth, however, remains the coastal and offshore deposits of hydrocarbons. Petroleum production from the original Miri field, in onshore Sarawak, has ceased, but discoveries off shore, made in the 1960s, have maintained production. At the end of 2003 Malaysia's crude and condensate petroleum reserves totalled 4,000m. barrels (compared with 5,500m. barrels in 1996), and natural gas reserves 84,900,000m. cu ft. Malaysia's production of crude petroleum was estimated at 38.8m. metric tons in 2003. Petroleum production in that year averaged 875,000 barrels per day from Malaysia's 38 oilfields.

Until the rise of petroleum, Malaysia's main economic resource was the agricultural potential of the peninsula. This derived not so much from the inherent superiority of its soils—indeed, those of Sarawak and Sabah are similar—but rather from its accessibility for commercial enterprise. Rubber and, subsequently, oil palm flourished in this environment. Sabah and Sarawak rely heavily upon their vast wealth in tropical timbers. However, the rate of extraction of timber has been so rapid since the late 1970s that serious efforts are now having to be made to conserve resources, particularly in the peninsula.

POPULATION AND CULTURE

The total population of Malaysia, including adjustments for underenumeration, was 23,274,690 at the census of July 2000 (compared with the August 1991 census total of 18,379,655), of whom 18,523,632 resided in Peninsular Malaysia (the most urbanized part of the country, where the average density was 140.7 per sq km), 2,679,552 were in Sabah and Labuan (36.4 per sq km) and 2,071,506 in Sarawak (16.6 per sq km). In Peninsular Malaysia the indigenous population, apart from some 50,000 or so aboriginal peoples, consists mainly of Muslim Malays, who, according to the census of 1991, form 57.0% of the total population, which also includes 28.7% Chinese and a further 9.3% Indians (an ethnic term which applies to people from India, Pakistan or Bangladesh). In Sabah and Sarawak, by contrast, Malays and other Muslim peoples are confined mainly to the coastal zone, while various other ethnic groups occupy the interior. There is also a large Chinese element, amounting to 27.7% of the population in Sarawak and 11.7% in Sabah in 1991. According to the census of July 2000, Muslim Malays comprised 65.1%, Chinese 26.0% and Indians 7.7% of the total population. In 2003, according to official estimates, the total population of Malaysia was 25,050,000.

History

ROBERT CRIBB

Based on an earlier article by IAN BROWN

The founding of the Melaka (Malacca) sultanate in c. 1400 AD conventionally marks the beginning of the modern history of the territory that constitutes present-day Malaysia. Until its capture by the Portuguese in 1511, Melaka was not only the dominant trading centre in the region, and arguably the greatest emporium in the Asia of that period, but also a vigorous centre of Malay culture, influential in shaping the political institutions and traditional culture of the Malays through the succeeding centuries.

The Portuguese were unable to maintain Melaka's dominance in regional trade and in 1641 they lost control of the city to the Dutch. The Dutch were principally interested in Java, however, and during the 17th and 18th centuries local and outside influences vied for power in the Malay peninsula. Johor (Johore) inherited some of the prestige of Melaka, but it was rivalled in influence by the Dutch, the Aceh and Siak in Sumatra, and the Siamese. Refugees and explorers from southern Sulawesi (in what is now Indonesia) took control of some regions, including Johor, and Minangkabau migrants from Sumatra became a strong presence in Negeri Sembilan (Negri Sembilan). By the mid-18th century the peninsula was divided into a number of small Malay states, which had various links with outside powers.

The United Kingdom began its territorial advance in the late 18th century, aiming to expand its commercial activities east of India, and, in particular, to secure the trade route to China through the Straits of Melaka. Pinang (Penang) was acquired in 1786 and Melaka in 1795, and a trading settlement was founded on the island of Singapore in 1819. In 1826 Pinang, Melaka and Singapore became a single administrative unit, the Straits Settlements, which remained under the authority of British India until 1867, when administrative responsibility was transferred to the Colonial Office in London, United Kingdom. British interest in the peninsula was further secured by an Anglo-Dutch treaty of 1824, which established the Straits as the boundary of their respective spheres of influence. The treaty contained no reference to Borneo, but in 1841 an English adventurer, James Brooke, was installed by the Sultan of Brunei as Raja of Sarawak, thus founding a dynasty of 'white rajas' which was to rule the territory for over a century (see History of Brunei); in 1847 Brunei ceded the small island of Labuan to the United Kingdom as a coaling station. In the north-east corner of Borneo concessions granted by the Sultans of Brunei and Sulu were eventually acquired by the British North Borneo Co, formed in 1881 as a chartered company under the British crown. Brunei itself was reduced to two small enclaves under British protection.

Through the middle decades of the 19th century, British policy (formulated in Calcutta (now Kolkata), in India, and in London) sought to avoid involvement in the Malay states, but important economic developments in the west coast states of the peninsula gradually drew the Straits Settlements into a closer commercial, and then political, relationship with its hinterland.

The discovery of major tin deposits in Larut (Perak) in the 1840s led to a substantial expansion of mining activity on the west coast, financed in large part by mercantile interests in the Straits, and dependent on a very considerable influx of Chinese immigrant labour. Rival Chinese secret societies, frequently in alliance with factions within the Malay ruling houses, fought for control of this important new source of wealth. By the late 1860s Perak and Selangor, in particular, were approaching anarchy, and the British administration in the Straits Settlements was urged by local merchants, Chinese as well as European, to intervene to restore order.

Eventually, in 1873, the British Government itself agreed to intervene, possibly because it feared that a rival European power, imperial Germany, might take advantage of the disorder in the peninsula to establish a base that would threaten the British strategic domination of the Straits, but more fundamentally in order to secure wider British commercial interests in the east. With the Pangkor Treaty of January 1874, British administration was accepted in Perak; it was then rapidly extended to Selangor, Negeri Sembilan and, in 1888, to Pahang. In 1896 these four states were brought together as the Federated Malay States (FMS), with the federal capital at Kuala Lumpur. In 1909 the four northern states of Kedah, Perlis, Kelantan and Terengganu (Trengganu), long within the influence of Siam, were transferred to British authority and accepted British advisers; the northern Malay states, including Pattani and Singgora (Songkhla), remained under Siamese hegemony. In 1914 a permanent adviser was appointed to Johor, the one remaining independent Malay state. However, these five states did not enter the centralized administration of the FMS, but became collectively known as the Unfederated Malay States (UMS). Thus, by 1914 British authority had been extended throughout the Malay Peninsula, except for the Siamese north, but with considerable variation in both constitutional form and administrative practice.

BRITISH RULE IN THE MALAY STATES

The basis of British administration in the Malay states, until the Japanese invasion in 1941, was a short phrase in the Pangkor Treaty which required the Sultan to accept a British Resident whose advice 'must be asked and acted upon on all questions other than those touching Malay religion and custom'. Consequently, as the residential system evolved in the FMS, government was in the name of the Sultan but executive authority lay very firmly with the Resident. Although there was a brief Malay uprising in Perak in 1875, in general the Malay ruling families found little difficulty in reconciling themselves to their loss of political powers. Acceptance was eased by the fact that the British administration, eager to sustain the fiction of Malay rule with British advice, not only maintained the full splendour of Malay court ceremonial but also treated the Malay rulers in public with the deference due to royalty. In the UMS, particularly in Johor and Kedah, the appointed British official, here called the Adviser, in general carried significantly less executive authority than his counterpart, the Resident, in the FMS. The more assertive Sultans here could exercise independence of British advice.

Malaya emerged as a major world producer of tin and, from the early 20th century, of rubber. Both the tin industry and the plantation rubber interests were heavily dependent on immigrant labour, from the southern coastal provinces of China in the case of the former, and from southern India in the case of the latter. By 1931 Chinese constituted 39% of the population of British Malaya, compared with 45% Malays. Moreover, Chinese clearly outnumbered Malays in the states of the west coast, and in the main urban centres: it was only in the four northern unfederated states that the Malays maintained their numerical superiority. Chinese (and Indians) were dominant in the production of tin and rubber for the world market, in internal trade and in money-lending. The Malays were predominantly subsistence rice farmers, although they did develop a considerable rubber smallholding production by the inter-war years, despite strong opposition from a colonial administration that sought to reinforce the self-sufficiency of their *kampung* economy and to encourage their cultivation of food crops for the immigrant labouring populations.

In the inter-war period Malay nationalism began to develop, fostered principally by a vocal generation of village teachers trained at the Malay-medium Sultan Idris Training College (founded in 1922), by English-educated Malay civil servants and by religious figures who drew on the Islamic reformist ideas then emanating from the Middle East. The political concerns of the Chinese and Indians in this period lay primarily outside Malaya. From the 1920s the Kuomintang Government in China sought political support and financial contributions from among the Malayan Chinese, and, as China itself fell victim to Japanese aggression in the following decade, there was an upsurge of patriotism among the community in Malaya, which manifested itself most clearly in a series of boycotts of Japanese trade. Similarly, the Malayan Indians focused their attention on the gathering independence struggle in the subcontinent. With the important exception of the Communist Party of Malaya (CPM), formally organized in 1930, no radical party emerged to secure significant support among the Malayan peoples. The colonial order, founded on accommodation with both the Malay aristocracy and the wealthy Chinese mercantile class, appeared unshakeable.

OCCUPATION AND INDEPENDENCE

The Japanese invasion and occupation of Malaya swept aside that order and made its full restoration impossible. The rapid collapse of British power in Malaya and Singapore in the early weeks of 1942 did much to destroy the myth of white superiority which had sustained colonial rule in Malaya, as elsewhere, while the three years of Japanese administration that followed the British surrender greatly heightened the political sensibilities and racial antagonisms of the diverse elements in the Malayan population. The Malay rulers, who had refused to evacuate with the retreating British administration, largely collaborated with the Japanese, but did not receive the deference that had been theirs in the pre-war colonial order. Moreover, through the organization of mass demonstrations, pan-Malayan conferences and paramilitary youth groups, the Japanese sought to weaken the established allegiance of the Malays to their individual state and sultan, and to encourage loyalty to a peninsula-wide entity. In August 1943, however, Japan transferred the four northern states, which were predominantly Malay, to Thailand, giving the immigrant communities overwhelming numerical dominance in the remaining states. The Indian community fared less well. Many Indian estate workers were forcibly conscripted for Japanese projects, including work on the Siam–Burma railway, and never returned. However, substantial enlistment into the Indian National Army, which was supported by Japan, helped to protect the community from the harshest treatment. The Chinese offered the only serious resistance to Japanese rule, for the anti-Japanese patriotism of the community had been strongly fuelled by the aggression of the 1930s, and they suffered most brutally during the occupation. Chinese comprised by far the largest component in the Malayan People's Anti-Japanese Army, the main resistance force, which was dominated by the CPM. When the Japanese administration suddenly collapsed in mid-August 1945, the CPM was left as the only effective political-military organization in the peninsula in advance of the returning British.

During the Japanese occupation, British officials in London, isolated from the Malayan reality, drew up proposals for major post-war constitutional reforms. Previous attempts to simplify the peninsula's complex constitutional and administrative structures had been largely unsuccessful, but the perceived need to promote rapid economic recovery in the post-war years gave greater urgency to centralized direction. It was proposed to incorporate the FMS, the UMS, Pinang and Melaka into a unified administrative unit, the Malayan Union, while making Singapore a separate crown colony. It was also proposed that citizenship of the Malayan Union be extended to all, irrespective of race or origin, and that all citizens would have equal rights. Sovereignty was to be transferred from the Malay rulers to the British crown. The opening of liberal citizenship provisions for Chinese and Indians reflected, in part, an awareness that both populations had now taken on the character of a predominantly settled community, and, in part, a recognition that they had, in general, remained loyal to the United Kingdom during the occupation. It also seriously challenged, however, the long-established privileged position and rights of the Malays, who considered themselves to be the indigenous people of the country, despite the long history of migration from other parts of South-East Asia.

When the Malayan Union was announced by the restored colonial administration in January 1946, and formally introduced in April, the usually quiescent Malay community was angered, not only by the actual provisions, but also by the overbearing manner in which the new constitutional form had been introduced. Malay opposition was brought together in a new political force, the United Malays National Organization (UMNO), inaugurated in May. The provisions of the Union could not be brought into effect in the face of determined Malay opposition. The British then began negotiations with the Malay rulers and UMNO for a new constitutional arrangement. This was the Federation of Malaya, which was eventually inaugurated in February 1948. The Federation maintained the Malayan Union concept of a unified administrative unit (embracing the FMS, the UMS, Pinang and Melaka), but it also reaffirmed the sovereignty of the Sultans, preserved the special privileges of the Malays, and introduced citizenship provisions which, for Chinese and Indians, were markedly more restrictive than those contained in the abortive Malayan Union.

Many Chinese saw in the Federation a betrayal of the loyalty that they had shown towards the colonial power during the Japanese invasion and occupation, and in this sense its introduction undoubtedly strengthened the position of those within the CPM who sought an armed confrontation with the British. For the three years from the re-establishment of British administration in 1945, the CPM had pursued the 'open and legal' struggle, which principally had involved organizing and radicalizing the labour movement in Malaya. The immediate post-war years thus saw a high level of labour unrest, notably on the estates and in the public services. In late 1947 the Government introduced stringent controls over the organization and structure of trade unions. With its legal position thus threatened, in early 1948 the CPM inevitably moved towards an armed struggle. A spate of murders of European planters signalled to the authorities the change in CPM strategy, and on 18 June 1948 the Government proclaimed a state of emergency throughout Malaya.

The communist insurrection posed a severe military challenge to the colonial Government. The guerrilla forces secured considerable initial success against European plantation and mining personnel, and in October 1951 they assassinated the British High Commissioner, Sir Henry Gurney. The CPM received only limited support from the Malayan population at large, for its overwhelmingly Chinese membership denied it access to the Malay community, while its ideology found little acceptance among wealthy Chinese. Moreover, from the early 1950s the colonial Government undertook a resettlement of the Chinese squatter communities which had sustained the CPM with food, information and recruits, relocating them into secured, military-protected 'new villages'. By the mid-1950s the communist insurrection had collapsed, although the state of emergency officially remained in force until 1960. The CPM finally abandoned its armed struggle in December 1989, when the long-term leader, Chin Peng, signed agreements with the Malaysian and Thai Governments.

The CPM's strength in the mid-1950s was also undermined by a British announcement that Malaya would move quickly to political independence. From the late 1940s the colonial administration had sought to encourage a non-communal Malayan leadership, and in this context placed its faith largely in Dato' Onn Ja'afar, President of UMNO; however, Onn's inability to persuade his party to open its membership to all races, and the subsequent electoral failure of his non-communal, but élite, Independence of Malaya Party, closed that avenue. An alternative approach to securing a viable political leadership for an independent Malaya emerged from within the Malayan communities themselves. In the Kuala Lumpur municipal elections of February 1952, the local branches of UMNO and the Malayan Chinese Association (MCA) successfully contested seats as a united front. From this grew a national alliance (into which the Malayan Indian Congress (MIC) was incorporated in 1954) in which each party retained its separate identity and policies, while the Alliance acted as a single organization in selecting the candidates and party to contest each particular seat. In the 1955 federal elections the UMNO-MCA-MIC Alliance secured 51 of the 52 contested seats and 81% of the vote.

As Malaya now approached self-government, a new Constitution was prepared, providing for a single nationality, with citizenship open to all those in Malaya who qualified either by birth or by fulfilling requirements of residence and language. In order to meet Malay unease that these provisions were too liberal, the Constitution also provided for an unusual elective monarchy. The Yang di-Pertuan Agong (paramount ruler) was to be selected for a five-year term by, and from among, the nine ruling Sultans of the Malay states. He held a special responsibility for safeguarding the privileged position of the Malays, who were designated *bumiputra* ('sons of the soil'). Islam was made the official religion, but the degree to which Islamic law might be implemented remained under the control of the states. Independence (*merdeka*) was proclaimed on 31 August 1957, with Tunku Abdul Rahman, President of UMNO, as the first Prime Minister.

FORMATION OF MALAYSIA

Singapore, excluded from the unified Malaya created in 1948, followed a separate path to political independence. Internal self-government was secured in 1959, with the prospect of full independence being achieved in 1963. This prospect, however, caused considerable concern in independent Malaya, for it was feared that Singapore, whose politics had been notably radical in the 1950s, might soon be in a position to encourage and aid the remnants of Chinese left-wing elements in the peninsula. It was therefore proposed from 1961, notably by Tunku Abdul Rahman, that an independent Singapore be incorporated into a federation with Malaya: this would not only enable the latter to restrain the more volatile political forces in Singapore, but would also reinforce the natural economic relationship between the island entrepôt and its peninsular hinterland. However, federation with overwhelmingly Chinese Singapore threatened the numerical superiority of the Malays, and it was therefore further proposed that the northern Borneo territories also be brought within the new alignment. Sarawak and North Borneo both had significant Chinese communities, but the Muslim Malays and the often Christian indigenous peoples (principally Dayak in Sarawak and Kadazan in North Borneo) together outnumbered the immigrant communities in Malaysia.

Since 1946 both Sarawak and North Borneo had been crown colonies, following the respective terminations of Brooke rule and the administration of the British North Borneo Co, while Brunei remained a British protectorate. In none of these territories, however, had the United Kingdom made significant preparations for self-government, and there were strong local doubts regarding the prospect of being ruled from distant Kuala Lumpur. The Sultan of Brunei considered himself senior to the Sultans of the peninsula and, in view of Brunei's petroleum resources, the federation had few economic attractions for the protectorate. A basic nationalist movement thus emerged, with the aim of establishing an independent northern Borneo. However, the United Kingdom exploited disunity within the movement and concern that neighbouring Indonesia might seek to annex a small independent state. The British Government consequently proceeded with the creation of the federation, although the Sultan of Brunei finally decided not to join the grouping. In September 1963 both Sarawak and North Borneo (now renamed Sabah) joined Singapore and the Federation of Malaya to form the independent Federation of Malaysia. Within this Federation, Sarawak and Sabah retained a higher degree of autonomy than the peninsular states, including the right to control immigration from the peninsula.

Indonesia, for reasons that derived primarily from its own internal political tensions, sought from the first to break up the Federation. In addition, the Philippines pursued a claim to the territory of Sabah. In September 1963 Indonesia and the Philippines broke off diplomatic relations with Malaysia, and Indonesia launched a series of military raids into Sarawak and Sabah from Indonesian Borneo (Kalimantan). However, this military challenge was successfully contained by Malaysian and Commonwealth forces and, following the fall from power of President Sukarno, a *rapprochement* between the two countries was achieved in August 1966.

An internal challenge to Malaysia arose from the determination of the People's Action Party (under Lee Kuan Yew), the governing party in Singapore, to campaign for the Chinese vote in peninsular elections, and so, in effect, oppose the MCA. This strategy threatened to undermine the Alliance consensus, and Malaysian leaders accused Singapore of provoking inter-communal animosity. In August 1965 Singapore was effectively expelled from Malaysia, despite protests from the island's leaders.

RESHAPING THE POLITICAL ORDER

The 1957 Constitution envisaged a multi-ethnic Malaysia in which the political and administrative dominance of the Malays would be balanced by the continuing economic pre-eminence of the Chinese and, to a lesser extent, the Indians. Within the parliamentary system, it was assumed, the interests of these groups would be represented by their respective élites, UMNO, the MCA and the MIC, who would negotiate policies aimed primarily at preserving the status quo and then at gradually achieving a convergence in the social standing of the three communities. Thus, it was expected that the Malays would achieve a steadily greater share of the economy, while the Chinese and Indians would gradually obtain more political influence, although there was no model for the kind of society this process might eventually produce and there was no prospect of any kind of ethnic or cultural fusion. The expulsion of Singapore from the Federation appeared to ensure that Malays would take part in this system from a position of numerical dominance, and would thus enjoy perpetual political pre-eminence, as long as they remained disciplined in their support of UMNO. The planners of the 1950s, however, had not allowed for the rapidity of social change in the new country. Amongst Malays, the impact of modernization contributed to a growing responsiveness to Islam and to political support for the Parti Islam se Malaysia (PAS—Islamic Party of Malaysia). The Islamic party had its political base amongst rural Islamic teachers and scholars, who were generally unsympathetic to the Malay aristocracy dominating UMNO. The Chinese for their part grew increasingly dissatisfied with political subordination, and in the federal election of May 1969 the MCA lost considerable ground among the Chinese voters to opposition parties, notably Gerakan Rakyat Malaysia (GERAKAN—Malaysian People's Movement) and the Democratic Action Party (DAP), which had demanded a more rapid end to Malay political predominance. Subsequent communal violence in the federal capital left many hundreds dead. To restore order, the Constitution was suspended and a national emergency declared.

The May 1969 riots forced a major readjustment both in the terms upon which communal interests and aspirations were accommodated, and in the manner by which the details of that accommodation were negotiated between the principal communities. The New Economic Policy (NEP), to be implemented over the 20 years to 1990, had as its primary objectives the eradication of poverty (irrespective of race) and of the identification of race with economic function. The latter objective implied securing for the Malays not only a far greater share in the wealth of the country but also a much wider range of educational and employment opportunities. It was intended that by 1990 30% of commercial and industrial share capital would be in *bumiputra* ownership (the ownership of Malays and other indigenous communities, or public enterprises acting on their behalf). Immediately after the 1969 riots, new legislation removed from public discussion such sensitive issues as the powers and status of the Sultans, Malay special rights, the status of Islam as the official religion and citizenship rights. At the same time the structure of coalition politics changed. The tripartite UMNO-MCA-MIC alliance was replaced by a broader coalition, registered in 1974 and called the Barisan Nasional (BN—National Front), which aimed to draw as many parties as possible into the Government under the leadership of UMNO. This strategy had the aim not only of giving the Government a broader electoral base but also of muting criticism by involving potential opposition parties in government policy. The BN has governed Malaysia ever since, and all significant parties except the DAP and Semangat '46 (see below, dissolved in October 1996) have joined it in government at least for a time.

A major complicating factor in Malaysian politics is the position of the state Governments. The states vary greatly in their ethnic composition, their socio-economic structure and even their constitutional relationship to the federal Government. Politics, for instance, has a very different character in Pinang (urbanized, mainly Chinese), Kelantan (rural, predominantly Malay and ruled by a Sultan) and Sabah (Malays, Chinese and mainly Christian indigenous Borneans in roughly equal proportions). The nine peninsular states with reigning Sultans have greater political influence in the Conference of Rulers than the four states that do not (Pinang, Melaka, Sabah and Sarawak), but the two Borneo states have maintained the distinct status that allows them to control immigration from the peninsula as well as to exercise greater economic and cultural autonomy than any peninsular state. Most of Malaysia's politicians, therefore, work from a power base within just one state, and build their national political careers by means of alliances with politicians from other regions. Movement between the state and federal legislatures and Governments is also quite common. Although the political role of the Sultans is constitutionally limited, individual Sultans have had considerable influence within their own states from time to time.

Under the rule of the BN, the idiom of politics changed from one of national coalition between ethnic groups to one of Malay dominance as *bumiputra* (indigenes). The term *bumiputra* has always been problematic: not only does the Malay community include the descendants of 17th and 18th century immigrants to the peninsula from Sumatra and Sulawesi in what is now Indonesia, but more recent migrants from Indonesia and the Philippines, and even Muslim Cham refugees from Cambodia, have sometimes found it possible to acquire 'indigenous' status, even though their ties to the country are much more recent than those of some Chinese communities. Whereas on the peninsula *bumiputra* status, moreover, is closely linked to Islam, the indigenous peoples of the Borneo states are mostly Christian. Because of the sedition laws governing discussion of the rights of Malays and the position of Islam, however, such issues have seldom received more than tangential mention in Malaysian political discourse. None the less, the growing influence of the Malays was unmistakable. Bahasa Malaysia was adopted as the language of education and administration at all levels, racial quotas gave advantages to Malays, especially in the education system, and the cultural heritages of Chinese and Indian Malaysians were given clearly subordinate standing. UMNO patronage became increasingly important in the allocation of government posts, and Chinese and Indians appeared to be losing ground in the administration. Malays began to take a greater share of the economy, both through private firms receiving significant support from the Government and through state-owned enterprises. A steady trickle of educated Chinese and Indian Malaysians to Singapore and other destinations contributed to Malay dominance. In 1995 official policy on the use of Bahasa Malaysia was partially reversed, with a decision to reintroduce English as a language of instruction at tertiary level for the sake of international business and technological connections. In July 2002 the Government announced that, from primary level, all schools would shift from the use of Bahasa Malaysia to English as the language of instruction, despite strong resistance from leaders of the Chinese community, who perceived the move as being an attack on Chinese culture. The Government later announced that the use of English would initially be limited to education in science and mathematics.

THE MAHATHIR YEARS 1981–2003

Within UMNO itself, however, important changes were in progress. Malaysia's first Prime Minister, Tunku Abdul Rahman (1963–70), was a royal prince, and his successors, Tun Abdul Razak (1970–76) and Dato' Hussein bin Onn (1976–81), were both Malay aristocrats. In 1981, however, Hussein Onn was succeeded by Dr Mahathir Mohamad, a professional and a commoner, who was widely seen as representing a new generation of modern Malays who were in thrall neither to the old aristocratic élite nor to the Islamic teachers of the countryside. Mahathir, who had earlier written a controversial defence of Malay interests (*The Malay Dilemma*, 1970), also represented a generation determined to consolidate the dominant position of the Malays in Malaysia. Mahathir's rule was characterized by a determined modernizing drive, based on both industrialization and the exploitation of raw materials (especially timber). In the early 1980s the Government was widely commended for its forceful insistence on efficiency and honesty, but by the late 1980s accusations of corruption and of abuse of power for economic advantage had become increasingly common. Business figures linked with UMNO appear to have prospered greatly from their connections, though legal restrictions have hampered detailed investigation of these allegations. In June 1991 the New Development Policy (NDP) was launched to succeed the NEP. The NDP shifted the emphasis from the NEP's racial economic restructuring to overall economic growth and the eradication of poverty. Racial quotas were retained, but no specific date was set for achieving the NEP's target of 30% *bumiputra* control of the country's corporate assets. An important part of this programme was the so-called 'Vision 2020', which Mahathir outlined in February 1991, embodying an ambition to turn Malaysia into a 'fully developed' country by the year 2020. Under this vision, Malaysia was not just to enjoy the material benefits of being fully developed (health care, public transport, education and general infrastructure) but was to be self-confident, harmonious, just, dynamic and democratic. Inevitably, however, progress towards the more material signs of development was easier to achieve than the less tangible goals.

The UMNO leadership faced two main challenges to its dominance over the Malays. PAS, whose main base was in rural Kelantan and Terengganu, also had some following amongst urban Malays who continued to resent Chinese economic influence and who wished to see Malaysia become more Islamic. The old Malay élite also resented Mahathir's drive for a more meritocratic society (at least amongst the Malays), and were shocked in 1983 by his proposals to restrict the power of the hereditary rulers by effectively removing the rulers' right to withhold assent from legislation. The Government and Sultans later reached a compromise on the proposals, but the conflict lingered on in a growing rift between Mahathir and his Minister of Trade and Industry, Tunku Razaleigh Hamzah, an aristocrat from Kelantan. Razaleigh narrowly failed in a challenge to Mahathir's party leadership in April 1987. Mahathir then removed Razaleigh supporters from most of the significant positions in both the Government and the party, but was given an unexpected opportunity to consolidate his dominance in February 1988, when the High Court ruled that, because of irregularities in the UMNO elections of 1987, the party was in breach of the Societies Act and was technically illegal. To regularize the party's position, Mahathir created a new party, UMNO Baru (New UMNO), later in February, which inherited the property of the 'old' UMNO. All members of the former party, however, were required to apply anew for registration. In this way, Razaleigh and other opponents were excluded without the difficulty of going through expulsion proceedings. A number of dissidents subsequently accepted Mahathir's dominance and joined UMNO Baru (henceforth referred to as UMNO), but in October 1989 Razaleigh announced the creation of a new movement called Semangat '46 (Spirit of 1946, the year of founding of the original UMNO).

Although the ideas and social groups represented by Razaleigh still had considerable support amongst Malays, the electoral prospects of the new party (which was registered in May 1990) were hampered by its lack of access to government patronage and by its need to seek allies amongst the forces outside the BN. Semangat '46 quickly developed an alliance with PAS (although PAS and the Razaleigh group had been bitter rivals in Kelantan state politics and Razaleigh himself had engineered the expulsion of PAS from the BN in 1978), under the name Angkatan Perpaduan Ummah (APU—Muslim Unity Movement). As the 1990 elections approached, however, Semangat '46 also developed an alliance, called the Gagasan Rakyat (People's Concept), with the overwhelmingly Chinese DAP. Although PAS and the DAP each shared interests and concerns with Semangat '46, they had virtually nothing in common with each other, and the coalition looked implausible as an alternative government. The elections of October 1990 thus returned the Mahathir Government. The BN won slightly fewer seats (127 of the 180 seats in the extended House of Representatives), but retained the two-thirds' majority required to amend the Constitution. The opposition did well in the DAP stronghold of Pinang and won overwhelmingly in Kelantan, but for the most part performed poorly elsewhere. Semangat '46 in fact won only six of the 61 seats it contested. This victory gave Mahathir a platform from which to pursue his campaign to reduce the authority of the Sultans. In July 1992 he oversaw the introduction of a code of conduct for the rulers, which restricted their ability to intervene in the political process. In January 1993 Parliament passed legislation removing the Sultans' personal legal immunity. When the rulers withheld assent from the bill, the Government orchestrated a campaign of public revelations of excesses and abuses of position on the part of various Sultans until the Sultans capitulated. The Government also rescinded the traditional privileged treatment that the rulers had enjoyed from government officials and which some of them had used to develop significant business interests. In May 1994 the House of Representatives approved a constitutional amendment definitively removing the rulers' power to block legislation by withholding assent. In September 1997 Mahathir continued his attack on state power by convening a conference to centralize the administration of Islamic law, previously a state matter. His target in this case, however, was not so much state politicians as state religious authorities whose conservative *fatwa*, he alleged, were impeding national development. At the UMNO assembly in September, he accused many Muslims of being more concerned with form (beards, clothing, etc.) than with the substance of Islamic teaching.

During the late 1980s the Mahathir Government came into increasing conflict with the judiciary after a number of cases in which government decisions were overturned, although the Malaysian courts have traditionally been rather conservative. In May 1987, just before one of the crucial court hearings over the legal status of UMNO, the Yang di-Pertuan Agong suspended the Lord President of the Supreme Court, Tun Salleh Abbas, from his post. A tribunal of judges then concluded he was guilty of 'misbehaviour' in the form of bias against the Government and he was dismissed in August. Five of the remaining Supreme Court judges were also suspended at this time, and two were dismissed, in connection with the legal proceedings surrounding the dismissal of Salleh Abbas. These measures were followed in March 1988 by constitutional amendments limiting the power of the judiciary to interpret laws. In 1994 the Government restyled the Supreme Court as the Federal Court and the Lord President as the Chief Justice, as well as introducing a mandatory code of ethics for judges.

The Government was also willing to use its extensive security powers against opposition groups. In October 1987 it had several politicians, journalists, lawyers and leaders of pressure groups imprisoned, including Lim Kit Siang, the Secretary-General of the DAP. (All were released by the following April.) Three newspapers were closed by government order and political rallies were banned. As the BN extended its dominance over the media, legislation was introduced imposing stringent penalties on editors and publishers if they published what the Government regarded as 'false' news, and empowering the Minister of Information to monitor all radio and television broadcasts, and to revoke the licence of any private broadcasting company not conforming with 'Malaysian values'. In June 1988 the Government introduced legislation removing the right of persons being detained under the Internal Security Act (ISA) to have recourse to the courts. At the end of December 1991 the High Court upheld a ruling by the Ministry of Home Affairs to ban the public sale of party newspapers. This was widely interpreted as an attempt to undermine further the potential effectiveness of the opposition parties, because it mainly affected opposition journals, including *The Rocket* (DAP), *Harakah* (PAS) and *Berita Rakyat* (Parti Rakyat Malaysia), all of which were henceforth allowed to be distributed only to party members.

The general election of 24–25 April 1995 approached, with few observers believing that the opposition Gagasan Rakyat had any chance of victory. Malaysia's rapid economic growth had led to a widespread feeling that the administration was performing well. The uneasy alliance between Semangat '46, the DAP and PAS, moreover, had been made still more uncomfortable by an announcement from the PAS state Government in Kelantan in 1992 that it intended to introduce the Islamic criminal code (*hudud*) in the state. Unlike the laws on Islamic religious practice that the state already applied to Muslims in Kelantan, this would have subjected non-Muslims to Islamic law. It also prescribed the death sentence for apostasy, although the implementation of the penalty required federal consent, which was not forthcoming. The legislation was adopted in 1993, despite claims that it contravened the Constitution's guarantee of freedom of religion. This and related issues sowed such discord in opposition ranks that the DAP withdrew from the Gagasan in January 1995. The opposition also had to cope with a lack of media coverage and an official campaign period limited by the Government to just nine days. In fact, the only issues that seemed likely to detract significantly from the Government's electoral support were the persistent allegations of corruption that emerged against its senior figures. In September 1994 the Chief Minister of Melaka and leader of the powerful UMNO youth wing, Tan Sri Datuk Rahim Thamby Chik, was forced to resign following allegations that he had had sexual relations with a minor, an offence constituting statutory rape. The charges were later abandoned. The DAP youth leader, Lim Guan Eng, was subsequently tried for sedition for remarks he had made suggesting that the authorities had not been even-handed in the case, and in April 1997 he was convicted and fined RM 15,000 (Lim was released in August 1999 following the failure of an appeal against his conviction—the judgment automatically disqualified him from holding a parliamentary seat).

The 1995 election results were a triumph for Mahathir. The BN's share of the vote rose to 64%, from 53% in 1990, and it won 162 of the 192 seats in the extended House of Representatives, as well as all the seats in the State Legislative Assemblies in Perlis and Johor, and a majority in every other state except Kelantan. The PAS-Semangat '46 coalition retained power in Kelantan and lost only two seats nationally, but the DAP lost 11 of its 20 seats in the House of Representatives and all but one of its 14 state seats in Pinang. The Parti Bersatu Sabah (PBS), led by Pairin Kitingan, won eight of Sabah's 20 federal seats. The election defeat led to a crisis of confidence in Semangat '46. The party's coalition with PAS in Kelantan began to break down and in October 1996 the party was formally dissolved, with its 200,000 members joining UMNO.

At the UMNO party congress in late 1993 there had been considerable speculation over likely successors to Mahathir, who was born in 1925. Attention focused then on the Minister of Finance and former UMNO youth leader, Dato' Seri Anwar Ibrahim, who was elected Deputy President of the party (normally a stepping stone to the leadership) and whose self-styled 'Vision Team' won all three vice-presidential positions. The 1995 election results briefly ended such speculation, but by the end of the year it had resurfaced as signs of jostling for power grew at all levels of the UMNO organization. Observers were far from certain that Mahathir supported Anwar as his successor, and in the aftermath of the election the Prime Minister appeared to shift the balance of power in the Cabinet slightly away from Anwar and his allies. In particular, Dato' Seri Najib Tun Razak was moved from the defence to the education portfolio, a significant post because the last three Malaysian Prime Ministers had all served in that department before acceding to power. Najib had been one of Anwar's 'Vision Team' vice-presidents, but was now seen as having shifted to the Mahathir camp. In November 1995 the annual UMNO assembly decided not to allow any challenge to Mahathir or Anwar at the triennial party congress scheduled for October 1996, thus postponing the battle for succession until 1999. The contest, however, continued by proxy over the several junior leadership positions that were at stake, although in July 1996 the party's Supreme Council banned open campaigning by or for any candidate. The return of Semangat '46 members to UMNO took place too late to affect the outcome of the congress, but most observers believed that their presence strengthened the position of Mahathir, whose relations with Razaleigh were surprisingly cordial. The vice-presidential elections saw the return to office of Najib and of Anwar loyalist Mohammad Taib, and the election of a Mahathir supporter, the Minister of Foreign Affairs, Abdullah Ahmad Badawi. On the other hand, two Mahathir allies who had been damaged by corruption accusations, Rafidah Aziz (head of the UMNO women's wing) and Rahim Thamby Chik, were defeated by Anwar supporters, Dr Siti Zaharah Sulaiman and Ahmad Zahid Hamidi. Both were thus victims, in part, of Mahathir's increasing outspokenness against what was commonly called 'money politics', which he described in October 1995 as the only factor that could destroy UMNO. Taib, a prominent Anwar supporter, fell victim to the same affliction in December 1996, when he was arrested at the airport in Brisbane, Australia, carrying the equivalent of more than RM 2.3m., thus contravening an Australian law requiring the declaration of amounts greater than about RM 10,000 being taken into or out of the country. Taib subsequently resigned as Chief Minister of Selangor, and in May 1997 he was forced to resign as Vice-President of UMNO; however, he was later acquitted of all charges. New Chief Ministers in Selangor and Melaka announced strict standards of probity for politicians and senior officials within their states, but appeared to find it difficult to distinguish clearly between legitimate business interests and corrupt behaviour. Throughout 1996–98 many lesser figures were charged or otherwise disciplined for alleged financial irregularities, but public opinion did not interpret these moves as a genuine attempt to eliminate corrupt practices. The weakening of Anwar's position that Taib's disgrace implied, however, was counterbalanced by a period of two months from mid-May 1997, during which Mahathir took leave to travel, write and promote Malaysia abroad, leaving Anwar as acting Prime Minister in his place.

Mahathir returned to office in time to face a growing political challenge arising from the Asian financial crisis. In mid-July 1997 the ringgit began to depreciate, placing sudden pressure on the many firms and investors who had borrowed in foreign currency, making foreign credits suddenly much more difficult to obtain and forcing down prices on the Malaysian stock exchange. The Malaysian crisis was precipitated by simple contagion from events in Thailand, but the markets soon focused on what were seen as structural problems in the Malaysian system, notably overexpenditure on prestige infrastructure projects and opacity in the economy, partly a result of corruption, partly a result of the formal policy of promoting the economic development of the Malay community. The ringgit's decline became catastrophic in October, when it lost 40% of its previous value within a month. Mahathir's public response, however, was to portray the crisis as a selfish, and perhaps malevolent, attack on Malaysia's economic achievements by a small group of foreign speculators and Western media, possibly in league with unfriendly governments. Remarks made by Mahathir, which were widely interpreted as showing that the Prime Minister did not understand the issues that concerned the financial markets, placed still further pressure on the ringgit.

Unlike Thailand, Indonesia and the Republic of Korea, Malaysia declined to accept IMF intervention, which would almost certainly have required the abandonment of most of the major infrastructure projects favoured by Mahathir and might have condemned the policy of preference for Malays. None the less, Anwar, as Minister of Finance, was able to persuade Mahathir to abandon several large projects. In early December 1997 all federal and state ministers took a 10% cut in salary and smaller reductions were imposed on parliamentarians and senior bureaucrats. Most observers agreed that Anwar's standing as Mahathir's likely successor had been strengthened by his measured response to the crisis, but he personally showed no keenness to displace Mahathir and thus to assume prime responsibility for the continuing difficulties.

The atmosphere of crisis in 1997 was exacerbated by a heavy pall of smoke (generally referred to as 'haze') from forest fires in Indonesia, which drifted over large areas of Peninsular Malaysia and Sarawak, reaching its worst in September. Air pollution levels were dangerous to public health for several weeks and hundreds of airline flights were cancelled, causing massive business losses, especially in the tourism industry. In November the Government banned the country's academics from making public statements about the smoke, on the grounds that commentators had been alarmist and had been 'manipulated' by foreign media.

In November 1997 Anwar moved a parliamentary motion of confidence in Mahathir in order to defuse the growing speculation about a possible change of leadership. Internal elections for divisional committee members in UMNO in March 1998 left most of Mahathir's supporters in place, leaving Anwar with no clear power base from which to launch a challenge. None the less, the resignation of the Indonesian President, Suharto, in May revived speculation that the economic crisis might also bring down Mahathir. In a reflection of the Indonesian resentment of the business interests of Suharto's children, there was increased public attention in Malaysia to the business activities of Mahathir's three sons. The Prime Minister's eldest son, Mirzan Mahathir, received government assistance to save his businesses from bankruptcy under circumstances that some Malaysians regarded as favouritism, although several other prominent business executives had also received assistance as part of a programme intended to keep the local corporate sector relatively intact until the economy improved.

Although Anwar was careful not to give any indication of a challenge to Mahathir, he spoke out publicly against corruption and political restriction in a way that clearly laid out a case for a change of leadership. Then, at the UMNO annual conference in June 1998, one of Anwar's supporters launched a strong attack on the party leadership over the same issues. Mahathir's supporters responded with a brochure entitled 'Fifty Reasons Why Anwar Cannot Become Prime Minister', in which Anwar was accused of conspiracy and sexual offences. Information on Anwar associates who appeared to have benefited from 'cronyism' was also released. During the following weeks, Mahathir systematically dismantled Anwar's power base, appointing Dato' Paduka Daim Zainuddin to oversee the economy, dismissing newspaper editors close to Anwar, and ordering the arrest of a close associate of Anwar on firearms charges carrying the death penalty under Malaysian law. Increasingly lurid accounts of Anwar's alleged sexual misdemeanours began to circulate, and Mahathir reportedly pressed his deputy to resign or to face dismissal and criminal charges. Anwar declined to resign and was dismissed on 2 September.

The rift between the two men appeared to be based partly on their very different responses to the economic crisis: Mahathir blamed foreign speculators and believed that Malaysia's economic salvation lay in protection and in continued economic expansion, whereas Anwar, as Minister of Finance, supported austerity measures close in spirit to IMF prescriptions. Anwar's dismissal was preceded by an announcement from Mahathir imposing tight currency and stock-market controls in direct opposition to global pressures, and Anwar's preference, for greater liberalization. Many observers believed that Mahathir wished to forestall the possibility that Anwar might succeed to the premiership and reverse such measures. Mahathir himself formally denied that policy divisions had prompted his action, and stated that Anwar's lifestyle made him unsuited to lead a conservative, religious country; further underlining this point, on 3 September 1998 the UMNO Supreme Council voted unanimously in favour of Anwar's expulsion from the party.

Anwar responded to his dismissal by denying the rumours of impropriety, alleging that there was a conspiracy against him and denouncing Mahathir as 'paranoid'. He forecast that a popular movement would bring Mahathir down, as had toppled President Suharto in Indonesia. His supporters held daily demonstrations, bearing portraits of Anwar and banners with the word *'reformasi'* (reform), which had been the rallying cry of the opposition to Suharto in April and May 1998. Although government officials were quick to accuse Anwar of hypocrisy in attacking the system of which, until recently, he had himself been a part, the authorities in turn had difficulty explaining why a minister so central to the Mahathir team should be suddenly so ferociously vilified. As well as arousing his own considerable following, Anwar's dismissal crystallized broader resentment of the autocratic methods of the Mahathir Government. As the demonstrations in support of Anwar gathered momentum, Anwar and 16 supporters were arrested by riot police on 20 September, under the ISA. Anwar's wife, Dr Wan Azizah Wan Ismail, was summoned for questioning by police. A restriction order was issued, banning her from holding rallies at her house, which had become a focus for the *reformasi* movement. Anwar appeared in court on 29 September and was charged with five counts of corruption and five of unnatural sexual acts. He appeared badly bruised and claimed to have been assaulted while in custody. Anwar's allegations provoked expressions of extreme concern from foreign governments—in particular from the Presidents of the Philippines and Indonesia—and prompted the UN Secretary-General, Kofi Annan, to urge the Malaysian Government to ensure humane treatment for Anwar. Although the Malaysian Government initially dismissed Anwar's claims of assault, it was subsequently announced that a special investigation was to be established. In December the Inspector-General of the Malaysian police force, Tan Sri Abdul Rahim Noor, resigned after an initial inquiry blamed the police for the injuries Anwar had sustained. Malaysia's Attorney-General publicly admitted in January 1999 that Anwar had been assaulted while in police custody, and an official inquiry completed in March found Tan Sri Rahim Noor to be personally responsible for the beating. He was later sentenced to two months' imprisonment and a fine of RM 2,000.

Meanwhile, popular demonstrations demanding the reform of the ISA, Mahathir's resignation and Anwar's release continued to take place, but were ignored by the local media. In October 1998, however, the influential Malaysian Bar unanimously adopted resolutions condemning detention without trial and demanding that independent inquiries be held into allegations of police brutality. It was subsequently announced that Anwar was no longer to be detained under the ISA, but would be transferred to a regular prison. (The last of the supporters of Anwar also detained under the ISA in September were released in November.) The trial of Anwar on four charges of corruption (which referred to efforts allegedly made by Anwar in 1997 to use his position to suppress an investigation of his alleged sexual misconduct) began on 2 November 1998. During the trial the credibility of the prosecution was undermined by a number of factors, and in January 1999 the charges against Anwar were unexpectedly amended, with the emphasis being shifted from sexual misconduct to abuse of power; the amendment meant that the prosecution would no longer have to prove that Anwar had committed sexual offences, but only that he had attempted to use his position to influence the police to quash the investigation into the allegations, effectively making it easier for a conviction to be obtained. The judge subsequently ruled that further testimony relating to the earlier allegations of sexual misconduct was irrelevant, leading the defence to claim that it had been denied the opportunity to refute the damaging and by now widely publicized accusations. (The ruling also meant that the defence would no longer be able to carry out its planned cross-examination of Mahathir and Daim Zainuddin on the issue of whether the allegations of sexual misconduct in particular were part of a political conspiracy against Anwar.) The trial, the longest-running criminal trial in Malaysia's history, was ended abruptly in late March, and the verdict was delivered in mid-April: Anwar was found guilty on each of the four charges of corruption and was sentenced to six years' imprisonment. (Under Malaysian law, a sentence of this length is automatically followed by a five-year period of disqualification from political office.)

Meanwhile, Mahathir sought to confirm his dominance by appointing a new Deputy Prime Minister, Abdullah Badawi, in a major reorganization of the Cabinet effected in January 1999. Also in the reshuffle, Mahathir relinquished the home affairs portfolio to Badawi and the finance portfolio to Daim Zainuddin, thus restoring a 'normal' distribution of portfolios in the Cabinet. Badawi's background as Minister of Foreign Affairs made him a useful foil to the continuing allegations of UMNO corruption, while his subdued political style meant that he was unlikely to become a political threat to Mahathir. Within UMNO, a wide-ranging purge removed many known Anwar supporters. With Mahathir preparing the ground for UMNO's campaign in the forthcoming general election (which was to be held by June 2000), Anwar's supporters countered by founding a new political party, the Parti Keadilan Nasional (PKN, National Justice Party), in early April under the leadership of Wan Azizah; the new party reportedly aimed to establish itself as a multi-ethnic and multi-religious party (although its initial membership appeared to be predominantly Muslim), and declared that, should it come to power, it would seek a royal pardon for Anwar. PAS membership reportedly grew by 25%, to 600,000, between September 1998 and mid-1999, while the PKN claimed to have 150,000 members. PAS also sought to broaden its appeal by opening a dialogue with non-Muslims, courting local businessmen and opening social-welfare centres, which provided legal and medical assistance to those Malaysians suffering the effects of the regional financial crisis. In June 1999 PAS, the PKN, the predominantly Chinese DAP and the Parti Rakyat Malaysia (PRM—Malaysian People's Party) formed a coalition, the Barisan Alternatif (BA—Alternative Front), to contest the elections, prompting Mahathir to accuse them of being supported by foreigners who wished to undermine the Malaysian economy. The BA was a disparate group united mainly by hostility to Mahathir, whom Wan Azizah described as a 'once respected leader who has lost all sense of perspective, all sense of right and wrong and all sense of reality'. The coalition subsequently selected Anwar as its prime ministerial candidate, but quarrelled over electoral strategy and experienced difficulty in its attempts to avoid misgivings over the differences in the long-term aims of the individual parties within the coalition.

The Government responded to this wave of dissent by accusing its opponents of deliberately seeking chaos and by issuing numerous civil lawsuits in which ruling politicians sued journalists and opposition figures for defamation. The Attorney-General warned that those who alleged selective prosecution would themselves be prosecuted for contempt of court.

In late April 1999 Anwar was further charged with one count of illegal homosexual activity, to which he pleaded not guilty; it was announced that four other similar charges and one additional corruption charge against him had been 'suspended'. The second trial of Anwar began on 7 June. On the first day of the trial the prosecution amended the wording of its charge, changing the month and year in which the alleged crimes were supposedly committed. Meanwhile, at the UMNO General Assembly in mid-May, Mahathir made it clear that there was no prospect of any reconciliation with Anwar. Anwar responded from prison with accusations that Mahathir and his colleagues were guilty of corruption and nepotism and claimed that he had been removed from the Government because of his commitment to ending such practices. (In October the Government responded with accusations that Anwar himself had accumulated RM 780m. in illicit funds.) In September the trial was adjourned and Anwar was sent for medical examination, following claims by the defence that Anwar had proven high levels of arsenic in his blood and was quite possibly the victim of deliberate poisoning. In the same month Anwar lost a defamation suit that he had filed against Mahathir following his arrest. In October doctors concluded that Anwar 'showed no clinical signs of acute or chronic arsenic poisoning' and his trial resumed. On 29 April the Court of Appeal upheld Anwar's original conviction on charges of abuse of power, and the former Deputy Prime Minister and Minister of Finance was finally convicted of sodomy in August, receiving a sentence of nine years' imprisonment to commence after the expiry of his earlier sentence, meaning that he could remain imprisoned until 2014. In May 2001 the prosecution abandoned the five remaining charges against Anwar. In July 2002 Anwar's appeal against his corruption conviction was finally rejected by the High Court; his appeal against the sodomy conviction suffered the same fate in April 2003, and he was also refused bail to travel to Germany for medical treatment.

In November 1999 the Government unexpectedly announced that a general election was to be held later the same month. The opposition expressed dissatisfaction at the limited period of time allowed for the election campaign. During this brief official campaign period, the opposition focused on what it claimed was corruption and arrogance on the part of the Government, while the BN attacked the character of Anwar and emphasized the incompatibility of the four opposition parties. Constituencies were threatened with the loss of federal government funds if they failed to return BN candidates at the election, and Mahathir warned of the prospect of an outbreak of violence if the Government were not returned to office.

The national elections held on 29 November 1999 returned the BN to government, the coalition winning 148 of the 193 seats in the House of Representatives. Although the BN retained the two-thirds' majority required to allow the Government to amend the Constitution, the coalition's share of the national vote declined from 64.1% (at the previous general election) to 56.5%, while UMNO's representation in the legislature decreased to 72 members. There was a substantial vote against UMNO in the states of Kedah and Pahang, and even Mahathir's winning margin in his own constituency fell from 17,000 to 10,000 votes. PAS, by contrast, increased its number of seats from seven to 27 and the PAS leader, Fadzil Nor, was officially appointed parliamentary leader of the opposition. The party also retained power in the state of Kelantan and took office in Terengganu. However, PAS's partners in the BA performed poorly. The DAP won 10 seats, but its Secretary-General, Lim Kit Siang, lost his seat, as did the party's Deputy Chairman, Karpal Singh (who was also Anwar's legal representative). Wan Azizah retained Anwar's former seat for the PKN, but the party won only four other seats, all of them in regions that otherwise would probably have returned a PAS candidate. The distribution of votes suggested that 70% of Chinese voters and 90% of Indians had voted for the BN, whereas only about 50% of Malays had done so. For the first time in a federal election, there were widespread allegations of malpractice.

Mahathir made few changes in his Cabinet after the election, although former Minister of Education, Najib Tun Razak, who only narrowly retained his seat in the House of Representatives, was demoted from the education portfolio to that of defence. There was much interest, however, in the policies of the new PAS state legislature in Terengganu. The new Chief Minister, Abdul Hadi Awang, had pledged to ban gambling, restrict alcohol sales and make Friday a public holiday, and had also suggested the introduction of both strict Islamic criminal law and a religious income tax, called *kharaj*, on non-Muslims. Once in office, however, Awang banned *karaoke* and gambling but postponed more radical proposals and concentrated on winning business confidence and establishing for himself a reputation for efficiency and honesty. Expensive prestige projects were cancelled, and a new openness was introduced into the allocation of government tenders and licences. In early September 2000 the federal Government, alarmed by Terengganu's successes, announced that it would no longer pay Terengganu a royalty percentage on petroleum production in the state, but would instead channel the revenue into a federally controlled development fund. (Royalty payments amounted to about RM 1,000m. per year and accounted for 90% of Terengganu state revenues.)

In January 2000 the authorities arrested five prominent opposition leaders, including Karpal Singh and the editor and publisher of the popular PAS newspaper, *Harakah*, most of them on charges of sedition. *Harakah* itself was threatened with closure, as a result of the newspaper's illegal sale to the public. In March the Ministry of Home Affairs announced that it would permit *Harakah* to be published only twice a month, rather than twice a week, but later emphasized Mahathir's insistence that Malaysian internet websites, including opposition news websites, such as that of *Harakah* itself, would not be censored. Other opposition figures also faced charges, which had apparently been brought with the aim of removing them from political life. In April and May 2001 16 opposition figures, most of them members of the PKN, were arrested under the ISA. Both courts and the Malaysian Human Rights Commission criticized the Government for using legislation that was intended to combat communist insurrection as a means of suppressing its political opponents. With increasing restrictions on the print media, an internet newspaper, *Malaysiakini*, established in mid-2000, became an increasingly important alternative source of information for the Malaysian public.

Contrary to many expectations, the BA did not disintegrate after the election, but continued to function as a coalition in opposition, creating a form of shadow cabinet in March 2000, maintaining pressure on the BN in parliamentary debates and continuing to minimize the fundamental policy differences between its members on the issue of the place of Islam in Malaysian politics and society. As PAS broadened its appeal to Muslim professionals and younger, urban voters, the party's Islamic character became steadily less intimidating to non-Malay groups. The PAS Government in Terengganu further sought the favour of non-Muslims by allowing a Chinese cultural festival to take place, and issuing permits for non-Muslim places of worship, which had been continually refused by the BN Government. The PKN, by contrast, grew increasingly uncertain of itself, divided over the extent to which it should continue to focus on the treatment of Anwar and over whether to promote public demonstrations against the Government, even though these actions appeared to be attracting fewer supporters. In June, however, the party achieved increased support in a by-election, although it still failed to win the seat, and was further boosted when the PRM agreed in principle to merge with it. In November 2000 the BA defeated the BN in a by-election for the Lunas seat in Kedah, which had been won by the BN one year earlier. The victory was significant, since Kedah was Mahathir's home state and since the result deprived the BN of its symbolic former two-thirds' majority in the state legislature. The UMNO leadership was also concerned by resistance within the party to a proposal to extend the term of UMNO officials from three to five years. In addition, the MCA faced corruption scandals. UMNO's response to this pressure was to make efforts to improve its Islamic credentials. In March 2000 the BN state Government of Perlis introduced legislation prescribing detention at a Faith Rehabilitation Centre for those found guilty of 'deviationism' and apostasy, while Johor activated 1997 legislation, whereby lesbianism, pre-marital sex, prostitution, pimping and incest would be punishable by whipping and terms of imprisonment. The federal Government envisaged legislation to make misuse of religion a criminal offence. In May 2000 the Government introduced a requirement for all Muslim civil servants to attend classes in religion twice weekly. The Government also continued to emphasize long-term reforms, such as the introduction of Islamic banking and insurance. A proposal, presented in late 1999, for the inclusion of a statement of religious affiliation on identity cards (which must be carried by all Malaysians), was, however, suspended. The measure was widely criticized as likely to exacerbate religious differences and to hinder the emergence of the shared Malaysian identity, which formed an integral part of Mahathir's 'Vision 2020'. In September 2000 the Government charged 29 members of the al-Ma'unah cult with treason, following a robbery of weapons in Perak in July, in which two hostages were killed. Government attempts to attribute the incident to PAS were received with considerable scepticism. In December 2001 19 of those charged were found guilty of treason; three were sentenced to death and the remainder to life imprisonment. A total of 10 others pleaded guilty to lesser charges and were given 10-year prison terms.

Controversy continued over the random destruction of Chinese cultural heritage in Malaysia. During 2000 the federal Government began to develop a new 10-year plan, provisionally known as the Vision Development Policy, which was to replace the NDP. Although the new plan maintained affirmative action policies for Malays, speculation that the extent of these policies would be reduced prompted Malay demonstrations against Chinese organizations, and fierce condemnation of alleged Chinese extremism. Chinese and Indian communities, in turn, criticized government plans for multi-ethnic so-called 'Vision Schools', which they viewed as a means of ending the separate Chinese and Indian educational systems.

As the result of a decision of the UMNO Supreme Council, the post of President, held by Mahathir, was not contested at the UMNO General Assembly on 11–13 May 2000, and Badawi stood as the sole candidate for the post of Deputy President, vacant since Anwar's dismissal. There was strong competition, however, for three positions of Vice-President. The Minister of Defence, Najib Tun Razak, an incumbent Vice-President, was re-elected, along with a former Chief Minister of Selangor, Mohammad Taib, and the Minister of Domestic Trade and Consumer Affairs, Muhyiddin Mohd Yassin. Although all three had been members of Anwar's 'Vision Team' in 1993, they were nominated by Mahathir's own Kubang Pasu faction in UMNO. Hints from the Prime Minister that he was not happy with the outcome of the ballot seemed to be ingenuous, as none of the defeated candidates, including three state Chief Ministers, had Mahathir's favour. At the conclusion of the Assembly, Mahathir announced that he had not decided when to retire but that he intended to relinquish greater responsibility to Badawi in preparation for the eventual transfer of office.

During 2001 Mahathir continued his public attacks on corruption within UMNO, urging party officials to declare their assets and suggesting that the very rich should not hold principal posts. Six division-level UMNO leaders were dismissed for having used bribery to win party elections in April, and the Minister of Information and UMNO Secretary-General, Khalil Yacoob, was charged with misuse of public funds. Many observers, however, considered the Minister of Finance, Daim Zainuddin, who was one of Malaysia's most wealthy men, to be Mahathir's main target. Daim took two months' leave of absence from his post from 19 April, and formally resigned in June. Mahathir assumed the finance portfolio, in addition to carrying out his duties as Prime Minister. Daim had been widely criticized for using public funds to assist Malay business associates, but was also known to have policy differences with Mahathir. Daim's former significant role in fund-raising for UMNO cast further uncertainty on the issue of whether Badawi would be able to defeat leadership challenges from Razaleigh and Najib Tun Razak, in the event of Mahathir's departure from politics.

During the second half of 2000 PAS became less concerned to downplay its intention to introduce an Islamic state, while the BN seemed to disengage from its campaign to match PAS in the promotion of Islam. In September 2000 Datuk Haji Nik Abdul Aziz Nik Mat, spiritual leader of PAS, declared, *inter alia*, that the revealing clothes worn by some women encouraged rape; in October he urged women to be banned from competitions in the reading of the Koran because their voices might seduce men. In July 2001 a senior Islamic leader, Hashim Yahya, ruled that a man might divorce his wife by sending her a text message from a mobile cellular telephone. Women's groups called on the Government to reject the finding and to insist that divorce cases continue to be heard in the courts. Fadzil Nor assured PAS's partners in the BA that his party would not seek to impose an Islamic state without fully discussing the topic with other groups; he was, however, unable to dissuade the DAP from leaving the BA over the issue in September of that year. The opposition was further weakened when the PBS, which had withdrawn from the BN in 1990, applied in late 2001 to rejoin the Government. In contrast, the continuation of serious infighting within the MCA, which had publicly divided into factions labelled 'Team A' and 'Team B', did not seem to affect the BN. The BA's Islamic character was reinforced when the uncompromising Chief Minister of Terengganu, Abdul Hadi Awang, became leader of PAS after Fadzil Nor's death in June 2002. Abdul Hadi Awang's pronounced hostility towards the US-led campaign to oust the regime of Saddam Hussain in Iraq in early 2003 garnered further public support for his party.

The Government responded to the developments of October 2000 by suspending proposed legislation that would have authorized punishment for Muslims who abandoned their faith. Addressing a conference on Islamic law, Mahathir condemned Islamic intolerance and urged flexibility and tolerance in the observance of Islam. The Government tightened its control over Islamic groups in August 2001, detaining Nik Adli Nik Abdul Aziz, son of Nik Abdul Aziz Nik Mat, and 16 others on charges of planning to overthrow the Government. They were ordered to be detained for two years without trial under the ISA. In the same month the Government banned public political meetings and passed legislation formally outlawing discrimination against women and making it possible for women to sue in cases where they had suffered discrimination. In September measures were also announced against religious leaders who raised political themes in sermons and, later in the same month, Mahathir declared that Malaysia was in fact already an Islamic state, apparently implying that further Islamization, as proposed by PAS, was unnecessary (although this declaration caused consternation in non-Muslim circles). Meanwhile, under a new Chief Justice, Tan Sri Dato' Mohamed Dziaddin bin Haji Abdullah, the High Court had significantly recovered its reputation for independence, and had opposed the Government in prominent cases in which contempt proceedings and defamation charges had been used against opposition figures.

The terrorist attacks of 11 September 2001 on the USA gave the Government a further opportunity to suppress Islamist extremism. In October six people were arrested under the ISA on suspicion of membership of the Kumpulan Mujahidin Malaysia (KMM), which was suspected of having connections with the al-Qa'ida international terrorist network. During 2002 a further 65 people were arrested, and in late 2002 the Government announced plans to tighten the ISA still further by removing all recourse to the courts for those arrested under its provisions. The Government implied openly that the KMM and the related Jemaah Islamiah terrorist group were connected with PAS. Nik Abdul Aziz Nik Mat encouraged these perceptions by demanding a *jihad* ('holy war') against the USA and urging Malaysians to support the Taliban regime of Afghanistan. During campaigning for a by-election in January 2002, propaganda produced by the Government explicitly linked PAS with the Taliban, drawing particular attention to the subordinate position of women under Taliban rule. In February 2001 UMNO had established Puteri UMNO, an organization aimed at women under the age of 35. This group was generally perceived to hold more modern views and to be opposed to radical Islamic practices. In late 2002 the head of Puteri UMNO, Azalina Othman Said, was publicly accused of lesbianism, but UMNO was unwilling to allow a repeat of the Anwar scandal and the case was dismissed. Increasingly, the BN Government was concerned over what it perceived to be the propagation of radical Islamic ideas in private schools for Muslim children, which comprise 80% of all schools in Kelantan. In October 2002 it announced that funding for these schools would be reduced and that the Government would introduce 'after-hours' religious education in government-controlled schools, with the aim of promoting a less radical view of Islam. This proposal, however, met with misgivings from non-Muslim communities, who feared that Islam would be entrenched further as the national religion.

In July 2002 the PAS-controlled Government of Terengganu followed Kelantan in formally introducing state Islamic law, which included penalties such as that of stoning for adultery and amputation for theft. As in Kelantan, the law could not be implemented because criminal law remained under federal control and, on this occasion, the police force announced publicly that it would not participate in the enforcement of the new laws. None the less, the legislation was widely seen to constitute a statement of PAS's intentions if it secured victory in the national elections due to be held in 2004. Women's groups were incensed that the new laws required the testimony of four male witnesses to prove that a rape had occurred. The law also prescribed the death penalty for apostasy and whipping for the consumption of alcohol.

On 22 June 2002 Mahathir threw Malaysian politics into confusion by abruptly announcing his resignation during a speech to the annual congress of UMNO. Within an hour, however, party officials announced that the Prime Minister had been persuaded to withdraw his resignation. Although opposition figures immediately described the incident as a pre-election ploy to rally support for Mahathir, others believed that the Prime Minister was increasingly keen to relinquish power after 21 years in office, and that the speech was intended to prepare the ground for an orderly transfer of power to his deputy and heir apparent, Minister of Home Affairs Abdullah bin Ahmad Badawi. At the congress Badawi had already sought to strengthen his position by adopting as a theme the cleansing of UMNO of corruption, which was widely regarded as a major reason for the party's electoral set-backs. During the approach to the congress, six senior party officials had been dismissed on corruption charges. Badawi's credentials as an Islamic scholar were widely seen to be an advantage in UMNO's political contest with PAS. A transfer of power within the faction-ridden MCA in June 2003, from the rival leaders of 'Team A' and 'Team B' to younger members, was generally perceived to constitute the first step in a wider process of generational renewal in Malaysia. In September 2002 Mahathir announced that he would not contest the general election due in 2004 and later scheduled his retirement for October 2003. Although there was speculation that Abdullah bin Ahmad Badawi might be challenged for the leadership by Najib Tun Razak or others, the UMNO conference in June 2003 passed without any indication of unrest. Badawi, however, pointedly refused to designate his preferred deputy.

In his final months in office, Mahathir demonstrated his concern for Malaysia's future by introducing a National Service Training Act in June 2003. Under the Act, 100,000 male and female school leavers (about one-fifth of all school leavers) were to be chosen at random for a three-month programme, which would include some military instruction, along with training in patriotism. The aim of the programme was to strengthen discipline and national commitment among young Malaysians.

The credibility of Malaysia's security forces was undermined in July 2003, when police identified an unregistered paramilitary movement, Pasukan Khas Persekutuan Malaysia (PKPM—Federal Special Forces of Malaysia), headed by a self-styled lieutenant-general, Nor Azami Ahmad Ghazali, who had formerly been a technician in the Draining and Irrigation Department. The PKPM allegedly planned to overthrow the Government and claimed 8,000 members, including civil servants and former army personnel. Its funds, however, appeared to come from the sale to its members of ranks and uniforms resembling those of the federal police. PKPM members held 'authority cards', supposedly issued by the Prime Minister's Office, authorizing them to arrest political leaders and ministers. The organization also raided shops selling pirated video compact discs. The existence of the PKPM highlighted dissatisfaction with the performance of the police itself, which was increasingly believed to be involved in corruption and abuse of power.

When Mahathir promised in June 2002 to step down from his post in the following year, he designated his deputy, Abdullah Badawi, as his successor. Owing to Mahathir's dominant position, however, and the fate of his previous deputies and presumed successors, especially Anwar Ibrahim and Datuk Musa Hitam (who had resigned as Deputy Prime Minister in February 1986, owing to 'irreconcilable differences' with Mahathir), an air of uncertainty surrounded the Badawi succession until the last moment. There was also widespread speculation that Badawi, reputed to have a mild character, might be little more than a tool for continued dominance by Mahathir from behind the scenes. In May 2003 Badawi, as Minister of Home Affairs, ordered the release of three senior PKN officials who had been detained under the ISA, on the grounds that the PKN no longer presented a security threat, but it was unlikely that Mahathir opposed the move. In August the PKN merged with the smaller PRM, forming the Parti Keadilan Rakyat Malaysia (PKR, People's Justice Party).

BADAWI'S ACCESSION TO POWER

Despite the speculation, Mahathir resigned as Prime Minister on 31 October 2003, and Badawi was sworn in on the following day. Mahathir's farewell speech included a condemnation of factionalism and 'money politics' and a call for the emergence of 'New Malays' who would be better able to compete with other communities. Preparations immediately began for the 2004 general election, which would be Badawi's first test of public approval. The number of parliamentary seats to be filled was increased from 193 to 219, and the deposit required from candidates was also raised. (This measure was intended to discourage frivolous candidatures, because the deposit is forfeit if a candidate does not achieve a certain percentage of the vote. It works, however, against poor parties, however popular, because they may be unable to afford many deposits.) As UMNO local branches were based on electorates, the increase in seats provided the UMNO leadership with an opportunity to reshape local branches and to tighten central control.

Three days before Mahathir's resignation, the PAS Government in Terengganu introduced new state regulations conforming to Islamic law, including the penalty of amputation for theft and stoning for adultery, although with criminal law the responsibility of the federal Government, it seemed impossible that the new laws would be implemented. The PAS action, however, enabled Badawi to signal his intentions, and the federal Attorney-General immediately launched a court case to have the law overturned. Badawi was less effective in making his own mark on politics in other ways. He delayed appointing a Deputy Prime Minister until 7 January 2004, but then chose the Minister of Defence, Najib Tun Razak, who had been publicly favoured for the post by Mahathir. Badawi's first Cabinet, too, largely retained Mahathir-era figures. A Royal Commission into the police force was established in December 2003, but its terms of reference excluded corruption and the abuse of power. On the other hand, Badawi cancelled a number of Mahathir's so-called 'mega-projects', and spoke many times in public about the need to develop Malaysia's democratic culture and to respect the independence of the judiciary. The new Prime Minister also promised to expand the use of parliamentary committees to examine draft legislation, and launched a high-profile campaign against corruption. In February 2004 police arrested a number of senior and middle-ranking figures, including the Minister of Land and Co-operative Development, Kasitah Gaddam, on corruption charges, sending a surge of apprehension though the Mahathir-era élite.

Badawi's new approach produced a dramatic improvement in the BN's performance at national elections held on 21 March 2004, delivering the ruling alliance 198 of the 219 seats in the House of Representatives. The elections proceeded peacefully, although fears of violence had led the authorities to reduce the campaigning period to a mere seven days. There was some evidence of irregularities and vote-rigging, but overall the result seemed to reflect widespread approval for the new Prime Minister. PAS performed poorly, its parliamentary representation declining from the 27 seats won in the 1999 general election to seven. The Islamic party also lost control of the state Government in Terengganu and retained Kelantan only by a narrow margin. Middle-class voters who had previously voted against the BN in protest at the treatment of Anwar were believed to account for much of the trend towards the party. Badawi's promises to reduce poverty and combat corruption were also popular, whereas the implementation of Islamic law by PAS state governments had proved to be irksome to many urbanized Muslims. Badawi, coming from a family of Muslim scholars and himself with a degree in Islamic studies, had put much effort into arguing that Islam was not a religion of social restriction, but rather a religion of development, which encouraged hard work and an appetite for science, technology and knowledge. The PKR lost four of the five seats previously held by the PKN, with only Anwar Ibrahim's wife, Wan Azizah, retaining her seat. By contrast, the DAP staged a modest recovery, increasing its number of seats from 10 to 12.

Badawi's progress in addressing corruption and undemocratic practices following the election seemed relatively modest. His Cabinet contained a number of figures linked with corruption allegations in the past, and he urged, rather than required, the systematic disclosure of assets by ministers. He prepared for the UMNO General Assembly in September by urging that there be no contest for senior posts. A National Integrity Plan, launched in April 2004, emphasized the development of civic and environmental consciousness, but did not seem to be directed at the causes of corruption and abuse of power.

On 2 September 2004, however, Badawi's promises of a more independent judiciary were vindicated, when a federal court overturned Anwar Ibrahim's conviction for sodomy and, since he had already served his sentence for corruption, ordered his release from prison. Anwar appeared in a neck brace, made necessary by the beating he had received from the police chief at the time of his original detention, and flew to Germany for medical treatment. Under Malaysian law, he would remain ineligible for election for a further five years because of his corruption conviction. None the less, Anwar indicated that he planned to resume political activity.

INTERNATIONAL RELATIONS

Malaysia effected a major realignment of its external alliances after it achieved independence in 1957. Whereas the country formerly enjoyed a strong relationship with the Western powers, notably the United Kingdom, it began, in the years following independence, to show a greater commitment towards its neighbours in South-East Asia and to develop closer ties with the emerging powers of eastern Asia.

In the early and middle years of the 1960s the newly formed Federation faced major challenges from Indonesia and the Philippines: in surmounting those challenges, Malaysia was assisted by Commonwealth military forces, composed mainly of British troops. Malaysia's pro-Western orientation was illustrated by its support of US involvement in Viet Nam, and by the absence of formal diplomatic relations with any communist countries, including the USSR and the People's Republic of China. The shift in Malaysia's foreign policy occurred in 1970, following the British Government's decision to withdraw British military forces 'east of Suez', including those in the Malaysia-Singapore area, and the reduction, and subsequent complete withdrawal, of US land forces from Viet Nam. The Western withdrawal was accompanied in the early 1970s by the emergence of China and Japan as major political and economic influences in South-East Asia. Diplomatic relations between Malaysia and the People's Republic of China were established in 1974, when the latter undertook not to interfere in Malaysia's internal affairs.

The most important development in this period was the formation of a strong regional grouping in South-East Asia in

1967, the Association of South East Asian Nations (ASEAN), the founder-members of which were Malaysia, Indonesia, the Philippines, Thailand and Singapore; Brunei joined in January 1984. ASEAN was primarily concerned with economic development, although co-operation in foreign-policy issues increased rapidly. The organization represented its members' political solidarity in opposing the Vietnamese occupation of Cambodia (then Kampuchea), and played a significant role in negotiations prior to the Vietnamese withdrawal from Cambodia, effected in 1989. Following the admission of Viet Nam to ASEAN in 1995, there was a general expectation that Laos, Cambodia and Myanmar (formerly Burma) would follow in due course. Initially, Malaysia preferred to make admission dependent on stable and reasonably democratic conditions in the candidate countries, but by mid-1996 it had come to favour rapid and unconditional admission, partly because Malaysian firms had become major investors in Myanmar. This policy led to verbal clashes with the USA and other Western countries, which wished to see Myanmar excluded until its observance of human rights improved. In the event, Myanmar and Laos were admitted to ASEAN in July 1997. Malaysia then pressed to have Myanmar accepted as a participant at the 1998 Asia-Europe Meeting (ASEM) in London, threatening to boycott the meeting if Myanmar were excluded, as the British Government announced would be the case. By the time the summit took place, however, the Asian financial crisis had intervened and the issue of Myanmar's participation was postponed. Cambodia joined ASEAN in 1999. In July 2002 Malaysia offered to host a permanent secretariat for 'ASEAN + 3', the informal grouping linking ASEAN with China, Japan and the Republic of Korea. At a meeting of ASEAN foreign ministers in Brunei, however, the proposal was rejected on the grounds that it might weaken ASEAN as an institution.

In January 1992, at the fourth ASEAN Summit, the member-states agreed to establish an ASEAN Free Trade Area (AFTA) within 15 years. (AFTA was formally established on 1 January 2002.) In July 1994 the first ASEAN Regional Forum (ARF) was convened in Bangkok, Thailand; the ARF was expected to provide a platform for the discussion of political and security issues. Malaysia's and Mahathir's role within ASEAN and within groupings of developing nations over trade and security issues continued to expand. Mahathir's articulation of his proposal to establish an East Asia Economic Caucus (EAEC), a trade group that was to exclude the USA, Australia and New Zealand, angered both the US Government (which was concerned to continue to promote the US-inspired Asia-Pacific Economic Co-operation—APEC) and other nations. In July 1993, at the ASEAN ministerial meeting, a consensus was reached on the EAEC. It was agreed, despite the continuing reluctance of Japan to participate, that the EAEC should operate as an East Asian interest group within APEC. The EAEC formally came into existence at an ASEAN informal summit in November 2000. Its economic significance was limited by the aftermath of the Asian economic crisis, but it provided an opportunity for Malaysia's proposal for the establishment of a trans-Asian railway, which would promote economic integration in the region.

In the early 1990s international attention focused on the issues surrounding the logging of the rain forest and its effect on the lives and land of indigenous communities. Prior to the UN Conference on Environment and Development in Rio de Janeiro, Brazil, in June 1992, this issue received much attention both within Malaysia and internationally. Mahathir became the effective international spokesperson for the developing countries against those of the industrialized North. In April 1992 he chaired a meeting on the environment attended by the developing nations, in Kuala Lumpur, in which he led demands for a withdrawal of international criticism on such issues as logging in developing nations, and appealed to the countries of the North to stop their 'imperialist agendas' and to adjust their own consumption and production patterns to avoid environmental pollution. This message, in the form of the Kuala Lumpur Declaration, which also included a rejection of the linking of development aid with human rights improvements, was taken to the Rio Conference. Prior to the 1995 election, Mahathir also offered sharp criticism of what he termed the continuing racism, hedonism and immorality of Western countries. Malaysia, however, also hosted an annual 'International Dialogue' between heads of state of developing countries and senior international business executives, which acted as a forum for seeking practical solutions to the problems of globalization. During 2000 Mahathir argued that the next meeting of the World Trade Organization (WTO) should be postponed until problems arising from existing agreements had been solved. He strongly attacked the West over what he called its abandonment of the Muslims in Bosnia and Herzegovina and sought, without conspicuous success, to co-ordinate a more vigorous response by the international Muslim community to the crisis. Although Mahathir formally wrote to the Indonesian President, Suharto, to protest against the smoke that disrupted commerce and daily life in much of Malaysia in September 1997, Malaysia's public criticism of Indonesia was muted, perhaps because its own environmental record was vulnerable to criticism.

From the mid-1990s relations with Singapore underwent some strain, in part owing to several incidents in which Singaporean officials were openly critical of Malaysia. For example, in mid-1996 Singapore's former Prime Minister, Lee Kuan Yew, stated that a reunification of Singapore and Malaysia might be possible and desirable if Malaysia abandoned its policy of preference for *bumiputra*. The first volume of Lee's memoirs, published in September 1998, also contained criticisms of Malaysian leaders of the 1950s and 1960s, provoking a new deterioration in relations. Relations were further strained in 1998 when Singapore unilaterally transferred its immigration officials from Tanjong Pagar railway station in Singapore to a modern facility near the Malaysian border. Other bilateral problems included Malaysia's failure to provide a formal agreement on water supply to Singapore and Singapore's refusal to allow Peninsular Malaysians to withdraw mandatory savings until the age of 55 even if they left Singapore, a rule that did not apply to other foreign workers. Most of these issues were resolved in a treaty signed on 4 September 2001. However, during 2002 the two sides clashed again over allegations that land reclamation work being carried out by Singapore was impeding access to Malaysian ports, and over rival plans for replacing the causeway linking the two countries with one or more bridges.

In May 1997 the Malaysian and Indonesian Governments agreed to refer to the International Court of Justice (ICJ) their conflicting claims to the sovereignty of two small islands off the coast of Borneo, Sipadan and Ligitan. The ICJ's judgment, handed down in December 2002, awarded the islands to Malaysia on the basis of occupation and use before the dispute had arisen. Another territorial claim being pursued through the ICJ was the dispute with Singapore over the island of Batu Putih (Pedra Branca). The two countries had originally agreed to refer the dispute to the ICJ in September 1994, but the formal agreement was signed only in February 2003; the ICJ hearing was expected in 2006. In March 1998 Malaysia began to construct a building on Investigator Shoal, a part of the disputed Spratly Islands to which Malaysia, the Philippines and China maintained overlapping claims; Malaysia's move was probably prompted by China's construction of substantial facilities on Mischief Reef further to the north. The discovery of new oil reserves off the northern Borneo coast in 2002 gave immediacy to an unresolved maritime boundary dispute with Brunei.

During the early 1990s, as economic growth fuelled the demand for labour, Malaysia became an increasingly attractive destination for illegal immigrants from nearby countries, especially the Philippines, Indonesia and Bangladesh. About 1m. illegal immigrants were believed to be living in Malaysia in the mid-1990s. By 1995 both the number of illegal immigrants and their treatment (by employers and, if they were apprehended, by the Government) had become an issue of increasing public concern. In June 1996 the Malaysian and Indonesian authorities agreed on joint measures to limit the flow of illegal workers into Malaysia. Malaysia repatriated the remaining Vietnamese refugees from its territory by the UN-imposed deadline of June 1996; some had been held in camps since 1975. The financial crisis led Malaysia to prepare plans for the large-scale repatriation of both illegal and now unemployed legal immigrant workers. By March 1998 17,000 Indonesians had been arrested as part of 'Operation Nyah (Go Away)' and naval patrols of the coastline were increased.

The restriction of illegal immigration to Malaysia became more stringent during 2001, and an estimated 120,000 illegal immigrants were expelled from the country. In early 2002 the Government ended the legal recruitment of labourers from Indonesia and began a campaign of bulldozing squatter settlements; sea patrols to interdict people smugglers also increased in number. The Malaysian authorities routinely expelled illegal immigrants from Myanmar across its northern border into Thailand, despite compelling evidence that some were being captured by criminal gangs and held for ransom, sold into slavery or smuggled back into Malaysia. In April the Government announced that it would take yet firmer action against illegal immigrants, who were estimated to number between 300,000 and 1m. The move was prompted by several factors. Malaysia's economic difficulties (see Economy) had led to increased rates of unemployment and public pressure for a 'Malaysians First' policy. Members of the public also perceived the illegal workers as being responsible for rising crime levels; the perception of the immigrants as being unruly was reinforced when clashes occurred between police and Indonesian illegal immigrants in a detention centre in December 2001, and in a textile factory in January 2002. Moreover, especially after the 11 September terrorist attacks on the USA, the Government feared that the immigrants might include political and religious militants. Under new legislation which entered into force from 1 August 2002, illegal immigrants faced punishments of fines, whipping or prison terms, although an amnesty was granted to those with confirmed tickets to leave the country by the end of the month. For the first time the employers of illegal immigrants also faced strict penalties. Many of the immigrants came from Indonesia and the Philippines, and both countries sent ships to assist in the repatriation of their citizens. Bangladeshis and Muslim Rohingyas from Myanmar were also affected. There was much controversy over the conditions in which illegal immigrants were held in Malaysian detention centres, and both Indonesia and the Philippines formally complained to Malaysia concerning the treatment of their nationals. Government officials estimated that about 300,000 people, mainly Indonesians, had left the country prior to 1 August. On the eve of the deadline imposed for departure, hundreds of illegal immigrants, mainly Acehnese from Indonesia and Rohingyas, sought certification of refugee status from the office of the United Nations High Commissioner for Refugees (UNHCR) in Kuala Lumpur, claiming that they would suffer discrimination if they returned to their own countries. Malaysia, however, was not a signatory of the 1951 UN Convention on Refugees, and seemed unlikely to accept any immigrants on these grounds. An estimated 1.7m. foreign workers remained legally in Malaysia, but by 2003 the departure of illegal workers had begun to cause labour shortages in some sectors, leading Malaysia to negotiate agreements with Indonesia and the Philippines to allow illegals to continue working under some circumstances.

Relations with Thailand were strained in early 1996 by Malaysia's construction of a wall along its border with Thailand to stem the arrival of illegal immigrants. Thailand accused Malaysia of arbitrarily expelling illegal immigrants to Thai territory and of making insufficient effort to prevent Muslim separatists in southern Thailand from using Malaysian territory as a sanctuary. However, in January 1998 three Thai Muslim separatists were arrested by the Malaysian authorities and deported to Thailand, and in April an agreement was signed jointly to develop a huge offshore gas and oilfield in a disputed area of the Gulf of Thailand.

The arrest of the former Deputy Prime Minister and Minister of Finance, Anwar Ibrahim, in September 1998 prompted critical reactions from President B. J. Habibie of Indonesia and President Joseph Estrada of the Philippines, both of whom had enjoyed good relations with Anwar; for the most part, however, Malaysia was able to invoke the ASEAN custom that implicitly prohibited interference in the internal affairs of fellow member countries. The Malaysian Government reacted indignantly in November when the US Vice-President, Al Gore, expressed support for the *reformasi* movement while attending an APEC summit meeting in Malaysia. Relations with the Philippines were further strained in 2000 after Muslim separatists from the southern Philippines kidnapped 21 tourists, 12 of them foreigners, from the island of Sipadan, off Sabah, in April. Malaysia complained that lack of security in the Philippines had damaged its interests. Meanwhile, Malaysia's relations with China became closer during May 1999, when the two countries signed a 12-point co-operation agreement. Mahathir and the Chinese Prime Minister, Zhu Rongji, exchanged state visits later in the year. The *rapprochement* partly reflected the countries' shared suspicion of economic liberalization. Criticism from Western countries over the conviction of Anwar on sodomy charges in August 2000, however, left Malaysia temporarily somewhat isolated in world affairs.

The terrorist attacks on the US cities of New York and Washington, DC, on 11 September 2001 initially prompted speculation that the USA would identify Malaysia as a country that harboured terrorists, especially after Mahathir described the attacks as a retaliation for Israel's 'state terrorism'. Evidence was also discovered revealing that two of those who had perpetrated the September attacks had stayed in the home of a former Malaysian army captain in 2000. Malaysia criticized the USA for killing innocent people during its attacks on the Taliban regime in Afghanistan. In the event, however, the Malaysian Government's swift suppression of the KMM allowed it to be perceived as clearly opposing Islamist extremism. Malaysia also promised to co-operate with the USA in tracing al-Qa'ida funds in Malaysia. In May 2002 Mahathir visited the USA to reinforce the improved bilateral relationship and to sign a formal agreement to co-operate in combating terrorism, but he warned the West against becoming 'impatient' with Malaysia over democratization and human rights issues, stating that a transition to democracy was likely to be slow. The two sides later announced plans for the establishment of a joint anti-terrorism training centre in Malaysia to serve the ASEAN countries. The Malaysian Government also initiated a major expansion of the country's military capacity, purchasing both submarines and F/A-18 fighter jets. Malaysia also signed an anti-terrorism agreement with Australia in August 2002, but relations were strained by official warnings against travel to Malaysia issued by the Australian Department of Foreign Affairs and Trade, which Malaysia considered to be unwarranted. In December the Malaysian Government also agreed to work closely with that of Indonesia. Mahathir was strongly critical of US policies on Iraq both before and after the US-led invasion in early 2003 had taken place, and the UMNO youth wing collected 2.5m. signatures for a petition against the offensive. Mahathir, however, retained surprisingly good relations with the USA, owing to his determination to prevent Malaysia from becoming a haven for groups identified as posing a terrorist threat.

Mahathir's last months in office were marked by several verbal clashes with neighbours and world powers. In a speech at the summit meeting of the Organization of the Islamic Conference (OIC), held in Malaysia just before his retirement, in October 2003, the Prime Minister accused Jews of ruling the world by proxy and of using others to fight and die for them. He was consequently condemned by the US Congress, while the US Senate voted to reduce military aid to Malaysia. In a speech at the APEC summit meeting in Bangkok shortly afterwards, he attacked globalization, the WTO and free trade, calling instead for fair trade. Meanwhile, in August of that year Mahathir signed a contract to replace the Malaysian half of the causeway to Singapore with a bridge that would enable shipping to pass through the strait separating the two countries. He also accused Singapore of profiteering in the complex arrangements by which Malaysia supplies untreated water to Singapore and receives some treated water in exchange. Malaysia's relations with Singapore were further strained by that country's programme of land reclamation, which produced changes in the median-line maritime boundary between the two countries. Malaysia came under international pressure, on the other hand, over its failure to recognize as refugees Acehnese who were fleeing the Indonesian military operations in Aceh.

Mahathir's successor as Prime Minister, Abdullah Badawi, soon softened Mahathir's dogmatic approach to international relations, seeking a *rapprochement* with both Singapore and Australia. Progress was made in resolving the long-running dispute with Singapore over water supplies, while negotiations began over a free-trade agreement with Australia. As Chairman of the OIC, Badawi sought to stress the commitment of Muslim states to suppressing Islamic terrorism, while insisting that

Islam itself was not a religion of extremism or violence. Relations with Thailand were strained when Thai troops reportedly killed 112 Islamic militants in southern Thailand in late April 2004, and Malaysia offered sanctuary to Muslim Thais seeking refuge. Thailand accused Malaysia of harbouring Islamic militants. In response, Malaysia agreed to closer co-operation with Thailand in its operations against Islamists in southern Thailand. In November 2003, however, Badawi had dismissed the editor of the daily newspaper *New Straits Times* over an article that accused the Government of Saudi Arabia of hypocrisy, profligacy and encouraging Islamic extremism. The Saudi Government had strongly criticized Malaysia and had sharply reduced Malaysia's quota for the *Hajj* (pilgrimage to Mecca). In April 2004 the Deputy Prime Minister and Minister of Defence, Najib Tun Razak, reacted sharply to a suggestion from the US naval commander in the Pacific, Adm. Thomas Fargo, that US warships might patrol the Melaka Strait to guard against terrorist attacks. Subsequently, however, he agreed to closer collaboration with the USA against terrorist groups.

Economy

CHRIS EDWARDS

Revised by ROKIAH ALAVI

INTRODUCTION

From the mid-1970s Malaysia had one of the most dynamic and fastest-growing economies in Asia, and between 1985 and 1995 it was the eighth fastest-growing economy in the world. In the 25 years following the adoption of the New Economic Policy (NEP) in 1971, Malaysia's gross domestic product (GDP) grew at an annual average rate of about 7%. Between 1985 and 1995 GDP growth averaged over 8% per annum, which, with the population expanding at a little over 2.5% annually, gave an average annual growth in per caput income of 5.7%. During the 1980s real GDP growth averaged 5% annually, but the growth rate was uneven. The year-by-year GDP growth rate was particularly spectacular until the mid-1990s. However, Malaysia's economy decelerated drastically in 1997, owing to the regional economic crisis, which started in July of that year. The impact of the crisis became more evident in 1998, in which year GDP contracted by 7.4%. However, various government policies helped the economy to recover to the extent that GDP grew by 6.1% in 1999 and by 8.9% in 2000. The performance of the Malaysian economy in 2001 was adversely affected by the global economic slowdown, particularly in the USA. This was reflected in the modest rate of economic expansion in that year; GDP grew by only 0.3%. Strong domestic demand and improved export performance accelerated economic growth in 2002, when a GDP growth rate of 4.1% was recorded. In 2003 GDP expanded by 5.3%, thus exceeding the official forecast of 4.5%.

The unemployment rate increased from 2.7% in 1997 to 3.2% in 1998. Official estimates indicated that the number of employees retrenched increased to 83,865 in 1998, compared with 18,863 workers in 1997 and 7,773 in 1996. As the economy strengthened, the unemployment rate declined to 3.4% in 1999, and to 3.1% in 2000. The rate increased slightly in 2001, to 3.7%, and declined only minimally in 2002, to 3.4%, owing to the slowdown in the economy. In 2003 the unemployment rate was 3.6%, but retrenchment decreased by 20% in that year (see below). The inflation rate, which showed a significant upward trend in late 1997 and in the first half of 1998 (owing to the impact of the depreciation of the ringgit), declined progressively after June 1998. For 1998 as a whole inflation rose by 5.3%. The recovery of the economy contributed to a decline in the rate of inflation to 2.7% in 1999, and further, to 1.5% in 2000. Inflationary pressures remained subdued, and the consumer price index (CPI) experienced modest increases of only 1.4% and 1.8% in 2001 and 2002, respectively. The rate of inflation remained both low and stable in 2003, at only 1.2%. According to the Bank Negara Report 2003, although demand had strengthened, excess capacity in selected sectors and the absence of wage cost pressures, together with improving labour productivity, helped to contain price pressures.

The Sectoral Growth Pattern

During 1971–90 the manufacturing sector registered the highest rate of output growth, expanding at an annual average of 10.3%. The sector grew at a higher rate of 11.3% per annum in the 1970s, compared with 9.4% annually in the 1980s. However, during the second half of the 1980s, the sector expanded at an average annual rate of 13.7%. Between 1990 and 1995 the sector experienced growth of an average annual rate of 13.3%. Manufacturing output subsequently grew at a slightly slower pace of 12.2% in 1996 and 12.5% in 1997. In 1998 the output of the manufacturing sector contracted by 13.4%, the first decline since 1985; this was due mainly to the economic slowdown in the Asia-Pacific region as a whole. However, the sector experienced strong positive growth of 11.7% in 1999. The sector grew further by 18.3% in 2000, as a result of significant increase in external demand as well as sustained strong domestic requirements. However, the slowdown in major industrial countries and the downturn in the global electronics cycle adversely affected the overall performance of Malaysia's manufacturing sector in 2001, leading to a contraction of 5.9%. Nevertheless, the sector recovered in 2002 and 2003, expanding by 4.1% and 8.3%, respectively. This was attributable to a strengthening of domestic demand and a recovery in exports. The sector's share of GDP increased significantly from 16.9% in 1975 to 33.7% in 1997; in 1998 its share increased further to 34.4%, despite the regional economic crisis, indicating the sector's considerable importance in the context of the economy as a whole. The manufacturing sector continued to contribute approximately one-third of annual GDP during 1999–2003.

The services sector also expanded rapidly, growing at an average annual rate of 7.6% during 1971–90, 9.3% between 1990 and 1995, and 9.7% and 8.7% in 1996 and 1997, respectively; however, as a result of the economic problems of 1997–98, this sector contracted by 0.4% in 1998. In accordance with the overall strong expansion of the economy, the services sector grew at a faster rate of 4.5% in 1999, 6.7% in 2000 and 6.0% in 2001. The sector expanded by 6.4% and 4.4% in 2002 and 2003, respectively. The services sector has gained in importance over time, expanding at an annual rate of 8.2% during 1990–2002, faster than the average 6.5% annual increase in overall GDP growth over the same period. The overall contribution of the services sector to GDP increased from 38.3% in 1970 to 46.8% in 1990, and was 47.4% in 2003. The intermediate services group (comprising transport, storage and communications and finance, insurance, real estate and business services) expanded more rapidly, at a rate of 5.5% in 2003, than the final services group (comprising electricity, gas and water, wholesale and retail trade, hotels and restaurants and government services), which grew by 3.8% in that year. The strong expansion in the transport, storage and communications sub-sector could primarily be attributed to the rapid growth of the mobile cellular telephone and internet services segment of the telecommunications industry and to the increase in the transhipment activities of local ports. In the final services group sector, the utilities sub-sector recorded a higher rate of growth owing to higher demand by domestic industries and the commercial sector. The construction sector had a lower growth rate during 1971–90 (6.4% annually) but this accelerated to 13.3% per annum during 1990–95, and 14.2% and 9.5% in 1996 and 1997, respectively. In 1998 the sector experienced a contraction in output of 24.5%, mainly as a result of lower aggregate demand. In 1999, however, the sector strengthened as a result of various policy measures introduced by the Government, recording a rate of contraction of only 4.4%. In 2000 and 2001, however, the sector recorded positive growth rates of 0.6% and 2.1%, respectively. In 2003 the construction sector expanded at a more moderate pace of 1.9% (compared with 2.3% in 2002), owing to a slowdown in the civil engineering sub-sector following the completion of several privatization projects. The agricultural sector, which traditionally provided the impetus for growth in the economy, experienced a continued deceleration in growth rates, from an average of 5% per annum in the 1970s to 3.8% in the 1980s, 2.0% per annum during 1990–95, 2.2% in 1996 and 3.5% in 1997. In 1998 the sector experienced a contraction of 2.8%, before recording a rate of growth of about 0.5% in 1999. In 2000 the sector expanded by 6.1%, before contracting by 0.6% in 2001, owing to slower growth in the production of crude palm oil. However, the agricultural sector enjoyed higher growth rates of 2.6% and 5.7% in 2002 and 2003, respectively, as higher production levels and prices, mainly of palm oil and rubber, propelled growth. The sector's contribution to GDP was reduced substantially, from 29% in 1970 to approximately 9.3% in 2003. The change in the sectoral growth pattern clearly demonstrates the change in development strategy, moving from an economy based on agriculture to one based on industry.

Regional and Rural–Urban Divisions

Regional development strategies implemented since independence have resulted in improved living standards and a higher quality of life for the inhabitants of all states, as well as the reduction of regional disparities. The population of Malaysia in 2003 was approximately 25.1m. An estimated 43% of the population resided in the less developed states, with Sabah and Sarawak populated by more than 2m. people. Based on the composite index of development in 2000, the states of Johor, Perak, Pulau Pinang (Penang), Melaka (Malacca), Negeri Sembilan (Negri Sembilan), Selangor and the Federal Territory of Kuala Lumpur were categorized as more developed states, while Kedah, Kelantan, Pahang, Perlis, Sabah, Sarawak and Terengganu (Trengganu) were categorized as less developed states. The average per caput income of those living in the more developed states was RM 17,410 in 2000, compared with RM 10,893 for the inhabitants of the less developed states. The national average per caput income was RM 14,592 or US $3,840 in 2003. However, the average growth of per caput GDP between 1995 and 2000 in the less developed states was slightly higher (6.3%) than in the more developed states (6.1%). This indicated that the income gap between the more developed and less developed states was narrowing. The overall incidence of poverty in Malaysia declined from 8.7% in 1995 to 7.5% in 1999. In the more developed states the incidence of poverty declined slightly, from 4.2% in 1995 to 3.9% in 1999. In the less developed states the incidence declined from 15.6% in 1995 to 13.2% in 1999, with Terengganu experiencing the most significant decline, from 23.4% in 1995 to 14.9% in 1999. In terms of access to infrastructure and social amenities, 100% of rural areas in Peninsular Malaysia had access to electricity supply, while Sabah and Sarawak had 80% and 70% access, respectively.

Malaysia is a rapidly urbanizing society. In 1980 just over one-third of the Malaysian population lived in urban areas, but by 1991 this had increased to 51% and by 2000 to 62%. In the decade to 1995 the urban population increased at an average annual rate of 4%, much faster than the 2.5% increase in the population as a whole. Between 1990 and 1995 the average incomes of urban households grew faster (at 8.1% per annum) than those of rural households (at 5.3%).

Employment and Income Distribution

The economically active population accounted for 62.9% of the whole in 2000. The Labour Force Survey indicated that 55.3% of the labour force was concentrated in urban areas. The creation of new jobs took place mainly in the services sector, where 217,300 persons found employment in 2000. The manufacturing sector registered a decline in the number of new jobs created in 2000, while the construction sector witnessed the establishment of 55,500 new jobs, owing to an increase in residential housing and public infrastructure projects.

Malaysia began to experience persistent labour shortage problems from the second half of the 1990s. The Eighth Malaysia Plan, for the 2001–05 period, estimated that the number of registered foreign workers in the labour force had almost doubled, from 650,000 in 1995 to 1,239,862 in 2003 (constituting about 12% of the total work-force), but these figures were widely thought to be underestimates. The number of foreign workers in Malaysia was probably close to 2m. in 2000. In connection with the Eighth Malaysia Plan, it was also calculated that, from a total of 749,200 foreign employees holding work permits in the year 2000, 31.3% were engaged in manufacturing, 22.9% in agriculture, 8.7% in construction and 7.4% in services, while 20.3% worked as maids. In 2003 the majority (70%) originated from Indonesia, while others mainly came from Nepal, Bangladesh, India and Viet Nam. About 16% of foreigners recruited in the first half of 2001 were skilled workers. India accounted for the largest number of approved skilled workers, followed by Japan and the People's Republic of China. In terms of occupational groups, directors accounted for the largest number, followed by managers and engineers.

In 2003 unemployment in Malaysia was estimated at 3.6% of the labour force, compared with a rate of 8.7% in 1987. The average rate conceals quite considerable disparities among states. For example, Terengganu and Kelantan, in the east of the peninsula, and Sabah and Sarawak all had unemployment rates of 4.5% or more, whereas the federal capital, Selangor and Pulau Pinang had rates of 1.1% and below.

The impact of the slowdown in economic activity was experienced by the labour market, and was most acutely reflected by the number of retrenched workers in the manufacturing sector. In 2001 a total of 38,116 workers were retrenched (compared with 25,236 in 2000); 75.6% of these had been engaged in manufacturing activities. The electronics and electrical products sub-sector accounted for almost one-half (45.7%) of the total number of workers made redundant. The main reason for the retrenchments was a decline in the demand for manufactured products. Local workers accounted for the majority (86.9%) of retrenched workers, and 96.9% of all retrenchments were made in West Malaysia. In 2003 the number of workers retrenched declined significantly, falling by 20% to 21,206 persons (compared with 26,452 in 2002), reflecting the improved economic climate. Retrenchment occurred mainly in the manufacturing and tourism-related sectors.

The avowed aims of the NEP were to bring about (over a 20-year period from 1971 to 1990) a redistribution of income and wealth, particularly in favour of the *bumiputra* (indigenous population, mostly Malays), since in 1971 Malays were disproportionately poor. In 1995 about 58% of the total population possessed *bumiputra* status. In terms of reducing absolute poverty, the NEP was successful since, whereas in 1970 almost one-half of the households in Peninsular Malaysia were classified as poor, by 1990 less than one-sixth of Malaysian households were so classified. Furthermore, by 1995 the proportion of households that were classified as poor had declined to less than 9%. In 1997 the overall incidence of poverty declined to 6.1%. However, the regional economic crisis led to an increase in this index to 8.5% in 1998, before the level declined to 7.5% in 1999. Both rural and urban households recorded reductions in poverty between 1995 and 1999. The incidence of rural poverty decreased fron 14.9% in 1995 to 12.4% in 1999, while urban poverty declined from 3.6% in 1995 to 3.4% in 1999. The incidence of poverty amongst agricultural workers was highest, at 16.4% in 1999. The proportion of households classified as very poor (defined as having less than one-half of the poverty line income and known as 'hardcore' poverty) has also fallen in recent years, from 3.9% in 1990 to 1.4% in 1999. The incidence of rural 'hardcore' poverty declined from 3.6% in 1995 to 2.4% in 1999, while in the urban areas it decreased from 0.9% to 0.5% in these years.

In terms of reducing relative poverty, the NEP was less successful. Between 1970 and 1999 the 'Gini coefficient' (a commonly used measure of inequality) had improved only slightly, and in 1999 the average monthly income of *bumiputra* households was RM 1,984, compared with an average for Chinese households of RM 3,456. Thus, in 1999 the gap in incomes between the two major ethnic groups remained substantial. A major cause of this was the higher proportion of *bumiputra* households living in rural areas. In both rural and urban areas the gap between the ethnic groups was less but, in general, rural households were worse off, with the average rural household income (RM 1,718) being less than the average urban household income (RM 3,103). Income distribution in Malaysia in general has remained unequal, with the 'Gini coefficient' of 0.446 in 1990 declining marginally to 0.443 in 1999. At the beginning of the 21st century the NEP's original objective to achieve 30% *bumiputra* ownership had yet to be realized. In 1999 the proportion of corporate equity owned by *bumiputra* was 19.1%, compared with 37.9% under Chinese ownership. The highest proportion of equity controlled by the *bumiputra* was in the transport (32.2%), construction (27.1%) and agriculture sectors (24.3%). In comparison, the proportion of equity of non-*bumiputra* controlled companies was high in almost all sectors, ranging from 41% to 57%, except in the manufacturing sector, which was 60% under foreign control.

In 1991 the NEP was replaced by the National Development Policy (NDP). Under the NDP, there is less rhetorical emphasis on *bumiputra* preferences, but the consensus is that there has been a continuation of the NEP's strategy of encouraging the emergence of a sizeable core of Malay entrepreneurs. Furthermore, there is a widespread belief that the beneficiaries of preferences consist disproportionately of prominent politicians and their close contacts.

AGRICULTURE, FORESTRY AND FISHING

Land Use

In 1995 it was officially reported that a little under 19m. ha of land in Malaysia was under forest, compared with about 20.5m. in 1980. This 19m. represented just under 60% of Malaysia's total land area (of 33m. ha). The strength of agricultural production in Malaysia lies in the large plantations of commercial crops such as rubber, oil palm, cocoa and pineapple. These crops occupy most of the arable land. The area used for cultivation of agricultural commodities in 2000 was 6m. ha or about 18% of the total land area, of which 8% was under oil palm cultivation, 5% rubber, 2% paddy and 1% each under cocoa, coconuts and fruit. An increase in the area under cultivation was recorded for oil palm, pepper, tobacco, vegetables and fruits. However, about 430,800 ha of land used for the cultivation of rubber and cocoa was converted to oil palm cultivation and other uses.

Contribution to Export Revenue

Agriculture and forestry's share of total export earnings have declined rapidly, from 53% in 1973 to 7% in 2003. The traditional dominance of rubber, which provided almost one-half of Malaysia's export earnings in 1961, continues to be eroded. In 2003 rubber accounted for only 0.7% of total exports and about 9% of total agricultural exports. The most significant export-earner of all agricultural commodities is palm oil, constituting about 56% of total agricultural exports in 2003.

Rubber

Rubber, a principal export of Malaysia since colonial times, has experienced a relative decline over the past few decades. Since the 1960s there has been considerable replanting of rubber land with other crops, especially oil palm. As a result, the area under rubber declined from 2m. ha in 1980 to 1.66m. in 1997 and further, to 1.52m., in 1999. Over 85% of the land under rubber is in Peninsular Malaysia and only 15% in East Malaysia. The contribution of rubber production to national GDP has declined substantially, from 5% in the early 1980s to 0.6% in 2002. However, rubber remains important in a socio-economic context, especially in terms of providing job opportunities. About 265,000 families were engaged in the rubber smallholdings sector in 2002. The sector also has strong linkages to domestic, downstream, rubber-based industries such as those manufacturing rubber gloves, tyres and tubes, as well as rubber footwear. Malaysia's most significant export product is rubber gloves; Malaysia is believed to control over 60% of the global rubber glove market, with manufacturers including Top Glove Corporation Bhd, Oon, Comfit, Profeel, Dermagrip, Supergloves and RadiaXon. Local producers of tyres and related products include Sime Darby Bhd and Silverstone Bhd. The main importers of Malaysian rubber and rubber products are the countries of the European Union (EU), the USA, the People's Republic of China and the Republic of Korea. In 2003 Malaysia's rubber exports totalled US $3,582m. Malaysia is the third largest producer and net exporter of natural rubber in the world after Thailand and Indonesia.

Palm Oil

Since the early 1960s, when the Government and the private sector began to diversify agriculture away from rubber, palm oil production has increased spectacularly, with an average annual growth rate in production of 19.6% between 1970 and 1980. As a result, Malaysia has become the largest producer and exporter of palm oil, accounting for 49% of world production in 2003, and for 58% of international trade in palm oil. An average annual increase in production of more than 8% over the period 1986–94 was due to improvements in planting, harvesting techniques, the increase in mature oil palm trees and the introduction of the Cameroon weevil to stimulate production. Representing the fastest-growing sub-sector within agriculture, the palm oil sub-sector contributed substantially to the value-added of the agricultural sector as a whole in the late 1990s, and in 2001 grew by 8.9% compared with the rate of growth of 2.5% of the agricultural sector overall.

The bulk of crude palm oil comes from Peninsular Malaysia, which accounted for an estimated 67.5% of the country's production (or 7.96m. metric tons) in 2001. Sarawak produced 0.54m. tons, or 4.6% of the total, in that year, while Sabah produced 3.29m. tons, or 27.9% of the total. The total area planted with oil palm in 2003 was estimated at 3.75m. ha, of which 3.1m. ha were matured holdings. In terms of ownership, the industry is dominated by private estates, which constituted 60% of the total planted in 2002. There are four large oil palm corporations in Malaysia—Kumpulan Guthrie Bhd, Golden Hope Plantations Bhd, IOI Corporation Bhd and Sime Darby Bhd. The equity of plantation companies is largely under Malaysian ownership and they are primarily owned by Perbadanan Nasional Berhad (PERNAS) and the Employee Provident Fund (EPF). Of the rest, smallholders organized under the Federal Land Development Authority (FELDA), the Federal Land Consolidation and Rehabilitation Authority (FELCRA), the Rubber Industry Smallholders Development Authority (RISDA), and various state land schemes accounted for 30.5%, with independent smallholders owning 9.5%.

After 1975 the Government attempted to change the emphasis of the industry from planting and exporting crude palm oil to producing, refining, manufacturing and selling oil-based products. Consequently, by the early 1990s there were 264 operational palm oil mills, 54 palm oil refineries and a growing oleochemical industry. By 1996 there were 290 mills in the country. However, labour shortages were becoming a serious problem as workers moved to urban areas for higher wages; on some estates migrant foreign labour accounted for up to 70% of the work-force. Malaysian plantation workers were, however, paid more than three times the amount earned by their Indonesian counterparts in 1993, and Indonesian labour remained a cheaper option despite a government levy payable for every foreign worker. Another way in which Malaysian companies were able to take advantage of cheaper Indonesian labour was by investing in projects in Indonesia. In recent years 27 joint-venture agreements and MoUs have been co-signed by Malaysian and Indonesian firms to develop 1.5m. ha of oil palm, but in early 1997 the Indonesian Government imposed a freeze on Malaysian investment in the oil palm sector, stating that it had reached an unacceptably high level. Other such joint-ventures are in India, the Philippines, Papua New Guinea, Latin America and Africa.

In 1980 exports of palm oil totalled RM 2,500m., but they had risen to RM 4,300m. by 1990 and to RM 20,200m. by 2003. Like rubber, however, palm oil faced a declining real price. In the 1980s the average price of crude palm oil in the Rotterdam market was US $459 per metric ton, whereas the average price in 2003 was $1,617 per ton. In May 2000 the Malaysian Palm Oil Board (MPOB) was established, following the merger of the Palm Oil Registration and Licensing Authority (PORLA) and the Palm Oil Research Institute of Malaysia (PORIM). The objective of establishing the MPOB was to strengthen the role of agricultural agencies; the Board was expected to undertake research and development, licensing and enforcement of regulations for the industry, and to disseminate information related to palm oil. Continued research by the MPOB resulted in the commencement of development of 174 technology products by the end of 2002, of which 47 were launched during the year. A further 35 new technologies and products were launched for commercialization in 2003.

Forestry

The production of timber is an important sector of the Malaysian economy. Total sawlog production in 1994 was 36m. cu m, but this declined to 21.4m. cu m in 2003. Output in Peninsular Malaysia and Sabah fell by 28.0% and 13.5%, respectively, and Sarawak accounted for over one-half of Malaysia's total production, reflecting the large-scale clearance of forests to make way for the Bakun hydroelectric dam (see below). The central bank reported that the decline was principally a result of weaker demand from traditional buyers (mainly from within the Asia-Pacific region). As well as adverse weather conditions, overall production was also affected by the Government's conservation policy and commitment to the achievement of sustainable forest management by the year 2000, in compliance with the requirements of the International Tropical Timber Organization.

In 2001 Sarawak was the principal producer of sawlogs, accounting for 62% of total production, with Sabah accounting for 12% and Peninsular Malaysia 26%. The principal markets for Malaysian logs in 2003 were India (24.9%), Japan (24.6%), China (24.3%), Taiwan (12.6%) and Hong Kong (1.9%). To ensure sustainable forest management, the Government initiated a National Forestry Policy in 1978, aimed at encouraging sustainable development of Malaysia's timber resources through forest conservation, the promotion of under-utilized species and the reduction of waste. Successive five-year plans reduced logging targets from 223,000 ha per year for the Fourth Plan (1981–86) to 152,140 ha per year for the 1986–90 period. The annual permissible logging area in permanent forests was reduced to 46,000 ha in Peninsular Malaysia, 60,000 ha in Sabah and 170,000 ha in Sarawak during the Seventh Malaysia Plan (1996–2000) period. In addition, logging companies in Sarawak were required to allocate 50% of their output to local processors. This was to increase the domestic value-added of the sector. Further evidence of the government policy of encouraging downstream activity in the forestry sector was apparent in 1993, when the Government introduced for the first time export levies on selected species of timber and veneer, in furtherance of its policy to provide adequate cheap resources for the expanding downstream wood-processing activities. From 1 January 1994 the export of rubberwood was banned to encourage local furniture and panel manufacturing activities.

As a result of these measures, the production of sawlogs declined by 5.6% per annum, from 31.8m. cu m in 1995 to about 24m. cu m in 2000. Accordingly, the production of sawn timber decreased from 7.5m. cu m in 1995 to 5.2m. cu m in 2000. However, rubberwood became a major source of raw material in the furniture and other wood-based industries, with total exports increasing from RM 1,900m. in 1995 to RM 4,300m. in 2000. Rubberwood furniture exports increased by 11.1%, to RM 3,900m., in 2001 (compared with RM 3,500m. in 2000).

Although the Government has promoted conservation, Malaysia has still attracted international and domestic criticism from anti-tropical hardwood campaigners for its forestry policies. In response to this, the Malaysian timber industry has established its own certification scheme, operated by the Malaysian Timber Certification Council (MTCC). By the end of 2003 seven states had been awarded the certificate for forest management, and 38 timber companies had been given permission to use the MTCC logo to provide assurance to buyers of Malaysian timber and timber products that they were from sustainable and legal sources of forest products. This effort was fruitful, and in 2003 Denmark endorsed the certification scheme as one of the accepted schemes in its Environmental Guidelines for Purchasing Tropical Timber. One of the most recent challenges to the timber industry was related to trade in ramin timber. In February 2004 the United Kingdom's Environmental Investigation Agency (EIA) and an Indonesian-based non-governmental organization (NGO), Telapak, released a report alleging that Malaysian officials had provided false documentation of the Convention on International Trade in Endangered Species (CITES) and 'knowingly allowed' the issuing of false certificates of origin from Malaysia in order to facilitate the 'laundering' of illegal ramin from Indonesia. In 2001 Indonesia had asked for ramin to be listed under CITES, which still allows the timber to be traded but requires CITES documentation, including a certificate of origin from the state of export.

Subsistence and Other Agriculture

The principal subsistence crop in Malaysia is rice. Malaysia was about 75% self-sufficient in rice in 1995, compared with about 80% self-sufficiency in 1990. The self-sufficiency level reduced further, to 71%, in 2000. In real terms, the world price of rice declined particularly sharply from the mid-1980s, so that in spite of government incentives and subsidies for the sector, production and area under paddy are declining. Poverty is high in the paddy-producing areas, and there is little prospect of reversing this through paddy production itself. The trend of youths leaving the paddy-producing areas is likely to continue, exacerbating the existing labour shortages. However, rice production exceeded 2.0m. metric tons in the three years 2000–02, before declining to 1.4m. tons in 2003. Malaysia exported approximately 2.5m. tons of rice in 2003, but imported 358,000 tons of rice in the same year. This provided an indication of Malaysia's increasing dependency on imported rice.

Cocoa production expanded rapidly during the late 1980s to become Malaysia's fourth most important agricultural subsector after palm oil, timber and rubber. Production increased from 88,000 metric tons in 1984 to 235,000 tons in 1991, making Malaysia the world's fourth largest producer of cocoa, accounting for around 7% in 1991, after Côte d'Ivoire, Brazil and Ghana (a position Malaysia was subsequently unable to maintain). In 1991 the Malaysian Cocoa Board introduced compulsory grading to control the quality of exports. However, low cocoa prices, the conversion of cocoa-cultivated land to other crops, increasing labour costs and the rising cost of agricultural inputs together resulted in a decrease in Malaysian cocoa production and exports throughout the 1990s. In 2003 total cocoa production was an estimated 36,240 tons (compared with 58,000 tons in 2001). Export revenue from cocoa declined from RM 110m. in 2002 to RM 83m. in 2003.

Pepper production, 99% of which is located in Sarawak, rapidly increased in the late 1980s to reach 31,000 metric tons in 1990. However, production declined by one-half to about 15,000 tons in 1995. The decline in output was partly due to substantial hectarage being replanted to replace old and diseased pepper plants, as well as a marginal decrease in the area under cultivation. Pepper production increased, however, to 28,500 tons in 2001, as a result of efforts to promote low-cost planting and enhanced efficient fertilizer application, as well as increased demand. Malaysia accounted for 11.1% of the world's supply (making it the fifth largest producer) in 2000.

Fruit production has expanded in Malaysia and has more than kept pace with consumption. As a result, Malaysia was 91.3% self-sufficient in fruits and 88.5% self-sufficient in vegetables in 2000; however, it is a net importer of both fruits and vegetables. The value of fruit and vegetable exports in 2000 was RM 701.2m., while the value of imports was RM 1,913.6m. A total of 15 different types of fruit were promoted for commercial cultivation, including banana, papaya, pineapple, watermelon, starfruit, mango, durian, rambutan, guava and citrus fruits. The principal local fruits are durians and bananas, and these account for over two-thirds of fruit output. Fruit production registered an average annual increase of 6.2%, from 1m. metric tons in 1995 to 1.4m. tons in 2000, while vegetable production recorded a growth rate of 7.2% per annum during the same period. In 2002 production of vegetables and fruits expanded by 19.9% and 4.2%, respectively; this was mainly attributable to an expansion in the areas under cultivation, including Permanent Food Production Parks (PFPP). By the end of 2002 more than 1,840 ha of land had been developed under the PFPP programme.

In 1990 Malaysia was more than self-sufficient in poultry, pork and egg production, and remained a net exporter in 1995, with the principal market being Singapore. However, the self-sufficiency level for pork had declined from 104% in 1995 to 80% in 2000. In beef and mutton production, Malaysia was 22.7% (compared to 19.2% in 1995) and 6.4% (compared with 6% in 1995) respectively self-sufficient in 2000.

In 1995 Malaysia was about 91% self-sufficient in fishery products, the same percentage as in 1990. However, the level had declined to 89% in 2000. In 2003 the country's total fish production amounted to 1.3m. metric tons (equivalent to RM 5,500m.) and a total value of RM 1,000m. of fish was imported in that year, reflecting insufficient domestic production to meet the increase in per caput consumption. Malaysia's per caput consumption of fish, totalling 49 kg in 1999, was one of the highest in the Association of South East Asian Nations (ASEAN). Although aquaculture production has expanded rapidly, more than 85% of the value of fisheries production comes from the sea. In 2000 1.3m. tons of fish were produced from capture fisheries and 151,800 tons from aquaculture. Given the larger potential of the aquaculture sub-sector, the Government has identified a total of 4,472 ha to be classified as aquaculture industrial zones. Under the Third National Agricultural Policy, aquaculture production was envisaged to increase fourfold, to 600,000 tons, by 2010.

MINING

Mineral output accounts for over 40% of Malaysia's industrial production and for about 7% of total GDP, with about four-fifths of this being accounted for by petroleum. The mining sector's contribution to GDP has remained consistent, at approximately 10%–11%, since 1980.

Tin

Tin production was once one of Malaysia's leading industries, but between 1980 and 2003 the annual output of tin-in-concentrates declined substantially, from 61,000 metric tons to 3,502 tons. In 1980 Malaysia was the largest producer in the world, accounting for 31% of global output. By contrast, in 1997 Malaysia was the eighth largest global producer of tin, providing only an estimated 2.4% of total output.

The decline in Malaysian production reflected a fall in the tin price. In 1980 the price was RM 35.7 per kg, but it had collapsed to RM 15.5 by 1986. Prices (and Malaysian production) experienced a mild recovery until 1989, as the Supply Rationalization Scheme (SRS) of the Association of Tin Producing Countries (ATPC) managed to limit the supply of tin through the imposition of export controls on member countries. However, the international recession, the political turmoil in Eastern Europe and the unpredictable release of tin stocks by the US Defence Logistics Agency caused a further decline in tin prices, from RM 23.1 per kg in 1989 to less than RM 15 in late 1996. In 2000 only 40 mines were in production, compared with 847 in 1980. In 2003 employment in the sector had fallen to 1,215 workers, compared with almost 50,000 in 1970 and 39,000 in 1980. At the end of 2003 there were 26 mines that were considered to be productive. However, these mines were operating near the end of leasing periods and the quality of tin obtained was of a lower grade. The cost of production was also high, as the tin could only be obtained from a greater depth.

Despite the sharp decline in tin production, in 1997 Malaysia remained the world's fourth largest centre of tin smelting. The Malaysia Smelting Corporation, the world's largest tin smelter, is the only tin smelter in Malaysia, following the closure of the Escoy tin smelter in March 1998. As domestic production of tin-in-concentrates declined, Malaysia was obliged to import supplies in order to keep its smelters working at near capacity. In 2003 imports of tin-in-concentrates totalled 7,661 metric tons. In the same year Malaysia exported 15,164 tons of processed tin.

Petroleum and Gas

Malaysian petroleum production is not significant in international terms: in 1997 Malaysia's production of crude petroleum accounted for just 1.0% of world output. However, Malaysia is more than self-sufficient in primary commercial energy, and the surplus over self-sufficiency increased from 176 petajoules in 1990 to 511 petajoules in 1995.

Average production of domestic crude petroleum declined from 663,000 barrels per day (b/d) in 1995 to 606,000 b/d in 2000. In 2003 Malaysia produced a total of 268.4m. barrels of oil. The bulk of the total output was from Peninsular Malaysia (about 63%), with Sarawak and Sabah accounting for 23.2% and 13.8%, respectively. At the end of 1998 there were 37 oilfields in production, including two new oilfields in Peninsular Malaysia and one in Sabah: of these 37 oilfields, 16 were in Peninsular Malaysia, while 13 were in Sabah and eight were in Sarawak. By 2000 the number of oilfields in production had declined to 33. However, by 2003 the number of oilfields had increased to 51.

Petroleum and gas production is the responsibility of Petroliam Nasional Bhd (PETRONAS), which operates mainly through production-sharing contracts. In 1995 PETRONAS was partially privatized by the Government. By the end of 1991 a total of 29 production-sharing contracts had been signed under the revised terms introduced in 1985.

In 1997 reserves of crude petroleum in Malaysia were estimated at 534,247,000 metric tons, compared with 397,260,000 tons in 1990. At 1 January 2004 Malaysia's crude oil reserves, including condensates, had increased by 6.6%, to 4,840m. barrels, compared with 4,540m. barrels at the beginning of 2003, as a result of new discoveries. At the current rate of production, Malaysia's oil reserves are expected to last for another 18 years. In 1980 the Government implemented the National Depletion Policy (NPD), in order to slow exploitation rates and hence conserve the country's reserves. However, during the recession of the early 1980s the Government became increasingly dependent on revenue from petroleum exports, and the policy was reportedly abandoned. However, the Seventh Malaysia Plan stated that production throughout the Sixth Plan period was in line with the NPD. Between 1993 and 2003 Malaysia's reserves of crude petroleum declined by 20%, from 5,000m. barrels to 4,000m. barrels. The decline was mainly due to sustained production and the maturity of existing fields. To increase reserves, PETRONAS ventured into upstream activities abroad by securing the rights to several exploration areas in Algeria, Angola, Chad, Gabon, Indonesia, Iran, Libya, Myanmar, Pakistan, Sudan, Syria, Tunisia, Turkmenistan and Viet Nam. In 2002 crude petroleum production increased to 597,000 b/d, almost attaining the production target of 600,000 b/d set for the year under the NDP. The increase arose from the commencement of production from five new oilfields during the year, as well as higher external demand. Domestic crude petroleum in Malaysia, which has a low sulphur content and is therefore considered to be of premium quality, was largely exported. In 2000 natural gas produced off the east coast state of Terengganu was processed to produce several components for use as fuel by petrochemical industries, while gas produced off the coast of Sabah was utilized by the methanol and hot briquette iron plants there. Gas from Sarawak was used to produce liquid natural gas (LNG) for export to Japan and the Republic of Korea. The LNG complex in Bintulu, Sarawak, is the world's largest gas field. Malaysia's oil and gas exports amounted to RM 25,500m. (US $6,700m.) in 2000, equivalent to 6.8% of the country's total export earnings. In the downstream sector, domestic crude petroleum-refining capacity increased by 72%, to 356,000 b/d, as a result of a 100,000 b/d-capacity PETRONAS refinery in Melaka coming on stream in 1994 and the expansion of existing refineries. In 1997 Malaysia had a refining capacity of 18.5m. tons, representing 0.5% of world refining capacity.

In 1995 Malaysia's reserves of natural gas were estimated at 2,200m. metric tons of oil equivalent (TOE), or 85,000,000m. cu ft. This constituted an increase of 700m. TOE compared with 1990. Extraction of natural gas in 1995 was 3,500m. cu ft per day, almost double the 1990 extraction level. At the end of 2003 natural gas reserves were estimated at 84,900,000m. cu ft, sufficient to sustain production at that year's level for another 45 years.

During the Sixth Plan period (1991–95) there was a marked increase in the utilization of gas in Malaysia, particularly for electricity generation. In 2002 demand from the power generation industry accounted for more than 70% of the country's total gas consumption. To diversify gas utilization in the country and to encourage greater use of cleaner transportation fuels, gas was promoted as a fuel for vehicles. Under the natural gas for vehicles (NGV) promotion programme, the fuel was exempted from excise duty, making its retail price at the pump half that of premium petrol. In addition, conversion kits that allowed petrol engines to use gas were exempted from import duty and sales tax. With the implementation of the NGV programme, a total of 18 public NGV refuelling stations and two private NGV outlets were built, while 3,700 vehicles were converted to operate on natural gas. In 1998 PETRONAS was given approval to import, in stages, 1,000 monogas taxis, of which 300 were already in operation. The Gas District Cooling (GDC) system was introduced to further diversify the use of gas. The GDC, which utilizes gas to produce chilled water for air-conditioning and waste heat for power generation, helps to lower peak load demand and reduces investment for peaking capacity. Three GDC plants—at the Kuala Lumpur City Centre (KLCC), Kuala Lumpur's international airport and Putrajaya—have started operations.

To ensure the sustainable development of gas resources, a long-term utilization limit of 2,000m. cu ft of processed gas was adopted for Peninsular Malaysia in 1993. Of this, 1,300m. cu ft was reserved for electricity generation, with most of the rest set aside for use as a feedstock by the petrochemical industries.

Other Minerals

The production of other minerals in Malaysia generally declined in the 1990s. In 2003 production of bauxite was an estimated 5,732 metric tons, compared with 376,418 tons in 1991. In 1996 production of copper totalled an estimated 87,220 tons, compared with 111,593 tons in 1992. Production of iron ore in 1996 increased by an estimated 60.7% compared with the previous year, but was still below production in 1991, which had totalled 375,869 tons. In 2003 iron ore production increased further, to 601,612 tons.

INDUSTRY AND INDUSTRIAL POLICY

Manufacturing

Following independence in 1957, Malaysia, like many developing countries, had only a rudimentary manufacturing sector, with the majority of national income coming from the exploitation of its primary resources. However, successive five-year plans focused on developing Malaysia's industrial base, producing a rapid expansion in manufacturing output, which culminated in manufacturing superseding agriculture as the largest contributor to GDP in 1987, providing 22.3%, compared with 8.7% in 1960. Since the Second Malaysia Plan (1971–75) an additional aim of the industrialization policy has been to increase the participation by indigenous Malays in non-agricultural activities. During the 1960s industrialization proceeded through an import substitution strategy, which focused on increasing domestic processing of natural products and on producing basic consumer goods for domestic consumption. However, during the 1970s the emphasis changed to export-orientated, labour-intensive industries such as textiles and electronics, which were predominantly financed by foreign investment. During the first half of the 1980s there was a second phase of import substitution focusing on heavy industries, such as steel and car production, and led by the public sector, particularly the Heavy Industries Corporation of Malaysia (HICOM). This was the period when the national car project (Perusahaan Otomobil Nasional—PROTON) was launched. However, these large investments were likely to be slow to develop, given the small size of the domestic market for industries that are subject to considerable economies of scale.

The manufacturing sector has played an important role in the economic growth of Malaysia since the country's independence. The sector's share of GDP grew from 8.6% in 1960 to 14.8% in 1970, and further to 21% in 1988; by 2003 the sector accounted for 29.7% of GDP. The contribution of manufactured exports to overall export earnings increased significantly from 12% in 1970 to 22% in 1980, and sharply to 59% in 1990 and 86.5% in 2002. However, this share fell to 79.4% in 2003. During the period 1988–99 both the value-added and exports of the manufacturing sector grew strongly, by an average annual rate of 13.9% and 24.3%, respectively.

Manufactured exports have been concentrated mainly in the electrical and electronics industrial sub-sector since the mid-1970s. Its share of total manufactured exports increased rapidly from 9% in 1970 to 48% in 1980, and further increased to 57% and 66% in 1990 and 1997, respectively; this share increased still more, to 66.7%, in 2003 (compared with 70% in 2002). In 2003 employment in the electrical and electronics sector totalled 355,077 workers, or 35% of total employment in the manufacturing sector. Of the total electrical and electronics industry work-force, 73.7% was employed in the semiconductor devices sub-sector. The major export destinations for electrical and electronics products were the USA, Singapore, Hong Kong, Japan and the People's Republic of China.

The second largest manufacturing sub-sector is the transport equipment industry. It comprises four major sub-sectors: the automotive, automobile parts, marine and aerospace industries. Malaysia is predominantly a passenger car market. In 2003 passenger cars accounted for a 76% share of overall production and approximately 80% of total sales of new vehicles. The first national car project, PROTON, was launched in 1983. It was established by a joint-venture agreement between the Mitsubishi Motor Corporation (MMC) and the Mitsubishi Corporation (which both owned 15% of the company's equity) and HICOM (which owned 70% of the company's equity). There are another three national automotive manufacturers, Perusahaan Otomobil Kedua Sdn Bhd (PERODUA), HICOM MTB and Inokom Corporation Sdn Bhd, each with a different technology partner. The second national car project, PERODUA, was launched in 1993 to produce mini passenger cars with Daihatsu (a subsidiary of Toyota Corporation). These manufacturers accounted for 80% of total commercial vehicles and passenger cars sold in 2003, and the combined market share of PROTON and PERODUA alone was about 76%. In 2003 there were 13 assemblers in Malaysia. The automotive industry in Malaysia is domestic-orientated, and about 85%–90% of PERODUA and PROTON's products are sold locally. Since the inception of PROTON, the industry has been protected from local and foreign competition by import duties ranging from 40% to 300% on imported vehicles and parts. PROTON was also awarded a preferential import duty rate of 13% on 'completely knocked down' (CKD) parts and a 50% exemption from excise duty. As a result of the highly protected environment and a rapid growth in domestic demand, the automobile industry grew at an average annual rate of 10% between 1991 and 2000.

The automobile parts manufacturing industry also expanded to serve the national car projects and assemblers. Approximately 200 Malaysian-controlled local companies, as well as 50 licensees and joint ventures of multinational automobile part manufacturers, were operating in the country in 2001. However, many of these local manufacturers were small and medium-sized operations, with only 32 Tier One manufacturers that supplied mainly to PROTON. Of these, 25 were majority foreign-owned companies. However, tariffs on manufactured goods, including cars and vehicle parts, were to be reduced to no more than 5% by the end of 2004 for most ASEAN members. Malaysia requested a delay in the implementation of the tariff cuts, and will reduce tariffs for imported cars to 20% from 1 January 2005. The 5% maximum tariff mandated by the ASEAN Free Trade Area (AFTA) was scheduled to take effect only in 2008 for Malaysia. As a result, the market for domestically-produced cars decreased substantially, in anticipation of Malaysia's compliance to AFTA regulations. PROTON's share of the market declined from 70% in the early 1990s to 50% in 2002. Analysts, however, cautioned that it would be difficult for the Malaysian Government to lift the protection given to local car manufacturers, owing to the political sensitivity of the issue. PROTON, for example, employs almost 10,000 Malaysians, mainly *bumiputra*, and is a significant tax payer. In addition, more than 300 component manufacturers in the country are heavily dependent upon these anchor companies. There has been speculation that PROTON has been seeking to forge strategic alliances with foreign car makers, although the Government wishes it to remain a Malaysian-controlled company. In preparing for the liberalization of the automotive sector, the Malaysian Government has relaxed its restrictions on the ownership of equity. Foreign partners of Malaysian automotive companies, such as Daihatsu, Ford Motors and Volvo, have increasingly taken advantage of the resultant opportunities to increase the size of their shares in their respective partner companies.

Foreign Direct Investment

Since independence in 1957 the Malaysian Government has striven to attract foreign direct investment (FDI) into the manufacturing sector. It was estimated that in 1970 foreign investors owned something like 60% of the manufacturing sector. The statistics on ownership in Malaysia are crude, but it is clear that the flow of FDI into Malaysia has been exceptionally high, amounting to just under 5% of GDP between 1967 and 1986—not as high as the flow into Singapore (at 7% of GDP), but well above the flows into Indonesia and the Republic of Korea over the same period (less than 1% of GDP). Between 1989 and 1994 inflows of FDI into Malaysia totalled US $22,500m., equivalent to 5% of GDP. The comparable figures for Singapore and Thailand were $26,300m. (7% of GDP) and $10,700m. (1%), respectively. Bank Negara reported that despite the currency crisis, net inflows of foreign investment remained strong at RM 19,200m. in 1997, compared with RM 19,400m. in 1996. Inflows of FDI declined to below 4% of GDP in 1998, owing to the financial crisis that adversely affected the region in 1997. In 1998 net inflows of FDI totalled RM 11,600m., and by 2001 FDI had fallen to just RM 2,100m. However, in 2002 the situation was reversed when net FDI inflows increased substantially, to RM 12,000m. In 2003 net inflows totalled RM 9,398m. There has also been a notable development in the nature of FDI inflows, which have become more broadly-based with greater shares of new inflows being directed towards the services sector. This sector received 38% of total FDI, while the manufacturing and gas sectors attracted 28% and 34%, respectively. During the 1990s most of the inflows of FDI were channelled to the manufacturing sector, which accounted for about 65% of total inflows. The oil and gas sector received 18% of foreign investment inflows, while services acquired only 10% and the property sector 7%. Prior to the crisis, the major investing countries were Japan, Singapore, the USA, Taiwan and the United Kingdom. During the period following the Asian financial crisis some significant changes in terms of major investing countries took place. Of the countries that increased their share of investment in Malaysia, the most notable was the USA, followed by Japan, Germany and Singapore.

Malaysia has continued to attempt to attract FDI on an ever-increasing scale, by improving investment conditions and by creating a more attractive destination for the relocation of overseas firms. Prior to 1994 companies that had been granted pioneer status were given tax exemption on 70% of statutory income for a period of five years. After that date companies granted the Income Tax Allowance (ITA) enjoyed a concession of 60% of qualifying capital expenditure, with the amount to be given exemption for any year of assessment not to exceed 70% of statutory income. These percentages were subsequently raised to 80% and 85%, respectively. As a result, during the Sixth Plan period (1991–95) the inflow of FDI continued. In this period approved private investment in the manufacturing sector totalled RM 116,000m., of which FDI accounted for 53%. Foreign investment was particularly important in petroleum products, textiles and electronic products. To encourage greater domestic investment, in 1993 the Government launched the Domestic Investment Initiative, and in that year approved domestic investment exceeded the approvals for foreign investment for the first time since 1987. The Government had taken steps to promote high technology, knowledge-based and capital-intensive industries, and foreign investment into these areas was highly encouraged. Among various initiatives undertaken by the Government was the establishment in 1996 of the Multimedia Super Corridor (MSC) project, in accordance with which companies with approved MSC status enjoy attractive benefits under a 10-point guarantee plan. At July 2000 there were 347 companies with MSC status. Approvals for foreign investment in the MSC almost trebled, to RM 9,400m., in 2000, compared with RM 3,200m. in 1999.

Overseas investment by Malaysian-owned companies had risen rapidly since 1991. However, overseas investment declined in 1997, owing to the uncertainty in the region. Net overseas investment declined to RM 8,100m. in 1998, compared with RM 11,500m. in 1997. In 2002 and 2003 overseas investment by Malaysian companies totalled RM 7,300m. and RM 5,204m., respectively. The significant decrease in overseas investment was attributable to various factors; these included the domestic economic deceleration, the Government's directive to defer overseas investments that do not have direct linkages with the domestic economy, the tightening of exchange control regulations on overseas investment since 1 September 1998 and liquidation of the overseas assets of Malaysian companies. Major overseas investments have been made by the state-controlled hydrocarbon company, PETRONAS (with investments in Algeria, Iran and Sudan), and by the private sector. Both the private and public corporations have invested particularly heavily in South Africa since 1994, and in early 1996 PROTON, the national car-maker, acquired a prestigious British car company, Lotus. The major recipient countries of Malaysian overseas investment in 1998 were Singapore (28%), the USA (22%), the United Kingdom (11%), Thailand (7%) and the Netherlands (4%). Investment in Singapore was concentrated principally in the finance and business services sector (mainly in investment holding companies) and in the manufacturing sector.

Regional Co-operation

Malaysia has pursued regional links at the sub-regional, South-East Asian and East Asian levels. At the sub-regional level, following the success of the first ASEAN growth triangle (the Southern Growth Triangle—SGT), initiated by Malaysia, Singapore and Indonesia in the late 1980s, other intra-ASEAN economic activities have been initiated. In July 1993 the Indonesia-Malaysia-Thailand Growth Triangle, encompassing northern Sumatra, southern Thailand and the northern states of Peninsular Malaysia, was established. The East ASEAN Growth Area (EAGA) was formed in March 1994, covering: Sarawak, Sabah and Labuan in East Malaysia; Brunei; East and West Kalimantan, together with North Sulawesi, in Indonesia; and Mindanao in the southern Philippines.

At country level, attempts have been made to promote integration through ASEAN, which celebrated its 30th anniversary in June 1997. In order to enhance intra-ASEAN trade, in 1992 ASEAN member countries endorsed the Framework Agreement on Enhancing ASEAN Economic Co-operation. A significant outcome of this agreement was the decision to establish AFTA by the year 2008. This was subsequently brought forward to the year 2003 and then to 2002, when AFTA was formally implemented. The main mechanism for the realization of AFTA was the Common Effective Preferential Tariff scheme, through which tariff barriers in ASEAN would be gradually reduced to 5% or less. The scheduled time frame for the realization of AFTA was 2002/03 for the original six member countries, 2006 for Viet Nam and 2008 for Laos and Myanmar. ASEAN has also agreed to eliminate duties on all products by 2010 for the original six members, and by 2015 for the newer members of the grouping. Total intra-ASEAN trade in 1999 accounted for 22.0% of total ASEAN global trade in that year. Malaysia accounted for 25.7% of total intra-ASEAN trade (RM 91,177m.) in 2002.

Development at the level of ASEAN has been slow, but integration at the wider level of East Asia has been even slower. There have been conflicts of interest concerning the Malaysian-inspired East Asia Economic Caucus (EAEC), a trade group that was to include the People's Republic of China and Japan but exclude the USA, Australia and New Zealand. Some governments regarded it as a potential counterweight to the EU and the North American Free Trade Agreement (NAFTA), whilst others, notably Japan, saw the EAEC as a way of integrating the region across the Pacific to North America and Australasia. China, Japan and certain members of ASEAN, however, were reluctant to jeopardize trade relations with the USA, which opposed the new grouping. In July 1993, at an ASEAN ministerial meeting, agreement was reached for the EAEC to operate as a regional grouping within the broader forum of Asia-Pacific Economic Co-operation (APEC).

The 1997–98 financial and economic crisis resulted in greater efforts towards regional co-operation. For example, in March 1999 Ministers of Finance of the Association's member nations established the ASEAN Surveillance Process, which was to monitor current economic and financial sector developments and their impact on the region, and required the submission of information by member states. Subsequently, in March 2000 ASEAN agreed to expand the foreign-exchange repurchasing arrangement (which had been initially established in August 1997 by five member countries) to include the remaining ASEAN countries. In addition, ASEAN, together with China, Japan and the Republic of Korea (designated 'ASEAN + 3'), agreed to enhance existing co-operation among the region's central banks and monetary authorities, with the pronouncement of the Chiang Mai Initiative (CMI) in May 2000. The initiative involved the establishment of a network of bilateral swap arrangements (BSA) among East Asian countries. In 2002 eight BSAs were concluded and signed—between Japan and the Republic of Korea, Thailand, the Philippines, and Malaysia and between China and Thailand, Japan, and Korea, as well as one between Korea and Malaysia—with a total value of US $20,000m. Negotiations on two BSAs, between Korea and Thailand and Korea and the Philippines, are already at an advanced stage, while negotiations for another four BSAs, between Japan and Indonesia, Japan and Singapore, China and Malaysia and China and the Philippines, have been initiated.

An informal summit meeting of ASEAN + 3 leaders, which took place in Singapore in November 2000, further demonstrated closer co-operation within the group and the region. At the meeting a 'Protocol Regarding the Implementation of the Common Effective Preferential Tariff (CEPT) Scheme Temporary Exclusion List' was approved. This agreement provided the mechanism for member countries to delay temporarily the transfer of products into the 'Inclusion List', or to suspend concessions on products already transferred onto the list. This protocol made it possible to accommodate Malaysia's request to defer the inclusion of automotive products into the CEPT Scheme. At the end of 2000 98.3% of all products were already on the Inclusion List, with 92.7% of the products having tariffs of less than 5%. The average ASEAN tariff rate for the original six countries was 3.5%, and it was expected to decline progressively to 2.4% by 1 January 2003.

ASEAN has concluded a number of Free Trade Agreements (FTAs) in recent years. An FTA between ASEAN and the People's Republic of China was signed in November 2002 in Phnom-Penh, Cambodia. FTAs were also concluded between ASEAN and both Japan and India in October 2003, and a Trans-Regional EU-ASEAN Trade Initiative was signed in April 2003. Bilateral FTAs initiated by Malaysia included: a Malaysia-Japan FTA, over which negotiations commenced in December 2003; and a Malaysia-USA Trade and Investment Framework Agreement (TIFA). In order to narrow the disparity within ASEAN and to enhance the competitiveness of the organization, there was an initiative whereby the more developed ASEAN member countries would assist the less developed through education and training projects and the construction and improvement of rail, road and air links. One of the projects approved was the construction of a Singapore–Kunming rail connection, which was initiated by Malaysia to link countries in the region. The route will connect eight countries: China, Thailand, Laos, Cambodia, Myanmar, Viet Nam, Malaysia and Singapore. The ASEAN Framework Agreement on Services (AFAS) was signed in December 1995 in Bangkok, Thailand. AFAS was intended to guide the liberalization of services and to promote co-operation between services suppliers in ASEAN. ASEAN services liberalization commitments existed in addition to any commitments made by its members to the World Trade Organization (WTO). By mid-2004 ASEAN had completed three stages of services liberalization. These stages involved progressive liberalization in the construction, telecommunication, business services, financial, air and maritime transport, and tourism sectors. This included the granting of preferential market access to other ASEAN countries in the establishment of services entities and employment of professionals.

TOURISM

Tourism was not identified as a potential area for growth until 1986. In 1987 the Ministry of Tourism was formed and in 1990 the Government sponsored 'Visit Malaysia Year 1990', which comprised an international promotion and advertising campaign. The approach was so successful that a similar campaign was launched in 1994, which led to an increase of 12% in tourist arrivals compared with the previous year. However, in 1996 the number of tourist arrivals declined for the first time since 1991, to 7.2m. (from 7.5m. in 1995). In 1997 the performance of the tourism industry was particularly affected by the prolonged haze in the South-East Asian region, and this problem of pollution persisted in 1998. Furthermore, the regional economic downturn exacerbated the situation. Tourist arrivals in 1997 and 1998 declined to 6.2m. and 5.6m., respectively. However, in 1999 there was a 41.0% increase in the number of tourist arrivals, to 7.9m., and in 2000 arrivals exceeded 10.2m. Despite the uncertainty faced by the regional tourism industry owing to the heightened fear of terrorism precipitated by the September 2001 suicide attacks on the USA and the bombings in Bali, Indonesia, in October 2002, tourist arrivals increased to 12.8m. and 13.3m. in 2001 and 2002, respectively. However, in 2003 tourist arrivals declined significantly, to 10.6m. To provide for these higher tourist arrivals, there has been much investment in hotel accommodation in recent years. The hotel occupancy rate in Kuala Lumpur increased by more than 55% in 2000, compared with 49% in 1999. Several tourist destinations in the island resorts of Langkawi, Pangkor and Redang and in the highland resorts also recorded a hotel occupancy rate of more than 70%. In 2000 70.3% of tourists came from ASEAN countries, Japan (4.5%) and China (4.2%). The most rapid increase in the number of tourist arrivals was from emerging markets such as India and China, with annual average growth rates of 34.6% and 32.4%, respectively. In 2002 tourists from East Asia, excluding Japan, accounted for about 81% of total tourist arrivals, and contributed about 72% of total tourism receipts. Domestic tourism expanded significantly in the late 1990s as a result of aggressive promotional activities such as 'Cuti-cuti Malaysia', 'Visit Perak Year', 'Visit Selangor Year' and other similar campaigns. In addition, the declaration of holidays for those employed in public services on the first Saturday of the month, effective from 1 January 1999, and also the third Saturday of the month, effective from 1 February 2000, had a positive effect on domestic tourism. The number of domestic tourist excursions increased by 89.9%, to 15.8m. trips, in 1999, compared with 8.32m. trips between August 1997 and July 1998.

INFRASTRUCTURE AND THE CONSTRUCTION INDUSTRY

Energy

In the early 1990s there was a shortage of electricity in Malaysia, and five independent power producers (IPPs) were licensed to build and operate electricity-generating plants. By 1996 the power shortage had been transformed into a significant surplus, as the IPPs established more than 4,000 MW of capacity. In March 1997, in this context of over-supply, the 77% state-owned monopoly, Tenaga Nasional Bhd (TNB), submitted proposals to the Government for the restructuring of its operations. It proposed a separation of its generating and distribution facilities, a diversification of investments into other sectors and a revised consumer tariff formula. By the end of 1996 a wholly owned subsidiary had been established, with responsibility for managing the 7,621 MW of installed capacity. The revised tariff formula proposed that more of the burden of buying high-priced electricity from the five IPPs be passed on to consumers. Between 1995 and 2000 the other two major utilities—Sabah Electricity Board (SEB) and Sarawak Electricity Supply Company (SESCO)—were also restructured. In 2000 TNB's share of total generation in Peninsular Malaysia was at 63%, with the remainder contributed by the IPPs. The IPPs contributed 40% in Sabah and 36% in Sarawak. The average electricity tariff for domestic consumers in Peninsular Malaysia increased by 17.3%, from 20.03 sen per kWh in 1995 to 23.5 sen per kWh in 2000. Sabah maintained its tariff at 24.4 sen per kWh during this period, while that in Sarawak was reduced to 27.1 sen from 28.5 sen per kWh. At the same time, TNB has been seeking a lower price for gas feedstock purchased from PETRONAS, and the IPPs are now bearing some of the cost of subsidizing supplies to rural areas. In 2000 rural electricity coverage in Malaysia was 93%. Malaysia was not expected to need any more new plants until 2004, although the excess supply was scheduled to decline from 50% in 1996 to 30% by 2000. In December 2000 a contract was signed between a local Malaysian IPP, GB3, and the French company ALSTOM for an additional 650-MW generation unit to be built at the site of the Lumut Power Plant. Meanwhile, in the same month German company Siemens AG was contracted to construct two 710-MW power plants, one at Teluk Gong for the IPP Powertek Sdn Bhd, and one in Sepang for the Malaysian Resources Corporation Bhd. No more IPPs were expected to be licensed in the near future.

In the longer term some 2,400 MW of additional capacity should be provided upon the completion of the controversial Bakun hydroelectric dam in Sarawak. The six shareholders in the company that is to manage the plant formally committed themselves to the project in April 1997. The main shareholder (with a 32% holding) is the private company Ekran. Five government institutions hold a further 43% of the shares. The huge project, which is expected to cost RM 13,600m. to complete, has been slow to attract foreign investors and was deferred in December 1997, owing to the economic crisis. At the end of the 1990s the Government revived the project; Sarawak Hidro Sdn Bhd, a company operating under the control of the Ministry of Finance, has been appointed to act as the project's implementing agency. The sub-sea transmission line concept has been abandoned, and the Malaysian Government is exploring the possibility of sales of electricity to Brunei and Indonesia.

Telecommunications

The telecommunications sector continues to develop, following its partial liberalization in mid-1994, which allowed for the establishment of several new operators. However, Telekom Malaysia (in which the Government holds a controlling interest) maintains a dominant share of the market, precisely because the liberalization remains partial, with the Government deciding to defer equal access for other operators to Telekom's network from July 1996 to January 1999. Competition has been keener in the more liberalized, and rapidly expanding, mobile telephone market. In 1996 this market grew by 49% to reach 1.5m. subscribers, with foreign investors gaining sizeable stakes (of between 20% and 30%) in three domestic companies. In February 1998 the Government decided to raise the permitted ceiling on foreign ownership of telecommunications companies from 30% to 49%, and increased it further to 61% in April. However, foreign companies were required to reduce their stake to 49% again within five years.

The communication sub-sector also grew strongly in 2000, benefiting from the increased level of economic activity, as well as the rapid expansion in the mobile telephone industry. In 2002 the number of cellular telephone subscribers had increased to 8.5m. (compared with 3.0m. in 1999). In 2003 the number of subscribers increased again, to 11.1m. New services were introduced over the cellular network, such as Short Messaging Services (SMS), Voice Mail and Calling Line Identification, as well as internet access with Wireless Application Protocol (WAP). Subscription for internet services also increased, with Jaring subscribers rising to 326,930 (compared with 255,100 in 1999), while TMNet's subscription base increased to 712,000 (compared with 405,300 subscribers in 1999). At the end of 2002 the number of internet subscribers had expanded to 2.6m.

Transport

In general, road, rail and port communications are well developed in Malaysia, and are more highly developed in Peninsular than in East Malaysia.

The total network of main roads in 1996 was an estimated 94,500 km, of which about three-quarters was paved. During the Sixth Plan period (1991–95) road development was based on a three-pronged strategy, to improve inter-urban links, to alleviate capacity constraints and to open new growth areas. As well as the 848-km North–South Highway down the western corridor of the peninsula (which opened in September 1994), three major projects were undertaken along the east–west corridor. In addition, various projects were initiated along the Klang valley between Kuala Lumpur and Port Klang, the main seaport. However, congestion in the major urban areas grew rapidly in the Sixth Plan period. It was particularly bad in Johor Bahru, Georgetown and Kuala Lumpur. Traffic entering the capital city increased by 17% per annum, and it was clear that new road improvements would do little to alleviate congestion. As a result, the Government embarked on the construction of an integrated public transport system, in which a major part was to be played by a Light Rail Transit (LRT) system. LRT System I (LRT STAR) began its commercial operations in December 1996, followed by LRT System II (LRT PUTRA) in September 1998. By June 1999 the LRT System in the Klang valley encompassed a total route length of 56 km. Use of the LRT STAR increased from an initial average of 46,853 passengers per day in 1997 to an average of 77,803 passengers per day by the end of 2000. The LRT PUTRA also experienced a high growth rate, with the number of passengers increasing from an average of 12,532 per day to 121,950 per day by the end of 2000. The KL Monorail and the Express Rail Link (ERL) to Kuala Lumpur International Airport (KLIA, see below) from Kuala Lumpur Sentral, a privatized intra-city light rail network of 8.6 km in length, have been in operation since April 2001. In addition, the eight existing Kuala Lumpur-based bus companies were amalgamated into two consortia in 1994.

The railway development programme during the Sixth Plan period was aimed at increasing haulage capacity, enhancing operational safety and improving commuter train services, but more ambitious railway schemes were announced in early 1997. In April the Government announced that approval had been given to a diversified transport group (DRB-HICOM) to build a RM 1,800m. high-speed tilting-train service between Rawang, on the outskirts of Kuala Lumpur, and Ipoh, 175 km to the north. It was also announced that the Government might authorize the construction of a RM 8,300m. high-speed rail link between Kuala Lumpur and Singapore, which has been proposed by Renong. The latter, a Malaysian conglomerate, has a 50% stake in the consortium (in which DRB-HICOM also has a 30% stake) which assumed control of Keretapi Tanah Melayu (KTP), the state railway corporation. Other projects that were completed included: a 180-km inter-city electrified double track railway linking Rawang and Ipoh; a 32-km rail link from Kempas to the Port of Tanjung Pelepas; a 14-km rail link to Port Klang; a 3-km rail link to the North Butterworth Container Terminal; and a rail link to the Segamat Inland Port.

Malaysia's five main seaports handled 112m. metric tons of cargo in 1996, an increase of 19% on 1995. Port Klang (the main port, west of Kuala Lumpur) handled 49m. tons, almost one-half of the total, while Johor Bahru handled just under 20% of the total. Port Klang is now one of the world's leading container terminals, having risen from 57th to 12th in the rankings between 1980 and 2000. As a result of the increase in the volume of cargo handled by the major ports in Malaysia (an increase of 9.5% in 1999), in 2000 various expansion plans were being undertaken by the Government to enhance port facilities and services. Bank Negara reported that at West Port in Pulau Indah, for example, work had begun on the construction of a new 600-m terminal which would increase the capacity of the port to 1.8m. 20-ft equivalent units (TEUs) annually. The construction of 12 new berths at the Kuantan Port Consortium (KPC) was expected to raise the port's capacity by 9m. tons. In the south a new port, the Port of Tanjung Pelepas (PTP), commenced operations in 2000. The port recorded notable success, attaining a capacity of 2.5m. TEUs in 2002. This was attributable mainly to the relocation of the operations of Maersk Sealand and Evergreen, two of the world's largest shipping companies, from Singapore to the port. Another port in the south, the Segamat Inland Port, which began operations in October 1998, was expected to attract shippers located mainly in southern Melaka and northern Johor. During the Seventh Malaysia Plan, more than 90% of Malaysia's international trade was conducted through seaports.

As part of the expansion of infrastructural facilities, in 1991 the Government laid out plans for a new airport at Sepang, 70 km south of Kuala Lumpur, to ease congestion at the existing main airport at Subang (renamed the Sultan Abdul Aziz Shah Airport in 1996). The first phase of the new KLIA, costing RM 9,000m., began operations in June 1998. The airport, consisting of two 4-km runways and a four-winged satellite building with monorail, was able to handle 25m. passengers a year initially, and ultimately 45m., compared with the previous airport's 14m. passengers. In 2002 passenger traffic at KLIA totalled 16.4m. passengers.

The major airline is Malaysia Airlines (formerly the Malaysian Airline System—MAS), which operates an extensive domestic and international service. In 1992 Malaysia Airlines announced a comprehensive aircraft modernization programme at an estimated eventual cost of US $5,000m. The number of international destinations to which it operated services increased from 66 in 1995 to 81 in 2000, while the number of domestic destinations was reduced from 36 to 33. In late 1994 a second national carrier, Air Asia Sdn Bhd, was launched to operate both domestic and regional flights. Pelangi Airways Sdn

Bhd and Transmile Air Sdn Bhd were established later and provide air services to selected domestic and regional destinations, particularly tourist resorts such as Tioman, Pangkor and Medan in Indonesia.

Megaprojects and the Construction Industry

There was considerable concern in Malaysia that the very large infrastructure projects initiated in the 1990s would overheat the economy. Major projects included the hydroelectric dam at Bakun in Sarawak (at an estimated cost of RM 13,600m.), the world's tallest twin towers in Kuala Lumpur (RM 2,000m.), the new airport (RM 9,000m.) and the new administrative capital at Putrajaya (RM 20,000m.), south of Kuala Lumpur. In 1997 several of the projects were deferred, in an attempt to reduce the trade deficit and restore confidence in the financial markets, following the significant decline in the value of the ringgit and concomitant turmoil in the stock exchange.

The construction sector experienced a contraction in output in 1998, principally as a result of lower aggregate demand. In the same year the value-added component of the sector declined by 24.5%, following growth of 9.5% in 1997. The contraction moderated in 1999, however, to just 5.6%. The growth impetus for this sub-sector was derived from the higher allocation of funds in the 1999 budget to infrastructure projects and the resumption of construction work on the new Pantai Expressway, the express rail link connecting central Kuala Lumpur and the new Kuala Lumpur International Airport, and the People-Mover Rapid Transit System. In 2000 growth in the construction industry emanated largely from four privatized road projects (the Kajang Ring Road, Ipoh–Lumut Highway, Guthrie Corridor Expressway and the Butterworth Outer Ring Road) and one independent power plant (the Technology Tenaga in Perlis). In 2001 the construction sector recovered to record growth of 2.3%, owing primarily to government spending under the fiscal stimulus programme, privatized infrastructure projects and residential housing development. In 2002 major ongoing public projects included the electrified double tracking of the railway track between Ipoh and Rawang.

Several measures were introduced in 1998 and 1999 to support the construction industry. Emphasis was placed upon the construction of low- and medium-cost houses, for which import content was low and sectoral linkages were high whilst the underlying demand remained strong. In May 1998 a RM 2,000m. fund for the Special Scheme for Low and Medium Cost Houses was established, of which RM 1,000m. was made available for bridging finance and another RM 1,000m. for end-financing. Other measures introduced included the liberalization of lending for the construction or purchase of residential properties costing RM 250,000 and below; the removal of the RM 100,000 levy on foreign purchases (effective from August 1997); and a reduction of the Real Property Gains Tax to 5%. With effect from late April 1998, foreigners were permitted to purchase any type of residential units, shop-houses, commercial and office space costing above RM 250,000 per unit, provided that the financing for any such purchase was not obtained from within Malaysia and also that purchases remained confined to newly completed projects or to those which were at least 50% completed. To help reduce excess stocks, the Government assisted the First and Second Home Ownership campaigns. Incentives offered during the campaigns included an exemption of stamp duties and a minimum price discount of 5% on properties costing RM 100,000 or less (and of 10% on properties costing above RM 100,000). As a result of these efforts, the value-added of the residential sub-sector grew by 32.8% in 1999, a marked improvement on 1998, in which year the sector contracted by 16.5%. However, the residential sub-sector continued to be faced with the problem of excess supply. At the end of June 1999 the number of unsold units was estimated at 93,599. In 2001 and 2002 unsold units numbered 40,977 and 52,419, respectively, mainly owing to an increase in the number of new units and to locational factors. Houses priced in the RM 50,000–RM 150,000 range constituted the majority (56%) of unsold units, while houses priced above RM 250,000 acccounted for only 6%. Low-cost houses accounted for 12% of total unsold residential property. Meanwhile, house prices continued their downward trend in 1999, with the Malaysian House Price Index having declined by 12% in the first half of that year following a contraction of 9.4% in 1998. Depressed house prices stimulated purchases by foreign buyers, which increased by 18.7% in 1999. However, from the second half of 1999 into 2000, prices of residential properties rose by 15.4%. In 2001 prices increased only marginally (by 0.9% in the first half of that year). The implementation of the Foreign Investment Committee (FIC) guidelines, effective April 2001, resulted in a substantial increase in the foreign purchase of residential properties in Malaysia. In 2002 total foreign purchases rose by 35%, to RM 590m. Meanwhile, the non-residential sub-sector remained weak, owing to the continued excess supply of office and retail space. The overall occupancy rate for purpose-built office space in the Klang valley declined from 79.8% in 1998 to 76.2% in 1999. However, the occupancy rate of the retail space sub-sector improved substantially (by 76.6%) in 1999, compared with an increase of 59.5% in 1998. According to Bank Negara, the main reasons for this improvement were improved business and consumer sentiments and lower levels of new supply. As at the end of September 2002 the average occupancy rates for office space and retail complexes had stabilized at 78% and 77%, respectively, due to a decline in supply.

PRIVATIZATION

In 1991 the Privatization Master Plan (PMP) was introduced, and during the Sixth Malaysia Plan period a total of 204 projects were privatized. It is in the area of transport and infrastructure that the main thrust of privatization has taken place in Malaysia. Principal projects included the LRT system in Kuala Lumpur, the National Sports Complex, PROTON, the ports of Klang and Johor, the Second Causeway Link to Singapore and TNB, the electricity supply company.

During the Seventh Malaysia Plan (1996–2000) 98 projects were privatized, of which 47 were existing projects and 51 new projects. Of the total number of new projects, 49 were in the construction, transport, and electricity and gas sectors. Five of the projects were completed, while 25 were in various stages of implementation and the remainder in the planning stage. The completed projects were: the Damansara–Puchong Highway; the upgrading of the Sungei Besi Road; and the three IPPs in Sandakan, Karambunai and Batu Sapi in Sabah. By 2002 the privatized road projects that had been completed were: the Kajang Ring Road; the Butterworth Outer Ring Road; the Guthrie Corridor Expressway; the new Pantai Highway; Package C of the SPRINT Expressway; the ERL; and the Kajang–Seremban Expressway.

Malaysia's water and healthcare services have also undergone privatization. The transfer to the private sector of national water utilities began in the 1990s, when deals were designed by states under guidelines determined by the Federal Government. There are currently more than 20 local companies providing water-supply related services in Malaysia, with three or four foreign enterprises also involved in the sector. Privatization of healthcare services also started in the early 1990s. In 1992 the Ministry of Health announced that haemodialysis clinics in 16 hospitals would be contracted to private companies. In the same year the Institut Jantung Negara (National Heart Institute) was separated from the Kuala Lumpur General Hospital and was corporatized as a fully government-owned referral heart centre. In 1994 Malaysia's drug distribution system, which was previously run by the Government's General Medical Store (GMS), was privatized. All state hospitals were required to procure their supplies from a new company, Southern Tasek Sdn Bhd (STSB), a subsidiary of Renong. Owing to various problems and to a generally poor performance, the Government awarded a 15-year renewable concession to another company, Remedi Pharmaceuticals, a subsidiary of United Engineers (Malaysia) Bhd (UEM). Food supplies were privatized in 1996. Meanwhile, under the Seventh Malaysian Plan (1996–2000), the Government had indicated that many health facilities, including hospitals and specialist units, would eventually be privatized.

TRADE, THE BALANCE OF PAYMENTS AND THE EXCHANGE RATE

Malaysia is an extremely open economy in the sense that in 1995 exports of goods and non-factor services were slightly greater than GDP. At the end of the 1980s the current account of the balance of payments was in surplus, but by the early 1990s it had moved sharply into deficit. According to Malaysian sources, the merchandise account remained in surplus. Exports of goods and services grew by an annual average of 9.2% during 1971–90, 18.4% during 1990–95 and 13.3% in 1997. In 2000 exports of manufactured goods recorded an increased growth of 17% (compared with 14.3% in 1999). However, exports of manufactured goods declined by 10.3% in 2001. Exports recovered in 2002, and growth of 8.3% was recorded in 2003. In 1997 the merchandise account registered a surplus of RM 11,300m., compared with a surplus of RM 933m. in 1996. The performance of the account was even better in 1998 and 1999, in which years it recorded a surplus of more than RM 69,300m. and RM 83,000m., respectively. However, the trade account surplus decreased steadily, to RM 51,000m. in 2002. In 2003 the merchandise account registered a surplus of RM 97,701m. Growth in gross exports, which had remained positive since 1997, declined by 10.4% in 2001. Imports also declined, by 9.9%. In 2002 gross exports recorded growth of 6.0%, largely owing to a significant acceleration in the export of semiconductors. In 2003 growth was 11.6%.

The high merchandise growth in exports was accompanied by rapid import growth. Imports grew by an annual average of 10.0% during 1971–90, compared with 19.7% during 1990–95, and 9.9% in 1999. In 2002 imports grew by 8.3%, outpacing the 6.0% growth of exports. The majority of imports were intermediate goods, which accounted for 71.3% of total imports in 2002, while imports of capital goods accounted for 15% and consumption goods 6.3% in the same year. The primary cause of the overall growth in imports was the increase in imports of intermediate goods (which recorded a rate of growth of 6.2% in 2000) and, to a lesser extent, the increase in imports of consumption and dual-use goods (imports of which increased by 15.9% in 2002). Imports of capital goods experienced an increase of 10.6% in 2002, which occurred mainly in the services, construction and mining sectors, while capital imports for the manufacturing sector declined as a result of excess capacity in some sub-sectors.

The services account has been in deficit since the late 1970s, and the deficit widened in the 1980s largely owing to high outflows of investment income, freight and insurance and other services. It was this rapid growth in the services deficit that caused the deficits in the current account from 1990. In the second half of the 1990s, however, the services-account deficit narrowed; this improvement was supported by the significant increase in tourism receipts and, to a lesser extent, by an improvement in the foreign-exchange earnings on airport and port-related services. Bank Negara reported that in terms of gross national income (GNI), the services-account deficit declined to 2.9% in the period 1996–2002, from 4.6% in 1991–95. In 2003 the services account experienced a deficit of RM 15,206m. According to the Bank Negara Report 1997, prior to 1980 Malaysia had not experienced prolonged periods of deficit on the current account of the balance of payments. The first period of sustained current-account deficit occurred in 1980–86, when the deficits averaged 6.5% of GNI. However, following voluntary structural adjustment policies undertaken by the Government, surpluses were recorded in the period 1987–89. The current-account deficit re-emerged during the period 1990–97, averaging 6.2% of GNI. There was a significant improvement in the current-account position in 1998, however, when a surplus of RM 36,100m. was recorded, following a deficit of RM 14,200m. in 1997; this constituted the first surplus on the current account of the balance of payments since 1989. In terms of GNI, the surplus increased to 13.7% of GNI in 1998 from 5.4% in 1997, surpassing the previous high of 8.9% achieved in 1987 after the last recession. The current account improved further in 1999, recording an increased surplus of RM 47,381m. or 16.9% of GNI (compared with a surplus of RM 36,800m. in 1998). However, the surplus declined to RM 31,959m. in 2000 and further, to RM 25,070m., in 2001. The surplus remained at approximately RM 27,000m. in 2001 and 2002. However, in 2003 the surplus expanded to RM 50,848m.

Large inflows of long-term capital into Malaysia, however, were more than sufficient to finance the current-account deficit. Net inflows of long-term capital were moderate in the 1960s and 1970s. In the late 1970s and early 1980s the Government relied on external loans to sustain public-sector investments, leading to higher net inflows in the long-term official capital account (reaching a peak of RM 6,300m. in 1983). The net long-term official capital account was in deficit during 1987–92, indicating repayment for the large external debt incurred in the early 1980s. However, increasing net long-term private capital inflows from 1988 onwards caused an overall surplus on the balance of payments. This was due to the new incentives introduced in 1986, combined with the buoyant domestic economy and the lower exchange rate of the ringgit. This led to large inflows of FDI mainly into the manufacturing and petroleum sectors, with smaller amounts accruing to the agricultural and property sector. It increased from RM 1,100m. in 1987 to total more than RM 10,000m. throughout the 1990s. In 1997 the net inflow of long-term capital increased by 27.7% in real terms to RM 18,800m. (compared with RM 13,600m. in 1996). This consisted of RM 13,200m. (RM 12,800m. in 1996) of net long-term private investment and RM 5,600m. (RM 800m. in 1996) of net official long-term capital. Thus, a surplus of RM 5,700m. was recorded on the capital account in 1997, compared with a surplus of RM 1,300m. in 1996.

In 1998 the surplus on the current account was higher than the surplus on the capital account, and from 1999 the current-account surplus began to offset the deficit on the capital account. In 1998 the surplus on the current account was RM 37,394m., while that on the long-term capital account declined to RM 10,900m. as a result of the weakened global economy and of a risk-averse attitude amongst investors. Since the first quarter of 2001 Malaysia's balance-of-payment accounts have been estimated in accordance with the guidelines prescribed by the fifth edition of the Balance-of-Payments Manual (BPM5). In this new format, the capital account has become known as the 'financial account', recording direct, portfolio and other investments. Using this format, the current-account surplus for 2000 was RM 31,959m., and for 2001 was RM 25,070m., while the financial account displayed deficits of RM 23,848m. and RM 17,948m. in 2000 and 2001, respectively. In 2002 the financial account recorded an improvement, registering a lower net outflow of RM 12,000m. However, the overall balance showed a significant deficit of RM 10,900m. in 1997 (compared with RM 6,200m. in 1996), which led to a decline in external reserves from RM 70,020m. to RM 59,120m. The large deficit was a result of substantial outflows of short-term portfolio investment from the country. Movement of private short-term capital only became significant in 1989. Net short-term capital inflows increased from RM 1,600m. in 1989 to RM 13,000m. in 1993, owing to an increase in the net external liabilities of the commercial banks and large inflows of portfolio funds into the booming stock market. In 1994 Malaysia experienced a high net outflow, totalling RM 7,900m., for speculative reasons. In 1995 the net inflow of short-term capital improved, and there was a small surplus of RM 734m. before increasing rapidly to RM 11,200m. in 1996. However, the short-term capital account recorded a large outflow of RM 14,000m. and RM 21,700m. in 1997 and 1998, respectively, in the wake of the currency turmoil in the region. The short-term capital account recorded a further substantial net outflow of RM 36,000m. in 1999. The bulk of the net outflow of funds in 1997 was due to the liquidation of portfolio investment as well as intervention to support the ringgit exchange rate. This resulted in a huge deficit in the overall balance of the balance of payments. In 1998 the overall balance of payments reverted to a surplus of RM 40,300m., following a deficit in 1997. Net international reserves of Bank Negara increased to RM 99,400m. by the end of 1998 from RM 59,100m. at the end of 1997; this level of reserves was sufficient to finance 5.7 months of retained imports. In 1999 larger outflows of short-term capital, together with exchange revaluation losses of RM 5,400m., led to a decline in the overall balance of payments to RM 17,819m. (reduced from RM 40,301m. in 1998). In the same year, external reserves increased to RM 117,244m., or US $30,854m. The level of reserves for 2000 and 2001 were

RM 113,541m. and RM 117,202m., or $29,900m. and $29,720m., respectively. In 2002 the level of reserves increased steadily, to RM 131,394m., mainly owing to the strong trade surplus.

Commentators argued that deficits in the 1990s were largely driven by the private sector, unlike the deficits in the 1980s, which were led by the government sector. Large-scale investment projects undertaken by the private sector in the 1990s resulted in deficits in the resource balance. Prior to the 1980s there had been no serious deficit in the savings-investment position. However, in the first half of the 1980s the savings-investment gap widened to 8.8%, mainly owing to large-scale investment by the Government. Following the voluntary adjustment programme undertaken by the Government, in the latter half of the 1980s savings were higher than investment. However, the national resource position reverted to a deficit in 1990. For the first half of the 1990s, the deficit widened further as investment growth continued to accelerate. The deficit in the resource balance in the 1990s was mainly due to a significant growth in private-sector investment activities, despite the high rate of domestic savings in the country by international standards. In terms of GNI, the share of gross national savings was 38.5% and 40.0% in 1996 and 1997, respectively. Meanwhile, the gross domestic capital formation as a share of GNI was 45.1% in 1997, compared with 43.6% in 1996. This led to a deficit in the savings-investment position of RM 14,200m. in 1997 (compared with RM 12,300m. in 1996). However, the situation was reversed in 1998, in which year gross national savings increased by 5.1% in spite of the ongoing regional financial crisis, and thereby increased its share of GNI to 41.2%. This increase, combined with a decline in gross capital formation of 38.4%, resulted in a surplus in the savings-investment position of RM 36,100m., or 13.7% of GNI. The country's resource balance position recorded an increased surplus of RM 47,900m. (or 17.1% of GNI) in 1999. A smaller total resource surplus of RM 31,200m., accounting for 10% of GNI, was recorded for 2000. In 2002 the stronger economic performance resulted in a slightly lower resource surplus of RM 27,400m., or 8.1% of GNI (compared with a surplus of RM 27,700m., or 9% of GNI, in 2001).

Bank Negara reported that total medium- and long-term external debt at the end of 1998 had increased by 3% on the previous year to RM 131,300m. In US dollar terms, the debt increased by 5.5%, to US $34,500m. in 1998 from $32,700m. in 1997. Owing to prudent management and an efficient debt-monitoring system, however, total external debt outstanding declined by 1.4% to RM 159,000m. in 1999. Short-term debt, comprising mainly the external borrowings of commercial banks and the non-bank private sector, had declined by 34% to RM 28,500m. at the end of 1998, and further by 29% to RM 22,800m. at the end of 1999. For the first time since 1989, in 1999 the private sector recorded a net repayment of external loans of RM 1,800m. However, this was offset by higher net borrowing by the Federal Government and non-financial public enterprises (NFPEs) totalling RM 6,800m. in 1999. The Federal Government's external debt, which accounted for 9.3% of total external debt in 1998, had increased by 15.2% to RM 14,900m. at the end of 1998 from RM 12,900m. in 1997. This debt had increased further, by 23%, to RM 18,400m. at the end of 1999, accounting for an increased share of 12.0% of total external debt. Meanwhile, the outstanding debt of the NFPEs increased by 10.2% to RM 58,600m. in the same year. Private-sector external debt declined by 37.4% to RM 9,600m. in 1998 from RM 15,400m. in 1997, owing mainly to a slowdown of the economy resulting from the financial crisis. Total outstanding external debt declined by 3% to RM 157,000m. at the end of 2000, principally owing to the reduction in short-term debt and, to a lesser extent, the revaluation gains resulting from the stronger ringgit. In 2001 the total debt was RM 169,800m. and the ratio of external debt to GNI was 55.4% (the ratio had been 64% in 1997 following the regional financial crisis). The new loans were mainly drawn by the public sector (RM 7,200m.), with new external borrowing totalling RM 2,800m.

FINANCE AND BANKING

Monetary policy has emerged as an important aspect of economic management in Malaysia, with the rapid expansion in the external reserves held by the central bank and the commercial banks, and with the growth in the money and capital markets. Since the inflationary problems experienced in the early 1980s, Bank Negara has attempted to exercise strict control over commercial and merchant banks, in order to control the growth in the money supply. However, following an apparently successful period of low inflation and high growth in the mid- to late 1980s, the inflation rate (up to 4.4% in 1991 and 4.8% in 1992, compared with less than 1% during 1985–87) and the growth in the money supply (11% in 1991, compared with 3.6% in 1987) increased during the early 1990s. The renewed inflationary pressures were a consequence of the strong domestic demand for credit and the sharp increase in liquidity caused by inflows of foreign capital, which were related to Malaysia's rapid economic growth. In August 1991 Bank Negara introduced a strict monetary policy to curb inflation and restrict the growth in the money supply. During 1992 high interest rates and restrictions on credit resulted in a slowing of consumer spending. These high interest rates attracted funds from abroad, however, accelerating the appreciation of the ringgit. In December 1992 the dollar notation was abandoned, and the official currency denomination became ringgit Malaysia (RM).

In the mid-1990s Bank Negara increasingly resorted to direct action to control the monetary sector, attempting to stabilize liquidity by changing reserve requirements, and imposing limits on the composition of lending. In early 1998 monetary policy was tightened in order to contain inflation and restore stability in the foreign-exchange market. There was some disagreement between the Prime Minister, Mahathir Mohamad, and the then Minister of Finance, Anwar Ibrahim, over the interest-rate policy: Mahathir believed that lowering the interest rate would stimulate the weakening economy and constituted a politically favourable option, while Anwar insisted on a tight monetary policy, arguing that it was necessary that the interest rate remain high in order to restore confidence in the economy and to avoid a deterioration of the exchange rate as well as capital flight. However, the appointment of Dato' Paduka Daim Zainuddin as Minister of Special Functions in charge of economic development by Mahathir in June 1998 severely undermined Anwar's position as Minister of Finance (a post from which he was subsequently removed), and signalled a change of direction in the interest-rate policy. In early August 1998 the three-month intervention rate was reduced from 11% to 10.5%. This rate was subsequently reduced further to 10% and then to 9.5%, and was reduced again to 8% in early September following the introduction of new exchange-control measures on 1 September. By early November the rate had been reduced to 7%. Through a combination of monetary and fiscal policy, together with a rapidly expanding economy, the Government was also quite successful in controlling inflation.

A more influential central role for Bank Negara was part of the wide-ranging Banking and Financial Institutions Act of 1989, which also aimed to bring greater competition to the money and capital markets. As part of the legislation, foreign banks operating in Malaysia were required to incorporate locally and transfer more than 50% of equity to local ownership by October 1994 or face expulsion.

In the 1990s the banking sector flourished. In 1995 its pre-tax profit was RM 6,900m. In 1996 the sector's pre-tax profits rose by 26%, to RM 8,700m., with commercial banks accounting for RM 6,200m. In March 1993 Bank Negara attracted strong criticism when it was revealed that it had incurred losses of about US $3,570m. in 1992, mostly in speculation on the foreign-exchange markets. This scandal was compounded by further losses of $2,100m. in foreign-currency trading in 1993. Since May 1998, however, the Government has placed priority on strengthening the financial system in an effort to stimulate the country's economic recovery. One of the measures adopted was a comprehensive plan to restructure the banking sector which included: a merger programme; the creation of an asset management company, Pengurusan Danaharta Nasional Berhad (Danaharta); and the establishment of a special-purpose vehicle (SPV), Danamodal Nasional Berhad (Danamodal), and of the Corporate Debt Restructuring Committee (CDRC). The merger programme for the domestic banking sector was completed with the merger of Arab-Malaysian Finance Bhd with MBf Finance Bhd in June 2002, and the acquisition of Bank Utama (Malaysia) Bhd by RHB Bank Bhd at the end of 2002. With the completion of these mergers, the domestic banking system has been transformed from one that was highly fragmented, with 71 institutions prior to the 1997 regional crisis, to 30 banking institutions under 10 domestic banking groups.

The CDRC, which started its operations in 1998, successfully resolved 48 cases involving debts of RM 52,000m., representing approximately 65% of the cases under its remit. Of these, 32 cases, with debts totalling RM 36,000m., were fully restructured. The implementation of the remaining 16 cases, with total debts of RM 16,600m., would reduce non-performing loans (NPLs) by two percentage points. The CDRC ceased operations on 15 August 2002, when Danaharta assumed responsibility for the implementation of recovery strategy through restructuring, settlement, foreclosure and other arrangements intended to maximize the recovery of the NPLs acquired. At the end of 2002 the average recovery rate was 57%. Danaharta was expected to finish its operations in 2005.

In 1990 the Government declared the island of Labuan an international offshore financial centre, as part of an attempt to stimulate Malaysia's financial sector. By 1993, however, only 13 banks were licensed to establish offshore units in Labuan. Its failure to attract significant investment was largely due to a lack of basic infrastructure and telecommunications systems. In mid-1993 Malaysia banned Singapore-based offshore banks from financing Malaysian business ventures in an attempt to promote Labuan. By September 1998 Labuan had issued 64 licences.

In November 1994 Bank Negara introduced a two-tier regulatory system, separating larger banks from those less well-capitalized. The 11 commercial banks in the Tier One league at the end of 1996 were required to have shareholders' funds of at least RM 1,000m. by the end of 1998, and a similar level of paid-up capital by the end of 2000 to retain the status. The Tier One designation entitled these banks to privileges that lesser banks do not enjoy, such as the rights to issue negotiable instruments, to participate in the derivatives market and to expand overseas. However, this system was abolished on 10 April 1999, and incentives that had previously been accorded only to Tier One banking institutions were made available to all institutions (subject to the approval of Bank Negara). In April 1996 it was announced that foreign state-owned banks would be permitted to establish offices in Malaysia, but a ban on new privately owned foreign banks (except in the offshore centre of Labuan) would remain in place to protect the financial markets. However, in the late 1990s the domestic financial system moved towards greater liberalization, as evidenced by the significant foreign presence in the Malaysian financial sector by 2000. In the banking sector at mid-2000 there were 13 wholly foreign-owned commercial banks, with foreign interests in the aggregate, accounting for about 30% of the total assets of the country's commercial banks. In the insurance sector, meanwhile, the foreign market share amounted to 74% of life insurance premiums and 35% of general insurance premiums. However, foreign banks are conditioned to operate as locally owned subsidiaries and are required to obtain 60% of their local credit from Malaysian banks.

Other developments since the mid-1980s include the establishment of an Islamic bank, the formation of a secondary mortgage market, the establishment of property unit trusts and, in 1991, of the country's first credit rating company, the Rating Agency of Malaysia. In April 1994 the Government announced the establishment of Khazanah Nasional to manage government assets. Khazanah, which was based on the highly successful state-run investment agency in Singapore, Temasek Holdings, unified all government corporate holdings, and was able to invest in industries considered important for the country's economic development.

Since the early 1980s the Government has also focused on decreasing its budget deficit, primarily through a reduction in public expenditure and the privatization of publicly owned companies. In every year between 1993 and 1996 there was a surplus on government finances, with the surplus of operating revenue over operating expenditure being more than RM 14,000m. in each of the three years 1994–96. Owing to the economic crisis, the 1997 budget aimed for an increase in the overall fiscal surplus to 1.6% of GNI in 1997, compared with the surplus of 0.8% of GNI achieved in 1996. A budgetary surplus of 2.7% of GNI was initially forecast for 1998. However, the severe deflationary impact of the regional financial crisis on the domestic economy resulted in a revision of policy: additional budgetary funds were allocated from mid-1998, mainly for infrastructure development, and this contributed to a fiscal deficit equivalent to 3.7% of GDP in 1998. According to the IMF, the budget deficit amounted to US $9,488m. in 1999 (equivalent to 3.2% of GDP). The budget deficits for 2000 and 2001 were estimated at 4.0% and 5.5% of GDP, respectively. The Government's budgetary operations remained expansionary in 2002, resulting in a deficit of approximately RM 20,300m., or 5.6% of GDP.

The Kuala Lumpur Stock Exchange (KLSE) was established in 1973 but assumed an independent existence only after the traditional links with the Stock Exchange of Singapore were severed in December 1989. Confidence in the Exchange was eroded in the first half of 1990, leading the Government to introduce higher capital requirements for licensed brokerage firms from June of that year, which caused many companies to merge. Owing to the Government's privatization policy, the KLSE's market capitalization was higher than that of its regional competitors in Bangkok, Thailand and Singapore. In 1995 the Government announced a series of measures that aimed to liberalize the capital market, including permission for the full foreign ownership of fund-management companies, in anticipation of a future role as a major regional capital market. In May 1996 the Government granted an operating licence for the country's second financial futures exchange, the Commodity and Monetary Exchange of Malaysia. An additional exchange, the Malaysian Exchange of Securities Dealings and Automated Quotations (MESDAQ), which was aimed at technology companies, began operating at the end of 1998. (In March 2002, however, MESDAQ merged with the KLSE and ceased to operate as a stock exchange.) In 1996 new funds tapped from the KLSE grew by 37%, to a record RM 15,900m., with rights issues accounting for about one-third of these funds and with initial public offers accounting for a further quarter. In 1996 the KLSE's daily turnover almost doubled to 268m. shares, worth RM 1,900m. In 1999 positive market sentiment, driven mainly by the country's strong economic performance and improved corporate earnings, led to an increase in the KLSE composite index by 38.6% (from a low of about 420 in July 1998, owing to the regional financial crisis) to end the year at 812.33 points. The composite index failed to return to its pre-crisis level, however. In 2000 the KLSE composite index fell by 16.3%, to end the year at 679.64 points, while the market capitalization declined by 19.6%, to RM 444,350m. The index improved by 2.4% in 2000, while the market capitalization rose by 4.6%, to RM 465,000m. The central bank reported that the KLSE performed quite favourably compared with the regional bourses. The contributing factors to this positive development were: strong economic fundamentals; the completion of the restructuring of several large corporations; and favourable changes in corporate governance.

In February 2000 multilateral trade negotiations conducted under the auspices of the WTO concerning the services sector commenced. At the Fourth WTO Ministerial Conference, which was held in Doha, Qatar, on 9–14 November 2001, it was agreed that member countries should submit initial requests for specific commitments by 30 June 2002, with initial offers to be entered by 31 March 2003. It was also agreed that negotiations relating to the services sector were to be completed by no later than 1 January 2005. Malaysia had committed, under the General Agreement on Trade in Services (GATS), to allow an aggregate maximum foreign equity limit in the domestic banking institutions of 30%. A commitment was also made to allow existing, original, foreign shareholders to own up to 51% of companies within the insurance sector. The AFAS was signed in 1995, with the intention of liberalizing financial services in ASEAN. The ASEAN finance ministers agreed that the commitment offer under the AFAS was preferable to existing commitments under the GATS. Foreign ownership in the financial sector is quite substantial. Of the 27 commercial banks (including two Islamic banks), 14 are wholly foreign-owned and foreigners own an average of 18% of the total equity of five domestically owned banks. Twenty-five out of 63 insurance companies are completely under foreign ownership.

DEVELOPMENT AND PLANNING

Malaysia has launched 10 development plans since independence: the First Malaya Plan (1956–60), the Second Malaya Plan (1961–65) and the eight five-year Malaysia Plans covering the period 1966–2005.

In the 1950s a basic feature of the Malaysian economy was the concentration of Malays and other indigenous people in the 'traditional', subsistence agriculture sector of the economy with other races, notably the Chinese, having the major role in the modern rural and urban sectors. In order to redress the balance, the Second Malaysia Plan and, to some extent, the Third Plan were based on the NEP, which was inaugurated in 1970 in the aftermath of the 1969 racial riots. The two main aims of the NEP (and of the Second Malaysia Plan) were to reduce and eventually eradicate poverty by increasing income levels and employment opportunities for all Malaysians, irrespective of race; and to accelerate the process of restructuring Malaysian society in order to correct economic and geographical imbalances, thereby reducing, and eventually eliminating, the identification of race with economic function. The period 1971–80 represented the first decade of the Outline Perspective Plan (OPP) 1971–90, within which the objectives of the NEP were to be realized.

Over the period covered by the NEP, the controversial target of 30% *bumiputra* equity ownership was not achieved. However, the *bumiputras'* share of the corporate sector increased from 2.4% to 20.3% between 1970 and 1990. Chinese and Indian equity ownership increased from 32% to 46% over the same period, while that of foreigners was reduced from 63% to 25%. The proportion not accounted for was held by nominee companies which were not classified by race, promoting conflicting claims about the accuracy of the declared share of *bumiputra* ownership. However, the transfer of state assets to *bumiputra* ownership resulted in the creation of a wealthy élite within the Malay community, where income disparity was now greater than in other ethnic groups.

In June 1991 the New Development Policy (NDP), the successor to the NEP, was announced. This formed the basis of the Prime Minister's 'Vision 2020', the date by which he intended Malaysia to become an industrialized country. The focus of the NDP differed slightly from that of the NEP; racially based economic and social engineering was relaxed in favour of national unity. The 30% target for *bumiputra* corporate ownership was retained, but there was no deadline set for its achievement. The emphasis was placed on the development of skills to promote the consolidation of *bumiputra* wealth. In the early 1990s there were too few Malays with relevant management qualifications; this shortage of suitable personnel hindered government efforts to broaden *bumiputra* participation at a high level. The NDP's macroeconomic and social targets were contained in the OPP for 1991–2000. Real GDP was projected to grow at an average rate of 7% per year, while development expenditure was to total RM 224,000m. Manufacturing's share of exports was projected to increase to more than 80%, while average annual growth in imports was to be restrained to 11.8%, compared with the average 15.7% increase that occurred under the NEP. The level of unemployment was to be reduced from 6% of the labour force in 1991 to 4% by the year 2000, and the poverty rate was to decrease from 17% to 7%.

In May 1996 the Seventh Malaysia Plan (1996–2000) was introduced. Under this, the average annual GDP growth target was set at 8.6%, with the manufacturing sector expected to grow at an annual average of just under 11%. Exports of manufactures were expected to increase at an average of 17% annually. The major thrust of the Seventh Malaysia Plan was to upgrade the economy and to achieve a strategic shift from low-skill, low value-added output to high-skill, high value-added output. A major aspect of this was an emphasis on improved vocational training, which was to be stimulated by the Human Resources Development Council, and the provision of greater opportunities within tertiary education under a commercialized and partly privatized university system.

However, one of the major obstacles to this restructuring, which the Seventh Plan did not address in detail, was the increasing reliance of the Malaysian economy on imported unskilled labour. There is also a growing caucus that is concerned about the increasing pollution and congestion problems, which are seen as being related to the rapid rate of growth that Malaysia achieved in the late 1980s and early 1990s and which was projected to continue under the Seventh Malaysia Plan. However, these concerns were superseded by the economic crisis of 1997–98, which transformed the economic situation and delayed by at least 10 years attainment of industrialized status, the Prime Minister's 'Vision 2020', put forward in 1991. ('Vision 2020' had predicted, on the basis of forecasts of the country's future average income per head, that Malaysia would be classified as a high-income or developed country by 2020.)

In early 2001 the Malaysian Government introduced the Third OPP, a 10-year development programme for 2001–10, which focused on creating a resilient and competitive economy. In addition, the Eighth Malaysia Plan (2001–05) was announced on 23 April 2001. The main aims of the plan were: to shift the growth strategy from one that is input-driven to one that is knowledge-driven; to accelerate structural transformation within the manufacturing and services sectors; and to strengthen socio-economic stability through the equitable distribution of income and wealth. The plan laid emphasis on nine principal strategies: maintaining macroeconomic stability; eradicating poverty and restructuring society; enhancing productivity-driven growth; increasing competitiveness in key sectors; expanding usage of information and communications technology; enhancing human resource development; achieving sustainable development; ensuring quality of life; and strengthening moral and ethical values. Under the plan, the Government aimed to achieve economic growth averaging 7.5% per year, with low inflation and sustainable budgetary and external accounts. It was envisaged that the manufacturing and services sectors would continue to provide the greatest contribution to the economic growth of Malaysia.

Statistical Survey

Sources (unless otherwise stated): Department of Statistics, Blok C6, Parcel C, Pusat Pentadbiran Kerajaan Persekutuan, 62514 Putrajaya; tel. (3) 88857000; fax (3) 88889248; e-mail jpbpo@stats.gov.my; internet www.statistics.gov.my; Departments of Statistics, Kuching and Kota Kinabalu.

Note: Unless otherwise indicated, statistics refer to all states of Malaysia.

Area and Population

AREA, POPULATION AND DENSITY

Area (sq km)	
Peninsular Malaysia	131,686
Sabah (incl. Labuan)	73,711
Sarawak	124,450
Total	329,847*
Population (census results)	
14 August 1991	18,379,655
5–20 July 2000	
Males	11,853,432
Females	11,421,258
Total	23,274,690
Population (official estimate at mid-year)	
2002	24,530,000
2003	25,050,000
Density (per sq km) at mid-2003	75.9

* 127,355 sq miles.

PRINCIPAL ETHNIC GROUPS
(at census of August 1991)*

	Peninsular Malaysia	Sabah†	Sarawak	Total
Malays and other indigenous groups	8,433,826	1,003,540	1,209,118	10,646,484
Chinese	4,250,969	218,233	475,752	4,944,954
Indians	1,380,048	9,310	4,608	1,393,966
Others	410,544	167,790	10,541	588,875
Non-Malaysians	322,229	464,786	18,361	805,376
Total	14,797,616	1,863,659	1,718,380	18,379,655

* Including adjustment for underenumeration.
† Including the Federal Territory of Labuan.

Mid-1997 (estimates, '000 persons): Malays 10,233.2; Other indigenous groups 2,290.9; Chinese 5,445.1; Indian 1,541.7; Others 685.7; Non-Malaysians 1,468.9; Total 21,665.5.

STATES
(census of 5–20 July 2000)

	Area (sq km)	Population*	Density (per sq km)	Capital
Johor (Johore)	18,987	2,740,625	144.3	Johore Bahru
Kedah	9,425	1,649,756	175.0	Alor Star
Kelantan	15,024	1,313,014	87.4	Kota Bahru
Melaka (Malacca)	1,652	635,791	384.9	Malacca
Negeri Sembilan (Negri Sembilan)	6,644	859,924	129.4	Seremban
Pahang	35,965	1,288,376	35.8	Kuantan
Perak	21,005	2,051,236	97.7	Ipoh
Perlis	795	204,450	257.2	Kangar
Pulau Pinang (Penang)	1,031	1,313,449	1,274.0	George Town
Sabah	73,619	2,603,485	35.4	Kota Kinabalu
Sarawak	124,450	2,071,506	16.6	Kuching
Selangor	7,960	4,188,876	526.2	Shah Alam
Terengganu (Trengganu)	12,955	898,825	69.4	Kuala Trengganu
Federal Territory of Kuala Lumpur	243	1,379,310	5,676.2	—
Federal Territory of Labuan	92	76,067	826.8	Victoria
Total	329,847	23,274,690	70.6	

* Including adjustment for underenumeration.

PRINCIPAL TOWNS
(population at 2000 census)

Kuala Lumpur (capital)*	1,297,526	Kuala Terengganu (Kuala Trengganu)	250,528
Ipoh	566,211	Seremban	246,441
Kelang (Klang)	563,173	Kota Baharu (Kota Bahru)	233,673
Petaling Jaya	438,084	Sandakan	220,000†
Shah Alam	319,612	Taiping‡	183,320
Kuantan	283,041	George Town (Penang)	180,573

* The new town of Putrajaya is now the administrative capital.
† Provisional.
‡ Excluding a part of Pondok Tanjong, which is in the District of Kerian.
Source: Thomas Brinkhoff, *City Population* (internet www.citypopulation.de).

Mid-2003 (UN estimate, incl. suburbs): Kuala Lumpur 1,352,057 (Source: UN, *World Urbanization Prospects: The 2003 Revision*).

BIRTHS AND DEATHS

	Registered live births		Registered deaths	
	Number	Rate (per 1,000)	Number	Rate (per 1,000)
1993	541,760	27.7	87,594	4.5
1994	537,611	26.7	90,051	4.5
1995	539,234	24.9	95,025	4.4
1996	540,866	25.5	95,520	4.5
1997	537,104	24.8	97,042	4.5
1998	554,573	25.0	97,906	4.4

Source: UN, mainly *Demographic Yearbook*.

1999 (rates per 1,000): Births 25.4; Deaths 4.6 (Source: Ministry of Health, Kuala Lumpur).

2000 (provisional, rates per 1,000): Births 24.5; Deaths 4.4 (Source: Ministry of Health, Kuala Lumpur).

2001 (provisional, rates per 1,000): Births 22.3; Deaths 4.4 (Source: Ministry of Health, Kuala Lumpur).

2002 (provisional, rates per 1,000): Births 21.8; Deaths 4.4 (Source: Ministry of Health, Kuala Lumpur).

Expectation of life (WHO estimates, years at birth): 72.0 (males 69.6; females 74.7) in 2002 (Source: WHO, *World Health Report*).

ECONOMICALLY ACTIVE POPULATION*
(sample surveys, ISIC major divisions, '000 persons aged 15 to 64 years)

	2000	2001	2002
Agriculture, forestry and fishing	1,712	1,497	1,422
Mining and quarrying	27	29	29
Manufacturing	2,126	2,155	2,071
Electricity, gas and water	48	57	48
Construction	799	849	907
Trade, restaurants and hotels	1,790	2,107	2,118
Transport, storage and communications	423	477	496
Finance, insurance, real estate and business services	462	582	639
Community, social and personal services	1,935	1,564	1,555
Private households with employed persons	—	210	267
Total employed	9,322	9,526	9,552
Unemployed	294	366	334
Total labour force	9,616	9,892	9,886

* Excluding members of the armed forces.

Source: IMF, *Malaysia: Statistical Appendix* (March 2004).

2003 ('000 persons): Total employed 9,869.7; Unemployed 369.8; Total labour force 10,239.6.

Health and Welfare

KEY INDICATORS

Total fertility rate (children per woman, 2002)	2.9
Under-5 mortality rate (per 1,000 live births, 2002)	8
HIV/AIDS (% of persons aged 15–49, 2003)	0.40
Physicians (per 1,000 head, 1996)	2.01
Health expenditure (2001): US $ per head (PPP)	345
Health expenditure (2001): % of GDP	3.8
Health expenditure (2001): public (% of total)	53.7
Human Development Index (2002): ranking	59
Human Development Index (2002): value	0.793

For sources and definitions, see explanatory note on p. vi.

Agriculture

PRINCIPAL CROPS
('000 metric tons)

	2000	2001	2002
Rice (paddy)	2,141	2,094	2,091
Maize*	65	67	70
Sweet potatoes†	41	41	41
Cassava (Manioc)†	380	380	370
Other roots and tubers†	45	44	44
Sugar cane†	1,600	1,600	1,600
Groundnuts (in shell)†	6	6	6
Coconuts	734*	700†	738†
Oil palm fruit†	56,600	58,950	58,390
Cabbages†	47	37	37
Tomatoes†	10	10	10
Pumpkins, squash and gourds†	14	14	14
Cucumbers and gherkins†	40	50	45
Other vegetables and melons†	250	245	240
Watermelons	86	96	91
Mangoes	20	20	20
Pineapples†	92	86	87
Papayas†	60	65	65
Bananas†	540	530	500
Oranges†	12	12	12
Other fruit (excl. melons)†	299	301	300
Coffee (green)†	18	20	22
Cocoa beans	70	58	48
Tea (made)	6*	5*	5†
Tobacco (leaves)	7	9	12
Natural rubber	615	546	589

* Unofficial figure(s).
† FAO estimate(s).

Source: FAO.

LIVESTOCK
('000 head, year ending September)

	2000	2001	2002
Cattle	724	742	748
Buffaloes	142	148	154
Goats	233	247	248
Sheep	157	129	118
Pigs	1,808	1,973	1,824
Chickens	123,650	149,586	160,843
Ducks*	13,000	13,000	13,000

* FAO estimates.

Source: FAO.

LIVESTOCK PRODUCTS
('000 metric tons)

	2000	2001	2002
Beef and veal*	18	19	21
Buffalo meat*	3	3	4
Pig meat	160	168	207
Poultry meat*	771	781	836
Cows' milk	30	32	37
Buffaloes' milk*	7	7	7
Hen eggs	391	411	432
Other poultry eggs	10	10	11
Cattle and buffalo hides*	3	3	4

* FAO estimates.

Source: FAO.

Forestry

ROUNDWOOD REMOVALS
('000 cubic metres, excl. bark)

	2000	2001	2002
Sawlogs, veneer logs and logs for sleepers	15,095	16,161	17,913
Fuel wood*	3,346	3,286	3,228
Total	18,441	19,447	21,141

* FAO estimates.

Source: FAO.

SAWNWOOD PRODUCTION
('000 cubic metres, incl. railway sleepers)

	2000	2001	2002
Total (all broadleaved)	5,590	4,696	4,594

Source: FAO.

Fishing

('000 metric tons, live weight)

	2000	2001	2002
Capture	1,289.2	1,234.7	1,275.6
Indian scad	84.2	77.4	90.3
Kawakawa	58.1	53.2	59.4
Yellowstripe scad	44.3	39.9	41.0
Indian mackerels	98.1	99.5	87.9
Prawns and shrimps	96.0	77.5	76.0
Squids	54.4	45.3	52.5
Aquaculture	151.8	158.2	165.1
Blood cockle	64.4	70.8	78.7
Total catch	1,441.0	1,392.9	1,440.7

Note: Figures exclude crocodiles, recorded by number rather than by weight. The number of estuarine crocodiles caught was: 559 in 2000; 375 in 2001; 122 in 2002. Also excluded are shells and corals. Catches of turban shells (FAO estimates, metric tons) were: 80 in 2000; 80 in 2001; 80 in 2002. Catches of hard corals (FAO estimates, metric tons) were: 4,000 in 2000; 4,000 in 2001; 4,000 in 2002.

Source: FAO.

Mining

PRODUCTION
(metric tons, unless otherwise indicated)

	2000	2001	2002*
Tin-in-concentrates	6,307	4,972	4,215
Bauxite	123,000	64,000	40,000
Iron ore†	259,000	376,000	404,000
Kaolin	233,885	364,458	258,273
Gold (kg)	4,026	3,965	4,289
Silver (kg)§	5	3	n.a.
Barytes	7,274	649	1,602
Hard coal	382,942	497,733	352,513
Crude petroleum ('000 barrels)	249,159	243,696	255,922
Natural gas (million cu m)‡	56,929	58,751	61,091
Ilmenite†	124,801	129,750	106,046
Zirconium†	3,642	3,768	5,293

* Data are preliminary.
† Figures refer to the gross weight of ores and concentrates.
‡ Including amount reinjected, flared and lost.
§ Includes by-product from a copper mine in Sabah, tin mines in Peninsular Malaysia and gold mines in Peninsular Malaysia and Sarawak.

Sources: Minerals and Geoscience Dept; US Geological Survey; UN, *Statistical Yearbook for Asia and the Pacific*.

Industry

SELECTED PRODUCTS
('000 metric tons, unless otherwise indicated)

	2001	2002	2003
Canned fish, frozen shrimps/prawns	40.1	42.2	47.6
Palm oil (crude)	11,804	11,908	n.a.
Refined sugar	1,210.4	1,404.8	1,424.1
Soft drinks ('000 litres)	501.6	440.2	521.4
Cigarettes (metric tons)	25,618	23,079	23,971
Woven cotton fabrics (million metres)	177.4	166.4	160.5
Veneer sheets ('000 cu metres)	1,173.2	1,089.6	956.3
Plywood ('000 cu metres)	3,937.5	3,972.3	4,171.4
Kerosene and jet fuel	3,293.1	3,170.7	3,056.6
Liquefied petroleum gas	2,308.0	2,945.4	3,188.2
Inner tubes and pneumatic tyres ('000)	26,391	27,107	27,873
Rubber gloves (million pairs)	12,256.3	12,207.7	15,072.2
Earthen brick and cement roofing tiles (million)	1,196.1	1,315.6	1,579.0
Cement	13,820	14,336	17,227
Iron and steel bars and rods	2,691.3	3,221.5	3,385.8
Refrigerators for household use ('000)	186.0	172.1	184.5
Television receivers ('000)	9,501.2	10,409.7	9,915.2
Radio receivers ('000)	28,839	21,735	27,640
Semiconductors (million)	13,524	15,036	15,932
Electronic transistors (million)	19,989	20,401	24,206
Integrated circuits (million)	17,457	19,916	23,269
Passenger motor cars ('000)*	383.6	418.8	342.7
Commercial vehicles('000)*	62.7	72.3	80.8
Motorcycles and scooters ('000)	235.9	242.4	256.3

* Vehicles assembled from imported parts.

Tin (smelter production of primary metal, metric tons): 28,913 in 1999; 26,228 in 2000; 32,566 (preliminary figure) in 2001 (Source: US Geological Survey).

Source (unless otherwise indicated): Bank Negara Malaysia, Kuala Lumpur.

Finance

CURRENCY AND EXCHANGE RATES

Monetary Units

100 sen = 1 ringgit Malaysia (RM—also formerly Malaysian dollar).

Sterling, US Dollar and Euro Equivalents (31 May 2004)

£1 sterling = RM 6.9719;
US $1 = RM 3.8000;
€1 = RM 4.6535;
RM 100 = £14.34 = US $26.32 = €21.49.

Exchange Rate

A fixed exchange rate of US $1 = RM 3.8000 has been in effect since September 1998.

FEDERAL BUDGET
(RM million)

Revenue	2001	2002	2003*
Tax revenue	61,492	66,860	64,891
Taxes on income and profits	42,097	44,351	43,016
Companies (excl. petroleum)	20,770	24,642	23,990
Individuals	9,436	9,889	7,984
Petroleum	9,858	7,636	8,466
Export duties	867	803	1,157
Import duties	3,193	3,668	3,919
Excises on goods	4,130	4,745	5,031
Sales tax	7,356	9,243	7,965
Service tax	1,927	2,214	2,038
Other revenue	18,076	16,655	27,913
Total	79,567	83,515	92,804

* Preliminary figures.

Expenditure	2000	2001	2002
Defence and security	6,958	8,310	9,030
Economic services	6,637	5,150	6,015
Agriculture and rural development	1,323	1,366	1,446
Trade and industry	3,761	1,870	1,838
Transport	1,286	1,672	2,069
Social services	18,784	21,757	24,798
Education	12,923	14,422	16,982
Health	4,131	4,680	5,152
Transfer payments	2,524	5,561	5,778
Debt service charges	9,055	9,634	9,670
Pensions and gratuities	4,187	4,711	5,134
General administration	8,401	8,636	8,274
Total	56,547	63,757	68,699

Source: Bank Negara Malaysia, Kuala Lumpur.

FEDERAL DEVELOPMENT EXPENDITURE
(RM million)

	2001	2002	2003
Defence and security	3,287	4,333	6,026
Social services	15,384	18,043	17,704
Education	10,363	12,436	10,194
Health	1,570	1,503	2,684
Housing	1,269	1,808	1,928
Economic services	12,725	12,433	13,799
Agriculture and rural development	1,394	1,364	1,621
Public utilities	1,092	1,808	920
Trade and industry	4,830	3,474	3,463
Transport	5,042	5,401	7,354
General administration	3,839	1,168	1,824
Total	35,235	35,977	39,353

Source: Bank Negara Malaysia, Kuala Lumpur.

INTERNATIONAL RESERVES
(US $ million at 31 December)

	2001	2002	2003
Gold*	51	56	61
IMF special drawing rights	125	151	178
Reserve position in IMF	764	790	871
Foreign exchange	29,585	33,280	43,466
Total	30,526	34,277	44,576

* Valued at SDR 35 per troy ounce.

Source: IMF, *International Financial Statistics*.

MONEY SUPPLY
(RM million at 31 December)

	2001	2002	2003
Currency outside banks	22,148	23,897	26,101
Demand deposits at commercial banks	57,791	63,892	75,083
Total money (incl. others)	83,882	91,932	105,602

Source: IMF, *International Financial Statistics*.

COST OF LIVING
(Consumer Price Index; base 2000 = 100)

	1999	2001	2002
Food	98.1	100.7	101.4
Beverages and tobacco	99.1	100.7	101.8
Clothing and footwear	101.8	97.4	95.2
Rent and other housing costs, heating and lighting	98.6	101.4	102.1
Furniture, domestic appliances, tools and maintenance	100.0	100.1	99.7
Medical care	98.0	102.9	105.4
Transport and communications	98.0	103.6	110.4
Education and leisure	99.4	99.9	100.1
Other goods and services	99.1	100.7	101.8
All items	98.5	101.4	103.2

Source: Bank Negara Malaysia, Kuala Lumpur.

NATIONAL ACCOUNTS
(RM million at current prices)

Expenditure on the Gross Domestic Product

	2001	2002	2003
Government final consumption expenditure	42,265	50,015	54,913
Private final consumption expenditure	150,644	159,506	172,366
Change in stocks	−3,339	2,217	−2,842
Gross fixed capital formation	83,345	83,764	87,089
Total domestic expenditure	272,915	295,502	311,526
Exports of goods and services	389,256	415,040	450,592
Less Imports of goods and services	327,765	348,918	367,920
GDP in purchasers' values	334,404	361,624	394,200
GDP at constant 1987 prices	211,227	219,988	231,674

Source: Bank Negara Malaysia, Kuala Lumpur.

Gross Domestic Product by Economic Activity

	2001	2002	2003
Agriculture, forestry and fishing	27,566	33,296	38,223
Mining and quarrying	33,935	33,960	40,957
Manufacturing	101,735	110,372	122,420
Electricity, gas and water	11,415	12,202	12,911
Construction	14,163	14,606	14,910
Trade, restaurants and hotels	48,726	50,937	52,604
Transport, storage and communications	23,636	25,005	26,788
Finance, insurance, real estate and business services	41,544	45,520	46,885
Government services	24,104	27,625	29,643
Other services	23,853	25,245	26,363
Sub-total	350,677	378,768	411,704
Import duties	5,653	6,605	6,386
Less Imputed bank service charges	21,925	23,749	23,891
GDP in purchasers' values	334,404	361,624	394,200

Source: Bank Negara Malaysia, Kuala Lumpur.

BALANCE OF PAYMENTS
(US $ million)

	2000	2001	2002
Exports of goods f.o.b.	98,429	87,981	93,383
Imports of goods f.o.b.	−77,602	−69,597	−75,248
Trade balance	20,827	18,383	18,135
Exports of services	13,941	14,455	14,878
Imports of services	−16,747	−16,657	−16,448
Balance on goods and services	18,020	16,182	16,565
Other income received	1,986	1,847	2,139
Other income paid	−9,594	−8,590	−8,734
Balance on goods, services and income	10,412	9,439	9,970
Current transfers received	756	537	661
Current transfers paid	−2,680	−2,689	−3,442
Current balance	8,488	7,287	7,190
Direct investment abroad	−2,026	−267	−1,905
Direct investment from abroad	3,788	554	3,203
Portfolio investment liabilities	−2,145	−666	−836
Other investment assets	−5,565	−2,702	−4,597
Other investment liabilities	—	−830	1,868
Net errors and omissions	−3,221	−2,394	−391
Overall balance	−1,009	999	3,657

Source: IMF, *International Financial Statistics*.

External Trade

PRINCIPAL COMMODITIES
(RM million)

Imports c.i.f.	2001	2002	2003
Mineral products	16,190	15,659	18,953
Chemical products	15,291	15,726	16,237
Plastics and rubber	10,105	11,476	12,140
Machinery, mechanical appliances and electrical equipment	160,645	176,567	185,170
Transportation equipment	9,141	11,650	10,613
Instruments, measuring, musical, etc.	9,330	9,365	10,143
Total (incl. others)	280,229	303,090	317,746

Exports f.o.b.	2001	2002	2003
Mineral products	32,779	31,087	40,682
Animal or vegetable fats	11,761	16,935	22,977
Chemical products	11,197	13,325	16,263
Plastics and rubber	13,433	14,920	18,294
Wood and wood products	10,714	11,222	11,919
Base metals and articles thereof	9,128	9,336	11,732
Machinery, mechanical appliances and electrical equipment	200,826	213,606	225,665
Total (incl. others)	334,284	357,430	398,882

Source: Asian Development Bank, *Key Indicators of Developing Asian and Pacific Countries.*

PRINCIPAL TRADING PARTNERS
(RM million)

Imports c.i.f.	2001	2002	2003
Australia	5,944	5,415	4,802
China, People's Republic	14,473	23,328	25,897
France	4,349	4,263	4,565
Germany	10,451	11,188	14,758
Hong Kong	7,064	8,809	8,233
India	2,934	2,442	2,561
Indonesia	8,536	9,683	11,196
Ireland	3,811	2,678	1,553
Italy	2,865	2,517	2,410
Japan	53,750	53,813	53,412
Korea, Republic	11,249	16,006	16,250
Philippines	6,987	9,862	9,928
Singapore	35,352	36,243	36,928
Taiwan	15,930	16,848	15,026
Thailand	11,120	11,982	14,215
United Kingdom	6,846	5,966	5,987
USA	44,881	51,679	46,823
Total (incl. others)	280,229	302,589	307,266

Exports f.o.b.	2001	2002	2003
Australia	7,795	8,011	9,607
China, People's Republic	14,683	19,961	24,467
France	3,569	5,170	6,662
Germany	7,766	7,962	9,140
Hong Kong	15,437	20,165	23,285
India	5,992	6,690	9,541
Indonesia	5,930	6,801	8,085
Japan	44,393	39,690	41,757
Korea, Republic	11,108	11,832	11,173
Netherlands	15,438	13,136	12,914
Philippines	4,892	5,071	5,264
Singapore	56,643	60,525	62,689
Taiwan	12,167	13,202	13,463
Thailand	12,756	15,087	17,394
United Kingdom	8,759	8,327	8,779
USA	67,618	69,293	68,786
Total (incl. others)	334,284	354,078	382,303

Source: Bank Negara Malaysia, Kuala Lumpur.

Transport

RAILWAYS
(traffic, Peninsular Malaysia and Singapore)

	1998	1999	2000
Passenger-km (million)	1,397	1,313	1,220
Freight ton-km (million)	992	908	917

Source: UN, *Statistical Yearbook.*

ROAD TRAFFIC
(registered motor vehicles at 31 December)

	1998*	1999*	2000
Passenger cars	3,517,484	3,852,693	4,212,567
Buses and coaches	45,643	47,674	48,662
Lorries and vans	599,149	642,976	665,284
Road tractors	286,898	304,135	315,687
Motorcycles and mopeds	4,692,183	5,082,473	5,356,604

* Source: International Road Federation, *World Road Statistics*; data for 1999 are estimates.

SHIPPING

Merchant Fleet
(registered at 31 December)

	2001	2002	2003
Number of vessels	882	915	972
Total displacement ('000 grt)	5,207.1	5,394.4	5,745.8

Source: Lloyd's Register-Fairplay, *World Fleet Statistics.*

Sea-borne Freight Traffic*
(Peninsular Malaysia, international and coastwise, '000 metric tons)

	1998	1999	2000
Goods loaded	35,206	39,755	42,547
Goods unloaded	48,314	54,854	56,537

* Including transhipments.

Source: UN, *Monthly Bulletin of Statistics.*

CIVIL AVIATION
(traffic on scheduled services)

	1997	1998	1999
Kilometres flown (million)	183	189	207
Passengers carried ('000)	15,592	13,654	14,985
Passenger-km (million)	28,698	29,372	33,708
Total ton-km (million)	3,777	3,777	4,431

Source: UN, *Statistical Yearbook.*

Tourism

TOURIST ARRIVALS BY COUNTRY OF RESIDENCE*

	2001	2002	2003
Brunei	309,529	256,952	215,634
China, People's Republic	453,246	557,647	350,597
Indonesia	777,449	769,128	621,651
Japan	397,639	354,563	213,527
Singapore	6,951,594	7,547,761	5,922,306
Taiwan	249,811	209,706	137,419
Thailand	1,018,797	1,166,937	1,152,296
United Kingdom	262,423	239,294	125,569
Total (incl. others)	12,775,073	13,292,010	10,576,915

* Including Singapore residents crossing the frontier by road through the Johore Causeway.

Tourism receipts (RM million): 24,221.5 in 2001; 25,781.1 in 2002; 21,291.1 in 2003.

Source: Malaysia Tourism Promotion Board.

Communications Media

	2001	2002	2003
Telephones ('000 main lines in use)	4,709.6	4,669.9	4,571.6
Mobile cellular telephones ('000 subscribers)	7,385.2	9,253.4	11,124.1
Personal computers ('000 in use)*	3,000	3,600	n.a.
Internet users ('000)*	6,346.6	7,840.6	8,692.1

Radio receivers ('000 in use, 1997): 9,100.

Television receivers ('000 in use, 2000): 3,900.

Facsimile machines ('000 in use): 175.0* in 1998.

Mobile cellular telephones ('000 subscribers): 8,500 in 2002.

Book production (1999): 5,084 titles (29,040,000 copies in 1996)†.

Newspapers (1996): 42 dailies (average circulation 3,345,000 copies); 44 non-dailies (average circulation 1,424,000 copies).

Periodicals (1992): 25 titles (average circulation 996,000 copies).

* Estimate(s).

† Including pamphlets (106 titles and 646,000 copies in 1994).

Sources: International Telecommunication Union; UNESCO, *Statistical Yearbook*; UN, *Statistical Yearbook*.

Education

(January 2004, unless otherwise indicated)

	Institutions	Teachers	Students
Primary	7,557	179,622	3,044,797
Secondary	1,962	130,372	2,093,645
Regular	1,751	116,006	1,979,526
Fully residential	53	3,135	29,813
Technical	88	7,224	34,710
Religious	55	3,053	37,732
Special	3	140	884
Special Model	10	688	10,437
Sports*	2	142	912
Tertiary†	48	14,960	210,724
Universities	9	7,823	97,103
Teacher training	31	3,220	46,019
MARA Institute of Technology	1	2,574	42,174

* 2003 figures.

† 1995 figures.

Source: Ministry of Education.

Pre-primary: 9,743 schools (1994); 20,352 teachers (1994); 459,015 pupils (1995) (Source: UNESCO, *Statistical Yearbook*).

Adult literacy rate (UNESCO estimates): 88.7% (males 92.0%; females 85.4%) in 2002 (Source: UN Development Programme, *Human Development Report*).

Directory

Note: in 2004 telephone and fax numbers in Malaysia remained in the process of change. See www.telekom.com.my for further details.

The Constitution

The Constitution of the Federation of Malaya became effective at independence on 31 August 1957. As subsequently amended, it is now the Constitution of Malaysia. The main provisions are summarized below.

SUPREME HEAD OF STATE

The Yang di-Pertuan Agong (King or Supreme Sovereign) is the Supreme Head of Malaysia.

Every act of government is derived from his authority, although he acts on the advice of Parliament and the Cabinet. The appointment of a Prime Minister lies within his discretion, and he has the right to refuse to dissolve Parliament even against the advice of the Prime Minister. He appoints the Judges of the Federal Court and the High Courts on the advice of the Prime Minister. He is the Supreme Commander of the Armed Forces. The Yang di-Pertuan Agong is elected by the Conference of Rulers, and to qualify for election he must be one of the nine hereditary Rulers. He holds office for five years or until his earlier resignation or death. Election is by secret ballot on each Ruler in turn, starting with the Ruler next in precedence after the late or former Yang di-Pertuan Agong. The first Ruler to obtain not fewer than five votes is declared elected. The Deputy Supreme Head of State (the Timbalan Yang di-Pertuan Agong) is elected by a similar process. On election the Yang di-Pertuan Agong relinquishes, for his tenure of office, all his functions as Ruler of his own state and may appoint a Regent. The Timbalan Yang di-Pertuan Agong exercises no powers in the ordinary course, but is immediately available to fill the post of Yang di-Pertuan Agong and carry out his functions in the latter's absence or disability. In the event of the Yang di-Pertuan Agong's death or resignation he takes over the exercise of sovereignty until the Conference of Rulers has elected a successor.

CONFERENCE OF RULERS

The Conference of Rulers consists of the Rulers and the heads of the other states. Its prime duty is the election by the Rulers only of the Yang di-Pertuan Agong and his deputy. The Conference must be consulted in the appointment of judges, the Auditor-General, the Election Commission and the Services Commissions. It must also be consulted and concur in the alteration of state boundaries, the extension to the federation as a whole, of Islamic religious acts and observances, and in any bill to amend the Constitution. Consultation is mandatory in matters affecting public policy or the special position of the Malays and natives of Sabah and Sarawak. The Conference also considers matters affecting the rights, prerogatives and privileges of the Rulers themselves.

FEDERAL PARLIAMENT

Parliament has two Houses—the Dewan Negara (Senate) and the Dewan Rakyat (House of Representatives). The Senate has a membership of 70, comprising 26 elected and 44 appointed members. Each state legislature, acting as an electoral college, elects two Senators; these may be members of the State Legislative Assembly or otherwise. The Yang di-Pertuan Agong appoints the other 44 members of the Senate; these include four Senators representing the three Federal Territories—Kuala Lumpur, Labuan and Putrajaya. Members of the Senate must be at least 30 years old. The Senate elects its President and Deputy President from among its members. It may initiate legislation, but all proposed legislation for the granting of funds must be introduced in the first instance in the House of Representatives. All legislative measures require approval by both Houses of Parliament before being presented to the Yang di-Pertuan Agong for the Royal Assent in order to become law. A bill originating in the Senate cannot receive Royal Assent until it has been approved by the House of Representatives, but the Senate has delaying powers only over a bill originating from and approved by the House of Representatives. Senators serve for a period of three years, but the Senate is not subject to dissolution. Parliament can, by statute, increase the number of Senators elected from each state to three. The House of Representatives consists of 219 elected members (see Amendments). Of these, 165 are from Peninsular Malaysia (including 11 from Kuala Lumpur and one from Putrajaya), 28 from Sarawak and 26 from Sabah (including one from Labuan). Members are returned from single-member constituencies on the basis of universal adult franchise. The term of the House of Representatives is limited to five years, after which time a fresh general election must be held. The Yang di-Pertuan Agong may dissolve Parliament before then if the Prime Minister so advises.

THE CABINET

To advise him in the exercise of his functions, the Yang di-Pertuan Agong appoints the Cabinet, consisting of the Prime Minister and an unspecified number of Ministers (who must all be Members of Parliament). The Prime Minister must be a citizen born in Malaysia and a member of the House of Representatives who, in the opinion of the Yang di-Pertuan Agong, commands the confidence of that House. Ministers are appointed on the advice of the Prime Minister. A number of Deputy Ministers (who are not members of the Cabinet) are also appointed from among Members of Parliament. The Cabinet meets regularly under the chairmanship of the Prime Minister to formulate policy.

PUBLIC SERVICES

The Public Services, civilian and military, are non-political and owe their loyalty not to the party in power but to the Yang di-Pertuan Agong and the Rulers. They serve whichever government may be in power, irrespective of the latter's political affiliation. To ensure the impartiality of the service, and its protection from political interference, the Constitution provides for a number of Services Commissions to select and appoint officers, to place them on the pensionable establishment, to determine promotion and to maintain discipline.

THE STATES

The heads of nine of the 13 states are hereditary Rulers. The Ruler of Perlis has the title of Raja, and the Ruler of Negeri Sembilan that of Yang di-Pertuan Besar. The rest of the Rulers are Sultans. The heads of the States of Melaka (Malacca), Pinang (Penang), Sabah and Sarawak are each designated Yang di-Pertua Negeri and do not participate in the election of the Yang di-Pertuan Agong. Each of the 13 states has its own written Constitution and a single Legislative Assembly. Every state legislature has powers to legislate on matters not reserved for the Federal Parliament. Each State Legislative Assembly has the right to order its own procedure, and the members enjoy parliamentary privilege. All members of the Legislative Assemblies are directly elected from single-member constituencies. The head of the state acts on the advice of the State Government. This advice is tendered by the State Executive Council or Cabinet in precisely the same manner in which the Federal Cabinet tenders advice to the Yang di-Pertuan Agong.

The legislative authority of the state is vested in the head of the state in the State Legislative Assembly. The executive authority of the state is vested in the head of the state, but executive functions may be conferred on other persons by law. Every state has its own Executive Council or Cabinet to advise the head of the state, headed by its Chief Minister (Ketua Menteri in Melaka, Pinang, Sabah and Sarawak and Menteri Besar in other states), and collectively responsible to the state legislature. Each state in Peninsular Malaysia is divided into administrative districts, each with its District Officer. Sabah is divided into four residencies: West Coast, Interior, Sandakan and Tawau, with headquarters at Kota Kinabalu, Keningua, Sandakan and Tawau, respectively. Sarawak is divided into five Divisions, each in charge of a Resident—the First Division, with headquarters at Kuching; the Second Division, with headquarters at Simanggang; the Third Division, with headquarters at Sibu; the Fourth Division, with headquarters at Miri; the Fifth Division, with headquarters at Limbang.

AMENDMENTS

From 1 February 1974, the city of Kuala Lumpur, formerly the seat of the Federal Government and capital of Selangor State, is designated the Federal Territory of Kuala Lumpur. It is administered directly by the Federal Government and returns five members to the House of Representatives.

In April 1981 the legislature approved an amendment empowering the Yang di-Pertuan Agong to declare a state of emergency on the grounds of imminent danger of a breakdown in law and order or a threat to national security.

In August 1983 the legislature approved an amendment empowering the Prime Minister, instead of the Yang di-Pertuan Agong, to declare a state of emergency.

The island of Labuan, formerly part of Sabah State, was designated a Federal Territory as from 16 April 1984.

The legislature approved an amendment increasing the number of parliamentary constituencies in Sarawak from 24 to 27. The amendment took effect at the general election of 20–21 October 1990. The total number of seats in the House of Representatives, which had increased to 177 following an amendment in August 1983, was thus expanded to 180.

In March 1988 the legislature approved two amendments relating to the judiciary (see Judicial System).

In October 1992 the legislature adopted an amendment increasing the number of parliamentary constituencies from 180 to 192. The Kuala Lumpur Federal Territory and Selangor each gained three seats, Johor two, and Perlis, Kedah, Kelantan and Pahang one. The amendment took effect at the next general election (in April 1995).

In March 1993 an amendment was approved which removed the immunity from prosecution of the hereditary Rulers.

In May 1994 the House of Representatives approved an amendment which ended the right of the Yang di-Pertuan Agong to delay legislation by withholding his assent from legislation and returning it to Parliament for further consideration. Under the amendment, the Yang di-Pertuan Agong was obliged to give his assent to a bill within 30 days; if he failed to do so, the bill would, none the less, become law. An amendment was simultaneously approved restructuring the judiciary and introducing a mandatory code of ethics for judges, to be drawn up by the Government.

In 1996 an amendment was approved, increasing the number of parliamentary constituencies from 192 to 193.

In July 2001 an amendment was approved banning all discrimination on grounds of gender.

From 1 February 2001 the city of Putrajaya, formerly part of Selangor State, was designated a Federal Territory.

In 2003 the legislature approved an amendment increasing the number of parliamentary constituencies from 193 to 219. The amendments took effect at the next general election (in March 2004).

The Government

SUPREME HEAD OF STATE

HM Yang di-Pertuan Agong: HM Tuanku SYED SIRAJUDDIN IBNI AL-MARHUM SYED PUTRA JAMALULLAIL (Raja of Perlis) (took office 13 December 2001).

Deputy Supreme Head of State

Timbalan Yang di-Pertuan Agong: HRH Sultan MIZAN ZAINAL ABIDIN (Sultan of Terengganu).

THE CABINET
(September 2004)

Prime Minister and Minister of Finance and of Internal Security: Dato' Seri ABDULLAH BIN Haji AHMAD BADAWI.

Deputy Prime Minister and Minister of Defence: Dato' Seri NAJIB BIN Tun Haji ABDUL RAZAK.

Minister of Housing and Local Government: Dato' ONG KAH TING.

Minister of Foreign Affairs: Datuk Seri SYED HAMID BIN SYED JAAFAR ALBAR.

Minister of Home Affairs: Datuk AZMI KHALID.

Minister of International Trade and Industry: Dato' Seri Paduka RAFIDAH BINTI AZIZ.

Minister of Domestic Trade and Consumer Affairs: Datuk SHAFIE APDAL.

Minister of Transport: Dato' CHAN KONG CHOY.

Minister of Energy, Water and Communications: Datuk Seri Dr LIM KENG YAIK.

Minister of Works: Dato' Seri S. SAMY VELLU.

Minister of Youth and Sports: Datuk AZALINA OTHMAN SAID.

Minister of Education: Datuk HISHAMMUDDIN Tun HUSSEIN.

Minister of Higher Education: Datuk Dr SHAFIE MOHD SALLEH.

Minister of Information: Datuk Paduka ABDUL KADIR Sheikh FADZIR.

Minister of Human Resources: Datuk Dr FONG CHAN ONN.

Minister of Natural Resources and the Environment: Datuk Seri ADENAN SATEM.

Minister of Plantation Industries and Commodities: Datuk PETER CHIN FAH KUI.

Minister of Arts, Culture and Heritage: Datuk Seri Dr RAIS YATIM.

Minister of Tourism: Datuk LEO MICHAEL TOYAD.

Minister of Science, Technology and Innovations: Datuk Dr JAMALUDDIN JARJIS.

Minister of Health: Datuk Dr CHUA SOI LEK.

Minister of Agriculture and Agro-Based Industry: Tan Sri Dato' Haji MUHYIDDIN BIN Haji MOHD YASSIN.

Minister of Rural and Regional Development: Datuk ABDUL AZIZ SHAMSUDDIN.

Minister of Federal Territories: Tan Sri MOHAMED ISA ABDUL SAMAD.

Minister of Entrepreneur and Co-operative Development: Datuk MOHAMED KHALED NORDIN.

Minister of Women, Family and Community Development: Datuk SHAHRIZAT BTE ABDUL JALIL.

Minister of Finance II: NOR MOHAMED YAKCOP.

Ministers in the Prime Minister's Department: Tan Sri BERNARD GILUK DOMPOK, Dato' Seri MOHAMAD NAZRI BIN ABDUL AZIZ, Datuk MUSTAPA BIN MOHAMED, Datuk Seri MOHD RADZI BIN Sheikh AHMAD, Prof. Datuk Dr ABDULLAH BIN MOHD ZIN, Datuk Dr MAXIMUS JOHNITY ONGKILI.

MINISTRIES

Prime Minister's Office (Jabatan Perdana Menteri): Blok Utama, Tingkat 1–5, Pusat Pentadbiran Kerajaan Persekutuan, 62502 Putrajaya; tel. (3) 88888000; fax (3) 88883424; e-mail ppm@pmo.gov.my; internet www.pmo.gov.my.

Ministry of Agriculture and Agro-Based Industry: Tingkat 4, Wisma Tani, Jalan Sultan Salahuddin, 50624 Kuala Lumpur; tel. (3) 26982011; fax (3) 26913758; e-mail admin@moa.my; internet www.agrolink.moa.my.

Ministry of Arts, Culture and Heritage: Menara Dato' Onn, 34th–36th Floors, POB 5–7, Putra World Trade Centre, 45 Jalan Tun Ismail, 50694 Kuala Lumpur; tel. (3) 26937111; fax (3) 26941146; e-mail mocat@tourism.gov.my; internet www.mocat.gov.my.

Ministry of Defence (Kementerian Pertahanan): Wisma Pertahanan, Jalan Padang Tembak, 50634 Kuala Lumpur; tel. (3) 26921333; fax (3) 26914163; e-mail cpa@mod.gov.my; internet www.mod.gov.my.

Ministry of Domestic Trade and Consumer Affairs (Kementerian Perdagangan Dalam Negeri Dan Hal Ehwal Pengguna): Tingkat 33, Menara Dayabumi, Jalan Sultan Hishamuddin, 50632 Kuala Lumpur; tel. (3) 22742100; fax (3) 22745260; e-mail menteri@kpdnhq.gov.my; internet www.kpdnhq.gov.my.

Ministry of Education (Kementerian Pendidikan): Blok J, Tingkat 7, Pusat Bandar Damansara, 50604 Kuala Lumpur; tel. (3) 2586900; fax (3) 2543107; e-mail webmaster@moe.gov.my; internet www.moe.gov.my.

Ministry of Energy, Water and Communications: Wisma Damansara, 1st Floor, Jalan Semantan, 50668 Kuala Lumpur; tel. (3) 20875000; fax (3) 20957901; e-mail webmaster@ktkm.gov.my; internet www.ktkm.gov.my.

Ministry of Entrepreneur and Co-operative Development: Tingkat 22–26, Medan MARA, Jalan Raja Laut, 50652 Kuala Lumpur; tel. (3) 26985022; fax (3) 26917623; e-mail nazri@kpun.gov.my; internet www.kpun.gov.my.

Ministry of Federal Territories (Kementerian Wilayah Persekutuan): Aras 3, Blok Barat, Bangunan Perdana Putra, Pusat Pentadbiran Kerajaan Persekutuan, 62502 Putrajaya.

Ministry of Finance (Kementerian Kewangan): Kompleks Kementerian Kewangan, Precinct 2, Pusat Pentadbiran Kerajaan Persekutuan, 62592 Putrajaya; tel. (3) 88823000; fax (3) 88823892; e-mail mk1@treasury.gov.my; internet www.treasury.gov.my.

Ministry of Foreign Affairs (Kementerian Luar Negeri): Wisma Putra, 1 Jalan Wisma Putra, 62602 Putrajaya; tel. (3) 88874000; fax (3) 88891717; e-mail webmaster@kln.gov.my; internet www.kln.gov.my.

Ministry of Health (Kementerian Kesihatan): Jalan Cenderasari, 50590 Kuala Lumpur; tel. (3) 26985077; fax (3) 26985964; e-mail CJM@moh.gov.my; internet www.moh.gov.my.

Ministry of Higher Education (Kementerian Pengajian Tinggi): Paras 2, Blok B (Utara), Pusat Bandar Damansara, 50604 Kuala Lumpur.

Ministry of Home Affairs (Kementerian Dalam Negeri): Blok D1, Parcel D, Pusat Pentadbiran Kerajaan Persekutuan, 62546 Putrajaya; tel. (3) 88868000; e-mail irg@kdn.gov.my; internet www.kdn.gov.my.

Ministry of Housing and Local Government (Kementerian Perumahan dan Kerajaan Tempatan): Paras 4 and 5, Blok K, Pusat Bandar Damansara, 50782 Kuala Lumpur; tel. (3) 2547033; fax (3) 2547380; e-mail menteri@kpkt.gov.my; internet www.kpkt.gov.my.

Ministry of Human Resources (Kementerian Sumber Manusia): Level 6–9, Blok D3, Parcel D, Pusat Pentadbiran Kerajaan Persekutuan, 62502 Putrajaya; tel. (3) 88865000; fax (3) 88893381; e-mail mhr@po.jaring.my; internet www.jaring.my/ksm.

Ministry of Information (Kementerian Penerangan): 5th Floor, Wisma TV, Angkasapuri, 50610 Kuala Lumpur; tel. (3) 2825333; fax (3) 2821255; e-mail webmaster@kempen.gov.my; internet www.kempen.gov.my.

Ministry of Internal Security (Kementerian Dalam Negeri): Blok D1, Parcel D, Pusat Pentadbiran Kerajaan Persekutuan, 62546 Putrajaya; tel. (3) 88868000; e-mail pro@kdn.gov.my; internet www.kdn.gov.my.

Ministry of International Trade and Industry (Kementerian Perdagangan Antarabangsa dan Industri): Blok 10, Kompleks Pejabat Kerajaan, Jalan Duta, 50622 Kuala Lumpur; tel. (3) 62033022; fax (3) 62031303; e-mail mitiweb@miti.gov.my; internet www.miti.gov.my.

Ministry of Natural Resources and the Environment (Kementerian Sumber Asli dan Alam Sekitar): Tingkat 11, Wismah Tanah, Jalan Semarak,50574 Kuala Lumpur; e-mail adenan@nre.gov.my; internet www.ktpk.gov.my.

Ministry of Plantation Industries and Commodities: Menara Dayabumi, 6th–8th Floors, Jalan Sultan Hishamuddin, 50654 Kuala Lumpur; tel. (3) 22747511; fax (3) 22745014; e-mail webeditor@kpu.gov.my; internet www.kpu.gov.my.

Ministry of Rural and Regional Development: Aras 5–9, Blok D9, Parcel D, Pusat Pentadbiran Kerajaan Persekutuan, 62606 Putrajaya; tel. (3) 88863500; fax (3) 88892096; e-mail info@kplb.gov.my; internet www.kplb.gov.my.

Ministry of Science, Technology and Innovations: Level 4, Blok C5, Parcel C, Federal Government Administrative Centre, 62662 Putrajaya; tel. (3) 88858000; fax (3) 88892980; e-mail mastic@mastic.gov.my; internet www.moste.gov.my.

Ministry of Tourism (Kementerian Pelancongan): Tingkat 27, Menara Dato' Onn, Pusat Dagangan Dunia Putra, 45 Jalan Tun Ismail, 50654 Kuala Lumpur; e-mail menteri@mocat.gov.my; internet www.mocat.gov.my.

Ministry of Transport (Kementerian Pengangkutan): Wisma Perdana, Level 5–7, Blok D5, Parcel D, Federal Government Administrative Centre, 62502 Putrajaya; tel. (3) 88866000; fax (3) 88892537; e-mail leelc@mot.gov.my; internet www.mot.gov.my.

Ministry of Women, Family and Community Development: Wisma Bumi Raya, Floors 19–21, Jalan Raja Laut, 50562 Kuala Lumpur; tel. (3) 26925022; fax (3) 26937353; e-mail adminkpn@kempadu.gov.my; internet www.kempadu.gov.my.

Ministry of Works (Kementerian Kerja Raya): Ground Floor, Blok A, Kompleks Kerja Raya, Jalan Sultan Salahuddin, 50580 Kuala Lumpur; tel. (3) 27111100; fax (3) 27116612; e-mail menteri@kkr.gov.my; internet www.kkr.gov.my.

Ministry of Youth and Sports (Kementerian Belia dan Sukan): Blok G, Jalan Dato' Onn, 50570 Kuala Lumpur; tel. (3) 26932255; fax (3) 26932231; e-mail webmaster@kbs.gov.my; internet www.kbs.gov.my.

Legislature

PARLIAMENT

Dewan Negara

(Senate)

The Senate has 70 members, of whom 26 are elected. Each State Legislative Assembly elects two members. The Supreme Head of State appoints the remaining 44 members, including four from the three Federal Territories.

President: Dr Abdul Hamid Pawanteh.

Dewan Rakyat

(House of Representatives)

The House of Representatives has a total of 219 members: 165 from Peninsular Malaysia (including 11 from Kuala Lumpur and one from the Federal Territory of Putrajaya), 28 from Sarawak and 26 from Sabah (including one from the Federal Territory of Labuan).

Speaker: (to be appointed).

Deputy Speaker: Dr Yusof Yacob.

General Election, 21 March 2004

Party	Seats
Barisan Nasional (National Front)	198
United Malays National Organization	109
Malaysian Chinese Association	31
Parti Pesaka Bumiputera Bersatu	11
Parti Gerakan Rakyat Malaysia	10
Malaysian Indian Congress	9
Parti Bansa Dayak Sarawak	6
Sarawak United People's Party	6
Parti Bersatu Sabah	4
Sabah Progressive Democratic Party	4
United Kadazan People's Organization	4
Sabah Progressive Party	2
Parti Bersatu Rakyat Sabah	1
Parti Progresif Pendukuk Malaysia	1
Democratic Action Party	12
Parti Islam se Malaysia	7
Parti Keadilan Rakyat	1
Independents	1
Total	219

The States

JOHOR
(Capital: Johor Bahru)

Sultan: HRH Tuanku MAHMOOD ISKANDAR IBNI AL-MARHUM Sultan ISMAIL.

Menteri Besar: Datuk Haji ABDUL GHANI OTHMAN.

State Legislative Assembly: 56 seats: Barisan Nasional 55; Parti Islam se Malaysia 1; elected March 2004.

KEDAH
(Capital: Alor Star)

Sultan: HRH Tuanku Haji ABDUL HALIM MU'ADZAM SHAH IBNI AL-MARHUM Sultan BADLISHAH.

Menteri Besar: Datuk SYED RAZAK SYED ZAIN BARAKBAH.

State Legislative Assembly: 36 seats: Barisan Nasional 31; Parti Islam se Malaysia 5; elected March 2004.

KELANTAN
(Capital: Kota Baharu)

Sultan: HRH Tuanku ISMAIL PETRA IBNI AL-MARHUM Sultan YAHAYA PETRA.

Menteri Besar: Tuan Guru Haji Nik ABDUL AZIZ BIN Nik MAT.

State Legislative Assembly: 45 seats: Parti Islam se Malaysia 24; Barisan Nasional 21; elected March 2004.

MELAKA (MALACCA)
(Capital: Melaka)

Yang di-Pertua Negeri: Tan Sri KHALIL YAAKOB.

Ketua Menteri: Datuk WIRA MOHAMED ALI RUSTAM.

State Legislative Assembly: 28 seats: Barisan Nasional 26; Democratic Action Party 2; elected March 2004.

NEGERI SEMBILAN
(Capital: Seremban)

Yang di-Pertuan Besar: Tuanku JA'AFAR IBNI AL-MARHUM Tuanku ABDUL RAHMAN.

Menteri Besar: Datuk MOHAMAD HASAN.

State Legislative Assembly: 36 seats: Barisan Nasional 34; Democratic Action Party 2; elected March 2004.

PAHANG
(Capital: Kuantan)

Sultan: HRH Haji AHMAD SHAH AL-MUSTA'IN BILLAH IBNI AL-MARHUM Sultan ABU BAKAR RI'AYATUDDIN AL-MU'ADZAM SHAH.

Menteri Besar: Dato' ADNAN BIN YAAKOB.

State Legislative Assembly: 41 seats: Barisan Nasional 40; Democratic Action Party 1; elected March 2004.

PERAK
(Capital: Ipoh)

Sultan: HRH Sultan Tuanku AZLAN MUHIBUDDIN SHAH IBNI AL-MARHUM Sultan YUSUF IZUDDIN GHAFARULLAH SHAH.

Menteri Besar: Dato' Seri DiRaja MOHAMAD TAJOL ROSLI.

State Legislative Assembly: Pejabat Setiausaha Kerajaan Negeri, Perak Darul Ridzuan, Bahagian Majlis, Jalan Panglima Bukit Gantang Wahab, 30000 Ipoh; tel. (5) 2410451; fax (5) 2552890; e-mail master@perak.gov.my; internet www.perak.gov.my; 59 seats: Barisan Nasional 52; Democratic Action Party 7; elected March 2004.

PERLIS
(Capital: Kangar)

Regent: Tuanku SYED FAIZUDDIN PUTRA IBNI Tuanku SYED SIRAJUDDIN PUTRA JAMALULAIL.

Menteri Besar: Dato' Seri SHAHIDAN KASSIM.

State Legislative Assembly: internet www.perlis.gov.my; 15 seats: Barisan Nasional 14; Parti Islam se Malaysia 1; elected March 2004.

PINANG (PENANG)
(Capital: George Town)

Yang di-Pertua Negeri: HE Datuk ABDUL RAHMAN Haji ABBAS.

Ketua Menteri: Tan Sri Dr KOH TSU KOON.

State Legislative Assembly: 40 seats: Barisan Nasional 38; Democratic Action Party 1; Parti Islam se Malaysia 1; elected March 2004.

SABAH
(Capital: Kota Kinabalu)

Yang di-Pertua Negeri: HE Datuk AHMADSHAH ABDULLAH.

Ketua Menteri: Datuk Seri MUSA AMAN.

State Legislative Assembly: Dewan Undangan Negeri Sabah, Aras 4, Bangunan Dewan Undangan Negeri Sabah, Peti Surat 11247, 88813 Kota Kinabalu; tel. (88) 427533; fax (88) 427333; e-mail pejduns@sabah.gov.my; internet www.sabah.gov.my; 60 seats: Barisan Nasional 59; Independents 1; elected March 2004.

SARAWAK
(Capital: Kuching)

Yang di-Pertua Negeri: HE Tun Datuk Patinggi Abang Haji MUHAMMED SALAHUDDIN.

Ketua Menteri: Datuk Patinggi Tan Sri Haji ABDUL TAIB BIN MAHMUD.

State Legislative Assembly: Bangunan Dewan Undangan Negeri, 93502 Petra Jaya, Kuching, Sarawak; tel. (82) 441955; fax (82) 440790; e-mail mastapaj@sarawaknet.gov.my; internet www.dun.sarawak.gov.my; 62 seats: Barisan Nasional 60; Democratic Action Party 1; Independents 1; elected September 2001.

SELANGOR
(Capital: Shah Alam)

Sultan: Tuanku IDRIS SALAHUDDIN ABDUL AZIZ SHAH.

Menteri Besar: MOHAMAD KHIR TOYO.

State Legislative Assembly: 56 seats: Barisan Nasional 54; Democratic Action Party 2; elected March 2004.

TERENGGANU
(Capital: Kuala Terengganu)

Sultan: HRH Sultan Tuanku MIZAN ZAINAL ABIDIN IBNI AL-MARHUM Sultan MAHMUD.

Menteri Besar: Datuk IDRIS JUSOH.

State Legislative Assembly: 32 seats: Barisan Nasional 28; Parti Islam se Malaysia 4; elected March 2004.

Political Organizations

Barisan Nasional (BN) (National Front): Suites 1–2, 8th Floor, Menara Dato' Onn, Pusat Dagangan Dunia Putra, Jalan Tun Ismail, 50480 Kuala Lumpur; tel. (3) 2920384; fax (3) 2934743; e-mail info@bn.org.my; internet www.bn.org.my; f. 1973; the governing multi-racial coalition of 14 parties; Chair. Dato' Seri ABDULLAH BIN Haji AHMAD BADAWI; Sec.-Gen. Dato' Datuk Sri MOHAMMED RAHMAT; Comprises:

Liberal Democratic Party: Tingkat 2, Lot 1, Wisma Jasaga, POB 1125, Sandakan, 90712 Sabah; tel. (89) 271888; fax (89) 288278; e-mail ldpkk@tm.net.my; Chinese-dominated; Pres. Datuk CHONG KAH KIAT; Sec.-Gen. Datuk ANTHONY LAI VAI MING.

Malaysian Chinese Association (MCA): Wisma MCA, 8th Floor, 163 Jalan Ampang, POB 10626, 50720 Kuala Lumpur; tel. (3) 21618044; fax (3) 21619772; e-mail info@mca.org.my; internet www.mca.org.my; f. 1949; c. 1,038,496 mems; Pres. Dato' Seri ONG KAH TING; Sec.-Gen. Tan Sri Dato' Seri Dr TING CHEW PEH.

Malaysian Indian Congress (MIC): Menara Manickavasagam, 6th Floor, 1 Jalan Rahmat, 50350 Kuala Lumpur; tel. (3) 4424377; fax (3) 4427236; internet www.mic.malaysia.org; f. 1946; 401,000 mems (1992); Pres. Dato' Seri S. SAMY VELLU; Sec.-Gen. Dato' G. VADIVELOO.

Parti Bansa Dayak Sarawak (PBDS) (Sarawak Native People's Party): 622 Jalan Kedandi, Tabuan Jaya, POB 2148, Kuching, Sarawak; tel. (82) 365240; fax (82) 363734; f. 1983 by fmr mems of Sarawak National Party; Pres. Datuk DANIEL TAJEM.

Parti Bersatu Rakyat Sabah (PBRS) (United Sabah People's Party): POB 20148, Luyang, Kota Kinabalu, 88761 Sabah; tel. and fax (88) 269282; f. 1994; breakaway faction of PBS; mostly Christian Kadazans; Leader Datuk JOSEPH KURUP.

Parti Bersatu Sabah (PBS) (Sabah United Party): Block M, Lot 4, 2nd and 3rd Floors, Donggongon New Township, 89500 Penampang, Sabah; tel. (88) 714891; fax (88) 718067; e-mail hq@pbs-sabah.org; internet www.pbs-sabah.org; f. 1985; multiracial party, left the BN in 1990 and rejoined in Jan. 2002; Pres. Datuk Seri JOSEPH PAIRIN KITINGAN; Sec.-Gen. Datuk RADIN MALLEH.

Parti Gerakan Rakyat Malaysia (GERAKAN) (Malaysian People's Movement): Tingkat 5, Menara PGRM, 8 Jalan Pudu Ulu Cheras, 56100 Kuala Lumpur; tel. (3) 92876868; fax (3) 92878866; e-mail gerakan@gerakan.org.my; internet www.gerakan.org.my; f. 1968; 300,000 mems; Pres. Dato' Seri Dr LIM KENG YAIK; Sec.-Gen. CHIA KWANG CHYE.

Parti Pesaka Bumiputera Bersatu (PBB) (United Traditional Bumiputra Party): Lot 401, Jalan Bako, POB 1953, 93400 Kuching, Sarawak; tel. (82) 448299; fax (82) 448294; f. 1983; Pres. Tan Sri Datuk Patinggi Amar Haji ABDUL TAIB MAHMUD; Dep. Pres. Datuk ALFRED JABU AK NUMPANG.

Parti Progresif Penduduk Malaysia (PPP) (People's Progressive Party): 27–29A Jalan Maharajalela, 50150 Kuala Lumpur; tel. (3) 2441922; fax (3) 2442041; e-mail info@ppp.com.my; internet www.jaring.my/ppp; f. 1953 as Perak Progressive Party; joined the BN in 1972; Pres. Datuk M. KAYVEAS.

Sabah Progressive Party (SAPP) (Parti Maju Sabah): Lot 23, 2nd Floor, Bornion Centre, 88300 Kota Kinabalu, Sabah; tel. (88) 242107; fax (88) 254799; e-mail sapp@po.jaring.my; internet www.sapp.org.my; f. 1994; non-racial; Pres. Datuk YONG TECK LEE; Sec.-Gen. RICHARD YONG WE KONG.

Sarawak Progressive Democratic Party (SPDP): Lot 4319–4320, Jalan Stapok, Sungai Maong, 93250 Kuching, Sarawak; tel. (82) 311180; fax (82) 311190; f. 2003 by breakaway faction of Sarawak National Party; Pres. Datuk WILLIAM MAWAN ANAK IKOM; Sec.-Gen. Agung Dr JUDSON SAKAI TAGAL.

Sarawak United People's Party (SUPP): 7 Jalan Tan Sri Ong Kee Hui, POB 454, 93710 Kuching, Sarawak; tel. (82) 246999; fax (82) 256510; e-mail supp@po.jaring.my; internet www.sarawak.com.my/supp/; f. 1959; Sarawak Chinese minority party; Pres. Datuk Dr GEORGE CHAN HONG NAM; Sec.-Gen. SIM KHENG HUI.

United Kadazan People's Organization (UPKO): Penampang Service Centre, Km 11, Jalan Tambunan, Peti Surat 420, 89507 Penampang, Sabah; tel. (88) 718182; fax (88) 718180; e-mail n4upko@yahoo.com; internet www.upko.org.my; f. 1994 as the Parti Demokratik Sabah (PDS—Sabah Democratic Party); formed after collapse of PBS Govt by fmr leaders of the party, represents mostly Kadazandusun, Rungus and Murut communities; Pres. Tan Sri BERNARD GILUK DOMPOK.

United Malays National Organization (Pertubuhan Kebangsaan Melayu Bersatu—UMNO Baru) (New UMNO): Menara Dato' Onn, 38th Floor, Jalan Tun Dr Ismail, 50480 Kuala Lumpur; tel. (3) 40429511; fax (3) 40412358; e-mail email@umno.net.my; internet www.umno.org.my; f. 1988 to replace the original UMNO (f. 1946) which had been declared an illegal organization, owing to the participation of unregistered branches in party elections in April 1987; Supreme Council of 45 mems; 2.5m. mems; Pres. Dato' Seri ABDULLAH BIN Haji AHMAD BADAWI; Sec.-Gen. Tan Sri MOHD KHALIL BIN YAACOB.

Angkatan Keadilan Insan Malaysia (AKIM) (Malaysian Justice Movement): f. 1994 by fmr members of PAS and Semangat '46; Pres. HAMBALI YAZID.

Barisan Alternatif (Alternative Front): Kuala Lumpur; f. June 1999 to contest the general election; opposition electoral alliance originally comprising the PAS, the DAP, the PKN and the PRM; the DAP left in Sept. 2001.

Barisan Jama'ah Islamiah Sa-Malaysia (Berjasa) (Front Malaysian Islamic Council—FMIC): Kelantan; f. 1977; pro-Islamic; 50,000 mems; Pres. Dato' Haji WAN HASHIM BIN Haji WAN ACHMED; Sec.-Gen. MAHMUD ZUHDI BIN Haji ABDUL MAJID.

Bersatu Rakyat Jelata Sabah (Berjaya) (Sabah People's Union): Natikar Bldg, 1st Floor, POB 2130, Kota Kinabalu, Sabah; f. 1975; 400,000 mems; Pres. Haji MOHAMMED NOOR MANSOOR.

Democratic Action Party (DAP): 24 Jalan 20/9, 46300 Petaling Jaya, Selangor; tel. (3) 7578022; fax (3) 7575718; e-mail dap.Malaysia@pobox.com; internet www.malaysia.net/dap; f. 1966; main opposition party; advocates multiracial society based on democratic socialism; 12,000 mems; Chair. LIM KIT SIANG; Sec.-Gen. M. KULA SEGARAN (acting).

Democratic Malaysia Indian Party (DMIP): f. 1985; Leader V. GOVINDARAJ.

Kongres Indian Muslim Malaysia (KIMMA): Kuala Lumpur; tel. (3) 2324759; f. 1977; aims to unite Malaysian Indian Muslims politically; 25,000 mems; Pres. AHAMED ELIAS; Sec.-Gen. MOHAMMED ALI BIN Haji NAINA MOHAMMED.

Malaysian Solidarity Party: Kuala Lumpur.

Parti Hisbul Muslimin Malaysia (Hamim) (Islamic Front of Malaysia): Kota Bahru, Kelantan; f. 1983 as an alternative party to PAS; Pres. Datuk ASRI MUDA.

Parti Ikatan Masyarakat Islam (Islamic Alliance Party): Terengganu.

Parti Islam se Malaysia (PAS) (Islamic Party of Malaysia): Pejabat Agung PAS Pusat, Lorong Haji Hassan, off Jalan Batu Geliga, Taman Melewar, 68100 Batu Caves, Selangor Darul Ehsan; tel. (3) 61895612; fax (3) 61889520; e-mail editor@parti-pas.org; internet www.parti-pas.org; f. 1951; seeks to establish an Islamic state; 700,000 mems; Pres. ABDUL HADI AWANG; Sec.-Gen. NASHARUDIN MAT ISA.

Parti Keadilan Masyarakat (PEKEMAS) (Social Justice Party): Kuala Lumpur; f. 1971 by fmr mems of GERAKAN; Chair. SHAHARYDDIN DAHALAN.

Parti Keadilan Rakyat (PKR) (People's Justice Party): 75A Jalan Lawan Pedang 13/27, Tadisma Business Park, 40000 Shah Alam, Selangor; tel. (3) 55133416; fax (3) 55129646; e-mail contact@partikeadilanrakyat.org; internet www.partikeadilanrakyat.org; f. Aug. 2003 following merger between Parti Keadilan Nasional (PKN) and Parti Rakyat Malaysia (PRM); Pres. WAN AZIZAH WAN ISMAIL; Sec.-Gen. ABDUL RAHMAN OTHMAN.

Parti Nasionalis Malaysia (NasMa): f. 1985; multiracial; Leader ZAINAB YANG.

Parti Rakyat Jati Sarawak (PAJAR) (Sarawak Native People's Party): 22A Jalan Bampeylde, 93200 Kuching, Sarawak; f. 1978; Leader ALI KAWI.

Persatuan Rakyat Malaysian Sarawak (PERMAS) (Malaysian Sarawak Party): Kuching, Sarawak; f. March 1987 by fmr mems of PBB; Leader Haji BUJANG ULIS.

Pertubuhan Bumiputera Bersatu Sarawak (PBBS) (United Sarawak National Association): Kuala Lumpur; f. 1986; Chair. Haji WAN HABIB SYED MAHMUD.

Pertubuhan Rakyat Sabah Bersatu (United Sabah People's Organization—USPO): Kota Kinabalu, Sabah.

Sabah Chinese Party (PCS): Kota Kinabalu, Sabah; f. 1986; Pres. Encik FRANCIS LEONG.

Sabah Chinese Consolidated Party (SCCP): POB 704, Kota Kinabalu, Sabah; f. 1964; 14,000 mems; Pres. JOHNNY SOON; Sec.-Gen. CHAN TET ON.

Sarawak National Party (SNAP): 304–305 Bangunan Mei Jun, 1 Jalan Rubber, POB 2960, 93758 Kuching, Sarawak; tel. (82) 254244; fax (82) 253562; f. 1961; deregistered Nov. 2002 but deregistration deferred indefinitely in April 2003 following appeal; Pres. EDWIN DUNDANG BUGAK; Sec.-Gen. STANLEY JUGOL.

Setia (Sabah People's United Democratic Party): Sabah; f. 1994.

United Malaysian Indian Party: aims to promote unity and economic and social advancement of the Indian community; Sec. KUMAR MANOHARAN.

Diplomatic Representation

EMBASSIES AND HIGH COMMISSIONS IN MALAYSIA

Afghanistan: 2nd Floor, Wisma Chinese Chamber, 258 Jalan Ampang, 50450 Kuala Lumpur; tel. (3) 42569400; fax (3) 42566400; e-mail afghemb@tm.net.my; internet www.afghanembassykl.org; Chargé d'affaires a.i. AMANULLAH JAYHOON.

Albania: 2952 Jalan Bukit Ledang, off Jalan Duta, 50480 Kuala Lumpur; tel. (3) 20937808; fax (3) 20937359; Chargé d'affaires a.i. HAJDAR MUNEKA.

Algeria: 5 Jalan Mesra, off Jalan Damai, 55000 Kuala Lumpur; tel. (3) 21488159; fax (3) 21488154; e-mail dzkl1@yahoo.com; Ambassador AMMAR BELLANI.

Argentina: 3 Jalan Semantan Dua, Damansara Heights, 50490 Kuala Lumpur; tel. (3) 20950176; fax (3) 20952706; e-mail emsia@pd.jaring.my; Ambassador ALFREDO MORELLI.

Australia: 6 Jalan Yap Kwan Seng, 50450 Kuala Lumpur; tel. (3) 21465555; fax (3) 21415773; e-mail info@australia.org.my; internet www.australia.org.my; High Commissioner JAMES WISE.

Austria: MUI Plaza, 7th Floor, Jalan P. Ramlee, POB 10154, 50704 Kuala Lumpur; tel. (3) 21484277; fax (3) 21489813; e-mail kuala-lumpur-ob@bmaa.gv.at; Ambassador Dr OSWALD SOUKOP.

Bangladesh: Block 1, Lorong Damai 7, Jalan Damai, 55000 Kuala Lumpur; tel. (3) 21487490; fax (3) 21413381; e-mail bddoot@pc.jaring.my; High Commissioner SHAFIE U. AHMAD.

Belgium: 8A Jalan Ampang Hilir, 55000 Kuala Lumpur; tel. (3) 42525733; fax (3) 42527922; e-mail kualalumpur@diplobel.org; internet www.diplomatie.be/kualalumpur/; Ambassador ROLAND VAN REMOORTELE.

Bosnia and Herzegovina: JKR 854, Jalan Bellamy, 50460 Kuala Lumpur; tel. (3) 21440353; fax (3) 21426025; e-mail hsomun@hotmail.com; Ambassador MUSTAFA MUJEZINOVIĆ.

Brazil: 22 Pesiaran Damansara Endah, Damansara Heights, 50490 Kuala Lumpur; tel. (3) 20948607; fax (3) 20955086; e-mail brazil@po.jaring.my; Ambassador MARCOS CARAMURU DE PAIVA.

Brunei: Tingkat 19, Menara Tan & Tan, Jalan Tun Razak, 50400 Kuala Lumpur; tel. (3) 21612800; fax (3) 2631302; e-mail bhckl@brucomkul.com.my; High Commissioner Dato' Paduka Haji ABDULLAH BIN Haji MOHAMMAD JAAFAR.

Cambodia: 46 Jalan U Thant, 55000 Kuala Lumpur; tel. (3) 42573711; fax (3) 42571157; e-mail reckl@tm.net.my; Ambassador KEO PUTH REASMEY.

Canada: POB 10990, 50732 Kuala Lumpur; tel. (3) 27183333; fax (3) 27183399; e-mail klmpr-td@dfait-maeci.gc.ca; internet www.dfait-maeci.gc.ca/kualalumpur; High Commissioner MELVYN L. MACDONALD.

Chile: 8th Floor, West Block 142-C, Jalan Ampang, Peti Surat 27, 50450 Kuala Lumpur; tel. (3) 21616203; fax (3) 21622219; e-mail eochile@ppp.nasionet.net; internet www.chileembassy-malaysia.com.my; Ambassador PATRICIO TORRES.

China, People's Republic: 229 Jalan Ampang, 50450 Kuala Lumpur; tel. (3) 21428495; fax (3) 21414552; e-mail cn@tm.net.my; Ambassador WANG CHUNGUI.

Colombia: Level 26, UOA Centre, 19 Jalan Pinang, 50450 Kuala Lumpur; tel. (3) 21645488; fax (3) 21645487; e-mail ekualalumpur@minrelext.gov.co; Chargé d'affaires a.i. YOBANI VELASQUEZ QUINTERO.

Croatia: 3 Jalan Menkuang, off Jalan Ru Ampang, 55000 Kuala Lumpur; tel. (3) 42535340; fax (3) 42535217; e-mail croemb@tm.net.my; Ambassador ZELJKO CIMBUR.

Cuba: 20 Lingkungan U Thant, off Jalan U Thant, 55000 Kuala Lumpur; tel. (3) 42516808; fax (3) 42520428; e-mail malacub@po.jaring.my; internet www.cubaemb.com.my; Ambassador PEDRO MONZÓN BARATA.

Czech Republic: 32 Jalan Mesra, off Jalan Damai, 55000 Kuala Lumpur; tel. (3) 21427185; fax (3) 21412727; e-mail kualalumpur@embassy.mzv.cz; internet www.mzv.cz/kualalumpur; Chargé d'affaires a.i. MILAN TOUS.

Denmark: POB 10908, 50728 Kuala Lumpur; tel. (3) 20322001; fax (3) 20322012; e-mail kulamb@un.dk; internet www.denmark.com.my; Ambassador LASSE REIMANN.

Ecuador: 10th Floor, West Block, Wisma Selangor Dredging, 142-C Jalan Ampang, 50450 Kuala Lumpur; tel. (3) 21635078; fax (3) 21635096; e-mail embecua@po.jaring.my; Ambassador Dr MARCO TULIO CORDERO ZAMORA.

Egypt: 28 Lingkungan U Thant, POB 12004, 55000 Kuala Lumpur; tel. (3) 42568184; fax (3) 42573515; e-mail egyembkl@tm.net.my; Ambassador MOHAMED AFIFI.

Fiji: Level 2, Menara Chan, 138 Jalan Ampang, 50450 Kuala Lumpur; tel. (3) 27323335; fax (3) 27327555; e-mail fhckl@pd.jaring.my; High Commissioner Adi SAMANUNU Q. TALAKULI CAKOBAU.

Finland: Level 5, Wisma Chinese Chamber, 258 Jalan Ampang, 50450 Kuala Lumpur; tel. (3) 42577746; fax (3) 42577793; e-mail sanomat.kul@formin.fi; Ambassador UNTO JUHANI TURUNEN.

France: 196 Jalan Ampang, 50450 Kuala Lumpur; tel. (3) 20535500; fax (3) 20535501; e-mail ambassade.kuala-lumpur-amba@diplomatie.gouv.fr; internet www.ambafrance-my.org; Ambassador JACQUES LAPOUGE.

Germany: 26th Floor, Menara Tan & Tan, 207 Jalan Tun Razak, 50400 Kuala Lumpur; tel. (3) 21709666; fax (3) 21619800; e-mail contact@german-embassy.org.my; internet www.german-embassy.org.my; Ambassador HERBERT GESS.

Ghana: 14 Ampang Hilir, off Jalan Ampang, 55000 Kuala Lumpur; tel. (3) 42526995; fax (3) 42578698; e-mail ghcomkl@tm.net.my; High Commissioner JOHN BENTUM-WILLIAMS.

Guinea: 5 Jalan Kedondong, off Jalan Ampang Hilir, Kuala Lumpur; tel. (3) 42576500; fax (3) 42511500; e-mail mwcnakry@sotelgui.net.gn; Ambassador MAMADOU TOURÉ.

Hungary: City Square Centre, 30th Floor, Empire Tower, Jalan Tun Razak, 50400 Kuala Lumpur; tel. (3) 21637914; fax (3) 21637918; e-mail huembkl@tm.net.my; Ambassador TAMAS TOTH.

India: 2 Jalan Taman Duta, off Jalan Duta, 50480 Kuala Lumpur; tel. (3) 20933510; fax (3) 20933507; e-mail highcomm@po.jaring.my; High Commissioner RAJAMANI LAKSHMI NARAYAN.

Indonesia: 233 Jalan Tun Razak, POB 10889, 50400 Kuala Lumpur; tel. (3) 21452011; fax (3) 21417908; e-mail kbrikl@po.jaring.my; internet www.kbrikl.org.my; Ambassador K. P. H. RUSDIHARDJO.

Iran: 1 Lorong U Thant Satu, off Jalan U Thant, 55000 Kuala Lumpur; tel. (3) 42514824; fax (3) 42562904; e-mail ir_emb@tm.net.my; internet www.iranembassy.com.my; Ambassador MOHAMMAD GHASEM MOHEB ALI.

Iraq: 2 Jalan Langgak Golf, off Jalan Tun Razak, 55000 Kuala Lumpur; tel. (3) 21480555; fax (3) 21414331; Chargé d'affaires a.i. Dr MAHMOUD AL-MSAFIR.

Ireland: Ireland House, The Amp Walk, 218 Jalan Ampang, POB 10372, 50450 Kuala Lumpur; tel. (3) 21612963; fax (3) 21613427; e-mail ireland@po.jaring.my; Ambassador DANIEL MULHALL.

Italy: 99 Jalan U Thant, 55000 Kuala Lumpur; tel. (3) 42565122; fax (3) 42573199; e-mail embassyit@italy-embassy.org.my; internet www.italy-embassy.org.my; Ambassador ANACLETO FELICANI.

Japan: 11 Pesiaran Stonor, off Jalan Tun Razak, 50450 Kuala Lumpur; tel. (3) 21427044; fax (3) 21672314; internet www.my.emb-japan.go.jp; Ambassador MASAKI KONISHI.

Jordan: 2 Jalan Kedondong, off Jalan Ampang Hilir, 55000 Kuala Lumpur; tel. (3) 42521268; fax (3) 42528610; e-mail jordanembassy@po.jaring.my; Ambassador MAZEN JUMA.

Kazakhstan: POB 21, Wisma Selangor Dredging, 3rd Floor, South Block, 142A Jalan Ampang, 50540 Kuala Lumpur; tel. (3) 21664144; fax (3) 21668553; e-mail klkazemb@po.jaring.my; Ambassador MUKHTAR TLEUBERDI.

Kenya: Kuala Lumpur Empire Tower Unit, 38C, 38th Floor, 182 Jalan Tun Razak, 50400 Kuala Lumpur; tel. (3) 21645015; fax (3) 21645017; e-mail kenya@po.jaring.my; High Commissioner DAVID GACHOKI NJOKA.

Korea, Democratic People's Republic: 4 Jalan Persiaran Madge, off Jalan U Thant, 55000 Kuala Lumpur; tel. (3) 42569913; fax (3) 42569933; Ambassador KIM WON HO.

Korea, Republic: Lot 9 and 11, Jalan Nipah, off Jalan Ampang, 55000 Kuala Lumpur; tel. (3) 42512336; fax (3) 42521425; e-mail korem-my@mofat.go.kr; Ambassador LEE YOUNG-JOON.

Kuwait: 229 Jalan Tun Razak, 50400 Kuala Lumpur; tel. (3) 21410033; fax (3) 21456121; e-mail kuwait@streamyx.com; Ambassador ABDULHAMID AL-FAILAKAWI.

Kyrgyzstan: 10 Lorong Damai 9, 55000 Kuala Lumpur; tel. (3) 21649862; fax (3) 21632024; e-mail kyrgyz@tm.net.my; Ambassador AKHBAR RYSKULOV.

Laos: 25 Jalan Damai, 55000 Kuala Lumpur; tel. (3) 21487059; fax (3) 21450080; Ambassador CHALEUNE WARINTHRASAK.

Libya: 6 Jalan Madge, off Jalan U Thant, 55000 Kuala Lumpur; tel. (3) 21411035; fax (3) 21413549; Ambassador Dr AHMAD MOHAMED ALI AL-HANESH.

Luxembourg: Menara Keck Seng Bldg, 16th Floor, 203 Jalan Bukit Bintang, 55100 Kuala Lumpur; tel. (3) 21433134; fax (3) 21433157; e-mail emluxem@po.jaring.my; Chargé d'affaires a.i. CHARLES SCHMIT.

Mauritius: Lot W17-B1 and C1, 17th Floor, West Block, Wisma Selangor Dredging, Jalan Ampang, 50450 Kuala Lumpur; tel. (3) 21636306; fax (3) 21636294; e-mail maur@tm.net.my; High Commissioner RAMJANALLY YOUSOUF MOHAMED (acting).

Mexico: Menara Tan & Tan, 22nd Floor, 207 Jalan Tun Razak, 50400 Kuala Lumpur; tel. (3) 21646362; fax (3) 21640964; e-mail embamex@po.jaring.my; internet www.embamex.org.my; Ambassador ALFREDO PÉREZ BRAVO.

Morocco: 3rd Floor, East Block, Wisma Selangor Dredging, 142B Jalan Ampang, 50450 Kuala Lumpur; tel. (3) 21610701; fax (3) 21623081; e-mail sifmakl@tm.net.my; Ambassador BADRE EDDINE ALLALI.

Myanmar: 12 Jalan Ru, off Jalan Ampang Hilir, 55000 Kuala Lumpur; tel. (3) 42560280; fax (3) 42568320; e-mail mekl@tm.net.my; Ambassador U HLA MAUNG.

Namibia: Suite 15-01, Level 15, 3 Jalan Kia Peng, 50450 Kuala Lumpur; tel. (3) 21646520; fax (3) 21688790; e-mail namhckl@po.jaring.my; High Commissioner NEVILLE MELVIN GERTZA.

Nepal: Suite 13A-01, 13th Floor, Wisma MCA, 163 Jalan Ampang, 40450 Kuala Lumpur; tel. (3) 21645934; fax (3) 21648659; Chargé d'affaires a.i. DEEPAK DHITAL.

Netherlands: The Amp Walk, 7th Floor, 218 Jalan Ampang, POB 10543, 50450 Kuala Lumpur; tel. (3) 21686200; fax (3) 21686240; e-mail nlgovkl@netherlands.org.my; internet www.netherlands.org.my; Ambassador J. C. F. VON MÜHLEN.

New Zealand: Menara IMC, 21st Floor, 8 Jalan Sultan Ismail, 50250 Kuala Lumpur; tel. (3) 20782533; fax (3) 20780387; e-mail nzhckl@po.jaring.my; High Commissioner GEOFF RANDAL.

Nigeria: 85 Jalan Ampang Hilir, 55000 Kuala Lumpur; tel. (3) 42517843; fax (3) 42524302; e-mail archives@nigeria.org.my; High Commissioner Dr WAHAB OLASEINDE DOSUNMU.

Norway: Suite CD, 53rd Floor, Empire Tower, Jalan Tun Razak, 50400 Kuala Lumpur; tel. (3) 21637100; fax (3) 21637108; e-mail emb.kualalumpur@mfa.no; Ambassador ARILD EIK.

Oman: 109 Jalan U Thant, 55000 Kuala Lumpur; tel. (3) 42577378; fax (3) 42571400; e-mail omanemb@po.jaring.my; Ambassador Sheikh GHAZI BIN SAID BIN ABDULLAH AL BAHR AL RAWAS.

Pakistan: 132 Jalan Ampang, 50450 Kuala Lumpur; tel. (3) 21618877; fax (3) 21645958; e-mail parepklumpur@po.jaring.my; internet www3.jaring.my/pakistanhc/; High Commissioner Maj.-Gen. (retd) TALAT MUNIR.

Papua New Guinea: 11 Lingkungan U Thant, off Jalan U Thant, 55000 Kuala Lumpur; tel. (3) 42575405; fax (3) 42576203; High Commissioner PETER P. MAGINDE.

Peru: Wisma Selangor Dredging, 6th Floor, South Block 142-A, Jalan Ampang, 50450 Kuala Lumpur; tel. (3) 21633034; fax (3) 21633039; e-mail info@embperu.com.my; internet www.embperu.com.my; Chargé d'affaires a.i. FRANCISCO GARCIA YRIGOYEN.

Philippines: 1 Changkat Kia Peng, 50450 Kuala Lumpur; tel. (3) 21484233; fax (3) 21483576; e-mail webmaster@philembassykl.org.my; internet www.philembassykl.org.my; Ambassador ROMUALDO ANOVER ONG.

Poland: 495 Bt 4½ Jalan Ampang, 68000 Ampang, Selangor; tel. (3) 42576733; fax (3) 42570123; e-mail polamba@tm.net.my; Ambassador EUGENIOSZ SAWICKI.

Romania: 114 Jalan Damai, off Jalan Ampang, 55000 Kuala Lumpur; tel. (3) 21423172; fax (3) 21448713; e-mail roemb@streamyx.com; Ambassador GABRIEL GAFITA.

Russia: 263 Jalan Ampang, 50450 Kuala Lumpur; tel. (3) 42567252; fax (3) 42576091; e-mail ruemvvl@tm.net.my; Ambassador VLADIMIR NIKOLAEVICH MOROZOV.

Saudi Arabia: Level 4, Wisma Chinese Chamber, 258 Jalan Ampang, 50450 Kuala Lumpur; tel. (3) 42579433; fax (3) 42578751; e-mail saembssy@tm.net.my; Ambassador HAMED MOHAMMED YAHYA.

Senegal: 5 Persiaran Ampang, 55000 Kuala Lumpur; tel. (3) 42567343; fax (3) 42563205; e-mail senamb_mal@yahoo.fr; Ambassador AMADOU FAYE.

Seychelles: 12th Floor, West Block, Wisma Selangor Dredging, POB 24, 142c Jalan Ampang, 50450 Kuala Lumpur; tel. (3) 21635726; fax (3) 21635729; e-mail seyhicom@po.jaring.my; High Commissioner LOUIS SYLVESTRE RADEGONDE.

Singapore: 209 Jalan Tun Razak, 50400 Kuala Lumpur; tel. (3) 21616277; fax (3) 21616343; e-mail shckl@po.jaring.my; internet www.mfa.gov.sg/kl; High Commissioner ASHOK KUMAR MIRPURI.

Slovakia: 11 Jalan U Thant, 55000 Kuala Lumpur; tel. (3) 21150016; fax (3) 21150014; e-mail slovemb@tm.net.my; Ambassador MILAN TANCAR.

South Africa: 12 Lorong Titiwangsa, Taman Tasik Titiwangsa, Setapak, 53200 Kuala Lumpur; tel. (3) 40244456; fax (3) 40249896; e-mail sahcpol@tm.net.my; internet www.afrikaselatan.com; High Commissioner Dr ABRAHAM SOKHAYA NKOMO.

Spain: 200 Jalan Ampang, 50450 Kuala Lumpur; tel. (3) 21484868; fax (3) 21424582; e-mail embespmy@mail.mae.es; Chargé d'affaires a.i. BLANCA LONDAIZ LABORDE.

Sri Lanka: 116 Jalan Damai, off Jalan Ampang, 55000 Kuala Lumpur; tel. (3) 21612199; fax (3) 21612219; e-mail slhicom@putra.net.my; internet www.kuala-lumpur.mission.gov.lk; High Commissioner (vacant).

Sudan: 2A Persiaran Ampang, off Jalan Ru, 55000 Kuala Lumpur; tel. (3) 42569104; fax (3) 42568107; e-mail sudanikuala@hotmail.com; Ambassador ABDEL RAHMAN HAMZAH ELRAYA.

Swaziland: Suite 22.03 and 03 (A), Menara Citibank, 165 Jalan Ampang, 50450 Kuala Lumpur; tel. (3) 21632511; fax (3) 21633326; e-mail swazi@tm.net.my; High Commissioner NEWMAN SIZWE NTSHANGASE (acting).

Sweden: Wisma Angkasa Raya, 6th Floor, 123 Jalan Ampang, POB 10239, 50708 Kuala Lumpur; tel. (3) 21485433; fax (3) 21486325; e-mail ambassaden.kuala-lumpur@foreign.ministry.se; internet www.embassyofswedenmy.org; Ambassador BRUNO BEIJER.

Switzerland: 16 Persiaran Madge, 55000 Kuala Lumpur; tel. (3) 21480622; fax (3) 21480935; e-mail vertretung@kua.rep.admin.ch; Ambassador Dr PETER A. SCHWEIZER.

Syria: Suite 23.03, 23rd Floor, Menara Tan & Tan, Jalan Tun Razak, 50400 Kuala Lumpur; tel. (3) 21634110; fax (3) 21634199; Chargé d'affaires a.i. ABDUL AZIZ ALI.

Thailand: 206 Jalan Ampang, 50450 Kuala Lumpur; tel. (3) 21488222; fax (3) 21486527; e-mail thaikul@mfa.go.th; Ambassador CHAISIRI ANAMARN.

Timor-Leste: 62 Jalan Ampang Hilir, 55000 Kuala Lumpur; tel. (3) 42562046; fax (3) 42562016; e-mail embaixada_tl_kl@yahoo.com; Ambassador DJAFAR AMUDE BIN ALCATIRI.

Turkey: 118 Jalan U Thant, 55000 Kuala Lumpur; tel. (3) 42572225; fax (3) 42572227; e-mail turkbe@tm.net.my; Ambassador KORAY TARGAY.

Ukraine: Suite 22.02, 22nd Floor, Menara Tan & Tan, 207 Jalan Tun Razak, 50400 Kuala Lumpur; tel. (3) 21669552; fax (3) 21664371; e-mail emb_my@mfa.gov.ua; Ambassador OLEKSANDR SHEVCHENKO.

United Arab Emirates: 1 Gerbang Ampang Hilir, off Persiaran Ampang Hilir, 55000 Kuala Lumpur; tel. (3) 42535221; fax (3) 42535220; e-mail uaemal@tm.net.my; Ambassador NASSER SALMAN ALABOODI.

United Kingdom: 185 Jalan Ampang, 50450 Kuala Lumpur; tel. (3) 21702200; fax (3) 21442370; e-mail political.kualalumpur@fco.gov.uk; internet www.britain.org.my; High Commissioner BRUCE CLEGHORN.

USA: 376 Jalan Tun Razak, POB 10035, 50700 Kuala Lumpur; tel. (3) 21685000; fax (3) 21422207; internet www.usembassymalaysia.org.my; Ambassador CHRISTOPHER J. LAFLEUR.

Uruguay: 6 Jalan 3, Taman Tun Abdul Razak, 68000 Ampang, Selangor Darul Ehsan; tel. (3) 42518831; fax (3) 42517878; e-mail urukual@po.jaring.my; Ambassador ROBERTO PABLO TOURINO TURNES.

Uzbekistan: 2 Jalan 12, Taman Tun Abdul Razak, 68000 Ampang, Selangor; tel. (3) 42532406; fax (3) 42535406; e-mail uzbekemb@streamyx.com; Ambassador AYBEK KHASANOV.

Venezuela: Suite 20-05, 20th Floor, Menara Tan & Tan, 207 Jalan Tun Razak, 50400 Kuala Lumpur; tel. (3) 21633444; fax (3) 21636819; e-mail venezuela@po.jaring.my; internet www.venezuela.org.my; Ambassador Maj.-Gen. (retd) NOEL ENRIQUE MARTÍNEZ OCHOA.

Viet Nam: 4 Jalan Persiaran Stonor, 50450 Kuala Lumpur; tel. (3) 21484036; fax (3) 21483270; e-mail daisevn@putra.net.my; Ambassador NGUYEN QUOC DUNG.

Yemen: 7 Jalan Kedondong, off Jalan Ampang Hilir, 55000 Kuala Lumpur; tel. (3) 42511793; fax (3) 42511794; e-mail yemenkl@tm.net.my; Ambassador Dr ABDUL NASSER ALI ABDO MUNIBARI.

Zimbabwe: 124 Jalan Sembilan, Taman Ampang Utama, 68000 Ampang, Selangor Darul Ehsan; tel. (3) 42516779; fax (3) 42517252; e-mail zhck@tm.net.my; Ambassador LUCAS PANDE TAVAYA.

Judicial System

The two High Courts, one in Peninsular Malaysia and the other in Sabah and Sarawak, have original, appellate and revisional jurisdiction as the federal law provides. Above these two High Courts is the Court of Appeal, which was established in 1994; it is an intermediary court between the Federal Court and the High Court. When appeals to the Privy Council in the United Kingdom were abolished in 1985 the former Supreme Court became the final court of appeal. Therefore, at that stage only one appeal was available to a party aggrieved by the decision of the High Court. Hence, the establishment of the Court of Appeal. The Federal Court (formerly the Supreme Court) has, to the exclusion of any other court, jurisdiction in any dispute between states or between the Federation and any state; and has special jurisdiction as to the interpretation of the Constitution. The Federal Court is headed by the Chief Justice (formerly the Lord President); the other members of the Federal Court are the President of the Court of Appeal, the two Chief Judges of the High Courts and the Federal Court Judges. Members of the Court of Appeal are the President and the Court of Appeal judges, and members of the High Courts are the two Chief Judges and their respective High Court judges. All judges are appointed by the Yang di-Pertuan Agong on the advice of the Prime Minister, after consulting the Conference of Rulers. In 1993 a Special Court was established to hear cases brought by or against the Yang di-Pertuan Agong or a Ruler of State (Sultans).

The Sessions Courts, which are situated in the principal urban and rural centres, are presided over by a Sessions Judge, who is a member of the Judicial and Legal Service of the Federation and is a qualified barrister or a Bachelor of Law from any of the recognized universities. Their criminal jurisdiction covers the less serious indictable offences, excluding those that carry the death penalty.

Civil jurisdiction of a Sessions Court is up to RM 250,000. The Sessions Judges are appointed by the Yang di-Pertuan Agong.

The Magistrates' Courts are also found in the main urban and rural centres and have both civil and criminal jurisdiction, although of a more restricted nature than that of the Sessions Courts. The Magistrates consist of officers from the Judicial and Legal Service of the Federation. They are appointed by the State Authority in which they officiate on the recommendation of the Chief Judge.

There are also Syariah (Shariah) courts for rulings under Islamic law. In July 1996 the Cabinet announced that the Syariah courts were to be restructured with the appointment of a Syariah Chief Judge and four Court of Appeal justices, whose rulings would set precedents for the whole country.

Prior to February 1995 trials for murder and kidnapping in the High Courts were heard with jury and assessors, respectively. The amendment to the Criminal Procedure Code abolished both the jury and the assessors systems, and all criminal trials in the High Courts are heard by a judge sitting alone. In 1988 an amendment to the Constitution empowered any federal lawyer to confer with the Attorney-General to determine the courts in which any proceedings, excluding those before a Syariah court, a native court or a court martial, be instituted, or to which such proceedings be transferred.

Federal Court of Malaysia

Bangunan Sultan Abdul Samad, Jalan Raja, 50506 Kuala Lumpur; tel. (3) 26939011; fax (3) 26932582; internet www.kehakiman.gov.my.

Chief Justice of the Federal Court: Tan Sri Dato' Sri AHMAD FAIRUZ BIN Dato' Sheikh ABDUL HALIM.

President of the Court of Appeal: (vacant).

Chief Judge of the High Court in Peninsular Malaysia: Dato' HAIDAR BIN MOHAMED NOOR.

Chief Judge of the High Court in Sabah and Sarawak: Datuk STEVE SHIP LIM KIONG.

Attorney-General: ABDUL GANI PATAIL.

Religion

Islam is the established religion but freedom of religious practice is guaranteed. Almost all ethnic Malays are Muslims, representing 53% of the total population in 1985. In Peninsular Malaysia 19% followed Buddhism (19% in Sarawak and 8% in Sabah), 7% were Christians (29% in Sarawak and 24% in Sabah), and Chinese faiths, including Confucianism and Daoism, were followed by 11.6%. Sikhs and other religions accounted for 0.5%, while 2%, mostly in Sabah and Sarawak, were animists.

Malaysian Consultative Council of Buddhism, Christianity, Hinduism and Sikhism (MCCBCHS): 8 Jalan Duku, off Jalan Kasipillai, 51200 Kuala Lumpur; tel. (3) 40414669; fax (3) 40444304; e-mail hsangam@po.jaring.my; f. 1981; a non-Muslim group.

ISLAM

President of the Majlis Islam: Datuk Haji MOHD FAUZI BIN Haji ABDUL HAMID (Kuching, Sarawak).

Jabatan Kemajuan Islam Malaysia (JAKIM) (Department of Islamic Development Malaysia): Aras 4–9, Block D7, Pusat Pentadbiran Persekutuan, 62502 Putrajaya; tel. (3) 88864000; e-mail faizal@islam.gov.my; internet www.islam.gov.my.

Istitut Kefahaman Islam Malaysia (IKIM) (Institute of Islamic Understanding Malaysia): 2 Langgak Tunku, off Jalan Duta, 50480 Kuala Lumpur; tel. (3) 62010889; fax (3) 62014189; internet www.ikim.gov.my.

BUDDHISM

Malaysian Buddhist Association (MBA): MBA Building, 113, 3¼ Miles, Jalan Klang, 58000 Kuala Lumpur; tel. (3) 7815595; e-mail mbapg@po.jaring.my; internet www.jaring.my/mba; f. 1959; the national body for Chinese-speaking monks and nuns and temples from the Mahayana tradition; Pres. Venerable CHEK HUANG.

Young Buddhist Association of Malaysia (YBAM): 10 Jalan SS2/75, 47300 Petaling Jaya, Selangor; tel. (3) 78764591; fax (3) 78762770; e-mail ybamhq@po.jaring.my; internet www.ybam.org.my; f. 1970.

Buddhist Missionary Society Malaysia (BMSM): 123 Jalan Berhala, off Jalan Tun Sambanthan, 50470 Brickfields, Kuala Lumpur; tel. (3) 22730150; fax (3) 22740245; e-mail president@bmsm.org.my; internet www.bmsm.org.my; f. 1962 as Buddhist Missionary Society; Pres. ANG CHOO HONG.

Malaysian Fo Kuang Buddhist Association: 2 Jalan SS3/33, Taman University, 47300 Petaling Jaya, Selangor.

Buddhist Tzu-Chi Merit Society (Malaysia): 24 Jesselton Ave, 10450 Pinang; e-mail mtzuchi@po.jaring.my; internet www.tzuchi.org.my.

Sasana Abhiwurdhi Wardhana Society: 123 Jalan Berhala, off Jalan Tun Sambanthan, 50490 Kuala Lumpur; f. 1894; the national body for Sri Lankan Buddhists belonging to the Theravada tradition.

CHRISTIANITY

Majlis Gereja-Gereja Malaysia (Council of Churches of Malaysia): 26 Jalan Universiti, 46200 Petaling Jaya, Selangor; tel. (3) 7567092; fax (3) 7560353; e-mail cchurchm@tm.net.my; internet www.ccmalaysia.org; f. 1947; 16 mem. churches; 8 associate mems; Pres. Most Rev. Datuk YONG PING CHUNG (Anglican Bishop of Sabah); Gen. Sec. Rev. Dr HERMEN SHASTRI.

The Anglican Communion

Malaysia comprises three Anglican dioceses, within the Church of the Province of South East Asia.

Primate: Most Rev. Datuk YONG PING CHUNG (Bishop of Sabah).

Bishop of Kuching: Rt Rev. MADE KATIB, Bishop's House, POB 347, 93704 Kuching, Sarawak; tel. (82) 240187; fax (82) 426488; e-mail bkg@pc.jaring.my; has jurisdiction over Sarawak, Brunei and part of Indonesian Kalimantan (Borneo).

Bishop of Sabah: Most Rev. Datuk YONG PING CHUNG, Rumah Bishop, Jalan Tangki, POB 10811, 88809 Kota Kinabalu, Sabah; tel. (88) 247008; fax (88) 245942; e-mail pcyong@pc.jaring.my.

Bishop of West Malaysia: Rt. Rev. Tan Sri Dr LIM CHENG EAN, Bishop's House, 14 Pesiaran Stonor, 50450 Kuala Lumpur; tel. (3) 20312728; fax (3) 20313225; e-mail diocese@tm.net.my.

The Baptist Church

Malaysia Baptist Convention: 2 Jalan 2/38, 46000 Petaling Jaya, Selangor; tel. (3) 77823564; fax (3) 77833603; e-mail mbcpj@tm.net.my; Chair. Dr TAN ENG LEE.

The Methodist Church

Methodist Church in Malaysia: 69 Jalan 5/31, 46000 Petaling Jaya, Selangor; tel. (3) 79541811; fax (3) 79541788; e-mail methmas@tm.net; 140,000 mems; Bishop Dr PETER CHIO SING CHING.

The Presbyterian Church

Presbyterian Church in Malaysia: Joyful Grace Church, Jalan Alsagoff, 82000 Pontian, Johor; tel. (7) 711390; fax (7) 324384; Pastor TITUS KIM KAH TECK.

The Roman Catholic Church

Malaysia comprises two archdioceses and six dioceses. At 31 December 2002 approximately 3.2% of the population were adherents.

Catholic Bishops' Conference of Malaysia, Singapore and Brunei

Xavier Hall, 133 Jalan Gasing, 46000 Petaling Jaya, Selangor; tel. and fax (3) 79581371; e-mail cbcmsb@pc.jaring.my.

Pres. Most Rev. NICHOLAS CHIA (Archbishop of Singapore).

Archbishop of Kuala Lumpur: Most Rev. MURPHY NICHOLAS XAVIER PAKIAM, Archbishop's House, 528 Jalan Bukit Nanas, 50250 Kuala Lumpur; tel. (3) 20788828; fax (3) 20313815; e-mail mpakiam@pd.jaring.my.

Archbishop of Kuching: Most Rev. JOHN HA TIONG HOCK, Archbishop's Office, 118 Jalan Tun Abang Haji Openg, POB 940, 93718 Kuching, Sarawak; tel. (82) 242634; fax (82) 425724; e-mail johnha@pd.jaring.my.

BAHÁ'Í FAITH

Spiritual Assembly of the Bahá'ís of Malaysia: 4 Lorong Titiwangsa 5, off Jalan Pahang, 53200 Kuala Lumpur; tel. (3) 40235183; fax (3) 40226277; e-mail nsa-sec@bahai.org.my; internet www.bahai.org.my; mems resident in 800 localities.

The Press

PENINSULAR MALAYSIA DAILIES

English Language

Business Times: Balai Berita 31, Jalan Riong, 59100 Kuala Lumpur; tel. (3) 22822628; fax (3) 22825424; e-mail bt@nstp.com

.my; internet www.btimes.com.my; f. 1976; morning; Editor Zainul Arifin; circ. 15,000.

Malay Mail: Balai Berita 31, Jalan Riong, 59100 Kuala Lumpur; tel. (3) 22822829; fax (3) 22821434; e-mail malaymail@nstp.com.my; internet www.mmail.com.my; f. 1896; afternoon; Editor Fauzi Omar; circ. 75,000.

Malaysiakini: 2–4 Jalan Bangsa-Utang 9, 59000 Kuala Lumpur; tel. (3) 22835567; fax (3) 22892579; e-mail editor@malaysiakini.com; internet www.malaysiakini.com; Malaysia's first online newspaper; English and Malay; Editor Steven Gan.

New Straits Times: Balai Berita 31, Jalan Riong, 59100 Kuala Lumpur; tel. (3) 2823322; fax (3) 2821434; e-mail news@nstp.com .my; internet www.nst.com.my; f. 1845; morning; Group Editor-in-Chief Kalimullah Masheerul Hassan; circ. 190,000.

The Edge: G501–G801, Levels 5–8, Block G, Phileo Damansara I, Jalan 16/11, off Jalan Damansara, 46350 Petaling Jaya, Selangor; tel. (3) 76603838; fax (3) 76608638; e-mail eeditor@bizedge.com; internet www.theedgedaily.com; f. 1996; weekly, with daily internet edition; business and investment news.

The Star: 13 Jalan 13/6, 46200 Petaling Jaya, POB 12474, Selangor; tel. (3) 7581188; fax (3) 7551280; e-mail msd@thestar.com.my; internet www.thestar.com.my; f. 1971; morning; Group Chief Editor Ng Poh Tip; circ. 192,059.

The Sun: Sun Media Corpn Sdn Bhd, Lot 6, Jalan 51/217, Section 51, 46050 Petaling Jaya, Selangor Darul Ehsan; tel. (3) 7946688; fax (3) 7952624; e-mail editor@sunmg.po.my; f. 1993; Editor-in-Chief Ho Kay Tat; Man. Dir Tan Boon Kean; circ. 82,474.

Chinese Language

China Press: 80 Jalan Riong, 59100 Kuala Lumpur; tel. (3) 2828208; fax (3) 2825327; circ. 206,000.

Chung Kuo Pao (China Press): 80 Jalan Riong, 59100 Kuala Lumpur; tel. (3) 2828208; fax (3) 2825327; f. 1946; Editor Poon Chau Huay; Gen. Man. Ng Beng Lye; circ. 210,000.

Guang Ming Daily: 19 Jalan Semangat, 46200 Petaling Jaya, Selangor; tel. (3) 7582888; fax (3) 7575135; circ. 87,144.

Kwong Wah Yit Poh: 19 Jalan Presgrave, 10300 Pinang; tel. (4) 2612312; fax (4) 2615407; e-mail editor@kwongwah.com.my; internet www.kwongwah.com.my; f. 1910; morning; Chief Editor Tan Aye Choo; circ. 72,158.

Nanyang Siang Pau (Malaysia): 1 Jalan SS7/2, 47301 Petaling Jaya, Selangor; tel. (3) 78726888; fax (3) 78726800; e-mail editor@nanyang.com.my; internet www.nanyang.com.my; f. 1923; morning and evening; Editor-in-Chief Cheng Khee Chien; circ. 180,000 (daily), 220,000 (Sunday).

Shin Min Daily News: 31 Jalan Riong, Bangsar, 59100 Kuala Lumpur; tel. (3) 2826363; fax (3) 2821812; f. 1966; morning; Editor-in-Chief Cheng Song Huat; circ. 82,000.

Sin Chew Jit Poh (Malaysia): 19 Jalan Semangat, POB 367, Jalan Sultan, 46200 Petaling Jaya, Selangor; tel. (3) 7582888; fax (3) 7570527; internet www.sinchew-i.com; f. 1929; morning; Chief Editor Liew Chen Chuan; circ. 227,067 (daily), 230,000 (Sunday).

Malay Language

Berita Harian: Balai Berita 31, Jalan Riong, 59100 Kuala Lumpur; tel. (3) 2822323; fax (3) 2822425; e-mail bharian@bharian.com.my; internet www.bharian.com.my; f. 1957; morning; Group Editor Ahmad Rejal Arbee; circ. 350,000.

Metro Ahad: Balai Berita 31, Jalan Riong, 59100 Kuala Lumpur; tel. (3) 2822328; fax (3) 2824482; e-mail metahad@nstp.com.my; internet www.metroahad.com.my; circ. 132,195.

Mingguan Perdana: 48 Jalan Siput Akek, Taman Billion, Kuala Lumpur; tel. (3) 619133; Group Chief Editor Khalid Jafri.

Utusan Malaysia: 46m Jalan Lima, off Jalan Chan Sow Lin, 55200 Kuala Lumpur; tel. (3) 2217055; fax (3) 2220911; e-mail corpcomm@utusan.com.my; internet www.utusan.com.my; Editor Abdul Aziz Ishak; circ. 239,385.

Watan: 23–1 Jalan 9a/55a, Taman Setiawangsa, 54200 Kuala Lumpur; tel. (3) 4523040; fax (3) 4523043; circ. 80,000.

Tamil Language

Malaysia Nanban: 11 Jalan Murai Dua, Batu Kompleks, off Jalan Ipoh, 51200 Kuala Lumpur; tel. (3) 6212251; fax (3) 6235981; circ. 45,000.

Tamil Nesan: 28 Jalan Yew, Pudu, 55100 Kuala Lumpur; tel. (3) 2216411; fax (3) 2210448; f. 1924; morning; Editor V. Vivekananthan; circ. 35,000 (daily), 60,000 (Sunday).

Tamil Osai: 19 Jalan Murai Dua, Batu Kompleks, Jalan Ipoh, Kuala Lumpur; tel. (3) 671644; circ. 21,000 (daily), 40,000 (Sunday).

Tamil Thinamani: 9 Jalan Murai Dua, Batu Kompleks, Jalan Ipoh, Kuala Lumpur; tel. (3) 66719; Editor S. Nachiappan; circ. 18,000 (daily), 39,000 (Sunday).

SUNDAY PAPERS

English Language

New Sunday Times: Balai Berita 31, Jalan Riong, 59100 Kuala Lumpur; tel. (3) 2822328; fax (3) 2824482; e-mail news@nstp.com .my; f. 1932; morning; Group Editor (vacant); circ. 191,562.

Sunday Mail: Balai Berita 31, Jalan Riong, 59100 Kuala Lumpur; tel. (3) 2822328; fax (3) 2824482; e-mail smail@nstp.com.my; f. 1896; morning; Editor Joachim S. P. Ng; circ. 75,641.

Sunday Star: 13 Jalan 13/6, 46200 Petaling Jaya, POB 12474, Selangor Darul Ehsan; tel. (3) 7581188; fax (3) 7551280; f. 1971; Editor David Yeoh; circ. 232,790.

Malay Language

Berita Minggu: Balai Berita 31, Jalan Riong, 59100 Kuala Lumpur; tel. (3) 2822328; fax (3) 2824482; e-mail bharian@bharian .com.my; f. 1957; morning; Editor Dato' Ahmad Nazri Abdullah; circ. 421,127.

Mingguan Malaysia: 11a The Right Angle, Jalan 14/22, 46100 Petaling Jaya; tel. (3) 7563355; fax (3) 7577755; f. 1964; Editor Mohd Hassan Mohd Noor; circ. 543,232.

Utusan Zaman: 11a The Right Angle, Jalan 14/22, 46100 Petaling Jaya; tel. (3) 7563355; fax (3) 7577755; f. 1939; Editor Mustafa Fadula Suhaimi; circ. 11,782.

Tamil Language

Makkal Osai: 11 Jalan Murai Dua, Batu Kompleks, off Jalan Ipoh, 51200 Kuala Lumpur; tel. (3) 6212251; fax (3) 6235981; circ. 28,000.

PERIODICALS

English Language

Her World: Berita Publishing Sdn Bhd, Balai Berita 31, Jalan Riong, 59100 Kuala Lumpur; tel. (3) 2824322; fax (3) 2828489; monthly; Editor Alice Chee Lan Neo; circ. 35,000.

Malaysia Warta Kerajaan Seri Paduka Baginda (HM Government Gazette): Percetakan Nasional Malaysia Berhad, Jalan Chan Sow Lin, 50554 Kuala Lumpur; tel. (3) 92212022; fax (3) 92220690; e-mail pnmb@po.jaring.my; fortnightly.

Malaysian Agricultural Journal: Ministry of Agriculture, Publications Unit, Wisma Tani, Jalan Sultan Salahuddin, 50624 Kuala Lumpur; tel. (3) 2982011; fax (3) 2913758; f. 1901; 2 a year.

Malaysian Forester: Forestry Department Headquarters, Jalan Sultan Salahuddin, 50660 Kuala Lumpur; tel. (3) 26988244; fax (3) 26925657; e-mail skthai@forestry.gov.my; f. 1931; quarterly; Editor Thai See Kiam.

The Planter: Wisma ISP, 29 33 Jalan Taman U Thant, POB 10262, 50708 Kuala Lumpur; tel. (3) 21425561; fax (3) 21426898; e-mail isphq@tm.net.my; internet www.isp.org.my; f. 1919; publ. by Isp Management (M); monthly; Editor W. T. Perera; circ. 4,000.

Young Generation: 11a The Right Angle, Jalan 14/22, 46100 Petaling Jaya, Selangor; tel. (3) 7563355; fax (3) 7577755; monthly; circ. 50,000.

Chinese Language

Mister Weekly: 2a Jalan 19/1, 46300 Petaling Jaya, Selangor; tel. (3) 7562400; fax (3) 7553826; f. 1976; weekly; Editor Wong Ah Tai; circ. 25,000.

Mun Sang Poh: 472 Jalan Pasir Puteh, 31650 Ipoh; tel. (5) 3212919; fax (5) 3214006; bi-weekly; circ. 77,958.

New Life Post: 80M Jalan SS21/39, Damansara Utama, 47400 Petaling Jaya, Selangor; tel. (3) 7571833; fax (3) 7181809; f. 1972; bi-weekly; Editor Low Beng Chee; circ. 231,000.

New Tide Magazine: Nanyang Siang Pau Bldg, 2nd Floor, Jalan 7/2, 47301 Petaling Jaya, Selangor; tel. (3) 76202118; fax (3) 76202131; e-mail newtidemag@hotmail.com; f. 1974; monthly; Editor Nellie Ooi; circ. 39,000.

Malay Language

Dewan Masyarakat: Dewan Bahasa dan Pustaka, Jalan Wisma Putra, POB 10803, 50926 Kuala Lumpur; tel. (3) 2481011; fax (3) 2484211; f. 1963; monthly; current affairs; Editor Zulkifli Salleh; circ. 48,500.

Dewan Pelajar: Dewan Bahasa dan Pustaka, Jalan Wisma Putra, POB 10803, 50926 Kuala Lumpur; tel. (3) 2481011; fax (3) 2484211; f. 1967; monthly; children's; Editor ZALEHA HASHIM; circ. 100,000.

Dewan Siswa: POB 10803, 50926 Kuala Lumpur; tel. (3) 2481011; fax (3) 2484208; monthly; circ. 140,000.

Gila-Gila: 38-1, Jalan Bangsar Utama Satu, Bangsar Utama, 59000 Kuala Lumpur; tel. (3) 22824970; fax (3) 22824967; fortnightly; circ. 70,000.

Harakah: Jabatan Penerangan dan Penyelidikan PAS, 28A Jalan Pahang Barat, Off Jalan Pahang, 53000 Kuala Lumpur; tel. (3) 40213343; fax (3) 40212422; e-mail hrkh@pc.jaring.my; internet www.harakahdaily.com; two a month; Malay; organ of the Parti Islam se Malaysia (PAS—Islamic Party of Malaysia); Editor ZULKIFLI SULONG.

Jelita: Berita Publishing Sdn Bhd, 16–20 Jalan 4/109E, Desa Business Park, Taman Desa, off Jalan Klang Lama, 58100 Kuala Lumpur; tel. (3) 76208111; fax (3) 76208114; e-mail jelita@beritapub.com.my; internet www.jelita.com.my; monthly; fashion and beauty magazine; Editor ROHANI PA' WAN CHIK; circ. 80,000.

Mangga: 11A The Right Angle, Jalan 14/22, 46100 Petaling Jaya, Selangor; tel. (3) 7563355; fax (3) 7577755; monthly; circ. 205,000.

Mastika: 11A The Right Angle, Jalan 14/22, 46100 Petaling Jaya, Selangor; tel. (3) 7363355; fax (3) 7577755; monthly; Malayan illustrated magazine; Editor AZIZAH ALI; circ. 15,000.

Utusan Radio dan TV: 11A The Right Angle, Jalan 14/22, 46100 Petaling Jaya, Selangor; tel. (3) 7363355; fax (3) 7577755; fortnightly; Editor NORSHAH TAMBY; circ. 115,000.

Wanita: 11A The Right Angle, Jalan 14/22, 46100 Petaling Jaya, Selangor; tel. (3) 7563355; fax (3) 7577755; monthly; women; Editor NIK RAHIMAH HASSAN; circ. 85,000.

Punjabi Language

Navjiwan Punjabi News: 52 Jalan 8/18, Jalan Toman, 46050 Petaling Jaya, Selangor; tel. (3) 7565725; f. 1950; weekly; Assoc. Editor TARA SINGH; circ. 9,000.

SABAH DAILIES

Api Siang Pau (Kota Kinabalu Commercial Press): 24 Lorong Dewan, POB 170, Kota Kinabalu; f. 1954; morning; Chinese; Editor Datuk LO KWOCK CHUEN; circ. 3,000.

Borneo Mail (Nountan Press Sdn Bhd): 1 Jalan Bakau, 1st Floor, off Jalan Gaya, 88999 Kota Kinabulu; tel. (88) 238001; fax (88) 238002; English; circ. 14,610.

Daily Express: News House, 16 Jalan Pasar Baru, POB 10139, 88801 Kota Kinabalu; tel. (88) 256422; fax (88) 238611; e-mail sph@tm.net.my; internet www.dailyexpress.com.my; f. 1963; morning; English, Bahasa Malaysia and Kadazan; Editor-in-Chief SARDATHISA JAMES; circ. 30,000.

Hwa Chiaw Jit Pao (Overseas Chinese Daily News): News House, 16 Jalan Pasar Baru, POB 10139, 88801 Kota Kinabalu; tel. (88) 256422; e-mail sph@tm.net.my; internet www.dailyexpress.com.my; f. 1936; morning; Chinese; Editor HII YUK SENG; circ. 30,000.

Merdeka Daily News: Lot 56, BDC Estate, Mile 1½ North Road, POB 332, 90703 Sandakan; tel. (89) 214517; fax (89) 275537; e-mail merkk@tm.net.my; f. 1968; morning; Chinese; Editor-in-Chief FUNG KON SHING; circ. 8,000.

New Sabah Times: Jalan Pusat Pembangunan Masyarakat, off Jalan Mat Salleh, 88100 Kota Kinabalu; POB 20119, 88758 Kota Kinabalu; tel. (88) 230055; fax (88) 241155; internet www.newsabahtimes.com.my; English, Malay and Kadazan; Editor-in-Chief EDDY LOK; circ. 30,000.

Syarikat Sabah Times: Kota Kinabalu; tel. (88) 52217; f. 1952; English, Malay and Kadazan; circ. 25,000.

Tawau Jih Pao: POB 464, 1072 Jalan Kuhara, Tawau; tel. (89) 72576; Chinese; Editor-in-Chief STEPHEN LAI KIM YEAN.

SARAWAK DAILIES

Berita Petang Sarawak: Lot 8322, Lorong 7, Jalan Tun Abdul Razak, 93450 Kuching; POB 1315, 93726 Kuching; tel. (82) 480771; fax (82) 489006; f. 1972; evening; Chinese; Chief Editor HWANG YU CHAI; circ. 12,000.

Borneo Post: 40 Jalan Tuanku Osman, POB 20, 96000 Sibu; tel. (84) 332055; fax (84) 321255; internet www.borneopost.com.my; morning; English; Man. Dir LAU HUI SIONG; Editor NGUOI HOW YIENG; circ. 60,000.

International Times: Lot 2215, Jalan Bengkel, Pending Industrial Estate, POB 1158, 93724 Kuching; tel. (82) 482215; fax (82) 480996; e-mail news@intimes.com; internet www.intimes.com.my; f. 1968; morning; Chinese; Editor LEE FOOK ONN; circ. 37,000.

Malaysia Daily News: 7 Island Rd, POB 237, 96009 Sibu; tel. (84) 330211; tel. (84) 320540; f. 1968; morning; Chinese; Editor WONG SENG KWONG; circ. 22,735.

Sarawak Tribune and Sunday Tribune: Lot 231, Jalan Nipah, off Jalan Abell, 93100 Kuching; tel. (82) 424411; fax (82) 420358; internet www.jaring.my/tribune; f. 1945; English; Editor FRANCIS SIAH; circ. 29,598.

See Hua Daily News: 40 Jalan Tuanku Osman, POB 20, 96000 Sibu; tel. (84) 332055; fax (84) 321255; f. 1952; morning; Chinese; Man. Editor LAU HUI SIONG; circ. 80,000.

United Daily News: internet www.uniteddaily.com.my; f. 2004 following merger between Chinese Daily News and Miri Daily News; morning; Chinese; Dep. Publr WONG KEH HUONG; circ. 35,000.

PERIODICALS

Pedoman Rakyat: Malaysian Information Dept, Mosque Rd, 93612 Kuching; tel. (82) 240141; f. 1956; monthly; Malay; Editor SAIT BIN HAJI YAMAN; circ. 30,000.

Pembrita: Malaysian Information Services, Mosque Rd, 93612 Kuching; tel. (82) 247231; f. 1950; monthly; Iban; Editor ALBAN JAWA; circ. 20,000.

Sarawak Gazette: Sarawak Museum, Jalan Tun Abang Haji Openg, 93566 Kuching; tel. (82) 244232; fax (82) 246680; e-mail museum@po.jaring.my; f. 1870; 2 a year; English; Chief Editor Datu Haji SALLEH SULAIMAN.

Utusan Sarawak: Lot 231, Jalan Nipah, off Jalan Abell, POB 138, 93100 Kuching; tel. (82) 424411; fax (82) 420358; internet www.tribune.com.my/tribune; f. 1949; Malay; Editor Haji ABDUL AZIZ Haji MALIM; circ. 32,292.

NEWS AGENCIES

Bernama (Malaysian National News Agency): Wisma Bernama, 28 Jalan 1/65A, off Jalan Tun Razak, POB 10024, 50700 Kuala Lumpur; tel. (3) 2945233; fax (3) 2941020; e-mail sjamil@bernama.com; internet www.bernama.com; f. 1968; general and foreign news, economic features and photo services, public relations wire, screen information and data services, stock market on-line equities service, real-time commodity and monetary information services; daily output in Malay and English; in June 1990 Bernama was given the exclusive right to receive and distribute news in Malaysia; Gen. Man. SYED JAMIL JAAFAR.

Foreign Bureaux

Agence France-Presse (AFP): 26 Hotel Equatorial, 1st Floor, Jalan Sultan Ismail, 50250 Kuala Lumpur; tel. (3) 26911906; fax (3) 21615606; Correspondent MERVIN NAMBIAR.

Associated Press (AP) (USA): Wisma Bernama, 28 Jalan 1/65A, off Jalan Tun Razak, POB 12219, Kuala Lumpur; tel. (3) 2926155; Correspondent HARI SUBRAMANIAM.

Inter Press Service (IPS) (Italy): 32 Jalan Mudah Barat, Taman Midah, 56000 Kuala Lumpur; tel. (3) 9716830; fax (3) 2612872; Correspondent (vacant).

Press Trust of India: 114 Jalan Limau Manis, Bangsar Park, Kuala Lumpur; tel. (3) 940673; Correspondent T. V. VENKITACHALAM.

United Press International (UPI) (USA): Room 1, Ground Floor, Wisma Bernama, Jalan 1/65A, 50400 Kuala Lumpur; tel. (3) 2933393; fax (3) 2913876; Rep. MARY LEIGH.

Reuters (United Kingdom) and Xinhua (People's Republic of China) are also represented in Malaysia.

PRESS ASSOCIATION

Persatuan Penerbit-Penerbit Akhbar Malaysia (Malaysian Newspaper Publishers' Asscn): Unit 706, Block B, Phileo Damansara 1, 9 Jalan 16/11, off Jalan Damansara, 46350 Petaling Jaya; tel. (3) 76608535; fax (3) 76608532; e-mail mnpa@macomm.com.my; Chair. ROSELINA JOHARI.

Publishers

KUALA LUMPUR

Arus Intelek Sdn Bhd: Plaza Mont Kiara, Suite E-06-06, Mont Kiara, 50480 Kuala Lumpur; tel. (3) 62011558; fax (3) 62018698; e-mail arusintelek@po.jaring.my; Man. Datin AZIZAH MOKHZANI.

Berita Publishing Sdn Bhd: Balai Berita, 31 Jalan Riong, 59100 Kuala Lumpur; tel. (3) 2824322; fax (3) 2821605; internet www.jelita.com.my; education, business, fiction, cookery; Man. ABDUL MANAF SAAD.

Dewan Bahasa dan Pustaka (DBP) (Institute of Language and Literature): POB 10803, 50926 Kuala Lumpur; tel. (3) 21481011; fax (3) 21444460; e-mail aziz@dbp.gov.my; internet www.dbp.gov.my; f. 1956; textbooks, magazines and general; Chair. Tan Sri Kamarul Ariffin Mohamed Yassin; Dir-Gen. Dato' Haji A. Aziz Deraman.

International Law Book Services: Lot 4.1, Wisma Shen, 4th Floor, 149 Jalan Masjid India, 50100 Kuala Lumpur; tel. (3) 2939864; fax (3) 2928035; e-mail gbc@pc.jaring.my; Man. Dr Syed Ibrahim.

Jabatan Penerbitan Universiti Malaya (University of Malaya Press): University of Malaya, Lembah Pantai, 50603 Kuala Lumpur; tel. (3) 79574361; fax (3) 79574473; e-mail hamedi@um.edu.my; internet www.um.edu.my/umpress; f. 1954; general fiction, literature, economics, history, medicine, politics, science, social science, law, Islam, engineering, dictionaries; Chief Editor Dr Hamedi Mohd Adnan.

Malaya Press Sdn Bhd: Kuala Lumpur; tel. (3) 5754650; fax (3) 5751464; f. 1958; education; Man. Dir Lai Wing Chun.

Pustaka Antara Sdn Bhd: Lot UG 10–13, Upper Ground Floor, Kompleks Wilayah, 2 Jalan Munshi Abdullah, 50100 Kuala Lumpur; tel. (3) 26980044; fax (3) 26917997; e-mail pantara@tm.net.my; textbooks, children's, languages, fiction; Man. Dir Datuk Abdul Aziz bin Ahmad.

Utusan Publications and Distributors Sdn Bhd: 1 and 3 Jalan 3/91A, Taman Shamelin Perkasa, Cheras, 56100 Kuala Lumpur; tel. (3) 9856577; fax (3) 9846554; e-mail rose@utusan.com.my; internet www.upnd.com.my; school textbooks, children's, languages, fiction, general; Exec. Dir Roselina Johari.

JOHOR

Penerbitan Pelangi Sdn Bhd: 66 Jalan Pingai, Taman Pelangi, 80400 Johor Bahru; tel. (7) 3316288; fax (7) 3329201; e-mail ppsb@po.jaring.my; internet www.pelangibooks.com; children's books, guidebooks and reference; Man. Dir Samuel Sum Kown Cheek.

Textbooks Malaysia Sdn Bhd: 49 Jalan Tengku Ahmad, POB 30, 85000 Segamat, Johor; tel. (7) 9318323; fax (7) 9313323; school textbooks, children's fiction, guidebooks and reference; Man. Dir Freddie Khoo.

NEGERI SEMBILAN

Bharathi Press: 166 Taman AST, POB 74, 70700 Seremban, Negeri Sembilan Darul Khusus; tel. (6) 7622911; f. 1939; Mans M. Subramania Bharathi, Bharathi Thasan.

PINANG

Syarikat United Book Sdn Bhd: 187–189 Lebuh Carnarvon, 10100 Pulau Pinang; tel. (4) 61635; fax (4) 615063; textbooks, children's, reference, fiction, guidebooks; Man. Dir Chew Sing Guan.

SELANGOR

Federal Publications Sdn Bhd: Lot 46, Subang Hi-Tech Industrial Park, Batu Tiga, 40000 Shah Alam, Selangor; tel. (3) 56286888; fax (3) 56364620; e-mail fpsb@tpg.com.my; f. 1957; computer, children's magazines; Gen. Man. Stephen K. S. Lim.

FEP International Sdn Bhd: 6 Jalan SS 4c/5, POB 1091, 47301 Petaling Jaya, Selangor; tel. (3) 7036150; fax (3) 7036989; f. 1969; children's, languages, fiction, dictionaries, textbooks and reference; Man. Dir Lim Mok Hai.

Mahir Publications Sdn Bhd: 39 Jalan Nilam 1/2, Subang Sq., Subang Hi-Tech Industrial Park, Batu Tiga, 40000 Shah Alam, Selangor; tel. (3) 7379044; fax (3) 7379043; e-mail mahirpub@tm.net.my; Gen. Man. Zainora binti Muhamad.

Minerva Publications (NS) Sdn Bhd: 51 Jalan SG 3/1, Tan Sri Gombak, Batu Caves, 68100 Selangor; tel. (3) 61882876; fax (3) 61883876; e-mail minerva@streamyx.com; internet www.minervaa.com; f. 1974; general, children's, reference, medical, law; Dir and Chief Editor Sujaudeen; Man. Dir Thanjudeen.

Pearson Education Malaysia Sdn Bhd: Lot 2, Jalan 215, off Jalan Templer, 46050 Petaling Jaya, Selangor; tel. (3) 77820466; fax (3) 77818005; e-mail inquiry@pearsoned.com.my; internet www.pearsoned.com.my; textbooks, mathematics, physics, science, general, educational materials; Dir Wong Wee Woon; Man. Wong Mei Mei.

Pelanduk Publications (M) Sdn Bhd: 12 Jalan SS 13/3E, Subang Jaya Industrial Estate, 47500 Subang Jaya, Selangor; tel. (3) 7386885; fax (3) 7386575; e-mail pelpub@tm.net.my; internet www.pelanduk.com; Man. Jackson Tan.

Penerbit Fajar Bakti Sdn Bhd: 4 Jalan U1/15, Sekseyen U1, Hicom-Glenmarie Industrial Park, 40150 Shah Alam, Selangor; tel. (3) 7047011; fax (3) 7047024; e-mail edes@pfb.po.my; school, college and university textbooks, children's, fiction, general; Man. Dir Edda de Silva.

Penerbit Pan Earth Sdn Bhd: 11 Jalan SS 26/6, Taman Mayang Jaya, 47301 Petaling Jaya, Selangor; tel. (3) 7031258; fax (3) 7031262; Man. Stephen Cheng.

Penerbit Universiti Kebangsaan Malaysia: Universiti Kebangsaan Malaysia, 43600 UKM, Selangor; tel. (3) 8292840; fax (3) 8254375; Man. Hasrom bin Haron.

Pustaka Delta Pelajaran Sdn Bhd: Wisma Delta, Lot 18, Jalan 51A/22A, 46100 Petaling Jaya, Selangor; tel. (3) 7570000; fax (3) 7576688; e-mail dpsb@po.jaring.my; economics, language, environment, geography, geology, history, religion, science; Man. Dir Lim Kim Wah.

Pustaka Sistem Pelajaran Sdn Bhd: Lot 17–22 and 17–23, Jalan Satu, Bersatu Industrial Park, Cheras Jaya, 43200 Cheras, Selangor; tel. (3) 9047558; fax (3) 9047573; Man. T. Thiru.

Sasbadi Sdn Bhd: 103A Jalan SS 21/1A, Damansara Utama, 47400 Petaling Jaya, Selangor; tel. (3) 7182550; fax (3) 7186709; Man. Law King Hui.

SNP Panpac (Malaysia) Sdn Bhd: Lot 3, Jalan Saham 23/3, Kawasan MIEL Phase 8, Section 23, 40300 Shah Alam, Selangor Darul Ehsam; tel. (3) 55481088; fax (3) 55481080; e-mail eastview@snpo.com.my; f. 1980; fmrly SNP Eastview Publications Sdn Bhd; school textbooks, children's, fiction, reference, general; Dir Chia Yan Heng.

Times Educational Sdn Bhd: 22 Jalan 19/3, 46300 Petaling Jaya, Selangor; tel. (3) 79571766; fax (3) 79573607; textbooks, general and reference; Man. Foong Chui Lin.

GOVERNMENT PUBLISHING HOUSE

Percetakan Nasional Malaysia Bhd (Malaysia National Printing Ltd): Jalan Chan Sow Lin, 50554 Kuala Lumpur; tel. (3) 2212022; fax (3) 2220690; fmrly the National Printing Department, incorporated as a company under govt control in January 1993.

PUBLISHERS' ASSOCIATION

Malaysian Book Publishers' Association: 306 Block C, Glomac Business Centre, 10 Jalan SS 6/1 Kelana Jaya, 47301 Petaling Jaya, Selangor; tel. (3) 7046628; fax (3) 7046629; e-mail mabopa@po.jaring.my; internet www.mabopa.com.my; f. 1968; Pres. Ng Tieh Chuan; Hon. Sec. Kow Ching Chuan; 95 mems.

Broadcasting and Communications

TELECOMMUNICATIONS

Celcom (Malaysia) Sdn Bhd: Wisma Telekom Semarak, 82 Jalan Raja Muda Abdul Aziz, 503000 Kuala Lumpur; tel. (3) 26873838; e-mail cpr@celcom.com.my; internet www.celcom.com.my; f. 1988; private co licensed to operate mobile cellular telephone service; merged with TM Cellular Sdn Bhd in 2003; Chair. Dr Mohamed Munir bin Abdul Majid; CEO Dato' Mohamed Yunus Ramli bin Abbas.

DiGi Telecommunications Sdn Bhd: Lot 30, Jalan Delima 1/3, Subang Hi-Tech Industrial Park, 40000 Shah Alam, Selangor; tel. (3) 57211800; fax (3) 57211857; internet www.digi.com.my; private co licensed to operate mobile telephone service; Chair. Tan Sri Dato' Seri Vincent Tan Chee Yioun; CEO Tore Johnsen.

Jabatan Telekomunikasi Malaysia (JTM) (Department of Telecommunications): c/o Ministry of Energy, Communications and Multimedia, Wisma Damansara, 3rd Floor, Jalan Semantan, 50668 Kuala Lumpur; tel. (3) 20875000; fax (3) 20957901; internet www.ktkm.gov.my; regulatory body for telecommunications industry.

Maxis Communications Bhd: Menara Maxis, Aras 18, Kuala Lumpur City Centre, 50088 Kuala Lumpur; tel. (3) 23307000; fax (3) 23300008; internet www.maxis.com.my; f. 1995; provides mobile, fixed line and multimedia services; approx. 3.25m. subscribers in 2003; CEO Jamaludin Ibrahim; Chair. Datuk Megat Zaharuddin bin Megat Mohamed Noor.

Telekom Malaysia Bhd: Level 51, North Wing, Menara Telekom, off Jalan Pantai Baru, 50672 Kuala Lumpur; tel. (3) 22401221; fax (3) 22832415; internet www.telekom.com.my; f. 1984; public listed co responsible for operation of basic telecommunications services; 74% govt-owned; 4.22m. fixed lines (95% of total); Chair. Haji Muhammad Radzi bin Haji Mansor; Chief Exec. Dr Md Khir Abdul Rahman.

Technology Resources Industries Bhd (TRI): Menara TR, 23rd Floor, 161B Jalan Ampang, 50450 Kuala Lumpur; tel. (3) 2619555; fax (3) 2632018; operates mobile cellular telephone service; Chair. and Chief Exec. Tan Sri Dato' Tajudin Ramli.

Time dotCom Bhd: Wisma Time, 1st Floor, 249 Jalan Tun Razak, 50400 Kuala Lumpur; tel. (3) 27208000; fax (3) 27200199; internet www.time.com.my; f. 1996 as Time Telecommunications Holdings Bhd; name changed as above in Jan. 2000; state-controlled co licensed to operate trunk network and mobile cellular telephone service; Chair. Dato' Wan Muhamad Wan Ibrahim; Man. Dir Tan See Yin.

BROADCASTING

Regulatory Authority

Under the Broadcasting Act (approved in December 1987), the Minister of Information is empowered to control and monitor all radio and television broadcasting, and to revoke the licence of any private company violating the Act by broadcasting material 'conflicting with Malaysian values'.

Radio Televisyen Malaysia (RTM): Dept of Broadcasting, Angkasapuri, Bukit Putra, 50614 Kuala Lumpur; tel. (3) 22825333; fax (3) 2824735; e-mail helpdesk@rtm.net.my; internet www.rtm.net.my; f. 1946; television introduced 1963; supervises radio and television broadcasting; Dir-Gen. Jaafar Kamin; Dep. Dir-Gen. Tamimuddin Abdul Karim.

Radio

Radio Malaysia: Radio Televisyen Malaysia (see Regulatory Authority), POB 11272, 50740 Kuala Lumpur; tel. (3) 2823991; fax (3) 2825859; f. 1946; domestic service; operates six networks; broadcasts in Bahasa Malaysia, English, Chinese (Mandarin and other dialects), Tamil and Aborigine (Temiar and Semai dialects); Dir of Radio Madzhi Johari.

Suara Islam (Voice of Islam): Islamic Affairs Division, Prime Minister's Department, Blok Utama, Tingkat 1–5, Pusat Pentadbiran Kerajaan Persekutuan, 62502 Putrajaya; internet www.smpke.jpm.my; f. 1995; Asia-Pacific region; broadcasts in Bahasa Malaysia on Islam.

Suara Malaysia (Voice of Malaysia): Wisma Radio, Angkasapuri, POB 11272, 50740 Kuala Lumpur; tel. (3) 22887824; fax (3) 22847594; f. 1963; overseas service in Bahasa Malaysia, Arabic, Myanmar (Burmese), English, Bahasa Indonesia, Chinese (Mandarin/Cantonese), Tagalog and Thai; Controller of Overseas Service Stephen Sipaun.

Radio Televisyen Malaysia—Sabah: Jalan Tuaran, 88614 Kota Kinabalu; tel. (88) 213444; fax (88) 223493; f. 1955; television introduced 1971; a dept of RTM; broadcasts programmes over two networks for 280 hours a week in Bahasa Malaysia, English, Chinese (two dialects), Kadazan, Murut, Dusun and Bajau; Dir of Broadcasting Jumat Engson.

Radio Televisyen Malaysia—Sarawak: Broadcasting House, Jalan P. Ramlee, 93614 Kuching; tel. (82) 248422; fax (82) 241914; e-mail pvgrtmsw@tm.net.my; f. 1954; a dept of RTM; broadcasts 445 hours per week in Bahasa Malaysia, English, Chinese, Iban, Bidayuh, Melanau, Kayan/Kenyah, Bisayah and Murut; Dir of Broadcasting Norhyati Ismail.

Rediffusion Sdn Bhd: Rediffusion House, 17 Jalan Pahang, 53000 Kuala Lumpur; tel. (3) 4424544; fax (3) 4424614; f. 1949; two programmes; 44,720 subscribers in Kuala Lumpur; 11,405 subscribers in Pinang; 6,006 subscribers in Province Wellesley; 20,471 subscribers in Ipoh; Gen. Man. Rosni B. Rahmat.

Time Highway Radio: Wisma Time, 10th Floor, Jalan Tun Razak, 50400 Kuala Lumpur; tel. (3) 27202993; fax (3) 27200993; e-mail chief@thr.fm; internet www.thr.fm; f. 1994; serves Kuala Lumpur region; broadcasts in English; CEO Abdul Aziz Hamdan.

Television

Measat Broadcast Network Systems Sdn Bhd: All Asia Broadcast Centre, Technology Park Malaysia, Lebuhraya Puchong, Simpang Besi, Bukit Jalil, 57000 Kuala Lumpur; tel. (3) 95434188; fax (3) 95437333; e-mail custcare@astro.com.my; internet www.astro.com.my; nation-wide subscription service; Malaysia's first satellite, Measat 1, was launched in January 1996; a second satellite was launched in October of that year; Chair. T. Ananda Krishnan.

Mega TV: Kuala Lumpur; internet www.megatv.com.my; subscription service; began broadcasting in November 1995; 5 foreign channels; initially available only in Klang Valley; 40%-owned by the Govt.

MetroVision: 33 Jalan Delima, 1/3 Subang Hi-Tech Industrial Park, 40000 Shah Alam, Selangor; tel. (3) 7328000; fax (3) 7328932; e-mail norlin@metrovision.com.my; internet www.metrovision.com.my; began broadcasting in July 1995; commercial station; operates only in Klang Valley; 44%-owned by Senandung Sesuria Sdn Bhd, 56%-owned by Metropolitan Media Sdn Bhd; Man. Dr Sabri Abdul Rahman.

Radio Televisyen Malaysia—Sabah: see Radio.

Radio Televisyen Malaysia—Sarawak: see Radio.

Sistem Televisyen Malaysia Bhd (TV 3): 3 Persiaran Bandar Utama, Bandar Utama, 47800 Petaling, Selangor Darul Ehsan; tel. (3) 77266333; fax (3) 77261333; internet www.tv3.com.my; f. 1983; Malaysia's first private television network, began broadcasting in 1984; Chair. Dato' Mohd Noor Yusof; Man. Dir Hisham Dato' Abd. Rahman.

Televisyen Malaysia: Radio Televisyen Malaysia (see Regulatory Authority); f. 1963; operates two national networks, TV1 and TV2; Controller of Programmes Ismail Mohamed Jah.

Under a regulatory framework devised by the Government, the ban on privately owned satellite dishes was ended in 1996.

Finance

(cap. = capital; auth. = authorized; res = reserves; dep. = deposits; m. = million; brs = branches; amounts in ringgit Malaysia)

BANKING

In January 2004 there were 46 domestic commercial banks, merchant banks and finance companies. In February 2000 the Government announced that it had approved plans for the creation of up to 10 banking groups to be formed through the merger of existing institutions. By August 2001 51 banks had merged under the terms of these plans. In February 2004 53 banks held offshore licences in Labuan.

Central Bank

Bank Negara Malaysia: Jalan Dato' Onn, 50480 Kuala Lumpur; tel. (3) 26988044; fax (3) 2912990; e-mail info@bnm.gov.my; internet www.bnm.gov.my; f. 1959; bank of issue; financial regulatory authority; cap. 100.0m., res 42,307.2m., dep. 123,101.0m. (Dec. 2003); Gov. Tan Sri Dato' Sri Zeti Akhtar Aziz; 6 brs.

Regulatory Authority

Labuan Offshore Financial Services Authority (LOFSA): Level 17, Main Office Tower, Financial Park Labuan, Jalan Merdeka, 87000 Labuan; tel. (87) 408188; fax (87) 413328; e-mail communication@lofsa.gov.my; internet www.lofsa.gov.my; regulatory body for the International Offshore Financial Centre of Labuan established in October 1990; Chair. Datuk Zeti Akhtar Aziz (Gov. of Bank Negara Malaysia); Dir-Gen. Rosnah Omar.

Commercial Banks

Peninsular Malaysia

ABN Amro Bank Bhd: Levels 25–27, MNI Twins, Tower II, 11 Jalan Pinang, POB 10094, 50704 Kuala Lumpur; tel. (3) 21626666; fax (3) 21625692; e-mail info@abnamro.com.my; internet www.abnamromalaysia.com; f. 1963.

Affin Bank Bhd: Menara AFFIN, 17th Floor, Jalan Raja Chulan, 50200 Kuala Lumpur; tel. (3) 20559000; fax (3) 20261415; e-mail head.ccd@affinbank.com.my; internet www.affinbank.com.my; f. 1975 as Perwira Habib Bank Malaysia Bhd; name changed to Perwira Affin Bank Bhd 1994; merged with BSN Commercial Bank (Malaysia) Bhd Jan. 2001, and name changed as above; cap. 1,017.3m., res 368.2m., dep. 17,523.0m. (Dec. 2003); Chair. Gen. Tan Sri Dato' Seri Ismail Haji Omar; Pres. and CEO Dato' Abdul Hamidy Abdul Hafiz; 110 brs.

Alliance Bank Malaysia Bhd: Menara Multi-Purpose, Ground Floor, Capital Sq., 8 Jalan Munshi Abdullah, 50100 Kuala Lumpur; POB 10069, 50704 Kuala Lumpur; tel. (3) 26948800; fax (3) 26946727; e-mail multilink@alliancebg.com.my; internet www.alliancebank.com.my; f. 1982 as Malaysian French Bank Berhad; name changed to Multi-Purpose Bank Bhd 1996; name changed as above Jan. 2001, following acquisition of six merger partners; cap. 596.5m., res 540.4m., dep. 13,535.1m. (March 2003); Chair. Tan Sri Abu Talib Othman; Chief Exec. Dir Ng Siek Chuan; 79 brs.

AmBank Bhd: 22nd Floor, Bangunan AmBank Group, 55 Jalan Raja Chulan, 50200 Kuala Lumpur; tel. (3) 20782633; fax (3) 20316453; e-mail customercare@ambg.com.my; internet www.ambg.com.my; f. 1994; fmrly Arab-Malaysian Bank Bhd; name changed as above 2002; cap. 505.5m., res –72.0m., dep. 9,036.3m. (March 2003); Chair. Tan Sri Dato' Azman Hashim; Man. Dir Kung Beng Hong.

Bangkok Bank Bhd (Thailand): 105 Jalan Tun H. S. Lee, 50000 Kuala Lumpur; tel. (3) 2324555; fax (3) 2388569; e-mail bbb@tm.net.my; f. 1958; cap. 100.0m., res 68.0m., dep. 541.9m. (Dec. 2002); Chair. Albert Cheok Saychuan; CEO Chalit Tayjasanant; 1 br.

Bank of America Malaysia Bhd: Wisma Goldhill, Jalan Raja Chulan, 50200 Kuala Lumpur; tel. (3) 20321133; fax (3) 20319087; internet www.bankofamerica.com.my; cap. 135.8m., res 177.2m., dep. 546.7m. (Dec. 2002); Chair. RICHARD LINEBAUGH.

Bank of Nova Scotia Bhd: POB 11056, Menara Boustead, 69 Jalan Raja Chulan, 50734 Kuala Lumpur; tel. (3) 21410766; fax (3) 21412160; e-mail bns.kualalumpur@scotiabank.com; internet www.scotiabank.com.my; f. 1973; cap. 122.4m., res 222.6m., dep. 1,806.1m. (Oct. 2002); Man. Dir RASOOL KHAN.

Bank of Tokyo-Mitsubishi (Malaysia) Bhd (Japan): 1 Leboh Ampang, 50100 Kuala Lumpur; ; tel. (3) 20789100; fax (3) 20708340; e-mail edpbtm@tm.net.my; f. 1996 following merger of the Bank of Tokyo and Mitsubishi Bank; cap. 200m., res 516.1m., dep. 1,832.5m. (Dec. 2002); Chair. YOSHIHIRO WATANABE; Pres. and CEO HIROYUKI KUDO.

Bank Pembangunan & Infrastruktur Malaysia Bhd: Menara Bank Pembangunan, POB 12352, Jalan Sultan Ismail, 50774 Kuala Lumpur; tel. (3) 26152020; fax (3) 26928520; e-mail bpimb-pr@bpimb.com.my; internet www.bpimb.com.my; f. 1973; cap. 1,200m., res 731.5m., dep. 4,534.7m. (Dec. 2002); Man. Dir and CEO Dato' Haji ABDUL RAHIM MOHAMED ZIN; Chair. Datuk MOHAMED ADNAN BIN ALI; 13 brs.

Bumiputra Commerce Bank Bhd: 6 Jalan Tun Perak, 50050 Kuala Lumpur; tel. (3) 26983022; fax (3) 26986628; internet www.bcb.com.my; f. 1999 following merger of Bank Bumiputra Malaysia Bhd with Bank of Commerce Bhd; cap. 2,064.0m., res 2,059.4m., dep. 52,559.3m. (Dec. 2002); Chair. Tan Sri RADIN SOENARNO AL-HAJ; Man. Dir and CEO Dr ROZALI MOHAMED ALI; 230 brs.

Citibank Bhd (USA): 165 Jalan Ampang, POB 10112, 50450 Kuala Lumpur; tel. (3) 2325334; fax (3) 2328763; internet www.citibank.com.my; f. 1959; cap. 121.7m., res 1,577.4m., dep. 17,132.8m. (Dec. 2002); Country Officer PIYUSH GUPTA; 3 brs.

Deutsche Bank (Malaysia) Bhd (Germany): 18–20 Menara IMC, 8 Jalan Sultan Ismail, 50250 Kuala Lumpur; tel. (3) 20536788; fax (3) 20319822; f. 1994; cap. 125.0m., res 226.9m., dep. 1,980.7m. (Dec. 2002); Man. Dir KUAH HUN LIANG.

EON Bank Bhd: Wisma Cyclecarri, 12th Floor, Jalan Raja Laut, 50350 Kuala Lumpur; tel. (3) 26941188; fax (3) 26949588; e-mail caf@eonbank.com.my; internet www.eonbank.com.my; f. 1963; fmrly Kong Ming Bank Bhd; merged with Oriental Bank Bhd, Jan. 2001; cap. 1,329.8m., dep. 20,910.2m. (Dec. 2003), res 278.5m. (Dec. 2002); Chair. Tan Sri Dato' Seri MOHAMED SALEH BIN SULONG; 96 brs.

Hong Leong Bank Bhd: Wisma Hong Leong, Level 3, 18 Jalan Perak, 50450 Kuala Lumpur; tel. (3) 21642828; fax (3) 27156365; internet www.hlb.com.my; f. 1905; fmrly MUI Bank Bhd; merged with Wah Tat Bank Bhd, Jan. 2001; cap. 1,435.0m., res 1,899.1m., dep. 23,746.3m. (June 2003); Chair. Tan Sri QUEK LENG CHAN; Man. Dir YVONNE CHIA; 167 local brs, 2 overseas brs.

HSBC Bank Malaysia Bhd (Hong Kong): 2 Leboh Ampang, POB 10244, 50912 Kuala Lumpur; tel. (3) 20700744; fax (3) 20702678; e-mail manager.public.affairs@hsbc.com.my; internet www.hsbc.com.my; f. 1860; fmrly Hongkong Bank Malaysia Bhd; adopted present name in 1999; cap. 114.5m., res 1,779.5m., dep. 25,423.9m. (Dec. 2003); Chair. MICHAEL SMITH.

Malayan Banking Bhd (Maybank): Menara Maybank, 14th Floor, 100 Jalan Tun Perak, 50050 Kuala Lumpur; tel. (3) 20708833; fax (3) 20702611; e-mail publicaffairs@maybank.com.my; internet www.maybank2u.com; f. 1960; acquired Pacific Bank Bhd, Jan. 2001; merged with PhileoAllied Bank (Malaysia) Bhd, March 2001; cap. and res 12,099m., dep. 111,046m. (June 2004); Chair. Tan Sri MOHAMED BASIR BIN AHMAD; Pres. and CEO Datuk AMIRSHAM A. AZIZ; 415 domestic brs, 81 overseas brs.

OCBC Bank (Malaysia) Bhd: Tingkat 1–8, Wisma Lee Rubber, Jalan Melaka, 50100 Kuala Lumpur; tel. (3) 26920344; fax (3) 26926518; internet www.ocbc.com.my; f. 1932; cap. 287.5m., res 1,465.2m., dep. 20,737.9m. (Dec. 2003); Chair. Tan Sri Dato' NASRUDDIN BAHARI; CEO ALBERT YEOH BEOW TIT; 25 brs.

Public Bank Bhd: Menara Public Bank, 146 Jalan Ampang, 50450 Kuala Lumpur; tel. (3) 21638888; fax (3) 21639917; internet www.publicbank.com.my; f. 1965; merged with Hock Hua Bank Bhd, March 2001; cap. 3,206.6m., res 4,320.3m., dep. 48,201.3m. (Dec. 2003); Chair. Tan Sri Dato' Dr TEH HONG PIOW; 216 domestic brs, 3 overseas brs.

RHB Bank Bhd: Towers Two and Three, RHB Centre, 426 Jalan Tun Razak, 50400 Kuala Lumpur; tel. (3) 92878888; fax (3) 92879000; e-mail md_ceo@rhbbank.com.my; internet www.rhbbank.com.my; formed in 1997 by a merger between DCB Bank Bhd and Kwong Yik Bank Bhd; acquired Sime Bank Bhd in mid-1999; merged with Bank Utama (Malaysia) Bhd, May 2003; cap. 3,318.1m., res 1,420.8m., dep. 38,994.4m. (June 2002); Chair. Dato' ALI BIN HASSAN; Man. Dir and CEO (vacant); 148 brs.

Southern Bank Bhd: Level 3, Menara Southern Bank, 83 Medan Setia Satu, Plaza Damamsara, Bukit Damansara, 50490 Kuala Lumpur; tel. (3) 2573000; fax (3) 3817200; e-mail info@sbbgroup.com.my; internet www.sbbgroup.com.my; f. 1963; merged with Ban Hin Lee Bank Bhd, July 2000; cap. 1,122.8m., res 1,131.6m., dep. 20,523.7m. (Dec. 2002); Chair. Tan Sri OSMAN S. CASSIM; CEO Dato' TAN TEONG HEAN; 105 domestic brs, 1 overseas br.

Standard Chartered Bank Malaysia Bhd: 1st Floor, 2 Jalan Ampang, 50450 Kuala Lumpur; tel. (3) 20726555; fax (3) 2010621; internet www.standardchartered.com.my.

United Overseas Bank (Malaysia) Bhd: Menara UOB, Jalan Raja Laut, POB 11212, 50738 Kuala Lumpur; tel. (3) 26927722; fax (3) 26981228; e-mail uobmtre@uob.com.my; internet www.uob.com.my; f. 1920; merged with Chung Khiaw Bank (Malaysia) Bhd in 1997 and with Overseas Union Bank (Malaysia) Bhd in 2002; cap. 470m., res 1,130.8m., dep. 15,174.1m. (Dec. 2002); Pres. WEE EE CHEONG; Dir and CEO FRANCIS LEE CHIN YONG; 25 brs.

Merchant Banks

Affin Merchant Bank Bhd: Menara Boustead, 27th Floor, 69 Jalan Raja Chulan, POB 1124, 50200 Kuala Lumpur; tel. (3) 2423700; fax (3) 2424982; e-mail general@affinmerchantbank.com.my; internet www.affinmerchantbank.com.my; f. 1970 as Permata Chartered Merchant Bank Bhd; name changed as above March 2001; cap. 187.5m., res 122.2m., dep. 1,519.3m. (Dec. 2001); Chair. Tan Sri YAACOB MOHAMED ZAIN; CEO Datin ZURAIDAH ATAN-SHARARIMAN.

Alliance Merchant Bank Bhd: Menara Multi-Purpose, 20th Floor, Capital Sq., 8 Jalan Munshi Abdullah, 50100 Kuala Lumpur; tel. (3) 26927788; fax (3) 26928787; e-mail ambb@alliancemerchant.com.my; internet www.alliancemerchant.com.my; f. 1974 as Amanah-Chase Merchant Bank Bhd; name changed as above Jan. 2001, following merger with Bumiputra Merchant Bankers Bhd; cap. 365.0m., res 54.3m., dep. 1,809.5m. (March 2003); Chair. Tan Sri ABU TALIB OTHMAN; Chief Exec. Dir T. JEYARATNAM.

AmMerchant Bank Bhd: 22nd Floor, Bangunan AmBank Group, 55 Jalan Raja Chulan, 50200 Kuala Lumpur; tel. (3) 20782644; fax (3) 20314891; e-mail customercare@ambg.com.my; internet www.ambg.com.my; f. 1975; fmrly Arab-Malaysian Merchant Bank Bhd; name changed as above 2002; cap. 300.0m., res 480.3m., dep. 13,871.2m. (March 2003); Chair. Tan Sri Dato' AZMAN HASHIM; Man. Dir CHEAH TEK KUANG; 4 brs.

Aseambankers Malaysia Bhd: Menara Maybank, 33rd Floor, 100 Jalan Tun Perak, 50050 Kuala Lumpur; tel. (3) 20591888; fax (3) 20784194; e-mail faudziah@aseam.com.my; internet www.aseam.com.my; f. 1973; cap. 50.1m., res 246.0m., dep. 2,791.1m. (June 2003); CEO and Dir AGIL NATT; Chair. Dato' MOHAMED BASIR AHMAD; 2 brs.

Commerce International Merchant Bankers Bhd: Bangunan CIMB, 10th Floor, Jalan Semantan, Damansara Heights, 50490 Kuala Lumpur; tel. (3) 2536688; fax (3) 2535522; e-mail info@cimb.com.my; f. 1974; cap. 319.2m., res 876.7m., dep. 6,774.6m. (Dec. 2002); Chair. Dato' MOHAMED NOR BIN MOHAMED YOUSOF.

Malaysian International Merchant Bankers Bhd: Wisma Cyclecarri, 21st Floor, 288 Jalan Raja Laut, 50350 Kuala Lumpur; tel. (3) 26910200; fax (3) 26948388; internet www.mimb.com.my; f. 1970; shareholders' funds 265m. (Dec. 2003); Chair. Dato' ZULKIFLI BIN ALI; 1 br.

Public Merchant Bank Bhd: 25th Floor, Menara Public Bank, 146 Jalan Ampang, 50450 Kuala Lumpur; tel. (3) 21669382; fax (3) 21669362; f. 1973 as Asian International Merchant Bankers Bhd; became Sime Merchant Bankers Bhd 1996; name changed as above 2000; cap. 165.0m., res –28.9m., dep. 1,086.6m. (Dec. 2002); Chair. Datuk ISMAIL BIN ZAKARIA.

RHB Sakura Merchant Bankers Bhd: Tower Three, 9th Floor, RHB Centre, 426 Jalan Tun Razak, 50400 Kuala Lumpur; tel. (3) 92873888; fax (3) 92878000; e-mail publicaffairs@rhb.com.my; internet www.rhb.com.my; f. 1974; cap. 338.6m., res 409.4m., dep. 2,735.4m. (June 2002); Chair. Datuk AZLAN ZAINOL; Man. Dir (vacant).

Southern Investment Bank Bhd: 11th Floor, Wisma Genting, Jalan Sultan Ismail, 50250 Kuala Lumpur; tel. (3) 20594188; fax (3) 20722964; e-mail sibb@sibb.com.my; internet www.southernbank.com.my; f. 1988; fmrly Perdana Merchant Bankers Bhd; cap. 77.9m., res –32.2m., dep. 216.7m. (Dec. 2000); Chair. Dato' Nik IBRAHIM KAMIL; CEO YAP FAT (acting).

Utama Merchant Bank Bhd: Central Plaza, 27th Floor, Jalan Sultan Ismail, 50250 Kuala Lumpur; POB 12406, 50776 Kuala Lumpur; tel. (3) 21438888; fax (3) 21430357; e-mail umbb@umbb.po.my; internet www.cmsb.com.my/ubg; f. 1975 as Utama Wardley Bhd, name changed 1996; cap. 223.0m., res 63.6m., dep. 992.6m.

(Dec. 1999); Chair. Nik HASHIM BIN Nik YUSOFF; CEO DONNY KWA SOO CHUAN; 1 br.

Co-operative Bank

Bank Kerjasama Rakyat Malaysia Berhad: Bangunan Bank Rakyat, Jalan Tangsi, Peti Surat 11024, 50732 Kuala Lumpur; tel. (3) 2985011; fax (3) 2985981; f. 1954; 83,095 mems, of which 823 were co-operatives (Dec. 1996); Chair. Dr YUSUF YACOB; Man. Dir Dato' ANUAR JAAFAR; 67 brs.

Development Banks

Bank Industri & Teknologi Malaysia Bhd (Industrial Development Bank of Malaysia): Level 28, Bangunan Bank Industri, Bandar Wawasan, 1016 Jalan Sultan Ismail, POB 10788, 50724 Kuala Lumpur; tel. (3) 26929088; fax (3) 26985701; e-mail pru@bankindustri.com.my; internet www.bankindustri.com.my; f. 1979; govt-owned; finances long-term, high-technology projects, shipping and shipyards and engineering (plastic, electrical and electronic); cap. 670.5m., res 10.1m., dep. 3,485.1m. (Dec. 2000); Chair. Tan Sri Dato' OTHMAN MOHD RIJAL; Man. Dir Encik MD NOOR YUSOFF.

Sabah Development Bank Bhd: SDB Tower, Wisma Tun Fuad Stephens, POB 12172, 88824 Kota Kinabalu, Sabah; tel. (88) 232177; fax (88) 261852; e-mail sdbank@po.jaring.my; internet www.borneo-online.com.my/sdb; f. 1977; wholly owned by State Government of Sabah; cap. 350m., res –117.5m., dep. 861.3m. (Dec. 2002); Chair. PETER SIAU; Man. Dir and CEO PETER LIM.

Islamic Banks

Bank Islam Malaysia Bhd: Darul Takaful, 14th Floor, Jalan Sultan Ismail, 50250 Kuala Lumpur; tel. (3) 26935842; fax (3) 26922153; e-mail bislam@po.jaring.my; internet www.bankislam.com.my; f. 1983; cap. 500m., res 613.1m., dep. 12,397.1m. (June 2003); Chair. Dato' MOHAMED YUSOFF BIN MOHAMED NASIR; Man. Dir Dato' Haji AHMAD TAJUDIN ABDUL RAHMAN; 76 brs.

Bank Muamalat Malaysia Bhd: Menara Bumiputra, 21 Jalan Melaka, 50913 Kuala Lumpur; tel. (3) 26988787; fax (3) 26910388; e-mail webmaster@muamalat.com.my; internet www.muamalat.com.my; f. 1999; CEO Tuan Haji MOHD SHUKRI HUSSIN; 40 brs.

'Offshore' Banks

ABN Amro Bank, Labuan Branch: Level 9 (A), Main Office Tower, Financial Park Labuan, Jalan Merdeka, 87000 Labuan; tel. (87) 423008; fax (87) 421078; Man. ANTHONY RAJAN.

Al-Hidayah Investment Bank (Labuan) Ltd: Level 7 (C), Main Office Tower, Financial Park Labuan, Jalan Merdeka, 87000 Labuan; tel. (87) 451660; fax (87) 583088.

AMInternational (L) Ltd: Level 12 (B), Block 4, Office Tower, Financial Park Labuan, Jalan Merdeka, 87000 Labuan; tel. (87) 413133; fax (87) 425211; e-mail felix-leong@ambg.com.my; internet www.ambg.com.my; CEO PAUL ONG WHEE SEN.

AmMerchant Bank Bhd, Labuan Branch: Level 12 (B), Block 4, Office Tower, Financial Park Labuan, Jalan Merdeka, 87000 Labuan; tel. (87) 413133; fax (87) 425211; Gen. Man. PAUL ONG WHEE SEN.

Bank Islam (L) Ltd: Level 15, Block 4, Office Tower Penthouse B, Financial Park Labuan, Jalan Merdeka, 87000 Labuan; tel. (87) 451802; fax (87) 451800; e-mail bislamln@tm.net.my; CEO MOHAMAD NAJIB SHAHARUDDIN.

Bank of America, National Trust and Savings Association, Labuan Branch: Level 13 (D), Main Office Tower, Financial Park Labuan, Jalan Merdeka, 87000 Labuan; tel. (87) 411778; fax (87) 424778; Gen. Man. Pengiran NUR FARHAH OOI ABDULLAH.

Bank of East Asia Ltd, Labuan Offshore Branch: Level 10 (C), Main Office Tower, Financial Park Labuan, Jalan Merdeka, 87000 Labuan; tel. (87) 451145; fax (87) 451148; e-mail bealbu@hkbea.com.my; Gen. Man. THOMAS WONG WAI YIP.

Bank Muamalat Malaysia Bhd, Labuan Branch: Level 15 (A1), Main Office Tower, Financial Park Labuan, Jalan Merdeka, 87000 Labuan; tel. (87) 412898; fax (87) 451164; e-mail fuad@muamalat.com.my; Gen. Man. ZAINOL RASHID KHAIRUDDIN.

Bank of Nova Scotia, Labuan Branch: Level 10 (C2), Main Office Tower, Financial Park Labuan, Jalan Merdeka, 87000 Labuan; tel. (87) 451101; fax (87) 451099; Man. KWAN SING HUNG.

Bank of Tokyo-Mitsubishi Ltd, Labuan Branch: Level 12 (A & F), Main Office Tower, Financial Park Labuan, Jalan Merdeka, 87000 Labuan; tel. (87) 410487; fax (87) 410476; e-mail pulaubtm@tm.net.my; Gen. Man. WATURU TANAKA.

Barclays Bank PLC: Level 5(A), Main Office Tower, Financial Park Labuan, Jalan Merdeka, 87000 Labuan; tel. (87) 425571; fax (87) 425575; e-mail barclay@tm.net.my; Man. MIAW SIAW LOONG.

Bayerische Landesbank Girozentrale, Labuan Branch: Level 14 (C), Block 4, Office Tower, Financial Park Labuan, Jalan Merdeka, 87000 Labuan; tel. (87) 422170; fax (87) 422175; e-mail blblab@tm.net.my; Exec. Vice-Pres., CEO and Gen. Man. LOUISE PAUL.

BNP Paribas, Labuan Branch: Level 9 (E), Main Office Tower, Financial Park Labuan, Jalan Merdeka, 87000 Labuan; tel. (87) 422328; fax (87) 419328, e-mail bnpkul@tm.net.my; Gen. Man. YAP SIEW YING.

Bumiputra Commerce Bank (L) Ltd: Level 14 (B), Main Office Tower, Financial Park Labuan, Jalan Merdeka, 87000 Labuan; tel. (87) 410302; fax (87) 410313; e-mail bumitrst@tm.net.my; Gen. Man. ASARAF ABU BAKAR.

Cathay United Bank, Labuan Branch: Level 3 (C), Main Office Tower, Financial Park Labuan, Jalan Merdeka, 87000 Labuan; tel. (87) 452168; fax (87) 453678; Gen. Man. YEH PIN HUNG.

CIMB (L) Ltd: Unit 11 (B1), Level 11, Main Office Tower, Financial Park Labuan, Jalan Merdeka, 87000 Labuan; tel. (87) 451608; fax (87) 451610; CEO ADHA AMIR ABDULLAH.

Citibank Malaysia (L) Ltd: Level 11 (F), Main Office Tower, Financial Park Labuan, Jalan Merdeka, 87000 Labuan; tel. (87) 421181; fax (87) 419671; Gen. Man. CLARA LIM AI CHENG.

City Credit Investment Bank Ltd: Level 11 (D1), Main Office Tower, Financial Park Labuan, Jalan Merdeka, 87000 Labuan; tel. (87) 582268; fax (87) 581268; Dir ABDUL RAHMAN ABDULLAH.

Commercial IBT Bank, Labuan Branch: 02-01, 2nd Floor, Wisma Lucas Kong Bldg, U0185 Jalan Merdeka, 87000 Labuan; tel. (87) 411868; fax (87) 416818; e-mail aong@cibtbank.com; Pres. Dir Dr ADRIAN ONG CHEE BENG.

Commerzbank AG, Labuan Branch: Level 6 (E), Main Office Tower, Financial Park Labuan, Jalan Merdeka, 87000 Labuan; tel. (87) 416953; fax (87) 413542; Prin. Officer HO KAH HENG.

Crédit Agricole Indosuez, Labuan Branch: Level 11 (C), Main Office Tower, Financial Park Labuan, Jalan Merdeka, 87000 Labuan; tel. (87) 425118; fax (87) 424998; Gen. Man. BOON EONG TAN.

Crédit Industriel et Commercial: Level 11 (C2), Main Office Tower, Financial Park Labuan, Jalan Merdeka, 87000 Labuan; tel. (87) 452008; fax (87) 452009; Gen. Man. YEOW TIANG HUI.

Crédit Lyonnais, Labuan Branch: Level 6 (B), Main Office Tower, Financial Park Labuan, Jalan Merdeka, 87000 Labuan; tel. (87) 408331; fax (87) 439133; Man. CLEMENT WONG.

Crédit Suisse First Boston, Labuan Branch: Level 10 (B), Main Office Tower, Financial Park Labuan, Jalan Merdeka, 87000 Labuan; tel. (87) 425381; fax (87) 425384; Gen. Man. RUDOLF ZAUGG.

Danaharta Managers (L) Ltd: Tingkat 10, Bangunan Setia 1, 15 Lorong Dungun, Bukit Damansara, 50490 Kuala Lumpur; tel. (3) 2531122; fax (3) 2534375; Gen. Man. (vacant).

Deutsche Bank AG, Labuan Branch: Level 9 (G2), Main Office Tower, Financial Park Labuan, Jalan Merdeka, 87000 Labuan; tel. (87) 439811; fax (87) 439866; Man. Dir KUAH HUN LIANG.

Development Bank of Singapore (DBS Bank) Ltd, Labuan Branch: Level 12 (E), Main Office Tower, Financial Park Labuan, Jalan Merdeka, 87000 Labuan; tel. (87) 423375; fax (87) 423376; Gen. Man. KEVIN WONG.

Dresdner Bank AG, Labuan Branch: Level 13 (C), Main Office Tower, Financial Park Labuan, Jalan Merdeka, 87000 Labuan; tel. (87) 419271; fax (87) 419272; Gen. Man. JAMALUDIN NASIR.

ECM Libra Investment Bank Ltd: Level 3 (I1), Main Office Tower, Financial Park Complex, Jalan Merdeka, 87000 Labuan; tel. (87) 408525; fax (87) 408527.

Hongkong & Shanghai Banking Corporation, Offshore Banking Unit: Level 11 (D), Main Office Tower, Financial Park Labuan, Jalan Merdeka, 87000 Labuan; tel. (87) 417168; fax (87) 417169; Man. PREM KUMAR.

ING Bank NV: Level 8 (B2), Main Office Tower, Financial Park Labuan, Jalan Merdeka, 87000 Labuan; tel. (87) 425733; fax (87) 425734; Gen. Man. MILLY TAN.

International Commercial Bank of China: Level 7 (E2), Main Office Tower, Financial Park Labuan, Jalan Merdeka, 87000 Labuan; tel. (87) 581688; fax (87) 581668; Gen. Man. TAI CHI-HSIEN.

J. P. Morgan Chase Bank, Labuan Branch: Level 5 (F), Main Office Tower, Financial Park Labuan, Jalan Merdeka, 87000 Labuan; tel. (87) 424384; fax (87) 424390; e-mail fauziah.hisham@chase.com; Gen. Man. LEONG KET TI.

J. P. Morgan Malaysia Ltd: Unit 5 (F), Level 5, Main Office Tower, Financial Park Labuan, Jalan Merdeka, 87000 Labuan; tel. (87) 459000; fax (87) 451328; Gen. Man. LEONG KET TI.

KBC Bank NV, Labuan Branch: Level 3 (B), Main Office Tower, Financial Park Labuan, Jalan Merdeka, 87000 Labuan; tel. (87) 581778; fax (87) 583787; Gen. Man. KONG KOK CHEE.

Lloyds TSB Bank PLC: Lot B, 11th Floor, Wisma Oceanic, Jalan OKK Awang Besar, 87007 Labuan; tel. (87) 418918; fax (87) 411928; e-mail labuan@lloydstsb.com.my; Dir and Gen. Man. BARRY FRANCIS LEA.

Macquarie Bank Ltd, Labuan Branch: Level 3 (A), Main Office Tower, Financial Park Labuan, Jalan Merdeka, 87000 Labuan; tel. (87) 583080; fax (87) 583088; Division Dir DARREN WOODWARD.

Maybank International (L) Ltd: Level 16 (B), Main Office Tower, Financial Park Labuan, Jalan Merdeka, 87000 Labuan; tel. (87) 414406; fax (87) 414806; e-mail millmit@tm.net.my; Gen. Man. LAM HEE.

Mizuho Corporate Bank Ltd, Labuan Branch: Level 9 (B and C), Main Office Tower, Financial Park Labuan, Jalan Merdeka, 87000 Labuan; tel. (87) 417766; fax (87) 419766; Gen. Man. ISAKU TANIMURA.

Natexis Banque Populaires: Level 9 (G), Main Office Tower, Financial Park Labuan, Jalan Merdeka, 87000 Labuan; tel. (87) 581009; fax (87) 583009; Gen. Man. RIZAL ABDULLAH.

National Australia Bank, Labuan Branch: Level 12 (C2), Main Office Tower, Financial Park Complex, Jalan Merdeka, 87008 Labuan, tel. (87) 426386; fax (87) 428387; e-mail natausm@po.jaring.my; Gen. Man. LIONEL LIM.

OSK Investment Bank (Labuan) Ltd: Lot 3B, Level 5, Wisma Lazenda, Jalan Kemajuan, Labuan; tel. (87) 581885; fax (87) 582885; Prin. Officer ONG LEONG HUAT.

Oversea-Chinese Banking Corporation Ltd, Labuan Branch: Level 8 (C), Main Office Tower, Financial Park Labuan, Jalan Merdeka, 87000 Labuan; tel. (87) 423381; fax (87) 423390; Gen. Man. BERNARD FERNANDO.

Public Bank (L) Ltd: Level 8 (A and B), Main Office Tower, Financial Park Labuan, Jalan Merdeka, 87000 Labuan; tel. (87) 411898; fax (87) 413220; Man. ALEXANDER WONG.

RHB Bank (L) Ltd: Level 15 (B), Main Office Tower, Financial Park Labuan, Jalan Merdeka, 87000 Labuan; tel. (87) 417480; fax (87) 417484; Gen. Man. TOH AY LENG.

RUSD Investment Bank, Inc.: Level 4–A1, Main Office Tower, Financial Park Labuan, Jalan Merdeka, 87000 Labuan; tel. (87) 452100; fax (87) 543100; Man. Dir NASEERUDDIN A. KHAN.

Société Générale, Labuan Branch: Level 11 (B), Main Office Tower, Financial Park Labuan, Jalan Merdeka, 87000 Labuan; tel. (87) 421676; fax (87) 421669; Gen. Man. RAMZAN ABU TAHIR.

Standard Chartered Bank Offshore Labuan: Level 10 (F), Main Office Tower, Financial Park Labuan, Jalan Merdeka, 87000 Labuan; tel. (87) 417200; fax (87) 417202; Gen. Man. EDWARD NG.

Sumitomo Mitsui Banking Corpn, Labuan Branch: Level 12 (B and C), Main Office Tower, Financial Park Labuan, Jalan Merdeka, 87000 Labuan; tel. (87) 410955; fax (87) 410959; Gen. Man. JUNICHI IKENO.

UBS AG, Labuan Branch: Level 5 (E), Main Office Tower, Financial Park Labuan, Jalan Merdeka, 87000 Labuan; tel. (87) 421743; fax (87) 421746; Man. ZELIE HO SWEE LUM.

UFJ Bank Ltd, Labuan Branch: Level 10 (D), Main Office Tower, Financial Park Labuan, Jalan Merdeka, 87000 Labuan; tel. (87) 419200; fax (87) 419202; Gen. Man. MASAYUKI KUNISHIGE.

United Overseas Bank Ltd, Labuan Branch: Level 6 (A), Main Office Tower, Financial Park Labuan, Jalan Merdeka, 87000 Labuan; tel. (87) 424388; fax (87) 424389; Gen. Man. HO FONG KUN.

United World Chinese Commercial Bank: Level 3 (C), Main Office Tower, Financial Park Labuan, Jalan Merdeka, 87000 Labuan; tel. (87) 452168; fax (87) 453678; Gen. Man. PIN HUNG YEH.

Banking Associations

Association of Banks in Malaysia (ABM): UBN Tower, 34th Floor, 10 Jalan P. Ramlee, 50250 Kuala Lumpur; tel. (3) 20788041; fax (3) 20788004; e-mail banks@abm.org.my; internet www.abm.org.my; f. 1973; Chair. Dr ROZALI BIN MOHAMED ALI; Exec. Dir WONG SUAN LYE.

Institute of Bankers Malaysia: Wisma IBI, 5 Jalan Semantan, Damansara Heights, 50490 Kuala Lumpur; tel. (3) 20956833; fax (3) 20952322; e-mail ibbm@ibbm.org.my; internet www.ibbm.org.my; Chair. Tan Sri Dato' Dr ZETI AKHTAR AZIZ.

Malayan Commercial Banks' Association: POB 12001, 50764 Kuala Lumpur; tel. (3) 2983991.

Persatuan Institusi Perbankan Tanpa Faedah Malaysia (Association of Islamic Banking Institutions Malaysia—AIBIM): Tingkat 9, Menara Tun Razak, Jalan Raja Laut, 50350 Kuala Lumpur; tel. (3) 26932936; fax (3) 26910453; e-mail secretariat@aibim.com.my; internet www.aibim.com.my.

STOCK EXCHANGES

Kuala Lumpur Stock Exchange (KLSE): Exchange Sq., Bukit Kewangan, 50200 Kuala Lumpur; tel. (3) 20267099; fax (3) 27102308; e-mail commsdept@klse.com.my; internet www.klse.com.my; f. 1973; in 1988 KLSE authorized the ownership of up to 49% of Malaysian stockbroking companies by foreign interests; 283 mems (April 2003); 921 listed cos (March 2004); merged with Malaysian Exchange for Securities Dealing and Automated Quotation Bhd (MESDAQ) in March 2002; Chair. MOHAMAD AZLAN HASHIM; Pres. Dato' MOHD SALLEH BIN ABDUL MAJID.

Malaysia Derivatives Exchange Bhd (MDEX): 10th Floor, Exchange Sq., Bukit Kewangan, 50200 Kuala Lumpur; tel. (3) 20708199; fax (3) 20702376; e-mail info@mdex.com.my; internet www.mdex.com.my; f. 2001 as a result of the merger of the Kuala Lumpur Options and Financial Futures Exchange Bhd (KLOFFE) and the Commodity and Monetary Exchange of Malaysia; multi-product futures exchange; Exec. Chair. Dato' ABDUL JABBAR BIN ABDUL MAJID; Gen. Man. RAGHBIR SINGH BHART.

Regulatory Authority

Securities Commission (SC): 3 Persiaran Bukit Kiara, Bukit Kiara, 50490 Kuala Lumpur; tel. (3) 62048000; fax (3) 62015078; e-mail cau@seccom.com.my; internet www.sc.com.my; f. 1993; Chair. Datuk ALI ABDUL KADIR.

INSURANCE

From 1988 onwards, all insurance companies were placed under the authority of the Central Bank, Bank Negara Malaysia. In 1997 there were 69 insurance companies operating in Malaysia; nine reinsurance companies, 11 composite, 40 general and life and two takaful insurance companies.

Principal Insurance Companies

Allianz General Insurance Malaysia Bhd: Wisma UOA II, Floors 23 and 23A, 21 Jalan Pinang, 50450 Kuala Lumpur; tel. (3) 21623388; fax (3) 21626387; e-mail partner@allianz.com.my; internet www.allianz.com.my; f. 2001; CEO WILLIAM MEI YORK LIANG; Chair. Tan Sri RAZALI ISMAIL.

Allianz Life Insurance Malaysia Bhd: Wisma UOA II, Floors 23 and 23A, 21 Jalan Pinang, 50450 Kuala Lumpur; tel. (3) 21623388; fax (3) 21626387; e-mail partner@allianz.com.my; internet life.allianz.com.my; fmrly MBA Life Assurance Sdn Bhd; Chief Financial Officer CHARLES ONG ENG CHOW.

Asia Insurance Co Ltd: Bangunan Asia Insurance, 2 Jalan Raja Chulan, 50200 Kuala Lumpur; tel. (3) 2302511; fax (3) 2323606; f. 1923; general.

Capital Insurance Bhd: 38 Jalan Ampang, POB 12338, 50774 Kuala Lumpur; tel. (3) 2308033; fax (3) 2303657; Gen. Man. MOHD YUSOF IDRIS.

Commerce Life Assurance Bhd: 338 Jalan Tunku Abdul Rahman, 50100 Kuala Lumpur; tel. (3) 26123600; fax (3) 26987035; internet www.xlife.com.my; f. 1992 as AMAL Assurance Bhd; name changed as above in 1999.

Great Eastern Life Assurance (Malaysia) Bhd: Menara Great Eastern, 303 Jalan Ampang, 50450 Kuala Lumpur; tel. (3) 42598888; fax (3) 42590500; e-mail wecare@lifeisgreat.com.my; internet www.lifeisgreat.com.my; Dir and CEO ALEX FOONG SOO HAH.

John Hancock Life Insurance (Malaysia) Bhd: Menara John Hancock, 12th Floor, 6 Jalan Gelenggang, Damansara Heights, 50490 Kuala Lumpur; tel. (3) 20948055; fax (3) 20935487; internet www.jhancock.com.my; f. 1963; life and non-life insurance; fmrly British American Life and General Insurance Bhd; Chair. Tan Sri Dato' Haji YAHYA BIN ABDUL WAHAB; Man. Dir KHOR HOCK SENG.

Hong Leong Assurance Sdn Bhd: Menara HLA, 26th Floor, Jalan Kia Peng, 50450 Kuala Lumpur; tel. (3) 76501818; fax (3) 27101735; internet www.hla.com.my; Chair. Tan Sri QUEK LENG CHAN.

ING Insurance Bhd: Menara ING, 84 Jalan Raja Chulan, POB 10846, 50927 Kuala Lumpur; tel. (3) 21617255; fax (3) 21610549; internet www.ing.com.my; f. 1987; fmrly Aetna Universal Insurance Bhd; Chair. Tengku ABDULLAH IBNI AL-MARHUM Sultan ABU BAKAR.

Jerneh Insurance Corpn Sdn Bhd: Wisma Jerneh, 12th Floor, 38 Jalan Sultan Ismail, POB 12420, 50788 Kuala Lumpur; tel. (3) 2427066; fax (3) 2426672; f. 1970; general; Gen. Man. GOH CHIN ENG.

Malaysia National Insurance Sdn Bhd: Tower 1, 26th Floor, MNI Twins, 11 Jalan Pinang, 50450 Kuala Lumpur; tel. (3) 21769000; fax (3) 21769090; internet www.mni.com.my; f. 1970; life and general; CEO MOHAMED NAJIB ABDULLAH.

Malaysian Co-operative Insurance Society Ltd: Wisma MCIS, Jalan Barat, 46200 Petaling Jaya, Selangor; tel. (3) 7552577; fax (3) 7571563; e-mail info@mcis.po.my; internet www.mcis.com.my/mcis; f. 1954; CEO L. MEYYAPPAN.

Mayban Assurance Bhd: Mayban Assurance Tower, Dataran Maybank, 1 Jalan Maarof, 50000 Kuala Lumpur; tel. (3) 22972888; fax (3) 22972828; e-mail mayassur@tm.nct.my; internet www.maybank2u.com.my; Chair. Tan Sri MOHAMED BASIR AHMAD.

MBf Insurans Sdn Bhd: Plaza MBf, 5th Floor, Jalan Ampang, POB 10345, 50710 Kuala Lumpur; tel. (3) 2613466; fax (3) 2613466; Man. MARC HOOI TUCK KOK.

MCIS Zürich Insurance Bhd: Wisma MCIS Zürich, Jalan Barat, 46200 Petaling Jaya, Selangor; tel. (3) 79552577; fax (3) 79574780; e-mail info@mciszurich.com.my; internet www.mciszurich.com.my; CEO Datuk L. MEYYAPPAN; Chair. Dato' MOHAMAD WAHIDUDDIN BIN ABDUL WAHAB.

Multi-Purpose Insurans Bhd: Menara Multi-Purpose, 9th Floor, Capital Square, 8 Jalan Munshi Abdullah, 50100 Kuala Lumpur; tel. (3) 26919888; fax (3) 26945758; e-mail info@mpib.com.my; fmrly Kompas Insurans Bhd; Senior Gen. Mans WONG FOOK WAH, VISWANATH A. L. KANDASAMY.

Overseas Assurance Corpn Ltd: Wisma Lee Rubber, 21st Floor, Jalan Melata, 50100 Kuala Lumpur; tel. (3) 2022939; fax (3) 2912288; Gen. Man. A. K. WONG.

Progressive Insurance Sdn Bhd: Plaza Berjaya, 9th, 10th and 15th Floors, 12 Jalan Imbi, POB 10028, 50700 Kuala Lumpur; tel. (3) 2410044; fax (3) 2418257; Man. JERRY PAUT.

RHB Insurance Bhd: Tower 1, 4th Floor, RHB Centre, Jalan Tun Razak, 50450 Kuala Lumpur; tel. (3) 9812731; fax (3) 9812729; Man. MOHAMMAD ABDULLAH.

Sime AXA Assurance Bhd: Wisma Sime Darby, 15th Floor, Jalan Raja Laut, 50350 Kuala Lumpur; tel. (3) 2937888; fax (3) 2914672; e-mail hkkang@simenet.com; Gen. Man. HAK KOON KANG.

South-East Asia Insurance Bhd: Tingkat 9, Menara SEA Insurance, 1008 Jalan Sultan Ismail, 50250 Kuala Lumpur; POB 6120 Pudu, 55916 Kuala Lumpur; tel. (3) 2938111; fax (3) 2930111; internet www.sea.com.my; CEO HASHIM HARUN.

UMBC Insurans Sdn Bhd: Bangunan Sime Bank, 16th Floor, Jalan Sultan Sulaiman, 50000 Kuala Lumpur; tel. (3) 2328733; fax (3) 2322181; f. 1961; CEO ABDULLAH ABDUL SAMAD.

United Oriental Assurance Sdn Bhd: Wisma UOA, 36 Jalan Ampang, 50450 Kuala Lumpur; tel. (3) 2302828; fax (3) 2324250; e-mail uoa@uoa.po.my; f. 1976; CEO R. NESARETNAM.

Trade and Industry

GOVERNMENT AGENCIES

Danamodal Nasional Bhd (Danamodal): 10th Floor, Bangunan Sime Bank, Jalan Sultan Sulaiman, 50000 Kuala Lumpur; tel. (3) 20312255; fax (3) 20310786; e-mail info@danamodal.com.my; internet www.bnm.gov.my/danamodal/main2bck.htm; f. 1998 to recapitalize banks and restructure financial institutions, incl. arranging mergers and consolidations; Man. Dir MARIANUS VONG SHIN TZOI; Chair. Raja Datuk ARSHAD Raja Tun UDA.

Federal Agricultural Marketing Authority (FAMA): Bangunan Fama Point, Lot 17304, Jalan Persiaran 1, Bandar Baru Selayang, 68100 Batu Caves, Selangor Darul Ehsan; tel. (3) 61389622; fax (3) 61365597; internet agrolink.moa.my/fama; f. 1965 to supervise, coordinate and improve marketing of agricultural produce, and to seek and promote new markets and outlets for agricultural produce; Chair. AZIZI MEOR NGAH; Dir-Gen. HARON A. RAHIM.

Federal Land Development Authority (FELDA): Jalan Maktab, 54000 Kuala Lumpur; tel. (3) 2935066; fax (3) 2920087; f. 1956; govt statutory body formed to develop land into agricultural smallholdings to eradicate rural poverty; 893,150 ha of land developed (1994); involved in rubber, oil palm and sugar-cane cultivation; Chair. Raja Tan Sri MUHAMMAD ALIAS; Dir-Gen. MOHAMED FADZIL YUNUS.

Khazanah Nasional: 21 Putra Place 100, Ilu Putra, 50350 Kuala Lumpur; e-mail knb@po.jaring.my; f. 1994; state-controlled investment co; assumed responsibility for certain assets fmrly under control of the Minister of Finance Inc.; holds 40% of Telekom Malaysia Bhd, 40% of Tenaga Nasional Bhd, 6.6% of HICOM Bhd and 17.8% of PROTON; Chair. Datuk Seri Dr MAHATHIR MOHAMAD.

Malaysia Export Credit Insurance Bhd: Bangunan Bank Industri, 17th Floor, Bandar Wawasan, 1016 Jalan Sultan Ismail, POB 11048, 50734 Kuala Lumpur; tel. (3) 26910677; fax (3) 26910353; e-mail mecib@mecib.com.my; internet www.mecib.com.my; f. 1977; wholly owned subsidiary of Bank Industri & Teknologi Malaysia Bhd; provides insurance, financial guarantee and other trade-related services for exporters of locally manufactured products and for banking community; cap. RM 150m., exports declared RM 1,077m. (2003); Gen. Man. EN. AMINURRASHID ZULKIFLY; 67 employees.

Malaysia External Trade Development Corpn (MATRADE): Wisma Sime Darby, Jalan Raja Laut, 50350 Kuala Lumpur; tel. (3) 26947259; fax (3) 26947362; e-mail info@hq.matrade.gov.my; internet www.matrade.gov.my; f. 1993; responsibility for external trade development and promotion; CEO MERLYN KASIMIR.

Malaysian Institute of Economic Research: Menara Dayabumi, 9th Floor, Jalan Sultan Hishamuddin, POB 12160, 50768 Kuala Lumpur; tel. (3) 22725897; fax (3) 22730197; e-mail Admin@mier.po.my; internet www.mier.org.my; Exec. Dir MOHAMED ARIFF.

Malaysian Palm Oil Board (MPOB): Lot 6, SS6, Jalan Perbandaran, 47301 Kelana Jaya, Selangor; tel. (3) 7035544; fax (3) 7033533; internet porla.gov.my; f. 2000 by merger of Palm Oil Registration and Licensing Authority and Palm Oil Research Institute of Malaysia; Dir-Gen. Dato' MOHD YUSOF BASIR.

Malaysian Timber Industry Board (Lembaga Perindustrian Kayu Malaysia): 13–17 Menara PGRM, Jalan Pudu Ulu, POB 10887, 50728 Kuala Lumpur; tel. (3) 92822235; fax (3) 92003769; e-mail info@mtib.gov.my; internet www.mtib.gov.my; f. 1973 to promote and regulate the export of timber and timber products from Malaysia; Chair. Tan Sri Datuk Dr ABDULLAH BIN MOHD TAHIR; Dir-Gen. MOHD NAZURI BIN HASHIM SHAH.

Muda Agricultural Development Authority (MADA): MADA HQ, Ampang Jajar, 05990 Alor Setar, Kedah; tel. (4) 7728255; fax (4) 7722667; internet www.mada.gov.my; Chair. Dato' Seri SYED RAZAK BIN SYED ZAIN.

National Economic Action Council: NEAC-MTEN, Office of the Minister of Special Functions, Level 2, Block B5, Federal Government Administrative Centre, 62502 Putrajaya, Selangor Darul Ehsan; tel. (88) 883333; internet www.neac.gov.my; Exec. Dir Dato' MUSTAPA MOHAMED.

National Information Technology Council (NITC): Kuala Lumpur; Sec. Datuk Tengku Dr MOHD AZZMAN SHARIFFADEEN.

National Timber Certification Council: Kuala Lumpur; Chair. CHEW LYE TENG.

Pengurusan Danaharta Nasional Bhd (Danaharta): Tingkat 10, Bangunan Setia 1, 15 Lorong Dungun, Bukit Damansara, 50490 Kuala Lumpur; tel. (3) 20931122; fax (3) 20934360; internet www.danaharta.com.my; f. 1998 to acquire non-performing loans from the banking sector and to maximize the recovery value of those assets; Man. Dir ZUKRI SAMAT.

Perbadanan Nasional Bhd (PERNAS): Kuala Lumpur; tel. (3) 2935177; f. 1969; govt-sponsored; promotes trade, banking, property and plantation development, construction, mineral exploration, steel manufacturing, inland container transportation, mining, insurance, industrial development, engineering services, telecommunication equipment, hotels and shipping; cap. p.u. RM 116.25m.; 10 wholly owned subsidiaries, over 60 jointly owned subsidiaries and 18 assoc. cos; Chair. Tunku Dato' SHAHRIMAN BIN Tunku SULAIMAN; Man. Dir Dato' A. RAHMAN BIN HAMIDON.

DEVELOPMENT ORGANIZATIONS

Fisheries Development Authority of Malaysia: 7th–11th Floors, Wisma PKNS, Jalan Raja Laut, 50784 Kuala Lumpur; tel. (3) 26177000; fax (3) 26911931; e-mail info@kim.moa.my; internet agrolink.moa.my/lkim; Dir-Gen. Dato' Sheikh AHMAD BIN Sheikh LONG.

Johor Corporation: 13th Floor, Menara Johor Corporation, Kotaraya, 80000 Johor Bahru; tel. (7) 2232692; fax (7) 2233175; e-mail pdnjohor@jcorp.com.my; internet www.jcorp.com.my; development agency of the Johor state govt; Chief Exec. Dato' H. MUHAMMAD ALI.

Kumpulan FIMA Bhd (Food Industries of Malaysia): Kompleks FIMA, International Airport, Subang, Selangor; tel. (3) 7462199; f. 1972; fmrly govt corpn, transferred to private sector in 1991; promotes food and related industry through investment on its own or by co-ventures with local or foreign entrepreneurs; oil palm, cocoa and fruit plantation developments; manufacturing and packaging, trading, supermarkets and restaurants; Man. Dir Dato' MOHD NOOR BIN ISMAIL; 1,189 employees.

Majlis Amanah Rakyat (MARA) (Trust Council for the People): Bangunan Medan MARA, 13th Floor, 21 Jalan Raja Laut, 50609 Kuala Lumpur; tel. (3) 26915111; fax (3) 26913620; internet www.mara.gov.my; f. 1966 to promote, stimulate, facilitate and undertake economic and social development; to participate in industrial and commercial undertakings and jt ventures; Chair. Tan Sri NAZRI AZIZ.

Malaysian Agricultural Research and Development Institute (MARDI): POB 12301, General Post Office, 50774 Kuala Lumpur; tel. (3) 89437111; fax (3) 89483664; e-mail saharan@mardi.my; internet www.mardi.my; f. 1969; research and development in food and tropical agriculture; Dir-Gen. Dr SAHARAN BIN Haji ANANG.

Malaysian Industrial Development Authority (MIDA): Wisma Damansara, 6th Floor, Jalan Semantan, POB 10618, 50720 Kuala Lumpur; tel. (3) 2553633; fax (3) 2557970; e-mail promotion@mida.gov.my; internet www.mida.gov.my; f. 1967; Chair. Tan Sri Datuk ZAINAL ABIDIN BIN SULONG; Dir-Gen. Dato' ZAINUN AISHAH AHMAD.

Malaysian Industrial Development Finance Bhd: Bangunan MIDF, 195A Jalan Tun Razak, 50400 Kuala Lumpur; tel. (3) 21610066; fax (3) 21615973; e-mail inquiry@midf.com.my; internet www.midf.com.my; f. 1960 by the Govt, banks, insurance cos; industrial financing, advisory services, project development, merchant and commercial banking services; Man. Dir Dato' MOHAMED SALLEHUDDIN BIN OTHMAN.

Malaysian Pepper Marketing Board: Tanah Putih, POB 1653, 93916 Kuching; tel. (82) 331811; fax (82) 336877; e-mail pmb@pepper.po.my; internet www.sarawakpepper.gov.my/sarawakpepper; f. 1972; responsible for the statutory grading of all Sarawak pepper for export, licensing of pepper dealers and exporters, trading and the development and promotion of pepper grading, storage and processing facilities; Gen. Man. GRUNSIN AYOM.

Pinang Development Corporation: 1 Pesiaran Mahsuri, Bandar Bayan Baru, 11909 Bayan Lepas, Pinang; tel. (4) 6340111; fax (4) 6432405; e-mail enquiry@pdc.gov.my; internet www.pdc.gov.my; f. 1969; development agency of the Pinang state government; Chair. Tan Sri Dr KOH TSU KOON.

Sarawak Economic Development Corpn: Menara SEDC, 6th–11th Floors, Sarawak Plaza, Jalan Tunku Abdul Rahman, POB 400, 93902 Kuching; tel. (82) 416777; fax (82) 424330; e-mail ssedc@pop1.jaring.my; internet www.sedc.com.my; f. 1972; statutory org. responsible for commercial and industrial development in Sarawak either solely or jtly with foreign and local entrepreneurs; responsible for the development of tourism infrastructure; Chair. Datuk Haji TALIB ZULPILIP.

Selangor State Development Corporation (PKNS): Persiaran Barat, off Jalan Barat, 46505 Petaling Jaya, Selangor; tel. (3) 79572955; fax (3) 79575250; e-mail general@pkns.gov.my; internet www.pkns.gov.my; f. 1964; partially govt-owned; Corporate Man. YUSOF OTHMAN.

CHAMBERS OF COMMERCE

The Associated Chinese Chamber of Commerce: Wisma Chamber, 4th Floor, Lot 214, Jalan Bukit Mata, 93100 Kuching; tel. (82) 428815; fax (82) 429950; e-mail kcjong@pc.jaring.my; f. 1965; Pres. TIONG SU KOUK; Sec. Gen. LEE KHIM SIN.

Associated Chinese Chambers of Commerce and Industry of Malaysia: 8th Floor, Office Tower, Plaza Berjaya, 12 Jalan Imbi, 55100 Kuala Lumpur; tel. (3) 2452503; fax (3) 2452562; e-mail acccim@mol.net.my; internet www.acccim.org.my; Pres. Dato' LIM GUAN TEIK; Exec. Sec. ONG KIM SENG.

Malay Chamber of Commerce Malaysia: Plaza Pekeliling, 17th Floor, Jalan Tun Razak, 50400 Kuala Lumpur; tel. (3) 4418522; fax (3) 4414502; f. 1957 as Associated Malay Chambers of Commerce of Malaya; name changed as above 1992; Pres. Tan Sri Dato' TAJUDIN RAMLI; Sec.-Gen. ZAKI SAID.

Malaysian Associated Indian Chambers of Commerce and Industry: 116 Jalan Tuanku Abdul Rahman, 2nd Floor, 50100 Kuala Lumpur; tel. (3) 26931033; fax (3) 26911670; e-mail klsicci@po.jaring.my; internet www.maicci.com; f. 1950; Pres. Dato' K. KENNETH ESWARAN; Hon. Sec.-Gen. MUTHUSAMY V. V. M. SAMY; 8 brs.

Malaysian International Chamber of Commerce and Industry (MICCI) (Dewan Perniagaan dan Perindustrian Antarabangsa Malaysia): C-8-8, 8th Floor, Block C, Plaza Mont' Kiara, 50480 Kuala Lumpur; tel. (3) 62017708; fax (3) 62017705; e-mail micci@micci.com; internet www.micci.com; f. 1837; brs in Pinang, Perak, Johor, Melaka and Sabah; 1,100 corporate mems; Pres. P. J. DINGLE; Exec. Dir STEWART J. FORBES.

National Chamber of Commerce and Industry of Malaysia: 37 Jalan Kia Peng, 50450 Kuala Lumpur; tel. (3) 2419600; fax (3) 2413775; e-mail nccim@po.jaring.my; internet www.nccim.org.my/nccim; f. 1962; Pres. Tan Sri TAJUDIN RAMLI; Sec.-Gen. Dato' ABDUL HALIM ABDULLAH.

Sabah Chamber of Commerce and Industry: Jalan Tiga, Sandakan; tel. (89) 2141; Pres. T. H. WONG.

Sarawak Chamber of Commerce and Industry (SCCI): POB A-841, Kenyalang Park Post Office, 93806 Kuching; tel. (82) 237148; fax (82) 237186; e-mail phtay@pc.jaring.my; internet www.cmsb.com.my/scci; f. 1950; Chair. Datuk Haji MOHAMED AMIN Haji SATEM; Dep. Chair. Datuk Abang Haji ABDUL KARIM Tun Abang Haji OPENG.

South Indian Chamber of Commerce of Sarawak: 37C India St, Kuching; f. 1952; Pres. HAJA NAZIMUDDIN BIN ABDUL MAJID; Vice-Pres. SYED AHMAD.

INDUSTRIAL AND TRADE ASSOCIATIONS

Federation of Malaysian Manufacturers: Wisma FMM, 3 Persiaran Dagang, PJU 9 Bandar Sri Damansara, 52200 Kuala Lumpur; tel. (3) 62761211; fax (3) 62741266; e-mail webmaster@fmm.org.my; internet www.fmm.org.my; f. 1968; 2,145 mems (Jan. 2004); Pres. JEN (B.) Dato' MUSTAFA MANSUR; CEO LEE CHENG SUAN.

Federation of Rubber Trade Associations of Malaysia: 138 Jalan Bandar, 50000 Kuala Lumpur; tel. (3) 2384006.

Malayan Agricultural Producers' Association: Kuala Lumpur; tel. (3) 42573988; fax (3) 42573113; f. 1997; 464 mem. estates and 115 factories; Pres. Tan Sri Dato' Haji BASIR BIN ISMAIL; Dir MOHAMAD BIN AUDONG.

Malaysian Iron and Steel Industry Federation: 28E, 30E, 5th Floor, Block 2, Worldwide Business Park, Jalan Tinju 13/50, Section 13, 40675 Shah Alam, Selangor; tel. (3) 55133970; fax (3) 55133891; e-mail misif@po.jaring.my; Chair. Tan Sri Dato' SOONG SIEW HOONG; 150 mems.

Malaysian Palm Oil Association (MPOA): Bangunan Getah Asli I, 12th Floor, 148 Jalan Ampang, 50450 Kuala Lumpur; tel. (3) 27105680; fax (3) 27105679; e-mail kay@mpoa.org.my; internet www.mpoa.org.my; f. 1999 as result of rationalization of plantation industry; secretariat for producers of palm oil; Chief. Exec. M. R. CHANDRAN.

Malaysian Pineapple Industry Board: Wisma Nanas, 5 Jalan Padi Mahsuri, Bandar Baru UDA, 81200 Johor Bahru; tel. (7) 2361211; fax (7) 2365694; e-mail mpib@tm.net.my; Dir-Gen. Tuan Haji ISMAIL BIN ABD JAMAL.

Malaysian Rubber Products Manufacturers' Association: 1 Jalan USJ 11/1J, Subang Jaya, 47620 Petaling Jaya, Selangor; tel. (3) 56316150; fax (3) 56316152; e-mail mrpma@po.jaring.my; internet www.mrpma.com; f. 1952; Pres. Tan Sri Datuk ARSHAD AYUB; 144 mems.

Malaysian Rubber Board: 148 Jalan Ampang, 50450 Kuala Lumpur; tel. (3) 92062000; fax (3) 21634492; e-mail dg@lgm.gov.my; internet www.lgm.gov.my; f. 1998; implements policies and development programmes to ensure the viability of the Malaysian rubber industry; regulates the industry (in particular, the packing, grading, shipping and export of rubber); Dir-Gen. Dato' ABDUL HAMID BIN SAWAL.

Malaysian Timber Certification Council (MTCC): 19th Floor, Menara PGRM, 8 Jalan Pudu Ulu, Cheras, 56100 Kuala Lumpur; tel. (3) 92005008; fax (3) 92006008; e-mail mtcc@tm.net.my; internet www.mtcc.com.my; f. 1999; operates a voluntary national timber certification scheme; CEO Dato' Dr FREEZAILAH BIN CHE YEOM.

Malaysian Wood Industries Association: 19B, 19th Floor, Menara PGRM, 8 Jalan Pudu Ulu, Cheras, 56100 Kuala Lumpur; tel. (3) 92821789; fax (3) 92821779; e-mail mwia@tm.net.my; f. 1957; Exec. Officer PANG SUET KUM.

National Tobacco Board Malaysia (Ibu Pejabat Lembaga Tembakau Negara): Kubang Kerian, POB 198, 15720 Kota Bharu, Kelantan; tel. (9) 7652933; fax (9) 7655640; e-mail ltnm@ltn.gov.my; Dir-Gen. TEO HUI BEK.

Northern Malaya Rubber Millers and Packers Association: 22 Pitt St, 3rd Floor, Suites 301–303, 10200 Pinang; tel. (4) 620037; f. 1919; 153 mems; Pres. HWANG SING LUE; Hon. Sec. LEE SENG KEOK.

Palm Oil Refiners' Association of Malaysia (PORAM): 801C/802A Blok B, Executive Suites, Kelana Business Centre, 97 Jalan SS7/2, 47301 Kelana Jaya, Selangor; tel. (3) 74920006; fax (3) 74920128; e-mail poram@po.jaring.my; internet www.poram.org.my; f. 1975 to promote the palm oil refining industry; Chair. ER KOK LEONG; 27 mems.

Rubber Industry Smallholders' Development Authority (RISDA): 4½ Miles, Jalan Ampang, 50450 Kuala Lumpur; tel. (3) 4564022; Dir-Gen. MOHD ZAIN BIN Haji YAHYA.

Tin Industry Research and Development Board: West Block, 8th Floor, Wisma Selangor Dredging, Jalan Ampang, POB 12560, 50782 Kuala Lumpur; tel. (3) 21616171; fax (3) 21616179; e-mail mcom@mcom.com.my; Chair. MOHAMED AJIB ANUAR; Sec. MUHAMAD NOR MUHAMAD.

EMPLOYERS' ORGANIZATIONS

Malaysian Employers' Federation: 3A06–3A07, Block A, Pusat Dagangan Phileo Damansara II, 15 Jalan 16/11, off Jalan Damansara, 46350 Petaling Jaya, Selangor; tel. (3) 79557778; fax (3)

79559008; e-mail mef-hq@mef.org.my; internet www.mef.org.my; f. 1959; Pres. JAFAR ABDUL CARRIM; private-sector org. incorporating 13 employer organizations and 3,745 individual enterprises, including:

Association of Insurance Employers: c/o Royal Insurance (M) Sdn Bhd, Menara Boustead, 5th Floor, 69 Jalan Raja Chulan, 50200 Kuala Lumpur; tel. (3) 2410233; fax (3) 2442762; Pres. NG KIM HUONG.

Commercial Employers' Association of Peninsular Malaysia: c/o The East Asiatic Co (M) Bhd, 1 Jalan 205, 46050 Petaling Jaya, Selangor; tel. (3) 7913322; fax (3) 7913561; Pres. HAMZAH Haji GHULAM.

Malayan Commercial Banks' Association: see Banking Associations, above.

Malaysian Chamber of Mines: West Block, Wisma Selangor Dredging, 8th Floor, Jalan Ampang, 50350 Kuala Lumpur; tel. and fax (3) 21616171; e-mail mcom@mcom.com.my; internet www.mcom.com.my; f. 1914; promotes and protects interests of Malaysian mining industry; Pres. MOHAMED AJIB ANUAR; Exec. Dir MUHAMAD NOR MUHAMAD.

Malaysian Textile Manufacturers' Association: Wisma Selangor Dredging, 9th Floor, West Block, 142C Jalan Ampang, 50450 Kuala Lumpur; tel. (3) 21621587; fax (3) 21623953; e-mail textile@po.jaring.my; internet www.fashion-asia.com; Pres. BAHAR AHMAD; Exec. Dir CHOY MING BIL; 230 mems.

Pan Malaysian Bus Operators' Association: 88 Jalan Sultan Idris Shah, 30300 Ipoh, Perak; tel. (5) 2549421; fax (5) 2550858; Sec. Datin TEOH PHAIK LEAN.

Sabah Employers' Consultative Association: Dewan SECA, No. 4, Block A, 1st Floor, Bandar Ramai-Ramai, 90000 Sandakan, Sabah; tel. and fax (89) 272846; Pres. E. M. KHOO.

Stevedore Employers' Association: 5 Pengkalan Weld, POB 288, 10300 Pinang; tel. (4) 2615091; Pres. ABDUL RAHMAN MAIDIN.

UTILITIES

Electricity

Energy Commission of Malaysia: Levels 15, 19 and 20, Menara Dato' Onn, Putra World Centre, Jalan Tun Ismail, Kuala Lumpur; internet www.st.gov.my; f. 2002; regulatory body supervising electricity and gas supply.

Tenaga Nasional Bhd: 129 Jalan Bangsar, POB 11003, 50732 Kuala Lumpur; tel. (3) 2825566; fax (3) 2823274; e-mail webadmin@tnb.com.my; internet www.tnb.com.my; f. 1990 through the corporatization and privatization of the National Electricity Board; 53% govt-controlled; generation, transmission and distribution of electricity in Peninsular Malaysia; generating capacity of 7,621 MW (63% of total power generation); also purchases power from 12 licensed independent power producers; Chair. JAMALUDDIN JARIS; CEO Dato' FUAD B. JAAFAR (acting).

Sabah Electricity Board (SEB): Wisma Lembaga Letrik Sabah, 88673 Kota Kinabalu; tel. (88) 211699; generation, transmission and distribution of electricity in Sabah.

Sarawak Electricity Supply Corpn (SESCO): POB 149, Kuching, Sarawak; tel. (82) 441188; fax (82) 444434; internet www.sesco.com.my; generation, transmission and distribution of electricity in Sarawak.

Gas

Energy Commission of Malaysia: see above.

Gas Malaysia Sdn Bhd: 5 Jalan Serendah 26/17, Seksyen 26, Peti Surat 7901, 40732 Shah Alam, Selangor Darul Ehsan; tel. (3) 51923000; e-mail ccu@gasmalaysia.com; internet www.gasmalaysia.com; f. 1992; Chair. Tan Sri Datuk Dr AHMAD TAJUDDIN ALI; CEO MUHAMAD NOOR HAMID.

Water

Under the federal Constitution, water supply is the responsibility of the state Governments. In 1998, owing to water shortages, the National Water Resources Council was established to co-ordinate management of water resources at national level. Malaysia's sewerage system is operated by Indah Water Konsortium, owned by Prime Utilities.

National Water Resources Council: c/o Ministry of Works, Jalan Sultan Salahuddin, 50580 Kuala Lumpur; tel. (3) 2919011; fax (3) 2986612; f. 1998 to co-ordinate management of water resources at national level through co-operation with state water boards; Chair. Dato' Seri Dr MAHATHIR BIN MOHAMAD.

Regulatory Authorities

Johor State Regulatory Body: c/o Pejabat Setiausaha Kerajaan Negeri Johor, Aras 1, Bangunan Sultan Ibrahim, Jalan Bukit Timbalan, 80000 Johor Bahru; tel. (7) 223850; Dir Tuan Haji OMAR BIN AWAB.

Kelantan Water Department: Tingkat Bawah Blok 6, Kota Darul Naim, 15503 Kota Bahru, Kelantan; tel. (9) 7475240; Dir Tuan Haji WAN ABDUL AZIZ BIN WAN JAAFAR.

Water Supply Authorities

Kedah Public Works Department: Bangunan Sultan Abdul Halim, Jalan Sultan Badlishah, 05582 Alor Setar, Kedah; tel. (4) 7334041; fax (4) 7341616; Dir Dr NORDIN BIN YUNUS.

Kelantan Water Sdn Bhd: 14 Beg Berkunci, Jalan Kuala Krai, 15990 Kota Bahru, Kelantan; tel. (10) 9022222; fax (10) 9022236; Dir PETER NEW BERKLEY.

Kuching Water Board: Jalan Batu Lintang, 93200 Kuching, Sarawak; tel. (82) 240371; fax (82) 244546; Dir DAVID YEU BIN TONG.

Labuan Public Works Department: Jalan Kg. Jawa, POB 2, 87008 Labuan; tel. (87) 414040; fax (87) 412370; Dir Ir ZULKIFLY BIN MADON.

LAKU Management Sdn Bhd: Soon Hup Tower, 6th Floor, Lot 907, Jalan Merbau, 98000 Miri; tel. (85) 442000; fax (85) 442005; e-mail chuilin@pd.jaring.my; serves Miri, Limbang and Bintulu; CEO YONG CHIONG VAN.

Melaka Water Corpn: Tingkat Bawah, 1 10–13, Graha Maju, Jalan Graha Maju, 75300 Melaka; tel. (6) 2825233; fax (6) 2837266; Ir ABDUL RAHIM SHAMSUDI.

Negeri Sembilan Water Department: Wisma Negeri, 70990 Seremban; tel. (6) 7622314; fax (6) 7620753; Ir Dr MOHD AKBAR.

Pahang Water Supply Department (Jabatan Bekalan Air Pahang): 9–10 Kompleks Tun Razak, Bandar Indera Mahkota, 25582 Kuantan, Pahang; tel. (9) 5721222; fax (9) 5721221; e-mail p-jba@pahang.gov.my; Dir Ir Haji ISMAIL BIN Haji MAT NOOR.

Pinang Water Supply Department: Level 29, KOMTAR, 10000 Pinang; tel. (4) 6505462; fax (4) 2645282; e-mail lyc@sukpp.gov.my; f. 1973; Gen. Man. Datuk Ir LEE YOW CHING.

Perak Water Board: Jalan St John, Peti Surat 589, 30760 Ipoh, Perak; tel. (5) 2551155; fax (5) 2556397; Dir Ir SANI BIN SIDIK.

Sabah Water Department: Wisma MUIS, Blok A, Tingkat 6, Beg Berkunci 210, 88825 Kota Kinabalu; tel. (88) 232364; fax (88) 232396; Man. Ir BENNY WANG.

SAJ Holdings Sdn Bhd: Bangunan Ibu Pejabat SAJ Holdings, Jalan Garuda, Larkin, POB 262, 80350 Johor Bahru; tel. (7) 2244040; fax (7) 2236155; e-mail support@saj.com.my; internet www.saj.com.my; f. 1999; Exec. Chair. Dir Dato' Haji HAMDAN BIN MOHAMED.

Sarawak Public Works Department: Wisma Seberkas, Jalan Tun Haji Openg, 93582 Kuching; tel. (82) 244041; fax (82) 429679; Dir MICHAEL TING KUOK NG.

Selangor Water Department: POB 5001, Jalan Pantai Baru, 59990 Kuala Lumpur; tel. (3) 2826244; fax (3) 2827535; f. 1972; Dir Ir LIEW WAI KIAT.

Sibu Water Board: Km 5, Jalan Salim, POB 405, 96007 Sibu, Sarawak; tel. (84) 211001; fax (84) 211543; e-mail swbs@swb.gov.my; Man. DANIEL WONG PARK ING.

Terengganu Water Department: Tkt 3, Wisma Negeri, Jalan Pejabat, 20200 Kuala Terengganu; tel. (9) 6222444; fax (9) 6221510; Ir Haji WAN NGAH BIN WAN.

MAJOR COMPANIES

The following are among the major industrial undertakings in Malaysia (cap. = capital; res = reserves; m. = million; amounts in ringgit Malaysia):

Aluminium Co of Malaysia Bhd: Lot 8, Jalan Universiti, 46200 Petaling Jaya, Selangor Darul Ehsan; tel. (3) 79561588; fax (3) 79564940; e-mail kok-heng.chan@alcan.com; f. 1960; mfrs of aluminium sheet, foil and extruded and fabricated products; cap. and res 210.4m., sales 280.0m. (2001); Man. Dir ANTONIO TADEU COELHO NARDOCCI; 921 employees.

AMDB Bhd Group: Bangunan AMDB, 20th Floor, 1 Jalan Lumut, 50400 Kuala Lumpur; tel. (3) 40432311; fax (3) 40430311; e-mail amdb@po.jaring.my; internet www.amdbgroup.com; f. 1965; fmrly Arab-Malaysian Development Bhd; mfrs and exporters of cotton and finished fabrics; cap. and res 378.5m., sales 188.0m. (2002/03); Chair. Tan Sri Dato' AZMAN HASHIM; CEO AZMI HASHIM; 2,175 employees.

Asiatic Development Bhd: Wisma Genting, 10th Floor, Jalan Sultan Ismail, 50250 Kuala Lumpur; tel. (3) 21612288; fax (3) 21616149; e-mail info@asiatic.com.my; internet www.asiatic.com.my; f. 1977; plantation and property development, food production; cap. and res 1,336.6m., sales 490.8m. (2003); Chair. Tan Sri MOHD

AMIN BIN OSMAN; Chief Execs Tan Sri LIM KOK THAY, Dato' BAHARUDDIN BIN MUSA; 3,259 employees.

Berjaya Group Bhd: 11th Floor, Menara Berjaya, KL Plaza, 179 Jalan Bukit Bintang, 55100 Kuala Lumpur; tel. (3) 29358888; fax (3) 29358043; e-mail corpcom@berjaya.com.my; internet www.berjaya .com.my; f. 1967; mfg, commercial and residential property, insurance and finance; cap. and res 1,498.2m., sales 7,195.6m. (2002/03); Chair. and CEO Tan Sri Dato' VINCENT TAN CHEE YIOUN; Man. Dir Dato' DANNY TAN CHEE SING; 22,300 employees.

British American Tobacco (Malaysia) Bhd: Virginia Park, Jalan Universiti, 46200 Petaling Jaya, Selangor Darul Ehsan; tel. (3) 79566899; fax (3) 79558416; internet www.batmalaysia.com; cigarette and other tobacco products mfrs; cap. and res 405.8m., sales 3,010m. (2001); Chair. Tan Sri ABU TALIB BIN OTHMAN; 1,530 employees.

Carlsberg Brewery Malaysia Bhd: 55 Persiaran Selangor, Section 15, 40000 Shah Alam, Selangor; POB 10617, Kuala Lumpur; tel. (3) 55191621; fax (3) 55191931; e-mail info@carlsberg.com.my; internet www.carlsberg.com.my; brewers of beer and stout; cap. and res 444.4m., sales 841.1m. (2001); Chair. MICHAEL IUUL; Man. Dir Dato' JORGEN BORNHOFT; 977 employees.

Cement Industries of Malaysia Bhd: Bukit Ketri, Mukim of Chuping, 02450 Kangar, Perlis; tel. (4) 9382006; fax (4) 9382722; e-mail coo@pc.jaring.my; internet www.cima.com.my; f. 1975; mfrs of cement and investment holding; cap. and res 644.8m., sales 354.6m. (2000); Chair. Dato' Dr YAHA BIN ISMAIL; Man. Dir LEE TUCK FOOK; 1,328 employees.

Chocolate Products (Malaysia) Bhd: Menara Lion, 46th Floor, 165 Jalan Ampang, 50450 Kuala Lumpur; tel. (3) 21622155; fax (3) 21623448; internet www.lion.com.my; mfrs of chocolate and related products, cocoa butter, cocoa powder and snacks, property development and investment holding; cap. and res 602.2m., sales 614.4m. (2000/01); Chair. Tan Sri WILLIAM H. J. CHENG; Man. Dir HEAH SIEU LAY; 7,165 employees.

CSM Corpn Bhd: Menara Cold Storage, 10th Floor, Jaya Shopping Centre, Jalan Semangat, 46100 Petaling Jaya, Selangor; tel. (3) 79588888; fax (3) 79581289; f. 1903; mfrs of ice cream, UHT and non-carbonated drinks, butter and dairy spreads, ghee, margarine, squashes, cordials, ice and meat products; operates retail supermarkets, pharmacies and shopping arcades; imports and distributes refrigerated and non-refrigerated foods, beverages and pharmaceuticals; cap. and res 322.8m., sales 218.7m. (1998); Chair. Dato' ABDUL RAHMAN BIN HAMZAH; Chief Exec. GAN GWO CHYANG; 700 employees.

Cycle and Carriage Bintang Bhd: Lot 9, Jalan 219, Federal Highway, 46100 Petaling Jaya, Selangor; tel. (3) 79572422; fax (3) 79560593; internet www.cyclecarriage.com.my; f. 1967; franchise holders for Mercedes Benz and Mazda commercial and passenger vehicles; cap. and res 575.2m., sales 740.8m. (2001); Chair. Tan Sri Dato' ABDUL BIN ALI; Man. Dir MOHAMAD HAJI HASAN; 2,209 employees.

DMIB Bhd: 4 Jalan Tandang, 46050 Petaling Jaya, POB 66, Selangor; tel. (3) 77818833; fax (3) 77825414; e-mail info@dmi.com .my; internet www.dmi.com.my; f. 1961; mfrs of a complete range of Dunlop tyres, chemical products, industrial gloves, mattresses and golf balls; cap. and res 299.5m., sales 580.4m., (2000/01); Chair. TUNKU TAN SRI DATO' SERI AHMAD BIN TUNKU YAHAYA; Man. Dir AHMAD ZUBAIR Haji MURSHID; 1,800 employees.

DRB-HICOM Bhd: Tingkat 5, Wisma DRB-HICOM, 2 Jalan Usahawan, 40150 Shah Alam, Selangor; tel. (3) 20528000; fax (3) 20528118; internet www.drb-hicom.com; f. 1996; involved in design and construction of Electrified Double Track Project between Rawang and Ipoh; cap. and res 2,150.1m., sales 5,073.3m. (2000/01); Chair. Tan Sri Dato' Seri MOHD SALEH SULONG; 17,395 employees.

Esso Malaysia Bhd: Menara ExxonMobil, Tingkat 16, Kuala Lumpur City Centre, 50088 Kuala Lumpur; tel. (3) 20533000; fax (3) 3803438; f. 1960; refines and markets all classes of petroleum products, lubricating oils, gas and ammonia; cap. and res 552.8m., sales 4,134.4m. (2001); Chair. and CEO RICHARD M. KRUGER; 529 employees.

FCW Holdings Bhd: 83 Jalan Segambut, 51200 Kuala Lumpur; tel. (3) 40439266; fax (3) 40436750; provision of management services and the trading of telecommunications equipment; cap. and res 153.5m., sales 55.0m. (2000/01); Chair. TAN HUA CHOON; 126 employees.

FFM Bhd: Wisma Jerneh, 16th Floor, 38 Jalan Sultan Ismail, 50250 Kuala Lumpur; tel. (3) 21424282; fax (3) 21456255; internet www.ffmb.com.my; f. 1962; fmrly Federal Flour Mills Bhd; flour-milling, soya bean-processing, maize, wheat, palm oil refining and animal feed; cap. and res 1,438.2m., sales 4,532.5m. (2001); Man. Dir Datuk OH SIEW NAM; Exec. Chair. OH SIEW NAM; 3,710 employees.

General Corpn Bhd: Plaza Ampang City, 19th Floor, 332A-19 Jalan Ampang, 50450 Kuala Lumpur; tel. (3) 42564599; fax (3) 42578197; quarrying, construction, property management and manufacturing; cap. and res 478.7m., sales 117.8m. (2000/01); Chair. Tun MOHAMED HANIF BIN OMAR; Man. Dir LOW KENG BOON.

Goodyear (Malaysia) Bhd: POB 7049, 40914 Shah Alam, Selangor; tel. (3) 55192411; fax (3) 55195729; e-mail vijaian.k@ goodyear.com; internet www.goodyear.com; mfrs of passenger car, truck and tractor tyres and tubes; Chair. Dato' SHAHRIMAN BIN TUNKU SULAIMAN; Man. Dir HARISH KHOSIA; 720 employees.

Guinness Anchor Bhd: Sungei Way Brewery, POB 144, 46710 Petaling Jaya, Selangor Darul Ehsan; tel. (3) 78414688; fax (3) 78740986; internet www.gab.com.my; mfrs of beer and stout; cap. and res 297.2m., sales 725.4m. (2002/03); Chair. Tan Sri SAW HUAT LYE; Man. Dir THEO DE ROND; 993 employees.

HICOM Diecastings Sdn Bhd: Wisma HICOM, 5th Floor, 2 Jalan Usahawan U1/8, 4050 Shah Alam, Selangor; tel. (3) 2028000; fax (3) 2028118; e-mail hicom@hicom.drb-hicom.com.my; internet www .drb-hicom.com; fmrly HICOM Holdings Bhd; industrial projects incl. building materials, commercial vehicles and welded pipes; cap. and res 2,082.0m., sales 1,703.3m. (1998/99); Chair. Dato' MOHD SALEH BIN SULONG.

Highlands and Lowlands Bhd: Wisma Guthrie, 21 Jalan Gelenggang, Bukit Damansara, 50490 Kuala Lumpur; tel. (3) 20941644; fax (3) 20957934; internet www.kumpulanguthrie.com; f. 1975; cultivation and processing of rubber, oil palm, coconut and cocoa; property investment and devt; cap. and res 2,497.4m., sales 290.2m. (2002); Chair. Dato' ABDUL KHALID BIN IBRAHIM.

Hume Industries (Malaysia) Bhd: Wisma Hong Leong, 9th Floor, 18 Jalan Perak, 50450 Kuala Lumpur; tel. (3) 21642631; fax (3) 21642514; e-mail HIMB@hongleong.com.my; internet www.himb .com.my; f. 1961; mfrs of asbestos cement products, steel and concrete pipes, pre-stressed concrete beams and piles, tanks, electrical conduits and other moulded products, pressure vessels, autoclaves and lift gates; cap. and res 555.2m., sales 1,903.3m. (2002/03); Chair. Tan Sri QUEK LENG CHAN; Pres. and CEO KWEK LENG SAN; 11,868 employees.

IOI Corpn Bhd: 2 IOI Sq., IOI Resort, 62505 Putrajaya; tel. (3) 89478888; fax (3) 89432266; e-mail corp@ioigroup.com; internet www.ioigroup.com; cultivation and processing of oil palm, rubber and cocoa, property devt, production of industrial and medical gases; cap. and res 3,530.2m., sales 3,907.9m. (2003); Exec. Chair. Tan Sri Dato' LEE SHIN CHENG.

Keck Seng (Malaysia) Bhd: 2-G Foh Chong Bldg, Jalan Ibrahim, Johor Baru, Johor; tel. (7) 3555866; fax (7) 3540827; f. 1958; cultivation and processing of oil palm, cocoa, housing devt, property investment; cap. and res 805.7m., sales 937.9m. (1998); Chair. HO KIAN GUAN; 1,500 employees.

Kemayan Corpn Bhd: 167 Jalan Glasier, Taman Tasek, 80200 Johor Baru, Johor; tel. (7) 2362390; fax (7) 2365307; f. 1965; cultivation and processing of oil palm and cocoa; cap. and res –848.5m., sales 72.5m. (2000/01); Chair. Dato' ONG KIM HOAY; 326 employees.

Kuala Lumpur Kepong Bhd: Wisma Taiko, 1 Jalan S. P. Seenivasagam, 30000 Ipoh; tel. (5) 2417844; fax (5) 2555466; internet www .klk.com.my; f. 1973; plantations cover 145,967 ha; cap. and res 3,775.8m., sales 3,473.5m. (2002/03); Chair. and CEO Dato' LEE OI HIAN; 30,892 employees.

Kumpulan Guthrie Bhd: Wisma Guthrie, 21 Jalan Gelenggang, Damansara Heights, 50490 Kuala Lumpur; tel. (3) 20941644; fax (3) 20957934; e-mail megat@guthrie.com.my; internet www .kumpulanguthrie.com; production, processing, export and distribution of rubber, palm oil, palm kernel, cocoa and coconut; cap. and res 2,803.6m., sales 3,077.9m. (2002); Chair. Tan Sri Dato' MUSA HITAM; 54,235 employees.

Lion Land Bhd: Menara Lion, 17th Floor, 165 Jalan Ampang, 50450 Kuala Lumpur; tel. (3) 21622155; fax (3) 21613166; internet www.lion.com.my; f. 1924; mfrs of computer components, property development, construction, steel-mill operations; cap. and res 836.9m., sales 938.4m. (2000/01); Chair. Tan Sri Dato' MUSA BIN HITAM; Man. Dir Datuk CHENG YONG KIM; 2,994 employees.

Malakoff Bhd: 8th Floor, Blok B, Wisma Semantan, 12 Jalan Gelenggang, Damansara Heights, 50490 Kuala Lumpur; tel. (3) 20923388; fax (3) 20922288; e-mail malakoff@malakoff.com.my; internet www.malakoff.com.my; f. 1975; cultivation and processing of natural rubber and oil palm, generation and sale of electrical energy and generating capacity; cap. and res 1,835.6m., sales 1,473.2m. (1999/2000); Chair. Tan Sri IBRAHIM MENUDIN; Man. Dir AHMAD JAUHARI YAHYA.

Malaysia Mining Corpn Bhd (MMCB): Menara PNB, 32nd Floor, 201A Jalan Tun Razak, 50400 Kuala Lumpur; tel. (3) 21616000; fax (3) 21612951; e-mail corpsec@mol.net.my; f. 1981 by merger of Malayan Tin Dredging Co and Malaysia Mining Corpn; the world's largest tin-mining group (active in the exploration, mining, smelting and marketing of tin) until April 1993, when it ceased tin-mining operations, owing to depressed tin prices; plantations and diamond exploration, property and financial services; cap. and res 1,592.9m. (1998/99), sales 276.0m. (1999/2000); Chair. Tan Sri Datuk IBRAHIM MENUDIN; 2,211 employees.

Malaysian Oxygen Bhd: 13 Jalan 222, 46100 Petaling Jaya, Selangor Darul Ehsan; tel. (3) 79554233; fax (3) 79566389; internet www.mox.com.my; f. 1960; mfrs of industrial and medical gases and electrodes, supplies welding, safety, marine, medical and fire-fighting equipment; cap. and res 491.2m., sales 434.7m. (1999/2000); Chair. Tun Dato' Haji OMAR YOKE LIN ONG; Man. Dir DAVID JOHN FULLER; 643 employees.

Malaysian Tobacco Co Bhd: Tingkat 21, MNI Twins, 11 Jalan Pinang, 50450 Kuala Lumpur; tel. (3) 21665855; fax (3) 21613623; f. 1956; cigarette mfrs; cap. and res 774.1m., sales 20.6m. (2001); Chair. Datuk UMAR BIN Haji ABU.

Maruichi Malaysia Steel Tube Bhd: B-8-7 Megan Phileo Promenade, 189 Jalan Tun Razak, 50400 Kuala Lumpur; tel. (3) 21616322; fax (3) 21610501; e-mail maruichi@tm.net.my; internet www.maruichi.com.my; f. 1969; steel pipes and tubes, steel wire, engineering services and share registration services; cap. and res 569.7m., sales 378.3m. (2000/01); Chair. ZAIN AZAHARI BIN ZAINAL ABIDIN; Man. Dir YANG YEN FANG; 526 employees.

Mitsubishi Electric (Malaysia) Sdn Bhd: Senai Industrial Area, Lot 32, Senai, 81400 Johor; tel. (607) 5996060; fax (607) 5996076; f. 1989; mfr of audio and video equipment; sales 2,259m. (1995); Man. Dir MICHOMI HIGUCHI; 1,500 employees.

Motorola Malaysia Sdn Bhd: 2 Jalan SS8/2, Free Industrial Zone Sungai Way, 47300 Petaling Jaya, Selangor Darul Ehsan; tel. (3) 78731133; fax (3) 78731055; e-mail mmscinfo@motorola.com; internet www.motorola.com/my; mfr of electronic components and telecommunications equipment; sales 3,310m. (1995); Man. Dir J. A. LEW; 8,300 employees.

Multi-Purpose Holdings Bhd: 38th Floor, Menara Multi-Purpose, Capital Square, 8 Jalan Munshi Abdullah, 50100 Kuala Lumpur; tel. (3) 26948333; fax (3) 26941380; e-mail info@mphb.com.my; internet www.mphb.com.my; financial services, property development and investment, gaming and leisure and utilities; cap. p.u. 781.6m., sales 548.6m. (2000); Chair. Dr CHAN CHIN CHEUNG.

Nylex (Malaysia) Bhd: Persiaran Selangor, Seksyen 15, Shah Alam Industrial Estate, 40200 Shah Alam, Selangor Darul Ehsan; tel. (3) 55191706; fax (3) 55100088; e-mail nylex@nylex.com; internet www.nylex.com; mfr of vinyl coated fabrics, calendered film, and sheeting and plastic products; cap. and res 194.8m., sales 365.9m. (May 2002); Chair. Raja Tun MOHAR BIN Raja BADIOZAMAN; Man. Dir Dato' SIEW KA WEI; 1,364 employees.

Shell Refining Co (FOM) Bhd: Bangunan Shell Malaysia, off Jalan Semantan, Damansara Heights, POB 11027, 50490 Kuala Lumpur; tel. (3) 20959144; fax (3) 20912957; f. 1960; refining and manufacture of all classes of petroleum products; cap. and res 744.3m., sales 4,601.4m. (2001); Chair. Datuk LIM HAW KUANG; Man. Dir SHAHUL HAMID BIN MOHD ISMAIL; 320 employees.

Silverstone Bhd: Lot 5831, Kawasan Perusahaan Kamunting II, POB 2, 34600 Kamunting, Taiping, Perak Darul Ridzuan; tel. (5) 8911077; fax (5) 8911079; e-mail silverstone@silverstone.com.my; internet www.silverstone.com.my; f. 1988; tyre manufacturer; Exec. Dir and CEO PHANG WAI YEEN.

Sime Darby Bhd: Wisma Sime Darby, 21st Floor, Jalan Raja Laut, 50350 Kuala Lumpur; tel. (3) 26914122; fax (3) 26987398; e-mail enquiries@simenet.com; internet www.simedarby.com; plantation management, manufacturing tyres and trucks, commodity trading, insurance services, oil and gas; cap. and res 8,424.7m., sales 14,903.5m. (2003/04); Chair. Dato' Seri AHMAD SARJI BIN ABDUL HAMID; Chief Exec. Dao' Haji AHMAD ZUBIR BIN Haji MURSHID; 26,842 employees.

Tan Chong Motor Holdings Bhd: 62–68 Jalan Ipoh, 51200 Kuala Lumpur; tel. (3) 40478888; fax (3) 40478636; e-mail tcmh@tanchong.com.my; internet www.nissan.com.my; f. 1972; assembly and distribution of motor vehicles, provision of after-sales services and related financial services; cap. and res 970.1m., sales 1,677.5m. (2003); Vice-Chair. EN AHMAD BIN ABDULLAH; Man. Dir Tan ENG SOON; 2,762 employees.

Tasek Corporation Bhd: Tingkat 5, Wisma Hong Leong, 18 Jalan Perak, 50450 Kuala Lumpur; tel. (3) 21641818; fax (3) 21643703; e-mail tasek@tasek.com.my; internet www.tasekcement.com; fmrly Tasek Cement Bhd; mfr of building materials; cap. and res 568.0m., sales 206.7m. (2000/01); Chair. Tan Sri QUEK LENG CHAN; 644 employees.

Top Glove Corpn Bhd: Lot 4969, Jalan Teratai, Batu 6, off Jalan Meru, 41050 Klang, Selangor; tel. (3) 33921992; fax (3) 33921291; e-mail top@topglove.com.my; internet www.topglove.com.my; cap. and res US $40m. (June 2004); manufacture and exporter of natural rubber gloves; Exec. Dir K. M. LEE; 3,500 employees.

United Engineers (Malaysia) Bhd: UE Complex, 5 Jalan 217, 46700 Petaling Jaya, Selangor Darul Ehsan; tel. (3) 76205010; fax (3) 76205030; internet www.uem.com.my; f. 1966; iron, steel and non-ferrous founders; mechanical, electrical, civil, structural and telecommunication engineers for contract and project schemes; cap. and res 3,370.4m., sales 2,692.0m. (2000); Chair. Tan Sri RADIN SOENARNO AL-HAJ; Man. Dir Dato' Dr RAMLI BIN MOHAMAD; 12,141 employees.

United Plantations Bhd: Jendarata Estate, 36009 Teluk Intan, Perak; tel. (5) 6411411; fax (5) 6411876; e-mail upnet@tm.net.my; cultivation and processing of oil palm and coconut; cap. and res 806.3m., sales 456.9m. (2003); Chair. Tan Sri Datuk Dr JOHARI BIN MAT; 7,460 employees.

MAJOR INVESTMENT HOLDING COMPANIES

The following are among the major investment holding concerns in Malaysia (cap. = capital; res = reserves; m. = million; amounts in ringgit Malaysia):

AMMB Holdings Bhd: Bangunan Arab-Malaysian, 22nd Floor, 55 Jalan Raja Chulan, 50200 Kuala Lumpur; tel. (3) 2382633; fax (3) 2382842; e-mail gpa@ambg.com.my; internet www.ambg.com.my; f. 1991; investment holding; cap. and res 940m., sales 4,372m. (March 1999); Chair. Tan Sri Dato' AZMAN HASHIM; 6,009 employees.

Amsteel Corpn Bhd: Menara Lion, 46th Floor, 165 Jalan Ampang, 50450 Kuala Lumpur; tel. (3) 21622155; fax (3) 21641036; internet www.lion.com.my; f. 1977; investment holding, mfrs of steel and steel products, assembly of motor cycle engines, etc.; cap. and res 225.6m., sales 4,564.2m. (2002/03); Chair. Tan Sri Dato' ZAIN HASHIM; Man. Dir LIM KANG SENG; 14,649 employees.

Antah Holdings Bhd: 9577 Jalan SS 16/1, Subang Jaya, 47500 Petaling Jaya, Selangor; tel. (3) 56328668; fax (3) 2558464; e-mail info@antah.com.my; internet www.antah.com.my; f. 1976; investment holding and provision of management services; cap. and res 470.8m., sales 614.1m. (2000/01); Chair. Tunku NAQUIYUDDIN IBNI Tuanku JA'AFAR.

Arab-Malaysian Corpn Bhd: Lot 271, 1st Floor, Jalan Dua, off Jalan Chan Sow Lin, 55200 Kuala Lumpur; tel. (3) 2228870; fax (3) 2217793; e-mail amcorp@amcorp.com.my; internet www.amcorp.com.my; investment holding, operation of rubber and oil palm plantations, management services; cap. and res 1,624.8m., sales 161.5m. (2002/03); Chair. Tan Sri Dato' AZMAN HASHIM; Man. Dir SOO KIM WAI.

Batu Kawan Bhd: Wisma Taiko, 1 Jalan S. P. Seenivasagam, 30000 Ipoh, Perak Darul Ridzuan; tel. (5) 2417844; fax (5) 2548054; e-mail bkawan@pd.jaring.my; f. 1965; investment holding and manufacture of chemicals; cap. and res 1,854.5m., sales 151.4m. (2002/03); Group Man. Dir Dato' LEE HAU HIAN; 500 employees.

Berjaya Capital Bhd: 11th Floor, Menara Berjaya, KL Plaza, 179 Jalan Bukit Bintang, 55100 Kuala Lumpur; tel. (3) 29358888; fax (3) 29358043; e-mail judytan@berjaya.com.my; internet www.berjaya.com.my; investment holding, property investment and development, hotels, development of resorts and recreational facilities, travel, gaming and lottery management; cap. and res 1,660.7m., sales 238.7m. (2002/03); Chair. Dato' Seri SULAIMAN BIN MOHAMED AMIN.

Cahya Mata Sarawak Bhd: Wisma Mahmud, 6th Floor, Jalan Sungai Sarawak, POB 2710, 93754 Kuching, Sarawak; tel. (82) 238888; fax (82) 333828; internet www.cmsb.com.my; investment holding, management services, property development; cap. and res 973.7m., sales 1,083.0m. (2003); Chair. Dato' Sri SULAIMAN ABDUL RAHMAN TAIB; CEO DAVID WILLIAM BERRY; 2,956 employees.

Chemical Co of Malaysia Bhd: 9th Floor, Wisma Sime Darby, 14 Jalan Raja Laut, 50350 Kuala Lumpur; tel. (3) 26919366; fax (3) 26919901; internet www.ccm.com.my; f. 1963; subsidiaries engaged in mfr of fertilizers, chlor-alkali products pharmaceuticals and healthcare products; cap. and res 472.0m., sales 497.6m. (2001); Man. Dir Dato' LIM SAY CHONG; 1,200 employees.

Commerce Asset-Holding Bhd: Tingkat 12, Commerce Square, Jalan Semantan, Damansara Heights, 50490 Kuala Lumpur; tel. (3) 20935333; fax (3) 20933335; internet www.commerz.com.my; f. 1924; investment holding; property management; cap. and res 5,505.2m. (2001); Chair. Dato' MOHD DESA PACHI; 10,381 employees.

Eng Teknologi Holdings Bhd: 69–70 Pesara Kampung Jawa, Bayan Lepas Industrial Zone, 11900 Penang; tel. (4) 6440122; fax (4) 6423430; e-mail info@engtek.com; internet www.engtek.com; f. 1992; investment holding; mfr and design of precision tools; cap. and res 135.6m., sales 202.0m. (2003); Exec. Chair. and CEO Dato' ALFRED TEH EONG LIANG; 1,771 employees.

Genting Bhd: Wisma Genting, 24th Floor, Jalan Sultan Ismail, 50250 Kuala Lumpur; tel. (3) 21612288; fax (3) 21615304; e-mail gbinfo@genting.com.my; internet www.genting.com.my; gaming operations, hotels and plantations, property development, manufacturing and trading in paper and related products, electricity generation and supply; cap. and res 5,736.8m., sales 3,148.4m. (2001); Exec. Chair. LIM KOK THAY; Man. Dir Dato' LIM KOK THAY; 15,300 employees.

Golden Hope Plantations Bhd: Menara PNB, 13th Floor, 201A Jalan Tun Razak, 50400 Kuala Lumpur; tel. (3) 21619022; fax (3) 21618221; internet www.goldenhope.com.my; f. 1976; production and processing of rubber, palm oil, palm kernels, cocoa and copra; cap. and res 3,833.7m., sales 1,317.4m. (2000/01); plantations cover 116,811 ha; Chair. Tan Sri Dato' Seri AHMAD SARJI BIN ABDUL HAMID; 21,078 employees.

Golden Plus Holdings Bhd: Level 6, Wisma E & C, 2 Lorong Dungun Kiri, Damansara Heights, 50490 Kuala Lumpur; tel. (3) 20923311; fax (3) 20947788; e-mail info@gplus.com.my; internet www.gplus.com.my; property development and construction; cap. and res 300.0m. (2003), sales 130.9m. (1999); Chair. Dato' Haji Dr ZAINAL ABIDIN BIN HAJI MOHD ALI; Man. Dir TEH SOON SENG.

Hong Leong Industries Bhd: Wisma Hong Leong, 9th Floor, 18 Jalan Perak, 50450 Kuala Lumpur; tel. (3) 21642631; fax (3) 21642514; internet www.hongleong.com; f. 1982; subsidiaries engaged in investment and property holding, property management, manufacture of mosaic and ceramic tiles, steel products, PVC flooring, office products; cap. and res 718.1m., sales 1,520.7m. (2000/01); Chair. Tan Sri QUEK LENG CHAN; Pres. and CEO QUEK LENG SENG; 14,052 employees.

Innovest Bhd: Suite 201, 20th Floor, Menara Haw Par, Jalan Sultan Ismail, 50250 Kuala Lumpur; tel. (3) 2633633; fax (3) 2633033; investment holding and management services; cap. and res 11.7m., sales 51.4m. (1999); Chair. Dato' AZRAT GULL BIN AMIRZAT GULL.

IOI Oleochemical Industries Bhd: 2411 Lorong Perusahaan Satu, Prai Industrial Complex, 13600 Prai, Pinang; tel. (4) 3906766; fax (3) 3900067; e-mail webmaster@acidchem.com.my; internet www.palmco-holdings.com; f. 1976; fmrly Palmco Holdings Bhd; name changed as above Oct. 2003; subsidiaries engaged in production of palm kernel oil, manufacture of fatty acids and glycerine, bulk cargo warehousing, property devt and oil palm plantations; cap. and res 860.9m., sales 1,099.8m. (2002/03); Exec. Chair. Tan Sri Dato' LEE SHIN CHENG; 860 employees.

Jaya Tiasa Holdings Bhd: 1–9 Pusat Suria Permata, 96000 Sibu, Sarawak; tel. (84) 213255; fax (84) 213855; e-mail jayatiasa@jthsibu.po.my; internet www.jayatiasa.com; investment holding and management services, manufacture and sales of veneer plywood and sawn timber; cap. and res 748.7m., sales 440.3m. (2001/02); Chair. Tan Sri ABDUL RAHMAN BIN ABDUL HAMID; Man. Dir TIONG CHIONG HOO.

KFC Holdings (Malaysia) Bhd: Tingkat 17, Wisma KFC, Jalan Sultan Ismail, 50250 Kuala Lumpur; tel. (3) 2063388; fax (3) 2387786; internet www.kfcholdings.com.my; f. 1993; investment holding; subsidiaries involved in restaurant operation and poultry breeding and processing; cap. and res 392m., sales 1,071m. (2001); Man. Dir Dato' Haji JOHARI BIN ABDUL GHANI; 10,991 employees.

Leader Universal Holdings Bhd: Wisma Leader, 8 Jalan Larut, 10050 Pinang; POB 923, 10810 Pinang; tel. (4) 2292888; fax (4) 2292333; internet www.leaderuniversal.com; mfr and sale of telecommunication and power cables, copper and aluminium rods and conductors, cable installation and engineering services, power generation and property development; cap. and res 430.1m., sales 1,137.2m. (2001); Chair. Tan Sri RAZALI ISMAIL; Dep. Chair and CEO Dato' Seri HING BOK SAN.

Malayan Cement Bhd: Level 12, Bangunan TH Uptown 3, 3 Jalan SS21/39, 47400 Petaling Jaya, Selangor Darul Ehsan; tel. (3) 77238200; fax (3) 77224100; e-mail mcem@bciplc.com; internet www.malayancement.com.my; mfg and marketing of cement, ready-mixed concrete and allied products; cap. and res 3,083m., sales 1,657m. (2001); Chair. Tunku ABDULLAH IBNI AL-MARHUM Tunku ABDUL RAHMAN; 2,615 employees.

Malayan United Industries Bhd: MUI Plaza, 14th Floor, Jalan P. Ramlee, 50250 Kuala Lumpur; tel. (3) 21482566; fax (3) 21445209; e-mail muipa@po.jaring.my; internet www.mui-global.com; activities include retailing, hotels, food and confectionery, financial services, property, and travel and tourism; cap. and res 1,048m., sales 1,394m. (2003); Chair. and Chief Exec. Tan Sri Dato' Dr KHOO KAY PENG; 11,000 employees.

Malaysian Pacific Industries Bhd: Wisma Hong Leong, 9th Floor, 18 Jalan Perak, 50450 Kuala Lumpur; tel. (3) 2642631; fax (3) 2644801; f. 1962; subsidiaries engaged in manufacture of cartons, semiconductors and electronic components; cap. and res 615.6m., sales 1,006.6m. (1998/99); Chair. Tan Sri QUEK LENG CHAN; Man. Dir DAVID E. COMLEY; 3,300 employees.

Malaysian Resources Corpn Bhd: Menara MRCB, 2 Jalan Majlis, Seksyen 14, 40000 Shah Alam, Selangor Darul Ehsan; tel. (3) 55138080; fax (3) 55122608; e-mail info@mrcb.com.my; internet www.mrcb.com.my; f. 1968; property development, construction and civil engineering, telecommunications; cap. and res 1,141.6m., sales 248.6m. (1999/2000); Man. Dir and CEO SHAHRIL RIDZA RIDZUAN; Chair. Dato' Seri SYED ANWAR JAMALULLAIL; 160 employees.

Metroplex Bhd: Level 10, Grand Seasons Ave, 72 Jalan Pahang, 53000 Kuala Lumpur; tel. (3) 26931828; fax (3) 26912798; investment holding, hotel and casino operations, property development, quarry operations; cap. and res 1,114.1m., sales 343.3m. (2000/01); Chair. LIM SIEW KIM; Man. Dir CHAN TEIK HUAT.

Minho (M) Bhd: 31C Jalan Satu Kaw 16, Berkeley Town Centre, off Federal Highway, Klang, 41300 Selangor; tel. (3) 3911300; fax (3) 3912100; activities include manufacture, export and dealing in moulded timber and timber products; cap. and res 146.2m., sales 308.1m. (1999); Chair. Tunku Tan Sri IMRAN IBNI JA'AFAR.

MNI Holdings Bhd: Tower 1, 26th Floor, MNI Twins, 11 Jalan Pinang, 50450 Kuala Lumpur; tel. (3) 21645000; fax (3) 21641010; e-mail administrator@mni.com.my; internet www.mni.com.my; investment holding, insurance, takaful, manufacturing and marketing of welding supplies, tin mining; cap. and res 878.7m., sales 618.8m. (1998/99); Chair. Tan Sri Dato' Seri Dr AHMAD SARJI ABDUL HAMID; CEO Dato' Dr SHAMSUDDIN BIN KASSIM.

Nestlé (Malaysia) Bhd: Nestlé House, 4 Lorong Pesiaran Barat, 46200 Petaling Jaya, Selangor; tel. (3) 79554466; fax (3) 79550992; internet www.nestle.com.my; subsidiaries engaged in manufacture and marketing of milk products and halal food and beverage products; cap. and res 500.5m. (2001), sales 2,657.0m. (2003); Chair. Tan Sri Dato' MOHD GHAZALI SETH; Man. Dir JOSÉ LOPEZ Y VARGAS; 3,281 employees.

Oriental Holdings Bhd: Wisma Pinang Garden, 1st Floor, 42 Jalan Sultan Ahmad Shah, 10050 Pinang; tel. (4) 2266363; fax (4) 2265860; subsidiaries engaged in manufacture of plastic articles, etc.; cap. and res 2,201.6m., sales 2,583.0m. (2001); Chair. Dato' LOH CHENG YEAN; Man. Dir Dato' WONG LUM KONG; 9,328 employees.

PacificMas Bhd: 2nd Floor, Wisma Genting, Jalan Sultan Ismail, 50250 Kuala Lumpur; tel. (3) 21761000; fax (3) 20266868; e-mail jlcphuah@pacificmas.com.my; internet www.pacbanc.com.my/pacific/; f. 1919; fmrly The Pacific Bank Bhd, acquired by Malayan Banking Bhd and became investment holding company in Jan. 2001; cap. 171.0m., res 633.1m. (Dec. 2001), dep. 913.1m. (Dec. 2000); Chair. CHOI SIEW HONG; Gen. Man. NG HON SOON.

Perusahaan Otomobil Kedua Sdn Bhd (PERODUA): Sungai Choh, Locked Bag 226, 48009 Rawang, Selangor; tel. (3) 67338888; e-mail ILHAM@perodua.com.my; internet www.perodua.com.my; f. 1993; manufacture, assembly and sale of motor vehicles; Man. Dir (vacant); 6,600 employees.

Perusahaan Otomobil Nasional Bhd (PROTON): Kawasan Perindustrian HICOM, Batu Tiga, 40000 Shah Alam, Selangor; tel. (3) 5111055; fax (3) 5111252; internet www.proton.com; f. 1983; manufacture, assembly and sale of motor vehicles; cap. and res 3,008.0m., sales 8,301.0m. (2000/01); Chair. Datuk ABU HASSAN BIN KENDUT; CEO Tan Sri Dr MAHALEEL BIN Tengku ARIFF; 10,067 employees.

Petroliam Nasional Bhd (PETRONAS): Tower 1, Petronas Twin Towers, Persiaran KLCC, 50088 Kuala Lumpur; tel. (3) 20515000; fax (3) 20265055; e-mail webmaster@petronas.com.my; internet www.petronas.com.my; f. 1974; national oil co engaged in exploration, production, refining and marketing; total assets 144,128m. (March 2002); Chair. Tan Sri Dato' Seri AZIZAN ZAINUL ABIDIN; Pres. and CEO Tan Sri Dato' MOHD HASSAN BIN MARICAN; 23,450 employees.

Petronas Dagangan Bhd: Tower 1, Petronas Twin Towers, Persiaran KLCC, 50088 Kuala Lumpur; tel. (3) 2065500; fax (3) 2065505; internet www.mymesra.com.my; f. 1982; domestic marketing of petroleum products, operation of service stations and distribution of lubricants; cap. and res 2,540.1m., sales 9,830.4m. (2003/04); Man. Dir and CEO IBRAHIM BIN MARSIDI.

Petronas Gas Bhd: Tower 1, Petronas Twin Towers, Persiaran KLCC, 50088 Kuala Lumpur; tel. (3) 20515000; fax (3) 2065885; separates natural gas into components, and stores, transports and distributes them; cap. and res 6,189.8m., sales 1,772.4m. (2000/01); Chair. Tan Sri Dato' MOHD HASSAN BIN MARICAN; CEO ABDUL HAMID BIN IBRAHIM; 1,628 employees.

PPB Group Bhd: Wisma Jerneh, 17th Floor, 38 Jalan Sultan Ismail, 50250 Kuala Lumpur; tel. (3) 21412077; fax (3) 21418242; e-mail ppb@po.jaring.my; internet www.ppbgroup.com; fmrly Perlis Plantations Bhd; sugar, flour and feed milling, film distribution, edible oils processing and marketing and computer services; cap. and res 2,775.5m., sales 5,629.1m. (2001); Chair. ONG LE CHEONG; Dep. Chair. OH SIEW NAM; 15,492 employees.

Promet Bhd: 3rd Floor, Plaza Kelanamas, 19 Lorong Dungun, Damansara Heights, 50490 Kuala Lumpur; tel. (3) 2521919; fax (3) 2521911; internet www.promet.com.my; steel fabrication, civil engineering and construction; cap and res –377.6m., sales 161.9m.

(1998/99); Chair. Tan Sri MOHD NGAH SAID; Pres. and CEO SOH CHEE WEN.

Rashid Hussain Bhd: Tower 1, 9th Floor, RHB Centre, Jalan Tun Razak, 50400 Kuala Lumpur; tel. (3) 92852233; fax (3) 92855522; internet www.rhb.com.my; f. 1987; commercial banking, securities, merchant banking, financial and management services, insurance; cap. and res 383.7m., sales 2,800.8m. (2000/01); Chair. Tan Sri Dato' ABDUL RASHID HUSSAIN; 9,554 employees.

Renong Bhd: Bangunan MCOBA, 2nd Floor, 42 Jalan Syed Putra, 50460 Kuala Lumpur; tel. (3) 22742166; fax (3) 22743979; internet www.renong.com.my; f. 1982; engineering, construction and infrastructure project procurement and management, strategic investment; cap. and res –4,618.9m., sales 757.1m. (1998/99); Chair. Tan Sri Dato' SERI HALIM SAAD; 17,000 employees.

Resorts World Bhd: Wisma Genting, 24th Floor, Jalan Sultan Ismail, 50250 Kuala Lumpur; tel. (3) 21612288; fax (3) 21615304; e-mail roomrsv@genting.po.my; internet www.mol.net.my/genting; hotels, restaurants, theme parks, gaming, time-share ownership, tours and travel-related services; cap. and res 4,025.8m., sales 2,178.5m. (1999); Chair. and Chief Exec. Tan Sri LIM GOH TONG; Man. Dir Dato' LIM KOK THAY.

RHB Capital Bhd: Level 8 Tower 3, RHB Centre, 426 Jalan Tun Razak, 50400 Kuala Lumpur; tel. (3) 92806777; fax (3) 92806507; internet www.rhb.com.my; f. 1994; investment holding, banking; cap. and res 3,608.2m., sales 3,238.2m. (2001/02); Exec. Chair. Tan Sri Dato' ABDUL RASHID HUSSAIN.

R. J. Reynolds Bhd: Menara John Hancock, 6th Floor, 6 Jalan Gelenggang, Damansara Heights, 50490 Kuala Lumpur; tel. (3) 2549011; fax (3) 2550230; manufacture, marketing and sale of tobacco products; cap. and res 406.5m., sales 583.0m. (2000); Chair. Dato' MOHD NADZMI BIN MOHD SALLEH; 931 employees.

Road Builder (M) Holdings Bhd: Menara John Hancock, Level 16, 6 Jalan Gelenggang, Damansara Heights, 50490 Kuala Lumpur; tel. (3) 20939888; fax (3) 20925498; e-mail info@rb.com.my; internet www.rb.com.my; f. 1992; building, civil construction, quarry operations; cap. and res 1,258.1m., sales 1,234.7m. (2002/03); Chair. Tengku Tan Sri Dato' Seri AHMAD RITHAUDEEN BIN Tengku ISMAIL; Man. Dirs Dato' SHAMSUDIN BIN MOHD DUBI, Dato' LOW KENG KOK; 1,707 employees.

Sarawak Enterprise Corpn Bhd (SECB): Custodev Twin Tower, 1st Floor, 2679 Rock Rd, 93200 Kuching, Sarawak; tel. (82) 244000; fax (82) 248588; e-mail info@secb.com.my; internet www.jaring.my/mphb/secb; fmrly Dunlop Estates Bhd; investment holding; power generation, transmission and distribution; property development and investment; manufacture and trading of plastic packaging products; cap. and res 2,865m., sales 157.5m. (2000); Chair. Dato' MOHAMAD TAHA BIN ARIFFIN; CEO Datuk WAN ALI TUANKU YUBI.

Sports Toto Malaysia Sdn Bhd: Menara Prime, Levels 8–10, 30 Jalan Sultan Ismail, 50250 Kuala Lumpur; tel. (3) 21489888; fax (3) 21419581; e-mail webmaster@sportstoto.com.my; internet www.sportstoto.com.my; investment holding, management services, betting services, property development; cap. and res 708.6m., sales 2,195.4m. (1998/99); Chair. Tan Sri Dato' Seri VINCENT TAN CHEE YIOUN.

Technology Resources Industries Bhd: Menara Technology Resources, 20th Floor, 161B Jalan Ampang, 50450 Kuala Lumpur; tel. (3) 21619555; fax (3) 21632018; internet www.tri.com.my; f. 1966; telecommunications and transport services; mfr of consumer electricals and electronics; cap. and res 520.4m., sales 2,121.8m. (2000); Chair. and Chief Exec. Tan Sri Dato' TAJUDIN RAMLI.

UMW Holdings Bhd: The Corporate, 3rd Floor, 15/7 Jalan Utas, POB 7052, 40915 Shah Alam, Selangor; tel. (3) 55191911; fax (3) 55193890; internet www.umw.com.my; automotive equipment, manufacturing and engineering, oil and gas; cap. and res 1,491,756m., sales 3,332,716m. (2001); Chair. Haji ASMAT BIN KAMALUDIN; Man. Dir and CEO Haji ABDUL HALIM BIN HARUN.

YTL Corpn Bhd: Yeoh Tiong Lay Plaza, 11th Floor, 55 Jalan Bukit Bintang, 55100 Kuala Lumpur; tel. (3) 21426633; fax (3) 21412703; e-mail ctrl@ytl.com.my; internet www.ytlcommunity.com; property devt, manufacture of industrial products; cap. and res 4,139.8m., sales 2,326.0m. (2000/01); Chair Tan Sri Dato' Dr YEOH TIONG LAY; Man. Dir Tan Sri Dato' FRANCIS YEOH SOCK PING; 3,016 employees.

TRADE UNIONS

In 1995 there were 502 trade unions, 56% of which were from the private sector. About 8.2% of the Malaysian work-force of 7.9m. belonged to unions.

Congress of Unions of Employees in the Public Administrative and Civil Services (CUEPACS): a nat. fed. with 53 affiliates, representing 120,150 govt workers (1994).

Malaysian Trades Union Congress: Wisma MTUC, 10–5, Jalan USJ 9/5T, 47620 Subang Jaya, Selangor; POB 3073, 46000 Petaling Jaya, Selangor; tel. (3) 80242953; fax (3) 80243224; e-mail mtuc@tm.net.my; internet www.mtuc.org.my; f. 1949; 241 affiliated unions; Pres. ZAINAL RAMPAK; Sec.-Gen. G. RAJASEKARAN.

Principal affiliated unions:

All Malayan Estates Staff Union: POB 12, 46700 Petaling Jaya, Selangor Darul Ehsan; tel. (3) 7249533; e-mail mes@po.jaring.my; 2,654 mems; Pres. TITUS GLADWIN; Gen. Sec. D. P. S. THAMOTHARAM.

Amalgamated Union of Employees in Government Clerical and Allied Services: 32A Jalan Gajah, off Jalan Yew, Pudu, 55100 Kuala Lumpur; tel. (3) 9859613; fax (3) 9838632; 6,703 mems; Pres. IBRAHIM BIN ABDUL WAHAB; Gen. Sec. MOHAMED IBRAHIM BIN ABDUL WAHAB.

Chemical Workers' Union: Petaling Jaya, Selangor; 1,886 mems; Pres. RUSIAN HITAM; Gen. Sec. JOHN MATHEWS.

Electricity Industry Workers' Union: 55-2 Jalan SS 15/8A, Subang Jaya, 47500 Petaling Jaya, Selangor; tel. (3) 7335243; 22,000 mems; Pres. ABDUL RASHID; Gen. Sec. P. ARUNASALAM.

Federation of Unions in the Textile, Garment and Leather Industry: c/o Selangor Textile and Garment Manufacturing Employees Union, 9D Jalan Travers, 50470 Kuala Lumpur; tel. (3) 2742578; f. 1989; four affiliates; Pres. ABDUL RAZAK HAMID; Gen. Sec. ABU BAKAR IBRAHIM.

Harbour Workers' Union, Port Kelang: 106 Persiaran Raja Muda Musa, Port Kelang; 2,426 mems; Pres. MOHAMED SHARIFF BIN YAMIN; Gen. Sec. MOHAMED HAYAT BIN AWANG.

Kesatuan Pekerja Tenaga Nasional Bhd: 30 Jalan Liku Bangsar, POB 10400, 59100 Kuala Lumpur; tel. (3) 2745657; 10,456 mems; Pres. MOHAMED ABU BAKAR; Gen. Sec. IDRIS BIN ISMAIL.

Kesatuan Pekerja-Pekerja FELDA: 2 Jalan Maktab Enam, Melalui Jalan Perumahan Gurney, 54000 Kuala Lumpur; tel. (3) 26929972; fax (3) 26913409; 2,900 mems; Pres. INDERA PUTRA Haji ISMAIL; Gen. Sec. MOHAMAD BIN ABDUL RAHMAN.

Kesatuan Pekerja-Pekerja Perusahaan Membuat Tekstil dan Pakaian Pulau Pinang dan Seberang Prai: 23 Lorong Talang Satu, Prai Gardens, 13600 Prai; tel. (4) 301397; 3,900 mems; Pres. ABDUL RAZAK HAMID; Gen. Sec. KENNETH STEPHEN PERKINS.

Malayan Technical Services Union: 3A Jalan Menteri, off Jalan Cochrane, 55100 Kuala Lumpur; tel. (3) 92851778; fax (3) 92811875; 6,500 mems; Pres. Haji MOHAMED YUSOP Haji HARMAIN SHAH; Gen. Sec. SAMUEL DEVADASAN.

Malaysian Rubber Board Staff Union: POB 10150, 50908 Kuala Lumpur; tel. (3) 4565102; 1,108 mems; Pres. JUDE MICHAEL; Gen. Sec. NG SIEW LAN.

Metal Industry Employees' Union: Metalworkers' House, 5 Lorong Utara Kecil, 46200 Petaling Jaya, Selangor; tel. (3) 79567214; fax (3) 79550854; e-mail mieum@tm.net.my; 15,491 mems; Pres. SAMUSUDDIN USOP; Gen. Sec. JACOB ENGKATESU.

National Union of Commercial Workers: Bangunan NUCW, 98A–D Jalan Masjid India, 50100 Kuala Lumpur; POB 12059, 50780 Kuala Lumpur; tel. (3) 2927385; fax (3) 2925930; f. 1959; 11,937 mems; Pres. TAIB SHARIF; Gen. Sec. C. KRISHNAN.

National Union of Plantation Workers: 428 A–B, Jalan 5/46, Gasing Indah, POB 73, 46700 Petaling Jaya, Selangor; tel. (3) 77827622; fax (3) 77815321; e-mail sangkara@tm.net.my; f. 1946; 34,338 mems; Pres. AWI BIN AWANG; Gen. Sec. Dato' G. SANKARAN.

National Union of PWD Employees: 32B Jalan Gajah, off Jalan Yew, 55100 Kuala Lumpur; tel. (3) 9850149; 5,869 mems; Pres. KULOP IBRAHIM; Gen. Sec. S. SANTHANASAMY.

National Union of Telecoms Employees: Wisma NUTE, 17A Jalan Bangsar, 59200 Kuala Lumpur; tel. (3) 2821599; fax (3) 2821015; 15,874 mems; Pres. MOHAMED SHAFIE B. P. MAMMAL; Gen. Sec. MOHD JAFAR BIN ABDUL MAJID.

Non-Metallic Mineral Products Manufacturing Employees' Union: 99A Jalan SS 14/1, Subang Jaya, 47500 Petaling Jaya, Selangor; tel. (3) 56352245; fax (3) 56333863; e-mail nonmet@tm.net.my; 10,000 mems; Pres. ABDULLAH ABU BAKAR; Sec. S. SOMAHSUNDRAM.

Railwaymen's Union of Malaya: Bangunan Tong Nam, 1st Floor, Jalan Tun Sambathan (Travers), 50470 Kuala Lumpur; tel. (3) 2741107; fax (3) 2731805; 5,500 mems; Pres. ABDUL GAFFOR BIN IBRAHIM; Gen. Sec. S. VEERASINGAM.

Technical Services Union—Tenaga Nasional Bhd: Bangunan Keselamatan, POB 11003, Bangsar, Kuala Lumpur; tel. (3) 2823581; 3,690 mems; Pres. RAMLY YATIM; Gen. Sec. CLIFFORD SEN.

Timber Employees' Union: 10 Jalan AU 5C/14, Ampang, Ulu Kelang, Selangor; 7,174 mems; Pres. ABDULLAH METON; Gen. Sec. MINHAT SULAIMAN.

Transport Workers' Union: 21 Jalan Barat, Petaling Jaya, 46200 Selangor; tel. (3) 7566567; 10,447 mems; Pres. NORASHIKIN; Gen. Sec. ZAINAL RAMPAK.

Independent Federations and Unions

Kongres Kesatuan Guru-Guru Dalam Perkhidmatan Pelajaran (Congress of Unions of Employees in the Teaching Services): Johor; seven affiliates; Pres. RAMLI BIN MOHD JOHAN; Sec.-Gen. KASSIM BIN Haji HARON.

Malaysian Medical Association: MMA House, 4th Floor, 124 Jalan Pahang, 53000 Kuala Lumpur; tel. (3) 40420617; fax (3) 40418187; e-mail mma@tm.net.my; internet www.mma.org.my; 10 affiliates; Pres. Datuk Dr N. ARUMUGAM.

National Union of Bank Employees: NUBE Bldg, 61 Jalan Ampang, POB 12488, 50780 Kuala Lumpur; tel. (3) 20789800; fax (3) 20703800; e-mail nubehq@pd.jaring.my; internet www.nube.org.my; 27,000 mems; Gen. Sec. J. SOLOMON.

National Union of Journalists: 30B Jalan Padang Belia, 50470 Kuala Lumpur; tel. (3) 2742867; fax (3) 2744776; f. 1962; 1,700 mems; Gen. Sec. ONN EE SENG.

National Union of Newspaper Workers: 11B Jalan 20/14, Paramount Garden, 46300 Petaling Jaya, Selangor; tel. (3) 78768118; fax (3) 78751490; e-mail nunwl@tm.net.my; 3,000 mems; Pres. GAN HOE JIAN; Gen. Sec. R. CHANDRASEKARAN.

Sabah

Sabah Banking Employees' Union: POB 11649, 88818 Kota Kinabalu; internet sbeukk@tm.net.my; 729 mems; Gen. Sec. LEE CHI HONG.

Sabah Civil Service Union: Kota Kinabalu; f. 1952; 1,356 mems; Pres. J. K. K. VOON; Sec. STEPHEN WONG.

Sabah Commercial Employees' Union: Sinsuran Shopping Complex, Lot 3, Block N, 2nd Floor, POB 10357, 88803 Kota Kinabalu; tel. (88) 225971; fax (88) 213815; e-mail sceu-kk@tm.net.my; f. 1957; 980 mems; Gen. Sec. REBECCA CHIN.

Sabah Medical Services Union: POB 11257, 88813 Kota Kinabalu; tel. (88) 242126; fax (88) 242127; e-mail smsu65@hotmail.com; 4,000 mems; Pres. KATHY LO NYUK CHIN; Gen. Sec. LAURENCE VUN.

Sabah Petroleum Industry Workers' Union: POB 1087, Kota Kinabalu; tel. (88) 720737; e-mail spiwu@sabah.org.my; internet www.sabah.org.my/spiwu; f. 1966; 168 mems; Pres. BONIFACIO OBON.

Sabah Teachers' Union: POB 10912, 88810 Kota Kinabalu; tel. (88) 420034; fax (88) 431633; f. 1962; 3,001 mems; Pres. KWAN PING SIN; Sec.-Gen. PATRICK Y. C. CHOK.

Sarawak

Kepak Sarawak (Kesatuan Pegawai-Pegawai Bank, Sarawak): POB 62, Bukit Permata, 93100 Kuching, Sarawak; tel. (19) 8549372; e-mail kepaksar@tm.net.my; bank officers' union; 1,430 mems; Gen. Sec. DOMINIC CH'NG YUNG TED.

Sarawak Commercial Employees' Union: POB 807, Kuching; 1,636 mems; Gen. Sec. SONG SWEE LIAP.

Sarawak Teachers' Union: 139A Jalan Rock, 1st Floor, 93200 Kuching; tel. (82) 245727; fax (82) 245757; e-mail swktu@po.jaring.my; internet www.geocities.com/swktu; f. 1965; 12,832 mems; Pres. WILLIAM GHANI BINA; Sec.-Gen. THOMAS HUO KOK SEN.

Transport

RAILWAYS

Peninsular Malaysia

The state-owned Malayan Railways had a total length of 1,672 km in Peninsular Malaysia in 1996. The main railway line follows the west coast and extends 782 km from Singapore, south of Peninsular Malaysia, to Butterworth (opposite Pinang Island) in the north. From Bukit Mertajam, close to Butterworth, the Kedah line runs north to the Thai border at Padang Besar where connection is made with the State Railway of Thailand. The East Coast Line, 526 km long, runs from Gemas to Tumpat (in Kelantan). A 21-km branch line from Pasir Mas (27 km south of Tumpat) connects with the State Railway of Thailand at the border station of Sungei Golok. Branch lines serve railway-operated ports at Port Dickson and Telok Anson as well as Port Klang and Jurong (Singapore). An express rail link connecting central Kuala Lumpur and the new Kuala Lumpur International Airport (KLIA) opened in 2001.

Keretapi Tanah Melayu Bhd (KTMB) (Malayan Railways): KTMB Corporate Headquarters, Jalan Sultan Hishamuddin, 50621 Kuala Lumpur; tel. (3) 22757142; fax (3) 27105706; e-mail pro@ktmb.com.my; internet www.ktmb.com.my; f. 1885; incorporated as a co under govt control in Aug. 1992; privatized in Aug. 1997; managed by the consortium Marak Unggal (Renong, DRB and Bolton); Chair. Tan Sri Dato' THONG YAW HONG.

Sabah

Sabah State Railway: Karung Berkunci 2047, 88999 Kota Kinabalu; tel. (88) 254611; fax (88) 236395; 134 track-km of 1-m gauge (1995); goods and passenger services from Tanjong Aru to Tenom, serving part of the west coast and the interior; diesel trains are used; Gen. Man. IR BENNY WANG.

ROADS

Peninsular Malaysia

Peninsular Malaysia's road system is extensive, in contrast to those of Sabah and Sarawak. In 1999 the road network in Malaysia totalled an estimated 65,877 km, of which 16,206 km were highways and 31,777 km secondary roads; 75.8% of the network was paved.

Sabah

Jabatan Kerja Raya (Public Works Department): Jalan Sembulan, 88582 Kota Kinabalu, Sabah; tel. (88) 244333; fax (88) 237234; e-mail pos@jkr.sabah.gov.my; internet www.jkr.sabah.gov.my; f. 1881; implements and maintains public infrastructures such as roads, bridges, buildings and sewerage systems throughout Sabah; maintains a network totalling 14,297 km, of which 1,230 km were trunk roads in 1997; the total included 5,011 km of sealed roads; Dir DAVID CHIU SIONG SENG.

Sarawak

Jabatan Kerja Raya (Public Works Department): Tingkat 11–18, Wisma Saberkas, Jalan Tun Abg Haji Openg, 93582 Kuching, Sarawak; tel. (82) 203100; fax (82) 429679; e-mail limkh@sarawaknet.gov.my; internet www.jkr.sarawak.gov.my; implements and maintains public infrastructures in Sarawak; road network totalling 10,979 km, of which 3,986 km were sealed roads; Dir HUBERT THIAN CHONG HUI.

SHIPPING

The ports in Malaysia are classified as federal ports, under the jurisdiction of the federal Ministry of Transport, or state ports, responsible to the state ministries of Sabah and Sarawak.

Peninsular Malaysia

The federal ports in Peninsular Malaysia are Klang (the principal port), Pinang, Johor and Kuantan.

Johor Port Authority: POB 66, 81707 Pasir Gudang, Johor; tel. (7) 2517721; fax (7) 2517684; e-mail jport@lpj.com.my; internet www.lpj.gov.my; f. 1973; Gen. Man. MOHD ROZALI BIN MOHD ALI.

Johor Port Bhd: POB 151, 81707 Pasir Gudang, Johor; tel. (7) 2525888; fax (7) 2522507; e-mail joport@silicon.net.my; internet www.joport.com.my; Exec. Chair. Dato' MOHD TAUFIK ABDULLAH.

Klang Port Authority: POB 202, Jalan Pelabuhan, 42005 Port Klang, Selangor; tel. (3) 31688211; fax (3) 31670211; e-mail pka_admin@pka.gov.my; f. 1963; Gen. Man. Datin Paduka O. C. PHANG.

Kuantan Port Authority: Tanjung Gelang, POB 161, 25720 Kuantan, Pahang; tel. (9) 5833201; fax (9) 5833866; e-mail lpk@po.jaring.my; internet www.lpktn.gov.my; f. 1974; Gen. Man. KHAIRUL ANUAR BIN ABDUL RAHMAN.

Penang Port Commission: POB 143, 10710 Pinang; tel. (4) 2633211; fax (4) 2626211; e-mail sppp@po.jaring.my; internet www.penangport.gov.my; f. 1956; Gen. Man. Dato' Capt. Haji ADBUL RAHIM ABDUL AZIZ.

Sabah

The chief ports are Kota Kinabalu, Sandakan, Tawau, Lahad Datu, Kudat, Semporna and Kunak and are administered by the Sabah Ports Authority. Many international shipping lines serve Sabah. Local services are operated by smaller vessels. The Sapangar Bay oil terminal, 25 km from Kota Kinabalu wharf, can accommodate oil tankers of up to 30,000 dwt.

Sabah Ports Authority: Bangunan Ibu Pejabat LPS, Jalan Tun Fuad, Tanjung Lipat, Locked Bag 2005, 88617 Kota Kinabalu, Sabah; tel. (88) 538400; fax (88) 223036; e-mail sabport@po.jaring.my; internet www.infosabah.com.my/spa; Gen. Man. RAMLI AMIR.

Sarawak

There are four port authorities in Sarawak: Kuching, Rajang, Miri and Bintulu. Kuching, Rajang and Miri are statutory ports, while Bintulu is a federal port. Kuching port serves the southern region of

Sarawak, Rajang port the central region, and Miri port the northern region.

Kuching Port Authority: Jalan Pelabuhan, Pending, POB 530, 93710 Kuching, Sarawak; tel. (82) 482144; fax (82) 481696; e-mail kuport@po.jaring.my; f. 1961; Gen. Man. CHOU CHII MING.

Rajang Port Authority: 96000 Sibu, Sarawak; tel. (84) 319004; fax (84) 318754; e-mail rajang@po.jaring.my; f. 1970; Gen. Man. Haji BAHRIN Haji ADENG.

Principal Shipping Companies

Achipelego Shipping (Sarawak) Sdn Bhd: Lot 267/270, Jalan Chan Chin Ann, POB 2998, 93758 Kuching; tel. (82) 412581; fax (82) 416249; Gen. Man. MICHAEL M. AMAN.

Malaysia Shipping Corpn Sdn Bhd: Office Tower, Plaza Berjaya, Suite 14C, 14th Floor, 12 Jalan Imbi, 55100 Kuala Lumpur; tel. (3) 21418788; fax (3) 21429214; Chair. Y. C. CHANG.

Malaysian International Shipping Corpn Bhd (National Shipping Line of Malaysia): Suite 3–8, Tingkat 3, Wisma MISC, 2 Jalan Conlay, 50450 Kuala Lumpur; tel. (3) 2428088; fax (3) 2486602; e-mail zzainala@miscnote1.miscbhd.com; internet www.misc-bhd.com; f. 1968; regular sailings between the Far East, South-East Asia, Australia, Japan and Europe; also operates chartering, tanker, haulage and warehousing and agency services; major shareholder, Petroliam Nasional Bhd (PETRONAS); Chair. Tan Sri Dato' MOHD HASSAN BIN MARICAN; Man. Dir Dato' Haji MOHD ALI BIN Haji YASIN.

Perbadanan Nasional Shipping Line Bhd (PNSL): Kuala Lumpur; tel. (3) 2932211; fax (3) 2930493; f. 1982; specializes in bulk cargoes; Chair. Tunku Dato' SHAHRIMAN BIN Tunku SULAIMAN; Exec. Dep. Chair. Dato' SULAIMAN ABDULLAH.

Persha Shipping Agencies Sdn Bhd: Bangunan Mayban Trust, Penthouse Suite, Jalan Pinang, 10200 Pinang; tel. (4) 2612400; fax (4) 2623122; Man. Dir MOHD NOOR MOHD KAMALUDIN.

Syarikat Perkapalan Kris Sdn Bhd (The Kris Shipping Co Ltd): 3A07 Block A, Kelana Centre Point, 3 Jalan SS7/19, Kelana Jaya; POB 8428, 46789 Petaling Jaya, Selangor; tel. (3) 7046477; fax (3) 7048007; domestic services; Chair. Dato' Seri SYED NAHAR SHAHABUDIN; Gen. Man. ROHANY TALIB; Dep. Gen. Man. THO TEIT CHANG.

Trans-Asia Shipping Corpn Sdn Bhd: Unit 715–718, Block A, Kelana Business Centre, 97 Jalan SS7/2, Kelana Jaya, 47301 Petaling Jaya, Selangor; tel. (3) 78802020; fax (4) 78802200; e-mail ahmad@tasco.com.my; internet www.tasco.com.my; Man. Dir LEE CHECK POH.

CIVIL AVIATION

The new Kuala Lumpur International Airport (KLIA), situated in Sepang, Selangor (50 km south of Kuala Lumpur) began operations in June 1998, with an initial capacity of 25m.–30m. passengers a year, rising to 45m. by 2020. It replaced Subang Airport in Kuala Lumpur (which was renamed the Sultan Abdul Aziz Shah Airport in 1996). An express rail link between central Kuala Lumpur and KLIA opened in early 2001. There are regional airports at Kota Kinabalu, Pinang, Johor Bahru, Kuching and Pulau Langkawi. In addition, there are airports catering for domestic services at Alor Star, Ipoh, Kota Bahru, Kuala Terengganu, Kuantan and Melaka in Peninsular Malaysia, Sibu, Bintulu and Miri in Sarawak and Sandakan, Tawau, Lahad Datu and Labuan in Sabah. There are also numerous smaller airstrips.

Department of Civil Aviation (Jabatan Penerbangan Awam Malaysia): Aras B1, 1, 2 and 3, Blok D5, Parcel D, Pusat Pentadbiran Kerajaan Persekutuan, 62502 Putrajaya; tel. (3) 88866000; fax (3) 88891541; internet www.dca.gov.my; Dir-Gen. Dato' KOK SOO CHON.

Air Asia Sdn Bhd: Wisma HICOM, 6th Floor, 2 Jalan Usahawan, U1/8, Seksyen Ul, 40150 Shah Alam, Selangor; tel. (3) 2028007; fax (3) 2028137; internet www.airasia.com; f. 1993; a second national airline, with a licence to operate domestic, regional and international flights; 85%-owned by HICOM.

Berjaya Air: Apprentice Training Bldg, 1st Floor, Mas Complex B (Hangar 1), Lapangan Terbang Sultan Abdul Aziz Shah, Shah Alam, Selangor, 47200 Kuala Lumpur; tel. (3) 7476828; fax (3) 7476228; e-mail berjayaa@tm.net.my; f. 1989; scheduled and charter domestic services; Pres. Dato TENGKU ADNAN MANSOR.

Malaysia Airlines: 32nd Floor, Bangunan MAS, Jalan Sultan Ismail, 50250 Kuala Lumpur; tel. (3) 21655140; fax (3) 21633178; e-mail grpcomm@malaysiaairlines.com.my; internet www.malaysiaairlines.com.my; f. 1971 as the Malaysian successor to the Malaysia Singapore Airlines (MSA); known as Malaysian Airline System (MAS) until Oct. 1987; services to 33 domestic points and to 79 international destinations; Chair. Tan Sri Datuk Seri AZIZAN ZAINUL ABIDIN.

Pelangi Airways Sdn Bhd: Kuala Lumpur; tel. (3) 2624453; fax (3) 2624515; internet www.asia123.com/pelangi/home.htm; f. 1988; domestic scheduled passenger services; Chair. Tan Sri SAW HWAT LYE.

Transmile Air Sdn Bhd: Wisma Semantan, Mezzanine 2, Block B, 12 Jalan Gelenggang, Bukit Damansara, 50490 Kuala Lumpur; tel. (3) 2537718; fax (3) 2537719; f. 1992; scheduled and charter regional and domestic services for pasengers and cargo; Chair. Tan Sri ZAINOL MAHMOOD.

Tourism

Malaysia has a rapidly-growing tourist industry, and tourism remains an important source of foreign-exchange earnings. In 2003 some 10.6m. tourists visited Malaysia, a decline of 20.4% compared with the previous year. In 2003 tourist receipts totalled RM 21,191.1m.

Malaysia Tourism Promotion Board: Menara Dato' Onn, 15th–18th, 24th–27th, 29th–30th Floors, Putra World Trade Centre, 45 Jalan Tun Ismail, 50480 Kuala Lumpur; tel. (3) 26158188; fax (3) 26935884; e-mail enquiries@tourism.gov.my; internet www.tourismmalaysia.gov.my; f. 1972 to co-ordinate and promote activities relating to tourism in Malaysia; Minister Datuk LEO MICHAEL TOYAD; Dep. Minister Datuk AHMAD ZAHID HAMIDI.

Sabah Tourist Association: Kota Kinabalu; tel. (88) 211484; internet www.sabahtourist.com.my; f. 1963; 55 mems; parastatal promotional org.; Chair. CLEMENT LEE.

Sarawak Tourism Board: Levels 6 and 7, Bangunan Yayasan Sarawak, Jalan Masjid, 93400 Kuching; ; tel. (82) 423600; fax (82) 416700; e-mail stb@sarawaktourism.com; internet www.sarawaktourism.com; f. 1995; CEO ABANG Haji KASHIM ABANG MORSHIDI.

Defence

In August 2003 the total strength of the armed forces was 104,000; army 80,000, navy 14,000, air force 10,000; military service is voluntary. Paramilitary forces included the General Operations Force of 18,000 and the People's Volunteer Corps of 240,000. Malaysia is a participant in the Five-Power Defence Arrangements with Singapore, Australia, New Zealand and the United Kingdom.

Defence Budget: The budget for 2003 allocated RM 7,700m. (excluding procurement allowance) to defence.

Chief of the Defence Forces: Gen. Tan Sri Dato' Seri MOHD ZAHIDI BIN Haji ZAINUDDIN.

Chief of Army: Gen. Datuk MOHAMAD AZUMI MOHAMED.

Chief of Navy Staff: Adm. MOHD ANWAR MOHD NOR.

Chief of Air Force Staff: Gen. Datuk Nik ISMAIL Nik MOHAMAD.

Education

Under the Malaysian education system, free schooling is provided at government-assisted schools for children between the ages of six and 18. There are also private schools, which receive no government financial aid. Education is compulsory for 11 years between the ages of six and 16 years. Government expenditure on education was budgeted at RM 29,931m. in 2003 (27.2% of projected expenditure). Total enrolment in primary and secondary education in 1997 was equivalent to 82% of children in the relevant age group (males 80%; females 85%). Scholarships are awarded at all levels and there are many scholarship-holders studying at universities and other institutes of higher education at home and abroad.

PRIMARY EDUCATION

The national language, Bahasa Malaysia, is the main medium of instruction, although English, Chinese and Tamil are also used. Two-thirds of the total primary school enrolment is in National Schools where Malay is used and the remainder in National-Type Primary Schools where Tamil or Chinese is used. A place in primary school is now assured to every child from the age of six onwards, and parents are free to choose the language of instruction. Enrolment in primary education was equivalent to 101% (males 101%; females 101%) of children in the relevant age group in 1997. The primary school course lasts for six years.

SECONDARY EDUCATION

Bahasa Malaysia is the main medium of instruction in secondary schools, while English is taught as a second language and Chinese and Tamil are taught as pupils' own languages. There are, however, private Chinese secondary schools. In January 2003 new legislation came into effect, requiring that all mathematics and science classes in schools be taught in English. Secondary education lasts for seven years, comprising a first cycle of three years and a second of four. In 1997 enrolment at the secondary level was equivalent to 64% of pupils of the relevant age (males 56%; females 69%).

HIGHER EDUCATION

In 1997 there were nine universities, including the International Islamic University. In the same year there were eight polytechnics and colleges including the MARA Institute of Technology. In 1995 enrolment in tertiary education was equivalent to 12% of those in the relevant age-group.

Malaysia's universities adhere to a quota system (55% of entrants should be Malays and 45% non-Malays), which, in practice, makes it considerably more difficult for non-Malays to gain a place at university. The Government was attempting to encourage foreign universities to establish campuses in Malaysia to improve standards and reduce the cost of sending Malaysian students abroad to study. In 1994 the Prime Minister announced that from 1995, contrary to previous policy, English would be used to teach scientific and technological subjects at university. In 1995 the Government announced that, from 1996, basic university degree courses would be reduced in length from four to three years, in order to increase the number of graduates entering the employment market.

Bibliography

GENERAL

Brown, Ian, and Ampalavanar, Rajeswary. *Malaysia.* World Bibliographical Series, Vol. 12. Oxford, Clio Press, 1986.

Hodder, B. W. *Man in Malaya.* London, University of London Press, 1959.

Purcell, V. *Malaysia.* London, Thames and Hudson, 1965.

Smith, T. E., and Bastin, J. *Malaysia.* London, Oxford University Press, 1967.

PEOPLES OF MALAYSIA

Ackerman, S. E., and Lee, R. *Heaven in Transition: Non-Muslim Religious Innovation and Ethnic Identity in Malaysia.* Kuala Lumpur, Forum, 1990.

Chandra, Muzaffar. *Islamic Resurgence in Malaysia.* Kuala Lumpur, Penerbit Fajar Bakti, 1987.

Hew Cheng Sim. *Women Workers, Migration and Family in Sarawak.* London, RoutledgeCurzon, 2002.

Hong, Evelyne. *The Natives of Sarawak.* Pinang Institut Masyarakat, 1987.

Ibrahim, Zawawi. *The Malay Labourer: By the Window of Capitalism.* Singapore, Institute of Southeast Asian Studies, 1998.

King, Victor T. *The Peoples of Borneo.* Oxford, Blackwell, 1993.

de Koninck, Rodolphe. *Malay Peasants Coping with the World: Breaking the Community Circle?* Singapore, Institute of Southeast Asian Studies, 1992.

Loh Kok, Francis, and Khoo Boo Teik. *Democracy in Malaysia. Discourses and Practices.* Richmond, Surrey, Curzon Press, 2001.

Pillai, P. *People on the Move—An Overview of Recent Immigration and Emigration in Malaysia.* Kuala Lumpur, Institute of Strategic and International Studies, 1992.

Sellato, Bernard. *Nomads of the Borneo Rainforest: The Economics, Politics and Ideology of Settling Down.* Honolulu, HI, University of Hawaii Press, 1994.

Wong, Diana. *Peasants in the Making: Malaysia's Green Revolution.* Singapore, Institute of Southeast Asian Studies, 1987.

HISTORY

Amin, Mohamad, and Caldwell, M. (Eds). *Malaya: The Making of a Neo-Colony.* Nottingham, Spokesman, 1979.

Andaya, Barbara Watson, and Andaya, Leonard. *A History of Malaysia.* London, Macmillan, 1982.

A History of Malaysia. 2nd edn. Honolulu, HI, University of Hawaii Press, 2001.

Black, Ian. *A Gambling Style of Government. The Establishment of the Chartered Company's Rule in Sabah, 1878–1915.* Kuala Lumpur, Oxford University Press, 1983.

Cheah Boon Kheng. *Red Star over Malaya. Resistance and Social Conflict during and after the Japanese Occupation of Malaya, 1941–46.* Singapore, Singapore University Press, 1983.

Malaysia: The Making of a Nation. Singapore, Institute of Southeast Asian Studies, 2002.

Comber, Leon. *13 May 1969: A Historical Survey of Sino-Malay Relations.* Singapore, Graham Brash, 2001.

Emerson, R. *Malaysia, a Study in Direct and Indirect Rule.* Kuala Lumpur, University of Malaya Press, 1964.

Goh Cheng Teik. *The May Thirteenth Incident and Democracy in Malaysia.* Kuala Lumpur, Oxford University Press, 1971.

Gullick, J. M. *Malay Society in the Late Nineteenth Century: The Beginnings of Change.* Singapore, Oxford University Press, 1987.

Rulers and Residents: Influence and Power in the Malay States 1870–1920. Singapore, Oxford University Press, 1992.

Jackson, Robert. *The Malayan Emergency: The Commonwealth's Wars 1948–66.* London, Routledge, 1991.

Kathirithamby-Wells, Jeyamalar. *Nature and Nation: Forests and Development in Peninsular Malaysia.* Honolulu, HI, University of Hawaii Press, 2004.

Kaur, Amarjit, and Metcalfe, Ian (Eds.). *The Shaping of Malaysia.* Basingstoke, Macmillan, 1999.

Kratoska, Paul H. *The Japanese Occupation of Malaya: A Social and Economic History.* London, Hurst, 1998.

Leigh, Michael B. *Mapping the Peoples of Sarawak.* Kota Samarahan, Sarawak, Institute of East Asian Studies, 2002.

Milner, Anthony. *The Invention of Politics in Colonial Malaya.* Cambridge, Cambridge University Press, 1995.

Parkinson, C. N. *British Intervention in Malaya 1867–77.* Kuala Lumpur, University of Malaya Press, 1960.

Pringle, R. *Rajahs and Rebels: The Iban of Sarawak under Brooke Rule 1841–1941.* London, Macmillan, 1971.

Rahman, Abdul Embong. *State-led Modernization and the New Middle Class in Malaysia.* New York, Palgrave Macmillan, 2001.

Rashid, Rehman. *A Malaysian Journey.* Kuala Lumpur, Rehman Rashid, 1993.

Reece, R. H. W. *The Name of Brooke. The End of White Rajah Rule in Sarawak.* Kuala Lumpur, Oxford University Press, 1982.

Riddell, Peter G. *Islam and the Malay-Indonesian World: Transmission and Responses.* Honolulu, HI, University of Hawaii Press, 2001.

Rimmer, Peter J., and Allen, Lisa M. (Eds). *The Underside of Malaysian History: Pullers, Prostitutes, Plantation Workers.* Singapore, Singapore University Press, 1990.

Roff, W. *The Origins of Malay Nationalism.* New Haven, CT, Yale University Press, 1967.

Sandhu, Kernial Singh, and Wheatley, Paul (Eds). *Melaka: The Transformation of a Malay Capital c. 1400–1980.* Kuala Lumpur, Oxford University Press, 1983.

Shennan, Margaret. *Out In The Midday Sun: The British in Malaya 1880–1960.* London, John Murray, 2000.

Short, A. *The Communist Insurrection in Malaya 1948–1960.* London, Frederick Muller, 1975.

Simandjuntak, B. *Malayan Federalism 1945–63.* Kuala Lumpur, Oxford University Press, 1969.

Stockwell, A. J. *British Policy and Malay Politics during the Malayan Union Experiment 1942–48.* Kuala Lumpur, Malaysian Branch of the Royal Asiatic Society, 1979.

British Documents on the End of Empire. (Series B, Vol. 3), *Malaya.* London, HMSO, 1995.

Stubbs, R. *Hearts and Minds in Guerilla Warfare: The Malayan Emergency 1948–1960.* Singapore, Eastern Universities Press, 2004.

Swettenham, F. A. *British Malaya.* London, Allen and Unwin, 1948.

Tregonning, K. G. *A History of Modern Sabah 1881–1963.* Oxford, Oxford University Press, 1965.

Turnbull, C. M. *The Straits Settlements 1826–67.* London, Athlone Press, 1972.

Verma, Vidhu. *Malaysia: State and Civil Society in Transition.* Boulder, CO, Lynne Rienner Publrs, 2002.

White, Nicholas. *Business, Government and the End of Empire: Malaya 1942–1957.* Kuala Lumpur, Oxford University Press, 1996.

British Business in Post-Colonial Malaysia, 1957–70: Neo-colonialism or Disengagement? London, RoutledgeCurzon, 2004.

Weiss, Meredith, and Saliha, Hassan (Eds). *Social Movements in Malaysia: From Moral Communities to NGOs.* London, RoutledgeCurzon, 2002.

Yong, C. F., and McKenna, R. F. *The Kuomintang Movement in British Malaya 1912–1949.* Singapore, Singapore University Press, 1990.

POLITICS

Abdullah, Kamarulnizam. *The Politics of Islam in Contemporary Malaysia.* Kuala Lumpur, Penerbit Universiti Kebangsaan Malaysia, 2002.

ALIRAN. *Reflections on the Malaysian Constitution.* Pinang, ALIRAN, 1987.

Atkins, William. *The Politics of Southeast Asia's New Media.* London, RoutledgeCurzon, 2002.

Bowie, Alasdair. *Crossing the Industrial Divide: State, Society, and the Politics of Economic Transformation in Malaysia.* New York, Columbia University Press, 1991.

CARPA. *Tangled Web: Dissent, Deterrence and the October 1987 Crackdown.* Australia, Collins/Angus and Robertson Publishers Australia, 1988.

Cheong Mei Sui, and Amin, Adibah. *Daim: The Man Behind the Enigma.* Petaling Jaya, Pelanduk Publications, 1995.

Clutterbuck, Richard. *Conflict and Violence in Singapore and Malaysia 1945–83.* Singapore, Graham Brash, 1985.

Crouch, Harold. *Government and Society in Malaysia.* Ithaca, NY, Cornell University Press, 1996.

Crouch, H., Lee, K. H., and Ong, M. *Malaysian Politics and the 1978 Election.* 1980.

Esman, M. J. *Administration and Development in Malaysia.* Ithaca, NY, Cornell University Press, 1972.

Goh Cheng Teik. *Malaysia: Beyond Communal Politics.* Petaling Jaya, Pelanduk Publications, 1994.

Gomez, Edward Terence. *Money Politics in the Barisan Nasional.* Kuala Lumpur, Forum, 1990.

UMNO's Corporate Investments. Kuala Lumpur, Forum, 1990.

Political Business: Corporate Involvement in Malaysian Political Parties. Townsville, James Cook University of North Queensland, 1994.

The 1995 Malaysian General Elections: A Report and Commentary. Singapore, Institute of Southeast Asian Studies, 1996.

Gomez, Edward Terence (Ed.). *The State of Malaysia: Ethnicity, Equity and Reform.* London, RoutledgeCurzon, 2004.

Gould, J. W. *The United States and Malaysia.* Cambridge, MA, Harvard University Press, 1969.

Hefner, Robert W. (Ed.). *The Politics of Multiculturalism: Pluralism and Citizenship in Malaysia, Singapore, and Indonesia.* Honolulu, HI, University of Hawaii Press, 2001.

Heng Pek Koon. *Chinese Politics in Malaysia: A History of the Malaysian Chinese Association.* Kuala Lumpur, Oxford University Press, 1988.

Hilley, John. *Malaysia: Mahathirism, Hegemony and the New Opposition.* London, Zed Books, 2001.

Hooker, Virginia, and Othman, Norani (Eds). *Malaysia: Islam, Society and Politics.* Singapore, Institute of Southeast Asian Studies, 2003.

Hua Wu Yin. *Class and Communalism in Malaysia: Politics in a Dependent Capitalist State.* London, Zed Books, 1983.

International Bar Association. *Justice in Jeopardy: Malaysia in 2000—Report of a Mission (17–27 April 1999) on behalf of the International Bar Association, the ICJ Centre for the Independence of Judges and Lawyers, the Commonwealth Lawyers' Association, the Union Internationale des Avocats.* London, International Bar Association, 2000.

In-Won Hwang. *Personalized Politics: The Malaysian State Under Mahathir.* Singapore, Institute of Southeast Asian Studies, 2003.

Kahn, Joel S., and Loh Kok Wah, Francis (Eds). *Fragmented Vision: Culture and Politics in Contemporary Malaysia.* Sydney, Asian Studies Association of Australia, 1993.

Khoo Boo Teik. *Paradoxes of Mahathirism: An Intellectual Biography of Mahathir Mohamad.* Kuala Lumpur, Oxford University Press, 1995.

Beyond Mahathir: Malaysian Politics and Its Discontents. London, Zed Books, 2003.

Lee, H. P. *Constitutional Conflicts in Contemporary Malaysia.* Kuala Lumpur, Oxford University Press, 1995.

Loh Kok Wah, Francis, and Saravanamuttu, Johan (Eds). *New Politics in Malaysia.* Singapore, Institute of Southeast Asian Studies, 2003.

Mahathir Mohamad. *The Malay Dilemma.* Singapore, Asia Pacific Press, 1970, republished 1981.

The Way Forward: Growth, Prosperity and Multiracial Harmony in Malaysia. London, Weidenfeld and Nicholson, 1998.

A New Deal for Asia. Kuala Lumpur, Pelanduk Publications, 1999.

Mandal, Sumit K. *Challenging Authoritarianism in Southeast Asia: Comparing Indonesia and Malaysia.* London, Routledge, 2003.

Milne, R. S., and Mauzy, Diane K. *Politics and Government in Malaysia.* Singapore, Federal Publications, 1978.

Malaysian Politics under Mahathir. London, Routledge, 1999.

Munro-Kua, Anne. *Authoritarian Populism in Malaysia.* New York, St Martin's Press, 1997.

Mutalib, Hussin. *Islam and Ethnicity in Malay Politics.* Singapore, Oxford University Press, 1990.

Nair, Shanti. *Islam in Malaysian Foreign Policy.* London, Routledge, 1997.

Pathmanathan, Murugesu, and Lazarus, David. *Winds of Change: The Mahathir Impact on Malaysia's Foreign Policy.* Petaling Jaya, Eastview Productions Sdn Bhd, 1984.

Peletz, Michael G. *Islamic Modern: Religious Courts and Cultural Politics in Malaysia.* Princeton, NJ, Princeton University Press, 2002.

Ratnam, K. J. *Communalism and the Political Process in Malaya.* Kuala Lumpur, University of Malaya Press, 1965.

Roff, M. C. *The Politics of Belonging: Political Change in Sabah and Sarawak.* London, Oxford University Press, 1975.

Salleh Abas, Tun, and Das, K. *May Day for Justice.* Kuala Lumpur, Magnus Books, 1990.

Shamsul, A. B. *From British to Bumiputera Rule: Local Politics and Rural Development in Peninsular Malaysia.* Singapore, Institute of Southeast Asian Studies, 2004.

Shome, Tony. *Malay Political Leadership.* Richmond, Surrey, Curzon Press, 2001.

Stewart, Ian. *The Mahathir Legacy: A Nation Divided, A Region at Risk.* St Leonards, NSW, Allen and Unwin, 2003.

Tan Liok Ee. *The Politics of Chinese Education in Malaya 1945–61.* Kuala Lumpur, Oxford University Press, 1998.

Tan, Andrew T. H. *Security Perspectives of the Malay Archipelago: Security Linkages in the Second Front in the War on Terrorism.* Cheltenham, Edward Elgar Publishing, 2004.

Teng, Fan Yew. *The UMNO Drama.* Kuala Lumpur, Egret Books, 1989.

Yeoh, Michael (Ed.). *21st Century Malaysia: Challenges and Strategies in Attaining Vision 2020.* London, ASEAN Academic Press, 2002.

ECONOMY

Alavi, Rokiah. *Industrialization in Malaysia: Import Substitution and Infant Industry Performance.* London, Routledge, 1996.

Ariffin, Jamilah. *Women and Development in Malaysia.* Petaling Jaya, Pelanduk Publications, 1992.

Athukorala, Prema-chandra. *Crisis and Recovery in Malaysia: The Role of Capital Controls.* Cheltenham, Edward Elgar Publishing, 2001.

Baginda, Abdul Razak. *Malaysia in Transition: Politics, Economics and Society.* London, ASEAN Academic Press, 2002.

Barlow, Colin. *The Natural Rubber Industry.* Kuala Lumpur, Oxford University Press, 1978.

Modern Malaysia in the Global Economy: Political and Social Change into the 21st Century. Cheltenham, Edward Elgar Publishing, 2001.

Barlow, Colin, and Loh Kok Wah, Francis (Eds). *Malaysian Economics and Politics in the New Century.* Cheltenham, Edward Elgar Publishing, 2003.

Bunnell, Tim. *Malaysia, Modernity and the Multimedia Super Corridor: A Critical Geography of Intelligent Landscapes.* London, RoutledgeCurzon, 2004.

Chai, H. C. *The Development of British Malaya 1896–1909.* Kuala Lumpur, Oxford University Press, 1964.

Cho, George. *The Malaysian Economy: Spatial Perspectives.* London, Routledge, 1990.

Cooke, Fadzillah M. *The Challenge of Sustainable Forests: Forest Resource Policy in Malaysia, 1970–1995.* St Leonards, NSW, Allen and Unwin, 1999.

Drabble, J. H. *Rubber in Malaya 1876–1922.* London, Oxford University Press, 1973.

An Economic History of Malaysia, c.1800–1990: The Transition to Modern Economic Growth. Basingstoke, Palgrave Macmillan, 2000.

Faaland, J., Parkinson, J. R., and Saniman R. *Growth and Ethnic Inequality; Malaysia's New Economic Policy*. London, Hurst and Company, 1990.

Fisk, E. K., and Osman-Rani, H. (Eds). *The Political Economy of Malaysia*. Kuala Lumpur, Oxford University Press, 1982.

Gill, Ranjit. *The Making of Malaysia Inc.: A 25-year Review of the Securities Industry of Malaysia and Singapore*. London, ASEAN Academic Press, 2003.

Gomez, E. T. *Political Business; Corporate Involvement of Malaysian Political Parties*. Townsville, James Cook University of North Queensland, 1994.

Chinese Business in Malaysia: Accumulation, Ascendance, Accommodation. Honolulu, HI, University of Hawaii Press, 1999.

Gomez, E. T., and Jomo, K. S. *Malaysia's Political Economy: Politics, Patronage and Profits*. Cambridge, Cambridge University Press, 1997.

Gullick, John. *Malaysia: Economic Expansion and National Unity*. London, Ernest Benn, 1981.

Hill, R. D. *Rice in Malaya: A Study in Historical Geography*. Kuala Lumpur, Oxford University Press, 1977.

Hoffman, Lutz, and Tan, Siew Ee. *Review of Industrial Growth, Employment and Foreign Investment in Peninsular Malaysia*. Kuala Lumpur, Oxford University Press, 1980.

Jesudason, J. *Ethnicity and the Economy; The State, Chinese Business and Multinationals in Malaysia*. Kuala Lumpur, Oxford University Press, 1989.

Jomo, K. S. *Growth and Structural Change in the Malaysian Economy*. London, MacMillan, 1990.

U-Turn? Malaysian Economic Development Policies after 1990. Townsville, James Cook University of North Queensland, 1994.

Jomo, K. S. (Ed.). *A Question of Class: Capital, the State and Uneven Development in Malaysia*. Singapore, Oxford University Press, 1986.

Mahathir's Economic Policies. Petaling Jaya, INSAN, 1989.

Industrializing Malaysia; Policy, Performance, Prospects. London, Routledge, 1993.

Privatizing Malaysia: Rents, Rhetoric and Reality. Boulder, CO, Westview Press, 1994.

Japan and Malaysian Development: In the Shadow of the Rising Sun. London, Routledge, 1995.

Rethinking Malaysia. Hong Kong, Asia 2000, 1999.

Malaysian Eclipse: Economic Crisis and Recovery. London, Zed Books, 2001.

Jomo, K. S., Chang, K. T., and Khoo, K. J. *Deforesting Malaysia: The Political Economy and Social Ecology of Agricultural Expansion and Commercial Logging*. London, Zed Books, 2004.

Jomo, K. S., Felker, Greg, and Rajah, Rasiah. *Industrial Technology Development in Malaysia*. London, Routledge, 1999.

Kanpathy, V. (Ed.). *Managing Industrial Transition in Malaysia*. Kuala Lumpur, Institute of Strategic and International Studies, 1995.

Kanpathy, V., and Ismail Salleh (Eds). *Malaysian Economy: Selected Issues and Policy Directions*. Kuala Lumpur, Institute of Strategic and International Studies, 1994.

Khera, H. S. *The Oil Palm Industry of Malaysia: An Economic Study*. Kuala Lumpur, University of Malaya Press, 1976.

King, Victor T. *Tourism in Borneo: Issues and Perspectives*. Williamsburg, Borneo Research Council, 1994.

Lee, H. L. *Household Saving in West Malaysia and the Problem of Financing Economic Development*. Kuala Lumpur, Faculty of Economics and Administration, University of Malaya Press, 1971.

Public Policies and Economic Diversification in West Malaysia, 1957–70. Kuala Lumpur, University of Malaya Press, 1978.

Lee Kiong Hock, and Shyamala, Nagaraj (Eds). *The Malaysian Economy beyond 1990*. Kuala Lumpur, Persatuan Ekonomi Malaysia, 1991.

Lim, T. G. *Peasants and their Agricultural Economy in Colonial Malaya 1874–1941*. Kuala Lumpur, Oxford University Press, 1977.

McGee, T. G. *The Urbanization Process in the Third World*. London, Bell, 1971.

Mahathir Mohamad. *The Malaysian Currency Crisis: How and Why It Happened*. New York, NY, Weatherhill Publishers, 2003.

Malaysia: Economic Planning Unit.

Second Malaysia Plan 1971–75. Kuala Lumpur, Government Press, 1971.

Third Malaysia Plan 1976–80. Kuala Lumpur, Government Press, 1976.

Fourth Malaysia Plan 1981–85. Kuala Lumpur, Government Press, 1981.

Fifth Malaysia Plan 1986–90. Kuala Lumpur, Government Press, 1986.

Sixth Malaysia Plan 1991–95. Kuala Lumpur, Government Press, 1991.

Seventh Malaysia Plan 1996–2000. Kuala Lumpur, Government Press, 1996.

Eighth Malaysia Plan 2001–05. Kuala Lumpur, Government Press, 2001.

Mehmet, Ozay. *Development in Malaysia: Poverty, Wealth and Trusteeship*. London, Croom Helm, 1986.

Navaratnam, Ramon V. *Managing the Malaysian Economy: Challenges and Prospects*. Petaling Jaya, Pelanduk Publications, 1997.

Malaysia's Economic Sustainability: Confronting New Challenges Amidst Global Realities. Philadelphia, PA, Coronet Books, 2002.

Malaysia's Socioeconomic Challenges: Debating Public Policy Issues. Petaling Jaya, Pelanduk Publications, 2003.

Nyland, Chris, Smith, Wendy, Smyth, Russell, and Vicziany, Marika (Eds). *Malaysian Business in the New Era*. Cheltenham, Edward Elgar Publishing, 2001.

Rao, V. V. Bhanoji. *Malaysia: Development Pattern and Policy, 1947–71*. Singapore, Singapore University Press, 1980.

Searle, Peter. *The Riddle of Malaysian Capitalism—Rent-seekers or Real Capitalists?* Honolulu, HI, University of Hawaii Press, 1998.

Sieh Lee Mei Ling. *Taking on the World: Globalization Strategies in Malaysia*. Kuala Lumpur, McGraw Hill, 2000.

Silcock, T. H., and Fisk, E. K. (Eds). *The Political Economy of Independent Malaya*. Singapore, Eastern Universities Press, 1963.

Snodgrass, D. R. *Inequality and Economic Development in Malaysia*. Kuala Lumpur, Oxford University Press, 1980.

Sundaram, Jomo (Ed.). *Malaysian Eclipse—Economic Crisis and Recovery*. London, Zed Books, 2001.

Vincent, Jeffrey R., and Ali, Rozali Mohamed. *Managing Natural Wealth: Environment and Development in Malaysia*. Washington, DC, Resources for the Future, 2004.

World Bank. *Malaysia: Selected Issues in Rural Poverty*. Washington, DC, World Bank, 1980.

Yip, Y. H. *The Development of the Tin Mining Industry in Malaya*. Singapore, University of Malaya Press, 1969.

Young, Kevin, Bussink, Willem C. F., and Hasan, Parvez (Eds). *Malaysia: Growth and Equity in a Multiracial Society*. Baltimore, MD, and London, Johns Hopkins University Press, 1980.

MONGOLIA

Physical and Social Geography

ALAN J. K. SANDERS

Mongolia occupies an area of 1,564,116 sq km (603,909 sq miles) in east-central Asia. It is bordered by only two other states: the Russian Federation, along its northern frontier (extending for 3,543 km according to a new survey completed in December 2001), and the People's Republic of China, along the considerably longer southern frontier (4,676 km).

PHYSICAL ENVIRONMENT

For the purpose of geographical description Mongolia may be divided into five regions. In the west is the Altai area, where peaks covered with eternal snow rise to more than 4,300 m above sea-level. To the east of this lies a great depression dotted with lakes, some of salt water and some of fresh. Some of these, such as Uvs *nuur* (3,350 sq km) and Khövsgöl *nuur* (2,760 sq km) reach a considerable size. The north-central part of the country is occupied by the Khangai-Khentii mountain complex, enclosing the relatively fertile and productive agricultural country of the Selenge-Tuul basin. This has always been the focus of what cultural life existed in the steppes of north Mongolia: the imperial capital of Karakorum lay here and the ruins of other early settlements are still to be seen. To the east again lies the high Mongolian plateau reaching to the Chinese frontier, and to the south and east stretches the Gobi or semi-desert.

Water is unevenly distributed. In the mountainous north and west of the country large rivers originate, draining into either the Arctic or the Pacific. A continental watershed divides Mongolia, and the much smaller rivers of the south drain internally into lakes or are lost in the ground.

CLIMATE

The climate shows extremes of temperature between the long, cold, dry winter and the short, hot summer during which most of the year's precipitation falls. In Ulan Bator (Ulaanbaatar) the July temperature averages 17.0°C and the January temperature -26.1°C. Annual precipitation is variable but light. Ulan Bator's average is 233 mm, with 72.6 mm of rain in July. Rain is liable to fall in sudden, heavy showers or more prolonged outbursts in mid-summer, with severe flooding and damage to towns and bridges. Average snowfall in Ulan Bator in October–March is 16.9 mm. The bitter winter weather is relieved by the almost continuous blue sky and sunshine. Mongolia occasionally suffers severe earthquakes, especially in mountainous regions, but the population is too widely scattered for heavy losses to be caused.

POPULATION

Mongolia is mostly sparsely inhabited. The January 2000 census indicated a population of 2,373,493, or 1.5 persons per sq km; the population was officially estimated at 2,504,000 at the end of 2003. It is not correct to regard the Mongols as essentially nomadic herdsmen, although stock-movement (*otor*), sometimes covering large distances, has been a regular feature of rural life. Of the total population, more than 58% live in towns, with more than one-half of these, an estimated 893,400 at the end of 2003, residing in the capital, Ulan Bator. There are over 300 rural settlements, inhabited by about 22% of the population. The population is, relatively speaking, homogenous. Some 90% of the people are Mongols, and of these the overwhelming majority belong to the Khalkha (Halh) group. The only important non-Mongol element in the population is that of the Kazakhs, a Turkic-speaking people dwelling mostly in the far west, and representing approximately 4.3% (in 2000) of the whole. The population has grown steadily over recent years: between 1963 and 1983 it increased by 74%, but the annual growth rate peaked at 2.6% in 1990 and has declined since then to between 1.2% and 1.4%. As a result, there is a preponderance of young people: according to the January 2000 census, 66% of the population were under 30 years of age. According to official figures, infant mortality declined from 64.4 per 1,000 births in 1990 to 23 in 2003. The official language is Mongol, written in a native vertical script or for everyday purposes in modified Cyrillic. Mongol is quite different from both Russian and Chinese, its geographic neighbours, but does show certain similarities, perhaps fortuitous, to Turkish, Korean and Japanese. Several Mongol dialects beside the dominant Khalkha are spoken, and in the Kazakh province of Bayan-Ölgiy the first language is Kazakh, most people being bilingual in Mongol.

History

ALAN J. K. SANDERS

EARLY HISTORY

Today only a minority of ethnic Mongols live in Mongolia, the sole independent Mongol state. Besides the related Buryat and Kalmyk peoples who are to be found within the Russian Federation in their own republics (near Lake Baikal and on the lower Volga, respectively), many true Mongols dwell outside Mongolia, most of them in the Inner Mongolia Autonomous Region of the People's Republic of China and adjacent areas—Heilongjiang, Jilin, Liaoning, Gansu, Ningxia and Xinjiang.

This division came about in the following way: in the early 17th century the Manchus, expanding southwards from Manchuria towards their ultimate conquest of all China, passed through what came to be called Inner Mongolia, which lay across their invasion routes. Many of the Mongol princes allied themselves with the Manchus (sometimes reinforcing such alliances by marriage), others submitted voluntarily to them, while yet others were conquered. In 1636, after the death of Ligdan Khan (the last Mongol Emperor), the subordination of these princes to the new, rising dynasty was formalized. The princes of Khalkha or Outer Mongolia maintained a tributary relationship towards the Manchus for a further half-century but, in their turn, lost their independence at the Convention of Dolonnor in 1691. The Manchus had, in 1688, entered Khalkha to expel Galdan, the ruler of the West Mongol Oirats, who was both terrorizing the Khalkhas and challenging the Manchus for supremacy in this area. With Galdan defeated, the three great princes of Khalkha and the Javzandamba Khutagt, or Living Buddha of Urga, the head of the lamaist church in Mongolia, had to accept Manchu overlordship. The 1728 Treaty of Kyakhta (Khiagt) defined the western borders between the Russian and Manchu empires, and confirmed Manchu rule in Outer Mongolia and Tannu Tuva (Uriankhai).

Outer Mongolia was administered by the Manchus as a separate area from Inner Mongolia. A fourth princedom (*aimag*) was created in addition out of the existing three in 1725, and soon afterwards the princedoms were renamed leagues and removed from the jurisdiction of the hereditary princes, to be administered instead by Mongol league heads, appointed by the imperial Government in Beijing. Within the league organization Mongolia was divided into about 100 banners and a number of temple territories, while the Living Buddha owned a huge number of widely dispersed serfs. This state structure survived the fall of the Manchus in 1911 and lasted until the foundation of the Mongolian People's Republic (MPR) in 1924. In spite of their dependence on Beijing, however, the Mongols always considered themselves allies of the Manchus, not subjects on the same level as the Chinese, and made good use of this distinction when the Manchu (Qing) dynasty lost the throne of China.

AUTONOMOUS MONGOLIA

The beginnings of the existence of modern Mongolia can be traced back to 1911. In that year the fall of the Manchus enabled the Mongols to terminate their association with China. With some political and military support from Russia, a number of leading nobles proclaimed Mongolia an independent monarchy, and the throne was offered to the Living Buddha. The new Government, in an unrealistic excess of euphoria, invited all Mongols everywhere to adhere to the new state, but this involved them in conflict with China, which retained control of Inner Mongolia. Nor did they obtain much useful support from Russia, which was bound by secret treaties with Japan not to obstruct the latter's interests in Inner Mongolia, and which in any case was reluctant to engage in a doubtful pan-Mongolist adventure. In 1915 Russian, Chinese and Mongol representatives, meeting at Kyakhta on the Russo-Mongol border, agreed to the reduction of Mongolia's would-be independence to autonomy under Chinese suzerainty. At this time autonomous Mongolia consisted more or less of the territory of present-day Mongolia, the only substantial difference being the accession of Dariganga in the south-east at the time of the 1921 revolution. Inner Mongolia, Barga and the Altai district of Xinjiang were to remain under Chinese control. Tannu Tuva, after a brief period of autonomy as a 'people's republic', was absorbed by the USSR in 1944.

Autonomous Mongolia was a theocratic monarchy and during the few years of its existence very little happened to change the conditions inherited from Manchu times. Russian advisers began to modernize the Mongol army and to bring some sort of order into the fiscal system. Several primary schools and a secondary school were opened, some children (including the future dictator, Khorloogiin Choibalsan) were sent to study in imperial Russia, and the first newspaper appeared; but the state structure, the feudal organization of society and the administration of justice remained more or less as they had been, while the Buddhist clergy managed to consolidate and enhance its position of privilege. While legally subject to Chinese suzerainty, Mongolia was, in fact, a Russian protectorate. When Russian power and prestige in Central Asia were sapped by the collapse of the tsarist regime and the outbreak of revolution in 1917, this dependence of Mongolia became very apparent, and China lost no time in reasserting its authority. By mid-1919 the abrogation of autonomy was being discussed by the Mongol Government and the Chinese resident in Urga, the capital, but the process was brutally accelerated by the arrival in Mongolia of Gen. Xu Shuzeng who, with a large military force at his disposal, forced the Mongols to relinquish all authority to the Chinese in February 1920.

THE REVOLUTIONARY MOVEMENT

Towards the end of 1919 two revolutionary groups had been founded in Urga; the next year these amalgamated to form the Mongolian People's Party (MPP). There was no long-standing revolutionary tradition in Mongolia, which perhaps explains how it was that the Mongol revolution fell so completely under Soviet control. The members of the groups were men of varied social origin, including lamas (such as Dogsomyn Bodoo, the premier of 1921, who was liquidated in 1922), government servants, workers, soldiers (such as Damdiny Sükhbaatar) and students who had returned from Russia (such as Khorloogiin Choibalsan). They had the sympathy of several prominent nobles through whom they were able to approach the King, while at the same time they acquired some knowledge of Marxism from their acquaintance with left-wing Russian workers in Urga.

The first real contacts with Soviet Russia took place in early 1920 when a Comintern agent, Sorokovikov, came to Urga to assess the situation. It is therefore not surprising to find that the aims of the revolutionaries were at this time fairly moderate. First of all they desired national independence from the Chinese, then an elective government, internal administrative reforms, improved social justice, and the consolidation of the Buddhist faith. With Sorokovikov's approval they planned to send a delegation to Russia to seek help against the Chinese. They obtained the sanction of the King and carried with them a letter authenticated with his seal. They were, in fact, authorized only to obtain advice from Russia, not to negotiate actual intervention.

In their absence from Mongolia the situation was complicated by the incursion into the country of White Russian forces under Baron von Ungern-Sternberg. At first the Mongol authorities and the people welcomed the White Russians who dislodged the oppressive Chinese, and with the help of Ungern, the King was restored to the throne. However, Ungern's brutalities soon turned the Mongols against him. More important, the Soviet agents dealing with the Mongol delegation were able to use Ungern's apparent ascendancy over the Urga regime to extract far-reaching concessions. They made the offer of help conditional upon the establishment in Urga later of a new government friendly to them.

In March 1921 the first Congress of the MPP was held at Kyakhta on Soviet territory, and a provisional revolutionary Government was formed there, in opposition to the legal authorities in Urga who had sponsored the delegates who now abandoned them. This provisional Government gathered a small band of partisans who, with substantial Soviet forces, entered Mongolia, defeated Ungern and then marched on Urga. Here, in July 1921, a new Government was proclaimed, under the restored King. The monarchy existed now, however, in name only. Mongolia came more and more under Soviet direction. A secret police force was set up and in 1922 the first of a long series of political purges took place. In 1924 the King died. The MPP was officially renamed the Mongolian People's Revolutionary Party (MPRP), and a People's Republic, with a Soviet-style Constitution, was proclaimed.

THE MONGOLIAN PEOPLE'S REPUBLIC

Mongolia was now, in name, a people's republic, the second socialist state in the world, but its primitive stage of development posed daunting problems. Buddhism, which commanded deep loyalty from the people, weighed heavily on the economy and was a powerful ideological opponent of communism. Local separatism, especially in the far west, took years to overcome and in some outlying parts local government could not be established until 1928 or 1929. Moreover, it was easy for disillusioned herdsmen to trek with their herds over the frontiers into China and there were considerable losses of population by emigration. There was widespread illiteracy and many of those who could read and write were lamas whose skill was in Tibetan rather than Mongol.

The country's economy depended exclusively on extensive animal herding. Trade and crafts were in the hands of foreigners, almost all of them Chinese. There was no banking system, no national currency, no industry and no medical service in the modern sense. Finally, most of those men who were politically experienced and capable of running the local administration were lamas or nobles, two classes at whose eventual annihilation the revolutionary regime aimed.

Thus, the stage of economic, social and intellectual development which Mongolia had reached was far below that of the USSR and its capacity for independent action was extremely limited against its one international partner, the immeasurably more powerful USSR. Mongolia was ineluctably involved with Soviet interests and developments, and its history over the next two decades shows the same progression of events as characterized Stalin's USSR.

At first, until 1928, there was some measure of semi-capitalist development, during which the privileges of the nobility and clergy were not seriously curtailed. In international contacts, too, the Mongols reached out to France and Germany. However, parallel with the rise of Stalin and the swing to the left in the USSR there developed in Mongolia what came to be known as the 'leftist deviation'. All foreign contacts other than with the USSR were terminated. The USSR monopolized Mongolia's trade, in which it had hitherto had only a modest share. Between 1929 and 1932 an ill-prepared programme of collectivization ruined the country's economy, stocks of cattle falling by at least one-third. A rigorous anti-religious campaign did much to turn people against the MPRP, and in 1932 uprisings broke out which, particularly in West Mongolia, reached the proportions of civil war and necessitated the intervention of the Soviet army. Thousands of Mongols deserted the country with their herds. This disastrous course was reversed only on the direct instructions of the Comintern in June 1932. Leaders who until then had been enthusiastic leftists, such as Peljidiin Genden, who became Chairman of the Council of Ministers (Prime Minister) and was later 'unmasked' as a 'Japanese spy' and liquidated by Choibalsan, now adopted a more moderate line; and, under what was termed the 'New Turn Policy', private ownership of cattle and private trade were again encouraged, and Buddhism was treated more leniently.

However, from 1936 onwards, Mongolia fell under the dictatorship of Marshal Choibalsan (died 1952), whose methods were indistinguishable from those of Stalin. Buddhism was largely destroyed, with much loss of life and property, and most of the former leadership of revolutionaries, politicians, high military officers and intellectuals were liquidated on charges, usually of treasonable plotting with the Japanese, which have since been acknowledged to have been quite false. Thus, Choibalsan declared in 1940 that Mongolia could begin the transition from the 'democratic stage' of the revolution to the socialist stage.

The progress made by 1940 had been mostly negative, consisting in the elimination of old social groupings and the redistribution of wealth confiscated from the former nobles, liquidated in and after 1929, and the clergy. A certain amount of reconstruction had been achieved, in the fields of education, medical services, communications and industry, but it was not until well after the Second World War that any extensive programme of modernization was to be attempted in Mongolia. One reason for this tardiness was the threat posed by the Japanese in Manchuria, which meant that most of the Soviet expenditure in Mongolia was devoted to a military build-up. It is significant that the only railway to be built in pre-war years served the town of Bayantümen, renamed Choibalsan, in eastern Mongolia. Only after the war was Mongolia's main economic region, the area around Ulan Bator, to be connected with the Trans-Siberian railway line.

POLITICAL DEVELOPMENTS SINCE 1945

Mongolia escaped the worst of the Second World War, though not without suffering some effects. The Japanese in Manchuria had for some years been probing the defences of Mongolia, and in mid-1939 they provoked a series of battles on the Khalkha River (Khalkhyn Gol) in which they were heavily defeated by Soviet and Mongol troops. From then on a truce reigned until August 1945, when Mongolia followed the USSR in declaring war on Japan. Mongol forces advanced as far as the Pacific coast of China, but were soon afterwards withdrawn, and the only advantage Mongolia drew from its belated participation in the war was the labour of a number of Japanese prisoners. Imports from the USSR almost ceased during the war years, and Mongolia made a heavy contribution to the Soviet war effort, although it was never at war with Germany. As a result, there was practically no economic progress.

Following the allied Powers' agreement in Yalta to preserve the *status quo* in Mongolia, a plebiscite in October 1945 confirmed the country's wish for independence, which was recognized by China in January 1946. However, Mongolia's international position of isolation, in sole dependence on the USSR, did not change until the communization of Eastern Europe and the success of the communists in China provided it with a new and ready-made field of diplomatic activity. Between October 1948 and March 1950 it exchanged diplomatic recognition with all the then existing communist states except Yugoslavia, and thereafter with a number of non-aligned countries such as India, Burma (Myanmar) and Indonesia. The United Kingdom was the first Western European state to recognize Mongolia (1963). Mongolia was admitted to the United Nations in 1961.

Mongolia continued to look mainly to the USSR for guidance and help in its affairs, in spite of its widening international contacts. In 1946 the traditional alphabet was abandoned in favour of a form of the Cyrillic script. Mongolia's alignment with the USSR in the Sino-Soviet dispute was predictable, and the official press continued to adopt an uncompromising anti-Beijing line. China was accused, among other things, of carrying out a colonialist policy in its minority areas, including Inner Mongolia, and of openly preparing for war with the USSR and Mongolia. Soviet troops were stationed in the MPR at the Mongolian Government's request, because of the 'real threat' of Chinese 'great-power expansion'.

By 1986, however, Mongolia's relations with China appeared to have improved significantly, with a visit to Ulan Bator by the Chinese Vice-Minister of Foreign Affairs and the subsequent signing of an agreement on consular relations between the two countries. Mongolia's position as a 'buffer' state between China and the USSR was illustrated in July, when the Soviet leader, Mikhail Gorbachev, offered to withdraw some of the Soviet troops stationed in Mongolia, as a step towards the normalization of relations between Moscow and Beijing. A partial withdrawal (of about 20% of the estimated total) took place between April and June 1987, and a second stage began in May 1989. Following a series of high-level Mongolian-Chinese negotiations, Mongolia and China subsequently declared the normalization of relations, and in May 1990 the Mongolian Head of State, Punsalmaagiin Ochirbat, paid a short visit to Beijing. The final stage of Soviet troop withdrawals was completed in September 1992.

The partial *détente* which was initiated in the USSR by Nikita Khrushchev in the 1950s and 1960s was imitated in Mongolia, where several of the leaders who had been executed in the 1930s were 'rehabilitated'. Contacts with non-communist foreigners were permitted, a small tourist industry was developed and controls on publications were slightly relaxed. A feature of this period was the reassertion of feelings of Mongolian nationalism, which for 20 years had been repressed. Since 1936 the existence of pre-revolutionary culture in Mongolia had been systematically denied. Nothing of ancient Mongol literature was taught in schools, no old books were reprinted and manuscripts considered to be contrary to contemporary ideology were destroyed. After 1956 this policy was modified. School curricula, while still insisting that children be given a communist education, were liberalized to the extent that they included the study of extracts from ancient literature once more. The Committee (later Academy) of Sciences was able to begin a programme of research and publication in the fields of literature, history and linguistics, and to organize in 1959 the First International Congress of Mongolists. This was the first occasion on which scholars from the Western world, the Soviet bloc and China conferred together in Mongolia.

This renascence of national sentiment was rebuffed from time to time when it clashed with Soviet requirements of greater international communist conformity, as in 1962 when the Mongols celebrated the 800th anniversary of the birth of Genghis Khan. The enthusiasm provoked in Mongolia was regarded by Moscow, and in more orthodox quarters in Mongolia itself, as manifesting excessive feelings of nationalism at the expense of 'proletarian internationalism' and the celebrations were abruptly cancelled. In early 1963 an ideological conference was held in Ulan Bator with the participation of a strong Soviet delegation, in order to reassert the correct political line. It was not until the early 1990s that Genghis Khan was officially rehabilitated; President Ochirbat referred to him in 1992 as 'a national hero and the pride of the country'.

In June 1974 Yumjaagiin Tsedenbal, the Chairman of the Council of Ministers since 1952, became Chairman of the Presidium of the People's Great Khural (Head of State), succeeding Jamsrangiin Sambuu, who had died in May 1972. The new

Chairman of the Council of Ministers, Jambyn Batmönkh, was a comparative newcomer to political life. Tsedenbal was concurrently First Secretary of the MPRP Central Committee from 1958 (from 1981 General Secretary), and had been General Secretary of the party in 1940–54.

In August 1984, following an extraordinary session of the MPRP Central Committee, Tsedenbal, who was on holiday in the USSR, was unexpectedly replaced as General Secretary of the Central Committee by Jambyn Batmönkh. He was also removed from the Politburo and relieved of the post of Chairman of the Presidium of the People's Great Khural—ostensibly owing to ill health and with his full agreement. In December 1984 Batmönkh was elected to the post of Chairman of the Presidium of the People's Great Khural, while Dumaagiin Sodnom was elected to membership of the Politburo and the post of Chairman of the Council of Ministers.

In late 1986 government ministries responsible for agriculture, water supply and construction were reorganized, and more extensive restructuring took place in late 1987–early 1988, with the aim of improving the efficiency and productiveness of the country's economy.

In November 1988 the MPRP Politburo, obliged to admit that economic renewal was not succeeding because of the need for social reforms, proposed wide-ranging improvements in procedures for elections to party and legislative offices, and other changes in the name of 'democratization', *il tod* (openness) and *öörchlön shinechlelt* (renewal). The proposals were reported to have received widespread public approval.

THE BIRTH OF DEMOCRACY

Between December 1989 and March 1990 there was a great upsurge in public political activity, as several newly formed opposition movements organized a series of peaceful demonstrations in Ulan Bator, demanding political and economic reforms. The most prominent of these groups was the Mongolian Democratic Association (MDA), which was founded in December 1989. In January 1990 dialogue was initiated between MPRP officials and representatives of the MDA, including its chief co-ordinator, Sanjaasürengiin Zorig (a lecturer at the Mongolian State University).

The emergence of further opposition groups, together with escalating public demonstrations (involving as many as 20,000 people), led to a crisis of confidence within the MPRP. At a party plenum, held in mid-March 1990, Batmönkh announced the resignation of the entire Politburo as well as of the Secretariat of the Central Committee. Gombojavyn Ochirbat, a former head of the Ideological Department of the Central Committee and a former Chairman of the Central Council of Mongolian Trade Unions, was elected the new General Secretary of the party, replacing Batmönkh. A new five-member Politburo was formed. Delegates at the plenum voted to expel the former MPRP General Secretary, Yumjaagiin Tsedenbal, from the party and to rehabilitate several prominent victims of Tsedenbal's purges of the 1960s.

At a session of the People's Great Khural, which was held shortly after the MPRP plenum, the senior positions in the Presidium were reorganized. The People's Great Khural also adopted amendments to the Constitution, including the removal of references to the MPRP as the 'guiding force' in Mongolian society, approved a new electoral law and brought forward the date of the next general election from 1991 to 1990.

In April 1990 an extraordinary congress of the MPRP was held, at which more than three-quarters of the membership of the Central Committee was renewed. General Secretary Gombojavyn Ochirbat was elected to the restyled post of Chairman of the party. The Politburo was renamed the Presidium, and a new four-member Secretariat of the Central Committee was appointed.

In May 1990 the People's Great Khural approved a law on political parties, which legalized the new 'informal' movements through official registration; it also adopted further amendments to the Constitution, introducing a presidential system with a standing legislature called the State Little Khural, elected by proportional representation of parties.

At the July 1990 general election and subsequent re-elections, 430 deputies were elected to serve a five-year term: 357 from the MPRP (in some instances unopposed), 16 from the Mongolian Democratic Party (MDP, the political wing of the MDA), nine from the Mongolian Revolutionary Youth League, six from the Mongolian National Progress Party (MNPP), four from the Mongolian Social-Democratic Party (MSDP) and 39 without party affiliation. Under the constitutional amendments adopted in May, the People's Great Khural was required to convene at least four times in the five years of its term.

In September 1990 the People's Great Khural elected Punsalmaagiin Ochirbat to be the country's first President, with a five-year term of office; the post of Chairman of the Presidium lapsed and Jambyn Gombojav was elected Chairman (Speaker). Radnaasümbereliin Gonchigdorj of the MSDP was subsequently elected Chairman of the State Little Khural, *ex officio* becoming Vice-President of Mongolia. Dashiin Byambasüren was appointed Prime Minister (formerly the post of Chairman of the Council of Ministers) and began consultations on the formation of a multi-party government. The newly restyled Cabinet was elected by the State Little Khural in September and October. Under the amended Constitution, the President, Vice-President and Ministers were not permitted to remain concurrently deputies of the People's Great Khural; therefore, re-elections of deputies to the People's Great Khural took place in mid-November.

The 20th Congress of the MPRP, which was held in February 1991, elected a new 99-member Central Committee, which, in turn, appointed a new Presidium. The Central Committee also elected a new Chairman, Büdragchaagiin Dash-Yondon, the Chairman of the Ulan Bator City Party Committee, who had become a Presidium member in November 1990.

A new Constitution was adopted by an almost unanimous vote of the Great Khural in January 1992 and entered into force in the following month. It provided for a unicameral Mongolian Great Khural, comprising 76 members, to replace the People's Great Khural following legislative elections, to be held in June. The State Little Khural was abolished. The country's official name was changed from the Mongolian People's Republic to Mongolia and the communist gold star was removed from the national flag.

At the elections to the Mongolian Great Khural in June 1992, a total of 293 candidates stood in 26 constituencies, comprising the 18 *aimag* (provinces), the towns of Darkhan and Erdenet, and Ulan Bator City (six). The constituencies had two, three or four seats, according to the size of the local electorate. The MPRP presented 82 candidates, compared with 51 put forward by an alliance of the MDP, the MNPP and the United Party (UP), and 30 by the MSDP; six other parties and another alliance also took part, although with fewer candidates.

A total of 1,037,392 voters (95.6% of the electorate) participated in the elections, although 62,738 ballots were declared invalid. Candidates were elected by a simple majority, provided that they obtained the support of at least 50% of the electorate in their constituency. The MPRP candidates received altogether 1,719,887 votes (some 57%), while the candidates of the other parties (excluding independents) achieved a combined total of 1,205,350 votes (40%), of which the MDP-MNPP-UP alliance won 521,883 votes and the MSDP 304,548. The outcome of the election was disproportionate, however, the MPRP taking 70 seats (71, if a pro-MPRP independent was included). The remaining seats went to the MDP (two, including an independent), the MSDP, the MNPP and the UP (one each).

The first session of the Mongolian Great Khural opened in July 1992 with the election of officers, the nomination of Puntsagiin Jasrai (who had been a Deputy Chairman of the Council of Ministers and a candidate member of the MPRP Politburo at the end of the communist period) to the post of Prime Minister, and the approval of his Cabinet. Natsagiin Bagabandi, a Vice-Chairman of the MPRP Central Committee, was elected Chairman of the Great Khural. Jambyn Gombojav (Chairman of the People's Great Khural from late 1990 to late 1991) was elected Vice-Chairman of the new Khural. Meanwhile, in June, a National Security Council was established, with the country's President as its Chairman, and the Prime Minister and Chairman of the Great Khural as its members.

In October 1992 the MDP, the MNPP, the UP and the Mongolian Renewal Party amalgamated to form the Mongolian National Democratic Party (MNDP), with a General Council headed by the MNPP leader, Davaadorjiin Ganbold, and including Sanjaasürengiin Zorig and other prominent opposition politicians. In the same month the MPRP Central Committee was renamed the MPRP Little Khural, and its membership was increased to 169 (subsequently to 198). The Presidium was replaced by a nine-member Party Leadership Council.

Political life in Mongolia during the first half of 1993 was dominated by the country's first direct presidential election, held on 6 June. Apparently dissatisfied with the increasingly independent line adopted by the incumbent President, Punsalmaagiin Ochirbat, and angered by presidential vetoes on legislation proposed by the Government, the MPRP Little Khural decided not to support Ochirbat, who had been an MPRP member, and nominated Lodongiin Tüdev as its candidate. Meanwhile, Ochirbat received the nomination of the organizationally weaker opposition coalition of the MNDP and the MSDP. The MPRP expected to win by imposing party discipline on its more numerous supporters, but miscalculated. The outcome of the election was a victory for Ochirbat: 57.8% of the vote, as against 38.7% for Tüdev.

Amendments to the Election Law introduced in early 1996 increased the number of constituencies for election to the Great Khural from 24 to 76; all would be single-seat constituencies, with representatives elected by the majority vote system. The parliamentary opposition parties, the MNDP and MSDP, supported by the Mongolian Green Party and the Mongolian Believers' (Buddhist) Democratic Party, formed a coalition, the Democratic Alliance, to contest the general election of June 1996.

THE END OF COMMUNIST RULE

In the general election held on 30 June 1996 the Democratic Alliance confounded most observers by winning 50 of the 76 seats in the Great Khural. The ruling MPRP took only 25 seats, while one seat went to a candidate of the pro-MPRP United Heritage Party (UHP). A total of 1,057,182 voters (officially 92.15% of the electorate) participated in the elections (47,022 ballots were spoiled). Although official nation-wide totals were not published, it can be calculated from constituency returns that the Democratic Alliance polled 469,586 votes (46.67%), the MPRP 408,977 (40.64%), and other parties and independents 127,684 (12.69%); the last figure included nearly 4,000 ballots which were blank but ruled as valid votes for no candidate.

The first session of the newly elected Great Khural opened in mid-July 1996 amidst confusion. The election of the MSDP leader, Radnaasümbereliin Gonchigdorj (who had been Vice-President of Mongolia during 1990–92), to the post of Chairman of the Great Khural passed without incident. The Democratic Alliance's choice of Prime Minister, Mendsaikhany Enkhsaikhan (head of the presidential secretariat), was nominated by President Ochirbat and voted into office by the Great Khural. However, the MPRP had issued a list of demands, including the allocation of the vice-chairmanship of the Khural and two important standing committee chairmanships to MPRP members, and when these demands were rejected by the Democratic Alliance, MPRP members walked out of the Great Khural, leaving it inquorate and unable to function.

The boycott of the Great Khural by the MPRP lasted three days. The Khural elected the MNDP leader, Tsakhiagiin Elbegdorj, to be its Vice-Chairman; thereafter, the political confrontation focused on the six standing committees, whose chairmen were all members of the Democratic Alliance. Moreover, the Democratic Alliance declared that it intended to remove MPRP officials from all important posts in the administration.

Eight members of Prime Minister Enkhsaikhan's Government were presented to the Great Khural on 30 July 1996 and voted into office. The selection process had been delayed by a ruling of the Constitutional Court (MPRP appointees) that no member of the Government could remain a member of the Great Khural; the ruling was later overturned by the Great Khural. In late July 1996 the MPRP Little Khural elected a new General Secretary, Nambaryn Enkhbayar, the former Minister of Culture, and a new Leadership Council.

In local government elections, held in October 1996, the Democratic Alliance won 208 seats at provincial council level, while the MPRP achieved 142. However, the Democratic Alliance failed to gain control of Ulan Bator City Council (MPRP 23 seats, MNDP nine, MSDP six, Mongolian Green Party one). At rural and urban district council level, the MPRP won 3,660 seats, the Democratic Alliance secured 3,169 and other parties 29 seats.

It became clear that the Democratic Alliance was losing support to the MPRP in its efforts to promote privatization and the development of a market economy in the face of industrial stagnation, increasing poverty and unemployment. For the presidential election of 18 May 1997 the Democratic Alliance nominated the incumbent President, Punsalmaagiin Ochirbat, as its candidate. The MPRP put forward the former Chairman of the Great Khural (1992–96), Natsagiin Bagabandi, who had been elected MPRP Chairman in February. The third party in the Mongolian Great Khural, the UHP, nominated Jambyn Gombojav, who had defected from the MPRP after Bagabandi's election, altering the balance of power in the Great Khural. The election was won by Bagabandi, with 60.8% of the votes, compared with 29.8% for Ochirbat and 6.6% for Gombojav, a result which many observers saw as an expression of popular dissatisfaction with the Democratic Alliance. Meanwhile, the chairmanship of the MPRP reverted to Nambaryn Enkhbayar, who, in mid-August, won the by-election for Bagabandi's seat in the Great Khural.

A CRISIS OF CONFIDENCE

In April 1998 the Democratic Alliance decided that the Cabinet would be headed by the Alliance's leader and, unlike the Enkhsaikhan Cabinet, would comprise members of the Great Khural. Thus, Tsakhiagiin Elbegdorj, leader of the MNDP, was appointed Prime Minister, although the formation of the new Cabinet took a month. The policies of the Elbegdorj Cabinet did not differ much from those of its predecessor, although it sought to project a reformist image.

Soon afterwards the Government became embroiled in a dispute over its amalgamation of the state-owned Reconstruction Bank (which had been declared bankrupt after having over-extended its credit) with the private Golomt Bank. Although various authorities, including the IMF and the Asian Development Bank (ADB), approved this amalgamation, some Democratic Alliance members of the Great Khural were unsure about it, while the opposition MPRP bitterly opposed it. Amidst accusations that Democratic Alliance leaders and ministers had obtained loans from the Bank just before it collapsed, the MPRP resorted to another boycott of the Great Khural. When the spring session closed (in August 1998), although the Khural had sat for 103 days, 20 working days had been lost, thus delaying important legislation, including finance bills.

Insisting that the Democratic Alliance Cabinet should resign, the MPRP refused to accept a government compromise but returned to the Great Khural to pursue a vote of 'no confidence' in the Government. The vote of 'no confidence' was carried on 24 July 1998 by 42 votes to 33, with 15 members of the Democratic Alliance 'crossing the floor'. The three-month-old Elbegdorj Government resigned, leaving the Democratic Alliance 30 days to choose a new Prime Minister.

Meanwhile, a new altercation arose over the management of the Mongolian-Russian copper-mining joint venture at Erdenet. In February 1998 the Enkhsaikhan Government had become embroiled in a dispute over the reappointment of the Mongolian director-general of the Erdenet enterprise, Shagdaryn Otgonbileg. For many years a member of the MPRP Little Khural, he had been in charge of the enterprise from its inception. The Russians wanted him to be maintained in the post, but Enkhsaikhan refused to reappoint him on the grounds that he had attempted to privatize part of the enterprise illegally. However, President Bagabandi intervened on the Russian side and Otgonbileg's contract was extended.

When Otgonbileg's contract finally ended, he refused to attend the ceremony marking the official hand-over to his successor—the former Minister of Defence, Dambyn Dorligjav—but sought a court ruling that his 'dismissal' was illegal. The court's decision in his favour was overturned on appeal by the State Property Committee.

The Elbegdorj Cabinet established a government commission and imposed a 'special regime' on the Erdenet enterprise. President Bagabandi then stated that the Government could not impose such a 'special regime' on the enterprise unilaterally, without consulting the Russian co-directors. The government commission accused Otgonbileg of criminal negligence and appealed to the President to withdraw his support for the Russian position. President Bagabandi retorted that the 'special regime' had been imposed in disregard of his opinion, and he issued a statement criticizing anti-Russian reports in the 'official media'. Various MPRP leaders joined him in condemning the 'politicization' of the Erdenet affair as harmful to relations with the 'northern neighbour'. Later in the year a meeting of the full board of Erdenet approved Dorligjav's appointment and the 'special regime' was terminated.

In mid-August 1998 the Democratic Alliance nominated as their choice for the next Prime Minister Davaadorjiin Ganbold, the Chairman of the Economic Standing Committee of the Great Khural, who had served as First Deputy Chairman of the Government in 1990–92 and President of the MNDP in 1992–96. However, President Bagabandi refused to accept Ganbold's nomination, on the grounds that he had done nothing as Chairman of the Economic Standing Committee to resolve the bank merger crisis. Ganbold was nominated a second time, but the President rejected him again. The Democratic Alliance protested that the President had no constitutional right to reject its nomination, but he did so yet again and put forward his own nominee, Dogsomyn Ganbold, whom the Democratic Alliance ignored. After Davaadorjiin Ganbold's nomination had been rejected a sixth time by the President, the Democratic Alliance presented a new nominee, the acting Minister of External Relations, Rinchinnyamyn Amarjargal. His nomination was accepted by President Bagabandi on 31 August 1998 but rejected by the Great Khural on the following day by a majority of only one vote. After a further period of delay, the Democratic Alliance nominated a new candidate, Galsangiin Gankhuyag, who was rejected by the President on the grounds that he might face charges of drunken driving. In late September Bagabandi rejected a further nominee, Erdeniin Bat-Üül, who had been replaced as Vice-President of the MNDP by Davaadorjiin Ganbold in August. On 2 October Sanjaasürengiin Zorig, the Minister of Infrastructure Development, was murdered at his home. Zorig, the founder of the Mongolian democratic movement, had been widely seen as a potential candidate for the post of Prime Minister, although he had not been nominated.

After Zorig's state funeral, President Bagabandi issued the names of six more candidates of his own, including Dogsomyn Ganbold and the Mayor of Ulan Bator, Janlavyn Narantsatsralt. The Democratic Alliance ignored the presidential list and, for the seventh time, nominated Davaadorjiin Ganbold. Although the nomination was supported by all 48 Democratic Alliance members of the Great Khural, the President again rejected him. Later the same month the political crisis was deepened by the Constitutional Court's latest ruling, reaffirming that members of the Great Khural could not serve concurrently in the Government. The ruling overturned an amendment to the Law on the Status of Great Khural Members adopted in January 1998. Two months later the Democratic Alliance finally nominated Bagabandi's candidate, Janlavyn Narantsatsralt, who was appointed Prime Minister in early December. The formation of his Government was completed with the appointment of the last four ministers in mid-January 1999.

Narantsatsralt's Government remained in power for just over six months. In July 1999 the Prime Minister was challenged in the Great Khural over a letter that he had written in January to Yurii Maslyukov, First Deputy Chairman of the Russian Government, in which he seemingly acknowledged Russia's right to privatize its share in the Erdenet joint venture without reference to the Mongolians. Unable to offer a satisfactory explanation, in late July Narantsatsralt lost a vote of confidence, in which MSDP members of the Great Khural voted with the opposition MPRP. The Democratic Alliance nominated Rinchinnyamyn Amarjargal for the post of Prime Minister, but the proposal was immediately challenged by President Bagabandi. The President insisted that, following the Constitutional Court ruling of late 1998, he could consider Amarjargal's suitability for nomination in the Great Khural only after the candidate had resigned his seat. After several days of arguments, representatives of the Democratic Alliance and the President adopted a formula that allowed the Great Khural's approval of the prime ministerial nomination and the nominee's resignation of his Great Khural seat to take place simultaneously. Amarjargal was elected Prime Minister at the end of July. The ministers of Narantsatsralt's Government remained in office in an acting capacity until early September, when all but one (the Minister of Law) were reappointed. In November Amarjargal replaced Narantsatsralt as President of the MNDP.

The 1992 Constitution was amended for the first time in December 1999 by a Mongolian Great Khural decree, supported by all three parliamentary parties, which *inter alia* simplified the procedure for the appointment of the Prime Minister and allowed members of the Great Khural to serve as government ministers while retaining their seats in the legislature. However, the President vetoed the decree, stating that the amendments (due to come into force from 15 July) could not be approved by the Great Khural alone, without public consideration of his opinion and that of the Constitutional Court. The presidential veto was rejected by the Great Khural in January 2000, but in March a five-member session of the Constitutional Court ruled that the decree had been unconstitutional. When the Great Khural opened its spring session in April, members rejected the ruling and refused to discuss it. The Constitutional Court's demand for a statement on the issue was disregarded by the Great Khural.

As the general election approached, party political activity increased dramatically. A breakaway grouping of the MNDP reconstituted the Mongolian Democratic Party, and a faction of the MSDP founded the Mongolian New Social Democratic Party. Sanjaasürengiin Oyuun, the sister of the murdered minister, Zorig, established the Civil Courage Party (CCP, or Irgenii Zorig Party) drawing away from the MNDP several more members of the Great Khural, and formed an electoral alliance with the Green Party. The MNDP, unable to reconstitute the previously-successful Democratic Alliance with the MSDP, therefore formed a new Democratic Alliance with the Mongolian Believers' Democratic Party. A grand alliance of nine non-parliamentary parties was quickly reduced to three—the Democratic Renewal Party, Mongolian Traditional United Party (UHP) and For Mongolia Party—following a decision by the Mongolian Democratic New Socialist Party (MDNSP) and the Mongolian Republican Party (MRP, headed by wealthy businessmen, a new feature of Mongolian politics) to present their own candidates.

THE RETURN TO POWER OF THE MPRP

At the election, held on 2 July 2000, three coalitions and 13 parties were represented by a total of 603 candidates, including 27 independents. The MPRP took 72 of the 76 seats in the Great Khural. Prime Minister Rinchinnyamyn Amarjargal and his entire Cabinet lost their seats. The MPRP received 50.2% of the votes cast, the level of participation being 82.4% of the electorate. The party's victory was attributed to widespread popular support for its policy of social welfare and poverty reduction, and to the disintegration of the MNDP-MSDP coalition. Moreover, in the more numerous rural constituencies, where its main support lay, the MPRP was widely seen as willing and able to put an end to the economic and social stagnation of the countryside.

The four seats not taken by the MPRP went to Sanjaasürengiin Oyuun (President of the CCP); Badarchiin Erdenebat (Chairman of the MDNSP); Lamjavyn Gündalai (Independent), a businessman from Khövsgöl Province; and ex-Prime Minister Janlavyn Narantsatsralt (MNDP). The Democratic Alliance, which presented 71 candidates, received 13% of the votes cast; the MDNSP, with 73 candidates, received 10.7% of the votes cast. The 67 MSDP candidates received 8.9% of votes cast, but won no seats.

When the new Great Khural opened, Lkhamsürengiin Enebish, the MPRP General Secretary, was elected to the post of Chairman (Speaker). However, the nomination of the MPRP Chairman, Nambaryn Enkhbayar, for the post of Prime Minister was rejected by President Natsagiin Bagabandi, on the grounds that priority be given to the constitutional amendments. After a week of discussion a compromise was reached whereby Enkhbayar's nomination was presented to the Great Khural, while the amendments remained in force pending a Great Khural debate and a full nine-member session of the

Constitutional Court. On 26 July 2000 the Great Khural approved Prime Minister Enkhbayar's appointment by 67 MPRP members' votes to three. It approved the membership of Enkhbayar's Cabinet on 9 August.

Reflecting the MPRP's emphasis on social issues, Enkhbayar divided the former Ministry of Health and Social Welfare into two separate ministries, the Ministry of Health and the Ministry of Social Protection and Welfare. The Ministry of Law became the Ministry of Justice and Home Affairs and took charge of the border troops. Furthermore, the MSDP alleged that the new Government was acting in violation of the spirit of the Constitution, by dismissing large numbers of civil servants because of their party affiliation. On 28 September 2000 Rinchinnyamyn Amarjargal, leader of the MNDP, and Radnaasümbereliin Gonchigdorj, leader of the MSDP, signed a joint declaration announcing that a conference of the parties would be held on 6 December, to formalize the merging of the two parties. This was done with a view to ensuring the necessary parliamentary basis for Gonchigdorj's nomination in the 2001 presidential election. The two parties formed a new Coalition of Democratic Forces (together with four smaller parties) to contest the provincial and Ulan Bator local government elections held on 1 October. None the less, the MPRP won 552 of the 695 seats available in the city and *aimag* (provincial) khurals, a voter participation rate of 60.75% being recorded.

At a conference held on 6 December 2000, altogether five democratic parties—the MNDP, MSDP, Mongolian Democratic Party, Believers' Democratic Party, and Democratic Renewal Party—resolved to disband themselves and form a new Democratic Party (DP). Dambyn Dorligjav, the former Minister of Defence and director of the Erdenet copper combine, was elected Chairman, while ex-Prime Minister Janlavyn Narantsatsralt and former Minister of the Environment Sonomtserengiin Mendsaikhan were elected Vice-Chairmen. When registered on 26 December the DP claimed a membership of 160,000. Lamjavyn Gündalai, elected to the Great Khural as an Independent, joined the DP. After the formation of the DP's primary organizations nationwide in February 2001, they elected the party's National Advisory Committee, comprising two members from each of the Great Khural's 76 constituencies. They included many of the former leaders of the MNDP and MSDP who in the Great Khural elections of July 2000 had failed to be elected. In July 2001 the DP's Secretary-General, Zandaakhüügiin Enkhbold, was released to attend a study course in the USA, and his duties were taken over by the former Mongolian ambassador to the United Kingdom, Tsedenjavyn Sükhbaatar. A new Secretary-General, Norovyn Altankhuyag, was elected in September 2001.

On 14 December 2000 the Great Khural readopted unchanged the decree of December 1999 amending the 1992 Constitution for immediate implementation. The decree was vetoed by the President, and his veto was rejected, but the Constitutional Court was unable to meet in full session because the election of replacements for time-expired members was delayed in the Great Khural. Finally, President Bagabandi sealed the amendments in May.

The MPRP's 23rd Congress was held at the end of February 2001. The Chairman, Nambaryn Enkhbayar, and General Secretary, Lkhamsürengiin Enebish, were re-elected. Two of the three party secretaries, Taukein Sultaan and Baldangiin Enkhmandakh, were released for diplomatic duties, while Sanjbegziin Tömör-Ochir and Luvsandagvyn Amarsanaa joined the secretariat, although Amarsanaa was soon nominated to be the next Mongolian ambassador to China. The membership of the Party Leadership Council was increased from 11 to 15.

The presidential election on 20 May 2001 was won by the MPRP's Natsagiin Bagabandi, who received 574,553 votes (57.95% of the total ballot), as against 362,684 votes for Radnaasümbereliin Gonchigdorj of the DP (36.58%) and 35,139 votes for Luvsandambyn Dashnyam of the Civil Courage Party (3.54%).

In January 2001, meanwhile, a member of the Great Khural and former director of the Erdenet copper combine, Shagdaryn Otgonbileg, was killed when an Mi-8 helicopter carrying 23 passengers and crew investigating the *zud* in western Mongolia crashed in Malchin district of Uvs *aimag*. The eight dead included several UN staff. Otgonbileg was given a state funeral, and in May his widow, Tuyaa, was elected unopposed as Great Khural MPRP member for his Zavkhan 22 constituency. The Chairman of the Great Khural, Lkhamsürengiin Enebish, died in September 2001, and was succeeded by Sanjbegziin Tömör-Ochir. He was replaced as General Secretary of the MPRP by Doloonjingiin Idevtkhen, and Byambajavyn Övgönhüü was elected for the MPRP in Enebish's constituency. In March 2002 Sanjaasürengiin Oyuun's Civil Courage Party merged with Bazarsadyn Jargalsaikhan's MRP to form the Civil Courage Republican Party under Oyuun's leadership. The new party began to disintegrate in June 2003 when the party expelled Jargalsaikhan. He took away the MRP banner and pledged to rebuild the party. It was rumoured that he would form a new alliance with the MPRP, which it was said had offered him help with the repayment of a large commercial debt.

A member of the Great Khural for the MPRP, Mendiin Zenee, died in June 2002. In a September by-election in Töv 36 constituency he was replaced by Sunduin Batbold (MPRP), who had been chairman of the *aimag* assembly. Also in June 2002 the Great Khural approved the Law on Land and the Law on Land Privatization, scheduled to enter into force during 2003. Although under 1% of the country's total territory was to be available for privatization, the laws generated a great deal of controversy. From November 2002 there were several demonstrations by tractor-driving farmers who were arrested for parking their vehicles on Sükhbaatar Square in Ulan Bator. Headed by Erdeniin Bat-Üül of the DP, President of the Movement for Justice in Land Privatization, they protested that the poor would be denied land by the 'oligarchy'.

An effective protest against the MPRP Government's 'gagging' of the opposition (the majority MPRP had prevented Great Khural minority members from addressing opening sessions of the Great Khural) was made by DP Great Khural member Lamjavyn Gündalai, who interrupted the Prime Minister's televised speech and disrupted the opening ceremony of the 2002 autumn session by displaying to television cameras a series of placards on which he demanded the right to speak, condemned the Government's media monopoly and criticized the land privatization programme. When the 2003 spring session opened at the beginning of April Gündalai again displayed a range of slogans; the President and the Prime Minister were unable to deliver their speeches, and the televised session was suspended. On 1 October Gündalai succeeded for a third time in disrupting the opening of the Great Khural (obstructing its autumn session), with the result that the Prime Minister was once again unable to deliver his report to the cameras.

Meanwhile, in the predominantly Kazakh *aimag* of Bayan-Ölgii, in western Mongolia, dissatisfaction with local leaders led to hunger strikes and demonstrations by up to 4,000 people in February and March 2003. Against the background of poverty and unemployment, the immediate causes seemed to be corruption and the abuse of human rights. In May 2003 the three Kazakh members of the Great Khural were refused permission to form a national minority group of Great Khural members. The Law on the State Language, adopted by the Great Khural in the same month, was interpreted to mean that Kazakh speakers in Bayan-Ölgii would have to deal with registration and other local government matters in the Mongolian language.

In April 2003 the Eagle (Bürged) television station closed down after a funding dispute, cutting off many viewers from an opposition voice and image of the outside world. According to a survey by the US-based non-governmental organization Freedom House, reported in April, the Mongolian press was 'half-free'. In July President Bagabandi vetoed a government resolution stipulating that daily newspapers should agree to publish government policies and decisions.

At the height of the season of Tourism Year 2003 the Ulan Bator authorities ordered that the city's central Sükhbaatar Square be dug up and rebuilt. There were plans to remove the remains of Sükhbaatar and Choibalsan from the mausoleum and rebury them (with those of President Jamsrangiin Sambuu, interred in the State Palace gardens) at Altan-Ölgii cemetery. The discovery of the remains of several hundred people at Khambyn Ovoo in the city suburbs in May reopened the debate about the political purges of the 1930s and the compensation of victims' relatives. Most of the remains were shown to be of Buddhist lamas shot in the head; it was decided to cremate their bones and build a memorial stupa.

In June 2003 the Mongolian Great Khural approved the National Latin Script Programme, which declared that drafting of standards for the 'Romanization of the letters of the Mongolian Cyrillic Alphabet' was already under way. The preparatory stage was to remain in force during 2003–04, while the rules for the new Latin script were formulated and plans drawn up for the publication of textbooks, dictionaries and other aids, to be followed by the implementation stage, which was to run during 2005–06. The National Human Rights Commission's 2003 annual report was highly critical of bureaucracy, corruption and cronyism. The police were accused of numerous cases of brutality; the right of detainees to contact a lawyer was widely abused, the report stated. The National Human Rights Programme was adopted in December. Also in 2003, Damirangiin Enkhbat, who had been suspected of the murder of the Minister of Infrastructure Development, Sanjaasürengiin Zorig, in 1998 and had since been resident in France, was reported to have been abducted by Mongolian secret agents and subsequently imprisoned in Mongolia. In early 2004 reports from Amnesty International, the human rights organization, suggested that he had been tortured during interrogation. In January 2004 the Government established a Chief Directorate for Disaster Relief under Püreviin Dash, former deputy head and chief of staff of the Border Troops.

THE 2004 ELECTION IMPASSE

In the spring of 2004 the political campaigning began for the Mongolian Great Khural election, which was held in late June. Huge placards went up all over Ulan Bator, praising the achievements and plans of the ruling MPRP. Mongolian opinion polls indicated that the MPRP was likely to win the election, with a small decrease in its overall majority. The new General Election Committee (GEC) incorporated many MPRP nominees, 'like Choibalsan's special plenipotentiary commission' (which ordered thousands of executions in the 1930s), one opposition spokesman remarked. The electorate was calculated to total 1,279,516 persons, but there was no provision to vote for the some 70,000 people who were resident abroad. The opposition 'Motherland-Democracy' election pact formed by Mendsaikhany Enkhsaikhan's DP and Badarchiin Erdenebat's 'Motherland' Mongolian Democratic New Socialist Party was joined by Sanjasurengiin Oyuun's Civil Courage Republican Party, minus the followers of Bazarsadyn Jargalsaikhan, who left to re-form the Republican Party.

After the registration of participating political parties and coalitions the GEC set about the registration of candidates. There was disquiet among 30 or so sitting MPRP members of the Great Khural when they learned that they had been deselected. The GEC examined the official lists, rejecting all Mongolian Youth Party candidates. Before polling day three candidates withdrew, leaving the final count at 241: 76 each for the MPRP and the Motherland Democracy coalition, 33 for the Republican Party, 23 for the National Solidarity Party, nine for the Mongolian Traditional United Party (also known as the United Heritage Party), five for the Green Party, four for the Liberal Party and 15 Independents.

The initial results of the election of 27 June 2004 (compiled as percentages of the total ballot in each constituency) dealt a major reverse to the MPRP and left the political scene in disarray: the MPRP and the Motherland Democracy coalition had each won about one-half of the seats, leaving neither with the necessary majority of 39 (one-half the Great Khural seats plus one seat). The three Independents elected, although all DP members, were ruled as not counting in this process. The Republican Party won one seat. The number of votes cast was 1,051,812 (82.2% of registered voters). In 25 constituencies there was a straight contest between the MPRP and the Motherland Democracy coalition. The MPRP and the Motherland Democracy coalition accused each other of bribery and electoral fraud in a number of constituencies, and efforts to form a government soon become embroiled in disputes at the GEC and the recently established City Administrative Court (which has powers of supervision over the GEC). On 1 July Lamjavyn Gündalai and other Motherland Democracy leaders and supporters forced their way into the Mongolian Radio and Television building and broadcast their account of the election results, protesting against the MPRP monopoly of the state media. The MPRP responded with its own late-night broadcasts on other television stations. A lawsuit charging Gündalai with 'illegally entering a special objective under state protection' was later dismissed by the court.

The GEC submitted the results in 74 of the 76 constituencies to President Bagabandi on 9 July 2004 at the first session of the newly elected Great Khural, which was boycotted by the MPRP. The President stated that it was right to convene the first session, even if two results had yet to be confirmed, because of the need to discuss the many issues that the legislature should address. The Motherland Democracy members were not allowed to take the oath, however. Meeting separately, 70 of the MPRP members elected in 2000 filed a lawsuit against the President on the grounds that he had contravened the Constitution and allowed the Great Khural to meet without a quorum (57 members being present) before the closing session of the Great Khural elected in 2000 had been held.

On 17 July 2004 the GEC held a new ballot in one of the two still-contested constituencies at only 12 hours' notice, giving a narrow victory to the MPRP candidate (the Minister of Defence, Jügderdemidiin Gürragchaa) which led to a storm of protest. The closing session of the 2000 Great Khural was held on 22 July. Among other decisions, it released the Great Hural's Deputy Chairman Jamsrangiin Byambadorj (who had lost his parliamentary seat in the recent election) to take up a vacant seat in the Constitutional Court and accused the President of acting unconstitutionally in convening the 9 July session. All these decisions were vetoed by President Bagabandi in late July as unconstitutional.

Postponed from 23 July after another MPRP boycott, the first plenary session of the 2004 Great Khural was held on 26 July, when 74 members were sworn in. The meeting was chaired by the senior member present, Damdingiin Demberel. The first business was the election of the new Chairman (Speaker) of the chamber. The MPRP group supported the candidature of the MPRP leader and acting Prime Minister, Nambaryn Enkhbayar, but protracted discussion of this proposal with the Motherland Democracy members of the Khural continued for days without resolution. In late July the MPRP members of the Great Khural proposed that the Motherland Democracy members should nominate the next Prime Minister, while the Motherland Democracy members proposed the formation of a joint working group to draw up a programme for a government of 'national accord'. Meanwhile, the GEC had still not declared the results in the remaining two constituencies, and the Motherland Democracy coalition threatened to boycott the Khural until it did so.

Eventually, at the end of August, former Prime Minister Nambaryn Enkhbayar of the MPRP was appointed Chairman of the Great Khural, and Tsakhiagiin Elbegdorj of the Motherland Democracy coalition was nominated as Prime Minister. The appointment of new government ministers was, however, further delayed. Although the newly elected members of the Mongolian Great Khural agreed on the formation, chairmanship and membership of the Khural's standing committees and subcommittees, discussion of the basic principles for the formation of a coalition government were protracted. In mid-September Prime Minister Elbegdorj forwarded to President Bagabandi the outlines of a Cabinet to include a Deputy Prime Minister and 13 ministers. Finally, after months of disagreement between the two main parliamentary groups, on 28 September a new Government was appointed under the leadership of Elbegdorj. It had been agreed that the terms of office of the Prime Minister and the newly created post of Deputy Prime Minister would be shared between the MPRP and the Motherland Democracy coalition, implying that Prime Minister Elbegdorj would be required to cede his position to an MPRP Prime Minister after two years.

EXTERNAL RELATIONS

The collapse of the USSR in December 1991 had far-reaching effects on Mongolia, which was obliged to negotiate separate treaties with the USSR's former constituent parts to ensure the continuation of aid and trade, upon which Mongolia remained largely dependent. In January 1993 President Ochirbat visited Moscow, where he signed with President Yeltsin a new 20-year

Mongolian-Russian Treaty of Friendly Relations and Co-operation to replace the defunct Mongolian-Soviet treaty of 1986. Ochirbat and Yeltsin also issued a joint statement expressing regret at the imprisonment and execution of Mongolian citizens in the USSR during the Stalinist purges. Similar treaties of friendly relations and co-operation were concluded with Kazakhstan and Ukraine.

Relations between Mongolia and China had improved by April 1994, when a new Treaty of Friendship and Co-operation was concluded during a visit to Ulan Bator by the Chinese Premier, Li Peng. An agreement on cultural, economic and technical co-operation was also signed.

In July 1994 two important documents outlining Mongolian foreign policy objectives were published, the *National Security Concept* and the *Foreign Policy Concept*. While emphasizing 'complete equality' in its co-operation with Russia and China, Mongolia also focused on the development of relations with its 'third neighbour'—primarily the USA, Japan, Western Europe, the Asia-Pacific region, the UN and international financial bodies.

In May 1995 the US Congress issued a statement of support for Mongolia, and in August President Bill Clinton authorized the provision of US military aid to Mongolia. His wife, Hillary Clinton, paid a brief visit to Mongolia in September and announced further aid of US $4.5m. In the same month President Ochirbat flew to Germany and then to the headquarters of the European Union in Brussels for aid talks. Ochirbat returned to Europe in April 1996 for official visits to France and the United Kingdom; in London he promoted bilateral trade, and was received by Queen Elizabeth. Prime Minister Jasrai had paid an official visit to China at the end of March.

Relations with China were consolidated in 1997 by the visits to Ulan Bator of Qiao Shi, the Chairman of the Standing Committee of the National People's Congress (in April), and of the Minister of Foreign Affairs, Qian Qichen (in August). The Malaysian Prime Minister, Mahathir Mohamad, visited Mongolia in September, but various aid and co-operation projects discussed during his visit were postponed owing to the financial crisis affecting Malaysia.

In July Mongolia was admitted to the Association of South East Asian Nations (ASEAN) Regional Forum (ARF) at its ministerial meeting in Manila. In March 2000 Prime Minister Amarjargal paid an official visit to the United Kingdom. He was followed a fortnight later by a delegation of 10 members of the Mongolian Great Khural. In May Amarjargal attended a conference in Riga, Latvia, where Mongolian membership of the European Bank for Reconstruction and Development (EBRD) was approved. In June 2000 Mongolia became the 24th member of the Parliamentary Union of the Countries of Asia and the Pacific.

In November 2000 Russian President Vladimir Putin made an overnight stop in Ulan Bator on his way to the Asia-Pacific Economic Co-operation (APEC) conference in Brunei. He was the first senior-level Moscow visitor to Mongolia since Soviet Communist Party General Secretary Leonid Brezhnev's visit in 1974. Presidents Putin and Bagabandi issued a joint declaration which pledged Mongolia and Russia 'not to join any military-political alliances against one another, nor conclude any treaty or agreement with third countries harmful to the interests of the other's sovereignty or independence. Neither side will allow its territory to be used by a third state for purposes of aggression or other acts of violence harmful to the sovereignty, security and public order of the other'. Russia confirmed its adherence to the five nuclear powers' declaration of guarantees for Mongolia's security in connection with its nuclear-weapons-free status.

Enkhbayar travelled abroad as Prime Minister for the first time to the Davos international economic forum in Switzerland in January 2001, then to Japan. Russian Premier Mikhail Kasyanov's brief visit to Mongolia in March 2002 raised once more the dispute between Moscow and Ulan Bator over Mongolia's repayment to Russia of the large debt for Soviet aid (see Economy).

The seventh British-Mongolian 'Round-Table' Conference in Ulan Bator in September 2002 was attended by a parliamentary secretary of the British Foreign and Commonwealth Office. The Dalai Lama visited Ulan Bator in November 2002, travelling via Japan after Russia refused him a visa and Korean Air, the national carrier of the Republic of Korea, banned him on the grounds that he posed a security threat. Although his visit was at the invitation of Mongolia's Buddhist leaders rather than the Government, the Chinese authorities indicated their displeasure by halting rail traffic on their mutual border for 36 hours. The UN Secretary-General, Kofi Annan, paid a brief visit to Mongolia in October 2002.

Hu Jintao, the Chinese President, visited Ulan Bator in early June 2003; he conducted talks with President Bagabandi and Prime Minister Enkhbayar, and addressed the Great Khural on the subject of 'neighbourly partnership of mutual trust', which, besides reiterating respect for each other's independence, sovereignty and territorial integrity, embodied the handling of bilateral relations 'in the spirit of consultation, co-operation and friendship'. China granted Mongolia 50m. yuan for the building of a road across the border between the two countries (from Zamyn-Üüd to Erlian) and offered loans, the exact use of which was to be determined at a later date.

A company of 170 Mongolian soldiers was dispatched to Iraq in September 2003; their duties mainly consisted of working on construction projects and guarding oil pipelines.

Economy

ALAN J. K. SANDERS

Between 1948 and 1990 the Mongolian economy was developed under a series of Five-Year Plans, with large-scale assistance from the USSR and other communist countries, principally those that formed the Council for Mutual Economic Assistance (CMEA, dissolved in mid-1991). The two salient features of development during this period were the completion of the transition to the socialist system of production and a rate of economic expansion very much faster than was achieved in the first 30 years of the republic.

In the early 1990s, parallel with the political developments taking place in Mongolia, the Government initiated a series of far-reaching reforms aimed at achieving a market economy and privatization. Mongolia was faced with a rapidly growing population and an increasing demand for foodstuffs and consumer goods. During 1980–91 the population increased by an annual average of 2.7%, but, with the population of working age rising by 3.4% per year, it was feared that unemployment would increase considerably. (The population rose by an annual average of 1.2% in 1992–2002.) A high rate of infant mortality was disclosed after decades of concealment. (According to the UN Development Programme (UNDP), this stood at 61 deaths per 1,000 births in 2001.) Poor living conditions were reported to be responsible for 50% of infant deaths in the 1980s, genetic and ecological conditions for 30% and inadequate medical services for 20%. During the 1980s there was also a severe shortage of convertible currency, necessary to acquire new technology, to stimulate production and export earnings, and to reduce Mongolia's foreign debt. In an attempt to revitalize the economy, Mongolia planned to promote the tourist industry and to secure a relatively small medium-term loan in convertible currency. This was to be used to buy small, mobile mining machinery, to exploit the country's deposits of gold and silver, and modern technology for the improvement of livestock-breeding and the associated processing industry.

In 1989, according to official sources, gross national income (GNI) per head was US $473. Reflecting the dislocation of the economy in the transition to a free market, GNI per head had declined to $112 by 1991. Large devaluations of the tögrög since then have made the exact figure uncertain, but the World Bank estimated GNI per head to be $390 in 1997 and $380 in 1998. In July 2001 the President of Mongolbank estimated average annual per caput income at $360–$380, although the World Bank estimated GNI per head to have increased to $400 for that year.

AGRICULTURE

After the catastrophe of the period of 'leftist deviation' (1929–32), animal herding, the mainstay of Mongolia's economy, reverted to private enterprise, and, apart from taking compulsory deliveries of produce during the period of the Second World War, the Government did not interfere with herding activity. Small-scale mutual help in carrying out certain tasks was practised, and some herdsmen joined together in small producers' associations, but under the 'New Turn Policy' the formation of co-operatives was discouraged. By 1952 the existing co-operatives contained only 280,000 animals out of a national stock of nearly 23m. However, collectivization of herding had by 1957 again become a matter of policy. Although collectivization was said to have been carried out voluntarily by the herdsmen, the initiative came from the MPRP, and propaganda and economic compulsion were widely used to persuade people to join. Thus, state loans were granted to newly formed co-operatives, discriminatory rates of taxation were imposed on individuals who owned large herds and similarly differential norms of compulsory deliveries of produce were set.

The collectivization programme reached its height in 1958 and by April 1959 all but a tiny minority of herdsmen had been collectivized. The herdsmen's associations (*negdel*) were quite different in character from the earlier producers' associations. They were of considerable size, and were units of local administration as well as economic units. Labour was regulated by means of work-books issued to each member, who received pay according to his work. All families were allowed to retain a certain number of private animals. Produce was purchased by the State which also granted loans to the co-operatives. Internally, each co-operative was organized into a number of permanent brigades, each with its own territory and headquarters and its own special tasks. The brigade in its turn contained a number of sections (*heseg*) comprising several bases, each of usually two households living in felt tents and looking after a number of animals.

An innovation in the Mongolian rural economy has been the development of large-scale agriculture. This did not affect any previous pattern of economic activity and from the start was organized as a direct state venture. Ten state farms existed in 1940 and 52 state farms in 1988. In and after 1959 the area under cultivation increased sharply as large tracts of virgin land were opened, but the sown area was smaller in 1970 than in 1965. By 1978 1m. ha of virgin land had been ploughed. It was reported that the 1985 grain harvest reached a record 889,400 metric tons, enabling Mongolia to meet its own grain requirements and to export surplus wheat to Siberia. Production amounted to 798,600 tons in 1989, but it subsequently declined annually and by 2002 totalled only 125,860 tons.

Increasing attention was paid to mechanization and the introduction of scientific methods of farming. The division of activity between the 255 *negdel* and the 52 state farms was not a strict one. Co-operatives also engaged in field work, especially fodder growing, while state farms were expected to supply good breeding animals. The principal crops produced by the state farms were cereals, potatoes and other vegetables. An apparently successful innovation was the establishment of 17 inter-co-operative production enterprises, in which neighbouring co-operatives combined resources to specialize in particular farm-related activities. A new law on co-operatives came into force in January 1990. This law controlled co-operative activity in small-scale industry, trade and services and transformed the *negdel* into proper co-operatives. In 1991, however, following the new Government's initiation of political and economic reforms, these were privatized and mostly divided into smaller units. All restrictions on private livestock ownership were removed in 1990. Some state farms became joint stock companies, others were broken up.

At the annual year-end livestock census in 1995, Mongolia had a record total of 28.6m. head of sheep, goats, horses, cows and camels. This was largely due to a 2m. head increase in the number of goats, encouraged by the growth in the cashmere industry. In 1996 the rise in livestock numbers continued, reaching 29.3m., but this was also a reflection of low levels of industrial consumption of meat and hides.

A survey of the herding community at the end of 1996 revealed that, of the 517,700 families in Mongolia, 170,100, or 32.8%, were herding families, with 395,400 herdsmen (16.8% of the population) engaged in livestock raising. Only 9.3% of these families had electricity. Mongolia's livestock herds increased in 1999 to a new record of 33.6m. (compared with 31.3m. in 1997). However, serious drought in late 1999 was followed in early 2000 by several months of severe cold, snow and frost. In consequence, by mid-2000 an estimated 2.5m. head of livestock had died from starvation and cold. In response to government appeals, international donors offered financial aid, and relief teams of the International Committee of the Red Cross (ICRC) were mobilized to take food and medicine to isolated herding families. Almost 2,500 families were believed to have lost all their animals by late June, when survivals of new-born stock numbered 8.4m. head, compared with 10.2m. in 1999.

The December 2000 livestock census recorded a fall in the herds to 30,096,400 head, and in the following months the weakened animals faced even more severe conditions, with deep snow and very low temperatures. It was reported at the end of June 2001 that 3,312,000 head had died, 26.7% more than the year before, and that 7,364 households had lost all their animals. The hardest-hit provinces were Zavkhan (28.9% of stock lost), Arkhangai (18.9%) and Khövsgöl (18.3%). Eastern provinces were affected by foot-and-mouth disease. Newborn stock thriving in the summer of 2001 numbered 7.6m., 762,800 fewer than in 2000. At the end of 2002 Mongolia's stock of sheep, goats, horses, cattle and camels totalled some 23,684,500, after further losses in that year. By the end of 2003 total livestock numbers had recovered to reach 25,427,700. As of the end of June 2004, the number of young livestock surviving from birth had reached 8.6m., 1.2m. more than in the same period in 2003.

Prime Minister Enkhbayar expressed concern about the state of the nomadic herding economy, stating that it was important to improve the quality of stock, raise yields and invigorate product processing. The key was the intensification of livestock raising to meet the growing needs of large towns, especially Ulan Bator. The capital's population was estimated at 893,400 in December 2003, with some 100,000–150,000 people commuting into the city daily.

There was a continuing decline in the numbers of herding households (from 191,526 in 2000 to 172,412 in 2003) and of herdsmen (from 421,392 in 2000 to 377,936 in 2003), especially of herdsmen aged under 35. Some 28.9% of herdsmen's households had an electricity supply in 2003 (a 5% increase on 2000). Although there was a decline in the number of agricultural specialists, the number of vets increased slightly, from 1,800 in 1989 to 2,000 in 2003. Over the same period the number of tractors declined by about two-thirds and of grain harvesters by more than one-half. The number of animal shelters roughly doubled between 1990 and 2003, while the number of wells in operation rose from 34,600 in 1995 to 40,900 in 2003.

INDUSTRY

Large-scale industry has developed only since the 1960s. Before the Revolution most manufactured requisites were imported or were made locally, chiefly by Chinese craftsmen. In the 1920s technicians from Western Europe were engaged to help develop Mongolia's infant industry. In particular they built a power station and a brickworks. After the swing to the left in 1929, however, only Soviet aid and expertise were welcome, and in the years prior to the Second World War industrial growth was slow. Only one enterprise of any size was commissioned, the Industrial Combine in Ulan Bator, which in 1934 commenced production of leather goods and felts. Industry developed in two channels. Co-operative industry had a much smaller output than state industry, producing many items needed for domestic use, as well as providing repair services. After the Second World War state-operated industry expanded rapidly and many new enterprises were commissioned.

Mongolia received enormous aid from its political allies—some 10,000m. roubles from the USSR alone between 1945 and 1990—without which its industrial advance, modest though it is in world terms, could not have been envisaged. For a time, in the 1950s, it seemed as if China was hoping to challenge the USSR's leading position in Mongolia, using the weapon of economic aid. A first gift of 160m. roubles in 1955 was followed by the dispatch of Chinese labourers to help Mongolia's inadequate and under-trained labour force. Exact numbers are not available, but in the peak years of 1959 and 1960 several thousand Chinese labourers were working on diverse projects, such as building apartment blocks, laying roads and installing irrigation systems. Many had their families with them. As the Sino-Soviet rift widened, the Chinese workers began to leave Mongolia, until by mid-1964 most had returned home. To some extent the loss of these workmen was made good by the supply of Soviet construction labourers and engineers (some 50,000 at their peak), working principally in Ulan Bator, the Darkhan area and Choibalsan, where a third industrial area was planned. However, the break with China had other adverse effects. Chinese consumer goods, in particular silk and cloth, which were plentiful in 1959, were, by 1968, no longer available. The drastic fall in railway through-goods traffic between the USSR and China also meant a considerable loss of state revenue.

The principal centres of industry are in the central economic region, at Ulan Bator and Darkhan, half-way between the capital and the Russian frontier. Both centres are situated in the area of densest population and have direct road and rail communication with Russia. Both have their own coal supplies also, Darkhan at an open-cast mine at Sharyn Gol, to which it is linked by a new rail spur, and Ulan Bator at a large, new, open-cast mine at Baganuur, linked to the Trans-Mongolian railway. The two centres account for most of Mongolia's production in terms of electric power, capital materials such as cement, bricks and wall panels, and consumer goods—food, drink, leather goods, china, sweets, soap and so on. A large new cement and lime complex at Khötöl, between Darkhan and Erdenet, was expected to produce 100,000 metric tons of lime and 500,000 tons of cement per year, enough to satisfy the country's total requirements. In the mid-1990s, however, the construction industry was working at only some 20% of its capacity. Many of Mongolia's industrial enterprises were not viable in market conditions and ceased production, being in effect bankrupt. A factory producing disposable syringes, financed by the Republic of Korea, began production in the mid-1990s. At the end of the 1990s there was an improvement in the output of the knitwear and garment industry.

One of the most important developments in Mongolia's economy has been the joint Soviet-Mongolian exploitation of copper and molybdenum deposits at Erdenet in Orkhon province. The deposits are located near Khangal *sum*, about two hours' journey by road to the west of Darkhan, and thus accessible to transport routes. The Salkhit–Erdenet railway line, linking the new complex with the main rail system, went into operation in October 1975, and the ore concentrator in December 1978. The combine attained its full capacity in November 1983. In the 1990s the Erdenet copper enterprise became Mongolia's largest source of foreign exchange, while Erdenet town emerged as an industrial centre in its own right, with a large carpet factory and other enterprises.

In 1995 copper concentrate exports were sufficient to finance more than 67% of Mongolia's imports. However, in 1995–96 fluctuating world copper prices and the declining metal content of exported concentrate underlined Mongolia's vulnerability as a one-asset economy. There was a continuing slow decline in the share of copper in Mongolia's total exports, falling from 53.0% in 1995 to 27.6% in 2002. In 2000–01 the Erdenet concern operated at a loss, unable to reduce production costs or to pay government taxes, which had been set too high on world copper price forecasts.

Output of fluorspar declined by 200,000 metric tons in 1994, when only 88,000 tons were exported, owing to the lack of a steady market; production increased again from 1995, reaching 597,100 tons in 1999. The most successful industry was that of gold-mining, with production rising from 4.5 tons in 1995 to 11.8 tons in 2000, and 13.7 tons in 2001, before declining to 12.1 tons in 2002 and to 11.1 tons in 2003. In mid-2003 the Canadian company Ivanhoe Mines continued to improve the prospects for developing the Oyuu Tolgoi deposit in the Gobi region, upgrading its estimate of copper and gold content and finding plentiful supplies of underground water. Finance was being sought for construction of a road from the deposit to the border with China. The Great Khural's examination in January 2004 of a draft decree which would allow the Government to alter the boundaries of special protected areas (nature reserves) when appropriate for the exploitation of mineral resources led to protests by the Mongolian offices of the World Wide Fund for Nature (WWF) and the UNDP. It was said that 3.1m. ha were under threat in the Daguur, Onon-Balj and Gobi Large and Small Reserves, including the habitat of the extremely rare Gobi bear. In April the Government issued a decree to stop illegal gold-mining in the Gobi Small Reserve by hundreds of 'ninjas', as the freelance miners were called.

In 1997 exploratory drilling at Tamsag, in eastern Mongolia, began to yield crude petroleum, at a rate of 1,500 barrels per day. The first consignment was delivered to China for analysis and refining. A clear sign of long-term improvement in government revenues appeared in July 1997, with income of 500m. tögrög from petroleum exploitation included for the first time. The number of wells was scheduled to reach eight in 1998 and 17 in 1999. Production of crude petroleum rose from 44,791 barrels in 1998 to 71,914 barrels in 1999, then fell slightly to 65,522 barrels in 2000. The SOCO oil company had suffered a severe fire at its Tamsag base in eastern Mongolia, but restarted extraction at its three production wells for export to China (73,700 barrels in 2001). Crude extraction amounted to 139,205 tons in 2002 and 183,047 tons in 2003.

The 1990s witnessed a resurgence of Chinese business activity in Mongolia. By 2001 companies with Chinese investment accounted for 35% of all companies with foreign investment, and companies benefiting from Hong Kong investment for a further 13.7%, while the Republic of Korea accounted for 23.7%.

Mongolia's hopes of benefiting from the oil and gas pipelines to be built from Russia to China finally ended in mid-2001 when it emerged that they would bypass Mongolia on the way to the refineries at Daqing in the Chinese province of Heilongjiang.

According to the Asian Development Bank (ADB), gross industrial production in 2002 totalled 289,728.1m. tögrög, an increase of 5.0% compared with 2002 production, in real terms. Real output of the mining sector decreased by 6.8% in 2002.

FOREIGN TRADE

Trade was traditionally almost entirely with the countries formerly constituting the 'socialist bloc': in 1989 93.1% of Mongolia's exports went to socialist countries (CMEA 90.3%), which were also the source of 95.6% of imports (CMEA 92.5%). However, the share of Mongolia's total foreign trade conducted with former socialist states had declined to some 74% by 1992. In 2000 the principal source of imports (supplying 33.6%) was Russia, followed by China (20.5%) and Japan (11.9%). The principal markets for exports were China (taking 51.2%), the USA (24.3%) and Russia (8.4%).

In 2002 Mongolian exports to China were worth US $212.2m. (42.4% of all exports) and imports from China $160.8m. (24.4%), compared with $238.3m. (45.7%) and $136.1m. (21.3%) the previous year. Exports to Russia and imports from Russia were worth $43.1m (8.6%) and $224.6m. (34.1%) respectively, as against $44.9m. (8.6%) and $225.9m. (35.4%) in 2001.

Mongolia exports mainly primary products and imports industrial goods and equipment. In 1989 some 42.8% of its exports consisted of fuels, minerals and metals, while raw materials, including foodstuffs, accounted for a further 35.7%. In that year industrial consumer goods accounted for 17.5% of total exports. In 1993 industrial goods accounted for 75.4% of total imports, while consumer goods comprised the remaining 24.6%. Fuel and petroleum accounted for 32.6% of industrial imports.

In 1993 59.9% of imports were conducted by general (convertible currency) trade, 29.5% by barter and 10.6% by other kinds of trade. The breakdown of exports was as follows: 51.6% general trade, 35.2% barter and 13.2% other kinds of trade. In 1994–95 Mongolia aimed to achieve a trade surplus for the first time, reducing petroleum imports and promoting the export of cashmere goods. Mongolia's international trade recorded a deficit of US $87.4m. in 1996. With world copper prices low, the trade deficit in 1998 was US $155.6m., compared with a surplus of $30.2m. in 1997. The trade deficit was estimated by the IMF at $90.2m. in 1999, $72.6m. in 2000, $100.6m. in 2001 and $156.2m. in 2002.

In 2001 the principal imports were mineral products (22.8%), machinery (17.8%), transport equipment (10.8%) and textiles (9.9%). The principal exports in 2001 were copper concentrate (28.4%), gold (14.3%), hides, skins and furs (11.3%) and dehaired cashmere (10.5%). Exports of copper concentrate were worth US $147.9m. and dehaired cashmere $54.9m., as against $160.3m and $54.5m. respectively in 2000. Renewed exports of frozen beef to Russia were worth $10.4m. (2.7%) in 2001, down slightly on 2000 ($14.3m.).

In 2002 Mongolia's exports were worth US $500.9m. and imports $659.0m. The main exports by value were copper concentrate ($138.0m.—27.6%) and dehaired cashmere ($30.1m.—6.0%), both mostly to China, textiles ($136.6m.—27.3%), hides, skins and furs ($43.3m.—8.6%) and gold ($117.6m.—23.5%). Copper concentrate exports decreased by 0.4% by volume and their value fell by 6.7%. Mineral products (essentially refined petroleum from Russia) accounted for $122.9m. (18.6%) of imports, and machinery $128.3m. (19.5%).

In the first half of 2003 exports were worth US $219.3m. and imports $360.9m., both registering considerable growth with European Union (EU) countries. Exports of copper ore decreased by 9.9% in terms of volume and by 0.9% in terms of value compared with the first half of 2002, world copper prices having increased by 10% over the period. In the first six months of 2004 exports totalled $307.2m. and imports $479.9m. Compared with the first half of 2003, mineral exports increased by $82.8m. and textile exports by $4.4m., while traditional exports like hides and skins declined by $12m.

The law on the legal status of the free economic zone at Zamyn-Üüd (a railway station on the Chinese border) was adopted in June 2003. The Altanbulag free economic zone (approved in June 2002) on the northern border with Russia, where there is an international road border crossing point, continues its development with infrastructure and trade service centres under construction. Another free economic zone, at Tsagaan Nuur on the western border with Russia, where there is also an international road border crossing point, was in the course of approval in 2004.

POWER AND TRANSPORT

Fuel of many types is used in Mongolia. At one end of the scale is the Ulan Bator No. 4 power station, built with Soviet assistance, which went into operation in 1985. Its capacity of 380 MW doubled the country's generating capacity. The station is fuelled by coal from Baganuur. Provincial (*aimag*) centres may have thermal power stations or diesel generators, and in rural centres small diesel generators are common. Domestic heating in apartment blocks in Ulan Bator is by central town-heating from the power station. Elsewhere wood, roots, bushes and dried animal dung are used for domestic firing. In 1998 plans were revealed for a US $120m. petroleum refinery project in the former coal-mining town of Nalaikh. Part of the projected annual output of 1.4m. metric tons of petroleum products was to be fuel oil for a new 70-MW power station nearby.

Transport shows a similar range of sophistication. Ulan Bator is linked with Beijing and Höhhot (China), Ulan-Ude and Irkutsk (Russian Federation), as well as provincial centres by An-24 turbo-prop aircraft. The airline's one Airbus flies to Moscow and Berlin and also to Seoul, two Boeing 727s also fly the shorter international routes, and a Boeing 737-800 was delivered in mid-2002. Mongolian railways connect the Trans-Siberian main line with the Chinese railway system, providing a direct route from Moscow to Beijing and an eastern link with the town of Choibalsan. Because the Mongolian and Chinese gauges are different, goods are transhipped at the border stations Zamyn-Üüd or Erenhot. In order to encourage foreign investment, Mongolia is planning long-term road improvements. To mark the millennium the Government launched the building of a road running from west to east across the country, linking Siberian Russia with northern China, and improving communications with Ulan Bator from outlying parts of Mongolia. The route was from Kosh-Agach in the Altai Republic to Ölgii, the Zavkhan valley and through Arkhangay to Ulan Bator, then via Öndörkhaan in Khentii to Sümber (Khalkhyn Gol) and into Inner Mongolia. The project was expected to take about 10 years to complete, but because of the cost the road was not planned to be hardtopped for the whole length. Meanwhile, long-distance transport is mainly cross-country by lorry or Soviet-built UAZ (jeep-type vehicle). Horse, ox and camel carts are still widely used, even in Ulan Bator, and camels are employed as beasts of burden.

In the mid-1990s the decline in rail and road freight traffic continued. Only 150,000 of the scheduled 670,000 metric tons of transit freight were carried from Russia to China in 1994 and 3,000 of the scheduled 300,000 tons from China to Russia. The amount of freight transported by rail rose steadily in the late 1990s, however, reaching almost 8.2m. tons in 1999. Rail passenger traffic also increased steadily. In September 1995 there were new concerns over the state of Mongolia's ageing aviation equipment after a MIAT Antonov An-24 twin turbo-prop crashed into a mountain 17 km from Mörön airport in Khövsgöl province, killing 41 people. Although pilot error was blamed, the civil aviation authority decided to acquire ground position indicators for the fleet. The crash of a MIAT Yu-12 in June 1997, which killed eight passengers, appeared to have been caused by freak weather conditions. In May 1998 a Chinese-built Yu-12 belonging to MIAT crashed in the mountains of southern Khövsgöl province, while on an internal flight from Erdenet to Tosontsengel, killing the 28 crew and passengers. Although the crash was attributed to 'pilot error', MIAT's remaining Yu-12 aircraft were to be returned to China.

REFORM AND RECENT DEVELOPMENTS

The regular pattern of Five-Year Plans and CMEA-orientated trade was disrupted in 1990–91 by the collapse of command economies in the USSR and Eastern Europe, by the transition to payments in convertible currencies, and by the first steps towards privatization and a market economy in Mongolia itself. There was a rapid decline in Mongolia's foreign trade, which, with traditional partners, was largely reduced, by the shortage of convertible currencies, to barter transactions. Petroleum, medicines, and some imported foodstuffs and consumer goods were in particularly short supply, owing to Mongolia's inability to pay for all its requirements at prevailing world prices. Industrial production also declined sharply. The rationing of basic foodstuffs, such as flour, pasta, sugar, tea, vodka and meat, was introduced for the urban population. In January 1991 the Government doubled wages and private savings in order to compensate for the expected doubling of retail prices.

The Government initiated a programme for the privatization of state property in two stages. The first, involving small-scale enterprises, began in May 1991 with the auctioning of several shops and restaurants in Ulan Bator to private individuals. The privatization of large state industrial enterprises began haltingly in early 1992. However, the Government prohibited the privatization of railways, roads, the airline, the state oil company, gold mines, hunting and forestry enterprises, and large irrigation systems. The State also retained a 51% interest in power stations and transmission lines, mines producing coal and metalliferous ores, communications links, some flour mills, meat-packing plants and cement works, motor transport depots, the brewery, distillery, etc. Although it had been claimed that the members of the *negdel* did not wish them to be dissolved, it was announced in mid-1991 that at least some of the members would leave with their stock, and an Association of Individual Herdsmen was established. The privatization of livestock farming and of internal trade and services had been 80% and 90% completed, respectively, by mid-1993.

In June 1991 a massive devaluation of the tögrög changed the official exchange rate from US $1 = 7.10 tögrög at the end of May to $1 = 40.00 tögrög. A devaluation of the currency on 1 January 1993 from $1 = 40.00 tögrög to 150.00 was followed at the end of

May by its flotation at $1 = 400.00 tögrög, stabilizing at 395.00. This policy had the support of the IMF, whose confidence in the Government's economic programme ensured the continuation of international aid for Mongolia. The first signs of economic growth and a sharp fall in inflation emerged in 1994. While the World Bank welcomed these developments, it estimated that Mongolia would still need aid worth $150m.–$200m. per year for the foreseeable future. Among the Government's immediate priorities were to minimize the social impact of the transition to a market economy, to reduce the government sector in agriculture and to improve the environment.

A UNDP report, completed in June 1994, stated that privatization in Mongolia had failed to create an environment in which the market mechanism could operate efficiently. Many supposedly privatized enterprises were still partially owned by the State. The report declared that 'the problem is not a low volume of credit but an allocation of credit that responds to political interference and personal private connections rather than to commercial criteria', and stated that the Mongolian Government should have promoted new private-sector enterprises, creating additional wealth, rather than simply redistributing assets through the voucher scheme. Substantial aid was forthcoming, the report added, but it was not to be used to sustain consumption rather than investment. The UNDP urged price stabilization, lower interest rates, bank regulation and a system of commercial law, as well as a mechanism for the enforcement of contracts.

In November 1994 an IMF report criticized state interference in the allocation of bank credits, which the Government had directed to agriculture and strategic public-sector projects without consultation with Mongolbank, the central bank. This practice had 'adverse consequences for monetary management', constraining private-sector activity. Prime Minister Jasrai's address to Mongolia's first convention of industrialists and business representatives, held in April 1995, was not successful. Delegates complained about government bureaucracy, and criticized middle- and low-ranking officials who were hindering the growth of private businesses with their excessive administrative procedures.

Following the establishment of a Securities Committee in November 1994, under the chairmanship of Jigjidsürengiin Yadamsüren, former Chairman of Ulan Bator City Council, the Mongolian Securities Exchange opened in late August 1995. The annual rate of inflation declined from 183% in 1993 to 66% in 1994. However, the tögrög declined from its 1993 flotation value of US $1=400 to $1=534 at the official rate of exchange by July 1996. New 5,000 and 10,000 tögrög notes and coins of 20, 50, 100 and 200 tögrög, recycled from older demonetized coins, were put into circulation. Growth in gross domestic product (GDP) in 1995 was estimated at 7.0%.

One of the first actions of the Enkhsaikhan Government, which took office in mid-1996, was to remove the ban on the export of raw cashmere. The ban had been imposed in March 1994, to protect the cashmere garment industry's 'added value'. The lifting of the ban was welcomed by the IMF, the World Bank and the ADB, which had opposed it on the grounds that it interfered with trade. The Government subsequently abolished customs duty on almost all imports, on the grounds that the imports were needed to boost the country's export production capacity. To protect domestic garment production, a new tax was levied on raw cashmere and camel wool exports.

Prime Minister Enkhsaikhan urged public support for the Democratic Alliance's proposals to transform the state executive and to reform government agencies so as to meet the demands of the market system. The Democratic Alliance's rationalization programme began with the establishment of 'sectoral ministries' which combined (and in some cases relocated and co-sited) ministries of the previous Government. The Democratic Alliance Government's first year in office produced no real achievements in its efforts to stimulate the free-market economic revival it had hoped for. With GDP growth at 2.4% in 1996, inflation, at 44.6%, was lower than in the previous year. Consumer prices increased at an average annual rate of 20.5% in 1997.

The restructuring of the social security 'net', without extra revenue, was not well received by those most in need of financial support, i.e. pensioners, the very poor and the unemployed. The number of registered unemployed reached 64,200 at the end of June 1997, but several estimates put the true figure at around 227,000, if school-leavers, ex-servicemen and others who had never been employed were included. The number of unemployed declined to 49,800 in 1998 and to 39,800 in 1999, the latter figure representing a rate of 4.6% of the economically active population. Unemployment continued to decline, to 38,600 in 2000 (4.6%), but increased to 40,300 in 2001, before again declining in 2002, to some 30,900 at the end of the year. It was reported at the end of December 2003 that the Poverty Reduction Programme, renamed the Sustainable Household Livelihood Programme, was considered a failure after 10 years' operation. Business investment was discouraged by licences, 'red tape', bribery and excessive taxation, the report said. The minimum wage was set in January 2004 at 236.68 tögrög per hour or 40,000 tögrög per month; the average civil service salary was 81,750 tögrög, and the minimum state pension was 32,000 tögrög.

It was disclosed in June 2004 that the IMF had become aware of the Bank of Mongolia and the Mongolian Government in late 2002 and early 2003 'conducting large quasi-fiscal and/or extra-budgetary operations'. The IMF had 'questioned the legality of these operations and cautioned that they could threaten debt sustainability, undermine fiscal transparency and accountability and give rise to misreporting'. The Mongolian authorities had now taken 'significant steps' to address these concerns, relating to the foreign financing of large infrastructure projects—two hydroelectric power stations and a section of the Millennium Road. At the same time, independent supervision of the Board of the Bank of Mongolia would be instituted.

Privatization continued to make slow progress, with attention focused on the free disposal of state housing stock and the auction of small businesses, rather than the dispersal of large state-owned enterprises, such as the department store in central Ulan Bator, for which there were no bidders. By the beginning of 1996 more than 91% of trade enterprises had been privatized, as well as 88% of agricultural assets, but over 50% of industrial enterprises remained wholly under state ownership. In September 1997 a radical privatization programme was revealed by the Government, under which it was proposed to sell by auction to the highest bidder the entire state sector except some assets such as the railway.

The Agricultural Bank was sold in February 2003 to a Japanese-financed company for US $6.8m. The 80% state share in the petroleum distributor NIK was sold for $7.32m. to East Oil International Consortium, including Anais Ltd, Irkutsk Oil Co (both of Russia) and a company registered in Cyprus. However, the Government cancelled this sale in September after 'negative reports' about the consortium, and NIK was put out to tender again. In February 2004 NIK was sold to the Mongolian distribution company Petrovis, its chief competitor; its director-general, Janchivyn Oyuungerel, had been the manager of NIK during 1990–96 and then a director representing the 20% of private shareholders. The Mongol Daatgal (insurance) company was successfully privatized in December 2003, but a bid for the Gobi cashmere company from Itochu and the Mongolian MCS company the same month failed on a technicality.

The value of the tögrög declined by 13% in the first half of 1997, falling to around $1 = 830 tögrög in March/April before stabilizing and standing at an average rate of $1 = 840.83 in 1998. In 1999, however, the average rate of exchange declined to $1 = 1,021.87 tögrög. The tögrög continued to decline during 2000–03, to $1 = 1,076.67 in 2000, $1 = 1,097.70 in 2001, $1 = 1,110.31 in 2002 and $1 = 1,146.54 in 2003.

Real GDP growth in 1998 was estimated at 3.5%, compared with 4.0% in 1997. Consumer prices, meanwhile, rose by only 6.0% in 1998. Industrial production increased slightly in terms of value but declined in terms of unit output in 1998. In 1999 GDP registered growth of 3.5%, to reach 873,700m. tögrög or 359,583 tögrög ($335) per caput. GDP growth was 1.4% in 2000 and 1.0% in 2001, increasing to 4.0% in 2002, according to the ADB. The IMF stated a growth rate of 5.0% for 2003, and forecast that growth in 2004 and 2005 would also be around 5.0%. The annual rate of inflation rose to 10.0% in 1999 and by mid-2000 had reached 17.4%, before declining to 12.6% in mid-2001 and 11.2% for the year as a whole. Annual inflation declined to 3.1% in 2002. According to the IMF, inflation in 2003 was 5.0%. The country's external currency reserves had increased to US $128m. by mid-1997, declining to $107m. at the end of the year. According to official sources, foreign-exchange reserves increased from $116.9m. (11.1 weeks' imports) at the end of 1999 to $140m. (12.6 weeks' imports) at mid-2000, and $173.3m. in May 2002. At the end of 2003 foreign exchange reserves amounted to $235.9m.

Substantial deficits were incorporated into the budgets for 1996–2002. The 2003 budget forecast revenue of 455,737.8m. tögrög and expenditure of 558,198.7m. tögrög. The budget deficit in 2002 increased to some 5.7% of GDP, having fallen from 12.9% in 1999 to 6.8% in 2000 and 5.3% in 2001. The budget for 2004 envisaged a deficit of 85,994.9m. tögrög.

According to the World Bank, Mongolia's foreign debt at the end of December 2001 amounted to US $885.0m, of which $823.7m. was long-term public debt. At the end of 2002 total debt rose to $1,037.2m. (of which $950.4m. was long-term public debt), reaching the equivalent of 92.6% of GNI.

In April 2003 Mongolia joined the 1973 London Convention on Shipping, as part of its preparations to instigate its own register of shipping, with help from a Singaporean company, Maritime Chain. It was reported that Russian Far Eastern ship owners had hurried to register, but there was some concern that the Mongolian authorities might not be able to regulate the safety and insurance of the ships they registered.

EXTERNAL AID

In the early 1990s, in response to an appeal by the Mongolian Government for foreign aid, the USA offered grain credits and, during his visit to Ulan Bator in July 1991, the US Secretary of State, James Baker, pledged further US aid to the value of US $10.6m. Japan also provided wheat, and in August, on the occasion of Prime Minister Kaifu's visit to Mongolia, the Japanese Government announced economic assistance worth $15m. and development aid of $7m. for improvements to communications. In September a conference of representatives of the IMF, the ADB (which two bodies Mongolia had joined in February 1991), the European Community (now European Union—EU) and Japan approved emergency aid totalling $155m. to Mongolia. Another donors' conference, held in Tokyo in September 1993, pledged $250m. to Mongolia for 1993–94.

The next international donors' conference, held in Tokyo in November 1994, pledged US $210m. to Mongolia in grants and credits for 1994–95. Apart from the international banks, Japan continued to be the biggest donor, with $82.5m. in grants and $87m. in credits pledged for 1993–95. The next donors' conference, held in Tokyo in February 1996, followed up with new pledges totalling $212.5m. The outgoing Jasrai Government was criticized by Japan for not taking up aid quickly enough. In 1997 there was continued support for Mongolia at the Tokyo international aid donors' conference, held in October, which pledged Mongolia $256.1m. (the ADB and Japan $60m. each, Russia $30m., the World Bank $25m., the Republic of Korea $21m., Germany $17m. and the IMF $15m.).

Convening in Ulan Bator for the first time, the aid donors at the seventh Mongolia Assistance Group Meeting, held in June 1999, pledged a record US $320m. in aid for banking-sector reform, poverty relief and privatization. The Assistance Group urged the Mongolian Government to strengthen the financial sector by increasing transparency and improving conditions for private enterprise, including the privatization of state-owned banks and a legal framework for other financial institutions. Future Assistance Group meetings were to take the form of Consultative Group meetings chaired by the World Bank, the venue rotating between Ulan Bator, Tokyo and Paris.

In February 2000, following several months of drought and continuing severe cold, an appeal by the Mongolian Government for international aid was supported by the UN. The climatic conditions of 1999–2000 were believed to have killed millions of livestock (see Agriculture) and directly or indirectly affected the livelihood of at least 500,000 people. Most of the US $5m. received in aid was donated by the Japanese, German, US and British Governments, and the World Bank pledged the balance of $1.3m. from its Poverty Alleviation Programme to Mongolia's relief measures. By mid-2000 a total of $7.4m. had been pledged in relief assistance. Mongolia's membership of the European Bank for Reconstruction and Development (EBRD) was approved in May 2000.

The eighth meeting of the Mongolia Consultative Group took place in Paris at the end of May 2001, attended by representatives of 11 countries, the EU, UN, ADB, IMF and World Bank, and three Mongolian ministers. Prime Minister Enkhbayar announced the outcome of the Paris meeting at a press conference in Ulan Bator: Mongolia would receive US $330m. in loans and donations for the coming year, compared with the 1999 meeting's provision of $320m. over 18 months.

On returning to Ulan Bator the Minister of Finance and Economics, Chültemiin Ulaan, explained that US $1,900m. of the $2,600m. of Mongolia's development aid 1991–2000 had been implemented. Interest was payable on 47.6% of this sum; 37% of the money had been allocated to infrastructure, 16.5% to social services, 10.8% to industry and agriculture, and 23% to financial and economic management sectors; 40% of the money came from the ADB, 26% from Japan, 18% from the World Bank, and 6% from Germany.

About 1,700 companies from 65 countries had invested around US $420m. in Mongolia as of June 2001. The leading investors were China, the Republic of Korea, Japan, the USA and Russia. The main sectors of investment were mining, light industry and agricultural processing. Foreign investment had created 50,000 jobs and accounted for 14.2% of Mongolia's tax revenue. Foreign investment increased by $125,280 in 2001, 38.3% more than the increase of $90,586 in 2000, mainly owing to renewed interest in Mongolia's copper and gold resources, especially from China, which had accounted for 37% of foreign investment during the preceding three years.

The ninth meeting of the Mongolia Consultative Group took place in Ulan Bator at the beginning of July 2002, attended by representatives of 11 countries, the EU, EBRD, ADB, IMF, World Bank and various UN bodies. The meeting pledged US $333m. in aid for the next year, but the IMF and other donors suspended some measures and complained about Mongolia's budget deficits, bureaucracy and the slow pace of reform. The Mongolian Deputy Minister of Finance and Economics, Luvsandagvyn Enkhtaivan, told the press that some $2,300m. of the $2,800m. in aid pledged by the previous eight meetings had been implemented, 83% of loans totalling $1,300m., and $1,100m. of grants totalling $1,400m. Most of the loans were for 30–40 years at an interest rate of 0.5%–0.75%.

The 10th Mongolia Consultative Group meeting held in Tokyo in November 2003 pledged US $335m. in grants and loans for the coming year. The donors praised Mongolia's 'effective utilization' of aid. Between 1991 and 2002 Mongolia had received $2,500m. in aid, 52.5% as grants worth $1,200m., 46% of which was for technical assistance, and 47.5% as loans worth $1,140m., 87.9% of which was spent on economic services, including 81.1% on projects and 12.1% on social services (the amount of loans implemented was $913.1m.). In 2002 Japan provided 25.4% of Mongolia's grants and loans, WHO 8.1%, Germany 5.7%, the ADB 5.4%, the EU 4.2% and the USA 3%; 23% of aid was spent on infrastructure, 12.5% on agriculture and 8.7% on health.

Russian Premier Mikhail Kasyanov's brief visit to Ulan Bator in March 2002 and the signing of routine co-operation agreements passed in the usual cordial atmosphere. However, the Russian media reported that he and Mongolian Prime Minister Enkhbayar had discussed Mongolia's repayment to Moscow of its Soviet-era aid debt 'amounting to US $11,600m.' Enkhbayar said later that only the terms of repayment had been discussed, not the exact amount, but the Mongolian media and opposition fiercely disputed the equivalence of the defunct transferable Soviet rouble to the US dollar. The issue remained unresolved, although it had been raised at many inter-governmental meetings over the past decade. Following the March 2002 Enkhbayar-Kasyanov meeting, Enkhbayar had been expected to visit Moscow in March 2003 for futher discussions, but the meeting was postponed. The Minister of Finance, Chültemiin Ulaan, was reported as saying that Moscow was seeking a 90%–95% discount of the debt, with repayment of the remaining $500m. payable over 40 years. However, when the two Prime Ministers did eventually meet in Moscow in early July, Mongolia's repayment of the Soviet-era debt was hardly discussed, and the major outcome of the talks was a new five-year agreement on the operation of the Erdenet copper enterprise that preserved Mongolia's 51% ownership of stock. Russia agreed that Mongolia had already repaid the cost of building the Erdenet plant. Otherwise, Mongolia's main concern was to reduce Russian taxes on imports of Mongolian goods. At the end of December 2003 Russia announced that it had received Mongolia's payment in settlement of the 'big debt' (as the Mongolians referred to the money Russia claimed it was owed for Soviet aid to Mongolia during 1947–91). Russia had waived 98% of the total debt of 11,400m. transferable roubles and accepted payment of US $250m. Prime Minister Nambaryn Enkhbayar celebrated a political and diplomatic victory for the MPRP Government, but the details of the settlement remained unclear. The two countries had been unable to agree terms for the previous 10 years, especially the equivalence of the now-defunct transferable rouble, but also because of Mongolian opposition pressure to offset the cost of damage done to the environment by Soviet military activity. Democratic Party leader and ex-Prime Minister Enkhsaikhan pointed out that under the Constitution international agreements were supposed to be approved by the Great Khural. Former Prime Minister Byambasüren revived charges of Soviet price-fixing, stating that from 1976 Mongolia had been paid only 0.39 roubles for every one rouble of the value of its exports to the USSR but had been charged 1.5 roubles for its imports of Soviet goods.

Statistical Survey

Unless otherwise indicated, revised by Alan J. K. Sanders

Area and Population

AREA, POPULATION AND DENSITY

Area (sq km)	1,564,116*
Population (census results)	
5 January 1989	2,043,400
5 January 2000	
Males	1,177,981
Females	1,195,512
Total	2,373,493
Population (official estimates at 31 December)	
2001	2,442,500
2002	2,475,400
2003	2,504,000
Density (per sq km) at 31 December	1.6

* 603,909 sq miles.

ADMINISTRATIVE DIVISIONS
(estimates at 31 December 2003)

Province (Aimag)	Area ('000 sq km)	Estimated population ('000)	Provincial centre
Arkhangai	55.3	96.1	Tsetserleg
Bayankhongor	116.0	83.2	Bayankhongor
Bayan-Ölgii	45.7	100.8	Ölgii
Bulgan	48.7	62.8	Bulgan
Darkhan-Uul	3.3	86.5	Darkhan
Dornod (Eastern)	123.6	74.4	Choibalsan
Dornogobi (East Gobi)	109.5	52.1	Sainshand
Dundgobi (Central Gobi)	74.7	50.5	Mandalgobi
Gobi-Altai	141.4	61.4	Altai
Gobi-Sümber	5.5	12.2	Choir
Khentii	80.3	71.1	Öndörkhaan
Khovd	76.1	87.5	Khovd
Khövsgöl	100.6	121.5	Mörön
Orkhon	0.8	75.1	Erdenet
Ömnögobi (South Gobi)	165.4	46.7	Dalanzadgad
Övörkhangai	62.9	113.2	Arvaikheer
Selenge	41.2	101.8	Sükhbaatar
Sükhbaatar	82.3	56.4	Baruun Urt
Töv (Central)	74.0	92.5	Zuun mod
Ulaanbaatar (Ulan Bator)*	4.7	893.4	(capital city)
Uvs	69.6	81.9	Ulaangom
Zavkhan	82.5	82.9	Uliastai
Total	1,564.1	2,504.0	

* Ulaanbaatar, including Nalaikh, and Bagakhangai and Baganuur districts beyond the urban boundary, has special status as the capital city.

ETHNIC GROUPS
(January 2000 census)

	Number	%
Khalh (Khalkha)	1,934,700	81.5
Kazakh (Khasag)	103,000	4.3
Dörvöd (Durbet)	66,700	2.8
Bayad (Bayat)	50,800	2.1
Buryat (Buriat)	40,600	1.7
Dariganga	31,900	1.3
Zakhchin	29,800	1.3
Uriankhai	25,200	1.1
Other ethnic groups	82,600	3.5
Foreign citizens	8,100	0.3
Total	2,373,500	100.0

PRINCIPAL TOWNS
(estimated population, December 1999 unless otherwise indicated)

Ulan Bator (capital)	893,400*	Erdenet	65,700
Darkhan	72,600	Choibalsan	40,900†

* December 2003.
† January 2000 census.

BIRTHS, MARRIAGES AND DEATHS

	Registered live births		Registered marriages*		Registered deaths	
	Number	Rate (per 1,000)	Number	Rate (per 1,000)	Number	Rate (per 1,000)
1996	51,806	22.9	14,188	11.1	17,550	7.7
1997	49,488	21.5	14,421	11.0	16,980	7.4
1998	49,256	21.2	13,908	10.3	15,799	6.8
1999	49,461	21.0	13,722	10.1	16,105	6.8
2000	48,721	20.4	12,601	9.0	15,472	6.5
2001	49,658	20.5	12,393	8.6	15,999	6.6
2002	46,922	19.1	13,514	9.2	15,857	6.4
2003	45,723	18.4	14,572	9.6	16,006	6.4

* Persons aged 18 years and over.

Source: *Mongolian Statistical Yearbook*.

Expectation of life (years at birth): 63.6 (males 60.8; females 66.5) in 2003.

EMPLOYMENT
('000 employees at 31 December)

	2001	2002	2003
Agriculture and forestry	402.4	391.4	387.5
Industry*	93.3	99.2	109.5
Transport and communications	35.1	38.8	39.5
Construction	20.4	25.5	35.1
Trade	90.3	104.5	129.7
Public administration	41.0	43.9	44.8
Education Science, research and development }	55.2	59.3	55.3
Health	33.0	34.5	36.8
Total (incl. others)	832.3	870.8	926.5

Source: *Mongolian Statistical Yearbook.*

Unemployed ('000 registered at 31 December): 40.3 in 2001; 30.9 in 2002; 33.3 in 2003.

* Comprising manufacturing (except printing and publishing), mining and quarrying, electricity, water, logging and fishing.

Source: Asian Development Bank, *Key Indicators of Developing Asian and Pacific Countries.*

Health and Welfare

KEY INDICATORS

Total fertility rate (children per woman, 2002)	2.4
Under-5 mortality rate (per 1,000 live births, 2003)	71
HIV/AIDS (% of persons aged 15–49, 2003)	<0.10
Physicians (per 1,000 head, 2003)	2.7
Hospital beds (per 1,000 head, 2003)	7.3
Health expenditure (2001): US $ per head (PPP)	122
Health expenditure (2001): % of GDP	6.4
Health expenditure (2001): public (% of total)	72.3
Access to water (% of persons, 2000)	60
Access to sanitation (% of persons, 2000)	30
Human Development Index (2002): ranking	117
Human Development Index (2002): value	0.668

For sources and definitions, see explanatory note on p. vi.

Agriculture

PRINCIPAL CROPS
(metric tons)

	2001	2002	2003
Cereals*	150,900	125,861	165,046
Potatoes	58,000	51,888	78,673
Other vegetables	44,500	39,721	59,610
Hay	831,490	767,000	840,700

* Mostly wheat, but also small quantities of barley and oats.

LIVESTOCK
(at 31 December)

	2001	2002	2003
Sheep	11,938,100	10,636,000	10,756,400
Goats	9,683,700	9,134,800	10,652,900
Horses	2,190,800	1,988,900	1,968,900
Cattle	2,069,600	1,884,300	1,792,800
Camels	285,500	253,000	256,700
Pigs	14,773	13,274	13,733
Poultry	54,367	61,603	90,591

LIVESTOCK PRODUCTS
('000 metric tons, unless otherwise indicated)

	2001	2002	2003
Meat	226.4	204.4	153.4
Beef	66.9	60.7	43.6
Mutton and goat meat	104.6	94.9	80.9
Sheep's wool	19.8	17.0	15.2
Cashmere	3.1	2.9	2.7
Milk	290.3	276.6	292.4
Eggs (million)	7.7	4.2	7.1

Source: *Mongolian Statistical Yearbook.*

Forestry

ROUNDWOOD REMOVALS
('000 cubic metres)

	2001	2002	2003
Total	609.9	568.3	576.5

Source: *Mongolian Statistical Yearbook.*

SAWNWOOD PRODUCTION
('000 cubic metres, incl. railway sleepers)

	2001	2002	2003
Total	39.7	28.3	55.0

Source: *Mongolian Statistical Yearbook.*

Fishing

(metric tons, live weight)

	2000	2001	2002
Total catch (freshwater fishes)	425	117	129

Source: FAO.

Mining

(metric tons, unless otherwise indicated)

	2001	2002	2003
Salt	877	680	281
Coal	5,134,200	5,544,400	5,666,100
Fluorspar concentrate	209,000	159,800	198,400
Copper concentrate*	381,400	376,300	372,200
Molybdenum concentrate*	3,028	3,384	3,836
Gold (kilograms)	13,674	12,097	11,118
Crude petroleum (barrels)	73,700	139,205	183,047

* Figures refer to the gross weight of concentrates. Copper concentrate has an estimated copper content of 25%, while the metal content of molybdenum concentrate is 47%.

Source: *Mongolian Statistical Yearbook.*

Industry

SELECTED PRODUCTS

	2001	2002	2003
Flour ('000 metric tons)	37.7	49.6	49.6
Bread ('000 metric tons)	23.3	21.7	21.7
Confectionery ('000 metric tons)	6.0	5.9	6.5
Vodka ('000 litres)	8,626.5	9,436.2	8,873.1
Beer ('000 litres)	4,267.8	3,375.3	3,027.6
Soft drinks ('000 litres)	11,082.7	12,907.3	24,561.1
Cashmere (combed) (metric tons)	608.4	622.1	396.9
Wool; scoured (metric tons)	2,089.7	1,179.6	500.0
Felt ('000 metres)	110.5	112.9	303.0
Camelhair blankets ('000)	43.1	38.2	27.4
Spun thread (metric tons)	45.6	55.9	55.1
Knitwear ('000 garments)	2,315.7	5,563.6	5,148.1
Carpets ('000 sq metres)	614.8	533.9	633.1
Leather footwear ('000)	16.7	9.5	4.6
Felt footwear ('000 pairs)	33.4	16.1	9.0
Surgical syringes (million)	17.7	22.8	25.3
Bricks (million)	21.0	13.2	22.9
Lime ('000 metric tons)	30.1	42.5	42.1
Cement ('000 metric tons)	67.7	147.6	162.3
Ferroconcrete ('000 cu metres)	17.0	11.9	6.9
Steel sheet and blanks ('000 metric tons)	17.2	26.3	60.0
Copper (metric tons)	1,475.9	1,500.0	1,341.1
Electricity (million kWh)	3,017.0	3,111.7	3,137.7

Sources: *Mongolian Statistical Yearbook*.

Finance

CURRENCY AND EXCHANGE RATES

Monetary Units

100 möngö = 1 tögrög (tughrik).

Sterling, Dollar and Euro Equivalents (31 May 2004)

£1 sterling = 2,126.4 tögrög;
US $1 = 1,159.0 tögrög;
€1 = 1,419.3 tögrög;
10,000 tögrög = £4.703 = $8.628 = €7.046.

Average Exchange Rate (tögrög per US $)

2001 1,097.70
2002 1,110.31
2003 1,146.54

BUDGET
(million tögrög)

Revenue	2001	2002	2003
Tax revenue	328,203.2	359,179.2	404,410.6
Income tax	64,504.5	72,433.9	81,811.8
Customs duty	53,330.0	51,321.3	58,914.6
Taxes on goods and services	166,415.9	178,605.1	190,341.4
Value-added tax	104,193.8	118,688.2	121,685.2
Other current revenue	101,748.2	110,569.4	121,958.0
Social insurance	53,956.7	54,397.6	63,623.2
Foreign aid (grants)	9,176.4	6,841.7	8,668.1
Capital revenue (privatization)	162.2	458.6	759.0
Total	439,290.0	477,049.0	535,795.7

Expenditure	2001	2002	2003
Goods and services	254,706.3	283,957.2	286,859.9
Wages and salaries	92,060.4	105,034.1	116,283.0
Interest payments	16,513.6	19,581.9	33,631.5
Subsidies and transfers	95,480.9	109,928.0	125,787.3
Capital expenditure	59,135.2	68,100.3	88,819.7
Lending (net)	63,894.4	67,071.9	81,438.5
Total (incl. others)	489,730.5	548,639.2	616,536.9

Source: *Mongolian Statistical Office*.

2004 (forecasts, million tögrög): Total revenue 553,332.7; Total expenditure 639,327.6 (Source: *Töriin Medeelel*).

INTERNATIONAL RESERVES
(US $ million at 31 December)

	2001	2002	2003
Gold*	50.91	49.79	6.65
IMF special drawing rights	0.02	0.04	0.04
Reserve position in IMF	0.08	0.12	0.14
Foreign exchange	205.60	349.50	235.90
Total	256.61	399.44	242.73

* Valued at 4,300 tögrög per gram.

Source: IMF, *International Financial Statistics*.

MONEY SUPPLY
(million tögrög at 31 December)

	2001	2002	2003
Currency outside banks	109,131	120,755	131,482
Demand deposits at deposit money banks	46,995	66,944	81,337
Total money	156,126	187,699	212,819

Source: IMF, *International Financial Statistics*.

COST OF LIVING
(Consumer Price Index at December; base: December 2000 = 100)

	2001	2002	2003
Foods	108.8	107.0	115.2
Clothing and footwear	104.5	110.4	110.3
Rent and utilities	121.7	127.6	125.3
All items (incl. others)	108.0	109.8	114.9

Source: *Mongolian Statistical Yearbook*.

NATIONAL ACCOUNTS
(million tögrög at current prices, unless otherwise indicated)

Expenditure on the Gross Domestic Product

	2001	2002	2003
Government final consumption expenditure	217,491.3	236,658.8	259,587.7
Private final consumption expenditure	834,662.4	957,789.8	1,072,724.6
Increase in stocks	50,614.5	38,689.6	48,077.2
Gross fixed capital formation	351,594.1	360,918.0	375,354.7
Total domestic expenditure	1,454,362.3	1,594,056.2	1,755,744.2
Exports of goods and services / *Less* Imports of goods and services	−211,310.0	−264,596.4	−291,170.6
Sub-total	1,243,052.3	1,329,459.8	1,464,573.6
Statistical discrepancy*	−127,410.9	−88,673.0	−102,046.9
GDP in purchasers' values	1,115,641.4	1,240,786.8	1,362,526.6
GDP at constant 1995 prices	639,013.0	664,868.3	701,756.6

* Referring to the difference between the sum of the expenditure components and official estimates of GDP, compiled from the production approach.

Source: Asian Development Bank, *Key Indicators of Developing Asian and Pacific Countries*.

Gross Domestic Product by Economic Activity

	2001	2002	2003
Agriculture	277,561.0	256,623.5	272,852.1
Mining	100,832.0	125,896.3	129,043.7
Manufacturing	90,144.3	77,974.9	81,561.7
Electricity, heating and water supply	32,955.3	46,812.1	48,637.7
Construction	21,931.9	29,013.7	33,365.8
Trade	297,831.9	344,010.5	391,881.0
Hotels and restaurants	14,259.3	15,413.4	16,646.5
Transport and communications	144,941.2	182,765.1	205,869.2
Finance	–55.3	4,020.3	6,257.1
Real estate	11,428.5	14,717.5	16,590.0
Public administration, defence and social security	48,179.5	55,959.4	62,423.5
Education and health, other services	75,631.8	87,580.1	97,398.3
GDP in purchasers' values	1,115,641.4	1,240,786.8	1,362,526.6

Source: *Mongolian Statistical Yearbook.*

BALANCE OF PAYMENTS
(US $ million)

	2000	2001	2002
Exports of goods f.o.b.	535.8	523.2	524.0
Imports of goods f.o.b.	–608.4	–623.8	–680.2
Trade balance	–72.6	–100.6	–156.2
Exports of services	77.7	113.5	183.9
Imports of services	–162.8	–205.4	–265.8
Balance on goods and services	–157.7	–192.5	–238.1
Other income received	13.0	14.8	14.1
Other income paid	–19.5	–16.8	–18.6
Balance on goods, services and income	–164.2	–194.5	–242.6
Current transfers received	25.0	40.3	126.9
Current transfers paid	–16.9	—	–42.3
Current balance	–156.1	–154.2	–158.0
Direct investment from abroad	53.7	43.0	77.8
Other investment assets	–44.3	–5.2	–32.1
Other investment liabilities	80.5	69.2	111.7
Net errors and omissions	–19.3	–32.2	14.1
Overall balance	–85.5	–79.4	–13.4

Source: IMF, *International Financial Statistics.*

External Trade

PRINCIPAL COMMODITIES
(US $ million)

Imports c.i.f.	2001	2002	2003
Vegetable products	51.2	56.3	34.5
Prepared foodstuffs; beverages, spirits and vinegar; tobacco and manufactured substitutes	53.2	54.5	57.9
Mineral products	145.4	122.9	151.5
Products of chemical or allied industries	33.8	33.3	40.9
Textiles and textile articles	63.2	83.7	81.5
Base metals and articles thereof	29.7	29.7	41.3
Machinery and mechanical appliances; electrical equipment; sound and television apparatus	113.4	128.3	158.9
Vehicles, aircraft, vessels and associated transport equipment	69.1	69.8	79.7
Total (incl. others)	637.7	659.0	787.3

Exports f.o.b.	2001	2002	2003
Live animals and animal products	26.4	25.6	18.3
Mineral products	175.2	165.5	199.7
Copper concentrate	147.9	138.0	n.a.
Fluorite concentrate	19.7	13.9	n.a.
Hides, skins and furs	59.0	43.3	48.7
Hides and skins	17.3	9.7	n.a.
Cashmere (dehaired)	54.9	30.1	n.a.
Textiles and textile articles	171.5	136.6	152.9
Knitwear	18.9	17.5	n.a.
Precious metals and jewellery	75.4	119.2	139.8
Gold	74.7	117.6	n.a.
Total (incl. others)	521.5	500.9	600.2

Sources: National Statistical Office, *Monthly Bulletin of Statistics.*

PRINCIPAL TRADING PARTNERS
(US $ million)

Imports c.i.f.	2001	2002	2003
Australia	1.4	11.3	19.6
China, People's Republic	136.1	167.7	196.3
Germany	30.3	30.4	38.0
Japan	55.9	42.8	63.4
Korea, Republic	58.3	86.3	67.7
Russia	225.9	237.6	265.4
USA	14.9	23.4	23.5
Total (incl. others)	637.7	690.8	801.0

Exports f.o.b.	2001	2002	2003
Australia	10.0	17.7	34.5
China, People's Republic	238.3	220.5	287.0
Japan	15.7	6.3	8.5
Republic of Korea	20.2	22.5	7.5
Russia	44.9	165.7	142.9
United Kingdom	12.4	17.5	26.1
USA	144.5	165.7	142.9
Total (incl. others)	521.5	524.0	615.9

Source: *Mongolian Statistical Yearbook.*

Transport

FREIGHT CARRIED
('000 metric tons)

	2001	2002	2003
Rail	10,147.7	11,637.0	12,284.7
Road	1,658.2	1,888.7	5,335.9
Air	2.9	2.4	2.2
Water	1.7	1.8	—
Total	11,810.5	13,529.9	17,622.8

Source: Mongolian Statistical Directorate.

PASSENGERS CARRIED
(million)

	2001	2002	2003
Rail	4.1	4.0	3.9
Road	94.1	101.4	163.7
Air*	0.3	0.3	0.3
Total	98.5	105.7	167.9

* MIAT only.

Source: *Mongolian Statistical Yearbook.*

RAILWAYS
(traffic)

	2001	2002	2003
Passengers carried ('000)	4,082.8	3,962.6	3,947.8
Freight carried ('000 metric tons)	10,147.7	11,637.0	12,284.7
Freight ton-km (million)	5,287.9	6,461.3	7,253.3

Source: *Mongolian Statistical Yearbook.*

ROAD TRAFFIC
(motor vehicles in use)

	2001	2002	2003
Passenger cars	53,198	63,224	68,458
Buses and coaches	10,187	10,841	9,834
Lorries, special vehicles and tankers	29,686	29,740	27,483

Source: *Mongolian Statistical Yearbook.*

CIVIL AVIATION
(traffic on scheduled services)

	2001	2002	2003
Passengers carried ('000)	254.5	269.5	251.9
Freight carried (tons)	2,890.8	2,394.2	2,230.7

Source: Mongolian Statistical Directorate.

Tourism

FOREIGN ARRIVALS BY NATIONALITY

Country	2001	2002	2003
China, People's Republic	67,360	92,657	91,934
France	2,764	2,891	2,751
Germany	5,388	6,856	4,999
Japan	11,565	13,708	7,757
Korea, Republic	10,098	14,536	17,205
Russia	66,415	71,368	53,330
United Kingdom	3,122	3,537	2,859
USA	6,653	6,860	5,570
Total (incl. others)	192,057	235,165	204,845

Source: *Mongolian Statistical Yearbook.*

Tourism receipts (US $ million): 36 in 2000; 39 in 2001; 130 in 2002 (Source: World Tourism Organization).

Communications Media

	2001	2002	2003
Television receivers ('000 in use)	173.1	200.0	220.0
Telephones ('000 main lines in use)	119.7	126.7	135.5
Mobile cellular telephones ('000 subscribers)	91.2	256.8	319.4
Internet users ('000)	10.3	10.0	11.2
Newspapers printed (million copies)	14.6	17.0	18.5

Book production (1994): 128 titles; 640,000 copies.

Non-daily newspapers (titles): 80 in 1998.

Other periodicals (titles): 24 in 1998.

Sources: mostly *Mongolian Statistical Yearbook.*

Education

(2003/04)

	Institutions	Teachers	Students
General education schools:			
Primary (grades 1–3)	72	7,200	232,400
Incomplete secondary (grades 4–8)	193	13,600 (grades 4–10)	232,000
Complete secondary (grades 9–10)	421		73,000
Vocational schools (incl. private)	32	n.a.	21,500
Higher education:			
Universities and colleges*:			
State-owned	46	n.a.	73,900
Private	137	n.a.	34,600

* Excluding 1,700 students studying abroad.

Pre-school institutions (2003): 687 kindergartens (incl. 155 with crèche facilities) attended by 90,215 infants, and a further 14 crèches attended by 3,700 infants.

Source: *Mongolian Statistical Yearbook.*

Adult literacy rate (estimates based on census data): 97.8% (males 98.0%; females 97.5%) in 2002 (Source: UN Development Programme, *Human Development Report*).

Directory

The Constitution

The Constitution was adopted on 13 January 1992 and came into force on 12 February of that year. It proclaims Mongolia (*Mongol Uls*), with its capital at Ulan Bator (Ulaanbaatar), to be an independent sovereign republic which ensures for its people democracy, justice, freedom, equality and national unity. It recognizes all forms of ownership of property, including land, and affirms that a 'multi-structured economy' will take account of 'universal trends of world economic development and national conditions'.

The 'citizen's right to life' is qualified by the death penalty for serious crimes, and the law provides for the imposition of forced labour. Freedom of residence and travel within the country and abroad may be limited for security reasons. The citizens' duties are to respect the Constitution and the rights and interests of others, pay taxes, and serve in the armed forces, as well as the 'sacred duty' to work, safeguard one's health, bring up one's children and protect the environment.

Supreme legislative power is vested in the Mongolian Great Khural (Assembly), a single chamber with 76 members elected by universal adult suffrage for a four-year term, with a Chairman and Vice-Chairman elected from amongst the members. The Great Khural recognizes the President on his election and appoints the Prime Minister and members of the Cabinet. A presidential veto of a decision of the Great Khural can be overruled by a two-thirds majority of the Khural. Decisions are taken by a simple majority.

The President is Head of State and Commander-in-Chief of the Armed Forces. He must be an indigenous citizen at least 45 years old who has resided continuously in Mongolia for the five years before election. Presidential candidates are nominated by parties with seats in the Great Khural; the winning candidate in general presidential elections is President for a four-year term.

The Cabinet is the highest executive body and drafts economic, social and financial policy, takes environmental protection measures, strengthens defence and security, protects human rights and implements foreign policy for a four-year term.

The Supreme Court, headed by the Chief Justice, is the highest judicial organ. Judicial independence is protected by the General Council of Courts. The Procurator General, nominated by the President, serves a six-year term.

Local administration in the 21 *aimag* (provinces) and Ulan Bator is effected on the basis of 'self-government and central guidance', comprising local khurals of representatives elected by citizens and governors (*zasag darga*), nominated by the Prime Minister to serve four-year terms.

The Constitutional Court, which guarantees 'strict observance' of the Constitution, consists of nine members nominated for a six-year term, three each by the Great Khural, the President and the Supreme Court.

The first amendments to the Constitution, adopted by the Mongolian Great Khural in December 2000, despite opposition over procedure from the Constitutional Court, were finally approved by President Bagabandi in May 2001. The main effects of the amendments were to clarify the method of appointment of Prime Ministers, enable decision-making by a simple majority vote, and shorten the minimum length of sessions of the Khural from 75 days to 50.

The Government

PRESIDENCY

President and Commander-in-Chief of the Armed Forces: NATSAGIIN BAGABANDI (elected President of Mongolia 18 May 1997; re-elected 20 May 2001).

Head of Presidential Secretariat: BADAMDORJIIN BATKHISHIG.

Director of the Presidential Information Service: DALANTAIN KHALIUN.

NATIONAL SECURITY COUNCIL

The President heads the National Security Council; the Prime Minister and the Chairman of the Mongolian Great Khural are its members. The Secretary is the President's national security adviser.

Chairman: NATSAGIIN BAGABANDI.

Members: NAMBARYN ENKHBAYAR, TSAKHIAGIIN ELBEGDORJ.

Secretary: DÜGERJAVYN GOTOV.

CABINET

(September 2004)

Prime Minister: TSAKHIAGIIN ELBEGDORJ.

Deputy Prime Minister: CHÜLTEMIIN ULAAN.

General Ministries

Minister of Foreign Affairs: TSENDIIN MÖNKH-ORGIL.

Minister of Finance and Economy: NOROVYN ALTANKHUYAG.

Minister of Justice and Home Affairs: TSENDIIN NYAMDORJ.

Sectoral Ministries

Minister of Nature and the Environment: ULAMBAYARYN BARSBOLD.

Minister of Defence: BADARCHIIN ERDENEBAT.

Minister of Education, Culture and Science: PUNTSAGIIN TSAGAAN.

Minister of Construction and Town Management: NYAMJAVYN BATBAYAR.

Minister of Road Transport, Travel and Tourism: GAVAAGIIN BATKHÜÜ.

Minister of Social Welfare and Labour: TSEVELMAAGIIN BAYARSAIKHAN.

Minister of Industry and Trade: SÜKHBAATARYN BATBOLD.

Minister of Fuel and Power: TÜVDENGIIN OCHIRKHÜÜ.

Minister of Food and Agriculture: DENDEVIIN TERBISHDAGVA.

Minister of Health: TÖGSJARGALYN GANDI.

Head of Government Affairs Directorate: SANGAJAVYN BAYARTSOGT.

Minister in charge of Professional Qualifications Control: ICHINKHORLOOGIIN ERDENEBAATAR.

Minister in charge of Disaster Relief: UKHNAAGIIN KHÜRELSÜKH.

MINISTRIES AND GOVERNMENT DEPARTMENTS

All Ministries and Government Departments are in Ulan Bator.

Prime Minister's Office: State Palace, Sükhbaataryn Talbai 1, Ulan Bator 12; fax (11) 328329; internet www.pmis.gov.mn/primeminister.

Ministry of Construction and Town Management: Ulan Bator.

Ministry of Defence: Government Bldg 7, Dandaryn Gudamj, Bayanzürkh District, Ulan Bator 61; tel. (11) 458495; fax (11) 451727; e-mail mef@mongolnet.mn; internet www.pmis.gov.mn/mdef/mongolian.

Ministry of Education, Culture and Science: Government Bldg 3, Baga Toiruu 44, Sükhbaatar District, Ulan Bator 210620; tel. (11) 322480; fax (11) 323158; internet www.med.pmis.gov.mn.

Ministry of Finance and Economy: Government Bldg 2, Negdsen Ündestnii Gudamj 5/1, Chingeltei District, Ulan Bator 210646; tel. (11) 320247; fax (11) 322712; internet www.mof.pmis.gov.mn.

Ministry of Food and Agriculture: Government Bldg 9, Enkh Taivny Örgön Chölöö 16A, Bayanzürkh District, Ulan Bator 210349; tel. (11) 450258; fax (11) 452554; e-mail mofa@mofa.pmis.gov.mn; internet www.pmis.gov.mn/food&agriculture.

Ministry of Foreign Affairs: Enkh Taivny Örgön Chölöö 7A, Sükhbaatar District, Ulan Bator 210648; tel. (11) 311311; fax (11) 322127; e-mail mongmer@magicnet.mn; internet www.extmin.mn.

Ministry of Fuel and Power: Ulan Bator.

Ministry of Health: Government Bldg 8, Olimpiin Gudamj 2, Sükhbaatar District, Ulan Bator; tel. (11) 323381; fax (11) 320916; e-mail moh@moh.mng.net; internet www.pmis.gov.mn/health/index.

Ministry of Industry and Trade: Block A, Government Bldg 2, Negdsen Ündestnii Gudamj 5/1, Chingeltei District, Ulan Bator 210646; tel. (11) 329222; fax (11) 322595; e-mail mit@mit.pmis.gov.mn; internet www.mit.pmis.gov.mn.

Ministry of Justice and Home Affairs: Government Bldg 5, Khudaldaany Gudamj 6/1, Chingeltei District, Ulan Bator 210646; tel. (11) 322383; fax (11) 325225; e-mail forel@moj.pmis.gov.mn; internet www.pmis.gov.mn/mjus.

Ministry of Nature and the Environment: Government Bldg 3, Baga Toiruu 44, Sükhbaatar District, Ulan Bator 210620; tel. (11) 320943; fax (11) 321401; e-mail mtt@magicnet.mn; internet www.pmis.gov.mn/men.

Ministry of Road Transport, Travel and Tourism: Ulan Bator.

Ministry of Social Welfare and Labour: Government Bldg 2, UN St 5, Ulan Bator 210646; tel. (11) 324918; fax (11) 328634; e-mail mswl@mongolnet.mn; internet www.mswl.gov.mn.

Government Affairs Directorate (Cabinet Secretariat): State Palace, Sükhbaataryn Talbai 1, Ulan Bator 12; tel. (11) 323501; fax (11) 310011; internet www.pmis.gov.mn/cabinet.

National Statistical Office: Government Bldg 3, Baga Toiruu 44, Sükhbaatar District, Ulan Bator 210620; tel. (11) 322424; fax (11) 324518; e-mail info@nso.mn; internet www.nso.mn.

President and Legislature

PRESIDENT

Office of the President: Ulan Bator; fax (11) 311121; internet www.pmis.gov.mn/president.

Election, 20 May 2001

Candidate	Votes	%
Natsagiin Bagabandi (MPRP)	574,553	57.95
Radnaasümbereliin Gonchigdorj (Democratic Party)	362,684	36.58
Luvsandambyn Dashnyam (Civil Courage Party)	35,139	3.54

MONGOLIAN GREAT KHURAL

Under the fourth Constitution, which came into force in February 1992, the single-chamber Mongolian Great Khural is the State's supreme legislative body. With 76 members elected for a four-year term, the Great Khural must meet for at least 50 working days in every six months. Its Chairman may act as President of Mongolia when the President is indisposed.

Chairman: NAMBARYN ENKHBAYAR.

Vice-Chairpersons: SANJAASÜRENGIIN OYUUN, DANZANGIIN LUNDEEJANTSAN.

General Secretary: NAMSRAIJAVYN LUVSANJAV.

General Election, 27 June 2004

Party	Seats
Mongolian People's Revolutionary Party (MPRP)	36
Motherland Democracy coalition*	34
Mongolian Republican Party	1
Independents†	3
Undeclared‡	2
Total	76

* Comprising the Democratic Party (DP), the 'Motherland' Mongolian Democratic New Socialist Party (MDNSP) and the Civil Courage Republican Party.
† All three candidates were DP members who stood independently.
‡ The final results of voting in two constituencies remained in dispute.

Political Organizations

Civil Courage-Republican Party (CC-RP): Rm 1, Altai College, Sükhbaatar District, Ulan Bator (CPO 13, Box 37); tel. and fax (11) 319006; e-mail oyun@mail.parl.gov.mn; f. 2002 by merger of the Civil Courage Party and Mongolian Republican Party; also known as the Citizens' Will (Republican Party); Chair. SANJAASÜRENGIIN OYUUN.

Democratic Party (DP): Chingisiin Örgön Chölöö 1, Ulan Bator; f. 2000 by amalgamation of the Mongolian National Democratic Party, Mongolian Social-Democratic Party, Mongolian Democratic Party, Mongolian Democratic Renewal Party and the Mongolian Believers' Democratic Party; c. 170,000 mems. (May 2002); Chair. MENDSAIKHANY ENKHSAIKHAN; Sec.-Gen. LÜIMEDIIN GANSÜKH.

Mongolian Democratic New Socialist Party (MDNSP): Erel Co, Bayanzürkh District, Ulan Bator; internet www.mongol.net/mdnsp; f. 1998; amalgamated with Mongolian Workers' Party 1999; 110,000 mems (May 2002); Chair. BADARCHIIN ERDENEBAT.

Mongolian Democratic Party (MDP): Ulan Bator; f. 1990; merged in Oct. 1992 with other parties to form the Mongolian National Democratic Party (MNDP); reconstituted in Jan. 2000, the party won no seats in the 2000 elections, and in Dec. 2000 most members merged with the MDNP and other parties to form the Democratic Party; a splinter group opposed to the merger tried unsuccessfully to challenge the legal status of the DP and then elected a new MDP leadership; Chair. DAMDINDORJIIN NINJ.

Mongolian Green Party: POB 51, Erkh Chölöönii Gudamj 11, Ulan Bator 38; tel. (11) 323871; fax (11) 458859; f. 1990; political wing of the Alliance of Greens; 5,000 mems (March 1997); Chair. DAVAAGIIN BASANDORJ.

Mongolian Liberal Party: f. 1999 as Mongolian Civil Democratic New Liberal Party, renamed 2004; ruling body Little Khural of 90 mems with Leadership Council of nine; Chair. D. BANZRAGCH.

Mongolian Liberal Democratic Party: POB 470, Ulan Bator 44; tel. (11) 315555; fax (11) 310076; e-mail mldp@magicnet.mn; internet www.mldp.mn; f. 1998 as Mongolian Socialist Democratic (Labour) Party; ruling body Political Council; 848 mems (1998); Chair. TÜVSHINBATYN TÖMÖRMÖNKH.

Mongolian National Solidarity Party: 'Ikh Zasag' Institute Bldg, 4th Horoo, Bayanzürkh District, Ulan Bator; tel. and fax (11) 455736; f. 1994 as Mongolian Solidarity Party; 15,480 mems (March 2002); Chair. NAMSRAIN NYAM-OSOR.

Mongolian New Social Democratic Party: f. 2000 by dissidents of the Mongolian Social-Democratic Party (MSDP); Chair. LANTUUGIIN DAMDINSÜREN.

Mongolian People's Party (MPP): Ulan Bator; tel. (11) 311083; f. 1991; forestalled any MPRP plans to revert to its original name, MPP; 2,000 mems (June 1995); in March 2000 some mems (led by Chairman Dembereliin Ölziibaatar) claimed to have merged the MPP with the MPRP; others reaffirmed the MPP's independence at a party congress in April; Chair. Lama DORLIGJAVYN BAASAN.

Mongolian People's Revolutionary Party (MPRP): Baga Toiruu 37/1, Ulan Bator 11; tel. (11) 322745; fax (11) 320368; f. 1921 as Mongolian People's Party; c. 122,000 mems (May 2002); ruling body Party Little Khural (244 mems in Feb. 2001), which elects the Leadership Council; Chair. NAMBARYN ENKHBAYAR; Gen. Sec. DOLOONJINGIIN IDEVKHTEN.

Mongolian Republican Party: c/o The Mongolian Great Hural, Ulan Bator; this party was re-registered in 2004 after the split in the leadership of the Civil Courage Republican Party (see above); Chair. BAZARSADYN JARGALSAIKHAN.

Mongolian Rural Development Party: f. 1995 as the Mongolian Countryside Development Party, reorganized in December 1999; Pres. L. CHULUUNBAATAR.

Mongolian Traditional United Party: Huvisgalchdyn Örgön Chölöö 26, Ulan Bator; tel. (11) 325745; fax (11) 342692; also known as the United Heritage (conservative) Party; f. 1993 as an amalgamation of the United Private Owners' Party and the Independence Party; 14,000 mems (1998); ruling body General Political Council; Chair. ÜRJINGIIN KHÜRELBAATAR.

Diplomatic Representation

EMBASSIES IN MONGOLIA

Bulgaria: Olimpiin Gudamj 8, Ulan Bator (CPO Box 702); tel. and fax (11) 322841; e-mail posolstvobg@magicnet.mn; Ambassador NIKOLAI MARIN.

China, People's Republic: Zaluuchuudyn Örgön Chölöö 5, Ulan Bator (CPO Box 672); tel. (11) 320955; fax (11) 311943; Ambassador GAO SHUMAO.

Cuba: Negdsen Ündestnii Gudamj 18, Ulan Bator (CPOB 710); tel. (11) 323778; fax (11) 327709; e-mail embacuba@mongol.net; Ambassador EDUARDO CASTELLANOS SOTO.

Czech Republic: POB 665, Olimpiin Gudamj 14, Ulan Bator; tel. (11) 321886; fax (11) 323791; e-mail czechemb@mongol.net; internet www.mzv.cz/ulaanbaatar; Ambassador JIŘI NEKVASIL.

France: Diplomatic Corps Bldg, Apartment 48, Ulan Bator 6 (CPO Box 687); tel. (11) 324519; fax (11) 329633; e-mail ambafrance@magicnet.mn; Ambassador NICOLAS CHAPUIS.

Germany: Negdsen Ündestnii Gudamj 7, Ulan Bator (CPO Box 708); tel. (11) 323325; fax (11) 323905; e-mail germanemb_ulanbator@mongol.net; Ambassador MICHAEL VORWERK.

Hungary: Enkh Taivny Gudamj 1, Ulan Bator (CPO Box 668); tel. (11) 323973; fax (11) 311793; e-mail huembuln@mongol.net; Ambassador (vacant).

India: Zaluuchuudyn Örgön Chölöö 10, Ulan Bator (CPO Box 691); tel. (11) 329522; fax (11) 329532; e-mail indembmongolia@magicnet.mn; Ambassador GAURI SHANKAR GUPTA.

Japan: Zaluuchuudyn Gudamj 12, Ulan Bator 13 (CPO Box 1011); tel. (11) 320777; fax (11) 313332; e-mail eojmongol@magicnet.mn; internet www.eojmongolia.mn; Ambassador TATSUO TODA.

Kazakhstan: Diplomatic Corps Bldg, Apartment 11, Chingeltei District, Ulan Bator (CPO Box 291); tel. (11) 312240; fax (11) 312204; e-mail kzemby@magicnet.mn; Ambassador J. S. KARYBDZHANOV.

Korea, Democratic People's Republic: Ulan Bator; Ambassador PAK JONG-DO.

Korea, Republic: Olimpiin Gudamj 10, Ulan Bator (CPO Box 1039); tel. (11) 321548; fax (11) 311157; Ambassador KEUM BYUNG-MOK.

Laos: Ikh Toiruu 59, Ulan Bator (CPO Box 1030); tel. (11) 322834; fax (11) 321048; e-mail laoemb@mongol.net; Ambassador SHAMPONG SOURIPOIMI.

Poland: Diplomatic Corps Bldg, 95, Apartment 66–67, Ulan Bator 13; tel. (11) 320641; fax (11) 320576; Ambassador KRZYSZTOF DEBNICKI.

Russia: Enkh Taivny Gudamj A-6, Ulan Bator (CPO Box 661); tel. (11) 327851; fax (11) 327018; e-mail embassy_ru@mongol.net; Ambassador OLEG DERKOVSKII.

Turkey: Enkh Taivny Örgön Chölöö 5, Ulan Bator (CPO Box 1009); tel. (11) 313992; fax (11) 313992; e-mail turkemb@mongol.net; Ambassador TANER KARAKAŞ.

United Kingdom: Enkh Taivny Gudamj 30, Ulan Bator 13 (CPO Box 703); tel. (11) 458133; fax (11) 458036; e-mail britemb@magicnet.mn; Ambassador RICHARD AUSTEN.

USA: Ikh Toiruu 59/1, Ulan Bator (CPO Box 1021); tel. (11) 329606; fax (11) 320776; e-mail webmaster@us-mongolia.com; Ambassador PAMELA JO SLUTZ.

Viet Nam: Enkh Taivny Örgön Chölöö 47, Ulan Bator (CPO Box 670); tel. (11) 458917; fax (11) 458923; Ambassador CHAN NGUYEN TICH.

Judicial System

Under the fourth Constitution, judicial independence is protected by the General Council of Courts, consisting of the Chief Justice (Chairman of the Supreme Court), the Chairman of the Constitutional Court, Procurator General, Minister of Law and others. The Council nominates the members of the Supreme Court for approval by the Great Khural. The Chief Justice is chosen from among the members of the Supreme Court and approved by the President for a six-year term. Civil, criminal and administrative cases are handled by Ulan Bator City court, the 21 *aimag* (provincial) courts, *sum* (rural district) and urban district courts, while the system of special courts (military, railway, etc.) is still in place. The Procurator General and his deputies, who play an investigatory role, are nominated by the President and approved by the Great Khural for six-year terms. The Constitutional Court safeguards the constitutional legality of legislation. It consists of nine members, three nominated each by the President, Great Khural and Supreme Court, and elects a Chairman from among its number.

Chief Justice: CHIMEDLKHAMYN GANBAT.

Procurator General: MONGOLYN ALTANKHUYAG.

Chairman of Constitutional Court: NAVAANPERENLEIN JANTSAN.

Religion

The 1992 Constitution maintains the separation of Church and State but forbids any discrimination, declaring that 'the State shall respect religion and religion shall honour the State'. During the early years of communist rule Mongolia's traditional Mahayana Buddhism was virtually destroyed, then exploited as a 'show-piece' for visiting dignitaries (although the Dalai Lama himself was not permitted to visit Mongolia until the early 1980s). The national Buddhist centre is Gandantegchinlen Khiid (monastery) in Ulan Bator, with about 100 lamas and a seminary; it is the headquarters of the Asian Buddhist Conference for Peace. In the early 1990s some 2,000 lamas established small communities at the sites of 120 former monasteries, temples and religious schools, some of which were being restored. These included two other important monasteries, Erdene Zuu and Amarbayasgalant. The Kazakhs of western Mongolia are nominally Sunni Muslims, but their mosques, also destroyed in the 1930s or closed subsequently, are only now being rebuilt or reopened. Traces of shamanism from the pre-Buddhist period still survive. In recent years there has been a new upsurge in Christian missionary activity in Mongolia. However, the Law on State-Church Relations (of November 1993) sought to make Buddhism the predominant religion and restricted the dissemination of beliefs other than Buddhism, Islam and shamanism. The law was challenged by human rights campaigners and Mongolian Christians as unconstitutional.

BUDDHISM

At the end of 2003 there were 214 Buddhist temples and monasteries in Mongolia (including 53 in Ulan Bator) with 2,091 lamas (monks) and 4,465 apprentices in religious schools, of whom 1,021 were under 16 years of age. It is estimated that about 70% of the adult population (975,600) are Buddhists, that is, some 39% of the total population.

Living Buddha: The Ninth Javzandamba Khutagt (Ninth Bogd), Jambalnamdolchoijinjaltsan (resident in Dharamsala, India).

Gandantegchinlen Monastery: Ulan Bator 58; tel. (11) 360023; Centre of Mongolian Buddhists; Khamba Lama (Abbot) DEMBERELIIN CHOIJAMTS.

Mongolian Buddhist Association: Pres. G. ENKHSAIKHAN.

Pethub Buddhist Institute: PO Box 105, Ulan Bator 38; tel. (11) 322366; fax (11) 320676; e-mail pethubmongolia@magicnet.mn; internet www.pethubmonastery.com; f. 2001 by Ven. Kushok Bakula Rinpoche (Indian Ambassador to Mongolia 1990–2000).

CHRISTIANITY

Roman Catholic Church

The Church is represented in Mongolia by a single mission. At April 2004 there were 187 Catholics in the country.

Catholic Mission: POB 694, Ulan Bator; tel. (11) 458825; fax (11) 458027; e-mail ccmvatican@magicnet.mn; f. 1992; Superior Bishop WENCESLAO PADILLA.

Cathedral of St. Peter and St. Paul: Bayanzürkh District, Ulan Bator.

Protestant Church

Association of Mongolian Protestants: f. 1990; Pastor M. BOLDBAATAR.

Mongolian Evangelical Alliance: Kseni Bldg, Baga Toiruu, Chingeltei District, Ulan Bator; tel. 312771; e-mail mea@magicnet.mn; f. 1998; a branch of the World Evangelical Alliance.

Russian Orthodox Church

Holy Trinity Church: Jukovyn Gudamj 55, Ulan Bator; opened in 1870, closed in 1930; services recommenced 1997 for Russian community; Father ANATOLII FESECHKO.

ISLAM

Muslim Society: f. 1990; Hon. Pres. K. SAIRAAN; Chair. of Central Council M. AZATKHAN.

BAHÁ'Í FAITH

Bahá'í Community: Ulan Bator; tel. 321867; f. 1989; Leader A. ARIUNAA.

The Press

PRINCIPAL NATIONAL NEWSPAPERS

State-owned publications in Mongolia were denationalized with effect from 1 January 1999, although full privatization could not proceed immediately.

MN-Önöödör (Today): Mongol News Co, Ikh Toiruu 20, Ulan Bator; tel. (11) 352504; fax (11) 352501; e-mail mntoday@mobinet.mn; internet www.mongolnews.mn; f. 1996; 304 a year; Editor-in-Chief TS. BALDORJ; circ. 5,500.

Mongolyn Medee (Mongolian News): Erel Group, 13th Sub-District, Bayanzürkh District, Ulan Bator; tel. (11) 453169; fax (11) 453816; e-mail news-of-mon@magicnet.mn; 256 a year; Editor-in-Chief D. SANDAGSÜREN; circ. 2,900.

Ödriin Sonin (Daily News): Ikh Toiruu, Ulan Bator 20; tel. 99263536; fax (11) 352499; e-mail daily_news@mbox.mn; internet www.dn.mongolmedia.com; f. 1924; restored 1990; fmrly Ardyn Erkh, Ardyn Ündesnii Erkh, Ündesnii Erkh and Ödriin Toli; 312 a year; Editor-in-Chief JAMBALYN MYAGMARSÜREN; circ. 14,200.

Ünen (Truth): Baga Toiruu 11, Ulan Bator; tel. (11) 323847; fax (11) 321287; e-mail unen@magicnet.mn; internet www.unen.mn; f. 1920; publ. by MPRP; 256 a year; Editor-in-Chief TSERENSODNOMYN GANBAT; circ. 8,330.

Zuuny Medee (Century's News): Amaryn Gudamj 1, Ulan Bator 20; tel. (11) 320940; fax (11) 321279; e-mail zuunii_medee@yahoo.com; internet www.zuuniimedee.mn; f. 1991 as *Zasgiin Gazryn Medee*; 312 a year; Editor-in-Chief TSERENDORJIIN TSETSEGCHULUUN; circ 8,000.

PRINCIPAL PERIODICALS

81-r Suvag (Channel 81): POB 367/13, Ulan Bator; tel. 99118359; e-mail channel81@mongolmedia.com; internet www.channel81.mongolmedia.com; publishes views of the Mongolian Newspaper Asscn; 40 a year; Editor-in-Chief JAMSRANGIIN BAYARJARGAL.

Anagaakh Arga Bilig (The Healthy Way of Yin and Yang): POB 1053/13, Ulan Bator; tel. (11) 321367; e-mail arslny7144@magicnet.mn; twice monthly; Editor YA. ARSLAN.

Ardyn Elch (People's Envoy): St. Petersburg Centre, 2nd Floor, Rm 36, Sükhbaataryn Örgön Chölöö; tel. 95252346; fax (11) 314580; monthly; Editor-in-Chief B. TSEDEN.

Bagsh (Teacher): Rm 106, Teachers' College, Ulan Bator; tel. 99183398; f. 1989 by Ministry of Education; 24 a year.

Business Times: Chamber of Commerce and Industry, Government Bldg 1, Rm 806, Erkh Chölöönii Talbai, Ulan Bator; tel. (11) 325374; fax (11) 324620; e-mail bis_times@usa.net; 48 a year; Editor B. SARANTUYAA.

Deedsiin Amidral (Elite's Life): Bldg 6, Rm 207, A. Amaryn Gudamj, POB 536/13, Ulan Bator; tel. 91189399; fax (11) 323847; e-mail elitslife@mongolmedia.com; internet www.elitslife.mongolmedia.com; Editor-in-Chief TSEGMIDIIN CHIMIDDONDOG.

Deedsiin Khüreelen (Elite's Forum): CPO Box 1114, Ulan Bator; tel. (11) 325687; 48 a year; Editor NYAMBUUGIIN BILGÜÜN; circ. 12,000.

Dorno-Örnö (East-West): POB 17/48, Ulan Bator; fax (11) 322613; f. 1978; publ. by Institute of Oriental and International Studies of Acad. of Sciences; scientific and socio-political journal; history, culture, foreign relations; articles in Mongolian; two a year; Editor-in-Chief Dr N. ALTANTSETSEG.

Erüül Mend (Health): Ulan Bator; fax (11) 321278; publ. by Ministry of Health; monthly; Editor-in-Chief SH. JIGJIDSÜREN; circ. 5,600.

Il Tovchoo (Openness): Mongol Nom Co Bldg, Rm 8, Ulan Bator; POB 234, Ulan Bator 46; tel. (11) 325517; e-mail iltovchoo@magicnet.mn; f. 1990; current affairs and cultural news; 36 a year; Editor-in-Chief G. AKIM.

Khani (Spouse): POB 600/49, National Agricultural Co-operative Members' Association Bldg, Rm 103, Enkh Taivny Gudamj 18A-1, Bayanzürkh District, Ulan Bator; women and family issues; 36 a year; Editor-in-Chief D. Otgontuyaa; circ. 64,920.

Khöldömör (Labour): Sükhbaataryn Talbai 3, Ulan Bator 210664; tel. (11) 323026; f. 1928; publ. by the Confederation of Mongolian Trade Unions; 36 a year; Editor-in-Chief Tsoodolyn Khulan; circ. 64,920.

Khökh Tolbo (Blue Spot): POB 306/24, Mon-Azi Co Bldg, Ulan Bator; tel. and fax (11) 313405; 48 a year; Editor-in-Chief Batyn Erdenebaatar; circ. 3,500.

Khuuli Züin Medeelel (Legal Information): Ministry of Justice and Home Affairs, Ulan Bator; f. 1990; 24 a year.

Khuviin Amidral (Private Life): POB 429/44, Ulan Bator 44; tel. (11) 329926; fax (11) 310181; 36 a year; Editor-in-Chief T. Amardavaa; circ. 2,900.

Khuviin Soyol (Personal Culture): CPOB 1254, Rm 2, Block 39, behind No. 5 School, Baga Toiruu, Ulan Bator; Editor Bekhbazaryn Bekhsüren.

Khümüün Bichig (People and Script): Montsame, Ulan Bator; tel. (11) 329486; fax (11) 327857; e-mail montsame@pop.mn; current affairs in Mongolian classical script; 36 a year; Editor T. Galdan; circ. 15,000.

Khümüüs (People): POB 411, Ulan Bator 46; tel. (11) 328732; fax (11) 318363; e-mail khumuus@mongol.net; 48 a year; Editor O. Mönkh-Erdene; circ. 21,500.

Khümüüsiin Amidral (People's Life): POB 411, Ulan Bator 46; tel. (11) 328732; fax (11) 318363; e-mail khumuus@mongol.net; 48 a year; Editor B. Khulan.

Mash Nuuts (Top Secret): POB 113/49, Ulan Bator 49; 24 a year; Editors N. Batdelger, Ts. Mönkhtuyaa.

Mongol Taims (Mongol Times): Jamyangiin Gudamj 9, Ulan Bator; tel. and fax (11) 458802; 48 a year; Editor Chonoin Kulanda.

Mongol Törkh (Mongolian Style): POB 418/46, MPRP Bldg, Rm 102, Sükhbaatar District, Ulan Bator; tel. and fax (11) 311406; 36 a year; Editor D. Badamgarav.

Mongoljin Goo (Mongolian Beauty): POB 717/44, Mongolian Women's Federation, Ulan Bator; tel. (11) 320790; fax (11) 367406; e-mail monwofed@magicnet.mn; f. 1990; monthly; Editor J. Erdenechimeg; circ. 3,000.

Mongolyn Anagaakh Ukhaan (Mongolian Medicine): CPOB 696/13, Ulan Bator 13; tel. (11) 112306; fax (11) 451807; e-mail nymadawa@hotmail.com; publ. by Scientific Society of Mongolian Physicians, Sub-assembly of Medical Sciences and Mongolian Academy of Sciences; quarterly; Editor-in-Chief Prof. Pagvajavyn Nyamdavaa.

Mongolyn Khödöö (Mongolian Countryside): Mongolian National University, Zaisan, Khan-Uul District, Ulan Bator; tel. 91164884; e-mail tulgaa_h2000@yahoo.com; publ. by Ministry of Agriculture and Industry, and Academy of Agricultural Sciences; monthly; Editor-in-Chief Sodnomdarjaagiin Battulga.

Montsame-giin Medee (Montsame News): CPOB 1514, Montsame, Ulan Bator; tel. (11) 314507; fax (11) 314511; daily news digest primarily for government departments; 264 a year; Editor B. Naminchimed.

Notstoi Medee (Important News): POB 359/20, Maximum Press Co, Ulan Bator; tel. 99133270; Editor B. Mönkhzul.

Oroin Medee (Evening News): Bayangol District, Ulan Bator; tel. 90151080; e-mail e_news@hotmail.com; Editor-in-Chief S. Badamragchaa.

Sain Baina Uu? (Hello!): CPOB 1085, Chingisiin Örgön Chölöö 1, Ulan Bator; tel. (11) 310757; fax (11) 372810; internet sain.mongolmedia.com; publ. by the Democratic Party; 48 a year; Editor Ch. Chuluunbaatar.

Serüüleg (Alarm Clock): POB 1094/13, Flat 2, Entrance 1, Block A4, 1-40 Myangat, Ulan Bator; tel. and fax (11) 318006; 40 a year; Editor-in-Chief Sengeegiin Bayarmönkh; circ. 28,600.

Setgüülch (Journalist): POB 600/46, Ulan Bator; tel. (11) 325388; fax (11) 313912; f. 1982; publ. by Union of Journalists; journalism, politics, literature, art, economy; quarterly; Editor Tsendiin Enkhbat.

Shar Sonin (Yellow Newspaper): POB 76/20A, Ulan Bator; tel. (11) 313984; 48 a year; Editor B. Odgerel.

Shinjlekh Ukhaany Akademiin Medee (Academy of Sciences News): Sükhbaataryn Talbai 3, Ulan Bator; tel. (11) 321993; e-mail MAS@ac.mn; f. 1961; publ. by Academy of Sciences; quarterly; Editor-in-Chief D. Regdel.

Shuurkhai Zar (Quick Advertisement): POB 151/46A, Ulan Bator Bank Bldg, Rm 104, 1st Floor, Ulan Bator; tel. (11) 313778; e-mail shirevger@mobinet.mn; Editor E. Tseyenkhorloo.

Tavan Tsagarig (Five Rings): National Olympic Committee, Ikh Toiruu 20, Ulan Bator; tel. (11) 352487; fax (11) 343541; e-mail t_ts_sport@yahoo.com; noc@olympic.mn; Editor-in-Chief Sodnomdarjaagiin Batbaatar.

Töriin Medeelel (State Information): Secretariat of the Mongolian Great Khural, State Palace, Ulan Bator 12; tel. (11) 329612; fax (11) 322866; e-mail elbegsaihan@maild.parl.gov.mn; internet www.parl.gov.mn; f. 1990; presidential and governmental decrees, state laws; 48 a year; circ. 5,000.

Tsenkher Delgets (Light Blue Screen): Ulan Bator; tel. (11) 312010; fax (11) 311850; e-mail bsnews@magicnet.mn; weekly guide to TV and radio programmes; Editor-in-Chief Galigaagiin Bayarsaikhan.

Tsonkh (Window): CPOB 1085, Chingisiin Örgön Chölöö 1, Ulan Bator; tel. (11) 310717; publ. by the Democratic Party's Political Department; monthly.

Üg (Word): CPOB 680, Rm 2, Block 8B, 1–40 Myangat, Ulan Bator; tel. (11) 321165; fax (11) 329795; e-mail ugsonin@mol.mn; fmrly the journal of the Mongolian Social Democratic Party (until 2000); Editor-in-Chief A. Ganbaatar.

Ulaanbaatar Times (Ulan Bator Times): A. Amaryn Gudamj 1, Ulan Bator 20; tel. and fax (11) 311187; f. 1990 as *Ulaanbaatar*; publ. by Ulan Bator City Govt; weekly; Editor-in-Chief Jünpereliin Saruulbuyan; circ. 2,000.

Utga Zokhiol Urlag (Literature and Art): Union of Writers, Sükhbaataryn Gudamj 11, Ulan Bator 46; tel. (11) 321863; f. 1955; monthly; Editor-in-Chief Ü. Khürelbaatar; circ. 3,000.

Zar Medee (Advertisement News): Government Bldg 5, 1st Floor, Rm 130, Chingeltei District, Ulan Bator; tel. and fax (11) 324885; e-mail advertisement-news@yahoo.com; personal and company advertisements; 102 a year; Editor D. Bayasgalan.

Zindaa (Ranking): Kyoküshyuzan Development Fund Bldg, Chingeltei District, Ulan Bator; tel. (11) 312008; internet www.zindaa.mongolmedia.com; wrestling news; 36 a year; Editor-in-Chief Kh. Mandakhbayar.

FOREIGN LANGUAGE PUBLICATIONS

Menggu Xiaoxi Bao (News of Mongolia): Montsame News Agency, CPOB 1514, Ulan Bator; e-mail mgxxbao@yahoo.com; f. 1929 as Ajilchny Zam; closed 1991, reopened 1998; 48 a year; in Chinese; Sec. B. Mönkhtuul.

Mongolian Magazine: Interpress Publishers, Ulan Bator; f. 2004; English-language monthly illustrated magazine about Mongolian history, culture, nature, life and customs.

Mongolia This Week: Ulan Bator; tel. and fax (11) 318339; e-mail mongoliathisweek@mobinet.mn; internet www.mongoliathisweek.mn; weekly in English, online daily; Editor-in-Chief D. Narantuyaa; English Editor Eric Mustafa.

Mongoliya Segodnya (Mongolia Today): Box 609, PO 46, Ulan Bator; tel. and fax (11) 324141; e-mail MNSegodnia@mongol.net; weekly; Editor-in-Chief Dünger-Yaichiliin Solongo.

The Mongol Messenger: Montsame News Agency, CPOB 1514, Ulan Bator 13; tel. (11) 325512; fax (11) 327857; e-mail monmessenger@magicnet.mn; internet www.mongolmessenger.mn; f. 1991; weekly newspaper in English; Editor-in-Chief B. Indra; circ. 2,000.

Montsame Daily News: CPOB 1514, Montsame, Ulan Bator; tel. (11) 314574; fax (11) 327857; daily English news digest for embassies, etc.

Novosti Mongolii (News of Mongolia): CPOB 1514, Montsame, Ulan Bator; tel. (11) 310157; fax (11) 327857; e-mail montsame@pop.magicnet.mn; f. 1942; weekly; in Russian; Editor-in-Chief Enkhbayaryn Ravdan.

The UB Post: Mongol News Co, Ikh Toiruu 20, Ulan Bator; tel. (11) 352487; fax (11) 352495; e-mail ubpost@yahoo.com; internet ubpost.mongolnews.mn; f. 1996; weekly; in English; Editor-in-Chief Namsrain Oyuunbayar; circ. 4,000.

NEWS AGENCIES

Montsame (Mongol Tsakhilgaan Medeenii Agentlag) (Mongolian News Agency): Jigjidjavyn Gudamj 8, Ulan Bator 13 (CPO Box 1514); tel. (11) 314502; fax (11) 327857; e-mail montsame@magicnet.mn; internet www.montsame.mn; f. 1921; govt-controlled; Gen. Dir Dorjiin Ariunbold (acting); Editor-in-Chief Givaandondogiin Pürevsambuu.

Mongolyn Medee (Mongolian News): operated by Mongolteleviz; Dir D. SANDAGSÜREN.

Foreign Bureaux

Informatsionnoye Telegrafnoye Agentstvo Rossii—Telegrafnoye Agentstvo Suverennykh Stran (ITAR—TASS) (Russia): Khudaldaany Gudamj 4, Rm 323, Ulan Bator; Bureau Chief V. B. IONOV; Correspondent N. A. KERZHENTSEV.

Rossiiskoye Informatsionnoye Agentstvo—Novosti (RIA—Novosti) (Russia): POB 686, Ulan Bator; tel. 327384; Correspondent ALEKSANDR ALTMAN.

Xinhua (New China) News Agency (People's Republic of China): Ulan Bator; tel. (11) 322718; Correspondent LI REN.

Publishers

Mongolian Book Publishers' Association: Ulan Bator; Exec. Dir S. TSERENDORJ.

Mongolian Free Press Publishers' Association: POB 306, Ulan Bator 24; tel. and fax (11) 313405; Pres. BATYN ERDENEBAATAR.

The Government remains the largest publisher, but the ending of the state monopoly has led to the establishment of several small commercial publishers, including Shuvuun Saaral (Ministry of Defence), Mongol Khevlel and Soyombo Co, Mongolpress (Montsame), Erdem (Academy of Sciences), Süülenkhüü children's publishers, Sudaryn Chuulgan, Interpress, Sükhbaatar Co, Öngöt Khevlel, Admon, Odsar, Khee Khas Co, etc. Newspaper publishing fell from 134.1m. copies in 1990 to 21.1m. copies in 1999. Book printing likewise declined, from 96.3m. printer's sheets (each of 16 pages) in 1990 to 19.2m. in 1993. In 1994 128 book titles appeared, with a total of 640,000 copies. The two main press, periodical and book subscription agencies are Mongol Shuudan and Gurvan Badrakh Co.

Broadcasting and Communications

TELECOMMUNICATIONS

Digital exchanges have been installed in Ulan Bator, Darkhan, Erdenet, Sükhbaatar, Bulgan and Arvaikheer, while radio-relay lines have been digitalized between: Ulan Bator–Darkhan–Sükhbaatar; Ulan Bator–Darkhan–Erdenet; and Dashinchilen–Arvaikheer. Mobile telephone companies operate in Ulan Bator and other central towns, in addition to Arvaikheer, Sainshand and Zamyn-Üüd.

Bodicom: Ulan Bator; tel. 325144; fax 318486; e-mail bodicom@mongolnet.mn; internet www.bodicom.mn.

Datakom: Negdsen Ündestnii Gudamj 49, Ulan Bator 46; tel. (11) 315544; fax (11) 320210; e-mail sales@datacom.mn; internet www.web.mn; service provider for Magicnet connection to internet; Dir DANGAASÜRENGIIN ENKHBAT.

MagicNet Co: Rm 104, Ground Fl., Science and Technology Information Centre, Ulan Bator; tel. (11) 312061; fax (11) 315668; e-mail support@magicnet.mn; internet www.magicnet.mn; internet service provider.

MCSCom: MCS Plaza, 3rd Fl., Ulan Bator; tel. (11) 327854; fax (11) 311323; e-mail mcscom@mcs.mn; internet www.mcscom.mn; internet service provider.

Micom: Mongol Tsakhilgaan Kholboo Co Bldg, 3rd Floor, Sükhbaataryn Talbai 1, Ulan Bator 210611; tel. (11) 313229; fax (11) 322473; e-mail info@micom.mng.net; internet www.micom.mng.net; Dir R. BATMEND.

MobiCom: Mobicom Corp. Bldg, Enkhtayvyny Örgön Chölöö 3/1, Ulan Bator 210620; tel. (11) 312222; fax (11) 324017; e-mail mobicom@mobicom-corp.com; internet www.mobicom.mn; mobile telephone service provider; Dir G. BATTÖR.

Moncom: 51 Post Office, Box 207, Ulan Bator; tel. (11) 329409; e-mail ch.enkhmend@hotmail.com; pager services.

Mongolia Telecom: Sükhbaataryn Talbai 1, Ulan Bator 210611 (CPO Box 1166); tel. (11) 320597; fax (11) 325412; e-mail mt@mtcone.net; internet www.mongol.net; 54.6% state-owned, 40.0% owned by Korea Telecom; Pres. and CEO OONOI SHAALUU.

Railcom: Business Services, Ulan Bator Railway Directorate, Ulan Bator; tel. (11) 942600; internet www.railcom.mn; telephone, TV and internet service provider.

Skytel: Skytel Plaza Centre, Chingisiin Örgön Chölöö, Ulan Bator; tel. (11) 319191; fax (11) 318487; e-mail skytel@mtcone.net; mobile telephone and voice mail service provider; Mongolia-Republic of Korea joint venture; Marketing Man. G. TÜVSHINTÖGS.

BROADCASTING

A 1,900-km radio relay line from Ulan Bator to Altai and Ölgii provides direct-dialling telephone links as well as television services for western Mongolia. New radio relay lines have been built from Ulan Bator to Choibalsan, and from Ulan Bator to Sükhbaatar and Sainshand. Most of the population is in the zone of television reception, following the inauguration of relays via satellites operated by the International Telecommunications Satellite Organization (INTELSAT).

All provincial centres receive two channels of Mongolian national television; and all district centres can receive television, although only one-third can receive Mongolian television.

Head of Mongolian Radio and Television Affairs Directorate: BAASANJAVYN GANBOLD.

Radio

Mongolradio: Khuvisgalyn Zam 3, Ulan Bator 11; tel. (11) 323096; f. 1934; operates for 17 hours daily on three long-wave and one medium-wave frequency, and VHF; programmes in Mongolian (two) and Kazakh; Dir BAARANGIIN PÜREVDASH.

Voice of Mongolia: Ulan Bator; e-mail radiomongolia@magicnet.mn; external service of Mongolradio; broadcasts in Russian, Chinese, English and Japanese on short wave; Dir B. NARANTUYAA.

The World Service of the British Broadcasting Corpn is relayed on 103.1 MHz in Ulan Bator.

AE and JAAG Studio: Amryn Gudamj 2, Ulan Bator 210620, POB 126, Ulan Bator; tel. (11) 310631; fax (11) 326545; e-mail aejaag@magicnet.mn; f. 1996; broadcasts for 4.5–5 hours daily; CEO Z. ALTAI.

FM 100.1: Orkhun Centre (East entrance), No.3 Ger District, Ulan Bator; tel. (11) 305388; Elgen Nutag (Homeland) radio.

FM 102.1: Ulan Bator; 'Homeland' Youth Radio; run by the Open Information Foundation.

FM 103.6: radio station run by TV-9.

FM 106.6: Ulan Bator; f. 2001; Voice of America news and information in Mongolian; English lessons and music.

FM 107 (New Century Radio): Ulan Bator; local entertainment; also relays Voice of America broadcasts in English and Russian; Dir M. BUYANBADRAKH.

Info Radio: Youth Palace, Ulan Bator; tel. (11) 329353; e-mail inforadio@mongol.net; broadcasts on 105.5 MHz.

Khökh Tenger (Blue Sky Radio): Ulan Bator; broadcasts for 12 hours Monday to Saturday and for shorter hours on Sundays on 100.9 MHz and 4850 kHz; Dir L. AMARZAYAA.

Puls-Misheel: Ulan Bator; f. 2001 by Mongolradio and Buryat Puls radio; Russian-language broadcaster on 102.5 MHz, 24 hours; hourly news in Russian and Mongolian.

Radio Ulaanbaatar: Ulan Bator; f. 1995; fmrly *Dörvön Uul* (Four Mountains); broadcasts on 101.6 MHz; Dir U. BULGAN.

There are seven long- and short-wave radio transmitters and 49 FM stations in 23 towns (13 in Ulan Bator).

Television

Mongolteleviz (MTV): Khuvisgalyn Zam 3, Ulan Bator 11 (CPO Box 365); tel. (11) 326663; fax (11) 327234; e-mail mrtv@magicnet.mn; f. 1967; daily morning and evening transmissions of locally-originated material relayed by land-line and via INTELSAT satellites; short news bulletins in English Mon., Wed. and Fri.; state-owned; Dir TSENDIIN ENKHBAT.

MN Channel 25: AE and JAAG Studio, Ulan Bator; broadcasts entertainment in the evening from Tuesday to Sunday; CEO Z. ALTAI.

Sansar (STV): Ikh Toiruu 46, Ulan Bator; tel. (11) 313752; fax (11) 313770; cable television equipment and services, reaches 220,000 households; f. 1994; Dir-Gen. A. ENKHBAT.

TV-5: evening broadcasts from 6 p.m., repeated the following morning.

TV-9: f. 2003; broadcaster for the Ulan Bator area, general entertainment, with 20% religious content; Dir N. DULAMSÜREN.

UBS: Ulaanbaatar Broadcasting System (operated by Ulan Bator City Government); evening broadcasts repeated the following morning except Mondays; Dir-Gen. L. BALKHJAV.

Cable TV companies (29 in total) operate in 19 towns. There are local TV stations in Ulan Bator (three), Darkhan, Sükhbaatar and Baganuur. ORT and RTR may be received via Russian satellites, and Kazakh television is received in Bayan-Ölgii. NHK, Deutsche Welle, RTF and Inner Mongolia (China) Television programmes are relayed in Ulan Bator on Channel 10.

Finance

(cap. = capital; res = reserves; dep. = deposits; m. = million; brs = branches; amounts in tögrög)

BANKING

Before 1990 the State Bank was the only bank in Mongolia, responsible for issuing currency, controlling foreign exchange and allocating credit. With the inauguration of market reforms in Mongolia in the early 1990s, the central and commercial functions of the State Bank were transferred to the newly created specialized commercial banks: the Bank of Capital Investment and Technological Innovation and the State Bank International. In May 1991 the State Bank became an independent central bank, and the operation of private, commercial banks was permitted. By the end of 1996 there were 15 commercial banks. The performance of these banks was poor, owing to high levels of non-performing loans, inexperienced management and weak supervision. Loss of confidence in the sector resulted in the implementation of extensive restructuring: in November 1996 amendments were made to banking legislation to improve the regulation and supervision of commercial banks, and two major insolvent banks were liquidated. Restructuring continued in the late 1990s and beyond.

Central Bank

Bank of Mongolia (Mongolbank): Baga Toiruu 9, Ulan Bator 46; tel. (11) 322169; fax (11) 311471; e-mail ad@mongolbank.mn; internet www.mongolbank.mn; f. 1924 as the State Bank of the Mongolian People's Republic; cap. 5,000m., res 27,335m. (July 2004); Gov. OCHIRBATYN CHULUUNBAT.

Other Banks

AG Bank—Agricultural Bank (Khaan Bank): Enkh Taivny Gudamj 51, Ulan Bator 240149 (POB 185, Ulan Bator 51); tel. and fax (11) 457880; fax (11) 458670; e-mail haanbank@magicnet.mn; f. 1991; purchased by H and S Securities (Japan) in Feb. 2003; cap. 9,270m., res 5,262m., dep. 83,027m. (June 2004); CEO J. PETER MORROW; 380 brs.

Anod Bank: Khudaldaany Gudamj 18, Ulan Bator 211238 (POB 361, Ulan Bator 13); tel. (11) 313315; fax (11) 313070; e-mail anod@magicnet.mn; internet www.anod.mn; cap. 10,971m., dep. 17,483m., dep. 104,142m. (June 2004); Dir-Gen. D. ENKHTÖR.

Avtozam Bank (Motor Roads Bank): Bridge-Building Office, Ulan Bator; tel. (11) 381744; fax (11) 368094; f. 1990; cap. 95m.; Dir-Gen. TS. SANGIDORJ.

Aziin Khöröngö Oruulaltyn Bank (Asian Capital Investment Bank): Khölög Group Bldg, Eastern entrance, 2nd Floor, Bayangol District, Ulan Bator; tel. (11) 367386; Dir-Gen. TSEDENGIIN BATBOLD.

Capital Bank (Capital Bank of Mongolia): Chingeltei District, Sambuugiin Gudamj 48, Ulan Bator 211238; tel. (11) 319247; fax (11) 310833; e-mail center@capitalbank.mn; internet www.capitalbank.mn; cap. 4,000m., res 400m., dep 8,024m. (June 2004); Chair. AGVAANJAMYN ARIUNBOLD; Exec. Dir ENKHBAZARIN AMGALAN; 10 brs.

Capitron Bank: Enkh Taivny Örgön Chölöö, Ulan Bator 210648; tel. (11) 327550; fax (11) 315635; e-mail info@capitronbank.mn; internet www.capitronbank.mn; f. 2001; cap. 4,000m., res 5,000m., dep 12,000m. (2002); Exec. Dir A. MÖNKHBAT.

Chingis Khan Bank: Chinggis Khan Bank Bldg, 5th Sub-District, Bayangol District, Ulan Bator; tel. (11) 633105; fax (11) 633185; e-mail chkhbank@magicnet.mn; f. 2001 by Millennium Securities Management Ltd and Coral Sea Holdings Ltd (British Virgin Islands); cap. US $31.4m.

Erel Bank: 'Erel' Co No 2 Bldg, Chingisiin Örgön Chölöö, 3rd Sub-District, Ulan Bator (POB 500, Ulan Bator 36); tel. and fax (11) 343567; e-mail info@erelbank.mn; f. 1997; cap. 3,000m., res 243.6m., dep. 1636.9m. (Oct. 2003); Owner BADARCHIIN ERDENEBAT; CEO GOMBOJAVYN DORJ.

Export Import Bank: Zamchny Gudamj 1, Bayangol District, Ulan Bator (CPO Box 28/52); tel. (11) 311693; fax (11) 311323; f. 1993 as Ulaanbaatar Bank; cap. 506.8m.; CEO L. SARANGEREL.

Golomt Bank of Mongolia: Bodi Tower, Sükhbaataryn Talbai 3, 4th Floor, Ulan Bator 210620; tel. (11) 311530; fax (11) 312307; e-mail psdc@golomtbank.com; internet golomtbank.com; f. 1995 by Mongolian-Portuguese IBH Bodi International Co Ltd; cap. 11,126m., res 51,043m., dep. 78,965m. (June 2004); Chair. LUVSANVANDANGIIN BOLD; Dir-Gen. D. BAYASGALAN; 10 brs.

Interbank of Mongolia: POB 1130, Interbank Bldg, Sükhbaatar District, Ulan Bator; tel. (11) 327403; fax (11) 328372; e-mail interbank@mongolnet.mn; internet www.interbank.mn; f. 2001; cap. 2,238.1m., dep. 10,554.8m., total assets 12,908.0m. (Dec. 2002); Gen. Dir KHOROLSÜRENGIIN CHINBAT.

Khadgalamjiin Bank (Savings Bank): Khudaldaany Gudamj 6, Ulan Bator 11; tel. (11) 312043; fax (11) 310621; e-mail savbank@magicnet.mn; internet www.savingsbank.mn; f. 1996 as Ardyn Bank; cap. 3,123.1m., res. 174.7m., dep. 46,368.7m. (Dec. 2002); Dir-Gen. G. TSERENPÜREV; 39 brs.

Kredit Bank: State Palace Bldg (east side), Sükhbaataryn Talbai, Ulan Bator; tel. (11) 310853; fax (11) 319039; e-mail creditbank@magicnet.mn; internet www.creditbank.mn; cap. 4,292m., res. 1,293m., dep. 2,828m. (June 2004); Exec. Dir N. BATKHISHIG.

Mongol Shuudan Bank (Post Bank): Kholboochdyn Gudamj 4, Ulan Bator 13 (POB 874); tel. (11) 310103; fax (11) 328501; e-mail post_bank@mongol.net; internet www.postbank.mn; f. 1993; cap. 4,932m., res 6,328m., dep. 32,300m. (June 2004); 100% in private ownership; Exec. Dir D. OYUUNJARGAL.

Teever Khögjliin Bank (Transport and Development Bank): Amarsanaagiin Gudamj 2, Bayanzürkh District, Ulan Bator; tel. (11) 458617; Exec. Dir D. ENKHTUYAA; cap. 1,072m.

Trade and Development Bank of Mongolia (Khudaldaa Khögjliin Bank): Khudaldaany Gudamj 6, Ulan Bator 11; tel. (11) 326289; fax (11) 311618; e-mail tdbank@tdbm.mn; internet www.tdbm.mn; f. 1991; carries out Mongolbank's foreign operations; cap. 15,876m., res. 8,383m., dep. 38,064m. (June 2004); 76% equity bought by Banca Commerciale (Lugano) and Gerald Metals (Stanford, CT), May 2002; Exec. Dir SIILEGMAAGIIN MÖNKHBAT.

Tülsh-Erchim Bank (Fuel and Power Bank): Baga Toiruu, Ulan Bator; tel. (11) 310605; fax (11) 310981; f. 1992; Dir M. MYAGMARSÜREN.

Ulaanbaatar Bank: Baga Toiruu 15, Ulan Bator (PO 370, Ulan Bator 46); tel. (11) 312155; fax (11) 311067; e-mail ubbank@magicnet.mn; f. 1998 by Capital City with assistance from the Bank of Taipei (Taiwan); cap. 4,800m; Dir-Gen. G. BUYANBAT.

XacBank: Yörönkhii said Amaryn Gudamj, Sükhbaatar District, Ulan Bator 210646 (POB 46/721); tel. (11) 318185; fax (11) 328701; e-mail bank@xacbank.org; internet www.xacbank.org; f. 2001; cap. 4,267m., res 2,626m., dep. 9,132m. (June 2004); Exec. Dir CHULUUNY GANKHUYAG; Chair. STEPHEN MITCHELL.

Zoos Bank: Baga Toiruu, 1st Sub-District, Chingeltei District, Ulan Bator (POB 314, Ulan Bator 44); tel. (11) 312107; fax (11) 329537; e-mail zoosbank@mongol.net; internet www.zoosbank.mn; f. 1999; cap. 5,360.6m., res 4.9m., dep. 19,259.8m. (Sept. 2003); Exec. Dir SH. CHUDANJII.

Banking Associations

Bankers' Association: Ulan Bator; Pres. D. NYAMTSEREN; Chair. of Bd S. ORGODOL.

Mongolian Bankers Association: Ulan Bator; e-mail monba@mongolnet.mn; Pres. LUVSANVANDANGIIN BOLD; Exec. Dir JIGJIDIIN ÜNENBAT.

STOCK EXCHANGE

In 2003 there were 402 listed companies, of which 68 were partly state-owned. In 250 trading days 8.5m. shares were traded (about half the number traded in 2001). The turnover was 25,600m. tögrög, (just over half the amount in 2002). Market capitalization rose from 35,800m. tögrög in 2002 to 49,500m. tögrög in 2003. The leading companies in market capitalization in 2003 were Mongolyn Tsakhilgaany Kholboo (telecommunications) with a market value of 18,100m. tögrög; Mongol Securities, 11,400m. tögrög; Gobi (cashmere), 3,900m. tögrög; NIK (petrol distribution), 2,500m. tögrög; Baganuur (mining), 2,300m. tögrög; and Shivee Ovoo (mining), 2,000m. tögrög.

Stock Exchange: Sükhbaataryn Talbai 2, Ulan Bator; tel. (11) 310501; fax (11) 325170; e-mail mse@mongol.net; internet www.mse.mn; f. 1991; Dir DULAMSÜRENGIIN DORLIGSÜREN.

INSURANCE

Mongol Daatgal (National Insurance and Reinsurance Co): Seoul St, Ulan Bator 210644; tel. (11) 313641; fax (11) 313647; e-mail insurance@mongoldaatgal.mn; internet www.mongoldaatgal.mn; f. 1934; sold Dec. 2003 to consortium formed by Angara-SKB and Chingis Khan Bank; Chair. and CEO B. ENKHBAT; Snr Vice-Pres. CH. BATTSOGT; Gen. Man. U. BYAMBASÜREN.

Trade and Industry

GOVERNMENT AGENCIES

Labour Co-ordination Directorate: Khuvisgalchdyn Gudamj 14, Ulan Bator; tel. and fax 327906; Dir D. JANTSAN.

Mineral Resources Authority: Government Bldg 5, Barilgachdyn Talbai 13, Ulan Bator 211238; tel. and fax 310370; e-mail mram@magicnet.mn; Dir O. CHULUUN.

Mongol Gazryn Tos: Üildverchnii Gudamj, Ulan Bator 37; tel. 61584; e-mail petromon@magicnet.mn; supervises petroleum exploration and development; Dir O. DAVAASAMBUU.

State Industry and Trade Control Service: POB 38/66, Barilgachdyn Talbai, Ulan Bator 38; tel. and fax 328049; e-mail chalkhaajavd@mongolnet.mn; f. 2000; enforces laws and regulations relating to trade and industry, services, consumer rights, and geology and mining; Dir DAMBADARJAAGIIN CHALKHAAJAV.

State Property Committee: Government Bldg 4, Ulan Bator 12; tel. 312460; fax 312798; internet odmaa@spc.gov.mn; supervision and privatization of state property; Chair. LKHANAASÜRENGIIN PÜREVDORJ.

DEVELOPMENT ORGANIZATIONS

Economics and Market Research Center: Government Bldg 11, J. Sambuugiin Gudamj 11, Ulan Bator 38; tel. 324258; fax 324620; e-mail emrc@mongolchamber.mn; internet www.mongolchamber.mn; Dir J. BOZKHÜÜKHEN.

Foreign Investment and Foreign Trade Agency (FIFTA): Government Bldg 1, J. Sambuugiin Gudamj 11, Ulan Bator; tel. 326040; fax 324076; e-mail fifta@investnet.mn; internet www.investnet.mn; Chair. BAASANKHÜÜGIIN GANZORIG.

Mongolian Business Development Agency: Yörönkhii said Amaryn Gudamj, Ulan Bator; CPO Box 458, Ulan Bator 13; tel. 311094; fax 311092; e-mail mbda@mongol.net; internet www.mbda-mongolia.org; f. 1994; Gen. Man. D. BAYARBAT.

CHAMBERS OF COMMERCE

Central Asian Chamber of Commerce: PO Box 470, Ulan Bator 44; tel. 38970; fax 311757; Chair. G. TÖMÖRMÖNKH.

Junior Chamber of Commerce: Youth Union Bldg, Ulan Bator; tel. 328694; Chair NATSAGDORJ.

Mongolian National Chamber of Commerce and Industry: Sambuugiin Gudamj 11, Ulan Bator 38; tel. 312501; fax 324620; e-mail chamber@mongolchamber.mn; internet www.mongolchamber.mn; f. 1960; responsible for establishing economic and trading relations, contacts between trade and industrial organizations, both at home and abroad, and for generating foreign trade; organizes commodity inspection, press information, and international exhbns and fairs at home and abroad; registration of trademarks and patents; issues certificates of origin and of quality; Sec. Gen. ENEBISHIIN OYUUNTEGSH; Chair. and CEO SAMBUUGIIN DEMBEREL.

INDUSTRIAL AND TRADE ASSOCIATIONS

Association of Exporters of Livestock Raw Materials and Semi-Processed Products: Ulan Bator; Exec. Dir B. TORMONH.

Association of Private Herders' Co-operatives: POB 21/787, Ulan Bator 211121; tel. 633601; fax 325935; e-mail mongolherder@magicnet.mn; f. 1991; Pres. R. ERDENE.

Mongolian Farmers and Flour Producers' Association: AgroPro Business Centre, 19th Sub-District, Bayangol District, Ulan Bator 24; tel. 300114; fax 362875; e-mail agropro@magicnet.mn; f. 1997; research and quality inspection services in domestic farming and flour industry; Pres. SHARAVYN GUNGAADORJ.

Mongolian Franchising Council of the Mongolian National Chamber of Commerce and Industry: Ulan Bator; tel. 327178; fax 324620; e-mail tecd@mongolchamber.mn; Pres. BAATARYN CHADRAA.

Mongolian Marketing Association: Ulan Bator; Pres. D. DAGVADORJ.

Mongolian National Mining Association: Sky Plaza Centre, Olimpiin Gudamj 14, PO Box 46/910, Ulan Bator 210646; tel. 314877; fax 314877; e-mail mongma@mobinet.mn; internet www.owc.org.mn/monnma; f. 1994; provides members of the mining sector with legal protection, and reflects their views in Government mining policy and mineral sector development; Pres. DUGARYN JARGALSAIKHAN; Exec. Dir NAMGARYN ALGAA.

Mongolian Printing Works Association: Ulan Bator; Pres. G. KHAVCHUUR.

Mongolian Wool and Cashmere Federation: Khan-Uul District, Ulan Bator; tel. 341871; fax 342814; Pres. G. YONDONSAMBUU.

EMPLOYERS' ORGANIZATIONS

Employers' and Owners' United Association: Rm 401, 4th Fl., Mongolian Youth Association 'B' Bldg, Ulan Bator; tel. (11) 326513; Exec. Dir. B. SEMBEEJAV.

Federation of Professional Business Women of Mongolia: Ulan Bator 20; tel. and fax 315638; e-mail mbpw@mongolnet.mn; f. 1992; provides education, training, and opportunities for women to achieve economic independence, and the running of businesses; Pres. OCHIRBATYN ZAYAA; 7,000 mems, 14 brs.

Free Labour Managers' Association: Ulan Bator; Pres. N. PÜREVDORJ.

Immovable Property (Real Estate) Business Managers' Association: Ulan Bator; Pres. J. BYAMBADORJ.

Mongolian Management Association: Ulan Bator; Chair. Exec. Council DAGVADORJIIN TSERENDORJ.

Private Business Owners' Association: Ulan Bator; Pres. T. NYAMDORJ.

Private Employers' Association: Ulan Bator; Pres. O. NATSAGDORJ.

Private Industry Owners' Association: Ulan Bator; f. 1990; with 39 mems; Pres. LUVSANBALDANGIIN NYAMSAMBUU.

UTILITIES

Electricity

TsShSG: Ulan Bator; tel. 41294; supervision of electric power network in Ulan Bator; Dir D. BASSAIKHAN.

Water

DShSG: Ulan Bator; tel. 343047; supervision of hot water district heating network in Ulan Bator; Dir SHARAVYN BAASANJAV.

USAG: Khökh Tengeriin Gudamj 5, Ulan Bator 49; tel. 455055; fax 450120; e-mail usag@magicnet.mn; supervision of water supply network in Ulan Bator; Chair. OSORYN ERDENEBAATAR.

IMPORT AND EXPORT ORGANIZATIONS

Agrotekhimpeks: Ulan Bator 32; imports agricultural machinery and implements, seed, fertilizer, veterinary medicines and irrigation equipment.

Altjin: Ulan Bator; company importing and distributing oil and oil products and also running distilleries, a spin-off from APU; Dir G. ALTAN.

Arisimpex: Ulan Bator 52; tel. 343007; fax 343008; exports hides and skins, fur and leather goods; imports machinery, chemicals and accessories for leather, fur and shoe industries; Pres. A. TSERENBALJID.

Avtoimpeks: Sonsgolon-2, Ulan Bator (POB 37, Ulan Bator 211137); tel. 331860; fax 331383; f. 1934; privatization pending 1999; imports of vehicles and spare parts, vehicle servicing; Exec. Dir TS. TOGTMOL.

Barter and Border: Khuvisgalchdyn Örgön Chölöö, Ulan Bator 11; tel. 324848; barter and border trade operations.

Böönii Khudaldaa: Songinokhairkhan District, Ulan Bator; wholesale trader; privately-owned; Dir-Gen. OCHBADRAKHYN BALJINNYAM.

Khorshoololimpeks: Tolgoit 37, Ulan Bator (PO Box 262); tel. 332926; fax 331128; f. 1964; exports sub-standard skins, hides, wool and furs, handicrafts and finished products; imports equipment and materials for housing, and for clothing and leather goods; Dir L. ÖLZIIBUYAN.

Kompleksimport: Enkh Taivny Gudamj 7, Ulan Bator; tel. and fax 382718; f. 1963; imports consumer goods, foodstuffs, sets of equipment and turnkey projects; training of Mongolians abroad; state-owned pending privatization in 1999; cap. 3,500m. tögrög.

Makhimpeks: 4th Sub-District, Songinokhairkhan District, Ulan Bator; tel. 63247; fax 632517; e-mail makhimpex@mongol.net; f. 1946; abattoir, meat processing, canning, meat imports and exports; 51% share privatized in 1999; cap. 7,800m. tögrög; Dir H. BATTULGA.

Materialimpex: Teeverchdiin Gudamj 35, Ulan Bator 35; tel. 365125; fax 367904; e-mail miehnco@magicnet.mn; f. 1957; exports cashmere, wool products, animal skins; imports glass, roofing material, dyes, sanitary ware, metals and metalware, wallpaper, bitumen, wall and floor tiles; privatized Feb. 1999 (most shares state-owned); Gen. Dir B. ZORIG; 126 employees.

Mongoleksport Co Ltd: Government Bldg 7, 8th Floor, Erkh Chölöönii Talbai, Ulan Bator 11; tel. 329234; fax 324848; exports wool, hair, cashmere, mining products, antler, skins and hides; Dir-Gen. D. CHIMEDDAMBAA.

Mongolemimpex: Ikh Toiruu 39, Ulan Bator 28; tel. 323961; fax 323877; e-mail moemim@magicnet.mn; procurement and distribution to hospitals and pharmacies of drugs and surgical appliances; Dir-Gen. R. BYAMBAA.

Mongolimpeks: Khuvisgalchdyn Örgön Chölöö, Ulan Bator 11; tel. 326081; exports cashmere, camels' wool, hair, fur, casings, powdered blood and horn, antler, wheat gluten, alcoholic drinks, cashmere and camels' wool knitwear, blankets, copper concentrate, souvenirs, stamps and coins; imports light and mining industry machinery, scientific instruments, chemicals, pharmaceuticals and consumer goods; state-owned; Dir-Gen. DORJPALAMYN DÖKHÖMBAYAR.

Mongol Safari Co: Ulan Bator 38; tel. 360267; fax 360067; e-mail monsafari@magicnet.mn; exports hunting products; imports hunting equipment and technology; organizes hunting and trekking tours; Dir-Gen. U. BUYANDELGER.

Monnoos: POB 450, Ulan Bator 36; tel. 343201; fax 342591; e-mail monnoos@mongolnet.mn; wool trade enterprise; Dir SANJIIN BAT-OYUUN.

Monos Cosmetics: Sonsgolon Toiruu 5, Songinokhairkhan District 20, POB 62, Ulan Bator 211137; tel. and fax 633257; e-mail info@monoscosmetics.mn; internet www.monoscosmetics.mn; f. 1990; production, export and import of cosmetics; Chair. and CEO BALDANDORJIIN ERDENEKHISHIG; 90 employees.

Monospharma Co Ltd: Chingünjavyn gudamj 9, Bayangol District 2, Ulan Bator 210526; tel. and fax 361419; e-mail monospharma@mongol.net; f. 1990; production, export and import of medicine, medical equipment and health food; Dir-Gen. LUVSANGIIN KHÜRELBAATAR; 280 employees.

Noosimpeks: Ulan Bator 52; tel. 342611; fax 343057; exports scoured sheep's wool, yarn, carpets, fabrics, blankets, mohair and felt boots; imports machinery and chemicals for wool industry.

Nüürs: Ulan Bator 21; tel. 327428; exports and imports in coal-mining field; Man. D. DÜGERJAV.

Packaging: Tolgoit, Ulan Bator; tel. 31053; exports raw materials of agricultural origin, sawn timber, consumer goods, unused spare parts and equipment, and non-ferrous scrap; imports machinery and materials for packaging industry, and consumer goods.

Petrovis: Ulan Bator; oil products importer and distributor; in February 2004 acquired the 80% state-owned shares in the country's biggest distributor NIK (Neft Import Kontsern) for $8.5m.; Exec. Dir J. OYUUNGEREL.

Raznoimpeks: 3rd Sub-District, Bayangol District, Ulan Bator 36; tel. 329465; fax 329901; f. 1933; exports wool, cashmere, hides, canned meat, powdered bone, alcoholic drinks, macaroni and confectionery; imports cotton and woollen fabrics, silk, knitwear, shoes, fresh and canned fruit, vegetables, tea, milk powder, acids, paints, safety equipment, protective clothing, printing and packaging paper; state-owned pending planned privatization of 51% share in 1999; cap. 6,100m. tögrög; Exec. Dir Ts. BAT-ENKH.

Tekhnikimport: Ulan Bator; tel. 32336; imports machinery, instruments and spare parts for light, food, wood, building, power and mining industries, road-building and communications; state-owned; Dir-Gen. ERDENIIN GANBOLD.

Tüshig Trade Co Ltd: Enkh Taivny Örgön Chölöö, Ulan Bator 44, PO Box 481; tel. 314062; fax 314052; e-mail d.ganbaatar@mongol.net; exports sheep and camel wool, and cashmere goods; imports machinery for small enterprises, foodstuffs and consumer goods; Dir-Gen. D. GANBAATAR.

CO-OPERATIVES

Central Association of Consumer Co-operatives: Ulan Bator 20; tel. and fax 329025; f. 1990; wholesale and retail trade; exports animal raw materials; imports foodstuffs and consumer goods; Chair. G. MYANGANBAYAR.

Individual Herdsmen's Co-operatives Association: Ulan Bator; Pres. R. ERDENE.

Mongolian Central Association of Savings and Credit Co-operatives: Ulan Bator; f. ; Pres. E. SANDAGDORJ.

Mongolian Co-operatives Development Centre: Ulan Bator; Dir DANZANGIIN RADNAARAGCHAA.

Mongolian Producer Co-operatives Central Association: Ulan Bator; Pres. CHOGSOMJAVYN BURIAD.

Mongolian United Association of Savings and Credit Co-operatives: Ulan Bator; Pres. TSERENDORJIIN GANKHUYAG.

National Association of Mongolian Agricultural Co-operatives: Enkh Taivny Gudamj 11, Ulan Bator; tel. and fax 358671; Pres. N. NADMID.

Union of Mongolian Production and Services Co-operatives: POB 470, Ulan Bator 210646; tel. 328446; fax 329669; f. 1990; Pres. of Supreme Council (vacant).

MAJOR COMPANIES

Ajnay (Mongol Savkhi): Chingis Khaany Örgön Chölöö, Khan-Uul District, Ulan Bator 52; tel. 343237; fax 342356; leather processing and garment manufacture; Dir-Gen. S. HÜNIIHÜÜ.

Altai Holdings: CPOB 513, Eronkhii Said Amaryn Gadamj, Ulan Bator; tel. and fax 358067; precious metal exploration and trading; subsidiaries include Altai Petroleum, Altai Mining, Altai Trading and Altai Travel; Pres. S. BATBOLD.

Altan Taria: Songinokhairkhan District, Ulan Bator 211125; tel. and fax 632057; e-mail altantaria@magicnet.mn; flour milling and retailing; 51% owned by Mongolian-Czech Credit Co; Dir-Gen. P. TSENGÜÜN.

APU: Chingis Khaany Örgön Chölöö 14, Khan-Uul District, Ulan Bator; tel. 342434; fax 343063; e-mail apu@magicnet.mn; vodka, beer and non-alcoholic drinks; privatized in 2002; Dir GAVAAGIIN BATKHÜÜ.

Baganuur: Baganuur town (130 km from Ulan Bator) Ulan Bator office; tel. 457717; fax 457715; coal-mining co, main supplier of Ulan Bator's power stations; 75% state-owned; Dir D. DAMBAPELJEE.

Bayangol Zochid Buudal: Chingis Khaany Örgön Chölöö, Ulan Bator; tel. 328869; fax 326880; 418-bed hotel and restaurants; CEO Ü. OTGONBAYAR.

Berkh-Uul: Berkh, Khentii Province; fluorspar mining and concentrating; part of Mongolrostsvetmet jt enterprise; state-owned.

Betonarmatur: Chingis Khaany Örgön Chölöö, Khan-Uul District, Ulan Bator 52; tel. and fax 342622; ferro-concrete structures.

Biokombinat: Khan-Uul District, Ulan Bator 210131; tel. and fax 326642; medicines, drugs, veterinary medicines; state-owned; Dir LUVSANGIIN DORJSAMBUU.

Bodi International: Sükhbaataryn Talbai 3, 4th Floor, Ulan Bator; POB 11/20A, Ulan Bator; tel. 311971; fax 329057; financial services, trade, property, gold-mining; Dir-Gen. M. ZORIGT.

Bolovsrol: POB 982, Surguuliin Gudamj 8, Ulan Bator 210646; tel. and fax 320674; e-mail bolovsrol@magicnet.mn; f. 1990; educational and medical services, food and industrial goods production; Dir-Gen. S. BYAMBADORJ; 100 employees.

Bor-Öndör: Bor-Öndör, Khentii Province; fluorspar mining enterprise under Mongolrostsvetmet; Dir B. A. KUTMIN.

Buligaar: Ulan Bator; tanning, shoemaking; exporting shoes to India and China; Exec. Dir OCHIRBATYN RAGCHAA.

Buyan: Khan-Uul District, Ulan Bator; tel. 325413; fax 326755; cashmere and camel-wool processing and garment manufacture; Dir-Gen. B. JARGALSAIKHAN.

Chinggis Khan Hotel: Tokiogiin Gudamj, Ulan Bator 49 (CPO Box 513, Ulan Bator); tel. 458076; fax 458067; e-mail altaiholdings@magicnet.mn; 186 rooms and 400 beds; trade, conference and shopping centres and a casino.

Darkhan Nekhii: POB 901, Darkhan Nekhii JSC, Industrial Zone, Darkhan-Uul Province; tel. (37) 227025; fax (37) 223149; f. 1972; cap. US $3,055,891; sheepskin processing and garment manufacture; Exec. Dir. E. BATSAIKHAN; 300 employees.

Em Hangamj: Barilgachdyn Talbai, Ulan Bator; tel. and fax 324504; distribution of pharmaceuticals to Ulan Bator retail pharmacies.

Emiin Üildver: Ulan Bator; state-owned pharmaceuticals co; Dir O. DAMBA.

Erchim Khüchnii Zasvaryn Üildver: next to No. 4 Power Station, Bayangol District, Ulan Bator; tel. and fax 332626; production, repair and service of equipment and spares for electric power generation; state-owned; Dir CH. TÖMÖR.

Erdenet: Erdenet, Orkhon Province; tel. (35) 73505; fax (35) 23002; e-mail ikhtashir@magicnet.mn; copper mining and concentrating; jt-stock co with Russia (24% owned by Russian State, 25% owned by Russian businesses), 51% state-owned by Mongolia; Jt Chair. of Board LKHANAASÜRENGIIN PÜREVDORJ (Mongolia), A. S. LADYGO (Russia); Dir-Gen. KHALZHÜÜGIIN NARANKHÜÜ; 7,420 employees.

Erdenet Hivs: Nairamdal District, Erdenet; tel. (35) 20111; fax (35) 21617; wool spinner and woollen carpet manufacturer; Dir-Gen. J. DELGER-TSETSEG.

Erdes: Mardai, Dornod Province; uranium mining; former Soviet concession, transferred to Mongolia June 1999; state-owned pending planned establishment of a Mongolian-US-Canadian jt venture.

Erdmin: POB 631 Erdenet, Orkhon Province; tel. (35) 72176; Mongolian-US jt venture f. 1994 between Erdenet Concern (51%) and Armada Copper (49%) production since 1997 of pure (A grade) copper cathodes by chemical leaching of low-grade ore from Erdenet mine's waste dump; most sold to Marubeni Corpn; Dir-Gen. J. Damdinjav.

Erel Group: POB 88, Ulan Bator 51; tel. 341714; fax 341739; mining, geological research, construction, banking, financial services, investment; Dir-Gen. Badarchiin Erdenebat.

Gobi: Khan-Uul District, Ulan Bator 52; tel. 342713; fax 343081; e-mail gobimon@magicnet.mn; f. 1981; cashmere processing and knitwear production; cashmere and camel-wool processing, yarn, knitted and woven production; jt stock company; Government withdrew tender in Dec. 2003 when response to sale of state-owned shares failed to meet expectations; Dir-Gen. Badarchiin Yondonjamts.

Gurvan Saikhan: Khairkhan, Dundgobi Province; uranium mining; jt venture by Mongolia, USA and Russia; Exec. Dir Aleksandr V. Rudchenko; Mongolian Dir Dashiin Bat-Erdene.

Jagar International: CPO Box 9, Ulan Bator 13; tel. 322823; fax 313289; e-mail jagar@magicnet.mn; transport and storage of wholesale household goods, foodstuffs and cosmetics, production of soft drinks; leasing of retail premises; Dir J. Erdenetsogt.

Khünstreyd: Songinokhairkhan District, Tolgoyt 37, Ulan Bator; tel. 631846; fax 631891; domestic and foreign food trader and wholesaler; state-owned; Dir-Gen. Ürdeegiin Tömörbaatar.

Monel: Science and Culture Bldg, 8th Floor, Sükhbaataryn Talbai, Ulan Bator 11; tel. 327546; electronics manufacturer and trader; 40% state-owned.

Monenzym: Research and Production Association of Enzymology and Microbiology, Tolgoyt 25, Ulan Bator; tel. 32431; research, testing and production of enzymes and microbiological products.

Mongol-Amikal: Ulan Bator; joint venture (USA 55%, Mongolia 45%) producing cashmere, goathair and camelhair; annual capacity 450 metric tons; Exec. Dir Ronny Lamb.

Mongol Energy Service: PO Box 193, Ulan Bator 28; tel. 320468; fax 327146; e-mail mes@magicnet.mn; installation and maintenance of electricity generation equipment and district heating pipelines; Mongolian-Russian joint venture; Dir-Gen. B. Shatar.

Mongolrostsvetmet: Jukovyn Gudamj, Bayanzürkh District, Ulan Bator 51; tel. 458072; fax 458380; e-mail mrtsvetmet@mongol.net; f. 1973; fluorspar mining at Bor Öndör, Airag and Örgön, gold mining at Zaamar Placer; joint venture with Russia; Joint Chair. of Board L. Pürevdorj (Mongolia), A. S. Ladygo (Russia); Dir-Gen. Khookhoryn Badamsüren; 3,985 employees.

Mongol Shuudan: CPO Box 1106, Ulan Bator 13; tel. 320137; fax 328413; e-mail shuudan@magicnet.mn; handling of letters, parcels, subscriptions to periodicals, book and newspaper retail sales, counter services, printing of postage stamps, rural transport services; Dir-Gen. Jasrayn Jantsan.

Mongol Tamga: Sükhbaatar District, Ulan Bator; tel. 310667; fax 310571; official seals and stamps; state-owned; Dir Col Sangidorjiin Sodnombaljir.

Mongolyn Alt Corpn: POB 287, Ulan Bator 21; tel. 750199; fax 758075; e-mail mglgold@magicnet.mn; gold-mining, coal mining, agriculture, geology exploration; Pres. B. Nyamtaishir.

Mongolyn Nüürs: POB 147, Ulan Bator 13; tel. and fax 682570; association of Mongolian coal-mines; state-owned; Pres. Dorjiin Dondov; Exec. Dir Toshoony Sambasanchir.

Mongolyn Ünet Tsaasny Khevlekh Üildver (Mongolian Securities Printing Works): POB 613, Ulan Bator 13; tel. 327101; fax 328555; e-mail mnspc@magicnet.mn; jt venture with British company De La Rue; Exec. Dir Ch. Chuluunbaatar.

Monmap: Sükhbaataryn Talbai 3, 4th Floor, Ulan Bator 46; tel. and fax 320728; e-mail monmap@magicnet.mn; surveying and remote sensing; Chair. M. Saandar.

Monnoos: Khan-Uul District, Ulan Bator; tel. 342591; fax 342592; wool-washing and spinning; Dir-Gen. Norov.

Monsam: Ulan Bator; tel. 351069; fax 351068; f. 1995; jt venture with Samsung; surgical syringes and needles; Dr B. Doyoddorj.

Monsan: Baga Toiruu 5, Chingeltei District, Ulan Bator; tel. 310775; fax 320810; electrical and plumbing installation; Dir M. Batbaatar.

Naran Trade Co Ltd: POB 46/568, Ulan Bator 46; tel. 322758; fax 320396; e-mail narantrade@mongol.net; f. 1990; retail and wholesale trade, media, real estate; Man. Dir Sereeteriin Boldkhet.

Petroleum Production Co: Orkhon Province; f. 2001 as Mongolia-Kyrgyzstan joint venture; oil refinery producing petrol, diesel and furnace oil from imported Russian gas condensate; annual capacity 50,000 metric tons.

Oyuuny Undraa: POB 867, Ulan Bator 46; tel. 325496; fax 325495; printing, sale and service of cars and aviation equipment; Dir-Gen. S. Otgonbayar.

SAPU: Ulan Bator; soft-drinks company; f. 2003 with South Korean partnership by ex-director of APU (see above); Dir Sandagdorjiin Jamyansüren.

Shijir Alt: Jukovyn Gudamj 18/6, Bayanzürkh District, POB 244, Ulan Bator 51; tel. 453521; f. 1995; gold mining in the Tuul river valley; CEO N. Janchiv.

Spirt Bal Buram: Mandal District, Züünkharaa, Selenge Province; tel. Züünkharaa 397; producer of alcohol and treacle; Dir. L. Eldev-Ochir.

Süü: Üildverchnii Gudamj 13, Songinokhairkhan District, Ulan Bator; tel. 331950; fax 331901; milk and dairy produce; Dir T. Bayarkhüü.

Talkh Chikher: Songinokhairkhan District, Ulan Bator 57; tel. 633383; fax 631580; bread, confectionery; state-owned; Exec. Dir D. Pürevsüren.

TAS Group: POB 142, Ulan Bator 46; tel. 312223; fax 311753; mining, trade and retail services, tourism, light and food industry, banking and investment services; Pres. J. Batbaatar.

Tsement: Darkhan; tel. (37) 4770; fax (37) 4570; output of Portland cement; Dir-Gen. I. Dorjgotov.

Tsement Shokhoi: Khötöl, Selenge Province; producer of cement and lime; state-owned; Dir-Gen. R. Shiilegjambal.

Tüshig Co: Natsagdorjiin gudamj, Sükhbaatar District, Ulan Bator; parent company of Tüshig Trade and other companies; Dir G. Battüshig.

Ulaanbaatar Barilga: POB 52, Ulan Bator; fax 342806; building materials, construction; Dir-Gen. B. Jügder.

Ulaanbaatar Khivs JSC: Khan-Uul District, Ulan Bator 36; tel. 342559; fax 343311; e-mail ubk@magicnet.mn; internet www.mongolia-carpet.com; woollen carpet manufacturer; Exec. Dir D. Munkhjargal.

Ulaanbaatar Zochid Buudal: Baga Toiruu, Ulan Bator; tel. 320230; fax 324485; 280-bed hotel and restaurants; Exec. Dir. Ch. Ochirsükh.

Ulsyn Ikh Delgüür: Enkh Taivny Gudamj, Chingeltei District, Ulan Bator; tel. 325720; fax 320792; department store retailing household goods, foodstuffs, clothing, etc; state-owned.

XL-TA Holding: POB 981, Ulan Bator 13; fax 310133; f. 1990; investment, tourism, banking, trade, security services; Pres. Batdelgeriin Batbold.

TRADE UNIONS

Confederation of Mongolian Trade Unions: Sükhbaataryn Talbai 3, Ulan Bator 11; tel. 327253; fax 322128; brs throughout the country; Chair. Nyamjavyn Sodnomdorj; International Relations Adviser Ts. Natsagdorj; 450,000 mems (1994).

'Khökh Mongol' (Blue Mongolia) Free Trade Unions: Ulan Bator; f. 1991 by Mongolian Democratic Union; Leader Sh. Tömörbaatar.

United Association of Free Trade Unions: f. 1990; Pres. Sh. Dorjpagma.

Transport

MTT (Mongol Transport Team): POB 582, Ulan Bator 21; tel. 689000; fax 684953; e-mail mtt@mtteam.mn; internet www.mtteam.mn; international freight forwarders by air, sea, rail and road; offices in Beijing, Berlin, Moscow and Prague.

Tuushin Co Ltd: Yörönkhii said Amaryn Gudamj 2, Ulan Bator 210620; tel. and fax 325909; e-mail tuushin@magicnet.mn; f. 1990; international freight forwarders; transport and forwarding policy and services, warehousing, customs agent; tourism; offices in Beijing, Moscow and Prague; Dir-Gen. N. Zorigt.

RAILWAYS

In 2004 the total track length was 2,083km., of which 1,815km. was the main-line track. In 1990 the system carried 2.6m. passengers and 14.5m. tons of freight (about 70% of total freight traffic). However, traffic by rail later declined, owing to fuel shortages; in 1991 rail freight carriage was 10.3m. tons, falling to 8.6m. tons in

1992. In 2003 the railways carried 3.9m. passengers and 12.4m. tons of freight (about 70% of all freight carriage). Freight turnover in 2003 was 7,253.3m. ton/km of which 5,470.7m. ton/km was international traffic, including 4,041.7 ton/km of transit freight traffic. Passengers carried on international routes totalled 141,100, while 3.8m. passengers were carried on local routes.

Ulan Bator Railway: POB 376, Ulan Bator 13; tel. (11) 944401; fax (11) 328360; e-mail slc-mr@magicnet.mn; internet www.mtz.mn; f. 1949; joint-stock co with Russian Federation; Dir RADNAABAZARYN RAASH; First Dep. Dir N. BATMÖNKH; Chair. A.I. KASYANOV.

External Lines: from the Russian frontier at Naushki/Sükhbaatar (connecting with the Trans-Siberian Railway) to Ulan Bator and on to the Chinese frontier at Zamyn-Üüd/Erenhot, connecting with Beijing (total length 1,110 km).

Branches: from Darkhan to Sharyn Gol coalfield (length 63 km); branch from Salkhit near Darkhan, westwards to Erdenet (Erdenet-iin-ovoo open-cast copper mine) in Bulgan Province (164 km); from Bagakhangai to Baganuur coal-mine, south-east of Ulan Bator (96 km); from Khar Airag to Bor-Öndör fluorspar mines (60 km); from Sainshand to Züünbayan oilfield (63 km).

Eastern Railway, linking Mongolia with the Trans-Siberian and Chita via Borzya; from the Russian frontier at Solovyevsk to Choibalsan (238 km), with branch from Chingis Dalan to Mardai uranium mine near Dashbalbar (110 km), possibly inactive.

IFFC (International Freight-forwarding Centre): 2/F Ulan Bator Railway Building, Ulan Bator, CPO Box 376; tel. (11) 312509; fax (11) 313165; e-mail iffc@magicnet.mn; international freight forwarding.

Mongolian Express Co: Khos Jürj Bldg, Chingisiin Örgön Chölöö, Ulan Bator; tel. (11) 318329; e-mail info@monex.mn; internet monexpress@magicnet.mn; international dispatching agency; Dir Gen. D. ENKHBAT.

Mongoltrans: Khan-Uul District, Ulan Bator; POB 373, Ulan Bator 211121; tel. (11) 682100; fax (11) 687517; e-mail mtc@mongoltrans.mn; internet www.mongoltrans.mn; rail freight forwarding; offices in Beijing and Moscow; Dir-Gen. B. NOROLKHOOSUREN.

Progresstrans: PO Box 345, Ulan Bator 210526; tel. (11) 633992; fax (11) 631924; e-mail ptrans@magicnet.mn; internet www.progresstrans.com; international despatch and freight forwarding.

ROADS

Main roads link Ulan Bator with the Chinese frontier at Zamyn üüd/Erenhot and with the Russian frontier at Altanbulag/Kyakhta. A road from Chita in Russia crosses the frontier in the east at Mangut/Onon (Ölzii) and branches for Choibalsan and Öndörkhaan. In the west and north-west, roads from Biisk and Irkutsk in Russia go to Tsagaannuur, Bayan-Ölgii aimag, and Khankh, on Lake Khövsgöl, respectively. The total length of the road network was 49,250 km in 2003, of which asphalted roads comprised 5,632 km and gravel and improved earth roads comprised 1,801 km. The first section of a hard-surfaced road between Ulan Bator and Bayankhongor was completed in 1975. The road from Darkhan to Erdenet was also to be surfaced. Mongolia divides its road system into state-grade and country-grade roads. State-grade roads run from Ulan Bator to provincial centres and from provincial centres to the border. Country-grade roads account for the remaining 41,817 km, but they are mostly rough cross-country tracks.

To mark the millennium, the Government decided to construct a new east–west road, linking the Chinese and Russian border regions via Ölgii, Lake Khar Us, Zavkhan and Arkhangai provinces, Dashinchilen, Lün, Ulan Bator, Nalaikh, Baganuur, Öndörkhaan and Sümber. Construction was expected to take about 10 years, but because of the cost, the road was not expected to be surfaced for the whole length.

At the end of 2003 Mongolia had 68,458 cars (of which 53,512 were privately owned), 9,834 buses, 22,975 lorries, 1,320 tankers and 3,188 specialized vehicles.

There are bus services in Ulan Bator and other large towns, and road haulage services throughout the country on the basis of motor transport depots, mostly situated in provincial centres. However, in some years services have been truncated, owing to fuel shortages.

Tav Co: 20th sub-district, Songinokhairkhan District, Ulan Bator; tel. 632741; fax 632737; e-mail avtotav@mongol.net; road freight transport from Ulan Bator to the Mongolian provinces.

CIVIL AVIATION

Civil aviation in Mongolia, including the provision of air traffic control and airport management, is the responsibility of the Civil Aviation Directorate of the Ministry of Infrastructure. It supervises the Mongolian national airline (MIAT) and smaller operators such as Khangarid and Tengeriin Ulaach, which operate local flights. There are scheduled services to Ulan Bator (Buyant-Ukhaa) by Aeroflot (Russia) and Air China. In 2003 Mongolian airlines carried 138,800 passengers on international routes and 113,000 on internal routes.

Director-General of Civil Aviation: MANJIIN DAGVA.

Aero Mongolia Co Ltd: Buyant-Ukhaa Airport, Ulan Bator (PO Box 105, Ulan Bator 34); fax (11) 379616; e-mail management@aeromongolia.mn; internet www.aeromongolia.mn; f. 2001; began operations June 2003, scheduled international flights to Irkutsk and Hohhot and scheduled internal flights to five provincial centres and Juulchin's South Gobi tourist camp by Fokker-50 aircraft.

Blue Sky Aviation: POB 932, Ulan Bator 13; tel. (11) 312085; fax (11) 322857; e-mail bsa@whizzmail.com; jt venture of Mission Aviation Fellowship and Exodus International; operates charter flights and medical emergency services; Dir PAAVO KILPI; Operations Man. JAVKHLANTÖGSIIN GANBAATAR.

Khangarid: Room 210, MPRP Bldg, Baga Toiruu 37/1, Ulan Bator; tel. (11) 379935; fax (11) 379973; domestic and international passenger and freight services; Dir L. SERGELEN.

Mongolian Civil Air Transport (MIAT): Buyant-Ukhaa, Ulan Bator; tel. (11) 379935; fax (11) 379919; internet www.miat.com; f. 1956; operated by Air Consulting International (Republic of Ireland) as part of preparations for privatization; scheduled services to Moscow, Irkutsk, Beijing, Seoul, Osaka, Berlin, Hohhot, and to some Mongolian provincial centres; Dir LUTYN SANDAG.

Tengeriin Ulaach Shine: Buyant-Ukhaa 34-17, Khan-Uul District, Ulan Bator; tel. (11) 983043; fax (11) 379765; internal transport for tourists and businesspeople; Dir L. TÖMÖR.

Tourism

A foreign tourist service bureau was established in 1954, but tourism is not very developed. There are 12 hotels for foreign tourists in Ulan Bator, with some 1,500 beds, and the outlying tourist centres (Terelj, South Gobi, Öndör-Dov and Khujirt) have basic facilities. The country's main attractions are its scenery, wildlife and historical relics. There were 204,845 foreign visitors to Mongolia in 2003, of whom 21,890 had tourist visas. Tourist receipts totalled US $28m. in 1999. Tourism profits in 2002 amounted to US $102.9m., or 10.2% of GDP.

Juulchin Tourism Corporation of Mongolia: Chingis Khaany Örgön Chölöö 5B, Ulan Bator 210543; tel. (11) 328428; fax (11) 320246; e-mail juulchin@Mongol.net; internet www.mongoljuulchin.mn; f. 1954; offices in Berlin, New Jersey, Beijing, Tokyo, Osaka and Seoul; tours, trekking, safaris, jeep tours, expeditions; Exec. Dir S. NERGÜI.

Mongolian Tourism Association: 3/F, Rm 318, Building of the Mongolian Trade Unions Confederation, Sükhbaataryn Talbai 11, Ulan Bator; tel. (11) 327820; e-mail info@travelmongolia.org; internet www.travelmongolia.org; Pres. TSEVELMAA BAYARSAIKHAN.

National Travel and Tourism Agency: Government Bldg 14, Sambuugiin Gudamj 11, Ulan Bator 38; tel. (11) 311102; fax (11) 318492; e-mail mtb@magicnet.mn; internet www.mongoliatourism.gov.mn; Dir M. ENKHBAYAR.

Defence

Under the 1992 Constitution, the President of Mongolia is *ex officio* Commander-in-Chief of the Armed Forces. The defence roles of the President, the Mongolian Great Khural, the Government and local administrations are defined by the Mongolian Law on Defence (November 1993). Mongolia's Military Doctrine, a summary of which was issued by the Great Khural in July 1994, defines the armed forces as comprising general purpose troops, air defence troops, construction troops and civil defence troops. The border troops and internal troops, which are not part of the armed forces, are responsible for protection of the borders and of especially important installations, respectively. The general purpose troops comprise rear services, communications, artillery, etc. The Ministry of Defence has departments in charge of personnel, foreign relations and contracts, etc. In August 2003, according to the International Institute for Strategic Studies, Mongolia's defence forces numbered 8,600, comprising an army of 7,500 (of whom 3,300 were thought to be conscripts), 800 air defence personnel and 300 construction troops. There was a paramilitary force of about 7,200, comprising 1,200 internal security troops and 6,000 border guards. Military service is for 12 months (for males aged 18–25 years), but a system of alternative service is being introduced, and only about 40% of conscripts are found fit for service. Besides deferment for family or educational reasons, a limited number of conscripts may buy themselves out of military service. There are financial inducements for regular service soldiers, especially those in the best trained and equipped 'élite' battalions, some of whom have seen service with coalition forces in Iraq. Training of officers and NCOs is concentrated at the Defence University in Ulan Bator. Transport aircraft and helicopters for support of the ground forces have been authorized to carry civilian passengers, thus reducing the distinction between military and civil air transport operations.

Defence Expenditure: Defence spending for 2004 was projected at 20,742.3m. tögrög (about 3.2% of total planned government expenditure for 2004).

Chief of Staff of the Mongolian Armed Forces: Maj.-Gen. TSEVEGSÜRENGIIN TOGOO.

Education

General education is entirely state-administered. Ten-year education is compulsory, and 11-year education is being introduced. In the 2003/04 school year there were 537,400 pupils and a teaching staff totalling 20,792. The 32 vocational schools, with 21,500 students in 2003/04, train personnel for the service industries, including electricians, drivers and machine operators.

In 2003/04 131,700 students were enrolled at state and private institutes of higher education, including those studying in Russia, Germany, Turkey, the USA and elsewhere. The Mongolian National University comprises four faculties (mathematics, natural sciences, physics and social sciences), four institutes (biology, economics, law and Mongol studies) and the School of Foreign Service (diplomatic training). In 2003/04 there were seven other state-owned universities and three private ones, as well as 38 state-owned and 134 private higher schools and colleges. About 34,600 students attended the private higher schools and colleges. In 2003/04 the number of students at higher schools, colleges and universities was 108,500.

The state budget allocation to the Ministry of Education, Culture and Science for 2004 was 138,382.7m. tögrög (21.6% of planned budgetary expenditure).

Bibliography

GENERAL AND ECONOMY

Allen, Benedict. *Edge of Blue Heaven—A Journey through Mongolia*. London, BBC Books, 1998.

Asian Development Bank. *Mongolia: A Centrally Planned Economy in Transition*. Manila, 1992.

Badarch, Dendeviin, Zilinskas, Raymond A., and Balint, Peter J. (Eds) *Mongolia Today—Science, Culture, Environment and Development*. London, RoutledgeCurzon, 2002.

Batbayar, Bat-Erdene (trans Ed. Kaplonski, C.). *Twentieth Century Mongolia*. Cambridge, White Horse Press, 1999.

Batbayar, Tsedendambyn (Ed.). *Renovation of Mongolia on the Eve of the XXI Century and Future Development Patterns*. Ulan Bator, Mongolian Development Research Centre, 2000.

Bawden, Charles R. *Mongolian–English Dictionary*. London, Kegan Paul, 1997.

Mongolian Traditional Literature: An Anthology. London, Kegan Paul, 2003.

Becker, Jasper. *The Lost Country: Mongolia Revealed*. London and Sydney, Hodder and Stoughton, 1992.

Blunden, Jane. *Mongolia: The Bradt Travel Guide*. Chalfont St Peter, Bradt, 2004.

Bruun, Ole, and Odgaard, Ole (Eds). *Mongolia in Transition: Old Patterns, New Challenges*. Richmond, Curzon Press, 1996.

Bulag, Uradyn E. *Nationalism and Hybridity in Mongolia*. Oxford, Clarendon Press, 1998.

Damdinsuren, H. (Ed.) *Ulaanbaatar Capital of Mongolia*. Ulan Bator, Mongolian Business Press and Information Bureau and Montsame News Agency, 2000.

Doing Business in Mongolia: A Guide for European Companies: Ulan Bator, Mongolian Business Development Agency and European Union Tacis Programme for Mongolia, 1996.

Enkhtuvshin, A. (Ed.) *Reference Book on Mongolian Investment (National Almanac 98–99)*. Ulan Bator, Board of Foreign Investment and Hamag Mongol National Advertising Agency, 1999.

Ganbold, D. (Ed.). *Facts about Mongolia*. Ulan Bator, Admon, 2000.

Goldstein, Melvyn C., and Beall, Cynthia M. *The Changing World of Mongolia's Nomads*. Hong Kong, The Guidebook Company, 1994.

Griffin, Keith (Ed.). *Poverty and the Transition to a Market Economy in Mongolia*. Basingstoke, Macmillan Press, 1995.

Hansen, Henny Harald. *Mongol Costumes*. London, Thames and Hudson, 1993.

Humphrey, Caroline, and Onon, Urgunge. *Shamans and Elders*. Oxford, Clarendon Press, 1996.

Jagchid, S., and Hyer, P. *Mongolia's Culture and Society*. Folkestone, Wm Dawson and Sons Ltd, 1979.

Janhunen, Juha (Ed.) *The Mongolic Languages*. London, Routledge, 2003.

Kaplonski, Christopher. *Truth, History and Politics in Mongolia—The Memory of Heroes*. London and New York, RoutledgeCurzon, 2004.

Kotkin, Stephen, and Elleman, Bruce A. (Eds). *Mongolia in the Twentieth Century: Landlocked Cosmopolitan*. London, M. E. Sharpe, 1999.

Krouchkin, Yuri. *Mongolia Encyclopedia*. Ulan Bator, Edmon Publishing House, 1998. (A business-orientated collection of laws and regulations.).

Kullman, Rita, and Tserenpil, D. *Mongolian Grammar*. Hong Kong, Jensco Ltd, 1996.

Lawless, Jill. *Wild East: Travels in the New Mongolia*. Toronto, ECW Press, 2000.

Man, John. *Gobi: Tracking the Desert*. London, Weidenfeld and Nicolson, 1997.

Marder, Aariimaa Baasanjav. *Mongolian Dictionary and Phrasebook*. New York, NY, Hippocrene Books, 2002.

Mayhew, Bradley. *Mongolia: Discover a Land without Fences*. Hawthorn, Vic, Lonely Planet, 2001.

Mongolian Business Directory. Ulan Bator, Mongolian Chamber of Commerce and Industry, 1994.

Mongolian Economy and Society in 1996. Ulan Bator, State Statistical Office, 1997.

Mongolia in a Market System 1989-2002. Ulan Bator, National Statistical Office, 2004.

Mongolian National Security and Defense Policy Handbook. USA, International Business Publications, 2002.

Mongolian Statistical Yearbook 2003. Ulan Bator, National Statistical Office, 2004.

MPR Academy of Sciences. *Information Mongolia*. Oxford, Pergamon, 1990.

Myagmarsuren, D. (Ed.). *Special Protected Areas of Mongolia*. Ulan Bator, Environmental Protection Agency, 2000.

Nixson, Frederick, Walters, Bernard, Suvd, B., and Luvsandorj, P. *The Mongolian Economy*. Cheltenham, Edward Elgar, 2000.

Nordby, Judith. *Mongolia*. World Bibliographical Series Vol. 156, Oxford, Clio Press, 1993.

Pegg, Carole. *Mongolian Music, Dance, and Oral Narrative*. Seattle and London, University of Washington Press, 2001.

Purev, Otgony. *The Religion of Mongolian Shamanism*. Ulan Bator, GENCO, 2002.

Purevsambuu, G. (Ed.). *Mongolia*. Ulan Bator, Montsame News Agency, 2003.

Sanders, A. J. K. *Mongolia: Politics, Economics and Society*. London, Frances Pinter, 1987.

Historical Dictionary of Mongolia. Lanham, MD, and London, Scarecrow Press, 2003 (2nd edition).

Sanders, A. J. K., and Bat-Ireedüy, J. *Colloquial Mongolian*. London, Routledge, 2002 (reprint).

Mongolian Phrasebook. Footscray, Vic, Lonely Planet, 2003 (reprint).

Sermier, Claire. *Mongolia: Empire of the Steppes*. Hong Kong, Odyssey/Airphoto International, 2002.

Severin, Tim. *In Search of Genghis Khan*. London and Sydney, Hutchinson, 1992.

Stewart, Stanley. *In the Empire of Genghis Khan—A Journey among Nomads*. London, HarperCollins, 2000.

Thevenet, Jacqueline. *Les Mongols de Genghis Khan et d'aujourd'hui*. Paris, Armand Colin, 1986.

Traders' Manual for Asia and the Pacific: Mongolia. New York, UN Economic and Social Commission for Asia and the Pacific, 1995.

Waugh, Louisa. *Hearing Birds Fly: A Nomadic Year in Mongolia.* London, Little, Brown, 2003.

Weatherford, Jack. *Genghis Khan and the Making of the Modern World.* New York, Crown Publishers, 2004.

Worden, Robert L., and Savada, Andrea Matles (Eds). *Mongolia: A Country Study.* Washington, DC, Library of Congress, 1991, 2nd edn.

HISTORY

Akiner, Shirin (Ed.). *Mongolia Today.* London, Kegan Paul International, 1992.

Allsen, Thomas T. *Mongol Imperialism.* University of California, 1987.

Barkmann, Udo. *Geschichte der Mongolei* (History of Mongolia). Bonn, Bouvier, 1999.

Bat-Ochir, Bold. *Mongolian Nomadic Society—A Reconstruction of the 'Medieval' History of Mongolia.* Richmond, Curzon Press, 1999.

Bawden, Charles R. *The Modern History of Mongolia.* (2nd edn, with afterword by A. J. K. Sanders). London, Kegan Paul International, 1989.

Biran, Michal. *Qaidu and the Rise of the Independent Mongolian State in Central Asia* Richmond, Surrey, Curzon Press, 1997.

Brown, W. A., and Onon, U. *History of the Mongolian People's Republic.* Cambridge, MA, Harvard University Press, 1976. English trans., with extensive footnotes, of Vol. III of Mongolia's official history.

Buell, Paul D. *Historical Dictionary of the Mongol World Empire.* Lanham and Oxford, Scarecrow Press, 2003.

Colvin, John. *Nomonhan.* London, Quartet Books, 1999.

Dashpurev, D., and Soni, S. K. *Reign of Terror in Mongolia 1920–1990.* New Delhi, South Asian Publishers, 1992.

Dashpurev, D., and Prasad, Usha. *Mongolia: Revolution and Independence 1911–1992.* New Delhi, Subhash and Associate, 1993.

de Rachewiltz, Igor. *Papal Envoys to the Great Khans.* London, Faber and Faber, 1971.

(Trans. and commentary) *The Secret History of the Mongols—A Mongolian Epic Chronicle of the Thirteenth Century.* Leiden and Boston, Brill, 2004.

Ewing, E. E. *Between the Hammer and the Anvil? Chinese and Russian Policies in Outer Mongolia 1911–21.* Bloomington, IN, Indiana University, 1980.

Marshall, Robert. *Storm from the East.* London, BBC Books, 1993.

Morgan, D. *The Mongols.* Oxford, Blackwell, 1990.

Musée National des Arts Asiatiques. *Trésors de Mongolie XVIIe-XIXe siècles.* Paris, Guimet, 1993.

Onon, Urgunge (Trans.). *Mongolian Heroes of the 20th Century.* New York, AMS Press, 1976.

The Golden History of the Mongols. London, Folio Society, 1993.

The Secret History of the Mongols—The Life and Times of Chinggis Khan. London, RoutledgeCurzon, 2001.

Onon, Urgunge, and Pritchatt, D. *Asia's First Modern Revolution.* Leiden, E. J. Brill, 1989.

Ratchnevsky, Paul (trans. Thomas Haining). *Genghis Khan: His Life and Legacy.* Oxford, Blackwell, 1991.

Rupen, R. A. *Mongols of the Twentieth Century, Vol. 1* (History), *Vol. 2* (Bibliography). The Hague, Indiana University and Mouton and Co, 1964 (reprinted).

Sabloff, Paula (Ed.). *Modern Mongolia: Reclaiming Genghis Khan.* Philadelphia, University of Pennsylvania Museum of Archaeology and Anthropology, 2001.

Sandag, Shagdariin, and Kendall, Harry H. *Poisoned Arrows: The Stalin-Choibalsan Mongolian Massacres 1921–1941.* Oxford, Westview Press, 2000.

Saunders, J. J. *The History of the Mongol Conquests.* London, Routledge and Kegan Paul, 1971.

The Mongolia Society Inc, of Indiana University, Bloomington, IN, publishes historical and contemporary studies (see Part Three, Regional Information—Research Institutes, USA).

MYANMAR
(BURMA)

Physical and Social Geography

HARVEY DEMAINE

The Union of Myanmar (until 1989 the Socialist Republic of the Union of Burma), which covers a total area of 676,552 sq km (261,218 sq miles), lies to the east of India and Bangladesh and to the south-west of the People's Republic of China, and has a long coastline facing the Bay of Bengal and the Andaman Sea. Much the greater part of its territory, lying between latitudes 28° 50′ and 16° N, forms a compact unit surrounded on three sides by a great horseshoe of mountains and focusing on the triple river system of the Ayeyarwady, Chindwinn and Sittoung (or Irrawaddy, Chindwin and Sittang, respectively). In addition, Tanintharyi (Tenasserim), consisting of a narrow coastal zone backed by steep mountains, extends south from the Gulf of Martaban to Victoria Point, only 10° N of the Equator.

PHYSICAL FEATURES

Structurally, the topography of Myanmar falls into three well-marked divisions, of which the first comprises the mid-Tertiary fold mountains of the west. These ranges, swinging in a great arc from the Hukwang valley to Cape Negrais, appear to represent a southward continuation of the eastern Himalayan series. From north to south these western ranges are known successively as the Patkai, Naga, and Chin Hills, and the Arakan Yoma, though hills is a misleading description of ranges with summits exceeding 3,650 m in the case of the Patkai and reaching 1,800 m–2,400 m in the case of the Chin and Naga Hills. Further south, in the Arakan Yoma, the summit levels gradually decrease to 900 m–1,500 m.

The second major structural unit consists of the eastern mountain ranges, of Mesozoic or earlier origin, which, beginning as a continuation of the Yunnan plateau of China across the Myanma border into the north-eastern corner of Kachin State, extend thence through the Shan and Kayinni (or Karenni) plateaux into more subdued but still rugged upland, which forms the divide between Tanintharyi and peninsular Thailand. In the far north, where this system adjoins the western mountain system, the general plateau level is of the order of 1,800 m, with higher ridges frequently attaining 3,000 m. The corresponding altitudes in the Shan area, however, are only about one-half as great, though here also the surface is dissected, with the main rivers, notably the great Thanlwin (Salween), rushing southwards in deeply incised gorges.

Between the two main mountain systems lies the third major structural unit, namely the vast longitudinal trough of central Myanmar, containing the great alluvial lowlands which form the cultural and economic heart of the country. Throughout the length of these lowlands the Ayeyarwady provides the central artery, both of drainage and of communication. To the north it is paralleled by its largest tributary, the Chindwinn, which joins it near the centre of the Dry Zone, and further south by the Sittoung, which flows separately to the sea on the opposite side of the recent volcanic uplands of the Pegu Yoma. Central Myanmar is a zone of crustal instability; a severe earthquake in July 1975 caused extensive damage. Altogether, the Ayeyarwady drains a total area of some 400,000 sq km, and its huge delta potentially provides one of the greatest 'rice bowls' of the world.

CLIMATE

Apart from the highest uplands in the far north of the country, the climate of practically the whole of Myanmar may be classified as tropical monsoonal, although important regional variations nevertheless occur within that overall category. In all parts of the country the main rains come during the period of the south-west monsoon, i.e. between May and October, and those areas, notably Rakhine (or Arakan) and Tanintharyi, that face the prevailing winds and are backed by steep and high ranges, receive some of the heaviest rainfall in the world (Sittwe, or Akyab, 5,180 mm annually; Kyaikkami, or Amherst, 4,980 mm). Moreover, even the flat and low-lying Ayeyarwady delta receives an annual rainfall of about 2,500 mm, and in all of these three areas mean annual temperatures are around 27°C, though the seasonal range varies from 6.5°C in Sittwe to 3.5°C in Kyaikkami.

A considerable portion of the interior of the central lowland constitutes a rain-shadow area relative to the south-west monsoon, and here the total annual precipitation is less than 1,000 mm. In some places it is even below 640 mm. Although in this Dry Zone the seasonal incidence is similar to that of other areas, the spectacular difference in total amount is reflected in a major change of vegetation, from the heavy tropical monsoon forest prevailing elsewhere, to a much more open cover and, in places, a mere thorny scrub. Moreover, the relative aridity is also responsible for a wider range of temperature, as is shown by Mandalay's 21°C in January and 32°C in April, immediately before the onset of the rains. Finally, in the eastern plateaux, rainfall is above that of the Dry Zone, but much less than along the western coastal margins, and this fact, combined with temperatures some 6°–8°C below those of the torrid plains, gives the Shan plateau the most equable climate of any part of the country.

NATURAL RESOURCES

Myanmar's natural resources are closely related to the salient features of its physical geography. Thus, the greatest wealth of the humid mountain slopes lies in timber, particularly teak, and while the young folded mountains of the west are not noted for mineral wealth, the older plateaux of the east have long been noted for a variety of metallic minerals, including the silver, lead and zinc of Bawdwin and the tungsten of Mawchi. Further south, Tanintharyi forms a minor part of the South-East Asian tin zone, though its resources in this respect are very small compared with those of Malaysia and Thailand. More important than any of these metals, or the sub-bituminous coal deposits at Kalewa, near the Chindwinn–Myittha confluence, are the petroleum and natural gas deposits that occur in the Tertiary structures underlying the middle Ayeyarwady lowlands. Prospects for the petroleum industry, which remained small by world standards, were much improved by the early 1990s, owing to foreign participation in onshore exploration, and to expectations of future foreign participation in offshore exploration.

It is in agricultural resources that Myanmar is potentially most richly endowed, and the Ayeyarwady delta should eventually fulfil its potential as a 'rice bowl' under the stimulus of higher prices, higher-yielding varieties and improved water control. The Dry Zone is also well suited for the production of oilseeds and cotton, especially under irrigation, and in Tanintharyi conditions are appropriate, though not ideal, for the cultivation of rubber and fruit crops.

POPULATION AND ETHNIC GROUPS

At mid-2003, according to UN estimates, Myanmar's population had reached 49,485,000: a density of only 73.1 per sq km. The greatest concentration of population occurs in the delta. In the lowlands, including Rakhine and Tanintharyi, the Bamar (Burmans) form the majority element in the population. The uplands are more sparsely inhabited by a series of minority groups at varying levels of development. The Bamar, whose ancestors came from the Sino-Tibetan borders and supplanted all but a fraction of the earlier Mon population of lowland Burma, formed some 68% of the total population of Myanmar in 1992. A further 7% were Shan, the ethnic kinsfolk of the Thai and Lao, who follow the Theravada form of Buddhism, like the Bamar and the Mon (2.3%), while the Arakanese constituted a further 3.8% of the population.

Of the non-Buddhist indigenous groups, who are often referred to collectively as hill peoples, the Kayin (Karen) are the most numerous (6% of the total population in 1992). Their homeland occupies the uplands between the Shan plateau and Tanintharyi, but many have migrated into the lowlands around Mawlamyine (Moulmein) and to the Ayeyarwady delta, and considerable numbers have discarded animism and adopted Christianity. Other upland peoples include the Kachin (2.3%), Chin (2.3%), Wa-Palaung, Lolo-Muhso and Naga, who are still mostly animists.

Myanmar has remained below the South-East Asian average in respect of urbanization, with Yangon (Rangoon), the capital, having a population of 2,513,023 at the census of 31 March 1983, and only seven other towns exceeding 100,000 inhabitants. By mid-2003, according to UN estimates, the population of Yangon had increased to 3,873,739.

History

ROBERT CRIBB

Based on an earlier article by JOSEF SILVERSTEIN

EARLY HISTORY

Modern Myanmar falls into three distinct geopolitical zones: the valley of the Ayeyarwady (Irrawaddy), the hill country that surrounds the valley, and the coastal areas of Rakhine (Arakan) and Tanintharyi (Tenasserim). The long, fertile valley of the Ayeyarwady, which is navigable for 1,400 km from Bhamo to the sea, has witnessed a succession of powerful kingdoms based on both agriculture and trade; only for brief periods, however, were those kingdoms able to unite the entire valley in a single polity. The earliest kingdoms were founded by the Mon, ethnically close to today's Cambodians, and the Pyus. However, from about the 10th century a Tibeto-Burmese people, the Burmans (now Bamars), entered the valley and eventually conquered the Mon and Pyus, absorbing many elements of those earlier cultures in the process. The hill country, which surrounds the valley in a broad horseshoe of territory, is inhabited by smaller ethnic groups—the Shan (who are closely related to the Thai), the Karen (now Kayin), the Kachin, the Chin and the Naga—which sometimes briefly established kingdoms extending into the valley, but which were more often subject or tributary to the kingdoms of the Mon and Burmans. The third zone, which comprises the coastal strips of Rakhine and Tanintharyi, has remained relatively isolated from the country's heartland and has a long history of maritime engagement with the outside world.

The first large Burman kingdom, Pagan, was founded in the 11th century and it was in the context of this kingdom that Theravada Buddhism was established as the religion of the Burmans. Pagan was destroyed by a Mongol invasion in 1287. Its successor kingdom, Ava, was weakened by fighting with the Mon and the Shan. In the mid-16th century the Toungoo dynasty created an empire which extended briefly from Rakhine in the west to Laos in the east. However, spread over too wide an area, the Toungoo empire was unable to defeat counter-attacks from Rakhine and Siam, and subsequently collapsed at the end of the 16th century. Only in the late 18th century did a new dynasty, the Konbaung, unite the country and resume Burma's expansion. In doing so, however, it came into conflict with British power in India, and in three wars between 1824 and 1885 Burma lost first Rakhine and Tanintharyi, then the lower reaches of the Ayeyarwady, and finally the north of the country. The British abolished the Burman monarchy and annexed Burma to British India, inheriting at the same time the Konbaung dynasty's contested claims to the hill regions. The Kachin region in the far north was not fully subdued until 1915, and the British also faced continuing resistance in the countryside of the Ayeyarwady valley.

BRITISH AND JAPANESE RULE

British rule (which lasted from 1896 to 1948) transformed Burma socially, economically and administratively. The power of the court was removed and that of the aristocracy greatly diminished, while village headmen were included in a new centralized administrative system, leaving the Buddhist monastic establishment (*sangha*) as the country's most powerful indigenous institution and a centre of hostility to foreign rule. Indians began to migrate to Burma in large numbers, as traders and labourers, to the extent that the capital, Rangoon (now Yangon), became a predominantly Indian city. The establishment of Rangoon University in 1920 made Western learning accessible to a new generation of Burmese and with the growth of exports of rice, oil, teak, tin, rubies and cotton, Burma became more integrated than ever before into the global economy. This integration led to serious hardship in Burma during the Depression, and discontent was expressed in the Saya San uprising of 1930–31.

As part of British India, Burma was included in the gradual development of democratic institutions. In 1923 the British introduced a partly elected Legislative Council and, under the system known as 'dyarchy', devolved political power in several fields to ministers responsible to the Council. In 1937 the administration of Burma was separated from that of India and Burmese cabinet ministers responsible to an elected parliament took over all areas of government except defence, foreign relations and monetary policy. The influence of Burman parliamentarians, however, was circumscribed by the existence of separate, more generously represented electorates for several minorities (Karen, Indians, Anglo-Indians, Chinese and Europeans), so that the four successive Governments that served in office between 1937 and 1942 were all coalitions between Burman and minority representatives. These constitutional arrangements, however, excluded most of the hill regions, which remained under direct British rule.

The modern Burmese nationalist movement that arose in the context of these developments was strongly dominated by ethnic Burmans. The Young Men's Buddhist Association (YMBA), founded in 1906, campaigned initially on religious issues, but subsequently developed into the General Council of Buddhist Associations (GCBA), which pursued a more political agenda. In contrast, the Thakin (master) movement, founded by members of the young intelligentsia in Rangoon in 1930, emphasized a more secular Burmese nationalism which incorporated Marxist elements but which drew strongly on the Burman language, culture and traditions. This movement provided the new generation of leaders who guided the nation through the war and into independence. Throughout the 1930s, while the British tried to lead Burma towards full self-government, the nationalist movement sought the country's independence. Believing that only armed force could overthrow the British, 30 Thakins, including Aung San, went to Japan in 1940 and returned in 1942 with invading Japanese forces and a small but symbolic Burma Independence Army (BIA).

The Japanese advanced rapidly, driving the British out of most of Burma by mid-1942; however, they were unable to press more than a few kilometres beyond Burma's borders into India.

The Japanese ruled Burma with a combination of conciliation and repression. In August 1943, for example, they granted 'puppet' independence to a state headed by the pre-war Burman leader, Dr Ba Maw, and also expanded the BIA into the Burma National Army (BNA). However, they demanded labour and materials for the war effort and were brutal in their repression of the growing resistance movement, especially amongst the Karen and other hill peoples. In August 1943 the Japanese transferred part of the Shan region to Thailand. In late 1944, with Japan in retreat on its Indian front, Aung San and leaders of the growing Communist movement founded the Anti-Fascist People's Freedom League (AFPFL), turning the BNA against the Japanese in March 1945 and launching a revolt which hastened the Allied victory in Burma.

The British announced in May 1945 that they were willing to grant Burma independence within the British Commonwealth. Following the end of Allied military administration in October, however, this promise was suspended in favour of a period of economic reconstruction. Concerned that such a delay would allow British economic domination to continue, Aung San and other nationalist leaders led a campaign of unrest which forced the British to appoint AFPFL members to the governor's Executive Council and to call elections in April 1947 for a constituent assembly to draft a new constitution. The AFPFL won 171 of the 182 seats contested, and the assembly drafted a constitution, which gave Burma full independence outside the Commonwealth. In July 1947 Aung San and six other members of the Executive Council were assassinated; however, the governor appointed Thakin Nu (later U Nu) as Aung San's successor and allowed the movement towards independence to continue. Burma thus became independent on 4 January 1948.

CONSTITUTIONAL GOVERNMENT, 1948–62

At independence in 1948, the governing party was the socialist AFPFL. Throughout independence negotiations, the United Kingdom had preferred to deal with the non-communist wing of the AFPFL. As a consequence, the communist wing, with strong popular support both in the Ayeyarwady delta and amongst former members of the BIA, perceived itself to be excluded from power in the new Government. In March 1948, after Government incursions against them, the communist element went into revolt, taking significant sections of the army and the militia with it. In 1949 the situation became more complicated when the Karen also launched a revolt. The independent Union of Burma was an unequally balanced federation: alongside the Burman heartland, it included the hill regions as states with varying degrees of autonomy. The Shan and Kayah states had the formal right to secede from the Union; the Kachin and Karen states had autonomy but no right to secede; the Chin Special Division had only limited autonomy, while a separate status for Arakan was only foreshadowed. The army was also constructed federally, with each unit drawn primarily from one or other of the ethnic groups. From the outset, however, the Government in Rangoon tended to concentrate power in its own hands, and the Karen, who were generally well-educated and who had prospered on the whole under British rule, grew increasingly discontented. In time, revolts were to break out amongst almost all the country's minorities. Also, during 1949, the eastern Shan region was occupied by Chinese Nationalist armies in retreat after the victory of the Chinese Communist Party.

For several months the central government controlled only Rangoon and a few other centres. Burma's formal unity survived partly because the rebels fought amongst themselves, partly because they had no significant external support, and partly because U Nu's Government responded vigorously to the crisis. U Nu reformed the army, placing Gen. Ne Win in charge. Ne Win subsequently centralized and enlarged the military, eliminating the ethnic units and placing Burman officers in most command positions. He also operated ruthlessly against the rebels, developing for the army a reputation for brutality against civilian opponents. Meanwhile, U Nu campaigned energetically in the country's heartland, winning support as a charismatic proponent of state Buddhism. By 1951 the Government was once again in control of the Burman heartland and was well-established in parts of the hill country, although perhaps 10% of the country remained outside its control.

The AFPFL won national elections in 1951–52 and 1956, with an overwhelming but decreasing share of the vote, and in 1958 the party split into 'Clean' and 'Stable' rival factions. To avert renewed civil war, U Nu invited Gen. Ne Win to form a 'caretaker' government to prepare the country for new elections which were to be held in 1960. The 18-month interlude of military rule was generally welcomed as a period of law and order, economic growth and the implementation of serious measures against corruption. Although the U Nu-led 'Clean' faction of the AFPFL, renamed the Union Party, won the elections, U Nu soon alienated minorities by declaring Buddhism as the state religion, which provoked a revolt in the Shan and Kachin regions and exacerbated discontent amongst the Mon and Karen. With corruption and administrative mismanagement plaguing the country once more, the armed forces under Gen. Ne Win carried out a *coup d'état* on 2 March 1962, arresting members of the Government, suspending the Constitution and appointing a Revolutionary Council (RC) to govern Burma by decree.

MILITARY RULE, 1962–74

Following the *coup d'état*, Gen. Ne Win quickly rescinded U Nu's declaration of Buddhism as the state religion; in other respects, however, his regime paid little attention to public opinion. The democratic institutions of the 1950s were dismantled and power was concentrated in the hands of the RC, which comprised a small group of senior officers led by Gen. Ne Win. In form and theory, Burma remained a federal state; however, for all practical purposes the new leadership treated the country as a unitary state. At the local level, authority was placed under the control of new Security and Administration Committees (SACs), headed by local military commanders. For the next decade, however, the Ne Win Government continued to fight ethnic and communist insurgents in almost all the hill country outside the Burman heartland. In 1969–71 U Nu tried unsuccessfully to create a coalition of rebel forces under his leadership; his lack of popularity among the ethnic minorities, however, rendered this impossible. A more serious threat to the Government came from the Communist Party of Burma (CPB), which received substantial support from the People's Republic of China, especially during the latter's Cultural Revolution. From 1970, however, relations between Burma and China improved, and Chinese support for the CPB was reduced. Ne Win's long-term political aim was to reshape Burma into a Socialist one-party state, and to this end he created the Burma Socialist Programme Party (BSPP, also known as Lanzin, from its Burmese name) in June 1962. The party adopted as its doctrine a manifesto issued two months earlier by the RC entitled *The Burmese Way to Socialism*, which took a Marxist approach to capitalism, but which incorporated doctrinal elements from Theravada Buddhism, in particular the concept of impermanence. At the same time, the Government sought to prevent the Buddhist *sangha* (order of monks) from becoming a focus of political opposition, and in 1965 an attempt was made by the Government to require all Buddhist monks to register and to carry identity cards. The manifesto adopted by the BSPP also included the notion that human greed needs to be kept under control by state intervention. The BSPP refrained from becoming involved in international conflicts, emphasizing the need for Burma to follow its own path and keeping the country remarkably isolated from the outside world. In 1966, in order to ensure that these ideas were widely disseminated throughout society, the RC launched a thorough reform of the education system, bringing schools (including monastic schools) under the closer control of the Ministry of Education and adding an ideological element to the curriculum. Ne Win's Government also sought to 'Burmanize' the economy, nationalizing land, the banking sector, oil wells, foreign trade, the insurance sector, shipping, wholesale trade, cinemas, and much of the publishing, mining and saw-milling industries in the years between 1963 and 1971. In so doing, they displaced the European, Indian and Chinese business interests which had previously dominated these sectors and caused an exodus of technical and managerial personnel. The economy gradually began to show signs of stagnation, including the emergence of a 'black market'. Formerly a major exporter of rice, Burma ceased to produce rice surpluses. Meanwhile, the reputation of the regime was not helped by growing indications of corruption and the abuse of power by members of the RC itself. The competence of the administration was further diminished by the exclusion of civilians from the Government: whereas Gen. Ne Win had made considerable use of highly qualified civilians in his administration during the 'caretaker' Government of 1958–60, the Government that held office from 1962 to 1974 was dominated by a small group of senior military officers. In those 12 years, only three civilians achieved ministerial rank, while the Council of Ministers—Burma's Cabinet—was reshuffled only twice.

In its early stages, the BSPP was a cadre party with a small membership: when the party held its first Congress in 1971 it had only 73,369 members, more than half of whom were drawn from the military. At the party's inaugural Congress Gen. Ne Win was formally elected as leader. The party identified its mission as twofold: to transform itself from a cadre party into a mass party in order to replace centralism with democratic centralism; and to transform the nation into a socialist democratic state under a new constitution. In July 1971 the military leaders announced their intention to draft a new constitution and to transfer power to a civilian government. A recruitment campaign for the party subsequently began in 1972, and within a decade its membership had expanded to 1.5m. Meanwhile, in April 1972, to accord with the move toward constitutionalism, Gen. Ne Win and 20 of his senior commanders retired from the army and became civilian members of the Government. The RC began to prepare the way for its own dissolution, proclaiming the end of the revolutionary Government and its replacement by the Government of the Union of Burma. Gen. Ne Win became Prime Minister, heading a Cabinet of nine retired officers, three serving officers and two civilians. This 'civilianization' of Gen. Ne Win's rule was completed in early 1974, in which year elections were held, the new Constitution came into effect and the RC was dissolved. Ne Win was elected Chairman of the Council of State and thus became President under the new Constitution. Eleven of the 29 members of the Council of Ministers were carried over from the now-defunct RC.

BSPP RULE, 1974–88

Civilian rule was not merely intended to make military dominance appear more acceptable. Ne Win hoped that the one-party state would provide a format for engaging the public in government and national development without, however, permitting the political turmoil and drift which had blighted the parliamentary system. In January 1978 and October 1981, therefore, the Government held national elections, although only BSPP candidates were permitted to stand. The BSPP Government also presented itself as the protector of Burmese national identity in 1982, by formalizing a citizenship law, which created three categories of citizen—national, associated and naturalized. Resident Indians and Chinese were consigned to the latter categories, under which they had restricted rights in politics and the economy and were not permitted to join the armed forces. The Government also reversed the RC's previous antagonism towards the Buddhist *sangha*: rather than confronting organized Buddhism, the regime provided it with financial and administrative support, establishing a Ministry of Religious Affairs and acknowledging the importance of Buddhist tradition to the history and culture of the nation. In May 1980 the regime sponsored a nation-wide congregation for the purification, propagation and perpetuation of Buddhism. At this convention, representatives created a centralized authority to control the *sangha* throughout the nation and finally approved the introduction of identification cards for monks and nuns. In 1981 the religious courts, which had been inactive for years, were revived. Two sects were brought before them and charged with teaching heretical doctrines. After an extended trial, both sects were found guilty and ordered to be dissolved. Monks belonging to the two orders were to recant publicly or cease to be monks. These measures seemed to eliminate the potential of the *sangha* to serve as a base for opposition, and gave the Government the confidence to declare an extensive amnesty for U Nu and his followers, who were permitted to return to the country. Thousands of political prisoners were also released.

In December 1974, less than one year after the new civilian Government had been installed, riots broke out in Rangoon and other centres over food shortages, corruption and generally declining economic conditions. (The immediate catalyst for the riots had, however, been the perceived lack of proper honour accorded by the Government to the funeral of U Thant, a former Secretary-General of the UN and a friend of U Nu.) Students in Rangoon were particularly angry at declining standards and conditions at the universities, and further demonstrations broke out in March 1976. The BSPP responded initially by modifying its socialist programme, approving the idea of foreign aid and investment. The complaints about official corruption were met by new regulations requiring leaders at all levels to disclose their assets, so that corruption amongst those in power could be more easily detected. These signs of independent initiative on the part of the BSPP appear to have alarmed Ne Win, who launched a series of purges of the party. In October 1976 some 50,000 party members were expelled, the party's structure was reformed and more than half the Central Committee resigned, followed by several members of the national Cabinet, including the Prime Minister, U Sein Win. Further purges took place in September 1977 and February 1978, leaving the upper ranks of the BSPP and much of the Cabinet under the control of serving or retired military officers. In 1977 the Pyithu Hluttaw (People's Assembly) adopted a Private Enterprises Law, which firmly rejected the economic liberalization proposed by the BSPP. The law prohibited private foreign investment in Burma, and only grudgingly permitted local private investment in sectors not yet taken over by the State or the co-operatives. In 1981 Ne Win, who had by that time taken the honorific title U, announced that he would retire as President of the Socialist Republic of the Union of Burma (as the country had been redesignated), while continuing to lead the party. Formal procedures for Ne Win's succession were put in place, and San Yu took over formally as President, although U Ne Win in fact continued to control Burmese politics from behind the scenes.

The return to socialism under a remilitarized BSPP led to continuing economic decline in Burma. Unfavourable weather, mismanagement in agricultural production and distribution, and growing inflation caused increasing hardship throughout the country. In August 1987 U Ne Win finally admitted publicly to 'failures and faults' in his Government's past management of the economy and announced major reforms. The Government freed the purchase, sale, transport and storage of basic food items from state control, hoping that farmers would increase production to meet local market demands. In early 1988 the Government also terminated its 25-year monopoly on rice exports. These measures were popular, but they were followed by a regulation demonetizing the 25, 35 and 75 kyat currency notes, intended to strike at the wealth of black marketeers. This action, which effectively confiscated money from the public, was deeply resented, and student protests broke out in Rangoon and elsewhere. In March 1988 the police responded with violence to the demonstrations. At least 50 people were killed, leading to mass public demonstrations against the regime. The Government closed the universities, and briefly quelled the dissent by promising to investigate the students' grievances. By mid-1988, however, public impatience with the Government had returned and the demonstrations resumed. In June security forces brutally suppressed a demonstration by about 5,000 students and others demanding the release of persons detained in March. The Government responded by imposing a curfew and closing universities and schools indefinitely. The regional cities of Tanggyi, Pegu and Prome were also placed under curfew as unrest spread throughout the country. In an atmosphere of crisis, U Ne Win called an extraordinary meeting of the BSPP in July, at which he and San Yu resigned from their posts. Sein Lwin, a close military associate of U Ne Win and a known advocate of harsh repressive measures, took over as both BSPP Chairman and State President. Although U Ne Win warned in his resignation speech that the army would shoot to kill if the demonstrations continued, the public sensed that the regime had been placed on the defensive, and preparations began for massive strikes and demonstrations to begin on the supposedly auspicious date, 8 August 1988 (8-8-88), at 8.08 a.m. The protests proceeded peacefully until late in the evening of that day, when security forces responded with violence to the demonstrations. During the following five days, some 2,000 to 3,000 demonstrators are believed to have been killed as the reform movement was violently suppressed. On 13 August, however, with unrest continuing, Sein Lwin resigned and was replaced by the more moderate Dr Maung Maung, hitherto the Attorney-General. Maung Maung ended martial law, released political prisoners and offered political reforms, which implied that the BSPP was willing to surrender power to the opposition in a peaceful transition. An emergency meeting of the BSPP held in September 1988 agreed to hold free elections within three months and decided that members of the armed forces, police and civil service could no longer be affiliated to a political party. New political groups—including the All Burma Students Union (ABSU) and the National United Front for Democracy, later renamed the National League for Democracy (NLD)—were permitted to form in preparation for the election. The NLD quickly emerged as the main opposition group, its leaders including Brig.-Gen. Aung Gyi (formerly an associate of U Ne Win, but subsequently an outspoken critic of the regime, who was briefly detained under Sein Lwin), Gen. U Tin Oo (a former chief of staff and Minister of Defence) and Aung San Suu Kyi, daughter of the assassinated nationalist leader Aung San, whose portraits had been particularly prominent during the demonstrations of 8 August. Even though she had been out of the country since 1960, Aung San Suu Kyi increasingly came to represent the peaceful, prosperous country which people imagined Burma might have become but for her father's death. Amid general expectations that the NLD would resoundingly win the coming elections, and with Buddhist monks already taking over municipal administrations in many parts of the country, the military under Gen. Saw Maung launched a coup on 18 September 1988, ostensibly to maintain public order in the approach to the elections. Saw Maung, however, immediately created a new ruling body, the State Law and Order Restoration Council (SLORC), comprising 19 senior military officers, with himself as Chairman. All state organs, including the Pyithu Hluttaw, the State Council and the Council of Ministers, were abolished, demonstrations were banned and a nation-wide dusk to dawn curfew was imposed. In the first days after the military coup, more than 1,000 demonstrators were killed by security forces, and thousands of students and others then fled to areas along the border with Thailand, where they sought the protection and help of ethnic insurgents.

THE SLORC, 1988–97

The SLORC announced the formation of a nine-member Government, with Saw Maung as Minister of Defence and of Foreign Affairs and subsequently also Prime Minister. It was widely believed, however, that U Ne Win, although in retirement, retained a controlling influence over the new leaders, all of whom, including Saw Maung, were known to be his supporters. The new Government changed the official name of the country to the Union of Burma (as it had been before 1974). The law maintaining the BSPP as the sole party was abrogated, and new parties were encouraged to register for the forthcoming elections. By February 1989 a total of 233 parties had registered. Most were small and based on ethnicity, religion or location. Beyond registering, the new parties had little influence; under the Government's martial law regulations, group gatherings were limited to five persons, a curfew existed, and there were restrictions on travel, publication and public meetings. The BSPP was re-established under a new name, the National Unity Party (NUP), with U Tha Kyaw, the former Minister of Transport, as Chairman. Although the NLD registered, it was uncertain whether it would contest elections, which, it asserted, could not be held fairly under military rule. In December 1988, owing to disagreements with Aung San Suu Kyi, Aung Gyi was expelled from the NLD and founded the Union National Democracy Party (UNDP). In May 1989 the Government ratified electoral legislation, which had been promulgated in March. It provided for the holding of multi-party elections in May 1990, and permitted campaigning only in the three months prior to the election date. Stringent campaign regulations were imposed by the SLORC, including restrictions on public rallies and official censorship of speeches. In June 1989 the SLORC reiterated that martial law was still in force, and that it would retain power after the elections until the resulting legislative assembly drafted a new constitution and formed a civilian government. In the same month 3,000 people demonstrated on the anniversary of the student protests of 1988. Troops fired on the crowd and one person was killed.

On 18 June 1989 the name of the country was changed to the Union of Myanmar (Myanma Naing-ngan). The SLORC explained that the change had been made to avoid the racial connotation of the previous name, Union of Burma, which implied that the population were all Burmans, while, in fact, it included many racial groups. The Roman transliteration of many towns, divisions, states, rivers and nationalities was changed (although the pronunciation remained the same in Burmese). Thus, Rangoon was changed to Yangon, Pegu to Bago and the Irrawaddy River became the Ayeyarwady. The race known as the Burmans was renamed the Bamars, and the Karen and Karenni were restyled Kayin and Kayinni, respectively.

Tension between the SLORC and the opposition groups increased in July 1989. In an attempt to crush political dissent, the SLORC established five military tribunals (and a further six in August) to try persons violating martial law regulations (e.g. by failing to observe the curfew or participating in gatherings of more than five people). The tribunals could impose penalties ranging from three years' imprisonment to death. On 19 July, the anniversary of the 1947 assassination of her father, Aung San Suu Kyi planned to lead a protest march through the capital with members of more than 100 of the country's political parties. Suu Kyi cancelled the rally, however, fearing that troops deployed by the Government would fire on the marchers, as they had done in September 1988. The next day she and Tin Oo (the Chairman of the NLD) were placed under house arrest. They were accused of attempting to create disunity within the army and 'nurturing public hatred for the military'. In December 1989 Tin Oo was tried and sentenced to three years' imprisonment, with hard labour. Former Prime Minister U Nu was placed under house arrest for refusal to disband a parallel government which he had proclaimed during the uprising in 1988. Other leaders, especially amongst the student parties, were either arrested or went 'underground'. When the NLD sought to nominate Suu Kyi as an election candidate, an opposition candidate challenged the nomination and the committee responsible for certifying candidates upheld the challenge on the grounds of her marriage to a British citizen. In March 1990 reports began to circulate abroad that the armed forces were forcibly moving people out of certain areas in Yangon, Mandalay and elsewhere to distant new living quarters. Since most of the people moved were believed to be supporters of the NLD, this was seen as an effort by the army to weaken support for the opposition in the cities.

The 1990 Election and its Aftermath

Despite SLORC efforts to weaken opposition leaders and eliminate dissidents, 93 parties presented a total of 2,297 candidates to contest the 492 constituencies. In every constituency there were at least two candidates. Prior to the election, it was widely believed that the armed forces favoured certain parties, particularly the NUP but also Aung Gyi's UNDP. No single party was expected to win a clear majority. Two days before the election, the SLORC unexpectedly issued visas to 61 foreign journalists to observe and report on proceedings. The voting, which took place on 27 May 1990, was orderly, quiet and free. The NLD won an overwhelming victory, taking 392 of the 485 seats that were, in the event, contested, while the NUP won only 10. The remainder of the seats were allocated to 23 other parties. Parties representing ethnic groups achieved considerable success: the Shan Nationalities League for Democracy (SNLD) won 23 seats, and the Arakan (Rakhine) League for Democracy 11. Following the election, nearly all the parties representing non-Bamar nationalities formed a coalition known as the United Nationalities League for Democracy (UNLD), including 65 elected representatives. The total representation in the assembly of anti-SLORC forces—the NLD, the UNLD, the Party for National Democracy (PND, led by Dr Sein Win, a cousin of Suu Kyi; which had secured three seats) plus one other elected deputy—was thus 461 of 485 seats, or some 95%.

Following the NLD's electoral victory, its leaders demanded immediate talks with the SLORC, and movement towards popular rule. However, the SLORC announced that the election was intended only to produce a constituent assembly, which was to draft a constitution under the direction of a national convention to be established by the SLORC. In July 1990 the SLORC issued Order 1/90, which stated: that the SLORC had international legitimacy because it was recognized by the UN and individual countries; that it was incumbent on the SLORC to prevent the disintegration of the Union and of national solidarity and to ensure the perpetuity of state sovereignty; and that the SLORC would continue as the *de facto* Government until a new constitution was accepted by all the races of Myanmar. The elected membership of the NLD responded (independently of their leadership) with the 'Gandhi Hall Declaration'. This proclaimed that the NLD was ready to hold discussions with the SLORC and to convene the national assembly, according to the electoral law; that the national assembly was the highest authority in the State and not simply a constituent assembly; that the NLD had drafted a provisional Constitution by which it could govern and that this provisional Constitution would bring about the transfer of power in accordance with the law, pending the drafting of a new Constitution; finally, it demanded the immediate restoration of democratic rights.

In early August 1990 at an anti-Government protest in Mandalay, commemorating the deaths of thousands of demonstrators at the hands of the armed forces in 1988, troops killed four protesters. This led to a decision by various Buddhist orders to withhold their services from members of the military. Later in the same month the NLD, with the support of other parties, announced plans to convene a national assembly in September. In early September, however, the SLORC arrested six members of the NLD, including its acting leader, Kyi Maung, and its acting General Secretary, Chit Hlaing, on charges of passing state secrets to unauthorized persons. Both were tried in the special tribunals and given 10-year prison sentences. Also in September, NLD representatives discussed plans to declare a provisional government in Mandalay, without the support of the party's central executive committee. Influential monks agreed to support the declaration which was to take place in October. However, the plan was abandoned after government troops surrounded the monasteries; the SLORC subsequently banned all but nine Buddhist sects and empowered military commanders to impose death sentences on rebellious monks. More than 50 senior members of the NLD were arrested, and in late October and early November members of all political parties were required to endorse Order 1/90. In acquiescing, the NLD effectively nullified its demand for an immediate transfer of power.

Leaders of other parties were also arrested, disappeared or went 'underground' to avoid arrest. Reports of political prisoners being tortured and dying while in detention circulated widely. In the face of the steady repression and threats to the elected members of the national assembly, many of those in the NLD, not under arrest, assembled secretly in Mandalay in November 1990 and agreed to send some of their members to the border to create a provisional government. In early December eight opposition politicians arrived at Manerplaw, the Kayin (Karen) headquarters, and entered into discussions with the Democratic Alliance of Burma (DAB—a 21-member organization uniting ethnic rebel forces with student dissidents and monks). They subsequently agreed to form the National Coalition Government of the Union of Burma (NCGUB). The aims of the NCGUB were: to wage war against the military rulers; to convene a national conference of all elected leaders, representatives of the DAB, democratic forces and other notable individuals; to draft a new federal constitution; and to form a true democratic government. On 18 December the NCGUB was constituted with Sein Win, the leader of the PND, as its President. The remainder of his cabinet consisted of six elected NLD representatives and one independent. Other elected members of the national assembly made their way to Manerplaw and gave their support to this rival government. The SLORC quickly denounced the NCGUB and compelled the NLD to do likewise.

Members of the NCGUB agreed to form a Supreme Council of Burma Democratic Forces, known as the Burma Front, in partnership with the minority party leaders. It was to be the policy-making body during the interim period. It consisted of six Bamars and five minority members under the leadership of Gen. Bo Mya, then leader of the Karen (Kayin) National Union (KNU). In February 1991 the NCGUB sent a small delegation to Switzerland to address the UN Commission on Human Rights during its consideration of the issue of human rights violations in Myanmar. Later, the delegation visited other European countries to discuss conditions inside Myanmar and possible measures to bring about change. Although the NCGUB did not receive any official recognition by foreign states, it received funds from non-governmental organizations in Canada, Switzerland and Norway.

In April 1991, following intense pressure from the army, the NLD central executive committee was restructured. Suu Kyi and Tin Oo, both still under house arrest, were deprived of their former posts of General Secretary and Chairman. Suu Kyi was replaced by U Lwin, a little-known political figure, and Tin Oo by the former Acting Chairman, U Aung Shwe. In the same month Lt-Gen. (later Senior Gen., then Field Marshal) Than Shwe, the Vice-Chairman of the SLORC and the Deputy Commander-in-Chief of the Armed Forces, officially ruled out a transfer of power to those elected in May 1990, condemning the political parties that had taken part in the elections as subversive. In May 1991 34 opposition politicians (25 of them elected representatives) were sentenced to prison terms of up to 25 years, for alleged treason and attempting to establish an alternative government. Suu Kyi, who was under increasing pressure from the army to leave the country, was systematically attacked in the state-controlled press. At the end of May a military tribunal extended the sentences of Kyi Maung and Chit Hlaing to 20 years' imprisonment.

In October 1991 Aung San Suu Kyi was awarded the Nobel Peace Prize. The award ceremony, which took place in Oslo, Norway, in December, was attended by the leader of the NCGUB, Sein Win. In Myanmar students staged the first demonstrations since 1989 to coincide with the ceremony. The students were dispersed, and universities and colleges, which had reopened in May 1991, were closed. The now-compliant NLD leadership expelled Suu Kyi from the party.

On 23 April 1992 the SLORC announced the resignation of its Chairman, Saw Maung, on the grounds of ill health (Saw Maung died in July 1997), and his replacement by Than Shwe (who had already replaced Saw Maung as Minister of Defence in March 1992). On the following day Than Shwe was named as Prime Minister, although Maj.-Gen. (later Lt-Gen.) Khin Nyunt, the First Secretary of the SLORC and the head of the intelligence service, was still widely regarded as the most powerful member of the SLORC, owing to U Ne Win's patronage. The SLORC permitted Suu Kyi to receive visits from her husband and sons. The SLORC also announced that political prisoners who were no longer a threat to the regime would be released. By mid-May about 100 prisoners had been released, including U Nu and numerous NLD representatives. Shortly thereafter, the SLORC announced that it would convene a co-ordinating meeting for a national convention (the first meeting between representatives of political parties, ethnic groups and members of the SLORC) to develop the principles for a new constitution and to determine who would participate in drawing up the future basic law. In June the co-ordinating meeting convened with 43 delegates, comprising 15 representatives from the SLORC, 15 from the NLD, six from the SNLD, three from the NUP, four from separate ethnic minority parties and one independent.

In August 1992 universities and colleges were reopened, and in September the night curfew, imposed four years previously, was repealed. In the same month a reorganization of the Cabinet included the appointment of two Deputy Prime Ministers and the promotion to ministerial posts of six regional commanders of the armed forces who were known to oppose Khin Nyunt. These appointments, which ended the first overt power struggle under the SLORC, strengthened Khin Nyunt's position by effectively depriving the commanders of their regional power. Following the government reshuffle, the SLORC revoked two martial law decrees, which had been in force for three years, although the ban on gatherings of more than five people remained in place. In late September further political prisoners were granted amnesties, bringing the total released to 534. It was estimated, however, that a further 1,600 prisoners remained in detention.

The National Convention

In January 1993 the National Convention finally assembled. From the outset the Convention was under the firm control of the SLORC: of the initial 702 delegates, only 93 were NLD members (elected as legislators in 1990), while some 80% of all delegates were appointed by the SLORC. The first session was swiftly adjourned, owing to the objections of the opposition members to the SLORC's demand that the armed forces be allocated a leading role in government under the new Constitution. When the National Convention reconvened at the beginning of February, the NLD issued a statement opposing military dominance and proposing a national referendum on whether it should be incorporated in the new Constitution. The Convention was again adjourned. The SLORC reacted to opposition intransigence by suspending its conciliatory gestures; many arrests were reported during January and February. Following several meetings and adjournments of the Convention between March and September, the Chairman of the National Convention's Convening Committee, U Aung Toe (the Chief Justice), subsequently announced (seemingly without grounds) that consensus existed in favour of the SLORC's demands, which included: the representation, in both the lower and upper chambers of a proposed parliament, of military personnel (to be appointed by the Commander-in-Chief of the Defence Services); the independent self-administration of the armed forces; and the right of the Commander-in-Chief to exercise state power in an emergency (effectively granting legitimate status to a future coup).

In September 1993 an alternative mass movement to the NUP (which had lost credibility through its election defeat) was formed to establish a civilian front through which the armed forces could exercise control. The Union Solidarity and Development Association (USDA), whose aims were indistinguishable from those of the SLORC, was not officially registered as a political party, thus enabling civil servants to join the organization, with the incentive of considerable privileges.

In January 1994 the National Convention's discussions on the draft Constitution resumed. In that month large numbers of people were coerced into joining USDA rallies to demonstrate support for the constitutional proposals presented by the SLORC. In February a delegation led by a member of the US Congress was granted permission to visit Suu Kyi (the first time that she had received visitors, other than family members, during the period of her detention). Suu Kyi sent an encouraging message to the democracy movement and appealed for a meeting with the SLORC, expressing her willingness to negotiate on all issues except her exile. On the following day the SLORC announced that Suu Kyi would be detained until at least 1995 (despite the legal maximum of five years under house arrest, whereby Suu Kyi would be released in July 1994), since her first year in detention had only been an 'arrest period'.

The National Convention adjourned in April 1994, having determined three significant chapters of the future Constitution. According to these, Myanmar was to be renamed the Republic of the Union of Myanmar (Pyidaungsu Thammada Myanmar Naingandaw). The Republic's territorial organization was to preserve the existing seven *taing* or divisions in central and southern Myanmar (Yangon, Bago, Ayeyarwady, Tanintharyi, Magway, Sagaing and Mandalay), inhabited mainly by Bamars, and seven *pyinay* or states, associated with minority ethnic groups (Rakhine, Chin, Kachin, Shan, Kayinni, Kayin and Mon). The Republic would be headed by an executive president, elected by the legislature for five years; a number of conditions, including one disqualifying any candidate with a foreign spouse or children, were clearly designed to prevent Suu Kyi from entering any future presidential election (her husband and children were British citizens).

The Convention, which reconvened in early September 1994, again stressed that the central role of the military (as 'permanent representatives of the people') be enshrined in the new Constitution. Proposals by six smaller ethnic minority groups—Naga, Wa, Pa-O, Danu, Kokang and Palaung—for their own self-administered 'national zones' were considered favourably, although comparable demands by several other minority groups were rejected. The establishment of the six 'national zones' was agreed in April 1995, by which time agreement had also been reached on key chapters of the Constitution, covering the legislature, judiciary and Government, at Union, state and regional level. It was determined that legislative power be shared between a bicameral Pyidaungsu Hluttaw (Union Assembly) and regional and state hluttaws, all of which were to include representatives of the military. The Pyidaungsu Hluttaw was to comprise the Pyithu Hluttaw (People's Assembly) and the Amyotha Hluttaw (Assembly of Nationalities). The former would comprise 330 elected deputies and 110 members of the Tatmadaw (armed forces) and would be elected for five years. The latter would comprise 224 members: 12 elected from each of the 14 divisions and states, as well as 56 members of the Tatmadaw.

In late November 1995 the National Convention reconvened. Under instructions from their party, all 86 NLD delegates boycotted this session and were expelled after two days, leaving only a few small elected parties, mainly representing minorities, among the 545 delegates present. The proportion of delegates representing elected parties had declined from over 20% at the start of the Convention to less than 10%; the rest were various categories of SLORC appointees. In late March 1996 30 delegates from five of these remaining parties, the largest of which was the SNLD, protested publicly about the proposed military appointments to the People's Assembly. Not surprisingly, the protest had no effect. The 104 'basic principles' remained in place, and more of the 15 draft chapters continued to be produced. When the NLD representatives who had been expelled in late 1995 asked to rejoin the convention in 1997, their requests were rejected. However, observers representing various ethnic groups who had signed truces with the SLORC were permitted to join. The Convention was described in *The Nation*, a major Thai newspaper, as 'the world's slowest-operating rubber-stamp body', but in fact it ceased to operate at all after going into recess in 1996, though official statements commonly implied that it would reassemble one day.

The Opening of Dialogue with Aung San Suu Kyi

In July 1994 Khin Nyunt announced what appeared to be a major change of policy: that the SLORC was prepared to hold talks with Suu Kyi. In the same month the Minister of Foreign Affairs, U Ohn Gyaw, attending a meeting of the Association of South East Asian Nations (ASEAN) in Bangkok, stated that his Government would accept the invitation of the UN Secretary-General, Dr Boutros Boutros-Ghali, to discuss issues of democratization, national reconciliation and human rights in Myanmar. Following mediation by a senior Buddhist monk between Suu Kyi and leading members of the SLORC, in mid-September Suu Kyi was permitted to leave her house for the first time during her five-year detention to meet Than Shwe and Khin Nyunt. Parts of the meeting were broadcast on state television, although no specific details of the talks were actually reported. In late October Suu Kyi was invited to a second meeting with senior members of the SLORC, and in November she was permitted to meet Tin Oo and Kyi Maung, who were still imprisoned. In the same month, however, a delegation led by the US Deputy Assistant Secretary of State for East Asian and Pacific Affairs was not permitted to visit Suu Kyi. In February 1995 the fourth round of talks was held in Yangon between leading members of the SLORC and UN Assistant Secretary-General Alvaro de Soto on the range of issues agreed in mid-1994. However, de Soto was prevented from visiting Suu Kyi. Nevertheless, in March 1995 the Government released 31 political prisoners, including Tin Oo and Kyi Maung.

A reorganization of the Cabinet was carried out in June 1995, in which a new Ministry of Immigration and Population was created specifically for Lt-Gen. Maung Hla, a member of the SLORC, who had led the successful campaign against the Kayins earlier in the year (see below). As in the 1992 reshuffle, several regional military commanders were appointed ministers, in an apparent attempt to prevent them from developing a local power base.

On 10 July 1995, after almost six years under house arrest, Suu Kyi was released by the SLORC. No dialogue had taken place between Suu Kyi and the military regime since October 1994, and thus her release had not been expected at this time. Some observers suggested that the SLORC had finally yielded to international pressure, while others indicated that the regime could afford to release Suu Kyi at a time when it considered its position stronger than before (largely owing to the defeat or surrender of virtually all the ethnic insurgent groups, see below). Suu Kyi immediately began to re-establish contact with those remaining NLD leaders who had not been imprisoned or exiled, as well as with the party leadership which had compromised with the SLORC. Despite the continuing official ban on gatherings of more than five persons, crowds of up to 1,000 supporters of Suu Kyi assembled daily outside her house in Yangon; they were not dispersed by the police. However, there was no announcement in the official media of Suu Kyi's release until 19 July, when state television covered the annual Martyrs' Day ceremony in Yangon, at which Suu Kyi was permitted to lay a wreath on her father's grave. Than Shwe and Khin Nyunt were noticeably absent from the ceremony.

In public statements and extensive interviews with the foreign media, Suu Kyi emphasized that her release was only an initial step in a very long process and that pressure for democratic reforms should be maintained. She appealed to the SLORC for the release of all political prisoners, the gradual easing of martial-law restrictions, the recognition of the 1990 election result, and the holding of talks on national reconciliation. In August 1995 de Soto revisited Myanmar, and urged the military leaders to initiate dialogue with Suu Kyi. In the following month Dr Madeleine Albright, then US Permanent Representative to the UN, visited Yangon (the most senior US official to do so since 1988). During talks with leading members of the SLORC she stated that any change in US policy towards Myanmar would depend on fundamental changes in the regime's treatment of Myanmar's people. Dr Albright also met Suu Kyi. By mid-November, however, there appeared to be no indication of willingness by the SLORC to resume dialogue with Suu Kyi. Indeed, in late October the authorities decreed illegal the NLD's reinstatement of Suu Kyi as its General Secretary (with Aung Shwe re-elected as Chairman, and Tin Oo and Kyi Maung as Vice-Chairmen).

In November 1995 23 retired political and military leaders, including Bohmu Aung (one of the 'Thirty Comrades' together with Gen. Aung San and Gen. Ne Win), issued an open letter urging dialogue between the SLORC and the NLD. They were severely reprimanded and threatened by the SLORC. Later that month the remaining NLD delegates boycotted the National Convention and were expelled. The SLORC continued to harass the NLD and arrest its members and supporters throughout 1996, and Suu Kyi's activities were progressively restricted. Although no longer under house arrest, Suu Kyi was often warned by the military surrounding her home not to go out 'for her own safety'. On one occasion in 1996, she was prevented from travelling to Mandalay when the train carriage that she had booked was suddenly found to be unserviceable and removed. NLD colleagues who assisted her visit to a Kayin New Year celebration were subsequently imprisoned, and several well-known comedians who told anti-SLORC jokes at an NLD rally were sent to a labour camp. In September 1996 the military blocked off the area around Suu Kyi's house in University Avenue, rendering impossible the delivery of her regular weekend speeches from her garden. In November the car in which Suu Kyi was travelling was attacked by a 200-strong mob and severely damaged. Also, in August, Win Htein, Suu Kyi's personal assistant, was arrested and sentenced to seven years' imprisonment (subsequently doubled) for allegedly conspiring with groups in India to destabilize the country. Many foreigners, especially journalists, politicians and any groups communicating with the NLD before arrival in the country, or with intentions to meet with Suu Kyi during their visit, found their visas refused or revoked, their aircraft inexplicably delayed and their other scheduled meetings cancelled, while most local people were unable even to get near her house, let alone to see or hear her.

In May 1996 the NLD held its first party congress. Prior to its opening, 262 delegates, including 238 of those elected in 1990, were arrested and held until the congress was over; up to 20 remained political prisoners in July. However, the congress proceeded with the attendance of hundreds of NLD members, as well as foreign diplomats and journalists; the main outcome of the congress was the adoption of a resolution to draft an alternative constitution. In early June 1996 the SLORC intensified pressure on the NLD by issuing an order authorizing the Ministry of Home Affairs to ban any organization holding unlawful gatherings or obstructing the drafting of a new constitution by the National Convention; it was announced that members of a proscribed party could be sentenced to between five and 20 years' imprisonment. Each time the NLD attempted to hold a large meeting, hundreds of those planning to attend were arrested or detained until after the proposed date.

Harassment, detention, arrests, imprisonment and deaths in prison of the middle and lower level NLD leadership also continued in 1997, while government propaganda vilifying the party grew more frequent. In July SLORC First Secretary Khin Nyunt held a meeting with the NLD Chairman, Aung Shwe; however, the leaders of the ruling military junta had not met Suu Kyi since her 'release' from house arrest in July 1995. A further meeting with the SLORC leadership scheduled to take place in September was cancelled by the NLD because the military junta refused to allow Suu Kyi to participate. The SLORC media continued to assert that the NLD (to which they referred as the 'notorious league for demons') was funded from overseas, although this appeared not to be the case.

THE SPDC

Following the circulation during 1997 of rumours anticipating changes in the membership of the SLORC, on 15 November the leadership unexpectedly announced the dissolution of the SLORC and its immediate replacement by a new State Peace and Development Council (SPDC). The four most senior members of the SLORC retained their positions at the head of the new council: Gen. Than Shwe, despite his deteriorating health, was appointed Chairman of the SPDC (and also remained Prime Minister and Minister of Defence); the Commander-in-Chief of the Army, Gen. Maung Aye, was nominated Vice-Chairman; Lt-Gen. Khin Nyunt, head of military intelligence, was appointed First Secretary and Lt-Gen. Tin Oo Second Secretary (as opposed to the former NLD chairman, also called Tin Oo). Lt-Gen. Win Myint, the Adjutant-General, was appointed Third Secretary. Unlike the SLORC, the 19-member SPDC was composed entirely of serving military commanders from central and regional levels. Moreover, only the SPDC Chairman, Gen. Than Shwe, concurrently held posts in the Cabinet, of which about half the members were new appointments, many of them younger civilians from the USDA. The formation of the SPDC therefore appeared to be part of the military's slowly developing strategy to outflank the NLD by creating 'safe' political institutions. The USDA in particular expanded rapidly under the sponsorship and funding of the Ministry of Education; its Secretary-General was the Minister of Education, Than Aung, while Gen. Than Shwe was its chief patron. By late 1996 the USDA claimed 5m. members (more than 10% of the population), with a further 1m. in affiliated youth organizations. (By mid-2002 the organization claimed 16m. members.) Government employees were under heavy pressure to join the USDA, and there were strong indications that the regime was seeking to model the organization on Indonesia's then state-supported Golkar party. Lt-Gen. Khin Nyunt subsequently emerged as the chief strategist in this process. He was identified as a protégé of the ageing Gen. Ne Win, and in 1994 had established an Office of Strategic Studies (OSS) within the Ministry of Defence which took over responsibility for formulating policy on problematic issues, such as the drugs trade and ethnic minorities, and on the delicate issue of corruption in the upper echelons of the political administration. (In November 1997 several ministers were arrested on corruption charges.) Khin Nyunt also appeared to be the influential force behind the work of a committee responsible to the National Convention which, in the late 1990s, was said by the Government to be close to completing the draft of a new Constitution; in September 1998 he created a Political Affairs Committee (PAC), including both civilian and military figures, to oversee this process. A cabinet reorganization effected in November 1998 also indirectly further strengthened the position of Khin Nyunt: in the reshuffle, Ohn Gyaw was replaced as Minister of Foreign Affairs by Win Aung, hitherto the Myanma ambassador to the United Kingdom and reportedly a close ally of Khin Nyunt.

Faced with the gradual stalling of their 1990 democratic mandate, the NLD held a congress in May 1998 and subsequently issued an ultimatum to the SPDC to convene the Pyithu Hluttaw, in accordance with the results of the 1990 election, by 21 August 1998, announcing that it would convene the Assembly itself if the SPDC failed to comply. The SPDC, however, both rejected the demand and began a series of arrests of NLD deputies and members. By November about 850 deputies and members were reported to be in detention; about half of these were released after they agreed not to take part in an NLD-convened Assembly, but some 270 were given prison sentences. By mid-1999 at least three NLD deputies, including NLD founder member Tin Shwe, had died in prison since 1990. The Government claimed that only 129 of the original 392 NLD parliamentarians still held an electoral mandate; the remainder were reported variously to have died, resigned, left the party or been disqualified. Furthermore, the regime allegedly began a systematic campaign of coercion and intimidation in an attempt to secure the involuntary resignation of vast numbers of NLD members and the closure of a number of regional party headquarters; by the end of 1998 some 40,000 members were reported to have left the party. The junta also began to orchestrate calls for the party to be declared illegal, thus exposing its members to the risk of imprisonment. Rather than convening the Pyithu Hluttaw under such circumstances, in September 1998 the NLD instead created a Committee Representing the People's Parliament (CRPP) to issue statements on behalf of the Pyithu Hluttaw and to develop policy. Meanwhile, speculation continued on the subject of whether there were elements within the regime that might be interested in opening a dialogue with the NLD. The SPDC Vice-Chairman, Gen. Maung Aye, was said to be the most vehement opponent of dialogue. It was reported in September 1998 that 15 senior military officers had been arrested for seeking contact with Suu Kyi. In late 1998 the UN reportedly raised the possibility of providing US $1,000m. in aid and assistance to Myanmar if the Government would release political prisoners and begin a dialogue with the NLD, which in turn was to be requested to abandon its demand that the Pyithu Hluttaw convene on the basis of the results of the 1990 election. However, when news of the proposal emerged, both the SPDC and the NLD angrily rejected the suggestion of any possible compromise. The establishment of genuine dialogue was rendered unlikely by the SPDC's continuation of the SLORC's campaign of attrition against Suu Kyi, and also by its refusal to allow her involvement in any discussions. The validity of Suu Kyi's Myanma nationality was repeatedly questioned on the grounds of her marriage to a British citizen, Dr Michael Aris, and in mid-1998 the regime repeatedly used road-blocks to prevent her from attending NLD meetings, leading to prolonged confrontations between Suu Kyi and the military. In early 1999 Dr Aris, who had been diagnosed with terminal cancer, was refused permission to enter Myanmar to see Suu Kyi for the last time; he died in March. (Although the junta had encouraged Suu Kyi to visit Aris in the United Kingdom, she declined to do so for fear that she would not be permitted to return to Myanmar.) In many other respects also, the regime seemed far from conciliatory. After student riots against the Government in December 1996, universities and some high schools were closed indefinitely. Further large-scale anti-Government demonstrations were staged by students in September 1998. In January 1999 four medical schools were reopened, but other areas of the higher education sector remained closed, and plans were announced to move several campuses to remote regions before they would be permitted to open again. In July 2000 the ruling junta reportedly allowed some 60,000 university students to resume their education in a further step in the phased reopening of Myanmar's universities and colleges; all returning students were reportedly required to sign an oath, pledging not to engage in political activity. It was estimated that many thousands more students, however, remained excluded from education. In August 2001 the Government announced plans for a major expansion in distance education, apparently intended to develop the human capital needed for Myanmar's economic development, while avoiding the political risks associated with campus-based education. In the late 1990s reports emerged from Arakan, the Kayin State, the Chin State and elsewhere of deliberate harassment of religious minorities by government forces, and of the destruction by soldiers, local government officials, and even Buddhist monks, of mosques and churches to make way for Buddhist temples. In urban areas, the cemeteries of non-Buddhists were reportedly targeted for redevelopment, and relatives were told to move remains of the deceased at short notice and at their own expense. In 2001 communal riots between Buddhists and Muslims were reported in Taungoo, north of Yangon, and in Rakhine.

Meanwhile, the army, or Tatmadaw, increased in size from 175,000 to 400,000 in 1999, according to one estimate (with perhaps an additional 100,000 in the police force and militia groups). The age of recruitment, officially 17, was reportedly lowered in order to increase numbers. Some opponents of the regime also reported that units of boys under 18 were sometimes given amphetamines and used for 'human wave' attacks across minefields against fortified positions, although such accounts cannot be confirmed. There is no doubt, however, that the army used local forced labour extensively for portering and logistical support work. In addition, army units were placed in charge of development projects, such as the building of roads, railways and dams, which again relied mainly on forced labour and the labour of prisoners. In principle, non-prison labour was remunerated, but often much of the payment was retained by the military officer in charge. A report published by the International Labour Organization (ILO) in August 1998 confirmed that the use of forced labour was 'pervasive' throughout Myanmar and accused the military junta of using beatings, torture, rape and murder in the implementation of its forced labour policy (see below).

In April 1999 Suu Kyi was reported to have sent a videotaped appeal to the Office of the UN High Commissioner for Human Rights in Geneva, Switzerland, requesting that the organization make a 'firm resolution' to protect human rights in Myanmar. In the same month the UN adopted a unanimous resolution deploring the escalation in the persecution of the democratic opposition. Nevertheless, the harassment and intimidation of NLD members continued throughout 1999 and early 2000, with the resignations from the party of nearly 300 members reported in July 1999; further resignations were reported in November and January 2000, and in December 1999 it was reported that an elected representative of the People's Assembly, U Maung Maung Myint, had been forced to resign by the ruling SPDC. The NLD's failure to make any progress against the regime in 1999 led to increasing tensions within the opposition itself and even, among some elements of the party, to growing dissatisfaction with the uncompromising policy of Suu Kyi. In April three senior NLD members wrote to the party Chairman, Aung Shwe, and to Khin Nyunt, urging both the NLD and the Government to be more conciliatory; Aung San Suu Kyi publicly rejected their suggestion, however, and the three were suspended from party membership. In early August a series of protests was staged by supporters of the democratic opposition to mark the anniversary of the massacre of thousands of pro-democracy demonstrators by the military Government in 1988. Local and overseas opposition groups attempted to force a confrontation with the Government on 9 September 1999 (9-9-99), designated an auspicious date by analogy with 8-8-88; government forces were well prepared for the protests, however, and there were no more than small disturbances in several centres. In October 1999 Myanmar's Supreme Court issued a decision rejecting a claim by the NLD that its activities had been 'continuously disrupted, prevented and destroyed' and that hundreds of its members had been illegally detained; the NLD, however, subsequently expressed its intention to resubmit the claim.

In February 2000 senior Buddhist monks called for renewed dialogue between the regime and the opposition, evidently irritating the Government and Suu Kyi to equal degrees. By September, however, all indications were that attitudes on both sides had hardened. In late August the regime engaged in a stand-off with Suu Kyi when she attempted to travel from her home in Yangon to conduct party business at other centres. After nine days, troops eventually forcibly removed Suu Kyi from an informal camp she had established at a road-block and returned her to her home. Following the conclusion of the stand-off, the ruling junta continued to restrict the movements of Suu Kyi and a number of other NLD leaders, holding them under effective house arrest at Suu Kyi's Yangon home and denying diplomatic access to Suu Kyi. Although the restrictions were lifted in mid-September, later the same month Suu Kyi was prevented from leaving Yangon by train. Tin Oo and eight other NLD workers, who were planning to accompany Suu Kyi, were detained in a 'government guest house', while Suu Kyi herself was again placed under house arrest. Meanwhile, the NLD lost its lease over the building housing its headquarters, while an estranged brother of Suu Kyi, resident in the USA, began legal proceedings to take possession of her home in Yangon. Both the NLD and its leader thus faced the prospect of homelessness. In September 2000 the NLD-dominated CRPP announced that it would proceed with developing its own constitution for the country. On the other hand, in October, Gen. Khin Nyunt held talks with Suu Kyi, although no results from the discussions were announced. In January 2001 the regime released from detention 85 NLD members, including Tin Oo, while a court dismissed the case brought by Suu Kyi's brother. Another NLD leader, Maung Wuntha, was released in June 2001, and both sides became noticeably more restrained in their comments about the other. Talks continued without publicity throughout 2001, mediated by the special envoy of the UN Secretary-General, Malaysian diplomat Tan Sri Razali Ismail. By the end of 2001 some 174 NLD detainees had been released and the party had been permitted to reopen some offices in Yangon. Following the downfall of Indonesia's President Suharto in 1998, there was speculation that the SPDC was seeking an alternative model for political stability and economic development, and that it was attracted by the example of Malaysia. Representatives of the SPDC and USDA attended the General Assembly of the dominant United Malays National Organization (UMNO) in Malaysia in June 2002. The Malaysian Prime Minister, Mahathir Mohamad, visited Myanmar in August 2002 and commented that a rapid transition to democracy would be destructive.

In a reorganization of the Cabinet announced on 29 October 1999 the Minister of Commerce, Maj.-Gen. Kyaw Than, and the Minister of Sports, Brig.-Gen. Sein Win, were both replaced. In early December the Minister at the Office of the Chairman of the SPDC, Brig.-Gen. Maung Maung, was forced to resign his portfolio and was also removed from his position as Chairman of the Myanmar Investment Commission; Maung Maung was reportedly dismissed as a result of the ruling junta's displeasure at the sharp decline in foreign investment in the country. In mid-August 2000 it was reported that the Deputy Minister of National Planning and Economic Development, Brig.-Gen. Zaw Tun, had been forced to resign from the SPDC as a result of his criticism of the Government's economic policies. It was further reported that a number of other senior government officials and military officers had recently been dismissed or forced into retirement; among these individuals was former Commander-in-Chief of the Navy and SPDC member, Vice-Adm. Nyunt Thein. A further set of retirements took place in November 2001, some of them allegedly prompted by private warnings from Malaysia's Mahathir Mohamad that the SPDC needed to take action against corruption if it was to create an attractive business climate in Myanmar. There was much speculation over the implications of these changes for the political fortunes of the rival factions headed by Gen. Maung Aye—widely perceived as a hardliner in the context of his dealings with the NLD—and Lt-Gen. Khin Nyunt, but there was no clear indication that either faction would predominate. In August 2002 Maung Aye was promoted to the military rank of deputy senior general, while Khin Nyunt became a full general.

In February 2001 the death in a helicopter crash of Lt-Gen. Tin Oo, who had been SPDC Second Secretary and Army Chief of Staff, led to speculation as to whether the crash was an accident or an assassination. Also ambiguous was the arrest in March 2002 of the son-in-law and three grandsons of U Ne Win, on charges of plotting a coup. The heads of the air force and the national police were both dismissed shortly before the arrests took place. Since there was little evidence of any coup-planning, speculation focused on whether the detentions were related to U Ne Win's hostility to any *rapprochement* with Suu Kyi, or whether they were prompted by his family's reputed involvement in a criminal gang. There was some suggestion that their real target was U Ne Win's unpopular daughter, Sandar Win. In September 2002 those arrested were convicted on charges of treason and sentenced to death. Ne Win himself died in December 2002 and was cremated in a private ceremony. The significance of these events in Yangon was uncertain, however, because of the increasing power of Myanmar's regional military commanders. In practice, the SPDC devolved much administrative and economic decision-making to these commanders, who were therefore able to develop strong regional economic interests and to build powerful local political bases, sometimes approaching those of regional warlords. In November 2001 10 of the 12 regional commanders were recalled to Yangon to assume senior positions in the central Government, but only two appeared to have accepted this removal from regional power.

On 6 May 2002 Aung San Suu Kyi was, at last, released unconditionally from detention. There was no expectation that the SPDC would recognize the 1990 election result and relinquish power to the NLD, but the release did lead to increasing speculation that members of the NLD might be included in a transitional government preceding the holding of new elections. Talks between Suu Kyi and the SPDC, however, ceased after she was freed and Razali, who was seen as having played a key role in negotiating her release, continued to seek a dialogue between the two sides on political progress. Suu Kyi was obliged to inform the authorities about her travel plans but otherwise appeared free to move about the country. A steady stream of 283 NLD prisoners was also released over the following months. This process of relative liberalization, however, ended abruptly on 30 May 2003, when Suu Kyi was again placed in detention following a clash between government and NLD supporters. The authorities claimed that she was being held in protective custody for her own safety, and she was not permitted visitors, apart from a brief meeting with UN envoy Razali Ismail in June. NLD offices around the country, which had begun to proliferate, were also closed down. In September the SPDC permitted a delegation from the International Committee of the Red Cross (ICRC) to visit Suu Kyi in order to quash claims by the USA that Suu Kyi had begun a hunger strike whilst in captivity.

In August 2003 a cabinet reshuffle took place, in which Khin Nyunt became Prime Minister, replacing Than Shwe. However, observers were divided over whether this appointment strengthened Khin Nyunt's position as, although his formal powers increased, the appointment removed him from direct military command, which had been an important source of his power. He was believed, however, to have retained his position as head of military intelligence. Than Shwe remained Chairman of the SPDC and Commander-in-Chief of the Defence Services, as well as Minister of Defence, while also taking on the functions of state President. Maung Aye became responsible for vice-presidential duties. Khin Nyunt was succeeded as First Secretary of the SPDC by Lt-Gen. Soe Win, former Second Secretary. Soe Win was generally believed to be a hardliner in dealings with the NLD. Five other ministers were also replaced during the reorganization. Following his appointment, the new Prime Minister outlined a 'road map', which he promised would lead to a new constitution and to 'free and fair' elections as part of a 'disciplined' democratic system. Absent from the 'road map', however, was any gesture towards reconciliation with the NLD. It seemed likely, rather, that the SPDC would simply reassemble the National Convention (suspended since March 1996) to put in place a system based on partnership between the military and the military-sponsored USDA. Khin Nyunt also neglected to offer a timetable for this process.

In late September 2003 the regime released Aung San Suu Kyi from detention and confined her under house arrest after she had undergone major surgery. There was no relaxation, however, of restrictions on the NLD. Although the SPDC released five members of the NLD's central committee in November 2003 and a further 29 NLD members in January 2004, it was difficult to detect other signs of any softening in the harsh treatment of its opponents. The NLD Chairman, Aung Shwe, and the party Secretary, U Lwin, remained under house arrest, while NLD Vice-Chairman U Tin Oo was held in prison, before being released into house arrest in February 2004. In December 2003 nine people, including the editor of a sports magazine, were arrested on charges of plotting to assassinate members of the SPDC in a series of bomb attacks.

On 17 May 2004 the SPDC reconvened the National Convention under the leadership of a senior regime figure, the SPDC Second Secretary, Lt-Gen. Thein Sein. More than 1,000 delegates were present, but the NLD, which had been allocated only 50 places, refused to participate without the release from house arrest of Suu Kyi and U Tin Oo. (It was reported that Khin Nyunt had favoured allowing Suu Kyi to attend the Convention, but that he had been overruled by Than Shwe.) The authorities made it clear that the Convention would proceed from the point at which it had been suspended in 1996, meaning that controversial constitutional provisions, notably the reservation of 25% of parliamentary seats for the military, would not be discussed further. Freedom of debate was also restricted by several rules, including provisions that delegates were not to challenge the role of the SPDC, were not to make 'anti-national' statements, were not to criticize the convention itself, or stage any form of protest against its proceedings, and were not to discuss proceedings with outsiders.

In September 2004 a minor cabinet reorganization took place, in which the Minister of Foreign Affairs, U Win Aung, was replaced by Maj.-Gen Nyan Win; the Ministers of Agriculture and Irrigation and of Transport were also replaced. In the following month the Prime Minister, Gen. Khin Nyunt, was reported to have been dismissed and placed under house arrest, allegedly owing to allegations of corruption that had been made against him. The First Secretary of the SPDC, Lt-Gen. Soe Win, was subsequently appointed Prime Minister, and was succeeded as First Secretary by Second Secretary Lt-Gen. Thein Sein.

INSURGENTS AND DISSIDENTS

The Communist Party of Burma (CPB), located on the Chinese border in the northern part of the Shan State, posed the most serious military challenge to the Burmese armed forces until 1989. The party was well armed and supported by anti-Government, pro-CPB broadcasts from its China-based secret radio transmitter, the 'Voice of the People of Burma'. In 1984 China's policy changed, and the supply of weapons to the CPB diminished. Faced with declining fortunes, the CPB allegedly began exporting illicit drugs through Viet Nam and Laos, using the proceeds to purchase Soviet weapons. In early 1989 the CPB was split by internal dissent; certain ethnic factions of the party, critical of its 'narrow racial policies', challenged the central leadership. In April the Wa hill tribesmen and Kokang Chinese, who had served as cadres for the party, mutinied and forced the ageing CPB leaders across the border into China. Thus, the CPB was effectively defunct; its various regional armies divided into ethnic groupings, the most prominent being the Wa, Shan, Kokang Chinese, and Kachin.

Apart from the CPB, several groups of ethnic insurgents have been engaged in low-level warfare against the authorities since 1948, when Burma's independence was achieved. Despite the Government's efforts to eliminate and destroy the various insurgent groups, they survived and united to form the National Democratic Front (NDF) in 1975. The political objective of the NDF was the creation of a truly federal union (as envisaged by the original Constitution of 1947), based on the principles of self-determination, equality and democracy. The NDF hoped to achieve its objective through negotiations with the central Government, but fighting would continue until such talks could be arranged. The NDF formed three commands (northern, central and southern) to co-ordinate the military effort. In 1988 the NDF numbered 10 groups, representing Karen (Kayin), Arakanese (Rakhine), Mon, Karenni (Kayinni), Shan, Kachin, Palaung, Pa-O and Wa. In May 1986 the NDF formed a political and military anti-Government alliance with the CPB, following an NDF conference at Pa Jau in January. This alliance was maintained despite opposition from the Karen (Kayin) National Union (KNU), one of the largest ethnic insurgent groups in the NDF. The Government initially responded with renewed attacks on Karen guerrillas, although this resulted in early victories for the combined forces. By May 1987, however, government forces had gained control of 60 km of border areas previously controlled by the CPB, and had reasserted control over 500 sq km of territory in the north-east. Also in the same month, the Government launched a major assault against the Kachin, temporarily capturing their military and political headquarters.

In 1988 the unity of the minorities was tested as the Karen and Mon openly fought over control of the 'black-market' trade at the Three Pagoda Pass area on the Thai-Burmese border. Through the mediation of the NDF, the fighting ended in August. In April a prominent human rights organization, Amnesty International, issued two long reports documenting human rights violations committed by the Burmese Army in its war against the minorities. These reports drew world attention, but were ignored by the Burmese Government.

The insurgent groups were sympathetic to the anti-Government movements in Rangoon and other major cities in 1988. After the armed forces seized power, an estimated 7,000–10,000 students fled to the border areas to seek refuge and arms in order to continue their struggle against the Government. Despite shortages of weapons and supplies, units of the NDF gave them refuge and some training. At the same time, units of the NDF and the CPB launched attacks on government outposts. During 1989 and 1990 the armed forces' assault against the KNU continued, with the capture of six river enclaves, although the KNU retained their headquarters at Manerplaw. During the same period, the Kachin continued to engage the armed forces in their area. The Kachin held most of the countryside in their state, while government forces controlled the two major cities, Bhamo and Myitkyina, and their rail and road connections with the rest of Myanmar.

In the light of the changes in Rangoon, the NDF created a new political grouping, the Democratic Alliance of Burma (DAB), to include Burman students, monks and expatriates in a broad coalition which would, it hoped, eventually incorporate the peoples in the Burmese heartland, on the model of the original post-war nationalist movement, the AFPFL. Maj.-Gen. Bo Mya, the KNU leader, was elected President of the DAB, and Brang Seng, the Kachin leader, was named First Vice-President. At its first convention, in November 1988, 23 separate groups participated. It declared as its main objectives the overthrow of the military Government, the establishment of democratic rule, the ending of the civil war, the restoration of internal peace and national reconciliation, and the creation of a genuine federal union. Student forces within the DAB united to form the All-Burma Student Democratic Front (ABSDF), which operated in border regions and received weapons and training from the KNU and from Kayin rebels. The DAB did not replace the NDF, however; the latter continued to exist and control its own areas and armed forces. In late 1988 it admitted its 11th member, the Chin National Front.

The collapse of the CPB in 1989 provided the SLORC with an opportunity to divide its ethnic opposition and cause some groups to end their participation in the civil war. The Government successfully approached first the Kokang Chinese, and later the Wa, and offered much-needed rice in exchange for support in the government campaign against the NDF and Chang Chi Fu, known as Khun Sa, the 'opium warlord' and leader of the rebel Mong Tai Army (MTA, formerly the Shan United Army), in the Shan State. It was reported that the armed forces promised, in addition to food, to allow the Wa to continue to trade opium, the major crop of their area. The SLORC also pledged to initiate a border development programme (including the construction of roads, hospitals and schools) in the former CPB areas. The SLORC also approached members of the NDF, and was successful in securing agreements with the Shan State Progressive Party in September 1989, the Pa-O National Organization in February 1991 and the Palaung State Liberation Organization in April. In July 1991, at its third Congress, the NDF responded by expelling these three organizations, thus reducing the NDF membership to eight.

Following the establishment of the 'parallel' government (the NCGUB, see above) in Manerplaw in December 1990, the SLORC intensified its attacks on KNU positions. In October 1991 the KNU surprised the government armed forces by infiltrating the Ayeyarwady delta area and launching an attack with local support at Bogale, 200 km south-west of Rangoon, now Yangon. Until the attack, the SLORC had considered the area secure. Government forces responded with air and land attacks, and, after nearly one month of fighting, they regained control. In December the SLORC launched a major military campaign to defeat the KNU, attacking Manerplaw and Kawmoora. It aimed to capture the KNU and DAB headquarters, which was also the seat of the NCGUB, before Armed Forces Day on 27 March 1992. Its failure either to defeat the KNU in the field or capture its headquarters (owing, in part, to Thailand's actions, see below) led directly to the announcement in April that it was halting its war against the KNU. In October, however, government troops resumed hostilities.

In February 1993 the Kachin Independence Organization (KIO) agreed to attend peace talks with the Government in the Kachin capital, Myitkyina. The discussions, which took place during February and March, were the result of pressure exerted by China and Thailand. The Kachin (who were the second largest ethnic insurgent force, after the KNU) reportedly demanded a nation-wide cease-fire and that further talks include other rebel groups. However, the KIO finally appeared to have agreed to a bilateral cease-fire with the Government, prior to comprehensive peace talks with all ethnic rebels. The KIO were suspended from the DAB in October for negotiating separately with the SLORC, and the DAB reiterated its conditions for discussions with the SLORC in a series of open letters. The DAB stipulations included: the recognition of the DAB as a single negotiating body (the SLORC insisted on meeting each ethnic group separately); an immediate end to the forcible mass relocation of villages; a new body to draft a constitution; and the release of all political detainees, beginning with Suu Kyi. The cease-fire agreement between the KIO and the Government was announced in October.

The process of reconciliation continued with the return to the 'legal fold' in May 1994 of the Karenni (Kayinni) National People's Liberation Front, the 11th insurgent group to conclude a cease-fire agreement with the SLORC. This was followed, in July, by the declaration of a cease-fire by the Kayan New Land Party, and in October by the Shan State Nationalities Liberation Organization. In late December government forces launched a new offensive against the KNU, capturing its headquarters at Manerplaw in January 1995 and forcing many hundreds of KNU fighters to retreat across the border into Thailand. The virtual defeat of the KNU forces, after almost 50 years of military resistance, was attributed to their reportedly severe lack of ammunition and funds and to the defection from the Christian-led KNU in December 1994 of a mainly Buddhist faction, which established itself as the Democratic Karen (Kayin) Buddhist Organization (DKBO). The DKBO, which had comprised an estimated 10% of the strength of the KNU's military wing, the Karen (Kayin) National Liberation Army (KNLA), allegedly supported the government forces in their offensive. In February 1995 the Myanma army captured another major KNU stronghold, at Kawmoora, and in the following month Bo Mya resigned as the commander-in-chief of the KNLA (although he remained the leader of the KNU and was re-elected to the post in August 1995). The SLORC and their proxies, the Democratic Karen (Kayin) Buddhist Army (DKBA, the military wing of the DKBO) continued to attack KNU bases as well as refugee camps inside Thai territory. Discussions between the KNU and the SLORC began in early 1994. Despite the KNU's participation in a series of formal cease-fire negotiations with the SLORC between late 1995 and late 1996, as well as its request for further discussions, the SLORC attacked the KNU in February 1997 and drove its forces from nearly all of their remaining fixed bases. The KNU 16th Battalion Commander, Thamu Hae, surrendered, and it was rumoured that he might be given some position of authority in the SLORC Government of the Karen State. Unlike the previous offensive, few, if any, DKBA Karen troops were used in 1997. Because the rains were very late in 1997, this offensive continued beyond the middle of the year. It also displaced most of the ABSDF and other forces who co-operate with the KNU. From 1995 the DKBA made frequent incursions into Thai territory, abducting KNU members, attempting to coerce refugees into returning, robbing, raping and destroying villages (including two large Karen refugee villages, Huay Kaloke and Mawker, in March 1998). In March 1999 it was reported that Kayin rebels had executed at least 13 government officials whom they had abducted the previous month.

In the Shan region, after extensive battles with government forces between late 1993 and late 1995, there was great tension within the 20,000-strong MTA between the genuine Shan nationalists among the field officers and the Chinese leaders, who were more interested in the drugs trade. In attempts to prevent a split, the Chinese-Shan leader Khun Sa, long a prominent figure in the drugs trade, proclaimed an independent Shan State in May 1994, and in mid-1995 established a new Shan State National Congress Party as the political wing of the MTA, with Khun Kan Chit, a Shan, as its nominal leader. This did not prevent a substantial revolt of some northern MTA units led by Karn Yord, who formed the Shan State National Army in July, taking nearly one-quarter of the MTA forces with him. Weakened by this division, as well as by major government attacks, desertions and, reportedly, by his own ill health, Khun Sa, who remained the real MTA leader, agreed to a cease-fire with the SLORC in January 1996. The MTA headquarters at Homong, some other MTA troops at Mong Hsat and a few other locations in the southern Shan State surrendered later in January. Most Shan commanders and soldiers of the MTA disapproved of the surrender and continued their resistance, linking up with other Shan groups. This alliance was formalized in September 1997 when three major groups, including the Shan State National Army (SSNA), remnants of the MTA, and the Shan State Peace Council (SSPC) and its military wing, the Shan State Army (SSA), joined together in an enlarged SSA. As a result, the truce between the Government and the SSPC/SSA broke down, and SSA elements resumed guerrilla activity in various parts of the Shan State. Khun Sa subsequently moved to Yangon, like the Kokang drugs-trafficker, Lo Hsing-han, and started a number of businesses, including a bus service and an overseas trading company. Having referred to Khun Sa as a 'narco-terrorist' until late 1995, the SLORC Government subsequently used the honorific prefix, U, before his name and refused to extradite him to the USA for prosecution on drugs-trafficking charges (he had been indicted by the USA in 1989 and 1992). Clashes between SSA units and government forces were reported in December. In March 2000, however, following the group's announcement that it wished to seek a peaceful settlement with the ruling junta, the SSA issued a statement outlining cease-fire terms.

The Karenni (Kayinni) National Progress Party (KNPP) cease-fire with the SLORC, which was signed in March 1995, collapsed almost immediately, and government offensives on the KNPP area continued. These attacks intensified from January 1996 after Khun Sa's surrender at Homong, to the north-east of the KNPP area. In that month the KNPP and the KNU agreed to fight together again against the SLORC and both groups continued to exist as guerrilla armies along much of the Myanma border with western Thailand. Following negotiations begun in December 1993, however, the New Mon State Party (NMSP) signed a cease-fire with the SLORC in June 1995. This truce was particularly significant as the NMSP had controlled part of the area where a major gas pipeline to Thailand was under construction. The last members of the NMSP surrendered in May 1997, and the KNU was virtually eliminated from this area.

The SLORC regularly reported that all ethnic groups except the KNU had ceased armed resistance; however, various small groups in other areas (which were not often mentioned) continued their struggle: several Rohingya Muslim and Rakhine Buddhist groups in the Rakhine State, the Chin National Army in the Chin State, the National Socialist Council of Nagaland (various factions) in the Naga areas of western Sagaing Division, and a number of breakaway groups from the MTA, which continued resistance after the MTA surrender in January 1996 and formed a substantial Shan alliance in June. Many of the groups with which truces continued to hold were restive, and clashes between SLORC forces and KIA forces were reported in early 1996. In December 1995 the United Wa State Party (UWSP) was reportedly very discontented by the replacement of MTA forces by Myanma army troops in some areas adjacent to them and by the restrictions that were placed on their movements from early 1996.

ABSDF students in areas controlled by various groups, such as the KIA and KNU, experienced difficulties following the various cease-fires and offensives; the SLORC also attacked them directly, capturing their bases in Myanmar. For years they were also divided by factionalism. With the development of closer relations between Thailand and Myanmar, the Thai security forces increased their harassment, detention, and deportation of the thousands of Myanma refugee students in Thailand; some were moved to a remote camp in western Thailand known as the 'Safe Area.' In September 1996 the ABSDF reunited, with Dr Naing Aung as Chairman and Moe Thee Zun as Vice-Chairman. However, it continued to lose ground militarily. In March 1998 the SPDC (which had replaced the SLORC) detained 40 members of the ABSDF whom it accused of planning terrorist attacks on government buildings and embassies and the assassination of national leaders; the allegations were rejected by the ABSDF.

Various other groups who had signed cease-fires with the SLORC also expanded their involvement in the drugs trade; these included the Wa UWSP and its armed wing, the United Wa State Army (UWSA, reported to have a major amphetamine factory near the Thai border, in addition to its opium and heroin interests) and other former CPB groups in the eastern Shan State, as well as several Chinese groups in the Kokang area of the north-eastern Shan State, notably the Eastern Shan State Army of Lin Min Shin (Sai Lin). The surrender of Khun Sa and the break-up of the MTA did not in any way reduce opium production. Despite the truces, many of the former members of the DAB continued to be in contact; a meeting was held in January 1996, and a DAB congress in April. According to the SLORC, all groups that had signed cease-fire agreements were invited to send observers to the National Convention; apparently few did so.

Following the cease-fires with most ethnic rebels, the Tatmadaw was able to move into many areas where it had had no presence for 20–30 years or more. In some areas, the Tatmadaw forced groups of ethnic soldiers to relocate, contrary to the agreed cease-fire terms. In many zones, especially around the remaining active ethnic rebellions, there was a massive forced movement of villages away from rebel areas and into relocation areas in the plains. Hundreds of thousands of people were displaced from their homes in the south-western Shan State, the eastern Kayah State and the eastern Karen State in 1996 and 1997. In May 1999 it was reported that at least 300,000 Shan had been forced from their villages into resettlement camps by government troops. This population movement severely affected the local support base of the Karen KNU, Karenni KNPP, Pa-O and former MTA, which did not participate in the surrender of early 1996. Many of the villagers who did not wish to relocate in the plains became refugees, especially in Thailand but also in Bangladesh, India and China: the Tatmadaw offensive in the Karen State which started in February 1997, for example, drove an estimated 20,000 additional Karen refugees into Thailand. In many cases, potential refugees were turned back at the Thai, Indian and Chinese borders; in others, people were persuaded to return to Myanmar. Most such people ended up in relocation areas, which some non-governmental organizations working along the Thai–Myanma border termed 'concentration camps'. In 1996 Myanmar announced that the local population would be removed from a large tract of land in Tanintharyi (Tenasserim) to create a major nature reserve. Since the gas pipeline to Thailand was to pass through this region, however, the announcement was widely interpreted as a security measure, rather than a gesture towards nature conservation. In 2001 it was reported that government agents were fomenting tension between Buddhists and Muslims in Rakhine.

On 12–14 December 1998 23 ethnic groups met to issue a statement of support for the NLD and for the convening of the elected National Assembly. The SSA and the KNPP announced that they would co-operate to fight the regime. In January 2000 Gen. Bo Mya was succeeded as the leader of the KNU by the former Secretary-General of the organization, Saw Ba Thin. Following his appointment as leader, Ba Thin announced that, while the KNU intended to continue its struggle against the ruling SPDC, the movement was prepared to negotiate a political settlement with the military regime. It was subsequently reported by both government and independent sources that an initial but inconclusive round of talks between the KNU and the SPDC was held in February, followed by further discussions in March.

The *rapprochement* between the SPDC and the NLD in 2001 aroused alarm among many ethnic groups, which feared that the two sides in Yangon would reach an agreement that gave no place to greater regional autonomy. In September 2001 several ethnic groups formed the National Solidarity and Co-operation Committee to press for a tripartite dialogue.

The breakdown of relations between the SPDC and the NLD in 2003 turned the minorities once more into a potential ally of the regime against the NLD. Particularly after Khin Nyunt became Prime Minister, the SPDC renewed efforts to sign cease-fires with ethnic insurgents and to collect promises from ethnic organizations to participate in the planned new National Convention. The largest group still fighting the Government, the KNU, took part in cease-fire negotiations with government representatives in December 2003 and January 2004. Once the National Convention had been reconvened in May 2004, however, and several minority organizations had refused to attend, hostilities between the Myanma army and the insurgents flared up again.

REFUGEES AND OTHER MIGRANTS

As a result of SLORC policies since 1988, large numbers of ethnic minority refugees have flooded into every neighbouring country. This includes at least 288,000 (or nearly one-half) of the Muslim Rohingya population of Rakhine State, who fled to Bangladesh between 1989 and 1991. In Thailand there were about 70,000 Kayins from the KNU areas, about 10,000 Kayinnis from the KNPP area, and some 18,000 Mon from the former NMSP areas in the south; most of these fled Myanmar in the early 1990s, but more Kayin and Kayinni continued to arrive from 1996 onwards. There were also many Kachin refugees (estimated at as many as 15,000) in China, and Chin and Naga refugees (up to 10,000) in north-eastern India. The most recent group of refugees were the Shan from the former MTA area, who arrived in Thailand following the MTA surrender at Homong in January 1996; their number was estimated at up to 100,000, but this was probably an exaggeration. These Shan refugees were not placed in camps but dispersed around northern Thailand. The office of the UN High Commissioner for Refugees (UNHCR) was involved in sending more than 200,000 Rohingya back into Myanmar between 1992 and 1997, as they did in 1976–78. In July 1999, however, about 20,000 Rohingya refugees remained in camps in Bangladesh, despite the expiry of the official deadline for their repatriation in August 1997. In April 2000 the International Federation of Human Rights Leagues (FIDH) issued a report condemning the treatment of Rohingya Muslims by the Myanma Government, including forced labour, punitive taxes and extrajudicial killings. The FIDH claimed that the regime was attempting to force the exodus of Rohingyas from their native Rakhine and criticized UNHCR for its effective complicity with the Myanma regime in designating the more recent refugees as economic migrants. The Mon were repatriated in early 1996. Japan offered humanitarian aid to support the repatriation of the Kayin on a similar basis. Returning refugees were often concentrated in camps rather than being allowed to return to their homes, or were unable to recover their homes and land.

In addition to these recognized refugees, there were many economic migrants working in Thailand, estimated by some Thai sources to number hundreds of thousands or even 1m.; they provided much of the menial and seasonal labour needs of northern, western and southern Thailand and supplied the Thai sex industry. There were also many more who were internally displaced, either because they had obeyed SLORC demands to move from remote areas and urban centres to relocation zones, or because they had moved away from SLORC-controlled areas. Nearly all such people were undergoing severe hardship, deprived of their traditional homes, land and crops.

Large numbers of people were conscripted into forced labour squads, used by the SLORC to complete infrastructure projects (roads, railways, irrigation systems, etc.) or to carry weapons and supplies for the army, thus also undergoing temporary displacement. The SLORC defended this kind of labour as traditional in Myanmar; but the scale of the works undertaken was the greatest in recent history, and inadequate provision was made for feeding and accommodating the workers, who were required to bring their own tools and sometimes even to feed themselves, while neglecting their own crops or other work for weeks or months at a time. It was usually possible to pay to avoid this labour, but this option was too expensive for many people and only further enriched the local military commanders, who also sometimes received payment for the work done. The mortality rate among such labourers was high, but lower than that for gangs of prisoners (criminal and some political) engaged in such work full time.

In the first half of 1996 an estimated 70,000–75,000 Kayah were relocated within the Kayah State, while about 80,000 Shan were relocated in the former MTA area of south-western Shan State. Similar numbers of Karen were relocated away from the KNU in the first half of 1997. With each relocation, thousands more refugees appeared at the Thai border. Although many, especially Shan, were absorbed into local villages and the local work-force in Thailand, Thai police conducted occasional raids and deported those without Thai documentation. Some, particularly Karen, were placed in refugee camps along the western border. The SLORC on a number of occasions sent DKBA soldiers to burn these camps and kidnap KNU leaders. After the Asian economic crisis began in Thailand in July 1997, Thai soldiers increasingly turned back newly arrived refugees, and also repatriated some existing refugees. Nevertheless, about 98,000 Karen remained in western Thailand, living in camps which they could no longer leave for work, as well as a similar number of more widely dispersed Shan refugees. In December 1999 Thailand expelled tens of thousands of refugees who had been working in the country illegally. In January 2000 it was reported that 800–1,000 Karen had fled across the border into Thailand following clashes between government forces and the KNU.

There has always been considerable population movement in both directions across Myanmar's long border with China, but during the 1990s many observers reported a substantial increase in the Chinese population of the Myanma border regions and in northern cities, including Mandalay.

FOREIGN RELATIONS

After attaining independence in 1948, Burma pursued a policy of neutrality and non-alignment in world affairs. Burma declined to join the British Commonwealth and was the first non-communist country to recognize the People's Republic of China in 1949. Cordial relations with China were consolidated by a 1960 agreement to settle outstanding border disagreements, but relations were suspended between 1967 and 1970 over what Burma perceived as attempts to promote the Cultural Revolution amongst Burmese of Chinese descent. Burma supported the UN, and in 1961 became a founder member of the Non-aligned Movement (it subsequently withdrew from the Movement for a number of years, but was readmitted in 1992). Burma's foreign policy following the military seizure of power in 1962 continued to emphasize independence and non-alignment. Although Ne Win's Government was initially rather isolationist and sought to exclude foreign influences, governments from the 1970s onwards were aware of the importance of foreign contacts and investment for development. They sought to open the country to the West to the extent that this could be done without undermining government authority. Western interest in Burma was prompted especially by the country's importance in the drugs trade. In 1976 Burma began to participate in drug enforcement programmes of the UN and the USA, in an attempt to suppress the cultivation of opium in north-eastern Burma which, together with northern Thailand and north-western Laos, was known as the 'Golden Triangle'. Substantial anti-narcotics assistance, including funds and equipment, failed to halt drug production, as many of the areas cultivated for opium were under the control of insurgent groups which relied on the proceeds of drugs-trafficking to support their anti-Government campaign. The military coup in 1988 prompted the USA to suspend all assistance, and the annual crop increased significantly. The Myanma authorities continued to express willingness to co-operate with the outside world in controlling the trade, arguing that international isolation over human rights issues made this co-operation unnecessarily difficult. In June 1989 Myanmar ratified the 1988 UN Vienna Convention against trafficking in illegal drugs. During the 1990s, however, Myanmar became increasingly important as a regional source of amphetamines and methamphetamines, which were traded especially across the border with China and increasingly through India and Sri Lanka, as well as along more traditional routes through Thailand and the countries of Indo-China. It has been estimated that 50% of the heroin consumed in the USA is of Myanma origin, while earnings from drugs exports are believed to exceed the value of all Myanmar's other exports. Weak supervision of the Myanma banking system facilitates this trade. Although allegations were persistently heard that the SLORC and later the SPDC were themselves involved in drugs-trafficking (with particular attention being given to claimed links between Gen. Khin Nyunt and Wa producers), regional military commanders appeared to be the most important local figures in the trade. Opium production appeared to increase in Myanmar in 2000 after the Taliban Government in Afghanistan restricted production there, but statistics were unreliable and the growing importance of the industrial production of amphetamines and methamphetamines meant that the cultivation of opium itself constituted only a small part of the problem. Meanwhile, the ready availability of heroin for injecting within Myanmar, and the social disruption caused by the civil wars, created an environment in which HIV/AIDS spread rapidly. A survey in Kachin state indicated that 90% of heroin users were HIV-positive, and observers have suggested that as many as 1m. people may be infected.

The assumption of power by the armed forces in September 1988, and the subsequent brutal suppression of demonstrations by the opposition movement, provoked widespread international censure. Many creditor nations, including Japan, the Federal Republic of Germany, the USA, the United Kingdom and Australia, suspended economic aid, pending an improvement in Burma's human rights record. Among Burma's immediate neighbours, India was most vocal in its criticism of the military repression; it closed Indian trade routes to Burma and established refugee camps near the border. In March 1993, however, the Indian Minister of External Affairs visited Myanmar and signed bilateral agreements on the suppression of separatist movements and drugs-trafficking along the common border. In 1995–96 two border crossings between India and Myanmar were opened and, in 2002, India agreed with Myanmar to begin the construction of a network of roads linking the two countries; there were also joint military operations against Naga rebel groups. This closer relationship was widely believed to be India's reaction to the growing influence of China in Myanmar. Relations with Bangladesh, with which Myanmar shares a short border, were strained both by the presence of an estimated 21,000 Muslim refugees from Myanmar in camps in Bangladesh and by Myanma plans to dam the shared Naf river.

In response to the events of 1988, the USA halted all non-humanitarian assistance, barred Burma from trade benefits under its Generalized System of Preferences, decertified Burma on narcotics and opposed loans from the World Bank, the IMF and other international financial institutions. It blocked all sales of arms from the USA to Burma and urged other nations to do the same. In 1990 the US Congress passed, and the President signed into law, the Customs and Trade Act, which empowered the President to restrict trade with Myanmar if it failed to comply with certain conditions, which were to be monitored twice annually. These were: the protection of human rights; an end to martial law and the introduction of democracy; and adherence to the requirements of the 1986 Narcotics Trade Control Act. In conformity with the law, in July 1991 the President invoked economic sanctions by declining to renew a bilateral textile agreement. (In 1990 textile exports to the USA amounted to nearly one-half of total Myanma exports.) The US administration retained its tough stance towards the SLORC, and subsequently the SPDC, throughout the 1990s, and continued to condemn the widespread abuse of human rights in Myanmar and to demand the release of all political detainees. In mid-1996 the US Congress passed a law prohibiting new investment and trade with Myanmar; an official ban was announced in April 1997. During 1998 the US administration criticized the military junta in Myanmar for its treatment of Suu Kyi and, once again, demanded the release of hundreds of political prisoners. In March 1999 the US Secretary of State, Madeleine Albright, publicly criticized the regime for taking insufficient action to combat the production of, and trade in, narcotics within Myanmar. During the 1990s the European Union (EU) also maintained a common position of refusing to deal with the Myanma authorities, and in October 1998 an existing ban on visas for members of the Myanma Government and their families was extended to cover Myanma citizens working in the tourism industry.

Japan was also critical of the ruling military junta in Myanmar, but like other donors it completed existing aid projects after 1988. Japanese investment and business activities never ceased, and intensified after a Japan Federation of Economic Organizations (Keidanren) visit in June 1994. Substantial new Japanese aid started in mid-1995 after the release of Suu Kyi, and in April 2001 Japan agreed to provide US $28m. for the rehabilitation of a hydroelectric dam. Other countries, including Australia, Norway and Switzerland, have provided humanitarian aid to refugees outside Myanmar, and more limited humanitarian aid for projects within the country. After resistance to its attempts to observe and reduce human rights abuses, the ICRC withdrew from Myanmar in July 1995. However, the ICRC subsequently regained permission to visit a limited number of prisons in Myanmar in 1999. Although she was initially critical of the ICRC for reaching an agreement with the SPDC, Suu Kyi was reported subsequently to have expressed her support for the Committee's work with political prisoners. Despite visits in 1995 and subsequently, the Asian Development Bank (ADB), IMF and World Bank had not resumed loans by mid-1998. Relations with the USA, the EU, Australia and many other non-Asian countries continue to be difficult. The US Secretary of State denounced Myanmar publicly for its human rights abuses and lack of progress towards democracy at the ministerial meeting that followed the ASEAN meeting in July 1997. Myanmar was vigorously defended, however, by Prime Minister Mahathir of Malaysia and, earlier in 1997, former Prime Minister Lee Kuan Yew of Singapore was widely reported as having said that Suu Kyi would not be capable of ruling Myanmar. Suu Kyi continued to advocate a diplomatic, economic and tourist boycott of Myanmar under the SLORC (and, subsequently, the SPDC).

Facing such consistent hostility from the West, the Government of Ne Win placed great emphasis on fostering its relations with its immediate neighbours, especially Thailand. From about 1988, during the boom years before the Asian economic crisis of 1997, Thailand became a major source of foreign investment in Myanmar. The SLORC granted licences to Thai business interests to exploit raw materials in Burma, especially teak and other timber, in return for much-needed foreign exchange, while the Thai Government offered technical training and other assistance. Plans have been made for a bridge across the Sai river, which would improve communications between Myanmar and Thailand, for a gas pipeline to Thailand from reserves off the Ayeyarwady delta and for a dam in Shan State to supply water to Thailand.

Although there was no announcement of any official Thai-Burmese agreement, offensives by government forces against rebel groups, particularly the KNU, achieved unprecedented success, with Myanma troops frequently entering Thai territory and attacking insurgent bases from the rear. Partly because of deep-seated historical suspicion between the two countries, these operations led to tension. In early 1992 the Thai Government warned Myanma soldiers not to cross into Thailand in their attempts to capture the KNU headquarters at Manerplaw. Following repeated border violations, there were clashes between Myanma and Thai forces in March as the Thai military forced hundreds of Myanma troops out of entrenched positions that they had taken up to attack the KNU base from the rear. In October, however, government troops resumed hostilities and made several incursions into Thai territory. In December the Thai and Myanma Governments reached agreement to relocate the Myanma armed forces, and in February 1993 they resolved to delineate their common border. However, incursions by Myanma forces into Thai territory continued in early 1995, leading to a further deterioration in bilateral relations. Tatmadaw shelling of, and incursions into, Thai territory in pursuit of KNU and KNPP rebels occurred frequently, and Thai soldiers, police, paramilitary force members and local Thai villagers were occasionally killed. In December 1998 two Thai naval officers were killed during a clash between two fishing vessels, and in January 1999 a confrontation between naval patrols from the two countries occurred near the Thai city of Ranong. Clashes between Myanma and Thai troops along Thailand's northern border took place between February and May 2001. Thailand also became increasingly concerned at the inflow of drugs, including methamphetamines, from Myanmar.

Myanmar's relations with Thailand were further strained by the large number of Myanma refugees in Thailand. Refugees began reaching Thailand in significant numbers following the military suppression of the pro-democracy movement in 1988, though Thailand was then reluctant to grant them refugee status and in 1990 forcibly repatriated 1,000 Myanma dissidents. By mid-1998 the number of Myanma refugees in Thailand had increased to an estimated 300,000–500,000 (including many economic refugees). Many of these, it was believed, were seeking to escape conscription into forced labour squads used by the Myanma Government for the construction of roads, railways and other infrastructure projects. Others were members of various ethnic or insurgent groups which had been defeated by, or had capitulated to, the SLORC. In early November 1999 Myanma government troops threatened to shoot at least one group of Myanma illegal labourers whom Thai officials were attempting to deport from Thailand. In October 1999 a group of armed Myanma student activists seized control of their country's embassy in Bangkok, demanding the release of all political prisoners in Myanmar and the opening of dialogue between the Government and the opposition. More than 30 hostages, including tourists as well as diplomats, were taken. They were released by the gunmen within 24 hours, in exchange for the Thai Government's provision of helicopter transport to the Thai–Myanma border. Myanmar closed its border with Thailand immediately after the incident. In January 2000 Thai troops shot dead 10 armed Myanma rebels who had taken control of a hospital in Ratchaburi, holding hundreds of people hostage. The rebels, who were reported by some sources to be linked to the Kayin insurgent group, God's Army (a small breakaway faction of the KNU—the KNU itself denied any connection with the rebels), had issued several demands, including that the shelling of their base on the Thai–Myanma border by the Thai military be halted, that co-operation between the Thai and Myanma armies against the Kayins should cease, and that Kayin tribespeople be allowed to seek refuge in Thailand. Whilst the Thai Government denied reports that the perpetrators had been summarily executed after handing over their weapons, the brutal resolution of the incident was praised by the military Government in Myanmar.

None the less, shared economic interests helped to keep Thailand's relations with Myanmar close, despite these strains. Border crossings, which had been closed due to fighting in 1995, were reopened in early 1996, including those at Tachilek and at Myawaddy. Senior Gen. (later Field Marshal) Than Shwe, the head of the SLORC, visited Thailand in December 1995 and his visit was reciprocated in 1996 and 1997. Thai Prime Minister Thaksin Shinawatra made a visit to Yangon in June 2001 during which both sides made it clear that they wished to improve relations, and in September 2001 Gen. Khin Nyunt paid a three-day visit to Thailand, which was hailed as establishing a new basis for closer co-operation. In particular, Thailand began to work closely with the Myanma authorities to control the flow of drugs by launching military operations against drug organizations and by promoting crop substitution programmes in production areas. In January 2002, at a meeting of a joint Thai-Myanma commission (the first to have been held since 1999), the two countries also agreed to establish a task force to aid in the repatriation of illegal workers. In April 2002 Gen. (later Dep. Senior Gen.) Maung Aye made a visit to Bangkok, during which he promised to help reduce the flow of drugs between the two countries. However, relations deteriorated in the following month when fighting began between Myanma government troops, allied with the UWSA, and the SSA, which had seized four posts on the border. Thai troops were alleged to have fired shells into the country, claiming that the fighting had encroached upon Thai territory. In response the Myanma Government accused Thailand of supporting the SSA. Shortly afterwards the Thai–Myanma border was closed in response to the escalating tensions, which were compounded by reports that the SPDC had expelled hundreds of Thai workers from the country. Border incursions continued throughout June 2002 as relations worsened. Early in that month masked gunmen, thought to be members of the KNU, opened fire on a Thai school bus travelling close to the border, resulting in the deaths of three children. The

KNU denied responsibility for the attack and Thai Prime Minister Thaksin subsequently stated that he did not believe the KNU or Myanma soldiers to have been responsible. However, in August the Thai Minister of Foreign Affairs, Surakiart Sathirathai, met with SPDC leaders in an effort to resolve the tensions. Talks were reported to have been fruitful, and in October 2002 the border was reopened. By early 2003 relations had improved further, and officials from both countries were engaged in ongoing discussions concerning co-operation over investment and the suppression of drugs-related activity.

From 1988, as Burma's international isolation intensified, China assumed an increasingly important role. In August 1988 the two countries signed a broad cross-border trade agreement, which opened the Burmese market and resources to Chinese exploitation. By 1990 China had become the major supplier of consumer goods, which had previously been purchased mainly from Thailand. In 1991 China also became the SLORC's chief arms supplier, with the sale of weapons valued at more than US $1,000m. It was believed that many of the fighter aircraft that Myanmar purchased from China in the early 1990s were used in the suppression of anti-Government rebel groups. China played a significant role in delaying the passage (both in the UN Commission on Human Rights in Geneva in 1989 and 1990 and at the UN General Assembly in 1990) of strong resolutions criticizing the SLORC's human rights violations. China also improved road and bridge links with northern Myanmar, and Chinese firms became important in the timber industry. In the late 1990s several border towns became the centre of a tourist industry based on gambling, drugs and prostitution. It was reported that in some towns the Myanma kyat had been displaced by the Chinese yuan for virtually all transactions. The Chinese President, Jiang Zemin, visited Myanmar in December 2001, and agreements for economic co-operation in several fields were concluded. During the late 1990s, however, Chinese authorities became increasingly concerned by the flow of drugs from Myanmar into the newly affluent coastal regions of China and began to co-operate more closely with the Myanma Government in attempting to control the trade. In 2000 the Myanma army reportedly began to develop closer ties with the army of Pakistan.

From the 1980s, relations with ASEAN were central to Myanmar's efforts to avoid international isolation. As relatively developed economies, both Malaysia and Thailand were interested in the emerging economic opportunities in Myanmar, while ASEAN as a whole, and Indonesia in particular, was keen to ensure that Myanmar did not fall into China's orbit. In 1991 ASEAN denied a US request to use its influence with Myanmar to help persuade it to end human rights violations and allow democratic government to be restored. ASEAN declared that it pursued a policy of 'constructive engagement' with the SLORC and that it would not interfere in Myanmar's internal affairs. Trade between Myanmar and ASEAN continued to expand, and in July 1994 U Ohn Gyaw, then Myanma Minister of Foreign Affairs, travelled to Bangkok to attend the opening and closing ceremonies of the annual meeting of ASEAN ministers responsible for foreign affairs (the first time that Myanmar had been invited to attend as a guest). Myanmar's links with ASEAN became steadily closer from this time. It signed a treaty of friendship and co-operation with ASEAN members in 1995, and was granted full observer status at the July 1996 ASEAN meeting in Jakarta, also becoming a full member of the ASEAN Regional Forum (ARF). Myanmar applied to join ASEAN in October 1996 and was admitted as a full member on 1 July 1997. It also joined the BIST-EC (Bangladesh-India-Sri Lanka-Thailand Economic Forum) in 1997. However, the country was not allowed to participate in the second Asia-Europe Meeting (ASEM) in early 1998, at the insistence of various European states. In early 1999 a meeting between ASEAN and the EU scheduled to take place in February was postponed indefinitely as a result of continued disagreement over the representation of Myanmar at the meeting. ASEAN objected that the EU should not attempt to dictate ASEAN membership and defended its approach of 'constructive engagement'. The Minister of Foreign Affairs for Singapore commented, 'In Asia, we marry first and expect the bride to adapt her behaviour after the marriage'. As EU governments would not permit Myanma senior officials to enter their territory, ASEAN hosted a two-day ministerial summit in Vientiane, Laos, in December 2000, but only junior ministers attended from the European side. The meeting ended with an agreement that an EU delegation would visit Myanmar for discussions with both the Government and Aung San Suu Kyi. The visit took place in January 2001, but produced no formal outcome. In April 1996 James Leander (Leo) Nichols, a businessman, honorary consul of Norway and Denmark and local representative for Sweden, was arrested and imprisoned for having an unauthorized fax machine and additional telephone lines. Despite many vigorous but polite requests for his release on the grounds of old age and ill health, including a visit by the Norwegian, Danish and Swedish diplomatic representatives in Thailand and Singapore, Nichols was kept at Insein prison, where he died in June. The SLORC autopsy reported the cause of death as a stroke; Insein prison is notorious for the ill-treatment and torture of prisoners, many of whom have died in custody. Observers believed that the real reason for Nichols' arrest was his close relationship with Suu Kyi. Strong protests about his death were made by Australia (where much of his family lived) as well as Norway, Denmark and Sweden, and Myanmar's hopes for better relations with the EU thus suffered a major set-back.

In September 1999 a diplomatic dispute began between Myanmar and the United Kingdom, after British consular staff were refused permission to visit two Britons, James Mawdsley and Rachel Goldwyn, being held in prison in Myanmar for their separate involvement in pro-democracy protest action. The British Government expressed its grave concern over the treatment of the pair. Rachel Goldwyn was released in November; James Mawdsley was sentenced to 17 years' imprisonment by the ruling junta in September for illegal entry into the country and sedition, however, and served more than 400 days in solitary confinement before being released in October 2000 as a result of pressure from the British and US Governments, the UN and other international bodies.

In mid-1997 the ILO began a major investigation of forced labour and the suppression of trade unions in Myanmar. The findings of the investigation were subsequently published in late August 1998: in its report, the ILO found the use of forced labour to be 'pervasive' throughout the whole of the country (and to include women, children, the sick and the elderly), and accused the regime of using beatings, torture, rape and murder in the enforcement of its forced labour policy, constituting a 'gross denial of human rights'. In June 1999 a resolution condemning Myanmar for its widespread use of forced labour was adopted by the member countries of the ILO, and the country was barred from participating in any ILO activities. At the organization's annual conference in Geneva on 14 June 2000, its members voted by an overwhelming majority to adopt measures against Myanmar if the military Government did not halt the practice of forced labour within four months. The sanctions came into effect in November 2000, though their implementation depended on the co-operation of individual countries. The Myanma Government responded by describing the sanctions as 'unfair', 'unreasonable' and 'unjust' and announcing that it would cease to co-operate with the ILO. In May 2001 ASEAN labour ministers issued a joint communiqué noting what they described as concrete actions by Myanmar to eradicate forced labour and calling on the ILO to end its campaign against Myanmar. The ILO sent a delegation to Myanmar in October 2001 to investigate the claims of improvement; it reported, however, that forced labour was still widespread in areas controlled by the military. In March 2002 the ILO established a liaison office in the country, and in 2003 talks were under way regarding the development of a plan of action for the elimination of forced labour in Myanmar.

Myanmar has also been condemned annually since 1991 by the UN for human rights violations within the country, with increasingly strong resolutions having been passed by the General Assembly each year. As a result, companies operating in Myanmar may find themselves subject to boycotts and other protest activity, such as that which persuaded Pepsi, Texaco and many others to withdraw completely from the country. Various national governments now encourage companies not to invest in Myanmar, and several local governments, such as that of New York City in the USA, have selective purchasing policies and will not do business with companies that operate in Myanmar.

(In June 1998, however, a US court ruled that state and local governments could not impose a boycott on Myanmar by law, since such actions infringed the right of the federal Government to conduct foreign policy.) In late November the Republic of Korea (South Korea) became the first Asian country to co-sponsor a UN General Assembly resolution criticizing human rights abuses in Myanmar. In January 2002 the European lingerie manufacturer Triumph decided to close its factory in Myanmar in response to a long campaign in Europe claiming that it was exploiting forced labour.

From late 1998, however, Myanmar's international isolation began, in some respects, to diminish, and Western diplomats began to speak approvingly of the possibility of effecting change in Myanmar through the employment of policies of engagement, rather than isolation, towards its Government. In October 1998 the British Government convened a low-key meeting at Chilston Park in the United Kingdom at which ministers of foreign affairs and ambassadors from several countries, together with UN and World Bank officials, began to develop a strategy of engaging with the Myanma Government. These meetings gave rise to the proposal, quickly rejected by the ruling junta, that Myanmar should receive US $1,000m. in aid in return for significant political concessions. A follow-up meeting in Seoul, Republic of Korea, in March 2000, nicknamed Chilston-2, failed to come up with further strategies. None the less, in February 2000 Myanmar hosted a major Interpol conference addressing drugs-related crime. Although the United Kingdom and the USA boycotted the meeting, the military junta convinced some of those attending that it was interested in attacking the trade. The Australian, Japanese and South Korean Governments all made official contact with the Myanma Government to explore possibilities for political change in Myanmar. In November 1999 the Japanese Prime Minister, Keizo Obuchi, met with Senior Gen. Than Shwe during an ASEAN summit meeting in Manila, the Philippines; the meeting was the first between the leader of a major world power and a senior member of the military Government since the junta's suppression of the democratic opposition in 1988. In April 2000 the EU announced the introduction of increased sanctions against Myanmar, but in the second half of 2001, in response to increasing indications that dialogue was taking place between the SPDC and Aung San Suu Kyi, it agreed to allow Myanmar to take part in future ASEAN-EU dialogues and to sponsor a variety of aid programmes, while keeping the sanctions in place. The removal of restrictions on Suu Kyi in May 2002 was welcomed in international circles, although it was widely described as being only the first step towards change. The first diplomatic benefit of the release came in August 2002, when the Japanese Minister of Foreign Affairs, Yoriko Kawaguchi, visited Myanmar. The USA also signalled its willingness to lift sanctions in return for political progress. In January 2003 the EU waived its ban on the granting of visas to members of the SPDC in order to allow the Deputy Minister of Foreign Affairs, Khin Maung Wien, to attend a meeting in Europe. By April, however, the EU was threatening to tighten sanctions unless overt progress towards democratization was made. In February the USA had made similar threats. Following the renewed detention of Suu Kyi in May, the USA expanded its sanctions against Myanmar, banning all imports and 'freezing' SPDC assets in the USA. The EU extended its visa restrictions. Japan, which was Myanmar's largest aid donor, announced that no new aid would be granted until Suu Kyi was freed.

To the surprise of many observers, ASEAN also condemned the detention of Suu Kyi and called for a resumption of political dialogue. The Malaysian Prime Minister, Mahathir Mohamad, even speculated that Myanmar might be expelled from the organization. ASEAN leaders appeared to be concerned that Myanmar's poor relations with the West might compromise the efforts of other ASEAN countries to maintain good relations and damage the interests of ASEAN as a whole, since Myanmar was due to chair the association in 2006. President Megawati Sukarnoputri of Indonesia urged all ASEAN governments to encourage political participation, and a parliamentary caucus to promote democracy in Myanmar was formed in Malaysia, presumably with at least the tolerance of the new Malaysian Prime Minister, Abdullah Badawi. None the less, ASEAN remained committed to a strategy of engagement with, rather than sanctions against, the regime. Despite private expressions of disappointment with Myanmar's progress towards democracy, the association limited itself publicly to a call to involve 'all strata of society' in the democratization process. Thailand was particularly keen to maintain good relations with the regime, in order to secure co-operation in its attempts to control the flow of drugs from Myanmar into Thailand. Accordingly, in mid-December 2003 Thailand was host to an international forum, called the 'Bangkok Process', at which the Myanma Minister of Foreign Affairs, U Win Aung, presented the 'road map' to representatives from Australia, Austria, the People's Republic of China, France, Germany, India, Indonesia, Italy, Japan and Singapore, along with Razali Ismail, the special envoy for Myanmar of UN Secretary-General Kofi Annan. U Win Aung announced at this meeting that the first step of the 'road map'—the reassembly of the National Convention—would take place in early 2004. Myanmar's other main neighbours, India and China, also in competition for influence there, remained reluctant to exert pressure on the Government. The most public pressure on the regime came from Kofi Annan, who urged that the transition to democracy foreshadowed in Khin Nyunt's 'road map' be completed by 2006. Annan and Razali Ismail emphasized the need for dialogue with Aung San Suu Kyi and the NLD, and Annan expressed dismay that Suu Kyi had not been released to take part in the National Convention in May 2004.

The EU's insistence on not permitting the three newest members of ASEAN (Cambodia, Laos and Myanmar) to join the ASEM forum was placed under strain by the Union's request that its own new members be permitted to participate. ASEAN's official response was that the new European members would be permitted to join only when Cambodia, Laos and Myanmar were admitted. Behind the scenes, however, there was a protracted search for a formula that might somehow permit ASEM to grow while saving face on both sides. Meanwhile, the SPDC also placed some hope in the new Bay of Bengal Initiative for Multi-Sectoral Technical and Economic Co-operation (BIMST-EC), formerly known as the Bangladesh, India, Myanmar, Sri Lanka and Thailand Economic Co-operation Forum (which had been established in 1997 following the admission of Myanmar into the BIST-EC). Gen. Khin Nyunt attended the first summit meeting of the grouping in Bangkok in July 2004 to discuss potential co-operation on public health, education, information technology and biotechnology. Plans for a BIMST-EC free-trade agreement were envisaged, along with an India–Myanmar–Thailand highway. Rumours also circulated that Myanmar was developing closer ties with the Democratic People's Republic of Korea (North Korea), although relations had not been formally restored since a bomb attack in 1983 by North Korean agents on a South Korean delegation visiting Myanmar. Myanmar was said to have obtained missiles and other strategic weapons from North Korea, and the possibility was raised that North Korea was secretly helping Myanmar to build a nuclear reactor.

Economy

RICHARD VOKES

Revised for this edition by MYA THAN

INTRODUCTION

In November 1997 Myanmar's military junta renamed the ruling State Law and Order Restoration Council (SLORC) the State Peace and Development Council (SPDC). Within a year several powerful regional commanders who had recently been promoted to the Council were replaced, and the Cabinet doubled in size to consolidate around two leading generals, Maung Aye and Khin Nyunt (see below), who competed for control, but who both favoured economic privatization. However, sanctions imposed by Western democracies severely hindered this policy, so attention was turned to agriculture, where the expansion of cultivated land became the focus of a revitalized Ministry of Agriculture and Irrigation. Whereas high-yielding rice and double cropping were emphasized during the period of economic expansion between 1993 and 1996, the priority shifted to expanding the area under cultivation by means of awarding benefits to new agri-corporations for the purchase and cultivation of unused land. The SPDC's goal was to achieve self-sufficiency in food grains and to re-enter the global rice market.

Myanmar's economic growth rate was negative or low throughout the late 1980s, but the situation improved in the 1990s, owing to a recovery in the level of exports resulting from foreign direct investment (FDI). In 1996/97, according to the Asian Development Bank (ADB), real gross domestic product (GDP) increased by 6.4%, compared with a growth rate of 6.9% in 1995/96, but GDP growth declined to 5.7% in 1997/98, before increasing to 5.8% in 1998/99. GDP growth was officially declared to be 10.9% in 1999/2000, 13.6% in 2000/01 and 10.5% in 2001/02. In the latter year the Economist Intelligence Unit (EIU) suggested that a growth rate of 5.4% was more realistic. This decrease in the rate of growth was forecast to continue in 2002/03 as a result of adverse global economic conditions, declining foreign investment, poor returns in the industry sector, the effect of sanctions and the sluggish performance of the agriculture sector as a whole. Furthermore, it was thought that power shortages would have a negative effect on manufacturing GDP. However, according to official statistics released by the Ministry of National Planning and Economic Development, GDP performed strongly in 2002/03, growing by 10.0%. Compared with the previous year, the GDP of the manufacturing sector increased by 5.0% in 1997/98, by 6.2% in 1998/99, by 14.5% in 1999/2000 and by 23.4% in 2000/01. According to official statistics, growth in manufacturing GDP fell sharply in 2001/02, to 5.0%.

While the Government attracted foreign investment in several natural resource sectors, particularly energy development, after the Foreign Investment Law came into effect in December 1988, by 1999 investment had declined to one-tenth of the 1992 level. Meanwhile, the agricultural sector grew at an average annual rate of 5.0% from 1992 to 1996, as foreign capital flowed into a number of large corporate farming ventures, but then entered a brief period of decline as the Ministry of Agriculture and Irrigation tried to manipulate the market by means of quotas and price controls. However, in 1999/2000 real growth in agriculture was reported to have exceeded 11%, following good weather and an extension of the area under cultivation, reversing the downward trend of the previous years. Between 1996 and 1999 overall FDI declined from US $2,814m. to a mere $29.5m. This sharp decline abated slightly in 1999/2000, when investment increased to $55.6m., owing to a shift in capital flow, according to Myanmar's Central Statistical Organization (CSO). The same source stated that FDI increased to $217.7m. in 2000/01. However, FDI fell significantly in 2001/02, to only $19.0m., before recovering to reach $86.9m. in 2002/03. The USA and France had invested heavily in petroleum and natural gas in the early 1990s, but US sanctions took effect in 1998, following which the source of capital flow shifted to Asian sources, primarily Japan, Thailand, Hong Kong, Singapore and the Republic of Korea. The six principal investors in Myanmar in 2001/02 remained: the Republic of Korea ($5m.); Japan ($4.7m.); the People's Republic of China ($3.3m.); Hong Kong ($1.5m.); Indonesia ($1.5m); and Malaysia ($1.5m.). By 2002 FDI in Myanmar from member countries of the Association of South East Asian Nations (ASEAN) had fallen to 17.1% ($3m.) of the total, from 39% in 2000. The decrease in FDI was a result of the impact of the 1997 Asian financial crisis, the terrorist attacks on the USA of 11 September 2001, the overall global economic slowdown and an unfavourable environment for domestic investment. According to official statistics, in the period between 1988 and March 2002 FDI reached a total of $7,400m. By the end of January 2003 it had increased to $7,467.6m.

Official data ignore a large influx of 'black money' earned from narcotics and smuggling, estimated at between US $1,000m. and $2,000m. annually. The highway between China and Mandalay provides extensive evidence of substantial unrecorded expenditures on elaborate new pagodas and monasteries, palatial houses, expensive cars and construction equipment, and upmarket resorts. A small but affluent new class of officials, merchants and farmers deals in this illicit hard currency. While urban property prices have decreased sharply in recent years, the volume of highway traffic remains unaffected; indeed, one multifaceted corporation owned by a retired trader in narcotics contracted to expand and maintain this key highway near China, to be financed by levying tolls. In addition, several small private airlines and upmarket bus companies apparently capitalized by 'black money' opened in the mid-1990s, hoping to serve Myanmar's new rich and an expected influx of foreign tourists. Energy shortages reflect unwillingness by the members of Myanmar's new class to invest their wealth in long-term projects such as hydroelectric dams or thermal plants, and a failure by the Government under the *Burmese Way to Socialism* (see below) and the military regimes since 1988, to implement a comprehensive programme of economic reform. In particular, at the official exchange rate, the currency continued to be highly overvalued; in September 2002 the unofficial exchange rate had risen sharply to 1,200 kyats = US $1, while the official rate remained at 6.9 kyats = US $1. However, by July 2004 it had fallen to 960 kyats = US $1, approximately 145 times the official exchange rate of 6.6 kyats = US $1, owing to the US ban on imports and the 'freezing' of US dollar remittances (see below). Moreover, the limited inflow of foreign assistance prevented major structural problems, such as the overvalued exchange rate, from being addressed. This failure undermined the pattern of growth recorded in the first half of the 1990s, as suggested by the deceleration between 1996/97 and 1998/99.

Meanwhile, a high annual rate of inflation continued to prevail, owing largely to the constantly expanding budget deficit and to a decline in the rice harvest in 2001. Although actual economic data are unavailable, it has been estimated that consumer price inflation averages around 20%–30% per annum, largely attributable to the growth in money supply (M2). According to the CSO, in 2002/03 the average rate of inflation exceeded 58% (see below), mainly as a result of higher food prices, as well as of the Government's imposition of limits upon cash withdrawals in order to avert a crisis in the banking sector. With no prospect of a substantial increase in external assistance, development opportunities remain poor. The resumption of both aid and investment depends on greater political stability, which in turn depends on some reasonable transition to democracy. By mid-2001 the SPDC had taken its first tentative steps towards liberalization, but combating poverty remained its most pressing problem. Per caput income was estimated at US $97 (at the free-market exchange rate) in 2001 and $90 in 2002, confirming that Myanmar remained one of the poorest countries in Asia. In 1997, according to the World Bank, some 23m. people, or almost one in four households, were living below minimum subsistence levels and an estimated 22.9% of the population were living in poverty.

ECONOMIC POLICY AND PLANNING

Until the reforms of 1988, Myanmar had been ruled consistently by Governments that espoused an essentially socialist path to development. The military coup in March 1962 and the subsequent publication, in April, of the new Government's economic policy document, *The Burmese Way to Socialism*, signified a turning-point in Myanmar's post-war economic development. The document outlined the military Government's commitment to the establishment of a 'socialist democratic state'. The four principal objectives of the Government's economic policy were: the elimination of foreign control of the economy; a reduction in the country's dependence on foreign markets; a restructuring of the economy away from its dependence on primary production towards a more balanced industrial condition; and, finally, the centralization of economic power in the hands of the State, in order to reduce the power of the private market. These objectives were to be realized through the nationalization of all vital means of production, including those in agriculture, industry, commerce, transport, communications and external trade. Ownership of the means of production was to be vested either in the State, or in co-operatives or unions. While the importance of the role of agriculture in the economy was recognized, priority was given to the development of industry and the improvement of social services.

Implementation of these policies initially brought some benefit to the population, especially in rural areas. However, the country soon faced a growing economic crisis. As a result of the priority given to building a modern industrial sector, agriculture was neglected and, coupled with low procurement prices, this led to a decline in the production of rice, one of the country's principal exports as well as its staple food, thus resulting in a severe shortage of foreign exchange. The industrial sector also failed to perform well, despite the priority it received in terms of investment. This was due to the bias towards continued investment in new plants, despite the fact that existing capacity was being used inefficiently, and to inadequate managerial capacity. Growth in the output of the state-owned industrial sector was insufficient to compensate for the decline in the private sector; as a result, shortages and rationing developed, leading to the growth of a 'black market', which continues to exist.

As the country's economic crisis worsened, the Government was forced to introduce its first programme of economic reform. This involved: increased priority for the production and export of primary products; the decentralization of management and greater use of commercial incentives, in an effort to improve efficiency within state enterprises; an enhanced role for the private sector; and a new willingness to seek and accept external assistance, in an effort to gain access to both capital and technology. After 1972 there was a rapid growth in foreign assistance, particularly from Japan, and from the World Bank and the ADB. Foreign assistance, which had totalled only US $21.7m. in 1970/71, grew rapidly, reaching a peak of $512m. in 1979/80. All these policies were to be implemented within the framework of a 20-Year Plan. This represented the country's first real attempt at long-term planning, and provided only broad guide-lines with respect to goals and priorities: it was to be divided up into five Four-Year Plans.

While economic performance did improve after 1972, and particularly after 1976, by 1982 the country was once more facing serious economic problems, caused by slow growth in the agricultural sector, a significant decline in the price of both teak and rice, the two principal exports, and growing debt-service payments. In response to the worsening crisis, further economic reforms were introduced by the Government of the Burma Socialist Programme Party (BSPP) between August 1987 and July 1988. The implementation of these reforms was disrupted by widespread political unrest, which resulted, finally, in the establishment of the SLORC in September 1988.

The economic reform programme followed since then, based on the earlier reforms introduced by the BSPP, implies the complete abandonment of the policy of self-sufficiency and the adoption of an 'open-door policy' on foreign investment, which is now accepted as crucial to the country's economic development. Under the Foreign Investment Law published in November 1988, 100% foreign-owned companies were allowed, in addition to joint ventures where foreign capital had to form at least 35% of total capital. The law also offered exemption from tax for at least three years and a range of other incentives, as well as guarantees against nationalization. In seeking foreign investment, priority was to be given to the promotion and expansion of exports, the exploitation of natural resources requiring large investment, the acquisition of advanced technology, employment generation, energy conservation and regional development. By 1991 the Government had had some success in attracting foreign investment, particularly into the energy sector. However, Myanmar's ability to absorb a major inflow of foreign investment was limited in the short term, most notably by its poor infrastructure, inefficient banking sector and shortages of skilled labour.

In 1996 the Government embarked upon its second Five-Year Plan. This was intended to achieve a 6% rise in the GDP growth rate. According to the SPDC, it had succeeded in reaching an annual growth rate of 8.4% by 2001. The previous Four-Year Plan (1992–96) had apparently achieved an average annual growth rate of 7.5%. Overall, economic growth had more than doubled since the end of the 1980s as the Government's short-term economic plans took effect, creating favourable conditions for Myanmar's economic development.

In March 2001 the Government implemented its third Five-Year Plan. The plan beginning in 2001 projected average GDP growth rates in excess of 8% until 2005/06. The main aims of the plan were: to promote the establishment of agro-based industries; to develop the power and energy sectors; to expand the agriculture, meat and fish sectors in order to meet local demand and provide a surplus for export; to establish forest reserves; to extend the health and education sectors; and to develop the rural areas. The plan was also considered to be the first Five-Year Plan for the decade 2001/02–2010/11. The objective of the decade plan was to double GDP growth within 10 years through the achievement of an average annual growth rate of 7.2% during the plan period. At the same time per caput GDP was projected to increase at an average annual rate of 5.1%. However, given Myanmar's recent economic performance and outlook, the decade plan seemed very ambitious.

AGRICULTURE, FORESTRY AND FISHING

Agriculture (including livestock, forestry and fishing) remains a major sector in the Myanma economy, contributing 57.2% of GDP in 2001/02, according to the ADB. In 2002/03 the sector contributed 53.3% of GDP. Approximately 70% of the population are engaged in farming. Exports from this sector accounted for more than three-quarters of total exports in 1995/96. Agriculture is thus central to the performance of the whole economy. The performance of the sector improved markedly after it was accorded increased priority in the economic reforms of 1971/72, although the sector has still frequently failed to achieve the targets set. The agricultural sector expanded at an average annual rate of 6.6% over the period 1973–83, but it subsequently advanced more slowly, achieving a growth rate of 4.7% under the Fourth Plan (1982/83–1985/86). The sector was seriously affected by the economic and political disruptions of 1988 and 1989, with production declining significantly. Between 1990/91 and 1995/96 agricultural output increased in all but one year, and in the years from 1992/93 to 1996/97 the sector averaged annual growth of 5.0%. However, whilst the improvement of irrigation systems has been a central part of the Government's economic policy, other factors (such as continued vulnerability to adverse weather and pests, a shortage of agricultural inputs such as fertilizers, and government rice procurement policies of doubtful wisdom) have continued to constrain the sector. Another major factor previously undermining agricultural development was the Government's failure to provide adequate incentives, in the form of higher prices and support services, to farmers, although these have been improved. In 1997/98, compared with the previous year, agricultural GDP expanded by 3.7%; the rate of growth slowly increased to 4.5% in 1998/99 before benefiting from favourable climatic conditions to reach 11.5% in 1999/2000 and 10.9% in 2000/01, according to the ADB. According to official figures, real growth in agriculture fell sharply, to 3.4%, in 2001/02, largely owing to limitations in the production process, such as shortages of fertilizer and other agricultural inputs. The EIU indicated a similar trend (in calendar years), stating that agricultural output expanded at an average annual rate of approximately 6% between 1990 and 2000, before growth slowed to 3.9% in 2001 and to 1.2% in 2002.

Until the mid-1970s growth in agricultural production was achieved largely through a gradual increase in the area under cultivation. Since then, greater emphasis has been given to raising productivity through the use of improved seeds and modern inputs. Owing to a lack of funds for maintenance and the shortage of fuel to operate pumps, the area under irrigation actually declined by 7.4% between 1986/87 and 1991/92. A further expansion in irrigation was vital to efforts to promote greater use of modern inputs, especially fertilizer and pesticides. Between 1988 and 2002, according to the Ministry of Agriculture and Irrigation, the Government built 127 new dams at a cost of US $140.2m., bringing to 265 the total number of dams in the country. Some 1.92m. ha (19%) of Myanmar's arable land is cultivated through irrigation.

In 1995/96 an estimated 9.1m. ha of land were under cultivation, representing about 13.1% of the land area, and slightly less than one-half of the land available for cultivation. By 2002 this figure had risen to 15.0m. ha. While this suggests that there is ample scope for expanding the land frontier, much virgin land is remote from the major centres of population and would require considerable investment in infrastructure to bring it into production. The increase of area under cultivation has not kept pace with population growth, resulting in a decline in the availability of cultivated land per caput, from 0.56 ha in 1940/41 to only 0.20 ha in 1992/93. By 2002/03 this area had increased marginally, to 0.21 ha. However, there has been a rapid increase in the cultivation of non-paddy crops, such as pulses and beans (which by 1997 occupied 14% of cultivable land), sesame (10%) and sugar (0.7%). By 2002/03 the area under pulses and beans had increased to 25.1% of total cultivated land, and that under sesame had also increased, to 13.5%.

Rice

Rice dominates Myanmar's economy. It is the main source of employment as well as an important export earner. It is also the staple food, providing the bulk of the calorie and protein intake for the population. The total area under paddy in 1995/96 was estimated at 6.2m. ha, 47.5% of all land under cultivation. Of this, some 1.3m. ha were planted with summer paddy, and 4.9m. ha with the monsoon crop. By 1999/2000 paddy production had increased to 20.1m. metric tons, compared with 14.1m. tons in 1986/87. According to official statistics, in 2001/02 Myanmar produced 22.3m. tons of rice. The Government intended to cultivate 6.8m. ha of paddy with the aim of producing 28.2m. tons of rice in 2002/03; rice exports were forecast to reach 1m. tons. However, according to the EIU, the country produced only 21.6m. tons of rice in that year.

The increase in production was particularly marked from the late 1970s, reflecting the success of government efforts to promote the use of high-yielding varieties (HYV). An experimental 'Whole Township Special High Yield Variety Paddy Cultivation Programme', introduced in 1975/76, was gradually extended, and by 1987 the programme covered 82 townships. While extension and other support services are concentrated in these areas, the use of HYV has also spread to areas outside the programme. Government data record the use of HYV on 60% of the total paddy area, providing 70% of production, in 1992/93. The average yield of HYV paddy in 1991/92 was 3,300 kg per ha, compared with 2,900 kg per ha for paddy as a whole; both figures, however, are low by regional standards. Paddy accounted for an estimated 74.8% of the irrigated area, 82.9% of fertilizer use in Myanmar and most of the pesticide use in 1995/96.

Until 1986 the State, working through the Agricultural and Farm Produce Trade Corporation (AFPTC—renamed the Myanma Agricultural Products Agency in 1989), was responsible for procurement and distribution of paddy and rice. After 1986 the Government began to liberalize the marketing of paddy and rice, first by allowing co-operative societies to operate alongside the AFPTC and then, in September 1987, by opening the trade to the private sector. This was followed in January 1988 by the ending of the State's monopoly on rice exports. These policies were intended to increase both production and exports. However, their introduction coincided with a decline in production and a period of rapidly rising rice prices that helped to fuel the growing unrest in urban areas during 1988. As a result, after 1988, state procurement, albeit in competition with the private sector, was reintroduced. Although state procurement prices were increased, they were still below those offered by private traders. However, because the state marketing organizations were able to provide fertilizer and credit at subsidized prices, they were still able to attract supplies from farmers. In 1990/91 it was estimated that the State purchased around 15% of total paddy production, while co-operatives purchased a further 7%. In 1997 a quota system, under which state agencies procured paddy from farmers at low prices, was still in operation. However, in late 1997, prices paid to farmers by the Government were raised, and a relaxation of the state monopoly on rice exports was attempted.

Given the country's acute shortage of foreign exchange, the SLORC sought to boost the level of rice exports, although priority was still being given to ensuring an adequate domestic supply of rice in an attempt to reduce pressure on rice prices. Owing to unreliable harvests, however, the Government achieved varying degrees of success in its attempt to raise rice exports. In 1994/95 rice exports exceeded their 30-year record level of 1m. tons, reaching almost 1.2m. tons, mainly owing to the expansion of rice production in the summer months. In seeking to expand its exports of rice, Myanmar was faced with strong competition from other exporters, especially Thailand and the USA and, from 1989, Viet Nam; Myanmar was also hampered by the low quality of its rice, although this had improved as a result of a programme, completed in 1985, to modernize the country's facilities for the storage and processing of rice. Exports of rice dwindled from late 1997 onwards, owing to a burgeoning population which consumed the majority of the domestic supply, and also because of ongoing smuggling through Yunnanese traders. According to the CSO, rice exports fell to a record low of 28,300 tons in 1997/98. In December 2000 rice prices declined to an estimated 50% of their 1999 value, owing to unfavourable market conditions, causing rice exports to reach only 55,000 tons in 1999/2000, before increasing to reach 939,500 tons in 2001/02. In late 2002 a massive increase in rice prices, precipitated by a production shortage, threatened to contribute to rising levels of starvation amongst the Myanma population. As a result, Myanmar exported only 793,500 tons of rice in 2002/03, owing to a decline in output and yield as a result of a lack of fertilizer and quality seeds and of flooding in mid-2002. Production of rice was expected to increase in 2003/04, since the Government announced in May 2003 that it intended to relax its 40-year-old paddy procurement policy.

Other Crops

Besides paddy, Myanmar produces a wide range of other crops, including maize, pulses, groundnuts, sesame seed and the so-called 'industrial crops', jute, cotton, sugar cane and tobacco. These are produced primarily for domestic use, although pulses and sesame seeds are important export commodities, and small quantities of maize, rubber, jute and tobacco have been exported in recent years. Both the area and production of these crops have increased steadily since the mid-1970s, with the exception of rubber. In an attempt to raise production levels, the Government increased the size of the area under rubber plantation from 81,000 ha in 1994 to 182,250 ha by the end of 2001. A total of 2.6m. ha are reportedly suitable for rubber cultivation. According to the UN Food and Agriculture Organization (FAO), rubber production was 27,000 metric tons in 2000/01 and 23,000 tons in 2001/02, most of which was exported. In 2001, according to official statistics, Myanmar exported 18,400 tons of raw rubber, a 29.2% decrease in comparison with 2000. Export earnings reached US $9.0m., a decline of 33.4% from 2000. In the same year the country imported rubber goods worth $34.8m., 13.0% less than 2000.

The Government has tried to encourage the use of improved varieties on all non-paddy crops; during the Fourth Plan, the

'Whole Township Programme' was extended to include 20 other crops, and particular attention was given to those crops with export potential, or to those crops which could be substituted for imports. The SPDC adopted the 'four pillars' policy in order to boost production of four key crops—rice, pulses, cotton and sugar cane. After paddy, sesame seed, groundnuts, maize, sugar cane and beans and pulses are the most important crops in terms of the area cultivated. The prospects for exports of maize, beans and pulses in the 1990s were favourable as demand for these commodities remained high in neighbouring countries and Japan. Jute production increased in the early 1990s, from 12,363 metric tons in 1990/91 to an estimated 43,000 tons in 1995/96. According to FAO, jute production totalled 33,000 tons in 1999/2000 and 2000/01, before increasing to 42,000 tons in 2001/02 and 2002/03. Tobacco is produced primarily for the local market, but small quantities of Virginia are exported. Production of tobacco was 47,000 tons in 2000/01 and 48,000 tons in 2001/02 and 2002/03. Since the Government does not seek to control the prices of non-rice crops, decisions with respect to the planted area of such crops have been determined more by relative prices on the private market than by government plans and targets; efforts are now being made, however, to design cropping patterns that are beneficial to both farmers and the State, as well as being suited to local conditions.

Livestock and Fisheries

The livestock and fisheries sector has increased rapidly in recent years, achieving an annual growth rate of 6.4% under the Fourth Plan. Following a poor performance between 1986/87 and 1990/91, the sector recorded a growth rate of 7.2% in 1991/92, and a further average annual increase of 5.8% between 1991/92 and 1995/96. In 1996/97 the sector achieved a growth rate of 9.7%. The rate of growth of the sector increased to 11.3% in 2001/02. Continued growth is regarded by the Government as an important means of expanding domestic consumption, thereby improving nutrition, and, in the case of fisheries, expanding exports. In addition, the livestock sector is vital to the development of agriculture, since the vast majority of farmers continue to rely on animal draught power. The number of draught animals was estimated at 6.8m. in 1995/96 and 13.7m. (cattle and buffalo) in 2001/02.

Fish is the major source of animal protein in the Myanma diet. In 1985/86 consumption per caput amounted to 11.02 viss (1 viss = 1.63 kg or 3.60 lb). By contrast, meat consumption was only 3.32 viss per caput. Owing to shortages, meat prices are high, and meat is therefore regarded as a 'luxury' food. There was, however, a particularly rapid increase in chicken and duck production from the late 1970s. Total meat production increased by an estimated 27.2% between 1992/93 and 1995/96, from 1.2m. viss to 1.5m. viss, while milk production increased by only 4% over the same period, having doubled between 1979/80 and 1986/87. In 1998/99 total meat production (including the production of poultry meat, output of which reached an estimated 144,000 metric tons) increased to an estimated 390,000 tons. In the same year the production of cows', buffaloes' and goats' milk totalled an estimated 588,000 tons. In 1999/2000 total meat production was approximately 437,000 tons (of which poultry meat constituted an estimated 177,000 tons) and milk output was 604,000 tons. In 2000/01, according to FAO estimates, 444,000 tons of meat were produced (including 198,800 tons of poultry meat), and 744,000 tons of milk. According to the CSO, in 2001/02 total meat production was 316.5m. viss, or about 515,957 tons, and total milk production (including buffalo milk) was 466.6m. viss, or 760,600 tons.

Since the mid-1970s significant efforts have been made to promote the fisheries sector, particularly offshore marine fisheries. Considerable donor assistance has been given to offshore fisheries, and production has increased significantly since 1976/77. In contrast to the livestock sector, a primary development objective of the fisheries sector is to increase exports. Export earnings from the fisheries sector have increased rapidly, and the prospects for further expansion in the sector are favourable. In 1993/94 government figures suggested that the total fish catch was 922,500 metric tons, of which 75% came from the sea. In 1994/95 output of marine products increased by 1.7%, compared with 1993/94, although export earnings from the fisheries sector increased by 57%. Although exports of fish and fish products declined in 1995/96, exports of fresh and dried prawns continued to show strong growth, reaching an estimated 392.3m. kyats (equivalent to 9.7% of the value of total exports). Exports of fish and fish products recovered in 1996/97, totalling an estimated 227.8m. kyats in that year. Exports of fresh and dried prawns increased from 518.7m. kyats in 2001/02 to 623.4m. kyats in 2002/03. Meanwhile, the value of exports of fish and fish products grew from 309.9m. kyats to 466.2m. kyats over the same period, according to the CSO. According to FAO, the total fish catch rose from 912,100 tons in 1997/98 to 1,010,500 tons in 1998/99 and 1,168,600 tons in 1999/2000. In 2001/02 the total fish catch was 890.7m. viss (or 1,451,866 tons), of which marine products accounted for 70.4%.

From early 1989 several agreements were signed allowing foreign trawlers, including those from Thailand, Malaysia, Japan and the Republic of Korea, access to Myanma waters. Future development prospects could now be adversely affected by overfishing. In mid-1993 the SLORC announced the cancellation of fishing rights granted to Thai firms. A fresh attempt to revitalize the fisheries sector was made by the Government in August 1994, when it announced the dissolution of the state-owned Myanma Fisheries Enterprise, whose work was assumed by the Department of Fisheries. The aim was to attract greater private-sector participation.

Forestry

Prior to the Second World War, teak was the third most important export commodity after rice and petroleum. Production was severely disrupted by the war, and, after the attainment of independence in 1948, by insurgency. Teak production in 1970/71, at 291,000 cu tons (each of 50 cu ft or 1.416 cu m), was still below pre-war levels (447,000 tons in 1939/40), while teak exports were almost one-half of their pre-war level. None the less, teak exports still accounted for 25% of export earnings in that year, and, with rice exports continuing to decline, the development of the forestry sector was identified as one of the most important means of increasing foreign-exchange earnings. Increased investment resulted in a rise in teak production to 436,000 tons in 1981/82, but production declined to 291,000 tons in 1988/89 before rising to a post-war peak of 440,000 tons in 1990/91. Owing to conservation measures (see below), teak production declined steadily thereafter, to an estimated 220,000 tons in 1996/97. Production of other hardwoods totalled 1,840,000 tons in 1992/93 (most of which was for domestic consumption), but subsequently declined to an estimated 1,109,000 tons in 1995/96. Forest products accounted for an estimated 22% of exports in 1994/95, compared with 32.1% in 1993/94. Nearly 70% of such exports were derived from teak. In 1994/95 forestry contributed around 5% to GDP. Since then the forestry sector has accounted for about 1% of GDP annually. The value of exports of teak and other hardwood was estimated to have declined from 1,060.9m. kyats in 1994/95 to 1,048.5m. kyats in 1995/96, but increased to an estimated 1,140.9m. kyats in 1996/97. However, having increased to 1,880m. kyats in 2001/02, the value of exports fell to 1,874m. kyats in 2002/03. Timber production declined by an annual average of 4.9% between 1992/93 and 1996/97. In 2001/02 total extraction of teak and hardwood increased by 17.6%, from 1,201,100 cu tons in 2000/01 to 1,411,980 cu tons. Extraction of teak grew from 250,500 cu tons in 2000/01 to 276,067 cu tons in 2001/02, while extraction of hardwood increased from 950,600 cu tons to 1,135,821 cu tons over the same period. In 1998 exports of roundwood were estimated at US $137.2m., compared with $98.1m. in 1997. Roundwood removals were estimated at 38.08m. cu m in 2000/01, at 39.37m. cu m in 2001/02 and at 40.94m. cu m in 2002/03.

The illegal felling of teak, and its smuggling into Thailand, has been a major problem. In an effort to ease the country's acute shortage of foreign exchange, the then BSPP Government announced an end to the 40-year ban on cross-border trade in teak in early 1988. From September 1988 onwards the SLORC made strenuous efforts to boost foreign-exchange earnings and its foreign reserves by signing a series of logging contracts with Thai companies. This resulted in a significant increase in the rate of felling, particularly for hardwoods other than teak, mainly in areas along the Myanmar–Thailand border. The increase in the rate of timber extraction gave rise to growing concern about over-felling and likely adverse environmental impacts. This led to the announcement in January 1990 of an 80% reduction in the level of timber exports allowed by the Thai companies and a statement that no new timber contracts would be allowed. In June 1990 export fees on teak and other hardwoods were increased by more than 100%. However, policing such regulations was difficult. Many of the concessions awarded to Thai companies were in areas affected by the long-running insurgency. Measures taken to prevent rapid deforestation were undermined by the SLORC's initiation of agreements with insurgent groups in border areas, which provided for the rebels' cross-border trade to be legalized in exchange for truce arrangements with government forces. A logging ban was reimposed in 1994. Deforestation was held to be partly responsible for the serious flooding that affected the country in mid-1997.

In the more medium term there is a crucial need to reduce the export of raw logs and to increase the value added from local processing, both as a means of increasing export earnings from forestry and ensuring a more sustainable development of the sector. At 1990 felling rates, it was estimated that the country's teak resources could be exhausted in as little as 15 years. In mid-1993 the SLORC cancelled 47 Thai timber-logging concessions in an attempt to slow the pace of deforestation. It also launched a new, three-year reafforestation programme (with an annual target of some 50,000 ha). In 1996/97 an estimated 48% of Myanmar was still under forest cover. The area of reserved forest increased from 10.2m. ha in 1992/93 to 10.4m. ha in 1995/96. In the early 21st century Myanmar's forest covered 50% of its total land area, 7% less than in 1962. To overcome teak shortages, Myanmar launched a special plantation plan in 1997 and has been able to grow more than 32,400 ha of teak. Of the total forest area, 18.6% was protected, with the target due to increase to 30%. Myanmar has since planted more than 32,400 ha of teak. According to its decade plan, the SPDC plans to extract 301,700 cu tons of teak and 1,044,500 cu tons of hardwood annually by 2010/11.

MINING AND PETROLEUM

Myanmar has a rich natural resource base which offers considerable potential for long-term developments. Mining was another sector accorded a higher priority under the 20-Year Plan. This primarily reflected the need to increase petroleum production to meet the country's development needs. Emphasis was, however, also given to expanding the supply of mineral raw materials for local industry, and to raising the level of mineral exports. Although production targets were not always achieved, the sector grew rapidly from the mid-1970s onwards, recording an annual growth rate of 8.2% under the Third Plan and 12.8% under the Fourth Plan (the latter representing the second-highest growth rate of any sector during the Plan period). In the final year of the Fourth Plan, output was provisionally estimated to have risen by 21.9%, but the increase was later adjusted to 9.9%. Since then the value of production has often declined in real terms, owing to a lack of capital and technology and the continuing shortage of fuel. In 1991/92 production was about 20% below the level of 1985/86. However, mining output has increased considerably since the introduction of the Myanmar Mining Law in 1994, which granted concessions to foreign mining companies. By mid-1998 12 foreign mining companies had signed joint-venture agreements, and the production of oil, gas, zinc, lead, gold and gems had increased substantially. In 1998/99 the mining sector provided 0.9% of GDP, decreasing to 0.5% in 1999/2000 before increasing marginally to 0.6% in 2000/01 and declining again, to 0.5%, in 2001/02. In 1997/98 the sector accounted for 0.7% of employment. In 1996/97 the output of the sector expanded by 13%, largely as a result of foreign investment (which totalled US $178.3m. in that year). By 2002 total foreign investment in Myanmar's mining sector had amounted to $522.5m. since 1988. The Lepadaung project in Monywa, the country's north-western Sagaing division, is expected to become one of the world's largest sources of copper production. US firm Ivanhoe, which has invested $150m. in a joint venture with the Myanma Mining Enterprise to operate the mine, estimates that Lepadaung is about five times larger than the other two deposits at Sabetaung and Kyinsintaung.

Although petroleum was a principal export commodity prior to the Second World War, the industry was all but destroyed by the hostilities. While efforts were made to rehabilitate the industry in the post-war period, these proceeded only slowly, and petroleum imports were still required during the 1970s. An independent study carried out in 1995 estimated that (contrary to official statistics) Myanmar is still able to meet only some 50% of domestic fuel requirements. Once it had acquired improved access to both capital and foreign technology, the Government was successful in increasing annual production from the 6.8m. barrels recorded in 1974/75, to 11m. barrels in 1979/80; in that year 1m. barrels of petroleum were exported to Japan. In 1995/96 production was estimated at only 6.9m. barrels, decreasing steadily to 3.39m. barrels in 1999/2000, before increasing to 3.54m. barrels in 2000/01. Production increased to 4.70m. barrels in 2001/02 and to an estimated 4.92m. barrels in 2002/03, according to the US Geological Survey. During 2001 petroleum imports totalled US $217.6m. By the early 2000s total foreign investment in the oil and gas sector since 1988 had reached $2,600m. from at least 34 contracts, which included exploration projects being conducted in 47 inland blocks and 15 joint-venture contracts in the same undertakings at 25 offshore blocks in the Mottama, Tanintharyi and Rakhine coastal areas. Growth continued in 2002; Myanmar produced a total of 820,000 barrels of petroleum in the first two months of the year, an increase of 12% compared with the corresponding period of 2001. The BSPP Government followed a policy of not allowing petroleum imports. However, this was only possible as a result of severe restrictions on consumption; petrol was rationed and there was a thriving 'black market', a situation that continued under the SLORC. The rationing of petrol was relaxed in August 1997, however, in an attempt to limit the 'black market'. Foreign involvement in the petroleum and natural gas sector has been problematic. The difficult operating environment, combined with the failure to locate commercially viable deposits, has prompted a number of foreign companies to cease exploration, notably the US oil company, Amoco, in March 1994. In 1995–96 new production-sharing contracts, both off shore and on shore, were awarded to Australian, Japanese and US companies, amongst others. However, the Myanma Oil and Gas Enterprise, a state corporation established to deal with foreign investors, proved to be a weak and inexperienced ally for the new shareholders, causing them to abandon their expectations of reasonable profits. Baker Hughes, the largest operator in central Myanmar, was among seven companies that withdrew from the country or became inactive in 2000.

Prospects for natural gas exports to Thailand were excellent throughout the 1990s until the sharp decline in the value of the kyat in 1998. In 1994/95 and 1995/96 natural gas production was 45,599m. cu ft and 69,540m. cu ft, respectively; in 1996/97 production declined slightly to 65,700m. cu ft. Production in 2000/01 reached 8,477m. cu m, while 5,608.6m. cu m of natural gas was exported, generating US $523.1m. in foreign exchange. In 2002/03 Myanmar produced an estimated 8,500m. cu m of natural gas. Most of the output came from wells on the west bank of the Ayeyarwady (Irrawaddy) river in central Myanmar. A further major impetus to natural gas production and exports of liquefied natural gas (LNG) was anticipated, following the discovery, in 1982 and 1983, of major offshore fields in the Gulf of Martaban. The country's production of liquefied petroleum increased from 4.3m. gallons in 1997/98 to 6.2m. gallons in 2002/03. In September 1994 the Myanma and Thai Governments concluded an agreement, under which Thailand was to purchase around $400m. worth of natural gas annually from Myanmar for a period of 30 years, starting in 1998. To this end construction of a 670-km pipeline to transport natural gas from the Gulf of Martaban to Thailand began in 1996. Partners in the project included Total of France (which held 31.24% of shares) and Unocal of the USA (28.26%), as well as the Petroleum Authority of Thailand (25.50%) and the Myanma Oil and Gas Enterprise (15.00%). In 1997 the project accounted for about one-third of all foreign investment committed in Myanmar. Production was scheduled to start in July 1998, but was delayed owing to environmental and human rights protests on the Thai side of the border. Production delays on the Thai side from 1998 delayed natural gas purchases until 2000, when the first gas began to flow into the Ratchaburi power station; sales were only $50m., instead of the expected level of $400m.

A major impetus to economic growth in 2002/03 was expected to be the additional exploitation of the Yadana and Yetagun offshore gas fields, which were approaching their targeted production levels in 2002. The main destinations for Myanmar's gas are neighbouring Thailand and other South-East Asian destinations. In February 2002 Daewoo International of the Republic of Korea announced that it would begin exploration on its Block A-1 offshore concession near the north-west coast, where estimates place reserves at 10,000,000m. cu ft of gas. In the following month the Petroleum Exploration and Production Company of Thailand (PTTEP) announced that its Yetagun II offshore gas project in the Andaman Sea's Gulf of Martaban would go on stream in October 2002 at approximately 215,000m. cu ft per day. A sub-sea pipeline was to carry the gas produced to Myanmar for processing. Yadana, a second Myanma gas project in the Andaman Sea to provide output to Thailand, was expected to produce 525,000m. cu ft per day within five years. In 2002 gas was Myanmar's major export earner. It generated US $872m. in export revenue and accounted for 29% of total exports, making it by far Myanmar's most important source of export revenue.

There was also scope for expanding onshore gas production and this was to receive increased priority. In the late 1990s total onshore reserves were estimated to be 1,100,000m. cu ft. To ease fuel shortages, the Government has sought to encourage greater use of natural gas as an alternative fuel, and production of compressed natural gas increased by over 50% between 1988/89 and 1991/92. According to official reports, Myanmar has about 400,000m. cu m of exploitable offshore natural gas reserves. Myanmar started using natural gas in a small number of government vehicles and buses in 1986. The consumption of gasoline (petrol) in Myanmar rose to 440,280 kilolitres in 2003 from 209,610 kilolitres in 1993, while the consumption of diesel fuel almost tripled, to 1,510,650 kilolitres, during the same period. Myanmar imported US $200m. of fuel from Malaysia in 2003 and planned to import a further $250m. of fuel from the same source in 2004. Although further foreign investment in this sector is crucial, its realization is not secure, particularly following the decision by the USA in 1997 to ban new investment by its nationals in the country. Indeed, in late 1997 the US company, Texaco, sold its gas interests in Myanmar, in response to shareholder and other pressure linked to the issue of human rights in the country. In late 1996 a lawsuit was brought against Unocal, the major company involved in pipeline construction in Yadana, because of its alleged direct responsibility for human rights abuses in Myanmar. In July 2000 a US federal judge agreed to hear the case against Unocal but, following a trial in September 2000, the company was exonerated of all charges of unlawful conduct. In September 2002 the British company, Premier Oil, announced that it was to dispose of all its interests in Myanmar. While claiming that its decision was made purely for commercial reasons, the company was thought to have been influenced by a campaign against it led by human rights activists. In November 2003 a consortium of South Korean and Indian companies, comprising Daewoo International (which held a 60% stake), the Korea Gas Corporation (10%) and the Indian public-sector firms the Oil and Natural Gas Commission (ONGC) Videsh Ltd and the Gas Authority of India Ltd (GAIL—30%), commenced exploration of the waters off the Arakan Coast and produced gas at a rate of 32m. cu ft per day by drill stem test. Production was to start in 2009 and Daewoo International predicted that at least $86.2m. in net annual profit would be generated for 20 years, commencing in 2010, from the exploitation of natural gas in the zone. A feasibility study on the construction of a gas pipeline through Bangladesh connecting Myanmar's gas fields with India has already been completed. In August 2004 the PTTEP was awarded a production-sharing contract permitting exploration of Blocks M-3 and M-4 in the Gulf of Martaban, about 200–300 km south of Yangon. The Blocks, in which the PTTEP has a 100% interest, are gas prone, sharing basic characteristics with the Yetagun gas-condensate field and the Yedana field, both of which are producing fields. The PTTEP has committed to spend $13m. over the first four years of the contract term and to conduct a geological and geophysical study, seismic surveys, and the drilling of at least one exploration well over the two tracts, which cover a total area of 18,000 sq km.

Myanmar possesses a range of other important minerals that are commercially exploited. These include coal, tin, tungsten, gold, copper, lead, zinc, silver, barytes and gypsum. Following increases during the early 1980s, the production of many of these minerals declined after 1985/86, again largely owing to the shortage of fuel, foreign exchange and new technology. Gold production, however, has increased significantly since 1988/89, following the completion of a new gold-mining and -processing centre at Bawdwin, north-east of Mandalay. Myanmar's refined gold production in 1999/2000 was 6,145 oz. This declined to 3,619 oz in 2000/01. Myanmar also possesses commercially exploitable quantities of jade and gems. In the first half of 2001 Myanmar's sapphire production reached 2.643m. carats, a decline of 5.2% compared with the corresponding period of 2000. Ruby production increased by 13.2%, to 1.184m. carats, during the same period, while production of jade reached 69.645 metric tons.

INDUSTRY AND ENERGY

In spite of the priority given to the development of the agricultural, forestry and mining sectors under the 20-Year Plan, the share of public investment allocated to the industrial sector actually increased after 1972. From an average allocation of around 13% during the 1960s, the share increased to around 30% in the 1970s, with the industrial sector receiving the largest single share in all but a few years. Although industry's share subsequently declined, it still continued to receive one of the largest allocations of public investment through the 1980s. However, since 1990/91 its share has declined significantly and in 1992/93 the sector received only 5.7% of estimated public investment expenditure, reflecting the increased priority being given by the Government to construction, transport and communications and social services. In contrast to earlier policies, which aimed to promote the development of heavy industry, after 1972 emphasis was placed on the development of agro-based and resource-based industries, ensuring an effective link between industry and primary production, and on light consumer goods industries. At the same time, greater emphasis was given to improving the rate of utilization and efficiency of existing plants. While these remain key elements of industrial policy, after the mid-1980s greater emphasis was given to the development of export-orientated manufacturing industries. The increased level of investment after 1972, as well as reforms in management and accounting practices, and the introduction of incentives, meant that the industrial sector grew at a far greater rate than in the period 1962–72 (when growth of 2.8% was recorded), although it has still generally failed to reach plan targets. Since the mid-1980s the performance of the sector has been seriously affected by the shortage of foreign exchange and energy (see below), production being further disrupted by the political unrest of 1987/88 and 1988/89. The estimated value of production in 1991/92 (in constant prices) was still some 18% below that of 1985/86. Between 1992/93 and 1996/97 production increased by an annual average of 9.8% per year. Compared with the previous year, industrial GDP rose by an estimated 6.1% in 1998/99, by 13.7% in 1999/2000, by 20.1% in 2000/01 and by 14.7% in 2001/02. In 2001/02, according to the ADB, the industrial sector (including mining and quarrying, manufacturing, utilities and construction) contributed an estimated 10.5% of GDP. In 2002/03, according to official statistics, the sector's contribution to GDP had risen to 12.7%.

Despite the nationalization of all major industries after 1962, the private sector retained an important role in industry, and this has been increasing since the reforms of 1988. In 1992/93 the private sector accounted for 69% of industrial output, 94% of all industrial establishments and as much as 80% of employment. However, the majority of these establishments (over 90%) are small enterprises employing fewer than 10 persons. In 1995/96, of a total of 474 factories employing more than 100 workers, 399 were state-owned, 68 were privately owned and seven were run by co-operatives. In 1995, according to the ADB, state-owned enterprises accounted for about one-third of public investment but contributed only one-fifth of government receipts. Subsequent assessments suggest that there has been little, if any, improvement in the management and productivity of state-owned enterprises.

The economic reforms of 1971 recognized the role that the private sector could play in the development of the industrial sector, and some efforts were subsequently made to encourage increased private investment. A Rights of Private Enterprise Law was passed in 1977, which provided for guarantees against nationalization, but only until 1994. In mid-1994 the SLORC introduced new procedures for private investment, particularly in the manufacturing sector. The publication of a foreign investment law in December 1988, under which both 100% foreign-owned and joint-venture companies were allowed, was expected to lead to a significant expansion in industrial investment and production in the medium term. The sector remained dominated by food and beverage processing, which accounted for an estimated 85.0% of gross manufacturing output in 1995/96. Other principal activities included the production of industrial raw materials (5.4%), mineral and petroleum products (2.5%) and the manufacture of textiles and clothing (2.3%). While growth in agro-based industry has slowed, the rate of growth of the manufacturing sector has increased. However, the development of the manufacturing sector was hindered by the shortage of foreign exchange (which restricted the imports required by the sector), power shortages, a lack of key imported inputs, import restrictions, the opening of border trade with China, Thailand and India, and the country's poor infrastructure. While deficiencies in the sector, along with international concern over the abuse of human rights in the country, have discouraged foreign investment to some extent, foreign investment in light manufacturing increased in 1996/97.

Shortages of energy supply are a major constraint on the long-term development of the industrial sector. While energy reserves were more than adequate in the late 1980s, supply fell well short of demand, despite a 20% increase in generating capacity between 1985/86 and 1991/92. The average annual growth of energy production was 3.9% in 1980–89, compared with an average annual increase in consumption of 4.2%. Although 18 new steam and gas turbines were installed during the Fourth Plan period, the Government has continued to stress the expansion of hydroelectric power, and in 1985/86 a 56-MW hydroelectric plant was commissioned at the Kinda dam near Mandalay, and seven small-scale hydroelectric plants were constructed. In June 1989 the Sedawgyi hydroelectric power plant near Mandalay was commissioned. Thailand's National Energy Administration proposed the construction of seven dams on rivers along the border with Myanmar. At the end of 1989 Thai and Myanma officials agreed on the construction of the first two of these dams which would provide water for irrigation, as well as increasing hydroelectric power output. In April 1990 two new power projects, including a 6,000-MW hydropower plant near Yangon, were announced, with assistance provided by the Japanese private sector. Following an agreement with Thailand in May 1994, a number of smaller hydropower plants were to be constructed, including one on the Mae Sai river.

In 1994/95 the state-owned Myanma Electric Power Enterprise (MEPE) had an installed capacity of 845 MW, of which 48% was generated by natural gas, 35% by hydroelectric power, 9% by diesel generators and the remaining 7% by thermal energy. Power shortages in 1991/92 were aggravated by renovation work on one of the generators at the Lawpita hydroelectricity station and by low water levels in a number of other major hydroelectric dams. Lawpita, built in the 1950s by the Japanese, provided some 40% of Myanmar's electricity in 1990. Shortages of power continued to be a major problem in the 1990s, with the result that industry rarely operated at more than 50% of full capacity. In 1996 hydroelectric power projects were estimated to have the potential to add 1,000 MW to installed capacity, but it was not clear when this would be realized and some projects were unlikely to receive financial support. In 1999/2000 total installed electrical capacity was 1,162 MW, compared with 1,055 MW in 1998/99 and 1,042 MW in 1997/98. According to the CSO, total electricity production increased from 5,674m. kWh in 2001/02 to 6,614m. kWh in 2002/03. Despite this, Myanmar suffers from a serious shortage of electricity, and this has a significant impact upon its economic development and ability to attract FDI. Moreover, it is difficult for many people to receive a normal power supply, and some have to rely on their own power generators for domestic purposes. In December 2000 63.00% of electricity was generated by gas, 21.27% from hydroelectric power, 14.49% from thermal energy and 1.24% from diesel generators. The increase in capacity was due mainly to the commissioning of three new hydroelectricity plants, designed to extend electrification to areas not previously served by the national grid. Electricity prices, however, were reportedly increased 10 fold in March 1999. In September 2000 the Export and Import Bank of China announced a US $120m. low-interest loan for the completion of a plant near Pyinmana, which had been partially built by MEPE. The Yunnan Machinery and Equipment Import and Export Corporation was to supply equipment, design and construction supervision for the project, which would have a generating capacity of 105 MW and would increase power generation by as much as 30% upon completion, scheduled for 2003.

A total of 2,225m. kWh of electricity was distributed in 1988/89. A total of 32 new power plants—26 hydroelectric power stations and six gas-fired and combined recycle power plants—have been built in Myanmar since 1988, and power generation in 2000/01 exceeded 5,024m. kWh of electricity. Since 1988 the country's total installed electrical power generating capacity (IGC) had increased by 509 MW. The nation's IGC, according to the MEPE, remained at 1,172 MW at the end of June 2001. However, it increased to 1,570 MW in 2002/03. In May 2002 the Japanese Government announced a US $5m. loan for the Baluchaung No. 2 hydroelectric power plant. In the same month the MEPE announced plans for the construction of its first coal-burning electric power station, a 120-MW unit to be built in partnership with the Myanma Mining Enterprise in Pinlaung. In 2002 the MEPE announced plans to build five hydroelectric power plants—three of them with Chinese assistance—at Paunglaung, Zaungtu, Mone, Thaphanseik and Maipan. These were expected to assist in the generation of a further 2,000 MW of electricity within the next five years.

TRANSPORT AND COMMUNICATIONS

The poor state of the country's transport infrastructure was another factor undermining development efforts in the late 1980s and 1990s. Roads provide the most important means of transport for both freight and passenger traffic. In 2001/02 there were 28,598 km of roads accessible to motor vehicles in Myanmar, compared with 21,816 km in 1988/89. Many more roads linking the northern and southern regions or eastern and western sectors of the nation were under construction from 2002. These included: the Mandalay–Myitkyina highway on the west bank of the Ayeyarwady River; the Mandalay–Bhamo highway on the east bank of the Ayeyarwady; the Myitkyina–Putao highway; the Pa-an–Zathabyin–Moulmein highway; the Taunggyi–Ywangan–Hanmyintmo highway; the Pyinmana–Pinlaung highway; the Sittwe–An–Minbu highway; the Kawthoung–Bokpyin–Tavoy–Moulmein highway; the Taungup–Maei–Kyaukpru highway; and the Monywa–Hkamti highway. Most road freight is handled by the private sector and there has also been a significant increase in the number of privately owned passenger buses since 1988/89. The number of private haulage trucks increased from 23,753 in 1992/93 to an estimated 29,318 in 1995/96, whereas state-owned trucks declined from 2,044 to an estimated 1,969. Similarly, private passenger buses increased from 108,815 in 1992/93 to an estimated 154,853 in 1995/96, while state-owned buses declined from 1,082 to 951. The total number of trucks (both light and heavy duty) declined from 54,180 in 1999/2000 to 53,892 in 2000/01 and then to 53,541 in 2001/02. Meanwhile, the number of buses declined from 16,998 in 1999/2000 to 16,866 in 2000/01, but increased to 17,552 in 2001/02.

In 1996/97 Myanma Railways operated a network of 3,955 km, most of which was single-track. In the same year the network carried an estimated 53.4m. passengers and 3.3m. long tons of freight. As a result of the acquisition of new locomotives and rolling stock from abroad, both passenger and freight traffic has expanded since 1988/89. In 2001/02 the railways transported 61.3m. passengers and 3.4m. long tons of freight. Meanwhile, the number of passenger-miles travelled increased from 2,800m. in 2001/02 to 2,900m. in 2002/03. At the same time freight ton-miles increased from 720m. to 723m. The SLORC oversaw some upgrading and expansion of the rail network, including the construction of a new 160-km line from Ye in Mon State to Tavoy in the south of Tanintharyi Division. The latter project received widespread condemnation by international human rights organizations, which claim that forced labour squads were being used in its construction. Financial aid from the People's Republic of China and the involvement in projects of Chinese engineers have been of great importance in various projects undertaken to repair and extend both railways and roads. By 2002 the Government had spent more than US $398m. on railways since 1988 and had constructed new rail routes linking the following towns: Prome, Aunglan and Taungdwingyi, passing through the central Yoma mountain range; Pakokku and Kyaw, cutting through the Pondaung–Ponnya mountain range; and Shwenyaung, Taunggyi and Hsaik-kaung, which traverses the Shan–Yoma mountain ranges. The length, in terms of route mileage, increased slightly, from 2,907 miles in 1998/99 to 2,974 miles in 2001/02.

Air transport has been particularly affected by the shortage of foreign exchange, necessary for the purchase of spare parts and aviation fuel. As a result, both passenger and freight traffic declined in the early 1980s. Since 1988/89 there has been some recovery, at least in passenger traffic. Work on upgrading the country's one international airport, near Yangon, began in 1988, but was suspended, owing to aid restrictions. However, a new international airport opened near Mandalay in 2000. In April 1993 Myanma Airways International (MAI) was formed to operate international routes, leaving Myanma Airways as the domestic carrier. In September a private consortium took control of 70% of MAI in Myanmar's first large-scale privatization. In 1995 Myanma Airways formed a joint-venture airline, Air Mandalay, with Air Mandalay Holdings of Singapore, to cater for the anticipated increase in foreign tourist demand. In 1995/96 the two airlines carried an estimated total of 719,000 and 138,000 passengers and 2,800 short tons and 3,000 short tons of freight on internal and external flights, respectively. In January 2004 United Myanmar Air (UMA), the second international carrier to be based in Myanmar, postponed the launch of its international services (to Hong Kong, Malaysia, Singapore and Thailand) until the end of that year, owing to financial difficulties. UMA was founded in July 2003 as a joint venture between the state-owned domestic Myanmar Airways and the Hong Kong-based Sunshine Strategic Investment Holdings and its two local partners. Meanwhile, a new international airline, Air Myanmar, was to become the country's second international airline following the commencement of operations at the end of 2004. It planned to fly two extended routes: Yangon to Sydney, Australia, via Singapore; and Yangon to Fukuoka, Japan, via Bangkok, Shanghai, China, and Seoul, Republic of Korea. The airline intended to use two types of aircraft, the Airbus A300-600 and A310-300. The Air Myanmar Company Ltd was a joint venture between Myanmar Airways, the Dawn Light Company, the Hong Kong-based Cathay Aviation and First Growth Associates Ltd.

Although freight ton-miles declined significantly, from 906,000 in 1997/98 to 435,000 in 2002/03, the number of passenger-miles flown increased from 130.7m. to 161.0m. during the same period. According to official statistics, the number of aircraft decreased from 20 in 1980/81 to just 10 in 2001/02. By January 2002 Myanmar had 12 airports capable of handling jet aircraft; a further six airports were under construction. In April China's Yunnan Airlines became the first foreign carrier to commence commercial flights from Kunming to Mandalay, an initiative designed to attract more Chinese visitors to Myanmar.

In the mid-1990s inland water transport continued to provide an important link in north–south communications. There are some 12,800 km of inland waterways in Myanmar, including approximately 8,000 km of navigable rivers, of which the Ayeyarwady, the Chindwin (Chindwinn) and the delta creeks are the principal arteries. By January 2002 nine bridges spanned the Ayeyarwady (at Magwe, Dedaye and Mandalay), the Chindwin (at Monywa), the Sittang (at Shwekyin-Madauk), and the Salween (at Moulmein). The state-owned Inland Water Transport carried an estimated 3.2m. tons of freight and 24.5m. passengers in 1995/96. However, like road transport, most river transport was controlled by the private sector. In an attempt to overcome transport delays and to improve internal distribution and export trade, controls on the transport of goods to ports were removed, with effect from 1 June 1988. As far as public, inland water transport is concerned, freight ton-miles increased from 335,422 in 1997/78 to 370,872 in 2002/03 and the number of passenger-miles travelled also increased, from 433,956 to 480,733 during the same period. In November 2002 a Chinese engineering company announced its decision to construct a four-lane highway bridge across the Ayeyarwady River, providing better vehicular access to Myanmar's second largest city, Mandalay.

Efforts were being made to upgrade other forms of communications, especially telecommunications. In 1992/93 there were an estimated 89,318 telephones in Myanmar, approximately equivalent to one for every 474 persons. This had increased to 197,026 telephones by 1996/97, and by 2001 there were 295,200 main line telephones in use. In 1996 cellphone equipment was installed in Mandalay and Yangon, and a digital telephone system was in place in a number of towns. The completion of a satellite ground station in March 1994 resulted in Myanmar having international direct dialling with 40 other countries. In 1996/97 there were 227 telex services and 1,503 licensed fax machines in Myanmar. In 2002 there were 351,000 telephones in operation in Myanmar (11,000 of them CDMA mobile cellular phones) with 569 exchange stations, compared with 67,000 in 1988. In March 2002 the China National Electronics Import and Export Shenzhen Company signed an agreement with Myanma Posts and Telecommunications (MPT) to construct digital auto-telephone exchanges with a capacity of 13,500 lines in 12 towns across the country over the following 18 months, at a cost of US $6.5m. According to the World Bank, there were eight fixed line and mobile telephones per 1,000 people and 25,000 internet users in 2002.

The transport, storage and communications sector contributed an estimated 5.4% of GDP in 2001/02. By 2002/03, according to the Ministry of National Planning and Economic Development, this share had increased to 7.0%. The Government has given increasing priority to the sector, which was allocated 13.3% of public investment expenditure under the 1989/90 plan, although this declined to only 9.5% in 1992/93. In that year the private sector owned 58% of the transport facilities, while 40% was controlled by the state sector and the balance by co-operatives.

TOURISM

The exploitation of tourism potential has been regarded in recent years as one way of increasing foreign-exchange earnings. During the unrest of 1988 the Government stopped the issuing of tourist visas. The new Government began issuing visas again in May 1989, allowing tourists on 'package' tours to remain for as long as two weeks in the country, compared with the previous one-week limit. However, tourist arrivals, estimated at 11,430 in 1992, were substantially fewer than in the period prior to 1988 (arrivals had totalled 41,904 in 1987), although foreign-exchange earnings from tourism were at a record high (nearly double the 1990 figure). This was due to the higher cost of package tours and the related higher incomes and expenditure of recent tourist arrivals. The SLORC's aim was to attract some 250,000 tourists in 1996/97, which was officially declared 'Visit Myanmar Year'; however, this target was widely regarded as over-ambitious, particularly in view of the opposition's demands for an international boycott of Myanmar. The actual number of tourist arrivals in 1996/97, as reported by Myanmar Hotels and Tourism Services, was 179,594. Foreign-exchange receipts from tourism during 'Visit Myanmar Year' totalled 400.6m. kyats, according to the CSO. According to the World Tourism Organization, tourist arrivals rose to 189,000 in 1997/98 and to 201,000 in 1998/99, before declining slightly, to 198,210, in 1999/2000. However, arrivals rose to 207,665 in 2000/01, before falling to 204,862 in 2001/02. Of these, 8.2% were from China and 6.6% from the USA. Between April 2003 and January 2004 258,832 tourists visited Myanmar. The country hopes eventually to raise the annual number of foreign tourist arrivals to roughly 500,000. Revenue from tourism totalled a reported US $35m. in 1999/2000. According to the Union of Myanmar Travel Association (UMTA), in 2003 tourist arrivals numbered 205,610, generating US $116m. in tourism receipts. Since the late 1980s a number of agreements for the construction of new 'five-star' hotels and tourist-related facilities have been concluded with foreign investors. However, in recent years a combination of Western sanctions, Asian financial problems, stringent visa procedures and restricted air access has caused a huge downturn in the hotel trade. Hotels have been forced to implement drastic price reductions in order to survive, and several planned developments have been abandoned. In early 2001, in an effort to combat the decline, the tourism industry announced a new advertising campaign with a 'Mystical Myanmar' theme and lobbied the Government to make Myanmar more accessible to visitors. Funding issues, however, continued to create problems. By 2002, according to the Ministry of Information, total foreign investment in the country's hotel projects had reached $585.3m. A total of 25 new hotel projects were completed between 1993 and 2001. These included 19 in Yangon, three in Mandalay, and one each in Bagan, Kawthaung and Tachilek. In 2004, according to the UMTA, there were 10,275 employees working to service a total of 17,039 rooms at 563 hotels in Myanmar.

DOMESTIC COMMERCE, FINANCE AND PRICES

Much of the country's internal trading network was nationalized after 1963 as part of the military Government's programme of 'Burmanization' of the economy. However, a considerable proportion of the country's internal trade remained officially in private ownership. In 1985/86 the State controlled 46.5% of the internal trade sector, while the private sector controlled 41.3% and co-operatives 12.2%. Owing to the moves to liberalize both internal and foreign trade begun by the BSPP in 1987 and 1988 and continued by the SLORC, the share of internal trade controlled by the private and co-operative sectors has been increasing, particularly since the legalization after 1988 of many of the trading activities (including border trade) previously associated with the 'black market'. Thus, while the volume of internal trade handled by the state sector more than doubled between 1988/89 and 1992/93, the State's share declined to only 14% in the latter year. In July 1994 the Government announced that 3,815 registered limited companies, 311 foreign companies and 974 'partnerships' had been registered since 1988, with the majority involved in the retail trade. By 2001/02 there were 16,267 registered limited companies, 1,450 foreign firms and 1,269 'partnerships'. The ongoing shortage of consumer goods, including some basic necessities, resulted in the continuation of some 'black market' activities, with prices based on the free-market exchange rate. One method that the Government periodically employed to combat the 'black market' was the demonetization of large-denomination currency notes. This occurred twice in the second half of the 1980s: in November 1985 and September 1987. While these actions did cause some temporary disruption to 'black market' activity, their main effect was to undermine still further the population's faith in the Myanma currency and to increase the degree of dissatisfaction with the Government's management of the economy. Indeed, the demonetization measures of September 1987, which resulted in 60% of the currency in circulation being declared worthless with almost no compensation, set off the chain of unrest that culminated in the military coup of September 1988. In an effort to restore confidence in the currency, the SLORC strongly emphasized that there would be no further demonetizations.

From January 1993 private banks were permitted to operate, in a move designed to facilitate private business. From April 1994 local private banks were permitted for the first time to deal in foreign exchange. Moreover, in the early 1990s the SLORC issued licences to foreign banks, particularly from Thailand and Singapore, to open representative offices in Myanmar. In 1996 there were about 20 foreign banks with representative offices in Yangon. However, many restrictions remained regarding their activities, with the Government giving priority to the development of domestic private banks. Most foreign representative offices are primarily involved with arranging short-term trade finance.

In February 1993 the Government introduced foreign-exchange certificates (FECs) for use by overseas visitors. From December 1995 Myanma citizens were legally allowed to hold these FECs, which could be exchanged for kyat at a rate close to the unofficial rate of around 350 kyats = US $1; by 2000 SPDC policy obliged most private companies to pay for imported goods in FECs, and confiscated US dollars for the settlement of government foreign accounts.

One of the most serious financial developments to have threatened Myanmar's economy in recent times was the banking crisis that began in early 2003. In February more than a dozen deposit-taking institutions in Myanmar collapsed. As the deposits at these banks were unsecured, panic ensued. Mass withdrawals occurred at many of Myanmar's largest private banks, including the Asia Wealth Bank, Yoma Bank, Myanmar May Flower Bank and Myanma Oriental Bank. By March most private banks had suspended cash withdrawals and payments on credit cards that they had issued, as they were facing severe liquidity problems. In a bid to restore confidence in the banking system, the Central Bank reportedly provided around 25,000m. kyats (US $3,900m. at the official exchange rate and about $25m. at the free market rate) in financial assistance in an attempt to prevent the collapse of three private banks, the Asia Wealth Bank, Yoma Bank and Kanbawza Bank. This, in turn, contributed to higher consumer prices. According to official statistics, the inflation rate rose from 34.5% in 2001/02 to 58.1% in 2002/03. This sharp increase could be attributed to the rising money supply, the panic buying of essential goods (owing to the restrictions imposed on cash withdrawal) and rising food prices. In 2003/04, however, the rate of inflation was reported to have decreased to 24.9%. The banking crisis exacerbated an already deteriorating domestic economic situation. It was thought that this would create problems for Myanmar's fledgling industry sector, as investment expenditure contracted.

Since 1962 the Government has placed considerable emphasis on the maintenance of price stability; procurement prices for crops, especially for paddy, have thus been kept low. Inflation was officially put at only 2% per year in the 1960s, but increased significantly in the 1970s. The Yangon consumer price index (CPI) more than doubled between 1972 and 1975. It was subsequently more stable, although after 1986, when inflation was officially around 10%, the rate of inflation rose significantly. Rice prices increased rapidly following the liberalization of rice marketing in November 1987, and the unrest during 1988 aggravated the already widespread shortages of food and consumer goods. On a new base of 1986 = 100, the Yangon CPI stood at 349.3 by 1992 and at 455.3 in 1993. Officially inflation was estimated at around 22% for 1992, but official figures took no account of price movements in the 'black market'. Inflation continued to be fuelled by the rapid growth in money supply to enable the SLORC to fund the budget deficit, which was estimated to be equivalent to about 8% of GDP in 1990/91. The budget deficit was subsequently reduced, to approximately 2.5% of GDP in 1994/95, although monetization of the deficit continued. The budget deficit increased to 3.3% of GDP in 1995/96, but declined to 2.2% and 1.0% of GDP in 1996/97 and 1997/98, respectively, before increasing to 4.0% of GDP in 1998/99. The deficit fell to 4.9% of GDP in 2001/02, from 8.3% of GDP in 2000/01. Inflation reached an annual average of 23.8% between 1992 and 1996. According to official data, inflation stood at 30%–35% in 1995/96, rising to 40% in 1996/97; some analysts, however, estimated it to be considerably higher. As a result of harvest failure and increases in the prices of fuel and electricity, inflationary pressure remained strong in the late 1990s. According to the ADB, consumer prices in Yangon increased by 29.7% in 1997/98 and by 30.1% in 1998/99, before prices steadied somewhat, registering a smaller increase of 21.0% in 1999/2000. In 2000/01 consumer prices were estimated by the ADB to have fallen by 0.1% in comparison with the previous year, although prices were estimated subsequently to have risen by 40.1% in 2001/02 and by 43.5% in 2002/03 (although official statistics declared the figure for the latter year to be 58.11%). The budget deficit for 1999/2000 was 20,200m. kyats. With total expenditure originally projected to decline by some 37% compared with the previous year, from 591,200m. kyats in 1998/99 to 430,400m. kyats in 1999/2000, the budget reflected the need for fiscal retrenchment in the context of the deceleration in economic growth.

FOREIGN TRADE

The adoption of a more liberal policy towards foreign investment and trade in 1987 and 1988 represented a major shift in the country's economic policy which, since independence, had been designed to reduce the country's dependence on foreign markets, in an effort to minimize its exposure to fluctuations in commodity markets and to protect domestic industry. After 1962 the military Government sought, with renewed determination, as far as possible to pursue a policy of self-sufficiency, with minimum foreign contacts. A government monopoly was extended to embrace all foreign trade. However, the neglect of the trade sector, and the resultant dramatic fall in export earnings, meant that the country's economic performance was subsequently undermined by a shortage of foreign exchange. Thus, from 1972, rather than seeking to minimize foreign trade, the Government was compelled to give it first priority, by placing particular emphasis on expanding the level of the country's traditional primary exports.

Measures introduced by the SLORC after September 1988 to liberalize foreign trade contributed to the recovery in export earnings. These measures included an end to the state's monopoly on foreign trade, except for teak, petroleum, natural gas, pearls and gems. In addition, the Government legalized much of the previous 'black market' border trade. In November 1988 official border trade was initiated with China, with three trading points opened. In 1988–90 exporters were permitted to retain an increasing share of their export earnings. In 1989–95 the cross-border trade carried out by certain groups of insurgents was legalized in exchange for peace agreements with the SLORC.

Rice exports, amounting to over 3m. metric tons in 1940/41, had declined to a mere 262,000 tons by 1972/73. Rice exports subsequently increased in line with the Government's policy of promoting traditional exports, totalling 646,000 tons in 1985/86. In the three years from 1994/95 rice exports decreased in volume terms from 675,000 tons to 250,000 tons. Revenue from rice exports, meanwhile, declined sharply from US $198m. to $47m. Teak and other hardwoods have constituted a major source of foreign exchange since 1985. In 1992/93 the sale of forest products temporarily eclipsed rice, accounting for 32% of the value of total exports. In 1994/95 rice and rice products accounted for 21.6% of total exports, while pulses and beans accounted for 14.8%, teak and other hardwood 19.6% and rubber 2.3%. However, the contribution to total exports made by rice and rice products declined steadily over the following years, to a mere 0.6% in 1997/98. Rice constituted 2.4% of total exports in 1998/99, 1.6% in 2000/01, 4.4% in 2001/02 and 3.2% in 2002/03. Meanwhile, the contribution of pulses and beans to total exports increased to 27.1% in 1995/96, but had declined to 11.1% by 2001/02 and to 8.0% by 2002/03. In 2001/02, according to official statistics, pulses were the second largest source of export revenue, generating $286m. Exports of both rubber and teak and other hardwood increased in 1995/96 (to 20.9% and 3.6% of total exports, respectively); the export of teak and other hardwood continued to increase in 1996/97 (representing 21.8% of total exports in that year) but declined to just 15.6% of total exports in 1997/98 and to 11.8% in 1998/99. Although teak and hardwood exports staged a modest recovery in 1999/2000 and 2000/01, in 2001/02 teak and hardwood exports constituted 11.0% of total exports. By 2002/03 this proportion had declined to 9.4%. The export of rubber declined in both 1996/97 (when it accounted for 3.2% of total exports) and 1997/98 (to 2.1% of total exports). Rubber exports fell by 13.6% in 2002/03, to 22,000m. tons from 25,000m. tons in 2001/02. In 2000/01 rice exports increased by 389%, to 692,700 tons. Moreover, the CSO stated that Myanmar exported 230,000 tons of rice in the first two months of 2002—an increase of more than 220% compared with the corresponding period of 2001. However, in 2002/03 rice exports decreased by 18.4%, to 793,500 tons. Timber, including teak and hardwood, has become Myanmar's third most significant export after gas and agricultural products, with the country accounting for some 85% of the world's teak. In 2001/02 Myanmar exported 283,707 cu m of teak, a decline of 8.1% from 308,753 cu m in the previous year. In 2002/03 teak exports registered a 214% increase, to 429,300m. tons. In 2001/02 Myanmar also exported 404,407 cu m of hardwood, a reduction of 13.2% from 466,101 cu m in 2000/01. However, exports of hardwood increased by 7.8%, to 308,000m. tons. Earnings from timber exports totalled $280m. In 2001/02 the export of teak and hardwood, according to official statistics, accounted for 8.9% of total exports, making the sector Myanmar's third largest export earner. The industry received a boost in May 2002 when Myanmar signed a $1m. agreement to export 1,000 tons of teak to India—the largest timber deal ever concluded.

In recent years garment exports have increased, although the USA's decision in 1996 to cease purchasing from Myanmar had a serious, negative impact on exports from within this sector. Export earnings of garments in 2002/03 fell to 2,973.2m. kyats, from 2,985.3m. kyats in 2001/02. The imposition of US sanctions on Myanmar's exports from August 2003 would affect garment exports from 2003/04 onwards. Export earnings from the garments and textiles sector rose significantly, from 435.9m. kyats in 1997/98 to 3,785.3m. kyats in 2000/01, before declining gradually, to 2,973.2m. kyats, in 2002/03. In recent years the USA has accounted for around 65% of Myanmar's exports of garments and textiles. Even if the USA had not imposed sanctions, exports of garments and textiles would automatically be stopped due to the withdrawal of preferential quotas granted to developing countries such as Cambodia, Laos and Myanmar. According to the EIU, it has been alleged that garments are finished in Myanmar before being sent to other Asian destinations to be relabelled and sold to other countries. Myanmar conducts the majority of its external trade with neighbouring Asian countries. Trade levels with neighbouring countries were expected to increase following an announcement by the Ministry of Commerce in July 2002 that it had opened a total of 10 border trade points with China (at Muse, Lweje and Laizar), Thailand (at Tachilek, Myawaddy, Kawthoung and Myeik), Bangladesh (at Maungtaw and Sittwe) and India (at Tamu). In 2003 Thailand was the principal destination (accounting for 30.7% of the total) for exports; other major markets were the USA, India, the People's Republic of China and Japan.

Despite the rising trend in the volume of exports after 1972, the country experienced acute problems in the trade sector from 1981/82, largely as a result of falling prices for export commodities, increased import prices and, more recently, the decline in export volumes. Government figures show that prices for rice, teak, hardwoods and minerals fell by over 30% between 1981/82 and 1986/87. Export prices continued to fall subsequently. In contrast, the prices of imports, which consisted mainly of capital goods, equipment and industrial raw materials, have risen sharply since 1981/82, and particularly since 1985/86, increasing by about 35% between 1985/86 and 1991/92. As a result, Myanmar suffered a steady decline in its terms of trade, with trade in 2001/02 reaching only 41.7% of its 1985/86 level. The structure of imports has changed since 1999/2000, with imports of consumer goods accounting for the largest share of total imports. In 2002/03 consumer goods accounted for 39.6% of the total value of imports, followed by intermediate goods (35.5%) and capital goods (24.9%). In 1994/95 the value of imports increased to an estimated US $1,414.2m., and by 1999/2000 this figure stood at $2,400m., according to official estimates. In 2003/04 total imports amounted to $1,980m., according to Myanmar's Customs Department. The EIU estimated that imports in 2002, 2003 and 2004 totalled $2,200m., $2,400m. and $2,500m., respectively. Official statistics revealed that the total cost of imports increased from 14,336.1m. kyats in 1997/98 to 18,377.7m. kyats in 2001/02, before declining for the first time in 2002/03, to 14,910m. kyats, mainly owing to trade restrictions, shortage of foreign exchange and sluggish domestic demand. The principal source of imports in 2003 was the People's Republic of China (accounting for 29.5% of all imports); other major suppliers were Singapore, Thailand, Malaysia and the Republic of Korea. In 1999/2000, according to the CSO, Myanmar's foreign trade, including cross-border legal trade, totalled $3,800m., having declined by 6.9% compared with 1998/99. In 1999/2000 imports were valued at $2,600m. (of which 69.6% was accounted for by the private sector). Exports in that year amounted to $1,180m. (the value of the private sector was 72.4% of the total). In 2002, according to official statistics, exports surged by an impressive 27.5%, from $2,360m. in 2001 to $3,008m. in 2002. In local currency terms, exports increased by 16.5%, from 1,730.7m. kyats in 2001/02 to 19,955.1m. kyats in 2002/03. Over the same period, imports declined by 18.8%, from $2,852m. in 2001 to $2,315m. in 2002. This helped Myanmar's current account to record a rare surplus. During this period, the import value of intermediate goods, capital goods and consumer goods accounted for 37.6%, 24.8% and 37.5% of total imports, respectively. According to official statistics, in local currency terms imports declined by 23.3%, from 18,377.7m. kyats in 2001/02 to 14,910.0m. kyats in 2002/03. According to the Customs Department, in 2003/04 total exports amounted to $2,400m., while total imports were valued at $1,900m. The strong growth in exports, together with the decline in imports in 2002/03 and 2003/04 helped to push the balance of trade into surplus for two successive years.

A major obstacle to increased exports remained the country's exchange rate, which favoured imports at the expense of exports. The Government indicated that the exchange rate would not be adjusted, as this would have an inflationary effect. In June 1996, in order to increase revenue from customs duties and to simplify the tariff system, a new customs valuation rate of 100 kyats = US $1 was introduced. However, from June 2004 the commercial tax on imports was raised to a flat rate of 25%, from a variable rate of between 2.5% and 20%, without any warning or reason being given. In addition, tax on all imports was to be calculated at 450 kyats = US $1, rather than 100 kyats = US $1 for consumer items, 150 kyats = US $1 for televisions and 180 kyats = US $1 for roofing sheets. Some items, such as medicines, computers, fertilizers, pesticides, diesel and petrol, were exempted from the tax. In January 1998, in an attempt to stabilize the currency, the Government revoked the licences of Myanmar's 30 foreign-exchange dealers. In March the Central Bank suspended the rights of the 10 part-state-owned and private banks to engage in foreign-exchange transactions. In mid-2000 the 'black market' rate was around 400 kyats = US $1, and by September 2002 the kyat reached an all-time low of 1,200 kyats = US $1, prompting the Government to begin targeting 'black market' currency traders. However, in 2003 the 'black market' rate was around 990 kyats = US $1 and in August 2004 it remained at 860 kyats = US $1.

Myanmar developed increasingly close relations with ASEAN during the 1990s, and in July 1997 the country was admitted as a full member of the Association; Myanmar's trading position appeared likely to improve as a result of its membership of ASEAN, although the country's external trade suffered as a result of the loss of trade privileges with a number of Western countries. In 2003/04 exports to ASEAN were estimated at US $1,300m. (54.2% of total exports) and imports from ASEAN totalled approximately $1,000m. (52.6% of total imports). Thailand was the largest ASEAN importer of Myanmar's exports and Singapore was the most significant ASEAN source of Myanmar's imports. While the Government had imposed new restrictions on imports in 1998, hoping thereby to curb the outflow of foreign currency from the country, by 2000 the SPDC had relaxed those controls but imposed the use of FECs in order to achieve the same objective.

FOREIGN AID, INVESTMENT AND DEBT

After the 1972 economic reforms, foreign assistance increased rapidly, from only US $21.7m. in 1971 to $511.8m. in 1979/80. Between 1980/81 and 1985/86 the Myanma Government found that it was increasingly difficult to maintain the flow of aid and concessionary loans, and this was a major factor contributing to balance-of-payments problems from 1981 onwards. In 1986 total foreign assistance amounted to $413.8m., with commitments of $306.1m. from bilateral donors and $107.7m. from multilateral donors. Japan was by far the largest bilateral donor, providing some 73% of all bilateral aid in 1986 in the form of both extensive project and programme aid. The World Bank and the ADB accounted for the bulk of multilateral lending, providing 41% and 25% of multilateral aid in 1986 respectively. At the Burma Aid Group meeting in Tokyo, Japan, in January 1986, donors agreed to provide approximately $500m. a year over the period of the Fifth Plan (1986/87–1989/90), with about one-half of this sum donated by Japan. Since the Government had come to rely heavily on foreign aid to maintain the level of imports and public investment, the economy was severely affected by the boycott on aid, introduced by Japan and Western donors, following the political unrest in 1988. While Japan did recognize the new military Government in February 1989, it indicated that aid payments would be made only for existing projects. This was still considerable, however, as Japan was involved in five grant projects, their cost totalling 9,200m. yen ($66m.), of which 65% had been disbursed, and 19 loan projects totalling 125,000m. yen, of which only 20% had been disbursed. As a result of the aid boycott, the country received virtually no other aid in 1989–95. Following the release of opposition leader Aung San Suu Kyi from house arrest in July 1995, Japan reinstituted some limited aid programmes; in 1998 the Japanese Government agreed partially to resume official development assistance to Myanmar. Japan's bilateral aid increased from $16.1m in 1998 to $34.2m in 1999 and to $51.8m. in 2000, in which year Japanese bilateral aid accounted for 76% of total bilateral aid. Total multilateral aid flows into the country came mainly from UN agencies. Multilateral aid increased from $31.3m. in 1998 to $37.8m. in 2000. Total net official development assistance (ODA) stood at $72.1m. in 1998, $81.1m. in 1999, $106.8m. in 2000 and $126.8m. in 2001, according to the EIU. According to the World Bank, in 2002 aid per person in Myanmar totalled only $2.50. In general, however, there were severe problems with regard to foreign investment in the country: according to the IMF, of the $3,200m. of investment approved since the first reforms of 1989, only $1,300m. had been invested by 31 March 2000. Net inflows of foreign investment declined from $317.8m. in 1998 to $210.3m. in 2001 and $128.7m. in 2002. At 1 January 2003 approved foreign investment in Myanmar was $7,467.56m., including investment from those donors that had already withdrawn. Realized investment has been estimated at around 40% of total approved investment.

Despite the aid boycott, the Government continued to receive private credit, which increased the external debt burden. Concessionary loans accounted for almost 80% of disbursed debt in 1986, reflecting the Government's reliance on official development assistance rather than commercial borrowing. Even so, the debt-service ratio rose to almost 60% in 1987; in 1965 it had been only 4.5%. The status of 'least developed country', granted at the end of 1987, brought some relief to Myanmar's debt problems; most of the new aid that was announced during the first six months of 1988 was in the form of grants rather than loans. Following the suspension of aid payments, Myanmar stopped repayments on debt to its major creditors, Japan and Germany. While arrears have accumulated, the debt-service ratio, based on actual as opposed to contractual debt-service payments, declined to 11.1% in 1991/92. By May 1989 Myanmar was 11,600m. yen in arrears on repayments to Japan. In March 1990 Myanmar made a debt repayment of 3,500m. yen to Japan, which Japan later returned to Myanmar in the form of a debt-relief grant. In May 1991 Myanmar repaid 3,000m. yen, which was also subsequently returned as a debt-relief grant, under which Myanmar could purchase raw materials, machinery or spare parts from any country. In June 1991 the French Government announced that it would waive debts incurred by the country's Government in 1976. Myanmar's debt to China at the end of 1993/94 totalled US $73m., excluding several hundred million dollars worth of credits that had helped pay for imports of military equipment. In 1995 the USA secured an agreement with 'Paris Club' donors that a debt settlement with Myanmar would have to be reached via a multilateral route on the basis of an agreement with the IMF. Total external debt was $5,063m. in 1997, compared with $5,184m. in 1996 and $5,771m. in 1995. In 1999 Myanmar's external debt rose to $5,999m., of which long-term public debt accounted for $5,333m. In 2000, according to the ADB, outstanding external debt amounted to $6,046.1m. In 2001 it fell modestly, to $5,691m., before rising again, to $6,135m., in 2002. According to the World Bank, by 2001 Myanmar's debt-service ratio had declined significantly, falling to 2.9% in that year from 5.0% in 1998.

The BSPP Government had been forced to deplete its foreign-exchange reserves to finance its balance-of-payments deficit. By the end of 1987, they had declined to less than US $30m., sufficient for only about three weeks' imports. By 1990 reserves had recovered and stood at $560.7m., owing mostly to the revenue gained from granting concessions for timber, fisheries and minerals. Foreign-exchange reserves reached a record $674m. in May 1995 but subsequently declined, falling to $229m. in December 1996 before increasing to $250m. in

December 1997. Foreign-exchange reserves rose from $314.6m. in December 1998 to $398.5m. at the end of 1999. In March 2000 reserves were reported to stand at $240m. By July 2000, however, as had been the case in September 1998, it was reported that foreign-exchange reserves were sufficient to finance only one month's imports. By 2002 foreign-exchange reserves stood at $469.9m. In the first quarter of 2003 foreign-exchange reserves (excluding gold) stood at $552.7m., sufficient to cover only 2.7 months of imports. An improvement in Myanmar's export performance and a resumption of aid from its Western donors remained vital to a long-term improvement in the balance of payments.

In January 2003 the Myanma Government secured a US $200m. 'package' of preferential loans from the Chinese authorities, a further indication of Beijing's increasing influence within Myanmar's economy. The two Governments also agreed to promote technical and economic co-operation and to initiate a number of new industrial projects, including the construction of roads, bridges, shipyards and telecommunications facilities. In June 2004 Thailand also agreed to provide Myanmar with a $100m. preferential loan, as part of an economic co-operation strategy that had been launched by Cambodia, Laos, Myanmar and Thailand in 2003. In July of that year India agreed to provide Myanmar with a $56m. loan to renovate the country's railway system.

The previous Government's sensitivity over foreign interference and its refusal to be bound by 'conditionality' were reasons why Myanmar made little use of the IMF in the past; however, a decline in export earnings and increasing debt-service obligations caused the Government to negotiate a US $250m. loan with the IMF in 1986. In addition, Myanmar was one of 62 poor, commodity-dependent countries with serious debt-repayment problems that were eligible for special concessionary loans under the IMF's Enhanced Structural Adjustment Facility, introduced in December 1987. However, more extensive reform, including a devaluation of the kyat, was likely to be a condition of any significant new assistance from the IMF. In 2000 the USA was still opposing Myanma requests for loans from the IMF, the World Bank and other international lending institutions, and supported the World Bank's unusual decision, in September 1998, to sever financial ties with Myanmar on account of the country's failure to make repayments on earlier loans, arrears being estimated at $14m. The International Labour Organization (ILO) passed a resolution regarding Myanmar in June 2000, which was expected to affect adversely the provision of international assistance to the country. However, the decision, made at the annual summit of the IMF and the World Bank in Prague, Czech Republic, in September 2000, to provide debt relief for debtor developing countries appeared likely to benefit Myanmar in the near future.

Foreign investment in Myanmar since the promulgation of enabling legislation in 1988 has been largely limited to petroleum, gas and mining concerns. South-East Asian investors have also tended to concentrate on US dollar-earning sectors, such as hotels and tourism. In 1996 cumulative foreign investment approvals totalled almost US $3,800m., although the disbursement rate is poor. Singapore was the largest investor in Myanmar in 1996/97, followed by the United Kingdom, France, Thailand and Malaysia. In the same year, FDI approvals totalled $2,800m.; however, in 1997/98 approved FDI declined to about $1,000m. In 1996 there was an increasing consumer boycott, particularly in the USA and Europe, of companies investing in Myanmar in protest at the SLORC's abuses of human rights. This led to the withdrawal from the country in that year of a number of large investors, including the drinks manufacturers PepsiCo, Carlsberg and Heineken; US garment manufacturers such as Oshkosh, Madison Avenue and Levi Strauss, have also ceased operations in Myanmar, although a number of other well-known companies (for example, the Australian brewing company Foster's and the German car manufacturer BMW) continued to operate and to expand in the country. Inflows have slowed further following the imposition by the USA in early 1997 of sanctions on new investment in Myanmar by US companies, although in June 1998 a US court ruled that state and local governments could not by law impose a boycott on Myanmar. While the Asian economic crisis of 1997–98 reduced the inflow of FDI into Myanmar, by mid-2000 China, Thailand and Singapore had committed themselves with renewed vigour to investments in both the public and private sectors. There was a gradual relaxation of international attitudes towards Myanmar from 2001. Various ASEAN members, most notably Malaysia, proclaimed their interest in making investments in Myanmar, and in April 2001 Japan agreed to fund a $28m. bilateral aid programme intended to facilitate the rehabilitation of a hydroelectric dam, a direct result of a liberalization of the SPDC stance. However, Japan later postponed the funding of the project indefinitely, owing to protests by human rights groups that it involved alleged human rights violations. In October 2001 the European Union (EU) voted to ease its sanctions, having extended them for a six-month period in April of that year, although most Western countries awaited more substantial concessions from the SPDC. However, the EU extended its sanctions in 2003 and 2004 owing to the lack of political reform in Myanmar. In 2002 FDI totalled only $45m., according to the IMF, compared with $208m. in the previous year. In February 2002 the US Government issued a report offering the possibility of an easing of sanctions, but only if the country made further discernible progress towards democracy. Following the release of Aung San Suu Kyi in May, it was thought that more international concessions would be forthcoming. However, a lack of further substantive political progress in the country during subsequent months, culminating in the rearrest of Suu Kyi in May 2003, meant that Myanmar continued to await full reintegration into the global economic and political community, particularly after the USA extended its sanctions against the country in mid-2003 and again in mid-2004. The US sanctions included not only a ban on imports and on the issuing of visas to high-ranking SPDC officials, but also a ban on US banks carrying out almost any financial transactions with Myanmar, thus effectively making it impossible to conduct trade in US dollars. In November 2003 the USA designated Myanmar and two of its private commercial banks as being of 'primary money-laundering concern'. However, the Myanma Government claimed to be combating money-laundering by promulgating legislation intended to prevent it.

Statistical Survey

Source (unless otherwise stated): Ministry of National Planning and Economic Development, 653–691 Merchant St, Yangon; tel. (1) 272009; fax (1) 282101.

Area and Population

AREA, POPULATION AND DENSITY

Area (sq km)	676,552*
Population (census results)	
31 March 1973	28,885,867
31 March 1983†	
Males	17,507,837
Females	17,798,352
Total	35,306,189
Population (UN estimates at mid-year)‡	
2001	48,205,000
2002	48,852,000
2003	49,485,000
Density (per sq km) at mid-2003	73.1

* 261,218 sq miles.

† Figures exclude adjustment for underenumeration. Also excluded are 7,716 Myanma citizens (males 5,704 males, females 2,012) abroad.

‡ Source: UN, *World Population Prospects: The 2002 Revision*.

PRINCIPAL TOWNS
(population at census of 31 March 1983)

Yangon (Rangoon)	2,513,023	Pathein (Bassein)	144,096
Mandalay	532,949	Taunggyi	108,231
Mawlamyine (Moulmein)	219,961	Sittwe (Akyab)	107,621
Bago (Pegu)	150,528	Manywa	106,843

Source: UN, *Demographic Yearbook*.

Mid-2003 (UN estimate, incl. suburbs): Yangon 3,873,739 (Source: UN, *World Urbanization Prospects: The 2003 Revision*).

BIRTHS AND DEATHS
(UN estimates, annual averages)

	1985–90	1990–95	1995–2000
Birth rate (per 1,000)	31.3	29.9	26.5
Death rate (per 1,000)	13.3	12.3	11.5

Source: UN, *World Population Prospects: The 2002 Revision*.

Birth rate (2001): 23.8 per 1,000 (Source: UN, *Statistical Yearbook for Asia and the Pacific*).

Death rate (2001): 6.2 per 1,000 (Source: UN, *Statistical Yearbook for Asia and the Pacific*).

Expectation of life (WHO estimates, years at birth): 58.9 (males 56.2; females 61.8) in 2002 (Source: WHO, *World Health Report*).

EMPLOYMENT
(estimates, '000 persons aged 15 to 59 years)

	1995/96	1996/97	1997/98
Agriculture, hunting, forestry and fishing	11,848	11,960	12,093
Mining and quarrying	116	132	121
Manufacturing	1,481	1,573	1,666
Electricity, gas and water	19	21	26
Construction	354	378	400
Trade, restaurants and hotels	1,715	1,746	1,781
Transport, storage and communications	441	470	495
Community, social and personal services (excl. government)	563	577	597
Administration and other services	776	835	888
Activities not adequately defined	274	272	270
Total employed	17,587	17,964	18,359

Unemployed ('000 persons aged 18 years and over): 425.3 registered in 1998/99.

Source: mainly UN, *Statistical Yearbook for Asia and the Pacific*.

Health and Welfare

KEY INDICATORS

Total fertility rate (children per woman, 2002)	2.9
Under-5 mortality rate (per 1,000 live births, 2002)	109
HIV/AIDS (% of persons aged 15–49, 2003)	1.20
Physicians (per 1,000 head, 1999)	0.30
Hospital beds (per 1,000 head, 1999)	0.30
Health expenditure (2001): US $ per head (PPP)	26
Health expenditure (2001): % of GDP	2.1
Health expenditure (2001): public (% of total)	17.8
Access to water (% of persons, 2000)	68
Access to sanitation (% of persons, 2000)	46
Human Development Index (2002): ranking	132
Human Development Index (2002): value	0.551

For sources and definitions, see explanatory note on p. vi.

Agriculture

PRINCIPAL CROPS
('000 metric tons)

	2000	2001	2002
Wheat	94	96	100†
Rice (paddy)	21,324	21,914	22,780†
Maize	365	532	660†
Millet	169	159	160*
Potatoes	255	319	319*
Sweet potatoes	38	57	57*
Cassava	77	97	97*
Sugar cane	5,449	5,894	6,333*
Other sugar crops*	242	242	242
Dry beans	1,285	1,467	1,467*
Dry peas	28	100†	100*
Chick-peas	84	119	119*
Dry cow peas	77	152†	152*
Pigeon peas	189	300†	300*
Soybeans (Soya beans)	99	110	115†
Groundnuts (in shell)	634	731	700†
Sunflower seed	160	268	265†
Sesame seed	296	426	225†
Seed cotton	176	153	168†
Coconuts	225	275†	275†
Dry onions	476	593	593*
Garlic	67	82	82*
Other vegetables (incl. melons)*	2,800	2,850	2,850*
Plantains	354	400*	400*
Other fruits (excl. melons)*	955	965	965
Areca (Betel) nuts	39	51	51*
Tea (made)	19†	19†	19*
Pimento and allspice*	40	40	40
Tobacco (leaves)	47	48	48*
Jute	33	42	42*
Cotton (lint)	59	51	56†
Natural rubber	27	36	35*

* FAO estimate(s).

† Unofficial figure.

Source: FAO.

LIVESTOCK
('000 head, year ending September)

	2000	2001	2002
Horses*	120	120	120
Cattle	10,982	11,243	11,551
Buffaloes	2,441	2,502	2,552
Pigs	3,974	4,261	4,499
Sheep	390	403	432
Goats	1,392	1,439	1,542
Chickens	44,755	55,080	57,128
Ducks	6,173	6,000*	6,100*
Geese*	500	500	500

* FAO estimate(s).

Source: FAO.

LIVESTOCK PRODUCTS
('000 metric tons)

	2000	2001	2002
Beef and veal*	102.0	104.4	108.0
Buffalo meat*	20.4	20.9	22.1
Goat meat	7.0	7.3	7.8
Pig meat*	112.9	121.0	122.7
Poultry meat	198.8	220.3	256.6
Cows' milk	498.4	510.0	525.1
Buffaloes' milk	111.0	113.7	116.0
Goats' milk	7.3	7.5	8.1
Butter*	11.0	11.2	11.6
Cheese*	31.3	32.0	32.9
Hen eggs	77.5	85.9	101.7
Other poultry eggs	10.1	10.5	11.0
Cattle hides (fresh)*	21.5	22.0	22.0

* FAO estimates.

Source: FAO.

Forestry

ROUNDWOOD REMOVALS
('000 cubic metres, excl. bark)

	2000	2001	2002
Sawlogs, veneer logs and logs for sleepers	2,337	2,662	2,662*
Other industrial wood	1,275	1,300	2,877
Fuel wood	34,471	35,403	35,403*
Total	38,083	39,365	40,942

* FAO estimate.

Source: FAO.

SAWNWOOD PRODUCTION
('000 cubic metres, incl. railway sleepers)

	2000	2001	2002
Total (all broadleaved)	545	671	381

Source: FAO.

Fishing

('000 metric tons, live weight, year ending 31 March)

	1999/2000	2000/01	2001/02
Capture*	1,069.7	1,166.9	1,312.6
Freshwater fishes	189.7	235.4	304.5
Marine fishes	849.0	900.5	975.1
Aquaculture	98.9	121.3*	121.3*
Roho labeo	93.9	85.8	76.2
Total catch*	1,168.6	1,288.1	1,433.9

* FAO estimate(s).

Source: FAO.

Mining

(metric tons, unless otherwise indicated)

	2000	2001	2002*
Coal and lignite	52,811	41,736	57,000
Crude petroleum ('000 barrels)	3,538	4,696	4,920
Natural gas (million cu m)†	8,477§	8,804	8,500
Copper ore‡	26,711	25,900	28,000
Lead ore*‡	1,200	900	500
Zinc ore‡	437	467	350
Tin concentrates‡	212	212	190
Chromium ore (gross weight)*	3,000	3,000	3,000
Tungsten concentrates‡	74	49	30
Silver ore (kilograms)‡	2,457	1,804	1,500
Gold ore (kilograms)‡§	250*	200*	200
Feldspar*§	12,000	10,000	10,000
Barite (Barytes)	30,370	31,015	18,000
Salt (unrefined, excl. brine)*	35,000	35,000	35,000
Gypsum (crude)	48,067	64,609	113,000
Rubies, sapphires and spinel ('000 metric carats)§	8,351	8,630	2,760
Jade ('000 kilograms)	8,318	87	33

* Estimated production.

† Marketed production.

‡ Figures refer to the metal content of ores and concentrates (including mixed concentrates).

§ Twelve months beginning 1 April of year stated.

Source: US Geological Survey.

Industry

SELECTED PRODUCTS
('000 metric tons, unless otherwise indicated)

	1999	2000	2001
Raw sugar*	53	60	103
Refined sugar†	43	60	n.a.
Cigarettes (million)†	2,270	2,512	2,650
Cotton yarn†	4.8	5.7	5.5
Plywood ('000 cu m)	8	14	n.a.
Printing and writing paper	16	17	20
Nitrogenous fertilizers‡	64	160	39
Petroleum refinery products ('000 barrels)§‖	5,605	5,536	5,286
Cement§	338,025	393,355	377,961
Tin—unwrought (metric tons)§	149	212	212
Electric energy (million kWh, net)	4,558	5,028	n.a

Beer (hectolitres): 13,000 in 1994.

Woven cotton fabrics (million sq m): 16 in 1995.

* Data from FAO.

† Production by government-owned enterprises only.

‡ Production in terms of nitrogen during twelve months ending 30 June of stated year.

§ Twelve months beginning 1 April of year stated. Data from US Geological Survey.

‖ Figure includes gasoline, jet fuel, kerosene, diesel, distillate fuel oil and residual fuel oil.

Source (unless otherwise specified): UN, *Statistical Yearbook for Asia and the Pacific*.

Finance

CURRENCY AND EXCHANGE RATES

Monetary Units
100 pyas = 1 kyat.

Sterling, Dollar and Euro Equivalents (31 May 2004)
£1 sterling = 10.6279 kyats;
US $1 = 5.7927 kyats;
€1 = 7.0938 kyats;
1,000 kyats = £94.09 = $172.63 = €140.97.

Average Exchange Rate (kyats per US $)
2001 6.6841
2002 6.5734
2003 6.0764

Note: Since January 1975 the value of the kyat has been linked to the IMF's special drawing right (SDR). Since May 1977 the official exchange rate has been fixed at a mid-point of SDR 1 = 8.5085 kyats. On 1 June 1996 a new customs valuation exchange rate of US $1 = 100 kyats was introduced. In September 2001 the free market exchange rate was $1 = 450 kyats.

CENTRAL GOVERNMENT BUDGET
(million kyats, year ending 31 March)

Revenue*	1998/99	1999/2000	2000/01
Tax revenue	56,653	60,294	75,727
Taxes on income, profits and capital gains	20,515	21,169	26,140
Domestic taxes on goods and services	30,748	33,750	44,101
General sales, turnover or VAT	22,720	24,576	32,961
Taxes on international trade and transactions	5,390	5,375	5,486
Other revenue	59,334	62,390	58,324
Entrepreneurial and property income	43,689	47,269	41,144
Administrative fees, nonindustrial and incidental sales	15,645	15,121	17,180
Capital revenue	79	211	257
Total	116,066	122,895	134,308

Expenditure†	1998/99	1999/2000	2000/01
Current expenditure	63,095	81,608	134,068
General public services, incl. public order	15,011	23,562	20,435
Defence	39,627	45,040	63,453
Education	9,572	12,132	31,345
Health	3,233	4,144	7,388
Social security and welfare	1,599	2,436	4,993
Recreational, cultural and religious affairs	2,114	2,013	2,668
Economic affairs and services	41,015	46,641	66,132
Agriculture, forestry, fishing and hunting	17,918	19,479	38,447
Transportation and communication	22,161	26,279	25,917
Other current expenditure	11,588	17,332	24,475
Capital expenditure	60,969	71,889	87,187
Total	124,064	153,497	221,255

* Excluding grants received from abroad (million kyats): 524 in 1998/99; 221 in 1999/2000; 242 in 2000/01.

† Excluding lending minus repayments (million kyats): –528 in 1998/99; 63 in 1999/2000; –127 in 2000/01.

Source: IMF, *Government Finance Statistics Yearbook*.

INTERNATIONAL RESERVES
(US $ million at 31 December)

	2001	2002	2003
Gold*	10.2	11.0	12.0
IMF special drawing rights	0.6	0.1	0.1
Foreign exchange	399.9	469.9	550.1
Total	410.7	481.0	562.2

* Valued at SDR 35 per troy ounce.

Source: IMF, *International Financial Statistics*.

MONEY SUPPLY
(million kyats at 31 December)

	2001	2002	2003
Currency outside banks	494,521	718,633	1,102,937
Demand deposits at deposit money banks	206,349	290,520	82,948
Total money (incl. others)	701,153	1,009,471	1,186,104

Source: IMF, *International Financial Statistics*.

COST OF LIVING
(Consumer Price Index for Yangon; base: 1990 = 100)

	1997	1998	1999
Food (incl. beverages)	548.2	834.6	997.0
Fuel and light	457.3	590.2	632.3
Clothing (incl. footwear)	366.0	620.3	700.4
Rent	405.1	568.9	624.1
All items (incl. others)	498.1	754.6	893.4

All items (base: 1990 = 100): 151.7 in 2000; 183.8 in 2001; 288.6 in 2002.

Source: ILO.

NATIONAL ACCOUNTS
(million kyats at current prices, year ending 31 March)

National Income and Product

	1996/97	1997/98	1998/99
Domestic factor incomes	773,940	1,087,997	1,534,089
Consumption of fixed capital	18,040	21,557	25,907
Gross domestic product (GDP) at factor cost	791,980	1,109,554	1,559,996
Indirect taxes, *less* subsidies	—	—	—
GDP in purchasers' values	791,980	1,109,554	1,559,996
Net factor income from abroad	–116	–69	–220
Gross national product	791,864	1,109,485	1,559,776
Less Consumption of fixed capital	18,040	21,557	25,907
National income in market prices	773,824	1,087,928	1,533,869

Source: UN, *National Accounts Statistics*.

Expenditure on the Gross Domestic Product

	1997/98	1998/99	1999/2000
Final consumption expenditure	987,513	1,419,709	1,906,136
Increase in stocks	–10,276	–7,604	48,325
Gross fixed capital formation	150,240	206,912	241,694
Total domestic expenditure	1,127,477	1,619,017	2,196,155
Exports of goods and services	6,290	7,700	9,394
Less Imports of goods and services	–14,258	–16,941	–15,248
GDP in purchasers' values	1,119,509	1,609,776	2,190,301
GDP at constant 1985/86 prices	75,123	79,460	88,134

Source: IMF, *International Financial Statistics*.

Gross Domestic Product by Economic Activity

	1999/2000	2000/01	2001/02
Agriculture, hunting, forestry and fishing	1,312,285	1,461,150	2,013,713
Mining and quarrying	10,842	15,032	16,119
Manufacturing	143,244	182,897	273,894
Electricity, gas and water	2,558	3,444	3,125
Construction	40,425	46,044	76,668
Wholesale and retail trade	524,403	613,686	855,864
Transport, storage and communications	105,669	153,371	188,776
Finance	2,215	2,641	3,231
Government services	16,505	39,354	44,261
Other services	32,174	35,114	47,864
GDP in purchasers' values	2,190,320	2,552,733	3,523,515

Source: Asian Development Bank, *Key Indicators of Developing Asian and Pacific Countries*.

BALANCE OF PAYMENTS
(US $ million)

	1999	2000	2001
Exports of goods f.o.b.	1,293.9	1,661.6	2,316.9
Imports of goods f.o.b.	−2,181.3	−2,165.4	−2,587.9
Trade balance	−887.3	−503.8	−271.1
Exports of services	512.2	477.9	423.6
Imports of services	−291.1	−328.1	−380.2
Balance on goods and services	−666.3	−354.0	−227.7
Other income received	51.6	35.5	35.9
Other income paid	−54.6	−168.8	−402.6
Balance on goods, services and income	−669.2	−487.3	−594.3
Current transfers received	384.8	289.7	298.5
Current transfers paid	−0.3	−14.1	−12.8
Current balance	−284.7	−211.7	−308.5
Direct investment from abroad	255.6	258.3	210.3
Other investment liabilities	−4.4	−45.4	188.8
Net errors and omissions	−12.4	−24.5	89.3
Overall balance	−45.9	−23.4	179.8

Source: IMF, *International Financial Statistics*.

External Trade

PRINCIPAL COMMODITIES*
(distribution by SITC, million kyats, year ending 31 March)

Imports c.i.f.	1999/2000	2000/01	2001/02
Food and live animals	620	586	838
Mineral fuels, lubricants, etc.	1,654	1,145	3,839
Animal and vegetable oils, fats and waxes	488	412	253
Chemicals and related products	1,871	1,924	1,787
Basic manufactures	4,125	4,401	4,548
Machinery and transport equipment	4,868	3,754	5,110
Miscellaneous manufactured articles	643	1,000	726
Total (incl. others)	16,265	15,073	18,378

Exports f.o.b.†	1999/2000	2000/01	2001/02
Food and live animals	2,237	3,206	3,774
Dried beans, peas, etc. (shelled)	1,179	1,658	1,898
Crude materials (inedible) except fuels	1,819	1,401	2,750
Teak and other hardwood	925	803	1,880
Basic manufactures	602	1,240	168
Miscellaneous manufactured articles	176	1,570	104
Total (incl. others)	8,947	12,736	17,131

* Totals include, but distribution by commodities excludes, border trade, mainly with the People's Republic of China, Thailand and Bangladesh.
† Excluding re-exports.

Source: Asian Development Bank, *Key Indicators of Developing Asian and Pacific Countries*.

PRINCIPAL TRADING PARTNERS
(US $ million)

Imports	2001	2002	2003
China, People's Republic	547.3	797.3	998.5
Germany	18.4	20.1	17.4
Hong Kong	70.1	69.9	48.4
India	58.1	63.1	72.0
Indonesia	75.9	59.8	66.8
Japan	205.3	126.9	116.0
Korea, Republic	255.3	157.8	180.0
Malaysia	216.7	263.1	300.1
Singapore	465.6	576.6	716.0
Thailand	390.5	355.9	483.4
Total (incl. others)	2,661.2	2,950.7	3,387.7

Exports	2001	2002	2003
China, People's Republic	122.0	124.5	154.1
France	72.1	79.5	56.9
Germany	100.3	73.2	96.9
India	179.8	195.2	222.6
Japan	92.8	100.3	123.0
Malaysia	71.1	69.8	79.6
Singapore	102.1	97.3	76.2
Thailand	735.4	831.2	831.7
United Kingdom	87.4	87.8	92.6
USA	456.2	345.4	268.6
Total (incl. others)	2,626.1	2,626.6	2,709.4

Source: Asian Development Bank, *Key Indicators of Developing Asian and Pacific Countries*.

Transport

RAILWAYS
(traffic, million)

	1998	1999	2000
Passenger-kilometres	3,948	4,112	4,451
Freight ton-kilometres	988	1,043	1,222

Source: UN, *Statistical Yearbook*.

ROAD TRAFFIC
(registered motor vehicles at 31 March)

	1994	1995	1996
Passenger cars	119,126	131,953	151,934
Trucks	39,939	36,728	42,828
Buses	19,183	14,624	15,639
Motorcycles	71,929	82,591	85,821
Others	8,377	6,251	6,611
Total	258,554	272,147	302,833

Source: Department of Road Transport Administration.

Passenger cars: 177,900 in 1997; 177,600 in 1998; 171,100 in 1999; 173,900 in 2000 (Source: UN, *Statistical Yearbook*).

Commercial vehicles: 74,800 in 1997; 75,900 in 1998; 83,400 in 1999; 90,400 in 2000 (Source: UN, *Statistical Yearbook*).

INLAND WATERWAYS
(traffic by state-owned vessels)

	1993/94	1994/95*	1995/96†
Passengers carried ('000)	36,003	26,582	24,491
Passenger-miles (million)	617	531	544
Freight carried ('000 metric tons)	3,172	3,194	3,158
Freight ton-miles (million)	353	346	351

* Provisional.
† Estimates.

SHIPPING

Merchant Fleet
(registered at 31 December)

	2001	2002	2003
Number of vessels	124	124	122
Displacement (grt)	379,819	402,159	433,039

Source: Lloyd's Register-Fairplay, *World Fleet Statistics*.

International Sea-Borne Traffic
(state-owned vessels)

	1993/94	1994/95*	1995/96†
Passengers carried ('000)	69	77	60
Passenger-miles (million)	23	26	20
Freight carried ('000 metric tons)	773	1,213	1,030
Freight ton-miles (million)	2,655	2,765	2,807

* Provisional.
† Estimates.

CIVIL AVIATION
(traffic on scheduled services)

	1997	1998	1999
Kilometres flown (million)	9	8	9
Passengers carried ('000)	575	522	537
Passenger-km (million)	385	345	355
Total ton-km (million)	46	40	40

Source: UN, *Statistical Yearbook*.

Tourism

TOURIST ARRIVALS BY COUNTRY OF NATIONALITY

	1999	2000	2001
Australia	3,642	4,120	4,442
China, People's Republic	12,148	14,336	16,788
France	13,594	13,313	12,461
Germany	9,039	9,920	11,450
India	5,083	5,605	5,572
Italy	6,925	6,852	6,618
Japan	25,319	21,930	20,118
Korea, Republic	5,885	7,423	7,581
Malaysia	7,583	9,938	11,296
Singapore	11,074	11,645	9,939
Taiwan	32,977	32,098	26,020
Thailand	19,392	19,070	17,123
United Kingdom	9,267	9,020	8,424
USA	10,270	12,669	13,524
Total (incl. others)	198,210	207,665	204,862

Source: World Tourism Organization, *Yearbook of Tourism Statistics*.

Tourism receipts (US $ million): 35 in 1999; 42 in 2000; 45 in 2001 (Source: World Bank).

Communications Media

	2001	2002	2003
Telephones ('000 main lines in use)	295.2	341.3	357.3
Mobile cellular telephones ('000 subscribers)	22.7	48.0	66.5
Personal computers ('000 in use)	55	250	n.a.
Internet users ('000)	10.0	25.0	28.0

Book production (1999): 227 titles.

Newspapers (1996): 5 dailies (average circulation 449,000).

Facsimile machines (number in use): 2,540 in 1999.

Radio receivers (1999): 3,157,000 in use.

Television receivers ('000 in use): 344.3 in 2000.

Sources: International Telecommunication Union; UNESCO, *Statistical Yearbook*; UN, *Statistical Yearbook*.

Education

(provisional, 1994/95)

	Institutions	Teachers	Students
Primary schools*	35,856	169,748	5,711,202
Middle schools	2,058	53,859	1,390,065
High schools	858	18,045	389,438
Vocational schools	86	1,847	21,343
Teacher training	17	615	4,031
Higher education	45	6,246	247,348
Universities	6	2,901	62,098

* Excluding 1,152 monastic primary schools with an enrolment of 45,360.

1997/98 (provisional): *Primary:* Institutions 35,877, Teachers 167,134, Students ('000) 5,145.4; *General Secondary:* Institutions 2,091, Teachers 56,955, Students ('000) 1,545.6; *Tertiary:* Institutions 923, Teachers 17,089, Students ('000) 385.3 (Source: UN, *Statistical Yearbook for Asia and the Pacific*).

Adult literacy rate (UNESCO estimates): 85.3% (males 89.2%; females 81.4%) in 2002 (Source: UN Development Programme, *Human Development Report*).

Directory

The Constitution

On 18 September 1988 a military junta, the State Law and Order Restoration Council (SLORC), assumed power and abolished all state organs created under the Constitution of 3 January 1974. The country was placed under martial law. The state organs were superseded by the SLORC at all levels with the Division, Township and Village State Law and Order Restoration Councils. The SLORC announced that a new constitution was to be drafted by the 485-member Constituent Assembly that was elected in May 1990. In early 1993 a National Convention, comprising members of the SLORC and representatives of opposition parties, met to draft a new constitution; however, the Convention was adjourned in March 1996 and remained in recess in early 2004. In November 1997 the SLORC was dissolved and replaced by the newly formed State Peace and Development Council (SPDC). In August 2003 the SPDC announced that it planned to reconvene the National Convention in 2004 in order that it could commence the drafting of a new constitution.

The Government

HEAD OF STATE

Chairman of the State Peace and Development Council: Field Marshal THAN SHWE (took office as Head of State 23 April 1992).

STATE PEACE AND DEVELOPMENT COUNCIL
(October 2004)

Chairman: Field Marshal THAN SHWE.

Vice-Chairman: Dep. Senior Gen. MAUNG AYE.

First Secretary: Lt-Gen. THEIN SEIN.

Second Secretary: (vacant).

Third Secretary: (vacant).

Other members: Rear-Adm. KYI MIN, Lt-Gen. KYAW THAN, Lt-Gen. AUNG HTWE, Lt-Gen. YE MYINT, Lt-Gen. KHIN MAUNG THAN, Maj.-Gen. KYAW WIN, Maj.-Gen. THURA SHWE MANN, Maj.-Gen. MYINT AUNG, Lt-Gen. MAUNG BO, Maj-Gen. THIHA THURA TIN AUNG MYINT OO, Lt-Gen. TIN AYE.

CABINET
(October 2004)

Prime Minister: Lt-Gen. SOE WIN.

Minister of Defence: Field Marshal THAN SHWE.

Minister of Military Affairs: Maj.-Gen. THIHA THURA TIN AUNG MYINT.

Minister of Agriculture and Irrigation: Maj.-Gen. HTAY OO.

Minister of Industry (No. 1): U AUNG THAUNG.

Minister of Industry (No. 2): Maj.-Gen. SAW LWIN.

Minister of Foreign Affairs: Maj.-Gen. NYAN WIN.

Minister of National Planning and Economic Development: U SOE THA.

Minister of Transport: Maj.-Gen. THEIN SHWE.

Minister of Labour and Minister at the Office of the Prime Minister: U TIN WIN.

Minister of Culture: Maj.-Gen. KYI AUNG.

Minister of Co-operatives: Col ZAW MIN.

Minister of Rail Transportation: Maj.-Gen. AUNG MIN.

Minister of Energy: Brig.-Gen. LUN THI.

Minister of Education: U THAN AUNG.

Minister of Health: Dr KYAW MYINT.

Minister of Commerce: Brig.-Gen. TIN NAING THEIN.

Minister of Communications, Posts and Telegraphs and of Hotels and Tourism: Brig.-Gen. THEIN ZAW.

Minister of Finance and Revenue: Maj.-Gen. HLA TUN.

Minister of Religious Affairs: Brig.-Gen. THURA MYINT MAUNG.

Minister of Construction: Maj.-Gen. SAW TUN.

Minister of Science and Technology: U THAUNG.

Minister of Immigration and Population and of Social Welfare, Relief and Resettlement: Maj.-Gen. SEIN HTWA.

Minister of Information: Brig.-Gen. KYAW HSAN.

Minister of Progress of Border Areas, National Races and Development Affairs: Col THEIN NYUNT.

Minister of Electric Power: Maj.-Gen. TIN HTUT.

Minister of Sports: Brig.-Gen. THURA AYE MYINT.

Minister of Forestry: Brig.-Gen. THEIN AUNG.

Minister of Home Affairs: Col TIN HLAING.

Minister of Mines: Brig.-Gen. OHN MYINT.

Minister of Livestock and Fisheries: Brig.-Gen. MAUNG MAUNG THEIN.

Minister at the Office of the Prime Minister: U THAN SHWE, U KO LAY, Brig.-Gen. PYI SONE.

MINISTRIES

Office of the Chairman of the State Peace and Development Council: 15–16 Windermere Park, Yangon; tel. (1) 282445.

Prime Minister's Office: Minister's Office, Theinbyu St, Botahtaung Township, Yangon; tel. (1) 283742.

Ministry of Agriculture and Irrigation: Thiri Mingala Lane, off Kaba Aye Pagoda Rd, Yangon; tel. (1) 665587; fax (1) 664493; e-mail dap.moai@mptmail.net.mm; internet www.myanmar.com/Ministry/agriculture.

Ministry of Commerce: 228–240 Strand Rd, Yangon; tel. (1) 287034; fax (1) 280679; e-mail com@mptmail.net.mm; internet www.myanmar.com/Ministry/commerce.

Ministry of Communications, Posts and Telegraphs: 361 Pyay Rd, nr Hanthawaddy Circus, Sanchaung Township, Yangon; tel. (1) 515034; internet www.mcpt.gov.mm.

Ministry of Construction: 39 Nawaday St, Dagon Township, Yangon; tel. (1) 283938; fax (1) 289531.

Ministry of Co-operatives: 259–263 Bogyoke Aung San St, Yangon; tel. (1) 277096; fax (1) 287919.

Ministry of Culture: 131 Kaba Aye Pagoda Rd, Bahan Township, Yangon; tel. (1) 543235; fax (1) 283794; internet www.myanmar.com/Ministry/culture.

Ministry of Defence: Ahlanpya Phaya St, Yangon; tel. (1) 281611.

Ministry of Education: Theinbyu St, Botahtaung Township, Yangon; tel. (1) 285588; fax (1) 285480.

Ministry of Electric Power: 197–199 Lower Kyimyindaing Rd, Ahlone Township, Yangon; tel. (1) 229366; fax (1) 221006.

Ministry of Energy: 23 Pyay Rd, Yangon; tel. (1) 221060; fax (1) 222964; e-mail myanmoe@mptmail.net.mm; internet www.energy.gov.mm.

Ministry of Finance and Revenue: 26(A) Setmu Rd, Yankin Township, Yangon; tel. (1) 284763; internet www.myanmar.com/Ministry/finance.

Ministry of Foreign Affairs: Pyay Rd, Dagon Township, Yangon; tel. (1) 222844; fax (1) 222950; e-mail mofa.aung@mptmail.net.mm; internet www.myanmar.com/Ministry/mofa.

Ministry of Forestry: Thirimingala Lane, Kaba Aye Pagoda Rd, Yangon; tel. (1) 289184; fax (1) 664459; internet www.myanmar.com/Ministry/Forest.

Ministry of Health: Theinbyu St, Botahtaung Township, Yangon; tel. (1) 277334; fax (1) 282834; internet www.myanmar.com/Ministry/health.

Ministry of Home Affairs: cnr of Saya San St and No. 1 Industrial St, Yankin Township, Yangon; tel. (1) 549208; internet www.myanmar.com/Ministry/Moha.

Ministry of Hotels and Tourism: 77–91 Sule Pagoda Rd, Kyauktada Township, Yangon; tel. (1) 282705; fax (1) 287871; e-mail mtt.mht@mptmail.net.mm; internet www.myanmar.com/Ministry/Hotel_Tour.

Ministry of Immigration and Population: cnr of Mahabandoola Rd and Theinbyu St, Botahtaung Township, Yangon; tel. (1) 249090; internet www.myanmar.com/Ministry/imm&popu.

Ministry of Industry (No. 1): 192 Kaba Aye Pagoda Rd, Yangon; tel. (1) 566066; internet www.myanmar.com/Ministry/MOI-1.

Ministry of Industry (No. 2): 56 Kaba Aye Pagoda Rd, Yankin Township, Yangon; tel. (1) 666134; fax (1) 666135; e-mail dmip@mptmail.net.mm; internet www.myanmar.com/Ministry/moi2.

Ministry of Information: 365–367 Bo Sung Kyaw St, Yangon; tel. (1) 245631; fax (1) 289274.

Ministry of Labour: Theinbyu St, Botahtaung Township, Yangon; tel. (1) 278320; fax (1) 256185.

Ministry of Livestock and Fisheries: Theinbyu St, Botahtaung Township, Yangon; tel. (1) 280398; fax (1) 289711.

Ministry of Military Affairs: Yangon.

Ministry of Mines: 90 Kanbe Rd, Yankin Township, Yangon; tel. (1) 577316; fax (1) 577455; internet www.mining.com.mm.

Ministry of National Planning and Economic Development: 653–691 Merchant St, Pabedan Township, Yangon; tel. (1) 241918; fax (1) 282101.

Ministry of Progress of Border Areas, National Races and Development Affairs: Minister's Office, Theinbyu St, Botahtaung Township, Yangon; tel. (1) 280032; fax (1) 285257.

Ministry of Rail Transportation: 88 Theinbyu St, Yangon; tel. (1) 292769.

Ministry of Religious Affairs: Kaba Aye Pagoda Precinct, Mayangone Township, Yangon; tel. (1) 665620; fax (1) 665728; internet www.myanmar.com/Ministry/religious.

Ministry of Science and Technology: 6 Kaba Aye Pagoda Rd, Yankin Township, Yangon; tel. (1) 667639; fax (1) 651026.

Ministry of Social Welfare, Relief and Resettlement: Theinbyu St, Botahtaung Township, Yangon; tel. (1) 665697; fax (1) 650002; e-mail social-wel-myan@mptmail.net.mm; internet www.myanmar.com/Ministry/social-welfare.

Ministry of Sports: Minister's Office, Theinbyu St, Botahtaung Township, Yangon; tel. (1) 553958.

Ministry of Transport: 363–421 Merchant St, Botahtaung Township, Yangon; tel. (1) 296815; fax (1) 296824; internet www.myanmar.com/Ministry/Transport.

Legislature

CONSTITUENT ASSEMBLY

Following the military coup of 18 September 1988, the 489-member Pyithu Hluttaw (People's Assembly), together with all other state organs, was abolished. A general election was held on 27 May 1990. It was subsequently announced, however, that the new body was to serve as a constituent assembly, responsible for the drafting of a new constitution, and that it was to have no legislative power. The next legislative election was provisionally scheduled for September 1997, but did not take place.

General Election, 27 May 1990

Party	% of Votes	Seats
National League for Democracy	59.9	392
Shan Nationalities League for Democracy	1.7	23
Arakan (Rakhine) League for Democracy	1.2	11
National Unity Party	21.2	10
Mon National Democratic Front	1.0	5
National Democratic Party for Human Rights	0.9	4
Chin National League for Democracy	0.4	3
Kachin State National Congress for Democracy	0.1	3
Party for National Democracy	0.5	3
Union Pa-O National Organization	0.3	3
Zomi National Congress		2
Naga Hill Regional Progressive Party		2
Kayah State Nationalities League for Democracy		2
Ta-ang (Palaung) National League for Democracy		2
Democratic Organization for Kayan National Unity		2
Democracy Party		1
Graduates' and Old Students' Democratic Association		1
Patriotic Old Comrades' League	12.8	1
Shan State Kokang Democratic Party		1
Union Danu League for Democracy Party		1
Kamans National League for Democracy		1
Mara People's Party		1
Union Nationals Democracy Party		1
Mro (or) Khami National Solidarity Organization		1
Lahu National Development Party		1
United League of Democratic Parties		1
Karen (Kayin) State National Organization		1
Independents		6
Total	100.0	485

Political Organizations

A total of 93 parties contested the general election of May 1990. By October 1995 the ruling military junta had deregistered all except nine political parties:

Kokang Democracy and Unity Party: Yangon.

Mro (or) Khami National Solidarity Organization: f. 1988; Leader U San Tha Aung.

National League for Democracy (NLD): 97B West Shwegondine Rd, Bahan Township, Yangon; f. 1988; initially known as the National United Front for Democracy, and subsequently as the League for Democracy; present name adopted in Sept. 1988; central exec. cttee of 10 mems; Gen. Sec. Daw Aung San Suu Kyi; Chair. U Aung Shwe; Vice-Chair. U Tin Oo, U Kyi Maung.

National Unity Party (NUP): 93C Windermere Rd, Kamayut, Yangon; tel. (1) 278180; f. 1962 as the Burma Socialist Programme Party; sole legal political party until Sept. 1988, when present name was adopted; 15-mem. Cen. Exec. Cttee and 280-mem. Cen. Cttee; Chair. U Tha Kyaw; Jt Gen. Secs U Tun Yi, U Than Tin.

Shan Nationalities League for Democracy: f. 1988; Leader Khun Htun Oo.

Shan State Kokang Democratic Party: 140 40 St, Kyauktada; f. 1988; Leader U Yankyin Maw.

Union Karen (Kayin) League: Saw Toe Lane, Yangon.

Union Pa-O National Organization: f. 1988; Leader U San Hla.

Wa National Development Party: Byuhar St, Yangon.

The following parties contested the general election of March 1990 but subsequently had their legal status annulled:

Anti-Fascist People's Freedom League: Bo Aung Kyaw St, Bahan Township, Yangon; f. 1988; assumed name of wartime resistance movement which became Myanmar's major political force after independence; Chair. Bo Kyaw Nyunt; Gen. Sec. Cho Cho Kyaw Nyein.

Democracy Party: f. 1988; comprises supporters of fmr Prime Minister U Nu; Chair. U Thu Wai; Vice-Chair. U Khun Ye Naung.

Democratic Front for National Reconstruction: Yangon; f. 1988; left-wing; Leader Thakin Chit Maung.

Lahu National Development Party: f. 1988; deregistered 1994; Leader U Daniel Aung.

League for Democracy and Peace: 10 Wingaba Rd, Bahan Township, Yangon; f. 1988; Gen. Sec. U Thein Sein.

Party for National Democracy: Yangon; f. 1988; Chair. Dr Sein Win.

Union National Democracy Party (UNDP): 2–4 Shin Saw Pu Rd, Sanchaung Township, Yangon; f. 1988 by Brig.-Gen. Aung Gyi (fmr Chair. of the National League for Democracy); Chair. U Kyaw Myint Lay.

United League of Democratic Parties: 875 Compound 21, Ledauntkan St, Sa-Hsa Ward, Thingangyun Township, Yangon; f. 1989.

United Nationalities League for Democracy: Yangon; an alliance of parties representing non-Bamar nationalities; won a combined total of 65 seats at the 1990 election.

Other deregistered parties included the Arakan (Rakhine) League for Democracy, the Mon National Democratic Front, the National Democratic Party for Human Rights, the Chin National League for Democracy, the Kachin State National Congress for Democracy, the Zomi National Congress, the Naga Hill Regional Progressive Party, the Kayah State Nationalities League for Democracy, the Ta-ang (Palaung) National League for Democracy, the Democratic Organization for Kayan National Unity, the Graduates' and Old Students' Democratic Association, the Patriotic Old Comrades' League, the Union Danu League for Democracy, the Kamans National League for Democracy, the Mara People's Party and the Karen (Kayin) State National Organization.

The following groups are, or have been, in armed conflict with the Government:

Burma Democratic Alliance (BDA): f. 2004; opposition alliance comprised of several dissident organizations; Leader Dr Naing Aung.

Chin National Army: Chin State.

Chin National Front: f. 1988; forces trained by Kachin Independence Army 1989–91; first party congress 1993; conference in March 1996; carried out an active bombing campaign in 1996–97, mainly in the Chin State; Pres. Thomas Tang No.

Communist Party of Burma (CPB): f. 1939; reorg. 1946; operated clandestinely after 1948; participated after 1986 in jt military operations with sections of the NDF; in 1989 internal dissent resulted in the rebellion of about 80% of CPB members, mostly Wa hill tribesmen and Kokang Chinese; the CPB's military efficacy was thus completely destroyed; Chair. of Cen. Cttee Thakin Ba Thein Tin (exiled).

Democratic Alliance of Burma (DAB): Manerplaw; f. 1988; formed by members of the NDF to incorporate dissident students, monks and expatriates; Pres. Maj.-Gen. Bo Mya; Gen. Sec. U Tin Maung Win; Remaining organizations include:

All-Burma Student Democratic Front (ABSDF): Dagwin; f. 1988; in 1990 split into two factions, under U Moe Thi Zun and U Naing Aung; the two factions reunited in 1993; Chair. Than Khe; Sec.-Gen. Myo Win.

Karen (Kayin) National Union (KNU): f. 1948; in process of negotiating peace agreement with SPDC in 2004; Chair. Saw Ba Thin; Vice-Chair. Maj.-Gen. Bo Mya; Sec.-Gen. Mahn Sha; Military wing: **Karen (Kayin) National Liberation Army (KNLA)**; c. 6,000 troops; Chief of Staff Gen. Tamalabaw.

Karenni (Kayinni) National Progressive Party: agreement with the SLORC signed in March 1995 but subsequently collapsed; resumed fighting in June 1996; Chair. Gen. Aung Than Lay; Military wing: **Karenni (Kayinni) Revolutionary Army**.

God's Army: breakaway faction of the KNU; Leaders Johnny Htoo, Luther Htoo (surrendered to the Thai authorities in Jan. 2001).

National Democratic Front (NDF): f. 1975; aims to establish a federal union based on national self-determination; largely defunct.

National Socialist Council of Nagaland: Sagaing Division; comprises various factions.

Shan State Army (SSA): enlarged in Sept. 1997 through formal alliance between the following:

Shan State National Army (SSNA): Shan State; f. 1995; breakaway group from Mong Tai Army (MTA); Shan separatists; 5,000–6,000 troops; Leaders KARN YORD, YEE.

Shan State Peace Council (SSPC): fmrly Shan State Progressive Party; Pres. HSO HTEN; Gen. Sec. KARN YORD; Gen. Sec. KARN YORD; cease-fire agreement signed in Sept. 1989, but broken by SSA elements following establishment of above alliance in Sept. 1997; Military wing: original **Shan State Army** (5,000 men); Leaders SAI NONG, KAI HPA, PANG HPA.

Other MTA remnants also participated in the alliance.

Vigorous Burmese Student Warriors: f. 1999.

Most of the following groups have signed cease-fire agreements, or reached other means of accommodation, with the ruling military junta (the date given in parentheses indicates the month in which agreement with the junta was concluded):

Democratic Karen (Kayin) Buddhist Organization: Manerplaw; breakaway group from the KNU; Military wing: **Democratic Karen (Kayin) Buddhist Army.**

Kachin Democratic Army: (Jan. 1991); fmrly the 4th Brigade of the Kachin Independence Army; Leader U ZAW MAING.

Kachin Independence Organization (KIO): (Oct. 1993); Chair. U LAMON TU JAI; Military wing: **Kachin Independence Army.**

Karen (Kayin) Solidarity Organization (KSO): f. 1997; fmrly All Karen Solidarity and Regeneration Front; breakaway group from the KNU; 21-mem. exec. cttee; advocates nation-wide cease-fire and the settlement of all national problems through negotiations; Pres. SAW W. P. NI; Sec.-Gen. MAHN AUNG HTAY.

Karenni (Kayinni) National People's Liberation Front: (May 1994); Leader U TUN KYAW.

Kayan National Guard: (Feb. 1992).

Kayan New Land Party: (July 1994); Leader U THAN SOE NAING.

Myanmar National Democracy Alliance: (March 1989).

National Democracy Alliance Army: (June 1989).

New Democratic Army: Kachin; (Dec. 1989).

New Mon State Party: (June 1995); Chair. (vacant); Military wing: **Mon National Liberation Army.**

Palaung State Liberation Organization: (April 1991); Military wing: **Paulung State Liberation Army**; 7,000–8,000 men.

Pa-O National Organization: (Feb. 1991); Chair. AUNG KHAM HTI; Military wing: **Pa-O National Army.**

Shan State Nationalities Liberation Organization: (Oct. 1994); Chair. U THA KALEI.

United Wa State Party: (May 1989); fmrly part of the Communist Party of Burma; Military wing: **United Wa State Army**(10,000–15,000 men); Leaders CHAO NGI LAI, PAO YU CHANG.

Since 1991 the National Coalition Government of the Union of Burma, constituted by representatives elected in the general election of 1990, has served as a government-in-exile:

National Coalition Government of the Union of Burma (NCGUB): Washington Office, 1319 F St NW, Suite 303, Washington, DC 20004, USA; tel. (202) 639-0639; fax (202) 639-0638; e-mail ncgub@ncgub.net; internet www.ncgub.net; Prime Minister Dr SEIN WIN.

Diplomatic Representation

EMBASSIES IN MYANMAR

Australia: 88 Strand Rd, Yangon; tel. (1) 251810; fax (1) 246159; internet www.embassy.gov.au/mn; Ambassador PAUL GRIGSON.

Bangladesh: 11B Thanlwin Rd, Yangon; tel. (1) 515275; fax (1) 515273; e-mail bdootygn@mptmail.mm; Ambassador A. B. MANJOOR RAHIM.

Brunei: 51 Golden Valley, Bahan, Yangon; tel. (1) 510422; fax (1) 512854; Ambassador Pehin Datu PEKERMA Dato' Paduka Haji HUSSIN BIN Haji SULAIMAN.

Cambodia: 34 Kaba Aye Pagoda Rd, Yangon; tel. (1) 546157; fax (1) 546156; e-mail recyangon@mptmail.net.mm; Ambassador HUL PHANY.

China, People's Republic: 1 Pyidaungsu Yeiktha Rd, Yangon; tel. (1) 221281; fax (1) 227019; e-mail chinaemb_mm@mfa.gov.cn; Ambassador LI JINJUN.

Egypt: 81 Pyidaungsu Yeiktha Rd, Yangon; tel. (1) 222886; fax (1) 222865; Ambassador MOHAMED EL MENEISSY.

France: 102 Pyidaungsu Yeiktha Rd, POB 858, Yangon; tel. (1) 212523; fax (1) 212527; Ambassador JEAN-MICHEL LACOMBE.

Germany: 9 Bogyoke Aung San Museum Rd, POB 12, Yangon; tel. (1) 548951; fax (1) 548899; e-mail post@botschaftrangun.net.mm; Ambassador Dr KLAUS WILD.

India: 545–547 Merchant St, POB 751, Yangon; tel. (1) 243972; fax (1) 254086; e-mail amb.indembygn@mptmail.net.mm; Ambassador RAJIV KUMAR BHATIA.

Indonesia: 100 Pyidaungsu Yeiktha Rd, POB 1401, Yangon; tel. (1) 254465; fax (1) 254468; e-mail kbriygn@indosat.net.id; Ambassador WYOSO PROJOWARSITO.

Israel: 15 Khabaung St, Hlaing Township, Yangon; tel. (1) 515115; fax (1) 515116; e-mail yangon@israel.org; Ambassador YAAKOV AVRAHAMI MERON.

Italy: 3 Inya Myaing Rd, Golden Valley, Yangon; tel. (1) 527100; fax (1) 514565; e-mail ambitaly@ambitaly.net.mm; internet www.embassyofitaly-yangon.org; Ambassador RAFFAELE MINIERO.

Japan: 100 Natmauk Rd, Yangon; tel. (1) 549644; fax (1) 549643; e-mail embassyofjapan@mptmail.net.mm; Ambassador NOBUTAKE ODANO.

Korea, Republic: 97 University Ave, Yangon; tel. (1) 515190; fax (1) 513286; e-mail hankuk@koremby.net.mm; Ambassador LEE KYUNG-WOO.

Laos: A1 Diplomatic Quarters, Franser Rd, Yangon; tel. (1) 222482; fax (1) 227446; Ambassador CHANTHAVY BODHISANE.

Malaysia: 882 Diplomatic Quarters, Pyidaungsu Yeiktha Rd, Yangon; tel. (1) 220249; fax (1) 221840; e-mail mwyangon@mweb.com.na; Ambassador Dato' CHEAH SAM KIP.

Nepal: 16 Natmauk Yeiktha Rd, POB 84, Tamwe, Yangon; tel. (1) 545880; fax (1) 549803; e-mail rnembygn@datseco.com.mm; Ambassador DIBYA DEV BHATTA.

Pakistan: A4 Diplomatic Quarters, Pyay Rd, Dagon Township, Yangon; tel. (1) 222881; fax (1) 221147; e-mail parepygn@myanmar.com.mm; Ambassador YOUSUF SHAH.

Philippines: 50 Saya San Rd, Bahan Township, Yangon; tel. (1) 558149; fax (1) 558154; e-mail phyangon@mptmail.net.mm; Ambassador PHOEBE ABAYA GOMEZ.

Russia: 38 Sagawa Rd, Yangon; tel. (1) 241955; fax (1) 241953; e-mail rusinmyan@mptmail.net.mm; Ambassador GLEB A. IVASHENTSOV.

Serbia and Montenegro: 114A Inya Rd, POB 943, Yangon; tel. (1) 515282; fax (1) 504274; e-mail yuamb@yangon.net.mm; Chargé d'affaires a.i. VLADIMIR STAMENOVIĆ.

Singapore: 238 Dhamazedi Rd, Bahan Township, Yangon; tel. (1) 559001; fax (1) 559002; e-mail singemb@seygn.com.mm; internet www.mfa.gov.sg/yangon/; Ambassador SIMON TENSING DE CRUZ.

Sri Lanka: 34 Taw Win Rd, POB 1150, Yangon; tel. (1) 222812; fax (1) 221509; e-mail srilankaemb@mpt.net.mm; Ambassador D. M. M. RANARAJA.

Thailand: 73 Manawhari St, Dagon Township, Yangon; tel. (1) 224647; fax (1) 225929; e-mail thaiygn@mfa.go.th; Ambassador SUPHOT DHIRAKAOSAL.

United Kingdom: 80 Strand Rd, POB 638, Yangon; tel. (1) 370863; fax (1) 289566; e-mail chancery.Rangoon@fco.gov.uk; Ambassador VICKY BOWMAN.

USA: 581 Merchant St, POB 521, Yangon; tel. (1) 379880; fax (1) 256018; e-mail info.rangoon@state.gov; internet rangoon.usembassy.gov; Chargé d'affaires CARMEN MARIA MARTINEZ.

Viet Nam: 317–319 U Wisara Rd, Sanchaung Township, Yangon; tel. (1) 524285; fax (1) 524658; e-mail vnembmyr@bertech.net.mm; Ambassador PHAM QUANG KHON.

Judicial System

A new judicial structure was established in March 1974. Its highest organ, composed of members of the People's Assembly, was the Council of People's Justices, which functioned as the central Court of Justice. Below this Council were the state, divisional, township, ward and village tract courts formed with members of local People's Councils. These arrangements ceased to operate following the imposition of military rule in September 1988, when a Supreme Court with five members was appointed. A chief justice, an attorney-general and a deputy attorney-general were also appointed. In March 2003 a deputy chief justice, four more justices and two further deputy attorney-generals were appointed.

Office of the Supreme Court
101 Pansodan St, Kyauktada Township, Yangon; tel. (1) 280751.

Chief Justice: U Aung Toe.

Attorney-General: Aye Maung.

Religion

Freedom of religious belief and practice is guaranteed. In 1992 an estimated 87.2% of the population were Buddhists, 5.6% Christians, 3.6% Muslims, 1.0% Hindus and 2.6% animists or adherents of other religions.

BUDDHISM

State Sangha Maha Nayaka Committee: c/o Dept of Promotion and Propagation of the Sasana, Kaba Aye Pagoda Precinct, Mayangone Township, Yangon; tel. (1) 660759.

CHRISTIANITY

Myanmar Naing-ngan Khrityan Athin-dawmyar Kaung-si (Myanmar Council of Churches): Myanmar Ecumenical Sharing Centre, 601 Pyay Rd, University PO, Yangon 11041; tel. (1) 537957; fax (1) 296848; e-mail oikom@yangon.net.mm; f. 1914 as Burma Representative Council of Mission; reconstituted as Burma Council of Churches in 1974; 13 mem. nat. churches, 9 mem. nat. Christian orgs; Pres. Rev. Saw Mar Gay Gyi; Gen. Sec. Rt Rev. Smith N. Za Thawng.

The Roman Catholic Church

Myanmar comprises three archdioceses and nine dioceses. At 31 December 2002 an estimated 1.1% of the total population were adherents.

Catholic Bishops' Conference of Myanmar
292 Pyi Rd, POB 1080, Sanchaung PO, Yangon 11111; tel. (1) 525868; fax (1) 527198; e-mail clspcbcm@mptmail.net.mm.

f. 1982; Pres. Most Rev. Charles Maung Bo (Archbishop of Yangon).

Archbishop of Mandalay: Most Rev. Paul Zingtung Grawng, Archbishop's House, 82nd and 25th St, Mandalay 05071; tel. (2) 36369.

Archbishop of Taunggyi: Most Rev. Matthias U Shwe, Archbishop's Office, Bayint Naung Rd, Taunggyi 06011; tel. (81) 21689; fax (81) 22164; e-mail matthias@myanmar.com.mm.

Archbishop of Yangon: Most Rev. Charles Maung Bo, Archbishop's House, 289 Theinbyu St, Botataung, Yangon; tel. (1) 246710.

The Anglican Communion

Anglicans are adherents of the Church of the Province of Myanmar, comprising six dioceses. The Province was formed in February 1970, and contained an estimated 45,000 adherents in 1985.

Archbishop of Myanmar and Bishop of Yangon: Most Rev. Samuel San Si Htay, Bishopscourt, 140 Pyidaungsu Yeiktha Rd, Dagon PO (11191), Yangon; tel. (1) 285379; fax (1) 251405.

Protestant Churches

Lutheran Bethlehem Church: 181–183 Theinbyu St, Mingala Taung Nyunt PO 11221, POB 773, Yangon; tel. (1) 246585; Pres. Rev. Jenson Rajan Andrews.

Myanmar Baptist Convention: 143 Minye Kyawswa Rd, POB 506, Yangon; tel. (1) 223231; fax (1) 221465; e-mail mbc<mbc@mptmail.net.mm; f. 1865 as Burma Baptist Missionary Convention; present name adopted 1954; 550,104 mems (1994); Pres. Rev. U Tha Din; Gen. Sec. Rev. Dr Simon Pau Khan En.

Myanmar Methodist Church: Methodist Headquarters, 22 Signal Pagoda Rd, Yangon; Bishop Zothan Mawia.

Presbyterian Church of Myanmar: Synod Office, Falam, Chin State; 22,000 mems; Rev. Sun Kanglo.

Other denominations active in Myanmar include the **Lisu Christian Church** and the **Salvation Army**.

The Press

DAILIES

Botahtaung (The Vanguard): 22–30 Strand Rd, Botahtaung PO, POB 539, Yangon; tel. (1) 274310; daily; Myanmar.

Guardian: 392–396 Merchant St, Botahtaung PO, POB 1522, Yangon; tel. (1) 270150; daily; English.

Kyehmon (The Mirror): 77 52nd St, Dazundaung PO, POB 819, Yangon; tel. (1) 282777; daily; Myanmar.

Myanmar Alin (New Light of Myanmar): 58 Komin Kochin Rd, Bahan PO, POB 21, Yangon; tel. (1) 250777; f. 1963; fmrly Loktha Pyithu Nezin (Working People's Daily); organ of the SPDC; morning; Myanmar; Chief Editor U Soe Myint; circ. 400,000.

New Light of Myanmar: 22–30 Strand Rd, Yangon; tel. (1) 297028; internet www.myanmar.com/nlm; f. 1963; fmrly Working People's Daily; organ of the SPDC; morning; English; Chief Editor U Kyaw Min; circ. 14,000.

PERIODICALS

A Hla Thit (New Beauty): 46 90th St, Yangon; tel. (1) 287106; international news.

Dana Business Magazine: 72 8th Street, Lanmadaw Township, Yangon; tel. and fax (1) 224010; e-mail dana@mptmail.net.mm; economic; Editor-in-Chief William Chen.

Do Kyaung Tha: Myawaddy Press, 184 32nd St, Yangon; tel. (1) 274655; f. 1965; monthly; Myanmar and English; circ. 17,000.

Gita Padetha: Yangon; journal of Myanma Music Council; circ. 10,000.

Guardian Magazine: 392/396 Merchant St, Botahtaung PO, POB 1522, Yangon; tel. (1) 296510; f. 1953; nationalized 1964; monthly; English; literary; circ. 11,600.

Kyee Pwar Yay (Prosperity): 296 Bo Sun Pat St, Yangon; tel. (1) 278100; economic; Editor-in-Chief U Myat Khine.

Moethaukpan (Aurora): Myawaddy Press, 184 32nd St, Yangon; tel. (1) 274655; f. 1980; monthly; Myanmar and English; circ. 27,500.

Myanma Dana (Myanmar's Economy): 210A, 36th St, Kyauktada PO, Yangon; tel. (1) 284660; economic; Editor-in-Chief U Thiha Saw.

Myanmar Morning Post: Yangon; f. 1998; weekly; Chinese; news; circ. 5,000.

Myanmar Times & Business Review: Level 1, 5 Signal Pagoda Rd, Dagon Township, Yangon; tel. (1) 242711; fax (1) 242669; e-mail myanmartimes@mptmail.net.mm; internet www.myanmar.com/myanmartimes; f. 2000; Editor-in-Chief Ross Dunkley.

Myawaddy Journal: Myawaddy Press, 184 32nd St, Yangon; tel. (1) 274655; f. 1989; fortnightly; news; circ. 8,700.

Myawaddy Magazine: Myawaddy Press, 184 32nd St, Yangon; tel. (1) 274655; f. 1952; monthly; literary magazine; circ. 4,200.

Ngwetaryi Magazine: Myawaddy Press, 184 32nd St, Yangon; tel. (1) 274655; f. 1961; monthly; cultural; circ. 3,400.

Pyinnya Lawka Journal: 529 Merchant St, Yangon; tel. (1) 283611; publ. by Sarpay Beikman Management Board; quarterly; circ. 18,000.

Shwe Thwe: 529 Merchant St, Yangon; tel. (1) 283611; weekly; bilingual children's journal; publ. by Sarpay Beikman Management Board; circ. 100,000.

Taw Win Journal (Royal Journal): 149 37th St, Yangon; news; Editor-in-Chief Soe Thein.

Teza: Myawaddy Press, 184 32nd St, Yangon; tel. (1) 274655; f. 1965; monthly; English and Myanmar; pictorial publication for children; circ. 29,500.

Thwe Thauk Magazine: Myawaddy Press, 185 48th St, Yangon; f. 1946; monthly; literary.

Ya Nant Thit (New Fragrance): 186 39th St, Yangon; tel. (1) 276799; international news; Editor-in-Chief U Chit Win Mg.

NEWS AGENCIES

Myanmar News Agency (MNA): 212 Theinbyu Rd, Botahtaung, Yangon; tel. (1) 270893; f. 1963; govt-controlled; Chief Editors U Zaw Min Thein (domestic section), U Kyaw Min (external section).

Foreign Bureaux

Agence France-Presse (AFP) (France): 12L Pyithu Lane, 7th Mile, Yangon; tel. (1) 661069; Correspondent U Khin Maung Thwin.

Agenzia Nazionale Stampa Associata (ANSA) (Italy): POB 270, Yangon; tel. (1) 511663; fax (1) 526400; e-mail mtgroup@myanmar.com.mm; Rep. (vacant).

Associated Press (AP) (USA): 283 U Wisara Rd, Sanchaung PO 11111, Yangon; tel. (1) 527014; Rep. Aye Aye Win.

Xinhua (New China) News Agency (People's Republic of China): 105 Leeds Rd, Yangon; tel. (1) 221400; Correspondent Zhang Yuhfei.

Reuters (United Kingdom) and **UPI** (USA) are also represented in Myanmar.

Publishers

Hanthawaddy Press: 157 Bo Aung Gyaw St, Yangon; f. 1889; textbooks, multilingual dictionaries; Man. Editor U Zaw Win.

Knowledge Publishing House: 130 Bo Gyoke Aung San St, Yegyaw, Yangon; art, education, religion, politics and social sciences.

Kyipwaye Press: 84th St, Letsaigan, Mandalay; tel. (2) 21003; arts, travel, religion, fiction and children's.

Myawaddy Press: 184 32nd St, Yangon; tel. (1) 285996; journals and magazines; CEO U Thein Sein.

Sarpay Beikman Management Board: 529 Merchant St, Yangon; tel. (1) 283611; f. 1947; encyclopaedias, literature, fine arts and general; also magazines and translations; Chair. Aung Htay.

Shumawa Press: 146 West Wing, Bogyoke Aung San Market, Yangon; mechanical engineering.

Shwepyidan: 12A Haiaban, Yegwaw Quarter, Yangon; politics, religion, law.

Smart and Mookerdum: 221 Sule Pagoda Rd, Yangon; arts, cookery, popular science.

Thu Dhama Wadi Press: 55–56 Maung Khine St, POB 419, Yangon; f. 1903; religious; Propr U Tin Htoo; Man. U Pan Maung.

Government Publishing House

Printing and Publishing Enterprise: 365/367 Bo Aung Kyaw St, Kyauktada Township, Yangon; tel. (1) 294645; Man. Dir Mya Mya.

PUBLISHERS' ASSOCIATION

Myanma Publishers' Union: 146 Bogyoke Market, Yangon.

Broadcasting and Communications

TELECOMMUNICATIONS

Posts and Telecommunications Department: Block 68, Ayeyar Wun Rd, South Dagon Township, Yangon; tel. (1) 591388; fax (1) 591383; e-mail dg.ptd@mptmail.net.mm; internet www.mcpt.gov.mm/ptd/; Dir-Gen. of Posts and Telecommunications Dept U Kyi Than.

Myanma Posts and Telecommunications (MPT): 43 Bo Aung Gyaw St, Kyauktada Township, Yangon; tel. (1) 297722; fax (1) 251911; internet www.mpt.net.mm; Man. Dir Col Maung Maung Tin.

BROADCASTING

Radio

Myanma TV and Radio Department (MTRD): 426 Pyay Rd, Kamayut 11041, Yangon; POB 1432, Yangon 11181; tel. (1) 531850; fax (1) 530211; radio transmissions began in 1937; broadcasts in Bamar, Arakanese (Rakhine), Shan, Karen (Kayin), Kachin, Kayah, Chin, Mon and English; Dir-Gen. U Kyi Lwin; Dir of Radio Broadcasting U Ko Ko Htway.

In 1992 the National Coalition Government of the Union of Burma (NCGUB) began broadcasting daily to Myanmar from Norway under the name Democratic Voice of Burma (DVB). In 1995 it was believed that the DVB was being operated by Myanma student activists from the Norway-Burma Council, without any formal control by the NCGUB.

Television

Myanma TV and Radio Department (MTRD): 426 Pyay Rd, Yangon; tel. (1) 531850; fax (1) 530211; f. 1946; colour television transmissions began in 1980; Dir-Gen. U Khin Maung Htay; Dir of Television Broadcasting U Phone Myint.

TV Myawaddy: Hmawbi, Hmawbi Township, Yangon; tel. (1) 620270; f. 1995; military broadcasting station transmitting programmes via satellite.

Finance

(cap. = capital; res = reserves; dep. = deposits; m. = million; brs = branches; amounts in kyats)

BANKING

In July 1990 new banking legislation was promulgated, reorganizing the operations of the Central Bank, establishing a state-owned development institution, the Myanma Agricultural and Rural Development Bank, and providing for the formation of private-sector banks and the opening of branches of foreign banks.

Central Bank

Central Bank of Myanmar: 26A Settmu Rd, POB 184, Yankin Township, Yangon; tel. (1) 543511; fax (1) 543621; e-mail cbm.ygn@mptmail.net.mm; f. 1968; bank of issue; cap. 350m., dep. 13,545m.; Gov. U Kyaw Kyaw Maung; 37 brs.

State Banks

Myanma Economic Bank (MEB): 1–19 Sule Pagoda Rd, Yangon; tel. (1) 289345; fax (1) 283679; provides domestic banking network throughout the country; Man. Dir U Kyaw Kyaw.

Myanma Foreign Trade Bank: 80–86 Maha Bandoola Garden St, POB 203, Kyauktada Township, Yangon; tel. (1) 284911; fax (1) 289585; e-mail mftb-hoygn@mptmail-net.mm; f. 1976; cap. 110m., res 483.9m., dep. 2,425.9m. (March 1999); handles all foreign exchange and international banking transactions; Chair. U Ko Ko Gyi; Sec. U Tin Maung Aye.

Development Banks

Myanma Agricultural Development Bank (MADB): 1–7 cnr of Latha St and Kanna Rd, Yangon; tel. (1) 226734; f. 1953 as State Agricultural Bank, reconstituted as Myanma Agricultural and Rural Development Bank 1990 and as above 1996; state-owned; cap. 60.0m., dep. 615.6m. (Sept. 1993); Man. Dir U Chit Swe.

Myanma Investment and Commercial Bank (MICB): 170/176 Bo Aung Kyaw St, Botataung Township, Yangon; tel. (1) 250509; fax (1) 256871; e-mail micbhoygn@mptmail.net.mm; f. 1989; state-owned; cap. 400m., res 786m., dep. 7,035m. (March 2000); Chair. and Man. Dir U Mya Than; 1 br.

Private Banks

In 2003 a crisis in Myanmar's private banking sector forced the closure of six of the country's 20 private banks. Following government intervention, three subsequently reopened in early 2004, having been cleared of committing banking irregularities.

Asia Wealth Bank: 638 cnr of Maha Bandoola St and 22nd St, Latha Township, Yangon; tel. (1) 243700; fax (1) 245456; e-mail customer@awb.com.mm; internet www.awb.com.mm; Chair. Win Maung; Vice-Chair. Aik Htun; 38 brs.

Asian Yangon Bank Ltd: 319/321 Maha Bandoola St, Botataung Township, Yangon; tel. (1) 245825; fax (1) 245865; f. 1994 as Asian Yangon International Bank Ltd; name changed as above in 2000; Gen. Man. Tun Nyunt.

Co-operative Bank Ltd: 334–336 Kanna Rd, Yangon; tel. (1) 272641; fax (1) 283063; Gen. Man. U Nyunt Hlaing.

First Private Bank Ltd (FPB): 619–621 Merchant St, Pabedan Township, Yangon; tel. (1) 289929; fax (1) 242320; e-mail fpb.hq@mptmail.net.mm; f. 1992 as the first publicly-subscribed bank; fmrly the Commercial and Development Bank Ltd; provides loans to private business and small-scale industrial sectors; cap. 765.35m. (March 2003); Chair. Dr Sein Maung; 15 brs.

Innwa Bank Ltd: cnr 35th St and Merchant St, Yangon.

Kanbawza Bank Ltd: 615/1 Pyay Rd, Kamayut Township, Yangon; tel. (1) 538075; fax (1) 538069; e-mail kbz@mptmail.net.mm; Chair. U Aung Ko Win; 22 brs.

Myanma Citizens Bank Ltd (MCB): 383 Maha Bandoola St, Kyauktada Township, Yangon; tel. (1) 273512; fax (1) 245932; f. 1991; Chair. U Hla Tin.

Myanma Oriental Bank Ltd: 166–168 Pansodan St, Kyauktada Township, Yangon; tel. (1) 246594; fax (1) 253217; e-mail mobl.ygn@mptmail.net.mm; Chair. U Myat Kyaw; Man. Dir and CEO U Win Myint.

Myanma Universal Bank: 81 Theinbyu Rd, Yangon; tel. (1) 297339; fax (1) 245449; f. 1995.

Myanmar Industrial Development Bank Ltd: 26–42 Pansodan St, Kyauktada Township, Yangon; tel. (1) 249536; fax (1) 249529; f. 1996; cap. US $335m.

Myanmar May Flower Bank Ltd: 1B Yadanar Housing Project, 9 Mile Pyay Rd, Mayangone Township, Yangon; tel. (1) 666112; fax (1) 666110; e-mail mmb-hq@mptmail.net.mm; Chair. U Kyaw Win.

Myawaddy Bank Ltd: 24–26 Sule Pagoda Rd, Kyauktada Township, Yangon; tel. (1) 283665; fax (1) 250093; e-mail mwdbankygn@mtpt400.stems.com; Gen. Mans U Tun Kyi, U Mya Min.

Tun Foundation Bank Ltd: 165–167 Bo Aung Gyaw St, Yangon; tel. (1) 270710; Chair. U Thein Tun.

Yadanabon Bank Ltd: 26th St, cnr of 84th and 85th Sts, Mandalay; tel. (2) 23577.

Yangon City Bank Ltd: 12–18 Sepin St, Kyauktada Township, Yangon; tel. (1) 289256; fax (1) 289231; f. 1993; auth. cap. 500m.; 100% owned by the Yangon City Development Committee; Chair. Col MYINT AUNG.

Yoma Bank Ltd: 1 Kungyan St, Mingala Taung Nyunt Township, Yangon; tel. (1) 242138; fax (1) 246548; Chair. SERGE PUN.

Foreign Banks

By November 2003 18 foreign banks had opened representative offices in Yangon.

STOCK EXCHANGE

Myanmar Securities Exchange Centre: 1st Floor, 21–25 Sule Pagoda Rd, Yangon; tel. (1) 283984; f. 1996; jt venture between Japan's Daiwa Institute of Research and Myanma Economic Bank; Man. Dir EIJI SUZUKI.

INSURANCE

At the end of November 2003 there were three representative offices of foreign insurance companies in Myanmar.

Myanma Insurance: 627–635 Merchant St, Yangon; tel. (1) 256244; fax (1) 250275; e-mail MYANSURE@mptmail.net.com; f. 1976; govt-controlled; Man. Dir Col THEIN LWIN.

Trade and Industry

GOVERNMENT AGENCIES

Inspection and Agency Service: 383 Maha Bandoola St, Yangon; tel. (1) 276048; fax (1) 284823; works on behalf of state-owned enterprises to promote business with foreign companies; Man. Dir U OHN KHIN.

Myanmar Investment Commission: Ministry of National Planning and Economic Development, 653–691 Merchant St, Pabedan Township, Yangon; tel. (1) 241918; fax (1) 282101; Chair. U THAUNG; Vice-Chair. Maj.-Gen. TIN HTUT.

Union of Myanmar Economic Holdings: 72–74 Shwadagon Pagoda Rd, Yangon; tel. (1) 78905; f. 1990; public holding co; auth. cap. 10,000m. kyats; 40% of share capital subscribed by the Ministry of Defence and 60% by members of the armed forces.

CHAMBER OF COMMERCE

Chamber of Commerce and Industry: 504–506 Merchant St, Kyauktada Township, Yangon; tel. (1) 243151; fax (1) 248177; e-mail umcci@mptmail.net.mm; f. 1919; Gen. Sec. ZAW MIN WIN.

UTILITIES

Electricity

Myanma Electric Power Enterprise (MEPE): 197–199 Lower Kyimyindine Rd, Yangon; tel. (1) 220918; fax (1) 221006; e-mail mepe@mptmail.net.mm; Man. Dir U YAN NAING.

Water

Mandalay City Development Committee (Water and Sanitation Dept): cnr of 26th and 72nd Sts, Mandalay; tel. (2) 36173; f. 1992; Head of Water and Sanitation Dept U TUN KYI.

Water Resources Utilization Department (WRUD): Ministry of Agriculture and Irrigation, Thiri Mingala Lane, off Kaba Aye Pagoda Rd, Yangon; tel. (1) 666359; fax (1) 667456; f. 1995.

Yangon City Development Committee (Water and Sanitation Dept): City Hall, Yangon; tel. (1) 204052; fax (1) 246016; e-mail priycdc@mptmail.net.mm; internet www.yangoncity.com.mm; f. 1992; Head of Water and Sanitation Dept U ZAW WIN.

CO-OPERATIVES

In 1993/94 there were 22,800 co-operative societies, with a turnover of 23,603m. kyats. This was estimated to have increased to 24,760 societies, with a turnover of 20,927m. kyats, in 1994/95.

Central Co-operative Society (CCS) Council: 334/336 Strand Rd, Yangon; tel. (1) 274550; Chair. U THAN HLANG; Sec. U TIN LATT.

Co-operative Department: 259–263 Bogyoke Aung San Rd, Yangon; tel. (1) 277096; Dir-Gen. U MAUNG HTI.

MAJOR COMPANIES

State Enterprises

Livestock Feedstuff and Milk Products Enterprise: Station Rd, Insein Township, Yangon; tel. (1) 642019; fax (1) 642023; Man. Dir U KHIN MAUNG AYE.

Myanma Agricultural Produce Trading (MAPT): 302/304 cnr Bogyoke St and Pansodan St, Kyauktada Township, Yangon; tel. (1) 254018; fax (1) 371734; e-mail mapt.hr@mptmail.net.mm; Man. Dir U MIN HLA AUNG.

Myanma Agriculture Service: Kanbe Rd, Yankin Township, Yangon; tel. (1) 663541; fax (1) 283651; f. 1972; fmrly Agriculture Corpn; name changed as above 1989; Man. Dir Dr MYA MAUNG.

Myanma Ceramic Industries: 192 Kaba Aye Pagoda Rd, Bahan Township, POB 11201, Yangon; tel. (1) 566077; fax (1) 578226; e-mail MMI1MCI@mtmail.net.mm; produces cement, glass, pottery, marble, asbestos sheets and bricks; Man. Dir U THAN SHWE.

Myanma Cotton and Sericulture Enterprise: Thiri Mingala Lane, off Kaba Aye Pagoda Rd, Yangon; tel. (1) 666067; fax (1) 666065; f. 1994; Man. Dir Dr THEIN HTAY.

Myanma Export-Import Services: 622–624 Merchant St, Yangon; tel. (1) 550661; fax (1) 289587; Man. Dir U THAUNG SEIN.

Myanma Farms Enterprise: Pyi Rd, 9th Mile, Yangon; tel. (1) 665631; f. 1957; Man. Dir Col U NYUNT MG.

Myanma Foodstuff Industries: 192 Kaba Aye Pagoda Rd, Bahan Township, POB 11201, Yangon; tel. (1) 561298; fax (1) 561802; e-mail mfi.mi1@mptmail.net.mm; produces foodstuffs, incl. soft drinks, biscuits, lagers and distilled spirits; Man. Dir U SOE HLAING.

Myanma Gems Enterprise: 66 Kaba Aye Pagoda Rd, Mayangone Township, Yangon; tel. (1) 660365; fax (1) 665092; f. 1976; Man. Dir U KHIN OO.

Myanma General and Maintenance Industries: 192 Kaba Aye Pagoda Rd, Yangon; tel. (1) 566845; fax (1) 566859; mfr of shoes, leather products, paper and corrugated cartons; Gen. Man U MYINT SWE.

Myanma Heavy Industries: 56 Kaba Aye Pagoda Rd, POB 370, Yangon; tel. (1) 662880; fax (1) 660465; f. 1960; mfr of vehicles, electrical appliances, electronic goods and agricultural machinery; Man. Dir U SOE THEIN.

Myanma Jute Industries: 257 Yangon-Insein Rd, Yangon; tel. (1) 578946; Man. Dir U MYINT MAUNG.

Myanma Oil and Gas Enterprise: 74–80 Minye Kyawswa Rd, Lanmadaw Township, Yangon; tel. (1) 222874; fax (1) 222964; fmrly Myanma Oil Corpn (previously Burma Oil Co); nationalized 1963; Man. Dir U PE KYI.

Myanma Paper and Chemical Industries: 192 Kaba Aye Pagoda Rd, PO 11201, Yangon; tel. (1) 565776; fax (1) 565817; Man. Dir U ZAW SEIN WIN.

Myanma Perennial Crops Enterprise: Thiri Mingala Lane, off Kaba Aye Pagoda Rd, Yangon; tel. (1) 666011; fax (1) 667446; f. 1994.

Myanma Petrochemical Enterprise: 23 Minye Kyawswa Rd, Yangon; tel. (1) 222822; fax (1) 221723; f. 1975; Man. Dir HLAING MYINT SAN.

Myanma Petroleum Products Enterprise: 74–80 Minye Kyawswa Rd, Lanmadaw Township, Yangon; tel. (1) 221098; Man. Dir U AUNG HLAING.

Myanma Pharmaceutical Industries: 192 Kabu Aye Pagoda Rd, Bahan Township, POB 11201, Yangon; tel. (1) 566740; fax (1) 566722; Man. Dir U TIN HLAING.

Myanma Sugarcane Enterprise: Thiri Mingala Lane, off Kaba Aye Pagoda Rd, Yangon; tel. (1) 666041; fax (1) 666107; f. 1994; Man. Dir U MYO MYINT .

Myanma Textile Industries: 192 Kaba Aye Pagoda Rd, Bahan Township, POB 11201, Yangon; tel. (1) 566050; fax (1) 573373; produces wide range of fabrics and yarns; sales 23,365,040m. kyats (1998); Man. Dir U SAN KYI.

Myanma Timber Enterprise: POB 206, Ahlone, Yangon; tel. (1) 220637; fax (1) 221816; f. 1948; extraction, processing, and main exporter of teak and other timber, veneers, plywood and other forest products; Gen. Man. U MYINT KYU PE.

No. 1 Mining Enterprise: 90 Kanbe Rd, Yankin, Yangon; tel. (1) 577453; fax (1) 577309; e-mail 1myn@mptmail.net.mm; development and mining of non-ferrous metals; Man. Dir U KO KO.

No. 2 Mining Enterprise: 90 Kanbe Rd, Yankin, Yangon; tel. (1) 544089; fax (1) 577309; e-mail ME-2@mptmail.net.mm; development and mining of tin, tungsten and antimony; Man. Dir Col TIN WIN.

No. 3 Mining Enterprise: 90 Kanbe Rd, Yankin, Yangon; tel. (1) 577444; fax (1) 566224; e-mail ME3Myanmar@mptmail.net.mm; govt-controlled; production of pig iron, steel billet, steel grinding balls, coal, barytes, gypsum, limestone, chromite, antimony, various industrial minerals, etc.; Man. Dir U San Tun.

Private Companies

Daewoo Electronics: 139 MHI Compound, Kaba Aye Pagoda Rd, POB 737, Yangon; tel. (1) 64886; fax (1) 62870; f. 1990; South Korean company in joint venture with Myanma Heavy Industries; mfr of televisions, refrigerators and audio systems; Man. Dir Kim Chang Hun.

Myanmar Inctech: Yangon; f. 1991; auth. cap. 50m. kyats; joint venture between Inctech (Singapore); import, export and leasing of finished and semi-finished construction-related products.

Myanmar International Hotels: 77/91 Sule Pagoda Rd, Yangon; tel. (1) 62857; joint venture between Strand Hotels International and Myanmar Hotels and Tourism Services; construction and renovation of hotels.

Myanmar Ivanhoe Copper Co Ltd: 70 I Bo Chein St, Pyay Rd, Yangon; tel. (1) 514194; fax (1) 514194; jt venture between Canadian Ivanhoe Mines Ltd and No. 1 Mining Enterprise; copper mining.

Myanmar Natsteel Hardware Centre: 262 Seikkantha St, Yangon; tel. (1) 84985; f. 1991; joint venture between Natsteel Trade International (Singapore) and the local Construction and Electrical Stores Trading; mfg and marketing of building materials and steel products.

WORKERS' AND PEASANTS' COUNCILS

Conditions of work are stipulated in the Workers' Rights and Responsibilities Law, enacted in 1964. Regional workers' councils ensure that government directives are complied with, and that targets are met on a regional basis. In January 1985 there were 293 workers' councils in towns, with more than 1.8m. members. They are co-ordinated by a central workers' organization in Yangon, formed in 1968 to replace trade union organizations which had been abolished in 1964. The Myanma Federation of Trade Unions operates in exile.

Peasants' Asiayone (Organization): Yangon; tel. (1) 82819; f. 1977; peasants' representative org.; Chair. Brig.-Gen. U Than Nyunt; Sec. U San Tun.

Workers' Unity Organization: Central Organizing Committee, 61 Thein Byu St, Yangon; tel. (1) 284043; f. 1968; workers' representative org.; Chair. U Ohn Kyaw; Sec. U Nyunt Thein.

Transport

All railways, domestic air services, passenger and freight road transport services and inland water facilities are owned and operated by state-controlled enterprises.

RAILWAYS

The railway network comprised 3,955 km of track in 1996/97, most of which was single track.

Myanma Railways: Bogyoke Aung San St, POB 118, Yangon; tel. (1) 280508; fax (1) 284220; f. 1877; govt-operated; Man. Dir U Aung Thein; Gen. Man. U Thaung Lwin.

ROADS

In 1996 the total length of the road network in Myanmar was an estimated 28,200 km, of which an estimated 3,440 km were paved. In 2001/02 the total length of road accessible to motor vehicles was 28,598 km.

Road Transportation Department: 375 Bogyoke Aung San St, Yangon; tel. (1) 284426; fax (1) 289716; f. 1963; controls passenger and freight road transport; in 1993/94 operated 1,960 haulage trucks and 928 passenger buses; Man. Dir U Ohn Myint.

INLAND WATERWAYS

The principal artery of traffic is the River Ayeyarwady (Irrawaddy), which is navigable as far as Bhamo, about 1,450 km inland, while parts of the Thanlwin and Chindwinn rivers are also navigable.

Inland Water Transport: 50 Pansodan St, Yangon; tel. (1) 222399; fax (1) 286500; govt-owned; operates cargo and passenger services throughout Myanmar; 36m. passengers and 3.1m. tons of freight were carried in 1993/94; Man. Dir U Khin Maung.

SHIPPING

Yangon is the chief port. Vessels with a displacement of up to 15,000 tons can be accommodated.

In 2003 the Myanma merchant fleet totalled 122 vessels, with a combined displacement of 433,039 grt.

Myanma Port Authority: 10 Pansodan St, POB 1, Yangon; tel. (1) 280094; fax (1) 295134; f. 1880; general port and harbour duties; Man. Dir U Tin Oo; Gen. Man. U Hlaing Soon.

Myanma Five Star Line: 132–136 Theinbyu Rd, POB 1221, Yangon; tel. (1) 295279; fax (1) 297669; e-mail mfsl.myr@mptmail.net.mm; f. 1959; cargo services to the Far East and Australia; Man. Dir U Khin Maung Kyi; Gen. Man. U Kyaw Zaw; fleet of 26 coastal and ocean-going vessels.

CIVIL AVIATION

Mingaladon Airport, near Yangon, is equipped to international standards. The newly built Mandalay International Airport was inaugurated in September 2000. In 2002 plans for the construction of the country's third international airport, Hanthawaddy International Airport in Bago (Pegu), were finally approved; the airport was scheduled to become operational in 2007.

Department of Civil Aviation: Mingaladon Airport, Yangon; tel. (1) 665144; fax (1) 665124; e-mail dca.myanmar@mpt.mail.net.mm; Dir-Gen. U Win Maung.

Air Mandalay: 146 Dhammazedi Rd, Bahan Township, Yangon; tel. (1) 525488; fax (1) 525937; e-mail info@airmandalay.com; internet www.airmandalay.com; f. 1994; Myanmar's first airline; jt venture between Air Mandalay Holding and Myanma Airways; operates domestic services and regional services to Chiang Mai and Phuket, Thailand, and Siem Reap, Cambodia; Chair. Dr Tun Chin; Man. Dir Eric Kang Tian Lye; 242 employees.

Air Myanmar: Yangon; f. 2004; jt venture between Myanmar Airways, Dawn Light Co, Singapore-based Fast Growth Associates Ltd and Hong Kong-based Cathay Aviation; international services to Sydney, Australia, via Singapore and to Fukuoka, Japan, via Bangkok, Shanghai and Seoul.

Myanmar Airways (MA): 123 Sule Pagoda Rd, Yangon; tel. (1) 80710; fax (1) 255305; e-mail 8mpr@maiair.com.mm; internet www.maiair.com; f. 1993; govt-controlled; internal network operates services to 21 airports; Chief Operating Officer Prithpal Singh.

Myanmar Airways International (MAI): 08–02 Sakura Tower, 339 Bogyoke Aung San Rd, Yangon; tel. (1) 255260; fax (1) 255305; e-mail management@maiair.com; internet www.maiair.com; f. 1993; govt-owned; established by Myanmar Airways in jt venture with Highsonic Enterprises of Singapore to provide international services; operates services to Bangkok, Dhaka, Hong Kong, Kuala Lumpur and Singapore; Man. Dir Gerard de Vaz.

United Myanmar Air: Summit Parkview Hotel, Yangon; f. 2003; jt venture between Myanmar Airways and Sunshine Strategic Investments Holdings of Hong Kong; international services to Bangkok, Hong Kong, Kuala Lumpur and Singapore; CEO Edward Tan.

Yangon Airways: MMB Tower, Level 5, 166 Upper Pansodan Rd, Mingalar Taungnyunt Township, Yangon; tel. (1) 383100; fax (1) 383109; e-mail ygnhq@yangonair.com.mm; internet www.yangonair.com; f. 1996; domestic services to 13 destinations.

Tourism

Yangon, Mandalay, Taunggyi and Pagan possess outstanding palaces, Buddhist temples and shrines. The number of foreign visitors to Myanmar declined severely following the suppression of the democracy movement in 1988. In the early 1990s, however, the Government actively promoted the revival of the tourism industry, and between 1995 and 1998 alone the number of hotel rooms almost doubled, reaching a total of nearly 14,000. In 2001 there were 204,862 foreign tourist arrivals (compared with only 5,000 in 1989). In 2001 revenue from tourism totalled an estimated US $45m.

Myanmar Hotels and Tourism Services: 77–91 Sule Pagoda Rd, Yangon 11141; tel. (1) 282013; fax (1) 254417; e-mail mtt.mht@mptmail.net.mm; govt-controlled; manages all hotels, tourist offices, tourist department stores and duty-free shops; Man. Dir U Kyi Htun.

Myanmar Tourism Promotion Board: 5 Signal Pagoda Rd, Yangon; tel. (1) 243639; fax (1) 245001; e-mail mtpb@mptmail.net.mm; internet www.myanmar-tourism.com.

Myanmar Travels and Tours: 77–91 Sule Pagoda Rd, POB 559, Yangon 11141; tel. (1) 374281; fax (1) 254417; e-mail mtt.mht@mptmail.net.mm; internet www.myanmars.net/mtt/; f. 1964; govt tour operator and travel agent; handles all travel arrangements for groups and individuals; Gen. Man. U Htay Aung.

Union of Myanmar Travel Association (UMTA): Bldg 69, No 609B, 6th Floor, Yuzana Condo Tower, cnr Shwegonedaing Rd and Kabaraye Pagoda Rd, Bahan Township, Yangon; tel. (1) 559673; fax

(1) 545707; e-mail UMTA@mptmail.net.mm; internet www.umtanet.com; f. 2002; organizes private travel agencies and tour operators; promotes Myanmar as a tourist destination; Chair. U KHIN ZAW.

Defence

In August 2003 the total strength of the armed forces was reported to be some 488,000; army some 350,000, navy 16,000, air force 15,000. Military service is voluntary. Paramilitary forces comprise a people's police force (72,000 men) and a people's militia (35,000 men). As Myanmar maintains a policy of neutrality and has no external defence treaties, the armed forces are engaged mainly in internal security duties.

Defence Expenditure: Budgeted at an estimated US $1,500m. for 2003.

Commander-in-Chief of the Defence Services: Field Marshal THAN SHWE.

Commander-in-Chief of the Army: Dep. Senior Gen. MAUNG AYE.

Commander-in-Chief of the Navy: Rear-Adm. SOE THEIN.

Commander-in-Chief of the Air Force: Maj.-Gen. MYAT HEIN.

Education

The organization and administration of education is the responsibility of the Ministry of Education. Education is compulsory for five years between five and 10 years of age. Pre-school education begins at four years of age. Primary education lasts for five years between the ages of five and 10. Education in lower secondary, or middle, schools lasts four years from the age of 10 to 14, when pupils take the external government examination. Upper secondary, or high, school lasts for a further two years. In 1995 the total enrolment at secondary level was equivalent to 32% of children in the relevant age-group. In the same year enrolment at primary and secondary levels was equivalent to 65% of the school-age population.

In 1994/95 there were 45 institutes of higher education and six universities. Student enrolment in tertiary education totalled 245,317 in that year (not including students enrolled in medical science courses), including 89,717 university students.

In 2000/01 government expenditure on education was 31,345m. kyats (14.2% of total expenditure).

Bibliography

GENERAL

Aung San Suu Kyi. *Freedom from Fear and Other Writings*. Harmondsworth, Penguin, 1991.

Aung San Suu Kyi: Letters From Burma. London, Penguin, 1997.

Aung San Suu Kyi: The Voice of Hope. Conversations with Alan Clements. London, Penguin, 1997.

Becka, Jan. *Historical Dictionary of Myanmar*. Metuchen, NJ, Scarecrow Press, 1995.

Bradley, David. *Myanmar, A Comparative Study*. Canberra, Australian Government Publishing Service, 1992.

Evans, Grant, Hutton, Chris, and Kuah Khun Eng (Eds). *Where China meets Southeast Asia: Social and Cultural Change in the Border Regions*. New York, NY, St Martin's Press, 2000.

Herbert, Patricia M. *Burma (World Bibliographical Series)*. Oxford and Denver, CO, Clio Press, 1991.

The Life of the Buddha. London, British Library Board, 1993.

Images Asia. *No Childhood At All: A Report about Child Soldiers in Burma*. Chiang Mai, Images Asia, 1996.

Mawdsley, James. *The Heart Must Break: The Fight for Democracy and Truth in Burma*. London, Century, 2001.

Mi Mi Khaing. *Burmese Family*. London, Longman, Green, 1946.

The World of Burmese Women. London, Zed Books, 1984.

Nash, Manning. *The Golden Road to Modernity: Village Life in Contemporary Burma*. Chicago, IL, University of Chicago Press, 1965.

Open Society Institute. *Country in Crisis: A Burma Handbook*. New York, NY, Soros Foundation, 1997.

Rodrigues, Yves. *Nat-Pwe: Burma's Supernatural Sub-Culture*. Gartmore, Kiscadale Publications, 1993.

Shulman, Frank Joseph. *Burma: An Annotated Bibliographical Guide to International Doctoral Dissertation Research 1898–1985*. Lanham, MD, University Press of America, 1986.

Shwe Yoe (Sir George Scott). *The Burman: His Life and Notions*. London, 1982.

Spiro, Melford E. *Buddhism and Society: A Great Tradition and its Burmese Vicissitudes*. London, Allen and Unwin, 1971.

Kinship and Marriage in Burma. Berkeley, CA, University of California Press, 1977.

Anthropological Other or Burmese Brother? Studies in Cultural Analysis. New Brunswick, NJ, Transaction Publishers, 1992.

Steinberg, David I. *Burma: A Socialist Nation of South-East Asia*. Boulder, CO, Westview Press, 1982.

Tinker, H. *The Union of Burma*. 4th Edn, London, Oxford University Press, 1967.

Tucker, Shelby. *Among Insurgents: Walking through Burma*. London, Flamingo, 2001.

Victor, Barbara. *The Lady: Aung San Suu Kyi: Nobel Laureate and Burma's Prisoner*. London, Faber & Faber, 1998.

Yegar, Moshe. *Between Integration and Secession: The Muslim Communities of the Southern Philippines, Southern Thailand, and Western Burma/Myanmar*. Lanham, MD, Lexington Books, 2002.

HISTORY

Allen, Louis. *Burma: The Longest War 1941–45*. London, Dent, 1985.

Apple, Betsy. *School for Rape: The Burmese Military and Sexual Violence*. Bangkok, EarthRights, 1998.

Aung San Suu Kyi. *Aung San of Burma*. Edinburgh, Kiscadale Publications, 1991.

Burma and India: Some Aspects of Intellectual Life Under Colonialism. Shimla, Indian Institute of Advanced Study, Allied Publishers Pvt Ltd, 1990.

Aung-Thwin, Michael. *Pagan: The Origins of Modern Burma*. Honolulu, HI, University of Hawaii Press, 1985.

Myth and History in the Historiography of Early Burma: Paradigms, Primary Sources, and Prejudices. Athens, OH, Ohio University Center for International Studies, 1998.

Aye Kyaw. *The Voice of Young Burma*. New York, NY, Cornell University Press, 1993.

Ba Maw. *Breakthrough in Burma*. New Haven, CT, Yale University Press, 1968.

Bachoe, Ralph, and Stothard, Debbie (Eds). *From Consensus to Controversy: ASEAN's Relationship with Burma's SLORC*. Bangkok, Altsean, 1997.

Badgley, John H. (Ed.) *Reconciling Burma / Myanmar: Essays on US Relations with Burma*. Seattle, WA, National Bureau of Asian Research, 2004.

Becka, J. *The National Liberation Movement in Burma during the Japanese Occupation Period (1941–45)*. Prague, Oriental Institute of Academia, Publishing House of the Czechoslovak Academy of Sciences, 1983.

Blackburn, Terence. *The British Humiliation of Burma*. Bangkok, Orchid Press, 2001.

Cady, J. F. A. *History of Modern Burma*. Ithaca, NY, Cornell University Press, 1958.

The United States and Burma. Cambridge, MA, Harvard University Press, 1976.

Clements, Alan. *Burma: The Next Killing Fields?* Berkeley, CA, Odonian Press, 1992.

Clements, Alan, and Kean, Leslie. *Burma's Revolution of the Spirit: The Struggle for Democratic Freedom and Dignity*. New York, NY, Aperture, 1995.

Falla, Jonathan. *True Love and Bartholomew: Rebels on the Burmese Border*. Cambridge, Cambridge University Press, 1991.

Fink, Christina. *Living Silence: Burma under Military Rule*. London, Zed Books, 2001.

Furnivall, John S. *The Fashioning of Leviathan: The Beginnings of British Rule in Burma*. Canberra, Department of Anthropology Occasional Paper, Research School of Pacific Studies, Australian National University, 1991.

Ghosh, Parimal. *Brave Men of the Hills: Resistance and Rebellion in Burma, 1825–1932.* London, Hurst & Co, 2000.

Gooden, Christian. *Three Pagodas: A Journey down the Thai–Burmese Border.* Halesworth, Jungle Books, 1996.

Hall, D. G. E. *Early English Intercourse with Burma 1587–1743.* London, Frank Cass, 1968.

Images Asia. *Nowhere To Go: A Report on the 1997 SLORC Offensive against Duplaya District (KNU 6th Brigade), Karen State, Burma.* Chiang Mai, Images Asia, 1997.

A Question of Security: A Retrospective on Cross-border Attacks on Thailand's Refugee and Civilian Communities along the Burmese Border since 1995. Chiang Mai, Images Asia, 1998.

All Quiet on the Western Front? The Situation in Chin State and Sagaing Division, Burma. Chiang Mai, Images Asia, 1998.

Khin Yi. *The Dobama Movement in Burma.* Vol. 2. Ithaca, NY, Cornell University Press, 1988.

Kin Oung. *Who Killed Aung San?* Bangkok, White Lotus, 1993.

Koenig, William J. *The Burmese Polity, 1752–1819: Politics, Administration, and Social Organization in the Early Kon-baung Period.* Ann Arbor, MI, Michigan Papers on South and Southeast Asia, Center for South and Southeast Asian Studies, University of Michigan, Number 34, 1990.

Lehman, F. K. (Ed.). *Military Rule in Burma since 1962.* Singapore, Institute of Southeast Asian Studies, 1981.

Liberman, Victor B. *Burmese Administrative Cycles: Anarchy and Conquest 1580–1760.* Princeton, NJ, Princeton University Press, 1984.

Lintner, Bertil. *Outrage: Burma's Struggle for Democracy.* Hong Kong, Review Publishing Co Ltd, 1989.

Land of Jade: A Journey Through Insurgent Burma. Edinburgh, Kiscadale Publications, 1990.

The Rise and Fall of the Communist Party of Burma. Ithaca, NY, Cornell University Press, 1990.

Burma in Revolt: Opium and Insurgency Since 1948. Boulder, CO, Westview Press, 1994.

The Kachin: Lords of Burma's Northern Frontier. Chiang Mai, Teak House, 1997.

Marshall, Andrew. *The Trouser People: A Story of Burma in the Shadow of the Empire.* Boulder, CO, Counterpoint Press, 2002.

Maung Htin Aung. *A History of Burma.* New York, NY, Columbia University Press, 1968.

Maung Maung. *Burmese Nationalist Movements: 1940–1948.* Honolulu, HI, University of Hawaii Press, 1990.

The 1988 Uprising in Burma. New Haven, CT, Yale University Southeast Asia Studies, 1999.

McRae, Alister. *Scots in Burma: Golden Times in a Golden Land.* Edinburgh, Kiscadale Publications, 1990.

Ministry of Information, Government of the Union of Myanmar. *The Conspiracy of Treasonous Minions Within the Myanmar Naing-ngan and Traitorous Cohorts Abroad.* Yangon, 1989.

Mirante, Edith T. *Burmese Looking Glass: A Human Rights Adventure.* New York, NY, Grove Press, 1992.

Naw, Angelene. *Aung San and the Struggle for Burmese Independence.* Chiang Mai, Silkworm Books, 2001.

Ni Ni Myint. *Burma's Struggle Against British Imperialism.* Yangon, Universities Press, 1983.

Nu, Thakin. *Burma under the Japanese.* London, Macmillan, 1954.

O'Brien, Harriet. *Forgotten Land: A Rediscovery of Burma.* London, Michael Joseph, 1991.

Phayre, Arthur P. *History of Burma: From the Earliest Time to the End of the First War with British India.* London, Routledge, 2000.

Pollak, Oliver B. *Empires in Collision: Anglo-Burmese Relations in the Mid-Nineteenth Century.* Westport, CT, Greenwood Press, 1979.

Rogers, Benedict. *A Land Without Evil: Stopping the Genocide of Burma's Karen People.* Crowborough, Monarch Publications, 2004.

Selth, Andrew. *Death of a Hero: The U Thant Disturbances in Burma, December 1974.* Australia-Asia Papers, No. 49, April 1989. Nathan, Qld, Griffith University.

Transforming the Tatmadaw: The Burmese Armed Forces since 1988. Canberra, Strategic and Defence Studies Centre, Australian National University, 1996.

Burma's Intelligence Apparatus. Canberra, Strategic and Defence Studies Centre, Australian National University, 1997.

The Burma Navy. Canberra, Strategic and Defence Studies Centre, Australian National University, 1997.

Burma's Secret Military Partners. Canberra, Strategic and Defence Studies Centre, Australian National University, 2000.

Burma's Armed Forces: Power Without Glory. Norwalk, CT, EastBridge, 2002.

Burma's China Connection and the Indian Ocean Region Canberra, Strategic and Defence Studies Centre, Australian National University, 2003.

Burma's North Korean Gambit: A Challenge to Regional Security? Canberra, Strategic and Defence Studies Centre, Australian National University, 2004.

Silverstein, Josef. *The Political Legacy of Aung San.* Ithaca, NY, Cornell University Press, 1993.

Singh, Balwant. *Independence and Democracy in Burma, 1945–1952: The Turbulent Years.* Ann Arbor, MI, University of Michigan Press, 1993.

Stewart, A. T. Q. *The Pagoda War: Lord Dufferin and the Fall of the Kingdom of Ava, 1885–86.* London, Faber & Faber, 1972.

Tatsuro, Izumiya. *The Minami Organ.* Yangon, Universities Press, 1981.

Thant, Myint U. *The Making of Modern Burma.* New York, NY, Cambridge University Press, 2001.

Tinker, Hugh (Ed.). *Burma: The Struggle For Independence 1944–1948.* 2 vols. London, HMSO, 1984.

Trager, Frank. *Burma: From Kingdom to Republic.* New York, NY, Praeger, 1966.

Trager, Frank, and Koenig, William. *Burmese Sit-tans 1764–1826: Records of Rural Life and Administration.* Tucson, AZ, University of Arizona Press, 1979.

Tucker, Shelby. *Burma: The Curse of Independence.* London, Pluto Press, 2001.

Varey, Bertram S. *The Chin Hills.* New Delhi, Gyan Publishing House, 2004.

Win Naing Oo. *Human Rights Abuse in Burmese Prisons.* Sydney, NSW, ACFOA, 1996.

ECONOMY AND POLITICS

Andrus, J. R. *Burmese Economic Life.* Stanford, CA, Stanford University Press, 1948.

Aung-Thwin, Michael. *Irrigation in the Heartland of Burma: Foundations of the Pre-Colonial Burmese State.* Occasional Paper No. 15. DeKalb, IL, Center for Southeast Asian Studies, 1990.

Ball, Desmond. *Burma and Drugs: The Regime's Complicity in the Global Drug Trade.* Canberra, Strategic and Defence Studies Centre, Australian National University, 1999.

Ball, Desmond, and Lang, Hazel. *Factionalism and the Ethnic Insurgent Organisations.* Canberra, Strategic and Defence Studies Centre, Australian National University, 2001.

Boucaud, André and Louis. *Burma's Golden Triangle: On the Trail of the Opium Warlords.* Bangkok, Asia Books, 1988.

Callahan, Mary P. *Making Enemies: War and State Building in Burma.* Ithaca, NY, Cornell University Press, 2003.

Carey, Peter (Ed.). *Burma: The Challenge of Change in a Divided Society.* Basingstoke, Macmillan, 1997.

Chakravarti, N. *The Indian Minority in Burma: The Rise and Decline of an Immigrant Community.* London, Oxford University Press, 1971.

Chao Tzang Yawnghwe. *The Shan of Burma; Memoirs of a Shan Exile.* Singapore, Institute of Southeast Asian Studies, 1987.

Cheng Siok-Hwa. *The Rice Industry of Burma, 1852–1940.* Kuala Lumpur, University of Malaya Press, and London, Oxford University Press, 1969.

Collignon, Stefan, and Taylor, Robert H. (Eds). *Burma: Political Economy Under Military Rule.* Basingstoke, Palgrave Macmillan, 2001.

Fleischmann, Klaus. *Die Kommunistische Partei Birmas: Von den Anfangen bis zur Gegenwart.* Hamburg, Mitteilungen des Instituts für Asienkunde, 1989.

Furnivall, J. S. *Colonial Policy and Practice: A Co-operative Study of Burma and Netherlands India.* London, Cambridge University Press, 1957.

The Governance of Modern Burma. New York, NY, Institute of Pacific Relations, 1960.

Gravers, Mikael. *Nationalism as Political Paranoia in Burma: An Essay on the Historical Practice of Power.* Richmond, Surrey, Curzon, 1999.

Houtman, Gustaaf. *Mental Culture in Burmese Crisis Politics: Aung San Suu Kyi and the National League for Democracy.* Tokyo, Tokyo University of Foreign Studies—Institute for the Study of Languages and Cultures of Asia and Africa, 1999.

International Commission of Jurists. *The Burmese Way to Where? Report of a Mission to Myanmar (Burma).* Geneva, 1991.

Ismael Khin Maung. *The Myanmar Labour Force: Growth and Change, 1973–83.* Singapore, Institute of Southeast Asian Studies, 1997.

Johnstone, W. C. *Burma's Foreign Policy: A Study in Neutralism.* Cambridge, MA, Harvard University Press, 1963.

Khin Maung Gyi. *Memoirs of Oil Industry in Burma: 905 AD—1980 AD.* Yangon, 1989.

Khin Maung Nyunt. *Foreign Loans and Aid in the Economic Development of Burma, 1974/75 to 1985/86.* Chulalongkorn University Paper, No. 46. Bangkok, Institute of Asian Studies, 1990.

Lang, Hazel J. *Fear and Sanctuary: Burmese Refugees in Thailand.* Ithaca, NY, Cornell University Press, 2002.

Leach, E. R. *Political Systems of Highland Burma—A Study of Kachin Social Structure.* Cambridge, MA, Harvard University Press, 1954.

Lehman, F. K. *The Structure of Chin Society.* Urbana, IL, University of Illinois, 1963.

Liang Chi-sha. *Burma's Foreign Relations: Neutralism in Theory and Practice.* New York, NY, Praeger, 1990.

Lissak, Moshe. *Military Roles in Modernization: Civil-Military Relations in Thailand and Burma.* Beverly Hills, CA, Sage Publications, 1976.

Mason, Jana M. *No Way Out, No Way In: The Crisis of Internal Displacement in Burma.* Washington, DC, US Committee for Refugees, 2000.

Maung Maung. *Burma's Constitution.* 2nd edn. The Hague, Nijhoff, 1961.

Burma and General Ne Win. London, Asia Publishing House, 1969.

Maung Maung Gyi. *Burmese Political Values: The Socio-Political Roots of Authoritarianism.* New York, NY, Praeger, 1983.

Mendelson, E. Michael, and Ferguson, John (Eds). *Sangha and State in Burma.* Ithaca, NY, Cornell University Press, 1975.

Mi Mi Khine. *The World of Burmese Women.* London, Zed Books, 1984.

Moscotti, Albert D. *Burma's Constitution and the Elections of 1974.* Singapore, Institute of Southeast Asian Studies, 1977.

Mya Maung. *The Burma Road to Poverty.* New York, NY, Praeger, 1991.

Totalitarianism in Burma: Prospects for Economic Development. New York, NY, Paragon House, 1992.

Mya Than. *Growth Pattern of Burmese Agriculture: A Productivity Approach.* Occasional Paper No. 81. Singapore, Institute of Southeast Asian Studies, 1988.

Myanmar's External Trade: An Overview in the Southeast Asian Context. Singapore, Institute of Southeast Asian Studies, 1992.

Mya Than and Joseph L. H. Tan. *Myanmar Dilemmas and Options: The Challenge of Economic Transition in the 1990s.* Singapore, Institute of Southeast Asian Studies, 1990.

Mya Than and Myat Thein (Eds.). *Financial Resources for Development in Myanmar: Lessons from Asia.* Singapore, Institute of Southeast Asian Studies, 1999.

Mya Than and Gates, Carolyn L. (Eds). *ASEAN Enlargement: Impacts and Implications.* Singapore, Institute of Southeast Asian Studies, 2001.

Myat Thein. *Economic Development of Myanmar.* Singapore, Institute of Southeast Asian Studies, 2004.

Nishizawa, Nobuyoshi. *Economic Development of Burma in Colonial Times.* Hiroshima, Institute for Peace Science, Hiroshima University, 1991.

Nu, U. *U Nu—Saturday's Son.* New Haven, CT, Yale University Press, 1975.

Pedersen, Morten B., Rudland, Emily, and May, Ronald J. *Burma-Myanmar: Strong Regime, Weak State?* Adelaide, SA, Crawford House Publishing, 2000.

Pradhan, M. V. *Burma, Dhamma and Democracy.* Mumbai, Mayflower Publishing House, 1992.

Sarkisyanz, E. *Buddhist Backgrounds of the Burmese Revolution.* The Hague, Martinus Nijhoff, 1965.

Saito, Teruko, and Lee Kin Kiong (Eds). *Statistics on the Burmese Economy: The 19th and 20th Centuries.* Singapore, Institute of Southeast Asian Studies, 1999.

Sakhong, Lian H. *In Search of Chin Identity: A Study in Religion, Politics and Ethnic Identity in Burma.* Copenhagen, NIAS Press, 2002.

Schendel, Willem van. *Three Deltas: Accumulation and Poverty in Rural Burma, Bengal and South India.* Newbury Park, CA, Sage Publications, 1991.

Seekins, Donald M. *The Disorder in Order: The Army-State in Burma since 1962.* Bangkok, White Lotus Press, 2002.

Shwe Lu Maung. *Burma—Nationalism and Ideology.* Dhaka University Press, 1989.

Silverstein, Josef. *Burma: Military Rule and the Politics of Stagnation.* Ithaca, NY, Cornell University Press, 1977.

Burmese Politics: The Dilemma of National Unity. New Brunswick, NJ, Rutgers University Press, 1980.

Independent Burma at Forty Years: Six Assessments. Ithaca, NY, Cornell Southeast Asia Program, 1989.

Smith, D. E. *Religion and Politics in Burma.* Princeton, NJ, 1965.

Smith, Martin. *Burma: Insurgency and the Politics of Ethnicity.* London, Zed Books, 1999.

State of Fear: Censorship in Burma. Article 19, Country Report. London, 1991.

South, Ashley. *Mon Nationalism and Civil War in Burma: The Golden Sheldrake.* London, RoutledgeCurzon, 2002.

Steinberg, David I. *Burma's Road toward Development: Growth and Ideology under Military Rule.* Boulder, CO, Westview Press, 1981.

The Future of Burma: Crisis and Choice in Myanmar. Lanham, MD, and New York, NY, Asia Society/University Press of America, 1990.

Crisis in Burma: Stasis and Change in a Political Economy in Turmoil. Bangkok, Institute of Security and International Studies, Chulalongkorn University, 1989.

Burma, the State of Myanmar. Washington, DC, Georgetown University Press, 2001.

Stewart, Whitney. *Aung San Suu Kyi: Fearless Voice of Burma.* Minneapolis, MN, Lerner Publications, 1996.

Taylor, Robert H. *Marxism and Resistance in Burma 1942–45: Thein Pe Myint's Wartime Traveler.* Athens, OH, Ohio University Press, 1984.

The State in Burma, 1987. London, Hurst & Co, 1988.

Taylor, Robert H. (Ed.). *Burma: Political Economy under Military Rule.* London, St Martin's Press, 2000.

Thanakha Team (Eds). *Burma: Voices of Women in the Struggle.* Bangkok, Altsean, 1998.

Thaw Han, Daw. *Common Vision: Burma's Regional Outlook.* Washington, DC, Institute for the Study of Diplomacy, 1988.

Walinsky, Louis. *Economic Development in Burma 1951–1960.* New York, NY, Twentieth Century Fund, 1962.

NEW ZEALAND

Physical and Social Geography

A. E. MCQUEEN

New Zealand lies 1,600 km south-east of Australia. It consists of two main islands, North Island with an area of 115,777 sq km (44,702 sq miles) and South Island with an area of 151,215 sq km (58,384 sq miles), plus Stewart Island (or Rakiura) to the south, with an area of 1,746 sq km (674 sq miles), and a number of smaller islands. North and South Islands are separated by Cook Strait, which is about 30 km wide at the narrowest point. The total area of New Zealand is 270,534 sq km (104,454 sq miles).

CLIMATE

There are three major factors affecting the climate of New Zealand, particularly as regards pasture growth. The first is the country's situation in the westerly wind belt which encircles the globe. The main islands lie between 34°S and 47°S, and are therefore within the zone of the eastward moving depressions and anti-cyclones within this belt. The second factor is the country's location in the midst of a vast ocean mass, which means that extremes of temperature are modified by air masses passing across a large expanse of ocean. It also means that abundant moisture is available by evaporation from the ocean, and rainfall is considerable and fairly evenly distributed throughout the year. The mean annual rainfall varies from 330 mm east of the Southern Alps to more than 7,500 mm west of the Alps, but the average for the whole country is between 600 mm and 1,500 mm. The third factor is more of local significance: the presence of a chain of mountains extending from south-west to north-east through most of the country. The mountains provide a barrier to the westward moving air masses, and produce a quite sharp climatic contrast between east and west. The rain shadow effect of the mountains produces in certain inland areas of the South Island an almost continental climate, although no part of New Zealand is more than 130 km from the sea.

The annual range of mean monthly temperatures in western districts of both islands is about 8°C, while elsewhere it is 9°C–11°C, except in inland areas of the South Island, where it may be as high as 14°C. The mean temperatures for the year vary from 15°C in the far north to 12°C about Cook Strait and 9°C in the south. With increasing altitude, mean annual temperatures fall about 1°C per 180 m. Snow is rare below 600 m in the North Island, and falls for only a few days a year at lower altitudes in the South Island. Rainfall and temperature combine to give a climate in which it is possible to graze livestock for all the year at lower altitudes in all parts of the country, and at higher altitudes, even in the South Island, for a considerable part of each year. Pasture growth varies according to temperature and season, but ranges from almost continual growth in North Auckland to between eight and 10 months in the South Island.

PHYSICAL FEATURES

Altitude and surface configuration are important features affecting the amount of land which is readily available for farming. Less than one-quarter of the land surface lies below 200 m; in the North Island the higher mountains, those above 1,250 m, occupy about 10% of the surface, but in the South Island the proportion is much greater. The Southern Alps (or Ka Tiriti o te Moana), a massive chain including 16 peaks over 3,100 m, run for almost the entire length of the island. The economic effect of the Southern Alps as a communications barrier has been considerably greater than that of the North Island mountains; their rain-shadow effect is also significant for land use, as the lower rainfall, while giving a reduced growth rate, also produces the dry summers of the east coast plains which are major grain-growing areas. In addition, the wide expanses of elevated open country have led to the development of large-scale pastoral holdings, and it is the South Island high country that produces almost all New Zealand's fine wools, particularly from the Merino sheep.

New Zealand rivers are of vital importance for hydroelectric power production, with their high rate of flow and reliable volume of ice-free water. Many of the larger lakes of both islands, most of which are situated at quite high altitudes, are also important in power production, acting as reservoirs for the rivers upon which the major stations are situated. Earthquakes are a particular risk along a zone west of the Southern Alps, through Wellington and thence north-east to Napier and Wairoa.

NATURAL RESOURCES

In general geological terms, New Zealand is part of the unstable circum-Pacific mobile belt, a region where, in parts, volcanoes are active and where the earth's crust has been moving at a geologically rapid rate. Such earth movements, coupled with rapid erosion, have formed the sedimentary rocks which make up about three-quarters of the country. New Zealand also includes in its very complex geology schist, gneiss, and other metamorphic rocks, most of which are hundreds of millions of years old, as well as a number of igneous rocks. In such a geologically mobile country the constant exposure of new rock has led to young and generally fertile soils.

Within this broad pattern a variety of minerals has been found, many of them, however, in only very small deposits. Non-metallic minerals, such as coal, clay, limestone and dolomite are today both economically and industrially more important than metallic ores; but new demands from industry, and a realization that apparently small showings of more valuable minerals may well indicate much larger deposits, have led to a surge of prospecting since the early 1960s. One of the most successful results so far has been the proving of ironsand deposits on the west coast of the North Island. This development, along with the discovery of a satisfactory method of processing the sands, has paved the way to the founding of an iron and steel industry which began production in 1969. However, knowledge of New Zealand's economic geology is still far from complete; in many respects the concentration upon farming had led to the belief that there were no economically attractive mineral resources. Exploration and prospecting carried out by both local and overseas firms has only recently shown that the nation has other worthwhile natural resources than its climate and soil. A notable discovery in 1970 was a large natural gas field, off the South Taranaki coast. The construction of pipelines to bring the gas to major centres, and of plants to convert this gas into synthetic petrol and methanol, had been completed by 1985.

When the first European settlers came to New Zealand they found two-thirds of the land's surface covered by forest. In 1995 about 29% of the area was forested, most of it kept as reserve or as national parks. The rest has been felled, much of it with little regard for land conservation principles. Of today's total forested area of 7.9m. ha, 6.4m. are still indigenous forest, much of it (some 4.9m. ha) unmillable protected forest. The 1.5m. ha of man-made exotic forest provide building timber and raw material for well-established pulp and paper and other forest product industries. These activities are based largely on the extensive exotic forests of the Bay of Plenty-Taupo region, near the centre of the North Island, but widespread planting has taken place in other regions for eventual use in newly developed processing plants.

POPULATION

The majority of the population is of European origin. At the census of March 1996, 523,374 persons (14.5% of the usually resident population) were enumerated as indigenous Maori; New Zealand's total population stood at 3,681,546. At the census of March 2001 the total population was enumerated at 3,737,277. In terms of international comparison, the population is small, the density (15.0 per sq km in mid-2004) and growth rates are low and the degree of urbanization is high. Some 85% of the population reside in cities, boroughs or townships with populations greater than 1,000.

Between 1926 and 1945 an average annual increase in population of 1.1% was recorded, one of the lowest rates in the country's history. Between 1945 and 1970, however, the rate of population growth was maintained at a high level—a trend supported by a significant level both of natural increase and of immigration. The high post-war birth rate reached a peak of 27.0 per 1,000 in 1961. The birth rate declined steadily to 15.7 per 1,000 in 1983, before gradually rising to 17.9 in 1990, then decreasing to 15.8 in 1995 and to 14.0 in 2003. This contrasted with death rates of 8.1 per 1,000 in 1983, 7.6 in 1995 and 7.0 in 2003, among the lowest in the world. The high rate of increase was complemented by a steady influx of immigrants, some of them assisted by the New Zealand Government. The total number of immigrants increased annually until 1952/53, since when there have been large variations in both the immigration and emigration rates. In 2003 immigrants totalled 92,660 and emigrants 57,754, resulting in a net increase of 34,906, compared with a net increase of 9,726 in 2001. In the year to June 1996 Taiwan displaced the United Kingdom as the leading source of immigrants.

The processes of economic development have dictated a steadily increasing concentration of population, farm products processing, and industrial output in the major cities, especially within the North Island. Larger towns contain more of the population than other urban areas. In 1996 69% of the population resided in urban areas with populations of over 30,000.

History

JEANINE GRAHAM

THE COLONIAL ERA

New Zealand's earliest migrant population was well established in coastal settlements by the 12th century. Oral tradition tells of voyages from Hawaiiki, the islands of eastern Polynesia, and of subsequent migrations within New Zealand waters. The Maori culture, which was to be of such interest to early European explorers and observers, developed in isolation as one variant of Polynesian culture.

Although the Dutch navigator, Abel Tasman, sailed in 1642–43 along part of the western coastline of the country that he called Staten Landt, it was the three scientific expeditions led by the British naval captain, James Cook (in 1769, 1773 and 1774), that brought New Zealand into prominence. During his first voyage Cook spent more than six months circumnavigating the islands and produced a chart remarkable for its detail and accuracy. Scientists and artists with the expedition recorded their impressions of the nature and habits of the indigenous Maori and New Zealand's unique flora and fauna. Voyages by other 18th and early-19th-century European explorers were peripheral. The traders, missionaries, scientists and settlers who followed in Cook's wake built on the secure foundations of his achievements.

Sealers, whalers and traders sought to exploit New Zealand's natural resources for economic advantage. Christian missionaries aimed to evangelize the *tangata whenua* (people of the land). The consequences of cultural encounter for the Maori people were varied. Against such diseases as influenza, measles, tuberculosis and whooping cough, the indigenous inhabitants had no immunity. Death rates remained high throughout the 19th century. Tribal receptiveness to new technology, literacy and Christianity was selective. Although the lifestyle of the Maori people was disrupted, it was not, especially in the case of those tribes in locations away from the coast, necessarily undermined. The early Europeans were, in fact, heavily dependent upon Maori co-operation and goodwill, a pattern that persisted during the first two decades after New Zealand became part of the British Empire in 1840.

The New Zealand Co and related organizations founded the settlements of Wellington, Nelson, Wanganui, New Plymouth, Christchurch and Dunedin. Until 1864, the colonial administration was based in Auckland, the character of this town being influenced by its strong links with the Australian colonies. A steady stream of assisted or independent immigrants flowed into New Zealand until the 1880s, with a dramatic influx during the gold rushes of the 1860s and large numbers of labouring-class settlers arriving during the era of public-works expansion in the 1870s. A minority in their own land by the late 1850s, the Maori population had declined to 42,000 by 1896, while Pakeha (non-Maori) settlers numbered some 700,000.

The insatiable demand of European settlers for land was a major cause of conflict between Maori and European during the 19th century. Settlers south of the Cook Strait region scarcely felt the impact of racial tension, for relatively few Maori lived there, but many North Island communities were affected. Sporadic and localized fighting in the 1840s was followed by a major outbreak in central and west-coast areas of the North Island in the early 1860s. The Treaty of Waitangi, signed on 6 February 1840 by the Lieutenant-Governor-elect, Capt. William Hobson, and subsequently by many—but by no means all—of the major Maori chiefs, had acknowledged Maori ownership of land, and accorded *tangata whenua* the status of British subjects. However, post-conflict confiscation of much tribal land and the introduction of a system of individualized land tenure, to replace the traditional communal form of ownership, caused major cultural upheaval during the last decades of the 19th century and effectively undermined the socio-economic base of those tribes most affected. Only since the 1970s have New Zealanders of European origins begun to appreciate that Maori understandings of the Treaty's guarantees had largely been ignored by successive Governments for more than a century.

British legislation provided a constitution for New Zealand in 1852. A bicameral General Assembly and six Provincial Councils were functioning by 1854, but responsible government was not granted until 1856. Improved internal and coastal communications, returns from wool and gold exports and the need to co-ordinate loan applications undermined the rationale for separate provincial development. The provincial governments were abolished in 1876 and were replaced by a plethora of local bodies which dominated local government for more than a century and, as with central government, took little cognizance of bicultural values. The granting of the vote to Maori men in 1867 and the establishment of four Maori seats in the House of Representatives in the same year were not sufficient measures to ensure that Maori interests were articulated at parliamentary level and that their needs were satisfied. Inter-tribal initiatives to set up complementary political forums met with no parliamentary support.

The economic crises of the 1880s caused many recent immigrants to leave New Zealand for the Australian colonies. Social distress and revelations of 'sweated labour' in the clothing industry fostered a sense of radicalism and protest against the apparently ineffectual, conservative 'continuous ministry'. Politicians espousing a liberal ideology came to form a coherent grouping and took office under the leadership of John Ballance in January 1891. During the ensuing 18 months, a hostile, nominated Legislative Council continually blocked reforms, a situation resolved only when the Governor was instructed by the British Secretary of State for the Colonies not to reject the advice of his responsible ministers in making new appointments. Ballance died in early 1893 and was succeeded as Premier by Richard John Seddon. For the next 13 years 'King Dick' Seddon dominated New Zealand life and politics. Assisted initially by some very able colleagues, Seddon carried through legislation that provided for industrial conciliation and arbitration, better factory conditions, accident compensation, shorter working hours and old-age pensions. Maori leaders found Seddon sympathetic to their people's plight, but the initiatives for urgent health reform came from a small group of Maori university graduates. Liberal policy towards Maori needs remained ambivalent, however: increased pressure for the further alienation of Maori land in the North Island resulted from measures that were implemented during the 1890s to foster the expansion of small farming, especially dairying. The technology of refrigeration had provided the key to resolving New Zealand's export difficulties. Such was the new sense of accomplishment and colonial pride that few in New Zealand were disposed to entertain seriously the idea of federation with the Australian colonies. New Zealanders entered the 20th century firmly committed to the ideal of the British Empire, a loyalty that cost the country and its people dearly during the First World War (1914–18).

THE INTER-WAR PERIOD

After 1918, under conditions of general prosperity and full employment, New Zealand resumed a selective, assisted migration policy, which had increased the population by nearly 70,000 by 1928, when the intake temporarily ceased. As trade conditions worsened, the Government increasingly intervened to control the nation's economy. Efforts to borrow in the face of the oncoming depression proved fruitless, and a coalition Government of United and Reform parties took office in September 1931. Orthodox financial solutions were tried. Railway construction was halted and expenditure on defence reduced. Emergency taxes were imposed. Incomes fell by an average of 20%. The depression worsened, and with it the social distress suffered by the growing number of unemployed. An estimated 80,000 people—12% of the labour force—were out of work. Employment schemes, including state forestry planning, provided relatively little relief. Labour unrest was rife throughout the country and short-lived bouts of rioting occurred in the major cities. A noted Maori politician, Sir Apirana Ngata, nevertheless succeeded in promoting a cultural revival amongst the Maori and addressing issues of Maori poverty through consolidation of fragmented land titles and the establishment of Maori-owned corporations to develop unproductive land.

The election of 1935 marked a radical political change. Small farmers combined with urban workers to return the Labour Party to power with a substantial majority. Under its leader, Michael Joseph Savage, the prices of essential commodities were fixed. Dairy farmers were given a guaranteed price for their butter and cheese, which were marketed by the Government both at home and abroad. Unemployment was tackled by the introduction of large-scale public works such as road and railway construction. The State offered the right to work at a fixed basic wage and made trade-union membership compulsory. Salary and wage cuts were restored, and a 40-hour week was introduced to increase the demand for labour. An ambitious housing programme was launched. These various undertakings were financed by means of increased taxation and public loans. A political alliance forged in 1936 between Savage and a Maori spiritual leader, Tahupotiki Wiremu Ratana, resulted in Ratana candidates being elected to all four Maori seats; the Ratana movement continued to support Labour until 1979.

In 1938 the Labour Government was re-elected with an increased majority and immediately began to build upon the social legislation of the 1890s. A Social Security Act provided for free general practitioner services, medicines, hospital treatment and maternity benefits, and for family allowances and increased old-age pensions. To finance such comprehensive insurance a special tax was levied on all incomes and supplemented by grants from ordinary revenue. Such measures were possible because these were years of prosperity and prices for exports were high. By 1939, however, heavy withdrawals to meet overseas commitments, together with the flight of private capital, forced the Government to restrict imports and control the export of capital.

The British declaration of war in 1939 was felt to be as binding on New Zealand as that of 1914, even though the two countries had differed publicly in their attitude to League of Nations' policy initiatives during the 1930s. A special war cabinet was formed on the basis of a coalition of all parties. The Prime Minister, Peter Fraser, concentrated economic policy around the objective of stabilization, and in this regard he was largely successful. Conscription was introduced for overseas service, essential commodities were rationed and the country's resources were mobilized for the war effort. However, the rationalizing of manpower in 1942 provoked a crisis in the coal-mining industry and led the Government to assume control of the mines. Incensed at what it regarded as the Government's capitulation to the coal miners, the National Party withdrew from the war cabinet. When Japan entered the Second World War, New Zealand responded to meet the threat of invasion, but, even after the attack on Pearl Harbor (the US naval base in Hawaii, now a state of the USA) in December 1941, it did not recall its troops from the Middle East. The 40-hour week was suspended, and in 1943 thousands of men were released from the armed forces in order to increase food and factory production and thereby to assist the USA in its Pacific campaign. Employment opportunities, together with the difficulty of providing for a growing population on fragmented landholdings, prompted significant numbers of rural Maori to leave their subsistence lifestyle and migrate to the cities.

THE POST-WAR YEARS

The Labour Party won the election held in November 1946, but by a narrow majority; shortages, rationing, high food prices, poor housing and a reaction against wartime bureaucratic control all undermined the Labour vote. The subsequent introduction of compulsory military training in peacetime further alienated Labour supporters. The National Party, led by Sidney (later Sir Sidney) Holland, capitalized on public dissatisfaction with government regulation, and won a decisive victory at the 1949 election.

With the full approval of the Labour opposition, the Government abolished the Legislative Council, which, as a parliamentary chamber of review, had gradually declined in importance. Subsidies on certain commodities were withdrawn, but attempts to 'freeze' prices failed to prevent inflation. In a climate marked by Cold War fears of communism, the Government severely restricted civil liberties during a prolonged strike by dock workers, and eventually dispatched troops to end the unrest. Among the public there were strong feelings concerning the Prime Minister's handling of the crisis, but in the election called at short notice in 1951 the Government secured an increased parliamentary majority. Growing dissatisfaction with the performance of both parties was reflected three years later in the 1954 election, when the newly formed Social Credit League won 11% of the votes cast. While support for its economic policies was always problematic, as a third party Social Credit continued to benefit from tactical and protest voting until the mid-1980s. Electoral legislation in 1956 entrenched the franchise and electoral system and made enrolment compulsory for Maori, as it had been for Europeans since 1924.

Two months before the 1957 election, Holland retired from office because of ill health. Under his successor, Keith (later Sir Keith) Holyoake, the National Party narrowly lost the election. Labour's marginal victory was attributed to its espousal of a new Pay As You Earn (PAYE) system of collecting income tax. Almost immediately the Labour Government, with 75-year-old Walter Nash as Prime Minister, was embarrassed by pressing economic problems. Falling prices for meat and dairy produce led to an adverse balance of trade and to an alarming depletion of the country's overseas reserves. The Government reimposed import licensing and exchange control and thereby raised over £30m. in London and Australia and another £20m. internally. These measures arrested the decline, but the Government's fate was sealed by the 'black budget' of 1958, which, while introducing 3% loans for housing, increasing pensions, countering a balance-of-payments problem and encouraging industry, also increased taxes on beer, spirits, cigarettes and petrol. At the election in 1960 the National Party, led by Holyoake, was returned to power. Among the more important changes that followed was the appointment in 1962 of a Parliamentary Commissioner (ombudsman) to inquire into public complaints arising from governmental administrative decisions. The National Party retained office in an election in 1963. Labour's new leader, Arnold Nordmeyer, had tried unsuccessfully to revive the Labour Party's popularity by denying the social need for class struggle. The formation in 1966 of the small Socialist Unity Party, many of the 60 adherents of which were former members of the New Zealand Communist Party, indicated the persistence of more radical views on the subject of class conflict. Meanwhile, in 1965 Nordmeyer was ousted as Labour's leader by a former Party President, Norman Kirk.

Although economic policy was the issue most frequently debated in the 1966 election, attitudes to New Zealand's involvement in Viet Nam cut across party lines and secured a further victory for the National Party. Holyoake's consensus style of politics continued to have electoral appeal, but he retired in February 1972, to be succeeded by John (later Sir John) Marshall, a leader whose urbane demeanour was challenged by ambitious and aggressive younger politicians within the party. Labour gained a sweeping victory in November 1972 in an election that had been contested by National, Labour, Social Credit and a newly formed Values Party, one of the world's first 'Green' parties, which, by emphasizing the importance of environmental issues, appealed to many younger voters. As Labour's leader, Kirk was vigorous and forthright, especially in the field of foreign policy, but his leadership was cut short by his untimely death in August 1974. His successor, Wallace (later Sir Wallace) Rowling, opposed by a new leader of the National Party, Robert (later Sir Robert) Muldoon, continued Kirk's policy, but conducted it less forcefully.

In the post-war world, New Zealand's first major step in pursuing a policy independent of British interests was made in 1944 with the signing of the Canberra Pact, a mutual security agreement with Australia. In 1947 New Zealand finally adopted the Statute of Westminster, which gave it complete autonomy and freedom of action in international affairs. As a strong supporter of the UN, in 1950 the Government sent troops to join the UN Command in the Republic of Korea. Despite official misgivings about the exclusion of the United Kingdom, in 1952 it signed the Australia, New Zealand and the USA (ANZUS) defence treaty, and two years later New Zealand became a member of the South-East Asia Treaty Organization (SEATO). New Zealand sent a military unit to Malaya (now Malaysia) in 1955, and combined with Australia in an ANZAC (Australia and New Zealand Army Corps) unit in Viet Nam. Along with Australia, its troops in Singapore and Malaysia combined with the remaining British forces to safeguard the political status quo of the region. In 1971 New Zealand joined Australia, Malaysia, Singapore and the United Kingdom in establishing new arrangements for the defence of Malaysia and Singapore, and maintained a presence there until the last troops were withdrawn in 1990.

New Zealand played an increasingly important role in the Pacific after joining the South Pacific Commission (now Pacific Community) as a founder member in 1947. It administered the UN Trust Territory of Western Samoa until 1962, when those islands became independent, and in 1965 it assisted the Cook Islands to achieve self-government in free association with New Zealand. Niue became an associated territory in 1974, while the tiny islands of Tokelau are now New Zealand's only remaining Pacific dependency. When Fiji became independent from the United Kingdom in 1970, New Zealand helped to establish the University of the South Pacific in Suva, and in 1971 it joined the South Pacific Forum (now Pacific Islands Forum), which was founded to promote economic and political co-operation in the region.

RECENT DEVELOPMENTS

Political Affairs

Playing on public fears of national economic disaster consequent on the inflationary effects of Labour's policy of heavy overseas borrowing, the opposition leader, Robert Muldoon, campaigned on the promise of a more attractive and egalitarian pensions system than that offered by the Government. At the November 1975 general election a revitalized National Party won 55 of the 87 seats in the House of Representatives. Labour lost 23 seats, including those of five cabinet members.

The National Party retained office in the 1978 election, although its share of the total vote fell, largely owing to the abrasive style of leadership practised by Muldoon as Prime Minister. The Labour Party received more votes than the National Party but won fewer seats, while the Social Credit Party secured only one seat. The Values Party, with no firm base of appeal, fared poorly (in 1989 it changed its name to the Green Party of Aotearoa). To stabilize the parliamentary representation of the South Island (as its share of the country's population in proportion to that of the North Island had been—and is still—declining), its number of seats was fixed at 25. In 1981 public opinion was highly polarized by the Government's failure to discourage a South African Springbok rugby tour of New Zealand. The National Party narrowly retained office in November. The Social Credit Party lost political credibility, but in August 1983 a new third party was founded: the right-wing New Zealand Party, led by a wealthy property-owner, Robert (later Sir Robert) Jones. Meanwhile, in February 1983, the Labour Party had replaced its much respected but publicly unimpressive leader, Wallace Rowling, with David Lange, a lawyer with a more vigorous style and greater charisma, who became Prime Minister in July 1984, when Labour capitalized on Muldoon's unpopularity and poor tactics in calling an early general election. The New Zealand Party gained no representation but attracted 12% of the total number of votes cast. The high turnout of eligible voters (86%) was an indication of public dissatisfaction with the Muldoon administration.

In November 1984 Muldoon was replaced as leader of the National Party by James McLay, the deputy leader and a former Attorney-General. The transfer was not effected smoothly, as Muldoon made no pretence of co-operation and maintained a high political profile, working to destabilize the leadership of his successor. In March 1986 McLay was ousted by his deputy leader, James (Jim) Bolger, a farmer from King Country.

Initially, the Labour Government benefited considerably from the opposition's disarray, for the programme of economic restructuring along free-market lines was alienating many of Labour's traditional urban-based supporters. The social cost of New Zealand's high level of inflation and worsening balance-of-payments difficulties, and of Labour's policy initiatives, bore most heavily upon low-income earners and welfare beneficiaries, in which group Maori and Pacific Islanders were heavily represented.

At the general election of August 1987, popular support for the Government's anti-nuclear defence policy, a stance then opposed (but later endorsed) by the National Party, enabled Lange's administration to become the first Labour Government since 1946 to be elected for two successive terms of office. Labour's political euphoria was short-lived, however. Policy disagreements emerged between finance minister Roger Douglas, the architect of Labour's deregulation and free-market economics, and David Lange, who wished to prevent the extension of market principles into social policy. Douglas lost his portfolio, but in August 1989 Lange resigned as Prime Minister, his position having been consistently undermined by the continued internal political intriguing of Douglas, and by the formation of the NewLabour Party (led by a former President of the New Zealand Labour Party, Jim Anderton), which aimed to appeal to erstwhile Labour supporters who had become disillusioned with the Government's continued advocacy of privatization. The new Deputy Prime Minister, Helen Clark, a former lecturer in political science, became the first woman to hold such office in New Zealand's parliamentary history. Geoffrey (later Sir Geoffrey) Palmer, previously a professor of law, came to the Prime Minister's position with a reputation for cautious and careful administration and did not find it easy to project a more personable image. In September 1990, less than eight weeks before the next election, Palmer resigned from his position. Michael Moore, the Minister of External Relations and Trade (who had contested the August 1989 leadership election), replaced Palmer as Prime Minister.

The National Party secured 67 of the 97 seats in the House of Representatives in the October 1990 election, the Labour Party won 29 seats, and the NewLabour Party one seat. The NewLabour Party provided one option for protest votes, while disenchanted voters with environmental concerns supported the Green Party. Non-voting increased again, and the incoming National Government actually received fewer votes than had Labour in 1987. Maori voters still had the choice of registering on either the general or the Maori electoral roll, but the Mana Motuhake party, founded in November 1979 by the former member for Northern Maori, Matiu Rata, did not have the resources to contest more than a few seats.

Jim Bolger thus became the country's fourth Prime Minister within 15 months, and the new National Party Government immediately signalled a series of cost-cutting measures (see Economy), designed essentially to reduce the welfare burden, which was to consume 60% of total government spending in the 1991/92 financial year. Under the direction of the Minister of Finance, Ruth Richardson, 'user pays' initiatives in health and education undermined long-cherished principles of a welfare state, while a crisis over pensions funding, with which the Minister of Social Welfare, Jenny Shipley, had to contend, was essentially a result of political decisions made by the National Party at the time of the 1975 general election. The passing of the Employment Contracts Act in 1991 promoted the rapid deregulation of the labour market and further undermined a trade-union movement already weakened by high unemployment. More than 10% of the labour force was out of work in 1992, with the Maori unemployment rate being significantly greater than that of Pakeha (those of predominantly British origin). By 1992 there was a new political alignment, the Alliance, made up of a coalition of minor parties, including the Greens, NewLabour, Mana Motuhake and the New Zealand Democratic Party (descended from the former Social Credit Party). It also included another new party, the Liberal Party, formed by two dissident ex-National Party MPs. The maverick political behaviour of the former National and subsequently independent member for Tauranga, Winston Peters, resulted in his expulsion from the National Party; he subsequently formed New Zealand First, a political grouping to which a number of prominent Alliance members defected, once it became apparent that Peters would not support the coalition.

Such political manoeuvrings were presented as an argument against electoral reform, for which strong support had been expressed by voters during a referendum in September 1992; but at a second, binding referendum, held in conjunction with the general election on 6 November 1993, the electorate voted in favour of the introduction of a mixed-member proportional representation (MMP) system. The close election results (National 50 seats, Labour 45, Alliance two, New Zealand First two) led to some weeks of political uncertainty, but Bolger continued to govern with his marginal majority because the Alliance rejected Labour proposals that it form a coalition to bring down the Government. The snap by-election held in August 1994 in the Canterbury constituency of Selwyn, following the resignation from politics of the former Minister of Finance, Ruth Richardson, confirmed National's position. Labour's poor third-place rating in the polls reflected public reaction to the internecine wrangling that dominated party affairs after Helen Clark replaced Michael Moore as Labour's leader following the November 1993 election.

As a result of opposition disunity, National was able to continue to govern effectively; no key policy issue faced a unified challenge, not even the ongoing sale of state assets. The growing appeal (in particular among Maori and the elderly) of New Zealand First, the leadership of which was openly critical of overseas investment, Asian immigration and government policies affecting the elderly, continued to erode support for both National and Labour. The increasing dogmatism of both individuals and political parties in the months preceding the 1996 election caused many New Zealanders, desirous of more responsive government, to view the forthcoming change to MMP with considerable scepticism (which was little assuaged by subsequent events).

At the general election held on 12 October 1996, the National Party won 44 of the 120 parliamentary seats, the Labour Party 37, New Zealand First 17 and the Alliance 13. An increased number of Maori and women were returned as MPs: 15 and 36, respectively. As no party had secured an outright majority, prolonged negotiations ensued, with both Labour and National endeavouring to establish an alliance with New Zealand First. Eventually, on 10 December, a coalition Government was formed comprising National and New Zealand First (which together represented 47% of the votes cast). Winston Peters, as well as being appointed Deputy Prime Minister, was assigned the post of Treasurer, giving him substantial control over economic policy. However, Bill Birch, a National MP, remained Minister of Finance.

Older voters, many of whom had supported New Zealand First in protest at National's health and pensions policies, were highly critical of the new coalition, which for National was essentially an arrangement of political convenience. The political inexperience of many New Zealand First MPs and factionalism within that party contributed to its steady decrease in popularity. The leadership change effected by Jenny Shipley and her supporters during Jim Bolger's absence overseas in November 1997 was an indication of National's growing concern that its own standing was being undermined by the reputation and performance of its coalition partner. Differences of economic and social policy, including the continuing privatization of state assets, culminated in Shipley's dismissal of Winston Peters as Deputy Prime Minister and Treasurer on 14 August 1998. The engagement of both parties in an agreed disputes-resolution process was a formality. The coalition Government was dissolved four days later. Shipley subsequently led a minority Government that was dependent for its survival upon dissident ex-New Zealand First MPs, independents and the right-wing ACT New Zealand, a party founded by Sir Roger Douglas and led by a former Labour cabinet minister, Richard Prebble. During 1999 Shipley's political reputation was not enhanced by her Government's handling of various controversies concerning substantial severance payments to senior officers within the departments of tourism, the fire service, and work and income support services: the appropriateness of trying to imbue the public service with a corporate ethos continued to be widely debated. In August the withdrawal of the IBM corporation from further software development of the much-vaunted new police computer system, already $NZ30m. beyond its original $NZ100m. budget and remaining incomplete after five years of effort, was a further political embarrassment for a Government that had pledged a strong policy on law enforcement but appeared to have left its 'front-line' police officers inadequately resourced to cope with the growing rates of drugs- and gang-related crime.

At the election conducted on 27 November 1999, the opposition Labour Party won the largest share of votes cast. A recount of votes in one constituency, where the Green Party candidate then unexpectedly took the seat from the incumbent National MP, combined with the incorporation of 'special votes' (which included those cast by New Zealanders overseas), led to a substantial modification of the initial results. Having secured 38.7% of the votes cast, the Labour Party was finally allocated 49 of the 120 seats in the House of Representatives, while the National Party, which had won 30.5% of the votes, received 39 seats. The Alliance was allocated 10 seats and ACT New Zealand nine seats. Under the recently introduced system of proportional representation, the Green Party's victory in the one constituency automatically entitled the movement to a further six seats in the legislature. New Zealand First's representation declined to five; the party's leader, Winston Peters, only narrowly retained his seat. United New Zealand took the one remaining seat. Having previously discounted any co-operation with the Green Party, the Labour Party was thus obliged to seek the support not only of the Alliance but also of the seven Green MPs. The leader of the Labour Party, Helen Clark, therefore became the first woman politician to win a New Zealand general election. On 4 April 2001 Dame Silvia Cartwright took office as Governor-General, leading to an unprecedented situation: five important public roles in the country—that of Prime Minister, Leader of the Opposition, Attorney-General, Chief Justice and Governor-General—were all occupied by women. In October, however, Bill English replaced Shipley as leader of the National Party.

Clark's minority Government, which incorporated several members of the Alliance, including Jim Anderton as Minister for

Regional Development, took office in December 1999, and in 2001 continued to function relatively harmoniously under the Prime Minister's strong control, although progress on initiatives to address the disparities between Maori and non-Maori communities, particularly in the socio-economic indicators of health, housing, income, education, criminality and domestic violence, was slow. In March 2001 the Government announced *ex gratia* payments of $NZ30,000 per person to survivors of military and civilian Japanese prison camps in the Second World War, or to surviving spouses. In conservation and environmental issues the political influence of the Green Party became apparent, notably in debates over research involving genetic modification.

The disintegration of the Alliance and the clear differences between Labour and the Greens over the continuation of a moratorium banning the commercial release of genetically modified organisms (due to expire in October 2003) contributed to Helen Clark's decision to call an early general election, a political strategy used only twice previously in the post-war era, in 1951 and 1984. The Government hoped to take advantage of its popularity and secure sufficient electoral support to function without the need for coalition partners. The growing volatility of the New Zealand electorate was apparent in the outcome. Only 77% of registered voters participated on election day, 27 July 2002. Labour polled 41% of the party votes and secured 52 of the 120 seats in the House of Representatives, three more than in 1999, but not sufficient for an outright majority. The Greens secured 7% of votes cast (nine seats) but the Alliance failed to win any parliamentary representation. Jim Anderton's Progressive Coalition party, a personal grouping formed after the disintegration of the Alliance, won two seats, despite securing only 1.7% of the party votes. The centre-right vote fragmented. New Zealand First received 10% of the votes cast (13 seats), a significant recovery in its political fortunes following adverse reaction to its contribution to the collapse of the 1996–98 coalition Government. ACT New Zealand, the most rightward leaning of the campaigning parties, won 7% (nine seats), and subsequently underwent a peaceful change of leadership in April 2004, when Richard Prebble resigned and was replaced with Rodney Hide. Centre-right support shifted to the centrist alliance of United Future New Zealand, which secured 7% of the votes cast (eight seats). The final outcome was disastrous for the National Party, which lost 12 of its sitting MPs and recorded the worst result of its 66-year history as a major force in New Zealand politics, winning just 21% of the party vote and only 27 seats.

Subsequent negotiations enabled Clark in August 2002 to announce that her minority coalition Government, comprising Labour and the Progressive Coalition, had enlisted United Future New Zealand's support on matters of confidence and supply. Formed in 1995 through an alliance of United with Future New Zealand, a small Christian-based party, United Future New Zealand, led by Peter Dunne, appealed to many middle-class voters because of its emphasis on 'family values'. By remaining outside the Government, Dunne hoped to avoid the fate of the junior coalition partners (New Zealand First and the Alliance, respectively) in the previous two Governments elected under the system of proportional representation. In meeting the expectations of its centre-left supporters, Labour needed to maintain good working relationships with the Greens, the co-leader of which, Jeanette Fitzsimons, lost her electorate seat but was returned to Parliament as a party-list member. After the general election, Helen Clark's firm and effective leadership contributed significantly to Labour's continued popularity in opinion polls. The National Party struggled to re-establish itself as a viable opposition party, while the Alliance virtually disintegrated as a political force. The Greens, however, continued to put pressure on the Government on environmental matters, especially those relating to genetic modification.

In October 2003 Don Brash, a former Governor of the Reserve Bank of New Zealand who had stood successfully as a candidate in the 2002 election, broke with normal National Party convention and publicly announced his challenge to the leadership of Bill English. Despite considerable criticism of both the strategy and his alleged political naivety, Brash won the caucus vote on 28 October and proceeded to transform National's political fortunes, as indicated by a dramatic rise in the public opinion polls. In a highly controversial speech addressed, as has become customary for National Party leaders, to the Orewa Rotary Club at the end of January, Brash strongly criticized the fundamental direction of government policy in dealing with Treaty of Waitangi-related issues, and thereby broke the political consensus that had evolved over past decades. The resultant popular debate, much of it ill-informed, revealed deep-seated disquiet over the Labour Government's affirmative action policies, even though many of these initiatives, in areas of health, education and employment especially, had been meeting with considerable success. Despite much of the current rhetoric, however, contemporary public discourse is far more concerned with avoiding separatism rather than promoting it.

The Labour Government's highly controversial legislation, which, in practice, proposes Crown ownership of the foreshore and seabed on behalf of all current and future New Zealanders, while giving recognition and protection to customary rights that Maori currently have under common law, has generated both protest and support across society. It has, however, seriously undermined Labour's traditional support (since the 1930s) amongst Maori voters. Thousands of Maori and Pakeha joined in a foreshore and seabed hikoi (protest walk), which progressed through the country in late April 2004 and reached Parliament on 5 May, one day before the proposed legislation survived its first reading in the House of Representatives and was sent to a select committee for public submissions. The Associate Minister of Maori Affairs,Tariana Turia, was dismissed from her portfolio because of her decision to vote against the measure; she also resigned from the Labour Party and forced a by-election in which she was returned, in July, as co-leader of a newly formed Maori Party, which pledged to field candidates in a number of constituencies in the general election due in 2005. With highly respected urban Maori leader and social reformer Dr Pita Sharples as the other co-leader, and with a goal of promoting Maori political unity, the Maori vote in 2005 was expected to be a significant but unpredictable factor affecting both major parties, especially as National's sole Maori female MP, Georgina Te Heuheu, was dismissed from her position as party spokesperson for Maori affairs and Treaty of Waitangi issues in February 2004 for criticizing the divisive nature of Brash's Orewa speech.

Other Domestic Affairs

Official immigration policies continue to be a source of electoral concern (exploited most obviously by New Zealand First). The 2001 national census revealed a continuing trend towards ethnic diversity in the country's population: Pacific Peoples comprised 6.5% of the total population, while those identifying as Asian totalled 6.6% (compared with 3% in 1991), despite the imposition in October 1996 of a $NZ20,000 fee on immigrants lacking competence in the English language. Appropriate support services were not always in place to assist new migrants. Highly skilled professionals, admitted to New Zealand because of their qualifications, were frequently unable to work in their area of expertise owing to language difficulties, administrative barriers or racial prejudice. More recent immigration selection policies, introduced in early 2004, target areas of skills shortages and aim to ensure that new arrivals will be gainfully employed in those fields. According to the 2001 census, Maori totalled 14.7% of the population, while those identifying as European accounted for 80%. An increase in South African and Croatian immigrants was noticeable within the European category. The median age varied widely across the different ethnic groups, that of Europeans being considerably older (36.8 years) than that of Maori (21.9) and Pacific Peoples (21.0). The ageing of the post-1945 'baby boomers' will begin to have a major impact on the social-services, health and superannuation policies by the end of the first decade of the 21st century.

More than 80% of Maori are urban dwellers, a situation that has caused many to become separated from their tribal background and traditions, though sustained and vigorous educational efforts are gradually arresting the decline in Maoritanga (Maori culture). In 1985 the Lange Government enabled the Waitangi Tribunal (established in 1975) to consider retrospectively Maori grievances dating back to 1840. The evidence presented by claimants to the Tribunal is enabling Maori perspectives on the colonial past to be aired in a public forum, and the findings of the Tribunal, while not binding on the Government, have already contributed significantly towards changing public perceptions on this issue. The strengthening cultural renaissance initially encountered a mixed reception from Pakeha. While the status of Maori as an official language and the establishment of the *Kohanga Reo* ('language nest') early-childhood programmes were soon accepted, the need for affirmative action to redress educational and social disadvantage, much of which is the unforeseen consequence of past policies, was more contentious. National Government promotion of a $NZ2,000m. 'fiscal envelope' scheme as a means of resolving outstanding claims was soundly rejected by Maori at a series of consultative *hui* (gatherings) held throughout the country in 1995. Direct negotiation with the Crown remains an option, however, and a 1995 agreement, which included the return of assets and a formal apology by the Crown, with the powerful North Island Tainui tribes met with a mixed reception from other claimants. In October 1996 a similar settlement was reached with the Ngai Tahu tribes of South Island. The government-initiated Sealord deal, whereby fishing quotas were allocated to coastal tribes, caused dissension about the fairest method of distributing returns within Maoridom. Urban Maori authorities argue that payments according to *iwi* (tribal) affiliation deny them a share of the resources to which their urban-born membership is entitled. Costly legal action has resulted, nevertheless, in an affirmation that an urban Maori authority cannot be defined as an *iwi*, even if it serves that purpose for its members: affiliation based on *whakapapa* (descent), not on residential location, must be the basis upon which payments are made. Efforts continue to resolve the issue without further redress to the courts. Disputed rights of ownership over the foreshore and sea-bed had also emerged as a contentious issue at mid-2003. Following a Court of Appeal decision that the Maori Land Court could consider an application concerning a tribal proposal for mussel farming in the Marlborough Sounds, and thereby make a ruling on the ownership question, the Labour Government sought to allay public concerns by declaring that the foreshore and seabed are for all New Zealanders. Debates over the recognition of customary title have continued. Public submissions were invited in the latter months of 2003, and proposed legislation proceeded to select committee hearings in mid-2004. A resolution satisfactory to all parties will not be easily attained, because of a public perception that the recreational interests of all New Zealanders could be affected. Government assurances that Maori customary rights will be upheld have not mollified protesters, angered at what they perceive to be a denial of Maori rights to follow due legal process by testing the Court of Appeal's ruling before the Maori Land Court.

Labour also aroused widespread public debate during 2004 over proposed legislation to address the legal inequalities affecting *de facto* and same-sex couples. MPs were permitted a conscience vote on the Civil Union Bill, which would affect the one in five New Zealanders who, according to the 2001 Census, were identified as living in a relationship but were not married. An accompanying Relationships (Statutory References) Bill aimed to remove the legal discrimination that affected all those established in some recognized form of personal partnership. The parliamentary select committees dealing with both issues were inundated with public submissions, and their successful resolution of the issue was a political priority for the Labour administration.

Increasing numbers of New Zealanders are becoming better-informed about their country's colonial history, and the overwhelming public response to Te Papa, the new national museum opened in Wellington in 1998, indicated a high level of interest. Yet for citizens adversely affected by recent social and economic policy, the problems of the present assume much greater significance than any knowledge of the past. The conspicuous rise in poverty, violent crime (much of it drugs-related), youthful offending, gang association and drugs-trafficking, and the persistence of youth unemployment, do not augur well for social harmony, particularly when Maori and Polynesian youths are over-represented in such statistics. High rates of teenage pregnancy, alcohol abuse, methamphetamine addiction and youth suicide are also matters of social concern, as are child obesity and child abuse. Despite political rhetoric concerning the need for a more skilled work-force in a knowledge-based economy, escalating costs for tertiary education have created extremely high levels of indebtedness amongst the student population, an increasing number of whom have been forced to rely upon bank loans to fund their training. Shortages of trained staff in the health, education and information-technology sectors especially reflect the trend for recent graduates to seek higher-paid employment abroad in an effort to reduce their burden of debt. Successive Governments have fostered the notion of community responsibility for social welfare, but have yet to provide adequate resourcing, particularly in mental health. The social cost of the market-led 'reforms' is no longer a matter of debate. The Labour Government has addressed the issue of market-value rents for state housing, a practice that contributed to the plight of urban low-income families, and to the poor health of their children, particularly in the South Auckland area, with its high concentration of unskilled and unemployed Pacific Island Polynesians. Evoking the precedent of the influential Maori Land March of 1975, in September 1998 the Anglican Church organized an ecumenical Hikoi of Hope, in which marchers from both northern and southern ends of the country converged on Parliament to protest against the trends in social policy. The closure of rural hospitals, increased privatization and an expensive restructuring of the health service have yet to be perceived by the public at large as improvements on the previous public health-care system. Campaigns against smoking, which initially had considerable success in the creation of smoke-free work environments, have still to meet targets amongst youth (young women and Maori, in particular); entrenched attitudes that equate heavy drinking with masculinity are proving to be difficult to change, despite the obvious link between alcohol consumption and a high incidence of driving accidents, domestic violence and homicides. Obesity, use of the drug pure methampetamine (also known as 'P'), a dramatic increase in diabetes, and the high incidence of Meningococcal group B disease are further areas of major social concern with serious financial implications for the health sector. A mass immunization programme against Meningococcal B, involving 1.15m. young people, commenced in South Auckland in August 2004, at a cost of $NZ200m. Many New Zealanders entered the 21st century with a strong conviction that the individualism and ideology of market forces that prevailed in the last two decades of the 20th century had not served the country and its people well, but a growing divergence of political viewpoints over social reform may hinder the implementation of policies designed to redress the obvious inequalities.

Two major natural disasters affected the country during 2004, with extensive and serious flooding in the lower North Island (Rangitikei and Manawatu regions) in February and millions of dollars of flood damage also occurring in the Bay of Plenty in July. Major environmental concerns have also been raised about the condition of many of the country's lakes and waterways, especially those closer to urban areas and affected by nutrient contamination, as is the case with Lake Rotorua. Meanwhile, efforts on the part of the Department of Conservation to save the endangered kakapo parrot were set back by the loss of three female birds through an infection spread by seabirds.

International Relations

Continuing the more independent foreign policy initiated by Norman Kirk, from 1984 the Lange Government made a determined effort to foster Pacific unity and to build closer relations with New Zealand's Asian neighbours. Strong diplomatic and trading links were established with Japan, the Republic of Korea (South Korea) and the People's Republic of China. In 1986 the Lange Government was also the first foreign power to give diplomatic recognition to the Aquino administration in the Philippines. Support for the policy of creating a nuclear-free zone in the Pacific—in 1972 the Kirk Government sent a naval frigate into the testing zone near Mururoa Atoll, in French Polynesia, in an effort to force French testing to be conducted underground—led to a strengthening of ties between New Zealand and the other island nations of the Pacific (see below for relations with France). Long-term strategies to promote Pacific unity were seriously impeded, however, by the successful military coups in Fiji in May and September 1987. During 1989 New Zealand endeavoured to give leadership on other issues of mutual concern to South Pacific nations. For example, at a meeting of the South Pacific Forum in Vanuatu in July, New Zealand stated its intention to ban fishing by drift gill-nets from its territorial waters and to refuse permission for vessels using such nets to enter its ports. In 1990 the Japanese decision to cease drift gill-net fishing in Pacific waters received general approval. There have been similar small-power initiatives concerning measures to protect the ozone layer in the atmosphere and to conserve Antarctica's special status, and in September 2002, at the major international forum on sustainable development held in South Africa, New Zealand officials strongly advocated policies limiting carbon-dioxide emissions. New Zealand's initiative to broker a peace agreement between Bougainville separatists and the Government of Papua New Guinea led to a cessation of hostilities and to the presence of a New Zealand peace-keeping mission on the island of Bougainville throughout 1998. In May 2000 the Labour Government strongly condemned the hostage-taking and eventual overthrow of the Indian-dominated elected Government of Fiji, and also in the early 2000s supported efforts to resolve the civil conflict in Solomon Islands.

In July 1985 the trawler *Rainbow Warrior*, the flagship of the anti-nuclear environmentalist group, Greenpeace (which was to have led a flotilla to Mururoa Atoll to protest against French testing of nuclear weapons in the Pacific), was blown up and sunk in Auckland harbour. The vessel's photographer died in the explosion. Alain Mafart and Dominique Prieur, two agents of the French external security division, the Direction générale de la sécurité extérieure (DGSE), were subsequently arrested and, after being convicted of manslaughter at a court hearing in November, were imprisoned in Auckland. In July 1986, after pressure from France (including the imposition of trade sanctions) that they be released or returned to France, the two agents were transferred to the atoll of Hao in French Polynesia, where they were scheduled to be detained for the remainder of their 10-year sentence. The French Government made a formal apology for the sabotage operation, and paid the New Zealand Government $NZ7m. in compensation. Relations deteriorated in late 1987, however, when Mafart was returned to Paris, ostensibly for medical treatment, and in May 1988 a pregnant Prieur was airlifted back to France immediately prior to the French presidential election. Prieur's promotion to the rank of major on the fourth anniversary of the sinking of the *Rainbow Warrior* and the public decoration of Mafart in July 1991, together with the continuation of French nuclear testing in the Pacific, showed that protests by New Zealand had little real impact on French policy. President Jacques Chirac's resumption of underground nuclear testing at Mururoa Atoll in September 1995, however, outraged world opinion and was vehemently opposed by countries in the South Pacific. In response to public pressure within New Zealand, the National Government sent a naval research vessel to Mururoa. Meanwhile, New Zealand's attempt to reopen its 1973 case against France at the International Court of Justice in The Hague, Netherlands, was unsuccessful.

New Zealand's relationship with Australia was strengthened by the implementation of the Closer Economic Relations (CER) agreement. Inaugurated on 1 January 1983 and taking full effect on 1 July 1990, the arrangement was not without local critics, who pointed to a disparity of access, maintaining that Australian manufacturers had a competitive advantage owing to their unrestricted entry into the open New Zealand market. Meanwhile, defence links between the two countries were seriously undermined by dissension that developed within the ANZUS alliance as a consequence of the Labour Government's opposition to nuclear testing and weaponry in the Pacific region. In February 1985 the Lange Government upheld its popular election manifesto by imposing a ban on the entry of any nuclear-capable ships into New Zealand ports. The US Government maintained that such a standpoint rendered the ANZUS treaty inoperable as, for security reasons, it was policy neither to confirm nor to deny that any US vessel was nuclear-propelled or nuclear-armed. New Zealand's access to intelligence-sharing and military co-operation was withheld, and in August 1986 security guarantees to New Zealand under the ANZUS treaty were suspended. In November Lange reaffirmed his Government's decision to impose the ban on nuclear-capable ships. The US Government responded with the announcement, in February 1987, of its decision not to renew a 1982 memorandum of understanding (due to be renegotiated in June of that year), whereby New Zealand was able to purchase military equipment from the USA at concessionary prices.

The New Zealand Government subsequently proposed a new defence strategy, based on increased self-reliance for the country's military forces, in conjunction with continuing co-operation with Australia and greater involvement in the affairs of the South Pacific region. US military aircraft, which, by the nature of their operations, did not need to carry nuclear missiles, were, nevertheless, permitted to continue using the US Antarctic supply base at Christchurch (income from which is some $NZ20m. annually). In June 1987 the ban on nuclear-capable vessels became law, with the enactment of the New Zealand Nuclear Free Zone, Disarmament and Arms Control Bill. The Labour Government's anti-nuclear policy continued to command widespread support within the country, although the National Government, which took office in 1990, appeared to be uncomfortable with such a commitment, as was evidenced by the diplomatic overtures extended towards the USA by the Minister of External Relations, Don McKinnon. By 1992 the Government was signalling that it wished to amend the anti-nuclear law to allow for the entry of nuclear-powered vessels, but the National Party could not risk its slender majority by initiating such a change. The ban on nuclear arms remains supported by all parties—only the removal of nuclear weapons from the vessels of the US Pacific fleet would enable the Government to contemplate visits by US warships, although improved relations (see below) later permitted a resumption of some military co-operation between the two countries.

Apart from the common heritage of Commonwealth membership, New Zealand had little diplomatic contact with the African continent but, following Lange's visit there in April 1985, a High Commission was established in Harare, Zimbabwe. Limited promises of aid were also made. In the mid-1990s the National Government sought to improve relations with South Africa, previously compromised by controversy over the apartheid system and the ill-conceived and government-supported Springbok rugby tour of New Zealand in 1981. The success of President Nelson Mandela's visit to New Zealand in November 1995 was consolidated when Bolger opened an embassy in Pretoria in August 1996. While in South Africa, New Zealand's Prime Minister acknowledged that the 1981 Tour had been a costly mistake. At the meeting of Commonwealth heads of government in March 2002, New Zealand officials joined in the condemnation of President Robert Mugabe's controversial policies of farm seizures and land redistribution in Zimbabwe. In South Asia, the offence caused to the Indian Government by the Muldoon administration's closure of the New Zealand High Commission in New Delhi was redressed in 1985 by the appointment of the world-famous mountaineer, Sir Edmund Hillary, as High Commissioner in India. (The 50th anniversary of Hillary and Norgay Tenzing's ascent of Mt Everest was celebrated in Nepal and New Zealand in 2003.)

McKinnon's successful campaign to obtain for New Zealand a seat on the UN Security Council for a two-year term from January 1993 increased the country's international profile and encouraged a greater domestic awareness of foreign policy issues, as did his accession to the position of Commonwealth Secretary-General. Members of the armed forces joined the UN peace-keeping force in Bosnia and Herzegovina in the first half of the 1990s, and the National Government agreed to continue involvement despite the escalation of the conflict in former Yugoslavia. New Zealand's UN role and increasing global support for an international nuclear test ban led to some easing of tension with the USA. Bolger was received in Washington, DC, by President Bill Clinton in early 1995. The US President joined other world leaders at the 'summit' meeting of the Asia-Pacific Economic Co-operation (APEC) forum held in Auckland in September 1999. President Clinton announced the end of the 14-year ban on New Zealand's participation in military exercises with the USA, in preparation for the dispatch of a multinational peace-keeping force to East Timor (then part of Indonesia), of which New Zealand troops formed a part throughout 2000. Subsequent—and controversial—government decisions concerning defence expenditure indicated, nevertheless, a continuing commitment to such roles in the future. Thus, in December 2001 New Zealand announced that, at the request of the UN transitional administration, its 660 peace-keeping troops would remain in East Timor (now the independent state of Timor-Leste) until November 2002. In mid-2003 New Zealand defence personnel joined an Australian-led regional intervention force deployed in Solomon Islands. The multinational South Pacific force, which also included troops from Fiji, Papua New Guinea, Tonga, Vanuatu and Samoa, was successful in restoring law and order and enforcing a firearms amnesty for illegal weapons. The rebel leader, Harold Keke, surrendered to the international force in August 2003. The military component of the intervention force was significantly reduced at the end of July 2004, although New Zealand was to maintain a contingent of police and support personnel in Solomon Islands for a further two years, and provide aid for the improvement of domestic infrastructure and public-sector reform. Substantial financial assistance was also given to Niue following the devastation caused by tropical cyclone Heta on the island in January 2004.

The need for export markets has been a major influence on New Zealand's recent foreign policy. New Zealand has sought access to the markets of the European Union (EU), Eastern Europe and the Middle East, and has also become a member of the Organisation for Economic Co-operation and Development (OECD). Responses to international crises have been tempered by trade considerations. As part of the general protest against the Soviet intervention in Afghanistan, New Zealand temporarily severed diplomatic relations with the USSR and supported the US appeal for a boycott of the Moscow Olympic Games in 1980. In 1982, with considerable swiftness, the Muldoon Government offered aid to the United Kingdom during the crisis over the Falkland Islands, and imposed a boycott on Argentine goods. However, the government-authorized massacre of students in Tiananmen Square, Beijing, in June 1989 received only diplomatic condemnation, as trade with the People's Republic of China, especially in wool, had assumed significant proportions since the late 1970s. An official visit to China by Prime Minister Helen Clark in April 2001 was intended to improve New Zealand's trading position with that country prior to its accession to the World Trade Organization (WTO). Prime Minister Bolger's state visits to the Republic of Korea and Japan in the 1990s, and the accompanying public insistence that New Zealand's trading future was to be found in Asia, led to economic initiatives that were to prove unsustainable by that market. In 1998 the tourism and forestry sectors were particularly badly affected by the Asian economic crisis, thus further exposing the social costs of New Zealand's vulnerable position within the global economy. Again, in 2003 the Severe Acute Respiratory Syndrome (SARS) epidemic affected tourism numbers to New Zealand, and also contributed to a significant decline in the number of international, fee-paying students from Asia attending New Zealand educational institutions (from primary and secondary through to tertiary level). The December 2003 release of the third and final film in *The Lord of the Rings* series continued the positive tourism-related publicity gained for New Zealand from Peter Jackson's filming of the trilogy in his home country.

Responses to world events during 2001 and 2002 reflected New Zealand's non-isolationist stance. The Government was quick to express support for the USA in the aftermath of the terrorist attacks of 11 September 2001 and committed troops from the special combat force to service in Afghanistan, without full parliamentary debate on the decision. In 2001, for humanitarian reasons, the Government accepted 141 of the Afghan refugees rescued at sea by a Norwegian vessel but denied access to the Australian mainland (see the chapter on the Australian territory of Christmas Island). The New Zealand Government has continued to take small numbers of these refugees, without public fanfare. The nature of US policies in the ongoing 'war on terror' was a source of debate and concern among many New Zealanders. In the build-up to the US-led invasion of Iraq, the New Zealand Government's stance in favour of action only through the UN was popularly endorsed, as were the Prime Minister's reputed criticisms of the US Government's approach. Relations with the US Administration were adversely affected, as evidenced by Australia's signing of a free-trade agreement with the USA, while New Zealand continued to press, unsuccessfully, for free-trade access to US markets. International negotiations at the WTO meeting in Geneva, Switzerland, in July–August 2004, however, resulted in both the EU and the USA agreeing to phase out all agricultural export subsidies. If implemented, this commitment could mean a significant rise in New Zealand's export earnings.

In July 2004 New Zealand's relations with Israel deteriorated when it was revealed that two Israelis, convicted on charges related to a fraudulent attempt to obtain a New Zealand passport, were acting on behalf of Israel's intelligence services. High-level visits and consulations between the two countries were immediately suspended, and Prime Minister Clark made it clear that she considered a formal apology to be appropriate for this breach of New Zealand sovereignty and international law.

The New Zealand Prime Minister's presence at a meeting of the Pacific Islands Forum held in Fiji in early 2002 indicated a renewal of closer relations with that country following its return to democracy. New Zealand expressed disquiet about the level of ongoing violence and civil disorder in Solomon Islands (see above), and about the political unrest that surrounded the general election in Papua New Guinea in mid-2002, concerns that were shared by other Pacific nations. New Zealand also welcomed Australia's increased level of commitment to political and economic development in the Pacific, reflected at the meeting of the Pacific Islands Forum hosted by New Zealand in August 2003. Prime Minister Clark attended the 2004 Pacific Islands Forum meeting held in Apia, Samoa, in August 2004, at which members acknowledged the urgent need for preventive measures against the further spread of HIV/AIDS in the region.

Economy

KENNETH E. JACKSON

Based on an earlier article by J. W. ROWE

New Zealand is best described as a small, open economy centred on two main islands. On 24 April 2003 its estimated population reached 4m., with a further 61,300 added by mid-2004. The country's renowned advocacy for, and implementation of, a regime of deregulation, free trade and liberalized investment remains in place, although the policy is no longer followed with the same zeal as previously. In fact, most of the liberalization efforts were concentrated in the mid-1980s and the early 1990s, during the first terms of the Labour administration (1984–87) and of the successor National Party (1990–93).

New Zealand is far distant from its traditional markets and relatively distant from its nearest major market, Australia. Export receipts continue to be concentrated upon a very narrow range of primary products. These are subject to considerable fluctuations in their prices, which are largely determined externally. Market access for agricultural products has increased, particularly since the successful conclusion of the 'Uruguay Round' of the General Agreement on Tariffs and Trade (GATT), in December 1993, and the subsequent establishment of the World Trade Organization (WTO). Expectations of improved agricultural trade conditions arose following the effective resumption of talks on the WTO agricultural agreement at the end of July 2004, although the eventual impact remains to be seen. Restrictions on New Zealand exports remain an occasional problem on the demand side, whilst drought and other constraints on output affect supply from time to time. The internal market is restricted by physical obstacles and by the low density of population over much of the land surface. A relatively high exchange rate by historical standards also affects export receipts.

Debate continued over the causes of New Zealand's relatively poor macroeconomic performance compared with the average of the other member countries of the Organisation for Economic Co-operation and Development (OECD) and with the Australian experience in particular. Some commentators, including the former Governor of the Reserve Bank of New Zealand (RBNZ), Don Brash, now leader of the opposition National Party, took issue with this assessment, arguing that from 1990 New Zealand's average annual per-head increase in real gross domestic product (GDP), of some 2.5%, had set the country behind only Ireland, Australia, Norway, the USA and the Netherlands. Judicious choice of other start and end dates of course gave an entirely different outcome. The average growth rate in the long run (since 1900) was of the order of 1.6% and more in line with Argentina than with the countries normally thought of as being in the same 'club', such as Australia, Canada or the United Kingdom. Australia generally outperforms New Zealand, despite its castigation from time to time in the media and elsewhere for its lack of microeconomic reform on the New Zealand pattern. New Zealand incomes had declined from their 1995 peak of 87% of the OECD average to 84% by 2001, when that organization estimated the Australian figure to be 108% of the OECD average. In recent years New Zealand's macroeconomic performance has improved, however, with GDP growth of 3.6% recorded in the year to March 2004 and of 4.4% in the previous year.

Migration figures moved from negative to positive in the 12 months to mid-2002 and remained positive for the years to June 2003 and 2004, with a net inflow of migrants of 42,500 and 22,000 for 2003 and 2004, respectively, compared with a net gain of 32,800 for 2002. The inflow largely settles in the Auckland area, fuelling a rise in house prices in that region. Such issues have led the central bank, the RBNZ, to pursue an aggressive monetary policy, with the official cash rate (OCR) set above the OECD average. Following the terrorist attacks on the USA on 11 September 2001, New Zealand interest rates generally declined, but by a smaller margin than elsewhere and with a slight delay. In mid-2002 rates rose again, to reach a peak of 5.75%, and by July 2004 interest rates were at 6%. How such relatively high rates will affect the Government's desire to encourage growth and innovation remains to be seen.

The development of the New Zealand economy since the initiation of the 1984 reform programme has incorporated a continuation of many of the previous characteristic features, such as relatively slow growth rates and comparatively high real interest rates, along with a propensity to import, which inclines the economy towards a deficit on the current account of the balance of payments. Unemployment, as indicated by conventional measures, was virtually non-existent in 1970, but subsequently increased to a significant level, with major peaks in 1983 and 1992. The unemployment rate subsequently fluctuated around a slowly declining trend, and was estimated at 4.0% of the labour force in mid-2004, the second lowest rate in OECD. Inflation also rose in the 1970s, but slowed to an annual average rate of 8.5% in 1995–98, was negligible by 1999, and in 2003 remained below the 3% upper limit of the target set by the RBNZ. As of June 2004 the annual inflation rate was estimated at 2.4%. Past difficulties were partly attributable to external factors, in the guise of prolonged periods of international recession. Restricted market access for some agricultural exports has continued from time to time, but the country's economic problems have been aggravated, even where not originally induced, by the slow pace of domestic adjustment to external market changes.

The New Zealand of the reform era has been described by Don Brash as 'a model for change' from an economic perspective. Outcomes, however, have failed to match expectations. Apart from the period of 1991–96, for most of the time since the late 1960s New Zealand has performed below the OECD average and fallen well behind not only Australia (see above), but also the United Kingdom and the USA in terms of GDP per head. This has occurred despite the post-1984 campaign for extensive deregulation of the economy, involving deregulation of financial institutions, liberalization of shopping hours and liquor-licensing laws, delicensing of the internal transport system and the meat-processing industry, and the removal of the last vestiges of import licensing. Farm subsidies were discontinued, and many state trading activities were corporatized if not privatized. The model still has its critics, but whether other models and prescriptions would have performed any better is a moot point.

The policy measures adopted from 1984, starting with the deregulation of financial and foreign-exchange sectors of the New Zealand economy, were part of a broader range of policies avowedly intended to promote more rapid economic adjustment. Not all the desirable aims were, in fact, achieved immediately. Fiscal balance and then a fiscal surplus were accomplished, but only in the longer run. The taxation system was radically altered with the introduction of a goods and services tax (GST) and the reduction of marginal income tax rates on higher incomes. The Government attempted to take a secondary role in economic matters, regarding its function as one of encouraging the efficient operation of markets, competition and an economy able to respond quickly and flexibly to changing circumstances and demands. Initially, a burgeoning fiscal deficit emerged, as taxation revenue stagnated while welfare spending rose. Restructuring involved excessive unemployment, at least in the short term. Regional, age and skill disparities appeared in the unemployment figures.

In October 1990, when the National Party returned to power, a second spate of enthusiasm for deregulation occurred. Greater emphasis was placed on decreasing the fiscal deficit and further reducing inflation. In addition, from early 1991 legislation was introduced with the aim of deregulating the labour market. The Employment Contracts Act (ECA) marked a major shift in negotiating power. It reintroduced voluntary unionism and enhanced the employer's bargaining power. Union membership halved. The ECA was claimed to have produced greater flexibility in the labour market, resulting in a significant stimulus to employment and production, notably for export. The exact outcome was difficult to assess, as the figures were confusing and the choice of benchmark for comparison was an important determinant of the result. Amendments to the employment law were made from October 2000 through the Employment Relations Act (ERA), which, among other measures, restored a greater role for collective agreements.

In 1991 the Government also moved to reduce the cost of the welfare state and to contain the rising bill for national superannuation. The ensuing outcry led to some reversal, but the main features of the cut-backs remained. Measures were also put in place to curb the increasing public costs of health and education, along with institutional changes intended to improve service delivery and accountability. For a while, in the early 1990s, external income and expenditure were approximately balanced for the first time in two decades, annual inflation had declined to around 1% and, temporarily, strong growth in production and employment was apparent. From 1993 to 1996 the economy continued to perform strongly, with unemployment declining and fiscal balance being achieved by mid-1994, largely owing to buoyant tax revenues, but also to spending restraint. A policy of allocating priority to the repayment of overseas public debt from these fiscal surpluses was then implemented, thus improving the country's international credit rating.

Economic problems re-emerged in 1996–97, albeit less serious than those of the 1970s and 1980s. The strength of the New Zealand dollar in the mid-1990s and high interest rates curbed the growth of exports and production generally, whilst inflation rose slightly. Some decline in the exchange rate gave exporters a boost. Parliamentarians conducted the new electoral system of mixed-member proportional representation (MMP, see History) in an unstable manner, and one not highly conducive to economic growth. Economic instability was exacerbated in 1998 by the adverse effects of the Asian financial crisis, which was particularly damaging to the tourism and forestry sectors. These factors, combined with New Zealand's high propensity to demand imports, resulted in a balance-of-payments deficit equivalent to some 7% of GDP. Any recovery in 1998 and 1999 was somewhat slow and hesitant, as well as uneven, both geographically and sectorally. A revision of some of the previous reforms emerged following the 1999 election of a new Government, with further reviews occurring since the general election of July 2002. The methods and targets of the RBNZ remained under some question, including demands for further rate reductions in 2003 from manufacturers, exporters and trade unions. The changes from a quantity-type control to an interest rate control developed during 1999 and 2000, with the original target band of price inflation being set at 0%–2%, and subsequently amended to 1%–3%. The major monetary control variable, the OCR, stood at 6.00% in August 2004, with a further rise expected in September. Fears of the economy overheating resulted from buoyancy in the housing market, low unemployment and companies' pricing intentions. The strong growth experienced in primary-product export receipts by New Zealand farmers in 2001 stabilized in 2002, and receipts declined in 2003, with milk and other primary-product prices falling from their earlier peaks. A continuing trend for the value of the New Zealand dollar to remain steady, if not increase, against the Japanese yen, the US dollar and the pound sterling caused disquiet amongst exporters, who complained about continuing rises in interest rates.

AGRICULTURE, FORESTRY AND FISHING

Agriculture

Farming and associated processing industries play a greater part in the New Zealand economy than in most developed countries. The historic concentration on pastoral products for export income arose partly from preferential access to a large market, the United Kingdom, but production was also favoured by a benign climate. Since soil fertility is not generally high, a scientific approach to the improvement of soil productivity and of stock-breeding and management has been necessary, together with a heavy reliance on (imported) phosphatic fertilizers. Diversification into manufacturing and service industries has taken place in recent decades, although traditionally these have been limited to the small domestic market, with its protection by distance and by government intervention. The agricultural sector remains of fundamental importance for external trade (although wool receipts are not as prominent a contributor to export earnings as previously), and, of necessity, farming is highly efficient by world standards. Capital intensity, in the guise of large farm size and a high degree of mechanization, has been a major factor in maintaining the international competitiveness of agricultural and pastoral production.

Dairying is now the most important sector of farming in New Zealand. Favourable factors have been product and market diversification, increasing average farm size and concentration of processing in a few very large plants. The New Zealand Dairy Board, which in earlier years had played a vital role in improving efficiency in dairying, both on-farm and off-farm, was replaced in 2001 by a new organization, Globalco, which subsequently began trading under the new name of Fonterra, one of the biggest such concerns in the world. Issues relating to fears of cross-subsidization of export production and long-term implications for efficiency and competition were expressed at the time of its introduction, but farmers largely supported the move. New Zealand domestic dairy prices generally have remained below world levels, rather than above them, allaying overseas competitors' suspicions of cross-subsidization, and some measures to ensure domestic competition have been put in place. Domestic dairy prices, however, have been rising overall since the new organization was established. Declines in the world price for dairy products in 2002–03 were not matched by similar falls in domestic prices, but the rise in the exchange rate accounted for some of the variation.

Pastoral farming is still of major significance and is also predominantly export-orientated. Problems of access to some overseas markets and limited progress in the processing of products have restricted profitability. In recent years livestock diversification has occurred, such as moves to expand the number of deer and goats, whilst reducing sheep numbers. Cropping, fruit growing and horticulture are also of increasing importance, although problems occur from time to time, such as those facing apple growers in 2004. Latterly there has been a major expansion of viticulture, with significant exports of wine. Fluctuating grape crops, however, affect the rate of growth of wine production. Consistent quantity production capacity problems have proven something of a difficulty for this industry in the past, but the development of major producers in newer regions has contributed to an improvement in the situation.

The agricultural sector faced major reorganization when support prices were abandoned in 1985/86 as part of the programme of economic reform. The sector withstood the restructuring surprisingly well. Despite diversification elsewhere in the economy, exports of farm products of all kinds still account for a substantial percentage of total exports and for much of the growth in exports witnessed since 1999. Meat and dairy products together accounted for 31.6% of total export earnings in 2003. The overall merchandise terms of trade moved favourably in early 2004, rising by 2.1% compared to the last quarter of 2003.

New Zealand welcomed the successful conclusion in December 1993 of the 'Uruguay Round' of GATT negotiations, and has actively participated in subsequent bilateral negotiations, which on the whole have been reasonably successful. Agricultural protectionism in Japan, the European Union (EU) and elsewhere remains a problem, however, despite optimism over the recommencement of previously stalled agricultural trade talks in 2004. Effective implementation of existing agreements remains as important an issue as the conclusion of new agreements.

Forestry

Most parts of New Zealand were still heavily forested when European settlers arrived, although substantial areas had been deforested during the centuries of Maori occupation. Today large tracts remain in indigenous forest, although mainly in rugged terrain. Large-scale clearance of forest cover was needed for pastoral expansion. Timber from forest clearance was originally used for ships' masts, then as inputs into farming through fencing and construction, a process that extended to urban construction and later export to Australia, to supply Melbourne in its late-19th-century building boom for dwellings. In the last decades of the 19th century, clearance was so rapid that much of the timber was burnt rather than used.

Today plantation-grown *Pinus radiata*, along with other introduced species, has largely replaced indigenous timber for production purposes. *Pinus radiata*, which in New Zealand produces logs suitable for milling within 25–30 years, also provides the raw material for a widening range of manufactured products—pulp and paper, paperboard and compressed boards of various sorts, as well as exports of sawlogs on a considerable scale. The forestry industry's contribution to export earnings has increased, and by 2003 the value of exports of wood and wood articles had reached $NZ2,385m., representing over 8% of total export earnings. In 2003 logs, wood and wood articles were the third major commodity contributor, behind dairy products and meat. Major plantings in the late 1960s and 1970s were reaching maturity in the early 21st century. A 74% increase in wood available for export was anticipated between 1996 and 2010, assuming 60,000 ha of new plantings every year, but processing expansion requires heavy investment. There has also been extensive planting of *Pinus radiata* for land stabilization and environmental, as well as for production, purposes. These developments point to forestry becoming markedly more important in the years ahead, with the amount of land used for farming continuing to decline, whilst that allocated to forestry production continues to increase.

Legislation relating to native forests has increased the level of protection of old-growth forests so that the felling of indigenous trees has been reduced, mainly for heritage and conservation reasons. Restrictions on indigenous forest use on the west coast of the South Island and the 2001 conflict over a proposed extension to gold-mining in the same area continue to cause tension between advocates of conservation and those espousing the economic exploitation of such resources. Farm forestry, combining both aesthetic and economic considerations, has reduced the vulnerability of pastoral farming to fluctuations in meat and wool prices. It has also made better use of marginal land and has been instrumental in protecting some hill farms from erosion problems. Forestry development is one of the few areas where tax advantages are to be found in New Zealand's deregulated, non-interventionist atmosphere, but the long-term nature of the investment tends to deter many would-be participants.

Fishing

New Zealand has a wide variety of fish in waters around its long coastline, and many of these are now recognized to be commercially valuable species. The Territorial Sea and Exclusive Economic Zone Act of 1978 gave the country control over the fisheries resources in more than 4m. sq km (1.3m. nautical sq miles) of sea surrounding New Zealand territory, or approximately 13 times the land area of New Zealand. Joint ventures with foreign partners have contributed substantially to the development of the industry, especially in the trawl fisheries of deeper waters. There has also been continuing growth in the domestic finfish industry. Exports of fish, crustaceans and molluscs were valued at $NZ1,167m. and accounted for 4.1% of total exports in 2003. Poaching and organized smuggling of paua (abalone) has emerged as a serious threat to the sustained development of such activity.

Essentially dating from the 1980s, fish and shellfish farming and processing has continued to develop, although disputes over property rights in the seabed and foreshore areas have created some uncertainty in the industry. Production is directed primarily at the export market. Trout, introduced from the northern hemisphere, have not so far been commercially exploited, but salmon are farmed, as well as fished recreationally, in parts of the country. Aquaculture, principally involving the production of mussels, salmon and Pacific oysters, has become a significant part of total fisheries output. By 2000 its value was already estimated at some $NZ200m. Transferable quota arrangements have been developed as a way of dealing with issues relating to the long-term sustainability of common property rights to fish stocks. Allocation of quota to Maori has also been undertaken as part of the Treaty of Waitangi process (see History).

MINING

Mining of metallic minerals is limited except for titanomagnetic ironsands from the west coast of the North Island, originally for export but later also for local steel production. Recent efforts to revive mining for precious metals have been remarkably successful, although they are capital-intensive and face problems with meeting environmental-protection requirements in many cases. The Resource Management Act (RMA) covering environmental requirements has made compliance a more expensive undertaking for the mining sector, as well as for many other industries.

Coal production is limited and mainly for use in thermal power stations. Household consumption is insignificant. New Zealand has substantial coal reserves, but their economic utilization has not been high. Japan, Chile, India and the People's Republic of China are the principal export markets. Natural gas has become increasingly important, especially since the onshore supply has been supplemented by offshore gas from the large Maui field, and a considerable amount is utilized in the production of synthetic petrol and in thermal power stations. For many years drilling companies failed to make any significant petroleum discoveries, but there are now several onshore producing wells, which together make a small, yet significant, contribution to meeting New Zealand's oil demands. In 2003 there was an energy crisis in terms of electricity generation, with low lake levels and concerns about the life expectancy of the Maui gas field.

MANUFACTURING AND CONSTRUCTION

Most of the country's larger industrial enterprises are export-orientated and process pastoral products (milk, livestock and wool), natural gas, logs or ironsand, but there is also a large aluminium smelter using imported alumina. Other large enterprises based on imported raw materials have faced increasing import competition, and some enterprises have closed down as border protection has been reduced or removed. The WTO, the Asia-Pacific Economic Co-operation (APEC) group and the Closer Economic Relations (CER) agreement with Australia have all played a part in this process, but successive New Zealand Governments have moved more generally towards a progressive reduction in tariffs. This has included a unilateral programme aimed at increasing the competitive environment in the domestic market. It has also led many manufacturers to relocate their operations off shore. Others have become more specialized, while diversification is evident in the expansion of non-traditional manufactured exports. There has recently been a renewed interest in New Zealand, as elsewhere in the world, in regional and bilateral trade arrangements, with talks conducted and actual agreements concluded with several countries.

Some manufacturers have called for a deceleration of the pace of liberalization. In 2000 the Government acted to curb the rate of tariff reduction, as well as to promote the need for more involvement in regional development. New Zealand already has fewer non-tariff barriers to imports than is the case in many other countries. Of equal importance are the issues of the exchange rate and the prevailing relatively high interest rates, which have tended to inhibit manufactured exports. In early 2004 the overall trade-weighted index fell slightly, but generally it has remained stable against the currencies of New Zealand's major trading partners. Greater trans-Tasman integration has continued, with common food regulations with Australia being implemented, and some calls for a common currency and stock market. Achieving easier access to the greater Australian market, combined with improved efficiency, remain the key factors for New Zealand's secondary industrial development.

The construction industry in New Zealand has three distinct sectors: the residential construction sector, traditionally timber-framed in single units; commercial and industrial construction, which also employs timber framing as well as steel and concrete; and civil engineering. Since the mid-1960s the industry as a whole, and especially its residential component, has been subject to marked cyclical fluctuations arising from demographic trends (including sharp fluctuations in migration) and economic vicissitudes. The reversal from a net migration outflow to a net inflow in 2001–02 led to some recovery in activity levels and prices in the residential sector in particular. More generally, the Building Activity Survey recorded strong increases during 2000 and into 2001. This trend still continued in 2002 and the first half of 2003, despite public concern about the so-called 'leaky building syndrome' and a resultant reversion to the use of treated timber for internal framing and other regulatory changes. By early 2004 there was still buoyancy in the sector, although expectations were for a slowing down of price rises in the near future. Shortages of skilled labour were being reported for all sectors of the construction industry.

INFRASTRUCTURE

Energy

New Zealand has substantial hydroelectric generating potential, although limited storage capacity is a weakness, revealed by substantial rises in the 'spot' price in the winter of early 2001 and calls for a voluntary savings campaign. The winter of 2002 passed without any such crisis. In 2003 there was another voluntary savings campaign and the prospect of power cuts as hydro-lake levels fell. By June the crisis had passed, but plans for extra generating capacity were brought forward. There were no significant electricity shortages in the year to September 2004. Electricity output is more than 112.40 petajoules per year, with hydroelectricity contributing approximately two-thirds of the total. The main thermal stations are fired by a variety of fuels; there are also two large geothermal stations, with a capacity comparable to that of both the main thermal stations. Two-way direct current cables link the two main islands, so virtually the whole country is served by the grid, even relatively remote areas, as the result of the previous subsidized extensions to rural networks.

In a major departure from tradition, generation of electricity was separated from the operation of the grid to encourage non-state generation. Until the 1980s such private involvement had been hampered, if not prohibited directly, by government regulation. These changes at the wholesale level were accompanied by the corporatization and some privatization of 'power boards', which used to handle the retail distribution of most electricity. With the development of a wholesale electricity market, electricity retail companies are now in apparent competition with one another, although the reality is that with existing technology this is difficult to make effective for small consumers. In an attempt to control the abuse of monopoly power in the retail market, separate transmission-line companies and retail energy companies have been formed. Further refinements to the regulatory system now allow some vertical integration, whilst attempting to ensure the separating out of the natural monopoly elements of the wholesale and retail distribution networks, from generation and retail supply.

Since 1973 considerable efforts have been made to reduce the country's dependence on petroleum products, both crude and refined, through measures such as the expansion and modification of the existing petroleum refinery, the conversion of natural gas to methanol as feed stock for the refinery and moves towards more efficient energy use. Such developments, together with increasing production from onshore and offshore gas and petroleum wells, and the switch from oil to coal and gas in thermal power stations, greatly increased energy self-sufficiency. Not all of the 1970s efforts made economic sense, with some developments failing to be subjected to adequate project appraisal in the rush to save on imported energy. More recently, energy policy debates have placed a greater emphasis on the reduction of carbon-dioxide emissions and their offset by large-scale afforestation in line with predictions of global warming. In 2003 discussion continued as to the merits or otherwise of ratifying the international Kyoto Protocol on the reduction of the emission of 'greenhouse gases', although some of these discussions were more focused on methane emissions from sheep and cattle than those from the energy sector, in a debate which at times appeared to move from the serious to the comic.

Transport and Communications

Historically, the central Government was heavily and directly involved in transport, as with other areas of infrastructure, through its ownership of New Zealand Rail (now Tranz Rail Ltd), Air New Zealand and the Shipping Corporation. All were privatized, although Air New Zealand was renationalized in October 2001, following the collapse of Ansett, its Australian subsidiary carrier. In 2002 the Government held an 82% stake in Air New Zealand. The Qantas operations in New Zealand were being managed by the airline directly, following the Australian carrier's own problems with an associated operator flying New Zealand domestic routes. Aggressive competition then emerged on the main domestic routes. Proposals for various forms of combination or agreement between Air New Zealand and Qantas continued to be opposed by the Commerce Commission in 2004.

The Government remains involved in transport regulation, notably through the Land Transport Safety Authority and Transit New Zealand, which have responsibility for road safety and main highways, respectively, and the latter, in part, for urban passenger subsidies. Since mid-1998 debates concerning how to apply 'user pays' principles to land transport, and how to restructure it, have continued. Proposals have involved active consideration of private-public partnership schemes and toll roads. In other areas of transport, major airports have been privatized or corporatized, with Auckland and Christchurch airports being the focus of investigations by the Commerce Commission over allegations of overcharging. The prospect of price controls over Auckland International Airport was raised, but not enacted. Reform of port operations was largely complete by the mid-1990s, a period during which transport became more efficient and competitive. In the early 21st century, however, shipping and air services across the Tasman Sea remained far from totally liberalized. The introduction of new carriers into the trans-Tasman service, such as Emirates (of the United Arab Emirates) in mid-2003 and, later, the low-cost airline Virgin Blue, as well as a change in Air New Zealand pricing structures, increased the competitive environment.

Equally dramatic changes have taken place in communications, with the privatization of telecommunications and the corporatization of postal services. Historically, the Post Office in New Zealand, as a government department, handled postal services and various collections on behalf of other government agencies, as well as telecommunications. It also operated a savings bank catering mainly for small depositors. The Government separated telecommunications services and later re-established banking operations in the guise of Kiwibank, while NZ Post remains a 'stand-alone', but still state-owned, enterprise, although its monopoly of mail services has been removed. National Post and other competitors initially appeared, offering household delivery and separate on-street mail collection boxes, but National Post subsequently withdrew from the market, following financial difficulties, and its owners were unable to find a buyer.

SERVICE INDUSTRIES

The increase in complexity of the economy has been accompanied by a steady growth in the proportion of the labour force in service industries, some of which render specialist services to primary and secondary industries as well as to the community in general. This growth has been at the expense of manufacturing in particular, but also of the service-like activities of building and construction and transport and communications (see above).

The tourism sector is a major component of economic activity, providing jobs for around 10% of the population. Economic expansion in industrial countries generated a significant increase in international tourism, from which New Zealand benefited considerably. In 1987/88 tourism, as officially defined, became the single largest source of foreign-exchange earnings, its expansion said to have been aided by the increased competitiveness of the industry following the freeing up of the labour market. As might be expected, given its proximity, Australia remains the main source of overseas visitors, followed by the United Kingdom, the USA and Japan. As a result of the Asian economic crisis, however, the total number of visitor arrivals declined from 1,551,341 in 1996/97 to 1,464,766 in 1997/98, the number of tourists from the Republic of Korea showing the sharpest decrease (44%). Total receipts from tourism declined to $NZ3,068m. in 1998. A recovery in the tourism sector subsequently became evident, with visitor numbers reaching record levels by mid-2001 and being only briefly affected by the repercussions of the terrorist attacks on the USA in September of that year. Growth subsequently continued. In the year to June 2003 the numbers of visitors increased to some 2,050,000, although 2003 again saw some some declines from Asian areas, owing to the epidemic of Severe Acute Respiratory Syndrome (SARS), and the average stay declined from 22 days to 21. By mid-2004 the industry was experiencing a strong recovery, with the figures for June 2004 over 20% above those for June 2003. The services sector, including the various components of the tourism industry, is notoriously difficult to assess in terms of labour productivity, but its expansion is an essential part of modern economic development, and New Zealand exhibits much the same trend as other developed economies.

PUBLIC SECTOR

Until 1991/92 the public sector's claim on national resources was inexorably rising. Within the public sector itself, the central Government was increasingly put under pressure to transfer resources to territorial and other local authorities, because the latter lack the automatically increasing flow of funds that a relatively fixed and progressive (personal) income-tax structure so conveniently provides in an era of rising wages and salaries. At the same time, there was a tendency for local or regional authorities to assume responsibility for a greater proportion of total public-sector expenditure. This was given further impetus by the extensive reform of local government undertaken in 1989. As in many other countries, deficit-financing was adopted in the early 1980s, deficits averaging 7% of GDP. Despite efforts by successive Governments, public-sector expenditure, including transfers, was equivalent to more than 40% of GDP by the end of 1990, when the National Party took office. By 2001 the proportion was approximately 35%, a level similar to that of the previous three–four years, although about one-third of payments represented transfers rather than direct expenditure. Between 1984 and 1990 the Labour Government had initiated wide-ranging reforms in the public sector, notably by corporatizing many enterprises and privatizing others. This process was continued by the National Party Government. Some of the apparent decline in the share of GDP represented by government expenditure was owing to the shifting of these enterprises off the government account. Fiscal surpluses were achieved in the early 1990s (even when the proceeds of asset sales were excluded). Net public-sector debt as a percentage of GDP also declined during this period. Household indebtedness, however, increased consistently through to 2003, and since then has maintained a level comparable with that of other developed countries. Mortgage debt alone rose by 17% during 2003. Overseas debt, at a level of approximately 75% of GDP, is now largely a private-sector obligation rather than a public one.

INTERNATIONAL TRADE

Exports and imports of goods and services are usually roughly equivalent, at a little under 30% of GDP, although a persistent balance-of-payments deficit was recorded in the 28 years to 2000/01, with the current-account deficit peaking at 6.6% of GDP in the year to March 2000 and subsequently narrowing to just over 2.0%. In 2003 the current-account deficit was some 4.2% of GDP, at around $NZ5,700m. Forestry, agriculture and pastoral exports remained a major component of foreign-exchange commodity earnings (some 65%), whilst fish are of increasing relative importance. Exports of manufactures (other than processed primary products) are also growing steadily, and earnings of New Zealand companies operating overseas are now becoming appreciable. About 50% of New Zealand's foreign exchange, however, still comes from the exports of just 30 companies, a reflection of the small size of the economy. Meanwhile, 96% of companies do not export.

Exports have been diversified, as regards both commodity composition and market destination. In 1950 the United Kingdom purchased nearly two-thirds of New Zealand's exports. By the 1980s it had been overtaken by Australia, the USA and Japan, which remained the three principal export markets in 2003. The People's Republic of China became the fourth largest recipient of New Zealand's exports in that year, just displacing the United Kingdom, which was followed by the Republic of Korea. For imports the pattern was similar, the principal suppliers being Australia, the USA and Japan, followed by the People's Republic of China and Germany, with the United Kingdom in sixth place. The increasing interdependence of the Australian and New Zealand economies led their respective Governments to implement the CER agreement, signed by the two countries on 14 December 1982, following three years of negotiations. The bilateral trade agreement took effect on 1 January 1983. Its main objective was to encourage the development of economically strong productive structures in both countries. Two basic mechanisms were set in place by the CER agreement: the phasing out of tariffs both ways; and the progressive elimination of import-licensing and tariff quotas between Australia and New Zealand. By and large, CER has benefited both countries, perhaps more so New Zealand, and mutual undertakings have generally been respected. The CER was larger in impact than the preceding agreement, which had required items for liberalization to be included in the list rather than the CER approach, which stipulates the listing of exceptions.

The New Zealand dollar's fixed relationship with the US dollar was ended in July 1973, when the authorities introduced a trade-weighted managed daily 'float' involving weighting against the currencies of New Zealand's main trading partners. In July 1984, immediately following the election of a Labour Government, the currency was devalued by 20%. In March 1985 the New Zealand dollar was freely 'floated', in response to fears of a renewed 'run' on the currency, which had been caused by a sharp upsurge in inflation following devaluation and the end of a 'freeze' of wages and prices. By mid-1986, contrary to the expectation of many people, the currency had remained very close to its average value at the time of the flotation, and it continued to fluctuate around this level until 1991, when the trade-weighted index fell appreciably, giving exporters a welcome boost. In 1993 the New Zealand dollar strengthened again, and in particular appreciated markedly *vis-à-vis* the Australian dollar. It also proved remarkably resilient in the face of turbulence in the exchange-rate mechanism of the EU. These phenomena reflected improvement in the economy as well as relatively high real interest rates. The new-found strength of the New Zealand dollar at that time no doubt also owed something to the independence enjoyed by the RBNZ and confidence in the fiscal stance of the National Party Government. Thereafter, the New Zealand dollar appreciated further, against the US dollar and the Australian dollar in particular. This helped to contain inflationary pressures, but rendered New Zealand's exports less competitive internationally. To the relief of exporters, the New Zealand dollar depreciated slightly against the US dollar in 1997. There was a continuation and increase in this trend through to early 2001, with a decline to the low 40s in terms of US cents per New Zealand dollar, although the cross rate against the Australian currency was rising. The New Zealand dollar stood at 0.82 Australian cents by August 2001. Relief in terms of additional export revenue was somewhat slower to appear than expected, and there were even calls for the introduction of a common trans-Tasman currency in order to obtain a reduction in transaction costs for traders. Further rises against both the US dollar (to 65 cents) and the Australian dollar (to 90 cents) occurred in the first half of 2004.

DOMESTIC ECONOMY

Throughout the 1950s and 1960s the average rate of growth of real GDP was more than 4% per year, which was not much less than the average for OECD countries. New Zealand's performance in the 1970s, particularly after the first oil shock with its sharp rises in international petroleum prices in 1973 and 1974, was comparatively worse than the OECD average, with the terms of trade deteriorating substantially. Between 1975 and 1982 there was virtually no real growth in GDP. Between 1982 and 1987 there was modest growth, but this improvement was not sustained. Large fiscal deficits emerged in the late 1970s and, together with increasing external imbalance, acted to create a mounting burden of debt in which the external component was increasingly important. By the early 1980s economic performance—as measured by a wide range of indicators including growth, unemployment, inflation, external deficit, budgetary imbalance and debt—had deteriorated substantially. In real terms, GDP grew by an average annual rate of 2.5% over the 1990s, but this covers a wide range of economic activities. In 1998 GDP contracted by 0.3%. Signs of a recovery from the recession had emerged by mid-1999, when GDP expanded by an estimated 3.5%. Growth of 1.8% was achieved in 2000, with a similar figure recorded in 2001, and budget estimates for GDP growth in 2003 were around 3.5%, comfortably above the OECD average of 2.2%.

In the early 1980s unemployment emerged for the first time since the 1930s as a serious problem. There had effectively been full employment in New Zealand throughout most of the post-Second World War period. It was only in 1978 that the unemployment rate first rose above 1% of the labour force. It then increased steadily, reaching a peak in 1983, and, despite declining in 1984 and 1985 owing to renewed growth in the economy, rose to 6% at the end of 1988, and to over 10% in 1992. Subsequently, unemployment levels appeared to stabilize, growth in employment having more than offset rising labour force participation. In the 1998 recession the average rate of unemployment increased to 7.5% of the work-force. Unemployment decreased to 6.8% of the labour force in 1999, with a further decline to 4.0% recorded at mid-2004. Employment growth had continued strongly over the previous year.

The average annual rate of inflation was 4.8% in 1985–95. Lower rates set in thereafter: consumer prices increased by an annual average of 2.3% in 1996, by only 1.2% in 1997 and by 1.3% for 1998. By 1999 some commentators had declared inflation to be defunct—in the first quarter of 1999 the consumer price index (CPI) was negative compared with 12 months previously. In June 2004, however, the CPI stood at an annual rate of 2.4%.

Until May 1991 wage-bargaining was based on national awards, covering broad occupational groups, and took little or no account of regional considerations or of the economic well-being of individual industries and enterprises. The wage-bargaining process had had a long history of inflexibility, regarded as out of place in the more competitive, less regulated environment of recent times.

In its efforts to liberalize the economy from 1984 to 1990s, the Labour Government floated the currency on international exchange markets, maintained a non-accommodating monetary policy and embarked on a programme of taxation reform in an effort to increase economic efficiency and competitiveness on a medium- to long-term basis. Debate on the moves towards increased deregulation and exposure to market forces focused mainly on the pace and sequence of change, rather than on the need for it. From 1990 the National Party Government maintained the momentum, particularly in containing spending and introducing a greater element of 'user pays' in health and education.

Attempts to reduce the budget deficit without increasing the money supply led to high real interest rates in the 1980s, and these, in turn, sustained the exchange rate. In accordance with a programme of tax reform that was initiated in 1984, a 10% tax on all goods and services became effective from late 1986 (the GST rate was raised to 12.5% in July 1989). This was followed by a reduction in the rate of corporate tax, from 45% to 33%, and the introduction of a basically two-tier rate of personal income tax at 24% and 33%, compared with the former 30% and 48%, respectively. Between mid-1996 and mid-1998 the lower rate of personal income tax was reduced further. From April 2000 this trend was reversed, with a higher rate of 39% being imposed on annual income in excess of $NZ60,000. Budget surpluses have in the past been used not only to fund tax reductions, but also to decrease public-debt levels. The 2004 budget estimated a current surplus in excess of $NZ5,700m. for 2004/05. Government operating expenses had been reduced by 2002/03 to just over 30% of GDP, down from the 40% of 1992/93. In addition, some pre-funding of future pension obligations was put in place. Net official sovereign debt was to approximate zero by 2007/08, as pension funds equalled and offset official debt levels.

The position in New Zealand with regard to social-welfare expenditure is similar to that in most other OECD countries. Benefit payments account for about two-fifths of the Government's total health, education and welfare budget, while health and education each account for about one-quarter. The rate of increase in expenditure in all sectors tends to outstrip economic growth. Following its election in late 1990, the National Party Government made a determined effort to curb welfare spending, as well as to reduce expenditure in other areas. At the same time it placed more emphasis on relevant education, recognizing that education and vocational training play a pivotal role in the drive to improve overall economic performance. The indigenous Maori population is making explicit claims for a greater influence over the way in which the country is administered, as well as for ownership of a higher proportion of its output. In 1999 the incoming Labour administration regarded a strong economy as necessary to achieve success in its varied goals. External economic problems and the changes currently taking place in New Zealand, together with major demographic and social developments (in common with many other countries), are being accompanied by changes in political philosophy and traditional affiliations. The July 2002 election resulted in the formation of a three-party coalition led by the Labour Party, to replace the previous two-party administration. The inclusion of the United Future New Zealand party in support of the coalition Government did not, however, result in any immediate or dramatic change in economic policy.

Statistical Survey

Source (unless otherwise stated): Statistics New Zealand, Aorangi House, 85 Molesworth St, POB 2922, Wellington 1; tel. (4) 495-4600; fax (4) 472-9135; e-mail info@stats.govt.nz; internet www.stats.govt.nz.

Area and Population

AREA, POPULATION AND DENSITY

Area (sq km)	270,534*
Population (census results)†	
5 March 1996	3,618,303
6 March 2001	
Males	1,823,004
Females	1,914,273
Total	3,737,277
Population (official estimates at mid-year)	
2002	3,939,100
2003	4,009,200
2004‡	4,061,300
Density (per sq km) at mid-2004	15.0

* 104,454 sq miles.

† Figures refer to the population usually resident. The total population (including foreign visitors) was: 3,681,546 in 1996; 3,820,749 in 2001.

‡ Provisional figure.

ADMINISTRATIVE REGIONS
(census of March 2001)

	Area (sq km)	Population	Density (per sq km)
North Island			
Northland	13,296	140,133	10.5
Auckland	5,048	1,158,891	229.6
Waikato	26,170	357,726	13.7
Bay of Plenty	11,428	239,412	21.0
Gisborne	8,355	43,974	5.3
Hawke's Bay Region	13,764	142,947	10.4
Taranaki	7,227	102,858	14.2
Manawatu-Wanganui	22,687	220,089	9.7
Wellington	8,056	423,765	52.6
Total North Island	116,031	2,829,798	24.4
South Island			
Tasman	14,538	41,352	2.8
Nelson	444	41,568	93.6
Marlborough	12,493	39,558	3.2
West Coast	23,351	30,303	1.3
Canterbury	45,845	481,431	10.5
Otago	31,476	181,542	5.8
Southland	25,392	91,005	3.6
Total South Island	153,540	906,753	5.9
Area outside regions	963	726	0.8
Total	270,534	3,737,277	13.8

PRINCIPAL CENTRES OF POPULATION
(population at census of 6 March 2001)

Auckland	1,074,507	Palmerston North	72,681
Wellington (capital)	339,747	Hastings	58,139
Christchurch	334,107	Napier	54,534
Hamilton	166,128	Nelson	53,685
Dunedin	107,088	Rotorua	52,608
Tauranga	95,694		

BIRTHS, MARRIAGES AND DEATHS

	Live births*		Marriages†		Deaths*	
	Number	Rate (per '000)	Number	Rate (per '000)	Number	Rate (per '000)
1996	57,280	15.4	20,453	5.5	28,255	7.6
1997	57,604	15.3	19,953	5.3	27,471	7.3
1998	57,251	15.1	20,135	5.3	26,206	6.9
1999	57,053	15.0	21,085	5.5	28,117	7.4
2000	56,605	14.7	20,655	5.4	26,660	7.0
2001	55,799	14.5	19,972	5.1	27,825	7.0
2002	54,021	13.7	20,690	5.3	28,065	7.1
2003	56,134	14.0‡	21,419	5.3‡	28,010	7.0‡

* Data for births and deaths are tabulated by year of registration rather than by year of occurrence.
† Based on the resident population concept, replacing the previous *de facto* concept.
‡ Provisional figure.

Expectation of life (WHO estimates, years at birth): 78.9 (males 76.6; females 81.2) in 2002 (Source: WHO, *World Health Report*).

IMMIGRATION AND EMIGRATION

	2001	2002	2003
Long-term immigrants*	81,094	95,951	92,660
Long-term emigrants†	71,368	57,753	57,754

* Figures refer to persons intending to remain in New Zealand for 12 months or more, and New Zealand citizens returning after an absence of 12 months or more.
† Figures refer to New Zealand citizens intending to remain abroad for 12 months or more, and overseas migrants departing after a stay of 12 months or more.

ECONOMICALLY ACTIVE POPULATION
('000 persons aged 15 years and over, excl. armed forces)

	2001	2002	2003
Agriculture, hunting, forestry	161.5	161.3	152.8
Fishing	3.9	3.6	3.9
Mining and quarrying	3.5	3.7	3.3
Manufacturing	289.1	290.8	278.7
Electricity, gas and water	10.1	9.8	8.5
Construction	112.1	120.7	138.6
Wholesale and retail trade; repair of motor vehicles, motorcycles and personal and household goods	314.1	316.8	347.7
Restaurants and hotels	94.5	106.9	94.8
Transport, storage and communications	112.4	113.3	111.0
Financial intermediation	52.2	53.7	54.1
Real estate, renting and business activities	180.7	190.2	196.8
Public administration and defence; compulsory social security	93.4	84.5	110.6
Education	137.3	146.7	151.3
Health and social work	158.0	172.8	173.4
Other community, social and personal service activities	88.7	92.2	89.9
Private households with employed persons	7.3	6.6	2.5
Extra-territorial organizations and bodies	0.9	1.0	—
Activities not adequately defined	4.0	2.2	3.2
Total employed	1,823.4	1,876.8	1,921.0
Unemployed	102.3	102.5	93.9
Total labour force	1,925.7	1,979.2	2,014.9
Males	1,050.1	1,079.7	1,093.1
Females	875.6	899.6	921.8

Source: ILO.

Health and Welfare

KEY INDICATORS

Total fertility rate (children per woman, 2002)	2.0
Under-5 mortality rate (per 1,000 live births, 2002)	6
HIV/AIDS (% of persons aged 15–49, 2003)	0.1
Physicians (per 1,000 head, 2000)	2.2
Hospital beds (per 1,000 head, 1998)	6.2
Health expenditure (2001): US $ per head (PPP)	1,724
Health expenditure (2001): % of GDP	8.3
Health expenditure (2001): public (% of total)	76.8
Human Development Index (2002): ranking	18
Human Development Index (2002): value	0.926

For sources and definitions, see explanatory note on p. vi.

Agriculture

PRINCIPAL CROPS
('000 metric tons)

	2000	2001	2002
Wheat	326	364	301
Barley	302	365	441
Maize	181	177	149
Oats	35	22	35
Potatoes	500	500	500
Dry peas	64	38	29
Cabbages	36	33	30
Lettuce	37	33	30
Tomatoes	87	87	87
Cauliflower	56	63	53
Pumpkins, squash and gourds	155	155*	156*
Green onions and shallots	240	245*	240*
Green peas	60	43	45
Carrots	77	78	57
Green corn	101	98*	98*
Other vegetables (excl. watermelons)*	153	150	149
Grapes	80	71	119
Apples	620	474	531
Pears	42	34	42
Kiwi fruit	262	271	248

* FAO estimate(s).

Source: FAO.

LIVESTOCK
('000 head at 30 June)

	2000	2001	2002
Cattle	9,015	9,281	9,637
Sheep	42,260	40,033	39,546
Goats	183	163	155
Pigs	369	354	341
Horses	73*	77	76
Chickens	13,000	13,000	17,128
Ducks*	180	170	180
Geese*	65	65	68
Turkeys*	60	70	70

* FAO estimate(s).

Source: FAO.

LIVESTOCK PRODUCTS
('000 metric tons)

	2000	2001	2002
Beef and veal*	571.8	590.4	576.3
Mutton and lamb*	533.0	562.2	520.9
Pig meat*	46.8	46.9	46.7
Poultry meat*	106.7	115.9	130.0
Game meat*	21.8	23.8	25.6
Cows' milk†	12,235.4	13,119.4	13,865.9
Butter†	343.8	391.1	414.6
Cheese†	296.7	280.6	307.0
Hen eggs	43.0	43.9	45.7
Other poultry eggs	2.4	2.4	2.4
Honey	9.6	9.1	4.7
Wool: greasy	257.2	236.7	228.3
Wool: scoured	202.4	198.7	195.3
Cattle hides (fresh)‡	48.3	50.0	51.0
Sheepskins (fresh)‡	99	103	98

* Twelve months ending 30 September of year stated.
† Twelve months ending 31 May of year stated.
‡ FAO estimate(s).

Source: FAO.

Forestry

ROUNDWOOD REMOVALS
('000 cubic metres, year ending 31 March, estimates)

	2000/01	2001/02	2002/03
Sawlogs	7,274	7,382	8,386
Pulp logs	3,566	3,504	3,288
Export logs and chips	6,421	7,838	8,435
Other	2,206	2,216	2,342
Total	19,287	20,940	22,451

Source: Forestry Statistics Section, Ministry of Agriculture and Forestry, Wellington.

SAWNWOOD PRODUCTION
('000 cubic metres, year ending 31 March)

Species	2000/01	2000/02	2002/03
Radiata pine	3,625	3,678	4,214
Other introduced pines	32	5	9
Douglas fir	136	124	164
Rimu and miro	17	13	5
Total (incl. others)	3,848	3,864	4,436

Source: Forestry Statistics Section, Ministry of Agriculture and Forestry, Wellington.

Fishing

('000 metric tons, live weight)

	2000	2001	2002
Capture*	549.3	561.9	559.3
Southern blue whiting	23.0	29.8	42.1
Blue grenadier (Hoki)	234.0	223.7	192.5
Pink cusk-eel	21.6	18.6	20.3
Orange roughy	17.9	14.0	18.0
Oreo dories	22.8	24.2	17.6
Jack and horse mackerels	22.5	28.5	32.3
Snoek (Barracouta)	21.9	25.2	23.1
Wellington flying squid	20.9	35.1	50.0
Aquaculture	86.5	76.0	86.6
New Zealand mussel	76.0	64.0	78.0
Total catch	635.8	638.0	645.9

Note: Figures exclude aquatic mammals (recorded by number rather than by weight) and sponges. The number of whales and dolphins caught was: 13 in 2000; 19 in 2001; 13 in 2002. The catch of sponges (in metric tons) was: 18 in 2000 (FAO estimate); 23 in 2001; 0 in 2002.

* Excluding catches made by chartered vessels and landed outside New Zealand.

Source: FAO.

Mining

('000 metric tons, unless otherwise indicated)

	2001	2002	2003
Coal (incl. lignite)	3,911	4,459	5,180
Gold (kg)	9,850	9,770	9,300
Crude petroleum ('000 barrels)	12,400	13,000*	8,711
Gross natural gas (million cu m)	5,000*	5,000*	4,926
Liquid petroleum gas ('000 barrels)*	2,000	2,100	193,953†
Iron sands	1,636	1,740	n.a.
Silica sand	36.0	59.4	48.4
Limestone	4,746	5,294	4,876

* Estimate(s).
† Official figure quoted in metric tons.

Sources: Ministry of Economic Development, Wellington; US Geological Survey.

Industry

SELECTED PRODUCTS
(metric tons, unless otherwise indicated)

	1999	2000	2001
Wine ('000 hectolitres)*	602	602	533
Beer (sales, '000 hectolitres)	3,148	2,980	3,070
Wool yarn (pure and mixed)	23,500	22,200	n.a.
Knitted fabrics†	3,700	3,600	3,500
Chemical wood pulp‡	645,032	753,885	744,991
Mechanical wood pulp‡	755,707	773,680	827,286
Newsprint‡	383,372	360,623	380,614
Other paper and paperboard‡	430,942	469,189	490,993
Fibre board (cu m)‡	600,673	744,879	801,493
Particle board (cu m)‡	169,569	188,054	204,524
Veneer (cu m)‡	285,825	378,282	401,590
Plywood (cu m)‡	192,445	239,947	243,702
Jet fuels ('000 metric tons)	839	838	n.a.
Motor spirit—petrol ('000 metric tons)	1,360	1,429	n.a.
Gas-diesel (Distillate fuel) oils ('000 barrels)	13,000	13,000§	13,000§
Residual fuel oils ('000 barrels)	3,500	3,500§	3,500§
Cement ('000 metric tons)§	960	950	950
Aluminium—unwrought ('000 metric tons)			
Primary	326.7	328.4	322.3
Secondary	21.4	21.5§	21.5§
Electric energy (million kWh)	34,152	34,700	35,252

* Data from New Zealand Wines Online.
† Twelve months ending 30 September of year stated.
‡ Twelve months ending 31 March of year stated. Source: Ministry of Agriculture and Forestry, Wellington.
§ Estimate(s).

2003 (year ending 31 March): Chemical wood pulp (metric tons) 684,187; Mechanical wood pulp (metric tons) 828,415; Newsprint (metric tons) 355,569; Other paper and paperboard (metric tons) 496,094; Fibre board (cu m) 883,117; Particle board (cu m) 209,977; Veneer (cu m) 601,600; Plywood (cu m) 321,655 (Source: Ministry of Agriculture and Forestry, Wellington).

2003: Wine ('000 hectolitres) 550; Electrical energy (million kWh) 36,256.

Sources (unless otherwise stated): UN, *Industrial Commodity Statistics Yearbook* and *Monthly Bulletin of Statistics*; US Geological Survey; New Zealand Wines Online.

Finance

CURRENCY AND EXCHANGE RATES

Monetary Units

100 cents = 1 New Zealand dollar ($NZ).

Sterling, US Dollar and Euro Equivalents (31 May 2004)

£1 sterling = $NZ2.9076;
US $1 = $NZ1.5848;
€1 = $NZ1.9407;
$NZ100 = £34.39 = US $63.10 = €51.53.

Average Exchange Rate (US $ per New Zealand dollar)

2001	0.4204
2002	0.4625
2003	0.5804

BUDGET

($NZ million, year ending 30 June)

Revenue	2000/01	2001/02	2002/03
Direct taxation	23,863	24,557	26,778
Indirect taxation	12,875	13,863	13,007
Compulsory fees, fines, penalties and levies	385	520	2,763
Operational revenue	2,369	2,702	14,479
Total	39,492	41,642	57,027

Expenditure	2000/01	2001/02	2002/03
Core government services	1,817	1,602	1,655
Defence	1,267	1,197	1,154
Law and order	1,560	1,755	1,911
Education	6,690	7,124	7,788
Health	7,342	7,713	7,412
Transport and communications	1,026	1,120	5,619
Social security and welfare	13,216	13,487	17,084
Economic and industrial services	1,141	1,157	4,280
Total (incl. others)	38,186	39,699	55,224

Source: New Zealand Treasury, Wellington.

INTERNATIONAL RESERVES

(US $ million at 31 December)

	2001	2002	2003
IMF special drawing rights	16	22	28
Reserve position in IMF	387	459	644
Foreign exchange	2,605	3,258	4,238
Total	3,009	3,739	4,910

Source: IMF, *International Financial Statistics*.

MONEY SUPPLY

($NZ million at 31 December)

	2001	2002	2003
Currency outside banks	2,241	2,451	2,597
Demand deposits at banking institutions	16,506	17,487	19,068
Total money (incl. others)	18,795	20,019	21,706

Source: IMF, *International Financial Statistics*.

COST OF LIVING

(Consumer Price Index; base: 1990 = 100)

	2000	2001	2002
Food (incl. beverages)	112.7	119.6	123.3
Fuel and light	151.8	152.7	160.4
Clothing (incl. footwear)	105.8	107.6	108.9
Rent	146.4	132.1	134.4
All items (incl. others)	119.1	122.2	125.5

Source: ILO, *Yearbook of Labour Statistics*.

NATIONAL ACCOUNTS

National Income and Product

($NZ million at current prices, year ending 31 March)

	2000/01	2001/02	2002/03
Compensation of employees	48,133	51,883	55,095
Gross operating surplus	52,372	56,722	58,086
Taxes on production and imports	14,697	15,486	16,787
Less Subsidies	360	355	389
GDP in market prices	114,842	123,736	129,579
Net factor income from abroad	−7,172	−6,611	−7,093
Gross national income	107,670	117,125	122,486
Less Consumption of fixed capital	15,504	16,431	17,153
Net national income	92,166	100,694	105,333
Other current transfers from abroad (net)	411	134	134
Net national disposable income	92,577	100,828	105,468

Expenditure on the Gross Domestic Product

($NZ million at current prices, year ending 31 March)

	2000/01	2001/02	2002/03
Government final consumption expenditure	20,236	21,720	22,746
Private final consumption expenditure	68,913	72,302	77,144
Change in inventories	1,383	2,010	674
Gross fixed capital formation	22,344	24,535	26,626
Total domestic expenditure	112,876	120,567	127,190
Exports of goods and services	41,269	43,595	41,836
Less Imports of goods and services	39,303	39,914	39,911
Statistical discrepancy	—	−513	464
GDP in market prices	114,842	123,736	129,579
GDP at constant 1995/96 prices	105,177	108,777	113,509

Gross Domestic Product by Economic Activity

($NZ million in constant 1995/96 prices)

	1998/99	1999/2000	2000/01
Agriculture	5,184	5,926	7,949
Hunting and fishing	309	335	364
Forestry and logging	1,073	1,109	1,242
Mining and quarrying	1,143	1,200	1,367
Manufacturing	16,039	16,745	18,018
Electricity, gas and water	2,631	2,896	2,834
Construction	4,203	4,847	4,752
Wholesale and retail trade	13,208	13,829	14,537
Hotels and restaurants	1,830	1,887	2,092
Transport, storage and communications	7,686	7,808	7,992
Financial intermediation (incl. insurance)	5,560	5,946	6,186
Property and business activities	13,820	14,906	15,239
Ownership of dwellings	8,284	8,307	8,332
Public administration and defence	4,627	4,712	4,867
Education	4,181	4,463	4,654
Health and community services	5,211	5,443	5,796
Cultural and recreational services	1,969	2,270	2,326
Personal and other services	1,471	1,588	1,672
Sub-total	98,428	104,218	110,215
Less Financial intermediation services indirectly measured	3,415	3,533	3,575
Gross value added at basic prices	95,013	100,685	106,640
Goods and services tax on production	6,840	7,312	7,569
Import duties	510	612	633
Other taxes on production (net)	102	−40	—
GDP in market prices	102,465	108,570	114,842

BALANCE OF PAYMENTS
(US $ million)

	2001	2002	2003
Exports of goods f.o.b.	13,920	14,517	16,828
Imports of goods f.o.b.	–12,448	–14,014	–17,219
Trade balance	1,471	503	–391
Exports of services	4,373	5,161	6,440
Imports of services	–4,287	–4,763	–5,637
Balance on goods and services	1,557	901	412
Other income received	596	1,077	1,379
Other income paid	–3,594	–4,037	–5,274
Balance on goods, services and income	–1,441	–2,058	–3,482
Current transfers received	578	616	792
Current transfers paid	–390	–496	–647
Current balance	–1,253	–1,938	–3,337
Capital account (net)	443	765	508
Direct investment abroad	1,144	–185	–299
Direct investment from abroad	775	738	2,438
Portfolio investment assets	–1,219	–935	–856
Portfolio investment liabilities	1,527	2,401	2,184
Other investment assets	–3,537	–1,086	318
Other investment liabilities	3,068	–103	–379
Net errors and omissions	–1,135	1,429	206
Overall balance	–187	1,086	782

Source: IMF, *International Financial Statistics.*

External Trade

Source: Statistics New Zealand, Overseas Trade, Wellington.

PRINCIPAL COMMODITIES
($NZ million)

Imports (value for duty)	2001	2002	2003
Food and live animals	1,945.1	2,016.2	1,881.1
Mineral fuels, lubricants, etc.	2,884.6	2,819.1	2,725.3
Petroleum, petroleum products, etc.	2,877.5	2,810.7	2,700.7
Chemicals and related products	3,815.7	3,633.7	3,438.5
Basic manufactures	4,022.5	4,142.9	3,972.9
Machinery and transport equipment	11,576.4	12,226.0	12,360.5
Machinery specialized for particular industries	1,127.3	1,282.0	1,110.0
General industrial machinery, equipment and parts	1,211.1	1,265.3	1,216.1
Office machines and automatic data-processing machines	1,516.1	1,542.1	1,518.3
Telecommunications and sound equipment	1,380.1	1,188.7	1,218.3
Other electrical machinery, apparatus, etc.	1,316.3	1,310.0	1,269.6
Road vehicles (incl. air-cushion vehicles) and parts*	3,448.9	4,146.2	4,387.3
Other transport equipment and parts*	1,066.4	932.1	1,056.7
Miscellaneous manufactured articles	4,099.2	4,180.6	4,204.0
Clothing and accessories (excl. footwear)	889.6	880.7	1,769.6
Total (incl. others)	29,612.3	30,328.6	29,817.0

* Excluding tyres, engines and electrical parts.

Total imports c.i.f. ($NZ million): 31,682.7 in 2001; 32,337.4 in 2002; 31,781.7 in 2003.

Exports f.o.b	2001	2002	2003
Food and live animals	14,961.6	13,905.8	12,835.9
Meat and meat preparations	4,380.3	4,343.9	4,204.4
Dairy products and birds' eggs	6,379.7	5,222.9	4,767.4
Fish (not marine mammals), crustaceans, molluscs, etc., and preparations	1,465.2	1,483.8	1,167.2
Vegetables and fruit	1,646.3	1,710.8	1,590.0
Crude materials (inedible) except fuels	4,166.8	4,095.9	3,512.1
Cork and wood	1,679.2	1,878.3	1,479.5
Chemicals and related products	3,023.9	2,263.4	1,769.1
Basic manufactures	4,176.7	4,034.0	3,507.5
Non-ferrous metals	1,221.8	1,084.2	883.5
Machinery and transport equipment	3,016.0	3,043.3	3,160.1
Miscellaneous electrical machinery, apparatus and appliances, and parts thereof	937.7	858.8	871.1
Miscellaneous manufactured articles	1,387.9	1,438.7	1,411.5
Total (incl. others)*	32,670.0	31,033.7	28,397.2

* Including re-exports ($NZ million): 1,094.1 in 2001; 1,213.2 in 2002; 1,090.7 in 2003.

PRINCIPAL TRADING PARTNERS
($NZ million)

Imports (value for duty)*	2001	2002	2003
Australia	6,563.9	6,966.6	6,818.5
Belgium	296.3	335.3	316.4
Canada	428.9	338.6	535.0
China, People's Republic	2,067.0	2,433.9	2,673.0
France	573.2	638.9	863.5
Germany	1,431.4	1,568.8	1,596.1
Indonesia	387.4	395.7	340.2
Italy	677.0	779.2	728.2
Japan	3,220.8	3,562.6	3,399.4
Korea, Republic	676.2	734.3	796.9
Malaysia	912.7	802.2	683.2
Netherlands	271.5	302.6	274.1
Oman	286.7	324.9	244.8
Saudi Arabia	363.2	354.2	226.9
Singapore	584.6	566.6	592.6
Sweden	327.0	324.1	317.4
Taiwan	625.9	617.0	654.8
Thailand	469.6	513.8	530.8
United Arab Emirates	260.2	321.9	419.5
United Kingdom	1,144.4	1,119.5	991.6
USA	4,777.7	4,146.7	3,519.9
Total (incl. others)	29,612.7	30,328.5	29,816.9

* Excluding specie and gold.

Exports*	2001	2002	2003
Australia	6,180.9	6,221.4	6,118.7
Belgium	546.5	626.4	631.6
Canada	631.4	648.5	547.7
China, People's Republic	1,349.4	1,430.1	1,376.4
France	367.3	384.1	372.6
Germany	838.7	908.9	745.9
Hong Kong	786.3	629.0	559.1
Indonesia	550.5	456.1	383.2
Italy	548.5	466.1	444.2
Japan	4,083.4	3,569.2	3,123.1
Korea, Republic	1,440.2	1,376.4	993.2
Malaysia	697.9	592.7	545.5
Mexico	535.9	466.2	411.4
Philippines	564.7	467.1	488.9
Singapore	398.6	387.4	306.7
Taiwan	710.9	680.2	625.7
Thailand	412.0	353.3	333.0
United Kingdom	1,558.2	1,494.7	1,363.1
USA	4,850.6	4,760.6	4,118.8
Viet Nam	371.2	129.1	158.3
Total (incl. others)	32,670.0	31,033.7	28,397.1

* Including re-exports. Excluding specie and gold.

Transport

RAILWAYS
(traffic, year ending 30 June)

	2000/01	2001/02	2002/03
Freight ('000 metric tons)	14,461	14,330	14,822
Passengers ('000)	12,714	12,521	12,300*

* Excludes passengers on the Tranz Scenic network.

Source: Tranz Rail Ltd, Wellington.

ROAD TRAFFIC
(vehicles licensed at 31 March)

	2000	2001	2002
Passenger cars	1,877,850	1,909,480	1,960,503
Goods service vehicles	368,964	364,928	366,639
Taxis	7,479	7,181	7,366
Buses and service coaches	12,264	12,667	13,339
Trailers and caravans	359,400	368,994	381,914
Motorcycles and mopeds	48,722	47,670	47,423
Tractors	20,526	21,377	22,914

Source: Ministry of Transport, Wellington.

SHIPPING

Merchant Fleet
(registered at 31 December)

	2001	2002	2003
Number of vessels	159	173	170
Displacement (grt)	174,942	180,435	205,188

Source: Lloyd's Register-Fairplay, *World Fleet Statistics*.

Vessels Handled
(international, '000 grt)

	1993	1994	1995
Entered	37,603	39,700	48,827
Cleared	35,128	37,421	42,985

Source: UN, *Statistical Yearbook*.

International Sea-borne Freight Traffic
('000 metric tons, year ending 30 June)

	2000/01	2001/02	2002/03*
Goods loaded	22,522	24,671	25,310
Goods unloaded	14,162	15,447	16,155

* Provisional.

CIVIL AVIATION
(domestic and international traffic on scheduled services)

	1997	1998	1999
Kilometres flown (million)	173	174	172
Passengers carried ('000)	9,435	8,655	8,892
Passenger-km (million)	20,983	19,014	19,322
Total metric ton-km (million)	2,816	2,700	2,746

Source: UN, *Statistical Yearbook*.

Tourism

VISITOR ARRIVALS

Country of residence	2001	2002	2003
Australia	630,549	632,470	702,162
China, People's Republic	53,174	76,534	65,989
Germany	52,482	48,951	52,534
Japan	149,085	173,567	150,851
Korea, Republic	87,167	109,936	112,658
United Kingdom	211,646	236,986	264,819
USA	187,381	205,289	211,624
Total (incl. others)	1,909,381	2,045,064	2,104,420

Tourism receipts ($NZ million, year ending 31 March): 5,544 in 2001/02; 6,192 in 2002/03; 6,313 in 2003/04.

Source: Tourism Research Council, Wellington.

Communications Media

	2000	2001	2002
Television receivers ('000 in use)	2,000	2,130	n.a.
Telephones ('000 main lines in use)	1,831	1,823	1,765
Mobile cellular telephones ('000 subscribers)	1,542	2,288	2,449
Personal computers ('000 in use)	1,380	1,500	1,630
Internet users ('000)	1,515	1,762	1,908

2003: Mobile cellular telephones ('000 subscribers) 2,599; Internet users ('000) 2,110.

Radio receivers ('000 in use, 1997): 3,750.

Facsimile machines ('000 in use, year ending 31 March 1996): 65.

Daily newspapers: 26 (circulation 850,000 copies, 2002).

Non-daily newspapers: 2 (circulation 311,380 copies, 2002).

Book production (1999): 4,800 titles.

Sources: partly International Telecommunication Union; UNESCO, *Statistical Yearbook*; UN, *Statistical Yearbook*.

Education

(July 2003)

	Institutions	Teachers (full-time equivalent)	Students
Early childhood services	4,280	11,485[1]	180,000[2]
Primary schools[3]	2,177	23,358[4]	456,782
Composite schools[5]	136	1,861[4]	44,782
Secondary schools[6]	333	15,596[4]	257,586
Special schools	47	763[4]	2,605
Polytechnics	20	4,223	95,782[7]
Colleges of education	4	491	10,788[7]
Universities	8	6,562	132,396[7]
Wananga[8]	3	781	27,535[7]
Private training establishments receiving government grants	522	4,177	53,385[7]

[1] Excludes 1,205 playcentre teaching staff and Te Kohanga Reo personnel (responsible for Maori 'Language Nests').

[2] Includes children on the regular roll of the Correspondence School, kindergartens, playcentres, Te Kohanga Reo, Early Childhood Development Unit funded playgroups, Early Childhood Development Unit funded Pacific Islands language groups, education and care centres (incl. home-based childcare).

[3] Primary schools include Full Primary Years 1–8, Contributing Years 1–6, Intermediate Years 7–8.

[4] Teachers employed in state schools at 1 March 2002.

[5] Composite schools provide both primary and secondary education (includes area schools and the Correspondence School).

[6] Secondary schools include Years 7–15, Years 9–15.

[7] 2002 figure.

[8] Tertiary institutions providing polytechnic and university level programmes specifically for Maori students, with an emphasis on Maori language and culture.

Source: Ministry of Education, Wellington.

Directory

The Constitution

New Zealand has no written constitution. The political system is closely modelled on that of the United Kingdom (with an element of proportional representation introduced to the legislature in 1996). As in the United Kingdom, constitutional practice is an accumulation of convention, precedent and tradition. A brief description of New Zealand's principal organs of government is given below:

HEAD OF STATE

Executive power is vested in the monarch and is exercisable in New Zealand by the monarch's personal representative, the Governor-General.

In the execution of the powers and authorities vested in him or her, the Governor-General must be guided by the advice of the Executive Council.

EXECUTIVE COUNCIL

The Executive Council consists of the Governor-General and all the Ministers. Two members, exclusive of the Governor-General or the presiding member, constitute a quorum. The Governor-General appoints the Prime Minister and, on the latter's recommendation, the other Ministers.

HOUSE OF REPRESENTATIVES

Parliament comprises the Crown and the House of Representatives. At the 1996 general election, a system of mixed member proportional representation was introduced. The House of Representatives comprises 120 members: 67 electorate members (five seats being reserved for Maoris) and 53 members chosen from party lists. They are designated 'Members of Parliament' and are elected for three years, subject to the dissolution of the House before the completion of their term.

Everyone over the age of 18 years may vote in the election of members for the House of Representatives. Since August 1975 any person, regardless of nationality, ordinarily resident in New Zealand for 12 months or more and resident in an electoral district for at least one month is qualified to be registered as a voter. Compulsory registration of all electors except Maoris was introduced at the end of 1924; it was introduced for Maoris in 1956. As from August 1975, any person of the Maori race, which includes any descendant of such a person, may enrol on the Maori roll for the particular Maori electoral district in which that person resides.

By the Electoral Amendment Act 1937, which made provision for a secret ballot in Maori elections, Maori electors were granted the same privileges, in the exercise of their vote, as general electors.

In local government the electoral franchise is the same.

The Government

Head of State: HM Queen ELIZABETH II (acceded to the throne 6 February 1952).

Governor-General and Commander-in-Chief: Dame SILVIA CARTWRIGHT (took office 4 April 2001).

CABINET

(September 2004)

A coalition of the Labour Party and the Progressive Coalition.

Prime Minister and Minister for Arts, Culture and Heritage: HELEN CLARK.

Deputy Prime Minister, Minister of Finance, Minister of Revenue and Leader of the House: Dr MICHAEL CULLEN.

Minister for Economic Development and Minister for Industry and Regional Development: JIM ANDERTON*.

Minister for Social Development and Employment, of Housing and of Broadcasting: STEVE MAHAREY.

Minister of Foreign Affairs and Trade, of Justice and of Pacific Island Affairs: PHIL GOFF.

Minister of Health and for Food Safety: ANNETTE KING.

Minister of Agriculture and Forestry, for Biosecurity and for Trade Negotiations: JIM SUTTON.

Minister of Education, of State Services and for Sport and Recreation: TREVOR MALLARD.

Minister of Energy, of Transport, of Research, Science and Technology and for Crown Research Institutes: PETE HODGSON.

Attorney-General and Minister of Commerce and in Charge of Treaty of Waitangi Negotiations: MARGARET WILSON.

Minister of Maori Affairs: PAREKURA HOROMIA.

Minister of Police, of Internal Affairs, of Civil Defence and of Veteran's Affairs: GEORGE HAWKINS.

Minister of Defence, for State Owned Enterprises and of Tourism: MARK BURTON.

Minister of Labour, of Immigration, of Communications, of Corrections and for Information Technology: PAUL SWAIN.

Minister for the Environment and for Disarmament and Arms Control: MARIAN HOBBS.

Minister for Accident Compensation Corporation, for Senior Citizens and of Women's Affairs: RUTH DYSON .

Minister of Youth Affairs, for Land Information and of Statistics: JOHN TAMIHERE.

Minister of Conservation and of Local Government: CHRIS CARTER.

Minister for Courts and of Customs: RICK BARKER.

Minister of Fisheries: DAVID BENSON-POPE.

* Denotes member of the Progressive Coalition.

In addition, there are six Ministers outside the Cabinet.

MINISTRIES AND GOVERNMENT DEPARTMENTS

Department of the Prime Minister and Cabinet: Executive Wing, Parliament Bldgs, Wellington; tel. (4) 471-9700; fax (4) 473-2508; internet www.dpmc.govt.nz.

Ministry of Agriculture and Forestry: POB 2526, Wellington; tel. (4) 474-4100; fax (4) 474-4111; internet www.maf.govt.nz.

Ministry of Civil Defence and Emergency Management: POB 5010, Wellington; tel. (4) 473-7363; fax (4) 473-7369.

Department of Conservation: POB 10-420, Wellington; tel. (4) 471-0726; fax (4) 471-1082; e-mail tsmith@doc.govt.nz; internet www.doc.govt.nz.

Ministry for Culture and Heritage: POB 5364, Wellington; tel. (4) 499-4229; fax (4) 499-4490; e-mail info@mch.govt.nz; internet www.mch.govt.nz.

Ministry of Defence: POB 5347, Wellington; tel. (4) 496-0999; fax (4) 496-0859; internet www.defence.govt.nz.

Ministry of Economic Development: POB 1473, 33 Bowen St, Wellington; tel. (4) 472-0030; fax (4) 473-4638; e-mail info@med.govt.nz; internet www.med.govt.nz.

Ministry of Education: Level 7, St Paul's Square, 45–47 Pipitea St, Thorndon, Wellington; tel. (4) 463-8000; fax (4) 463-8001; e-mail enquiries.national@minedu.govt.nz; internet www.minedu.govt.nz.

Ministry for the Environment: POB 10-362, Wellington; tel. (4) 917-7400; fax (4) 917-7523; e-mail library@mfe.govt.nz; internet www.mfe.govt.nz.

Ministry of Fisheries: POB 1020, Wellington; tel. (4) 470-2600; fax (4) 470-2601; internet www.fish.govt.nz.

Ministry of Foreign Affairs and Trade: Private Bag 18901, Wellington; tel. (4) 439-8000; fax (4) 472-9596; e-mail enquiries@mfat.govt.nz; internet www.mfat.govt.nz.

Ministry of Health: POB 5013, Wellington; tel. (4) 496-2000; fax (4) 496-2340; internet www.moh.govt.nz.

Ministry of Housing: 256 Lambton Quay, Wellington; tel. (4) 472-2753; fax (4) 499-4744; e-mail info@minhousing.govt.nz; internet www.minhousing.govt.nz.

Department of Internal Affairs: POB 805, Wellington; tel. (4) 495-7200; fax (4) 495-7222; e-mail webmaster@dia.govt.nz; internet www.dia.govt.nz.

Ministry of Justice: POB 180, Wellington; tel. (4) 918-8800; fax (4) 918-8820; e-mail reception@justice.govt.nz; internet www.justice.govt.nz.

Department of Labour: POB 3705, Wellington; tel. (4) 915-4444; fax (4) 915-0891; e-mail info@dol.govt.nz; internet www.dol.govt.nz.

Ministry of Maori Development (Te Puni Kokiri): POB 3943, Wellington 6015; tel. (4) 922-6000; fax (4) 922-6299; e-mail info@tpk.govt.nz; internet www.tpk.govt.nz.

Ministry of Pacific Island Affairs: POB 833, Wellington; tel. (4) 473-4493; fax (4) 473-4301; e-mail contact@minpac.govt.nz; internet www.minpac.govt.nz.

Ministry of Research, Science and Technology: POB 5336, Wellington; tel. (4) 917-2900; fax (4) 471-1284; e-mail talk2us@morst.govt.nz; internet www.morst.govt.nz.

Ministry of Social Development: Bowen State Bldg, Bowen St, POB 12136, Wellington; tel. (4) 916-3300; fax (4) 918-0099; e-mail information@msd.govt.nz; internet www.msd.govt.nz.

State Services Commission: POB 329, Wellington; tel. (4) 495-6600; fax (4) 495-6686; e-mail commission@ssc.govt.nz; internet www.ssc.govt.nz.

Statistics New Zealand: POB 2922, Wellington; tel. (4) 931-4600; fax (4) 931-4610; e-mail info@stats.govt.nz; internet www.stats.govt.nz.

Ministry of Tourism: POB 5640, Wellington; tel. (4) 498-7440; fax (4) 498-7445; e-mail info@tourism.govt.nz; internet www.tourism.govt.nz.

Ministry of Transport: POB 3175, Wellington; tel. (4) 472-1253; fax (4) 473-3697; e-mail reception@transport.govt.nz; internet www.transport.govt.nz.

Treasury: POB 3724, Wellington 6001; tel. (4) 472-2733; fax (4) 473-0982; e-mail information@treasury.govt.nz; internet www.treasury.govt.nz.

Ministry of Women's Affairs: POB 10-049, Wellington; tel. (4) 473-4112; fax (4) 472-0961; e-mail mwa@mwa.govt.nz; internet www.mwa.govt.nz.

Ministry of Youth Affairs: POB 10–300, Wellington; tel. (4) 471–2158; fax (4) 471–2233; e-mail info@youthaffairs.govt.nz; internet www.youthaffairs.govt.nz.

Legislature

HOUSE OF REPRESENTATIVES

Speaker: JONATHAN HUNT.

General Election, 27 July 2002

Party	Number of votes	% of votes	Party seats	List seats	Total seats
NZ Labour Party . . .	838,219	41.26	45	7	52
NZ National Party . .	425,310	20.93	21	6	27
New Zealand First . .	210,912	10.38	1	12	13
ACT New Zealand . .	145,078	7.14	—	9	9
Green Party	142,250	7.00	—	9	9
United Future NZ . .	135,918	6.69	1	7	8
Progressive Coalition .	34,542	1.70	1	1	2
Total (incl. others) . .	2,031,617	100.00	69	51	120

Political Organizations

In March 2003 20 parties were registered.

ACT New Zealand: Old Mercury Bldg, Level 1, Block B, Nuffield St, POB 99-651, Newmarket, Auckland; tel. (9) 523-0470; fax (9) 523-0472; e-mail info@voteact.org.nz; internet www.act.org.nz; f. 1994; supports free enterprise, tax reform and choice in education and health; Pres. CATHERINE JUDD; Leader RODNEY HIDE.

Green Party of Aotearoa—New Zealand: POB 11-652, Wellington; tel. (4) 801-5102; fax (4) 801-5104; e-mail greenparty@greens.org.nz; internet www.greens.org.nz; f. 1989; fmrly Values Party, f. 1972; Co-Leaders ROD DONALD, JEANETTE FITZSIMONS.

Mana Motuhake o Aotearoa (New Zealand Self-Government Party): Private Bag 68-905, Newton, Auckland; f. 1979; pro-Maori; promotes bicultural policies; Leader WILLIE JACKSON.

Maori Party: PO Box 20683, Glen Eden, Waitakere City; tel. (9) 813-9204; fax (9) 813-9206; e-mail enquiry@maoriparty.com; internet www.maoriparty.com; f. 2004; Co-Leader Dr PITA SHARPLES; Co-Leader Hon. TARIANA TURIA; Co-Leader Dr WHATARANGI WINIATA.

The New Zealand Democratic Party Inc: 414 Glenfield Rd, POB 40364, Glenfield, North Shore City; tel. (9) 442-2364; fax (9) 442-2438; e-mail nzdp.inc@xtra.co.nz; internet www.democrats.org.nz; f. 1953 as Social Credit Political League; adopted present name 1985; liberal; Pres. JOHN PEMBERTON; Leader STEPHNIE DE RUYTER.

New Zealand First: c/o House of Representatives, Wellington; fax (4) 472-8557; e-mail nzfirst@parliament.govt.nz; f. 1993 by fmr National Party mems; Leader WINSTON PETERS; Pres. DOUG WOOLERTON.

New Zealand Labour Party: 160–162 Willis St, POB 784, Wellington; tel. (4) 384-7649; fax (4) 384-8060; e-mail nzlpho@labour.org.nz; internet www.labour.org.nz; f. 1916; advocates an organized economy guaranteeing an adequate standard of living to every person able and willing to work; Pres. MIKE WILLIAMS; Parl. Leader HELEN CLARK; Gen. Sec. MIKE SMITH.

New Zealand New National Party: Willbank House, 14th Floor, 57 Willis St, POB 1155, Wellington 60015; tel. (4) 472-5211; fax (4) 478-1622; e-mail hq@national.org.nz; internet www.national.org.nz; f. 1936; centre-right; supports private enterprise and competitive business, together with maximum personal freedom; Pres. JUDY KIRK; Parl. Leader DON BRASH.

Progressive Coalition: c/o House of Representatives, Wellington; f. 2002 to contest the general election; Leader JIM ANDERTON.

United Future New Zealand (UNZ): c/o House of Representatives, Wellington; e-mail united.nz@parliament.govt.nz; internet www.united.org.nz; f. 1995 by four mems of National Party, two mems of Labour Party and leader of Future New Zealand; joined by New Zealand Conservative Party in 1998; Pres. GRAEME REEVES.

Other parties that contested the 2002 election included the Alliance, the Aotearoa Legalise Cannabis Party, the Christian Heritage Party, Mana Maori Movement, One New Zealand Party and Outdoor Recreation New Zealand.

Diplomatic Representation

EMBASSIES AND HIGH COMMISSIONS IN NEW ZEALAND

Argentina: Sovereign Assurance House, 14th Floor, 142 Lambton Quay, POB 5430, Lambton Quay, Wellington; tel. (4) 472-8330; fax (4) 472-8331; e-mail enzel@arg.org.nz; internet www.arg.org.nz; Ambassador ENRIQUE J. DE LA TORRE.

Australia: 72–78 Hobson St, Thorndon, POB 4036, Wellington; tel. (4) 473-6411; fax (4) 498-7118; e-mail nzinbox@dfat.gov.au; internet www.australia.org.nz; High Commissioner Dr ALLAN HAWKE.

Brazil: Wool House, Level 9, 10 Brandon St, Wellington; tel. (4) 473-3516; fax (4) 473-3517; e-mail brasemb@brazil.org.nz; internet www.brazil.org.nz; Ambassador SÉRGIO BARBOSA SERRA.

Canada: 61 Molesworth St, POB 12049, Wellington 1; tel. (4) 473-9577; fax (4) 471-2082; High Commissioner VALERIE RAYMOND.

Chile: 19 Bolton St, POB 3861, Wellington; tel. (4) 471-6270; fax (4) 472-5324; e-mail embchile@ihug.co.nz; Ambassador CARLOS APPELGREN BALBONTÍN.

China, People's Republic: POB 17257, Karori, Wellington; tel. (4) 472-1382; fax (4) 499-0419; internet www.chinaembassy.org.nz; Ambassador CHEN MINGMING.

Fiji: 31 Pipitea St, Thorndon, POB 3940, Wellington; tel. (4) 473-5401; fax (4) 499-1011; e-mail viti@paradise.net.nz; internet www.fiji.org.nz; High Commissioner BAL RAM.

France: Open Networks Bldg, 12th Floor, 34–42 Manners St, POB 11-343, Wellington; tel. (4) 384-2555; fax (4) 384-2577; e-mail amba.france@actrix.gen.fr; internet www.ambafrance-nz.org; Ambassador JEAN-MICHEL MARLAUD.

Germany: 90–92 Hobson St, POB 1687, Wellington; tel. (4) 473-6063; fax (4) 473-6069; e-mail germanembassywellington@xtra.co.nz; Ambassador ERICH RIEDLER.

Greece: 5–7 Willeston St, 10th Floor, POB 24-066, Wellington; tel. (4) 473-7775; fax (4) 473-7441; e-mail info@greece.org.nz; Ambassador EVANGELOS DAMIANAKIS.

Holy See: Apostolic Nunciature, 112 Queen's Drive, Lyall Bay, POB 14-044, Wellington 6041; tel. (4) 387-3470; fax (4) 387-8170; e-mail nuntius@ihug.co.nz; Apostolic Nuncio Most Rev. PATRICK COVENEY (Titular Archbishop of Satriano).

India: 180 Molesworth St, POB 4045, Wellington 1; tel. (4) 473-6390; fax (4) 499-0665; e-mail hicomind@xtra.co.nz; internet www.hicomind.org.nz; High Commissioner K. P. RAM (acting).

Indonesia: 70 Glen Road, Kelburn, POB 3543, Wellington; tel. (4) 475-8699; fax (4) 475-9374; e-mail kbriwell@ihug.co.nz; internet www.indonesianembassy.org.nz; Ambassador PRIMO ALUI JOELIANTO.

Iran: The Terrace, POB 10-249, Wellington; tel. (4) 386-2983; fax (4) 386-3065; e-mail embassy.of.iran@xtra.co.nz; Ambassador KAMBIZ SHAYKH HASANI.

Italy: 34–38 Grant Rd, Thorndon, POB 463, Wellington 1; tel. (4) 473-5339; fax (4) 472-7255; e-mail ambwell@xtra.co.nz; internet www.italy-embassy.org.nz/em_well.html; Ambassador ROBERTO PALMIERI.

Japan: Majestic Centre, Levels 18–19, 100 Willis St, POB 6340, Wellington 1; tel. (4) 473-1540; fax (4) 471-2951; Ambassador MASAKI SAITO.

Korea, Republic: ASB Bank Tower, Level 11, 2 Hunter St, POB 11-143, Wellington; tel. (4) 473-9073; fax (4) 472-3865; e-mail korembec@world-net.co.nz; Ambassador SHIN JUNG-SEUNG.

Malaysia: 10 Washington Ave, Brooklyn, POB 9422, Wellington; tel. (4) 385-2439; fax (4) 385-6973; e-mail mwwelton@xtra.co.nz; High Commissioner Dato' ZULKIFLY ABDUL RAHMAN.

Mexico: Perpetual Trust House, Level 8, 111–115 Customhouse Quay, Manners St, POB 11-510, Wellington; tel. (4) 472-0555; fax (4) 496-3559; e-mail mexico@xtra.co.nz; internet www.mexico.org.nz; Ambassador MARÍA ANGÉLICA ARCE MORA.

Netherlands: Investment House, 10th Floor, Cnr Featherston and Ballance Sts, POB 840, Wellington; tel. (4) 471-6390; fax (4) 471-2923; e-mail wel@minbuza.nl; internet www.netherlandsembassy.co.nz; Ambassador A. E. DE BIJLL NACHENIUS.

Papua New Guinea: 279 Willis St, POB 197, Wellington; tel. (4) 385-2474; fax (4) 385-2477; e-mail pngnz@globe.net.nz; High Commissioner LUCY BOGARI.

Peru: Cigna House, Level 8, 40 Mercer St, POB 2566, Wellington; tel. (4) 499-8087; fax (4) 499-8057; e-mail embassy.peru@xtra.co.nz; internet www.embassyofperu.org.nz; Ambassador JAVIER LEÓN.

Philippines: 50 Hobson St, Thorndon, POB 12-042, Wellington; tel. (4) 472-9848; fax (4) 472-5170; e-mail embassy@wellington-pe.co.nz; Chargé d'affaires a. i. LOLITA CAPCO.

Poland: 17 Upland Rd, Kelburn, POB 10211, Wellington; tel. (4) 475-9453; fax (4) 475-9458; e-mail polishembassy@xtra.co.nz; internet poland.org.nz; Ambassador LECH MASTALERZ.

Russia: 57 Messines Rd, Karori, Wellington; tel. (4) 476-6113; fax (4) 476-3843; e-mail eor@netlink.co.nz; Ambassador GENNADII SHABANNIKOV.

Samoa: 1A Wesley Rd, Kelburn, POB 1430, Wellington; tel. (4) 472-0953; fax (4) 471-2479; e-mail samoahicom@clear.net.nz; High Commissioner FEESAGO SIAOSI FEPULEA'I.

Singapore: 17 Kabul St, Khandallah, POB 13140, Wellington; tel. (4) 470-0850; fax (4) 479-4066; e-mail shcwlg@xtra.co.nz; internet www.mfa.gov.sg/wellington; High Commissioner SEETOH HOY CHENG.

Switzerland: Panama House, 22 Panama St, POB 25004, Wellington; tel. (4) 472-1593; fax (4) 499-6302; e-mail vertretung@wel.rep.admin.ch; Ambassador SYLVIE MATTEUCCI.

Thailand: 2 Cook St, Karori, POB 17-226, Wellington; tel. (4) 476-8618; fax (4) 476-3677; e-mail thaiembassynz@xtra.co.nz; Ambassador ANUCHA OSATHANOND.

Turkey: 15–17 Murphy St, Level 8, POB 12-248, Wellington; tel. (4) 472-1292; fax (4) 472-1277; e-mail turkem@xtra.co.nz; Ambassador (vacant).

United Kingdom: 44 Hill St, POB 1812, Wellington; tel. (4) 924-2888; fax (4) 473-4982; e-mail bhc.wel@xtra.co.nz; internet www.britain.org.nz; High Commissioner RICHARD FELL.

USA: 29 Fitzherbert Terrace, POB 1190, Wellington; tel. (4) 462-6000; fax (4) 472-3537; internet usembassy.state.gov/wellington; Ambassador CHARLES SWINDELLS.

Judicial System

The Judicial System of New Zealand comprises a Court of Appeal, a High Court and District Courts, all of which have civil and criminal jurisdiction, and the specialist courts, the Employment Court, the Family Court, the Youth Court and the Maori Land Court. Final appeal is to the Judicial Committee of the Privy Council in the United Kingdom. In civil cases parties have an appeal to the Privy Council from the Court of Appeal as a matter of right if the case involves $NZ5,000 or more; leave to appeal is not automatically granted for criminal cases. The system of final appeal to the Privy Council was scheduled to be replaced by a local Supreme Court during 2004.

The Court of Appeal hears appeals from the High Court and from District Court Jury Trials, although it does have some original jurisdiction. Its decisions are final, except in cases that may be appealed to the Privy Council. Appeals regarding convictions and sentences handed down by the High Court or District Trial Courts are by leave only.

The High Court has jurisdiction to hear cases involving crimes, admiralty law and civil matters. It hears appeals from lower courts and tribunals, and reviews administrative actions.

District Courts have an extensive criminal and civil law jurisdiction. They hear civil cases up to $NZ200,000, unless the parties agree to litigate a larger sum (up to $NZ62,500 per year for rent and up to $NZ500,000 for real estate). Justices of the Peace can hear minor criminal and traffic matters if less than $NZ5,000. The Family Court, which is a division of the District Courts, has the jurisdiction to deal with dissolution of marriages, adoption, guardianship applications, domestic actions, matrimonial property, child support, care and protection applications regarding children and young persons, and similar matters.

The tribunals are as follows: the Employment Tribunal (administered by the Department of Labour), Disputes Tribunal, Complaints Review Tribunal, Residential Tenancies Tribunal, Waitangi Tribunal, Environment Court, Deportation Review Tribunal and Motor Vehicles Disputes Tribunal. The Disputes Tribunal has the jurisdiction to hear civil matters involving sums up to $NZ7,500. If the parties agree, it can hear cases involving sums up to $NZ12,000.

In criminal cases involving indictable offences (major crimes), the defendant has the right to a jury. In criminal cases involving summary offences (minor crimes), the defendant may elect to have a jury if the sentence corresponding to the charge is three months or greater.

Attorney-General: MARGARET WILSON.

Chief Justice: Dame SIAN ELIAS.

THE COURT OF APPEAL

President: THOMAS MUNRO GAULT.

Judges: Dame SIAN ELIAS (*ex officio*), NOEL CROSSLEY ANDERSON, JOHN J. MCGRATH, Sir KENNETH KEITH, PETER BLANCHARD, ANDREW PATRICK CHARLES TIPPING, SUSAN GLAZEBROOK.

THE HIGH COURT

Permanent Judges: Chief Justice Dame SIAN ELIAS (*ex officio*), Justice ELLIS, Justice JOHN ANTHONY DOOGUE, Justice JAMES BRUCE ROBERTSON, Justice ROBERT GRANT HAMMOND, Justice LOWELL GODDARD, Justice J. WARWICK GENDALL, Justice WILD, Justice E. T. DURIE, Justice RONALD YOUNG, Justice FRANCE, Justice NEAZOR, D. I. GENDALL, Justice ROBERT LLOYD FISHER, Justice DAVID STEWART MORRIS, Justice HUGH WILLIAMS, Justice DAVID BARAGWANATH, Justice PETER M. SALMON, Justice BARRY J. PATERSON, Justice JUDITH M. POTTER, Justice JOHN A. LAURENSON, Justice TONY RANDERSON, Justice NICHOLSON, Justice CHAMBERS, Justice RODNEY HANSEN, Justice O'REGAN, Justice PRIESTLEY, Justice HARRISON, Justice HEATH, Justice TOMPKINS, Justice SMELLIE, Justice PENLINGTON, ANNE GAMBRILL, J. A. FAIRE, G. L. LANG, Justice JOHN HANSEN, Justice PANCKHURST, Justice CHISOLM, Justice WILLIAM YOUNG, Justice FRASER, G. J. VENNING.

Religion

CHRISTIANITY

Conference of Churches in Aotearoa New Zealand: POB 22-652, Christchurch; tel. and fax (3) 377-2703; fax (3) 377–2702; e-mail admin@ccanz.net.nz; internet www.ccanz.net.nz; f. 1987 to replace the National Council of Churches in New Zealand (f. 1941); 11 mem. churches.

Te Runanga Whakawhanaunga i Nga Hahi o Aotearoa (Maori Council of Churches in New Zealand): Private Bag 11903, Ellerslie, Auckland, Aotearoa-New Zealand; tel. (9) 525-4179; fax (9) 525-4346; f. 1982; four mem. churches; Administrator TE RUA GRETHA.

The Anglican Communion

The Anglican Church in Aotearoa, New Zealand and Polynesia comprises Te Pihopatanga o Aotearoa and eight dioceses (one of which is Polynesia). In 1996 the Church had an estimated 631,764 members in New Zealand.

Primate of the Anglican Church in Aotearoa, New Zealand and Polynesia, and Bishop of Auckland: Rt Rev. JOHN CAMPBELL PATERSON, POB 37-242, Parnell, Auckland; tel. (9) 302-7201; fax (9) 377-6962.

General Secretary and Treasurer of the Anglican Church in Aotearoa, New Zealand and Polynesia: ROBIN NAIRN, POB 885, Hastings; tel. (6) 878-7902; fax (6) 878-7905; e-mail gensec@hb.ang.org.nz; internet www.anglican.org.nz.

The Roman Catholic Church

For ecclesiastical purposes, New Zealand comprises one archdiocese and five dioceses. At 31 December 2002 there were an estimated 465,635 adherents.

Bishops' Conference

New Zealand Catholic Bishops' Conference, POB 1937, Wellington; tel. (4) 496-1747; fax (4) 496-1746; e-mail adickinson@nzcbc.org.nz. f. 1974; Pres. Most Rev. DENIS BROWNE (Bishop of Hamilton); Sec. Bishop JOHN DEW (Auxiliary Bishop of Wellington); Executive Officer ANNE DICKINSON.

Archbishop of Wellington: Cardinal THOMAS STAFFORD WILLIAMS, POB 1937, Wellington 6015; tel. (4) 496-1795; fax (4) 486-1728; e-mail archbishop@wn.catholic.org.nz.

Other Christian Churches

Baptist Churches of New Zealand: 8 Puhinui Rd, Manukau City, POB 97543, South Auckland; tel. (9) 278-7494; fax (9) 278-7499; e-mail info@baptist.org.nz; internet www.baptist.org.nz; f. 1882; 22,456 mems; Pres. GRAHAM CREAHAN; Nat. Leader Rev. BRIAN WINSLADE.

Congregational Union of New Zealand: 10 The Close, 42 Arawa St, New Lynn, Auckland; tel. and fax (9) 827-3708; e-mail cunz@xtra.co.nz; f. 1884; 786 mems, 15 churches; Gen. Sec. BOB FRANKLYN; Chair. BARBARA KENNETT.

Methodist Church of New Zealand: Connexional Office, POB 931, Christchurch; tel. (3) 366-6049; fax (3) 366-6009; e-mail info@methodist.org.nz; 22,524 mems; Gen. Sec. Rev. JILL VAN DE GEER.

Presbyterian Church of Aotearoa New Zealand: 100 Tory St, POB 9049, Wellington; tel. (04) 801-6000; fax (04) 801-6001; e-mail aes@presbyterian.org.nz; internet www.presbyterian.org.nz; 50,000 mems; Moderator Rt Rev. MICHAEL THAWLEY; Assembly Exec. Sec. Rt Rev. GARRY MARQUAND.

There are several Maori Churches in New Zealand, with a total membership of over 30,000. These include the Ratana Church of New Zealand, Ringatu Church, Church of Te Kooti Rikirangi, Absolute Maori Established Church and United Maori Mission. The Antiochian Orthodox Church, the Assemblies of God, the Greek Orthodox Church of New Zealand, the Liberal Catholic Church and the Society of Friends (Quakers) are also active.

BAHÁ'Í FAITH

National Spiritual Assembly of the Bahá'ís of New Zealand: POB 21-551, Henderson, Auckland 1231; tel. (9) 837-4866; fax (9) 837-4898; e-mail natsec@nsa.org.nz; internet www.bahai.org.nz; CEO SUZANNE MAHON.

The Press

NEWSPAPERS AND PERIODICALS

Principal Dailies

In 2001 there were 25 daily newspapers in New Zealand (seven morning, 18 evening).

Bay of Plenty Times: 108 Durham St, Private Bag 12002, Tauranga; tel. (7) 577-7770; fax (7) 578-0047; e-mail news@bopp.co.nz; internet www.mytown.co.nz/bayofplenty; f. 1872; evening; Mon.–Sat.; Gen. Man. ROD HALL; Editor CRAIG NICHOLSON; circ. 23,285.

The Daily News: Currie St, POB 444, New Plymouth; tel. (6) 758-0559; fax (6) 758-6849; e-mail editor@dailynews.co.nz; f. 1857; morning; Gen. Man. KEVIN NIELSEN; Editor (vacant); circ. 27,316.

The Daily Post: 1143 Hinemoa St, POB 1442, Rotorua; tel. (7) 348 6199; fax (7) 346 0153; e-mail daily@dailypost.co.nz; f. 1885; evening; Gen. Man. MIKE FLETCHER; Editor KARYN SCHERER; circ. 12,100.

Dominion Post: Dominion Post House, 40 Boulcott St, POB 1297, Wellington; tel. (4) 474-0000; fax (4) 474-0350; e-mail editor@dompost.co.nz; internet www.wnl.co.nz; f. 2002 following merger of *Evening Post* and *The Dominion*; morning; Gen. Man. DON CHURCHILL; Man. Editor TIM PANKHURST; circ. 101,600.

Evening Standard: POB 3, Palmerston North; tel. (6) 356-9009; fax (6) 350-9836; e-mail editor@msl.co.nz; f. 1880; evening; Gen. Man. MIKE HELLEUR; Editor TONY CURRAN; circ. 20,860.

Gisborne Herald: 64 Gladstone Rd, POB 1143, Gisborne; tel. (6) 868-6655; fax (6) 867-8048; e-mail editorial@gisborneherald.co.nz; f. 1874; evening; Man. Dir M. C. MUIR; Editor IAIN GILLIES; circ. 9,587.

Hawke's Bay Today: 113 Karamu Rd, POB 180, Hastings; tel. (6) 873-0800; fax (6) 873-0805; e-mail hb_news@hbtoday.co.nz; f. 1999; evening; conservative; Gen. Man. RON D. HALL; Editor LOUIS PIERARD; circ. 33,000.

Marlborough Express: 62–64 Arthur St, POB 242, Blenheim; tel. (3) 577-2950; fax (3) 577-2953; e-mail laurab@marlexpress.co.nz; internet www.marlboroughexpress.co.nz; f. 1866; Gen. Man. ROGER G. ROSE; Editor LAURA BASHAM; circ. 10,431.

The Nelson Mail: 15 Bridge St, POB 244, Nelson; tel. (3) 548-7079; fax (3) 546-2802; f. 1866; evening; Business Man. MARK HUGHES; Editor DAVID J. MITCHELL; circ. 18,555.

New Zealand Herald: 46 Albert St, POB 32, Auckland; tel. (9) 379-5050; fax (9) 373-6421; internet www.nzherald.co.nz; f. 1863; morning; Man. Dir TIM MURPHY; Editor KEN STEINKE; circ. 215,000.

The Northern Advocate: 36 Water St, POB 210, Whangarei; tel. (9) 438-2399; fax (9) 430-5669; e-mail daily@northernadvocate.co.nz; internet www.wilsonandhorton.co.nz; f. 1875; evening; Gen. Man. J. E. P. HENTON; Editor A. VERDON; circ. 15,112.

Otago Daily Times: Lower Stuart St, POB 517, Dunedin; tel. (3) 477-4760; fax (3) 474-7422; e-mail odt.editor@alliedpress.co.nz; internet www.odt.co.nz; f. 1861; morning; Man. Dir JULIAN C. S. SMITH; Editor ROBIN CHARTERIS; circ. 44,500.

The Press: Cathedral Sq., Private Bag 4722, Christchurch; tel. (3) 379-0940; fax (3) 364-8492; e-mail editorial@press.co.nz; internet www.press.co.nz; f. 1861; morning; Gen. Man. CHRIS JAGUSCH; Editor PAUL THOMPSON; circ. 92,000.

Southland Times: 67 Esk St, POB 805, Invercargill; tel. (3) 218-1909; fax (3) 214-9905; e-mail editor@stl.co.nz; f. 1862; morning; Gen. Man. BLAIR BURR; Editor FRED TULETT; circ. 33,300.

Timaru Herald: 52 Bank St, POB 46, Timaru; tel. (3) 684-4129; fax (3) 688-1042; e-mail editor@timaruherald.co.nz; f. 1864; morning; Man. CHRIS MCAUSLIN; Editor DAVE WOOD; circ. 14,141.

Waikato Times: Private Bag 3086, Hamilton; tel. (7) 849-6180; fax (7) 849-9603; f. 1872; evening; independent; Gen. Man. P. G. HENSON; Editor VENETIA SHERSON; circ. 42,000.

Wanganui Chronicle: 59 Taupo Quay, POB 433, Wanganui; tel. (6) 349-0710; fax (6) 349-0722; e-mail wc_editorial@wilsonandhorton.co.nz; f. 1856; morning; Gen. Man. R. A. JARDEN; Editor J. MASLIN; circ. 14,000.

Weeklies and Other Newspapers

Best Bets: POB 1327, Auckland; fax (9) 366-4565; Sun. and Thur.; horse-racing, trotting and greyhounds; Editor MIKE BROWN; circ. 10,000.

Christchurch Star: 293 Tuam St, POB 1467, Christchurch; tel. (3) 379-7100; fax (3) 366-0180; e-mail bob_cotton@christchurchstar.co.nz; f. 1868; 2 a week; Chief Reporter BOB COTTON; circ. 118,170.

MG Business: 8 Sheffield Crescent, Christchurch 8005; tel. (3) 358-3219; fax (3) 358-4490; f. 1876; fmrly *Mercantile Gazette*; fortnightly; Mon.; economics, finance, management, stock market, politics; Editor BILL HORSLEY; circ. 16,300.

The National Business Review: Bank of New Zealand Tower, Level 26, 125 Queen St, POB 1734, Auckland; tel. (9) 307-1629; fax (9) 307-5129; e-mail editor@nbr.co.nz; f. 1970; weekly; Editor NEVIL GIBSON; circ. 14,328.

New Truth and TV Extra: News Media (Auckland) Ltd, 155 New North Rd, POB 1327, Auckland; tel. (9) 302-1300; fax (9) 307-0761; e-mail editor@truth.co.nz; f. 1905; Friday; local news and features; TV and entertainment; sports; Editor CLIVE NELSON; circ. 22,000.

New Zealand Gazette: Dept of Internal Affairs, POB 805, Wellington; tel. (4) 470-2930; fax (4) 470-2932; e-mail gazette@parliament.govt.nz; internet www.gazette.govt.nz; official government publication; f. 1840; weekly; Man. JANET GOOTJES; circ. 1,000.

North Shore Times: POB 33-235, Takapuna, Auckland; tel. (9) 489 4189; fax (9) 486-1950; e-mail janet.ainsworth@snl.co.nz; 3 a week; Man. JANET AINSWORTH; Editor IVAN DUNN; circ. 64,000.

Sunday Star-Times: POB 1409, Auckland; tel. (9) 302-1300; fax (9) 309-0258; f. 1994 by merger; Editor CATE BRETT; circ. 210,510.

Sunday News: POB 1327, Auckland; tel. (9) 302 1300; fax (9) 358 3003; e-mail editor@sunday-news.co.nz; internet www.sundaynews.co.nz; Editor CLIVE NELSON; circ. 113,422.

Taieri Herald: POB 105, Mosgiel; tel. (3) 489-7123; fax (3) 489-7668; f. 1962; weekly; Man. Editor LEE HARRIS; circ. 10,700.

Waihi Gazette: Waihi; weekly; Editor FRITHA TAGG; circ. 8,650.

Wairarapa News: POB 87, Masterton; tel. (6) 370-5690; fax (6) 379-6481; f. 1869; Editor ERIC TURNER; circ. 18,200.

Other Periodicals

AA Directions: AA Centre, Cnr Albert and Victoria Sts, Auckland; tel. (9) 966-8800; fax (9) 966-8975; e-mail editor@nzaa.co.nz; internet www.aa.co.nz/Online; quarterly; official magazine of the The New Zealand Automobile Association; Editor JOHN CRANNA; circ. 547,912.

Air New Zealand Inflight Magazine: Private Bag 47-920, Ponsonby, Auckland; tel. (9) 379-8822; fax (9) 379-8821; e-mail nzsales@pol.net.nz; monthly; in-flight magazine of Air New Zealand; circ. 61,000.

Architecture New Zealand: AGM Publishing Ltd, Private Bag 99-915, Newmarket, Auckland; tel. (9) 846-4068; fax (9) 846-8742; e-mail johnw@agm.co.nz; f. 1987; every 2 months; Editor JOHN WALSH; circ. 10,000.

Australian Women's Weekly (NZ edition): Private Bag 92-512, Wellesley St, Auckland; tel. (9) 308-2735; fax (9) 302-0667; monthly; Editorial Dir LOUISE WRIGHT; circ. 100,550.

Computer Buyer New Zealand: 246 Queen St, Level 8, Auckland; tel. (9) 377-9902; fax (9) 377-4604; every 2 months; Man. Editor DON HILL; circ. 120,968.

Dairying Today: POB 3855, Auckland; tel. (9) 307-0399; fax (9) 307-0122; e-mail rural_news@clear.net.nz; monthly; Editor ADAM FRICKER; circ. 25,979.

Fashion Quarterly: ACP Media Centre, Private Bag 92-512, Auckland; tel. (9) 308-2735; fax (9) 302-0667; e-mail fq@acpmedia.co.nz; f. 1982; 5 a year; Editor LEONIE BARLOW; circ. 31,500.

Grapevine: Private Bag 92-124, Auckland; tel. (9) 624-3079; fax (9) 625-8788; monthly; family magazine; Editor JOHN COONEY; circ. 149,658.

Info-Link: AGM Publishing Ltd, Private Bag 99-915, Newmarket, Auckland; tel. (9) 846-4068; fax (9) 846-8742; e-mail pengelly@agm.co.nz; internet www.info-link.co.nz; quarterly; Editors SALLY LINDSAY, REBECCA WOOD; circ. 22,000.

Landfall: University of Otago Press, POB 56, Dunedin; tel. (3) 479-8807; fax (3) 479-8385; e-mail landfall@otago.ac.nz; f. 1947; 2 a year; literary; Editor JUSTIN PATON; circ. 1,200.

Mana Magazine: POB 1101, Rotorua; tel. (7) 349-0260; fax (7) 349-0258; e-mail editor@manaonline.co.nz; internet www.manaonline.co.nz; Maori news magazine; Editor DEREK FOX.

Management: Wellesley St, POB 5544, Auckland; tel. (9) 630-8940; fax (9) 630-1046; e-mail editor@management.co.nz; internet www.profile.co.nz; f. 1954; monthly; business; Editor REG BIRCHFIELD; circ. 12,000.

New Idea New Zealand: 48 Greys Ave, 4th Floor, Auckland; tel. (9) 979-2700; fax (9) 979-2721; weekly; women's interest; Editor ANATHEA JOHNSTON; circ. 59,039.

New Truth: 155 New North Rd, Auckland 1; tel. (9) 302-1300; fax (9) 307-0761; e-mail editor@truth.co.nz; weekly; Editor CLIVE NELSON; circ. 22,000.

New Zealand Dairy Exporter: POB 299, Wellington; tel. (4) 499-0300; fax (4) 499-0330; e-mail lance@dairymag.co.nz; f. 1925; monthly; Man. Editor L. MCELDOWNEY; circ. 22,739.

The New Zealand Farmer: Great South Rd, POB 4233, Auckland; tel. (9) 520-9451; fax (9) 520-9459; f. 1882; weekly; Editor HUGH STRINGLEMAN; circ. 18,000.

New Zealand Forest Industries: POB 5544, Auckland; tel. (9) 630-8940; fax (9) 630-1046; e-mail info@nzforest.co.nz; internet www.nzforest.co.nz; f. 1960; monthly; forestry; Publr REG BIRCHFIELD; circ. 2,500.

New Zealand Gardener: POB 6341, Wellesley St, Auckland; tel. (4) 293-4495; f. 1944; monthly; Editor P. MCGEORGE; circ. 77,077.

New Zealand Horse and Pony: POB 12965, Auckland; tel. (9) 634-1800; fax (9) 634-2918; e-mail editor@horse-pony.co.nz; f. 1959; monthly; Editor JOAN E. GILCHRIST; circ. 9,252.

New Zealand Medical Journal: Department of Surgery, Christchurch Hospital, POB 4345, Christchurch; tel. (3) 364-1277; fax (3) 364-1683; e-mail frank.frizelle@cdhb.govt.nz; internet www.nzma.org.nz; 2 a month; Editor Prof. FRANK A. FRIZELLE; circ. 5,000.

New Zealand Science Review: POB 1874, Wellington; fax (4) 389-5095; e-mail mberridge@malaghan.org.nz; internet www.rsnz.govt.nz/clan/nzmss; f. 1942; 4 a year; reviews, policy and philosophy of science; Editor M. V. BERRIDGE.

New Zealand Woman's Day: Wellesley St, Private Bag 92-512, Auckland; tel. (9) 308-2700; fax (9) 357-0978; e-mail wdaynz@acp.nz.co.nz; weekly; Editor in Chief LOUISE WRIGHT; circ. 143,420.

New Zealand Woman's Weekly: NZ Magazines Ltd, POB 90-119, Auckland Mail Centre, Auckland; tel. (9) 360-3820; fax (9) 360-3826; e-mail editor@nzww.co.nz; internet www.on-line.co.nz; f. 1932; Mon.; women's issues and general interest; Editor NICKY PELLEGRINO; circ. 130,706.

Next: Level 4, Cnr Fanshawe and Beamont Sts, Westhaven, Private Bag 92-512, Auckland 1036; tel. (9) 308-2773; fax (9) 377-6725; e-mail next@acpmedia.co.nz; f. 1991; monthly; home and lifestyle; Editor LIZ PARKER; circ. 63,000.

North & South: 100 Beaumont St, Wellesley St, Private Bag 92-512, Auckland; tel. (9) 308-2700; fax (9) 308-9498; e-mail northsouth@acpmedia.co.nz; f. 1986; monthly; current affairs and lifestyle; Editor ROBYN LANGWELL; circ. 35,959.

NZ Catholic: POB 147-000, Ponsonby, Auckland 1034; tel. (9) 378-4380; fax (9) 360-3065; e-mail catholic@iconz.co.nz; internet www

.nzcatholic.org.nz; f. 1996; fortnightly; Roman Catholic; Editor PAT McCARTHY; circ. 7,000.

NZ Home and Entertaining: ACP Media Centre, Cnr Fanshawe and Beaumont Sts, Private Bag 92-512, Auckland; tel. (9) 308–2700; e-mail cmccall@acpmedia.co.nz; f. 1936; bi-monthly, design, architecture, lifestyle; Editor CLARE McCALL; circ. 25,000.

NZ Listener: NZ Magazines Ltd, POB 90-119, Auckland Mail Centre, Auckland; tel. (9) 360-3820; fax (9) 360-3831; f. 1939; weekly; general features, politics, arts, TV, radio; Editor FINLAY MACDONALD; circ. 96,000.

NZ Turf Digest: Fairfax Sunday Newspapers Ltd, 155 New North Rd, POB 1327, Auckland; tel. (9) 302-1310; fax (9) 366-4565; Sun./Thur.; Editor MIKE BROWN; circ. 7,500.

Otago Southland Farmer: POB 805, Invercargill; tel. (3) 211-1051; fax (3) 214-3713; fortnightly; Editor REX TOMLINSON; circ. 19,000.

Pacific Wings: NZ Wings Ltd, Harewood, POB 39-099, Christchurch; tel. (3) 359-0256; fax (3) 359-0471; e-mail editor@nzwings.co.nz; internet www.nzwings.co.nz; f. 1932; monthly; Editor CALLUM MACPHERSON; circ. 20,000.

Prodesign: AGM Publishing Ltd, Private Bag 99-915, Newmarket, Auckland; tel. (9) 846-4068; fax (9) 846-8742; e-mail agm.publishing@xtra.co.nz; f. 1993; every 2 months; Man. Editor STEVE BOHLING; circ. 9,000.

PSA Journal: PSA House, 11 Aurora Terrace, POB 3817, Wellington 1; tel. (4) 917-0333; fax (4) 917-2051; e-mail enquiries@psa.org.nz; internet www.pas.org.nz; f. 1913; 8 a year; journal of the NZ Public Service Asscn; circ. 52,000.

Reader's Digest: Auckland; f. 1950; monthly; Editor TONY SPENCER-SMITH; circ. 105,000.

RSA Review: 181 Willis St, POB 27248, Wellington; tel. (4) 384-7994; fax (4) 385-3325; e-mail rsareview@paradise.net.nz; every 2 months; magazine of the Royal New Zealand Returned Services' Asscn; circ. 106,000.

Rural News: POB 3855, Auckland; tel. (9) 307-0399; fax (9) 307-0122; e-mail rural_news@clear.net.nz; fortnightly; Editor ADAM FRICKER; circ. 85,813.

SHE Magazine: Wellesley St, Private Bag 92-512, Auckland; tel. (9) 308-2735; fax (9) 302-0667; e-mail she@acpmedia.co.nz; Editor LEONIE BARLOW.

Spanz: POB 9049, Wellington; tel. (4) 801-6000; fax (4) 801-6001; e-mail commsmanager@presbyterian.org.nz; internet www.presbyterian.org.nz; f. 1987; bi-monthly; magazine of Presbyterian Church; circ. 21,500.

Straight Furrow: c/o Rural Press, POB 4233, Auckland; tel. (9) 376-9786; fax (9) 376-9780; e-mail straightfurrow@ruralpress.com; internet www.straightfurrow.co.nz; f. 1933; fortnightly; Editor SUSAN TOPLESS; circ. 85,000.

Time New Zealand: Hopetoun St, Level 8, Newton; fax (9) 366-4706; internet www.timepacific.com; weekly; circ. 35,467.

TV Guide (NZ): POB 1327, Auckland; tel. (9) 302-1300; fax (9) 373-3036; e-mail editor@tv-guide.co.nz; f. 1986; weekly; Editor ANTHONY PHILLIPS; circ. 225,648.

UNANewZ: UN Asscn of NZ, POB 12324, Wellington; tel. (4) 473-0441; fax (4) 473-2339; e-mail unanz@xtra.co.nz; internet www.converge.org.nz/unanz; f. 1945; quarterly; Editor (vacant).

NEWS AGENCIES

New Zealand Press Association: Newspaper House, 93 Boulcott St, POB 1599, Wellington; tel. (4) 472-7910; fax (4) 473-7480; e-mail news@nzpa.co.nz; internet www.newsquest.co.nz; f. 1879; non-political; Chair. JULIAN SMITH.

South Pacific News Service Ltd (Sopacnews): Lambton Quay, POB 5026, Wellington; tel. and fax (3) 472-8329; f. 1948; Man. Editor NEALE McMILLAN.

Foreign Bureaux

Agence France-Presse (AFP): Manners St, POB 11-420, Wellington; tel. (021) 688438; fax (021) 471085; e-mail afpnz@clear.net.nz; Correspondent MICHAEL FIELD.

Reuters New Zealand Ltd (United Kingdom): POB 11-744, Wellington; tel. (4) 473-4746; e-mail wellington.newsroom@reuters.com; Correspondent GYLES BECKFORD.

United Press International (UPI) (USA): Press Gallery, Parliament Bldgs, Wellington; tel. (4) 471-9552; fax (4) 472-7604; Correspondent BRENDON BURNS.

Xinhua (New China) New Agency (People's Republic of China): 136 Northland Rd, Northland, Wellington; tel. (4) 475-7607; fax (4) 475-7607; e-mail xinhua@ihug.co.nz; Correspondent ZHOU CIPU.

PRESS COUNCIL

New Zealand Press Council: The Terrace, POB 10879, Wellington; tel. (4) 473-5220; fax (4) 471-1785; e-mail presscouncil@asa.co.nz; internet www.presscouncil.org.nz; f. 1972; Chair. Sir JOHN JEFFRIES; Sec. M. E. MAJOR.

PRESS ASSOCIATIONS

Commonwealth Press Union (New Zealand Section): POB 1066, Wellington; tel. (4) 472-6223; fax (4) 471-0987; Chair. G. ELLIS; Sec. L. GOULD.

Newspaper Publishers' Association of New Zealand (Inc): Newspaper House, 93 Boulcott St, POB 1066, Wellington 1; tel. (4) 472-6223; fax (4) 471-0987; e-mail npa@npa.co.nz; f. 1898; 31 mems; Pres. J. SANDERS; CEO L. GOULD; Corporate Affairs Man. H. SOUTER.

Publishers

Auckland University Press: Private Bag 92-019, University of Auckland, Auckland; tel. (9) 373-7528; fax (9) 373-7465; e-mail aup@auckland.ac.nz; internet www.auckland.ac.nz/aup; f. 1966; scholarly; Dir ELIZABETH P. CAFFIN.

Christchurch Caxton Press Ltd: 113 Victoria St, POB 25-088, Christchurch 1; tel. (3) 366-8516; fax (3) 365-7840; f. 1935; human and general interest, local and NZ history, tourist publs; Man. Dir BRUCE BASCAND.

Dunmore Press Ltd: POB 5115, Palmerston North; tel. (6) 358-7169; fax (6) 357-9242; e-mail dunmore@xtra.co.nz; internet www.dunmore.co.nz; f. 1975; non-fiction, educational; Publrs MURRAY GATENBY, SHARMIAN FIRTH.

HarperCollins Publishers (New Zealand) Ltd: 31 View Rd, Glenfield, Auckland; tel. (9) 443-9400; fax (9) 443-9403; f. 1888; general and educational; Man. Dir BARRIE HITCHON.

Hodder Moa Beckett Publishers Ltd: POB 100-749, North Shore Mail Centre, Auckland 1330; tel. (9) 478-1000; fax (9) 478-1010; e-mail admin@hoddermoa.co.nz; f. 1971; Man. Dir KEVIN CHAPMAN.

Learning Media Ltd: POB 3293, Wellington; tel. (4) 472-5522; fax (4) 472-6444; e-mail info@learningmedia.co.nz; internet www.learningmedia.com; f. 1947 as School Publications; became Crown-owned company in 1993; general and educational books, audio cassettes, videos and computer software in English, Spanish, Maori, etc.; International Man. TRISH STEVENSON.

Legislation Direct: c/-Securacopy, Thorndon Quay, POB 12 357, Wellington; tel. (4) 495-2882; fax (4) 495-2880; e-mail idenquiries@legislationdirect.co.nz; internet www.legislationdirect.co.nz; general publishers and leading distributor of government publs; fmrly Govt Printing Office/GP Publications; Publications Man. WENDY CAYLOR.

LexisNexis NZ Ltd: 205–207 Victoria St, POB 472, Wellington; tel. (4) 385-1479; fax (4) 385-1598; e-mail Customer.Relations@lexisnexis.co.nz; internet www.lexisnexis.co.nz; legal; Country Man. STEPHEN DUNN.

McGraw-Hill Book Co, New Zealand Ltd: Westfield Tower, 2nd Floor, Westfield Shopping Town, Manukau City, POB 97082, Wiri; tel. (9) 262-2537; fax (9) 262-2540; e-mail cservice_auckland@mcgraw-hill.com; f. 1974; educational; Man. FIRGAL ADAMS.

New Zealand Council for Educational Research: POB 3237, Wellington; tel. (4) 384-7939; fax (4) 384-7933; e-mail sales@nzcer.org.nz; internet www.nzcer.org.nz; f. 1934; scholarly, research monographs, educational, academic, periodicals; Chair. Prof. RUTH MANSELL; Dir ROBYN BAKER.

Pearson Education New Zealand Ltd: Private Bag 102-908, North Shore Mail Centre, Glenfield, Auckland 10; tel. (9) 444-4968; fax (9) 444-4957; e-mail rosemary.stagg@pearsoned.co.nz; f. 1968; fmrly Addison Wesley Longman; educational; Dirs ROSEMARY STAGG, A. HARKINS, P. FIELD.

Penguin Group (NZ) Ltd: Cnr Airborne and Rosedale Rds, Albany, Private Bag 102-902, North Shore Mail Centre, Auckland; tel. (9) 415-4700; fax (9) 415-4701; e-mail geoff.walker@penguin.co.nz; internet www.penguin.co.nz; f. 1973; Publ. Dir GEOFF WALKER; Man. Dir TONY HARKINS.

Wendy Pye Ltd: Private Bag 17-905, Greenlane, Auckland; tel. (9) 525-3575; fax (9) 525-4205; e-mail admin@sunshine.co.nz; children's fiction and educational; Man. Dir WENDY PYE.

Random House New Zealand Ltd: Private Bag 102-950, North Shore Mail Centre, Glenfield, Auckland; tel. (9) 444-7197; fax (9)

444-7524; e-mail admin@randomhouse.co.nz; internet www.randomhouse.co.nz; f. 1977; general; Man. Dir MICHAEL MOYNAHAN.

Reed Publishing (NZ) Ltd: Private Bag 34-901, Birkenhead, Auckland 10; tel. (9) 441 2960; fax (9) 480-4999; e-mail info@reed.co.nz; internet www.reed.co.nz; children's and general non-fiction; Heinemann Education primary, secondary tertiary and library; Man. Dir ALAN L. SMITH.

University of Otago Press: POB 56, Dunedin; tel. (3) 479-8807; fax (3) 479-8385; e-mail university.press@otago.ac.nz; f. 1958; academic, trade publs; Man. Editor WENDY HARREX.

Whitcoulls Ltd: 210 Queen St, 3rd Floor, Private Bag 92-098, Auckland 1; tel. (9) 356-5410; fax (9) 356-5423; NZ, general and educational; Gen. Man. S. PRESTON.

PUBLISHERS' ASSOCIATION

Book Publishers' Association of New Zealand Inc: POB 36-477, Northcote, Auckland 1309; tel. (9) 480-2711; fax (9) 480-1130; e-mail bpanz@copyright.co.nz; internet www.bpanz.org.nz; f. 1977; Pres. ELIZABETH CAFFIN.

Broadcasting and Communications

TELECOMMUNICATIONS

Asia Pacific Telecom (NZ) Ltd: POB 331214, Takapuna, Auckland.

Compass Communications Ltd: POB 2533, Auckland.

Global One Communications Ltd: Phillips Fox Tower, Level 15, 209 Queen St, Auckland; tel. (9) 357-3700; fax (9) 357-3737.

NewCall Communications Ltd: Symonds St, POB 8703, Auckland; Man. Dir NORMAN NICHOLLS.

Singtel Optus Ltd: ASB Centre, Level 14, 135 Albert St, Auckland; tel. (9) 356-2660; fax (9) 356-2669.

Telecom Corpn of New Zealand Ltd: Telecom Networks House, 68 Jervois Quay, POB 570, Wellington; tel. (4) 801-9000; fax (4) 385-3469; internet www.telecom.co.nz; Chair. RODERICK DEANE; Chief Exec. THERESA GATTUNG.

Telecom Mobile Telecommunications: Midcity Tower, 141 Willis St, Wellington 6001; tel. (4) 801-9000; internet www.telecom.co.na; f. 1987; Gen. Man. LORRAINE WITTEN.

TelstraClear: TelstraClear Centre, Cnr Northcote and Taharoto Rds, Takapuna, Private Bag 92-143, Auckland; tel. (9) 912-4200; fax (9) 912-4442; e-mail webmaster@clear.co.nz; internet www.clear.co.nz; f. 1990 as Clear Communications Ltd; business solutions, local and toll services, enhanced internet, etc.; principal shareholders include Telstra Corpn and Austar United; Chair. JOHN B. EDE; CEO ROSEMARY HOWARD; Man. Dir STEVE BURDON.

Telstra Saturn Ltd: Telstra Business Centre, Level 9, 191 Queen St, Auckland; tel. (9) 980-8800; fax (9) 980-8801; e-mail enquiries@telstra.co.nz; internet www.telstra.co.nz; f. 2000 through merger of Telstra New Zealand and Saturn Communications.

Vodafone New Zealand Ltd: 21 Pitt St, Private Bag 92-161, Auckland; tel. (9) 357-5100; fax (9) 357-4836; internet www.vodafone.co.nz; fmrly Bell South; cellular network; CEO LARRY CARTER; Man. Dir GRAHAME MAHER.

WorldxChange Communications Ltd: 55 Shortland St, POB 3296, Auckland; tel. (9) 308-1300.

Regulatory Authority

Telecommunications Policy Section, Ministry of Economic Development: 33 Bowen St, POB 1473, Wellington; tel. (4) 472-0030; fax (4) 473-7010; e-mail info@med.govt.nz; internet www.med.govt.nz/pbt/telecom.html.

BROADCASTING

In December 1995 Radio New Zealand Commercial (RNZC) and New Zealand Public Radio Ltd (NZPR) became independent entities, having assumed responsibility for, respectively, the commercial and non-commercial activities of Radio New Zealand. RNZC was sold to the New Zealand Radio Network Consortium in 1996, and NZPR, which remained a Crown-owned company, assumed the name of its now-defunct parent company, to become Radio New Zealand Ltd. In late 1999 there were 190 radio stations broadcasting on a continuous basis, of which 170 were operating on a commercial basis.

Radio

Radio Broadcasters' Association (NZ) Inc: POB 3762, Auckland; tel. (9) 378-0788; fax (9) 378-8180; e-mail rba@xtra.co.nz; represents commercial radio industry; Exec. Dir D. N. G. INNES; Sec. JANINE BLISS; 142 mems.

Radio New Zealand Ltd: RNZ House, 155 The Terrace, POB 123, Wellington; tel. (4) 474-1999; fax (4) 474-1459; e-mail rnz@radionz.co.nz; internet www.radionz.co.nz; f. 1936; Crown-owned entity, operating non-commercial national networks: National Radio and Concert FM; parliamentary broadcasts on AM Network; Radio New Zealand News and Current Affairs; the short-wave service, Radio New Zealand International; and archives; Chief Exec. SHARON CROSBIE.

The Radio Network of New Zealand Ltd: 54 Cook St, Private Bag 92-198, Auckland; tel. (9) 373-0000; e-mail enquiry@radionetwork.co.nz; operates 95 commercial stations, reaching 1.3m. people; Chief Exec. JOHN MCELHINNEY.

Television

Television New Zealand (TVNZ) Ltd: Television Centre, Cnr Hobson and Victoria Sts, POB 3819, Auckland; tel. (9) 916-7000; fax (9) 379-4907; internet www.tvnz.co.nz; f. 1960; the television service is responsible for the production of programmes for two TV networks, TV One and 2; networks are commercial all week and transmit in colour; both channels broadcast 24 hours a day, seven days a week, and reach 99.9% of the population; Chair. Dr ROSS ARMSTRONG; CEO RICK ELLIS; Gen. Man. ALAN BROOKBANKS.

Private Television

Auckland Independent Television Services Ltd: POB 1629, Auckland.

Bay Satellite TV Ltd: Hastings; tel. (6) 878-9081; fax (6) 878-5994; Man. Dir JOHN LYNAM.

Broadcast Communications Ltd: POB 2495, Auckland; tel. (9) 916-6400; fax (9) 916-6404; internet www.bclnz.co.nz; principal shareholder is Television New Zealand (TVNZ) Ltd; Man. Dir GEOFF LAWSON.

Sky Network Television Limited: 10 Panorama Rd, POB 9059, Newmarket, Auckland; tel. (9) 579-9999; fax (9) 579-0910; internet www.skytv.co.nz; f. 1990; UHF service on six channels, satellite service on 18 channels; Chair. T. MOCKRIDGE; CEO J. FELLET.

TV3 Network Services Ltd: Symonds St, Private Bag 92-624, Auckland; tel. (9) 377-9730; fax (9) 366-5999; internet www.tv3.co.nz; f. 1989; owned by CanWest Global Corpn; Man. Dir RICK FRIESEN.

TV4 Network Ltd: Symonds St, Private Bag 92-624, Auckland; tel. (9) 377-9730; fax (9) 366-5999; e-mail di.winks@tv3.co.nz; internet www.ctv4.co.nz; 'free-to-air' entertainment channel; Man. Dir RICK FRIESEN.

Finance

(cap. = capital; res = reserves; dep. = deposits; m. = million; amounts in New Zealand dollars)

BANKING

Central Bank

Reserve Bank of New Zealand: 2 The Terrace, POB 2498, Wellington; tel. (4) 472-2029; fax (4) 473-8554; e-mail rbnz-info@rbnz.govt.nz; internet www.rbnz.govt.nz; f. 1934; res 411.5m., dep. 8,147.3m. (June 2002); Gov. ALAN BOLLARD; Dep. Gov. Dr R. M. CARR.

Registered Banks

As a result of legislation which took effect in April 1987, several foreign banks were incorporated into the domestic banking system.

ANZ Banking Group (New Zealand) Ltd: Private Bag 92-210 AMSC, Auckland; tel. (9) 374-4040; fax (9) 374-4038; internet www.anz.co.anz; f. 1979; subsidiary of Australia and New Zealand Banking Group Ltd of Melbourne, Australia; cap. 406.0m., dep. 23,451m. (Sept. 2001); Chair. R. S. DEANE; Man. Dir Dr M. J. HORN; 143 brs and sub-brs.

ASB Bank Ltd: ASB Bank Centre, Cnr Wellesley and Albert Sts, POB 35, Auckland 1; tel. (9) 377-8930; fax (9) 358-3511; e-mail helpdesk@asbbank.co.nz; internet www.asbbank.co.nz; f. 1847 as Auckland Savings Bank, name changed 1988; cap. 323.1m., res 496.6m., dep. 18,762.8m. (June 2001); Chair. G. J. JUDD; Man. Dir G. HUGH BURRETT; 119 brs.

Bank of New Zealand (BNZ): BNZ Centre, 1 Willis St, POB 2392, Wellington; tel. (4) 474-6999; fax (4) 474-6861; internet www.bnz.co.nz; f. 1861; owned by National Australia Bank; cap. 1,791m., dep. 32,165m. (Sept. 2001); Chair. T. K. MCDONALD; Man. Dir P. THODEY; 241 brs and sub-brs.

Citibank NA (USA): 23 Customs Street East, POB 3429, Auckland; tel. (9) 307-1902; fax (9) 307-1983; internet www.citibank.com; CEO Andrew Au; 2 brs.

Deutsche Bank AG: Wellesley St, POB 6900, Auckland; tel. (9) 351-1000; fax (9) 351-1001; internet www.deutsche-bank.co.nz; f. 1986; fmrly Bankers Trust New Zealand; Chief Country Officer Brett Shepherd.

Hongkong and Shanghai Banking Corporation Ltd (Hong Kong): 1 Queen St, Level 9, POB 5947, Auckland; tel. (9) 308-8888; fax (9) 308-8997; e-mail hsbcplb@clear.net.nz; internet www.asiapacific.hsbc.com; CEO S. Edwards; 6 brs.

Kiwibank Ltd: Private Bag 39888, Wellington; tel. (4) 462-7900; fax (4) 462-7996; internet www.kiwibank.co.nz; f. 2001; 100% New Zealand-owned; savings bank for small depositors; Chair. Jim Bolger; Chief Exec. Sam Knowles.

National Bank of New Zealand Ltd: 1 Victoria St, POB 1791, Wellington 6000; tel. (4) 494-4000; fax (4) 494-4023; internet www.nationalbank.co.nz; f. 1873; merged with Countrywide Banking Corpn in 1999; cap. 472m., res 1,705m., dep. 33,757m. (Dec. 2001); Chair. Sir Wilson Whineray; Dir and CEO Sir John Anderson; 160 brs.

Rabobank (New Zealand): POB 1069, Auckland; tel. (9) 375-3700; fax (9) 375-3710; f. 1996; full subsidiary of Rabobank Nederland.

TSB Bank Ltd: POB 240, New Plymouth; tel. (6) 968-3700; fax (6) 968-3768; internet www.tsbbank.co.nz; f. 1850; dep. 1,800m. (March 2004); Man. Dir K. W. Rimmington; 15 brs.

Westland Bank Ltd: 99 Revell St, POB 103, Hokitika; tel. (3) 755-8680; fax (3) 755-8277; full subsidiary of ASB Bank Ltd; cap. and res 6,176m., dep. 108m. (June 1992); Man. Dir K. J. Beams; 9 brs.

WestpacTrust: 318 Lambton Quay, POB 691, Wellington; tel. (4) 498-1000; fax (4) 498-1350; acquired Trust Bank New Zealand; New Zealand division of Westpac Banking Corpn (Australia); Chief Exec. D. T. Gallagher.

Other Banks

Postbank: 58–66 Willis St, Wellington 1; tel. (4) 729-809; f. 1987; owned by Australia and New Zealand Banking Corpn; 550 brs.

Association

New Zealand Bankers' Association: POB 3043, Wellington; tel. (4) 472-8838; fax (4) 473-1698; e-mail acopland@nzba.org.nz; internet www.nzba.org.nz; f. 1891; Chief Exec. Alan Yates.

STOCK EXCHANGES

Dunedin Stock Exchange: POB 298, Dunedin; tel. (3) 477-5900; Chair. E. S. Edgar; Sec. R. P. Lewis.

New Zealand Exchange Ltd: ASB Tower, Level 9, 2 Hunter St, POB 2959, Wellington 1; tel. (4) 472-7599; fax (4) 473-1470; e-mail info@nzx.com; internet www.nzx.com; Chair. Simon Allen; Man. Dir M. Weldon.

Supervisory Body

New Zealand Securities Commission: POB 1179, Wellington; tel. (4) 472-9830; fax (4) 472-8076; e-mail seccom@sec-com.govt.nz; internet www.sec-com.govt.nz; f. 1979; Chair. Jane Diplock.

INSURANCE

ACE Insurance NZ Ltd: POB 734, Auckland; tel. (9) 377-1459; fax (9) 303-1909; e-mail grant.simpson@ace-ina.com; Gen. Man. Grant Simpson.

AMI Insurance Ltd: 29–35 Latimer Sq., POB 2116, Christchurch; tel. (3) 371-9000; fax (3) 371-8314; e-mail amichch@es.co.nz; f. 1926; Chair. K. G. L. Nolan; CEO John B. Balmforth.

ANZ Life Assurance Co Ltd: POB 1492, Wellington; tel. (4) 496-7000; fax (4) 470-5100; Man. R. A. Dean.

BNZ Life Insurance Ltd: POB 1299, Wellington; tel. (4) 382-2577; fax (4) 474-6883; e-mail rodger-murphy@bnz.co.nz; Gen. Man. R. J. Murphy.

Commercial Union General Insurance Co Ltd: 142 Featherston St, POB 2797, Wellington; tel. (4) 473-9177; fax (4) 473-7424; fire, accident, travel, engineering; Gen. Man. T. Wakefield.

Farmers' Mutual Group: 68 The Square, POB 1943, Palmerston North 5330; tel. (6) 356-9456; fax (6) 356-4603; e-mail enquiries@fmg.co.nz; internet fmg.co.nz; comprises Farmers' Mutual Insurance Asscn, Farmers' Mutual Insurance Ltd and other cos; fire, accident, motor vehicle, marine, life; CEO G. Smith.

Gerling NCM EXGO: POB 3933, Wellington 6015; tel. (4) 472-4142; fax (4) 472-6966; e-mail info@gerlingEXGO.co.nz; internet www.gerlingEXGO.co.nz; trade credit insurance services; division of Gerling NZ Ltd since December 2001; Gen. Man. Arthur C. Davis.

Metropolitan Life Assurance Co of NZ Ltd: Sovereign Assurance House, 33–45 Hurstmere Rd, Takapuna, Auckland 1; tel. (9) 486-9500; fax (9) 486-9501; f. 1962; life; Man. Dir Peter W. Fitzsimmons.

New Zealand Insurance: NZI House, 151 Queen St, Private Bag 92130, Auckland 1; tel. (9) 309-7000; fax (9) 309-7097; internet www.nzi.co.nz; Chief Exec. M. Hannan.

New Zealand Local Government Insurance Corporation Ltd (Civic Assurance): Local Government Bldg, 114–118 Lambton Quay, POB 5521, Wellington; tel. (4) 470-0037; fax (4) 471-1522; e-mail rod.mead@civicassurance.co.nz; internet www.civicassurance.co.nz; f. 1960; fire, motor, all risks, accident; Chair. K. N. Sampson; Gen. Man. R. D. Mead.

Promina Group: 48 Shortland St, Auckland; tel. (9) 363-2222; internet www.promina.co.nz; f. 1878; fmrly Royal & SunAlliance; comprises AA Insurance, Vero Insurance New Zealand Ltd and other cos; fire, accident, marine, general; CEO Michael Wilkins.

QBE Insurance (International) Ltd: POB 44, Auckland; tel. (9) 366-9920; fax (9) 366-9930; e-mail gevans@qbe.co.nz; f. 1890; Gen. Man. Graeme Evans.

Sovereign Ltd: Sovereign House, 33–45 Hurstmere Rd, Private Bag Sovereign, Auckland; tel. (9) 486-9500; fax (9) 486-9501; e-mail emailus@sovereign.co.nz; internet www.sovereign.co.nz; f. 1989; life insurance and investment.

State Insurance Ltd: Microsoft House, 3–11 Hunter St, POB 5037, Wellington 1; tel. (4) 496-9600; fax (4) 476-9664; internet www.state.co.nz; f. 1905; mem. NRMA Insurance Group; Man. Dir T. C. Sole.

Tower Insurance Ltd: 67–73 Hurstmere Rd, POB 33-144, Takapuna, Auckland; tel. (9) 486-9340; fax (9) 486-9368; internet www.towerlimited.com; f. 1873; fmrly National Insurance Co of New Zealand; Man. Dir P. R. Hunt.

Associations

Insurance Council of New Zealand: POB 474, Wellington; tel. (4) 472-5230; fax (4) 473-3011; e-mail icnz@icnz.org.nz; internet www.icnz.org.nz; CEO Christopher Ryan.

Investment Savings and Insurance Association of New Zealand Inc: POB 1514, Wellington; tel. (4) 473-8730; fax (4) 471-1881; e-mail isi@isi.org.nz; f. 1996; from Life Office Asscn and Investment Funds Asscn; represents cos that act as manager, trustee, issuer, insurer, etc. of managed funds, life insurance and superannuation; Chief Exec. Vance Arkinstall; Exec. Officer Deborah Keating.

Trade and Industry

GOVERNMENT AGENCY

New Zealand Trade and Enterprise: POB 2878, Wellington; tel. (4) 910-4300; fax (4) 910-4309; e-mail info@nzte.govt.nz; internet www.nzte.govt.nz; f. 2003 to combine the functions of Trade New Zealand and Industry New Zealand; charged with helping New Zealand's businesses achieve success at home and in the global marketplace; CEO Tim Gibson; Chair. Phil Lough.

CHAMBERS OF COMMERCE

Auckland Regional Chamber of Commerce and Industry: POB 47, Auckland; tel. (9) 309-6100; fax (9) 309-0081; e-mail mbarnett@chamber.co.nz; internet www.chamber.co.nz; CEO Michael Barnett; Pres. David Truscott.

Canterbury Employers' Chamber of Commerce: 57 Kilmore St, POB 359, Christchurch; tel. (3) 366-5096; fax (3) 379-5454; e-mail info@cecc.org.nz; internet www.cecc.org.nz; CEO Peter Townsend.

Otago Chamber of Commerce and Industry Inc: Westpac Trust Bldg, Level 7, 106 George St Dunedin; tel. (3) 477-0341; fax (3) 643-0341; e-mail office@otagochamber.co.nz; internet www.otagochamber.co.nz; f. 1861; CEO J. A. Christie; Pres. N. Smith.

Wellington Regional Chamber of Commerce: 109 Featherston St, 9th Floor, POB 1590, Wellington; tel. (4) 914-6500; fax (4) 914-6524; e-mail info@wgtn-chamber.co.nz; internet www.wgtn-chamber.co.nz; f. 1856; Chief Exec. Philip Lewin; Pres. Peter Steel; 1,000 mems.

INDUSTRIAL AND TRADE ASSOCIATIONS

Canterbury Manufacturers' Association: POB 13-152, Armagh, Christchurch; tel. (3) 353-2540; fax (3) 353-2549; e-mail cma@cma.org.nz; internet www.cma.org.nz; f. 1879; CEO John Walley; 500 mems.

Employers' and Manufacturers' Association (Central Inc): Federation House, 95–99 Molesworth St, POB 1087, Wellington; tel. (4) 473-7224; fax (4) 473-4501; e-mail ema@emacentral.org.nz; f. 1997 by merger of Wellington Manufacturers' Asscn and Wellington Regional Employers' Asscn; Pres. R. KERR-NEWELL; 2,200 mems.

Employers' and Manufacturers' Association (Northern Inc): 159 Khyber Pass Rd, Grafton, Private Bag 92066, Auckland; tel. (9) 367-0900; fax (9) 367-0902; e-mail ema@ema.co.nz; internet www.ema.co.nz; f. 1886; fmrly Auckland Manufacturers' Asscn; Pres. T. ARNOLD; 5,000 mems.

ENZAFRUIT: POB 279, Hastings; tel. (6) 878-1898; fax (6) 878-1850; e-mail info@enza.co.nz; fmrly New Zealand Apple and Pear Marketing Board; Man. Dir MICHAEL DOSSOR; Chair. B. BIRNIE.

Federated Farmers of New Zealand (Inc): Agriculture House, 6th Floor, 12 Johnston St, Wellington 1; POB 715, Wellington; tel. (4) 473-7269; fax (4) 473-1081; e-mail wellingtonoffice@fedfarm.org.nz; internet www.fedfarm.org.nz; f. 1946; Pres. ALISTAIR POLSON; CEO TONY ST CLAIR; 16,000 mems.

Kiwifruit New Zealand: POB 4246, Mt Maunganui South; tel. (7) 574-7139; fax (7) 574-7149; Chair. PETER TRAPSKI.

National Beekeepers' Association of New Zealand (Inc): POB 234, Te Kuiti 2500; tel. and fax (7) 878-7193; e-mail waihon@actrix.co.nz; internet www.nba.org.nz; f. 1913; 450 mems; Pres. JANE LORIMER; Sec. PAULINE BASSETT.

New Zealand Animal By-Products Exporters' Association: 11 Longhurst Terrace, POB 12-222, Christchurch; tel. (3) 332-2895; fax (3) 332-2825; 25 mems; Sec. J. L. NAYSMITH.

New Zealand Berryfruit Growers' Federation (Inc): POB 10-050, Wellington; tel. (4) 473-5387; fax (4) 473-6999; e-mail berryfcd@xtra.co.nz; 530 mems; Pres. JOHN GARELJA; Executive Officer PETER ENSOR.

New Zealand Council of Wool Exporters Inc: POB 2857, Christchurch; tel. (3) 353-1049; fax (3) 374-6925; e-mail cwe@woolexport.net; internet www.woolexport.net; f. 1893; Exec. Man. R. H. F. NICHOLSON; Pres. JOHN HENDERSON.

The New Zealand Forest Industries Council: POB 2727, Wellington; tel. (4) 473-9220; fax (4) 473-9330; e-mail info@nzfic.org.nz; internet www.nzfic.org.nz; CEO STEPHEN JACOBI.

The New Zealand Forest Owners' Association: POB 1208, Wellington; tel. (4) 473-4769; fax (4) 499-8893; e-mail robmcl@nzfoa.org.nz; internet www.nzfoa.nzforestryco.nz; CEO ROB MCCLAGAN.

New Zealand Fruitgrowers' Federation: Huddart Parker Bldg, POB 2175, Wellington, 6015; tel. (4) 472-6559; fax (4) 472-6409; e-mail hans@fruitgrowers.org.nz; internet www.fruitgrowers.org.nz; f. 1928; 5,000 mems; CEO P. R. SILCOCK.

New Zealand Fruit Wine and Cider Makers: POB 912, New Plymouth, Taranaki; tel. and fax (6) 757-8049; e-mail admin@fruitwines.co.nz; internet www.fruitwines.co.nz; 30 mems; Chair. BRIAN SHANKS; Exec. Officer CHRIS GARNHAM.

New Zealand Meat Board: Wool House, 10 Brandon St, POB 121, Wellington; tel. (4) 473-9150; fax (4) 474-0800; e-mail help@meatnz.co.nz; internet www.meatnz.co.nz; f. 1922; Chair. JEFF GRANT; Sec. A. DOMETAKIS; 13 mems.

New Zealand Pork Industry Board: POB 4048, Wellington; tel. (4) 385-4229; fax (4) 385-8522; e-mail info@pork.co.nz; internet www.pork.co.nz; f. 1937; Chair. C. TRENGROVE; CEO A DAVIDSON.

New Zealand Seafood Industry Council: Private Bag 24-901, Wellington; tel. (4) 385-4005; fax (4) 384-2727; e-mail info@seafood.co.nz; internet www.seafood.co.nz; CEO OWEN SYMMANS; Chair. DAVID SHARP.

New Zealand Timber Industry Federation: 2–8 Maginnity St, POB 308, Wellington; tel. (4) 473-5200; fax (4) 473-6536; e-mail enquiries@nztif.co.nz; internet www.nztif.co.uk; 350 mems; Exec. Dir W. S. COFFEY.

New Zealand Vegetable and Potato Growers' Federation (Inc): POB 10232, Wellington 1; tel. (4) 472-3795; fax (4) 471-2861; e-mail information@vegfed.co.nz; internet www.vegfed.co.nz; 4,000 mems; Pres. B. GARGIULO; CEO P. R. SILCOCK.

New Zealand Wool Board: 10 Brandon St, Box 3225, Wellington; tel. (4) 472-6888; fax (4) 473-7872; e-mail info@woolboard.co.nz; internet www.woolboard.co.nz; assists research, development, production and marketing of NZ wool; Chair. B. C. MUNRO; CEO MARK O'GRADY.

NZMP: 25 The Terrace, POB 417, Wellington; tel. (4) 462-8096; fax (4) 471-8600; internet www.nzmp.com; f. 2001; through merger of New Zealand Dairy Board and Fonterra Co-operative.

Registered Master Builders' Federation (Inc): 234 Wakefield St, Level 6, POB 1796, Wellington; tel. (4) 385-8999; fax (4) 385-8995; internet www.masterbuilder.org.nz; Chief Exec. C. PRESTON.

Retail Merchants' Association of New Zealand (Inc): Willbank House, 8th Floor, 57 Willis St, Wellington; tel. (4) 472-3733; fax (4) 472-1071; e-mail bhellberg@retail.org.nz; f. 1920 as the NZ Retailers' Fed. (Inc); present name adopted in 1987, following merger with the NZ Wholesale Hardware Fed.; direct membership 4,600; Pres. TERENCE DELANEY; CEO JOHN ALBERTSON.

EMPLOYERS' ORGANIZATION

Business New Zealand: Lumley House, Level 6, 3–11 Hunter St, Wellington; tel. (4) 496-6555; fax (4) 496-6550; e-mail admin@businessnz.org.nz; internet www.businessnz.org.nz; Chief Exec. SIMON CARLAW.

UTILITIES

Energy Efficiency and Conservation Authority (EECA): NGC Bldg, Level 1, 44 The Terrace, Wellington; tel. (4) 470-2200; fax (4) 499-5330; e-mail eecainfo@eeca.govt.nz; internet www.eeca.govt.nz.

Electricity

Bay of Plenty Electricity Ltd: 52 Commerce St, POB 404, Whakatane 3080; tel. (7) 922-2700; fax (7) 307-0922; e-mail bopelec@bopelec.co.nz; internet www.bopelec.co.nz; f. 1995; generation, purchase and supply of electricity and natural gas; Commercial Man. CHRIS POWER; CEO DAVID BULLEY.

Contact Energy Ltd: Harbour City Tower, Level 1, 29 Brandon St, Wellington; tel. (4) 449-4001; fax (4) 499-4003; e-mail investor.centre@mycontact.co.nz; internet www.contactenergy.co.nz; f. 1996; generation of electricity, wholesale and retail of energy; transferred to the private sector in 1999; Chair. PHIL PRYKE; CEO and Man. Dir STEPHEN BARRETT.

The Marketplace Co Ltd (M-CO): Wool House, Level 2, 10 Brandon St, POB 5422, Wellington; tel. (4) 473-5240; fax (4) 473-5247; e-mail info@m-co.co.nz; internet www.m-co.co.nz; administers wholesale electricity market; Chief Exec. CHRIS RUSSELL.

ON Energy Ltd: 10 Hutt Rd, Petone, Wellington; tel. (4) 576-8700; fax (4) 576-8600; e-mail transalta@transalta.co.nz; internet www.transalta.co.nz; f. 1993; fmrly Transalta New Zealand Ltd; retailer of electricity and gas; Chair. R. H. ARBUCKLE; CEO MURRAY NELSON; Chair. D. A. JOHNSTON.

Powerco: 84 Liardet St, Private Bag 2061, New Plymouth; tel. (6) 759-6200; fax (6) 758-6818; internet www.powerco.co.nz; operates power networks in Wairarapa, Manawatu, Wanganui and Taranaki; Chair. BARRY R. UPSON; CEO STEVEN R. BOULTON.

Transpower New Zealand Ltd: Unisys House, 56 The Terrace, POB 1021, Wellington; tel. (4) 495-7000; fax (4) 495-7100; internet www.transpower.co.nz; manages national grid; Chair. Sir COLIN MAIDEN; Chief Exec. RALPH CRAVEN.

TrustPower Ltd: Private Bag 12-023, Tauranga, Auckland; tel. (7) 574-4800; fax (7) 574-4825; e-mail trustpower@trustpower.co.nz; internet www.trustpower.co.nz; f. 1920 as Tauranga Electric Power Board; independent generator; Chair. HAROLD TITTER.

Vector Ltd: POB 99-882, Newmarket, Auckland; tel. (9) 978-7788; fax (9) 978-7799; internet www.vector-network.co.nz; fmrly Mercury Energy Ltd; Chair. WAYNE BROWN; CEO Dr PATRICK STRANGE.

Wairarapa Electricity Ltd: 316–330 Queen St, Masterton, Auckland; tel. (6) 377-3773; fax (6) 378-2347; f. 1993; owned by South Eastern Utilities Ltd; distribution of electric power; Chair. JOHN MATHESON; CEO A. LODGE.

Gas

Bay of Plenty Electricity Ltd: see Electricity, above.

Natural Gas Corpn: 10 Hutt Rd, Petone, Private Bag 39980, Wellington Mail Centre, Wellington; tel. (4) 576-8700; fax (4) 576-8600; internet www.natgas.co.nz; f. 1992; purchase, processing and transport of natural gas; wholesale and retail sales; Chair. G. J. W. MARTIN; CEO RICHARD J. BENTLEY.

Nova Gas Ltd: POB 10-141, Wellington; tel. (4) 472-6263; fax (4) 472-6264; e-mail hansv@novagas.co.nz; internet www.novagas.co.nz.

Water

Waste Management NZ Ltd: 86 Lunn Ave, Mt Wellington, Auckland; tel. (9) 527-8044; fax (9) 570-5595; f. 1985; water and waste water services; Chair. J. A. JAMIESON; Man. Dir K. R. ELLIS.

MAJOR COMPANIES

Construction and Cement

Golden Bay Cement Co Ltd: POB 1359, 2nd Floor, 308 Great South Rd, Greenlane, Auckland; tel. (9) 523-3050; fax (9) 520-2163; f. 1909; mfrs of cement; Gen. Man. C. BADGER; employees: 210.

Milburn New Zealand Ltd: POB 6040, Christchurch; tel. (3) 348-8509; fax (3) 348-3442; internet www.milburn.co.nz; f. 1888; sales $NZ195.3m. (1998), operating revenue $NZ249.9m. (2000/01); mfrs of cement and concrete, aggregate lime and associated products; Chair. J. C. R. MAYCOCK; Man. Dir R. R. WILLIAMS; 751 employees.

Food and Drink

DB Breweries Ltd: Citibank Centre, 23 Customs St East, Auckland 1; tel. (9) 377-8990; fax (9) 377-3871; e-mail db@db.co.nz; internet www.db.co.nz; f. 1966; fmrly DB Group Ltd; cap. and res $NZ128.1m., sales 278.4m. (2000/01); mfrs and distributors of beer, brewers, and bottlers; Chair. DAVID G. SADLER; Man. Dir BRIAN JAMES BLAKE; 500 employees.

Fonterra Co-operative Group Ltd: 103 Leonard Isitt Drive, Auckland; tel. (9) 256-5400; fax (9) 256-5419; internet www.fonterra.com; f. 2001; New Zealand Dairy Group (f. 1918) was replaced by a new org., Globalco, in 2001, which subsequently began trading under the above name; sales $NZ5,604m. (2001/02); collecting and processing of milk, production of dairy products, etc.; Chair. JOAN RADLEY; CEO CRAIG NORGATE; 20,000 employees.

Lion Nathan Ltd: Level 5, Tower 2, Shortland Centre, 55–65 Shortland St, Auckland; tel. (9) 303-3388; fax (9) 303-3307; internet www.lion-nathan.co.nz; f. 1923; cap. and res $NZ–334m., sales $NZ1,612m. (1999/2000); brewers, bottlers, wine and spirits merchants; Chair. DOUGLAS MYERS; CEO GORDON CAIRNS; Exec. Dir HIROTAKI KOBAYASHI; 3,792 employees.

Sanford Ltd: 22 Jellicoe St, Freeman's Bay, Auckland 1001; tel. (9) 379-4720; fax (9) 309-1190; e-mail info@sandford.co.nz; internet www.sanford.co.nz; f. 1904; cap. and res $NZ404.3m., sales $NZ354.9m. (2000/01); processing and distribution of seafood; Chair W. D. GOODFELLOW; Man. Dir E. F. BARRATT; 1,500 employees.

Forestry, Pulp and Paper

Carter Holt Harvey Ltd: Private Bag 92-106, 640 Great South Rd, Manukau City, Auckland; tel. (9) 262-6000; fax (9) 262-6099; internet www.chh.com; f. 1971; cap. and res $NZ4,700m., sales $NZ3,755m. (2000/01); activities include sawmilling, pulp forestry, tissue, paperboard and plastic packaging; Chair. W. J. WHINERAY; CEO CHRIS LIDDELL.

Steel

BHP-New Zealand Steel Ltd: Private Bag 92 121, Auckland 1; tel. (9) 375-8999; fax (9) 375-8959; e-mail colorsteel@bhp.com.au; internet www.bhp.com.au; f. 1966; cap. p.u. $NZ650m. (1998); suppliers of iron-sand concentrate; mfrs of hot- and cold-rolled steel, hollow sections, galvanized flat sheet and coil and pre-painted flat sheet and coil; share cap. $NZ464.7m., sales $NZ349.6m. (2000/01); Man. Dir J. HOWARD; 1,670 employees.

Steel and Tube Holdings Ltd: 15–17 Kings Cres., Lower Hutt; tel. (4) 570-5000; fax (4) 569-4218; internet www.steelandtube.co.nz; f. 1953; cap. and res $NZ130.3m., sales $NZ390.0m. (2000/01); holding company; 20 subsidiary companies; Chair. Dr R. L. EVERY; CEO N. CALAVRIAS; 1,042 employees.

Miscellaneous

Bendon Ltd: POB 1134, Auckland; tel. (9) 273-0850; fax (9) 273-0851; e-mail info@bendon.co.nz; internet www.bendon.co.nz; f. 1987; fmrly Ceramco Corpn Ltd; name changed as above 2000; cap. and res $NZ52.5m., sales $NZ89.2m. (1998/99); mfrs of underwear and lingerie; CEO STEFAN PRESTON; 744 employees.

Cavalier Corporation Ltd: 7 Grayson Ave, Papatoetoe, POB 97–040, South Auckland Mail Centre; tel. (9) 277-6000; fax (9) 278-7417; internet www.carcorp.co.nz; cap. and res $NZ63.0m., sales $NZ193.0m. (2002/03); carpet mfrs; Chair. A. M. JAMES; CEO and Man. Dir W. K. CHEUNG.

Donaghys Ltd: 179 Pages Rd, POB 15007, Aranui, Christchurch; tel. (3) 349-1735; fax (3) 349-1736; e-mail enquiries@donaghys.co.nz; internet www.donaghys.co.nz; cap. and res $NZ55.5m., sales $NZ116.4m. (1998/99); building materials; Chair. GRAEME J. MARSH; CEO JEREMY SILVA; Man. Dir ROSS A. CALLON; 650 employees.

Fernz Corporation Ltd: Level 38, Coopers and Lybrand Tower, 23–29 Albert St, Auckland; tel. (9) 379-3001; fax (9) 366-1394; internet www.fernz.com; f. 1916; fmrly NZ Farmers' Fertilizer Co; cap. and res $NZ283.4m., sales $NZ1,363.8m. (1997/98); mfrs and distributors of fertilizers, sulphuric acid, sulphate of alumina, chrome sulphate, agricultural chemicals, animal health products; 16 subsidiaries; Chair. W. WILSON; Man. Dir KERRY M. HOGGARD; 1,194 employees.

Fisher & Paykel Industries Ltd: POB 58550, Greenmount, Auckland; tel. (9) 273-0600; fax (9) 273-0538; e-mail customer.care@fp.co.nz; internet www.fisherpaykel.co.nz; f. 1934; cap. and res $NZ384.0m., sales $NZ932.8m. (2000/01); electricals; Chair. Sir COLIN MAIDEN; Man. Dir and CEO G. A. PAYKEL; 2,700 employees.

ICI Industrial Group: 8 Pacific Rise, Mt Wellington, Auckland 1006; tel. (9) 573-2700; fax (9) 573-2710; f. 1935; manufacturers and suppliers of industrial chemicals and explosives, synthetic resins, pharmaceuticals, plastics, dyestuffs, agricultural chemicals, polythene films and packaging, industrial and domestic paints; Chair. T. SILVERSON; CEO J. MORGAN; Man. Dir D. P. PIKE; 840 employees.

Montana Group (New Zealand) Ltd: Level 15, Tower II, The Shortland Centre, 55-65 Shortland St, Auckland; tel. (9) 307-2660; fax (9) 366-1579; e-mail admin@montanagroup.co.nz; internet www.montanawines.co.nz; fmrly Corporate Investments Ltd; cap. and res $NZ278.1m., sales $NZ449.6m. (1999/2000); production and export of wine; Exec. Chair. PETER H. MASFEN; 1,188 employees.

New Zealand Refining Co Ltd: Marsden Point, Private Bag 9024, Whangarei; tel. (9) 432-3811; fax (9) 432-8035; e-mail gsmith@nzrc.co.nz; internet www.nzrc.co.nz; f. 1961; refines petrol, diesel oils, fuel oils and bitumen; cap. and res $NZ326.8m., sales $NZ176.7m. (2001); Chair. IAN P. FARRANT; CEO ALAN P. DAVEY; 248 employees.

Richina Pacific Ltd: 11th Floor, 385 Queen St, Auckland; tel. (9) 375-2188; fax (9) 375-2104; e-mail richina@richinapacific.co.nz; internet www.richinapacific.com.nz; f. 1969; cap. and res $NZ69.8m., sales $NZ647.1m. (2002); construction, leather; Chair. Sir ALASTAIR MCCORMICK; CEO RICHARD C. L. YAN; 1,750 employees.

Sealord Group Ltd: Sealord House, Trafalgar St, Nelson 7001; tel. (3) 548-3069; fax (3) 546-0966; e-mail corporate@sealord.co.nz; internet www.sealord.co.nz; f. 1973; 50% owned by Treaty of Waitangi Fisheries Commission, 50% owned by Nippon Suisan Kaisha (Nissui) of Japan; catching, processing and marketing of seafood; sales $NZ581m. (2001); Chair S. JONES; 1,600 employees.

Tenon Ltd: Private Bag 92 036, Auckland 1030; tel. (9) 571-9800; fax (9) 571-9801; e-mail info@tenon.co.nz; internet www.tenon.co.nz; f. 1981 as Fletcher Challenge Ltd; name changed as above in 2004; reorganized as holding co in 1996; cap. $NZ558m., sales $NZ678m. (2002/03); operations in building, energy, forestry and paper; manufacture and distribution of building materials, incl. concrete, steel, plasterboard, wood-based panel products and aluminium extrusion, also commercial, industrial, civil and residential construction; petroleum exploration, production and distribution; plantation forest ownership and management, wood processing and distribution; manufacture of communications papers and speciality pulps; international company, with major operations in New Zealand, Australia, Canada, the USA, South America, Asia; Chair. TONY GIBBS; CEO JOHN DELL.

The Warehouse Group Ltd: 26 The Warehouse Way, Akoranga Drive, Northcote, Auckland; tel. (9) 489-7000; fax (9) 489-7444; e-mail investor@twl.co.nz; internet www.thewarehouse.co.nz; f. 1982; cap. and res $NZ346.7m., sales $NZ2,035m. (2002/03); general merchandise retailing; Chair. KEITH R. SMITH; Man. Dir IAN MORRICE; 8,005 employees.

TRADE UNIONS

In December 2000 a total of 134 unions were in operation; 318,519 workers belonged to a union.

New Zealand Council of Trade Unions: Education House, West Block, 178–182 Willis St, POB 6645, Wellington 1; tel. (4) 385-1334; fax (4) 385-6051; e-mail ctu@nzctu.org.nz; internet www.union.org.nz; f. 1937; present name since 1987; affiliated to ICFTU; 34 affiliated unions with 250,000 mems; Pres. ROSS WILSON; Sec. CAROL BEAUMONT.

Principal Affiliated Unions

Association of Staff in Tertiary Education (ASTE)/Te hau Takitini O Aotearoa: POB 27141, Wellington; tel. (4) 801-5098; fax (4) 385-8826; e-mail national.office@aste.ac.nz; 3,500 mems; Nat. Sec. SHARN RIGGS.

Association of University Staff (AUS): POB 11-767, Wellington; tel. (4) 915-6690; fax (4) 915-6699; e-mail national.office@aus.ac.nz; internet www.aus.ac.nz; 6,000 mems; Gen. Sec HELEN KELLY.

Central Amalgamated Workers Union (AWUNZ): POB 27-291, 307 Willis St, Wellington; tel. (4) 384-4049; fax (4) 801-7306; e-mail centralawunz@xtra.co.nz; Sec. JACKSON SMITH.

FinSec Finance and Information Workers Union: POB 27-355, Wellington; tel. (4) 385-7723; fax (4) 385-2214; e-mail union@finsec

.org.nz; internet www.finsec.org.nz/; Pres. SUE BORASTON; Sec. ANDREW CASIDY.

Maritime Union of New Zealand: POB 27004, Wellington; tel. (4) 385-0792; fax (4) 384-8766; e-mail trevor.hanson@munz.org.nz; 2,800 mems; Sec. TREVOR HANSON.

Meat and Related Trades Workers Union of Aotearoa: POB 17056, Greenlane, Auckland; tel. (9) 520-0034; fax (9) 523-1286; e-mail meat.union@xtra.co.nz; Sec. GRAHAM COOKE.

New Zealand Dairy Workers Union, Inc: POB 9046, Hamilton; tel. (7) 839-0239; fax (7) 838-0398; e-mail nzdwu@nzdwu.org.nz; internet www.nzdwu.org.nz; f. 1992; 6,400 mems; Sec. JAMES RITCHIE; Pres. JOHN SMITH.

New Zealand Educational Institute (NZEI)Te Riu Roa (Te Riu Roa): POB 466, Wellington; tel. (4) 384-9689; fax (4) 385-1772; e-mail nzei@nzei.org.nz; internet www.nzei.org.nz; f. 1883; Pres. COLIN TARR; Sec. LYNNE BRUCE.

New Zealand Engineering, Printing & Manufacturing Union (EPMU): POB 31-546, Lower Hutt; tel. (4) 568-0086; fax (4) 576-1173; e-mail andrew.little@epmu.org.nz; internet www.epmu.org .nz; Sec. ANDREW LITTLE.

New Zealand Meat Workers and Related Trades Union: POB 13-048, Armagh, Christchurch; tel. (3) 366-5105; fax (3) 379-7763; e-mail nzmeatworkersunion@clear.net.nz; 8,439 mems; Pres. J. REID; Gen. Sec. D. W. EASTLAKE.

New Zealand Nurses' Organisation: POB 2128, Wellington; tel. (4) 385-0847; fax (4) 382-9993; e-mail nurses@nzno.org.nz; internet www.nzno.org.nz; 32,000 mems; CEO GEOFF ANNALS.

New Zealand Post Primary Teachers' Association: POB 2119, Wellington; tel. (4) 384-9964; fax (4) 382-8763; e-mail gensec@ppta .org.nz; internet www.ppta.org.nz; Pres. JEN McCUTCHEON; Gen. Sec. KEVIN BUNKER.

NZ PSA (New Zealand Public Service Association): PSA House, 11 Aurora Terrace, POB 3817, Wellington 1; tel. (4) 917-0333; fax (4) 917-2051; e-mail enquiries@psa.org.nz; internet www.psa.org.nz; 40,000 mems; Pres. IAN BAMBER.

Rail & Maritime Transport Union Inc: POB 1103, Wellington; tel. (4) 499-2066; fax (4) 471-0896; e-mail admin@rmtunion.org.nz; internet rmtunion.org.nz; 3,497 mems; Pres. J. KELLY; Sec. W. BUTSON.

Service and Food Workers' Union: Private Bag 68-914, Newton, Auckland; tel. (9) 375-2680; fax (9) 375-2681; e-mail darien.fenton@ sfwu.org.nz; internet www.sfwu.org.nz; 23,000 mems; Pres. DAELE O'CONNOR; Sec. DARIEN FENTON.

Other Unions

Manufacturing & Construction Workers Union: Manners St, POB 11-123, Wellington; tel. (4) 385-8264; fax (4) 384-8007; e-mail M .C.Union@TradesHall.Org.NZ; Gen. Sec. GRAEME CLARKE.

National Distribution Union (NDU): 120 Church St, Private Bag 92-904, Onehunga, Auckland; tel. (9) 622-8355; fax (9) 622-8353; e-mail ndu@nduunion.org.nz; internet www.nduunion.org.nz; 18,500 mems; Pres. BILL ANDERSEN; Sec. MIKE JACKSON; Vice Pres. DENNIS DAWSON.

New Zealand Building Trades Union: Manners St, POB 11-356, Wellington; tel. (4) 385-1178; fax (4) 385-1177; e-mail nzbtu@ tradeshall.org.nz; Pres. P. REIDY; Sec. ASHLEY RUSS.

New Zealand Seafarers' Union: Marion Square, POB 9288, Wellington; tel. (4) 385-9288; fax (4) 384-9288; e-mail admin@seafarers .org.nz; f. 1993; Pres. DAVE MORGAN.

Transport

RAILWAYS

There were 3,912 km of railways in New Zealand in 2001, of which more than 500 km were electrified.

Tranz Rail Ltd: Wellington Station, Private Bag, Wellington 1; tel. (4) 498-3000; fax (4) 498-3259; e-mail info@tranzrail.co.nz; internet www.tranzrail.co.nz; f. 1990; transferred to private ownership in 1993; fmrly New Zealand Rail; 3,904 km of railway; also provides road and ferry services; Man. Dir MICHAEL BEARD.

ROADS

In June 2001 there were a total of 92,188 km of maintained roads in New Zealand, including 10,775 km of state highways and motorways.

Transfund New Zealand: BP House, 20 Customhouse Quay, POB 2331, Wellington; tel. (4) 916-4220; fax (4) 916-0028; e-mail reception@transfund.govt.nz; internet www.transfund.govt.nz; f. 1996; Crown entity responsible for allocating resources to ensure an integrated, safe, responsive and sustainable land transport system; Chair. Dr JAN WRIGHT; Chief Exec. WAYNE DONNELLY.

Transit New Zealand: 20–26 Ballance St, POB 5084, Wellington; tel. (4) 499-6600; fax (4) 496-6666; e-mail deborah.willett@transit .govt.nz; internet www.transit.govt.nz; Crown agency responsible for management and development of the state highway network; Chair. DAVID STUBBS; CEO Dr RICK VAN BARNEVELD.

SHIPPING

There are 13 main seaports, of which the most important are Auckland, Tauranga, Wellington, Lyttleton (the port of Christchurch) and Port Chalmers (Dunedin). In December 2002 the New Zealand merchant fleet comprised 173 vessels, with a total displacement of 180,435 grt.

Principal Companies

P & O Nedlloyd Ltd: 2–10 Customhouse Quay, POB 1699, Wellington; tel. (4) 462-4000; fax (4) 462-4162; e-mail g.f.quirke@ponl .com; internet www.ponl.com; world-wide shipping services; Man. Dir GARY F. QUIRKE.

Sofrana Unilines NZ Ltd: 396–404 Queen St, POB 3614, Auckland; tel. (9) 356-1407; fax (9) 356-1407; internet www.sofrana.co.nz; Chair DIDIER LEROUX; Man. Dir BENOIT MARCENAC.

Other major shipping companies operating services to New Zealand include Blue Star Line (NZ) Ltd and Columbus Line, which link New Zealand with Australia, the Pacific Islands, South-East Asia and the USA.

CIVIL AVIATION

There are international airports at Auckland, Christchurch and Wellington.

Civil Aviation Authority of New Zealand: Aviation House, 10 Hutt Rd, Petone, POB 31441, Lower Hutt; tel. (4) 560-9400; fax (4) 569-2024; e-mail info@caa.govt.nz; internet www.caa.govt.nz; Dir of Civil Aviation JOHN JONES.

Principal Airlines

Air Nelson: Private Bag 32, Nelson 7030; tel. (3) 547-8700; fax (3) 547-8788; e-mail john.hambleton@airnewzealand.co.nz; internet www.airnewzealand.com; f. 1979; owned by Air New Zealand; changed to present name 1986; operates services throughout New Zealand; Gen. Man. JOHN HAMBLETON.

Air New Zealand: Quay Tower, 29 Customs St West, Private Bag 92007, Auckland 1; tel. (9) 366-2400; fax (9) 366-2401; e-mail investor@airnz.co.nz; internet www.airnz.co.nz; f. 1978 by merger; privatized in 1989, but returned to state sector in late 2001; services to Australia, the Pacific Islands, Asia, Europe and North America, as well as regular daily services to 26 cities and towns in New Zealand; Chair. JOHN PALMER; Man. Dir and CEO RALPH NORRIS.

Tourism

New Zealand's principal tourist attractions are its high mountains, lakes, forests, volcanoes, hot springs and beaches. In 2003 New Zealand received 2,104,420 visitors. Receipts from tourism totalled $NZ6,141m. in 2002.

Tourism New Zealand: POB 95, Wellington; tel. (4) 917-5400; fax (4) 915-3817; e-mail reception@tnz.govt.nz; internet www .newzealand.com; f. 1901; responsible for marketing of New Zealand as a tourism destination; offices in Auckland, Wellington and Christchurch; 13 offices overseas; Chair. WALLY STONE; Chief Exec. GEORGE HICKTON.

Defence

In August 2003 the total strength of the regular forces was 8,610: army 4,430, navy 1,980 and air force 2,200. In addition, there were 8,600 regular reserves (army 4,420, navy 1,980, air force 2,200) and 2,660 territorial reserves (army 2,070, navy 370, air force 220). Military service is voluntary. New Zealand is a participant in the Five-Power Defence Arrangements with Australia, Malaysia, Singapore and the United Kingdom.

Defence Expenditure: Budgeted at $NZ1,202m. for 2002/03.

Chief of Defence Force: Air-Marshal BRUCE FERGUSON.

Chief of Army: Maj.-Gen. JERRY MATEPARAE.

Chief of Navy: Rear-Adm. DAVID IAN LEDSON.

Chief of Air Force: Air Vice-Marshal JOHN H. S. HAMILTON.

Commander Joint Forces New Zealand: Maj.-Gen. MARTYN J. DUNNE.

Education

Education in New Zealand is free and secular in state schools. It is compulsory for all children aged six to 16 years, although in practice almost 100% start at the age of five years. Budgetary expenditure on education by the central Government in 2000/01 was $NZ6,097m., and was projected at $NZ7,041m for 2001/02.

In October 1989 the central administrative body, the Department of Education, and the education boards were disestablished and replaced by six agencies, assigned responsibility for the administration of early childhood and compulsory education: the Early Childhood Development Unit, the Education Review Office, the Ministry of Education, the Special Education Service, the Parent Advocacy Council (later disestablished) and the Teacher Registration Board. In July 1990 three agencies to administer the post-compulsory sector were established: the Education and Training Support Agency, the New Zealand Qualifications Authority and Quest Rapuara, the Career Development and Transition Education Service. The most significant change in the delivery and administration of primary and secondary education consists in the participation of parents and the community, in partnership with teachers. Responsibility for the administration of primary and secondary schools, previously controlled by education boards and the regional offices of the Department of Education, was decentralized to boards of trustees of individual schools. Boards of trustees are accountable for attaining the objectives declared in their charter, and for expenditure made from bulk grants received from Government to administer institutions. Boards are required to report to the Education Review Office, which, in turn, reports directly to the Minister of Education.

PRE-SCHOOL PRIMARY AND SECONDARY EDUCATION

Local early childhood centres are maintained and controlled by voluntary associations to which the Government gives substantial assistance including grants and subsidies. Since 1 July 1990 all early childhood centres wishing to receive government funding have been required to hold a charter. Those not holding a charter must still be licensed. In July 2001 there were 171,333 children enrolled in early childhood education services.

In July 2001 there were 449,491 pupils enrolled in primary classes at 2,209 schools. Pupils may complete the eight-year primary course at a full primary school (or, in rural areas, an area school) or, as is the case for most pupils, they may proceed to an intermediate school for the final two years. An intermediate school is a centrally-located school, usually instructing 300–600 pupils between the ages of 11 and 13 years. In July 2000 58,852 pupils were enrolled at 133 of these schools, which, because of economies of scale, can provide specialist teachers and facilities not normally within the reach of primary schools.

All children are entitled to free secondary education until the end of their 19th year. In mid-2000 a total of 239,481 pupils received secondary education. The majority of pupils leave school at the end of their fourth year. The School Certificate examination is taken at the end of the third or fourth year of secondary school. Pupils who pass subjects in this examination may go on to a year in the sixth form. At the end of the sixth-form year they may obtain the Sixth Form Certificate. Pupils intending to go on to university usually spend a further year in the seventh form to obtain a Higher School Certificate which is awarded without examination and provides a higher bursary. They may also sit examinations for university bursaries and entrance scholarships.

RURAL EDUCATION

In order to give children in country districts the advantage of special equipment and the more specialized teaching of larger schools, the consolidation of the smaller rural schools has been undertaken wherever this is practicable. In certain cases boarding allowances are granted to pupils living in areas where there are no convenient transport services enabling them to attend school. In small rural districts area schools provide primary and secondary education for all pupils in the immediate vicinity, and education from the first to the seventh form is provided in larger districts in separate schools.

CORRESPONDENCE SCHOOL

This school, which had an enrolment of 8,119 in July 2000, serves students who cannot attend school because they live in remote areas, because of illness or for other reasons. It also provides courses for pupils who wish to study subjects not offered at their local school, and adults who do not have access to secondary-school classes.

POLYTECHNICS AND TEACHER TRAINING

Since the early 1980s vocational education and training has moved away from the secondary to the continuing education sector, with training formerly provided by technical high schools now provided by polytechnics. Polytechnics provide a diverse range of vocational education resources and cover an increasing number of subjects at various levels of specialization.

In 2001 there were 22 polytechnics, including the Open Polytechnic of New Zealand. This last is the largest polytechnic in the country, with 17,322 students in July 1998. In mid-2001 there were a total of 3,999 full-time equivalent staff and 87,965 students in the polytechnic system.

In mid-2001 there were 10,884 full-time and part-time students undergoing kindergarten, primary and secondary teacher training in four colleges of education. Most pre-school primary teacher trainees follow a three-year course. University graduates take a course lasting one or two years.

UNIVERSITIES

The eight universities are autonomous bodies with their own councils. Eighty per cent of the universities' funds are state-provided. About 20% of pupils leaving secondary school go to university. In mid-2001 there were 125,668 students enrolled at the universities. There was a teaching staff of 6,103 full-time equivalent teachers.

EDUCATION OF MAORI AND PACIFIC ISLAND STUDENTS

There has been a steady increase in the number of Maori children attending pre-school institutions, and Maori students make up about 20% of the primary and secondary school roll. The period of stay of Maori secondary pupils, however, remains considerably shorter than that for other pupils. In July 2000 there were 146,913 Maori children receiving primary and secondary education. Recognized play centres provided for 1,832 Maori and 324 Pacific Island children out of a total enrolment of 15,808 in July 2000, and kindergartens for 7,048 Maori and 3,437 Pacific Island children out of a total of 45,869. In addition, 8,921 Maori and 3,890 Pacific Island children regularly attended licensed childcare centres. Te Kohanga Reo ('Language Nests') were established by Maori in the early 1980s to provide an educational environment in which children can learn Maori language and Maori cultural values. Kohanga Reo are special-purpose childcare centres under the control of Te Kohanga Reo Trust (Inc.), and licences are approved by the trust. Small non-profit-making pre-school groups and Pacific Island play-groups receive special funding support from the Government, but are exempt from the need to hold a charter or licence. Pacific Island communities operate early childhood centres with an emphasis on language development, both in their indigenous language and in English, increasing parental knowledge in early childhood care and education and assisting a child's transition into the formal education system. Home-based programmes are also linked to these centres. In July 1999 there were 11,859 pupils enrolled at licensed Te Kohanga Reo, 524 children at developing Te Kohanga Reo and 2,948 at developing Pacific Island language groups.

There is a renaissance in the learning and teaching of Maori language and culture, in recognition of the reality that New Zealand is the only place in the world where the Maori culture can be preserved. In-service training courses are run for teachers of Maori students, universities and colleges of education offer Maori Studies courses, and advisers and teachers throughout the country are developing Maori Studies programmes for all pupils. There are increasing numbers of bilingual classes in primary and secondary schools, and 'total immersion' Maori language schools (Kura Kaupapa Maori) have also been established. By 1998 the Kura Kaupapa Maori totalled 60. The three wananga, tertiary institutions providing polytechnic and university level programmes with an emphasis on Maori language and culture, had an enrolment of 11,278 students in mid-2001.

In 1961 the Maori Education Foundation was set up to provide financial assistance to Maori students, and it has also been active in other areas, notably pre-school education. A Pacific Island Polynesian Education Foundation was also established in 1972.

Bibliography

GENERAL

The Dictionary of New Zealand Biography, Vols 1–5, 1769–1960 1990–2000; internet www.dnzb.govt.nz.

Dunn, Michael. *New Zealand Painting: A Concise History*. Auckland, Auckland University Press, 2003.

Jackson, Keith, and McRobie, Alan. *Historical Dictionary of New Zealand.* Auckland, Longman, 1996.

Kirkpatrick, Russell. *Bateman Contemporary Atlas: New Zealand: The Shapes of Our Nation.* Auckland, Bateman, 1999.

McKinnon, Malcolm (Ed.). *New Zealand Historical Atlas: Papatuanuku e Takoto Nei.* Wellington, David Bateman in association with Historical Branch, Department of Internal Affairs, 1997.

McLintock, A. H. (Ed.). *An Encyclopaedia of New Zealand.* Wellington, Government Printer, 1966.

Patterson, Brad and Kathryn (Eds). *World Bibliographical Series, Vol. 18, New Zealand.* Oxford, Clio Press, 1998.

Reed, A. W. *An Illustrated Encyclopedia of Traditional Maori Life*, revised edn. by Mikaere, Buddy. London, New Holland Publrs, 2003.

Statistics New Zealand. *New Zealand Official Yearbook*, 1990, 1993 and 2000 editions especially. Wellington, annually.

Sturm, Terry (Ed.) *The Oxford History of New Zealand Literature in English*, 2nd edn, Auckland, Oxford University Press, 1998.

Thorns, David, and Sedwick, Charles. *Understanding Aotearoa/New Zealand: Historical Statistics.* Palmerston North, Dunmore Press, 1997.

HISTORY

Bassett, Michael, and King, Michael. *Tomorrow Comes the Song: A Life of Peter Fraser.* Auckland, Penguin, 2000.

Belich, James. *Making Peoples: a History of the New Zealanders from Polynesian Settlement to the End of the Nineteenth Century.* Auckland, Allen Lane, the Penguin Press, 1996.

Paradise Reforged: A History of the New Zealanders from the 1880s to the Year 2000. Auckland, Allen Lane, Penguin Press, 2001.

Binney, Judith, Bassett, Judith, and Olssen, Erik. *The People and the Land: Te Tangata me Te Whenua: an illustrated history of New Zealand, 1820–1920.* Wellington, Allen and Unwin, 1990.

Binney, Judith (Ed.). *The Shaping of History: Essays from the New Zealand Journal of History 1967–1999.* Wellington, Bridget Williams Books, 2001.

Byrnes, Giselle. *Boundary Markers: Land Surveying and the Colonisation of New Zealand.* Wellington, Bridget Williams Books, 2002.

Chapman, Robert. *New Zealand Politics and Social Patterns.* Victoria University Press, 2001.

Gustafson, Barry. *His Way: A Biography of Robert Muldoon.* Auckland, Auckland University Press, 2000.

Ihimaera, Witi (Ed.). *Growing Up Maori.* Auckland, Tandem Press, 1998.

Ip, Manying (Ed.). *Unfolding History, Evolving Identity: the Chinese in New Zealand.* Auckland, Auckland University Press, 2003.

Kawharu H. (Ed.). *Waitangi: Contemporary Maori and Pakeha Perspectives on the Treaty.* Auckland, Oxford University Press, 1989.

King, Michael. *Death of the Rainbow Warrior.* Harmondsworth, Penguin, 1986.

The Penguin History of New Zealand. Auckland, Penguin, 2003.

Wrestling with the Angel: A Biography of Janet Frame. Auckland, Penguin, 2000.

McGibbon, Ian (Ed.). *The Oxford Companion to New Zealand Military History.* Auckland, Oxford University Press, 2000.

Mulgan, Richard. *Politics in New Zealand.* (3rd edn, updated by Peter Aimer.) Auckland, Auckland University Press, 2004.

Oliver, W. H. *Claims to the Waitangi Tribunal.* Wellington, Daphne Brassell Associates, 1991.

Orange, C. *The Treaty of Waitangi.* Wellington, Allen and Unwin/Port Nicholson Press, 1987.

Owen, Alwyn. (Ed.). *Snapshots of the Century.* Auckland, Tandem Press, 1998.

Reader's Digest. *New Zealand Yesterdays.* Auckland, David Bateman, 2001.

Rice, G. R. (Ed.). *The Oxford History of New Zealand*, 2nd edn. Auckland, Oxford University Press, 1992.

Rollo, Arnold. *New Zealand's Burning.* Wellington, Victoria University Press, 1994.

Salmond, Anne *Two Worlds: first meetings between Maori and Europeans 1642–1772.* Auckland, Viking, 1991.

*Between Worlds: Early Exchanges between Maori and Europeans 1773–1815.*Auckland, Viking, 1997.

Sharp, Andrew (Ed.). *Histories, Power and Loss: Uses of the Past—A New Zealand Commentary*. Bridget Williams Books, 2002.

Sinclair, Keith (Ed.). *The Oxford Illustrated History of New Zealand.* Auckland, Oxford University Press, 1990.

Sinclair, Keith. *A History of New Zealand*, revised edn with additional material by Raewyn Dalziel. Auckland, Penguin, 2000.

Trapeznik, Alexander. *Common Ground: Heritage and Public Places in New Zealand.* University of Otago Press, 2000.

Vowles, Jack, Aimer, Peter, Banducci, Susan, Karp, Jeffrey, and Miller, Raymond. (Eds). *Voters' Veto: The 2002 Election in New Zealand and the Consolidation of Minority Government.* Auckland, Auckland University Press, 2004.

Walker, Ranginui. *Ka Whawhai Tonu Mataou: Struggle without end.* Auckland, Penguin Books, 1990.

He Tipuna: The Life and Times of Sir Apirana Ngata. Auckland, Viking, 2001.

Ward, Alan. *An Unsettled History: Treaty Claims in New Zealand Today.* Wellington, Bridget Williams Books, 1999.

Wilson, J. (Ed.). *From the Beginning: the Archaeology of the Maori.* Auckland, Penguin Books, 1987.

Websites: www.nz.history.net.nz; and www.timeframes.natlib.govt.nz.

ECONOMY

Birks, Stuart. *The New Zealand Economy, Issues and Policies*, 3rd edn. Palmerston North, Dunmore Press, 1997.

Dalziel, Paul, and Lattemore, Ralph. *New Zealand Macroeconomy*, 3rd edn. Auckland, Oxford University Press, 1999.

Davey, Judith (Ed.). *Economic Policy, Social Aspects.* Wellington, New Zealand Institute of Policy Studies.

Douglas, Roger. *Unfinished Business.* Auckland, Random House, 1993.

Duncan, Ian, and Bollard, Alan. *Corporatisation and Privatisation: Lessons from New Zealand.* Auckland, Oxford University Press, 1994.

Easton, Brian. *In Stormy Seas—The New Zealand Economy since 1945.* Dunedin, University of Otago Press, 1997.

The Whimpering of the State: Policy after MMP. Auckland, Auckland University Press, 1999.

Fryer, Glenda, and Oldfield, Yvonne. *New Zealand Employment Relations.* Auckland, Addison Wesley Longman New Zealand Ltd, 1994.

Goldfinch, Shaun. *Remaking New Zealand and Australian Economic Policy: Ideas, Institution and Policy Communities.* Georgetown University Press, 2000.

Hawke, G. R. *The Making of New Zealand, An Economic History.* Cambridge University Press, 1985.

Hazledine, Tim. *Taking New Zealand Seriously: The Economics of Decency.* Auckland, HarperCollins, 1998.

Henderson, David. *Economic Reform, New Zealand in International Perspective.* Wellington, New Zealand Business Roundtable, 1996.

Kelsey, Jane. *The New Zealand Experiment—A World Model for Structural Adjustment.* Auckland, Auckland University Press/Bridget Williams Books, 1995.

Reclaiming the Future: New Zealand and the Global Economy. University of Toronto Press, 2000.

Massey, Patrick. *New Zealand—Market Liberalisation in a Developed Economy.* New York, St Martin's Press, 1995.

McKinnon, Michael. *Treasury: A History of the New Zealand Treasury.* Auckland, Auckland University Press, 2003.

Organisation for Economic Co-operation and Development. *New Zealand*. Paris, OECD, annual country surveys, 1985 and subsequent years.

Roper, B., and Rudd, Chris (Eds.). *State and Economy in New Zealand (Oxford Readings in New Zealand Politics)*. Oxford University Press, 1993.

Sandrey, R., and Reynolds, R. *Farming Without Subsidies*. Wellington, Government Printer, 1990.

Savage, J., and Bollard, A. *Turning It Around: Closure and Revitalization in New Zealand Industry*. Auckland, Oxford University Press, 1990.

THE PACIFIC ISLANDS

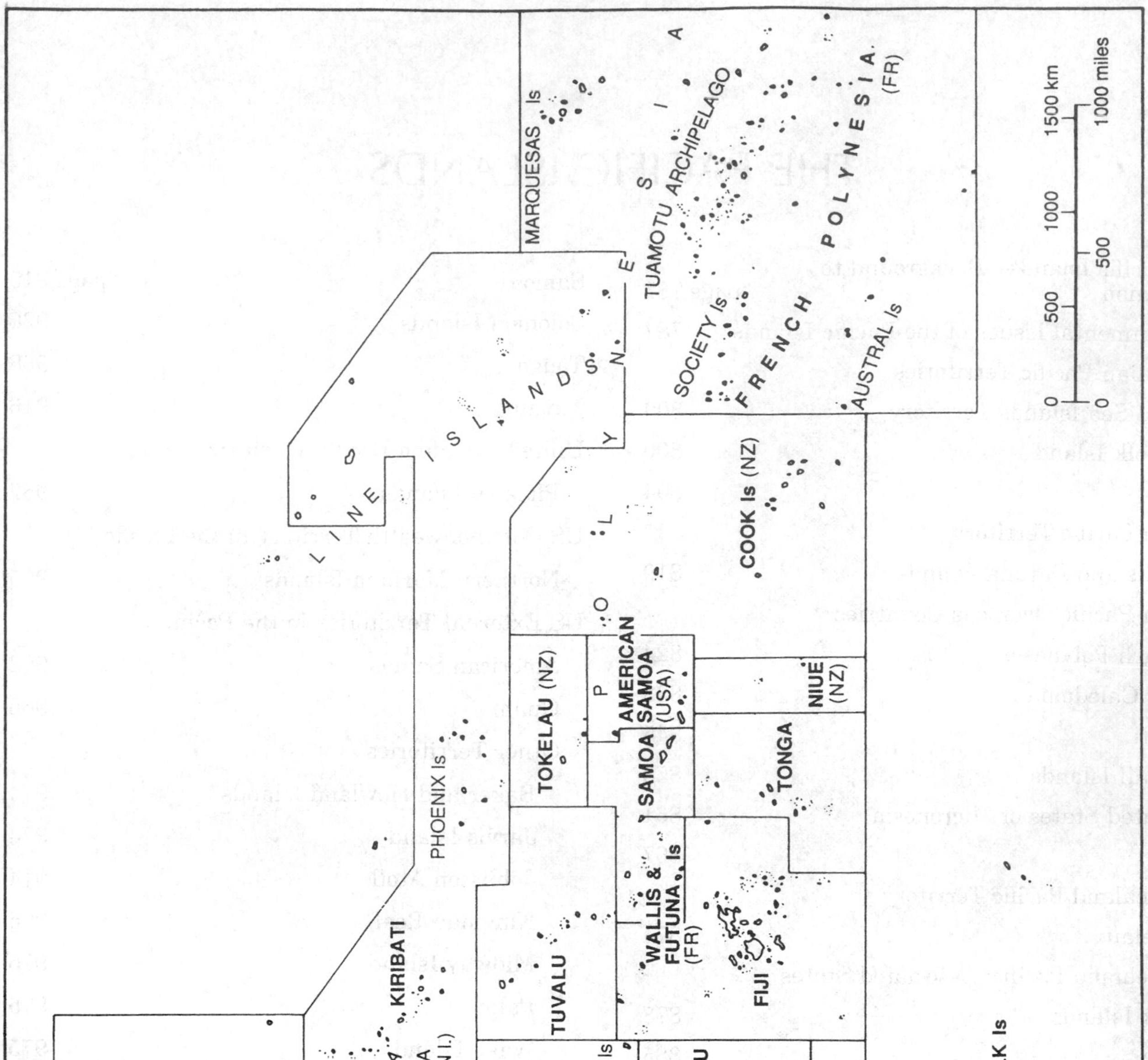

The Pacific Islands

THE PACIFIC ISLANDS

BACKGROUND TO THE REGION

BRYANT J. ALLEN

With subsequent revisions by JENNY J. BRYANT and the editorial staff

PHYSICAL AND SOCIAL GEOGRAPHY

The Pacific Ocean occupies one-third of the earth's surface. Within it are located many thousands of islands, more than in all the rest of the world's seas combined. The large number of the Pacific islands, and their widespread distribution, gives rise to a great variety of physical, social and economic environments. Their location relative to the continents and larger islands that border the Pacific, which include North and South America, Japan, China, the Philippines, Indonesia, Australia and New Zealand, continues to influence political and economic conditions in them. Their small size and physical isolation have rendered them vulnerable to influences from the rest of the world. Rapid and often traumatic ecological, social, economic and political changes have occurred throughout the Pacific following penetration by European and Asian explorers and colonists, a process that is still under way, as improved communications and the neo-colonialism of mining, investment and tourism bring the area closer to the modern world.

A number of broad classifications of Pacific islands exist. The islands may be divided into continental islands, high islands, low islands and atolls. The people of the Pacific may be divided into Melanesians, Polynesians and Micronesians. Melanesians occupy the larger islands in the south-west, Papua (formerly Irian Jaya) and Papua New Guinea, Solomon Islands, Vanuatu, Fiji and New Caledonia. Polynesians live on islands that are located over an immense area from Hawaii in the north to Easter Island in the south-east to New Zealand in the south-west. In the central Pacific, Polynesians occupy the major groups of Tonga, Samoa, the Society Islands including Tahiti, and the Cook Islands, as well as numerous small atolls. The Micronesians live in the north, central and west Pacific in the Mariana, Caroline, Marshall, Gilbert (Tungaru), Phoenix and Line groups.

Physical Features

No agreement exists on the origin of the Pacific basin. One hypothesis suggests that the basin is the result of shrinkage and depression of the earth's crust, while another argues that the gradual expansion of the earth, the emergence of new rocks on the ocean floors and the drift of continental land masses on great plates, has created the great oceans of the world, of which the Pacific is the largest.

The physical features of the Pacific basin are better known. The basin is 4–6 km deep and roughly circular in shape. The boundary is, in most places, the continental margin, but elsewhere it is obscured in a jumble of island arcs and fragmented continental blocks. The northern half of the basin forms one relatively deep unit measuring 5-6 km deep, and the southern half another shallower one. The north is characterized by a number of enormous volcanoes and numerous clusters of smaller ones. The crust here is broken by very long faults. The south is deformed by a series of very long broad arches or rises with associated block and wrench faulting. Island arcs and deep trenches occur along the margins of the basins and parallel to them, archipelagos of volcanic islands and clusters of submarine volcanoes occur in all parts of the basins but most are in the west and south-west.

These structures give rise to a number of characteristic island types. West of the so-called 'andesite-line', representing the furthest eastward limit of the continental blocks of Asia and Australia, are islands formed on the broken edges of the continental blocks. These continental islands have foundations of ancient folded and metamorphosed sediments which have been intruded by granites. Vulcanism has overlaid these rocks with lavas, tuff and ash, and transgressions by the ocean have laid down softer and younger marine sediments. Erosion has resulted in plains, deltas and swamps along the modern coastline. New Guinea (comprising Papua New Guinea, Papua and surrounding islands) is the best example of these continental islands; it is dominated by a massive central cordillera within which lie dissected and flat-floored montane valleys. The highest peak in the island is over 5,000 m. Active volcanoes exist along the north coast and in the New Guinea islands. North and south of the central mountains are broken hills and vast swamps. The coastal pattern is one of small coastal plains alternating with low river terraces, high marine terraces, coastal hills and steep mountain slopes plunging straight into the sea. The largest rivers of all the Pacific islands are found here. The River Fly is navigable by motorized vessels for about 800 km and the Sepik for 500 km. Other continental islands are Fiji, Solomon Islands, Vanuatu and New Caledonia.

The high islands of the central Pacific are composed almost entirely of volcanic materials, together with reef limestone and recent sediments. The islands are the peaks of the largest volcanoes in the world. The Hawaiian volcano of Mauna Loa, for example, rises 9 km from the ocean floor and is over 200 km in diameter. Characteristic landforms of the high islands are striking peak and valley forms, with old volcanic cores often eroded to form fantastic skylines. Waterfalls, cliff faces and narrow beaches, with fringing coral reefs, complete the pattern. High islands in the Pacific include the Samoas, Tahiti and the Marquesas, Rarotonga in the Cook Islands, Pohnpei in the Eastern Carolines and the Northern Mariana Islands.

Low islands are of two types: some are volcanic islands which have been eroded, while others are raised atolls, which resemble sea-level reefs, but which are now elevated above modern sea-level. Caves and sinkholes occur widely. Small pockets of soil occur within the limestone rocks. Surface water is uncommon. Examples of low volcanic islands include Aitutaki, in the Cook Islands, and Wallis Island. Raised coral islands include some of the islands of the Tuamotu, Society, Cook, Line, Tokelau, Marshall, Caroline and Kiribati groups. Low islands with raised reefs are also common. One of the best examples is Mangaia in the Southern Cook group, which has a central core of volcanic rock 180 m high surrounded by an unbroken 1 km-wide band of coral limestone raised 70 m above the present sea-level. A new fringing reef now surrounds the island.

The atolls are the fourth island form, roughly circular reefs of coral limestone, partly covered by sea water, on which there are small islands made up of accumulations of limestone debris, and within each of which there occurs a lagoon of calm water. Atoll islets are commonly less than 3 m above the high-tide level. It is generally agreed that atolls have developed on the tops of volcanoes, which now no longer protrude above sea-level. Atolls vary in size from Rose Island (American Samoa), an atoll about 3 km by 3 km, to the Kwajalein Atoll in the Marshall Islands which is over 60 km long. Sources of fresh water are rain and a freshwater lens which is frequently found floating on salt groundwater beneath the islets. Hurricanes and typhoons frequently sweep over atolls, partially or completely submerging islets.

Climate

Five atmospheric circulation regions have been identified in the Pacific. A middle latitude area is characterized by the occurrence of extra-tropical cyclones with characteristic distinctive frontal weather systems. The Marianas sometimes receive this type of weather in the northern winter. The trade winds regions, where at least 60% of prevailing winds are from the north-west in the northern hemisphere and the south-east in the southern hemisphere, lie in an arc from the west coast of Mexico through Hawaii to the Marshall Islands in the north, and from the west coast of South America across the Marquesas and Tuamotus to the Society Islands. In these areas distinct wet and dry zones appear on larger islands. The monsoon area occurs to the far west and influences few of the Pacific islands. The weather of Papua New Guinea is influenced, however, and a wet season and a dry season are distinguishable, although they are by no means as sharp as the term 'monsoon' implies. A doldrums area occurs in a poorly-defined band south of the Equator in an arc extending east from Solomon Islands to the Phoenix Islands. Finally, a hurricane zone exists in the northern Pacific in an arc extending west from Panama and including the Marshall, Caroline and Mariana Islands. A similar zone occurs in the south extending from the Tuamotus west across the Cooks, the Samoas, Tonga and Fiji to the north-east Australian coast. Serious cyclonic storm damage is sometimes incurred. Cyclone Ofa, one of the most severe of the century, occurred in February 1990, causing devastation in Niue, Samoa, Tokelau and Tuvalu.

Rainfall in the Pacific is geographically most variable; some islands are semi-arid while others are very wet. In the northern Pacific, for example, Midway receives a mean annual rainfall of 1,194 mm and Honolulu 550 mm. Further south Yap receives 3,023 mm, Palau 3,900 mm, Pohnpei 4,700 mm and Fanning 2,054 mm. Islands in the eastern Pacific near the Equator are frequently barren. Rainfall decreases from west to east along the Equator: Nauru receives 2,050 mm, Ocean Island (Banaba) 1,930 mm and Christmas Island (Kiritimati) 950 mm. Further south in an arc extending east from New Guinea to the Society Islands in the Central Pacific average annual rainfall varies from between 3,500 mm in the west to 2,000 mm in the east. In Papua New Guinea altitude and local relief influence climate. Areas exposed to the north-west and south-east winds receive over 5,000 mm of rain, while inland areas cut off from moist air masses may receive less than 1,500 mm. On the south-eastern coast, east and west of Port Moresby, average annual rainfall is less than 1,000 mm.

The predictions of climate change and the possibility of rising sea-levels could have devastating consequences for the low-lying islands of the Pacific. Even a half-a-metre rise in sea-level would mean the flooding of territories such as the Tokelau group of atolls, Tuvalu, the Marshall Islands and parts of Kiribati. On high islands, arable low-lying land would also be lost. The effects of this on population migration, both within and outside the Pacific, need to be planned for. In addition, the salinization of soil and drinking water, the changing environment for the practice of subsistence agriculture, the likelihood of more frequent and intense cyclones, the pressure on land, and consequently the political implications for the region, constitute major threats to the Pacific islands.

Soils, Minerals, Vegetation

Geology, soil, altitude, land-forms, location and climate are all combined in the creation of the widely varying physical environments of Pacific islands. Continental islands exhibit the widest range of environments, from high alpine grasslands, through montane forest and lowland rain forest to savannah and mangrove swamps. They also contain the richest deposits of minerals: nickel, chrome and manganese are mined in New Caledonia, gold and copper in New Guinea. The high volcanic islands contain no known minerals of commercial value. A rich source of minerals with potential for future exploitation are the deep-sea manganese nodules (containing copper, nickel and cobalt), which have been found in the north and south-west Pacific. Terrain is frequently a limit to cultivation, although soils are in general heavily leached and of low fertility, with low mineral and humus content. Raised coral islands lack groundwater and soils are shallow and often scattered in pockets. Phosphate deposits are mined from the coralline limestone on Nauru and Banaba (Ocean Island), and were formerly mined on Makatea in the Society Islands. The atolls contain only sparse resources. Soil development is often nil, fresh water difficult to obtain and foodplants other than coconuts and pandanus nuts difficult to cultivate. Special techniques are used to cultivate taro, but storms or high sea-levels frequently destroy gardens. The atolls provide the most tenuous existence for man in the Pacific.

The flora and fauna of the Pacific islands are unbalanced in comparison with those of the continents in that many major categories of plants have not reached the islands. A few ancestral immigrants have given rise to the entire endemic biota. Lack of ecological competition appears to have resulted in genera developing many more species than plants on continental land masses have been able to produce.

Prehistory, Culture and Early Society

The continental and oceanic Pacific islands were never linked by land bridges to the Asian continent and the Indonesian islands east of Bali and west of New Guinea form a frontier zone between a realm of placental mammals and marsupial mammals, the Wallace Line. Many palaeanthropologists argue therefore that man, a placental mammal, is an intruder in the Pacific. The first men to immigrate across the Wallace Line are believed to have been *Homo sapiens* approaching the modern form. The people of the interior of New Guinea are classified as Australoid populations, which are thought to have begun moving into the area more than 40,000 years ago. Recent archaeological discoveries in Papua New Guinea have been dated as 38,000 years old, and evidence of agricultural activity more than 9,000 years ago exists in the central highlands. Archaeological evidence indicates that the Pacific islands east of the Bismarck Archipelago were devoid of human settlement until about 3000 BC. Between 4,000 and 2,000 years ago, people who are thought to have lived in north-eastern Indonesia and the Philippines, and who had descended from a Mongoloid stock, spread into the Pacific and along the coasts of the continental islands, intermarrying with the existing Australoid populations of eastern Indonesia and New Guinea. Modern Melanesians, Polynesians and Micronesians are thus, to varying degrees, the outcome of the mixing of these early Australoid and Mongoloid stocks.

Therefore, the Melanesians who inhabit the island chains from Solomon Islands eastwards to Fiji are basically an Australoid group, while the Fijians are a more intermediate group. Polynesians tend towards the Mongoloid end of the continuum and Micronesians more so. The actual pattern, however, is far more complex than this simple description.

Origins of the three groups may also be evidenced in their cultures. The Polynesians are culturally and linguistically the most homogenous. Polynesian societies are basically patrilineal and genealogically ranked, with elaborate hierarchical systems of rank and class, best developed on the Hawaiian, Tongan and Society Islands. Micronesian societies are mainly matrilineal, with the exception of those in Yap and Kiribati. Melanesia is culturally the most diverse area of all. Hereditary ranking occurs in Fiji, but in many areas, especially in Papua New Guinea, status is achieved rather than inherited. Most groups are patrilineal, but matrilineal societies occur in New Guinea, Solomon Islands and Vanuatu.

Throughout the Pacific, the pre-contact subsistence economy was based on the vegetative propagation of root and tree crops, together with fishing and some pig husbandry and hunting. The only domesticated animals were dogs, pigs and fowls, but all three were not present everywhere in the region. The major root crops, taro and yam, have Asian origins, but one, the sweet potato, which was grown in New Guinea and in the Hawaii, Marquesas, Society and Easter Island groups prior to European contact, has a South American origin. Shifting cultivation was the main agricultural technique in most areas, although the intensity of land use and the periodicity of cycles varied widely in relation to population densities. In New Caledonia and parts of Polynesia, notably Hawaii, Tahiti and the Cook Islands, taro was cultivated in relatively elaborate, terraced, irrigated gardens.

Short-distance ocean voyaging was well-established in Polynesia and Micronesia before European contact, with large double-hulled canoes and navigation based on stars, wave patterns, bird flights and inherited geographical knowledge. Large ocean-going and coastal canoes were also used in Papua New Guinea and Fiji.

About 1,200 different languages, and many more dialects, are spoken in the Pacific islands, some 750 being found in Papua New Guinea alone. They belong to two groups, the non-Austronesian phyla found in Papua New Guinea (and in scattered pockets in Indonesia), and the Austronesian phyla, which are spoken in coastal Papua New Guinea, most of island Melanesia, all of Polynesia and Micronesia (as well as in parts of Indonesia, the Philippines, South-East Asia and Madagascar).

To summarize, existing evidence suggests that Papua New Guinea was already settled over 30,000 years ago by ancestral Australoid populations who were followed about 3,000 years ago by Austronesian speakers of Mongoloid stock who probably brought pottery, horticulture and pigs to Papua New Guinea. Intermixing occurred, followed by further movements east to New Caledonia and Vanuatu. Fiji was then settled by people who carried with them a pottery technology previously established in Papua New Guinea and islands to the east, and further movements into the Pacific Ocean took place. During the last 2,500 years further intermixing has occurred in Melanesia while Polynesian and Micronesian populations have had less interaction.

Population and Socio-Economic Characteristics

The 22 countries and territories of the region encompassed by the Pacific Community had an estimated population of 8.6m. at mid-2004; some 67% of that total were found in Papua New Guinea, which includes the eastern part of New Guinea, the second largest island in the world. The islands of Melanesia together constituted some 86.4% of the region's total population, while Polynesia forms 7.4% and Micronesia 6.2%. The combined population of the seven smallest island countries and territories totalled some 51,800 in mid-2004, equivalent to no more than the joint population of Vanuatu's two towns, Port Vila and Luganville. It was predicted that the population of the region would double between 2004 and 2032 if current growth rates continued. The 22 countries and territories occupy about 550,000 sq km of land in 30m. sq km of ocean.

The rate of increase in the population of Pacific nations is near the average for developing countries (annual average growth rates of 2.7% were recorded in 1990–2003), but there are variations among states. For example, during the period 1990–2003 the population of the Northern Mariana Islands increased by an annual average of 4.4%, while on Niue a decline of 2.4% was recorded. In many other islands with more moderate population growth rates, natural increase is partly offset by out-migration to the Pacific rim nations of New Zealand, Australia and the USA. A feature of the Polynesian Pacific, in particular, is that about 300,000 islanders reside in the metropolitan rim countries, where they seek employment, education and generally better standards of living.

Pacific countries are becoming increasingly urbanized. In the early 2000s more than 50% of the population in 11 out of 22 island countries and territories lived in urban areas, with urban population growth rates of 8.2% recorded in both American Samoa and the Marshall Islands. The associated problems of overcrowding, housing shortages, the widening economic gap between rural and urban areas, environmental pressures and the need for economic diversification faced all Pacific nations at the beginning of the 21st century.

Many of the Pacific countries and territories have a narrow resource base and limited arable land. Those that do possess more diverse resources, such as minerals and the possibilities of plantation agriculture, have frequently suffered over-exploitation of resources, political turmoil and dislocation of the populations. The susceptibility of the small island nations to cyclones, droughts or the possible impacts of climatic change and rising sea-levels make them particularly vulnerable. Future sustainable and carefully-managed development of the Pacific nations will need to involve a considerable degree of regional co-operation and co-ordinated policy, particularly as regards access to land and resources. Cordial relations with the metropolitan powers on the Pacific rim, to mitigate the impact of international economic and political changes, will also need to be maintained if the countries are to succeed as economically-viable nations. Although many of the countries are very small, some have extremely large exclusive economic zones (EEZs). For example, the EEZ of Kiribati, with a population of about 88,800 in 2003, covers 3,550,000 sq km of ocean, which, although used for fishing by local people, is also subject to major and lucrative fishing agreements with Pacific rim powers. Throughout the Pacific the ocean floor is being explored for mineral resources (fuel and manganese).

Agriculture and Employment

Most Pacific countries have depended on mixed subsistence cultivation of coconuts, root crops (taro, sweet potatoes and cassava) and a large number of other tree and leaf crops. Cash-cropping has become a major part of the agricultural sector, with the introduction of oil palm, coffee, cocoa, rice, sugar cane, pumpkins, ginger and nuts, as well as commercial exploitation of crops that are also grown for subsistence purposes, such as taro, pineapples and papayas (pawpaws). Pacific island nations are increasingly dependent upon imported foodstuffs, and the declining nutritional levels of the population reflect this change.

The majority of the Pacific islands' labour force continues to be employed in the agricultural sector, with the next largest employer being the government/administrative sector. Service employment is increasing, particularly in Fiji, where tourism has overtaken sugar in the provision of the nation's gross domestic product. Manufacturing employs only a small proportion of the Pacific labour force, although this is increasing as countries move to industrialize.

Income, Education and Literacy

The average annual income per caput of Pacific islanders in the mid-1990s was around US $1,000. The phosphate island of Nauru, however, had an average of some US $7,000 during this period (although it subsequently declined significantly), while in Kiribati the annual income per caput was only about US $800 by the early 2000s. French and US territories are generally more prosperous and well provided with services, although they are experiencing increasing social problems (annual average income per head in French Polynesia and New Caledonia was some US $16,000 in 2000). Although levels of primary school enrolment and literacy appear to be high in the Pacific (with more than 90% of the adult population being literate and more than 80% of the relevant group attending primary school in many island countries and territories), there is evidence that these ratios have been declining in recent years. Moreover, Papua New Guinea, the most populous country in the region, had an adult literacy rate of only 65% in 2002, while in Vanuatu the figure was as low as 34%. It was believed that the increasing inequalities could, in part, be a result of rapid urban development, decreases in world commodity prices, the shift towards more industrial-based economies and the declining ability of governments to satisfy demands for more urban services. The economy of the region as a whole was forecast to increase by 2.5% in 2003 and by 2.7% in 2004.

POLITICAL SITUATION

The islands may be divided into politically dependent and independent states. Dependent states governed wholly or partially by colonial administrations include the Pitcairn Islands (the United Kingdom); and Wallis and Futuna (France). Other dependent states are internally self-governing and include Tokelau, the Cook Islands and Niue (New Zealand); Norfolk Island (Australia); and the Northern Mariana Islands, Guam and American Samoa (USA).

Former dependencies that have achieved full political independence are Samoa, formerly Western Samoa (from New Zealand in 1962); Nauru (from the UN and Australia in 1968); Tonga (from the United Kingdom in 1970); Fiji (from the United Kingdom in 1970); Papua New Guinea (from Australia in 1975); Tuvalu (from the United Kingdom in 1978); Solomon Islands (from the United Kingdom in 1978); Kiribati (from the United Kingdom in 1979); and Vanuatu (from the United Kingdom and France in 1980). The Federated States of Micronesia and the Marshall Islands achieved independence from the USA when the Compact of Free Association took effect in 1991. Palau, the last remaining component of the Trust Territory of the Pacific Islands, achieved independence under a similar agreement in October 1994. The political future of the region's dependent territories was considered in a UN report on decolonization, compiled in 1996, which particularly focused on the status of New Caledonia, Tokelau and Guam. In the late 1990s there were moves towards full autonomy by both New Caledonia and French Polynesia. In 1998 France and New Caledonia signed the Nouméa Accord, which provided for the gradual transfer of powers to local authorities over the next 15 to 20 years. As part of these changes, in 1999 New Caledonia acquired the status of an 'overseas country'. French Polynesia acquired similar status in 2004. Papua (formerly Irian Jaya), the western half of the island of New Guinea, is a province of Indonesia and is not considered in this section; Hawaii, a state of the USA, New Zealand and the Philippines are also excluded.

REGIONAL ORGANIZATIONS

In 1947 the South Pacific Commission (SPC) was established by the United Kingdom, France, the USA, New Zealand, Australia and the Netherlands. Its aim was to 'promote the economic and social welfare and advancement of the peoples of the region', although commentators feel that it was more a move by Australia and New Zealand to extend their power and to deter others from obtaining a foothold, and it was obvious that organizations such as the SPC served the interests of the metropolitan powers.

In 1950 the South Pacific Conference was established as a separate body for the island nations. The moves towards independence for the island nations in the 1960s and 1970s challenged the role of the South Pacific Conference, which was still controlled by the metropolitan powers. This was most evident in the debates over nuclear testing in the Pacific during 1970, when the French delegation walked out of a meeting of the South Pacific Commission. In 1971 the South Pacific Forum (SPF) was established.

In May 1997 members of the South Pacific Commission voted to change the name of the organization to reflect the expanded membership of the group. The change to Pacific Community was approved at a meeting of the South Pacific Conference which took place in October 1997, and became effective in February 1998.

The Pacific Islands Forum

The Pacific Islands Forum (PIF—as the SPF was restyled in 2000) has become the most important regional organization in the Pacific region, enabling Pacific nations to work towards and foster regional responses to a number of economic, social and political issues which, because of the isolation, distance and small populations of the Pacific nations, might otherwise be ineffectively managed. The PIF comprises the heads of government of the independent and self-governing island countries, as well as Australia and New Zealand. The Forum is a political body, which can raise and discuss any issue. The organization has an administrative, research and development arm, the Pacific Islands Forum Secretariat (called the South Pacific Bureau for Economic Co-operation (SPEC) until 1988 and the South Pacific Forum Secretariat until 2000).

The 31st summit meeting in October 2000 in Kiribati saw the official launch of the organization's new title of 'Pacific Islands Forum', which aimed to reflect the group's expanding membership. The 34th summit meeting, held in Auckland, New Zealand, in August 2003, was overshadowed by problems in Solomon Islands and the imminent deployment of an Australian-led regional intervention force to that country. At the Auckland summit an Australian national, Greg Urwin, was elected to the post of secretary-general for the first time in the organization's history. His appointment provoked controversy among some commentators in the region, particularly given Australia's role in the intervention in Solomon Islands and its Government's stated willingness to take similar action where it believed its own or the region's security to be under threat. The 35th summit meeting was to be held in Niue in August 2004, but following the devastation wreaked on the island by Cyclone Heta in January of that year the venue was changed to Apia, Samoa. (See Regional Organizations for further details of PIF membership and activities.)

Other Agreements

In 1981 a trade agreement was signed by Australia, New Zealand and the SPF countries. Known as the South Pacific Regional Trade and Economic Co-operation Agreement (SPARTECA), it was designed to ease import restrictions on island goods into Australia and New Zealand, in an attempt to adjust the massive trade imbalance that exists. On 1 January 1987 new terms of trade came into effect, whereby all duties and quotas on island imports were abolished, with the exception of those on garments, footwear, sugar, steel and motor vehicles. The disparities in size and economic power have presented an obstacle to SPARTECA. Fiji is generally considered to be benefiting from regional trade agreements to a greater extent than other, smaller nations which have fewer export resources.

In 1986 a five-year fisheries treaty was signed between the USA and members of the South Pacific Forum Fisheries Agency, bringing to a close a long-running dispute which had involved the capture and confiscation of US fishing boats by Papua New Guinea and Solomon Islands. The treaty guaranteed royalties of some US $60m., to be paid over a five-year period, and was renegotiated for a further 10 years in 1992. The fisheries treaty developed out of the lack of return Pacific countries were receiving for tuna taken by distant nations from their waters. The total access fees paid to all countries by the fishing nations were on average less than 3% of the annual market value of tuna caught in the region (which in the early 1990s was valued at some US $1,200m.). In June 1997 the members of the South Pacific Forum Fisheries Agency, together with representatives from several major fishing nations, signed the Majuro Declaration, which contained a commitment to introduce measures for the conservation and management of the region's migratory fish stocks. The South Pacific region was thought to be the only major fishing area in the world not to have adopted such an agreement already. A report by the World Bank, published in 1999, claimed that overfishing posed a serious threat to many Pacific island countries. A new organization, the Western and Central Pacific Fisheries Commission, was established in June 2004, following several years of regional negotiations. The commission introduced a new tuna conservation convention, which aimed to promote the responsible fishing of shared tuna stocks and to end unregulated fishing on the high seas. the convention had been ratified by 14 countries by mid-2004.

During 1986 and 1987 continued discussions were held on the implications of the South Pacific Nuclear-Free Zone (SPNFZ) treaty, signed in 1985 by Australia, New Zealand, the Cook Islands, Fiji, Kiribati, Niue, Tuvalu and Western Samoa (now Samoa). (Vanuatu and Solomon Islands had refused to sign, on the grounds that the treaty's provisions were not far-reaching enough.) The treaty designates a 'nuclear-free zone' in the South Pacific, and imposes a ban on the manufacture, testing, storage and use of nuclear weapons, and the dumping of nuclear waste, in the region. Both the USSR and the People's Republic of China signed protocols attached to the treaty, but in 1987 the Governments of the USA and the United Kingdom rejected the South Pacific signatories' request for their endorsement of the treaty, a decision that provoked expressions of anger and concern from member countries of the SPF. Widespread protests against the French Government's programme of nuclear testing on Mururoa Atoll, in French Polynesia, continued in the early 1990s. In 1993 the French Government decided to continue the moratorium on testing, introduced in the previous year, for an indefinite period. However, shortly after his election in May 1995, President Chirac announced that France would resume nuclear testing, with a programme of eight tests (subsequently curtailed to six) commencing in September 1995. The decision was condemned for its apparent disregard for regional opinion, as well as for undermining the considerable progress made by Western nations towards a world-wide ban on nuclear testing. After its final test in January 1996, France signed the SPNFZ treaty, together with the United Kingdom and the USA, and was reinstated as a dialogue partner to the SPF (following its earlier suspension).

The establishment of the Joint Commercial Commission in Hawaii in 1992 created greater opportunities for private-sector trade between the USA and the countries of the Pacific region.

At a meeting in Hawaii in early 1996 representatives of the countries and territories of Micronesia formed the Council of

Micronesian Government Executives. The new body aimed to facilitate discussion on economic developments in the region and to examine possibilities for reducing the considerable cost of shipping essential goods between the islands. In early 1997 the Council established a technical working group on fisheries management in the region.

Under the Cotonou Agreement, concluded in June 2000 by the EU and ACP states as a successor arrangement to the Lomé Convention, the eight Pacific ACP countries and the EU were to commence negotiations in September 2002 on a regional economic partnership agreement; this was to enter into effect by 2008 at the latest and would allow the EU increased access to the Pacific markets. However, representatives of the island nations were concerned that this could have implications for their countries' economies, as many of them are dependent on customs duties for their internal revenue.

In mid-2003 the Australian Minister for Foreign Affairs, Alexander Downer, announced the initiation of discussions with regional leaders, aimed at the eventual establishment of a free-trade agreement and possible common market. A report published by the ANZ Bank in February 2004 expressed strong support for the establishment of a free-trade area in the region. The Pacific Island Countries Trade Agreement, launched shortly before the report, provided for the gradual introduction of such an area over the following 10 years.

SPREP

Of increasing significance in the region is the South Pacific Regional Environment Programme (SPREP). SPREP grew out of concern over environmental problems in the region voiced by the South Pacific Conference in the 1960s. In 1982 the South Pacific Commission, with support from the United Nations Environment Programme's Regional Seas Programme and the regional Governments, established SPREP at an intergovernmental conference held in Rarotonga, Cook Islands. SPREP operates on the basis of a Convention and several protocols, which required several rounds of meetings and conferences before they were adopted and ratified. In 1990 the SPREP Convention came into force when ratified by 10 countries. The Convention addresses the threat to marine and coastal environments from all types of pollution (including mining, oil spillages, dumping of toxic wastes and hazardous materials) and the need for co-operation to ensure sustainable resource management. The island Governments have remained firm in their demand for a prohibition of nuclear dumping and testing, but the final convention excludes a prohibition clause which is dealt with in the separate Nuclear-Free Zone Treaty (NFZT) banning radioactive waste dumping and testing. The SPREP Convention then allowed for full regional participation, including that of France and its territories.

The ratification of the Programme's Convention meant that SPREP became the recipient of a great deal of aid and assistance from bilateral funding agreements. The regional Governments participating in SPREP took the environmental concerns of the Pacific to the UN Conference on Environment and Development (UNCED), held in Brazil in June 1992, and to the third Conference of the Parties (COP) to the UN Framework Convention on Climate Change (UNFCCC), held in Kyoto, Japan, in December 1997. Prior to the fourth COP of the UNFCCC, held in Buenos Aires, Argentina, in November 1998, the SPREP published a report detailing the effects of global warming and climate change in the Pacific. The South Pacific delegations at the conference emphasized the need to ensure compliance with the Kyoto Protocol on the reduction of the emission of 'greenhouse gases'. A Pacific Islands Conference on Climate Change, Variability and Sea Level Rise was convened by SPREP in April 2000 at Rarotonga; the participating governments aimed to develop a regional strategic framework for action on climate change.

OTHER RECENT DEVELOPMENTS

Major social and political changes swept through the Pacific in the late 1980s and 1990s. In 1987 and 1988 violence in New Caledonia, by Melanesian Kanaks seeking independence from France, and two successful military coups in Fiji suggested fundamental political changes in the future of the Pacific. Racial tensions in Fiji, exacerbated by the success of an ethnic-Indian-led Government, culminated in violence and a coup in April 2000. In March 1988 Papua New Guinea, Solomon Islands and Vanuatu formed a 'Melanesian Spearhead Group', stressing that this was in no way a threat to the SPF. It has been seen, however, as a reaction to both the informal formation of a 'Polynesian Community', which has received overt and covert support from France, and to Australian and New Zealand disapproval of Vanuatu's links with Libya, now discontinued. The Melanesian group also supported more radical efforts on the part of the SPF towards the decolonization of French territories in the Pacific. In 1996 a fourth country, Fiji, was admitted to the group. In April 1991 Western Samoa (now Samoa) held its first election based on universal suffrage. In Papua New Guinea charges of major corruption in the forestry industry and the closure of the Bougainville copper mine, in addition to a major social crisis manifested in a breakdown of law and order, have made the future of the country as a resource-rich nation less stable. In Tonga, where demands for political reform were increasing, in the nine years to 1991 the country earned millions of dollars from the sales of its passports. Some of the sales proved to be illegal, and the money from those sales was missing.

Political and economic events in the region in the 1990s were increasingly influenced by environmental concerns. A sudden increase in unsustainable logging activity in the islands (most notably in Solomon Islands, Papua New Guinea and Vanuatu) prompted widespread expressions of concern. At the SPF summit meeting in 1994 the exploitation of the region's natural resources, particularly its tropical forests, was the principal topic of debate and in that year the World Bank joined various environmental and political organizations in criticizing the Pacific islands, particularly Papua New Guinea and Solomon Islands, for placing insufficient importance on the conservation of their environment, most notably their tropical forests.

Controversy over Japan's hunting of whales for the purposes of 'scientific research' continued. At a meeting of the International Whaling Commission (IWC) in mid-2004, Japan succeeded for the sixth consecutive year in defeating a proposal to establish a South Pacific Whale Sanctuary. A number of PIF members, however, including the Cook Islands, Niue and Papua New Guinea, had meanwhile established their own whale sanctuaries within their respective EEZs. It seemed likely that Tonga might also establish a whale sanctuary within its waters, following a statement made in mid-2003.

Proposals to establish a large-scale nuclear waste storage facility in the Marshall Islands provoked considerable controversy among environmentalists and neighbouring Pacific islands during 1994. Moreover, Japanese shipments of plutonium, which passed within the maritime boundaries of several Pacific islands during the 1990s, caused widespread concern in the region. Most controversial, however, was President Chirac's decision in mid-1995 to resume France's nuclear-testing programme on Mururoa Atoll, French Polynesia. The announcement led to angry demonstrations throughout the region, which intensified when French commandos violently seized *Rainbow Warrior II,* the flagship of the environmentalist group Greenpeace, and its crew, which had been protesting peacefully near the test site. Following the sixth and final test in early 1996, the French Government, which had attracted almost universal criticism for its actions, initiated a series of measures aimed at improving relations with the Pacific islands, including the doubling of its aid budget to the region.

The South Pacific Sea Level and Climate Monitoring Project was launched in 1991 with Australian aid and aimed to allow 11 Pacific island nations to monitor the effects of climate change on their territory. In 2000 it was announced that the programme would be extended until 2006. The increasing impact of the 'greenhouse effect' (the heating of the earth's atmosphere as a consequence of pollution) on the Pacific islands has continued to cause concern in the region. As well as the rise in sea-levels, with its implications for the continued habitability of many low-lying atolls, the phenomenon was also thought to be responsible for a dramatic increase in the frequency of cyclones, tidal waves and other associated natural disasters in the region. The issue dominated the 28th SPF summit meeting in August 1997 and provoked an unprecedented rift within the group when Australia refused to support recommended reductions in the emission of pollutant gases known to contribute to the 'greenhouse effect'. In mid-1999 the SPREP announced that two uninhabited islands in Kiribati had disappeared beneath the sea as a result of rising sea-levels. Following the USA's decision in April 2001 to reject the Kyoto Protocol, in March 2002 Kiribati and Tuvalu announced their decision to take legal action against the USA. Tuvalu reported in 2003 that, following the disappearance of an islet and the increasing amount of seawater seeping into its crop plantations, it had appealed to Australia to be allowed to purchase some uninhabited islands in the Torres Strait in order to resettle its population. The request was rejected by the Australian Government, as was another to allow the resettlement of Tuvaluans in Queensland. A spokesperson for the Australian Government stated only that his country would be part of an international response to deal with any emergency situation on Tuvalu should it arise. Niue, which for many years had suffered from a declining population, officially invited the residents of Tuvalu in 2003 to resettle on its territory. The feasibility of this offer, however, was in question following the devastating effects of Cyclone Heta, which struck Niue in January 2004.

Tourism remains vitally important to the Pacific islands' economy. In 1993, when 745,000 tourists visited the region, the industry earned US $658.4m. (equivalent to 24.6% of the value of total regional exports) and employed some 50,000 people directly and indirectly. By 1998 the annual total of tourists visiting the region had increased to 967,600, and tourism earned the region some US $1,100m. In 2003 tourist arrivals to the region were estimated at 1,075,000. However, the growth of the industry on many islands is restricted by limited infrastructure and inadequate transport facilities, whereas in those countries that have experienced a rapid expansion in tourist arrivals, damage to the environment has often resulted. Consequently, in many islands the development of 'eco-tourism', which encourages the conservation and appreciation of the natural environment, is being explored as a more sustainable alternative to traditional tourism. The South Pacific Tourism Organization (formerly the Tourism Council of the South Pacific) has focused attention on the problems affecting the industry, including the shortage of hotel accommodation in the region and the paucity of regular air services. The industry was seriously affected by the Asian economic crisis of the late 1990s and again by the repercussions of the terrorist attacks on the USA in September 2001.

The economic development of much of the region, particularly the smaller islands, has been impeded by the lack of efficient communications. Regional airlines have often suffered from problems relating to the islands' small size and populations, their remoteness and the frequent inability of their Governments to subsidize unprofitable routes. In an attempt to address this problem, SPF member nations commissioned a report in 1995 on the possible rationalization of regional air services and established a regular Forum Aviation Policy Meeting. At a meeting of the group in May 1998 representatives adopted a policy framework to manage the region's airspace as a unified area and to replace the numerous bilateral agreements between countries with an integrated regional aviation plan, with the possible establishment of a single aviation market among SPF members.

The development of 'offshore' financial services throughout the region during the 1990s represented an important new source of revenue for the islands. However, repeated accusations that these services were being targeted by criminal organizations for the processing of funds from their illegal activities (most significantly the illicit drugs trade) led to problems in the establishment of the industry and prompted the introduction of new legislation throughout much of the region from 2000. In mid-2004 Nauru and the Cook Islands remained on the list of non co-operative countries and territories drawn up by the Paris-based Financial Action Task Force (FATF) on Money Laundering (established in 1989 on the recommendation of the Group of Seven (G-7) industrialized nations) or on the list of un-co-operative tax havens maintained by the Organisation for Economic Co-operation and Development (OECD). The licensing of internet services and the sale of website domain names had also become an important source of revenue for the Pacific islands by the early 2000s.

Australia's decision to involve the Pacific islands in its policy towards people seeking asylum was the focus of international attention in September 2001 when a Norwegian freighter, the MV *Tampa*, was refused permission to dock in the Australian territory of Christmas Island and most of the 460 predominantly Afghan nationals on board were diverted to Nauru. This initiative, which subsequently became known as the 'Pacific Solution', involved the temporary housing in detention camps of up to 800 refugees on Nauru in return for increased aid payments of $A20m. In December 2001 Nauru signed a new agreement to accommodate 1,200 asylum-seekers in return for a further $A10m. An additional agreement to accept up to 1,500 people and to extend the duration of the camps' operations was signed in December 2002. The policy attracted widespread criticism for its apparent disregard for the welfare of the asylum-seekers (many of whom were reported to have been detained for extended periods in overcrowded and inadequate conditions and, in many cases, separated from their family members), as well as for its apparent unwillingness to address the issue of asylum in Australia. However, the 'Pacific Solution' was extended to include detention facilities on Manus Island in Papua New Guinea (accommodating up to 1,000 people) and on the Australian territory of Christmas Island (with facilities expected to house several thousand immigrants). In late 2003 the camp on Manus Island was closed but a large number of asylum-seekers remained on Nauru and Christmas Island.

A deepening social and political crisis, together with the virtual disintegration of law and order in Solomon Islands led the Government of that country to request international assistance in 2003. Consequently, in July of that year an Australian-led regional intervention force, comprising more than 2,000 troops and other security personnel, began to deploy in the country with the aim of restoring peace and security to the islands. The intervention, which represented the largest military undertaking in the region since the Second World War, also involved an economic programme including increased development assistance and the appointment of numerous Australian officials to senior positions in the Solomon Islands public and financial sectors. The Australian Government subsequently expressed its intention to take a more active role in the affairs of the region as a whole, claiming that 'failed states' could be exploited by international criminal networks or terrorists, and could thereby present a threat to the security of the entire region. Papua New Guinea was identified as requiring particularly urgent attention, and, consequently, in early 2004 some 250 Australian security personnel were dispatched with the aim of restoring law and order to troubled parts of the country. Moreover, during 2004 some 70 Australian officials took up senior public roles in areas including finance, law, immigration and border security. Although the Papua New Guinea Government had been reluctant to accept Australia's new role in its affairs, under the so-called Enhanced Co-operation Program, Australian Prime Minister John Howard stated that future aid payments would be dependent on the country's co-operation. Howard announced a further initiative, the Pacific Governance Support Program, at the Pacific Islands Forum summit meeting in August 2004, which was to involve the establishment of links between Australian government departments and agencies and their counterparts in the Pacific islands.

ENVIRONMENTAL ISSUES OF THE PACIFIC ISLANDS

ROSS STEELE

RICH PHYSICAL AND HUMAN DIVERSITY WITH SHARED ECONOMIC AND ENVIRONMENTAL VULNERABILITY

The Pacific Islands encompass a vast region that stretches from Papua New Guinea in the west to French Polynesia and Pitcairn in the east and from New Caledonia in the south to the Northern Mariana Islands in the north. On the basis of political jurisdiction, this essay excludes the Hawaiian Islands (USA) to the north and Easter Island (Chile) to the east. The whole region is referred to as the Pacific Community. It comprises 22 countries and territories with a total land area of more than 551,000 sq km and a vast 30m. sq km of the Pacific Ocean, an area more than three times larger than the USA or Australia. The islands of the region number almost 30,000, of which about 1,000 are inhabited. They range in size from the largest in the west to the smallest in the east. In 2004 the region was home to an estimated 8.8m. people but, if the nation of Papua New Guinea with the largest landmass and population is excluded, the land area decreases to less than 89,000 sq km and the total population to under 3.0m. (see Table 1). The region has a wide variety of peoples and cultures, with three commonly recognized sub-regional ethnographic groupings—Melanesians, Polynesians and Micronesians, who among them speak more than 2,000 different languages. The Melanesian peoples have strong Australoid origins and occupy the larger islands to the south-west including Papua New Guinea, Solomon Islands, Vanuatu, Fiji and New Caledonia. Like the Melanesians, Polynesians are believed to have descended from intermarriage between the early Australoid peoples and Mongoloid peoples who had migrated into the Pacific from north-eastern Indonesia and the Philippines, but they tend to be more similar to their Mongoloid ancestors, with the Micronesians even more so. Polynesians inhabit islands spread over a huge area of the Pacific Ocean, stretching from Hawaii in the north to Easter Island in the south-east to New Zealand in the south-west. They occupy the central island groups of Samoa, Tonga, the Society Islands (including Tahiti) and the Cook Islands as well as many small atolls. The Micronesians inhabit an arc of islands to the north of the Melanesians and Polynesians in the western and central Pacific including the Mariana, Caroline, Marshall, Gilbert (Tungara), Phoenix and Line groups.

To many city dwellers exposed to alluring travel brochures portraying beautiful coral reefs, Pacific atolls, sheltered lagoons and pristine sandy beaches fringed with coconut palms, or the dramatic volcanic peaks and turquoise seascapes of Bora Bora (French Polynesia), the word 'Pacific Island' projects images of a paradise in the South Pacific where it is possible to escape from the cares of modern city life and enjoy the pleasures of an unspoilt natural landscape and the 'simple life'. This image is reinforced by tourists who make a brief visit and limit their explorations to the exotic tourist enclaves developed by major resort chains and, as far as the tourism industry of the developed world is concerned, that is 'the only image' of the Pacific Islands. Unfortunately, the reality for most of the residents of the Pacific Islands in the early 21st century is very different. Globalization and the modern economic realities that accompany it have long since caught up with this most pleasant region of the world, with metropolitan powers persistently exploiting the mineral, agricultural, physical and human resources of the islands for the last 150 years, leading to serious environmental consequences for the local populations. In recent decades rapid population growth and increased urbanization, rising aspirations, increased consumerism and consumption have all combined to create large foreign-debt burdens for many of the countries and territories of the Pacific Islands. This burden, and the consequent need to maximize hard-currency revenue, has led to a 'fast-tracking' of tourism, mining and timber extraction development in many of the Pacific Islands. It has even enticed Nauru and Papua New Guinea to accede to requests from Australia to provide camps for asylum-seekers (the so-called 'Pacific Solution'), whom Australia itself did not wish to accept. Similar economic and political pressures have forced other islands to act as nuclear-testing sites, missile sites and as dumps for the toxic wastes of regional and metropolitan powers. The ever-present Pacific Ocean influences all aspects of life in the Pacific Island communities, from the development of natural hazards such as tropical cyclones (referred to as hurricanes in the northern hemisphere) over its warm waters, to tsunamis and volcanic eruptions caused by earthquakes and the movement of continental plates and landslides in its deeps, to extreme isolation within its huge expanse. The Pacific Islands have also shared a traditional dependence on marine resources for their daily needs, foods, tools, transport and waste disposal despite new technologies and lifestyle changes. The waters of the Pacific Ocean contain the highest level of marine diversity in the world and in the words of the 1999 report of the United Nations Environment Programme (UNEP) 'represent almost the sole opportunity for substantial economic development for nations such as the Marshall Islands, Kiribati and Tuvalu'. The small-island developing status of their economies has also created a common vulnerability to external economic shocks, a vulnerability reinforced by their dependence on a limited and fragile resource base, an over-reliance on a few key economic activities, relatively high labour costs (often caused by pay rates developed for an expatriate public sector in colonial times), very expensive transport costs and an under-resourced and unresponsive system of governance.

The islands of the Pacific have been classified by Bryant Allen and others into three or four major groupings based on their physiography and geological origins (see the article on Background to the Region). The populations of the countries and territories within the region also exhibit marked diversity, but they are all small in a global sense, making their composition and growth rates sensitive to trends in international migration. A growth rate of 4.4% annually over the years 1990–2003 in the Northern Mariana Islands and a decline of 2.4% per year over the same period in Niue's population are extreme and emphasize this point. Over the period 1990–2003 the Pacific Islands' population grew at the rapid rate of 2.69% per annum, although growth rates in the last few years appear to have decreased to 2.2%–2.3% annually as fertility levels have declined or stabilized. Nevertheless, if current trends continue, the region's population will surpass 15m. by 2030, making already problematic goals of sustainable economic development, adequate levels of welfare and service provision, and the amelioration of serious environmental issues even more difficult to overcome in future decades. Population growth rates in some Pacific Island nations such as the Marshall Islands and Solomon Islands are considered very high and to some extent environmentally unsustainable. Indeed, the resource depletion that may occur on these islands and in their marine environments, as a result of continued population growth, has the potential to cause serious depopulation almost on the scale of Easter Island, if the 'safety valve' of emigration is not available. As well as high population growth rates, the Pacific Island nations share other common demographic characteristics such as high but declining or stable fertility rates, increased life expectancy, and higher urban than rural population growth. The regions with the most rapid rates of population growth have the fastest growing urban populations. Parents are leaving inadequate school and medical facilities in the more isolated outer islands or rural villages to migrate to provincial or national urban seats of government in order to provide their children with a better future. According to a background paper on demographic issues prepared for the UN Economic and Social Commission for Asia and the Pacific (ESCAP) Ministerial Conference on Environment and Development in 2000, the Marshall Islands and American Samoa both had urban population growth rates of 8.2% in the late 1990s (a doubling time of under nine years), Vanuatu 7.3%, Solomon Islands 6.2%, the Northern Mariana Islands 5.6% and Papua New Guinea 4.2%. Growth rates of this magnitude far exceeded the capacity of governments to provide housing and essential urban services such as potable water, sewage disposal and adequate health and dental care. The result has been the creation of crowded shanty towns on the edges of the major urban centres such as Suva, Nadi and Lautoka (in Fiji), Port Moresby and Lae (Papua New Guinea), Port Vila (Vanuatu), Honiara (Solomon Islands), Nouméa (New Caledonia) and Nuku'alofa (Tonga). In these shanty towns crime, homelessness, underemployment and unemployment, intestinal diseases, ear and eye infections, and insect borne diseases such as dengue fever are prevalent. Population densities in some of the capitals of the atoll nations are very high and comparable to metropolitan cities such as Hong Kong. Tarawa (Kiribati) and Ebeye (Marshall Islands) are examples, with Ebeye having a density of over 23,000 persons per sq km. High fertility rates are still prevalent in many of the Pacific Islands, with one-half of the 22 countries or territories in the late 1990s still having total fertility rates in excess of twice the replacement rate of 2.1 children per woman. Clearly in the late 1990s these 11 countries were still in the early stages of their fertility transition. Because of their continued high but decreasing fertility and their recent decline in infant mortality brought about by external medical interventions in the last several decades, the population in many of the Pacific Islands is very youthful, with most having median ages of below 25 years. This youthful age structure will maintain the momentum of high population growth in the region and sustain the substantial demand for health and education services and for employment in the short to medium term, even if fertility rates continue to decline

The relatively rapid transformation from traditional active lifestyles where islanders were involved in growing, hunting and fishing to provide a diet of predominantly fresh fish, root crops such as taro, breadfruit, coconuts and leafy vegetables, all of which were rich in fibre and complex carbohydrates, to a more modern and sedentary lifestyle, where much of the food is imported, has created serious health problems for many island people. This modern diet includes more cereal-based foods, animal products, fats, oils and high density energy foods with high sugar and salt content. This change in diet and lifestyle, with increased smoking and drinking, has led to a rise in non-communicable diseases such as hypertension, heart disease, diabetes, cancer, stroke and obesity and undernutrition in the form of Vitamin A and iron deficiencies. The problem has been compounded by the Pacific Islanders' genetic predisposition to store fat for times of scarcity (thrift gene phenotype). Pacific Islanders have paid a hefty cost for this change. They have experienced not only an increase in ill-health, but a significant decrease in their life expectancy. A Suva-based study of deaths among mostly male life-insurance policy-holders in Fiji, found a life expectancy of just 49 years, and although this is not a representative study of all Fijians, or of other Pacific Islanders, it does highlight the problems caused by a change in nutrition and lifestyle and the potential threat that they pose to the region's human resource base.

Population pressures vary dramatically among different island groups. In many of the Polynesian and Micronesian countries emigration to the Pacific Rim countries (Australia, New Zealand and the USA) and to France with which they have political associations, is an important population 'safety valve' and has kept population growth rates at moderate levels despite continued high fertility. Examples are Tonga, the Cook Islands,

Table 1
(2003, unless otherwise indicated)

Country/Territory	Population ('000)*	Land area (sq km)	Density (per sq km)	Net migration rate (per '000)†	Urban population (%)	Food production index‡§	GDP per capita ($ US)‡	Forest area (% of land area)‖
Melanesia	7,623	539,537	14.1	–0.4	—	—	—	—
Fiji	847	18,376	46.1	–3.7	50.2	96.9	2,169	39.3
New Caledonia	233	18,575	12.5	2.4	60.3	132.3	15,750	19.8
Papua New Guinea	5,836	462,840	12.6	—	27.6	121.4	541	67.4
Solomon Is	491	27,556	17.8	—	20.0	152.3	661	87.0
Vanuatu	217	12,190	17.8	—	22.1	87.6	1,127	36.5
Micronesia	535	3,227	165.8	–1.2	—	—	—	—
Micronesia, Fed. States	110	700	157.1	–5.9	28.6	n.a.	2,165	21.7
Guam	165	549	300.5	–10.8	38.0	129.1	11,465	38.2
Kiribati	89	811	109.7	—	43.5	130.5	609	38.4
Marshall Is	54	181	298.3	–20.0	66.0	54.6	1,938	—
Nauru	13	21	619.0	–0.2	n.a.	105.6	2,500	—
Northern Mariana Is	83	457	181.6	33.5	89.5	n.a.	n.a.	30.4
Palau	21	508	41.3	8.1	69.5	n.a.	12,051	76.1
Polynesia	643	8,682	74.1	–8.4	—	—	—	—
American Samoa	63	201	313.4	2.7	n.a.	96.2	n.a.	60.0
Cook Islands	18	237	75.9	–21.7	67.7	96.8	4,388	91.3
Niue	2	263	7.6	–43.2	33.5	106.8	7,470	23.1
Pitcairn	0.05	36	1.4	n.a.	n.a.	n.a.	n.a.	n.a.
French Polynesia	248	4,167	59.5	—	53.0	101.2	13,891	27.3
Samoa	180	2,831	63.6	–17.6	22.0	104.2	1,509	35.5
Tokelau	2	12	166.7	–24.9	n.a.	n.a.	n.a.	n.a.
Tonga	105	748	140.4	–15.1	32.6	97.5	1,348	4.1
Tuvalu	11	26	423.1	–5.2	47.0	100.4	1,342	n.a.
Wallis and Futuna	15	161	93.2	–8.7	n.a.	n.a.	n.a.	n.a.
Total	8,801	551,446	16.0	–1.1	—	—	—	—
Excluding Papua New Guinea	2,965	88,606	33.5	n.a.	—	—	—	—

* Estimates at mid-2004 (Source: UN, *World Population Prospects: The 2002 Revision)*.
† Refers to the migration rate in the late 1990s obtained from Secretariat of the Pacific Community, *Oceania Population 2000 Poster*.
‡ 2002 figures.
§ Base 1989–91 = 100.
‖ 2000 figures. The Forest Resources Assessment 2000 did not identify natural forest cover separately, therefore this was calculated by subtracting 'Forest Plantations' from 'Total Forest' (Table 2, FAO, 2003) as suggested by Emily Matthews in Understanding the FRA 2000, p. 4 *Forest Briefing No. 1*, World Resources Institute, 2001.

Sources: Secretariat of the Pacific Community, Population Programme, *Oceania Population 2000 Poster*. Nouméa, New Caledonia; UN Economic and Social Commission for Asia and the Pacific (ESCAP), *Population Data Sheet 2004* and *Asia-Pacific in Figures*. Bangkok, Thailand.

Kiribati, Tokelau and Niue, where remittances comprise a major portion of foreign-exchange income. On other islands such as the Marshalls, the Northern Marianas, Nauru and Palau, populations have grown rapidly. Over the period 1990–2003 the Melanesian countries had the highest population growth rates at 2.9% annually, followed by Micronesia at a more modest 1.8% and Polynesia with the lowest rate of 1.5%. One reason for this is that the Melanesian people have been more reluctant to emigrate and populations are rapidly increasing on the Melanesian islands of Papua New Guinea, Solomon Islands, Vanuatu and New Caledonia. Fiji would have experienced similar growth rates, but a racially inspired coup in 1987 encouraged the emigration of many of Fiji's Indian population.

The UNEP *Pacific Islands Environment Outlook* report of 1999 identified the major environmental problems that are placing pressure on the natural resources, lifestyles and economic development of this region as: loss of biological diversity, threats to freshwater resources, degradation of coastal environments, climate change and sea-level rise, and land- and sea-based pollution. Before discussing these and other environmental problems in detail, however, it needs to be stressed that different areas within this huge region have unique combinations of environmental issues that are an inextricable product of the ecology, natural resources, history, population and ethnic composition of that part of the region. The high islands of the Western Pacific such as Papua New Guinea, Solomon Islands, Vanuatu, New Caledonia and Fiji are rich in mineral and forestry resources and have the largest human populations. Papua New Guinea alone accounted for an estimated 5.8m. (66%) of the region's total population of 8.8m. in 2004. The three countries of Papua New Guinea, Solomon Islands and Vanuatu had at least 70% of their people living in rural settlements as subsistence farmers and most enjoyed an almost undiluted Melanesian culture. The per caput income levels in these three countries were low, as were their rankings on the UN Human Development Index, and each had very high levels of fertility and infant mortality. The French Overseas Country of New Caledonia has an ethnically diverse population of 233,000, comprising peoples of Melanesian, Asian and European backgrounds. The 847,000 population of Fiji is also ethnically mixed and comprises Melanesians, Polynesians, Indians and Europeans. Both countries have more stable and diversified economies than their Melanesian neighbours, are more urbanized and more developed economically.

The main environmental problems of all of these high islands were land degradation, unsustainable deforestation, water pollution from mining, invasion of exotic species, local depletion of coastal fisheries and, except for Fiji and New Caledonia, rapid population growth. These countries also face the problems of rapidly growing urban populations and their attendant problems of unemployment, underemployment, poverty, sanitation and housing. They also regularly endure the natural hazards brought by drought, fire, volcanic eruptions, earthquakes and tsunamis. The mid-sized islands of Polynesia (Tonga, Samoa, and French Polynesia) and Micronesia (Palau and the Federated States of Micronesia) and the high island Territories of the United States (Guam, American Samoa and the Commonwealth of the Northern Mariana Islands) have limited land resources, few commercial forests and no exploitable mineral deposits. Most of the people of Polynesia and Micronesia are agrarian and rural; however, there is considerable variation. For example, Guam is dominated by urban-based, ethnically mixed communities, and Tonga and Samoa receive large remittances from expatriate island communities in Pacific Rim countries. French Polynesia is an Overseas Country of France, while Guam, the Northern Mariana Islands and American Samoa are Territories of the USA. These Territories of metropolitan powers enjoy a high standard of living as a result of subsidies, and have few of their own tradeable resources and no manufacturing capabilities. The major environmental problems that they face are: a growing shortage of land, especially of good quality agricultural land that is suitable for the production of food crops, loss of remaining native forest and associated loss of biodiversity, decline of coastal fishery resources, coral reef degradation, invasion of exotic species, solid waste disposal and pollution of ground water and coastal areas by agricultural chemicals and sewage. The small coral islands and especially the atoll states (Cook Islands, Kiribati, Tuvalu, the Federated States of Micronesia, the Marshall Islands, Niue and Nauru) have very limited land resources, but are spread over vast areas of the Pacific Ocean. As illustrated in Table 1, 54,000 Marshall Islanders live on 181 sq km of coral islands, giving each person only 0.33 ha of land, with the situation being exacerbated by high population growth rates, particularly in the urban areas. In all of the atoll states the situation was slightly better at 0.8 ha per person. On the other hand, each person had economic control over 41.4 sq km of ocean. These low islands are the most vulnerable places on earth to the adverse effects of climate change and sea-level rise, with some of the islands in Tuvalu, Tonga, the Federated States of Micronesia and the Marshall islands likely to be submerged completely. Coastal erosion, often related to inadequately planned construction work, has caused serious damage to many islands, and two low-lying islands have already eroded below sea-level. The most serious environmental problems on these small low islands are: vulnerability to storms and droughts, fresh water availability and the pollution of groundwater with sewage and salt, scarce agricultural land, solid waste disposal, food security and rapid urban population growth.

LAND DEGRADATION, DEFORESTATION AND LOSS OF BIODIVERSITY

Land Degradation and Commercial Agriculture

Given the recent nature of population growth in the majority of the islands of the Pacific and their relatively well watered and lush landscapes, most of the region has not experienced the scale of land degradation witnessed in more populous and climatically marginal habitats such as north-western China. However, there is evidence that local land degradation did occur on some islands where population densities exceeded local resource-carrying capacities, or where destructive land-clearing activities were used. Examples include Easter Island, and the fernlands of Viti Levu in Fiji, and Wallis and Futuna, where excessive burning of land for agriculture denuded soils to such an extent that only ferns could grow. However, for many environmentalists, modern commercial agriculture is regarded as the most environmentally destructive human activity in the Pacific Islands region. The all-pervasive impacts of commercial agriculture result from its disruption of existing ecosystems: biodiversity loss is caused by clearing of the natural vegetation and its replacement by a monoculture of commercial and plantation crops such as sugar cane, drainage of wetlands and accelerated erosion of topsoils and by destruction of soil structure, pollution of surface and ground waters with agricultural chemicals, and its contribution to global warming through the loss of trees and the production of methane. Commercial agriculture has also been a contributor to landlessness, as village land traditionally devoted to subsistence cropping was converted to plantation and other commercial crops. These crops were then processed or exported and the end result was that in many of the Pacific Islands local food production levels in 2002 were lower or similar to 1989–91 levels, despite significant population increases over the last decade or so. For example, in Fiji, Vanuatu, American Samoa, French Polynesia, Samoa, Tonga and Tuvalu local food production per head of population would appear to be less than a decade ago (Table 1), meaning that commercial agriculture has not only caused severe environmental problems, but contributed to the health and nutritional problems currently evident in Pacific Island communities, as they have become increasingly dependent on imported foods (and their associated additives of fat and salt), rather than traditional and healthier locally produced fresh foods.

Traditional Pacific Island agricultural systems were highly sustainable owing to their discontinuous and limited areal extent amid natural vegetation, their genetic diversity, mixed cropping and minimal tillage practices and long fallow periods that permitted the replenishment of soil fertility. The ESCAP background paper on agriculture of 2000 cited the example of Vanuatu, where the steep slopes of Pentecoast and Ambae were traditionally cultivated for a variety of crops, including commercial kava plantations, but because of their small size and traditional practices these gardens did not contribute significantly to soil erosion and degradation. Similarly in Tonga, traditional shifting cultivation with mixed cropping under a canopy of up to

100 associated tree species, allowed ample fallow time and the regeneration of soils, reduced pest problems and prevented erosion for more than 3,000 years. These environmentally sustainable traditional farming systems decreased in importance as farmers entered the cash cropping system. Small productive mixed gardens with plentiful tree cover were either burned or bulldozed. The cleared areas were tilled by tractor and exposed to leaching and erosion from the heavy rainfall and seasonal cyclones. The cleared land was replanted with a treeless single species monoculture of densely planted cash crops that inevitably resulted in outbreaks of pests and the need to apply poisons to control them. The yields were boosted with the applications of chemical fertilizers, but pressing commercial imperatives, the growing demand for consumer goods and the cash to buy them have combined with growing population pressures to lead to a shortening of fallow times and further environmental pressures on the limited land resources. Such trends are particularly serious on smaller islands, especially atolls with limited land, poor soils and few other land resources. Land degradation is most evident where populations and economic activity are spatially concentrated together, particularly around towns, and where timber and mineral resources are overexploited. Village cash-cropping is carried out in the same land use and tenure context as subsistence production, but the land requirements for larger plantation ventures compete with those required for the expansion of food production. Crop production is then forced to expand into marginal land that is more susceptible to soil erosion or has lower natural fertility. In Fiji and Samoa, for example, there is already severe pressure on good arable land, and subsistence gardening has been pushed to increasingly marginal soil types and slopes where land has been degraded by water and wind erosion, particularly during cyclonic storms or prolonged drought, such as the intense El Niño event of 1997–98. In Fiji, burning to clear land and remove sugar cane debris altered the characteristics of soils and made them more susceptible to erosion. Clear felling of forests for kava plantations has also reduced the forest habitat in Fiji, a habitat needed for yam and other wild foods that were traditionally important staples during food shortages. The costs of erosion are difficult to estimate, but in Fiji the on-site cost from ginger farming was estimated by ESCAP in 1995 at US $0.4–$1.2m. per year, owing to the loss of 27,000–81,000 metric tons of soil. In other countries such as Papua New Guinea and Solomon Islands, where only a small proportion of land is currently used for agriculture, increased production mostly comes from an intensification of land use within areas already in production and some movement upslope onto terrain more susceptible to erosion. In these instances, when the systems of land use and their traditional management practices are no longer ecologically suited to the more intensive production levels demanded from them or to their new more marginal sites, land degradation also results. Land degradation by its very nature is usually a gradual and somewhat subtle phenomenon, and so it is perhaps understandable that few Pacific Island countries have developed, and even fewer have implemented, sustainable land-management policies essential to their long-term food self-sufficiency.

Deforestation

In most of the region, deforestation and forest degradation were insignificant prior to European contact because populations were small, commercial activities were absent and steel tools were unknown. This all changed following European contact in the mid- to late 19th century, when deforestation and forest degradation accelerated and coastal and lowland forests were converted to large-scale commercial coconut, cocoa and banana plantations on many islands. The introduction of steel tools and mechanized transport encouraged the conversion of forest lands to agricultural uses, and in recent decades the introduction of chainsaws combined with increased commercial pressures have encouraged even more rapid removal of forest, with the spread of timber logging and the development of commercial agriculture. Few data are available on the rates at which natural forest was removed in recent decades. In Samoa the UNEP estimated rates of deforestation as high as 2% per year in the mid-1990s. In part of the Federated States of Micronesia only 15% of the land was undisturbed forest in 1995 compared with 42% in 1976. The UNEP report of 1999 cited forest cover as a percentage of total land area in 1998 at 50% in Fiji, 64% in Niue, 86% in Papua New Guinea, 65% in Samoa, 85% in Solomon Islands and 74% in Vanuatu. Estimates for 2000 based on the *State of the World's Forests* 2003 report are shown in Table 1, with the area of forest plantations being subtracted from the total forest cover to give an estimate of natural forest cover. These more recent estimates are generally much lower in the case of Papua New Guinea (67.4%), Fiji (39.3%), Niue (23.1%), Samoa (35.5%) and Vanuatu (36.5%), implying accelerated deforestation since 1998, but the impossibly large decrease in Niue and the small and equally improbable increase in forest cover in Solomon Islands imply definitional problems and serious shortcomings in the comparability of the estimates. There is little doubt, however, that in recent years forest and tree cover has continued to decrease as a result of a combination of population pressures, loss of traditional village-level controls, shifting cultivation and shorter fallow periods, owing to increased population and commercial pressures, pasture development, mining and logging activities.

In recent years the environmental impact of logging in Papua New Guinea and Solomon Islands has received unfavourable publicity, and it is also an issue in Vanuatu, Fiji, Niue, Samoa and Tonga. The loss of forest as a result of agro-deforestation is also significant on those islands with substantial population densities such as the Cook Islands, the Federated States of Micronesia, Kiribati, Marshall Islands, Tokelau, Tonga, Tuvalu and Samoa with significant areas of forest lost annually to fire in Fiji, the Federated States of Micronesia, Cook Islands, Samoa and Guam. The loss of forest cover has had severe adverse impacts on soil conservation and productivity as well as on the duration of stream flow and its quality and on the incidence of flash floods. Loss of forest habitat and degradation have a negative impact on wildlife, reduce the availability of medicinal plants and gathered foodstuff, and can have an adverse effect on human nutrition and increase women's workloads. Large- and medium-scale logging operations can also accelerate breakdowns of traditional systems of social sanctions and an increase in alcohol consumption. All of the four larger countries (i.e. Papua New Guinea, Fiji, Solomon Islands and Vanuatu) have developed national codes of logging practice, and Fiji and Papua New Guinea have attempted to implement them in an effort to minimize the adverse effects of logging on the environment. However, analysis of logging programmes in Solomon Islands and Papua New Guinea illustrates the difficulties in trying to implement sustainable forest management policies within the context of the political and economic imperatives facing Pacific Island nations.

Logging for export began in 1961 in Solomon Islands and has accelerated since the early 1990s. In 1999 UNEP claimed that deforestation rates were so high that harvesting could not be sustained for more than another eight years (i.e. until 2007) and that its cessation would cause a significant decline in national income. An earlier Asian Development Bank (ADB) assessment in 1994 estimated sustainable annual yields of 270,000 cu m, yet logging licences had been granted for up to 1.4m. cu m per year. The ADB suggested that in the more carefully logged areas reforestation or regeneration of the forests may take 30–40 years, but that the widespread damage to residual forests and forest site productivity may mean that regeneration or reforestation may take anything from 45 to 200 years. In addition, it was reported that there had been significant under-reporting of the volumes and value of the timber exported and that this had eroded the royalties and export tax returns to Solomon Islands.

With over 75% of its original forest cover still untouched by commercial logging, Papua New Guinea's forests represent one of the largest remaining tracts of virgin tropical forest in the world. These forests play a central role in the social, economic and cultural life of most of the country's population who, owing to colonial land policies, still retain traditional ownership over 97% of the country. Small-scale timber cutting began in Papua New Guinea in the 1930s; by the end of the decade nine small sawmills had been established and walnut logs and lengths of timber had begun to be exported to the USA. However, except for the clearance of forests for plantation development, from 1930 until the mid-1960s the Australian colonial administration attempted strictly to control the forestry sector to conserve timber resources for future generations. Despite these efforts, by 1960 the industry employed more than 2,000 workers, representing one-half of the total factory employment of the islands. However, by the early 1960s the Australian colonial administration was concerned with the continued economic dependence of its Territory of Papua New Guinea and requested a review of its economic potentialities by the International Bank for Reconstruction and Development (World Bank). The report, completed in 1964, recommended accelerated development of the forestry sector and large-scale logging and milling operations. The Department of Forests almost immediately increased its acquisition of Timber Rights Purchase (TRP) areas from indigenous landowners and began the tendering of large-scale timber concessions. The area under TRP agreements increased from 317,000 ha in 1962/63 to over 1.1m. ha by 1967/68, and the volume of logs harvested increased by 230% over the same period.

The granting of independence to Papua New Guinea in 1975 led to the reversal of some of these changes. The newly independent Government restricted the export of sawlogs and required timber producers to establish domestic value-adding processing facilities. By the end of the 1970s, the realities of a high-cost and under-performing economy had forced a reassessment, and for the first time forest policies were put in place that emphasized log exports as the nation's principal means of revenue generation and offered large log export quotas to timber developers in exchange for their construction of public infrastructure and the development of follow-up agricultural projects. The next eight years witnessed rapid and largely uncontrolled logging of Papua New Guinea's forests. However, by 1986 accusations that foreign logging companies had engaged in widespread tax evasion, transfer-pricing and abuse of local villagers forced the Government of Papua New Guinea to establish a Royal Commission of Inquiry into the operations of the industry, led by Justice Barnett. Barnett documented in detail a history of systemic fraud, corruption, exploitation and abuse and implicated senior members of the Papua New Guinea Parliament in these practices. The Commission estimated that transfer payments and lost company taxes on hidden profits reached nearly US $32m. during 1986/87, or around 15% of the nation's total earnings from the timber industry in that year. Public outrage forced the Government to invite the World Bank to recommend reforms to the industry, and in mid-1989 both Barnett and the World Bank delivered their findings. A two-year moratorium was immediately placed on new timber permits, and in the next five years forestry policy was completely revised under a new Forestry Act to focus on balancing growth with the needs of future generations through the practice of sustained-yield harvesting and the establishment of a comprehensive national forest inventory. The forest bureaucracy was restructured and oversight of the industry was placed under the control of a National Forest Board. Unfortunately, the impact of the changes remains in doubt, with the interests of indigenous customary landowners being largely excluded from most aspects of forest planning and management. In particular, it was argued that the key interrelationships that tie Papua New Guinea's traditional communities to their forests were ignored, just as they were in colonial times, and that there was a failure to appreciate the immediate physical impact that logging operations have on the subsistence economies and environment of village communities, in particular on the supply of wild foodstuffs, water sources, building materials and medicinal plants that sustain these communities. The socio-cultural disruptions caused by the loss of the male work-force to logging activities and the subsequent decline in subsistence gardening and the loss of important cultural and spiritual resources found within the forests were also ignored. Instead, the new policies, like their predecessors, viewed forestry as an industrial production-orientated enterprise controlled by the central Government and its bureaucracy, which excluded landowners from planning and management decisions very early in the production cycle. Like the old, the new policy encouraged the state to play multiple, conflicting roles where it is meant equitably to balance the needs of landowners, but at the same time promote the development of the industry for the revenue that it depends on, work closely with logging contractors, and monitor and regulate logging operations. The net effect was that the new policy virtually guaranteed logging contractors more influence over decision-making processes than it did landowners, while at the same time it transferred the greatest costs of logging onto the traditional communities living within forest areas.

A recent review of the state of Papua New Guinea's forest resources suggests that, despite all these attempts at reform, the logging industry is still not sustainable. The study found that in 2002 Papua New Guinea had approximately 26m. ha of forest, of which about 11m. ha was suitable for commercial logging, and of this 7m. ha had already been allocated for large-scale logging. In the 1970s and early 1980s the large-scale log export industry was focused on the islands of New Britain and New Ireland because of high stocking densities of commercial species and easy access to wharf facilities built along the coastline. Revenue was high and costs low, ensuring substantial profits for the logging companies. The area under concession quadrupled between 1982 and 1991, but there was no corresponding increase in reported log export volumes, suggesting widespread fraud and illegality, as noted by the Barnett Royal Commission. With the exhaustion of these island resources, logging companies were forced onto the mainland where lower densities of high-value species and poor access necessitated that they acquire ever-larger concessions, if they were to remain profitable. The export of logs from Papua New Guinea reached a peak in the mid-1990s at 3m. cu m per year, but since then there has been a decline to around one-half of that figure in 2001, simply because the best commercial forests have already been logged out. It is claimed by that the maximum sustainable harvest from the 11m. ha of forest suitable for large-scale exploitation is only 2m. cu m per year. However, this estimate assumes a 40-year cutting cycle that has no scientific basis, and many commentators suggest that the forests will take 70, rather than 40, years to regenerate. Using the 'optimistic' 40-year assumption, some analysts assert that only one-quarter of logging concessions in Papua New Guinea were cutting at or below the sustainable rate and the remainder were cutting at more than twice the sustainable level. They claim that although timber concessions were intended to last 35–40 years, most were exhausted within 11 or 12 years. They conclude that the future looks bleak for the Papua New Guinea forestry sector, with exports predicted to fall to 400,000 cu m in 2004 and 100,000 cu m by 2010, if no new concessions are granted. Exports could be stabilized at 1.5m. cu m until 2015 if 10 new concessions covering 2.5m. ha of forest previously identified as suitable for commercial logging are brought into production, but after 2015 they would rapidly decline. These projections illustrate the failure of Papua New Guinea, the nation with the largest share of forestry resources in the Pacific Islands region, to manage its forest industry in a sustainable manner over the last 35 years. Similar deficiencies in forestry management are widespread among the Pacific Island nations, which have already lost, or are in the process of losing, most of their high-value commercial forests. The subsistence communities of the islands are now experiencing the costs of this loss to their traditional lifestyles, societies and cultures, while their environment is being degraded at an accelerating rate. Commenting on a Malaysian logging company's impact in the Vanimo Timber Area in 1999, an environmental and social impact assessment stated that few things had changed since 1989 when Judge Barnett stated: 'In many cases, the timber industry has made life harder for the landowners at all levels. Not only do they have to face destruction of their environment, but they face the destruction of their society'.

Loss of Biodiversity

The Pacific region is one of the world's most biologically diverse regions, the western Pacific alone being considered to have the highest marine diversity in the world, with up to 3,000 species being found on a single reef. The many thousands of islands are surrounded by rich coastal ecosystems comprising mangroves, seagrass beds and estuarine lagoons, along with complex coral reef systems. The extreme isolation of most of the islands and the evolution of island biogeography have led to a high endemism (i.e. unique to that area) in terrestrial species (in excess of 80% on some islands), especially on the larger islands. Ironically, this treasure trove of biodiversity now ranks as one of the most endangered in the world, with perhaps up to 50% of its total biodiversity at risk because of habitat destruction and alien invasive species, according to the South Pacific Regional Environment Programme (SPREP). Isolation and their dependence on very specialized micro-habitats make these species especially vulnerable to rapid extinction from fire or deforestation. The destruction began with early human settlement from the Philippines, South America or both some 3,000 years ago. This early settlement resulted in major changes to the biodiversity of the region as forests were cleared and converted to settlements or agricultural land, forest and lagoon resources were exploited and alien species introduced. For example, there is evidence that a number of bird and other animal species were hunted to extinction by the early Pacific Islanders who had a profound effect on biodiversity even with relatively low population densities and pre-modern technologies. The threats to biodiversity increased enormously with the arrival of European colonization and the introduction of new post-industrial technologies and exploitative commercial values.

In New Caledonia, for example, 75% of the flora and fauna evolved on the islands and several plant species, unique in the world, are limited to only a small area of one mountain. This rich and diverse genetic heritage is of such scientific importance that New Caledonia has been listed as one of only 10 locations in the world where the primary forest is at once exceptional and endangered. The bird life of New Caledonia is the most diverse in the south-west Pacific, with 68 species, of which 22 are endemic to the country. This bird fauna was even more exceptional before the 18th century, when a giant, flightless bird, like the famous (and also extinct) New Zealand moa, was still common. The extinction of these birds began with the arrival of Melanesians about 900 years ago and was probably caused by fire, slash-and-burn agriculture, and hunting. The arrival of European settlers exacerbated the rate of loss, through a combination of logging and mining; furthermore, natural drought conditions resulted in massive fires that destroyed a majority of the habitats on the southern part of the main island. Unfortunately for the biodiversity of the main island, it has the world's largest-known deposits of nickel in the world, which today generate about 90% of New Caledonia's foreign exchange. The impact of the open-cast mining has been devastating, causing large expanses of deforestation and habitat destruction, leaving behind bare slopes and waste heaps. The erosion of the mined areas has caused siltation of streams, pollution of water supplies and the destruction of offshore coral reefs. The intentional and accidental introduction of alien species for food or recreational purposes has also been extremely damaging, as these can often out-compete and replace many of the original species. It is estimated by Conservation International Biodiversity Hotspots that there are now approximately 800 alien plant species, 400 alien invertebrates and 36 alien vertebrates established on the islands. An international market for rare species of birds and marine animals such as the Ouvéa (Uvea) parakeet, the horned parakeet and an endemic cephalopod, poses another threat to the islands' rare and unique species. The experience of New Caledonia was repeated throughout the Pacific. In the Marquesas, Polynesian settlers exterminated eight of the 20 species of sea birds and 14 of the 16 land birds. On Easter Island, the early settlers denuded the entire island of trees and exterminated 22 species of sea birds and all six species of land birds. Given recent high rates of population growth, current trends in land degradation and the excessive exploitation of near-shore and offshore marine resources, the high rates of extinctions experienced in the Pacific Islands are expected to continue, despite a growing recognition of the value of biological diversity among the peoples of the region.

A dramatic example of the effects on biodiversity of the accidental introduction of an alien species is found on Guam where the venomous brown tree snake (boiga irregularis) was accidentally introduced (probably from Solomon Islands) after the Second World War. The snake lacked natural predators in its new home and, with an ample supply of prey, its population increased dramatically. By the 1970s it was found all over the island, and it is now so common that the snakes cause over 100 power supply breakdowns per year when they climb a power pole and short-circuit the lines. In the 2002 SPREP report the snakes were estimated to number an incredible 18m. Their impact has been devastating on Guam's native species, especially the birds, its preferred diet, but they also feed on rats, shrews and lizards. Not having evolved with a night-time arboreal (tree-climbing) predator, Guam's native birds had no physical defences against the snake. By the mid-1980s nine of Guam's 11 native forest birds had become extinct and the survival of the others was very tenuous, despite active protection programmes. It now appears that the Micronesian starling is the only native bird that may ultimately survive the introduction of the snakes to Guam. Of Guam's six native geckos, two are now extinct, three have been severely reduced in number and only the house gecko is still common. All of Guam's native skinks are now either rare or extinct, and the only common species is the carlia fusca, which was introduced from Papua New Guinea and probably evolved with such predators. Two of the island's three species of bats are now believed to be extinct and because of their loss and the loss of birds, insect populations have increased, causing problems for local agricultural industries and outbreaks of insect-carried disease such as dengue fever. The Guam experience illustrates the extreme dangers to local biodiversity posed by the introduction of just one alien species. Numerous sightings of the snake have been made on other islands, and an incipient population is probably already established on Saipan, in the Northern Mariana Islands. With Guam continuing its role as a transportation hub, it has the potential to cause similar devastation to the biodiversity of many other Pacific islands, including Hawaii, where repercussions from its introduction could cost over US $100m. Some of the many other uninvited alien species that continue to cause great damage to the local ecosystems are the Merremia vine, which is rapidly smothering Pacific forests; the black rat, which has already eradicated many unique species of birds and lizards; and the giant African snail, which is threatening agriculture and native plants with its voracious eating habits.

THREATS TO FRESHWATER RESOURCES, DEGRADATION OF COASTAL ENVIRONMENTS, CORAL REEFS AND COASTAL AND OCEANIC FISHERIES

Threats to Freshwater Resources

Occasional water shortages during droughts have been a long-standing problem throughout the Pacific, with the most severe water shortages being on the atolls and raised limestone islands where there are no streams and thus inhabitants must rely on the groundwater lens floating on top of the sea water for their water needs. As early as 1992, two-thirds of SPREP members reported water supply and storage problems and an even higher number reported groundwater pollution. The region-wide drought in 1998 highlighted problems of water shortage in Papua New Guinea and Samoa and the need to reduce wastage and consumption. Groundwater pollution levels have not been researched in detail, but both solid and liquid waste disposal systems are inadequate and poorly planned. Waste flows are increasing with population growth, local industrial development and tourism, and pollution problems are likely to worsen. Localized pollution and excessive sedimentation, resulting from uncontrolled watershed development, are common problems in Fiji, Samoa and Solomon Islands. In some atoll communities where the water lens used for cooking and drinking water has been polluted, health problems such as diarrhoea and hepatitis are prevalent, along with occasional outbreaks of typhoid, and in Kiribati, Tuvalu and the Marshall Islands cholera has been diagnosed. As the 1999 UNEP report cautions: 'Pumping from the freshwater lens needs to be carefully monitored and controlled in order to provide warning of impending saltwater intrusion and to test water quality for bacteria counts, chemical residues and total dissolved salts'. Freshwater lenses are governed by the interaction of rainfall volume and periodicity, tidal fluctuations, seepage, hydraulic conductivity and abstraction rates. Once the lens, which is in a dynamic state of equilibrium, is contaminated by saltwater intrusion, for example, it may take years to re-establish. If the contamination is land-based from pesticides or leachate, the recovery time may be much longer. For these reasons, detailed research is required into the sustainability of freshwater resources and provision made for greater investment in appropriate water supplies and sanitation services, before new developments such as large tourism resorts are approved by government authorities. There is a growing consensus that global warming may lead to increased energy in the hydrological cycle and consequent greater intensity and frequency of extreme events such as drought and floods which, combined with sea-level rise, may further increase the threats to freshwater resources.

Degradation of Coastal Environments

As the coastal zone and its immediate hinterland is home to most of the region's inhabitants and the focus of critical economic activities—such as fishing, coastal shipping, port and harbour development, water-based recreational activities, road and urban development, infrastructure development, sewage treatment and disposal, rubbish dumping and coastal protection—it is the zone where the environment is most heavily threatened. Imminent threats to the zone include nutrients derived from sewage, soil erosion and chemical fertilizers, solid waste disposal, sedimentation from land clearance and increased erosion, physical alteration caused by destruction of fringing reefs, beaches, wetlands and mangroves for coastal development and by sand and coral extraction, and over-exploitation of coastal food fisheries, particularly through the use of destructive fishing methods such as explosives. Land reclamation and natural erosion owing to wave action are seen as imminent threats to the marine environment by the Cook Islands, Samoa and American Samoa. In many islands sewage discharge has reduced water quality, reef fish are being over-exploited, rubbish is being dumped along the foreshore and sea turtles no longer find the foreshore attractive for nesting. Marine invasive species are also an issue in some ports and coastal habitats, along with ship-sourced marine pollution. Overall the coastal zone is extremely vulnerable, and many of the impacts such as the destruction of seagrasses, mangroves and reef habitats and the attendant wave erosion are irreversible.

Coral Reefs

The coral reefs of the region are among the most biologically diverse ecosystems on the planet, with species diversity in corals decreasing from west to east across the Pacific. Unfortunately, many of these unique ecosystems have been destroyed by human activities. It was estimated that in 1998 10% of the region's reefs were under a high risk of destruction (19% in Fiji, 12% in Papua New Guinea and 8% in Solomon Islands), 31% were under medium risk (48% Fiji, 42% Solomon Islands and 38% Papua New Guinea), with the remaining reefs (59%) being under a low risk. Assessment criteria included proximity to coastal development, marine pollution, over-exploitation, destructive fishing, and inland pollution and erosion. The less developed regions such as the Marshall Islands and French Polynesia appear to have some of the more pristine reef ecosystems. During the Second World War more than 1,800 ships were sunk in the Asia-Pacific area, many of them on these reef systems, along with a vast load of armaments. On Chuuk, in the Federated States of Micronesia, people have removed explosives from munition dumps to stun and kill fish, and in 1994 it was reported that blasting had killed 10% of the reefs in the lagoon. However, even if these threats are averted, the coral reefs of the region are being heavily stressed by the abnormally high sea-water temperatures caused by global warming and severe and frequent El Niño events, which cause widespread coral bleaching and often coral death (see Part One, Environmental Issues of Asia).

Coastal Fisheries

Subsistence fisheries are of vital importance to the local people, with some 83% of the coastal households in Solomon Islands fishing for home consumption and 35%, 50%, 87% and 99% of the rural households in Vanuatu, Samoa, Marshall Islands and Kiribati, respectively, also fishing primarily for their own consumption. Prior to the 1980s it was believed that the fish stocks used for subsistence fishing were underfished and thus offered an important route to economic development. A 1999 World Bank report on coastal management revealed that this view was wrong and that overfishing, aggravated by the use of diving equipment and gill nets, poses a major threat to national food security in many Pacific Islands. However, the report also found that the most serious threats to coastal fish stocks were not only from overfishing, but also from pollution and habitat loss, which depressed the ability of inshore stocks to recover even once extraction rates and fishing practices were controlled. The seagrass beds, coral reefs, mangroves and the sea surface micro layer are all critical nursery habitats for marine plants and animals, and serious damage has been done to these habitats by siltation, pesticides, hazardous chemicals, domestic waste, logging and mining. Mangrove habitats have been especially undervalued in the islands of the Pacific where they have been frequently used as municipal dump sites and filled for housing. Destructive fishing practices such as dynamiting and fish poisoning have also posed major problems. Fish stocks of shallow water coral reefs are in good condition away from population centres but, like stocks of important key species that are commercially valuable, they have been overfished near villages and urban centres. Valuable commercial species such as giant clams, lobsters, and most large reef fish have been heavily fished even on the more remote reefs. Many countries in the region received financial assistance to increase their fishing fleet for deep-bottom fishing of deep reefs and sea mounts from 100 m to 500 m below the surface, but it was then discovered that the large fish at these depths took decades to reach maturity and that replenishment was so slow that they could not be successfully fished again for many years. Depletion of the sea mounts in Tonga was particularly rapid. Fishers then moved to more distant sea mounts, but increased fuel costs and problems of refrigeration reduced their profits and many loans could not be repaid. Unfortunately, many administrations have excluded the impact of sport fishing, diving, agricultural, forestry, and mining practices from the fisheries development process, and most communities in the Pacific have had limited capacity to control their fish catches and pollution sources. Since the early 1990s the Secretariat of the Pacific Community (SPC) has helped shift national fishing priorities from increasing commercial fishing capabilities to sustainable management of fisheries resources, and to an increased awareness of the need to involve local communities in self-management programmes.

Oceanic Fisheries

The commercial exploitation of oceanic marine resources in the Pacific Islands region is dominated by the high-technology fishing fleets of distant fishing nations engaged in harvesting migratory tuna. The tuna harvest is worth some US $3,000m. annually, with the tuna fisheries producing more than nine times the amount of fish of all other fisheries in the region combined. In 2004 the environmental group Greenpeace estimated that 95% of the profits from the tuna fisheries go to distant water fishing nations (DWFNs—nations with boats fishing a long way from home, usually because they have overfished their own waters) such as Japan, the Republic of Korea, Taiwan, China, the USA, the Philippines and the European Union. Only 10% of the total catch is caught by Pacific Islanders. Despite this unequal distribution of profits and catch, many Pacific island nations rely on income from access fees paid by foreign vessels for the right to fish their Exclusive Economic Zones (EEZs) that cover 74% of the region's water surface. In 2001 the annual tuna catch was equivalent to 11% of the combined gross domestic product of all countries in the region and accounted for one-third of all exports from the region. The catch was also a major source of employment for Pacific Island nationals, with 10,000 Pacific Islanders being employed on tuna vessels and in tuna-processing plants. The total direct and indirect employment was estimated to be between 21,000 and 31,000 people, equivalent to 5%–8% of all wage employment in the region. The Pacific remains one of the world's last healthy fisheries, supplying two-thirds of the global tuna catch of 4m. metric tons a year, of which the western and central Pacific contributed an estimated 2m. metric tons in 2002. The 2002 catch was the second highest annual catch ever recorded after the catch of 1998, and was proof of a 400% increase in the region's tuna harvest over the previous 30 years. The most dramatic increase in the tuna harvest began with a doubling of the catch in the 1980s, as the new highly efficient purse-seine fishing vessels that round up schools of tuna in a single net began cruising into the central and western Pacific from the American coast and East Asia. In the period between 1997–98 and 2001–02 the number of registered fishing boats increased from 827 to 1,233, placing even more pressure on fish stocks. In 2002 the catch taken by purse-seine boats was 1.16m. metric tons, equivalent to 58% of the total catch, with 17% being taken by pole-and-line methods, 11% by longlines and the remaining 14% taken by troll gear and a variety of artisanal gears. In April 2004 Japan warned that a new 'jumbo' class of fishing boat used mainly by Taiwanese operators was threatening the long-term future of the richest tuna stocks in the Pacific Ocean. These new 'super' purse-seine vessels work with sophisticated fish-finding technology and larger nets, and can take up to 20,000 metric tons of tuna a year. The 2002 catch featured a high skipjack tuna catch (66% of the total tonnage), with yellowfin tuna comprising a further 22%, bigeye tuna 6% and albacore tuna 6%. A 2002 survey of the status of tuna stocks indicated that stocks of some individual tuna species were exhibiting the early warning signs of overexploitation. Fishing of skipjack tuna has reduced the stock biomass by 20%–25% in recent years, but the authors of the report maintained that: 'current levels of stock biomass are high and recent catch levels are easily sustainable under current stock productivity levels'. The report also stated that the yellowfin tuna stock was not being overfished; however, it warned that it was nearing full exploitation in 2002, and that any future increases in the catch would not result in sustainable increases in yield and may move the species to an overfished state. The equatorial regions appeared fully exploited, especially the Indonesian fisheries, whereas the temperate regions were underexploited, suggesting the need for different management plans in different regions. There was considerable uncertainty about the bigeye tuna stock; although it may not be generally overfished, overfishing was occurring and the 2002 level of exploitation was not considered sustainable. South Pacific albacore tuna stock had declined moderately since the early 1980s, but there was a slight recovery in the mid-1990s, apparently as a result of changes in oceanographic conditions, but the impact of the fishery on stock levels was small and higher levels of catch were thought to be sustainable.

Since the early 1990s the Forum Fisheries Agency (FFA) has been representing the small island developing states of the South Pacific in seeking the adoption of an agreement under international law to limit development of tuna fisheries to sustainable levels in the western central Pacific. The aim has been to balance the sovereign rights of coastal states to set catch limits against the need to co-operate to ensure that the collective catch taken within the EEZs of coastal states and on the high seas does not exceed a sustainable agreed level. In June 2004 the Convention on the Conservation and Management of Highly Migratory Fish Stocks in the Western and Central Pacific Ocean (WCP Fisheries Convention) came into force. Fourteen countries of the Pacific Region have ratified the agreement, which gives signatory nations under the UN Fish Stocks Agreement the right, for the first time, to board and inspect on the high seas within the Convention Area fishing vessels flagged by a state that has also ratified the UNFSA. Although both Australia and New Zealand are among the ratifying countries and both have provided naval and air force resources to monitor the Convention, its ultimate success will depend on the willingness of major DWFNs such as the USA, Japan, China, Taiwan, the Republic of Korea and the countries of the European Union to accept its regulations and honour their responsibilities. In the mean time, the threat to the region's fisheries from illegal fishing activities and bilateral licensing agreements between small island nations and powerful DWFNs continues, with European Union vessels joining the more than 1,000 Asian and American boats fishing in the EEZs of the region.

CLIMATE CHANGE AND ITS CONSEQUENCES

The evidence of global warming, its causes and general consequences, and the progress and effectiveness of the Kyoto Protocol have been summarized in Part One, Environmental Issues of Asia, and will not be elaborated upon in this essay. From the viewpoint of the peoples of the Pacific Islands and their Governments whose well-being is so dependent on coastal ecosystems, there are four anticipated consequences of climate change that will be critical to their future: global warming and sea-level rise; an increase in climate-related natural disasters (storms, cyclones, floods and droughts); disruption to agriculture owing to changes in temperature, rainfall and winds; and increased health hazards.

Dangers of Sea-Level Rise

The Intergovernmental Panel on Climate Change (IPCC) in its 2001 Assessment Report predicted a range of global sea-level rises with a central value of 48cm by the year 2100. Although changes of this magnitude may appear minor on a global scale, for a region with a large number of low-lying atolls and some atoll-based countries such as Tuvalu, Kiribati and the Marshall Islands, they would be very threatening indeed. However, there is still considerable uncertainty about sea-level rise predictions. Examination of long-term tide-gauge records gave estimated positive relative sea-level trends of 1.07 and 0.8mm per year, but the records were of poor quality, there were likely datum shifts and gaps in the records and the estimates were not adjusted for land movement. A more sophisticated tide-gauge array has now been deployed at 12 sites to monitor sea-level in the region with Australian assistance, but the new SEAFRAME gauges have only been in place for 10 years, not enough accurately to estimate long-term trends in sea-level change. As Bill Mitchell, the director of Australia's National Tidal Facility at Flinders University, has cautioned, 'We are yet to see the acceleration of sea-levels that the climatologists have predicted'. Nevertheless, the 10-year record does show a dramatic drop in sea-level associated with the 1997/98 El Niño event. Tuvalu experienced a drop in sea-level of almost 40 cm, whereas other countries such as Nauru experienced a sea-level rise in the year preceding El Niño, giving a total sea-level variation of over 50 cm from one year to the next. These dramatic changes are likely to have an impact on coastal geomorphic processes, and may cause a movement of sediments toward the lagoon on low-lying atoll islands, rather than a simple readjustment of the beach profile to the new sea-level. This is an important consideration for island Governments trying to manage their vulnerability. Certainly any rise in the mean sea-level would increase the zone subject to flooding and intensify the impact of high-tide events and storm waves and cyclones. Coastal erosion would increase and the erosion of fringing reefs would disturb lagoon ecology, mangrove habitats would be damaged and fishing nurseries reduced. Tourism facilities, human settlements and infrastructure would be threatened and possibly damaged. The likely increased frequency of wave overwash owing to rising sea-levels would also increase the salinity of the freshwater lens found on most atolls above 1.5 ha in size and reduce its potability, thereby threatening the habitability of many atolls. Another factor complicating the measurement of long-term trends in mean sea-level is that much of the monitoring has been conducted in areas affected by human interference such as causeway construction, dredging or reclamation. In Kiribati, for example, international print media highlighted the disappearance of Bikeman islet within Tarawa lagoon in the early 1990s as evidence of the damage already being wrought by climate change, but other commentators suggested its disappearance was more likely to have resulted from the construction of a major causeway between two nearby islands that affected current and sand flows.

Other media coverage has been similarly alarmist and lacking in any scientific caution. For example, in February 2002 a routine spring high tide in Funafuti, Tuvalu, was described as proof of inundation caused by global warming, and in the same year a publication in the USA claimed that the nine coral atolls of Tuvalu had already been drowned by the rising waters of the Pacific Ocean. Unfortunately, in the case of Tuvalu the more alarmist and emotive views have been adopted by the local political élite, an understandable stance given that the country's highest point is no more than 5 m above sea-level and most areas are well below that. Tuvalu's domestic development options are few, with the small independent state having a total land area of just 26 sq km on three small reef islands and six coral atolls that have poor soils and limited resources, and are frequently exposed to environmental stresses associated with cyclones and droughts. Nevertheless, the aspirations of its estimated 11,000 people have increased considerably as many have experienced wage employment in Nauru or New Zealand. At the same time emigration has increased, and Tuvalu's economic future has become more dependent on remittances from the 20% of the country's work-force now employed overseas. Internal migration from the outer islands to the main island of Funafuti in search of wage and salary employment has placed enormous environmental stress on the main island, the population of which rose from 871 in 1973 to over 4,000 in the late 1990s, leading to attendant overcrowding, poor sanitation and inadequate fresh water supplies. In effect, the natural hazard of global warming and associated sea-level rise has been embraced by local political leaders as a means of overcoming the challenges of economic development in a very small state, and in recent times even more strident appeals have been made for imminent relocation to Australia or New Zealand, along with threats of litigation against the major emitters of greenhouse gases such as Australia and the USA. The problem with this approach is that emotion and rapid environmental degradation have overwhelmed scientific detachment and with it the need for policy changes in major greenhouse-gas-emitting states. At the same time, the critical need to develop appropriate environmental management policies within the atoll states themselves, such as mitigation strategies to combat storm surges, cyclonic damage and groundwater pollution, have been ignored as the islanders see their future as being somewhere else.

Climate-Related Natural Hazards, Agricultural Production and Health

Another predicted consequence of global warming is an increased incidence of more extreme weather events coupled with El Niño, which would have a severe impact on agricultural productivity through drought and water shortages. The *Pacific Islands Environment Outlook* claimed that as early as 1999 there was growing evidence of these impacts. The report cited water shortages and drought in Papua New Guinea, the Marshall Islands, the Federated States of Micronesia, American Samoa, Samoa and Fiji, and floods in New Zealand. It also cited research from New Zealand's National Institute of Water and Atmospheric Research that climate had changed from the mid-1970s in the following regions: Kiribati, the northern Cook Islands, Tokelau and northern French Polynesia had become wetter; New Caledonia, Fiji and Tonga had become drier; Samoa, eastern Kiribati, Tokelau and north-east French Polynesia had become warmer and cloudier; New Caledonia, Fiji, Tonga, the southern Cook Islands and south-west French Polynesia had become warmer and sunnier; Western Kiribati and Tuvalu had become sunnier.

Increases in the incidence of heavy rainfall and flooding are predicted to raise the rates of water-borne diseases, while higher temperatures and humidity may increase the incidence of vector-borne diseases, such as dengue fever, malaria and yellow fever. Research has shown a strong link between the incidence of El Niño events and dengue fever outbreaks. In 1997/98, during the drought associated with the El Niño event, a dengue fever epidemic affected 24,000 people and killed 13, at a cost of US $3–$6m. Drought and water shortages also tend to increase the likelihood of diarrhoea, eye and skin diseases, poor nutrition and a general low level of health, particularly in poorer areas such as urban squatter settlements where clean water supplies and sanitation services are inadequate.

The UNEP report also claims that cyclones have become more frequent. Tokelau for example, had only three major storms since 1846, but two cyclones struck in the early 1990s alone. Tuvalu experienced an average of three cyclones per decade from the 1940s until the 1970s, but eight struck the island nation in the 1980s. There is also concern that tropical cyclones may become more intense, which could increase storm surge height. These concerns were underlined in January 2004 when the island of Niue was struck by Cyclone Heta, a Category 5 cyclone (a once-in-1,000-year event), with winds of up to 300 km an hour and a storm surge that brought waves up over 18 m cliffs, leaving 200 of the island's 1,600 residents homeless. Niue's close association with New Zealand and Australia guaranteed that in this case international emergency aid was provided very quickly, but the small size of most of the Pacific Islands, their remoteness and limited financial resources, combined with a steady degrading of traditional coping measures, have made the islanders and the island ecosystems more vulnerable to disasters than in earlier times. Poor farming practices and logging on steep slopes mean that in recent cyclonic conditions massive erosion usually occurs, which in turn pollutes water supplies and deposits massive loads of silt on the coral reefs. Small gardens that were once protected by trees are today large unprotected gardens that can be totally destroyed by cyclonic downpours. In Pohnpei, in the Federated States of Micronesia, for example, after a severe cyclone in 1997 massive landslides caused loss of life, destruction of commercial kava plantations on the upper slopes and damaged coastal reef communities. The cyclone precipitated this disaster, but the large-scale clearing of upland forest for commercial kava plantations had established the ideal pre-conditions.

LAND- AND SEA-BASED POLLUTION

It is a sad irony that Governments in an area of the world renowned for the unspoilt beauty of its lagoons and beaches now list the prevention of pollution as their major environmental concern. Indeed, pollution is now so serious that it is recognized as one of the major threats to the sustainable development of the Pacific Islands region, causing damage to its tourism and trade, food supplies, public health and environment. The main types of pollution within the region have been identified by SPREP as shipping-related pollution, hazardous chemicals and hazardous wastes, and inadequate solid waste management and disposal. The region's marine resources are increasingly being threatened by introduced marine species, shipwrecks, marine accidents and spills and ships' waste. There are also increasing quantities of solid waste and a poor control of chemicals imported into the region, and a lack of capacity to manage waste disposal, let alone develop programmes to recycle it. Much of the rubbish slowly breaks down and leaches into the soil and into drinking water supplies, but in the mean time it is dumped on the nearest available government land where there is a proliferation of plastics, paper, glass, metal and even drums of hazardous chemicals. Foul-smelling organic wastes attract disease carrying pests such as mosquitoes, rats and flies, increasing the incidence of vector-borne disease. In the absence of publicly funded waste disposal systems, piles of household rubbish collect on beaches and in mangrove swamps, detracting from the reputation of the region's famed beaches and waterways. The economic vulnerability and political dependence of many of the countries in the region has exposed their people to the pollution and hazards of nuclear testing, dumping of toxic and radioactive wastes and their shipment, as well as to pollution from mining activities desperately required to bolster meagre foreign-exchange earnings.

Land-Based Pollution and Waste Management

For most of the last century the impacts of pollution were small and there was little need for waste management in the Pacific Island countries because most waste products were biodegradable and populations were dispersed. Wastes were usually disposed of through individual dumping in lagoons and rivers or on unused land close to villages. The growth of urban populations and increased imports of non-biodegradable materials and chemicals related to agricultural and manufacturing activities created environmental health problems and the need to manage waste and toxic and hazardous substances. The small size of many Pacific Islands and their dependence on marine resources and a limited land resource base make them very vulnerable to contamination by toxic and hazardous wastes and chemicals and radioactive materials. Although biodegradable material such as vegetable and putrescible waste and garden waste still dominated the household waste stream in most Pacific Island states in 1994, during the 1990s there was a rapid increase in the importation of packaged consumer goods. These have added to the growing amount of non-biodegradable waste such as plastics, glass, cardboard, paper and metals. Increased urbanization placed greater demands on inadequate sanitation systems, resulting in high coliform contamination of surface waters and groundwater near urban centres. Public health has been affected, with diarrhoea the third most common cause of hospitalization in the Pacific Islands. In Kiribati, cholera, diarrhoea and other water-related diseases were the major cause of death. In Ebeye lagoon, in the Marshall Islands, pollution rates reached levels 25,000 times higher than World Health Organization safe levels, and epidemics of gastroenteritis were almost impossible to control. Toxins from industrial waste, effluent from abattoirs, fish canneries or other food-processing plants, leachate from sawmills and copper-chrome arsenic chemicals used in the preservation treatment of wood have also caused problems. The continued use of chemicals in agriculture and in manufacturing, such as in fibreglass fabrication and the manufacture of plastic packaging, has exacerbated the amount of pollution from toxic and hazardous substances, the presence of which can usually not even be monitored by the inadequate laboratory facilities existing in most Pacific Island countries. A 1999 SPREP/AusAID study of persistent organic pollutants in the region found that considerable stockpiles existed in some countries and that a number of sites had been contaminated through past disposal or storage of these chemicals. For example, even without data from Papua New Guinea, French and US Territories, which were not included in the study, Pacific Island countries had an estimated 130 metric tons of agricultural chemicals (including DDT) that were polluting sites, 220,000 litres of potentially polychlorinated biphenyls (PCB) contaminated transformer oil and another 21 pesticide-contaminated sites. Particular attention is required at the national level to strengthen the capacity of island countries to minimize and prevent pollution, but the small size of local markets and the limited land areas mean that increased international assistance will be required to tackle disposal issues, and to develop programmes to reuse, recycle and reduce wastes and to give effect to the Waigani Convention and the Global Programme of Action for the Protection of the Marine Environment from Land-Based Activities.

Sea-Based Pollution

Sea-based pollution is also a major problem for the Pacific Islands region. The dangers have come from two major sources: shipwrecks sunk in the Pacific Theatre in the Second World War and their unexploded ordnances still on board, and from spills of oil and other chemicals from ships visiting or traversing the region. The Pacific was the scene of some of the largest and most famous naval battles of the Second World War such as Pearl Harbour, Midway, Truk lagoon, Bismarck Sea, Guadalcanal (Iron Bottom Sound) and the Coral Sea. During these and other battles in the early 1940s more than 1,800 ships were sunk in the Asia-Pacific area. The shipwrecks include 23 large aircraft carriers, 213 destroyers, 22 battle ships, hundreds of Japanese planes and submarines and 50 oil tankers. It is the oil tankers that pose the biggest environmental threat. Now, over 50 years since their sinking, these rusting hulks are beginning to discharge their potentially deadly cargo. One tanker alone holds some 19m. litres of fuel oil. Some ships are lying in open water, but others lie within the fringing reef so that any leak is trapped within the island's lagoon. However, the lack of major land-barriers throughout the Pacific, combined with the complex pattern of transoceanic currents means that in terms of water circulation the Pacific Ocean is perhaps the most highly connected and continuous ocean. These characteristics of the Pacific region compound the seriousness of marine pollution within the region. Pollution incidents in one area have potentially serious implications for other areas. The organization SPREP has developed a comprehensive programme to address marine pollution from ship-based sources called the Pacific Ocean Pollution Prevention Programme (PACPOL). The danger of these wrecks to the marine environment was highlighted in July 2001 when a tropical storm disturbed the wreck of the *USS Mississinewa*, an auxiliary oiler that had been sunk in Ulithi lagoon, Yap State (Federated States of Micronesia), in 40 m of water by a manned suicide torpedo in November 1944. The *USS Mississinewa* sank with an estimated 9.6m. litres of oil on board, and after the storm of July 2001 it began to leak oil into one of the largest turtle hatcheries in the region. The US Navy acted quickly to seal the leaks, but not before up to 91,000 litres of oil had escaped into the lagoon and caused its closure to fishing. The hull soon began leaking again, however, and the oil eventually had to be pumped out and salvaged by the US Navy, which paid the US $5m. clean-up bill. By 2002 SPREP through PACPOL had set up a regional database that held details on more than 1,500 wrecks across the Pacific region. The next steps will be to identify a generic risk-assessment model to classify the sites, agree on the interventions for each risk category, assess each site and then begin active interventions. Other sea-based threats to the marine environment include anti-fouling paints, introduced rodents or organisms in ballast waters, nutrient enrichment from rusting steel and dispersion of fishing gear that may entangle marine life.

Globalization, the Dumping of Wastes and Nuclear Testing in the Pacific

With the outbreak of the Pacific War, the region assumed a strategic importance to the major powers, which it has not relinquished since. Unfortunately, this role has come at great environmental cost and in the five decades since the Second World War the region's huge distances, political dependence and relative isolation from the homelands of the major powers, such as the USA, the United Kingdom and France, have encouraged its use as a nuclear-testing site, a missile-testing zone, and as a storage and disposal site for chemical weapons. Nuclear testing by the USA began in the region at Bikini and Enewetak atolls in the Marshall Islands, some 4,800 km west of Honolulu and 7,700 km from the West Coast of the USA, in June 1946 and April 1948, respectively. By the time that testing stopped at Bikini atoll in 1958, it had been the site for 23 atmospheric atomic bomb tests. The world's first thermonuclear detonation, a prototype of the hydrogen bomb with a yield of 10.4 megatons (an explosive force 693 times more powerful than the atomic bomb that annihilated Hiroshima in 1945), was detonated at Enewetak atoll in November 1952. The explosion vaporized the island of Elugelab and left behind a 3-km-wide crater and a deeply fractured reef platform that cleaved away and plummeted into the ocean depths after an adjacent thermonuclear test in 1958. This was followed up by the even more powerful Bravo test of a hydrogen bomb at Bikini atoll in February 1954, which at 15 megatons was the most powerful bomb ever exploded by the USA (being 1,000 times more powerful than the atomic bomb dropped on Hiroshima). The force of this bomb was so great that it vaporized three islands and threw radioactive debris over nearly 50,000 sq miles. It also covered the atolls of Rongelap and Utirik and their residents with radioactive fall-out. By the time the USA ceased its nuclear-testing programme in the Marshall Islands in 1958, after 12 years of testing, it had detonated 67 nuclear bombs in and around the land, air and water of the territory. The bombs had a total yield of 108,496 kilotons, over 7,200 times more powerful than the atomic weapons used during the Second World War. The testing totally destroyed six small islands and hundreds of people had been irradiated. The Bikini and Enewetak islanders had been evacuated before the explosions, but peoples of Rongelap, Bikini, Enewetak, Utirik, Ailuk, Likrip and other Marshallese continue to suffer from cancer, miscarriages, and tumours resulting from the radioactive fall-out. Some 84% of those who lived on Rongelap and were aged less than 10 years at the time of the explosions have required surgery for thyroid tumours.

Almost 50 years since the cessation of testing, the Bikinians, Rongelapese and the Enjebi community from Enewetak are still 'nuclear nomads', unable to return to their native atolls and suffering hardships on islands less hospitable than their original homelands. The Enewetakese were unable to return home for 33 years, during which time they were exiled on desolate Ujelang atoll where the US Government acknowledged that they suffered grave deprivations. On the return of some islanders in 1980, they found a vastly different Enewetak, only 43% of the land area was habitable and some 8% or 154.4 acres had been vaporized. Other tests were carried out in the region on Kiritimati (Christmas) Island (one of the Line Islands and previously part of the British colony of the Gilbert and Ellice Islands —now respectively Kiribati and Tuvalu) by the United Kingdom from 1957 until 1958, and these were resumed again in 1962 at Kiritimati by the United Kingdom and the USA, in a successful attempt to force the USSR to agree to an atmospheric test ban treaty. All testing ceased at Kiritimati in 1962. Tests were also conducted at Johnston Atoll, a US possession 1,900 km to the north-west from 1958 to 1962. The French Government had also begun atmospheric nuclear testing in the Pacific at Mururoa atoll in the south-east corner of the Tuamotu archipelago in French Polynesia in 1966, and underground tests were begun in 1974. Over the period 1966–92 a total of 41 atmospheric tests and 138 underground tests were conducted. The testing was halted as part of the Nuclear Non-Proliferation

Treaty, in 1992, but then in the face of outrage from Pacific Island countries resumed for six more underground tests between 1995 and 1996 when they were finally halted. Although the atmospheric tests at Mururoa, Kiritimati and Johnston increased the risks to humans and aquatic life, they were smaller in yield than those at Bikini and Enewetak. They were also exploded at high altitude, unlike the larger ground-level tests and appear to have had a less harmful impact on the inhabitants who were only partly evacuated by the British, US and French authorities. A 1978 study of rats at Enewetak found possible inherited genetic effects caused by radiation, and there were some fears that genetic mutations could have occurred among marine life at Mururoa where radiation was reputed to have seeped from underground fissures. Certainly, French evidence has shown that years of testing cracked the atoll and probably altered the land plates. Of all the testing nations, the USA has probably been the most generous, not surprisingly given the massive scale of its testing and the damage caused to the Marshall Islands and their people. Since 1946 the USA has paid at least US $749m. in nuclear-related compensation to the Marshallese, but medical, clean-up and resettlement costs have continued to rise, and the US Congress has resisted requests for more assistance.

Although nuclear testing ceased in most of the Pacific Islands region in 1962 and at Mururoa in 1996, chemical weapons began to be stockpiled at Johnston Atoll in 1971. Incineration of these weapons in 'burnships' began in 1990 and was completed in 2000, with more than 400,000 rockets, projectiles, bombs, mortars and mines being destroyed. A clean-up of the atoll was commenced, and in June 2004 all military personnel departed and the island was transferred to the control of the US Fish and Wildlife Service. There is considerable concern, however, that pollution from the plutonium landfill could be absorbed by fish and the threat carried elsewhere. The radioactive rubble left behind has been contained by a sea wall, but the life of that sea wall is predicted to be less than 50 years. The threat of further contamination, however, continues. In the late 1980s, for example, there were at least 10 attempts to use the Pacific Islands as a place to install hazardous waste dumps, incineration sites or storage areas. These proposals were sophisticated and contained a multitude of financial incentives. The independent Governments of the Pacific Island countries rejected the proposals on environmental grounds. The Basel Convention came into force to ensure the safe shipment of hazardous wastes in 1992 and it was strengthened in 1994, to outlaw the export of hazardous waste from countries of the Organisation for Economic Co-operation and Development (OECD) to non-OECD countries. In 1995 the South Pacific Forum (now the Pacific Islands Forum) presented to its members the Waigani Convention to ban the import of hazardous and radioactive waste into the territory of its members and to control the trans-boundary movement of these wastes. Parties to the Waigani Convention are: Australia, the Cook Islands, the Federated States of Micronesia, Fiji, Kiribati, Nauru, New Zealand, Papua New Guinea, Samoa, Solomon Islands and Tuvalu.

IMPACT OF MINING ACTIVITIES

Mining has had a significant impact on the environment of many of the Pacific Islands, and, in the extreme case of Nauru, decades of phosphate mining have made most of the island a lunar landscape that is quite incapable of supporting its population. There are four kinds of mining in the Pacific Islands: mineral extraction (nickel, gold, silver, copper, iron, uranium and titanium); coal mining; construction mining (for fill, building stone and cement); and petroleum and gas extraction. Papua New Guinea, New Caledonia, Solomon Islands and Fiji are the major mineral mining centres, and Papua New Guinea also produces petroleum and natural gas, mostly from off shore. Mining in these countries has resulted in unavoidable localized damage to the environment from mine tailings, processing fumes and siltation of streams. In the mountainous areas of New Caledonia and Papua New Guinea, strip mining has been particularly damaging. For example, the Panguna copper mine on Bougainville was developed without strict environmental controls and dumped enormous quantities of tailings that turned the fertile Jabs and Kawerong river valleys into wastelands, made the water undrinkable and destroyed coastal fisheries. This caused enormous resentment among local villagers which, combined with Rio Tinto's failure to pay compensation to the local landowners, contributed to a civil war that led to the closure of the mine in 1989. The Ok Tedi gold and copper mine was set up with numerous controls to avoid the problems of Bougainville, and the agreement even contained clauses on how to close a mine after its resources were exhausted. However, when settlement ponds were destroyed in an earthquake and the mine was allowed to continue operations, the results were much the same, with sediments severely polluting the Fly and other nearby rivers and killing local gardens and fisheries. Tens of thousands of metric tons of tailings containing copper, zinc, cadmium and lead were dumped directly into the Fly and Ok Tedi rivers every day for two decades. Environmental damage was so severe that local landowners attempted to sue the mine owner for $A4,000m. in compensation. The case was dismissed on a technicality, and the Papua New Guinea landholders reluctantly accepted an out-of-court settlement. However, concern about the huge costs involved in the development of a comprehensive waste management scheme forced BHP Billiton to consider closing the mine, and in early 2002 it divested its shares to a local company. Given the importance of the mine to the Papua New Guinea economy (it accounted for 18% of the country's foreign-exchange earnings in 2000) and the fact that it employed, both directly and indirectly, 3,500 local villagers in 2002, the Program Company and the Papua New Guinea Government had little choice but to continue to operate the mine until its reserves run out in about 2007. Both the Panguna and Ok Tedi mines are examples of the dependence of even larger Pacific Island economies on major mining ventures and their limited ability to enforce strict environmental controls in the face of highly mobile international capital.

NATURAL HAZARDS—TSUNAMIS, EARTHQUAKES AND VOLCANIC ERUPTIONS

In addition to the threat of damage from severe cyclones generated in the tropical waters of the Pacific, many of the islands in the region are exposed to natural hazards such as volcanic eruptions, earthquakes and tsunamis because of their location on the so called 'Pacific Rim of Fire', the zone of collision of many of the earth's major tectonic plates. Some of the islands of the Pacific located on this 'Rim of Fire' are: New Britain, Solomon Islands, New Caledonia and Vanuatu, and as a result they contain dozens of active volcanoes and countless extinct craters and calderas. An example is along the north coast of Papua New Guinea where the Australian Plate is colliding with the subducting Pacific Plate to its north, along a major east–west line of subduction and tectonic activity. In September 1994 two active volcanoes in the Rabaul Caldera on the north-east tip of the island of New Britain in Papua New Guinea erupted and covered the city of Rabaul and its port in 1m–2m of ash and pumice. The eruptions were accompanied by earthquakes and tremors that were felt throughout East New Britain. A timely warning and rapid evacuation of the 30,000 inhabitants of the city saved thousands of lives and only two deaths were reported. A decade later the town had been largely abandoned, with many of the surrounding plantations and village gardens so badly damaged that it had been impossible to restore them to production. Another example of the localized but severe damage that can be caused by natural hazards associated with this zone of tectonic plate activity occurred in 1998. In July of that year a series of tsunamis (commonly, but misleadingly, termed tidal waves) struck the north-west coast of the Papua New Guinea mainland at Sissano. The tsunami struck the shoreline about 20 minutes after a nearby magnitude 7 earthquake shook the area. More than 25 km of coastline were swept by the waves that killed more than 2,200 people and reached a maximum measured water height of 15 m above sea-level, making it the largest tsunami related to a magnitude 7 earthquake in the 1990s. Subsequent research has indicated that the tsunami was generated by a massive underwater land slump located off shore, with the energy of the tsunami being amplified by wave refraction along two submarine canyons directly off Sissano lagoon. Unfortunately, offshore tectonic activity is likely to produce future earthquakes in this area, which will probably be accompanied by submarine subsidence that could once again expose the nearby low-lying coastline to tsunamis. The only policy approach is to attempt to minimize likely future damage and loss of life by encouraging villagers to build their houses in the more sheltered sites along this shoreline.

CONCLUSION

This brief survey of the environmental issues and problems of the Pacific Islands region has emphasized the very strong nexus between the geography of the region, its economic and political dependence, and its extreme vulnerability to environmental degradation and pollution. Although the geographic isolation and marginal political importance of most Pacific Island nations should at first sight protect these countries from the acute environmental pressures faced by the more economically developed and populous countries of the world, the reality has been quite different. Instead of being protected by these factors, the economic weakness of their economies, their extreme geographic isolation, their very limited natural resources and their historical links with major colonial powers have conspired to exacerbate their environmental vulnerability. In this context, and in view of rising expectations among their people, the Pacific Islands have been forced to become economically dependent on a few key industries, such as mining, logging and tourism. Their relative vulnerability to outside pressures has led to the exploitation of their marine and land resources and to the use of the region to test nuclear weapons and to store and dispose of chemical weapons. Even their geographical isolation has conspired to maximize the threats to their biological diversity. Many of them are located on the 'Pacific Rim of Fire', which has exposed them to the natural hazards associated with earthquake and volcanic activity and tsunamis. Their tropical waters have also been the breeding ground for major cyclones, which in recent years have devastated many of their island habitats. However, there now appears to be the beginnings of a new era of co-operation and self-help among the Pacific Island nations that, if fostered by genuine financial, technical and political assistance from nearby industrialized countries, may be able to minimize or even overcome their environmental problems.

BIBLIOGRAPHY

General

Allen, Bryant J. 'The Pacific Islands: Background to the region', in *The Far East and Australasia 2004*, 35th Edition. London, Europa Publications, pp.751–757, 2003.

Nukuro, E. 'A Glimpse at the Pacific Island Countries' Population, Development and Environmental Issues'. Nadi, Fiji Islands. Presented at the Millennium Media Conference on the Environment, 24–28 July, 2000.

Steele, Ross 'Environmental Issues of Asia and the Pacific', in *The Far East and Australasia 2005*, 36th Edition. London, Europa Publications.

Thistlethwaite, R., and Votaw, G. *Environment and Development: A Pacific Island Experience*. Manila, Asian Development Bank, 1992.

UN Economic and Social Commission for Asia and the Pacific (ESCAP). *Review of the State of the Environment of the Pacific Islands*. Kitakyushu, Japan, Ministerial Conference on Environment and Development in Asia and the Pacific 2000, September 2000; internet www.unescap.org/mced2000/pacific/SoE-pacific.htm; also www.unescap.org/mced2000/pacific/index.htm; link to 'Background information' for background papers on the major environmental issues in the Pacific Islands.

UN Environment Programme (UNEP). *Pacific Islands Environment Outlook*. London, United Nations Environment Programme, 1999; internet www.grid.unep.ch/geo2000/region/pieo.pdf; June 2004.

Demographic and Health Issues

Keith-Reid, R. 'Region's not Starving. But There are Worries', Pacific Magazine and Islands Business, September 2001; internet www.pacificislands.cc/pm92001/pmdefault.php?urlarticleid=0018.

Ragogo, M. 'Islanders 'Unhealthy' And Trend Worsening: Need to act now, or else...', Pacific Magazine and Islands Business, January 2004; internet www.pacificislands.cc/pm12004/pmdefault.php?urlarticleid=0031.

Secretariat of the Pacific Community, Population Programme. Population Poster. Nouméa, New Caledonia, March 2003; internet www.spc.org.nc/demog/; link to 'Recent Statistics'.

UN Economic and Social Commission for Asia and the Pacific (ESCAP). *Review of the State of the Environment of the Pacific Islands*. Kitakyushu, Japan, Ministerial Conference on Environment and Development in Asia and the Pacific 2000, September 2000; internet www.unescap.org/mced2000/pacific/SoE-pacific.htm; also www.unescap.org/mced2000/pacific/index.htm; links to 'Background information' and 'Demographic Issues'.

UN Economic and Social Commission for Asia and the Pacific (ESCAP), Statistics Division. Asia-Pacific in Figures Database; internet www.unescap.org/stat/data/apif/results_ind.asp; particular tabulations available for downloading.

Land Degradation, Deforestation and Loss of Biodiversity

Barnett, T. E. *Report of the Commission of Inquiry into Aspects of the Forest Industry*. Unpublished report to the Government of Papua New Guinea.

Brunton, B. D. *'Underlying Causes of Deforestation and Forest Degradation, Oceania and Pacific: Forest Loss in Papua New Guinea', Compendium of Discussion Papers in the Oceania Region*, September 1998; internet www.wrm.org.uy/deforestation/Oceania/Papua.html.

Clark, W. C. and Thaman, R. R. (Eds). *Agro-Forestry in the Pacific Islands: Systems for Sustainability*. Tokyo, United Nations University Press, 1993.

Conservation International. *Biodiversity Hotspots: New Caledonia*; internet www.biodiversityhotspots.org/xp/Hotspots/new_caledonia/; also information on Polynesia and Micronesia.

Earth Crash Earth Spirit (ECES). *Invasive Species: Brown tree snake*; internet eces.org/gallery/000442.php.

Forests Monitor Limited and Individual and Community Rights Advocacy Forum Inc. (ICRAF). *Environmental and Social Impact Assessment of Logging Operations in the Vanimo Timber Area, Sandaun Province, Papua New Guinea*. Forests Monitor Limited and ICRAF, May 1999.

Montagu, A. S. *Reforming Forest Planning and Management in Papua New Guinea, 1991–94: Losing People in the Process*, Journal of Environmental Planning and Management, Vol. 44, No. 5, pp. 649–662, 2001.

Forest Planning and Management in Papua New Guinea, 1884 to 1995: A Political Ecological Analysis, Planning Perspectives, Vol. 17, pp. 21–40, 2002.

Nunn, P. *Oceanic Islands*. Oxford, Blackwell, 1994.

PNG Forest Watch. *Far Less than the Sum of its Parts: An analysis of recommendations made for remediation in individual projects as part of the Independent Forestry Review*. PNG Forest Watch, November 2001; PNG Eco-Forestry Forum; internet www.ecoforestry.org.pg/publications.html.

Shearman, P., and Cannon, J. *PNG Forest Resources and the Log Export Industry* The Papua New Guinea Eco-Forestry Forum, Working Paper, 14 pages, April 2002; internet www.ecoforestry.org.pg/publications.html.

Steadman, D. W. *Prehistoric Extinctions of Pacific Island Birds: Biodiversity Meets Zooarchaeology*, Science, Vol. 267, pp. 1123–31.

United Nations Development Programme (UNDP). *Pacific Human Development Report: Putting People First*. Provisional edn, Suva, Fiji, 1994.

UN Economic and Social Commission for Asia and the Pacific (ESCAP). *Review of the Environment and Development Trends in the South Pacific*. ESCAP, Port Vila, 1995.

Freshwater Resources, Coastal Degradation, Coral Reefs, Coastal and Oceanic Fisheries

Gillett, R., McCoy, M., Rodwell, L. and Tamate, J. *Tuna: A Key Economic Resource in the Pacific Islands* Asian Development Bank and the Forum Fisheries Agency, Manila, 2001.

Langley, A., Hampton, J., and Williams, P. *The Western and Central Pacific Tuna Fishery: 2002 Overview and Status of Stocks*, Tuna Fisheries Assessment Report No. 5, Oceanic Fisheries Programme, Secretariat of the Pacific Community, Nouméa, 2004.

South Pacific Regional Environment Programme (SPREP). *South Pacific Regional Environment Programme Annual Report 1996/97*. SPREP, Apia, 1998.

Sustaining Pacific resources and development: SPREP Annual Report 2002. SPREP, Apia, 2003; internet www.sprep.org.ws/publication/pub_top.asph; 2002 annual report.

Pacific Islands Forum Fisheries Agency. *MCS Newsletter*. Forum Fisheries Agency, Honiara, 2004; internet www.ffa.int/docs/MCS.newsletter.v6.i2.pdf.

Climate Change and its Consequences for the Pacific Region

Barnett, J. *Adapting to Climate Change in Pacific Island Countries: The Problem of Uncertainty*, World Development, Vol. 29, pp. 977–993, 2001.

Connell, J. *Losing ground? Tuvalu, the greenhouse effect and the garbage can*, Asia Pacific Viewpoint, Vol. 44, No. 2, pp. 89–107, August 2003.

Harvey, N., and Mitchell, B. *Monitoring sea-level change in Oceania*, Tiempo, Issue 50, pp. 1–6, December 2003.

Inter-governmental Panel on Climate Change (IPCC). *IPCC Third Assessment Report: Contributions of IPCC Working Groups. Climate Change 2001: The Scientific Basis, Chapter 11*; internet www.grida.no/climate/ipcc_tar/wg1/429.htm.

McMichael, A., Woodruff, R., Whetton, P., Hennessy, K., Nicholls, N., Hales, S., Woodward, A. and Kjellstrom, T. *Human Health and Climate Change in Oceania: A Risk Assessment 2002*. Canberra. Australian Department of Health, 2003; internet www.health.gov.au/pubhlth/strateg/envhlth/climate/.

Land- and Sea-Based Pollution

Nawadra, S., Polglaze, J., Le Provost, I., Hayes, T., Lindsay, S., Pasisi, B., Hillman, S. and Hilliard, R. *Improving Ships' waste management in Pacific Islands Ports*. Apia, Samoa. South Pacific Regional Environment Programme (SPREP), 2002.

South Pacific Regional Environment Programme (SPREP). *Pollution in the Pacific*; internet www.sprep.org.ws/topic/pollution_p.htm.

A Regional Strategy to Address Marine Pollution from World War II Wrecks. Majuro, Marshall Islands. The Pacific Ocean Pollution Prevention Programme, 2002.

SPREP *Annual Report 2002: Sustaining Pacific resources and development*. Apia, Samoa. SPREP, 2003; internet www.sprep.org.ws/.

SPREP *Annual Report 2003: Conserving the environment for the peoples of the Pacific*. A Apia, Samoa. SPREP, 2004; internet www.sprep.org.ws/.

South Pacific Regional Environment Programme and Australian Agency for International Development (SPREP/AusAID). *Assessment of Persistent Organic Pollution in Pacific Island Countries*. Canberra. SPREP/AusAID, 1999.

UN Economic and Social Commission for Asia and the Pacific (ESCAP). *Pollution and Waste*. Kitakyushu, Japan, Ministerial Conference on Environment and Development in Asia and the Pacific 2000, Background paper, September 2000; internet www.unescap.org/mced2000/pacific/background/pollution.htm.

Globalization, the Dumping of Wastes and Nuclear Testing in the Pacific

Anon. *Britain's Nuclear Weapons: British Nuclear Testing*; internet nuclearweaponarchive.org/Uk/UKTesting.html.

Anon. *Johnston Island History*; internet www.janeresture.com/johnston/; 7 July 2004.

Anon. *Nuclear Weapons: US Atmospheric Nuclear Tests Page*; internet www.zvis.com/nuclear/usnuks.shtml.

Keever, B. D. *Un-Remembered Origins of 'Nuclear Holocaust': World's First Thermonuclear Explosion of Nov. 1, 1952*. Originally published by the *Honolulu Weekly*, 30 October, 2002; internet www.nuclearfiles.org/kinuclearweapons/25_keever_origins-nuclear-holocaust.htm.

Republic of the Marshall Islands. *The Republic of the Marshall Islands and the United States: A Strategic Partnership*; internet www.rmiembassyus.org/ updated 2002; link to 'Nuclear' to enter obtain details of the impact of nuclear testing on the Marshall Islands and their people.

Salvador, R. N. *The Nuclear History of Micronesia and the Pacific*. Waging Peace: Website of the Nuclear Age Peace Foundation. August 1999; internet www.nyu.edu/globalbeat/asia/Salvador0899.html.

Impact of Mining Activities

UN Economic and Social Commission for Asia and the Pacific (ESCAP). *Review of the State of the Environment of the Pacific Islands*. Kitakyushu, Japan, Ministerial Conference on Environment and Development in Asia and the Pacific 2000, September 2000; internet www.unescap.org/mced2000/pacific/SoE-pacific.htm; also www.unescap.org/mced2000/pacific/index.htm; links to 'Background information' and 'Pollution and Waste'.

Natural Hazards—Earthquakes, Tsunamis and Volcanoes

González, F. I. *Tsunami*. Scientific American, 18 May, 1999.

Watts, P., Matsuyama, M. and Tappin, D.R. *Potential landslide tsunamis near Aitape, Papua New Guinea* ITS (International Thermoelectric Society) 2001 Proceedings, Session 2, Nos 2–11, 2001.

United States Geological Survey. *Descriptive Model of the July 17, 1998 Papua New Guinea Tsunami*; internet walrus.wr.usgs.gov/tsunami/PNG.html.

AUSTRALIAN PACIFIC TERRITORIES

There are two external dependencies or Territories of the Commonwealth of Australia in the Pacific Ocean: the Coral Sea Islands Territory and Norfolk Island, the latter being self-governing. The Australian Minister for Territories is responsible for the administration of the dependencies, which lie within the jurisdiction of the Commonwealth Government.

Head of State: HM Queen ELIZABETH II (succeeded to the throne 6 February 1952).

Governor-General: Maj.-Gen. MICHAEL JEFFERY (assumed office 11 August 2003).

Department of Transport and Regional Services with responsibility for Territories and Local Government: GPOB 594, Canberra, ACT 2601, Australia; tel. (2) 6274-2111; fax (2) 6274-7706; e-mail publicaffairs@dotrs.gov.au; internet www.dotrs.gov.au.

Minister for Local Government, Territories and Roads: JIM LLOYD.

CORAL SEA ISLANDS TERRITORY

The Coral Sea Islands became a Territory of the Commonwealth of Australia under the Coral Sea Islands Act of 1969. The Territory lies east of Queensland, between the Great Barrier Reef and longitude 156° 06′E, and between latitude 12°S and 24°S, and comprises several islands and reefs. The islands are composed largely of sand and coral, and have no permanent fresh water supply, but some have a cover of grass and scrub. The area has been known as a notorious hazard to shipping since the 19th century, the danger of the reefs being compounded by shifting sand cays and occasional tropical cyclones. The Coral Sea Islands have been acquired by Australia by numerous acts of sovereignty since the early years of the 20th century.

Spread over a sea area of approximately 780,000 sq km (300,000 sq miles), all the islands and reefs in the Territory are very small, totalling only a few sq km of land area. They include Cato Island, Chilcott Islet in the Coringa Group, and the Willis Group. In 1997 the Coral Sea Islands Act was amended to include Elizabeth and Middleton Reefs. A meteorological station, operated by the Commonwealth Bureau of Meteorology and with a staff of four, has provided a service on one of the Willis Group since 1921. The other islands are uninhabited. There are eight automatic weather stations (on Cato Island, Flinders Reef, Frederick Reef, Holmes Reef, Lihou Reef, Creal Reef, Marion Reef and Gannet Cay) and several navigation aids distributed throughout the Territory.

The Act constituting the Territory did not establish an administration on the islands, but provides means of controlling the activities of those who visit them. The Lihou Reef and Coringa-Herald National Nature Reserves were established in 1982 to provide protection for the wide variety of terrestrial and marine wildlife, which include rare species of birds and sea turtles (one of which is the largest, and among the most endangered, of the world's species of sea turtle). The Australian Government has concluded agreements for the protection of endangered and migratory birds with Japan and the People's Republic of China. The Governor-General of Australia is empowered to make ordinances for the peace, order and good government of the Territory and, by ordinance, the laws of the Australian Capital Territory apply. The Supreme Court and Court of Petty Sessions of Norfolk Island have jurisdiction in the Territory. The Territory is administered by a parliamentary secretary appointed by the Minister for Local Government, Territories and Roads. The area is visited regularly by the Royal Australian Navy.

NORFOLK ISLAND

Physical and Social Geography

Norfolk Island lies off the eastern coast of Australia, about 1,400 km east of Brisbane, to the south of New Caledonia and 640 km north of New Zealand. The Territory also comprises uninhabited Phillip Island and Nepean Island, 7 km and 1 km south of the main island respectively. Norfolk Island is hilly and fertile, with a coastline of cliffs and an area of 34.6 sq km (13.4 sq miles). It is about 8 km long and 4.8 km wide. The climate is mild and subtropical, and the average annual rainfall is 1,350 mm, most of which occurs between May and August. The resident population numbered 2,037 at the census of August 2001, and consists of 'islanders' (descendants of the mutineers from HMS *Bounty*, evacuated from Pitcairn Island, who numbered 824 in 1996) and 'mainlanders' (originally from Australia, New Zealand or the United Kingdom). Of the 2,037 residents, 1,574 were living permanently on Norfolk Island in 2001. English is the official language, but a Norfolk Island dialect (a mixture of old English and Tahitian) is also spoken. The capital of the Territory is Kingston.

History

The island was uninhabited when discovered in 1774 by a British expedition, led by Capt. James Cook. Norfolk Island was used as a penal settlement from 1788 to 1814 and again from 1825 to 1855, when it was abandoned. In 1856 it was resettled by 194 emigrants from Pitcairn Island, which had become overpopulated. Norfolk Island was administered as a separate colony until 1897, when it became a dependency of New South Wales. In 1913 control was transferred to the Australian Government.

Under the Norfolk Island Act 1979, Norfolk Island is progressing to responsible legislative and executive government, enabling it to manage its own affairs to the greatest practicable extent. Wide powers are exercised by the nine-member Legislative Assembly and by the Executive Council, comprising the executive members of the Legislative Assembly who have ministerial-type responsibilities. The Act preserves the Australian Government's responsibility for Norfolk Island as a Territory under its authority, with the Minister for Territories as the responsible minister. The Act indicated that consideration would be given within five years to an extension of the powers of the Legislative Assembly and the political and administrative institutions of Norfolk Island. In 1985 legislative and executive responsibility was assumed by the Norfolk Island Government for public works and services, civil defence, betting and gaming, territorial archives and matters relating to the exercise of executive authority. In 1988 further amendments empowered the Legislative Assembly to select a Norfolk Island Government Auditor (territorial accounts were previously audited by the Commonwealth Auditor-General). The office of Chief Minister was replaced by that of the President of the Legislative Assembly. David Ernest Buffett was reappointed to this post following the May 1992 general election. A lack of consensus among members of the Executive Council on several major issues prompted early legislative elections in April 1994. The newly elected Legislative Assembly was remarkable in having three female members. Following elections in April 1997, in which 22 candidates contested the nine seats, George Smith was appointed President (subsequently reverting to the title of Chief Minister) of the eighth Legislative Assembly. At legislative elections in February 2000 three new members were elected to the Assembly, and Ronald Nobbs was subsequently appointed Chief Minister. Geoffrey Gardner, hitherto Minister for Health, replaced Nobbs as Chief Minister following the elections of November 2001. The incoming Assembly included four new members.

In December 1991 a referendum took place at which a proposal by the Australian Government to include Norfolk Island within the Australian federal electorate was overwhelmingly rejected by the islanders. The outcome of the poll led the Australian Government, in June 1992, to announce that it had abandoned the plans. Similarly, in late 1996 a proposal by the Australian Government to combine Norfolk Island's population with that of Canberra for record-keeping purposes was strongly opposed by the islanders.

In late 1997 the Legislative Assembly debated the issue of increased self-determination for the island. Pro-independence supporters argued that the Territory could generate sufficient income by exploiting gas- and oilfields in the island's exclusive economic zone.

In August 1998 a referendum proposing that the Norfolk Island electoral system be integrated more closely with that of mainland Australia (initiated by the Minister for Regional Development, Territories and Local Government) was rejected by 78% of the Territory's electorate. A similar referendum in May 1999 was opposed by 73% of voters.

Frustration with the Australian Government's perceived reluctance to facilitate the transfer of greater powers to the Territory (as outlined in the Norfolk Island Act of 1979, see above) led the island's Legislative Assembly in mid-1999 to vote by seven members to one in favour of full internal self-government. Negotiations regarding the administration of crown land on the island, which continued in 2000, were seen as indicative of the islanders' determination to pursue greater independence from Australia.

In April 2002 a hotel worker from Sydney, Australia, was murdered. This was the first murder to occur on Norfolk Island in more than 150 years. As part of its investigation, the police force took fingerprints of virtually the entire adult population in August. The Australian authorities stated in April 2003 that they were no closer to solving the crime and in July announced that they had contacted more than 400 Australian and New Zealand nationals, who were visiting Norfolk Island at the time of the murder, to request fingerprints. In March 2004 the reward for information about the murder was tripled, and in May a further inquest was launched.

At a referendum held in August 2002 almost two-thirds of those who participated voted not to allow the operation of mobile telephones on the island. Furthermore, in March 2004 the Territory approved legislation to grant a monopoly on satellite internet connections to the government-owned Norfolk Telecom. The legislation, which was expected to put the Territory's private internet service provider, Norfolk Island Data Services, out of business, was criticized by the Australian Government.

The inhabitants of Norfolk Island were profoundly shocked when Deputy Chief Minister and long-standing political figure Ivens Buffett was shot dead in his office in mid-July 2004. A local man was arrested at the scene of the crime, although no motive was immediately apparent for the killing. It was subsequently revealed that the suspect was the politician's son.

A total of 14 candidates contested the legislative election held on 20 October 2004. Geoffrey Gardner, the Chief Minister, regained his seat in the nine-member Legislative Assembly, as did Speaker David Buffett. Two long-serving members, however, were defeated in the poll.

Legislation was approved in late March 2003 to amend the requirements to vote in Norfolk Island elections. Under the new system Australian, New Zealand and British citizens were to be allowed to vote after a residency period of 12 months (reduced from 900 days). The amendments, which followed a series of occasionally acrimonious discussions with the Australian Government, provoked concern among islanders who feared that succumbing to Australian pressure to reform the Norfolk Island Act would result in the effective removal of authority over electoral matters from island control. Moreover, a report by an Australian parliamentary committee published in July 2003 was critical of Norfolk Island's Government and public services. Many residents believed the report constituted a further attempt by Australia to undermine their autonomy. The Chief Minister refuted the committee's claims stating that he would refuse any offers of financial assistance from Canberra or any moves to introduce income tax in return for access to services from the mainland.

Economy

Despite the island's natural fertility, agriculture is no longer the principal economic activity. About 400 ha of land are arable. The main crops are Kentia palm seed, cereals, vegetables and fruit. Cattle and pigs are farmed for domestic consumption. Development of a fisheries industry is restricted by the lack of a harbour. Some flowers and plants are grown commercially. The administration is increasing the area devoted to Norfolk Island pine and hardwoods. Seed and seedlings of the Norfolk Island pine are exported. Potential oil- and gas-bearing sites in the island's waters may provide a possible future source of revenue. A re-export industry has developed to serve the island's tourist industry. In early 1999 the Norfolk Island Legislative Assembly announced plans to seek assistance from the Australian Government to establish an 'offshore' financial centre on the island. The Government is the most important employer and there were 151 public-sector workers in June 2003.

In 2000/01 the cost of the island's imports totalled almost $A41.3m. (compared with $A59.1m. in 1999/2000), while revenue from exports amounted to $A2.7m. (compared with almost $A2.8m. in the previous year). Norfolk Island's trade is conducted mainly with Australia and New Zealand. Imports from Australia in 2000/01 cost $A19m. (compared with almost $A23m. in the previous year). In 1999/2000 exports to Australia earned $A1m. but they were negligible in 2000/01. In 1999/2000 imports from New Zealand totalled $A9.3m. (compared with $A9.0m. in 1998/99). The authorities receive revenue from customs duties (some $A3.9m., equivalent to 34.3% of total revenue in 2000/01), mail order services and the sale of postage stamps, but tourism is the island's main industry, with departure fees providing a major source of revenue. In 2002/03 there were 37,671 tourist arrivals on the island. In 1985 and 1986 the Governments of Australia and Norfolk Island jointly established the 465-ha Norfolk Island National Park. This was to protect the remaining native forest, which is the habitat of several unique species of flora (including the largest fern in the world) and fauna (such as the Norfolk Island green parrot, the guavabird and the boobook owl). Conservation efforts include the development of Phillip Island as a nature reserve.

Statistical Survey

Source: The Administration of Norfolk Island, Administration Offices, Kingston, Norfolk Island 2899; tel. 22001; fax 23177; internet www.norfolk.gov.nf.

AREA AND POPULATION

Area: 34.6 sq km (13.3 sq miles).

Population: 2,181, including 409 visitors, at census of 6 August 1996; 2,601 (males 1,257, females 1,344), including 564 visitors, at census of 7 August 2001.

Density (2001): 75.2 per sq km.

Births, Marriages and Deaths (2002/03): Live births 13; Marriages 35; Deaths 10.

Economically Active Population (persons aged 15 years and over, 2001 census): 1,609 (males 849, females 760).

FINANCE

Currency and Exchange Rates: Australian currency is used.

Budget (year ending 30 June 2003): Revenue $A23,251,781; Expenditure $A19,440,903.

EXTERNAL TRADE

2000/01 (year ending 30 June): *Imports:* $A41,260,213, mainly from Australia and New Zealand. *Exports:* $A2,708,120.

Trade with Australia ($A million, 2000/01): *Imports* 19.

Trade with New Zealand ($NZ million, 2000/01): *Imports* 6.7; *Exports* 0.04.

TOURISM

Visitors (year ending 30 June): 40,221 in 2000/01; 33,619 in 2001/02; 37,671 in 2002/03.

COMMUNICATIONS MEDIA

Radio Receivers (1996): 2,500 in use.

Television Receivers (1996): 1,200 in use.

Telephones (2002/03): 2,374 main lines in use.

Internet Users (2002/03): 494.

Non-daily Newspaper (2002): 1 (estimated circulation 1,400).

EDUCATION

Institution (2003): 1 state school incorporating infant, primary and secondary levels.

Teachers (2002/03): 22.

Students (1999/2000): Infants 79; Primary 116; Secondary 119.

Directory

The Constitution

The Norfolk Island Act 1979 constitutes the administration of the Territory as a body politic and provides for a responsible legislative and executive system, enabling it to administer its own affairs to the greatest practicable extent. The preamble of the Act states that it is the intention of the Australian Parliament to consider the further extension of powers.

The Act provides for an Administrator, appointed by the Australian Government, who shall administer the government of Norfolk Island as a territory under the authority of the Commonwealth of Australia. The Administrator is required to act on the advice of the Executive Council or the responsible Commonwealth Minister in those matters specified as within their competence. Every proposed law passed by the Legislative Assembly must be effected by the assent of the Administrator, who may grant or withhold that assent, reserve the proposed law for the Governor-General's pleasure or recommend amendments.

The Act provides for the Legislative Assembly and the Executive Council, comprising the executive members of the Assembly who have ministerial-type responsibilities. The nine members of the Legislative Assembly are elected for a term of not more than three years under a cumulative method of voting: each elector is entitled to as many votes (all of equal value) as there are vacancies, but may not give more than four votes to any one candidate. The nine candidates who receive the most votes are declared elected.

The Government

The Administrator, who is the senior representative of the Commonwealth Government, is appointed by the Governor-General of Australia and is responsible to the Minister for Regional Services, Territories and Local Government. A form of responsible legislative and executive government was extended to the island in 1979, as outlined above.

Administrator: ANTHONY J. MESSNER (assumed office on 4 August 1997).

EXECUTIVE COUNCIL
(September 2004)

Chief Minister and Minister for Intergovernment Relations: GEOFFREY R. GARDNER.

Minister for Finance: GRAEME DONALDSON.

Minister for Tourism and Community Services: DAVID E. BUFFETT.

Minister for Land and the Environment: (vacant).

MINISTRIES

All Ministries are located at: Old Military Barracks, Quality Row, Kingston, Norfolk Island 2899; tel. 22003; fax 23378; e-mail executives@assembly.gov.nf.

GOVERNMENT OFFICES

Office of the Administrator: New Military Barracks, Norfolk Island 2899; tel. 22152; fax 22681.

Administration of Norfolk Island: Administration Offices, Kingston, Norfolk Island 2899; tel. 22001; fax 23177; e-mail ljohnson@admin.gov.nf; internet www.norfolkisland.gov.nf; all govt depts; CEO LUKE JOHNSON.

Legislature

LEGISLATIVE ASSEMBLY

Nine candidates are elected for not more than three years. The most recent general election was held on 20 October 2004.

Speaker: DAVID E. BUFFETT.

Deputy Speaker: to be appointed.

Other Members: RONALD C. NOBBS, GEOFFREY R. GARDNER, NEVILLE C. CHRISTIAN, TIMOTHY J. SHERIDAN, LORRAINE C. BOUDAN, TIMOTHY J. BROWN, VICKY JACK, JOHN T. BROWN.

Judicial System

Supreme Court of Norfolk Island

Kingston.

Appeals lie to the Federal Court of Australia.

Chief Magistrate: RON CAHILL.

Judges: BRYAN ALAN BEAUMONT (Chief Justice), MURRAY RUTLEDGE WILCOX.

Religion

The majority of the population professes Christianity (66.6%, according to the census of 2001), with the principal denominations being the Church of England (34.9%), the Catholic Church (11.7%) and the Uniting Church (11.2%).

The Press

Norfolk Island Government Gazette: Kingston, Norfolk Island 2899; tel. 22001; fax 23177; internet www.info.gov.nf/gazette; weekly.

Norfolk Islander: Greenways Press, POB 150, Norfolk Island 2899; tel. 22159; fax 22948; e-mail news@islander.nf; f. 1965; weekly; Co-Editors TOM LLOYD, TIM LLOYD; circ. 1,350.

Broadcasting and Communications

TELECOMMUNICATIONS

Norfolk Telecom: New Cascade Rd, Kingston; internet www.telecom.gov.nf; Man. KIM DAVIES.

BROADCASTING

Radio

Norfolk Island Broadcasting Service: New Cascade Rd, POB 456, Norfolk Island 2899; tel. 22137; fax 23298; e-mail 2niradio@ni.net.nf; govt-owned; non-commercial; broadcasts 112 hours per week; relays television and radio programmes from Australia; Broadcast Man. ROGER NEWMAN.

Radio VL2NI: New Cascade Rd, POB 456, Norfolk Island 2899; tel. 22137; fax 23298; e-mail 2niradio@ni.net.nf; internet www.users.nf/nfradio/.

Television

Norfolk Island Broadcasting Service: see Radio.

Norfolk Island Television Service: f. 1987; govt-owned; relays programmes of Australian Broadcasting Corpn, Special Broadcasting Service Corpn and Central Seven TV by satellite.

TV Norfolk (TVN): locally operated service featuring programmes of local events and information for tourists.

Finance

BANKING

Commonwealth Bank of Australia (Australia): Taylors Rd, Norfolk Island 2899; tel. 22144; fax 22805.

Westpac Banking Corpn Savings Bank Ltd (Australia): Burnt Pine, Norfolk Island 2899; tel. 22120; fax 22808.

Trade

Norfolk Island Chamber of Commerce Inc: POB 370, Norfolk Island 2899; tel. 22317; fax 23221; e-mail photopress@ni.net.nf; f. 1966; affiliated to the Australian Chamber of Commerce; 60 mems; Pres. GARY ROBERTSON; Sec. MARK MCGUIRE.

Norfolk Island Gaming Authority: POB 882, Norfolk Island 2899; tel. 22002; fax 22205; e-mail secgameauth@norfolk.net.nf; internet www.gamingauthority.nlk.nf; Dir KEVIN LEYSHON.

Transport

ROADS

There are about 100 km of roads, including 85 km of sealed road.

SHIPPING

Norfolk Island is served by the three shipping lines, Neptune Shipping, Pacific Direct Line and Roslyndale Shipping Company Pty Ltd. A small tanker from Nouméa (New Caledonia) delivers petroleum products to the island and another from Australia delivers liquid propane gas.

CIVIL AVIATION

Norfolk Island has one airport, with two runways (of 1,900 m and 1,550 m), capable of taking medium-sized jet-engined aircraft. Air New Zealand operates a twice-weekly direct service between Christchurch and Norfolk Island (via Auckland). Charter flights from Lord Howe Island and occasionally from New Caledonia also serve the island. The cessation of scheduled services from Australia by Ansett

Australia in 1997 had an adverse effect on the island's important tourist industry. As a consequence, Norfolk Jet Express was established to provide a weekly service to Australia. In February 1999 Air Nauru began to operate a weekly charter flight service to Norfolk Island from Sydney, under contract with Norfolk Jet Express. In mid-2002 services from Brisbane and Sydney were also being provided by Alliance Airlines. The latter were increased to three times per week in September 2003.

Tourism

Visitor arrivals totalled 37,671 in 2002/03, of whom some 81% came from Australia, with most of the remainder travelling from New Zealand. In that year tourist accommodation totalled 1,551 beds.

Norfolk Island Visitors Information Centre: Taylors Rd, Burnt Pine, POB 211, Norfolk Island 2899; tel. 22147; fax 23109; e-mail info@norfolkisland.com.au; internet www.norfolkisland.com.au; Gen. Man. BRUCE WALKER.

Education

Education is free and compulsory for all children between the ages of six and 15. Pupils attend the one government school from infant to secondary level. In 2002/03 a total of 187 pupils were enrolled at infant and primary levels and 118 at secondary levels. Students wishing to follow higher education in Australia are eligible for bursaries and scholarships. The budgetary allocation for education was $A2.12m. in 2002/03 (of which $A1.93m. was recurrent expenditure).

FIJI

Physical and Social Geography

The Republic of Fiji lies in the south-west Pacific Ocean, south of the Equator, 1,770 km north of Auckland (New Zealand) and 2,730 km north-east of Sydney (Australia). To the west lies Melanesia: Solomon Islands in the north-west, Vanuatu and New Caledonia. East of Fiji is Tonga and, in the north-east, other Polynesian islands, those of Wallis and Futuna and Western Samoa. Tuvalu is to the north. The Fiji group comprises four main islands, Viti Levu (where 70% of the population lives), Vanua Levu, Taveuni and Kadavu and some 840 smaller islands, atolls and reefs, of which fewer than 100 are inhabited. The island of Rotuma, 386 km (240 miles) north of Vanua Levu, and the eight smaller islands of the group also constitute part of the Republic. The total area of the Republic of Fiji is 18,376 sq km (7,095 sq miles). The climate is tropical, with temperatures ranging from 16° to 32°C (60° to 90°F). Rainfall is heaviest between November and April, but is more constant on the windward side.

Fiji is characterized by racial diversity. The indigenous Fijian population declined sharply during the 1850s, owing to epidemics of measles and influenza in which thousands died, and only in the 1950s did it begin to rise. The Indian population was originally brought to Fiji as labour for the canefields from 1879. The population at the census of August 1986 was 715,375, of whom 48.7% were Indians and 46.1% Fijians. Following the coups of 1987, there was emigration on a large scale, particularly from among the Indian community. In 1989 official statistics claimed that ethnic Fijians again formed the largest part of the population; and by 1996 it was estimated that ethnic Fijians comprised 51.1% of the population, Indians 43.6% and others 5.3%. In 1986 53% of the population were Christians (mainly Methodists), 38% were Hindus and 8% Muslims. Fiji's population totalled 819,000 in mid-2002. English is the official language, but Fijian (the principal dialect being Bauan) and Hindi (the locally developed dialect being known as Hindustani) are widely spoken. The capital is Suva on Viti Levu.

History

The Fijian islands were settled some 3,500 years ago, by Melanesian and Polynesian peoples. The first Europeans to settle on the islands were sandalwood traders, missionaries and shipwrecked sailors. Under their influence, local fighting and jealousies reached unprecedented heights until, by the 1850s, one Ratu (chief), Thakombau, had gained a tenuous influence over the whole of the western islands. Thakombau ran foul of US interests during the 1850s and turned to the British for assistance, unsuccessfully at first, but in 1874 Britain agreed to a second offer of cession, and Fiji was proclaimed a British possession. The island of Rotuma and its dependencies were added to the territory in 1881. Fiji became independent, within the Commonwealth, on 10 October 1970.

The racial diversity, compounded by actions of the past colonial administrations, presents Fiji with one of its most difficult problems. The colonial Government consistently favoured the Fijian population, protecting them from exploitation and their land from alienation, but allowed the importation of foreign labour. Approximately 80% of the islands were owned by Fijian communities, but over 90% of the sugar crop, Fiji's largest export, was produced by Indians, usually on land leased from Fijians. Until the mid-20th century Indians were poorly represented politically, while Fijians had their own administrative and judicial systems. The army is comprised almost entirely of ethnic Fijians.

After independence, Fijian politics were, for a long time, dominated by Ratu Sir Kamisese Mara, leader of the Alliance Party (AP). In a general election in 1977, the Indian-led opposition won a majority of seats in the House of Representatives, but failed to form a government because its leaders were uncertain whether Fijians would accept an Indian leadership. At elections in 1982 the AP won 28 seats, while the National Federation Party (NFP) won 22 seats.

Following tension between the Government and its opposition in Parliament (supported by the labour movement), a meeting of union leaders in May 1985 represented the start of discussions which culminated in the founding of the Fiji Labour Party (FLP), officially inaugurated in Suva in July 1985. Sponsored by the Fijian Trades Union Conference (FTUC), and under the presidency of Dr Timoci Bavadra, the new party was formed with the aim of presenting a more effective parliamentary opposition, and announced the provision of free education and a national medical scheme to be among its priorities.

At a general election in April 1987 a coalition of the FLP and the NFP won 28 seats in the House of Representatives, thereby defeating the ruling AP, which won 24 seats. Bavadra subsequently took office as Prime Minister, leading the first Fijian Government to be dominated by MPs of ethnic Indian, rather than ethnic Fijian (i.e. Melanesian), origin. Bavadra, a commoner, announced the formation of a new Cabinet, comprising seven Indians and five Fijians (including Bavadra himself). However, on 14 May 1987 the Government was removed from power by a military coup, led by Lt-Col (later Maj.-Gen.) Sitiveni Rabuka, who forcibly abducted and imprisoned Bavadra and the other 27 members of the coalition Government, seeking to justify his unconstitutional action by claiming that the militant ethnic Fijian Taukei Movement had been planning to attack Bavadra and his Government. Rabuka immediately formed a 17-member interim ruling Council (which included Mara, the former Prime Minister), of which he declared himself the Chief Minister. The Governor-General, Ratu Sir Penaia Ganilau, refused to recognize Rabuka's administration (although he subsequently swore in Rabuka as its Chief Minister), and declared a state of emergency. In an attempt to resolve the crisis, Ganilau appointed a 19-member Advisory Council, comprising Rabuka and seven other members of his intended Council of Ministers, together with members of Fiji's Great Council of Chiefs or Bose Levu Vakaturaga (a traditional body comprising every hereditary chief, or Ratu, of a Matagali, or Fijian clan) and a number of public servants. Bavadra, one other member of the elected Government (all of whose members had been released from detention), and another Indian were offered posts on the Advisory Council, but declined to participate, on the grounds that the Council was unconstitutional and biased in its composition. Of the remaining 16 appointed members, most were supporters of Mara's AP and only one was an ethnic Indian.

Widespread racial violence, continuing demonstrations of protest against the interim administration and public demands for Bavadra's reinstatement as Prime Minister led to increased civil tension and political uncertainty. In July 1987 the Great Council of Chiefs approved proposals for constitutional reform, including an increase in the number of seats in the House of Representatives, more than one-half of which were to be held by ethnic Fijians. According to the proposals, the Prime Minister was invariably to be of Melanesian origin, and was to be appointed only by Fijian MPs. Negotiations, between delegations led by Bavadra and Mara, took place in September, despite violence perpetrated (allegedly by the radical Taukei Movement) against one of Bavadra's spokesmen, and it was announced that the two factions had agreed to form an interim bipartisan Government.

The implementation of this compromise plan was, however, forestalled on 25 September 1987 by a second *coup d'état*, again led by Rabuka, who announced his intention of declaring a republic. Ganilau, who refused to recognize Rabuka's seizure of power, sought to reconcile the opposing factions. Negotiations between Ganilau, Rabuka, Bavadra and Mara took place on 29–30 September, but collapsed on 1 October, when Rabuka formally revoked the Constitution and deposed Queen Elizabeth II as Head of State. Further efforts were made to seek a new arrangement whereby Ganilau would remain in office at the head of a constitutional government, but these were abandoned on 7 October, when Rabuka installed an interim Council of Ministers, of which more than one-half of the members were drawn from the Taukei Movement. Rabuka, who became Minister for Home Affairs and the Public Service (with control of the armed forces), stated that the Council would remain in power for at least one year, during which time a new constitution, guaranteeing the dominance of ethnic Fijians, would be promulgated. On 15 October Ganilau, who had refused to accept the presidency of a Republic of Fiji, resigned as Governor-General, and Fiji was deemed to have left the Commonwealth. The Chief Justice and senior judges, who had opposed the coup, were removed from office, and Rabuka declared himself Head of State. While the new Government received recognition from France, the Governments of Australia and New Zealand condemned the Rabuka regime. On 6 December Rabuka resigned as Head of State, and Ganilau, the former Governor-General, was appointed the first President of the Fijian Republic, although he had earlier refused to accept the post. Mara was reappointed Prime Minister, and Rabuka became Minister of Home Affairs. The new Cabinet contained no member of Bavadra's deposed Government.

In February 1988 Rotuma (the only Polynesian island in the country), which lies to the north-west of Vanua Levu, attempted to declare independence from Fiji, announcing that it did not recognize Fiji's newly declared status as a republic and affirming its continued loyalty to the Commonwealth. However, Fijian troops were dispatched to the island and soon quelled the dissent. In the constitutional settlement of 1990, however, Rotumans received a special status, with one seat in each of the houses of Parliament.

A new draft Constitution was approved by the interim Government in September 1988. A constitutional committee, which was multiracial but included no member of the former Bavadra Government, received submissions during the following year and published a revised draft in September 1989. Bavadra and the FLP-NFP coalition, however, continued to condemn the proposals.

In January 1989 statistical information, released by the interim Government, indicated that the islands' ethnic Fijians were the largest group in the population for the first time since 1946. Some 9,500 ethnic Indians and 2,800 others had emigrated since the coup of May 1987.

In November 1989 Bavadra died and was succeeded as leader of the FLP by his widow, Adi (a female honorific corresponding to the chiefly title of Ratu) Kuini Bavadra. In October Mara agreed to remain as Prime Minister only if Rabuka left the Cabinet and returned to his military duties. Rabuka and two other army officers duly left the Cabinet in January 1990.

In June 1990 the Great Council of Chiefs approved the draft Constitution. It also initiated the formation of a new party, the Soqosoqo ni Vakavulewa ni Taukei (SVT) or Fijian Political Party, to advocate the cause of ethnic Fijians. The new Constitution was promulgated on 25 July 1990 by President Ganilau. This was reported to have been prompted by fears of another coup. The FLP-NFP coalition immediately condemned the Constitution and announced that it would not participate in any elections held under it. The opposition criticized the legislative majority of ethnic Fijians, who were reserved 37 of the 70 elective seats, compared with 27 Indian seats. Furthermore, the Great Council of Chiefs was to nominate ethnic Fijians to 24 of the 34 seats in the Senate and to appoint the President of the Republic.

In July 1991 Rabuka resigned as Commander of the Armed Forces in order to join the Cabinet as Deputy Prime Minister and Minister of Home Affairs, but later that year he relinquished the post to assume the leadership of the SVT.

Disagreements between the Government and the FTUC re-emerged at the beginning of 1991. In February a strike by more than 900 members of the Fijian Miners' Union over union recognition, pay and poor working conditions led to the dismissal of some 400 of the workers. In May the Government announced a series of reforms to the labour laws, including the abolition of the minimum wage, restrictions on strike action and derecognition of unions that did not represent at least two-thirds of the work-force. A significant announcement by the Government in late 1992 was the official recognition of the FTUC (withheld since 1986) as the sole representative of workers in Fiji.

At legislative elections held in May 1992 the SVT secured 30 of the 37 seats reserved for ethnic Fijians, while the NFP won 14 and the FLP 13 of the seats reserved for Indian representatives. Following the election, the FLP agreed to participate in Parliament and to support Rabuka in his campaign for the premiership, in return for a guarantee from the SVT of a full review of the Constitution and of trade union and land laws. Rabuka was, therefore, appointed Prime Minister, and formed a coalition Government (consisting of 14 members of the SVT and five others).

In July 1992 a report was published, detailing the findings of a corruption inquiry, undertaken following the military coups of 1987. Rabuka aroused some controversy by ordering that the report remain 'classified'. Nevertheless, in December Rabuka formally invited the opposition leaders, Jai Ram Reddy of the NFP and Mahendra Chaudhry of the FLP, to form a government of national unity. The move was largely welcomed, but Indian politicians expressed reluctance to take part in a government whose political control remained fundamentally vested with ethnic Fijians. Rabuka was criticized equally by nationalist extremists of the Taukei Solidarity Movement, who accused him of conceding too much political power to Fijian Indians. Following the appointment of a new Cabinet in June 1993, all 13 of the FLP members began an indefinite boycott of Parliament, in protest at Rabuka's failure to implement the reforms, which he had agreed to carry out in return for their support for his election to the premiership in June 1992.

In December 1993 President Ganilau died, following a long illness, and was replaced by Ratu Sir Kamisese Mara, who took office in January 1994 (re-elected on 18 January 1999). At legislative elections held in February 1994 the SVT increased the number of its seats in the House of Representatives to 31, while the Fijian Association Party (FAP, formed in January by former members of the SVT) secured only five seats, of a total of 37 reserved for ethnic Fijians. Of the 27 seats reserved for ethnic Indian representatives, 20 were secured by the NFP. The SVT subsequently formed a governing coalition with the General Voters' Party (GVP) and an independent member, under the premiership of Rabuka, who announced the formation of a new Cabinet composed entirely of ethnic Fijians. In response to international concern regarding the continued existence of Fiji's racially biased Constitution, Rabuka announced in June 1994 that the Constitutional Review Commission had been established, which, it was hoped, would have completed a review of the Constitution by 1997.

In January 1995 the Government announced that it was to recommend that Parliament vote to repeal the Sunday observance law (imposed after the coups of 1987), which prohibited work and organized entertainment and sport on that day. The announcement aroused intense opposition from nationalist politicians and Methodist church leaders, who organized demonstrations in three cities, attended by more than 12,000 people. In February, however, the House of Representatives voted in favour of removing the regulations. The Senate narrowly rejected the proposal (by 15 votes to 14), thus effectively delaying the implementation of any changes. The Sunday observance law was finally repealed in November 1995.

The issue of independence for the island of Rotuma was raised again in September 1995 with the return of the King of Rotuma from exile in New Zealand. King Gagaj Sa Lagfatmaro, who had fled to New Zealand following death threats made against him during the military coups of 1987, appeared before the Constitutional Review Commission to petition for the island's independence within the Commonwealth, reiterating his view that Rotuma remained a British colony rather than a part of Fiji.

In September 1995 the Government decided to transfer all state land (comprising some 10% of Fiji's total land area), hitherto administered by the Government Lands Department, to the Native Land Trust Board. The decision was to allow the allocation of land to indigenous Fijians on the basis of native custom. However, concern among the Fijian Indian population increased following reports in early 1996 that many would not be able to renew their land leases (most of which were due to expire between 1997 and 2024 under the Agricultural Landlords and Tenants Act—ALTA). The reports were strongly denied by the Government, despite statements by several Fijian land-owning clans that Indians' leases would not be renewed. Moreover, a recently formed sugar cane growers' association solely for ethnic Fijians, the Taukei Cane-Growers' Association, announced its intention to campaign for ethnic Fijian control of the sugar industry, largely by refusing to renew land leases to ethnic Indians (who held some 85% of sugar farm leases). In mid-1999 a government survey indicated that at least 70% of leases were likely to be renewed under ALTA, although there were continuing reports that ethnic Fijian landowners were unwilling to sign new contracts under the act. Concern was expressed that mounting tensions between landowners and tenants might lead to violence, and that the situation was affecting investor confidence. The Government announced that it was considering alternative proposals, including the transferral of land to the Native Land Trust Board, under which leases were subject to more frequent renewal than under ALTA.

Meanwhile racial tension intensified in October 1995, following the publication of the SVT's submission to the Constitutional Review Commission. In its report the party detailed plans to abandon the present multiracial form of government, recommending instead the adoption of an electoral system based on racial representation, in which each ethnic group selects its own representatives. The expression of numerous extreme anti-Indian sentiments in the document was widely condemned as offensive. Josefata Kamikamica of the FAP was one of several political leaders to describe the submission as disgraceful and insulting to Fijian, as well as to Indian, sensibilities.

A rift within the GVP in early 1996, which resulted in two of the four GVP members of the House of Representatives withdrawing their support for the Government, prompted Rabuka to seek alternative coalition partners from among the opposition, in an attempt to establish a more secure majority. However, the Prime Minister was unsuccessful in persuading parliamentary members of the FAP to join the Government. The administration's troubles during 1996 contributed to the defeat of the SVT in virtually every municipality at local elections, which took place in September.

Existing divisions within the Government were further exacerbated by the presentation to the House of Representatives, in September 1996, of the Constitutional Review Commission's report. The report included recommendations to enlarge the House of Representatives to 75 seats, with 25 seats reserved on a racial basis (12 for ethnic Fijians, 10 for Fijian Indians, two for General Electors and one for Rotuma Islanders), and also proposed that the Prime Minister should be a Fijian of any race, while the President should continue to be an indigenous Fijian. Rabuka and Mara both endorsed the findings of the report, while several nationalist parties, including the Vanua Independent Party, the Fijian Nationalist United Front Party (FNUFP) and the Taukei Solidarity Movement, expressed extreme opposition to the proposals, and formed a coalition in an attempt to further their influence within Parliament. In addition, a number of SVT members of the House of Representatives aligned themselves with the nationalists, and in early 1997 were reported to be responsible for a series of political manoeuvres within the Cabinet, aimed at undermining Rabuka's position. The parliamentary committee reviewing the report agreed on a majority of the 700 recommendations, but proposed that the House of Representatives be enlarged to only 71 seats, with 46 seats reserved on a racial basis (23 for ethnic Fijians, 19 for Indians, three for General Electors and one for Rotuma Islanders) and 25 seats open to all races. The committee's modified proposals were presented in May to the Great Council of Chiefs, which endorsed the recommendations on constitutional change, but demanded that the number of lower chamber seats reserved for ethnic Fijians be increased from 23 to 28. However, the Council approved the proposal to reduce the number of nominated senators to 15 (from 24). The reforms, as proposed by the committee, were officially endorsed by the Great Council of Chiefs in early June. The Constitution Amendment Bill was approved unanimously by the House of Representatives and the Senate in the following month. Rabuka was anxious to reassure extremist nationalist Fijians, who had vociferously opposed the reforms throughout the debate, that their interests would be protected under the amended Constitution and that indigenous Fijians would continue to play a pre-eminent role in the government of the country.

Despite opposition from both the FLP and the nationalist parties, Fiji was readmitted to the Commonwealth at a meeting of member states in October 1997. In the same month Rabuka was granted an audience with Queen Elizabeth II in London, at which he formally apologized for the military coups of 1987. (Rabuka subsequently apologized to the Indo-Fijian community, in April 1999, for the events of 1987.)

Events in early 1998 were dominated by political reaction to the reformed Constitution. An extremist nationalist group of former SVT supporters, including senior church, military and police officials were rumoured to be planning to overthrow Rabuka's Government. Meanwhile, it was reported that opponents of the Constitution were discussing the establishment of several new political parties, the most significant of which was the Christian Fellowship Party, formed in March under the leadership of Rev. Manasa Lasaro. Nevertheless, the new Constitution came into effect on 27 July 1998.

Following a period of severe drought, nation-wide water restrictions were introduced in early 1998. The sugar industry was drastically affected by the drought, resulting in an annual crop some 60% lower than predicted, with exporters unable to meet commitments to foreign buyers. By September of that year more than one-quarter of the population were dependent on emergency food and water rations. Devastating floods in the west of the country in early 1999, in which seven people died, destroyed the drought rehabilitation programme and that year's sugar crop.

A dispute between tribal landowners and the Government over compensation payments for land flooded by the Monosavu hydroelectric power station erupted into violence in July 1998. Landowners, who had been demanding compensation worth some $A30m. since the plant was constructed in 1983, seized control of the station (which supplies 90% of Fiji's electricity) and carried out a series of arson attacks on Fiji Electricity Authority property. In October the Government agreed to pay the landowners compensation totalling some $A12m., although many involved in the dispute rejected the offer and announced their intention of pursuing a legal claim against the Government. In late 1998 the Government was also believed to be considering a claim for compensation by landowners whose land was compulsorily acquired during the Second World War and now occupies the site of Nadi airport.

The prospect of a general election, to be held in early May 1999, prompted reports of political manoeuvring as parties sought to increase their influence by forming alliances. In addition, a number of changes in the country's political organizations occurred in late 1998. Following the death of Josefata Kamikamica, Adi Kuini Speed (widow of ex-President Bavadra) was elected leader of the Fijian Association Party. Meanwhile, the General Voters' Party and the General Electors' Party merged to form the United General Party under the leadership of the Minister for Tourism, Transport and Civil Aviation, David Pickering, and Rabuka was re-elected leader of the SVT (despite the party being required to amend its constitution in order for this to be possible). The formation of a new party, the Veitokani ni Lewenivanua Vakarisito (Christian Democratic Alliance—VLV), by several senior church and military leaders and former members of the nationalist Taukei Movement, was widely criticized for its extremist stance and refusal to accept the newly formed, multiracial Constitution. In April 1999 Poesci Bune defeated former army commander Ratu Epeli Ganilau to become leader of the party.

At legislative elections held on 8–15 May 1999, the first to be held under the new Constitution, Rabuka's coalition Government was defeated by Mahendra Chaudhry, leader of the Indian-dominated FLP, who thus became Fiji's first ethnic Indian Prime Minister. Chaudhry's broad-based Government (a coalition of the FLP, FAP and the Party of National Unity—PANU) seemed initially threatened by the reluctance of FAP members to serve under an Indian Prime Minister. The leaders were persuaded to remain in the coalition in the interests of national unity, after the intervention of President Mara. However, in June there were renewed demands for the FAP to withdraw from the coalition because of the alleged discontent of party members at the lack of representation afforded to the party in the Senate nominations put forward by Chaudhry. Political stability after the elections was marred by demands for Chaudhry's resignation by the Fijian Nationalist Vanua Takolavo Party, and by a number of arson attacks, allegedly linked to the outgoing SVT party (although these allegations were denied by Rabuka).

Following the SVT's decisive defeat in the elections, Rabuka resigned as party leader; he was replaced by Ratu Inoke Kubuabola, the former Minister for Communications, Works and Energy. Rabuka was subsequently appointed the first independent Chairman of the newly autonomous Great Council of Chiefs. In August 1999 three bombs exploded in Suva. A member of the Nationalist Vanua Takolavo Party was subsequently arrested. In the same month a parliamentary vote of 'no confidence' against Prime Minister Chaudhry was overwhelmingly defeated. In the latter half of 1999 there were persistent demands by various nationalist groups (including the SVT) that Chaudhry be replaced by a leader of indigenous Fijian descent, and a number of demonstrations were organized expressing disillusionment with the Government. In October the opposition coalition was successful in gaining control of a majority of councils in local elections, while the FLP won control of four municipal councils (having previously been in charge of none).

The Government's decision to disband the Fiji Intelligence Service from December 1999 was criticized by the opposition as 'foolish' and racially motivated. An announcement by the Chief Justice in December of a project to amend a number of laws that did not comply with the terms of the new Constitution, together with reports that the Government was planning to withdraw state funds previously provided to assist indigenous Fijian business interests, prompted further accusations of racism by the Government against ethnic Fijians. Consequently, the SVT and the VLV held talks to discuss ways of consolidating the ethnic Fijian political base. Proposed legislation, which would alter the distribution of power between the President and the Prime Minister, attracted further criticism from the opposition, and in February 2000 a faction of the FAP announced its withdrawal from the governing coalition, citing dissatisfaction with Chaudhry's leadership. Of greater significance, however, was the announcement in April that the extreme nationalist Taukei Movement (which had been inactive for several years) had been revived with the sole intention of removing the Prime Minister from office. The group subsequently publicized a campaign of demonstrations and civil disobedience, prompting the army to issue a statement distancing itself from any anti-Government agitation and pledging loyalty to Chaudhry. The campaign attracted considerable popular support, however, and culminated in a march through Suva by some 5,000 people in early May.

On 19 May 2000 a group of armed men, led by businessman George Speight, invaded the parliament building and ousted the Government, taking hostage Chaudhry and 30 other members of the governing coalition. President Mara condemned the coup and declared a state of emergency as Speight's supporters rampaged through the streets of Suva, looting and setting fire to Indian businesses. Speight declared that he had reclaimed Fiji for indigenous people and had dissolved the Constitution. Moreover, he threatened to kill the hostages if the military intervened. On 22 May Mara formally invited Rabuka, in his role as chairman of the Great Council of Chiefs, to seek a resolution of the crisis. In the following days the Great Council of Chiefs convened to discuss the situation and proposed the replacement of Chaudhry's Government with an interim administration, an amnesty for Speight and the rebels, and the amendment of the Constitution. Speight rejected the proposals, demanding that Mara also be removed from office. Meanwhile, violent clashes erupted at the headquarters of Fiji Television when the rebels stormed the building following the broadcast of an interview with an opponent of the coup. A police officer was shot dead, television equipment was destroyed and the station's employees were taken hostage. On 29 May Mara resigned, and the Commander of the Armed Forces, Frank Bainimarama, announced the imposition of martial law and a curfew, in an attempt to restore calm and stability to the country. In an expression of his apparent reluctance to assume the role, Bainimarama gave Mara a whale's tooth, a traditional Fijian symbol of regret.

Negotiations between the Military Executive Council and the Great Council of Chiefs continued throughout June 2000. Failure to reach a conclusive outcome seemed to be the result of inconsistencies in Speight's demands and an ambivalent attitude on the part of the military towards the coup. Regular patrols by the security forces curbed rioting in Suva, although outbreaks of violence in rural areas (mostly in the form of attacks on Indian Fijians, the looting and burning of Indian-owned farms and the occupation of several tourist resorts) were reported. On 25 June the four female hostages were released from the parliament building. The Military Executive Council announced its intention to appoint an interim government without consulting Speight and demanded that the rebel leader release the remaining hostages. Speight reiterated his threat to kill all those held if any rescue attempts were made.

An interim administration of 19 indigenous Fijians led by Laisenia Qarase (the former managing director of the Merchant Bank of

Fiji) was sworn in on 4 July 2000. Minutes after the ceremony a gun battle erupted outside the parliament building in which four civilians and one rebel were injured; the rebel subsequently died. Speight announced that he did not recognize the interim authority, and most of Fiji's mainstream political parties similarly denounced it, although the Methodist Church declared its support for the body. On 12 July a further nine hostages were released, and on the following day the remaining 18, including Chaudhry, were liberated. In accordance with Speight's wishes, Ratu Josefa Iloilo, hitherto the First Vice-President, was then installed as President. In the same month Chandrika Prasad, a farmer, quickly brought a legal challenge to the abrogation of the 1997 Constitution in the High Court of Fiji. Chaudhry launched an international campaign to reinstate both the Constitution and the People's Coalition Government.

Incidents of civil unrest (including the occupation of the hydroelectric dam at Monasavu and of the army barracks on Vanua Levu) continued throughout July 2000 as Speight sought to manipulate existing grievances, particularly disputes over land ownership, in order to mobilize additional support. On 29 July, however, Speight was finally arrested, along with dozens of his supporters, for breaking the terms of his amnesty by refusing to relinquish weapons. Armed rebels responded violently to the arrest, and in Labasa Indian Fijians were rounded up and detained in army barracks by supporters of Speight. In early August more than 300 rebels appeared in court on a variety of firearms and public order offences. Speight was similarly charged with several minor offences. On 11 August Speight and 14 of his supporters were formally charged with treason. On 15 November the High Court ruled that the existing Constitution remained valid and that the elected Parliament, ousted in the coup, remained Fiji's legitimate governing authority. Laisenia Qarase responded by declaring that the interim authority, of which he was leader, would continue as the country's national government until new elections could be organized and a new constitution drafted within 18 months.

In mid-December 2000 Chaudhry's campaign to re-establish his Government suffered a significant reversal when the ministers of foreign affairs of Australia and New Zealand announced that they were abandoning their appeal for its reinstatement, although they would continue to support a return to the 1997 Constitution. However, within Fiji supporters of a return to democracy formed the Fiji First Movement, which aimed to consolidate opposition to the post-coup regime. The group organized a series of protests across the country in late 2000 and early 2001.

In February 2001 an international panel of judges at the Court of Appeal began the hearing against the November 2000 ruling, which found the abrogation of the 1997 Constitution to be illegal. In its final judgment the court ruled that the 1997 Constitution remained the supreme law of Fiji, that the interim civilian government could not prove that it had the support of a majority of Fijian people and was therefore illegal and that, following Mara's resignation, the office of President remained vacant. The ruling was welcomed by many countries in the region, including Australia and New Zealand, and appeared to be accepted by the interim authority, which announced that it would organize elections as soon as possible. However, on 14 March Iloilo informed Chaudhry in a letter that he had been dismissed as Prime Minister, claiming that by advising Iloilo to dissolve the authority in preparation for elections he had accepted that he no longer had the mandate of Parliament. Chaudhry rejected the decision as unconstitutional and unlawful. Ratu Tevita Momoedonu—Labour and Industrial Relations Minister in Qarase's Government—was appointed Prime Minister on the same day. On the following day, however, Iloilo dismissed Momoedonu, on the advice of the Great Council of Chiefs, and reinstated Laisenia Qarase as head of the interim authority, with a mandate to prepare the country for general elections. The Great Council of Chiefs had reconfirmed Iloilo as President and agreed to accept the 1997 Constitution. It was announced that a general election would be held in August–September 2001, and would be conducted under the preferential voting system, similar to that of Australia, as used in Fiji's 1999 election.

There followed a period of factionalism and fragmentation among Fiji's political parties. George Speight had already been appointed President of the new Matanitu Vanua (MV—Conservative Alliance Party) party, despite facing the charge of treason for his part in the 2000 coup. On 9 May 2001 Qarase formed the Soqosoqo Duavata ni Lewenivanua (SDL—Fiji United Party), a new contender for the indigenous Melanesian vote, thus rivalling the established SVT. Another indigenous party, the Bai Kei Viti, was launched on 28 June. In the same month Tupeni Baba, former Deputy Prime Minister in Chaudhry's Government, left the FLP and formed the New Labour United Party. The election took place between 25 August and 1 September. Qarase's SDL was victorious, but failed to obtain an overall majority. The SDL secured 31 seats in the House of Representatives (increasing to 32 of the 71 seats after a by-election on 25 September). The FLP won 27 seats, the MV six seats and the NLUP two seats. International monitors were satisfied that the election had been contested fairly.

Following the election, however, by refusing to allow the FLP any representation in his new Cabinet, Qarase was accused of contravening a provision of the Constitution whereby a party winning more than 10% of the seats in the House of Representatives was entitled to a ministerial post. Two members of George Speight's MV were included in the Cabinet. Qarase claimed that Mahendra Chaudhry had not accepted that the Government should be based fundamentally on nationalist Fijian principles. In October 2001, when members of the House of Representatives were sworn in, Chaudhry refused to accept the position of Leader of the Opposition, a title that consequently fell to Prem Singh, leader of the NFP. In December Parliament approved the Social Justice Bill, a programme of affirmative action favouring Fijians and Rotumans in education, land rights and business-funding policies.

The Prime Minister defended himself against demands for his resignation in January 2002. The former Minister for Agriculture had alleged that Qarase had broken the Electoral Act after it was revealed that more than F$25m. had been misused by the interim Government's Ministry of Agriculture during the 2001 election campaign. In February 2002, furthermore, an appeal court ruled that the Prime Minister had violated the Constitution by failing to incorporate any member of the FLP in his Cabinet. Qarase had previously declared that he would resign if the legal challenge against him were to be successful. In September, in advance of the ruling by the Supreme Court on the issue of the inclusion of the FLP in the Cabinet, the Prime Minister effected a ministerial reorganization, assuming personal responsibility for a number of additional portfolios. In April 2003 the Commander of the Armed Forces, Frank Bainimarama, intervened in the ongoing dispute, stating that if the judicial ruling went against Qarase then he should resign. Meanwhile, rumours circulated that a further coup might be attempted if the Government were ordered to include FLP members in the Cabinet. The Supreme Court finally delivered its ruling on 18 July, finding in favour of Chaudhry and declaring that, in order to uphold the Constitution, Qarase should form a new cabinet including eight members of the FLP. Qarase responded by proposing to retain his current 22-member Cabinet and to add 14 FLP members. Both the opposition and the SVT leader, Sitiveni Rabuka, criticized the proposal, which would result in more than one-half of all members of the House of Representatives serving as cabinet ministers, and would give Fiji the largest cabinet in the world, in proportion to its population. Chaudhry expressed dissatisfaction with the suggestion, claiming that the positions offered to his party were too junior. However, Qarase remained defiant, his intransigence resulting in several more weeks of political impasse. At the end of August Qarase formally nominated a cabinet which included 14 FLP members (although Chaudhry was not among those named). The opposition, however, continued to resist the proposal, claiming that they should be consulted over the composition of the Cabinet. The ransacking of the FLP office in Lautoka at the beginning of September, in which records and property were destroyed, was believed by many observers to be an act of intimidation against the party. In May 2004 the President of Qarase's SDL party, Ratu Kalokalo Loki, issued a statement denouncing democracy as unsuited to Fiji's needs and added that the government of national unity was unworkable. Qarase reinforced this view in the following month when he appealed for the 1997 Constitution to be changed, claiming that it had failed, owing to its provision for multi-party politics. The remarks of both men were criticized by the opposition. In July the Supreme Court delivered its long-awaited ruling on the composition of the Cabinet, but its judgment contained no clear instruction on the numbers of seats to be awarded to the SDL and FLP. Both parties were advised to resolve the issue by negotiation. Qarase, however, asserted that there would be no further discussions on the matter and that thorough negotiations had already been conducted. He stated that his initial offer of cabinet positions to the FLP remained but that if the party rejected this his Cabinet would continue in its current form. The impasse appeared likely to remain when Chaudhry appealed to the Prime Minister to dismantle his Cabinet and renegotiate the terms of its composition.

Meanwhile, in June 2002 the Prime Minister and the FLP leader co-operated briefly in addressing the issue of expiring land leases that were threatening Fiji's sugar industry. A committee, comprising members of both the SDL and FLP, was established to try to negotiate land leases that would satisfy both Indo-Fijian tenants and their predominantly ethnic Fijian landowners. Most of the 30-year leases drawn up under the ALTA (see above) were expiring, and both tenants and the FLP were opposed to its replacement by the Native Land Trust Act (NLTA), which they saw as disproportionately favouring landowners. Two parliamentary bills had been approved by the Senate in April, reducing the land under state control to around 1% of the total and increasing the amount under the Native Land Trust Board to over 90%. During 2003 more than 1,100 tenants on Vanua Levu were evicted following the expiry of their land leases.

In August 2002, however, the FLP abandoned a second round of land lease discussions and announced that it would boycott most of the proceedings in the current session of Parliament. Chaudhry accused the Government of attempting to accelerate the passage of six bills through Parliament without regard for the mandatory 30 days' notice of a bill being tabled. He also complained that the Government had not given the FLP the full details of the proposed NLTA. The Prime Minister protested that this would compel the Government to accept the decision of the Great Council of Chiefs regarding the leases. Tensions between the ruling SDL and the FLP and, furthermore, between ethnic Fijians and Indo-Fijians had been further exacerbated by anti-Indian comments made by the Minister for Women, Social Welfare and Land Resettlement, Asenaca Caucau, which the Prime Minister had not denounced. In September Qarase effected a reorganization of cabinet portfolios in which he assumed direct responsibility for the reform of the sugar industry and restated his commitment to resolve the long-standing issue of land leases. In April 2004 Qarase and Chaudhry agreed on a plan to restructure the sugar industry devised by a team of technical experts from India. The Government of India was expected to finance the programme, which was estimated to cost some US $58m. initially and a further $170m. to complete.

Justice for both the perpetrators and the victims of Fiji's military coup of 2000 remained a slow process. The trial of George Speight and his accomplices on charges of treason opened in May 2001. (Speight was refused bail that would have enabled him to occupy the seat that he won in the legislative election later in the year.) All the accused pleaded guilty to their involvement in the coup of May 2000, and at the conclusion of the trial in February 2002 Speight was sentenced to death. Within hours of the verdict, however, President Iloilo signed a decree commuting the sentence to life imprisonment (Fiji being in the process of abolishing the death penalty). Prison sentences of between 18 months and three years were imposed on 10 of Speight's accomplices, the charges of treason having been replaced by lesser charges of abduction. The trial of two other defendants began in July, following the rejection of protests from the defendants that they were protected by an Immunity Decree promulgated by the Commander of the Armed Forces, Frank Bainimarama. Both were found guilty in March 2003, when a further 23 people were arrested on charges relating to the coup. In April 2004 an additional 61 soldiers were charged with various crimes concerning their involvement in the coup. Moreover, in the following month Vice-President Ratu Jope Seniloli was charged with having attempted to usurp President Mara during the coup, in addition to the existing charges against him. Also in May George Speight made a further court appearance on charges of hostage-taking during the 2000 coup. The leader of the FLP, Mahendra Chaudhry, and member of Parliament, Ganesh Chand, were claiming US $3.6m. in compensation for the 56 days they were held hostage in the government buildings by Speight and his accomplices. In July 2004 a group of 21 soldiers (the second such group) was found guilty of mutiny for their role in the plot to assassinate military commander Bainimarama and to liberate coup leader George Speight. In the following month Seniloli was found guilty of treason and sentenced to four years' imprisonment.

In June 2003 the High Chief of a district of Vanua Levu publicly apologized for his involvement in the coup during a public reconciliation ceremony and announced a ban, in his locality, on words that differentiate ethnic Fijians from their Indian Fijian neighbours. Almost one year later the Great Council of Chiefs issued a historic public apology to all Indian Fijians for injustices committed against them during the coups of 1987 and 2000. Moreover, when, in July 2004, Adi Litia Cakobau (a senator appointed by the Great Council of Chiefs) proposed a motion in the upper house to outlaw the use of the terms Indian Fijian and Indo-Fijian, claiming that they were insulting to ethnic Fijians, the Great Council expressed outrage at her position.

A national security alert was issued in January 2004 following intelligence reports that suggested unrest within the military and a possible destabilization of the country. Rumours of a rift within the army over support for Commander Frank Bainimarama were compounded when a number of senior military officers were dismissed and replaced with individuals known to be loyal to the military leader. Moreover, a military exercise near the government buildings in June 2004 prompted rumours that Bainimarama had been arrested.

Fears that Fiji was becoming a haven for criminal organizations proved well founded in June 2004 when drugs and the chemicals associated with their production, valued at an estimated US $500m., were discovered during a raid by Fijian police on three warehouses near the capital, Suva. The warehouses were believed to be the largest laboratory producing the illegal drug methamphetamine in the southern hemisphere. Seven people were arrested during the raid, of whom six were Chinese. In the same month the Pacific Transnational Crime Centre opened in Suva in response to an increase in crime in the region, in particular the activities of criminal networks from outside the Pacific region, which were believed to be establishing operations in the Pacific islands.

In late September 2004 the Minister for National Reconciliation and Multi-Ethnic Affairs and the Cabinet's only ethnic Indian member, George Shiu Raj, resigned following accusations that he had overspent government money while on an official trip to India during 2003.

Fiji's foreign relations had been severely compromised by the coups of 1987. Its traditional links with the Commonwealth, particularly India, Australia and New Zealand, deteriorated markedly, most notably in terms of military co-operation, which Fiji sought instead from France and several Asian nations. India remained a consistent critic of the new regime, and tension increased after the fire-bombing of Indian places of worship in Fiji in October 1989. Comments by the Indian ambassador resulted in his expulsion from Fiji and the redesignation of the embassy as a consulate. In May 1990 the Fijian Government ordered the complete closure of the diplomatic post. In May of the following year the Secretary-General of the Commonwealth stated that Fiji would not be readmitted to the organization until it changed its Constitution. Following the approval of major constitutional reforms in July 1997, therefore, Fiji applied for readmission to the Commonwealth and rejoined the organization at a meeting of member states in October 1997. Diplomatic relations were resumed between Fiji and India in 1998, and in May 1999 India reopened its High Commission in Suva. However, Fiji's relations with the international community suffered a major reversal following the coup of May 2000, which was condemned by the UN, the Commonwealth, the United Kingdom, Australia, New Zealand and several other nations in the region. In June Fiji was partially suspended from the Commonwealth, and a delegation of ministers of foreign affairs from the organization visited the islands to demand the reinstatement of the 1997 Constitution. Following the democratic elections of August–September 2001, both Australia and New Zealand removed the bilateral sanctions they had imposed. The sanctions imposed by the European Union (EU) remained in place until early 2002. In December 2001 the Commonwealth Ministerial Action Group on the Harare Declaration (CMAG) recommended that Fiji be readmitted to meetings of the Commonwealth. In late 2003 the EU announced the resumption of development aid to Fiji (suspended since 2000) and stated its willingness to discuss further co-operation with the Fijian Government. The Fijian High Commission in India was reopened in April 2004.

The decision by the French Government in June 1995 to resume nuclear testing in the region was widely condemned by the Pacific island nations and, in protest at the announcement, Fiji cancelled its annual military exercise with France. A demonstration by some 5,000 people in July against the tests typified the reaction of Pacific islanders throughout the region. In June 1999 the Pacific Concerns Resources Centre announced that it had filed a case with the European Court of Human Rights to establish British responsibility for the alleged exposure of Fijian soldiers to radioactive material on Christmas Island (now part of Kiribati) in the 1950s. Earlier that year the Fijian Government announced that it had decided to grant pensions to Fijian soldiers and sailors involved in the United Kingdom's nuclear testing in the Pacific.

Economy

Traditionally the Fijian economy is agricultural and dominated by the sugar industry, although both tourism and exports of garments have also become valuable sources of foreign exchange. In 2002, according to estimates by the World Bank, Fiji's gross national income (GNI), measured at average 2000–02 prices, was US $1,775m., equivalent to US $2,160 per head (or US $5,310 on an international purchasing-power parity basis). In 1990–2002 the population increased by an annual average of 0.9%, while GDP per head was estimated to have increased by an annual average of 1.8%. According to figures from the Asian Development Bank (ADB), Fiji's gross domestic product (GDP) increased, in real terms, by an average of an estimated 2.7% during 1990–2003 . Compared with the previous year, real GDP was estimated to have increased by 3.0% in 2001, by 4.1% in 2002 and by 10.1% in 2003. Real GDP was forecast to increase by 4.1% in 2004.

Agriculture (including forestry and fishing) contributed an estimated 13.9% of GDP in 2003. According to FAO, agriculture engaged 39.6% of the economically active population in 2001. During 1990–2003, according to ADB estimates, agricultural GDP rose by an average annual rate of 0.2%. The sector's GDP was provisionally estimated to have increased by 0.4% in 2003. According to FAO, in 1999 less than 16% of Fiji's land area was used for permanent crops or as arable land.

Sugar cane is the principal cash crop. Following a successful crop rehabilitation programme in the late 1990s, sugar cane production increased by 81% in 1999. However, the stability of the industry was disrupted by the May 2000 coup, and by the consequences of ongoing uncertainty surrounding the expiry of land leases. In 2003 sugar accounted for an estimated 18.1% of total export earnings. Sugar production declined by 25.9% in 2001 but increased by 41.7% in the following year. In 2003 production of the crop again declined owing to the effects of a prolonged drought and the continued issue of the non-renewal of land leases.

Other important export crops are coconuts and ginger, although production levels of both crops declined in the early 1990s. Vegetables and fruit are grown for domestic consumption, and paddy rice has been particularly encouraged as a subsistence crop. The most important livestock products are beef and poultry meat.

Fiji has extensive timber reserves, and forestry became important as an export trade from the mid-1980s. Pulpwood and pine timber accounted for most of the increase in log production. Overall timber production increased by 4.3% in 2001. Lumber accounted for 4.2% of exports in 2000, worth $F44.9m.

Fishing is an important activity in Fiji, both for export and for domestic consumption. Commercial fish-farming became a significant sector in the 1990s. In 2003 fish products earned some $F85m. in export revenue (6.7% of export receipts). Fiji also benefits from the sale to foreign interests of licences to fish in the islands' exclusive economic zone. Some 50 vessels operated in 1996, providing $F80m. in licence fees. However, plans to increase the number of licences permitted from 80 to 150 promised to augment revenue from this source.

Industry (including mining, manufacturing, power and construction) contributed an estimated 25.0% of GDP in 2003. In 1998 the sector engaged 33.8% of those in paid employment. The GDP of the industrial sector, according to the ADB's figures, increased at an average rate of 3.5% per year during 1990–2003. The industrial sector's GDP was estimated to have contracted by 7.4% in 2000, before increasing by 7.7% in 2001. Output from the sector remained constant in 2002 but increased by an estimated 6.8% in 2003.

Mining and quarrying contributed an estimated 2.1% of GDP in 2003 and employed 1.8% of the paid labour force in 1999. Gold and silver are the major mineral exports. Gold exploration activity increased in the mid-1990s (with the number of prospecting licences totalling 49 in late 1998). A large copper-mining project in Namosi, central Viti Levu, began operations in 1997. Fiji's production of gold totalled 3,517 kg in 2003, when exports earned an estimated $F76.5m., equivalent to 6.0% of total export earnings. The mining sector's GDP, however, remained almost constant in real terms (increasing at an average annual rate of less than 0.05%) between 1990 and 2002, a sharp decrease of 14.1% being recorded in 2000. Other mineral reserves include marble. In addition, 20 potential petroleum-bearing sites were identified in early 1993.

Manufacturing contributed an estimated 13.8% of GDP in 2003. The sector employed 26.0% of those in paid employment in 1999. Manufacturing GDP increased at an average rate of 3.3% per year during 1990–2003. Compared with the previous year, manufacturing GDP rose by 11.6% in 2001 and by 3.9% in 2002. The most important activity is food processing, notably sugar, molasses, coconut oil and fish-canning. The bottling of mineral water for export was becoming increasingly important in the early 2000s. Manufacturing has received particular encouragement since 1987, when the Government established a register for factories which, provided 95% of their products were exported, would be exempt from any taxation. By 1997 there were 156 tax-free factories in operation (compared with 44 in 1988), including 96 garment factories. Garment exports reached $F332.9m. in 2000, before declining to $F245.4m. in 2002, in which year they represented 20.5% of total export earnings. However, exports from the garment industry increased to some $F252.7m. in 2003, equivalent to 19.8% of total export earnings.

Hydroelectricity is the principal source of power, providing some 90% of Fiji's electricity in the late 1990s. Electricity, gas and water contributed an estimated 4.4% of GDP in 2003. Mineral fuels accounted for 14.7% of total imports in 2003.

The service industries contributed an estimated 61.1% of GDP in 2003, and in 1996 engaged 66.8% of those in paid employment. During 1990–2003 the sector's GDP rose by an average annual rate of 2.9%, according to the ADB. The ADB estimated that the GDP of the services sector had decreased by an estimated 1.8% in 2000 before rising by 3.3% in 2001, by 6.5% in 2002 and by 5.4% in 2003. Tourist arrivals declined sharply to 294,070 in 2000, owing to the repercussions of the coup in May of that year, but increased to 348,014 in 2001 and further, to 397,859, in the following year. In 2003 an estimated 430,800 tourists arrived in Fiji, providing revenue of $F615.4m. A further increase in visitors was expected in 2004 with the introduction of budget airline services from Australia and New Zealand. Most visitors are from Australia (31.1% of total arrivals in 2002), followed by New Zealand (17.2%), the USA (14.8%), the United Kingdom (10.9%) and Japan (6.6%). The trade, restaurants and hotels sector contributed 18.5% of GDP in 2002.

Fiji consistently records a trade deficit. According to the ADB, this amounted to US $340m. in 2003, when the country also recorded a deficit of US $171m. on the current account of the balance of payments. Export earnings increased from $F1,194.8m. in 2002 to $F1,273.1 in 2003. Imports rose from $F1,953.2m. in 2002 to $F2,214.6 in 2003. The principal sources of imports in 2003 were Australia (34.9%), New Zealand (17.1%) and Singapore (19.5%). The principal markets for exports were Australia (19.6%), the USA (23.7%) and the United Kingdom (13.6%). The principal imports in 2003 were machinery and transport equipment (30.0% of total costs in that year). Fiji's principal domestic exports are clothing, sugar, gold and fish. Fiji also re-exports mineral fuels (including bunkers for ships and aircraft).

In 2004 there was a projected budgetary deficit equivalent to 3.9% of GDP, compared with a similar figure for 2003. According to the ADB, Fiji's total external debt was US $258m. at the end of 2003 . In that year external debt was equivalent to 4.2% of GDP and the cost of debt-servicing was equivalent to 2.5% of the revenue from goods and services. The average annual rate of inflation was 3.4% in 1990–2000, decreasing to 1.1% in 2000. According to the ADB, the inflation rate increased to 4.3% in 2001, but stood at 0.8% in the following year, before increasing to 4.1% in 2003. Of the total labour force, some 14.1% was unemployed in 2002. Since 1987 Fiji has suffered a very high rate of emigration, particularly of skilled and professional personnel. Although the rate of emigration of professional and technical workers decreased by 2.8% in 2001, the number of emigrants from the services sector increased by 28.2%. An increase of some 6% in the number of emigrants was recorded in 2003, a disproportionate percentage of whom were professionals from the health and education sectors.

The overthrow of the Government during a coup in May 2000 had a severe impact on Fiji's economy, resulting in the first contraction in GDP since 1997. In the second half of 2000 it was estimated that 6,700 people lost their jobs, the overwhelming majority of these having been engaged either in tourism-related activities or in manufacturing. A significant amount of that year's excellent sugar crop was not harvested, and a large proportion of tourist bookings was cancelled. In addition, the islands lost a considerable sum of revenue owing to cancellation of foreign aid and investment, and the Government was unable to collect taxes worth more than $F90m.

In response to the economic difficulties following the coup, the Government introduced an emergency interim budget in July 2000, including austerity measures which were successful in keeping government debt at a manageable level. The 2001 budget reduced business taxes as a way of stimulating investment in the economy. In 2002 the Government established the Fiji Investment Corporation to manage and promote investment in tourism, forestry, fisheries and inter-island shipping services. In that year the sale of Amalgamated Telecom Holdings earned over $F64m., the first of the Government's proposed major privatization plans. Economic growth in 2003 resulted largely from a significant increase in tourist arrivals partly, prompted by the country's recent hosting of the South Pacific Games, and by an expansion of the construction sector, following the launch of several large private-sector projects and a number of important public infrastructure programmes. A slower rate of growth was predicted for 2004 owing to ongoing problems around the issue of land-ownership, the increasing shortage of skilled workers and the collapse of the sugar industry. The end of preferential access to the Australian market for Fijian garment exports in 2005 was also expected to have a negative impact on the economy. In mid-2003 the Fiji Manufacturers' Association protested that the Melanesian Free Trade Agreement (of which Fiji was a member) was harming the country's economy by restricting the export of certain goods from Fiji to other member countries.

In April 2003 the Government announced that the restructuring of Fiji's important sugar industry would be postponed for at least a further year because of disagreement among stakeholders. It was hoped that sugar farmers would be given assistance to diversify their activities before 2007 when the European Union (EU) protocol governing sugar prices (under which Fiji had been guaranteed the sale of its sugar at preferential rates) was due to expire.

Fiji is a member of the Pacific Islands Forum (formerly the South Pacific Forum), the Pacific Community (formerly the South Pacific Commission), the UN's Economic and Social Commission for Asia and the Pacific (ESCAP), the Asian Development Bank, the Colombo Plan, and the International Sugar Organization. Fiji is a signatory to the South Pacific Regional Trade and Economic Co-operation Agreement (SPARTECA) and also to the Lomé Conventions and successor Cotonou Agreement with the EU. In 1996 Fiji was admitted to the Melanesian Spearhead Group.

Statistical Survey

Sources (unless otherwise stated): Bureau of Statistics, POB 2221, Government Bldgs, Suva; tel. 315144; fax 303656; internet www.spc.int/prism/country/fj/stats ; Reserve Bank of Fiji, POB 1220, Suva; tel. 313611; fax 301688; e-mail rbf@reservebank.gov.fj; internet www.reservebank.gov.fj.

AREA AND POPULATION

Area (incl. the Rotuma group): 18,376 sq km (7,095 sq miles). Land area of 18,333 sq km (7,078 sq miles) consists mainly of the islands of Viti Levu (10,429 sq km—4,027 sq miles) and Vanua Levu (5,556 sq km—2,145 sq miles).

Population: 715,375 at census of 31 August 1986; 775,077 (males 393,931, females 381,146) at census of 25 August 1996; 826,281 (official estimate) at 31 December 2002.

Density (31 December 2002): 45.0 per sq km.

Principal Towns (population at 1996 census): Suva (capital) 77,366; Lautoka 36,083; Nadi 9,170; Labasa 6,491; Ba 6,314. *Mid-2003* (UN estimate, incl. suburbs): Greater Suva 210,472 (Source: UN, *World Urbanization Prospects: The 2003 Revision).*

Ethnic Groups (1996 census): Fijians 393,575; Indians 338,818; Part-European 11,685; European 3,103; Rotuman 9,727; Chinese 4,939; Other Pacific Islanders 10,463; Others 2,767; Total 775,077. *2002* (official estimates at 31 December): Fijians 441,092; Indians 327,667; Others 57,522; Total 826,281.

Births, Marriages and Deaths (registrations, 1998): Live births 17,944 (birth rate 22.5 per 1,000); Marriages 8,058 (marriage rate 10.1 per 1,000); Deaths 5,241 (death rate 6.6 per 1,000). Source: UN, *Demographic Yearbook.*

Expectation of Life (WHO, estimates years at birth): 67.3 (males 64.6; females 70.3) in 2002. Source: WHO, *World Health Report.*

Economically Active Population (persons aged 15 years and over, census of 31 August 1986): Agriculture, hunting, forestry and fishing 106,305; Mining and quarrying 1,345; Manufacturing 18,106; Electricity, gas and water 2,154; Construction 11,786; Trade, restaurants and hotels 26,010; Transport, storage and communications 13,151; Financing, insurance, real estate and business services 6,016; Community, social and personal services 36,619; Activities not adequately defined 1,479; *Total employed* 222,971 (males 179,595, females 43,376); Unemployed 18,189 (males 10,334, females 7,855); *Total labour force* 241,160 (males 189,929, females 51,231). *1999* (total labour force, provisional): 330,800 (unemployed 25,100). *2000* (total labour force, provisional): 341,700 (unemployed 41,700).

HEALTH AND WELFARE

Key Indicators

Total Fertility Rate (children per woman, 2002): 2.9.

Under-5 Mortality Rate (per 1,000 live births, 2002): 21.

HIV/AIDS (% of persons aged 15–49, 2003): 0.1.

Physicians (per 1,000 head, 1997): 0.48.

Health Expenditure (2001): US $ per head (PPP): 224.

Health Expenditure (2001): % of GDP: 4.0.

Health Expenditure (2001): public (% of total): 67.1.

Access to Water (% of persons, 2000): 47.

Access to Sanitation (% of persons, 2000): 43.

Human Development Index (2002): ranking: 81.

Human Development Index (2002): value: 0.758.

For sources and definitions, see explanatory note on p. vi.

AGRICULTURE, ETC.

Principal Crops (mostly FAO estimates, '000 metric tons, 2002 unless otherwise indicated): Sugar cane 3,300; Coconuts 170; Copra 13 (2000); Cassava 33; Rice (paddy) 15; Sweet potatoes 6; Bananas 7; Yams 5; Taro 38. Source: FAO.

Livestock (FAO estimates, '000 head, year ending September 2002): Cattle 340; Pigs 138; Goats 247; Horses 44; Chickens 3,700. Source: FAO.

Livestock Products (FAO estimates, metric tons, 2002): Poultry meat 8,420; Beef and veal 9,120; Goat meat 960; Pig meat 3,915; Hen eggs 2,668; Cows' milk 57,500; Butter and ghee 1,770; Honey 103. Source: FAO.

Forestry ('000 cubic metres, 2002): *Roundwood Removals* (excl. bark): Sawlogs and veneer logs 200; Fuel wood 37; Pulpwood 273; Total 510. *Sawnwood Production* (incl. sleepers): 79. Source: FAO, *Yearbook of Forest Products.*

Fishing (FAO estimates, '000 metric tons, live weight, 2002): Capture 42.8 (Groupers 1.5; Snappers 1.7; Emperors 2.9; Barracudas 2.8; Mullets 2.9; Narrow-barred Spanish mackerel 2.1; Albacore 8.0; Freshwater molluscs 5.2; Anadara clams 2.9; Other marine molluscs 3.2); Aquaculture 1.7; *Total catch* 44.5. Figures exclude aquatic plants ('000 metric tons): 0.8 (capture 0.5, aquaculture 0.3). Also excluded are trochus shells (150 metric tons), pearl oyster shells (10 tons) and corals (1,000 tons). Source: FAO.

MINING

Production (kg, 2003): Gold 3,517; Silver 1,247.

INDUSTRY

Production (metric tons, 2001, unless otherwise indicated): Sugar 496,000 (2003); Molasses 142,000 (1997); Coconut oil 8,000 (2003); Flour 64,000 (2003); Soap 3,000 (2003); Cement 100,000 (2003); Paint ('000 litres) 2,789; Beer ('000 litres) 18,000; Soft drinks ('000 litres) 24,160; Cigarettes 442; Matches ('000 gross boxes) 128; Electric energy (million kWh) 545 (2000 estimate); Ice cream ('000 litres) 1,257; Toilet paper ('000 rolls) 20,225. Source: UN, *Industrial Commodity Statistics Yearbook*, and ADB, *Key Indicators of Developing Asian and Pacific Countries.*

FINANCE

Currency and Exchange Rates: 100 cents = 1 Fiji dollar ($F). *Sterling, US Dollar and Euro Equivalents* (31 May 2004): £1 sterling = $F3.2473; US $1 = $F1.7699; €1 = $F2.1674; $F100 = £30.79 = US $56.50 = €46.14. *Average Exchange Rate* ($F per US $): 2.2766 in 2001; 2.1869 in 2002; 1.8958 in 2003.

General Budget ($F million, 2002): *Revenue:* Current revenue 949.4 (Taxes 796.5, Non-taxes 152.9); Capital revenue 80.7; Total 1,030.1. *Expenditure:* General public services 149.5; Defence 56.2; Education 168.4; Health 82.0; Social security and welfare 3.4; Housing and community amenities 7.53; Economic services 73.7 (Agriculture 6.5, Industry 31.4, Electricity, gas and water 11.3, Transport and communications 24.5); Other purposes 632.5; Total 1,173.2 (Current 971.8, Capital 201.4). Source: Asian Development Bank, *Key Indicators of Developing Asian and Pacific Countries. 2003:* Total current revenue 1,079.1; Total current expenditure 1,083.4.

International Reserves (US $ million at 31 December 2003): Gold (valued at market-related prices) 0.35; IMF special drawing rights 7.70; Reserve position in IMF 22.57; Foreign exchange 393.35; Total 423.97. Source: IMF, *International Financial Statistics.*

Money Supply ($F million at 31 December 2003): Currency outside banks 226.2; Demand deposits at commercial banks 666.0; Total money (incl. others) 900.6. Source: IMF, *International Financial Statistics.*

Cost of Living (Consumer Price Index; base: 1993 = 100): 124.7 in 2001; 125.6 in 2002; 130.9 in 2003.

Expenditure on the Gross Domestic Product ($F million at current prices, 2001): Government final consumption expenditure 655.4; Private final consumption expenditure 2,188.0; Increase in stocks 40.0; Gross fixed capital formation 516.8; *Total domestic expenditure* 3,400.2; Exports of goods and services 2,136.0; *Less* Imports of goods and services 2,327.0; Statistical discrepancy 626.64; *GDP in purchasers' values* 3,835.8. Source: Asian Development Bank, *Key Indicators of Developing Asian and Pacific Countries.*

Gross Domestic Product by Economic Activity ($F million at constant 1989 prices, 2003, preliminary): Agriculture, forestry and fishing 335.4; Mining and quarrying 50.6; Manufacturing 333.3; Electricity, gas and water 105.8; Building and construction 111.3; Wholesale and retail trade 448.6; Transport and communications 339.2; Finance, real estate, etc. 254.9; Public administration 409.9; Other services 17.8; *Sub-total* 2,406.8; *Less* Imputed bank service charges 129.2; *GDP at factor cost* 2,277.6. Source: Asian Development Bank, *Key Indicators of Developing Asian and Pacific Countries.*

Balance of Payments (US $ million, 1999): Exports of goods f.o.b. 537.7; Imports of goods f.o.b. –653.3; *Trade balance* –115.6; Exports of services 525.1; Imports of services –389.8; *Balance on goods and services* 19.7; Other income received 47.3; Other income paid –82.8; *Balance on goods, services and income* –15.8; Current transfers received 42.7; Current transfers paid –14.2; *Current balance* 12.7; Capital account (net) 14.0; Direct investment abroad –53.0; Direct investment from abroad –33.2; Other investment assets –62.2; Other investment liabilities 44.4; Net errors and omissions 32.5; *Overall balance* –44.9. Source: IMF, *International Financial Statistics*.

EXTERNAL TRADE

Principal Commodities ($F million, 2002): *Imports c.i.f.* (distribution by SITC): Food and live animals 297.0; Mineral fuels, lubricants, etc. 336.0; Chemicals and related products 131.8; Basic manufactures 95.8; Machinery and transport equipment 446.8; Miscellaneous manufactured articles 31.1; Total (incl. others) 1,953.2. *Exports f.o.b.:* Food and live animals 137.09; Beverages and tobacco 339.9; Basic manufactures 79.2; Machinery and transport equipment 42.7; Total (incl. others) 1,194.8. Source: Asian Development Bank, *Key Indicators of Developing Asian and Pacific Countries*. *2003:* Total imports c.i.f. 2,214.6; Total exports f.o.b. 1,273.1.

Principal Trading Partners (US $ million, 2003): *Imports c.i.f.:* Australia 341.0; China, People's Republic 28.6; Hong Kong 25.8; India 18.7; Indonesia 17.8; Japan 44.0; New Zealand 167.1; Singapore 190.0; Thailand 21.9; USA 21.6; Total (incl. others) 976.1. *Exports:* Australia 139.8; Japan 45.3; Kiribati 11.5; New Zealand 27.9; Samoa 38.3; Tonga 18.6; United Kingdom 96.9; USA 168.7; Total (incl. others) 712.6. Source: Asian Development Bank, *Key Indicators of Developing Asian and Pacific Countries*.

TRANSPORT

Road Traffic (motor vehicles registered at 31 December 2003): Passenger cars 66,028; Goods vehicles 38,025; Buses 2,019; Taxis 3,729; Rental vehicles 5,511; Motorcycles 4,670; Tractors 5,619; Total (incl. others) 130,878.

Shipping: *Merchant Fleet* (registered at 31 December 2003): Vessels 49; Total displacement ('000 grt) 23.1 (Source: Lloyd's Register-Fairplay, *World Fleet Statistics*). *International Freight Traffic* ('000 metric tons, 1990): Goods loaded 568; Goods unloaded 625 (Source: UN, *Monthly Bulletin of Statistics*).

Civil Aviation (traffic on scheduled services, 1999): Passengers carried 525,000; Passenger-kilometres 2,159 million; Total ton-kilometres 218 million. Source: UN, *Statistical Yearbook*.

TOURISM

Foreign Visitors by Country of Residence (excluding cruiseship passengers, 2002): Australia 123,606; Canada 9,802; Japan 26,382; New Zealand 68,293; Pacific Islands 24,051; United Kingdom 43,393; USA 58,815; Total (incl. others) 397,859. *2003:* Total arrivals 430,800.

Tourism Receipts ($F million): 413.5 in 2000; 495.5 in 2001; 615.4 in 2003.

COMMUNICATIONS MEDIA

Radio Receivers (1999): 545,000 in use*.

Television Receivers (2000): 92,000 in use†.

Telephones (2003): 102,000 main lines in use†.

Facsimile Machines (1999): 2,815 in use†.

Mobile Cellular Telephones (2003): 109,900 subscribers†.

Personal Computers (2002): 40,000 in use†.

Internet Users (2003): 55,000†.

Book Production (1980): 110 titles (84 books, 26 pamphlets); 273,000 copies (229,000 books, 44,000 pamphlets).

Daily Newspapers (2001): 3 (estimated combined circulation 49,124)‡.

Non-daily Newspapers (provisional, 1988): 7 (combined circulation 99,000)*.

* Source: UNESCO, *Statistical Yearbook*.
† Source: International Telecommunication Union.
‡ Source: Audit Bureau of Circulations, Australia.

EDUCATION

Primary (2002): 709 schools (1995); 5,107 teachers; 142,102 pupils.

General Secondary (2002): 146 schools (1995); 4,142 teachers; 67,212 pupils.

Vocational and Technical (2000): 33 institutions (1998); 1,024 teachers (including special schools); 1,730 students.

Teacher Training (2000): 4 institutions (1999); 97 teachers; 1,003 students.

Medical (1989): 2 institutions; 493 students.

University (1999): 1 institution; 277 teachers (1991); 9,208 students.

Source: mainly UNESCO, *Statistical Yearbook*, and UN, *Statistical Yearbook for Asia and the Pacific*.

Adult Literacy Rate (UNESCO estimates): 92.9% (males 94.5%, females 91.4%) in 2002. Source: UN Development Programme, *Human Development Report*.

Directory

The Constitution

On 1 March 2001 President Iloilo reinstated the 1997 Constitution, after the Great Council of Chiefs (Bose Levu Vakaturaga—a traditional body, with some 70 members, consisting of every hereditary chief or Ratu of each Fijian clan) had approved the draft. The Constitution Amendment Bill that was approved in July 1997 included provisions to ensure of a multi-racial Cabinet. The following is a summary of the main provisions:

The Constitution, which declares Fiji to be a sovereign, democratic republic, guarantees fundamental human rights, a universal, secret and equal suffrage and equality before the law for all Fijian citizens. Citizenship may be acquired by birth, descent, registration or naturalization and is assured for all those who were Fijian citizens before 6 October 1987. Parliament may make provision for the deprivation or renunciation of a person's citizenship. Ethnic Fijians, and the Polynesian inhabitants of Rotuma, receive special constitutional consideration. The Judicial and Legal Services Commission, the Public Service Commission and the Police Service Commission are established as supervisory bodies.

THE GREAT COUNCIL OF CHIEFS

The Great Council of Chiefs (Bose Levu Vakaturaga) derives its authority from the status of its members and their chiefly lineage. The Great Council appoints the President of the Republic and selects the 14 nominees for appointment to the Senate, the upper chamber of the Parliament.

The Great Council became fully independent of the Government in mid-1999.

THE EXECUTIVE

Executive authority is vested in the President of the Republic, who is appointed by the Great Council of Chiefs, for a five-year term, to be constitutional Head of State and Commander-in-Chief of the armed forces. The Presidential Council advises the President on matters of national importance. The President, and Parliament, can be empowered to introduce any necessary measures in an emergency or in response to acts of subversion which threaten Fiji.

In most cases the President is guided by the Cabinet, which conducts the government of the Republic. The Cabinet is led by the Prime Minister, who is a Fijian of any ethnic origin and is appointed by the President from among the members of Parliament, on the basis of support in the legislature. The Prime Minister selects the other members of the Cabinet (the Attorney-General, the minister responsible for defence and security and any other ministers) from either the House of Representatives or the Senate on a multi-party and multiracial basis. The Cabinet is responsible to Parliament.

THE LEGISLATURE

Legislative power is vested in the Parliament, which comprises the President, the appointed upper house or Senate and an elected House of Representatives. The maximum duration of a parliament is five years.

The Senate has 32 members, appointed by the President of the Republic for the term of the Parliament. A total of 14 senators are nominated by the Great Council of Chiefs, nine are appointed on the advice of the Prime Minister, eight on the advice of the Leader of the Opposition, and one on the advice of the Rotuma Island Council. The Senate is a house of review, with some powers to initiate legislation, but with limited influence on financial measures. The Senate is important in the protection of ethnic Fijian interests, and its consent is essential to any attempt to amend, alter or repeal any provisions affecting ethnic Fijians, their customs, land or tradition.

The House of Representatives has 71 elected members, who themselves elect their presiding officials and the Speaker from outside the membership of the House, and Deputy Speaker from among their number (excluding ministers). Voting is communal, with universal suffrage for all citizens of the Republic aged over 21 years. Seats are reserved on a racial basis: 23 for ethnic Fijians, 19 for Indians, three for other races (General Electors), one for Rotuma Islanders and 25 open seats. Elections must be held at least every five years and are to be administered by an independent Supervisor of Elections. An independent Boundaries Commission determines constituency boundaries.

THE JUDICIARY

The judiciary is independent and comprises the High Court, the Fiji Court of Appeal and the Supreme Court. The High Court and the Supreme Court are the final arbiters of the Constitution. The establishment of Fijian courts is provided for, and decisions of the Native Lands Commission (relating to ethnic Fijian customs, traditions and usage, and on disputes over the headship of any part of the Fijian people, with the customary right to occupy and use any native lands) are declared to be final and without appeal.

The Government

HEAD OF STATE

President: Ratu Josefa Iloilo (appointed 12 July 2000 by an interim authority established following the coup of 19 May 2000).

Vice-President: Ratu Jope Seniloli (convicted in August 2004 on charges of involvement in the coup of 2000).

THE CABINET
(September 2004)

Prime Minister, Minister for Fijian Affairs, Culture and Heritage, National Reconciliation and Unity, Multi-Ethnic Affairs and Reform of the Sugar Industry: Laisenia Qarase.

Minister for Foreign Affairs and External Trade: Kaliopate Tavola.

Minister for Finance and National Planning : Ratu Jone Kubuabola.

Attorney-General and Minister for Justice: Qoriniasi Bale.

Minister for Lands and Mineral Resources: Ratu Naiqama Lalabalavu.

Minister for Works and Energy: Savenaca Draunidalo.

Minister for Education: Ro Teimumu Vuikaba Kepa.

Minister for Tourism: Pita Nacuva.

Minister for Transport and Civil Aviation: Josefa Vosanibola.

Minister for Labour, Industrial Relations and Productivity: Kenneth Zinck.

Minister for Commerce, Business Development and Investment: Tomasi Vuetilovoni.

Minister for Information, Communications and Media Relations: Simione Kaitani.

Minister for Local Government, Housing, Squatter Settlement and Environment: Mataiasi Vave Ragigia.

Minister of Health: Solomone Naivalu.

Minister for Agriculture, Sugar and Land Resettlement: Jonetani Galuinadi.

Minister for Women, Social Welfare and Poverty Alleviation: Adi Asenaca Caucau.

Minister for Youth, Employment Opportunities and Sports: Isireli Leweniqila.

Minister for Home Affairs and Immigration: Joketani Cokanasiga.

Minister for Fisheries and Forests: Konisi Yabaki.

Minister for Public Enterprises and Public Sector Reform: Irami Matairavula.

MINISTRIES

Office of the President: Government Bldgs, POB 2513, Suva; tel. 3314244; fax 3301645.

Office of the Prime Minister: Government Bldgs, POB 2353, Suva; tel. 3211201; fax 3306034; e-mail pmsoffice@is.com.fj; internet www.fiji.gov.fj.

Office of the Attorney-General and Ministry of Justice: Government Bldgs, Victoria Parade, POB 2213, Suva; tel. 3309866; fax 3305421.

Ministry of Agriculture, Sugar and Land Resettlement: PMB, POB 358, Raiwaqa; tel. 3384233; fax 3385048; e-mail maffinfo@is.com.fj.

Ministry of Commerce, Business Development and Investment: Government Bldgs, POB 2131, Suva; tel. 3305411; fax 3302617.

Ministry of Education: Marela House, Thurston St, PMB, Suva; tel. 3314477; fax 3303511.

Ministry of Finance and National Planning: Government Bldgs, POB 2212, Suva; tel. 3307011; fax 3300834.

Ministry of Fisheries and Forests: POB 2218, Government Bldgs, Suva; tel. 3301611; fax 3301595; e-mail forestry-hq@msd.gov.fj.

Ministry of Foreign Affairs and External Trade: Government Bldgs, POB 2220, Suva; tel. 3309628; fax 3301741; e-mail info@foreignaffairs.gov.fj; internet www.foreignaffairs.gov.fj.

Ministry of Health: Government Bldgs, POB 2223, Suva; tel. 3306177; fax 3306163; e-mail info@health.gov.fj.

Ministry of Home Affairs and Immigration: Government Bldgs, POB 2349, Suva; tel. 3211401; fax 3317721; e-mail inforhomeaff@govnet.gov.fj.

Ministry of Information, Communications and Media Relations: Government Bldgs, POB 2225, Suva; tel. 3211700; fax 3303146; e-mail info@fiji.gov.fj.

Ministry of Labour, Industrial Relations and Productivity: Government Bldgs, POB 2216, Suva; tel. 3211640; fax 3304701; e-mail minlabour@is.com.fj.

Ministry of Lands and Mineral Resources: Government Bldgs, POB 2118, Suva; tel. 3211556; fax 3302730; e-mail mbaravilala@lands.gov.fj.

Ministry of Local Government, Housing, Squatter Settlement and Environment: *National Planning*: Government Bldgs, POB 2351, Suva; *Local Government*: Government Bldgs, POB 2131; *Environment*: Government Bldgs, POB 2131; tel. 3313411; fax 3304809; tel. 3211310; fax 3303515; tel. 3311069; fax 3312879; e-mail msovaki@govnet.gov.fj.

Ministry of Information, Communications and Media Relations: Government Bldgs, POB 2225, Suva; tel. 3211700; fax 3303146; e-mail info@fiji.gov.fj.

Ministry of Public Enterprise and Public Sector Reform: Government Bldgs, POB 2278, Suva; tel. 3315577; fax 3315035; e-mail mshareem@govnet.gov.fj; internet www.publicenterprises.gov.fj.

Ministry of Tourism: Government Bldgs, POB 2278, Suva; tel. 3312788; fax 3302060; e-mail infodesk@fijivb.gov.fj; internet www.bulafiji.com.

Ministry of Transport and Civil Aviation: Private Mail Bag, Samabula; tel. 3316866; fax 3307839; e-mail infodesk@fijifvb.gov.fj; internet www.bulafiji.com.

Ministry of Women, Social Welfare and Poverty Alleviation: POB 14068, Suva; tel. 3312199; fax 3303829; e-mail slomaloma@govnet.gov.fj.

Ministry of Works and Energy: POB 2493, Government Bldgs, Suva; tel. 3384111; fax 3383198.

Ministry of Youth, Employment Opportunities and Sports: Government Bldgs, POB 2448, Suva; tel. 3315960; fax 3305348.

Legislature

PARLIAMENT

Senate

The Senate is also known as the House of Review. The upper chamber comprises 32 appointed members (see The Constitution).

House of Representatives

The lower chamber comprises 71 elected members: 23 representing ethnic Fijians, 19 representing ethnic Indians, three representing other races (General Electors), one for Rotuma Islanders and 25 seats open to all races.

Speaker: Ratu Epeli Nailatikau.

General Election, 25 August–1 September 2001

	Communal Seats				
	Fijian	Indian	Other*	Open Seats	Total Seats
Fiji United Party (SDL)	18	—	1	13	32
Fiji Labour Party (FLP)	—	19	—	8	27
Conservative Alliance Party (MV)	5	—	—	1	6
New Labour Unity Party (NLUP)	—	—	1	1	2
National Federation Party (NFP)	—	0	—	1	1
United General Party (UGP)	—	—	1	—	1
Independents	—	—	1	1	2
Total	23	19	4	25	71

* One Rotuman and three General Electors' seats.

Notes: The total includes one seat won by the Fiji United Party in a by-election held on 25 September 2001. In September 2002, following a court ruling, the one NFP seat was awarded to the FLP, thus bringing the latter's representation to 28 seats.

Political Organizations

Bai Kei Viti: Suva; f. 2001; Sec. Ratu TEVITA MOMOEDONU.

Fiji Indian Congress: POB 3661, Samabula, Suva; tel. 3391211; fax 3340117; f. 1991; Gen. Sec. VIJAY RAGHWAN.

Fiji Indian Liberal Party: Rakiraki; f. 1991; represents the interests of the Indian community, particularly sugar-cane farmers and students; Sec. SWANI KUMAR.

Fiji Labour Party (FLP): POB 2162, Government Bldgs, Suva; tel. 3305811; fax 3305317; e-mail flp@connect.com.fj; internet www.flp.org.fj; f. 1985; Pres. JOKAPECI KOROI; Sec.-Gen. MAHENDRA PAL CHAUDHRY.

Fijian Association Party (FAP): Suva; f. 1995 by merger of Fijian Association (a breakaway faction of the SVT) and the multiracial All Nationals Congress; Leader Adi KUINI SPEED; Pres. Ratu INOKE SERU.

Fijian Conservative Party: Suva; f. 1989 by former mems of the Fiji Nationalist Party and the Alliance Party; Leader ISIRELI VUIBAU.

Fijian Nationalist United Front Party (FNUFP): POB 1336, Suva; tel. 3362317; f. 1992 to replace Fijian Nationalist Party; seeks additional parliamentary representation for persons of Fijian ethnic origin, the introduction of other pro-Fijian reforms and the repatriation of ethnic Indians; Leader SAKEASI BAKEWA BUTADROKA.

Janata Party: Suva; f. 1995 by former mems of NFP and FLP.

Matanitu Vanua (MV) (Conservative Alliance Party): c/o House of Representatives, Suva; f. 2001; Leader GEORGE SPEIGHT.

National Federation Party (NFP): POB 13534, Suva; tel. 3305811; fax 3305317; f. 1960 by merger of the Federation Party, which was multiracial but mainly Indian, and the National Democratic Party; Leader ATTAR SINGH; Pres. JOGINDRA SINGH.

Nationalist Vanua Takolavo Party (NVTLP): c/o House of Representatives, Suva; Leader SAULA TELAWA.

New Labour Movement: Suva; Gen. Sec. MICHAEL COLUMBUS.

New Labour United Party (NLUP): Suva; f. 2001 ; Leader DR TUPENI BABA; Pres. RATU MELI VESIKULA.

Party of National Unity (PANU): Ba; f. 1998 to lobby for increased representation for the province of Ba; Leader MELI BOGILEKA.

Soqosoqo Duavata ni Lewenivanua (SDL) (Fiji United Party): c/o House of Representatives, Suva; f. 1990; Leader LAISENIA QARASE; Gen. Sec. JALE BABA.

Soqosoqo ni Vakavulewa ni Taukei (SVT) (Fijian Political Party): Suva; f. 1990 by Great Council of Chiefs; supports constitutional dominance of ethnic Fijians but accepts multiracialism; Pres. Ratu SITIVENI RABUKA; Gen. Sec. EMA DRUAVESI.

Taukei Movement: POB 505, Lautoka; f. 1987, following merger of Taukei Liberation Front and Domo Ni Taukei group; right-wing indigenous Fijian nationalist group; Pres. Ratu TEVITA BOLOBOLO; Vice-Pres. Ratu INOKE KUBUABOLA; Gen. Sec. APISAI TORA.

United General Party (UGP): Suva; f. 1998 by the merger of the General Electors' Party and the General Voters' Party (fmrly the General Electors' Association, one of the three wings of the Alliance Party—AP, the ruling party 1970–87); represents the interests of the minority Chinese and European communities and people from other Pacific Islands resident in Fiji, all of whom are classed as General Electors under the 1998 Constitution; Pres. DAVID PICKERING.

Vanua Independent Party: Leader ILIESA TUVALOVO; Sec. URAIA TUISOVISOVI.

Veitokani ni Lewenivanua Vakarisito (VLV) (Christian Democratic Alliance): c/o House of Representatives, Suva; f. 1998 in opposition to constitutional reforms and to defend Christian and Melanesian interests; Leader POESCI WAQALEVU BUNE; Sec. TANIELA TABU.

Other minor parties that contested the 2001 election included the Justice and Freedom Party, the Dodonu ni Taukei Party, the Girmit Heritage Party, the General Voters Party, the Lio'on Famor Rotuma Party and the Party of the Truth.

Supporters of secession are concentrated in Rotuma.

Diplomatic Representation

EMBASSIES AND HIGH COMMISSIONS IN FIJI

Australia: 37 Princes Rd, POB 214, Suva; tel. 3382211; fax 3382065; e-mail public-affairs-suva@dfat.gov.au; High Commissioner JENNIFER RAWSON.

China, People's Republic: 147 Queen Elizabeth Dr., PMB, Nasese, Suva; tel. 3300215; fax 3300950; e-mail chinaemb@is.com.fj; Ambassador ZHANG JUNSAI.

France: Dominion House, 7th Floor, Thomson St, Suva; tel. 3312233; fax 3301894; e-mail eugene.berg@diplomatie.gouv.fr; Ambassador EUGENE BERG.

India: POB 471, Suva; tel. 3301125; fax 3301032; e-mail hicomindsuva@is.com.fj; High Commissioner ISHWAR SINGH CHAUHAN.

Japan: Dominion House, 2nd Floor, POB 13045, Suva; tel. 3304633; fax 3302984; e-mail eojfiji@is.com.fj; Ambassador HISATO MURAYAMA.

Korea, Republic: Vanua House, 8th Floor, PMB, Suva; tel. 3300977; fax 3303410; Ambassador LIM DAE-TAEK.

Malaysia: Pacific House, 5th Floor, POB 356, Suva; tel. 3312166; fax 3303350; e-mail mwsuva@is.com.fj; High Commissioner MOHAMMED TAKWIR DIN.

Marshall Islands: 41 Borron Rd, Government Bldgs, POB 2038, Suva; tel. 3387899; fax 3387115; Ambassador MACK KAMINAGA.

Micronesia, Federated States: 37 Loftus St, POB 15493, Suva; tel. 304566; fax 3300842; e-mail fsmsuva@sopacsun.sopac.org.fj; Ambassador KODARO MARTIN GALLEN.

Nauru: Ratu Sukuna House, 7th Floor, Government Bldgs, POB 2420, Suva; tel. 3313566; fax 3302861; High Commissioner Dr LUDWIG KEKE.

New Zealand: Reserve Bank of Fiji Bldg, 10th Floor, Pratt St, POB 1378, Suva; tel. 3311422; fax 3300842; e-mail nzhc@connect.com.fj; High Commissioner ADRIAN SIMCOCK.

Papua New Guinea: Credit Corporation House, 3rd Floor, Government Bldgs, POB 2447, Suva; tel. 3304244; fax 3300178; e-mail kundufj@is.com.fj; High Commissioner ALEXIS OAENGO MAINO.

Tuvalu: 16 Gorrie St, POB 14449, Suva; tel. 3301355; fax 3308479; High Commissioner TAUKELINA FINIKASO.

United Kingdom: Victoria House, 47 Gladstone Rd, POB 1355, Suva; tel. 3311033; fax 3301406; e-mail publicdiplomacysuva@fco.gov.uk; internet www.britishhighcommission.fco.gov.uk/fiji; High Commissioner CHARLES F. MOCHAN.

USA: 31 Loftus St, POB 218, Suva; tel. 3314466; fax 3300081; e-mail usembsuva@is.com.fj; internet www.amembassy-fiji.gov; Ambassador DAVID LYON.

Judicial System

Justice is administered by the Supreme Court, the Fiji Court of Appeal, the High Court and the Magistrates' Courts. The Supreme Court of Fiji is the superior court of record presided over by the Chief Justice. The 1990 Constitution provided for the establishment of Fijian customary courts and declared as final decisions of the Native Lands Commission in cases involving Fijian custom, etc.

Supreme Court

Suva; tel. 3211524; fax 3300674.

Chief Justice: DANIEL FATIAKI.

President of the Fiji Court of Appeal: JAI RAM REDDY.

Director of Public Prosecutions: PETER RIDGWAY.

Solicitor-General: NAINENDRA NAND.

Religion

CHRISTIANITY

Most ethnic Fijians are Christians. Methodists are the largest Christian group, followed by Roman Catholics. In the census of 1986 about 53% of the population were Christian (mainly Methodists).

Fiji Council of Churches: POB 2300, Government Bldgs, Suva; tel. (1) 3313798; f. 1964; seven mem. churches; Pres. Rev. Apimeleki Qilio; Gen. Sec. Rev. Ifereimi Cama; (acting).

The Anglican Communion

In April 1990 Polynesia, formerly a missionary diocese of the Church of the Province of New Zealand, became a full and integral diocese. The diocese of Polynesia is based in Fiji but also includes Wallis and Futuna, Tuvalu, Kiribati, French Polynesia, Cook Islands, Tonga, Samoa and Tokelau.

Bishop of Polynesia: Rt Rev. Jabez Leslie Bryce, Bishop's Office, 8 Desvoeux Rd, Suva; e-mail episcopus@connect.com.fj; tel. 3304716; fax 3302687.

The Roman Catholic Church

Fiji comprises a single archdiocese. At 31 December 2002 there were an estimated 85,518 adherents in the country.

Bishops' Conference: Episcopal Conference of the Pacific Secretariat (CEPAC), 14 Williamson Rd, POB 289, Suva; tel. 3300340; fax 3303143; e-mail cepac@connect.com.fj; f. 1968; 17 mems; Pres. Most Rev. Apuron Anthony Sablan (Archbishop of Agaña, Guam); Admin. Officer Thomas Tavutonivalu.

Regional Appeal Tribunal for CEPAC

14 Williamson Rd, POB 289, Suva; tel. 300340; fax 3303143; e-mail cepac@is.com.fj.

f. 1980; 17 mems; Judicial Vicar Rev. Theo Koster.

Archbishop of Suva: Most Rev. Petero Mataca, Archdiocesan Office, Nicolas House,35 Pratt St, POB 109, Suva; tel. 3301955; fax 3301565; e-mail cabajog@yahoo.com.

Other Christian Churches

Methodist Church in Fiji (Lotu Wesele e Viti): Epworth Arcade, Nina St, POB 357, Suva; tel. 3311477; fax 3303771; f. 1835; autonomous since 1964; 211,550 mems (2003); Gen. Sec. Rev. Jione Langi; Pres. Rev. Laisiasa Ratabacaca.

Other denominations active in the country include the Assembly of God (with c. 7,000 mems), the Baptist Mission, the Congregational Christian Church and the Presbyterian Church.

HINDUISM

Most of the Indian community are Hindus. According to the census of 1986, 38% of the population were Hindus.

ISLAM

In 1993 some 8% of the population were Muslim. There are several Islamic organizations:

Fiji Muslim League: POB 3990, Samabula, Suva; tel. 3384566; fax 3370204; e-mail fijimuslim@is.com.fj; f. 1926; Nat. Pres. Hafizud Dean Khan; Gen. Sec. Nisar Ahmad Ali; 26 brs and 3 subsidiary orgs.

SIKHISM

Sikh Association of Fiji: Suva; Pres. Harkewal Singh.

BAHÁ'Í FAITH

National Spiritual Assembly: National Office, POB 639, Suva; tel. 3387574; fax 3387772; e-mail nsafijiskm@suva.is.com.fj; mems resident in 490 localities; national headquarters for consultancy and co-ordination.

The Press

NEWSPAPERS AND PERIODICALS

Coconut Telegraph: POB 249, Savusavu, Vanua Levu; f. 1975; monthly; serves widely-scattered rural communities; Editor Lema Low.

Fiji Calling: POB 12095, Suva; tel. 3305916; fax 3301930; publ. by Associated Media Ltd; every 6 months; English; Publr Yashwant Gaunder.

Fiji Cane Grower: POB 12095, Suva; tel. 3305916; fax 3305256.

Fiji Daily Post: 10–16 Toorak Rd, POB 2071, Govt Bldgs, Suva; f. 1987 as *Fiji Post* , daily from 1989; English; 100% govt-owned since Sept. 2003; Chair. Malakai Naiyaga; Editor Fred Wesley.

Fiji Magic: POB 12095, Suva; tel. 3305916; fax 3302852; e-mail fijimagic@fijilive.com; internet www.fijilive.com/fijimagic; publ. by Associated Media Ltd; monthly; English; Publr Yashwant Gaunder; circ. 15,000.

Fiji Republic Gazette: Printing Dept, POB 98, Suva; tel. 3385999; fax 3370203; f. 1874; weekly; English.

Fiji Sun: 12 Amra St, Walubay, Suva; tel. 3307555; fax 3311455; e-mail fijisun@sun.com.fj; re-established 1999; daily; Editor Leone Cabenatabua; Publr Michael Richards.

Fiji Times: 20 Gordon St, POB 1167, Suva; tel. 3304111; fax 3301521; f. 1869; publ. by Fiji Times Ltd; daily; English; Man. Dir Tony Yianni; Editor Samisoni Kakaiualu; circ. 34,000.

Fiji Trade Review: The Rubine Group, POB 12511, Suva; tel. 3313944; monthly; English; Publr George Rubine; Editor Mabel Howard.

Islands Business Magazine: 46 Gordon St, POB 12718, Suva; tel. 3303108; fax 3301423; e-mail editor@ibi.com.fj; internet www.pacificislands.cc; fmrly *Pacific Magazine*; regional monthly news and business magazine featuring the Fiji Islands Business supplement; English; Publr Robert Keith-Reid; Editor-in-Chief Laisa Taga; circ. 8,500.

Na Tui: 422 Fletcher Rd, POB 2071, Govt Bldgs, Suva; f. 1988; weekly; Fijian; Publr Taniela Bolea; Editor Samisoni Bolatagici; circ. 7,000.

Nai Lalakai: 20 Gordon St, POB 1167, Suva; tel. 3304111; fax 3301521; e-mail fijitimes@is.com.fj; f. 1962; publ. by Fiji Times Ltd; weekly; Fijian; Editor Samisoni Kakaivalu; circ. 18,000.

Pacific Business: POB 12095, Suva; tel. 3305916; fax 3301930; publ. by Associated Media Ltd; monthly; English; Publr Yashwant Gaunder.

Pacific Telecom: POB 12095, Suva; tel. 3300591; fax 3302852; e-mail review@is.com.fj; publ. by Associated Media Ltd; monthly; English; Publr Yashwant Gaunder.

Pactrainer: PMB, Suva; tel. 3303623; fax 3303943; e-mail pina@is.com.fj; monthly; newsletter of Pacific Journalism Development Centre; Editor Peter Lomas.

PINA Nius: Pacific Islands News Association, 46 Gordon St, PMB, Suva; tel. 3303623; fax 3303943; e-mail pina@is.com.fj; internet www.pinarius.org; monthly newsletter of Pacific Islands News Association; Editor Nina Ratulele.

The Review: POB 12095, Suva; tel. 3305916; fax 3301930; e-mail review@is.com.fj; publ. by Associated Media Ltd; monthly; English; Publr Yashwant Gaunder.

Sartaj: John Beater Enterprises Ltd, Raiwaqa, POB 5141, Suva; f. 1988; weekly; Hindi; Editor S. Daso; circ. 15,000.

Shanti Dut: 20 Gordon St, POB 1167, Suva; f. 1935; publ. by Fiji Times Ltd; weekly; Hindi; Editor Nilam Kumar; circ. 12,000.

Top Shot: Suva; f. 1995; golf magazine; monthly.

Volasiga: 10–16 Toorak Rd, POB 2071, Suva; f. 1988; weekly; Fijian; Gen. Man. Anura Bandara (acting); Editor Samisoni Bolatagici.

The Weekender: 2 Dension Rd, POB 15652, Suva; tel. 3315477; fax 3305346; publ. by Media Resources Ltd; weekly; English; Publr Josefata Nata.

PRESS ASSOCIATIONS

Fiji Islands Media Association: c/o Vasiti Ivaqa, POB 12718, Suva; tel. 3303108; fax 3301423; national press asscn; operates Fiji Press Club and Fiji Journalism Training Institute; Sec. Nina Ratulele.

Pacific Islands News Association: 46 Gordon St, PMB, Suva; tel. 3303623; fax 3303943; e-mail pina@is.com.fj; internet www.pinarius.org; regional press asscn; defends freedom of information and expression, promotes professional co-operation, provides training and education; Administrator Nina Ratulele; Pres. Johnson Honimae.

Publishers

Fiji Times Ltd: POB 1167, Suva; tel. 3304111; fax 3302011; e-mail tyianni@fijitimes.com.fj; f. 1869; Propr News Corpn Ltd; largest newspaper publr; also publrs of books and magazines; Man. Dir Tony Yianni.

Lotu Pasifika Productions: POB 2401, Suva; tel. 3301314; fax 3301183; f. 1973; cookery, education, poetry, religion; Gen. Man. SERU L. VEREBALAVU.

University of the South Pacific: University Media Centre, POB 1168, Suva; tel. 3313900; fax 3301305; e-mail austin_l@usp.ac.fj; f. 1986; education, natural history, regional interests.

Government Publishing House

Printing and Stationery Department: POB 98, Suva; tel. 3385999; fax 3370203.

Broadcasting and Communications

TELECOMMUNICATIONS

Fiji International Telecommunicatons Ltd (FINTEL): 158 Victoria Parade, POB 59, Suva; tel. 3312933; fax 3300750; e-mail prichards@fintelfiji.com; 51% govt-owned; 49% C&W plc; CEO SAKARAIA TUILAKEPA.

Telcom Fiji Ltd: Private Mail Bag, Suva; tel. 3304019; fax 3301765; internet www.tfl.com.fj/; Chair. LIONEL YEE; CEO WINSTON THOMPSON.

Vodafone Fiji Ltd: Private Mail Bag, Suva; tel. 3312000; fax 3312007; e-mail aslam.khan@vodafone.com.fj; 51% owned by Telecom Fiji, 49% by Vodafone International Holdings BV; Man. Dir ASLAM KHAN.

BROADCASTING

Radio

Fiji Broadcasting Commission—FBC (Radio Fiji): Broadcasting House, POB 334, Suva; tel. 3314333; fax 3301643; f. 1954; statutory body; jointly funded by govt grant and advertising revenue; Chair. DANIEL WHIPPY; CEO SIRELI KINI.

Radio Fiji 1 broadcasts nationally on AM in English and Fijian;

Radio Fiji 2 broadcasts nationally on AM in English and Hindi;

Radio Fiji Gold broadcasts nationally on AM and FM in English;

104 FM and Radio Rajdhani 98 FM, mainly with musical programmes, broadcast in English and Hindi respectively, but are received only on Viti Levu;

Bula FM, musical programmes, broadcasts in Fijian, received only on Viti Levu.

Communications Fiji Ltd: 231 Waimanu Rd, PMB, Suva; tel. 3314766; fax 3303748; e-mail cfl@fm96.com.fj; f. 1985; operates three commercial stations; Man. Dir WILLIAM PARKINSON; Gen. Man. IAN JACKSON.

FM 96, f. 1985, broadcasts 24 hours per day, on FM, in English;

Navtarang, f. 1989, broadcasts 24 hours per day, on FM, in Hindi;

Viti FM, f. 1996, broadcasts 24 hours per day, on FM, in Fijian.

Radio Light/Radio Naya Jiwan: Shop 11A & B, Pacific Harbour Culture Centre, POB 319, Pacific Harbour; tel. and fax 3450007; fax 3450096; e-mail radiolight@connect.com.fj; internet www.radiolight.org; f. 1990; non-profit religious organization; broadcasts in English (Radio Light) and Hindi (Radio Naya Jiwan)on FM 106, FM 93.6 and FM 94.8; Station Man. and Programmes Dir DOUGLAS ROSE.

Radio Pasifik: POB 1168, University of the South Pacific, Suva; tel. 3313900; fax 3312591; e-mail schuster@usp.ac.fj; Gen. Man. ALFRED SCHUSTER.

Television

Film and Television Unit (FTU): c/o Ministry of Information and Communications, Govt Bldgs, POB 2225, Suva; tel. 3314688; fax 3300196; video library; production unit established by Govt and Hanns Seidel Foundation (Germany); a weekly news magazine and local documentary programmes.

Fiji Television Ltd: 20 Gorrie St, POB 2442, Govt Bldgs, Suva; tel. 3305100; fax 3305077; e-mail fijitv@is.com.fj; internet www.fijitv.com; f. 1994; operates two services, Fiji 1, a free channel, and Sky Fiji, a three-channel subscription service; Chair. OLOTA ROKOVUNISEI; CEO KEN CLARK; Head of Programming RICHARD BROADBRIDGE.

In 1990 two television stations were constructed at Suva and Monsavu, with aid from the People's Republic of China. A permanent television station became operational in July 1994.

Finance

In 1996 the Ministry of Finance announced that it had secured financial assistance for the undertaking of a study to investigate the possibility of developing an 'offshore' financial centre in Fiji.

BANKING

(cap. = capital; res = reserves; dep. = deposits; m. = million; brs = branches; amounts in Fiji dollars)

Central Bank

Reserve Bank of Fiji: 1 Pratt St, PMB, Suva; tel. 3313611; fax 3301688; e-mail rbf@reservebank.gov.fj; internet www.reservebank.gov.fj; f. 1984 to replace Central Monetary Authority of Fiji; bank of issue; administers Insurance Act; cap. 2.0m., res 81.5m., dep. 241.6m. (Dec. 2002); Chair. and Gov. SAVENACA NARUBE.

Commercial Bank

Colonial National Bank : 3 Central St, POB 1166, Suva; tel. 3314400; fax 3302190; f. 1974 as National Bank of Fiji; 51% acquired from Fiji Govt by Colonial Ltd in 1999; 51% owned by Commonwealth Bank of Australia, 49% owned by Govt of Fiji; cap. 15.0m., res 0.3m., dep. 251.4m. (June 2002), dep. 317.2m. (June 2004); Chair. MALAKAI NAIYAGA; Gen. Man. MIKE UPPERTON; 15 brs; 45 agencies.

Development Bank

Fiji Development Bank: 360 Victoria Parade, POB 104, Suva; tel. 3314866; fax 3314886; f. 1967; finances the development of natural resources, agriculture, transportation and other industries and enterprises; statutory body; cap. 50.8m., res 14.0m., dep. 182.6m. (June 1993); Chair. CHARLES WALKER; 9 brs.

Merchant Banks

Merchant Finance and Investment Company Ltd: 91 Gordon St, POB 14213, Suva; tel. 3314955; fax 3300026; e-mail merchantfi@connect.com.fj; f. 1986; fmrly Merchant Bank of Fiji Ltd; owned by Fijian Holdings Ltd (80%), South Pacific Trustees (20%); Man. Dir S. WELEILAKEBA; 3 brs.

National MBf Finance (Fiji) Ltd: Burns Philp Bldg, 2nd Floor, POB 13525, Suva; tel. 302232; fax 3305915; e-mail mbf@is.com.fj; f. 1991; 51% owned by the National Bank of Fiji, 49% by MBf Asia Capital Corpn Holding Ltd (Hong Kong); Chief Operating Officer SIEK KART; 4 brs.

Foreign Banks

Agence Française de Développement (ADF) (France): Suva; licensed to operate in Fiji in 1997.

Australia and New Zealand (ANZ) Banking Group Ltd: ANZ House, 25 Victoria Parade, POB 179, Suva; tel. 3213000; fax 3312527; bought Bank of New Zealand in Fiji (8 brs) in 1990; Gen. Man. (Fiji) JOHN VELEGRINIS; 17 brs; 9 agencies.

Bank of Baroda (India): Bank of Baroda Bldg, Marks St, POB 57, Suva; tel. 3311400; fax 3302510; e-mail bobsuva@is.com.fj; f. 1908; CEO S. K. BAGCHI; 7 brs; 2 agencies.

Bank of Hawaii (USA): 67–69 Victoria Parade, POB 273, Suva; tel. 3312144; fax 3312464; f. 1993; Gen. Man. BRIAN BLISS; 3 brs.

Habib Bank (Pakistan): Narsey's Bldg, Renwick Rd, POB 108, Suva; tel. 3304011; fax 3304835; e-mail hblfibas@is.com.fj; Chief Man. (Fiji) ABDUL MATIN; licensed to operate in Fiji 1990; 3 brs.

Westpac Banking Corporation (Australia): 1 Thomson St, Suva; tel. 3300666; fax 3301813; Chief Man. (Pacific Islands region) TREVOR WISEMANTEL; 12 brs; 9 agencies.

STOCK EXCHANGE

South Pacific Stock Exchange: Level 2, Plaza One, Provident Plaza, 33 Ellery St, POB 11689, Suva; tel. 3304130; fax 3304145; e-mail suvastockex@is.com.fj; internet www.suvastockex.com; formerly Suva Stock Exchange; name changed as above in 2000; Chair. FOANA T. NEMANI; Man. MESAKE NAWARI.

INSURANCE

Blue Shield (Pacific) Ltd: Parade Bldg, POB 15137, Suva; tel. 3311733; fax 3300318; Fijian co; subsidiary of Colonial Mutual Life Assurance Society Ltd; medical and life insurance; Chief Exec. SIALENI VUETAKI.

Colonial Mutual Life Assurance Society Ltd: Colonial Bldg, PMB, Suva; tel. 314400; fax 3303448; f. 1876; inc in Australia; life; Gen. Man. SIMON SWANSON.

Dominion Insurance Ltd: Civic House, POB 14468, Suva; tel. 3311055; fax 3303475; partly owned by Flour Mills of (Fiji) Ltd; general insurance; Man. Dir GARY S. CALLAGHAN.

FAI Insurance (Fiji) Ltd: Suva.

Fiji Reinsurance Corpn Ltd: RBF Bldg, POB 12704, Suva; tel. 3313471; fax 3305679; 20% govt-owned; reinsurance; Chair. Ratu JONE Y. KUBUABOLA; Man. PETER MARIO.

Fijicare Mutual Assurance: 41 Loftus St, POB 15808, Suva; tel. 3302717; fax 3302119; f. 1992; CEO JEFF PRICE.

Insurance Trust of Fiji: Loftus St, POB 114, Suva; tel. 3311242; fax 3302541; Man. SAMUEL KRISHNA.

National Insurance Co of (Fiji) Ltd: McGowan Bldg, Suva; tel. 3315955; fax 3301376; owned by New Zealand interests; Gen. Man. GEOFF THOMPSON.

New India Assurance Co Ltd: Harifam Centre, POB 71, Suva; tel. 3313488; fax 3302679; Man. MILIND A. KHARAT.

Queensland Insurance (Fiji) Ltd: Queensland Insurance Center, Victoria Parade, POB 101, Suva; tel. 3315455; fax 3300285; e-mail info.fiji@qbe.com; internet www.qbe.com/asiapacific; owned by Australian interests; Gen. Man. JOHN HUNT.

There are also two Indian insurance companies operating in Fiji.

Trade and Industry

GOVERNMENT AGENCIES

Fiji National Training Council (FNTC): Beaumont Rd, POB 6890, Nasinu; tel. 3392000; fax 3340184; e-mail gen-enq@fntc.ac.fj; internet www.tpaf.ac.fj; Dir-Gen. JONE USAMATE.

Fiji Trade and Investment Board: Civic House, 6th Floor, Victoria Parade, Suva; tel. 3315988; fax 3301783; e-mail ftibinfo@ftib.org.fj; internet www.ftib.org.fj; f. 1980, restyled 1988, to promote and stimulate foreign and local economic development investment; Chair. JAMES DATTA; CEO JESONI VITUSAGAVULU.

Mineral Resources Department: Private Mail Bag, Suva; tel. 3381611; fax 3370039; e-mail director@mrd.gov.fj; internet www.mrd.gov.fj/index.html.

DEVELOPMENT ORGANIZATIONS

Fiji Development Company Ltd: POB 161, FNPF Place, 350 Victoria Parade, Suva; tel. 3304611; fax 3304171; e-mail hfc@is.com.fj; f. 1960; subsidiary of the Commonwealth Development Corpn; Man. F. KHAN.

Fiji-United States Business Council: CI-FTIB; POB 2303; Suva; f. 1998 to develop and expand trade links between the two countries; Pres. RAMENDRA NARAYAN.

Fijian Development Fund Board: POB 122, Suva; tel. 3312601; fax 3302585; f. 1951; funds derived from payments of $F20 a metric ton from the sales of copra by indigenous Fijians; deposits receive interest at 2.5%; funds used only for Fijian development schemes; Chair. Minister for Fijian Affairs; CEO VINCENT TOVATA.

Land Development Authority: c/o Ministry for Agriculture, Sugar and Land Resettlement, POB 5442, Raiwaqa; tel. 33384900; fax 33384058; f. 1961 to co-ordinate development plans for land and marine resources; Chair. JONETANI GALUINADI.

CHAMBERS OF COMMERCE

Ba Chamber of Commerce: POB 99, Ba; tel. 6670134; fax 6670132; Pres. DIJENDRA SINGH.

Labasa Chamber of Commerce: POB 992, Labasa; tel. 8811467; fax 8813009; Pres. SUBRAIL PRASAD.

Lautoka Chamber of Commerce and Industry: POB 366, Lautoka; tel. 6661834; fax 6662379; e-mail vaghco@connect.com.fj; Pres. NATWARLAL VAGH.

Levuka Chamber of Commerce: POB 85, Levuka; tel. 3440248; fax 3440252; Pres. ISHRAR ALI.

Nadi Chamber of Commerce: POB 2735, Nadi; tel. 6701704; fax 6702314; e-mail arunkumar@is.com.fj; Pres. VENKAT RAMANI AIYER.

Nausori Chamber of Commerce: POB 228, Nausori; tel. 3478235; fax 3400134; Pres. ROBERT RAJ KUMAR.

Sigatoka Chamber of Commerce: POB 882, Sigatoka; tel. 6500064; fax 6520006; Pres. NATWAR SINGH.

Suva Chamber of Commerce and Industry: 37 Viria Rd, Vatuwaqa Industrial Estate, POB 337, Suva; tel. 3380975; fax 3380854; f. 1902; Joint Sec. VERONIKA HANSRAJ; 150 mems.

Tavua-Vatukoula Chamber of Commerce: POB 698, Tavua; tel. 6680390; fax 6680390; Pres. SOHAN SINGH.

INDUSTRIAL AND TRADE ASSOCIATIONS

Fiji Forest Industries (FFI): Suva; Deputy Chair. Ratu SOSO KATONIVERE.

Fiji Kava Council: Suva; Chair. Ratu JOSATEKI NAWALOWALO.

Fiji National Petroleum Co Ltd: Suva; f. 1991; govt-owned, distributor of petroleum products.

Fiji Sugar Corporation Ltd: Western House, 3rd Floor, Cnr of Bila and Vidilo St, PMB, Lautoka; tel. 6662655; fax 6664685; nationalized 1974; buyer of sugar-cane and raw sugar mfrs; Chair. HAFIZUD D. KHAN; Man. Dir JOHN MCFADDEN.

Fiji Sugar Marketing Co Ltd: Dominion House, 5th Floor, Thomson St, POB 1402, Suva; tel. 3311588; fax 3300607; Man. Dir JONETANI GALUINADI.

Mining and Quarrying Council: 42 Gorrie St, Suva; tel. 33313188; fax 3302183; e-mail employer@is.com.fj; Chief Exec. K. A. J. ROBERTS.

National Trading Corporation Ltd: POB 13673, Suva; tel. 3315211; fax 3315584; f. 1992; a govt-owned body set up to develop markets for agricultural and marine produce locally and overseas; processes and markets fresh fruit, vegetables and ginger products; CEO APIAMA CEGUMALINA.

Native Lands Trust Board: POB 116, Suva; e-mail info@nltb.com.fj; internet www.nltb.com.fj; manages holdings of ethnic Fijian landowners; Gen. Man. KALIVATI BAKANI.

Pacific Fishing Co: Suva; fish-canning; govt-owned.

Sugar Cane Growers' Council: Canegrowers' Bldg, 3rd Floor, 75 Drasa Ave, Lautoka; tel. 6650466; fax 6650624; e-mail canegrower@is.com.fj; f. 1985; aims to develop the sugar industry and protect the interests of registered growers; CEO JAGANNATH SAMI; Gen. Man. BALA DASS; Chair. RUSIATE MUSUDROKA.

Sugar Commission of Fiji: POB 5993, Lautoka; tel. 6664866; fax 6664051; e-mail scof@is.com.fj; Chair. GERALD BARRACK.

EMPLOYERS' ORGANIZATIONS

Fiji Employers' Federation: 42 Gorrie St, POB 575, Suva; tel. 3313188; fax 3302183; e-mail employer@fef.com.fj; represents 230 major employers; Pres. HAFIZ U. KHAN; CEO KENNETH A. J. ROBERTS.

Fiji Manufacturers' Association: POB 1308, Suva; tel. 9212223; fax 3302567; e-mail volau-m@usp.ac.fj; internet www.fijibusiness.com; f. 1902; Pres. DESMOND WHITESIDE; 55 mems.

Local Inter-Island Shipowners' Association: POB 152, Suva; fax 3303389; e-mail consortship@connect.com.fj; Pres. GEORGE PATTERSON; Sec. LEO B. SMITH.

Textile, Clothing and Footwear Council: POB 10015, Nabua; tel. 3384777; fax 3370446; Sec. R. DUNSTAN.

UTILITIES

Electricity

Fiji Electricity Authority (FEA): PMB, Suva; e-mail ceo@fea.com.fj; tel. 3311133; fax 3311882; f. 1966; govt-owned; responsible for the generation, transmission and distribution of electricity throughout Fiji; CEO ROKOSERU NABALARUA.

Water

Water and Sewerage Section: Public Works Dept, Ministry of Works and Energy, Nasilivata House, Kings Rd, PMB, Samabula; tel. 3384111; fax 3383013; e-mail rsshandil@fijiwater.gov.fj; Dir RAM SUMER SHANDIL.

MAJOR COMPANIES

BPT (South Sea) Co Ltd: POB 355, Suva; tel. 3384888; fax 3370309; f. 1920; importers and distributors of motor vehicles, machinery and outboard motors; Gen. Man. ALMA MAHARAJ.

Carlton Brewery (Fiji) Ltd: POB 696, Suva; f. 1957; a subsidiary of Foster's Brewing Group Ltd (Australia); Gen. Man. J. PICKERING.

Carter Holt Harvey (Fiji) Ltd: POB 427, Suva; tel. 3410011; fax 3410808; f. 1965; mfrs of polythene bags and paper products; Gen. Man. C. D. BOSSLEY.

Central Manufacturing Co Ltd: POB 560, Suva; tel. 3381144; fax 3370080; f. 1955; mfrs and distributors of cigarettes and tobacco; Gen. Man. JOHN J. NELSON.

Crest Chicken Ltd: POB 83, Nausori; tel. 3478400; fax 3400061; f. 1965; mfrs of animal food; Gen. Man. DON MACLELLAN.

Eddie Hin Industries Ltd: Marine Drive, POB 98, Lautoka; tel. 6661433; fax 6665886; f. 1947; mfrs of non-alcoholic beverages; Man. Dir EDDIE WONG.

Emperor Gold Mining Co Ltd: Vatukoula; tel. 6680477; fax 6680779; f. 1935; gold mining and processing; assoc. cos: Jubilee Mining and Koula Mining; 2,000 employees; Chair. GEORGE DRYSDALE.

Fiji Motor Works Ltd: Namaka, Nadi; f. 1997; assembly of cars and small commercial vehicles; Man. Dir RICHARD BRETNAL.

Fiji National Petroleum Co Ltd: Suva; f. 1991; govt-owned; distributor of petroleum products.

Fiji Pine Ltd: Vakabuli Rd, Drasa, Lautoka; tel. 6661388; fax 6661561; e-mail fijipine@connect.com.fj; internet www.fijipine.com; f. 1991 to replace Fiji Pine Commission; owned by Govt and Fiji Pine Trust; Chair. NAVITALAI NAISORO; CEO ALEC CHANG.

Flour Mills of Fiji Ltd: POB 977, Suva; fax 3300944; f. 1971; Man. Dir HARI PUNJA; CEO SANJAY PUNJA.

Homecentres (Fiji) Ltd: GPOB 15278, Rodwell Rd, Suva; e-mail hcmerchandise@courts.com.fj; Dir of Merchandise DHARMA NAND.

Ika Corporation Ltd: POB 1371, Suva; tel. 3361922; fax 3351194; incorporated 1990; govt-owned; tuna-fishing co; exporter of fresh fish; leasing of vessels; Gen. Man. MITIELI BALEIVANUALALA.

Natural Waters of Viti Ltd: Civic House, Ground Floor, POB 14128, Suva; tel. 3302654; fax 3302714; mineral water bottling.

P. A. Lal & Co Ltd: POB 1242, Suva; f. 1946; builders of buses, trucks, trailers, coaches, furniture and fibreglass work; Dir RICHARD LAL.

Pacific Fishing Company (PAFCO): 2nd Floor, Marks St, POB 1371, Suva; tel. 3304405; fax 3301904; e-mail info@pafcofiji.com; internet www.pafcofiji.com; f. 1963; Fiji's largest fish-processing factory.

Pacific Green Industries (Fiji) Ltd: POB 832, Sigatoka; tel. 6500055; fax 6520014; e-mail pgfij@connect.com.fj; internet www.pacificgreenfiji.com; exports furniture made from coconut wood; Gen. Man. RAVIN CHANDRA; Man. Dir BRUCE DOWSE.

Rewa Co-operative Dairy Company: Ratu Mara Rd, Suva; processes, packages and supplies dairy produce; Chair RAM CHAND.

Shell Fiji Ltd: POB 168, Suva; tel. 3313933; fax 3302279; f. 1928; petrol stations and distribution of petroleum products; Gen. Man. KEVIN DWYER.

Shipbuilding Fiji Ltd: POB 16695, Sannergren Drive, Korovou, Walu Bay, Suva; tel. 3314699; fax 3303500; e-mail sfl@is.com.fj; 49% govt-owned, 51% owned by MCI/Carpenters; builds and services ocean-going vessels; CEO WAYNE SHAW.

South Pacific Distilleries Ltd of Fiji: Lautoka; Gen. Man. JOE RODAN; Chair. NUNO D'AQUINO.

Voko Industries: Suva; fish canners.

TRADE UNIONS

Fiji Trades Union Congress (FTUC): 32 Des Voeux Rd, POB 1418, Suva; tel. 3315377; fax 3300306; e-mail ftucl@is.com.fj; f. 1951; affiliated to ICFTU and ICFTU—APRO; 35 affiliated unions; more than 42,000 mems; Pres. DANIEL URAI; Gen. Sec. FELIX ANTHONY.

Principal affiliated unions:

Association of USP Staff: POB 1168, Suva; tel. 3313900; fax 3301305; f. 1977; Pres. GANESH CHAND; Sec. D. R. RAO.

Federated Airline Staff Association: Nadi Airport, POB 9259, Nadi; tel. 6722877; fax 6720068; e-mail fasa@ats.com.fj; Sec. DEWAN CHAND.

Fiji Aviation Workers' Association: FTUC Complex, 32 Des Voeux Rd, POB 5351, Raiwaqa; tel. 3303184; fax 3311805; Pres. VALENTINE SIMPSON; Gen. Sec. ATTAR SINGH.

Fiji Bank and Finance Sector Employees' Union: 101 Gordon St, POB 853, Suva; tel. 3301827; fax 3301956; e-mail fbeu@connect.com.fj; Nat. Sec. DIWAN C. SHANKAR.

Fiji Electricity and Allied Workers Union: POB 1390, Lautoka; tel. 6666353; e-mail feawu@is.com.fj; Pres. LEONE SAKETA; Sec. J. A. PAUL.

Fiji Garment, Textile and Allied Workers' Union: c/o FTUC, Raiwaqa; f. 1992.

Fiji Nursing Association: POB 1364, Suva; tel. 3305855; e-mail fna@connect.com.fj; Gen. Sec. KUINI LUTUA.

Fiji Public Service Association: 298 Waimanu Rd, POB 1405, Suva; tel. 3311922; fax 3301099; e-mail fpsa@is.com.fj; f. 1943; 3,434 mems; Pres. AISEA BATISARESARE; Gen. Sec. RAJESHWAR SINGH.

Fiji Sugar and General Workers' Union: 84 Naviti St, POB 330, Lautoka; tel. 6660746; fax 664888; 25,000 mems; Pres. SHIU LINGAM; Gen. Sec. FELIX ANTHONY.

Fiji Teachers' Union: 1–3 Berry Rd, Govt Bldgs, POB 2203, Suva; tel. 3314099; fax 3305962; e-mail ftu@is.com.fj; f. 1930; 3,200 mems; Pres. BALRAM; Gen. Sec. AGNI DEO SINGH.

Fijian Teachers Association: POB 14464, Suva; tel. 3315099; fax 3304978; e-mail fta@.com.fj; Pres. TEVITA KOROI; Gen. Sec. MAIKA NAMUDU.

Insurance Officers' Association: POB 71, Suva; tel. 3313488; Pres. JAGDISH KHATRI; Sec. DAVID LEE.

Mineworkers' Union of Fiji: POB 876, Tavua; f. 1986; Pres. HENNESY PETERS; Sec. KAVEKINI NAVUSO.

National Farmers' Union: POB 522, Labasa; tel. 8811838; 10,000 mems (sugar-cane farmers); Pres. DEWAN CHAND; Gen. Sec. M. P. CHAUDHRY; CEO MOHAMMED LATIF SUBEDAR.

National Union of Factory and Commercial Workers: POB 989, Suva; tel. 3311155; 3,800 mems; Pres. CAMA TUILEVEUKA; Gen. Sec. JAMES R. RAMAN.

National Union of Hotel and Catering Employees: Nadi Airport, POB 9426, Nadi; tel. 670906; fax 6700181; Pres. EMOSI DAWAI; Sec. TIMOA NAIVAHIWAQA.

Public Employees' Union: POB 781, Suva; tel. 3304501; 6,752 mems; Pres. SEMI TIKOICINA; Gen. Sec. FILIMONE BANUVE.

Transport and Oil Workers' Union: POB 903, Suva; tel. 3302534; f. 1988 following merger of Oil and Allied Workers' Union and Transport Workers' Union; Pres. J. BOLA; Sec. MICHAEL COLUMBUS.

There are several independent trade unions, including Fiji Registered Ports Workers' Union (f. 1947; Pres. JIOJI TAHOLOSALE).

Transport

RAILWAYS

Fiji Sugar Corporation Railway: Rarawai Mill, POB 155, Ba; tel. 6674044; fax 670505; for use in cane-harvesting season, May–Dec.; 595 km of permanent track and 225 km of temporary track (gauge of 600 mm), serving cane-growing areas at Ba, Lautoka and Penang on Viti Levu and Labasa on Vanua Levu; Gen. Man. ADURU KUVA.

ROADS

At the end of 1996 there were some 3,440 km of roads in Fiji, of which 49.2% were paved. A 500-km highway circles the main island of Viti Levu.

SHIPPING

There are ports of call at Suva, Lautoka, Levuka and Savusavu. The main port, Suva, handles more than 800 ships a year, including large passenger liners. Lautoka handles more than 300 vessels and liners and Levuka, the former capital of Fiji, mainly handles commercial fishing vessels. In 1996 a feasibility study into the possible establishment of a free port at Suva was commissioned. In May 1997 the Government approved 14 new ports of entry in the northern, western and central eastern districts of Fiji.

Maritime and Ports Authority of Fiji (MPAF): Administration Bldg, Princes Wharf, POB 780, Suva; tel. 3312700; fax 3300064; corporatized in 1998; Chair. DANIEL ELISHA; Port Master Capt. GEORGE MACOMBER.

Ports Terminals Ltd: POB S13, Suva; tel. 3304725; fax 3304769; e-mail herbert@suv.ptl.com.fj; f. 1998; stevedore, pilotage, storage and warehousing; CEO H. HAZELMAN; Port Manager E. KURUSIGA.

Burns Philp Shipping (Fiji) Ltd: Rodwell Rd, POB 15832, Suva; tel. 3315444; fax 3302754; e-mail info@pacshipfiji.com.fj; shipping agents, customs agents and international forwarding agents, crew handling; Gen. Man. DANNY REES.

Consort Shipping Line Ltd: Muaiwalu Complex, Rona St, Walubay, POB 152, Suva; tel. 3313344; fax 3303389; CEO HECTOR SMITH; Man. Dir JUSTIN SMIT.

Fiji Maritime Services Ltd: c/o Fiji Ports Workers and Seafarers Union, 36 Edinburgh Drive, Suva; f. 1989 by PAF and the Ports Workers' Union; services between Lautoka and Vanua Levu ports.

Inter-Ports Shipping Corpn Ltd: 25 Eliza St, Walu Bay; POB 152, Suva; tel. 3313638; f. 1984; Man. Dir JUSTIN SMITH.

Transcargo Express Fiji Ltd: POB 936, Suva; f. 1974; Man. Dir LEO B. SMITH.

Wong's Shipping Co Ltd: Suite 647, Epworth House, Nina St, POB 1269, Suva; tel. 3311867.

CIVIL AVIATION

There is an international airport at Nadi (about 210 km from Suva), a smaller international airport at Nausori (Suva) and 15 other airfields. Nadi is an important transit airport in the Pacific and in 1990 direct flights to Japan also began.

Airports Fiji Ltd: Nadi International Airport, Nadi; tel. 6725777; fax 6725 161; e-mail info@afl.com.fj; Chair. VILIAME LEQA; CEO Ratu SAKIUSA TUISOLIA.

Air Fiji Ltd: 219 Victoria Parade, POB 1259, Suva; tel. 3314666; fax 3300771; internet www.airfiji.net; operates 65 scheduled services daily to 15 domestic destinations; daily service to Tonga and Tuvalu and direct flights to Auckland and Sydney commenced in 1999; charter operations, aerial photography and surveillance also conducted; partly owned by the Fijian Govt, which was expected to sell a majority of its shares in 2001; Chair. DOUG HAZARD; CEO KEN MACDONALD.

Air Pacific Ltd: Air Pacific Centre, POB 9266, Nadi International Airport, Nadi; tel. 6720777; fax 6720512; internet www.airpacific.com; f. 1951 as Fiji Airways, name changed in 1971; domestic and international services from Nausori Airport (serving Suva) to Nadi and international services to Tonga, Solomon Islands, Cook Islands, Vanuatu, Samoa, Japan, Australia, New Zealand and the USA; 51% govt-owned, 46.05% owned by Qantas (Australia); Chair. GERALD BARRACK; Man. Dir MICHAEL MCQUAY; CEO JOHN CAMPBELL.

Fijian Airways International: POB 10138, Nadi International Airport, Nadi; tel. 6724702; fax 6724654; f. 1997; service to London via Singapore and Mumbai (India) planned; Chair. NEIL UNDERHILL; CEO ALAN LINDREA.

Hibiscus Air Ltd: Nadi International Airport, Nadi; domestic airline operating charter and non-scheduled flights around Fiji.

Sunflower Airlines Ltd: POB 9452, Nadi International Airport, Nadi; tel. 6723555; fax 6720085; e-mail sun@is.com.fj; internet www.fiji.to; f. 1980; scheduled flights to domestic destinations, also charter services; Man. Dir DON IAN COLLINGWOOD.

Vanua Air Charters: Labasa; f. 1993; provides domestic charter and freight services; Proprs Ratu Sir KAMISESE MARA, CHARAN SINGH.

Tourism

Scenery, climate, fishing and diving attract visitors to Fiji, where tourism is an important industry. The number of foreign tourist arrivals increased from 348,014 in 2001 (excluding cruise-ship passengers who numbered 6,858 in 2001) to 397,859 in 2002 and to an estimated 430,800 in 2003. In 2002 some 31.1% of visitors came from Australia, 17.2% from New Zealand, 14.8% from the USA, 10.9% from the United Kingdom and 6.6% from Japan. A total of 5,542 rooms in 187 hotels were available in 2001. Receipts from tourism increased from $F495.5m. in 2001 to an estimated $F615.4m. in 2003. The South Pacific Tourism Organization (formerly the Tourism Council of The South Pacific) is based in Suva. In 1998 the Government announced its intention further to develop the tourist industry in Fiji through the establishment of the Fiji Tourism Development Plan: 1998–2005. In April 1999 construction began on a new luxury resort at Korotoga, situated half-way between Suva and Nadi. The industry was severely affected by the coup of May 2000. However, tourism expanded significantly during 2002–03 and was expected to contribute substantially to high economic growth rates in the following years.

Fiji Hotel Association (FHA): 42 Gorrie St, GPOB 13560, Suva; tel. 3302980; fax 3300331; e-mail fha@connect.com.fj; represents 85 hotels; Pres. DIXON SEETO; Chief Exec. OLIVIA PARETI.

Fiji Visitors' Bureau: POB 92, Suva; tel. 3302433; fax 3300986; e-mail infodesk@fijifvb.gov.fj; internet www.bulafiji.com; f. 1923; Chair. SITIVENI WELEILAKEBA; Chief Exec. VILIAME GAVOKA; Dir of Tourism ERONI LUVENIYALI.

Defence

The Fiji Military Forces consist of men in the regular army, the Naval Squadron, the conservation corps and the territorials. The defence budget for 2002 was an estimated $F61m. At 1 August 2003 the total armed forces numbered 3,500 (3,200 in the army and 300 in the navy). As the country is a member of the Commonwealth, Fijians are entitled to work in the British armed forces and in late 2003 some 2,000 Fijian soldiers were serving in Iraq with the British army, while a further 520 were working as mercenaries in the region. In the following year a private security firm sought to recruit a further 500 Fijians for deployment in Iraq. The Government, however, refused a request by the Government of the USA to send a further 700 troops to Iraq in mid-2004 because of difficulties in funding the deployment. The Republic of Fiji has co-operated militarily with France, Taiwan, the People's Republic of China and Malaysia. Defence co-operation with New Zealand and Australia, which ceased following the coups of 1987, was resumed in 1992 and 1993. Collaboration was again interrupted by the coup of May 2000.

Commander-in-Chief: President of the Republic.

Commander of the Armed Forces: Cdre FRANK BAINIMARAMA.

Deputy Commander of the Armed Forces: Col IOANE NAIVALARUA.

Education

Education in Fiji is not compulsory, but in 1992 about 97% of children of school age were enrolled at the country's schools, and the Government provided free education for the first eight years of schooling. Primary education begins at six years of age and lasts for six years. Secondary education, beginning at the age of 12, lasts for a further six years. State subsidies are available for secondary and tertiary education in cases of hardship. In 2002 the Government announced plans to extend the duration of compulsory free education. In 1995 there were 709 state primary schools (with a total enrolment of 142,912 pupils in 2000) and 146 state secondary schools (with an enrolment of 66,905 pupils in 2000). In 1998 there were 33 vocational and technical institutions (with 1,730 enrolled students in 2000). In 1999 Fiji had four teacher-training colleges (with 1,003 students in 2000). There were 200 holders of Fiji government scholarships at the University of the South Pacific in Fiji in 1989. In 1999 university students on campus totalled about 4,014 and extension students totalled 5,194. Fiji has experienced a shortfall in the number of trained teachers, as a result of increased emigration prompted by the political upheavals. Budgetary proposals for 2002 allocated an estimated $F168.4m. for education, equivalent to 14.4% of total budgetary expenditure. The adult literacy rate in 2000 averaged an estimated 92.9% (males 94.9%; females 90.8%).

FRENCH PACIFIC TERRITORY

Following the change in status of French Polynesia in 2004, there is one remaining French Overseas Territory, territoire d'outre-mer, in the Pacific Ocean, namely Wallis and Futuna (the second being the French Southern and Antarctic Territories). They are integral parts of the French Republic. Each is administered by a High Commissioner or Chief Administrator, who is the representative of the State and appointed by the French Government. Each permanently inhabited Territory has a Territorial Assembly or Congress (elected by universal adult suffrage) and representation in the Assemblée Nationale (National Assembly) and Sénat (Senate) in Paris. They have varying degrees of autonomy. French policy in the Pacific region, particularly as regards development and relations with other Pacific countries and territories, has been mooted by the Council of the South Pacific (Conseil du Pacifique Sud). This body, chaired by the President of the Republic, was established in 1985 and revived in 1990, following a period of inactivity. It comprises French ambassadors in the region and representatives of the French Government, of the French State High Commissioners, etc. The French Republic has membership of the Pacific Community (formerly the South Pacific Commission) in respect of its Pacific possessions. The French Government includes a Minister of the Overseas Possessions.

Head of State: President JACQUES CHIRAC (took office 17 May 1995; re-elected 5 May 2002).

Ministry of the Overseas Possessions: 27 rue Oudinot, 75005 Paris, France; tel. 1-53-69-20-00; internet www.outre-mer.gouv.fr.

Minister of the Overseas Possessions: BRIGITTE GIRARDIN.

WALLIS AND FUTUNA ISLANDS

Physical and Social Geography

The self-governing French Overseas Territory of the Wallis and Futuna Islands comprises two groups: the Wallis Islands, including Uvea (Wallis Island) and 19 islets (*motu*) on the surrounding reef, and, 230 km to the south-west, the volcanic and mountainous Futuna (Horn) Islands, comprising the two small islands of Futuna and Alofi (the latter being uninhabited owing to lack of water). The islands are located some 600 km to the north-east of Fiji, south-east of Tuvalu and west of Samoa. The total area is 160.5 sq km. Uvea rises to 151 m at Mt Lulu, and Futuna to 524 m at Mt Puke. Temperatures are generally between 23°C (73°F) and 30°C (86°F), and there is a cyclone season between December and March.

The inhabitants, who are nearly all nominally Roman Catholic, are mainly Polynesian. French is an official language and widely spoken, but the indigenous languages of Uvean (Wallisian) and Futunian are generally used. In 2000 the population of the islands totalled 14,600, the majority living on Uvea. The capital is the town of Mata'Utu on Uvea in the Wallis Islands.

History

The Wallis and Futuna Islands were originally settled by proto-Polynesian peoples, Wallis from Tonga in around 1200 BC, and Futuna from Samoa and northern Fiji in around 800 BC. Futuna was subsequently discovered by the Dutch navigators, Schouten and Le Maire, in 1616, and Uvea was discovered by Samuel Wallis in 1767. Three kingdoms later emerged, and in 1837 the first Marist missionaries arrived. By 1842 the majority of the population of the two archipelagos had been converted to Christianity. In April 1842 the authorities in Wallis requested French protection, coinciding with a similar proclamation in Tahiti (now French Polynesia). In 1851 a *fakauvea* war between the Catholic majority and Methodist minority assisted by Tonga (which had commenced in 1843) came to an end when 500 Wallisians left for Tonga. Protectorate status was formalized in 1887 for Wallis and in 1888 for the two kingdoms of Futuna, but domestic law remained in force. The islands were never formally annexed; French law or representative institutions were never introduced, although Wallis and Futuna were treated as a dependency of New Caledonia. The economy was based on family-run agriculture and the export of copra to Fiji. During the Second World War, Wallis was used as an airforce base by the USA. The mountainous Futuna remained isolated. After the war, many of its youth began to migrate to New Caledonia and the New Hebrides. In 1959 the traditional Kings and chiefs requested integration into the French Republic. The islands formally became an Overseas Territory in July 1961, following a referendum in December 1959, in which 94.4% of the electorate requested this status (almost all the opposition was in Futuna, which itself recorded dissent from only 22.2% of the voters; Wallis was unanimous in its acceptance). Government is conducted by the Chief Administrator, the representative of the French State, with the advice of the Territorial Council (including the three Kings) and the Territorial Assembly, elected for the first time in 1961. The Territory also has representation in the French National Assembly.

Although there was no movement in Wallis and Futuna seeking secession of the Territory from France (in contrast with the situation in the other French Pacific Territories, French Polynesia and New Caledonia), in November 1983 the two Kings whose kingdoms share the island of Futuna first requested that the island groups of Wallis and Futuna become separate Overseas Territories of France. It was argued that the administration and affairs of the Territory had become excessively concentrated on Uvea (Wallis Island).

At elections to the 20-member Territorial Assembly in March 1982, the conservative Rassemblement pour la République (RPR) and its allies won 11 seats, while the remaining nine went to candidates affiliated to the Union pour la Démocratie Française (UDF), a metropolitan, moderate right-wing party. Later that year one member of the Lua kae tahi, a group affiliated to the metropolitan UDF, defected to the RPR group. In November 1983, however, three of the 12 RPR members joined the Lua kae tahi, forming a new majority. In the subsequent election for President of the Territorial Assembly, this 11-strong block of UDF-associated members supported the ultimately successful candidate, Falakiko Gata, even though he had been elected to the Territorial Assembly in 1982 as a member of the RPR.

In 1985 the RPR's majority in the Territorial Assembly was restored, following Falakiko Gata's formation of a new political party, the Union Populaire Locale (UPL), a conservative party, which was committed to giving priority to local, rather than metropolitan, issues.

In 1987 a dispute broke out between two families both laying claim to the throne of Sigave, the northern kingdom on the island of Futuna. The conflict arose following the deposition of the former King, Sagato Keletaona, and his succession by Sosefo Vanaï. The intervention of the island's administrative authorities, who attempted to ratify Vanaï's accession to the throne, was condemned by the Keletaona family as an interference in the normal course of local custom, according to which such disputes are traditionally settled by a fight between the protagonists.

At elections to the Territorial Assembly, held in March 1987, the UDF together with affiliated parties) and the RPR each won seven seats. However, by forming an alliance with the UPL, the RPR was able to maintain its majority, and Falakiko Gata was subsequently re-elected President. At the elections for the French National Assembly in June 1988, Benjamin Brial was re-elected Deputy. However, when the result was contested by an unsuccessful candidate, the Futunian and left-wing Kamilo Gata, the election was investigated by the French Constitutional Council and the result declared invalid, owing to electoral irregularities. When the election was re-contested in January 1989, Kamilo Gata, with the help of Michel Hoatau and Gaston Lutui, was elected Deputy, obtaining 57.4% of the total vote. Brial was dismissed from the Territorial Assembly.

Statistical information, gathered in 1990, showed that the emigration rate of Wallis and Futuna islanders had risen to over 50%. In October of that year 13,705 people (of whom 97% were Wallisians and Futunians) lived in the Territory, while 14,186 were resident in New Caledonia. At the 1996 census the number of Wallisians and Futunians resident in New Caledonia had increased to 17,563. According to the results, a proportion of the islanders had chosen to emigrate to other French Overseas Possessions or to metropolitan France. The principal reason for the increase was thought to be the lack of employment opportunities in the islands.

At elections to the Territorial Assembly in March 1992, the newly founded Taumu'a Lelei secured 11 seats, while the RPR won nine. The new Assembly was remarkable in being the first since 1964 in which the RPR, directed now by Clovis Logologofalau and Basile Tui, did not hold a majority. At elections to the French National

Assembly in March 1993, Kamilo Gata was re-elected Deputy, obtaining 52.4% of total votes cast to defeat Clovis Logologofolau. In the same month an earthquake on Futuna resulted in the deathsof five people and the destruction of many buildings and roads. In 1994 the socialist French Government nominated Clovis Logologofalau of the RPR as the islands' the Social and Economic Adviser.

In June 1994 the Union Locale Force Ouvrière, directed by Soane Uhila, organized a general strike in protest at the increasing cost of living in the Territory and the allegedly inadequate education system. It was reported that demonstrations continued for several days, during which the Territorial Assembly building was damaged in an arson attack.

In October 1994 it was reported that the King of Sigave, Lafaele Malau, had been deposed by a unanimous decision of the chiefs of the northern kingdom of Sigave on Futuna. The action followed the appointment of two customary leaders to represent the Futunian community in New Caledonia, which had led to unrest among the inhabitants of Sigave. He was succeeded by Esipio Takasi.

At elections to the Territorial Assembly in December 1994, the RPR secured 10 seats, while the coalition group of Kamilo Gata, Union Populaire pour Wallis et Futuna (UPWF), won seven, and independent candidates three. Mikaele Tauhavili of the RPR was subsequently elected President of the Assembly.

The refusal by 10 of the 20 members of the Territorial Assembly to adopt budgetary proposals, in January 1996, led to appeals for the dissolution of the Territorial Assembly by France and the holding of new elections. The US $20m. budget, some $4.5m. lower than the previous year, aroused opposition for its apparent lack of provision for development funds, particularly for the islands' nascent tourist industry.

Elections to the Territorial Assembly took place on 16 March 1997. A participation rate of 87.2% was recorded at the poll, in which RPR candidates secured 14 seats and left-wing candidates (including independents and members of various political groupings) won six seats. Victor Brial, a representative of the RPR, was elected President of the Territorial Assembly. At elections to the French National Assembly in May–June 1997 Victor Brial defeated the socialist candidate, Kamilo Gata, obtaining 3,241 votes in the second round, equivalent to 51.3% of the total.

Allegations that electoral irregularities had occurred in the elections to the Territorial Assembly of March 1997 were investigated and upheld for 11 of the seats. As a result, new elections were organized for the 11 seats on 6 September 1998, at which the RPR's representation in the Assembly was reduced to 11 seats overall, while the left-wing UPWF and the independents increased their share of the seats to six and three respectively. Also in that month, in a second round of voting, Fr Robert Laufoaulu was elected to the French Senate, defeating Kamilo Gata in a vote by the Territorial Assembly. Laufoaulu, a priest and director of Catholic education in the islands, stood as a left-wing candidate, nominated by Victor Brial in preference to other RPR candidates, and was elected with the reluctant support of right-wing politicians.

In late March 1999 festivities were held to commemorate the 40th anniversary of the accession of the King of Wallis Island, Lavelua Tomasi Kulimoetoke. An estimated 10,000 people paid homage to the King, who was crowned in 1959. In late 1999 Kamilo Gata was appointed Social and Economic Adviser by the French Government. From March 2000 delegations from Wallis and Futuna made regular visits to New Caledonia to discuss mutual arrangements concerning bilateral free trade and employment rights.

In January 2001 two candidates of the RPR contested the presidency of the Territorial Assembly. Patalione Kanimoa was elected by the majority of the RPR (eight votes) and of the UPWF (four votes). Soane Muni Uhila, the previous President of the Territorial Assembly, then formed a new party, La Voix des Peuples Wallisiens et Futuniens, along with five other RPR dissidents. The new majority RPR-UPWF grouping elected Albert Likuvalu (of the UPWF) president of the permanent commission.

In June 2001 senior officials from Wallis and Futuna and from New Caledonia agreed on a project to redefine their bilateral relationship under the Nouméa Accord on greater autonomy, signed in 1998. Following decades of migration, the population of Wallis and Futuna was 15,000, while the number of migrants and descendants from the islands in New Caledonia had risen to 20,000. In exchange for controlling immigration, New Caledonia stated that it would make a financial contribution to economic development in Wallis and Futuna. The Nouméa Accord also called for a separate arrangement providing for open access to New Caledonia for residents of Wallis and Futuna. The French State was to address the issue of financial aid before the two territories' assemblies, under the arrangement.

In January 2002 a delegation from Wallis and Futuna met President Jacques Chirac in Paris to discuss the status of members of their community living in New Caledonia. Under the Nouméa Accord, New Caledonia was to have signed a separate agreement with Wallis and Futuna better to define their status, with particular regard to the job market.

A general election was held on 10 March 2002 for the 20 seats of the Territorial Assembly. The RPR won 12 of the seats, whilst socialist candidates, or affiliated independents, won eight. Patalione Kanimoa of the RPR was re-elected President of the Assembly. An unprecedented 82.7% of some 9,500 registered voters cast their vote; there were 32 lists with 134 candidates. The election campaign was the first to give parties coverage on television and radio, provided by the national broadcasting company

At elections to the National Assembly in June 2002, Victor Brial was re-elected as the Territory's Deputy to the French legislature, winning 50.4% of the votes cast and thus defeating Penisio Tialetagi, a local merchant. For the first time, the extreme right-wing Front National presented a candidate in the islands, Gaston Lutui, a Wallisian socialist politician and former member of the RPR. Meanwhile, Christian Job replaced Alain Waquet as the islands' Chief Administrator in August 2002. In December the Constitutional Council ruled that the result of the National Assembly election in Wallis and Futuna was invalid, as certain ballot papers had been improperly marked; Brial subsequently won the by-election in March 2003. Also in December 2002 the French Senate approved a bill providing for a constitutional amendment that would allow Wallis and Futuna (along with French Polynesia) to become an Overseas Country; both chambers of the legislature in Paris ratified the amendments to the Constitution in March 2003. In July of that year, during an official visit to the neighbouring Overseas Country of New Caledonia, President Chirac received a delegation from Wallis and Futuna.

In December 2003 the President of New Caledonia signed a special accord governing relations between France, New Caledonia and Wallis and Futuna. The signing of the accord, which had been agreed two years previously in an attempt to address the situation of the 20,000 Wallis and Futuna islanders permanently resident in New Caledonia, had been delayed by the continuing ethnic tensions in New Caledonia. Under the agreement, Wallis and Futuna and New Caledonia were henceforth to deliver separate public services. Concerns had been raised by the former's increasing debt (estimated to total 2,500m. francs CFP) to the Government of New Caledonia, a major creditor being the New Caledonian hospital. It was therefore hoped that Wallis and Futuna would become more self-sufficient in the areas of health and secondary education and that the islanders would be encouraged to remain on Wallis and Futuna, while those already settled in New Caledonia would become more integrated.

Between February and April 2003 the postal service was affected by strike action, which was resolved in favour of the striking workers. During May–June 2004 a strike halted the operations of the territory's main television and radio broadcasting company, and again the dispute was resolved in favour of the striking employees, whose demands included the resignation of the station manager.

Economy

In 1995 it was estimated that Wallis and Futuna's gross domestic product (GDP) was US $28.7m., equivalent to some $2,000 per head. Most monetary income on the islands is derived from government employment (60% in 2000) and remittances sent home by islanders employed in New Caledonia and also in Metropolitan France.

Agricultural activity is of a subsistence nature. Yams, taro, bananas, cassava and other food crops are also cultivated. Tobacco is grown for local consumption. In 1998 almost all the cultivated vegetation on the island of Wallis, notably the banana plantations, was destroyed by a cyclone. In response to the cyclone damage, the French Government provided exceptional aid of 80m. francs CFP to alleviate the situation. An estimated 25,000 pigs a year are reared on the islands. Three units rear 800, 500 and 450 hens a year respectively, which are used principally for eggs, and meet an estimated 80% of the territory's commercial needs. Apiculture was revived in 1996, and in 2000 honey production was sufficient to meet the demands of the local market. Copra, which formerly provided the main cash income for the islands, has been seriously affected by rhinoceros beetle. Fishing activity in the Territory's exclusive economic zone increased during the 1990s; the total annual catch was estimated at 300 metric tons between 1999 and 2002, compared with 70 tons in 1991. The Territorial Assembly accorded Japan, and also the Republic of Korea, deep-water fishing rights to catch 3,000 tons of fish a year in the islands' exclusive economic zone, a broad area of 200 miles (370 km) around Wallis and Futuna over which France exerts sovereignty.

The only industry is the manufacture of handicrafts (traditional artefacts). In 2000 the cost of the islands' imports reached 4,735.7m. francs CFP. Road vehicles, parts and accessories accounted for 10.0% of the total value of imports, followed by mineral fuels and products (9.8%), electrical machinery and sound and television apparatus (7.4%) and meat products (6.4%). Exports totalled only 22.4m. francs CFP. Traditional food products, mother of pearl (from the trochus shell) and handicrafts are the only significant export commodities. The principal market is New Caledonia and the large

Wallisian community there. Exports of copra from Wallis ceased in 1950, and from Futuna in the early 1970s. The principal sources of imports in 2000 were France, which supplied 28.7% of the total, and Australia (22.6%). Most of the islands' exports were purchased by Italy. In August 2001 the frequency of supplies to Wallis and Futuna was significantly improved when the Sofrana shipping company, based in Auckland, began operating a new route linking New Zealand, Tonga and the Samoas to Wallis and Futuna.

Mineral fuels are the principal source of electric energy, although it is hoped that hydroelectric power can be developed, especially on Futuna. There is a 4,000-kW thermal power station on Wallis, and a 2,600-kW thermal power station was built on Futuna in 2000. The hydroelectric power station on Futuna (Vainifao) provided 10% of the production needed. Total electricity output in 2003 reached almost 4.8m. kWh.

There were 291 businesses operating in the Territory in 2000, of which 24 were in the industrial and artisanal sector, 68 in construction and 199 in the service and commercial sectors; 47 of those businesses were located on Futuna. In 2002 a new commercial centre was to open in Wallis. The tourism sector is very limited. In 2002 Wallis had four hotels and Futuna two. In September 2004 the Territory established its own airline, Air Wallis, as a joint venture between government and local private investors. Previously the country had been served only by Aircalin, and it was hoped that competition would lead to a reduction in the price of air fares. The first flight was expected to operate to New Caledonia in January 2005; other destinations would include Vanuatu, Fiji and French Polynesia.

The islands' principal source of revenue is taxation. French aid to Wallis and Futuna totalled 7,048m. francs CFP in 1999, increasing to 10,329m. francs CFP in 2002. The islands' budgetary expenditure in 2001 totalled an estimated US $25.0m. Budgetary aid from France was to rise from €83,178m. in 2001 to €86,610m. in 2002. The territorial budget in 2003 totalled 3,087.3m. francs CFP

The annual rate of inflation in 1989–2002 averaged 1.5%. The rate of inflation in 2003 reached 1.8%. In July 2001 the Chief Administrator, Alain Waquet, acknowledged that the minimum wage (of 58,178 francs CFP per month) had not been revised since July 1998. He therefore announced that the monthly minimum wage was to be raised to 70,000 francs CFP by July 2002; the increase was to be implemented in three stages. A rise in the cost of living was expected to follow.

In early 1995 the Territory signed a convention with the French Government, which provided for 2,800m. francs CFP in funding for development projects over five years. Projects to be undertaken included the renovation of school buildings and the construction of a technical college on Uvea, a training scheme for young agricultural workers, the rebuilding of the wharf at Sigave on Futuna, a road-improvement scheme and the construction of low-cost housing.

During his visit to Paris in June 1985, the President of the Territorial Assembly informed the French Government of his belief that the policies relating to agricultural and fisheries development since 1960 had failed completely. It was hoped that these areas of the economy could be improved through new administrative arrangements, whereby development funding would be channelled through traditional chiefs. In August 1989 the visiting French Prime Minister, Michel Rocard, announced an additional 55m. francs CFP in aid, for social and economic development, and in June 1991 Clovis Logologofolau, the President of the Territorial Assembly, and Kamilo Gata, the Territory's Deputy to the French National Assembly, travelled to Paris to renegotiate funding for further development projects. The Territory is very dependent upon metropolitan financial assistance.

In 2001 a delegation from the Institut de Recherche pour le Développement was sent to the islands to assess the possibilities of self-development. In January 2002 an economic orientation document was signed between the French Government and the Territorial Assembly, with the participation of the three kingdoms. Meanwhile, discussions continued regarding reform of the 1961 statute, to give increased responsibilities to local representatives; particular emphasis was placed on issues of youth employment. In mid-2003 the National Assembly approved the Overseas Territories Development Bill, which would provide support for economic and social development in Wallis and Futuna (together with French Polynesia and New Caledonia) by attracting foreign investment, and among other benefits allow for overseas French residents to travel to mainland France to take advantage of free education.

Although France is also a member, the Territory has membership in its own right of the Pacific Community (formerly the South Pacific Commission), which is based in New Caledonia.

Statistical Survey

Source (unless otherwise indicated): Service Territorial de la Statistique et des Etudes Economiques, Centre de Havelu, Mata'Utu, 98600 Uvea, Wallis Islands, Wallis and Futuna, via Nouméa, New Caledonia; tel. and fax 722403; e-mail stats@wallis.co.nc; internet www.wallis.co.nc/stats.

AREA AND POPULATION

Area (sq km): 160.5. *By Island:* Uvea (Wallis Island) 77.5; Other Wallis Islands 18.5; Futuna Island 45; Alofi Island 19.5.

Population: Total population 14,166 (males 6,984, females 7,182) at census of 1996; 17,563 Wallisians and Futunians resided in New Caledonia. *July 2003* (provisional census figures): Wallis Islands 10,071; Futuna Island 4,873 (Alo 2,993, Sigave 1,880); Total 14,944.

Density (provisional, census of July 2003): 93.1 per sq km.

Principal Town (provisional, census of July 2003): Mata'Utu (capital), population 1,191.

Births and Deaths (estimates, 2000): Birth rate 21.7 per 1,000; Death rate 5.6 per 1,000.

Expectation of Life (estimates, years at birth): 73.8 (males 73.2; females 74.4) in 1998.

Economically Active Population (salaried workers, 1996): 2,465 (males 1,397, females 1,068). *Unemployment* (persons actively seeking work, 2002): 456 (males 259, females 197). In addition, 46 persons were enrolled as trainees at the national employment training agency.

AGRICULTURE, ETC.

Principal Crops (FAO estimates, '000 metric tons, 2002): Cassava 2.4; Taro (coco yam) 1.6; Yams 0.5; Other roots and tubers 1.0; Coconuts 2.3; Vegetables 0.5; Bananas 4.1; Other fruits (excl. melons) 4.6. Source: FAO.

Livestock (FAO estimates, '000 head, year ending September 2002): Pigs 25; Goats 7; Chickens 63. Source: FAO.

Livestock Products (FAO estimates, metric tons, 2002): Pig meat 315; Goat meat 15; Poultry meat 46; Cows' milk 30; Hen eggs 33; Honey 11. Source: FAO.

Fishing (FAO estimates, metric tons, live weight, 2002): Total catch 300 (Marine fishes 294). Note: Figures exclude trochus shells (FAO estimate, metric tons); 25. Source: FAO.

INDUSTRY

Electricity Production ('000 kWh, net production, 2003): 4,778.

FINANCE

Currency and Exchange Rates: 100 centimes = 1 franc de la Communauté française du Pacifique (franc CFP or Pacific franc). *Sterling, Dollar and Euro Equivalents* (31 May 2004): £1 sterling = 178.78 francs CFP; US $1 = 97.45 francs CFP; €1 = 119.33 francs CFP; 1,000 francs CFP = £5.593 = $10.262 = €8.380. *Average Exchange Rate* (francs CFP per US $): 133.35 in 2001; 126.80 in 2002; 105.73 in 2003. Note: Until 31 December 1998 the value of the franc CFP was fixed at 5.5 French centimes (1 French franc = 18.1818 francs CFP). Since the introduction of the euro, on 1 January 1999, an official exchange rate of 1,000 francs CFP = €8.38 (€1 = 119.332 francs CFP) has been in operation. Accordingly, the value of the franc CFP has been adjusted to 5.4969 French centimes (1 French franc = 18.1920 francs CFP), representing a 'devaluation' of 0.056%.

Territorial Budget (million francs CFP): 3,271.1 in 2001; 3,388.6 in 2002; 3,087.7 in 2003 (provisional).

Public Expenditure (million francs CFP, 2003): Operational expenditure 6,616.0 (Agriculture and fisheries department 95.5, Education 4,629.3, Economy, finance and industry department 265.7, Interior and decentralization department 99.2, Justice 39.5, Overseas affairs 864.1, Transport and communications 10.9, Maritime department 0.8, Youth and sports department 49.1, Health and welfare 501.6, Labour 58.6, Environment 1.7); Investment expenditure 603.4; Special Treasury accounts 24.0; Extra-budgetrary expenditure 16.1; Total expenditure 7,259.6.

Aid from France (million francs CFP, 2002): 10,329. Source: Institut d'Emission d'Outre-Mer.

Money Supply (million francs CFP at 30 Sept. 2001): Currency in circulation 759; Demand deposits 1,595; Total money 2,355. Source: Institut d'Emission d'Outre-Mer.

Cost of Living (Consumer Price Index; base: July–Sept. 1989 = 100): All items 120.7 in 2001; 123.6 in 2002; 125.8 in 2003.

EXTERNAL TRADE

Principal Commodities (million francs CFP): *Imports c.i.f.* (2000): Meat and edible meat offal 302.1; Preparations of meat, of fish or of crustaceans, molluscs or other aquatic invertebrates 178.2; Beverages, spirits and vinegar 199.0; Mineral fuels, mineral oils and products of their distillation; bituminous substances; mineral waxes 462.5; Articles of iron or steel 153.9; Boilers, machinery, mechanical appliances and parts 281.2; Electrical machinery, equipment and parts; sound and television apparatus 350.4; Road vehicles, parts and accessories 474.2; Total (incl. others) 4,735.7. *Exports f.o.b.* (2001): Preparations of molluscs and other aquatic invertebrates 0.3; Coral and shells 5.5; Braids and mats of vegetable material 0.9; Total 5.6.

Principal Trading Partners (million francs CFP): *Imports c.i.f.* (2003): Australia 1,051.7; Fiji 215.5; France (incl. Monaco) 1,371.4; Japan 161.3; New Caledonia 325.1; New Zealand 433.8; Spain 130.7; USA 103.6; Total (incl. others) 4,377.8. *Exports f.o.b.* (2001): Samoa 5.5; Japan 0.3; New Caledonia 0.9; Total 5.6.

TRANSPORT

Shipping: *Merchant Fleet* (31 December 2003): Vessels registered 4; displacement ('000 grt) 113.7. Source: Lloyd's Register-Fairplay, *World Fleet Statistics*.

Civil Aviation (2001): aircraft arrivals and departures 1,310; freight handled 131.1 metric tons; passenger arrivals and departures 32,445; mail loaded and unloaded 58.3 metric tons.

TOURISM

Visitors: 400 in 1985.

COMMUNICATIONS MEDIA

Telephones (2003): 1,683 main lines in use.

Facsimile Machines (1993): 90 in use (Source: UN, *Statistical Yearbook*).

Internet Users (2003): 384.

EDUCATION

Pre-primary (2002/03): 3 institutions.

Primary (2002/03): 14 institutions; 2,938 students (incl. pre-primary); 271 teachers.

Secondary (2002/03): 7 institutions (2 vocational); 2,293 students; 191 teachers.

Source: Vice-Rectorat de Wallis et Futuna.

Directory

The Constitution

The Territory of the Wallis and Futuna Islands is administered according to a statute of 1961, and subsidiary legislation, under the Constitution of the Fifth Republic. The Statute declares the Wallis and Futuna Islands to be an Overseas Territory of the French Republic, of which it remains an integral part. The Statute established an administration, a Council of the Territory, a Territorial Assembly and national representation. The administrative, political and social evolution envisaged by, and enacted under, the Statute is intended to effect a smooth integration of the three customary kingdoms with the new institutions of the Territory. The Kings are assisted by ministers and the traditional chiefs. The Chief Administrator, appointed by the French Government, is the representative of the State in the Territory and is responsible for external affairs, defence, law and order, financial and educational affairs. The Chief Administrator is required to consult with the Council of the Territory, which has six members: three by right (the Kings of Wallis, Sigave and Alo) and three appointed by the Chief Administrator upon the advice of the Territorial Assembly. This Assembly assists in the administration of the Territory; there are 20 members elected on a common roll, on the basis of universal adult suffrage, for a term of up to five years. The Territorial Assembly elects, from among its own membership, a President to lead it. The Territory elects national representatives (one Deputy to the National Assembly, one Senator and one Economic and Social Councillor) and votes for representatives to the European Parliament in Strasbourg. In 2003 both the National Assembly and Senate in Paris ratified amendments to the Constitution providing for Wallis and Futuna (along with French Polynesia) to become an Overseas Country.

The Government

(September 2004)

Chief Administrator (Administrateur Supérieur): CHRISTIAN JOB (appointed 2002).

CONSEIL DU TERRITOIRE

Chair: Chief Administrator.

Members by Right: King of Wallis, King of Sigave, King of Alo.

Appointed Members: MIKAELE HALAGAHU (Faipule), ATOLOTO UHILA (Kulitea), KELETO LAKALAKA (Sous réserves).

GOVERNMENT OFFICE

Government Headquarters: Bureau de l'Administrateur Supérieur, Havelu, BP 16, Mata'Utu, 98600 Uvea, Wallis Islands, Wallis and Futuna; tel. 722727; fax 722324; all departments.

Legislature

ASSEMBLÉE TERRITORIALE

The Territorial Assembly has 20 members and is elected for a five-year term. The most recent general election took place on 10 March 2002, at which RPR candidates secured a total of 12 seats and various left-wing candidates (including independents) won eight seats.

President: PATALIONE KANIMOA (UMP).

Territorial Assembly: Assemblée Territoriale, Havelu, BP 31, Mata'Utu, 98600 Uvea, Wallis Islands, Wallis and Futuna; tel. 722504; fax 722054.

PARLEMENT

Deputy to the French National Assembly: VICTOR BRIAL (UMP).

Representative to the French Senate: Fr ROBERT LAUFOAULU (UMP).

Social and Economic Adviser: KAMILO GATA (UPWF).

The Kingdoms

WALLIS

(Capital: Mata'Utu on Uvea)

Lavelua, King of Wallis: TOMASI KULIMOETOKE.

Council of Ministers: Prime Minister (Kivalu) and five other ministers; Kivalu KAMALIELE MULILOTO.

The Kingdom of Wallis is divided into three districts (Hihifo, Hahake, Mua), and its traditional hierarchy includes three district chiefs (Faipule) and 20 village chiefs (Pule).

SIGAVE

(Capital: Leava on Futuna)

Keletaona, King of Sigave: PASILIO KELETAONA.

Council of Ministers: six ministers, chaired by the King.

The Kingdom of Sigave is located in the north of the island of Futuna; there are five village chiefs.

ALO

(Capital: Ono on Futuna)

Tu'i Agaifo, King of Alo: SAGATO ALOFI.

Council of Ministers: five ministers, chaired by the King.

The Kingdom of Alo comprises the southern part of the island of Futuna and the entire island of Alofi. There are nine village chiefs.

Political Organizations

Taumu'a Lelei (Bright Future): c/o Assemblée Territoriale; f. 1992; Leader SOANE MUNI UHILA.

Union pour la Démocratie Française (UDF): c/o Assemblée Territoriale; centrist; based on Uvean (Wallis) support.

Union Populaire pour Wallis et Futuna (UPWF): c/o Assemblée Territoriale; affiliated to the Parti Socialiste of France in 1998; f. 1994; Leader KAMILO GATA.

Union pour un Mouvement Populaire (UMP): c/o Assemblée Territoriale; f. 2002 by members of the former Rassemblement pour la République and Démocratie Libérale parties; Territorial Leader Clovis Logologofolau.

La Voix des Peuples Wallisiens et Futuniens: c/o Assemblée Territoriale; f. 2001 by dissident RPR mems.

Religion

Almost all of the inhabitants profess Christianity and are adherents of the Roman Catholic Church.

CHRISTIANITY

The Roman Catholic Church

The Territory comprises a single diocese, suffragan to the archdiocese of Nouméa (New Caledonia). The diocese estimated that there were 14,700 adherents (98% of the population) at 31 December 2002. The Bishop participates in the Catholic Bishops' Conference of the Pacific, currently based in Fiji.

Bishop of Wallis and Futuna: Mgr Lolesio Fuahea, Evêché Lano, BP G6, 98600 Mata'Utu, Uvea, Wallis Islands, Wallis and Futuna; tel. 722932; fax 722783; e-mail eveche.wallis@wallis.co.nc.

The Press

Te-Fenua Fo'ou: BP 435, 98600 Mata'Utu, Uvea, Wallis Islands, Wallis and Futuna; tel. 721746; e-mail tff@wallis.co.nc; f. 1995; weekly; French, Wallisian and Futurian; ceased publication in April 2002.

Broadcasting and Communications

TELECOMMUNICATIONS

France Telecom (FCR): Télécommunications extérieures de Wallis et Futuna, BP 54, 98600 Mata'Utu, Uvea, Wallis Islands, Wallis and Futuna; tel. 722436; fax 722255; e-mail admin@wallis.co.nc.

Service des Postes et Télécommunications: Administration Supérieure des Iles Wallis et Futuna, BP 00, 98600 Mata'Utu, Uvea, Wallis Islands, Wallis and Futuna; tel. 722121; fax 722500; e-mail adminspt.get@wallis.co.nc.

BROADCASTING

Radio and Television

Radio Wallis et Futuna: BP 102, 97911 Mata'Utu, Uvea, Wallis Islands, Wallis and Futuna; tel. 722020; fax 722346; e-mail rfo.wallis@wallis.co.nc; internet www.rfo.fr; fmrly Radiodiffusion Française d'Outre-mer (RFO); transmitters at Mata'Utu (Uvea) and Alo (Futuna); programmes broadcast 24 hours daily in Uvean (Wallisian), Futunian and French; a television service on Uvea, transmitting for 12 hours daily in French, began operation in 1986; a television service on Futuna was inaugurated in December 1994; satellite television began operation in March 2000; Regional Dir Joseph Caihé; Chief Editor Lusia Kavakava.

Finance

BANKING

Bank of Issue

Institut d'Emission d'Outre-Mer: 98600 Mata'Utu, Uvea, Wallis Islands, Wallis and Futuna; tel. 722505; f. 1998.

Other Banks

Agence Française de Développement: 98600 Mata'Utu, Uvea, Wallis Islands, Wallis and Futuna; tel. 722505; fmrly Caisse Française de Développement; development bank.

Banque de Wallis et Futuna: BP 59, 98600 Mata'Utu, Uvea, Wallis Islands, Wallis and Futuna; tel. 722124; fax 722156; e-mail bnp.nc@bnpparibas.com; internet www.bnp.nc/bwf/identite.htm; f. 1991; 51% owned by BNP Paribas (New Caledonia).

Paierie de Wallis et Futuna: 98600 Mata'Utu, Uvea, Wallis Islands, Wallis and Futuna, via Nouméa, New Caledonia.

Trade and Industry

GOVERNMENT AGENCY

Economie Rurale et Pêche: BP 05, 98600 Mata'Utu, Uvea, Wallis Islands, Wallis and Futuna; Antenne de Futuna, BP 05, 98620 Sigave, Futuna, Wallis and Futuna; tel. 720400; fax 720404; e-mail ecoru@wallis.co.nc; tel. 723214; fax 723402.

UTILITIES

Electricité et Eau de Wallis et Futuna (EEWF): BP 28, 98600 Mata'Utu, Uvea, Wallis Islands, Wallis and Futuna; tel. 721500; fax 721196; e-mail eewf@wallis.co.nc; Dir F. Hijltsch; 32.3% owned by the territory and 66.64% owned by EEC of New Caledonia.

TRADE UNIONS

Union Interprofessionnelle CFDT Wallis et Futuna (UI CFDT): BP 178, 98600 Mata'Utu, Uvea, Wallis Islands, Wallis and Futuna; tel. 721880, Sec. Gen. Kalolo Hanisi.

Union Territoriale Force Ouvrière: BP 325, Mata-Utu, 98600 Wallis and Futuna; tel. 721732; fax 721732; Sec.-Gen. Christian Vaamei.

Transport

ROADS

Uvea has a few kilometres of road, one route circling the island, and there is also a partially surfaced road circling the island of Futuna; the only fully surfaced roads are in Mata'Utu.

SHIPPING

Mata'Utu serves as the seaport of Uvea and the Wallis Islands, while Sigave is the only port on Futuna. A total of 45 general freight ships and 17 gas and petroleum tankers docked in Wallis in 1999.

Services des Douanes et Affaires Maritimes: Aka'aka, 98600 Mata'Utu, Uvea, Wallis Islands, Wallis and Futuna; tel. 722571; fax 722986.

AMACAL (General Agent): POB 1080, Nouméa, New Caledonia; tel. 232910; fax 287388; e-mail amb@amb.nc.

CIVIL AVIATION

There is an international airport in Hihifo district on Uvea, about 5 km from Mata'Utu. Air Calédonie International (Aircalin—New Caledonia) operates five flights a week from Wallis to Futuna, one flight a week from Wallis to Tahiti (French Polynesia) and two flights a week from Wallis to Nouméa (New Caledonia). The airport on Futuna is at Vele, in the south-east, in the Kingdom of Alo. The Compagnie Aerienne de Wallis et Futuna (Air Wallis) was established in 2004, as a joint venture between the Government and local business interests, and was scheduled to commence operations in January 2005.

Tourism

Tourism remains undeveloped. There are four small hotels on Uvea, Wallis Islands. In 1985 there were some 400 tourist visitors, in total, to the islands. There are two small guest-houses for visitors on Futuna.

Education

In 2002/03 there were 19 state-financed primary schools (including three pre-primary, 14 primary and two homecraft centres), and seven secondary schools (including two vocational schools) in Wallis and Futuna. Primary and pre-primary pupils totalled 2,938 and secondary students numbered 2,293. In 1999/2000 361 students attended various universities overseas.

FRENCH PACIFIC OVERSEAS COUNTRIES

French Polynesia and New Caledonia were formerly Overseas Territories, territoires d'outre-mer, of France. The Nouméa Accord, concluded in 1998, provided for a gradual transfer of powers to New Caledonia. Under its terms, the Territory became an Overseas Country (Pays d'outre-mer), within the French Republic, in 1999. Amendments to the French Constitution, which were approved by Parlement (Parliament) in March 2003 and formed part of the French Government's programme of decentralization, were not to apply to New Caledonia, uniquely among the overseas possessions of the French Republic. Legislation promulgated in February 2004 granted autonomy to French Polynesia, which was henceforth also to have the status of an Overseas Country within the French Republic. The High Commissioner of French Polynesia is also responsible for Clipperton Island, although it is not part of the Overseas Country.

Head of State: President JACQUES CHIRAC (took office 17 May 1995; re-elected 5 May 2002).

FRENCH POLYNESIA

Physical and Social Geography

French Polynesia is an Overseas Country of France, covering an area of 4,167 sq km (1,609 sq miles—land area of 3,521 sq km), including 35 volcanic islands and about 183 low-lying coral atolls in five archipelagos: Society, Austral (or Tubuai), Tuamotu, Gambier and Marquesas. The Society Islands, in the west of the Territory, comprise a Windward Islands group (Iles du Vent—including Tahiti and Moorea) and, about 160 km to the north-west, a Leeward Islands group (Iles Sous le Vent—including the islands of Raiatea and Bora Bora). The five Austral Islands proper lie some 160 km to the south of Tahiti and include Tubuai itself; 770 km to the south-east of Tubuai is the separate island of Rapa. The Tuamotu Archipelago is a chain of about 80 atolls stretching from the north of Tahiti, south-east for about 1,500 km, from the islands around Rangiroa (the largest in the Territory) towards those beyond Mururoa Atoll, which is over 1,000 km from Tahiti. The chain is continued in the south-east by the small group of the Gambier Islands, including Mangareva, which is some 1,700 km from Tahiti. In the north-east, 1,500 km from Tahiti, are the Marquesas Islands, which comprise a southern and a northern group around the chief island of Nuku Hiva. The Territory's nearest neighbours are Kiribati to the north-west and the Cook Islands to the west. The small, uninhabited island of Clipperton, located far to the north-east of the Territory, some 600 km off the coast of Mexico, is administered from French Polynesia, being under the direct jurisdiction of the High Commissioner.

The average monthly temperature throughout the year varies between 20°C (68°F) and 29°C (84°F), and most rainfall occurs between November and April, the average annual precipitation being 1,625 mm (64 ins). The more mountainous volcanic islands (notably the Society Islands) receive the most rainfall, and the south of the Territory is cooler than the north.

The official languages are French and Tahitian. Polynesian languages are spoken by the indigenous population. Christianity is the principal religion, and the Protestant Evangelical Church the largest denomination. The population, which is predominantly Polynesian, includes minorities of French and other Europeans, Chinese (a small but economically significant community) and 'Demis' (persons of mixed race—Polynesian with others). In 1990 it was estimated that 82.8% of the population were Polynesian or of Polynesian descent and 11.9% were European. At the census of 7 November 2002 the population totalled 245,516. Some 87% of the inhabitants live in the Society Islands, which constitute about one-half of the land area. The population of the capital, Papeete (on Tahiti), was 26,181 in 2002.

History

The islands of French Polynesia were already inhabited by the Maohi, a Polynesian people, when first visited by Spanish explorers during the 16th century. Dutch, French and British explorers followed during the 1700s. Descriptions of Tahiti and other Society Islands by Wallis, who first visited in 1767, Bougainville in 1768 and Capt. James Cook in 1769 gave rise in Europe to a Utopian concept of the South Pacific, a romantic view which drew Europeans to the islands. In fact, European discovery dealt the Tahitian and other island groups' populations a severe blow. Disease caused rapid declines in population, and many more were killed in inter-island and inter-group warfare.

The Tahitian-based kingdom of the Pomare monarchs was made a French protectorate in 1842 and a colony in 1880. All the islands that now constitute French Polynesia had been annexed by the end of the 19th century. Clipperton Island, which is administered as a distinct entity, was first claimed by France in 1857. After disputes with the USA and Mexico, the island was restored to France by the arbitration of the King of Italy in February 1931.

The islands were governed from France under a decree of 1885 until 1946, when the French signed a new decree allowing the Polynesians to manage their domestic affairs while maintaining institutional control through a Governor in Papeete, the capital, on Tahiti. The Territorial Assembly was established in 1957 to assist the Governor. Moves towards increased local autonomy began in 1977, and new statutes creating a fully elected local executive were approved in Paris in May 1977.

Following elections to the Territorial Assembly in May 1982, the Gaullist Tahoeraa Huiraatira party (led by Gaston Flosse) formed successive ruling coalitions, first with the Ai'a Api party, and in September with the Pupu Here Ai'a Te Nunaa Ia Ora party, which advocates territorial autonomy. Seeking self-government, especially in economic matters, elected representatives of the Assembly held discussions with the French Government in Paris in 1983, and in September 1984 a new statute was approved by the French National Assembly. This allowed the Territorial Government greater powers, mainly in the sphere of commerce and development; the Council of Government was replaced by a Council of Ministers, whose President was to be elected from among the members of the Territorial Assembly. Gaston Flosse became the first President of the Council of Ministers.

At elections held in March 1986, the Tahoeraa Huiraatira gained the first outright majority to be achieved in the Territory, winning 24 of the 41 seats in the Territorial Assembly. Leaders of opposition parties subsequently expressed dissatisfaction with the election result, claiming that the Tahoeraa Huiraatira victory had been secured only as a result of allocating a disproportionately large number of seats in the Territorial Assembly to one of the five constituencies. The constituency at the centre of the dispute comprised the Mangareva and Tuamotu islands, where the two French army bases at Hao and Mururoa constituted a powerful body of support for Flosse and the Tahoeraa Huiraatira, which, in spite of winning a majority of seats, had obtained a minority of individual votes in the election (30,571, compared with the opposition parties' 43,771). At the concurrent elections for French Polynesia's two seats in the National Assembly in Paris, Flosse and Alexandre Léontieff, the candidates of the Rassemblement pour la République (RPR—to which party the Tahoeraa Huiraatira is affiliated), were elected, Flosse subsequently ceding his seat to Edouard Fritch. Flosse was later appointed as Secretary of State for Pacific Affairs in the French Council of Ministers.

In April 1986 Flosse was re-elected President of the Territory's Council of Ministers. However, he faced severe criticism from leaders of the opposition for his allegedly inefficient and extravagant use of public funds. He was accused, in particular, of corrupt electoral practice, having distributed government-financed gifts of construction materials, food and clothing, in an attempt to influence voters during his election campaign. Finally, in February 1987 Flosse resigned as President of the Council of Ministers, and was replaced by Jacques Teuira.

Unrest among dock-workers in Papeete led to serious rioting in October 1987 and the declaration of a state of emergency by the authorities. In December a coalition of opposition parties and a breakaway faction of Tahoeraa Huiraatira, the Te Tiarama party, led by Alexandre Léontieff, took power from Tahoeraa. The Léontieff Government survived several challenges in the Territorial Assembly to its continuation in office during 1988 and 1989. Amendments to the Polynesian Constitution, which were enacted by July 1990, augmented the powers of the President of the Territorial Government and increased the competence of the Territorial Assembly. Furthermore, five consultative Archipelago Councils were established, comprising Territorial and municipal elected representatives. The main purpose of these amendments was to clarify the areas of responsibility of the State, the Territory and the judiciary, which was considered particularly necessary following various disputes about the impending single market of the European Community (EC, now European Union—EU). In June 1989, in protest, 90% of the electorate refused to vote in the elections to the European Parliament.

At territorial elections in March 1991 the Tahoeraa Huiraatira won 18 of the 41 seats. Flosse then formed a coalition with the Ai'a Api, resulting in a majority of 23 seats in the Territorial Assembly. Emile Vernaudon, leader of the Ai'a Api, was elected President of the Assembly, and Flosse was elected President of the Council of Ministers. In September Flosse announced the end of the coalition between his party and the Ai'a Api, accusing Vernaudon of disloyalty, and signed a new alliance with the Pupu Here Ai'a Te Nunaa Ia Ora, led by Jean Juventin (a former political rival of Flosse).

In April 1992 Flosse was found guilty of fraud (relating to an illegal sale of government land to a member of his family), and there were widespread demands for his resignation. In November Juventin and Léontieff were charged with 'passive' corruption, relating to the construction of a golf course by a Japanese company. In the following month the French Court of Appeal upheld the judgment against Flosse, who received a six-month suspended prison sentence. The case provoked a demonstration by more than 3,000 people in January 1993, demanding the resignation of Flosse and Juventin. In September 1994 Flosse succeeded in having the conviction overruled, in a second court of appeal, on a procedural matter. In October 1997 Léontieff was found guilty of accepting substantial bribes in order to facilitate a business tenure and was sentenced to three years in prison (half of which was to be suspended). In May 1998 Léontieff was sentenced to a further three years' imprisonment (two of which were to be suspended) for corruption.

French presidential elections took place in April–May 1995. During the second round of voting in the Territory, the socialist candidate, Lionel Jospin, received 39% of the total votes, while the RPR candidate, Jacques Chirac, won 61%. (Chirac was elected to the presidency with 52.6% of votes cast throughout the republic.)

In November 1995 the Territorial Assembly adopted a draft statute of autonomy, which proposed the extension of the Territory's powers to areas such as fishing, mining and shipping rights, international transport and communications, broadcasting and the offshore economic zone. France, however, would retain full responsibility for defence, justice and security in the islands. Advocates of independence for French Polynesia criticized the statute for promising only relatively superficial changes, while failing to increase the democratic rights of the islanders. The statute was approved by the French National Assembly in December and came into force in April 1996.

On 13 May 1996 the Gaullist Tahoeraa Huiraatira achieved an outright majority at the territorial elections, although the principal pro-independence party, Tavini Huiraatira/Front de Libération de la Polynésie (FLP), made considerable gains throughout the Territory (largely owing to increased popular hostility towards France since the resumption of nuclear-weapons tests at Mururoa Atoll—see below). Tahoeraa Huiraatira secured 22 of the 41 seats in the Territorial Assembly, with 38.7% of total votes cast, while Tavini Huiraatira won 10 seats, with 24.8% of votes. Other anti-independence parties won a total of eight seats and an additional pro-independence grouping secured one seat. Flosse defeated the independence leader, Oscar Temaru, by 28 votes to 11 to remain as President of the Council of Ministers later in the month, and Justin Arapari was elected President of the Territorial Assembly. Allegations of voting irregularities led to legal challenges, which overturned the results in 11 constituencies. Following by-elections in May 1998 for the 11 seats, Tahoeraa Huiraatira increased its representation by one seat. Tavini Huiraatira again claimed that the elections had not been fairly conducted.

At elections for French Polynesia's two seats in the French National Assembly in May 1997, Michel Buillard and Emile Vernaudon, both supporters of the RPR, were elected, with 51.6% and 58.9% of total votes cast, respectively. However, the pro-independence leader, Oscar Temaru, was a strong contender for the western constituency seat, securing 41.6% of the votes. Flosse was elected as the Territory's representative to the French Senate in September 1998. In March 1999 proposals to increase French Polynesia's autonomy, as part of constitutional reforms, were announced in Paris. The proposed changes would allow the Territory to draft its own laws and negotiate its own international treaties. These proposals followed an initial agreement between the Territory and the French Government in late 1998, on the future of French Polynesia. In October 1999 the French Senate adopted a constitutional amendment granting French Polynesia a greater degree of autonomy. According to the bill (which had also been approved by the National Assembly in June), the status of the islands was to be changed from that of Overseas Territory to Overseas Country, and a new Polynesian citizenship was to be created. Although France was to retain control over areas such as foreign affairs, defence, justice and electoral laws, French Polynesia would have the power to negotiate with other Pacific countries, and sign its own international treaties. The constitutional amendment was presented to a joint session of the French Senate and National Assembly for final ratification in late January 2000.

In November 1999 Flosse was found guilty of corruption, on charges of accepting more than 2.7m. French francs in bribes from the owner of an illegal casino, allegedly to help fund his party. Flosse was sentenced to a two-year suspended prison term, a large fine, and a one-year ban on seeking office. Demonstrations, organized by the pro-independence FLP, took place in Tahiti, in protest at Flosse's refusal to resign from his post as President of the Territorial Council of Ministers. In October Flosse lodged an appeal with the High Court, which reversed the ruling in May 2001. In November 2002 the Court of Appeal in Paris announced that Flosse should be pardoned.

In December 2000 some 2,000 workers went on strike in Papeete, protesting against low wages, and demanding that their pay be raised to a level commensurate with the prosperous state of the Territory's economy. In that month provision was made for the number of seats in the Territorial Assembly to be increased from 41 to 49, in an attempt to reflect demographic changes in the Territory more accurately. On 6 May 2001 elections to the Assembly took place. Tahoeraa Huiraatira won 28 seats, securing a fifth successive term in office. The pro-independence Tavini Huiraatira took 13 seats. Gaston Flosse was subsequently re-elected President of the Territorial Assembly, and a government reorganization ensued.

In December 2001 the President of the Employers Council made a complaint about the lack of consultation over the Government's budget for 2002, particularly criticizing the significant increase in fiscal pressures on citizens, companies and employees (see Economic Affairs). Following the Territorial Assembly's approval of the controversial budget, trade unions and employers joined forces to stage a demonstration on 11 December; shops were closed, and 4,000 people participated in a protest march in Tahiti.

In January 2002 representatives of state and local government met to review the first five years of the Restructuring Fund, an agreement implemented in 1996 to further the economic autonomy of Polynesia and to regulate financial subsidies to the Territory following the cessation of nuclear testing. The President and Prime Minister of France took part in further such meetings in June and July. It was agreed that funding would be extended for a further 10 years after 2006. The French delegation supported the proposal for a new autonomy statute to grant more powers of self-government to French Polynesia. The Territory's status, however, was to remain unchanged. In June 2002 elections for the Territory's two seats in the French National Assembly were won by the Tahoeraa Huiraatira candidates Michel Buillard and Béatrice Vernaudon. The FLP did not present a candidate.

In December 2002 the French Senate approved a bill providing for a constitutional amendment that would allow French Polynesia (along with Wallis and Futuna) to become an Overseas Country; both houses of the French legislature ratified the amendments to the Constitution in March 2003. In July the Territorial Assembly and the Government ratified the amendments. The approval of the French State Council (the highest administrative court in France), the French Government and the Constitutional Council was expected later that year. Meanwhile, in May it was announced that French Polynesia (together with New Caledonia) would be allocated one additional seat in the French Senate, probably effective from the next senatorial elections in 2007. The final text of the autonomy statute was approved by the National Assembly in January 2004 and, with minor amendments, by the Constitutional Council in mid-February. In March, when President Chirac signed the requisite decree, French Polynesia formally became an Overseas Country of France, henceforth being permitted to 'govern itself freely and democratically'. French Polynesia was thus granted greater authority over matters such as labour law, civil aviation and regional relations, while France retained control of law and order, defence and money supply.

In April 2004 the local legislature was dissolved, in preparation for elections to a new 57-member Assembly, which were held on 23 May. The Union pour la Démocratie (UPD) coalition, led by Oscar Temaru of Tavini Huiraatira, won a total of 29 seats in the new Assembly, thus ending the long-standing dominance of the Tahoeraa Huiraatira; some 78% of the electorate voted, 10% more than at the elections of 2001. In June Antony Géros, a close ally of Temaru, was elected as President of the Assembly. The French State Council and the Court of Administrative Law of French Polynesia both rejected attempts by Gaston Flosse to have nullified the result of the election of the President of the Assembly, and of other officers, and also the decisions taken by the new legislature. Temaru replaced Gaston Flosse as President of the Council of Ministers, and a new Government was appointed. Meanwhile, it emerged that, at the prospect of a pro-independence government coming to power, the French Minister of the Overseas Possessions, Brigitte Girardin, had reportedly threatened to discontinue funding to French Polynesia.

In June 2004 the UPD coalition gained an extra seat in the Assembly following the defection of Jean-Alain Frébault from the opposition Tahoeraa Huiraatira party; in July another member left to sit as an independent. In September a new grouping was reported to have been formed by dissident legislators—from within the ruling coalition and the Tahoeraa Huiraatira—in protest over budgetary proposals, which they deemed to contravene election promises, and over the allocation of ministerial portfolios. Antonio Perez, a former member of Fe'tia Api, announced the founding of the Te' Avei'a (Te Ara) party. Later that month Frébault and another member both left the UPD to sit as independents, reducing the coalition's representation to 28 seats. Meanwhile, in the same month, the new Minister for Outer Islands Development and Decentralization, Emile Vernaudon, was convicted of misusing local government resources while serving as a mayor between 1992 and 1999; he received a suspended prison sentence and was fined US $30,000. In early October separate motions of no confidence were filed against the Government by the Tahoeraa Huiraatira and Te' Avei'a (Te Ara) parties. A total of 29 Assembly members voted against the Government, precipitating a constitutional crisis; Temaru contested the legitimacy of the vote and condemned the French Government's decision not to permit new legislative elections. Gaston Flosse was subsequently re-elected, by a narrow parliamentary majority, as the President of French Polynesia. Flosse's election was immediately disputed by Temaru.

The testing of nuclear devices by the French Government at Mururoa Atoll, in the Tuamotu Archipelago, began in 1966 at the Centre d'Expérimentation du Pacifique (CEP). In July 1985 the trawler, *Rainbow Warrior*, the flagship of the anti-nuclear environmentalist group, Greenpeace, which was to have led a flotilla to Mururoa Atoll to protest against the French nuclear-test programme, was blown up and sunk in Auckland Harbour, New Zealand. One member of the crew was killed. Two agents of the French secret service, the Direction Générale de la Sécurité Extérieure (DGSE), were found responsible, and relations between France and New Zealand were seriously affected by the resulting dispute, especially over the treatment of the two agents.

France, however, continued to perform tests in the islands despite growing local opposition and the protests of the South Pacific Forum nations. Between 1975 and 1992 France conducted 135 underground and 52 atmospheric nuclear tests. In May 1991, during a visit to New Zealand, the French Prime Minister, Michel Rocard, formally apologized for the bombing of the *Rainbow Warrior*. However, in July 1991 tension between France and the region was exacerbated by the French Government's decision to award a medal for 'distinguished service' to one of the agents convicted for his role in the bombing.

In April 1992 the French Government announced that nuclear tests would be suspended until the end of the year. Although the decision was welcomed throughout the South Pacific, concern was expressed in French Polynesia over the economic implications of the move, because of the Territory's dependence on income received from hosting the nuclear-test programme. Similarly, it was feared that unemployment resulting from the ban would have a serious impact on the economy. A delegation of political leaders subsequently travelled to Paris to voice its concerns, and in January 1993 accepted assistance worth 7,000m. francs CFP in compensation for lost revenue and in aid for development projects. It was subsequently confirmed that the suspension of tests would continue for an indefinite period.

However, shortly after his election in May 1995, President Chirac announced that France would resume nuclear testing, with a programme of eight tests between September 1995 and May 1996. The decision provoked almost universal outrage in the international community, and was condemned for its apparent disregard for regional opinion, as well as for undermining the considerable progress made by Western nations towards a world-wide ban on nuclear testing. Scientists also expressed concern at the announcement; some believed that further explosions at Mururoa could lead to the collapse of the atoll, which had been weakened considerably. Large-scale demonstrations and protest marches throughout the region were accompanied by boycotts of French products and the suspension of several trade and defence co-operation agreements. Opposition to the French Government intensified in July 1995, when French commandos violently seized the flagship of Greenpeace, *Rainbow Warrior II*, together with its crew, who had been protesting peacefully near the test site. Chirac continued to defy mounting pressure from the EU, Japan and Russia, as well as Australia, New Zealand and the South Pacific community, all of which urged him to reverse the decision to carry out the tests.

French Polynesia became the focus of world attention when the first test was carried out in September 1995. The action attracted further statements of condemnation from Governments around the world, and provoked major demonstrations in many countries. In Tahiti hitherto peaceful protests soon developed into full-scale riots, as several thousand demonstrators, enraged by the French authorities' intransigent stance, rampaged through the capital demanding an end to French rule. Meanwhile, violent clashes with police, and the burning of dozens of buildings in Papeete during the riots, left much of the capital in ruins. In defiance of world opinion, a further five tests were carried out, the sixth and final one being conducted in January 1996. In September 1998 the trial of more than 60 people charged with offences relating to the riots and protests of September 1995 began in Papeete. Hiro Tefaare, a pro-independence member of the Territorial Assembly and former police-officer, was found guilty of instigating the riots and sentenced to three years' imprisonment (of which 18 months were to be suspended). Furthermore, in September 1999 the French Government was ordered by the Administrative Tribunal to pay 204m. francs CFP in compensation for failing to maintain law and order.

Meanwhile, in early 1996 the French Government confirmed reports by a team of independent scientists that radioactive isotopes had leaked into the waters surrounding the atoll, but denied that they represented a threat to the environment. However, following the election of a new socialist administration in France in mid-1997, the French Minister of the Environment demanded in August 1998 that the matter be investigated further, stating that she had not been reassured by the initial reports. Work to dismantle facilities at the CEP test site began in 1997 and was completed in July 1998.

In early 1999 a study by the French Independent Research and Information Commission reported that there was serious radioactive leakage into underground water, lagoons and the ocean at Mururoa and Fangataufa atolls. These claims were dismissed by a New Zealand scientist who had taken part in an earlier study by the International Atomic Energy Agency, which had claimed that radiation levels were nearly undetectable. In May 1999 a French government official admitted that fractures had been found in the coral cone at the Mururoa and Fangataufa nuclear testing sites. The reports by Greenpeace that the atoll was in danger of collapsing had always been previously denied by France. However, France's claim that no serious long-term damage had been done was contested by Greenpeace, which also suggested the need for an urgent independent study of the test sites. In January 2000, in what was considered to be a significant development, the commander of the French armed forces in French Polynesia admitted that there were significant cracks in the coral reef surrounding Mururoa, and that these could lead to the occurrence of a tsunami.

In July 2003 some 2,000 islanders, led by Oscar Temaru, demonstrated in favour of self-determination during an official visit by President Chirac to Papeete. Chirac attempted to assure islanders that the nuclear tests had created no danger to health and pledged that France would assume responsibility should any evidence to the contrary emerge. At the beginning of Chirac's visit, some 200 members of an association of those formerly employed at Mururoa and Fangataufa staged a demonstration to demand that France recognize the existence of a connection between nuclear testing and the subsequent health of those involved, as Australia and the USA had done. Following his election as President, in May 2004, Temaru agreed to establish a body to investigate the impact of the French nuclear programme on the health of the population and the environment. He also expressed his support for the Nuclear Veterans' Association, which was seeking compensation from the French Government: according to the group around 30% of some 15,000 former nuclear workers were either suffering—or had died—from cancers or related diseases; the French military estimated the number of workers to be only 4,700. In late September a pre-trial investigation began in France in relation to a case filed in November 2003 by the French Nuclear Veterans' Association and Moruroa e Tatou, the French Polynesian association of former test workers, against an unidentified defendant.

The French Government is responsible for conducting the external affairs of French Polynesia although, by the constitutional settlement of 1990, the Territorial Government was enabled to enter into treaties with other Pacific countries and territories concerning matters in which it had competence. This competence was enhanced by the new status granted in 2004. An agreement settling the delimitation of the conflicting exclusive economic zones claimed by the Cook Islands and French Polynesia was signed in August 1990. In early August 2004 French Polynesia was granted observer status at the Pacific Islands Forum; Oscar Temaru used the occasion to announce that he would be seeking the support of the Forum members to have French Polynesia returned to the UN list of territories to be decolonized. Later that month it was announced that a High Council would be established in French Polynesia, in an advisory capacity, to decide how best to divide administrative responsibilities between Papeete and Paris.

Economy

The economy of French Polynesia is dominated by income from the French State and especially, from 1966, by the presence of the Commission d'Energie Atomique (CEA) and, until 1998, the Centre d'Expérimentation du Pacifique (CEP—see Recent History). In 2000, according to World Bank estimates, French Polynesia's gross national income (GNI), measured at average 1998–2000 prices, was US $3,794m., equivalent to $16,150 per head (or $24,360 per head on an international purchasing-power parity basis). During 1990–2000, it was estimated, GNP per head increased, in real terms, at an average annual rate of 0.5%. French Polynesia's gross domestic product (GDP) increased, in real terms, at an average annual rate of 2.3% in 1990–2000. Real GDP increased by 4.0% in both 1999 and 2000. One of French Polynesia's main problems has been the rapid rate of population growth (although this has shown signs of declining because of a fall in the birth rate), which, between 1990 and 2000, was estimated at an annual average of 1.8%. At the census of 1996, 43% of the population were estimated to be under 20 years of age.

Agriculture, forestry and fishing contributed only 4.7% of GDP in 2000, but provided most of French Polynesia's exports. The sector engaged 14.6% of the employed labour force in 1996; this had decreased to an estimated 3.6% by 2003. Coconuts are the principal cash crop, and in 2002 the estimated harvest was 88,000 metric tons. Most coconut trees are grown in the Tuamotu-Gambier region. Monoï oil is produced by macerating tiaré flowers in coconut oil, and in 2001 5,000 metric tons were exported. Vegetables, fruit (especially citrus fruit) and coffee are also cultivated. In 2004 the Government introduced a programme to boost annual vanilla production to 62.5 metric tons by 2009. The most important livestock products are dairy produce, eggs and honey. The forestry sector is still being developed.

Most commercial fishing, principally for tuna, is conducted, under licence, by Japanese and Korean fleets. The total catch by French Polynesian vessels in 2002 was 15,600 metric tons. The aquaculture sector produced only 65 metric tons in 2002. A US $40m. tuna-farming project in the Tuamotu Archipelago—previously the site of the CEP (see above)—was expected to begin operation in June 2004; it was hoped that the project would produce 1,000 tons of tuna in its first year of activity, rising to 2,000 tons within five years. The most important of the marine industries (which include the farming of shrimps and mussels) is the production of cultured black pearls, of which the quantity exported increased from about 50 kg in 1978 to 8,182 kg in 1999; it accounted for 54.8% of total exports in 2001. In the latter year exports of black pearls earned some 14,244m. francs CFP (compared with 17,328m. francs CFP and 20,934m. francs CFP in 1999 and 2000 respectively). During the mid-1990s French Polynesia was estimated to have produced more than 95% of the world's cultured black pearls. Japan is the biggest importer of black pearls from the country, and in 1998 purchased an estimated 65% of black pearls produced in that year (worth 9,600m. francs CFP). In 2000 French Polynesia's pearl producers embarked on a 'clean-up' operation aimed at preserving high standards and deterring the sale of cheaper, poor-quality pearls. However, the black pearl industry continued to decline in 2001. Auctions early in the year raised less revenue than expected, with a marked reduction in the number of buyers from Japan in particular, largely owing to the downturn in that country's economy, as well as in Europe and the USA. In December the industry lowered prices to maintain the level of sales, while at the same time it formed a centralized buying syndicate, to ensure minimum prices and quality control. Meanwhile, the largest producer of black pearls had suspended operations at five farms and reduced staff levels, in the hope that pearl prices would rise. In early 2003 the Government allocated some 150m. francs CFP for the promotion of the black pearl industry overseas.

Industry (including construction) engaged 16.9% of the employed labour force in 2003 and provided 15.5% of GDP in 1997. There is a small manufacturing sector, which is heavily dependent on agriculture. Coconut oil and meal or oilcake (for animal feed) are produced from copra, as are soap and monoï. There are breweries, fruit juice and soft drinks factories and some textile and handicraft manufacturers. The mining and manufacturing sector engaged 7.2% of all employees in 2003 and contributed 7.0% of GDP in 1997. Construction is an important industrial activity, contributing 5.3% of GDP in 1997 and engaging 8.9% of the employed labour force in 2003.

Hydrocarbon fuels are the main source of energy in the country, with the Papeete thermal power station providing about three-quarters of the electricity produced. Hydroelectric dams with the capacity to generate 36% of Tahiti's electricity requirements have been constructed. Solar energy is also increasingly important, especially on the less-populated islands. In 2000 Electricité de Tahiti's installed capacity amounted to 181,494 MW, supplying 60,925 customers.

The services sector engaged 79.5% of the employed labour force in 2003 and provided 80.4% of GDP in 1997. Tourism remains the primary source of revenue, and receipts from tourism totalled an estimated 49,900m. francs CFP in 2000. French Polynesia received some 212,692 tourists in 2003. As hotel occupancy was low in late 2001, the prices of various services were reduced. In June 2002 P&O agreed to operate two luxury cruise liners, which had been abandoned in Papeete harbour when their former owners, Renaissance Cruises, went bankrupt in 2001. Corsair, a French airline company, announced in August 2002 that it was to cease its twice-weekly service from Paris to Papeete, which had carried some 61,000 passengers to and from French Polynesia in 2001. The President of French Polynesia, Gaston Flosse, and the Chairman of Air Tahiti Nui attempted to persuade Corsair to continue the service until April 2003.

In 2000, according to the Institut d'Émission d'Outre-Mer (IEOM—the French overseas reserve bank), French Polynesia recorded a visible trade deficit of 81,052m. francs CFP. The country's recurring deficit is partly offset by receipts from tourism, but mainly by transfers from the French treasury. On the current account of the balance of payments there was a surplus of 45,660m. francs CFP, equivalent to 11% of GDP and an increase of 75.8% compared with the previous year. In 2003 imports reached 165,033m. francs CFP and exports totalled 15,818m. francs CFP (a decline of 25% compared with the previous year). France is traditionally the principal trading partner, accounting for 41.8% of imports and 11.5% of exports in 2003. The USA was next in importance that year, supplying 8.9% of imports and receiving 14.6% of exports. Japan, Thailand, New Zealand and Italy are also significant trading partners. The principal imports in 2003 included road vehicles (10.0% of the total) and machinery and mechanical appliances (14.7%). The principal commodity exports in 2003 were cultured black pearls (which provided 66.8% of total export revenue). Coconut oil, vanilla, fish and cut flowers were also important exports.

A budgetary deficit of 38,314m. francs CFP was projected in 2001 (the recorded deficit in 2000 was 44,579m. francs CFP). In 2002 expenditure by the French State in French Polynesia totalled 131,626m. francs CFP (25.8% of which was on the military budget), compared with 128,586m. francs CFP in 2001. In May 2004 the incoming Government of Oscar Temaru inherited a deficit from the previous administration of around US $140m. The total external debt was estimated at US $390m. in 1992. The annual rate of inflation averaged 1.3% in 1990–2002, rising to 2.9% in 2002. In 2003, however, the inflation rate declined to 0.4%.

The country suffers from a high rate of unemployment, recorded at 13.2% of the labour force in 1996. Problems are accentuated by the marked predominance of young people in the population, as well as by the migration from the countryside and the outer islands to the urban centres since the 1960s, and the consequent housing difficulties and loss of subsistence skills.

French Polynesia is an Overseas Country in association with the European Union (EU). The country is part of the Franc Zone (an association of countries with currencies linked to the French franc), and has membership in its own right of the Pacific Community (formerly the South Pacific Commission). French Polynesia is also an associate member of the UN's Economic and Social Commission for Asia and the Pacific (ESCAP).

French Polynesia enjoys a very high standard of living, compared with its island neighbours. However, despite one of the region's highest incomes per head in statistical terms, French Polynesia has considerable disparities of wealth. Perhaps the most significant, if distorting, economic factor has been the presence of the CEP and the CEA. The changes brought about by the nuclear-testing programme effectively transformed the islands from a state of self-sufficiency to one of import-dependency within less than a generation. Prior to the establishment of the CEP and the CEA and the start of nuclear testing in 1966, French Polynesia exported commodities to the value of 80% of its imports and the primary sector provided 54% of employment (1962). It was estimated that, in 1989, the CEP alone provided 12.5% of local jobs and accounted for 55% of all French Polynesia's external financial aid, 22% of GDP and 28% of total imports. With the help of revenues from the programme of nuclear testing, the economy of French Polynesia has been dominated by the large number of semi-public companies—such as the Société de Financement du Développement de la Polynésie Française and Air Tahiti Nui.

In 1993 the Governments of French Polynesia and Metropolitan France concluded an agreement, the 'pacte de progrès', which was to provide the country with aid worth some 26,000m. francs CFP over five years. This financial arrangement was extended following the conclusion of a further series of nuclear tests to some 28,300m. francs CFP annually over 10 years (1996–2006). It was hoped that the arrangement would enable French Polynesia to establish an economy that would be more reliant upon local resources and would consequently create greater employment, thereby enhancing the country's potential for durable independence. During a visit to French Polynesia in mid-2003, President Chirac of France confirmed that the annual grant of €151m. would be continued indefinitely. In May 2004 the new Temaru administration revealed that the French Government owed French Polynesia payments in arrears amounting to US $250m. In an attempt to supplement revenue provided by the Contribution de Solidarité Territoriale (an income tax introduced in 1993), the Territorial Government announced the introduction of a value-added tax (VAT) from October 1997. French Polynesia's steady economic growth resulted partly from the development of the services sector, notably in hotel construction and other tourism-related services, which led to significant employment creation. Other sectors of the economy, such as pearl farming, however, have not expanded as rapidly, principally because of regional economic conditions (notably the recession in Japan, one of the largest importers of black pearls). In 2001 customs duties were decreased, while VAT rates were maintained at similar levels, as the Government continued the anti-inflationary policy that it had instigated in 1996. The principal aim of the 2002 budget, however, was to avoid recession and to counter the repercussions of the deteriorating global economic situation. VAT rates were increased by an average of 3.3%, and total VAT receipts for the year were expected to reach 36,600m. francs CFP. The 2002 budget also introduced new taxes on alcohol, soft drinks, sugar and new road vehicles. The Government defended the introduction of these new taxes by citing the 'economic uncertainties' faced by French Polynesia. Of total budgetary expenditure of 139,000m. francs CFP, some 50,000m. francs CFP were allocated for investment purposes. The 2003 budget allocated more than 5% of total expenditure to social housing programmes. In mid-2003 the National Assembly approved the Overseas Territories Development Bill, which would provide support for economic and social development in French Polynesia (together with New Caledonia and Wallis and Futuna) by attracting foreign investment and, among other benefits, allow for overseas French residents to travel to mainland France to take advantage of free education. The tourism sector, meanwhile, proved vulnerable in the aftermath of the terrorist attacks on the USA on 11 September 2001. The number of visitor arrivals declined sharply in late 2001, and by early 2002 many of French Polynesia's hotels were reporting occupancy rates as low as 35%. In 2002 the total number of visitors declined by 17% compared with the previous year, but in 2003 visitor numbers increased by 11%.

Statistical Survey

Source (unless otherwise indicated): Institut Statistique de la Polynésie Française, Immeuble UUPA, rue Edouard Ahne, BP 395, 98713 Papeete; tel. 473434; fax 427252; e-mail ispf@ispf.pf; internet www.ispf.pf.

AREA AND POPULATION

Area: Total 4,167 sq km (1,609 sq miles); Land area 3,521 sq km (1,359 sq miles).

Population: 188,814 at census of 6 September 1988; 219,521 (males 113,934, females 105,587) at census of 3 September 1996; 245,516 at census of 7 November 2002. *2003* (official estimate): 246,000 at 1 January.

Population by Island Group (2002 census): Society Archipelago 214,445 (Windward Islands 184,224, Leeward Islands 30,221); Marquesas Archipelago 8,712; Austral Islands 6,386; Tuamotu-Gambier Islands 15,973.

Density (provisional figure, November 2002 census): 58.9 per sq km.

Ethnic Groups (census of 15 October 1983): Polynesian 114,280; 'Demis' 23,625 (Polynesian-European 15,851, Polynesian-Chinese 6,356, Polynesian-Other races 1,418); European 19,320; Chinese 7,424; European-Chinese 494; Others 1,610; Total 166,753. *1988 Census* ('000 persons): Polynesians and 'Demis' 156.3; Others 32.5.

Principal Towns (population at 2002 census, provisional figures): Faaa 28,339; Papeete (capital) 26,181; Punaauía 23,706; Pirae 14,499; Moorea-Maiao 14,550; Mahina 13,334; Paea 12,276. *Mid-2003* (UN estimate, incl. suburbs): Papeete 126,193 (Source: UN, *World Urbanization Prospects: The 2003 Revision*).

Births, Marriages and Deaths (2002, provisional figures): Registered live births 4,843 (birth rate 19.6 per 1,000); Marriages (1997) 1,176 (marriage rate 5.3 per 1,000); Registered deaths 1,090 (death rate 4.5 per 1,000).

Expectation of Life (years at birth, 2000): Males 72.5; Females 77.2. Source: UN, *Statistical Yearbook for Asia and the Pacific*.

Economically Active Population (persons aged 14 years and over, 1996 census): Agriculture, hunting, forestry and fishing 10,888; Mining and manufacturing 6,424; Electricity, gas and water 459; Construction 4,777; Trade, restaurants and hotels 9,357; Transport, storage and communications 3,788; Financial services 1,482; Real estate 383; Business services 3,710; Private services 9,033; Education, health and social welfare 10,771; Public administration 13,475; *Total employed* 74,547 (males 46,141, females 28,406); Persons on compulsory military service 1,049 (all males); Unemployed 11,525 (males 6,255, females 5,270); *Total labour force* 87,121 (males 53,445, females 33,676). *December 2003:* Agriculture, hunting, forestry and fishing 2,294; Mining and manufacturing 4,532; Electricity, gas and water 488; Construction 5,591; Trade and repairs 9,712; Hotels and restaurants 6,403; Transport, storage and communications 5,658; Financial services 1,488; Real estate, renting and business services 4,478; Domestic services 1,419; Education, health, collective services and social welfare 6,615; Public administration 14,275; Total employed 62,953.

HEALTH AND WELFARE

Key Indicators

Physicians (per 1,000 head, 2001): 1.8.

For definitions, see explanatory note on p. vi.

AGRICULTURE, ETC.

Principal Crops (metric tons, 2002): Roots and tubers 12,400 (FAO estimate); Vegetables and melons 7,210 (FAO estimate); Pineapples 3,500 (FAO estimate); Other fruit 5,520 (FAO estimate); Coconuts 88,000 (FAO estimate); Vanilla 35 (unofficial figure); Coffee (green) 18 (FAO estimate). Source: FAO.

Livestock (FAO estimates, year ending September 2002): Cattle 10,800; Horses 2,200; Pigs 30,000; Goats 16,500; Sheep 440; Chickens 240,000; Ducks 33,000. Source: FAO.

Livestock Products (metric tons, 2002): Beef and veal 200 (FAO estimate); Pig meat 1,051; Goat meat 75 (FAO estimate); Poultry meat 693 (FAO estimate); Cows' milk 1,200 (FAO estimate); Hen eggs 1,600 (FAO estimate); Other poultry eggs 85 (FAO estimate); Honey 32. Source: FAO.

Fishing ('000 metric tons, live weight, 2002): Capture 15.5 (Common dolphinfish 0.6; Skipjack tuna 1.5; Albacore 4.7; Yellowfin tuna 1.2; Bigeye tuna 0.7; Marlins and sailfishes 0.5; Sharks, rays, skates, etc. 1.1); Aquaculture 0.1; *Total catch* 15.6. Note: Figures exclude pearl oyster shells (kilograms): 1,289,490. Source: FAO.

INDUSTRY

Production: Copra 8,262 metric tons (sales, 2001); Coconut oil 5,000 metric tons (2001)*; Oilcake 2,500 metric tons (2001)*; Beer 129,000 hectolitres (1992); Printed cloth 200,000 m (1979); Japanese sandals 600,000 pairs (1979); Electric energy 506.7m. kWh (2002).
*FAO estimate.

FINANCE

Currency and Exchange Rates: see Wallis and Futuna Islands.

Territorial Budget (million francs CFP, 2002): *Revenue*: Current 111,351 (Indirect taxation 64,346). *Expenditure*: Current 107,103, Capital 42,562; Total 149,665.

French State Expenditure (million francs CFP, 1996): Civil budget 61,706 (Current 56,564, Capital 5,142); Military budget 46,119 (Current 37,160, Capital 8,959); Pensions 10,458; Total (incl. others) 123,774 (Current 109,060, Capital 14,714). *1998* (million francs CFP): 121,788 (incl. military budget 37,982). *1999* (million francs CFP): 120,631 (incl. military budget 34,343). *2000* (million francs CFP): 124,800 (incl. military budget 35,400). *2001* (million francs CFP): 128,586 (incl. military budget 36,774). *2002* (million francs CFP): 131,626 (incl. military budget 33,926).

Money Supply (million francs CFP at 31 December 2001): Currency in circulation 9,366; Demand deposits 100,617; Total money 109,983. Source: Institut d'Emission d'Outre-Mer.

Cost of Living (Consumer Price Index; base: 1990 = 100): All items 114.0 in 2001; 117.3 in 2002; 117.8 in 2003. Source: UN, *Monthly Bulletin of Statistics*.

Gross Domestic Product (million francs CFP at current prices): 378,501 in 1997; 404,886 in 1998; 412,100 in 1999; 453,460 in 2000.

Expenditure on the Gross Domestic Product (million francs CFP at current prices, 1993): Government final consumption expenditure 126,127; Private final consumption expenditure 202,563; Increase in stocks –536; Gross fixed capital formation 53,494; *Total domestic expenditure* 381,648; Exports of goods and services 34,523; *Less* Imports of goods and services 86,905; *GDP in purchasers' values* 329,266. Source: UN, *National Accounts Statistics*.

Gross Domestic Product by Economic Activity (million francs CFP at current prices, 1997): Agriculture, forestry and fishing 15,534; Manufacturing 26,360*; Electricity, gas and water 12,221*; Construction 20,104; Trade 81,854; Transport and telecommunications 27,832. Other private services 96,714; Government services 97,238; Domestic services 646; *GDP in purchasers' values* 378,503.

* Manufacturing of energy-generating products is included in electricity, gas and water. Source: UN, *National Accounts Statistics*.

EXTERNAL TRADE

Principal Commodities (million francs CFP, 2003): *Imports c.i.f.:* Live animals and animal products 9,975.2 (Meat and edible meat offal 6,065.0); Prepared foodstuffs; beverages, spirits and vinegar; tobacco and manufactured substitutes 26,660.3; Mineral products 14,086.4 (Mineral fuels, mineral oils and products of their distillation; bituminous substances; mineral waxes 10,459.2); Products of chemical or allied industries 11,272.8 (Pharmaceutical products 4,872.2); Plastics, rubber and articles thereof 5,431.9; Paper-making materials; paper and paperboard and articles thereof 4,265.3; Textiles and textile articles 2,041.1; Base metals and articles thereof 9,288.6; Machinery and mechanical appliances; electrical equipment; sound and television apparatus 24,323.1 (Nuclear reactors, boilers, machinery, mechanical appliances and parts 15,551.1, Electrical machinery, equipment, etc. 10,339.6); Vehicles, aircraft, vessels and associated transport equipment 46,574.0 (Aircraft and aeronautical craft 21,392.2; Road vehicles, parts and accessories 16,472.3); Miscellaneous manufactured articles 5,270.4; Total (incl. others) 165,032.7. *Exports f.o.b.:* Animal products 946.4 (Fish and crustaceans, molluscs and other aquatic invertebrates 656.1); Prepared foodstuffs; beverages, spirits and vinegar; tobacco and manufactured substitutes 2,024.7 (Preparations of vegetables, fruit, nuts or other parts of plants 725.6); Natural or cultured pearls, precious or semi-precious stones, precious metals and articles thereof; imitation jewellery; coin 10,566.8; Vehicles, aircraft, vessels and associated transport equipment 1,633.6 (Aircraft, spacecraft and parts 678.0); Total (incl. others) 15,817.9.

Principal Trading Partners (million francs CFP, 2003): *Imports:* Australia 18,809.1; Belgium 2,746.2; China, People's Republic 6,123.7; France 68,917.8; Germany 5,333.3; Italy 4,858.5; Japan 5,033.4; Korea, Republic 2,473.7; New Zealand 6,123.7; Romania 2,526.4; Spain 2,012.6; Thailand 3,573.3; United Kingdom 2,734.1; USA 14,767.0; Total (incl. others) 165,032.7. *Exports:* Belgium 270.5; Canada 194.1; France 1,817.4; Italy 210.1; Hong Kong 3,241.4; Japan 4,789.9; New Caledonia 451.4; New Zealand 209.1; Thailand 1,280.9; USA 2,304.5; Total (incl. others) 15,817.9.

TRANSPORT

Road Traffic (1987): Total vehicles registered 54,979; (1996 census): Private cars 47,300.

Shipping (1990): *International Traffic*: Passengers carried 47,616; Freight handled 642,314 metric tons. *Domestic Traffic*: Passengers carried 596,185; Freight handled 261,593 metric tons. (2002): Goods unloaded 960,297 metric tons; Goods loaded 29,626 metric tons.

Civil Aviation (2002): *International Traffic*: Passengers carried 572,961; Freight handled 8,766 metric tons. *Domestic Traffic*: Passengers carried 803,255; Freight handled 2,935 metric tons.

TOURISM

Visitors (excluding cruise passengers and excursionists): 227,658 in 2001; 188,998 in 2002; 212,692 in 2003.

Tourist Arrivals by Country of Residence (2003): Australia 7,506; Canada 5,523; France 48,177; Italy 9,213; Japan 22,882; New Zealand 6,106; United Kingdom 7,204; USA 77,768; Total (incl. others) 212,692.

Tourism Receipts (US $ million): 359 in 1997; 354 in 1998; 394 in 1999. Source: World Tourism Organization.

2000: 49,900m. francs CFP (Source: Ministère du Tourisme).

COMMUNICATIONS MEDIA

Radio Receivers (1997): 128,000 in use*.

Television Receivers (2000): 44,000 in use†.

Telephones (2002): 52,500 main lines in use†.

Facsimile Machines (1998): 3,000 in use†.

Mobile Cellular Telephones (subscribers, 2002): 90,000†.

Internet Users (2002): 35,000†.

Personal Computers (number in use, 2002): 70,000.

Daily Newspapers (2000): 2.

* Source: UNESCO, *Statistical Yearbook*.
† Source: International Telecommunication Union.

EDUCATION

Pre-primary (2000/01): 54 schools; 408 teachers (1996/97); 13,720 pupils.

Primary (2000/01): 173 schools; 2,811 teachers (1996/97); 26,249 pupils.

General Secondary (2000/01): 2,035 teachers (1998/99); 24,743 pupils.

Vocational (1992): 316 teachers; 3,730 students.

Tertiary (1993): 34 teachers; 892 students.

Source: partly UNESCO, *Statistical Yearbook*.

Directory

The Constitution

The constitutional system in French Polynesia was established under the aegis of the Constitution of the Fifth French Republic and specific laws of 1977, 1984 and 1990. The French Polynesia Statute 1984, the so-called 'internal autonomy statute', underwent amendment in a law of July 1990. A further extension of the Territory's powers under the statute was approved by the French National Assembly in Paris in December 1995. In January 2000 a constitutional amendment granting French Polynesia a greater degree of autonomy was presented to a joint session of the French Senate and National Assembly for final ratification (see History). In March 2003 both the National Assembly and Senate ratified amendments to the Constitution providing for French Polynesia (along with Wallis and Futuna) to become an Overseas Country. In July the Territorial Assembly and the Government of French Polynesia ratified the amendments. The approval of the French State Council, the French Government and the Constitutional Council duly followed, and in March 2004 French Polynesia became an Overseas Country.

French Polynesia is declared to be an Overseas Country of the French Republic, of which it remains an integral part. The High Commissioner, appointed by the French Government, exercises the prerogatives of the State in matters relating to defence, foreign relations, the maintenance of law and order, communications and citizenship. The head of the local executive and the official who represents the Overseas Country is the President of the Government, who is elected by the French Polynesian Assembly from among its own members. The President appoints and dismisses the Council of Ministers and has competence in international relations as they affect French Polynesia and its exclusive economic zone, and is in control of foreign investments and immigration. The Assembly, which has financial autonomy in budgetary affairs and legislative authority within French Polynesia, is elected for a term of up to five years on the basis of universal adult suffrage. Following the elections of May 2004, it comprised 57 members: 37 elected by the people of the Windward Islands (Iles du Vent—Society Islands), eight by the Leeward Islands (Iles Sous le Vent—Society Islands), three by the Gambier Islands-East Tuamotu Archipelago, three by the West Tuamotu Archipelago, three by the Austral Islands and three by the Marquesas Islands. The Assembly elects a Permanent Commission from among its members, and itself meets for two ordinary sessions each year and upon the demand of the majority party, the President or the High Commissioner. Local government is conducted by the municipalities. There is an Economic, Social and Cultural Council (composed of representatives of professional groups, trade unions and other organizations and agencies which participate in the economic, social and cultural activities of French Polynesia), an Audit Office and a judicial system which includes a Court of the First Instance, a Court of Appeal and an Administrative Court. The Overseas Country, as a part of the French Republic, also elects two deputies to the National Assembly and one member of the Senate, and may be represented in the European Parliament.

The Government

(September 2004)

High Commissioner: MICHEL MATHIEU (appointed October 2001).

Secretary-General: JACQUES MICHAUT.

COUNCIL OF MINISTERS

President and Minister for International Relations, the Development of the Municipalities and Urban Affairs: OSCAR TEMARU (removed from office in a parliamentary vote of no confidence on 8 October 2004).

Vice-President, Government Spokesperson and Minister for Tourism and the Environment: JACQUI DROLLET.

Minister for Economy and Finance: EMILE VANFASSE.

Minister for Land Affairs, Planning and Housing: GILLES TEFAATAU.

Minister for Education and Culture: JEAN-MARIUS RAAPOTO.

Minister for Utilities, Land and Sea Transport: JAMES SALMON.

Minister for the Promotion of Natural Resources: KEITAPU MAAMAATUAIAHUTAPU.

Minister for Health, Welfare and Solidarity: MARIE-LAURE VANIZETTE.

Minister for Outer Islands Development and Decentralization: EMILE VERNAUDON.

Minister for Social Dialogue: FRANCIS STEIN.

GOVERNMENT OFFICES

Office of the High Commissioner of the Republic: Bureau du Haut Commissaire, ave Bruat, BP 115, 98713, Papeete; tel. 468597; fax 468556; e-mail bcom@haut-commissariat.pf; internet www.polynesie-francaise.gouv.fr.

Office of the President of the Government: BP 2551, 98713 Papeete; tel. 472000; fax 419781; e-mail presid@mail.pf; internet www.presidence.pf.

Government of French Polynesia: BP 2551, Papeete; 28 blvd Saint-Germain, 75005 Paris, France; tel. 472000; fax 419781; e-mail presid@mail.pf; tel. 426510; fax 426409.

Economic, Social and Cultural Council (CESC): ave Bruat, BP 1657, 98716 Papeete; tel. 416500; fax 419242; e-mail cesc@cesc.gov.pf; internet www.cesc.pf; f. 1977 as the Economic and Social Council; name changed as above in 1984; Pres. PIERRE FRÉBAULT; Sec.-Gen KATIA TESTARD.

Ministry of Agriculture: Fare Ute, BP 100, 2444-001 Papeete; tel. 428144; fax 420831; e-mail secretariat@ressources.gov.pf; internet www.agriculture.gov.pf.

Ministry of Craft Industry: BP 4451, 98713 Papeete; tel. 423225; fax 436478; e-mail maraea.teissier@broche.pf; internet www.artisanat.gov.pf.

Ministry of Culture: BP 543, 98713 Papeete; tel. 432695; fax 425251; e-mail sce@culture.gov.pf; internet www.culture.gov.pf.

Ministry of the Economy and Finance: BP 2551, 98713 Papeete; tel. 468086; fax 450050; e-mail hinathunot@yahoo.fr; internet www.finances.gov.pf.

Ministry of Education: BP 2551, 98713 Papeete; tel. 501501; fax 424285; e-mail titaua.peu@education.gov.pf; internet www.education.gov.pf.

Ministry of Environment: Quartier Broche, rue Dumont d'Urville, BP 2551, 98713 Papeete; tel. 472000; fax 472005; e-mail dircab-cab@presidence.pf; internct www.environnement.gov.pf.

Ministry of Fisheries: BP 20, 98713 Papeete; tel. 422440; fax 451880; e-mail secretariat@ressources.gov.pf; internet www.mer.gov.pf.

Ministry of Health: BP 611, 98713 Papeete; tel. 460002; fax 433974; e-mail dircab.msa@sante.gov.pf; internet www.sante.gov.pf.

Ministry of Housing: Ave Bruat—Gouvernement, Rez de chaussée, BP 2551, 98713 Papeete; tel. 549575; fax 454343; e-mail dircab.mla@logement.gov.pf; internet www.logement.gov.pf.

Ministry of Land Affairs: BP 114, 98713 Papeete; tel. 549575; fax 454343; e-mail dircab.maa@foncier.gov.pf; internet www.foncier.gov.pf.

Ministry of the Pearl Industry: BP 2551, 98713 Papeete; tel. 472295; fax 472287; e-mail secretariat@ressources.gov.pf; internet www.perle.gov.pf.

Ministry of Public Works and Harbours: Bâtiment A1, rue du Commandant Destremeau, BP 85, 98713 Papeete; tel. 468019; fax 483792; e-mail contact.met@equipement.gov.pf; internet www.equipement.gov.pf.

Ministry of Solidarity and Family Affairs: Service des Affaires Sociales, BP 1707, 98713 Papeete; tel. 465846; fax 438920; e-mail dir.das@mail.pf; internet www.solidarite.gov.pf.

Ministry of Tourism and Transportation: BP 2551, 98713 Papeete; tel. 472152; fax 472005; e-mail maraea.teissier@broche.pf; internet www.tourisme.gov.pf.

Ministry of Youth: BP 2551, 98713 Papeete; tel. 501501; fax 424285; e-mail titaua.peu@education.gov.pf; internet www.jeunesse.gov.pf.

Legislature

ASSEMBLÉE

President: ANTONY GÉROS.

Assembly: Assemblée de la Polynésie Française, BP 28, 98713 Papeete; tel. 416100; fax 416149; internet www.assemblee.pf.

Election, 23 May 2004

Party	Seats
Tahoeraa Huiraatira	28
Tavini Huiraatira*	27
Fe'tia Api*	1
No Oe E Te Nunaa*	1
Total	57

* Campaigned as part of the Union pour la démocratie (UPD) coalition, which also included candidates from Ai'a Api, Here Ai'a and Heiura-Les Verts Polynésiens.

PARLEMENT

Deputies to the French National Assembly: MICHEL BUILLARD (Tahoeraa Huiraatira/UMP), BÉATRICE VERNAUDON (Tahoeraa Huiraatira/UMP).

Representative to the French Senate: GASTON FLOSSE (Tahoeraa Huiraatira/UMP).

Political Organizations

Ai'a Api (New Land): BP 11055, Mahina, Tahiti; tel. 481135; f. 1982; after split in Te E'a Api; Leader EMILE VERNAUDON.

Fe'tia Api (New Star): c/o Assemblée de la Polynésie Française, BP 140 512, Arue; Leader PHILIP SCHYLE.

Heiura-Les Verts Polynésiens: BP 44, Borabora; tel. and fax 677174; e-mail heiura@mail.pf; Pres. JACKY BRYANT.

Ia Mana Te Nunaa (Power to the People): rue du Commandant Destrémau, BP 1223, Papeete; tel. 426699; f. 1976; advocates 'socialist independence'; Sec.-Gen. JACQUI DROLLET.

No Oe E Te Nunaa (This Country is Yours): c/o Assemblée de la Polynésie Française, BP 28, Papeete; favours autonomy; Leader NICOLE BOUTEAU.

Pupu Here Ai'a Te Nunaa Ia Ora: BP 3195, Papeete; tel. 420766; f. 1965; advocates autonomy; 8,000 mems.

Taatiraa No Te Hau: BP 2916, Papeete; tel. 437494; fax 422546; f. 1977; Pres. ROBERT TANSEAU.

Tahoeraa Huiraatira (People's Rally): rue du Commandant Destrémeau, BP 471, Papeete; tel. 429898; fax 450004; f. 1958; supports links with France, with internal autonomy; Pres. GASTON FLOSSE; Hon. Pres. JACQUES TEUIRA.

Tapura Amui No Te Faatereraa Manahune-Tuhaa Pae: c/o Assemblée de la Polynésie Française, BP 140 512, Arue; represents the Austral Islands; Leader CHANTAL FLORES.

Tavini Huiraatira no te ao Ma'ohi /Front de Libération de la Polynésie (Polynesian People's Servant): c/o Assemblée de la Polynésie Française, BP 140 512, Arue; f. 1977; independence movement; anti-nuclear; Leader OSCAR TEMARU.

Te' Avei'a (Te Ara): c/o Assemblée de la Polynésie Française, BP 140 512, Arue; f. 2004 by fmr mems of Fe'tia Api and Tavini Huiraatira; Pres. ANTONIO PEREZ.

Te Henua Enana Kotoa: Papeete; Leader LOUIS TAATA.

Te Hono e Tau I te Honaui (Link Between Generations): Papeete; f. 2002; Leader STANLEY CROSS.

Judicial System

Court of Appeal: Cour d'Appel de Papeete, ave Bruat, BP 101, 98713 Papeete; tel. 415546; fax 453767; e-mail rgi.sar.ca-papeete@justice.fr; Pres. OLIVIER AIMOT; Attorney-General FRANÇOIS DEBY; Clerk of the Court DENISE BIANCONI.

Court of the First Instance: Tribunal de Première Instance de Papeete, ave Bruat BP 101, Papeete; tel. 415500; fax 454012; internet www.polynesie-francaise.gouv.fr; e-mail tapapeete@mail.pf; Pres. GUY RIPOLL; Procurator JEAN BIANCONI; Clerk of the Court DENISE DANGLADE.

Court of Administrative Law: Tribunal Administratif, ave Bruat, BP 4522, Papeete; tel. 509025; fax 451724; e-mail tadelapolynesiefrancaise@mail.pf; Pres. ALFRED POUPET; Cllrs RAOUL AUREILLE, MARIE-CHRISTINE LUBRANO, ALAIN LEVASSEUR, HÉLÈNE ROULAND, DONA GERMAIN.

Religion

About 54% of the population are Protestants and 38% are Roman Catholics.

CHRISTIANITY

Protestant Church

Maohi Protestant Church: BP 113, Papeete; tel. 460600; fax 419357; e-mail eepf@mail.pf; f. 1884; autonomous since 1963; fmrly L'Eglise évangélique en Polynésie française (Etaretia Evaneria i Porinetia Farani); c. 95,000 mems; Pres. of Council Rev. JACQUES TERAI IHORAI; Sec.-Gen. (vacant).

The Roman Catholic Church

French Polynesia comprises the archdiocese of Papeete and the suffragan diocese of Taiohae o Tefenuaenata (based in Nuku Hiva, Marquesas Is). At 31 December 2002 there were an estimated 89,000 adherents in the Territory. The Archbishop and the Bishop participate in the Episcopal Conference of the Pacific, based in Fiji.

Archbishop of Papeete: Most Rev. HUBERT COPPENRATH, Archevêché, BP 94, Vallée de la Mission, 98713 Papeete; tel. 502351; fax 424032; e-mail catholic@mail.pf.

Other Churches

There are small Sanito, Church of Jesus Christ of Latter-day Saints (Mormon), and Seventh-day Adventist missions.

The Press

La Dépêche de Tahiti: Société Océanienne de Communication, BP 50, Papeete; tel. 464343; fax 464350; e-mail journalistes@france-antilles.pf; internet www.ladepechedetahiti.com; f. 1964; daily; French; Editor-in-Chief THIERRY DURIGNEUX; Dir-Gen. CHRISTOPHE RUET; circ. 15,000.

Les Nouvelles de Tahiti: place de la Cathédrale, BP 629, Papeete; tel. 508100; fax 508109; f. 1956; daily; French; Editor MURIEL PONTAROLLO; Publr PHILIPPE HERSANT; circ. 6,500.

Le Semeur Tahitien: BP 94, 98713 Papeete; tel. 502350; e-mail catholic@mail.pf; f. 1909; bi-monthly; French; publ. by the Roman Catholic Church.

Tahiti Beach Press: BP 887, 98713 Papeete; tel. 426850; fax 423356; e-mail tahitibeachpres@mail.pf; f. 1980; monthly; English; Publr G. WARTI; circ. 10,000.

Tahiti Pacifique Magazine: BP 368, Maharepa, Moorea; tel. 562894; fax 563007; e-mail tahitipm@mail.pf; internet www.tahiti-pacifique.com; monthly; French; Dir and Editor ALEX DU PREL; circ. 5,000.

Tahiti Rama: Papeete; weekly.

La Tribune Polynésienne: place du Marché, BP 392, Papeete; tel. 481048; fax 481220; weekly; Dir LOUIS BRESSON.

Ve'a Katorika: BP 94, 98713 Papeete; f. 1909; monthly; publ. by the Roman Catholic Church.

Ve'a Porotetani: BP 113, Papeete; tel. 460623; fax 419357; e-mail eepf@mail.pf; f. 1921; monthly; French and Tahitian; publ. by the Evangelical Church; Dir IHORAI JACQUES; circ. 5,000.

NEWS AGENCIES

Tahitian Press Agency (ATP): Papeete; f. 2001; bilingual French and English news service providing pictures and radio reports.

Foreign Bureaux

Agence France-Presse (AFP): BP 629, Papeete; tel. 508100; fax 508109; e-mail international@france-antilles.pf; Correspondent CHRISTIAN BRETAULT.

Associated Press (AP) (USA): BP 912, Papeete; tel. 437562; Correspondent AL PRINCE.

Reuters (UK): BP 50, Papeete; tel. 475298; fax 464390; e-mail danielpardon@mail.pf; Correspondent DANIEL PARDON.

Publishers

Haere Po No Tahiti: BP 1958, Papeete; fax 582333; f. 1981; travel, history, botany, linguistics and local interest.

Au Vent des Iles: BP 5670, 98716 Pirae; tel. 509595; fax 509597; e-mail mail@auventdesiles.pf; internet www.tahiti-books.com; f. 1992; Gen. Man. CHRISTIAN ROBERT.

Government Printer

Imprimerie Officielle: 43 rue des Poilus-Tahitiens, BP 117, 98713 Papeete; tel. 500580; fax 425261; e-mail compta.clients@imprimerie.gov.pf; f. 1843; printers, publrs; Dir (acting) CLAUDINO LAURENT.

Broadcasting and Communications

TELECOMMUNICATIONS

France Câbles et Radio (FCR): Télécommunications extérieures de la Polynésie française, BP 99, Papeete; tel. 415400; fax 437553.

Office des Postes et Télécommunications: 8 rue de la Reine Pomare IV, Papeete; tel. 414242; fax 436767; internet www.opt.pf; Chair. ALPHONSE TERIIEROOITERAI; Dir-Gen. MATAHI BROTHERS.

BROADCASTING

Radio

RFO Polynésie: 410 rue Dumont d'Urville, POB 60, Pamatai; tel. 861616; fax 861611; e-mail jrbodin@mail.pf; internet www.rfo.fr.html; public service radio and television station operated by Réseau France Outre-Mer (RFO), Paris; daily programmes in French and Tahitian; Area Man. WALLES KOTRA; Communications Man. JEAN-RAYMOND BODIN.

Private Stations

NRJ Radio: Papeete; tel. and fax 464346; e-mail nrj@mail.pf; French.

Radio Bleue: Papeete; tel. 483436; fax 480825; e-mail redaction@radiobleue.pf; affiliated to the political party Ai'a Api; French.

Radio Maohi: Maison des Jeunes, Pirae; tel. 819797; fax 825493; e-mail tereo@mail.pf; owned by the political party Tahoeraa Huiraatira; French and Tahitian.

Radio One: Fare Ute, BP 3601, Papeete; tel. 434100; fax 423406; e-mail infos@radio1.pf; Editor-in-chief YVES FORTUNET.

Radio Tefana: Papeete; tel. 819797; fax 825493; e-mail tereo@mail.pf; affiliated to the political party Tavini Huiraatira; French and Tahitian.

Radio Tiare: Fare Ute, Papeete; tel. 434100; fax 423406; e-mail contact@tiarefm.pf.

Television

Radio-Télé-Tahiti: see Radio.

Canal Plus Polynésie: Immeuble Pomare, blvd Paofai, BP 20051, 98713 Papeete; tel. 540754; fax 540755; e-mail serge.lamagnere@canal-caledonie.com; privately owned; CEO SERGE LAMAGNÈRE.

TNTV (Tahiti Nui Television): BP 348 98713, Papeete; tel. 473601; fax 532721; e-mail tntv@tntv.pf; internet www.tntv.pf; broadcasts in French and Tahitian; Pres. JEAN-PAUL BARRAL; Gen. Man. AHITI ROOMATAAROA.

TNS (Tahiti Nui Satellite): see TNTV; French.

Finance

(cap. = capital; res = reserves; dep. = deposits; m. = million; brs = branches; amounts in CFP francs)

BANKING

Commercial Banks

Banque de Polynésie SA: 355 blvd Pomare, BP 530, 9713 Papeete; tel. 466666; fax 466664; e-mail sec.gen@sg-bdp.pf; internet www.sg-bdp.pf; f. 1973; 80% owned by Société Générale (France); cap. 1,380.0m.,res 3,509.0m., dep. 94,102.9m. (Dec. 2002); Pres. and Chair. RONALD GEORGES; Gen. Man. JEAN PERRE; 14 brs.

Banque de Tahiti SA: rue François Cardella, BP 1602, Papeete; tel. 417000; fax 423376; e-mail dirgene@bt.pf; internet www.banque-tahiti.pf; f. 1969; owned by Financière Océor (95.4%); merged with Banque Paribas de Polynésie, Aug. 1998; cap. 1,336.5m., res 4,787.8m., dep. 85,986.8m. (Dec. 2002); Chair., Pres. and Gen. Man. GILLES THERRY; Gen. Sec. SYLVAIN FAURE; 17 brs.

Banque SOCREDO—Société pour le Crédit et le Développement en Océanie: 115 rue Dumont d'Urville, BP 130, 98713 Papeete; tel. 415123; fax 433661; internet www.socredo.pf; f. 1959;

public body; affiliated to Banque Nationale de Paris (France); total assets 182.9m. (Dec. 2003); Pres. JEAN VERNAUDON; Dir ERIC POMMIER; 24 brs.

Trade and Industry

DEVELOPMENT ORGANIZATIONS

Agence pour l'Emploi et la Formation Professionnelle: BP 540, Papeete; tel. 426375; fax 426281.

Agence Française de Développement (ACFD): Immeuble Hokule'a, 2 rue Cook–Paofai, BP 578, Papeete; tel. 544600; fax 544601; e-mail afdpapeete@pf.groupe-afd.org; internet www.afd.fr; public body; development finance institute.

Service de l'Artisanat Traditionnel: BP 4451, Papeete 98713; tel. 423225; fax 436478; Dir TEURA IRITI.

Service du Développement de l'Industrie et des Métiers (SDIM): Bâtiment des Affaires Economiques à Fare Ute, 1er étage, BP 9055, 98713 Papeete; tel. 502880; fax 412645; e-mail secretariat .sdim@industrie.gov.pf; internet www.creation-entreprises.pf; f. 1988; industry and small business development administration.

Société pour le Développement de l'Agriculture et de la Pêche: BP 1247, Papeete; tel. 836798; fax 856886; agriculture and marine industries.

SODEP (Société pour le Développement et l'Expansion du Pacifique): BP 4441, Papeete; tel. 429449; f. 1961 by consortium of banks and private interests; regional development and finance co.

CHAMBERS OF COMMERCE

Chambre de Commerce, d'Industrie, des Services et des Métiers de Polynésie Française (CCISM): 41 rue du Docteur Cassiau, BP 118, Papeete 98713; tel. 472700; fax 540701; e-mail info@cci .pf; internet www.ccism.pf; f. 1880; 36 mems; Pres. JULES CHANGUES; Sec. ADRIEN BEAUMONT.

Chambre d'Agriculture et de la Pêche Lagonaire: route de l'Hippodrome, BP 5383, Pirae; tel. 425393; fax 438754; f. 1886; 10 mems; Pres. CLAUDE HAUATA.

Jeune Chambre Economique de Tahiti: BP 2576, Papeete; tel. 454542; fax 466070; internet www.jce.pf; Pres. CATHY GOURBAULT.

EMPLOYERS' ORGANIZATIONS

Association Française des Banques/Comité en Polynésie Française (AFB/CPF): BP 530, 98713 Papeete; fax .

Association des Transporteurs Locaux de Polynésie Française (ATAL): BP 314, 98713 Papeete.

Chambre Syndicale des Entrepreneurs du Bâtiment et des Travaux Publics (CSEBTP): BP 2218, 98713 Papeete; tel. 541040; fax 423237; Pres. GEORGES TRAMINI.

Chambre Syndicale des Métiers du Génie Civil (CSMGCTP): BP 51120, 98716 Pirae.

Confédération Générale des Petites et Moyennes Entreprises de Polynésie Française (CGPME): BP 1733, 98713 Papeete; tel. 426333; fax 835608; e-mail courrier@cgpme.pf; internet www.cpme .pf.

Conseil des Employeurs: Immeuble Farnham, rue Clappier, BP 972, Papeete; tel. 541040; fax 423237; internet cepf@cepf.pf; f. 1983; Pres. DANIEL DE MARIGNY; Sec.-Gen. CÉDRIC VIDAL.

Conseil des Professionnels de l'Hotellerie (CPH): BP 972, 98713 Papeete.

Fédération Générale du Commerce (FGC): angle rue Albert Leboucher et rue Clappier, BP 1607, 98713 Papeete; e-mail fgc@ mail.pf ; internet www.fgc.pf; tel. 541042; fax 422359; Pres. GILLES YAU; Sec. PATRICIA LO MONACO.

Syndicat des Industriels de Polynésie Française (SIPOF): tel. ; fax ; e-mail ; internet ; f. 1974; represents workers in industry, engineering, manufacturing and printing; 2,000 mems in 53 cos; Pres. ALAIN SIU; Sec. BRUNO BELLANGER.

Syndicat Professionnel des Producteurs de Perles: BP 850, 98713 Papeete.

Union Polynésienne de l'Hôtellerie (UPHO): 76 rue Wallis, BP 9179 Motu Uta, Papeete; tel. 429553.

Union Patronale de Polynésie Française (UPPF): BP 317, 98713 Papeete; tel. 541040; fax 423237; f. 1948; 63 mems; Pres. DIDIER CHOMER.

UTILITES

Electricity

Electricité de Tahiti (EDT): route de Puurai, BP 8021, Faaa-Puurai; tel. 867777; internet www.edt.pf; subsidiary of Suez; c. 61,000 customers (2000); Pres. JOËL ALLAIN; Dir-Gen. CHRISTIAN LEKKIEFRE.

Water

Société Polynésienne des Eaux et Assainissements: BP 25795, 98713 Papeete, Tahiti; fax 421548; e-mail spea@spea.pf.

Syndicat Central de l'Hydraulique: Tahiti.

TRADE UNIONS

Confédération A Tia I Mua: Immeuble Ia Ora, 1er étage, ave Georges Clemenceau, BP 4523, Papeete; tel. 544010; fax 450245; e-mail atiaimua@ifrance.com; affiliated to CFDT (France); Gen. Sec. TU YAN.

Confedetation des Syndicats des Travailleurs de Polynésie/Force Ouvrière (CSTP/FO): Immeuble Farnham, 1er étage, BP 1201, 98713 Papeete; tel. 426049; fax 450635; Pres. COCO TERAIEFA CHANG; Sec.-Gen. PIERRE FRÉBAULT.

Confédération des Syndicats Independants de la Polynésie Française (CSIP): Immeuble Allegret, 1er étage, ave du Prince Hinoi, BP 468, 98713 Papeete.

Confédération O Oe To Oe Rima: Immeuble Brown, 1er étage, Papeete, BP 52866, 98716; tel. 483445; fax 483445; Gen. Sec. RONALD TEROROTUA.

Confédération Otahi: Immeuble Allegret, 1er étage, ave du Prince Hinoi BP148, 98713 Papeete; tel. 450654; fax 451327.

Transport

ROADS

French Polynesia has 792 km of roads, of which about one-third are bitumen-surfaced and two-thirds stone-surfaced.

SHIPPING

The principal port is Papeete, on Tahiti.

Port Authority: Port Autonome de Papeete, Motu Uta, BP 9164, 98715 Papeete; tel. 505454; fax 421950; e-mail portppt@mail.pf; internet www.portdepapeete.com; Harbour Master CLAUDE VIGOR; Port Man. BÉATRICE CHANSIN.

Agence Maritime Internationale de Tahiti: BP 274, Papeete, Tahiti; tel. 428972; fax 432184; e-mail amitahiti@mail.pf; services from Asia, the USA, Australia, New Zealand, American Samoa and Europe.

CGM Tour du Monde SA: 80 rue du Général de Gaulle, BP 96, Papeete; tel. 420890; fax 436806; shipowners and agents; freight services between Europe and many international ports; Dir HENRI C. FERRAND.

Compagnie Française Maritime de Tahiti: Immeuble Importex No. 45 Fare Ute, POB 368, 98713 Papeete; tel. 426393; fax 420617; e-mail taporo@mail.pf; Man. MORTON GARBUTT.

Compagnie Polynésienne de Transport Maritime: BP 220, Papeete; tel. 426240; fax 434889; e-mail aranui@mail.pf; internet www.aranui.com; Dir JEAN WONG.

EURL Transport Maritime des Tuamotu Ouest: BP 1816, 98713 Papeete; tel. 422553; fax 422557; Dir SIMÉON RICHMOND.

Leprado Valere SARL: POB 3917, Papeete; tel. 450030; fax 421049.

SARL Société de Transport Insulaire Maritime (STIM): BP 635, 98713 Papeete; tel. 452324; fax 452444; Dir ROLAND PAQUIER.

Société de Navigation des Australes: BP 1890, Papeete; tel. 509609; fax 420609.

CIVIL AVIATION

There is one international airport, Faaa airport, 6 km from Papeete, on Tahiti, and there are about 40 smaller airstrips. International services are operated by Air France, Air New Zealand, LAN-Chile, Air Outre Mer, Air Calédonie International, Corsair and Hawaiian Airlines (USA).

Service d'État de l'Aviation Civile: BP 6404, 98702, Faaa, Papeete; tel. 861000; fax 861009; e-mail buchwalter_axelle@seac.pf; internet www.seac.pf; Dir GUY YEUNG.

Air Moorea: BP 6019, Faaa; tel. 864100; fax 864269; e-mail freddy .chanseau@airmoorea.pf; internet www.airmoorea.com; f. 1968;

operates internal services between Tahiti and Moorea Island and charter flights throughout the Territory; Pres. MARCEL GALENON; Dir-Gen. FREDDY CHANSEAU.

Air Tahiti: BP 314, 98713 Papeete; tel. 864000; fax 864009; e-mail direction.generale@airtahiti.pf; internet www.airtahiti.pf; f. 1953; Air Polynésie 1970–87; operates domestic services to 39 islands; 718,728 passengers (2003); Chair. and CEO CHRISTIAN VERNAUDON; Gen. Man. MARCEL GALENON.

Air Tahiti Nui: Rue Paul Gauguin, BP 1673, 98713 Papeete; tel. 460202; fax 460290; e-mail fly@airtahitinui.pf; internet www.airtahitinui.com; f. 1996; commenced operations 1998; scheduled services to the USA, France, Japan and New Zealand; Chair. and CEO NELSON LEVY.

Tourism

Tourism is an important and developed industry in French Polynesia, particularly on Tahiti, and 212,692 people visited the Territory in 2003. In that year some 36.6% of arrivals were from the USA, 22.7% from France and 10.8% from Japan. There were a total of 3,650 hotel rooms in Tahiti in 2000. In that year the tourism industry earned an estimated 49,900m. francs CFP.

GIE Tahiti Tourisme: Immeuble Paofai, bâtiment D, blvd Pomare, BP 65, 98713 Papeete; tel. 505700; fax 436619; e-mail tahiti-tourisme@mail.pf; internet www.tahiti-tourisme.pf; f. 1966 as autonomous public body, transformed into private corpn in 1993; tourist promotion; CEO DANY PANERO.

Service du Tourisme: Fare Manihini, blvd Pomare, BP 4527, Papeete; tel. 476200; fax 476201; govt dept; manages Special Fund for Tourist Development; Dir CLARISSE GODEFROY.

Syndicat d'Initiative de la Polynésie Française: BP 326, Papeete; Pres. PIU BAMBRIDGE.

Defence

In August 2003 France maintained a force of 2,700 military personnel in the Territory, as well as a gendarmerie of 600. France began testing nuclear weapons at Mururoa Atoll, in the Tuamotu Archipelago, in 1966. The military presence has been largely connected with the Centre d'Expérimentation du Pacifique (CEP) and the Commission d'Energie Atomique (CEA). An indefinite suspension of tests was announced in mid-1993. In June 1995, however, the French Government announced its decision to resume nuclear testing at Mururoa Atoll. The final test was conducted in January 1996. The military budget for 2002 was 33,926m. francs CFP.

Education

Education is compulsory for eight years between six and 14 years of age. It is free of charge for day pupils in government schools. Primary education, lasting six years, is financed by the territorial budget, while secondary and technical education are supported by state funds. In 2002 there were 45 kindergartens, 182 primary schools and 24 centres for young adolescents; there were 15,136 children enrolled at kindergarten and 28,988 pupils in the primary sector. Secondary education is provided by 10 public lycées, 24 public high schools and 11 private or church schools. In the same year a total of 29,466 pupils attended general secondary schools in , while 3,032 secondary pupils were enrolled at vocational institutions. The Territorial Government assumed responsibility for secondary education in 1988. Technical and professional education includes eight technical institutions, a tourism training programme, preparation for entrance to the metropolitan Grandes Ecoles, a National Conservatory for Arts and Crafts and training centres for those in the construction industry, health services, traditional handicrafts, primary school teaching and social work. The French University of the Pacific was established in French Polynesia in 1987. In 1999 it was divided into two separate branches, of which the University of French Polynesia is now based in Papeete. In 1995 the Papeete branch had 50 teachers. In 2003/04 some 2,343 students were enrolled at the University. Total public expenditure on education in the Territory was 40,300m. francs CFP in 1999.

NEW CALEDONIA

Physical and Social Geography

New Caledonia is a French Overseas Country located in the Pacific Ocean, lying south and slightly west of Vanuatu, some 1,500 km east of Australia and 1,700 km north of New Zealand. It comprises one large island and several smaller islands, with a total land area of 18,575 sq km (7,172 sq miles). The main island, New Caledonia (la Grande-Terre), is long and narrow, and has a total area of 16,372 sq km (6,321 sq miles); its island chain is continued by the Bélep Archipelago at the north end of the island, the Isle of Pines and the Huon Islands, a group of four barren coral islets, in the south. The coralline Loyalty Islands (including Lifou, Maré and Ouvéa) lie to the north-east of Nouméa in a chain that runs parallel to the main island. The third main group of islands is the uninhabited, well-wooded Chesterfield Islands, which lie about 400 km north-west of the main island. The isolated island of Walpole, a limestone coral island, is to the west of the main island of New Caledonia and the Loyalty Islands. The sporadically active volcanic Matthew and Hunter Islands (also claimed by Vanuatu) lie east of Walpole Island. Rugged mountains, rising to 1,628 m at Mt Panie, divide the west of the main island from the east, and there is little flat land. The climate is generally a mild one, with an average temperature of about 24°C (75°F) and a rainy season between December and March. The average rainfall in the east of the main island is about 2,000 mm (80 ins) per year, and in the west about 1,000 mm (40 ins). French is the official language and the mother tongue of the Caldoches (French settlers). There are some 28 Melanesian languages spoken by the indigenous Kanaks, divided into four main groups. The four most widely spoken and taught forms are Drehu (Lihou), Nengone (Maré), Aji'e (Houaïlou) and Paicï (Poindimié). Other immigrants speak Polynesian and Asian languages.

The population was 196,836 at the 1996 census, comprising Melanesians (44%), Europeans, mainly French (43%), Wallisians and Tahitians (Polynesian) (9%) and others, mainly Indonesians and Vietnamese (3%). In 1996 some 68% of the population lived in the Province of the South, 21% in the Province of the North and 11% in the Province of the Loyalty Islands. At the 1996 census there were 341 tribes (which have legal status under a high chief—living in reserves which covered 21% of total land), representing about 28.7% of the population. An estimated 55% of the population are Roman Catholics, and there is a substantial Protestant minority. The capital is Nouméa, on the main island of New Caledonia. The total population was estimated at 220,467 at mid-2003.

History

New Caledonia became a French possession in 1853, when the island was annexed as a dependency of Tahiti. In 1860 a separate administration was established, and in 1885 a Conseil Général was elected to defend the local interests before metropolitan France. From 1887 two separate administrations existed, for Melanesian Kanaks and for expatriates, until New Caledonia became an Overseas Territory of the French Republic in 1946. A Territorial Assembly was established in 1956, comprising 30 members elected by universal adult suffrage. There is a governor, the High Commissioner, who retains control of all French national departments and the gendarmerie, but otherwise there is substantial self-government.

Early European settlers on New Caledonia quickly set about acquiring Kanak land, which involved iniquitous legislation and numerous rebellions by Kanaks, the last of which took place in 1917. Further seizures of land followed as punishment. Cattle grazing, practised by the Europeans, disrupted indigenous agriculture, and today large areas of formerly irrigated taro terraces lie abandoned.

The French Government dismissed the Council of Government in March 1979, following its failure to support a proposal for a 10-year contract between France and New Caledonia because the plan did not acknowledge the possibility of New Caledonian independence. The Territory was then placed under the direct authority of the High Commissioner. A new electoral law, recommended by the French Minister for Overseas Departments and Territories, ensured that minor political parties were not represented in the Assembly following the July 1979 general election. Consequently, the numerous small pro-independence groups with largely Kanak supporters were excluded, and the elections resulted in the two new 'national' federations of parties loyal to France (Rassemblement pour la Calédonie dans la République—RPCR—and Fédération pour une Nouvelle Société Calédonienne—FNSC) winning 22 of the 36 seats in the Assembly.

Following the election of François Mitterrand as President of France, tension grew sharply after the assassination in September 1981 of Pierre Declercq, Secretary-General of the pro-independence party, Union Calédonienne (UC). In December, recognizing the need for major reforms, the French authorities outlined their most immediate aims, including equal access for all New Caledonians to positions of authority, land reforms and the fostering of Kanak cultural institutions. To help effect these reforms, the French Government simultaneously announced that it would rule by decree for a period of a year. In 1982 the FNSC joined with the recently established Front Indépendantiste (FI) in an attempt to form a government that was more favourable to the proposed reforms. In November 1983 the French Government proposed a five-year programme, which provided for an Act of self-determination in 1989, with independence as one of the options to be offered. The French proposals were opposed in New Caledonia, both by parties in favour of independence and by those against, and it was rejected by the Territorial Assembly in April 1984. Sittings of the Territorial Assembly had meanwhile been boycotted by the FI, rendering the Assembly unable to function. Nevertheless, a Statute of Autonomy was approved by the French National Assembly in September 1984. Under the provisions of the Statute, the Territorial Council of Ministers was responsible for many internal matters of government, its President henceforth being an elected member instead of the French High Commissioner; a second legislative chamber, with the right to be consulted on development planning and budgetary issues, was created at the same time.

All of the main parties seeking independence (except the Libération Kanake Socialiste—LKS—party, which left the FI) boycotted elections for a new Territorial Assembly in November 1984. Following the subsequent dissolution of the FI, these parties formed a new movement, the Front de Libération Nationale Kanake Socialiste (FLNKS); its congress instituted a 'provisional government', headed by Jean-Marie Tjibaou (of the UC, the dominant party in the FLNKS), on 1 December. The elections to the Territorial Assembly attracted only 50.1% of the electorate, and the anti-independence RPCR won 34 of the 42 seats. An escalation of violence by both Kanaks and Caldoches (French settlers) began in November. In December this culminated in the burning of farms belonging to French settlers, the murder of three settlers by pro-independence activists and the deaths of 10 Kanak activists (two of whom were Tjibaou's brothers), who were shot dead by *métis* settlers at Hienghène, in the far north of the Territory.

In January 1985 Edgard Pisani, the new High Commissioner, presented a plan by which the Territory might become independent 'in association with' France, subject to the result of a referendum. Kanak groups opposed the plan, insisting that the indigenous population be allowed to determine its own fate. At the same time, the majority of the population, which supported the RPCR, demonstrated against the plan and in favour of remaining within the French Republic. A resurgence of violence followed Pisani's announcement, and a state of emergency was declared after two incidents in which a leading member of the FLNKS was shot dead by the Groupe d'Intervention de la Gendarmerie Nationale (GIGN) and the son of a French settler was killed by Kanak activists.

In April 1985 the French Prime Minister, Laurent Fabius, announced a new plan for the future of New Caledonia, whereby the referendum on independence, formerly scheduled for July 1985, was deferred until an unspecified date not later than the end of 1987. Meanwhile, the Territory was to be divided into four Regions, each to be governed by its own elected autonomous Council, which would have extensive powers in the spheres of planning and development, education, health and social services, land rights, transport and housing. The elected members of all four Councils together would serve as regional representatives in a Territorial Congress (to replace the Territorial Assembly). The new set of proposals (known as the 'Fabius plan') was well received by the FLNKS, which voted in favour of participating in the regional elections. The RPCR, however, condemned the plan, and the proposals were rejected by the predominantly anti-independence Territorial Assembly in May 1985. Nevertheless, the necessary legislation was approved by the National Assembly in Paris in July, and the Fabius plan came into force.

The parties opposing independence eventually agreed to participate in the elections, which were held in September 1985. As expected, only in the Southern Region around Nouméa, where most of the population are non-Melanesian, was an anti-independence majority returned. However, the Kanaks, in spite of their majorities in the three non-urban Regions, would be in a minority in the Territorial Congress.

The FLNKS boycotted the French legislative elections of March 1986, in which only about 50% of the New Caledonia electorate participated. In May the French Council of Ministers approved a draft law providing for a referendum to be held in New Caledonia within 12 months, whereby a choice would be offered between independence and a further extension of regional autonomy. In December the French Government announced that all those who had been resident in New Caledonia for at least three years would be eligible to participate in the referendum. This concession was rejected by FLNKS leaders, who maintained that, unless eligibility were confined to the Kanak community, the result of such a referendum would be unfairly weighted in favour of the French and other non-indigenous groups (who together outnumbered the Melanesians), and would be unrepresentative of the wishes of the indigenous population.

In December 1986, despite vigorous French diplomatic opposition, the UN General Assembly voted to reinstate New Caledonia on the UN list of non-self-governing territories, thereby affirming the population's right to self-determination. In January 1987 the French Government expelled the Australian Consul-General from New Caledonia, in retaliation against the allegedly central role that he had played in determining the UN decision.

In March 1987 the UN Committee on Decolonization issued a declaration in support of the FLNKS, urging France to make preparations for New Caledonia 'freely and truly' to exert its right to independence. However, in April, despite continuing opposition from President Mitterrand, the French Government's legislation providing for a referendum was approved in the National Assembly. In May, Jean-Marie Tjibaou denounced the referendum and urged a boycott of it. The South Pacific Forum also condemned the referendum, maintaining that it did not satisfy the UN code of principles and practices for decolonization. In June the leader of the LKS, Nidoïsh Naisseline (a high chief of Maré island), stated that his party had also decided to recommend that its supporters boycott the referendum.

The referendum was held on 13 September 1987. Votes cast in favour of New Caledonia's continuation as part of the French Republic numbered 48,611 (98.3% of the total), while those in favour of independence numbered only 842 (1.7%). The level of participation, at 58.9% of the registered electorate, was higher than had been expected. The rate of abstention in constituencies inhabited by a majority of Kanaks (where, in most cases, at least 90% of the electorate abstained) was significantly higher than in those inhabited chiefly by the Caldoches.

In October 1987 seven pro-French loyalists were acquitted on a charge of murdering 10 Kanak separatists in 1984 (see above). Jean-Marie Tjibaou, who reacted to the ruling by declaring that his followers would have to abandon their stance of pacifism, and his deputy, Yeiwéné Yeiwéné, were indicted for 'incitement to violence'. In April 1988 four gendarmes were killed, and 27 held hostage in a cave on the island of Ouvéa (Uvéa), by supporters of the FLNKS. Two days later Kanak separatists prevented about one-quarter of the Territory's polling stations from opening, when local elections, scheduled to coincide with the French presidential election, were held. The elections were boycotted by the FLNKS. Although 12 of the gendarmes taken hostage were subsequently released, six members of a French anti-terrorist squad were also captured. When negotiations to release those held were unsuccessful, French security forces stormed the cave, freeing the hostages but leaving 19 Kanaks and two gendarmes dead. In the aftermath of the siege, it was alleged that three Kanaks had been executed or left to die after being arrested.

In June 1988 talks were convened between the French Prime Minister, Michel Rocard, the leader of the RPCR, Jacques Lafleur, and Jean-Marie Tjibaou, as the leader of the FLNKS. Agreement was reached to transfer the administration of New Caledonia to Paris for 12 months and to separate the Territory into three Provinces, prior to a territorial plebiscite to be held in 1998. This agreement became known as the Matignon Accord. The programme was presented to the French electorate in a referendum, held on 6 November 1988, and approved by a majority (of 80%), although an abstention rate of 63% was recorded. The programme was approved by a 57% majority in New Caledonia, where the rate of abstention was 37%. In November, under the terms of the agreement, 51 separatists were released from prison, including 26 Kanaks implicated in the incident on Ouvéa.

An increase in support for the pro-independence parties was demonstrated in the municipal elections of March 1989, in which they gained control of 20 out of 32 municipal districts (communes). On 4 May, however, Jean-Marie Tjibaou and Yeiwéné Yeiwéné were assassinated by Djubelly Wéa, a former pastor and a member of the Front Uni de Libération Kanake (FULK), a party that had opposed the participation of the FLNKS in the Matignon Accord with France. In June elections were held for the three provincial councils. The FLNKS gained control of the North Province and the Province of the Loyalty Islands, while the RPCR won control in the South Province. In accordance with the Matignon Accord, direct rule by France over

New Caledonia was discontinued on 14 July 1989 and the Provincial Councils gained a degree of limited autonomy.

In April 1991 the LKS announced its intention to withdraw from the Matignon Accord, accusing the French Government, as well as several Kanak political leaders, of seeking to undermine Kanak culture and tradition. The RPCR's policy of encouraging the immigration of skilled workers from mainland France and other European countries continued to be a source of conflict between the conservative coalition and the FLNKS.

At elections for the Representative to the French Senate in September 1992, the RPCR's candidate, Simon Loueckhote, narrowly defeated Roch Wamytan, the Vice-President of the FLNKS.

Debate concerning the political future of the Territory continued in 1994. In October Jacques Lafleur, proposed that New Caledonia abandon the planned 1998 referendum on self-determination, in favour of a 30-year agreement with France, similar to the Matignon Accord, but with provision for greater autonomy in judicial matters. The UC, however, rejected the proposal and reiterated its demand for a gradual transfer of power from France to New Caledonia, culminating in a return to sovereignty in 1998.

French presidential elections took place in April–May 1995. During the second round of voting in the Territory (in which only 35% of the electorate participated), the socialist candidate, Lionel Jospin, received 25.9% of the total votes, while the candidate of the Gaullist Rassemblement pour la République (RPR), Jacques Chirac, won 74.1%. (Chirac was elected to the presidency with 52.6% of votes cast throughout the republic.)

At provincial elections in July 1995 the RPCR remained the most successful party, although its dominance was reduced considerably. The FLNKS remained in control of the Provinces of the North and the Loyalty Islands, while the RPCR retained a large majority in the Province of the South. The RPCR retained an overall majority in the Territorial Congress, while the FLNKS remained the second largest party. Considerable gains were made by a newly formed party led by a Nouméa businessman, Une Nouvelle-Calédonie pour Tous (UNCT), which secured seven seats in the Territorial Congress and seven seats in the Provincial Assembly of the South. An estimated 67% of the electorate participated in the elections. However, a political crisis subsequently arose as a result of the UNCT's decision to align itself with the FLNKS, leaving the RPCR with a minority of official positions in the congressional committees. Lafleur would not accept a situation in which the UNCT appeared to be the dominant party in the chamber, and Pierre Frogier, the RPCR's President of Congress, refused to convene a congressional sitting under such circumstances. The deadlock was broken only when the FLNKS released a statement in October, reiterating the importance of the relationship between the FLNKS and the RPCR as signatories of the Matignon Accord, and proposing the allocation of congressional positions on a proportional basis.

Negotiations between the French Prime Minister, Alain Juppé, the Minister for Overseas Territories and delegations from the FLNKS and the RPCR were held in Paris in late 1995. It was agreed that further discussions would take place in early 1996, involving representatives from numerous interest groups in the Territory, to examine the possibility of achieving a consensus solution on the future of the islands. Thus, the major political groups in New Caledonia sought to achieve a consensus solution on the Territory's future, which could be presented to the electorate for approval in the 1998 referendum. It was widely believed that this was preferable to a simple 'for' or 'against' vote on independence, which would necessarily polarize the electorate and create a confrontational political climate.

Elections to the French National Assembly in May–June 1997 were boycotted by the pro-independence FLNKS and LKS, resulting in a relatively low participation rate among the electorate. Jacques Lafleur and Pierre Frogier, both candidates of the RPCR, were elected to represent the Territory.

Intensive negotiations involving the RPCR, the FLNKS and the French Government took place throughout early 1996. The process, however, was disrupted by a dispute over the disclosure of confidential information regarding the talks to the French press (responsibility for which was later admitted by Jacques Lafleur) and the belief by pro-independence leaders that France had apparently reneged on its promise to consider all available options for the Territory's political future by discounting the possibility of outright independence. France's refusal to grant final approval for a large-scale nickel smelter project in the North Province (see Economy) until the achievement of consensus in the discussions on autonomy prompted accusations of blackmail from several sources within the Territory and fuelled suspicions that metropolitan France would seek to retain control of the islands' valuable mineral resources in any settlement on New Caledonia's future status. The issue proved to be a serious obstacle in the negotiations and resulted in the virtual cessation of discussions between the two sides during the remainder of 1996. The FLNKS argued that the smelter project should be administered by local interests, consistent with the process of reallocating responsibility for the economy from metropolitan France to the Territory as advocated in the Matignon Accord. Their demands were supported by widespread industrial action in the mining sector during late 1996.

In February 1997 the French Minister for Overseas Territories travelled to the Territory in an attempt to achieve an exchange agreement on nickel between the Société Minière du Sud-Pacifique (SMSP), controlled by the North Province, and the French-owned company, with numerous interests in the islands, Société Le Nickel (SLN). The minister failed to resolve the dispute during his visit; however, at the end of the month, in a complete reversal of its previous position, the French Government announced its decision not to compensate SLN for any losses incurred. The decision provoked strong criticism from SLN and Eramet, the French mining conglomerate, of which SLN is a subsidiary, and attracted protests from shareholders and employees of the company. During March large-scale demonstrations were held by the UC and the pro-independence trade union Union Syndicale des Travailleurs Kanak et des Exploités (USTKE), in support of the SMSP's acquisition of the smelter. Meanwhile, another trade union, Union des Syndicats des Ouvriers et Employés de Nouvelle-Calédonie (USOENC—which represented a high proportion of SLN employees), organized a protest rally against the unequal exchange of mining sites. Frustrated at SLN's seemingly intransigent position in the negotiations, the FLNKS organized protests and blockades at all the company's major mining installations. Supporters of the pro-independence organization also restricted shipments of ore around the Territory. Consequently, four mines were forced to close, while a 25% reduction in working hours was imposed on 1,500 mine workers, prompting protests by SLN employees and demands from USOENC that the blockades be removed. In January 1998 Roch Wamytan, now leader of the FLNKS, urged the French Prime Minister, Lionel Jospin, to settle the dispute by the end of the month in order that official negotiations on the political future of the Territory, in preparation for the referendum, might begin. The position of the FLNKS had been somewhat undermined by the decision, in the previous month, of a breakaway group of pro-independence politicians (including prominent members of the UC, the Parti de Libération Kanak (PALIKA), the UMP and the LKS) to begin negotiations with the RPCR concerning the dispute. These moderate supporters of independence formed the Fédération des Comités de Coordination des Indépendantistes (FCCI) in 1998. A draft agreement on the exchange of deposits was signed by SLN and the SMSP on 1 February, and in April a deal between the SMSP and a Canadian company for the establishment of the nickel smelter was concluded. Unless construction of the smelter had begun by the end of 2005, control of the nickel deposits transferred to the SMSP was to revert to SLN. Meanwhile, the French Government agreed to pay compensation of about 1,000m. French francs to Eramet for the reduction in the company's reserves. Following the signing of the nickel agreement, tripartite talks on the constitutional future of New Caledonia resumed in Paris in late February 1998. Discussions between representatives of the French Government, the FLNKS and the RPCR continued in March, despite a temporary boycott of the talks by the RPCR delegation, which requested the inclusion of various other minor political groups at the negotiations, including the FCCI. On 21 April, following a final round of talks in Nouméa, an agreement was concluded by the three sides. The agreement, which became known as the Nouméa Accord, postponed the referendum on independence for a period of between 15 and 20 years but provided for a gradual transfer of powers to local institutions. The document also acknowledged the negative impact of many aspects of French colonization on the Territory and emphasized the need for greater recognition of the importance of the Kanak cultural identity in the political development of the islands. The Nouméa Accord was signed on 5 May during the two-day inauguration of the Centre Culturel Tjibaou, a centre for Kanak art.

On 6 July 1998 the French Parliament (the National Assembly and the Senate) voted in favour of adopting the proposed changes regarding the administration of New Caledonia, which were to be incorporated in an annex to the French Constitution. In the following month the French Minister for Overseas Territories returned to New Caledonia for discussions on draft legislation for the devolution process. In September a new political group, called the Comité Provisoire pour la Défense des Principes Républicains de la Nouvelle-Calédonie Française, was formed in opposition to the Nouméa Accord, with support from members of the Front National and other right-wing parties. The UNCT, which was dissatisfied with several aspects of the Accord, also urged its supporters to vote against the agreement.

The Nouméa Accord was presented to the electorate of New Caledonia in a referendum on 8 November 1998 and was decisively approved, 71.9% of votes cast being in favour of the agreement. A participation rate of 74.2% was recorded. The Province of the North registered the strongest vote in favour of the accord (95.5%), while the Province of the South recorded the most moderate level of approval (62.9%). In late December the French National Assembly unanimously approved draft legislation regarding the definitive adoption of the accord. The Senate similarly approved the legislation in February 1999. In March of that year, however, the French Constitutional Council declared its intention to allow any French person who had resided in New Caledonia for 10 years or more to vote in provincial elections. This decision was criticized by Roch Wamytan as well as by politicians in the French National Assembly and Senate, who claimed that this was in breach of the Nouméa Accord, whereby only those residing in New Caledonia in 1998 would be permitted to vote in provincial elections. Pro-independence groups threatened to boycott the elections (to be held in May). In response to this, the French Government announced that the Accord would be honoured, claiming that the Constitutional Council had breached the Nouméa Accord, and stating that this contravention would be rectified. In June the French Council of Ministers announced that it had drafted legislation restricting eligibility for voting in provincial elections and in any future referendums on sovereignty, to those who had been eligible to vote in the November 1998 referendum on the Nouméa Accord, and to their children upon reaching the age of majority. This decision was condemned by the right-wing Front National and by the RPCR. Constitutional amendments on the matter were still to be passed into law.

In compliance with the Nouméa Accord, an agreement was concluded in February 1999 to enable the transfer of 30% of SLN's share capital (and 8% of Eramet's capital) to a newly created company representing local interests, the Société Territoriale Calédonienne de Participation Industrielle (STCPI), to be owned by the development companies of the three New Caledonian provinces. In this way, it was hoped, the islands' people would be more closely associated with the management of New Caledonia's principal source of wealth. However, agreement on the establishment of the STCPI was repeatedly delayed, owing to disagreements between the FLNKS and the RPCR. Finally, in July 2000, after two years of negotiations, New Caledonia's political leaders signed an agreement on the formation of the STCPI. The new company was to be owned equally by PROMOSUD (representing the South Province) and NORDIL (combining the interests of the North Province and the Loyalty Islands). The French Government welcomed the agreement and announced its commitment of 1,040m. French francs to fund the transfer of shares in SLN and Eramet by means of a loan to the provinces from the Agence Française de Développement. In August the assemblies of the three provinces endorsed the plan, and in early September the STCPI was formally established. At its first meeting the new company confirmed Raphaël Pidjot, the managing director of the SMSP, as chairman. Later in September shares in SLN and Eramet were transferred to the STCPI, thus reducing Eramet's interest in SLN from 90% to 60%. Representatives of the STCPI joined SLN's board of directors in October 2000. In November, however, Raphaël Pidjot and other officials of the SMSP and Falconbridge mining companies died in a helicopter accident, but these companies were to continue their joint project for a nickel operation in the North, to open in 2004.

At the general election held on 9 May 1999, no party gained an absolute majority. However, Jacques Lafleur's anti-independence RPCR won 24 of the 54 seats in the Congress, and formed a coalition with the recently established FCCI and, on an informal level, with the Front National, thus creating an anti-independence block of 31 seats in the Congress. The pro-independence FLNKS won 18 seats. Simon Loueckhote was re-elected as President of the Congress in late May. Results of the elections in the Loyalty Island Province were officially challenged by the moderate pro-independence LKS and FCCI parties, as well as by the RPCR, following the issue by the electoral commissioner for the Province of a report claiming that a large number of irregularities had occurred. A new election was held in June 2000, at which a coalition of the RPCR, FCCI, LKS and FULK obtained 44.8% of votes and six seats. The FLNKS obtained 37.3% of votes and six seats, and PALIKA 17.8% of votes and two seats. The composition of the Congress therefore remained unchanged. Robert Xowie was re-elected as President of the Province.

On 28 May 1999 the Congress elected Jean Lèques as the first President of the Government of New Caledonia, under the increased autonomy terms of the Nouméa Accord. The new Government was elected on the basis of proportional representation and replaced the French High Commissioner as New Caledonia's executive authority. The election of Léopold Jorédié, leader of the FCCI, as Vice-President, was denounced by the FLNKS, which argued that, as the second largest party in Congress and as joint negotiators in the Nouméa Accord, the post should have gone to its leader, Roch Wamytan. In the formation of the Government the RPCR-FCCI coalition was awarded seven positions and the FLNKS four.

In October 1999 Wamytan threatened to withdraw from the Government, in protest at the lack of co-operation among parties. He claimed that sections of the Nouméa Accord requiring power to be distributed among the various political parties had not been observed (see above). In December Vice-President Léopold Jorédié received a one-year suspended prison sentence following accusations of misuse of public funds. Jorédié was charged with illegally obtaining grants totalling an estimated 5.5m. francs CFP, for the benefit of his son.

In July 2000 concern was expressed by the French Government over the implementation of the 'collegiality' clause in the Nouméa Accord, which provides for greater political co-operation among parties. Repeated threats by the FLNKS to withdraw from the Government because of its discontent with the RPCR's lack of power-sharing led to the establishment of an agreement between New Caledonia and the French State detailing the role of the two Governments in areas such as education and foreign policy; the role of the traditional chiefs in legal matters was also specified. In August Jacques Lafleur threatened to resign from his seat in the French National Assembly, following the upholding of a ruling convicting him of slander against Bruno Van Peteghem, an activist opposing construction plans for a complex near Nouméa. However, in early September Lafleur retracted his threat following pleas by RPCR members.

In August 2000 the Government introduced a 'social pact', aimed at addressing social issues and ongoing industrial conflicts. It proposed an increase in the minimum wage and suggested local employment priorities, subject to ratification by employers' unions in September.

In November 2000 Roch Wamytan was re-elected President of the FLNKS, at the party's annual conference, and was also narrowly re-elected leader of the UC. At the same time, the UC and the FLNKS both officially recognized a new pro-independence party, the group UC du Congrès (formed by a breakaway faction of the UC).

In March 2001 the municipal elections confirmed the predominance of the RPCR in the South, when it took 39 of the 49 seats in Nouméa. However, in the country as a whole the RPCR controlled only 14 of the 33 municipalities in New Caledonia, against the 19 held by pro-independence parties, principally the UC, PALIKA, LKS and FLNKS. The FLNKS won a majority in the North and took all three communes in the Loyalty Islands. Jean Lèques resigned as President in early April and was replaced by fellow RPCR politician, Pierre Frogier. Déwé Gorodey of the FLNKS was elected Vice-President. The election to the two most senior posts took place after the Congress had elected an 11-member Government, consisting of seven RPCR-FCCI coalition members, three from the FLNKS and one from the UC.

In June 2001, as a result of the failure to resolve industrial action over the dismissal of 12 employees from public works company Lefèbvre Pacifique, the USTKE extended its strike to 24-hour blockades of supermarkets, petrol stations, state radio and television companies, schools, the port and airport. In July 2001 a 100,000 francs CFP monthly minimum wage (as provided for in the 'social pact' brokered by Jacques Lafleur in September 2000) was implemented. Further strike action affected the tourist industry when, in December, USTKE launched a strike and occupation of the Château Royal complex, following Club Med's announcement of the Nouméa holiday resort's closure. The USTKE was demanding compensation for the 92 members of staff who were to be made redundant.

In October 2001 the French Council of State ruled that the 11th seat in the New Caledonian Government had been incorrectly allocated to the FLNKS following the local elections of April 2001. As a result, FCCI leader Raphaël Mapou replaced Aukusitino Manuohalalo of the FLNKS as Minister for Social Security and Health; Roch Wamytan threatened to resign from the Government in protest. In the same month, however, Wamytan was replaced as President of the UC by his deputy, Pascal Naouna; many members believed that Wamytan's dual role as President of both the UC and FLNKS was weakening the party. Then, in November, Wamytan lost the presidency of the FLNKS following a leadership struggle between its two main factions, the UC and PALIKA. The political bureau of the FLNKS was to lead the party until its internal disputes were resolved. Wamytan was subsequently replaced as Minister for Customary Affairs and Relations with the Senate by Mapou, necessitating a minor government reshuffle; Manuohalalo returned as Minister for Social Security and Health.

Meanwhile, a long-standing dispute over the ownership of a Wallisian settlement—Ave Maria, near Nouméa—had led to intense fighting between Wallisian and Kanak communities from December 2001, resulting in one Kanak fatality in February 2002. The Kanak community in neighbouring Saint-Louis had demanded the departure of all Wallisians from Ave Maria by March. The French High Commissioner mediated at several meetings in January 2002 in an attempt to resolve the issue, leading to the Wallisian spokesman's suggestion that his community might be prepared to leave Ave Maria, provided it was offered an alternative 25 ha of land on which to resettle. Four working groups were established in April to rehouse Wallisians, to improve the area's public facilities and to help reintegrate youths who had abandoned the education system. In early June a Wallisian man was ambushed by gunmen and shot dead in Ave Maria. Of a total of 127 families, some 60 had been relocated to social housing by June. In November the Kanak and Wallisian communities signed an accord, mediated by the High Commissioner, to end the violence, which had claimed three lives in total. About one-half of the Wallisian families present in Saint-Louis had left by the end of 2002. Under the terms of the accord, the remainder were to be resettled in 2003. In June of that year some 100 Wallisians were forced to flee Ave Maria and take up residence in temporary housing, following another outbreak of racial violence with the Kanak community. By mid-September the remaining 30 Wallisian families had left the area to be resettled in and around Nouméa. The leader of the opposition Alliance pour la Calédonie, Didier Leroux, condemned the failure of the French Government to intervene as tantamount to 'ethnic cleansing'.

In the first round of the presidential election, held on 21 April 2002, Jacques Chirac obtained 48.4% of the vote on the islands, followed by Jospin, who won 22.4%, and was eliminated nationally. In the second round, held on 5 May, Chirac overwhelmingly defeated Jean-Marie Le Pen of the extreme right-wing Front National, winning 80.4% of New Caledonian votes. Elections to the French National Assembly were held in June 2002. The UC and PALIKA had yet to agree upon their choice of President for the FLNKS. The UC therefore refused to take part in the elections and urged its supporters to abstain from the poll, thereby depriving the President of PALIKA, Paul Néaoutyine, of any chance of election to the National Assembly. Jacques Lafleur was thus re-elected as New Caledonia's representative in the French legislature, winning 55.74% of the votes cast in the second round of polling, and Pierre Frogier, who won 55.71%, was thus returned as New Caledonia's second deputy to the National Assembly in Paris. In December Lafleur announced his 'progressive retirement' from politics, although he gave no precise date.

Following two resignations in July 2002, a series of ministerial reorganizations took place. The Minister for Labour, Public Service and Educational Training resigned, citing personal reasons, and was replaced by a fellow RPCR member, Georges Naturel. In late July Raphaël Mapou resigned, following his outspoken criticism of the tendering of mining and prospecting rights in the Southern Province to a Canadian mining company. He was replaced by Corinne Fuluhea of the RPCR. Shortly afterwards, her portfolio was altered to that of Professional Training. In November the sole UC member of the Government, Gerald Cortot, resigned, prompting the immediate dissolution of the Government, as stipulated in the Nouméa Accord. Later that month the size of the Council of Ministers was reduced to 10 members, and the Congress appointed a new administration, with Frogier reappointed as President; the incoming Government contained seven members of the RPCR-FCCI coalition, two from the FLNKS and one from the UC. In May 2003 it was announced that New Caledonia (together with French Polynesia) would be granted one additional seat in the French Senate, probably effective from the next senatorial elections in 2007. Also in May 2003, at a congress of the FLNKS, officials of the party reportedly claimed that the terms of the Nouméa Accord were not being fully observed by the French Government. In July the USTKE called a general strike to coincide with a visit to New Caledonia by President Jacques Chirac. A rally organized by the mainly Kanak trade union was attended by 2,000 protesters. Chirac's four-day visit also provoked demonstrations by several hundred members of the UC. Police used tear gas in an attempt to disperse the pro-independence protesters, who had gathered near an official ceremony in honour of the visiting head of state. Kanak representatives, meanwhile, expressed dismay at their exclusion from meetings with the French President. During his visit Chirac refrained from making any direct comment on the status of New Caledonia.

Legislative elections took place on 9 May 2004. The RPCR lost its majority in both the Congress of New Caledonia and in the Province of the South. The results demonstrated an increase in support for the pro-independence movements, notably the recently formed Avenir Ensemble. Each Provincial Assembly in turn elected its President: Philippe Gomès of the Avenir Ensemble became President of the southern province, replacing Jacques Lafleur; Paul Néaoutyine of the UNI-FLNKS was re-elected in the northern province; and Néko Hnépeune, also of the UNI-FLNKS, was elected President in the Loyalty Islands. At its inaugural meeting in late May the incoming Congress elected Harold Martin of the Avenir Ensemble as its President and began the process of appointing a new Government. In early June Marie-Noëlle Thémereau of the Avenir Ensemble and Déwé Gorodey of the UNI-FLNKS were respectively nominated President and Vice-President. However, the Government collapsed within hours following the resignation of three UMP cabinet ministers, who claimed that they were entitled to four seats under power-sharing terms set out in the Nouméa Accord. Congress granted the UMP the seats but with the result that the decision of the Council of Ministers on its leadership reached a stalemate. In late June, following negotiations that also involved the French High Commissioner in New Caledonia and discussions in Paris between Harold Martin and the French Government, a new vote returned Thémereau and Gorodey to their elected posts. The incoming Government endorsed a code of conduct that emphasized the importance of consensus-building among the parties and 'collegiality' in the decision-making process. In her inaugural address to Congress in mid-August President Thémereau announced the Government's commitment to economic and social reform. Her administration would seek to raise the minimum wage, replace the existing import tax with value-added tax (VAT), and promote local business and indigenous Kanak culture.

In late July 2004 USTKE members blockaded the harbour wharf in protest at moves by stevedoring companies to consolidate cargo-handling activities, which it was believed would lead to redundancies. The police obtained a legal injunction to lift the blockade after five days, but in response the USTKE mobilized some 5,000 workers in a 'roving' general strike: pickets disrupted freight services at the wharf and airport; refuse collections were prevented and the RFO radio and television network was unable to broadcast. Nickel-mining companies were affected by the action, as was the principal flour mill, leading to a shortage of bread. Blockades of fuel depots forced the closure of service stations and brought school bus services to a standstill. After three days the cargo-handling agreement was reversed, and the USTKE withdrew its pickets in early August. However, in an unrelated strike over contractual terms, the USTKE forced RFO off the air again in mid-August; the strike action continued in September, during which time it spread to affect *Les Nouvelles Calédoniennes*, halting production of the newspaper for over two weeks.

In September 2004 politicians from across the political spectrum were urging a boycott of a forthcoming census on the grounds that it would not record individuals' ethnic identity. During his visit to New Caledonia in July 2003 President Chirac had declared his opposition to the inclusion of such details, taking the position that France recognized French citizens without regard to specific ethnicity. In September 2004 Congress voted to extend the census period by one week to accommodate an additional survey of ethnicity at a cost of some US $1.5m.; the UC refused to participate in the vote, stating that the survey was a waste of public funds.

In August 2001 the Government hosted the 14th Melanesian Spearhead Group (MSG) summit. The organization was established in 1988 as a support movement for the independence of New Caledonia and has developed into a body promoting economic and cultural exchange in the Melanesian archipelago; members are Fiji, Papua New Guinea, Solomon Islands, Vanuatu and the FLNKS. Discussions were held regarding eventual New Caledonian membership.

Economy

In 2000, according to World Bank estimates, New Caledonia's gross national product (GNP) at average 1998–2000 prices totalled US $2,989m., equivalent to $14,050 per head (or $21,960 per head on an international purchasing-power parity basis). During 1990–2000 it was estimated that GNP per head increased, in real terms, at an average annual rate of 0.7%. Over the same period the population rose by an average annual rate of 2.4%. During 1990–2000 New Caledonia's gross domestic product (GDP) increased, in real terms, at an average annual rate of 1.6%. GDP declined by 3.2% in 1998, but increased by 0.9% in 1999 and by 2.1% in 2000.

Since the first local discovery of nickel ore in 1864, and of cobalt and chromium ore in 1875, New Caledonia has been strongly influenced by the presence of large-scale mining enterprises. Large numbers of New Hebrideans (ni-Vanuatu), Japanese, Vietnamese and Javanese were imported to work the mines. After 1946 the indigenous Melanesian Kanaks began to move into the towns and to the mines, where heavy capitalization increased productivity. Further expansion absorbed more labour from rural areas, and the high level of earnings by nickel workers threatened all forms of rural production and eliminated chromium and cobalt mining, removing almost all forms of non-urban employment, particularly in the northern part of the island. Chromium mining was revived in the mid-1980s at Tiebaghi in the North Province; however, the exhaustion of reserves resulted in the closure of the mine in 1990. The discovery of large deposits of chromium in the South Province in 1992 prompted suggestions of a revival of the industry.

New Caledonia possesses the fifth largest nickel deposits in the world, accounting for about 8% of global reserves. It has the fourth largest nickel production in the world and was the largest ferro-nickel producer in 2002. In 1999 some 56,481 metric tons of ferro-nickel and nickel matte were produced and these products, together with nickel ore, accounted for 89.3% of New Caledonia's export income in 1999 (56.2% of which was provided by ferro-nickel alone), the main customers being France and Japan. It was estimated that in 2002 the nickel industry as a whole contributed some 80% of foreign export earnings. Production of nickel ore (metal content) declined from 123,500 tons in 2000 to 58,973 tons in 2001 and again to 59,867 in 2002. Exports of ferro-nickel and nickel matte reached 65,604 tons in 2003. France has declared nickel a strategic material and maintains strict control over mining. Mining and quarrying accounted for some 10.0% of GDP in 2002 (compared with 24% in 1989). Considerable growth was experienced in the industry in the mid-1990s. In 1998, however, the industry was adversely affected by the Asian financial crisis, which resulted in a dramatic decline in demand for nickel, a period of unrest among workers at the most severely affected mines and the consequent announcement by Société Le Nickel (SLN) of the loss of some 300 jobs.

Compared with the previous year, metallurgical production increased by 3.9% in 2001. The introduction of the single European currency, effected in January 2002, provided a boost to the nickel industry. (Both the one- and two-euro coins use New Caledonian nickel.) In November 2001, however, in response to a decline in world nickel prices, the Société Minière du Sud Pacifique (SMSP) announced plans to reduce its mining activity by 50%, entailing the redundancy of 600 workers.

A major nickel mine at Kopeto, near Nepoui, in the North Province, which had been closed since the 1970s, was reopened in mid-1994. Moreover, in 1999 the Canadian company Inco inaugurated a pilot smelter at Goro, in the South Province, to test new technology for nickel processing. A full-scale commercial plant producing nickel oxide and cobalt was developed subsequently, at a cost of an estimated US $1,400m.; the project, which envisaged the construction of a deep-water port, international airport, power station and associated infrastructure, was expected to provide 2,000 jobs. Construction of the plant was suspended in December 2002, owing to an increase in the overall cost of developing the site to around $2,000m. However, by redesigning the plant project Inco anticipated being able to reduce the overall cost to $1,850m.; projected annual production was 60,000 metric tons of nickel, some 10,000 tons more than initially calculated, and up to 5,100 tons of cobalt. In mid-August 2004 a public inquiry raised concerns about the project's environmental impact but recommended that it proceed. A final decision (expected by mid-October) was dependent on the French Government providing some US $25m. either in the form of tax rebates or finance towards the construction of a power station to supply the plant (see below).

Two further new nickel mines were being developed in 2003: one at Koniambo in the North Province by the Canadian company, Falconbridge, in association with the SMSP; the other in the South Province by SLN, a subsidiary of Eramet of France. Falconbridge expected to reach a final decision on the Koniambo project by the end of 2004. The overall cost of the project was estimated at US $1,150m. and would require the construction of a new town and port; annual nickel production was projected at 30,000 metric tons over 60 years. SLN planned to increase its nickel output to 75,000 tons by 2006, through the development of a new mine at Tiébaghi and a second smelter at its Doniambo site, on the outskirts of Nouméa. Overall, New Caledonia's annual nickel production capacity was expected to rise to 200,000 tons by 2007.

In late 1999 a joint French and Australian research mission made an offshore discovery of what was believed to be the world's largest gas deposit, measuring an estimated 18,000 sq km. It was hoped that this indicated the presence of considerable petroleum reserves.

Agriculture and fishing contributed only 1.9% of GDP in 1999, and the sector engaged 4.8% of the employed labour force in 2001. Some 94% of agriculturally productive land was used for pasture or fodder in 1992, mainly for cattle and pigs. New Caledonia also began to export deer, principally to Thailand, in 1994. In 1998 an Australian company announced major investment in a chicken-farming project in the North Province, increasing the number of chickens from 1,000 in 1995 to 330,000 in 2000. Maize, yams, sweet potatoes and coconuts have traditionally been the principal crops. However, pumpkins (squash) became an exclusive export crop for the Japanese market during the 1990s, and 2,200 metric tons were exported in 2001. Lychees became a new, albeit as yet minor, export in 2001, with Japan, New Zealand and French Polynesia each purchasing 8 metric tons in that year. The main fisheries products are tuna and shrimps (most of which are exported to Japan). A total of 1,815 metric tons of shrimps were produced in farms in 2002 (compared with 632 tons in 1993), and a giant-clam project was undertaken in 1996. Exports of fisheries and aquaculture products were worth 2,973m. francs CFP in 2001, some 5.0% of total exports.

The manufacturing sector provided 4.0% of GDP in 1997 and (together with electricity, gas and water services) accounted for 11.0% of total employment in 2001. The sector consists mainly of small and medium-sized enterprises, most of which are situated around the capital, Nouméa, producing building materials, furniture, salted fish, fruit juices and perishable foods.

Electric energy is provided by thermal power stations (78.5% in 2001) and by hydroelectric plants. Solid fuels and mineral products accounted for 14.8% of total imports in 2000. In 1996 construction began on a plant producing wind-generated electricity near Nouméa, at a cost of US $7m., which provides some 4.5m. kWh of energy per year. In 2001 nine windmills were installed on Lifou, one of the Loyalty Islands, to deliver more than 800,000 kWh through the island's electricity network, as part of the Government's attempt to reduce expensive imports of diesel fuel. Plans were announced in early 2003 to install a further 31 windmills in the South Province. In July 2004 Prony Energies announced that it would begin construction of a 50-MW coal-fired power station at Goro, in a joint venture with the local energy company Enercal and Elyo of France; a decision on a second 50-MW station, to supply the Inco nickel plant, was expected at the end of September.

Service industries together contributed 73.7% of GDP in 1999 and engaged 67.1% of the employed labour force in 2001. The tourism sector in New Caledonia, however, has failed to achieve expansion similar to that experienced in many other Pacific islands. Tourist arrivals were seriously affected by political unrest and outbreaks of violence in the late 1980s between the Caldoches and Kanaks. Several luxury tourist resorts were under construction in the mid-1990s in an attempt to attract more visitors to the islands. Having declined from 109,587 in 2000 to 100,515 in 2001, the number of tourist arrivals rose to 103,933 in 2002, but declined again to 101,983 in 2003. The sector aimed to increase the number of visitors to 200,000 by 2007; between 1984 and 2002 tourist arrivals had risen by fewer than 10,000. Receipts from tourism declined from US $112m. in 1999 to $93m. in 2001. Air France ceased operations in New Caledonia in March 2003; the route to Tokyo, where Air France would provide connections from Japan to Paris, was taken over by Air Calédonie International (Aircalin), which was given a tax concession by the French Government on 75% of the cost of two new aircraft. In mid-August 2004 Australian airline Virgin Blue was granted landing rights in New Caledonia, and it was hoped that tourist numbers would begin to rise once the company began flights in October. However, in late September the company postponed its inaugural flight as a result of disagreement over service charges at the airport and what it described as a 'boycott' by local travel agencies closely linked to Aircalin.

New Caledonia's external trade is dominated by its principal commodity export, nickel. New Caledonian commerce benefits from its status as a French Overseas Country in association with the EU. In 2000 New Caledonia recorded a visible trade deficit of some 42,490m. francs CFP, and there was a surplus of 25,242m. francs CFP, equivalent to 7.2% of that year's GDP, on the current account of the balance of payments (a deficit of 3,000m. francs CFP was registered in 1999). In 2001, however, the trade deficit increased to an estimated 64,700m. francs CFP. The principal imports in 1999 were mineral fuels, foodstuffs, and machinery and transport equipment. France is the main trading partner, providing some 40.3% of imports and purchasing 26.1% of exports in 2001. Other major trading partners in that year included Japan (which purchased 23.1% of exports), Taiwan (17.6% of exports), Australia and Singapore.

The budget for 2002 envisaged revenue and expenditure balancing at 81,337m. francs CFP. Budgetary expenditure for 2002 was set at 86,400m. francs CFP. In 1998 the islands received official development assistance totalling US $338.4m., nearly all of which was from France. French budgetary aid totalled €755,317m. in 2001, increasing to €777,965 in 2002. The average annual rate of inflation in Nouméa was 1.9% in 1990–2003, standing at 1.9% in 2002 and 1.2% in 2003. Some 16.0% of the labour force were unemployed in 2001. A total of 10,325 were registered as unemployed in June 2002.

New Caledonia is a member of the Franc Zone (an association of countries and territories with currencies linked to the French franc), an associate member of the UN's Economic and Social Commission for Asia and the Pacific (ESCAP) and, in its own right, of the Pacific Community (formerly the South Pacific Commission—based in Nouméa). Following the adoption of the Nouméa Accord in mid-1998 (see History), New Caledonia obtained observer status at the South Pacific Forum (now Pacific Islands Forum) in September 1999.

New Caledonia's economy is vulnerable to factors affecting the islands' important nickel industry, which included political unrest during the 1980s and fluctuations in international prices for the commodity in the 1990s. The Nouméa Accord, approved by referendum in November 1998 (see History), aimed to improve the economic conditions of the Kanak population and to increase their participation in the market economy and in public administration. Despite previous attempts to redress the balance of New Caledonian society (most importantly in the Matignon Accord of 1988), the indigenous population remained largely excluded from New Caledonia's economic and political administration, and a considerable proportion continued to experience economic hardship in the late 1990s. In mid-1999 the RPCR leader, Jacques Lafleur, revealed proposals for the creation of an inter-provincial committee for economic development, which would allow the comparatively wealthy South Province to assist the economic development of the other Provinces. Moreover, it was hoped that the development of those sectors generally considered not to have reached their full potential (notably tourism, aquaculture, fishing and farm agriculture) would alleviate the economic uncertainty created by the need to reduce dependency on France. It was reported that the French Government estimated its financial assistance to New Caledonia in 2003 at US $1,450m. In an effort to encourage investment in New Caledonia, legislation to replace the system of tax deductions with one of tax credits was expected to be adopted in 2002. In the tourism sector, visitor arrivals declined by 8.3% in 2001 compared with the previous year, but rose by 3.4% in 2002, before decreasing by 1.9% in 2003. The effect of the decline in world nickel prices during 2001 was compounded by the downturn in the aviation industry that followed the terrorist attacks in the USA in September (aeronautical companies normally being significant consumers of special nickel alloys). Nickel prices, however, recovered in 2002 and rose strongly during 2003. In mid-2003 the National Assembly approved the Overseas Territories Development Bill, which would provide support for economic and social development in New Caledonia (together with French Polynesia and Wallis and Futuna) by attracting foreign investment, and among other benefits allow for overseas French nationals to travel to mainland France to take advantage of free education.

Statistical Survey

Source (unless otherwise stated): Institut Territorial de la Statistique et des Etudes Economiques, BP 823, 98845 Nouméa; tel. 275481; fax 288148; internet www.itsee.nc.

AREA AND POPULATION

Area (sq km): New Caledonia island (Grande-Terre) 16,372; Loyalty Islands 1,981 (Lifou 1,207, Maré 642, Ouvéa 132); Isle of Pines 152; Belep Archipelago 70; Total 18,575 (7,172 sq miles).

Population: 164,173 (males 83,862, females 80,311) at census of 4 April 1989; 196,836 (males 100,762, females 96,074) at census of 16 April 1996; 220,467 (official estimate) at 1 July 2003. *Population By Province* (1996 census): Loyalty Islands 20,877; North Province 41,413; South Province 134,546.

Density (1 July 2003): 11.9 per sq km.

Ethnic Groups (census of 1996): Melanesians 86,788; French and other Europeans 67,151; Wallisians and Futunians (Polynesian) 17,763; Indonesians 5,003; Tahitians (Polynesian) 5,171; Others 14,960.

Principal Town (1 January 2003): Nouméa (capital), population 83,266. *Mid-2003* (UN estimate, incl. suburbs) Nouméa 139,798 (Source: UN, *World Urbanization Prospects: The 2003 Revision*.

Births, Marriages and Deaths (2001): Registered live births 4,326 (birth rate 20.2 per 1,000); Registered marriages 925 (marriage rate 4.3 per 1,000); Registered deaths 1,131 (death rate 5.3 per 1,000). *2003:* Birth rate 18.6 per 1,000; Marriage rate 4.0 per 1,000; Death rate 5.1 per 1,000.

Expectation of Life (years at birth, 2001): Males 70.5; Females 76.1.

Employment (2001): Agriculture, hunting, forestry and fishing 2,490; Mining and quarrying 1,772; Manufacturing and Electricity, gas and water 5,720; Construction 7,031; Trade, restaurants and hotels 9,799; Transport, storage and communications 2,861; Financing, insurance, real estate and business services 5,461; Community, social and personal services 11,165; Activities not adequately defined 5,488; *Total employed* 51,787; Unemployed 9,860; *Total labour force* 61,647. Source: UN, *Statistical Yearbook for Asia and the Pacific*.

AGRICULTURE, ETC.

Principal Crops ('000 metric tons, 2002): Maize 3.9; Potatoes 1.9; Sweet potatoes 3.0; Cassava 2.8 (FAO estimate); Taro (Coco yam) 2.3 (FAO estimate); Yams 11.0 (FAO estimate); Coconuts 16.0 (FAO estimate); Vegetables 4.0; Bananas 0.7; Other fruits 2.9 (unofficial figure). Source: FAO.

Livestock ('000 head, year ending September 2002): Horses 11.5; Cattle 111.3 (FAO estimate); Pigs 25.4; Sheep 2.3; Goats 8.1; Poultry 510 (FAO estimate). Source: FAO.

Livestock Products ('000 metric tons, 2002): Beef and veal 4.0; Pig meat 1.4 (FAO estimate); Poultry meat 0.7; Cows' milk 3.6 (FAO estimate); Hen eggs 1.6 (FAO estimate); Cattle hides (fresh) 0.5 (FAO estimate). Source: FAO.

Forestry ('000 cubic metres, 1994): *Roundwood Removals*: Sawlogs and veneer logs 2.8; Other industrial wood 2.0 (FAO estimate); Total 4.8. *Sawnwood Production*: 3.3 (all broadleaved). *1995–2002*: Annual output as in 1994 (FAO estimates). Source: FAO.

Fishing (metric tons, live weight, 2002): Capture 3,417 (Albacore 1,165; Yellowfin tuna 572; Bigeye tuna 189; Other tuna-like fishes 265; Sea cucumbers 450 (FAO estimate)); Aquaculture 1,882 (Shrimps and prawns 1,815); Total catch 5,299. Source: FAO.

MINING

Production ('000 metric tons): Nickel ore (metal content) 123.5 in 2000; 60.0 in 2001; 59.9 in 2002 (provisional figure). Source: US Geological Survey.

INDUSTRY

Production (2003): Ferro-nickel and nickel matte 61,523 metric tons (nickel content); Electric energy 1,758 million kWh.

Cement: 100,080 metric tons in 2002 (provisional figure). Source: US Geological Survey.

FINANCE

Currency and Exchange Rates: see Wallis and Futuna Islands.

Territorial Budget (million francs CFP, 2001): *Revenue*: 81,337; *Expenditure*: 81,337. French government transfers 86,344.

Aid from France (US $ million): 336.6 in 1997; 336.3 in 1998; 314.9 in 1999. Source: UN, *Statistical Yearbook for Asia and the Pacific*.

Money Supply (million francs CFP at 31 December 2001): Currency in circulation 10,473; Demand deposits 94,525; Total money 104,998. Source: Institut d'Emission d'Outre-Mer.

Cost of Living (Consumer Price Index for Nouméa; base: December 1992 = 100): 111.4 in 1999; 113.1 in 2000; 115.7 in 2001.

Gross Domestic Product (million francs CFP at current prices): 347,303 in 2000. Source: *Bank of Hawaii, An Update on New Caledonia*.

Expenditure on the Gross Domestic Product (million francs CFP at current prices, 1992): Government final consumption expenditure 94,770; Private final consumption expenditure 159,514; Increase in stocks 416; Gross fixed capital formation 66,818; *Total domestic expenditure* 321,518; Exports of goods and services 47,246; *Less* Imports of goods and services 88,403; *Sub-total* 280,361; Statistical discrepancy 1,066; *GDP in purchasers' values* 281,427. Source: UN, *National Accounts Statistics*.

Gross Domestic Product by Economic Activity (million francs CFP at current prices, 1997): Agriculture and food processing 12,835; Energy, mining and metallurgy 37,232; Other manufactures 13,815; Construction and public works 17,447; Transport and communication 23,415; Services and commerce 152,659; Salaries 91,857; *Total* 349,260. Source: Bank of Hawaii, *An Update on New Caledonia*.

Balance of Payments (million francs CFP, 2000): Trade balance −42,490; Services (net) −5,161; *Balance on goods and services* −47,651; Other income (net) 33,394; *Balance on goods, services and income* −14,257; Current transfers (net) 39,500; *Current balance* 25,242; Capital account (net) 179; Direct investment (net) −5,492; Portfolio investment (net) −15,367; Other capital (incl. charges in reserves) −4,738; Net errors and omissions 176. Source: Institut d'Emission d'Outre-Mer.

EXTERNAL TRADE

Principal Commodities (million francs CFP, 1999): *Imports:* Prepared foodstuffs, beverages, spirits and vinegar, tobacco and manufactured substitutes 18,233; Mineral products 10,579; Products of chemical or allied industries 8,860; Plastics, rubber and articles thereof 4,526; Paper-making material, paper and paperboard and articles thereof 3,115; Textiles and textile articles 3,967; Base metals and articles thereof 8,048; Machinery and mechanical appliances, electrical equipment, sound and television apparatus 22,617; Vehicles, aircraft, vessels and associated transport equipment 17,613; Total (incl. others) 112,808. *Exports:* Ferro-nickel 29,445; Nickel matte 8,764; Nickel ore 8,583; Prawns 1,868; Total (incl. others) 52,388. *2003* (million francs CFP): Total imports 163,869; Total exports 82,152 (Ferro-nickel and matte 65,604; Nickel ore 9,288; Fish products 2,532).

Principal Trading Partners (million francs CFP, 2001): *Imports:* Australia 20,666.6; Belgium 1,659.9; China, People's Republic 2,977.9; France (metropolitan) 50,046.6; Germany 3,996.6; Italy 3,619.7; Japan 3,868.5; Korea, Republic 1,445.0; New Zealand 6,131.5; Singapore 7,551.4; Spain 2,485.9; United Kingdom 2,812.1; USA 4,910.6; Total (incl. others) 124,037.8. *Exports:* Australia 4,492.4; Belgium 1,236.2; Finland 1,422.2; France (metropolitan) 15,523.2; Italy 1,370.8; Japan 13,733.0; Korea, Republic 2,612.7; Spain 4,355.3; Taiwan 10,455.8; USA 1,446.4; Total (incl. others) 59,511.9. Source: Direction Régionale des Douanes.

TRANSPORT

Road Traffic (motor vehicles in use, 2001): Total 85,499.

Shipping (2001): Vessels entered 478; Goods unloaded 1,298,829 metric tons, Goods loaded 210,197 metric tons. *Merchant Fleet* (vessels registered, '000 grt, at 31 December 1992): 14.

Civil Aviation (La Tontouta airport, Nouméa, 2001): Passengers arriving 173,913, Passengers departing 172,854; Freight unloaded 3,661 metric tons, Freight loaded 1,401 metric tons.

TOURISM

Foreign Tourist Arrivals (arrivals by air): 100,515 in 2001; 103,933 in 2002; 101,983 in 2003.

Tourist Arrivals by Country of Residence (2003): Australia 15,957; France 29,440; Japan 28,490; New Zealand 6,030; Total (incl. others) 101,983.

Receipts (US $ million): 112 in 1999; 110 in 2000; 93 in 2001.
Source: UN, *Statistical Yearbook for Asia and the Pacific*.

COMMUNICATIONS MEDIA

Radio Receivers (1997): 107,000 in use*.

Television Receivers (2000): 106,000 in use†.

Telephones (2001): 50,700 main lines in use†.

Facsimile Machines (1994): 2,200 in use‡.

Mobile Cellular Telephones (2002): 80,000 subscribers†.

Internet Users (estimate, 2002): 25,000†.

Daily Newspapers (1999): 1.

* Source: UNESCO, *Statistical Yearbook*.
† Source: International Telecommunication Union.
‡ Source: UN, *Statistical Yearbook*.

EDUCATION

Pre-primary (2000): 81 schools; 13,033 pupils.

Primary (2001): 289 schools; 1,837 teachers (incl. pre-primary); 36,996 pupils.

Secondary (2001): 64 schools; 2,371 teachers; 29,036 pupils.

Higher (2000): 55 teachers; 2,069 students.
Source: Vice-Rectorat de Nouvelle-Calédonie.

2003: *Pre-primary and Primary:* 36,995 pupils; *Secondary:* 30,940 pupils/students; *Higher:* 2,597 students.

Adult Literacy Rate (1989): Males 94.0%; Females 92.1% .

Directory

The Constitution

The constitutional system in New Caledonia is established under the Constitution of the Fifth French Republic and specific laws, including those enacted in July 1989 in accordance with the terms agreed by the Matignon Accord and the Transitional Provisions appended by legislation on 20 July 1998. A referendum on the future of New Caledonia (originally expected to be conducted in 1998) was postponed for a period of between 15 and 20 years while a gradual transfer of power from metropolitan France to local institutions is effected under the terms of the Nouméa Accord, concluded in 1998. Under the Nouméa Accord, the islands are declared to be an Overseas Country of the French Republic, of which they remain an integral part. The High Commissioner is the representative of the State in New Caledonia and is appointed by the French Government. The High Commissioner is responsible for external relations, defence, law and order, finance and secondary education. New Caledonia is divided into three Provinces, of the South, the North and the Loyalty Islands. Each is governed by a Provincial Assembly, which is elected on a proportional basis and is responsible for local economic development, land reform and cultural affairs. Members of the Assemblies (40 for the South, 22 for the North and 14 for the Loyalty Islands) are subject to re-election every five years. A proportion of the members of the three Provincial Assemblies together form the Congress of New Caledonia (32 for the South, 15 for the North and seven for the Loyalty Islands), which is responsible for the New Caledonian budget and fiscal affairs, infrastructure and primary education. The Assemblies and the Congress each elect a President as leader. The Government of New Caledonia is elected by the Congress, and comprises between seven and 11 members. Under the terms of the Nouméa Accord, it replaces the French High Commissioner as New Caledonia's executive authority. Provision is also made for the maintenance of Kanak tradition: there are eight custom regions, each with a Regional Consultative Custom Council. These eight Councils, with other appropriate authorities, are represented on the Customary Senate, which is composed of 16 members (two elected from each regional council for a six-year period); the Senate is consulted by the Congress and the Government. Local government is conducted by 33 communes. New Caledonia also elects two deputies to the National Assembly in Paris and one Senator, on the basis of universal adult suffrage. One Economic and Social Councillor is also nominated. New Caledonia may be represented in the European Parliament.

The Government

(September 2004)

STATE GOVERNMENT

High Commissioner: DANIEL CONSTANTIN (took office August 2002).

Secretary-General: ALAIN TRIOLLE.

Deputy Secretary-General: LOUIS LE FRANC.

LOCAL GOVERNMENT

Secretary-General: ALAIN SWETSCHKIN.

Deputy Secretary-General: MARTINE MICHEL.

Deputy Secretary-General: LÉON WAMYTAN.

COUNCIL OF MINISTERS

President, responsible for Social Affairs and Solidarity: MARIE-NOËLLE THÉMEREAU (AE).

Vice-President and Minister for Culture, Women's Affairs and Citizenship: EPÉRI 'DÉWÉ' GORODEY (UNI-FLNKS).

Minister for Agriculture, Fisheries and Livestock: ERIC BABIN (AE).

Minister for Education and Research: CHARLES WASHETINE (UNI-FLNKS).

Minister for Vocational Training, Employment and Public Services: ALAIN SONG (AE).

Minister for Transport, Road Safety, Infrastructure and Energy: GÉRALD CORTOT (UC).

Minister for Foreign Trade, External Affairs and Relations with the European Union: PIERRE FROGIER (UMP).

Minister for Finance and the Budget: JEAN-CLAUDE BRIAULT (UMP).

Minister for the Economy, Taxes, Sustainable Development, Mines, Aviation and Communication: DIDIER LEROUX (AE).

Minister for Health and the Disabled: MARIANNE DEVAUX (UMP).

Minister for Youth and Sports: MAURICE PONGA (UMP).

GOVERNMENT OFFICES

Office of the High Commissioner: Haut-commissariat de la République en Nouvelle-Calédonie, 1 ave Maréchal Foch, BP C5, 98844 Nouméa Cédex; tel. 266300; fax 272828; e-mail haussariat@nouvelle-caledonie.gouv.fr; internet www.nouvelle-caledonie.gouv.fr.

New Caledonian Government: *Présidence du Gouvernement:* 8 route des Artifices, BP M2, 98849 Nouméa Cédex; *Congrès de la Nouvelle-Calédonie:* 1 blvd Vauban, BP 31, 98845 Nouméa Cédex; tel. 246565; fax 246550; tel. 273129; fax 277020; e-mail cellule.communication@gouv.nc; e-mail courrier@congres.nc.

Office of the Secretary-General of the Government of New Caledonia: Immeuble administratif Jacques Iekawe, 18 ave Paul Doumer, BP M2, 98844 Nouméa Cédex; tel. 256000; fax 286848; e-mail lmoprini@gouv.nc; internet www.gouv.nc.

Government of the Province of the Loyalty Islands: Gouvernement Provincial des Iles Loyauté, BP 50, 98820 Wé, Lifou, Loyalty Islands; tel. 455100; fax 455100; e-mail loyalty@loyalty.nc; internet www.loyalty.nc.

Government of the Province of the North: Gouvernement Provincial du Nord, BP 41, 98860 Koné, Grande-Terre; tel. 477100; fax 355475.

Government of the Province of the South: Hôtel de la Province Sud, 9 Route des Artifices Artilleries, Port Moselle, BP 4142, 98846 Nouméa Cédex; tel. 258000; fax 274900; internet www.province-sud.nc.

GOVERNMENT DEPARTMENTS

Department of Administrative Services and Computer Technology (SMAI): 3 rue Gustave Flaubert, Baie de l'Orphelinat, BP 8231, 98807 Nouméa Cédex; tel. 275888; fax 281919; e-mail smai@gouv.nc; internet www.gouv.nc.

Department of Civil Aviation: 179 rue Gervolino, BP H01, 98849 Nouméa Cédex; tel. 265200; fax 265202; e-mail dac-nc@aviation-civile.gouv.fr; internet www.dgac.fr.

Department of the Budget and Financial Affairs (DBAF): 18 ave Paul Doumer, BP M2, 98849 Nouméa Cédex; tel. 256082; fax 283133; e-mail dbaf@gouv.nc.

Department of Culture: Mission aux Affaires Culturelles, Haut-Commisariat de la Republique en Nouvelle Calédonie, 56 Rue Bataille, BP C5, Nouméa; tel. 242181; fax 242180; e-mail jj.garnier@nouvelle-caledonie.gouv.fr; Dir JEAN-JACQUES GARNIER.

Department of Education (DENC): Immeuble Foch, 19 ave du Maréchal Foch, BP 8244, 98807 Nouméa Cédex; tel. 239600; fax 272921; e-mail denc@gouv.nc; internet www.prim.edu.gouv.nc.

Department of Employment (DT): 12 rue de Verdun, BP 141, 98845 Nouméa Cédex; tel. 275572; fax 270494; e-mail dt@gouv.nc; internet www.dtnc.gouv.nc .

Department of Economic Affairs (DAE): 7 rue du Général Gallieni, BP 2672, 98846 Nouméa Cédex; tel. 279200; fax 272675; e-mail dae@gouv.nc; internet www.gouv.nc.

Department of Fiscal Affairs (DSF): Fiscaux Hôtel des Impôts, 13 rue de la Somme, BP D2, 98848 Nouméa Cédex ; tel. 257500; fax 251166; e-mail dsf@gouv.nc; internet www.dsf.gouv.nc.

Department of Human Resources and Civil Service (DRHFPT): 18 ave Paul Doumer, BP M2, 98849 Nouméa Cédex; tel. 256000; fax 274700; e-mail drhfpt@gouv.nc; internet www.drhfpt.gouv.nc.

Department of Infrastructure, Topography and Transport (DITTT): 1 bis rue Edouard Unger, 1ère Vallée du Tir, BP A2, 98848 Nouméa Cédex; tel. 280300; fax 281760; e-mail dittt@gouv.nc; internet www.gouv.nc.

Department of Maritime Affairs: 2 bis rue Félix Russeil, BP 36, Nouméa Cédex; tel. 272626; fax 287286; e-mail affmar@gouv.nc; internet www.gouv.nc.

Department of Health and Social Services (DASS): 5 rue Général Galliéni, BP 3278, 98846 Nouméa Cédex; tel. 243700; fax 243702; e-mail dtass@gouv.nc; internet www.dass.gouv.nc.

Department of Industry, Mines and Energy (DIMENC): 1 ter rue Edouard Unger, 1ère Vallée du Tir, BP 465, 98845 Nouméa Cédex; tel. 273944; fax 272345; internet www.gouv.nc.

Department of Veterinary, Food and Rural Affairs (DAVAR): 209 rue Auguste Bénébig, BP 180, 98845 Nouméa Cédex; tel. 255100; fax 255129; e-mail davar@gouv.nc; internet www.davar.gouv.nc; f. 2001; fmrly Dept of Rural Economy.

Department of Vocational Training (DFPC): 19 ave du Maréchal Foch, BP 110, 98845 Nouméa Cédex; tel. 246622; fax 281661; e-mail dfpc@gouv.nc.

Department of Youth and Sports (DJS): 23 rue Jean Jaurès, BP810, 98845 Nouméa Cédex; tel. 252384; fax 254585; e-mail djsnc@gouv.nc; internet www.djs.gouv.nc.

Legislature

ASSEMBLÉES PROVINCIALES

Members of the Provincial Assemblies are elected on a proportional basis for a five-year term. Each Provincial Assembly elects its President. A number of the members of the Provincial Assemblies sit together to make up the Congress of New Caledonia. The Assembly of the Northern Province has 22 members (including 15 sitting for the Congress), the Loyalty Islands 14 members (including seven for the Congress) and the Southern Province has 40 members (including 32 for the Congress).

Election, 9 May 2004 (results by province)

Party	North	South	Loyalty Islands
Le Rassemblement-Union pour un Mouvement Populaire (UMP)	3	16	2
L'Avenir Ensemble	1	19	—
Union Nationale pour l'Indépendance-Front de Libération Nationale Kanak Socialiste (UNI-FLNKS)	11	—	2
Union Calédonienne (UC)	7	—	4
Front National (FN)	—	5	—
Fédération des Comités de Coordination des Indépendantistes (FCCI)	—	—	2
Libération Kanak Socialiste (LKS)-KAP Identité	—	—	2
Unir et Construire dans le Renouveau (UCR)	—	—	2
Total	22	40	14

Province of the North: President PAUL NÉAOUTYINE (UNI-FLNKS).

Province of the South: President PHILIPPE GOMÈS (Avenir Ensemble).

Province of the Loyalty Islands: President NÉKO HNÉPEUNE (UNI-FLNKS).

CONGRÈS

A proportion of the members of the three Provincial Assemblies sit together, in Nouméa, as the Congress of New Caledonia. There are 54 members (out of a total of 76 sitting in the Provincial Assemblies).

President: HAROLD MARTIN.

Election, 9 May 2004 (results for the Territory as a whole)

Party	Votes	%	Seats
Le Rassemblement-UMP	21,880	24.43	16
L'Avenir Ensemble	20,338	22.71	16
UNI-FLNKS	12,556	14.02	8
UC	10,624	11.86	7
FN	6,684	7.46	4
FCCI-FULK	2,864	3.20	1
LKS-KAP Identité	2,572	2.87	1
UC Renouveau	1,587	1.77	1
Others	10,467	11.69	—
Total	89,572	100.00	54

PARLEMENT

Deputies to the French National Assembly: JACQUES LAFLEUR (RPCR), PIERRE FROGIER (RPCR).

Representative to the French Senate: SIMON LOUECKHOTE (RPCR).

Political Organizations

L'Avenir Ensemble (AE): 19 Bd Extérieur, Faubourg Blanchot, Nouméa; tel. 870371; fax 870379; e-mail courier@avenirensemble.nc; internet www.avenirensemble.nc; f. 2004; combined list including fmr mems of Rassemblement pour la Calédonie dans la République and Alliance pour la Calédonie; Pres. HAROLD MARTIN.

Fédération des Comités de Coordination des Indépendantistes (FCCI): f. 1998 by breakaway group from FLNKS; includes Front du Développement des Iles Loyauté and Front Uni de Libération Kanak; Leaders LÉOPOLD JORÉDIÉ, RAPHAËL MAPOU, FRANÇOIS BURCK.

Front Calédonien (FC): extreme right-wing; Leader M. SARRAN.

Front de Libération Nationale Kanak Socialiste (FLNKS): Nouméa; tel. 272599; f. 1984 (following dissolution of Front Indépendantiste); pro-independence; Pres. (vacant); a grouping of the following parties:

Groupe UC du Congrès: Nouméa; f. 2000 by breakaway faction of the UC.

Parti de Libération Kanak (PALIKA): f. 1975; 5,000 mems; Leader PAUL NÉAOUTYINE.

Rassemblement Démocratique Océanien (RDO): Nouméa; f. 1994 by breakaway faction of UO; supports Kanak sovereignty; Pres. ALOISIO SAKO.

Union Nationale pour l'Indépendence (UNI): Leader PAUL NÉAOUTYINE.

Union Progressiste Mélanésienne (UPM): f. 1974 as the Union Progressiste Multiraciale; 2,300 mems; Pres. VICTOR TUTUGORO; Sec.-Gen. RENÉ POROU.

Front National (FN): BP 4198, 98846 Nouméa; tel. 258068; fax 258064; e-mail george@province-sud.nc; right-wing; Leader GUY GEORGE.

Génération Calédonienne: f. 1995; youth-based; aims to combat corruption in public life; Pres. JEAN-RAYMOND POSTIC.

Libération Kanak Socialiste (LKS): Maré, Loyalty Islands; moderate, pro-independence; Leader NIDOÏSH NAISSELINE.

Le Rassemblement-UMP: 13 rue de Sebastopol, BP 306, 98845 Nouméa; tel. 282620; fax 284033; e-mail rassemblement@bluepack.nc; f. 1977; affiliated to the metropolitan Union pour un Mouvement Populaire (UMP); includes fmr mems of Rassemblement pour la Calédonie dans la République (RPCR); in favour of retaining the status quo in New Caledonia; Leader JACQUES LAFLEUR; a coalition of the following parties:

Centre des Démocrates Sociaux (CDS): f. 1971; Leader JEAN LÈQUES.

Parti Républicain (PR): Leader PIERRE MARESCA.

Union Calédonienne (UC): f. 1952; left FLNKS coalition prior to elections of 2004; 11,000 mems; Pres. PASCAL NAOUNA; Sec.-Gen. DANIEL YEIWÉNÉ.

Union Océanienne (UO): Nouméa; f. 1989 by breakaway faction of RPCR; represents people whose origin is in the French Overseas Territory of Wallis and Futuna; conservative; Leader MICHEL HEMA.

Other political organizations participating in the elections of May 2004 included: Avance, Calédonie Mon Pays, Construire Ensemble l'Avenir, Le FLNKS pour l'Indépendance, Mouvement Chiraquien des Démocrates Chrétiens UMP, Mouvement des Citoyens Calédoniens, Patrimoine et Environnement avec les Verts, Rassemblement Océanien dans la Calédonie (ROC) and Union Calédonienne Renouveau.

Judicial System

Court of Appeal: Palais de Justice, BP F4, 98848 Nouméa; tel. 279350; fax 269185; e-mail pp.ca-noumea@justice.fr; First Pres. GÉRARD FEY; Procurator-Gen. GÉRARD NÉDELLEC.

Court of the First Instance: 2 blvd Extérieur, BP F4, 98848 Nouméa; fax 276531; e-mail p.tpi-noumea@justice.fr; Pres. JEAN PRADAL; Procurator of the Republic ROBERT BLASER; There are two subsidiary courts, with resident magistrates, at Koné (Province of the North) and Wé (Province of the Loyalty Islands).

Customary Senate of New Caledonia: Conseil Consultatif Coutumier, 68 ave J. Cook, POB 1059, Nouville; tel. 242000; fax 249320; f. 1990; consulted by Local Assembly and French Govt on matters affecting land, Kanak tradition and identity; mems: 40 authorities from eight custom areas; Pres. Chief PAUL JEWINE; Sec.-Gen.. SARIMIN BOUENGHI.

Religion

The majority of the population is Christian, with Roman Catholics comprising about 55% of the total in 2002. About 3% of the inhabitants, mainly Indonesians, are Muslims.

CHRISTIANITY

The Roman Catholic Church

The Territory comprises a single archdiocese, with an estimated 110,000 adherents in 2002. The Archbishop participates in the Catholic Bishops' Conference of the Pacific, based in Fiji.

Archbishop of Nouméa: Most Rev. MICHEL-MARIE-BERNARD CALVET, Archevêché, BP 3, 4 rue Mgr-Fraysse, 98845 Nouméa; tel. 265353; fax 265352; e-mail archeveche@ddec.nc.

The Anglican Communion

Within the Church of the Province of Melanesia, New Caledonia forms part of the diocese of Vanuatu (q.v.). The Archbishop of the Province is the Bishop of Central Melanesia (resident in Honiara, Solomon Islands).

Protestant Churches

Eglise évangélique en Nouvelle-Calédonie et aux Iles Loyauté: BP 277, Nouméa; f. 1960; Pres. Rev. SAILALI PASSA; Gen. Sec. Rev. TELL KASARHEROU.

Other churches active in the Territory include the Assembly of God, the Free Evangelical Church, the Presbyterian Church, the New Apostolic Church, the Pentecostal Evangelical Church and the Tahitian Evangelical Church.

The Press

L'Avenir Calédonien: 10 rue Gambetta, Nouméa; organ of the Union Calédonienne; Dir PAÏTA GABRIEL.

La Calédonie Agricole: BP 111, 98845 Nouméa Cédex; tel. 243160; fax 284587; every 2 months; official publ. of the Chambre d'Agriculture; Pres. ANDRÉ MAZURIER; Man. GEORGES ROUCOU; circ. 3,000.

Eglise de Nouvelle-Calédonie: BP 3, 98845 Nouméa; fax 265352; f. 1976; monthly; official publ. of the Roman Catholic Church; circ. 450.

Les Nouvelles Calédoniennes: 41–43 rue de Sébastopol, BP G5, 98848 Nouméa; tel. 272584; fax 281627; e-mail lnc@canl.nc; internet www.nouvelles-caledoniennes.nc; f. 1971; daily; Publr ROBERT HERSANT; Gen. Man. THIERRY MASSÉ; Editor MICHEL MEKKI; circ. 18,500.

Télé 7 Jours: route de Vélodrome, BP 2080, 98846 Nouméa Cédex; tel. 284598; weekly.

NEWS AGENCY

Agence France-Presse (AFP): 15 rue Docteur Guégan, 98800 Nouméa; tel. 263033; fax 278699; Correspondent FRANCK MADOEUF.

Publishers

Editions d'Art Calédoniennes: 3 rue Guynemer, BP 1626, Nouméa; tel. 277633; fax 281526; art, reprints, travel.

Editions du Santal: 5 bis rue Emile-Trianon, 98846 Nouméa; tel. and fax 262533; history, art, travel, birth and wedding cards; Dir PAUL-JEAN STAHL.

Grain de Sable: BP 577, 98845 Nouméa; tel. 273057; fax 285707; e-mail lokisa@canl.nc; internet www.pacific-bookin.com; literature, travel.

Ile de Lumière: BP 8401, Nouméa Sud; tel. 289858; history, politics.

Savannah Editions SARL: BP 3086, 98846 Nouméa; e-mail savannahmarc@hotmail.com; f. 1994; sports, travel, leisure.

Société d'Etudes Historiques de la Nouvelle-Calédonie: BP 63, 98845 Nouméa; tel. 767155.

Broadcasting and Communications

TELECOMMUNICATIONS

France Câbles et Radio (FCR): Télécommunications extérieures de la Nouvelle-Calédonie, BP A1, 98848 Nouméa Cédex; tel. 266600; fax 266666; Dir CLAUDE BONNEAU.

Offices des Postes et Télécommunications: Le Waruna, 2 rue Monchovet, Port Plaisance, 98841 Nouméa Cédex; tel. 268217; fax 262927; e-mail direction@opt.nc; internet www.opt.nc; Dir JEAN-YVES OLLIVAUD.

BROADCASTING

Radio

NRJ Nouvelle-Calédonie: 41–43 rue Sébastopol, BP G5, 98848 Nouméa; tel. 279446; fax 279447.

Radio Djiido: 29 rue du Maréchal Juin, BP 1671, 98803 Nouméa Cédex; tel. 253515; fax 272187; f. 1985; community station; broadcasts in French; pro-independence.

Radio Nouvelle-Calédonie: Radio Télévision Française d'Outre Mer (RFO), 1 rue Maréchal Leclerc, Mont Coffyn, BP G3, 98848 Nouméa Cédex; tel. 239999; fax 239975; e-mail internet.nc@rfo.fr; f. 1942; fmrly Radiodiffusion Française d'Outre-mer (RFO); 24 hours of daily programmes in French; Gen. Man. FRANÇOIS GUILEBEAU; Editor-in-Chief FRANCIS ORNY.

Radio Rythme Bleu: BP 578, Nouméa; tel. 254646; fax 284928; e-mail RRB@lagoon.nc; Dir CHRISTIAN PROST.

Television

RFO-Télé Nouvelle-Calédonie: Radio Télévision Française d'Outre Mer (RFO), 1 rue Maréchal Leclerc, Mont Coffyn, BP G3, 98848 Nouméa Cédex; tel. 239999; fax 239975; e-mail internet.nc@rfo.fr; internet www.rfo.fr; f. 1965; transmits 10 hours daily; Gen. Man. GÉRALD PRUFER.

Canal Calédonie: 30 rue de la Somme, BP 1797, 98845 Nouméa; CEO SERGE LAMAGNÈRE; subscription service.

Canal Outre-mer (Canal+): Nouméa; f. 1995; cable service.

Finance

(cap. = capital; res = reserves; dep. = deposits; m. = million; brs = branches; amounts in CFP francs unless otherwise stated)

BANKING

Agence Française de Développement: 5 rue Barleux, BP JI, 98849 Nouméa Cédex; tel. 282088; fax 282413.

Banque de Nouvelle-Calédonie: 25 ave de la Victoire, BP L3, 98849 Nouméa Cédex; tel. 257400; fax 274147; internet www.bnc.nc; f. 1974; adopted present name Jan. 2002; 95.8% owned by Financière Océor; cap. 3,711.1m., res 1,252.6m., dep. 44,640.4m. (Dec. 2002); Pres. JEAN-CLAUDE CLARAC; Gen. Man. JEAN-PIERRE FLOTAT; 7 brs.

Banque Calédonienne d'Investissement (BCI): 54 ave de la Victoire, BP K5, 98846 Nouméa; tel. 256565; fax 274035; e-mail dg@bci.nc; internet www.bci.nc; cap. 7,500m. (Dec. 2003); Pres. DANIEL LABORDE; Gen. Man. ALAIN CELESTE.

BNP Paribas Nouvelle-Calédonie (France): 37 ave Henri Lafleur, BP K3, 98849 Nouméa Cédex; tel. 258400; fax 258459; e-mail bnp.nc@bnpparibas.com; f. 1969 as Banque Nationale de Paris; present name adopted in 2001; cap. €28.0m., res €315.6m. (Dec. 2001); Pres. GÉRARD HAYAUD; Gen. Man. JEAN-CLAUDE DANG; 10 brs.

Société Générale Calédonienne de Banque: 44 rue de l'Alma, Siège et Agence Principale, BP G2, 98848 Nouméa Cédex; tel. 256300; fax 276245; e-mail svp.sgcb@canl.nc; f. 1981; cap. 1,068.4m., res 5,053.5m., dep. 81,547.6m. (Dec. 2001); Gen. Man. DOMINIQUE POIGNON; Chair. JEAN-LOUIS MATTEI; 12 brs.

Trade and Industry

DEVELOPMENT ORGANIZATIONS

New Caledonia Economic Development Agency (ADECAL): 15 rue Guynemer, BP 2384, 98846 Nouméa Cédex; tel. 249077; fax 249087; e-mail adecal@offratel.nc; internet www.adecal.nc; f. 1995; promotes investment within New Caledonia; Dir JEAN-MICHEL ARLIE.

Agence de Développement de la Culture Kanak: Centre Tjibaou, rue des Accords de Matignon BP 378, 98845 Nouméa Cédex; tel. 414545; fax 414546; e-mail adck@adck.nc; internet www.adck.nc; Pres. MARIE-CLAUDE TJIBAOU; Dir-Gen. OCTAVE TOGNA; Dir of Culture EMMANUEL KASARHEROU.

Agence de Développement Rural et d'Aménagement Foncier (ADRAF): 1 rue de la Somme, BP 4228, 98847 Nouméa Cédex; tel. 258600; fax 258604; e-mail dgadraf@offratel.nc; f. 1986, reorganized 1989; acquisition and redistribution of land; Chair. DANIEL CONSTANTIN; Dir-Gen. LOUIS MAPOU.

Agence pour l'Emploi de Nouvelle Calédonie: 3 rue de la Somme, BP 497, 98845 Nouméa Cédex; tel. 281082; fax 272079; internet www.apenc.nc.

Conseil Economique et Social: 14 ave Georges Clemenceau, BP 4766, 98847 Nouméa Cédex; tel. 278517; fax 278509; e-mail ces@gouv.nc; represents trade unions and other organizations that are involved in economic, social and cultural life; Pres. BERNARD PAUL; Gen. Sec. YOLAINE ELMOUR.

Institut Calédonien de Participation: Nouméa; f. 1989 to finance development projects and encourage the Kanak population to participate in the market economy.

CHAMBERS OF COMMERCE

Chambre d'Agriculture: 3 rue A. Desmazures, BP 111, 98845 Nouméa Cédex; tel. 243160; fax 284587; e-mail canc-gr@canl.nc; f. 1909; 33 mems; Pres. ANDRÉ MAZURIER.

Chambre de Commerce et d'Industrie: 15 rue de Verdun, BP M3, 98849 Nouméa Cédex; tel. 243100; fax 243131; e-mail cci@cci.nc; internet www.cci.nc; f. 1879; 29 mems; Pres. MICHEL QUINTARD; Dir-Gen. MICHEL MERZEAU.

Chambre des Métiers: 10 ave James Cook, BP 4186, 98846 Nouméa Cédex; tel. 282337; fax 282729.

STATE-OWNED INDUSTRIES

Société Minière du Sud Pacifique (SMSP): Nouméa; dirgen@smsp.nc; 87% owned by SOFINOR, 5% owned by SODIL; nickel-mining co; subsidiaries: Nouméa Nickel, Nickel Mining Corporation (NMC), Bienvenue, Sam 3; Man. Dir ANDRÉ DANG VAN NHA.

Société Le Nickel (SLN): Doniambo, Nouméa; tel. 245300; fax 275989; 60% owned by the Eramet group, privatized in 1999; nickel mining, processing and sales co; CEO YVES RAMBAUD; Man. Dir PHILIPPE VECTEN.

EMPLOYERS' ORGANIZATION

MEDEF Nouvelle-Calédonie/Fédération Patronale: Immeuble Jules Ferry, 1 rue de la Somme, BP 466, 98845 Nouméa Cédex; tel. 273525; fax 274037; e-mail medefnc@medef.nc; f. 1936; represents the leading companies of New Caledonia in the defence of professional interests, co-ordination, documentation and research in socio-economic fields; Pres. PHILLIPE VECTEN; Sec.-Gen. FRANÇOIS PERONNET.

TRADE UNIONS

CGT Force Ouvrière: 13 rue Jules Ferry, BP R2, 98851 Nouméa Cédex; tel. 274950; fax 278202; e-mail utfonc@lagoon.nc; f. 1982; Sec.-Gen. JEAN-CLAUDE NÈGRE.

Confédération des Travailleurs Calédoniens: Nouméa; Sec.-Gen. R. JOYEUX; grouped with:

Fédération des Fonctionnaires: Nouméa; Sec.-Gen. GILBERT NOUVEAU.

Syndicat Général des Collaborateurs des Industries de Nouvelle-Calédonie: Sec.-Gen. H. CHAMPIN.

Union Syndicale des Travailleurs Kanak et des Exploités (USTKE): BP 4372, Nouméa; tel. 277210; fax 277687; Pres. GÉRARD JODARD; Sec.-Gen CLAUDE WEMA.

Union des Syndicats des Ouvriers et Employés de Nouvelle-Calédonie (USOENC): BP 2534, Vallée du Tir, 98846 Nouméa; tel. 259640; fax 250164; e-mail usoenc@canl.nc; Sec.-Gen. DIDIER GUÉNANT-JEANSON.

Transport

ROADS

In 1983 there was a total of 5,980 km of roads on New Caledonia island; 766 km were bitumen-surfaced, 589 km unsealed, 1,618 km stone-surfaced and 2,523 km tracks in 1980. The outer islands had a total of 470 km of roads and tracks in 1980.

SHIPPING

Most traffic is through the port of Nouméa. Passenger and cargo services, linking Nouméa to other towns and islands, are regular and frequent. There are plans to develop Nepoui, in the Province of the North, as a deep-water port and industrial centre.

Port Autonome de la Nouvelle-Calédonie: BP 14, 98845 Nouméa Cédex; tel. 255000; fax 275490; e-mail noumeaportnc@canl.nc; Port Man. PHILIPPE LAFLEUR; Harbour Master EDMUND MARTIN.

Compagnie Wallisienne de Navigation: BP 1080, Nouméa; tel. 232910; fax 287388; e-mail jyboileau@amb.nc; Chair. ARMAND BALLANDE; Man. Dir JEAN-YVES BOILEAU.

Moana Services: BP 2099, 98846 Nouméa; tel. 273898; fax 259315; e-mail moana@canl.nc; fmrly Somacal.

CIVIL AVIATION

There is an international airport, Tontouta, 47 km from Nouméa, and an internal network, centred on Magenta airport, which provides air services linking Nouméa to other towns and islands. Air France, which formerly operated a service between Nouméa and Paris via Tokyo, Japan, ceased operations in New Caledonia in early 2003. The route to Tokyo was taken over by Air Calédonie International (Aircalin). AOM operates flights from Paris to Nouméa via Sydney. Other airlines providing services to the island include Air New Zealand, Air Vanuatu and Qantas.

Air Calédonie: BP 212, Nouméa; tel. 250300; fax 254869; e-mail directoire@air-caledonie.nc; internet www.air-caledonie.nc; f. 1955; services throughout New Caledonia; 301,000 passengers carried in 2000/01; Pres. and CEO OLIVIER RAZAVET.

Air Calédonie International (Aircalin): 8 rue Frédéric Surleau, BP 3736, 98846 Nouméa Cédex; tel. 265546; fax 272772; e-mail aircalin@aircalin.nc; internet www.aircalin.nc/aci.htm; f. 1983; ; services to Sydney and Brisbane (Australia), Auckland (New Zealand), Nadi (Fiji), Papeete (French Polynesia), Wallis and Futuna Islands, Port Vila (Vanuatu) and Osaka and Tokyo (Japan); Chair. CHARLES LAVOIX; CEO JEAN-MICHEL MASSON.

Tourism

An investment programme was begun in 1985 with the aim of developing and promoting tourism. A total of 2,388 hotel rooms were available in 1999. In 2003 there were 101,983 visitors to New Caledonia, of whom 28.3% came from France, 27.9% from Japan and 15.6% from Australia. The industry earned US $93m. in 2001.

New Caledonia Tourism South: Galerie Nouméa Centre, 20 rue Anatole France; BP 688, 98845 Nouméa Cédex; tel. 242080; fax 242070; e-mail info@nctps.com; internet www.nctps.com; f. 2001; international promotion of tourism in the Southern Province of New Caledonia; Chair. CHRISTINE GAMBEY; Gen. Man. JEAN-MICHEL FOUTREIN.

Defence

In August 2003 France was maintaining a force of 2,700 military personnel in New Caledonia.

Education

Education is compulsory for 10 years between six and 16 years of age. Schools are operated by both the State and churches, under the supervision of three Departments of Education: the Provincial department responsible for primary level education, the New Caledonian department responsible for primary level inspection, and the State department responsible for secondary level education. The French Government finances the state secondary system. Primary education begins at six years of age, and lasts for five years; secondary education, beginning at 11 years of age, comprises a first cycle of four years and a second, three-year cycle. In 2000 there were 81 pre-primary schools, and in 2001 there were 289 primary schools and 64 secondary schools. In 1999 there were 19 technical and higher institutions (of which 11 were private). Some 400 students attend universities in France. In 1987 the French University of the Pacific (based in French Polynesia) was established, with a centre in Nouméa, and divided into two universities in 1999. In 2000 the University of New Caledonia had 60 teachers and 1,600 students. Several other vocational tertiary education centres exist in New Caledonia. In 1989 the rate of adult illiteracy averaged 6.9% (males 6.0%, females 7.9%). According to UNESCO, total public expenditure on education in 1993 was 1,652m. French francs.

KIRIBATI

Physical and Social Geography

The Republic of Kiribati (pronounced 'Kir-a-bas') comprises one island and 32 atolls, in three principal groups, scattered over about 5m. sq km of ocean along the Equator and extending about 3,780 km from east to west and 2,050 km from north to south. The island of Banaba (Ocean Island), a solid coral outcrop 306 km to the east of Nauru, and the 16 Gilbert Islands lie in the west of Kiribati, with Tuvalu to the south and the Marshall Islands to the north. The eight Phoenix Islands, which are largely uninhabited, lie some 1,300 km south-east of the Gilbert group and to the north of Tokelau. In the east, Kiribati also comprises eight of the Line Islands (three others are uninhabited dependencies of the USA), including Kiritimati (Christmas Island), the largest coral atoll in the world, which covers 388 sq km (150 sq miles), or more than half of the total 717 sq km (277 sq miles) of dry land in Kiribati. Tahiti, French Polynesia, lies some 900 km to the south-east of the Line Islands.

Most of Kiribati has a maritime equatorial climate, the northern and southern islands being in the tropical zone. Temperature varies very little through the year, and the mean annual temperature ranges from 29°C (84°F) in the southern Gilberts to 27°C (81°F) in the Line Islands. There is a season of north-westerly trade winds from March to October and a season of rains and gales from October to March. Average annual rainfall varies greatly, averaging about 3,000 mm in the islands north of Tarawa, about 1,500 mm on Tarawa and for most of the Gilberts group, and 700 mm in the Line Islands. All the islands are prone to severe drought.

The population, which is mainly Micronesian and Christian, is concentrated in the Gilbert Islands. The principal languages are I-Kiribati (Gilbertese) and English. At the census of November 2000 there were 84,494 people in Kiribati. The total population was estimated at 88,800 at mid-2003. The capital is located on Bairiki island in Tarawa.

History

The I-Kiribati people first settled the islands of the Gilberts (or Tungaru) group between 1000 and 1300 AD. European contact began in the 16th century. In 1892 the United Kingdom established a protectorate over the 16 atolls of the Gilbert Islands and the nine Ellice Islands (now Tuvalu). The two groups were administered together under the jurisdiction of the Western Pacific High Commission (WPHC), which was based in Fiji until its removal to the British Solomon Islands (now Solomon Islands) in 1953. Phosphate-rich Ocean Island (Banaba), west of the Gilbert group, was annexed by the United Kingdom in 1900. The Gilbert and Ellice Islands were annexed in 1915, effective from January 1916, when the protectorate was declared a colony. Later in 1916 the new Gilbert and Ellice Islands Colony (GEIC) was extended to include Ocean Island and two of the Line Islands, far to the east. Christmas Island (now Kiritimati), another of the Line Islands, was added in 1919 and the eight Phoenix Islands (then uninhabited) in 1937. A joint British-US administration for two of the Phoenix group, Kanton (Canton) and Enderbury, was agreed in April 1939.

During the Second World War the GEIC was invaded by Japanese forces, who occupied the Gilbert Islands in 1942–43. Tarawa Atoll, in the Gilbert group, was the scene of some of the fiercest fighting in the Pacific between Japan and the USA.

As part of the British Government's programme of developing its own nuclear weapons, the first test of a British hydrogen bomb was conducted near Christmas Island in May 1957, when a device was exploded in the atmosphere. Two further tests took place in the same vicinity later that year.

Preparations for self-government in the GEIC began in 1963, and there were new Constitutions in 1967 (when an assembly containing elected representatives was first introduced) and in 1970. In January 1972 a Governor of the GEIC was appointed to assume almost all the functions previously exercised in the colony by the High Commissioner. At the same time the five uninhabited Central and Southern Line Islands, previously administered directly by the High Commissioner, became part of the GEIC. In May 1974 a House of Assembly, with 28 elected members and three official members, replaced the previous legislature. The House elected Naboua Ratieta as Chief Minister.

On 1 October 1975 the Ellice Islands were allowed to separate from the GEIC to form a distinct territory, named Tuvalu. The remainder of the GEIC was renamed the Gilbert Islands and the House of Assembly's membership was reduced.

In 1975 the British Government refused to recognize as legitimate a demand for independence by the people of Banaba, or Ocean Island, who had been in litigation with the British Government since 1971 over revenues derived from exports of phosphate. The discovery of the guano deposits on the 600 ha island was a prime motive in Britain's annexation of the island. Between 1920 and 1979 the British Phosphate Commission (BPC), a consortium of the British, Australian and New Zealand Governments, mined phosphate for use as a fertilizer in Australia and New Zealand. Open-cast mining so adversely affected the island's environment that most Banabans were removed from the island during the Second World War and resettled on Rabi Island, 2,600 km away in the Fiji group, becoming citizens of Fiji in 1970. Banabans remain the landowners of Banaba.

The Banabans rejected the British Government's argument that phosphate revenues should be distributed over the whole Gilbert Islands group and in 1973, despite winning 50% of the revenues, continued with litigation. They claimed unpaid royalties from the British Government and damages for the destruction of the island's environment against both the Government and the BPC. The Banabans took these two cases to the British High Court in London. In 1976, after a lengthy hearing, their claim for royalties was dismissed but that for damages upheld. In May 1977 the British Government offered them an *ex gratia* payment of $A10m. without admitting liability for damages and on condition that no further judicial appeal would be made. The offer was not accepted.

The Gilbert Islands obtained internal self-government on 1 January 1977. Later that year the number of elected members in the House of Assembly was increased to 36. This was subsequently adjusted to 35, with the remaining seat to be filled by a nominee of the Rabi Council of Leaders. Following a general election in February 1978, Ieremia Tabai, Leader of the Opposition in the previous House of Assembly, was elected Chief Minister in March. On 12 July 1979 the Gilbert Islands became an independent republic, within the Commonwealth, under the name of Kiribati. The country did not become a member of the UN (owing to financial considerations). The House of Assembly was renamed the Maneaba ni Maungatabu, and Ieremia Tabai became the country's first President (Beretitenti). In September Kiribati signed a treaty of friendship with the USA, which relinquished its claim to the Line and Phoenix Islands, including Kanton and Enderbury. In April 1981 the Banaban community on Rabi accepted the British Government's earlier *ex gratia* offer of $A10m. in compensation together with the interest accrued ($A14.58m. in total), although they continued to seek self-government.

The first general election since independence began in March 1982, with a second round of voting in April. The members of the new Maneaba all sat as independents. President Tabai was re-elected in May. However, the Government resigned in December, following the Maneaba's second rejection of proposals to increase salaries for civil servants. As a result, the Maneaba was dissolved, and another general election took place in January 1983. President Tabai was again re-elected in February 1983. Following a general election in March 1987, Tabai was re-elected for a fourth term in May. At a general election in May 1991 eight incumbent members of the Maneaba lost their seats. A presidential election was held in July, when the former Vice-President, Teatao Teannaki, narrowly defeated Roniti Teiwaki to replace Tabai, who had served the maximum period in office permitted by the Constitution.

In January 1992 the Maneaba approved an opposition motion urging the Government to seek compensation from Japan for damage caused during the Second World War. The motion, which had been presented to previous sessions of the legislature, was thought to have been prompted by the success of a similar claim by the Marshall Islands. The intention to pursue a compensation claim against the Japanese was reiterated by President Teburoro Tito in late 1994.

In May 1994 the Government was defeated on a motion of confidence, following opposition allegations that government ministers had misused travel allowances. As a result of the vote, the Maneaba was dissolved, and at legislative elections in July five cabinet ministers lost their seats. Thirteen of the newly elected members were supporters of the opposition Maneaban Te Mauri, while only eight were known to support the previously dominant National Progressive Party grouping. At the presidential election in September Teburoro Tito was the successful candidate, receiving 51.1% of the total votes. The new President declared that reducing Kiribati's dependence on foreign aid would be a major objective for his Government. He also announced his intention to pursue civil and criminal action against members of the previous Government for alleged misuse of public funds while in office.

In 1995 a committee was created with the aim of assessing public opinion regarding possible amendments to the Constitution. In March 1998 more than 200 delegates attended a Constitutional Review Convention in Bairiki to consider the committee's recommendations, which included equalizing the status of men and women regarding the citizenship rights of foreigners marrying I-Kiribati and changing the structure of the Council of State. Leaders of the Banaban community in Rabi, Fiji, were also consulted during 1998 as part of the review process.

A general election, held in September 1998 and contested by a record 191 candidates, failed to produce a conclusive result, necessitating a second round of voting, at which the Government and opposition each lost seven seats. At a presidential election in November Tito was re-elected with 52.3% of total votes cast, defeating Dr Harry Tong, who obtained 45.8% of votes, and Ambreroti Nikora, with 1.8%.

In late 1999 concerns were expressed by a Pacific media organization after a New Zealand journalist working for Agence France-Presse was banned from entering Kiribati. The Kiribati Government claimed that a series of articles by the correspondent, unfavourable to Kiribati, which had been published in a regional magazine, were biased and sensationalist. In December former President Ieremia Tabai and Atiera Tetoa, a former member of the Maneaba, were fined, having been convicted of importing telecommunications equipment without a permit. They had launched Newair FM, an independent commercial radio station, 12 months previously; it had been immediately suspended and a criminal investigation was instigated by the police. Tab'ai subsequently established Kiribati's first private newspaper, the *Kiribati Newstar*, in an attempt to reduce the Government's control over the media in the islands. Its first published edition appropriately coincided with Media Freedom Week in May 2000.

In November 2000 the Vice-President and Minister for Home Affairs and Rural Development, Tewarika Tentoa, collapsed while addressing the Maneaba and died. The post of Vice-President was subsequently combined with the cabinet portfolio of finance and economic planning.

In mid-October 2002 Tito's administration suffered a rare defeat when a government bill aimed at amending the Constitution was rejected, after failing to gain the two-thirds' support necessary. The proposed legislation, which received 20 parliamentary votes in favour and 13 against, included measures to replace the Chief Justice with the Attorney-General on the Council of State and amendments to the number of members of Parliament from each island. Opposition members criticized the proposals, which they claimed would favour the Government in the forthcoming general election.

Campaigning for the general election during November 2002 was characterized by numerous allegations of improper conduct. Observers noted that officials from the Chinese embassy in Tarawa, accompanied by government candidates, had been donating gifts to the local community in the weeks preceding the election. (The Government had recently amended the Elections Act to allow gifts to be distributed to the public by candidates during their electoral campaigns, a practice that had been banned hitherto.) The opposition, which had stated its intention to close the Chinese satellite-tracking station (based on South Tarawa—see below) if elected, claimed that this action constituted a clear attempt to influence voters. Moreover, under a newly amended Newspaper Registration Act, Tito ordered police to seize opposition election pamphlets in November. Further allegations that the Government was attempting to stifle freedom of expression were made by former President Ieremia Tabai, whose private radio station was finally granted a licence to broadcast in December, following delays totalling almost four years in issuing the permit.

A total of 176 candidates contested the general election on 29 November 2002. The Government suffered significant losses, with 14 of its members (including seven ministers) failing to retain their seats. The presidential election was postponed from its original date and finally took place on 25 February 2003. At the poll Tito received 14,160 votes, while the opposition candidate, Taberannang Timeon, secured 13,613. Tito was sworn in for his third term as President on 28 February, and many of his former opponents in the legislature were expected to cross the floor to support him. However, in late March Tito was narrowly defeated on a motion of 'no confidence' and his Government was replaced by an interim Council of State (comprising the Speaker, the Chief Justice and the Public Service Commissioner). In accordance with the Constitution, another general election took place on 9 and 14 May, at which supporters of Tito secured a majority of seats. A presidential election took place in early July at which the opposition candidate, Anote Tong, narrowly defeated his brother, Harry Tong. Anote Tong's electoral campaign, which had focused on his pledge to review the lease of the Chinese satellite-tracking station on South Tarawa, had been characterized by a series of personal attacks on his brother.

In early November 2003 President Anote Tong announced the establishment of diplomatic relations with Taiwan. The Government's decision to switch its allegiance from the People's Republic of China to Taiwan caused considerable controversy within Kiribati and the region. Several hundred people staged a protest in Tarawa against the Government's decision, claiming that it had been made in return for Taiwanese funding of Anote Tong's electoral campaign. The President strongly refuted this allegation, but did, however, state that Taiwan had offered extensive development funds to Kiribati for adopting its position. By late November it was reported that Chinese technicians were dismantling the satellite-tracking station, which had played an important role in China's recent first manned space flight. The Chinese embassy, however, remained open while China requested that Kiribati reconsider its decision. Despite its efforts, which, according to many commentators, were motivated largely by the islands' strategic importance to China, in late November that country suspended diplomatic relations with Kiribati. In the same week the police in Kiribati announced the launch of an investigation into the death threats received by President Tong, believed to be from a Chinese source. A Taiwanese embassy was opened in Tarawa in January 2004. The continued presence of Chinese officials in Kiribati (three diplomats remained as caretakers of the Chinese embassy building in Tarawa in mid-2004) caused the authorities some concern. President Tong expressed fears that the representatives were remaining on the islands in order to influence a change of government.

Reports that Palmyra Atoll (an uninhabited US territory some 200 km north of Kiribati's northern Line Islands) was to be sold and used by a US company for the storage of nuclear waste led the Maneaba to pass a unanimous resolution in May 1996 urging the Government to convey the islanders' concerns over the proposals to the US Government. The islanders' anxieties centred largely on Palmyra's geographical proximity to Kiribati, coupled with the belief that the atoll's fragility and porous structure make it an unstable environment for the storage of highly toxic materials. In June Kiribati formally requested that the US Government veto the proposals. In November 2000 Palmyra was purchased by The Nature Conservancy, a conservation group that planned to preserve the natural state of the atoll.

In April 1999 the Kiribati Government announced its desire to acquire Baker, Howland, and Jarvis Islands (see p. 974) from the USA, because of the economic value of their fish resources. In mid-1999 there was some controversy over the renamed Millennium Island (previously Caroline Island). The island had been renamed in 1997 in an attempt to market it as a tourist destination for the coming year 2000. In 1994 Kiribati moved the international date-line to incorporate the Line and Phoenix Island groups (including Millennium Island) in the same time zone as the Gilbert group, thus creating a large eastward projection in the date-line. Millennium Island's position as the first place to celebrate the New Year was subsequently confirmed. In early 2000, however, opposition politicians severely criticized the Government for failing to attract the predicted number of tourists to the islands' millennium celebrations, despite expenditure of more than $A1m.

In 1995 Kiribati suspended its diplomatic relations with France in protest at the French Government's decision to renew nuclear-weapons testing at Mururoa Atoll. Kiribati was admitted to the UN in September 1999.

Owing to the territory's high rate of population growth (more than 2% per year) and, in particular, the situation of over-population on South Tarawa and its associated social and economic problems, it was announced in 1988 that nearly 5,000 inhabitants were to be resettled on outlying atolls, mainly in the Line Islands. The migration began in 1989. A further programme of resettlement from South Tarawa to five islands in the Phoenix group was initiated in 1995.

A Chinese satellite-tracking station was opened on South Tarawa in late 1997 but was dismantled in late 2003 following the suspension of diplomatic relations between the two countries (see above). Sea Launch, a US consortium led by the Boeing Commercial Space Company, also announced plans to undertake a rocket-launching project from a converted oil-rig near the islands. The facility began operations in March 1999, when a prototype satellite was launched. Kiribati was not, however, expected to benefit financially from the proposed project, as the consortium had controversially sought to carry out its activities in international waters near the outer limits of the islands' exclusive economic zone. The US authorities dismissed environmental concerns about the negative impact of the site (particularly the dumping of large quantities of waste fuel in the islands' waters), which were expressed both by the Government of Kiribati and the South Pacific Regional Environmental Programme in mid-1998. These fears were compounded in March 2000 after a rocket launched from the site crashed and, furthermore, Sea Launch refused to disclose where it had landed. In November 1999 it was announced that the Government of Kiribati and the National Space Development Agency of Japan had reached agreement on the proposed establishment of a space-vehicle launching and landing facility on Kiritimati. In the following year the Japanese organization was also given permission by the Kiribati Government to use land and runway facilities on the island, free of charge, until 2020 and to construct a 100–150-room hotel.

Kiribati is a member of the Pacific Islands Forum (formerly the South Pacific Forum), which expressed concern about the threat of the 'greenhouse effect' (the heating of the earth's atmosphere and a resultant rise in sea-level, as a consequence of pollution) in 1988. In 1989 a UN report on the 'greenhouse effect' listed Kiribati as one of the countries that would completely disappear in the 21st century unless drastic action were taken. None of the land on the islands is more than two metres above sea-level. A rise in sea-level would not only cause flooding, but would also upset the balance between sea and fresh water (below the coral sands), rendering water supplies undrinkable. In mid-1999 it was announced that two uninhabited coral reefs had sunk beneath the sea as a result of the 'greenhouse effect'. In late 1997 President Tito strongly criticized the Australian Government's refusal, at the Conference of the Parties to the Framework Convention on Climate Change (see Regional Organizations, UN Environment Programme) in Kyoto, Japan, to reduce its emission of gases known to contribute to the 'greenhouse effect'.

As a result of discussions held at the Pacific Islands Forum summit meeting in Kiribati in October 2000, Japan announced that it was willing to negotiate compensation claims with the islanders for damage caused during the Second World War. Furthermore, a six-day visit by President Tito to Japan in February 2001 resulted in a number of informal agreements aimed at enhancing relations between the two countries. These included a decision to try to resolve a dispute over tuna fishing, caused by Japan's refusal to sign a convention aimed at protecting tuna stocks in the central and western Pacific Ocean and agreements to address their differences over whaling and nuclear-fuel shipments. Tito also appeared to modify his position on nuclear energy following the visit to Japan, stating that emissions of harmful 'greenhouse gases' could be reduced by replacing fossil with nuclear fuel. In April 2001 the USA's decision to reject the Kyoto Protocol to the UN's Framework Convention on Climate Change was widely criticized. In March 2002 Kiribati, Tuvalu and the Maldives announced their decision to take legal action against the USA for its refusal to sign the Kyoto Protocol.

Economy

Until 1979 phosphate rock, derived from rich deposits of guano, was mined on Banaba by the British Phosphate Commission for export to Australia and New Zealand, where it was used for fertilizer. The ending of phosphate production had a devastating effect on the economy, as receipts from phosphates had accounted, on average, for about 80% of total export earnings and 50% of government taxation revenue. The investment of some phosphate earnings from 1956 did result in a considerable fund of foreign reserves, which have been maintained since 1979. In December 2003 the value of the Revenue Equalization Reserve Fund (RERF) was valued at $A400m. The RERF has provided the Government with investment income equivalent to around 33% of gross domestic product (GDP) per year, and in 2001 the Asian Development Bank (ADB) estimated that the value of the RERF was such that Kiribati had sufficient foreign reserves for 10 years' worth of imports. It was estimated that, owing to the cessation of phosphate mining, the country's GDP per head was halved between the late 1970s and early 1980s.

In 2002, according to estimates by the World Bank, Kiribati's gross national income (GNI), measured at average 2000–02 prices, was US $77m., equivalent to US $810 per head. During 1990–2002, it was estimated, GDP per head increased, in real terms, at an average annual rate of 0.3%. Over the same period the population increased by an average of 2.3% per year. According to figures from the Asian Development Bank (ADB), Kiribati's overall GDP increased, in real terms, by an estimated average of 4.1% per year in 1990–2002. Real GDP expanded by 0.9% in 2002 and by 3.3% in 2003. The ADB forecast an increase of 1.8% in real GDP for 2004.

Most islanders participate in subsistence activities, although, according to FAO, agriculture and fishing engaged only 26% of the economically active population in 2002. Apart from Banaba, Kiribati is composed of coral atolls with poor quality soil. Most of them are covered with coconut palms, which provide the only agricultural export in the form of copra. A government-owned company operates a coconut plantation on Kiritimati, and there are commercial plantations on two other atolls in the Line Islands. Production of copra, however, declined to 6,133 metric tons in 2000 from 11,776 tons in 1999. In the latter year, according to the ADB, exports of the commodity earned almost $A9.0m. (equivalent to 63.9% of total domestic export revenue), although this figure decreased to $A2.5m. (23.4% of exports) in 2000. According to the ADB, agriculture (including fishing) contributed an estimated 14.2% of monetary GDP in 2002. Bananas, screw-pine (*Pandanus*), breadfruit and papaya are cultivated as food crops. The cultivation of seaweed began on Tabuaeran in the mid-1980s: seaweed provided an estimated 15.9% of domestic export earnings in 2000. Pigs and chickens are kept.

As a result of various development projects, the fishing sector's contribution to the country's GDP trebled between 1979 and 1984. The significance of fishing to the economy, however, drastically declined thereafter, and the closure of the state fishing company was announced in 1991. Fish provided only 1.8% of export earnings in 2000 (compared with 46.2% in 1990). Nevertheless, pet fish have become a significant commodity, contributing 13.6% of export earnings in 1999 (although this figure declined to 1.9% in 2000). The sale of fishing licences to foreign fleets (notably from South Korea, Japan, Taiwan and the USA) provides an important source of income. Revenue from the sale of fishing licences increased from 1996, when the Government increased their cost in response to the continued illegal operations of foreign vessels, which was noticeable in the following year's fishing licence revenues of $A28m. Income from this source rose from an estimated $A12.8m. in 1999 (uncharacteristically low as a result of unfavourable climatic conditions) to reach $A52m. in 2001 before declining to $A32m. in 2002. According to figures from the ADB, the GDP of the agricultural sector decreased at an average annual rate of 0.4% in 1990–2002. The sector's GDP remained constant in 2001 but decreased by 4.1% in 2002.

Industry (including manufacturing, construction and power) contributed an estimated 10.9% of monetary GDP in 2002. The sector employed 38.9% of the work-force in 1995. According to figures from the ADB, industrial GDP increased by an average of 8.6% per year in 1990–2002. Compared with the previous year, industrial GDP was estimated by the ADB to have expanded by 25.5% in 2001, but declined by 10.1% in 2002. The production of solar-evaporated salt for export to Hawaii, Fiji, Guam, American Samoa and Samoa (for use on fishing vessels with brine refrigeration systems) began on Kiritimati in 1985. A small industrial centre was established on Betio, with British aid, in 1990. Manufacturers based there produce garments, footwear, furniture, leather goods and kamaimai (a toddy condensed from coconut). Construction of a new copra mill, near Betio port, began in March 2002. The project, in partnership with Australian company Techso, was to extend copra production to the outer islands and produce oils and soaps for both local and international markets. Manufacturing contributed an estimated 0.8% of monetary GDP in 2002. Manufacturing GDP increased by an annual average of 5.4% in 1990–2002. Compared with the previous year, the GDP of the manufacturing sector expanded by an estimated 18.6% in 2001 but contracted by 3.0% in 2002.

Production of electrical energy declined from 14.5m. kWh in 2000 to 12.5m. kWh in 2001. Mineral fuels accounted for an estimated 8.8% of total import costs in 2000. In August 2001 the EU announced that it planned to fund the introduction of 1,500 new solar energy systems, valued at more than $A6m., to Kiribati. Moreover, in May 2003 the Government announced the completion of a Japanese-funded programme to construct a new power station, to install two new generating units and to upgrade 16 km of power lines. The project, which cost a total of US $11m. to implement, was expected to ensure a power supply sufficient to meet Kiribati's increasing demand.

Services provided 75.0% of monetary GDP in 2002. Tourism makes a significant contribution to the economy: the trade and hotels sector provided an estimated 16.6% of GDP in 2000. The GDP of the trade and hotels sector increased by an annual average of 4.2% in 1990–98. The services sector employed 87.4% of the work-force in 1995. Kiribati's tourist sector was expected to expand from 2001 when, in February, the Government signed an agreement with a major cruise line, which agreed to include the island in its cruise ship itinerary. Up to 10 ships per year were expected to visit the island, with each visit generating around US $15,000 for the local economy. Tourist arrivals at Tarawa and Kiritimati, however, declined from 4,842 in 2001, in which year receipts from tourism reached $A3.0m., to 4,831 in 2002 and to 4,288 in 2003. The GDP of the services sector increased at an annual average rate of 4.7% in 1990–2002. According to estimates from the ADB, the services sector's GDP decreased by 0.7% in 2001 but increased by 3.0% in 2002.

Kiribati recorded a trade deficit of an estimated US $27.6m., and a surplus of US $1.75m. on the current account of the balance of payments in 2001. In that year Kiribati's trade deficit narrowed to 58.5% of GDP, as a decline in income from copra exports was more than offset by a fall in imports. By 2003 the deficit on the current account of the balance of payments was estimated by the ADB to be equivalent to 18.7% of GDP. The major imports in 2000 were food and live animals, machinery and transport equipment, manufactures, mineral fuels, beverages and tobacco, and chemicals. The major domestic exports were copra, seaweed and shark fins. In 2003 the value of exports declined to US $32.2m., while imports increased to US $54.6m. In that year the principal sources of imports were Australia (39.8%) and Fiji (23.0%). The principal recipients of exports in that year were Japan (59.7%) and the Republic of Korea (19.9%).

In 2000 Kiribati recorded an overall budget deficit of $A3.0m. Budgetary expenditure for 2002 was projected at $A77.9m., 15% less than the revised estimates for 2001, and required a drawdown of

$A16.7m. from the RERF. In 2002, according to the ADB, the government surplus increased to reach the equivalent of 21.4% of GDP, but a deficit of 13.4% was projected for 2003. In April 2003 the interim administration enacted a budget providing for expenditure of $A32m. over the next five months. Kiribati's total external debt in 2003 was estimated by the ADB at US $16m., and the cost of debt-servicing was equivalent to 1.9% of revenue from exports of goods and services. The annual rate of inflation averaged 3.1% in 1990–2000. Consumer prices increased by an annual average of 6.0% in 2001, by 3.2% in 2002 and by 1.8% in 2003. About 1.6% of the labour force were unemployed in 2000. Only around 8,600 people, equivalent to less than 20% of the working-age population, were formally employed in 2001.

Kiribati became a member of the ADB in 1974. In 1986 Kiribati joined the IMF. Following negotiations held in the same year at the Government's request, Kiribati was placed on the UN list of least developed countries, a status that attracts concessional loans from the World Bank and the IMF, and renders the country's exports eligible for special tariff rates. Kiribati is an associate member of the UN Economic and Social Commission for Asia and the Pacific (ESCAP). The republic is also a member of the South Pacific Forum and of the Pacific Community (formerly the South Pacific Commission). It is a signatory of the South Pacific Regional Trade and Economic Co-operation Agreement (SPARTECA) and of the Lomé Conventions and successor Cotonou Agreement with the EU. In early 1996 Kiribati joined representatives of the other countries and territories of Micronesia at a meeting in Hawaii, at which a new regional organization, the Council of Micronesian Government Executives, was established. The new body aimed to facilitate discussion on economic developments in the region and to examine possibilities for reducing the considerable cost of shipping essential goods between the islands.

According to UN criteria, Kiribati is one of the world's least-developed nations. The islands' vulnerability to adverse climatic conditions was illustrated in early 1999 when a state of national emergency was declared following a prolonged period of severe drought. Kiribati's extremely limited export base and dependence on imports of almost all essential commodities result in a permanent (and widening) trade deficit, which is in most years only partially offset by revenue from fishing licence fees, interest earned on the RERF and remittances from I-Kiribati working overseas. The majority of the latter group is comprised of seamen working on foreign ships, particularly Japanese fishing boats and cruise liners. In 2000 there were some 2,000 of these overseas workers, whose remittances totalled $A10.4m. in that year. Although the value of the RERF had trebled within 10 years, the fund then declined in value in 2001 as a result of the downturn in world stock markets (its assets being invested in offshore markets). The country is reliant on foreign assistance for its development budget. Official development assistance declined from a total of US $20.9m. in 1999 to US $17.9m. in 2000. In 2003/04 New Zealand provided $NZ3.14m. and Australia $A11.4m. in development assistance. In early 2003 Kiribati established a non-governmental body to secure US $840,000 in EU funding under the Cotonou Agreement, which would be spent on projects over the next five years. Development finance was expected to be received from Taiwan following the establishment of diplomatic relations in late 2003. Dependence on external finance, however, is widely regarded as having left Kiribati vulnerable to foreign exploitation. Moreover, concern has been expressed that, although foreign companies specializing in advanced technology (particularly telecommunications and satellite systems) are seeking to establish operations in the islands, Kiribati will not benefit significantly from the major investment involved in such projects. In 2000 the Government generated some US $400,000 of revenue through the sale of I-Kiribati passports to investors in the islands. The scheme was introduced largely to encourage people from overseas, in particular Chinese nationals, to establish business interests in Kiribati. Further passport sales in 2001 were worth US $375,000 and, combined with sales of Kiribati Residential Permits, produced more than $A2.5m. in revenue. However, following international pressure in the aftermath of the terrorist attacks on the USA, in September 2001, the Government announced in 2004 that it would discontinue the scheme. The Government's policy of subsidizing copra producers following a fall in world prices of the commodity, however, had a negative impact on the economy; in 2001 these subsidies totalled $A2m. The economy recovered somewhat in 2001, largely as a result of a substantial increase in recurrent government expenditure, an improvement in copra production and the implementation of various development projects. A modest rate of economic growth was maintained in 2002, prior to stronger expansion in 2003. The Government's National Development Strategy for 2000–03 sought to reform the public sector and promote private-sector development. The new proposals were intended to encourage greater investment in the country's private sector, which would allow the creation of jobs and the broadening of the islands' narrow base of exports.

Statistical Survey

Source (unless otherwise stated): Statistics Office, Ministry of Finance and Economic Planning, POB 67, Bairiki, Tarawa; tel. 21082; fax 21307.

AREA AND POPULATION

Area: 810.5 sq km (312.9 sq miles). *Principal Atolls* (sq km): Banaba (island) 6.29 Tarawa 31.02 (North 15.26, South 15.76); Abemama 27.37; Tabiteuea 37.63 (North 25.78, South 11.85); Total Gilbert group (incl. others) 285.52; Kanton (Phoenix Is) 9.15; Teraina (Fanning) 33.73; Kiritimati (Christmas—Line Is) 388.39.

Population: 77,658 at census of 7 November 1995; 84,494 (males 41,646, females 42,848) at census of 7 November 2000; 88,800 (estimate) at mid-2003. *Principal Atolls* (2000): Banaba (island) 276; Abaiang 5,794; Tarawa 41,194 (North 4,477, South—including Bairiki, the capital—36,717); Tabiteuea 4,582 (North 3,365, South 1,217); Total Gilbert group (incl. others) 78,158; Kanton (Phoenix Is) 61; Kiritimati 3,431; Total Line and Phoenix Is (incl. others) 6,336. *Mid-2003* (UN estimate, incl. suburbs) South Tarawa (including Bairiki, the capital) 41,530 (Source: UN, *World Urbanization Prospects: The 2003 Revision)*.

Density (mid-2003): 109.6 per sq km.

Ethnic Groups (census of 2000): Micronesians 83,452; Polynesians 641; Europeans 154; Others 247; Total 84,494.

Principal Towns: (population in '000, 1990): Bairiki (capital) 18.1; Bikenibeu 5.1; Taburao 3.5; Butaritari 3.2. Source: Stefan Helders, *World Gazetteer* (internet www.world-gazetteer.com). *Mid-2003* (UN estimate, incl. suburbs) South Tarawa (including Bairiki, the capital) 41,530 (Source: UN, *World Urbanization Prospects: The 2003 Revision)*.

Births, Marriages and Deaths: Registered live births (1996) 2,299 (birth rate 29.5 per 1,000); Marriages (registrations, 1988) 352 (marriage rate 5.2 per 1,000); Death rate (estimate, 1995) 7 per 1,000.

Expectation of Life (WHO estimates, years at birth): 64.1 (males 61.8; females 66.7) in 2002. Source: WHO, *World Health Report*.

Employment (paid employees, 1995, provisional): Agriculture, hunting, forestry and fishing 487; Manufacturing 104; Electricity, gas and water 182; Construction 215; Trade, restaurants and hotels 1,026; Transport, storage and communications 710; Financing, insurance real estate and business services 349; Community, social and personal services 4,778; *Total employed* 7,848. Source: UN, *Statistical Yearbook for Asia and the Pacific*. *Mid-2002* (estimates, '000 persons employed): Agriculture, etc. 10; Total 38 (Source: FAO).

HEALTH AND WELFARE

Key Indicators

Total Fertility Rate (children per woman, 2002): 4.1.

Under-5 Mortality Rate (per 1,000 live births, 2002): 69.

Physicians (per 1,000 head, 1998): 0.30.

Hospital Beds (per 1,000 head, 1990): 4.27.

Health Expenditure (2001): US $ per head (PPP): 143.

Health Expenditure (2001): % of GDP: 8.6.

Health Expenditure (2001): public (% of total): 98.8.

Access to Water (% of persons, 2000): 47.

Access to Sanitation (% of persons, 2000): 48.
For sources and definitions, see explanatory note on p. vi.

AGRICULTURE, ETC.

Principal Crops (FAO estimates, '000 metric tons, 2002): Taro (Coco yam) 1.7; Other roots and tubers 7.2; Coconuts 96; Vegetables 5.6; Bananas 4.6; Other fruits 1.2.

Livestock (FAO estimates, '000 head, year ending September 2002): Pigs 12; Chickens 450. Source: FAO.

Livestock Products (FAO estimates, metric tons, 2002): Pig meat 892; Poultry meat 450; Hen eggs 230. Source: FAO.

Fishing ('000 metric tons, live weight, 2002): Capture 31.0 (Snappers 0.9; Flyingfishes 1.6; Jacks and crevalles 3.7; Skipjack tuna 3.8; Emperors 1.4; Clupeoids 3.6; Marine molluscs 2.3); Aquaculture 0.0; Total catch 31.0. Figures exclude aquatic plants ('000 metric tons): 12.6 (all aquaculture). Source: FAO.

INDUSTRY

Copra Production (metric tons): 5,853 in 2000; 6,741 in 2001; 5,903 in 2002.

Electric Energy (million kWh): 12.86 in 1999; 14.48 in 2000; 12.46 in 2001.
Source: Asian Development Bank, *Key Indicators of Developing Asian and Pacific Countries*.

FINANCE

Currency and Exchange Rates: Australian currency: 100 cents = 1 Australian dollar ($A). *Sterling, US Dollar and Euro Equivalents* (31 May 2004): £1 sterling = $A2.5685; US $1 = $A1.4000; €1 = $A1.7144; $A100 = £38.93 = US $71.43 = €58.33. *Average Exchange Rate* (US $ per Australian dollar): 1.9334 in 2001; 1.8406 in 2002; 1.5419 in 2003.

Budget ($A million, 2000): *Revenue*: Tax revenue 18.7 (Corporate tax 4.7, Import duties 13.9); Other current revenue 42.6 (Fishing licence fees 31.2, Fees and incidental sales 5.8); Capital revenue 0.0; Total 61.3. *Expenditure*: General public services 7.0; Public order and safety 4.7; Education 12.8; Health 8.8; Welfare and environment 1.1; Community amenities 1.5; Agriculture and fishing 2.3; Construction 1.9; Communication 1.7; Commerce 0.7; Labour affairs 1.7; Others 20.0; Total 64.3 (Current 53.9, Capital 10.4).

Cost of Living (Consumer Price Index for Tarawa; base: 1996 = 100): 114.3 in 2001; 118.0 in 2002;120.2 in 2002.

Expenditure on the Gross Domestic Product ($A '000 at current prices, 1992): Government final consumption expenditure 25,039; Private final consumption expenditure 31,592; Increase in stocks 250; Gross fixed capital formation 25,750; *Total domestic expenditure* 82,631; Exports of goods and services 5,798; *Less* Imports of goods and services 52,625; Statistical discrepancy 10,456; *GDP in purchasers' values* 46,260. Source: UN, *Statistical Yearbook for Asia and the Pacific*.

Gross Domestic Product by Economic Activity ($A '000 at current factor cost, 2002): Agriculture (incl. fishing) 11,605; Manufacturing 695; Electricity, gas and water 1,000; Construction 7,224; Trade, hotels and bars 11,632; Transport and communications 9,834; Finance (incl. imputed bank charges) 680; Public administration 34,640; Others (incl. owner-occupied dwelling) 4,680; *Sub-total* 81,990; Indirect taxes, *less* subsidies 16,223; *GDP in market prices* 98,213. Source: Asian Development Bank, *Key Indicators of Developing Asian and Pacific Countries* .

Balance of Payments (estimates, US $ million, 2001): *Trade balance* −27.6; Exports of services and income 39.3; Imports of services and income −21.5; *Balance on goods, services and income* −9.8; Current transfers received 13.0; Current transfers paid −1.4; *Current balance* 1.7; Capital account (net) 3.6; Direct investment (net) −0.5; Portfolio investment −5.7; *Overall balance* −0.9. Source: Asian Development Bank, *Key Indicators of Developing Asian and Pacific Countries*.

EXTERNAL TRADE

Principal Commodities (estimates, $A million, 2000): *Imports f.o.b.*: Food and live animals 19.5; Beverages and tobacco 4.6; Mineral fuels, lubricants, etc. 6.0; Chemicals 3.1; Basic manufactures 11.1; Machinery and transport equipment 15.9; Miscellaneous manufactured articles 5.4; Total (incl. others) 67.9. *Exports f.o.b.*: Copra 2.5; Seaweed 1.7; Pet fish 0.2; Shark fins 0.4; Total (incl. others) 10.7 .

Principal Trading Partners (US $ million, 2003): *Imports*: Australia 21.7; Fiji 12.6; Japan 4.8; New Zealand 4.4; Total (incl. others) 54.6. *Exports* (incl. re-exports): Australia 1.8; Japan 19.2; Korea, Republic 6.4; Poland 1.0; Thailand 1.3; USA 1.7; Total (incl. others) 33.5. Source: Asian Development Bank, *Key Indicators of Developing Asian and Pacific Countries*.

TRANSPORT

Road Traffic (motor vehicles registered on South Tarawa, 2000): Motor cycles 702; Passenger cars 477; Buses 10; Trucks 267; Minibuses 392; Others 13; Total 1,861.

Shipping: *Merchant Fleet* (registered, at 31 December 2003): 8 vessels; total displacement 4,198 grt. (Source: Lloyd's Register-Fairplay, *World Fleet Statistics*). *International Sea-borne Freight Traffic* ('000 metric tons, 1990): Goods loaded 15; Goods unloaded 26 (Source: UN, *Monthly Bulletin of Statistics*).

Civil Aviation (traffic on scheduled services, 1998): Passengers carried 28,000; Passenger-km 11 million; Total ton-km 2 million. Source: UN, *Statistical Yearbook*.

TOURISM

Foreign Tourist Arrivals (at Tarawa and Kiritimati): 4,842 in 2001; 4,831 in 2002; 4,288 in 2003.

Tourism Receipts ($A million): 2.1 in 1999; 2.2 in 2000; 3.0 in 2001.

COMMUNICATIONS MEDIA

Radio Receivers (1997): 17,000 in use.

Television Receivers (1997): 1,000 in use.

Telephones (main lines in use, 2001): 3,600.

Facsimile Machines (1996): 200 in use.

Mobile Cellular Telephones (subscribers, 2001): 500.

Personal Computers ('000 in use, 2001): 2.

Internet Users ('000, 2001): 2.

Non-daily Newspapers (2002): 2; estimated combined circulation 3,600.
Sources: UNESCO, *Statistical Yearbook*; UN, *Statistical Yearbook*; International Telecommunication Union; Australian Press Council.

EDUCATION

Primary (2001): 88 schools; 16,096 students; 627 teachers.

Junior Secondary (2001): 19 schools; 5,743 students; 324 teachers.

Secondary (2001): 14 schools; 5,743 students; 324 teachers.

Teacher-training (2001): 198 students; 22 teachers.

Vocational (2001): 1,303 students; 17 teachers.

Adult Literacy Rate (UNESCO estimates): 92.5% (males 93%; females 92%) in 2001. Source: UNESCO, *Assessment of Resources, Best Practices and Gaps in Gender, Science and Technology in Kiribati*.

Directory

The Constitution

A new Constitution was promulgated at independence on 12 July 1979. The main provisions are as follows:

The Constitution states that Kiribati is a sovereign democratic Republic and that the Constitution is the supreme law. It guarantees protection of all fundamental rights and freedoms of the individual and provides for the determination of citizenship.

The President, known as the Beretitenti, is Head of State and Head of the Government and presides over the Cabinet which consists of the Beretitenti, the Kauoman-ni-Beretitenti (Vice-President), the Attorney-General and not more than eight other ministers appointed by the Beretitenti from an elected parliament known as the Maneaba ni Maungatabu. The Constitution provided that the pre-independence Chief Minister became the first Beretitenti, but that in future the Beretitenti would be elected. After each general election for the Maneaba, the chamber nominates, from among its members, three or four candidates from whom the Beretitenti is elected by universal adult suffrage. Executive authority is vested in the Cabinet, which is directly responsible to the Maneaba ni Maungatabu. The Constitution also provides for a Council of State consisting of the Chairman of the Public Services Commission, the Chief Justice and the Speaker of the Maneaba.

Legislative power resides with the single-chamber Maneaba ni Maungatabu, composed of 40 members elected by universal adult suffrage for four years (subject to dissolution), one nominated member (see below) and the Attorney-General as an *ex-officio* member if he is not elected. The Maneaba is presided over by the Speaker, who is elected by the Maneaba from among persons who are not members of the Maneaba.

One chapter makes special provision for Banaba and the Banabans, stating that one seat in the Maneaba is reserved for a nominated member of the Banaban community. The Banabans' inalienable right to enter and reside in Banaba is guaranteed and, where any right over or interest in land there has been acquired by the Republic of Kiribati or by the Crown before independence, the Republic is required to hand back the land on completion of phosphate extraction. A Banaba Island Council is provided for, as is an independent commission of inquiry to review the provisions relating to Banaba.

The Constitution also makes provision for finance, for a Public Service and for an independent judiciary (see Judicial System).

The Government

HEAD OF STATE

President (Beretitenti): ANOTE TONG (elected 4 July 2003).

Vice-President (Kauoman-ni-Beretitenti): TEIMA ONORIO.

THE CABINET
(September 2004)

President and Minister for Foreign Affairs: ANOTE TONG.

Vice-President and Minister for Education, Youth and Sport Development: TEIMA ONORIO.

Minister for Commerce, Industry and Co-operatives: IOTEBA REDFERN.

Minister for Communications, Transport and Tourism Development: NAATAN TEEWE.

Minister for Environment, Lands and Agricultural Development: MARTIN TOFINGA.

Minister for Finance and Economic Development: NABUTI MWEMWENIKARAWA.

Minister for Health and Medical Services: NATANERA KIRATA.

Minister for Human Resources Development: BAURO TONGAAI.

Minister for Internal Affairs and Social Development: AMBEROTI NIKORA.

Minister for the Line and Phoenix Islands: TAWITA TEMOKU.

Minister for Natural Resources Development: TETABO NAKARA.

Minister for Public Works and Utilities: JAMES TAOM.

MINISTRIES

Office of the President (Beretitenti): POB 68, Bairiki, Tarawa; tel. 21183; fax 21145.

Ministry of Commerce, Industry and Co-operatives: POB 510, Betio, Tarawa; tel. 26158/26157; fax 26233; e-mail commerce@tskl.net.ki.

Ministry of Communications, Transport and Tourism Development: POB 487, Betio, Tarawa; tel. 26003/26435; fax 26193.

Ministry of Education, Youth and Sport Development: POB 263, Bikenibeu, Tarawa; tel. 28091/28033; fax 28222.

Ministry of the Environment, Lands and Agricultural Development: POB 234, Bikenibeu, Tarawa; tel. 28211/28071; fax 28334; e-mail ps@melad.gov.ki.

Ministry of Finance and Economic Development: POB 67, Bairiki, Tarawa; tel. 21802/21805; fax 21307.

Ministry of Foreign Affairs: POB 68, Bairiki, Tarawa; tel. 21342; fax 21466; e-mail mfa@tskl.net.ki.

Ministry of Health and Medical Services: POB 268, Bikenibeu, Tarawa; tel. 28100; fax 28152.

Ministry of Human Resources and Development: POB 69, Bairiki, Tarawa; tel. 21068/21071; fax 21452.

Ministry of Internal Affairs and Social Development: POB 75, Bairiki, Tarawa; tel. 21092; fax 21133; e-mail homeaffairs@tskl.net.ki.

Ministry of Line and Phoenix Islands: Kiritimati Island; tel. 21449/81213; fax 81278.

Ministry of Natural Resources Development: POB 64, Bairiki, Tarawa; tel. 21099; fax 21120.

Ministry of Public Works and Utilities: POB 498, Betio, Tarawa; tel. 26192; fax 26172.

President and Legislature

PRESIDENT

Election, 4 July 2003

Candidate	Votes
Anote Tong	13,556
Harry Tong	12,457

A third candidate, Banuera Berina, secured a small number of votes in the election.

MANEABA NI MAUNGATABU
(House of Assembly)

This is a unicameral body comprising 40 elected members (most of whom formally present themselves for election as independent candidates), and one nominated representative of the Banaban community. A general election was held on 29 November 2002. However, in late March 2003, following its defeat on a motion of 'no confidence', the Government was replaced by an interim authority. A further general election was held on 9 and 14 May 2003, at which supporters of former President Teburoro Tito secured a majority of seats.

Speaker: TAOMATI IUTA.

Political Organizations

There are no organized political parties in Kiribati. However, loose groupings of individuals supporting similar policies do exist, the most prominent being the Maneaban Te Mauri (Protect the Maneaba), led by Teburoro Tito, the National Progressive Party, led by Teatao Teannaki, the Liberal Party, led by Tewareka Tentoa, and the Boutokan Te Koaua (Pillars of Truth), led by Dr Harry Tong.

Diplomatic Representation

EMBASSY AND HIGH COMMISSIONS IN KIRIBATI

Australia: POB 77, Bairiki, Tarawa; tel. 21184; fax 21904; e-mail AHC_Tarawa@dfat.gov.au; High Commissioner JUREK JUSZCZYK.

China (Taiwan): Bairiki, Tarawa; Ambassador SAMUEL CHEN.

New Zealand: POB 53, Bairiki, Tarawa; tel. 21400; fax 21402; e-mail nzhc@tskl.net.ki; High Commissioner JOHN GOODMAN.

United Kingdom: POB 5, Bairiki, Tarawa; tel. 22501; fax 22505; e-mail ukrep@tskl.net.ki; internet www.ukinthepacific.bhc.org.fj; High Commissioner CHARLES F. MOCHAN (resident in Suva, Fiji).

Judicial System

There are 24 Magistrates' Courts (each consisting of one presiding magistrate and up to eight other magistrates) hearing civil, criminal and land cases. When hearing civil or criminal cases, the presiding magistrate sits with two other magistrates, and when hearing land cases with four other magistrates. A single magistrate has national jurisdiction in civil and criminal matters. Appeal from the Magistrates' Courts lies, in civil and criminal matters, to a single judge of the High Court, and, in matters concerning land, divorce and inheritance, to the High Court's Land Division, which consists of a judge and two Land Appeal Magistrates.

The High Court of Kiribati is a superior court of record and has unlimited jurisdiction. It consists of the Chief Justice and a Puisne Judge. Appeal from a single judge of the High Court, both as a Court of the First Instance and in its appellate capacity, lies to the Kiribati Court of Appeal, which is also a court of record and consists of a panel of three judges.

All judicial appointments are made by the Beretitenti (President).

High Court

POB 501, Betio, Tarawa; tel. 26007; fax 26149; e-mail highcourt@tskl.net.ki.

Chief Justice: ROBIN MILLHOUSE.

Judges of the Kiribati Court of Appeal: ROBIN MILLHOUSE (President), Sir MAURICE CASEY, Sir MICHAEL HARDIE-BOYS, Sir DAVID TOMPKINS, PETER PENLINGTON.

Religion

CHRISTIANITY

Most of the population are Christians: 53.4% Roman Catholic and 39.2% members of the Kiribati Protestant Church, according to the 1990 census.

The Roman Catholic Church

Kiribati forms part of the diocese of Tarawa and Nauru, suffragan to the archdiocese of Suva (Fiji). At 31 December 2002 the diocese contained an estimated 48,908 adherents. The Bishop participates in the Catholic Bishops' Conference of the Pacific, based in Suva (Fiji).

Bishop of Tarawa and Nauru: Most Rev. PAUL EUSEBIUS MEA KAIUEA, Bishop's House, POB 79, Bairiki, Tarawa; fax 21401; e-mail cathchurch@tskl.net.ki.

The Anglican Communion

Kiribati is within the diocese of Polynesia, part of the Anglican Church in Aotearoa, New Zealand and Polynesia. The Bishop in Polynesia is resident in Fiji.

Protestant Church

Kiribati Protestant Church: POB 80, Bairiki, Tarawa; tel. 21195; fax 21453; f. 1988; Moderator Rev. BAITEKE NABETARI; Gen. Sec. Rev. TIAONTIN ARUE; 29,432 mems in 1998.

Other Churches

Seventh-day Adventist, Church of God and Assembly of God communities are also represented, as is the Church of Jesus Christ of Latter-day Saints (Mormon).

BAHÁ'Í FAITH

National Spiritual Assembly: POB 269, Bikenibeu, Tarawa; tel. and fax 28074; e-mail emi@tskl.net.ki; 2,400 mems resident in 100 localities in 1995.

The Press

Butim'aea Manin te Euangkerio: POB 80, Bairiki, Tarawa; tel. 21195; e-mail kpc@tskl.net.ki; f. 1913; Protestant Church newspaper; weekly; a monthly publication Te Kaotan te Ota is also produced; Editor Rev. TOOM TOAKAI.

Kiribati Business Link: Bairiki, Tarawa; English.

Kiribati Newstar: POB 10, Bairiki, Tarawa; tel. 21652; fax 21671; e-mail newstar@tskl.net.ki; internet www.users.bigpond.com/kiribati_newstar; f. 2000; independent; weekly; English and I-Kiribati; Editor-in-Chief NGAUEA UATIOA.

Te Itoi ni Kiribati: POB 231, Bikenibeu, Tarawa; tel. 28138; fax 21341; f. 1914; Roman Catholic Church newsletter; monthly; circ. 2,300.

Te Uekera: Broadcasting and Publications Authority, POB 78, Bairiki, Tarawa; tel. 21162; fax 21096; f. 1945; weekly; English and I-Kiribati; Editor TIBWERE BOBO; circ. 5,000.

Broadcasting and Communications

TELECOMMUNICATIONS

Telecom Kiribati Ltd: Bairiki, Tarawa; Gen. Man. ENOTA INGINTAU.

Telecom Services Kiribati Ltd: POB 72, Bairiki, Tarawa; tel. 21446; fax 21424; e-mail ceo@tskl.net.ki; internet www.tski.net.ki; owned by Govt of Kiribati; Gen. Man. STUART EASTWARD; CEO CLIFF MACALPINE.

BROADCASTING

Regulatory Authority

Broadcasting and Publications Authority: POB 78, Bairiki, Tarawa; tel. 21187; fax 21096.

Radio

Radio Kiribati: Broadcasting and Publications Authority, POB 78, Bairiki, Tarawa; tel. 21187; fax 21096; f. 1954; statutory body; station Radio Kiribati broadcasting on SW and MW transmitters; programmes in I-Kiribati (90%) and English (10%); some advertising; Gen. Man. TANIERI TEIBUAKO.

Television

Television Kiribati: Broadcasting and Publications Authority, POB 78, Bairiki, Tarawa; tel. 21187; fax 21096; in process of establishing services.

Finance

(cap. = capital; dep. = deposits; res = reserves)

BANKING

The Bank of Kiribati Ltd: POB 66, Bairiki, Tarawa; tel. 21095; fax 21200; e-mail bankofkiribati@tskl.net.ki; f. 1984; 75% owned by ANZ Bank, 25% by Govt of Kiribati; dep. $A42.8m., res $A1.3m., total assets $A46.3m. (Sept. 1999); Chair. R. GOUDSWAARD; Pres. and Man. Dir N. OLDHAM; 3 brs.

Development Bank of Kiribati: POB 33, Bairiki, Tarawa; tel. 21345; fax 21297; e-mail dbk@tskl.net.ki; f. 1986; took over the assets of the National Loans Board; identifies, promotes and finances small-scale projects; auth. cap. $A2m.; Gen. Man. KIETAU TABWEBWEITI; 5 brs.

A network of lending entities known as 'village banks' operates throughout the islands, as do a number of credit unions under the management of the Credit Union League. In August 1995 there were 26 credit unions operating in Tarawa and seven in the outer islands with a total membership of 1,808 people.

INSURANCE

Kiribati Insurance Corpn: POB 38, Bairiki, Tarawa; tel. 21260; fax 21426; e-mail kirins@tskl.net.ki; f. 1981; govt-owned; only insurance co; reinsures overseas; Gen. Man. TEAIRO TOOMA.

Trade and Industry

GOVERNMENT AGENCIES

Kiribati Housing Corporation: Bairiki, Tarawa; tel. 21092; operates the Housing Loan and Advice Centre; Chair. TOKOREAUA KAIRORO.

Kiribati Provident Fund: POB 76, Bairiki, Tarawa; tel. 21300; fax 21186; f. 1977; total equity $A56.8m. (Dec. 1998); Gen. Man. TOKAATA NIATA.

CHAMBER OF COMMERCE

Kiribati Chamber of Commerce: POB 550, Betio, Tarawa; tel. 26351; fax 26351; Pres. WAYSANG KUM KEE; Sec.-Gen. TIARITE KWONG.

UTILITIES

Public Utilities Board: POB 443, Betio, Tarawa; tel. 26292; fax 26106; e-mail pub@tskl.net.ki; f. 1977; govt-owned; provides electricity, water and sewerage services in Tarawa; CEO TOKIA GREIG.

Solar Energy Company: Tarawa; e-mail sec@tskl.net.ki; a co-operative administering and implementing solar-generated electricity projects in North Tarawa and the outer islands.

CO-OPERATIVE SOCIETIES

Co-operative societies dominate trading in Tarawa and enjoy a virtual monopoly outside the capital, except for Banaba and Kiritimati.

The Kiribati Copra Co-operative Society Ltd: POB 489, Betio, Tarawa; tel. 26534; fax 26391; f. 1976; the sole exporter of copra; seven cttee mems; 29 mem. socs; Chair. RAIMON TAAKE; CEO RUTIANO BENETITO.

Bobotin Kiribati Ltd: POB 485, Betio, Tarawa; tel. 26092; fax 26224; replaced Kiribati Co-operative Wholesale Society; govt-owned; Gen. Man. AKAU TIARE.

MAJOR COMPANIES

Abamakoro Trading Ltd: POB 492, Betio, Tarawa; tel. 26568; fax 26415; e-mail abamakoro@tskl.net.ki; importer and wholesaler of general merchandise.

Atoll Seaweed Co Ltd: POB 528, Betio, Tarawa; tel. 26442; fax 26442; e-mail atoll.seaweed@tskl.net.ki; CEO KEVIN ROUATU.

Betio Shipyard Ltd: POB 468, Betio, Tarawa; tel. 26282; fax 26064; e-mail shipyard@tskl.net.ki; Gen. Man. TEMAIA EREATA.

Coconut Products Ltd: POB 280, Betio, Tarawa; manufacturer of coconut oil soap and cosmetics.

Kiribati Oil Co Ltd: Betio, Tarawa.

Kiribati Supplies Co Ltd: POB 71, Bairiki, Tarawa; tel. 21185; fax 21104; e-mail kscl@tskl.net.ki.

Marine Export Ltd: Betio, Tarawa.

Tarawa Biscuit Co Ltd: Betio, Tarawa.

Te Mautari Ltd: POB 508, Betio, Tarawa; govt-owned; fishing co; operates five vessels.

TRADE UNIONS

Kiribati Trades Union Congress (KTUC): POB 502, Betio, Tarawa; tel. 26277; fax 26257; f. 1982; unions and asscns affiliated to the KTUC include the Fishermen's Union, the Co-operative Workers' Union, the Seamen's Union, the Teachers' Union, the Nurses' Asscn, the Public Employees' Asscn, the Bankers' Union, Butaritari Rural Workers' Union, Christmas Island Union of Federated Workers, the Pre-School Teachers' Asscn, Makim Island Rural Workers' Org., Nanolelei Retailers' Union, the Plantation Workers' Union of Fanning Island and the Overseas Fishermen's Union (formed in 1998); 2,500 mems; Pres. TATOA KAITEIE; Gen. Sec. TAMARETI TAAU.

Transport

ROADS

Wherever practicable, roads are built on all atolls, and connecting causeways between islets are also being built as funds and labour permit. A programme to construct causeways between North and South Tarawa was completed in the mid-1990s. Kiribati has about 670 km of roads that are suitable for motor vehicles; all-weather roads exist in Tarawa and Kiritimati. In 2000 there were about 1,468 motor vehicles registered in the islands, of which some 48% were motorcycles.

SHIPPING

A major project to rehabilitate the port terminal and facilities at Betio, with finance totalling some US $22m. from Japan, was completed in May 2000. There are other port facilities at Banaba, Kanton and English Harbour.

Kiribati Shipping Services Ltd: POB 495, Betio, Tarawa; tel. 26195; fax 26204; e-mail kssl@tskl.net.ki; operates three passenger/freight vessels on inter-island services and one landing craft; govt-owned; Gen. Man. Capt. ITIBWINNANG AIAIMOA.

MATS Shipping and Transport: POB 413, Betio, Tarawa; tel. 26355; operates a fortnightly passenger and cargo service to the outer islands and occasional longer journeys.

CIVIL AVIATION

There are five international airports (Bonriki on South Tarawa, Cassidy on Kiritimati, Antekana on Butaritari, as well as others on Kanton and Tabuaeran) and several other airfields in Kiribati. The airport at Bonriki was enlarged in the early 1990s, using a loan from the Bank of China. Air Nauru and Air Marshall Islands also operate international services to Tarawa, and Aloha Airlines operates a charter flight service between Kiritimati Island and Honolulu, Hawaii. In December 2001 the Government announced its decision to lease a prop-jet, which it intended to operate from Tarawa to the Marshall Islands, Tuvalu and Fiji.

Air Kiribati Ltd: POB 274, Bonriki, Tarawa; tel. 28088; fax 28216; e-mail airkiribati.admin@tsklnet.ki; f. 1977; fmrly Air Tungaru; national airline; operates scheduled services to 15 outer islands; Chair. TAKEI TAOABA; CEO TANIERA TEIBUAKO.

Tourism

Previous attempts to establish tourism have been largely unsuccessful, owing mainly to the remoteness of the islands. There were 14,211 visitor arrivals in 1998 (of whom fewer than 40% were tourists). Of total tourist arrivals in 1998, some 58.3% came from Asia and Oceania, and 26.9% came from the Americas. The number of tourist arrivals at Tarawa and Kiritimati airports rose from 3,112 in 1999 to 4,829 in 2000 and totalled 4,288 in 2003. In 2000 the industry earned some $A2.2m. In 1996 there were 201 hotel rooms in the islands. In 1989 the Government adopted a plan to develop hotels in the Line Islands and to exploit sites of Second World War battles. A further Tourism Development Action Plan was introduced in 1997. Game-fishing and 'eco-tourism', particularly bird-watching, were promoted in the late 1990s in an attempt to increase tourist arrivals to Kiritimati. Kiribati also exploited the location of some of its islands in the Line and Phoenix group by marketing the area as a destination for tourists wishing to celebrate the year 2000. In 1997 Caroline Island, situated close to the recently realigned international date-line, was renamed Millennium Island in an attempt to maximize its potential for attracting visitors. In late 2000 an agreement was signed with the Norwegian Shipping Line company allowing large cruise ships to make weekly calls to the Line Islands from the end of 2001.

Kiribati National Tourism Office: Ministry of Communications Transport and Tourism Development, POB 487, Betio, Tarawa; tel. 26003; fax 26193; e-mail sto@mict.gov.ki; internet www.spto.com; Sec. TEBWE IETAAKE; Senior Tourist Officer TARATAAKE TEANNAKI.

Education

Education is compulsory for children between six and 15 years of age. This generally involves six years at a primary school and at least three years at a secondary school. Every atoll is provided with at least one primary school. An estimated 92% of children aged six to 12 receive primary education. In 2001 there were 88 primary schools, three government and six private secondary schools. There were 16,096 primary students and 5,743 general secondary students enrolled in 2001. The tertiary sector is based on Tarawa, except for one of the two private colleges which is based on Abemama. The Government administers a technical college and training colleges for teachers, nurses and seamen. There were 198 students enrolled in teacher-training and 1,303 in other vocational training in 2001. An extra-mural centre of the University of the South Pacific (based in Fiji) is also located on South Tarawa. In mid-1999 plans for the establishment of a college of advanced education were announced. In 2000 government expenditure on education totalled $A12.8m. (equivalent to 19.9% of total recurrent budgetary expenditure).

THE MARSHALL ISLANDS

Physical and Social Geography

The Republic of the Marshall Islands consists of two groups of islands, the Ratak ('sunrise') and Ralik ('sunset') chains, comprising 29 atolls (some 1,225 islets) and covering about 180 sq km (70 sq miles) of land. The islands lie within the area of the Pacific Ocean known as Micronesia, some 3,200 km south-west of Hawaii (USA) and about 2,100 km south-east of Guam. The nearest neighbours are Kiribati to the south and the Federated States of Micronesia to the west. Rainfall decreases from south to north, with January, February and March being the driest months, although seasonal variations in rainfall and temperature are generally small. The indigenous population comprises a Micronesian people, whose language is known as Marshallese. Both Marshallese and English are official languages. The traditional society of the Marshall Islands consists of a complex system of matrilineal clans and social stratification. Most of the population are Christians. According to the November 1999 census, the population totalled 50,848, with 23,682 living on Majuro Atoll, where the capital, Dalap-Uliga-Darrit Municipality, is situated.

History

The ancestors of the Micronesians settled the Marshall Islands some 4,000 years ago. The islands were ruled by several, warring, feudal states. The high chiefs, or Iroij, continued to exercise power, following expeditions under the authority of the Spanish Crown (Fernão de Magalhães—Ferdinand Magellan, the Portuguese navigator—in 1521 and Miguel de Saavedra in 1529) and the formal claim to the Marshall Islands by Spain in 1592. The islands received their name from the British explorer, John Marshall, who visited them at the end of the 18th century. Spanish sovereignty over the Marshall Islands was recognized in 1886 by the Papal Bull of Pope Leo XIII, which also gave Germany trading rights there (German trading companies had been active in the islands from the 1850s). In 1899 Germany also acquired, by purchase from Spain, the Caroline Islands and the Northern Mariana Islands (i.e. the Mariana Islands excluding Guam, which was ceded, by Spain, to the USA in 1898). In 1914, at the beginning of the First World War, Japan occupied all of German Micronesia, and received a formal mandate for its administration from the League of Nations in 1920. The territory was intensively colonized. In 1944 the islands were occupied by the US military, and in 1947 the United Nations included the Marshall Islands in the Trust Territory of the Pacific Islands, which was to be administered by the USA. (For the history of the Trust Territory, see the chapter on Palau.)

Following the rejection in a plebiscite, in July 1978, of participation in a federal Micronesian state, the Marshall Islands drafted its own Constitution, which came into effect on 1 May 1979. Negotiations on a Compact of Free Association with the USA continued, and draft agreements were initialled in October and November 1980. The USA signed the Compact of Free Association with the Republic of Palau in August 1982, and with the Marshall Islands and the Federated States of Micronesia in October of that year. The trusteeship of the islands was due to end after the principle and terms of the Compacts had been approved by the respective peoples and legislatures of the new countries, by the US Congress and by the UN Security Council. Under the Compacts, the four countries (including the Northern Mariana Islands) would be independent of each other and would manage their internal and foreign affairs separately, while the USA would be responsible for defence and security. Moreover, Marshallese citizens were granted the right to live and work in the USA. The Compacts with the Federated States of Micronesia and the Marshall Islands were approved in plebiscites in June and September 1983, respectively. Under the Compact with the Marshall Islands, the USA was to retain its military bases in the Marshall Islands for at least 15 years and, over the same period, was to provide annual aid of US $30m. The Compact came into effect on 21 October 1986. The Republic of the Marshall Islands ceased to be part of the Trust Territory, the final dissolution of which occurred on 1 October 1994, when the Compact between the Republic of Palau and the USA came into effect. In December 1990 the UN Security Council finally voted to ratify formally the termination of the Trusteeship Agreement. A revised Compact (see below) was signed by the Governments of the two countries on 1 May 2003; it was ratified by the US Congress in November 2003, and signed by US President George W. Bush in December of that year. The Marshall Islands became a member of the UN in 1991.

Bikini and Enewetak Atolls, in the Marshall Islands, were used by the USA for experiments with nuclear weapons, Bikini in 1946–58 and Enewetak in 1948–58. A total of 67 such tests were carried out during this period. The indigenous inhabitants of Enewetak were evacuated before tests began, and were allowed to return to the atoll in 1980, after much of the contaminated area had supposedly been rendered safe. The inhabitants of Bikini campaigned for similar treatment, and in 1985 the entire population of Rongelap Atoll (which had been engulfed in nuclear fall-out from the tests at Bikini in 1954) was resettled on Mejato Atoll, after surveys suggested that levels of radiation there remained dangerous. Subsequently, the USA agreed to decontaminate Bikini Atoll over a period of between 10 and 15 years. Under the terms of the Compact, the US Administration consented to establish a US $150m. trust fund to settle claims against the USA resulting from the testing of nuclear devices in the Marshall Islands during the 1940s and 1950s. Accordingly, the Marshall Islands Nuclear Claims Tribunal was established in 1988, with jurisdiction to 'render final determination upon all claims past, present and future, of the Government, citizens and nationals of the Marshall Islands' in respect of the nuclear testing programme. In 1989 the US Supreme Court ruled that the Compact of 1983 and subsequent US legislation in 1986 prevented the islanders from suing the USA for any additional compensation, but in 1990 further compensation was approved for the islanders. A compensation programme was implemented in 1991 for personal injuries deemed to have resulted from the testing programme. Following an approach defined in legislation adopted by the US Congress in 1990, which established a 'presumptive' programme of compensation for specified diseases contracted by US civilian and military personnel who had been physically present in what was termed the 'affected area' during periods of atmospheric testing in Nevada, the Marshall Islands Nuclear Claims Tribunal initially identified 25 diseases for which credible evidence demonstrated a significant statistical relationship between exposure to radiation and subsequent development of a disease; in response to the findings of later studies, the Tribunal's list had by 2003 been extended to include 11 further conditions.

By the end of 2003 compensation awards totalling US $83m. had been made to, or on behalf of, 1,865 individuals who had contracted one or more of those conditions identified by the Tribunal. Additionally, an award of some $578m. had been ordered in May 2000 in respect of a class action brought by the people of Enewetak for loss of and damage to property; and an award of $563m. had been made in March 2002 in settlement of a class action brought by the peoples of Bikini Atoll; settlements of similar class actions by the peoples of Rongelap and Utrik Atolls were being finalized, while a new class action had been submitted by the people of Ailuk Atoll. However, only $45.8m. had been made available for actual payment of awards decided by the Tribunal; furthermore, less than $6m. remained of the original value of the fund. In view of the inadequacy of the fund to meet the compensation awards made by the Tribunal, in September 2000 the Marshall Islands Government had formally petitioned the US Congress for a renegotiation of the settlement agreed under the Compact; the basis of the petition, which sought additional compensation amounting to $2,000m., was an article of the agreement providing for what were termed 'Changed Circumstances'. After much delay by the US authorities, in mid-June 2004 the President of the Marshall Islands, Kessai Note, met with several influential members of the US Congress who expressed their strong support for holding a formal congressional hearing on the 'Changed Circumstances' petition. In early August 2004 the Tribunal declared a deadline for islanders' compensation claims of the end of that month. The reason given for the deadline was the diminution of the Tribunal's funds, which not only put into doubt future compensation payments but also the existence of the Tribunal itself.

In January 1994, meanwhile, several senior members of the Marshall Islands' legislature, the Nitijela, demanded that the US authorities release detailed information on the effects of its nuclear-testing programme in the islands. In July documentation released by the US Department of Energy gave conclusive evidence that Marshall Islanders had been deliberately exposed to high levels of radiation in order that its effects on their health could be studied by US medical researchers. Further evidence emerged during 1995 that the USA had withheld the medical records of islanders involved in radiation experiments (which included tritium and chromium-51 injections and genetic and bone-marrow transplant experiments).

Despite the publication of a study conducted by US scientists (in 1992) into contamination levels on Bikini Atoll, which suggested that radiation levels there remained dangerous, in February 1997 a group of Bikini islanders returned for the first time since 1946 to assist in the rehabilitation of the atoll for resettlement. The operation was to involve the removal of radioactive topsoil (although the matter of its disposal presented a serious problem) and the saturation of the remaining soil with potassium, which was believed to inhibit the absorption of radioactive material by root crops. In early 1999 the Nuclear Claims Tribunal demanded the adoption of US Environmental Protection Agency standards in the rehabilitation of contaminated islands, claiming that Marshall Islanders deserved to receive the same treatment as US citizens would in similar circumstances. The US Department of Energy, however, expressed strong resistance to the suggestion. In February 2001 a report published by an eminent Japanese scientist stated that radiation levels on Rongelap Atoll, according to research conducted in 1999, had now declined to such a level that human habitation of the island was again possible. In early 2004 the Marshall Islands protested that a reduction, decided upon by the US Department of Energy without consultation with local representatives, of some US $740,000 in congressional funding allocated to nuclear test-related studies would result in the closure of a centre on Bikini Atoll used to support scientific studies at the former test site.

Another atoll in the Marshall Islands, Kwajalein, has been used since 1947 as a target for the testing of missiles fired from California, USA. The Compact as ratified in 1986 committed the US Government to provide an estimated US $170m. in rent over a period of 30 years for land used as the site of a missile-tracking station, and a further $80m. for development projects. The inhabitants of Kwajalein Atoll were concentrated on the small island of Ebeye, adjacent to the US base on Kwajalein Island, before a new programme of weapons-testing began in 1961. Consequent overcrowding reportedly led to numerous social problems on Ebeye. In 1989 the Marshall Islands Government agreed that the USA could lease a further four islands in the atoll, for five years, for the purpose of military tests. A further lease agreement was signed in 1995 for the use of Biken Island, in Aur Atoll, and Wake Island to be used in the missile-testing programme. In January 2003 it was announced that the Marshall Islands Government and the USA had reached agreement, under the amended Compact (see below), on new terms extending the lease of the Kwajalein site, previously to end in 2066. The renegotiated terms envisaged the payment of some $3,100m. for use of the site during this period, with an option to extend the lease by a further 20 years. It was confirmed that the USA would continue to provide an annual $1.9m. in social funding for the residents of Ebeye. However, Kwajalein landowners, who deemed the new terms unacceptable, asserted that the new arrangement was invalid, since they had not consented, as constitutionally required, to its terms.

The first President of the Republic of the Marshall Islands was Iroijlaplap (paramount chief) Amata Kabua, who was re-elected for further four-year terms of office in 1984, 1988, 1992 and 1995. Following legislative elections in November 1995, at which eight incumbent members of the 33-seat Nitijela were defeated, Kabua had been re-elected for a fifth presidential term. The President died in December 1996. Iroijlaplap Imata Kabua, a cousin of the late President, was elected as his successor on 13 January 1997.

In 1996 the Nitijela approved legislation allowing for the introduction of gambling in the islands, in order to provide an additional source of revenue. However, income earned from the venture did not fulfil expectations and aroused controversy for attracting more local than tourist interest. Moreover, a vociferous campaign by local church leaders in the islands to revoke the legislation led to fierce debate in the Nitijela in early 1998. Divisions within the Cabinet ensued, with three members supporting the President's pro-gambling stance and four others expressing support for the church's position. In April of that year the Nitijela voted to repeal the law legalizing gambling, following the disqualification from the vote of several influential politicians (including Imata Kabua) known to have major gambling interests. A second bill containing further measures to ensure the prohibition of all gambling activity in the islands was narrowly approved. In August a cabinet reorganization was effected, in which three ministers who had supported the anti-gambling legislation were dismissed. In the following month one of the dismissed ministers proposed a motion of 'no confidence' in Kabua, the first such motion to be presented in the islands' legislature. The President and his supporters boycotted subsequent sessions of the Nitijela, effectively preventing the vote from taking place. The lack of a quorum also resulted in a delay in the approval of the budget for the imminent new financial year. Meanwhile, the opposition claimed that Kabua's continued absence from the Nitijela violated the terms of the Constitution. The motion of 'no confidence' in Kabua was eventually defeated in October, by a margin of one vote.

At legislative elections held on 15 November 1999 the opposition United Democratic Party (UDP) secured a convincing victory over the incumbent administration, winning 18 of the 33 seats in the Nitijela (including four of the five seats in Majuro). A total of five senior members of the outgoing Government were defeated, among them the Ministers of Finance and of Foreign Affairs and Trade—both of whom had played a prominent role in the establishment of diplomatic relations with Taiwan in 1998 (see below). The former Nitijela Speaker, Kessai Note, was elected President on 3 January 2000 (the islands' first non-traditional leader to assume the post). The UDP Chairman, Litokwa Tomeing, became Speaker of the legislature. Note subsequently appointed a 10-member Cabinet, and reiterated his administration's intention to pursue anti-corruption policies. In May 2000 a task-force was established by the Government for the purpose of investigating misconduct and corruption, with a view to rendering government more accountable.

In November 2000 it was reported that finance officials had discovered that Imata Kabua had used funds granted to the Marshall Islands under the terms of the Compact of Free Association to pay off a personal loan, although the former President denied any wrongdoing. In mid-January 2001 Imata Kabua and former ministers in his Government, including the former Minister of Education, Justin DeBrum, presented a 'no confidence' motion against President Note to the Nitijela. Although it was suggested that the vote had been intended to delay the publication of a report into mismanagement and corruption on the part of the former Government, DeBrum stated that the motion resulted from a number of failings by the Note Government, including its unwillingness to renegotiate land rental payments with the USA for the use of the military base on Kwajalein Atoll and also the development of an economic relationship between the Note Government and Rev. Sun Myung Moon, the founder of the Unification Church (known as the 'Moonies'). However, the Government was successful in defeating the vote by a margin of 19 to 14.

In September 2000 the Nitijela approved legislation to ensure the closer regulation of the banking and financial sector. In May of that year the Group of Seven industrialized nations (G-7) had expressed its view that the Marshall Islands had become a significant centre for the 'laundering' of money generated by international criminal activity, and, in June the Marshall Islands was one of more than 30 countries and territories criticized by the Organisation for Economic Co-operation and Development (OECD) for the provision of inappropriate 'offshore' financial establishments. OECD threatened to implement sanctions against 'non-co-operative tax havens' unless reforms were introduced before July 2001. Following a commendation from the IMF on a series of new measures to combat fraud, including specific legislation and the establishment of a Domestic Financial Intelligence Unit, in October 2002 the Financial Action Task Force on Money Laundering removed the Marshall Islands from its list of countries judged to be unhelpful in the combating of international financial crime.

Negotiations began between the US and Marshall Islands Governments in July 2001 to renew the provisions of the Compact of Free Association ratified in 1986, which was due to expire at the end of September 2001. A two-year extension was permitted while negotiations were under way, during which time annual assistance to the Marshall Islands was to increase by some US $5.5m. An agreement was originally scheduled for early 2002 in order to allow adequate time for the US Congress to review it and to approve the requisite legislation (by 1 October 2003), but the procedure was postponed until early May 2002 after the Marshall Islands Government submitted a proposal seeking financing of more than $1,000m. over 15 years. The Government had also objected to being allocated 25%–30% less in US grant assistance per caput than that apportioned to the Federated States of Micronesia since the year 2000. In a further attempt to increase the national income, the Government sought to raise significantly the level of taxes levied on the Kwajalein base (see above). In early November 2002 the USA and the Marshall Islands Government announced a programme of direct funding of $822m., to be disbursed over 20 years, in addition to the expansion of many US government services in the islands. It was envisaged that the Marshall Islands would receive some $30.5m. per year; furthermore, a trust fund would be established, to which the USA would contribute $7m. annually in order to provide a means of income after the termination of direct US assistance in 2023. The amended Compact of Free Association was signed by the Governments of the Marshall Islands and the USA in May 2003, at which time a number of issues—including a requirement for an initial US payment into a Kwajalein landowners' trust fund, and consideration of the 'Changed Circumstances' petition—remained to be resolved. Under the new Compact, Marshall Islanders would for the first time require passports in order to enter the USA. They would, however, retain the right to enter the USA to live, work and study, and would no longer be required to obtain work authorization documentation before taking up employment in the USA. Other than the issue of the Kwajalein lease, a principal obstacle to the negotiation of Compact amendments had been that of immigration: the USA, increasingly preoccupied by issues of homeland security, had been notably concerned to prevent future sales of Marshallese passports (a controversial programme of which had been implemented in the 1990s, although this had officially been suspended in 1997). The US authorities had also sought to close a loophole in the original Compact provisions that had effectively allowed the unmonitored adoption of Marshallese children by US citizens. In December 2003 the Nitijela approved legislation introducing a compulsory scheme for adoption agencies operating in the Marshall Islands. Another principal objective of the amended Compact was the economic self-reliance of the Marshall Islands by the end of the financial year 2023. Final terms, including the restoration of some rights of access to US health-care and education programmes, were approved by the US Congress in November 2003, and ratified by President George W. Bush in December.

At a general election held on 17 November 2003 the UDP returned 20 Senators to the 33-member Nitijela. The opposition grouping Ailin Kein Ad (Our Islands), which had campaigned against the terms of the renewed Compact and which received particularly strong support from Marshall Islanders resident in the USA, secured 10 seats (the re-election of the incumbent Ailin Kein Ad Senator for Ailinglaplap Atoll was decided following a recount of votes conducted in late January 2004). Note was re-elected for a second presidential term in a vote held in the Nitijela on 4 January 2004, defeating Justin DeBrum, the candidate of Ailin Kein Ad, by 20 votes to nine. He and his new Cabinet were sworn in on 12 January.

In 1989 a UN report on the 'greenhouse effect' (heating of the earth's atmosphere) predicted a possible resultant rise in sea-level of some 3.7 m by 2030, which would completely submerge the Marshall Islands. The islands' Government strongly criticized the Australian Government's refusal, at the conference on climate change in Kyoto, Japan, in December 1997, to reduce its emission of pollutant gases known to contribute to the 'greenhouse effect'. Furthermore, in early 2002 the Intergovernmental Panel on Climate Change (IPCC) projected that during the 21st century global sea-level rises would submerge over 80% of Majuro Atoll. However, the Marshall Islands Government has itself caused regional concerns regarding pollution, notably with regard to the possible establishment of large-scale facilities for the storage of nuclear waste. Criticism by the US Government of the plans, announced in 1994, was strongly denounced by the Marshall Islands authorities, which claimed that the project constituted the only opportunity for the country to generate sufficient income for the rehabilitation of contaminated islands and the provision of treatment for illnesses caused by the US nuclear-test programme. In mid-1997 President Imata Kabua announced the indefinite suspension of the project (despite the initiation of a feasibility study into the development of a nuclear waste storage facility). None the less, the Government approved plans for a new feasibility study on the subject in April 1998. The new Minister of Foreign Affairs appointed by President Note in January 2000 was known to be strongly opposed to the scheme.

In November 1998 the Marshall Islands established full diplomatic relations with Taiwan. The action was immediately condemned by the People's Republic of China, which in December severed diplomatic ties with the islands, closing its embassy in Majuro and suspending all intergovernmental agreements. The Marshall Islands Government insisted that it wished to maintain cordial relations with both Governments. The Note administration, which took office in January 2000, emphasized its commitment to the maintenance of diplomatic relations with Taiwan. In February 2001 a proposed visit by a flotilla of Taiwanese naval vessels to the Marshall Islands was vetoed by the USA, on the grounds that the defence protocol of the Compact of Free Association prohibited such a visit. In August 2004 the Chinese Vice-Minister of Foreign Affairs, Zhou Wenzhong, expressed China's willingness to normalize relations, under the condition that the Marshall Islands withdraw its diplomatic recognition of Taiwan. In September it was announced that Taiwan was to contribute more than US $40m. over a 20-year period to the Marshall Islands trust fund established earlier in the year (see above).

Economy

In 2002, according to estimates by the World Bank, the Marshall Islands' gross national income (GNI), measured at average 2000–02 prices, was US $125m., equivalent to $2,350 per head. During 1990–2002, it was estimated, the population increased at an average annual rate of 1.2%, while in 1990–2001 gross domestic product (GDP) per head declined, in real terms, at an average rate of 2.5% per year. Overall GDP declined by 0.9% in 1990–2002. According to the Asian Development Bank (ADB), GDP decreased by 1.5% in 2001, but was estimated to have increased by 4.0% in 2002 and by 3.0% in 2003.

Agriculture (including fishing and livestock-rearing) contributed an estimated 10.5% of GDP in the financial year ending September 2001, and according to FAO engaged some 6,000 people in 2002, including forestry and fishing. According to estimates by the ADB, compared with the previous year the GDP of the agricultural sector increased by 15.5% in 2000 and by 4.2% in 2001. The sector is mainly on a subsistence level, the principal crops being coconuts, cassava and sweet potatoes. Coconuts are processed into copra (dried coconut meat), some of which is then pressed for coconut oil. In 2002 some 2,653 short tons of copra were produced (compared with 5,256 tons in 2001 and 4,273 tons in 2000). In 2000 exports of coconut oil and copra accounted for 27.0% of the total value of exports. Low prices, transport problems and an ageing tree stock adversely affected copra production in the late 1990s. There is a commercial tuna-fishing industry, including a tuna-canning factory and transhipment base on Majuro. The cultivation of seaweed was developed extensively in 1992, and in 1994 a project to cultivate blacklip pearl oysters on Arno Atoll was undertaken with US funding. The sale of fishing licences is an important source of revenue and earned the islands some US $3m. in 2000/01. All shark-fishing licences, however, were revoked in August 2004 owing to low stocks. In 2001 the Japanese Government funded the construction of a commercial fishing base at Jaluit Atoll, scheduled for completion in 2002. The Marshall Islands expected to receive annual revenues of some $21m. following the renewal of a treaty between the USA and the Forum Fisheries Agency (FFA) group of Pacific island nations in 2003.

Industry (including mining, manufacturing, construction and power) contributed some 19.8% of GDP in 2000/01. Between 1990 and 1999 industrial GDP declined by an average annual rate of 1.5%. According to the ADB, compared with the previous year the sector's GDP rose by 12.0% in 1999, 11.7% in 2000 and by 2.9% in 2001. There are few mineral resources, although high-grade phosphate deposits exist on Ailinglaplap Atoll. Manufacturing activity, which provided an estimated 4.6% of GDP in 2001, and which engaged a total of 800 workers in 2000, consists mainly of the processing of agricultural products. Tobolar, the Marshall Islands' leading copra-processing company, was unable to process any further supplies of the commodity in 2001, owing to the size of its stockpile of coconut oil and the lack of any additional storage capacity. In April 2002 the Taiwanese Government invested US $1.2m., enabling Tobolar to sell its stockpile, to begin processing again and to open a new oil refinery (hitherto, the Marshall Islands had produced only unrefined oil). Construction activity, which is largely associated with the islands' developing tourist industry, provided some 11.5% of GDP in 200001. An estimated 24.4% of the employed population worked in production, transport and labour in 1999, according to census information.

The services sector provided an estimated 69.7% of GDP in 2000/01. The development of the tourist industry has been hindered by the difficulty of gaining access to the islands and a lack of suitable facilities. There were 5,444 visitors to the islands in 2001, a 3.8% increase compared with 2000. A short-term tourism development programme focusing on special-interest tourism was established in 2000. Tourist receipts totalled US $4m. in 2002. According to the ADB, compared with the previous year the GDP of the services sector contracted by 0.5% in 1999, by 2.9% in 2000, and by 2.3% in 2001.

The international shipping registry experienced considerable expansion following the political troubles in Panama in 1989, and continued to expand in the mid-1990s (largely as a result of US ships reflagging in the islands). The Marshallese merchant fleet numbered 515 vessels at the end of 2003, with a tonnage greater than that of the USA. The shipping industry also benefited from the construction of a floating dry-dock in Majuro in 1995. Nevertheless, it was reported in mid-2003 that despite a 29% increase in ship registrations in the year to July, registration payments were only 42% of their total at the corresponding time of the previous year.

In the year to September 2003, according to ADB estimates, the Marshall Islands recorded a trade deficit of US $47.1m., but a surplus of $22.2m. on the current account of the balance of payments. Domestic exports in 2000 consisted mainly of coconut products and pet fish. Re-exports of diesel fuel totalled $6.6m., their value exceeding that of domestic exports. The principal imports included mineral fuels and lubricants, food and live animals and machinery and transport equipment. In 2003 the principal sources of imports were the USA (which provided 6.8% of total imports), Australia and Japan.

A budgetary surplus equivalent to 6.2% of GDP was forecast for 2004, compared with 14.1% in the previous year. Financial assistance from the USA (at an annual rate of more than US $65m.), in accordance with the terms stipulated in the Compact of Free Association, contributed a large part of the islands' external revenue. In 2000/01 budgeted aid from the USA amounted to $20.7m. (35% of which was provided under the Compact agreement). Aided by an increase in this support, estimates for 2001/02 envisaged a rise in budgetary expenditure to $74m. (compared with $66m. in the previous year). The amended Compact (signed in May 2003) provided for a further $3,500m. to be paid to the Marshall Islands and the Federated States of Micronesia by the end of 2023, by which time, it is hoped, both countries will have achieved economic self-reliance.

The islands' external debt was estimated by the ADB at US $58m. in 2002. In that year the debt-service ratio was equivalent to 6.1% of the value of exports of goods and services, a figure considerably lower than the equivalent 60.4% for 2001. The Marshall Islands received $499 of aid per caput in 2000, which increased to $551 budgeted for 2003. The US aid budget for 2002 included a grant of $2.5m. to the Marshall Islands for an extension of the Military Use and Operating Rights Agreement (in addition to its mandatory annual payments of support for Enewetak Atoll and the Compact of Free Association). Aid has also been provided by Japan and Taiwan. The annual rate of inflation in Majuro averaged 4.9% in 1990–2001. According to the ADB, consumer prices increased by an annual average of 1.8% in 2001 and by an estimated 1.7% in 2002 and 1.8% in 2003. The unemployment rate stood at 30.0% of the economically active population in 2000; a survey prepared by the ADB in 2003 recorded that the more inaccessible islands suffered from the most extreme levels of poverty and unemployment.

The introduction, from the mid-1990s, of retrenchment measures in the public sector was welcomed by several international financial organizations and supported by the ADB. However, it was subsequently observed that reform of the public sector, which until the recession of the mid-1990s had employed up to one-half of the economically active population, had been accompanied by a decline in employment in the private sector; business employment declined by 7.4% during 1995–99, leading to a very high rate of unemployment and suggesting that the private sector in the Marshall Islands existed largely as a secondary sector. In 1999 the Government reduced import duties by more than 50% on many items, in an attempt to revitalize the local economy. Later that year reforms to promote the private sector were announced, in the hope that private businesses would assume responsibility for some of the services hitherto administered by the Government. Plans to transfer a number of state-owned companies to the private sector were also under way. A significant component of the new Compact of Free Association signed in May 2003 was an attempt to stimulate the private sector by reducing the islands' widely acknowledged dependence on aid from the USA. This was to be achieved through a phased annual decline in direct transfers to the Marshallese Government. In the same month, a report published on behalf of the ADB claimed that the reluctance of Marshallese financial institutions to lend money for private-sector development was damaging the country's prospects for economic growth. The document also alleged that an 'atmosphere of tension and mistrust' existed between the Government and the wider private sector, owing to the continuing dominance of the public sector in the Marshallese economy. Moreover, the Government's widespread failure to collect taxes, and reports that the Marshall Islands Development Bank had illegally used US aid funds for commercial purposes, further diminished investors' confidence in the local economy. The widely acknowledged need for reform of taxation and tax collection has met with considerable obstruction from those benefiting from the status quo. The 2004 budget provided for increased public spending, especially in health and education.

During the 1990s the Marshall Islands sought to diversify its economic relations with the international community. This policy achieved most notable success with the People's Republic of China, prior to the Marshall Islands' decision in 1998 to recognize Taiwan (see History). The Marshall Islands subsequently benefited from numerous economic agreements with Taiwan, worth an estimated US $20m., which financed many projects, including the construction of roads, the acquisition of boats and the development of the agricultural sector. In mid-1999 the two countries signed a commercial cooperation agreement based primarily on the islands' potential as a site for fisheries investment. However, concern was expressed by the ADB in that year that reliance on external aid (notably from the USA and Taiwan) was hampering economic reform in the Marshall Islands. (In 2000 bilateral and multilateral aid to the Marshall Islands totalled $57.2m.) The ADB itself approved a $12m. low-interest loan in June 2001, urging the Government to use it to improve budgeting and accounting practices. The Japanese Government, however, announced in 2002 that it was to reduce aid by 10%, owing to its own financial difficulties (since 1996 Japan has provided more than $2.2m. in small grants and millions of dollars in infrastructural development assistance). In June 2002 the European Union (EU) rejected a country support strategy plan designed to qualify the Marshall Islands for a five-year $1.5m. annual funding programme. (The proposed strategy focused on education and human-resource development; the EU, however, also required emphasis on alternative-energy installations.) In September 2002 the Marshallese Government's audit of its own 2001 accounts identified a possible misallocation of some $4.4m. in US aid funds, and criticized accounting practices prior to the Note Government's period in office. In December 2002 the ADB approved a further total of $7.25m. in loans and grants in order to improve the country's transport infrastructure. In September 2004 it was announced that the Taiwanese Government had agreed to make a substantial contribution to the Marshall Islands government trust fund (see History), which had been established in May with an initial investment of $25m. and a deposit of $7m. from the USA. Taiwan planned to transfer a total of $1m. to the fund each year until 2009 and thereafter, until 2023, an annual sum of $2.4m.

The lack of internationally marketable natural resources and the remote location of the islands present major challenges for the Marshall Islands Government in its efforts to revitalize and expand the economy. Attempts to circumvent these difficulties, including efforts to introduce gambling and 'offshore' financial services (see History) have generated political controversy, both domestically and internationally. As allegations of money-laundering activities continued, in early 2002 the Marshall Islands remained on the Financial Action Task Force (FATF) list of Non Co-operative Countries and Territories. However, the IMF commended the islands on progress in implementing a comprehensive framework against such fraud, including specific legislation, the establishment of a Domestic Financial Intelligence Unit and joining various international organizations. The Marshall Islands was removed from the list in October of the same year, since its inclusion had been based only on the absence of sufficient preventative measures. None the less, despite this apparent progress, it was reported in mid-2003 that financial 'shell' companies registered in the Marshall Islands had been involved in substantial international frauds originating in Canada and Serbia.

The sharp decline in external debt repayments in 2002 led to the release of substantial resources, which the Government had planned to channel into the Marshall Islands Intergenerational Trust Fund. In 2001 the Government had transferred a total of US $14m. to the Fund, to which, it was envisaged, future contributions from the USA and financial assistance from other donors would be allocated. The new Compact of Free Association, signed in May 2003, projected that within 20 years the Fund might total $800m. and in due course provide an annual income of $30m.–$40m. However, the subsequent downturn in the US financial markets was believed to have significantly affected the value of the Fund.

The Marshall Islands is a member of the Pacific Community (formerly the South Pacific Commission), the Pacific Islands Forum (formerly the South Pacific Forum) and the Asian Development Bank (ADB). In 1991 the islands were admitted to the UN (in addition to the UN's Economic and Social Commission for Asia and the Pacific—ESCAP), and in 1992 the Marshall Islands became a full member of the IMF and the World Bank. In early 1996 the Marshall Islands joined representatives of the other countries and territories of Micronesia at a meeting in Hawaii, USA, at which a new regional organization, the Council of Micronesian Government Executives, was established. The new body aimed to facilitate discussion of economic developments in the region and to examine possibilities for reducing the considerable cost of shipping essential goods between the islands. In August 2000 it was announced that the Marshall Islands was to participate, together with other Pacific nations, in the establishment of a commission to manage regional tuna stocks, which was to be based in the Federated States of Micronesia.

Statistical Survey

AREA AND POPULATION

Area: 181.4 sq km (70.0 sq miles) (land only); two island groups, the Ratak Chain (88.1 sq km) and the Ralik Chain (93.3 sq km).

Population: 43,380 at census of 13 November 1988; 50,848 (males 26,034, females 24,814) at census of June 1999. *Mid-2003* (estimate): 58,800 (Source: ADB, *Key Indicators of Developing Asian and Pacific Countries*). *By Island Group* (1999): Ratak Chain 30,932 (Majuro Atoll 23,682); Ralik Chain 19,916 (Kwajalein Atoll 10,903).

Density (mid-2003, land area only): 324.1 per sq km.

Births and Deaths (estimates, 2001): Birth rate 29.6 per 1,000; Death rate 5.2 per 1,000. Source: UN Economic and Social Commission for Asia and the Pacific, *Statistical Yearbook for Asia and the Pacific*.

Expectation of Life (WHO estimates, years at birth): 62.7 (males 61.1; females 64.6) in 2002. Source: WHO, *World Health Report*.

Economically Active Population (2000): Agriculture, forestry and fishing 2,100; Manufacturing 800; Activities not adequately defined 7,400; *Total employed* 10,300; Unemployed 4,500; *Total labour force* 15,000 (Source: Asian Development Bank, *Key Indicators of Developing Asian and Pacific Countries*). *Mid-2002* (estimates): Agriculture, etc. 6,000; Total labour force 23,000 (Source: FAO).

HEALTH AND WELFARE

Key Indicators

Total Fertility Rate (children per woman, 2002): 5.5.

Under-5 Mortality Rate (per 1,000 live births, 2002): 66.

Physicians (per 1,000 head, 1996): 0.42.

Health Expenditure (2001): US $ per head (PPP): 343.

Health Expenditure (2001): % of GDP: 9.8.

Health Expenditure (2001): public (% of total): 64.7.
For sources and definitions, see explanatory note on p. vi.

AGRICULTURE, ETC.

Principal Crops, Livestock and Livestock Products: see the chapter on the Federated States of Micronesia.

Fishing (metric tons, live weight): Total catch 38,742 in 2002. Source: FAO.

INDUSTRY

Electric Energy (million kWh, Majuro only): 74 in 2001; 80 in 2002; 81 in 2003. Source: Asian Development Bank, *Key Indicators of Developing Asian and Pacific Countries*.

FINANCE

Currency and Exchange Rates: United States currency is used: 100 cents = 1 United States dollar (US $). *Sterling and Euro Equivalents* (31 May 2004): £1 sterling = US $1.8347; €1 = US $1.2246; US $100 = £54.50 = €81.66.

Budget (estimates, US $ million, year ending 30 September 2002): *Revenue:* Recurrent 24.8 (Tax 18.6, Non-tax 6.2); Grants 58.8; Total 83.6. *Expenditure:* Recurrent 58.7; Capital (incl. net lending) 15.3; Total 74.0. *2003* (estimates, US $ million, year ending 30 September): Total revenue and grants 84.8; Total expenditure 67.2. Source: Asian Development Bank, *Key Indicators of Developing Asian and Pacific Countries*.

Cost of Living (Consumer Price Index for Majuro; base: Oct.–Dec. 1982 = 100): All items 200.6 in 2001; 204.0 in 2002 (estimate); 207.6 in 2003 (estimate). Source: Asian Development Bank, *Key Indicators of Developing Asian and Pacific Countries*.

Gross Domestic Product by Economic Activity (estimates, US $ million at current prices, year ending 30 September 2001): Agriculture 10.3; Mining 0.3; Manufacturing 4.5; Electricity, gas and water 3.4; Construction 11.3; Trade 16.9; Transport and communications 5.0; Finance 15.5; Public administration 13.0; Other services 18.0; *Sub-total* 98.3; Import duties 7.2; *Less* Imputed bank service charges 6.3; *GDP in purchasers' values* 99.2. Source: Asian Development Bank, *Key Indicators of Developing Asian and Pacific Countries*.

Balance of Payments (estimates, US $ million, year ending 30 September 2001): Merchandise exports f.o.b. 8.3; Merchandise imports c.i.f. -60.5; *Trade balance* -52.2; Exports of services 35.0; Imports of services -13.2; *Balance on goods and services* -30.4; Private unrequited transfers (net) 0.7; Official unrequited transfers (net) 44.1; *Current balance* 14.4; Capital account (net) -22.0; Net errors and omissions 9.3; *Overall balance* 1.7. Source: Asian Development Bank, *Key Indicators of Developing Asian and Pacific Countries*.

EXTERNAL TRADE

Principal Commodities (estimates, US $ million, 2000): *Imports:* Food and live animals 5.0, Beverages and tobacco 6.0; Crude materials, inedible, except fuels 2.6; Mineral fuels, lubricants and related materials 20.4; Animal and vegetable oils and fats 2.4; Chemicals 0.1; Basic manufactures 3.0; Machinery and transport equipment 8.2; Miscellaneous manufactured articles 1.4; Goods not classified by kind 5.8; Total 54.7. *Exports:* Coconut oil (crude) 1.1; Copra cake 1.4; Pet fish 0.5; Total (incl. others) 9.1. Source: Asian Development Bank, *Key Indicators of Developing Asian and Pacific Countries*.

Principal Trading Partners (estimates, US $ million): *Imports* (2003): Australia 10.1; Hong Kong 2.5; Japan 3.7; New Zealand 2.5; USA 49.2; Total (incl. others) 72.3. *Exports* (2000): USA 5.2; Total (incl. others) 9.1. Source: mainly Asian Development Bank, *Key Indicators of Developing Asian and Pacific Countries*.

TRANSPORT

Road Traffic (vehicles registered, 1999): Trucks 64; Pick-ups 587; Sedans 1,404; Jeeps 79; Buses 75; Vans 66; Scooters 47; Other motor vehicles 253.

Shipping: *Merchant Fleet* (at 31 December 2003): Vessels 515; Displacement ('000 grt) 17,628 (Source: Lloyd's Register-Fairplay, *World Fleet Statistics*). *International Sea-borne Freight Traffic* (estimates, '000 metric tons, 1990):* Goods loaded 29; Goods unloaded 123 (Source: UN, *Monthly Bulletin of Statistics*).
* Including the Northern Mariana Islands, the Federated States of Micronesia and Palau.

Civil Aviation (traffic on scheduled services, 1999): Passengers carried 19,000; Passenger-km 12 million. Total ton-km 1 million. Source: UN, *Statistical Yearbook*.

TOURISM

Tourist Arrivals: 4,622 in 1999; 5,246 in 2000; 5,444 in 2001.

Arrivals by Country (2001): Australia 190; Fiji 94; Japan 996; Kiribati 283; Federated States of Micronesia 203; Philippines 222; USA 2,039; Total (incl. others) 5,444.

Tourism Receipts (US $ million): 4 in 2000; n.a. in 2001; 4 in 2002. Sources: World Tourism Organization, *Yearbook of Tourism Statistics*, and Marshall Islands Visitor Authority.

COMMUNICATIONS MEDIA

Telephones (main lines in use): 4,500 in 20031*.

Mobile Cellular Telephones (subscriptions): 600 in 2003*.

Facsimile Machines (number): 160 in 1996†.

Personal Computers: 3,000 in 2002*.

Internet Users: 1,400 in 2003*.

Non-daily Newspaper: 1 (average circulation 10,000 copies) in 1996‡.
* Source: International Telecommunication Union.
† Source: UN, *Statistical Yearbook*.
‡ Source: UNESCO, *Statistical Yearbook*.

EDUCATION

Primary (1998): 103 schools; 548 teachers; 12,421 pupils enrolled.

Secondary (1998): 16 schools; 162 teachers; 2,667 pupils enrolled.

Higher (1994): 1 college; 25 teachers; 1,149 students enrolled.

Directory

The Constitution

On 1 May 1979 the locally drafted Constitution of the Republic of the Marshall Islands became effective. The Constitution provides for a parliamentary form of government, with legislative authority vested in the 33-member Nitijela. Members of the Nitijela are elected by a popular vote, from 25 districts, for a four-year term. There is an advisory council of 12 high chiefs, or Iroij. The Nitijela elects the President of the Marshall Islands (who also has a four-year mandate) from among its own members. The President then selects members of the Cabinet from among the members of the Nitijela. On 25 June 1983 the final draft of a Compact of Free Association was signed by the Governments of the Marshall Islands and the USA, and the Compact was effectively ratified by the US Congress on 14 January 1986. An amended Compact was signed by the Governments of the two countries on 1 May 2003; final terms were ratified by the US Congress in November, and signed by the US President in December of that year. By the terms of the Compact, free association recognizes the Republic of the Marshall Islands as an internally sovereign, self-governing state, whose policy concerning foreign affairs must be consistent with guide-lines laid down in the Compact. Full responsibility for defence lies with the USA, which undertakes to provide regular economic assistance. The economic and defence provisions of the Compact are renewable after 15 years, but the status of free association continues indefinitely.

The Government

HEAD OF STATE

President: KESSAI H. NOTE (took office 10 January 2000; re-elected by the Nitijela 4 January 2004).

THE CABINET
(September 2004)

Minister in Assistance to the President: WITTEN T. PHILIPPO.

Minister of Education: WILFRED I. KENDALL.

Minister of Finance: BRENSON S. WASE.

Minister of Transportation and Communication: MICHAEL M. KONELIOS.

Minister of Health and Environment: ALVIN T. JACKLICK.

Minister of Public Works: MATTLAN ZACKHRAS.

Minister of Internal Affairs: REIN MORRIS.

Minister of Justice: DONALD CAPELLE.

Minister of Resources and Development: JOHN M. SILK.

Minister of Foreign Affairs and Trade: GERALD ZACKIOS.

MINISTRIES

Office of the President: Govt of the Republic of the Marshall Islands, POB 2, Majuro, MH 96960; tel. (625) 3213; fax (625) 4021; e-mail presoff@ntamar.net.

Ministry of Education: POB 3, Majuro, MH 96960; tel. (625) 5262; fax (625) 7735/3861; e-mail secmoe@ntamar.net.

Ministry of Finance: POB D, Majuro, MH 96960; tel. (625) 7420; fax (625) 3607; e-mail secfin@ntamar.net.

Ministry of Foreign Affairs and Trade: POB 1349, Majuro, MH 96960; tel. (625) 3181; fax (623) 4979; e-mail mofatadm@ntamar.net.

Ministry of Health and Environment: POB 16, Majuro, MH 96960; tel. (625) 3355; fax (625) 3432; e-mail mipamohe@ntamar.net.

Ministry of Internal Affairs: POB 18, Majuro, MH 96960; tel. (625) 8240/8718; fax (625) 5353; e-mail rmihpo@ntamar.net.

Ministry of Justice: c/o Office of the Attorney General, Majuro, MH 96960; tel. (625) 3201/8245; fax (625) 3323; e-mail agoffice@ntamar.net.

Ministry of Public Works: POB 1727, Majuro, MH 96960; tel. (625) 8911; fax (625) 3005; e-mail rndadm@ntamar.net.

Ministry of Resources and Development: POB 1727, Majuro, MH 96960; tel. (625) 3206/3277; fax (625) 5447; e-mail rndsec@ntamar.net.

Ministry of Transportation and Communication: POB 1079, Majuro, MH 96960; tel. (625) 3129/8869; fax (625) 3486; e-mail rmimotc@ntamar.net.

STATE TRIBUNAL

Nuclear Claims Tribunal: POB 702, Majuro, MH 96960; tel. (625) 3396; fax (625) 3389; e-mail nctmaj@ntamar.net; internet www.nuclearclaimstribunal.com; f. 1988; authorized under Section 177 of the first Compact of Free Association between the Government of the Marshall Islands and the Government of the USA to decide all claims arising from the nuclear testing programme conducted by the USA in the Marshall Islands in 1946–1958; Chair. JAMES H. PLASMAN; Defender of the Fund PHILIP A. OKNEY; Public Advocate BILL GRAHAM.

Legislature

THE NITIJELA

The Nitijela (lower house) consists of 33 elected Senators. The most recent national election was held on 17 November 2003, as a result of which the United Democratic Party held 20 seats and Ailin Kein Ad 10.

Speaker: Sen. LITOKWA TOMEING.

THE COUNCIL OF IROIJ

The Council of Iroij is the upper house of the bicameral legislature, comprising 12 tribal chiefs who advise the Presidential Cabinet and review legislation affecting customary law, land tenure or any traditional practice.

Chairman: Iroij KOTAK LOEAK.

Political Organizations

Ailin Kein Ad (Our Islands): Majuro; f. 2002; opposed to President Note; Leader TONY DEBRUM.

United Democratic Party: Majuro; Chair. LITOKWA TOMEING.

Diplomatic Representation

EMBASSIES IN THE MARSHALL ISLANDS

China (Taiwan): A5-6, Lojkar Village, Long Island, POB 1229, Majuro, MH 96960; tel. (247) 4141; fax (247) 4143; e-mail eoroc@ntamar.net; Ambassador LIEN-GENE CHEN.

Japan: A-1 Lojkar Village, POB 300, Majuro, MH 96960; tel. (247) 7463; fax (247) 7493; Ambassador KENRO IINO.

USA: POB 1379, Majuro, MH 96960; tel. (247) 4011; fax (247) 4012; e-mail publicmajuro@state.gov; internet usembassy.state.gov/majuro; Ambassador GRETA N. MORRIS.

Judicial System

The judicial system consists of the Supreme Court and the High Court, which preside over District and Community Courts, and the Traditional Rights Court.

Supreme Court of the Republic of the Marshall Islands: POB 378, Majuro, MH 96960; tel. (625) 3201; fax (625) 3323; e-mail jutrep@ntamar.com; Chief Justice DANIEL CADRA.

High Court of the Republic of the Marshall Islands: Majuro; Chief Justice CARL INGRAM.

District Court of the Republic of the Marshall Islands: Majuro, MH 96960; tel. (625) 3201; fax (625) 3323; Presiding Judge BOKEPOK HELAI.

Traditional Rights Court of the Marshall Islands: Majuro, MH 96960; customary law only; Chief Judge RAILEY ALBERILTAR.

Religion

The population is predominantly Christian, mainly belonging to the Protestant United Church of Christ. The Roman Catholic Church, Assembly of God, Bukot Nan Jesus, Seventh-day Adventists, the Church of Jesus Christ of Latter-day Saints (Mormons), the Full Gospel and the Bahá'í Faith are also represented.

CHRISTIANITY

The Roman Catholic Church

The Apostolic Prefecture of the Marshall Islands included 4,601 adherents at 31 December 2002.

Prefect Apostolic of the Marshall Islands: Rev. Fr JAMES C. GOULD, POB 8, Majuro, MH 96960; tel. (625) 6675; fax (625) 5520; e-mail catholic@ntamar.com.

Protestant Churches

The Marshall Islands come under the auspices of the United Church Board for World Ministries (475 Riverside Drive, New York, NY 10115, USA); Sec. for Latin America, Caribbean and Oceania Dr PATRICIA RUMER.

BAHÁ'Í FAITH

National Spiritual Assembly: POB 1017, Majuro, MH 96960; tel. (247) 3512; fax (247) 7180; e-mail nsamarshallislands@yahoo.com; internet www.mh.bahai.org; mems resident in 50 localities; Sec. Dr IRENE J. TAAFAKI.

The Press

Kwajalein Hourglass: POB 23, Kwajalein, MH 96555; tel. (355) 3539; e-mail jbennett@kls.usaka.smdc.army.mil; internet www.smdc.army.mil/KWAJ/Hourglass/Hourglass.html; f. 1954; 2 a week; Editor JIM BENNETT; circ. 2,300.

Marshall Islands Gazette: monthly; government publ.

Marshall Islands Journal: POB 14, Majuro, MH 96960; tel. (625) 8143; fax (625) 3136; e-mail journal@ntamar.com; f. 1970; weekly; Editor GIFF JOHNSON; circ. 3,700.

Broadcasting and Communications

TELECOMMUNICATIONS

National Telecommunications Authority (NTA): POB 1169, Majuro, MH 96960; tel. (625) 3852; fax (625) 3952; e-mail aefowler@ntamar.net; internet www.ntamar.com; privatized in 1991; sole provider of local and long-distance telephone services and internet communications in the Marshall Islands; Chair. ALEX C. BING; Pres. and Gen. Man. ALAN FOWLER.

BROADCASTING

Radio

Radio Marshalls V7AB: POB 3250, Majuro, MH 96960; tel. (625) 3411; fax (625) 5353; govt-owned; commercial; programmes in English and Marshallese; Station Man. ANTARI ELBON.

Marshall Islands Broadcasting Co: POB 19, Majuro, MH 96960; tel. (625) 3250; fax (625) 3505; privately-owned; Chief Information Officer PETER FUCHS.

Television

AFN (American Forces Network): Kwajalein Atoll; e-mail comments@abs.army.mil; internet http://abs-afn.army.mil/index1.htm; information and entertainment network for US service personnel.

Alele Museum Foundation: POB 629, Majuro, MH 96960; tel. and fax (625) 3226; broadcasts educational programmes.

Marshalls Broadcasting Co Television: POB 19, Majuro, MH 96960; tel. (625) 3413; privately-owned; Chief Information Officer PETER FUCHS.

Finance

(cap. = capital; res = reserves; dep. = deposits; amounts in US dollars)

BANKING

Bank of Guam (USA): POB C, Majuro, MH 96960; tel. (625) 3322; fax (625) 3444; internet www.bankofguam.com; Man. ROMY A. ANGEL; brs in Ebeye, Kwajalein and Majuro.

Bank of the Marshall Islands: POB J, Majuro, MH 96960; tel. (625) 3636; fax (625) 3661; e-mail bankmar@ntamar.com; internet www.angelfire.com/ms/bankofMI; f. 1982; 40% govt-owned; dep. 47.5m., total assets 61.8m. (Dec. 2003); Chair. GRANT LABAUN; Gen. Man. PATRICK CHEN; brs in Majuro, Ebeye and Santo.

Marshall Islands Development Bank: POB 1048, Majuro, MH 96960; tel. (625) 3230; fax (625) 3309; f. 1989; total assets 19.5m. (Dec. 1992); lending suspended in 2003; Man. Dir AMON TIBON.

INSURANCE

Majuro Insurance Company: POB 60, Majuro, MH 96960; tel. (625) 8885; fax (625) 8188; Man. LUCY RUBEN.

Marshalls Insurance Agency: POB 113, Majuro, MH 96960; tel. (625) 3366; fax (625) 3189; Man. TOM LIKOVICH.

Moylan's Insurance Underwriters (Marshall) Inc: POB 727, Majuro, MH 96960; tel. (625) 3220; fax (625) 3361; e-mail marshalls@moylansinsurance.com; internet www.moylansinsurance.com; Pres. JOEL PHILLIP.

Trade and Industry

DEVELOPMENT ORGANIZATIONS AND STATE AUTHORITIES

Marshall Islands Environmental Protection Authority: POB 1322, Majuro, MH 96960; tel. (625) 3035; fax (625) 5202; e-mail eparmi@ntamar.net; Dir ABRAHAM HICKIN (acting).

Marshall Islands Development Authority: Majuro, MH 96960; Gen. Man. DAVID KABUA.

Marshall Islands Marine Resources Authority: Majuro, MH 96960; Dir DANNY WASE.

Kwajalein Atoll Development Authority (KADA): POB 5159, Ebeye Island, Kwajalein, MH 96970; Dir JEBAN RIKLON.

Tobolar Copra Processing Authority: POB G, Majuro, MH 96960; tel. (625) 3494; fax (625) 7206; e-mail tobolar@ntamar.com; Gen. Man. MIKE SLINGER.

CHAMBER OF COMMERCE

Majuro Chamber of Commerce: POB 3279, Majuro, MH 96960; tel. (625) 2525; fax (625) 2500; e-mail emihess@yahoo.com; Pres. KIRTLEY PINHO; Vice-Pres. DON HESS.

UTILITIES

Electricity

Marshalls Energy Company: POB 1439, Majuro, MH 96960; tel. (625) 3829; fax (625) 3397; e-mail meccorp@ntamar.net; Gen. Man. WILLIAM F. ROBERTS.

Kwajalein Atoll Joint Utility Resource (KAJUR): POB 5819, Ebeye Island, Kwajalein, MH 96970; tel. (329) 3799; fax (329) 3722.

Water

Majuro Water and Sewage Services: POB 1751, Majuro, MH 96960; tel. (625) 8934; fax (625) 3837; e-mail mwsc@ntamar.net; internet www.omip.org/majuro.html; Man. TERRY MELLAN.

CO-OPERATIVES

These include the Ebeye Co-op, Farmers' Market Co-operative, Kwajalein Employees' Credit Union, Marshall Is Credit Union, Marshall Is Fishermen's Co-operative, and the Marshall Is Handicraft Co-operative.

Transport

ROADS

Macadam and concrete roads are found in the more important islands. In 1996 there were 152 km of paved roads in the Marshall Islands, mostly on Majuro and Ebeye. Other islands have stone and coral-surfaced roads and tracks. In 1997 the Marshall Islands received a grant of some US $0.5m. from Japan for a road-improvement project on Majuro. The project was to form part of an extensive programme costing US $15m., and was completed in 1999.

SHIPPING

The Marshall Islands operates an 'offshore' shipping register. In mid-2003 the merchant fleet comprised 569 vessels, with a combined displacement of some 17.5m. grt.

Vessel Registry:

Marshall Islands Maritime and Corporate Administrators Inc: 11495 Commerce Park Drive, Reston, VA 20191-1507, USA; tel. (703) 620-4880; fax (703) 476-8522; e-mail info@register-iri.com; internet www.register-iri.com.

The Trust Company of the Marshall Islands Inc: Trust Company Complex, Ajeltake Island, POB 1405, Majuro, MH 96960; tel. (247) 3018; fax (247) 3017; e-mail tcmi@ntamar.net; Pres. GUY EDISON CLAY MAITLAND.

CIVIL AVIATION

In 1995 the Marshall Islands, Kiribati, Nauru and Tuvalu agreed to begin discussions on the establishment of a joint regional airline. In 1997 the Marshall Islands signed a bilateral agreement on international air transport with the Federated States of Micronesia. Continental Micronesia operates three flights a week from Honolulu and Guam; Air Marshall Islands provides a daily domestic service; and Aloha Airlines provides a weekly service from Honolulu to Kwajalein and to Majuro.

Air Marshall Islands (AMI): POB 1319, Majuro, MH 96960; tel. (625) 3731; fax (625) 3730; e-mail amihdq@ntamar.net; internet www.airmarshallislands.com; f. 1980; internal services for the Marshall Islands; international operations ceased in early 1999; also charter, air ambulance and maritime surveillance operations; agency for Aloha airlines; Chair. KUNIO LAMARI; CFO NEIL ESCHERRA.

Continental Micronesia: POB 156, Majuro; tel. (625) 3209; fax (625) 3730; international flights between Majuro, the Federated States of Micronesia, Guam and Honolulu; also internal services between Majuro and Kwajalein; based in Hagåtña, Guam; Man. LEO SION.

Tourism

Tourism, which has been hindered by the difficulty of gaining access to the islands and a lack of transport facilities, was expected to develop significantly from the late 1990s, owing to the establishment of major resort complexes on Majuro and on Mili Atoll, funded at an estimated cost of US $1,000m. by South Korean investors. Tourism receipts totalled some $4m. in 2002. There were 5,444 tourist arrivals in 2001. In that year some 37.5% of visitors came from the USA and 18.3% from Japan. The islands' attractions include excellent opportunities for diving, game-fishing and the exploration of sites and relics of Second World War battles. The Marshall Islands Visitor Authority has implemented a short-term tourism development programme focusing on special-interest tourism markets. In the longer term the Visitor Authority planned to promote the development of small-island resorts throughout the country. Following consultations with other members of the Pacific Asia Travel Association (PATA) in June 2004, the Marshall Islands Visitor Authority requested an increase of 150% in its government funding.

Marshall Islands Visitor Authority: POB 5, Majuro, MH 96960; tel. (625) 6482; fax (625) 6771; e-mail tourism@ntamar.net; internet www.visitmarshallislands.com; f. 1997; Chair. KIRT PINHO.

Defence

Defence is the responsibility of the USA, which maintains a military presence on Kwajalein Atoll. The US Pacific Command is based in Hawaii.

Education

There is a school system, based on that of the USA, operated by the state. However, only 25% of students continue their education beyond a primary level, owing to limited resources and inadequate instruction. In 1999 there were 103 primary schools, with a total enrolment of 12,421 pupils, but only 16 secondary schools, with a total of 2,667 pupils enrolled. The College of the Marshall Islands, which became independent from the College of Micronesia in 1993, is based on Majuro. (However, the College faced a severe financial crisis in 2003 following the discovery of widespread misappropriation of its funds.) In 1994 there were 1,149 students enrolled at the College, and in 1991 118 students were enrolled at colleges and universities in the USA. In early 1995 the islands received a grant of US $6m. from Japan to fund projects to improve secondary education, as well as a loan of US $2m. from the People's Republic of China to finance the construction of a new secondary school on Majuro. Government expenditure on education in 1997/98 was budgeted at US $10.0m., equivalent to 23.5% of total budgetary spending.

THE FEDERATED STATES OF MICRONESIA

Physical and Social Geography

The Federated States of Micronesia forms (with Palau, q.v.) the archipelago of the Caroline Islands, about 800 km east of the Philippines. The Federated States of Micronesia includes (from west to east) the states of Yap, Chuuk (formerly Truk, renamed in January 1990), Pohnpei (formerly Ponape, renamed in November 1984) and Kosrae, and consists of some 607 islands and atolls, extending for some 2,900 km. The islands of Yap are in the Western Caroline Islands and are about 870 km south-west of Guam. The remaining islands are in the Eastern Carolines and the easternmost island of Kosrae lies some 4,587 km south-west of Hawaii. The islands are subject to heavy rainfall, although precipitation decreases from east to west. January, February and March are the driest months, although seasonal variations in rainfall and temperature are generally small. Average annual temperature is 27°C (81°F). The indigenous population, which is predominantly Micronesian, consists of various ethno-linguistic groups, the principal ones being Yapese, Ulithian-Woleaian, Chuukese, Pohnpeian, Kosraean and Kapingimarangi-Nukuoroan. English is widely understood. The traditional social structure is based on matrilineal (except in the Polynesian islands of Kapingimarangi and Nukuoro) clans. A considerable degree of social stratification occurred, particularly in Yap, much of which remains. In Pohnpei social status was gained by complex competition for bestowed titles. The erosion of the traditional way of life is a source of considerable social strain. Most of the population, which totalled 107,008 at the census of 2000 according to provisional results, are Christian. The federal capital is at Palikir, on Pohnpei.

History

The Caroline Islands were first settled by the ancestors of the Micronesians some 4,000 years ago. The islands that now constitute the Federated States of Micronesia were then ruled by numerous clan chieftains. A form of tributary empire emerged at one point, initially based in Kosrae, but eventually succumbing to an economic and religious empire dominated by Yap. In 1525 Portuguese navigators first came upon the islands of Yap and Ulithi. Spanish expeditions subsequently reached the other Caroline Islands, and Spain only renounced its sovereignty in 1899, when the Carolines (including Palau) and the Mariana Islands (except the southernmost island of Guam) were sold to Germany, which had already secured the Marshall Islands. In 1914 this territory of German Micronesia was occupied by the Japanese, who acquired a League of Nations mandate to administer it in 1920. The USA conquered the territory in the Second World War and, in 1947, agreed to administer it, on behalf of the UN, as the Trust Territory of the Pacific Islands (TTPI). (For the history of the TTPI, see the chapter on Palau.)

Until 1979 the four districts of Yap, Truk (formerly Hogoleu, now Chuuk), Ponape (now Pohnpei) and Kosrae were governed by a local Administrator, appointed by the High Commissioner of the TTPI. However, on 10 May 1979 the four districts ratified a new Constitution to become the Federated States of Micronesia. The districts of Palau and the Marshall Islands had rejected participation in the federation.

The USA signed a Compact of Free Association with the Federated States of Micronesia in October 1982. The Compact with the Federated States of Micronesia was approved in a plebiscite in June 1983, and the Congress of the Federated States of Micronesia ratified the decision in September. The Compact of Free Association came into effect on 3 November 1986, and the territory was deemed to be subject to the Trusteeship no longer. In November President Ronald Reagan issued a proclamation which formally ended US administration in Micronesia. The final dissolution of the TTPI (which, in effect, consisted only of Palau) occurred on 1 October 1994, when the Compact between Palau and the USA came into effect. The UN Security Council formally terminated the Trusteeship Agreement in December 1990. Ponape was renamed Pohnpei in November 1984, when its Constitution came into effect. Truk was renamed Chuuk in January 1990, following the implementation of its new Constitution.

The Federated States of Micronesia has established diplomatic relations with numerous countries world-wide, and was admitted to the UN in September 1991.

The incumbent President, John Haglelgam, was replaced by Bailey Olter, a former Vice-President, in May 1991. At congressional elections in March 1995, Olter was re-elected to the Pohnpei Senator-at-Large seat, and in early May was re-elected to the presidency unopposed. Similarly, Jacob Nena was re-elected as Vice-President. Allegations that financial mismanagement by the Governor of Chuuk, Sasao Gouland, had resulted in state debts of some US $20m. led to his resignation in June 1996, in order to avoid impeachment proceedings. In July Olter suffered a stroke. Jacob Nena served as acting President during Olter's absence from office, and in May 1997 was sworn in as President of the country. Olter died in February 1999.

Congressional elections took place in early March 1997 for the 10 Senators elected on a two-yearly basis, at which all of the incumbents were returned to office. A referendum held concurrently on a proposed amendment to the Constitution (which envisaged increasing the allocation of national revenue to the state legislatures from 50% to 80% of the total budget) was approved in Chuuk and Yap, but rejected in Pohnpei and Kosrae.

Allegations of government interference in the media became widespread when the editor of the country's principal newspaper, *FSM News*, was refused permission to re-enter the islands in June 1997. The Government had sought to deport the editor (who was a Canadian national) following publication in the periodical of reports on government spending, which the authorities claimed were false and malicious. It was also thought that by enforcing the exclusion order, the Government hoped to suppress the publication in the newspaper of information relating to alleged corruption among public officials. The newspaper ceased publication in late 1997.

In February 1998 Congress approved proposals to restructure and reorganize the Cabinet. Several ministerial portfolios were consequently merged or abolished, with the aim of reducing government expenditure. Congressional elections took place on 2 March 1999, at which President Nena was re-elected to the Kosrae Senator-at-Large seat and Vice-President Leo Falcam to the Pohnpei Senator-at-Large seat. On 11 May Congress elected Falcam as President and the Chuuk Senator-at-Large, Redley Killion, as Vice-President.

A first round of renegotiations of the Compact of Free Association (certain terms of which were due to expire in 2001) was completed in late 1999. The USA and the Federated States of Micronesia pledged to maintain defence and security relations. It was also agreed that the USA would continue to provide economic aid to the islands and assist in the development of the private sector, as well as in promoting greater economic self-sufficiency. In July 2001 the USA offered annual assistance of US $61m. and a trust fund of $13m., and expressed concern that the $2,600m. it had given to Micronesia and the Marshall Islands since 1986 had been mismanaged. The Compact's funding terms for Micronesia were originally due to expire on 3 November 2001, but negotiations regarding a new Compact were not completed by this time.

Funding was, nevertheless, continued at the Compact's 15-year average level while negotiations remained in progress. Following a proposal by the USA in April 2002 to extend economic assistance for a period of 20 years, a new draft funding structure was agreed, and in March the US budget projections for 2004 granted Micronesian citizens access to private health-care resources in the USA as part of the Federated States' continued entitlement to US federal programmes. On 1 May 2003 the amended Compact of Free Association was signed by representatives of the two countries in Pohnpei. The new Compact envisaged direct annual grants of US $76.2m. in 2004, in addition to a further $16m. annually, which was to be paid into a Trust Fund for Micronesia. From 2007, direct grants were to decrease by some $800,000, with this amount being transferred to the Trust Fund. (The total amount to be paid prior to the expected termination of US assistance in 2023 amounted, in 2004 terms, to some $1,760m.) Furthermore, the Micronesian Government also undertook to provide frequent, strictly monitored audit information on all US funding in order to ensure greater accountability. In October 2003 final agreement was reached on some outstanding security and immigration issues, and the US Congress approved the amended Compact in November. US President George W. Bush approved the pact in December, and representatives of both Governments signed a document of implementation in June 2004. Nevertheless, there remained widespread concern in the Federated States of Micronesia that the new Compact represented a substantial overall reduction in annual income over the long term. Moreover, the formula for the distribution of Compact funds to each of Micronesia's states and the removal of certain US subsidies remained the subject of considerable controversy. In August representatives of Micronesia and the USA met to review the management of Compact funds in the first session of the Joint Economic Management Committee.

In late 2002 unrest occurred on the Faichuk islands, part of Chuuk, where the Faichuk Commission for Statehood continued its campaign to secede from Chuuk and gain equal status within the Federation. The secessionists believe that independence would bring to the islands more goods, services, medical treatment and capital improvement projects. Local dissatisfaction worsened in September following allegations of electoral manipulation against the village mayor of Udot island, with a large crowd appearing to support attempts by local security forces to prevent the mayor's arrest. (The issue of Faichuk's status remained unresolved in 2004.)

In September 2002, meanwhile, a referendum was held on a number of proposed amendments to the Constitution. The proposed changes included the direct election of presidential candidates, the right of islanders to hold dual citizenship and changes to the distribution formula for Compact of Free Association funds. However, the measures did not receive the required three-quarters' majority of votes and were thus rejected.

At congressional elections held on 6 March 2003 President Leo Falcam unexpectedly failed to achieve re-election to a further four-year term as Senator-at-Large for Pohnpei. In mid-May Congress appointed the Senator-at-Large for Yap, Joseph J. Urusemal, to the presidency. The elections were the subject of some controversy, as it appeared that elected officials had disbursed a portion of the 2002 US funding for Micronesia in order to enhance their electoral prospects. The alleged misallocation of funds was reportedly a significant factor in the worsening fiscal positions of Chuuk, Pohnpei and Kosrae. Moreover, perceptions of official accountability continued to deteriorate in 2003; in November three serving congressmen were indicted for their role in an alleged fraud involving some US $1.2m. in public funds. In January 2004 the National Congress approved a resolution to dismiss the judge assigned to the case. President Urusemal lodged a petition against the dismissal on the grounds that it infringed the constitutionally guaranteed separation of powers. In August the Supreme Court ruled in favour of the petition and overturned the judge's dismissal. In January, meanwhile, national congressmen attempted to introduce legislation effectively absolving public officials from corruption allegations relating to Compact of Free Association funds. The proposals aroused widespread public hostility, and several representatives of state legislatures threatened to secede from the federation unless the measure were withdrawn. In March the so-called 'amnesty bill' was returned to a congressional sub-committee for further discussion.

In December 1996 a state of emergency was declared after a typhoon wreaked devastation on the islands of Yap. Flooding and mudslides in Pohnpei in April 1997, as a result of heavy rain, caused the deaths of more than 20 people; another typhoon in mid-1997 left several people dead. Furthermore, in early 1998 a state of emergency was declared, owing to severe drought throughout the islands, believed to have been caused by El Niño (a periodic warming of the tropical Pacific Ocean). In July 2002 at least 49 people died and extensive damage was caused to crops and buildings on Chuuk during a severe tropical storm. In December the islands suffered further damage (albeit without loss of life) when a typhoon struck. US President George W. Bush declared Micronesia a federal disaster area and ordered emergency US funding and resources to be allocated to the relief effort. A further typhoon which struck Yap in April 2004 left 1,200 people homeless; the US Government offered to assume 75% of the cost of the recovery effort.

In late 2000 marine biologists issued a warning regarding the erosion of the islands' coastlines, caused by the destruction of the coral reefs by pollution, overfishing and increasing sea temperatures. Furthermore, in late 2003 concerns over environmental pollution increased, owing to environmental damage caused by former US and Japanese military equipment that was submerged in Micronesian waters. A former US Navy oil tanker submerged off the remote Ulithi Atoll was reported at this time to be leaking. Meanwhile, in September 2003 President Urusemal urged the UN General Assembly to work towards halting climate change and its consequent effects on sea-levels and weather systems.

In May 2000 a state of emergency was declared on Pohnpei, following an outbreak of cholera. In August the Government announced a vaccination scheme for the entire population over two years of age. Import restrictions were introduced on the surrounding islands. Pohnpei was officially declared free of cholera on 16 February 2001. During the epidemic some 20 people had died and a further 3,525 were estimated to have been infected.

In November 2002, following a conference in Manila, the Philippines, attended by Pacific nations and those industrialized countries with significant fishing interests in the region, Micronesia was awarded the contract to establish a headquarters for a new body to manage migratory fish stocks in the Central and Western Pacific, to be known as the Tuna Commission. The remit for the organization included the management of waters outside each nation's 200-mile exclusive economic zone, in line with the framework established under the 1995 UN Fish Stocks Agreement. In February 2003 Pohnpei hosted the first Summit of Micronesian Leaders. At the second summit meeting, held in Koror, Palau, in March 2004, President Urusemal and the leaders of Palau, the Northern Mariana Islands and Guam undertook to increase co-operation among the Pacific island states in the areas of tourism and the environment.

Economy

In 2002, according to estimates by the World Bank, gross national income (GNI) in the Federated States of Micronesia, measured at average 2000–02 prices, was US $242m., equivalent to $1,980 per head. Gross domestic product (GDP) per head decreased at an average annual rate of 0.5% in 1990–2002. Over the same period the population increased by an annual average of 2.0%. Overall GDP increased, in real terms, at an average annual rate of 1.6% in 1990–2002. According to the Asian Development Bank (ADB), growth in 2003 was 0.1%, compared with 0.9% in the previous year. The ADB projected a contraction of 1.5% in 2004.

Agriculture is mainly on a subsistence level, although its importance is diminishing. The principal agricultural crops are coconuts (from which some 500 short tons of copra were produced in 2001), bananas, citrus fruit, betel-nuts, cassava and sweet potatoes. White peppercorns are produced on Pohnpei. The sector (including forestry and fishing) contributed 19.1% to GDP in 1996 and engaged 55.3% of the employed labour force in 2000. Exports of bananas accounted for 1.2% of export earnings in 1999, while exports of marine products accounted for 91.9% of total export revenue in that year. A dramatic increase in the re-export of fish to Japan by foreign vessels operating in the islands' waters in 1994 was responsible for growth of almost 300% in earnings from the fishing industry. In 1996 the islands' share of earnings under the Multilateral Fisheries Treaty with the USA increased from 8% to 20%. In 1999/2000 fishing access fees, mainly from Japanese fleets, totalled US $16.8m. In 1997 the Government received a technical assistance grant, worth some $0.9m., for the development and implementation of a national fisheries plan aimed at ensuring the efficient management and conservation of the islands' marine resources. The National Fisheries Corporation maintains a specialized air freight service to export tuna to Japan. In November 2002 Micronesia was confirmed as the location of the headquarters for a new multilateral agency to manage migratory fish stocks in the region, to be known as the Tuna Commission; the establishment of the commission had been announced in mid-2000.

The tourist industry is a significant source of foreign exchange. It was hoped that several projects aimed at improving communications would stimulate tourism, hitherto hindered by the territory's remote situation. The industry has been identified by the Asian Development Bank (ADB) as having the greatest potential for development and thus contribution to the islands' economic growth. Some 15,265 non-resident tourists arrived on the islands in 2001, and tourism receipts for that year totalled some US $13m.

In the year ending 30 September 2003 there was a visible trade deficit of US $88.8m., but a surplus of $17.7m. on the current account of the balance of payments. The value of merchandise exports, including re-exports, was estimated at $20.1m. in 2002/03, when imports cost some $108.9m. The principal imports in 1999 were food and live animals (24.8% of the total), mineral fuels and lubricants (20.3%), machines and transport equipment (19.5%) and basic manufactures (18.9%). Fish is the major export commodity, mainly in the form of re-exports to Japan. Other significant exports are garments and buttons (8.4%), handicrafts, bananas, copra, trochus shells and pepper. The USA provided 60.1% of total imports in 2002, Japan 11.7% and Australia 7.3%. Japan was the principal market for exports in 1999, taking 83.9% of the total.

In 2002/03 there was an estimated budget surplus of US $4.1m., compared with a estimated surplus of $5.5m. in the previous financial year. The 2003 national budget envisaged expenditure of $37.8m. for central government operations. The Federated States of Micronesia relies heavily on financial assistance, particularly from the USA. Under the terms of the amended Compact of Free Association (signed in May 2003), gross US financial contributions were to be $104.9m. annually from September 2005. It is hoped that Micronesia will have achieved financial self-sufficiency by the year 2023. Under the amended Compact, instead of general budgetary grants, US aid will be targeted more towards specific projects and government departments. Also, aid will be conditional upon the efficiency of its management and use, to be reviewed annually by the Joint Economic Management and Financial Accountability Committee (JEMFAC). This Committee first convened in August 2004, in Honolulu, Hawaii. In April 2001, meanwhile, Japan donated US $7m. to improve the road network in the state of Yap.

At the end of the 2003 financial year the islands' total external debt was estimated at US $54.3m. In that year the cost of debt-servicing was equivalent to just 6.1% of the value of exports of goods and services, compared with a ratio of 65.7% in 2000 and 32.8% in 2001. According to the ADB, the rate of inflation decreased from an estimated annual average of 6.0% in 1993 to 1.3% in 2001. For both the years 2002 and 2003 there was deflation of 0.2%, mainly attributable to the fall in food prices. Inflation of around 0.5% was anticipated in 2004. Some 2.6% of the labour force was unemployed in 2000.

The islands are vulnerable to adverse climatic conditions, as was illustrated in late 1997 and early 1998, when a prolonged drought caused problems throughout the islands, and by a series of tropical storms in 2002–04. An extremely high rate of natural increase in the population has exacerbated certain economic problems, but is partially offset by an annual emigration rate of more than 2%. In 1997 the ADB approved a loan of US $17.7m. for the funding of its so-called Private Sector Development programme, a major structural adjustment of the economy, in preparation for the cessation of US assistance under the Compact of Free Association, the original terms of which expired in September 2001. The reform programme comprised measures to attract new sources of foreign aid and private investment, fiscal reform and the strengthening of the private sector, as well as severe reductions in the number of public-sector employees. In late 1999 an Economic Policy Implementation Council (EPIC) was established in an attempt to monitor the reform process. While commending the successful implementation of certain policies, notably the retrenchment in government expenditure and the development of the private sector, concerns were expressed by the ADB regarding the Federated States of Micronesia's overdependence on foreign aid, and the adverse impact of this on the establishment of economic self-sufficiency. In December 2000 the ADB approved an $8m. loan to fund a six-year reform programme of the health and education sectors. In December 2001 the ADB granted a further loan of $13m. targeted at job creation, increased production for both domestic and export markets and the development of a competitive services sector. No further ADB loans were approved in 2003, although three technical assistance projects, totalling $1.2m., were approved. In August 2004 a loan of $19.2m. to help improve infrastructure in Micronesia was under appraisal. Cumulative ADB lending to the Federated States of Micronesia, as of 31 December 2003, was $56.1m.

The uncertainty surrounding the future of the public sector, meanwhile, had continued to deter investment in the private sector. Furthermore, the private sector was constrained by the disproportionately high cost of domestic labour, rates of pay in the public sector having risen substantially in recent years as a result of the large external inflows (although public-sector salaries were 'frozen' under the 2004 budget). In February 2003 an IMF report noted that facilities for the private sector remained underdeveloped, notwithstanding the authorities' largely positive oversight of the banking sector. Further criticism was attached to the private sector's effective role as a provider of services to the public sector, and the latter's tendency to operate in unequal competition with private-sector interests. Concerns also continued as to the relative lack of progress in restructuring the economy in preparation for the potentially dramatic impact of the decline and eventual withdrawal of direct US aid in 2023. (According to the UN, bilateral and multilateral aid to Micronesia totalled US $101.6m. in 2000, the latter accounting for only $5.0m. of the total.) The Trust Fund for Micronesia established to alleviate such pressures was expected to remain vulnerable to external factors, the majority of this capital being invested in US stock markets. The 2001/02 fiscal deficit, (i.e. excluding the increase in US aid payments in the year, which produced an overall surplus) of approximately $11m.–$12m. (equivalent to some 5% of GDP), was reportedly due to the continued financial difficulties of the National Government and Chuuk state. None the less, a number of apparently positive measures were adopted in 2002, including an agreement in June between the Micronesian Government and the ADB to prepare a National Poverty Reduction Strategy. Following four consecutive years of deficit, the country's budget returned to surplus in 2002. The repayment of substantial portions of Micronesia's external debt (see above) during 2001 was expected to improve public finances in the long term. In the 12 months to September 2002 the total debt outstanding decreased by 19% compared with the previous financial year (before rising slightly in 2002/03). However, the islands' remote position remained an obstacle to the development of future sources of earnings such as tourism, and the fishing sector continued to decline.

In early 1996 the Federated States of Micronesia joined representatives of the other countries and territories of Micronesia at a meeting in Hawaii, at which a new regional organization, the Council of Micronesian Government Executives, was established. The new body aimed to facilitate discussion of economic developments in the region and to examine possibilities for reducing the considerable cost of shipping essential goods between the islands. The Federated States is a member of, *inter alia*, the Pacific Community (formerly the South Pacific Commission), the Pacific Islands Forum (formerly the South Pacific Forum), the South Pacific Regional Trade and Economic Co-operation Agreement (SPARTECA) and the UN Economic and Social Commission for Asia and the Pacific (ESCAP).

Statistical Survey

AREA AND POPULATION

Area: 700 sq km (270.3 sq miles): Chuuk (294 islands) 127 sq km; Kosrae (5 islands) 110 sq km; Pohnpei (163 islands) 344 sq km; Yap (145 islands) 119 sq km.

Population: 105,506 (53,923 males, 51,583 females) at census of 18 September 1994; 107,008 (males 54,191, females 52,817) at 2000 census (provisional). *By State* (2000): Chuuk 53,595; Kosrae 7,686; Pohnpei 34,486; Yap 11,241. *Mid-2002* (UN estimate): 119,551 (Source: UN, *Population and Vital Statistics Report*).

Density (mid-2002): 170.8 per sq km.

Births and Deaths (2000, official estimates): Birth rate 27.1 per 1,000; Death rate 6.0 per 1,000.

Expectation of Life (WHO estimates, years at birth): 66.5 (males 64.9; females 68.1) in 2002. Source: WHO, *World Health Report*.

Economically Active Population ('000 persons, 2000): Agriculture, forestry and fishing 17.25; Total employed (incl. others) 31.21; Unemployed 0.82; Total labour force 32.02. Source: Asian Development Bank, *Key Indicators of Developing Asian and Pacific Countries*.

HEALTH AND WELFARE

Key Indicators

Total Fertility Rate (children per woman, 2002): 3.8.

Under-5 Mortality Rate (per 1,000 live births, 2002): 24.

Physicians (per 1,000 head, 1999): 0.57.

Hospital Beds (per 1,000 head, 1989): 3.47.

Health Expenditure (2001): US $ per head (PPP): 319.

Health Expenditure (2001): % of GDP: 7.8.

Health Expenditure (2001): public (% of total): 72.0 .

For sources and definitions, see explanatory note on p. vi.

AGRICULTURE, ETC.

Principal Crops* (FAO estimates, '000 metric tons, 2002): Coconuts 140; Cassava 12; Bananas 2; Sweet potatoes 3. Source: FAO.

Livestock* (FAO estimates, '000 head, year ending September 2002): Pigs 32; Cattle 14; Goats 4; Poultry 185. Source: FAO.

Livestock Products* ('000 metric tons, 2002): Beef and veal 245; Pig meat 873; Poultry meat 135; Hen eggs 175; Cattle hides 45. Source: FAO.

* Including the Northern Mariana Islands, the Marshall Islands and Palau.

Fishing ('000 metric tons, live weight, 2002): Skipjack tuna 14.0; Yellowfin tuna 3.9; Bigeye tuna 1.0 ; Total capture (incl. others) 20.4. Source: FAO.

FINANCE

Currency and Exchange Rates: United States currency is used: 100 cents = 1 United States dollar (US $). *Sterling and Euro Equivalents* (31 May 2004): £1 sterling = US $1.8347 ; €1 = US $1.2246; US $100 = £54.50 = €81.66.

Budget (estimates, US $ million, year ending 30 September 2003): *Revenue:* Current 48.4 (Tax 28.5, Non-tax 19.9); Grants 115.3; Total 163.7. *Expenditure:* Current 130.7; Capital 28.9; Total 159.6. Source: Asian Development Bank, *Key Indicators of Developing Asian and Pacific Countries*.

Gross Domestic Product (US $ million, year ending 30 September, current prices): 220.6 in 2001; 222.1 in 2002; 221.8 in 2003.

GDP by Economic Activity (US $ million at current prices, 1996): Agriculture, forestry and fishing 34.7; Mining and quarrying 0.7; Manufacturing 2.6; Electricity, gas and water 1.9; Construction 1.9; Wholesale and retail trade 39.5; Hotels and restaurants 4.1; Transport, storage and communications 8.5; Finance, real estate and business services 5.5; Government services 76.5; Other services 5.6; *GDP in purchasers' values* 181.5. Source: IMF, *Federated States of Micronesia: Recent Economic Developments* (August 1998).

Balance of Payments (estimates, US $ million, year ending 30 September 2003): Merchandise exports (incl. re-exports) f.o.b. 20.1; Merchandise imports f.o.b. –108.9; *Trade balance* –88.8; Exports of services 37.5; Imports of services –52.3; *Balance on goods and services* –103.6; Private unrequited transfers 2.3; Official unrequited transfers 119.0; *Current balance* 17.7; Capital transfers (net) 37.5; Net errors and omissions –26.0; *Overall balance* 29.2. Source: Asian Development Bank, *Key Indicators of Developing Asian and Pacific Countries*.

EXTERNAL TRADE

Principal Commodities (estimates, US $'000, 1999): *Imports:* Food and live animals 3,053; Beverages and tobacco 738; Crude materials (inedible) except fuel 52; Mineral fuels, lubricants, etc. 2,503; Chemicals 534; Basic manufactures 2,326; Machinery and transport equipment 2,406; Miscellaneous manufactured articles 701; Total (incl. others) 12,328. *Exports:* Fish 1,956; Bananas 25; Total (incl. others) 2,128. Source: Asian Development Bank, *Key Indicators of Developing Asian and Pacific Countries*.

Principal Trading Partners (estimates, US $'000, 2002): *Imports:* Australia 6,810; Japan 10,823; USA 55,795; Total (incl. others) 92,859. *Exports* (1999): Japan 1,785; Marshall Islands 23; Total (incl. others) 2,128. Source: Asian Development Bank, *Key Indicators of Developing Asian and Pacific Countries*.

TRANSPORT

Shipping: *Merchant Fleet* (registered at 31 December 2003): Vessels 21; Total displacement ('000 grt) 18.3. Source: Lloyd's Register-Fairplay, *World Fleet Statistics*.

TOURISM

Foreign Tourist Arrivals: 16,140 in 1999; 20,051 in 2000; 15,265 in 2001. Source: World Tourism Organization, *Yearbook of Tourism Statistics*.

Tourist Arrivals by Country of Residence (2001): Australia 516; Japan 3,118; Other Asia (including Philippines, Republic of Korea, People's Republic of China, Taiwan, etc.) 9,183; USA 6,903; Total (incl. others) 15,265. Source: World Tourism Organization, *Yearbook of Tourism Statistics*.

COMMUNICATIONS MEDIA

Telephones ('000 main lines in use, 2001): 10.0*.

Facsimile Machines (number in use, 1998): 539*.

Radio Receivers (1996): 22,000 in use.

Internet Users ('000, 2000): 4.0*.

Television Receivers (1996): 19,800 in use.

* Source: International Telecommunication Union.

EDUCATION

Primary (1995): 174 schools; 1,051 teachers (1984); 27,281 pupils.

Secondary (1995): 24 schools; 314 teachers (1984); 6,898 pupils.

Tertiary (1994): 1,461 students.
Source: UN, *Statistical Yearbook for Asia and the Pacific*.

Directory

The Constitution

On 10 May 1979 the locally-drafted Constitution of the Federated States of Micronesia, incorporating the four states of Kosrae, Yap, Ponape (formally renamed Pohnpei in November 1984) and Truk (renamed Chuuk in January 1990), became effective. Each of the four states has its own Constitution, elected legislature and Governor. The Constitution guarantees fundamental human rights and establishes a separation of the judicial, executive and legislative powers. The federal legislature, the Congress of the Federated States of Micronesia, is a unicameral parliament with 14 members, popularly elected. The executive consists of the President, elected by the Congress, and a Cabinet. The Constitution provides for a review of the governmental and federal system every 10 years.

In November 1986 the Compact of Free Association was signed by the Governments of the Federated States of Micronesia and the USA. By the terms of the Compact, the Federated States of Micronesia is an internally sovereign, self-governing state, whose policy concerning foreign affairs must be consistent with guide-lines laid down in the Compact. Full responsibility for defence lies with the USA, and the security arrangements may be terminated only by mutual agreement. Furthermore, the Compact guaranteed exclusivity to US military forces in Micronesia's waters. The Governments of the Federated States of Micronesia and the USA signed an amended Compact on 1 May 2003, whereby its terms where renewed until 2023. The agreement was approved by the US Congress in November 2003 and ratified by President George W. Bush in December. The status of free association continues indefinitely.

The Government

HEAD OF STATE

President: Joseph J. Urusemal (took office 14 July 2003).

Vice-President: Redley Killion.

THE CABINET
(September 2004)

Secretary of the Department of Finance and Administration: Nick L. Andon.

Secretary of the Department of Foreign Affairs: Sebastian L. Anefal.

Secretary of the Department of Economic Affairs: Ishmael Lebehn (acting).

Secretary of the Department of Health, Education and Social Services: Dr Jefferson B. Benjamin.

Secretary of the Department of Justice: Harry Seymour (acting).

Secretary of the Department of Transportation, Communications and Infrastructure: Akallino H. Susaia.

Public Defender: Beauleen C. Worswick.

Postmaster-General: Bethwel Henry.

GOVERNMENT OFFICES

Office of the President: POB PS-53, Palikir, Pohnpei, FM 96941; tel. 320-2228; fax 320-2785.

Department of Economic Affairs: POB PS-12, Palikir, Pohnpei, FM 96941; tel. 320-2646; fax 320-5854; e-mail fsmrd@mail.fm; e-mail invest@fsminvest.fm; internet www.fsminvest.fm.

Department of Finance and Administration: POB PS-158, Palikir, Pohnpei, FM 96941; tel. 320-2640; fax 320-2830.

Department of Foreign Affairs: POB PS-123, Palikir, Pohnpei, FM 96941; tel. 320-2641; fax 320-2933; e-mail foreignaffairs@mail.fm.

Department of Health, Education and Social Services: POB PS-70, Palikir, Pohnpei, FM 96941; tel. 320-2872; fax 320-2933.

Department of Justice: POB PS-105, Palikir, Pohnpei, FM 96941; tel. 320-2644; fax 320-2234.

Department of Transportation, Communications and Infrastructure: POB PS 123 Palikir, Pohnpei, FM 96941; tel. 320-2613; fax 320-2933.

Office of the Public Defender: POB PS-174, Palikir, Pohnpei, FM 96941; tel. 320-2648; fax 320-5775.

Public Information Office: POB PS-34, Palikir, Pohnpei, FM 96941; tel. 320-2548; fax 320-4356; e-mail fsmpio@mail.fm; internet www.fsmpio.fm.

The Legislature

CONGRESS OF THE FEDERATED STATES OF MICRONESIA

The Congress comprises 14 members (Senators), of whom four are elected for a four-year term and 10 for a two-year term.

Speaker: Peter M. Christian.

STATE LEGISLATURES

Chuuk State Legislature: POB 189, Weno, Chuuk, FM 96942; tel. 330-2234; fax 330-2233; Senate of 10 mems and House of Representatives of 28 mems elected for four years; Gov. Ansito Walter.

Kosrae State Legislature: POB 187, Tofol, Kosrae, FM 96944; tel. 370-3002; fax 370-3162; e-mail kosraelc@mail.fm; unicameral body of 14 mems serving for four years; Gov. Rensley A. Sigrah.

Pohnpei State Legislature: POB 39, Kolonia, Pohnpei, FM 96941; tel. 320-2235; fax 320-2505; internet www.fm/pohnpeileg/; 27 representatives elected for four years (terms staggered); Gov. Johnny P. David.

Yap State Legislature: POB 39, Colonia, Yap, FM 96943; tel. 350-2108; fax 350-4113; 10 mems, six elected from the Yap Islands proper and four elected from the Outer Islands of Ulithi and Woleai, for a four-year term; Gov. Vincent A. Figir.

Diplomatic Representation

EMBASSIES IN MICRONESIA

Australia: POB S, Kolonia, Pohnpei, FM 96941; tel. 320-5448; fax 320-5449; e-mail australia@mail.fm; internet www.australianembassy.fm; Ambassador BRENDAN F. DORAN.

China, People's Republic: POB 530, Kolonia, Pohnpei, FM 96941; tel. 320-5575; fax 320-5578; e-mail chinaemb@mail.fm; Ambassador ZHANG BINHUA.

Japan: Pami Bldg, 3rd Floor, POB 1847, Kolonia, Pohnpei, FM 96941; tel. 320-5465; fax 320-5470; Chargé d'affaires a.i. HYOSUKE YASUI.

USA: POB 1286, Kolonia, Pohnpei, FM 96941; tel. 320-2187; fax 320-2186; e-mail usembassy@mail.fm; internet www.fm/usembassy; Ambassador SUZANNE HALE.

Judicial System

Supreme Court of the Federated States of Micronesia: POB PS-J, Palikir Station, Pohnpei, FM 96941; tel. 350-2159; fax 320-2756; Chief Justice ANDON L. AMARAICH.

State Courts and Appellate Courts have been established in Yap, Chuuk, Kosrae and Pohnpei.

Religion

The population is predominantly Christian, mainly Roman Catholic. The Assembly of God, Jehovah's Witnesses, Seventh-day Adventists, the Church of Jesus Christ of Latter-day Saints (Mormons), the United Church of Christ, Baptists and the Bahá'í Faith are also represented.

CHRISTIANITY

The Roman Catholic Church

The Federated States of Micronesia forms a part of the diocese of the Caroline Islands, suffragan to the archdiocese of Agaña (Guam). The Bishop participates in the Catholic Bishops' Conference of the Pacific, based in Fiji. At 31 December 2002 there were 77,733 adherents in the diocese.

Bishop of the Caroline Islands: Most Rev. AMANDO SAMO, Bishop's House, POB 939, Weno, Chuuk, FM 96942; tel. 330-2399; fax 330-4585; e-mail diocese@mail.fm; internet www.diocesecarolines.org.

Other Churches

United Church of Christ in Pohnpei: Kolonia, Pohnpei, POB 864, FM 96941; tel. 320-2271; fax 320-4404; Pres. BERNELL EDWARD.

Liebenzell Mission: Rev. Roland Rauchholz, POB 9, Weno, Chuuk, FM 96942; tel. 330-3869; e-mail missions@liebenzellusa.org; internet www.liebenzellusa.org/mcrnesia.htm.

The Press

Chuuk News Chronicle: POB 244, Wenn, Chuuk, FM 96942; f. 1983; Editor MARCIANA AKASY.

FSM News: Pohnpei, FM 96941; tel. 320-4256; fax 320-4256; f. 1994; monthly; US-funded.

The Island Tribune: Pohnpei, FM 96941; f. 1997; fortnightly.

Micronesia Focus: Pohnpei, POB 627, FM 96941; tel. 320-4672; f. 1993; Editor KETSON JOHNSON.

The National Union: FSM Public Information Office, POB 490, Kolonia, Pohnpei, FM 96941; tel. 320-2548; fax 320-2785; f. 1979; 2 a month; Public Information Officer KETSON JOHNSON; circ. 500.

Yap State Bulletin: Dept of Youth and Civic Affairs, POB 430, Colonia, Yap 96943; tel. 350-2168; fax 350-3898; e-mail bigj@mail.fm; f. 1989; fortnightly; circ. 500.

Broadcasting and Communications

TELECOMMUNICATIONS

FSM Telecommunication Corporation: POB 1210, Kolonia, Pohnpei, FM 96941; tel. 320-2740; fax 320-2745; e-mail takinaga@mail.fm; internet www.telecom.fm; provides domestic and international services; Gen. Man. TAKURO AKINAGA.

BROADCASTING

Radio

Federated States of Micronesia Public Information Office: POB PS-34, Palikir, Pohnpei, FM 96941; tel. 320-2548; fax 320-4356; e-mail fsmpio@mail.fm; internet www.fsmpio.fm; govt-operated; four regional stations, each broadcasting 18 hours daily; Information Officer KESTER JAMES.

Station V6AH: POB 1086, Kolonia, Pohnpei, FM 96941; programmes in English and Ponapean; Man. DUSTY FREDERICK.

Station V6AI: POB 117, Colonia, Yap, FM 96943; tel. 350-2174; fax 350-4426; programmes in English, Yapese, Ulithian and Satawalese; Man. PETER GARAMFEL.

Station V6AJ: POB 147, Tofol, Kosrae, FM 96944; tel. 370-3040; fax 370-3880; e-mail v6aj@.mail.fm; programmes in English and Kosraean; Man. MCDONALD ITTU.

Station V6AK: Wenn, Chuuk, FM 96942; tel. 330-2596; programmes in Chuukese and English; Man. JOE COMMOR.

WSZA Yap: Dept of Youth and Civic Affairs, POB 30, Colonia, Yap 96943; tel. 350-2174; Media Dir PETER GARAMFEL.

WSZD Pohnpei: Kolonia, Pohnpei 96943; tel. 320-2296; Man. FRANCIS A. ZARRED.

Television

Island Cable TV—Pohnpei: POB 1628, Pohnpei, FM 96941; tel. 320-2671; fax 320-2670; e-mail ictv@mail.fm; f. 1991; Pres. BERNARD HELGENBERGER; Gen. Man. DAVID O. CLIFFE.

TV Station Chuuk (TTKK): Wenn, Chuuk, FM 96942; commercial.

TV Station Pohnpei (KPON): Central Micronesia Communications, POB 460, Kolonia, Pohnpei, FM 96941; f. 1977; commercial; Pres. BERNARD HELGENBERGER; Tech. Dir DAVID CLIFFE.

TV Station Yap (WAAB): Colonia, Yap, FM 96943; tel. 350-2160; fax 350-4113; govt-owned; Man. LOU DEFNGIN.

Finance

BANKING

Regulatory Authority

Federated States of Micronesia Banking Board: POB 1887, Kolonia, Pohnpei, FM 96941; tel. 320-2015; fax 320-5433; e-mail fsmbb@mail.fm; f. 1980; Chair. LARRY RAIGETAL; Commissioner WILSON F. WAGUK.

Banks are also supervised by the US Federal Deposit Insurance Corporation.

Commercial Banks

Bank of the Federated States of Micronesia: POB BF, Tofol, Kosrae FM 96944; tel. 370-3225; e-mail bofsmhq@mail.fm; brs in Kosrae, Yap, Pohnpei and Chuuk.

Bank of Guam (USA): POB 367, Kolonia, Pohnpei, FM 96941; tel. 320-2550; fax 320-2562; e-mail bogpohn@mail.fm; internet www.bankofguam.com; Man. VIDA B. RICAFRENTE; Pres. ANTHONY A. LEON GUERRERO; brs in Chuuk and Pohnpei.

Kosrae Credit Union: POB KU, Tofol, Kosrae; tel. 370-3195; fax 370-3213.

Yap Credit Union: POB 610, Colonia, Yap; tel. 350-2142.

Development Bank

Federated States of Micronesia Development Bank: POB M, Kolonia, Pohnpei, FM 96941; tel. 320-2840; fax 320-2842; e-mail fsmdb@mail.fm; f. 1979; Chair. IHLEN JOSEPH; Pres. ANNA MENDIOLA; 4 brs.

Banking services for the rest of the islands are available in Guam, Hawaii and on the US mainland.

INSURANCE

Actouka Executive Insurance: POB 55, Kolonia, Pohnpei; tel. 320-5331; fax 320-2331; e-mail mlamar@mail.fm.

Caroline Insurance Underwriters: POB 37, Chuuk; tel. 330-2705; fax 330-2207.

FSM Insurance Group: Kosrae; tel. 370-3788; fax 370-2120.

Islands Insurance: POB K, Kolonia, Pohnpei; tel. 320-3422; fax 320-3424.

Moylan's Insurance Underwriters: POB 1448, Kolonia, Pohnpei; tel. 320-2118; fax 320-2519; e-mail moylan90@mail.fm.

Oceania Insurance Co: POB 1202, Weno, Chuuk, FM 96942; tel. 330-3036; fax 330-2334; e-mail oceanpac@mail.fm; also owns and manages Pacific Basin Insurance.

Pacific Islands Insurance Underwriters: POB 386, Colonia, Yap; tel. 350-2340; fax 350-2341.

Transpacific Insurance: POB 510, Kolonia, Pohnpei; tel. 320-5525; fax 320-5524.

Yap Insurance Agency: POB 386, Colonia, Yap; tel. 350-2340; fax 350-2341; e-mail tachelioyap@mail.fm.

Trade and Industry

GOVERNMENT AGENCIES

Coconut Development Authority: POB 297, Kolonia, Pohnpei, FM 96941; tel. 320-2892; fax 320-5383; e-mail fsmcda@mail.fm; responsible for all purchasing, processing and exporting of copra and copra by-products in the islands; Gen. Man. NAMIO NANPEI.

Pohnpei Economic Development Authority: POB 738, Kolonia, Pohnpei, FM 96941; tel. 320-2298; fax 320-2775; e-mail eda@mail.fm; Chair. President JOSEPH J. URUSEMAL (*ex officio*); Exec. Dir SHELTEN NETH.

FSM National Fisheries Corporation: POB R, Kolonia, Pohnpei, FM 96941; tel. 320-2529; fax 320-2239; e-mail nfcairfreight@mail.fm; internet www.fsmgov.org/nfc/; f. 1984; established in 1990, with the Economic Devt Authority and an Australian co, the Caroline Fishing Corpn (three vessels); promotes fisheries development; Pres. PETER SITAN.

Micronesian Fisheries Authority: PS122, Palikir, Pohnpei State, FM 96941; tel. 320-2700; fax 320-2383; e-mail fsmfish@mail.fm.

National Oceanic Resource Management Authority (NORMA): POB PS-122, Palikir, Pohnpei, FM 96941; tel. 320-2700; fax 320-2383; e-mail norma@mail.fm; fmrly Micronesian Fisheries Authority; name changed 2002; responsible for conservation, management and development of tuna resources and for issue of fishing licences; Exec. Dir BERNARD THOULAG; Deputy Dir EUGENE PANGELINAN.

UTILITIES

Chuuk Public Works (CPW): POB 248, Weno, Chuuk, FM 96942; tel. 330-2242; fax 320-4815; e-mail chkpublicworks@mail.fm.

Kosrae Utility Authority: POB 277, Tofol, Kosrae, FM 96944; tel. 370-3799; fax 370-3798; e-mail KUA@mail.fm; corporatized in 1994.

Pohnpei Utility Corporation: POB C, Kolonia, Pohnpei, FM 96941; tel. 320-5606; fax 320-2505; f. 1992; provides electricity, water and sewerage services.

Yap Public Services Corporation: POB 621, Colonia, Yap, FM 96943; tel. 350-2175; fax 350-2331; f. 1996; provides electricity, water and sewerage services.

CO-OPERATIVES

Chuuk: Chuuk Co-operative, Faichuk Cacao and Copra Co-operative Asscn, Pis Fishermen's Co-operative, Fefan Women's Co-operative.

Pohnpei: Pohnpei Federation of Co-operative Asscns (POB 100, Pohnpei, FM 96941), Pohnpei Handicraft Co-operative, Pohnpei Fishermen's Co-operative, Uh Soumwet Co-operative Asscn, Kolonia Consumers' and Producers' Co-operative Asscn, Kitti Minimum Co-operative Asscn, Kapingamarangi Copra Producers' Asscn, Metalanim Copra Co-operative Asscn, PICS Co-operative Asscn, Mokil Island Co-operative Asscn, Ngatik Island Co-operative Asscn, Nukuoro Island Co-operative Asscn, Kosrae Island Co-operative Asscn, Pingelap Consumers' Co-operative Asscn.

Yap: Yap Co-operative Asscn, POB 159, Colonia, Yap, FM 96943; tel. 350-7028; fax 350-4114; e-mail yca@mail.fm; f. 1952; Pres. JAMES GILMAR; Gen. Man. TONY GANNGIYAN; 1,200 mems.

Transport

ROADS

Macadam and concrete roads are found in the more important islands. Other islands have stone and coral-surfaced roads and tracks.

SHIPPING

Pohnpei, Chuuk, Yap and Kosrae have deep-draught harbours for commercial shipping. The ports provide warehousing and transhipment facilities.

Caroline Fisheries Corporation (CFC): POB 7, Kolonia, Pohnpei, FM 96941; tel. 320-5791; fax 320-4733; e-mail cfc@mail.fm; Gen. Man. MILAN KAMBER.

Pacific Shipping Agency: POB 154, Lelu, Kosrae FM 96944; tel. 370-2912; Gen. Man. THEODORE SIGRAH.

Pohnpei Transfer & Storage, Inc.: POB 340, Kolonia, Pohnpei FM 96941; tel. 320-2552; fax 320-2389; e-mail fsmlinejv@mail.fm; Gen. Man. JOE VITT.

Truk Transportation Company (TRANSCO): POB 99, Weno, Chuuk FM 96942; tel. 330-2143; fax 330-2726; e-mail transco@mail.fm; Gen. Man. LINDA MORI HARTMAN.

Waab Transportation Company.: POB 177, Colonia, Yap FM 96943; tel. 350-2301; fax 350-4110; e-mail waabtrans@mail.fm; agents for PM & O Lines (USA); Gen. Man. CYRIL CHUGRAD.

CIVIL AVIATION

The Federated States of Micronesia is served by Continental Micronesia, Air Nauru and Continental Airlines (USA). Pacific Missionary Aviation, based in Pohnpei and Yap, provides domestic air services. There are international airports on Pohnpei, Chuuk, Yap and Kosrae, and airstrips on the outer islands of Onoun and Ta in Chuuk.

Tourism

The tourist industry is a significant source of revenue, although it has been hampered by the lack of infrastructure. Visitor attractions include excellent conditions for scuba-diving (notably in Chuuk Lagoon), Second World War battle sites and relics (many underwater) and the ancient ruined city of Nan Madol on Pohnpei. In 1990 there was a total of 362 hotel rooms. The number of tourist arrivals totalled 15,265 in 2001, and tourist receipts totalled some US $13m. in that year.

Federated States of Micronesia Visitors Board: Dept of Economic Affairs, National Government, PO Box PS-12, Palikir, Pohnpei, FM 96941; tel. 320-5133; fax 320-3251; e-mail fsminfo@visit-fsm.org; internet www.visit-fsm.org.

Chuuk Visitors Bureau: POB FQ, Weno, Chuuk, FM 96942; tel. 330-4133; fax 330-4194; e-mail cvb@mail.fm.

Kosrae Visitors Bureau: POB 659, Tofol, Kosrae, FM 96944; tel. 370-2228; fax 370-2187; e-mail kosrae@mail.fm; internet www.kosrae.com.

Pohnpei Department of Tourism and Parks: POB 66, Kolonia, Pohnpei, FM 96941; tel. 320-2421; fax 320-6019; e-mail tourismparks@mail.fm; Deputy Chief BUMIO SILBANUZ.

Pohnpei Visitors Bureau: POB 1949, Kolonia, Pohnpei, FM 96941; tel. 320-4851; fax 320-4868; e-mail pohnpeiVB@mail.fm; internet www.visit-pohnpei.fm.

Yap Visitors Bureau: POB 36, Colonia, Yap, FM 96943; tel. 350-2298; fax 350-2571; e-mail yvb@mail.fm; internet www.visityap.com.

Defence

Defence and security are the responsibility of the USA. The US Pacific Command is based in Hawaii.

Education

Primary education, which begins at six years of age and lasts for eight years, is compulsory. Secondary education, beginning at 14 years of age, comprises two cycles, each of two years. The education system is based on the US pattern of eight years' attendance at an elementary school and four years' enrolment at a high school. In 1995 there were a total of 27,281 pupils enrolled at 174 primary schools and 6,898 pupils at 24 secondary schools. In 1994 some 1,461 students were involved in tertiary education.

NAURU

Physical and Social Geography

The Republic of Nauru is a small island in the central Pacific Ocean, lying about 4,000 km north-east of Sydney, Australia, and 306 km west of Banaba (Ocean Island), in Kiribati, its nearest neighbour. Covering an area of 21.3 sq km (8.2 sq miles), Nauru is a low-lying island (its highest point is 65 m—213 ft), comprising a narrow, coastal strip of fertile land surrounding coralline cliffs rising to a plateau of phosphatic rock which covers more than three-fifths of the land area. Nauru has a tropical climate, with a westerly monsoon season from November to February, during which time most rainfall occurs. Annual rainfall averages 2,060 mm (80 ins), but there are marked variations from year to year. Indigenous Nauruans are of mixed Polynesian, Micronesian and Melanesian descent, but are predominantly Polynesian. Nauruan is the official language, although English is widely used and generally understood. At the 1992 census the population was 9,919. In July 2001 the population was estimated at 12,088. There is no capital as such, but Parliament House and most government offices are in Yaren district.

History

The first European to discover Nauru was Capt. John Fearn, whose whaling ship, *Hunter*, reached the island in 1798. The territory, which he named Pleasant Island, was described as being inhabited by relatively large numbers of predominantly Polynesian people, organized in 12 clans. The arrival of traders during the 19th century and the subsequent introduction of firearms and alcohol on to the island, however, led to unrest between the tribes and precipitated 'The Ten-Year War': a civil war in which more than one-third of the population were killed. When the island was eventually annexed by Germany in 1888, its inhabitants numbered little more than 900 in total. In 1914, shortly after the outbreak of the First World War, the island was captured by Australian forces. It continued to be administered by Australia under a League of Nations mandate (granted in 1920) which also named the United Kingdom and New Zealand as co-trustees. Between 1942 and 1945 Nauru was occupied by the Japanese, who deported 1,200 islanders to Truk (now Chuuk), Micronesia, where some 500 died as a result of starvation and bombing. In 1947 the island was placed under UN Trusteeship, with Australia as the administering power on behalf of the Governments of Australia, New Zealand and the United Kingdom. The UN Trusteeship Council proposed in 1964 that the indigenous people of Nauru be resettled on Curtis Island, off the Queensland coast. This offer was made in anticipation of the progressive exhaustion of the island's phosphate deposits. The Nauruans, however, elected to remain on the island, and studies were initiated in 1966 for the shipping of soil to the island to replace the phosphate rock. Nauru received a considerable measure of self-government in January 1966, with the establishment of Legislative and Executive Councils, and proceeded to independence on 31 January 1968 (exactly 22 years after the surviving Nauruans returned to the island from exile in Micronesia). In early 1998 Nauru announced its intention to seek UN membership and full Commonwealth membership. The decision was largely based on the islanders' desire to play a more prominent role in international policies relating to issues that affect them, most notably climate change (see below). Nauru attained full membership of the Commonwealth in May 1999. The country's bid to seek UN membership was challenged by the People's Republic of China because of Nauru's links with Taiwan. Nevertheless, Nauru became a member of the UN in September 1999.

The Head Chief of Nauru, Hammer DeRoburt, was elected President in May 1968 and re-elected in 1971 and 1973. Dissatisfaction with his increasingly personal rule led to the election of a new President, Bernard Dowiyogo (leader of the informal Nauru Party), in 1976. Dowiyogo was re-elected President after a general election in late 1977. DeRoburt's supporters, however, adopted obstructive tactics in Parliament, and in January 1978 Dowiyogo resigned, in response to a parliamentary impasse over budgetary legislation; he was re-elected shortly afterwards, but was again forced to resign in April, following the defeat of a legislative proposal concerning phosphate royalties. Lagumot Harris, another member of the Nauru Party, succeeded him, but resigned three weeks later, when Parliament rejected a finance measure, and DeRoburt was again elected to the presidency. He was re-elected in December 1980 and again in May and December 1983.

In September 1986, following the Government's defeat on a parliamentary motion proposing an amendment to the annual budget legislation, DeRoburt resigned, and was replaced as President by Kennan Adeang, who was elected in Parliament by nine votes to DeRoburt's eight. However, after holding office for only 14 days, Adeang was, in turn, defeated in a parliamentary vote expressing 'no confidence' and DeRoburt resumed the presidency. Following a general election in December 1986, Adeang was narrowly elected President. However, he was subsequently ousted by another vote of 'no confidence', and DeRoburt was again reinstated as President. The atmosphere of political uncertainty, generated by the absence of a clear majority in Parliament, led DeRoburt to dissolve Parliament in preparation for another general election in January 1987, at which DeRoburt was re-elected to the presidency by 11 votes to six.

In February 1987 Adeang announced the establishment of the Democratic Party of Nauru, thus giving formal status to the first new political grouping in Nauru since 1976. Eight of the 18 members of Parliament subsequently joined the new party, which declared that its aim was to curtail the extension of presidential powers and to promote democracy. In August 1989 a motion of 'no confidence' in DeRoburt (proposed by Adeang) was approved by 10 votes to five, and Kenai Aroi was subsequently elected President. Following a general election, held in December, Bernard Dowiyogo was re-elected President, defeating his only opponent, DeRoburt, by 10 votes to six. The next presidential election, held shortly after a general election in November 1992, resulted in victory for Dowiyogo, who defeated Buraro Detudamo by 10 votes to seven. Following a general election in November 1995, in which all cabinet members were re-elected, Nauru's newly elected Parliament voted, by nine votes to eight, to replace Dowiyogo with Lagumot Harris in the presidency.

The resignation of the Chairman of Air Nauru, following allegations of misconduct, prompted Parliament to vote on a motion of 'no confidence' in the Government in November 1996. The motion was narrowly approved, and Harris was replaced by Dowiyogo as President. Later that month, however, Dowiyogo's new Government was itself defeated in a parliamentary vote of 'no confidence', and Kennan Adeang was elected to the presidency. A widespread perception that the new Government lacked experience was thought to be a major factor prompting a further motion of 'no confidence' in December, at which Adeang was similarly removed from office. At a subsequent presidential contest Reuben Kun, a former Minister for Finance, defeated Adeang by 12 votes to five, on the understanding that his administration would organize a general election. An election duly took place on 8 February 1997, at which four new members were elected to Parliament, following an apparent agreement between the supporters of Harris and those of Dowiyogo to end the political manoeuvring that had resulted in several months of instability in Nauru. At the election to the presidency on 13 February, Kinza Clodumar (who had been nominated by Dowiyogo) defeated Harris by nine votes to eight.

In early 1998 five members of Parliament (including former President Lagumot Harris) were dismissed by Adeang, the Speaker, for refusing to apologize for personal remarks about him that had been published in an opposition newsletter in late 1997. At the resultant by-elections held in late February 1998 three of the five members were re-elected. A motion expressing 'no confidence' in the President was approved in June, and Dowiyogo was consequently elected to replace Clodumar. In a further vote of 'no confidence', in late April 1999, Dowiyogo was defeated by 10 votes to seven; his replacement was Rene Harris, previously Chairman of the Nauru Phosphate Corporation. Former President Lagumot Harris died in September 1999.

Following legislative elections held on 8 April 2000, Rene Harris was re-elected President, narrowly defeating Dowiyogo by nine votes to eight. Ludwig Scotty was elected Speaker of Parliament. On 18 April, however, Scotty and his deputy, Ross Cain, subsequently resigned, stating only that they were unable to continue under the 'current political circumstances'. Harris tendered his resignation and was replaced by Dowiyogo on 19 April, whereupon Scotty and Cain were re-elected to their posts in the legislature. Observers attributed the manoeuvring to shifting political allegiances within the legislature.

In early 2001, in another reversal to Dowiyogo's leadership, Anthony Audoa, the Minister for Home Affairs, Culture, Health and Women's Affairs, resigned and requested that Parliament be recalled. He claimed that Dowiyogo had squandered Nauru's wealth during his various tenures as President and that in promoting the island as a tax haven he had allowed Nauru to be used by Russian criminal gangs to 'launder' their illegal funds, prompting speculation that he intended to mount a challenge for the presidency. In late March 2001 Dowiyogo was ousted from the presidency in a parliamentary vote of 'no confidence' while he was undergoing hospital treatment in Australia. The motion, which was passed by two votes, led to Rene Harris regaining the presidency. In October, however, Harris was flown to Australia for emergency medical treatment for a diabetes-related illness, during which time Remy Namaduk performed the role of Acting President.

Allegations that Nauru's 'offshore' financial centre was being used extensively by Russian criminal organizations for 'laundering' the proceeds of their illegal activities, led Dowiyogo to order a full review of the industry in March 1999. The Government subsequently stated that it intended to modernize legislation governing the island's financial sector. In early 2000 President Rene Harris announced that Nauru was to suspend its 'offshore' banking services and improve the accountability of existing banks on the island, as part of the Government's efforts to bring Nauru's financial services regulations into conformity with international standards. Dowiyogo similarly reaffirmed his commitment to reform the 'offshore' sector, following his election in April 2000. However, in February 2001 11 members of Nauru's 18-member legislature signed a petition requesting that Dowiyogo attend a special session of Parliament to answer questions relating to the island's alleged role in 'laundering' significant funds from Russian criminal organizations. The allegations originated in claims by Russia's central bank that some US $70,000m. of illegal funds had been processed in 'offshore' banks in Nauru. It was estimated that 400 such banks existed on the island in early 2001. The Government introduced new legislation designed to deter money laundering in August 2001, but the Financial Action Task Force (FATF—established in 1989 on the recommendation of the Group of Seven (G-7) industrialized nations) found that the new laws contained several deficiencies, and imposed sanctions in December 2001. Following the suicide attacks on New York and Washington, DC, on 11 September, for which the USA held the fundamentalist Islamist al-Qa'ida (Base) organization to be principally responsible, Nauru's financial system was subject to international scrutiny amid suspicion that it might have been used as a conduit for the terrorists' funds. The Government announced new anti-money-laundering legislation in December, and was considering legal action against the FATF in early 2002. However, in June 2002 Nauru was again listed by the FATF as one of 15 countries that refused to take adequate measures to prevent money laundering and the financing of terrorist organizations. The FATF was particularly concerned that the existence of about 400 'shell' banks, which had no physical presence in the country, was an unacceptable money-laundering risk.

Following the annual review of the FATF in February 2004, it was decided that Nauru should remain on the group's list of 'non-co-operative' countries and territories, owing to the island's failure to implement effective measures to combat the practice of international 'money laundering'. As a result of continued US pressure, however, and on the same day as the FATF meeting, Nauru's Parliament approved a new law to address the problem of 'money laundering', along with legislation to close down the country's offshore banks. In December 2003, meanwhile, following the island's commitment to improve transparency and to exchange information on tax matters with other countries, Nauru had been removed from a list of unco-operative tax havens, issued in April 2002 by the Organisation for Economic Co-operation and Development (OECD)

On 8 January 2003 President Rene Harris was defeated in a motion of 'no confidence' by eight votes to three and was replaced by Bernard Dowiyogo. The vote followed a political crisis resulting from the defeat of the Government's budget proposals at the end of December 2002, as well as reports of increasing dissatisfaction with Harris's alleged economic mismanagement of the country. Nauru's deteriorating financial situation, in addition to the Government's decision to accept more than 1,000 asylum-seekers in return for aid from Australia, was believed to be a major factor in the loss of confidence in Harris, which had led to the defection to the opposition of two cabinet ministers, two backbenchers and the Speaker in late 2002. However, Harris applied to the Supreme Court, and on 10 January 2003 an injunction was issued against Dowiyogo accepting the presidency. This decision had been based on the fact that only 11 of the 18 members of Parliament had attended the session when the vote took place, thus rendering it invalid; Harris and his Cabinet had staged a boycott of Parliament when the motion was to be proposed. Despite the injunction, Dowiyogo maintained his position and appointed a new Cabinet. Several days of confusion and political instability ensued. Finally, Harris was reinstated as President, following the intervention of Nauru's Melbourne-based Chief Justice. However, he resigned from the presidency the following day. In the resultant contest Dowiyogo defeated Kinza Clodumar by nine votes to eight to become the new President on 20 January. A lack of support for Dowiyogo within Parliament continued to create problems, however, amid appeals for an early election to resolve the impasse.

Meanwhile, a complete collapse of Nauru's telecommunications system in early January 2003 increased the problems experienced by the island. Nauru, thus, effectively became cut off from the rest of the world, with external contact only possible when ships equipped with satellite telephones were calling. A speech by Dowiyogo claiming that Nauru was on the verge of bankruptcy, unable to pay its public servants or to send its sick citizens to Australia for treatment, and appealing to donor countries for emergency assistance, could not be transmitted for almost a month. Telecommunications services were restored in early March following a visit from a technician supplied by Australia's government aid agency, AusAID.

In early March 2003 Dowiyogo travelled to Washington, DC, at the request of the US Government, which had threatened to impose harsh economic sanctions on the island and to repossess Air Nauru's only aircraft, if Nauru did not discontinue its 'offshore' banking services. The Administration of George W. Bush was reported to have been angered by the possibility that individuals with links to terrorist organizations might have used the island's financial services to 'launder' their funds; some 400 'offshore' banks were registered on the island in the early 2000s. Consequently, Dowiyogo agreed to sign executive orders not to renew any banking licences or to issue any further so-called 'investor passports'. Shortly after the meeting, however, Dowiyogo collapsed and, following emergency heart surgery, died on 9 March. Derog Gioura was appointed acting Head of State and on 20 March was elected President by nine parliamentary votes to seven. Legislation providing for the expiry of most 'offshore' banking licences within 30 days (and for the remainder within six months) was passed by Parliament in late March, in accordance with the agreement that Dowiyogo had signed in the USA. At the end of March acting President Derog Gioura himself suffered a heart attack and was flown to Australia for treatment.

A general election took place on 3 May 2003, at which six new members were elected to the legislature. However, the new Speaker resigned one day after his election, and with no further nominations for the position, Parliament was unable to proceed to a presidential election. The impasse was resolved when a Speaker was finally elected in late May and Ludwig Scotty won the subsequent presidential election, defeating Kinza Clodumar with 10 parliamentary votes to seven. Scotty, who named a new six-member Cabinet in June, stated his Government's intention to focus on 'prudent management and financial stability'. However, on 8 August Scotty was ousted from office by a 'no confidence' motion and replaced by Rene Harris, who became the fourth President of 2003. The reasons for Scotty's removal were not clear, although concerns had been expressed about his plans to close recently opened embassies in Washington, DC, and Beijing, and there had been speculation that he intended to switch Nauru's diplomatic allegiance from the People's Republic of China back to Taiwan (see below).

In August 2003 workers at the Nauru Phosphate Corporation began a strike in support of demands for almost six months of unpaid salaries. Opposition politicians claimed that the dispute was an indication of more widespread dissatisfaction with corruption and mismanagement in the industry and within the Government.

In January 2004 President Rene Harris was flown to Australia amid rumours that he had suffered a physical collapse and was in a poor state of health. Officials declined to respond to queries surrounding the President, merely stating that Derog Gioura would be acting President in his absence. In the following month the Minister for Justice resigned, precipitating a vote of 'no confidence' in the President. The country faced a further political crisis when the motion received an equal number of votes in favour and against. Moreover, when Parliament was unable to agree on the election of a new Speaker, following the resignation of the incumbent in early April, the resulting impasse meant that Parliament could not be formally convened. As the country's financial crisis deepened, President Harris travelled to Australia in mid-April to request assistance in averting imminent bankruptcy for the island. On his return to Nauru Harris faced angry demonstrations by hundreds of government employees protesting at the hardship imposed on them as a result of their salaries having been unpaid for 12 months. Reports indicated that government employees (who constituted the majority of paid employees on the island) were surviving on subsistence diets of fish and coconuts. In the same month receivers were appointed to manage the assets of the Nauru Phosphate Royalties Trust (including Nauru's extensive property portfolio in Australia) which was unable to pay off debts of some $A230m. to US interests. Nauruans working in government-owned buildings in Melbourne and Sydney were served with eviction notices. Meanwhile, with neither the Government nor the opposition willing to nominate a Speaker from among its members (thereby giving the other side a majority in Parliament) the legislature was unable to produce a budget. In May, however, during another of Harris's overseas trips, the opposition elected one of its members as Speaker and immediately approved legislation making it illegal for a government to operate without a budget. In the following month, and before the Supreme Court had ruled on the matter, Kinza Clodumar, the Minister of Finance, crossed the floor, thereby allowing the opposition to approve a motion of 'no confidence' in Harris. Ludwig Scotty was subsequently elected to the presidency. In late July Australian Treasury official Peter Depta arrived in Nauru to take up the post of

Financial Secretary, effectively assuming control of the country's finances.

Nauru's precarious political situation deteriorated during September 2004, and by the end of the month President Ludwig Scotty had dissolved Parliament and declared a state of emergency. His action had been prompted by the Speaker Russell Kun's suspension of the Minister of Health, Kieron Keke, on the grounds that he held dual Nauruan and Australian nationality. Keke's suspension had resulted in the loss of the Government's one-seat majority and a consequent stalemate in Parliament, during which budget legislation had been unable to be approved. Scotty assumed sole responsibility over the government of the country until a general election took place on 23 October. At the election all nine members of Scotty's Government retained their seats in the legislature, while seven of the nine opposition members of parliament were not re-elected. The result thus gave President Scotty an ample majority in Parliament.

In September 2001 Nauru agreed to accept 310 of 460 predominantly Afghan asylum-seekers who were on board a Norwegian freighter, the MV *Tampa*, unable to disembark on Christmas Island as Australia refused to grant them entry into its territory (see the chapter on Christmas Island). The Australian Government agreed to fund the processing of the asylum-seekers and to pay an undisclosed sum to Nauru, which was to house the asylum-seekers for three months while their claims for asylum were assessed. Those found to be genuine refugees were to be given sanctuary in various nations, including Australia, New Zealand and Norway. Following the interception of another boat carrying asylum-seekers in Australian waters later the same month, Nauru received a pledge of $A20m. from the Australian Government for agreeing to host 800 asylum-seekers. In December 2001 Nauru signed an agreement with Australia's Minister for Foreign Affairs to accommodate a total of 1,200 at any one time, in return for a further $A10m. of aid, to be allocated to education, health and infrastructure programmes. During a brief visit to Australia in January 2002, the President of Nauru declared that the bilateral agreement should end as originally envisaged. However, local residents and owners of the land upon which the camps were located expressed concern over the delays in processing the asylum-seekers' claims. Processing was due to be completed by July 2002, and in May Australia offered monetary assistance to Afghan asylum-seekers as an incentive to return to their homeland. In June the President announced that he anticipated that all the asylum-seekers would have left Nauru within six months. However, in December some 700 people remained in the camps, despite the deportation of more than 100 Afghan asylum-seekers to Kabul. In the same month the President signed a new agreement with Australia's Minister for Foreign Affairs to extend the duration of the camps' operations and to accommodate up to 1,500 asylum-seekers.

In mid-2003 the Australian Government was accused of cruelty for detaining some 100 children in the camps on Nauru and for failing to reunite families held in separate camps for extended periods. The Nauru Government was subject to further criticism when the visa of a Catholic priest and prominent human rights activist was withdrawn hours before he was due to visit refugees detained on the island. It was believed that Nauruan officials had been instructed to cancel the visa by the Australian Government. Concerns about the conditions at the camps and the welfare of the detainees increased during 2003, and in December some of those held began a hunger strike in order to attract attention to their situation. A reported 40 asylum-seekers participated in the strike, some of whom stitched their lips together. The Nauruan Government's subsequent appeals for medical assistance from Australia to care for the hunger strikers, many of whom required hospital treatment during the following weeks, were refused. The emerging rift between the two countries widened when Nauru's Minister of Finance, Kinza Clodumar, condemned remarks made by Australia's Minister for Immigration, Amanda Vanstone, who had said that the health of the hunger strikers was of no concern to Australia. Clodumar, who stated that his country's limited medical resources could not cope with a problem of this scale, accused Australia of failing to recognize its obligations in continuing to ignore the plight of the asylum-seekers on Nauru. The strike ended about one month after it had begun, and in mid-January 2004 the Australian Government sent a delegation to inspect medical facilities available to asylum-seekers on the island. The resultant report, however, which found services for those held in detention to be adequate, was widely regarded as flawed, as it had failed to examine any of the detainees and had been compiled solely by Australian government officials. In late June 2004 the first of 146 Afghan refugees cleared to enter Australia left the detention centre.

In July 2002 a political crisis emerged after President Harris decided unilaterally to recognize the People's Republic of China, thus ending 22 years of diplomatic relations with Taiwan. Several cabinet ministers opposed the shift in policy, and the controversy increased after the President immediately accepted US $60m. in aid and US $77m. in debt forgiveness from the People's Republic of China. Following the switch in allegiance, Taiwan announced that it would take legal action to recover a loan of US $12.5m. which it had arranged to make available to Nauru.

In February 2003 two diplomatic missions were opened in Washington, DC, and Beijing primarily to address US concerns over money laundering and international terrorism (fears regarding the latter increased when, in the same month, two members of al-Qa'ida were found travelling on Nauruan passports). However, in July President Scotty announced plans to close the missions, citing economic constraints and his belief that they were not serving their intended purpose. Representatives in China and the USA expressed surprise at the announcement and queried the President's motives, in particular his commitment to ending the lucrative sale of Nauruan passports.

In February 2004 Nauru announced the establishment of diplomatic relations with Iceland. The two countries cited their common reliance on the sea as a unifying factor between them.

In July 2004 President Ludwig Scotty held emergency discussions with the Presidents of Kiribati and Tuvalu regarding some US $2m. in outstanding salary payments owed to their nationals employed in Nauru by the Nauru Phosphate Corporation.

In February 1987 the British, Australian and New Zealand Governments officially terminated the functions of the British Phosphate Commissioners, who from 1919 until 1970 had managed the mining of Nauru's phosphate deposits. President DeRoburt subsequently expressed concern over the distribution of the Commissioners' accumulated assets, which were estimated to be worth $A55m. His proposal that part of this sum be spent on the rehabilitation of areas of the island that had been mined before Nauru gained independence in 1968 was rejected by the three Governments involved. The ensuing dispute became known as the 'Matter of Rehabilitation'. In August 1987 DeRoburt established a government commission of inquiry to report on proposals for rehabilitation, and in the following year proposed that the three Governments each provide one-third of the estimated rehabilitation costs of $A216m. (As a result of extensive phosphate mining, 80% of the surface of the island is uninhabitable and impossible to cultivate.) In 1989, because of Australia's refusal to contribute to the rehabilitation of former phosphate mining areas, Nauru appealed to the International Court of Justice (ICJ), at The Hague, Netherlands, for compensation for damage to its environment. (New Zealand and the United Kingdom deposited instruments with the ICJ that prevented them from being sued by Nauru.) Australia agreed to comply with the eventual ruling of the ICJ. However, in 1993, following negotiations between President Dowiyogo and the Australian Prime Minister, Paul Keating, a Compact of Settlement was signed, under which the Australian Government was to pay a total of $A107m. to Nauru. New Zealand and the United Kingdom subsequently agreed to contribute $A12m. each towards the settlement. In 1995 a report commissioned by the Government was published, which gave details of a rehabilitation programme extending over the next 20–25 years and costing $A230m. The success of the rehabilitation scheme, however, was dependent on the co-operation of landowners, some of whom were expected to continue to allow areas to be mined for residual ore once phosphate reserves had been exhausted. In 1997 Parliament approved the Nauru Rehabilitation Corporation (NRC) Act, providing for the establishment of a corporate body to manage the rehabilitation programme. The programme, which began in May 1999, was expected to transform the mined areas into sites suitable for agriculture and new housing. The project, however, was hampered considerably by delays, and in early 2004 the chair of the NRC resigned, reportedly in frustration at problems regarding the implementation of a feasibility study into the mining of residual phosphate.

Nauru was persistently critical of France's use of the South Pacific region for nuclear-weapons testing, and was one of the most vociferous opponents of the French Government's decision in mid-1995 to resume its nuclear-testing programme. Diplomatic relations between the two countries, suspended in 1995, were formally resumed in early 1998.

In early 2001 Nauru voiced strong opposition to the US Government's plans to develop a missile defence system, in which missiles are deployed to shoot down other missiles in flight. Government officials in Nauru expressed fears that testing of the system in the region could result in missile debris landing on the Pacific islands.

The President of Nauru had discussions with the Cuban Minister of Foreign Affairs in November 2001 at a UN meeting, where they agreed to establish diplomatic relations between their two countries and discussed a proposed technical and economic co-operation agreement whereby Nauru would be provided with health experts from Cuba.

In 1989 a UN report on the 'greenhouse effect' (the heating of the earth's atmosphere and a resultant rise in sea-level, as a consequence of pollution) listed Nauru as one of the countries that might disappear beneath the sea in the 21st century, unless drastic action were taken. Another issue of environmental concern was raised in late 1992, when it was announced that a vessel carrying plutonium would pass close to Nauru *en route* to Japan from France. Dowiyogo travelled to Japan to protest at the proposal. Furthermore, in early 1997 President Kinza Clodumar urged the Japanese Government to ensure that a similar ship (travelling from France to Japan with a cargo of radioactive waste) remain outside its exclusive economic zone, particularly as the vessel had illegally entered the waters of several other countries. The Government of Nauru strongly criticized Australia's refusal, at the December 1997 Conference of the Parties to the Framework Convention on Climate Change, in Kyoto, Japan, to reduce its emission of pollutant gases known to contribute to the 'greenhouse effect'.

In August 2001 Nauru hosted the Pacific Islands Forum summit meeting, despite a problematic shortage of accommodation, caused by the presence of contingents of officials from Australia, refugee agencies, the UN and Eurest (the company subcontracted to operate Nauru's refugee camp). Fiji had been expected to perform this role, but its participation had been opposed owing to its failure to reinstate democratic rule following the coup of the previous year. Japan pledged financial aid for Nauru's hosting of the summit meeting. During the meeting the Nauruan President urged solidarity among South Pacific states in opposing international demands to restrict financial transactions.

Economy

In 1999, according to UN estimates, Nauru's gross domestic product (GDP), measured at current prices, was US $34m., equivalent to US $2,830 per head. In 1991–99, it was estimated, GDP decreased, in real terms, at an average annual rate of 5.9%. The population increased by an annual average of 2.0% per year in 1990–2000. The UN estimated that GDP declined, in real terms, by 1.9% in 1998 and by 1.9% in 1999. According to official estimates, Nauru's GDP grew in real terms by only 0.8% in 2000 (following growth of 3.2% in 1999). This was in part due to adverse climatic conditions, which affected coffee production, and low international prices for Nauru's coffee, copra and palm oil output. Real GDP growth of around 3% was predicted for 2001.

Agricultural activity comprises mainly the small-scale production of tropical fruit, vegetables, and livestock, although the production of coffee and copra for export is increasingly significant. According to FAO, agriculture engaged some 20% of the economically active population in 2001. Coconuts are the principal crop. Bananas, pineapples and the screw-pine (*Pandanus*) are also cultivated as food crops, while the islanders keep pigs and chickens. However, almost all Nauru's requirements (including most of its drinking water) are imported. Increased exploitation of the island's marine resources was envisaged following the approval by Parliament of important fisheries legislation in the late 1990s. Funding for a new harbour for medium-sized vessels was secured from the Government of Japan in 1998, and in 1999 the Marshall Islands Sea Patrol agreed to provide assistance in the surveillance of Nauru's exclusive economic zone. Revenue from fishing licence fees totalled $A8.5m. in 2000.

Until the early 1990s the island's economy was based on the mining of phosphate rock, derived from rich deposits of guano, which constituted about four-fifths of the island's surface area. Phosphate extraction has been conducted largely by indentured labour (with the majority of workers originating from Kiribati and Tuvalu). The annual output of phosphate rock totalled 1,181,000 metric tons in 1989, but declined to 510,000 tons in 1996. Although Nauru accounts for only a small proportion of world phosphate production, the revenue accruing to the island is, in relation to its size, very high; it was estimated to have provided about $A100m.–$A120m. annually since independence. Phosphate exports decreased to an annual average of 0.51m. tons in 1990–97 (compared with 1.58m. tons per year in the 1980s), mainly owing to the collapse of the Australian market. As a result of the Asian financial crisis, exports of phosphate declined by almost 18% in 1998 compared with the previous year. Mining production, meanwhile, decreased by 7.9% in 2000. Primary deposits of phosphate were expected to be exhausted by 2005. Feasibility studies have been conducted into the mining of secondary and residual deposits, although this activity would be less profitable. An Australian company was expected to undertake a detailed survey of the island's potential for secondary phosphate mining during 2004.

The revenue from phosphate exports has been shared among the Government (which takes about 50% of the profits), Nauruan landowners, a long-term trust fund (the Nauru Phosphate Royalties Trust—NPRT) and the Nauru Local Government Council. The value of the NPRT was estimated at some $A300m. in 2003, compared with $A1,300m. in 1991. Investments by Nauru, using phosphate revenues, have been made in Australia (extensively in Melbourne), the United Kingdom, the Philippines, New Zealand, Fiji, Guam, Samoa and the USA (Honolulu, Houston, Washington, DC, and Portland, Oregon). Mainly as a result of its phosphate revenues, the Government has been able to provide an extensive welfare system and to maintain an international airline. However, not all investments have been successful, and some Pacific hotel ventures, as well as the staging of a lavish musical production in London, United Kingdom, have been described as disastrous. Criticism regarding the inefficient management of Nauru's funds intensified following the conviction, in 1994, of an accountant employed by the NPRT, for the theft of $A1.2m. from the Trust. In the following year a further eight people were charged with, and later convicted of, grand larceny and fraud over the theft of some $A40m. from the NPRT. Moreover, in March 2000 an Australian businessman was found guilty of conspiring to defraud the Trust of some $A100m.

The Government is an important employer, engaging 1,138 people in 1989, although this number declined significantly as a result of a series of austerity measures introduced during the 1990s. The rate of unemployment among those aged 15–19 in the late 1990s was estimated by the Asian Development Bank (ADB) to be 33% of males and 52% of females. The national budget for 1998/99 envisaged total revenue of $A38.7m. and expenditure of $A37.2m. (dramatically reduced from expenditure of $A61.8m. in the previous year). According to the ADB, the budgetary deficit was projected at the equivalent of 10.6% in 2000/01. Deficits of some $A15.2m. and $A22m. were envisaged for 2002/03 and 2003/04 respectively, although the actual budget deficits were expected to be considerably higher. In 2000 overseas development assistance was estimated at US $4.0m., of which US $2.9m. was bilateral aid. Development assistance from Australia amounted to $A3.3m. in 1999/2000. Nauru's external debt was estimated at $A280m. in 2000. In that year the cost of debt-servicing was equivalent to an estimated 13% of total revenue from the exports of goods and services, while total public debt was equivalent to 60.8% of GDP. Consumer prices increased by 4.0% in 1998, by 6.7% in 1999 and by 17.9% in 2000; the annual rate of inflation averaged 4.0% in 2001.

The country's trade balance deteriorated significantly in 2001: imports decreased by 16.3%, while exports declined by 61.5% compared with the previous year, resulting in a trade deficit of US $18.4m. The principal imports are food and live animals (which comprised 83.7% of total imports in 1994, while beverages accounted for a further 4.1%), non-metallic mineral manufactures (4.9%) and machinery and transport equipment (2.8%). Phosphates are the most important export, earning $A38.1m. in 1995; exports of crude fertilizers to Australia totalled $A8.5m. in 2001. The principal export markets in 2001 were New Zealand (which purchased 30.6% of the total), Australia (25.3%), Thailand (15.8%) and the Republic of Korea (12.3%). The principal sources of imports were Australia (supplying 49.4%), the USA (16.9%) and Indonesia (7.9%).

After gaining independence in 1968, Nauru benefited from sole control of phosphate earnings and, as a result, its income per head was among the highest in the world. This, however, had serious repercussions for the country, which became excessively dependent on imported labour, foreign imports and convenience foods, precipitating severe social problems. According to a report on mortality published in 1989, one-third of the adult Nauruan population suffers from diabetes (the worst incidence of the disease anywhere in the world), while obesity, heart-disease, alcoholism and other illnesses attributable to social and dietary problems are also prevalent. By early 1996 the decline in phosphate sales had begun to cause financial difficulties, which led the Government to announce plans to sell a number of its overseas properties and to initiate a corporatization programme in an attempt to reduce external debts of some $A200m. From 1996, furthermore, the Bank of Nauru experienced an ongoing liquidity crisis. The Government's initial failure to address the problem led to a severe crisis of confidence in the bank.

A review of Nauru's economy by the ADB in late 1998 led the organization to devise a programme for major economic and structural adjustment, which aimed to ensure the island's financial security in view of the dramatic decline in revenue from phosphates and from many of its traditional investments. In June 1999 the Government announced that it had completed the first phase of the reform programme, funded by the ADB, involving some 434 redundancies in the civil service with a view to reducing personnel expenditure by 35% (although the retrenchment programme was impeded in 2000 by the appointment of some 500 additional employees to the state mining operation).

Measures to reform the financial sector, in response to allegations that Nauru's 'offshore' banking services were being abused for the purpose of money laundering, were announced in early 2000. However, serious allegations of money laundering re-emerged in early 2001, when the Financial Action Task Force (FATF—see History) declared Nauru to be one of the worst international offenders. The FATF imposed counter-measures, prescribing increased monitoring, surveillance and transparency in financial transactions. The Government subsequently attempted to implement further stringent reforms of the island's financial regulations. Despite these measures, however, the US Administration of George W. Bush obliged Nauru's President in March 2003 to sign papers effectively ending the island's 'offshore' banking industry or face damaging economic sanctions. The country remained on the FATF list in 2004, following the organization's annual review in February. In December 2003, meanwhile, Nauru was removed from a list of unco-operative tax havens issued in April 2002 by the Organisation for Economic Co-operation and Development (OECD).

Nauru received significant financial assistance in late 2001, in exchange for co-operating with Australia in its 'Pacific Solution' to the problem of asylum-seekers. By the end of 2001, in addition to meeting the costs of detaining the refugees in camps, the Australian Government had committed total aid exceeding $A30m. The aid was allocated to various programmes, including $A10m. towards fuel to power the island's electricity generators and $A3m. towards the purchase of new generators.

Nauru's economic outlook at the beginning of the 21st century remained poor. It remained unclear how the substantial budgetary deficits envisaged for the early 2000s would be fully financed; the Government had hitherto relied upon loans from official bilateral sources, overseas corporations or funds from the NPRT, the assets of which had been seriously depleted by 2001. (By mid-2004 Nauru's remaining assets were in the possession of receivers—see History). The Bank of Nauru, meanwhile, another source of budgetary support and financing of phosphate royalty payments to landowners, was believed to have become practically insolvent. During the early 2000s, furthermore, the repeated suspensions of Air Nauru's operations, owing to lack of funds, had led to serious disruptions in the provision of food, fuel and other essential supplies to the island. By early 2002 government payrolls and payments to several creditors had been severely delayed. In January 2003 President Dowiyogo broadcast an appeal for emergency aid, claiming that the island was on the verge of bankruptcy. A report published by the ADB in April stated that Nauru's economic situation was very serious and deteriorating. Furthermore, the report claimed that development assistance given to the country in the past had had little impact because of a lack of political commitment to change and a shortage of skills among the working population. In late 2003 the Australian Government announced that it was considering the possibility of offering Australian citizenship to Nauruans when revenue from the island's role as an immigration detention centre (which since 2001 had constituted Nauru's most important source of income) was no longer available. In April 2004 receivers acting on behalf of creditors in the USA seized Nauru's property portfolio, including five buildings in Sydney and Melbourne. The Nauruan Government had been given until early May to repay a debt of more than US $200m. to a US finance corporation. In February 2004 the Governments of Australia and Nauru signed a memorandum of understanding providing for $A22.5m. in assistance for the period July 2003–June 2005. In July 2004 Australian officials arrived in Nauru to assume management of the country's finances.

Nauru is a member of the Pacific Islands Forum (formerly the South Pacific Forum), the Pacific Community (formerly the South Pacific Commission) and the UN's Economic and Social Commission for Asia and the Pacific (ESCAP). (Nauru was admitted to the UN itself in September 1999.) In addition, Nauru became a member of the Asian Development Bank (ADB) in September 1991.

Statistical Survey

Source (unless otherwise stated): General Statistician, Nauru Government Offices, Yaren.

AREA AND POPULATION

Area: 21.3 sq km (8.2 sq miles).

Population: 8,042 (Nauruan 4,964, Other Pacific Islanders 2,134, Asians 682, Caucasians—mainly Australians and New Zealanders—262) at census of 13 May 1983; 9,919 (5,079 males, 4,840 females) at census of 17 April 1992; 11,845 (official estimate) at mid-2000.

Density (mid-2000): 556 per sq km.

Principal Towns (population, 1992 census): Aiwo (capital) 600; Anetan 427; Anabar 320; Anibare 165. Source: Thomas Brinkhoff, *City Population* (internet www.citypopulation.de).

Births, Marriages and Deaths (1995): Registered live births 203 (birth rate 18.8 per 1,000); Registered marriages 57 (marriage rate 5.3 per 1,000); Registered deaths 49 (death rate 4.5 per 1,000).

Expectation of Life (WHO estimates, years at birth): 62.7 (males 59.7; females 66.5) in 2002. Source: WHO, *World Health Report*.

Economically Active Population (census of 30 June 1966): 2,473 (Administration 845, Phosphate mining 1,408, Other activities 220). *Mid-2002* (estimates): Agriculture, etc. 1,000; Total labour force 6,000 (Source: FAO).

HEALTH AND WELFARE

Key Indicators

Total Fertility Rate (children per woman, 2002): 3.9.

Under-5 Mortality Rate (per 1,000 live births, 2002): 30.

Physicians (per 1,000 head, 1995): 1.57.

Health Expenditure (2001): US $ per head (PPP): 1,015.

Health Expenditure (2001): % of GDP: 7.5.

Health Expenditure (2001): public (% of total): 88.7.

For sources and definitions, see explanatory note on p. vi.

AGRICULTURE, ETC.

Principal Crop and Livestock (FAO estimates, 2002): Coconuts 1,600 metric tons; Pigs 2,800 head; Chickens 5,000 head. Source: FAO.

Fishing (FAO estimates, metric tons, live weight of capture, 2002): Total catch 21. Source: FAO.

MINING

Phosphate Rock ('000 metric tons): 487 in 1998; 600 (estimate) in 1999; 500 (estimate) in 2000. The phosphoric acid content (in '000 metric tons) was: 185 in 1998; 230 (estimate) in 1999; 195 (estimate) in 2000. Source: US Geological Survey.

INDUSTRY

Electric Energy (million kWh): 32 in 1998; 33 in 1999; 33 in 2000. Source: UN, *Industrial Commodity Statistics Yearbook*.

FINANCE

Currency and Exchange Rates: Australian currency: 100 cents = 1 Australian dollar ($A). *Sterling, US Dollar and Euro Equivalents* (31 May 2004): £1 sterling = $A2.5685; US $1 = $A1.4000; €1 = $A1.7144; $A100 = £38.93 = US $71.43 = €58.33. *Average Exchange Rate* (US $ per Australian dollar): 1.9334 in 2001; 1.8406 in 2002; 1.5419 in 2003.

Budget (estimates, $A '000, year ending 30 June 1999): *Revenue:* 38,700; *Expenditure:* 37,200.

EXTERNAL TRADE

Principal Commodities ($A '000, year ending 30 June 1994): *Imports:* Food and live animals 38,420; Beverages 1,890; Non-metallic mineral manufactures 2,268; Non-electrical machinery 758; Transport equipment 534; Total (incl. others) 45,906. *Exports:* Total 45,111. *1995* ($A '000): Total exports 38,081.

Principal Trading Partners (US $ million, 2001): *Imports:* Australia 13.51; India 1.31; Indonesia 2.17; New Zealand 1.05; USA 4.62; Total (incl. others) 27.37. *Exports:* Australia 2.26; Japan 0.62; Republic of Korea 1.10; New Zealand 2.74; Thailand 1.41; Total (incl. others) 8.95. Source: UN, *Statistical Yearbook for Asia and the Pacific*.

TRANSPORT

Road Traffic (1989): 1,448 registered motor vehicles.

Shipping: *Merchant Fleet* (displacement, '000 grt at 31 December): 15 in 1991 (at 30 June); 5 in 1992; 1 in 1993. Source: Lloyd's Register of Shipping. *International Freight Traffic* (estimates, '000 metric tons, 1990): Goods loaded 1,650; Goods unloaded 59. Source: UN, *Monthly Bulletin of Statistics*.

Civil Aviation (traffic on scheduled services, 1999): Kilometres flown (million) 2; Passengers carried ('000) 143; Passenger-km (million) 254; Total ton-km (million) 25. Source: UN, *Statistical Yearbook*.

COMMUNICATIONS MEDIA

Radio Receivers (1997): 7,000 in use*.

Television Receivers (1997): 500 in use*.

Telephones (main lines, 2000): 1,800 in use†.

Mobile Cellular Telephones (2000): 1,200 subscribers†.

* Source: UNESCO, *Statistical Yearbook*.

† Source: International Telecommunication Union.

EDUCATION

Pre-primary (2002): 6 schools; 46 teachers; 634 pupils.

Primary (2002): 5 schools; 64 teachers; 1,566 pupils.

Secondary (2002): 4 schools; 40 teachers; 609 pupils.

Vocational (2001): 6 teachers; 38 students.
Source: Department of Education, Yaren, Nauru.
Nauruans studying at secondary and tertiary levels overseas in 2001 numbered 85.

Directory

The Constitution

The Constitution of the Republic of Nauru came into force at independence on 31 January 1968, having been adopted two days previously. It protects fundamental rights and freedoms, and vests executive authority in the Cabinet, which is responsible to a popularly elected Parliament. The President of the Republic is elected by Parliament from among its members. The Cabinet is composed of five or six members, including the President, who presides. There are 18 members of Parliament, including the Cabinet. Voting is compulsory for all Nauruans who are more than 20 years of age, except in certain specified instances.

The highest judicial organ is the Supreme Court and there is provision for the creation of subordinate courts with designated jurisdiction.

There is a Treasury Fund from which monies may be taken by Appropriation Acts.

A Public Service is provided for, with the person designated as the Chief Secretary being the Commissioner of the Public Service.

Special mention is given to the allocation of profits and royalties from the sale of phosphates.

The Government

HEAD OF STATE

President: LUDWIG SCOTTY (elected 22 June 2004).

CABINET
(September 2004)

President, Minister of Civil Aviation, Minister of Customs and Immigration and Minister of Public Service: LUDWIG SCOTTY.

Minister of Foreign Affairs, Minister of Internal Affairs, Minister of Good Governance, Minister of Justice, Minister of Fisheries and Natural Resources, Minister of Police and Prisons and Minister assisting the President: DAVID ADEANG.

Minister of Finance and Minister of the Environment: KINZA CLODUMAR.

Minister of Health, Minister of Culture and Tourism, Minister of Shipping and Minister of Women's Affairs: KIERAN KEKE.

Minister of Economic Development and Industry and Minister of Sports: MARCUS STEPHEN.

Minister of Education and Vocational Training, Minister of Public Works, and Minister of Youth Affairs : DOGABE JEREMIAH.

MINISTRIES

Office of the President: Yaren, Nauru.

Ministry of Education: Yaren, Nauru; tel. 444-3130; fax 444-3718.

Ministry of Health and Youth Affairs: Yaren, Nauru; tel. 444-3166; fax 444-3136.

Ministry of Finance: Aiwo, Nauru; tel. 444-3140; fax 555-4477.

Ministry of Justice: Yaren, Nauru; tel. 444-3160; fax 444-3108.

Ministry of Works and Community Services: Yaren, Nauru; tel. 444-3177; fax 444-3135.

Legislature

PARLIAMENT

Parliament comprises 18 members. The most recent general election took place on 23 October 2004.

Speaker: (to be appointed).

Political Organizations

Democratic Party of Nauru: c/o Parliament House, Yaren, Nauru; f. 1987; revival of Nauru Party (f. 1975); Leader KENNAN ADEANG.

Naoero Amo (Nauru First): c/o Parliament House, Yaren, Nauru; e-mail visionary@naoeroamo.com; internet www.naoeroamo.com; f. 2001; Co-Leaders DAVID ADEANG, KIERAN KEKE.

Diplomatic Representation

EMBASSY IN NAURU

China, People's Republic: Yaren; Ambassador XU SHIGUO.

Judicial System

The Chief Justice presides over the Supreme Court, which exercises original, appellate and advisory jurisdiction. The Resident Magistrate presides over the District Court, and he also acts as Coroner under the Inquests Act 1977. The Supreme Court is a court of record. The Family Court consists of three members, one being the Resident Magistrate as Chairman, and two other members drawn from a panel of Nauruans. The Chief Justice is Chairman of the Public Services Appeals Board and of the Police Appeals Board.

SUPREME COURT

tel. 444-3163; fax 444-3104.

Chief Justice: BARRY CONNELL (non-resident).

DISTRICT COURT

Resident Magistrate: G. N. SAKSENA.

FAMILY COURT

Chairman: G. N. SAKSENA.

Religion

Nauruans are predominantly Christians, adhering either to the Nauruan Protestant Church or to the Roman Catholic Church.

Nauruan Protestant Church: Head Office, Nauru; Moderator (vacant).

Roman Catholic Church: POB 16, Nauru; tel. and fax 444-3708; Nauru forms part of the diocese of Tarawa and Nauru, comprising Kiribati and Nauru. The Bishop resides on Tarawa Atoll, Kiribati.

The Press

Central Star News: Nauru; f. 1991; fortnightly.

Nasero Bulletin: Nauru; tel. 444-3847; fax 444-3153; e-mail bulletin@cenpac.net.nr; fortnightly; English; local and overseas news; Editor SEPE BATSIUA; circ. 500.

The Nauru Chronicle: Nauru; Editor RUBY DEDIYA.

Broadcasting and Communications

TELECOMMUNICATIONS

Nauru Telecommunications Service: Nauru; tel. 444-3324; fax 444-3111; Dir EDWARD W. R. H. DEYOUNG.

BROADCASTING

Radio

Nauru Broadcasting Service: Information and Broadcasting Services, Chief Secretary's Department, POB 77, Nauru; tel. 444-3133; fax 444-3153; e-mail ntvdirector@cenpac.net.nr; f. 1968; state-owned and non-commercial; expected to be corporatized in the late 1990s; broadcasts in the mornings in English and Nauruan; operates Radio Nauru; Station Man. RIN TSITSI; Man. Dir GARY TURNER.

Television

Nauru Television (NTV): Nauru; tel. 444-3133; fax 444-3153; e-mail ntvmanager@cenpac.net.nr; began operations in June 1991; govt-owned; broadcasts 24 hrs per day on 3 channels; most of the programmes are supplied by foreign television companies via satellite or on videotape; a weekly current affairs programme is produced locally; Man. MICHAEL DEKARUBE; Dir of Media GARY TURNER.

Finance

(cap. = capital; res = reserves; dep. = deposits; m. = million; amounts in Australian dollars unless otherwise stated)

BANKING

State Bank

Bank of Nauru: Civic Centre, POB 289, Nauru; tel. 444-3238; fax 444-3203; f. 1976; state-owned; cap. 12.0m., res 123.0m., dep. 141.0m. (Dec. 1994); Chair. MARCUS STEPHEN.

INSURANCE

Nauru Insurance Corporation: POB 82, Nauru; tel. 444-3346; fax 444-3731; f. 1974; sole licensed insurer and reinsurer in Nauru; Chair. NIMES EKWONA.

Trade and Industry

GOVERNMENT AGENCIES

Nauru Agency Corporation: POB 300, Aiwo, Nauru; tel. 555-4324; fax 444-3730; e-mail nrugrp@cenpac.net.nr; functions as a merchant bank to assist entrepreneurs in the registration of holding and trading corporations and the procurement of banking, trust and insurance licences; chaired by the President of Nauru; Gen. Man. S. B. HULKAR.

Nauru Corporation: Civic Centre, Yaren, Nauru; f. 1925; operated by the Nauru Council; the major retailer in Nauru; Gen. Man. A. EPHRAIM.

Nauru Fisheries and Marine Resources Authority: POB 449, Nauru; tel. 444-3733; fax 444-3812; e-mail nfmra@cenpac.net.nr; f. 1997.

Nauru Phosphate Corporation: Aiwo, Nauru; tel. 444-3839; fax 444-2752; f. 1970; operates the phosphate industry and several public services of the Republic of Nauru (including provision of electricity and fresh water) on behalf of the Nauruan people; responsible for the mining and marketing of phosphate; Gen. Man. JOSEPH HIRAM; Chair. ALI AMWANO.

Nauru Phosphate Royalties Trust: Nauru; e-mail nprtnau@cenpac.net.nr; statutory corpn; invests phosphate royalties to achieve govt revenue; extensive international interests, incl. hotels and real estate; Sec. NIRAL FERNANDO.

Nauru Rehabilitation Corporation (NRC): Nauru; f. 1999; manages and devises programmes for the rehabilitation of those parts of the island damaged by the over-mining of phosphate; Chair. ALI AMWANO.

UTILITIES

Nauru Phosphate Corporation: Aiwo, Nauru; tel. 555-6481; fax 555-4111; operates generators for the provision of electricity and supplies the island's water; Chair. (vacant); Gen. Man. JOSEPH HIRAM.

Transport

RAILWAYS

There are 5.2 km of 0.9-m gauge railway serving the phosphate workings.

ROADS

A sealed road, 16 km long, circles the island, and another serves Buada District. There were 1,448 registered vehicles in 1989.

SHIPPING

As Nauru has no wharves, passenger and cargo handling are operated by barge. In late 1998 finance was secured from the Japanese Government for the construction of a harbour in Anibare district. Work on the project began in 1999.

Nauru Pacific: Government Bldg, Yaren, Nauru; tel. 444-3133; f. 1969; operates cargo charter services to ports in Australia, New Zealand, Asia, the Pacific and the west coast of the USA; Man. Dir (vacant).

CIVIL AVIATION

Air Nauru: Directorate of Civil Aviation, Government of Nauru Offices, POB 40, Yaren, Nauru; tel. 444-3274; fax 444-3705; e-mail write2us@airnauru.com.au; internet www.airnauru.com.au; f. 1970; corporatized in 1996 and moved to Australian aviation register in mid-1997; operates passenger and cargo services to Kiribati, Fiji, New Caledonia, Solomon Islands, Guam, Palau, the Philippines, the Federated States of Micronesia, Hawaii (USA), Australia and New Zealand; Chair. KEN McDONALD; CEO GEOFFREY BOWMAKER.

Defence

The Republic of Nauru has no defence forces. Under an informal agreement, Australia is responsible for the defence of the island.

Education

Education is free and compulsory for children between the ages of six and 16. In 2002 the island had six pre-primary schools, with 634 pupils, five primary schools, with 1,566 pupils, and four secondary schools with 609 pupils. In 2002 Nauruans studying overseas at secondary and tertiary levels numbered 85. An extension centre of the University of the South Pacific, based in Suva, Fiji, was opened in Nauru in the late 1980s.

NEW ZEALAND PACIFIC TERRITORY

Tokelau is the only remaining Island Territory of New Zealand, of which it is an integral part. New Zealand is a member of the Pacific Islands Forum (formerly the South Pacific Forum), both in its own right and in respect of its Pacific dependencies. The cabinet minister responsible is the Minister of Foreign Affairs and Trade, of Justice and of Pacific Island Affairs (responsibility falling within the remit of the portfolio of foreign affairs and trade).

Head of State: HM Queen ELIZABETH II (succeeded to the throne 6 February 1952).

Governor-General: Dame SILVIA CARTWRIGHT (took office 4 April 2001).

Ministry of Foreign Affairs and Trade: Private Bag 18901, Wellington, New Zealand; tel. (4) 494-8500; fax (4) 472-9596; e-mail enquiries@mfat.govt.nz; internet www.mfat.govt.nz.

Minister of Foreign Affairs and Trade, of Justice and of Pacific Island Affairs: PHIL GOFF.

TOKELAU

Physical and Social Geography

Tokelau is located in the central Pacific Ocean, about 480 km north of Apia in Samoa, its nearest neighbour. Tuvalu lies to the west and Kiribati to the north and east. The Territory consists of three atolls with a total area of 10 sq km (3.9 sq miles). The central atoll of Nukunonu is the largest; Atafu, the smallest, lies 64 km to the north-west and Fakaofo lies 92 km to the south-east of Nukunonu. Each atoll consists of a number of reef-bound islets, or *motu*, encircling a lagoon. The *motu* vary in size but are never wider than 200 m or higher than 5 m, although they can be up to 6 km in length. The average annual temperature is 28°C (82°F), July being the coolest month and May the warmest. Rainfall is heavy but inconsistent, and occasionally there are severe storms. The indigenous inhabitants are a Polynesian people, and Tokelauan is the official language, although English is widely spoken. The population is almost entirely Christian, being either Protestant (just over two-thirds) or Roman Catholic. The total population was 1,487 at the 1996 census, and was estimated at 1,445 in July 2001. Tokelau has no official capital, each atoll having its own administrative centre. However, the seat of Government is recognized as 'the capital' and is rotated on a yearly basis among the three atolls.

History

The islands now comprising Tokelau were inhabited by Polynesians, closely related to the people of Samoa, before becoming a British protectorate in 1877. At the request of the inhabitants, the United Kingdom annexed the territory, then known as the Union Islands, in 1916 and included it within the Gilbert and Ellice Islands Colony. In 1925 the British Government transferred administrative control to New Zealand. In 1946 the group was officially designated the Tokelau Islands, and in 1948 sovereignty was transferred to New Zealand. From 1962 until the end of 1971 the High Commissioner for New Zealand in Western Samoa (now Samoa) was also the Administrator of the Tokelau Islands. In November 1974 the administration of the Tokelau Islands was transferred to the Ministry of Foreign Affairs in New Zealand. In 1976 the Tokelau Islands were officially redesignated Tokelau.

New Zealand has undertaken to assist Tokelau towards increased self-government and economic self-sufficiency. The Territory was visited by the UN Special Committee on Decolonization in 1976 and 1981, but on both occasions the missions reported that the people of Tokelau did not wish to change the nature of the existing relationship between Tokelau and New Zealand. This opinion was reiterated by an emissary of the General Fono, the Territory's highest advisory body, in 1987, and by the Official Secretary in 1992. A report by the UN Special Committee in 2002 listed Tokelau as one of 16 dependent territories it was seeking to encourage towards independence. However, a UN decolonization mission which visited the islands in September of that year was told that the majority of Tokelauans wanted to remain part of New Zealand and that the Territory was far too dependent on that country to change its status.

A programme of constitutional change, agreed in 1992 and formalized in January 1994, provided for a more defined role for Tokelau's political institutions, as well as for their expansion. A process of relocating the Tokelau Public Service (hitherto based in Apia, Western Samoa, now Samoa) to the Territory began in 1994, and by 1995 all government departments, except Transport and Communications and part of the Administration and Finance Department, had been transferred to Tokelauan soil. The Tokelau Apia Liaison Office (formerly the Office for Tokelau Affairs) was, however, to remain in Western Samoa (now Samoa), owing to the country's more developed communications facilities.

The development of Tokelau's institutions at a national level has prompted renewed interest in the islands' prospects for greater internal autonomy. In June 1994 the General Fono adopted a National Strategic Plan, which gave details of Tokelau's progression (over the next five to 10 years) towards increased self-determination and, possibly, free association with New Zealand. The executive and administrative powers of the Administrator were formally transferred, in that year, to the General Fono and, when the Fono is not in session, to the Council of Faipule. A draft Constitution was subsequently drawn up. In May 1996 the New Zealand House of Representatives approved the Tokelau Amendment Bill, granting the General Fono the power to enact legislation, to impose taxes and to declare public holidays, effective from 1 August 1996 (although New Zealand was to retain the right to legislate for Tokelau). A visit to the islands by the Prime Minister of Tuvalu in mid-1996, for the signing of a mutual co-operation agreement (covering shipping, trade and fisheries), was widely interpreted as an indication of Tokelau's increased autonomy. A further co-operation agreement was established in March 2003, following a five-day visit to the islands by the Prime Minister of Samoa. Tokelau's traditional leaders agreed a framework for annual meetings with the Samoan Government to discuss issues of concern and mutual benefit in what was regarded as a sign of the growing relationship between the two parties.

Elections took place in January 1999, at which two of the Territory's Faipule were re-elected, while the remaining Faipule and three Pulenuku posts were secured by new candidates. Following electoral reforms, delegates were, for the first time, elected to the General Fono for a three-year term; they had previously been nominated by each Taupulega (Island Council or Council of Elders). As part of the same reform process, the number of delegates to the General Fono was reduced from 27 to 18. At elections in January 2002, all three incumbent Faipule and one Pulenuku were re-elected to office; two new Pulenuku were elected.

Mounting fears among islanders that, despite their wishes, New Zealand was seeking to loosen its ties with Tokelau led the New Zealand Minister of Foreign Affairs to state in April 2000 that his country would not impose independence on the Territory and that any change in its political status would only occur with the consent of Tokelauans. In early 2001 the head of the Tokelau Public Service Commission, Aleki Silau, reiterated the islanders' reluctance to renounce New Zealand citizenship, and emphasized that both sides have until 2010 to reach a decision. In accordance with legislation approved in 1999, management of the islands' public service was formally transferred to Tokelau in July 2001. A mission from the UN Special Committee on Decolonization visited the islands in September 2002 (see above). In late November 2003 New Zealand's Governor-General, Dame Sylvia Cartwright, made an official visit to Tokelau to sign the Principles of Partnership agreement. The document was described by New Zealand's Minister of Foreign Affairs, Phil Goff, as a step closer to decolonization for the islands. Goff reiterated that the final decision on Tokelau's future would be made by its inhabitants, although he also confirmed that he expected the islands to adopt a system of self-government in free association with New Zealand, similar to that existing in Niue and the Cook Islands. In March 2004 it was announced that new powers were to be granted to the three atolls' Taupulega, giving them greater control over local affairs. In May a senior government member reiterated the view that, despite the ambition of New Zealand and the UN for Tokelau to achieve self-determination, the islanders themselves were extremely reluctant to change their status.

In mid-August 2004 New Zealand's Prime Minister Helen Clark made an official visit to Tokelau (the first such visit in more than 20 years). During the visit Clark announced the provision of a grant worth some US $0.3m. towards improvements for boat access to the islands and a review of the islands' communications infrastructure. The Prime Minister also expressed her confidence that the islanders would vote in favour of free association with New Zealand when the issue was finally put to a referendum

New Zealand is responsible for the external relations of Tokelau. Strong ties are maintained with Samoa, to the people of which the Tokelauans are closely related. There is considerable co-operation in health and education matters.

In December 1980 New Zealand and the USA signed a treaty whereby the US claim to Tokelau, dating from 1856, was relinquished. At the same time, New Zealand relinquished a claim, on behalf of Tokelau, to Swains Island (Olohenga), administered by the USA since 1925 as part of American Samoa. The treaty was ratified in August 1983, although there was some dissent in Tokelau.

Tokelau is a member of the Pacific Community (formerly the South Pacific Commission) and, as a dependent territory of New Zealand, is represented by that country in the South Pacific Forum and other international organizations. In response to local concern, New Zealand outlawed drift-net fishing within Tokelau's exclusive economic zone (extending to 200 nautical miles (370 km) from the islands' coastline) from November 1989. In 1996 Tokelau signed a memorandum of understanding with Tuvalu which aimed to facilitate co-operation in shipping, trade and fisheries.

A further environmental threat is posed by the possible consequences of the 'greenhouse effect' (the heating of the earth's atmosphere as a result of pollution). A UN report on this subject, published in 1989, listed Tokelau as one of the island groups likely to be completely submerged during the 21st century, owing to a rise in sea-level.

Economy

Tokelau's size, isolation and lack of land-based resources limit economic development. The principal domestic sources of revenue are fees from shipping, fishing licences, radio and telegram excises and customs duties, and exports of copra, handicrafts, postage stamps and souvenir coins (these last are legal tender, although New Zealand currency is in general use). According to estimates by the UN Development Programme, in 1982 Tokelau's gross national product (GNP) was US $1.2m., equivalent to US $760 per head. Gross domestic product (GDP) was estimated at US $1.5m. in 1993 and GDP per head at US $1,000 in that year.

Tokelau's soil is thin and infertile. Apart from some copra production, agriculture is of a basic subsistence nature. In the year to March 1987 Tokelau exported some 107 metric tons of copra to Western Samoa (now Samoa). Exports were subsequently affected by cyclone damage to the coconut trees. Apart from coconuts, food crops include bananas, breadfruit, pulaka, papaya (pawpaw), and the edible fruit of the screw-pine (*Pandanus*). Livestock consists of pigs, ducks and other poultry. Ocean and lagoon fish and shellfish are staple constituents of the islanders' diet. There is little commercial fishing, but the sale of licences to fish in Tokelau's exclusive economic zone earns some revenue. Tokelau benefited from development assistance and revenue of some US $62,500 per year from the USA, under the treaty negotiated by the Forum Fisheries Agency in 1987. In 2000/01 fisheries licensing provided $NZ1.2m. in revenue, although the Government projected a more modest income of $NZ0.9m. in 2001/02. In 2001 $NZ0.68m. of the income from fisheries licensing was used to establish the Tokelau Trust Fund. In early 2004 the Pacific Community Secretariat produced a fisheries management plan for Tokelau. The plan, which was to be implemented in mid-2004, focused on community-based activities and included the increased exploitation of the islands' giant clam resources.

The only industries of significance are copra production and the manufacture of handicrafts. However, a factory on Atafu, processing highly priced yellowfin tuna, opened in 1990, providing another important source of income. The factory is supplied by local residents, appointed by Atafu's village council to fish on certain days to provide tuna for the plant. The principal markets for the product are New Zealand and Japan. Construction, mainly public works, is carried out with the help of New Zealand and foreign aid and, often, using the local labour levy. Electricity is provided by diesel generators based in the village of each atoll. Trading relations are dominated by New Zealand and Samoa; there are co-operative stores on each atoll. Most requirements are imported. Imports to the value of $NZ67,000 were purchased from New Zealand in 1997/98.

Since 1982 the General Fono has levied a tax on the salaries of public servants who are unavailable for the community service labour levy. In the mid-1990s this levy, which was equivalent to 6%–12% of their salaries, was collected from about 160 civil servants. In 1996/97 almost US $0.7m. of total budgetary expenditure (estimated at some US $4m.) was invested in village development projects. Measures were taken during 1996 to augment locally generated revenue, including the expansion of the islands' tax base and the introduction of charges for health and education services, which resulted in an increase in Tokelau's contribution to total budgetary revenue from 17% in 1995 to 25% in the following year. The proportion, however, declined to 13.8% in 1997/98. In 1999/2000 there was a budgetary deficit of $NZ0.9m. Tokelau's budget for 2001/02 was to include at least $NZ4.2m. from New Zealand, while an estimated $NZ1.7m. was to be obtained from local revenues such as fisheries licensing, duty, taxes, philatelic sales, freight charges and interest earned. Official development assistance from New Zealand totalled $NZ8.6m. in 2003/04. In addition to its links to New Zealand, Tokelau maintains a bilateral development assistance plan with Australia, centred upon human resource development. Australia provides scholarships for Tokelauan students to study in Australia or at regional academic institutions. Development assistance from Australia in 2001/02 was projected at a total of $A100,000.

In 1987 the General Fono decided to delay work on the construction of airstrips and to concentrate resources on the development of shipping links instead. The improvement of these links in the late 1980s and early 1990s, although of general economic benefit, resulted in an increase in imported foods, which, according to a report published in 1992, has contributed to serious social and dietary problems in the islands. In September 2002 the Faipule of Nukunonu, Pio Tuia, announced renewed proposals for the construction of wharves and improved access for shipping to facilitate the export of fish, and of an airport to enable tourists to visit the islands more easily (fewer than 30 tourists visited the islands in 2001). It was hoped that Tokelau would derive greater benefits from its fisheries resources as a result of a plan initiated by the Pacific Community Secretariat in 2004 (see above).

Remittances from Tokelauans working abroad, mainly in New Zealand, are also an important source of private income. At the March 1991 census in New Zealand, it was reckoned that there were 2,802 Tokelauans in that country. The New Zealand budget for 1989 allowed the payment of 50% of retirement income (under a national scheme) to people residing abroad. It was hoped that this would encourage Tokelauans to return to the islands. There is no unemployment as such in the communal system of Tokelau. Emigration has, however, relieved population pressure on the atolls, particularly at the village of Fakaofo (on the *motu* of Fale). A new settlement has been established on the larger *motu* of Fenua Fale nearby. The people of Atafu all live in one village, on the *motu* of Vao. This name is shared by the *motu* on which Nukunonu village is located (a bridge connects it to another islet where some families have settled). There is least population pressure on Nukunonu.

Statistical Survey

Source (unless otherwise indicated): Tokelau Apia Liaison Office, POB 805, Apia, Samoa; tel. 20822; fax 21761.

AREA AND POPULATION

Area: Atafu 3.5 sq km; Nukunonu 4.7 sq km; Fakaofo 4.0 sq km; Total 12.2 sq km (4.7 sq miles).

Population (census of March 1991): Total 1,577 (Atafu 543, Nukunonu 437, Fakaofo 597). *Tokelauans Resident in New Zealand:* 2,802. *Census of 1996* (incl. temporary visitors): Total 1,507 (Atafu 499, Fakaofo 578, Nukunonu 430); Total resident population 1,487.

Density (resident population, 1996): 121.9 per sq km.

Births and Deaths (1996): Birth rate 33.1 per 1,000; Death rate 8.2 per 1,000.

Expectation of Life (official estimates, years at birth, 1996): Males 68; Females 70. Source: Ministry of Foreign Affairs and Trade, Wellington.

AGRICULTURE, ETC.

Crop Production (FAO estimates, metric tons, 2002): Coconuts 3,000; Copra 45; Roots and tubers 300; Bananas 15; Other tropical fruits 46. Source: FAO.

Livestock (FAO estimates, year ending September 2002): Pigs 1,000; Chickens 5,000. Source: FAO.

Fishing (FAO estimates, metric tons, live weight, 2002): Total catch 200. Source: FAO.

INDUSTRY

Production (estimate, 1990): Electric energy 300,000 kWh.

FINANCE

Currency and Exchange Rates: New Zealand currency is legal tender. Tokelau souvenir coins have also been issued. New Zealand currency: 100 cents = 1 New Zealand dollar ($NZ); *Sterling, US Dollar and Euro Equivalents* (31 May 2004): £1 sterling = $NZ2.9076; US $1 = $NZ1.5848; €1 = $NZ1.9407; $NZ100 = £34.39 = US $63.10 = €51.53. *Average Exchange Rate* (US $ per $NZ): 2.3788 in 2001; 2.1622 in 2002; 1.7229 in 2003.

Budget ($NZ, year ending 30 June 1998): *Revenue:* Local 734,950; New Zealand subsidy 4,600,000; Total 5,334,950. *Expenditure:* Total 5,208,449.

Overseas Aid (projection, $NZ '000, 2002/03): Official development assistance from New Zealand 8,100 (of which Budget support 4,750, Projects and training 2,650). *2003/04:* Total development assistance from New Zealand $NZ8.6m. Source: Ministry of Foreign Affairs and Trade, Wellington.

EXTERNAL TRADE

Principal Commodities: *Imports:* Building materials WS $104,953 (1983/84); *Exports:* Copra $NZ43,542 (1982/83); Handicrafts WS $10,348 (1983/84). Note: Foodstuffs and fuel are also important import commodities. *1997/98:* Imports from New Zealand totalled $NZ67,000.

COMMUNICATIONS MEDIA

Radio Receivers (estimate, 1997): 1,000 in use.

EDUCATION

Schools (1999): 3 (one school for all levels on each atoll).

Teachers (1990): Qualified 43; Aides 8; Adult Learning Centre Co-ordinators 3. (1999): Qualified 30.

Pupils (1991): Pre-primary 133; Primary 361; General secondary 113.

Students Overseas (1990): Secondary 42; Technical and vocational 36; Teacher training 3; University 16. (1999): Secondary 22; Tertiary 20.

Directory

The Constitution

Tokelau is administered under the authority of the Tokelau Islands Act 1948 and subsequent amendments and regulations. The Act declared Tokelau (then known as the Tokelau Islands) to be within the territorial boundaries of New Zealand. The Administrator is the representative of the Crown and is responsible to the Minister of Foreign Affairs and Trade in the New Zealand Government. The office of Administrator is normally held conjointly with that of New Zealand's Secretary of Foreign Affairs and Trade, but provision is made for the offices to be held separately. Most of the powers of the Administrator are delegated to the Tokelau Apia Liaison Office, the General Fono and the Council of Faipule. The chief representative of the Administrator (and the Crown) on each atoll is the highest elected official, the Faipule, who exercises executive, political and judicial powers. The three Faipule, who hold ministerial portfolios and form the Council of Faipule, act as the representatives of the Territory in dealings with the administration and at international meetings, and choose one of their number to hold the title Ulu-O-Tokelau (Head of Tokelau) for a term of one year. The Ulu-O-Tokelau chairs sessions of the territorial assembly, the General Fono. The General Fono is a meeting of 18 delegates (including the Faipule and the Pulenuku—Village Mayor—from each atoll), representing the entire Territory. There are two or three meetings each year, which may take place on any of the atolls. The General Fono is the highest advisory body and the administration must consult it about all policy affecting the Territory. The assembly has responsibility for the territorial budget and has the power to enact legislation, to impose taxes and to declare public holidays. There are a number of specialist committees, such as the Budget Committee and the Law Committee.

Tokelau is an association of three autonomous atoll communities. Local government consists of the Faipule, the Pulenuku and the Taupulega (Island Council or Council of Elders). The Faipule, the Pulenuku and delegates to the General Fono are elected every three years on the basis of universal adult suffrage (the age of majority being 21). The Faipule represents the atoll community, liaises with the administration and the Tokelau Public Service, acts as a judicial commissioner and presides over meetings of the Taupulega. The Pulenuku is responsible for the administration of village affairs, including the maintenance of water supplies and the inspection of plantations, and, in some instances, the resolution of land disputes (practically all land is held by customary title, by the head of a family group, and may not be alienated to non-Tokelauans). The Taupulega is the principal organ of local government. The Taupulega also appoints the Failautuhi (Island Clerk), to record its meetings and transactions. The Taupulega in Atafu consists of the Faipule, the Pulenuku and the head of every family group; in Nukunonu it consists of the Faipule, the Pulenuku, the elders of the community and the nominated heads of extended families; in Fakaofo it consists of the Faipule, the Pulenuku and the elders (meetings of all the heads of family groups take place only infrequently).

The Government

(September 2004)

Administrator: Lindsay Watt (took office March 1993).

FAIPULE

At elections in January 2002, all three incumbent Faipule were re-elected to office.

Faipule of Fakaofo: Kolouei O'Brien.

Faipule of Nukunonu: Pio Tuia.

Faipule of Atafu: Kuresa Nasau.

PULENUKU

At elections in January 2002, two new Pulenuku were elected to office and one was re-elected.

Pulenuku of Fakaofo: Keli Neemia.

Pulenuku of Nukunonu: Panapa Sakaria.

Pulenuku of Atafu: Paulo Kitonia.

GOVERNMENT OFFICES

Tokelau Apia Liaison Office/Ofiha o Fehokotakiga Tokelau Ma Apia: POB 865, Apia, Samoa; tel. 20822; fax 21761; responsible for transport, accounting and consular functions; Dir Zac Patalesio.

The Tokelau Public Service has seven departments, divided among the three atolls, with a supervising administrative official located in each village. Two departments are established on each atoll, while the seventh department, the Office of the Council of Faipule, rotates on a yearly basis in conjunction with the position of Ulu-O-Tokelau. Management of the Tokelau Public Service was formally transferred to Tokelau in July 2001.

Judicial System

Tokelau's legislative and judicial systems are based on the Tokelau Islands Act 1948 and subsequent amendments and regulations. The Act provided for a variety of British regulations to continue in force and, where no other legislation applies, the law of England and Wales in 1840 (the year in which British sovereignty over New Zealand was established) was to be applicable. New Zealand statute law applies in Tokelau only if specifically extended there. In 1986 legislation formalized the transfer of High Court civil and criminal jurisdiction from Niue to New Zealand. Most cases are judged by the Commissioner established on each atoll, who has limited jurisdiction in civil and criminal matters. Commissioners are appointed by the New Zealand Governor-General, after consultation with the elders of the atoll.

Commissioner of Fakaofo: Lui Kelekolio.

Commissioner of Nukunonu: Atonio Egeliko.

Commissioner of Atafu: Maka Toloa.

Religion

On Atafu almost all inhabitants are members of the Tokelau Congregational Christian Church, on Nukunonu all are Roman Catholic, while both denominations are represented on Fakaofo. In the late 1990s some 70% of the total population adhered to the Congregational Christian Church, and 30% to the Roman Catholic Church.

CHRISTIANITY

Roman Catholic Church

The Church is represented in Tokelau by a Mission, established in 1992. There were an estimated 500 adherents at 31 December 2002.

Superior: Mgr Patrick Edward O'Connor, Catholic Mission, Nukunonu, Tokelau (via Apia, Samoa); tel. 4160; fax 4236; e-mail dr .tovite@clear.net.n3.

Broadcasting and Communications

Each atoll has a radio station to broadcast shipping and weather reports. Radio-telephone provided the main communications link with other areas until the late 1990s. A new telecommunications system established at a cost of US $2.76m. (US $1m. of which was provided by New Zealand) and operating through an earth station, linked to a communications satellite, on each atoll, became operational in 1997. A new radio station, broadcasting information and music, was to commence operations in 2002.

TELECOMMUNICATIONS

TeleTok: Fenuafala, Fakaofo.

Finance

In 1977 a savings bank was established on each atoll; commercial and other banking facilities are available in Apia, Samoa.

Trade and Industry

A village co-operative store was established on each atoll in 1977. Local industries include copra production, woodwork and plaited craft goods, and the processing of tuna. Electricity is provided by diesel generators based in the village on each atoll.

Transport

There are no roads or motor vehicles. Unscheduled inter-atoll voyages, by sea, are forbidden because the risk of missing landfall is too great. Passengers and cargo are transported by vessels that anchor off shore, as there are no harbour facilities. A scheme to provide wharves (primarily to facilitate the export of fish) was proposed in September 2002. Most shipping links are with Samoa, but a monthly service from Fiji was introduced in 1986. The vessel *Forum Tokelau*, operated by Pacific Forum Line, began a monthly service between Tokelau and Apia, Samoa, in mid-1997. A New Zealand-funded inter-atoll vessel commenced service in 1991, providing the first regular link between the atolls for 40 years. Plans to construct an airstrip on each atoll were postponed in 1987 in favour of the development of shipping links. In late 2002, however, proposals for the construction of an airport were again under consideration.

Education

Education is provided free of charge, and attendance is almost 100%. Kindergarten facilities are available for children from the age of three years, while primary education takes place between the ages of five and 14. The provision of an additional year of schooling, for those aged 15, is rotated among the Territory's three schools every five years. There were 30 qualified teachers, complemented by teacher aides, on the islands in 1999. Pupil enrolment in 1991 at primary and secondary levels totalled 474. In the financial year ending 30 June 1999, government expenditure on education totalled $NZ0.80m., or 18.0% of total budgetary expenditure. The New Zealand Department of Education provides advisory services and some educational equipment. The Education Department of Samoa organizes daily radio broadcasts. Scholarships are awarded for secondary and tertiary education, and for vocational training, in New Zealand, Australia and other Pacific countries. Link arrangements exist between Tokelau and the University of the South Pacific, based in Fiji. There were 97 Tokelauan students at overseas institutions in 1990.

NEW ZEALAND PACIFIC: ASSOCIATED STATES

The Associated States of New Zealand, the Cook Islands and Niue, were formerly Island Territories and integral parts of New Zealand. They now enjoy full self-government but continue in free association with New Zealand. New Zealand remains responsible for defence and represents the dependencies at the UN and in whichever external relations not conducted by the local Government.

Head of State: HM Queen ELIZABETH II (succeeded to the throne 6 February 1952).

Governor-General: Dame SILVIA CARTWRIGHT (took office 4 April 2001).

COOK ISLANDS

Physical and Social Geography

The 13 inhabited and two uninhabited islands of the Cook Islands are located in the southern Pacific Ocean. The territory lies between American Samoa, to the west, and French Polynesia, to the east. The total area of all the islands is 237 sq km (91.5 sq miles), but they extend over about 2m. sq km (more than 750,000 sq miles) of ocean, and form two groups: the Northern Cooks, which are all atolls and include Pukapuka (Danger Islands), Rakahanga (Rierson Island) and Manihiki (Humphrey Island), and the Southern Cooks, including Aitutaki, Mangaia and Rarotonga, which are all volcanic islands. From December to March the climate is warm and humid, with the possibility of severe storms; from April to November the climate is mild and equable. The average annual rainfall on Rarotonga is 2,134 mm (84 ins). The official language is English, but Polynesian languages are also spoken. The principal religion is Christianity, with about 58% of the population adhering to the Cook Islands Christian Church. The population was 18,027 in December 2001, according to the provisional results of the census; of these, 12,900 were estimated to be permanent residents. The capital is Avarua, on Rarotonga.

History

The islands were already settled by Polynesian (Maori) clans when the first Europeans, Spaniards, visited the territory. The Cook Islands are named after the leader of a British expedition of 1773, Capt. James Cook. The first islands were proclaimed a British protectorate in 1888. The first British Resident, in 1891, established an Elective Federal Parliament and a Federal Executive Council (the latter comprising Arikis or hereditary chiefs). This system was dissolved when the Cook Islands were annexed to New Zealand in 1901. Subsequent legislation developed government and representative institutions, and a Legislative Assembly was established in 1957. In 1962 the New Zealand Government presented the Assembly with four choices for constitutional development.

Following negotiations on the details and the enactment of the Constitution, the Cook Islands became a self-governing territory in free association with New Zealand on 4 August 1965. The people are New Zealand citizens. Sir Albert Henry, leader of the Cook Islands Party (CIP), was elected Premier in 1965 and re-elected in 1968, 1971, 1974 and March 1978. However, in July 1978, following an inquiry into alleged electoral malpractice, the Chief Justice disallowed votes cast in the elections to the Legislative Assembly by Cook Islands expatriates who had been flown from New Zealand, with their fares paid from public funds. The amended ballot gave a majority to the Democratic Party (DP), and its leader, Dr (later Sir) Thomas Davis, was sworn in as Premier by the Chief Justice. In August 1979 Sir Albert Henry was convicted of conspiracy to defraud, and was stripped of his knighthood.

In May 1981 an Amendment Bill made several changes in the Cook Islands' Constitution, increasing the membership of the Parliament, from 22 to 24, and extending the parliamentary term from four to five years.

In March 1983 Sir Thomas Davis lost power to the CIP, under Geoffrey (later Sir) Henry, cousin of the former Premier. However, with one seat already subject to re-election, Henry's majority of three was eroded by the death of one CIP member of Parliament and the transfer of support to the DP by another, Tupui Henry, son of the late Albert Henry. Geoffrey Henry resigned in August, and a general election in November returned the DP to power under Davis. In August 1984 Davis announced a major reorganization of cabinet portfolios, with three of the seven posts going to members of the CIP, to form a coalition Government, with Geoffrey Henry as Deputy Prime Minister. In mid-1985, however, Davis dismissed Geoffrey Henry, who had endorsed an unsuccessful motion expressing 'no confidence' in the Government, and the CIP withdrew from the coalition. Henry's successor as Deputy Prime Minister was Dr Terepai Maoate, one of four CIP members who continued to support the Davis Government. In July 1987 Davis was ousted as Prime Minister by a vote expressing 'no confidence' in him, following protracted controversy over contentious budget proposals. He was succeeded by Dr Pupuke Robati, a member of the Cabinet and a leading figure in the DP. Geoffrey Henry again became Prime Minister following a general election victory in January 1989.

In mid-1990 the defection of an MP from the DP to the CIP provided the CIP with 15 seats in Parliament and thus the minimum two-thirds' majority support necessary to amend the Constitution. In August 1991 a constitutional amendment was passed that increased the number of members of Parliament to 25, and at an election to the newly created seat a CIP candidate was successful. The amendment also provided for an increase in the number of cabinet members from seven to nine (including the Prime Minister).

In mid-1992 Norman George (parliamentary whip of the DP) was dismissed from the party, following a dispute over government spending, and in October he formed the Alliance Party.

At a general election in March 1994 the CIP increased its majority, winning 20 seats in Parliament, while the DP secured three and the Alliance Party two. Davis, who failed to win a seat, subsequently resigned as leader of the DP. A referendum, held simultaneously, revealed that a majority of the electorate favoured retaining the current name (69.8% of voters), national anthem (80.2%) and flag (48.5%) of the Cook Islands. (At subsequent by-elections the CIP lost two seats and the DP and Alliance Party each gained one seat.)

A financial scandal was narrowly averted following reports that during 1994 the Government had issued loan guarantees for foreign companies worth more than $NZ1,200m. (The islands' revenue for 1994/95 was estimated to total $NZ50m.) An investigation into the affair by the New Zealand Reserve Bank found that the Government had not been guilty of fraud, but rather had been coerced into the activity by unscrupulous foreign business interests. However, the affair led many investors to remove their funds from the islands, provoking a financial crisis which resulted in Henry's decision, in mid-1995, to withdraw the Cook Islands dollar from circulation, and to implement a programme of retrenchment measures. The crisis deepened during 1995 as new allegations emerged and Henry's Government was severely criticized by New Zealand for failing to co-operate with an official inquiry into accusations of fraud and tax evasion by several New Zealand companies. Henry maintained that the islands' bank secrecy laws prevented the disclosure of information relating to financial transactions. The situation de- teriorated further when it was revealed that the Government had defaulted on a debt of some US $100m. to an Italian bank. In response to pressure from New Zealand and in an attempt to restore a degree of financial stability to the islands. Henry announced a severe restructuring programme in April 1996. The measures included a 50% reduction in the pay of public-sector workers, the closure of almost all diplomatic missions overseas, a 60% reduction in the number of government departments and ministries and the privatization of the majority of government-owned authorities. A marked increase in 1995/96 in the emigration rate and a decline in the number of Cook Islanders returning to the islands following a period of residency overseas was attributed to the austere economic conditions created by the financial crisis.

Criticism of the Government's management of the austerity programme continued both from within the Government and the opposition in late 1996 and early 1997. However, aid donors praised the Government for the success of the reform programme at a consultative meeting in July 1997.

In August 1997 Parliament approved the Outer Islands Local Government Act, providing for a new budgetary system to allocate funds for projects in the outer islands and for increased powers for local authorities, was expected to result in a significant reduction in central government administration of the outer islands. As part of the plan, three new government bodies were elected in April 1998.

Henry's administration continued to attract controversy, with the announcement in December 1997 of the closure of the Ministry of Public Works, Survey, Housing, Water Supply and Environment Services for exceeding its budget. The minister responsible, Tihina Tom Marsters, resigned in protest at the closure, which resulted in the loss of more than 100 public servants' jobs, problems with the supply of utilities (particularly water) and the suspension of several development projects.

In November 1997 the northern Cook Islands were devastated by Cyclone Martin, which killed at least eight people and destroyed virtually all crops and infrastructure. The islands' important black pearl industry suffered severe losses as a result of extensive damage on Manihiki Atoll.

In April 1999 Chief Justice Sir Peter Quilliam ruled as unconstitutional an electoral amendment, put forward by the ruling CIP, to ban any electoral activity, including the formation of political parties. The amendment was contested by the Democratic Alliance Party (DAP, a grouping including the DP) and NAP, which claimed that it contravened fundamental human rights and freedoms.

At legislative elections in June 1999 the CIP won 11 of the 25 seats in Parliament, the DAP 10 seats, and the New Alliance Party (NAP, formerly the Alliance Party) four seats. Sir Geoffrey Henry of the CIP was reappointed Prime Minister and formed a new Cabinet, following the establishment of a political coalition with the NAP. The leader of the NAP, Norman George, became Deputy Prime Minister. With a total of 15 seats in Parliament, it was hoped that the new coalition's accession to power would result in greater political stability. However, three members of the CIP subsequently left the party to form a coalition with the DAP, in protest at the alliance with the NAP, and at the end of July Henry resigned and was replaced by a rebel CIP member, Dr Joe Williams. Williams was confirmed as the new Prime Minister by 13 votes to 12 in a vote of confidence by the Parliament. Williams' appointment provoked a public protest on Rarotonga, exacerbated by general discontent at the nomination of a Prime Minister whose parliamentary constituency was outside the Cook Islands. Electors also voted in a referendum on whether the parliamentary term should be reduced from five years to four. The shorter term was favoured by 63% of voters, and therefore failed to receive the support of the two-thirds' majority required to amend the Constitution. The result of the contest for the Pukapuka seat was challenged by the DAP. The matter was taken to the Court of Appeal, which subsequently declared the result invalid, stripping the Government of its one-seat majority. A by-election was held in late September to decide the Pukapuka seat; however, the result was again deemed to be invalid, and a further by-election was held in November 2000 (although the result of this poll was being disputed in the High Court in late 2000). The Government became a minority administration in mid-October when the Prime Minister dismissed his deputy, Norman George, along with the Minister of Education, following their defection to the opposition. Despite the appointment of three new ministers, Williams failed to regain a majority in Parliament. In November Williams resigned, shortly before a vote of 'no confidence' was to be tabled against him by Dr Terepai Maoate, the leader of the opposition DAP. Maoate won the vote by 14 votes to 11 and was appointed Prime Minister, forming a new coalition Government with the NAP. He subsequently reappointed Norman George to the post of Deputy Prime Minister.

In February 2001 Maoate removed the transport portfolio from among the responsibilities of Dr Robert Woonton (whose other duties included that of Minister of Foreign Affairs). Maoate underwent medical treatment in March, prompting speculation that he might stand down as Prime Minister later in the year. In July Maoate dismissed Norman George, alleging that he was working to undermine him; this was the second time that George had lost the position of Deputy Prime Minister. He was replaced by Dr Robert Woonton. Woonton, however, strongly criticized Maoate's leadership in the same month.

In late 2001 the rift between the Prime Minister and his Cabinet widened. Woonton announced his resignation, which Maoate refused to accept. This led to a motion of 'no confidence' in the Prime Minister, which he only narrowly survived. In February 2002 Maoate's leadership was again challenged: 15 of the 25 Members of Parliament voted against him in a second motion of 'no confidence'. He was therefore replaced by Robert Woonton. In an extensive ministerial reorganization Sir Geoffrey Henry returned to the Cabinet as Deputy Prime Minister. Woonton declared his priorities to be the encouragement of emigrant workers to return to the islands through income tax incentives and an ambitious redevelopment plan for the capital, Avarua.

The Government's decision not to name a senior civil servant in the Prime Minister's office who had been arrested and charged with fraud in August 2002 led to accusations of secrecy by the opposition. Criticism of the Government's position increased in the following month with the introduction of new media laws (which many observers believed would serve to suppress opposition to government policy), particularly when a government official stated his desire for 'Zimbabwe-style legislation to stop inaccurate reporting'.

In November 2002 Parliament approved a constitutional amendment abolishing the requirement for electoral candidates to reside in the islands for a qualifying period of three months. This action was widely interpreted as a way of retaining the overseas parliamentary seat for the CIP leader, Dr Joe Williams, who lived permanently in New Zealand. In the same month the CIP and the DP formed a coalition government (the fifth such coalition since the previous election) which left Norman George, who had been recently dismissed from his position in the Cabinet, as the sole opposition member of Parliament. The Government's action prompted a demonstration outside the parliament building by some 150 people, organized by the recently formed Group for Political Change. The protesters claimed that the virtual absence of an opposition constituted an erosion of democracy and appealed to the Prime Minister to commit to an early general election. However, in late January 2003 the CIP was ousted from the coalition. The continued political manoeuvring was widely denounced, particularly among the business community, for creating a climate of instability in the islands. Public dissatisfaction with the situation resulted, in March, in the presentation to the Government of a petition signed by a significant percentage of the population. The petition called for a number of political reforms, including a reduction in the number of members of Parliament, the introduction of a shorter parliamentary term and the abolition of the overseas seat. Moreover, businessman Teariki Heather announced the formation of a new political party, the Cook Islands National, in the same month. In September legislation was approved providing for the abolition of the overseas seat and for a referendum (to be held concurrently with the next general election) on a proposal to shorten the parliamentary term from five years to four.

In early November 2003 the Deputy Prime Minister, Terepai Maoate, and the Minister of Justice, Tangata Vavia, resigned following an unsuccessful attempt by Maoate to propose a motion of 'no confidence' in the Government. The Government was the focus of further criticism in December when about 200 people marched through Avarua to protest at the granting of a residency permit to New Zealander Mark Lyon. The protesters claimed that Lyon, a wealthy businessman with recent convictions for weapons possession and a reputation for behaviour deemed disrespectful to island traditions, was not a suitable candidate for residency in the islands. Moreover, the demonstrators demanded the resignation of the Prime Minister and his chief adviser, Norman George, over the matter, stating that the islands should not be obliged to host undesirable characters, regardless of their ability to finance important business projects.

The reputation of the public service suffered a reversal in November 2003 when the former Chief of Staff in the Office of the Prime Minister, Edward Drollet, was convicted of seven charges of receiving secret commissions and one of forgery and was sentenced to more than two years' imprisonment.

At a general election in September 2004 the DP won 14 seats, the CIP secured nine and an independent candidate won the remaining seat. Prime Minister Robert Woonton regained his seat by only four votes. In a referendum held concurrently, 82.3% of participating voters indicated their support for the shortening of the parliamentary term from five years to four.

The rate of emigration, meanwhile, continued to increase. In late 2000 it was announced that some 1,400 residents had left the islands during that year (compared with 641 in the previous year). This resulted in a reduction in the estimated population of the islands to 14,300 (the lowest level in more than 50 years) and prompted the Government to initiate campaigns in Australia and New Zealand to encourage former residents to return to their homeland. Private-sector businesses, many of which had experienced difficulties in recruiting workers in sufficient numbers, were also involved in the campaign. The resident population was estimated to number only 12,900 in December 2001, and the Government feared that it could decrease to just 10,000 by 2006, the majority of the loss being from the outer islands. In August 2002 the Government announced that it would allocate US $23,350 for the campaign.

In early 2000 the islands of Penrhyn, Pukapuka, Rakahanga and Manihiki expressed their desire to become fully devolved and to take sole control over areas such as administration, public expenditure and justice. In response, the Government pledged gradually to phase out the Minister of Outer Islands Development, as well as the post of Government Representative in the outer islands. In December of that year an additional US $2m. in funding under the Cotonou Agreement with the European Union (EU) was designated for projects on the outer islands.

A reported increase in the number of Russian nationals opening bank accounts in the Cook Islands led to allegations in early 1999 that the islands' 'offshore' financial centre was being used extensively by criminal organizations for 'laundering' the proceeds of their activities. The claims were vigorously denied by officials in the sector. However, in June 2000 the islands were named by both the Paris-based Organisation for Economic Co-operation and Development (OECD) and by the Financial Action Task Force (FATF) on Money Laundering as one of a number of countries that had failed to co-operate in international efforts to combat 'harmful' tax havens and the practice of 'money-laundering'. This led to increased pressure on the Government to implement stricter controls over its 'offshore' financial centre. Consequently, legislation was approved in August of that year providing for the creation of the Money Laundering Authority and the introduction of new regulations aimed at reducing criminal activity in the sector. However, in mid-2004 the Cook Islands remained on the FATF list of non-co-operative countries and territories.

New Zealand is ultimately responsible for the defence and foreign relations of the Cook Islands, although the territory has progressively assumed control over much of its foreign policy (a Ministry of Foreign Affairs was established in 1983). The Cook Islands is a member of the Pacific Islands Forum (formerly the South Pacific Forum) in its own right. In August 1985 eight members of the South Pacific Forum, including the Cook Islands, signed a treaty on Rarotonga, designating a nuclear-free zone in the South Pacific. The treaty imposed a ban on the manufacture, testing, storage and use of nuclear weapons, and the dumping of nuclear waste, in the region.

In January 1986, after the virtual disintegration of the ANZUS military alliance linking Australia, New Zealand and the USA, Sir Thomas Davis declared the Cook Islands a neutral country, because he considered that New Zealand was no longer in a position to defend the islands. In 1989 and 1990 the Henry Government sought to improve links with the neighbouring territory of French Polynesia (Cook Islanders and Tahitians are related) and secured French co-operation in the policing of the Cook Islands' exclusive economic zone (EEZ). In 1990 an agreement was signed that settled the exact delimitation of the EEZs of the Cook Islands and French Polynesia (the two claims overlapped). In 1991 the Cook Islands signed a treaty of friendship and co-operation with France. The establishment of closer relations with France was widely regarded as an expression of the Cook Islands' Government's dissatisfaction with existing arrangements with New Zealand. However, relations deteriorated considerably when the French Government resumed its programme of nuclear-weapons testing at Mururoa Atoll in September 1995. Henry was fiercely critical of the decision and dispatched a vaka (traditional voyaging canoe) with a crew of Cook Islands' traditional warriors to protest near the test site. The tests were concluded in January 1996, and diplomatic relations between the two countries were established in early 2000.

The islands established diplomatic relations at ambassadorial level with the People's Republic of China in July 1997. In November 1998 Henry made an official visit to China, during which the two countries signed a bilateral trade agreement and each conferred the status of 'most favoured nation' on the other. Henry stated that the move constituted a further attempt by his Government to reduce the islands' dependence on New Zealand. During her visit to the islands in June 2001, Helen Clark, the New Zealand Prime Minister, stated that if the Cook Islands desired complete independence and membership of international organizations, the process would not be obstructed by New Zealand. Cook Islanders would then, however, be obliged to renounce their New Zealand citizenship.

Economy

In 2002, according to the Asian Development Bank (ADB), the Cook Islands' gross domestic product (GDP), measured at current prices, totalled an estimated $NZ220.9m. GDP increased, in real terms, at an average annual rate of 1.9% in 1990–2001. In 2003 GDP per head was estimated at $NZ14,377, representing an increase of 9.7% compared with the previous year. According to ADB figures, GDP rose by 3.9% in 2002 and by 3.1% in 2003. A growth rate of 2.7% was forecast for 2004.

Agriculture has more than subsistence importance on the southern islands, with their fertile volcanic soil. According to estimates by the ADB, agriculture (including forestry and fishing) contributed 11.6% of GDP in 2002. In that year the sector engaged some 38% of the economically active population, according to FAO estimates. In 2000 only 35% of Rarotonga households were classified as being agriculturally active, compared with 74% in the southern outer islands. According to figures from the ADB, the real GDP of the agricultural sector was estimated to have contracted by 3.3% per year in 1990–2001. Compared with the previous year, the sector's GDP decreased by 2.9% in 2001 but increased by 9.4% in 2002. In that year the sector provided 32.6% of export earnings (compared with 5.3% in the previous year). Papaya is the Cook Islands' most important export crop, and accounted for 24.9% of total export earnings in 1996. Papaya exports were worth some $NZ250,000 in 1998. Most of the produce is sold to New Zealand. Other important cash crops are coconuts and tropical fruits such as mangoes, pineapples and bananas. Aitutaki Island is important for the production of bananas. Cassava, sweet potatoes and vegetables are cultivated as food crops. Pigs and poultry are kept. Plans to develop cattle-ranching, ostrich farming and the cultivation of vanilla, taro, coffee and arrowroot on Atiu, Mangaia and Mauke have been discussed.

Exploitation of the Cook Islands' maritime exclusive economic zone (EEZ) became a major earner of foreign exchange in the 1980s. The Cook Islands' EEZ, extending to 200 nautical miles (370 km) from the territory's shores, is monitored under Australia's regional fisheries programme, and also by the French Government. The sale of fishing licences to foreign fleets provides an important source of income. However, illegal fishing in the islands' exclusive economic zone increased during the 1990s, and in March 2003 the Government announced harsher penalties to deal with this problem. Aquaculture, in the form of giant clam farming and pearl oyster farming, was developed during the 1980s. The pearl industry expanded considerably during the 1990s. Pearl oyster farming at Manihiki and Penrhyn Island was the islands' most important industry and pearls were the most important export commodity by 2000, when they contributed 92.1% of total export earnings. The industry was adversely affected, however, by Cyclone Martin which devastated Manihiki Atoll in late 1997 and by a bacterial pearl shell disease in 2000. In response to the infection, the pearl industry agreed to a number of measures designed to protect the environment and reduce overfarming. In 2002 pearl exports earned $NZ6.4m., equivalent to 58.6% of total earnings. In early 2001 an Australian fishing company announced plans to invest some US $60m. (including the purchase of 25 long-line fishing vessels and the construction of a fish-processing facility) in a major fishing project in the islands. Revenue from the export of fresh and chilled fish was estimated at more than $NZ12m. in 2003, almost double the earnings of the previous year. Some 44 fishing vessels had been in operation during 2003, compared with 19 in 2002.

According to ADB estimates, industry (comprising mining and quarrying, manufacturing, construction and power) provided 8.1% of GDP in 2002. The sector engaged 12.1% of employees in 1993. Industrial GDP increased, in real terms, at an average rate of 2.4% per year during 1990–2001. Compared with the previous year, the industrial sector's GDP increased by some 13.3% in 2001 and by 5.0% in 2002.

Manufacturing contributed 4.1% of GDP in 1995, and engaged 4.5% of employees in 1993. The manufacturing and mining sectors together accounted for an estimated 3.3% of GDP in 2002. The real GDP of manufacturing and mining declined at an average rate of 0.2% per year during 1990–2001. Compared with the previous year, however, the two sectors' GDP increased by 3.5% in 2001, before decreasing by 2.6% in 2002. The most important industrial activities are fruit-processing, brewing, the manufacture of garments and handicrafts. Construction contributed an estimated 3.3% of GDP in 2002 and engaged 3.4% of the employed labour force in 1993.

The islands depend on imports for their energy requirements. Mineral fuels accounted for 21.9% of total imports in 2003. In September 1997 the Government signed an agreement with a consortium of Norwegian companies to mine cobalt, nickel, manganese and copper by extracting mineral-rich nodules found in the islands' EEZ between Aitutaki and Penrhyn. It was estimated that the deep-sea mining project, which was expected to begin in 2003/04, could earn the islands up to US $15m. per year and US $600m. in total. Trial operations began in 1999.

Service industries contributed an estimated 80.2% to GDP in 2002 and engaged 80.8% of the employed labour force in 1993. According to the ADB, the service sector's GDP expanded by 15.4% in 2000, by an 5.0% in 2001 and by 2.8% in 2002. Tourism expanded considerably in the late 1980s and early 1990s, and earned an estimated $NZ93.9m. in 2000/01. Visitor arrivals rose from 72,781 in 2002 to 78,328 in the following year. The trade, restaurants and hotels sector contributed 38.2% of GDP in 2002 and engaged 20.9% of the employed labour force in 1993. The withdrawal of Canada 3000's weekly flight from Toronto (following the airline's financial failure in late 2001) and a reduction in Air New Zealand's capacity, combined with the repercussions of the terrorist attacks on the USA in September 2001, had an adverse effect on the Cook Islands' tourism industry. Nevertheless, the islands were expected to retain their appeal as a relatively safe destination.

'Offshore' banking, introduced to the islands in 1982, expanded rapidly, with more than 2,000 international companies registered by 1987. In 1992 the islands were established as an alternative domicile for companies listed on the Hong Kong Stock Exchange. The financial and business services sector provided 8.1% of GDP in 2002 and engaged 3.6% of the employed labour force in 1993. A significant proportion of the islands' revenue is provided by remittances from migrants (who outnumber the residents of the islands).

The Cook Islands suffer from a persistent trade deficit, which increased to more than 70% of GDP in the 1990s . In 2002 the islands recorded a trade deficit of $NZ91.2m. In the same year there was a surplus on the current account of the balance of payments equivalent to 6.3% of GDP. This figure declined slightly to an estimated 5.9% in 2003. In 2003 the principal imports were machinery and transport equipment (which accounted for 32.4% of total imports), food and live animals (20.7%) and basic manufactures (16.9%), while food and live animals were the main export, accounting for 74.9% of total exports. In that year New Zealand provided 78.2% of the islands' total imports, while Japan and the USA were the major purchasers of the islands' exports, accounting for 32.4% and 27.3% respectively. Despite the large trade deficit, the islands consistently record a surplus on the current account of the balance of payments. This is primarily due to the role of tourism in generating income for the islands. Remittances sent back by migrant workers account for a large proportion of the islands' income.

According to official estimates, in the financial year ending June 2003 there was an overall budgetary deficit of $NZ17.2m. In mid-2003 the Government announced a budget that emphasized reduced spending, as well as the establishment of a reserve trust fund with US $185,000. Development assistance from New Zealand totalled $NZ6.2m. in 1999/2000, and was to remain at this level for the following three financial years (New Zealand is the guarantor of the Cook Islands' borrowing from the ADB, which had reached a total of $NZ25m. at 31 October 2001). Aid from New Zealand totalled $NZ6.24m. in 2003/04. Development assistance from Australia totalled $A10.4m. in 2002/03. The islands were also to receive $NZ1m. per year between 2003 and 2008 (to be spent on education, health and outer islands development) under the Cotonou Agreement with the European Union. It was estimated at the end of 2003 that the islands' external debt amounted to some $NZ109.7m., equivalent to 48.6% of GDP. The annual rate of inflation averaged 3.1% in 1990–2001. Average consumer prices rose by 3.9% in 2002 and by 2.4% in 2003. The ADB estimated the unemployment rate to be 12.7% in 1998. The unemployment rate had fallen to 9.0% by 2001, however, owing to increased demand for workers in the tourism and retail sectors.

During the 1990s and early 2000s development plans sought to expand the economy by stimulating investment in the private sector and developing the islands' infrastructure. In mid-1995 it was announced that the New Zealand dollar was to become the sole legal currency of the islands, following a financial crisis that led the Government to withdraw the Cook Islands dollar from circulation and to introduce a retrenchment programme. Between 1996 and 1998 the number of public-sector workers was reduced from 3,350 to 1,340, and the number of ministries from 52 to 22. (By early 2004, however, there were renewed concerns for the islands' financial welfare when it was reported that the number of public-sector employees had increased to some 1,800 and public payroll expenditure had risen by 30% in only two years.)

Formal discussions, organized by the ADB, took place between the Cook Islands and its creditors in September 1998, at which a preliminary agreement to reschedule the islands' external debts was reached. The Government of France (a major creditor to the islands) refused to participate in the arrangement. The Government subsequently adopted a policy of accumulating reserves for the purposes of debt-servicing as an annual budgetary allocation. Plans to restructure the 'offshore' sector, announced in 1998, were expected to increase revenue. However, the islands' financial regulations were severely criticized by the Financial Action Task Force (FATF—see History), which in late 2004 continued to include the Cook Islands on its list of countries and territories, the banking systems of which were allegedly being used for the purpose of 'money-laundering'.

Large-scale emigration remained a serious and deepening concern for the Cook Islands' economy in the early 2000s. In August 2002 the Government announced that it would allocate US $23,350 to the campaign to encourage islanders resident abroad to return. Meanwhile, the success of the tourist industry and the dramatic increase in arrivals to the islands led to expressions of concern that Rarotonga, in particular, was unable to sustain the growth. Reports indicated that waste disposal and energy provision were inadequate in relation to the demands of large numbers of visitors and that pollution of the lagoon was occurring as a result. In January 2003 the Cook Islands Tourism Corporation appealed for a moratorium on all new tourist projects following an independent report into the environmental and social impact of the industry on the islands. However, a project to expand the runway at Rarotonga airport, allowing for the arrival of larger aircraft, was expected to be completed by 2005.

In addition to holding membership of the Pacific Islands Forum (formerly the South Pacific Forum), the Cook Islands is a member of the Pacific Community (formerly the South Pacific Commission), the Asian Development Bank and an associate member of the UN's Economic and Social Commission for Asia and the Pacific (ESCAP). The Cook Islands has the status of observer to the Lomé Convention and successor Cotonou Agreement with the European Union.

Statistical Survey

Sources (unless otherwise stated): Statistics Office, POB 125, Rarotonga; tel. 29390; Prime Minister's Department, Government of the Cook Islands, Avarua, Rarotonga; tel. 29300; fax 22856.

AREA AND POPULATION

Area: 237 sq km (91.5 sq miles).

Population: 19,103 (males 9,842, females 9,261) at census of 1 December 1996; 18,027 (males 9,303, females 8,724) at census of December 2001 (provisional). *Resident Population:* 18,034 (males 9,297, females 8,737) at 1996 census; 14,990 (males 7,738, females 7,252) at 2001 census. *Mid-2002* (estimate): 18,400. *By Island* (resident population, 1996 census): Rarotonga (including the capital, Avarua) 10,337; Aitutaki 2,272; Atiu 942; Mangaia 1,083; Manihiki 656; Mauke 643; Mitiaro 318; Nassau 99; Palmerston (Avarua) 49; Penrhyn (Tongareva) 604; Pukapuka 778; Rakahanga 249; Suwarrow 4. *Cook Island Maoris Resident in New Zealand* (census of 6 March 2001): 52,569.

Density (resident population, mid-2002): 77.6 per sq km.

Principal Town (UN population estimate at mid-2003, incl. suburbs): Avarua (capital) 12,507. Source: UN, *World Urbanization Prospects: The 2003 Revision.*

Births and Deaths (2002): Registered live irths: 224 (birth rate 21.2 per 1,000); Registered deaths: 72 (death rate 3.9 per 1,000). Source: UN, *Population and Vital Statistics Report.*

Expectation of Life (WHO estimates, years at birth): 71.6 (males 69.2; females 74.2) in 2001. Source: WHO, *World Health Report.*

Employment (September 1993): Agriculture, hunting, forestry and fishing 457; Mining and quarrying 16; Manufacturing 290; Electricity, gas and water 254; Construction 215; Trade, restaurants and hotels 1,338; Transport, storage and communications 770; Financing, insurance, real estate and business services 231; Community, social and personal services 2,835; *Total employees* 6,406 (males 4,069, females 2,337) (Source: ILO, *Yearbook of Labour Statistics*). *1996* (rounded figures): Agriculture, hunting, forestry and fishing 600; Manufacturing, mining and quarrying 300; Activities not adequately defined 4,400; Total employed 5,200; Unemployed 800; Total labour force 6,000 (Source: Asian Development Bank, *Key Indicators of Developing Asian and Pacific Countries*).

HEALTH AND WELFARE

Key Indicators

Total Fertility Rate (children per woman, 2002): 3.2.

Under-5 Mortality Rate (per 1,000 live births, 2002): 23.

Physicians (per 1,000 head, 1997): 0.90.

Health Expenditure (2001): US $ per head (PPP): 598.

Health Expenditure (2001): % of GDP: 4.7.

Health Expenditure (2001): public (% of total): 67.6.

Access to Water (% of persons, 2000): 100.

Access to Sanitation (% of persons, 2000): 100.

For sources and definitions, see explanatory note on p. vi.

AGRICULTURE, ETC.

Principal Crops (mainly FAO estimates, metric tons, 2002): Cassava 3,000; Sweet potatoes 1,400; Coconuts 5,000; Vegetables and melons 1,802; Tomatoes 500; Mangoes 2,700; Papayas 908; Pineapples 50; Bananas 230; Avocados 150. Source: FAO.

Livestock (mainly FAO estimates, head, year ending September 2002): Pigs 40,000; Goats 2,500; Poultry 80,000; Horses 300. Source: FAO.

Livestock Products (metric tons, 2002): Hen eggs 250; Goatskins 2; Pigmeat 600; Poultry meat 80. Source: FAO.

Fishing (FAO estimates, metric tons, live weight, 2002): Total catch 1,768 (Groupers 60; Flyingfishes 30; Jacks and crevalles 40; Marlins and sailfishes 62; Swordfish 62; Sharks, rays and skates 20; Octopuses 30; Other marine molluscs 120). Figures exclude trochus shells: 15 metric tons and aquatic plants 50 metric tons (FAO estimates). Source: FAO.

INDUSTRY

Electric Energy (million kWh): 26 in 2001; 28 in 2002; 29 in 2003.

FINANCE

Currency and Exchange Rates: New Zealand currency is legal tender. In mid-1995 it was announced that the Cook Islands dollar (formerly the local currency, at par with the New Zealand dollar) was to be withdrawn from circulation. New Zealand currency: 100 cents = 1 New Zealand dollar ($NZ); for details of exchange rates, see Tokelau.

Budget ($NZ '000, year ending 30 June 2003): *Revenue:* Total revenue 70,976 (Tax 61,865, Non-tax 2,932, Capital 6,179). Excludes grants (10,000). *Expenditure:* Total expenditure 88,128 (Current 82,200, Capital 5,928).

Overseas Aid ($NZ '000): Official development assistance from New Zealand 6,200 in 2001/02 and in 2002/03; 6,240 in 2003/04. Source: Ministry of Foreign Affairs and Trade, Wellington.

Cost of Living (Consumer Price Index for Rarotonga; base: 1998 = 100): 113.4 in 2001; 117.2 in 2002; 119.5 in 2003.

Gross Domestic Product by Economic Activity (estimates, $NZ '000 in current prices, 2002): Agriculture, forestry and fishing 26,299; Mining, quarrying and manufacturing 7,477; Electricity, gas and water 3,406; Construction 7,585; Trade, restaurants and hotels 86,578; Transport and communications 32,387; Financial and business services 18,362; Public administration 25,577; Other services 18,964; *Sub-total* 226,635; *Less* Imputed bank service charge 5,739; *GDP in purchasers' values* 220,896. Source: Asian Development Bank, *Key Indicators of Developing Asian and Pacific Countries*.

EXTERNAL TRADE

Principal Commodities (distribution by SITC, $NZ '000, 2003): *Imports c.i.f.:* Food and live animals 25,077; Mineral fuels, lubricants, etc. 5,495; Chemicals 7,513; Basic manufactures 20,458; Machinery and transport equipment 39,220; Miscellaneous manufactured articles 15,697; Total (incl. others) 121,021. *Exports f.o.b.:* Food and live animals 10,917; Basic manufactures 2,843; Miscellaneous manufactured articles 656; Total (incl. others) 14,579. Source: Asian Development Bank, *Key Indicators of Developing Asian and Pacific Countries*.

Principal Trading Partners ($NZ '000, 2003): *Imports:* New Zealand 94,665; Australia 7,954; USA 2,956; Japan 3,444; Fiji 5,873; Total (incl. others) 121,021. *Exports:* Japan 4,722; New Zealand 3,056; Australia 1,010; USA 3,973; Total (incl. others) 14,579. Source: Asian Development Bank, *Key Indicators of Developing Asian and Pacific Countries*.

TRANSPORT

Road Traffic (registered vehicles, April 1983): 6,555. *New Motor Vehicles Registered* (Rarotonga, 2001): 1,698.

Shipping: *Merchant Fleet* (registered at 31 December 2003): 42 vessels, displacement 17,600 grt (Source: Lloyd's Register-Fairplay, *World Fleet Statistics*); *International Sea-borne Freight Traffic* (estimates, '000 metric tons): Goods unloaded 32.6 (2001); Goods loaded 9; Goods unloaded 32 (1990) (Source: UN, *Monthly Bulletin of Statistics*).

Civil Aviation (2001): *Aircraft Movements*: 559. *Freight Traffic* (metric tons): Goods loaded 290; Goods unloaded 659.

TOURISM

Foreign Tourist Arrivals: 74,575 in 2001; 72,781 in 2002; 78,328 in 2003.

Tourist Arrivals by Country of Residence (2000): Australia 12,128; Canada 5,992; Germany 5,232; New Zealand 22,020; United Kingdom 12,392; USA 6,734; Total (incl. others) 72,994. Source: World Tourism Organization, *Yearbook of Tourism Statistics*.

Visitor Expenditure (estimates, $NZ '000): 73,753 in 1999/2000; 93,929 in 2000/01.

COMMUNICATIONS MEDIA

Radio Receivers (1997): 14,000 in use*.

Television Receivers (1997): 4,000 in use*.

Telephones (main lines, 2000): 5,680 in use†.

Mobile Cellular Telephones (2000): 552 subscribers†.

Facsimile Machines (1990): 230 in use‡.

Daily Newspaper (1996): 1; circulation 2,000*.

Non-daily Newspaper (1996): 1; circulation 1,000*.

* Source: UNESCO, *Statistical Yearbook*.
† Source: International Telecommunication Union.
‡ Source: UN, *Statistical Yearbook*.

EDUCATION

Pre-primary (1998): 26 schools; 30 teachers; 460 pupils.

Primary (1998): 28 schools; 140 teachers; 2,711 pupils.

Secondary* (1998): 23 schools; 129 teachers; 1,779 pupils.

Higher (1980): 41 teachers; 360 pupils†.

* Includes high school education.
† Source: UNESCO, *Statistical Yearbook*.

Directory

The Constitution

On 5 August 1965 a new Constitution was proclaimed, whereby the people of the Cook Islands have complete control over their own affairs in free association with New Zealand, but they can at any time move into full independence by a unilateral act if they so wish.

Executive authority is vested in the British monarch, who is Head of State, and exercised through an official representative. The New Zealand Government also appoints a representative (from 1994 redesignated High Commissioner), resident on Rarotonga.

Executive powers are exercised by a Cabinet consisting of the Prime Minister and between five and seven other ministers including a Deputy Prime Minister. The Cabinet is collectively responsible to Parliament.

Legislation approved in September 2003 resulted in the abolition of the seat for one member elected by voters living overseas and consequently Parliament consists of 24 members elected by universal suffrage and presided over by the Speaker. Moreover, as a result of a referendum held concurrently with the general election of September 2004, the parliamentary term was shortened from five years to four. The House of Ariki comprises up to 15 members who are hereditary chiefs; it can advise the Government, particularly on matters relating to land and indigenous people but has no legislative powers. The Koutu Nui is a similar organization comprised of sub-chiefs, which was established by an amendment in 1972 of the 1966 House of Ariki Act.

Each of the main islands, except Rarotonga (which is divided into three tribal districts or *vaka*), has an elected mayor and a government representative who is appointed by the Prime Minister. In January 2000 it was announced that the post of Government Representative in the outer islands was to be phased out over two years.

The Government

Queen's Representative: Fred Goodwin.

New Zealand High Commissioner: Sir Thomas Davis.

THE CABINET
(September 2004)

Prime Minister, Minister of Foreign Affairs and Immigration, Tourism, House of Ariki, Airport Authority, Ports Authority, Telecom and Information Services and Broadcasting and Information: Dr Robert Woonton.

Deputy Prime Minister, Minister of Transport, Civil Aviation and Shipping, Parliamentary Services, National Research and Development Institute, National Development Council, Crown Law and Attorney-General: Ngamau Munokoa.

Minister of Agriculture, Justice, Works and Marine Resources: Robert Wigmore.

Minister of Education, Culture, Public Service Commission, Natural Heritage, Ombudsman and Human Resources: Jim Marurai.

Minister of Finance, Police, National Disaster Management, Energy, Outer Islands' Development, National Superannuation, Public Expenditure Review and Audit, Bank of the Cook Islands, Revenue Management and Development Investment Board: Tapi Taio.

Minister of Health, Internal Affairs, Youth, Sports and Recreation, Non-Government Organizations, Information Technology and Commerce Commission: Peri Vaevae Pare.

GOVERNMENT OFFICES

Office of the Queen's Representative: POB 134, Titikaveka, Rarotonga; tel. 29311.

Office of the Prime Minister: Government of the Cook Islands, Avarua, Rarotonga; tel. 21150; fax 23792; e-mail coso@pmoffice.gov.ck.

Office of the Public Service Commissioner: POB 24, Rarotonga; tel. 29421; fax 21321; e-mail pscom@oyster.net.ck.

Department of Justice: POB 111, Rarotonga; tel. 29410; fax 29610; e-mail justice@cookislands.gov.ck.

Department of Tourism and Transport: POB 61, Rarotonga; tel. 28810; fax 28816; e-mail tourism@cookislands.gov.ck.

New Zealand High Commission: 1st Floor, Philatelic Bureau Bldg, Takuvaine Rd, POB 21, Avarua, Rarotonga; tel. 22201; fax 21241; e-mail nzhraro@oyster.net.ck.

Ministries

Ministry of Agriculture: POB 96, Rarotonga; tel. 28711; fax 21881; e-mail cimoa@oyster.net.ck.

Ministry of Cultural Development: POB 8, Rarotonga; tel. 20725; fax 23725; e-mail culture1@oyster.net.ck; internet www.cinews.co.ck/culture/index.htm.

Ministry of Education: POB 97, Rarotonga; tel. 29357; fax 28357; e-mail dieducat@oyster.net.ck.

Ministry of Energy: POB 72, Rarotonga; tel. 24484; fax 24485.

Ministry of Finance and Economic Management: POB 120, Rarotonga; tel. 22878; fax 23877; e-mail finsec@oyster.net.ck.

Ministry of Foreign Affairs and Immigration: POB 105, Rarotonga; tel. 29347; fax 21247; e-mail secfa@foraffairs.gov.ck.

Ministry of Health: POB 109, Rarotonga; tel. 22664; fax 23109; e-mail aremaki@oyster.net.ck.

Ministry of Internal Affairs: POB 98, Rarotonga; tel. 29370; fax 23608; e-mail sec1@intaff.gov.ck.

Ministry of Marine Resources: POB 85, Rarotonga; tel. 28721; fax 29721; e-mail rar@mmr.gov.ck.

Ministry of Outer Islands' Development: POB 383, Rarotonga; tel. 20321; fax 24321.

Ministry of Works and Physical Planning: POB 102, Rarotonga; tel. 20034; fax 21134; e-mail herman@mow.gov.ck.

HOUSE OF ARIKI

House of Ariki

POB 13, Rarotonga; tel. 26500; fax 21260.

President: Tou Trauel Ariki.

Vice-President: Tamatoa Ariki.

KOUTU NUI

Koutu Nui

POB 13, Rarotonga; tel. 29317; fax 21260; e-mail nvaloa@parliament.gov.ck.

President: Tetika Mataiapo Dorice Reid.

Legislature

PARLIAMENT

Parliamentary Service

POB 13, Rarotonga; tel. 26500; fax 21260; e-mail nvaloa@parliament.gov.ck.

Speaker: Sir Pupuke Robati.

Clerk of Parliament: Nga Valoa.

General Election, 7 September 2004

Party	Seats
Democratic Party (DP)	14
Cook Islands Party (CIP)	9
Independent	1
Total	24

Political Organizations

Cook Islands Labour Party: Rarotonga; f. 1988; anti-nuclear; Leader Rena Ariki Jonassen.

Cook Islands National: Rarotonga; f. 2003; Leader Teariki Heather.

Cook Islands Party (CIP): Rarotonga; f. 1965; Gen. Sec. Tihina Tom Marsters; Leader Dr Joe Williams.

Democratic Party (DP): POB 73, Rarotonga; tel. 21224; f. 1971; Leader Makiuti Tongia; Pres. Fred Goodwin.

Democratic Tumu Party (DTP): POB 492, Rarotonga; tel. 21224; fax 22520; split from Democratic Party in 1985; Leader Vincent A. K. T. Ingram.

Judicial System

Ministry of Justice

POB 111, Avarua, Rarotonga; tel. 29410; fax 28610; e-mail offices@justice.gov.ck.

The judiciary comprises the Privy Council, the Court of Appeal and the High Court.

The High Court exercises jurisdiction in respect of civil, criminal and land titles cases on all the islands, except for Mangaia, Pukapuka and Mitiaro, where disputes over land titles are settled according to custom. The Court of Appeal hears appeals against decisions of the High Court. The Privy Council, sitting in the United Kingdom, is the final appellate tribunal for the country in civil, criminal and land matters; since May 1993 the Court of Appeal has been the final appellate tribunal for disputes concerning chiefly titles.

Attorney-General: Norman George.

Solicitor-General: Janet Grace Maki.

Chief Justice of the High Court: Laurence Murray Greig.

Judges of the High Court: Glendyn Carter, Norman Smith, David Williams, Heta Hingston.

Religion

CHRISTIANITY

The principal denomination is the Cook Islands (Congregational) Christian Church, to which about 58% of the islands' population belong, according to figures recorded in the census conducted in 1996.

Religious Advisory Council of the Cook Islands: POB 31, Rarotonga; tel. 22851; fax 22852; f. 1968; four mem. churches; Pres. Kevin Geelan; Gen. Sec. Tungane Pokura.

The Roman Catholic Church

The Cook Islands form the diocese of Rarotonga, suffragan to the archdiocese of Suva (Fiji). At 31 December 2001 the diocese contained an estimated 2,396 adherents. The Bishop participates in the Catholic Bishops' Conference of the Pacific, based in Suva.

Bishop of Rarotonga: Rt Rev. Stuart France O'Connell, Catholic Diocese, POB 147, Rarotonga; tel. 20817; fax 29817; e-mail sbish@oyster.net.ck.

The Anglican Communion

The Cook Islands are within the diocese of Polynesia, part of the Church of the Province of New Zealand. The Bishop of Polynesia is resident in Fiji.

Protestant Churches

Cook Islands Christian Church: Takamoa, POB 93, Rarotonga; tel. 26452; 11,193 mems (1986); Pres. Rev. Tangimetua Tangatatuta; Gen. Sec. Willie John.

Seventh-day Adventists: POB 31, Rarotonga; tel. 22851; fax 22852; e-mail umakatu@oyster.net.ck; 732 mems (1998); Pres. Uma Katu.

Other churches active in the islands include the Assembly of God, the Church of Latter-day Saints (Mormons), the Apostolic Church, the Jehovah's Witnesses and the Baptist Church.

BAHÁ'Í FAITH

National Spiritual Assembly: POB 1, Rarotonga; tel. 20658; fax 23658; e-mail nsacooks@bahai.org.ck; mems resident in eight localities; Sec. Johnny Frisbie.

The Press

Cook Islands Herald: POB 126, Tutakimoa, Rarotonga; e-mail bestread@ciherald.co.ck; internet www.ciherald.co.ck; weekly; Publr George Pitt.

Cook Islands News: POB 15, Avarua, Rarotonga; tel. 22999; fax 25303; e-mail editor@cookislandsnews.com; internet www.cookislandsnews.com; f. 1954 by Govt, transferred to private ownership 1989; daily; mainly English; Man. Dir Phil Evans; Editor Moana Moeka'a; circ. 1,800.

Cook Islands Star: POB 798, Rarotonga; tel. 29965; e-mail jason@oyster.net.ck; fortnightly; Chief Reporter Jason Brown.

Cook Islands Sun: POB 753, Snowbird Laundry, Arorangi, Rarotonga; f. 1988; tourist newspaper; twice a year; Editor WARREN ATKINSON.

Broadcasting and Communications

TELECOMMUNICATIONS

Telecom Cook Islands Ltd: POB 106, Rarotonga; tel. 29680; fax 20990; e-mail stu@telecom.co.ck; internet www.telecom.co.ck; CEO STUART DAVIES.

BROADCASTING

Radio

Cook Islands Broadcasting Corpn (CIBC): POB 126, Avarua, Rarotonga; tel. 29460; fax 21907; f. 1989 to operate new television service, and radio service of former Broadcasting and Newspaper Corpn; state-owned; Gen. Man. EMILE KAIRUA.

Radio Cook Islands: tel. 20100; broadcasts in English and Maori 18 hours daily.

KC Radio: POB 521, Avarua, Rarotonga; tel. 23203; f. 1979 as Radio Ikurangi; commercial; operates station ZK1ZD; broadcasts 18 hours daily on FM; Man. Dir and Gen. Man. DAVID SCHMIDT.

Television

Cook Islands Broadcasting Corpn (CIBC): see Radio.

Cook Islands TV: POB 126, Rarotonga; tel. 20101; fax 21907; f. 1989; broadcasts nightly, in English and Maori, from 5 p.m. to 10.15 p.m.; 10 hours of local programmes per week; remainder provided by Television New Zealand.

In early 1999 the French Société Nationale de Radio-Télévision Française d'Outre-Mer agreed to operate two or three new television channels in the islands.

Finance

Commission of Offshore Financial Services: Rarotonga; tel. 20798; fax 21798; e-mail comm@oyster.net.ck; f. 2003; Commissioner MATHILDA UHRLE; Chair. TREVOR CLARKE.

Cook Islands Monetary Board: POB 594, Rarotonga; tel. 20798; fax 21798; f. 1981; exercises control of currency; controlling body for trade and industry as well as finance; registers companies, financial institutions, etc.; Sec. M. BROWN.

Trustee Companies Association (TCA): Rarotonga; controlling body for the 'offshore' financial sector; Sec. LOU COLVEY.

BANKING

Development Bank

Bank of Cook Islands (BCI): POB 113, Avarua, Rarotonga; tel. 29341; fax 29343; e-mail bci@oyster.net.ck; f. July 2001; when Cook Islands Development Bank merged with Cook Island Savings Bank; finances development projects in all areas of the economy and helps islanders establish small businesses and industries by providing loans and management advisory assistance; Gen. Man. UNAKEA KAUVAI; brs on Rarotonga and Aitutaki.

Commercial Banks

Australia and New Zealand (ANZ) Banking Corpn: 1st Floor, Development Bank Bldg, POB 907, Avarua, Rarotonga; tel. 21750; fax 21760; e-mail lancaster@gatepoly.co.ck; Man. GARY RUNCIMAN.

The Wall Street Banking Corp Ltd: POB 3012, CITC House, Avarua, Rarotonga; tel. 23445; fax 23446; e-mail info@wallbank.co.ck; internet www.wallstreetbankingcorp.com; f. 1992; 100% owned by Natar Holdings Co Ltd; cap. US $15.0m., dep. US $74.9m. (Mar. 2001); Exec. Dir RIAZ PATEL.

Westpac Banking Corpn (Australia): Main Rd, POB 42, Avarua, Rarotonga; tel. 22014; fax 20802; e-mail bank@westpac.co.ck; Man. TERRY SMITH.

Legislation was adopted in 1981 to facilitate the establishment of 'offshore' banking operations.

INSURANCE

Cook Islands Insurance: POB 44, Rarotonga.

Trade and Industry

GOVERNMENT AGENCIES

Cook Islands Development Investment Board: Private Bag, Avarua, Rarotonga; tel. 24296; fax 24298; e-mail cidib@cidib.gov.ck; internet www.cookislands-invest.com; f. 1996 as replacement for Development Investment Council; promotes, monitors and regulates foreign investment, promotes international trade, advises the private sector and Government and provides training in business skills; CEO MARK SHORT.

Cook Islands Investment Corporation: Rarotonga; tel. 29391; fax 29381; e-mail ciic@oyster.net.ck; f. 1998; manages government assets and shareholding interests; Chair. JOHN SHORT; CEO TARITA HUTCHINSON.

Cook Islands Public Service Commission: POB 24, Rarotonga; tel. 29421; fax 21321; e-mail pscom@oyster.net.ck; Commissioner MAINE BROWNE.

Cook Islands Trading Corporation: Avarua, Rarotonga; fax 20857.

CHAMBER OF COMMERCE

Chamber of Commerce: POB 242, Rarotonga; tel. 20925; fax 20969; f. 1956; Pres. EWAN SMITH.

INDUSTRIAL AND TRADE ASSOCIATIONS

Pearl Federation of the Cook Islands, Inc: Manihiki; tel. and fax 43363; f. 1995 following the dissolution of the govt-owned Cook Islands Pearl Authority; oversees the activities and interests of pearl-producers in the northern Cook Islands.

Pearl Guild of the Cook Islands: Rarotonga; e-mail trevon@oyster.net.ck; f. 1994; monitors standards of quality within the pearl industry and develops marketing strategies; Pres. TREVON BERGMAN.

UTILITIES

Electricity

Te Aponga Uira O Tumutevarovaro (TAUOT): POB 112, Rarotonga; tel. 20054; fax 21944.

Water

Water Supply Department: POB 102, Arorangi, Rarotonga; tel. 20034; fax 21134.

TRADE UNIONS

Airport Workers Association: Rarotonga Int. Airport, POB 90, Rarotonga; tel. 25890; fax 21890; f. 1985; Pres. NGA JESSIE; Gen. Sec. (vacant).

Cook Islands Industrial Union of Waterside Workers: Avarua, Rarotonga.

Cook Islands Workers' Association (CIWA): POB 403, Avarua, Rarotonga; tel. 24422; fax 24423; largest union in the Cook Islands; Pres. MIRIAMA PIERRE; Gen. Sec. NGAMETUA ARAKUA.

Transport

ROADS

On Rarotonga a 33-km sealed road encircles the island's coastline. A partly sealed inland road, parallel to the coastal road and known as the Ara Metua, is also suitable for vehicles. Roads on the other islands are mainly unsealed.

SHIPPING

The main ports are on Rarotonga (Avatiu), Penrhyn, Mangaia and Aitutaki. The Cook Islands National Line operates a three-weekly cargo service between the Cook Islands, Tonga, Samoa and American Samoa. In August 2002 the Government approved proposals to enlarge Avatiu Harbour. The project, work on which was due to start in September 2002, received additional funding from the Ports Authority and from New Zealand.

Apex Maritime: POB 378, Rarotonga; tel. 27651; fax 21138.

Cook Islands National Line: POB 264, Rarotonga; tel. 20374; fax 20855; 30% govt-owned; operates three fleet cargo services between the Cook Islands, Niue, Samoa, Norfolk Island, Tonga and New Zealand; Dirs CHRIS VAILE, GEORGE ELLIS.

Cook Islands Shipping Ltd: POB 2001, Arorangi, Rarotonga; tel. 24905; fax 24906.

Ports Authority: POB 84, Rarotonga and Aitutaki; tel. 21921; fax 21191; Chair. DON BEER.

Reef Shipping Company: Rarotonga; operates services between Rarotonga and Aitutaki.

Taio Shipping Ltd: Teremoana Taio, POB 2001, Rarotonga; tel. 24905; fax 24906.

Triad Maritime (1988) Ltd: Rarotonga; fax 20855.

CIVIL AVIATION

An international airport was opened on Rarotonga in 1974. Polynesian Airlines and Air New Zealand operate services linking Rarotonga with other airports in the region. Air Pacific (Fiji) began a twice-weekly service between Nadi and Rarotonga in June 2000, and in August of that year Air New Zealand began a direct service from Rarotonga to Los Angeles, USA.

Airport Authority: POB 90, Rarotonga; CEO JOE NGAMATA.

Air Rarotonga: POB 79, Rarotonga; tel. 22888; fax 23288; e-mail bookings@airraro.co.ck; f. 1978; privately owned; operates internal passenger and cargo services and charter services to Niue and French Polynesia; Man. Dir EWAN F. SMITH.

Tourism

Tourism is the most important industry in the Cook Islands, and there were 78,328 foreign tourist arrivals in 2003, compared with 25,615 in 1984. Of total visitors in 2000, 30.2% came from New Zealand, 16.6% from Australia and 17.0% from the United Kingdom. There were 1,874 beds available at hotels and similar establishments in the islands in 1999. Most of the tourist facilities are to be found on Rarotonga and Aitutaki, but the outer islands also offer attractive scenery. Revenue from tourism was estimated at some $NZ93.9m. in 2000/01.

Cook Islands Tourism Corporation: POB 14, Rarotonga; tel. 29435; fax 21435; e-mail tourism@cookislands.gov.ck; internet www.cook-islands.com; CEO CHRIS WONG.

Education

Free secular education is compulsory for all children between six and 15 years of age. In 1998 there were 28 primary schools, with a total of 2,711 pupils, while the 18 secondary schools had a total enrolment of 1,779 pupils. Under the New Zealand Training Scheme, the New Zealand Government offers overseas scholarships in New Zealand, Fiji, Papua New Guinea, Australia and Samoa for secondary and tertiary education, career-training and short-term in-service training. There is an extension centre of the University of the South Pacific (based in Fiji) in the Cook Islands. Budgetary expenditure on education was estimated at $NZ11.67m. in 2003/04, equivalent to 13.2% of total expenditure.

NIUE

Physical and Social Geography

Niue is a coral island of 262.7 sq km (101 sq miles) located about 480 km east of Tonga and 930 km west of the southern Cook Islands. The island is mainly covered with bush and forest and, because of the rocky and dense nature of the terrain, fertile soil is not plentiful. Agriculture is further made difficult because there are no running streams or surface water. Rainfall occurs predominantly during the hottest months, from December to March, when the average temperature is 27°C (81°F). Average annual rainfall is 7,715 mm (298 ins). The restricted nature of local resources has led many islanders to migrate to New Zealand. The population declined from 5,194 in September 1966 to 2,088 at the census of August 1997 and to 1,788 at the census of September 2001. The population was officially estimated to total 2,033 in December 2003. The official languages are Niuean (a Polynesian language of the indigenous inhabitants) and English. Both are widely spoken. Most of the population are Christian, mainly Protestant. The capital and administrative centre (with a population of some 689 in 2003) is Alofi, on the west coast.

History

The first Europeans to discover Niue, which was inhabited by a Polynesian people related to the Tongans and Samoans, were members of a British expedition, led by Capt. James Cook, in 1774. Missionaries visited the island throughout the 19th century. In 1876 the clans and families of Niue elected a king, and in 1900 the island was declared a British protectorate. In 1901 Niue was formally annexed to New Zealand as part of the Cook Islands, but in 1904 it was granted a separate administration.

In October 1974 Niue attained the status of 'self-government in free association with New Zealand'. Niueans retain New Zealand citizenship, and 14,556 Niueans were resident in New Zealand in 1991. Robert (from 1982, Sir Robert) Rex, who had been Niue's political leader since the early 1950s, was the island's Premier when it became self-governing, and retained the post following three-yearly general elections in 1975–90.

The migration of Niueans to New Zealand has been a cause of increasing concern, and in 1985 the Government of New Zealand announced its intention to review its constitutional relationship with Niue, with the specific aim of preventing further depopulation of the island. In 1987 a six-member committee, comprising four New Zealanders and two Niueans, was formed to examine Niue's economic and social conditions, and to consider the possibility of the island's reverting to the status of a New Zealand-administered territory. It was hoped that the replacement of national superannuation by Guaranteed Retirement Income (GRI) in the July 1989 New Zealand budget would encourage the return of the Niueans resident in New Zealand, since all those eligible for GRI would immediately be able to receive 50% of the entitlement if they resided overseas for more than six months of the year.

At a general election held in March 1987, all except three of the 20 members of the Niue Assembly were re-elected. The newly founded Niue People's Action Party (NPAP), secured one seat. The NPAP, Niue's only political party, criticized the Government's economic policy, and, in particular, its apparent inability to account for a substantial amount of the budgetary aid received from New Zealand. A declared aim of the party has been to persuade Niueans residing in New Zealand to invest in projects on Niue.

In 1989 Rex survived a vote of 'no confidence' in the Assembly. The dissension was prompted by the New Zealand Auditor-General's report, which was highly critical of the Niuean Government's use of monetary aid from New Zealand. Opponents of the Government accused it of favouring public servants in the allocation of grants, and of insufficient long-term planning. In that year Niue's legislature proposed 37 changes to the island's Constitution. These included the replacement of the New Zealand Governor-General by a Niuean citizen as the British monarch's representative. The implication for relations with New Zealand, however, created a controversy that led the Assembly to withdraw the proposals in November of that year, before a planned referendum could take place.

At a general election in April 1990, the candidates of the NPAP and its sympathizers won 12 of the 20 seats. Earlier disagreements in the NPAP leadership, however, allowed Rex to secure the support of four members previously opposed to his Government. Rex therefore remained Premier.

In September 1990 disagreement within the Cabinet concerning reconstruction policy (following Cyclone Ofa, which had struck the island in February) led two ministers to support a proposal for a change of Premier. Rex dismissed the two, appointing Young Vivian, leader of the NPAP and of the unofficial opposition, and another member of the opposition to the Government.

The announcement in mid-1991 by the New Zealand Government that it was to reduce its aid payments to Niue by about $NZ1m. (a decrease of some 10% on the average annual allocation) caused considerable concern on the island. More than one-quarter of the paid labour force on Niue were employed by the Government, and following the reduction in aid about 150 (some 25%) lost their jobs. Members of the Government subsequently travelled to New Zealand to appeal against the decision and to request the provision of redundancy payments for the dismissed employees. Their attempts failed, however, with the New Zealand Government reiterating its claim that aid had been inefficiently used in the past.

In December 1992 Sir Robert Rex died, and Young Vivian (who had been serving as acting Premier at the time of Rex's death) was unanimously elected Premier by the Government. Legislative elections took place in February 1993, and in the following month the Niue Assembly elected Frank Lui, a former cabinet minister, as Premier. Lui, who defeated Young Vivian by 11 votes to nine, announced a new Cabinet following the election; among the new Premier's stated objectives were the development of tourism and further plans to encourage Niueans resident in New Zealand to return to the island.

In March 1994 Vivian proposed an unsuccessful motion of 'no confidence' in the Government, and a further attempt by the opposition to propose a similar motion was invalidated in the High Court in October on a procedural matter. However, during the ensuing debate, the Minister for National Planning and Economic Development, Sani Lakatani, resigned in order to join the opposition as its deputy leader, thus leaving the Government with only 10 official supporters in the Assembly. Subsequent opposition demands for the intervention of the Governor-General of New Zealand in dissolving the legislature, in preparation for a general election, were rejected, and, despite Lui's assurance that an early election would take place, in order to end the atmosphere of increasing political uncertainty, elections were not held until 16 February 1996. The Premier and his three cabinet ministers were re-elected to their seats, although support among the electorate for candidates of the Niue People's Party (NPP, formerly the NPAP) and independents appeared fairly equally divided; in one village the result was decided by the toss of a coin when both candidates received an equal number of votes. Frank Lui was re-elected by the Niue Assembly as Premier, defeating Robert Rex, Jr (son of Niue's first Premier) by 11 votes to nine.

The issue of Niue's declining population continued to cause concern, particularly when provisional census figures, published in late 1997, revealed that the island's population was at its lowest recorded level. The Government expressed disappointment that its policy of encouraging Niueans resident in New Zealand to return to the island had failed and announced its intention to consider introducing more lenient immigration laws in an attempt to increase the population.

At a general election on 19 March 1999 Lui lost his seat and subsequently announced his retirement from politics. The Minister for Finance also failed to be re-elected. On 29 March Sani Lakatani, leader of the NPP, was elected Premier by the new Assembly, defeating O'Love Jacobsen by 14 votes to six. Lakatani's stated priority as Premier was to increase Niue's population to at least 3,000; he claimed that the sharp decline in the number of residents constituted a threat to the island's self-governing status.

It was reported in May 1999 that New Zealand was to phase out aid to Niue by 2003; New Zealand's aid programme to the island had been reduced by $NZ0.25m. annually over the previous five years. However, doubts were expressed over the legality of the New Zealand Government's action, and it was suggested that New Zealand was required by law to provide financial assistance under the 1974 act that established Niue as a self-governing state.

In December 1999 a motion of 'no confidence' in Lakatani was proposed by a number of opposition ministers, in protest at the Government's plans to fund a new national airline (Coral Air Niue). The result of the vote was inconclusive, with an equal number of votes cast for and against the motion. The proposed airline did not materialize, and the Government lost $NZ400,000 of its initial investment in the project. The New Zealand Government subsequently criticized officials in Niue for failing to secure a business plan or feasibility study for the proposed airline.

In late 1999 allegations made by a foreign news agency that Niue was being used by criminal organizations for 'laundering' the proceeds of their illegal activities were strongly denied by Lakatani. However, in a report published in June 2000, the Financial Action Task Force (FATF) on Money Laundering (established in 1989 on the recommendation of the Group of Seven (G-7) industrialized nations) named the island as one of a number of countries and territories that had 'failed to co-operate' in regional efforts to combat money-laundering. As a result the Government suspended the issue of any further 'offshore' banking licences until stricter regulations governing the financial sector had been introduced. In early 2001 the USA imposed sanctions on Niue (including a ban on transactions with US banks), claiming that the island had not implemented all the recommendations of the report. Lakatani appealed directly to President George W. Bush to end the embargo, which he said was having a devastating effect on Niue's economy. The Government stressed its commitment to meeting international requirements in its financial sector but claimed that it was having difficulty doing so, given its limited legal resources. Moreover, the Premier expressed strong disapproval that a nation as powerful as the USA should choose to inflict such hardship on a small, economically vulnerable island, and urged other Pacific islands targeted by the report to unite against the forceful tactics of developed nations. In June 2001 the Government engaged a US law firm in an attempt to persuade two banks, Chase Manhattan and Bank of New York, to remove their bans on the transfer of some $NZ1m. to Niue via a business registry in Panama that the Government used for 'offshore' tax activity. Having failed to meet an FATF deadline in August 2001, in February 2002 the Government announced proposals to repeal the 'offshore' banking legislation. Premier Sani Lakatani was also considering closing down international business registrations based in Niue. The FATF announced in April 2002 that, in view of the island's commitment to improving the transparency of its tax and regulatory systems, the organization was to remove Niue from its list of non-co-operative territories; the decision was duly implemented in October. The bank-licensing legislation was repealed in June.

In December 2001 Frank Lui returned to politics to form a new organization, the Alliance of Independents, in order to contest the forthcoming general election. The party's spokesperson, O'Love Jacobsen, announced that the Alliance would campaign for a direct air link to New Zealand and for increased spending on public health.

At the general election, held on 20 April 2002, all 20 incumbent members were returned to the Niue Assembly. Independent candidate Toke Talagi received the highest number of votes (445), but overall the NPP was victorious. However, despite having polled the second highest number of votes (428), Sani Lakatani did not command the general support of his party, and faced a leadership challenge from his deputy, Young Vivian. Following several days of lobbying within the NPP, Vivian was chosen as Premier. Vivian announced that the party had the support of 10 elected members, having formed a coalition with several independents associated with Toke Talagi. Lakatani was appointed Deputy Premier; however, later that year and following a period of ill health, the former Premier indicated that he might withdraw from politics.

In July 2003 Niue's only formal political party, the Niue People's Party, was dissolved as a result of ongoing disagreement among its membership and the failure of several projects (the most prominent of which was the attempted establishment of Coral Air Niue, see above). Opposition member Terry Coe expressed satisfaction with the news, stating that he hoped that party politics would cease henceforth in Niue. Observers also commented that Robert Rex (Niue's widely respected first Premier) had strongly opposed party politics, believing it to cause rifts in families and communities.

In September 2003 the opposition expressed concern that too many government members were travelling overseas on business, and that a significant amount of public money was being used to fund these trips. At the time of the statement seven of the Niue Assembly's 20 members (including two cabinet ministers) were absent on engagements overseas. In the same month it was announced that Niue was to receive US $90,000 from the People's Republic of China in order to build new accommodation for the 300 delegates and visitors who were expected to visit the island for the Pacific Islands Forum summit meeting in 2004. The meeting was relocated, however, following the widespread devastation of the island by a cyclone in early 2004 (see below).

In October 2003 Niue's Premier issued a statement inviting the residents of Tuvalu (whose continued existence on those islands was increasingly threatened by rising sea-levels) to migrate to Niue. The Government of Tuvalu subsequently requested that Niue produce a memorandum of understanding giving formal details of this invitation and of the rights that Tuvaluans would enjoy on Niue.

In early January 2004 Niue was devastated by Cyclone Heta. Damage caused by the storm, which was described as the worst in the island's recent history, included the destruction of many buildings, the loss of most food crops, the death of one person and serious injury of several others and extensive damage to Niue's infrastructure, communications and coral reef. Relief supplies were sent from New Zealand as part of an initial aid programme worth some US $3.5m. It was estimated that US $25m. would be needed for a rebuilding programme to be carried out over a five-year period. The destruction of Alofi was so severe that the Government announced plans to relocate the island's capital to Fonuakula on the upper plateau. However, fears for the continued feasibility of the island were expressed by some observers, who suggested that many Niueans might exercise their right to take up residency in New Zealand, leaving the community on Niue unviable. As work began to repair or rebuild some 300 homes under the Government's New Niue or Niue Foou recovery plan, it was announced that efforts would be made to attract expatriate Niueans back to the island. A fish-processing plant due to open later in the year as well as a number of new agricultural projects were expected to provide some employment opportunities

In March 2000 a Niue-New Zealand joint consultative committee met, for the first time, in Alofi to consider the two sides' future constitutional relationship. Later that year the committee proposed to conduct a survey of islanders' views and to consider all options, from reintegration with New Zealand to full independence. A meeting of the joint committee took place in March 2001 in Wellington at which the issues of New Zealand aid and reciprocal immigration laws were discussed, as well as options for Niue's future constitutional status. In early 2001 Hima Takelesi was appointed Niue's first High Commissioner to New Zealand. New Zealand remained committed to annual assistance of $NZ6.3m. in the years 2001–03. At New Zealand's 2001 census, a total of 20,148 Niueans were recorded as resident in New Zealand. Discussions took place in Wellington in March 2003 between the New Zealand Prime Minister, Helen Clark, and Niue's Premier, Young Vivian. Topics debated included budgetary assistance, a review of the island's development plan and the continued migration of islanders from the territory to New Zealand.

Following almost 10 years of technical and political consultations, Niue and the USA signed a maritime boundary treaty in May 1997, delineating the precise boundary between the territorial waters of Niue and American Samoa.

Economy

Niue's economic development has been adversely affected by inclement weather, inadequate transport services and the annual migration of about 10% of the population to New Zealand. Two-thirds of the land surface is uncultivable, and marine resources are variable. Until the early 1990s more than 80% of the working population were employed in the public sector on social services and economic development projects, mainly in the agricultural sector. However, significant reductions in aid and budgetary support from New Zealand, which funded the majority of these projects, resulted in a sharp decline in the number of people employed in the public sector.

Agricultural activity is mostly of a subsistence nature, the main crops being coconuts, taro, yams, cassava (tapioca) and sweet potatoes (kumara). Honey is also produced for export, and, until the closure of a processing factory in 1989, coconut cream was the country's leading export. A taro export scheme was introduced in the early 1990s, and production of the crop increased by more than 500% in 1993. Plans to increase the production of vanilla as an export crop were discussed in 2003, but the promising harvest was destroyed by the cyclone of early 2004. The reintroduction of vanilla cultivation, as well as that of organic nonu, for export was initiated during 2004 as part of the Government's post-cyclone recovery programme. Pigs, poultry and beef cattle are reared mainly for local consumption. Fishing is also primarily for local consumption, although some fish and coconut crab are exported to Niueans living in New Zealand.

Preliminary tests in 1990 revealed the presence of uranium on the island. Further exploration, however, to establish the viability of mining the mineral was delayed by the financial problems of the Australian company concerned. There is no manufacturing industry on Niue, although a fish-processing factory at Amanau was expected to begin operations in September 2004. It was estimated that the new plant could generate some $NZ9m. annually in revenue.

An increase in the frequency of flights between Niue and Auckland in 1992 enhanced prospects for the island's nascent tourist industry, as did a project to construct 60 hotel rooms and to extend the runway at Hanan airport. A total of 2,010 people visited the island by air in 2000, almost 50% of whom were from New Zealand. Tourism earned some US $1.0m. in 1998. The tourism industry, however, was seriously affected by Royal Air Tonga's decision in March 2001 to suspend its twice-weekly service to the island. Niue therefore was forced to rely upon charter flights operated by Air Fiji. The Government of Niue subsequently agreed to subsidize the service and flights resumed in June. In November Kiribati announced plans to operate a 70-seat Airbus service between Niue and Fiji. More significantly, in October 2002, Samoa's Polynesian Airlines signed a five-year agreement to provide a direct weekly flight between Niue and Auckland, New Zealand. This was increased to a twice-weekly service in 2003, in which year visitor arrivals increased by 44% to 2,758. However, following the cyclone of January 2004 (see above) experts suggested that the island's tourist industry was unlikely to recover within less than four years.

Niue records an annual trade deficit, with imports in 1993 exceeding exports by around 1,300%. In that year New Zealand, Niue's main trading partner, provided 86.1% of imports. The principal exports in that year were root crops (which provided 87.1% of total export earnings), coconuts (1.9%), honey and handicrafts. The principal imports were foodstuffs (which cost 28.0% of total imports), electrical goods (11.8%), motor vehicles (10.6%) and machinery (5.4%). In 1999 imports totalled $NZ5.2m., while exports were worth only $NZ0.3m. Taro, honey and vanilla were Niue's most significant exports. The cost of imports from New Zealand totalled an estimated $NZ3.7m. in 2002/03, while exports to that country earned some $NZ0.2m.

In 1991 the New Zealand Government announced that budgetary aid to Niue was to be reduced considerably, following criticism by the New Zealand Auditor-General that aid had not been used efficiently. By 1993 aid provided by New Zealand was some 30% lower than in 1990, and as a result severe reductions in Niue's public services were made. In mid-1999 the New Zealand Government announced its intention to phase out aid to Niue by 2003. Development assistance from New Zealand totalled $NZ8.28m. in 2003/04 (compared with $NZ10.0m. in 1994/95). The Government aimed to reduce the budget deficit to $NZ1.7m. in the 2003/04 financial year, when total budgetary expenditure was projected at $NZ17.8m. Budgetary expenditure for the following financial year was projected at $NZ19m. with a deficit of some $NZ1m. The annual rate of inflation averaged 2.6% in 1990–2002; consumer prices increased by an average of 6.8% in 2001 and by 2.7% in 2002.

In an attempt to diversify Niue's aid sources, efforts were made to improve relations with Australia in the early 1990s. Premier Frank Lui announced further measures in 1993 aimed at encouraging the return to the island of Niueans resident in New Zealand. It was hoped that, by increasing the resident population, Niue's economy could be stimulated, and the island's prospects for self-sufficiency improved. However, the population continued to decline. The island's population had fallen to 1,788 at the census of September 2001, compared with 2,321 at the 1994 census. (In December 2003, however, the population was officially estimated at 2,033.) In all, 20,148 Niueans were resident in New Zealand at that country's March 2001 census. In late 2003 the Government extended an invitation to the residents of Tuvalu to move to Niue (see above).

In 1994 the Niue Assembly approved legislation allowing the island to become an 'offshore' financial centre. By mid-1996 the 'offshore' centre was believed to have attracted some US $280,000. However, following the threat of financial sanctions from the Paris-based Financial Action Task Force (FATF, see History), Niue declared its intention to repeal its 'offshore' banking legislation, despite fears that this would result in annual revenue losses of some US $80,000 in bank licence fees and more than US $500,000 in company registration fees. In June 2002 the Niue Assembly voted to end the issuing of banking licences. Further attempts to secure additional sources of revenue in Niue included the leasing of the island's telecommunications facilities to foreign companies for use in specialist telephone services. However, this enterprise caused considerable controversy when it was revealed that Niue's telephone code had been made available to companies offering personal services considered indecent by the majority of islanders. In addition, the island earned some US $0.5m. between 1997 and 2000 from the sale of its internet domain name '.nu', although similar controversy ensued when a report, published in July 2004, claimed that the island was hosting some 3m. pages of pornographic material via its .nu domain. (According to the report, Niue's internet suffix was the fourth largest host of pornography in the world.) In mid-2003 Niue became the first location in the world to have a national wireless internet system, allowing internet access from anywhere by means of solar-powered aerials attached to coconut palms. The system was financed largely from profits generated by the sale of the island's domain name. The imposition of harsh economic sanctions by the US Government in 2001, following accusations that the island was still allowing criminal organizations to 'launder' their funds through the territory's 'offshore' financial centre, led Niue's Government to investigate various alternative activities for generating revenue. In November 2001 Premier Sani Lakatani had announced that the Government was negotiating a deal with a US company interested in using Niue as the call-centre of a satellite service. Other large-scale projects included a scheme by a Korean religious sect to build a US $200m. holy walled city on Niue, which was abandoned in early 2003 following strong opposition from local residents.

In 1999 Premier Lakatani announced that he was seeking to gain membership of the Asian Development Bank (ADB) for Niue. It was hoped that Niue might receive a low-interest loan from the ADB if the New Zealand Government proceeded with its decision to withdraw aid to the island. In 2001, however, Niue's application was obstructed by the USA. In March 2002 Niue was in the process of concluding a 20-year development programme with the European Union (EU). Initial assistance of €2.6m., to be released in 2003, was to be used to finance renewable energy projects such as wind power generation. A five-year island development plan announced in 2003 included proposals to develop Niue's fishing industry by employing a fleet of used Korean fishing vessels, to expand the agricultural sector with increased vanilla cultivation, to encourage tourism given the availability of direct flights from New Zealand and to increase taro exports to New Zealand owing to a new regular shipping service which was expected to reduce freight costs significantly.

Niue's entire economy was severely affected by Cyclone Heta which struck the island in January 2004 causing extensive damage to housing, crops and infrastructure (see above).The subsequent recovery programme, known as New Niue or Niue Foou, emphasized rebuilding works and a major fish-processing project. The Government was also hoping to decrease the working hours of the island's 430 public servants, in an attempt to reduce salary expenses and to encourage more people to participate in both subsistence agriculture and the cultivation of vanilla and nonu for export.

Niue is a member of the Pacific Islands Forum (formerly the South Pacific Forum) and the Pacific Community (formerly the South Pacific Commission) and an associate member of the UN's Economic and Social Commission for Asia and the Pacific (ESCAP). In 2000 Niue became a signatory of the Cotonou Agreement with the EU. In September 2002 Niue joined the Pacific Agreement on Closer Economic Relations, which had earlier been ratified by the Governments of Australia, the Cook Islands, Fiji, New Zealand, Samoa and Tonga.

Statistical Survey

Source (unless otherwise stated): Economics, Planning, Development and Statistics Unit, POB 40, Alofi; tel. and fax 4219.

AREA AND POPULATION

Area: 262.7 sq km (101.4 sq miles).

Population: 2,088 at census of 17 August 1997; 1,788 (males 897, females 891) at census of 7 September 2001. An estimated 20,145 Niueans lived in New Zealand at the time of the 2001 census. *2003* (official estimate at 31 December): 2,033.

Density (at 31 December 2003): 7.7 per sq km.

Ethnic Groups (2001 census, declared ethnicity): Niueans 1,3999; Caucasian 81; Pacific Islander 182; Niuean/Caucasian 28; Niuean/Pacific Islander 42; Asian 4.

Religion (2001 census, declared adherence): Ekalesia Niue 1,093; Church of Jesus Christ of Latter-day Saints 158; Roman Catholic 128; Jehovah's Witnesses 43; Seventh-day Adventist 25; Other 151.

Principal Towns (population, 2001 census): Alofi (capital) 614; Hakupu 227. Source: Thomas Brinkhoff, *City Population* (internet www.citypopulation.de). *Mid-2003* (UN estimate, incl. suburbs): Alofi 689 (Source: UN, *World Urbanization Prospects: The 2003 Revision*).

Births, Marriages and Deaths (2001 census): Crude birth rate 18.5 per 1,000; Death rate 7.8 per 1,000. *2002:* Live births 25; Marriages 15; Deaths 13.

Expectation of Life (WHO estimates, years at birth): 70.3 (males 67.6; females 73.3) in 2002. Source: WHO, *World Health Report*.

Immigration and Emigration (2002): Arrivals 3,155; Departures 3,017.

Economically Active Population (2001 census, persons aged 15 years and over): Agriculture, forestry and fishing 60; Mining 17; Manufacturing 19; Electricity, gas and water 27; Construction 72; Trade 48; Restaurants and hotels 29; Transport 64; Finance 35; Real estate, etc. 3; Public administration 96; Education 63; Health, etc. 72; *Total employed* (incl. others) 633; Unemployed 21; *Total labour force* 654. Note: Figures exclude 63 subsistence workers.

HEALTH AND WELFARE

Key Indicators

Total Fertility Rate (children per woman, 2001): 2.6.

Physicians (per 1,000 head, 1996): 1.30.

Health Expenditure (2000): US $ per head (PPP): 1,111.

Health Expenditure (2000): % of GDP: 7.6.

Health Expenditure (2000): public (% of total): 96.2.

Access to Water (% of persons, 2000): 100.

Access to Sanitation (% of persons, 2000): 100.

For sources and definitions, see explanatory note on p. vi.

AGRICULTURE, ETC.

Principal Crops (FAO estimates, metric tons, 2002): Taro 3,200; Other roots and tubers 406; Coconuts 2,500; Bananas 70; Lemons and limes 110. Source: FAO.

Livestock (FAO estimates, 2002): Cattle 112; Pigs 1,700; Chickens 10,000. Source: FAO.

Livestock Products (FAO estimates, year ending September 2002): Pigmeat 49; Poultry meat 14; Hens eggs 12; Honey 6. Source: FAO.

Forestry (cu m, 1985): Roundwood removals 613; Sawnwood production 201.

Fishing (FAO estimates, metric tons, 2002): Total catch 200. Source: FAO.

INDUSTRY

Production (1998): Electric energy 3 million kWh (estimate).

FINANCE

Currency and Exchange Rates: 100 cents = 1 New Zealand dollar ($NZ). For details, see Tokelau.

Budget ($NZ '000, year ending 30 June 2003): Recurrent revenue 13,263 (New Zealand budgetary support 3,750, Internal revenue 9,513); Other 3,425; *Total revenue* 16,689; Recurrent 12,458; Capital 486; Corporations 5,437; *Total expenditure* 18,381. *2003/04:* Forecast expenditure $NZ17.8m.

Overseas Aid ($NZ '000, 2003/04): Official development assistance from New Zealand 8,275 . Source: Ministry of Foreign Affairs and Trade, Wellington.

Cost of Living (Consumer Price Index; base: March 1990 = 100): 123.4 in 2000; 131.9 in 2001; 137.2 in 2002.

Gross Domestic Product ($NZ '000 in current prices): 14,693 in 1998; 14,199 in 1999; 14,210 in 2000.

GDP by Economic Activity ($NZ '000 in current prices, 2000): Agriculture, forestry and fishing 2,934; Mining and quarrying 28; Manufacturing 204; Electricity, gas and water 402; Construction 546; Trade, restaurants and hotels 1,676; Transport and communications 1,741; Financial and business services; real estate, etc. 1,245; Public administration 5,241; Other community and personal services 518; *Sub-total* 14,535; *Less* Imputed bank service charge 360; *Gross value added in basic prices* 14,175; Indirect taxes 1,014; *Less* Subsidies 980; *GDP in purchasers' values* 14,210.

EXTERNAL TRADE

Principal Commodities ($NZ '000, 2002): *Imports c.i.f.:* Total 3,246. *Exports f.o.b.:* Total 135. *2003* (Exports only): Taro 199; Coconuts 16; Total 215 .

Principal Trading Partners ($NZ '000, 1993): *Imports c.i.f.:* Australia 101.7; Fiji 140.9; Japan 358.3; New Zealand 5,993.8; Samoa 47.2; USA 197.2; Total (incl. others) 6,962.1. *Exports f.o.b.:* Total 543.2. *2002/03* ($NZ '000): Trade with New Zealand: *Imports c.i.f.:* 3,684; *Exports f.o.b.:* 223 (Source: Ministry of Foreign Affairs and Trade, Wellington).

TRANSPORT

Road Traffic (2001 census): Passenger cars 323; Motorcycles 134; Vans 170; Trucks 74; Pick ups 76; Buses 11.

International Shipping: *Ship Arrivals* (1989): Yachts 20; Merchant vessels 22; Total 42. *Freight Traffic* (official estimates, metric tons, 1989): Unloaded 3,410; Loaded 10.

Civil Aviation: *Passengers* (1992): Arrivals 3,500; Departures 3,345; Transit n.a. *Freight Traffic* (metric tons, 1992): Unloaded 41.6; Loaded 15.7.

TOURISM

Foreign Tourist Arrivals (by air): 2,069 in 2001; 1,632 in 2002; 2,758 in 2003.

Tourist Arrivals by Country of Residence (2000): Australia 172; Fiji 58; New Zealand 1,000; Tonga 58; United Kingdom 32; USA 145; Total (incl. others) 2,010.

Tourism Receipts (US $ million): 1 in 1996; 2 in 1997; 1 in 1998. Source: mainly World Tourism Organization, *Yearbook of Tourism Statistics*.

COMMUNICATIONS MEDIA

Mobile Cellular Telephones (2001 census): 225 units in use*.

Radio Receivers (2001 census): 605 in use†.

Television Receivers (2001 census): 451 in use.

Personal Computers (2001 census): 77 in use.

Non-daily Newspaper (1996, estimate): 1, circulation 2,000†.

* Source: International Telecommunication Union.

† Source: UNESCO, *Statistical Yearbook*.

EDUCATION

Pre-primary (1998): 1 facility; 54 pupils.

Primary (2002, unless otherwise indicated): 1 school (1998); 17 teachers; 251 pupils.

Secondary and Post-secondary (2002, unless otherwise indicated): 1 school (1998); 29 teachers; 491 pupils.

Sources: mainly Department of Education, Niue, and UN, *Statistical Yearbook for Asia and the Pacific*.

Directory

The Constitution

In October 1974 Niue gained self-government in free association with New Zealand. The latter, however, remains responsible for

Niue's defence and external affairs and will continue economic and administrative assistance. Executive authority in Niue is vested in the British monarch as sovereign of New Zealand but exercised through the government of the Premier, assisted by three ministers. Legislative power is vested in the Niue Assembly or Fono Ekepule, which comprises 20 members (14 village representatives and six elected on a common roll), but New Zealand, if requested to do so by the Assembly, will also legislate for the island. There is a New Zealand representative in Niue, the High Commissioner, who is charged with liaising between the Governments of Niue and New Zealand.

The Government

New Zealand High Commissioner: SANDRA LEE.

Secretary to Government: BRADLEY PUNU.

THE CABINET
(September 2004)

Premier and Minister for External Affairs, International Relations and Aid Co-ordination, the Niue Public Service Commission, the Crown Law Office, Community Affairs, Arts, Culture and Village Councils, Religious Affairs, Population Development and Niueans Abroad, Women's Affairs, Youth and Sport, and Private Sector Development: YOUNG VIVIAN.

Deputy Premier and Minister for Agriculture, Forestry and Fishing, Planning, Economic Development and Statistics, the Niue Development Bank, Shipping, Investment and Trade, Civil Aviation, Police, Immigration, Disaster Management, Development and Training, Public Works, Water and Utilities, Port Services, and the Niue Broadcasting Corporation: BILL MOTUFOOU.

Minister for Education and Language Development, Health, Environment and Biodiversity, Justice, Finance, Customs and Revenue, Tourism, Post, Telecommunications and Information, Computer Technology Development, Philatelic Bureau and Numismatics: TOKE TALAGI.

GOVERNMENT OFFICES

All ministries are in Alofi.

Office of the New Zealand High Commissioner: POB 78, Tapeu, Alofi; tel. 4022; fax 4173.

Office of the Secretary to Government: POB 40, Alofi; tel. 4200; fax 4232; e-mail secgov.premier@mail.gov.nu.

Legislature

ASSEMBLY

The Niue Assembly or Fono Ekepule has 20 members (14 village representatives and six members elected on a common roll). The most recent general election was held on 20 April 2002.

Speaker: ATAPANA SIAKIMOTU.

Political Organization

Alliance of Independents: Alofi; f. 2001; Leader FRANK LUI; Spokesperson O'LOVE JACOBSEN.

Judicial System

The Chief Justice of the High Court and the Land Court Judge visit Niue quarterly. In addition, lay justices are locally appointed and exercise limited criminal and civil jurisdiction. Appeals against High Court judgments are heard in the Court of Appeal of Niue (created in 1992).

The High Court: exercises civil and criminal jurisdiction.

The Land Court: is concerned with litigation over land and titles.

Land Appellate Court

Hears appeals over decisions of the Land Court.

Chief Justice: NORMAN F. SMITH.

Religion

About 66% of the population belong to the Ekalesia Niue, a Protestant organization, which had 1,487 adherents at the time of the 1991 census. Within the Roman Catholic Church, Niue forms part of the diocese of Tonga. The Church of Jesus Christ of Latter-day Saints (Mormon), the Seventh-day Adventists, the Jehovah's Witnesses and the Church of God of Jerusalem are also represented.

Ekalesia Niue: Head Office, POB 25, Alofi; tel. 4195; fax 4352/4010; e-mail ekalesia.niue@niue.nu; f. 1846 by London Missionary Society, became Ekalesia Niue in 1966; Pres. Rev. MATAGI VILITAMA; Gen. Sec. Rev. ARTHUR PIHIGIA.

The Press

Niue Economic review: POB 91, Alofi; tel. 4235; monthly.

Niue Star: POB 151, Alofi; tel. 4207; weekly; Niuean and English; publ. by Jackson's Photography and Video; circ. 600.

Broadcasting and Communications

TELECOMMUNICATIONS

Director of Posts and Telecommunications: Alofi; tel. 4002.

Niue Telecom: Alofi; tel. 4000; internet www.niuenet.com; Man. RICHARD HIPA.

BROADCASTING

Radio

Broadcasting Corporation of Niue: POB 68, Alofi; tel. 4026; fax 4217; operates television service and radio service; govt-owned; Chair. PONI KAPAGA; CEO TREVOR TIAKIA; Gen. Man. PATRICK LINO.

Radio Sunshine: broadcasts in English and Niuean between 6am and 10pm Mon.-Sat.

Television

Broadcasting Corporation of Niue: see Radio.

Television Niue: broadcasts in English and Niuean, six days a week from 5 p.m. to 11 p.m.

Trade and Industry

GOVERNMENT AGENCIES

Business Advisory Service: Alofi; tel. 4228.

Office of Economic Affairs, Planning and Development, Statistics and Trade and Investment: POB 42, Alofi; tel. 4148; e-mail business.epdsu@mail.gov.nu; responsible for planning and financing activities in the agricultural, tourism, industrial sectors, business advisory and trade and investment.

UTILITIES

Niue Power Corporation: POB 198, Alofi; tel. 4119; fax 4385; e-mail gm.npc.@mail.gov.nu.

MAJOR COMPANY

NU Domain Ltd: Alofi; internet www.nunames.nu; f. 1997; responsible for the sale of Niue's internet domain name; Admin. Man. STAFFORD GUEST; Tech. Man. RICHARD ST CLAIR.

TRADE UNION

Public Service Association: Alofi.

Finance

DEVELOPMENT BANK

Fale Tupe Atihake Ha Niue (Development Bank of Niue): POB 34, Alofi; tel. 4335; fax 4010; f. 1993; began operations July 1994; Gen. Man. TERAI MCFADZIEN.

COMMERCIAL BANK

Westpac Banking Corpn: Main St, Alofi; tel. 4221; fax 4043; e-mail westpacniue@sin.net.nu; Man. R. J. COX.

Transport

ROADS

There are 123 km of all-weather roads and 106 km of access and plantation roads. A total of 655 motor vehicles were registered in 1992. The road network was extensively damaged by Cyclone Heta in January 2004. In mid-2004 it was estimated that some 48 km of sealed roads were clear and in good condition.

SHIPPING

The best anchorage is an open roadstead at Alofi, the largest of Niue's 14 villages. Work to extend a small wharf at Alofi began in mid-1998 with US assistance. The New Zealand Shipping Corporation operates a monthly service between New Zealand, Nauru and Niue. Fuel supplies are delivered by a tanker (the *Pacific Explorer*) from Fiji. In December 2002 the Government signed an agreement with Reef Shipping Ltd to provide a service to New Zealand every three–four weeks.

CIVIL AVIATION

Hanan International Airport has a total sealed runway of 2,350 m, following the completion of a 700-m extension in 1995, with New Zealand assistance. Air links were seriously affected by the cessation of the Air Nauru service in 1989. In 1995 Royal Tongan Airlines began operating a weekly service between Niue and Tonga, with connections to Auckland, New Zealand, and in 1996 the airline established a twice-weekly direct service between Niue and Tonga. However, Royal Tongan Airlines discontinued its service to Niue in October 2002. Air Rarotonga (Cook Islands) operates occasional charter flights to Niue. In November 2001 Kiribati announced plans to operate a flight between Niue and Fiji. In October 2002 the Niue Government concluded negotiations with Polynesian Airlines to introduce direct weekly flights from Niue to Auckland, New Zealand, for a five-year period.

Niue Airways Ltd (NAL): Hanan International Airport; f. 1990; registered in New Zealand; Dir RAY YOUNG.

Tourism

Niue has a small but significant tourism industry (specializing in holidays based on activities such as diving, rock-climbing, caving and game fishing), which was enhanced by an increase in the frequency of flights between the island and New Zealand in the early 1990s. A new 24-room hotel opened in 1996, increasing Niue's tourist accommodation to 158 beds. However, this resort was closed in April 2001 and its 20 staff were made redundant, following numerous cancelled bookings resulting from the cessation of the Royal Tongan Airlines service to the island. A total of 2,758 people arrived by air (about 50% of whom were from New Zealand) to visit Niue in 2003. The industry earned about US $1m. in 1998. The prospects for the island's tourist industry were severely hampered by the extensive damage caused by Cyclone Heta in January 2004. Experts believed that it would be several years before a recovery could be achieved. However, the island's Matavai resort was operating in mid-2004 and offering 33 rooms for visitors.

Niue Tourist Office: POB 42, Alofi; tel. 4224; fax 4225; e-mail niuetourism@mail.gov.nu; internet www.niueisland.com; Dir of Tourism IDA TALAGI HEKESI.

Education

Education is free and compulsory between six and 16 years of age (the school-leaving age having been raised from 14 in 1998). In 1987 the island's seven village primary schools were closed and a single national primary school was opened at Halamahaga. In 1998 this bilingual (Niuean/English) primary school had with 12 teachers and an enrolment of 282 pupils . There was one secondary school at Paliati, with a teaching staff of 27 and a total enrolment of 304 pupils in 1991. Post-secondary students were estimated to number 50 in March 1991. Higher education takes place at the Niue Extension Centre of the University of the South Pacific (based in Fiji), on government training schemes or by correspondence. Some study overseas, in the Pacific region and New Zealand. A private medical school opened in Niue in 2000 but closed in the following year. A private university offering online business and information technology courses opened in late 2003. The Government's budget for 1998 allocated $NZ1,190,386 to education.

PALAU

Physical and Social Geography

The Republic of Palau (also known as Belau) consists of eight principal and 252 smaller islands, in a chain about 650 km long, stretching from the small groups of islands around Tobi and Sonsorol, north-east to the main group, which extends from Angaur to Kayangel atoll. This latter group includes the main island of Babeldaob (Babelthuap), the second largest island in Micronesia (after Guam). Sometimes referred to by its Japanese name of the *hunto*, it is of volcanic origin and rises to 239 m. Palau lies about 7,150 km south-west of Hawaii and about 1,160 km south of Guam. The territory's nearest neighbour is Yap to the east, one of the Federated States of Micronesia. The Philippines lie to the west and Indonesia to the south and south-west. With the Federated States of Micronesia (q.v.), Palau forms the archipelago of the Caroline Islands. Palau is subject to heavy rainfall, and seasonal variations in precipitation and temperature are generally small. The indigenous population is Micronesian, most of whom speak Palauan, a language with little variation in dialect, although some linguists class Sonsorolese-Tobian (from south-west Palau) as a separate language. English is also an official language, and many people still speak Japanese. Most of the population are Christian, the main denomination being Roman Catholic, but many traditional beliefs persist, even among adherents of the Christian churches. The Modignai church is an indigenous, non-Christian religion. At the April 2000 census the population totalled 19,129, of whom more than 70% resided in Koror, on Koror Island, which is the provisional capital (the Constitution provides for the construction of a new official capital on the less-developed island of Babeldaob, in Melekeok state).

History

The earliest settlement of the islands of Palau occurred some 4,500 years ago, probably from Indonesia. A complex society of warring, matrilineal clans emerged. There was a considerable degree of social stratification, with every individual born to a definite rank in society. The clans came to be grouped into two loose confederations, that in the north being presided over by a high chief known as the Reklai and that in the south led by the Ibedul. Palau, as part of the Carolines, was in the Spanish sphere of influence from the 16th century, but was not formally annexed to the Spanish Crown until 1886. Until that date, from the late 18th century, the British had dominated trade with the islands. European contacts resulted in the population being devastated by dysentery and influenza. In 1899, following its defeat in the Spanish–American War of 1898, Spain sold the Caroline Islands (including Palau) and the Northern Mariana Islands (i.e. all the Marianas except Guam) to Germany. This area, together with the Marshall Islands, became known as German Micronesia (because of German trading rights in the region) until the First World War began in 1914, when Japan occupied the territory.

Japan's formal administration began in 1920, under a mandate from the League of Nations. The islands were colonized and greatly developed, with over 100,000 permanent Japanese settlers, compared with some 40,000 indigenous inhabitants. In Palau alone, by 1940, there were some 35,000 people, of whom only about 7,500 were Palauan. Micronesia was conquered by the USA in 1944 and the former Japanese territory, removed of its settlers, became the only strategic trusteeship of the 11 trusteeships established by the United Nations in 1947. The USA was named the administering authority of the Trust Territory of the Pacific Islands, which included the Marshall Islands, Ponape (now Pohnpei), Truk (now Chuuk), Yap, Palau and the Northern Mariana Islands; it was also a member of the Trusteeship Council (all permanent members of the UN Security Council, any administering countries and other non-administering countries elected by the UN General Assembly for three-year terms). Within the US federal government, the Secretary of the Navy was responsible for the administration of the Trust Territory until 1951, when it was transferred to the jurisdiction of the Secretary of the Interior.

From 1965 there were increasing demands for local autonomy within the Trust Territory. In that year the Congress of Micronesia was formed, and in 1967 a commission to examine the future political status of the islands was established. In 1970 it declared Micronesians' rights to sovereignty over their own lands, self-determination, the right to devise their own constitution and the right to revoke any form of free association with the USA. From the beginning of the negotiations it was clear that the aspirations of the people of the Northern Mariana Islands were different from those of the rest of the Trust Territory. The islands were separated administratively in 1976, as part of a process leading towards the achievement of incorporation within the USA as the Commonwealth Territory of the Northern Mariana Islands (q.v.). In May 1977 President Jimmy Carter announced that his Administration intended to adopt measures to terminate the Trusteeship Agreement by 1981. In 1978, however, despite the recommendation of both the USA and the UN, the districts of Palau and the Marshall Islands rejected participation in a single federated Micronesian state. Only Yap, Truk (now Chuuk), Ponape (now Pohnpei) and Kosrae proceeded to form the Federated States of Micronesia. The three jurisdictions that emerged in the Trust Territory, excluding the Northern Mariana Islands, all favoured a similar status of 'free association', a concept with no precise definition in international law. Although recognized under the constitutional treaty or Compact of Free Association between the USA and the respective governments, full sovereignty is generally considered to be precluded by the reservation of defence and security arrangements to the USA.

In the Palau District a referendum in July 1979 approved a proposed local Constitution, which came into effect on 1 January 1981, when the district became known as the Republic of Palau. The USA signed a Compact of Free Association with the Republic of Palau in August 1982, and with the Marshall Islands and the Federated States of Micronesia in October. The trusteeship of the islands was due to end after the principle and terms of the Compacts had been approved by the respective peoples and legislatures of the new countries, by the US Congress and by the UN Security Council. Under the Compacts, the three countries (excluding the Northern Mariana Islands) would be independent of each other and would manage both their internal and foreign affairs separately, while the USA would be responsible for defence and security. In addition, the USA was to allocate some US $3,000m. in aid to the islands.

More than 60% of Palauans voted in February 1983 to support their Compact, but fewer than the required 75% approved changing the Constitution to allow the transit and storage of nuclear materials. A revised Compact, which contained no reference to nuclear issues, was approved by 66% of votes cast in a plebiscite in September 1984. A favourable majority of 75% of the votes cast, however, remained necessary for the terms of the Compact to override the provisions of the Palau Constitution in the event of a conflict between the two.

In June 1985 President Haruo Remeliik of Palau was assassinated. (Relatives of a rival candidate in the 1984 presidential election, Roman Tmetuchl, were convicted of the murder, but remained at liberty pending an appeal; this was upheld, on the grounds of unreliable evidence, by Palau's Supreme Court in August 1987.) Alfonso Oiterong assumed office as acting President until elections in September 1985, when he was defeated in the presidential contest by Lazarus Salii.

In January 1986 representatives of the Palauan and US administrations reached a preliminary agreement on a new Compact, whereby the USA consented to provide US $421m. in economic assistance to the islands. The new Compact was approved by only 72% of Palauan voters at a referendum in February, but both Salii and President Ronald Reagan of the USA argued that the majority vote was sufficient for the Compact's approval, as the USA had guaranteed that it would observe the constitutional ban on nuclear material.

In May 1986 the UN Trusteeship Council endorsed the US Administration's request for the termination of the existing Trusteeship Agreement with the islands. However, a writ was subsequently submitted to the Palau High Court, in which it was claimed that approval of the Compact with the USA was unconstitutional because it had failed to obtain the requisite 75% of votes. The High Court ruled in favour of the writ, but the Palauan Government appealed against the ruling. In October the Compact was approved by the US Congress, but, at a new plebiscite in December, only 66% of Palauans voted in favour of the Compact. Its ratification, therefore, remained impossible.

The USA's Compacts with the Marshall Islands and with the Federated States of Micronesia came into effect in October and November 1986, respectively. In November President Reagan proclaimed the end of US administration in Micronesia and the end of the trusteeship: the Northern Mariana Islands achieved full commonwealth status within the USA, and the Republic of the Marshall Islands and the Federated States of Micronesia became Associated States. With the failure to implement the Compact with the Republic of Palau, the USA agreed to continue to administer it as the only remaining part of the Trust Territory of the Pacific Islands.

A fifth plebiscite on Palau's proposed Compact with the USA, held in June 1987, failed to secure the 75% vote in favour required by the Constitution. Under alleged physical intimidation by pro-nuclear supporters of the Compact, the House of Delegates (the lower house of the Palau National Congress) agreed to a further referendum in August of that year. In this referendum an amendment to the Constitution was approved, ensuring that a simple majority would be sufficient to approve the Compact. This was duly achieved in a further referendum in the same month. However, a writ was entered with the Supreme Court challenging the result. The ensuing violence and allegations of corruption and intimidation on the part of the Palau Government resulted, in February 1988, in the opening of an investigation by the US General Accounting Office (GAO). In April Palau's Supreme Court declared invalid the procedure by which the Compact had been approved in the previous August.

Three government employees, including Salii's personal assistant, were imprisoned in April 1988, having been found guilty of firing on the home of Santos Olikong, Speaker of Palau's House of Delegates. The attack was widely considered to have been prompted by Olikong's public opposition to the proposed Compact. In August Salii, who had been the principal subject of bribery allegations, was found dead from a gunshot wound. Rumours to the effect that Salii had committed suicide were confirmed in September.

At elections in November 1988 Ngiratkel Etpison, an advocate of the proposed Compact, narrowly defeated the anti-nuclear candidate, Roman Tmetuchl. Furthermore, Kuniwo Nakamura, another opponent of the Compact, was elected to the vice-presidency by a substantial majority, and the elections also returned opposition majorities in both chambers of the Olbiil era Kelulau (legislature).

A seventh referendum on the issue of the Compact was held in February 1990. The level of participation was low, and only 60% of voters approved the Compact. In July the US Department of the Interior declared its intention to impose stricter controls on the administration of Palau, particularly in financial matters. In the following year the leader of Palau's Council of Chiefs (a presidential advisory body), Yutaka Gibbons, initiated proceedings to sue the US Government. His claim centred on demands for compensation for the extensive damage caused to Palau's infrastructure by US forces during the Second World War and the subsequent retardation of the economy, allegedly as a result of the US administration of the islands.

During 1991 the US authorities reopened investigations into the assassination of Remeliik in 1985. In March 1992 the Minister of State, John Ngiraked, his wife, Emerita Kerradel, and Sulial Heinrick (already serving a prison sentence for another killing) were charged with Remeliik's murder. In April 1993 Ngiraked and Kerradel were found guilty of aiding and abetting the assassination of the President, while Heinrick was acquitted.

Legislative and presidential elections were held in November 1992. The electoral system had been modified earlier in the year to include primary elections, at which two presidential candidates are selected. At secondary elections the incumbent Vice-President, Kuniwo Nakamura, narrowly defeated Johnson Toribiong to become President. A referendum was held concurrently, proposing that in future polls a simple majority be sufficient to approve the adoption of the Compact of Free Association. Some 62% of voters were in favour of the proposal, which was approved in 14 of the territory's 16 states. A further referendum on the proposed Compact took place in November 1993. Some 68.3% of participating voters approved the proposed Compact, giving the Government a mandate to proceed with its adoption. Nevertheless, opposition to the changes remained fierce, and in January 1994 two legal challenges were mounted that questioned the validity of the amendments and stated that the Compact's approval had been procured by coercion. The challenges were not successful, however, and on 1 October 1994 Palau achieved independence under the Compact of Free Association. At independence celebrations, Nakamura appealed to opponents of the Compact to support Palau's new status. He announced that his Government's principal concern was the regeneration of the Palauan economy, which he aimed to initiate with an economic programme financed by funds from the newly implemented Compact. Palau was admitted to the UN in December 1994, and became a member of the IMF in December 1997.

At a preliminary round of voting in the presidential election in September 1996, Nakamura secured 52.4% of total votes, Toribiong received 33.5% and Yutaka Gibbons 14.2%. Nakamura and Toribiong were, therefore, expected to proceed to a second election due to take place in November. However, in late September a serious crisis struck Palau when the bridge linking the islands of Koror and Babeldaob collapsed, killing two people and injuring several others. The collapse of the bridge left the capital isolated from the international airport on Babeldaob, with disastrous economic repercussions for Palau, which relies on the route for all domestic and international communications. A state of emergency was declared as the authorities sought to re-establish water and power supplies and to reconnect telephone lines. It was subsequently revealed that major repairs had recently been carried out on the bridge, at a cost of US $3.2m., and several reports implied that inappropriate changes made to its structure were responsible for the disaster. Toribiong was harshly critical of Nakamura, who had commissioned the repair work, and demanded his resignation along with those of the public works officials involved. However, following the revelation that Toribiong's running-mate in the election for the vice-presidency was one of the officials involved in the work, Toribiong withdrew his candidacy from the second round of the presidential election. A new bridge linking Koror to Babeldaob opened in January 2002. Its construction was funded by the Japanese Government, at a cost of some $25m.

Legislative elections and the second round of presidential voting were held on 5 November 1996. Nakamura was re-elected with 62.0% of total votes, defeating Yutaka Gibbons (who had re-entered the contest, following the withdrawal of Toribiong in October). Legislation providing for the establishment of an 'offshore' financial centre in Palau was approved by the Senate (the upper house of the Palau National Congress) in October 1998, despite strong opposition by President Nakamura, who believed that the new status might attract criminal organizations seeking to 'launder' the proceeds of their illegal activities and that it might compromise the country's ability to enforce corporate law. In December 1999, following discussions with US government officials, President Nakamura signed an executive order establishing a National Banking Review Commission, with the aim of maintaining a legally responsible banking environment in Palau: both the Bank of New York (of the USA) and Deutsche Bank (of Germany) had alleged that Palau's 'offshore' banks were facilitating money-laundering. The new body was given wide-ranging powers to examine banking operations in the country and to evaluate current banking regulations.

At the presidential election held on 7 November 2000, Thomas E. Remengesau, Jr, hitherto Palau's Vice-President, was elected with 52% of the votes cast, defeating Senator Peter Sugiyama. Sandra Pierantozzi was elected as Vice-President. Remengesau was officially inaugurated on 19 January 2001, pledging to pursue Nakamura's policies of economic expansion and transparency in government. In July 2001 Remengesau introduced a formal resolution proposing the reduction of the legislature to a single chamber, to replace the existing House of Delegates and the Senate, claiming that a unicameral legislature would reduce bureaucracy. He also proposed that henceforth the President and Vice-President should base their election campaign on a common policy, to ensure efficient and cohesive government. However, owing to a lack of legislative progress on the necessary constitutional changes, in early 2004 Remengesau endorsed a congressional resolution to conduct a popular referendum. The poll, which would first require the signatures of 25% of the electorate, was scheduled to take place in November. The proposals included the creation of a unicameral legislature and restrictions on legislators' terms of office; furthermore, it was proposed that Palauans resident in the USA be offered the opportunity of dual citizenship.

Despite President Remengesau's inauguration pledges to improve transparency in public office, there were several instances of apparent official corruption in the early 2000s. In December 2002 the Speaker of the House of Delegates, Mario Gulibert, was arrested on charges relating to the alleged misuse of travel expenses. (The charges were subsequently abandoned on condition of repayment of the funds, but he was dismissed on unrelated charges in March 2004.) In February 2003 the former Governor of Ngardmau state, Albert Ngirmekur, was fined and sentenced to six months' imprisonment, following his impeachment on allegations of theft and misuse of public funds. In March, however, the appellate division of the Supreme Court reversed an earlier ruling that had found the Governor of Ngiwal, Elmis Mesubed, along with six state legislators, liable for irregularities relating to their inauguration expenses in 2000. In late 2003 President Remengesau was obliged to veto an attempt by legislators to eliminate the Office of the Special Prosecutor, which had conducted a number of investigations into alleged misuse of public funds. In February 2004, however, members of the National Congress under investigation for alleged misuse of expenses agreed to pay some US $250,000, on condition that the cases against them be withdrawn. Meanwhile, international concern was prompted by renewed attempts in the National Congress to modify stringent banking reforms enacted in 2001 (see Economy). In November, the Senate overturned President Remengesau's veto of a bill to ease licensing restrictions, although this decision was reversed in December following the threat of international financial sanctions.

In mid-1999 a delegation from Solomon Islands visited Palau to discuss the possibility of allowing Solomon Islanders to work in Palau, in an attempt to resolve the latter's severe labour shortage. In August 2001, however, the Government imposed a ban on the hiring of Indian and Sri Lankan workers, citing rising tensions and disputes with local employers, which the Government claimed were largely due to religious differences. The ban was to remain in place until legislation to create official recruitment agencies in Palau had been approved. In July 2002 the Senate passed a bill that would amend the islands' immigration law in order to give greater authority over immigration affairs to the President of Palau. (Non-Palauan nationals were estimated to represent about 33% of the islands' population in 2002.) In March 2003 the Government introduced a measure to extend employment permits for foreign workers, which allowed a new maximum extension of two years. In April some 200 Chinese migrant employees of a failed clothing business were stranded on Palau and placed under house arrest. However, the migrants were subsequently repatriated, following diplomatic intervention from the People's Republic of China.

Diplomatic relations were established, at ambassadorial level, with Taiwan in late 1999; the first Taiwanese Ambassador to Palau was formally appointed in April 2000. Reports in December 2000 that the Palau Government was considering establishing diplomatic relations with the People's Republic of China were denied, and President-elect Remengesau reaffirmed Palau's diplomatic ties with Taiwan. In early 2000, meanwhile, Palau and Taiwan signed an agreement pledging to develop bilateral projects in a number of areas, including agriculture, fisheries and tourism.

In October 2001 the Government sought to establish diplomatic relations with Malaysia and Indonesia, in an attempt to facilitate the resolution of disputes over overlapping territorial boundaries, amid concern over increasing instances of illegal fishing in Palau waters. Palau introduced more stringent regulations concerning illegal fishing in May 2002, and introduced a comprehensive marine protection law in September 2003.

Palau maintains strong diplomatic links with Japan, which was consistently a leading source of tourist revenue throughout the 1980s and 1990s. The Japanese Government provided US $25m. for the construction of the Koror bridge (see above), and has also funded projects including the construction of the terminal building at Palau's new international airport, which opened in May 2003. In June 2002 President Remengesau undertook his first state visit to the Republic of Korea, his itinerary incorporating visits to a number of infrastructural development projects.

Economy

In 2002, according to estimates by the World Bank, Palau's gross national income (GNI), measured at average 2000–02 prices, totalled US $141.9m., equivalent to $7,140 per head. During 1999–2001, it was estimated, the population increased at an average annual rate of 2.6%, while gross domestic product (GDP) per head decreased, in real terms, by an average of 1.6%. In 1990–2002 overall GDP increased, in real terms, at an average annual rate of 1.0%; growth in 2002 was 3.0%. The Asian Development Bank (ADB) estimated that GDP contracted by 4.7% in 2002 and by 0.1% in 2003.

Agriculture (including fishing) is mainly on a subsistence level, the principal crops being coconuts, root crops and bananas. Pigs and chickens are kept for domestic consumption. Eggs are produced commercially, and the introduction of cattle-ranching on Babeldaob was under consideration from the late 1990s. The agricultural sector, together with mining, engaged 2.0% of the employed labour force in 2003. In 2003 it was estimated that agricultural activities employed just 261 people. Agriculture and fisheries alone provided 4.0% of GDP in 2001, according to the IMF. Fishing licences are sold to foreign fleets, including those of Taiwan, the USA, Japan and the Philippines. Palau is a signatory of the Multilateral Fisheries Treaty, concluded by the USA and member states of the South Pacific Forum (now Pacific Islands Forum) in 1987. The islands, however, are believed to lose significant amounts of potential revenue through illegal fishing activity. Accordingly, the Palauan authorities have enforced stricter measures against illegal fishing in recent years. Revenue from the sale of fishing licences totalled US $39,000 in 1999/2000 and an estimated $76,000 in 2000/01, a considerable decline from the total of $230,000 recorded in 1994/95. Fish have traditionally been a leading export, accounting for some $13m. of exports in 1995, but earnings declined to an estimated $7m. in 2001, reportedly owing to adverse weather conditions. The output of the islands' tuna industry declined by more than 35% between 1993 and 1997.

The industrial sector (including mining and quarrying, manufacturing, construction and utilities) provided an estimated 12.6% of GDP in 2001. In June 2004 controversial legislation was approved by Congress that established a statutory framework for petroleum exploration and subsequent exploitation. The only manufacturing activity of any significance is a factory complex producing garments, which, in 1997, employed some 300 (mostly non-resident) workers. In 2001 manufacturing accounted for only 1.5% of GDP. Construction is the most important industrial activity, contributing an estimated 7.8% of GDP in 2001 and engaging 20.2% of the employed labour force in 2003. The strength of the sector reflects considerable investment in infrastructural projects in the early 2000s. Road building (undertaken with foreign financial aid) has been ongoing, and a new official capital was also under construction on the island. Electrical energy is produced by two power plants, at Aimeliik and Malakal. Construction of an innovative thermal energy generator and desalination plant, scheduled to begin in 2003, was to be partly funded by the Government of Japan, with technical assistance provided by a Japanese university.

Service industries dominate Palau's economy, providing an estimated 83.3% of GDP in 2001 and (with utilities) engaging 77.8% of the employed labour force in 2003. The Government is an important employer within the sector, with public administration engaging some 33.0% of the total employed labour force in 2003. Tourism is an important source of foreign exchange, with hotels and restaurants contributing an estimated 10.5% of GDP in 2001 and employing 9.4% of paid workers in 1995. In 2002/03 a total of 65,772 visitors arrived in the islands, the majority coming from Taiwan, Japan and the USA. Expenditure by tourists totalled an estimated US $58m. in that year. A $16m. terminal building at Palau's new international airport, on Babeldaob, was opened in May 2003. In August 2004 Palau's first locally-owned airline, Palau Micronesia Air, made its inaugural flight to Manila, in the Philippines.

In the year ending September 2003 the visible trade deficit was estimated at US $81.8m. The deficit on the current account of the balance of payments totalled $5.3m. in that year. The principal sources of imports in 2002 were the USA (which supplied 42.1% of the total), Guam (12.1%) and Japan (9.8%). The principal imports in 2002 were manufactured goods, machinery and transport equipment, and food and live animals.

The islands record a persistent budget deficit, projected at US $9.6m. in the year ending 30 September 2004. Financial assistance from the USA contributes a large part of the islands' external revenue. Furthermore, upon implementation of the Compact of Free Association with the USA in 1994, Palau became eligible for an initial grant of $142m. and for annual aid of $23m. over a period of 14 years. Assistance from the USA (including non-Compact funding) totalled $20.8m. in 2000/01 and an estimated $21.4m. in 2001/02, a respective 39.1% and 41.8% of total budgetary revenue in those years. Palau's external debt was relatively low in 2000/01, totalling $20.0m., largely owing to the role of foreign grants in funding the country's capital-intensive development projects. According to the ADB, however, in December 2003 the external debt stood at $32m., equivalent to 26.9% of GDP. The annual rate of inflation was thought to average between 1% and 3% in the mid-1990s. The inflation rate was estimated by the ADB at 0.1% in 2002, compared with deflation of 0.6% in the previous financial year, rising to 1.3% in 2003. In 2000 2.3% of the total labour force were unemployed.

A period of recession in the early 1990s (caused largely by reductions in government expenditure in an attempt to offset decreases in revenue) was alleviated by substantial aid payments from the US Government, following the implementation of the Compact of Free Association, and by the dramatic expansion of tourism in the country. Subsequent policies to reduce the size of the public sector, to encourage the return of Palauans resident overseas and to attract foreign investment aimed to stimulate economic growth. Particularly strong commercial ties were forged with Taiwan during the mid-1990s, and the establishment of diplomatic relations between Palau and Taiwan in late 1999 has resulted in substantial increases in Taiwanese investment. Trade between Palau and the Philippines also increased following the signing of a bilateral agreement in early 1998. The development of tourism is expected to underpin economic growth following the scheduled termination of direct US aid under the Compact of Free Association in 2009. The Government announced plans in 1998 for a major expansion of the industry, which had declined in the late 1990s following the Asian financial crisis, and expressed particular interest in developing the country as a speciality destination for visitors of above-average wealth. The Government has remained opposed to the development of mass tourism in the islands, and in June 2002 President Thomas Remengesau vetoed a bill to legalize casinos. In August of that year, none the less, President Remengesau announced a new five-year economic plan, a principal aim of which was to develop basic infrastructure in order to permit the further expansion of the tourist sector. However, concern has been expressed that large-scale development of tourism (especially the diving sector) could lead to ecological damage. Consequently, consideration was being given in 2004 to diversification in the tourist market, to include activities such as golf.

An IMF report issued in mid-2002 assessed that the country's investment laws (which prohibit foreign ownership of land or businesses) were contributing to a lack of transparency in financial dealings. Meanwhile, following allegations that Palau's 'offshore' banking system was being used for the purposes of money-laundering (see History), various new banking laws were approved, in an effort to restore the confidence of the international financial community. Although a number of concerns were raised by the Paris-based Financial Action Task Force on Money Laundering (the FATF having been established in 1989 on the recommendation of the Group of Seven industrialized nations), Palau was not included on the FATF's list of non-co-operative countries and territories. In June 2002 the legislature approved new measures to regulate the financial sector, including the creation of a Financial Institutions Commission. A subsequent IMF report in June 2004 judged the Palauan authorities 'largely compliant' with its recommendations on combating money-laundering, but 'materially non-compliant' with its recommendations on accounting standards.

Palau is a member of the Pacific Community (formerly the South Pacific Commission) and of the Pacific Islands Forum (formerly the South Pacific Forum); it is also an associate member of the UN's Economic and Social Commission for Asia and the Pacific (ESCAP). In early 1996 Palau joined representatives of the other countries and territories of Micronesia at a meeting in Hawaii, at which a new regional organization, the Council of Micronesian Government Executives, was established. The new body aimed to facilitate discussion of economic developments in the region. In December 2002 Palau requested admission to the World Trade Organization. The country joined the Asian Development Bank (ADB) in December 2003.

Statistical Survey

AREA AND POPULATION

Area: 508 sq km (196 sq miles); Babeldaob (Babeldaop, Babelthuap) island 409 sq km (158 sq miles).

Population: 17,225 (males 9,213, females 8,012) at census of 9 November 1995; 19,129 at census of 15 April 2000.

Density (2000): 37.7 per sq km (97.6 per sq mile).

Births and Deaths (1999): Live births 250 (birth rate 13.2 per 1,000); Deaths 131 (death rate 6.9 per 1,000). Source: UN, *Population and Vital Statistics Report*.

Expectation of Life (WHO estimates, years at birth): 68.5 (males 66.4; females 70.9) in 2002. Source: WHO, *World Health Report*.

Economically Active Population (2003): Agriculture, fishing and mining 261; Construction 2,678; Manufacturing 6; Transport, communications and utilities 526; Trade, hotels and restaurants 2,639; Finance, insurance and real estate 169; Public administration 4,374; Other services 2,595; *Total employed* 13,248. Source: IMF, *Republic of Palau: Selected Issues and Statistical Appendix* (April 2004).

HEALTH AND WELFARE

Key Indicators

Total Fertility Rate (children per woman, 2002): 2.4.

Under-5 Mortality Rate (per 1,000 live births, 2002): 29.

Physicians (per 1,000 head, 1998): 1.10.

Health Expenditure (2001): US $ per head (PPP): 426.

Health Expenditure (2001): % of GDP: 9.2.

Health Expenditure (2001): public (% of total): 92.0.

Access to Water (% of persons, 2000): 79.

Access to Sanitation (% of persons, 2000): 100.

For sources and definitions, see explanatory note on p. vi.

AGRICULTURE, ETC.

Principal Crops, Livestock and Livestock Products: see Chapter on the Federated States of Micronesia.

Fishing (FAO estimates, metric tons, live weight, 2002): Total catch 1,010 (Marine fishes 1,008). Source: FAO.

INDUSTRY

Production (estimate, 1999): Electric energy 210 million kWh. Source: UN, *Statistical Yearbook for Asia and the Pacific*.

FINANCE

Currency and Exchange Rates: United States currency is used: 100 cents = 1 United States dollar (US $). *Sterling and Euro Equivalents* (31 May 2004): £1 sterling = US $1.8347; €1 = US $1.2246; US $100 = £54.50 = €81.66.

Budget (official forecasts, US $'000, year ending Sept. 2004): *Revenue:* Domestic revenue 36,400 (Tax 29,600, Other current revenues 5,300, Local trust fund 1,500,); Grants 38,100 (Compact funds 14,100); Total 74,600. *Expenditure:* Current expenditure 62,500 (Wages and salaries 30,600, Purchase of goods and services 26,700, Subsidies and other transfers 4,100, Interest payments and investment fees 1,100); Capital expenditure 21,700; Total 84,200. Source: IMF, *Republic of Palau: Selected Issues and Statistical Appendix* (April 2004).

Gross Domestic Product by Economic Activity (US $'000 at current prices, 2001): Agriculture 1,399; Fishing 3,372; Mining and quarrying 240; Manufacturing 1,774; Electricity, gas and water 3,741; Construction 9,181; Trade 24,337; Hotels and restaurants 12,419; Transport and communications 10,855; Finance and insurance 4,647; Real estate and business services 5,036; Public administration 30,860; Other services 10,381; *Sub-total* 118,243; Import duties 3,842; *Less* Imputed bank service charges 1,250; *GDP in purchasers' values* 120,835. Source: IMF, *Republic of Palau: Selected Issues and Statistical Appendix* (April 2004).

Balance of Payments (estimates, US $ million, year ending Sept. 2003): Exports of goods f.o.b. 11.9; Imports of goods f.o.b. –93.7; *Trade balance* –81.8; Exports of services 57.9; Imports of services –8.7; *Balance on goods and services* –32.6; Other income (net) 3.9; *Balance on goods, services and income* –28.7; Private current transfers (net) –1.3; Official current transfers (net) 24.7; *Current balance* –5.3; Capital grants received 22.8; Loan repayments –0.6; Private direct investment 1.0; Net errors and omissions –21.3; *Overall balance* –3.4. Source: IMF, *Republic of Palau: Selected Issues and Statistical Appendix* (April 2004).

EXTERNAL TRADE

Principal Commodities (US $'000): *Imports f.o.b.* (2002): Food and live animals 14,381; Beverages and tobacco 7,004; Mineral fuels, lubricants, etc. 12,430; Chemicals 4,860; Basic manufactures 23,647; Machinery and transport equipment 21,902; Miscellaneous manufactured articles 9,921; Total (incl. others) 95,874. *Exports* (2001/02): 11,900 (including trochus, tuna, copra and handicrafts) (Source: mainly IMF, *Republic of Palau: Selected Issues and Statistical Appendix*, April 2004).

Principal Trading Partners (US $'000): *Imports* (2002): Australia 760; China, People's Republic 190; Guam 11,558; Hong Kong 3,457; Japan 9,364; Korea, Republic 8,523; Philippines 4,083; Singapore 7,439; Taiwan 5,439; USA 40,356; Total (incl. others) 95,874. *Exports* (2001/02): Total 11,900. Source: IMF, *Republic of Palau: Selected Issues and Statistical Appendix* (April 2004).

TRANSPORT

International Shipping (1995): *Ship Arrivals:* 280. *Freight Traffic* (metric tons): Goods unloaded 64,034.

TOURISM

Tourist Arrivals: 55,586 in 2000/01; 54,797 in 2001/02; 65,772 in 2002/03.

Tourist Arrivals by Country of Residence (2002/03): Japan 21,620; Philippines 3,409; Taiwan 26,813; USA 9,449.
Source: IMF, *Republic of Palau: Selected Issues and Statistical Appendix* (April 2004).

COMMUNICATIONS MEDIA

Radio Receivers (1997): 12,000 in use.

Television Receivers (1997): 11,000 in use.

EDUCATION

Elementary (1990): 22 government schools; 172 teachers (1998); 2,125 pupils at government schools; 369 pupils at private schools.

Secondary (1990): 1 government school; 5 private (church-affiliated) schools; 60 teachers (1998); 610 pupils at government schools; 445 pupils at private schools (1989).

Tertiary (1986): 305 students.

2004: 23 government schools (22 elementary; 1 secondary) with 3,145 pupils enrolled in 2001/02.

Directory

The Constitution

In October 1994 Palau, the last remaining component of the Trust Territory of the Pacific Islands (a United Nations Trusteeship administered by the USA), achieved independence under the Compact of Free Association. Full responsibility for defence lies with the USA, which undertakes to provide regular economic assistance.

From 1986 the three polities of the Commonwealth of the Northern Mariana Islands, the Republic of the Marshall Islands and the Federated States of Micronesia ceased, *de facto*, to be part of the Trust Territory. In December 1990 the United Nations Security Council agreed formally to terminate the Trusteeship Agreement for all the territories except Palau. The agreement with Palau was finally terminated in October 1994.

The islands became known as the Republic of Palau when the locally-drafted Constitution came into effect on 1 January 1981. The Constitution provides for a democratic form of government, with executive authority vested in the directly-elected President and

Vice-President. Presidential elections are held every four years. Legislative power is exercised by the Olbiil era Kelulau, the Palau National Congress, which is an elected body consisting of the Senate and the House of Delegates. The Senators represent geographical districts, determined by an independent reapportionment commission every eight years, according to population. There are currently 14 Senators (four from the northern part of Palau, nine from Koror and one from the southern islands). There are 16 Delegates, one elected to represent each of the 16 states of the Republic. The states are: Kayangel, Ngerchelong, Ngaraard, Ngardmau, Ngaremlengui, Ngiwal, Melekeok, Ngchesar, Ngatpang, Aimeliik, Airai, Koror, Peleliu, Angaur, Sonsorol and Tobi. Each state elects its own Governor and legislature.

The Government

HEAD OF STATE

President: THOMAS E. REMENGESAU, Jr (took office 19 January 2001).

Vice-President: SANDRA S. PIERANTOZZI.

THE CABINET
(September 2004)

Vice-President and Minister of Health: SANDRA S. PIERANTOZZI.

Minister of Commerce and Trade: OTOICHI BESEBES.

Minister of Resources and Development: FRITZ KOSHIBA.

Minister of Education: MARIO KATOSANG.

Minister of Justice: MICHAEL ROSENTHAL.

Minister of Community and Cultural Affairs: ALEXANDER R. MEREP.

Minister of State: TEMMY L. SHMULL.

Minister of Administration: ELBUCHEL SADANG.

COUNCIL OF CHIEFS

The Constitution provides for an advisory body for the President, comprising the 16 highest traditional chiefs from the 16 states. The chiefs advise on all traditional laws and customs, and on any other public matter in which their participation is required.

Chairman: Ibedul YUTAKA GIBBONS (Koror).

GOVERNMENT OFFICES AND MINISTRIES

Office of the President: POB 100, Koror, PW 96940; tel. 488-2403; fax 488-1662; e-mail roppresoffice@palaunet.com.

Department of the Interior, Office of Insular Affairs (OIA): OIA Field Office, POB 6031, Koror, PW 96946; tel. 488-2601; fax 488-2649; Field Rep. J. VICTOR HOBSON, Jr; Co-ordinator HAURO WILLTER.

Ministry of Commerce and Trade: POB 1471, Koror, PW 96940; tel. 488-4343; fax 488-3207.

Ministry of Health: POB 6027, Koror, PW 96940; tel. 488-2552; fax 488-1211; e-mail minister@palau-health.net.

Ministry of Resources and Development: POB 100, Koror, PW 96940; tel. 488-2701; fax 488-3380; e-mail mrd@palaunet.com.

All national government offices are based in Koror. Each state has its own administrative headquarters.

President and Legislature

PRESIDENT

At the presidential election held on 7 November 2000, Thomas E. Remengesau, Jr, who won 52% of the votes cast, narrowly defeated Sen. Peter Sugiyama.

OLBIIL ERA KELULAU
(Palau National Congress)

President of the Senate: SEIT ANDRES.

Vice-President of the Senate: HARRY R. FRITZ.

Speaker of the House of Delegates: (vacant).

Vice-Speaker of the House of Delegates: WILLIAM NGIRAIKELAU.

Political Organizations

(There are currently no active political parties in Palau)

Palau Nationalist Party: c/o Olbiil era Kelulau, Koror, PW 96940; inactive; Leader JOHNSON TORIBIONG.

Ta Belau Party: c/o Olbiil era Kelulau, Koror, PW 96940; inactive; Leader KUNIWO NAKAMURA.

Diplomatic Representation

EMBASSIES IN PALAU

China (Taiwan): POB 9087, Koror, PW 96940; tel. 488-8150; fax 488-8151; Ambassador CLARK K. H. CHEN.

Japan: POB 6050, Palau Pacific Resort, Arakebesang, Koror, PW 96940; tel. 488–6455; fax 488-6458; Ambassador KENRO IINO.

Philippines: 2nd Flr, M. Ueki Bldg., Iyebukel Hamlet, Koror; tel. 488-5077; fax 488-6310; e-mail philkor@palaunet.com; Ambassador LEONCIO R. PARUNGAO.

USA: POB 6028, Koror, PW 96940; tel. 488-2920; fax 488-2911; e-mail usembassy@palaunet.com; Chargé d'affaires RONALD A. HARMS.

Judicial System

The judicial system of the Republic of Palau consists of the Supreme Court (including Trial and Appellate Divisions), presided over by the Chief Justice, the National Court (inactive), the Court of Common Pleas and the Land Court.

Supreme Court of the Republic of Palau: POB 248, Koror, PW 96940; tel. 488-2482; fax 488-1597; e-mail cjngiraklsong@palaunet.com; Chief Justice ARTHUR NGIRAKLSONG.

Religion

The population is predominantly Christian, mainly Roman Catholic. The Assembly of God, Baptists, Seventh-day Adventists, the Church of Jesus Christ of Latter-day Saints (Mormons), and the Bahá'í and Modignai (or Modeknai) faiths are also represented.

CHRISTIANITY

The Roman Catholic Church

Palau forms part of the diocese of the Caroline Islands, suffragan to the archdiocese of Agaña (Guam). The Bishop, who is resident in Chuuk, Eastern Caroline Islands (see the Federated States of Micronesia), participates in the Catholic Bishops' Conference of the Pacific, based in Suva, Fiji.

MODIGNAI FAITH

Modignai Church: Koror, PW 96940; an indigenous, non-Christian religion; also operates a high school.

The Press

Palau Gazette: POB 100, Koror, PW 96940; tel. 488-3257; fax 488-1662; e-mail roppresoffice@palaunet.com; newsletter publ. by Govt; monthly.

Palau Horizon: POB 487, Meketii, Koror, PW 96940; tel. 488-4588; fax 488-4565; e-mail hprinting@palaunet.com; twice weekly; f. 1998; Pub. ABED E. YOUNIS; circ. 1,500.

Roureur Belau: POB 477, Koror, PW 96940; tel. 488-6365; fax 488-4810; e-mail myu@palaunet.com; weekly; Publr CLIFFORD 'SPADE' EBAS.

Tia Belau (This is Palau): POB 477, Koror, PW 96940; tel. 488-6365; fax 488-4810; e-mail myu@palaunet.com; f. 1992; fortnightly; English and Palauan; Editor RAOUL G. BRIONES; Publr MOSES ULUDONG; circ 1,500.

Rock Islander: POB 1217, Koror, PW 96940; tel. 488-1461; fax 488-1614; e-mail jerome@palaunet.com; quarterly; articles about Palau; Publr JACKSON HENRY.

Broadcasting and Communications

BROADCASTING

Radio

KRFM: Sure Save Store, Koror, PW 96940; tel. 488-1359; e-mail rudimch@palaunet.com.

Palau National Communications Corpn (PNCC): Bureau of Domestic Affairs, POB 279, Koror, PW 96940; tel. 587-9000; fax 587-1888; e-mail pncc@palaunet.com; internet www.palaunet.com; f. 1982; mem. of the Pacific Islands Broadcasting Asscn; operates station WSZB; broadcasts American, Japanese and Micronesian music; 18 hrs daily; Gen. Man. ED CARTER.

T8AA (Eco Paradise): POB 279, Koror, PW 96940; tel. 488-2417; fax 488-1932; broadcasts news, entertainment and music; govt.

WWFM: POB 1327, Koror, PW 96940; tel. 488-4848; fax 488-4420; e-mail wwfm@palaunet.com; internet www.brouhaha.net/palau/wwfm.html; Man. ALFONSO DIAZ.

WSZB Broadcasting Station: POB 279, Koror, PW 96940; tel. 488-2417; fax 488-1932; Station Man. ALBERT SALUSTIANO.

KHBN: POB 66, Koror, PW 96940; tel. 488-2162; fax 488-2163; e-mail hamadmin@palaunet.com; f. 1992; broadcasts religious material; CEO JACKIE MITCHUM YOCKEY.

Television

STV-TV Koror: POB 2000, Koror, PW 96940; tel. 488-1357; fax 488-1207; broadcasts 12 hrs daily; Man. DAVID NOLAN; Technical Man. RAY OMELEN.

Island Cable Television, Inc: POB 39, Koror, PW 96940; tel. 488-1490; fax 488-1499; e-mail nor_ictv@palaunet.com; owned by the Palau National Communications Corpn.

Finance

(cap. = capital; res = reserves; amounts in US dollars)

BANKING

Bank of Guam: POB 338, Koror, PW 96940; tel. 488-2696; fax 488-1384; internet www.bankofguam.com; Man. KATHRINE C. LUJAN.

Bank of Hawaii (USA): POB 340, Koror, PW 96940; tel. 488-2428; fax 488-2427; internet www.boh.com.

Bank Pacific: POB 1000, Koror, PW 96940; tel. 488-5635; fax 488-4752; Man. JOSEPH KOSHIBA.

National Development Bank of Palau: POB 816, Koror, PW 96940–0816; tel. 488-2578; fax 488-2579; e-mail ndbp@palaunet.com; f. 1982; cap. and res 11.5m. (Sept. 2002); Pres. KALEB ADUI, Jr.

Pacific Savings Bank: POB 399, Koror, PW 96940; tel. 488-1859; fax 488-1858; e-mail bank@palaunet.com; Pres. TIM TAUNTON.

In 1990 there were also 22 registered credit unions, with 1,025 members, but they were severely affected by the financial crisis in Palau.

INSURANCE

Century Insurance Co: POB 318, Koror, PW 96940; tel. 488-8580; fax 488-8632; e-mail knakamura@palaunet.com; internet www.tanholdings.com/cic.asp.

NECO Insurance Underwriters Ltd: POB 129, Koror, PW 96940; tel. 488-2325; fax 488-2880; e-mail necogroup@palaunet.com.

Poltalia National Insurance: POB 12, Koror, PW 96940; tel. 488-2254; e-mail psata@palaunet.com.

Trade and Industry

CHAMBER OF COMMERCE

Palau Chamber of Commerce: POB 1742, Koror, PW 96940; tel. 488-3400; fax 488-3401; e-mail pcoc@palaunet.com; f. 1984; Exec. Dir LEILANI NGIRTURONG.

CO-OPERATIVES

These include the Palau Fishermen's Co-operative, Palau Boatbuilders' Asscn and the Palau Handicraft and Woodworkers' Guild. In 1990, of the 13 registered co-operatives, eight were fishermen's co-operatives, three consumers' co-operatives (only two in normal operation) and two farmers' co-operatives (one in normal operation).

Transport

ROADS

Macadam and concrete roads are found in the more important islands. Other islands have stone and coral-surfaced roads and tracks. The Government is responsible for 36 km (22 miles) of paved roads and 25 km (15 miles) of coral- and gravel-surfaced roads. Most paved roads are located on Koror and are in a poor state of repair. A major project to construct a new 85-km (53-mile) road around Babeldaob began in 1999 and is scheduled for completion in 2005. The project is being funded with US $150m. from the Compact of Free Association.

SHIPPING

Most shipping in Palau is government-organized. However, the Micronesia Transport Line operates a service from Sydney (Australia) to Palau. A twice-weekly inter-island service operates between Koror and Peleliu. There is one commercial port at Malakal Harbor, which is operated by the privately-owned Belau Transfer and Terminal Company.

CIVIL AVIATION

There is an international airport on Babeldaob. A new terminal building, construction of which was funded by a grant of US 16m. from Japan, opened in May 2003. Domestic airfields (former Japanese military airstrips) are located on Angaur and Peleliu. Continental Micronesia (Northern Mariana Islands and Guam) provides daily flights to Koror from Guam, and twice-weekly flights from Manila (Philippines). Cebu Pacific operates direct flights from Davao (Philippines) to Koror. A civil aviation agreement between Palau and Taiwan was signed in 1997, and direct charter flights between the two countries began in the following year, operated by the Far Eastern Air Transport Corpn. Palau Trans Pacific also operates from Taiwan. In August 2004 Palau's first locally-owned airline, Palau Micronesia Air, made its inaugural flight to Manila. Its routes were to include the Federated States of Micronesia, the Philippines and Australia.

Palau Micronesia Air: POB 9048, Ernguul Rd, Topside, Koror , PW 96940; tel. 488-1111; fax 488-8830; e-mail info@palau-air.com; internet www.palau-air.com; serving Chuuk, Yap, Saipan, Manila (Philippines) and Darwin (Australia); Pres. and CEO ALAN SEID.

Palau Paradise Air, Inc: POB 488, Palau Int. Airport, Airai State, Babeldaob, PW 96940; tel. and fax 488-2348; fax 488-2348; internal services between Koror, Peleliu and Angaur.

Rock Island Airlines: managed by Aloha Airlines (Hawaii).

Tourism

Tourism is becoming increasingly important in Palau. The islands are particularly rich in their marine environment, and the Government has taken steps to conserve and protect these natural resources. The myriad Rock Islands, now known as the Floating Garden Islands, are a noted reserve in the lagoon to the west of the main group of islands. There were 1,049 hotel rooms in 2002. In that year there were 58,560 visitor arrivals, of whom 40.1% were from Japan, 27.0% from Taiwan, 14.5% from the USA and 5.8% from the Philippines. In the year ending September 2003 the number of arrivals had risen to 65,772. Tourist expenditure rose from US $53.2m. in 1999/2000 to an estimated $66.1m. in 2002.

Belau Tourism Association: POB 9032, Koror, PW 96940; tel. 488-4377; fax 488-1725; e-mail bta@palaunet.com.

Palau Visitors' Authority: POB 256, Koror, PW 96940; tel. 488-2793; fax 488-1453; e-mail info@visit-palau.com; internet www.visit-palau.com; Man. Dir MARY ANN DELEMEL.

Defence

The USA is responsible for the defence of Palau, according to the Compact of Free Association implemented in October 1994, and has exclusive military access to Palau's waters, as well as the right to operate two military bases on the islands. The US Pacific Command is based in Hawaii.

Education

The educational system is similar to that in the USA. The Government of Palau is now responsible for the state school system, which most children attend. Education is free and compulsory between the ages of six and 14, and secondary education may be obtained at the public High School or one of the five private ones. In 1990 there were 22 public elementary schools with 2,125 pupils (and 369 pupils attending private elementary schools). In the same year there were 610 students enrolled at government high schools, and 445 at private high schools. In 2001/02 3,145 pupils attended 19 public elementary schools and one government high school. The Micronesian Occupational College, based in Koror, provides two-year training programmes. Construction of Palau's first university, as part of a major development project (to include a housing complex for 5,000 people, a commercial centre and leisure facilities), was expected to begin in the late 1990s. The project was to be funded by Taiwanese interests. Government expenditure on education totalled US $9.1m. in 1999/2000, equivalent to 10.7% of total budgetary expenditure.

PAPUA NEW GUINEA

Physical and Social Geography

Papua New Guinea lies east of Indonesia and north of the north-eastern extremity of Australia. It comprises the eastern part of the island of New Guinea, the western section of which, Papua (formerly Irian Jaya), is part of Indonesia, and some smaller islands including the Bismarck Archipelago (mainly New Britain, New Ireland and Manus) and the northern part of the Solomon Islands (mainly Bougainville and Buka). It covers a total area of 462,840 sq km (178,704 sq miles). The climate is hot and humid throughout the year, with an average maximum temperature of 33°C (91°F) and an average minimum of 22°C (72°F).

The census of July 2000 recorded a population of 5,130,365. The UN estimated the population to be 5,711,000 at mid-2003.

History

New Guinea was visited by European navigators from the early 16th century onwards, but exploration and colonial settlement did not begin until the mid-19th century. The western part of New Guinea was administered until 1949 as part of the Netherlands East Indies and from 1949 until 1962 as the Nederlands Nieuw Guinea. In 1963, after military action by Indonesia, the territory was redesignated Daerah Irian Barat and declared a part of Indonesia by an 'act of free choice' in August 1962. It is now known as Papua, a province of Indonesia formerly known as Irian Jaya (or West Papua).

The southern part of eastern New Guinea became British New Guinea in 1906, following the establishment of a British protectorate in 1884 and annexation in 1886. Australia administered what became the Territory of Papua until 1949, when it was joined, under a unified administration, with the Trust Territory of New Guinea (see below).

In 1884 the northern part of eastern New Guinea came under German administration as Schutzgebiet Kaiser-Wilhelmsland und Bismarck-archipel, later becoming known as German New Guinea. In 1914 the Germans were removed from the territory by Australian troops, and Australia subsequently administered the area under a League of Nations mandate until 1942, when much of it fell under Japanese occupation. In 1945 the territory returned to Australian administration under UN trusteeship arrangements. In December 1973 the Territory of Papua New Guinea became internally self-governing, and on 16 September 1975 became the independent nation of Papua New Guinea.

A House of Assembly (renamed the National Parliament at independence) was elected in February 1972. It had 102 members elected by universal adult suffrage. There were 20 ministers including the Prime Minister, Michael Somare, and an inner cabinet of 10 ministers. The Government was formed from a loose coalition of three main political parties with the opposition made up of one main party and a number of independents supporting both sides.

Although broad cultural similarities occur among groups in the Papuan Coastal, New Guinea Highlands, New Guinea Coastal, New Guinea Islands and Bougainville regions, local group sympathies are strong and in rural areas where approximately 80% of the indigenous population live, there is little interest in national issues.

The Papua New Guinea Government inherited from Australia a highly bureaucratic, centralized administration unsuited to a country in which transport is so difficult and national awareness so low. Strong pressure from some provinces during 1975 and 1976 resulted in the Government reintroducing into the Constitution provisions for decentralized provincial governments. In 1976 leaders of a group styling itself the 'Independent Republic of the North Solomons' on Bougainville island threatened to secede from Papua New Guinea. After several months of negotiations, Bougainville was granted provincial government status and became self-governing in July 1976. By 1978 all 20 provinces had been granted provincial government status.

The third national election since independence was held in June 1982. The Pangu Pati won 41 seats in the 102-member National Parliament, which in August elected Michael Somare as Prime Minister. Sir Julius Chan's People's Progress Party (PPP) won only 12 seats. In 1983 the Somare Government effected a constitutional change to provide the central authorities with greater control of the provincial governments as a means of preventing abuse of their powers. As a result, Somare was able to suspend the Enga provincial government in 1984. Several other provincial governments were subsequently suspended for financial mismanagement.

In March 1985 a motion expressing 'no confidence' in Somare's Government was introduced in Parliament by Sir Julius Chan, who nominated Paias Wingti (hitherto Deputy Prime Minister and a member of Somare's Pangu Pati) as alternative Prime Minister. Somare quickly formed a coalition, comprising the ruling Pangu Pati, the National Party (NP) and the Melanesian Alliance (MA), and the 'no confidence' motion was rejected. Fourteen members of Parliament who had supported the motion were expelled from the Pangu Pati, and subsequently formed a new political party, the People's Democratic Movement (PDM), under the leadership of Wingti.

In August 1985 the NP withdrew from Somare's coalition Government, and in November Sir Julius Chan presented another motion of 'no confidence', criticizing Somare's handling of the economy. Somare was defeated, and Wingti took office as Prime Minister, at the head of a new five-party coalition Government (comprising the PDM, the PPP, the NP, the United Party (UP) and the MA), with Chan as Deputy Prime Minister.

At the mid-1987 general election to the National Parliament Somare's Pangu Pati won 26 of the 109 elective seats and Wingti's PDM obtained 18. However, by forming a coalition with four minor parties (the PPP, the UP, the People's Action Party (PAP) and the Papua Party) together with a group of independents, Wingti succeeded in securing a parliamentary majority and was re-elected Prime Minister. In July 1988 Wingti was defeated in a further 'no confidence' motion, and Rabbie Namaliu, who had replaced Somare as leader of the Pangu Pati, took office as Prime Minister.

In June 1990 Namaliu's position was strengthened when Utula Samana, leader of the Melanesian United Front (MUF), and four other members of the party transferred their support to the Government, although they remained members of the MUF. This development gave Namaliu the two-thirds' majority in Parliament which he required to pass constitutional amendments. Allegations that Namaliu had paid over K400,000 to four ministers dismissed in a government reorganization in June, in order to secure their support for an aborted 'no confidence' motion, provoked a demonstration of 6,000–7,000 people outside Parliament, calling for its dissolution and for Namaliu's resignation. The increasingly frequent use of the 'no confidence' motion in Parliament prompted the proposal of an amendment to the Constitution, granting incoming Prime Ministers a minimum period of 18 months before such a motion could be presented. In August the proposal was approved by Parliament and was incorporated into the Constitution in July 1991.

Apart from the continued unrest on Bougainville (see below), the principal cause for concern in Papua New Guinea's domestic affairs from the 1980s was the increase in serious crime on the islands. In 1991 the National Parliament approved a programme of severe measures to combat crime, including the introduction of the death penalty and the tattooing of the foreheads of convicted criminals.

In September 1991 a leadership tribunal found Ted Diro, the leader of the PAP, guilty of 81 charges of misconduct in office. However, the Governor-General, Sir Serei Eri, refused to ratify the tribunal's decision and reinstated Diro as Deputy Prime Minister, despite recommendations that he be dismissed. A constitutional crisis subsequently arose, during which a government envoy was sent to London to request that the Queen dismiss Eri. However, on 1 October the resignation of the Governor-General was announced and was followed shortly afterwards by that of Diro.

In 1992 the Government continued to be troubled by allegations of corruption and misconduct (notably bribery and misuse of public funds). The general election campaign began amid serious fighting among the various political factions, which led to rioting, in April 1992, by some 10,000 supporters of rival candidates. At the election, held in June, a total of 59 members of the legislature (including 15 ministers) lost their seats. The final result gave the Pangu Pati 22 of the 109 elective seats, while the PDM secured 15. Independent candidates won a total of 31 seats. Paias Wingti of the PDM was subsequently elected Prime Minister; he later formed a coalition of PDM, PPP and League for National Advancement (LNA) members, as well as several independents. As part of an anti-corruption policy, Wingti suspended six provincial governments for financial mismanagement in October, and announced plans to abolish the entire local government system.

In early 1993 a resurgence of tribal violence, mainly in the Enga province, resulted in the deaths of more than 100 people. This development, together with a continued increase in violent crime throughout the country (despite the severe measures introduced in May 1991), prompted the National Parliament to approve a new Internal Security Act in May 1993. The legislation provided the Government with greatly increased powers, which permitted the introduction of a system of national registration, the restriction of freedom of movement within the country and the erection of permanently policed gates on all major routes into the capital, Port

Moresby. Most significantly, the legal system was to be changed so that defendants accused of serious crimes would be required to prove their innocence, rather than be proved guilty. The measures were criticized by the opposition as oppressive. In May 1994 the Supreme Court nullified six of the 26 sections of the act (most of which concerned the extension of police powers) as unconstitutional.

In September 1993 Prime Minister Wingti announced his resignation to Parliament. The Speaker immediately requested nominations for the premiership, and Wingti was re-elected unopposed with 59 of the 109 votes. According to the Constitution (as amended in July 1991), a motion of 'no confidence' in the Prime Minister could not be presented for at least 18 months, and Wingti claimed that his action had been necessary in order to secure a period of political stability for the country. Opposition members, however, described the events as an abuse of the democratic process, and several thousand demonstrators gathered in the capital to demand Wingti's resignation.

In August 1994 the Supreme Court declared Wingti's re-election in September 1993 invalid. Wingti did not contest the ensuing parliamentary vote for a new Prime Minister, in which Sir Julius Chan of the PPP defeated the Speaker, Bill Skate, by 66 votes to 32. Chris Haiveta, the leader of the Pangu Pati, was appointed Deputy Prime Minister.

In September 1994 a state of emergency was declared following major volcanic eruptions and earthquakes around Rabaul in East New Britain Province. Some 100,000 inhabitants were evacuated from the area, and a programme of rehabilitation was announced in early 1995.

The abolition of the directly elected provincial government system, as proposed by Wingti, was rejected by Chan's Government. However, there was still considerable support among the opposition for the planned changes, and, as a result, it was decided that Parliament would vote on a series of motions to amend the Constitution accordingly; the first of these was approved in March 1995. However, in June the Pangu Pati withdrew its support for the reforms, and considerable opposition to the proposals was expressed by the provincial governments, particularly in Morobe. Nevertheless, the motion was approved later that month on condition that Chan lent his support to several opposition amendments to be considered later in the year. Chan subsequently dismissed six ministers for failing to vote for the legislation and effected a major reorganization of portfolios. The new regional authorities, comprising national politicians and selected local councillors and led by appointed governors, were appointed in that month. Wingti resigned as leader of the opposition in order to assume the post of Governor in the Western Highlands Province, and was replaced by another PDM member, Roy Yaki.

Meanwhile, the country continued to experience serious problems relating to crime and tribal violence, particularly in Port Moresby and in the form of banditry on the Highlands highway. In November 1996 the Government imposed a nation-wide curfew to combat the increasing lawlessness.

The mercenary affair on Bougainville in early 1997 (see below) led to a period of extreme instability throughout the country, culminating in Chan's resignation and the appointment of John Giheno as acting Prime Minister for about two months. The atmosphere of political uncertainty was exacerbated by serious outbreaks of violence in the weeks preceding the general election (which had been set for mid-June). The Government imposed a dusk-to-dawn curfew and a nation-wide ban on the sale of alcohol, and dispatched security personnel across the country in an attempt to quell the politically motivated disturbances. Many senior politicians failed to be re-elected in the general election, including Chan, Giheno and Wingti. Outbreaks of violence were reported in several Highlands constituencies as the results were declared. A period of intense political manoeuvring followed the election, as various members sought to form coalitions and groupings in an attempt to achieve a majority in Parliament. On 22 July Bill Skate of the People's National Congress (PNC), who was the former Speaker and Governor of the National Capital District, was elected Prime Minister, defeating Sir Michael Somare (who had established a new party—the National Alliance in 1996) by 71 votes to 35. Skate was supported by a coalition of the PNC, PDM, PPP, Pangu Pati and independent members. A new Government was appointed in late July, and extensive changes in the functional responsibilities of ministries were announced.

In September 1997 the Government declared a national disaster following a prolonged period of drought believed to have been caused by El Niño (a periodic warming of the tropical Pacific Ocean). By December more than 1,000 people had died, and as many as 1.2m. were threatened by starvation, as a result of the drought, which was the most severe in the country's recorded history. Several countries and organizations that had provided relief funds during the disaster were highly critical of the Government's management of the aid it received. The Minister for Finance, Roy Yaki, was dismissed, in part for his role in the affair, and responsibility for the administration of relief funds was subsequently transferred from the Department of Finance to the Department of Provincial Affairs. In March 1998 (when the drought was deemed to have ended following heavy rainfall) it was revealed that less than one-half of the relief aid received had been deployed to help the victims of the disaster.

On 14 November 1997 Silas (later Sir Silas) Atopare was appointed Governor-General, defeating Sir Getake Gam, head of the Evangelical Lutheran Church, by 54 votes to 44 in the legislature.

A serious political scandal erupted in late November 1997, following allegations of corruption against Skate. The accusations centred on a videotape broadcast on Australian television, which appeared to show Skate arranging bribes and boasting of his strong connections with criminal elements in Port Moresby. The Prime Minister dismissed his recorded comments (and the resultant allegations) saying that he had been drunk at the time of filming. Several senior politicians, including Somare, demanded his resignation over the affair. Meanwhile, Skate dismissed the leaders of the Pangu Pati and the PPP (Haiveta and Baing), his coalition partners, accusing them of conspiring against him. The situation intensified with the resignation of seven Pangu Pati members from the Government in early December, and the announcement that the Pangu Pati and the PPP would join the opposition. However, the PPP rejoined the Government shortly afterwards, having voted to replace Baing as leader of the party with Michael Nali. Similarly, four Pangu Pati ministers rejoined the Government, thereby restoring Skate's majority in the National Parliament. A major ministerial reorganization was subsequently announced, in which Nali was appointed Deputy Prime Minister. In April 1998 Skate announced the formation of a new political grouping, the Papua New Guinea First Party (which absorbed the PNC, the Christian Country Party and several other minor parties), and effected a cabinet reorganization. The Government's majority was subsequently undermined, however, following a series of decisions by the Court of Disputed Returns during mid-1998 which declared the election of seven government MPs (at the 1997 general election) null and void. Furthermore, in June the Pangu Pati officially joined the opposition, thereby reducing the Government's representation to 61 members in the National Parliament.

Rumours of a motion expressing 'no confidence' in the Prime Minister prompted the formation of a new pro-Skate coalition in the National Parliament in late July 1998. In the same month Skate announced a number of major reforms in the structure of the Government, including the merging of several ministerial portfolios, the establishment of new departments for Private Enterprise and Rural Development, a series of new appointments to various public bodies and an extensive ministerial reorganization.

In July 1998 a state of national disaster was declared following a series of tsunami (huge tidal waves caused by undersea earthquakes) which obliterated several villages on the north-west coast of the country and killed an estimated 3,000 people. An estimated K10m. was subsequently received for disaster relief operations, although widespread concern at the apparently inefficient distribution of the funds was expressed in the following months.

In October 1998 the PPP left the governing coalition and, in a subsequent ministerial reorganization, the party's leader, Michael Nali, was replaced as Deputy Prime Minister by Iairo Lasaro.

In late 1998 there was a series of scandals relating to the various serious misdemeanours of a number of provincial governors. Moreover, outbreaks of tribal fighting continued to cause problems. A serious conflict in the Eastern Highlands in early 1999 involved villagers using rocket launchers and grenades, and resulted in numerous deaths. In mid-1999 a state of emergency was declared in the Southern Highlands, following serious disturbances provoked by the death of a former provincial Governor, Dick Mune, in a road accident.

The establishment of a new political party, the PNG Liberal Party (PNGLP), in May 1999 by the Speaker, John Pundari (relaunched in June as the Advance PNG Party—APP), encouraged rumours of a forthcoming vote of 'no confidence' against the Prime Minister. In early June, as part of a government reorganization, Skate dismissed the PDM leader, Sir Mekere Morauta, and three other PDM ministers, replacing them with four Pangu Pati members, including Haiveta. Later that month both the PDM and the United Resource Party (URP) announced their decision to withdraw from the coalition Government, following the resignation of nine government ministers. Both parties were expected to support the APP, in an attempt to subject Skate to a vote of 'no confidence'. On 7 July, however, Skate unexpectedly resigned, but declared that he would remain in power, in an acting capacity, until the appointment of a new Prime Minister. The opposition alliance announced Morauta as their candidate for the premiership. On the day before the National Parliament was due to reconvene, Skate claimed that the APP had pledged support for the Government, thereby ensuring that it would have a sufficient majority to defeat any motion of 'no confidence'. At the opening session of the National Parliament in mid-July, Lasaro, the Prime Minister's nominee, defeated the opposition candidate, Bernard Narokobi of the PDM, by 57 votes to 45 to become Speaker. However, on the next day Morauta was elected Prime Minister with an overwhelming majority, following his nomination by Pundari, who had transferred his allegiance from Skate, having refused to accept the latter's nomination of himself as candidate for the post of Prime Minister. Lasaro immediately resigned as Speaker, and Narokobi was elected unopposed to the position. Morauta subsequently appointed a new Government, with Pundari as Deputy Prime Minister.

In early August 1999 Morauta, a former Governor of the central bank, presented a 'mini-budget', in an attempt to combat various economic problems which, he claimed, were a consequence of the previous Government's mismanagement. As part of a series of measures aimed at stabilizing the political situation in the country, legislation was drafted in October 1999 to prevent ministers transferring political allegiances. In early December Prime Minister Morauta expelled the APP from the coalition Government and dismissed Pundari from his post as Deputy Prime Minister, claiming that this constituted a further move towards the restoration of political stability. (Pundari was rumoured to have conspired with the opposition leader, Skate, to oust the Prime Minister.) Pundari was replaced by the former Minister of Works and deputy leader of the PDM, Mao Zeming. Following Pundari's sudden dismissal, four small political parties within the governing coalition—the National Alliance, the People's National Party, the Melanesian Alliance and the Movement for Greater Autonomy—joined to form the People's National Alliance. Also in that month, Skate (who was faced with a charge of attempted fraud during his term in office as Governor of the National Capital District Commission) announced that his party, the People's National Congress (PNC—which had 10 MPs), was to join the coalition Government; in May 2000 he assumed the leadership of the party. Later in December 1999, the resignation of the Minister of Agriculture and Livestock, Ted Diro, prompted Morauta to carry out a government reshuffle, including the appointment of three new ministers.

In March 2000 Morauta dismissed three ministers on the grounds that they had allegedly conspired to introduce a parliamentary motion of 'no confidence' in the Prime Minister. The parliamentary strength of the ruling coalition was increased to 76 MPs in April, following the readmission of the APP to government, including the appointment of Pundari to the post of Minister of Lands and Physical Planning. In 2000 Morauta introduced further measures designed to ensure future governmental stability. In August he proposed legislation that would limit the ability of MPs to change their party allegiance within a parliamentary session; he also announced the adjournment of Parliament between January and July 2001, the only period during which votes of 'no confidence' could be tabled. (The Constitution forbids such votes in the 18 months following the election of a Prime Minister, and in the 12 months before a general election.) In November, however, a revolt by 25 government MPs (including six cabinet ministers) prevented a vote on the so-called Political Parties Integrity Bill. The rebellion prompted a major cabinet reshuffle, in which all of the ministers involved in the revolt, including the Deputy Prime Minister, Mao Zeming, were dismissed. Further changes to the composition of the National Executive Council were made in December, notably the dismissal of Sir Michael Somare, the Minister for Foreign Affairs and Trade, following which Morauta secured the parliamentary approval of the bill. The Prime Minister claimed that the introduction of the new legislation represented the most important constitutional change in Papua New Guinea since independence and would greatly enhance the political stability of the country, as it required members of Parliament who wished to change party allegiance to stand down and contest a by-election.

Allegations of corruption and mismanagement resulted in the suspension of four provincial governments (Western Province, Southern Highlands, Enga and the National Capital District—NCD) in late 2000 and early 2001, with the central Government claiming that a failure to deliver services had resulted from the misuse of public funds. Moreover, in January 2001 the Minister for Provincial and Local Government, Iairo Lasaro, was arrested for the alleged misappropriation of public funds, and in mid-March Bill Skate was charged with the same crime, having been acquitted earlier in the month of conspiring to defraud an insurance company (the trial was to be abandoned in December 2001 owing to lack of evidence).

A Commonwealth report into the Papua New Guinea Defence Force published in January 2001 recommended reducing the number of army personnel by one-third. Subsequent plans by the Government to make more than 2,000 soldiers redundant (equivalent to some 50% of the entire Defence Force) resulted in a revolt at the Port Moresby barracks in March. It was believed that senior officers had helped to distribute arms to the rebels, who demanded the resignation of the Prime Minister and the transfer of power to a caretaker administration. The rebellion ended some two weeks later with an amnesty for the soldiers involved, during which hundreds of looted weapons were surrendered. In March 2002 another rebellion by soldiers protesting against the proposed reductions in defence personnel took place at the Moem barracks on the northern coast. The leader of the rebellion, Nebare Dege, was sentenced to 15 years' imprisonment in December 2002.

In April 2001 the Advance PNG Party, led by John Pundari, was dissolved and merged with the PDM, led by the Prime Minister. In May, following a minor ministerial reorganization in March, Morauta expelled the National Alliance from the ruling coalition and dismissed the party's ministers (including Bart Philemon, Minister for Foreign Affairs) from the National Executive Council, accusing Somare, its leader, of attempting to destabilize the Government.

In late June, following several days of protests against the Government's economic reforms, police used tear gas to disperse hundreds of demonstrators outside the Prime Minister's office in Port Moresby. In a separate incident, four students were killed and several injured when riot police allegedly entered the premises of the University of Papua New Guinea and opened fire. A temporary curfew was imposed, and a Commission of Inquiry was established; in December relatives of one of the dead students began legal action against the police force.

In October 2001 the Prime Minister effected another reallocation of ministerial portfolios. At the end of the month, furthermore, the Minister for Foreign Affairs, John Pundari, was dismissed. His removal from office followed his criticism of the Government's participation in Australia's 'Pacific Solution', whereby 216 refugees, who had attempted to enter Australia illegally, were being housed in a former military prison on Manus Island. In December negotiations with Australia were under way to accommodate a further 1,000 asylum-seekers, and Australia had requested the Government to hold the refugees for an additional six months, although no formal agreement was made. In January 2002 the Government decided to accept 784 additional asylum-seekers. The agreement with Australia was extended for a further 12 months in October 2002 and was to provide accommodation for an additional 1,000 asylum-seekers. The camp remained open, despite the Australian Government's announcement in August 2003 that it intended to close the facility. In early 2004 it was revealed that the sole occupant of the camp was a Palestinian refugee who had been held in solitary confinement for almost seven months at a cost of more than US $3m. Opposition politicians in Australia noted that the amount of money spent maintaining the camp on Manus Island and its one occupant would fund unemployment benefit payments for all the asylum-seekers currently detained in Australian centres.

Meanwhile, a general election was held in the latter part of June 2002. Voting commenced on 15 June and was to extend over a two-week period. Many polling stations, however, failed to open as scheduled, amid reports of the theft of ballot papers and subsequent strike action by electoral staff. Some 25 people were killed and dozens injured in election-related violence, much of which occurred in the Highlands provinces as a result of disputes between clan-based candidates. A Commonwealth inquiry was subsequently planned to investigate events surrounding the election, which was described as the worst in the country's history. In the final results announced in August six seats in the Highlands provinces remained empty where voting had been unable to proceed. Of the 103 seats declared, Somare's National Alliance won 19 and Morauta's PDM secured 12. Independent candidates secured 17 seats with the remainder divided among a large number of minor parties, many of which were formed specifically to contest the election.

On 5 August 2002 Sir Michael Somare was elected Prime Minister, with 88 parliamentary votes. No candidates stood against him, and Morauta and his supporters abstained from the vote. Bill Skate, who had played an important part in Somare's campaign, was elected Speaker. After appointing a 28-member Cabinet (which was dominated by 19 newly elected MPs), Somare began his third term as Prime Minister, pledging to restore stability, halt the privatization programme and reduce expenditure.

In April and May 2003 elections were held for the six Highlands constituencies where voting had failed to take place in the previous year. In June the Government submitted proposals for a number of changes to the political system, including a mandatory general election following the approval of a 'no confidence' motion and stricter rules governing the switching of party allegiances among members of the National Parliament. In late 2003 the Government proposed legislation that would extend the period that a new government should be exempt from votes of 'no confidence' from 18 months to three years. Both attempts to introduce these constitutional changes were unsuccessful, despite government inducements of some US $25,000 for members of Parliament to support the proposals.

On 18 September 2003 Sir Albert Kipalan defeated Sir Paulias Matane by a single vote in the final poll to elect a new Governor-General. Kipalan was expected to take up the position on 13 November when the term of the incumbent Sir Silas Atopare expired. However, questions concerning possible flaws in the procedure arose in the following week, casting doubt on the validity of the election, and the Supreme Court subsequently declared it to be defective and invalid. Speaker Bill Skate was appointed acting Governor-General until a new election could be arranged. At the poll, which was duly held on 4 December, Sir Pato Kakaraya was the successful candidate. However, Kipalan, who had also contested the election, challenged the result and, owing to a number of objections, was granted a legal injunction against Kakaraya's swearing-in, which had been due to take place in late January 2004. Moreover, in early March Bill Skate was obliged to resign briefly as acting Governor-General to stand trial on charges of misappropriating public funds. He resumed the position a few days later, having been cleared of the charges. In late March the Supreme Court declared the election of Sir Pato Kakaraya to be null and void. Somare consequently ordered the recall of the National Parliament in order that a new election to the post could be organized. Following four rounds of voting Sir Paulias Matane, a former Minister of Foreign Affairs, was narrowly elected Governor-General in late May and, when the Supreme Court had dismissed an injunction against his appointment by Kakaraya, was duly sworn in in the following month.

In May 2004 Somare dismissed Deputy Prime Minister Moses Maladina together with all the PNC members of the Cabinet who had refused to support proposed government legislation and who were believed to be planning a vote of 'no confidence' in the Government. Meanwhile, ongoing fears about increasing gun ownership were highlighted in early August when the National Parliament was adjourned for a period of three months after both the Government and the opposition expressed concern over the number of members of Parliament who were reported to be bringing firearms into Parliament.

In December 2003, following the apparent success of the intervention in Solomon Islands earlier in the year, Australia announced the deployment of some 250 of its security personnel to different regions of Papua New Guinea, as part of a five-year operation, aimed at restoring order to troubled areas of the country. Moreover, the Australian Government was to send up to 70 officials to Papua New Guinea to take up senior public roles in finance, law, public-sector management, immigration, border security and transport safety. The agreement, known as the Enhanced Co-operation Program, was officially concluded in June 2004 and incorporated a total aid 'package' worth US $690m. Somare had initially been reluctant to accept Australia's expanded role in the country, fearing that it would erode national sovereignty, but had been obliged to reconsider when the Australian Prime Minister, John Howard, implied that aid payments to Papua New Guinea were dependent on the country's co-operation.

Tribal conflicts, principally between the Ujimap and Wagia tribes, broke out near Mendi, the capital of Southern Highlands province, in December 2001. It was alleged that national and local politicians, along with tribal leaders, had made little effort to end the fighting, which had originated in a dispute over the governorship of the province in 1997; furthermore, many were dissatisfied at the election of Tom Tomiape as Governor, in late November 2001, and at the widespread political corruption and deteriorating public services in the province. Tomiape's election was ruled invalid by the Supreme Court in December, and Wambi Nondi was appointed acting Governor. A brief cease-fire was brokered in early January 2002 and an independent peace commission was formed in February, headed by Francis Awesa, a local businessman. By early 2002 more than 120 people had been killed since the onset of the fighting.

Ethnic and tribal violence continued in several regions of the country during 2003. Eight people were killed in fighting near Port Moresby in May, and 17 died during inter-clan violence involving more than 1,000 people in Enga province in July. Further killings led to rioting in the following month and to an appeal by the Governor of the province for those responsible to be hanged. In early 2004 eight villagers were killed in further tribal conflicts in Enga province and by mid-2004 ethnic violence in Morobe, Chimbu and East Highlands provinces had resulted in many more deaths, including those of several children. Concerns regarding the conduct of the national police force in tackling such conflicts intensified following an incident in August when a confrontation between security personnel and a group of youths in Enga province resulted in police setting fire to an entire village, killing domestic animals and shooting dead a villager and injuring several others. Earlier in the year the former Commander-in-Chief of the Defence Force, Jerry Singirok, stated that he believed that the marked increase in ownership of illegal firearms, particularly in the Highlands region, was the single most significant problem in Papua New Guinea.

In February 2000 it was announced that about 25,000 inhabitants of the Duke of York Islands, which are situated in East New Britain Province, were expected to be resettled by the Government, following the publication of a UN report indicating that the islands were becoming uninhabitable owing to rising sea-levels resulting from the 'greenhouse effect' (the heating of the earth's atmosphere as a consequence of pollution).

Fears that rising sea-levels would have very serious consequences for Pacific islanders increased in May 2003 when an emergency operation was undertaken to save the inhabitants of Carteret and Mortlock islands near Bougainville. Food supplies were sent to the islands, the 2,000 inhabitants of which were reported to be suffering from starvation and health problems related to poor diet, since the failure of their crops, which had been flooded by sea water. The islanders, however, were reluctant to accept a government proposal to relocate them to Bougainville, fearing the loss of their distinct Polynesian culture and way of life.

The country was beset by a series of severe floods in late 2003 and early 2004, which affected some 5,000 people in Morobe province and more than 10,000 in West Highlands province. Food crops, livestock and entire villages were destroyed by the floods.

The status of the province of Bougainville was increasingly questioned in the late 1980s, a problem that developed into civil unrest and a long-term national crisis. In April 1988 landowners on the island of Bougainville submitted compensation claims, amounting to K10,000m., for land mined by Bougainville Copper Ltd at Panguna since 1972. When no payments ensued, acts of sabotage were perpetrated in late 1988 by the Bougainville Revolutionary Army (BRA), led by Francis Ona, a former miner and surveyor, and the mine was obliged to suspend operations for an initial period of eight days. However, the mine's owners refused to resume operations for fear of further attack, and, following increased violence in January 1989, a curfew was imposed. Operations at the mine resumed but the violence continued. The BRA's demands increasingly favoured secession for the island of Bougainville (and North Solomons Province) from Papua New Guinea, together with the closure of the mine until their demands for compensation and secession had been met. In May, as the violent campaign on the island intensified, the mine was forced once more to suspend production, and in June the Papua New Guinea Government declared a state of emergency on Bougainville, dispatching some 2,000 security personnel to the island. Ted Diro, having been acquitted on a charge of perjury by the Supreme Court but still facing impending charges of corruption, was reinstated to the National Executive Council in May as Minister of State, and was accorded special responsibility for overseeing the Bougainville crisis. In September a minister in the Bougainville provincial Government, who had been negotiating an agreement with landowners to provide them with financial compensation, was shot dead. The signing of the accord, due to take place the following day, was postponed indefinitely, and Diro responded by offering a reward for the capture or killing of Ona and seven of his deputies, including the BRA's military commander, Sam Kauona.

In January 1990 the owners of the Bougainville mine made redundant 2,000 of the remaining 2,300 staff. The escalation of violence on Bougainville prompted the Australian Government to announce plans to send in military forces to evacuate its nationals trapped on the island, and to withdraw the remaining 300 personnel of Bougainville Copper Ltd. Growing criticism of the Government's failure to resolve the Bougainville dispute, the rising death toll among the security forces, rebels and civilians and the worsening economic crisis led the Government to negotiate a cease-fire with the BRA, with effect from the beginning of March 1990. The Government undertook to withdraw its security forces and to release 80 detainees. Bougainville came under control of the BRA in mid-March after the sudden departure of the security forces. The premature withdrawal of the troops was seen as an attempt by Paul Tohian, the police commissioner who had been in charge of the state of emergency on Bougainville, to disrupt the peace process, and was followed by an abortive coup attempt, allegedly led by Tohian, who was summarily dismissed from his post.

In March 1990 the Government imposed an economic blockade on Bougainville; this was intensified in May, when banking, telecommunications and public services on the island were suspended. On 17 May the BRA, allegedly in response to the Government's implementation of economic sanctions, proclaimed Bougainville's independence, renaming the island the Republic of Bougainville. The unilateral declaration of independence, made by Ona, who also proclaimed himself interim President, was immediately dismissed by Namaliu as unconstitutional and invalid. In July negotiations between the BRA and the Government finally began on board a New Zealand naval vessel, *Endeavour*. In the resulting 'Endeavour Accord', the BRA representatives agreed to defer implementation of the May declaration of independence and to hold further discussions on the political status of the island. The Government agreed to end the blockade, and to restore essential services to the island. However, despite assurances from Namaliu that troops would not be sent to the island, the first two ships that left for Bougainville with supplies were found to be carrying 100 security personnel, intending to disembark on the island of Buka, north of Bougainville. The BRA accused the Government of violating the accord, and a week later the security forces and the two ships, together with the supplies, withdrew. In mid-September the Government sent armed troops to take control of Buka, stating that this was in response to a petition for help from Buka islanders. Violent clashes ensued between the BRA and the armed forces on Buka, in which many people were reported to have been killed.

In January 1991 further negotiations took place, in Honiara (the capital of Solomon Islands), between representatives of the Papua New Guinea Government and of Bougainville, which resulted in the 'Honiara Accord'. The agreement stated that the Papua New Guinea

Government would not station its security forces on Bougainville if the islanders agreed to disband the BRA and to surrender all prisoners and weapons to a multinational peace-keeping force. The Bougainville secessionists were guaranteed immunity from prosecution. The agreement, however, made no provision for any change in the political status of Bougainville, and by early March it appeared to have failed.

Government troops launched a further attack on Bougainville in April 1991. In June Col Leo Nuia was dismissed from his post as Commander of the Papua New Guinea Defence Force, after admitting that his troops had committed atrocities during fighting on Bougainville in early 1990. Further allegations of human rights abuses and summary executions of BRA members and sympathizers by government troops prompted Namaliu to announce plans for an independent inquiry into the claims.

Fighting continued throughout 1991 and the situation deteriorated further when in early 1992, in an attempt to force the Government to end its economic blockade of the island, the BRA intercepted and burnt a supply ship, and held its crew hostage. As a result, all shipping and air services to Bougainville were suspended.

In October 1992 government troops began a major offensive against rebel-held areas of Bougainville and, later in the month, announced that they had taken control of the main town, Arawa. The BRA, however, denied the claim and began a campaign of arson against government offices and public buildings in Arawa. Violence on the island intensified in early 1993, and allegations of atrocities and violations of human rights, by both sides, were widely reported.

Talks between government representatives and secessionists in Honiara during 1994 led to the signing of a cease-fire agreement in September. Under the terms of the agreement, a regional peace-keeping force, composed of troops from Fiji, Vanuatu and Tonga, was deployed in October (with the Governments of Australia and New Zealand in a supervisory role) and the economic blockade of the island was lifted. In the following month Chan and a group of non-BRA Bougainville leaders signed the Charter of Mirigini, which provided for the establishment of a transitional Bougainville government. The BRA declared its opposition to the proposed authority, reiterating its goal of outright secession. However, in April 1995, following the suspension of the Bougainville provincial government, 27 members of the 32-member transitional administration were sworn in at a ceremony on Buka Island, attended by the Prime Minister and several foreign dignitaries. Theodore Miriong, a former legal adviser to Ona, was elected Premier of the authority. However, the three seats reserved for the BRA leaders, Ona, Kauona and Joseph Kabui (a former Premier of North Solomons Province), remained vacant, as the rebels urged their supporters to reject the new administration and to continue the violent campaign for independence. An amnesty, declared in May by the transitional administration and the Government, for all who had committed crimes during the conflict, was rejected by the BRA. Violence escalated during July and August, and a campaign of arson against public buildings was conducted by BRA rebels. Meanwhile, the Government denied any involvement in a series of attacks on BRA leaders, including an assassination attempt on Kabui. The murder of several more members of the security forces in March 1996 led the Government to abandon all talks with the secessionists, and troops to reimpose a military blockade on Bougainville. An escalation of violence in mid-1996 culminated in a major military offensive against rebel-held areas in June. Civilians on the island were encouraged to seek refuge in government 'care centres', and by August it was estimated that some 67,000 Bougainvilleans were being accommodated in 59 such centres. In the same month defence forces arrested Miriong, accusing him of incitement regarding the killing of 13 government soldiers at an army camp. The incident, in which Miriong was removed from Bougainville and kept under surveillance on Buka, caused the Government considerable embarrassment, particularly as it followed a number of similar cases, in which the security forces had chosen to act independently of government policy. On 12 October Miriong was assassinated at his home in south-west Bougainville by unidentified gunmen. The following month an official inquiry concluded that a group of government soldiers was responsible for the killing, assisted by pro-Government civilians (known as 'resistance fighters'). Meanwhile, Gerard Sinato was elected as the new Premier of the Bougainville Transitional Government.

An apparent deterioration in the situation on Bougainville in late 1996 was characterized by an increase in BRA attacks against civilian targets and a similar escalation in violence by government troops and 'resistance fighters'. The Red Cross temporarily suspended its operations on the island following an attack on one of its vehicles, and human rights organizations repeated demands that observers be allowed to monitor incidents on the island, following a series of attacks on civilians (for which both sides denied responsibility). A report commissioned by the Government and published in late 1996 recommended a thorough reorganization of the country's armed forces. The report identified a number of problems that had contributed to a lack of cohesion and discipline and to a marked decline in morale among troops, which had resulted in many soldiers refusing to serve in Bougainville.

In February 1997 unofficial reports suggested that the Government was planning to engage the services of a group of mercenaries on Bougainville. Chan reacted angrily to the reports, which he claimed were inaccurate; however, he confirmed that a company based in the United Kingdom, Sandline International (a subsidiary of Executive Outcomes, a notorious supplier of private armed forces in Africa), had been commissioned to provide military advice and training for soldiers on Bougainville. Subsequent reports of mercenary activity on the island and of the large-scale purchase of military equipment and weapons provoked expressions of condemnation from numerous interests in the region, including the British High Commission in Port Moresby, and, in particular, from the Government of Australia. The situation developed into a major crisis on 16 March when the Commander of the Defence Force, Brig.-Gen. (later Maj.-Gen.) Jerry Singirok, announced on national radio and television that the country's armed forces were refusing to co-operate with the mercenary programme and demanded the immediate resignation of Chan. He explained that the mercenaries (most of whom were from South Africa) had been captured by the armed forces and were being detained while arrangements were made for their deportation. Singirok denied that his actions constituted a coup attempt. The following day Chan dismissed Singirok, replacing him with Col Alfred Aikung. However, the armed forces rejected the new leadership, remaining loyal to Singirok. Popular support for the army's stance became increasingly vocal as several thousand demonstrators rampaged through the streets of the capital, looting and clashing with security forces. Armed forces in Australia were reported to be prepared for deployment to Papua New Guinea in the event of any worsening of the situation. In view of the escalation in civil unrest, Chan announced the suspension of the contract with Sandline International on 20 March, pending an inquiry into the affair. On the following day the remaining mercenaries left the country, and Aikung was replaced as Commander of the Defence Force by Col Jack Tuat. Despite these attempts at conciliation, however, demands for Chan's resignation intensified, with military, political and religious leaders, as well as the Governor-General, urging him to leave office. Moreover, four government ministers resigned from their posts in an attempt to increase the pressure on Chan to do likewise. On 25 March the National Parliament voted on a motion of 'no confidence' in the Prime Minister. When the vote was defeated protesters laid siege to the parliament building, effectively imprisoning more than 100 members inside the building until the next day, while an estimated 15,000 demonstrators marched on Parliament from across the capital. On the following day Chan announced his resignation, along with that of the Deputy Prime Minister and the Minister for Defence. John Giheno, the erstwhile Minister for Mining and Petroleum, was subsequently elected acting Prime Minister by the Cabinet.

In April 1997 an inquiry was initiated into the mercenary affair. The chief executive of Sandline International, Col (retd) Tim Spicer, was questioned over his alleged acceptance of bribes and in connection with various firearms offences. During the inquiry it was revealed that the company had requested part-ownership of the Panguna copper mine as payment for its military services. Criminal charges against Spicer were withdrawn within several days. The inquiry concluded in early June that Chan had not been guilty of misconduct in relation to the mercenary affair and, as a result (despite Giheno's stated intention to continue as acting Prime Minister until a general election had taken place), Chan announced his immediate resumption of his former position. Shortly after resuming office, Chan again provoked controversy by appointing Col Leo Nuia to the position of Commander of the Defence Force. (Nuia had been dismissed from the post in 1991, following an admission that his troops had committed atrocities during fighting on Bougainville.)

Following the election of a new Government, it was announced in August 1997 that a second inquiry into the mercenary affair, based on broader criteria, would be conducted. Meanwhile, government soldiers reacted angrily to the prosecution of military leaders involved in the operation to oust Sandline mercenaries from Bougainville in March. Nuia was imprisoned in his barracks by members of the Defence Force, while Maj. Walter Enuma, who was being held while awaiting trial on charges of 'raising an illegal force', was freed by rebel soldiers. The second inquiry concluded in September 1998 that Chris Haiveta (the former Deputy Prime Minister) had been the beneficiary of corrupt payments from Sandline and upheld the first inquiry's finding that Chan had not been guilty of any wrongdoing in the affair. In the same month an international tribunal ruled that the Government owed $A28m. to Sandline in outstanding payments under the mercenary contract. In May 1999 the Government agreed to pay this debt and the two parties undertook to end all legal action against each other. The Papua New Guinea Defence Force was to be allowed to retain Sandline military equipment stored on Bougainville.

In July 1997 talks were held in New Zealand (at the Burnham army base) between secessionists and representatives of the Bougainville Transitional Government. As a result of the negotiations, Sinato and Kabui signed the 'Burnham Declaration', which recommended the withdrawal of government troops from Bougainville and the deployment of a neutral peace-keeping force. Persistent reports of internal divisions within the BRA were refuted by the leadership, despite Ona's declared opposition to the 'Burnham Declaration' and to the decision to release the five hostages. However, hopes for a significant improvement in the political climate on Bougainville were encouraged by an official visit by the newly elected Prime Minister, Bill Skate, in August (the first such visit since 1994) and by the resumption of talks in New Zealand in the following month. Negotiations concluded on 10 October 1997 with the signing of the 'Burnham Truce', in which representatives from both sides agreed to a series of interim measures, which included refraining from acts of armed confrontation pending a formal meeting of government and secessionist leaders in early 1998. Ona (who appeared to be becoming increasingly marginalized within the BRA) refused to be a party to the truce, however, claiming that similar agreements had not been honoured by government troops and 'resistance fighters' in the past, and indirectly threatened the members of an unarmed group of regional representatives, established to monitor the truce. The Prime Ministers of both Papua New Guinea and Solomon Islands made an extended visit to Bougainville in December 1997 to demonstrate their united support for the truce. Further talks held at Lincoln University in Christchurch, New Zealand, in January 1998 resulted in the 'Lincoln Agreement', providing for an extension to the truce, the initiation of a disarmament process and the phased withdrawal of government troops from the island. Skate also issued a public apology for mistakes made by successive administrations during the conflict, which was estimated to have resulted in the deaths of some 20,000 people and to have cost a total of K200m.

In accordance with the provisions of the 'Burnham Truce', a permanent cease-fire agreement was signed in Arawa on 30 April 1998. The occasion was attended by senior government members from Australia, New Zealand, Solomon Islands and Vanuatu, as well as Papua New Guinea government representatives and secessionist leaders. Ona declined to take part in the ceremony, reiterating his opposition to the peace agreement. In the following month he attracted statements of condemnation from Kabui and Kauona for issuing a 'shoot-to-kill' order against the peace-keeping troops on Bougainville to the small band of rebels who remained loyal to him. In June government troops were withdrawn from Arawa under the terms of the agreement and reconstruction projects on Bougainville, financed by funds from various sources, including an aid package of $NZ1m. from the New Zealand Government, were initiated. An increase in crime and arson attacks, however, was reported in the demilitarized zones in the following months.

In August 1998 more than 2,000 representatives from different groups in Bougainville met in Buin (in the south of the island) to discuss their response to the 'Burnham Truce'. The resultant 'Buin Declaration' stated that the islanders were united in their aspiration for independence through peaceful negotiation. In October the National Parliament voted to amend the Constitution to allow the Bougainville Reconciliation Government to replace the Bougainville Transitional Government. The new authority came into existence on 1 January 1999, following the renewed suspension of the Bougainville provincial government, and at its first sitting elected Sinato and Kabui as its co-leaders. In April an agreement signed by Bougainville and Papua New Guinea government representatives (although not acknowledged by the BRA, which threatened to leave the peace process as a result), known as the Matakana and Okataina Understanding, reaffirmed both sides' commitment to the cease-fire, while undertaking to discuss options for the political future of the island.

Elections to the Bougainville People's Congress (BPC), formerly the Bougainville Reconciliation Government) were held in early May 1999, and were reported to have proceeded smoothly. At the first sitting of the BPC, Kabui was elected President by an overwhelming majority, securing 77 of the 87 votes, defeating his former co-leader, Sinato. Kabui subsequently appointed 29 members to the Congressional Executive Council, and Linus Konukong was elected Speaker. Ona refused Kabui's offer to join the BPC, stating that he did not wish to co-operate with the Government. At a subsequent session of the Council, Kabui announced his intention to campaign for independence for Bougainville. In response, Prime Minister Bill Skate stated that, although there was no possibility of independence (as this was not provided for in the Constitution), Parliament would consider terms for greater autonomy for the island. Plans for the disposal of weapons on Bougainville were expected to be drafted by the UN under terms agreed upon at a meeting with the Peace Process Consultative Committee in June. In August the suspension of the Bougainville provincial government was extended for a further six months in an attempt to find a peaceful resolution to the issue of autonomy for Bougainville; the decision was confirmed at a parliamentary session the following month. Following a visit to Bougainville, the Minister for Bougainville Affairs and for Foreign Affairs, Sir Michael Somare (as he had become), stated that the Government was willing to grant the island a greater degree of autonomy. Somare proposed that Bougainville be self-governing in all matters except foreign affairs, defence and policing, all of which would remain the responsibility of the central Government. The proposal was welcomed as an important development by members of the BPC, although they also announced their intention not to surrender their weapons until the Government had agreed to the holding of a referendum on independence. Following further talks between Somare and Kabui, the former reiterated Skate's earlier comments that there was no provision for a referendum on independence in the Constitution.

In October 1999 a Supreme Court ruling declared the suspension of the Bougainville provincial government illegal on technical grounds. On 9 December the provincial government was formally recognized, in theory, despite protests by members of the BPC and concerns that this development would hinder the peace process. In effect, however, the provincial government comprised only four members—the Bougainville Regional Member of Parliament, John Momis, who thus became Governor-elect (although he agreed not to exercise his powers for the mean time) and the three other Bougainville parliamentarians. Following talks in mid-December between Somare and members of the BPC, the BRA and elders of the island, Somare declared, in the consequent agreement, that he would consider the possibility of a referendum on independence for Bougainville. The agreement, known as the Hutjena Accord, stated that the highest possible degree of autonomy should be granted to the island. Further talks regarding the future status of Bougainville commenced in early March 2000. However, following the rejection of an initial proposal on the island's autonomy by secessionist leaders, the talks were suspended. Negotiations resumed in mid-March, and on 23 March an agreement, known as the Loloata Understanding, allowing for the eventual holding of a referendum on independence, once full autonomy had been implemented, and the formal establishment of the Bougainville Interim Provincial Government (BIPG), composed of an Executive Council and a 25-member Provincial Assembly (including the four original members), was signed by Somare and Kabui.

On 30 March 2000 the Provincial Assembly and the Executive Council were sworn in by the Governor-General. Another six seats were left vacant in the Provincial Assembly for other Bougainville officials such as Kabui and Ona; Kabui, however, stated that, rather than joining the BIPG, he would wait until the establishment of a fully autonomous government (which he hoped would be in place by early 2001). Ona and his supporters, who numbered around 8,000, refused to participate in any government that was not completely independent. In May 2000 the Office of Bougainville Affairs was renamed the Office of Peace and Reconstruction. In July Somare approved an allocation of K200,000, as part of a scheme co-ordinated by the United Nations Development Programme (UNDP), to facilitate the collection and disposal of weapons held by dissident groups. Although an initial deadline of 15 September to determine the form and date of a referendum was not met, a further round of negotiations that month was presented as a significant advance in the peace process. Uncertainties regarding the time-scale for the island's progression to autonomy continued to delay fulfilment of the peace process; Somare's statement in May that the referendum would be held 15 years hence contradicted a previously stated deadline of December 2000. Further concerns about renewed Australian funding of the Papua New Guinea Defence Force troops stationed on Bougainville Island, and the failure of the BRA in particular to begin disarmament before the implementation of political autonomy, further impeded progress. In October the BRA commander, Gen. (self-styled) Ishmael Toroama, threatened to abandon the group's cease-fire. Somare was absent from a further, 'final' round of peace talks in October because of illness and a subsequent visit to Australia. At these discussions the conflicting demands for the holding of a referendum on the question of independence, and for prior disarmament, continued to delay the signing of an agreement to implement autonomy in Bougainville. Following a further round of peace talks in November, Kabui confirmed that he had secured an agreement from Somare that a future referendum would include a legally-binding option of independence, although disagreement over how soon the vote should be held persisted. In the following month Somare was abruptly dismissed from his ministerial position and replaced by Bart Philemon; Morauta had apparently believed that Somare was impeding the progress of the Political Parties Integrity Bill.

In February 2001 agreement on the terms of the referendum was finally reached following the intervention of the Australian Minister for Foreign Affairs. The agreement stated that the referendum would be held in 10–15 years' time and would contain the option of independence. In the interim the provincial government was to be granted increased autonomy and the BRA would be expected to disarm. Despite a temporary breakdown in negotiations, in early May commanders of the BRA and the Bougainville Resistance Force (a militia that was allied to the Government during the civil conflict on Bougainville) signed an agreement to surrender their weapons.

In the same month, in Port Moresby, the Government and the Provincial Assembly held negotiations on autonomy for Bougainville. In late June Moi Avei, who had replaced Philemon as Minister for Bougainville Affairs in May, announced that further talks had resulted in a comprehensive agreement on autonomy for the island, with the Government ceding to the Provincial Assembly's demands that Bougainville be accorded its own system of criminal law and an autonomous police force. It was also agreed that the Papua New Guinea Defence Force's jurisdiction on the island would be strictly limited. On 30 August the Government and island leaders signed the Bougainville peace agreement in Arawa. Although he signed the accord, which was still to be approved by the National Parliament, Toroama stated that the BRA would campaign for the referendum to be held in three to five years' time, and would continue to seek full independence. Francis Ona did not attend the signing ceremony.

The weapons disposal process was threatened in late November 2001 when Henry Kiumo, a former commander of the pro-Government Bougainville Resistance Force (BRF) militia, was murdered. However, weapons disposal by the BRA and BRF began in early December, and the UN Observer Mission on Bougainville (UNOMB) formally acknowledged the Bougainville Peace Agreement later in that month. In January 2002 the National Parliament unanimously endorsed the Organic Law enacting the Bougainville peace agreement, along with a bill containing the requisite constitutional amendment. A second vote, held in late March, ratified the legislation, and the Papua New Guinea Defence Force began its withdrawal from Bougainville. The withdrawal of troops was completed in late December. In the same month, however, the peace process was jeopardized by the theft of 360 weapons from containers being used in the weapons disposal process.

In January 2003 it was announced that the BIPG and the BPC would be merged to form the Bougainville Constituent Assembly. The new body, which was expected to be composed of some 90 members, would debate and approve a proposed constitution for the island and complete the weapons disposal process. In late June an Australian-led group of regional representatives, who had been monitoring the truce on Bougainville since its signing in 1997, officially left the island. Joseph Kabui used the opportunity of their departure to appeal to ex-combatants not to endanger the peace process. Meanwhile, members of a group calling itself the Me'ekamui Fighters, loyal to Francis Ona, reiterated their disillusionment with the peace process and stated that they were fighting for independence rather than autonomy. Regional concern for the future of the island was highlighted when, in August, the Pacific Islands Forum urged the UN Security Council to remain involved in the peace process on Bougainville. In the same month the Government of Papua New Guinea formally announced its plan for autonomy in Bougainville, allowing provincial authorities to proceed with the establishment of a constitution and the eventual organization of elections for an autonomous government. In March 2004 the BIPG passed legislation providing for the establishment of the Bougainville Constituent Assembly (BCA), which was finally convened later that month to consider the third and final draft of the proposed constitution. In June the UN agreed to extend the term of its Observer Mission on the island by six months, but stressed that there would be no further extensions and urged the BCA to organize elections by the end of 2004. In the following month 50 newly trained officers began duty in their role as Bougainville's first police force. They were to be supported by a small team of Australian police officers under the terms of that country's Enhanced Co-operation Program with Papua New Guinea.

In September 2000 a group of landowners from Bougainville initiated legal action in a US court against Rio Tinto, the operator of the Panguna copper mine between 1972 and 1988. The group was reported to be suing the company for the environmental and social damage caused by its activities, including health problems experienced by workers and islanders living near the mine. Moreover, their case alleged that the company had effectively transformed the Papua New Guinea Defence Force into its own private army and was therefore responsible for the deaths of some 15,000 civilians in military action and a further 10,000 as a result of the economic blockade on the island.

In 1984 more than 9,000 refugees crossed into Papua New Guinea from the Indonesian province of Irian Jaya (officially known as Papua from January 2002), as a consequence of operations by the Indonesian army against Melanesian rebels of the pro-independence Organisasi Papua Merdeka (OPM—Free Papua Movement). For many years, relations between Papua New Guinea and Indonesia had been strained over the conflict in Irian Jaya, not least because the independence movement drew sympathy from many among the largely Melanesian population of Papua New Guinea. A new border treaty was signed in October 1984, and attempts were made to repatriate the refugees, based on assurances by the Indonesian Government that there would be no reprisals against those who returned. However, in 1985 and 1986 more refugees were still crossing into Papua New Guinea than returning to Irian Jaya. By mid-1987 more than 15,000 refugees were living in camps near the border in Papua New Guinea's Western Province, and only a few hundred had taken advantage of the programme of voluntary repatriation. A treaty signed by the two Governments in October 1995, providing for the settlement of disputes by consultation and arbitration, provoked strong criticism among opposition politicians in Papua New Guinea, who claimed that it effectively precluded the censure of any violation of human rights in Irian Jaya. In 1988 Somare (the then Minister for Foreign Affairs) condemned the numerous incursions by Indonesian soldiers into Papua New Guinea, and the resultant violence and killings, as a breach of the treaty, affirming that Papua New Guinea would not support Indonesia in its attempt to suppress the OPM. In late 1995 the Indonesian consulate in Vanimo was attacked by OPM rebels and a subsequent increase in Indonesian troops along the border was reported. Violent confrontations resulted in the killing of several rebels and security personnel and the kidnapping by OPM activists of some 200 villagers. Australia urged the Papua New Guinea Government to accept several thousand refugees living in camps along the border, and in May 1996 the Government announced that 3,500 Irian Jayans would be allowed to remain in Papua New Guinea on condition that they were not involved in OPM activities. In mid-1998, on an official visit to Indonesia, the Prime Minister, Bill Skate, signed a memorandum of understanding on bilateral relations, which was expected to provide the basis for closer co-operation in political, economic and defence matters. Under the terms of the agreement, troop numbers along the border were increased in March 1999. In May of that year 11 Javanese people were taken hostage and three people were reported to have been killed by OPM rebels in the West Sepik province of Papua New Guinea. The hostages were later released as a result of the intervention of the Papua New Guinea security forces. The OPM subsequently demanded an inquiry into the operation leading to the release of the hostages, after its communications director was allegedly shot dead at a border post. During 2000 hundreds of refugees were voluntarily repatriated to Indonesia (including more than 600 in a major operation in early September), although approximately 7,000 were believed to remain in Papua New Guinea in September. In late 2000 it was reported that Indonesian security forces had made some 400 incursions into Papua New Guinea in the previous two months while pursuing separatists. In December the border with Irian Jaya was officially closed, while Morauta reaffirmed his recognition of Indonesia's sovereignty over the region.

In January 2003 the Papua New Guinea Government denied a number of security breaches, which were reported to have occurred along its border with Indonesia. Concerns grew following a dramatic increase in the numbers of Indonesian troops along the border, from some 150 in late 2002 to a reported 1,500 in early 2003. In late January the Government of Papua New Guinea ordered its security forces to arrest OPM rebels believed to be staying in refugee camps along the border, following complaints from the Indonesian authorities that rebels based there were carrying out operations in the province of Papua. The Government also confirmed in April that it had been unable to persuade more than 400 Papuans living in a camp near the border town of Vanimo to return to Indonesia, despite assurances that their safety would not be at risk. In July 2004 it was announced that Vanuatu would host a series of negotiations between the Indonesian authorities and Papuan separatists.

Since 1990 relations with Solomon Islands have been overshadowed by the conflict on Bougainville. Solomon Islands (whose inhabitants are culturally and ethnically very similar to Bougainville islanders) has protested against repeated incursions by Papua New Guinea defence forces into Solomon Islands' territorial waters, while the Papua New Guinea Government has consistently accused Solomon Islands of harbouring members of the BRA and providing them with supplies. Despite several attempts during the 1990s to improve the situation between the two countries (including an agreement by Solomon Islands in 1993 to close the BRA office in Honiara), relations remained tense. In April 1997 the Solomon Islands Government announced that it was considering the initiation of proceedings against the Papua New Guinea Government concerning its attempted use of mercenaries on Bougainville. In June 1997 Papua New Guinea and Solomon Islands concluded a maritime border agreement, following several years of negotiations. The purpose of the agreement (which came into effect in January 1998) was not only to delineate the sea boundary between the two countries but also to provide a framework for co-operation in matters of security, natural disaster, customs, quarantine, immigration and conservation. In March 2000 relations between Papua New Guinea and the Solomon Islands were further strengthened following the opening of a Solomon Islands High Commission in Port Moresby. A bilateral agreement covering various areas of mutual concern was signed by the two countries in July 2004.

Concerns were expressed in mid-2003 that the rebel Solomon Islands leader Harold Keke (responsible for a campaign of terror in Solomon Islands) was recruiting new members for his Guadalcanal Liberation Force in Bougainville and strengthening his links with the BRA. These fears were exacerbated by sightings of Keke in Buka and Bougainville and attacks against journalists who had reported on the activities of Keke's supporters (including the stockpiling of weapons) along the border between Solomon Islands and Papua New Guinea. However, in August the Governor of Bougainville, John Momis, officially stated that he did not believe that any link existed between Keke and the BRA.

In March 1988 Papua New Guinea signed an agreement with Vanuatu and Solomon Islands to form the 'Melanesian Spearhead Group' (MSG), dedicated to the preservation of Melanesian cultural traditions and to the achievement of independence for the French Overseas Territory of New Caledonia. In 1989 Papua New Guinea increased its links with South-East Asia, signing a Treaty of Amity and Co-operation with ASEAN, and entering into negotiations with Malaysia on the creation of a Port Moresby Stock Exchange, which started trading in April 1999. The friendly relations between Papua New Guinea and the People's Republic of China which had been jeopardized by a briefly implemented decision to establish full diplomatic relations with Taiwan in mid-1999, were reconfirmed in 2000 when the Government of Papua New Guinea gave permission for China to construct a new building for the Ministry of Foreign Affairs in Port Moresby, despite fears that this could lead to a breach of national security.

The increasing exploitation of Papua New Guinea's natural resources in the 1990s led to considerable concern over the impact of the activities of numerous foreign business interests on the country's environment. Activity in the forestry sector increased dramatically in the early 1990s, and an official report, published in 1994, indicated that the current rate of logging was three times the sustainable yield of the country's forests. Attempts to introduce new regulations to govern the industry, however, were strongly opposed by several Malaysian logging companies with operations in the country. The exploitation of Papua New Guinea's extensive mineral resources had resulted in considerable environmental damage. In 1989 the Government gave the operators of the Ok Tedi gold and copper mine permission to discharge 150,000 metric tons of toxic waste per day into the Fly River. In 1994 6,000 people living in the region began a compensation claim against the Australian company operating the mine for damage caused by the resultant pollution. A settlement worth some $A110m. was reached in an Australian court in 1995 (and a further settlement worth $A400m. for the establishment of a containment system was concluded in mid-1996), despite opposition from the Papua New Guinea Government, which feared that such action might adversely affect the country's prospects of attracting foreign investment in the future. Similar claims were initiated by people living near the Australian-controlled Porgera gold mine in early 1996 for pollution of the Strickland River system, as well as by landowners near the Kutubu oilfield. The results of an independent study into the effects of mining activities by the Ok Tedi mine, published in mid-1999, showed that damage to the environment might be greater than originally believed, casting doubts over the future of the mine. The Government, however, announced that it would await the findings of an independent review, carried out by the World Bank. The results of this report, which were published in March 2000, found that closure of the mine was necessary on environmental grounds. However, it also emphasized the potentially damaging impact of the mine's early closure on both the local economy and world copper markets. In October 2001 BHP Billiton, the Australian operator, announced its withdrawal from the Ok Tedi mine and in early 2002 transferred its 52% stake to a development fund called the PNG Sustainable Development Programme Ltd.

Economy

In 2002, according to estimates by the World Bank, Papua New Guinea's gross national income (GNI), measured at average 2000–02 prices, was US $2,823m., equivalent to $530 per head (or $2,080 per head on an international purchasing-power parity basis). During 1990–2002, it was estimated, the population increased by an annual average of 2.5%. The average annual increase in gross domestic product (GDP) was 3.6% in 1990–2002. According to estimates by the Asian Development Bank (ADB), GDP decreased by 0.8% in 2002 but increased by 2.0% in 2003. A growth rate of 2.8% was projected for 2004.

More than 70% of the working population are engaged in subsistence agriculture, growing mainly roots and tubers. The principal cash crops are coffee (which accounted for an estimated 3.8% of export earnings in 2003), cocoa (3.3%), coconuts (for the production of copra and coconut oil), palm oil (5.4%), rubber and tea. Vanilla was also becoming an important export crop in the early 2000s. In 2000–01 agricultural exports were constrained by unusual climatic conditions, which particularly affected the coffee harvest, and by low international commodity prices; coffee and copra prices declined by some 30% on the world market in 2001. None the less, cocoa production increased in 2000 as a result of better management practices and earnings from the crop also rose as a result of higher world market prices for the commodity. Palm oil production also increased in the early 2000s. The sale of fishing licences to foreign fleets provides a significant source of revenue (estimated at US $15m. in 1998). However, a report published in 2004 claimed that the tuna industry alone could generate revenue of some $1,000m. per year if the catch was processed locally; some 95% of tuna was estimated to be caught by foreign fleets and shipped directly overseas.

In 2002 the agricultural sector contributed 34.1% of GDP and in 2001 it engaged an estimated 73.6% of the employed labour force. During 1990–2002 agricultural GDP rose, in real terms, by an annual average of 3.1%. According to the ADB, it decreased by an estimated 5.3% in 2001, but increased by 7.3% in 2002 and by 1.6% in 2003.

Forestry is an important activity. Papua New Guinea is one of the world's largest exporters of unprocessed tropical timber. In 1998 exports of timber provided 13.0% of total export earnings. Low prices for tropical hardwoods on the international market, together with the Asian financial crisis, however, led to a contraction in the industry in the late 1990s. By 2003 timber exports provided only 5.3% of total export earnings, despite large volumes of the commodity being exported. There is serious concern about the environmental damage caused by extensive logging activity in the country, much of which is illegal. Concern regarding the unsustainability of the industry increased in 1999, when tax incentives for logging operators led to a dramatic rise in production. In late 1999 the Government reinstated higher taxes on logging and introduced a moratorium on all new forestry licences in an attempt to control logging activity. However, a report published in 2004 claimed that legislation attempting to ensure the sustainability of the industry was largely ignored by most logging companies, and revealed that the forest resources of eight of the country's 12 provinces were almost entirely depleted.

The real GDP of the industrial sector (including mining and quarrying, manufacturing, construction and utilities) increased by an annual average of 5.6% in 1990–2002. According to the ADB, industrial GDP decreased by 10.8% in 2002, but increased by 3.8% in 2003. The sector contributed an estimated 36.5% of GDP in 2002. The manufacturing sector is limited by the small internal market, the low purchasing power of the population and the lack of an integrated transport network. Manufacturing provided 10.9% of GDP in 2002. Measured by the value of output, the principal branches of manufacturing are food products, beverages, wood products, metal products, machinery and transport equipment. Manufacturing GDP increased by an annual average of 4.0% in 1990–2002. Several fish canneries were established in the 1990s following the signing of a new quota agreement with the European Union (EU) in March 2003. A major new fish processing plant to enable the export of live fish and lobsters to China began operations in 2004 following the receipt of funding for the project of some US $30m. from a Chinese company. Exports of fresh, frozen and canned fish increased by some 40% to $109m. in 2003, representing the highest revenue ever earned by the industry.

High rainfall and relief give Papua New Guinea the potential to generate large amounts of hydroelectricity, which is estimated to account for more than 50% of electricity supplies. The sugar industry also provides the raw material for the production of ethyl alcohol (ethanol) as an alternative fuel. In the development of all Papua New Guinea's natural energy resources, foreign investment is of paramount importance.

A comparatively new development is the exploitation of Papua New Guinea's extensive mineral resources, mainly copper, gold and silver, and, more recently, substantial deposits of chromite, cobalt, nickel and quartz have been discovered. Papua New Guinea also has large reserves of natural gas and petroleum. The main source of copper and gold is the island of Bougainville, where copper-mining began in 1972. Operations at the mine ceased indefinitely in 1989, owing to civil unrest (see History). Copper is normally the country's principal export, accounting for 58% of export earnings (K702.8m.) in 1988 (although this figure had fallen to 18.0% by 2003). The development of a new mine in 1984 at Ok Tedi, in the Star mountains on the mainland, placed Papua New Guinea among the world's major gold-producing countries. Production of copper concentrates at Ok Tedi began in 1986, and in 1991 the mine produced the country's entire output of copper; it employed a total of 1,920 people in 1992. In 2000, however, the mine was threatened with closure because of a report indicating that damage to the environment caused by mining activities might be greater than originally believed (see History). The mine, on average, contributes 10% of the country's annual GNI and 20% of export revenues. Large deposits of gold were discovered at several sites during the 1980s (including the largest known deposit outside South Africa at Lihir, which began operations in 1997), and annual gold production subsequently rose at a steady rate, reaching an estimated 74 metric tons in 2000, and totalling 68 tons in 2003. Two new projects announced in mid-2003 were expected to increase gold production at Lihir by more that 30% over three years. A new medium-sized gold-mining project at Kainantu in the Eastern Highlands was expected to begin production in mid-

2005. Moreover, the important Ramu cobalt-nickel project was launched in February 2003. Exports of gold provided 35.9% of total export earnings in 2003. Mining and quarrying provided 16.6% of GDP in 2002 and 74.7% of total export earnings in 2003. Mining GDP declined by 15.9% in 2002, according to ADB figures.

Reserves of natural gas and petroleum were assessed in West Papua New Guinea in the mid-1980s. Recent geological surveys have indicated the presence of substantial gas reserves. In 1999 plans were under way for the construction of a US $3,500m. gas pipeline from the Southern Highlands to Queensland, Australia, although the agreement between landowners and the oil company involved to begin the project was not signed until June 2002. Continued problems with the project resulted in further delays and it was announced that the pipeline was not expected to begin operating until 2007. In 1987 potential recoverable petroleum reserves of an estimated 500m. barrels were discovered at the Iagifu oilfield in the Southern Highlands. Production of petroleum at the Hedinia and nearby Agogo oilfields in the Kutubu joint venture began in 1992, and by 1993 had reached some 130,000 barrels per day (b/d), although by early 2000 output had decreased to around 36,400 b/d, owing to natural depletion of the field, and an increase in gas production. Production from two new major oilfields (Moran and Gobe), which began in early 1998, was expected to double the volume of exported petroleum. However, production declined by 20.2% in 2000. The country's largest petroleum refinery at Napa Napa, was expected to begin production in mid-2004 and to process some 33,000 b/d. Exports of crude petroleum provided 20.8% of total export earnings in 2003.

The services sector contributed an estimated 29.4% of GDP in 2002. Tourism is an expanding sector, although political instability and reports of widespread crime have had a detrimental effect on the industry. During the late 1980s and early 1990s annual foreign tourist arrivals barely exceeded 40,000, but the total increased significantly, to 67,465, in 1998. By 2003 the number of visitors to Papua New Guinea had declined to 52,623. Tourism receipts were worth an estimated US $101m. in 2001. The GDP of the services sector expanded by an average annual rate of 3.6% in 1990–2002. According to the ADB, services GDP decreased by 1.2% in 2002 but increased by 2.3% in 2003.

Government finance is derived from three main sources: internal revenue, loans and overseas aid. Papua New Guinea has received substantial long-term loans from the ADB and from the World Bank. In 2003/04 aid for Papua New Guinea from Australia was projected at $A333.6m. A budgetary deficit equivalent to 1.7% of GDP was estimated in 2003 (compared with around 4% in the previous two years) and a further deficit of K300m. was forecast for 2004.

In 2003, according to the ADB, Australia provided 44.7% of Papua New Guinea's imports and took 29.3% of exports (including gold). Japan supplied 3.6% of imports and purchased 7.4% of exports in that year. Other significant trading partners are the People's Republic of China (6.1% of exports in 2003), New Zealand (7.3% of imports), Germany (3.7% of exports), Singapore (20.6% of imports in that year) and the United Kingdom. In 2003 there was a visible trade surplus of US $1,017m., and a surplus of $140m. on the current account of the balance of payments. Papua New Guinea's total external debt stood at $2,485.4m. at the end of 2002, of which $1,487.7m. was long-term public debt. In the previous year the cost of long-term debt-servicing had been equivalent to 12.7% of exports of goods and services. The annual inflation rate averaged 9.7% in 1990–2002; consumer prices increased by an average of 9.3% in 2001, by 11.8% in 2002 and by a similar figure in 2003.

Papua New Guinea remains dependent on aid from international donors for its economic development. Foreign investment is sought and encouraged, while fears of neo-colonialism and domination of the economy by overseas interests are frequently expressed. A marked rural–urban migration pattern has developed. The prolonged closure of the Bougainville copper mine from 1989 and the high cost of security operations there coincided with a period of low prices for Papua New Guinea's key agricultural commodities. The country is also vulnerable to adverse climatic conditions, as illustrated in 1997 when a serious drought resulted in the suspension of operations at the Porgera and Ok Tedi mines (owing to the low water-level in rivers essential for the transport of minerals) as well as in disastrous harvests for cash and subsistence crops. Similarly, a series of tsunami which struck the north-west of the country in 1998 had serious consequences for the country (see History), as did an earthquake and tsunami off the north coast, near Wewak, in September 2002 and severe floods in Madang in early 2004 which destroyed crops, infrastructure and entire villages. Growing ethnic and environmental problems (see above), widespread crime and concern over foreign exploitation also continued to threaten Papua New Guinea's economic success. A report published in late 1998 claimed that crime (particularly urban crime, the incidence of which was believed to have increased 20-fold since the 1970s) constituted Papua New Guinea's most serious economic problem. A total of 592 murders were reported in 2003 and this figure appeared likely to increase in 2004.

A 'recovery' budget for 2000 included the privatization of government assets, the proceeds of which were to be used to repay a proportion of the national debt. A joint funding programme, valued at US $500m., was agreed upon by the IMF and the World Bank to assist the Government in the implementation of the economic reforms, and Australia pledged a total of $1,000m. in programme aid over a period of three years. In late 2000 some 1,500 public servants were made redundant as part of ongoing government retrenchment measures. However, in early 2001 the World Bank expressed concern over the continued high incidence of official corruption in the country, and in February its resident representative (who had been monitoring the implementation of economic reforms) was deported for allegedly interfering in domestic politics. In April 2002 it was announced that, as part of the ongoing public-sector reforms, more than 3,000 civil servants were to be made redundant by the end of the year. However, the Government suffered a conclusive defeat at a general election held in June 2002, and the newly elected Prime Minister, Sir Michael Somare, declared that one of his Government's most urgent priorities was the immediate cessation of the World Bank-endorsed privatization programme. A marked increase in the country's debt, as well as a continued decline in the value of the kina (which fell by some 10% between October and November) resulted in serious financial difficulties for the new Government, particularly when a request to the Australian Government for help in restructuring $180m. of debt was refused. Expenditure in the 2003 budget was significantly reduced, although a large proportion of this was allocated to major infrastructure projects, such as the Highlands Highway (which, it was hoped, would stimulate the economy, particularly the agricultural sector). Despite the success of Papua New Guinea's agricultural exports (notably coffee, cocoa and palm oil) in 2002, owing to improved commodity prices and the low value of the kina, it seemed likely that additional revenue would need to be raised in 2003 through a resumption of the asset sales programme. Similarly, the 2004 budget was to include a temporary tax on imports and a reduction in personal tax rebates, in an attempt to control the predicted deficit of $100m. New agricultural developments were also to receive corporate tax concessions in the budget in order to encourage several important projects in the sector, although fears were expressed that export revenue from some of the country's most important commodities might decline when a trade agreement allowing free access to the EU market expires in 2007. Moreover, in August 2004 the Government announced plans to reduce the number of government employees by 10% by 2007. The Government's success in restoring a measure of economic stability to the country, including a significant reduction in the budget deficit and a lower rate of inflation, earned it praise from the IMF in mid-2004.

Papua New Guinea is a member of the Asian Development Bank, the Asia-Pacific Economic Co-operation (APEC) group, the Pacific Community (formerly the South Pacific Commission), the Pacific Islands Forum (formerly the South Pacific Forum), the International Cocoa Organization and the International Coffee Organization.

Statistical Survey

Source (unless otherwise stated): Papua New Guinea National Statistical Office, Post Office, Wards Strip, Waigani, NCD; tel. 3011229; fax 3251869; e-mail nsuvulo@nso.gov.pg; internet www.nso.gov.pg.

Area and Population

AREA, POPULATION AND DENSITY

Area (sq km)	462,840*
Population (census results)	
11 July 1990†	3,607,954
9 July 2000	
Males	2,661,091
Females	2,469,274
Total	5,130,365
Population (UN estimates at mid-year)‡	
2001	5,460,000
2002	5,586,000
2003	5,711,000
Density (per sq km) at mid-2003	12.3

* 178,704 sq miles.

† Excluding North Solomons Province (estimated population 154,000).

‡ Source: UN, *World Population Prospects: The 2002 Revision*.

PRINCIPAL TOWNS
(census of 9 July 2000)

Port Moresby (capital)	254,158	Mount Hagen	27,782
Lae	78,038	Madang	27,394
Arawa	36,443	Kokopo/Vunamami	20,262

Source: Thomas Brinkhoff, *City Population* (internet www.citypopulation.de).

Mid-2003 (UN estimate, incl. suburbs): Port Moresby 274,872 (Source: UN, *World Urbanization Prospects: The 2003 Revision*).

BIRTHS AND DEATHS (2002): Live births 175,824; Deaths 6,915. (Source: UN, *Population and Vital Statistics Report*). *1996* (estimates): Birth rate 32.6 per 1,000; Death rate 10.1 per 1,000.

Expectation of life (WHO estimates, years at birth): 59.8 (males 58.4; females 61.5) in 2002 (Source: WHO, *World Health Report*).

ECONOMICALLY ACTIVE POPULATION*
(estimates based on 1980 census results)

Agriculture, hunting, forestry and fishing†	564,500
Mining and quarrying	4,300
Manufacturing	14,000
Electricity, gas and water	2,800
Construction	21,600
Trade, restaurants and hotels	25,100
Transport, storage and communications	17,400
Financing, insurance, real estate and business services	4,500
Community, social and personal services	77,100
Others (incl. activities not stated or not adequately described)	1,500
Total labour force†	732,800

* Figures refer to citizens of Papua New Guinea only.
† Excluding persons solely engaged in subsistence agriculture, hunting, forestry and fishing.

1990 census (citizens aged 10 years and over, excl. North Solomons Province): Total employed 1,582,518 (Agriculture, hunting, forestry and fishing 1,269,744); Unemployed 132,812; Total labour force 1,715,330 (males 1,002,891, females 712,439).

Mid-2002 (estimates in '000): Agriculture, etc. 1,944; Total 2,657 (Source: FAO).

Health and Welfare

KEY INDICATORS

Total fertility rate (children per woman, 2002)	4.1
Under-5 mortality rate (per 1,000 live births, 2002)	94
HIV/AIDS (% of persons aged 15–49, 2003)	0.6
Physicians (per 1,000 head, 1998)	0.07
Hospital beds (per 1,000 head, 1990)	4.02
Health expenditure (2001): US $ per head (PPP)	144
Health expenditure (2001): % of GDP	4.4
Health expenditure (2001): public (% of total)	89.0
Access to water (% of persons, 1999)	42
Access to sanitation (% of persons, 1999)	82
Human Development Index (2002): ranking	133
Human Development Index (2002): value	0.542

For sources and definitions, see explanatory note on p. vi.

Agriculture

PRINCIPAL CROPS
('000 metric tons)

	2000	2001	2002
Maize*	7	8	8
Sorghum*	3	4	4
Sweet potatoes	480†	490*	490*
Cassava (Manioc)*	120	125	125
Yams*	230	232	232
Taro (Coco yam)*	172	174	174
Other roots and tubers*	277	281	281
Sugar cane*	383	450	425
Pulses*	3	3	3
Coconuts	1,032†	553†	513*
Oil palm fruit*	1,200	1,175	1,066
Vegetables and melons*	474	487	487
Pineapples*	12	14	14
Bananas*	710	725	725
Other fruit (excl. melons)*	512	524	524
Coffee (green)*	83†	63†	63
Cocoa beans†	47	39	45
Tea (made)*	9	9	9
Natural rubber (dry weight)*	7	4	4

* FAO estimate(s).
† Unofficial figure(s).

Source: FAO.

LIVESTOCK
(FAO estimates, '000 head, year ending September)

	2000	2001	2002
Horses	1.8	2	2
Cattle	87	88	89
Pigs	1,550	1,650	1,650
Sheep	6.3	6.5	6.5
Goats	2.3	2.4	2.4
Chickens	3,700	3,700	3,800

Source: FAO.

LIVESTOCK PRODUCTS
(FAO estimates, '000 metric tons)

	2000	2001	2002
Beef and veal	3	3	3
Pig meat	44	47	47
Poultry meat	5	5	6
Other meat	19	19	18
Poultry eggs	5	5	5

Source: FAO.

Forestry

ROUNDWOOD REMOVALS
('000 cubic metres, excluding bark)

	2000	2001	2002*
Sawlogs, veneer logs and logs for sleepers	2,064	1,611	1,611
Pulpwood	120	97	97
Fuel wood*	5,533	5,533	5,533
Total	7,717	7,241	7,241

* FAO estimates.

Source: FAO.

SAWNWOOD PRODUCTION
('000 cubic metres, including railway sleepers)

	2000	2001	2002
Coniferous (softwood)	0	0	10
Broadleaved (hardwood)	40	40	60
Total	40	40	70

Source: FAO.

Fishing

('000 metric tons, live weight of capture)

	2000	2001	2002
Mozambique tilapia*	2.3	2.3	2.3
Other freshwater fishes*	6.7	6.7	6.7
Sea catfishes*	1.9	1.9	1.9
Skipjack tuna	52.3	64.4	89.9
Yellowfin tuna	14.7	25.8	27.9
Total catch*	96.4	122.5	148.7

* FAO estimates.

Note: Figures exclude crocodiles, recorded by number rather than weight. The number of estuarine crocodiles caught was: 8,336 in 2000; 10,677 in 2001; 9,332 in 2002. The number of New Guinea crocodiles caught was: 16,018 in 2000; 20,668 in 2001; 18,798 in 2002.

Source: FAO.

Mining

	2000	2001	2002†
Petroleum, crude ('000 barrels)	24,967	20,423	20,000
Copper ('000 metric tons)*	203.1	203.8	204.0
Silver (metric tons)*	79.2	69.4	73.0
Gold (metric tons)*	74.5	67.0	70.0

* Figures refer to metal content of ore.

† Preliminary figures.

Source: US Geological Survey.

Industry

SELECTED PRODUCTS

	2000	2001	2002
Palm oil ('000 metric tons)*	336	329	316
Raw sugar ('000 metric tons)*	46	54	51

* Unofficial estimates.

Beer ('000 litres): 39,029 in 1997; 34,826 in 1998.

Electric energy (million kWh): 2,180 in 2000.

Sources: FAO, *Production Yearbook*; UN, *Industrial Commodity Statistics Yearbook*; and Papua New Guinea National Statistical Office.

Finance

CURRENCY AND EXCHANGE RATES

Monetary Units

100 toea = 1 kina (K).

Sterling, Dollar and Euro Equivalents (31 May 2004)

£1 sterling = 5.909 kina;
US $1 = 3.221 kina;
€1 = 3.944 kina;
100 kina = £16.92 = $31.05 = €25.36.

Average Exchange Rate (US $ per kina)

2001 0.2658
2002 0.2488
2003 0.3000

Note: The foregoing information refers to the mid-point exchange rate of the central bank. In October 1994 it was announced that the kina would be allowed to 'float' on foreign exchange markets.

BUDGET

(million kina)*

Revenue†	1999	2000‡	2001‡
Taxation	1,662	1,917	2,476
Taxes on income, profits, etc.	818	910	1,342
Individual	524	565	650
Corporate	293	345	692
Excises	171	169	185
Taxes on international trade and transactions	551	666	771
Import duties	472	542	636
Export duties	79	124	135
Non-tax revenue	1,148	600	383
Entrepreneurial and property income	1,066	483	111
Total	2,810	2,517	2,859

Expenditure§	1997	1998	1999
Recurrent expenditure	1,768	1,895	2,020
National and statutory	941	994	1,091
National	820	858	970
Salaries and wages	367	384	393
Goods and services	453	475	428
Statutory authorities	120	136	121
Provincial departments	530	566	537
Salaries and wages	288	307	384
Goods and services	73	125	89
Conditional grants	169	135	65
Interest payments	298	335	393
Domestic	206	228	261
External	91	107	132
Development expenditure	428	596	786
National and statutory	318	495	698
Provinces	110	101	88
Total	2,196	2,491	2,806

* Figures refer to the operations of the General Budget and of superannuation and retirement funds. The transactions of other central government units with individual budgets are excluded.

† Excluding grants received from abroad (million kina): 477 in 1999; 555 (estimate) in 2000; 551 (estimate) in 2001.

‡ Estimates.

§ Excluding net lending (million kina): –3 in 1997; –5 in 1998; –5 in 1999.

Source: partly IMF, *Papua New Guinea: Recent Economic Developments* (October 2000).

2000 (million kina): *Revenue* Recurrent revenue 2,910 (Tax revenue 2,678; Non-tax revenue 232); Grants 490; Total (incl. grants) 3,400. *Expenditure:* Recurrent expenditure 2,726; Capital expenditure 836; Total 3,562 (Source: Asian Development Bank, *Key Indicators of Developing Asian and Pacific Countries*).

INTERNATIONAL RESERVES

(US $ million at 31 December)

	2001	2002	2003
Gold (national valuation)	7.42	21.89	25.83
IMF special drawing rights	8.71	6.06	3.68
Reserve position in IMF	0.38	0.49	0.59
Foreign exchange	413.56	314.96	489.90
Total	430.06	336.85	520.00

Source: IMF, *International Financial Statistics*.

MONEY SUPPLY

(million kina at 31 December)

	2001	2002	2003
Currency outside banks	308.98	379.94	417.93
Demand deposits at deposit money banks	1,104.17	1,241.13	1,474.95
Total money (incl. others)	1,416.95	1,629.66	1,896.79

Source: IMF, *International Financial Statistics*.

COST OF LIVING
(Consumer Price Index; base: 1977 = 100)

	1998	1999	2000
Food	390.2	456.5	518.5
Clothing	286.1	324.8	375.9
Rent, fuel and power	216.5	225.6	242.2
All items (incl. others)	411.0	472.4	546.1

Source: National Statistical Office, *Statistical Digest 2000*.

All items (base: 2000 = 100): 109.3 in 2001; 122.2 in 2002; 140.2 in 2003 (Source: IMF, *International Financial Statistics*).

NATIONAL ACCOUNTS
(million kina at current prices)

National Income and Product

	1997	1998	1999
Compensation of employees	1,778.3	1,751.6	1,852.4
Operating surplus	3,963.2	4,804.6	5,519.4
Domestic factor incomes	5,741.5	6,556.2	7,371.8
Consumption of fixed capital	535.2	539.6	526.5
Gross domestic product (GDP) at factor cost	6,276.7	7,095.8	7,898.3
Indirect taxes / *Less* Subsidies	787.0	692.6	882.4
GDP in purchasers' values	7,063.7	7,788.5	8,780.8
Net factor income from abroad	–310.4	–343.3	–360.1
Gross national product	6,753.3	7,445.2	8,420.7
Less Consumption of fixed capital	535.2	539.6	526.5
National income in market prices	6,218.1	6,905.6	7,894.2

Expenditure on the Gross Domestic Product

	2000	2001	2002
Government final consumption expenditure	1,740.9	1,821.8	1,929.3
Private final consumption expenditure	6,462.0	8,450.9	9,881.5
Increase in stocks	143.7	186.3	206.1
Gross fixed capital formation	2,145.2	2,374.1	2,443.3
Total domestic expenditure	10,491.8	12,833.1	14,460.2
Exports of goods and services	4,720.3	4,974.8	5,188.8
Less Imports of goods and services	4,462.2	6,049.6	6,274.0
GDP in purchasers' values	10,749.9	11,758.3	13,375.0
GDP at constant 1983 prices	8,591.8	8,823.3	9,002.7

Sources: National Statistical Office, Bank of Papua New Guinea and Asian Development Bank.

Gross Domestic Product by Economic Activity

	2000	2001	2002
Agriculture, hunting, forestry and fishing	3,306.9	3,577.5	4,428.0
Mining and quarrying	2,463.5	2,401.1	2,149.6
Manufacturing	1,026.6	1,285.3	1,421.4
Electricity, gas and water	139.6	167.2	194.8
Construction	450.2	687.0	973.0
Wholesale and retail trade	902.5	1,078.6	1,441.1
Transport, storage and communications	439.4	560.9	620.8
Finance, insurance, real estate and business services*	395.6	406.7	407.0
Community, social and personal services (incl. defence)	1,297.3	1,240.6	1,351.8
Sub-total	10,421.6	11,404.9	12,987.5
Import duties, *less* subsidies	299.4	303.3	294.6
Statistical discrepancy	28.9	50.1	92.9
GDP in purchasers' values	10,749.9	11,758.3	13,375.0

* After deducting imputed bank service charge.

Source: Asian Development Bank, *Key Indicators of Developing Asian and Pacific Countries*.

BALANCE OF PAYMENTS
(US $ million)

	1999	2000	2001
Exports of goods f.o.b.	1,927.4	2,094.1	1,812.9
Imports of goods f.o.b.	–1,071.4	–998.8	–932.4
Trade balance	856.0	1095.3	880.5
Exports of services	247.5	242.7	285.1
Imports of services	–727.9	–772.3	-662.0
Balance on goods and services	375.6	565.7	503.7
Other income received	18.6	32.0	20.0
Other income paid	–291.1	–242.1	–250.2
Balance on goods, services and income	103.1	355.6	273.4
Current transfers received	60.3	62.4	75.9
Current transfers paid	–68.7	–72.7	–67.3
Current balance	94.7	345.3	282.0
Direct investment from abroad	296.5	95.9	62.5
Portfolio investment assets	89.0	–123.8	–72.7
Other investment assets	10.7	–41.0	–66.9
Other investment liabilities	–380.2	–185.2	–74.7
Net errors and omissions	14.3	13.1	-1.6
Overall balance	125.0	104.5	128.6

Source: IMF, *International Financial Statistics*.

External Trade

PRINCIPAL COMMODITIES

Imports f.o.b. (US $ million)*	1998	1999	2000
Food and live animals	220.2	128.7	168.6
Meat and meat preparations	44.0	25.4	41.6
Fresh, chilled or frozen meat	40.9	24.2	38.0
Cereals and cereal preparations	105.3	61.1	73.5
Rice	75.0	42.1	46.3
Milled or semi-milled rice	49.1	25.5	39.1
Chemicals and related products	130.5	86.1	71.0
Basic manufactures	275.7	198.0	159.5
Iron and steel	44.1	38.7	31.3
Iron or steel structures and parts	53.1	3.4	3.6
Machinery and transport equipment	529.1	481.9	305.3
Power-generating machinery and equipment	47.8	33.3	18.0
Machinery specialized for particular industries	71.7	82.6	43.0
General industrial machinery and equipment	105.2	58.9	53.9
Electrical machinery, apparatus and appliances, etc.	49.9	39.1	26.6
Road vehicles and parts†	145.1	172.7	119.6
Motor vehicles for goods transport	66.8	79.8	51.5
Other transport equipment and parts†	56.3	38.3	24.4
Miscellaneous manufactured articles	111.9	87.6	64.2
Total (incl. others)	1,360.1	1,086.4	1,035.1

* Figures include migrants' and travellers' dutiable effects, but exclude military equipment and some parcel post.

† Data on parts exclude tyres, engines and electrical parts.

Source: UN, *International Trade Statistics Yearbook*.

Exports f.o.b. (million kina)	2001	2002	2003
Coffee	247	277	299
Logs	332	414	416
Copper	859	1,019	1,415
Petroleum	1,889	1,431	1,632
Palm oil	245	390	421
Gold	2,115	2,295	2,811
Cocoa	121	226	258
Total (incl. others)	6,105	6,387	7,842

Source: Asian Development Bank, *Key Indicators of Developing Asian and Pacific Countries*.

PRINCIPAL TRADING PARTNERS
(US $ million)

Imports f.o.b. (excl. gold)	2001	2002	2003
Australia	574.1	589.1	603.3
China, People's Republic	21.0	30.0	67.0
Hong Kong	9.8	10.3	n.a.
Indonesia	31.2	33.2	37.1
Japan	51.6	50.5	48.8
Malaysia	31.1	35.7	40.7
New Zealand	45.0	52.5	99.1
Singapore	212.9	225.0	277.9
USA	24.5	25.6	33.2
Total (incl. others)	1,123.8	1,196.5	1,351.1

Exports f.o.b. (excl. gold)	2001	2002	2003
Australia	646.5	645.5	1,005.4
China, People's Republic	111.3	145.2	210.1
Germany	106.1	77.0	126.9
Japan	279.2	252.5	254.0
Korea, Republic	78.4	29.3	33.4
Philippines	59.4	12.8	43.5
United Kingdom	54.3	67.3	89.9
USA	39.5	87.9	64.0
Total (incl. others)	2,626.6	2,669.1	3,434.1

Source: Asian Development Bank, *Key Indicators of Developing Asian and Pacific Countries*.

Transport

ROAD TRAFFIC
(licensed vehicles)

	1986	1987	1988
Cars and station wagons	16,574	17,121	17,532
Commercial vehicles	26,989	26,061	29,021
Motorcycles	1,246	1,232	1,204
Tractors	1,287	1,313	1,414

1996 (estimates): Passenger cars 31,000; Lorries and vans 85,000.

Source: IRF, *World Road Statistics*.

SHIPPING

Merchant Fleet
(registered at 31 December)

	2001	2002	2003
Number of vessels	114	111	112
Total displacement ('000 grt)	77.0	72.4	73.3

Source: Lloyd's Register-Fairplay, *World Fleet Statistics*.

International Sea-borne Freight Traffic

	1997	1998	1999
Cargo unloaded ('000 metric tons)	2,208.6	2,209.0	2,062.8
Cargo loaded ('000 metric tons)	735.9	823.3	788.1

Source: Papua New Guinea Harbours Board, *Monthly Shipping Register Form*.

CIVIL AVIATION
(traffic on scheduled services)

	1997	1998	1999
Kilometres flown (million)	15	15	12
Passengers carried ('000)	1,114	1,110	1,102
Passenger-km (million)	735	736	641
Total ton-km (million)	86	87	80

Source: UN, *Statistical Yearbook*.

Tourism

FOREIGN TOURIST ARRIVALS

Country of origin	1998	1999	2000
Australia	35,403	33,818	29,285
Canada	912	923	1,762
Japan	1,834	2,427	3,244
New Zealand	3,661	3,712	2,648
United Kingdom	3,087	3,067	2,279
USA	6,101	5,619	5,429
Total (incl. others)	67,465	67,357	58,448

Receipts from tourism (US $ million): 76 in 1999; 92 in 2000; 101 in 2001.

Source: World Tourism Organization.

Communications Media

	2000	2001	2002
Television receivers ('000 in use)	100	110	n.a.
Telephones ('000 main lines in use)	64.8	62.0	62.0
Mobile cellular telephones ('000 subscribers)	8.6	10.7	15.0
Personal computers ('000 in use)	280	300	321
Internet users ('000)	45.0	50.0	75.0

Daily newspapers (1996): 2 (combined circulation 65,000 copies).

Radio receivers (1997): 410,000 in use.

Facsimile machines (1994): 795 in use.

Sources: UNESCO, *Statistical Yearbook*; International Telecommunication Union.

Education

(1999, unless otherwise indicated)

	Institutions*	Teachers	Students
Pre-primary	29	3,816	119,147
Primary	2,790	16,297	594,444
Secondary	n.a.	3,046	74,042
Tertiary	n.a.	815	13,761

* 1995 figures.

Sources: National Department of Education, *The state of education in PNG* (March 2001); UNESCO, *Statistical Yearbook*.

Adult literacy rate (UNESCO estimates): 64.6% (males 71.1%; females 57.7%) in 2001 (Source: UN Development Programme, *Human Development Report*).

Directory

The Constitution

The present Constitution came into effect on 16 September 1975, when Papua New Guinea became independent. The main provisions of the Constitution are summarized below:

PREAMBLE

The national goals of the Independent State of Papua New Guinea are: integral human development, equality and participation in the development of the country, national sovereignty and self-reliance, conservation of natural resources and the environment and development primarily through the use of Papua New Guinean forms of social, political and economic organization.

BASIC HUMAN RIGHTS

All people are entitled to the fundamental rights and freedoms of the individual whatever their race, tribe, place of origin, political opinion, colour, creed or sex. The individual's rights include the right to freedom, life and the protection of the law, freedom from inhuman treatment, forced labour, arbitrary search and entry, freedom of conscience, thought, religion, expression, assembly, association and employment, and the right to privacy. Papua New Guinea citizens also have the following special rights: the right to vote and stand for public office, the right to freedom of information and of movement, protection from unjust deprivation of property and equality before the law.

THE NATION

Papua New Guinea is a sovereign, independent state. There is a National Capital District which shall be the seat of government.

The Constitution provides for various classes of citizenship. The age of majority is 19 years.

HEAD OF STATE

Her Majesty the Queen of the United Kingdom of Great Britain and Northern Ireland is Queen and Head of State of Papua New Guinea. The Head of State appoints and dismisses the Prime Minister on the proposal of the National Parliament and other ministers on the proposal of the Prime Minister. The Governor-General and Chief Justice are appointed and dismissed on the proposal of the National Executive Council. All the privileges, powers, functions, duties and responsibilities of the Head of State may be exercised or performed through the Governor-General.

Governor-General

The Governor-General must be a citizen who is qualified to be a member of Parliament or who is a mature person of good standing who enjoys the respect of the community. No one is eligible for appointment more than once unless Parliament approves by a two-thirds' majority. No one is eligible for a third term. The Governor-General is appointed by the Head of State on the proposal of the National Executive Council in accordance with the decision of Parliament by simple majority vote. He may be dismissed by the Head of State on the proposal of the National Executive Council in accordance with a decision of the Council or of an absolute majority of Parliament. The normal term of office is six years. In the case of temporary or permanent absence, dismissal or suspension he may be replaced temporarily by the Speaker of the National Parliament until such time as a new Governor-General is appointed.

THE GOVERNMENT

The Government comprises the National Parliament, the National Executive and the National Judicial System.

National Parliament

The National Parliament, or the House of Assembly, is a single-chamber legislature of members elected from single-member open or provincial electorates. The National Parliament has 109 members elected by universal adult suffrage. The normal term of office is five years. There is a Speaker and a Deputy Speaker, who must be members of Parliament and must be elected to these posts by Parliament. They cannot serve as government ministers concurrently.

National Executive

The National Executive comprises the Head of State and the National Executive Council. The Prime Minister, who presides over the National Executive Council, is appointed and dismissed by the Head of State on the proposal of Parliament. The other ministers, of whom there shall be not fewer than six nor more than a quarter of the number of members of the Parliament, are appointed and dismissed by the Head of State on the proposal of the Prime Minister. The National Executive Council consists of all the ministers, including the Prime Minister, and is responsible for the executive government of Papua New Guinea.

National Judicial System

The National Judicial System comprises the Supreme Court, the National Court, Local Courts and Village Courts. The judiciary is independent.

The Supreme Court consists of the Chief Justice, the Deputy Chief Justice and the other judges of the National Court. It is the final court of appeal. The Chief Justice is appointed and dismissed by the Head of State on the proposal of the National Executive Council after consultation with the minister responsible for justice. The Deputy Chief Justice and the other judges are appointed by the Judicial and Legal Services Commission. The National Court consists of the Chief Justice, the Deputy Chief Justice and no fewer than four nor more than six other judges.

The Constitution also makes provision for the establishment of the Magisterial Service and the establishment of the posts of Public Prosecutor and the Public Solicitor.

THE STATE SERVICES

The Constitution establishes the following State Services which, with the exception of the Defence Force, are subject to ultimate civilian control.

National Public Service

The Public Service is managed by the Department of Personnel Management which is headed by a Secretary, who is appointed by the National Executive Council on a four-year contract.

Police Force

The Police Force is subject to the control of the National Executive Council through a minister and its function is to preserve peace and good order and to maintain and enforce the law.

Papua New Guinea Defence Force

The Defence Force is subject to the superintendence and control of the National Executive Council through the Minister of Defence. The functions of the Defence Force are to defend Papua New Guinea, to provide assistance to civilian authorities in a civil disaster, in the restoration of public order or during a period of declared national emergency.

The fourth State Service is the Parliamentary Service.

The Constitution also includes sections on Public Finance, the office of the Auditor-General, the Public Accounts Commission and the declaration of a State of National Emergency.

The Government

HEAD OF STATE

Queen: HM Queen ELIZABETH II.

Governor-General: Sir PAULIAS MATANE (sworn in 29 June 2004).

NATIONAL EXECUTIVE COUNCIL
(September 2004)

Prime Minister: Sir MICHAEL SOMARE.

Minister for Agriculture and Livestock: MATHEW SIUNE.

Minister for Communications and Information: BEN SEMRI.

Minister for Correctional and Institutional Services: POSI MENAI.

Minister for Culture and Tourism: DAVID BASUA.

Minister for Defence: MATHEW GUBAG.

Minister for Education, Science and Technology: BRIAN PULAYASI.

Minister for Environment and Conservation: WILLIAM DUMA.

Minister for Finance and Treasury: BART PHILEMON.

Minister for Fisheries: BEN SIMRI.

Minister for Foreign Affairs and Immigration: Sir RABBIE NAMALIU.

Minister for Forestry: PATRICK PRUAITCH.

Minister for Health: MELCHIOR PEP.

Minister for Housing and Urban Resettlement: ATUMENG BUHUTE.

Minister for Inter-Government Relations: Sir PETER BARTER.

Minister for Internal Security: BIRE KIMISOPA.

Minister for Justice: MARK MAIPAKAI.

Minister for Labour and Industrial Relations: TOM TOMIAPE.

Minister for Lands and Physical Planning: PETRUS THOMAS.

Minister for Mining: SAM AKOITAI.

Minister for National Planning and Monitoring: (vacant).

Minister for Petroleum and Energy: Sir MOI AVEI.

Minister for Public Services: (vacant).

Minister for Science and Technology: ROY BIYAMA.

Minister for State Enterprises: Dr PUKA TEMU.

Minister for Trade and Industry: PAUL TIENSTEN.

Minister for Transport and Civil Aviation: DON POLYE.

Minister for Welfare and Social Development: Lady CAROL KIDU.

Minister for Works: GABRIEL KAPRIS.

GOVERNMENT DEPARTMENTS AND OFFICES

Office of the Prime Minister: POB 639, Waigani, NCD; tel. 3276544; fax 3277380; e-mail primeminister@pm.gov.pg; internet www.pm.gov.pg/pmsoffice/PMsoffice.nsf.

Department of Agriculture and Livestock: Spring Garden Rd, POB 417, Konedobu 125, NCD; tel. 3231848; fax 3230563; internet www.agriculture.gov.pg.

Department of the Attorney-General: POB 591, Waigani, NCD; tel. 3230138; fax 3230241.

Department of Commerce and Industry: Central Government Offices, Kumul Ave, Post Office, Wards Strip, Waigani, NCD; tel. 3271115; fax 3271750.

Department of Corrective Institution Services: POB 6889, Boroko, NCD; tel. 3214917; fax 3217686.

Department of Defence: Murray Barracks, Free Mail Bag, Boroko 111, NCD; tel. 3242480; fax 3256117; internet www.defence.gov.pg.

Department of Education: POB 446, Waigani, NCD 131, Fin Corp Haus; tel. 3013555; fax 3254648; internet www.education.gov.pg.

Department of Environment and Conservation: POB 6601, Boroko 111, Kamul Ave, Waigani, NCD; tel. 3011607; fax 3011691.

Department of Family and Church Affairs: Ori Lavi Haus, Nita St, POB 7354, Boroko, NCD; tel. 3254566; fax 3251230; internet www.datec.com.pg/government/famchurch/default.htm.

Department of Finance and Treasury: POB 710, Waigani, Vulupindi Haus, NCD; internet www.treasury.gov.pg.

Department of Fisheries and Marine Resources: POB 2016, Port Moresby; tel. 3271799; fax 3202074; internet www.fisheries.gov.pg.

Department of Foreign Affairs: Central Government Offices, Kumul Ave, Post Office, Wards Strip, Waigani, NCD; tel. 3271311; fax 3254467.

Department of Health: POB 3991, Boroko, Aopi Centre, Waigani Drive, Waigani; tel. 3254648; fax 3013555; internet www.health.gov.pg.

Department of Housing and Urban Resettlement: POB 1550, Boroko, NCD; tel. 3247200; fax 3259918.

Department of Inter-Government Relations: Somare Haus, Independence Drive, Waigani, POB 1287, Boroko, NCD; tel. 3011002; fax 3250553; e-mail paffairs@dalton.com.pg; internet www.dplga.gov.pg.

Department of Justice: Port Moresby; internet www.justice.gov.pg.

Department of Labour and Industrial Relations: POB 5644, Boroko, NCD; tel. 3217408; fax 3214085.

Department of Lands and Physical Planning: POB 233, Boroko, NCD; tel. 3013175; fax 3013299; e-mail lubena@lands.gov.pg; internet www.datec.com.pg/government/lands.

Department of Mineral Resources: PMB PO, Konedobu, Port Moresby 121; tel. 3227600; fax 3213701; internet www.mineral.gov.pg.

Department of Mining: Private Mailbag, Port Moresby Post Office; tel. 3227670; fax 3213958.

Department of Personnel Management: POB 519, Wards Strip, Waigani, NCD; tel. 3276422; fax 3250520; e-mail perrtsiamalili@dpm.gov.pg.

Department of Petroleum and Energy: POB 1993, Port Moresby, NCD; tel. 3224200; fax 3224222; e-mail joseph_gabut@petroleum.gov.pg; internet www.petroleum.gov.pg.

Department of Planning and Implementation: POB 710, Waigani 131, Vulupindi Haus, Port Moresby, NCD; tel. 3288302; fax 3288375.

Department of Police: Police Headquarters, POB 85, Konedobu, NCD; tel. 3226100; fax 3226113; internet www.police.gov.pg.

Department of Private Enterprise: Central Government Offices, Kumul Ave, Post Office, Wards Strip, Waigani, NCD.

Department of Social Welfare and Development, Youth and Women: Maori Kiki Bldg, 2nd Floor, POB 7354, Boroko, NCD; tel. 3254967; fax 3213821.

Department of State: Haus To Makala, 5th Floor, Post Office, Wards Strip, Waigani, NCD; tel. 3276758; fax 3214861.

Department of Trade and Industry: Heduru Haus, Waigani Drive, POB 375, Waigani 131, NCD; tel. 3255311; fax 3254482; internet www.trade.gov.pg.

Department of Transport and Civil Aviation: POB 1489, Port Moresby; tel. 3222580; fax 3200236; e-mail dotdirectro@datec.com.pg.

National Forest Authority: POB 5055, Boroko, Frangipani St, Hohola, Port Moresby; tel. 3277800; fax 3254433; internet www.datec.com.pg/government/forest/default.htm.

Office of Peace and Reconstruction (fmrly Office of Bougainville Affairs): Morauta Haus, POB 343, Waigani 131, NCD; tel. 3276760; fax 3258038; internet www.oba.gov.pg.

Office of Civil Aviation: POB 684, Boroko 111, NCD; tel. 3257077; fax 3251919; e-mail dgoca@datec.net.pg; internet www.oca.gov.pg.

Office of Information and Communication: Port Moresby; internet www.communication.gov.pg.

Office of Works: POB 1489, Port Moresby 121, cnr Champion Parade and Musgrave St; tel. 3222500; fax 3200236; internet www.works.gov.pg.

Legislature

NATIONAL PARLIAMENT

The unicameral legislature has 109 elective seats: 89 representing open constituencies and 20 representing provincial constituencies. There is constitutional provision for up to three nominated members.

Speaker: JEFFREY NAPE (acting).

General Election, 15 June–29 July 2002

Party	Seats
National Alliance	19
People's Democratic Movement	12
People's Progress Party	8
Pangu Pati	6
People's Action Party	5
People's Labour Party	4
Christian Democratic Party	3
Melanesian Alliance	3
National Party	3
United Party	3
National Transformation Party	2
Pan-Melanesian Congress Party	2
People's National Congress	2
People's Solidarity Party	2
Pipol First Party	2
Rural Pipol's Pati	2
Others and undeclared	14
Independents	17
Total	109

Note: In many cases party affiliations were subject to review in the immediate aftermath of the election.

Autonomous Region

BOUGAINVILLE INTERIM PROVINCIAL GOVERNMENT—BIPG

The BIPG was formally established following the signing of the 'Loloata Understanding' on 23 March 2000, and is responsible for the island's budget, judicial system and local administration. It is composed of an Executive Council and a 25-member Provincial Assembly, which was appointed by the National Executive Council on the recommendation of the four Bougainville Members of Parliament (who themselves joined the Assembly).

BOUGAINVILLE PEOPLE'S CONGRESS

Prior to the establishment of the BIPG, the Bougainville People's Congress (BPC) officially represented the island of Bougainville. Elections to the Congress were held over several days starting on 3 May 1999. Joseph Kabui was elected President, winning 77 of the 87 votes. A Congressional Executive Council, comprising 29 members, was subsequently appointed by Kabui. Although the BPC unanimously endorsed the 'Loloata Understanding', it did not actively participate in the newly established BIPG.

President: JOSEPH KABUI.

Vice-President: JAMES TANIS.

Speaker: LINUS KONUKONG.

BOUGAINVILLE CONSTITUENT ASSEMBLY

In late March 2004 the BIPG approved legislation to establish the Bougainville Constituent Assembly (BCA). The organization, which was composed of the members of both the BIPG and the BPC, was convened for the first time on 30 March 2004 to finalize the third and final draft of the proposed constitution for Bougainville. Following its consideration by the Bougainville Constitutional Commission in August, the BCA was expected to recommend the proposed constitution to the National Government in late 2004.

Political Organizations

Bougainville Revolutionary Army (BRA): demands full independence for island of Bougainville; Leaders FRANCIS ONA, ISMAEL TOROAMA.

League for National Advancement (LNA): POB 6101, Boroko, NCD; f. 1986; Leader JOHN NILKARE.

Melanesian Labour Party: Port Moresby; Leader PAUL MONDIA.

Melanesian United Front: Boroko, NCD; f. 1988; fmrly Morobe Independent Group; Leader UTULA SAMANA.

National Alliance: c/o National Parliament, Port Moresby; f. 1996 to combat corruption in public life; Leader Sir MICHAEL SOMARE.

National Party (NP): Private Bag, Boroko, NCD; f. 1979; fmrly People's United Front; Leader MICHAEL MEL.

Pangu (Papua New Guinea Unity) Pati: POB 289, Waigani, NCD; tel. 3277628; fax 3277611; f. 1968; urban- and rural-based; Leader CHRIS HAIVETA; Pres. PATE WAMP; Sec.-Gen. MOSES TAIAN.

People's Action Party (PAP): Boroko, NCD; tel. 3251343; f. 1985; Leader TED DIRO.

People's Democratic Movement (PDM): POB 972, Boroko, NCD; f. 1985; merged with Advance PNG Party in 2001; Leader Sir MEKERE MORAUTA.

People's National Alliance (PNA): c/o National Parliament, Port Moresby; f. 1999 by merger of following:

Melanesian Alliance (MA): Port Moresby; tel. 3277635; f. 1978; socialist; Chair. Fr JOHN MOMIS; Gen. Sec. FABIAN WAU KAWA.

Movement for Greater Autonomy: Manus Province; Leader STEPHEN POKAWIN.

People's National Party: Port Moresby.

People's National Congress (PNC): c/o National Parliament, Port Moresby; Leader BILL SKATE.

People's Progress Party (PPP): POB 6030, Boroko, NCD; f. 1970; Leader ANDREW BAING; Nat. Chair. GLEN KUNDIN.

People's Unity Party: c/o National Parliament, Port Moresby; Leader ALFRED KAIABE.

United Resource Party (URP): Port Moresby; f. 1997; aims to secure greater representation in Government for resource owners; Leader MASKET IANGALIO; Chair. PITA IPATAS.

Diplomatic Representation

EMBASSIES AND HIGH COMMISSIONS IN PAPUA NEW GUINEA

Australia: POB 129, Waigani, NCD; tel. 3259333; fax 3259183; High Commissioner MICHAEL POTTS.

China, People's Republic: POB 1351, Boroko, NCD; tel. 3259836; fax 3258247; e-mail chnempng@daltron.com.pg; Ambassador LI ZHENGJUN.

Fiji: Defence House, 4th Floor, Champion Parade, Port Moresby, NCD; tel. 3211914; fax 3217220; e-mail ert_rakai@yahoo.co,uk; High Commissioner RATU INOKE KUBUABOLA.

France: Defens Haus, 6th Floor, Cnr of Hunter St and Champion Parade, POB 1155, Port Moresby; tel. 3215550; fax 3215549; e-mail ambfrpom@global.net.pg; Ambassador THIERRY BERNADAC.

Holy See: POB 98, Port Moresby; tel. 3256021; fax 3252844; e-mail nunciaturepng@datec.net.pg; Apostolic Nuncio Archbishop ADOLFO TITO YLLANA.

India: Port Moresby; tel. 3254757; fax 3253138; e-mail hcipom@datec.net.pg; High Commissioner H. V. S. MANRAL.

Indonesia: 1-2/410 Kiroki St, Sir John Guise Dr., Waigani, NCD; tel. 3253116; fax 3253544; e-mail kbripom@daltron.com.pg; Ambassador (vacant).

Japan: Cuthbertson House, Cuthbertson St, POB 1040, Port Moresby; tel. 3211800; fax 3217906; Ambassador (vacant).

Korea, Republic: Pacific View Apts, Lot 1, sec. 84, Pruth St, Korobosea, Port Moresby; tel. 3254755; fax 3259996; Ambassador Dr SEO HYUN-SEOP.

Malaysia: POB 1400, Pacific View Apts, Units 1 and 3, 2nd floor, Pruth St, Kovobosea, Port Moresby; tel. 3252076; fax 3252784; e-mail mwpom@datec.com.pg; High Commissioner MOHAMAD FADZIL AYOB.

New Zealand: Embassy Drive, POB 1051, Waigani, NCD; tel. 3259444; fax 3250565; e-mail nzhcpom@dg.com.pg; High Commissioner LAURIE MARKES.

Philippines: POB 5916, Boroko, NCD; tel. 3256577; fax 3231803; e-mail pomphpem@datec.com.pg; Ambassador BIENVENIDO TEJANO.

Solomon Islands: Port Moresby; e-mail sihicomm@daltron.com.pg; High Commissioner PHILIP KAPINI.

United Kingdom: Kiroki St, POB 212, Waigani 131, NCD; tel. 3251677; fax 3253547; e-mail bhcpng@datec.net.pg; internet www.britishhighcommission.gov.uk/papuanewguinea; High Commissioner DAVID GORDON-MACLEOD.

USA: Douglas St, POB 1492, Port Moresby; tel. 3211455; fax 3213423; internet www.altnews.com.au/usembassy; Ambassador ROBERT FITTS.

Judicial System

The Supreme Court is the highest judicial authority in the country, and deals with all matters involving the interpretation of the Constitution, and with appeals from the National Court. The National Court has unlimited jurisdiction in both civil and criminal matters. All National Court Judges (except acting Judges) are Judges of the Supreme Court. District Courts are responsible for civil cases involving compensation, for some indictable offences and for the more serious summary offences, while Local Courts deal with minor offences and with such matters as custody of children under the provision of Custom. There are also Children's Courts, which judge cases involving minors. Appeal from the District, Local and Children's Courts lies to the National Court. District and Local Land Courts deal with disputes relating to Customary land, and Warden's Courts with civil cases relating to mining. In addition, there are other courts with responsibility for determining ownership of government land and for assessing the right of Customary landowners to compensation. Village Courts, which are presided over by Magistrates with no formal legal qualification, are responsible for all Customary matters not dealt with by other courts.

Supreme Court of Papua New Guinea

POB 7018, Boroko, NCD; tel. 3245700; fax 3234492.

Chief Justice: Sir MARI KAPI.

Attorney-General: MICHAEL GENE.

Religion

The belief in magic or sorcery is widespread, even among the significant proportion of the population that has adopted Christianity (nominally 97% in 1990). Pantheism also survives. There are many missionary societies.

CHRISTIANITY

Papua New Guinea Council of Churches: POB 1015, Boroko, NCD; tel. 3259961; fax 3251206; f. 1965; seven mem. churches; Chair. EDEA KIDU; Gen. Sec. SOPHIA W. R. GEGEYO.

The Anglican Communion

Formerly part of the Province of Queensland within the Church of England in Australia (now the Anglican Church of Australia), Papua New Guinea became an independent Province in 1977. The Anglican Church of Papua New Guinea comprises five dioceses and had 246,000 members in 2000.

Archbishop of Papua New Guinea and Bishop of Aipo Rongo: Most Rev. JAMES AYONG, POB 893, Mount Hagen, Western Highlands Province; tel. 5421131; fax 5421181; e-mail acpnghgn@global.net.pg.

General Secretary: MARTIN GARDHAM, POB 673, Lae, Morobe Province; tel. 4724111; fax 4721852; e-mail acpng@global.net.pg.

The Roman Catholic Church

For ecclesiastical purposes, Papua New Guinea comprises four archdioceses and 15 dioceses. At 31 December 2001 there were 1,625,355 adherents.

Catholic Bishops' Conference of Papua New Guinea and Solomon Islands

POB 398, Waigani, NCD; tel. 3259577; fax 3232551; e-mail cbc@online.net.pg; internet www.catholicpng.org.pg.

f. 1959; Pres. Most Rev. KARL HESSE (Archbishop of Rabaul); Gen. Sec. LAWRENCE STEPHENS.

Archbishop of Madang: Most Rev. WILLIAM KURTZ, Archbishop's Residence, POB 750, Madang; tel. 8522946; fax 8522596; e-mail Kurtz_caom@global.net.pg.

Archbishop of Mount Hagen: Most Rev. MICHAEL MEIER, Archbishop's Office, POB 54, Mount Hagen, Western Highlands Province; tel. 5421285; fax 5422128; e-mail archdios@online.net.pg.

Archbishop of Port Moresby: Most Rev. Sir BRIAN BARNES, Archbishop's House, POB 1032, Boroko, NCD; tel. 3251192; fax 3256731; e-mail archpom@daltron.com.pg.

Archbishop of Rabaul: Most Rev. KARL HESSE, Archbishop's House, POB 357, Kokopo, East New Britain Province; tel. 9828369; fax 9828404; e-mail abkhesse@online.net.pg.

Other Christian Churches

Baptist Union of Papua New Guinea Inc: POB 705, Mount Hagen, Western Highlands Province; tel. 5522364; fax 5522402; e-mail bupng@global.net.pg; f. 1976; Gen. Sec. JOHN KAENKI; 48,000 mems.

Evangelical Lutheran Church of Papua New Guinea: Bishop Rt Rev. WESLEYKIGASUNG POB 80, Lae, Morobe Province; tel. 4723711; fax 4721056; e-mail bishop@elcpng.org.pg; f. 1956; Sec. REUBEN KURE; 815,000 mems.

Gutnius Lutheran Church of Papua New Guinea: Bishop Rev. DAVIDP.PISO POB 111, 291 Wabag, Enga Province; tel. 5471002; f. 1948; Gen. Sec. RICHARD R. MOSES; 95,000 mems.

Papua New Guinea Union Mission of the Seventh-day Adventist Church: POB 86, Lae, Morobe Province 411; tel. 4721488; fax 4721873; Pres. Pastor WILSON STEPHEN; Sec. Pastor BRADLEY RICHARD KEMP; 200,000 adherents.

The United Church in Papua New Guinea: POB 1401, Port Moresby; tel. 3211744; fax 3214930; f. 1968 by union of the Methodist Church in Melanesia, the Papua Ekalesia and United Church, Port Moresby; Moderator Rev. SAMSON LOWA; 600,000 mems; Gen. Sec. DEMAS TONGOGO APELIS.

BAHÁ'Í FAITH

National Spiritual Assembly: Private Mail Bag, Boroko, NCD; tel. 3250286; fax 3236474; e-mail nsapng@datec.net.pg.

ISLAM

In 2000 the Muslim community in Papua New Guinea numbered about 1,500, of whom approximately two-thirds were believed to be expatriates. The religion was introduced to the island in the 1970s. The first mosque there was opened in late 2000 at Poreporena Highway, Hohola, Port Moresby; Imam KHALID ARAI (acting).

The Press

There are numerous newspapers and magazines published by government departments, statutory organizations, missions, sporting organizations, local government councils and regional authorities. They are variously in English, Tok Pisin (Pidgin), Motu and vernacular languages.

Ailans Nius: POB 1239, Rabaul, East New Britain Province; weekly.

Foreign Affairs Review: Dept of Foreign Affairs, Central Government Offices, Kumul Ave, Post Office, Wards Strip, Waigani, NCD; tel. 3271401; fax 3254886.

Hailans Nius: Mount Hagen, Western Highlands Province; weekly.

Lae Nius: POB 759, Lae, Morobe Province; 2 a week.

The National: POB 6817, Boroko, NCD; tel. 3246888; fax 3246868; e-mail national@thenational.com.pg; internet www.thenational.com.pg; f. 1993; daily; Editor BRIAN GOMEZ; circ. 20,000.

Niugini Nius: POB 3019, Boroko, NCD; tel. 3252177; e-mail niusedita@pactok.net; internet pactok.net.au/docs/nius/; daily.

Papua and New Guinea Education Gazette: Dept of Education, PSA Haus, POB 446, Waigani, NCD; tel. 3272413; fax 3254648; monthly; Editor J. OBERLENTER; circ. 8,000.

Papua New Guinea Post-Courier: POB 85, Port Moresby; tel. 3091000; fax 3212721; e-mail postcourier@ssp.com.pg; internet www.postcourier.com.pg; f. 1969; daily; English; published by News Corpn; Gen. Man. TONY YIANNI; Editor OSEAH PHILEMON; circ. 25,044.

Sunkamap Times: Bougainville, North Solomons Province; monthly; f. 2004; community newsletter.

Wantok (Friend) Niuspepa: POB 1982, Boroko, NCD; tel. 3252500; fax 3252579; e-mail word@global.net.pg; f. 1970; weekly in New Guinea Pidgin; mainly rural readership; Publr ANNA SOLOMON; Editor YAKAM KELO; circ. 10,000.

Publishers

Gordon and Gotch (PNG) Pty Ltd: POB 107, Boroko, NCD; tel. 3254855; fax 3250950; e-mail ggpng@online.net.pg; f. 1970; books, magazines and stationery; Gen. Man. PETER G. PORTER.

Scripture Union of Papua New Guinea: POB 280, University, Boroko, NCD; tel. and fax 3253987; f. 1966; religious; Chair. RAVA TAVIRI.

Word Publishing Co Pty Ltd: POB 1982, Boroko, NCD; tel. 3252500; fax 3252579; e-mail word@global.net.pg; f. 1982; 60% owned by the Roman Catholic Church, 20% by Evangelical Lutheran, 10% by Anglican and 10% by United Churches; Gen. Man. JEREMY BURGESS.

Broadcasting and Communications

TELECOMMUNICATIONS

Office of Information and Communication: POB 639, Waigani; tel. 3256853; fax 3250412; internet www.communication.gov.pg.

Papua New Guinea Telecommunication Authority (Pangtel): POB 8444, Boroko, NCD; tel. 3258633; fax 3256868; internet www.pangtel.gov.pg; f. 1997; CEO PHILIP AEAVA.

Telikom PNG Pty Ltd: POB 7395, Boroko, NCD; tel. 3005000; fax 3259582; internet www.telikompng.com.pg; Man. Dir GEREA AOPI; Chair. FLORIAN GUBON.

BROADCASTING

Radio

National Broadcasting Corporation of Papua New Guinea: POB 1359, Boroko, NCD; tel. 3255233; fax 3230404; e-mail md.nbc@global.net.pg; f. 1973; commercial and free govt radio programmes services; broadcasting in English, Melanesian, Pidgin, Motu and 30 vernacular languages; Chair. CHRIS RANGATIN; Man. Dir Dr KRISTOFFA NINKAMA.

Kalang Service (FM): POB 1359, Boroko, NCD; tel. 3255233; commercial radio co established by National Broadcasting Commission; Chair. CAROLUS KETSIMUR.

Nau FM/Yumi FM: POB 774, Port Moresby; tel. 3201996; fax 3201995; internet www.naufm.com.pg; f. 1994; Gen. Mans MARK ROGERS, JUSTIN KILI.

Television

EM TV: POB 443, Boroko, NCD; tel. 3257322; fax 3254450; e-mail emtv@emtv.com.pg; internet www.emtv.com.pg; f. 1988; operated by Media Niugini Pty Ltd; CEO STEPHEN SMITH.

Media Niugini Pty Ltd: POB 443, Boroko, NCD; tel. 3257322; fax 3254450; e-mail emtv@emtv.com.pg; internet www.emtv.com.pg; f. 1987; owned by Nine Network Australia; CEO STEPHEN SMITH.

Finance

(cap. = capital; res = reserves; dep. = deposits; m. = million; brs = branches; amounts in kina unless otherwise stated)

BANKING

Central Bank

Bank of Papua New Guinea: Douglas St, POB 121, Port Moresby; tel. 3227200; fax 3211617; e-mail bpng@datec.com.pg; internet www.bankpng.gov.pg; f. 1973; bank of issue since 1975; sold to Bank of South Pacific in 2002; cap. 62.0m., res 680.2m., dep. 1,040.0m. (Dec. 2002); Gov. WILSON KAMIT.

Commercial Banks

Australia and New Zealand Banking Group (PNG) Limited: Defens Haus, 3rd Floor, cnr of Champion Parade and Hunter St, POB 1152, Port Moresby; tel. 3223333; fax 3223306; f. 1976; cap. 4.7m., res 1.1m., dep. 494.1m. (Sept. 1998); Chair. R. G. LYON; Man. Dir ALLAN MARLIN; 8 brs.

Bank of Hawaii (PNG) Ltd: Burns House, Champion Parade, POB 1390, Port Moresby; tel. 3213533; fax 3213115; e-mail bohpng@datec.com.pg; f. 1983 as Indosuez Niugini Bank Ltd, name changed in 1997; the bank was expected to cease operations in 2001/02; cap.

3.0m., res 0.8m., dep. 121.1m. (Dec. 1997); Chair. MARC DUMETZ; Man. Dir JEAN-PHILIPPE AUDUBERT; 2 brs.

Bank of South Pacific Ltd: Douglas St, POB 173, Port Moresby; tel. 3212444; fax 3200053; e-mail service@bsp.com.pg; internet www.bsp.com.pg; f. 1974; acquired from National Australia Bank Ltd by Papua New Guinea consortium (National Investment Holdings, now BSP Holdings Ltd) in 1993; cap. 182.4m., res 8.0m., dep. 1,554.0m. (Dec. 2002); Chair. NOREO BEANGKE; Man. Dir NOEL R. SMITH; 8 brs and 2 sub-brs.

Maybank (PNG) Ltd: Waigani, NCD; f. 1995.

MBf Finance (PNG) Ltd: Elsa Beach Towers, Ground Floor, cnr of Musgrave St, POB 329, Port Moresby; tel. 3213555; fax 3213480; f. 1989.

Papua New Guinea Banking Corporation: cnr of Douglas and Musgrave Sts, POB 78, Port Moresby; tel. 3229700; fax 3211683; f. 1974; corporatized in 2001; cap. 11.3m., res 72.9m., dep. 978.8m. (Dec. 1997); Chair. ROGER PALME; Man. Dir HENRY FABILA; 34 brs.

Westpac Bank—PNG—Ltd: Mogoru Motu Bldg, 5th Floor, Champion Parade, POB 706, Port Moresby; tel. 3220800; fax 3213367; e-mail westpacpng@westpac.com.au; f. 1910 as Bank of New South Wales, present name since 1982; 90% owned by Westpac Banking Corpn, Australia; cap. 5.8m., res 6.1m., dep. 554.7m. (Sept. 1999); Chair. ALAN WALTER; Man. Dir SIMON MILLETT; 15 brs.

Development Bank

Rural Development Bank of Papua New Guinea: Somare Crescent, POB 686, Waigani, NCD; tel. 3247500; fax 3259817; f. 1967 as Agriculture bank of Papua New Guinea; name changed as above in 1994; cap. 32.6m., res 17.1m. (Dec. 1992); statutory govt agency; Chair. RUPA MULINA; Man. Dir ANDREW NAGARI; 10 brs.

Savings and Loan Societies

Registry of Savings and Loan Societies: Financial System Supervision Dept, POB 121, Port Moresby; tel. 3227200; fax 3214548; 101 savings and loan societies; 125,306 mems (2001); total funds 163.5m., loans outstanding 90.0m., investments 21.4m. (June 2001); CEO ELIZABETH GIMA.

STOCK EXCHANGE

Port Moresby Stock Exchange (POMSoX) Ltd: Level 4, Defens Haus, POB 1531, Port Moresby; tel. 3201980; fax 3201981; e-mail pomsox@datec.com.pg; internet www.pomsox.com.pg; f. 1999; Chair. Sir ANTHONY SIAGURU; Gen. Man. EMILY GEORGE TAULE.

INSURANCE

Niugini Insurance Corporation Ltd: POB 331, Port Moresby; tel. 3214077; fax 3217898; f. 1978; govt-owned; Chair. JACOB POPUNA; Man. Dir DARRYL G. NATHAN.

Pan Asia Pacific Assurance (PNG) Pty Ltd (PAPA): POB 3757, Boroko, NCD ; tel. 3202344; fax 3203443; e-mail pap@daltron.com.pg; f. 1993; Gen. Man. SOMASHEKAR MIRLE.

There are branches of several Australian and United Kingdom insurance companies in Port Moresby, Rabaul, Lae and Kieta.

Trade and Industry

GOVERNMENT AGENCIES

Investment Corporation of Papua New Guinea: Hunter St, POB 155, Port Moresby; tel. 3212855; fax 3211240; f. 1971 as govt body to support local enterprise and to purchase shares in foreign businesses operating in Papua New Guinea; partially transferred to private ownership in 1993.

Investment Promotion Authority (IPA): POB 5053, Boroko, NCD 111; tel. 3217311; fax 3212819; e-mail iepd@ipa.gov.pg; internet www.ipa.gov.pg; f. 1992 following reorganization of National Investment and Development Authority; a statutory body responsible for the promotion of foreign investment; the first contact point for foreign investors for advice on project proposals and approvals of applications for registration to conduct business in the country; contributes to planning for investment and recommends priority areas for investment to the Govt; also co-ordinates investment proposals; Man. Dir SIMON PETER (acting).

Privatization Commission: Port Moresby; f. 1999 to oversee transfer of state-owned enterprises to private ownership; Exec. Chair. BEN MICAH.

DEVELOPMENT ORGANIZATIONS

CDC Capital Partners Ltd: CDC Haus, 2nd Floor, POB 907, Port Moresby; tel. 3212944; fax 3212867; e-mail png@cdc.com.pg; internet www.cdcgroup.com; fmrly Commonwealth Development Corpn; Man. Dir ASHLEY EMBERSON-BAIN.

Industrial Centres Development Corporation: POB 1571, Boroko, NCD; tel. 3232913; fax 3231109; promotes foreign investment in non-mining sectors through establishment of manufacturing facilities.

CHAMBERS OF COMMERCE

Papua New Guinea Chamber of Commerce and Industry: POB 1621, Port Moresby; tel. 3213057; fax 3210566; e-mail pngcci@global.net.pg; Pres. MICHAEL MAYBERRY; Vice-Pres. PHILIP FRANKLIN.

Papua New Guinea Chamber of Mines and Petroleum: POB 1032, Port Moresby; tel. 3212988; fax 3217107; e-mail ga@pngchamberminpet.com.pg; internet www.pngchamberminpet.com.pg; Exec. Dir GREG ANDERSON; Pres. PETER BOTTEN.

Port Moresby Chamber of Commerce and Industry: POB 1764, Port Moresby; tel. 3213077; fax 3214203; Pres. DAVID CONN.

Lae Chamber of Commerce and Industry: POB 265, Lae, Morobe Province; tel. 4722340; fax 4726038; e-mail lcci@global.net.pg; internet www.lcci.org.pg; Pres. ALAN MCLAY.

INDUSTRIAL AND TRADE ASSOCIATIONS

Cocoa Board of Papua New Guinea: POB 1165, Madaney; tel. 8823253; fax 8822198; f. 1974; Chair. JIMMY SIMITAB; CEO LAUATU TAUTEA.

Coffee Industry Corpn Ltd: POB 137, Goroka, Eastern Highlands Province; tel. 721266; fax 721431; e-mail webmaster@gka.coffeecorp.org.pg; internet www.coffeecorp.org.pg; CEO RICKY MITIO.

Fishing Industry Association (PNG) Inc: POB 2340, Boroko, NCD; e-mail netshop1@daltron.com.pg; Chair. MAURICE BROWNJOHN.

Forest Industries Association: POB 5055, Boroko, NCD; Pres. STANIS BAI; CEO ROBERT TATE.

Higaturu Oil Palms Pty Ltd: POB 28, Popondetta, Oro Province; tel. 3297177; fax 3297137; f. 1976; jtly owned by the Commonwealth Development Corpn (UK) and the Papua New Guinea Govt; major producer of palm oil and cocoa; Gen. Man. RICHARD CASKIE.

Kopra Indasrti Korporesen (KIK): POB 81, Port Moresby; tel. 3211133; fax 3214257; e-mail infor@kik.com.pg; regulates and markets all copra and cocnut products in Papua New Guinea; consists of a chair. and mems representing producers; formerly known as the Copra Marketing Board of Papua New Guinea; Chair. Sir JOHN MIDDLETON; CEO TAND WAIM (acting).

Manufacturers' Council of Papua New Guinea: POB 598, Port Moresby; tel. 3259512; fax 3230199; e-mail pngmadecouncil@datec.com.pg; internet www.pngmade.org.pg; Chair. WAYNE GOLDING; CEO MARYANNE MCDONALD.

Mineral Resources Development Corporation: POB 1076, Port Moresby; tel. 3255822; fax 3252633; e-mail info@mrdc.com.pg; Man. Dir FRANCIS KAUPA.

National Contractors' Association: Port Moresby; formed by construction cos for the promotion of education, training and professional conduct in the construction industry; Pres. ROY THORPE.

National Fisheries Authority: POB 2016, Port Moresby; tel. 3212643; fax 3202074; e-mail nfa@fisheries.gov.pg; internet www.fisheries.gov.pg; Man. Dir TONY LEWIS; Dep. Man. Dir MICHAEL BATTY.

National Housing Corpn (NHC): POB 1550, Boroko NCD; tel. 3247000; fax 3254363; Man. Dir GABRIEL TOVO.

New Britain Palm Oil Ltd: POB 389, Kimbe, West New Britain; tel. 9852177; fax 9852003; e-mail nbpol@nbpol.com.pg; internet www.nbpol.com.pg; f. 1967; 80% owned by Kulim (Malaysia), 20% owned by Govt, employees and local producers; major producer of palm oil, coffee trader and exporter, supplier of high quality oil palm seed; Man. Dir NICK THOMPSON; Sec. HIMSON WANINARA.

Niugini Produce Marketing Pty Ltd: Lae, Morobe Province; f. 1982; govt-owned; handles distribution of fruit and vegetables throughout the country.

Palm Oil Producers Association: Port Moresby; Exec. Sec. ALLAN MAINO.

Papua New Guinea Forest Authority: POB 5055, Boroko, NCD; tel. 3277800; fax 3254433; Man. Dir THOMAS NEN; Chair. VALENTINE KAMBORI.

Papua New Guinea Growers Association: POB 14, Rabaul, East New Britain ; tel. 9829123; fax 9829264; e-mail davidloh@global.net.pg; Pres. ROBIN MERIBA; Exec. Dir DAVID LOH.

Papua New Guinea Holdings Corpn: POB 131, Port Moresby; fax 3217545; f. 1992; responsible for managing govt privatization programme; Chair. MICHAEL MEL; Man. Dir PETER STEELE.

Papua New Guinea Log Carriers Association: f. 1993.

Pita Lus National Silk Institute: Kagamuga, Mount Hagen, Western Highlands Province; f. 1978; govt silk-producing project.

Rural Industries Council: Chair. PETER COLTON.

UTILITIES

Electricity

PNG Electricity Commission (Elcom): POB 1105, Boroko, NCD; tel. 3243200; fax 3214051; plans to privatize the organization were announced in 1999; Chair. PAUL AISA; Chief Exec. SEV MASO.

Water

Eda Ranu (Our Water): POB 1084, Waigani, NCD; tel. 3122100; fax 3122190; e-mail enquiries@edaranu.com.pg; internet www.edaranu.com.pg; fmrly Port Moresby Water Supply Company; Gen. Man. BILLY IMAR.

PNG Waterboard: POB 2779, Boroko, NCD; tel. 3235700; fax 3236317; e-mail pamini@pngwater.com.pg; f. 1986; govt-owned; operates 12 water supply systems throughout the country; Man. Dir PATRICK AMINI.

MAJOR COMPANIES

General

Coconut Products Ltd: POB 94, Rabaul, East New Britain Province; tel. 9839310; internet 9829305; e-mail camohe@global.net.pg; coconut oil and copra; Gen. Man. G. DOWNS.

Collins and Leahy Holdings Ltd: ANZ House, Second Floor, Central Avenue, Lae; tel. 4722644; fax 7321946; f. 1970; Chair. Sir MICHAEL BROMLEY; Sec. C. I. CUNNIMGHAME.

Kagamuga Natural Products Co Pty Ltd: POB 74, Mount Hagen, Western Highlands Province; tel. 551225; fax 551329; pyrethrum extract and vegetable seeds.

Pacific Helicopters Ltd: POB 342, Goroka, Eastern Highlands Province; tel. 7321226; fax 7321503; e-mail pachelf@datec.com.pg; f. 1975; helicopter services, hotel, real estate, engine overhaul, eggs, construction; Gen. Man. STEVEN MASON.

Steamships Trading Co Ltd: POB 1, Port Moresby; tel. 3220222; fax 3213595; internet www.steamships.com.pg; hotels, department stores, food and hardware, wholesale and retail; quarrying, manufactures building materials, fibreglass products and UHT products; stevedoring, shipping; road transport, automotive distributors and car rentals; property management; Chair. CHRIS D. PRATT; Man. Dir G. J. DUNLOP.

Thiess Watkins PNG Ltd: POB 1393, Boroko, NCD; tel. 253466; building contractors, civil engineers, property developers and property management; operates throughout Papua New Guinea.

Food, Drink and Tobacco

Angco Ltd: POB 136, Goroka, Eastern Highlands Province; tel. 7321677; fax 7322154; e-mail angco@global.net.pg; coffee and cocoa exporters; coffee and cocoa plantation managers and developers; CEO ARTHUR JONES; Gen. Man. CRAIG MCCONAGHTY.

Associated Mills Ltd: POB 1906, Lae, Morobe Province; tel. 4723555; fax 4723424; flour milling; Gen. Man. JIM GREGG.

W. R. Carpenter & Co Estates: POB 94, Mount Hagen, Western Highlands Province; tel. 5422700; fax 5421616; e-mail sales@wrcarpenters.com.pg; internet www.wrcarpenters.com.pg; coffee, cocoa and tea processing; Gen. Man. MIKE JACKSON.

Kongo Coffee Ltd: POB 338, Kundiawa, Simbu Province; tel. 7323069; fax 7323068; e-mail kongo@global.net.pg; coffee exporters; Man. Dir JERRY KAPKA.

Niugini Coffee, Tea and Spice Co Pty Ltd: POB 2531, Lae, Morobe Province; tel. 4725633; fax 4725614; e-mail niuginicoffee@global.net.pg; Gen. Man. BRIAN STEVENSON.

Ramu Sugar Ltd: Port Moresby; sugar processing; Gen. Man. ERROL JOHNSTON; Chair. PETER COLTON.

SP Brewery Ltd: POB 6550, Boroko, NCD; tel. 3128200; fax 3250656; f. 1951; beer; Chair. J. TAUVASA; Gen. Man. HEIN VAN DORT.

Star-Kist PNG Pty Ltd: POB 1341, Rabaul, East New Britain Province; fish processing.

Tanubada Dairy Products Pty Ltd: POB 6203, Boroko, NCD; tel. 3212522; fax 3212774; reconstituted milk, ice cream, ice confection, orange drinks; Gen. Man. JOHN GOODWIN.

Tokua Plantation Pty Ltd: POB 65, Kokopo; tel. 9839370; fax 9839370; copra and cocoa production; Gen. Man. JOHN CARROLL.

Wills (PNG) Ltd: Modilon Rd, POB 678, Madang; tel. 8523788; fax 8523667; cigarettes and tobacco.

Minerals and Heavy Engineering

Barracuda Pty Ltd: POB 1139, Port Moresby, NCD; tel. 3212633; e-mail 3212847; exploration for petroleum; Man. IAN TREVITT.

Barclay Bros (NG) Pty Ltd: POB 1180, Boroko, NCD; tel. 255711; fax 250094; all types of civil engineering and building construction.

Bishop Bros Engineering Pty Ltd: POB 9081, Hohola; tel. 252900; all types of civil engineering.

Chevron Niugini Pty Ltd: Credit House, Cuthbertson St, Port Moresby, NCD; tel. 3211088; fax 3225566; e-mail irta@chevron.com; wholly owned by ChevronTexaco Corpn (USA); eploration for and production of petroleum; sale and distribution of fuel; Man. Dir LARRY BARTHOLD; Dir MOSELEY MORAMORO; 1,000 employees.

Hebou Constructions (PNG) Pty Ltd: POB 6207, Boroko, NCD; tel. 3253077; fax 3253441; e-mail info@constant.com.pg; building, civil construction, sand and gravel, timber, hotels.

Highlands Pacific Ltd: Pacific Place, Level 9, cnr Champion Parade, Port Moresby; tel. 3217633; fax 3217551; e-mail info@highlandspacific.com; internet www.highlandspacific.com; exploration for gold, nickel, cobalt, copper and other minerals; Chair. ROBERT BRYAN; Man. Dir IAN R. HOLZBERGER.

Hornibrook NGI Pty Ltd: Spring Garden Rd, Hohola, Port Moresby, NCD; tel. 3253099; fax 3250387; e-mail hngi@datec.com.pg; civil engineering and steel manufacture, pre-fabricated buildings, bridges, tubular fabrications, plant hire; f. 1991; Man. Dir MALCOLM LEWIS.

Lihir Gold Ltd: POB 789, Port Moresby; tel. 3217711; fax 3214705; e-mail mark.laurie@lihir.com.pg; internet www.lihir.com.pg; mining; gold mining on Lihir Island; Chair. ROSS GARNAUT; CEO NEIL SWAN.

Metal Refining Operations Pty Ltd: POB 3980, Boroko, NCD; tel. 3252647; fax 3252959; e-mail metalsrefining@mail.com; internet www.metalsrefining.com.pg; gold mining; Man. Dir TOM HUNTER; Gen. Man. GEOFF WHEBELL.

Mobil Oil New Guinea Ltd: 5th Floor, Credit House, Cuthbertson St, Port Moresby; tel. 3212055; fax 3222100; sale and distribution of fuel; operation of service stations; Area Man. NAMAR T. MAWASON.

Monier (PNG) Ltd: POB 328, Port Moresby; tel. 3253344; fax 3253389; concrete pipes, masonry blocks and roofing tiles, cement supplies, pre-cast concrete, sand and gravel; paints; fibreglass and plastic products; Gen. Man. DUNCAN FRASER.

Naco (NG) Pty Ltd: POB 707, Port Moresby; aluminium windows and sliding doors, louvre windows and metal blades.

Niugini Mining Ltd: developers of the Lihir Island gold project; Chair. GEOFF LOUDON.

Oil Search Ltd: MMI House, POB 1031, Port Moresby; tel. 3213177; fax 3214379; e-mail gmarsden@osl.com.au; internet www.oilsearch.com.au; sales US $90.8m., cap. and res US $308.8m. (1998); oil and gas exploration; Chair. BRIAN HORWOOD; Man. Dir PETER R. BOTTEN.

Ok Tedi Mining Ltd: POB 1, Tabubil, Western Province; tel. 5483311; fax 5489199; e-mail info@oktedi.com; internet www.oktedi.com; gold and copper mining; 52.6% owned by BHP (Australia), 17.4% owned by Inmet Mining Corpn (Canada), 30% govt-owned; Man. Dir KEITH FAULKNER; Exec. Man. PAUL JOHNSON.

Orogen Minerals Ltd: Musgrave St, POB 2151, Port Moresby; tel. 3217600; fax 3202209; e-mail sling@datec.com.pg; internet www.orogen.com.au; merged with Oil Search Ltd in March 2002; 18% govt-owned, 82% privately-owned; Chair. LINDSAY MACALISTER; Man. Dir FRANCIS KAUPA.

Parker Drilling Co (PNG): POB 478–7F ANG House, Hunter St, Port Moresby, NCD; tel. 3213322; fax 3213213; e-mail pom-ops@daltron.com.pg; internet www.parkerdrilling.com; owned by Parker Drilling Co (USA); drilling contractor; Operations Mans MARK ANDREWS, HERB VIGEANT.

PNG Oil Refinery Pty Ltd: POB 1071, Boroko, NCD; e-mail hunterd@syd.egis.com; Port Moresby; Chair. SAM PEPENA.

Pacrim Energy Ltd: c/o Sinton Spence Chartered Accountants, POB 6861,Boroko, NCD; tel. 3257611; fax 3259389; e-mail info@pacrimenergy.com.au; internet www. pacrimenergy.com.au; f. 1988; owned by Pacarc NL; petroleum exploration; Chair. Sir BARRY HOLLOWAY.

Pangpang Development Corpn Pty Ltd: f. 1992 by Lihir Island landowners to participate in business generated by the Lihir gold project.

Uni Group: Boroko, NCD; tel. 260433; civil engineering and building contractors.

United Pacific Drilling (PNG) Pty Ltd: Fikus St, POB 108, Madang; tel. 8522411; fax 8522830; e-mail wkh1@orica.com.au; internet www.upd.com.pg; mineral exploration; Gen. Man. WILLIAM HUGHES.

Timber and Palm Oil

ANG Timbers Pty Ltd: POB 1984, Boroko, NCD; tel. 253966; timber and timber products.

Hargy Oil Palms Pty Ltd: POB 21, Bialla, West New Britain Province; tel. 9831005; fax 9831191; e-mail harg@online.net.pg; production of palm oil; Gen. Man. DAVID MATHER.

Higaturu Oil Palms Pty Ltd: see Industrial and Trade Associations, above.

Jant Pty Ltd: POB 714, Madang; wood chips, milled timber.

Milne Bay Estates Pty Ltd: POB 36, Alotau; tel. 6411211; fax 3252959; e-mail oilpalm@mbe.com.pg; internet www.mbe.com.pg; palm oil production; Gen. Man. JOHN CHESTER.

New Britain Palm Oil Development Ltd: see Industrial and Trade Associations, above.

Open Bay Timber Pty Ltd: POB 1020, Rabaul, East New Britain Province; tel. 9821633; fax 9821220; e-mail obterry@global.net.pg; milled timber, mouldings, scantlings, log and sawn timber export, reafforestation; Gen. Man. TOSHIHARU SHINOHARA; Man. Dir TERRY SAKAKI.

PNG Forest Products Pty Ltd: POB 89, Lae, MP411; tel. 4745322; fax 4745365; e-mail pngfp@global.net.pg; plywood, mouldings, milled timber, furniture, kitset buildings; Gen. Man. TONY HONEY.

Pasis Manua Inland Timber Resources Pty Ltd: f. 1992 by landowners in Kandrian district.

Stettin Bay Lumber Co Pty Ltd: POB 162, Kimbe, West New Britain Province; tel. 9854066; fax 9854028; timber merchants; Man. Dir S. W. LONG.

Taway Timbers Pty Ltd: POB 515, Madang; tel. 8523517; fax 8523384; Gen. Man. EDDIE FITZGERALD.

Unevulg Development Pty Ltd (UDPL): West New Britain Province; logging; plans to diversify into coconut oil production.

Chemicals

BOC Gases Papua New Guinea: POB 93, Lae, Morobe Province; tel. 4722377; fax 4726177; industrial gases, oxygen, dissolved acetylene, nitrogen, argons, medical gases, pestigas, insectigas and deodour gas; cutting and welding equipment and consumables; Man. Dir BARRY BURKE.

Boral Energy (PNG) Pty Ltd: POB 1468, Boroko, NCD; tel. 3214248; fax 3211570; e-mail william-lamur@boral.com.pg; distribution of LPG; Man. Dir ROY VEERHUIS.

Colgate Palmolive (PNG) Ltd: POB 981, Lae, Morobe Province; tel. 4723166; fax 4726280; soap; Gen. Man. MICHAEL BUBB.

Shell Papua New Guinea Ltd: POB 169, Port Moresby; tel. 3228700; fax 3211840; aviation fuel, other fuels, lubricants and industrial chemicals; Gen. Man. DON MANOA.

TRADE UNIONS

The Industrial Organizations Ordinance requires all industrial organizations that consist of no fewer than 20 employees or four employers to register. In 1977 there were 56 registered industrial organizations, including a general employee group registered as a workers' association in each province and also unions covering a specific industry or profession.

Papua New Guinea Trade Unions Congress (PNGTUC): POB 254, Boroko, NCD; tel. 3212132; fax 3212498; e-mail tucl@daltron.com.pg; Pres. GASPER LAPAN; Gen. Sec. JOHN PASKA; 52 affiliates, 76,000 mems.

The following are among the major trade unions:

Bougainville Mining Workers' Union: POB 777, Panguna, North Solomons Province; tel. 9958272; Pres. MATHEW TUKAN; Gen. Sec. ALFRED ELISHA TAGORNOM.

Central Province Building and Construction Industry Workers' Union: POB 265, Port Moresby.

Central Province Transport Drivers' and Workers' Union: POB 265, Port Moresby.

Employers' Federation of Papua New Guinea: POB 490, Port Moresby; tel. 3214772; fax 3214070; f. 1963; Pres. G. J. DUNLOP; Exec. Dir TAU NANA; 170 mems.

National Federation of Timber Workers: Madang; f. 1993; Gen. Sec. MATHIAS KENUANGI (acting).

Papua New Guinea Communication Workers' Union: Pres. GASPER LAPAN; Gen. Sec. EMMANUEL KAIRU.

Papua New Guinea National Doctors' Association: Pres. Dr BOB DANAYA; 225 mems.

Papua New Guinea Teachers' Association: POB 1027, Waigani, NCD; tel. 3262588; f. 1971; Pres. TAINA DAI; Gen. Sec. LEONARD JONLI; 13,345 mems.

Papua New Guinea Waterside Workers' and Seamen's Union: Port Moresby; f. 1979; an amalgamation of four unions; Sec. DOUGLAS GADEBO.

Police Association of Papua New Guinea: POB 903, Port Moresby; tel. 3214172; f. 1964; Pres. A. AVIAISA; Gen. Sec. (vacant); 4,596 mems.

Port Moresby Council of Trade Unions: POB 265, Boroko, NCD; Gen. Sec. JOHN KOSI.

Port Moresby Miscellaneous Workers' Union: POB 265, Boroko, NCD.

Printing and Kindred Industries Union: Port Moresby.

Public Employees' Association: POB 965, Boroko, NCD; tel. 3252955; fax 3252186; f. 1974; Pres. NAPOLEON LIOSI; Gen. Sec. JACK N. KUTAL; 28,000 mems.

Transport

There are no railways in Papua New Guinea. The capital city, Port Moresby, is not connected by road to other major population centres. Therefore, air and sea travel are of particular importance.

ROADS

In 1999 there were an estimated 19,600 km of roads in Papua New Guinea, of which 3.5% were paved. In 1995 the Australian Government announced that it was to donate $A155m. over five years for major road-building projects in Papua New Guinea. In 1998 the European Union (EU) committed €24m. over two years to upgrade the Lae–Madang Highway. Japan offered a grant-in-aid in 1998 of 940m. yen for the reconstruction of a bridge on the Highlands Highway. In October 2000 it was announced that this highway was also to be upgraded over six years at a cost of 26.3m. kina financed through a loan negotiated with the Asian Development Bank (ADB).

Papua New Guinea Roads Authority: Lae, Morobe Province; f. 2004; statutory authority established to maintain the road network, particularly the Highlands Highway; Chair. ALLAN MCLAY; Deputy Chair. ALPHONSE NIGGINS.

SHIPPING

Papua New Guinea has 16 major ports and a coastal fleet of about 300 vessels. In early 1999 a feasibility study was commissioned to investigate the possible relocation of port facilities in Port Moresby.

Papua New Guinea Harbours Board: POB 671, Port Moresby; tel. 3211400; fax 3211546; Chair. TIMOTHY BONGA.

Port Authority of Kieta: POB 149, Kieta, North Solomons Province; tel. 9956066; fax 9956255; Port Man. SAKEUS GEM.

Port Authority of Lae: POB 563, Lae, Morobe Province; tel. 4422477; fax 4422543; Port Man. JOSHUA TARUNA.

Port Authority of Madang: POB 273, Madang; tel. 8523381; fax 8523097; Port Man. WILLIE WANANGA.

Port Authority of Port Moresby: POB 671, Port Moresby; tel. 211400; fax 3211546; Gen. Man. T. AMAO.

Port Authority of Rabaul: POB 592, Rabaul, East New Britain Province; tel. 9821533; fax 9821535.

Shipping Companies

Coastal Shipping Pty Co Ltd: Sulphur Creek Rd, POB 423, Rabaul, East New Britain Province; tel. 9828518; fax 9828519.

Lutheran Shipping: POB 1459, Lae, Morobe Province; tel. 4722066; fax 4725806.

Morehead Shipping Pty Ltd: POB 1908, Lae, Morobe Province; tel. 4423602.

New Guinea Australia Line Pty Ltd: POB 145, Port Moresby; tel. 3212377; fax 3214879; e-mail ngal@daltron.com.pg; f. 1970; operates regular container services between Australia, Papua New Guinea, Singapore, Indonesia, Vanuatu, Tuvalu and Solomon Islands; Chair. (vacant); Gen. Man. GEOFFREY CUNDLE.

P & O PNG Ltd (trading as Century Shipping Agencies): MMI House, 3rd Floor, Champion Parade, POB 1403, Port Moresby; tel. 3229200; fax 3229251; e-mail cgcpom@popng.com.pg; owned by P & O (Australia); Gen. Man. ANDREW CRIDLAND.

Papua New Guinea Shipping Corporation Pty Ltd: POB 634, Port Moresby; tel. 3220290; fax 3212815; e-mail shipping@steamships.com.pg; f. 1977; owned by Steamships Trading Co Ltd; provides a container/break-bulk service to Australia and the Pacific islands; Chair. CHRISTOPHER PRATT; Man. Dir JOHN DUNLOP.

South Sea Lines Proprietary Ltd: POB 5, Lae, Morobe Province; tel. 4423455; fax 4424884; Man. Dir R. CUNNINGHAM.

Western Tug & Barge Co P/L: POB 175, Port Moresby; tel. 3212099; fax 3217950; shipowning arm of P & O PNG; operates 24 vessels.

CIVIL AVIATION

There is an international airport at Port Moresby, Jackson's Airport, and there are more than 400 other airports and airstrips throughout the country. International services from Lae and Mount Hagen airports began in March 1999. A programme to upgrade eight regional airports over three years, with finance of $A30m. from the Australian Government, was initiated in mid-1997. New domestic and international terminal buildings were opened at Jackson's Airport in 1998, following a 13-year project financed with K120m. from the Japanese Government. A project to redevelop Tari airport was announced in late 1999. It was expected that, following redevelopment, the airport would receive international flights.

Air Niugini: POB 7186, Boroko, NCD; tel. 3259000; fax 3273482; e-mail airniugini@airniugini.com.pg; internet www.airniugini.com.pg; f. 1973; govt-owned national airline (plans to privatize the airline were announced in 1999); operates scheduled domestic cargo and passenger services within Papua New Guinea and international services to Australia, Solomon Islands, Philippines, Singapore and Japan; Chair. JOSEPH PHILLIP KAPAL; CEO ROD NELSON; Man. Dir MICHAEL BULEAU (acting).

MBA Pty Ltd: POB 170, Boroko, NCD; tel. 3252011; fax 3252219; e-mail mba@mbapng.com; internet www.mbapng.com; f. 1984; operates domestic scheduled and charter services; Chair. and CEO JOHN R. WILD; Gen. Man. SIMON D. WILD.

Tourism

Despite Papua New Guinea's spectacular scenery and abundant wildlife, tourism makes only a small contribution to the economy. In 1998 there were 4,280 hotel beds. In 2003 visitor arrivals totalled some 52,623. The industry earned an estimated US $101m. in 2001.

PNG Tourism Promotion Authority: POB 1291, Port Moresby; tel. 3200211; fax 3200223; e-mail tourismpng@dg.com.pg; internet www.paradiselive.org.pg; CEO JOHN KAMBOWA.

Defence

In March 1975 the Papua New Guinea Government assumed responsibility for defence from the Australian Government. In August 2003 the fully integrated Papua New Guinea defence force had a total strength of some 3,100 (army 2,500, navy 400 and air force 200). Military service is voluntary. Australian training forces stationed in the country totalled 38. In addition, the 55-strong Bougainville Peace Monitoring Group included 35 Australian troops, the others being from New Zealand and Fiji. Government expenditure on defence in 2003 was estimated at K68m.

Commander-in-Chief of Papua New Guinea Defence Force: Col CARL MALPO.

Education

Education from pre-school to tertiary level is available in Papua New Guinea, although facilities remain inadequate and unevenly distributed. In 1995 there were 2,790 primary schools, with 594,444 pupils in 1999. In 1986 there were 116 secondary schools and 103 technical and vocational schools. Secondary-school pupils totalled 74,042 in 1999. There are two universities.

Children attend school from seven years of age. At the age of 13 they move from community schools to provincial high schools for a further three years and are then eligible to spend another two years at the national high schools, where they are prepared for entrance to tertiary education. Originally schooling was free, but in recent years fees and charges for equipment have been introduced.

In 1995 the total enrolment at public primary schools was equivalent to 82% of that school-age population (males 88%; females 75%). In the same year secondary enrolment was equivalent to only 14% of children in the relevant age-group (boys 17%; girls 11%). In some areas, such as East New Britain and Port Moresby, almost all eligible children attend primary schools, whereas in others, such as the Highlands provinces, attendance is as low as 34%. Access to secondary education ranges from 7% in the Eastern Highlands to almost 50% in East New Britain. In 2002, according to UNESCO estimates, adult illiteracy averaged 64.6% (males 71.1%; females 57.7%).

Budgetary expenditure on education by the central Government in 2002 totalled an estimated K150.0m.

SAMOA

Physical and Social Geography

The Independent State of Samoa (formerly Western Samoa) comprises the two large islands of Savai'i and Upolu and seven small islands, of which five are uninhabited. Their total area is 2,831 sq km (1,093 sq miles). These high volcanic islands, with rugged interiors and little flat land except along the coasts, lie in the South Pacific, about 2,400 km north of New Zealand. The country's nearest neighbour is American Samoa, to the east. The climate is tropical, with temperatures generally between 23°C (73°F) and 30°C (86°F). The rainy season is from November to April.

At the census of November 2001 the population of Samoa totalled 176,710. About 75% of the population resided on the island of Upolu. The population of Apia, the capital, totalled an estimated 39,666 in mid-2003.

History

The islands are populated by Polynesians and are thought to have been the origin of many of the people who now occupy islands further east. The Samoan language is believed to be the oldest extant form of Polynesian speech. Samoan society developed an intricate hierarchy of graded titles comprising titular chiefs and orator chiefs. One of the striking features of modern Samoa is the manner in which these titles and the culture prior to European contact remain a dominant influence. Most of the population have become Christians.

The Samoan islands were first visited by Europeans in the 1700s, but it was not until 1830 that missionaries from the London Missionary Society settled there. The eastern islands (now American Samoa) were ceded to the USA in 1904 but Western Samoa (as it was known until July 1997), a former German colony, was occupied by New Zealand in 1914 and the League of Nations granted a mandate over the territory to New Zealand in 1920. In 1946 the UN assumed responsibility for the Territory of Western Samoa through its Trusteeship Council, with New Zealand as the administering power. From 1954 measures of internal self-government were gradually introduced, culminating in the adoption of an independence Constitution in October 1960. This was approved by a UN-supervised plebiscite in May 1961 and the islands became independent on 1 January 1962. The office of Head of State was to be held jointly by two of the paramount chiefs but, upon the death of his colleague in April 1963, Malietoa Tanumafili II became sole Head of State for life.

Samoa has had a Legislative Assembly (Fono) since 1947. Since independence the islands have been governed under a parliamentary system, with a Prime Minister and Cabinet. Until 1991, only two of the 47 seats in the Fono were decided by universal suffrage, the rest being decided by the Matai (elected clan chiefs). The 1973 elections were won by Fiame Mata'afa Fuamui Mulinuu, who was first elected Prime Minister in 1959 but who lost the position in 1970 to Tupua Tamasese Lealofi. Fiame Mata'afa Mulinuu became Prime Minister again in 1973, remaining in office until his death in 1975, when Tupua Tamasese Lealofi was recalled to complete the term of office. Tupuola Taisi Efi was elected Prime Minister in March 1976 and re-elected in February 1979. Elections in February 1982 resulted in a victory for the opposition Human Rights Protection Party (HRPP), which won 24 of the 47 seats in the Fono. Va'ai Kolone was elected Prime Minister but was dismissed in September when a court ruled that he had won his seat improperly. Efi was restored as Prime Minister, but resigned in December 1982 after the Fono had rejected his budget. He was replaced by the new HRPP leader, Tofilau Eti Alesana. At elections in February 1985, the HRPP won 31 of the 47 seats, increasing its majority in the Fono from one to 15 seats. In December Tofilau Eti resigned, following the rejection of the proposed budget by the Fono, and Va'ai Kolone was again appointed Prime Minister. Tupua Tamasese Efi replaced Va'ai Kolone as leader of the ruling coalition in February 1988, immediately prior to a general election, at which both the HRPP and the Samoa National Development Party (SNDP) coalition initially secured 23 seats, with one constituency being tied. When two recounts proved inconclusive, a third was presided over by a New Zealand judge who declared the Christian Democratic Party (CDP) candidate the winner. Before a new government could be formed, however, a member of the SNDP, newly elected to the Fono, transferred allegiance to the HRPP. In April Tofilau Eti was re-elected Prime Minister and a new Government, comprising HRPP members, was formed.

Legislation proposed in early 1990 that would permit local village councils to fine or impose forced labour or exile on individuals accused of offending communal rules was widely perceived as a government attempt to gain the support of the Matai (elected clan chiefs, see above) for the next general election. In the late 1980s and early 1990s the number of Matai titles increased significantly, and this was perceived to have greatly undermined the system of chiefly leadership. A referendum was conducted in October 1990, at which voters narrowly accepted government proposals for the introduction of universal suffrage. A second proposal, to create an upper legislative chamber composed of the Matai, was rejected. A bill to implement universal adult suffrage was approved by the Fono in December 1990, despite strong opposition from the SNDP.

A general election was held in April 1991 (postponed from February, owing to the need to register an estimated 80,000 newly enfranchised voters). In the following weeks election petitions were filed with the Supreme Court against 11 newly elected members of the Fono who were accused of corrupt or illegal electoral practices. Moreover, subsequent political manoeuvring resulted in the HRPP increasing its parliamentary representation from an initial 26 to 30 seats, while the SNDP ultimately secured only 16 seats in the Fono, and the remaining seat was retained by an independent. At the first meeting of the new Fono, convened in early May, Tofilau Eti was re-elected for what, he later announced, would be his final term of office as Prime Minister.

In November 1991 the Fono approved legislation to increase the parliamentary term from three to five years and to create an additional two seats in the Fono. These seats were contested in early 1992 and won by the HRPP.

In December 1991 the islands were struck by a devastating cyclone (Cyclone Val), which caused 13 deaths and damage estimated at 662m. tala. This was the second major cyclone to hit Western Samoa in two years, the first being Cyclone Ofa in February 1990, which left an estimated 10,000 islanders homeless. The dramatic increase in the incidence of cyclones occurring in the region was widely attributed to climatic change caused by the 'greenhouse effect' (the heating of the earth's atmosphere as a consequence of pollution).

The introduction of a value-added tax on goods and services in January 1994 (which greatly increased the price of food and fuel in the country) provoked a series of demonstrations and protest rallies, as well as demands for the resignation of the Prime Minister. As a result of overwhelming opposition to the new regulations, the Government agreed, in March, to amend the most controversial aspects of the tax. Meanwhile, four members of the Fono (including three recently expelled HRPP members), who had opposed the financial reforms, established a new political organization, the Samoa Liberal Party, under the leadership of the former Speaker, Nonumalo Leulumoega Sofara.

In May 1994 treasury officials warned the Government that the financial crisis at the national airline, Polynesian Airlines, was threatening the country's economic stability. It was estimated that the company's debts totalled more than 45m. tala. A report by the Chief Auditor, Tom Overhoff, accused the Government of serious financial mismanagement relating to a series of decisions to commit public funds to the airline, and charged seven cabinet ministers with fraud and negligence in their handling of government resources. An inquiry into the allegations conducted in late 1994 cleared the ministers in question of all the charges, although its findings were harshly criticized by Overhoff, who claimed that the inquiry had been neither independent nor impartial.

Protests against the value-added tax on goods and services continued in early 1995, following the Government's decision to charge two prominent members of the Tumua ma Pule group of traditional leaders and former members of the Fono with sedition, for organizing demonstrations against the tax during 1994. In March 1995 3,000 people delivered a petition to the Prime Minister, bearing the signatures of 120,000 people (some 75% of the population), that demanded that the tax be revoked. The Prime Minister questioned the authenticity of the signatures and appointed a 14-member committee to investigate the matter. In late June the case against the two members of Tumua ma Pule, which had attracted attention from several international organizations (including the World Council of Churches and Amnesty International) was dismissed on the grounds of insufficient evidence.

In December 1995 the HRPP unanimously re-elected Tofilau Eti as the leader of the party, despite concern over the Prime Minister's deteriorating health, as well as a previous declaration that he would retire from politics upon completion of his current term in office.

In March 1996 one of the two female members of the Fono, Matatumua Naimoaga, left the HRPP in order to form the Samoa All-People's Party. The formation of the new party, in preparation for the forthcoming general election, was reportedly a result of dissatisfaction with the Government's alleged mismanagement of public assets together with concern over corruption. Legislation was introduced in April, which attempted to distinguish between the traditional Samoan practice of exchanging gifts and acts of bribery, amid numerous reports that voters were demanding gifts and favours from electoral candidates in return for their support.

A general election took place on 26 April 1996. The opposition was highly critical of the delay in the counting of votes (which took some three weeks in total), claiming that the length of time involved allowed the HRPP to recruit successful independent candidates in an attempt to secure a majority of seats in the Fono. It was eventually announced in mid-May that the HRPP had secured a total of 28 seats (with the recruitment of several independent members to their ranks), the SNDP had won 14 seats and independent candidates had secured seven. Tofilau Eti was subsequently re-elected as Prime Minister, defeating the Leader of the Opposition, Tuiatua Tupua Tamasese, with 34 votes to 14.

The issue of government involvement in the media became prominent in early 1997, following Tuiatua Tupua Tamasese's decision to be interviewed on Australian radio. In the interview he urged overseas donors to withhold aid from Western Samoa, alleging widespread mismanagement and corrupt practice within the Government. The opposition leader responded to criticism of his action by denouncing the government policy that denies opposition politicians access to the state-controlled media, and by urging New Zealand to persuade the Western Samoan Government to reform the system.

The strength of traditional religious beliefs among Samoans and resistance to foreign influences on their culture was illustrated in early 1997. In March a member of the Church of Jesus Christ of Latter-day Saints (Mormon), the fastest growing denomination in the country, was tied to a stake surrounded by kindling when he refused to obey a banishment order for criticizing the village council's decision to prevent the construction of a Mormon church in the village. The intended victim was rescued by police-officers and church leaders, who negotiated his release from the villagers. A request by the Samoa Council of Churches to impose a ban on the establishment of new churches in the country was refused by the Government as incompatible with the principle of freedom of religion, as guaranteed in the Constitution. The organization was also highly critical of a US religious television channel due to begin operating in the islands in April, claiming that the evangelical style of religion depicted would undermine traditional Samoan Christianity. In mid-2003 US officials arrived in Samoa to investigate the suspicious burning of a Mormon church.

In May 1997 the Prime Minister proposed a constitutional amendment in the Fono to change the country's name to Samoa. (The country has been known simply as Samoa at the UN since it was admitted to the organization in 1976.) On 3 July the Fono voted by 41 votes to one to approve the change, which came into effect on the next day when the legislation was signed by the Head of State. The neighbouring US territory of American Samoa, however, expressed dissatisfaction with the change (which was believed to undermine the Samoan identity of its islands and inhabitants), and in September introduced legislation to prohibit the recognition of the new name within the territory. In March 1998 the House of Representatives in American Samoa voted against legislation that proposed not to recognize Samoan passports (thereby preventing Samoans from travelling to the territory), but decided to continue to refer to the country as Western Samoa and to its inhabitants as Western Samoans. Nevertheless, in January 2000 Samoa and American Samoa signed a memorandum of understanding, increasing co-operation in areas including health, trade and education.

A series of reports in *The Samoa Observer* in mid-1997 alleged that a serious financial scandal involving the disappearance of some 500 blank passports, and their subsequent sale to Hong Kong Chinese for up to US $26,000 each, had occurred. The Government refused to comment on the newspaper's allegations, stating only that several senior immigration officials had been suspended pending the outcome of an investigation into the affair. Moreover, the Government subsequently brought charges of defamatory libel against the editor, Savea Sano Malifa, for publishing a letter criticizing the Prime Minister (who was reported to have told the Fono of his intention to change legislation governing business licences, such that publications could have their licences withdrawn for publishing dissenting material). The regional organization, the Pacific Islands News Association, also condemned the Prime Minister's comments as an attack on freedom of information and expression. The continued existence of the newspaper was placed in jeopardy when Savea Sano Malifa was found guilty of defaming the Prime Minister in two libel cases in July and September 1998, and was ordered to pay a total of some $17,000 in costs. The newspaper had alleged that public funds had been used to construct a hotel owned by the Prime Minister and had criticized the allocation of $0.25m. in the 1998 budget for Tofilau Eti's legal costs. The Government's increasingly autocratic style, its apparent intolerance of dissent and the perceived lack of accountability of its members, coupled with its poor economic record, resulted in frequent expressions of popular discontent during 1997. These culminated in a series of protest marches in late 1997 and early 1998, organized by the Tumua ma Pule group of chiefs and attended by several thousand people, which aimed to increase pressure on the Prime Minister to resign.

In November 1998 Tofilau Eti Alesana resigned as Prime Minister, owing to ill health. He was replaced by the Deputy Prime Minister, Tuila'epa Sailele Malielegaoi, and, at the same time, the Cabinet was reshuffled. Tofilau Eti Alesana died in March 1999.

In July 1999 the Minister of Public Works, Luagalau Levaula Kamu, was shot dead while attending an event commemorating the 20th anniversary of the foundation of the ruling HRPP. Speculation followed that the killer's intended target had been the Prime Minister, but this was denied both by Tuila'epa Sailele and by the New Zealand police officers sent to the island to assist in the investigation. A man identified as Eletise Leafa Vitale (son of the Minister of Women's Affairs, Leafa Vitale) was arrested and charged with the murder. Eletise Leafa Vitale was convicted of the murder and sentenced to death (subsequently commuted to life imprisonment). Leafa Vitale was also subsequently charged with the murder of Kamu, together with the former Minister of Telecommunications, Toi Akuso, who faced additional charges of incitement to murder Kamu and the Prime Minister, Tuila'epa Sailele. Both of the men were found guilty in April 2000 and were also sentenced to death (which was similarly commuted to life imprisonment; no death sentence had been carried out since Samoa's independence). It later emerged that Kamu had been killed in an attempt to prevent him from uncovering incidences of corruption and bribery in which the two ministers had become involved.

Meanwhile, in November 1999 the ruling HRPP increased its number of seats in the Fono to 34 (out of a possible 49) following the defection of an independent candidate to the HRPP. By-elections for the two imprisoned former ministers' seats were held in June 2000, and HRPP candidates were successful in both constituencies.

In August 2000 a supreme court ruling ordered the Government to allow opposition politicians access to the state-controlled media. For several years the opposition had been denied free access to the media (see above).

At a general election on 2 March 2001 the HRPP won 22 seats, the SNDP secured 13 seats and independent candidates won 14 seats. On 16 March Tuila'epa Sailele won 28 votes in the Fono, after securing the support of six independents, to be re-elected Prime Minister. However, the opposition mounted a number of legal challenges to his election. In August eight elected members of Parliament, including the Deputy Prime Minister, the Minister of Health and the Minister of Internal Affairs, Women's Affairs and Broadcasting, faced charges of electoral malpractice in the Supreme Court. None of the Cabinet Ministers was found guilty, and by-elections for the vacant parliamentary seats were held in October and November. The HRPP won all four contested seats.

An Electoral Commission, established shortly after the March elections, published its recommendations in October 2001, urging the replacement of the two Individual Voters Roll seats with two Urban Seats and that government employees who wished to stand for Parliament should first be obliged to resign from their offices.

In mid-2003 members of the medical profession expressed alarm at the increasing numbers of qualified medical staff choosing to emigrate or to leave the profession altogether. It was estimated that one-third of Samoa's nurses had left the profession between 2002 and 2003, while half of all doctors' positions remained vacant. Low pay was blamed for the situation: doctors' salaries were reported to be five times higher in American Samoa and three times higher in New Zealand. Nurses in Samoa, who had not received a pay increase since 1999, complained of similar conditions.

In January 2004 the Fono passed legislation, introduced by the Prime Minister, that abolished the death penalty in Samoa. Capital punishment had not been carried out in the islands for 52 years, and the motion was widely supported by government and opposition members alike.

Samoa was struck by Cyclone Heta in early January 2004. The storm killed one person and caused considerable damage to housing, roads and power supplies on the islands.

Samoa maintains strong links with New Zealand, where many Samoans now live and where many others receive their secondary and tertiary education. An appeal against attempts to curb the high level of migration to New Zealand led the Privy Council in London to rule in July 1982 that all Western Samoans born between 1924 and 1949, and their male children, were entitled to New Zealand citizenship. The ruling was estimated to affect about 100,000 Western Samoans. However, in August 1982 the New Zealand and Western Samoan Governments agreed to annul the ruling, declaring in its stead that illegal immigrants already in New Zealand would be allowed to apply for citizenship, and that a quota of 1,100 migrants per year would be accepted into New Zealand. In June 2002 New Zealand formally apologized for the mistakes it had committed while administering Samoa during 1914–62. These injustices included New Zealand's poor handling of the 1918 influenza pandemic (which had killed 22% of Samoa's population within a fortnight, the virus having been brought in on a ship from New Zealand); the murder of a Samoan paramount chief and independence leader, Tupua Tamasese Lealofi III, and the killing of nine other supporters of the pacifist Mau movement during a non-violent protest in 1929; and the banishment of native leaders, who were also stripped of their chiefly titles. The apology, while accepted, drew mixed reactions from Samoans, many of whom were more concerned with the issue of the restoration of their rights to New Zealand citizenship, as upheld by the Privy Council in 1982. Moreover, in March 2003 large protest marches took place in Samoa and New Zealand demanding the repeal of the 1982 law, which ended Samoans' automatic right to New Zealand citizenship. In May 2004 a parliamentary select committee rejected a 100,000-signature petition seeking a repeal of the law and upheld the principles of the 1982 ruling.

Also in March 2003, in common with many other island nations in the region, large demonstrations were organized in Samoa to protest against the US-led military assault against Iraq. The Samoan Government expressed support for the demonstrations, which were attended by some 2,000 people.

The announcement in mid-1995 by the French Government that it was to resume nuclear weapons testing in the South Pacific provoked large-scale demonstrations in Apia and statements of condemnation by the Government, which introduced an indefinite ban on visits to the islands by French warships and aircraft. The tests were concluded in January 1996.

The first visit by a member of the Japanese royal family in Samoa's history took place in October 2003. During their stay, Prince and Princess Akishino launched a Japanese-funded port expansion project in Apia.

In September 2004 the Samoan Government announced that it was seeking to alter its maritime boundary with American Samoa, owing to a number of recent, unspecified incidents. Discussions were scheduled to take place in late November.

Economy

In 2002, according to estimates by the World Bank, Samoa's gross national income (GNI), measured at average 2000–02 prices, stood at US $250m., equivalent to $1,420 per head (or $5,350 per head on an international purchasing-power parity basis). During 1990–2002, it was estimated, the population increased at an annual average rate of 0.8%. During 1994–2002 gross domestic product (GDP) per head increased, in real terms, by an annual average of 3.3%. According to the Asian Development Bank (ADB), it was estimated that GDP had increased by 1.5% in 2002 and by 3.5% in 2003. A growth rate of 4.0% was projected for 2004.

Agriculture, forestry and fishing engaged around 33% of the labour force in 2002, but provided only 13.5% of GDP in 2003. Between 1995 and 2002, according to figures from the ADB, agricultural GDP decreased, in real terms, at an average annual rate of 2.2%, but decreased by 4.0% in 2003. The principal cash crop is coconut. In total, coconut oil, cream and copra accounted for 15.0% of exports in 2000. In 2003 coconut cream accounted for 6.8% of export revenue. Meanwhile, the country's coconut oil mill had remained closed for much of 2000. Another important cash crop has been taro, which is also the country's primary staple food. Sales of taro provided 58% of all domestic export earnings in 1993, but an outbreak of taro leaf blight devastated the crop in 1994 and reduced exports to almost nil in that year and subsequently. A campaign to revive the taro industry was launched in mid-2000; exports of taro accounted for 3.0% of total exports in 2003. Small quantities of timber are also exported. Breadfruit, yams, maize, passion fruit and mangoes are cultivated as food crops. Exports of breadfruit and papaya were expected to increase significantly following the installation in early 2003 of a treatment facility to eradicate fruit fly from the produce. Pigs, cattle, poultry and goats are raised, mainly for local consumption. Some 1,800 head of cattle were brought to Samoa from Australia in early 2003, in an attempt to boost stocks and increase beef production. The country's commercial fishing industry expanded considerably in the late 1990s, with export revenues rising from US $4.8m. (33.0% of domestic export earnings) in 1997 to $29.0m. (62.7%) in 2002. Exports of fresh fish declined in the following year to contribute only 35.6% of total export earnings. The fishing industry contributed 7.6% of GDP in 2000. Significant reductions in fishing taxes announced by the Government in March 2003 were expected to stimulate the country's fishing industry.

Industry (comprising mining, manufacturing, construction and utilities) provided 27.9% of GDP in 2003 and employed 5.5% of the labour force in 1986. Between 1995 and 2002 industrial GDP increased by an annual average of 1.3%; sectoral GDP grew by 2.8% in 2003. During the same period manufacturing GDP remained virtually constant, in real terms, increasing at an average annual rate of only 0.01%. Manufacturing provided 17.2% of GDP in 2003 and (with mining) engaged 3.5% of the labour force in 1986. Traditionally, the principal manufactures have been beverages (beer—which accounted for 8.6% of exports in 2003—and soft drinks), coconut-based products and cigarettes. The manufacturing sector expanded considerably in the early 1990s with the establishment of a Japanese-owned factory, producing electrical components for road motor vehicles. The Yazaki Samoa factory engaged about 2,500 workers in 1996, making it the largest private-sector employer in the country. The factory assembles wire harnessing systems that are exported to car-manufacturing plants in Australia. Shipments from the Yazaki factory are generally excluded from official statistics for visible (merchandise) exports. Instead, the value added by the plant is recorded as part of 'invisible' trade, in the services account of the balance of payments. Net receipts from the harness assembly trade increased from US $5.7m. in 1995 to $8.2m. in 1996 and to $9.7m. in 1997; however, in 1997 the factory's operations were restructured, with output subsequently reduced. Net receipts in the first nine months of 1998 were about $3.6m., only one-half of the total in the corresponding period of the previous year. Nevertheless, the initial success of the venture, and the availability of government incentives for export processing activities, encouraged further investment in the manufacturing sector. A chocolate factory, processing locally grown cocoa, has been built with assistance from foreign aid funding. The clothing industry has also expanded, and in 2003 garments accounted for 30.1% of total export earnings.

With the opening of a 3,500-kW hydroelectric power station at Sauniatu in 1985, the country's dependence on imported diesel fuel for electric power generation was reduced from 80% of the total to about 40%. A grant of US $0.3m. was received from the ADB in early 2003 to establish a hydroelectric project on Savai'i island. However, the Government was forced to seek an alternative location for the power plant following the refusal of local chiefs to allow its construction in their village. Chiefs representing the people of Sili village rejected the scheme, which they claimed would pollute local water supplies and harm the environment. Imports of petroleum accounted for 12.4% of the value of total imports in 2001.

Services engaged 29.5% of the labour force in 1986 and accounted for 58.6% of GDP in 2003. Between 1995 and 2002, according to ADB figures, the sector's GDP increased at an average annual rate of 7.6%. Compared with the previous year, the GDP of the services sector expanded by 7.4% in 2002 and by 8.2% in 2003. Measures to encourage the development of tourism were undertaken during the 1980s, including the expansion of hotel facilities and improvements to the road network and airport. However, financial difficulties within the national airline, Polynesian Airlines, led to the disruption of services in 1994. The tourism sector performed well in 2000, as a result of the completion of new tourist facilities. The sector also benefited from political instability in Fiji and Solomon Islands. Tourist revenues, including the proportion of international travel credited to carriers based in Samoa, totalled an estimated 133.1m. tala in 2000, compared with 125.8m. tala in 1999. The number of tourist arrivals rose from 88,960 in 2002 to 92,313 in 2003.

In 2003 the country recorded a visible trade deficit of US $120.8m., and there was a surplus of $19.1m. on the current account of the balance of payments. In that year Australia was Samoa's principal trading partner, providing 17.7% of imports and purchasing 75.6% of exports, while the USA provided 5.5% of imports and purchased 5.9% of exports. Other important trading partners were New Zealand (21.6% of imports), Fiji (19.2% of imports) and Japan (11.5% of imports). The principal exports are fish, garments, coconut products and beer, and the main imports are food and beverages, industrial supplies and fuels.

In the year ending 30 June 2003 there was an overall budget deficit of 5.5m. tala (equivalent to 0.6% of GDP), reflecting an increase in development spending financed by external borrowing and a decrease in lending to the domestic banking system. At the end of 2002 the country's total external debt stood at US $234.4m., of which $156.8m. was long-term public debt. In 2000 the cost of debt-servicing was equivalent to 6.9% of total revenue from exports of goods and services. Remittances from Samoans overseas are important to the economy and totalled 150.7m. tala in 2000, equivalent to some 19.5% of GDP in that year (more than three times the value of merchandise exports and, for the first time, exceeding tourist remittances). Official development assistance from Australia totalled $A16.3m. in 2003/04. Aid from New Zealand amounted to $NZ8.3m in that year. In addition Samoa received US $4m. in aid from Japan in 2002, as well as US $6.5m. for a project to expand the port facilities in Apia. The annual rate of inflation averaged 3.3% in 1990–2000; consumer prices increased by 8.1% in 2002 and by 0.1% in 2003. According to the census conducted in late 2001, a total of 2,618 people were unemployed, compared with 1,175 in late 1991.

A programme of economic reforms, initiated by the Government in the mid-1990s, won the approval of the ADB, the World Bank and other international financial organizations. Strong growth characterized the economy during the mid-1990s, but this slowed considerably in 1997 and 1998, owing to restructuring by the Japanese-owned manufacturing plant, the completion of cyclone reconstruction projects and drought. In May 1999 it was announced that Samoa's Post Office and Telecommunications Department was to be privatized as part of a public-sector reform programme. In 2000–01 high rates of GDP growth were maintained (with the construction sector recording particularly strong expansion) and various items of legislation were approved, aiming to raise standards of fiscal and corporate governance and to improve regulation and supervision in the financial sector. The tourism sector, meanwhile, continued to perform well, following the completion of new facilities, and was also well placed to benefit from the political instability in neighbouring Fiji and Solomon Islands. The Government reaffirmed its policies in the Strategy for the Development of Samoa 2002–2004, and in April 2003 the number of government departments was reduced from 27 to 14 (largely through mergers) as part of the reforms. The document stressed the importance of opportunities for all through sustained economic growth and improvements in health and education. The islands were adversely affected by Cyclone Heta in January 2004, which caused damage to buildings, infrastructure and agriculture. Economic growth of around 4% was predicted for 2004 and was expected to result largely from increased activity in the construction industry, including rehabilitation projects in the aftermath of Cyclone Heta, continued private-sector hotel development and construction of facilities for the South Pacific Games, which Samoa was expected to host in 2007.

Samoa is a member of the Pacific Islands Forum (formerly the South Pacific Forum), the Pacific Community (formerly the South Pacific Commission), the UN's Economic and Social Commission for Asia and the Pacific (ESCAP), and the Asian Development Bank (ADB), and is a signatory of the Lomé Conventions and the successor Cotonou Agreement with the European Union. Samoa was expected to become a member of the World Trade Organization in 2005.

Statistical Survey

AREA AND POPULATION

Area: Savai'i and adjacent small islands 1,708 sq km, Upolu and adjacent small islands 1,123 sq km; Total 2,831 sq km (1,093 sq miles).

Population: 161,298 (males 84,601, females 76,697) (Savai'i 45,050, Upolu and adjacent small islands 116,248) at census of 3 November 1991; 176,710 (males 92,050, females 84,660) at census of 5 November 2001. *Mid-2003* (UN estimate): 178,000 (Source: UN, *World Population Prospects: The 2002 Revision*).

Density (mid-2003): 62.9 per sq km.

Principal Town: Apia (capital), population 34,126 at census of 3 November 1991. *Mid-2003* (UN estimate, incl. suburbs): Apia 39,666 (Source: UN, *World Urbanization Prospects: The 2003 Revision*).

Births and Deaths (2000): Birth rate 29.1 per 1,000; Death rate 5.5 per 1,000. Source: UN, *Statistical Yearbook for Asia and the Pacific*.

Expectation of Life (WHO estimates, years at birth): 68.2 (males 66.8; females 69.7) in 2002. Source: WHO, *World Health Report*.

Economically Active Population (census of 3 November 1986): Agriculture, hunting, forestry and fishing 29,023; Manufacturing and mining 1,587; Electricity, gas and water 855; Construction 62; Trade, restaurants and hotels 1,710; Transport, storage and communications 1,491; Financing, insurance, real estate and business services 842; Community, social and personal services 9,436; Activities not adequately defined 629; *Total labour force* 45,635 (males 37,054, females 8,581). *1991 Census* (persons aged 15 years and over, excluding armed forces): Total labour force 57,142 (males 38,839, females 18,303) (Source: ILO, *Yearbook of Labour Statistics*). *Mid-2002* (estimates): Agriculture, etc. 21,000; Total labour force 63,000 (Source: FAO).

HEALTH AND WELFARE

Key Indicators

Total Fertility Rate (children per woman, 2002): 4.2.

Under-5 Mortality Rate (per 1,000 live births, 2002): 25.

Physicians (per 1,000 head, 1996): 0.34.

Health Expenditure (2001): US $ per head (PPP): 199.

Health Expenditure (2001): % of GDP: 5.8.

Health Expenditure (2001): public (% of total): 82.2.

Access to Water (% of persons, 2000): 99.

Access to Sanitation (% of persons, 2000): 99.

Human Development Index (2002): ranking 75.

Human Development Index (2002): value 0.769.

For sources and definitions, see explanatory note on p. vi.

AGRICULTURE, ETC.

Principal Crops (FAO estimates, '000 metric tons, 2002): Taro 17; Yams 3; Other roots and tubers 3; Coconuts 140; Copra 11 (2001); Bananas 22; Papayas 4; Pineapples 5; Mangoes 4; Avocados 1; Other fruits 8; Cocoa beans 0.5. Source: FAO.

Livestock (FAO estimates, '000 head, year ending September 2002): Pigs 201; Cattle 28; Horses 1.8; Chickens 450. Source: FAO.

Livestock Products (FAO estimates, '000 metric tons, 2002): Beef and veal 1; Pigmeat 4. Source: FAO.

Forestry (FAO estimates, '000 cubic metres, 2002): *Roundwood Removals* (excl. bark): Sawlogs and veneer logs 58; Other industrial roundwood 3; Fuel wood 70; Total 131. *Sawnwood Production* (incl. sleepers): 21. Source: FAO.

Fishing (metric tons, live weight, 2002): Albacore 4,360; Yellowfin tuna 388; Total catch (incl. others, all capture) 12,392. Source: FAO, *Yearbook of Fishery Statistics*.

INDUSTRY

Electric Energy (million kWh): 91 in 2000; 105 in 2001; 124 in 2002. Source: Asian Development Bank, *Key Indicators of Developing Asian and Pacific Countries*.

FINANCE

Currency and Exchange Rates: 100 sene (cents) = 1 tala (Samoan dollar). *Sterling, US Dollar and Euro Equivalents* (31 May 2004): £1 sterling = 5.2181 tala; US $1 = 2.8441 tala; €1 = 3.4829 tala; 100 tala = £19.16 = US $35.16 = €28.71. *Average Exchange Rate* (US $ per tala): 0.2816 in 2001; 0.3109 in 2002; 0.3600 in 2003.

Budget (provisional, million tala, year ending 30 June 1998): *Revenue:* Tax revenue 138.2 (Income tax 29.5, Excise tax 25.8, Taxes on international trade 47.1, Value-added gross receipts and services tax (VAGST) 32.2, Other taxes 3.6); Other revenue 33.2 (Fees, service charges, etc. 9.7, Departmental enterprises 19.3, Rents, royalties and international investments 4.2); Total 171.4, excl. external grants received (60.5). *Expenditure:* Current expenditure 142.1 (General administration 39.0, Law and order 11.3, Education 32.0, Health 24.8, Social security and pensions 6.8, Agriculture 9.0, Public works 18.6, Land survey 6.2, Other economic services 4.6, Interest on public debt 4.0, Other purposes (residual) –0.7, Sub-total 155.6, *Less* VAGST payable by government 13.5); Development expenditure 71.3; Total 213.4, excl. net lending (8.2). *2002/03* (provisional, million tala, year ending 30 June): *Revenue:* Tax revenue 206.5; Other revenue 22.8; Total 229.4, excl. external grants received (74.2). *Expenditure:* Current expenditure 202.2; Capital expenditure 86.9; Total 289.1 (Source: Asian Development Bank, *Key Indicators of Developing Asian and Pacific Countries*).

International Reserves (US $ million at 31 December 2003): IMF special drawing rights 3.57; Reserve position in IMF 1.03; Foreign exchange 79.31; Total 83.91. Source: IMF, *International Financial Statistics*.

Money Supply (million tala at 31 December 2003): Currency outside banks 35.73; Demand deposits at banks 82.48; Total money 118.21. Source: IMF, *International Financial Statistics*.

Cost of Living (Consumer Price Index, excluding rent; base: 1995 = 100): 121.0 in 2001; 130.8 in 2002; 131.0 in 2003. Source: IMF, *International Financial Statistics*.

Gross Domestic Product by Economic Activity (million tala at current prices, 2003): Agriculture and fishing 122.9; Manufacturing 156.2; Electricity and water 38.4; Construction 58.3; Trade 205.5; Transport and communications 115.4; Finance 82.1; Public administration 72.9; Other services 55.4; *Sub-total* 907.1; *Less* Imputed bank service charges 9.4; *Total* 897.7. Source: Asian Development Bank, *Key Indicators of Developing Asian and Pacific Countries*.

Balance of Payments (estimates, US $ million, 2003): Exports of goods f.o.b. 14.7; Imports of goods f.o.b. –135.6; *Trade balance* –120.8; Services and other income (net) 55.6; *Balance on goods, services and income* –65.2; Private remittances (net) 63.5; Official transfers (net) 20.9; *Current balance* 19.1; Capital transactions (incl. net errors and omissions) –10.7; *Overall balance* 8.4. Source: Asian Development Bank, *Key Indicators of Developing Asian and Pacific Countries*.

EXTERNAL TRADE

Principal Commodities ('000 tala, 2003): *Imports c.i.f.:* Total 406,970. *Exports f.o.b.:* Coconut cream 3,005; Fish 15,761; Beer 3,812; Taro 1,314; Garments 13,318; Total (incl. others) 44,271. Source: Asian Development Bank, *Key Indicators of Developing Asian and Pacific Countries*. Note: The trade data above exclude purchases and sales by the Yazaki car components factory.

Principal Trading Partners (US $ million, 2003): *Imports:* Australia 38.73; Fiji 42.02; Indonesia 6.71; Japan 25.09; Republic of Korea 6.85; New Zealand 47.19; Singapore 11.34; USA 12.10; Total (incl. others) 218.95. *Exports:* American Samoa 2.06; Australia 64.05; Germany 1.02; Japan 1.19; New Zealand 1.97; USA 5.00; Total (incl. others) 84.68. Source: Asian Development Bank, *Key Indicators of Developing Asian and Pacific Countries*.

TRANSPORT

Road Traffic (motor vehicles registered, 1998): Private cars 2,100; Pick-ups 2,080; Taxis 804; Trucks 261; Buses 159; Motorcycles 40; Tractors 6; Total (incl. others) 5,813.

International Shipping (freight traffic, '000 metric tons, 2001): Goods loaded 3; Goods unloaded 16 (Source: UN, *Monthly Bulletin of Statistics*). *Merchant Fleet* (total displacement, '000 grt at 31 December 2003): 10.2; vessels 9 (Source: Lloyd's Register-Fairplay, *World Fleet Statistics*).

Civil Aviation (traffic on scheduled services, 1999): Passengers carried 92,000; Passenger-kilometres 244 million; Total ton-kilometres 23 million. Source: UN, *Statistical Yearbook*.

TOURISM

Visitor Arrivals: 88,263 in 2001; 88,960 in 2002; 92,313 in 2003.

Visitor Arrivals by Country (2000): American Samoa 30,063; Australia 10,954; Fiji 2,032; Germany 1,784; New Zealand 22,818; United Kingdom 2,092; USA 9,032. Source: World Tourism Organization, *Yearbook of Tourism Statistics*.

Tourism Receipts (million tala*): 115.2 in 1998; 125.8 in 1999; 133.1 in 2000.

* Includes the proportion of international travel credited to Samoan carriers.

COMMUNICATIONS MEDIA

Telephones (2002): 11,800 main lines in use*.

Facsimile Machines (1999): 500 in use*.

Personal Computers (2002): 1,000*.

Internet Users (2002): 4,000*.

Mobile Cellular Telephones (2002): 2,700 subscribers*.

Radio Receivers (1997): 410,000 in use†.

Television Receivers (2001): 26,000 in use*.

Non-daily Newspapers (1988): 5 (estimated circulation 23,000)†.

* Source: International Telecommunication Union.

† Source: UNESCO, *Statistical Yearbook*.

EDUCATION

Primary (1999): 155 schools (1996); 1,233 teachers; 35,749 pupils.

Intermediate (1983): 8,643 pupils.

General Secondary (1996): 665 teachers; 12,672 pupils.

Universities, etc. (1983): 11 teachers; 136 students.

Other Higher (1996): 71 teachers; 674 students.

Source: partly UNESCO, *Statistical Yearbook*.

Adult Literacy Rate (UNESCO estimates): 98.7% (males 98.9%; females 98.4%) in 2002. Source: UN Development Programme, *Human Development Report*.

Directory

The Constitution

A new Constitution was adopted by a constitutional convention on 28 October 1960. After being approved by a UN-supervised plebiscite in May 1961, the Constitution came into force on 1 January 1962, when Western Samoa became independent. A constitutional amendment adopted in July 1997 shortened the country's name to Samoa. The main provisions of the Constitution are summarized below:

HEAD OF STATE

The office of Head of State is held (since 5 April 1963, when his coruler died) by HH Malietoa Tanumafili II, who will hold this post for life. After that the Head of State will be elected by the Fono (Legislative Assembly) for a term of five years.

EXECUTIVE

Executive power lies with the Cabinet, consisting of the Prime Minister, supported by the majority in the Fono, and ministers selected by the Prime Minister. Cabinet decisions are subject to review by the Executive Council, which is made up of the Head of State and the Cabinet.

LEGISLATURE

The Fono consists of 49 members. It has a five-year term and the Speaker is elected from among the members. Beginning at the election of 5 April 1991, members are elected by universal adult suffrage: 47 members of the Assembly are elected from among the Matai (elected clan leaders) while the remaining two are selected from non-Samoan candidates.

The Government

HEAD OF STATE

O le Ao o le Malo: HH MALIETOA TANUMAFILI II (took office as joint Head of State 1 January 1962; became sole Head of State 5 April 1963).

CABINET
(September 2004)

Prime Minister and Minister of Foreign Affairs: TUILA'EPA SAILELE MALIELEGAOI.

Deputy Prime Minister and Minister of Finance: MISA TELEFONI.

Minister of Internal Affairs, Women's Affairs and Broadcasting: TUALA AINIU IUSITINO.

Minister of Education: FIAME NAOMI MATA'AFA.

Minister of Lands, Survey and Environment: TUALA SALE TAGALOA.

Minister of Health and Labour: MULITALO SIAFAUSA.

Minister of Public Works: FAUMUI LIUGA.

Minister of Agriculture, Forestry, Fisheries and Meteorological Services: TUISUGALETAUA SOFARA AVEAU.

Minister of Trade, Commerce and Industry: HANS JOACHIM KEIL.

Minister of Transport: PALUSALUE FAAPO II.

Minister of Youth, Sports and Culture: ULU VAOMALO KINI.

Minister for the Legislative Assembly: TINO GAINA.

Minister of Justice: (vacant).

MINISTRIES AND MINISTERIAL DEPARTMENTS

Prime Minister's Department: POB L 1861, Apia; tel. 63122; fax 21339; e-mail pmdept@ipasifika.net.

Ministry of Agriculture, Forestry, Fisheries and Meteorology: POB 1874, Apia; tel. 22561; fax 24576.

Broadcasting Department: POB 200, Apia; tel. 21420.

Customs Department: POB 44, Apia; tel. 21561.

Economic Affairs Department: POB 862, Apia; tel. 20471.

Education Department: POB 1869, Apia; tel. 21911; fax 21917.

Ministry of Foreign Affairs: POB L 1859, Apia; tel. 25313; fax 21504; e-mail mfa@mfa.gov.ws.

Health Department: Private Bag, Apia; tel. 21212.

Inland Revenue Department: POB 209, Apia; tel. 20411.

Justice Department: POB 49, Apia; tel. 22671; fax 21050.

Lands, Survey and Environment Department: Private Bag, Apia; tel. 22481; fax 23176.

Public Works Department: Private Bag, Apia; tel. 20865; fax 21927; e-mail pwdir@lesamoa.net.

Statistics Department: POB 1151, Apia; tel. 21371; fax 24675.

Department of Trade, Industry and Commerce: POB 862, Apia; tel. 20471; fax 21646; e-mail tipu@tci.gov.ws; internet www.tradeinvestsamoa.ws.

Ministry of Transport: POB 1607, Apia; tel. 23701; fax 21990; e-mail mvnofo@mot.gov.ws.

Treasury Department: Private Bag, Apia; tel. 34333; fax 21312; e-mail treasury@samoa.net.

Ministry of Youth, Sports and Culture: Apia; tel. 23315.

Legislature

FONO

(Legislative Assembly)

The Assembly has 47 Matai members, representing 41 territorial constituencies, and two individual members. Elections are held every five years. At a general election on 2 March 2001, the Human Rights Protection Party (HRPP) won 22 seats, the Samoa National Development Party won 13 seats and independent candidates secured 14 seats. Four by-elections held during October and November 2001 were won by the HRPP, as was a further by-election in August 2003.

Speaker: TALEAFOA FAISI.

Political Organizations

Human Rights Protection Party (HRPP): c/o The Fono, Apia; f. 1979; Western Samoa's first formal political party; Leader TUILA'EPA SAILELE MALIELEGAOI; Gen. Sec. LAULU DAN STANLEY.

Samoa All-People's Party: Apia; f. 1996; Leader MATATUMUA NAIMOAGA.

Samoa Liberal Party: Apia; f. 1994; Leader NONUMALO LEULUMOEGA SOFARA.

Samoa Mo Taeao (Samoans for a Better Tomorrow): Apia; Chair. TUIFA'ASISINA MEAOLE KEIL.

Samoa National Development Party (SNDP): POB 1233, Apia; tel. 23543; fax 20536; f. 1988 following general election; coalition party comprising the Christian Democratic Party (CDP) and several independents; Leader Hon. LEMAMEA R. MUALIA; Sec. VALASI TAFITO.

Samoa National Party: Apia; f. 2001; Sec. FETU TIATIA.

Diplomatic Representation

EMBASSIES AND HIGH COMMISSIONS IN SAMOA

Australia: Beach Rd, POB 704, Apia; tel. 23411; fax 23159; e-mail peterhooton@dfat.gov.au; High Commissioner PETER HOOTON.

China, People's Republic: Private Bag, Vailima, Apia; tel. 22474; fax 21115; Ambassador GU SICONG.

New Zealand: Beach Rd, POB 1876, Apia; tel. 21711; fax 20086; High Commissioner JOHN ADANK.

USA: POB 3430, Apia; tel. 21631; fax 22030; e-mail usembassy@samoa.ws; Chargé d'affaires CHARLES SWINDELL.

Judicial System

Attorney-General: BRENDA HEATHER.

The Supreme Court

Is presided over by the Chief Justice. It has full jurisdiction for both criminal and civil cases. Appeals lie with the Court of Appeal.

Chief Justice: TIAVAASUE FALEFATU MAKA SAPOLU.

Secretary for Justice: FAAITAMAI P. F. MEREDITH.

The Court of Appeal: consists of the President (the Chief Justice of the Supreme Court), and of such persons possessing qualifications prescribed by statute as may be appointed by the Head of State. Any three judges of the Court of Appeal may exercise all the powers of the Court.

The District Courts

Replaced the Magistrates' Court in 1998.

Judges: LESATELE RAPI VAAI, TAGALOA ENOKA FERETI PUNI.

The Land and Titles Court

Has jurisdiction in respect of disputes over Samoan titles. It consists of the President (who is also a judge of the Supreme Court) and three Deputy Presidents, assisted by Samoan judges and Assessors.

President of The Land and Titles Court: TIAVAASUE FALEFATU MAKA SAPOLU.

Religion

Almost all of Samoa's inhabitants profess Christianity.

CHRISTIANITY

Fono a Ekalesia i Samoa (Samoa Council of Churches): POB 574, Apia; f. 1967; four mem. churches; Sec. Rev. EFEPAI KOLIA.

The Anglican Communion

Samoa lies within the diocese of Polynesia, part of the Church of the Province of New Zealand. The Bishop of Polynesia is resident in Fiji, while the Archdeacon of Tonga and Samoa is resident in Tonga.

Anglican Church: POB 16, Apia; tel. 20500; fax 24663; Rev. PETER E. BENTLEY.

The Roman Catholic Church

The islands of Samoa constitute the archdiocese of Samoa-Apia. At 31 December 2002 there were an estimated 42,473 adherents in the country. The Archbishop participates in the Catholic Bishops' Conference of the Pacific, based in Fiji.

Archbishop of Samoa-Apia: Cardinal ALAPATI L MATA'ELIGA, Cardinal's Residence, Fetuolemoana, POB 532, Apia; tel. 20400; fax 20402; e-mail archdiocese@samoa.ws.

Other Churches

Church of Jesus Christ of Latter-day Saints (Mormon): Samoa Apia Mission, POB 1865, Apia; tel. 64210; fax 64222; f. 1888; Pres. RENDAL V. BROOMHEAD; f. 1888; 65,000 mems.

Congregational Christian Church in Samoa: Tamaligi, POB 468, Apia; tel. 22279; fax 20429; e-mail cccsgsec@lesamoa.net; f. 1830; 100,000 mems; Gen. Sec. Rev. MAONE F. LEAUSA.

Congregational Church of Jesus in Samoa: 505 Borie St, Honolulu, HI 96818, USA; Rev. NAITULI MALEPEAI.

Methodist Church in Samoa (Ekalesia Metotisi i Samoa): POB 1867, Apia; tel. 22282; f. 1828; 36,000 mems; Pres. Rev. SIATUA LEULUAIALII; Sec. Rev. FAATOESE AUVAA.

Seventh-day Adventist Church: POB 600, Apia; tel. 20451; f. 1895; covers Samoa and American Samoa; 5,000 mems; Pres. Pastor SAMUELU AFAMASAGA; Sec. UILI SOLOFA.

BAHÁ'Í FAITH

National Spiritual Assembly: POB 1117, Apia; tel. 23348; fax 21363.

The Press

Newsline: POB 2441, Apia; tel. 24216; fax 23623; twice a week.

Samoa News: POB 1160, Apia; daily; merged with the weekly *Samoa Times* (f. 1967) in Sept. 1994; Publr LEWIS WOLMAN.

The Samoa Observer: POB 1572, Apia; tel. 21099; fax 21195; f. 1979; five times a week; independent; English and Samoan; Editor AUMA'AGAOLU ROPETA'ALI; circ. 4,500.

Samoa Weekly: Saleufi, Apia; f. 1977; weekly; independent; bilingual; Editor (vacant); circ. 4,000.

Savali: POB L1861, Apia; publ. of Lands and Titles Court; monthly; govt-owned; Samoan edn f. 1904; Editor FALESEU L. FUA; circ. 6,000; English edn f. 1977; circ. 500; bilingual commercial edn f. 1993; circ. 1,500; Man. Editor (vacant).

South Seas Star: POB 800, Apia; tel. 23684; weekly.

Broadcasting and Communications

TELECOMMUNICATIONS

Samoa Communications Ltd: Apia; tel. 23456; fax 24000; corporatized in July 1999; telecommunications and postal services provider; CEO (vacant).

BROADCASTING

Radio

Samoa Broadcasting Service: Broadcasting Department, POB 1868, Apia; tel. 21420; fax 21072; f. 1948; govt-controlled with commercial sponsorship; operates Radio 2AP; broadcasts on two channels in English and Samoan for 24 hours daily; Dir J. K. BROWN.

Magik 98 FM: POB 762, Apia; tel. 25149; fax 25147; e-mail magic98fm@samoa.net; f. 1989; privately-owned; operates on FM wavelengths 98.1 and 99.9 MHz; Man. COREY KEIL.

Radio Graceland: Apia; broadcasts gospel music.

Television

Televise Samoa Corporation: POB 3691, Apia; tel. 26641; fax 24789; e-mail ceotvsamoa@samoa.net; f. 1993; govt-owned national television broadcasting service; locally-produced programmes and programmes supplied by Australian Television (ATV); in mid-2003 the Govt announced plans to merge its television and radio stations after failing to sell Televise Samoa Corpn; CEO LEOTA UELESE PETAIA.

Finance

(cap. = capital; res = reserves; dep. = deposits; m. = million; brs = branches; amounts in tala, unless otherwise indicated)

BANKING

Central Bank

Central Bank of Samoa: Private Bag, Apia; tel. 34100; fax 20293; e-mail cbs@lesamoa.net; internet www.cbs.gov.ws; f. 1984; cap. 10.0m., res 16.6m., dep. 40.6m. (Dec. 2001); Gov. and Chair. PAPALI'I TOMMY SCANLAN.

Commercial Banks

ANZ Bank (Samoa) Ltd: Beach Rd, POB L 1885, Apia; tel. 22422; fax 24595; e-mail anz@samoa.ws; internet www.anz.com/samoa; f. 1959 as Bank of Western Samoa, name changed 1997; owned by ANZ Banking Group Ltd; cap. 1.5m., res 28.4m., dep. 246.1m. (Sept. 2001); Dir R. G. LYON; Man. Dir G. R. TUNSTALL; 1 br.

Industrial Bank Inc: POB 3271, Lotemau Centre, Vaea St, Apia; tel. 21878; fax 21869; f. 1995; owned by Industrial Pacific Investments Ltd; cap. US $0.3m., res US $1.9m. (1995); Chair. and Pres. IAN BYSTROV.

International Business Bank Corporation Ltd: Chandra House, Convent St, Apia; tel. 20660; fax 23253; e-mail ibb@samoa.net; f. 1991; 46.7% owned by ELECS Investment Ltd, 22.5% by Tidal Funds Co Ltd; cap. US $25.5m., res US $0.8m., dep. US $22.2m. (Jan. 1997); Chair. ILIA KARAS; Exec. Dir SERGUEI GREBELSKI.

National Bank of Samoa: POB 3047L, Apia; tel. 26766; fax 23477; e-mail info@nationalbanksamoa.com; internet www.nationalbanksamoa.com; f. 1995; owned by consortium of private interests in Samoa, American Samoa and the USA; Chair. TERENCE BETHAM; CEO ANNE BONISCH; 5 agencies; 1 sub-br.

Samoa Commercial Bank: Apia; tel. 31233; fax 30250; e-mail info@scbl.ws; internet www.scbl.ws; f. 2003; CEO RAY AH LIKI.

Westpac Bank Samoa Ltd: Beach Rd, POB 1860, Apia; tel. 20000; fax 22848; e-mail samoa@westpac.com.ws; f. 1977 as Pacific Commercial Bank Ltd, current name adopted 2001; first independent bank; 93.5% owned by Westpac Banking Corpn (Australia); cap. 1.2m., res 5.5m., dep. 96.5m. (Sept. 2002); Chair. ALAN WALTER; Gen. Man. STEVE BAKER; 3 brs.

Development Bank

Development Bank of Samoa: POB 1232, Apia; tel. 22861; fax 23888; f. 1974 by Govt to foster economic and social development; cap. 12.9m. (1992); Gen. Man. FALEFA LIMA.

INSURANCE

National Pacific Insurance Ltd: NPF Bldg, Private Bag, Apia; tel. 20481; fax 23374; f. 1977; 30% govt-owned; Gen. Man. RICKY WELCH.

Progressive Insurance Company: POB 620, Lotemau Centre, Apia; tel. 26110; fax 26112; e-mail progins@samoa.ws; f. 1993; Gen. Man. I. O. FILEMU.

Western Samoa Life Assurance Corporation: POB 494, Apia; tel. 23360; fax 23024; f. 1977; Gen. Man. A. S. CHAN TING.

Trade and Industry

CHAMBER OF COMMERCE

Chamber of Commerce and Industry: Level one, Lotemau Centre, Convent St, POB 2014, Apia; tel. 31090; fax 31089; e-mail info@samoachamber.com; internet www.samoachamber.com; f. 1938; Pres. NORMAN WETZELL; Vice-Pres. SALA EPA'TUIOTI; Sec. JOHN F. BOYLE.

INDUSTRIAL AND TRADE ASSOCIATIONS

Samoa Coconut Products: Apia.

Samoa Forest Corporation: Apia.

UTILITIES

Electricity

Electric Power Corporation: POB 2011, Apia; tel. 22261; fax 23748; e-mail epcgm@samoa.ws.

Water

Samoa Water Authority: POB 245, Apia; tel. 20409; fax 21298; e-mail taputoa@swa.gov.ws; Man.Dir MAFAA'UO TAPUTOA TITIMAEA.

MAJOR INDUSTRIAL COMPANIES

Desico Samoa Ltd: Apia; Gen. Man. TAIMANG JENSEN.

Wilex CCP Ltd: Apia; operates chocolate factory; Gen. Man. EDDIE WILSON.

Yazaki Samoa: Apia; manufacturers of automotive components.

TRADE UNIONS

Journalists' Association of Samoa: Apia; Pres. APULU LANCE POLU.

Samoa Manufacturers' Association (SMA): Apia; Pres. EDDIE WILSON.

Samoa Nurses' Association (SNA): POB 3491, Apia; Pres. FAAMANATU NIELSEN; 252 mems.

Samoa Trade Union Congress (STUC): POB 1515, Apia; tel. 24134; fax 20014; f. 1981; affiliate of ICFTU; Pres. FALEFATA TUANIU PETAIA; Dir MATAFEO R. MATAFEO; 5,000 mems.

Transport

Public Works Department: see under The Government; Dir of Works ISIKUKI PUNIVALU.

ROADS

In 1983 there were 396 km of main roads on the islands, of which 267 km were bitumen surfaced; 69 km of urban roads, of which 32 km were bitumen surfaced; 440 km of unsealed secondary roads and about 1,180 km of plantation roads. In mid-2004 the Government announced a programme of road-building, including new roads from Apia to the airport and to the inter-island wharves. New four-lane roads were also to be constructed in the capital in response to increased vehicle numbers, estimated to have risen by some 7% annually in the previous 10 years.

SHIPPING

There are deep-water wharves at Apia and Asau. A programme of improvements to port facilities at Apia, funded by Japanese aid, was completed in 1991. Regular cargo services link Samoa with Australia, New Zealand, American Samoa, Fiji, New Caledonia, Solomon Islands, Tonga, US Pacific coast ports and various ports in Europe.

Samoa Ports Authority: POB 2279, Apia; tel. 23552; fax 25870; e-mail spa@lesamoa.net; f. 1999.

Samoa Shipping Services Ltd: POB 1884, Apia; tel. 20790; fax 20026; e-mail sss@lesamoa.net.

Samoa Shipping Corporation Ltd: Private Bag, Shipping House Matautu-tai, Apia; tel. 20935; fax 22352; e-mail ssc@samoa.net; internet www.samoashippingcorporation.com; Gen. Man. OLOIALII KOKI TUALA.

CIVIL AVIATION

There is an international airport at Faleolo, about 35 km from Apia and an airstrip at Fagali'i, 4 km east of Apia Wharf, which receives light aircraft from American Samoa. In mid-1999 US $19.4m. was allocated by the World Bank to improve facilities at Faleolo airport.

Polynesian Airlines (Holdings) Ltd: NPF Bldg, Beach Rd, POB 599, Apia; tel. 21261; fax 20023; e-mail enquiries@polynesianairlines.co.nz; internet www.polynesianairlines.co.nz; f. 1959; govt-owned and -operated; international services to American Samoa, Rarotonga (Cook Islands), Nadi (Fiji), Tonga, Sydney and Melbourne (Australia), Auckland and Wellington (New Zealand) and Hawaii and Los Angeles (USA); domestic services between islands of Upolu and Savai'i; Chair. TUILA'EPA SAILELE MALIELEGAOI; CEO JOHN FITZGERALD.

Samoa Air: tel. 22901; operates local shuttle services between Pago Pago and Apia; CEO ANDRE LAVIGNE.

Tourism

Samoa has traditionally maintained a cautious attitude towards tourism, fearing that the Samoan way of life might be disrupted by an influx of foreign visitors. The importance of income from tourism has, however, led to some development, including the expansion of hotel facilities and improvements to the road network and airport. Some 87,688 tourists arrived in 2000, and revenue from the tourist industry totalled 133.1m. tala in that year. In 2000 34.3% of tourists came from American Samoa, 26.0% from New Zealand, 12.5% from Australia and 10.3% from the USA. The principal attractions are the scenery and the pleasant climate.

Samoa Visitors' Bureau: POB 2272, Apia; tel. 26500; fax 20886; e-mail samoa@samoa.net; internet www.samoa.co.nz; f. 1986; Gen. Man. SONJA HUNTER; Marketing Man. ALISE FAULALO-STUNNENBERG.

Defence

In August 1962 Western Samoa (as it was then known) and New Zealand signed a Treaty of Friendship, whereby the New Zealand Government, on request, acts as the sole agent of the Samoan Government in its dealings with other countries and international organizations. In the year ending June 2003 expenditure on defence and security amounted to 16.9m. tala (equivalent to 8.4% of total budgetary expenditure).

Education

The education system is divided into primary, intermediate and secondary and is based on the New Zealand system. Legislation was passed in 1992 which made education compulsory until the age of 14. In 1997 there were 35,649 pupils at primary schools, and in 1996 12,672 pupils undergoing secondary-level education. Teaching staff at primary and intermediate levels numbered 1,479 in 1996; in that year there were 665 secondary-school teachers. There are also a trades training institute, a teacher-training college and a college for tropical agriculture. About 97% of the adult population are literate in Samoan. The National University of Samoa was founded in 1988, and had an initial intake of 328 students. A new campus for the university, providing accommodation for up to 8,000 students and built at a cost of US $14.4m. with assistance from the Government of Japan, was opened in September 1997. Samoa had joined other governments in the area in establishing the regional University of the South Pacific, based in Fiji, in 1977. Current government expenditure on education in the year ending 30 June 2003 was an estimated 45.2m. tala (22.3% of total current expenditure).

SOLOMON ISLANDS

Physical and Social Geography

Solomon Islands is a scattered Melanesian archipelago covering a land area of 27,556 sq km (10,639 sq miles) in the south-western Pacific Ocean, east of Papua New Guinea and north of Vanuatu. The country includes most of the Solomon Islands (those to the north-west being part of Papua New Guinea), Ontong Java Islands (Lord Howe Atoll), Rennell Island and the Santa Cruz Islands, about 500 km to the east. There are 21 large islands and numerous small ones. The principal islands, all in the main group, are Choiseul, Santa Isabel (Boghotu), New Georgia, Malaita, Guadalcanal and San Cristobal (Makira). The climate is equatorial, with small seasonal variations governed by the trade winds. Much of the country is mountainous and of volcanic origin, with steep terrain which remains under dense tropical rain forest; extensive tracts of native and introduced grassland cover the northern plains of Guadalcanal. The smaller islands are mainly coralline.

Most of the population are Melanesian, and they speak about 80 dialects and languages. Pidgin English (Pijin—much of the vocabulary is derived from standard English, but used in a Melanesian grammatical form and with different intonations) is the lingua franca and is widely understood, but standard English is the official language. Some 95% of the population are Christian, the largest denomination being the Church of Melanesia (Anglican). The total population at the 1999 census was 409,042, compared with 285,176 at the census of 1986 when some 94% of the population were Melanesian, 4% Polynesian (mainly from the 'outliers' such as Ontong Java), 1% Micronesian (originally resettled from the Gilbert Islands, now Kiribati) and some Europeans and Chinese. Solomon Islands has an extremely high rate of population growth, which averaged an annual rate of 2.8% in 1990–2002. The capital is Honiara, on the island of Guadalcanal.

History

Solomon Islands, long settled by Melanesian peoples, was named by a Spanish navigator, Alvaro de Mendaña, in 1568. European contacts were intermittent from then, and made little impact on the islanders' lives. It was not until the 19th century that traders, whalers and missionaries began to establish outposts on the main islands. Forcible recruiting of labour ('blackbirding') spread from the New Hebrides (now Vanuatu) to the Solomon Islands during the 1860s. The northern Solomon Islands became a German Protectorate in 1885 and the southern Solomons a British Protectorate in 1893. Rennell Island and the Santa Cruz Islands were added to the British Protectorate in 1898 and 1899. Germany ceded most of the northern Solomons and Ontong Java Islands to the United Kingdom between 1898 and 1900. The whole territory, known as the British Solomon Islands Protectorate, was placed under the jurisdiction of the Western Pacific High Commission (WPHC), with its headquarters in Fiji. The High Commissioner for the Western Pacific was represented locally by a Resident Commissioner.

The Solomon Islands were invaded by Japan in 1942 but, after a fierce battle on Guadalcanal, most of the islands were recaptured by US forces in 1943. After the Second World War the Protectorate's capital was moved from Tulagi, on Ngella (Florida Islands), to Honiara, on Guadalcanal, which was near to a major war-time airfield. In January 1953 the headquarters of the WPHC also moved to Honiara. Meanwhile, elected local councils were established on most of the islands and by 1966 almost the whole territory was covered by such councils. The introduction of responsible local government was initially prompted by the challenge of the Maa'sina (brotherhood) Ruru movement, also known by the anglicized form 'Marching Rule'. This originated in Malaita in 1927, but grew during and after the Second World War. It favoured strictly controlled, custom-based communities living in large villages and practising a communal agricultural economy and opposed close co-operation with the colonial administration or the dominant churches. The WPHC at first attempted to accommodate the movement, but its influence continued to spread and its suppression took place between 1948 and 1950.

Under a new Constitution, introduced in October 1960, a Legislative Council and an Executive Council were established for the Protectorate's central administration. Initially, all members of both bodies were appointed but from 1964 the Legislative Council included elected members, and the elective element was gradually increased as successive legislative and administrative bodies were created by new Constitutions in March 1970 and April 1974. The Constitution of April 1974 instituted a single Legislative Assembly with 24 members who chose a Chief Minister with the right to appoint his own Council of Ministers. A new office of Governor of the Protectorate was also created to assume almost all the functions previously exercised in the territory by the High Commissioner for the Western Pacific. Solomon Mamaloni, leader of the newly founded People's Progressive Party (PPP), was appointed the first Chief Minister in August 1974. The territory was officially renamed the Solomon Islands in June 1975, although it retained protectorate status.

In January 1976 the Solomon Islands received internal self-government, with the Chief Minister presiding over the Council of Ministers in place of the Governor. In June elections were held for an enlarged Legislative Assembly and in July the Assembly elected one of its new members, Peter (later Sir Peter) Kenilorea, to the position of Chief Minister. Following a constitutional conference in London in September 1977, Solomon Islands (as it was restyled) became an independent state, within the Commonwealth, on 7 July 1978. The Legislative Assembly became the National Parliament and designated Kenilorea the first Prime Minister.

The main political issue confronting the country was the proposed decentralization of authority to the regions, support for which was particularly strong in the Western District, the most commercially developed part of the country. In 1979 the PPP merged with the Rural Alliance Party to form the People's Alliance Party (PAP), with Solomon Mamaloni as its leader. The first general election since independence was held in August 1980. Independent candidates won more seats than any of the three parties. Parliament again elected Kenilorea as Prime Minister by an overwhelming majority. In August 1981, however, Parliament approved a motion expressing 'no confidence' in Kenilorea, and chose Mamaloni to succeed him as Prime Minister.

After elections to the National Parliament in October 1984, a majority among the minor parties and independent members was decisive in electing Kenilorea to the post of Prime Minister. The new Government consisted of a coalition of nine members of Kenilorea's United Party (UP), three from the newly formed Solomone Ano Sagufenua (SAS) and three independents. The five provincial ministries, established by Mamaloni, were abolished in line with Kenilorea's declared policy of restoring to central government control some of the powers held by the provincial governments. In October 1985 a new political party, the Nationalist Front for Progress (NFP), was formed, under the leadership of Andrew Nori. However, the SAS subsequently withdrew its support from the coalition, and Kenilorea formed a new Cabinet, comprising nine members of the UP, three of the NFP and three independents. In November 1986 three cabinet ministers (all members of the NFP) resigned, in reaction to allegations that Kenilorea had secretly accepted French aid, amounting to about $A70,000, to repair cyclone damage to his home village. Faced with a third motion expressing 'no confidence' (the previous two having been defeated), Kenilorea himself resigned, and in December he was replaced as Prime Minister by Ezekiel Alebua, the former Deputy Prime Minister. Alebua retained Kenilorea's Cabinet almost in its entirety, with the latter becoming Deputy Prime Minister.

In March 1988 a constitutional review committee, chaired by Mamaloni, recommended that Solomon Islands become a federal republic within the Commonwealth. The next general election was held in February 1989 and the PAP won 11 of the 38 seats, remaining the largest single party by a considerable margin. It had also sponsored several candidates who stood as independents and secured the support of others. Mamaloni was appointed Prime Minister after defeating Bartholomew Ulufa'alu (leader of the Solomon Islands Liberal Party and the parliamentary faction of the Coalition for National Unity) in a parliamentary ballot. The new Government included Sir Baddeley Devesi, the former Governor-General (who had been succeeded by Sir George Lepping in July 1988), and was described as the first since independence solely to comprise the members of a single party.

Dissatisfaction with Mamaloni's leadership was expressed throughout 1990, and, in October, he resigned as leader of the PAP, one week before a party convention. Mamaloni declared that he would remain as an independent Prime Minister and dismissed five members of the Cabinet, replacing them with four members of the opposition and a PAP back-bencher. Persistent demands for Mamaloni's resignation by more than one-half of the country's MPs were defied by the Prime Minister. The PAP subsequently asked the remaining 10 ministers to resign their posts in the interests of party unity, but they refused, and in February 1991 were expelled from the party. By establishing a coalition Government and dividing the opposition and the ruling party, Mamaloni's action was widely interpreted as a return to the political traditions of Solomon Islands, based on personalities rather than on organized parties. The country's economic situation deteriorated throughout 1991, and in November prompted the Solomon Islands Council of Trade Unions to issue an ultimatum demanding Mamaloni's resignation, in order to avert mass industrial action.

At legislative elections in May 1993 the Group for National Unity and Reconciliation, led by Mamaloni, won 21 of the 47 seats in the recently enlarged National Parliament. As Mamaloni's party had failed to achieve a majority the main opposition parties and independents agreed to form the National Coalition Partners. At elections to the premiership in June the independent member, Francis Billy Hilly (supported by the newly formed alliance), defeated Mamaloni by a single vote.

In October 1994 a constitutional crisis resulted in several weeks of political confusion, following attempts by the Governor-General to dismiss Hilly on the grounds that he no longer held a parliamentary majority. Hilly remained in office, however, with the support of a High Court ruling, and confusion intensified when the Governor-General appointed the opposition leader, Solomon Mamaloni, to the position of Prime Minister. Hilly finally resigned on 31 October, and the post was declared vacant. In a parliamentary election to the premiership on 8 November Mamaloni defeated the former Governor-General, Sir Baddeley Devesi, by 29 votes to 18. Among the new Government's stated objectives, including several proposed political and constitutional changes, Mamaloni expressed his intention to conduct a thorough review of the country's current logging policy. Consequently, during late 1994 it was announced that many of the regulations introduced by the previous Government, in an attempt to conserve the islands' forestry resources, were to be repealed or relaxed (see Economy).

In April 1995 security forces were dispatched to Pavuvu Island (some 50 km north-west of Honiara), following angry protests by islanders who were resisting a compulsory resettlement programme, which the Government had agreed to implement in return for the sale of logging rights (for some 1m. cu m of timber on the island) to a Malaysian company. Logging began in May, despite continuing protests by the islanders, as well as opposition demands to abandon the programme and a warning from the Solomon Islands Central Bank that the Government's policy of allowing virtually unrestrained logging would seriously undermine the country's long-term economic prospects. An opposition motion to reverse the decision, presented to the National Parliament in July, was, however, defeated by 15 votes to 12. The vote prompted accusations that the Mamaloni Government favoured the protection of foreign business interests over the welfare of its own people.

In late 1995 and early 1996 Mamaloni's Government suffered a series of allegations of corruption and misconduct. In December seven ministers appeared in court accused of accepting payments and other benefits from foreign logging companies between 1993 and 1995. The Prime Minister gave assurances that none of the ministers would be dismissed from their positions, and in February 1996 all were acquitted. In May, moreover, a serious financial scandal at the Ministry of Finance led to the suspension of 25 officials pending an investigation into the disappearance of funds valued at some US $10m.

Meanwhile, the Government's continued reluctance to implement regulations to restrain logging activity in the islands remained a source of controversy. Extensive logging on Mono Island in the Western Province, from early 1996, provoked a series of protests by islanders and landowners. In frustration at the continuation of virtually unchecked logging in the islands, the Australian Government announced that it was to reduce aid to the country by $A2.2m. annually.

Controversy arose in July 1996 when it was reported that members of the National Parliament had begun to present gifts to their constituencies (mostly in the form of canoes and outboard motors) in preparation for the general election scheduled for 1997. This followed the reinstatement of the controversial Constituency Development Fund, which entitled each member to US $66,000, and which had been widely used by members at the previous general election to secure re-election by purchasing gifts for voters in the constituency. Fears that corruption and bribery in the country had reached unacceptable levels were confirmed in September by the regional trades union organization, SPOCTU, which cited the problem as the greatest obstacle to the islands' development, and stated that, with the worst incidence of corruption in the region, investors and aid donors would remain reluctant to make financial commitments to Solomon Islands.

In early August 1996 the National Parliament approved legislation to reform the provincial government system. Under the new system, the legislative and administrative powers of the nine provincial governments were transferred to 75 area assemblies and councils, with financial control vested wholly in the central Government. In February 1997, however, the legislation, which had been vehemently opposed by the larger provinces, was declared invalid by the High Court.

Concern over foreign exploitation of Solomon Islands resources continued during 1996, and in early 1997 landowners near the proposed Gold Ridge gold mine on Guadalcanal initiated legal proceedings for compensation and greater provision for the protection of the environment against the Australian company involved in the project. Furthermore, incidents in which local people damaged or destroyed equipment belonging to foreign logging companies were reported to have increased throughout the islands during 1997.

In May 1997 Mamaloni announced his intention to hold an early general election, following which he would resign as leader of the Group for National Unity and Reconciliation. Meanwhile, the opposition leader, Ezekiel Alebua, resigned, following accusations of misconduct from fellow opposition members, and was replaced by Edward Hunuehu. A general election took place on 6 August to a legislature that had recently been enlarged to 50 seats (following the creation of three new constituencies). A total of 19 incumbent and 23 new members were successful in the poll, with the Group for National Unity and Reconciliation winning 24 seats and a coalition grouping, Alliance for Change, securing the remainder. A period of intense political manoeuvring followed the election, as groups of successful candidates sought to secure a majority in Parliament. On 27 August Bartholomew Ulufa'alu was elected Prime Minister, defeating the newly elected leader of the Group for National Unity and Reconciliation, Danny Philip, by 26 votes to 22. The new Government announced a programme of extensive structural reforms (approved in October) which aimed to address many of the country's ongoing problems. The proposals included a review of the public service (involving an expected loss of some 1,000 public servants' jobs) and measures to expand the role of the private sector in the economy and to encourage greater participation of non-governmental organizations in the country's socio-economic development. The measures aimed to restore a degree of economic stability to Solomon Islands and to attract increased foreign investment. Legislation proposing that a politician seeking to change party allegiance would automatically lose his or her seat and be subject to a by-election was similarly intended to increase political stability. The Government also announced plans for major structural changes in the forestry industry, including a sustainable harvesting policy with reduced quotas, greater involvement of local landowners in logging operations and a moratorium on new logging licences (implemented in April 1998).

In early 1998 a shipment of weapons (including a helicopter gunship, two military aircraft and smaller armaments) ordered by the previous administration for use in defending the maritime border with Papua New Guinea, arrived in the country from the USA. Ulufa'alu requested assistance from Australia in impounding the weapons, following widespread concern that unofficial organizations (particularly the Bougainville secessionist rebels within Papua New Guinea) might intercept the shipment. A dispute subsequently arose between Ulufa'alu and the former Prime Minister, Solomon Mamaloni, who had ordered the weapons. Mamaloni accused Ulufa'alu of treason for surrendering sovereign property to a foreign government, while the Prime Minister claimed that serious irregularities had occurred in procuring the arms and cited missing files and apparent overpayment as evidence of this.

In April 1998 Job Dudley Tausinga, whose Coalition for National Advancement (CNA) constituted the largest group outside the Government, was appointed Leader of the Opposition. Following the defection of six government members in July–August 1998, the opposition claimed to hold a majority (with 25 of the 49 sitting members) in the National Parliament and consequently sought to introduce a motion of 'no confidence' in the Prime Minister. The Government effectively delayed the vote for several weeks by lodging an unsuccessful appeal against it in the High Court. A vote took place on 18 September but was deemed to have been defeated as an equal number of votes were cast for and against the motion. Three opposition members subsequently defected to the Government, thus increasing their representation to 27 members. The climate of political instability resulting from the persistent political manoeuvring during mid-1998 prompted the Government to reiterate its proposal for legislation to restrict the rights of elected members to change party allegiance within the National Parliament. In late September Solomon Mamaloni was elected Leader of the Opposition. Mamaloni died in January 2000 and was succeeded by Manasseh Sogavare, under whose leadership the CNA reverted to its original name of the PPP.

From April 1998 violent unrest in Honiara was attributed to ethnic tensions, mainly between the inhabitants of Guadalcanal and Malaita provinces. The previously latent inter-ethnic tensions on Guadacanal island were attributed to the large-scale migration of ethnic Malaitans there in the Second World War period, and their subsequent achievement of status and wealth disproportionate to their numbers, to the perceived detriment of those indigenous to Guadacanal (Isatambu). A specific catalyst of the violent unrest from 1998 was the alienation of land by the Government since independence. Title to some land thus expropriated had been returned to Guadalcanal in late November, but this had not allayed a widespread feeling of resentment in the province at the financial burden imposed by hosting the capital. It was reported that a group styling itself the 'Guadalcanal Revolutionary Army' (GRA) had begun a campaign of militancy to force the Government to relocate the capital. The unrest intensified in early 1999, prompting the Government to establish a peace committee for the province. In mid-1999 talks between the Premier of Guadalcanal Province, Ezekiel Alebua, and the Solomon Islands Prime Minister failed to resolve the ethnic tensions in the province. Riots broke out in the capital, Honiara, and as many as 80 Malaitan immigrants were evacuated following threats by armed GRA militants. Alebua continued to pledge his full commitment to the peace process, and rumours that he was supporting the GRA led him to offer his resignation in an attempt to disprove any connection. Following Ulufa'alu's demands that peace be restored to the province before the implementation of any further measures, the GRA ordered an immediate halt to its activities. The Guadalcanal Provincial Assembly subsequently declared that it had accepted an initial payment of SI $500,000 in compensation for accommodating the national capital. A reconciliation ceremony was held between the two parties, during which Alebua apologized for the recent persecutions of Malaitan immigrants and appealed to the GRA to lay down its arms. However, the Malaitans subsequently demanded that they too be compensated, to the sum of US $600,000, for damage to their property by the GRA militants.

Following an increase in violence during which three Malaitan immigrants were reported to have been killed, and an estimated 10,000 forced to flee their villages, on 15 June 1999 a state of emergency was declared in Guadalcanal, and restrictions on media reporting of the conflict were tightened. A leader of the Isatabu Freedom Movement (IFM, formerly the GRA and also known in 1999 as the Isatabu Freedom Fighters—IFF), Andrew Te'e, announced that the movement was willing to surrender in return for a full amnesty for the militants; however, this was rejected by Ulufa'alu. The former Prime Minister of Fiji, Sitiveni Rabuka, was appointed Commonwealth Special Envoy, following a request for assistance by Ulufa'alu. After meeting with the parties concerned, Rabuka announced on 28 June that a peace agreement had been reached. A UN delegation and police officers from Fiji and Vanuatu were sent to monitor the conflict and help implement the peace plan. As part of the Honiara Peace Accord, the Solomon Islands Government agreed to pay SI $2.5m. into a Reconciliation Trust Fund, which was to be jointly administered by Guadalcanal Province and the national Government, to compensate the victims of the unrest. In return, the IFM agreed to disarm and to abandon their demands for a full amnesty for their supporters. Rabuka also urged the Government to lift the state of emergency and the media restrictions. In August, however, Rabuka and the UN monitoring team returned to the province amid speculation that the Honiara Peace Accord was on the verge of collapse, after four members of the IFM were reported to have been shot by police near Honiara.

On 12 August 1999 a new peace agreement was signed by Rabuka, Alebua and others. The agreement, known as the Panatina Agreement, allowed for a reduction in police activity in the province followed by the eventual revocation of the state of emergency, in return for the surrender of weapons by the IFM. Despite the extension of the disarmament deadline into September, the state of emergency was ended, and media restrictions were lifted in mid-October; following negotiations in the capital, Honiara, and the signing of an agreement in Fiji, a multinational peace-monitoring group from Fiji and Vanuatu, jointly funded by Australia and New Zealand, arrived in Guadalcanal later that month to monitor the disarmament process. In early December the peace-keeping force's mandate was extended into January 2000, and subsequently further extended by three months, following renewed outbreaks of violence and the emergence of a new guerrilla group, the Malaita Eagle Force (MEF), demanding US $40m. in compensation payments for loss of property incurred by Malaitans as a result of the conflict. Among their spokesmen was the former Minister for Finance, Andrew Nori.

Following further outbreaks of violence in the province that led to the death of four people, including two policemen, in February 2000, a decree was issued by the Governor-General outlawing membership of both the IFM and the MEF. Nevertheless, further clashes between the two rebel groups were reported in early March, and later that month riots took place in the capital, Honiara, during which Malaitan immigrants stoned the headquarters of the Guadalcanal Provincial Government. Peace talks took place in May without members of the IFM and the MEF, both of which refused to attend in protest at the decision to outlaw them earlier that year. However, a document known as the Buala Peace Communiqué was issued as the outcome of these talks on 5 May. In an attempt to advance the peace process, the order that outlawed the groups was suspended on 12 May. However, on 5 June members of the MEF, armed with weapons obtained in raids on police armouries, seized control of Honiara, placing Ulufa'alu under arrest. The rebels demanded the immediate resignation of the Prime Minister, claiming that he, himself an ethnic Malaitan, had failed to compensate displaced Malaitans within the established deadline (allegedly set for that day) and also demanded the appointment of a new Commissioner of Police, the *de facto* head of National Security. In renewed outbreaks of violence, it was alleged that up to 100 people had been killed. The MEF, meanwhile, claimed to have gained control of the Police Force, 98% of the military-style weapons in the territory, broadcasting services and the telecommunications infrastructure. Ulufa'alu was released four days later, following an agreement between the MEF and government negotiators that a special parliamentary sitting would be convened during which Ulufa'alu would be subject to a motion of 'no confidence'. A 14-day cease-fire was also called to guarantee the safe passage of a Commonwealth monitoring team. On 14 June Ulufa'alu resigned, one day before the scheduled 'no confidence' vote; however, he agreed to remain as caretaker Prime Minister for a 14-day transitional period during which negotiations between the MEF, the IFM and a Commonwealth Special Envoy, Prof. Ade Adefuye of Nigeria, were to take place. However, negotiations collapsed following the MEF's refusal to hand over its weapons, pending the appointment of a new Prime Minister. An extraordinary parliamentary session at which a new Prime Minister was to be appointed was set for 28 June. As this session failed to raise the necessary quorum, the election of a new Prime Minister was delayed until 30 June, when the leader of the Opposition, Manasseh Sogavare, defeated Rev. Leslie Boseto, incumbent Minister for Lands and Housing, by 23 votes to 21. (The third candidate, former Premier Francis Billy Hilly, withdrew in support of Boseto.) Sogavare declared that he would seek to establish peace without making significant changes to the policy of the previous Government. None the less, Sogavare announced a comprehensive reallocation of ministerial posts and restructuring of ministries. A new Ministry for National Unity, Reconciliation and Peace was established, and plans to create a Solomon Islands Defence Force and Ministry of Defence were announced. Previously defence issues had fallen within the remit of the police service, which had become compromised because of allegations of MEF infiltration. Sogavare also pledged to examine the issue of an alleged ethnic imbalance within the Solomon Islands Police Force; it was claimed that about 75% of the 900 officers were ethnic Malaitans, and that all 30 of the ethnic Isatambu were deployed outside of their home island of Guadacanal.

New cease-fire negotiations commenced on the Australian naval vessel, Tobruk, in early August 2000, resulting in the declaration of a 90-day cease-fire between the IFM and MEF. However, as occurred during the June cease-fire, intermittent violent disorder related to inter-ethnic tension continued across Guadalcanal. In late August IFM dissidents kidnapped the brother of Deputy Prime Minister Allan Kemakeza, demanding SI $6.5m. in compensation for the displaced persons of Guadalcanal. He was released unharmed after 10 days on 31 August without the payment of ransom. Logistical problems delayed the onset of further peace talks on the New Zealand navy frigate, *Te Kana*, until early September. These negotiations produced a communiqué outlining methods by which a lasting peace might be achieved. Meanwhile, in mid-September a breakaway group from the IFM, which reverted to the former name of the GRA and was led by Harold Keke, held an airline pilot hostage with a demand for SI $2m.; he was released unharmed without the demands being acceded to. A further round of peace talks took place in Queensland, Australia, which on 15 October led to the signing of a peace treaty known as the Townsville Agreement. Those party to the agreement were the MEF, the IFM, the Solomon Islands Government, and the Provincial Governments of Malaita and Guadalcanal. The agreement allowed for an amnesty giving immunity to all involved in crimes associated with the ethnic conflict, on condition of a surrender of weaponry within 30 days. It also envisaged the creation of an international peace-monitoring team, and the repatriation to their home villages of all MEF and IFM soldiers at the expense of the Solomon Islands Government. Infrastructure and services in the two provinces would also be restored and developed, within three months of these repatriations. Furthermore, a greater degree of autonomy was to be granted to the two provinces, by devolution of power or the implementation of constitutional amendments, and Malaita Province was to receive additional funding to reflect the demands placed on the region by the influx of 20,000 displaced persons from Guadalcanal. Public displays of forgiveness, reconciliation and confession were to be organized, and a Peace and Reconciliation Committee was to be established.

The implementation of the peace process, overseen by monitors from Australia and New Zealand, was threatened in mid-November 2000, when four people were killed in a shooting in Gizo, Western Province. Among the dead were two members of Papua New Guinea's Bougainville Revolutionary Army. Moreover, delays in the disarmament process involving the MEF and the IFM caused the deadline for the surrender of arms to be extended from 15 November until the end of that month, and then further until 15 December. In mid-November, following an announcement that the Guadalcanal Provincial Government headquarters, which had been occupied by members of the MEF since June, was to be rehabilitated as a symbol of national unity, arsonists, believed to be linked with the MEF, attacked the building. Despite a series of ceremonies in late November and early December, in which members of the IFM and

MEF surrendered weapons, it was believed that at least 400 illegally-held weapons remained in circulation at the end of December. The legislation granting immunity to those who had committed crimes during the conflict was passed by Parliament in mid-December, and was criticized by the prominent human rights organization, Amnesty International. In late December one man was wounded in an attack on a motel, in Honiara, in which disarmed former IFM rebels recruited to join the police were resident. Responsibility for the attack was attributed to a group calling itself the Marau Eagle Force, from the eastern Marau region of Guadalcanal, which was subject to separate peace negotiations. The breakaway GLF led by Harold Keke also announced that it had not accepted the cease-fire. In early January 2001 concerns were expressed that, as a result of the ongoing economic crisis and a larger number of rebels having opted to join the police than had been anticipated, the Government might be unable to pay the former rebels, thus precipitating further concerns about security.

In early December 2000 new legislation had permitted the appointment of two further ministers, a Minister for Rehabilitation, Reconstruction and Redirection, and a Minister for Economic Reform and Structural Adjustment. In early January 2001 it was reported that the Government was contemplating an extension of the amnesty. Further violence broke out between Guadalcanal militants and members of the Marau Eagle Force. A peace agreement between the group and the IFM was signed in early February, although the Peace Monitoring Council observed that the infrastructure of the Marau region had been almost entirely destroyed since 1998 and that there was little immediate prospect of recovery. In early March 2001 peace in the Marau region appeared to be threatened following an incident in which police officers, in connection with a drink-driving charge, opened fire on a vehicle owned by a former commander of the Marau Eagle Force. It was also reported that two other former Marau commanders had demanded SI $100,000 from the Government, and that other Marau militants would refuse to surrender their arms, in conformity with the agreement signed in February, unless the GLF were also disarmed. In mid-March tensions heightened following a security operation against the GLF leader, Keke, in which a government patrol boat fired at villages on the western coast of Guadalcanal. In the same month police disarmed and arrested a group of villagers from Munda, Western Province. Further incidents in early April, in which a boat fired at coastal targets in southern Guadalcanal, led the international peace monitoring force from Australia to state that the cease-fire arrangement reached under the Townsville Agreement might have been breached. The Assistant Commissioner of Police Operations protested that the Townsville Agreement did not require the police to surrender their guns, although a report published in April claimed that some of the weapons used in the security operation should have been surrendered the previous year. In June the Premier of Guadalcanal, Ezekiel Alebua, was shot and seriously injured in an assassination attempt, apparently carried out on the orders of former leaders of the IFM. Harold Keke met the Deputy Prime Minister and the secretary of the Peace Monitoring Council in June and, in October, he voiced his support for the imminent general elections. However, the failure of former militia groups to surrender their guns was the main obstacle to a lasting peace throughout 2001, and officials estimated that there remained some 500 high-powered weapons in the community, in addition to hand-made weapons. The Government declared an 'arms amnesty' in April 2002. Militants surrendered some 2,000 guns with impunity. Weapons disposal began in June, coinciding with the International Peace Monitoring Team's departure from the country.

Meanwhile, in the second half of 2000 Western, Choiseul, and Temutu Provinces all declared themselves to be semi-autonomous states within Solomon Islands; on 1 September the legislature of the latter, representing 20,000 inhabitants, approved a bill allowing for a referendum on the province's proposed independence. The worsening economic situation caused by the conflict in Guadalcanal was believed to be a determining factor; additionally, movements in Guadalcanal and Makira provinces demanding greater autonomy were reported to have gained strength at this time. In late November the Minister for Provincial Government and Rural Development, Nathaniel Waena, announced that legislation to amend the Constitution would be submitted in 2001, in order to institute a federal system of government. Controversy arose following an announcement by the Government, in mid-March 2001, that it intended to introduce legislation to extend the life of Parliament for a further year, to expire in August 2002, stating that the social and economic position of the islands would not facilitate the holding of elections as scheduled in August 2001. This proposal attracted widespread opposition, including that of churches, trade unions, all provincial premiers, and the principal overseas aid donors to Solomon Islands. When it became apparent that he would not possess a parliamentary majority to support the legislation, Sogavare withdrew the bill from Parliament in early May; later that month it was announced that the elections would be held as scheduled, funded wholly from overseas. In September the National Parliament approved a bill to increase national general election fees by 150%; Walter Folotalu, former member of the National Parliament for Baegu, challenged the proposed legislation in the High Court in October. In the same month the High Court also began hearing the case of former Prime Minister Bartholomew Ulufa'alu; he had challenged the legality of the Government, seeking a ruling that the coup and subsequent election that ousted him were unconstitutional. However, Ulufa'alu's challenge was unsuccessful.

In October 2001 public services deteriorated as nation-wide power cuts resulted from the inability of the government-owned Solomon Islands Electricity Authority (SIEA) to pay for its supplies of diesel fuel. The Government announced the introduction of health charges in order to maintain medical services, and was forced to appeal to Australia and New Zealand for assistance in policing, as violent crime became endemic. A storage container holding weapons relinquished under the Townsville Peace Agreement was broken into and, in November, the revelation that compensation totalling SI $17.4m. had been paid to former members of the MEF for alleged property damage precipitated a violent demonstration by protesters demanding similar recompense. The Prime Minister was prevented from leaving his office (to attend a session of the UN General Assembly in New York, USA), while the house of the Deputy Prime Minister was vandalized.

Accusations and rumours of bribery and intimidation were rife during the campaigning that preceded the general election, held on 5 December 2001. The electoral coalition of the Solomon Islands Alliance for Change Coalition (SIACC) won 12 seats, the PAP secured nine and the PPP six, while 22 seats were won by independent candidates. In the absence of a clear SIACC leader, 11 elected members of the coalition convened with 11 elected independents to decide upon a satisfactory premier. Later that month, following various shifts in allegiances, Sir Allan Kemakeza (as he had become), leader of the PAP and former Deputy Prime Minister, was declared Prime Minister. Despite having been previously accused of misappropriating state funds, the former Minister for Finance, Snyder Rini, was appointed Deputy Prime Minister and Minister for National Planning.

The new Government, however, was unable to improve the increasingly desperate political and economic situation on the islands. A peace summit, organized by the Peace Monitoring Council, scheduled to be held in March 2002 was, in February, postponed until June, and in early March, as the security situation in the country deteriorated further, numerous international peace monitors began withdrawing from the islands. Later in March the New Zealand Deputy High Commissioner to Solomon Islands, Bridget Nichols, was discovered at her home suffering from severe knife wounds, and died shortly afterwards in hospital, in what was initially believed to have been an incident linked to the recent escalation of violence. However, subsequent official investigations by both the New Zealand and Solomon Islands authorities concluded that the diplomat's death had been accidental. Also in March Kemakeza dismissed Michael Maina, the Minister for Finance, after Maina failed to consult the Cabinet prior to announcing a number of drastic budgetary measures, the most significant of which was his decision to devalue the national currency by 25%. Maina was replaced by Laurie Chan, who in early April reversed the devaluation.

There were reports of further disturbances on the western coast of Guadalcanal in early June 2002, during which it was claimed that 11 Malaitans, who were part of a force attempting to capture the GLF leader, Harold Keke, had been killed. In mid-July water and electricity supplies to the capital were interrupted after the SIEA was once again unable to purchase fuel to power its generators, and the Solomon Islands Water Authority had failed to pay rental arrears owed to the landowners of the Kongulai water source. Later that month the Prime Minister effected a reorganization of the Cabinet in which, most notably, Nollen Leni, hitherto Minister for Provincial Government and Rural Development, replaced Alex Bartlett as Minister for Foreign Affairs and Trade Relations; Bartlett was allocated the tourism and aviation portfolio.

The country was further adversely affected by a series of strikes in mid-August 2002 by public sector workers in protest at the non-payment of their salaries. Later that month the Minister for Youth and Sports and Women's Affairs, Rev. Augustin Greve, was assassinated. It was subsequently reported that Harold Keke had claimed responsibility for the murder. In September a church deacon was found beheaded on a beach on Guadalcanal's southern coast, an area considered to be a stronghold of the GLF. In a separate incident in the same month a woman and three children were shot dead and 11 others wounded in an assault which local chiefs attributed to Keke's organization. In late September Kemakeza made a formal request to the UN for a peace-keeping force to address the increasing state of lawlessness in the country, and in the following month a delegation of four UN officials visited the islands to assess the situation.

In mid-October 2002 the Cabinet approved the establishment of the National Peace Council, an interim body, to replace the Peace Monitoring Council, the mandate of which expired on 15 October under the terms of the Townsville Peace Agreement. A permanent body was to be established after 16 January 2003. It was confirmed in the same month, however, that more weapons were in circulation in the country than when the peace agreement was signed in October 2000. The National Peace Council expressed a belief that many of the weapons surrendered under the arms amnesty of April 2002 had been removed for use by police officers in their campaign to capture Keke.

In mid-December 2002 the Minister for Finance, Laurie Chan, resigned. The Government claimed that this action was in response to criticism of his budget, which had recently been approved by the Cabinet. However, Radio New Zealand reported that his resignation was in protest at the Government's decision to pay unscheduled allowances to the 'special constables' (former militants from the ethnic conflict who were allowed to join the police force as part of the peace agreement) who had demanded the payments with threats and violence. The latter had included an incident in which gunshots were fired at the Prime Minister's residence by a group of 'special constables'. The Government faced a serious challenge in mid-December when six independent members of Parliament resigned because of what they described as a 'leadership problem' within the coalition. However, the Government defeated an opposition motion of 'no confidence' by 28 votes to 17. Meanwhile, several financial advisers fled the country, reportedly as a result of the threats they had received.

In late December 2002 the remote islands of Tikopia and Anuta in the province of Temotu were struck by Cyclone Zoe and lost contact with the rest of the country. The islands suffered widespread devastation with the loss of all crops, buildings and water supplies, and heavy casualties were feared. The dispatch of a boat containing relief supplies was delayed by several days because of the Government's inability to pay for a crew or fuel for the vessel. When aid finally reached the islands, it was found that, contrary to all expectations, there had been no fatalities among the combined population of some 4,000, despite the severity of the storms. Islanders were believed to have used their knowledge of tunnels and caves on the islands to shelter from the cyclone.

The country's precarious peace process suffered a major reversal in February 2003 when a leading member of the National Peace Council, Sir Frederick Soaki, was assassinated. The motive for the killing was not immediately apparent, although there was speculation that it might have been connected to his involvement in a 'demobilization' programme for 'special constables'. A report by the human rights organization, Amnesty International, in March 2003 claimed that the country's 'special constables' had tortured and killed numerous people in the operation against Harold Keke along the southern coast of Guadalcanal. It was reported that a total of 861 'special constables' had been 'demobilized' by mid-July 2003 under the programme. In April three of Keke's close associates deserted him and reported that they had witnessed him carry out nine murders in recent weeks. The killings were mostly thought to have involved his own supporters whom he had suspected of collaborating with the police. In May it was reported that Keke had taken six missionaries hostage in southern Guadalcanal, all of whom were subsequently killed. In a separate incident on Malaita, in the same month, an Australian missionary was beheaded.

In early June 2003 Kemakeza travelled to Canberra for a meeting with the Australian Prime Minister, John Howard, at which he requested direct foreign intervention in order to halt the islands' worsening law and order crisis. (Former Prime Minister Manasseh Sogavare had made a similar appeal for Australia to send troops to the country at the time of the signing of the Townsville Peace Agreement in October 2000, but the request had been rejected.) In mid-June senior officials from Australia and New Zealand arrived in Solomon Islands to assess the possibility of mounting a regional intervention in the country. At a meeting of the Pacific Islands Forum in Sydney, Australia, later that month, delegates from the 16 nations agreed unanimously to send a multinational intervention force to Solomon Islands; eight island members stated their intention to commit personnel to the force. The proposed action was to constitute the largest armed intervention in the South Pacific since the Second World War.

Meanwhile, reports of violence and intimidation by Harold Keke and his rebel forces continued during June 2003. At least 23 people (including 11 members of a religious order) were taken hostage and several more were killed. An estimated 1,000 villagers fled their homes after Keke took control of a police post, thereby expanding the area under his control. The rebel leader also burned two villages where he believed that local people had informed the authorities of his activities. Furthermore, in the same month his militia forced some 1,200 people at gun-point to stand along several stretches of beach in order to serve as a human shield and thereby prevent a planned police landing in the area. Atrocities continued in the following month when another settlement (of some 500 people) was burned on Keke's orders and a number of people (including several children) were beaten to death or beheaded.

Discussions continued throughout July 2003 regarding the nature of the proposed regional intervention force and its role in the country. The Australian Government presented a document to Kemakeza stating its requirements for the intervention to proceed, which included unhindered access to the country's financial records and the appointment of up to 100 foreign nationals to senior positions in the islands' public service and government sectors. The economic component of the proposed intervention would also provide for the payment of Solomon Islands' domestic and foreign debts through increased aid from Australia and New Zealand. Displaced villagers (estimated to total 20,000 since ethnic violence intensified in 2000) were to receive a specific allocation of $A100,000 in aid from Australia. On 10 July the Solomon Islands Government unanimously approved legislation to allow the Australian-led force into the country. The Australian Cabinet approved the deployment on 22 July, and two days later the warship HMAS *Manoora* arrived in Honiara with 400 personnel. Howard had said, in relation to the intervention, that it was 'not in Australia's interests to have a number of failed states in the Pacific', which led some commentators to speculate that Australia was acting out of fears for its own security rather than through concern for the people of Solomon Islands. In the following days a total of 2,225 troops and police arrived in the islands from various other countries including New Zealand, Papua New Guinea, Fiji, Tonga, Vanuatu and Samoa.

In early August 2003 the UN Secretary-General, Kofi Annan, commended the Pacific island countries for their efforts to support Solomon Islands and stated that the UN was prepared to contribute actively to any future peace process in the country. On 13 August Harold Keke surrendered and was arrested, along with 10 of his close associates. Thousands of villagers gathered to watch as Keke handed over a large number of firearms and was taken away on HMAS *Manoora* for his own safety. Two days later the Malaita Eagle Force surrendered around 100 weapons in a decommissioning ceremony presided over by Nick Warner, the senior diplomat in charge of the multinational force. Later that month a 17-member 'economic assistance team' arrived in Solomon Islands, some of whom were to assume key roles in government and the public service in order to implement major reforms. The Australian Prime Minister, John Howard, also made a one-day visit to the islands. By early September the regional forces announced that they had collected a total of 3,850 weapons since the start of a firearms amnesty a month earlier. Australia began to withdraw its troops in late October and by early December more than one-half of its personnel had left the islands. It was announced in late August that some 100 Australian troops would remain in the country to support the work of the local civilian police force.

The capture and arrest of members of various militia groups, including two individuals who had been signatories to the Townsville Peace Agreement, continued during late 2003. Two senior members of the Malaita Eagle Force, including Police Superintendent Mannaseh Maelanga, who had served as its Supreme Commander, were arrested in November. Maelanga was sentenced to one year's imprisonment in March 2004. Meanwhile, Andrew Te'e, former Supreme Commander of the Isatabu Freedom Movement, was arrested and charged with the murder of three people. Moreover, the trial of Keke and three of his closest associates on multiple charges of murder and abduction began in February 2004.

The Government was subject to considerable embarrassment in December 2003 when the Minister for Communication, Aviation and Meteorology, Daniel Fa'afunua, was arrested and charged with the assault of his wife and of a female police officer, with being drunk and disorderly and with demanding money with menaces. The former minister was found guilty in February 2004 and received a lengthy prison sentence.

In June 2004 Nathaniel Waena, the former Minister for National Unity, Reconciliation and Peace, was elected Governor-General with 27 parliamentary votes and in the following month was duly sworn in.

In 1990 relations between Solomon Islands and Papua New Guinea deteriorated, following allegations by the Solomon Islands' Government that patrol boats from Papua New Guinea were interfering with the traditional crossing between Bougainville Island (Papua New Guinea) and the Shortland Islands, while the Papua New Guinea Government accused Solomon Islands of harbouring members of the rebel Bougainville Revolutionary Army (BRA) and of providing them with supplies. Despite the signing that year of an agreement on joint border surveillance and arrangements to host peace negotiations between the BRA and the Papua New Guinea Government, relations worsened considerably in 1992 when Papua New Guinea forces carried out several unauthorized incursions into the Shortland Islands, in which a fuel depot was destroyed and two Solomon Islanders were killed. Alleging Australian involvement in the incursions, Mamaloni suspended surveillance flights by the Australian air force over its territory, and relations between the two countries deteriorated significantly. Despite the initiation of discussions between Solomon Islands and Papua New Guinea in January 1993, further incursions were reported in April. Following the election of a new Government in May in Solomon Islands, however, relations appeared to improve, and subsequent negotiations between the two countries resulted in an agreement to close the BRA office in Honiara. Tensions with Papua New Guinea, however, increased in 1996 as violence on Bougainville intensified. Numerous incursions by Papua New Guinea defence forces into Solomon Islands' waters were reported, while the Papua New Guinea Government repeated accusations that Solomon Islands was harbouring BRA activists. However, in June 1997 Papua New Guinea and Solomon Islands concluded a maritime border agreement, following several years of negotiations. The purpose of the agreement was not only to delineate the sea boundary between the two countries but also to provide a framework for co-operation in matters of security, natural disaster, customs, quarantine, immigration and conservation. In December 1997 the Prime Ministers of the two countries paid an extended visit to Bougainville to express support for the recently established truce agreement. Furthermore, the Governor of Bougainville was sympathetic to the problem of increasing numbers of Solomon Islanders from the Western Province crossing to Bougainville in late 2001. Many were trading goods in Bougainville in exchange for food and services. Discussions regarding the border of Papua New Guinea were held in April 2002; both Governments were concerned about the increase of weapons trafficking from Bougainville to the Western Province. Renewed border discussions took place between the two countries in June 2003.

In March 1988 the 'Spearhead Group' was formed after talks between the Governments of Papua New Guinea, Vanuatu and Solomon Islands. The principal aims of the new group were to preserve Melanesian cultural traditions and to campaign for independence for the French Overseas Territory of New Caledonia. In March 1990, at a meeting in Honiara, the Melanesian Spearhead Group (MSG) admitted the FLNKS (the main Kanak, or Melanesian, political group in New Caledonia). In mid-1994 the group concluded an agreement regarded as the first step towards the establishment of a free-trade area by the three countries. Fiji was admitted to the group in mid-1996. Solomon Islands announced its commitment to further economic integration between the MSG countries in late 1997. In 1995 Solomon Islands became a signatory to the Federated States of Micronesia Agreement on Regional Fisheries Access. In early 1997 Solomon Islands and Vanuatu agreed to undertake negotiations on the maritime boundaries between the two countries in an attempt to clarify uncertainty regarding fishing rights.

Solomon Islands is a member of the Pacific Islands Forum (formerly the South Pacific Forum) and supported moves by that organization in 1988 and 1989 to ban drift-net fishing in the region, particularly by Japanese, Taiwanese and Korean fleets. Japan ceased its drift-net operations in 1992.

Solomon Islands attracted international criticism in mid-2003 for its practice of capturing and exporting large numbers of live dolphins. Police mounted a large-scale security operation around Honiara airport to prevent journalists from filming some 200 dolphins being loaded into a cargo aeroplane bound for Mexico. Several reports of harassment and violence against foreign journalists were received. It was believed that, owing to the high price commanded by the sale of the animals, senior Solomon Islands officials were likely to be involved. The trade in wild dolphins has been prohibited by most developed countries under a Convention on International Trade in Endangered Species of Wild Flora and Fauna. The trade was not only a cause for concern among foreigners, however, and in late 2003 one of the country's important traditional houses of chiefs, the Gela, expressed disquiet over the treatment that the animals had received.

A seven-member parliamentary delegation, including five cabinet ministers, travelled to Taiwan in September 2004 to discuss mutual co-operation. One of the most important issues to be debated was the review of Taiwan's annual aid programe of US $10.8m. to Solomon Islands.

Economy

In 2002, according to estimates by the World Bank, Solomon Islands' gross national income (GNI), measured at average 2000–02 prices, was US $254m., equivalent to US $570 per head (or US $1,520 on an international purchasing-power parity basis). During 1990–2002, it was estimated, the population increased at an annual average rate of 2.8%, while gross domestic product (GDP) per head decreased, in real terms, at an average annual rate of 2.9%. Overall GDP decreased, in real terms, at an average annual rate of 0.2% in 1990–2002. According to the Asian Development Bank (ADB) GDP declined by 14.3% in 2000, mainly because of the climate of political uncertainty following an upsurge in civil unrest and ethnic tension. According to revised figures from the ADB, GDP remained constant in 2001, but rose by 6.6% in 2002 and by 11.8% in 2003. It was projected that GDP would increase by 4.2% in 2004 and by a similar figure in 2005.

Agriculture (including hunting, forestry and fishing) contributed 47.7% of GDP (measured at constant 1985 prices) in 2002. The ADB estimated that, compared with the previous year, agricultural GDP declined by 43.0% in 2000 and by 12.6% in 2001. In 2002 an estimated 72.3% of the working population were involved in agriculture. The main subsistence crops are coconuts, sweet potatoes, taro, yams, cassava, rice, garden vegetables and fruit. Livestock development has concentrated on the strengthening of support for cattle, pig and poultry farmers. Spices, particularly chillies, are cultivated for export on a small scale, while in the early 1990s the production of honey became important.

The principal commercial agricultural product was traditionally copra, which was for many years the country's main export. Earnings from copra exports, however, declined from SI $34.7m. in 2000 to only SI $0.4m. in 2001, although revenue recovered somewhat to reach SI $7.8m in 2003. The operations of the islands' largest agricultural business, Solomon Islands Plantations Ltd (in which the Government purchased a stake in the late 1970s), were suspended in 2000 as a result of the civil disturbances, seriously affecting the country's palm oil exports. High prices on the world market for cocoa in the early 2000s, however, increased the value of exports of that commodity, and in April 2003 the Government announced plans to reopen the palm oil plant.

Besides being a traditional subsistence activity, the fishing sector accounted for 16.7% of total export earnings in 2003 (compared with 11.7% in 2000). By the late 1970s the Solomon Islands Government had been enabled to buy shares in the Japanese skipjack tuna fishing, freezing and canning company, Solomon Taiyo Ltd, operating in the islands. The Government also had its own fishing project, the National Fisheries Development Co (this was sold to Canadian interests in 1990). In 1997 the sale of licences to foreign fishing vessels (mainly from Japan and Taiwan) amounted to some SI $9.5m. Other maritime industries include the culture of marine prawns, seaweed and sea shells. The country's first shipment of seaweed was exported to France in early 2004. Giant clam farming also became an important activity in the mid-1990s, and by December 1996 a ban on the sale and export of giant clams was introduced in an attempt to conserve stocks. The fishing sector was badly affected from the late 1990s by the consequences of civil unrest on the islands and by low international prices for the product.

The forestry sector is an important source of export revenue. However, a dramatic increase in the production of timber in the early 1990s prompted several international organizations, including the World Bank, to express alarm at the current rate of logging in the country. By the late 1990s Solomon Islands was one of the few remaining countries in the world to allow the export of round logs. However, a dramatic decline in the price of timber on the international market, together with a decrease in demand for round logs, led to a crisis in the industry in 1997, which the recently elected Government aimed to address with a new logging policy (see History). Roundwood removals were estimated at 714,000 cu m in 2003, compared with 550,000 cu m in the previous year. Export earnings from forestry amounted to SI $371.4m. in 2003 (contributing 66.7% of total export earnings). It was estimated in the early 2000s that, if logging continued at current levels, all remaining virgin forest in Solomon Islands would be removed before 2015.

Industry (including mining, manufacturing, construction and power) contributed 8.2% of GDP in 2002 (at 1985 prices), and employed 13.7% of wage-earners in 1993. The GDP of the industrial sector expanded by an average annual rate of 5.0% in 1990–2000. The ADB estimated that industrial GDP declined by 49.2% in 2000 and by 33.2% in 2001.

A heavily mineralized area at Betilonga and in the Sutakiki valley, on Guadalcanal, has been investigated for gold, silver and copper, and there have been surveys of phosphate deposits, estimated at 10m. metric tons, on Bellona Island and of deposits of asbestos at Kumboro, on Choiseul, and high-grade bauxite on Rennell and Vaghena Islands. Deposits of lead, zinc and cobalt, as well as nickel deposits of some 25m. tons of ore on San Jorge and Isabel islands, have also been discovered. Small amounts of gold (which is mostly produced by alluvial mining) are exported and earned SI $1.3m., for 50 kg, in 1991, although this declined to SI $0.6m. for 2 kg in 1997. The discovery of significant deposits of gold on Vangunu Island, and of further deposits on Guadalcanal, were reported in 1996. The gold-mining project at Gold Ridge (Guadalcanal) began operations in September 1998, with a projected yield of 100,000 oz per year for 13 years. The forecast output for 1999/2000 was 130,000 oz. However, IFM dissidents (see History) occupied the mine in June 2000, forcing operations to be suspended. Further prospecting for gold took place in the late 1990s at Mase, New Georgia.

The islands' transport and energy facilities are seriously inadequate, which hampers agricultural and economic development. Energy is derived principally from hydroelectric power, although solar energy is fairly widely, and increasingly, utilized, particularly in rural areas. Solomon Islands policy is to reduce dependence upon imported mineral fuels, which nevertheless accounted for 17.3% of total import costs in 2003. During the 1990s exploratory projects revealed several potential petroleum-bearing sites in the islands.

Service industries contributed 44.1% of GDP (at 1985 prices) in 2002. The ADB estimated that the services sector contracted by 26.0% in 2000 and by 4.6% in 2001. Tourism remains relatively undeveloped in Solomon Islands. In 2003 an estimated 1,718 tourists visited the islands. Earnings from the sector were estimated at some US $13m. in 1998 and declined to US $6m. in 1999. Training programmes, investment, airport expansion and the proposed designation of Marovo (New Georgia) and Rennell Island as World Heritage sites (see under UNESCO) were expected to benefit the tourist industry. Most important, however, was the Australian-led intervention in 2003 (see History) which resulted in the restoration of a measure of stability to troubled areas of the country and which was expected to have a marked beneficial effect on tourism in the islands.

In 2003 Solomon Islands recorded a visible trade surplus of US $4.2m., and a surplus of US $27.2m. on the current account of the balance of payments. In 2003 the principal source of imports was Australia (providing 28.1% of the total), while the principal market for exports was the People's Republic of China (purchasing 25.5% of the total). Other major trading partners are Japan (18.3% of exports), Singapore (23.6% of imports), the Republic of Korea (14.2% of exports), Thailand and the Philippines. The principal exports in 2003 were timber, fish and other marine products, and cocoa. The principal imports were foodstuffs, mineral fuels, machinery and transport equipment and basic manufactures.

In 2002 there was an estimated budgetary deficit of SI $139.1m. According to the ADB, the fiscal deficit declined from the equivalent of 11.1% of GDP in 2002 to a projected 0.3% in 2003. Budgetary expenditure of SI $480m. was approved for 2004, of which about one-quarter was to be financed by grants from Australia and New Zealand. Overseas aid totalled US $68.4m. in 2000. Aid from Australia totalled $A37.4m. in 2003/04. In the same year financial assistance from New Zealand totalled $NZ14.0m. The total external debt was US $180.4m. in 2002, of which US $175.5m. was long-term public debt. In 2000 the cost of debt-servicing was equivalent to 6.9% of the value of exports of goods and services. The annual rate of inflation in Honiara averaged 9.6% in 1990–2002. The rate of inflation in Honiara was estimated by the ADB at 9.3% in 2002 and at 10.0% in 2003.

The economic development of Solomon Islands has been impeded by inadequate transport facilities, inclement weather and by fluctuations in prices on the international market for the major agricultural exports. Increasing environmental concern over the exploitation of the country's natural resources and over its vulnerability to unscrupulous foreign operators has continued to threaten the islands' economic stability and the country's ability to sustain itself in the future. Furthermore, the dependence of Solomon Islands upon forestry—logging being the prime constituent of both taxation and export revenues—rendered the economy vulnerable to the low international prices for round logs prevailing in 2001. The country's logging industry attracted renewed international attention in early 2004 when a chain of islands in the Western Solomons (one of which was the site of a World Bank-funded conservation project) was devastated by intensive logging. A number of foreign companies were believed to have arrived in the islands and begun operations without undertaking any environmental impact studies or negotiations with villagers and landowners. Reports indicated that some islands had been completed deforested within a matter of weeks and that coral reefs had been destroyed by heavy machinery in the area. The economy was severely affected by ethnic unrest in Guadalcanal, which intensified in 1999. Subsequent housing and welfare payments to displaced victims of the unrest became a severe drain on domestic finance, forcing the Government into increased foreign borrowing and unsustainable drawdowns from the Central Bank of Solomon Islands. Moreover, it was estimated that the conflict had cost the country a very significant amount of revenue in lost tourism receipts. During 2000 a total of 8,000 workers, or 15% of the total work-force, were estimated to have lost their jobs or to have been dispatched on unpaid leave. Youth unemployment, meanwhile, remained a particular problem.

In February 2001 the Central Bank of Solomon Islands reportedly declared that the country faced economic and social disaster, stating that the country's GDP had contracted by 19% within one year, and that exports had declined by 40% in six months, while domestic borrowing, particularly by the Government, had notably increased. The fiscal position deteriorated sharply in 2001. Payments to public servants, along with the transfer of funds to provincial health and education services, were delayed and the Government defaulted on various external and domestic debts. Furthermore, the dramatic rise in crime, which in part had occurred as a result of the alleged infiltration of the police force by rebel groups (see History), appeared to be a deterrent to growth and investment. In January 2002 the Central Bank announced that the country's total debt amounted to some US $200m., while its foreign reserves had fallen to less than US $18m. The Government subsequently announced plans to reduce expenditure by up to 50%, which would reportedly result in about one-third of public employees losing their jobs. In October the Government reduced the number of ministries from 20 to 10. In July 2003 an Australian-led regional intervention force arrived in the country to restore law and order (see History). The intervention, which had been requested by the Government of Solomon Islands, included an economic recovery programme involving increased assistance from Australia (totalling some $A1,000m. over 10 years) and the appointment of numerous Australian officials to key posts within the Government and public service and finance sectors. It was hoped that the country's potential for prolonged political stability, improved inter-ethnic relations and consequently enhanced social and economic conditions would be greatly assisted by the measures agreed with Australia.

Solomon Islands, in addition to being a member of the Pacific Islands Forum (formerly the South Pacific Forum—see above), has membership of the Pacific Community (formerly the South Pacific Commission) and the UN's Economic and Social Commission for Asia and the Pacific (ESCAP). The country is a member of the Asian Development Bank (ADB), and is a signatory of the Lomé Conventions and of the successor Cotonou Agreement with the European Union.

Statistical Survey

Source (unless otherwise stated): Statistics Office, POB G6, Honiara; tel. 23700; fax 20392.

AREA AND POPULATION

Area: 27,556 sq km (10,639 sq miles).

Population: 285,176 at census of 23–24 November 1986; 409,042 (males 211,381, females 197,661) at census of 21–22 November 1999. *Mid-2003* (UN estimate): 477,000 (Source: UN, *World Population Prospects: The 2002 Revision*).

Density (mid-2003): 17.3 per sq km.

Ethnic Groups (census of November 1986): Melanesians 268,536; Polynesians 10,661; Micronesians 3,929; Europeans 1,107; Chinese 379; Others 564.

Principal Towns (population in '000, 2000): Honiara (capital) 50.1; Gizo 7.0; Auki 5.0. Source: Stefan Helders, *World Gazetteer* (internet www.world-gazetteer.com). *Mid-2003* (UN estimate, incl. suburbs) Honiara 56,339 (Source: UN, *World Urbanization Prospects: The 2003 Revision*).

Births and Deaths (2001): Birth rate 38.3 per 1,000; Death rate 4.8 per 1,000. Source: UN, *Statistical Yearbook for Asia and the Pacific*.

Expectation of Life (WHO estimates, years at birth): 65.4 (males 63.6; females 67.4) in 2002. Source: WHO, *World Health Report*.

Employment (employees only, June 1993): Agriculture, hunting, forestry and fishing 8,106; Manufacturing (incl. mining and quarrying) 2,844; Electricity and water 245; Construction 977; Trade, restaurants and hotels 3,390; Transport, storage and communications 1,723; Finance, insurance, real estate and business services 1,144; Community, social and personal services 11,148; *Total* 29,577. *1996:* Total employed 34,200 (Source: UN, *Statistical Yearbook for Asia and the Pacific*). *Mid-2002* (estimates, '000): Agriculture, etc. 172; Total labour force 238 (Source: FAO).

HEALTH AND WELFARE

Key Indicators

Total Fertility Rate (children per woman, 2002): 4.5.

Under-5 Mortality Rate (per 1,000 live births, 2002): 24.

Physicians (per 1,000 head, 1995): 0.14.

Hospital Beds (per 1,000 head, 1992): 2.75.

Health Expenditure (2001): US $ per head (PPP): 133.

Health Expenditure (2001): % of GDP: 5.0.

Health Expenditure (2001): public (% of total): 93.5.

Access to Water (% of persons, 2000): 71.

Access to Sanitation (% of persons, 2000): 34.

Human Development Index (2002): ranking: 124.

Human Development Index (2002): value: 0.624.

For sources and definitions, see explanatory note on p. vi.

AGRICULTURE, ETC.

Principal Crops (FAO estimates, '000 metric tons, 2002): Coconuts 330; Oil palm fruit 155; Rice 5; Cocoa beans 3; Sweet potatoes 86; Yams 29; Taro 38; Vegetables and melons 8; Fruit 19. Source: FAO.

Livestock (FAO estimates, '000 head, year ending September 2002): Cattle 13; Pigs 68. Source: FAO.

Livestock Products (FAO estimates, '000 metric tons, 2002): Beef and veal 0.6; Pig meat 2.2. Source: FAO.

Forestry (FAO estimates, '000 cu m, 2002): *Roundwood Removals* (excl. bark): Industrial wood 554; Fuel wood 138; Total 692; *Sawnwood Production:* 12 (all broadleaved, incl. railway sleepers). Source: FAO.

Fishing (FAO estimates, metric tons, live weight, 2002): Skipjack tuna 13,897; Yellowfin tuna 3,913; Total catch (all capture, incl. others): 31,003. Source: FAO.

MINING

Production (kilograms, 2000): Gold 338; Silver 200 (estimate). Source: US Geological Survey.

INDUSTRY

Production (metric tons, 2002, unless otherwise stated): Copra 15,000 (2003); Coconut oil 15,000; Palm oil 34,000 (FAO estimates); Frozen fish 9,200 (2001); Electric energy 57 million kWh.

FINANCE

Currency and Exchange Rates: 100 cents = 1 Solomon Islands dollar (SI $). *Sterling, US Dollar and Euro Equivalents* (31 December 2003): £1 sterling = SI $13.3685; US $1 = SI $7.4906; €1 = SI $9.4607; SI $100 = £7.48 = US $13.35 = €10.57. *Average Exchange Rate* (SI $ per US $): 5.2780 in 2001; 6.7488 in 2002; 7.5059 in 2003.

Budget (SI $ million, 2003): *Revenue*: Taxes 340.4; Other current revenue 33.0; Total 373.4; *Expenditure*: Total 682.8 (Current 564.2; Capital 118.6). Source: Asian Development Bank, *Key Indicators of Developing Asian and Pacific Countries*.

Official Development Assistance (US $ million, 2000): Bilateral 22.1; Multilateral 46.3; Total 68.4 (Grants 69.7, Loans –1.3). Source: UN, *Statistical Yearbook for Asia and the Pacific*.

International Reserves (US $ million at 31 December 2003): IMF special drawing rights 0.00; Reserve position in IMF 0.82; Foreign exchange 36.39; Total 37.20. Source: IMF, *International Financial Statistics*.

Money Supply (SI $ million at 31 December 2003): Currency outside banks 101.43; Demand deposits at deposit money banks 230.09; *Total money*331.52. Source: IMF, *International Financial Statistics*.

Cost of Living (Consumer Price Index for Honiara; base: 1992 = 100): 223.3 in 2001; 244.2 in 2002; 268.5 in 2003. Source: IMF, *International Financial Statistics*.

Gross Domestic Product (SI $ million at current prices): 1,453.4 in 2000; 1,453.9 in 2001; 1,527.6 in 2002 (Source: IMF, *International Financial Statistics*).

GDP by Economic Activity (official estimates, SI $ million at constant 1984/85 prices, 2002): Agriculture 124.2; Mining 0.1; Manufacturing 10.6; Electricity, gas and water 4.6; Construction 6.1; Trade 27.2; Transport and communications 11.2; Finance 15.2; Other services (including public administration) 61.2; *GDP at factor cost* 260.4. Source: Asian Development Bank, *Key Indicators of Developing Asian and Pacific Countries*.

Balance of Payments (US $ million, 2003): Exports of goods f.o.b. 74.17; Imports of goods f.o.b. –69.97. *Trade balance* 4.20; Exports of services and income 28.77; Imports of services and income –59.07; *Balance on goods, services and income* –26.09; Current transfers received 72.13; Current transfers paid –18.87; *Current balance* 27.17; Capital account (net) 12.52; Net financial account (obtained as residual) –17.29; Net errors and omissions –3.67; *Overall balance* 18.73. Source: Asian Development Bank, *Key Indicators of Developing Asian and Pacific Countries*.

EXTERNAL TRADE

Principal Commodities (SI $ '000, 2003): *Imports c.i.f.*: Food and live animals 112,297; Beverages and tobacco 6,937; Crude materials (inedible) except fuels, mineral fuels, lubricants, etc., and animal and vegetable oils and fats 87,885; Chemicals 14,102; Basic manufactures 54,051; Machinery and transport equipment 64,459; Miscellaneous manufactured articles and other commodities and transactions 167,272; Total 557,013. *Exports f.o.b.*: Fish 92,869; Copra 7,821; Timber 371,394; Cocoa 53,186; Total (incl. others) 507,004.

Principal Trading Partners (US $ million, 2003): *Imports*: Australia 33.8; Fiji 5.3; Hong Kong 2.2; Japan 3.6; New Zealand 6.1; Papua New Guinea 5.1; Singapore 28.4; USA 2.1; Total (incl. others) 120.1. *Exports*: China, People's Republic 29.7; India 3.1; Japan 21.3; Korea, Republic 16.5; Malaysia 2.5; Philippines 11.1; Thailand 7.4; Total (incl. others) 116.4.

Source: Asian Development Bank, *Key Indicators of Developing Asian and Pacific Countries*.

TRANSPORT

Road Traffic (motor vehicles in use at 30 June 1986): Passenger cars 1,350; Commercial vehicles 2,026.

Shipping (international traffic, '000 metric tons, 1990): Goods loaded 278; Goods unloaded 349 (Source: UN, *Monthly Bulletin of Statistics*). *Merchant Fleet* (registered at 31 December 2003): Vessels 27; Total displacement ('000 grt) 7.4 (Source: Lloyd's Register-Fairplay, *World Fleet Statistics*).

Civil Aviation (traffic on scheduled services, 1999): Passengers carried 98,000; Passenger-km 80 million. Source: UN, *Statistical Yearbook*.

TOURISM

Foreign Tourist Arrivals: 1,245 in 2001; 1,658 in 2002; 1,718 in 2003. Source: Solomon Islands Visitors' Bureau.

Tourist Arrivals by Country (1999, incl. cruise-ship passengers): Australia 4,979; China, People's Republic 1,347; Fiji 729; Japan 887; Korea, Republic 1,237; New Zealand 925; Papua New Guinea 1,457; Philippines 2,102; Taiwan 478; United Kingdom 893; USA 1,589; Total (incl. others) 21,318. Source: World Tourism Organization, *Yearbook of Tourism Statistics*.

Tourism Receipts (US $ million): 16 in 1997; 13 in 1998; 6 in 1999. Source: World Tourism Organization.

COMMUNICATIONS MEDIA

Non-daily Newspapers (1996): 3; estimated circulation 9,000.

Radio Receivers (1997): 57,000 in use.

Television Receivers (2001): 12,000 in use.

Telephones (2002): 6,600 main lines in use.

Mobile Cellular Telephones (2002): 1,000 subscribers.

Personal Computers (2002): 18,000 in use.

Internet Users (2002): 2,200.

Facsimile Machines (1999): 764 in use.

Sources: UNESCO, *Statistical Yearbook*; International Telecommunication Union.

EDUCATION

Pre-primary (1994): 12,627 pupils.

Primary (1994): 523 schools (1993); 2,514 teachers; 60,493 pupils.

Secondary: 23 schools (1993); 618 teachers (1994); 7,981 pupils (1995).

Overseas Centres (1988): 405 students.

Source: mainly UNESCO, *Statistical Yearbook*.

Adult Literacy Rate (estimate based on census data): 76.6% in 2002. Source: UN Development Programme, *Human Development Report*.

Directory

The Constitution

A new Constitution came into effect at independence on 7 July 1978.

The main provisions are that Solomon Islands is a constitutional monarchy with the British sovereign (represented locally by a Governor-General, who must be a Solomon Islands citizen) as Head of State, while legislative power is vested in the unicameral National Parliament composed of 50 members (increased from 47 in 1997), elected by universal adult suffrage for four years (subject to dissolution), and executive authority is exercised by the Cabinet, led by the Prime Minister. The Governor-General is appointed for up to five years, on the advice of Parliament, and acts in almost all matters on the advice of the Cabinet. The Prime Minister is elected by and from members of Parliament. Other ministers are appointed by the Governor-General, on the Prime Minister's recommendation, from members of Parliament. The Cabinet is responsible to Parliament. Emphasis is laid on the devolution of power, and traditional chiefs and leaders have a special role within these arrangements. Legislation approved in August 1996 provided for the abolition of the provincial government system and the transfer of legislative and administrative powers from the nine provincial governments to 75 area assemblies and councils controlled by central Government.

The Constitution contains comprehensive guarantees of fundamental human rights and freedoms, and provides for the introduction of a 'leadership code' and the appointment of an

Ombudsman and a Public Solicitor. It also provides for 'the establishment of the underlying law, based on the customary law and concepts of the Solomon Islands people'. Solomon Islands citizenship was automatically conferred on the indigenous people of the islands and on other residents with close ties with the islands upon independence. The acquisition of land is reserved for indigenous inhabitants or their descendants.

In mid-1999 it was announced that two review committees had been established to amend the Constitution. They were expected to examine ways in which the traditions of the various ethnic groups could be better accommodated.

The Government

HEAD OF STATE

Monarch: HM Queen ELIZABETH II.

Governor-General: NATHANIEL WAENA (sworn in 7 July 2004).

THE CABINET
(September 2004)

Prime Minister: Sir ALLAN KEMAKEZA.

Deputy Prime Minister and Minister for National Planning and Human Resource Development: SNYDER RINI.

Minister for State, assisting the Prime Minister: JOHN GARO.

Minister for Agriculture and Livestock: ALEX BARTLETT.

Minister for Commerce, Employment and Trade: WALTON NAEZON.

Minister for Economic Reform and Structural Adjustment: (vacant).

Minister for Education and Training: MATHIAS TARO.

Minister for Finance: FRANCIS ZAMA.

Minister for Fisheries and Marine Resources: PAUL MAENU'U.

Minister for Foreign Affairs and Trade Relations: LAURIE CHAN.

Minister for Forestry, Environment and Conservation: DAVID HOLISIVI.

Minister for Health and Medical Services: BENJAMIN UNA.

Minister for Home Affairs: NELSON KILE.

Minister for Lands and Surveys: SIRIAKO USA.

Minister for Mines and Energy: STEPHEN PAENI.

Minister for National Unity, Reconciliation and Peace: AUGUSTINE TANEKO.

Minister for Police, Justice and National Security: MICHAEL MAINA.

Minister for Provincial Government and Rural Development: CLEMENT ROJUMANA.

Minister for Tourism and Culture: TREVOR OLAVAE.

Minister for Transport, Works and Communication: BERNARD GIRO.

Minister for Youth and Sports and Women's Affairs: (vacant).

MINISTRIES

Office of the Prime Minister: POB G1, Honiara; tel. 21867; fax 26088.

Ministry of Agriculture and Lands: POB G13, Honiara; tel. 21327; fax 21955; e-mail drsteve@solomon.com.sb.

Ministry of Education and Human Resource Development: POB G28, Honiara; tel. 23900; fax 20485.

Ministry of Finance and Treasury: POB 26, Honiara; tel. 22535; fax 20392; e-mail finance@welkam.solomon.com.sb.

Ministry of Foreign Affairs, Commerce and Tourism: POB G10, Honiara; tel. 2476; fax 20351; e-mail commerce@commerce.gov.sb; internet www.commerce.gov.sb.

Ministry of Health and Medical Services: POB 349, Honiara; tel. 20830; fax 20085.

Ministry of Infrastructure Development: POB G30, Honiara; tel. 38255; fax 38259; e-mail kudu@mnpd.gov.sb.

Ministry of Natural Resources: POB G24, Honiara; tel. 25848; fax 21245; e-mail kdfmp@welkam.solomon.com.sb.

Ministry of Police, Justice and National Security: POB 1723/404, Honiara; tel. 22208; fax 25949.

Ministry of Provincial Government, Home Affairs, National Unity, Reconciliation and Peace: POB G35, Honiara; tel. 21140; fax 21289.

Legislature

NATIONAL PARLIAMENT

POB G19, Honiara; tel. 21751; fax 23866.

Speaker: Sir PETER KENILOREA.

General Election, 5 December 2001

Party	Seats
Solomon Islands Alliance for Change	12
People's Alliance Party	9
People's Progressive Party	6
Labour Party	1
Independents	22
Total	50

Political Organizations

Parties in the National Parliament can have a fluctuating membership and an influence disproportionate to their representation. There is a significant number of independents who are loosely associated in the amorphous, but often decisive, 'Independent Group'. The following parties represent the main groupings:

National Action Party of Solomon Islands (NAPSI): Honiara; f. 1993; Leader FRANCIS SAEMALA.

National Party: Honiara; f. 1996; Pres. EZEKIEL ALEBUA.

People's Alliance Party (PAP): Honiara; f. 1979 by a merger of the People's Progressive Party (f. 1973) and the Rural Alliance Party (f. 1977); advocates the establishment of a federal republic; Leader Sir ALLAN KEMAKEZA; Sec. EDWARD KINGMELE.

People's Progressive Party (PPP): Honiara; f. 1973; latterly Coalition for National Advancement, reverted to original name in January 2000; Leader MANASSEH SOGAVARE; Pres. JOB DUDLEY TAUSINGA.

Solomon Islands Alliance for Change Coalition (SIACC): Honiara; f. 1997; formed from the previous Solomon Islands Alliance for Change (SIAC); Co-Leaders PATTERSON OTI, FRANCIS BILLY HILLY.

Solomon Islands Labour Party: Honiara; f. 1988; formed from the Solomon Islands Trade Union movement; Leader JOSES TUHANUKU; Gen. Sec. TONY KAGOVAI.

Solomon Islands Liberal Party (SILP): Honiara; f. 1976 as the National Democratic Party (NADEPA); present name adopted in 1986; Leader BARTHOLOMEW ULUFA'ALU.

United Democratic Party (UDP): f. 1980; fmrly United Solomon Islands Party; present name adopted in 2000; Pres. PETER KENILOREA; Chair. JOHN MAETIO.

Diplomatic Representation

EMBASSIES AND HIGH COMMISSIONS IN SOLOMON ISLANDS

Australia: Hibiscus Ave, POB 589, Honiara; tel. 21561; fax 23691; High Commissioner ROBERT DAVIS.

China (Taiwan): Pantina Plaza, POB 586, Honiara; tel. 38050; fax 38060; Ambassador PEI Y. TENG.

Japan: National Provident Fund Bldg, Mendana Ave, POB 560, Honiara; tel. 22953; fax 21006; Chargé d'affaires YUTAKA HIRATA.

New Zealand: Mendana Ave, POB 697, Honiara; tel. 21502; fax 22377; e-mail nzhicom@.solomon.com.sb; High Commissioner BRIAN SANDERS.

Papua New Guinea: POB 1109, Honiara; tel. 20561; fax 20562; High Commissioner PONABE YUWA.

United Kingdom: Telekom House, Mendana Ave, POB 676, Honiara; tel. 21705; fax 21549; e-mail bhc@solomon.com.sb; High Commissioner BRIAN BALDWIN.

Judicial System

The High Court is a Superior Court of Record with unlimited original jurisdiction and powers (except over customary land) as prescribed by the Solomon Islands Constitution or by any law for the time being in force in Solomon Islands. The Judges of the High Court are the Chief Justice, resident in Solomon Islands and employed by its Government, and the Puisne Judges (of whom there are usually three). Appeals from this Court go to the Court of Appeal, the members of which are senior judges from Australia, New Zealand

and Papua New Guinea. The Chief Justice and judges of the High Court are *ex officio* members of the Court of Appeal.

In addition there are Magistrates' Courts staffed by qualified and lay magistrates exercising limited jurisdiction in both civil and criminal matters. There are also Local Courts staffed by elders of the local communities, which have jurisdiction in the areas of established native custom, petty crime and local government by-laws. In 1975 Customary Land Appeal Courts were established to hear land appeal cases from Local Courts, which have exclusive original jurisdiction over customary land cases.

Office of the Registrar

High Court and Court of Appeal, POB G21, Honiara; tel. 21632; fax 22702; e-mail chetwynd@welkam.solomon.com.sb.

President of the Court of Appeal: Lord GORDON SLYNN.

Chief Justice of the High Court: ALBERT ROCKY PALMER.

Puisne Judges: ALBERT ROCKY PALMER, JOHN RODNEY BROWN, FRANK KABUI.

Registrar and Commissioner of the High Court: DAVID CHETWYND.

Chief Magistrate: DAVID CHETWYND (acting).

Attorney-General: PRIMO AFEAU.

Director of Public Prosecutions: FRANCIS MWANESALUA.

Solicitor-General: RANJIT HEWEGAMA.

Auditor-General: AUGUSTINE FATAI.

Public Solicitor: KENNETH HALL AVERRE.

Chair of Law Reform Commission: (vacant).

Religion

More than 95% of the population profess Christianity, and the remainder follow traditional beliefs. According to the census of 1976, about 34% of the population adhered to the Church of Melanesia (Anglican), 19% were Roman Catholics, 17% belonged to the South Seas Evangelical Church, 11% to the United Church (Methodist) and 10% were Seventh-day Adventists. Most denominations are affiliated to the Solomon Islands Christian Association. In many areas Christianity is practised alongside traditional beliefs, especially ancestor worship.

CHRISTIANITY

Solomon Islands Christian Association: POB 1335, Honiara; tel. 23350; fax 26150; e-mail essica@solomon.com.sb; f. 1967; five full mems, seven assoc. mem. orgs; Chair. Most Rev. ADRIAN SMITH; Gen. Sec. EMMANUEL IYABORA.

The Anglican Communion

Anglicans in Solomon Islands are adherents of the Church of the Province of Melanesia, comprising eight dioceses: six in Solomon Islands (Central Melanesia, Malaita, Temotu, Ysabel, Hanuato'o and Central Solomons, which was established in May 1997) and two in Vanuatu (one of which also includes New Caledonia). The Archbishop is also Bishop of Central Melanesia and is based in Honiara. The Church had an estimated 180,000 members in 1988.

Archbishop of the Province of Melanesia: Most Rev. Sir ELLISON POGO, Archbishop's House, POB 19, Honiara; tel. 21892; fax 21098; e-mail epogo@comphq.org.sb.

General Secretary: GEORGE KIRIAU, Provincial Headquarters, POB 19, Honiara; tel. 21892; fax 21098; e-mail gkiriau@comphq.oeg.sb.

The Roman Catholic Church

For ecclesiastical purposes, Solomon Islands comprises one archdiocese and two dioceses. At 31 December 2002 there were an estimated 89,371 adherents in the country. The Bishops participate in the Bishops' Conference of Papua New Guinea and Solomon Islands (based in Papua New Guinea).

Archbishop of Honiara: Most Rev. ADRIAN THOMAS SMITH, Holy Cross, POB 237, Honiara; tel. 21943; fax 26426; e-mail ahonccsi@solomon.com.sb.

Other Christian Churches

Assembly of God: POB 928, Honiara; tel. and fax 25512; f. 1971; Gen. Supt Rev. JERIEL OTASUI.

Christian Fellowship Church: Church, Paradise, Munda, Western Province; f. 1960; over 5,000 mems in 24 villages; runs 12 primary schools in Western Province.

Seventh-day Adventist Mission: POB 63, Honiara; tel. 21191; over 9,000 mems on Guadalcanal and over 6,800 on Malaita (Oct. 2000); Pres. of Western Pacific Region NEIL WATTS; Sec. Pastor J. PIUKI TASA.

South Seas Evangelical Church: POB 16, Honiara; tel. 22388; fax 20302; Pres. ERIC TAKILA; Gen. Sec. CHARLES J. RAFEASI.

United Church in Solomon Islands: POB 82, Munda, Western Province; tel. 61125; fax 61143; e-mail ucsihq@solomon.com.sb; a Methodist church; Bishop of Solomon Islands Region Rev. PHILEMON RITI; Gen. Sec. GINA TEBULU.

BAHÁ'Í FAITH

National Spiritual Assembly: POB 245, Honiara; tel. 22475; fax 25368; e-mail bahainsa@welkam.solomon.com.sb.

ISLAM

Solomon Islands Muslim League: POB 219, Honiara; tel. 21773; fax 24243; Gen. Sec. Dr MUSTAPHA RAMO; 66 mems.

The Press

Agrikalsa Nius (Agriculture News): POB G13, Honiara; tel. 21211; fax 21955; f. 1986; monthly; Editor ALFRED MAESULIA; circ. 1,000.

Citizens' Press: Honiara; monthly.

Link: Solomon Islands Development Trust, POB 147, Honiara; tel. 21130; fax 21131; pidgin and English; 3 or 4 per year.

Solomon Nius: POB 718, Honiara; tel. 22031; fax 26401; monthly; Dept of Information publication; Editor-in-Chief THOMAS KIVO; monthly; circ. 2,000.

Solomon Star: POB 255, Honiara; tel. 22913; fax 21572; f. 1982; daily; English; Dir JOHN W. LAMANI; Editor OFANI EREMAI (acting); circ. 4,000.

Solomon Times: POB 212, Honiara; tel. 39197; fax 39197; weekly; Chief Editor and Man. Dir EDWARD KINGMELE.

Solomon Voice: POB 1235, Honiara; tel. 20116; fax 20090; f. 1992; daily; circ. 3,000; Editor CAROL COLVILLE.

Broadcasting and Communications

TELECOMMUNICATIONS

Telekom (Solomon Telekom Company Ltd): POB 148, Honiara; tel. 21576; fax 23642; e-mail sales@telekom.com.sb; internet www.solomon.com.sb; 51% owned by Solomon Islands National Provident Fund, 41.9% by Cable and Wireless plc., 7.1% by Govt; operates national and international telecommunications links; Chair. DAVID QUAN; Gen. Man. MARTYN ROBINSON.

BROADCASTING

Radio

Solomon Islands Broadcasting Corporation: POB 654, Honiara; tel. 20051; fax 23159; e-mail sibcnews@solomon.com.sb; internet www.sibconline.com.sb; f. 1976; daily transmissions in English and Pidgin; broadcasts total 112 hours per week; Chair. FRANK PULE; Broadcast Operations Man. DAVID PALAPU; Gen. Man. JOHNSON HONIMAE; Editor WALTER NALANGU.

Finance

The financial system is regulated and monitored by the Central Bank of Solomon Islands. There are three commercial banks and a development bank. Financial statutory corporations include the Home Finance Corpn (which took over from the Housing Authority in 1990), the Investment Corporation of Solomon Islands (the state holding company) and the National Provident Fund. At the end of 1996 there were 142 credit unions, with some 17,000 members and total assets estimated at SI $18m.

BANKING

(cap. = capital; res = reserves; dep. = deposits; brs = branches; amounts in Solomon Islands dollars)

Central Bank

Central Bank of Solomon Islands: POB 634, Honiara; tel. 21791; fax 23513; e-mail cbsi-it@welkam.solomon.com.sb; internet www.cbsi.com.sb; f. 1983; sole bank of issue; cap. 2.6m., res 55.1m., dep. 83.2m. (Dec. 2002); Gov. RICK HOUWENIPWELA; Deputy Gov. JOHN KAITU.

Development Bank

Development Bank of Solomon Islands: POB 911, Honiara; tel. 21595; fax 23715; e-mail dbsi@welkam.solomon.com.sb; f. 1978; cap. 8.6m., res 4.6m. (Dec. 1998); in September 2004 the bank was declared insolvent and was expected to be liquidated by February 2005; Chair. JOHN MICHAEL ASIPARA; Man. Dir LUKE LAYMAN ETA; 4 brs; 5 sub-brs.

Commercial Banks

Australia and New Zealand Banking Group Ltd (Australia): Mendana Ave, POB 10, Honiara; tel. 21835; fax 22957; Gen. Man. TAIT JENKIN; (acting).

National Bank of Solomon Islands Ltd: Mendana Ave, POB 37, Honiara; tel. 21874; fax 24358; e-mail nbsi@welkam.solomon.com.sb; f. 1982; 51% owned by the Bank of Hawaii; 49% owned by Solomon Islands National Provident Fund; cap. 2.0m., res 17.4m., dep. 183.8m. (Dec. 2001); Chair. M. BAUER; Gen. Man. R. CANNOLES; 11 brs and 6 agencies.

Westpac Banking Corporation (Australia): 721 Mendana Ave, POB 466, Honiara; tel. 21222; fax 23419; e-mail gtaviani@westpac.com.au; Man. GIAN TAVIANI.

INSURANCE

About 10 major British insurance companies maintain agencies in Solomon Islands. In mid-1995 the Government announced a joint venture with an Australian insurance company to establish the Solomon Islands Insurance Company.

Trade and Industry

GOVERNMENT AGENCY

Investment Corporation of Solomon Islands: POB 570, Honiara; tel. 22511; fax 21263; holding company through which the Government retains equity stakes in a number of corporations; Chair. THOMAS KO CHAN.

DEVELOPMENT ORGANIZATION

Solomon Islands Development Trust (SIDT): POB 147, Honiara; tel. 21130; fax 21331; e-mail sidt@welkam.solomon.com.sb; f. 1982; development org; Chief Officer ABRAHAM BALANISIA.

CHAMBER OF COMMERCE

Solomon Islands Chamber of Commerce and Industry: POB 650, Honiara; tel. 39542; fax 39544; e-mail chamberc@solomon.com.sb; 86 member cos (Sept. 2003); Chair. LOYLEY NGIRA; Gen. Sec. SALLY ZIKY.

INDUSTRIAL AND TRADE ASSOCIATIONS

Association of Mining and Exploration Companies: c/o POB G24, Honiara; f. 1988; Pres. NELSON GREG YOUNG.

Commodities Export Marketing Authority: POB 54, Honiara; tel. 22528; fax 21262; e-mail cema@solomon.com.sb; sole exporter of copra; agencies at Honiara and Yandina; Chair. HUGO RAGOSO; Gen. Man. MOSES PELOMO.

Livestock Development Authority: Honiara; tel. 29649; fax 29214; f. 1977; privatized 1996; Man. Dir WARREN TUCKER.

Solomon Islands Business Enterprise Centre: POB 972, Honiara; tel. 26651; fax 26650; e-mail simbec@solomon.com.sb.

Solomon Islands Forest Industries Association: POB 1617, Honiara; tel. 26026; fax 20267; Chair. and Sec. KAIPUA TOHI.

EMPLOYERS' ORGANIZATIONS

Chinese Association: POB 1209, Honiara; tel. 22351; fax 23480; asscn of businessmen from the ethnic Chinese community.

Federation of Solomon Islands Business: POB 320, Honiara; tel. 22902; fax 21477.

Solomon Islands Farmers' Association: POB 113, Honiara; tel. 27508; fax 27509; founded a credit union in 1990; Technical Adviser Sri RAMON JUN QUITALES.

UTILITIES

Electricity

Solomon Islands Electricity Authority (SIEA): POB 6, Honiara; tel. 39442; fax 39472; e-mail mike@siea.com.sb; f. 1961; Chair. DANNY BULA; CEO M. L. NATION.

Water

Solomon Islands Water Authority (SIWA): POB 1407; Honiara; tel. 23985; fax 20723; f. 1994; Gen. Man. DONALD MAKINI.

CO-OPERATIVE SOCIETIES

In 1986 there were 156 primary co-operative societies, working mostly outside the capital. There are two associations running and aiding co-operative societies in Solomon Islands:

Central Co-operative Association (CCA): Honiara.

Salu Fishing Cooperative Association: POB 1041, Honiara; tel. 26550.

Solomon Islands Consumers Co-operative Society Ltd: Honiara; tel. 21798; fax 23640.

Solomon Islands Farmers and Producers Cooperative Association Ltd: Honiara; tel. 30908.

Western General Co-operative Association (WGCA): Gizo, Western Province.

MAJOR COMPANIES

Evergreen Forest Industries: POB 771, Honiara; tel. 30778.

Gold Ridge Mining: POB 1556, Honiara; tel. 25807; fax 25872; e-mail grml@solomon.com.sb; f. 1998; managed by Delta Gold (Australia), gold mining on Guadalcanal and New Georgia; operations suspended June 2000 due to ethnic tension; Man. DICK BRAIMBRIDGE.

Goodman Fielder (SI) Ltd: Honiara; tel. 30146; fax 30399; e-mail fielders@welkam.solomon.com.sb; flour millers, bakers and biscuit manufacturers; 17% Australian-owned; Gen. Man. PATRICK COTTER.

Honiara Timber Exporters: POB 959, Honiara; tel. 39200; fax 39199.

Kolombangara Forest Products Ltd: POB 382, Honiara; tel. 60230; fax 60020; e-mail admin@kfpl.com.sb; internet www.kfpl.com.sb; sustainable forestry; mahogany, teak, eucalyptus, gmelina; jt venture of govt of Solomon Islands and Commonwealth Development Corpn; FSC-certified company.

Linkali Timber Development Co Ltd: f. 1988; jtly owned by Xing Ling Timber Co Ltd and the Kalikoqu tribe.

National Fisheries Developments Ltd: Honiara; tel. 21506; fax 21459; f. 1977; sold by Govt in 1990; Singaporean-owned; operates fishing vessels and exports tuna; Gen. Man. PHIL ROBERTS.

Pacific Timbers: POB 201, Honiara; tel. 31100; fax 30062; e-mail movers@welkam.solomon.com.sb.

Russell Islands Plantation Estates Ltd: PO Yandina, Russell Islands, Central Province; tel. 21779; fax 21785; e-mail ripelyan@solomon.com.sb; coconut products; cocoa; livestock.

Solomon Islands Plantations Ltd: POB 350, Honiara; tel. 31121; fax 31188; e-mail sipl@solomon.com.sb; f. 1971; jt venture between Commonwealth Development Corpn, Investment Corpn of Solomon Islands and local landowners; major producer and exporter of palm oil and palm kernels; operations suspended mid-2000; Gen. Man. MIKE WORKMAN.

Solomon Islands Tobacco Co Ltd: Kukum Highway, Ranandi, Honiara; tel. 30127; fax 30463; part-owned by British American Tobacco; Gen. Man. ALLAN FAULDS.

Solomon Soaps Ltd: POB 326, Honiara; Gen. Man. EWALD TISCHLER.

Solomon Taiyo Ltd: POB 965, Honiara; tel. 21664; fax 23462; e-mail gm@stl.com.sb; f. 1972; 51% owned by Govt, 49% by Japanese co; offshore operations suspended Aug. 2000 following hijacking of vessel; Gen. Man. HIDEOTOSHI ITO.

Solrice: Honiara; 100% owned by Ricegrowers' Co-operative of Australia; Gen. Man. (vacant).

Waterking Marine: Gizo Hospital, Gizo, Western Province; tel. 60224; fax 60142; fresh and filleted marine products; Gen. Man. ISACC DAKEI.

TRADE UNIONS

There are 14 registered trade unions in Solomon Islands.

Solomon Islands Council of Trade Unions (SICTU): National Centre for Trade Unions, POB 271, Honiara; tel. 22566; fax 23171; f. 1986; Pres. DAVID P. TUHANUKU; Sec. BENEDICT ESIBAEA; the principal affiliated unions are:

Media Association of Solomon Islands (MASI): POB 654, Honiara; tel. 20051; fax 23300; e-mail sibcnews@welkam.solomon.com.sb; Pres. ROBERT IROGA.

Solomon Islands Medical Association: Honiara.

Solomon Islands National Teachers' Association (SINTA): POB 967, Honiara; f. 1985; Pres. K. SANGA; Gen. Sec. BENEDICT ESIBAEA.

Solomon Islands National Union of Workers (SINUW): POB 14, Honiara; tel. 22629; Pres. DAVID P. TUHANUKU; Gen. Sec. TONY KAGOVAI.

Solomon Islands Post and Telecommunications Union: Honiara; tel. 21821; fax 20440; Gen. Man. SAMUEL SIVE.

Solomon Islands Public Employees' Union (SIPEU): POB 360, Honiara; tel. 21967; fax 23110; Pres. MARTIN KARANI; Sec.-Gen. CLEMENT WAIWORI.

Solomon Islands Seamen's Association: POB G32, Honiara; tel. 24942; fax 23798.

Transport

ROADS

There are about 1,300 km of roads maintained by the central and provincial governments; in 1976 main roads covered 455 km. In addition, there are 800 km of privately maintained roads mainly for plantation use. Road construction and maintenance is difficult because of the nature of the country, and what roads there are serve as feeder roads to the main town of an island.

Honiara has a main road running about 65 km each side of it along the north coast of Guadalcanal, and Malaita has a road 157 km long running north of Auki and around the northern end of the island to the Lau Lagoon, where canoe transport takes over; and one running south for 35 km to Masa. On Makira a road links Kira Kira and Kakoranga, a distance of 35 km. Before it abandoned mining investigations in 1977, the Mitsui Mining and Smelting Co built 40 km of road on Rennell Island.

SHIPPING

Regular shipping services (mainly cargo) exist between Solomon Islands and Australia, New Zealand, Hong Kong, Japan, Singapore, Taiwan and European ports. In 1994 internal shipping was provided by 93 passenger/cargo ships, 13 passenger-only ships, 61 fishing vessels and 17 tugs. The four main ports are at Honiara, Yandina, Noro and Gizo. The international seaports of Honiara and Noro are controlled by the Solomon Islands Ports Authority. A new wharf, constructed with EU funding, was opened in Gizo in mid-2004. Six further wharves were to be constructed in other parts of the country with similar funding.

Solomon Islands Ports Authority: POB 307, Honiara; tel. 22646; fax 23994; e-mail ports@solomon.com.sb; f. 1956; responsible for the ports of Honiara and Noro; Chair. NELSON BOSO; Gen. Man. N. B. KABUI.

Sullivans (SI) Ltd: POB 3, Honiara; tel. 21643; fax 23889; e-mail shipserv@solomon.com.sb; shipping agents, importers, wholesalers.

Tradco Shipping Ltd: POB 114, Honiara; tel. 22588; fax 23887; e-mail tradco@solomon.com.sb; f. 1984; shipping agents; Man. Dir GERALD STENZEL.

CIVIL AVIATION

Two airports are open to international traffic and a further 25 serve internal flights. Air Niugini (Papua New Guinea), Air Nauru and Qantas (Australia) fly to the principal airport of Honiara International Airport—Henderson Field (renamed in September 2003 and formerly known as simply Henderson), 13 km from Honiara. In March 1998 a new terminal building, built with US $15.5m. of funds from Japan, was opened at Henderson airport (as it was then known).

Director of Civil Aviation: DEMETRIUS T. PIZIKI.

Solomon Airlines Limited: POB 23, Honiara; tel. 20031; fax 20232; e-mail gzoleveke@solair.com.sb; internet solomonairlines .com.au; f. 1968; govt-owned; international and domestic operator; scheduled services between Honiara and Port Moresby (Papua New Guinea), Nadi (Fiji), Brisbane (Australia) and Port Vila (Vanuatu); Chair. STEPHEN TONAFALEA; Gen. Man. GIDEON ZOLEVEKE Jr.

Tourism

Tourism is hindered by the relative inaccessibility of the islands and the inadequacy of tourist facilities. Tourist arrivals declined from 25,127 in 2000 to just 1,718 in 2003. The industry earned US $6m. in 1999, compared with $13m. in the previous year.

Solomon Islands Visitors Bureau: POB 321, Honiara; tel. 22442; fax 23986; e-mail visitors@solomon.com.sb; internet www.commerce .gov.sb/tourism/index.htm; f. 1980; Gen. Man. MORRIS OTTO NAMOGA; Marketing Man. ANDREW NIHOPARA.

Defence

Before the coup of June 2000, a unit within the Police Force, the Police Field Force, received technical training and logistical support from Australia and New Zealand. The force had two patrol boats and undertook surveillance activities in Solomon Islands' maritime economic zone. Additionally, in 1999–2000 a multinational force consisting largely of officers from Fiji and Vanuatu were deployed to quell inter-ethnic violence. (The officers from Fiji departed in April 2000.) The Solomon Islands Peace Plan of July 2000 (see History) foresaw the creation of separate Ministries of Defence and of Police, Justice and Legal Affairs. The Ministry of Defence would initially be responsible for introducing legislation allowing for the establishment of a Solomon Islands Defence Force. The Solomon Islands National Reconnaissance and Surveillance Force, founded in 1995, would also answerable to this ministry, and it was announced that an air surveillance unit would be established. In July 2003 an Australian-led force of more than 2,500 troops and police from numerous countries in the region arrived in Solomon Islands to intervene in the deteriorating situation in Guadalcanal and to restore law and order to the worst affected areas (see History). By August 2004 some 100 Australian troops remained in the islands, assisting the local civilian police force, and platoons of 33 troops from New Zealand, Fiji and Papua New Guinea were to remain on a three-month rotational basis until July 2006.

Commissioner of Police: WILLIAM MORRELL.

Commissioner of National Reconnaissance and Surveillance Force: Maj. MICHAEL WHEATLEY.

Education

About two-thirds of school-age children receive formal education, mainly in state schools. About 30% of the children who complete a primary school education receive secondary schooling, either in one of eight national secondary schools (at least one of which is run by the Government and the remainder by various Churches) or in one of 12 provincial secondary schools, which are run by provincial assemblies. The provincial secondary schools provide curricula of a practical nature, with a bias towards agriculture, while the national secondary schools offer more academic courses.

In 1993 there were 523 primary schools, with a total of 60,493 pupils in 1994. Also in 1993 there were 23 secondary schools, with a total of 7,981 pupils in 1995. In 2003 the Church of Melanesia announced plans to establish a new secondary school in each of the country's six dioceses. There are two teacher-training schools and a technical institute. According to the 1999 census, 57% of children aged five–14 attended school. Scholarships are available for higher education at various universities overseas. In 1977 the Solomon Islands Centre of the University of the South Pacific (based in Fiji) opened in Honiara. Central government expenditure on education in 1991 was SI $24m., or 7.9% of total spending.

TONGA

Physical and Social Geography

The Kingdom of Tonga, which is located in the central South Pacific about 650 km east of Fiji and south of Samoa, comprises 173 islands, totalling 748 sq km (289 sq miles) in area. The islands consist of two chains, those to the west being volcanic and those to the east being coral islands. They are divided into three groups: Vava'u in the north, Ha'apai and Tongatapu in the south. Only 36 of the islands are permanently inhabited. The climate is mild, with temperatures generally in the range 16°C–21°C (61°F–71°F) for most of the year, although usually hotter (27°C, or 81°F) in December and January.

The population is predominantly Polynesian, and Tongan is the language of the indigenous inhabitants. English is also widely spoken and is the language of education and administration. Most of the population are Christians, and the leading denomination is Wesleyan. The population of Tonga was estimated at 101,400 in mid-2003. About two-thirds of the population are resident on the largest island, Tongatapu, where the capital, Nuku'alofa, is situated.

History

From about the 10th century Tongan society developed a lineage of sacred chiefs, who gradually became effective rulers. Since European contact the chiefs have become known as kings. The Kingdom of Tonga adopted its first Constitution in 1875, during the reign of King George Tupou I, who had reunited the islands. As a result of increasing civil unrest, Tonga negotiated a treaty with the United Kingdom in 1900, whereby it came under British protection.

Queen Salote Tupou III came to the throne in 1918 and ruled Tonga until her death in December 1965. Her son, Prince Tupouto'a Tungi, who had been Prime Minister since 1949, succeeded her. He took the title of King Taufa'ahau Tupou IV and appointed his brother, Prince Fatafehi Tu'ipelehake, to be Prime Minister. In 1958 a treaty of friendship was signed between Tonga and the United Kingdom, providing for the appointment of a British commissioner and consul to be responsible to the Governor of Fiji, who held the office of British Chief Commissioner for Tonga. Tonga gained increased control over internal affairs in 1967 and became fully independent, within the Commonwealth, on 4 June 1970. Tonga was formally admitted to the UN in September 1999.

Elections held in May 1981 resulted in an unexpected defeat for the People's Representatives group, and the new Assembly became dominated by traditionalist conservatives. Further elections to the Legislative Assembly took place in May 1984. Following elections held in February 1987, new members, who were reported to include some of the Government's harshest critics, replaced six of the nine People's Representatives in the Legislative Assembly.

In July 1988 'Akilisi Pohiva, the editor of a local independent magazine, was awarded $T26,500 by the Supreme Court, after the Government had been found guilty of unfairly dismissing him in 1985 from his job in the Ministry of Education, for reporting on controversial issues. The court ruling intensified opposition demands for the abolition of perceived feudal aspects within Tongan society. The King, however, expressed his opposition to majority rule, on the grounds that the monarchy and royal-dominated Government could react to the needs of the people more quickly than the government of a parliamentary democracy. In 1989 the nine commoner members of the Legislative Assembly boycotted the Assembly, leaving it without a quorum, in protest at the absence of the Minister of Finance, Cecil Cocker, when they wished to question him. Upon resuming their seats in the Assembly, in September 1989, the commoners tabled a motion demanding reform of the Assembly to make it more accountable to the people. The motion proposed increasing popular representation to 15 and the reduction of noble representation to three seats (the Cabinet has 12 members who sit in the Assembly). The commoners indicated that, if they were re-elected in the general election of February 1990, they would consider the results as a mandate for change and as a sign of public approval for their attempts to achieve more equitable representation for the Tongan people. Pohiva, the leader of the pro-reform commoners, was the candidate who received the most votes and, with five colleagues, achieved an overwhelming victory. In September 1990 the newly constituted Appeal Court, comprising judges from other Pacific countries rather than the Privy Council, dismissed allegations of bribery and corruption made against Pohiva by some conservatives. In October 1990 Pohiva initiated a court case against the Government, claiming that its controversial sale of passports to foreign citizens was unconstitutional and illegal. The passports were sold mainly in Hong Kong to citizens of the territory and of the People's Republic of China, for as much as US $30,000 each, allowing the purchasers, in theory, to avoid travel restrictions imposed on Chinese passport-holders. In February 1991, however, a constitutional amendment, to legalize the naturalization of the new passport-holders, was adopted by an emergency session of the Legislative Assembly, and the case was therefore dismissed. In March a large demonstration was held in protest at the Government's actions, and a petition urging the King to invalidate the 426 passports in question, and to dismiss the Minister of Police (who was responsible for their sale), was presented by prominent commoners and church leaders. In the following month the Government admitted that the former President of the Philippines, Ferdinand Marcos, and his family had been given Tongan passports as gifts, after his fall from power in 1986. The events that ensued from the sale of passports were widely viewed as indicative of the growing support for reform and for greater accountability in the government of the country. By 1996, however, most of the 6,600 passports sold under the scheme had expired, as holders had failed to renew them.

In August 1991 the Prime Minister, Prince Fatafehi Tu'ipelehake, retired from office, owing to ill health, and was succeeded by the King's cousin, Baron Vaea, who had previously held the position of Minister of Labour, Commerce and Industries. Tu'ipelehake died in April 1999.

Plans by campaigners for democratic reform to establish a formal political organization were realized in November 1992, when the Pro-Democracy Movement was founded. The group, led by Fr Seluini 'Akau'ola (a Roman Catholic priest), organized a constitutional convention in the same month, at which options for the introduction of democratic reform were discussed. The Government, however, refused to recognize or to participate in the convention, prohibiting any publicity of the event and denying visas to invited speakers from abroad. Nevertheless, the pro-democracy reformists appeared to be enjoying increased public support, and, at elections in February 1993, won six of the nine People's Representative seats in the Legislative Assembly. However, Pohiva's position was undermined when, in December 1993 and February 1994, he lost two defamation cases in the Supreme Court, following the publication of allegations of fraudulent practice in his journal, *Kele'a*. In August 1994 Tonga's first political party was formed when the Pro-Democracy Movement launched its People's Party, under the chairmanship of a local businessman, Huliki Watab.

Elections took place in January 1996, at which pro-democracy candidates retained six of the nine People's Representative seats in the Legislative Assembly.

In July 1996 the Legislative Assembly voted to resume the sale of Tongan passports to Hong Kong Chinese, despite the controversy caused by a similar scheme in the early 1990s. As many as 7,000 citizenships were to be made available for between $T10,000 and $T20,000, granting purchasers all the rights of Tongan nationality, except ownership of land.

In September 1996 a motion to impeach the Minister of Justice and Attorney-General, Tevita Topou, was proposed in the Legislative Assembly. The motion alleged that Topou had continued to receive his daily parliamentary allowance during an unauthorized absence from the Assembly. Moreover, the publication of details of the impeachment motion, which had been reported to *The Times of Tonga* by Pohiva, before it had been submitted to the Legislative Assembly, resulted in the imprisonment of the latter and of the newspaper's editor, Kalafi Moala, and deputy editor, Filakalafi Akau'ola, for contempt of parliament. The three were subsequently released, although in October Moala was found guilty on a further charge of contempt, for critical remarks concerning the justice system in Tonga. In the same month the King closed the Legislative Assembly (which had been expected to sit until mid-November) until further notice. He denied that he had taken this decision in order to prevent further impeachment proceedings against Topou. In September 1997, however, the Legislative Assembly voted to abandon the impeachment proceedings.

The Government rejected accusations in early 1997 from journalists in Tonga and from media organizations throughout the region, that it was attempting to force the closure of *The Times of Tonga*, despite forbidding Moala (who resided in Auckland, New Zealand) to enter Tonga without written permission from the Government and issuing a ban on all government-funded advertising in the publication and on all government employees giving interviews to its journalists.

Parliament reopened in late May 1997. In June Akau'ola was arrested once again and charged with sedition for publishing a letter in *The Times of Tonga* that questioned government policy. However, in the same month the Government suffered a significant reversal when the Court of Appeal ruled that the imprisonment of Pohiva and the two journalists in late 1996 had been unlawful. In August the Prime Minister of New Zealand, James Bolger, paid an official visit to Tonga. Following the visit, Pohiva, who had unsuccessfully sought a meeting with Bolger, criticized New Zealand's relationship with Tonga, claiming that financial assistance from the country hindered democratic reform.

In mid-September 1998 the King closed the Legislative Assembly in response to a petition, signed by more than 1,000 people, which sought the removal from office of the Speaker, Eseta Fusitu'a. A parliamentary committee was forced to conduct an inquiry into the activities of Fusitu'a, who was accused of misappropriating public funds and of abusing his position. Pohiva and other pro-democracy activists commended the King for his decisive action in response to the petition. Legislation presented to the legislature later that month, which proposed that in future the Legislative Assembly should appoint cabinet ministers (a responsibility hitherto reserved for the King) was seen as further evidence of the increasing influence of the pro-democracy lobby in the political life of the country.

At the general election held on 11 March 1999 five members of the reformist Human Rights and Democracy Movement (formerly the People's Party) were returned to the Legislative Assembly, compared with six at the previous election. In April Veikune was appointed as Speaker and Chairman of the Legislative Assembly, replacing Eseta Fusitu'a, who had lost his seat at the general election. In late 1999 the former Minister of Lands, Survey and Natural Resources, Fakafanua, appeared in court on charges of forgery and bribery.

On 3 January 2000 the King appointed Prince 'Ulukalala-Lavaka-Ata as Prime Minister, replacing Baron Vaea, who had in 1995 announced his desire to retire. It had been expected that Crown Prince Tupouto'a would take up the position, but his support for constitutional reform in Tonga (notably the abolition of life-time terms for the Prime Minister and Ministers) contrasted with the King's more conservative approach. In March of that year a report published by the US State Department claimed that Tonga's system of government, whereby the Legislative Assembly is not directly elected, was in breach of UN and Commonwealth human rights guide-lines. The report was welcomed by 'Akilisi Pohiva, the leader of the Human Rights and Democracy Movement, who reiterated the party's demands for greater democracy, set out in a draft constitution completed in late 1999.

In November 2000 the Human Rights and Democracy Movement condemned the Government's decision to co-operate with an Australian biotechnology company wishing to carry out research into the genetic causes of a range of diseases. Tonga's extended family structures and the genetic isolation of its ethnically homogenous Polynesian population were cited as providing a rare opportunity for research. In return for its co-operation in the project, Tonga was to receive a share of the royalties from the sale of any drugs developed.

In January 2001 the Prime Minister announced a reallocation of cabinet positions in which he assumed responsibility for the newly created telecommunications portfolio. In the same month claims which appeared in a newspaper report that members of the pro-democracy movement had been involved in a plot to assist a prison break-out, to seize weapons from the army and to assassinate a cabinet minister were vehemently denied by representatives of the group. A spokesperson for the movement stated that the allegations were merely the latest in a series of claims aimed at discrediting the organization. Meanwhile, ongoing concerns for the freedom of the media in Tonga were renewed following the arrest of the deputy editor of *The Times of Tonga* on charges of criminal libel.

In February 2001 the Human Rights and Democracy Movement launched a public petition to amend Tonga's Nationality Act to allow Tongans who had taken up citizenship in other countries to retain Tongan citizenship. Their stated objective in undertaking the petition was to acknowledge Tonga's dependence on remittances from Tongans living overseas.

In September 2000, meanwhile, protests took place in Nuku'alofa, prompted by concerns that Chinese immigrant businesses, encouraged by the Government to establish themselves in Tonga, were creating unfavourable economic conditions for Tongan businesses. The Human Rights and Democracy Movement appealed to the Government to cease issuing work permits to foreign (predominantly Chinese) business people and to end the sale of Tongan passports, which had raised an estimated US $30m. during the 1980s. In order to protect this money from his ministers, whom he feared would squander it on unsuitable public works projects should it enter Tonga, the King requested that the funds be placed in the Tonga Trust Fund, held in a cheque account at the San Francisco, CA, branch of the Bank of America. In June 1999 Jesse Bogdonoff, a Bank of America employee, successfully sought royal approval to invest the money in a company in Nevada called Millennium Asset Management, where Bogdonoff was named as the Fund's Advising Officer. Bogdonoff later claimed that he had made a profit of about US $11m.; this so impressed the King that he appointed Bogdonoff as Court Jester.

The balance of the Fund and the interest it had accrued (altogether some US $40m.) was due to be returned to Tonga on 6 June 2001. Instead the money appeared to have vanished—along with Millennium Asset Management, which had ceased to exist. In September Princess Pilolevu, acting as Regent in the absence of the King and the Crown Prince, dismissed Kinikinilau Tutoatasi Fakafanua, currently Minister of Education, but who at the time of the incident had been Minister of Finance, along with the Deputy Prime Minister and Minister of Justice, Tevita Tupou, both of whom were trustees of the fund. She appointed the Minister of Police, Fire Services and Prisons, Clive Edwards, as acting Deputy Prime Minister. Parliament created a committee to investigate the crime and tabled a motion to impeach Fakafanua and Tupou. There were no apparent reprisals, or legal action, against Bogdonoff, who claimed that he had been deliberately misled as to the value of the funds. In October the acting Deputy Prime Minister denied that the Privy Council had directed the transfers from the Fund or that any ministers were implicated. He stressed that at least US $2.1m. of the fund, invested in Tongan banks, was duly accounted for. In June 2002, however, the Government admitted that some US $26m. had been lost as a result of Bogdonoff's actions and that legal proceedings had been initiated in the USA. It was announced in February 2003 that the case would be heard in December of that year; also in February the Legislative Assembly began impeachment proceedings against the ministers implicated in the matter. In February 2004 it was reported that the Government had agreed to settle out of court with Bogdonoff, who was to pay just US $1m. in compensation to the Tongan authorities.

In January 2002 Pohiva published allegations that the King held a secret offshore bank account containing US $350m., some of which was believed to be the proceeds of gold recovered from an 18th century shipwreck. He claimed to possess a letter written to the King from within the palace referring to the account. The Government dismissed the letter as a forgery. Nevertheless, Pohiva was briefly held in custody in February, whilst police searched the offices of the Human Rights and Democracy Movement and confiscated computer equipment in an attempt to find the source of the allegations. Later that month New Zealand's Minister for Foreign Affairs and Trade condemned Tonga as being endemically corrupt, implying that New Zealand's annual $NZ6m. profited the élite rather than the Tongan people as a whole. In February the King admitted that he did possess an overseas account, with the Bank of Hawaii, but declared that it contained the profits of vanilla sales from his own plantation. Nevertheless, Pohiva was formally charged with the use and publication of a forged document. In May 2002 a court began a preliminary hearing of the charges against him.

A general election was held on 6–7 March 2002, at which 52 candidates competed for the nine commoners' seats in the Legislative Assembly. The Human Rights and Democracy Movement won seven seats.

In late July 2002 it was reported that Pohiva, Moala and Akau'ola were seeking damages from the Tongan Government for wrongful imprisonment, following their incarceration in 1996 on charges of contempt of parliament.

Concerns for the freedom of the media re-emerged in February 2003, when the Government declared *The Times of Tonga*, the twice-weekly newspaper printed in New Zealand, to be a 'prohibited import', describing it as a foreign publication with a political agenda. The ban was challenged, and in April Tonga's Chief Justice, Gordon Ward, ruled that it was both illegal and unconstitutional. However, within hours of the ruling the Government ordered a new ban under different legislation. Ward overruled the ban in the following month, describing it as 'an ill-disguised attempt to restrict the freedom of the press'. However, on arrival from New Zealand 2,000 copies of the paper were seized by the Tongan customs authorities. Moreover, a few days later legislation was proposed which aimed to limit the power of the Supreme Court by excluding laws and ordinances approved by the Legislative Assembly and the Privy Council from judicial review. In early June the Supreme Court granted an injunction to *The Times of Tonga* ordering the Government to allow its distribution. The Government once again defied the order, stating that it would appeal. In mid-June the newspaper (which had been banned in Tonga since February) was finally allowed to go on sale within the country, after Gordon Ward had threatened each individual member of the Cabinet with contempt of court if they did not permit its distribution.

One of the most vocal critics of the affair was the King's nephew, Prince 'Uluvalu Tu'ipelehake, who in June 2003 expressed concern that the attempts to legislate against the freedom of the press and to restrict the right to seek judicial review would bring Tonga into disrepute. In mid-July it was reported that the Government had abandoned its proposals to limit the powers of the Supreme Court as a result of strong opposition to the plans. However, later that month the Legislative Assembly approved the Media Operators' Bill, which introduced restrictions on the involvement of foreign nationals in Tonga's media. The legislation was widely viewed as a further attempt to ban *The Times of Tonga*, the editor of which resided in

New Zealand. In early October Tonga's Roman Catholic bishop led a march by some 8,600 people to the government buildings in Nuku'alofa. The demonstration, which was the largest of its kind in Tonga's history, aimed to persuade the Government not to introduce any further restrictions on media freedom in the country. However, a few days later the Legislative Assembly voted in favour of a constitutional amendment allowing for greater control of the media, and shortly afterwards introduced the Newspaper Act, which gave the Government increased powers to regulate the content of newspapers in the country. It was reported that the legislation had been approved by 16 parliamentary votes to 11, with virtually all of the elected members voting against the proposals and nearly all of the appointed members voting in favour.

The introduction of legislation restricting media freedom in the country in late 2003 threatened to jeopardize Tonga's international relations, and in November the New Zealand Government announced that it was to review its relationship with Tonga as a result of these developments. The Tongan Government reacted angrily to the announcement, summoning New Zealand's ambassador to the islands to register its displeasure. However, in January 2004 Prince 'Uluvalu Tu'ipelehake appealed to Australia's Minister for Foreign Affairs, Alexander Downer, to encourage political reform and to support attempts to increase democratic representation in Tonga.

In February 2004 the King of Tonga's second son, Ma'atu Fatafehi Alaivahamama'o Tuku'aho, died of a heart attack at 48 years of age. Some 6,000 people attended his funeral and the King and Queen announced that they would undertake a period of mourning lasting 100 nights.

Serious and ongoing financial problems with the national airline, Royal Tongan Airlines, culminated in April 2004 with the grounding and repossession of the company's passenger jet in Auckland, New Zealand, and the consequent suspension of all international flights. Domestic flights continued to be operated by the airline, but in the following month the entire company ceased all operations when its last remaining domestic aircraft broke down and it was announced that no funds were available to repair it. The collapse of the airline caused considerable disquiet throughout Tonga, particularly among those involved in the islands' increasingly important tourist industry, which had relied heavily on the carrier's international services. Tourism officials estimated an almost immediate loss of bookings worth 1.2m. pa'anga , in addition to the loss of 261 jobs and a possible further 585. At the opening session of the Legislative Assembly in late May seven of the nine elected representatives staged a boycott, demanding the resignation of the Prime Minister, who they claimed had allowed millions of pa'anga to be wasted during his tenure as chairman of the airline.

A legal challenge to the constitutional amendments governing media freedom, introduced by the Government in 2003, was begun in the Supreme Court in August 2004. Shortly before the legal hearing began the Prime Minister announced the dismissal of three cabinet ministers, including the Minister of Police, Fire Services and Prisons, Clive Edwards, who had been one of the principal proponents of the constitutional amendments. No reasons were given for the dismissals, however, prompting expressions of concern for the stability of the Government and an appeal from the royalist Kotoa Movement for the Prime Minister to resign on the grounds that he had become too isolated from the people. At a public meeting in early September the King and Queen led several hundred Tongans in communal prayers for the guidance of their leaders.

A friendship treaty signed with the USA, in July 1988, provided for the safe transit of US nuclear-capable ships within Tongan waters. Tonga was virtually alone in the region in failing to condemn the French Government for its decision to resume nuclear weapons tests in the South Pacific in mid-1995. However, in May 1996 it was announced that Tonga was finally to accede to the South Pacific Nuclear-Free Zone Treaty (the Treaty of Rarotonga).

In September 1990 Tonga caused some controversy by reserving for itself the last 16 satellite positions in space that are suitable for trans-Pacific communications. Despite the protests of leading member-nations of Intelsat (an international consortium responsible for most of the world's satellite services) and allegations that the move was a profit-making ploy prompted by a US entrepreneur, the International Telecommunication Union was obliged to approve the claim. A dispute with Indonesia, concerning that country's use of satellite positions reserved by Tonga in 1990, was resolved by the signing of an agreement in December 1993. In May of the following year the two countries established diplomatic relations at ambassadorial level.

In January 1997 Tonga pledged its full support for New Zealand's efforts to combat illegal drugs-trafficking in the region. The Government's offer to co-operate with the New Zealand police followed the seizure in late 1996 of a shipment of cocaine which had been smuggled out of Tonga in hollowed yams, prompting fears that the country was becoming an *entrepôt* for drugs-traffickers operating in the South Pacific region.

In November 1998 Tonga announced that it had decided to terminate its diplomatic links with Taiwan and to establish relations with the People's Republic of China.

In January 2002, in the Red Sea, Israeli commandos seized a ship that was allegedly transporting weapons to Palestinian activists. The ship was flying a Tongan flag of convenience and had been registered in the Kingdom, although its ownership was uncertain. 'Akilisi Pohiva criticized Tonga's policy of international ship registration, which generated income through sales of flags of convenience. (The issue of international shipping registration had never been debated in Parliament.) Following the seizure, the registration system was closed in June. However, in September another Tongan-registered ship was seized off the coast of Italy and its crew arrested on suspicion of plotting an al-Qa'ida-sponsored terrorist attack in Europe. Furthermore, it was reported in early October that the Greek businessman in charge of the Tongan International Registry of Ships, Pelopidas Papadopoulos, had absconded with the proceeds of the operation owed to the Tongan Government, totalling some US $300,000.

Economy

In 2002, according to estimates by the World Bank, Tonga's gross national income (GNI), measured at average 2000–02 prices, was US $143m., equivalent to US $1,410 per head (or US $6,340 per head on an international purchasing-power parity basis). During 1990–2002, it was estimated, the population increased by an annual average of 0.4%, while gross domestic product (GDP) per head increased, in real terms, by an average 2.2% per year. Overall GDP increased, in real terms, at an annual average rate of 2.6% in 1990–2002. According to the Asian Development Bank (ADB), GDP increased by 2.1% in the 2001/02 financial year and by 2.9% in 2002/03.

The majority of the islands have an inherently fertile soil. Agriculture (with forestry and fishing) accounted for some 28.5% of GDP in 2001/02 and engaged 33.8% of the employed labour force in 1996. According to the ADB, agricultural GDP increased at an average annual rate of 0.8% in 1996–2001, rising by 10.6% in 2000, before declining by 1.1% in 2001. Sectoral growth was estimated at 2.1% in 2002 and at 1.4% in the following year. Coconuts and bananas have traditionally accounted for the bulk of Tonga's exports. However, squash have become increasingly important, and exports of the crop provided 45% of total export earnings in 2000/01, although this figure declined to 10.8% in the calendar year 2002, owing to over-production, which resulted in the depression of the crop's price on the important Japanese market. Production of vanilla has declined in recent years. Vanilla accounted for just 3.3% of export earnings in 2000/01, compared with 13.4% in 1995/96. Root crops, breadfruit, watermelons and citrus fruits are also cultivated. A significant increase in the number of farmers cultivating aloe vera was recorded in 2003. Livestock farming consists of pigs, cattle, goats, poultry and bees. The islands are not, however, self-sufficient, and imports of food accounted for some 30.2% of total import costs in 2001.

The fishing industry was strengthened in 1984 by the opening of a dockyard complex in the Ha'apai group and the construction of a port at Nuku'alofa, together with a fleet of up to 60 fishing-boats. Revenues from fishing improved steadily during the 1990s, and represented 10.8% of export income in 2002. In early 2001 the Australian Government agreed to fund a four-year fishing project on the islands, which was to include the provision of equipment to small-scale fishing operators, improved marketing opportunities and the development of commercial long-line tuna fishing.

Industry (including mining, manufacturing, construction and power) provided some 15.1% of GDP in 2001/02, and engaged 26.4% of the employed labour force in 1996. Industrial GDP was estimated by the ADB to have increased by 5.5% in 2002 and by 4.0% in 2003. Manufacturing contributed 5.6% of GDP in 2001/02, and (with mining) employed 22.8% of the labour force in 1996. Food products and beverages accounted for 56.8% of total manufactured goods produced in 1999/2000. The food and textile sectors of manufacturing registered the largest increases in 1999/2000, raising the value of their production by 14.6% and 56.8% respectively. The most important industrial activities are the production of concrete blocks, small excavators, clothing, furniture, handicrafts, leather goods, sports equipment (including small boats), brewing and coconut oil. There is also a factory for processing sandalwood. In early 2003 permission was granted for a Chinese businessman to establish a cigarette factory in the islands.

Service industries contributed some 56.3% of GDP in 2001/02. According to the ADB, the GDP of the services sector expanded by 3.4% in 2001, by 0.6% in 2002 and by 2.4% in 2003. Tourism makes a significant contribution to the economy and earned an estimated US $9m. in 2002. The trade, restaurants and hotels sector contributed 13.7% of GDP in 2001/02, and engaged 8.5% of the employed labour force in 1996. Visitor arrivals totalled 40,110 in 2003 (compared with 30,883 in 1999). However, the collapse of the national airline in mid-2004 (see History) resulted in severe losses for the industry and fears for its wider impact on the economy.

The country's visible trade deficit, which stood at US $57m. in 2003, is offset by income from tourism and by remittances from the large number of Tongans working overseas. The current account of the balance of payments recorded a deficit of US $5m. in the same year. Among Tonga's principal trading partners is New Zealand, accounting for 43.5% of total imports and 3.0% of exports in 2003. In the same year the USA provided 6.3% of imports and purchased 43.1% of exports and Japan accounted for 39.3% of exports earnings. Other important trading partners include Australia (which provided 11.7% of imports in 2003) and Fiji (20.6% of imports in that year). The principal exports in 2003 were foodstuffs (some 98% of total exports). The principal imports were foodstuffs (13.3% of total import costs), beverages and tobacco (13.9%), mineral fuels (19.1%) and machinery and transport equipment (15.1%).

Principal donors of aid to Tonga are the United Kingdom, New Zealand, Australia, Germany, Japan, the ADB and the European Union (EU). Official development assistance totalled an estimated US $18.8m. in 2000. Aid from Australia totalled $A11.7m. in 2003/04, in which year assistance from New Zealand totalled $NZ5.7m.

Unemployment and inflation are major problems, which have led to massive temporary migration: between 1974 and 1977 about 10,000 Tongans obtained entry visas to New Zealand. In 1995–2001 the average annual rate of inflation was 4.6%, and stood at 10.3% in 2002 and 11.7% in 2003. The rate was projected by the ADB to reach 11.0% in the following year. Unemployment was 13.3% of the labour force at the time of the 1996 census.

In 2001/02 there was an estimated overall budget deficit of 1.4m. pa'anga. The budgetary deficit was projected to be the equivalent of 3.1% of GDP in 2003. According to ADB figures, Tonga's total external debt was US $73.5m. at the end of 2002, of which US $72.6m. was long-term public debt. In that year the cost of debt-servicing was equivalent to 5.9% of the total revenue from exports of goods.

During recent years Tonga's economic development has been adversely affected by inclement weather, inflationary pressures, a high level of unemployment, large-scale emigration and over-reliance on the agricultural sector. Consequently the country has attempted to diversify its sources of income and establish an efficient source of energy, to enable a reduction in fuel imports. In August 2001, following legislative amendments, the country was removed from a list of non-co-operative countries and territories, drawn up by the Paris-based Financial Action Task Force (FATF), in efforts to combat the practice of international 'money laundering'. The repercussions of the suicide attacks on the USA in September 2001 not only depressed Tonga's tourism industry in the short term, but also jeopardized overseas remittances to the islands, because many emigrant workers were employed in the airline industry in the USA. It was reported in early 2002 that the Tongan Government had reached an agreement with a US company to develop the island of 'Eua as a rocket-launching site for the purposes of space tourism. The first launch was expected to take place in 2005. Meanwhile, the sharp rise in the Government's unbudgeted expenditure continued to cause concern in 2002. The Tongan economy was further weakened in 2003 by the continuing decline in the value of the pa'anga (which had lost some 54% against the New Zealand dollar since 2001), by high rates of inflation and by falling prices for squash on the international market. Supported by a loan, worth US $10m. over 24 years (the first tranche of which was released in 2002) and technical assistance from the ADB, the Economic Public Sector Reform Programme (EPSRP) aimed to improve the performance of the civil service and of public enterprises while decreasing costs, to implement a comprehensive reform of the country's tax system and reduce opportunities for tax evasion, and to develop better investment conditions.

Tonga signed a bilateral trade agreement with the People's Republic of China in 1999, and a trade delegation from that country visited the islands in the following year. It was thought that China was considering using Tonga as a base for the production of export goods (including garments and agricultural products) for the Australian and New Zealand markets. A delegation of senior Chinese trade officials visited Tonga in August 2004 to assess investment opportunities in agriculture, information technology and the power supply industry.

Tonga is a member of the Asian Development Bank (ADB), of the Pacific Islands Forum (formerly the South Pacific Forum), of the Pacific Community (formerly the South Pacific Commission) and of the UN's Economic and Social Commission for Asia and the Pacific (ESCAP). The Kingdom is also a signatory of the Lomé Conventions and of the successor Cotonou Agreement with the EU, and benefits from their provisions.

Statistical Survey

Source (unless otherwise indicated): Tonga Government Department of Statistics, POB 149, Nuku'alofa; tel. 23300; fax 24303; e-mail statdept@tongatapu.net.to.

AREA AND POPULATION

Area: 748 sq km (289 sq miles).

Population: 94,649 at census of 28 November 1986; 97,784 (males 49,615, females 48,169) at census of 30 November 1996; 101,400 (official estimate) at mid-2003. *By Group* (1996 census, provisional): Tongatapu 66,577; Vava'u 15,779; Ha'apai 8,148; 'Eua 4,924; Niuas 2,018.

Density (mid-2003): 135.6 per sq km.

Principal Towns (population in '000, 1986): Nuku'alofa (capital) 21.3; Mu'a 4.1; Neiafu 3.9. Source: Stefan Helders, *World Gazetteer* (internet www.world-gazetteer.com). *Mid-2003* (UN estimate, incl. suburbs): Nuku'alofa (capital) 34,654. Source: UN, *World Urbanization Prospects: The 2003 revision*.

Births, Marriages and Deaths (1998): Registered live births 2,737 (birth rate 27.0 per 1,000 at 1996 census); Registered marriages 736 (marriage rate 7.7 per 1,000); Registered deaths 498 (death rate 6.5 per 1,000 at 1996 census).

Expectation of Life (WHO estimates, years at birth): 70.7 (males 70.0; females 71.4) in 2002. Source: WHO, *World Health Report*.

Economically Active Population (persons aged 15 years and over, 1996): Agriculture, forestry and fishing 9,953; Mining and quarrying 43; Manufacturing 6,710; Electricity, gas and water 504; Construction 500; Trade, restaurants and hotels 2,506; Transport, storage and communications 1,209; Financing, insurance, real estate and business services 657; Public administration and defence 3,701; Education 1,721; Health and social work 510; Other community, social and personal services 1,320; Extra-territorial organizations 72; *Total employed* 29,406 (males 18,402, females 11,004); Unemployed 4,502 (males 3,293, females 1,209); *Total labour force* 33,908 (males 21,695, females 12,213). *Mid-2002* (estimates): Agriculture, etc. 12,000; Total labour force 37,000 (Source: FAO).

HEALTH AND WELFARE

Key Indicators

Total Fertility Rate (children per woman, 2002): 3.8.

Under-5 Mortality Rate (per 1,000 live births, 2002): 20.

Physicians (per 1,000 head, 1997): 0.44.

Health Expenditure (2001): US $ per head (PPP): 223.

Health Expenditure (2001): % of GDP: 5.0.

Health Expenditure (2001): public (% of total): 61.6.

Human Development Index (2002): ranking 63.

Human Development Index (2002): value 0.787.

AGRICULTURE, ETC.

Principal Crops (FAO estimates, '000 metric tons, 2002): Sweet potatoes 6; Cassava 9; Taro 4; Yams 4; Copra 4,109*; Coconuts 58; Pumpkins, squash and gourds 17,000*; Bananas 1; Plantains 3; Oranges 1; Lemons and limes 3*; Other fruits 3*.

* Unofficial figure.

Livestock (FAO estimates, '000 head, year ending September 2002): Pigs 81; Horses 11; Cattle 11; Goats 13; Chickens 300. Source: FAO.

Livestock Products (FAO estimates, metric tons, 2002): Pig meat 1,496; Other meat 684; Hen eggs 28; Honey 12; Cattle hides 40; Cow's milk 370. Source: FAO.

Forestry (FAO estimates, '000 cu m, 2002): *Roundwood Removals* (excl. bark): 2; *Sawnwood Production:* 2. Source: FAO.

Fishing (metric tons, live weight, 2002): Total catch 4,804 (Albacore 1,199; Marine crustaceans 337). Figures exclude aquatic plants (metric tons): 12,600 (all aquaculture). Source: FAO.

INDUSTRY

Production (2003): Electric energy 36 million kWh.

FINANCE

Currency and Exchange Rates: 100 seniti (cents) = 1 pa'anga (Tongan dollar or $T). *Sterling, US Dollar and Euro Equivalents* (31 March 2004): £1 sterling = $T3.5976; US $1 = $T1.9612; €1 = $T2.3973; $T100 = £27.80 = US $51.00 = €41.71. *Average Exchange Rate* (pa'anga per US $): 2.1236 in 2001; 2.1952 in 2002; 2.1420 in 2003.

Budget (estimates, million pa'anga, year ending 30 June 2003): *Revenue:* Taxation 76.2; Other current revenue 20.7; Capital receipts 0.4; Total 97.3 (excl. grants received from abroad 3.5). *Expenditure:* Current expenditure 91.0; Capital expenditure 10.1; Total 101.1 (excl. net lending –0.6). Source: Asian Development Bank, *Key Indicators of Developing Asian and Pacific Countries*.

International Reserves (US $ million at 31 December 2003): IMF special drawing rights 0.32; Reserve position in the IMF 2.54; Foreign exchange 39.76; Total 42.63. Source: IMF, *International Financial Statistics*.

Money Supply ('000 pa'anga at 31 December 2003): Currency outside banks 14,322; Demand deposits at deposit money banks 47,764; Total money 62,086. Source: IMF, *International Financial Statistics*.

Cost of Living (Consumer Price Index, excl. rent; base: 2000 = 100): 108.3 in 2001; 119.5 in 2002; 133.4 in 2003. Source: IMF, *International Financial Statistics*.

Gross Domestic Product (million pa'anga at current prices, year ending 30 June): 260.6 in 1999/2000; 270.3 (preliminary) in 2000/01; 296.5 (estimate) in 2001/02. Source: IMF, *Tonga: Statistical Appendix* (February 2003).

Gross Domestic Product by Economic Activity (preliminary figures, million pa'anga at current prices, year ending 30 June 2002): Agriculture, forestry and fishing 70.2; Mining and quarrying 1.1; Manufacturing 13.7; Electricity, gas and water 5.5; Construction 16.9; Trade, restaurants and hotels 33.8; Transport, storage and communications 19.7; Finance and real estate 25.8; Public administration and other 59.3; *Sub-total* 246.0; Indirect taxes, *less* subsidies 50.6; *GDP in purchasers' values* 296.6. Source: Asian Development Bank, *Key Indicators of Developing Asian and Pacific Countries*.

Balance of Payments (US $ million, year ending 30 June 2002): Exports of goods f.o.b. 24; Imports of goods f.o.b. –157; *Trade balance* –133; Exports of services and income 62; Imports of services and income –69; *Balance on goods, services and income* –140; Current transfers (net) 160; *Current balance* 20; Direct investment 1; Portfolio investment 0; Other long-term capital 26; Other short-term capital –1; Net errors and omissions –33; *Overall balance* 14. Source: Asian Development Bank, *Key Indicators of Developing Asian and Pacific Countries*.

EXTERNAL TRADE

Principal Commodities ('000 pa'anga, 2003): *Imports:* Food and live animals 26,881; Beverages and tobacco 28,021; Crude materials (inedible) except fuels 22,732; Mineral fuels 38,518; Chemicals 9,693; Basic manufactures 15,652; Machinery and transport equipment 30,395; Miscellaneous manufactured articles 8,119; Total (incl. others) 201,678. *Exports:* Food and live animals 36,576; Manufactured goods 437; Total (incl. others) 37,301.

Principal Trading Partners (US $ million, 2003): *Imports:* Australia 11.6; China, People's Repub. 2.3; Fiji 20.4; Indonesia 1.2; Japan 1.9; New Zealand 43.0; United Kingdom 1.9; USA 6.3; Total (incl. others) 98.9. *Exports:* Australia 0.4; Fiji 0.4; Italy 0.6; Japan 12.5; Lithuania 0.9; New Zealand 1.0; USA 13.7; Total (incl. others) 31.8. Source: Asian Development Bank, *Key Indicators of Developing Asian and Pacific Countries*.

TRANSPORT

Road Traffic (registered vehicles in use, 1998): Passenger cars 6,419; Buses and coaches 171; Lorries and vans 9,018.

Shipping (international traffic, '000 metric tons, 1998): Goods loaded 13.8; Goods unloaded 80.4. 1991: Vessels entered ('000 net registered tons) 1,950 (Source: UN, *Statistical Yearbook*). *Merchant Fleet* (registered at 31 December 2003): Vessels 99; Total displacement ('000 grt) 170.0 (Source: Lloyd's Register-Fairplay, *World Fleet Statistics*).

Civil Aviation (traffic on scheduled services, 1999): Passengers carried 91,000; Passenger-km 19 million; Total ton-km 2 million. Source: UN, *Statistical Yearbook*.

TOURISM

Foreign Tourist Arrivals: 32,386 in 2001; 36,585 in 2002; 40,110 in 2003.

Tourist Arrivals by Country (2000/01): Australia 5,453; New Zealand 10,674; Pacific islands 2,797; USA 7,053; Total (incl. others) 33,722.

Tourism Receipts (million pa'anga): 11.6 in 1998/99; 16.5 in 1999/2000; 12.7 in 2000/01.
Source: IMF, *Tonga: Statistical Appendix* (February 2003).

COMMUNICATIONS MEDIA

Radio Receivers (1997): 61,000 in use.

Television Receivers (1997): 2,000 in use.

Telephones ('000 main lines, 2002): 11.2 in use.

Mobile Cellular Telephones (2002): 3,400 subscribers.

Personal Computers (2002): 2,000.

Internet Users (2002): 2,900.

Facsimile Machines (1996): 250 in use.

Daily Newspapers (1996): 1; estimated circulation 7,000.

Non-daily Newspapers (2001): 2; estimated circulation 13,000.
Sources: UNESCO, *Statistical Yearbook*; UN, *Statistical Yearbook*; Audit Bureau of Circulations, Australia; and International Telecommunication Union.

EDUCATION

Primary (1999): 117 schools; 745 teachers; 16,206 pupils.

General Secondary (1999): 39 schools; 961 teachers; 13,987 pupils.

Technical and Vocational (1999): 4 colleges; 45 teachers (1990); 467 students.

Teacher-training (1999): 1 college; 22 teachers (1994); 288 students.

Universities, etc. (1985): 17 teachers; 85 students.

Other Higher Education (1985): 36 teachers (1980); 620 students.
In 1990 230 students were studying overseas on government scholarships.

Directory

The Constitution

The Constitution of Tonga is based on that granted in 1875 by King George Tupou I. It provides for a government consisting of the Sovereign; a Privy Council, which is appointed by the Sovereign and consists of the Sovereign and the Cabinet; the Cabinet, which consists of a Prime Minister, a Deputy Prime Minister, eight other ministers and the Governors of Ha'apai and Vava'u; a Legislative Assembly and a Judiciary. Limited law-making power is vested in the Privy Council and any legislation passed by the Executive is subject to review by the Legislative Assembly. The unicameral Legislative Assembly comprises the King, the Cabinet, nine hereditary nobles (chosen by their peers) and nine representatives elected by all adult Tongan citizens. Elected members hold office for three years.

The Government

HEAD OF STATE

The Sovereign: HM King Taufa'ahau Tupou IV (succeeded to the throne 15 December 1965).

CABINET
(September 2004)

Prime Minister and Minister of Defence, Civil Aviation, responsible for Telecommunications, Works, Marines and Ports: HRH Prince 'Ulukalala-Lavaka-Ata.

Deputy Prime Minister and Minister of Internal Affairs: Cecil Cocker.

Minister of Foreign Affairs: Tu'a Taumoepeau Tupou.

Minister of Finance: Siosiua 'Utoikamanu.

Minister of Education: Paula Sunia Bloomfield.

Minister of Defence: 'Aloua Fetu'utolu Tupou.

Attorney-General and Minister of Justice: Siaosi Taimani 'Aho.

Minister of Lands, Survey and Natural Resources: Fielakepa.

Minister of Health: Dr Viliami Ta'u Tangi.

Minister of Agriculture, Forestry and Fisheries: S. M. Tuita.

Governor of Ha'apai: Malupo.

Governor of Vava'u: 'AKAU'OLA.

GOVERNMENT MINISTRIES AND OFFICES

Office of the Prime Minister: POB 62, Taufa'ahau Rd, Kolofo'ou, Nuku'alofa; tel. 24644; fax 23888; e-mail fttuita@pmo.gov.to; internet www.pmo.gov.to.

Palace Office: Salote Rd, Kolofo'ou, Nuku'alofa; tel. 21000; fax 24102.

Ministry of Agriculture and Fisheries: Administration Office, Vuna Rd, Kolofo'ou, Nuku'alofa; tel. 23038; fax 23039; e-mail maf-holo@candw.to.

Ministry of Civil Aviation: POB 845, Salote Rd, Nuku'alofa; tel. 24144; fax 24145; e-mail info@mca.gov.to; internet www.mca.gov.to.

Ministry of Education: POB 61, Vuna Rd, Kolofo'ou, Nuku'alofa; tel. 23511; fax 23596; e-mail moe@kalianet.to.

Ministry of Finance: Treasury Building, POB 87, Vuna Rd, Kolofo'ou, Nuku'alofa; tel. 23066; fax 21010; e-mail minfin@candw.to.

Ministry of Foreign Affairs: National Reserve Bank Building, Salote Rd, Kolofo'ou, Nuku'alofa; tel. 23600; fax 23360; e-mail secfo@candw.to.

Ministry of Health: POB 59, Taufa'ahau Rd, Tofoa, Nuku'alofa; tel. 23200; fax 24921.

Ministry of Internal Affairs: POB 110, Salote Rd, Fasi-moe-afi, Nuku'alofa; tel. 23688; fax 23880.

Ministry of Justice: POB 130, Railway Rd, Kolofo'ou, Nuku'alofa; tel. 21055; fax 23098.

Ministry of Lands, Survey and Natural Resources: POB 5, Vuna Rd, Kolofo'ou, Nuku'alofa; tel. 23611; fax 23216.

Ministry of Marine and Ports: Fakafanua Centre, Vuna Rd, Ma'ufanga, Nuku'alofa; tel. 22555; fax 26234.

Ministry of Telecommunications: Nuku'alofa.

Ministry of Works and Disaster Relief Activities: 'Alaivahamama'o Rd, Vaololoa, Nuku'alofa; tel. 23100; fax 23102; e-mail mowtonga@kalianet.to.

Department of Civil Aviation: Salote Rd, Fasi-moe-afi; tel. 24144; fax 24145.

Legislative Assembly

The Legislative Assembly consists of the Speaker, the members of the Cabinet, nine nobles chosen by the 33 Nobles of Tonga, and nine representatives elected by all Tongans over 21 years of age. There are elections every three years, and the Assembly is required to meet at least once every year. The most recent election was held on 6–7 March 2002, when seven of the nine elected representatives were members of the reformist Tonga Human Rights and Democracy Movement.

Speaker and Chairman of the Legislative Assembly: TU'IVAKANO.

Political Organizations

Kotoa Movement: f. 2001; campaigns in support of monarchy.

Human Rights and Democracy Movement in Tonga (HRDMT): POB 843, Nuku'alofa; tel. 25501; fax 26330; e-mail demo@kalianet.to; f. 1992 as Tonga Pro-Democracy Movement; campaigns for democratic reform and increased parliamentary representation for the Tongan people; Chair. Rev. SIMOTE VEA; Sec. 'AKILISI POHIVA.

Diplomatic Representation

EMBASSY AND HIGH COMMISSIONS IN TONGA

Australia: Salote Rd, Private Bag, Nuku'alofa; tel. 23045; fax 23243; e-mail ahctonga@kalianet.to; High Commissioner COLIN HILL.

China, People's Republic: Vuna Rd, POB 877, Nuku'alofa; tel. 24554; fax 24595; e-mail chinaton@kalianet.to; Ambassador GAO SHANHAI.

New Zealand: cnr Taufa'ahau and Salote Rds, POB 830, Nuku'alofa; tel. 23122; fax 23487; e-mail nzhcnuk@kalianet.to; High Commissioner WARWICK A. HAWKER.

United Kingdom: Vuna Rd, POB 56, Nuku'alofa; tel. 24285; fax 24109; e-mail britcomt@kalianet.to; High Commissioner PAUL NESSLING.

Judicial System

There are eight Magistrates' Courts, the Land Court, the Supreme Court and the Court of Appeal.

Appeal from the Magistrates' Courts is to the Supreme Court, and from the Supreme Court and Land Court to the Court of Appeal (except in certain matters relating to hereditary estates, where appeal lies to the Privy Council). The Chief Justice and Puisne Judge are resident in Tonga and are judges of the Supreme Court and Land Court. The Court of Appeal is presided over by the Chief Justice and consists of three judges from other Commonwealth countries. In the Supreme Court the accused in criminal cases, and either party in civil suits, may elect trial by jury. In the Land Court the judge sits with a Tongan assessor. Proceedings in the Magistrates' Courts are in Tongan, and in the Supreme Court and Court of Appeal in Tongan and English.

Supreme Court

POB 11, Nuku'alofa; tel. 23599; fax 22380; e-mail cj_tonga@kalianet.to.

Chief Justice: ROBIN M. WEBSTER.

Puisne Judge: TONY FORD.

Chief Registrar: MANAKOVI PAHULU.

Religion

The Tongans are almost all Christians, and about 36% of the population belong to Methodist (Wesleyan) communities. There are also significant numbers of Roman Catholics (15%) and Latter-day Saints (Mormons—15%). Anglicans (1%) and Seventh-day Adventists (5%) are also represented. Fourteen churches are represented in total.

CHRISTIANITY

Kosilio 'ae Ngaahi Siasi 'i Tonga (Tonga National Council of Churches): POB 1205, Nuku'alofa; tel. 23291; fax 27506; e-mail tncc@kalianet.to; f. 1973; three mem. churches (Free Wesleyan, Roman Catholic and Anglican); Chair. Rt Rev. SOANE LILO FOLIAKI; Gen. Sec. Rev. SIMOTE M. VEA.

The Anglican Communion

Tonga lies within the diocese of Polynesia, part of the Church of the Province of New Zealand. The Bishop of Polynesia is resident in Fiji.

Archdeacon of Tonga and Samoa: The Ven. SAM KOY, The Vicarage, POB 31, Nuku'alofa; tel. 22136.

The Roman Catholic Church

The diocese of Tonga, directly responsible to the Holy See, comprises Tonga and the New Zealand dependency of Niue. At 31 December 2002 there were an estimated 15,767 adherents in the diocese. The Bishop participates in the Catholic Bishops' Conference of the Pacific, based in Fiji.

Bishop of Tonga: Dr SOANE LILO FOLIAKI, Toutai-mana Catholic Centre, POB 1, Nuku'alofa; tel. 23822; fax 23854; e-mail cathbish@kalianet.to.

Other Churches

Church of Jesus Christ of Latter-day Saints (Mormon): Mission Centre, POB 58, Nuku'alofa; tel. 26007; fax 23763; 40,000 mems; Pres. DOUGLAS W. BANKS.

Church of Tonga: Nuku'alofa; f. 1928; a branch of Methodism; 6,912 mems; Pres. Rev. FINAU KATOANGA.

Free Constitutional Church of Tonga: POB 23, Nuku'alofa; tel. 23966; fax 24458; f. 1885; 15,941 mems (1996); Pres. Rev. SEMISI FONUA; brs in Australia, New Zealand and USA.

Free Wesleyan Church of Tonga (Koe Siasi Uesiliana Tau'ataina 'o Tonga): POB 57, Nuku'alofa; tel. 23522; fax 24020; e-mail fwc@kalianet.to; f. 1826; 36,500 mems; Pres. Rev. Dr 'ALIFALETI MONE.

Tokaikolo Christian Fellowship: Nuku'alofa; f. 1978 as breakaway group from Free Wesleyan Church; 5,000 mems.

BAHÁ'Í FAITH

National Spiritual Assembly: POB 133, Nuku'alofa; tel. 21568; fax 23120; e-mail nsatonga@oceanoflight.to; mems resident in 142 localities.

The Press

Eva, Your Guide to Tonga: POB 958, Nuku'alofa; tel. 25779; fax 24749; e-mail vapress@kalianet.to; internet www.matangitonga.to; f. 1989; 4 a year; Editor PESI FONUA; circ. 4,500.

Ko e Kele'a (Conch Shell): POB 1567, Nuku'alofa; tel. 25501; fax 26330; f. 1986; monthly; activist-oriented publication, economic and political; Editor TAVAKE FUSIMALOHI; circ. 3,500.

Lali: Nuku'alofa; f. 1994; monthly; English; national business magazine; Publr KALAFI MOALA.

Lao and Hia: POB 2808, Nuku'alofa; tel. 14105; weekly; Tongan; legal newspaper; Editor SIONE HAFOKA.

Matangi Tonga: POB 958, Nuku'alofa; tel. 25779; fax 24749; e-mail vapress@kalianet.to; internet www.matangitonga.to; f. 1986; monthly; national news magazine, licence suspended in Feb. 2004; Editor PESI FONUA; circ. 2,000.

'Ofa ki Tonga: c/o Tokaikolo Fellowship, POB 2055, Nuku'alofa; tel. 24190; monthly; newspaper of Tokaikolo Christian Fellowship; Editor Rev. LIUFAU VAILEA SAULALA.

Taumu'a Lelei: POB 1, Nuku'alofa; tel. 27161; fax 23854; e-mail tmlcath@kalianet.to; f. 1931; monthly; Roman Catholic; Editor Dr SOANE LILO FOLIAKI.

The Times of Tonga/Koe Taimi'o Tonga: POB 880, Hala Velingatoni, Kolomotu'a; tel. 23177; fax 23292; e-mail times@kalianet.to; internet www.tongatimes.com; f. 1989; twice-weekly; English edition covers Pacific and world news, Tongan edition concentrates on local news; licence suspended in Feb. 2004; Publr KALAFI MOALA; Editor MATENI TAPUELUELU; circ. 8,000.

Tohi Fanongonongo: POB 57, Nuku'alofa; tel. 26533; fax 24020; e-mail fwctf@kalianet.to; monthly; Wesleyan; Editor Rev. TEVITA PAUKAMEA TIUETI.

Tonga Chronicle/Kalonikali Tonga: POB 197, Nuku'alofa; tel. 23302; fax 23336; e-mail chroni@kalianet.to; internet www.netstorage.com/kami/tonga/news; f. 1964; govt-sponsored; weekly; Editor MATEAKI-KIHE-IOTU HEIMULI; circ. 6,000 (Tongan and English).

Publisher

Vava'u Press Ltd: POB 958, Nuku'alofa; tel. 25779; fax 24749; e-mail vapress@kalianet.to; internet www.matangitonga.to; f. 1980; books and magazines; Pres. PESI FONUA.

Broadcasting and Communications

TELECOMMUNICATIONS

Tonga Communications Corporation: Private Bag 4, Nuku'alofa; tel. 26700; fax 26701; internet www.tcc.to; responsible for domestic and international telecommunications services; Gen. Man. MICHAEL W. SKINNER.

Tongasat—Friendly Islands Satellite Communications Ltd: POB 2921, Nuku'alofa; tel. 24160; fax 23322; e-mail kite@tongasat.com; 80% Tongan-owned; private co but co-operates with Govt in management and leasing of orbital satellite positions; Chair. Princess PILOLEVU TUITA; Man. Dir SIONE KITE; Sec. CLIVE EDWARDS.

BROADCASTING

Radio

Tonga Broadcasting Commission: POB 36, Tungi Rd, Fasi-moe-afi, Nuku'alofa; tel. 23555; fax 24417; independent statutory board; commercially operated; manages two stations, A3Z Radio Tonga 1 and Radio Tonga 2, with programmes in Tongan and English; Gen. Man. TAVAKE FUSIMALOHI.

93FM: Pacific Partners Trust, POB 478, Nuku'alofa; tel. 23076; fax 24970; broadcasts in English, Tongan, German, Mandarin and Hindi.

A3V The Millennium Radio 2000: POB 838, Nuku'alofa; tel. 25891; fax 24195; e-mail a3v@tongatapu.net.to; broadcasts on FM; musical programmes; Gen. Man. SAM VEA.

Tonga News Association: Nuku'alofa; Pres. PESI FONUA.

Television

The introduction of a television service has been mooted since 1984. Oceania Broadcasting Inc started relaying US television programmes in 1991. The Tonga Broadcasting Commission was expected to launch the country's first television service in July 2000.

Oceania Broadcasting Network: POB 91, Nuku'alofa; tel. 23314; fax 23658.

Finance

(cap. = capital; res = reserves; dep. = deposits; m. = million; amounts in Tongan dollars)

BANKING

Australia and New Zealand Banking Group Ltd: Cnr of Salote and Railway Rds, POB 910, Nuku'alofa; tel. 24944; fax 23870; internet www.candw.to/banks; Gen.Man. PAUL PELZER.

Bank of Tonga: POB 924, Nuku'alofa; tel. 23933; fax 23634; e-mail bot-gm@kalianet.to; f. 1974; owned by Govt of Tonga (40%) and Westpac Banking Corpn (60%); cap. 3.0m., res 10.6m., dep. 88.6m. (Sept. 2003); Chair. ALAN WALTER; Gen. Man. MISKA TU'IFA; 4 brs.

MBf Bank Ltd: POB 3118, Nuku'alofa; tel. 24600; fax 24662; e-mail mbfbank@kalianet.to; 93.35% owned by MBf Asia Capital Corpn Holdings Ltd, 4.75% owned by Crown Prince Tupouto'a, 0.95% owned by Tonga Investments Ltd, 0.95% owned by Tonga Co-operative Federation Society.

National Reserve Bank of Tonga: POB 25, Post Office, Nuku'alofa; tel. 24057; fax 24201; e-mail nrbt@reservebank.to; internet www.reservebank.to; f. 1989 to assume central bank functions of Bank of Tonga; issues currency; manages exchange rates and international reserves; cap. 1.0m., res 0.7m., dep. 61.9m. (June 2002); Gov. SIOSI C. MAFI; Chair. Prince 'ULUKALALA-LAVAKA-ATA.

Tonga Development Bank: Fatafehi Rd, POB 126, Nuku'alofa; tel. 23333; fax 23775; e-mail tdevbank@tdb.to; internet www.tdb.to; f. 1977 to provide credit for developmental purposes, mainly in agriculture, fishery, tourism and housing; cap. 10.5m., res 5.927m. (Dec. 2003); Man. Dir 'OTENIFI AFU'ALO MATOTO; 5 brs.

Westpac Banking Corporation: c/o Bank of Tonga, Railway Rd, Nuku'alofala; tel. 23933; fax 24048; Gen. Man. MISKA TU'IFUA.

Trade and Industry

DEVELOPMENT ORGANIZATIONS

Tonga Investments Ltd: POB 27, Nuku'alofa; tel. 24388; fax 24313; f. 1992 to replace Commodities Board; govt-owned; manages five subsidiary companies; Chair. Baron VAEA OF HOUMA; Man. Dir ANTHONY WAYNE MADDEN.

Tonga Association of Small Businesses: Nuku'alofa; f. 1990 to cater for the needs of small businesses; Chair. SIMI SILAPELU.

CHAMBER OF COMMERCE

Tonga Chamber of Commerce and Industries: Tungi Arcade, POB 1704, Nuku'alofa; tel. 25168; fax 26039; e-mail chamber@kalianet.to; Pres. AISAKE EKE.

TRADE ASSOCIATIONS

Tonga Kava Council: Nuku'alofa; to promote the development of the industry both locally and abroad; Chair. TOIIMOANA TAKATAKA.

Tonga Squash Council: Nuku;alofa; promotes the development of the industry; introduced a quota system for exports in 2004; Pres. TSUTOMU NAKAO; Sec. STEVEN EDWARDS.

UTILITIES

Shoreline Power Group: POB 47, Taufa'ahau Rd, Kolofo'ou, Nuku'alofa; tel. 23311; fax 23632; provides electricity via diesel motor generation, took over operations from the Tonga Electric Power Board in 2004; CEO SOANE RAMANLAL; Chair Prince TUPOUTO'A.

Tonga Water Board: POB 92, Taufa'ahau Rd, Kolofo'ou, Nuku'alofa; tel. 23298; fax 23518; operates four urban water systems, serving about 25% of the population; Man. SAIMONE P. HELU.

MAJOR COMPANIES

Royal Beer Co Ltd: POB 20 Nuku'alofa.

Sea Star Fishing Co Ltd: Faua Wharf Basin, Ma'ufanga, Nuku'alofa; tel. 25458; fax 24779; e-mail seastar@kalianet.to; Financial Man. TEVITA VEIKOSO.

CO-OPERATIVES

In April 1990 there were 78 registered co-operative societies, including the first co-operative registered under the Agricultural Organization Act.

Tonga Co-operative Federation Society: Tungi Arcade, Nuku'alofa.

TRADE UNIONS

Association of Tongatapu Squash Pumpkin Growers: Nuku'alofa; f. 1998.

Tonga Nurses' Association and Friendly Islands Teachers' Association (TNA/FITA): POB 150, Nuku'alofa; tel. 23200; fax 24291; Pres. FINAU TUTONE; Gen. Sec. 'ANA FOTU KAVAEFIAFI.

Transport

ROADS

Total road length was estimated at 680 km in 1996, of which some 27% were all-weather paved roads. Most of the network comprises fair-weather-only dirt or coral roads.

SHIPPING

The chief ports are Nuku'alofa, on Tongatapu, and Neiafu, on Vava'u, with two smaller ports at Pangai and Niuatoputapu.

Shipping Corporation of Polynesia Ltd: Vuna Rd, POB 453, Nuku'alofa; tel. 23853; fax 22334; e-mail shipcorp@kalianet.to; regular inter-islands passenger and cargo services; Chair. Prince 'ULUKALALA-LAVAKA-ATA; Gen. Man. Capt. VOLKER PAHL.

Uata Shipping Lines: 'Uliti Uata, POB 100, Nuku'alofa; tel. 23855; fax 23860.

Warner Pacific Line: POB 93, Nuku'alofa; tel. 21088; services to Samoa, American Samoa, Australia and New Zealand; Man. Dir MA'AKE FAKA'OSIFOLAU.

CIVIL AVIATION

Tonga is served by Fua'amotu International Airport, 22 km from Nuku'alofa, and airstrips at Vava'u, Ha'apai, Niuatoputapu, Niuafo'ou and 'Eua. The country's international airline, Royal Tongan Airlines, collapsed in May 2004 (see History). A new domestic carrier was established in June 2004 to provide limited internal air services (see below).

Peau 'o Vava'u: Vava'u; f. 2004; began operating domestic flights in June 2004 following the collapse of Royal Tongan Airlines in May 2004; operates a fleet of two DC-3s and one 10-seat Islander aircraft with technical support from Pion Air of New Zealand; Man. MOSIKAKA MOENGANGONGO.

Tourism

Tonga's attractions include scenic beauty and a mild climate. There were 33,868 visitors to the islands in 1999/2000 (an increase of more than 20% compared with the previous year). In addition 5,151 cruise-ship passengers visited the islands in 1999/2000. In 2003 some 40,110 tourists visited Tonga. In 2002 revenue from the industry earned US $9m. (an increase of some 14% compared with the previous year). The majority of tourists were from New Zealand, the USA and Australia. The collapse of the national airline Royal Tongan Airlines in mid-2004 was expected to have a detrimental effect on the tourist industry.

Tonga Tourist Association: POB 74, Nuku'alofa; tel. 23344; fax 23833; e-mail royale@kalianet.to; Pres. JOSEPH RAMANLAL.

Tonga Visitors' Bureau: Vuna Rd, POB 37, Nuku'alofa; tel. 25334; fax 23507; e-mail tvb@kalianet.to; internet www.tongaholiday.com; f. 1978; Dir VA'INGA PALU.

Defence

Tonga has its own defence force, the Tonga Defence Services, consisting of both regular and reserve units. The island also has a defence co-operation agreement with Australia. Projected government expenditure on defence in 1999/2000 was $T3.3m. (5.0% of total budgetary expenditure); in the same financial year estimated expenditure on law and order was $T4.6m. (6.9% of total current expenditure).

Education

Free state education is compulsory for children between five and 14 years of age, while the Government and other Commonwealth countries offer scholarship schemes enabling students to go abroad for higher education. In 1999 there were 117 primary schools, with a total of 16,206 pupils, and there was a total of 13,987 pupils in 39 secondary schools. There were also four technical and vocational colleges in 1999, with a total of 467 students, and one teacher-training college, with 288 students. In 1990 there were 230 Tongans studying overseas. Some degree courses are offered at the university division of 'Atenisi Institute. A new establishment offering higher education, the 'Unuaki 'o Tonga Royal Institute (UTRI), was expected to open in 2004. Recurrent government expenditure on education in 1999/2000 was an estimated $T14.8m. (equivalent to 12.9% of total recurrent budgetary expenditure).

TUVALU

Physical and Social Geography

Tuvalu is a scattered group of nine small atolls, (five of which enclose sizeable lagoons), extending 560 km from north to south and covering a land area of 26 sq km (10 sq miles) in the western Pacific Ocean. Its nearest neighbours are Fiji to the south, Kiribati to the north and Solomon Islands to the west. The climate is warm and pleasant, with a mean annual temperature of 29°C (84°F), and there is very little seasonal variation. The average annual rainfall is 3,000 mm. The inhabitants are a Polynesian people who speak Tuvaluan and English. Almost all profess Christianity, and about 98% are Protestants. The population was estimated to be 11,020 at mid-2003. The capital is on Funafuti Atoll.

History

Tuvalu was formerly known as the Ellice (or Lagoon) Islands. Between about 1850 and 1875 many of the islanders were captured by slave-traders and this, together with European diseases, reduced the population from about 20,000 to 3,000. In 1877 the United Kingdom established the Western Pacific High Commission, with its headquarters in Fiji, and the Ellice Islands and other groups were placed under its jurisdiction. In 1892 a British protectorate was declared over the Ellice Islands and the group was linked administratively with the Gilbert Islands to the north. In 1916 the United Kingdom annexed the protectorate, which was renamed the Gilbert and Ellice Islands Colony (GEIC). During the Japanese occupation of the Gilbert Islands in 1942–43, the administration of the GEIC was temporarily moved to Funafuti in the Ellice Islands. (For more details of the history of the GEIC, see the chapter on Kiribati.)

A series of advisory and legislative bodies prepared the GEIC for self-government. In May 1974 the last of these, the Legislative Council, was replaced by the House of Assembly, with 28 elected members (including eight Ellice Islanders) and three official members. A Chief Minister was elected by the House and chose between four and six other ministers, one of whom had to be from the Ellice Islands.

In January 1972 the appointment of a separate GEIC Governor, who assumed most of the functions previously exercised by the High Commissioner, increased the long-standing anxiety of the Ellice Islanders over their minority position as Polynesians in the colony, dominated by the Micronesians of the Gilbert Islands. In a referendum held in the Ellice Islands in August and September 1974, over 90% of the voters favoured separate status for the group, and in October 1975 the Ellice Islands, under the old native name of Tuvalu ('eight standing together', which referred to the eight populated atolls), became a separate British dependency. The Deputy Governor of the GEIC took office as Her Majesty's Commissioner for Tuvalu. The eight Ellice representatives in the GEIC House of Assembly became the first elected members of the new Tuvalu House of Assembly. They elected one of their number, Toaripi Lauti, to be Chief Minister. Tuvalu was completely separated from the GEIC administration in January 1976. The remainder of the GEIC was renamed the Gilbert Islands and achieved independence, under the name of Kiribati, in July 1979.

Tuvalu's first separate elections were held in August 1977, when the number of elective seats in the House of Assembly was increased to 12. An independence Constitution was finalized at a conference in London in February 1978. After five months of internal self-government, Tuvalu became independent on 1 October 1978, with Lauti as the first Prime Minister. The pre-independence House of Assembly was redesignated Parliament. Tuvalu is a 'special member' of the Commonwealth, taking part in functional activities but not represented at meetings of Heads of Government. Tuvalu was admitted to the UN on 5 September 2000.

In February 1979 Tuvalu signed a Treaty of Friendship with the USA, which renounced its claim, dating from 1856, to the four southernmost atolls. This treaty was ratified in August 1983. Meanwhile, following elections to the Parliament in September 1981, Dr Tomasi Puapua became Prime Minister.

In a general election in 1985 Puapua was re-elected Prime Minister. In February 1986 a nation-wide poll was conducted to establish public sentiment as to whether Tuvalu should remain a constitutional monarchy, with the British monarch at its head, or become a republic. On only one atoll did the community appear to favour republican status. Under a revised Constitution that took effect on 1 October 1986, the Governor-General's ability to veto government measures was abolished. Meanwhile, in March, Tupua (later Sir Tupua) Leupena, a former Speaker of Parliament, had become Governor-General, replacing Sir Penitala Teo, who had occupied the post since independence in 1978.

In September 1989 a general election was held, and in the following month Bikenibeu Paeniu was sworn in as the new Prime Minister. In October 1990 Toaripi Lauti, the former Prime Minister, succeeded Sir Tupua Leupena as Governor-General. Legislation approved by Parliament in mid-1991, which sought to prohibit all new religions from the islands and to establish the Church of Tuvalu as the State Church, caused considerable controversy and extensive debate. A survey showed the eight constituencies to be almost equally divided over the matter, although Paeniu firmly opposed the motion, describing it as incompatible with basic human rights.

In 1991 the Government announced that it was to prepare a compensation claim against the United Kingdom for the allegedly poor condition of Tuvalu's economy and infrastructure at the time of the country's achievement of independence in 1979. Moreover, Tuvalu was to seek additional compensation for damage caused during the Second World War when the United Kingdom gave permission for the USA to build airstrips on the islands (some 40% of Funafuti is uninhabitable because of the large pits created by US troops during the construction of an airstrip on the atoll). Relations with the United Kingdom deteriorated further in late 1992 when the British Government harshly criticized the financial policy of Paeniu's Government. Paeniu defended his Government's policies, and stated that continued delays in the approval of aid projects from the United Kingdom meant that Tuvalu would not be seeking further development funds from the British Government.

In mid-1992 a member of Parliament for Funafuti proposed a motion to establish Tuvalu as a republic. It was subsequently reported, however, that (as in 1986) only one of the eight parliamentary constituencies supported the proposal.

At a general election held in September 1993 three of the 12 incumbent members of Parliament lost their seats. At elections to the premiership held in the same month, however, Paeniu and Puapua received six votes each. When a second vote produced a similar result, the Governor-General dissolved Parliament, in accordance with the Constitution. Paeniu and his Cabinet remained in office until the holding of a further general election in November. At elections to the premiership in the following month Kamuta Latasi defeated Paeniu by seven votes to five. Puapua, who had agreed not to challenge Paeniu in the contest in favour of supporting Latasi, was elected Speaker of Parliament. In June 1994 Latasi removed the Governor-General, Toomu Malaefono Sione, from office, some seven months after he had been appointed to the position, and replaced him with Tulaga (later Sir Tulaga) Manuella. Latasi alleged that Paeniu's appointment of Sione had been politically motivated.

In December 1994, in what was widely regarded as a significant rejection of its political links with the United Kingdom, the Tuvaluan Parliament voted to remove Britain's union flag from the Tuvalu national flag. A new design was selected and the new flag was inaugurated in October 1995. Speculation that the British monarch would be removed as Head of State intensified during 1995, following the appointment of a committee to review the Constitution. The three-member committee was to examine the procedure surrounding the appointment and removal of the Governor-General, and, particularly, to consider the adoption of a republican system of government.

In late 1996 the Deputy Prime Minister, Otinielu Tausi, and the parliamentary Speaker, Dr Tomasi Puapua, both announced their decision to withdraw their support for Latasi's Government, thereby increasing the number of opposition members in Parliament from five to seven. This reversal appeared to be in response to increasing dissatisfaction among the population with Latasi. This had been perceived firstly with his unpopular initiative to replace the country's national flag, and was exacerbated by revelations that the leasing of Tuvalu's telephone code to a foreign company had resulted in the use of the islands' telephone system for personal services considered indecent by the majority of islanders. (It was announced by the Government in October 2000 that the lease was to cease by the end of the year.) Opponents of the Prime Minister submitted a parliamentary motion of 'no confidence' in his Government in December, which was approved by seven votes to five. Paeniu subsequently defeated Latasi, by a similar margin, to become Prime Minister, and a new Cabinet was appointed. The new premier acted promptly to restore the country's original flag, by proposing a parliamentary motion in February 1997, which was approved by seven votes to five.

A total of 35 candidates contested a general election on 26 March 1998. The period prior to the election had been characterized by a series of bitter disputes between Paeniu and Latasi, in which both had made serious accusations of sexual and financial misconduct against the other. Five members of the previous Parliament were returned to office, although Latasi unexpectedly failed to secure re-election. Paeniu was subsequently re-elected Prime Minister by 10 votes to two. In June the new Government announced a series of development plans and proposals for constitutional reform, including the introduction of a code of conduct for political leaders and the creation of an ombudsman's office. Paeniu stated that his administration intended to consult widely with the population before any changes were implemented. Also in 1998, Puapua was appointed Governor-General, replacing Manuella.

On 13 April 1999 Paeniu lost a parliamentary vote of confidence and was forced to resign. Later in the month Ionatana Ionatana, hitherto the Minister for Health, Education, Culture, Women and Community Affairs, was elected by Parliament as the new Prime Minister. On his appointment Ionatana immediately effected a reshuffle of the Cabinet.

Chronic water shortage on Funafuti led to a state of emergency being declared on the atoll in August 1999. It was hoped that a desalination plant, provided by Japan, would help alleviate the problem.

Potentially the most significant new source of revenue for many years was established in September 1998, when the Government signed an agreement to lease the country's national internet suffix '.tv' to a Canadian information company. The company, which defeated several other business interests to secure the deal, was expected to market the internet address to international television companies. Tuvalu was to receive an initial fee of US $50m. from the arrangement in addition to annual revenue of up to US $100m. In July 1999 it was announced that the deal had been abandoned after the company failed to meet its initial agreed payment of US $50m. In February 2000, however, it was announced that a US $50m. deal on the sale of the '.tv' suffix had been concluded with a US company. The sale was expected to generate some US $10m. in revenue annually. The funds generated from the sale enabled Tuvalu officially to join the UN and participate in the 55th annual UN General Assembly Meeting, held in September 2000.

In March 2000 18 schoolgirls and their supervisor were killed in a fire in a school on Vaitupu atoll. A government inquiry was established into the disaster, which was reportedly the worst in independent Tuvalu's history.

In December 2000 Prime Minister Ionatana Ionatana died unexpectedly. The Deputy Prime Minister, Lagitupu Tuilimu, was immediately appointed as interim Prime Minister, pending the election of a replacement. In February 2001 Parliament elected as Prime Minister the Minister for Internal Affairs and Rural and Urban Development, Faimalaga Luka; he assumed responsibility for the additional portfolios of foreign affairs, finance and economic planning, and trade and commerce, and immediately named a new Cabinet.

A vote of 'no confidence' was upheld against Luka in December 2001, while he was in New Zealand for a medical examination. Kolaoa Talake, a former Minister for Finance, was elected Prime Minister in the same month, winning eight of the 15 votes cast (the number of parliamentary seats having been increased from 12 to 15). He appointed an entirely new Cabinet.

Talake announced in March 2002 that lawyers were preparing evidence for legal action against the United Kingdom, seeking compensation for the alleged inequality of the division of assets between Tuvalu and Kiribati when the two nations had achieved independence in the late 1970s. Following the general election held on 25 July 2002, Saufatu Sopoanga, a former Minister for Finance, defeated Amasone Kilei, the opposition candidate, by eight votes to seven to become the new Prime Minister. Sopoanga subsequently announced his intention to hold a referendum on the adoption of a republican system of government in Tuvalu.

In May 2003 two by-elections resulted in the loss of the Government's one-seat majority. The Government's subsequent refusal to convene Parliament (allegedly in order to evade a vote of 'no confidence') was strongly criticized by the opposition. The Government maintained that it would regain its majority with an imminent defection from the opposition and would then convene Parliament. In July, however, the situation remained unchanged and the opposition consequently sought a court order obliging Sopoanga to convene Parliament. The appointment, in early September, of Faimalaga Luka, hitherto Speaker and a member of Parliament, as the country's new Governor-General necessitated an additional by-election which resulted in a further delay to Parliament's being convened. However, following the success of its candidate at the by-election, and (as anticipated) the defection of an opposition member, the Government regained its majority in mid-October. Sopoanga subsequently announced that Parliament would finally be convened in early November.

In April 2004 the Government announced that a team of officials was touring the outer islands to canvas opinion on the adoption of republican status for Tuvalu. If islanders indicated sufficient support for the proposal, it was thought that a referendum might be held by mid-2005.

In August 2004 the Prime Minister, Saufatu Sopoanga, was ousted by nine parliamentary votes to five in a vote of 'no confidence' after a member of the Government crossed the floor to vote with the opposition and was joined by the Speaker. The election of a new Prime Minister, however, was delayed by Sopoanga's decision to relinquish his seat, thus necessitating the organization of a by-election before Parliament could select a premier. Deputy Prime Minister Maatia Toafa assumed the role of acting Prime Minister in the interim. Sopoanga regained his seat in a by-election in early October, and on 11 October acting Prime Minister Maatia Toafa was elected to the premiership, defeating Sopoanga by eight votes to seven.

Tuvalu is a member of the Pacific Islands Forum (formerly the South Pacific Forum). The 19th Forum, meeting in September 1988, discussed the threat posed to low-lying countries, such as Tuvalu, by the predicted rise in sea-level as a result of the 'greenhouse effect' (the heating of the earth's atmosphere as a consequence of pollution); the highest point of Tuvalu's nine islands is no more than five metres above sea-level. A report published by the UN Environment Programme in late 1989 warned that Tuvalu was one of five island groups that were particularly threatened, and which, unless drastic measures were taken, could be completely submerged by the mid-21st century. At the UN World Climate Conference, held in Geneva in November 1990, Paeniu appealed for urgent action by developed nations to combat the environmental changes caused by the 'greenhouse effect', which were believed to include a 10-fold increase in cyclone frequency (from two in 1940 to 21 in 1990), an increase in salinity in ground water and a considerable decrease in the average annual rainfall. The Government remained critical, however, of the inertia with which it alleged certain countries had reacted to its appeal for assistance, and reiterated the Tuvaluan people's fears of physical and cultural extinction. The subsequent Prime Minister, Kamuta Latasi, was similarly critical of the industrial world's apparent disregard for the plight of small island nations vulnerable to the effects of climate change, particularly when Tuvalu was struck by tidal waves in 1994 (believed to be the first experienced by the islands). Attempts during the mid-1990s to secure approval for resettlement plans for Tuvaluans to other countries, including Australia and New Zealand, were largely unsuccessful. The Government of Tuvalu was strongly critical of Australia's refusal to reduce its emission of pollutant gases (known to contribute to the 'greenhouse effect') at the Conference of the Parties to the Framework Convention on Climate Change (see UN Environment Programme) in Kyoto, Japan, in late 1997. In July 2001 Australia did, however, adopt the Kyoto Protocol, which urged industrialized nations to reduce carbon dioxide emissions by 5.2% from 1990 levels by 2012. In March 2001 Tuvalu, Kiribati and the Maldives announced their decision to take legal action against the USA for its refusal to sign the Kyoto Protocol. In August Tuvalu was one of six states at the Pacific Islands Forum to demand a meeting with US President George W. Bush, in an effort to enlist his support for the Kyoto Protocol; the USA produced nearly one-third of the industrialized countries' carbon dioxide emissions and was the only country in the world not to adopt the protocol.

The installation of a new sea-level monitoring station began in December 2001 as part of the South Pacific Sea Level and Climate Monitoring Project administered by the Australian aid agency, AusAID. In January 2002 it was reported that the Government had engaged a US law firm to prosecute the USA and other nations for failing to meet their commitments to the UN Framework Convention on Climate Change (UNFCC). At the September 2002 World Summit for Sustainable Development in Johannesburg, South Africa, government delegates from Tuvalu attempted to persuade representatives of other Pacific islands to join their campaign against the USA and Australia. In September 2003 Tuvalu's Prime Minister addressed the 58th session of the UN General Assembly in New York, USA, and appealed for collective action to mitigate the impact of climate change and rising sea-levels on the islands. He once again urged all industrialized nations, particularly the USA, to sign the Kyoto Protocol. In the following month the Government of Niue, which had for many years suffered from a decline in population owing to the migration of its inhabitants to New Zealand, officially invited the residents of Tuvalu to resettle on Niue. The feasibility of this offer, however, was in question following the devastation caused to Niue by a cyclone in January 2004.

Tuvalu was subject to considerable international criticism in mid-2004 regarding its decision to join the International Whaling Commission. Environmental and animal welfare groups accused the Tuvaluan Government of accepting financial incentives from Japan in return for agreeing to use its vote to support a removal of the ban on commercial whaling at the commission's annual meeting in Italy in July 2004.

Economy

In 1999, according to UN estimates, Tuvalu's gross domestic product (GDP), measured at current prices, was US $16m., equivalent to US $1,556 per head. According to provisional estimates by the Asian Development Bank (ADB), in 2001 GDP totalled US $13m., equivalent to US $1,260 per head. Overall GDP increased, in real terms, by an annual average rate of 4.9% in 1990–2000. Compared with the previous year, GDP rose by an estimated 5.9% in 2001. GDP was estimated to have expanded by 1.3% in 2002 and by 7.0% in 2003. A growth rate of 3.0% was projected for 2004.

Tuvalu is composed of coral atolls with poor-quality soil. Most of the land is covered with coconut palms, which provide the only export in the form of copra. Exports of copra were worth $A7,376 for 32 metric tons in 2001. Agriculture (including fishing), which is, with the exception of copra production, of a basic subsistence nature, contributed some 16.6% of GDP in 2002. In that year, according to FAO, the sector engaged some 25% of the labour force. According to ADB figures, the GDP of the agricultural sector declined by an average annual rate of 1.2% in 1990–2002. Compared with the previous year, agricultural GDP decreased by 9.4% in 2002. Pulaka, taro, papayas, the screw-pine (*Pandanus*) and bananas are cultivated as food crops, and honey is produced. There is also subsistence farming of pigs, goats and poultry. Fishing is carried out on a small scale but, with the introduction of an exclusive economic zone covering about 1.3m. sq km (500,000 sq miles) of sea, exploitation of fish resources could be developed to form the basis of the economy. The sale of fishing licences to foreign fleets is an important source of revenue and earned $A11.8m. in 2001 (compared with $A3.6m. in 1997), equivalent to 50.3% of current revenue. Revenue from this source, however, declined in subsequent years.

Industry (including mining, manufacturing, construction and utilities) accounted for 14.8% of GDP in 2002. In 1990–2002, according to ADB figures, industrial GDP expanded at an average annual rate of 5.9%. Compared with the previous year, the sector's GDP increased by 6.5% in 2002. Manufacturing is confined to the small-scale production of coconut-based products, soap and handicrafts. The manufacturing sector contributed some 3.7% of GDP in 2002.

Energy is derived principally from a power plant (fuelled by petroleum) and, on the outer islands, solar power. In 1989 mineral fuels accounted for almost 13% of total import costs.

The Government is an important employer (engaging 1,185 people in 2001, equivalent to about one-half of the labour force) and consequently the services sector makes a relatively large contribution to Tuvalu's economy (providing some 68.6% of GDP in 2002). In 1990–2002, according to ADB figures, the GDP of the services sector increased at an average annual rate of 5.4%. Compared with the previous year, the sector's GDP expanded by 2.2% in 2002. The islands' remote situation and lack of amenities have hindered the development of a tourist industry. In the mid-1990s there was only one hotel (on Funafuti), with 17 rooms. Visitor arrivals totalled 1,496 in 2003 (compared with 770 in 1999). An important source of revenue has been provided by remittances from Tuvaluans working abroad. In the early 1990s some 1,200 Tuvaluans were working overseas, principally in the phosphate industry on Nauru, although many of these workers returned to Tuvalu during the late 1990s, as Nauruan phosphate reserves became exhausted. Some 300 Tuvaluans were working on Nauru in 2003, although the Government subsequently repatriated the majority of these workers, who had been suffering considerable hardship since Nauru's financial crisis resulted in the widespread non-payment of salaries. Remittances from some 450 Tuvaluan seafarers employed on foreign (predominantly German) merchant ships were estimated at more than $A5m. in 2003 (equivalent to some 20% of GDP). In 2001 receipts from the leasing of the islands' internet domain address reached US $1.6m., while revenue from telecommunication licence fees totalled US $0.31m.

In 2003 Tuvalu recorded a visible trade deficit of US $20.71m. The cost of imports reached US $22.16m., while export revenue totalled only US $1.45m. In 2003 imports totalling US $10.62m. were purchased from Fiji (which supplied 47.9% of the total) and US $3.42m. from Australia (15.4%). The principal markets for exports were the United Kingdom, which purchased 50.3% of the total, and Australia, which provided 13.8% of export earnings. In 2001 copra was the sole domestic export of any significance. The principal imports in 2000 were foodstuffs (some 35% of the total), mineral fuels (about 8%) and construction materials (some 5%). In 2003 total export earnings were equivalent to only 6.5% of the value of imports.

In 1986 the Government drafted plans to establish a $A27m. trust fund to finance Tuvalu's budget requirements. In 1987 the British, Australian and New Zealand Governments signed an agreement whereby the United Kingdom pledged to contribute $A8.5m. to the Tuvalu Trust Fund, Australia $A8m. and New Zealand $A8.2m. The market value of the Fund stood at some $A75.8m. in September 2003. Official development assistance declined from US $10.1m in 1997 to US $5.2m. in 1998. In 2003/04 New Zealand provided bilateral assistance worth $NZ2.05m., while aid from Australia totalled $A3.8m. The 2003 budget allowed for operating expenditure of $A25.1m. (some 38% higher than expenditure in the previous year) and for special development spending of $A1.8m. Revenue was projected to total $A20.7m. in the same year (some 45% lower than in 2002). An overall budgetary deficit of $A4.4m. was projected, equivalent to 16.3% of GDP (compared with a surplus equivalent to 47% in 2002). The annual rate of inflation averaged 3.2% in 1990–99. The average rate was 3.9% in 2000, 1.5% in 2001, 5.0% in 2002 and 2.9% in 2003, with a slightly lower rate anticipated in 2004.

According to UN criteria, Tuvalu is one of the world's least developed countries, a status that attracts concessionary loans from the World Bank and the IMF, and renders Tuvalu's exports eligible for special tariff rates under provisions of the General Agreement on Tariffs and Trade (GATT, superseded by the World Trade Organization—WTO). Its economic development has been adversely affected by inclement weather (as illustrated by Cyclone Kelo, which devastated the islands of Niulakita and Nukulaelae in mid-1997) and inadequate infrastructure. Tuvalu's vulnerability to fluctuations in the price of copra on the international market and the country's dependence on imports have resulted in a persistent visible trade deficit; it has also remained reliant on foreign assistance for its development budget. In August 1999 a US $4m. loan was secured to establish an outer islands development fund. The Island Development Programme aimed not only to decentralize administration but also to raise the standards of local public services and to encourage the development of small businesses. The capital assets of the Falekaupule Trust Fund, which was established in July 1999 and charged with promoting sustainable increases in funding for the development of the outer islands, reached US $8.2m. in 2001. In February 2000, meanwhile, the sale of the '.tv' internet suffix (see History) substantially increased the island's GDP (revenue from the sales totalled $A24.9m. in 2000, although this declined to some $A3m. in the following year). Government revenues in 2000 were almost twice the projected budgeted amount. Proceeds from the sale were to be used to develop the country's infrastructure and were channelled largely into improving roads and the education system, also allowing the Government to investigate the possibility of buying land in Fiji, should the resettlement of Tuvalu's population become necessary. In an attempt to ensure the continuity of flights to Tuvalu, the Government committed itself to the purchase of majority shares in Air Fiji in March 2002, using a loan from the National Bank of Tuvalu. A survey of the islands published by the Pacific Economic Bulletin (a publication of the Australian National University) claimed that Tuvalu's economy had been hampered significantly by the prolonged period of political instability that the islands had undergone during 2003. Meanwhile, the economy continued to suffer from the increasing impact of climate change on the islands, with high tides flooding homes, government buildings and the airport, and causing damage to agricultural produce in February 2004.

Tuvalu is a member of the UN's Economic and Social Commission for Asia and the Pacific (ESCAP), of the Pacific Community (formerly the South Pacific Commission) and of the Pacific Islands Forum (formerly the South Pacific Forum). Tuvalu was admitted to the Asian Development Bank (ADB) in May 1993.

Statistical Survey

AREA AND POPULATION

Land Area: 26 sq km (10 sq miles).

Population: 9,043 at census of 17 November 1991; 9,561 (males 4,729, females 4,832) at census of 1 November 2002. *By Atoll* (1996): Funafuti 3,836; Vaitupu 1,205; Niutao 749; Nanumea 818; Nukufetau 756; Nanumaga 644; Nui 608; Nukulaelae 370; Niulakita 75. *Mid-2003* (UN estimate, incl. suburbs): Funafuti 5,847 (Source: UN, *World Urbanization Prospects: The 2003 Revision*).

Density (2002 census): 367.7 per sq km.

Principal Town (population at mid-2000): Vaiaku (capital) 4,590.

Births and Deaths (2000): Birth rate 21.8 per 1,000; Death rate 7.7 per 1,000. Source: UN, *Statistical Yearbook for Asia and the Pacific*.

Expectation of Life (WHO estimates, years at birth): 60.6 (males 60.0; females 61.4) in 2002. Source: WHO, *World Health Report*.

Economically Active Population: In 1979 there were 936 people in paid employment, 50% of them in government service. In 1979 114 Tuvaluans were employed by the Nauru Phosphate Co, with a smaller number employed in Kiribati and about 255 on foreign ships. At the 1991 census the total economically active population (aged 15 years and over) stood at 2,383 (males 1,605, females 778). *Mid-2002* (estimates): Agriculture, etc. 1,000; Total labour force 4,000 (Source: FAO).

HEALTH AND WELFARE

Key Indicators

Total Fertility Rate (children per woman, 2002): 2.9.

Under-5 Mortality Rate (per 1,000 live births, 2002): 52.

Health Expenditure (2001): US $ per head (PPP): 673.

Health Expenditure (2001): % of GDP: 5.4.

Health Expenditure (2001): public (% of total): 53.4.

AGRICULTURE, ETC.

Principal Crops (FAO estimates, metric tons, 2002): Coconuts 1,500; Fruit (excl. melons) 650. Source: FAO.

Livestock (FAO estimate, '000 head, 2002): Pigs 13. Source: FAO.

Livestock Products (FAO estimates, metric tons, 2002): Poultry meat 41; Pig meat 90; Hen eggs 21; Honey 2. Source: FAO.

Fishing (FAO estimates, metric tons, live weight, 2002): Total catch 500 (Skipjack tuna 300; Yellowfin tuna 20). Source: FAO.

FINANCE

Currency and Exchange Rates: Australian and Tuvaluan currencies are both in use. Australian currency: 100 cents = 1 Australian dollar ($A). *Sterling, US Dollar and Euro Equivalents* (31 May 2004): £1 sterling = $A2.5685; US $1 = $A1.4000; €1 = $A1.7144; $A100 = £38.93 = US $71.43 = €58.33. *Average Exchange Rate* (US $ per Australian dollar): 1.9334 in 2001; 1.8406 in 2002; 1.5419 in 2003.

Budget (provisional, $A '000, 2002): Revenue 43,280 (Revenue from taxation 5,341, Non-tax revenue 37,939); Expenditure 35,040 (Current expenditure 25,656, Capital expenditure 9,384). Source: Asian Development Bank, *Key Indicators of Developing Asian and Pacific Countries*.

Official Development Assistance (US $ million, 2001): Bilateral 3.8; Multilateral 0.2; Total 4.0 (all grants). Source: UN, *Statistical Yearbook for Asia and the Pacific*.

Cost of Living (Consumer Price Index for Funafuti; base: July–Sept. 2003 = 100): 92.0 in 2001; 96.6 in 2002; 99.4 in 2003. Source: Asian Development Bank, *Key Indicators of Developing Asian and Pacific Countries*.

Gross Domestic Product by Economic Activity ($A '000 at current prices, 2002): Agriculture 4,565; Mining 237; Manufacturing 1,016; Electricity, gas and water 1,433; Construction 1,370; Trade, restaurants and hotels 3,700; Transport, storage and communications 3,429; Finance and real estate 4,055; Public administration 7,188; Community and personal services, *less* imputed bank charges 500; *Total* 27,490. Source: Asian Development Bank, *Key Indicators of Developing Asian and Pacific Countries*.

Balance of Payments ($A '000, 1996): Exports of goods f.o.b. 361; Imports of goods f.o.b. −10,740; *Trade balance* −10,379; Exports of services and other income 10,502; Imports of services and other income −8,758; *Balance on goods, services and income* −8,635; Unrequited transfers (net) 9,082; *Current balance* 447; Capital account (net) 2,088; Net errors and omissions −55; *Overall balance* 2,480. Source: Asian Development Bank, *Key Indicators of Developing Asian and Pacific Countries*.

EXTERNAL TRADE

Principal Commodities (US $ million, 1999): *Imports*: Food and live animals 2.1 (Meat and preparations 0.6; Cereals and cereal preparations 0.6); Beverages and tobacco 0.4; Refined petroleum products 0.7 (Gas oils 0.4); Chemicals and related products 0.4; Basic manufactures 1.2 (Metal manufactures 0.5); Machinery and transport equipment 2.0 (Industrial machinery and equipment 0.5; Road vehicles 0.4); Miscellaneous manufactured articles 0.8; Total (incl. others) 8.1 (Source: UN, *International Trade Statistics Yearbook*). *Exports* ($A '000, 1990): Copra 29.0; Handicrafts 2.0; Stamps 147.0; Total 178.0. *1997:* Copra 6.0.

Principal Trading Partners (US $ million, 2003): *Imports*: Australia 3.42; Fiji 10.62; Germany 1.71; Japan 2.41; New Zealand 1.82; Total (incl. others) 22.16. *Exports*: Australia 0.2; Fiji 0.11; Germany 0.05; Italy 0.08; Poland 0.06; United Kingdom 0.73; Total (incl. others) 1.45. Source: Asian Development Bank, *Key Indicators of Developing Asian and Pacific Countries*.

TRANSPORT

Shipping: *Merchant Fleet* (registered at 31 December 2003): Vessels 12; Total displacement ('000 grt) 60.6. Source: Lloyd's Register-Fairplay, *World Fleet Statistics*.

TOURISM

Tourist Arrivals: 976 in 2001; 1,236 in 2002; 1,496 in 2003. Source: South Pacific Tourism Organisation.

Tourist Arrivals by Country of Residence (2001): Australia 445; Fiji/Kiribati 861; Germany 68; Japan 317; New Zealand 232; United Kingdom 67; USA 101; Total (incl. others) 2,813. Source: *Tuvalu 2002 Economic and Public Sector Review*.

COMMUNICATIONS MEDIA

Non-daily Newspapers (1996): 1; estimated circulation 300*.

Telephones (main lines, 2000): 660 in use†.

Radio Receivers (1997): 4,000 in use*.

Facsimile Machines (1993): 10 in use‡.

* Source: UNESCO, *Statistical Yearbook*.

† Source: International Telecommunication Union.

‡ Source: UN, *Statistical Yearbook*.

EDUCATION

Primary (2001): 9 government schools, 1 private school; 102 teachers; 1,798 pupils.

General Secondary (2001): 1 government school; 32 teachers; 558 pupils. A maritime school offers training for 60 merchant seamen per year, with vocational, technical and commerce-related courses. The University of the South Pacific has an extension centre in Funafuti offering diploma and vocational courses and the first two years of degree courses (the latter requiring completion in Suva, Fiji).

Directory

The Constitution

A new Constitution came into effect at independence on 1 October 1978. Its main provisions are as follows:

The Constitution states that Tuvalu is a democratic sovereign state and that the Constitution is the Supreme Law. It guarantees protection of all fundamental rights and freedoms and provides for the determination of citizenship.

The British sovereign is represented by the Governor-General, who must be a citizen of Tuvalu and is appointed on the recommendation of the Prime Minister. The Prime Minister is elected by Parliament, and up to four other ministers are appointed by the Governor-General from among the members of Parliament, after consultation with the Prime Minister. The Cabinet, which is directly responsible to Parliament, consists of the Prime Minister and the other ministers, whose functions are to advise the Governor-General upon the government of Tuvalu. The Attorney-General is the principal legal adviser to the Government. Parliament is composed of 15 members directly elected by universal adult suffrage for four years, subject to dissolution, and is presided over by the Speaker (who is elected by the members). The Constitution also provides for the operation of a Judiciary (see Judicial System) and for an independent Public Service. Under a revised Constitution that took effect on 1 October 1986, the Governor-General no longer has the authority to reject the advice of the Government.

The Government

HEAD OF STATE

Sovereign: HM Queen ELIZABETH II.

Governor-General: FAIMALAGA LUKA (took office 9 September 2003).

CABINET

(October 2004)

Prime Minister, Minister for Works, Communications and Transport: MAATIA TOAFA.

Deputy Prime Minister: SAUFATU SOPOANGA.

Minister for Foreign Affairs and Labour: TINE LEULU.

Minister for Education, Sports and Culture, and Health: ALESANA KLIES SELUKA.

Minister for Finance and Economic Planning and and Industries: BIKENIBEU PAENIU.

Minister for Home Affairs and Rural Development: OTINIELU T. TAUSI.

Minister for Natural Resources, Energy, the Environment and Tourism: SAMUELU P. TEO.

MINISTRIES

The majority of ministries are situated on Vaiaku, Funafuti; the remainder are on Vaitupu.

Ministry of Tourism, Trade and Commerce: PMB, Vaiaku, Funafuti; tel. 20184; fax 20829; e-mail mttc@tuvalu.tv.

Ministry of Works, Communications and Energy: PMB, Vaiaku, Funafuti; tel. 20055; fax 20772.

Legislature

PARLIAMENT

Parliament has 15 members, who hold office for a term of up to four years. A general election was held on 25 July 2002. There are no political parties.

Speaker: OTINIELU TAUTELEIMALAE TAUSI.

Diplomatic Representation

There are no embassies or high commissions in Tuvalu. The British High Commissioner in Fiji is also accredited as High Commissioner to Tuvalu. Other Ambassadors or High Commissioners accredited to Tuvalu include the Australian, New Zealand, US, French and Japanese Ambassadors in Fiji.

Judicial System

The Supreme Law is embodied in the Constitution. The High Court is the superior court of record, presided over by the Chief Justice, and has jurisdiction to consider appeals from judgments of the Magistrates' Courts and the Island Courts. Appeals from the High Court lie with the Court of Appeal in Fiji or, in the ultimate case, with the Judicial Committee of the Privy Council in the United Kingdom.

There are eight Island Courts with limited jurisdiction in criminal and civil cases.

Chief Justice: Sir GAVEN DONNE (non-resident).

Attorney-General: IAKOBA ITALELEI TAEIA.

Religion

CHRISTIANITY

Te Ekalesia Kelisiano Tuvalu (The Christian Church of Tuvalu): POB 2, Funafuti; tel. 20755; fax 20651; f. 1861; autonomous since 1968; derived from the Congregationalist foundation of the London Missionary Society; some 98% of the population are adherents; Pres. Rev. ETI KINE; Gen. Sec. Rev. FILOIMEA TELIFO.

Roman Catholic Church: Catholic Centre, POB 58, Funafuti; tel. and fax 20527; e-mail cathcent@tuvalu.tv; 117 adherents (31 Dec. 2002); Superior Fr CAMILLE DESROSIERS.

Other churches with adherents in Tuvalu include the Church of Jesus Christ of Latter-day Saints (Mormons), the Jehovah's Witnesses, the New Apostolic Church and the Seventh-day Adventists.

BAHÁ'Í FAITH

National Spiritual Assembly: POB 48, Funafuti; tel. 20860; mems resident in 8 localities.

The Press

Tuvalu Echoes: Broadcasting and Information Office, Vaiaku, Funafuti; tel. 20138; fax 20732; f. 1984; fortnightly; English; Editor MELAKI TAEPE; circ. 250.

Te Lama: Ekalesia Kelisiano Tuvalu, POB 2, Funafuti; tel. 20755; fax 20651; quarterly; religious; Pres. Rev. ETI KINE; Editor Rev. KITIONA TAUSI; circ. 1,000.

Broadcasting and Communications

TELECOMMUNICATIONS

Telecom Tuvalu: Vaiaku, Funafuti; tel. 20010; fax 20002; f. 1994.

BROADCASTING

Tuvalu Media Corporation: PMB, Vaiaku, Funafuti; tel. 20731; fax 20732; e-mail media@tuvalu.tv; f. 1999; govt-owned; Chief Broadcasting and Information Officer PUSINELLI LAAFAI.

Radio

Radio Tuvalu: Broadcasting and Information Office, PMB, Vaiaku, Funafuti; tel. 20138; fax 20732; f. 1975; daily broadcasts in Tuvaluan and English, 43 hours per week; Programme Producer RUBY S. ALEFAIO.

Finance

BANKS

Development Bank of Tuvalu: PMB 9, Vaiaku, Funafuti; tel. 20199; fax 20850; f. 1993 to replace the Business Development Advisory Bureau.

National Bank of Tuvalu: POB 13, Vaiaku, Funafuti; tel. 20803; fax 20802; e-mail nbt@tuvalu.tv; f. 1980; commercial bank; govt-owned; ($A '000) cap. 471.0, res 2,467.5, dep. 24,101.7 (Dec. 2001); Chair. SEVE PAENIU; Gen. Man. IONATANA PEIA; brs on all atolls.

Trade and Industry

GOVERNMENT AGENCY

National Fishing Corporation of Tuvalu (NAFICOT): POB 93, Funafuti; tel. 20724; fax 20800; fishing vessel operators; seafood processing and marketing; agents for diesel engine spare parts, fishing supplies and marine electronics; Gen. Man. SEMU SOPOANGA TAAFAKI.

Tuvalu Philatelic Bureau: POB 24, Funafuti; tel. 20224; fax 20712; e-mail philatelic@tuvalu.tv.

CHAMBER OF COMMERCE

Tuvalu Chamber of Commerce: POB 27, Vaiaku, Funafuti; tel. 20917; fax 20646; e-mail tpasefika@hotmail.com; Chair. MATANILE IOSEFA; Sec. TEO PASEFIKA.

UTILITIES

Electricity

Tuvalu Electricity Corporation (TEC): POB 32, Funafuti; tel. 20350; fax 20351; e-mail thomas@tuvalu.tv.

CO-OPERATIVES

Tuvalu Co-operative Society Ltd: POB 11, Funafuti; tel. 20747; fax 20748; f. 1979 by amalgamation of the eight island socs; controls retail trade in the islands; Gen. Man. MONISE LAAFAI; Registrar SIMETI LOPATI.

Tuvalu Coconut Traders Co-operative: Contact TAAI KATALAKE.

TRADE UNION

Tuvalu Overseas Seamen's Union (TOSU): POB 99, Funafuti; tel. 20609; fax 20610; e-mail tosu@tuvalu.tv; Gen. Sec. TOMMY ALEFAIO.

Transport

ROADS

Funafuti has some impacted-coral roads; elsewhere, tracks exist.

SHIPPING

There is a deep-water lagoon at the point of entry, Funafuti, and ships are able to enter the lagoon at Nukufetau. Irregular shipping services connect Tuvalu with Fiji and elsewhere. The Government operates an inter-island vessel.

CIVIL AVIATION

In 1992 a new runway was constructed with EU aid to replace the grass landing strip on Funafuti. Air Marshall Islands operates a three-weekly service between Funafuti, Nadi (Fiji) and Majuro (Marshall Islands). In June 1995 Tuvalu, Kiribati, the Marshall Islands and Nauru agreed to begin discussions on the establishment of a joint regional airline. The Government of Tuvalu planned to purchase a majority shareholding in Air Fiji in 2002.

Tourism

In 1994 there was one hotel, with 17 rooms, on Funafuti. There were 1,496 tourist arrivals in 2003. On average a similar number of people visit the islands on official business each year. The majority of visitors are from Fiji, Kiribati, Australia, Japan and New Zealand.

Ministry of Tourism, Trade and Commerce: see under Ministries.

Tuvalu Tourism Office: PMB, Funafuti; tel. 20184; fax 20829.

Education

Education is provided by the Government, and is compulsory between the ages of six and 15 years. In 2001 there were nine government and one private primary schools, with a total of 1,798 pupils and 102 teachers . There was one secondary school in 2001. There were 558 secondary pupils and 32 teachers in 2001. The only tertiary institution is the Maritime Training School at Amatuku on Funafuti. About 60 people graduate from the school annually. Further training or vocational courses are available in Fiji and Kiribati. The University of the South Pacific (based in Fiji) has an extension centre on Funafuti. A programme of major reforms in the education system in Tuvalu, begun in the early 1990s, resulted in the lengthening of primary schooling (from six to eight years) and a compulsory two years of secondary education, as well as the introduction of vocational, technical and commerce-related courses at the Maritime Training School. Total government expenditure on education in 2001 was equivalent to some 35% of total budgetary expenditure.

UNITED KINGDOM PACIFIC TERRITORY

There is only one British Dependent Territory remaining in the Pacific, which is the Crown Colony of Pitcairn, Henderson, Ducie and Oeno Islands. Until the end of 1995 the United Kingdom maintained membership of the South Pacific Commission (now Pacific Community) in respect of the Pitcairn Islands. The British Minister responsible for overseas possessions is the Secretary of State for Foreign and Commonwealth Affairs.

Head of State: HM Queen ELIZABETH II (succeeded to the throne 6 February 1952).

Foreign and Commonwealth Office: Whitehall, London, SW1A 2AH, United Kingdom; tel. (20) 7270-3000; internet www.fco.gov.uk.

Secretary of State for Foreign and Commonwealth Affairs: JOHN (JACK) STRAW.

PITCAIRN ISLANDS

Physical and Social Geography

The British Dependent Territory of Pitcairn, Henderson, Ducie and Oeno Islands is commonly known as Pitcairn (after the one inhabited island) or the Pitcairn Islands. Pitcairn Island is situated at 25°04′ S and 130°06′ W, about midway between Panama and New Zealand and 2,172 km (1,350 miles) east-south-east of Tahiti (French Polynesia). It is a rugged and fertile island of volcanic origin, which rises to a height of some 330 m (1,100 ft) and has an area of 4.35 sq km (1.75 sq miles). Even at the only landing place, access from the sea is difficult. The climate is equable, with mean monthly temperatures ranging from 19°C (66°F) in August to 24°C (75°F) in February, and average annual rainfall of 2,000 mm. The resident population has been in decline since 1937 (when it peaked at 233), and in September 2004 numbered 47. There is a large population of some 1,500 descendants of Pitcairn Islanders, most of whom live in New Zealand (an estimated 171 in 2002) and Australia. It was estimated in 1990, however, that fewer than 500 people had been born on Pitcairn since 1790. The official language is English, but most of the islanders use Pitcairnese or Pitkern, a dialect based on 18th-century seafarers' English and Tahitian. The islanders are adherents of the Seventh-day Adventist Church. The chief settlement is Adamstown on Pitcairn Island.

The other three islands of the Territory are uninhabited, although the islanders regularly visit Henderson and Oeno Islands, the former being a large, raised atoll of 30 sq km (11.6 sq miles), 169 km east-north-east of Pitcairn, which provides miro wood, and the latter an atoll of less than 1 sq km and 121 km north-west of Pitcairn. Ducie Island (471 km east of Pitcairn) is the smallest of the four islands, and is largely inaccessible.

History

Pitcairn Island was discovered in 1767, when it was uninhabited, although there is evidence of previous occupation by Polynesian peoples. The island was first settled by the British in 1790, when it was occupied by nine mutineers of HMS Bounty (led by Fletcher Christian), accompanied by 12 women and six men from Tahiti. Despite the violence of the first decade (by 1800 the only surviving adult male was John Adams, who led the community until his death in 1829), the population increased steadily. Concern about the size of the population led to a temporary evacuation to Tahiti in 1831 and, when numbers reached 194 in 1856, the entire community was evacuated to a new home, provided by the British Government, on Norfolk Island. By 1864 43 Pitcairners had returned to the island, which has been permanently settled ever since. Pitcairn officially became a British settlement in 1887. In 1893 a parliamentary form of government was adopted, and in 1898 responsibility for administration was assumed by the High Commissioner for the Western Pacific. Pitcairn came under the jurisdiction of the Governor of Fiji in 1952, and, from 1970 onwards, of the British High Commissioner in New Zealand acting as Governor, in consultation with the Island Council, presided over by the Island Magistrate (who is elected triennially) and comprising one *ex-officio* member (the Island Secretary), five elected and three nominated members.

In 1987 the British High Commissioner in Fiji, acting on behalf of Pitcairn, the United Kingdom's last remaining dependency in the South Pacific, joined representatives of the USA, France, New Zealand and six South Pacific island states in signing the South Pacific Regional Environment Protection Convention, the main aim of which is to prevent the disposal of nuclear waste in the region.

In 1989 uninhabited Henderson Island was included on the UNESCO 'World Heritage List'. The island, 169 km east-north-east of Pitcairn, was to be preserved as a bird sanctuary. As well as many endemic plants, there are five species of bird unique to the island: the flightless rail or Henderson chicken, the green Henderson fruit dove, the Henderson crake, the Henderson warbler and the Henderson lorikeet. However, concern was expressed in 1994, following claims by scientists studying the island that its unique flora and fauna were threatened by the accidental introduction of foreign plant species by visitors, and by an increase in the rat population.

In April 1993, during an official visit to Pitcairn, the Governor was presented with a document expressing dissatisfaction with British policy towards the islands, and raising the question of a transfer of sovereignty.

Following some structural changes in the local government of Pitcairn, Steve Christian was elected to the position of Mayor in December 1999, presiding over the Island Council (a role previously fulfilled by the Island Magistrate).

In early 2000 British detectives began an investigation into an alleged rape case on the island. The British team was joined by the New Zealand police force in early 2001, when the case was widened to include 15 alleged sexual assaults, amid reports claiming that sexual abuse, particularly of children, was commonplace on the island. The trial, which was expected to take place in Auckland, New Zealand, would represent the first significant criminal case on Pitcairn since a murder trial in 1897. In November 2000 Auckland's Crown Solicitor, Simon Moore, was appointed Pitcairn's first Public Prosecutor, with the task of deciding whether to bring charges against 20 Pitcairn islanders. His decision was delayed by the fact that many of the complainants now lived in New Zealand and by the logistical problems of a trial that could potentially involve the entire populace. However, in April 2003 a judicial delegation of eight people visited Pitcairn and nine men on the island were charged with a total of 64 offences, some of which dated back about 40 years. Many islanders expressed serious concern that their community would not be able to sustain itself if the men (who constituted virtually the entire male work-force) were extradited to New Zealand to stand trial. In early June a further four men, all now resident in New Zealand, were charged with a total of 32 offences, including 10 charges of rape, which were alleged to have taken place on Pitcairn between five and 40 years previously. In April 2004 the Supreme Court of Pitcairn (sitting for only the second time in its history, at a special session in Auckland) rejected the accused men's application to be tried on Pitcairn, reiterating that they should stand trial in New Zealand. However, in late June Pitcairn's Court of Appeal (sitting for the first time in its history) overruled this decision and stated that, despite the logistical problems involved, the trial would be conducted on Pitcairn. Meanwhile, lawyers for the accused men argued that the trial should be abandoned as the islanders were all descendants of the Bounty mutineers, who had renounced all allegiance to the British Crown. Their claims that Pitcairn islanders were not subject to British jurisdiction appeared to be supported by historical documents discovered in London. Moreover, a group of women from Pitcairn issued a public statement claiming that sexual relations between men and girls below the British legal age of consent were commonplace on the island, and not considered to be a criminal offence by either party or by the community as a whole. However, in late September 2004 the trial of seven of the defendants began in converted school premises on Pitcairn. Some 25 lawyers, police officers and journalists travelled to the island, and witnesses in New Zealand gave evidence via live satellite video link. The trial concluded in late October. Six of the seven defendants were found guilty of many of the 51 charges against them, some of the assaults reportedly having taken place 40 years previously. One islander was acquitted of the charges against him. Among those convicted was the Mayor of Pitcairn, Steve Christian, who was found guilty of five counts of rape. The convicted men were expected to appeal.

The British Overseas Territories Act, which entered into effect in May 2002, granted citizenship rights in the United Kingdom to residents of the Overseas Territories, including Pitcairn. The legislation also entitled Pitcairn Islanders to hold British passports and to work in the United Kingdom and elsewhere in the European Union.

Economy

The economy has been based on subsistence gardening, fishing, handicrafts and the sale of postage stamps. Attempts to increase revenue from the island's agricultural output by producing dried fruits (notably bananas, mangoes and pineapples) began in 1999. Diversification of this sector to include production of jam, dried fish and coffee was subsequently under consideration. In early 1999 the Pitcairn Island police and customs office requested that no honey or beeswax be sent to the Island in order to protect from disease the island's growing honey industry, which was being developed as a source of foreign exchange. Pitcairn honey, which was pronounced to be 'exceptionally pure' by the New Zealand Ministry of Agriculture, began to be exported, largely through internet sales, in late 1999. A new stamp was issued to commemorate the launch of the industry.

A reafforestation scheme, begun in 1963, concentrated on the planting of miro trees, which provide a rosewood suitable for handicrafts. In 1987 the Governor of the islands signed a one-year fishing agreement with Japan, whereby the Japan Tuna Fisheries Cooperative Association was granted a licence to operate vessels within Pitcairn's EEZ. The agreement was subsequently renewed but lapsed in 1990. In 1992 an exclusive economic zone (EEZ), designated in 1980 and extending 370 km (200 nautical miles) off shore, was officially declared.

In early 1992 it was reported that significant mineral deposits, formed by underwater volcanoes, had been discovered within the islands' EEZ. The minerals, which were believed to include manganese, iron, copper, zinc, silver and gold, could (if exploited) dramatically affect the Territory's economy.

New Zealand currency has been in everyday use since 1967. There is no taxation, except for small licensing fees on guns and vehicles, and government revenue has been derived from philatelic sales (one-half of current revenue in 1992/93), and from interest earned on investments. In 1998/99 revenue totalled $NZ492,000 and expenditure $NZ667,000. In the early 2000s revenue was estimated to total about $NZ415,000 annually. Capital assistance worth an average of £100,000 annually is received from the United Kingdom. In 2002/03 exports from Pitcairn to New Zealand were worth an estimated $NZ38,000, while imports from New Zealand totalled $NZ100,000. Exports to Australia in 2000/01 totalled $A21,000 and imports $A7,000, consisting mainly of food. In 2001 Pitcairn's imports from the USA totalled US $5.5m., while exports to the USA were worth US $0.2m.

Hopes that Pitcairn might find an additional source of revenue through the sale of website addresses were boosted in early 2000 when the island won a legal victory to gain control of its internet domain name suffix, '.pn'.

Development projects have been focused on harbour improvements, power supplies, telecommunications and road-building. Pitcairn's first radio-telephone link was established in 1985, and a modern telecommunications unit was installed in 1992. A new health clinic was established, with British finance, in 1996.

A steady decline in the population, owing mainly to emigration to New Zealand, is the island's main problem. In March 2001 a New Zealand company, Wellesley Pacific, expressed an interest in acquiring development rights on Pitcairn with the aim of establishing fishing and tourism projects. The company claimed that if the islands achieved self-sufficiency through the proposals within five years, then Pitcairn could eventually become an independent state within the Commonwealth. In May 2002 the company's director reportedly announced that the development would begin within the next 12 months, describing plans for a lodge on Pitcairn Island and for a floating hotel off Oeno Island. The British High Commission in Wellington, however, emphasized that any such developments remained subject to the approval of Pitcairn Island Council, but that the company was welcome to submit proposals.

In August 2004 the British Government announced the provision of US $6.5m. in emergency assistance for the Territory in order to avert a financial crisis. The aid programme included a grant (partly financed by the European Union) to fund improvements to the road between Bounty Bay and Adamstown and to investigate the potential for ecotourism as a possible source of future revenue.

Statistical Survey

Source: Office of the Governor of Pitcairn, Henderson, Ducie and Oeno Islands, c/o British Consulate-General, Pitcairn Islands Administration, Private Box 105-696, Auckland, New Zealand; tel. (9) 366-0186; fax (9) 366-0187; e-mail pitcairn@iconz.co.nz.

AREA AND POPULATION

Area: 35.5 sq km. *By Island*: Pitcairn 4.35 sq km; Henderson 30.0 sq km; Oeno is less than 1 sq km and Ducie is smaller.

Population: 47 (September 2004).

Density (Pitcairn only, September 2004): 10.8 per sq km.

Employment (able-bodied men, 2002): 9.

FINANCE

Currency and Exchange Rates: 100 cents = 1 Pitcairn dollar. The Pitcairn dollar is at par with the New Zealand dollar ($NZ). New Zealand currency is usually used.

Budget ($NZ, 1998/99): Revenue 492,000; Expenditure 667,000.

EXTERNAL TRADE

Trade with New Zealand (($NZ '000, year ending 30 June)): *Imports:* 32 in 2000/01; 342 in 2001/02; 100 in 2002/03 *Exports:* 134 in 2000/01; 85 in 2001/02; 38 in 2002/03.

TRANSPORT

Road Traffic (motor vehicles, 2002): Passenger vehicles 29 (two-wheeled 1, three-wheeled 6, four-wheeled 23); Tractors 3; Bulldozer 1; Digger 1.

Shipping: *Local Vessels* (communally-owned open surf boats, 2000): 3. *International Shipping Arrivals* (visits by passing vessels, 1996): Ships 51; Yachts 30.

COMMUNICATIONS

Telephones (2002): a party-line service with 15 telephones in use; 2 public telephones; 2 digital telephones. Most homes also have VHF radio.

Directory

The Constitution and Government

Pitcairn is a British settlement under the British Settlements Act 1887, although the islanders reckon their recognition as a colony from 1838, when a British naval captain instituted a Constitution with universal adult suffrage and a code of law. That system served as the basis of the 1904 reformed Constitution and the wider reforms of 1940, effected by Order in Council. The Constitution of 1940 provides for a Governor of Pitcairn, Henderson, Ducie and Oeno Islands (who, since 1970, is concurrently the British High Commissioner in New Zealand), representing the British monarch. A Mayor is elected every three years to preside over the Island Council. The Local Government Ordinance 1964 constituted an Island Council of 10 members: in addition to the Mayor, five members are elected annually; three are nominated for terms of one year (the Governor appoints two of these members at his own discretion); and the Island Secretary is an *ex-officio* member. In addition to the Island Council there is an Island Magistrate who presides over the Magistrate's Court of Pitcairn, and is appointed by the Governor. Liaison between the Governor and the Island Council is conducted by a Commissioner, usually based in the Office of the British Consulate-General in Auckland, New Zealand.

Customary land tenure provides for a system of family ownership (based upon the original division of land in the 18th century). Alienation to foreigners is not forbidden by law, but in practice this is difficult. There is no taxation, and public works are performed by the community.

Governor of Pitcairn, Henderson, Ducie and Oeno Islands: RICHARD FELL (British High Commissioner in New Zealand—took office December 2001).

Office of the Governor of Pitcairn, Henderson, Ducie and Oeno Islands: c/o British High Commission, 44 Hill St, POB 1812, Wellington, New Zealand; tel. (4) 924-2888; fax (4) 473-4982; e-mail ppa.mailbox@fco.gov.uk; internet www.britain.org.nz; Gov. RICHARD FELL; Deputy Gov. MATTHEW FORBES; Commissioner LESLIE JAQUES.

Pitcairn Islands Administration: Private Box 105-696, Auckland, New Zealand; tel. (9) 366-0186; fax (9) 366-0187; e-mail admin@pitcairn.gov.pn; internet www.government.pn.

ISLAND COUNCIL
(September 2004)

Mayor: STEVE CHRISTIAN.

Island Secretary (ex officio): BETTY CHRISTIAN.

Government Treasurer: OLIVE CHRISTIAN.

Island Auditor: MIKE CHRISTIAN.

Chairman of Internal Committee: JAY WARREN.

Other Members: LEA BROWN, BRENDA CHRISTIAN, TOM CHRISTIAN, ROBERT MCDONALD, JOHN O'MALLEY, MERALDA WARREN, NOLA WARREN.

Elections to the Island Council take place each December. Meetings are held at the Court House in Adamstown.

Office of the Island Secretary: The Square, Adamstown.

Judicial System

Chief Justice: CHARLES BLACKIE.

Island Magistrate: LEA BROWN.

Public Prosecutor: SIMON MOORE.

Public Defender: PAUL DACRE.

Religion

CHRISTIANITY

Since 1887 many of the islanders have been adherents of the Seventh-day Adventist Church.

Pastor: JOHN O'MALLEY, SDA Church, The Square, POB 24, Adamstown; fax 872-7620/9763.

The Press

Pitcairn Miscellany: monthly four-page mimeographed news sheet; f. 1959; edited by the Education Officer; circulation 1,400 in 2002; Editor P. FOLEY.

Finance, Trade and Industry

There are no formal banking facilities. A co-operative trading store was established in 1967. Industry consists of handicrafts, honey and dried fruit.

Transport

ROADS

There are approximately 14 km (9 miles) of dirt road suitable for two-, three- and four-wheeled vehicles. In 2002 Pitcairn had one conventional motor cycle, six three-wheelers and 22 four-wheeled motor cycles, one four-wheel-drive motor car, three tractors, a five-ton digger and a bulldozer; traditional wheelbarrows are used occasionally. In 1995 a total of £79,000 was received from individual donors for work to improve the road leading to the jetty at Bounty Bay. Additional funding from the United Kingdom and the European Union was suspended in 2001, and the project remained incomplete.

SHIPPING

No passenger ships have called regularly since 1968, and sea communications are restricted to cargo vessels operating between New Zealand and Panama, which make scheduled calls at Pitcairn three times a year, as well as a number of unscheduled calls. There are also occasional visits by private yachts. The number of cruise ships calling at Pitcairn increased in the late 1990s, and 10 such vessels visited the island in 2000 (compared with just two or three annually in previous years). Bounty Bay, near Adamstown, is the only possible landing site, and there are no docking facilities. In 1993 the jetty derrick was refitted with an hydraulic system. The islanders have three aluminium open surf boats.

Education

Free primary education is provided on the island under the direction of a qualified schoolteacher, recruited in New Zealand. Scholarships, provided by the Pitcairn Government, are available for post-primary education or specialist training in New Zealand. In February 2004 there were seven children (four primary and three secondary) being educated on Pitcairn.

UNITED STATES COMMONWEALTH TERRITORY IN THE PACIFIC

There are two US Commonwealth Territories, the Northern Mariana Islands, in the Pacific Ocean, and Puerto Rico, in the Caribbean Sea. A Commonwealth is a self-governing incorporated territory that is an integral part of, and in full political union with, the USA. The Secretary of the Interior, in the federal Government, is responsible for relations with the Government of the Northern Mariana Islands. Within the US Department of the Interior, the Assistant Secretary for Policy, Management and Budget is responsible for the Office of Insular Affairs and exercises authority on behalf of the Secretary in all matters pertaining to the insular governments and territories. The USA maintains membership of the Pacific Community (formerly the South Pacific Commission), the Colombo Plan and the UN's Trusteeship Council (the operations of which were suspended in 1994) and its Economic and Social Commission for Asia and the Pacific (ESCAP).

Head of State: President GEORGE W. BUSH (took office 20 January 2001).

Department of the Interior, Office of Insular Affairs: 1849 C St, NW, Washington, DC 20240, USA; tel. (202) 208-6816; fax (202) 501-7759; internet www.doi.gov/oia.

Deputy Assistant Secretary of the Interior for Insular Affairs: DAVID B. COHEN.

Director, Office of Insular Affairs: NIKALAO A. PULA.

Resident Representative to the Government of the Commonwealth of the Northern Mariana Islands: JUAN A. BABAUTA.

THE NORTHERN MARIANA ISLANDS

Physical and Social Geography

The Commonwealth of the Northern Mariana Islands comprises 16 islands (all the Mariana group except Guam), most being of volcanic origin, lying in the western Pacific Ocean. The territory has a land area of 457 sq km (177 sq miles) and is situated about 5,300 km west of Honolulu (Hawaii) and due south of Japan, some 2,300 km from Tokyo. Its nearest neighbours are Guam and the Federated States of Micronesia. The climate is tropical and there is little seasonal variation in temperature, the mean annual temperature being some 28°C (82°F). Mean annual rainfall is some 2,120 mm (84 ins), the driest months being January–May. The Mariana Islands can be affected by monsoons between August and November. English, Chamorro and Carolinian are the official languages of the Commonwealth. Since the evacuation in 1990 of Anatahan, owing to a volcanic eruption, only five islands, including the three largest (Saipan, Tinian and Rota), have remained inhabited; the chief settlements, and the administrative centre of Capitol Hill, are on Saipan. The population totalled 69,221 at the census of April 2000.

History

The islands, already settled for more than 2,000 years by a Polynesian people now known as the Chamorros, were claimed for Spain by Fernão de Magalhães (Ferdinand Magellan) in 1521. They were known as the Ladrone Islands until 1668, when they were named the Mariana Islands in honour of Mariana of Austria, widow of Philip IV of Spain. Much of the Chamorro population was transferred to Guam in 1698, not to return to the Northern Marianas until the early 19th century. It was only during the Spanish period that the traditional matrilineal organization of Micronesian society changed, for the Chamorros, to a patrilineal system. Meanwhile, some Carolinian peoples, from southern Micronesia, settled in the islands. The Carolinians were primarily a fishing people, the Chamorros mainly agricultural. Spain sold the Northern Mariana Islands to Germany in 1899. During the 20th century the islands were administered successively by Germany, Japan and the USA, from 1947 as part of the UN Trust Territory of the Pacific Islands. The islands voted for separate status as a US Commonwealth Territory in June 1975. In March 1976 the US President, Gerald Ford, signed the 'Covenant to Establish a Commonwealth of the Northern Mariana Islands (CNMI) in Political Union with the USA'. In October 1977 President Jimmy Carter approved the Constitution of the Northern Mariana Islands, which provided that, from January 1978, the former Marianas District would become internally self-governing. The CNMI formally entered political union with the USA on 3 November 1986, when President Ronald Reagan issued a proclamation fully effecting the Covenant, after the ending of the Trusteeship in the district. Consequently, the residents of the CNMI became citizens of the USA.

Elections for a bicameral legislature, a Governor and a Lieutenant-Governor were first held in December 1977. Pedro Tenorio, the candidate of the Republican Party, was elected Governor in November 1981, and re-elected in 1985. With a constitutional limit of two consecutive terms as Governor, he did not seek re-election in November 1989, when a Republican, Lorenzo 'Larry' Guerrero, was elected Governor. Democrats retained control of the House of Representatives, but a Republican candidate, Juan Babauta, was elected as the Resident Representative in Washington, DC.

In December 1990 the UN Security Council voted to end the Trusteeship of the Northern Marianas, as well as that of two other Pacific Trust Territories. Although the decision to terminate the relationship had been taken in 1986, voting had been delayed. Guerrero, however, opposed the termination and had requested that the vote be postponed. The new relationship would, he argued, leave the islands subject to US law while remaining unrepresented in the US Congress; moreover, several important sovereignty issues (such as local control of marine resources) would remain unresolved.

At elections to the House of Representatives (which had been enlarged by three seats) in November 1991 Republicans regained a majority. Similarly, the party increased the number of its senators to eight. Republicans retained their majority at elections to the House of Representatives in November 1993. However, in the gubernatorial election a Democrat, Froilan Tenorio, was successful. Similarly, a Democratic candidate, Jesus Borja, was elected as Lieutenant-Governor, while Juan Babauta remained as Washington Representative. The Republicans increased their representation in both the Senate and the House of Representatives in elections in November 1995.

In January 1995 the minimum wage was increased by 12.2%, in accordance with the Governor's stated objectives; however, five days later Tenorio signed legislation that effectively reversed the decision. This prompted several members of the House of Representatives to demand that the US Federal Bureau of Investigation (FBI) conduct an investigation into the incident and the allegations of bribery surrounding the Governor's actions. The territory's reputation deteriorated further in April when the Government of the Philippines introduced a ban on its nationals accepting unskilled employment in the islands, because of persistent reports of abuse and exploitation of immigrant workers. Meanwhile, the US Congress announced that it was to allocate US $7m. towards the enforcement of the islands' labour and immigration laws, following the publication of a report in late 1994, which alleged the repeated violations of these regulations, as well as widespread corruption among immigration officials and business leaders.

At legislative elections in November 1995 Republican candidates won a convincing majority, securing 14 seats in the House of Representatives and six seats in the Senate.

Many of the territory's social and economic problems have been attributed to the dramatic increase in the Northern Marianas' population (from some 17,000 in 1979 to more than 69,000 in 2000). Acknowledging this situation, a US government report, published in mid-1996, recommended the expansion of prison and detention facilities on the islands, and offered further assistance with immigration control.

In May 1997 the US President, Bill Clinton, informed Tenorio of his intention to apply US immigration and minimum wage laws to the territory, stating that labour practices in the islands were inconsistent with US values. Clinton also criticized officials in the territory for failing to address the persistent problems of an inadequate minimum wage and reports of improper treatment of alien workers. In the previous month Democrat Congressman George Miller had proposed legislation in the US House of Representatives (the Insular Fair Wage and Human Rights Act) that would equalize the minimum wage level in the islands with that of the US mainland by 1999. The territory's Government, which denied many of the claims of exploitation of immigrant workers, responded to the proposed legislation by initiating a public relations campaign (at a cost of US $1m.) aimed at persuading the Republican majority in the US House of Representatives to oppose the bill. A further $0.5m. was spent funding a visit to the islands by seven Republican members of the US Congress, in an attempt to consolidate their support for Tenorio's stance. The issue dominated the election campaign during late 1997, with Willie Tan (a major entrepreneur who had been fined several million dollars for failing to comply with US Labor Department health and safety standards in his garment factories) providing considerable financial support towards Froilan Tenorio's efforts to remain in office.

At legislative elections on 1 November 1997 Republican candidates won 13 of the 18 seats in the House of Representatives and eight of the nine seats in the Senate. At the gubernatorial election, held concurrently, Pedro Tenorio was successful, securing 46% of total votes. The incumbent Froilan Tenorio secured some 27% of the vote and Jesus Borja (previously the Lieutenant-Governor) won 26%. Opponents of Pedro Tenorio subsequently initiated a legal challenge to his election on the grounds that it constituted his third term as Governor, thereby violating the Constitution, which states that a maximum of two gubernatorial terms may be served by any one individual (although the Constitution had been amended to include this provision only during Pedro Tenorio's second term in office). Following the legislative elections of 6 November 1999, Democrat candidates held two of the nine seats in the Senate and six of the 18 seats in the House of Representatives.

In January 1999 the Office of Insular Affairs (OIA) published a report in which it concluded that the Government's attempts to eradicate abuses of labour and immigration laws had been unsuccessful. In particular, it had failed to reduce the territory's reliance on alien workers, to enforce US minimum wage laws and to curb evasions of trade legislation governing the export of garments to the USA. In the same month former employees of 18 US clothing retailers initiated legal action against the companies, which were accused of failing to comply with US labour laws in Saipan. In April 2000 a settlement was reached with the garment manufacturers, providing some US $8m. in compensation for the workers. The companies also agreed to conform to regulations established by an independent monitoring system in Saipan. In August 2001 the Saipan Garment Manufacturers' Association (SGMA) denied a claim, made in a report by the US Department of Justice, that factory owners were giving employees methamphetamine to increase productivity. In March 2004 the Garment Oversight Board (established after a $20m. industry settlement in 2003) decertified one of the 26 participating garment manufacturers.

In February 2000, meanwhile, the US Senate approved a bill granting permanent residency in the Northern Marianas to some 10,000 immigrant workers. However, the bill also included provisions for limiting the stay of all future guest workers. In December 2000 Governor Tenorio announced that he was to oppose the decision by the US Government to bring the Northern Marianas' labour and immigration laws under federal control; Tenorio argued that this could have a negative impact on the Commonwealth's economy. In May 2001, following intense lobbying by the Northern Marianas Government, the US Congress abandoned the bill.

Legislative elections were held on 3 November 2001, at which the Republican Party secured 12 seats in the House of Representatives, the Democratic Party won five and the Covenant Party took one. The Republican Party won six seats in the Senate, the Democratic Party two and the Covenant Party one. Gubernatorial elections were held concurrently, at which Juan Nekai Babauta, the Republican Party candidate and former Representative to Washington, won a convincing victory, securing 42.8% of the votes cast. Benigno Fitial, of the Covenant Party, received 24.4%. Babauta was inaugurated as Governor in January 2002, while Diego Benavente, his running mate and the former Speaker of the House of Representatives, became Lieutenant-Governor.

In January 2002 the Supreme Court suspended deportation proceedings against an immigrant labourer working in the Northern Marianas illegally, after he appealed to the office of the UN High Commissioner for Refugees. The Court warned the Government that it might not be able to order the deportation of up to 10,000 of the Chinese, Sri Lankan and Bangladeshi workers in the Northern Marianas. However, in April 2004 more than 400 garment workers were referred to the Division of Immigration for probable deportation as a result of their non-co-operation with a Fair Labor Standards Act civil action against garment manufacturers.

In May 2002 the issue of the Northern Marianas' working conditions was raised again by US Senator Edward Kennedy, who proposed legislation seeking to increase the minimum wage incrementally. The first increase would occur 60 days after enactment, should the bill be successful. In late September 2002 seven further major US clothing retailers agreed to pay US $11.25m. in compensation to employees alleged to have suffered intolerable working conditions for poor rates of pay. The funds were also to be used for sponsorship of independent monitoring of labour conditions in the islands, and a new body was created in June 2003 in order to supervise the administration of some $20m. in compensation funds. In mid-2004 the minimum wage in the Northern Marianas remained some 40% less than in the USA. Nevertheless, a delegate from the US Commission on Civil Rights concluded in May of that year that the situation of garment workers on Saipan appeared to be improving.

In late 2002 the Government successfully resisted an attempt by the US Administration to place the Northern Mariana Islands' immigration and labour legislation under direct federal control. However, in July 2003 a US government report classified the Northern Mariana Islands as a 'high risk security area' owing to a reported increase in the activities of international criminal groups, and recommended that the territory's immigration laws be brought within the jurisdiction of the US Department of Justice. In September the Northern Mariana Islands announced a new immigration co-operation agreement with the US Department of the Interior, which removed the right of overseas political refugees in the territory to seek asylum in the USA.

In May 2002 the Government 'froze' the assets of the Bank of Saipan, pending auditing of its accounts, after the institution's former Chairman was arrested for allegedly attempting to defraud the bank of more than US $6.6m. The bank was placed in receivership and remained closed for 11 months. Following the bank's reopening in April 2003, with assets of some $12m., customers were permitted to retrieve a limited monthly quota of savings deposits. Three defendants were convicted in relation to the case in June 2003.

In September 2002 Babauta proposed reforms aimed at reducing government expenditure—reportedly, together with immigration issues, the cause of tension between the Governor and the Speaker of the House of Representatives, Heinz Hofschneider. In November 2002 the Government announced plans for a US $40m. bond issue to cover the cost of compensating traditional landowners for the loss of property expropriated for government use. However, in August 2003 an application to issue the bonds was still pending, and credit ratings agencies expressed concerns that the Government's other outstanding debts, in particular those to the Northern Mariana Islands' Retirement Fund, would prevent full repayment of the bonds.

Several instances of corruption in public office were reported in 2003–04; in April 2003 Senator Ricardo S. Atalig was found guilty of illegally employing relatives of another Senator, José M. de la Cruz. (De la Cruz was suspended from office following his own conviction in July.) Meanwhile, in May an investigation was launched into the issue of allegedly fraudulent expenses claimed by employees of the Northern Marianas' Liaison Office in Hawaii. In August 2004 the Superior Court convicted the chief financial officer of Tinian municipality, Romeo Atalig Diaz, in the first public corruption case lodged by the new anti-corruption unit of the Attorney General's Office. However, the sentence imposed (fines totalling US $2,800) was considered derisory by the local press.

Several typhoons badly damaged property and infrastructure in 2002–04. The cost of repairs was met largely by assistance from the US Federal Emergency Management Agency.

In late 2002, following the suspension of Northern Mariana islanders' access to medical facilities in Hawaii, islanders became legally obliged to contribute to the rising costs of their health care. In March 2003, however, the Bush Administration pledged US $15m. to assist the development of adequate health-care infrastructure in the US Pacific territories. None the less, the health and welfare of the population remained a cause for concern in 2002 and 2003, following the publication of independent surveys claiming that the Northern Marianas registered the world's third highest incidence of diabetes and a child poverty rate of some 38%.

The federal Government of the USA is responsible for defence and foreign relations. In 1990 a cause of tension between the federal and insular Governments was removed when the Administration of George Bush, Sr, finally appointed its Special Representative, a negotiator on bilateral relations. Co-operation between the Northern Marianas and the USA has increased in recent years. In early 2004 the Northern Marianas' Representative in Washington, Pete Tenorio, requested authorization for a non-voting delegate to the US Congress from the Northern Marianas. In May four members of the original team of Covenant negotiators considered publicly the possibility of the Northern Marianas becoming the 51st state of the USA.

In late 1999 it was announced that legal action was to be taken against the US Government following revelations that traces of a toxic chemical had been found in a village near the capital, Saipan. It was alleged that the chemicals had been abandoned by the US army in the 1960s. It was subsequently announced, in November 2000, that the Government of the Northern Marianas was to claim

US $1m. from the USA, to repay the costs of a screening programme, established to test the villagers for traces of the chemicals. Nevertheless, there were frequent reports of chemical contamination in 2002 and 2003. A US court order filed by the Center for Biological Diversity, an environmental lobby group, forced military training on the uninhabited island of Farallon de Medinilla to be suspended for 30 days in mid-2002. Despite environmentalists' concerns about the impact and legality of US military activity, in particular the testing of ordnance, upon the island's wildlife, the Northern Marianas Chamber of Commerce expressed fears that the substantial revenue generated by the visiting US armed forces might be jeopardized.

Economy

The economy of the Northern Marianas is dominated by the services sector, particularly tourism. In 1999 the Commonwealth's gross national income (GNI) was estimated by the Bank of Hawaii (BOH) to be US $696.3m. GNI per head was estimated at $8,582. The population increased at an estimated average annual rate of 7.2% in 1990–99. According to the BOH, the territory's gross domestic product (GDP) totalled $557.0m. in 2002.

Agriculture has not contributed significantly to the economy since the devastation of the sugar industry in 1944, during the Second World War. (In the late 1980s some 426 acres were used for arable farming, compared with 40,000 acres before the War.) Agriculture is now based on smallholdings, important crops being coconuts, breadfruit, tomatoes and melons. Arable farming is particularly important on Rota. Large-scale cattle-ranching is practised on Tinian. Vegetables, beef and pork are important exports. The sector engaged 1.5% of the employed labour force according to the census of 2000 and its commercial value is minimal. In 2002 the sector accounted for only 0.1% of gross business revenues (total values generated by business transactions; the Government does not calculate gross domestic product figures).

There is little commercial fishing based on Saipan or the other islands (the total catch was 193 metric tons in 1999); there is, however, a major tuna transhipment facility at Tinian harbour.

Before the Second World War Japan mined phosphate on the islands, but it is estimated that only about 50,000 short tons of low-grade guano phosphate remain on Rota. There is a possibility that mineral resources in Northern Marianan waters (particularly cobalt-rich manganese crusts) might be exploited in the future.

Industry (including manufacturing and construction) engaged 47.2% of the employed labour force in 2000. Manufacturing alone engaged 40.7% of workers, while construction employed 6.5%. The principal manufacturing activity is in the garment factories. Although the garment manufacturing industry was not established until the mid-1980s, it soon accounted for the largest amount of commodity exports from the islands. The manufacturers benefit from US regulations allowing duty-free imports of textiles to the USA from the Commonwealth, and advantageous local regulations allow manufacturers to certify garments as having been 'made in the USA'. Earnings from garment exports to the USA increased from an estimated US $419m. in 1995 to almost $1,000m. in 1998. In 2003 about 30 garment factories, officially employing some 16,000 Chinese and Filipino workers, were in operation. Garment manufacturing accounted for 34.7% of gross business revenues in 2000. Overall exports of garments were worth $925.7m. in 2001, compared with $1,017m. in 2000. However, earnings of only $660m. were envisaged for 2003, owing to weak demand in the USA. The terminal decline of this industry in the Northern Marianas was being widely predicted by 2004, owing to increased domestic regulation and taxation, the World Trade Organization (WTO) moratorium on garment quotas, which was due to come into force in January 2005, and stronger global competition (not least since the People's Republic of China joined the WTO in 2001). Other small-scale manufacturing industries include handicrafts and the processing of fish and copra.

The construction industry in the 1980s benefited from the rapid expansion in tourism and government development of the islands' infrastructure. However, the number of building permits sold, both commercial and residential, declined each year during 1997–2001.

In 2000 the services sector (including utilities) engaged 51.3% of the employed labour force, while accounting for 29.2% of gross business revenues. The islands' association with the USA, the availability of foreign investment and the opening of direct air links with Japan encouraged rapid economic development, mainly after 1978. By 1998 there were 3,942 hotel rooms available.

In mid-1995 a US company opened the Commonwealth Territory's first casino on Tinian. The islands were expected to receive some US $12m. annually in revenue from the casino, which was to be invested in infrastructure projects and health services. However, the Asian financial crisis of the late 1990s had a major impact on the industry, with many hotels reducing charges and employee numbers and others facing closure as a result of the decline in tourist arrivals. Continental Micronesia also significantly reduced its flights to Japan in response to the situation. In September 1998 the hotel and casino complex on Tinian sought the support of several senior politicians in an attempt to secure a loan for US $30m. from a foreign bank. The business, which relied heavily on visitors from Japan, Taiwan and the Republic of Korea, had been particularly badly affected by the Asian economic problems.

Visitor arrivals increased by 5.3% in 2000, when a total of 528,597 visitors travelled to the islands. However, the suicide attacks on the mainland USA on 11 September 2001 adversely affected the tourism industry in the Northern Marianas. The number of visitor arrivals by air decreased sharply, while the hotel occupancy rate declined to only 35% in December 2001. Of the total arrivals of 444,284 in 2001, more than 75% were from Japan. In 2002, however, visitor arrivals rose to 475,547. Tourism receipts were worth an estimated $225m. in 2002. Remittances from overseas workers and investments totalled $76.7m. in 2001.

The Northern Marianas is very dependent on imports, the value of which totalled US $836.2m. in 1997. The principal imports in that year were foodstuffs (9.6%), petroleum products (8.2% of the total), clothing (7.1%), automobiles and parts (5.0%) and construction materials (4.1%). In 1991 there was a visible trade deficit of $126.9m. The annual rate of inflation averaged 3.6% in 1990–98. Consumer prices decreased by 0.3% in 1999, increased by 2.0% in 2000, declined by 0.8% in 2001 and rose by 0.1% in 2002. Under the Covenant between the Commonwealth of the Northern Mariana Islands and the USA, the islands receive substantial annual development grants. Budget estimates predicted total revenue of $297.2m. in the financial year ending 30 September 2000. Total expenditure of $225.5m. was estimated for that year. Loans from the Commonwealth Development Authority totalled more than $0.5m. in 2001.

Rapid economic expansion in the 1980s led to a shortage of local labour and an increase in non-resident alien workers. The number of non-resident alien workers increased by 655% between 1980 and 1989, and by the early 1990s exceeded the permanent population of the islands. Owing to this increase in immigrant workers, wages remained relatively low, and there were widespread complaints of poor working conditions. Increasing concern over the alleged exploitation of foreign workers in the islands' garment factories led to the proposal in April 1997 of legislation in the US House of Representatives to increase the minimum wage to parity with the mainland level by 1999. In early 1999 a lawsuit was filed against garment manufacturers in Saipan in a further attempt to equalize conditions. In its verdict in September 2002 the court found against the garment manufacturers; it awarded compensation to the affected workers and provided for independent monitoring of labour practices in the Northern Marianas (see History). In total, 22,560 permits were issued to non-resident workers in the islands in 1994, 67.2% of which were for Filipino nationals. In 2003 the Government estimated that around 90% of all workers in the islands' garment industry were non-resident aliens.

The territory received a total of some US $13m. of US federal funding and development assistance in 2001/02. The islands' Republican administration, inaugurated in 2002, pledged to prioritize economic reform by promoting free, competitive markets with a minimum of government interference. It outlined an environmental rehabilitation programme in an effort to raise the territory's tourism profile. In May 2002 the assets of the Bank of Saipan were temporarily 'frozen' following the arrest of its chief executive officer on charges of fraud. Meanwhile, in March the Government introduced tax incentives for new businesses and developers, worth a potential 100% abatement of local taxes, or a 95% rebate of federal taxes. However, despite improved expenditure controls, the continued recession in the islands was estimated to have led to a deterioration in the Government's fiscal position in 2003; the 2004 budget envisaged increasing revenues and allocated some $4.2m. towards debt repayments. Moreover, in response to declining demand among tourists for air travel to the islands, Korean Air suspended direct services to Saipan in March 2003. In the following month a major development project to generate 10 MW of electric power per day using thermal technology appeared to have been cancelled, following a dispute between investors and a member of the Northern Mariana Islands Senate. During mid-2003 a series of electricity failures intensified debate over the planned partial privatization of the Commonwealth Utilities Corporation.

The main restraint on development is the need to expand the islands' infrastructure, combined with the problem of a labour shortage and the dependence on foreign workers. Many economic problems have been largely attributed to the dramatic increase in the islands' population (from some 17,000 in 1979 to more than 69,000 by 2000). The islands benefit from their political association with the USA and their relative proximity to Japan. The Commonwealth of the Northern Mariana Islands is a member of the Pacific Community (formerly the South Pacific Commission) and also an associate member of the UN's Economic and Social Commission for Asia and the Pacific (ESCAP).

Statistical Survey

Source: (unless otherwise stated): Department of Commerce, Central Statistics Division, POB 10007, Saipan MP, 96950; tel. 664-3000; fax 664-3001; internet www.commerce.gov.mp.

AREA AND POPULATION

Area: 457 sq km (176.5 sq miles). *By Island*: Saipan 120 sq km (46.5 sq miles); Tinian 102 sq km (39.2 sq miles); Rota 85 sq km (32.8 sq miles); Pagan 48 sq km (18.6 sq miles); Anatahan 32 sq km (12.5 sq miles); Agrihan 30 sq km (11.4 sq miles); Alamagan 11 sq km (4.4 sq miles); Asuncion 7 sq km (2.8 sq miles); Aguijan (Goat Is) 7 sq km (2.7 sq miles); Sarigan 5 sq km (1.9 sq miles); Guguan 4 sq km (1.5 sq miles); Farallon de Pajaros 3 sq km (1.0 sq miles); Maug 2 sq km (0.8 sq miles); Farallon de Medinilla 1 sq km (0.4 sq miles).

Population: 43,345 at census of 1 April 1990; 69,221 (males 31,984, females 37,237) at census of 1 April 2000. *By Island*: Saipan 62,392; Rota 3,283; Tinian 3,540; Northern Islands 6. *2002* (estimates): Saipan 67,011; Total 74,151.

Density (2002 census): 162 per sq km.

Ethnic Groups (2000 census): Filipino 18,141; Chinese 15,311; Chamorro 14,749; part-Chamorro 4,383; Total (incl. others) 69,221.

Principal Towns (2000 census): San Antonio 4,741; Garapan (capital) 3,588; Koblerville 3,543; San Vincente 3,494; Tanapag 3,318; Chalan Kanoa 3,108; Kagman 3,026. Source: Thomas Brinkhoff, *City Population* (internet www.citypopulation.de).

Births and Deaths (2001): Registered live births 1,451 (birth rate 20.4 per 1,000); Registered deaths 150 (death rate 2.1 per 1,000).

Employment (2000 census, persons aged 16 years and over): Agriculture, forestry, fisheries and mining 623; Manufacturing 17,398; Construction 2,785; Transport, communication and utilities 1,449; Trade, restaurants and hotels 9,570; Financing, insurance and real estate 1,013; Community, social and personal services 9,915; *Total employed* 42,753 (males 19,485, females 23,268); Unemployed 1,712 (males 888, females 824); *Total labour force* 44,465 (males 20,373, females 24,092). *Mid-2002* (estimates): Agriculture, etc. 7,000; Total labour force 33,000 (Source: FAO).

AGRICULTURE, ETC.

Livestock (1997): Cattle 1,789; Pigs 831; Goats 249; Poultry birds 29,409.

Fishing (metric tons, live weight, 2002): Total catch 198 (Skipjack tuna 73). Source: FAO, .

FINANCE

Currency and Exchange Rates: 100 cents = 1 United States dollar (US $). *Sterling and Euro Equivalents* (31 May 2004): £1 sterling = US $1.8347; €1 = US $1.2246; US $100 = £54.50 = €81.66.

Federal Direct Payments (US $ million, 2002, rounded figures): Social security 10.3; Retirement and disability 6.3; Veterans 0.5; Other direct payments 3.2; Total 21.0.

Budget (estimates, US $ million, year ending 30 September 2002): Total revenue 199.7 (Taxes 166.8, Service fees 28.7, Operating transfers 4.3); Total expenditure 212.1 (Wages, salaries and benefits 108.9; Other expenditure 103.2). Source: Bank of Hawaii, *Commonwealth of the Northern Mariana Islands Economic Report* (October 2003).

Cost of Living (Consumer Price Index for Saipan; base: 1990 = 100): 137.1 in 2000; 136.0 in 2001; 136.2 in 2002. Source: ILO.

Gross Domestic Product (US $ million in current prices, estimate): 557.0 in 2002. Source: Bank of Hawaii, *Commonwealth of the Northern Mariana Islands Economic Report* (October 2003).

EXTERNAL TRADE

Principal Commodities (f.o.b., US $ million, 2000): *Imports*: Food items 0.3; Construction materials 7.4; Automobiles (incl. parts) 5.9; Petroleum products n.a.; Clothing 73.7; Beverages other than dairy and juice 7.6; Total (incl. others) 267.2. *2001:* Total imports 214.4. *Exports* (1991): Total 258.4. Source: partly UN, *Statistical Yearbook for Asia and the Pacific*.

Principal Trading Partners (US $ million, 1997): *Imports*: Guam 298.0; Hong Kong 200.5; Japan 118.3; Korea, Republic 80.6; USA 63.3; Total (incl. others) 836.2.

TRANSPORT

Shipping: *Registered Fleet* (2001): 1,029 vessels (791 fishing vessels); *Traffic* ('000 short tons, 1997): Goods loaded 184.1; Goods unloaded 425.9.

Civil Aviation (Saipan Int. Airport, year ending September 1999): 23,853 aircraft landings; 562,364 boarding passengers. Source: Commonwealth Ports Authority.

Road Traffic (registered motor vehicles, 2001): 17,900.

TOURISM

Visitor Arrivals: 528,608 in 2000; 444,284 in 2001; 475,547 in 2002.

Visitor Arrivals by Country (2002): Japan 326,735; Korea, Republic 90,324; USA (incl. Guam) 35,858; China, People's Republic 10,471.

Tourism Receipts (US $ million): 407 in 1999; 430 in 2000; 225 in 2002 (approximate figure). Source: Bank of Hawaii, *Commonwealth of the Northern Mariana Islands Economic Report* (October 2003).

COMMUNICATIONS MEDIA

Radio Receivers (households with access, census of 2000): 10,684.

Television Receivers (estimate, 1995): 15,460 in use.

Telephones (main lines in use, 2001): 25,306.

Mobile Cellular Telephones (2000): 3,000 subscribers*.

TeleStations (1996): 1,200 in use†.

* Source: International Telecommunication Union.

† Source: UN, *Statistical Yearbook*.

EDUCATION

Pre-primary (2001/02, state schools, Headstart programme): 12 schools; 42 teachers; 523 pupils.

Primary (2001/02, state schools): 12 schools; 304 teachers; 5,828 students.

Secondary (2001/02, state schools): 6 schools; 208 teachers; 4,074 students.

Higher (2000/01): 1 college; 2,383 students (full- and part-time students).

Private Schools (2001/02): 18 schools; 216 teachers; 2,801 students.

Directory

The Government

(September 2004)

Governor: Juan Nekai Babauta (took office January 2002).

Lieutenant-Governor: Diego Tenorio Benavente.

GOVERNMENT OFFICES

Office of the Governor: Caller Box 10007, Capitol Hill, Saipan, MP 96950; tel. 664-2276; fax 664-2290; e-mail gov.frosario@saipan.com; internet www.saipan.com/gov/index.htm.

Office of the Resident Representative to the USA, Commonwealth of the Northern Mariana Islands: 2121 R St, NW, Washington, DC 20008; tel. (202) 673-5869; fax (202) 673-5873; e-mail rep@resrep.gov.mp; the Commonwealth Govt also has liaison offices in Hawaii and Guam.

Department of the Interior, Office of Insular Affairs (OIA): Field Office of the OIA, Dept of the Interior, POB 2622, Saipan, MP 96950; tel. 234-8861; fax 234-8814; e-mail jeff.schorr@saipan.com; OIA representation in the Commonwealth; Field Representative Jeffrey Schorr.

Legislature

Legislative authority is vested in the Northern Marianas Commonwealth Legislature, a bicameral body consisting of the Senate and the House of Representatives. There are nine senators, elected for four-year terms, and 18 members of the House of Representatives, elected for two-year terms. The most recent elections took place on 3 November 2001, following which the Republicans held six of the nine seats in the Senate, the Democratic Party two seats and the Covenant Party one seat. The Republican Party held 12 of the 18 seats in the House of Representatives, the Democratic Party five seats and the Covenant Party one seat.

Senate President: JOAQUIN G. ADRIANO.

Speaker of the House: BENIGNO R. FITIAL.

Commonwealth Legislature: Capitol Hill, Saipan, MP 96950; tel. 664-7757; fax 322-6344.

Political Organizations

Covenant Party: c/o Commonwealth Legislature, Capitol Hill, Saipan MP 96950; Leader BENIGNO R. FITIAL; Chair. ELOY INOS.

Democratic Party of the Commonwealth of the Northern Mariana Islands, Inc: POB 500676, Saipan, MP 96950-0676; tel. 234-7497; fax 233-0641; Pres. Dr CARLOS S. CAMACHO; Chair. LORENZO CABRERA.

Republican Party: c/o Commonwealth Legislature, Capitol Hill, Saipan MP 96950; Pres. JUAN REYES.

Judicial System

The judicial system in the Commonwealth of the Northern Mariana Islands (CNMI) consists of the Superior Court, the Commonwealth Supreme Court (which considers appeals from the Superior Court) and the Federal District Court. Under the Covenant, federal law applies in the Commonwealth, apart from the following exceptions: the CNMI is not part of the US Customs Territory; the federal minimum wage provisions do not apply; federal immigration laws do not apply; and the CNMI may enact its own taxation laws.

Religion

The population is predominantly Christian, mainly Roman Catholic. There are small communities of Episcopalians (Anglicans—under the jurisdiction of the Bishop of Hawaii, in the USA) and Protestants.

CHRISTIANITY

The Roman Catholic Church

The Northern Mariana Islands comprise the single diocese of Chalan Kanoa, suffragan to the archdiocese of Agaña (Guam). The Bishop participates in the Catholic Bishops' Conference of the Pacific, based in Suva, Fiji. At 31 December 2002 there were 43,000 adherents, including temporary residents, in the Northern Mariana Islands.

Bishop of Chalan Kanoa: Most Rev. TOMAS AGUON CAMACHO, Bishop's House, Chalan Kanoa, POB 500745, Saipan, MP 96950; tel. 234-3000; fax 235-3002; e-mail tcamacho@vzpacifica.net.

The Press

The weekly *Focus on the Commonwealth* is published in Guam, but distributed solely in the Northern Mariana Islands.

Marianas Observer: POB 502119, Saipan, MP 96950; tel. 233-3955; fax 233-7040; weekly; Publr JOHN VABLAN; Man. Editor ZALDY DANDAN; circ. 2,000.

Marianas Review: POB 501074, Saipan, MP 96950; tel. 234-7160; f. 1979 as *The Commonwealth Examiner*; weekly; English and Chamorro; independent; Publr LUIS BENAVENTE; Editor RUTH L. TIGHE; circ. 1,700.

Marianas Variety News and Views: POB 500231, Saipan, MP 96950; tel. 234-6341; fax 234-9271; e-mail mvariety@vzpacifica.net; internet www.mvariety.com; Mon.–Fri.; English and Chamorro; independent; f. 1972; Publrs ABED E. YOUNIS, PAZ YOUNIS; Editor ZALDY DANDAN; circ. 7,500.

North Star: Chalan Kanoa, POB 500745, Saipan, MP 96950; tel. 234-3000; fax 235-3002; e-mail north.star@saipan.com; internet www.cnmicatholic.org/north.htm; weekly; English and Chamorro; Roman Catholic; Publr BISHOP TOMAS A. CAMACHO; Editor RUY VALENTE M. POLISTICO.

Pacific Daily News (Saipan bureau): POB 500822, Saipan, MP 96950; tel. 234-6423; fax 234-5986; Publr LEE WEBBER; circ. 5,000.

Pacific Star: POB 505815 CHRB, Saipan, MP 96950; tel. 288-0746; fax 288-0747; weekly; Operational Man. NICK LEGASPI; circ. 3,000.

Pacifica: POB 502143, Saipan, MP 96950; monthly; Editor MIKE MALONE.

Saipan Tribune: POB 10001, PMB 34, Saipan, MP 96950-8901; tel. 235-8747; fax 235-3740; e-mail editor.tribune@saipan.com; internet www.saipantribune.com; 2 a week; Editor ALDWIN FAJARDO; Publr REX I. PALACIOS; circ. 3,500.

Broadcasting and Communications

TELECOMMUNICATIONS

Micronesian Telecommunications Corpn (MTC/GTE Pacifica): Saipan, MP 96950; internet www.gtepacifica.net/; owned by Verizon Corpn (USA).

BROADCASTING

Radio

Inter-Island Communications Inc: POB 500914, Saipan, MP 96950; tel. 234-7239; fax 234-0447; f. 1984; commercial; station KCNM-AM, or KZMI-FM in stereo; Gen. Man. HANS W. MICKELSON; Programme Dir KEN WARNICK; CEO ANGEL OCAMPO.

Far East Broadcasting Co: POB 500209, Saipan, MP 96950; tel. 322-9088; fax 322-3060; e-mail saipan@febc.org; internet www.febc.org; f. 1946; non-commercial religious broadcasts; Exec. Dir ROBERT L. SPRINGER; Pres. GREGG HARRIS.

KSAI-AM: tel. 234-6520; fax 234-3428; e-mail ksai@febc.org; f. 1978; local service; mainly religious broadcasts; Station Man. HARRY BLALOCK.

KFBS-SW: tel. 322-9088; e-mail saipan@febc.org; international broadcasts in Chinese, Indonesian, Russian, Vietnamese, Burmese, Mongolian; Chief Engineer ROBERT SPRINGER.

Power 99: POB 10000, Saipan, MP 96950; tel. 235-7996; fax 235-7998; e-mail cdancoe@spbguam.com; internet www.radiopacific.com; Man. CURTIS DANCOE.

The Rock 97.9: POB 10000 Saipan, MP 96950; tel. 235-7996; fax 235-7998; e-mail cdancoe@spbguam.com; internet www.radiopacific.com; Man. CURTIS DANCOE.

Station KHBI-SW: POB 501837, Saipan, MP 96950; tel. 234-6515; fax 234-5452; fmrly KYOI; non-commercial station owned by the *Christian Science Monitor* (USA); Gen. Man. DOMINGO VILLAR.

Station KPXP: PMB 415, Box 10000, Saipan, MP 96950; tel. 235-7996; fax 235-7998; e-mail power99@saipan.com; internet www.radiopacific.com/p99/; f. 1981; acquired Station KRSI in 2000, which also broadcasts on Saipan; Gen. Man. CURTIS DANCOE.

Television

Marianas CableVision: Saipan, MP 96950; tel. 235-4628; fax 235-0965; e-mail mcv@saipan.com; internet www.kmcv.co.mp; 55-channel cable service provider, broadcasting US and Pacific Rim programmes.

Tropic Isles Cable TV Corpn: POB 501015, Saipan, MP 96950; tel. 234-7350; fax 234-9828; 33-channel commercial station, with 5 pay channels, broadcasting 24 hours a day; US programmes and local and international news; 5,000 subscribers; Gen. Man. FRED LORD.

KMCV-TV: POB 501298, Saipan, MP 96950; tel. 235-6365; fax 235-0965; f. 1992; 52-channel commercial station, with 8 pay channels, broadcasting 24 hours a day; US programmes and local and international news; 5,650 subscribers; Gen. Man. WAYNE GAMBLIN.

Finance

BANKING

Bank of Guam (USA): POB 678, Saipan, MP 96950; tel. 234-6467; fax 234-3527; internet www.bankofguam.com; Gen. Man. MERCED TOMOKANE; brs on Tinian and Rota.

Bank of Hawaii: Bank of Hawaii Bldg, El Monte Ave, Garapan, POB 500566, Saipan, MP 96950; tel. 236-8450; fax 322-4210; Vice-Pres. DAVID BUEHLER; 2 brs.

Bank of Saipan: POB 500690, Saipan, MP 96950; tel. 234-6260; fax 235-1802; e-mail bankofsaipan@gtepacifica.net.

City Trust Bank: Qualo Rai, POB 501867, Saipan, MP 96950; tel. 234-7701; fax 234-8664; e-mail citytrustbank@saipan.com; Asst Vice-Pres. and Acting Man. MARIA LOURDES JOHNSON.

First Hawaiian Bank: Oleai Centre, Beach Rd, Chalan Laulau, Saipan 96950; tel. 234-6559; fax 236-8936; internet www.fhb.com; Gen. Man. KEN KATO.

First Savings and Loan Asscn of America (USA): Beach Rd, Susupe, POB 500324, Saipan, MP 96950; tel. 234-6617; Man. SUZIE WILLIAMS.

Guam Savings and Loan Bank: POB 503201, Saipan, MP 96950; tel. 233-2265; fax 233-2227; Gen. Man. GLEN PEREZ.

HSBC Ltd: Middle Rd, Garapan, Saipan, MP 96950; tel. 234-2468; fax 234-8882.

INSURANCE

Aon Insurance: Aon Insurance Micronesia (Saipan) Inc. POB 502177, Saipan, MP 96950; tel. 234-2811; fax 234-5462; e-mail rod.rankin@aon.com.au; internet www.aon.com.

Allied Insurance/Takagi and Associates, Inc: PPP 602 Box 10000 Saipan, MP 96950; tel. 233-2554; fax 670-2553; Gen. Man. PETER SIBLY.

Associated Insurance Underwriters of the Pacific Inc: POB 501369, Saipan, MP 96950; tel. 234-7222; fax 234-5367; Gen. Man. MAGGIE GEORGE.

General Accident Insurance Asia Ltd (Microl Insurance): POB 502177, Saipan, MP 96950; tel. 234-2811; fax 234-5462; Man. Dir ANDREW M. HOWLETT.

Century Insurance: Century Insurance PMB 193, POB 10000, Saipan, MP 96950; tel. 234-0560; fax 234-1845; e-mail Budgetmp@itecnmi.com.

Marianas Insurance Co Ltd: POB 502505, Saipan, MP 96950-2505; tel. 234-5091; fax 234-5093; e-mail mic@itecnmi.com; Gen. Man. ROSALIA S. CABRERA.

Mitsui Sumitomo Insurance Co Ltd (Japan): POB 502505, Saipan, MP 96950-2505; tel. 234-5091; fax 234-5093; Gen. Man. ANDREW M. HOWLETT.

Moylan's Insurance Underwriters (Int.) Inc: POB 500658, Saipan, MP 96950; tel. 234-6571; fax 632-3788; e-mail saipan@moylans.net; Gen. Man. TAMARA HUNTER.

Pacifica Insurance Underwriters Inc: POB 500168, Saipan, MP 96950; tel. 234-6267; fax 234-5880; e-mail norten@netpci.com; internet www.pacificains.com; Pres. NORMAN T. TENORIO.

Primerica Financial Services: POB 500964, Saipan, MP 96950; tel. 235-2912; fax 235-7910; Gen. Man. JOHN SABLAN.

Staywell: POB 2050, Saipan, MP 96950; tel. 235-4260; fax 235-4263; Gen. Man. LARRY LAVEQUE.

Trade and Industry

GOVERNMENT AGENCIES

Commonwealth Development Authority: POB 502149, Wakins Bldg, Gualo Rai, Saipan, MP 96950; tel. 234-7145; fax 234-7144; e-mail administration@cda.gov.mp; internet www.cda.gov.mp; govt lending institution; funds capital improvement projects and private enterprises; offers tax incentives to qualified investors; Chair. SIXTO K. IGISOMAR.

Office of Public Lands, Board of Public Lands Management: POB 500380, Saipan, MP 96950; tel. 234-3751; fax 234-3755; e-mail opl@vzpacifica.net; manages public land, which constitutes 82% of total land area in the Commonwealth (14% on Saipan).

CHAMBER OF COMMERCE

Saipan Chamber of Commerce: Chalan Kanoa, POB 500806 CK, Saipan, MP 96950; tel. 233-7150; fax 233-7151; e-mail saipanchamber@saipan.com; internet www.saipanchamber.com; Pres. ALEXANDER A. SABLAN; Exec. Dir CARLENE REYES-TENORIO.

EMPLOYERS' ASSOCIATIONS

Association of Commonwealth Teachers (ACT): POB 5071, Saipan, MP 96950; tel. and fax 256-7567; e-mail cnmiteachers@netscape.net; supports the teaching profession and aims to improve education in state schools.

Saipan Garment Manufacturers' Association (SGMA): Saipan, MP 96950; e-mail sgmaemy@vzpacifica.net; internet www.sgma-saipan.org; Chair. RICHARD A. PIERCE.

UTILITIES

Commonwealth Utilities Corpn: POB 501220, Saipan, MP 96950; tel. 235-7025; fax 235-6145; e-mail cucedp@gtepacifica.net; scheduled for privatization.

TRADE UNION AND CO-OPERATIVES

International Brotherhood of Electrical Workers: c/o Micronesian Telecommunications Corpn, Saipan, MP 96950; Local 1357 of Hawaii branch of US trade union based in Washington, DC.

The Mariana Islands Co-operative Association, Rota Producers and Tinian Producers Associations operate in the islands.

Transport

RAILWAYS

There have been no railways operating in the islands since the Japanese sugar industry railway, on Saipan, ceased operations in the Second World War.

ROADS

In 1991 there were 494 km (307 miles) of roads on the islands, 320 km (199 miles) of which are on Saipan. First grade roads constitute 135 km (84 miles) of the total, 99 km (62 miles) being on Saipan. There is no public transport, apart from a school bus system.

SHIPPING

The main harbour of the Northern Mariana Islands is the Port of Saipan, which underwent extensive renovation in the mid-1990s. There are also two major harbours on Rota and one on Tinian. Several shipping lines link Saipan, direct or via Guam, with ports in Japan, Asia, the Philippines, the USA and other territories in the Pacific.

Commonwealth Ports Authority: POB 501055, Saipan, MP 96950; tel. 664-3500; fax 234-5962; e-mail cpa.admin@saipan.com; internet www.cpa.gov.mp; Exec. Dir CARLOS H. SALAS; Chair. RAMÓN S. PALACIOS.

Mariana Express Lines: POB 501937,CTS Building, Saipan, MP 96950; tel. 322-1690; fax 323-6355; e-mail ctsispn@aol.com; services between Saipan, Guam, Japan and Hong Kong.

Saipan Shipping Co Inc (Saiship): Saiship Bldg, Charlie Dock, POB 500008, Saipan, MP 96950; tel. 322-9706; fax 322-3183; e-mail saiship.general@vzpacifica.net; weekly barge service between Guam, Saipan and Tinian; monthly services to Japan and Micronesia; Gen. Man. DARLENE CABRERA.

Westpac Freight: POB 2048, Puerto Rico, Saipan, MP 96950; tel. 322-8798; fax 322-5536; e-mail westpac@gtepacifica.net; internet www.westpacfreight.com; services between Saipan, Guam and the USA; Man. MICHIE CAMACHO.

CIVIL AVIATION

Air services are centred on the main international airport, Isley Field, on Saipan. There are also airports on Rota and Tinian.

Continental Micronesia: POB 508778, A.B. Won Pat International Airport, Tamuning, MP 96911; tel. 647-6595; fax 649-6588; internet www.continental.com; f. 1968, as Air Micronesia, by Continental Airlines (USA); name changed 1992; subsidiary of Continental Airlines; hub operations in Saipan and Guam; services throughout the region and to destinations in the Far East and the mainland USA; Pres. MARK ERWIN.

Freedom Air: POB 500239 CK, Saipan, MP 96950; tel. and fax 288-5663; scheduled internal flights.

Pacific Island Aviation: PMB 318, POB 10,000, Saipan, MP 96950; tel. 234-3600; fax 234-3604; e-mail piasaipan@saipan.com; internet www.pacificislandaviation.com; f. 1987; scheduled services to Rota and Guam in partnership with Northwest Airlines (USA); repair station; flight instruction; Pres. and CEO ROBERT F. CHRISTIAN; Exec. Vice-Pres. PAZ PABALINAS.

Tourism

Tourism is one of the most important industries in the Northern Mariana Islands, earning US $430m. in 2001. In 2002 there were 4,313 hotel rooms. Most of the islands' hotels are Japanese-owned, and in 2002 68.7% of tourists came from Japan. The Republic of Korea and the USA are also important sources of tourists. The islands received a total of 475,547 visitors in 2002 (an increase of some 7.0% on the figure for the previous year, but 10% fewer than in 2000). The islands of Asuncion, Guguan, Maug, Managaha, Sariguan and Uracas (Farallon de Pajaros) are maintained as uninhabited reserves. Visitors are mainly attracted by the white, sandy beaches and the excellent diving conditions. There is also interest in the *Latte* or *Taga* stones (mainly on Tinian), pillars carved from the rock by the ancient Chamorros, and relics from the Second World War.

Hotel Association of the Northern Mariana Islands: POB 501983, Saipan, MP 96950; tel. 234-3455; fax 234-3411; e-mail rds@itecnmi.com; internet www.marianashotels.org; f. 1985; Chair. RONALD D. SABLAN.

Marianas Visitors Authority: POB 500861 CK, Saipan, MP 96950; tel. 664-3200; fax 664-3237; e-mail mva@saipan.com; internet www.mymarianas.com; f. 1976; responsible for the promotion and development of tourism in the Northern Mariana Islands; Man. Dir PERRY JOHN TENORIO.

Defence

The USA is responsible for the defence of the Northern Mariana Islands. The US Pacific Command is based in Hawaii (USA).

Education

School attendance is compulsory from six to 16 years of age. In 2000/01 there were 12 state primary schools, with a total of 5,828 pupils enrolled, and there were six state secondary schools, with a total enrolment of 4,074 pupils. There was a total of 18 private schools, with a total enrolment of 2,801 pupils. There was one college of further education in 1993/94, with 3,051 students. Budgetary expenditure on education totalled US $49.6m. in 2000, equivalent to 22.0% of total government expenditure.

UNITED STATES EXTERNAL TERRITORIES IN THE PACIFIC

Most of the external or unincorporated territories of the USA (except the US Virgin Islands and uninhabited Navassa Island, in the Caribbean Sea) are located in the Pacific Ocean. There are two territories with differing degrees of self-government. The Trust Territory of the Pacific Islands was terminated on 1 October 1994, when the Republic of Palau achieved independence from the USA under the Compact of Free Association. Most of the seven remaining island jurisdictions were annexed to the USA by the Guano Act of 1856, together with other, now-relinquished claims. In the federal Government, the Secretary of the Navy is ultimately responsible for Kingman Reef and for Johnston Atoll, and the Secretary of the Air Force for Wake Island. Otherwise, the Secretary of the Interior is the responsible authority for the territories, although the administration may be exercised by another federal agency. Within the Department of the Interior, the Assistant Secretary for Policy, Management and Budget is responsible for the Office of Insular Affairs and acts on behalf of the Secretary in all matters relating to the insular governments and territories. The USA maintains membership, in its own right and in respect of its territories, of the Pacific Community (formerly the South Pacific Commission), the Colombo Plan and the UN's Economic and Social Commission for Asia and the Pacific (ESCAP).

Head of State: President George W. Bush (took office 20 January 2001).

Department of the Interior, Office of Insular Affairs: 1849 C St, Washington, DC 20240, USA; tel. (202) 208-6816; fax (202) 501-7759; internet www.doi.gov/oia.

Deputy Assistant Secretary of the Interior for Insular Affairs: David B. Cohen.

Director, Office of Insular Affairs: Nikalao A. Pula.

Department of Defense: The Pentagon, Washington, DC 20301, USA; tel. (703) 697-5737; fax (703) 695-1149; internet ww.defenselink.mil; for details of service departments, see under relevant insular possessions.

AMERICAN SAMOA

Physical and Social Geography

American Samoa comprises the five islands of Tutuila, Ta'u, Olosega, Ofu and Aunu'u, and the atolls of Swains Island (Olohenga) and uninhabited Rose Island. The islands lie in the South Central Pacific along latitude 14°S at about longitude 170°W, some 3,700 km south-west of Hawaii (USA). Swains Island, administered as part of American Samoa since 1925, lies 340 km to the north-west of the main group. The territory's nearest neighbour is Samoa (formerly known as Western Samoa). The five principal islands are high and volcanic with rugged interiors and little flat land except along the coasts. The area of the islands is 201 sq km (77.6 sq miles).

At the 2000 census the population was 57,291, of whom the majority lived on Tutuila, where the main town and capital, Pago Pago, is situated (the officially designated seat of government is the village of Fagatogo). Approximately 91,000 American Samoans are resident in the USA.

The islands are peopled by Polynesians and are thought to have been the origin of many of the people who now occupy islands further east. The Samoan language is believed to be the oldest extant form of Polynesian speech. Samoan society developed an intricate hierarchy of graded titles comprising titular chiefs and orator chiefs. The basis of Samoan society remains the aiga, or extended family unit, headed by a chiefly matai. One of the striking features of modern Samoa is the manner in which this system and the culture of the islands prior to European contact remains a dominant influence. It is referred to as *fa'a Samoa*, the Samoan Way. Most of the population are Christians.

History

The Samoan islands were first visited by Europeans in the 1700s, but it was not until 1830 that missionaries from the London Missionary Society settled there. In 1872 the Kingdom of Samoa, then an independent state, ceded Pago Pago harbour to the USA as a naval coaling station. The United Kingdom and Germany were also interested in the islands, but a treaty signed in 1889 by these powers and the USA guaranteed the neutrality of Samoa. The British withdrew in 1899, and in the same year, following internal conflicts between rival chiefs, the kingship was abolished and a further tripartite treaty left the Western islands for Germany to govern, while the Eastern islands passed under US influence. In 1900 the high chiefs, or Matai, of Tutuila formally ceded the islands of Tutuila and Aunu'u to the USA and, in 1904, the chiefs of the islands of Ta'u, Olosega, Ofu and Rose (Manu'a District) followed suit. These deeds of cession and American Samoa's annexation as a US territory, were enacted by Congress in the 1920s. The President of the USA was empowered to provide for the executive, legislative and judicial administration of the territory, this responsibility being vested in the Secretary of the Navy until 1951, and subsequently in the Secretary of the Interior. Since 1925 Swains Island has also been administered by the USA as part of American Samoa. In December 1980 the USA and New Zealand signed a treaty (ratified in 1983) whereby New Zealand, on behalf of the territory of Tokelau, relinquished a claim to Swains Island (Olohenga). At the same time, the USA relinquished its claim to Tokelau, which lies to the north of American Samoa.

From 1951 until 1978 American Samoa was administered by a Governor appointed by the US Department of the Interior, and a legislature comprising a Senate and a House of Representatives. In November 1977 the first popular vote to elect a Governor was held. In January 1978 the successful candidate, Peter Coleman, was inaugurated as Governor. He was re-elected to a second term in the election of November 1980, after three years in office instead of four (the normal term prescribed in the Constitution), to allow synchronization with US elections in that year. Coleman was ineligible to hold office for three consecutive terms and, at elections in November 1984, A. P. Lutali was elected Governor, and Eni Hunkin (who subsequently adopted the use of his chiefly name, Faleomavaega) was elected Lieutenant-Governor.

In October 1986 a constitutional convention completed a comprehensive rewriting of the American Samoan Constitution. The draft revision, however, had yet to be submitted to the US Congress in the early 2000s. In April 2004 it was announced that the Political Status Study Commission would study the status of the territory and submit its recommendations in time to call another constitutional convention in 2005.

In July 1988 the territory's Delegate to the US House of Representatives, Fofō Sunia, announced that he would not seek re-election, as he was then under federal investigation for alleged financial mismanagement. In October of the same year he was sentenced to a period of between five and 15 months in prison for fraud. Elections to choose a new Delegate took place in November and were won by Faleomavaega, the Lieutenant-Governor. At the same time, Peter Coleman was elected Governor and Galea'i Poumele was elected Lieutenant-Governor.

In November 1990 a proposal that the Governor's powers of veto over legislative affairs should be restricted was defeated by a 75% majority in a referendum. Following the vote, Coleman announced that he favoured autonomy for American Samoa, and that he would seek negotiations on the matter. At gubernatorial elections in November 1992 Coleman was defeated by A. P. Lutali. At elections to the House of Representatives in November 1994 about one-third of those members who sought re-election was defeated. Faleomavaega was re-elected as non-voting Delegate.

In late 1994 and early 1995 American Samoa was among a number of Pacific islands to express concern at the proposed passage through their waters of regular shipments of plutonium, *en route* from Europe to Japan. Moreover, the decision by France to resume nuclear testing in the South Pacific in September 1995 was fiercely criticized by Lutali, who described the action as an affront to the entire Pacific community.

A court action initiated by the Government of American Samoa against a US insurance company for its alleged failure to pay adequate compensation following a devastating hurricane in late 1991 was concluded in September 1995. The company (which had responded to a claim for US $50m. in 1991 with a payment of only $6.1m.) was ordered by the court to pay a total of $86.7m. to the islands in compensation.

Gubernatorial and legislative elections took place in November 1996. An estimated 87% of eligible voters participated in the election, at which only eight of the 18 members seeking re-election to the Fono were successful. At a second round of voting the incumbent Lieutenant-Governor, the Democrat Tauese Sunia, was elected Governor with 51.3% of the votes, defeating Leala Peter Reid, Jr, who secured 48.7% of votes. Faleomavaega was re-elected as Delegate to the US House of Representatives, with 56.5% of the votes, defeating Gus Hannemann, who won 43.5% of votes.

The decision in July 1997 by the Government of neighbouring Western Samoa to change the country's name to simply Samoa caused some controversy in the territory. Legislation approved in March 1998 in the House of Representatives stated that American Samoa should not recognize the new name, which was viewed by many islanders as serving to undermine their own Samoan identity. Similarly, legislation prohibiting citizens of Samoa (formerly Western Samoa) from owning land in American Samoa was approved in response to the change. Nevertheless, moves towards *rapprochement* have been ongoing. In August 2004 a joint Samoa-American Samoa trade show was planned. Also, on the agenda of the fifth round of inter-Samoan talks, scheduled for November 2004, were negotiations over the contested maritime boundary between the two countries.

In September 1998 a motion to impeach the Governor was introduced in the House of Representatives. Tauese Sunia was alleged to have used public funds to pay personal debts, to have sold government vehicles without following recognized procedure and to have used funds allocated for school building repairs to purchase and install a sauna at his home.

In November 1998 Faleomavaega was re-elected as Delegate to the US House of Representatives. In the same month, following a dispute lasting several years, a legal settlement was reached between the five leading US tobacco companies and 46 US states and several of its territories, including American Samoa. It was agreed that between 1998 and 2025 American Samoa was to receive a total of some US $29m. in compensation for the harmful effects of cigarette smoking suffered by its inhabitants. In April 2000 the House of Representatives rejected a bill proposing the legalization of gambling, on the grounds that this would foster social problems.

At elections held on 7 November 2000, Governor Tauese Sunia was narrowly re-elected, receiving 50.7% of votes cast, compared with 47.9% for opposition candidate, Senator Lealaifuaneva. However, it was later announced that a re-count of votes was to take place, although no date was set. At congressional elections, held concurrently, no candidate received the necessary 50% majority, Faleomavaega winning 45.7% of votes cast, compared with 30.3% for Hanneman. A second round of voting took place on 21 November, when Faleomavaega won a reported 61.1% of the votes and Hanneman received 38.9%. Faeomavaega thus returned to the US House of Representatives for a seventh two-year term.

The issue of US jurisdiction in the territory was raised once more in September 2001, when the Senate approved a resolution urging discussion with the US Government regarding the conferral of limited federal jurisdiction to the High Court of American Samoa, the only US territory without a sitting federal judge. A congressional public survey, conducted in the same month, registered wide popular support for a Federal Court and Public Prosecutor. At a meeting of the UN General Assembly in January 2002, the UN accepted American Samoa's proposal of May 2001 that it be removed from the list of colonized territories. The Governor had sent a resolution to the UN Committee on Decolonization affirming American Samoa's wish to remain a US territory.

In mid-September 2002 the territory's immigration procedures were amended to give the Attorney-General, rather than the Immigration Board, the ultimate authority to grant permanent resident status to aliens. The House of Representatives also approved a resolution to repeal legislation automatically conferring US citizenship in the territory to foreign parents. Meanwhile, in early June 2002 a former president of the Amerika Samoa Bank (now ANZ Amerika Samoa Bank), Will Cravens, was convicted of fraud in a scheme whereby a total of US $75m. of investors' money had been misappropriated. Further allegations of corruption in the banking sector followed in 2004. The government-owned Development Bank of American Samoa was being investigated in April of that year by an anti-corruption senate committee, after records suggested favouritism in the approval of loans.

The authorities were obliged to introduce fuel rationing in October 2002 after a shipment was turned away from Pago Pago when it was discovered that its contents did not meet the specifications of the Environmental Protection Agency. Emergency supplies were delivered from neighbouring Samoa, pending the arrival of the next shipment.

In elections for American Samoa's Delegate to the US Congress, held on on 6 November 2002, Faleomavaega won 41.3% of votes cast, Fagafaga Daniel Langkilde secured 32.1% and Aumua Amata Coleman won 26.6%. Since none of the candidates secured the 50% of votes required, Faleomavaega and Langkilde entered a second round of voting on 19 November, at which Faleomavaega secured an eighth term in office.

In mid-December 2002 legislation was introduced that prohibited nationals of 23 countries thought to present a terrorist threat from entering the territory, unless they were granted special permission. Opposition groups in Fiji notably expressed their displeasure at their country's inclusion on the list of countries, owing to its large Muslim population. In March 2003 airline representatives also complained that the strict security measures introduced at the airport were damaging business. Also in March, Tauese Sunia died while travelling to Hawaii for medical treatment. Lieutenant-Governor Togiola Tulafono became Governor and appointed Ipulasi Aitofele Sunia as his deputy.

In late January 2003 the Senate approved a motion to begin expulsion hearings against Senator Faamausili Pola. Members had already voted to remove Faamausili Pola in September 2002 when it was alleged that he had not been elected according to Samoan tradition. However, an American Samoan High Court ruling ordered Pola to be reinstated. In March 2003 it was reported that the South Korean owner of the Daewoosa Samoa clothing factory had been convicted by a US court of human trafficking, in the largest case of its kind in US history. The mainly Vietnamese and Chinese factory employees had received very low wages and suffered appalling working conditions. In March 2004 Governor Tulafono announced plans for legislation that would allow prosecution of human-trafficking offences in the territory. In July it was announced that a US senate committee would hold a hearing on human trafficking at the site of the closed garment factory.

During 2003 several allegations of official corruption were made against American Samoan government officials. In January 2004 Governor Tulafono announced plans for legislation that would expedite the investigation of corruption in the Government and companies that hold government contracts. In February the US Federal Bureau of Investigation (FBI) began a wide-ranging investigation into the alleged misuse of US government funds. In May Tafua Faau Seumanutafa, the former Chief Procurement Officer, pleaded guilty in the Hawaii Federal District Court to one count of conspiracy to defraud the US Government, in a case that also implicated senior officials from the Department of Health and Social Security and the Department of Education. Also in May the Senate Select Investigative Committee issued subpoenas to several senators to testify about alleged corruption in the allocation of government contracts. In early September the Committee recommended that, while under investigation for alleged corruption, Lieutenant-Governor Ipulasi Aitofele Sunia be placed on leave and have his name removed from the list of gubernatorial candidates for the 2004 election. (National presidential and legislative elections were also scheduled for November 2004.) Governor Tulafono claimed the charges were politically motivated.

In May 2003 severe flooding and landslides led to the deaths of five people, and in the following month President George W. Bush declared the territory a federal disaster area. In January 2004 a state of emergency was declared after a cyclone caused at least one fatality and major damage to some 4,000 homes, according to aid agencies. The USA allocated US $12.5m. towards the relief effort.

The federal Government of the USA remains responsible for defence and foreign affairs.

Economy

In 1985, according to estimates by the World Bank, American Samoa's gross national product (GNP), measured at average 1983–85 prices, was about US $190m., equivalent to $5,410 per head. GNP per head was estimated at some $8,000 in 1992. Between 1973 and 1985, it was estimated, GNP increased, in real terms, at an average rate of 1.7% per year, with real GNP per head rising by only 0.1% per year. In 1990–98 the population increased by an average of 4.2% per year. An estimated 91,000 American Samoans live on the US mainland or in Hawaii.

Agriculture, hunting, forestry and fishing engaged only 3.1% of the employed labour force in 2000. Agricultural production provides little surplus for export. Major crops are coconuts, bananas, taro, pineapples, yams and breadfruit. A state of disaster was declared in mid-1993, following the outbreak of taro-leaf blight in the islands, and in mid-1996 imports of root crops from Fiji were banned in an attempt to prevent the spread of a highly destructive beetle. Moreover, in 1998 the banana crop was severely reduced by aphid activity.

Tuna-canning plants at Pago Pago process fish from US, Taiwanese and South Korean vessels. The StarKist Samoa and COS Samoa Packing plants are among the largest in the world. Canned tuna constituted almost 97% of total export revenue in 1998/99, when earnings reached US $334.2m. Fish-canning is the dominant industrial activity, engaging more than 42% of the employed workforce in the early 1990s; in 2001 70% of employees were guest workers from Samoa (formerly Western Samoa). The industry is an important source of government revenue, with corporate taxes from the StarKist facility alone totalling $3m. in 2003. In 2002 the tuna-canning companies expressed concern that an impending US free-trade agreement with Central American countries, which included the removal of tariffs on canned tuna, would threaten their exports to the USA. Nevertheless, in March 2003 the Senate approved a bill to levy a 20% tax on foreign tuna meat sold to the territory's canneries, prompting criticism from cannery interests. In February and May 2004 Eni Faleomavaega, the American Samoan Delegate to the US Congress, made representations on behalf of the territory's cannery interests to the US International Trade Committee during negotiations over the Andean Trade Preferences Act (between the USA and Bolivia, Colombia, Ecuador and Peru) and the Thailand Free Trade Agreement. Faleomavaega opposed the abolition of tariffs on imported tuna from these countries, arguing that the consequences would be catastrophic for the American Samoan economy. In August the StarKist tuna-canning factory suspended activities for a week owing to a decline in the supply of tuna meat.

Other activities include meat-canning, handicrafts, dairy farming, orchid farming and the manufacture of soap, perfume, paper products and alcoholic beverages. A garment factory began operations at Tafuna in 1995, and in the following year employed more than 700 people (although almost one-half of these employees were foreign workers). However, the islands' garment manufacturing industry was adversely affected following a riot between local workers and Vietnamese employees at a clothing factory, during which a number of people were injured. The incident was widely condemned, and Governor Tauese Sunia stated that the future of the garment industry in American Samoa would require consideration. Exports of finished garments declined from US $15.4m. in 1997/98 to $4.6m. in 1998/99. In 2002 the management of the Daewoosa clothing manufacturer was ordered by the authorities to pay compensation to its employees owing to the poor working conditions at the factory, and in early 2003 the factory's South Korean owner was convicted of human trafficking. In July 2004 it was announced that a US senate committee would hold a hearing on human trafficking at the site of the closed Daewoosa factory. The construction sector engaged 6.4% of the employed labour force in 2000.

Service industries engage a majority of the employed labour force in American Samoa (55.2% in 2000). The Government alone employs almost one-third of workers, although in the mid-1990s a series of reductions in the number of public-sector employees was introduced. The tourist industry is developing steadily, and earned some US $10m. in 1998. Tourist arrivals rose from 35,672 in 1998 to 41,287 in 1999 and 44,158 in 2000.

The visible trade deficit rose from US $107.5m. in the year ending September 1999 to $159.6m. in 2000, when imports (including items imported by the Government and goods by the fish-canning sector) totalled $505.9m. and exports reached $346.3m. Most of American Samoa's trade is conducted with the USA. Other trading partners in 1999/2000 included Australia, which supplied 6.5% of total imports, and New Zealand (4.9 %). The UN estimated that, during 1990–2000, the territory's exports increased at an average annual rate of 2.4%, while imports increased at a rate of 3.2% a year. In 2001 there was a reported trade surplus of $86.0m., equivalent to some 37% of the value of imports. In March 2003 preliminary figures indicated that there had been a $121m. surplus on the balance of payments in 2002.

In late 1989 American Samoa experienced a severe financial crisis, in response to which, in early 1990, the Governor imposed pay cuts on 3,800 government workers and introduced the emergency 'freezing' of local prices. At the end of the year the budgetary deficit was estimated to have reached US $17.7m. (compared with $7.5m. in the previous year). The territory's total debt had reached some $60m. by September 1992. Assistance from the USA totalled more than $40m. in 1994. In 1996/97 the budget deficit was estimated at $7.5m. In 2001 a fiscal surplus of $200,000 was recorded, the territory's first budgetary surplus in more than 20 years. Government revenues in 2001 surpassed budgeted expectations by more than $4.2m. Annual inflation averaged 2.8% in 1990–96. Consumer prices rose by 1.7% in 1998 and by 0.9% in 1999. Following increases of 1.9% in 2000 and of 1.4% in 2001, an annual inflation rate of 5.2% was recorded in the third quarter of 2003. An estimated 10.5% of the total labour force were unemployed in 2000.

American Samoa's minimum wage structure has caused considerable controversy. As stipulated in the US Fair Labor Standards Act, American Samoa is exempt from the general minimum wage of mainland USA and, instead, applies industry-specific rates set by a special industry commission. In October 2003 the minimum rates of hourly pay for the garment and cannery sectors were, respectively, US $2.68 and $3.26. The situation (in which American Samoa has considerably lower minimum hourly rates of pay than the rest of the USA) has been largely attributed to the presence of the two tuna-canning plants on the islands, which provide much of the manufacturing employment on the islands, and consequently exert substantial influence over the setting of wage levels. A US recommendation for modest increases in wage levels, implemented in mid-1996, was strongly opposed by the American Samoan Government, which argued that, with higher costs, the territory's tuna-canning industry would be unable to compete with other parts of the world. In May 2001 the American Samoan Government waived the provision requiring foreign workers to perform domestic duties for a year before being allowed to work. This move was welcomed by the tuna canneries, which were having difficulties recruiting new workers.

American Samoa's continued financial problems have been compounded by the high demand for government services from an increasing population, the limited economic and tax base, and natural disasters. Attempts to reduce the persistent budget deficit and achieve a greater measure of financial security for the islands have included severe reductions in the number of public-sector employees, increased fees for government services and plans to diversify the economy by encouraging tourism and expanding manufacturing activity. Economic development, however, is hindered by the islands' remote location, their vulnerability to adverse climatic conditions, the limited infrastructure and lack of skilled workers.

In February 1998 the American Samoa Economic Advisory Commission was established; it was hoped that the Commission would help the territory attain self-sufficiency and reduce its budget deficit. Its task was facilitated in November of that year, when the so-called 'tobacco settlement' (see History) was reached. In late 1999 the Government opted to negotiate a loan of US $18.6m. from the US Department of the Interior, which was to be repaid from the proceeds of the 'tobacco settlement'. The money allowed the Government to pay outstanding debts—settlement of health-care and utility bills were a condition stipulated by the USA—and to launch a programme of fiscal reforms. This was bolstered by an insurance settlement for damages sustained during Hurricane Val, in 1991, worth $47.9m. In April 2002 the American Samoa Economic Advisory Commission published a report, further amended in July of that year, outlining proposals for the development of a more diverse market economy in the territory. The Commission identified four sectors—fisheries and agriculture, telecommunications and information technology, manufacturing and tourism—as growth industries with the most potential for development. The plan aimed to implement initiatives over a period of five to 10 years, leading ultimately to the creation of new enterprises and more small and medium-sized businesses.

American Samoa is a member of the Pacific Community (formerly the South Pacific Commission) and is an associate member of the UN's Economic and Social Commission for Asia and the Pacific (ESCAP). In mid-1986 the Governments of Western Samoa (now Samoa) and American Samoa signed a memorandum of understanding, in accordance with which a permanent committee was established to ensure mutual economic development.

Statistical Survey

Source (unless otherwise indicated): Statistics Division, Department of Commerce, Pago Pago, AS 96799; tel. 633-5155; fax 633-4195; internet www.amsamoa.com.

AREA AND POPULATION

Area: 201 sq km (77.6 sq miles); *By Island* (sq km): Tutuila 137; Ta'u 46; Ofu 7; Olosega 5; Swains Island (Olohenga) 3; Aunu'u 2; Rose 1.

Population: 46,773 at census of 1 April 1990; 57,291 (males 29,264, females 28,027) at census of 1 April 2000. *By Island* (2000): Tutuila 55,400; Manu'a District (Ta'u, Olosega and Ofu islands) 1,378; Aunu'u 476; Swains Island (Olohenga) 37.

Density (2000): 285.0 per sq km.

Ethnic Groups (2000 census): Samoan 50,545; part-Samoan 1,991; Asian 1,631; Tongan 1,598; Total (incl. others) 57,291.

Principal Towns (population at 2000 census): Tafuna 8,409; Nu'uuli 5,154; Pago Pago (capital) 4,278; Leone 3,568; Fagatogo 2,096.

Births, Marriages and Deaths (2001): Registered live births 1,655; Registered marriages 254; Registered deaths 239.

Expectation of Life (years at birth, 1995): Males 68.0; Females 76.0. Source: UN Economic and Social Commission for Asia and the Pacific.

Economically Active Population (persons aged 16 years and over, 2000 census): Agriculture, hunting, forestry, fishing and mining 517; Manufacturing 5,900; Construction 1,066; Trade, restaurants and hotels 2,414; Transport, storage, communications and utilities 1,036; Financing, insurance, real estate and business services 311; Community, social and personal services 5,474; *Total employed* 16,718 (males 9,804, females 6,914); Unemployed 909 (males 494, females 415); *Total labour force* 17,627 (males 10,298, females 7,329). Source: US Department of Commerce, *2000 Census of Population and Housing*. *2003 estimates*: Total employed 14,319; Unemployed 1,681; Total labour force 16,000 (Source: US Department of State).

AGRICULTURE, ETC.

Principal Crops (FAO estimates, '000 metric tons, 2002): Coconuts 5; Taro 2; Bananas 1. Source: FAO.

Livestock (FAO estimates, 2002): Pigs 10,500; Cattle 103; Chickens 38,000. Source: FAO.

Fishing (metric tons, live weight, 2002): Total catch 6,963 (Albacore 5,944; Yellowfin tuna 484). Source: FAO.

INDUSTRY

Production: Electric energy 171.10 million kWh (year ending September 2001). Source: American Samoa Department of Commerce.

FINANCE

Currency and Exchange Rates: United States currency is used. For details, see section on the Northern Mariana Islands.

Budget (US $ '000, year ending September 1997): *Revenue:* Local revenue 47,151 (Taxes 34,105, Charges for services 7,099); Grants from US Department of Interior 97,287 (Operating grant 23,056, Capital projects 4,015, Special revenue 70,216); Total 144,438. *Expenditure:* General government 21,615; Public safety 10,558; Public works 19,580; Health and recreation 41,706; Education and culture 43,533; Economic development 9,344; Capital projects 5,221; Debt-servicing 1,356; Total 152,912.

Cost of Living (Consumer Price Index; annual averages; base: 1990 = 100): All items 123.4 in 1999; 125.7 in 2000; 127.5 in 2001. Source: UN, *Statistical Yearbook for Asia and the Pacific*.

EXTERNAL TRADE

Principal Commodities (US $ million, year ending September 2001): *Imports:* Food 66.3 (Fish 27.4); Fuel and oil 25.8 (Diesel fuel 16.3); Textiles and clothing 7.3; Machinery and transport equipment 23.7 (Road motor vehicles and parts 12.7); Miscellaneous manufactured articles 85.4 (Tin plates 59.8); Construction materials 19.3; Total (incl. others) 231.0. *Exports:* Canned tuna 273.7; Total (incl. others) 317.0.

Principal Trading Partners (US $ million, year ending September 2000): *Imports:* Australia 33.1; Fiji 12.7; Japan 3.8; Korea, Republic 5.4; New Zealand 24.7; Samoa 7.0; USA 125.7; Total (incl. others) 505.9. *Exports:* Total 346.3 (almost entirely to the USA). Source: UN, *Statistical Yearbook for Asia and the Pacific*.

TRANSPORT

Road Traffic ('000 registered motor vehicles, 2001): Passenger cars 6.5; Total 7.4.

International Sea-borne Shipping (estimated freight traffic, '000 metric tons, 2001): Goods loaded 257; Goods unloaded 272.

Civil Aviation (Pago Pago Int. Airport, 2001): Flights 7,805; Passengers (excl. transit) 156,212 (Boarding 81,669, Disembarking 74,543, Transit 2,040); Freight and mail ('000 lb) 4,520 (Loaded 1,626, Unloaded 2,894).

TOURISM

Tourist Arrivals: 35,672 in 1998; 41,287 in 1999; 44,158 in 2000.

Tourist Arrivals by Country (2000): Australia 717; China 351; Fiji 685; France 498; Japan 93; New Zealand 3,916; Philippines 304; Tonga 519; United Kingdom 84; USA 7,521; Total (incl. others) 44,158. Figures exclude arrivals from Samoa (formerly Western Samoa). Source: World Tourism Organization, *Yearbook of Tourism Statistics*.

Tourism Receipts (US $ million): 9 in 1996; 10 in 1997; 10 in 1998. Source: World Tourism Organization.

COMMUNICATIONS MEDIA

Daily Newspapers (1996): 2; estimated circulation 5,000*.

Non-daily Newspapers (1996): 1; estimated circulation 3,000*.

Radio Receivers (1997): 57,000* in use.

Television Receivers (1999): 15,000* in use.

Telephones (2001): 12,587 main lines in use.

Facsimile Machines (2001): 694 subscribers.

Mobile Cellular Telephones (2001): 2,156 subscribers.

* Source: UNESCO, *Statistical Yearbook*.

EDUCATION

Pre-primary: 61 schools (2001); 123 teachers (1991/92); 1,893 pupils (2001).

Primary: 32 schools (2001); 524 teachers (1991/92); 11,343 pupils (2001).

Secondary: 10 high schools (2001); 266 teachers (1991/92); 4,217 pupils (2001).

Higher (2001): American Samoa Community College 1,178 students.

Sources: American Samoa Department of Education.

Directory

The Constitution

American Samoa is an Unincorporated Territory of the USA. Therefore, not all the provisions of the US Constitution apply. As an unorganized territory it has not been provided with an organic act by Congress. Instead the US Secretary of the Interior, on behalf of the President, has plenary authority over the Territory and enabled the people of American Samoa to draft their own Constitution.

According to the 1967 Constitution, executive power is vested in the Governor, whose authority extends to all operations within the Territory of American Samoa. The Governor has veto power with respect to legislation passed by the Fono (Legislature). The Fono consists of the Senate and the House of Representatives, with a President and a Speaker presiding over their respective divisions. The Senate is composed of 18 members, elected, according to Samoan custom, from local chiefs, or Matai, for a term of four years. The House of Representatives consists of 20 members who are elected by popular vote for a term of two years, and a non-voting delegate from Swains Island. The Fono meets twice a year, in January and July, for not more than 45 days and at such special sessions as the Governor may call. The Governor has the authority to appoint heads of government departments with the approval of the Fono. Local government is carried out by indigenous officials. In August 1976 a referendum on the popular election of a Governor and a Lieutenant-Governor resulted in an affirmative vote. The first gubernatorial elections took place on 8 November 1977 and the second occurred in November 1980; subsequent elections were to take place every four years.

American Samoa sends one non-voting Delegate to the US House of Representatives, who is popularly elected every two years.

The Government

(September 2004)

Governor: TOGIOLA TULAFONO.

Lieutenant-Governor: IPULASI AITOFELE SUNIA.

GOVERNMENT OFFICES

Governor's Office: Executive Office Building, Third Floor, Utulei, Pago Pago, AS 96799; tel. 633-4116; fax 633-2269; e-mail governorsoffice@asg-gov.com.

Department of the Interior, Office of Insular Affairs (OIA): Field Office of the OIA, Dept of the Interior, POB 1725, Pago Pago, AS 96799; tel. 633-2800; fax 633-2415; internet www.doi.gov/oia/Islandpages/asgpage; Field Representative LYDIA FALEAFINE NOMURA.

Office of the Representative to the Government of American Samoa: Amerika Samoa Office, 1427 Dillingham Blvd, Suite 210, Honolulu HI 96817, USA; tel. (808) 847-1998; fax (808) 847-3420; Representative SOLOALI'I FAALEPO, Jr.

Department of Administrative Services: American Samoa Government, Executive Office Building, Utulei, Pago Pago, AS 96799; tel. 633-4156; fax 633-1814; internet www.asg-gov.com/departments/as.asg; Dir LAUTI SIMONA.

Department of Agriculture: American Samoa Government, Executive Office Building, Utulei, Pago Pago, AS 96799; tel. 699-1497; fax 699-4031; internet www.asg-gov.com/departments/doa.asg; Dir APEFA'I TAIFANE.

Department of Commerce: American Samoa Government, Executive Office Building, 2nd Floor, POB 1147, Utulei, Pago Pago, AS

96799; tel. 633-5155; fax 633-4195; e-mail asgdoc@amsamoa.com; internet www.amsamoa.com; Dir ALI'IMAU H. SCANLAN, JR.

Department of Education: American Samoa Government, Executive Office Building, Utulei, Pago Pago, AS 96799; tel. 633-5237; fax 633-4240; internet www.doe.as; Dir Dr SILI SATAUA.

Department of Health: American Samoa Government, Pago Pago, AS 96799; tel. 633-4606; fax 633-5379; internet www.asg-gov.com/departments/doh.asg; Dir JOSEPH TUFA.

Department of Human Resources: American Samoa Government, Executive Office Building, Utulei, Pago Pago, AS 96799; tel. 633-4485; fax 633-1139; internet www.asg-gov.com/departments/dhr.asg; Dir PUNI PENEI SEWELL.

Department of Human and Social Services: American Samoa Government, Pago Pago, AS 96799; tel. 633-1187; fax 633-7449; internet www.asg-gov.com/departments/dhss/dhss.asg; Dir PATOLO MAGEO (acting).

Department of Legal Affairs: American Samoa Government, Executive Office Building, Utulei, Pago Pago, AS 96799; tel. 633-4163; fax 633-1838; internet www.asg-gov.com/departments/dla.asg; Dir Attorney-General FITI SUNIA.

Department of Local Government (Office of Samoan Affairs): American Samoa Government, Pago Pago, AS 96799; tel. 633-5201; fax 633-5590; internet www.asg-gov.com/departments/osa.asg; Gen. Sec. SOTOA SAVALI.

Department of Marine and Wildlife Resources: American Samoa Government, Executive Office Building, Utulei, Pago Pago, AS 96799; tel. 633-4456; fax 633-5590; internet www.asg-gov.com/departments/dmwr.asg; Dir UFAGAFA RAY TULAFONO.

Department of Parks and Recreation: American Samoa Government, Pago Pago, AS 96799; tel. 699-9614; fax 699-4427; internet www.asg-gov.com/departments/dpr.asg; Dir LAAU SEUI SR.

Department of Port Administration: American Samoa Government, Pago Pago, AS 96799; tel. 633-4251; fax 633-5281; internet www.asg-gov.com/departments/dpa.asg; Dir SEUGOGO SCHRIMER.

Department of Public Safety: American Samoa Government, Pago Pago, AS 96799; tel. 633-1111; fax 633-7296; internet www.asg-gov.com/departments/dps; Dir TUITELELEAPAGA PESETA FUE IONAE.

Department of Public Works: American Samoa Government, Executive Office Building, Utulei, Pago Pago, AS 96799; tel. 633-4141; fax 633-5958; internet www.asg-gov.com/departments/dpw.asg; Dir PUNAOFO TILEI.

Department of Treasury: American Samoa Government, Executive Office Building, Utulei, Pago Pago, AS 96799; tel. 633-4155; fax 633-4100; internet www.asg-gov.com/departments/dtr.asg; Dir AITOFELE SUNIA.

Department of Youth and Women's Affairs: American Samoa Government, Executive Office Building, Utulei, Pago Pago, AS 96799; tel. 633-2835; fax 633-2875; internet www.asg-gov.com/departments/dywa; Dir FIASILI PUNI E. HALECK.

Environmental Protection Agency: American Samoa Government, Office of the Governor, Pago Pago, AS 96799; tel. 633-2304; fax 633-5590; e-mail ppeshut@yahoo.com; Dir PETER PESHUT (acting).

Legislature

FONO

Senate

The Senate has 18 members, elected, according to Samoan custom, from local chiefs, or Matai, for a term of four years.

President: (vacant).

House of Representatives

The House has 20 members who are elected by popular vote for a term of two years, and a non-voting delegate from Swains Island.

Speaker: MATAGI MAILO RAY MCMOORE.

CONGRESS

Since 1980 American Samoa has been able to elect, for a two-year term, a Delegate to the Federal Congress, who may vote in committee but not on the floor of the House of Representatives. Elections to the post took place in November 2002.

Delegate of American Samoa: ENI F. H. FALEOMAVAEGA, US House of Representatives, 2422 Rayburn House Office Bldg, Washington, DC 20515, USA; tel. (202) 225-8577; fax (202) 225-8757; e-mail faleomavaega@mail.house.gov.

Judicial System

The judicial system of American Samoa consists of the High Court, presided over by the Chief Justice and assisted by Associate Justices (all appointed by the Secretary of the Interior), and a local judiciary in the District and Village Courts. The judges for these local courts are appointed by the Governor, subject to confirmation by the Senate of the Fono. The High Court consists of three Divisions: Appellate, Trial, and Land and Titles. The Appellate Division has limited original jurisdiction and hears appeals from the Trial Division, the Land and Titles Division and from the District Court when it has operated as a court of record. The Trial Division has general jurisdiction over all cases. The Land and Titles Division hears cases involving land or Matai titles.

The District Court hears preliminary felony proceedings, misdemeanours, infractions (traffic and health), civil claims less than US $3,000, small claims, Uniform Reciprocal Enforcement of Support cases, and *de novo* trials from Village Courts. The Village Courts hear matters arising under village regulations and local customs.

Chief Justice: MICHAEL KRUSE.

Associate Justice: LYLE L. RICHMOND.

Attorney-General: FITI SUNIA.

High Court

Office of the Chief Justice, High Court, Pago Pago, AS 96799; tel. 633-1261; e-mail Hcourt@samoatelco.com.

Judge of the District Court: JOHN L. WARD II, POB 427, Pago Pago, AS 96799; tel. 633-1101.

Judge of the Village Court: FAISIOTA TAUANU'U, Pago Pago, AS 96799; tel. 633-1102.

Religion

The population is largely Christian, more than 50% being members of the Congregational Christian Church and about 20% being Roman Catholics.

CHRISTIANITY

American Samoa Council of Christian Churches: c/o CCCAS Offices, POB 1637, Pago Pago, AS 96799; f. 1985; six mem. churches; Pres. Cardinal PIO TAOFINU'U (Roman Catholic Archbishop of Samoa-Apia and Tokelau); Gen. Sec. Rev. ENOKA L. ALESANA (Congregational Christian Church in American Samoa).

The Roman Catholic Church

American Samoa comprises the single diocese of Samoa-Pago Pago, suffragan to the archdiocese of Samoa-Apia and Tokelau. At 31 December 2002 there were 12,000 adherents in the islands. The Bishop participates in the Catholic Bishops' Conference of the Pacific, based in Suva, Fiji.

Bishop of Samoa-Pago Pago: Rev. JOHN QUINN WEITZEL, Diocesan Pastoral Center, POB 596, Fatuoaiga, Pago Pago, AS 96799; tel. 699-1402; fax 699-1459; e-mail quinn@samoatelco.com.

The Anglican Communion

American Samoa is within the diocese of Polynesia, part of the Church of the Province of New Zealand. The Bishop of Polynesia is resident in Fiji.

Protestant Churches

Congregational Christian Church in American Samoa (CCCAS): POB 1537, Pago Pago, AS 96799; tel. 699-9810; fax 699-1898; e-mail sampoga@aol.com; internet www.cwmission.org.uk/about/view_church.cfm?ChurchID=6; Gen. Sec. Rev SAMUEL TIALAVEA; 39,000 mems (incl. congregations in New Zealand and USA) in 2003.

Other active Protestant groups include the Baptist Church, the Christian Church of Jesus Christ, the Methodist Church, Assemblies of God, Church of the Nazarene and Seventh-day Adventists. The Church of Jesus Christ of Latter-day Saints (Mormons) is also represented.

The Press

News Bulletin: Office of Public Information, American Samoa Government, Utulei; tel. 633-5490; daily (Mon.–Fri.); English; non-commercial; Editor PHILIP SWETT; circ. 1,800.

Samoa Journal and Advertiser: POB 3986, Pago Pago, AS 96799; tel. 633-2399; weekly; English and Samoan; Editor MICHAEL STARK; circ. 3,000.

Samoa News: POB 909, Pago Pago, AS 96799; tel. 633-5599; fax 633-4864; e-mail samoanews@samoatelco.com; internet www.samoanews.com; 6 a week; English and Samoan; Publr VERA M. ANNESLEY; circ. 4,500.

Broadcasting and Communications

TELECOMMUNICATIONS

American Samoa Telecommunications Authority: Box M, Pago Pago, AS 96799; tel. 633-1211; e-mail info@samoatelco.com; internet www.samoatelco.com.

BROADCASTING

Radio

KSBS-FM: POB 793, Pago Pago, AS 96799; tel. 633-7000; fax 622-7839; internet www.ksbsfm.com; commercial; Gen. Man. ESTHER PRESCOTT.

WVUV: POB 4894, Pago Pago, AS 96799; tel. 688-7397; fax 688-1545; fmr govt-administered station leased to Radio Samoa Ltd in 1975; commercial; English and Samoan; 24 hours a day; Man. VINCENT IULI.

Television

KVZK-TV: Office of Public Information, POB 3511, Pago Pago, AS 96799; tel. 633-4191; fax 633-1044; f. 1964; govt-owned; non-commercial; English and Samoan; broadcasts 18 hours daily on two channels; Gen. Man. VAOITA SAVALI; Technical Dir CARL SEVERA.

Finance

(cap. = capital; dep. = deposits; m. = million; amounts in US dollars)

BANKING

Commercial Banks

ANZ Amerika Samoa Bank: POB 3790, Pago Pago, AS 96799; tel. 633-5053; fax 633-5057; e-mail decourtb@samoatelco.com; internet www.anz.com.au/americansamoa; f. 1979; fmrly Amerika Samoa Bank; joined ANZ group in April 2001; Pres. GARY AYRE; 4 brs.

Bank of Hawaii (USA): POB 69, Pago Pago, AS 96799; tel. 633-4226; fax 633-2918; f. 1897; Man. BRENT A. SCHWENKE; 3 brs.

Development Bank

Development Bank of American Samoa: POB 9, Pago Pago, AS 96799; tel. 633-4031; fax 633-1163; f. 1969; govt-owned and non-profit-making; cap. 6.4m. (1989); Chair. EUGENE G. C. H. REID; Pres. MANUTAFEA E. MEREDITH.

INSURANCE

American International Underwriters (South Pacific) Ltd: Pago Pago, AS 96799; tel. 633-4845.

Mark Solofa, Inc.: POB 3149 Pago Pago, AS 96799; tel. 684-5902; fax 684-5904; e-mail marksalofainc@yahoo.com.

National Pacific Insurance Ltd: Centennial Bldg, POB 1386, Pago Pago, AS 96799; tel. 633-4266; fax 633-2964; e-mail npi@samoatelco.com; f. 1977; Man. PETER MILLER.

Oxford Pacific Insurance Management: POB 1420, Pago Pago, AS 96799; tel. 633-4990; fax 633-2721; e-mail progressive_oxford@yahoo.com; f. 1977; represents major international property and life insurance cos; Pres. GREGG F. DUFFY.

South Seas Financial Services Corporation: POB 1448, Pago Pago, AS 96799; tel. 633-7896; fax 633-7895; e-mail ssfs@samoatelco.com.

Trade and Industry

DEVELOPMENT ORGANIZATIONS

American Samoa Development Corporation: Pago Pago, AS 96799; tel. 633-4241; f. 1962; financed by private Samoan interests.

American Samoa Economic Advisory Commission: Pago Pago; Chair. JOHN WAIHEE.

Department of Commerce: see Government Offices; Dir ALIIMAU H. SCANLAN, Jr.

UTILITIES

American Samoa Power Authority: Pago Pago, AS 96799; tel. 644-2772; fax 644-5005; e-mail abe@aspower.com; internet www.aspower.com; supplies water and electricity throughout the islands; also manages sewer and solid waste collection; CEO UTU ABE MALAE.

MAJOR COMPANIES

BP Oil South-West Pacific: POB 5620, Pago Pago, AS 96799; tel. 633-7386; fax 633-7389; internet www.bp.com; petroleum suppliers.

COS Samoa Packing: POB 957, Pago Pago, AS 96799; tel. 644-5272; fax 644-2290; e-mail cos@samoatelco.com; US-owned; tuna-canning; Human Resources Man. ALFONSO PETE GALEA'I.

Origin Energy: POB 159, Pago Pago, AS 96799; tel. 699-2948; fax 699-1852; e-mail origin@samoatelco.com.

StarKist Samoa, Inc.: Pago Pago, AS 96799; tuna-canning; owned by Del Monte Corpn (USA); f. 1963; Gen. Man. PHIL THIRKEL; 2,858 employees.

Transport

ROADS

There are about 150 km (93 miles) of paved and 200 km (124 miles) of secondary roads. Non-scheduled commercial buses operate a service over 350 km (217 miles) of main and secondary roads. There were an estimated 5,900 registered motor vehicles in the islands in 1996.

SHIPPING

There are various passenger and cargo services from the US Pacific coast, Japan, Australia (mainly Sydney) and New Zealand, that call at Pago Pago, which is one of the deepest and most sheltered harbours in the Pacific. Inter-island boats provide frequent services between Samoa and American Samoa.

Hamburg Süd: Samoan Sports Building, No. 1 Main St, Fagatogo, POB 1417, Pago Pago, AS 96799; tel. ; tel. 633-4665; fax 699-8110; e-mail spsi@samoatelco.com; internet www.hamburgsud.com.

PM&O Line: POB 3710, Pago Pago, AS 96799; e-mail paige@blueskynet.as.

Polynesia Shipping: POB 1478, Pago Pago AS 96799; tel. 633-1211; fax 633-1265; e-mail polyship@samoatelco.com.

CIVIL AVIATION

There is an international airport at Tafuna, 11 km (7 miles) from Pago Pago, and smaller airstrips on the islands of Ta'u and Ofu. International services are operated by Hawaiian Airlines and Polynesian Airlines.

Samoa Aviation: POB 280, Pago Pago Int. Airport, Pago Pago, AS 96799; tel. 699-9106; fax 699-9751; e-mail sales@samoair.com; internet www.samoaair.com; f. 1986; operates service between Pago Pago and Samoa, Tonga and Niue; Pres. ANDRE LAVIGNE.

Tourism

The tourist industry is encouraged by the Government, but suffers from the cost and paucity of air services linking American Samoa with its main sources of custom, particularly the USA. Pago Pago is an important mid-Pacific stop-over for large passenger aircraft. A total of 44,158 tourists visited the islands in 2000. In that year 17.0% of tourists came from the USA and 8.9% from New Zealand. In 1992 there were some 542 hotel beds available in the islands. The industry earned an estimated US $10m. in 1998.

Office of Tourism: Convention Center, POB 1147, Pago Pago, AS 96799; tel. 633-1092; fax 633-2092; e-mail amsamoa@amsamoa.com; e-mail amsamoa@amerikasamoa.info; internet www.amsamoa.com/tourism; Deputy Dir VIRGINIA SAMUELU.

Pago Pago Visitors Association (PPVA): f. 2004; Pres. TOM DRABBLE.

Defence

The USA is responsible for the defence of American Samoa. The US Pacific Command is based in Hawaii, but the territory receives regular naval visits and assistance in surveillance of its waters.

Education

Education is compulsory for children between six and 18 years of age. The education system is based on the US pattern of eight years' attendance at an elementary school and four years' enrolment at a high school. In 2001 there were 32 consolidated elementary or primary schools, 10 high schools and a community college. The Government's early childhood education division provides facilities for all children between three and five years of age. In 1996 pre-school facilities were available in 59 village centres. In 2001 total enrolment in primary and secondary schools was 15,560 pupils, and the community college had 1,178 students. Budgetary expenditure on education (and culture) in 1996/97 was US $43.5m., equivalent to 28.5% of total expenditure.

GUAM

Physical and Social Geography

Guam, the largest and southernmost of the Mariana Islands, covers an area of 549 sq km (212 sq miles). The island comprises a northern, coralline, limestone plateau and a mountainous area in the south, of volcanic origin, rising to some 395 m. It is situated about 2,400 km east of the Philippines, and its nearest neighbour is the Northern Mariana Islands, to the north. Guam has a tropical climate, with a mean annual temperature of 30°C (85°F), the hottest months being May and June. Most rain falls from July to October, and the island is sometimes subject to tropical storms and typhoons. The population is multiracial: some 45% are the indigenous Chamorro, 25% Filipino and 15% US immigrants. Most inhabitants profess the Christian religion, the largest denomination being Roman Catholic. Chamorro and, particularly, English are widely spoken and are both official languages. The population in 2000 was recorded by the US Census Bureau as 154,805, excluding members of the US armed forces stationed on the island, and was officially estimated at 166,090 in mid-2004. The capital is Hagåtña (formerly Agaña).

History

The ancestors of the Micronesians first settled in Guam and the other Mariana Islands some 4,000 years ago. A society of matrilineal clans evolved. Members of a Spanish expedition, under the Portuguese navigator Fernão Magalhães (Ferdinand Magellan), were the first Europeans to discover Guam, visiting the island in 1521, during a voyage that accomplished the first circumnavigation of the globe. The island was claimed by Spain in 1565 and the first Jesuit missionaries arrived in 1668. The native Micronesian population is estimated to have fallen from 100,000 in 1521 to fewer than 5,000 in 1741, owing largely to a combination of massacres by the Spaniards and exposure to imported diseases. The present-day Chamorro population are the descendants of the Micronesians, mingled with Spaniards and immigrant Filipinos. Following the Spanish–American War of 1898, Spain ceded Guam to the USA, and then sold the other Mariana Islands to Germany. Japan obtained a League of Nations mandate over the German islands, including the Caroline and Marshall Islands, in 1919. Japanese forces seized Guam in 1941, but the island was recaptured by US troops in 1944.

Guam is an Unincorporated Territory of the USA: the population are US citizens, but do not take part in US presidential elections, and not all the provisions of the US Constitution apply. The territory has been organized, however, under a constitutional or organic act. The island, previously the responsibility of the US Navy, was transferred to the Department of the Interior in 1950, when the territory was granted its Constitution, and formally placed under the Secretary of the Interior's full jurisdiction in 1951. In 1970 the island elected its first Governor, and in 1972 a new law gave Guam one Delegate to the US House of Representatives. The Delegate may vote only in committee. In November 1986 a Democratic majority was elected in the Legislature, but the Republican candidates were elected as Delegate and Governor. The latter, Joseph Ada, was re-elected as Governor in November 1990.

In 1987 the previous Governor of Guam, Ricardo Bordallo, was found guilty on charges of bribery, extortion and conspiracy to obstruct justice. He claimed that the prosecution had been politically motivated, and, in October 1988, Bordallo won an appeal and his sentence of imprisonment was overturned. Bordallo was, however, convicted on two subsidiary charges. He and his family remained popular: in November 1987 his wife, Madeleine, was elected to the Guam Legislature, and was a strong, though unsuccessful, contender as the Democratic candidate at gubernatorial elections in November 1990, following Bordallo's suicide in January of that year.

At elections held in November 1988 and November 1990 the Democrats retained their majority in the unicameral Legislature. Ben Blaz, a Republican, was re-elected as Guam's non-voting Delegate to the US Congress on both occasions.

In September 1976, meanwhile, an island-wide referendum decided that Guam should maintain close ties with the USA, but that negotiations should be held to improve the island's status. In a further referendum in 1982, in which only 38% of eligible voters participated, the status of commonwealth was the most favoured of six options, attracting 48% of the votes cast. In August 1987, in a referendum held on the provisions of a draft law aimed at conferring the status of Commonwealth on the Territory, voters approved the central proposal, while rejecting articles empowering the Guam Government to restrict immigration and granting the indigenous Chamorro people the sole right to determine the island's future political status. In a further referendum, later that year, both outstanding provisions were approved. Negotiations between the Guam Commission for Self Determination and the USA continued throughout the late 1980s and early 1990s.

At legislative elections in November 1992 the Democrats increased their representation to 14 seats, while the Republicans secured only seven. Robert Underwood was elected as the new Democrat Delegate to the US House of Representatives, replacing the Republican, Ben Blaz.

The issue of immigration re-emerged in early 1993, when a pressure group campaigning for the rights of indigenous people, 'Chamoru Nation', appealed for stricter control to be introduced. Members of the group expressed concern that the increased numbers of immigrants entering Guam would threaten the Chamorro culture as well as the political and social stability of the island.

In January 1994 the US Congress approved legislation providing for the transfer of 3,200 acres of land on Guam from federal to local control. This was a significant achievement for the Government of Guam, which had campaigned consistently for the return of 27,000 acres (some 20% of Guam's total area), appropriated by US military forces after the Second World War. Chamorro rights activists, however, opposed the move, claiming that land should not be transferred to the Government of Guam, but rather to the original landowners.

At gubernatorial elections in November 1994 the Democrat Carl Gutierrez defeated his Republican opponent, Tommy Tanaka, winning 54.6% of total votes cast, while Madeleine Bordallo (also a Democrat) was elected to the position of Lieutenant-Governor. Legislative elections held concurrently also resulted in a Democratic majority, with candidates of the party securing 13 seats, while the Republicans won eight. Robert Underwood was re-elected unopposed as Delegate to the US House of Representatives.

Reports that Chamorro rights activists had initiated a campaign for independence from the USA were denied by the Governor in July 1995. In September, however, at the Fourth World Conference on Women in Beijing, People's Republic of China, the Lieutenant-Governor, Madeleine Bordallo, expressed her support for the achievement of full autonomy for Guam and for the Chamorro people's desire for decolonization.

It was reported in 1995 that the US President, Bill Clinton, had appointed a team of commonwealth negotiators to review the draft Guam Commonwealth Act. In late 1999 the proposals were still awaiting a full hearing in the US Congress, despite considerable pressure from the Democrat Governor, Carl Gutierrez, for a serious debate on Guam's political status. However, evidence of the US Government's renewed interest in the Act was reported in early 1997, although officials denied suggestions that this apparent increase in co-operation was as a result of the Territory's substantial financial contribution to Clinton's re-election campaign in 1996. Guam's self-styled Commission on Decolonization was established in 1997, headed by the Governor. A plebiscite on the political status of Guam was originally scheduled for November 2000, but was deferred to coincide with planned legislative and presidential elections of November 2004. However, in May 2004 legislation was being drafted that would postpone the referendum until the number of Chamorro voters registered by the Guam Elections Committee had reached a requisite 50%.

At elections in November 1996 Republican members regained a majority in the Legislature, winning 11 seats, while Democrat candidates secured 10 seats. In a referendum held concurrently voters approved a proposed reduction in the number of Senators from 21 to 15 (effective from November 1998), and plans to impose an upper limit (2.5% of total budgetary expenditure) on legislative expenses. However, a proposal to restrict the number of terms that Senators can seek to serve in the Legislature was rejected by voters. Guam's Delegate to the US House of Representatives, Robert Underwood, was re-elected unopposed.

In March 1998 delegates from several Pacific island nations and territories met in Hawaii to discuss possible methods of controlling the increasing population of brown tree snakes in Guam. The reptile, which was accidentally introduced to the island from New Guinea after the Second World War, has been responsible for frequent power cuts, as a result of climbing electricity lines, as well as major environmental problems (including the decimation of native bird, rodent and reptile populations); it was feared that, if left unchecked, the snakes could spread to other islands in the region. By mid-2004 eight brown tree snakes had been found in Hawaii; all were believed to have been brought over in shipments from Guam. The US Geological Survey has estimated the density of tree snakes on Guam to be roughly 13,000 per sq mile of forested land. Until September 2004 a team of 25 full-time snake trappers and nine dog handlers captured an estimated 6,000 tree snakes annually at Guam's five main ports (air and sea). However, in that month plans were announced to reduce funding to this programme by nearly one-half. In October, however, the US Senate authorized expenditure of US $77m., to be divided among Guam, Hawaii and other islands for the purposes of snake eradication programmes.

In June 1998 US authorities approved changing the name of the capital from Agaña to Hagåtña. The change was effected in order to reflect more accurately the original Chamorro language name for the town.

Robert Underwood was re-elected, with 70.2% of total votes, as Guam's Delegate to the US House of Representatives on 3 November 1998, defeating Manuel Cruz, who secured 20.1% of the votes. Concurrent gubernatorial and legislative elections resulted in the return to office, as Governor, of Carl Gutierrez and, as Lieutenant-Governor, of Madeleine Bordallo. (In December former Governor Joseph Ada alleged that Gutierrez's victory had been achieved by fraudulent means, and later in the month a Guam court invalidated the election result and ordered a new poll to be held. In February 1999, however, the Superior Court of Guam found that no electoral malpractice had taken place and confirmed the appointments of Gutierrez and Bordallo.) In elections to the Legislature, where the number of seats had been reduced from 21 to 15, 12 candidates of the Republican Party and three candidates of the Democratic Party were returned as Senators.

US President Bill Clinton visited Guam in November 1998. During his visit President Clinton approved the transfer of some 1,300 ha of federal land to local administration and undertook to accelerate the transfer of a further 2,020 ha of former US air-force and navy land. In August 2001 the US army announced plans to move combat weaponry and equipment stored in Europe to bases in Guam, as well as in Taiwan and Hawaii. Also, in mid-2004 the US military was considering plans to construct a hangar for B-52 and B-2 bombers, at an estimated cost of US $32m. The increased military presence on Guam has been widely welcomed as a crucial source of new investment and employment.

In mid-1999 it was announced that Guam was no longer able to support the growing numbers of illegal immigrants from the People's Republic of China, some 500 having arrived between April and June alone, seeking refugee status. The immigrants were to be diverted to and detained in the nearby Northern Mariana Islands, the immigration laws of which differ from those of Guam.

At elections held on 7 November 2000, Underwood was re-elected as Guam's Delegate to the US House of Representatives, winning 78.1% of votes cast. At legislative elections, held concurrently, eight candidates of the Republican Party and seven candidates of the Democratic Party were returned as senators.

Robert Underwood secured the Democratic nomination for the gubernatorial elections of 5 November 2002, but was defeated by the Republican candidate, Felix Camacho, who received 55.4% of the votes cast. (Madeleine Bordallo was designated Guam's Delegate to the US Congress.) In the Legislature, nine Democrat candidates were elected, while the Republican Party's representation decreased to six seats. Governor Camacho pledged to halt the widely perceived rise in corruption and misallocation of public funds on Guam, and to improve public finances (see Economy).

Numerous investigations into alleged official corruption were conducted in the early 2000s. In July 2002 Marilyn Manibusan, a former Senator, and another former government official were indicted by a federal grand jury. They were charged with conspiring to extort real estate developers; both were convicted in May. In April 2003 a former Republican Senator and gubernatorial candidate of 1994, Tommy Tanaka, pleaded guilty to charges of misprision of felony, which related to his alleged involvement in a fraudulent government contract. In the same month Joseph Mafnas, a former chief of the Guam police department, was indicted on forgery charges. The case against him collapsed in September 2004, however, because the key witness for the prosecution was absent from the island. In March 2004 former Governor Carl Gutierrez was tried for his involvement in a private property development scheme, which allegedly involved the misappropriation of public funds. All but two minor charges against him were dismissed by a Superior Court judge in late April. However, in late August charges ranging from theft by deception to official misconduct were filed against the former Governor.

A War Claims Review Commission was established by the US Government in September 2003 to examine claims for compensation for the treatment received by residents of Guam under Japanese occupation during the Second World War. While claims for compensation were ongoing as of mid-2004, the Commission had acknowledged the hardship and suffering of the people of Guam, their 'courageous loyalty to the USA' and the inequality in compensation payments as regards similar claims. The Commission recommended that the US Government provide additional compensation to those injured and to the immediate families of those who had died as a direct result of the occupation. In May the US State Department's Radiation Exposure Compensation Program declared Guam eligible for compensation for the effects of nuclear tests carried out in the Pacific region during the long period of mutual hostility between the USA and the Soviet Union known as the 'Cold War'.

The island remained vulnerable to the impact of extreme weather formations during the early 2000s. President George W. Bush declared the island a federal disaster area on two occasions, following Typhoon Chata'an in July 2002 and Supertyphoon Pongsona in December. Although no loss of life was reported as a direct consequence of the severe storms, some 35,000 islanders remained homeless in early 2003. The US Government granted some US $10m. towards the recruitment of 4,000 largely unemployed islanders to assist in the recovery effort, and Governor Camacho retained several members of the former administration in order to ensure the provision of essential services during his transition to office. In July, in response to the high costs incurred by the recovery effort, the island's Delegate to the US Congress, Madeleine Bordallo, introduced a bill to amend the Organic Act of Guam. The legislation attempted to empower the US Secretary of the Interior to waive Guam's outstanding federal debt, in order to offset social costs caused by migration to Guam from Compact of Free Association countries. The measure was defeated, although the amended Compact of Free Association signed in December 2003 approved some $14m. to offset Guam's migration costs and erased $157m. of debt owed to the US federal Government. The financial cost of migration since 1986 from associated countries was estimated to be $269.3m. ($178.3m. for education and $91m. for health, safety, welfare and labour) in a report submitted by Governor Camacho to the US Department of the Interior in April 2004.

Economy

In 2000, according to estimates by the Bank of Hawaii, Guam's gross national income (GNI), at current prices, was US $2,772.8m., equivalent to $16,575 per head. Between 1988 and 1993, it was estimated, GNI increased, in real terms, at an average rate of some 10% per year. Real GNI rose by 3.9% in 1994. In 1990–2002, according to World Bank figures, the population increased at an average annual rate of 1.5%.

Guam's economy is based on tourism and the export of fish and handicrafts. The fishing industry expanded greatly during the late 1980s, and in 1997 there were 1,098 vessels operating from Guam (mainly Japanese, Korean and Taiwanese, fishing in the waters of the Federated States of Micronesia). The fishing catch totalled an estimated 464 metric tons in 2002. Fruit and vegetables, including watermelons, coconuts, cucumbers, bananas, runner beans, aubergines, squash, tomatoes and papaya are grown mainly for local consumption. Pigs, which numbered some 5,000 in 2002, are an important source of meat. Agriculture (including forestry, fishing and mining) engaged only 0.5% of the employed labour force in 2000.

Industrial enterprises, including a petroleum refinery (no longer operational) and textile and garment firms, were established in the early 1970s, in addition to existing smaller-scale manufacturing of soft drinks, confectionery and watches. Boat-building is also a commercial activity on Guam. The industrial sector accounted for some 15% of gross domestic product (GDP) in 1993. Construction engaged 6.1% and manufacturing (including textile and garment production and boat-building) 2.8% of the employed labour force in 2002. The Government has attempted to diversify Guam's economy by attracting increased foreign investment, principally from Asian manufacturers. Guam is a duty-free port and an important distribution point for goods destined for Micronesia. Re-exports constitute a high proportion of Guam's exports, major commodities being petroleum and petroleum products, iron and steel scrap and eggs.

In 2002 some 1,058,700 tourists (of whom 74% were Japanese) visited the island, a decline of almost 9% compared with the previous year. Tourism earned some US $1,908m. in 1999. It accounted for more than 10% of government revenue and 18% of GDP in 1994. The services sector as a whole engaged 90.1% of the employed labour force in 2002. The federal and territorial Governments alone employed 27.9% of workers in that year.

The territory consistently records a visible trade deficit. Total exports were valued at US $75.7m. in 1999, compared with $47.9m. in 1994, in which year imports of petroleum and petroleum products alone cost $163.7m. In 1999 total imports (excluding petroleum and petroleum products) were valued at $38.7m. In 1999 some 53.9% of Guam's exports were purchased by Japan, 24.5% by the Federated States of Micronesia, 6.2% by Palau and 2.0% by Taiwan. In 1991 the majority of imports were provided by the USA (44.6%) and Japan (31.6%).

In 2001 total budgetary operational expenditure stood at US $335.6m. and revenue at $446.8m. In 2002 federal expenditure on the territory totalled $1215.8m., comprising military spending of $561.7m., non-defence expenditure of $552.2m. and other federal assistance of $101.9m. The Government's total debt stood at $440m. in early 2002. Unemployment declined to 1.9% of the labour force in mid-1989, the lowest rate in the USA and its territories, but rose to 11.4% by 2002, although some estimates placed unemployment at around 20%. The average annual rate of inflation was 10.1% in 1990–95, and stood at 5.3% in 1995. Consumer prices increased by 3.2% in 1999, by 0.4% in 2000, but decreased in 2001 by 1.7%.

In the 1990s Guam's economy benefited from increased foreign investment, notably from Japan and the Republic of Korea. Much of this investment resulted in the rapid expansion of the tourist industry from the 1980s. Guam continues to receive considerable financial support from the USA (direct federal grants totalled US $143m. in 1997) in addition to revenue from US military installations on the islands (which accounts for some $400m. annually). In July 1999 it was announced that Guam was to receive a further $5m. in funding from the USA. Considerable interest in establishing an 'offshore' financial centre on Guam was expressed in 1992. The development of such a centre, however, was dependent upon Guam's achievement of commonwealth status, which would allow the introduction of new tax laws. Guam's economy, particularly its important tourist industry, was seriously affected by the Asian financial crisis in the late 1990s. The tourism sector was badly affected by the repercussions of the suicide attacks on the mainland USA in September 2001, compounded by the continued economic difficulties in Japan (the principal source of visitors). The extreme weather formations that struck Guam in July and December 2002 also damaged the island's tourism infrastructure. However, in mid-August 2004 it was reported that more than 1m. people had visited the island since the beginning of the financial year 2003/04. The Guam Visitor Bureau projected an increase of 240,000 visitors compared with the previous financial year.

In early 2003 the consequences of the conflict in Iraq, combined with the outbreak of the illness identified as Severe Acute Respiratory Syndrome (SARS) in East Asia and then elsewhere, led to a further deterioration in economic conditions, with the authorities reporting a significant net outflow of workers from the island. Despite an increase in US military spending on the island in 2002 and early 2003 (when US bomber aircraft were deployed on Guam), falling accommodation prices led to a sharp decline in the Government's revenues from tourism. In mid-2003 the Camacho administration announced several measures to improve its fiscal position; in April Camacho successfully sought the Legislature's authorization to issue some US $246m. in new debt, which was largely intended to refinance the Government's substantial obligations to the public pensions system. In the following months, the administration also announced initiatives to increase taxes, reduce the public-sector payroll and privatize the Guam Telephone Authority. However, by the latter part of 2003 the legality of the Government's measures to refinance its debts remained in doubt, and a budget deficit of some $200m. was forecast for the year. Nevertheless, in mid-2004 it was hoped that measures to rehabilitate and diversify the island's attractions would both stimulate and be stimulated by the recovery in the region's major tourist markets. Also at this time the Government was considering proposals from companies and consortia for the privatization of the Guam Telephone Authority.

Guam is a member of the Pacific Community (formerly the South Pacific Commission) and an associate member of the UN's Economic and Social Commission for Asia and the Pacific (ESCAP). In early 1996 Guam joined representatives of the other countries and territories of Micronesia at a meeting in Hawaii, at which a new regional organization, the Council of Micronesian Government Executives, was established. The new body aimed to facilitate discussion of economic developments in the region and to examine possibilities for reducing the considerable cost of shipping essential goods between the islands.

Statistical Survey

Sources (unless otherwise stated): Department of Commerce, Government of Guam, 102 M St, Tiyan, GU 96913; tel. 475-0321; fax 477-9031; e-mail commerce@ns.gov.gu; internet www.admin.gov.gu/commerce/economy.

AREA AND POPULATION

Area: 549 sq km (212 sq miles).

Population: 133,152 at census of 1 April 1990; 154,805 (males 79,181, females 75,624) at census of 1 April 2000. *Mid-2004* (official estimate): 166,090.

Density (mid-2004): 302.5 per sq km.

Ethnic Groups (2000 census): Chamorro 57,297; Filipino 40,729; White 10,509; Other Asian 9,600; part-Chamorro 7,946; Chuukese 6,229; Total (incl. others) 154,805.

Principal Towns (population at 2000 census): Tamuning 10,833; Mangilao 7,794; Yigo 6,391; Astumbo 5,207; Barrigada 4,417; Hagåtña (capital) 1,122.

Births, Marriages and Deaths (2003, unless otherwise indicated): Registered live births 3,298 (birth rate 20.2 per 1,000); Registered marriages (2000) 1,499; Registered deaths 700 (death rate 4.3 per 1,000). Source: mostly UN, *Population and Vital Statistics Report*.

Expectation of Life (UN estimates, years at birth, 2000): Males 75.5; Females 80.4. Source: UN, *Statistical Yearbook for Asia and the Pacific*.

Economically Active Population (persons aged 16 years and over, excl. armed forces, 2002 estimates): Agriculture, forestry, fishing and mining 290; Manufacturing 1,570; Construction 3,420; Transport, storage and utilities 4,590; Wholesale and retail trade 12,690; Finance, insurance and real estate 2,450; Public administration 16,500; Education, health and social services 14,510; *Total employed* 56,020; Unemployed 7,070; *Total labour force* 63,090. Source: Guam Department of Labor.

AGRICULTURE, ETC

Principal Crops (metric tons, 2002): Coconuts 52,000; Cucumbers and gherkins 260; Roots and tubers 1,500; Watermelons 2,200; Bananas 230. Source: FAO.

Livestock (head, 2002): Chickens 200,000; Pigs 5,000; Cattle 100; Goats 680; Horses 15; Buffaloes 65. Source: FAO.

Livestock Products (metric tons, 2002): Poultry meat 43; Hen eggs 700; Pig meat 140. Source: FAO.

Fishing (metric tons, live weight, 2002): Capture 231 (Common dolphinfish 36; Wahoo 22; Skipjack tuna 47); Aquaculture (estimate) 233 (Mozambique tilapia 100; Philippine catfish 20; Milkfish 80; Banana prawn 25); *Total catch* 464. Source: FAO.

INDUSTRY

Electric Energy (million kWh, estimates): 825 in 1997; 830 in 1998; 830 in 1999. Source: UN, *Industrial Commodity Statistics Yearbook*.

FINANCE

Currency and Exchange Rates: US currency is used. For details, see section on the Northern Mariana Islands.

Budget (estimates, US $ million, 2001): *Revenue*: Taxes 392.2; Federal contributions 49.1; Total (incl. others) 446.8. *Expenditure*: General government 38.3; Public order 59.6; Community services 5.7; Recreation 3.8; Public health 19.0; Public education 157.8; Transportation 2.8; Environmental protection 0.4; Economic development 4.9; Transfers to persons 20.7; Total (incl. others) 335.6.

Cost of Living (Consumer Price Index; base: July–September 1996 = 100): 100.5 in 1999; 104.4 in 2000; 102.6 in 2001. Source: Cost of Living Office, Economic Research Center, Department of Commerce, Government of Guam.

Gross Domestic Product (US $ million at current prices): 2,718.5 in 1999; 2,772.8 in 2000, 2,772.8 in 2001. Source: Office of Insular Affairs, US Department of the Interior.

EXTERNAL TRADE

Principal Commodities (US $ '000, 1999): *Imports*: Travel goods, handbags, etc. 7,607; Meat, prepared or preserved 1,043; Bakery products 804; Watches 1,558. Jewellery articles 1,392; Meat and edible offals (beef, pork and poultry) 2,475; Fruits (incl. nuts) and vegetables 1,335; Passenger motor cars 2,767; Non-alcoholic beverages 1,304; Beer 1,516; Cigars and cigarettes 830; Total (incl. others) 38,709. *Exports*: Fish (chilled, fresh, frozen, dried and salted) 40,720; Tobacco, cigars, etc. 5,123; Petroleum oils and gases 15,183; Perfumes and toilet waters 2,726; Passenger motor cars 2,367; Total (incl. others) 75,748. Note: Figures for imports do not include petroleum and petroleum products (totalling US $163,685,000 in 1994). *2001* (US $ million): Total imports 575.2; Total exports 55.6 (Source: UN, *Statistical Yearbook for Asia and the Pacific*).

Principal Trading Partners (US $ '000): *Imports (1991)*: Hong Kong 7,977.3; Japan 42,357.4; Korea, Republic 5,357.5; Philippines 3,281.3; Singapore 1,622.0; USA 59,837.8; Total (incl. others) 134,231.0. *Exports (2001)*: China, People's Repub. 1.5; Hong Kong 4.5; Japan 30.4; Federated States of Micronesia 5.6; Palau 5.7; Singapore 2.7; Switzerland 2.3; Taiwan 2.9; Thailand 1.5; USA 0.2. Total (incl. others) 60.8. Sources: UN, *Statistical Yearbook for Asia and the Pacific*, and Economic Research Center, Department of Commerce, Government of Guam.

TRANSPORT

Road Traffic (registered motor vehicles, 1999): Private cars 65,887; Taxis 537; Buses 627; Goods vehicles 26,220; Motorcycles 633; Total (incl. others) 99,618. Source: Department of Revenue and Taxation, Government of Guam.

International Sea-borne Shipping (estimated freight traffic, '000 metric tons, 1991): Goods loaded 195.1; Goods unloaded 1,524.1; Goods transhipped 314.7. *Merchant Fleet* (total displacement, '000 grt at 31 December 1992): 1 (Source: Lloyd's Register-Fairplay, *World Fleet Statistics*).

Air Cargo ('000 lb): 36,691 in 1998.

TOURISM

Foreign Tourist Arrivals ('000): 1,286.8 in 2000; 1,159.1 in 2001; 1,058.7 in 2002.

Tourist Arrivals by Country of Residence ('000, 2002): Japan 786.9; Korea, Republic 128.3; Taiwan 19.5; USA 41.5; Total (incl. others) 1,058.7. Source: Guam Visitors Bureau.

Tourism Receipts (US $ million): 2,361 in 1998; 1,908 in 1999. Source: World Tourism Organization.

COMMUNICATIONS MEDIA

Radio Receivers (1997): 221,000 in use*.

Television Receivers (1997): 106,000 in use*.

Telephones (2000): 80,300 main lines in use†.

Mobile Cellular Telephones (2000): 27,200 subscribers†.

Internet Users (estimate, 1999): 5,000†.

Daily Newspapers (1997): 1 (circulation 24,457).

Non-daily Newspapers (1988): 4 (estimated circulation 26,000)*.

* Source: UNESCO, *Statistical Yearbook*.

† Source: International Telecommunication Union.

EDUCATION

Institutions (public schools only, 2003): Elementary 26; Middle school 7; Senior high 4; Business colleges (private) 1; Guam Community College; University of Guam.

Teachers (public schools only, 1998/99): Elementary 1,063; Secondary 1,010.

Enrolment (public schools only, 1998/99): Elementary 16,102; Middle school 7,205; Senior high 8,364; Guam Community College 189; University of Guam 3,748.

Source: Department of Education, Government of Guam, and University of Guam.

Directory

The Constitution

Guam is governed under the Organic Act of Guam of 1950, which gave the island statutory local power of self-government and made its inhabitants citizens of the United States, although they cannot vote in presidential elections. Their Delegate to the US House of Representatives is elected every two years. Executive power is vested in the civilian Governor and the Lieutenant-Governor, first elected, by popular vote, in 1970. Elections for the governorship occur every four years. The Government has 48 executive departments, whose heads are appointed by the Governor with the consent of the Guam Legislature. The Legislature consists of 15 members elected by popular vote every two years (members are known as Senators). It is empowered to pass laws on local matters, including taxation and fiscal appropriations.

The Government

(September 2004)

Governor: Felix Camacho (Republican—took office January 2003).

Lieutenant-Governor: Kaleo Moylan.

GOVERNMENT OFFICES

Government offices are located throughout the island.

Office of the Governor: POB 2950, Hagåtña, GU 96932; tel. 472-8931; fax 477-4826; e-mail governor@mail.gov.gu; internet ns.gov.gu.

Department of the Interior, Office of Insular Affairs (OIA): Hagåtña, GU 96910; tel. 472-7279; fax 472-7309; Field Representative Keith A. Parsky.

Department of Administration: POB 884, Hagåtña, GU 96932; tel. 475-1101; fax 475-6788; e-mail doa@ns.gov.gu.

Department of Agriculture: 192 Dairy Rd, Mangilao, GU 96913; tel. 734-3942; fax 734-6569; e-mail guamagriculture@yahoo.com.

Department of Commerce: 102 M St, Tiyan, GU 96913; tel. 475-0321; fax 477-9031; e-mail commerce@mail.gov.gu; internet www.admin.gov.gu/commerce.

Department of Corrections: POB 3236, Hagåtña, GU 96932; tel. 734-4668; fax 734-4990; e-mail doc@mail.gov.gu.

Customs and Quarantine Agency: 13–16A Mariner Ave, Tiyan, Barrigada, GU 96913; tel. 475-6202; fax 475-6207; internet www.guamjustice.net/custom.

Department of Education: POB DE, Hagåtña, GU 96932; tel. 475-0461; fax 472-5003; e-mail rtainato@doe.edu.gu; internet www.doe.edu.gu.

Department of Labor: POB 9970, Tamuning, GU 96931; tel. 475-0101; fax 477-2988; e-mail labor@ns.gov.gu; internet www.labor.gov.gu.

Department of Land Management: POB 2950, Hagåtña, GU 96932; tel. 475-5252; fax 477-0883; e-mail dlm@mail.gov.gu; internet www.gov.gu/dlm.

Department of Law: Suite 2-200E, Judicial Ctr. building, 120 West O'Brien Drive, Hagåtña, GU 96910; tel. 475-3324; fax 475-2493; e-mail law@ns.gov.gu; internet www.justice.gov.gu/dol.

Department of Parks and Recreation: POB 2950, 13-8 Seagull Ave, Tiyan, Hagåtña, GU 96932; tel. 475-6296; fax 472-9626; e-mail dpr@gov.gu; internet www.gov.gu/dpr.

Department of Public Health and Social Services: POB 2816, Hagåtña, GU 96932; tel. 735-7102; fax 734-5910; e-mail director@dphss.govguam.net; internet www.admin.gov.gu/pubhealth.

Department of Public Works: 542 North Marine Drive, Tamuning, GU 96911; tel. 646-4388; fax 649-6178; e-mail dpwdir@ns.gov.gu.

Department of Revenue and Taxation: Bldg 13-1 Mariner Ave Tiyan Barrigada, GU 96913 ; tel. 475-5000; fax 472-2643; e-mail revtax@mail.gov.gu; internet www.admin.gov.gu/revtax.

Department of the Treasury: PDN Bldg, Suite 404, 238 Archbishop Flores St, Hagåtña, GU 96910.

Department of Youth Affairs: POB 23672, Guam Main Facility, GU 96921; tel. 734-2597; fax 734-7536; e-mail ddell@ns.gov.gu.

Office of Civil Defense (Guam Homeland Security): 221-B Chalan Palasyo, Agana Heights, GU 96910; tel. 475-9600; fax 477-3727; internet www.guamhs.org; f. 1999; Administrator Charles H. Ada, II (acting).

Legislature

GUAM LEGISLATURE

The Guam Legislature has 15 members, directly elected by popular vote for a two-year term. Elections took place in November 2002, when the Democratic Party won nine seats and the Republican Party six.

Speaker: Vicente 'Ben' C. Pangelinan.

CONGRESS

Guam elects a Delegate to the US House of Representatives. An election was held in November 2002, when the Democratic candidate, Madeleine Z. Bordallo, was elected as Delegate.

Delegate of Guam: MADELEINE Z. BORDALLO, Cannon House Office Bldg, 427, Washington, DC 20515–5301, USA; tel. (202) 225-1188; fax (202) 226-0341; e-mail madeleine.bordallo@mail.house.gov; internet www.house.gov/bordallo.

Judicial System

Attorney-General: DOUGLAS B. MOYLAN.

US Attorney: LEONARDO M. RAPADAS.

Supreme Court of Guam: Suite 300, Guam Judicial Center, 120 West O'Brien Drive, Hagåtña, GU 96910; tel. 475-3162; fax 475-3140; e-mail justice@guamsupremecourt.com; internet www.guamsupremecourt.com; Chief Justice F. PHILIP CABULLIDO.

District Court of Guam

4th floor, US Courthouse, 520 West Soledad Ave, Hagåtña, GU 96910; internet www.gud.uscourts.gov.

Judge appointed by the President of the USA. The court has the jurisdiction of a Federal district court and of a bankruptcy court of the United States in all cases arising under the laws of the United States. Appeals may be made to the Court of Appeals for the Ninth Circuit and to the Supreme Court of the United States.

Presiding Judge: (vacant).

Magistrate Judge: JOAQUIN V. E. MANIBUSAN, Jr.

Superior Court of Guam

120 West O'Brien Drive, Hagåtña, GU 96910; tel. 475-3250; internet www.guamjustice.net/superior/superior.htm.

Judges are appointed by the Governor of Guam for an initial eight-year term and are thereafter retained by popular vote. The Superior Court has jurisdiction over cases arising in Guam other than those heard in the District Court.

Presiding Judge: ALBERTO C. LAMORENA, III.

There are also Probate, Traffic, Domestic, Juvenile and Small Claims Courts.

Religion

About 81% of the population are Roman Catholic, but there are also members of the Episcopal (Anglican) Church, the Baptist churches and the Seventh-day Adventist Church. There are small communities of Muslims, Buddhists and Jews.

CHRISTIANITY

The Roman Catholic Church

Guam comprises the single archdiocese of Agaña. The Archbishop participates in the Catholic Bishops' Conference of the Pacific, based in Suva, Fiji, and the Federation of Catholic Bishops' Conferences of Oceania, based in Wellington, New Zealand.

At 31 December 2002 there were 131,430 adherents in Guam.

Archbishop of Agaña: Most Rev. ANTHONY SABLAN APURON, Chancery Office, Cuesta San Ramón 196B, Hagåtña, GU 96910; tel. 472-6116; fax 477-3519; e-mail arch@ite.net; internet www.archdioceseofagana.com.

BAHÁ'Í FAITH

National Spiritual Assembly: POB Box BA, Hagåtña, GU 96931; tel. 828-8639; fax 828-8112; mems resident in 19 localities in Guam and 10 localities in the Northern Mariana Islands.

The Press

NEWSPAPERS AND PERIODICALS

Bonita: POB 11468, Tumon, GU 96931; tel. 632-4543; fax 637-6720; f. 1998; monthly; Publr IMELDA SANTOS; circ. 3,000.

Directions: POB 27290, Barrigada, GU 96921; tel. 635-7501; fax 635-7520; f. 1996; monthly; Publr JERRY ROBERTS; circ. 3,800.

Drive Guam: POB 3191, Hagåtña, GU 96932; tel. 649-0883; fax 649-8883; e-mail glimpses@kuentos.guam.net; internet www.glimpsesofguam.com; f. 1991; quarterly; Publr STEPHEN V. NYGARD; circ. 20,000.

Guam Business News: POB 3191, Hagåtña, GU 96932; tel. 649-0883; fax 649-8883; e-mail glimpses@kuentos.guam.net; internet www.glimpsesofguam.com; f. 1983; quarterly; Publr STEPHEN V. NYGARD; Editor MAUREEN MARATITA; circ. 2,600.

Guam Tribune: POB EG, Hagåtña, GU 96910; tel. 646-5871; fax 646-6702; Tue. and Fri.; Publr MARK PANGILINAN; Man. Editor ROBERT TEODOSIO.

Hospitality Guahan: POB 8565, Tamuning, GU 96931; tel. 649-1447; fax 649-8565; e-mail ghra@ghra.org; internet www.ghra.org; f. 1996; quarterly; circ. 3,000.

Pacific Daily News and Sunday News: POB DN, Hagåtña, GU 96932; tel. 477-9711; fax 472-1512; e-mail news@guampdn.com; internet www.guampdn.com; f. 1944; Publr LEE P. WEBBER; Exec. Editor RINDRATY LIMTIACO; circ. 28,520 (weekdays), 26,237 (Sunday).

The Pacific Voice: POB 2553, Hagåtña, GU 96932; tel. 472-6427; fax 477-5224; f. 1950; Sunday; Roman Catholic; Gen. Man. TEREZO MORTERA; Editor Rev. Fr HERMES LOSBANES; circ. 6,500.

TV Guam Magazine: 237 Mamis St, Tamuning, GU 96911; tel. 646-4030; fax 646-7445; f. 1973; weekly; Publr DINA GRANT; Man. Editor EMILY UNTALAN; circ. 15,000.

NEWS AGENCY

United Press International (UPI) (USA): POB 1617, Hagåtña, GU 96910; tel. 632-1138; Correspondent DICK WILLIAMS.

Broadcasting and Communications

TELECOMMUNICATIONS

Guam Educational Telecommunication Corporation (KGTF): POB 21449, Guam Main Facility, Barrigada, GU 96921; tel. 734-2207; fax 734-5483; e-mail kgtfl2@ite.net.

Guam Telephone Authority: POB 9008, Tamuning, GU 96931; tel. 646-1427; fax 649-4821; e-mail gtagm@ite.net; internet www.privatize.gtaguam.com; proposals for privatization under consideration in mid-2004; Chair. PEDRO R. MARTINEZ; Gen. Man. RALPH TAITANO.

BROADCASTING

Radio

K-Stereo: POB 20249, Guam Main Facility, Barrigada, GU 96921; tel. 477-9448; fax 477-6411; operates on FM 24 hours a day; Pres. EDWARD H. POPPE; Gen. Man. FRANCES W. POPPE.

KGUM/KZGZ: Suite 800, 111 Chalan Santo, Hagåtña, GU 96910; tel. 477-5700; fax 477-3982; e-mail jontalk@k57.com; internet www.radiopacific.com; internet www.k57.com; Chair. and CEO REX SORENSEN; Pres. JON ANDERSON.

KOKU-FM: 508 West O'Brien Drive, Hagåtña, GU 96910; tel. 477-5658; fax 472-7663; e-mail bill@hitradio100.com; operates on FM 24 hours a day; Pres. KURT S. MOYLAN; Gen. Man. ROLAND R. FRANQUEZ.

KPRG FM: KPRG Public Radio for Guam UOG Station Mangilao, GU 96923; tel. 734-8930; fax 734-2958; e-mail kprg@kprg.org; internet www.kprg.org; operated by the University of Guam; news and music.

Radio Guam (KUAM): POB 368, Hagåtña, GU 96910; tel. 637-5826; fax 637-9865; e-mail kellykel@94jam2.com; internet www.kuam.com; f. 1954; operates on AM and FM 24 hours a day; Pres. PAUL M. CALVO; Gen. Man. JOEY CALVO.

Trans World Radio Pacific (TWR): POB CC, Hagåtña, GU 96932; tel. 477-9701; fax 477-2838; e-mail ktwr@twr.org; internet www.guam.net/home/twr; f. 1975; broadcasts Christian programmes on KTWR and one medium-wave station, KTWG, covering Guam and nearby islands, and operates five short-wave transmitters reaching most of Asia, Africa and the Pacific; Pres. THOMAS LOWELL; Station Dir MICHAEL DAVIS.

Television

Guam Cable TV: 530 West O'Brien Drive, Hagåtña, GU 96910; tel. 477-7815; fax 477-7847; f. 1987; Pres. LEE M. HOLMES; Gen. Man. HARRISON O. FLORA.

KGTF—TV: POB 21449 Guam Main Facility, Barrigada, GU 96921; tel. 734-3476; fax 734-5483; e-mail kgtf12@kgtf.org; internet www.kgtf.org; f. 1970; cultural, public service and educational programmes; Gen. Man. GERALDINE 'GINGER' S. UNDERWOOD; Operations Man. BENNY T. FLORES.

KTGM—TV: 692 Marine Dr, Tamuning 96911; tel. 649-8814; fax 649-0371.

KUAM—TV: 600 Harmon Loop, Dededo, Hagåtña, GU 96912; tel. 637-5826; fax 637-9865; e-mail rich@kuam.com; internet www.kuam.com; f. 1956; operates NTSC colour service channel 8; News Dir SABRINA SALAS.

Finance

(cap. = capital; res = reserves; = dep. = deposits; m. = million; brs = branches; amounts in US dollars)

BANKING

Commercial Banks

Allied Banking Corpn (Philippines): Suite 104, Bejess Commercial Bldg, 719 South Marine Drive, Tamuning, GU 96913; tel. 649-5001; fax 649-5002; e-mail abcguam@kuentos.guam.net; Man. MARIO R. PALISOC; 1 br.

Bank of Guam: POB BW, 134 W. Soledad Ave, Hagåtña, GU 96932; tel. 472-5300; fax 477-8687; e-mail customerservice@bankofguam.com; internet www.bankofguam.com; f. 1972; total assets $704.6m. (2003); Chair. ANTHONY A. LEON GUERRERO; Exec. Vice-Pres. WILLIAM D. LEON GUERRERO; 23 brs.

Bank of Hawaii (USA): PO Box BH, Hagåtña, GU 96932; tel. 479-3500; fax 479-3777; Vice-Pres. RODNEY KIMURA; 3 brs.

BankPacific, Ltd: 151 Aspinall Ave, Hagåtña, GU 96910; tel. 472-8160; fax 477-1483; e-mail philipf@bankpacific.com; internet www.bankpacific.com; f. 1954; Pres. and CEO PHILIP J. FLORES; Exec. Vice-Pres. MARK O. FISH; 4 brs in Guam, 1 br. in Palau; 1 br. in Northern Mariana Islands.

Citibank NA (USA): 402 East Marine Drive, POB FF, Hagåtña, GU 96932; tel. 477-2484; fax 477-9441; internet www.citibank.com/guam; Vice-Pres. RASHID HABIB; 2 brs.

Citizens Security Bank (Guam) Inc: POB EQ, Hagåtña, GU 96932; tel. 479-9000; fax 479-9090; Pres. and CEO DANIEL L. WEBB; 4 brs.

First Commercial Bank (Taiwan): POB 2461, Hagåtña, GU 96932; tel. 472-6864; fax 477-8921; Gen. Man. YAO DER CHEN; 1 br.

First Hawaiian Bank (USA): Compadres Mall 562, Harmon Loop Rd, Dededo GU 96912; tel. 475-7900; fax 637-9686; Regional Man. (Guam and Saipan) JOHN K. LEE; 2 brs.

First Savings and Loan Association (FSLA): 140 Aspinal Street, Hagåtña, GU 96910; affiliated to Bank of Hawaii; 6 brs.

HSBC Ltd: POB 27C, Hagåtña, GU 96932; tel. 647-8588; fax 646-3767; CEO GUY N. DE B. PRIESTLEY; 2 brs.

Metropolitan Bank and Trust Co (Philippines): 665 South Marine Drive, Tamuning, Guam 96911; tel. 649-9555; fax 649-9558; e-mail mbguam@metrobank.com.ph; f. 1975; Sen. Man. BENNETH A. REYES.

Union Bank of California (USA): 194 Hernan Cortes Ave, POB 7809, Hagåtña, GU 96910; tel. 477-8811; fax 472-3284; Man. KINJI SUZUKI; 2 brs.

CHAMBER OF COMMERCE

Guam Chamber of Commerce: Ada Plaza Center, Suite 102, 173 Aspinall Ave, POB 283, Hagåtña, GU 96932; tel. 472-6311; fax 472-6202; e-mail gchamber@guamchamber.com.gu; internet www.guamchamber.com.gu; f. 1924; Pres. ELOISE R. BAZA; Chair. JAMES L. ADKINS.

Trade and Industry

DEVELOPMENT ORGANIZATION

Guam Economic Development and Commerce Authority (GEDCA): Guam International Trade Center Bldg, Suite 511, 590 South Marine Drive, Tamuning, GU 96911; tel. 647-4332; fax 649-4146; e-mail ptudela@guameda.net; internet www.investguam.com; f. 1965; Administrator GERALD S. A. PEREZ.

EMPLOYERS' ORGANIZATION

Guam Employers' Council: 718 North Marine Drive, Suite 201, East-West Business Center, Upper Tumon, GU 96913; tel. 649-6616; fax 649-3030; e-mail scruz@ecouncil.org; internet www.ecouncil.org; f. 1966; private, non-profit assen providing management development training and advice on personnel law and labour relations; Exec. Dir BILL BORJA.

UTILITIES

Electricity

Guam Energy Office: 1504 East Sunset Boulevard, Tiyan, GU 96913; tel. 477-0538; fax 477-0589; e-mail guamenergy@kuentos.guam.net.

Guam Power Authority: POB 2977, Hagåtña, GU 96932; tel. 649-6818; fax 649-6942; e-mail gpa@ns.gov.gu; internet www.guampowerauthority.com; f. 1968; autonomous government agency; supplies electricity throughout the island.

Water

Guam Waterworks Authority: POB 3010, Hagåtña, GU 96932; tel. 479-7823; fax 649-0158; Gen. Man. DAVID CRADDICK (acting).

Public Utility Agency of Guam: Hagåtña, GU 96910; supplies the majority of water on the island.

TRADE UNIONS

Many workers belong to trade unions based in the USA such as the American Federation of Government Employees and the American Postal Workers' Union.

Guam Federation of Teachers (GFT): Local 1581, POB 2301, Hagåtña, GU 96932; tel. 735-4390; fax 734-8085; e-mail gft@netpci.com; f. 1965; affiliate of American Federation of Teachers; Pres. JOHN T. BURCH; Vice-Pres. BARBARA M. BLAS; 2,000 mems.

Guam Landowners' Association: Hagåtña; Sec. RONALD TEEHAN.

Transport

ROADS

There are 885 km (550 miles) of public roads, of which some 675 km (420 miles) are paved. A further 685 km (425 miles) of roads are classified as non-public, and include roads located on federal government installations.

SHIPPING

Apra, on the central western side of the island, is one of the largest protected deep-water harbours in the Pacific.

Port Authority of Guam: 1026 Cabras Highway, Suite 201, Piti, GU 96925; tel. 477-5931; fax 477-2689; e-mail pag4@netpci.com; internet www.netpci.com/~pag4; f. 1975; Government-operated port facilities; Gen. Man. JOSEPH F. MESA.

Ambyth, Shipping and Trading Inc: PAG Bldg, Suite 205, 1026 Cabras Highway, Piti, GU 96925; tel. 477-7250; fax 472-1264; agents for all types of vessels and charter brokers; Pres. LAM AKY; Dir ANDREW MILLER.

Atkins, Kroll Inc: 443 South Marine Drive, Tamuning, GU 96911; tel. 646-1866; f. 1914; Pres. ALBERT P. WERNER.

COAM Trading Co Ltd: PAG Bldg, Suite 110, 1026 Cabas Highway, Piti, GU 96925; tel. 477-1737; fax 472-3386.

Guam Shipping Agency: PO Box GD Hagåtña, GU 96932; tel. 477-7381; fax 477-7553; Gen. Man. H. KO.

Interbulk Shipping (Guam) Inc: Bank of Guam Bldg, Suite 502, 111 Chalan Santo Papa, Hagåtña, GU 96910; Man. S. GYSTAD.

Maritime Agencies of the Pacific Ltd: Piti, GU 96925; tel. 477-8500; fax 477-5726; e-mail rehmapship@kuentos.guam.net; f. 1976; agents for fishing vessels, cargo, dry products and construction materials; Pres. ROBERT E. HAHN.

Pacific Navigation System: POB 7, Hagåtña, GU 96910; f. 1946; Pres. KENNETH T. JONES, Jr.

Seabridge Micronesian, Inc: 1026 Cabras Highway, Suite 114, Piti, Guam 96925; tel. 477-7345; fax 477-6206; Gen. Man. J. L. CRUZ.

Sea-Land Service, Inc: 1010 Cabras Highway, Commercial Port, Apra Harbor, Guam; tel. 477-7861; internet www.horizon-lines.com; Gen. Man. RICK AUGUSTIN.

Tucor Services: 180 Guerrero St, Harmon Industrial Park, POB 6128, Tamuning, GU 96931; tel. 646-6947; fax 646-6945; e-mail boll@tucor.com; general agents for numerous dry cargo, passenger and steamship cos; Pres. MICHELLE BOLL.

CIVIL AVIATION

Guam is served by A. B. Won Pat International Airport.

Guam International Airport Authority: POB 8770, Tamuning, GU 96931; tel. 646-0300; fax 646-8823; e-mail rolendal@guamcell.net; internet www.airport.guam.net.

Continental Micronesia Airlines: POB 8778, Tamuning, GU 96931; tel. 649-6594; fax 649-6588; internet www.continental.com; f.

1968, as Air Micronesia, by Continental Airlines (USA); hub operations in Guam and Saipan (Northern Mariana Islands); services throughout the region and to destinations in the Far East and the mainland USA; Pres. and CEO MARK ERWIN.

Freedom Air: POB 1578, Hagåtña, GU 96932; tel. 472-8009; fax 4728080; e-mail freedom@ite.net; f. 1974; Man. Dir JOAQUIN L. FLORES, Jr.

Tourism

Tourism is the most important industry on Guam. In 2002 there were 1,058,700 visitor arrivals, compared with 876,700 in 1992. In 2002 some 74.3% of arrivals by air were visitors from Japan, 12.1% from the Republic of Korea, 3.9% from the USA and 1.8% from Taiwan. Most of Guam's hotels are situated in, or near to, Tumon, where amenities for entertainment are well-developed. Numerous sunken wrecks of aircraft and ships from Second World War battles provide interesting sites for divers. There were 7,879 hotel rooms on Guam at July 2003. A total of US $16.2m. was collected in hotel occupancy taxes in 2002. The industry as a whole earned some $1,908m. in 1999.

Guam Visitors Bureau: 401 Pale San Vitores Rd, Tumon, GU 96913; tel. 646-5278; fax 646-8861; e-mail guaminfo@visitguam.org; internet www.visitguam.org; Chair. DAVID B. TYDINGCO; Gen. Man. ALBERTO 'TONY' A.C. LAMORENA, V.

Defence

Guam is an important strategic military base for the USA, with about 1,580 members of the Air Force, 1,850 naval personnel and 30 army personnel stationed there in August 2002. In April 1992 it had been reported that the USA was to expand its military installations on the island considerably to compensate for the closure of a major naval base in the Philippines, despite continued requests by the Governor that the bases be returned to civilian use. However, the US Naval Air Station, Brewer Field, was returned to the Government of Guam for civilian use in 1995.

Education

School attendance is compulsory from six to 16 years of age. There were 26 public elementary schools, seven junior high and four senior high schools, as well as a number of private schools operating on the island in 2003. Total secondary enrolment in public schools in 1998/99 was 15,758 students. Some 3,748 students were enrolled in tertiary education in that year. In 2000 the rate of adult illiteracy was estimated at 1.0%. Government expenditure on education was US $157.8m. in 2001 (equivalent to 47% of total expenditure).

OTHER TERRITORIES

Baker and Howland Islands

The uninhabited Baker and Howland Islands lie about 60 km apart in the Central Pacific Ocean, just north of the Equator and some 2,575 km south-west of Honolulu, Hawaii (USA). Both are low-lying coral atolls without lagoons, some 2.5 km long, surrounded by narrow, fringing reefs. The islands were sighted by American vessels in the early 19th century, although there is evidence of previous Polynesian settlement on Howland. Baker Island (also then known as New Nantucket and Phoebe Island) was named by an American, Michael Baker, who visited the island in 1832 and 1839. Baker is sparsely vegetated. It was mined for guano in the late 19th century and a settlement, known as Meyerton, was established by the USA in 1935. Howland Island, to the north and a little west of Baker, has an area of some 162 ha and is more thickly vegetated, with several trees; it was reputedly named by the American whaler, Capt. George Netcher, after the look-out who first sighted the atoll. Howland was also mined for guano in the late 19th century and an American settlement, known as Itascatown, was established in 1935. An airfield (which is now unusable) was completed in 1937. In that year the aviatrix Amelia Earhart and her navigator were attempting to reach this airfield when their aircraft mysteriously disappeared. The lighthouse on the island is known as the Amelia Earhart Light. Both islands were evacuated in 1942, owing to Japanese air attacks. Baker and Howland Islands were claimed by the USA in 1857 and were made the responsibility of the Secretary of the Interior in 1936. Actual administrative authority was transferred to the US Fish and Wildlife Service as from 27 June 1974; both islands are national wildlife refuges. In 1990 legislation before Congress proposed that the islands be included within the boundaries of the State of Hawaii. Permission to land is required from the US Department of the Interior, US Fish and Wildlife Service, 1849 C St, NW, Washington, DC 20240; tel. (202) 208-3171; fax (202) 208-6965; internet www.doi.gov.

Jarvis Island

The uninhabited Jarvis Island lies just south of the Equator, about 2,090 km south of Hawaii (USA) and 160 km east of Baker and Howland Islands. The nearest inhabited territory is Christmas Island or Kiritimati, some 320 km to the north-east, in Kiribati. Jarvis is a low, basin-shaped coral island, with a diameter of about 2 km. There is a narrow fringing reef, but sparse vegetation, owing to the limited rainfall. Although known previously, under a variety of names, the island was officially discovered by a British sailor, Capt. Brown. Like Baker and Howland, Jarvis Island was claimed for the USA in 1857 (pursuant upon the Guano Act of 1856), mined for guano and then abandoned in 1879. The United Kingdom annexed the island in 1889 and some further guano mining took place. The island was reclaimed for the USA when a group of US settlers landed in 1935 and founded the village of Millersville and a weather station for the benefit of trans-Pacific aviation. This settlement was evacuated in 1942. Legislation before Congress in 1990 proposed that the island be included within the State of Hawaii. Jarvis Island was placed under the jurisdiction of the US Department of the Interior in 1936, but, as a national wildlife refuge, has since 1974 been administered by the US Fish and Wildlife Service, 1849 C St, NW, Washington, DC 20240; tel. (202) 208-3171; fax (202) 208-6965; internet www.doi.gov.

Johnston Atoll

Johnston Atoll lies in the Pacific Ocean, about 1,319 km west-south-west of Honolulu, Hawaii (USA), and comprises Johnston Island (population 327 in 1980), Sand Island (uninhabited) and two man-made islands, North (Akua) and East (Hikina). It has an area of 2.6 sq km (1.0 sq mile). The group was discovered by a passing British vessel in 1807. In 1858 conflicting claims to sovereignty were lodged by the USA and the Kingdom of Hawaii, and these remained unresolved until Hawaii became a US territory in 1898. Following a period of unsuccessful attempts to establish guano mining operations, Johnston Atoll was placed, in 1926, under the supervision of the US Department of Agriculture as a refuge and breeding ground for native birds. For strategic reasons, however, administrative control was transferred in 1934 to the US Department of the Navy, with the proviso that the islands' use as a bird sanctuary should continue. Operational control was transferred to the Department of the Air Force in 1948, but subsequently, in 1973, the Defense Nuclear Agency (DNA) assumed responsibility. In 1985 construction of a chemical weapons disposal facility began on the atoll and by 1990 it was fully operational. The building of the facility cost US $0.5m. In 1989 the US Government agreed to remove artillery shells containing more than 400 metric tons of nerve gas from the Federal Republic of Germany to the facility on Johnston Atoll. The shells were moved in late 1990, after an accelerated environmental review, and were due for destruction. This plan was a cause for considerable concern in the region, and prompted a variety of unsuccessful legal and diplomatic attempts to prevent the transfer of the weapons from Europe. In January 1995, following a series of fires and accidental chemical releases during the previous year, the US Army announced that it was to seek an extension to the 30 August 1995 deadline, the date by which the 400,000 weapons stored on the island were due to be destroyed. A facility capable of performing atmospheric nuclear-weapons tests remains operational on the atoll. A hurricane that struck Johnston Island in August 1994 forced the evacuation of 1,105 civilian and military personnel and resulted in damage estimated at some $15m. In early 1996 the weapons destruction programme was accelerated, and by May of that year it was reported that all nerve gases stored on the Atoll had been destroyed. However, 1,000 tons of chemical agents remained contained in land-mines, bombs and missiles at the site. In December 2000 it was announced that the destruction of the remaining stock of chemical weapons had been completed. The closure and decontamination of the facility was due for completion in 2004 and, in June of that year, all military personnel left and control of the Atoll was transferred to the US Fish and Wildlife Service. Johnston Atoll has been designated a Naval Defense Sea Area and Airspace Reservation, and is closed to public access. In January 2004 the Atoll had an estimated population of some 200 people, including personnel of the US Air Force, the US Fish and Wildlife Service and civilian contractors. Johnston Atoll falls under the jurisdiction of the Department of the Interior, US Fish and Wildlife Service (details as above, under Baker and Howland Islands). Operational control is the responsibility of the Defense Threat Reduction Agency (DTRA), Office of the General Counsel, 6801 Telegraph Rd, Room 109, Alexandria, VA 22310-3398; tel. (703) 325-7681. Permission to land on Johnston Island must be obtained from the DTRA. The residing military commander of Johnston Island acts as the agent for the DTRA.

Kingman Reef

Kingman Reef, located about 1,500 km south-west of Honolulu, Hawaii (USA), comprises a triangular, atoll-like reef and shoal, measuring about 8 km long and 15 km wide. Although the reef was first discovered in 1798, its precise location remained uncharted until 1853. In 1856 it was claimed for the USA. Kingman Reef posed a considerable hazard to the increasing volume of Pacific shipping during the second half of the 19th century, and was the site of notable shipwrecks in 1874, 1888 and 1893. The US Government formally took possession of the Reef in 1922, and in 1934 it was placed under the jurisdiction of the US Department of the Navy. Declared a National Defense Area in 1941, Kingman Reef now forms a Naval Defense Sea Area and Airspace Reservation, and public access to within a 5-km limit is forbidden. Legislation before the US Congress in 1990 proposed the inclusion of Kingman Reef within the boundaries of the State of Hawaii. The area is administered by the US Department of Defense, Department of the Navy, The Pentagon, Washington, DC 20530; tel. (202) 695-9020; internet www.navy.mil.

Midway Island

Midway Island is a coral atoll formed on a volcanic sea-mount at one end of the Hawaiian chain of islands. Midway was discovered in 1859 and was formally declared a US possession in 1867. Previously known as Brooks Island, the US Navy renamed it in recognition of its geographical location on the route between the USA and Japan. The group consists of Sand Island, Eastern Island and several small islets within the reef, and lies 1,850 km north-west of Hawaii. Midway has an area of about 5 sq km (2 sq miles) and in 1983 had a population of 2,200, although by 1990 the total had declined to 13. Since the transfer of the islands' administration from the US Department of Defense to the Department of the Interior in October 1996, limited tourism is permitted. A national wildlife refuge was established on the atoll of Midway Island, under an agreement with the US Fish and Wildlife Service. The territory is home to many species of birds, notably the frigate or gooney bird. Legislation before Congress in 1990 proposed the inclusion of the territory within the State of Hawaii. The atoll is administered by the US Department of the Interior, US Fish and Wildlife Service, 1849 C St, NW, Washington, DC 20240; tel. (202) 208-3171; fax (202) 208-6965; internet www.doi.gov.

Palmyra

Palmyra is a privately-owned atoll of islands, usually uninhabited, located some 1,600 km south of Honolulu, Hawaii (USA), and about midway between there and American Samoa. The territory consists of about 50 islets and a total area of 100 ha, thick with vegetation, but never more than 2 m above sea level. The Kingdom of Hawaii annexed the atoll in 1862 and it was formally annexed by the USA, with Hawaii, in 1898. The United Kingdom had annexed the atoll in 1889, and renewed British interest resulted in the USA dispatching a naval vessel to claim formal possession in 1912. Palmyra was, however, excluded from the boundaries of the State of Hawaii in 1959, and remains an unorganized and unincorporated territory of the USA, under the civil administration of the Department of the Interior since 1961. A Judge Cooper of Honolulu purchased the atoll in 1911 and 1912, and his heirs remain in possession of two islets, known as the Home Islets, following the sale of the bulk of the atoll to the Fullard-Leo family. During the Second World War the US Navy used and adapted the atoll and, until 1961, the US Air Force maintained the now unserviceable airstrip. The islands remain uninhabited. In 1990 legislation before Congress proposed the inclusion of Palmyra within the boundaries of the State of Hawaii. In mid-1996 it was announced that the owners (the Fullard-Leo family in Hawaii) were to sell the atoll to a US company, which, it was believed, planned to establish a nuclear waste storage facility in the Territory. The Government of Kiribati expressed alarm at the proposal, owing to Palmyra's proximity to the islands, and reiterated its intention to seek the reinclusion of the atoll within its own national boundaries. However, in June one of the Hawaiian Representatives to the US Congress proposed legislation in the US House of Representatives to prevent the establishment of such a facility, and a US government official subsequently announced that the atoll would almost certainly not be used for that purpose. In May 2000 it was announced that the island was to be purchased by The Nature Conservancy (internet www.tnc.org); the sale, for some US $30m., was confirmed in November of that year. The lagoons and surrounding waters within the 12-nautical mile zone of US territorial seas were transferred to the US Department of the Interior, US Fish and Wildlife Service (1849 C St, NW, Washington, DC 20240; tel. (202) 208-3171; fax (202) 208-6965; internet www.doi.gov) in January 2001, and designated a National Wildlife Refuge.

Wake Island

Wake Island is a coral atoll consisting of three islets, Wake, Wilkes and Peale. The atoll, situated on a submerged volcano, lies in the Pacific on the direct route from Hawaii to Hong Kong, about 3,200 km west of Hawaii and 2,060 km east of Guam. The group is 7.2 km long and 2.4 km wide, and covers less than 8 sq km (3 sq miles). Wake Island may have been sighted by a Spanish expedition in 1568, but, following its formal discovery by a British vessel (commanded by Capt. William Wake, for whom the island is named) in 1796, its exact location was lost and not re-established until 1841 by a US naval expedition. The US Government took formal possession in 1899. In 1935 Wake Island was placed under the control of the US Department of the Navy, and in the same year a commercial seaplane base was established to service trans-Pacific passenger flights. The island was occupied by Japanese forces during the Second World War. Administrative responsibility passed to the US Department of the Interior in 1962, but the territory is actually administered by the US Air Force. In 1972 the responsibility for civil administration was delegated to the General Counsel of the Air Force, the agent being the Military Commander of Wake Island. In 1983 the population was 1,600, and in 1988 was estimated to be almost 2,000. These numbers have declined in recent years to an estimated 200 people (all civilian contractors) in October 2001. In 1990 legislation before the US Congress proposed the inclusion of Wake Island within the boundaries of the territory of Guam. However, the Republic of the Marshall Islands, some 500 km to the south of Wake, then decided to exert its claim to the atoll, known as Enenkio to the Micronesians. President Amata Kabua of the Marshall Islands declared that Wake Island was a site of great importance to the traditional chiefly rituals of his islands and that the Marshall Islands, having achieved a measure of independence from the USA, could now claim the territory. Plans by a US company, announced in 1998, to establish a large-scale facility for the storage of nuclear waste on the atoll were condemned by environmentalists and politicians in the region. The group is administered by the US Department of Defense, Department of the Air Force (Pacific/East Asia Division), The Pentagon, Washington, DC 20330; tel. (202) 694-6061; fax (703) 696-7273; internet www.af.mil.

VANUATU

Physical and Social Geography

The Republic of Vanuatu (formerly the New Hebrides) comprises an archipelago of some 80 islands covering a land area of 12,190 sq km (4,707 sq miles), including the Banks and Torres Islands, stretching from south of Solomon Islands to Hunter and Matthew Islands, east of New Caledonia, 900 km in all. The islands range in size from 12 ha to 3,600 sq km. The islands have rugged mountainous interiors, with narrow coastal strips where most of the inhabitants dwell. Three islands have active volcanoes on them. The climate is tropical. Temperatures in Port Vila, the capital, range from 16°C (61°F) to 33°C (92°F). There is a rainy season between November and April, and the islands are vulnerable to cyclones during this period; south-east trade winds blow between May and October. The population was estimated to be 212,000 at mid-2003. Most of the inhabitants (approximately 95%) are Melanesians, and there are small numbers of Europeans, Micronesians and Polynesians. The national language is Bislama (a Pidgin English), and English and French are also official languages. Most of the population profess Christianity, the largest denomination being Presbyterian. The capital is Port Vila, which is located on the island of Efate. The capital's population increased dramatically during the late 1980s and early 1990s, and was estimated to total 33,987 by mid-2003.

History

The New Hebrides were governed until 1980 by an Anglo-French condominium, which was established in 1906. Under this arrangement there were three elements in the structure of administration: the British national service, the French national service and the condominium (joint) departments. Each power was responsible for its own citizens and other non-New Hebrideans who chose to be *ressortissant* of either power. Indigenous New Hebrideans were not permitted to claim either British or French citizenship. The result of this was two official languages, two police forces, three public services, three courts of law, three currencies, three national budgets, two resident commissioners in Port Vila (the capital) and two district commissioners in each of the four districts.

After the Second World War New Hebridean concern regarding the alienation of native land (more than 36% of the New Hebrides was owned by foreigners) prompted local political initiatives. Na-Griamel, one of the first political groups to emerge, had its source in cult-like activities. In 1971 Na-Griamel leaders petitioned the UN to prevent more land sales at a time when territory was being sold to US interests for development as tropical tourist resorts. In 1972 the New Hebrides National Party (NHNP) was formed with support from Protestant missions and covert support from British interests. French interests established the Union des communautés néo-hébridaises in 1974. In the same year discussions in the United Kingdom culminated in the replacement of the Advisory Council, established in 1957, by a Representative Assembly. Of the Assembly's 42 members, 29 were directly elected; this did not, however, fulfil nationalist aspirations.

The Representative Assembly was dissolved in early 1977 following a boycott by the NHNP, which had changed its name to the Vanuaaku Pati (VP) in 1976. However, the VP succeeded in reaching an agreement with the condominium powers for new elections for the Representative Assembly to be held, based on universal suffrage for all seats.

In July 1977 it was announced, at a conference of New Hebridean, British and French delegates, that the New Hebrides would become independent in 1980, following a referendum and elections. The VP, demanding immediate independence, boycotted this conference, refused to participate in the November 1977 elections and declared a 'people's provisional government'. A smaller (39-member) Assembly was, none the less, elected, and a degree of self-government was introduced in early 1978 with the creation of a Council of Ministers and of the office of Chief Minister, together with the inauguration of a single New Hebrides public service to replace the French, British and condominium services. In December 1978 a Government of National Unity was formed, with Fr Gérard Leymang, a Roman Catholic priest, as Chief Minister.

In November 1979 new elections resulted in victory for the VP, which secured 26 of the Assembly's 39 seats. The outcome provoked rioting on Espiritu Santo by Na-Griamel supporters, who threatened non-Santo 'foreigners'. In late November Fr Walter Lini, the President of the VP, was elected Chief Minister.

In June 1980 Jimmy Stevens, the leader of Na-Griamel, declared Espiritu Santo independent of the rest of the New Hebrides, renaming the island the 'Independent State of Vemarana'. Members of his movement, armed with bows and arrows (and allegedly assisted by French *colons* and supported by private US business interests), imprisoned government officers and police, who were later released and allowed to leave the island, together with other European and indigenous public servants. Later in the same month a peace-keeping force comprising about 200 British troops was dispatched to Espiritu Santo; this was strongly criticized by the French, who would not permit Britain's unilateral use of force on Espiritu Santo.

In mid-July 1980, however, agreement was reached between the two condominium powers and Lini, and the New Hebrides became independent within the Commonwealth, under the name of Vanuatu, on 30 July 1980. The former Deputy Chief Minister, George Kalkoa, who adopted the surname Sokomanu ('leader of thousands'), assumed the largely ceremonial post of President. Lini became Prime Minister. Shortly after independence, the Republic of Vanuatu signed a defence pact with Papua New Guinea, and in August units of the Papua New Guinea defence force replaced the British and French troops on Espiritu Santo and arrested the Na-Griamel rebels.

In February 1981 the French ambassador to Vanuatu was expelled, following the deportation from New Caledonia of the VP Secretary-General, who had been due to attend an assembly of the New Caledonian Independence Front. France immediately withdrew aid to Vanuatu but it was subsequently restored, and a new French ambassador was appointed. In June 1982 Vanuatu laid claim to the small, uninhabited islands of Matthew and Hunter, lying about 200 km south-east of Vanuatu's southern island of Aneityum, thus greatly increasing the size of the country's exclusive economic zone. France disputed the claim.

At a general election in November 1983 the VP retained a majority in Parliament. George Sokomanu resigned as President in February 1984, after pleading guilty in court to the late payment of road taxes, but was re-elected in the following month.

In mid-1986 the Government announced that it was to establish diplomatic relations with the USSR and Libya, both of which had hitherto been without diplomatic representation in the South Pacific region. This development, together with the Government's continued support for the Kanak National Liberation Front in its efforts to secure independence for New Caledonia, alarmed many of Vanuatu's Western trading partners (notably Australia, New Zealand and the USA), and attracted criticism of Lini's non-aligned foreign policy.

In October 1986 Vanuatu was one of a group of South Pacific island states to sign a five-year fishing agreement with the USA, whereby the US tuna fleet was granted a licence to operate vessels within Vanuatu's exclusive fishing zone. In January 1987, after protracted negotiations (and despite vigorous opposition from the leaders of other political parties), Lini concluded a one-year fishing agreement with the USSR. This agreement caused considerable disquiet among Australian and US officials, who expressed concern over the growing Soviet presence in the South Pacific. Furthermore, the proposed opening of a Libyan diplomatic mission in Vanuatu provoked censure from the Governments of Australia and New Zealand, and in May of that year it was announced that the establishment of diplomatic relations with Libya was to be postponed indefinitely.

In October 1987 the Government expelled the French ambassador, who, it was alleged, had provided 'substantial financial assistance' to the opposition Union of Moderate Parties (UMP). Elections to an enlarged legislature were held in December, at which the VP secured 26 seats and the UMP took the remaining 20 seats. Following the election, the Secretary-General of the VP, Barak Sope, unsuccessfully challenged Lini for the party presidency, but later accepted a portfolio in the Council of Ministers.

Rioting broke out in Port Vila in May 1988, at the time of a demonstration to protest against a government decision to abolish a local land corporation. Lini accused Sope of provoking the riots, and dismissed his rival from the Council of Ministers. Australia, New Zealand and Papua New Guinea guaranteed military support to Lini's Government in the case of a deterioration in the political situation. In July Sope and four colleagues resigned from the VP and were subsequently dismissed from Parliament at Lini's behest. Accusing Lini and the parliamentary Speaker of acting unconstitutionally, 18 members of the UMP boycotted successive parliamentary sessions and were in turn dismissed. (The Supreme Court subsequently upheld Lini's action.) In September Sope and his colleagues announced the formation of a new political party, the Melanesian Progressive Pati (MPP), and in the following month the VP expelled 128 of its members for allegedly supporting the new organization.

In October 1988 the Court of Appeal ruled as unconstitutional the dismissal from Parliament of Sope and his colleagues but upheld the expulsion of the 18 members of the UMP. In the following month Sope resigned from Parliament. By-elections for the vacated parliamentary seats were held in December, at which the VP increased its majority, but there was a low level of electoral participation; Sokomanu dissolved Parliament and announced that Sope would act as an interim Prime Minister, pending a general election to be held in February 1989. Lini immediately denounced Sokomanu's actions, and the Governments of Australia, New Zealand and Papua New Guinea refused to recognize the interim Government. The islands' police force remained loyal to Lini, and later in December Sokomanu, Sope and other members of the interim Government were arrested and charged with treason. Fred Timakata, the former Minister of Health, assumed the presidency in January 1989. In an attempt to ensure impartiality, a judge from Solomon Islands presided over trials, which were held in March: Sokomanu was sentenced to six years' imprisonment, while Sope and the leader of the parliamentary opposition, Maxime Carlot Korman, received five-year prison sentences, after having been convicted of seditious conspiracy and incitement to mutiny. However, representatives of the International Commission of Jurists, who had been present at the trials, criticized the court's rulings, and in April the Court of Appeal reversed the sentences, citing insufficient evidence for the convictions.

Major changes were effected within the Government in 1990, with Lini assuming the functions of several ministries. There was diminishing support for his premiership within the Council of Ministers, and in August 1991 a motion of 'no confidence' in Lini's leadership was approved at the VP's congress. Donald Kalpokas, the Secretary-General of the VP, was unanimously elected to replace Lini as President of the party. In September a motion of 'no confidence' in Lini was narrowly approved in Parliament, and Kalpokas was elected Prime Minister. Lini stated that he would challenge his opponents at the general election scheduled for December 1991, and formed the National United Party (NUP).

Following the election of Kalpokas as Prime Minister, further defections to Lini's newly formed NUP resulted in a narrowing of the VP's majority to two seats. At the election in December 1991 the UMP obtained 19 seats, while the VP and NUP each won 10 seats, the MPP four and the Tan Union, Fren Melanesia and Na-Griamel one each. The leader of the UMP, Maxime Carlot Korman, was appointed Prime Minister, and, in an unexpected move, a coalition Government was formed between his party and Lini's NUP.

The election of Carlot, Vanuatu's first francophone Prime Minister, helped to improve the country's uneasy relationship with France. However, the new Government reaffirmed its support for the Kanak independence movement in the French territory of New Caledonia, at a meeting with Kanak leaders in July 1992. The statement followed threats by Lini to withdraw from the Government unless Carlot moderated his pro-French policies.

The NUP-UMP coalition was beset by internal problems during 1993. A disagreement over the composition of the Council of Ministers led Lini to declare the coalition invalid, causing several NUP members to resign from government office, while several others expressed support for Carlot. During the following 30 days (the period required formally to dissolve the coalition) and while Carlot was overseas, the UMP President, Serge Vohor, negotiated an agreement with Lini to re-establish the two parties in government. Upon his return, however, Carlot rejected the agreement, deciding instead to form an alliance with the 'breakaway' members of the NUP, who had remained in their ministerial posts.

At a presidential election on 14 February 1994 neither the UMP's candidate, Fr Luc Dini, nor Fr John Bani, who was supported by the opposition, attained the requisite two-thirds of total votes cast. The election was rescheduled for March, and in the intervening period an atmosphere of political uncertainty prevailed, as parties sought to form alliances and several politicians changed party allegiances. The VP subsequently agreed to vote with the ruling UMP, in return for a guaranteed role in a future coalition government. As a result of this agreement, the UMP's candidate, Jean-Marie Leye, was elected to the presidency with 41 votes (Bani won only five.) The VP subsequently withdrew its support for the UMP when the Prime Minister refused to offer the party more than one ministerial post; the VP had requested three.

In May 1994 the 'breakaway' members of the NUP who had remained in their ministerial posts, Sethy Regenvanu, Edward Tabisari and Cecil Sinker, were expelled from the party. They subsequently formed the People's Democratic Party (PDP), and later that month signed an agreement with the UMP to form a new coalition Government, the third since the election of December 1991. The UMP-PDP coalition held a total of 26 legislative seats.

In August 1994 Parliament approved legislation providing for the introduction of a new system of local government, and in mid-September 11 local councils were dissolved and replaced by six provincial governments. Elections to the newly formed provincial authorities took place in the following month, at which the Unity Front (an opposition coalition comprising the VP, the MPP, Tan Union and Na-Griamel), the NUP and the UMP each won control of two councils. In October 1995 the election results in two councils were declared invalid, following evidence of irregularities in the voting, and a third authority was suspended.

Widespread controversy concerning a series of decisions taken by Leye led, in October 1994, to the granting of a restraining order to the Government by the Supreme Court against further actions by the President. Members of the Government and the judiciary had become increasingly alarmed by Leye's exercise of his presidential powers, which had included orders to free 26 criminals (many of whom had been convicted of extremely serious offences) and to appoint a convicted criminal to the position of Police Commissioner (on the recommendation of the Prime Minister). The Government was granted the restraining order pending the hearing of its application to the Supreme Court to overrule several of the President's recent decisions. In May Leye's decision to release the 26 criminals was ruled to have been unconstitutional and therefore invalid.

The issue of press freedom was prominent in late 1994 and early 1995, following allegations concerning the censorship of news reports at the government-controlled Vanuatu Broadcasting and Television Corporation. In April 1995 Carlot attracted severe criticism from the Vanuatu-based regional news agency, Pacnews, when he dismissed two senior government officials for making comments critical of the Government; journalists who reported the comments were threatened with dismissal. The Prime Minister's increasing reputation for intolerance of criticism was compounded by allegations that, as part of his Government's policy of reducing the number of employees in the public service, civil servants believed to be opposition sympathizers were among the first to lose their jobs.

In mid-1995 the Carlot administration was virtually alone in the region in not condemning France's decision to resume nuclear weapons testing in French Polynesia. Within Vanuatu, the Government's stance was criticized by the opposition, who claimed that it did not reflect the views of the vast majority of ni-Vanuatu. Moreover, the country's reputation for media censorship was further damaged by a government ban on all reports of the tests that had not been approved by the Government for broadcast or publication.

A general election took place in November 1995. The Unity Front coalition (now comprising the VP, the MPP and Tan Union) won 20 of the 50 seats in the enlarged Parliament, the UMP secured 17 seats, the NUP won nine, Fren Melanesia and Na-Griamel each obtained one, and independent candidates won two seats. A period of intense political manoeuvring ensued, as the Unity Front and UMP each sought to form coalitions with other parties in an attempt to secure a parliamentary majority. The situation was compounded by the emergence of two factions within the UMP, one comprising the supporters of Carlot and another led by Serge Vohor (the party's President). The Carlot faction of the UMP and the Unity Front both sought the political allegiance of the NUP. The latter's decision to accept the offer of a coalition with the UMP effectively excluded the Unity Front (the grouping with the largest number of seats) from government, and, in protest, its members boycotted the opening of Parliament in December, thus preventing the holding of a vote on the formation of a new government. At a subsequent parliamentary session Vohor was elected as Prime Minister. In January 1996, however, the Unity Front filed petitions in the Supreme Court against 12 members of the Government (among them Vohor and Carlot), alleging irregularities in their election—including the transfer of substantial funds from France to UMP activists for use in local projects aimed at securing support for the party shortly before the election.

In February 1996 seven dissident UMP members of Parliament proposed a motion of 'no confidence' in Vohor, supported by 22 other opposition members. However, Vohor announced his resignation as Prime Minister, thus preventing the vote taking place. A parliamentary session to elect a new premier was abandoned as a result of a boycott by supporters of Vohor, who was reported to have retracted his resignation. However, in a further sitting, Carlot was elected as Prime Minister with 30 parliamentary votes. A climate of instability persisted, and 25 elected members boycotted the opening of Parliament in March, following a declaration by the Chief Justice that rejected Vohor's application for reinstatement as Prime Minister and upheld the election of Carlot to the position. At the opening session President Jean-Marie Leye recommended an urgent review of the islands' Constitution, which he believed to contain inadequate provision for the avoidance of prolonged political turmoil such as that which had followed the inconclusive general election.

In July 1996 a report published by the national ombudsman revealed a serious financial scandal involving the issuing of 10 bank guarantees with a total value of US $100m. The Minister of Finance, Barak Sope, who had issued the guarantees in April (allegedly against the advice of the Attorney-General), had been persuaded by an Australian financial adviser, Peter Swanson, that the scheme could earn the country significant revenue. Swanson, who left Vanuatu after securing the guarantees, was subsequently traced and charged with criminal offences relating to his dealings with

Sope. (In February 1998 the Supreme Court found Swanson guilty on seven charges arising from the scandal, sentencing him to 18 months' imprisonment.) Carlot, meanwhile, rejected demands for his resignation for his compliance with the scheme and resisted considerable pressure to dismiss Sope and the Governor of the Reserve Bank of Vanuatu. In the following month, however, Sope was dismissed, following his defection to the opposition after Carlot had transferred him to the Ministry of Commerce, Trade and Industry.

In September 1996 a motion of 'no confidence' in the Government was approved by 28 of the 50 members of Parliament. Vohor was again elected Prime Minister and appointed Sope as his deputy. His coalition Government comprised the pro-Vohor faction of the UMP, the NUP, the MPP, Tan Union and Fren Melanesia.

A dispute by members of the 300-strong paramilitary Vanuatu Mobile Force (VMF) over unpaid allowances, dating from 1993, led to the cancellation of the annual Constitution Day celebrations in October 1996. When the Government failed to resolve the dispute, members of the VMF briefly abducted the President and Deputy Prime Minister to demand a settlement. Both Leye and Sope expressed sympathy for the VMF members. In the same month the Chief Justice, Charles Vaudin d'Imecourt, who had been the subject of numerous allegations of misconduct and partiality since 1995, was dismissed (and deported as an 'undesirable alien'), following his decision to issue arrest warrants against several members of the VMF. Similarly, Sope was replaced as Deputy Prime Minister by Donald Kalpokas, and Fr Walter Lini was appointed Minister of Justice. Following a further incident in November in connection with their pay dispute, in which an official from the Department of Finance was abducted and allegedly assaulted, Lini ordered the arrest of more than half of the members of the VMF (although most were later released). Vohor claimed that his Government's decision to arrest the VMF personnel had averted a military coup, and revealed that documentary evidence detailing the force's plans to seize power and to install a military administration had been discovered. In June 1999 18 VMF members were charged with the alleged kidnapping of a number of government officers in 1996. One was found guilty and was referred to a court martial for sentencing.

In March 1997 a memorandum of agreement was signed between the VP, the NUP and the UMP, the three parties of the newly formed governing coalition. The defection in May, however, of five NUP members of Parliament, including two government ministers, to the VP (thus increasing the VP's representation to 20 seats) led to the party's expulsion from the Government. As a result, a new coalition, comprising the UMP, the MPP, Tan Union and Fren Melanesia, was formed. The subsequent designation of a new Council of Ministers was controversial for the appointment of Barak Sope to the position of Deputy Prime Minister and Minister of Commerce, Trade and Industry. Sope had been described in January by the national ombudsman, Marie-Noëlle Ferrieux-Patterson, in a further report on the financial scandal of the previous year, as unfit for public office. Ferrieux-Patterson also recommended the resignation or dismissal of the newly appointed Minister of Finance, Willie Jimmy, who, together with ex-Prime Minister Carlot, had in 1993 made illegal 'compensation' payments to the 23 members dismissed from Parliament following their boycott of the legislature in 1988. In July 1997 legal action was initiated to recover the estimated US $300,000 of public funds paid to the members of Parliament. In the same month it was reported that the Government was preparing to amend legislation governing the powers of the ombudsman, in an attempt to contain increasingly vociferous criticism of certain public figures. In August the Tonga-based Pacific Islands News Association awarded its annual Freedom of Information prize to Ferrieux-Patterson. The decision was criticized by the Government as an interference in the country's affairs.

In November 1997 Parliament approved a private member's bill seeking the repeal of the Ombudsman Act. (Although the office of the ombudsman is guaranteed under the Constitution, the repeal of the Act, promulgated in 1995, would effectively curb the influence of Ferrieux-Patterson at a time when she was about to release a number of reports apparently damaging to several prominent public figures.) However, President Leye refused to sign the legislation, referring the bill to the Supreme Court for adjudication on its constitutionality.

The instability that had characterized political life in Vanuatu since the inconclusive general election of 1995 culminated in a constitutional crisis in late November 1997, when Carlot filed a parliamentary motion of 'no confidence' in Vohor's administration. In order to prevent debate of the motion, the Government withdrew all legislation from Parliament, whereupon the Speaker, Edward Natapei, declared the legislative session closed. In response, Leye announced the dissolution of Parliament, citing the need to restore institutional stability, and set new elections for January 1998. The sponsors of the 'no confidence' motion, protesting that their constitutional rights had been infringed, appealed to the Supreme Court, which overturned the President's dissolution order and also countermanded the dismissal, during the crisis, of Barak Sope and the Minister of Lands, Energy and Natural Resources, Sato Kilman. The issue of the dissolution was further referred to the Court of Appeal, which in early January ruled that Leye's order to dissolve Parliament had been within his constitutional competence. Parliament was thus dissolved, and an election set for 6 March.

Despite the vote to repeal the Ombudsman Act, Ferrieux- Patterson continued to publish reports condemning the activities of numerous prominent political figures, and, as the election approached, the Government increasingly criticized the ombudsman for interfering in politics. In addition to a series of reports issued in late 1997 and early 1998 that sought the prosecution of Vohor and several of his ministers for their alleged involvement with a South Korean business executive in a scheme to sell 80,000 passports to Asian nationals, at a cost of 40,000m. vatu, Ferrieux-Patterson accused the Government of appointing foreign nationals as honorary consuls and trade representatives in return for financial contributions. Furthermore, the revelation, in late December 1997, that Jimmy, as Minister of Finance, had appointed a number of his associates to the board of the Vanuatu National Provident Fund (VNPF), to which all workers are obliged to make pension contributions, and that public figures had obtained housing loans at preferential rates from the fund, prompted public outcry as workers demanded the return of their savings. In mid-January 1998 there was rioting and looting in the capital as police used tear gas to disperse protesters who had gathered outside the offices of the VNPF. The violence in Port Vila and also in Luganville prompted the Government to impose a state of emergency (which remained in force in some areas until mid-February). Almost 500 arrests were made during this time: among those detained were reportedly members of Parliament, government officials and civil servants who had joined the rioting. Meanwhile, a new board of directors of the VNPF was appointed, and it was announced that contributors would be permitted to withdraw their savings from the fund. (The subsequent increase in liquidity arising from such withdrawals by some 28,000 members resulted in considerable monetary instability in the weeks preceding the election.)

Following a campaign dominated by allegations of corruption, the general election, on 6 March 1998, resulted in a loss of power for the UMP. According to the official results, the VP won 18 seats, the UMP 12, the NUP 11, the MPP six and the Tanna-based John Frum Movement two. Carlot was the only member of the Vanuatu Republikan Pati, formed by the former Prime Minister in January, to be elected, the other two seats being won by independents. In mid-March the VP and NUP reached agreement on the formation of a coalition, and at the end of the month Parliament formally approved the appointment of Kalpokas as Prime Minister (also Minister of Foreign Affairs, the Comprehensive Reform Programme—CRP—and the Public Service), with Lini as Deputy Prime Minister and Minister of Justice and Internal Affairs.

In June 1998, despite strong opposition notably by the UMP, Parliament approved a new 'leadership code'. Regarded as a key element in ensuring greater accountability and transparency in public life in accordance with the CRP, the code defined clear guidelines for the conduct of state officials (including a requirement that all public figures submit an annual declaration of assets to Parliament), and laid down strict penalties for those convicted of corruption. Shortly beforehand, the Supreme Court upheld the repeal of the Ombudsman Act, as approved by Parliament in November 1997.

In February 1998 the Supreme Court found Peter Swanson guilty on seven charges arising from the bank guarantees scandal revealed in 1996, sentencing him to 18 months' imprisonment. Convictions on five of these charges, together with the custodial sentence, were upheld by the Court of Appeal in July 1998. In late September the Court of Appeal ruled that the dismissal of Charles Vaudin d'Imecourt as Chief Justice in 1996 had been inadmissible; a further ruling on compensation was expected.

In October 1998 Kalpokas expelled the NUP from the governing coalition, following reports that Fr Walter Lini, the party's leader, had organized a series of meetings with prominent members of the opposition, with the aim of forming a new coalition government which would exclude the VP. Fr Walter Lini died in February 1999. At elections in March 1999 Fr John Bani was chosen by the electoral college to succeed Jean-Marie Leye as President of Vanuatu.

The governing coalition was threatened in August 1999 by a decision by the National Council of the UMP to oppose participation in the Government. The Council issued a directive to the 17 members of Jimmy's faction to resign from the Kalpokas administration. Following their refusal to comply with the directive, the National Council acted to suspend the members in October. However, later that month the suspensions were overruled by the Supreme Court. The two factions of the UMP had previously attempted reunification, but negotiations had stalled over the demands of Vohor's faction for two of the four ministerial positions held by Jimmy's faction. The UMP achieved reunification in November 2000.

At the end of August 1999 four by-elections took place, three of which were won by opposition parties, thus eliminating the Government's majority in Parliament. In November the Government staged a boycott of Parliament to avoid a proposed vote of 'no confidence' by opposition parties against the Kalpokas administration (at which time both the Government and the opposition controlled 26 seats in Parliament). However, the subsequent defection of two members to the opposition forced the resignation of Kalpokas prior to a 'no confidence' motion on 25 November. The Speaker, Edward Natapei, announced his resignation shortly afterwards, and Paul Rentari of the NUP was elected as his replacement. The leader of the MPP, Sope, was elected to lead a new Government, defeating Natapei (the newly appointed President of the VP) with 28 votes to 24. Sope formed a five-party coalition Government, comprising the MPP, the NUP, the Vohor faction of the UMP, the VRP and the John Frum Movement. The composition of the new Council of Ministers was swiftly announced; it included Vohor as Minister of Foreign Affairs and Carlot as Minister of Lands and Mineral Resources, both of whom, with Sope, had been the subject to critical reports by the ombudsman. The new Government pledged to reduce the country's dependence on foreign advisers, review the recently introduced value-added tax and ensures that adequate services were delivered to rural communities.

In May 2000 Parliament approved controversial legislation (the Public Services Amendment Bill and the Government Amendment Bill) giving the Government direct power to appoint and dismiss public servants. The opposition criticized the changes, claiming that they contravened the principles of the CRP (a programme of economic measures supported by the Asian Development Bank—ADB). President John Bani subsequently referred both pieces of legislation to the Supreme Court, which, in August, ordered that he approve them. The ADB reacted angrily to the development, arguing that it allowed for political bias in the public sector, and threatened to withhold further funds from Vanuatu. The bank's stance served to perpetuate an ongoing dispute between the organization and the Vanuatu Government, which had often expressed the view that the bank imposed harsh conditions in return for its finance.

An incident in August 2000 in which the Deputy Prime Minister, Reginald Stanley, was allegedly involved in the serious assult of two people and in causing criminal damage to property while drunk in a bar in Port Vila led to Stanley's dismissal from the post. He was replaced by Minister of Trade Development, James Bule. In October, however, the ombudsman recommended his complete dismissal from the Council of Ministers (he had retained the portfolios of infrastructure and public utilities).

In September 2000 opposition leader, Edward Natapei, invited the Vohor faction of the UMP to join the opposition and form a new government. Vohor declined the offer, saying that his priority was the stability of the current Government. The resignation of a VP member in the following month prevented the success of a 'no confidence' motion proposed in the Prime Minister. A further planned motion of 'no confidence' in the Prime Minister was withdrawn in December. In October the Government was forced to defend a controversial plan, which allowed a Thai company, Apex, to pay off a portion of government debt, allegedly in return for tax-haven privileges. The President of Apex, Amarendra Nath Ghosh, was also Vanuatu's recently appointed honorary consul to Thailand. In January 2001 the Government deported Mark Neil-Jones, the publisher of the independent newspaper, *Trading Post*, on the grounds of instigating instability in the country. (The *Trading Post* had recently published several critical articles about the Government, including reports on the Government's financial dealings with Nath Ghosh.) However, the Supreme Court reversed the decision, declaring that the deportation order was illegal, and Neil-Jones was allowed to return to the country. An investigation was subsequently launched into the circumstances of the deportation.

In late January 2001 three members of the UMP resigned from the party, further reducing the Government's majority. The Government's problems intensified in March after the withdrawal of the UMP from the ruling coalition. Opposition attempts to vote on a motion of 'no confidence' were delayed as Sope initiated legal action against the motion, and the Speaker, Paul Ren Tari, refused to allow the vote while legal action was pending. The Chief Justice, however, ordered the vote to proceed, and after further postponements by the Speaker, the Sope Government was voted out of office on 13 April. A new Government, led by Edward Natapei of the VP, was elected. The incoming administration, a coalition of the VP and UMP, was sworn in on 17 April and pledged to continue the reform programme and to restore investor confidence. One of the Government's first acts was to remove Nath Ghosh from his diplomatic position. Vohor was appointed Deputy Prime Minister. In early May Parliament held an extraordinary session to debate a motion to remove Ren Tari as Speaker because of his conduct during the political crisis. Ren Tari responded by suspending Natapei, along with five other members of Parliament who held government posts, for breaching parliamentary procedure. Despite an order by the Chief Justice that they be allowed to return to Parliament to continue the extraordinary session, Ren Tari refused to open the legislature while he appealed to the Supreme Court against the order. In response, Ren Tari and his two deputies were arrested and charged with sedition. In May a new Speaker, Donald Kalpokas, President of the UMP, was elected, thus reducing Natapei's majority in Parliament to one. In September 2001 opposition leader Sope tabled a motion of 'no confidence' against the ruling coalition but was defeated.

In November 2001 Sope was ordered to appear in court to answer charges that he had forged two government-supported Letters of Guarantee, worth US $23m., while he was Prime Minister. It was reported that police were also investigating the activities of the former Minister of Finance, Morkin Steven, whose signature appeared on the documents. Meanwhile, Sope was under investigation for possible involvement in the illegal issuing of diplomatic passports and in connection with allegations of bribery. He was, however, able to contest the general election due to take place later in that year as the preliminary hearings into the forgery charges, originally scheduled to take place in February 2002, were postponed.

In March 2002 Parliament was dissolved after the Supreme Court ruled that its four-year term had expired. Prime Minister Natapei remained in charge of a 'caretaker' Government until the general election, scheduled for May 2002. Also in March it was announced that the newly established People's Progressive Party (PPP) and Fren Melanesia were to form a coalition with the ruling NUP to contest the forthcoming election. The announcement contradicted a statement issued earlier in the same month by the NUP, in which the party had declared that it would stand independently. Meanwhile, the Government dismissed the board of the Vanuatu Broadcasting and Television Corporation (VBTC), claiming that the organization had failed to consult with it over a number of important decisions and that it had incurred unnecessary costs. Later in March a total of 27 VBTC employees returned to work, having been dismissed in October 2001 for participating in a strike. They claimed to have been reinstated by the newly appointed government board—a claim denied by the board—and refused to leave. The individuals alleged to be responsible for the reinstatement subsequently resigned. Negotiations to settle the ongoing disputes began in April, amid allegations of political interference into the affairs of the company by government appointees. It was reported that the general election was to be monitored by an independent group of observers. The future of the Comprehensive Reform Programme was widely perceived to be the main issue at stake in the electoral campaign.

The general election was held on 2 May 2002. A total of 327 candidates contested the 52 seats available in Parliament. A record 136 candidates stood as independents, prompting Natapei to comment prior to the election that if Vanuatu were to attain political stability the electoral constituency should vote only for party candidates. Bad weather and problems with ballot papers delayed voting in some constituencies. After some initial uncertainty, it was announced that the UMP had won 15 seats in the new Parliament and that the VP had secured 14. The NUP won eight seats and the MPP, the Green Party and the VRP secured three apiece. The remaining seats went to independent candidates. In accordance with the terms of the coalition agreement, however, the VP was permitted to nominate the next Prime Minister. The new Government was formed on 3 June 2002, with Natapei duly re-elected Prime Minister and Henry Taga appointed as Speaker of Parliament.

In July 2002 former Prime Minister Barak Sope was convicted of fraud by the Supreme Court and sentenced to three years in prison. It was alleged soon afterwards that New Zealand had interfered in Vanuatu's internal affairs by funding the investigation that had led to Sope's conviction. In early August, following the controversial appointment of Mael Apisai as Vanuatu's new Police Commissioner, disaffected police officers staged a raid during which they arrested Apisai, Attorney-General Hamilton Bulu and 14 other senior civil servants on charges of seditious conspiracy. Following an investigation, the charges were abandoned owing to a lack of evidence. Prime Minister Natapei subsequently assumed the police and VMF portfolios from the Minister of Internal Affairs, Joe Natuman, in what was thought to be an attempt to distance Natuman from some members of the police force, with whom he had reportedly become too closely involved. Later in the month members of the VMF surrounded the police headquarters in Port Vila to serve arrest warrants on 27 of those who had been involved in the raid, including the acting police commissioner, Holis Simon, and the commander of the VMF, Api Jack Marikembo. Shortly afterwards, in an attempt to bring an end to the hostilities, the Government signed an agreement with representatives from the police department and the VMF during a traditional Melanesian reconciliation ceremony. The police officers involved, who had been suspended from their posts, were reinstated, the police and the VMF pledged to make no further arrests and it was agreed that Apisai's appointment would be reviewed by a newly appointed police services commission. At the same time it was decided that the allegations of conspiracy that had been brought against the 15 officials initially arrested would be considered by the judicial authorities. A new acting police commissioner, Lt-Gen. Arthur Coulton, was then appointed.

In early October 2002 it was announced that the charges against 18 of those arrested in connection with the August raid would be abandoned, leaving eight senior officers to face trial on charges of mutiny and incitement to mutiny; the trial was scheduled for early 2003. Meanwhile, the UMP urged the removal of the Australian High Commissioner, Steve Waters, while reportedly accusing Australia of interference into Vanuatu's internal affairs and of destabilizing the coalition Government by communicating solely with the VP at the expense of the UMP. Deputy Prime Minister Serge Vohor alleged that the Australian Federal Police (AFP) had been engaged in the surveillance of government ministers and other officials in Vanuatu. The AFP denied the charges. In early December four of the eight senior police officers tried were found guilty by the Supreme Court on charges of mutiny, incitement to mutiny, kidnapping and false imprisonment and were given suspended two-year prison sentences. In the same month police intervened to prevent former Prime Minister Barak Sope from reclaiming his seat in the National Assembly, asserting that this was nullified by his conviction for fraud in July, despite receiving a pardon in November from President Bani. The pardon, which Bani stated he had made on the grounds of Sope's poor health, provoked widespread public opposition and led the Government to announce the appointment of a commission of inquiry to investigate the President's decision. However, in November 2003, at a by-election for the seat vacated by his conviction for fraud, Barak Sope was re-elected.

In April 2003 the election of Ham Lini (brother of Walter Lini, the late founder of the NUP) as President of the NUP, led to the signing of a memorandum of understanding inviting that party to join the coalition Government. The NUP was expected to assume three ministerial portfolios (including finance) in a development that observers believed could signify the possible reunification of the NUP with the VP (which had been a single political organization until their split in 1991). However, in late April the NUP rejected the VP's offer and announced its intention to remove the Government in a motion of 'no confidence'. A further attempt to propose a motion of 'no confidence' in the Government led Natapei to remove the UMP from the ruling coalition in a reorganization of cabinet portfolios in November. The party was replaced in the coalition by members of the NUP, the PPP, the Green Party and independents. Continued instability within the governing coalition prompted three further cabinet reorganizations during the first three months of 2004.

In April 2004 Alfred Maseng Nalo was sworn in as Vanuatu's new President following a lengthy and closely contested election. Maseng defeated 31 other candidates during a total of four rounds of voting. However, the validity of the election was questioned when it was revealed, shortly after his appointment, that Maseng was serving a suspended prison sentence having been convicted of misappropriation and receiving property dishonestly. In May, only four weeks after his appointment to the presidency, the Supreme Court ruled that Maseng should be removed from office. The likelihood that the Government, which held a minority of seats in Parliament, would be removed in an imminent vote of 'no confidence' led to a decision by the Council of Ministers to dissolve Parliament in June.

A general election was held on 6 July 2004 at which no single party won an overall majority and 25 new members were elected, including many independent candidates. The validity of the election, however, was jeopardized by an incident on Tanna in which ballot boxes *en route* to Port Vila for counting were ambushed and burnt. More than 40 people were arrested in connection with the incident, including the acting Minister of Finance, Jimmy Nickelim of the VP. In late July Serge Vohor was elected Prime Minister defeating Ham Lini by 28 votes to 24. A Council of Ministers composed of five political groups and several independents was appointed shortly afterwards. However, the stability of the new administration was threatened by rumours of shifting allegiances and reports that several members were being persuaded to cross the floor of Parliament. Despite these suggestions, an opposition motion of 'no confidence' in the new Government, proposed in September, was defeated, with Vohor's administration securing the support of 31 of the 52 members. Further doubts over Vohor's ability to continue as Prime Minister were cast when the Police Commissioner, Robert di Niro, attempted to arrest him on charges of contempt, following comments made in Parliament accusing the Chief Justice, Vincent Lunabeck, of being a 'pikinini blong white man' and therefore unduly influenced by a desire to please foreign interests in the country. The Supreme Court, however, dismissed the charges against Vohor during an appeal in late September.

Questions surrounding the freedom of the media in Vanuatu were raised in September 2003 when a court ruling was issued prohibiting the *Vanuatu Daily Post* from publishing reports concerning the chairman of the Vanuatu Maritime Authority. The organization was the subject of an investigation into alleged mismanagement and its chairman, Christophe Emele, claimed that the newspaper was printing defamatory material about himself and his family. The publishers of the newspaper announced their intention to appeal against the injunction.

In March 1988 Vanuatu signed an agreement with Papua New Guinea and Solomon Islands to form the 'Spearhead Group', which aimed to preserve Melanesian cultural traditions and to lobby for independence for New Caledonia. In 1994 the group concluded an agreement regarded as the first step towards the establishment of a free-trade area between the three countries. Fiji joined the group in early 1996.

Economy

In 2002, according to estimates by the World Bank, Vanuatu's gross national income (GNI), measured at average 2000–02 prices, was US $221m., equivalent to $1,080 per head, or $2,770 per head on an international purchasing-power parity basis. During 1990–2002, it was estimated, the population increased at an average annual rate of 2.8%, while gross domestic product (GDP) per head decreased, in real terms, by an average of 0.8% per year. Overall GDP increased, in real terms, at an average annual rate of 3.4% in 1990–2003; GDP contracted by 2.8% in 2002, but increased by 1.6% in 2003.

Agriculture (including forestry and fishing) contributed an estimated 15.6% of GDP in 2002, compared with some 40% in the early 1980s. According to the Asian Development Bank (ADB), the GDP of the agricultural sector was estimated to have increased at an average annual rate of 1.5% in 1990–2002. Compared with the previous year, the sector's GDP was estimated to have increased by 1.7% in 2002 and by 8.7% in 2003. About 35% of the employed labour force were engaged in agricultural activities in 2002. The performance of the agricultural sector in 2001 was severely affected by the devastation caused by two cyclones that struck Vanuatu early in the year. The principal cash crop is coconut. Fluctuations in levels of production of copra (the dried coconut meat that is the source of coconut oil) exemplify the vulnerability of Vanuatu's agricultural economy to adverse climatic conditions. Moreover, international prices for copra have tended to fluctuate widely. Revenue from copra accounted for 8.7% of total domestic export earnings in 2003, compared with 30.3% in 2000. Efforts have been made to diversify the agricultural sector, and the cultivation of cocoa is also significant. Exports of the crop accounted for 9.1% of total domestic earnings in 2003. Coffee is cultivated for export. More than 20,000 households are involved in the cultivation of kava, from which a narcotic drink is produced; the country's first commercial extraction plant was inaugurated in early 1998. However, it was feared that a European ban on kava imports, imposed from 2001 owing to concerns that its consumption might damage health, could affect the country's kava industry. In the early 1990s it was estimated that some 75% of the population were involved in subsistence cultivation, the principal staple crops being yams, taro, manioc, sweet potatoes, breadfruit and coconuts. In 2003 exports of beef contributed 8.8% of domestic exports. A project to breed goats for meat and cheese production was initiated in 2000. The Government derives substantial amounts of revenue from the sale of fishing rights to foreign fleets: sales of licences to Taiwanese and South Korean vessels contributed more than $A136,000 in 1997.

In 1993 it was announced that the Government had granted a Malaysian group of companies a licence to log 70,000 cu m of timber annually on the islands of Erromango, Malekula and Espiritu Santo (compared with previous licences for all operators of 5,000 cu m). The agreement caused extreme controversy, both for its lack of environmental provision and for the apparently unscrupulous terms under which it was negotiated. Following the publication of a report on the industry, financed by Australia, it was announced that from mid-1994 the export of round logs would be banned completely and logged wood would be restricted to an annual total of 25,000 cu m. These regulations were modified in late 1994, although the ban on round log exports was to be maintained. Timber exports contributed 7.7% of total export earnings in 2003.

The industrial sector contributed about 9.0% of GDP in 2002. Only 3.5% of the employed labour force were engaged in industrial activities in 1989. Compared with the previous year, according to the ADB, industrial GDP declined by 5.9% in 2002 and by 1.1% in 2003. In 1990–2002 the GDP of the manufacturing sector decreased at an average annual rate of 2.2%. Manufacturing contributed about 3.5% of GDP in 2002. Most activities are agro-industrial: the processing of coconuts (to produce copra), fish and beef, for example. Governments have sought to promote the development of export-orientated industries, rather than of import-substitution.

Commercial extraction of manganese on Efate ceased in 1976 when the French mine operators withdrew from the island in anticipation of independence. However, following investigations in the early 1990s, it was found that some 2m. metric tons of high-grade ore remained in the area, and in August 2003 the Government announced plans to reopen the mine, predicting substantial potential revenue from manganese exports. In addition, an aerial geophysical survey of the islands, conducted in late 1994 (with Australian aid), identified several possibilities for gold- and copper-mining, as well as large deposits of petroleum around the islands of Malekula and Espiritu Santo. Electricity generation is thermal; in 2003 consumption of electricity totalled 41m. kWh. Imports of mineral fuels accounted for 14.5% of the value of total domestic imports in 2003. Funding to construct hydroelectric power stations on Espiritu Santo and Malekula was secured in the mid-1990s. In August 2000 Shofa province signed an agreement with an Australian company to establish an oil refinery on Efate.

The services sector contributed 75.4% of GDP in 2002. According to the ADB, the GDP of the services sector declined by 2.3% in 2001, by 3.6% in 2002 and increased by 0.1% in 2003. Tourism has become an important source of revenue, and it was estimated that as many as 5,000 ni-Vanuatu were engaged in tourism-related activities (the tourist industry is the largest generator of private-sector employment). Some 50,400 foreign tourists arrived in Vanuatu in 2003, compared with 49,463 in the previous year. Revenue from tourism was estimated at US $58m. in 2000 and at $46m. in 2001. The announcement in 2003 that a new international air terminal was to be built on Espiritu Santo was expected to attract increased numbers of overseas visitors once completed.

Personal income taxes and taxes on company profits are not currently levied by the Government of Vanuatu (domestic tax revenue being derived principally from import duties). The country has thus become attractive as an 'offshore' financial centre and 'tax haven', earnings from which provided about 12% of annual GDP in the late 1980s. However, a restructuring of the tax base (effective from August 1998), as part of the Comprehensive Reform Programme (CRP—see below), included the introduction of a 12.5% value-added tax (VAT), offset by reductions in import duties and the abolition or adjustment of certain other levies. The Government has vehemently denied suggestions that Vanuatu's strict banking secrecy laws might conceal irregular transactions. The country's status as an 'offshore' financial centre has, however, aroused international controversy, and in June 2000 the Paris-based Organisation for Economic Co-operation and Development (OECD) listed Vanuatu as one of a number of countries and territories operating as unfair tax havens. It was claimed that the country was being used to 'launder' the proceeds of illegal activities of the Russian mafia and drug cartels. Sanctions were threatened if Vanuatu failed to take action to prevent both 'money-laundering' and international tax evasion. This contributed to the introduction of anti 'money-laundering' legislation in August 2000. In 2001 the Minister of Finance rejected demands to provide information on the country's revenue from its international tax haven facility, and in 2002 the Government stated that it would not co-operate with an international initiative intended to eliminate tax evasion. However, the country's subsequent commitment to implement transparent tax and regulatory systems by 2005 led to its removal from the OECD's list of unco-operative tax havens in May 2003. Vanuatu also operates an 'offshore' shipping register: 352 ships were registered in December 2003.

Vanuatu consistently records a visible trade deficit. However, such deficits are usually counterbalanced by receipts from tourism and the financial services sector and by official transfers from abroad. In 2003 there was a visible trade deficit of US $64.5m. and a deficit of $31.2m. on the current account of the balance of payments. The principal exports are copra, cocoa, beef and timber, and the most significant markets for Vanuatu's export commodities were India (which took 31.9% of exports in 2003) and Thailand (27.5%). The principal imports in 2003 were machinery and transport equipment (which accounted for 20.6% of the value of total imports), food and live animals (19.6%) and mineral fuels (14.5%). Significant suppliers of goods to Vanuatu are Australia (which supplied 22.9% of Vanuatu's imports in 2003) and Singapore (16.5%).

Budget estimates for 2003 projected an overall surplus of 86m. vatu, compared with deficits of 2,374m. vatu in 2000, 1,254m. vatu in 2001 and 496m. vatu in 2002. Australia, New Zealand, France, the United Kingdom, Canada and Japan are significant suppliers of development assistance. In 2003/04 Australia provided aid of some $A22.7m. and development assistance from New Zealand totalled $NZ5.86m. in the same period. In February 2002 the European Union (EU) allocated Vanuatu 2,000m. vatu of aid to be disbursed over the next five years from the ninth European Development Fund; the funding was to be used principally for the development of education and human resources training. In 2002 Vanuatu's external debt totalled US $83.7m., of which $69.7m. was long-term public debt. In that year the cost of debt-servicing was equivalent to 16.7% of the value of exports of goods and services (compared with 2.1% in 2001). The annual rate of inflation averaged 2.9% in 1991–2001. Consumer prices increased by an average of 2.0% in 2002 and by 3.0% in 2003. Vanuatu's unit of currency, the vatu, replaced the New Hebrides franc in March 1982.

Vanuatu's economic development has been impeded by its dependence on the agricultural sector, particularly the production and export of copra, which is vulnerable to adverse weather conditions and fluctuations in international commodity prices. Successive administrations, therefore, have attempted to encourage the diversification of the country's economy, notably through the development of the tourism sector. It was hoped that a campaign, initiated in 1995, to establish Vanuatu as an important petroleum producer would significantly enhance the country's economic prospects. Economic development, however, remained inhibited by a shortage of skilled indigenous labour, a weak infrastructure and frequent foreign exploitation. In addition, political instability and a series of financial scandals (see History) prompted fears that foreign interests would be deterred from investing in the country.

Implementation of the CRP, sponsored by the ADB and with additional support from the country's bilateral and multilateral creditors, was regarded as crucial to Vanuatu's future economic prosperity. The programme's key aims included what was termed a 'right-sizing' of the public sector, involving a reduction of 10%–15% in civil service personnel, together with measures aimed at ensuring greater accountability and transparency in all areas of public administration. However, the political instability of late 1997, followed by rioting in early 1998 (see History), had a severe negative impact on the economy. In 1999 the ADB reported that despite a number of difficulties (mainly the impact of the Asian economic crisis on certain sectors of the economy and a decline in output from the agricultural sector) the Government had shown a strong level of commitment to the CRP: the number of ministries had been reduced from 34 to nine, and the number of civil service personnel had been decreased by 7%. Other measures included the introduction of VAT and the restructuring of the National Bank and the Development Bank. Development plans announced in 2000 aimed to encourage growth in the rural sector with the establishment of an Agricultural Development Bank and in tourism with the construction of a major new airport. Moreover, legislation, approved in August 2000, facilitated the establishment of internet gambling services—an important additional source of revenue. In November 2001, after five years of negotiations, the World Trade Organization (WTO) offered Vanuatu membership. However, the Government rejected the offer on the grounds that it wished to delay its entry while it negotiated a more favourable tariff agreement. The economy experienced a deceleration in 2001, largely owing to the detrimental effect of adverse weather conditions upon the performance of the agricultural sector (and particularly upon exports of copra, which decreased by nearly 50%) and to the effects of the European ban on kava imports. Moreover, the repercussions of the 11 September 2001 terrorist attacks on the USA also contributed to a decline in tourism in the final months of the year. GDP declined further in 2002 but registered a modest increase in 2003. Further increases of 2.1% and 2.6% were projected for 2004 and 2005 and were expected to be the result of growth in the agricultural sector, particularly in beef and cocoa production. However, as in previous years, these projections were dependent on the absence of adverse weather conditions (such as Cyclone Ivy, which struck the islands in February 2004) and on the stability of world commodity prices.

Vanuatu is a member of numerous regional and international organizations, including the UN's Economic and Social Commission for Asia and the Pacific (ESCAP), the Pacific Community (formerly the South Pacific Commission), the Pacific Islands Forum (formerly the South Pacific Forum), the Asian Development Bank (ADB) and the Asian and Pacific Coconut Community. Vanuatu is a signatory to the South Pacific Regional Trade and Economic Agreement (SPARTECA) and to the Lomé Conventions and successor Cotonou Agreement with the EU.

Statistical Survey

Source (unless otherwise indicated): Vanuatu Statistics Office, PMB 19, Port Vila; tel. 22110; fax 24583; e-mail stats@vanuatu.com.vu; internet www.spc.int/stats/vanuatu.

AREA AND POPULATION

Area: 12,190 sq km (4,707 sq miles); *By Island*: (sq km) Espiritu Santo 4,010; Malekula 2,024; Efate 887; Erromango 887; Ambrym 666; Tanna 561; Pentecost 499; Epi 444; Ambae 399; Vanua Lava 343; Gaua 315; Maewo 300.

Population: 142,419 at census of 16 May 1989; 186,678 (males 95,682, females 90,996) at census of 16–30 November 1999; 212,000 (UN estimate) at mid-2003 (Source: UN, *World Population Prospects: The 2002 Revision*). *By Island* (mid-1999, official estimates): Espiritu Santo 31,811; Malekula 19,766; Efate 43,295; Erromango 1,554; Ambrym 7,613; Tanna 26,306; Pentecost 14,837; Epi 4,706; Ambae 10,692; Vanua Lava 2,074; Gaua 1,924; Maewo 3,385.

Density (mid-2003): 17.4 per sq km.

Principal Town (mid-2003, UN estimate, incl. suburbs): Port Vila (capital) 33,987. Source: UN, *World Urbanization Prospects: The 2003 Revision*.

Births and Deaths (estimates, 1995–2000): Birth rate 33.7 per 1,000; Death rate 6.1 per 1,000. Source: UN, *World Population Prospects: The 2002 Revision*.

Expectation of Life (WHO estimates, years at birth): 67.7 (males 66.4; females 69.1) in 2002. Source: WHO, *World Health Report*.

Economically Active Population (census of May 1989): Agriculture, forestry, hunting and fishing 40,889; Mining and quarrying 1; Manufacturing 891; Electricity, gas and water 109; Construction 1,302; Trade, restaurants and hotels 2,712; Transport, storage and communications 1,030; Financing, insurance, real estate and business services 646; Community, social and personal services 7,891; Activities not adequately defined 11,126; *Total labour force* 66,597 (males 35,692, females 30,905). *Mid-2002* (estimates): Agriculture, etc. 32,000; Total labour force 91,000 (Source: FAO).

HEALTH AND WELFARE

Key Indicators

Total Fertility Rate (children per woman, 2002): 4.2.

Under-5 Mortality Rate (per 1,000 live births, 2002): 42.

Physicians (per 1,000 head, 1997): 0.12.

Health Expenditure (2001): US $ per head (PPP): 107.

Health Expenditure (2001): % of GDP: 3.8.

Health Expenditure (2001): public (% of total): 59.2.

Access to Water (% of persons, 2000): 88.

Access to Sanitation (% of persons, 2000): 100.

Human Development Index (2002): ranking: 129.

Human Development Index (2002): value: 0.570.

For sources and definitions, see explanatory note on p. vi.

AGRICULTURE, ETC.

Principal Crops (FAO estimates, '000 metric tons, 2002): Coconuts 180; Roots and tubers 40; Vegetables and melons 10; Bananas 13; Other fruit 8; Groundnuts (in shell) 3; Maize 1; Cocoa beans 1. Source: FAO.

Livestock (FAO estimates, 2002): Cattle 130,000; Pigs 62,000; Goats 12,000; Horses 3,100; Chickens 340,000. Source: FAO.

Livestock Products (FAO estimates, metric tons, 2002): Beef and veal 3,000; Pig meat 2,805; Cows' milk 2,900; Hen eggs 320; Cattle and buffalo hides 479; Goatskins 6. Source: FAO.

Forestry ('000 cu m, 2002): *Roundwood Removals* (excl. bark): Sawlogs and veneer logs 28; Fuel wood 91; Total 119. *Sawnwood Production* (all broadleaved, incl. railway sleepers): Total 28. Source: FAO.

Fishing (FAO estimates, metric tons, live weight, 2002): Skipjack tuna 6,230; Yellowfin tuna 5,808; Bigeye tuna 2,457; Total catch (incl. others) 17,139. Source: FAO.

FINANCE

Currency and Exchange Rates: Currency is the vatu. *Sterling, Dollar and Euro Equivalents* (31 May 2004): £1 sterling = 209.82 vatu; US $1 = 114.36 vatu; €1 = 140.05 vatu; 1,000 vatu = £4.77 = $8.74 = €7.14. *Average Exchange Rate* (vatu per US $): 145.31 in 2001; 139.20 in 2002; 122.19 in 2003.

Budget (estimates, million vatu, 2003): *Revenue:* Tax revenue 5,749; Non-tax revenue 667; Total 6,416, excluding grants from abroad (381). *Expenditure:* Current expenditure 6,625; Capital expenditure 86; Total 6,711. Source: Asian Development Bank, *Key Indicators of Developing Asian and Pacific Countries*.

International Reserves (US $ million at 31 December 2003): IMF special drawing rights 1.32; Reserve position in IMF 3.71; Foreign exchange 38.79; Total 43.82. Source: IMF, *International Financial Statistics*.

Money Supply (million vatu at 31 December 2003): Currency outside banks 2,108; Demand deposits at banks 10,067; Total money (incl. others) 12,271. Source: IMF, *International Financial Statistics*.

Cost of Living (Consumer Price Index for Port Vila; base: 2000 = 100): 105.6 in 2001; 107.9 in 2002; 111.2 in 2003. Source: Asian Development Bank, *Key Indicators of Developing Asian and Pacific Countries*.

Expenditure on the Gross Domestic Product (million vatu at current prices, 2002): Government final consumption expenditure 7,488; Private final consumption expenditure 19,968; Increase in stocks 10; Gross fixed capital formation 6,890; Statistical discrepancy 1,693; *Total domestic expenditure* 36,049; Exports of goods and services 18,130; *Less* Imports of goods and services 21,222; *GDP in purchasers' values* 32,957. Source: Asian Development Bank, *Key Indicators of Developing Asian and Pacific Countries*.

Gross Domestic Product by Economic Activity (million vatu at current prices, 2002): Agriculture, forestry and fishing 5,128; Manufacturing 1,167; Electricity, gas and water 701; Construction 1,096; Wholesale and retail trade, restaurants and hotels 9,846; Transport, storage and communications 3,962; Finance 2,765; Public administration 4,825; Other services 3,467; *GDP in purchasers' values* 32,957. Source: Asian Development Bank, *Key Indicators of Developing Asian and Pacific Countries*.

Balance of Payments (estimates, US $ million, 2003): Exports of goods f.o.b. 26.85; Imports of goods f.o.b. –91.37; *Trade balance*–64.52; Exports of services and income 113.64; Imports of services and income –84.85; *Balance on goods, services and income* –35.73; Current transfers received 16.71; Current transfers paid –12.20; *Current balance* –31.22; Capital account (net) –3.91; Direct investment from abroad 19.37; Portfolio investment assets 2.06; Other long-term capital 19.01; Net errors and omissions –4.48; *Overall balance* 0.83 Source: Asian Development Bank, *Key Indicators of Developing Asian and Pacific Countries*.

EXTERNAL TRADE*

Principal Commodities (million vatu, 2003): *Imports:* Food and live animals 2,490; Beverages and tobacco 713; Mineral fuels, lubricants, etc. 1,846; Chemicals 1,454; Basic manufactures 1,658; Machinery and transport equipment 2,621; Miscellaneous manufactured articles 1,354; Total (incl. others) 12,703. *Exports:* Cocoa 296; Copra 282; Beef 287; Timber 249; Total (incl. others) 3,248. Source: Asian Development Bank, *Key Indicators of Developing Asian and Pacific Countries*.

Principal Trading Partners (US $ million, 2003): *Imports:* Australia 31.1; Fiji 10.1; France 3.2; India 7.5; Japan 4.2; New Caledonia 4.6; New Zealand 13.4; Singapore 16.5; Total (incl. others) 135.6. *Exports:* Australia 4.3; India 29.5; Indonesia 5.6; Japan 3.8; Korea, Republic 9.4; Thailand 25.4; Total (incl. others) 92.3. Source: Asian Development Bank, *Key Indicators of Developing Asian and Pacific Countries*.

* Figures refer to domestic imports and exports only.

TRANSPORT

Road Traffic (estimates, '000 motor vehicles in use): 9.2 (Passenger cars 7.4) in 1995; 5.9 (Passenger cars 2.7) in 1996; 6.2 (Passenger cars 2.7) in 1997. Source: UN, *Statistical Yearbook*.

Shipping: *Merchant Fleet* (registered at 31 December 2003): Vessels 352; Total displacement ('000 grt) 1,618 (Source: Lloyd's Register-Fairplay, *World Fleet Statistics*). *International Sea-borne Freight Traffic* (estimates, '000 metric tons, 1990): Goods loaded 80; Goods unloaded 55 (Source: UN, *Monthly Bulletin of Statistics*).

Civil Aviation (traffic on scheduled services, 1999): Kilometres flown (million) 3; Passengers carried ('000) 86; Passenger-km (million) 178; Total ton-km (million) 18. Source: UN, *Statistical Yearbook*.

TOURISM

Foreign Tourist Arrivals: 53,203 in 2001; 49,463 in 2002; 50,400 in 2003.

Tourist Arrivals by Country of Residence (2000): Australia 36,751; New Caledonia 4,114; New Zealand 7,985; Total (incl. others) 57,364.

Tourism Receipts (US $ million): 56 in 1999; 58 in 2000; 46 in 2001. Source: World Tourism Organization.

COMMUNICATIONS MEDIA

Radio Receivers (1997): 62,000 in use*.

Television Receivers (1999): 2,000 in use†.

Telephones (2003): 6,500 main lines in use‡.

Facsimile Machines (1996): 600 in use†.

Mobile Cellular Telephones (2003): 7,800 subscribers‡.

Internet Users (2003): 7,500‡.

Personal Computers (2002): 3,000 in use‡.

Non-daily Newspapers (1996): 2 (estimated circulation 4,000)*.

* Source: UNESCO, *Statistical Yearbook*.
† Source: UN, *Statistical Yearbook*.
‡ Source: International Telecommunication Union.

EDUCATION

Pre-primary (1992): 252 schools; 49 teachers (1980); 5,178 pupils.

Primary (1995): 374 schools; 852 teachers (1992); 32,352 pupils.

Secondary (General) (1992): 27 schools (1995); 220 teachers; 4,269 students.

Secondary (Vocational): 50 teachers (1981); 444 students (1992).

Secondary (Teacher Training): 1 college (1989); 13 teachers (1983); 124 students (1991).

Source: mainly UNESCO, *Statistical Yearbook*.

Directory

The Constitution

A new Constitution came into effect at independence on 30 July 1980. The main provisions are as follows:

The Republic of Vanuatu is a sovereign democratic state, of which the Constitution is the supreme law. Bislama is the national language and the official languages are Bislama, English and French. The Constitution guarantees protection of all fundamental rights and freedoms and provides for the determination of citizenship.

The President, as head of the Republic, symbolizes the unity of the Republic and is elected for a five-year term of office by secret ballot by an electoral college consisting of Parliament and the Presidents of the Regional Councils.

Legislative power resides in the single-chamber Parliament, consisting of 39 members (amended to 46 members in 1987, to 50 in 1995 and further to 52 in 1998) elected for four years on the basis of universal franchise through an electoral system that includes an element of proportional representation to ensure fair representation of different political groups and opinions. Parliament is presided over by the Speaker elected by the members. Executive power is vested in the Council of Ministers which consists of the Prime Minister (elected by Parliament from among its members) and other ministers (appointed by the Prime Minister from among the members of Parliament). The number of ministers, including the Prime Minister, may not exceed a quarter of the number of members of Parliament.

Special attention is paid to custom law and to decentralization. The Constitution states that all land in the Republic belongs to the indigenous custom owners and their descendants. There is a National Council of Chiefs, composed of custom chiefs elected by their peers sitting in District Councils of Chiefs. It may discuss all matters relating to custom and tradition and may make recommendations to Parliament for the preservation and promotion of the culture and languages of Vanuatu. The Council may be consulted on any question in connection with any bill before Parliament. Each region may elect a regional council and the Constitution lays particular emphasis on the representation of custom chiefs within each one. (A reorganization of local government was initiated in May 1994, and resulted in September of that year in the replacement of 11 local councils with six provincial governments.)

The Constitution also makes provision for public finance, the Public Service, the Ombudsman, a leadership code and the judiciary (see Judicial System).

The Government

HEAD OF STATE

President: KALKOT MATASKELEKELE (appointed 16 August 2004).

COUNCIL OF MINISTERS

(September 2004)

A coalition of various parties.

Prime Minister: SERGE VOHOR.

Deputy Prime Minister and Minister for Home Affairs: HAM LINI.

Minister for Foreign Affairs: BARAK SOPE MAAUTAMATE.

Minister of Agriculture, Forestry and Fisheries: STEVEN KALSAKAU.

Minister for Health: KEASIPAI SONG.

Minister of Education: JOE NATUMAN.

Minister for Finance and Economic Development: MOANA CARCASSES KALOSIL.

Minister of Public Utilities: WILLIE JIMMY.

Minister of Lands, Geology and Mines: CHARLOT SALWAI.

Minister of Trade and Business Development: JAMES BULE.

Minister of the Comprehensive Reform Programme: PIPITE MARCELLINO.

Minister of Ni-Vanuatu Business Development: MORKIN STEVEN.

Minister of Youth and Sports: PIERRE TORE.

MINISTRIES AND DEPARTMENTS

Prime Minister's Office: PMB 053, Port Vila; tel. 22413; fax 22863; internet www.vanuatu.gov.vu.

Deputy Prime Minister's Office: PMB 057, Port Vila; tel. 22750; fax 27714.

Ministry of Agriculture, Livestock, Forestry and Fisheries: POB 39, Port Vila; tel. 23406; fax 26498.

Ministry of Civil Aviation, Meteorology, Postal Services, Public Works and Transport: PMB 057, Port Vila; tel. 22790; fax 27214.

Ministry of the Comprehensive Reform Programme: POB 110, Port Vila.

Ministry of Culture, Home Affairs and Justice: PMB 036, Port Vila; tel. 22252; fax 27064.

Ministry of Education, Youth and Sports: PMB 028, Port Vila; tel. 22309; fax 24569; e-mail andrews@vanuatu.com.vu.

Ministry of Energy, Lands, Mines and Rural Water Supply: PMB 007, Port Vila; tel. 27833; fax 25165.

Ministry of Finance: PMB 058, Port Vila; tel. 23032; fax 27937.

Ministry of Foreign Affairs, External Trade and Telecommunications: PMB 074, Port Vila; tel. 27045; e-mail depfa@vanuatu.com.vu.

Ministry of Health: PMB 042, Port Vila; tel. 22545; fax 26113.

Ministry of Trade, Industry, Co-operatives and Commerce: PMB 056, Port Vila; tel. 25674; fax 25677.

Ministry of Women's Affairs: PMB 028, Port Vila; tel. 25099; fax 26353.

Legislature

PARLIAMENT

Speaker: JOSIAS MOLI.

General Election, 6 July 2004

	Seats
National United Party	10
Union of Moderate Parties	8
Vanuaaku Pati	8
People's Progressive Party	4
Vanuatu Republikan Pati	4
Green Party	3
Melanesian Progressive Pati	3
National Community Association	2
Namangi Aute	1
People's Action Party	1
Independents	8
Total	52

Political Organizations

Efate Laketu Party: Port Vila; f. 1982; regional party, based on the island of Efate.

Green Party: Port Vila; f. 2001 by breakaway group of the UMP; Leader GERARD LEYMANG.

Independence Front: Port Vila; f. 1995 by breakaway group of the UMP; Chair. PATRICK CROWBY.

Melanesian Progressive Pati (MPP): POB 39, Port Vila; tel. 23485; fax 23315; f. 1988 by breakaway group from the VP; Chair. BARAK SOPE; Sec.-Gen. GEORGES CALO.

National Democratic Party (NDP): Port Vila; f. 1986; advocates strengthening of links with France and the UK; Leader JOHN NAUPA.

National United Party (NUP): Port Vila; f. 1991 by supporters of Walter Lini, following his removal as leader of the VP; Pres. HAM LINI; Sec.-Gen. WILLIE TITONGOA.

New People's Party (NPP): Port Vila; f. 1986; Leader FRASER SINE.

People's Democratic Party (PDP): Port Vila; f. 1994 by breakaway faction of the NUP.

People's Progressive Party (PPP): Port Vila; f. 2001; formed coalition with National United Party (NUP) and Fren Melanesia to contest 2002 elections; Pres. SATO KILMAN.

Tu Vanuatu Kominiti: Port Vila; f. 1996; espouses traditional Melanesian and Christian values; Leader HILDA LINI.

Union of Moderate Parties (UMP): POB 698, Port Vila; f. 1980; Pres. SERGE VOHOR; the UMP is divided into two factions, one led by SERGE VOHOR and the other by WILLIE JIMMY.

Vanuaaku Pati (VP) (Our Land Party): POB 472, Port Vila; tel. 22584; f. 1971 as the New Hebrides National Party; advocates 'Melanesian socialism'; Pres. EDWARD NATAPEI; First Vice-Pres. IOLU ABBIL; Sec.-Gen. SELA MOLISA.

Vanuatu Independent Alliance Party (VIAP): Port Vila; f. 1982; supports free enterprise; Leaders THOMAS SERU, GEORGE WOREK, KALMER VOCOR.

Vanuatu Independent Movement: Port Vila; f. 2002; Pres. WILLIE TASSO.

Vanuatu Labour Party: Port Vila; f. 1986; trade-union based; Leader KENNETH SATUNGIA.

Vanuatu Republikan Pati (VRP): Port Vila; f. 1998 by breakaway faction of the UMP; Leader MAXIME CARLOT KORMAN.

The **Na-Griamel** (Leader FRANKLEY STEVENS), **Namaki Aute, Tan Union** (Leader VINCENT BULEKONE) and **Fren Melanesia** (Leader ALBERT RAVUTIA) represent rural interests on the islands of Espiritu Santo and Malekula. The **John Frum Movement** represents interests on the island of Tanna.

Diplomatic Representation

EMBASSIES AND HIGH COMMISSIONS IN VANUATU

Australia: KPMG House, POB 111, Port Vila; tel. 22777; fax 23948; e-mail australia_vanuatu@dfat.gov.au; internet www.vanuatu.embassy.gov.au; High Commissioner STEPHEN WATERS.

China, People's Republic: PMB 071, Rue d'Auvergne, Nambatu, Port Vila; tel. 23598; fax 24877; e-mail trade@chinese-embassy.com.vu; Ambassador WU ZURONG.

France: Kumul Highway, POB 60, Port Vila; tel. 22353; fax 22695; Ambassador JEAN GARBE.

New Zealand: BDO House, Lini Highway, POB 161, Port Vila; tel. 22933; fax 22518; e-mail kiwi@vanuatu.com.vu; High Commissioner BRIAN SMYTHE.

United Kingdom: POB 567, Port Vila; tel. 23100; fax 23651; e-mail bhcvila@vanuatu.com.vu; internet www.britishhighcommission.gov.uk/vanuatu; High Commissioner MICHAEL HILL.

Judicial System

The Supreme Court has unlimited jurisdiction to hear and determine any civil or criminal proceedings. It consists of the Chief Justice, appointed by the President of the Republic after consultation with the Prime Minister and the leader of the opposition, and three other judges, who are appointed by the President of the Republic on the advice of the Judicial Service Commission.

The Court of Appeal is constituted by two or more judges of the Supreme Court sitting together. The Supreme Court is the court of first instance in constitutional matters and is composed of a single judge.

Magistrates' Courts have limited jurisdiction to hear and determine any civil or criminal proceedings. Island Courts have been established in several Local Government Regions, and are constituted when three justices are sitting together to exercise civil or criminal jurisdiction, as defined in the warrant establishing the court. A magistrate nominated by the Chief Justice acts as Chairman. The Island Courts are competent to rule on land disputes.

In late 2001 legislation was introduced to establish a new Land Tribunal which was intended to expedite the hearing of land disputes. The tribunal was to have three levels, and no cases were to go beyond the tribunal and enter either the Supreme Court or the Island Courts. The tribunal was to be funded by the disputing parties.

In 1986 Papua New Guinea and Vanuatu signed a memorandum of understanding, under which Papua New Guinea Supreme Court judges were to conduct court hearings in Vanuatu, chiefly in the Court of Appeal.

Supreme Court of Vanuatu

PMB 041, rue de Querios, Port Vila; tel. 22420; fax 22692.

Attorney-General: HAMILTON BULU (acting).

Chief Justice: VINCENT LUNABECK.

Chief Prosecutor: HEATHER LINI LEO.

Religion

Most of Vanuatu's inhabitants profess Christianity. Presbyterians form the largest Christian group (with about one-half of the population being adherents), followed by Anglicans and Roman Catholics.

CHRISTIANITY

Vanuatu Christian Council: POB 13, Luganville, Santo; tel. 03232; f. 1967 as New Hebrides Christian Council; five mem. churches, two observers; Chair. Rt Rev. MICHEL VISI; Sec. Rev. JOHN LIU.

The Roman Catholic Church

Vanuatu forms the single diocese of Port Vila, suffragan to the archdiocese of Nouméa (New Caledonia). At 31 December 2002 there were an estimated 29,100 adherents in the country. The Bishop participates in the Catholic Bishops' Conference of the Pacific, based in Fiji.

Bishop of Port Vila: Rt Rev. MICHEL VISI, Evêché, POB 59, Port Vila; tel. 22640; fax 25342; e-mail catholik@vanuatu.com.vu.

The Anglican Communion

Anglicans in Vanuatu are adherents of the Church of the Province of Melanesia, comprising eight dioceses: Vanuatu (which also includes New Caledonia), Banks and Torres and six dioceses in Solomon Islands. The Archbishop of the Province is the Bishop of Central Melanesia, resident in Honiara, Solomon Islands. In 1985 the Church had an estimated 16,000 adherents in Vanuatu.

Bishop of Vanuatu: Rt Rev. HUGH BLESSING BOE, Bishop's House, POB 238, Luganville, Santo; tel. 37065; fax 36026.

Bishop of Banks and Torres: Rt Rev. NATHAN TOME, Bishop's House, POB 19, Toutamwat, Torba Province.

Protestant Churches

Presbyterian Church of Vanuatu (Presbitirin Jyos long Vanuatu): POB 150, Port Vila; tel. 23008; fax 26480; f. 1948; 56,000 mems (1995); Moderator Pastor BANI KALSINGER; Assembly Clerk Pastor FAMA RAKAU.

Other denominations active in the country include the Apostolic Church, the Assemblies of God, the Churches of Christ in Vanuatu and the Seventh-day Adventist Church.

BAHÁ'Í FAITH

National Spiritual Assembly of the Bahá'ís of Vanuatu: POB 1017, Port Vila; tel. 22419; e-mail nsavanuatu@vanuatu.com.vu; f. 1953; Sec. CHARLES PIERCE; mems resident in 199 localities.

The Press

Hapi Tumas Long Vanuatu: POB 1292, Port Vila; tel. 23642; fax 23343; quarterly tourist information; in English; Publr MARC NEIL-JONES; circ. 12,000.

Logging News: Port Vila; environment and logging industry.

Pacific Island Profile: Port Vila; f. 1990; monthly; general interest; English and French; Editor HILDA LINI.

Port Vila Presse: 1st Floor, Raffea House, POB 637, Port Vila; tel. 22200; fax 27999; e-mail marke@presse.com.vu; internet www.presse.com.vu; f. 2000; daily; English and French; Publr MARKE LOWEN; Editor RICKY BINIHI.

Vanuatu Daily Post: POB 1292, Port Vila; tel. 23111; fax 24111; e-mail tpost@vanuatu.com.vu; internet www.vanuatudaily.com.vu; daily; English; Publr MARC NEIL-JONES; Editor LEN GARAE; circ. 2,000.

Vanuatu Weekly: PMB 049, Port Vila; tel. 22999; fax 22026; e-mail vbtcnews@vanuatu.com.vu; f. 1980; weekly; govt-owned; Bislama, English and French; circ. 1,700.

Viewpoints: Port Vila; weekly; newsletter of Vanuaaku Pati; Editor PETER TAURAKOTO.

Wantok Niuspepa: POB 1292, Port Vila; tel. 23642; fax 23343.

Broadcasting and Communications

TELECOMMUNICATIONS

Freedom Telecommunications Company (USA): Santo; f. 2000; provides services on Santo and islands in the north of Vanuatu.

Telecom Vanuatu Ltd (TVL): POB 146, Port Vila; tel. 22185; fax 22628; e-mail telecom@tvl.net.vu; internet www.vanuatu.com.vu; f. 1989 as a joint venture between the Government of Vanuatu, Cable & Wireless Ltd and France Câbles et Radio; operates all national and international telecommunications services in Vanuatu; Man. Dir RICHARD HALL.

BROADCASTING

Radio

Vanuatu Broadcasting and Television Corporation (VBTC): PMB 049, Port Vila; tel. 22999; fax 22026; e-mail vbtcnews@vanuatu.com.vu; internet www.vbtc.com.vu; fmrly Government Media Services, name changed in 1992; Gen. Man. JOE BOMAL CARLO; Chair. GODWIN LIGO; Dir of Programmes A. THOMPSON.

Radio Vanuatu: PMB 049, Port Vila; tel. 22999; fax 22026; f. 1966; govt-owned; broadcasts in English, French and Bislama; Dir JOE BOMAL CARLO.

Television

Vanuatu Broadcasting and Television Corporation (VBTC): see Radio.

Television Blong Vanuatu: PMB 049, Port Vila; f. 1993; govt-owned; French-funded; broadcasts for four hours daily in French and English; Gen. Man. CLAUDE CASTELLY; Programme Man. GAEL LE DANTEC.

Finance

(cap. = capital; res = reserves; dep. = deposits; amounts in vatu unless otherwise indicated)

Vanuatu has no personal income tax nor tax on company profits and is therefore attractive as a financial centre and 'tax haven'.

BANKING

Central Bank

Reserve Bank of Vanuatu: POB 62, Port Vila; tel. 23333; fax 24231; e-mail resrvbnk@vanuatu.com.vu; internet www.rbv.gov.vu; f. 1981 as Central Bank of Vanuatu; name changed as above in 1989; govt-owned; cap. 100.0m., res 668.2m., dep. 2,880.8m. (Dec. 2002); Gov. SERGE VOHOR.

Development Banks

Development Bank of Vanuatu: rue de Paris, POB 241, Port Vila; tel. 22181; fax 24591; f. 1979; govt-owned; cap. 315m. (Nov. 1988); Man. Dir AUGUSTINE GARAE.

Agence Française de Développement: Kumul Highway, La Casa d'Andrea Bldg, BP 296, Port Vila; tel. 22171; fax 24021; e-mail afd.arep@vanuatu.com.vu; fmrly Caisse Française de Développement; provides finance for various development projects; Man. BERNARD SIRVAIN.

National Bank

National Bank of Vanuatu: POB 249, Air Vanuatu House, rue de Paris, Port Vila; tel. 22201; fax 27227; e-mail nationalbank@vanuatu.com.vu; f. 1991; when it assumed control of Vanuatu Co-operative Savings Bank; govt-owned; cap. 600m., dep. 2,826m. (Dec. 2002); Chair. JOHN AHURUHI; Man. Dir BOB HUGHES; 19 brs.

Foreign Banks

ANZ Bank (Vanuatu) Ltd: Lini Highway, POB 123, Port Vila; tel. 22536; fax 23950; e-mail anzvanuatu@anz.com; internet www.anz.com/vanuatu; f. 1971; cap. 3.7m., res 317.7m., dep. 24,734.2m. (Sept. 2003); Man. Dir MICHAEL FLOWER; brs in Port Vila and Santo.

European Bank Ltd (USA): International Bldg, Lini Highway, POB 65, Port Vila; tel. 27700; fax 22884; e-mail info@europeanbank.net; internet www.europeanbank.net; f. 1972, obtained a full banking licence in 1995; cap. US $0.8m., res US $1.3m., dep. US $38.4m. (Dec. 2003); 'offshore' and private banking; Chair. THOMAS MONTGOMERY BAYER; Pres. ROBERT MURRAY BOHN.

Westpac Banking Corporation (Australia): Kumul Highway, POB 32, Port Vila; tel. 22084; fax 24773; e-mail westpacv@vanuatu.com.vu; Man. R. B. WRIGHT; 2 brs.

Financial Institution

The Financial Centre Association: POB 1128, Port Vila; tel. 27272; fax 27272; e-mail fincen@vanuatu.com.vu; f. 1980; group of banking, legal, accounting and trust companies administering 'offshore' banking and investment; Chair. MARK STAFFORD; Sec. ALISTAIR RODGERS.

INSURANCE

Pacific Insurance Brokers: POB 229, Port Vila; tel. 23863; fax 23089.

QBE Insurance (Vanuatu) Ltd: La Casa D'Andrea Bldg, POB 186, Port Vila; tel. 22299; fax 23298; e-mail info.van@qbe.com; Gen. Man. GEOFFREY R. CUTTING.

Trade and Industry

GOVERNMENT AGENCY

Vanuatu Investment Promotion Authority: PMB 9011, Port Vila; tel. 24096; fax 25216; internet www.investinvanuatu.com; fmrly the Vanuatu Investment Board, name changed as above in 2000; CEO JOE LIGO.

CHAMBER OF COMMERCE

Vanuatu Chamber of Commerce and Industry: POB 189, Port Vila; tel. 27543; fax 27542; e-mail vancci@vanuatu.com.vu; Pres. JOSEPH JACOBE.

MARKETING BOARD

Vanuatu Commodities Marketing Board: POB 268, Luganville, Santo; e-mail vcmb@vanuatu.com.vu; f. 1982; sole exporter of major commodities, including copra, kava and cocoa; Gen. Man. GEORGE CALO.

UTILITIES

Electricity

Union Electrique du Vanuatu (Unelco Vanuatu Ltd): POB 26, rue Winston Churchill, Port Vila; tel. 22211; fax 25011; e-mail unelco@unelco.com.vu; private organization contracted for the generation and supply of electricity in Port Vila, Luganville, Tanna and Malekula, and for the supply of water in Port Vila; Dir-Gen. JEAN FRANÇOIS BARBEAU.

Village generators provide electricity in rural areas. The Department of Infrastructure, Utilities and Public Works provides recirculated water supplies to about 85% of urban and 30% of rural households.

CO-OPERATIVES

During the early 1980s there were some 180 co-operative primary societies in Vanuatu and at least 85% of goods in the islands were distributed by co-operative organizations. Almost all rural ni-Vanuatu were members of a co-operative society, as were many urban dwellers. By the end of that decade, however, membership of co-operatives had declined, and the organizations' supervisory body, the Vanuatu Co-operative Federation, had been dissolved, after having accumulated debts totalling some $A1m.

MAJOR COMPANIES

South Pacific Fishing Co: POB 237, Santo; tel. 36319; state-controlled exporters of fish.

Vanuatu Brewing Ltd: POB 169, Port Vila; tel. 22435; fax 22152; e-mail tusker@vanuatu.com.vu; f. 1990; 50% owned by Pripps Bryggerier (Sweden), 25% state-owned, 25% owned by Provident Fund of Vanuatu; production of Pripps Lager and Vanuatu Tusker Beer; Gen. Man. MURRAY PARSONS.

Vanuatu Maritime Services Ltd: POB 102, Port Vila; tel. 22454; fax 22884; e-mail security@vila.net; provides registry and administration of maritime services.

Vanuatu Registries Ltd: Port Vila; marketing and promotion of Vanuatu as a jurisdiction for company registration and ship registration; br. in Singapore.

TRADE UNIONS

Vanuatu Council of Trade Unions (VCTU): PMB 89, Port Vila; tel. 24517; fax 23679; e-mail synt@vanuatu.com.vu; Pres. OBED MASINGIOW; Sec.-Gen. EPHRAIM KALSAKAU.

National Union of Labour: Port Vila.

The principal trade unions include:

Oil and Gas Workers' Union: Port Vila; f. 1984.

Vanuatu Airline Workers' Union: Port Vila; f. 1984.

Vanuatu Public Service Association: Port Vila.

Vanuatu Teachers' Union: Port Vila; Gen. Sec. CHARLES KALO; Pres. OBED MASSING.

Vanuatu Waterside, Maritime and Allied Workers' Union: Port Vila.

Transport

ROADS

There are about 1,130 km of roads, of which 54 km, mostly on Efate Island, are sealed. In early 1998 Japan granted more than 400m. vatu to finance the sealing of the main road around Efate. In January 2002 an earthquake caused significant damage to the transport infrastructure around Efate. Several foreign donors, including New Zealand, contributed to funding urgent repairs to roads and bridges in the area. Two bridges on Efate were rebuilt with Japanese finance in 2004.

SHIPPING

The principal ports are Port Vila and Luganville.

Vanuatu Maritime Authority: POB 320, Marine Quay, Port Vila; tel. 23128; fax 22949; e-mail vma@vanuatu.com.vu; domestic and international ship registry, maritime safety regulator; Commissioner of Maritime Affairs JOHN T. ROOSEN; Chair. LENNOX VUTI.

Ports and Marine Department: PMB 046, Port Vila; tel. 22339; fax 22475; Harbour Master Capt. PAUL PETER.

Burns Philp (Vanuatu) Ltd: POB 27, Port Vila.

Ifira Shipping Agencies Ltd: POB 68, Port Vila; tel. 22929; fax 22052; f. 1986; Man. Dir CLAUDE BOUDIER.

Sami Ltd: Kumul Highway, POB 301, Port Vila; tel. 24106; fax 23405.

South Sea Shipping: POB 84, Port Vila; tel. 22205; fax 23304; e-mail southsea@vanuatu.com.vu.

Vanua Navigation Ltd: POB 44, Port Vila; tel. 22027; f. 1977 by the Co-operative Federation and Sofrana Unilines; Chief Exec. GEOFFREY J. CLARKE.

The following services call regularly at Vanuatu: Compagnie Générale Maritime, Kyowa Shipping Co, Pacific Forum Line, Papua New Guinea Shipping Corpn, Sofrana-Unilines, Bank Line, Columbus Line and Bali Hai Shipping. Royal Viking Line, Sitmar and P & O cruises also call at Vanuatu.

CIVIL AVIATION

The principal airports are Bauerfield (Efate, for Port Vila) and Pekoa (Espiritu Santo, for Luganville). There are airstrips on all Vanuatu's principal islands and an international airport at White Grass on Tanna was completed in 1998. The Civil Aviation Corporation Act, approved by Parliament in August 1998, provided for the transfer of ownership of Vanuatu's airports to a commercially-run corporation, in which the Government is to have a majority shareholding. In early 2000 it was announced that a further three airports were to be built on the islands of Pentecost, Malekula and Tanna. Major improvements providing for the accommodation of larger aircraft at both Bauerfield and Pekoa airports began in mid-2000. Moreover, in September 2000 plans for a new international airport at Teouma (Efate) were announced, with finance from a private Thai investor. In late 2003 plans for the construction of an international air terminal at Pekoa (Espiritu Santo) costing US $1.6m. were announced.

Air Vanuatu (Operations) Ltd: POB 148, Du Vanuatu House, rue de Paris, Port Vila; tel. 23838; fax 23250; e-mail service@airvanuatu.com.vu; internet www.pacificislands.com/airlines/vanuatu.html; f. 1981; govt-owned national carrier since 1987; regular services between Port Vila and Sydney, Brisbane and Melbourne (Australia), Nadi (Fiji), Nouméa (New Caledonia), Auckland (New Zealand), and Honiara (Solomon Islands); the frequency of flights from Auckland, Sydney, Brisbane and Honiara to Port Vila was increased in mid-2004; Man. Dir and CEO JEAN-PAUL VIRELALA.

Vanair: rue Pasteur, PMB 9069, Port Vila; tel. 22643; fax 23910; e-mail vias@vanuatu.com.vu; internet www.islandsvanuatu.com/vanair.htm; operates scheduled services to 29 destinations within the archipelago; CEO Capt. YVES CHEVALIER.

Dovair: Port Vila; privately-owned; operates domestic services.

Tourism

Tourism is an important source of revenue for the Government of Vanuatu. Visitors are attracted by the islands' unspoilt landscape and rich local customs. The establishment of regular air services from Australia and New Zealand in the late 1980s precipitated a significant increase in the number of visitors to Vanuatu. In 2003 there were an estimated 50,400 foreign visitor arrivals in Vanuatu, compared with 49,463 in the previous year. In 1997 passengers on cruise ships visiting the islands numbered 30,530. In 2000 some 64% of visitors were from Australia and 14% were from New Zealand. There was a total of 1,641 hotel beds at the end of 1996. Receipts from tourism totalled some US $58m. in 2000. The development of the tourist industry has hitherto been concentrated on the islands of Efate, Espiritu Santo and Tanna; however, the promotion of other islands as tourist centres is beginning to occur.

Vanuatu Hotel and Resorts Association: POB 215, Port Vila; tel. 22040; fax 27579; Pres. JOHN GROCOCK.

Vanuatu Tourism Office: Lini Highway, POB 209, Port Vila; tel. 22685; fax 23889; e-mail tourism@vanuatu.com.vu; internet www.vanuatutourism.com; Gen. Man. LINDA KALPOI.

Defence

Upon Vanuatu's achievement of independence in 1980, a defence pact with Papua New Guinea was signed. A 300-strong paramilitary force, the Vanuatu Mobile Force, exists. In 2003 Vanuatu sent 50 defence personnel to join the Australian-led regional intervention force in Solomon Islands. In 2002 the Government allocated 766m. vatu for defence, equivalent to 7.4% of total recurrent budgetary expenditure.

Education

The abolition of nominal fees for primary education following independence resulted in a significant increase in enrolment at that level. Thus, at the beginning of the 1990s it was estimated that about 85% of children between the ages of six and 11 were enrolled at state-controlled primary institutions (which numbered 267 in 1995). Secondary education begins at 12 years of age, and comprises a preliminary cycle of four years and a second cycle of three years. In 1992 4,269 pupils attended the country's secondary schools (which numbered 27 in 1995). Vocational education and teacher-training are also available. The relatively low level of secondary enrolment is a cause of some concern to the Government, and a major programme for the expansion of the education system, costing US $17.8m., was inaugurated in 1989. The programme aimed to double secondary enrolment by 1996. Literacy rates are another cause of concern; it was estimated in 2002 that 66% of the population was illiterate.

An extension centre of the University of the South Pacific was opened in Port Vila in May 1989. Students from Vanuatu can also receive higher education at the principal faculties of that university (in Suva, Fiji), in Papua New Guinea or in France.

In July 2004 the country's first pre-school (the Vila North Model Pre-School) was opened. The pre-school, which had been built with funding of 5.6m. vatu from the EU, was to have a staff of four teachers and was expected to serve as a centre for training other pre-school teachers.

The 2002 budget allocated an estimated 2,062m. vatu to education (20.1% of total recurrent expenditure by the central Government).

Bibliography

Akram-Lodhi, A. Haroon (Ed.). *Confronting Fiji Futures*. Canberra, Australian National University, 2000.

Aldrich, Robert, and Connell, John. *France's Overseas Frontier*. Cambridge, Cambridge University Press, 1992.

Aldrich, Robert. *The French Presence in the South Pacific, 1842–1940*. London, Macmillan, 1990.

France and the South Pacific since 1940. London, Macmillan, 1993.

Alkire, William H. *An Introduction to the Peoples and Cultures of Micronesia*. Menlo Park, CA, Cummings Publishing Co, 1977.

American Samoa Economic Advisory Commission. *Transforming the Economy of American Samoa: A Report to the President of the United States of America through the U. S. Department of the Interior*. Pago Pago, April 2002 (revised July 2002).

Australian National University. *Pacific History Bibliography and Comment, 1979–1987, Journal of Pacific History Bibliography –97, Journal of Pacific History*. Canberra, Australian National University.

Angleviel, Frédéric, Coppell, William, and Charleux, Michel (Eds). *Bibliographie des Thèses sur le Pacifique*. Bordeaux, CRET, CEGET, Université de Bordeaux, 1991.

Angleviel, Frédéric, Atoloto, Malau, and Atonio, Takasi (Eds). *101 Mots Pour Comprendre Wallis et Futuna*. Nouméa, Ile de Lumière, 1999.

Angleviel, Frédéric (Ed.). *La Nouvelle-Calédonie, Terre des Recherches—Bibliographie Analytique des Thèses et Memoires*. Nouméa, Association Thèse-Pac, 1995.

101 Mots Pour Comprendre l'Histoire de la Nouvelle-Calédonie. Nouméa, Ile de Lumière, 1997.

Antheaume, Benoît, and Bonnemaison, Joël. *Atlas des Iles et Etats du Pacifique Sud*. Montpellier-Paris, GIP Reclus/Publisud, 1988.

Australian Joint Parliamentary Committee on Foreign Affairs, Defence and Trade. *Australia's Relations with Papua New Guinea*. Canberra, 1991.

Bachimon, Philippe. *Tahiti Entre Mythes et Réalités: Essai d'Histoire Geographique*. Paris, Comité des Travaux Historiques et Scientifiques, 1989.

Ballard, J. A. (Ed.). *Policy-Making in a New State: Papua New Guinea 1972–77*. Melbourne, Oxford University Press, 1981.

Banks, Glenn, Bonnell, Susanne, and Filer, Colin (Eds). *Dilemmas of Development: the social and economic impact of the Porgera gold mine 1989–1994*. Canberra, Asia Pacific Press, Australian National University, 2000.

Barillot, Bruno. *L'héritage de la bombe: Sahara, Polynésie (1960–2002): les faits, les personnels, les populations*. Lyon, Centre de documentation et de recherche sur la paix et les conflits, 2002.

Bates, S. *The South Pacific Island Countries and France: A Study in International Relations*. Canberra, Australian National University, 1990.

Bayliss-Smith, T., Bedford, R., Brookfield, H., Latham, M., with Brookfield, M. *Islands, Islanders and the World: The Colonial and Post-Colonial Experience of Eastern Fiji*. Cambridge, Cambridge University Press, 1988.

Bellwood, Peter S. *The Polynesians: Prehistory of an island People*. London, Thames and Hudson, 1987.

Bellwood, Peter S. (Ed.). *The Austronesians: Historical and Comparative Perspectives*. Canberra, Australian University Press, 1995.

Bennett, Judith A. *Wealth of the Solomons*. Honolulu, HI, University of Hawaii Press, 1986.

Bensa, Alban, and Leblic, Isabelle. *En Pays Kanak*. Paris, Edition de la Maison des Sciences de L'Homme, Mission du Patrimoine Ethnologique, Collection Ethnologie de la France, Cahier 14, 2000.

Bensa, Alban. *Nouvelle-Calédonie, un paradis dans la tourmente*. Paris, Gallimard, 1990.

Bird, I. *The Hawaiian Archipelago*. London, Picador, 1998.

Birkett, Dea. *Serpent in Paradise*. New York, NY, Doubleday, 1997.

Bonnemaison, J., Huffman, K., Kauffmann, C., and Tryon, D. *Arts of Vanuatu*. Bathurst, NSW, Crawford House Press, 1997.

Borofsky, Robert (Ed.). *Remembrance of Pacific Pasts: an invitation to remake history*. Honolulu, HI, University of Hawaii, 2000.

Brookfield, H. C. *Melanesia, a Geographical Interpretation of an Island World*. London, Methuen, 1971.

Colonialism, Development and Independence; the Case of the Melanesian Islands in the South Pacific. London, Methuen, 1972.

The Pacific in Transition: a Geographical Perspective on Adaptation and Change. Canberra, Australian National University, 1973.

Browne, C., and Scott, D. A. *Economic Development in Seven Pacific Island Countries*. Washington, DC, International Monetary Fund, 1989.

Bullard, Alice. *Exile to Paradise: Savagery and Civilization in Paris and the South Pacific, 1790–1900*. Stanford, CA, Stanford University Press, 2000.

Burt, Ben, and Clerk, Christian (Eds). *Environment and Development in the Pacific Islands*. Canberra, Australian National University, 1994.

Cameron, I. *Lost Paradise: the Exploration of the Pacific*. Massachusetts, Salem House, 1987.

Campbell, Ian. *Island Kingdom: Tonga Ancient and Modern*. Christchurch, University of Canterbury, 1992.

Worlds Apart: A New History of the Pacific Islands. Christchurch, Canterbury University Press, 2003.

Campbell, Ian C., and Latouche, Jean-Pierre. *Les insulaires du Pacifique—histoire et situation politique*. Paris, Puf, 2001.

Capie, D. *Under the Gun: The Small Arms Challenge in the Pacific*. Wellington, Victoria University Press, 2003.

Carano, P., and Sanchez, P. C. *A Complete History of Guam*. Rutland, VT, Charles E. Tuttle.

Carrier, James. *History and Tradition in Melanesian Anthropology*. Berkeley, CA, University of California, 1992.

Chappell, David. *Double Ghost: Oceanian Voyagers on Euro-American Ships*. London, Armonk, 1997.

Chesneaux, Jean, and Maclellan, Nic. *La France dans le Pacifique, De Bougainville à Moruroa*. Paris, La Découverte, 1992.

After Mururoa—France in the South Pacific. Melbourne, Ocean Press, 1998.

Clark, R. S., and Sann, M. (Eds). *The Case against the Bomb: Marshall Islands, Samoa and Solomon Islands before the International Court of Justice*. Mansfield, OH, Book Masters, 1996.

Clifford, James. *Person and Myth: Maurice Leenhardt in the Melanesian World*. Los Angeles, University of California Press, 1982.

Cochrane, S. *Contemporary Art in Papua New Guinea*. Sydney, Craftsman House, 1997.

Cole, Rodney V., and Cuthbertson, S. *Population Growth in the South Pacific Island States—Implications for Australia*. Canberra, Bureau of Immigration, 1995.

Cole, Rodney V. (Ed.). *Pacific 2010—Challenging the Future*. Canberra, Development Studies Centre, Australian National University, 1993.

Cole, Rodney V., and Parny, T. G. (Eds). *Selected Issues in Pacific Island Development*. Canberra, Development Studies Centre, Australian National University, 1986.

Connell, John. *New Caledonia or Kanaky? The Political History of a French Colony*. Canberra, Development Studies Centre, Australian National University, 1987.

Papua New Guinea: The Struggle for Development. London, Routledge, 1997.

Connell, John et al. *Encyclopedia of the Pacific Islands*. Canberra, Australian National University, 1999.

Connell, John, and Lea, John (Eds). *Planning the Future: Melanesian Cities in 2010*. Canberra, Development Studies Centre, Australian National University, 1993.

Urbanisation in Polynesia. Canberra, Development Studies Centre, Australian National University, 1996.

Urbanization in the Island Pacific. London, Routledge, 2002.

Connell, John, and McCall, G. (Eds). *A World Perspective on Pacific Islander Migration: Australia, New Zealand and the USA*. Centre for Pacific Studies, University of New South Wales, 1992.

Coutau-Bégarie, H. *Géostratégie du Pacifique*. Paris, Economica, 1987.

Craig, Robert D. *Historical Dictionary of Polynesia*. Metuchen, NJ, Scarecrow Press, 1994.

Craig, Robert, and King, Frank (Eds). *Historical Dictionary of Oceania*. London, Greenwood Press, 1981.

Crocombe, Ron. *The South Pacific*. Suva, Institute of Pacific Studies, University of the South Pacific, 1989 (revised 2001).

The Pacific Islands and the USA. Suva, Institute of Pacific Studies, 1995.

Crocombe, Ron, and Ali, A. *Politics in Melanesia*. Suva, Institute of Pacific Studies, University of the South Pacific, 1982.

Daniel, P., and Sims, R. *Foreign Investment in Papua New Guinea: Policies and Practices*. Canberra, Development Studies Centre, Australian National University, 1986.

Davidson, J. W. *Samoa Mo Samoa: the Emergence of the Independent State of Western Samoa*. Melbourne, Oxford University Press, 1967.

Davis, T. *Island Boy, an Autobiography*. Suva, Institute of Pacific Studies, University of the South Pacific, 1992.

Daws, G. *Shoal of Time. A History of the Hawaiian Islands*. Honolulu, HI, University of Hawaii Press, 1968.

A Dream of Islands. Queensland, Jacaranda Press, 1980.

Debsky, R. *The Organization of Development in the South Pacific*. Honolulu, Pacific Islands Studies Program, Center for Asian and Pacific Studies, 1986.

Decentralisation in the South Pacific: Local, Provincial and State Government in Twenty Countries. Suva, University of the South Pacific, 1986.

De Deckker, Paul, and Kuntz, Laurence. *La Bataille de la Coutume*. Paris, L'Harmattan, 1997.

De Deckker, Paul, and Tryon, Darell (Eds). *Identités en Mutation dans le Pacifique à l'Aube du Troisième Millénaire, Iles et Archipels No. 26*. Cret, Coll., Université de Bordeaux III, 1998.

De Deckker, Paul. *Le Peuplement du Pacifique et de la Nouvelle-Calédonie au X1Xè siècle*. Paris, L'Harmattan & U.F.P., 1994.

De Deckker, Paul, and Faberon, Jean-Yves (Eds). *Custom and the Law*. Canberra, Asia Pacific Press 2001.

De Deckker, Paul et al. *L'outre-mer français dans le Pacifique (Nouvelle-Calédonie, Polynésie française, Wallis-et-Futuna)*. Paris-Nouméa, L'Harmattan-CDP, 2003.

Delgado, James P. *Ghost fleet: The Sunken Ships of Bikini Atoll*. Honolulu, HI, University of Hawaii Press, 1996.

Denoon, Donald, Firth, S., Linnekin, J., Meleisea, M., and Nero, K. *The Cambridge History of the Pacific Islanders*. Cambridge, Cambridge University Press, 1998.

Denoon, Donald, and Mein-Smith, Philippa. *A History of Australia, New Zealand and the Pacific*. Malden, Blackwell, 2000.

Diamond, J. *Guns, Germs and Steel—The Fate of Human Societies*. New York, W. W. Norton, 1997.

Diaz, Vincente M. *Native Pacific Cultural Studies on the Edge*. Honolulu, HI, University of Hawaii Press, 2001.

Dibblin, Jane. *Day of Two Suns: US Nuclear Testing and the Pacific Islanders*. New York, NY, New Amsterdam Books, 1990.

Dinnen, Sinclair, and Ley, Alison (Eds). *Reflections on Violence in Melanesia*. Canberra, Asia Pacific Press, Australian National University, 2000.

Dinnen, Sinclair, May, Ron, and Regan, Anthony J. (Eds). *Challenging the State: The Sandline Affair in Papua New Guinea*. Canberra, Australian National University, 1998.

Dinnen, Sinclair. *Law and Order in a Weak State—Crime and Politics in Papua New Guinea, Pacific Islands Monograph Series, No.17*. Honolulu, HI, The University of Hawaii Press, 2001.

Diolé, P. *The Forgotten People of the Pacific*. London, Cassell, 1976.

Docherty, James. *Historical Dictionary of Australia, No.1*. Metchuen, NJ, Scarecrow Press, 1992.

Dorney, Sean. *The Sandline Affair: Politics and Mercenaries and the Bougainville Crisis*. ABC Books, 1998.

Dorrance, John C. *The United States and the Pacific Islands*. Westport, CT, Praeger, 1992.

Douglas, Bronwen. *Across the Great Divide—Journeys in History and Anthropology*. Amsterdam, Harwood Academic Publishers, 1998.

Douglas, Ngaire, and Norman, Douglas (Eds). *Pacific Islands Yearbook*. Suva, Fiji Times Ltd.

Doumenge, F. *L'Homme dans le Pacifique Sud; Etude Géographique*. Paris, Société des Océanistes, 1966.

Doumenge, J.P. *Du Terroir à la Ville, les Mélanésiens et leurs espaces en Nouvelle-Calédonie*. Bordeaux, CRET, CEGET, Université de Bordeaux 1982.

Dubois, Marie-Joseph. *Les Chefferies de Maré*. Paris, Librarie, H. Champion, 1977.

Dunmore, John. *Who's Who in Pacific Navigation*. Honolulu, HI, University of Hawaii Press, 1991.

Visions and Realities—France in the Pacific, 1695–1995. Auckland, Heritage Press, 1997.

Duroy, L. Hienghène, le désespoir calédonien. Barrault, 1988.

Dye, Bob (Ed.). *World War Two in Hawaii from the pages of 'Paradise of the Pacific'*. Honolulu, HI, University of Hawaii Press, 2001.

Emberson-Bain, A. *Labour and Gold in Fiji*. Cambridge, Cambridge University Press, 1994.

Emberson-Bain, A. (Ed.). *Sustainable Development of Malignant Growth? Perspectives of Pacific Island Women*. Suva, Marama Publications, 1995.

Ernst, Manfred. *Winds of Change*. Suva, Pacific Conference of Churches, 1994.

Fairbairn, T. I. J. *Island Economies: Studies From the South Pacific*. Suva, Institute of Pacific Studies, University of the South Pacific, 1985.

Fairbairn, T. I. J., Morrison, C., Baker, R., and Groves, S. *The Pacific Islands*. Honolulu, HI, University of Hawaii Press, 1991.

Figiel, S. *Where We Once Belonged*. Auckland, Pasifika Press, 1997.

Finnegan, Ruth, and Orbell, Margaret. *South Pacific Oral Traditions*. Bloomington, IN, Indiana University Press, 1995.

Fischer, Steven Roger. *A History of the Pacific Islands*. Basingstoke, Palgrave Global Publishing, 2002.

Fleming, Euan, and Hardaker, Brian. *Pacific 2010: Strategies for Melanesian Agriculture for 2010: Tough Choices*. Canberra, National Centre for Development Studies, Australian National University, 1994.

Foerstel, L., and Gilliam, A. (Eds). *Confronting the Margaret Mead Legacy: Scholarship, Empire and the South Pacific*. Philadelphia, Temple University Press, 1992.

Forbes, David W. (Ed.). *Hawaiian National Bibliography, 1780–1900, Vol. 2: 1831–1850*. Honolulu, HI, University of Hawaii Press, 2001.

Foster, Robert (Ed.). *Nation-making Emergent Identities in Post-colonial Melanesia*. Ann Arbor, University of Michigan Press, 1995.

Franceschi, M. *La Démocratie Massacrée: Consensus ou Mystification à Nouméa?* Paris, Pygmalion, 1998.

Freeman, D. *Margaret Mead and Samoa: The Making and Unmaking of an Anthropological Myth*. Harmondsworth, Penguin, 1983.

Fry, Gerald, and Rufino, Mauricio (Eds). *Pacific Basin and Oceania, World Bibliographical Series, Vol. 70*. Oxford. Clio Press, 1987.

Galipaud, Jean-Christophe, and Lilley, Ian (Eds). *Le Pacifique de 5000 à 2000 avant le présent—Suppléments a L'Histoire d'une Colonisation*. Paris, Edition de l'Institut de Recherche pour le Développement, 1999.

Gannicott, Ken. *Pacific 2010—Women's Education and Economic Development in Melanesia*. Canberra, National University Centre for Development Studies, 1994.

Garrett, John. *Island Exiles*. Melbourne, ABC Books, 1996.

To live among the stars—Christian origins in Oceania. Suva, University of the South Pacific, 1982.

Footsteps in the Sea—Christianity in Oceania Since World War II. Suva, University of the South Pacific, 1992.

Where Nets Were Cast—Christianity in Oceania Since World War II. Suva, University of the South Pacific, 1997.

Geddes, W. H., et al. *Atoll Economy: Social Change in Kiribati and Tuvalu, Islands on the Line Team Report No. 1*. Canberra, Australian National University Press, 1982.

Ghai, Y. *Law, Politics and Government in Pacific Island States*. Suva, University of the South Pacific, 1988.

Gill, W. W. *Cook Islands Customs*. Cook Islands, University of the South Pacific and the Ministry of Education, 1979.

Gille, Bernard et Toullelan, Pierre Yves. *De la Conquête à l'Exode—Histoire des Océaniens et de leurs Migrations dans le Pacifique. Tome 1. Les Migrations Contraintes en Océénia, Terres de Colonisation et D'immigration*. Papeete, Tahiti, Au Vent des Îles, 1999.

Goetzfridt, Nicholas. *Indigenous Literature of Oceania—A Survey of Criticism and Interpretation*. Westport, CT, Greenwood Press, 1995.

Goodman, R., Lepani, C., and Morawetz, D. *The Economy of Papua New Guinea: An Independent Review*. Canberra, Development Studies Centre, Australian National University, 1985.

Gorman, G. E., and Mills, J. J. (Eds). *Fiji, World Bibliography, Volume 173*. Oxford, Clio Press, 1992.

Gostin, O. *Cash Cropping, Catholicism and Change: Resettlement among the Kuni of Papua*. Canberra, Development Studies Centre, Australian National University, 1986.

Grynberg, Roman. *Rules of Origin Issues in Pacific Island Development*. Canberra, Asia Pacific Press, Australian National University, 1998.

Gunson, W. N. (Ed.). *The Changing Pacific: Essays in Honour of H. E. Maude*. Melbourne, Oxford University Press, 1978.

Hanlon, David, and White, Geoffrey. *Voyaging through the Contemporary Pacific*. New York, NY, Rowman, Littlefield and Publishers, 2000.

Harding, Thomas, Lockwood, Victoria, and Wallace, Ben (Eds). *Contemporary Pacific Societies: Studies in Development and Change*. Englewood Cliffs, Prentice Hall, 1993.

Hau'ofa, E., Naidu, V., and Waddell, E. (Eds). *A New Oceania: Rediscovering Our Sea of Islands*. Suva, University of the South Pacific, 1993.

Hayes, P., Zarskey, L., and Bello, W. *American Lake: Nuclear Peril in the Pacific*. Harmondsworth, Penguin, 1986.

Henningham, Stephen. *France and the South Pacific: A Contemporary History*. Sydney, Allen and Unwin, 1992.

Henningham, Stephen, and May, R. J. *Resources, Development and Politics in the Pacific Islands*. Bathurst, Crawford House Press, 1992.

Heyerdahl, Thor. *Green was the Earth on the Seventh Day*. London, Little Brown, 1997.

Hezel, F. X. *Strangers in Their Own Land: A Century of Colonial Rule in the Caroline and Marshall Islands*. Honolulu, HI, University of Hawaii Press, 1995.

The New Shape of Old Island Cultures—A Half Century of Social Change in Micronesia. Honolulu, HI, University of Hawaii Press, 2001.

Hintjens, Helen M., and Newitt, M. D. (Eds). *The Political Economy of Small Tropical Islands: The Importance of Being Small*. Exeter, University of Exeter Press, 1992.

History of Micronesia. Québec, Levesque Publications.

Hongo, G. *Volcano: A Memoir of Hawai'i*. Washington, Knopf, 1995.

Hooper, Antony (Ed.). *Culture and Sustainable Development in the Pacific*. Canberra, Australian National University, 2000.

Howard, Michael. *Mining, Politics and Development in the South Pacific*. Boulder, Westview Press, 1991.

Howe, Kerry, Kiste, Robert, and Lal, Brij (Eds). *Tides of History: The Pacific Islands in the Twentieth Century*. Sydney, Allen and Unwin, 1994.

Howe, Kerry. *The Loyalty Islands—A History of Culture Contacts 1840–1900*. Honolulu, HI, University of Hawaii Press, 1977.

Where the waves fall: A new South Seas islands history from first settlement to colonial rule. Honolulu, HI, University of Hawaii Press, 1984.

Nature, Culture and History—The 'Knowing' of Oceania. Honolulu, HI, University of Hawaii Press, 2000.

Huffer, Elise, and So'o, Asofou (Eds). *Governance in Samoa*. Canberra, Asia Pacific Press, Australian National University, 2000.

Huntsman, Judith, and Hooper, Antony. *Tokelau, A Historical Ethnography*. Honolulu, HI, University of Hawaii Press, 1997.

Institute of Pacific Studies, University of the South Pacific. *Micronesian Politics*. Suva, 1988.

Institute of Pacific Studies, University of the South Pacific, and Ministry of Education, Training and Culture, Kiribati. *Kiribati: Aspects of History*. 1979.

Jackson, Keith, and McRobie, Alan. *Historical Dictionary of New Zealand, No.5*. Metchuen, NJ, Scarecrow Press, 1996.

Jones, Peter D. *From Bikini to Belau: The Nuclear Colonization of the Pacific*. London, War Resisters International, 1988.

Kaeppler, Adrienne, and Nimmo, Arlo H. (Eds). *Directions in Pacific Traditional Literature*. Honolulu, HI, Bishop Museum Press, 1976.

Kamisese, Ratu. *The Pacific Way: a Memoir*. Honolulu, HI, University of Hawaii Press, 1997.

Kaul, M. M. *Pearls in the Ocean: Security Perspectives in the South-West Pacific*. New Delhi, UBSPD, 1993.

Keating, Elizabeth. *Power Sharing—Language, Rank, Gender and Social Space in Pohnpei, Micronesia*. Oxford University Press, 1999.

Kernahan, M. *White Savages in the South Seas*. London, Verso, 1996.

King, D., and Ranck, S. *Papua New Guinea Atlas*. Robert Brown and University of Papua New Guinea, 1982.

Kirch, Vinton. P. (Ed.). *Island Societies—Archeological Approaches to Evolution and Transformation*. Cambridge, Cambridge University Press, 1986.

Kirch, Vinton P. *On the Road of the Wind—An Archeological History of the Pacific Islands before the European Contact*. Berkeley, CA, University of California Press, 2000.

Kirch, Vinton P., and Hunt, Terry L. (Eds). *Historical Ecology in the Pacific Islands: Prehistoric Environmental and Landscape Change*. Yale University Press, 1997.

Klein II, D. A. (Ed.). *Marshall Islands Legends and Stories*. Honolulu, HI, Bess Press, 2003.

Kluge, P. F. *The Edge of Paradise: America in Micronesia*. New York, NY, Random House, 1991.

Koburger, Charles W. *Pacific Turning Point: The Solomon Islands Campaign, 1942–1943*. London, Greenwood Publishing, 1995.

Kramer, Anthony. *The Samoa Islands*. Auckland, Pasifika Press, 1994.

Kramer, Augustin. *The Samoa Islands: Volume 1: Constitution, Pedigrees and Traditions*. Honolulu, HI, University of Hawaii, 2000.

Krieger, M. *Conversations with Cannibals: the End of the Old Pacific*. Ecco Press, 1994.

Lal, Brij V. *Girmitiyas: the Origin of Fiji Indians*. Canberra, Journal of Pacific History, 1983.

Politics in Fiji: Studies in Contemporary History. Hawaii, Brigham Young University, Institute for Polynesian Studies, 1986.

Power and Prejudice: The Making of the Fiji Crisis. Wellington, New Zealand Institute of International Affairs, 1989.

A Vision for Change: A. D. Patel and the Politics of Fiji. Canberra, Asia Pacific Press, Australian National University, 1997.

Another way: the politics of constitutional reform in post-coup Fiji. Canberra, Australian National University, 1998.

Lacey, Rod. *'Whose voices are heard? Oral history and the Decolonisation of History' in Emerging from Empire?—Decolonisation in the Pacific*. Canberra, Australian National University, 1997, pp. 180–186.

Lal, Brij V., and Fortune, K. (Eds). *The Pacific Islands: An Encyclopedia*. Honolulu, HI, University of Hawaii Press, 2000.

Larmour, Peter (Ed.). *Governance and Reform in the South Pacific*. Canberra, Development Studies Centre, Australian National University, 1997.

Latukefu, S. *Church and State in Tonga*. Canberra, Australian National University, 1974.

Laux, Claire. *Les Théocraties Missionnaires en Polynésie (Tahiti, Hawaii, Cook, Tonga, Gambier, Wallis et Futuna) au XIVè siècle. Des Cités de Dieu dans les Mers du Sud?* Paris, L'Harmattan, 2000.

101 mots pour comprendre l'Océanie. Nouméa, Ile de Lumière 2002.

Lawson, Stephanie. *Tradition versus Democracy in the South Pacific: Fiji, Tonga and Western Samoa*. Cambridge, Cambridge University Press, 1996.

Leckie, Jacqueline. *To Labour with the State: The Fiji Public Service Association*. Dunedin, University of Otago Press, 1997.

Leibowitz, A. *Defining Status: A Comprehensive Analysis of the United States' Territorial Relations*. Hingham, MA, Kluwer Academic Publishers, 1989.

Levantis, Theodore. *Papua New Guinea: employment, wages and economic development*. Canberra, Australian National University, 2000.

Levy, Neil M. *Micronesia Handbook*. Emeryville, CA, Moon Publications, 1999.

Lewis, D. *The Voyaging Stars—Secrets of the Pacific Island Navigators*. Sydney and London, Collins, 1978.

Lieber, M. D. (Ed.). *Exiles and Migrants in Oceania*. Honolulu, HI, University of Hawaii Press, 1977.

Lindstrom, Lamont. *Knowledge and Power in a South Pacific Society*. Washington, DC, Smithsonian Institute Press, 1990.

Lobban, Christopher, and Schefter, Maria. *Tropical Pacific Island Environments*. Hagåtña, University of Guam Press, 1998.

Löffler, E. *Geomorphology of Papua New Guinea*. Canberra, Australian National University, 1977.

Lummis, Trevor. *Life and Death in Eden: Pitcairn Island and the Bounty Mutineers*. London, Phoenix, 2000.

Lynch, John, and Mugler, France. *Pacific Languages in Education*. Suva, University of the South Pacific, 1996.

Maclellan, Nic, and Chesneaux, Jean. *After Moruroa: France in the South Pacific*. Melbourne, Ocean Press, 1998.

Mageo, Jeanette, M. *Cultural Memory—Reconfiguring History and Identity in the Postcolonial Pacific*. Honolulu, HI, University of Hawaii Press, 2001.

Marchal, H (Ed.). *De Jade et de Nacre—Patrimonie Artistique Kanak*. Paris, Réunion des Musées Nationaux, 1990.

Marshall, M. *Beyond the Reef: The Transformation of a Micronesian Community*. Boulder, CO, Westview Press, 2004.

McConnel, Frasier (Ed.). *Papua New Guinea. World Bibliography, Volume 90*. Oxford, Clio Press, 1990.

Macdonald, Barrie. *Cinderellas of the Empire: Towards a History of Kiribati and Tuvalu*. Suva, Institute of Pacific Studies, University of the South Pacific, 2002.

McInnes, D. (Ed.). *Encyclopaedia of Papua New Guinea*. Mount Waverley, Dellasta Pacific, 1996.

McKnight, Tom Lee. *Oceania—The Geography of Australia, New Zealand and the Pacific Islands*. Englewood, Prentice Hall, 1995.

Meleisea, M. *Lalaga: A Short History of Western Samoa*. Suva, University of the South Pacific, 1987.

Meller, Norman. *Constitutionalism in Micronesia*. Honolulu, HI, University of Hawaii Press, 1986.

Miles, J., and Shaw, E. *The French Presence in the South Pacific 1838–1990*. Auckland, Greenpeace, 1990.

Miles, John. *Infectious Diseases—Colonising the Pacific?* Dunedin, University of Otago Press, 1997.

Mills, Peter R. *Hawaii's Russian Adventure: A New Look at Old History*. Honolulu, HI, University of Hawaii Press, 2002.

Moala, Kalafi. *Island Kingdom Strikes Back: The Story of an Independent Island Newspaper*. Auckland, Pacmedia Publishers, 2003.

Moorehead, A. *The Fatal Impact: The Invasion of the South Pacific 1767–1840*. London, Hamish Hamilton, 1966.

Murray, Spencer. *Pitcairn Island: The First 200 Years*. La Canada, CA, Bounty Sagas.

Naepels, Michel. *Histoires de Terrres Kanakes—Conflits Fonciers et Rapports Sociaux dans la Région de Houaïlou (Nouvelle Calédonie)*. Paris, Belin, 1998.

Narakobi Bernard Mullu. *The Melanesian Way*. Suva, Institute of Pacific Studies University, 1980.

Neemia, U. *Cooperation and Conflict: Costs, Benefits and National Interests in Pacific Regional Cooperation*. Suva, University of the South Pacific, Institute of Pacific Studies, 1986.

Newbury, C. *Tahiti Nui*. Honolulu, HI, University of Hawaii Press, 1980.

New Politics in the South Pacific. Suva, Institute of Pacific Studies, 1994.

Nicolson, Robert. *The Pitcairners*. Honolulu, HI, University of Hawaii Press, 1997.

Nordyke, E. C. *The Peopling of Hawaii*. Honolulu, HI, University of Hawaii Press, 1977.

Nunn, Patrick, and Waddell Eric (Eds). *The Margin Fades—Geographical Itineraries on a World of Islands*. Suva, University of the South Pacific, 1994.

O'Callaghan, Mary-Louise. *Enemies Within: Papua New Guinea, Australia and the Sandline Crisis*. Doubleday Australia, 1998.

Oliver, Douglas L. *Native Cultures of the Pacific Islands*. Honolulu, HI, University of Hawaii Press, 1989.

The Pacific Islands. Honolulu, HI, University of Hawaii Press, 1989.

Black Islanders—A Personal Perspective of Bougainville 1937–1991. Melbourne, Hyland House Publishing, 1991.

Polynesia in Early Historic Times. Honolulu, HI, University of Hawaii Press, 2002.

Otto, Ton, and Borsboom, Ad. *Cultural Dynamics of Religious Change in Oceania*. Leiden, Netherlands, Kitlv Press, 1997.

Overton, John, and Scheyrens (Eds). *Strategies for Sustainable Development—Experiences from the Pacific*. Sydney, University of New South Wales Press, 1999.

Parmentier, R. J. *The Sacred Remains: Myth, History and Polity in Belau*. Chicago, University of Chicago Press, 1987.

Peterson, G. *Ethnicity and Interest at the 1990 Federated States of Micronesia Constitutional Convention*. Canberra, Australian National University, 1994.

Pitts, Maxine. *Crime, Corruption and Capacity in Papua New Guinea*. Canberra, Asia Pacific Press, Australian National University, 2002.

Poirine, Bernard. *Les Petites Economies Insulaires: Théories et Stratégies de Développement*. Paris, L'Harmattan, 1995.

Pollock, Nancy J. *These Roots Remain: Food Habits in Islands of the Central and Eastern Pacific since Western Contact*. Honolulu, HI, University of Hawaii Press, 1992.

Poyer, Lin, Falgout, Suzanne, and Carucci, Laurence, M. (Eds). *The Typhoon of War, Micronesian Experiences of the Pacific War*. Honolulu, HI, University of Hawaii Press, 2001.

Quadling, P. *Bougainville—The Mine and the People*. Sydney, Centre for Independent Studies, 1991.

Quanchi, M. *Atlas of the Pacific Islands*. Honolulu, HI, Bess Press, 2003.

Quanchi, M., and Adams, R. (Eds). *Culture Contact in the Pacific: Essays on Contact Encounter and Response*. New York, Cambridge University Press, 1992.

Rainier, Chris, and Taylor, Meg. *Where Masks Still Dance: New Guinea*. London, Little Brown, 1997.

Rannels, J. *PNG—A Modern Fact Book on Papua New Guinea*. Melbourne, Oxford University Press, 1995.

Rapaport, Moshe (Ed.). *The Pacific Islands—Environment and Society*. Hawaii, HI, University of Hawaii Press, 1999.

Ravuvu, A. D. *Development or Dependence: The Pattern of Change in a Fijian Village*. Suva, Institute of Pacific Studies, University of the South Pacific, 1988.

Revue Tiers-Monde. *Le Pacifique Insulaire—Nations, Aides, Espaces, Numéro Special de la Revue Tiers-Monde, Tome XXXVIII, No. 149*. Paris, Université de Paris, 1997.

Ridgell, Reilly. *Pacific Nations and Territories*. Honolulu, HI, University of Hawaii Press, 1998.

Robert, Craig, and King, Frank. *Historical Dictionary of Oceania*. London, Greenwood Press, 1981.

Robie, D. *Eyes of Fire: The Last Voyage of the Rainbow Warrior*. Auckland, Lindon Publishing, 1986.

Robie, D. (Ed.). *Tu Galala—Social Change in the Pacific*. Wellington, Bridget Williams Books, 1992.

Robillard, A. B. (Ed.). *Social Change in the Pacific Islands*. London, Kegan Paul International, 1992.

Rodman, Margaret C. *Houses Far From Home—British Colonial Space in the New Hebrides*. Honolulu, HI, University of Hawaii Press, 2001.

Rogers, R. F. *Destiny's Landfall: A History of Guam*. Honolulu, HI, University of Hawaii Press, 1996.

Rose, R. G. *Hawaii: The Royal Isles*. Honolulu, HI, Bishop Museum, 1980.

Ross, K. *Regional Security in the South Pacific: the Quarter-Century 1970–95*. Canberra, Strategic and Defence Studies Centre, Australian National University, 1993.

Rubinstein, H. J., and Zimmet, P. *Phosphate, Wealth and Health in Nauru: a study of lifestyle change*. Caulfield, Brolga Press, 1993.

Rumsey, Alan, and Weiner, James. *Emplaced Myth—Space, Narrative and Knowledge in Aboriginal Australia and Papua New Guinea*. Honolulu, HI, The University of Hawaii Press, 2001.

Sacks, O. *The Island of the Colorblind*. London, Picador, 1997.

Saffu, Y. (Ed.). *The 1992 Papua New Guinea Election—Change and Continuity in Electoral Politics*. Canberra, Australian National University, 1996.

Sahlins, M. *How 'Natives' Think: About Captain Cook, For Example*. Chicago, University of Chicago, 1996.

Sand, Christophe. *Le Temps d'Avant—La Préhistoire de la Nouvelle Calédonie*. Paris, L'Harmattan, 1995.

Scarr, Deryck. *A History of the Pacific Islands: Passages through Tropical Time*. London, Curzon Press, 2000.

Schoeffel, Penelope. *Sociocultural Issues and Economic Development in the Pacific Islands*. Manila, Asian Development Bank, Pacific Studies Series, 1996.

Segal, G. *Rethinking the Pacific*. London, Oxford University Press, 1990.

Seward, Robert. *Radio Happy Isles: Media and Politics at Play in the Pacific*. Honolulu, HI, University of Hawaii Press, 1998.

Sharphan, J. *Rabuka of Fiji: the Authorised Biography of Major-General Sitiveni Rabuka*. Queensland, Central Queensland University Press, 2000.

Shineberg, Dorothy. *They came for Santalwood—A Study of the Santalwood Trade in the South-West Pacific, 1830–1865*. Melbourne, Melbourne University Press, 1967.

The People Trade—Pacific Island Laborers and New Caledonia, 1865–1930, Pacific Island Monograph series No.16. Honolulu, HI, University of Hawaii Press, 1999.

Short, F. G. *Sinners and Sandalwood*. North Leura, NSW, Jomaru Press, 1997.

Short, Iaveta, Crocombe, Ron, and Herrmann, John. *Reforming the Political System of the Cook Islands: Preparing for the Challenges of the 21st Century*. Suva, Institute of Pacific Studies, University of the South Pacific, 1998.

Siagura, Sir Anthony. *In-House in Papua New Guinea with Anthony Seguera*. Canberra, Asia Pacific Press, Australian National University, 2002.

Sissons, J. *Nation and Destination: Creating a Cook Islands Identity*. Suva, Institute of Pacific Studies, University of South Pacific, 1999.

Skully, M. T. (Ed.). *Financial Institutions and Markets in the South-west Pacific*. London, Macmillan, 1985.

Skully, M. T., and Fairbairn, T. I. J. *Private Sector Development in the South Pacific: Options for Donor Assistance*. Centre for South Pacific Studies, University of New South Wales, 1992.

Smith, Bernard. *European Vision and the South Pacific*. New Haven, CT, Yale University Press, 1989.

Smith, Gary. *Micronesia: Decolonization and US Military Interests in the Trust Territories of the Pacific Islands*. Canberra, Australian National University, 1991.

Smith, R. A., and Meehl, G. A. *Pacific Legacy*. New York, Abbeville Press, 2003.

Somare, M. T. *Sana, an Autobiography*. Port Moresby, Niugini Press, 1975.

So'o, Asofou. *Universal Suffrage in Western Samoa: the 1991 General Elections*. Canberra, Australian National University, 1994.

Spate, O. H. K. *The Pacific Since Magellan*. Canberra, Australian National University. Vol. 1: The Spanish Lake, 1981; Vol. 2: Monopolists and Freebooters, 1983.

Speiser, F. *Ethnology of Vanuatu*. Bathurst, NSW, Crawford House Press, 1991.

Stanley, D. *South Pacific Handbook*. Emeryville, CA, Moon Publications, 1989.

Tahiti—Polynesia Handbook. Emeryville, CA, Moon Publications, 1999.

Fiji Islands Handbook. Emeryville, CA, Moon Publications, 2001.

Tahiti, including the Cook Islands. Emeryville, CA, Moon Publications, 2003.

Subramani. *South Pacific Literature—From Myth to Fabulation*. Suva, University of the South Pacific, 1992.

Swain, Tony, and Trompf, Gary. *The Religions of Oceania*. London, Routledge, 1995.

Syed, Saifullah, and Mataio, Ngatokorua. *Agriculture in the Cook Islands—New Directions*. Institute of Pacific Studies, and University of the South Pacific, 1993.

Tait, M. (Ed.). *The Papua New Guinea Handbook*. Canberra, Australian National University, irregular.

Taylor, M. (Ed.). *Fiji: Future Imperfect?* Sydney, Allen and Unwin, 1987.

Temu, Ila (Ed.). *Papua New Guinea: a 20/20 vision*. Canberra, Australian National University, 1997.

Thakur, R. (Ed.). *The South Pacific: Problems, issues and prospects*. Macmillan and the University of Otago, 1991.

Thawley, John (Ed.). *Australasia and South Pacific Islands Bibliography, Area Bibliographies No. 12*. London and Lanham, MD, Scarecrow, 1997.

Thomas, Nicholas. *In Oceania: Visions, Artifacts, Histories*. Durham, NC/London, Duke University Press, 1997.

Marquesan Societies—Inequality and Political Transformation in Eastern Polynesia. Oxford, Clarendon Press, 1990.

Tjibaou, Jean-Marie, and Missotte, P. *Kanak, Melanésien de Nouvelle Calédonie*. Papeete, Tahiti, Les Editions du Pacifique, 1976.

Kanaké—the Melanesian Way. Papeete,Les Editions du Pacifique, with Suva, University of the South Pacific, 1982.

Tjibaou, Jean-Marie. *La Presence Kanak*. Paris, O. Jacob, 1996.

Tolron, Francine. *La Nouvelle-Zélande—Histoire et Représentations*. Avignon, Université d'Avignon, 2000.

Toohey, John. *Captain Bligh's Portable Nightmare*. Fourth Estate, 1998.

Treadgold, M. L. *Bounteous Bestowal: The Economic History of Norfolk Island*. Canberra, Development Studies Centre, Australian National University, 1988.

Tudor, J. (Ed.). *Pacific Islands Year Book and Who's Who*. Sydney, Pacific Publications, Ltd.

Turner, Ann (Ed.). *Historical Dictionary of Papua New Guinea No.4*. Metuchen, NJ, Scarecrow Press, 1994.

University of the South Pacific. *South Pacific Bibliography 1981, 1982, 1983, 1984, 1985, 1988, 1989-90, 1991, 1992-3, 1994-5*. Suva, Pacific Information Centre, University of the South Pacific.

University of the South Pacific. *South Pacific Bibliography 1996–97*. Suva, Pacific Information Centre, University of the South Pacific, 1998.

Uriam, K. K. *In Their Own Words—History and Society in Gilbertese Oral Tradition*. Canberra, Journal of Pacific History, 1996.

Usher, L. *Letters from Fiji 1987–1990*. Suva, Fiji Times Ltd, 1992.

Viviani, N. *Nauru*. Canberra, Australian National University, 1970.

Waiko, John Dademo. *A Short History of Papua New Guinea*. Oxford, Oxford University Press, 1995.

Ward, R. G. (Ed.). *Man in the Pacific Islands: Essays on Geographical Change in the Pacific Islands*. Oxford, Clarendon Press, 1972.

Ward, R. G., and Proctor, A. W. *South Pacific Agriculture: Choices and Constraints. South Pacific Agricultural Survey 1979*. Canberra, Australian National University, and Asian Development Bank, 1980.

Weeramantry, C. *Nauru—Environmental Damage Under International Trusteeship*. Melbourne, Oxford University Press, 1992.

Weightman, B. *Agriculture in Vanuatu: A Historical Review*. Cheam, British Friends of Vanuatu, 1989.

Wenkam, R., and Baker, B. *Micronesia*. Honolulu, HI, University of Hawaii Press, 1971.

White, Geoffrey M., and Lindstrom, Lamont (Eds). *Chiefs Today—Traditional Pacific Leadership and the Postcolonial State*. Cambridge, Cambridge University Press, 1998.

Wilson, Lynn. *Speaking to Power: Gender and Politics in the Western Pacific*. New York, Routledge, 1995.

Winchester, S. *The Pacific*. London, Hutchinson, 1991.

Wittersheim, Eric. *Melanesian Elites and Modern Politics in New Caledonia and Vanuatu, National Centre for Development Studies, Discussion and Policy Papers, No.3*. Canberra, Australian University Press, 1998.

Wurm, S. A., and Hattori, S. *Language Atlas of the Pacific Area. Part 1: New Guinea Area, Oceania, Australia*. Canberra, Australian Academy of the Humanities, and Tokyo, Japan Academy, Pacific Linguistics Series C, No. 66, 1981.

THE PHILIPPINES

Physical and Social Geography

HARVEY DEMAINE

The combined land area of the 7,100 islands that constitute the Republic of the Philippines amounts to 300,000 sq km (115,831 sq miles). With the intervening seas, most of which rank as Philippines territorial waters, the country extends over a considerably larger area, from above 18°N to below 6°N, lying between the South China Sea and the Pacific Ocean.

Of its multitudinous islands, some 880 are inhabited and 462 have an area of 2.6 sq km (1 sq mile) or more. The two largest, namely Luzon in the north, covering 104,688 sq km, and Mindanao in the south, with an area of 94,630 sq km, account for 66.4% of its territory, and this figure is raised to 92.3% if the next nine largest (Samar, Negros, Palawan, Panay, Mindoro, Leyte, Cebu, Bohol and Masbate) are also included.

PHYSICAL FEATURES

Structurally, the Philippines forms part of the vast series of island arcs that fringe the East Asian mainland and also include Japan, the Ryukyus and Taiwan to the north, and extend into Sulawesi, Papua (formerly Irian Jaya) and other Indonesian territories to the south. Two main and nearly parallel lines of Tertiary folding run roughly north–south through Luzon, swing approximately north-west–south-east through the smaller islands surrounding the Sabayan, Visayan and Mindoro seas, and resume a north–south trend in Mindanao. In addition to these two, a less pronounced north-east–south-west pair extend from the central Philippines through Panay and the smaller islands of the Sulu archipelago, ultimately linking up with the similar Tertiary structures of the north-eastern tip of Borneo.

These major lines of folding largely determine the broad pattern of relief throughout the country. Over most of the islands Tertiary sediments and Tertiary-Quaternary eruptives predominate, and more than a dozen major volcanoes are still active. Nearly all the larger islands have interior mountain ranges, typically attaining heights of 1,200 m–2,400 m above sea-level, but, apart from narrow strips of coastal plain, few have any extensive lowlands. This is the country's greatest natural liability. The central plain of Luzon, which represents the only significant exception, has therefore assumed a dominant role.

CLIMATE

Because of its mountainous character and its alignment across the south-west monsoon and the north-east trade winds, the Philippines shows considerable regional variation in both the total amount and the seasonal incidence of rainfall. Thus, in general, the western side of the country receives most of its rain during the period of the south-west monsoon (late June–late September) whereas on most of the eastern side the wettest period of the year is from November to March when the influence of the north-east trades is at its greatest, though here, in contrast to the west, there is no true dry season. These differences can be seen by comparing Manila (on the west side of Luzon), which, out of an annual total of 2,100 mm, receives 1,100 mm in July–September and only 150 mm in December–April, with Surigao (in the north-east of Mindanao), which receives an annual total of 3,560 mm, 2,230 mm between the months of November and March inclusive, but with no monthly total falling below the August figure of 120 mm. In some sheltered valleys, however, totals may be as low as 1,020 mm, which, in association with mean annual sea-level temperatures that are rarely much below 26.7°C (80°F) anywhere in the country, makes farming distinctly precarious. On the other hand, a different kind of climatic hazard affects many of the more exposed parts of the country as a result of their exposure to typhoons, which are most common in the later months of the year and tend to be most severe in eastern Luzon and Samar.

NATURAL RESOURCES

The central lowlands of Luzon provide by far the best major food-producing region within the country, and although many of the smaller lowlands are also intensively cultivated, their soils are, in most cases, of only average fertility. Since the 1940s the only substantial areas of lowland offering scope for any important extension of cultivation have been in the southern island of Mindanao. Once described as the frontier of the Philippines, this island has become more densely populated in recent years, and the once extensive resources of tropical hardwoods have been disappearing.

While, as elsewhere in South-East Asia, rice forms the most important single item in the country's agricultural system, its predominance is less marked than in other parts of the region, and indeed in several of the islands, partly because of their relatively low rainfall, and partly because of the close cultural link with Latin America, maize is the leading food crop. So far as export crops are concerned, the emphasis has hitherto been mainly on coconuts, bananas, pineapples and sugar.

The Philippines has a fairly wide range of metallic mineral deposits, the most important of which are copper (with reserves mainly found on Cebu and at Marinduque), chromite and nickel. The country has been, on the other hand, sadly deficient in energy minerals. In the late 1990s more than one-half of energy requirements (mostly petroleum) were imported. The increasing costs of petroleum imports resulted in a frantic search for domestic energy supplies. Petroleum was discovered off the island of Palawan in 1977 and new reserves were located in 1990, but domestic production remained relatively insignificant. However, proven reserves of coal and lignite were estimated at 369m. metric tons, with potential reserves of 1,700m. tons. Hydroelectric power, which contributes about 8% of domestic energy requirements, and geothermal energy, which provides about 5%, are both the product of the youthful landscape and unstable geological structure of the archipelago.

POPULATION AND CULTURE

With an average annual growth rate of about 2.4% in 1995–2000, the population of the Philippines was enumerated at 76,504,077 at the May 2000 census, compared with the 68,616,536 recorded at the September 1995 census. At September 1995 the country had an average population density of 228.7 per sq km, which was nearly double the South-East Asian average and, in the region, exceeded only by that of Singapore. By May 2000 population density had increased to 255.0 per sq km. According to official estimates, the total population at mid-2004 had increased further, to 84,237,476, with an average density of 280.8 per sq km. The shortage of lowland means that much the greater part of the population is concentrated in a relatively small area, and, particularly in the lowlands of central Luzon, the resultant pressure is now a serious problem and likely to become increasingly severe, owing to the exceptionally high rate of population growth.

Despite the existence of several regional languages spoken by the lowland Filipinos, the latter, who form the great majority of the population, share a basically common culture, which is much influenced by Roman Catholicism. In recent decades considerable progress has been made in developing Tagalog, the language of central Luzon, as a national language (Filipino), although, particularly among the largely *mestizo* élite elements, English is widely used.

Other than the Christian Filipinos, the only large indigenous group comprises the Muslim Moros inhabiting the southern and south-western peripheries of the country, who form approximately 5% of the total population. There are also a number of much smaller communities of animist hill peoples, principally in the remoter parts of Luzon and Mindanao, who together form perhaps 6% of the total. The Chinese population in the Philippines is very small, accounting for only about 1% of the total.

Largely because of its long history of colonial rule and its archipelagic nature, the Philippines now has numerous and widespread small administrative and market towns. Manila is the nation's capital, with a population of 1,581,082 at the 2000 census. However, this figure is misleading, since the boundaries of the administrative unit of Metropolitan Manila encompass a large number of former towns and districts, including Quezon City (2000 population of 2,173,831), and contain a total population of almost 7m. In mid-2003, according to UN estimates, the population of Metropolitan Manila was 10,352,249. Other macro-regions of the country also have their urban focuses, however, with Davao City (1,147,116 in 2000) in south-east Mindanao, Cebu City (718,821) in the Visayas and Zamboanga City (601,794) in south-west Mindanao being the largest.

History

MICHAEL PINCHES

Based on an earlier article by IAN BROWN

HISTORICAL BACKGROUND

In March 1521 a Spanish expedition, led by the Portuguese navigator Fernão Magalhães (Ferdinand Magellan), which had sailed west from Spain in September 1519, in search of the Spice Islands, landed at Samar in the central Philippine archipelago. In April 1565 the first Spanish settlement was established in Cebu by Miguel López de Legaspi. By the early 1570s Spanish control extended over Cebu, Leyte, Panay, Mindoro and central Luzon; the capital of the new Spanish dominion was Manila, captured in May 1571. The extremely fragmented nature of indigenous political organization at that time made large-scale co-ordinated resistance impossible. Political authority was focused on the *barangay*, commonly a kinship community which contained between 30 and 100 families. These restricted communities were frequently at war with each other, but no nation-wide state organization emerged from these conflicts. Thus, the pre-colonial archipelago did not sustain the great indigenous empire, or powerful traditional kingdom, that provided a foundation for the modern nation-state in the rest of South-East Asia.

The principal aim of Spanish rule in the Philippines was religious conversion. In this, Spain exerted a profound influence over the indigenous population, although the animists of northern Luzon and, more particularly, the Muslims of Mindanao and Sulu effectively resisted Roman Catholicism; even in regions of widespread conversion, animist beliefs were by no means expelled. Spanish rule was financed primarily by the galleon trade; silver dollars and bullion were brought to Manila from Acapulco (Mexico) to be exchanged for Chinese silks, porcelain, bronzes and jade. The Spanish made little attempt to exploit the economic resources of the archipelago, and, by restriction and monopoly, prevented any other Europeans from doing so—at least until the beginning of the 19th century.

In 1834 the port of Manila was opened to unrestricted commerce, irrespective of nationality. Between 1855 and 1873 six provincial ports, including Iloilo and Cebu, were also opened, for the Spanish were now permitting other Westerners to engage in agriculture and manufacturing on the islands. From these initiatives there emerged the production and export of the crops—sugar, coconuts, Manila hemp (abaca) and tobacco—that were to become the foundation of the economy for the remaining decades of colonial rule and into the post-independence period. These economic changes had important repercussions, both socially and politically. An increase in agricultural production, from the middle of the 19th century onwards, encouraged the settlement of a vigorous Chinese entrepreneurial class in the regions. As time passed, a *mestizo*, or half-caste, élite emerged (sons of Chinese fathers and native mothers, brought up as natives and Roman Catholics). The *mestizo* derived their wealth and influence from land ownership, and frequently sent their children, known as *ilustrados* (enlightened ones), to universities in Manila and Europe. The *ilustrados* became the channels for Spanish culture and liberal thought in the Philippines. By the close of the 19th century they were also to challenge the repressive orthodoxy and exclusiveness of Spanish rule in the islands, as exposed in the writings of Dr José Rizal (1861–96). The *ilustrados* sought reform through a so-called 'Propaganda Movement', but the mass of rural and urban Filipinos, impoverished by loss of land, exploitative labouring wages and unemployment, were moving towards armed revolt against the colonial power.

The Philippine revolution erupted in late August 1896, launched by a secret society, the Katipunan (Association of Sons of the People), under the leadership of Andres Bonifacio, who had been born in the Tondo slum district of Manila. Although Rizal played no part in the revolt, he was subsequently seized by the Spanish, and tried on charges of rebellion, sedition and illicit association. In December 1896 he was executed by a firing squad. Rizal's death provoked a major upsurge in revolutionary fervour. The most important initial military successes against the Spanish were achieved in the province of Cavite, south of Manila, under the leadership of Emilio Aguinaldo, the mayor of the town of Kawit. That success ensured a shift in the leadership of the revolution, away from Bonifacio and his dispossessed followers, to Aguinaldo and the provincial landed élite, the *principalia*; divisions also emerged in the revolutionary movement at this time. In May 1897 Bonifacio was executed, having been sentenced to death by a military court that Aguinaldo had appointed. The struggle then turned against the Filipinos; and in December 1897 Aguinaldo and the *principalia* ceased hostilities, in return for exile in Hong Kong and a financial settlement. In the Philippines itself, the struggle was maintained.

In late April 1898 the USA declared war on Spain, following the sinking of the *USS Maine* in the harbour at Havana, Cuba. In May the US Asiatic Squadron sailed into Manila Bay and destroyed the Spanish fleet. Aguinaldo, with US encouragement and assistance, returned to the Philippines, and on 12 June 1898 proclaimed Philippine independence. However, from this time onwards, a rift developed between the USA and its Filipino 'allies'. The US armed forces colluded with the Spanish colonial administration to stage a mock battle for Manila, so that the capital would be surrendered solely to US forces. On 14 August 1898 the Spanish capitulated; in the following December Spain ceded the Philippines to the USA for the payment of US $20m., under the terms of the Treaty of Paris.

On 23 January 1899 the Philippine Republic was inaugurated at Malolos, and Aguinaldo was sworn in as President. Early in the following month skirmishes occurred between the republican army and US occupying forces, leading rapidly to open conflict. For a year the new Republic fiercely resisted the US forces, but by early 1900 Aguinaldo's troops had been forced into retreat, in the mountains of northern Luzon. When Aguinaldo was captured in March 1901, major resistance was at an end. In April Aguinaldo took his oath of allegiance to the USA, and urged his fellow Filipinos to do the same. However, sporadic fighting by guerrillas, still motivated by the vision of *kalayaan* (freedom) which had inspired the initial outbreak of revolution in late 1896, continued to occur during the following decade.

US RULE

Under US dominance, the Philippines became even more heavily dependent on agricultural exports, and the rural masses found no relief from rigorously exploitative conditions of tenancy. Superficially, the US political administration did appear to constitute a decisive break with the past. While the Spanish had denied Filipinos even the prospect of political advancement, the new US administration moved rapidly to make political and bureaucratic positions accessible to them. Municipal elections were instituted in 1901; provincial government elections in 1902; an elective Lower-House Legislature in 1907; and an Upper House in 1916. By 1903 Filipinos occupied 49% of bureaucratic positions, although mainly at the lower levels; by 1913 they accounted for 71% of such positions; and by 1928 virtually the whole colonial Government (from clerks to cabinet ministers) was controlled by Filipinos. Crucially, however, these appointees were drawn almost exclusively from the ranks of the *ilustrados*, who, as soon as the leadership of the revolution had been wrested from the dispossessed in early 1897, had begun to collaborate closely with the new colonial power. The USA had strongly reinforced the dominant position of the *ilustrados* within Filipino society, in the interests of effective colonial administration. It was primarily for this reason that no significant attempt was made to reduce the extremely high levels of tenancy in the major regions of export production, for land reform would have threatened to destroy the foundation of *ilustrado* wealth and influence. Even colonial education policy, which was characterized by a major expansion of educational opportunity in the islands, reinforced the position of the *ilustrados* and their US promoters: the use of English as the medium of instruction in schools, and the introduction of a US-style school curriculum, brought an influx of US cultural values into the Philippines, which, in time, dissipated the intense indigenous identity that had animated the Filipino masses during the revolution and the resistance to US occupation.

Filipino politics were dominated in the period of US administration by the Nacionalista Party, led firstly by Sergio Osmeña, and then by Manuel L. Quezon; on numerous occasions the Nacionalista ranks were divided by the fierce personal ambitions of its leadership. In public, Nacionalista leaders fiercely demanded 'immediate, complete and absolute independence', but they were fully aware that they largely owed their power and privileged position to their relationship with the USA. As the prosperity of the islands increasingly came to depend on the free entry of its agricultural exports to the US market, dependence on the colonial power was reinforced. By the 1930s, whatever their public stance, the élite regarded the prospect of independence with increasing unease.

In the event, Philippine independence was secured less by Filipino agitation than by the USA's domestic rejection of its colonial role. Important elements in US society found it difficult to reconcile the proud heritage of independence in the USA with the acquisition of empire. As the USA plunged deeply into depression at the end of the 1920s, these elements were joined by the farming and labour lobbies in demanding protection against the free import of Philippine agricultural commodities—or, in effect, Philippine independence.

Under the Tydings-McDuffie Act, signed by President Franklin Roosevelt on 24 March 1934 and subsequently accepted by the Philippine legislature, the Philippines was to achieve full independence on 4 July 1946. This was to be preceded by a 10-year transitional period of internal self-government, during which the Philippine Commonwealth would remain under US control as far as foreign relations and national defence were concerned; the President of the USA would also be empowered to veto any constitutional amendment or any legislation affecting currency, coinage, imports or exports during this period. In July 1934 a commission was elected to draft a Philippine Constitution. The Constitution was approved by the US President and ratified by the Filipino electorate in May 1935. The Philippine Commonwealth was formally inaugurated on 15 November 1935, with Manuel L. Quezon as President and Sergio Osmeña as Vice-President.

During the 1920s and 1930s abuses on the part of landowners provoked an upsurge in peasant unrest. This period saw the emergence of large peasant unions, principally in central Luzon; the founding of the Partido Komunista ng Pilipinas (PKP—the Communist Party of the Philippines) in 1930; and the establishment of the *Sakdal* (Accusation) movement, which campaigned against the inequitable distribution of property, excessive taxes, and the large landholdings of the Roman Catholic Church, and demanded complete and immediate independence. In early May 1935 the *Sakdalistas*, drawn from the discontented peasantry in the provinces around Manila, began an uprising, but it was rapidly crushed. President Quezon responded to growing agrarian unrest by introducing a programme of 'Social Justice'. This involved much constructive legislation, establishing, for example, minimum wages for industrial and agricultural labourers, and written contracts between landowners and tenants. Quezon's legislative programme, however, proved virtually impossible to enforce against landed interests (including members of the President's own circle), who dominated both local administration and national politics.

In December 1941 Japanese forces landed in the Philippines. Manila was occupied in early January 1942. The political and bureaucratic élite remained largely intact during the occupation, except for Quezon, who escaped to Australia. On 14 October 1943 the Japanese declared the Philippines an independent Republic within the 'Greater East Asia Co-Prosperity Sphere'; the new President of the Republic was José P. Laurel. Japan's occupation did not go unchallenged. Particularly important here was the *Hukbalahap*, or *Huks* (People's Army Against Japan), formed in March 1942, and drawing its forces, in particular, from the militant peasant movement that had been approaching revolt in the late 1930s. Operating in central and southern Luzon, the *Huks* continuously harried Japanese forces; but they also strongly opposed the indigenous landlord class, and those Filipinos whom they saw as collaborating with the Japanese. In October 1944 US forces, under Gen. Douglas MacArthur, landed on Leyte island; US troops entered Manila in early February 1945 and operations continued for several months thereafter. In many areas, the US military advance was prepared by guerrilla groups, under the leadership of former officers of the US Armed Forces in the Far East and the *Huks*.

THE PHILIPPINE REPUBLIC

With a major part of the country's physical capital destroyed by the occupation and liberation, with the export economy in urgent need of rehabilitation, and with Filipino political life riven with charges and denials of wartime collaboration with the Japanese, the Philippines moved towards the attainment of full political independence on 4 July 1946. In the mean time, interim Commonwealth rule was re-established on 27 February 1945, under President Sergio Osmeña, who had succeeded Quezon as Head of the Government-in-exile on the latter's death in the USA in August 1944. In the presidential election of April 1946, Osmeña was narrowly defeated by Manuel Roxas (from the liberal wing of the Nacionalista Party), who had been a cabinet minister during the occupation, and a director of the wartime rice procurement agency which supplied the Japanese army, but who was now strongly supported by Gen. MacArthur. On his assumption of the presidency, Roxas's principal concern was to secure the Philippines' economic relationship with the USA. In March 1947 the USA and the new Republic concluded an agreement on military bases, whereby the USA received a 99-year lease on 23 bases in the islands. The lease was shortened to 25 years in subsequent negotiations. The principal military bases were Subic Bay Naval Base and Clark Air Base, both situated on the main island of Luzon. This agreement gave the US authorities extensive legal jurisdiction over Filipinos living within the bases, and, for this reason, it soon gave rise to major controversy.

Despite their resistance to the Japanese, the *Huks* were regarded with suspicion: the politicians who had collaborated with the Japanese during the war were aware that the *Huks'* open resistance could further expose them; landowners knew that, during the occupation, the *Huks* had been as ruthless with their class as with the Japanese; and the US administration, as well as the Philippine government élite, recognized the strong communist inspiration of the *Huks*. As the war came to a close, these forces combined in a strenuous attempt to suppress the movement. *Huk* units were disarmed, usually by force. In mid-1945 a group of *Huks*, on their way to Pampanga, were seized and executed by government forces, with the knowledge of the US military police in the area; a major part of the leadership was arrested, but subsequently released. In the immediate post-war months the *Huks*, working principally through the opposition Democratic Alliance (DA), sought entry into formal politics. In the April 1946 presidential and congressional elections, the *Huks* supported President Osmeña, while seven DA candidates (including the *Huk* leader, Luis Taruc) were elected to the legislature by the central Luzon peasantry. In mid-1946 Taruc offered to negotiate a cease-fire agreement with the Government of President Roxas, but the negotiations broke down over the demand that *Huks* surrender their weapons. Roxas then engineered the exclusion of successful DA candidates from the legislature, while Philippine army and police units launched a fierce and brutal campaign against the *Huks*. On 6 March 1948 Roxas declared the *Hukbalahap* to be an illegal organization.

Roxas died suddenly in April 1948. He was succeeded by his Vice-President, Elpidio Quirino. The new President immediately sought a truce with the *Huks*: indeed, agreement for an amnesty and surrender of arms was reached, and Luis Taruc, the leader of the *Hukbalahap*, was finally able to take his seat in the Philippine Congress. In mid-August 1948, however, Taruc suddenly left Manila, accusing the authorities of duplicity, primarily over the surrender of arms. The *Huks* were now in open revolt. Taruc announced that he had become a member of the PKP and in April 1949 he publicly declared, for the first time, that the *Hukbalahap* sought the overthrow of the Government.

Quirino sought re-election in 1949 as the Liberal candidate for the presidency, against the nominee of the Nacionalista Party, José Laurel, the wartime collaboration leader. After an election marred by fraud, violence and intimidation, Quirino was returned to office. He faced a deepening economic crisis and an intensified campaign of insurgency by the *Huks*. Many incompetent and corrupt army officers were dismissed, and a new Secretary of National Defense, Ramon Magsaysay, was appointed. Magsaysay pursued a dual policy. The excesses of the armed forces (which had done much to secure support for the *Huks* in rural districts) were sharply curtailed, and a programme for the rehabilitation of rebels who surrendered was instituted. In October 1950 the central committee of the *Hukbalahap* was seized in a number of army and police raids across Manila. The removal of a major part of the leadership, coupled with the Government's more calculated military strategy and more constructive response to rural grievances, which had been at the core of the revolt, including the first stage of a tenancy reform programme in Luzon, resulted in the elimination of the *Huks* as a political entity. On 17 May 1954 Luis Taruc, the *Huks'* leader, surrendered to the authorities. The urgency of demands for genuine agrarian reform waned and little further progress was made.

The disintegration of the *Hukbalahap* secured Magsaysay's prominence in Quirino's Cabinet. When it became clear that Quirino would seek re-election in 1953, Magsaysay left the Liberal Party and became the presidential candidate of the Nacionalistas. After a comparatively free and non-violent election campaign, Magsaysay won a convincing majority. During his presidency, Magsaysay reinforced the Philippines' relationship with the USA; in 1954 the Philippines became a founder-member of the US-inspired South East Asia Treaty Organization (SEATO), and in 1956 US parity with Filipinos in the exploitation of the Philippines' natural resources was extended to all economic activities until 1974, under the Laurel-Langley Agreement. In domestic policy, Magsaysay strongly emphasized his desire to improve the lot of the *tao* (common man), by amending land-tenure arrangements, providing agricultural credit and technical assistance, instituting community development projects and distributing some public lands to the landless. When Magsaysay died in an air accident in March 1957 he was succeeded by his Vice-President, Carlos P. Garcia, who won the presidential election held later that year. Garcia's presidency was initially characterized by austerity measures, and by a policy of economic nationalism (the 'Filipino First' policy) that was intended to lessen the Philippines' economic dependence on the USA, in particular. As Garcia's term of office came to a close, however, his Government foundered on rampant corruption, graft and financial scandal. In the 1961 presidential election Garcia was defeated by the Liberal Party nominee, Diosdado Macapagal. The new President instituted far-reaching economic reforms, encouraged by the USA and the IMF. Macapagal's Government established a Philippine claim to Sabah, on the eve of its inclusion in the new Federation of Malaysia in August 1963. This act caused the Philippine Government to sever diplomatic relations with Malaysia in September; relations were not resumed until June 1966.

THE MARCOS YEARS

In the presidential election of November 1965, Macapagal was defeated by Ferdinand E. Marcos, the Senate President, who had earlier left the Liberal Party to become the Nacionalista candidate. During his first term there was an increase in government expenditure on infrastructure, financed by foreign loans. By the end of the 1960s the principal and interest payments fell due and the country faced a balance-of-payments crisis, which was aggravated by massive election spending in 1969. Marcos was re-elected by a large majority. The late 1960s and early 1970s saw the emergence of student activism, demanding social justice and national sovereignty, and an increasing challenge to central authority from the Moro National Liberation Front (MNLF), a Muslim secessionist movement based on the southern islands of Mindanao and Sulu. Dissension within the communists' ranks led in 1968 to a split in the PKP. A Maoist breakaway group, led by Jose Maria Sison, formed the Communist Party of the Philippines (CPP), whose military wing, the New People's Army (NPA), frequently clashed with government military forces. An umbrella organization, the National Democratic Front (NDF), which included the CPP, was formed in the early 1970s. The PKP, meanwhile, condemned the use of violence and continued to operate within the parliamentary framework.

Acts of political intimidation and violence, frequently perpetrated by private armies under the control of political figures, increased alarmingly after 1969. The number and size of student demonstrations escalated and they were violently dispersed, leaving scores of students dead. As the political and social fabric of the Republic disintegrated, a constitutional convention began the task of preparing a new constitution. Reports of corruption by Marcos and his wife, Imelda, were widespread and included allegations of bribery of members of the convention to extend Marcos's term of office, which, by the terms of the Constitution, had to end in 1973. In August 1971 an opposition rally was bombed; many died and a number of senatorial candidates were seriously injured. Marcos suspended the writ of habeas corpus. In September 1972, claiming a conspiracy between right-wing oligarchs and Maoist revolutionaries, Marcos declared martial law. Leading left- and right-wing opponents to Marcos, including both politicians and journalists, were arrested and detained, among them a former senator, Benigno S. Aquino, Jr. The new Constitution, which was drafted as martial law was declared, received ratification in January 1973 from Citizens' Assemblies throughout the Philippines. The new Constitution conferred the offices of both President and Prime Minister on Marcos, who was also empowered to determine when the interim National Assembly would be convened. Martial law continued under the new Constitution. The first elections to the interim Assembly were held in April 1978; local elections followed, in January 1980. On both occasions, the ruling Kilusan Bagong Lipunan (KBL—New Society Movement—formed in 1978 by Marcos and other former members of the Nacionalista Party) was successful, although there were claims by the opposition of widespread electoral malpractice. Marcos followed his electoral success with a period of slight political relaxation; Benigno Aquino was released in May 1980, enabling him to undergo medical treatment in the USA. Martial law was lifted in January 1981, although Marcos's powers remained largely intact. In the following April a national plebiscite approved constitutional amendments which, in effect, replaced the parliamentary form of regime with a mixed presidential-parliamentary Government, and César Virata was appointed Prime Minister.

In the mid-1970s the Muslim secessionist movement in the southern Philippines provoked a military confrontation between government troops and the MNLF. Although the secessionists were initially successful, the movement was neutralized in the late 1970s by the Tripoli Agreement, sponsored by Col Muammar al-Qaddafi of Libya. Marcos defeated the MNLF by ostensibly granting regional autonomy to the Muslim areas and by allegedly paying huge sums of money to some Moro leaders.

In August 1983 Benigno Aquino returned to the Philippines from exile in the USA. As he came down the steps of his aircraft, he was assassinated; the alleged assassin, Rolando Galman, was immediately shot by military guards. Benigno Aquino's death precipitated a major political crisis. Large-scale street demonstrations and labour strikes followed, demanding the resignation of Marcos. This pressure led to the appointment of an independent commission of inquiry into the assassination. In its report, published in October 1984, the commission concluded that Galman was not the assassin, and that Benigno Aquino had been the victim of a military conspiracy. It indicted 25 military officers (including the Chief of Staff of the Philippine Armed Forces, Gen. Fabian Ver), and a trial followed in February 1985.

It was widely believed that the President's immediate circle of associates, if not the President himself and his wife, Imelda, were implicated in Benigno Aquino's death. Rumours of the vast personal wealth that the Marcos family and their associates had accrued through questionable practices, and of the gross extravagance of Imelda Marcos, now began to circulate more openly. Human rights violations, including torture and killings, were documented by both the human rights organization Amnesty International and the US Congress. The communists and the legal opposition co-operated in daily demonstrations against the beleaguered regime. At the same time, the massively indebted national economy was on the verge of collapse. Domestic political instability caused foreign creditors to suspend the short-term loans that had sustained the economy from the late 1970s onwards. New investment was curtailed, and capital fled the country. The Government of the USA brought pressure to bear on Marcos, urging him to submit his regime to electoral test.

On 3 November 1985 Marcos announced that the next presidential election, scheduled for May 1987, would take place in early 1986. At first Salvador Laurel, the President of the United Nationalist Democratic Organization (UNIDO—an alliance of opposition groups formed in 1982), was the only other declared contender. He was soon challenged by Eva Kalaw, one of the leaders of the Liberal Party, and also by Jovito Salonga, a former senator. Corazon Aquino (the widow of Benigno Aquino, Jr) was chosen by some cause-orientated groups and also by Lakas ng Bayan (the People's Power Movement, founded by Benigno Aquino in 1978) and the Pilipino Democratic Party (PDP), which had formed an alliance (PDP-Laban) in 1983, and which formed part of UNIDO. The fragmentation of the opposition groups continued until the end of November 1985, when efforts by the Roman Catholic primate, Cardinal Jaime Sin (the Archbishop of Manila), and the US embassy in Manila, to encourage unity among opposition groups, resulted in a compromise: Laurel agreed to become the vice-presidential candidate, while Corazon Aquino consented to become the presidential candidate of UNIDO.

During the election campaign Gen. Ver and other military escorts of Benigno Aquino were acquitted of murder, in a trial that appeared to have been manipulated by Marcos. Confusion followed the election, which was held on 7 February 1986, when the two monitoring bodies, the National Movement for Free Elections (Namfrel), financed by the US-based National Endowment for Democracy, and the Government's Commission on Elections (Comelec), each alleged that the other had manipulated the enumeration of votes. Namfrel, basing its result on 70% of total votes, declared that Corazon Aquino had won a majority, while Comelec recorded Marcos in the lead from the outset, and completed the count, whereupon it declared Marcos's victory. A few days later, Marcos was recognized by the *Batasang Pambansa* (National Assembly) as the winner of the elections.

On 14 February 1986 a pastoral statement, issued by the Roman Catholic Church, declared that the election had been 'a fraud unparalleled in history'. Two days later Aquino announced a campaign of non-violent civil disobedience in protest at the outcome of the election. On the evening of 21 February Juan Ponce Enrile, the Minister of National Defense, retreated to his office at Camp Aguinaldo, where he decided to resist his arrest by Marcos, following news that a coup plot by members of the Rebolusyonaryong Alyansang Makabayan (RAM—Nationalist Revolutionary Alliance—also known as the Reform the Armed Forces Movement), a group of right-wing reformist military officers, had been discovered. He was joined at Camp Aguinaldo by Lt-Gen. (later Gen.) Fidel Ramos, the Deputy Chief of Staff of the armed forces, and on 22 February both men announced that they had withdrawn their support from Marcos. Following an appeal by Cardinal Sin for the populace to come to the aid of the two men, thousands of Filipinos flocked into the streets, effectively blocking the way of Marcos's troops. On 24 February the Government of the USA declared that it supported Enrile and Ramos. What had begun as a military revolt against Marcos had developed into a 'snap revolution', in which the strength of civilian support rendered Marcos's troops helpless. Enrile later declared that it was only because of the strength of civilian support for Corazon Aquino, manifested in 'people power', that he and Ramos had relinquished their own intention to form a government, including some civilians, and decided to ally with her. In spite of these events, Marcos insisted on a formal inauguration on 25 February, before being flown to safety in Hawaii, together with his family and associates, in two aircraft supplied by the US air force.

THE AQUINO GOVERNMENT

Corazon Aquino was inaugurated as President of the Philippines in a separate ceremony, also on 25 February 1986. Among those appointed to her Cabinet were Enrile, who retained the position of Minister of National Defense, and Ramos, who became the new Chief of Staff of the armed forces, replacing Ver. Continuity in economic policy between the Marcos administration and the new Aquino Government was established through Jose Fernandez, who was reappointed Governor of the Central Bank, and through Jaime Ongpin, who became the new Minister of Finance. Two new commissions were created: the Presidential Commission on Good Government, which was given the onerous task of recovering Marcos's ill-gotten wealth, and the Presidential Commission on Human Rights, which was formed to investigate allegations of violations of human rights under the Marcos regime, and to seek redress for the victims.

More than 500 political detainees were released within days of Aquino's inauguration (the armed forces and the US Government were alarmed that communist leaders, including Jose Maria Sison, were released with other political prisoners); the right of habeas corpus was restored in early March 1986; censorship of the media was abolished; and elderly pro-Marcos military officers were retired. However, there was some concern that President Aquino's powers were virtually dictatorial: in March she abolished the legislature, under an interim Constitution, and, at the same time, she demanded the resignation of several Supreme Court judges, and dismissed local officials. She also abolished the position of Prime Minister, within a few days of conferring it on Laurel. Disagreement emerged between supporters of UNIDO and PDP-Laban, especially over those appointed to local government positions. At the end of May a 50-member constitutional commission was convened. A new Constitution was presented to the Government in early September and approved by Aquino's Cabinet on 12 October.

By mid-1986 the new Government still awaited the results of its negotiations with creditor institutions, and no major economic initiative had been implemented. There was also growing doubt concerning the President's ability to produce a cohesive policy from the widely varying opinions represented within her administration. Civil unrest was a serious problem and street rioting by pro-Marcos supporters frequently required armed intervention by the authorities. Marcos himself maintained daily contact with his Manila supporters, by telephone from his base in Hawaii, where he directed anti-Aquino operations. There were continuous rumours of impending political and military coups.

In June 1986 Aquino announced the beginning of formal negotiations with the outlawed NDF, which would aim to establish agreement for a cease-fire in the 17-year campaign by the NPA. However, the insurgents' apparent reluctance to observe a temporary cease-fire caused Enrile and Ramos to resume the government forces' anti-communist offensive while Aquino was out of the country. On her return, she came under strong pressure from members of the Cabinet and the armed forces, notably Enrile and Ramos, to adopt a tougher stance. In November a 60-day cease-fire was agreed, to be effective from 10 December. The pessimistic attitude of both the armed forces and the NDF, however, doomed the peace initiatives to failure and the cease-fire was broken.

In addition, the new Government sought to end the 16-year secessionist war being conducted by the MNLF in the southern Philippines. In August 1986 Agapito Aquino, the President's brother-in-law, and Nur Misuari, the Chairman of the MNLF, met in Saudi Arabia and, as a result, the Aquino Government agreed to grant autonomy to four mainly Muslim-populated provinces in Mindanao. A month later, Misuari returned to the Philippines and agreed to a truce, although there was doubt whether this would be observed by some factions within the MNLF. The initiatives were attacked by another Muslim secessionist grouping, the Moro Islamic Liberation Front (MILF), which was not included in the negotiations but claimed a wider membership than the MNLF. No agreement was achieved.

In July 1986, while Aquino was on a visit to Mindanao, Arturo Tolentino, Marcos's vice-presidential candidate in the February election, proclaimed himself acting president from the Manila Hotel, which had been seized by 300 rebel troops. He claimed to be acting on the instructions of Marcos and appointed a cabinet which included Enrile, the Minister of National Defense in Aquino's Government. Enrile immediately dissociated himself from the rebels, and ordered troops loyal to Aquino to surround the hotel; Tolentino and his supporters surrendered. Over the next two months, however, disagreement within the new Government intensified. By mid-October Enrile and Laurel, with some support from Ramos, were insisting upon the holding of a new presidential election and the adoption of a more forceful policy against the insurgents. In early November, following his appearance at an anti-Aquino rally, there were rumours of an impending coup by Enrile and his army supporters. Tensions increased with the assassination of Rolando Olalia, a trade union leader and prominent opponent of Enrile. Subsequent investigations implicated officers close to Enrile. On the night of 22–23 November there was an ill-co-ordinated uprising by officers at several military camps. Aquino reacted swiftly, dismissing Enrile as Minister of National Defense and demanding the resignation of the rest of the Cabinet.

The new Constitution, approved by Aquino's Cabinet in October 1986, was subject to a national plebiscite on 2 February 1987. In essence, it returned the Philippines to the constitutional form that had operated (except for the war years) from 1935 to 1973, with a US-style bicameral legislature and an executive presidency. The House of Representatives was to have 200 directly elected members and up to 50 presidential 'appointees' chosen from party lists, interest groups (including labour and women's organizations) and tribal minorities. The members of the 24-seat Senate were to be directly elected in national elections. The Constitution provided for an extension of the term of the Presidential Commission on Good Government, which aimed to recover large quantities of Marcos's hidden wealth, and formally recognized 'People Power' organizations. Under the new Constitution, Aquino and her Vice-President, Salvador Laurel, were given a mandate to govern until 30 June 1992. Subsequent presidents would serve terms of six years, and would not be eligible for re-election. The Constitution did, however, empower the President to declare martial law in times of national emergency. The Constitution also stipulated that there should be no installation of nuclear weapons in the Philippines, thus directly threatening the maintenance and future of the US military bases.

The weeks immediately prior to the holding of the constitutional referendum were unusually tense. On 22 January 1987 some 10,000–15,000 demonstrators, gathered at the Mendiola Bridge in Manila in support of demands for accelerated land reforms, were attacked by government security forces, and at least 20 people were killed. The 'Mendiola Massacre' seriously damaged the international reputation of the Aquino Government and also reduced domestic support for the President. The NDF immediately withdrew from peace talks with the Government. Civil unrest intensified during the following days, and on 27 January troops belonging to secret organizations within the armed forces attacked military and civilian targets in Manila. Troops loyal to Aquino quashed the insurrection within hours, although a complete surrender by the rebels did not take place until four days had elapsed. On 28 January US officials thwarted an attempt by Marcos to return to Manila, from Hawaii, in order to take advantage of the disorder resulting from the attempted coup. Thirteen officers and 359 troops were subsequently detained by Ramos, pending their trial by court martial.

On 2 February 1987 Aquino's position was considerably strengthened, both in her negotiations with the communists and in her relationship with the armed forces, when the new Constitution was approved in a national referendum by 76% of voters. In the same month all members of the armed forces swore an oath of allegiance to the new Constitution; this was followed by an order disbanding military fraternities, such as the RAM, because they 'encouraged divisiveness'. Congressional elections were announced for 11 May. A Grand Alliance for Democracy (GAD) was formed by some anti-Government and anti-communist opposition groups, including the Nacionalista Party and the KBL, to contest the elections. Left-of-centre groups also formed a coalition, the Alliance for New Politics; its programme included the removal of US bases from the Philippines, as well as more radical land reform. The election resulted in a victory for President Aquino's Lakas ng Bayan alliance, which secured 180 of the 200 elective seats in the House of Representatives and 22 of the 24 senatorial seats.

In early 1987 it had become evident that the Government's armed forces lacked the ability to fight both the MNLF and the NPA at the same time. Talks between the Government and MNLF leaders in late 1986, under the auspices of the Organization of the Islamic Conference (OIC), resulted in an undertaking by MNLF leaders, in January 1987, to relinquish demands for complete independence in Mindanao and to accept autonomy. In western Mindanao, however, fierce fighting ensued, instigated by members of the MILF. Although the new Constitution provided for the eventual granting of autonomy to Muslim provinces in Mindanao, MNLF leaders claimed that this would not be sufficient to meet their demands, and in March the MILF reiterated that it would not recognize any agreement between the Government and the MNLF concerning the future of Mindanao. Talks continued, if sporadically, throughout April and May, when the Government announced that MNLF leaders were willing to form a joint commission to draft an autonomy settlement for Mindanao; following a deadlock in talks regarding autonomy provisions, however, the Government announced, in August, that it would no longer consult MNLF leaders.

In April 1987 government forces overcame an abortive coup by rebel soldiers. In July Marcos was forbidden to leave Hawaii by US officials, following the discovery of a pro-Marcos coup plot in Manila; and in the same month the Philippine Government initiated legal proceedings against Marcos, his family and associates (including Enrile and Jose Fernandez, the Governor of the Central Bank) in a Manila court, claiming damages of US $22,500m. for the sufferings of the Filipino people as a result of the Marcos Government's alleged corruption. Following the assassination of Jaime Ferrer, the Secretary of Local Government, by the NPA in August, there was an outbreak of strikes and violence in Manila. On 28 August Ramos and troops loyal to President Aquino averted a fifth attempted coup, when rebel officers, led by Col Gregorio Honasan (a former leader of the RAM and closely associated with Enrile), occupied the army headquarters. In subsequent fighting 53 people were killed. On the next day Honasan and his supporters fled, successfully evading capture until December.

In September 1987 Salvador Laurel resigned as the Secretary (formerly Minister) of Foreign Affairs, owing to differences of opinion regarding government policy in combating insurgency: he was replaced by Senator Raul Manglapus; however, he retained the post of Vice-President. In October Laurel announced that he and Enrile had agreed to form a new anti-Aquino alliance. During late 1987 the Government was accused of both 'leftism' and 'rightism' by its various opponents, and its reputation was damaged by allegations of abuses of human rights, owing to mass detentions and the creation (with government approval) of armed vigilante groups to combat insurgency in urban areas.

In January 1988 Ramos was appointed Secretary of National Defense. The government campaign against communist insurgents was immediately intensified, although the President continued to resist demands from the armed forces that a state of emergency be declared. Among numerous CPP officials captured by the army were its General Secretary, Rafael Baylosis, and the Commander-in-Chief of the NPA, Romulo Kintanar (who escaped from custody in November 1988). Although there were almost 2,000 deaths in insurgency-related violence during the first six months of 1988, Aquino was criticized by the military for not dealing sufficiently harshly with rebel elements.

The politically sensitive issue of US military bases in the Philippines came to the fore in 1988. Under an agreement signed in 1983 by the US and Philippine Governments, the continued use of six military bases, including Clark Air Base and Subic Bay Naval Base (both on Luzon island), had been permitted until 1989. In March 1986 Aquino had confirmed that the agreement would be respected, although the 1987 Constitution stated that foreign military bases would not be allowed in the country after 1991, unless under the provisions of a treaty approved by the Senate and ratified by voters in a referendum. Despite the strong nationalist feelings aroused by this issue, the bases had proved a useful source of employment and government revenue. Negotiations for a further two-year extension to 1991 were opened in April 1988 and were eventually concluded in October. Under the new agreement, the Philippine Government was to receive annual economic and military aid totalling US $481m., representing almost 2.5 times the previous level of receipts. The USA also secured the right for its military aircraft and warships to use the Philippines' facilities without a contingent obligation to declare the presence of nuclear weaponry.

In June 1988 the Government introduced a Comprehensive Agrarian Reform Programme (CARP), providing for an extensive redistribution of agricultural lands, over a 10-year period, to untenured farmers and farm workers. However, the extended timetable, together with numerous exemptions and complicated provisions for legal appeals, seriously impeded its implementation. These provisions were included in the programme as a result of the numerical domination of Congress, particularly the House of Representatives, by landowners. Also in June the PDP and Lakas ng Bansa formed a new pro-Aquino alliance, the Laban ng Demokratikong Pilipino (LDP).

Vice-President Laurel reasserted his opposition to Aquino's coalition in August 1988 by demanding the President's resignation and announcing his leadership of a broadly based opposition front, the Union for National Action (UNA). Laurel also initiated a campaign against alleged corruption in the Aquino Government and among members of the President's family. In July 1988 the Presidential Commission on Good Government, which had been established to recover funds illegally acquired by Marcos, came under suspicion itself. The Head of the Commission was obliged to resign and a large-scale reorganization was ordered. In May 1989 Marcos's Nacionalista Party was revived, with Laurel elected as President and Enrile appointed Secretary-General.

In February 1989 the NDF offered to begin peace talks with the Government, if Aquino agreed not to renew the lease on US bases after 1991. After the Government's refusal of this offer, attacks were made by the NPA on US personnel and facilities. During the year the Government responded to accusations of human rights abuses by exposing atrocities allegedly committed by the NPA. In June revelations of irregularities in the administration of the CARP gave further embarrassment to the Government. Aquino's ensuing attempts to appoint as Secretary for Agrarian Reform Miriam Defensor Santiago (who had previously been in charge of the Commission on Immigration and Deportation and who had a reputation for combating corruption) were repeatedly rejected by members of the Commission on Appointments, influenced by landowning interests.

In September 1989 the Philippine Government began the first of 35 planned civil suits against former President Marcos, in his absence, with the aim of regaining embezzled funds. At the end of the month, however, Marcos died in Hawaii. The Supreme Court prohibited (in the interests of national security) the return of Marcos's body for burial in the Philippines, as demanded by his supporters. (The trial of Imelda Marcos, on charges of fraud and of illegal transfer of stolen funds into the USA, began in New York in April 1990: she was acquitted of all charges in July.)

In November 1989 a plebiscite was held in 13 provinces and nine cities in Mindanao, on proposed legislation that envisaged the autonomy of these provinces and cities, with direct elections to a unicameral legislature in each province; this contrasted with the MNLF's demand for autonomy in 23 provinces, to be granted without a referendum. The MNLF appealed, with considerable success, to the Muslim population (about 28% of the inhabitants of the region) to abstain from participating in the referendum. The Christian inhabitants (about 66% of the regional population) were largely opposed to autonomy. Four provinces (Lanao del Sur, Maguindanao, Tawi-Tawi and Sulu) voted in favour of the government proposal and formed the Autonomous Region of Muslim Mindanao (ARMM).

In December 1989 an abortive coup was staged by members of two élite military units, the Marines and the Scout Rangers, in collusion with the illicit RAM and officers loyal to Marcos. Rebel soldiers captured an armed forces base and the air-force headquarters, which enabled them to launch air attacks on government strongholds and on the presidential palace. The progress of the coup prompted Aquino to request US air support to deter further aerial attacks by the rebels. This aim was achieved by the mere presence in the air of units of the US Air Force, which did not fire on rebel troops. Most of the rebel strongholds were captured within 24 hours. The coup was thought to involve many officers who had graduated with Honasan. Aquino subsequently accused Laurel and Enrile (who were both included in an eight-member provisional junta named by the rebels) of involvement in the coup attempt. Suspicion also fell on her estranged cousin, Eduardo Cojuangco, who had fled to the USA with Marcos in 1986 and returned secretly from exile shortly before the coup.

In February 1990 Enrile was arrested, on charges of 'rebellion complexed with murder'. He was also charged with harbouring Honasan, prior to the coup attempt. In March Rodolfo Aguinaldo, who had been suspended as Governor of Cagayan in January for supporting the attempted coup, launched an unsuccessful rebellion, following an attempt to arrest him. Arrests of those implicated in the attempted coup continued. In April Lt-Commdr Bilbastro Bibit, a suspected leader of the coup, was freed from prison by masked rebel soldiers. The soldiers claimed to be members of a military movement called the Young Officers' Union (YOU), which was alleged to have played an important role in the December 1989 coup attempt. In June the Supreme Court ordered the charge of 'rebellion complexed with murder' against Enrile and 22 other alleged rebels to be amended to 'simple rebellion' or 'illegal possession of firearms'. The court ruled that the original charge had been removed by Aquino from the statute book in 1986 by a presidential decree. As a result of the Supreme Court decision, Aguinaldo surrendered and was formally charged.

In May 1990 negotiations were resumed with the US Government on the future of US military bases in the Philippines after the expiry of the current agreement in 1991. There were violent confrontations between demonstrators, opposed to the retention of the bases, and police. In November 1990 it was announced that all US fighter aircraft and more than 1,800 military personnel would be withdrawn from the Philippines by the end of 1991. It was also announced that the two sides were to negotiate a new agreement to replace the existing security treaty, drafted in 1947.

In September 1990, following a trial lasting more than three years, a special court convicted 16 members of the armed forces of the murder of Benigno Aquino and Rolando Galman, the alleged assassin, and acquitted 20 defendants. It was recognized, however, that those responsible for planning the assassination had not been brought to justice.

In early October 1990 Col Alexander Noble, a former officer in Aquino's personal security corps, and allegedly an associate of Honasan, led a revolt on the island of Mindanao. The insurrection, which was launched in Cagayan de Oro and Butuan with 400 men, was ostensibly aimed at achieving independence for Mindanao. The Government swiftly destroyed the rebels' headquarters and communication building in an air strike, and Noble subsequently surrendered. However, the rebellion was widely assumed to be a further attempt to destabilize Aquino's

Government, by forcing the transfer of troops from Manila to Mindanao.

The NPA killed 563 members of the armed forces between January and April 1991 and maintained an active presence in 55 of the 73 provinces. The fighting was most intense in the far north of the country, although clashes had occurred on the southern island of Mindanao. Following the eruption in June of the 1,780-m volcano Mt Pinatubo, situated on the boundary of the provinces of Tarlac, Zambales and Pampanga, the NDF offered to hold discussions with the Government, proposing a cease-fire in areas damaged by the volcano. Aquino rejected the offer, claiming that the NDF could not control the actions of certain members of the NPA. In July the armed forces captured the NPA's most important base, Camp Venus in Sagada, Mountain Province, representing a major success in the struggle against the communist insurgency. In early August Romulo Kintanar, the Commander-in-Chief of the NPA, became the 10th communist leader to be captured in two weeks.

In late August 1991, following negotiations extending over 14 months, the Philippines and the USA provisionally agreed a military bases treaty providing for a new 10-year lease on Subic Bay Naval Base, with generous US compensation for the Philippines. Following the agreement, Aquino, the armed forces and many business groups campaigned to gain the constitutionally necessary support of a two-thirds' majority in the Senate. In September the NDF pledged an immediate cease-fire, pending a commitment by the Senate to reject the treaty. This commitment was subsequently given and the unilateral cease-fire duly implemented. Despite widespread popular support for the US installations, the Senate rejected the treaty by 12 votes to 11. The Government, however, rescinded the formal notice on termination of the previous lease, which had been served in May 1990, effectively extending the lease for another year.

In September 1991 Lt-Gen. Lisandro Abadio, the Chief of Staff of the armed forces since April, suspended all operations against Honasan and other military rebels in a concerted attempt to achieve national reconciliation. Abadio met Honasan twice in August and released 68 rebel soldiers as a gesture of goodwill. The Government was also attempting to negotiate an end to the communist insurgency.

At the end of July 1991 Aquino announced that Imelda Marcos and her family would be permitted to return to the Philippines to stand trial on charges of fraud and tax evasion. The decision to lift the travel ban on Imelda Marcos was prompted by a ruling of the Swiss Supreme Court in December 1990, which stated that the transfer of funds held in Swiss bank accounts to the Philippines was conditional on Imelda Marcos being brought to trial for fraud in a Philippine court before December 1991. In October Aquino relaxed the ban on the return of Marcos's remains, but stipulated that the burial should take place in his home province of Ilocos Norte, rather than in Manila, with no state or military honours. (Marcos's funeral finally took place in July 1993.) In November 1991 Imelda Marcos returned to the Philippines, and in December she pleaded 'not guilty' to charges of tax evasion. During the ensuing months she faced more than 80 civil and criminal charges and, in July 1993, she was convicted of corruption and sentenced to 18 years' imprisonment (although she remained at liberty pending an appeal).

During 1991 there was much political manoeuvring with a view to the presidential election, to be held in May 1992. In April 1991 Ramos officially joined the ruling LDP to compete with the Speaker of the House of Representatives, Ramon Mitra, for the party nomination. Although Ramos's candidacy elicited much support from the commercial sector, it was Mitra who was selected as the LDP's presidential candidate in November. Ramos then resigned from the LDP and formed a new party, the Partido Lakas Tao (People Power Party—PPP).

In January 1992 President Aquino endorsed the presidential candidacy of Ramos, thus ending speculation that (despite repeated denials) she might seek re-election. In the same month Ramos discarded the PPP and registered a new party called EDSA-LDP, with the support of 25 former LDP members of Congress. (EDSA was the popular acronym for Epifanio de los Santos Avenue, the main site of the uprising of February 1986.) The party subsequently altered its title to Lakas ng EDSA (Power of EDSA), and formed an electoral alliance with Raul Manglapus's National Union of Christian Democrats (NUCD) and the Partido Demokratiko Sosyalista ng Pilipinas (Philippine Democratic Socialist Party). The grouping was subsequently known as Lakas-NUCD.

In February 1992 Comelec authorized the presidential candidacies of eight of a total of 78 nominees. They were: Eduardo Cojuangco; Joseph Estrada (a senator and former film actor); Salvador Laurel; Imelda Marcos (whose candidacy was supported by what remained of the KBL); Ramon Mitra (who received the endorsement of the Archbishop of Manila, Cardinal Sin); Fidel Ramos; Jovito Salonga; and Miriam Defensor Santiago (who was supported by the newly formed People's Reform Party). Enrile had withdrawn his candidacy earlier in the month, giving the support of his faction of the Nacionalista Party to Mitra. In March Estrada withdrew his presidential candidacy in order to compete for the vice-presidency, supporting Cojuangco.

On 11 May 1992 elections took place to select the President, Vice-President, 24 senators, 200 members of the House of Representatives and 17,014 local government councillors. Although more than 100 people were killed in election-related violence, the contest was regarded as relatively orderly, in comparison with previous occasions. Accusations of widespread electoral malpractice (voiced, in particular, by Santiago and Cojuangco) appeared to be largely unsubstantiated. Because of the protracted vote-counting procedure, the result of the presidential election was not proclaimed by Congress until 22 June. The victor was Ramos, with Estrada as the successful vice-presidential candidate. As a result of the country's 'first past the post' electoral system, Ramos had won with only 23.6% of the votes. His closest rivals were Santiago (with 19.7%) and Cojuangco (18.2%). The success of Ramos and the high degree of support for Santiago (whose electoral campaign had emphasized the need for eradicating corruption) were regarded as a rejection of traditional party politics, since neither candidate was supported by a large-scale party organization. In the legislative elections, however, the LDP (the only party that had local bases in every province) won 16 of the 24 seats in the Senate, and 89 of the 200 elective seats in the House of Representatives.

THE RAMOS GOVERNMENT

Ramos's ascent to the presidency was supported by three important political groups: retired yet still powerful military officers; business executives and technocrats, who were identified as economic and political reformers, some of whom had served under the Aquino Government; and traditional politicians from Lakas-NUCD. The principal appointees to Ramos's Cabinet, established after his inauguration on 30 June 1992, consequently came from these groups. Ramos declared that his administration's priorities were to restore political and civil stability through national reconciliation and an amnesty programme; to deregulate the economy and encourage foreign investment by dismantling monopolies; and to curb criminality and corruption.

In July 1992 Lakas-NUCD gained leadership in Congress by expanding its membership through defections from the LDP and other parties and by forging a 'Rainbow Coalition' with the Liberal Party, Cojuango's Nationalist People's Coalition (NPC) and the KBL to form a new majority. The Secretary-General of Lakas-NUCD, Jose de Venecia, a close ally of Ramos, assumed the post of Speaker of the House of Representatives. The LDP, however, retained control of the Senate, with Neptali Gonzales remaining as Senate President. In the same month Ramos submitted to Congress four initiatives to support his reconciliation programme: the repeal of the Anti-Subversion Law, which would in effect legalize the CPP and similar organizations; the granting of an amnesty to 4,485 former rebels; the creation of a National Unification Commission (NUC) to help frame the Government's peace and amnesty strategy; and the review of all cases of rebels under detention or serving a sentence. (Congress formally approved these initiatives seven months later, in February 1993.) Ramos announced that his administration would pursue peace talks on three fronts: with the communist insurgents; the Muslim secessionists; and the military rebels. In July 1992 Ramos appointed a member of the House of Representatives, Jose Yap, as an official emissary to the NDF, thus giving government sanction to informal talks which were already taking place between Yap and rebel representatives. Also in July Ramos formed the Presidential Anti-Crime Commission (PACC) and named the popular Vice-President, Estrada, as its Head in an effort to curb the rising incidence of crime, especially kidnapping for ransom, which together with the acute energy crisis threatened the Government's plan to attract more foreign investment.

In mid-August 1992 Ramos authorized Yap to start exploratory talks with the expatriate leaders of the NDF, Luis Jalandoni and Jose Maria Sison, in the Netherlands. Ramos ordered the temporary release from prison of five communist leaders, including NDF negotiator Saturnino Ocampo and Romulo Kintanar, as well as 16 military rebels. Government and NDF emissaries held open but preliminary discussions in the Netherlands on 31 August and 1 September (the first since the failure of the peace talks in 1987). They agreed to recommend to their respective principals the resumption of formal talks, including the discussion of socio-political issues, in an attempt to secure a negotiated settlement of the insurgency. The Chairman of the NDF, Manuel Romero, immediately accepted the proposal, while Ramos promised to submit it to the NUC, which was created later in September. Ramos appointed Haydee Yorac, an elections commissioner and human rights lawyer, as Head of the NUC, and selected members from the legislature (including Yap), the Cabinet and the Catholic and Protestant churches. The NUC began its series of public consultations in early October. Meanwhile, a breakthrough was made in the negotiations with the MNLF: a member of the NUC, Eduardo Ermita, met Nur Misuari in Tripoli, Libya, where the latter indicated his willingness to return home from self-exile to hold peace talks with the Government. Two weeks later Ramos agreed to negotiate with the MNLF, with the OIC acting as an observer. At the end of October, following a series of separate talks with rebel leaders Honasan and Jose Maria Zumel, Yorac announced that the military rebels had submitted a framework for negotiations.

On 30 September 1992 the USA started to vacate Subic Bay Naval Base. In November Ramos called for a review of the 1951 Mutual Defense Treaty between the Philippines and the USA, declaring that it was not clear about the terms on which the USA would come to the aid of the Philippines in the event of an external attack. In the same month the USA completed its withdrawal from Subic Bay Naval Base and formally transferred command and control of the 56,000-ha facility to the Philippine Government. The Subic Bay Metropolitan Authority was established to oversee the conversion of the area to commercial use.

In January 1993 Edgardo Angara of the LDP was elected as Senate President, replacing Neptali Gonzales (also of the LDP) after realignments in the chamber brought together LDP and Lakas-NUCD senators in a new pro-Ramos majority. In another show of political dominance, the gubernatorial candidate supported by Ramos, Liningding Pangandaman, won 72% of the votes cast at elections in the ARMM on 25 March 1993.

The rift within the CPP, which had started in 1992 as a result of a Maoist purge initiated by the Chairman, Sison, worsened in 1993. In July the CPP's Metro Manila-Rizal organization declared its secession from the central leadership. Five months earlier, in November 1992, the party's Executive Committee had ordered the dissolution of the region's leading committee and its armed unit, the Alex Boncayao Brigade, purportedly for factionalism and military excesses. In October 1993 four communist leaders, including Romulo Kintanar and Arturo Tabara, the Secretary-General of the Central Philippine Command of the NPA, were expelled from the CPP and the NPA for refusing to recognize the authority of Sison.

In April 1993, following an extensive review of the Philippine National Police (PNP), in which corruption was found to be widespread, Ramos ordered the dismissal of hundreds of personnel, including 63 of the 194 senior officers. In July Ramos established a 60-day period during which the security forces were to disarm the 560 private armies in the Philippines, which were controlled mostly by provincial politicians and wealthy landowners. The September deadline for the dissolution of the private militias was, however, subsequently extended until the end of November, but even by this date little progress had been made.

Despite progress in early 1993, which had facilitated the reintegration of mutinous soldiers into active service, in February 1994 the military rebels announced that peace negotiations between the Government, the RAM and the YOU would continue to be suspended, pending the release of six military detainees. In the following month Ramos proclaimed a general amnesty for all rebels and for members of the security forces charged with offences committed during counter-insurgency operations, as recommended by the NUC. Crimes of torture, arson, massacre, rape and robbery were, however, exempted from the pardon. The RAM rejected the amnesty, on the grounds that it failed to address the causes of the rebellion, and reiterated its commitment to electoral, military, social and economic reform. Saturnino Ocampo also dismissed the proclamation, claiming that it was biased against communist rebels, since 80% of them had been charged with common crimes not covered by the amnesty.

In January 1994 an increase in petroleum prices took effect following the imposition of a levy by Ramos in September 1993, which had initially been charged to the Oil Price Stabilization Fund. A group of autonomous socialist organizations led by elements of the CPP, called the Philippine Left, organized a popular campaign in protest at price increases, creating the first crisis of Ramos's term of office. The Trade Union Congress of the Philippines threatened a general strike, and the still-active Alex Boncayao Brigade initiated bomb attacks on the country's three leading petroleum companies. At the end of February Ramos rescinded the levy, and established a committee to study alternative methods of raising government revenue.

The first formal negotiations between the Government and the MNLF took place in October 1993 in Jakarta, Indonesia, following two phases of exploratory talks earlier in the year. The MNLF demanded the creation of an autonomous Islamic state in the south, as provided for under the 1976 Tripoli Agreement. In November the two sides signed a memorandum of understanding and an interim cease-fire was agreed. In response to the progress of the negotiations, further violence occurred in the south, including an assault on a Roman Catholic cathedral in Davao City, followed by retaliatory attacks on three mosques by Christian extremists. In January 1994 the terms of the cease-fire were agreed, stipulating that both government and rebel forces should remain in place and refrain from provocative action. Nur Misuari subsequently gave an assurance that MNLF demands for a transitional government in Mindanao would not affect the terms of office of the officials of the existing ARMM, due to expire in 1996. The second round of formal peace negotiations between the Government and the MNLF took place in Jakarta in April 1994.

In February 1994 a federal court in Hawaii awarded US $1,200m. in punitive damages to 10,000 Filipinos who had been tortured under President Marcos's administration. This followed a court ruling in October 1992 that victims of human rights violations under the Marcos regime could sue his estate for compensation. Imelda Marcos announced that she would appeal against the decision. Lawyers acting for the Philippine Government also filed counter-claims to the money, citing an Aquino decree that stipulated that all properties recovered from the Marcoses should be used only to finance the Government's agrarian reform programme. In January 1995 the Hawaiian court ordered the estate of Marcos to pay a further $774m. in compensatory damages to the 10,000 victims. At the end of that month the Philippine Supreme Court restored a 'freeze' on the assets of more than 500 companies controlled by former associates of Marcos, which had been revoked on technical grounds in 1991. In August 1995 the Swiss authorities approved the transfer of about $475m. from Swiss bank accounts held by Marcos to an account in the Philippines.

In June 1994 the Government launched an offensive against Abu Sayyaf, a Muslim secessionist group, which was responsible for terrorist activities, including kidnapping, on the southern islands of Jolo and Basilan. In retaliation for the military assault, Abu Sayyaf planted a bomb in Zamboanga City, injuring 28 civilians. On the following day the armed forces captured the group's headquarters on Jolo Island, prompting Abu Sayyaf's seizure of 74 Christian hostages on Basilan. Many hostages were swiftly released, but 15 were killed in retaliation for government 'summary executions'. In mid-June 20 of the remaining 21 hostages were released following the alleged payment of a ransom by the local authority and the intercession of the MNLF. Abu Sayyaf, however, retained one hostage, a Roman Catholic priest, in order to protect themselves from government attacks. Government troops subsequently captured the group's main headquarters in Basilan; the priest was later released unharmed.

In May 1994 Ramos succeeded in enacting a law amending the system of value-added tax (VAT) in order to offset the losses in revenue arising from the rescission of the levy on petroleum prices. Popular opposition to the extended VAT system, however, gathered momentum during June, and Ramos's attempts to diminish dissent by granting exemptions for newspapers and school textbooks by executive order served only to confirm the public perception that the legislation was flawed. The VAT legislation provided a rallying point for opposition parties, which organized broad-based coalitions to oppose the law. Senators who initially supported the bill began to advocate its abrogation. At the end of June the Supreme Court issued an injunction on the legislation, preventing it from being implemented at the beginning of July, pending rulings on seven suits claiming that the legislation and its enactment violated the Constitution. In August the Supreme Court ruled that the VAT legislation was constitutional, but it still could not be implemented pending an appeal against the ruling by opposition groupings.

In July 1994 the majority leader of the House of Representatives, Ronaldo Zamora, defected from Lakas-NUCD to the NPC. A few days later he and 19 other members of the NPC resigned from the ruling coalition. However, in August Lakas-NUCD and the LDP formed an alliance, known as Lakas-Laban; the ruling coalition, which also comprised the Nacionalista Party, the Liberal Party and PDP-Laban, thus controlled 158 of the 215 seats in the House of Representatives and 19 of the 23 occupied seats in the Senate. The priority of Lakas-Laban was to present a common list for the 12 seats in the Senate that were to be contested in 1995. Ramos would, however, also benefit from support in the Senate to expedite the passage of vital legislation.

In August 1994 the RAM agreed to support government programmes that would benefit the population. Later that month it signed an agreement with the ruling administration on electoral reform. In July 1995 negotiations between the Government, the RAM and the YOU resumed, following a three-month hiatus owing to the elections. The military groupings pledged that they would refrain from organizing any further coup attempts.

In October 1994 negotiations resumed between the Government and the NDF in the Netherlands, but collapsed owing to disagreements over the NDF's insistence on granting immunity to government negotiators for future talks to be held in areas of the Philippines claimed by the NDF to be under its control. In May 1995 discussions scheduled for June were threatened by the arrest of a leading NDF member, Sotero Llamas. The peace talks did take place in Belgium in June, but were suspended by the Government following NDF demands for Llamas's release to enable him to join the negotiations. The Government subsequently announced that it would guarantee safe passage to all members of the NDF panel for future negotiations in the Philippines, but that it was no longer prepared to attend meetings overseas, as the discussions frequently collapsed owing to the unreasonable demands made by the insurgents. The NDF rejected the new conditions, and Sison announced that peace negotiations would be postponed until the expiry of Ramos's term of office in 1998.

In December 1994 clashes took place in North Cotabato between government forces and the MILF. The MILF, which was covered by the government cease-fire but was not a party to the peace talks, was widely suspected of having taken advantage of the continuing negotiations between the MNLF and the Government to strengthen its position, both by an accumulation of arms and the recruitment of young militants disaffected with the compliance of the MNLF. Negotiations between the Government and the MNLF proceeded in Mindanao in January 1995, at the end of which it was announced that the Government had agreed to the establishment of an MNLF provisional government in Mindanao, subject to congressional approval and the holding of a referendum. In April an attack by about 200 guerrillas on the largely Christian town of Ipil in Mindanao, in which more than 50 civilians were massacred, was attributed to Abu Sayyaf. Members of an MNLF splinter group, the Islamic Command Council, were alleged to have taken part in the attack on Ipil, which was denounced by Misuari as an attempt to sabotage the peace talks. Ramos, however, pledged to continue negotiations with the MNLF.

On 8 May 1995 elections were held to contest 12 of the 24 seats in the Senate and the 204 elective seats in the House of Representatives; 76 provincial governorships and more than 17,000 local government positions were also contested. Prior to the elections, in which an estimated 80% of all eligible voters participated, more than 80 people were killed in campaign violence. In Mindanao widespread violence caused voting to be postponed until 27 May, when 30,000 troops were dispatched to the island to ensure security. During the campaign Ramos declared that the elections should be regarded as a referendum on his three years in office and emphasized his administration's progress towards economic liberalization. Lakas-NUCD formed an electoral alliance with a small Mindanao-based moderate Islamic party for the election to the House of Representatives, in order to attract broad national support. The opposition parties, dominated by the NPC, focused on the Government's failure to prevent both the execution in Singapore of a Filipino domestic servant, Flor Contemplacion, who had been convicted of two murders, and the Muslim separatist attack on Ipil, emphasizing the Government's apparent inability to protect its nationals' interests at home and abroad.

In the event the ruling coalition won the vast majority (about 70%) of seats in the House of Representatives and Lakas-Laban won nine of the 12 seats in the Senate. One of the three opposition seats was secured by Honasan, despite a campaign by Aquino against his candidacy. In his opening address to Congress in July 1995 Ramos urged the adoption of radical legislation to reduce public bureaucracy and eliminate corruption in the police force and in the judicial system.

In October 1995 Ramos decided to assume personal responsibility for the campaign against organized crime, reducing the role of the PACC to that of a co-ordinating agency. Estrada was widely regarded as having failed in his position as head of the agency, despite his alleged sanctioning of extra-judicial executions to combat violent crime. The continuing increase in serious offences, often with the alleged involvement of members of PNP, adversely affected public perceptions of the Ramos administration, particularly in view of Ramos's past affiliation to the PNP. (In April 1996, however, the PACC's powers, and one-third of its former strength, were restored.)

In January 1996 the LDP withdrew from its alliance with Lakas-NUCD, leaving Ramos with a minority in the Senate, which threatened to undermine his reform programme, but still with a substantial majority in the House of Representatives. Ramos's administration had become increasingly unpopular as a result of rising crime, a national rice shortage (owing largely to mismanagement) and increasing inflation. Furthermore, the extended VAT legislation finally took effect in January, and in February an increase in the price of petrol was announced. Angara, the President of the LDP, claimed that the party had received no concessions in return for helping the administration to pass unpopular legislation, and that, contrary to agreement, Lakas-NUCD had contested LDP seats in the local government elections in May 1995. The LDP finally withdrew from the alliance, however, in protest against proposed anti-terrorist legislation, which it described as the effective introduction of martial law, designed to facilitate Ramos's extension of his term of office.

In 1995 formal peace negotiations continued between the Government and the MNLF in Jakarta regarding the proposed establishment of an expanded autonomous region in Mindanao. In August Misuari agreed for the first time to a referendum (as stipulated under the Constitution) prior to the establishment of an autonomous zone, but demanded the immediate establishment of a provisional MNLF government to ensure that the referendum was conducted fairly. (The MNLF had previously objected to a plebiscite, since a number of predominantly Christian regions were likely to vote against autonomy.) In April 1996 two bomb explosions in Zamboanga City were widely attributed to Abu Sayyaf or to other groups opposed to the peace negotiations. In June it was announced that the MNLF and the Government had finally reached agreement on a proposal by Ramos for the establishment of a transitional administrative council, to be known as the Southern Philippines Council for Peace and Development (SPCPD), which was to derive powers from the Office of the President. The five-member SPCPD, which was to be headed by Misuari, was to co-ordinate peace-keeping and development efforts in 14 provinces and 10 cities in Mindanao, with the assistance of an 81-member Consultative Assembly and a religious advisory council. After a period of three years a referendum was to be conducted in each province and city to determine whether it would join the existing ARMM. (The MNLF had abandoned its demands for autonomy in 23 provinces in Mindanao.) Also in June it was announced that negotiations between the Government and the NDF had resumed in the Netherlands.

The announcement of the peace agreement between the Government and the MNLF prompted criticism from a number of Christian politicians and a series of protests by Christians in Mindanao, which culminated during an official visit by Ramos to the region, in July 1996, in an attempt to secure public support for the planned establishment of the SPCPD. The Government subsequently warned that opponents of the peace agreement could be charged with sedition, after a number of demonstrators displayed inverted Philippine flags (symbolizing a state of war). In the same month government officials announced that, under the peace agreement, Muslims were to be allocated one cabinet post, and were to be granted representation in state-owned companies and constitutional commissions. In addition, Ramos offered to support the candidacy of Misuari in the forthcoming gubernatorial election in the ARMM.

In early September 1996 the Government and the MNLF signed a final draft of the peace agreement in Jakarta; discussions continued regarding the integration of some 7,500 members of the MNLF's military wing into the national army and security forces (this commenced in March 1997), the establishment of a regional security force in Mindanao and the administrative structure of the SPCPD. The MILF refused to endorse the settlement, continuing to demand separatism for 23 provinces in Mindanao. On 9 September 1996 elections took place peacefully in Muslim Mindanao for the region's Governor and Assembly; Misuari, who, as agreed, contested the gubernatorial election with the support of Lakas-NUCD, was elected unopposed. (Critics of Misuari subsequently maintained that he was not permitted by law concurrently to chair the SPCPD and hold office as Governor of Muslim Mindanao.) Following his election as Governor, Misuari demanded reforms in the Senate whereby the number of seats would be increased from 24 to 72 (with 24 senators elected from Mindanao). In October he was officially appointed Chairman of the SPCPD.

MILF and government officials had met for preliminary peace discussions for the first time in August 1996, following the MILF's rejection of the peace agreement concluded by the MNLF. Clashes between government and MILF forces, which resulted in several deaths, were reported in October in North Cotabato and Maguindanao. Nevertheless, peace talks reconvened in January 1997, resulting in the announcement of a temporary cease-fire, but the high incidence of kidnappings by separatist rebels in Mindanao in early 1997 threatened to hinder the progress of the negotiations. In February members of Abu Sayyaf, who remained opposed to any peace agreement between the Government and the Muslim separatists, were responsible for the assassination of a Catholic bishop in Jolo. The abduction, allegedly by MILF rebels, of some 40 workers from the Philippine National Oil Co in mid-June resulted in the collapse of the peace talks and a renewed outburst of fierce fighting between government and MILF forces in Mindanao, with many thousands of civilians forced to evacuate the area. In an attempt to encourage the resumption of peace negotiations and to allow the citizens to return to their homes, President Ramos ordered a suspension of military offensives by the armed forces in early July. Prospects for peace appeared brighter following a meeting between MILF leaders and Misuari, at which the MILF reiterated its demands for the withdrawal of government forces from the region, and a new cease-fire agreement was subsequently signed by the Government and the MILF. Talks resumed later in July, with the negotiators aiming to reach a substantive agreement by the end of 1997.

During 1996–97 Ramos continued publicly to dissociate himself from efforts by his supporters to amend the constitutional stipulation that restricted the President to a single term in office. The business community criticized his prevarication, claiming that it was profoundly affecting investors' confidence in the Philippine economy. Ramos's attempt to transform Congress into a constituent assembly to revise the 1987 Constitution in early September prompted Cardinal Sin and Aquino to organize a protest demonstration on 21 September, the 25th anniversary of Marcos's declaration of martial law. Recognizing the support that the protest movement was already commanding (in the event, some 500,000 people took to the streets of Manila), Ramos issued a statement several hours before the demonstration was due to commence, declaring that he still believed that the Constitution should be amended, but that it should be changed after the 1998 presidential election, and that he was categorically not intending to offer his candidature.

The unsuccessful moves to introduce constitutional changes that would have enabled President Ramos to stand for a second term greatly weakened the capacity of the ruling Lakas-NUCD to contest the presidential election. There was widespread apprehension within Lakas-NUCD, and more widely within the élite and business community, that Ramos was the only candidate able to defeat the populist Vice-President, Estrada, who had been nominated as the candidate of his own small party, the Partido ng Masang Pilipino (Party of the Filipino Masses). Moreover, Lakas-NUCD was divided over whom to support as party candidate: none of the contenders had wide popular appeal and any choice threatened to split the party. Contrary to early expectations, in December 1997 Ramos finally endorsed the Speaker of the House of Representatives and the most influential figure in the party, Jose de Venecia, in preference to the former Secretary of National Defense, Gen. Renato de Villa, a close associate of Ramos. De Villa promptly left the ruling party and declared his candidature for the Partido para sa Demokratikong Reporma—Lapiang Manggagawa Coalition (Party for Democratic Reform). In early 1998 a former senator, Gloria Macapagal Arroyo, the only candidate seriously to rival Estrada in opinion polls, withdrew from the presidential race, and from her own party, to become Lakas-NUCD's vice-presidential candidate. In June 1997 Estrada's party formed a coalition with two leading opposition parties, the LDP and the NPC. The resultant party, Laban ng Makabayang Masang Pilipino (LaMMP—the Struggle of Nationalist Filipino Masses), endorsed Estrada as its presidential candidate in December.

As the election approached, the only presidential candidates with significant popular followings were Estrada, de Venecia, Senator Raul Roco, the mayor of Manila, Alfredo Lim, and the former Governor of Cebu, Emilio Osmeña. Imelda Marcos, who did not command wide popular support, withdrew her candidature in support of Estrada. Only de Venecia and Estrada benefited from significant party organizations, while Lim had the endorsement of Cardinal Sin, the head of the Catholic Church, and the former President, Aquino.

The election on 11 May 1998 was one of the country's most peaceful (there were 53 deaths) and fair. As the opinion polls had predicted, the presidency was won by the former film actor, Estrada, with 39.9% of the votes cast; his nearest rival was de Venecia, who secured 15.9%. Estrada incorporated his family name into his professional name, becoming President Joseph Ejercito Estrada, and the LaMMP, which gained 66 seats in the House of Representatives (compared with 106 for Lakas-NUCD), was renamed Laban ng Masang Pilipino (LMP, Struggle of Filipino Masses). Exit polls showed that Estrada's strongest vote came from among lower socio-economic groups. Unlike all the other candidates, Estrada had built his public reputation and his campaign around his identity with the *masa* (masses). Despite a comfortable educated middle-class background, he prided himself on his failure to complete his education. Estrada had decades of political experience as a former mayor, senator and Vice-President, but, unlike many other politicians in the post-Marcos era, had not been stigmatized as a 'traditional politician'. Indeed, the 1998 elections continued a post-Marcos trend, the rise of populist politicians and a decline in what had always been a weak party system. Along with

Estrada, several other film actors and celebrities were elected to positions of national and local authority. While Lakas-NUCD candidates won a majority of congressional seats, over one-half defected to Estrada's LMP soon after the election results were announced, as had occurred with the LDP after Ramos's election in 1992. Although a high proportion of those in Congress were younger and reputedly more idealistic than was the case in the past, Estrada was likely to encounter significant problems in trying to fulfil his plans to abolish 'pork barrel funds', which had hitherto supplied politicians with the means to reward their particular constituencies.

THE ESTRADA GOVERNMENT

President Estrada's Cabinet and coterie of advisers were reminiscent of the 'rainbow coalition' that marked the early period of Aquino's presidency. Among them were wealthy, mainly ethnic Chinese business executives, former Marcos cronies and political allies, friends from Estrada's mayoral days in San Juan, and activists formerly linked with the communist movement and left-wing non-governmental organizations. In a post-election climate of reconciliation, Ramos accepted a position as Senior Adviser, modelled on that of Singapore's former Prime Minister, Lee Kuan Yew. One of the most powerful figures to re-emerge alongside Estrada was a former close ally of Marcos, Eduardo Cojuangco, to whom Estrada was 'running mate' in 1992. In July 1998 Cojuangco became Chairman and CEO of the San Miguel corporation, the position he held during the Marcos years, and it appeared that the Government was to abandon its legal cases against him and another Marcos crony, Lucio Tan. Estrada announced his intention to disband the Presidential Commission on Good Government and also wanted to reach a compromise with the Marcos family over government claims to Marcos's Swiss bank accounts. However, Estrada prompted a public outcry when he agreed to allow the body of former President Marcos to be buried in the Cemetery of National Heroes and was forced to back down. While Estrada adopted a more conciliatory stance towards the leaders of the Marcos era than his two predecessors, he was also attempting to reassure the business community that he would continue the economically successful policies of liberalization, deregulation and privatization put in place under Ramos. Led by the Secretary of Education, Eduardo Angara (a former corporate lawyer), Estrada's coterie of advisers and cabinet secretaries on economic matters were people widely identified with these policies. Partly in response to the interests of the business community, Estrada also promised to address the problem of crime, and to create a more streamlined and transparent state bureaucracy.

What remained unclear, particularly in a climate of regional economic crisis, was how the Estrada administration would reconcile its free-market economic programme with its widely publicized promises of poverty alleviation, food security and agricultural development. With an impending budget deficit and substantial potential problems in reforming the tax system, the new regime would be struggling to fund the infrastructural and land reform programme it had promised. Attempting to implement some of the pro-poor policies would be the new Secretary for Agrarian Reform and the former head of the proscribed NDF, Horacio Morales. Morales claimed that he would complete the CARP land reform programme, started more than a decade earlier, within the next four years. While some former communists were participating in the Estrada Government, the leadership of what remained of the CPP-NDF condemned the new regime, much as it did that of Aquino and Ramos. Not long after taking office, on 30 July 1998 President Estrada approved a human rights agreement that had been negotiated with the NDF under Ramos, and invited the exiled NDF leadership to the Philippines for a resumption of peace negotiations. With the NDF demanding a reversal of the Government's liberalization policies, the next round of negotiations on social and economic reforms was likely to be difficult.

With continued Muslim secessionist pressures in western Mindanao, one of the foremost bills in Estrada's legislative agenda was an amendment to the Organic Act of the ARMM. A plebiscite was scheduled for the southern Philippines in 1999, with the intention of widening the ARMM beyond the four provinces it presently covers. However, with continuing disquiet over the implementation of the Organic Act and Nur Misuari's governorship of the ARMM, there were even threats from two provincial Governors that they would withdraw from the regional body. Alternating hostilities and short-term cease-fire agreements continued to characterize relations between the national Government and the secessionist MILF over the latter half of 1997 and 1998. Despite the 1996 settlement with the MNLF, this period witnessed the largest deployment of military personnel to western Mindanao since the mid-1970s. Nevertheless, both sides seemed committed to ongoing peace talks.

Over the course of his first year in office, President Estrada proved himself a much more adept leader than many of his critics had given him credit for. Not only did he maintain his mass popularity, but he also won the grudging acceptance of many opponents, notably from within the business community. Estrada's coalition party, the LMP (relaunched as the Lapian ng Masang Pilipino—LaMP, Party of the Filipino Masses—in August 1999), continued to hold the great majority of seats in both houses of Congress. While hardly spectacular, the Philippine economy continued its recovery under Estrada, and political stability was maintained. Where Ramos was noted for his work ethic and attention to detail, Estrada proved his skill in cultivating and drawing upon a range of experienced, generally well-respected advisers, who assumed key positions in the government and state bureaucracy. While factional tensions developed between some of the most powerful of these figures—notably the Executive Secretary, Ronaldo Zamora, and the Chief of the Presidential Management Staff, Leonora Vasquez-de Jesus—Estrada was able to negotiate a balance.

Some claimed that this system of particularistic authority had allowed Estrada to reduce bureaucratic inefficiency, but it also exposed him to accusations of cronyism. Rather than pursuing his predecessor's policies of liberalization, some critics argued that Estrada had overseen the creation of new monopolies, for instance in the petrochemical industry and port services, as well as a reconcentration of power in the area of telecommunications. Remaining at the centre of these accusations were former Marcos cronies, Eduardo Cojuangco and Lucio Tan, the majority owner of the heavily indebted Philippine Airlines. Estrada was also censured for his relative inaction in seeking the recovery of the Marcos family's ill-gained wealth, although the cases against the Marcoses continued to be dealt with through the courts. In March 1999 the Supreme Court upheld a 1990 decision by the Bureau of Internal Revenue against the Marcos heirs for US $600m. in unpaid inheritance taxes. Later in the year an anti-graft court blocked an earlier $150m. damages agreement between the Marcos family and about 10,000 victims of human rights abuses during the Marcos incumbency. The victims had agreed to abandon further claims in exchange for this settlement, but the court ruled against the release of funds from the $590m. sequestered by the Government from Swiss bank accounts.

Just as Estrada's record on economic liberalization was uneven and even contradictory, so too was the stance of the mainstream political opposition. Under the leadership of the former President, Aquino, and the head of the Catholic Church, Cardinal Sin, this opposition continued to define itself as the defender of the democratic gains made in the wake of the overthrow of the Marcos regime. Amidst mounting concerns that Estrada's presidency was beginning to undermine these gains, a 'Rally for Democracy' was held on 20 August 1999, demanding an end to cronyism, presidential interference in the print media and government moves to amend the 1987 Constitution. As was the case at the end of the Ramos presidency, the campaign against constitutional change was largely aimed at retaining those provisions limiting an incumbent in political office to a maximum period of six years. Yet the campaign was also directed against Estrada's stated plans to extend economic deregulation by removing those provisions limiting foreign ownership of land and public utilities. Although further rallies were promised, Estrada had the support of most major business associations to continue with these constitutional reforms.

For many, the biggest expectations of the Estrada Government centred on its 'pro-poor' platform. Much was made of this in government rhetoric and in the creation of various state instrumentalities, like the National Anti-Poverty Commission, but the results were disappointing, an outcome that could only partly be attributed to the economic crisis. The principal aim was 'food security', which was to be achieved through agricultural modernization. To this end, the Agricultural and Fisheries Modernisation Act was passed with an increased budget allocation to fund such projects as rural credit and the improvement of irrigation, roads and other infrastructure. To complement this, the enclave development of Subic Bay was reorientated better to provide for its depressed Luzon hinterland. The dramatic increase in agricultural output in the early part of 1999 was, however, largely a result of the natural disasters that marred the previous year, rather than a measure of successful rural reform. There also appeared to have been little progress in the Government's plans to redistribute the remaining 1.3m. ha of land set aside under CARP. One of the few achievements under the pro-poor programme by mid-1999 was the rehousing of some 25,000 urban poor in Manila under a new state-sponsored mortgage scheme.

While Estrada still enjoyed wide popular support, his Government's continued failure to put its pro-poor rhetoric into practice threatened to bring significant political consequences. The communist movement remained small and fragmented, and, although its influence did not compare with that of a decade previously, it was recruiting growing numbers of people disillusioned with the mainstream political process and the social inequalities that continued to affect Philippine society. The NDF withdrew from peace negotiations with the Government in May 1999, following the Senate's ratification of a defence treaty with the USA (see below); Estrada, who had previously expressed dissatisfaction with the progress of the negotiations, and had imposed a deadline of December for a settlement, then assumed a position of outright hostility towards the movement. Nor did the Estrada Government succeed in placating the rural poor who were the major constituency of the MILF in the southern Philippines. Indeed, Estrada's stance towards the MILF vacillated between peace negotiations and promises of agricultural development and declarations of war. In early 1999, for example, the armed forces launched a major offensive on the MILF's main training camp in Maguindanao, with many killed or wounded, and thousands displaced. Peace negotiations resumed, but the national Government's standing among Muslim Filipinos remained tenuous. The national Government continued to delay the ARMM plebiscite it hoped would extend its authority over the region. Though the Estrada Government made little progress in dealing with the communist and Muslim secessionist movements, it achieved success in reducing crime. Most notably, it presided over a sharp reduction in the number of kidnappings, in large part a consequence of measures jointly formulated by Estrada and the Chinese business community.

In the latter months of 1999 and during most of 2000 the Estrada Government came under mounting pressure and criticism. Opponents of planned constitutional change, led by Cardinal Sin and Corazon Aquino, organized a rally involving some 80,000 protesters in Manila on 21 September 1999 to coincide with the anniversary of the declaration of martial law in 1972. This was followed, in October, by a march on the presidential palace by farmers protesting against proposed constitutional reforms that would open up land ownership to foreigners. Eventually, in January 2000, Estrada announced that constitutional change had been put on hold. While Estrada thus placated some opponents, others emerged within the trade union movement, whose wage claims Estrada rejected, and among teachers, whose promised bonuses were withdrawn. Opposition was also mounting among powerful sectors of the business community, represented by the Makati Business Club, which were critical of Estrada's tardiness in pushing further liberalization and privatization reforms through Congress, despite the ruling party occupying 80% of seats in the legislature, many more than those commanded by former President Ramos. To a degree this criticism was addressed in March 2000, when legislation was passed that allowed foreigners to enter the retail industry, and in May, when similar legislation gave foreigners the right to buy into domestic banks. In April the regime even oversaw a significant improvement in what had long been an abysmal record of tax collection.

Yet criticism and disappointment did not abate. Allegations of indecisiveness and inaction also surfaced increasingly in relation to Estrada's 'pro-poor' programme. World Bank funds intended for development projects, ostensibly aimed at benefiting the poor, were withdrawn in December 1999 because they had not been used, and almost a year later, in September 2000, similar funds from the Asian Development Bank (ADB) and the Japan Bank for International Development were similarly unused, apparently because of government inaction. Earlier, in October 1999, the head of Estrada's mass housing programme, Karina Constantino-David, resigned over the slow pace of implementing housing reform and, in particular, over the growing influence of developers lobbying against it. Despite Estrada's continued populist rhetoric, dissatisfaction was growing among the rural and urban poor, his main constituency. In March 2000, for example, oil price rises, which were affecting the whole region, resulted in widespread protests among Manila jeepney drivers, and the Government's continued failure to implement significant agricultural land reform also resulted in growing rural discontent. From a low of 6,000 armed guerrillas in 1994, the communist NPA had, by mid-2000, expanded its numbers to an estimated 9,500, mainly drawn from disaffected rural and urban youth.

Persistent accusations of corruption and favouritism also contributed to Estrada's growing unpopularity, particularly among the middle classes and the élite. Especially damaging, both to Estrada's claims to even-handedness and good economic management, was the accusation that he had intervened to protect one of his business associates from the charge of insider trading in March 2000, the result of which had been the near collapse of the Philippine Stock Exchange. The Estrada Government's suspension of flights to Taiwan in October 1999, in order to protect Philippine Airlines from what it regarded as unfair competition from two Taipei airlines, was also interpreted by some as a favour to its owner, reputed crony Lucio Tan. With the resultant loss in exports, overseas contract work and Taiwanese investment, other critics simply interpreted the move as economic mismanagement and a retreat from the promised liberalization agenda. While Estrada reshuffled his Cabinet and dismissed several of his closest advisers in January 2000, partly in an effort to address accusations of favouritism and indecision, these allegations continued to plague his Government.

By April 2000 opinion polls showed that Estrada's net approval rating had fallen to 5%, compared with 65% less than one year earlier, and rumours were circulating that he would not complete his term of office and that plans of a coup or an assassination plot were afoot. Yet the principal formal opposition party, Lakas, lacked unity, occupying only a small minority of seats in Congress, and its leader, Vice-President Gloria Macapagal Arroyo, remained in Estrada's Cabinet and declared her continued support for the President. To some degree Lakas's opposition to the Government seemed to coalesce around former President Fidel Ramos, once a senior adviser to Estrada and subsequently one of his most outspoken critics.

Of all the crises faced hitherto by the Philippines under the Estrada regime, the most serious and costly, but also the one that regained some popularity for Estrada, involved the sharp escalation of hostilities in the Muslim south beginning in the early months of 2000. Despite promises of ongoing dialogue and the provision of greater development assistance, the Government's attitude towards the MILF increasingly assumed the profile of a military offensive. In February government aircraft bombed the MILF's Camp Omar and a secondary camp was seized, the military claiming that this was in retaliation for an earlier ambush. In May, after weeks of fierce fighting, government troops took control of an MILF-controlled section of highway near its headquarters in Camp Abubakar, Maguindanao, and continued their attack on MILF positions despite the offer of a cease-fire. By now hundreds had been killed, mostly on the MILF side, and well over 300,000 civilians, mainly Muslim peasants, had been forced to flee to evacuation centres. In May bombs exploded in two Manila shopping malls, killing one person and wounding others. Though widely attributed to Muslim terrorists, they were just as likely to have been the work of military elements seeking further public support for their war in the south.

In July 2000, after further offensives into MILF-controlled territory, government soldiers finally captured and destroyed much of Camp Abubakar. Now in retreat, and with no major camps left intact, the MILF leadership issued a call for a *jihad* (holy war) against the Philippine state. While government officials and the Catholic Church had long been anxious not to represent the war in the south in religious terms, a dangerous shift seemed to be taking place as more self-proclaimed Christian vigilantes had begun killing and terrorizing ordinary Muslims. Estrada announced plans to turn Camp Abubakar into a special economic zone for the mainly Muslim population of the region and, after initially putting a price on their heads, offered an amnesty to the MILF leadership. Rejecting these moves, but with most of its approximately 15,000-strong fighting force now fragmented, the MILF turned, once again, to guerrilla warfare, its leadership unbowed in its quest for an independent Islamic state.

In the Sulu archipelago to the west of Mindanao, the separate and much smaller secessionist group, Abu Sayyaf, resumed its campaign of terrorism with the kidnapping, in March 2000, of some 30 students and teachers, threatening to kill them unless various demands were met. A few weeks later, in April, as government troops were closing in, one Abu Sayyaf unit took the more dramatic step of kidnapping 21 tourists and staff, many of them European, from the island resort of Sipadan in eastern Malaysia, not far from Abu Sayyaf strongholds in Basilan and Jolo. In further bizarre developments, foreign journalists and a group of Filipino Christian evangelists who travelled to Jolo, with the aim of ending the crisis through prayer, were also taken hostage. In May 15 of the original Filipino hostages were freed by the military, but four others were found dead. By the end of September, after months of periodic military assaults, numerous civilian casualties and intermittent negotiation, nearly all of the hostages had been released. Most were freed in exchange for millions of dollars in ransom, much of it paid by the Libyan Government as 'development aid'. While Abu Sayyaf made massive gains in wealth, international publicity and new recruitment, it emerged from the drama less as a group of dedicated revolutionaries than a loose collection of brigands squabbling among themselves over spoils.

In October 2000 the Philippines was engulfed by a political crisis that arose as a result of allegations, made to an investigative committee of the Senate by the Governor of Ilocos Sur, Luis Singson, that President Estrada had accepted large sums of money as bribes from illegal gambling businesses and from provincial tobacco taxes. Despite the President's denial of these allegations in a televised statement to the nation, opposition parties announced their intention to begin, in late October, the process of impeaching the President. Earlier in the month Vice-President Gloria Macapagal Arroyo had announced her resignation from the Cabinet, in which she served as Secretary for Social Welfare and Development. The fact that she did not relinquish the vice-presidency at the same time led to speculation that she was preparing to succeed President Estrada in the event of his forced removal from office. On 13 November Estrada was impeached on charges of bribery, corruption, betrayal of public trust and culpable violation of the Constitution. The Speaker of the House of Representatives, Manuel Villar, ordered the transmittal of the articles of impeachment to the Senate without a vote, on the grounds that more than the required one-third of members had signed a petition endorsing impeachment. Villar and the President of the Senate, Franklin Drilon, were both replaced in their positions as they had resigned from the ruling coalition and urged Estrada's removal. A majority of two-thirds of the Senate, 15 of the 22 members, would be required to remove Estrada (who refused to resign) from office.

While the impeachment trial promised a balanced set of formal procedures to resolve the political uncertainties surrounding the corruption charges levelled against the President, a more volatile, extra-parliamentary process unfolded. The anti-Estrada street protests, which had begun with Singson's accusations, continued as growing numbers of Estrada allies withdrew their support, and public figures, like former President Fidel Ramos, repeated their calls for Estrada's resignation. On 30 December 2000, two weeks after the trial had started, Manila was rocked by five separate bomb explosions, resulting in 22 deaths and many more injured. While government sources blamed Estrada's political opponents or Muslim separatists, others attributed blame to Estrada's supporters, arguing that such violence was intended either to intimidate witnesses or to create an opportunity for the Government to declare martial law. However, the trial continued and damning testimony against Estrada emerged, notably that of Equitable PCI Bank's Senior Vice-President, Clarissa Ocampo, who alleged that she had been instructed to prevent the discovery of a bank account containing millions of dollars controlled by the President. Despite such testimony, the pro-Estrada majority in the Senate blocked a move by the prosecutors to open an envelope said to contain evidence of Estrada's corrupt banking practices.

With prosecutors resigning in protest, the trial came to an abrupt halt, more of Estrada's allies abandoned him, and the centre of political gravity moved to the street protests taking place at Manila's People Power monument, the scene of Marcos's overthrow in 1986. Covert lobbying by political, church and business leaders led the military and the police force to withdraw their support from Estrada in favour of Vice-President Gloria Macapagal Arroyo. Finally, on 20 January 2001, Estrada departed from the presidential palace and Macapagal Arroyo was sworn in as his successor.

THE MACAPAGAL ARROYO GOVERNMENT

Named People Power II by its supporters, the transition was welcomed by many, but it also aroused significant criticism because Estrada still appeared to have majority support, principally among his main constituency of urban and rural poor, and the transition created another political precedent, weakening the institutions of electoral democracy and the stability of the new regime. Although Estrada maintained his claim to the presidency and, hence, to immunity from prosecution, in March 2001 the Supreme Court ruled against both claims, reaffirmed Macapagal Arroyo as President and opened the way for charges of corruption, bribery and economic plunder to be brought against Estrada. In late April Estrada was imprisoned, awaiting trial.

Shortly afterwards, on 1 May 2001, mass protests were again staged in Manila, this time by Estrada's supporters. These disturbances were violent, resulting in several deaths and many injuries. President Macapagal Arroyo responded by declaring a week-long 'state of rebellion', and arrest orders, in response to an alleged aborted coup, were issued for Senators Juan Ponce Enrile and Gregorio Honasan, and for Estrada's former national police chief, Panfilo Lacson. Although Macapagal Arroyo withstood this challenge, doubt remained over her authority: it appeared likely that the President's coalition party, Lakas, would fare badly in elections for 13 of the 24 seats in the Senate, to the House of Representatives and to local administrations, scheduled for 14 May. In the event, however, Macapagal Arroyo emerged from the elections with a majority in both houses of Congress. While the veteran senator, Juan Ponce Enrile, lost his seat, both Honasan and Lacson, who remained in hiding, were returned to the Senate.

While Estrada's removal had much to do with the distaste he aroused among the élite and middle class, these represented President Macapagal Arroyo's principal, but not her only, basis of support: as the winning vice-presidential candidate in 1998 she had demonstrated a broad appeal, gaining the single largest number of votes of any candidate and thereby repeating an earlier performance in 1995, when she had received the most votes in elections to the Senate. In part she was able to take advantage of the fact that her father, Diosdado Macapagal, had been President before Marcos. However, she was also an experienced politician in her own right, having served two senatorial terms before becoming Vice-President; and as Under-Secretary of Trade and Industry in the Aquino administration. In addition to her élite social and political credentials, Macapagal Arroyo (who had a doctorate in economics) was admired by many in the business community, having drafted many bills aimed at liberalizing the economy during the era of President Ramos. Indeed, figures such as Ramos and organizations such as the powerful Makati Business Club, which favoured ongoing liberalization of the economy, were among Macapagal Arroyo's staunchest supporters. Like former President Aquino, but in contrast to Estrada, Macapagal Arroyo also had a reputation as a devout and conservative Catholic, which gained her strong support from the upper echelons of the Catholic Church.

Following her inauguration in January 2001, President Macapagal Arroyo pledged to return the Philippines to the course that had been planned under the presidencies of Aquino and Ramos: more stable democratic institutions, a campaign against official corruption, a war on poverty and steady economic growth based on policies of increased liberalization. In fulfilling these promises, however, Macapagal Arroyo was confronted by major problems. One was the difficult task of accommodating within her administration the disparate interests and personalities who had supported her in the overthrow of Estrada: the Catholic Church, the business élite, the military, militant trade unions, a range of left-orientated non-governmental organizations and a number of former Estrada cronies. Policies ranging from economic liberalization to family planning aroused opposing views within this broad coalition, with potential impediments to the implementation of the reform agenda that many regarded as necessary in the Philippines. Many of those appointed to Macapagal Arroyo's Cabinet were associates of Aquino and Ramos, such as the former Secretary of the Budget under Aquino and new Secretary of Finance, Alberto Romulo, and the former Secretary of National Defense under Ramos and incoming Chief Security Adviser and Executive Secretary, Gen. (retd) Renato de Villa. However, Macapagal Arroyo was also obliged to rely on many former supporters of Estrada who had changed their allegiance in the final days of his presidency. Notably, Luis Singson, whose exposure of Estrada's bribe-taking prompted the impeachment trial, was appointed as the new Government's gambling consultant. Such appointments raised doubts about the new President's capacity to eradicate corruption, a problem compounded by accusations that her husband, Mike Arroyo, had misused public lottery funds. However, the Macapagal Arroyo administration persisted in bringing several corruption charges against Estrada, the most serious being the embezzlement of US $80m. in state funds, a charge punishable by death. The trial began in October 2001.

The politicized class divide that seemed evident during the transition from Estrada to Macapagal Arroyo appeared to fade in the aftermath of the May 2001 elections, but the new President was still faced with the seemingly impossible task of implementing the liberal economic reforms of the Ramos era while, at the same time, addressing the deteriorating circumstances of the rural and urban poor. The Government's commitment to further liberalization appeared to be confirmed by the enactment, on 4 June 2001, of legislation, pending for five years, privatizing electric power supply. With regard to measures to combat poverty, President Macapagal Arroyo made a number of promises, undertaking fully to implement the CARP (see above), in particular its land reform provisions. The President also pledged to implement a housing and livelihood programme for the urban poor, and stated that the Government would grant loans and scholarships to the needy. However, Macapagal Arroyo's administration continued to be impeded by the high budget deficit inherited from the Estrada regime, by a corrupt and inadequate taxation system, by the high cost of servicing the country's foreign debt, and by a national economy severely weakened, both by domestic political turmoil and by the impact of the global economic downturn on major trading partners, such as the USA and Japan.

While investor confidence was stimulated by the change of government in early 2001, crime and rebellion continued to pose problems for the Macapagal Arroyo administration. The incidence of kidnapping for ransom increased sharply. The abductions were usually carried out by Manila-based and other urban syndicates, which mainly preyed on the wealthy Filipino-Chinese community; and by Abu Sayyaf, which targeted both wealthy Western tourists and Christian Filipinos, notably students and workers. The response of the Government was to intensify military intervention, in the case of Abu Sayyaf, and to establish a National Anti-Crime Commission. Despite these measures and a formal policy of non-negotiation, lucrative ransoms continued to be collected, while hostages were frequently killed, particularly by Abu Sayyaf. President Macapagal Arroyo continued to deploy the armed forces in attempts to free hostages held by Abu Sayyaf, but with only limited success.

While the Government continued efforts to combat Abu Sayyaf activities, it virtually abandoned the pursuit of a military solution to the conflicts with the MILF and the NPA. Between 1995 and 2000, in a context of widening wealth differences, the NPA was able almost to double its strength to more than 11,000 armed personnel. In March 2001 President Macapagal Arroyo declared a month-long cease-fire with the NPA and in April, in Norway, commenced formal peace talks with its leadership. Although the armed strength of the MILF had declined as a result of military action by the Estrada regime, it remained a potent force, with a strong recruitment base among ordinary Muslim Filipinos struggling with poverty and prejudice. In July, in Tripoli, Libya, a cease-fire between the Government and the MILF was concluded, and in August, after subsequent negotiations in Malaysia and a state visit to that country by President Macapagal Arroyo, a peace accord was reached. However, sporadic fighting between government and MILF forces continued. Moreover, the plebiscite held on 14 August in Mindanao, which the Government had hoped would increase the authority of the ARMM, was marked by a low turn-out and an emphatic 'no' vote, except in Basilan and Marawi City.

By mid-2002 President Macapagal Arroyo had fulfilled, or made substantial progress in implementing, many of the numerous pledges presented in her first State of the Nation address. Most surprisingly, in view of her own and her predecessor's respective bases of support, she had delivered more material benefits to the poor than had President Estrada. Notably, her Government was reported to have distributed more than 200,000 ha of land under the agrarian reform programme, to have provided security of tenure to over 180,000 urban poor families, and to have constructed some 150,000 houses for workers and others amongst the rural and urban poor. In addition, it had initiated a number of stores selling basic commodities at subsidized prices, and had established a national health insurance system (in which some 2.2m. had enrolled), which was largely designed to benefit the urban poor. The Macapagal Arroyo administration had also maintained its success, under new legislation, in promoting private-sector involvement in infrastructural development, most notably in rural electrification, although excessive increases in power prices prompted widespread protests. The encouragement of a more liberalized, privatized economy continued in a number of sectors, and attracted broad support from the local business community.

Despite these successes, government efforts continued to be impeded by a large budgetary deficit, inadequate tax collection and a series of social and political problems. One of these was the continued widespread support enjoyed by detained former President Joseph Estrada, whose trials on charges of corruption and economic sabotage proceeded extremely slowly, protracted by appeals to the Supreme Court and the dismissal of trial judges and defence lawyers, amidst claims by Estrada that he remained the legitimate President. In the mean time, from his military hospital prison, Estrada campaigned for the holding of an immediate presidential election, delivered an alternative State of the Nation speech to that of Macapagal Arroyo, and applied to leave the country for medical treatment. Although the charges against him appeared to be overwhelming, they were not expected to be resolved for some considerable time. While there was no serious prospect of Estrada returning to the presidency, Macapagal Arroyo's authority was, nevertheless, hampered by the manner in which she came to office and this was likely to continue until the next presidential election, scheduled for 2004, an event for which several prospective candidates, including Macapagal Arroyo, appeared to be preparing. Popular opinion polls had generally been unfavourable to her, although in August 2002 her national approval rating in one survey had increased to 55%.

President Macapagal Arroyo's Lakas-NUCD party continued to command a majority in the lower house, but its control of the Senate remained tenuous. Indeed, as a consequence of the defection of one senator in June 2002, and the absence abroad of another, proceedings in the Senate, and consequently the Government's legislative agenda, were suspended for nearly two months. Macapagal Arroyo's presidency also suffered through the resignation at about the same time of two of her leading cabinet members, the Secretary of Foreign Affairs, Teofisto Guingona, and the Secretary of Education, Culture and Sports, Raul Roco, over the government decision to allow the deployment of US troops in the Philippines, and corruption charges, respectively. Both politicians claimed that Macapagal Arroyo's management of these issues was also as much the cause of their resignations, reflecting a criticism that had been levelled more widely at the President's technocratic style of leadership. Some observers, however, argued that these resignations were to be viewed in terms of the political manoeuvring, prior to the 2004 elections. Local government elections, which took place in July 2002, appeared to have little immediate bearing on national politics. Although a number of candidates, officials and others were killed, the authorities announced that the elections had been conducted in a relatively peaceful manner.

The main areas in which President Macapagal Arroyo had little success were in curbing official corruption and escalating crime, particularly on the part of kidnap-for-ransom syndicates. Both posed a growing threat to local and foreign business confidence in the country. In her second State of the Nation address in mid-2002, Macapagal Arroyo identified these problems for special attention. Among other measures, she established a new taskforce (replacing two earlier bodies for combating corruption and crime), transferred the police force to the authority of local government, and authorized the establishment of citizen self-defence units, but with little apparent impact. Corruption was curtailed in the highest political echelons, but continued to prove extremely difficult to eradicate within the state bureaucracy, while law enforcement agencies made little progress against organized crime.

Meanwhile, a further series of peace talks began between the Government and the MILF in October 2001, and in May 2002 an agreement was reached, whereby the Government would provide reparations for damaged MILF property and establish a state-funded aid body, to be administered by the MILF, in order to assist the many thousands of Muslims who had been displaced by fighting in Mindanao. This agreement even included plans to relinquish control of the MILF stronghold at Abubakar for non-military purposes, but these plans were abandoned owing to strong opposition from within the military and Congress. Moreover, despite the cease-fire agreement negotiated in 2001, sporadic military encounters continued, and by late 2002 the pace of peace discussions had slowed. Indeed, the future of relations between the Government and the MILF, and the Macapagal Arroyo doctrine of a non-military solution, came under increasing pressure, following repeated claims that the MILF, rather than Abu Sayyaf, had the strongest links to international terrorist groups.

The CPP and NPA had also become more active under the Macapagal Arroyo presidency. Apart from conventional guerrilla warfare, the NPA was believed to have supported some 400 candidates in the July 2002 local elections. It was also reported that NPA militias had been instructed to open fire on US troops at the start of the Balikatan campaign (see below), although this did not, in effect, transpire. The exiled leader of the NPA, Joma Sison, was rumoured to have urged Filipino Muslims to join his movement. In August 2002, following the renewed campaign against Abu Sayyaf, most of the Filipino troops stationed in the south were redeployed for a further campaign in areas known to contain NPA strongholds. At the same time the US authorities declared the CPP and the NPA to be international terrorist organizations, thereby subjecting their supporters in the USA to the same legal sanctions as those associated with the al-Qa'ida (Base) organization of Osama bin Laden. Soon afterwards, the Dutch, British and Canadian authorities followed the US Government's example in freezing the assets of NPA groups and of their leaders, including Sison and many others who were exiled in the Netherlands. Although there had been ongoing efforts on both sides to reactivate peace talks between the Philippine Government and the CPP, Sison maintained that discussions would not resume while Macapagal Arroyo remained President. At the same time as these developments, various left-wing groups that had broken away from the CPP in the years following the overthrow of Marcos announced that they were to unite as the Partido ng Manggagawang Pilipino (PMP—Filipino Workers' Party).

Concerns over corruption, lawlessness and political instability continued to shape political and economic life in the Philippines throughout 2003. Indeed, President Macapagal Arroyo used the January 2003 anniversary of her assumption of the presidency to announce a national campaign against corruption, allocating public funds to such projects as the computerization of vote counting and reforms to the Bureau of Internal Revenue and the Customs Office. A number of kidnap-for-ransom syndicates, some involving renegade police and military officers, were apprehended, and four officials from the Bureau of Internal Revenue were dismissed. In May the Supreme Court upheld the Government's earlier cancellation of the contract for Manila's new international air terminal, on the grounds that it had been fraudulently awarded to Filipino and German joint-venture contractors under the Government's build-operate-transfer (BOT) policy. In July the Supreme Court also deemed that the US $683m. held in Marcos's Swiss bank accounts belonged to the Philippine Government, a ruling later upheld by the Swiss authorities. Finally, in September Congress passed a number of economic reform bills, providing, among other measures, for a better and more efficient system of tax collection.

Despite these and other moves, the success of which was partly assessed through an increase in state revenues, corruption continued to undermine government promises further to alleviate poverty, to attract foreign investment and, in other ways, to pursue its programme of national development. In particular, it was feared that the cancellation of the contract for the air terminal could prove particularly costly, not only in terms of the resulting compensation payments and the fact that the terminal remained almost complete but unused, but also in the effect that the case could have upon the confidence of other potential investors. In August 2003, in a survey of 66 countries conducted by the ADB, the Philippines ranked second in terms of the detrimental impact of corruption upon business in the country.

Rival allegations of corruption also came to dominate political debate at the expense of arguments over development policy. The corruption trial of former President Estrada continued to proceed, slowly, in the courts; Estrada himself filed impeachment complaints against eight justices of the Supreme Court. Meanwhile, Estrada's supporters threatened to impeach President Macapagal Arroyo on charges of corruption, while the former national police chief under Estrada, Senator Panfilo Lacson, and Macapagal Arroyo's husband, Mike Arroyo, continued to clash over allegations of 'money-laundering', racketeering and connections to drug syndicates.

While opposition efforts to undermine the credibility of President Macapagal Arroyo had only a limited impact, they appeared to have been part of a wider campaign of destabilization on the part of forces associated with Estrada. In late July 2003 the most immediate threat was presented by an attempted coup involving over 300 members of the Philippine armed forces, who took control of a shopping mall complex in the Manila business district of Makati and wired it with explosives. Reported to be part of a more serious plot discovered one week earlier, the mutiny was quickly quashed, in part owing to the lack of wider support that had been expected by some of its leaders. The soldiers involved were detained, but not before they had publicly aired their grievances, most notably the claim that high-ranking military personnel were guilty of systematic corruption in selling arms to, and otherwise profiting through collusion with, Muslim rebels in the southern Philippines. Secretary of National Defense Gen. (retd) Angelo Reyes was forced to resign, and several known associates of Joseph Estrada were charged with rebellion, including the former military coup leader Senator Gregorio Honasan. In contrast to the relaxed treatment of military mutineers under the Aquino administration, the soldiers involved in the coup attempt had criminal charges filed against them and were expected to be tried by court-martial. The state of rebellion called by the Government during the mutiny remained in force two months later, amid rumours of further attempts at political destabilization.

Concerns over corruption and instability were fuelled by the activities of various parties seeking to gain an advantage in the period prior to the May 2004 presidential election. In an apparent effort to limit the negative impact that such efforts were having on government, in December 2002 President Macapagal Arroyo announced that she would not stand as a candidate at the election and would focus instead on implementing the reforms that she had promised during her remaining period in office. However, by September 2003 the President was reconsidering her decision, buoyed by respectable economic growth, a reported drop in poverty levels, and an improvement in her popular opinion ratings; the latter was partly attributed to the firm manner in which she was perceived to have dealt with the coup attempt of July.

A number of potential candidates for the presidency were named, among them Senator Lacson and other former Estrada allies, including Senator Honasan and Danding Cojuangco. A third figure, usually identified as a liberal, was Raul Roco, a former cabinet member under President Macapagal Arroyo who had performed better than the President in several popular opinion polls. Lacson, a self-designated candidate, relied on his uncompromising reputation as former national police chief as his chief asset for the presidency during a period of increasing lawlessness, although his opponents accused him of human rights abuses, including murder and kidnapping. According to a respected national opinion poll conducted in September 2003, the most popular potential candidates were former television and radio broadcaster Noli de Castro, closely followed by Roco and Macapagal Arroyo. They, in turn, were followed by a well-known film actor, Fernando Poe, Jr, and Lacson, with Cojuangco and Honasan receiving only minimal support. The question of which parties candidates would choose to align themselves with continued to be the focus of much conjecture and manoeuvring as the election approached. Following the passage of legislation in February 2003, absentee Filipino overseas workers would, for the first time, be able to vote in the election, although only a small number had registered. One important figure who would be absent from the next election campaign, however, would be the head of the Roman Catholic Church, Cardinal Sin; he retired in September 2003 after serving for 30 years as the Archbishop of Manila. It remained to be seen whether the new Archbishop, Gaudencio Rosales, would take such an active role in national politics as his predecessor.

Beyond the political challenges posed by corruption, military coup attempts and the forthcoming presidential election, the Philippines continued to be affected by the ongoing rebellion of the NPA and Muslim separatists in the Southern Philippines. The NPA continued to increase in size and to engage government troops and local authorities in many parts of the country. It assumed responsibility for a number of attacks on government and private infrastructural installations, including the bombing of several mobile telephone relay stations, which was claimed by the Government to be a response to the failure of private companies to pay 'revolutionary taxation'. In an effort to reopen dialogue, in September 2003 plans were mooted for the holding of exploratory talks in Oslo, Norway, between Netherlands-based CPP officials and Philippine government representatives.

Government claims in 2002 that it had effectively immobilized the Abu Sayyaf group seemed to have been borne out by an apparent decline in hostilities involving Abu Sayyaf personnel. However, periodic violence continued to surface between government soldiers and Muslim separatists, and several damaging bomb attacks took place. In February 2003 key transmission pylons were destroyed, leaving the whole of Mindanao without power, and in the following month 21 civilians were killed and 134 injured when bombs were detonated outside Davao Airport. These and other attacks were thought to have been carried out by Abu Sayyaf or other breakaway factions of the MNLF or the MILF, although the leadership of the latter consistently denounced attacks on civilian targets. In June the MILF declared a unilateral cease-fire, and in July the Philippine Government asked Malaysia to host peace talks amidst pressures from the USA to have the MILF designated an international terrorist organization, a move against which President Macapagal Arroyo had campaigned. In July the Chairman of the MILF, Hashim Salamat, died, apparently of natural causes; his replacement was reported to be a moderate who continued to insist on the MILF's local orientation and organizational independence.

During late 2003 and the first half of 2004 political life in the Philippines, particularly in Manila, was dominated by the forthcoming elections, which were scheduled for 10 May, to select the President, Vice-President, 12 members of the Senate and 212 members of the House of Representatives, as well as more than 17,000 local officials. After the initial speculation and manoeuvring, five candidates were left to contend the presidential election, the incumbent President Macapagal Arroyo, who had reversed her earlier decision not to stand, Fernando Poe, Jr, Senators Lacson and Roco, and the leader of the Jesus is Lord Church, evangelist Eddie Villanueva. For a number of reasons only Arroyo and Poe emerged as serious contenders, with Poe leading in popular opinion polls for much of the campaign period. Like his friend Joseph Estrada, he had been a film idol, a former supporter of the Marcos regime and had failed to complete his secondary education. He was, if anything, a more popular public figure than Estrada, but had never been a politician. For his opponents, particularly within the intelligentsia and business community, this and his limited formal education were a major deficiency in his candidature; for his supporters, these qualities made him a 'man of the people', untarnished by the dishonest behaviour commonly associated with traditional politicians. Unlike Estrada, Poe had a reputation for marital fidelity and clean living, making him much less of a target for the Catholic Church than Estrada had been. Indeed, under the leadership of Cardinal Rosales, the Catholic Church played a less prominent role in the election campaign than had been the case under Cardinal Sin. Popular support for Poe was particularly evident in Manila, but more generally he assumed Estrada's supporter base, the members of which were still angered by the way in which their president had been removed from office and by his continued detention. Poe also advanced a similar pro-poor policy platform, although it was less pronounced and less developed during this campaign than in that of 1998. Indeed, the campaign in general was bereft of serious policy debate and centred rather on personality and manoeuvres such as the unsuccessful attempt by the Arroyo camp to have Poe excluded from the election because of his alleged alien citizenship.

Owing to the Government's failure to deliver on its promise of introducing a computerized counting system, and the fact that this helped precipitate various charges of fraud by a number of losing candidates, the outcome of the elections were not made public until six weeks after the vote, by which time more than 100 people had been killed in electoral violence, more than in 1998, but most of it associated with the simultaneous local elections. Although Poe warned of another 'people power' backlash when it became clear he would lose the presidential election, the street demonstrations that followed were relatively small and were not sustained, and even many of his supporters thought his campaign had been poorly managed. In part, Arroyo's victory, with 40.0% of the votes cast to Poe's 36.5%, was a consequence of her party's ability to secure as her running mate the highly popular celebrity politician Noli de Castro and the effective block vote of the Iglesia ni Cristo church, which had formerly supported Estrada.

Alhough the election was close, Macapagal Arroyo's victory gave her a mandate she did not formerly enjoy, having not originally been elected into the presidency. Moreover, the immediate dangers of a military coup, rumoured throughout most of her last incumbency, seemed to have significantly diminished, her party, Lakas-Christian Muslim Democrats (Lakas-CMD, as Lakas-NUCD had restyled itself in 2003), formally enjoyed majority support in both houses of Congress, and her 10-point plan for the nation, articulated in her inauguration and State of the Nation speeches, won broad endorsement from the principal political parties. Although she still faced a legal challenge to her electoral victory, as did many other successful candidates in the local elections, this had become a regular feature of Philippine politics and, in Arroyo's case, was likely to be rejected. More serious challenges were the major socio-economic and fiscal problems affecting the Philippines: high unemployment and an increasingly dangerous reliance on overseas contract work, a decline in foreign investment, a severe budget deficit resulting from inadequate state revenues and entrenched corruption, and a mounting foreign debt. While there was broad acknowledgement of these issues and the need to address them, vested interests in Congress threatened to hinder, for example, the introduction of new taxation measures. President Arroyo promised to address state inefficiency, not only by tightening and downsizing the bureaucracy, but also by replacing congressional government with a parliamentary system. This, too, had broad support among legislators, but would require constitutional change (planned for 2005), something that neither of her predecessors could achieve in the face of broad public opposition.

Plans to introduce a parallel federal system of government, in order to devolve more of the inordinate power concentrated in Congress and in Manila, were similarly likely to prove difficult to implement, as were constitutional amendments aimed at attracting more foreign investment into industries such as mining.

The other major problem confronting the new Arroyo Government was domestic rebellion, in large part a reflection of the need for social, legislative and constitutional reform. In November 2003 two men, one a former aviation chief, occupied the control tower of the Ninoy Aquino International Airport in Manila in order to draw attention to the national consequences of ceaseless political feuding. Although the siege was brief and the men were shot dead by the police, this event, like the earlier hotel siege by disaffected military officers, indicated the level of discontent felt among many Filipinos and their preparedness to challenge state authority openly. The NPA continued to engage government troops in battle in many parts of the country, with over 200 killed in just six months of fighting to June 2004. As a consequence of the CPP's shortage of income, following its branding as an international terrorist organization by the USA, the NPA further increased its imposition of 'revolutionary taxation' on candidates during the May elections, as well as on various business interests, notably cellular telephone companies, a number of their communication towers being bombed. Peace negotiations between the CPP and the Philippine Government had commenced in Oslo in February, but were suspended by the party in August, with a promise that attacks on the new Arroyo Government would be intensified.

Peace negotiations also continued between the Government and the MILF, but the change in leadership and continued efforts by Indonesian and Malaysian members of the regional terrorist organization Jemaah Islamiah to recruit and establish training camps in Mindanao created tensions within the MILF and, according to some sources, were leading to its fragmentation. In February 2004 Abu Sayyaf re-emerged in a bombing incident that led to the sinking of a new super-ferry, resulting in 116 deaths. This was reported to have been perpetrated in conjunction with the Manila-based Rajah Solaiman Movement, an organization of Muslim converts, who were also reported to have been planning a bomb attack on a shopping centre in the capital. In August 17 members of Abu Sayyaf, found guilty of kidnapping offences, were sentenced to death.

RECENT FOREIGN RELATIONS

Just as the Estrada Government sought to reassure local business that it would continue with Ramos's economic reforms, so it seemed intent on winning international support by promising continuity of policy. Estrada retained Domingo Siazon, Jr, as his Secretary of Foreign Affairs, and announced that he would encourage the Senate to ratify the Visiting Forces Agreement (VFA) with the USA, concluded under Ramos's administration. While Siazon sided with Thailand at the Association of South East Asian Nations (ASEAN) meeting in July 1998, in trying to reverse the regional association's policy of non-interference in member countries' internal affairs, Estrada was understood to oppose such a change.

Negotiations over the disputed sovereignty of the Spratly Islands, in the South China Sea, took place in the early 1990s; all parties agreed to reach a settlement by peaceful means and to develop jointly the area's natural resources. In December 1993 the Philippines and Malaysia agreed to co-operate on fishing rights for the Spratly Islands in the area not claimed by the other four countries. Following a visit to Viet Nam (the first official visit by a Philippine Head of State to that country), Ramos appealed for the six countries with claims to the Spratly Islands (the Philippines, the People's Republic of China, Taiwan, Viet Nam, Brunei and Malaysia) to remove all armed forces from the area. In May 1994 the Philippine Government granted a permit to a US company to explore for petroleum off the coast of the south-western island of Palawan, which covered part of the disputed area of the Spratly Islands. The People's Republic of China reaffirmed sovereignty over the region and lodged an official complaint. In June 1994 the Speaker of the House of Representatives, Jose de Venecia, defused the issue by successfully inviting China to be a partner in the project.

In February 1995 it was revealed that Chinese forces had occupied an area of the Spratly Islands claimed by the Philippines (the first time that China had seized territory in the South China Sea from a country other than Viet Nam). The Philippine Government lodged a formal diplomatic protest with the Chinese Government, which had established several permanent structures on a reef (Mischief Reef) about 200 km from the Philippine island of Palawan. Although the Philippines ruled out the use of force to regain the territory, air and sea patrols in the region were increased, 62 Chinese fishermen and five Chinese vessels were detained in the area and Chinese territorial markers were destroyed. Bilateral discussions in March in Beijing failed to resolve the dispute. Following two days of consultations in August, however, the Chinese Government agreed for the first time to settle disputes in the South China Sea according to international law, rather than insisting that historical claims should take precedence. The Chinese and Philippine Governments issued a joint statement agreeing on a code of conduct to reduce the possibility of a military confrontation in the area. A similar agreement was signed with Viet Nam in November 1995. In March 1996 the Philippines and China agreed to co-operate in combating piracy, which was also a source of tension between the two countries since pirate vessels in the area often sailed under Chinese flags. In May, however, the Philippines, together with other nations in the region, expressed concern after the Chinese Government announced new territorial claims.

Relations between the Philippines and China improved slightly in late 1996 when the Chinese President, Jiang Zemin, visited the Philippines in November and the two countries agreed to exchange military attachés. However, the construction of an airstrip by the Philippines on one of the Spratly Islands was criticized by China. The entry of Chinese warships into the waters around the Spratly Islands in April 1997 prompted a formal protest from the Philippine Government. Tension was heightened in May, when the Chinese flag, which had been raised on one of the islands, was replaced by the Philippine flag. Several Chinese fishermen, who had entered Philippine waters, were arrested by the Philippine authorities, and in late May negotiations were held between the two countries to resolve the situation. In August it was reported that the Philippine and Chinese Governments had agreed to put aside their territorial dispute and to concentrate on strengthening economic co-operation.

In March 1994 the East ASEAN Growth Area (EAGA), encompassing the southern Philippines, the eastern Malaysian states of Sabah and Sarawak, Brunei, and the Indonesian islands of Sulawesi and the Moluccas (Maluku), was formally conceived.

The escalating conflict over the Spratly Islands constituted the most urgent international issue confronting the Philippines during Estrada's first year of office. In November 1998 the Philippines protested to China over its construction of fort-like buildings on Mischief Reef, which the Philippines claims lies within its exclusive economic zone. The Philippine navy responded by arresting Chinese fishermen for illegal fishing in nearby waters, and impounded their boats. China, however, refused to dismantle the structures. Following unproductive talks between the two Governments in April 1999, relations deteriorated further in May when China claimed one of its fishing vessels had sunk after being deliberately rammed by a Philippine navy boat. A further Chinese fishing vessel sank in a collision with a Philippine navy vessel in July. In the mean time, Philippine relations with fellow ASEAN claimants of the Spratly Islands varied. Joint troop operations in the islands were agreed with Viet Nam, while in mid-1999 Malaysia stepped up its dispute with the Philippines with the construction of a two-storey building on Investigator Shoal, a reef claimed by both countries. The Philippines and Viet Nam drafted a new code of conduct for claimants of the Spratlys, which was approved by ASEAN.

As a result of China's expansionist policy in the South China Sea, relations between the Philippines and the USA improved considerably. Despite protests by the Catholic Church and the nationalist left, the May 1999 session of the Philippine Senate ratified the VFA between the Philippines and the USA, thereby allowing joint military operations to take place in the Philippines. Estrada campaigned strenuously for the ratification, despite his earlier support for the dismantling of the US military bases in 1992. Rationalized as an integral element to the modernization of the Philippine armed forces, the VFA also appeared to be part of the Government's strategy for dealing with its problems in the Spratly Islands and internal conflicts with the MILF and the communist movement.

Notwithstanding the ratification of the VFA, most of President Estrada's activities in foreign affairs focused on the region. All eight of his international visits, undertaken by mid-1999, were to neighbouring countries. While these generally worked at consolidating and building on existing alliances and economic agreements, Estrada also departed from ASEAN convention, raising the ire of the Malaysian Prime Minister, Mahathir Mohamad, by criticizing Malaysia over the arrest and beating of the former Deputy Prime Minister, Anwar Ibrahim.

The Estrada Government's largely inept handling, from March 2000, of the hostage crisis (see above), together with its fruitless suspension of flights to Taiwan (which resumed in October), left it with a somewhat tarnished international reputation. On a more positive note, the Philippines did participate successfully in the UN peace-keeping force in East Timor (now Timor-Leste), after the latter had opted for independence from Indonesia. Another achievement was a reduction in tensions with China over the Spratly Islands in the South China Sea. In Manila, in November 1999, China had rejected the code, prepared by ASEAN, for international conduct in the South China Sea, but it subsequently softened its hardline stance by circulating its own modified code in early 2000. During a state visit to China in May 2000 President Estrada concluded a number of agreements and issued a joint statement with China's President Jiang in which they undertook to seek the peaceful resolution of regional disputes. President Estrada subsequently visited the USA, but was forced to abandon his visit owing to the hostage crisis in the Philippines. Earlier, in January 2000, the first joint military exercises between the USA and the Philippines for five years took place, amid local protests, under the new VFA. Although it was not publicly acknowledged, US military personnel were reported to have been directly involved in attempts to resolve the Abu Sayyaf hostage crisis.

During the remainder of the Estrada period and beyond, under the presidency of Gloria Macapagal Arroyo, Moro separatist activities continued to be at the forefront of international relations. During the latter part of the Estrada period, the MILF was reported to have been training radical Muslim fighters from Indonesia, while the bombing, in August 2000, of the Philippine ambassador's residence in Jakarta was believed by some to have been carried out by Indonesian supporters of the MILF. Conversely, under Macapagal Arroyo, who assumed the presidency in early 2001, cease-fire agreements with the MILF were brokered through Libya and Malaysia. During this period President Macapagal Arroyo made her first state visit to Malaysia. Abu Sayyaf had continued to move in and out of Sabah with impunity, and in September 2000 had kidnapped three Malaysians from the resort of Pandanan, adding to the others it still held captive. The military freed a US hostage in April 2001, but in May Abu Sayyaf members took three more US citizens hostage, along with 17 Filipinos, from a luxury resort in Palawan. Abu Sayyaf later claimed to have beheaded one of the US captives. What most drew international attention to the activities of Abu Sayyaf was the hunt for members of al-Qa'ida, following attacks, allegedly by that organization, on New York and Washington, DC, in September 2001. It was well known that Abu Sayyaf had been assisted by this group, some of whose members from the Middle East had spent time in the Philippines. Declaring her support for the US-led alliance against the al-Qa'ida network, President Macapagal Arroyo agreed to grant the USA access to Philippine military bases and air space, while the Congress adopted legislation aimed at combating 'money-laundering'.

A major factor affecting Philippine internal politics and international relations from late 2001 was undoubtedly the US-led 'war on terror'. Not only was there continuing concern over Abu Sayyaf links to the al-Qa'ida terrorist network, but poor law enforcement in the Philippines appeared to make the country a potential haven for international terrorists. However, anti-Islamic sentiment among the majority Christian population and President Macapagal Arroyo's strident support for the US Government's position also made the Philippines the initial focus of the new US military presence in South-East Asia. In December the USA pledged US $100m. in military aid. In February 2002 660 US troops were deployed to the south as part of the six-month Balikatan campaign with the Philippine military to suppress Abu Sayyaf operations, although their role was to be limited to training, advice, surveillance and infrastructure development. While their presence attracted sharp criticism from the left, and from some nationalist politicians, including Vice-President Guingona, surveys indicated widespread popular support, at least among the Christian population. Despite concern that the US military presence would turn local secessionist sentiments in the Muslim south into popular support for the cause of al-Qa'ida, this did not appear to have transpired, in large part since US troops did not engage in direct combat and were concentrated in Basilan, Abu Sayyaf's stronghold, rather than in areas under the influence of the wider secessionist movement, led by the MILF. At the end of the six months the Balikatan campaign was declared a success, with the total of active Abu Sayyaf members reported to have declined from 4,000 to about 100. A number of Filipino military officers were expected to be court-martialled for collusion with Abu Sayyaf, but overall the government forces stationed in the south were believed to have improved in fighting power and discipline. While some hostages held by Abu Sayyaf were released during the campaign, at least two were killed, one US national and one Filipino, and other hostages were seized just before and soon after the campaign's completion. Indeed, two Filipinos were beheaded shortly after the departure of US forces. At the height of the campaign, US troops in the south were reinforced to number about 1,000, while a further 17,000 were engaged in military manoeuvres in the north. In October several people were killed and some 150 injured in two bomb attacks in Zamboanga City, which were widely attributed to Abu Sayyaf. A further three people were subsequently killed when a bomb exploded on a bus in Manila. Five suspected members of Abu Sayyaf were arrested later that month.

The Macapagal Arroyo Government's co-operation with the US anti-terrorist effort also extended to the closer scrutiny of possible 'front' organizations involved in covert al-Qa'ida operations in the Philippines. In May 2002 the Government signed an agreement with Indonesia and Malaysia, which was aimed at preventing cross-border terrorist activity, with particular focus on members of Jemaah Islamiah, an organization reported to be furthering al-Qa'ida objectives in South-East Asia. In January, and again in March, the Philippine authorities arrested several Indonesian nationals suspected of producing or transporting bombs, while at the same time arrests of foreign nationals were made elsewhere in the region. In November 2001 the Malaysian Government arrested, and then transferred to the Philippine authorities, the former MNLF leader, Nur Misuari, who earlier had led a violent, but unsuccessful, attack on a military outpost in Jolo, after being forced to relinquish his position as Governor of the ARRM. Malaysia also continued to take a significant role in the peace negotiations between the Philippine Government and the MILF.

While regional inter-government co-operation was prompted by the need to address the increasing threat of transnational terrorist activity, the latter also intensified some long-standing disputes. In an attempt to exercise tighter control over its population and borders, the Malaysian Government introduced stringent new immigration laws in August 2002, aimed at the estimated 80,000 illegal Filipino workers and residents, most of them Muslims, who had long been settled in Sabah. As thousands of Filipinos were assembled in detention centres and deported to the Philippines, reports emerged of their abuse and mistreatment by the Malaysian authorities. Also confronted with the difficult prospect of accommodating these large numbers, mainly in Mindanao, the Philippine authorities protested to Malaysia, while some Filipino politicians urged a revival of the Philippines' long-standing claims on Sabah. Following the intervention of President Macapagal Arroyo and the protests of employers and human rights activists in Malaysia, Prime Minister Mahathir suspended the mass deportations in early September.

As US forces escalated their 'war on terror' in early 2003 by leading an invasion of Iraq with the intention of ousting the regime of Saddam Hussain, President Macapagal Arroyo remained the USA's staunchest ally in South-East Asia. In February, in response to US pressure and citing intelligence reports on links between Iraqi agents and Abu Sayyaf, the Philippine Government expelled a number of Iraqi diplomats, embassy employees and businessmen. Meanwhile, in December 2002 the Philippines signed a five-year military agreement with the USA under which the Philippines would be provided with heavy equipment, training and logistical support in its efforts to address local and international terrorism. During a state visit to the USA by President Macapagal Arroyo in May 2003, the Philippines was granted US $356m. in US military aid, and, in September, about 2,000 US marines arrived in the Philippines to conduct joint military exercises. Regional security measures were also intensified in September 2003 with the holding of two regional summits: one of regional police chiefs in Manila, and one of army chiefs in Kuala Lumpur. It was expected that regional security measures would be further boosted with the appointment of the Philippines to the UN Security Council.

The focus of these international measures was the growth of the regional terrorist movement Jemaah Islamiah, which reputedly used the Philippines as its principal training ground. Included among those of its agents who had trained in the southern Philippines were several men arrested in Indonesia for the October 2002 bombing of a tourist resort on the Indonesian island of Bali, which killed over 200 people, mainly Indonesians and Australians. The penetration of the Philippines by international terrorists was also cited as the reason for the temporary closure of the Australian and Canadian embassies in Manila in November 2002. While the MILF steadfastly denied any support for, or organizational linkage to, Jemaah Islamiah, its leadership acknowledged that it might have unintentionally trained Jemaah Islamiah personnel. In May 2003 Saifullya Yunos, a former MILF member, was arrested on the basis of army intelligence reports that he had formed a Jemaah Islamiah 'cell' in the Philippines with the alleged manufacturer of the bombs used in the Bali attack, Rohman al-Ghoza, who was already in custody. Together with the alleged leader of Jemaah Islamiah, the Indonesian Hambali, who was later arrested in Bangkok, Thailand, Yunos and al-Ghoza were charged with having been responsible for a number of bombings in Manila and Mindanao. Al-Ghoza had already been convicted on one of these charges and was sentenced to a 10-year prison term. However, in July, on the same day that the Australian Prime Minister arrived in Manila to sign an anti-terrorist agreement, al-Ghoza escaped from custody, along with two Abu Sayyaf prisoners, apparently by bribing prison officials. Following a number of other escapes from police custody of high-profile terrorist suspects, the incident was not only an embarrassment to the Philippine authorities, but supported the claim, made in some quarters, that the Philippines provided a safe haven to international terrorists. In June the Japanese ambassador was forced to apologize after calling into question the ability of the Philippine authorities to address local law and order problems. In October al-Ghoza was located and killed by Philippine soldiers in Mindanao. This allayed some fears over the Government's ability to control the entry of foreign terrorists into the Philippines, as did the relatively trouble-free, albeit short, visit to Manila of US President George W. Bush on his way to a summit meeting of the Asia-Pacific Economic Co-operation (APEC) forum in Bangkok later in the same month.

However, US concerns over the Philippines stance on the 'war on terror' escalated in July 2004, when President Arroyo agreed to bring home, slightly ahead of schedule, a small police and military contingent from Iraq in response to threats by kidnappers to execute a Filipino truck driver. Reacting to strong domestic pressure to save the life of a man who had come to symbolize the 1.3m. Filipino civilian workers employed in the Middle East, including hundreds in Iraq, Arroyo's decision drew considerable local approval. While this support was invaluable, given Arroyo's recent, closely fought election victory, the decision was sharply condemned by fellow 'allies of the willing', the USA and Australia. In the same month a US envoy to the Philippines warned of ongoing concerns over intelligence reports noting the continued establishment of Jemaah Islamiah training camps in Mindanao. Shortly afterwards, US and Filipino troops commenced 'training exercises' in North Cotabato, amid claims of increased US surveillance activities in Mindanao. At the same time, US troops arrived for joint naval exercises in the northern Philippines. Despite the short period of diplomatic tension between the Philippines and its allies, President Arroyo and other government officials continued to pronounce their support for the 'war on terror'. Arroyo herself was reported to have been saved from a thwarted Jemaah Islamiah terrorist attack on her inauguration ceremony on 30 June. The Philippines was also forced to tighten security at its foreign embassies following threats, issued in Sweden, from sources identified with al-Qa'ida.

The Spratly Islands continued to arouse regional tensions in 2003–04, despite a 'declaration on the conduct of parties in the South China Sea' signed by China and the relevant ASEAN countries in November 2002, but, unusually, the Philippines was only mildly critical of its neighbours' transgressions. Moreover, a state visit by Arroyo to China in September 2004 resulted in the two countries signing a Defence Co-operation Agreement.

Economy

EDITH HODGKINSON

Since independence in 1946 the overall record of the economy of the Philippines has been one of underperformance—both relative to other countries in the region and in terms of its potential for growth. Initially, its rate of development was close to that of Japan, and it shared many characteristics with the countries of the Far East that graduated to newly-industrialized status in the decades after the Second World War (Taiwan, the Republic of Korea and Singapore), with the added advantage of a plentiful natural resource base and a potentially enormous domestic market. Yet it only intermittently matched the rapid economic expansion recorded by the 'tiger' economies of the region, both the initial group and the second group of the 1980s and 1990s (comprising Thailand, Malaysia and Indonesia). The less rapid progress of the Philippines' economy was a result of major and basic structural deficiencies. Among these were: a dependence on imported intermediate and capital goods, following a period of protectionism; a grossly inequitable distribution of wealth; a tendency for wealth to be acquired through ownership of monopoly privileges, derived from connections with influential persons, rather than entrepreneurship; and pervasive corruption, which distorted the design and implementation of economic policy. The Government that came to power after the overthrow of Ferdinand Marcos in 1986 espoused policies that would mitigate these features, and they were pursued with greater vigour and success by the administration of Gen. Fidel Ramos, who succeeded President Corazon Aquino in mid-1992. The economy was developed through the dismantling of domestic monopolies and the liberalization of the terms of entry of foreign goods and capital (notably in the banking sector). With investment rising, boosted by inflows of funds from the Philippine community overseas, and strong external markets for the country's manufactures, growth in gross national income (GNI) increased steadily in the mid-1990s, reaching 7.2% in 1996, and the Philippines at last seemed set to match the very rapid growth of other economies in the region.

This advance, however, was interrupted by the regional financial and currency crisis in 1997. The Philippines was forced to abandon the *de facto* pegging of the peso to the US dollar (which had made investment in Philippine bonds attractive and helped reduce inflation). The consequent depreciation of the currency could be restrained only by a steep increase in domestic interest rates. The much higher cost of credit, accompanied by tighter supply, as banks sought to reduce their exposure to bad debts, had a severe adverse effect on investment. While economic growth was sustained through late 1997 (the rise in GNI was reduced to 5.3% over the full year) and into early 1998 by continued dynamic foreign demand for Philippine manufactures, the drought associated with El Niño (a warm current, which periodically appears along the Pacific coast of South America, disrupting the usual weather patterns over a large area) caused a sharp fall in agricultural output in 1998. Finally, the growing economic difficulties in Japan, a major export market and leading source of capital, meant that GNI growth reached only 0.4% in 1998. An improvement in agriculture raised GNI growth to 3.7% in 1999, and the pace of economic recovery quickened in the following two years, with growth of 4.8% in 2000 and 5.6% in 2001. The improvement in 2001 occurred despite a marked deterioration in the external environment, with a sharp slowdown in the growth of demand in the USA in that year and a cyclical downturn in the information technology (IT) market resulting in a 16.2% fall in Philippine export earnings. The economy's resilience was largely attributable to the strong performance of agriculture and the high level of remittances from the Filipino community overseas. Economic growth was maintained at 4.5% in 2002, with a slower rate of agricultural expansion offset by a moderate recovery in exports as the IT market rebounded. The recovery of exports slowed sharply in 2003, with growth of only 1.5%, as the Philippines lost ground in the IT market, but the rate of GNI growth increased to 5.6% in that year, owing to another upturn in the agricultural sector and a resultant expansion of domestic consumer demand. These two factors, together with a strengthening in the export recovery as demand from Japan increased, pushed GNI growth to 6.2% in the first three months of 2004. The rate of growth for the rest of 2004 was expected to be around one point below this level.

GROWTH WITHOUT REFORM, 1946–85

On independence in 1946 the Philippines maintained the broad character of its economic policies as a US colony: freedom of trade with the USA, preferential entry for Philippine commodities into the US market and preferential treatment for US investment in the Philippines. However, with the vast increase in demand after the ravages of war, and despite reconstruction aid from the USA, import spending rose to levels far beyond the country's financing capacity. Consequently, import controls were imposed in 1949, and the country adopted the industrialization by import-substitution policy, which was to become the pattern among newly-independent developing countries. Very high tariffs on manufactured goods and quantitative controls were combined with a preferential exchange rate for the import of intermediate and capital goods, and interest 'ceilings', which underpriced capital. This set of measures served as a strong stimulus to manufacturing production, which grew by 12% a year in the 1950s, making the Philippines both the most industrialized country in South-East Asia by the end of the decade and the fastest-growing economy. The misuse of resources engendered by the policy of import-substitution, notably the recourse to capital-intensive processes in a country where labour was skilled and low-cost, was not immediately perceived. This was due to strong international prices for the country's agricultural exports, while the low population-to-land ratio allowed an expansion in the area under cultivation and, therefore, an increase in production. The Philippines was thus able to earn the foreign exchange needed to fund its dependence on imports.

However, at the beginning of the 1960s the prospects for a sustained growth in export earnings faded as a land constraint emerged. With rising imports threatening to create a balance-of-payments crisis, the Philippine Government took two significant steps towards stimulating export-manufacturing based on the country's factor endowment. The peso was devalued by around 100% and exchange controls were eliminated. However, other protectionist features were retained: high tariff rates, quantitative controls, and regulated and low interest rates. Thus, there was little incentive to switch from capital-intensive, domestic-orientated activities to labour-intensive export manufacture.

It was only after the introduction of martial law in 1972 that the Government implemented an effective policy, mainly through tax and tariff incentives, of encouraging export manufacture and attracting foreign investment. Coinciding with a period of buoyant world demand, the policy proved highly successful, increasing earnings from non-traditional manufactures by 25%–30% a year in 1973–78. With commodity prices booming in the 1970s, the rate of economic growth accelerated to an average 6.9% a year in 1973–79.

This record of success, and the ready availability of foreign funds as commercial banks sought to recycle 'petrodollars', prompted the Government to embark on a programme to widen the country's industrial base. The initiation of large-scale, mainly heavy industrial, projects was essentially aimed at enhancing self-sufficiency (only one was directed at export markets), but required massive inputs of foreign capital and technology. As a result, the foreign debt rose eightfold between 1975 and 1982. The debt-service payments required proved unsustainable when the international recession of the early 1980s brought a slowing in the growth in demand for Philippine manufactures. As a result, two-fifths of the country's foreign-exchange earnings were needed to cover the debt repayment and interest bill in 1982.

The deterioration in external economic conditions brought into sharp relief one of the characteristics of economic management under President Marcos: the concentration of ownership and control among members of the President's family and a group of close associates, in a system that came to be dubbed 'crony capitalism'. Marketing monopolies were set up for the coconut and sugar industries, while subsidies and special privileges (including preferential access to bank credit and government guarantees for foreign borrowing) were awarded to companies owned by 'crony' interests. The failings of this pattern of economic management were masked when the economy was expanding rapidly, but as growth slowed sharply in the early 1980s, to an average 2.4% a year in 1981–83, many of the 'crony' companies, including commercial banks, found themselves in severe financial difficulties. In August 1983 the assassination of President Marcos's leading opponent, Benigno Aquino, Jr, precipitated a downturn in the economy. It immediately produced a collapse of confidence both domestically, prompting a flight of capital, and among the country's foreign creditors, who refused to renew short-term financing. The already fragile balance-of-payments situation thus tipped into crisis. As the economy moved into deep recession in 1984 and 1985, with GNI contracting by almost one-quarter, a number of companies collapsed. This in turn undermined the viability of the big government banks, the portfolios of which included a high and growing proportion of non-performing assets and which consequently recorded massive losses in these two years.

The first set of measures introduced to address the crisis were emergency ones, designed to halt the collapse of foreign-exchange reserves. A moratorium was declared on the repayment of external debts, rigorous exchange controls were imposed and the peso was devalued by almost one-quarter. This devaluation, which had long been urged by the IMF and the World Bank, allowed negotiations to begin on a rescheduling of foreign debt and the provision of new funds by the IMF and the commercial banks. The Philippines was required to implement a severe austerity programme, with harsh reductions in government spending, restrictions on the liquidity of banks and a steep rise in interest rates.

In addition to its short-term stabilization objective, the programme agreed with the IMF had a long-term aim: the development of a more efficient economy, based on the exploitation of the country's factor endowment, in labour and in agricultural resources. The distortions represented by an overvalued currency, high import tariffs and quantitative controls, and interest subsidies were to be removed, and the role of market forces enhanced through the reduction of government intervention in the economy, specifically through the ending of the agricultural monopolies and the restructuring of government financial institutions.

A PERIOD OF INCOMPLETE RESTRUCTURING

Economic reform was only intermittently implemented by President Marcos, since it weakened the power system he operated. It was adopted with greater commitment after his overthrow in February 1986. Within months the coconut and sugar marketing monopolies were dismantled and a wide range of tax exemptions eliminated. Non-performing assets were removed from the portfolios of the government banks, their operations rationalized and their special privileges removed to place them on an equal footing with private banks. Government corporate holdings were put up for sale.

Supported by the country's international creditors (see Foreign Debt, below), the new Government was able to embark on a rapid restimulation of the economy, mainly through an increase in spending on infrastructure and on an emergency rural employment programme. The boost from the fiscal side was reinforced by the sharp improvement in world coconut prices in both 1986 and 1987. With demand for Philippine export manufactures strong, and the Government committed to stimulating the private sector, in the late 1980s investment began to improve. The economy entered into another period of strong expansion, with GNI growth averaging 6% a year in 1987–89.

This process was abruptly halted at the end of 1989. The economy encountered the familiar problem of a burgeoning deficit on the current account of the balance of payments, as import growth continued to exceed export expansion. From a modest US $390m. in 1988, the deficit had risen to $2,695m. in 1990. While debt restructuring by the country's commercial creditors, and new funds from official sources in 1989–90 (see below), reduced both total debt and the cost of servicing it, the balance of payments was adversely affected by the massive rise in the price of petroleum imports, following the Iraqi invasion of Kuwait in August 1990. The most serious reverse, however, was the severe shortfall of power in Luzon, the country's industrial centre (see below). The situation was compounded by the austere budgetary stance, which was the condition of IMF support, and GNI growth declined to about 1%–2% in 1991–93. It was only after electricity-generating capacity was rapidly expanded to meet the supply deficit that economic growth improved, increasing to just over 5% in both 1994 and 1995, and rising to 7.2% in 1996.

This surge in growth was in part due to the major structural reforms that were implemented by the Ramos administration. These included the removal of nearly all foreign-exchange controls, the termination of monopolies in telephones, aviation and inter-island shipping and the ending of the 45-year ban on the entry of foreign banks. Privatization, which had been a major feature of the Aquino administration (30% of the country's largest bank, the government-owned Philippine National Bank—PNB, was transferred to the private sector, and the national airline was privatized), was extended to petroleum-refining and -marketing, and steel manufacture. Moreover, the private sector was brought in to relieve one of the most serious constraints on economic growth, the inadequacy of the physical infrastructure, which could not be corrected by the public sector while budget finances were still precarious. The 'fast track' programme of capacity expansion that resolved the electricity supply crisis in 1993 employed a system of build-operate-transfer (BOT) schemes, under which private firms finance the installation of new facilities and contract to operate them for a fixed period, at the end of which they are transferred to government ownership. The same mechanism (or its development, the build-operate-own contract) was extended to other sectors, such as roads and commuter railways. Not all the liberalization targets had been achieved by the end of the Ramos administration in mid-1998, but the economy had been transformed from its state in 1986 when President Marcos left office. Economic liberalization continued under the next President, Joseph Estrada, although at a slower pace. The ban on foreign participation in retail trade was lifted in 2000, but the long-mooted privatization of the state-owned electricity utility, the National Power Corporation (NAPOCOR), remained mired in congressional debate. One significant privatization—the disposal of the Government's residual equity in the PNB—was conducted in a way that confirmed growing unease about a recrudescence of 'crony capitalism'. Major supporters of the President's election campaign were repeatedly accorded favourable treatment. One former Marcos crony had voting control over sequestered equity returned to him, allowing him to take over the country's leading food conglomerate, while another had a massive tax evasion case abandoned and his investment in the former state airline underpinned by the Government's suspension of the country's aviation agreement with Taiwan. A new feature was the incoherence and inconsistency of policy implementation, as the President's detached style of government allowed personal rivalries to develop unchecked within the administration. Hopes of an improvement in policy effectiveness after Gloria Macapagal Arroyo assumed the presidency in January 2001 were initially borne out. By June Congress had approved legislation to liberalize the electricity sector, including the privatization of the state electricity utility. More significantly, the Government achieved its principal policy target—the resumption of control over budget finances after the sharp deterioration under the Estrada administration—by holding the fiscal deficit in 2001 to within 2% of its target. However, the momentum was not maintained during the last two years of the first Macapagal Arroyo administration. The planned sale of the electricity transmission network was delayed by the failure of Congress to approve the transfer of the franchise and fiscal control was lost in 2002, when the deficit exceeded its target by two-thirds. Budget management improved once more in 2003 and the first half of 2004. However, the narrow margin by which Macapagal Arroyo secured victory at the May 2004 presidential election, and the challenging of the result by the opposition, blighted the prospect of further substantial reforms under the re-elected President.

POPULATION AND EMPLOYMENT

The rate of population growth has slowed in recent decades, from an annual average of 3.1% in the 1960s to 2.3% in the late 1990s. This reflected a decline in the birth rate, owing in part to birth-control programmes first implemented under the Marcos Government. However, in the absence of government policies supporting a reduction in the birth rate, population growth has steadied in recent years.

The rise in population numbers was not equalled by an increase in employment opportunities, and the constraint of land caused a significant migration to the urban areas, which failed to provide enough new jobs to absorb the addition to the labour force. The rate of unemployment fluctuated between 9% and 12% from the late 1990s until 2004, while underemployment was estimated to affect about one-sixth of the population. Unemployment would be greater were it not for the high level of emigration. Some 60,000–65,000 people a year emigrated officially in the first half of the 1990s, with the total declining to an annual average of some 51,500 in 2000–01. Much higher numbers (not classified as emigrants) undertake contract work abroad; in the 1990s overseas placements of Filipino workers averaged about 700,000 a year, with the number rising to 867,000 a year in 2000–02, equivalent to nearly 3% of the labour force. A total of some 8m. Filipinos were estimated by the Government to be living abroad in 2004.

AGRICULTURE, FORESTRY AND FISHING

In 2003 the agriculture, forestry and fishing sector contributed 14.4% of gross domestic product (GDP) and in 2004 it engaged 35.5% of the employed labour force. Its share of GDP has been exceeded by that of manufacturing since the 1980s, and as early as the mid-1970s the agricultural sector had lost its primacy as export earner. The farming system is extremely diverse and includes a large number of rice, maize and coconut holdings that are farmed by agricultural tenants or workers, as well as sugar *haciendas* and large plantations, devoted mainly to non-traditional export crops such as bananas and pineapples.

Agriculture experienced a period of relatively strong growth in the 1970s, averaging annual expansion of 5%, stimulated by measures to achieve self-sufficiency in food grains as well as by the rise in the area under crops, particularly through the clearing of virgin forest in Mindanao. The pace of expansion then slowed to an average of only around 2% a year in the 1980s through to the mid-1990s, as a land constraint emerged, government investment in infrastructure favoured urban areas, and the marketing monopolies operated under the Marcos administration depressed producer prices. Rural income was supported by measures introduced by the Aquino Government, such as the removal of export taxes on agricultural commodities and the dismantling of monopolies, while government investment in rural infrastructure and services for farmers has tended to rise, making the sector less vulnerable to adverse weather conditions. Agricultural growth has thus strengthened, averaging 3.9% a year in 2000–03.

Rice is the principal food crop, grown on around one-quarter of the cultivated area. The introduction of higher-yielding strains, together with an expansion in supplies of fertilizer and pesticides, has made the country broadly self-sufficient in rice (except in years of unfavourable weather conditions) since the late 1970s.

Coconuts are the most important cash crop, and the Philippines is the world's leading exporter of coconut products. Between 1975 and 1982 there was a rapid rise in annual output, from 1.9m. metric tons (copra equivalent) to 3.4m. tons. After 1983 output declined, largely owing to the ageing of trees, which reduced overall yields. A major replanting and rehabilitation programme was initiated in 1988, supported by World Bank funds, and began to have an impact in the mid-1990s, with output returning to 2m.–3m. tons a year.

Sugar, which was once a leading export crop, is now of minor importance. For many years a guaranteed share of the US market, at a fixed price, shielded the Philippine sugar industry from the effects of a long-term decline in productivity, but a lengthy period of world price weakness prompted a switch to higher-value crops and fish farming in the 1980s. With rising domestic demand, imports of sugar have become necessary to supply the premium US market. Sugar has been far exceeded as a dollar earner by bananas and pineapples, mainly grown on plantations developed in Mindanao by multinational companies. The production of mangoes and rubber has also increased in significance.

Agrarian Reform

One of the major requirements for sustained and broadly-based economic growth in the Philippines is land reform. Since independence, pressure for such a redistribution of resources has been continually resisted by the landed élite, who also constitute the political élite. In 1972 President Marcos launched a limited programme of agrarian reform covering land under rice and maize in holdings of 7 ha or more. By 1986 more than one-half of the 600,000-ha area covered by the programme still remained to be distributed. The Comprehensive Agrarian Reform Programme (CARP), introduced by the Aquino Government in 1988, provided for the redistribution of 4.5m. ha, or about 55% of all agricultural land, in three stages over a period of 10 years. The first phase involved unredistributed land covered by the Marcos programme and land held by the state—a total of some 1.09m. ha. The second phase, begun in 1989, involved private landholdings exceeding 50 ha, and the third phase, covering the vast majority of land to be redistributed, began in 1992. Landowners were allowed to retain 5 ha and their direct heirs 3 ha, but large estates could remain intact if they were made into corporate holdings with stock transfers substituting for land. Owing to obstruction by vested interests and reductions in government funding for landholder compensation, redistribution has remained well below target, and the completion date has been extended to 2008.

Fishing

The Philippines has extensive fishing resources, both marine and inland. Production increased rapidly to account for about 5% of GDP by the late 1980s. The fishing sector became a major source of foreign-exchange earnings, principally through the export of shrimps and prawns to Japan. While both freshwater ponds and most of the marine waters have not been fully developed, productivity in some areas has deteriorated because of pollution of coastal waters as the result of population growth, mining activities and destructive methods of exploitation. The infrastructure remains highly inadequate. Nevertheless, the total catch has increased gradually since the early 1990s, reaching 2.5m. metric tons in 2002.

Forestry

Forests were in the past one of the country's major resources, but suffered very severe depletion as the result of population pressure, shifting cultivation, illegal logging and inadequate reafforestation. In 1945 there were 15m. ha of virgin forest, but by 1999 the area was estimated at only 700,000 ha. A ban on logging in virgin forest, introduced in 1991, has proved largely ineffectual, and there is a prospect of its total elimination by 2010.

Government policy in the 1980s was to phase out exports of hardwood logs in order to stimulate the development of the local processing industry. An export ban has been in place since 1986, but substantial quantities of logs are believed to be illegally exported, with earnings put at about US $800m. a year in the late 1980s, more than three times the level of total officially recorded lumber export earnings at that time. The latter have fallen sharply, to an average of $32m. per annum in 2000–03, reflecting the overall decline in output as the resource base has narrowed.

MINING

The Philippines has extensive deposits of gold, silver, copper, nickel, lead and chromium. Lesser, but still important, minerals include zinc, cobalt and manganese. However, around one-quarter of the land area remains to be surveyed and some of the richest deposits are unexploited. During the 1970s the Government gave high priority to the development of minerals, which resulted in a rapid growth of the sector. Since the mid-1980s, however, mining has been in overall decline, owing to weak prices and an unstable system of taxes and incentives. By 2003 output of copper, the country's leading mineral product, had fallen to less than one-10th of its 1980 level, at 20,400 metric tons. The Philippines is also a significant producer of gold. Largely a by-product of copper mining, its output tends to reflect trends in the copper sector as well as in world prices. The non-metallic sector has performed better, with the output of coal stimulated by energy conversion in the cement and mining industries. Averaging some 1.2m. tons per year in the 1990s, coal production surged to a record 2m. tons in 2003.

Investment in the sector remains depressed by bureaucratic delay and legal challenges, with the improved access for foreign investment accorded by mining legislation introduced in 1995 ruled unconstitutional by the Supreme Court in January 2004. The sector thus remains underdeveloped, accounting for only 1% of GDP in 2003 and engaging 0.4% of the employed labour force in 2004.

ENERGY

In 1973, at the time of the first major increase in international petroleum prices, the Philippines was dependent on imported oil for 95% of its energy supply. This level had fallen to 50%–55% by the late 1990s and early 2000s, reflecting the development of various domestic sources of energy, notably coal, hydroelectric power, geothermal steam, and non-conventional sources (mainly bagasse, agricultural waste and dendrothermal). Dependence has declined further with the beginning of production, in 2001, at offshore gasfields.

The contribution of domestic petroleum production has been very limited. Output from petroleum deposits off the island of Palawan reached a peak in its first year, in 1979, of 42,000 barrels per day (b/d). Despite the entry into production of other commercial oilfields in the 1980s and early 1990s, production had declined to about 1,000 b/d in 2000–01. Some recovery was expected from the development of oil reserves at the offshore gasfield at Malampaya. However, in early 2004 the leading partners in the project decided not to develop the estimated 25m.–30m. barrels of oil reserves at the field, which they deemed not commercially viable. Natural gas reserves at this field are estimated at more than 3,000,000m. cu ft, and production commenced in September 2001, with full commercial operations starting in 2002, serving three power plants with a total capacity of 2,700 MW.

The exploitation of geothermal resources has been actively pursued: in 2001 the Philippines had an installed capacity of 1,931 MW (a total exceeded only by the USA). Considerable resources remain to be exploited: some estimates suggest that the total potential exceeds 35,000 MW.

To reduce the country's dependence on imported petroleum, a nuclear power station, with a generating capacity of 620 MW, was constructed in Bataan, Luzon. Due to enter operation in 1986–87, this highly controversial project (it was located near a seismic fault) was abandoned by the incoming Aquino Government. The failure of the Aquino Government to replace the Bataan scheme with other additions to power capacity exacerbated the emerging power shortage on Luzon island (the three island grids are not connected) as accelerating economic growth increased consumption. Interruptions in electricity supply in the Manila region in 1990–91 were estimated to have reduced GNI by close to 1%. The incoming Ramos administration embarked on a 'fast track' programme to expand electricity-generating capacity, which had achieved its aim of ending daily blackouts in Luzon by late 1993. This investment in capacity (some of it on terms which have since been questioned) proved excessive, and installed capacity was 2,000 MW higher than the required level in 2002. However, in the absence of substantial new investment in the sector—currently delayed owing to uncertainties concerning the pace of sectoral privatization and the regulatory framework—a supply shortage was expected to re-emerge by the end of the current decade.

MANUFACTURING

Philippine manufacturing developed relatively early, reaching 22%–25% of GDP by the 1960s, a share it has since broadly maintained. (Manufacturing contributed 22.9% of GDP in 2003 and engaged 9.8% of the employed labour force in 2004.) The sector was supported initially by exchange controls and, from the early 1960s, by tariffs and import quotas, which tended to promote production of consumer goods for the domestic market. Manufacturing for the export market was stimulated from the 1970s by the introduction of tax and duty exemptions for export producers and the establishment of four export-processing zones where 100% foreign ownership was permitted and companies were allowed to pay below the minimum wage. As a consequence there was a rapid growth of labour-intensive manufacturing, mainly textile products and electronic components, produced in many cases by Filipino enterprises working for multinationals on a subcontracting basis (these industries remained heavily dependent on imported inputs).

In the early 1980s the Government implemented a programme to develop the country's intermediate and heavy industrial base through 11 capital-intensive projects. Only four of these projects became operational: a copper smelter, a coco-chemical manufacturing project, a phosphate fertilizer project and the manufacture of diesel-engine components. Nevertheless, despite these developments and the elimination of import quotas in the late 1980s and extensive reductions in tariffs since, Philippine manufacturing remains orientated towards the provision of consumer goods for the domestic market. Its performance has thus been very responsive to movements in domestic demand. The external market for Philippine manufacturers is, nevertheless, significant, with the export-processing zone model introduced at four government-owned sites in the late 1970s replicated at centres established by the private sector since the early 1990s. In 2002 exports from these zones reached US $22,775m., representing two-thirds of the country's export receipts. The privately-operated zones and estates, which had for some time been attracting most investment, and which numbered 91 in mid-2004, accounted for the bulk of these exports. A major contribution to manufacturing, as well as to services such as tourism, transhipment and ship-repairing, is due to be made by the economic zone and free port at the former US naval base at Subic Bay. Another former US base, at Clark, is being developed similarly, and is centred on its airport facilities. However, the cyclical downturn in the market for electronics (the major activity in this sector) and competition from other, lower-cost countries have depressed the level of new investment since 2000. In 2003 approved investments in the export-processing zones (excluding Subic Bay and Clark) were valued at only $578m., compared with $1,585m. in 2001 and the peak of $5,423m. recorded in 1997. However, in early 2004 investment in the zones was recovering strongly, with total approvals increasing by 177% in January–May compared with the same period in the previous year. This growth owed much to a significant level of investment in call centres, the capacity of which was estimated at 20,000 seats over approximately 60 sites. The Government hoped for a doubling in the capacity of these centres in both 2004 and 2005, to a total of 80,000 seats. Employment at the zones was close to 966,000 in April 2004, representing around 5% of non-agricultural employment.

INFRASTRUCTURE

The country's physical infrastructure is characterized by marked regional disparity, which both reflects and reinforces the concentration of modern economic activity in Metropolitan Manila and regions immediately adjacent. Its efficiency was in decline from the mid-1980s as the result of reductions in budget expenditure under the austerity programme of that period. While considerable improvements were made in the 1990s as BOT contracts were used to mobilize private funds (both domestic and foreign) for investment in the transport infrastructure, overall the poor state of physical infrastructure has remained a major constraint on economic growth.

The road network in the Philippines accommodates about 60% of freight and 80% of passenger traffic. In 2000 there were 201,994 km of roads, of which 30,013 km were highways and 49,992 km were secondary roads; an estimated 42,419 km of the network were paved. Feeder roads are in poor condition. However, new urban highways are being built in Metropolitan Manila and adjacent areas.

The rail system is limited, with 740 km of single-line track in Luzon, and also in a poor state of repair. The network carried only 240,000 passengers in 2003. A mass transit system began operation in Manila in 1984, and has been undergoing extensive expansion since the late 1990s; 113m. passenger journeys were registered in 2003.

The seaport network services about 40% of freight and carries 10% of passenger traffic. The most important ports are Manila (which handled 58.7m. metric tons of cargo in 1996) and Cebu (which handled 11.1m. tons in 1995); both have container facilities. The inter-island fleet is old, safety regulations are poor and maritime navigational aids inadequate.

There are about 90 national airports; international airports include Manila, Cebu and General Santos. The former US military facility at Clark Air Base is being developed as a joint principal airport for the capital, with the Ninoy Aquino International Airport.

GOVERNMENT FINANCE

Fiscal imbalance has been endemic in the Philippines for decades, with budget income very rarely enough to cover spending and never sufficient to allow for adequate levels of investment in physical and social infrastructure. The tax effort—the ratio of tax revenue to GDP—has always been low, at around 15% in the late 1990s and declining in every year since 1998, reaching only 12.5% in 2002 and 2003. In large part this reflected a massive failure in tax collection. The Department of Finance has estimated that revenue agencies receive only 73% of personal income tax due, 40% of corporate income tax and 49% of value-added tax (VAT). Owing to congressional resistance to tax increases, and with sales of assets constituting only a limited option, the spending side has borne the brunt of the deficit reduction effort. Under President Marcos there was a steady rise in government expenditure, associated with ambitious programmes of public investment, often in urban-based capital-intensive projects, in an effort to sustain growth. Tax revenues failed to keep pace with the expanded expenditure, largely because of the proliferation of exemptions, poor compliance, inefficiency in tax administration and widespread evasion, generating persistent and rising fiscal deficits. Although expenditure was curtailed in the final years of the Marcos regime, as a condition of IMF support during the crisis years, the contraction in the economy in that period depressed revenue, so that by 1986 the budget deficit was still slightly above the 1982 result, at 4.2% of GNI. The new Government undertook to reduce the deficit, both absolutely and as a percentage of GNI. With occasional backsliding because of unforeseeable pressures on the spending side—such as the oil price subsidy to counter the surge in world prices in 1990—this objective was attained. Income was temporarily boosted by asset sales, such as the highly successful privatization of the country's national airline in 1992.

Under the administration of President Ramos, who took office in mid-1992, the budgetary balance moved into surplus for the first time, reaching 0.9% of GNI in 1994; this surplus was sustained, if at declining levels, in the following three years. This was due to a combination of higher than expected receipts from privatization and lower than projected interest (owing to falling interest rates and some reduction in government debt). The slowing in economic growth and the rise in interest payments in the wake of the currency and financial crisis in 1997 again resulted in a budget deficit in 1998, equivalent to 1.8% of GNI. The need to stimulate weak domestic demand meant that budget spending in 1999 was forecast to rise faster than revenue, but the deficit was even higher than projected, reaching 111,658m. pesos (3.6% of GNI), owing to internal tax receipts falling far below expectations. The budget balance deteriorated further in 2000, with the deficit, at 134,212m. pesos, more than double the target. This was in part the result of worsening investor sentiment because of the deepening political crisis, which meant that privatization proceeds were minimal, while interest outgoings were pushed up by the steep fall in the peso's value in October and the sharp rise in domestic interest rates to contain that fall. However, the most serious problem remained the shortfall in internal tax revenue, which was about 43,000m. pesos below forecast.

The incoming Macapagal Arroyo Government did not expect to reduce the budget deficit in 2001, but, given the record of its predecessors, near-attainment of the relatively ambitious target of 145,000m. pesos (the outturn was a deficit of 147,023m. pesos) was a major achievement. Revenue receipts were increased by a reinforced tax collection effort, including a tax amnesty scheme, while expenditure was aided by lower than expected interest rates. The deficit as a proportion of GNI was unchanged from the level of 3.8% in 2000. For 2002 the Government hoped to reduce the deficit in both nominal terms and as a percentage of GNI, with a target of 130,000m. pesos. This was to be achieved by faster growth in revenue (10.8%) than in expenditure (6.1%). In the event, revenue fell 9% below the level forecast, after Congress failed to approve proposed tax increases and tax collection remained poor. Meanwhile, spending was 3% above target despite another year of lower than scheduled interest costs. The 2002 deficit consequently exceeded the target by almost two-thirds, at 210,741m. pesos, equivalent to 5% of GNI. This forced the Government to shift its target for achievement of a zero balance from 2006 to 2009, with the deficit in 2003 expected to show only a marginal reduction, at 202,000m. pesos. With a renewed tax collection effort, which brought internal revenue in on target for the first time since 1996, and lower than expected interest costs, the budget deficit was held to 199,868m. pesos in 2003. A further, marginal reduction in the deficit was planned for 2004, with a faster rise in revenue than in spending yielding a balance of 197,800m. pesos. At mid-year the budget deficit was very close to target, aided by above-programme receipts from customs, but the full-year figure was forecast to exceed that scheduled unless Congress approved several new taxation measures, notably an increase in the levies on tobacco and alcohol.

The persistent and substantial deficits on the Government's budget and the drain from government-owned or -controlled corporations (GOCCs) operating at a loss have resulted in a major accumulation of public-sector debt. In January 2004 public-sector debt was estimated at 5,160,000m. pesos, equivalent to 112% of GNI. With the deficit of the GOCCs expected to reach 116,800m. pesos in 2004, a 79% increase from the previous year, no prospect of a major reduction until the privatization of NAPOCOR (a process which will involve the Government assuming up to 500,000m. pesos in debt liabilities), and against a background of continuing fiscal imbalance, the public-sector debt was certain to continue to rise until the end of the current decade. The debt was officially forecast to be 7,070,000m. pesos in 2010, a level likely to be only slightly below GNI in that year. At these levels, the Philippines is extremely vulnerable to adverse movements in interest rates or in economic output.

The Financial Sector

The banking and financial sector is still relatively undeveloped for an economy as large as that of the Philippines. Of the 42 commercial banks in operation at the end of 2003, 17 had assets of less than 20,000m. pesos, and a number were still family-owned. The sector has, however, been developing rapidly since the mid-1990s, stimulated by its liberalization, which has included the privatization of what was then the largest commercial bank, the PNB, the removal in May 1994 of the ban on the establishment of foreign banks, and legislation raising the maximum foreign-ownership level of local banks. At the end of 2002 13 foreign banks were in operation in addition to the four already in place since the 1950s. The banking system in the Philippines came under severe pressure as the result of the regional financial and currency crisis in 1997. However, it had relatively low exposure to the collapsing property sector, and the ratio of 'bad' loans in commercial banks' portfolios, while it had risen, was still relatively low at the end of 2003, at 14.1%. It was hoped that the quality of banks' asset portfolios would improve under legislation that took effect in early 2003 providing incentives for the sale of commercial banks' non-performing assets to asset management companies. However, no major disposals of such assets had been finally agreed by July 2004, just two months before the deadline for their registration with the central bank.

The local securities market is still small and speculative, but has been developing rapidly since the early 1990s. The market capitalization of the Philippine Stock Exchange (PSE) was US $80,645m. at the end of 1996, six times the level at the end of 1992. This reflected new listings, enhanced by the privatization programme, and growing foreign confidence in the Philippines. This process was reversed in 1997, as the regional crisis affected both foreign fund inflows and domestic corporate results, causing a decline of 61% (in dollar values) in the PSE's capitalization, to $31,270m. at the end of the year. Market capitalization had recovered to $48,099m. at the end of 1999, and after a renewed decline in early 2000 owing to the deterioration in investors' perceptions of political risk in the Philippines and the narrowing interest differential with dollar assets, its value recovered to $51,562m. by the end of the year. Capitalization fell again in 2001–03, in line with international trends, ending 2003 at $23,538m.

TRADE AND THE BALANCE OF PAYMENTS

Until the final years of the 1990s merchandise trade was constantly in deficit, with the shortfall extremely sensitive to variations in GDP. Sudden surges in economic growth led to unsustainable increases in imports, and domestic demand was such that the trade deficit was never entirely eliminated. Thus, as economic growth gained momentum in the mid-1990s, the trade deficit rose nearly every year, reaching a record of US $11,342m. in 1996 (equivalent to 13% of GNI in that year). The deficit declined slightly in 1997 as a sharp fall in the value of the peso in the second half of the year and a steep rise in the cost of credit began to suppress import demand. However, it was the downturn in the economy in mid-1998 that reduced the deficit to the negligible level of $28m. in that year. Contrary to the traditional pattern, the recovery in the economy in 1999 did not give rise to a surge in the trade deficit. Rather, a surplus, of $4,959m., was registered. This was the result of unexpectedly rapid growth in exports, as an earthquake in Taiwan boosted demand for the Philippines' leading export, electronic goods, and a continuing fall in expenditure on imports. The trend began to reverse in 2000 as export growth slackened to 9% (from 16% in 1999), owing to a sharp deceleration in the rise in electronics earnings, reflecting the downturn in the world IT market, while imports showed a moderate (9%) recovery. The trade surplus was consequently reduced by around $1,100m., to $3,814m. A much more severe deterioration followed in 2001. Continuing depression in the electronics sector (Philippine sales fell by one-quarter), compounded by the slackening in the US economy, caused exports to contract by 16.2%; this decline was only partly offset by a fall in import spending linked to weaker demand for inputs for electronics manufacturing. Trade consequently recorded a deficit of $743m. A surplus was re-established in 2002, although only a modest $408m., as exports recovered at a slightly faster pace than imports. This surplus proved short-lived. Exports performed sluggishly in 2003, as electronic exports declined once more, while import spending was boosted by stockpiling of (higher-priced) oil and production inputs in advance of the US-led invasion of Iraq in March. The trade gap consequently widened to $1,253m. The first five months of 2004 saw a strengthening recovery in exports as sales of electronics improved, matched by a rise in imports reflecting sustained, and accelerating, growth in domestic demand. The balance remained in deficit, having reached $1,059m. at the end of May. The USA and Japan continued to dominate Philippine international trade, with the level of transactions sustained by aid and private investment inflows at very important volumes. Changes in market shares have occurred among second-rank partners, as trade with other members of the Association of South East Asian Nations (ASEAN) has grown in importance as it has been liberalized. Their share of exports and imports increased from only 2% in 1973 to 17.6% and 16.3%, respectively, in 2003.

Exports and Imports

The commodity composition of the Philippines' export trade has been transformed since the 1970s, when four primary commodities—coconut products, sugar, timber and copper—accounted for about one-half of the total. The general decline in world prices of these commodities and/or the fall in the volume of production coincided with the development of the export manufacturing sector and the diversification of agricultural production, with the result that in 2003 the four traditional exports accounted for only 3% of the total, while manufactures accounted for 89%. One category alone, electrical and electronic equipment and components, accounted for 66% of all export earnings in 2003.

Changes in imports have been less marked, although the fall in world prices and some changes in energy use reduced the proportion represented by crude petroleum to only 6% in 1996–98 from the 23% recorded in 1983. Subsequent rises in world petroleum prices still left the share at 7.5% in 2000–03. The bulk of import spending is on capital goods (39% of the total in 2000–03) and semi-processed raw materials and intermediates (35% in the same period). Both categories are highly sensitive: the former to trends in foreign demand for Philippine electronics, since materials for their manufacture account for more than half the spending in this category, and the latter to trends in domestic investment. Thus imports of capital goods fell by 12% in the recession year 1998, and spending on intermediates declined in 2001, when exports of electronics fell by one-quarter, and improved markedly in 2002, as this sector staged a partial recovery.

Invisible Transactions

Invisible transactions (services, investment income and unrequited transfers) have always registered a significant surplus, owing primarily to inflows of remittances from Filipinos working abroad. These have tended to rise, reflecting the dependence of Philippine households on foreign wages and the strength of foreign demand for Filipino labour. In 2000–03 remittances from abroad averaged US $6,728m. per year, equivalent to one-fifth of merchandise export earnings. They serve as a significant stabilizer of growth when other economic sectors are under pressure. Tourism makes a useful, if much smaller, contribution, yielding a gross $1,740m. in 2003. Tourist numbers have fluctuated moderately. During much of the 1980s this was in response to political instability in the Philippines, while for most of the following decade the overall trend was steadily upward, rising from 951,365 tourist arrivals in 1991 to the highest level of 2.22m. in 1997. Numbers have since declined in most years. Arrivals from Asian countries were adversely affected by the regional economic crisis of the late 1990s, by the aftermath of the September 2001 terrorist attacks, which significantly reduced visits by overseas Filipinos (who normally account for one-12th of arrivals), and by the outbreak of Severe Acute Respiratory Syndrome (SARS) in early 2003, which affected tourism throughout the region. Total arrivals reached 1.91m. in 2003.

The major outflow on the invisible account has been interest payments, reflecting the size of the foreign debt. However, as a consequence of the stabilizing of the debt (in dollar terms) in the 1990s and the reduction in interest rates, owing in large part to measures of debt relief, outflows declined from their peak of US $2,164m. in 1989 to $1,518m. in 1993, before rising again, to average $2,972m. a year in 2000–02.

With the balance on merchandise trade turning positive in 1998–2000, the surplus on invisible transactions, which previously had served to offset some of the trade deficit, became incremental, moving the current account into very substantial surplus in both 1999 and 2000, at US $7,219m. and $6,258m., respectively, equivalent to 8%–9% of GNI. As merchandise trade slipped back into deficit in 2001, the current-account surplus fell dramatically, to $1,323m. It recovered once more in 2002, to $4,383m., as the trade surplus reappeared, but then eased in 2003, to $3,347m., as the rise in the surplus on invisible transactions (largely due to workers' remittances) was more than offset by the deterioration in the trade balance.

Capital Transactions

Traditionally, the capital account has been in surplus, more than offsetting the deficit on current payments in most years. This was largely the result of foreign borrowing. Although long-term loans remained the most significant inflow, the contribution of foreign investment, both direct and indirect, increased in the mid-1990s in response to the liberalization of the investment environment, political stability and sustained economic growth. Portfolio investment was by far the most important component, attracted by the strength of the peso and high domestic interest rates, but direct investment also rose rapidly, stimulated by the opportunities represented by strong external demand for Philippine manufactures. Like other long-term trends this was interrupted in 1997, when the sharp fall in portfolio investment inflows nearly halved the surplus on the capital account, to US $6,498m., which produced a deficit on the overall balance of payments of $3,094m., compared with the surplus of $4,338m. in 1996. While net capital inflows contracted sharply in 1998, in large part owing to the fall in foreign borrowing and in short-term inflows, and then became negative in 1999, the surplus on the current account in these years pushed overall payments back into surplus in both years, at $3,586m. in 1999. Political uncertainties caused another sharp contraction on the portfolio account in 2000, which increased the net outflow on capital transactions to $4,119m. and moved the overall balance of payments back into deficit, at $5,130m. The capital-account deficit narrowed in 2001, to $1,080m., as the political environment stabilized and the Central Bank and the Government embarked on major international borrowing. However, overall payments remained narrowly in deficit, at $192m. The capital-account balance deteriorated once more in 2002 and 2003, reaching a record level of $5,319m. as the result of much lower inflows of investment. While the rise in the surplus on the current account in 2002 more than compensated for the slightly higher deficit on the capital account, pushing payments into a surplus of $663m., the fall in the current-account surplus in 2003 meant that the overall payments surplus fell to only $111m.

CURRENCY

Ever since it was floated in February 1970, the peso has been in long-term decline, reflecting the economy's underlying external deficit. From 1991 to mid-1997, however, the currency's value against the US dollar fluctuated within a narrow band, with the average in 1996 5% greater than the value five years earlier. This strength, during a period of persistently large trade imbalances, reflected rising remittances from the Filipino community overseas and inflows of portfolio investment. The regional financial crisis in 1997 brought a sharp correction, with the value of the peso falling by over one-third between July 1997 and January 1998, to US $1 = 44 pesos. Consistently good export performance, continued high inflows of remittances and reduced demand for dollars as imports fell resulted in the value of the peso increasing throughout the remainder of 1998 and the first half of 1999, to an average of US $1 = 39 pesos in the 18-month period. The peso then entered a time of weakness, initially owing to the rise in US federal fund rates, which lessened demand for peso-denominated assets, and subsequently to a marked deterioration in investor sentiment, after the stock-exchange scandal of January 2000 prompted a year of political uncertainty, culminating in the removal of President Estrada on 20 January 2001.

Just before he was ousted the currency had reached a new low of US $1 = 55 pesos. It then strengthened and stabilized at around US $1 = 51 pesos through to mid-2002, as the new administration became more entrenched and basic economic indicators remained sound. However, the currency came under pressure in the second half of 2002 as the budget financing requirement far exceeded targets. Despite the reimposition of control on the budget and steady economic fundamentals, the weakening continued through the first half of 2003, to an average of US $1 = 54 pesos. Initially, this reflected fears about the effects of the invasion of Iraq on oil prices and remittances, and a persistent nervousness about the domestic security situation. However, from mid-2003 onwards the pressure resulted almost entirely from political events (see History): the attempted military coup in July 2003; the announcement in November 2003 by the populist film actor, Fernando Poe, Jr, of his candidacy for the May 2004 presidential election; the closeness of the presidential contest; and the challenge to the election result. The peso fell to an all-time low of US $1 = 56.4 pesos in mid-June 2004, before recovering slightly in July and August as fears of unrest abated.

FOREIGN DEBT

Following the removal of Ferdinand Marcos, the Philippines was left with massive foreign indebtedness (equivalent to 96.4% of GNI at the end of 1986, with a high short-term component) and an unsustainable debt-service burden (34.6% of earnings from exports of goods and services in the same year). Thus, both the Aquino and the Ramos administrations were obliged to seek regular rounds of debt relief, the price of which was compliance with IMF requirements on structural change and fiscal austerity. The debt relief took the form of rescheduling of both official and commercial credits, with grace periods when only interest payments were due. A new feature in 1990 was the 'buy-back' of debt, under which commercial banks sold US $1,340m. in Philippine debt to the Government at the secondary market rate, with the IMF, the World Bank and the USA providing funding. At the same time, the commercial banks agreed to provide new money, by buying 15-year government bonds at a reduced interest margin. A similar arrangement was agreed in 1992, when the consortium of the country's commercial bank creditors agreed to another 'buy-back' of $1,260m. and the conversion of $3,300m. to long-term, reduced-interest bonds. Against this background, the Government was able to return to the international capital market in 1993, with its first sovereign bond issue since before the 1983 payments crisis. Foreign debt, while still substantial at $40,146m. at the end of 1996, represented much less of a burden. It was then equivalent to 47% of GNI, while long-term debt to private creditors was predominantly in bonds rather than loans.

The reduced level of debt (in terms of the size of the economy), the stabilization of the current-account deficit, the increase in inflows of direct investment and the substantial growth in foreign-currency reserves (to over 10 times their end-1990 level by the end of 1996, at US $9,902m.) were all indicative of the improvement in the Philippines' payments position. However, the onset of the regional financial crisis in 1997 and its persistence throughout 1998 meant that the Philippines had to maintain a funding relationship with the IMF. A 'precautionary stand-by arrangement' agreed in March 1998 afforded access to $1,370m. over a two-year period. This pledge of IMF support, the policy implications of a continuing relationship with the Fund, and the promise of policy continuation under the new Estrada administration helped to maintain foreign confidence in the Philippines during an uncertain period for the whole regional economy. The Philippines was therefore able to return to the international bond market in 1999, with issues of $1,000m. in that year and $1,130m.in the first half of 2000. However, as foreign sentiment against the Philippines hardened, owing to increasingly serious doubts about both the quality of the Estrada administration and the feasibility of its budgetary targets, the Government had to curb its foreign borrowing. With investment confidence stimulated by the accession of President Macapagal Arroyo, and the spread on Philippine debt consequently narrowing, the Government could once more embark on a programme of international borrowing to help fund the budget deficit. The Central Bank, meanwhile, was active in 2001–03 in raising funds on the international capital market to prepay higher-cost debt and to maintain foreign-reserve levels during a period of relative weakness in external markets for Philippine goods. Thus foreign indebtedness, which had steadied at $57,459m. in 2000, had increased to $59,342m. by the end of 2002 and was an estimated $63,200m. at the end of 2003, equivalent to 74% of GNI.

Statistical Survey

Source (unless otherwise stated): National Statistics Office, Solicarel 1, Magsaysay Blvd, cnr Ampil St, POB 779, Metro Manila; tel. (2) 7160807; fax (2) 610794; internet www.census.gov.ph.

Area and Population

AREA, POPULATION AND DENSITY

Area (sq km)	300,000*
Population (census results)	
1 September 1995	68,616,536
1 May 2000	
Males	38,524,267
Females	37,979,810
Total	76,504,077
Population (official estimates at mid-year)	
2002	80,429,769
2003	82,311,608
2004	84,237,476
Density (per sq km) at mid-2004	280.8

* 115,831 sq miles.

REGIONS
(population at 2000 census)

	Area (sq km)	Population	Density (per sq km)
National Capital Region	636.0	9,932,560	15,617
Ilocos	12,840.2	4,200,478	327
Cagayan Valley	26,837.7	2,813,159	105
Central Luzon	18,230.8	8,030,945	441
Southern Tagalog	46,924.0	11,793,655	251
Bicol	17,632.5	4,686,669	266
Western Visayas	20,223.2	6,211,038	307
Central Visayas	14,951.5	5,706,953	382
Eastern Visayas	21,431.7	3,610,355	173
Western Mindanao	15,997.3	3,091,208	193
Northern Mindanao	14,033.0	2,747,585	196
Southern Mindanao	27,140.7	5,189,335	263
Central Mindanao	14,372.7	2,598,210	179
Cordillera Administrative Region	18,293.7	1,365,412	75
Autonomous Region of Muslim Mindanao	11,608.3	2,412,159	211
Caraga	18,847.0	2,095,367	111
Total*	300,000.3	76,504,077†	255

* Total includes a statistical adjustment.
† Including Filipinos in Philippine embassies, consulates and missions abroad (2,851 persons).

PRINCIPAL TOWNS
(population at 2000 census)

Manila (capital)*	1,581,082	Gen. Santos City	411,822
Quezon City*	2,173,831	Marikina City	391,170
Caloocan City*	1,177,604	Muntinlupa City*	379,310
Davao City	1,147,116	Iloilo City	365,820
Cebu City	718,821	Pasay City*	354,908
Zamboanga City	601,794	Iligan City	285,061
Pasig City*	505,058	Mandaluyong City*	278,474
Valenzuela City	485,433	Butuan City	267,279
Las Piñas City	472,780	Angeles City	263,971
Cagayan de Oro City	461,877	Mandaue City	259,728
Parañaque City	449,811	Baguio City	252,386
Makati City*	444,867	Olongapo City	194,260
Bacolod City	429,076	Cotabato City	163,849

* Part of Metropolitan Manila.

Mid-2003 (UN estimates, incl. suburbs): Metropolitan Manila 10,352,249; Davao City 1,254,388 (Source: UN, *World Urbanization Prospects: The 2003 Revision*).

BIRTHS, MARRIAGES AND DEATHS*

	Registered live births		Registered marriages		Registered deaths	
	Number	Rate (per 1,000)	Number	Rate (per 1,000)	Number	Rate (per 1,000)
1991	1,643,296	25.8	445,526	7.0	298,063	4.7
1992	1,684,395	25.8	454,155	7.0	319,579	4.9
1993	1,680,896	25.1	474,407	7.1	318,546	4.8
1994	1,645,011	24.0	490,164	7.2	321,440	4.7
1995	n.a.	n.a.	n.a.	7.4	324,737	4.7
1996	1,608,468	22.9	525,555	7.5	344,363	4.9
1997	1,653,236	23.1	562,808	7.9	339,400	4.7
1998	1,632,859	22.3	549,265	7.5	352,992	4.8

2000 (estimates, adjusted for under-registration): Birth rate 25.7; Death rate 5.77.

2001 (estimates, adjusted for under-registration): Birth rate 26.2; Death rate 5.83.

* Registration is incomplete. According to UN estimates, the average annual rates were: births 31.6 per 1,000 in 1990–95, 28.4 in 1995–2000; deaths 6.3 per 1,000 in 1990–95, 5.5 in 1995–2000 (Source: UN, *World Population Prospects: The 2002 Revision*).

Expectation of life (WHO estimates, years at birth): 68.3 (males 65.1; females 71.7) in 2002 (Source: WHO, *World Health Report*).

ECONOMICALLY ACTIVE POPULATION*
('000 persons aged 15 years and over, January)

	2002	2003	2004
Agriculture, hunting, forestry and fishing	11,006	11,150	11,174
Mining and quarrying	115	99	123
Manufacturing	2,830	2,781	3,104
Electricity, gas and water	120	117	110
Construction	1,531	1,585	1,713
Wholesale and retail trade, restaurants and hotels	6,194	6,318	6,653
Transport, storage and communications	2,167	2,248	2,436
Financing, insurance, real estate and business services	836	847	1,024
Public administration and defence	1,424	1,432	1,478
Community, social and personal services	2,167	2,222	2,181
Other services	1,316	1,316	1,548
Extraterritorial organizations and bodies	n.a.	5	2
Total employed	29,705	30,119	31,546
Unemployed	3,393	3,559	3,896
Total labour force	33,098	33,678	35,442

* Figures refer to civilians only and are based on annual household surveys (excluding institutional households).

Health and Welfare

KEY INDICATORS

Total fertility rate (children per woman, 2002)	3.2
Under-5 mortality rate (per 1,000 live births, 2002)	38
HIV/AIDS (% of persons aged 15–49, 2003)	<0.10
Physicians (per 1,000 head, 1996)	1.23
Hospital beds (per 1,000 head, 1993)	1.07
Health expenditure (2001): US $ per head (PPP)	169
Health expenditure (2001): % of GDP	3.3
Health expenditure (2001): public (% of total)	45.2
Access to water (% of persons, 1999)	87
Access to sanitation (% of persons, 1999)	83
Human Development Index (2002): ranking	83
Human Development Index (2002): value	0.753

For sources and definitions, see explanatory note on p. vi.

Agriculture

PRINCIPAL CROPS
('000 metric tons)

	2000	2001	2002
Rice (paddy)	12,389	12,955	13,271
Maize	4,511	4,525	4,319
Potatoes	64	66	68
Sweet potatoes	554	545	549
Cassava (Manioc)	1,766	1,652	1,626
Yams	26	24	26
Taro	96	97	97*
Sugar cane	24,491	24,962	25,835†
Dry beans	27	28	28*
Other pulses*	29	28	28
Groundnuts (in shell)	27	26	27†
Coconuts	12,995	13,208	13,683
Oil palm fruit*	212	212	212
Cabbages	88	90	90*
Tomatoes	148	146	146*
Pumpkins, squash and gourds	70	70*	70*
Aubergines (Eggplants)	166	170	180
Dry onions	84	83	82*
Green peas	28	29*	29*
Other vegetables (incl. melons)*	4,158	4,262	4,362
Watermelons	59	58	58*
Mangoes	848	884	880*
Avocados	38	37	37*
Pineapples	1,560	1,620	1,636
Papayas	76	77	78*
Bananas	4,930	5,061	5,264
Oranges	25†	29*	30*
Tangerines, mandarins, clementines and satsumas	49†	53*	53*
Lemons and limes	45†	52*	52*
Grapefruit and pomelos	46	43	44*
Other fruits*	3,040	3,265	3,365
Garlic	14	15	15*
Coffee (green)	126	132	131†
Cocoa beans	7	8†	7†
Tobacco (leaves)	49	48	56†
Natural rubber†	67	68	69

* FAO estimate(s).
† Unofficial figure(s).

Source: FAO.

LIVESTOCK
('000 head, year ending 30 June)

	2000	2001	2002
Cattle	2,479	2,496	2,548
Pigs	10,713	11,063	11,653
Buffaloes	3,024	3,066	3,122
Horses*	230	230	230
Goats*	6,245	6,197	6,250
Sheep*	30	30	30
Chickens	115,187	115,606	125,730
Ducks*	12,300	12,500	12,600

* FAO estimates.

Source: FAO.

LIVESTOCK PRODUCTS
('000 metric tons)

	2000	2001	2002
Beef and veal	190	183	183
Buffalo meat	72	72	76
Pig meat	1,008	1,064	1,332
Poultry meat	556	611	651
Cows' milk	10	11	11
Hen eggs*	445	445	495
Other poultry eggs*	73	74	74
Cattle and buffalo hides*	25	24	24

* FAO estimates.

Source: FAO.

Forestry

ROUNDWOOD REMOVALS
('000 cubic metres, excl. bark)

	2000	2001	2002
Sawlogs, veneer logs and logs for sleepers	384	260	288
Pulpwood	400	130	106
Other industrial wood	2,295	2,295	2,295*
Fuel wood	40,950	41,699	13,328*
Total	44,029	44,384	16,017

* FAO estimate.

Source: FAO.

SAWNWOOD PRODUCTION
('000 cubic metres, incl. railway sleepers)

	2000	2001	2002
Total	151	199	154

Source: FAO.

Fishing

('000 metric tons, live weight)

	2000	2001	2002
Capture	1,896.6	1,949.0	2,030.5
Scads (Decapterus)	261.0	291.8	283.6
Sardinellas	298.5	282.6	244.2
'Stolephorus' anchovies	79.6	100.9	74.1
Frigate and bullet tunas	112.2	111.7	163.1
Skipjack tuna	113.0	105.5	110.0
Yellowfin tuna	90.8	83.9	100.3
Freshwater molluscs	85.6	69.0	52.6
Aquaculture	393.9	434.7	443.3
Nile tilapia	77.6	89.5	104.4
Milkfish	210.0	225.3	232.0
Total catch	2,290.5	2,383.7	2,473.9

Note: Figures exclude aquatic plants ('000 metric tons): 707.4 (capture 0.4, aquaculture 707.0) in 2000; 786.4 (capture 0.4, aquaculture 785.8) in 2001; 895.6 (capture 0.7, aquaculture 894.9) in 2002.

Source: FAO.

Mining

('000 metric tons, unless otherwise indicated)

	1999	2000	2001†
Coal	1,300	1,300	1,500
Crude petroleum ('000 barrels)	400	400†	400
Chromium ore (gross weight)	19.6	20.9	2.6‡
Copper ore*	34.6	129.8	20.3‡
Salt (unrefined)	704.3	589.5	600.0
Nickel ore*	20.7	17.4	27.4‡
Gold (metric tons)*	31.1	36.5	33.8‡
Silver (metric tons)*	18.2	23.5	33.6
Limestone	16,738	22,244	23,000

* Figures refer to the metal content of ores and concentrates.
† Estimate(s).
‡ Reported figure.

Source: US Geological Survey.

Industry

SELECTED PRODUCTS
('000 metric tons, unless otherwise indicated)

	1999	2000	2001
Raw sugar*†	1,682	1,676	1,868
Plywood ('000 cubic metres)*	268	326	348
Mechanical wood pulp*‡	28	28	28
Chemical wood pulp*	145†	147†	147‡
Paper and paperboard*	1,010	1,107	1,056
Nitrogenous fertilizers (a)§	184	n.a.	n.a.
Phosphate fertilizers (b)§	193	n.a.	n.a.
Jet fuels ('000 barrels)‖	6,500	6,500	7,000‡
Motor spirit—petrol	1,460	n.a.	n.a.
Kerosene ('000 barrels)‖	4,500	4,500	5,000‡
Distillate fuel oils ('000 barrels)‖	40,000	40,000	40,000‡
Residual fuel oils ('000 barrels)‖	47,000	47,000	47,000‡
Liquefied petroleum gas‖	5,500	5,500	6,000‡
Cement‖	12,566	11,959	8,653‡
Smelter (unrefined) copper‖	162	n.a.	n.a.
Electric energy (million kWh)‡	41,337	n.a.	n.a.

* Source: FAO.
† Unofficial figure(s).
‡ Estimate(s).
§ Production of fertilizers is in terms of (a) nitrogen or (b) phosphoric acid.
‖ Source: US Geological Survey.

Sources: mainly UN, *Industrial Commodity Statistics Yearbook*; UN, *Statistical Yearbook for Asia and the Pacific*.

Finance

CURRENCY AND EXCHANGE RATES

Monetary Units
100 centavos = 1 Philippine peso.

Sterling, Dollar and Euro Equivalents (31 May 2004)
£1 sterling = 102.44 pesos;
US $1 = 55.84 pesos;
€1 = 68.38 pesos;
1,000 Philippine pesos = £9.76 = $17.91 = €14.62.

Average Exchange Rate (pesos per US $)

2001	50.993
2002	51.604
2003	54.203

GENERAL BUDGET
(million pesos)

Revenue*	2001	2002	2003
Tax revenue	489,859	496,372	537,361
Taxes on net income and profits	223,417	226,501	243,857
Excise tax	58,698	57,001	56,894
Sales taxes and licences	87,044	89,950	101,340
Import duties and taxes	96,232	96,250	106,092
Non-tax revenue	71,882	69,717	88,071
Bureau of the Treasury income	46,413	47,194	56,657
Fees and other charges	24,296	21,932	18,635
Total	561,741	566,089	625,432

Expenditure	2001	2002	2003
Current expenditure	648,893	n.a.	n.a.
Personnel services	190,893	n.a.	n.a.
Maintenance and operating subsidy	155,622	n.a.	n.a.
Allotment to local government units	118,179	140,540	145,502
Interest payments	174,834	185,861	226,408
Capital outlay	57,434	n.a.	n.a.
Infrastructure	15,895	n.a.	n.a.
Equity investment	484	1,486	2,623
Net lending	3,944	2,626	5,620
Total	710,755	777,882	826,498

* Excluding grants received (million pesos): 1,991 in 2001; 1,052 in 2002; 1,198 in 2003.

Source: Bureau of the Treasury.

INTERNATIONAL RESERVES
(US $ million at 31 December)

	2001	2002	2003
Gold*	2,216	3,036	3,408
IMF special drawing rights	14	10	2
Reserve position in IMF	110	119	130
Foreign exchange	13,319	13,015	13,331
Total	15,659	16,180	16,871

* Valued at market-related prices.

Source: IMF, *International Financial Statistics*.

MONEY SUPPLY
(million pesos at 31 December)

	2001	2002	2003
Currency outside banks	194,670	220,040	238,610
Demand deposits at commercial banks	191,940	252,770	274,920
Total money (incl. others)	392,250	478,480	519,840

Source: IMF, *International Financial Statistics*.

COST OF LIVING
(Consumer Price Index; base: 1994 = 100)

	2000	2001	2002
Food (incl. beverages and tobacco)	145.5	151.4	154.4
Fuel, light and water	146.6	164.4	172.3
Clothing (incl. footwear)	140.2	145.4	149.2
Rent	174.3	186.1	195.3
Services	182.8	203.8	213.9
Miscellaneous	122.9	129.3	131.8
All items	152.3	161.6	166.6

NATIONAL ACCOUNTS

National Income and Product
(million pesos at current prices)

	1998	1999	2000
Compensation of employees	184,967	222,793	281,672
Operating surplus	2,241,651	2,499,941	2,746,549
Domestic factor incomes	2,426,618	2,722,734	3,028,221
Consumption of fixed capital	238,442	254,170	274,368
Gross domestic product (GDP) at factor cost	2,665,060	2,976,904	3,302,589
Factor income from abroad	246,701	272,989	359,872
Less Factor income paid abroad	109,629	113,725	171,327
Gross national product (GNP)	2,802,132	3,136,168	3,491,134
Less Consumption of fixed capital	238,442	254,170	274,368
National income in market prices	2,563,690	2,881,998	3,216,766
Other current transfers from abroad	146,932	66,027	7,736
Less Other current transfers paid abroad	13,169	6,457	5,129
National disposable income	2,697,453	2,941,568	3,219,373

Expenditure on the Gross Domestic Product
('000 million pesos at current prices)

	2001	2002	2003
Government final consumption expenditure	444.8	488.7	498.4
Private final consumption expenditure	2,565.0	2,750.9	2,985.2
Increase in stocks	37.8	2.1	29.2
Gross fixed capital formation	720.7	774.1	786.8
Statistical discrepancy	19.6	27.5	167.1
Total domestic expenditure	3,787.9	4,043.3	4,466.7
Exports of goods and services	1,785.2	1,968.5	2,104.0
Less Imports of goods and services	1,899.4	1,989.1	2,211.7
GDP in purchasers' values	3,673.7	4,022.7	4,359.0
GDP at constant 1985 prices	987.4	n.a.	n.a.

Source: IMF, *International Financial Statistics*.

Gross Domestic Product by Economic Activity
('000 million pesos at current prices)*

	2001	2002	2003
Agriculture, hunting, forestry and fishing	548.7	592.1	629.6
Mining and quarrying	21.7	33.5	43.7
Manufacturing	831.6	915.2	998.2
Electricity, gas and water	116.3	124.1	137.0
Construction	222.1	235.4	230.7
Wholesale and retail trade, restaurants and hotels	517.5	556.3	603.3
Transport, storage and communications	247.6	276.7	313.2
Finance, insurance, real estate and business services	160.1	170.1	187.8
Public administration and defence	337.7	380.9	404.4
Other services	670.3	738.0	808.8
GDP in purchasers' values	3,673.7	4,022.7	4,359.0

* Totals may not be equal to sum of component parts, owing to rounding.

Source: Asian Development Bank, *Key Indicators of Developing Asian and Pacific Countries*.

BALANCE OF PAYMENTS
(US $ million)

	2000	2001	2002
Exports of goods f.o.b.	37,295	31,243	34,383
Imports of goods f.o.b.	–33,481	–31,986	–33,975
Trade balance	3,814	–743	408
Exports of services	3,972	3,148	3,056
Imports of services	–6,402	–5,198	–4,320
Balance on goods and services	1,384	–2,793	–856
Other income received	7,804	7,152	7,931
Other income paid	–3,367	–3,483	–3,381
Balance on goods, services and income	5,821	876	3,694
Current transfers received	552	517	594
Current transfers paid	–115	–70	–91
Current balance	6,258	1,323	4,197
Capital account (net)	38	–12	–19
Direct investment abroad	108	160	–85
Direct investment from abroad	1,345	982	1,111
Portfolio investment assets	–812	–399	–369
Portfolio investment liabilities	1,019	1,449	2,281
Other investment assets	–15,313	–13,898	–13,214
Other investment liabilities	9,611	11,312	7,536
Net errors and omissions	–2,630	–433	–1,432
Overall balance	–376	484	6

Source: IMF, *International Financial Statistics*.

External Trade

PRINCIPAL COMMODITIES
(distribution by SITC, US $ million)

Imports c.i.f.	2000	2001	2002
Food and live animals	2,252.6	2,310.8	2,276.0
Crude materials (inedible) except fuels	1,088.7	1,036.0	805.8
Mineral fuels, lubricants, etc.	4,102.9	3,593.9	3,281.6
Petroleum, petroleum products, etc.	3,683.7	3,244.4	3,002.1
Crude petroleum oils, etc.	3,170.8	2,822.9	2,262.8
Chemicals and related products	2,857.6	2,678.5	2,467.7
Basic manufactures	3,992.8	3,811.2	3,148.4
Textile yarn, fabrics, etc.	1,250.6	1,153.0	812.6
Iron and steel	991.6	928.1	977.8
Machinery and transport equipment	17,723.2	16,351.0	13,624.2
Machinery specialized for particular industries	1,164.8	885.9	809.6
Office machines and automatic data-processing equipment	2,428.4	3,103.2	3,274.4
Parts and accessories for office machines, etc.	2,202.5	2,912.1	3,129.8
Telecommunications and sound equipment	1,949.1	1,891.5	1,268.0

Imports c.i.f.— *continued*	2000	2001	2002
Other electrical machinery, apparatus, etc.	9,347.4	8,039.3	6,073.9
Thermionic valves, tubes, etc.	7,604.7	6,264.3	4,619.5
Electronic microcircuits	1,603.7	1,036.0	906.9
Road vehicles and parts (excl. tyres, engines and electrical parts)	1,086.7	1,048.4	1,060.8
Miscellaneous manufactured articles	1,470.8	1,287.8	986.7
Total (incl. others)	33,807.4	31,357.9	35,426.5

Source: UN, *International Trade Statistics Yearbook*.

2003 (US $ million): Food and live animals 2,112; Crude materials (inedible) except fuels 832; Mineral fuels, lubricants, etc. 3,766; Chemicals and related products 2,880; Basic manufactures 3,245; Machinery and transport equipment 15,175; Miscellaneous manufactured articles 1,060; Total (incl. others) 37,497.

Source: Asian Development Bank, *Key Indicators of Developing Asian and Pacific Countries*.

Exports f.o.b.	2000	2001	2002
Food and live animals	1,285.0	1,301.9	1,380.1
Basic manufactures	1,415.9	1,262.0	1,009.7
Machinery and transport equipment	28,989.5	23,870.7	14,089.6
Office machines and automatic data-processing equipment	7,208.1	7,040.7	6,020.5
Automatic data-processing machines and units	4,643.8	4,134.4	4,495.5
Complete digital central processing units	0.1	0.1	2,011.2
Digital central storage units, separately consigned	3.2	2,227.5	2,143.7
Peripheral units	3,966.6	462.2	333.4
Parts and accessories for office machines, etc.	2,517.1	2,815.5	1,480.0
Telecommunications and sound equipment	1,266.9	1,139.7	1,039.0
Other electrical machinery, apparatus, etc.	19,251.0	14,384.8	5,876.0
Switchgear, resistors, printed circuits, switchboards, etc.	1,622.4	960.9	315.1
Thermionic valves, tubes, etc.	16,662.8	12,569.9	4,771.6
Diodes, transistors, etc.	1,070.2	809.6	768.9
Electronic microcircuits	14,937.6	11,060.3	3,549.3
Miscellaneous manufactured articles	4,365.6	4,149.5	2,401.5
Clothing and accessories (excl. footwear)	2,575.9	2,422.9	1,366.1
Total (incl. others)	38,078.3	32,149.9	35,208.2

Source: UN, *International Trade Statistics Yearbook*.

2003 (US $ million): Food and live animals 1,528; Animal and vegetable oils, fats and waxes 535; Basic manufactures 1,145; Machinery and transport equipment 15,353; Miscellaneous manufactured articles 2,429; Total (incl. others) 36,298.

Source: Asian Development Bank, *Key Indicators of Developing Asian and Pacific Countries*.

PRINCIPAL TRADING PARTNERS
(US $ million)

Imports f.o.b.	2000	2001	2002
Australia	815.6	645.4	575.4
China, People's Republic	767.9	975.0	1,251.7
Finland	375.3	350.9	98.8
France (incl. Monaco)	347.7	296.2	290.7
Germany	734.2	792.4	708.0
Hong Kong	1,217.0	1,335.0	1,583.2
India	n.a.	248.2	428.4
Indonesia	692.7	759.8	764.8
Iran	795.6	758.5	433.3
Ireland	n.a.	278.1	372.6
Israel	328.3	375.1	97.0
Japan	6,027.4	6,633.1	7,232.6
Korea, Republic	2,350.8	2,081.6	2,754.2
Malaysia	1,141.7	1,080.0	1,293.2
Netherlands	323.9	246.1	234.0
Saudi Arabia	1,048.1	887.1	999.9
Singapore	2,115.0	2,072.9	2,311.1
Taiwan	1,948.0	1,969.5	1,782.7
Thailand	846.0	924.6	1,052.1
United Arab Emirates	877.2	605.8	368.5
United Kingdom	355.3	462.3	463.7
USA	5,323.3	6,410.7	7,285.7
Total (incl. others)	31,387.4	33,057.2	35,426.5

Source: National Statistical Coordination Board.

2003 (US $ million): China, People's Republic 3,140; Germany 1,064; Hong Kong 2,270; Japan 8,291; Korea, Republic 3,066; Malaysia 1,493; Saudi Arabia 1,263; Singapore 3,322; Thailand 1,687; USA 8,435; Total (incl. others) 44,939.

Source: Asian Development Bank, *Key Indicators of Developing Asian and Pacific Countries*.

Exports f.o.b.	2000	2001	2002
Canada	n.a.	281.6	377.9
China, People's Republic	663.3	792.8	1,355.8
Germany	1,328.6	1,323.1	1,386.1
Hong Kong	1,907.3	1,579.8	2,358.5
Japan	5,608.7	5,057.4	5,293.3
Korea, Republic	1,172.5	1,044.4	1,338.8
Malaysia	1,372.4	1,111.7	1,652.6
Netherlands	2,982.5	2,976.4	3,054.9
Singapore	3,124.2	2,307.5	2,471.7
Taiwan	2,861.3	2,127.4	2,484.9
Thailand	1,206.5	1,358.1	1,083.4
United Kingdom	1,506.3	997.3	946.3
USA	11,365.3	8,979.6	8,683.3
Total (incl. others)	38,078.3	32,150.2	35,208.2

Source: National Statistical Coordination Board.

2003 (US $ million): China, People's Republic 5,029; Germany 1,950; Hong Kong 3,329; Japan 6,063; Korea, Republic 1,456; Malaysia 1,983; Netherlands 2,079; Singapore 2,593; United Kingdom 1,047; USA 8,856; Total (incl. others) 41,921.

Source: Asian Development Bank, *Key Indicators of Developing Asian and Pacific Countries*.

Transport

RAILWAYS
(traffic)

	1994	1995	1996
Passengers ('000)	427	598	291
Passenger-km (million)	106	164	70
Freight ('000 metric tons)	13	14	6
Freight ton-km ('000)	3,080	4,000	1,476

Source: Philippine National Railways.

Passenger-km (million): 172 in 1997; 181 in 1998; 171 in 1999; 123 in 2000 (Source: UN, *Statistical Yearbook*).

ROAD TRAFFIC
(registered motor vehicles)

	2000	2001	2002
Passenger cars	763,834	726,044	745,795
Utility vehicles	1,387,878	1,488,998	1,652,036
Buses	33,886	31,686	33,915
Trucks	248,345	253,568	257,747
Motorcycles and mopeds*	1,236,236	1,337,980	1,470,376
Trailers	26,589	23,699	23,734

* Includes tricycles.

Source: Land Transportation Office, Manila.

SHIPPING

Merchant Fleet
(registered at 31 December)

	2001	2002	2003
Number of vessels	1,697	1,686	1,703
Total displacement (grt)	6,029,876	5,319,573	5,115,708

Source: Lloyd's Register-Fairplay, *World Fleet Statistics*.

International Sea-borne Shipping
(freight traffic)

	1994	1995	1996
Vessels ('000 net registered tons):			
entered	53,453	61,298	n.a.
cleared	53,841	61,313	n.a.
Goods ('000 metric tons):			
loaded	14,581	16,658	15,687
unloaded	38,222	42,418	51,830

CIVIL AVIATION
(traffic on scheduled services)

	1997	1998	1999
Kilometres flown (million)	96	40	53
Passengers carried ('000)	7,475	3,944	5,004
Passenger-km (million)	16,392	7,503	10,292
Total ton-km (million)	2,086	925	1,303

Source: UN, *Statistical Yearbook*.

Tourism

FOREIGN TOURIST ARRIVALS

Country of residence	2000	2001	2002
Australia	75,706	68,541	70,735
Canada	61,004	54,942	54,563
Germany	51,131	40,605	39,103
Hong Kong	146,858	134,408	155,964
Japan	390,517	343,840	341,867
Korea, Republic	174,966	207,957	288,468
Malaysia	42,067	30,498	31,735
Singapore	50,276	44,155	57,662
Taiwan	75,722	85,231	103,024
United Kingdom	74,507	60,147	48,478
USA	445,043	392,099	395,323
Total (incl. others)*	1,992,169	1,796,893	1,932,677

* Including Philippine nationals resident abroad (150,386 in 2000), not distributed by country.

Tourism receipts (US $ million): 2,134 in 2000; 1,723 in 2001; 1,741 in 2002.

Source: World Tourism Organization.

Communications Media

	2000	2001	2002
Television receivers ('000 in use)	11,000	n.a.	n.a.
Telephones ('000 main lines in use)	3,061.4	3,315.1	3,310.9
Mobile cellular telephones ('000 subscribers)	6,454.4	12,159.2	15,201.0
Personal computers ('000 in use)	1,480	1,700	2,200
Internet users ('000)	2,000	2,000	3,500

Source: International Telecommunication Union.

Radio receivers ('000 in use): 11,500 in 1997.

Facsimile machines (estimated number in use): 50,000 in 1995.

Telephones ('000 main lines in use): 3,338.9 in 2002.

Mobile cellular telephones ('000 subscribers): 14,216 in 2002.

Book production (titles, excluding pamphlets): 1,380 in 1999.

Daily newspapers: 47 (with average circulation of 5,700,000 copies) in 1996.

Non-daily newspapers: 243 (with average circulation of 153,000 copies) in 1995.

Sources: UN, *Statistical Yearbook*; UNESCO, *Statistical Yearbook*.

Periodicals (1990): 1,570 (estimated circulation 9,468,000).

Education

(1998/99)

	Institutions	Teachers	Pupils
Pre-primary	8,647	9,644*	n.a.
Primary schools	39,011	328,517	12,474,886
Secondary schools	7,021	108,981	5,066,190
University level	1,316	66,876†	2,481,809
Other tertiary level institutions	1,033	n.a.	4,134‡

* 1990/91 figure.
† 1993/94 figure.
‡ 1995/96 figure.

Sources: Department of Education, Culture and Sports; UNESCO, *Statistical Yearbook*.

2001/02: *Primary schools:* Institutions 40,763; Teachers 331,448; Pupils 12,826,218. *Secondary schools:* Institutions 7,683; Teachers 112,210; Pupils 5,813,879.

2002/03 (preliminary figures): *Primary schools:* Institutions 41,267; Teachers 337,082; Pupils 12,962,745. *Secondary schools:* Institutions 7,893; Teachers 119,235; Pupils 6,032,440.

Source: Department of Education, Culture and Sports.

Adult literacy rate (UNESCO estimates): 92.6% (males 92.5%; females 92.7%) in 2002 (Source: UN Development Programme, *Human Development Report*).

Directory

The Constitution

A new Constitution for the Republic of the Philippines was ratified by national referendum on 2 February 1987. Its principal provisions are summarized below:

BASIC PRINCIPLES

Sovereignty resides in the people, and all government authority emanates from them; war is renounced as an instrument of national policy; civilian authority is supreme over military authority.

The State undertakes to pursue an independent foreign policy, governed by considerations of the national interest; the Republic of the Philippines adopts and pursues a policy of freedom from nuclear weapons in its territory.

Other provisions guarantee social justice and full respect for human rights; honesty and integrity in the public service; the autonomy of local governments; and the protection of the family unit. Education, the arts, sport, private enterprise, and agrarian and urban reforms are also promoted. The rights of workers, women, youth, the urban poor and minority indigenous communities are emphasized.

BILL OF RIGHTS

The individual is guaranteed the right to life, liberty and property under the law; freedom of abode and travel, freedom of worship, freedom of speech, of the press and of petition to the Government are guaranteed, as well as the right of access to official information on matters of public concern, the right to form trade unions, the right to assemble in public gatherings, and free access to the courts.

The Constitution upholds the right of habeas corpus and prohibits the intimidation, detention, torture or secret confinement of apprehended persons.

SUFFRAGE

Suffrage is granted to all citizens over 18 years of age, who have resided for at least one year previously in the Republic of the Philippines, and for at least six months in their voting district. Voting is by secret ballot.

LEGISLATURE

Legislative power is vested in the bicameral Congress of the Philippines, consisting of the Senate and the House of Representatives, with a maximum of 274 members. All members shall make a disclosure of their financial and business interests upon assumption of office, and no member may hold any other office. Provision is made for voters to propose laws, or reject any act or law passed by Congress, through referendums.

The Senate shall be composed of 24 members; Senators are directly elected for six years by national vote, and must be natural-born citizens, at least 35 years of age, literate and registered voters in their district. They must be resident in the Philippines for at least two years prior to election, and no Senator shall serve for more than two consecutive terms. One-half of the membership of the Senate shall be elected every three years. No treaty or international agreement may be considered valid without the approval, by voting, of at least two-thirds of members.

A maximum of 250 Representatives may sit in the House of Representatives. Its members may serve no more than three consecutive three-year terms. Representatives must be natural-born citizens, literate, and at least 25 years of age. Each legislative district may elect one representative; the number of legislative districts shall be determined according to population and shall be reapportioned following each census. Representatives must be registered voters in their district, and resident there for at least one year prior to election. In addition, one-fifth of the total number of representatives shall be elected under a party list system from lists of nominees proposed by indigenous, but non-religious, minority groups (such as the urban poor, peasantry, women and youth).

The Senate and the House of Representatives shall each have an Electoral Tribunal which shall be the sole judge of contests relating to the election of members of Congress. Each Tribunal shall have nine members, three of whom must be Justices of the Supreme Court, appointed by the Chief Justice. The remaining six members shall be members of the Senate or of the House of Representatives, as appropriate, and shall be selected from the political parties represented therein, on a proportional basis.

THE COMMISSION ON APPOINTMENTS

The President must submit nominations of heads of executive departments, ambassadors and senior officers in the armed forces to the Commission on Appointments, which shall decide on the appointment by majority vote of its members. The President of the Senate shall act as ex-officio Chairman; the Commission shall consist of 12 Senators and 12 members of the House of Representatives, elected from the political parties represented therein, on the basis of proportional representation.

THE EXECUTIVE

Executive power is vested in the President of the Philippines. Presidents are limited to one six-year term of office, and Vice-Presidents to two successive six-year terms. Candidates for both posts are elected by direct universal suffrage. They must be natural-born citizens, literate, at least 40 years of age, registered voters and resident in the Philippines for at least 10 years prior to election.

The President is Head of State and Chief Executive of the Republic. Bills (legislative proposals) that have been approved by Congress shall be signed by the President; if the President vetoes the bill, it may become law when two-thirds of members in Congress approve it.

The President shall nominate and, with the consent of the Commission on Appointments, appoint ambassadors, officers of the armed forces and heads of executive departments.

The President is Commander-in-Chief of the armed forces and may suspend the writ of habeas corpus or place the Republic under martial law for a period not exceeding 60 days when, in the President's opinion, public safety demands it. Congress may revoke either action by a majority vote.

The Vice-President may be a member of the Cabinet; in the event of the death or resignation of the President, the Vice-President shall become President and serve the unexpired term of the previous President.

THE JUDICIARY

The Supreme Court is composed of a Chief Justice and 14 Associate Justices, and may sit *en banc* or in divisions comprising three, five or seven members. Justices of the Supreme Court are appointed by the President, with the consent of the Commission on Appointments, for a term of four years. They must be citizens of the Republic, at least 40 years of age, of proven integrity, and must have been judges of the lower courts, or engaged in the practice of law in the Philippines, for at least 15 years.

The Supreme Court, sitting *en banc*, is the sole judge of disputes relating to presidential and vice-presidential elections.

THE CONSTITUTIONAL COMMISSIONS

These are the Civil Service Commission and the Commission on Audit, each of which has a Chairman and two other Commissioners, appointed by the President (with the approval of the Commission on Appointments) to a seven-year term; and the Commission on Elections, which enforces and administers all laws pertaining to elections and political parties. The Commission on Elections has seven members, appointed by the President (and approved by the Commission on Appointments) for a seven-year term. The Commission on Elections may sit *en banc* or in two divisions.

LOCAL GOVERNMENT

The Republic of the Philippines shall be divided into provinces, cities, municipalities and barangays. The Congress of the Philippines shall enact a local government code providing for decentralization. A region may become autonomous, subject to approval by a majority vote of the electorate of that region, in a referendum. Defence and security in such areas will remain the responsibility of the national Government.

ACCOUNTABILITY OF PUBLIC OFFICERS

All public officers, including the President, Vice-President and members of Congress and the Constitutional Commissions, may be removed from office if impeached for, or convicted of, violation of the Constitution, corruption, treason, bribery or betrayal of public trust.

Cases of impeachment must be initiated solely by the House of Representatives, and tried solely by the Senate. A person shall be convicted by a vote of at least two-thirds of the Senate, and will then be dismissed from office and dealt with according to the law.

SOCIAL JUSTICE AND HUMAN RIGHTS

The Congress of the Philippines shall give priority to considerations of human dignity, the equality of the people and an equitable distribution of wealth. The Commission on Human Rights shall investigate allegations of violations of human rights, shall protect human rights through legal measures, and shall monitor the Government's compliance with international treaty obligations. It may advise Congress on measures to promote human rights.

AMENDMENTS OR REVISIONS

Proposals for amendment or revision of the Constitution may be made by:

i) Congress (upon a vote of three-quarters of members);

ii) A Constitutional Convention (convened by a vote of two-thirds of members of Congress);

iii) The people, through petitions (signed by at least 12% of the total number of registered voters).

The proposed amendments or revisions shall then be submitted to a national plebiscite, and shall be valid when ratified by a majority of the votes cast.

MILITARY BASES

Foreign military bases, troops or facilities shall not be allowed in the Republic of the Philippines following the expiry, in 1991, of the Agreement between the Republic and the USA, except under the provisions of a treaty approved by the Senate, and, when required by Congress, ratified by the voters in a national referendum.

The Government

HEAD OF STATE

President: GLORIA MACAPAGAL ARROYO (assumed office 20 January 2001; inaugurated for second term 30 June 2004).

Vice-President: NOLI DE CASTRO.

THE CABINET
(September 2004)

Executive Secretary: EDUARDO ERMITA.

Secretary of Agrarian Reform: RENE VILLA.

Secretary of Agriculture: ARTHUR YAP.

Secretary of the Budget and Management: EMILIA T. BONCODIN.

Secretary of Education, Culture and Sports: FLORENCIO ABAD.

Secretary of Energy: VICENTE S. PEREZ, Jr.

Secretary of the Environment and Natural Resources: MICHAEL DEFENSOR.

Secretary of Finance: JUANITA AMATONG.

Secretary of Foreign Affairs: ALBERTO G. ROMULO.

Secretary of Health: MANUEL M. DAYRIT.

Secretary of the Interior and Local Government: ANGELO REYES.

Secretary of Justice: RAUL GONZALES.

Secretary of Labor and Employment: PATRICIA A. SANTO THOMAS.

Secretary of National Defense: AVELINO J. CRUZ, Jr.

Secretary of Public Works and Highways: FLORANTE M. SORIQUEZ.

Secretary of Science and Technology: ESTRELLA F. ALABASTRO.

Secretary of Social Welfare and Development: NOLI DE CASTRO.

Secretary of Tourism: JOSEPH ACE DURANO.

Secretary of Trade and Industry: CESAR A. V. PURISIMA.

Secretary of Transportation and Communications: LEANDRO MENDOZA.

Director-General of the National Economic and Development Authority: ROMULO NERI.

Press Secretary: MILTON A. ALINGOD.

MINISTRIES

Office of the President: New Executive Bldg, Malacañang Palace Compound, J. P. Laurel St, San Miguel, Metro Manila; tel. (2) 7356047; fax (2) 7358006; e-mail opnet@ops.gov.ph; internet www.opnet.ops.gov.ph.

Office of the Vice-President: PICC, 2nd Floor, CCP Complex, Roxas Blvd, Pasay City, Metro Manila; tel. (2) 8312658; fax (2) 8312614; e-mail gma@easy.net.ph.

Department of Agrarian Reform: DAR Bldg, Elliptical Rd, Diliman, Quezon City, Metro Manila; tel. (2) 9287031; fax (2) 9292527; e-mail nani@dar.gov.ph; internet www.dar.gov.ph.

Department of Agriculture: DA Bldg, 4th Floor, Elliptical Rd, Diliman, Quezon City, Metro Manila; tel. (2) 9288741; fax (2) 9285140; e-mail dnotes@da.gov.ph; internet www.da.gov.ph.

Department of the Budget and Management: DBM Bldg, Gen. Solano St, San Miguel, Metro Manila; tel. (2) 7354807; fax (2) 7354927; e-mail dbmbiss@dbm.gov.ph; internet www.dbm.gov.ph.

Department of Education, Culture and Sports: Deped Complex, Meralco Ave, Pasig City, 1600 Metro Manila; tel. (2) 6321361; fax (2) 6320805; internet www.deped.gov.ph.

Department of Energy: Energy Center, Merritt Rd, Fort Bonifacio, Taguig, Metro Manila; tel. (2) 8441021; fax (2) 8442495; e-mail v_perez@doe.gov.ph; internet www.doe.gov.ph.

Department of the Environment and Natural Resources: DENR Bldg, Visayas Ave, Diliman, Quezon City, 1100 Metro Manila; tel. (2) 9296626; fax (2) 9204352; e-mail sechta@denr.gov.ph; internet www.denr.gov.ph.

Department of Finance: DOF Bldg, Roxas Blvd, cnr Pablo Ocampo St, Metro Manila; tel. (2) 5234955; fax (2) 5212950; e-mail camacho@dof.gov.ph; internet www.dof.gov.ph.

Department of Foreign Affairs: DFA Bldg, 2330 Roxas Blvd, Pasay City, Metro Manila; tel. (2) 8344000; fax (2) 8321597; e-mail webmaster@dfa.gov.ph; internet www.dfa.gov.ph.

Department of Health: San Lazaro Compound, Rizal Ave, Santa Cruz, 1003 Metro Manila; tel. (2) 7438301; fax (2) 7431829; e-mail mmdayrit@doh.gov.ph; internet www.doh.gov.ph.

Department of the Interior and Local Government: A. Francisco Gold Condominium II, Epifanio de los Santos Ave, cnr Mapagmahal St, Diliman, Quezon City, 1100 Metro Manila; tel. (2) 9250349; fax (2) 9250386; e-mail dilgmail@dilg.gov.ph; internet www.dilg.gov.ph.

Department of Justice: Padre Faura St, Ermita, Metro Manila; tel. (2) 5213721; fax (2) 5211614; e-mail sechbp@info.com.ph; internet www.doj.gov.ph.

Department of Labor and Employment: DOLE Executive Bldg, 7th Floor, Muralla Wing, Muralla St, Intramuros, 1002 Metro Manila; tel. (2) 5272131; fax (2) 5273494; e-mail osec@dole.gov.ph; internet www.dole.gov.ph.

Department of National Defense: DND Bldg, 3rd Floor, Camp Aguinaldo, Quezon City, Metro Manila; tel. (2) 9113300; fax (2) 9116213; e-mail osnd@philonline.com; internet www.dnd.gov.ph.

Department of Public Works and Highways: DPWH Bldg, Bonifacio Drive, Port Area, Metro Manila; tel. (2) 3043000; fax (2) 5275635; e-mail pid@dpwh.gov.ph; internet www.dpwh.gov.ph.

Department of Science and Technology: DOST Compound, Gen. Santos Ave, Bicutan, Taguig, Metro Manila; tel. (2) 8372071; fax (2) 8372937; e-mail wgp@sunl.dost.gov.ph; internet www.dost.gov.ph.

Department of Social Welfare and Development: Batasang Pambansa, Constitution Hills, Quezon City, Metro Manila; tel. (2) 9317916; fax (2) 9318191; e-mail dinky@mis.dswd.gov.ph; internet www.dswd.gov.ph.

Department of Tourism: Rm 317, DOT Bldg, Teodoro F. Valencia Circle, Rizal Park, Metro Manila; tel. (2) 5251805; fax (2) 5256538; e-mail oti@tourism.gov.ph; internet www.wowphilippines.com.ph.

Department of Trade and Industry: Industry and Investments Bldg, 385 Sen. Gil J. Puyat Ave, Buendia, Makati City, 3117 Metro Manila; tel. (2) 8953611; fax (2) 8956487; e-mail mis@dti.gov.ph; internet www.dti.gov.ph.

Department of Transportation and Communications: Columbia Tower, Ortigas Ave, Pasig City, 1555 Metro Manila; tel. (2) 7267106; fax (2) 7267104; e-mail dotc@i-next.net; internet www.dotcmain.gov.ph.

National Economic and Development Authority (NEDA—Department of Socio-Economic Planning): NEDA-sa-Pasig Bldg, 12 Blessed Josemaria Escriva St, Pasig City, 1605 Metro Manila; tel. (2) 6313747; fax (2) 6313282; e-mail info@neda.gov.ph; internet www.neda.gov.ph.

Philippine Information Agency (Office of the Press Secretary): PIA Bldg, Visayas Ave, Diliman, Quezon City, Metro Manila; tel. (2) 9247703; fax (2) 9204347; e-mail pia@ops.gov.ph; internet www.pia.ops.gov.ph.

President and Legislature

PRESIDENT

Election, 10 May 2004

Candidate	Votes	% of votes
Gloria Macapagal Arroyo (Lakas-CMD)	12,905,808	39.99
Fernando Poe, Jr (KNP)	11,782,232	36.51
Panfilo Lacson (LDP)	3,510,080	10.88
Raul S. Roco (Aksyon Demokratiko)	2,082,762	6.45
Eduardo Villanueva (Bangon Pilipinas)	1,988,218	6.16
Total*	32,269,100	100.00

* Total may not be equal to sum of components, owing to rounding.

THE CONGRESS OF THE PHILIPPINES

Senate

President of the Senate: Franklin M. Drilon.

Elections for 13 of the 24 seats in the Senate were held on 14 May 2001. The People Power Coalition (PPC) won eight seats and the Laban ng Demokratikong Pilipino-Puwersa ng Masa (LDP-PnM) won four. The remaining seat was taken by an independent candidate. Elections for 12 of the 24 seats took place on 10 May 2004. The Koalisyon ng Katapatan at Karanasan sa Kinabukasan (K-4) won seven seats and the Koalisyon ng Nagkakaisang Pilipino (KNP) won five. The result thus gave President Macapagal Arroyo a majority in the upper house.

House of Representatives

Speaker of the House: Jose de Venecia.

General Election, 10 May 2004

	Seats
Lakas ng EDSA-Christian Muslim Democrats (Lakas-CMD)	93
Nationalist People's Coalition (NPC)	54
Liberal Party (LP)	34
Laban ng Demokratikong Pilipino (LDP)	11
Nacionalista Party (NP)	5
Kabalikat ng Malayang Pilipino (KAMPI)	3
Partido ng Masang Pilipino (PMP)	3
Koalisyon ng Nagkakaisang Pilipino (KNP)	2
PDP-Laban Party	2
Independent	1
Others	4
Total*	235

* Total includes 23 members of minority and cause-orientated groups allocated seats in the House of Representatives under the party list elections, which also took place on 10 May 2004.

Autonomous Region

MUSLIM MINDANAO

The Autonomous Region of Muslim Mindanao (ARMM) originally comprised the provinces of Lanao del Sur, Maguindanao, Tawi-Tawi and Sulu. The Region was granted autonomy in November 1989. Elections took place in February 1990, and the formal transfer of limited executive powers took place in October. In August 2001 a plebiscite was conducted in 11 provinces and 14 cities in Mindanao to determine whether or not they would become members of the ARMM. The city of Marawi and the province of Basilan subsequently joined the Region. The regional Assembly, for which elections were held in November 2001, comprises 21 legislative seats.

Governor: Farouk Hussein (elected 26 November 2001).

Vice-Governor: Mahid Mutilan.

Political Organizations

Akbayan (Citizens' Action Party): 101 Matahimik St, Teacher's Village West, Quezon City, 1101 Metro Manila; tel. (2) 4336933; fax (2) 9252936; e-mail secretariat@akbayan.org; internet www.akbayan.org; f. 1998; left-wing party list; Pres. Ronald Llamas; Sec.-Gen. Arlene Santos.

Aksyon Demokratiko (Democratic Action Party): 16th Floor, Strata 2000 Bldg, Emerald Ave, Ortigas Centre, Pasig City, 1600 Metro Manila; tel. (2) 6385381; fax (2) 6343073; e-mail senator@raulroco.com; internet www.raulroco.com; f. 1997 to support presidential candidacy of Raul Roco; joined Alyansa ng Pag-asa in 2003 to contest 2004 elections; Chair. Jaime Galvez Tan; Pres. Raul Roco.

Alayon: c/o House of Representatives, Metro Manila.

Alyansa ng Pag-asa (AP) (Alliance of Hope): c/o House of Representatives, Metro Manila; f. 2003 to support presidential candidacy of Raul Roco; coalition of Aksyon Demokratiko, PROMDI and Reporma.

Bangon Pilipinas (Rise Philippines): 8th Floor, Dominion Bldg, 833 Arnaiz Ave, Legaspi Village, Makati City, 1200 Metro Manila; tel. (2) 8113355; fax (2) 8111110; e-mail feedback@bangonpilipinas.org; internet www.broeddie.com; supported candidacy of Eduardo Villanueva in 2004 presidential election; Pres. Eduardo Villanueva.

Bayan Muna: c/o House of Representatives, Metro Manila; e-mail bayanmuna@bayanmuna.net; internet www.bayanmuna.net; f. 1999; Pres. Satur Ocampo; Chair. Dr Reynaldo Lesaca, Jr.

Gabay ng Bayan (Nation's Guide): Metro Manila; fmrly Grand Alliance for Democracy; Leader Francisco Tadad.

Kabalikat ng Malayang Pilipino (KAMPI): c/o House of Representatives, Metro Manila; f. 1997; Chair. Margarita Cojuangco; Pres. Ronaldo Puno.

Kilusang Bagong Lipunan (KBL) (New Society Movement): Metro Manila; f. 1978 by Pres. Marcos and fmr mems of the Nacionalista Party; Sec.-Gen. Vicente Mellora.

Kilusan para sa Pambansang Pagpapanibago (BAGO): Metro Manila; f. 1997 to support presidential candidacy of Santiago F. Dumlao, Jr.

Koalisyon ng Katapatan at Karanasan sa Kinabukasan (K-4) (Coalition of Dedication and Experience for the Future): c/o House of Representatives, Metro Manila; f. 2003 to support presidential candidacy of GLORIA MACAPAGAL ARROYO; coalition of Lakas-CMD, Liberal Party (LP) and People's Reform Party (PRP); Leader GLORIA MACAPAGAL ARROYO.

Koalisyon ng Nagkakaisang Pilipino (KNP) (Coalition of the United Filipino): c/o House of Representatives, Metro Manila; f. Dec. 2003 to support presidential candidacy of FERNANDO POE, Jr; coalition of Angara faction of Laban ng Demokratikong Pilipino (LDP), PDP-Laban Party and Puwersa ng Masang Pilipino (PMP); Exec. Chair. EDGARDO ANGARA.

Laban ng Demokratikong Pilipino (LDP) (Fight of Democratic Filipinos): c/o House of Representatives, Metro Manila; f. 1987; reorg. 1988 as an alliance of Lakas ng Bansa and a conservative faction of the PDP-Laban Party; mem. of Lapian ng Masang Pilipino (LAMP) until Jan. 2001; split into two factions, led by EDGARDO ANGARA and AGAPITO AQUINO, to contest 2004 elections; Angara faction joined Koalisyon ng Nagkakaisang Pilipino (KNP) in Dec. 2003 to support presidential candidacy of FERNANDO POE, Jr; Aquino faction supported presidential candidacy of PANFILO LACSON; Pres. EDGARDO ANGARA; Sec.-Gen. AGAPITO AQUINO.

Lakas ng EDSA (Power of EDSA)-Christian Muslim Democrats (Lakas-CMD): c/o House of Representatives, Metro Manila; f. 1992 as alliance to support the presidential candidacy of Gen. FIDEL V. RAMOS; formed alliance with UMDP to contest 1998 and 2001 elections; fmrly Lakas-National Union of Christian Democrats (Lakas-NUCD); name changed as above in 2003; joined Koalisyon ng Katapatan at Karanasan sa Kinabukasan (K-4) in 2003; Pres. JOSE DE VENECIA; Sec.-Gen. HEHERSON ALVAREZ.

Lapian ng Masang Pilipino (LAMP) (Party of the Filipino Masses): Metro Manila; f. 1997 as Laban ng Makabayang Masang Pilipino (LaMMP, Struggle of Nationalist Filipino Masses), renamed Laban ng Masang Pilipino (LMP, Struggle of Filipino Masses) in 1998, present name adopted in 1999; originally coalition of Laban ng Demokratikong Pilipino (LDP), Nationalist People's Coalition (NPC) and Partido ng Masang Pilipino (PMP) until split in Jan. 2001.

Liberal Party (LP): 4th Floor, J & T Bldg, Magsaysay Blvd, Sta. Mesa, 1016 Metro Manila; tel. (2) 7168187; fax (2) 7168210; e-mail liberal@tri-isys.com; internet www.liberalparty.ph; f. 1946; represents centre-liberal opinion of the fmr Nacionalista Party, which split in 1946; joined Koalisyon ng Katapatan at Karanasan sa Kinabukasan (K-4) in Jan. 2004 to contest 2004 elections; Pres. FLORENCIO ABAD; Chair. FRANKLIN DRILON.

Nacionalista Party (NP): Metro Manila; tel. (2) 854418; fax (2) 865602; Pres. ARTURO TOLENTINO; Sec.-Gen. RENE ESPINA.

Nationalist People's Coalition (NPC): Metro Manila; f. 1991; breakaway faction of the Nacionalista Party led by EDUARDO COJUANGCO; mem. of Lapian ng Masang Pilipino (LAMP) from 1997 until Jan. 2001.

New National Alliance: Metro Manila; f. 1998; left-wing; Dep. Sec.-Gen. TEDDY CASINO.

Partido Bansang Marangal (PBM): Metro Manila; f. 1997 to support the presidential candidacy of MANUEL L. MORATO.

Partido Demokratiko Sosyalista ng Pilipinas (PDSP) (Philippine Democratic Socialist Party): Metro Manila; f. 1981 by mems of the Batasang Pambansa allied to the Nacionalista (Roy faction), Pusyon Visaya and Mindanao Alliance parties; joined People Power Coalition (PPC) in Feb. 2001; Leader NORBERTO GONZALES.

Partido Komunista ng Pilipinas (PKP) (Communist Party of the Philippines): f. 1930; Pres. FELICISIMO MACAPAGAL.

Partido Nacionalista ng Pilipinas (PNP) (Philippine Nationalist Party): Metro Manila; f. 1986 by fmr mems of KBL; Leader BLAS F. OPLE.

Partido ng Bayan (New People's Alliance): f. May 1986 by JOSE MARIA SISON (imprisoned in 1977–86), the head of the Communist Party of the Philippines (CPP); militant left-wing nationalist group.

Partido ng Manggagawang Pilipino (PMP) (Filipino Workers' Party): f. 2002 by fmr supporters of the CPP (see below).

Partido para sa Demokratikong Reporma-Lapiang Manggagawa Coalition (Reporma): c/o House of Representatives, Metro Manila; joined People Power Coalition (PPC) in Feb. 2001; Leader RENATO DE VILLA.

PDP-Laban Party: c/o House of Representatives, Metro Manila; f. February 1983 following merger of Pilipino Democratic Party (f. 1982 by fmr mems of the Mindanao Alliance) and Laban (Lakas ng Bayan—People's Power Movement, f. 1978 and led by BENIGNO S. AQUINO, Jr, until his assassination in August 1983); centrist; formally dissolved in Sept. 1988, following the formation of the LDP, but a faction continued to function as a political movement; Pres. JEJOMAR BINAY; Chair. AQUILINO PIMENTEL.

People's Reform Party (PRP): c/o House of Representatives, Metro Manila; f. 1991 by MIRIAM DEFENSOR SANTIAGO to support her candidacy in the 1992 presidential election; joined Koalisyon ng Katapatan at Karanasan sa Kinabukasan (K-4) to contest 2004 elections to Senate; Pres. MIRIAM DEFENSOR SANTIAGO.

Probinsya Muna Development Initiatives (PROMDI): 7 Pasteur St, Lahug, Cebu City; tel. (32) 2326692; fax (32) 2313609; e-mail emro@cebu.pw.net.ph; internet www.col.net.ph/emro; f. 1997; joined People Power Coalition (PPC) to contest 2001 elections; Leader EMILIO ('LITO') OSMEÑA.

Puwersa ng Masa (PnM): c/o House of Representatives, Metro Manila; f. 2001 by ex-President JOSEPH EJERCITO ESTRADA; formed an alliance with Laban ng Demokratikong Pilipino (LDP) to contest the 2001 elections to the Senate.

Puwersa ng Masang Pilipino (PMP): Metro Manila; mem. of Lapian ng Masan Pilipino (LAMP) from 1997 until Jan. 2001; joined Koalisyon ng Nagkakaisang Pilipino (KNP) to contest 2004 elections; Leader JOSEPH EJERCITO ESTRADA; Pres. HORACIO MORALES, Jr.

United Muslim Democratic Party (UMDP): Mindanao; moderate Islamic party; formed an electoral alliance with Lakas-NUCD for the election to the House of Representatives in May 2001.

United Negros Alliance (UNA): Negros Occidental.

The following organizations are, or have been, in conflict with the Government:

Abu Sayyaf (Bearer of the Sword): Mindanao; radical Islamic group seeking the establishment of an Islamic state in Mindanao; breakaway grouping of the MILF; est. strength 1,500 (2000); Leader KHADAFI JANJALANI.

Alex Boncayao Brigade (ABB): communist urban guerrilla group, fmrly linked to CPP, formed alliance with Revolutionary Proletarian Party in 1997; est. strength 500 (April 2001); Leader NILO DE LA CRUZ.

Islamic Command Council (ICC): Mindanao; splinter group of MNLF; Leader MELHAM ALAM.

Maranao Islamic Statehood Movement: Mindanao; f. 1998; armed grouping seeking the establishment of an Islamic state in Mindanao.

Mindanao Independence Movement (MIM): Mindanao; claims a membership of 1m.; Leader REUBEN CANOY.

Moro Islamic Liberation Front (MILF): Camp Abubakar, Lanao del Sur, Mindanao; aims to establish an Islamic state in Mindanao; comprises a faction that broke away from the MNLF in 1978; its armed wing, the Bangsa Moro Islamic Armed Forces, est. 10,000 armed regulars; Chair. Al-Haj MURAD.

Moro Islamic Reform Group: Mindanao; breakaway faction from MNLF; est. strength of 200 in 2000.

Moro National Liberation Front (MNLF): internet www.mnlf.org; seeks autonomy for Muslim communities in Mindanao; signed a peace agreement with the Govt in Sept. 1996; its armed wing, the Bangsa Moro Army, comprised an est. 10,000 mems in 2000; Chair. and Pres. of Cen. Cttee (vacant); Sec.-Gen. MUSLIMIN SEMA.

Moro National Liberation Front—Islamic Command Council (MNLF—ICC): Basak, Lanao del Sur; f. 2000; Islamist separatist movement committed to urban guerrilla warfare; breakaway faction from MNLF; Cmmdr DATU FIJRODIN, Cmmdr MUDS BAIRODIN, Cmmdr CALEB BEN MUHAMAD, Cmmdr ZAIDA BULLH, Cmmdr OMEN.

National Democratic Front (NDF): a left-wing alliance of 14 mem. groups; Chair. MARIANA OROSA; Spokesman GREGORIO ROSAL.

The NDF includes:

Communist Party of the Philippines (CPP): f. 1968; a breakaway faction of the PKP; legalized Sept. 1992; in July 1993 the Metro Manila-Rizal and Visayas regional committees, controlling 40% of total CPP membership (est. 15,000 in 1994), split from the Central Committee; Chair. JOSE MARIA SISON; Gen. Sec. BENITO TIAMZON.

New People's Army (NPA): f. 1969 as the military wing of the CPP; based in central Luzon, but operates throughout the Philippines; est. strength 9,500; Leader JOVENCIO BALWEG; Spokesman GREGORIO ROSAL.

Revolutionary Proletarian Party: Metro Manila; f. 1996; comprises mems of the Metro Manila-Rizal and Visayas regional committees, which broke away from the CPP in 1993; has a front organization called the Bukluran ng Manggagawang Pilipino (Association of Filipino Workers); Leader ARTURO TABARA.

Diplomatic Representation

EMBASSIES IN THE PHILIPPINES

Argentina: 8th Floor, Liberty Center, 104 H. V. de la Costa St, Salcedo Village, Makati City, 1262 Metro Manila; tel. (2) 8453218; fax (2) 8453220; e-mail embarfil@eastern.com.ph; Ambassador ISMAEL MARIO SCHUFF.

Australia: 23rd Floor, Tower II, RCBC Plaza, 6819 Ayala Ave, Makati City, 1200 Metro Manila; tel. (2) 7578100; fax (2) 7578268; e-mail public-affairs-MNLA@dfat.gov.au; internet www.australia.com.ph; Ambassador RUTH PEARCE.

Austria: Prince Bldg, 4th Floor, 117 Rada St, Legaspi Village, Makati City, 1200 Metro Manila; tel. (2) 8179191; fax (2) 8134238; e-mail manila-ob@bmaa.gv.at; Ambassador Dr CHRISTIAN KREPELA.

Bangladesh: Universal-Re Bldg, 2nd Floor, 106 Paseo de Roxas, Legaspi Village, Makati City, Metro Manila; tel. (2) 8175001; fax (2) 8164941; e-mail bdoot.manila@pacific.net.ph; Ambassador MUNIR UZ ZAMAN.

Belgium: Multibancorporation Centre, 9th Floor, 6805 Ayala Ave, Makati City, Metro Manila; tel. (2) 8451869; fax (2) 8452076; e-mail manila@diplobel.org; Ambassador CHRISTIAAN TANGHE.

Brazil: 16th Floor, Liberty Center, 104 H. V. de la Costa St, Salcedo Village, Makati City, 1229 Metro Manila; tel. (2) 8453651; fax (2) 8453676; e-mail brascom@info.com.ph; Ambassador CLAUDIO MARIA HENRIQUE DO COUTO LYRA.

Brunei: Bank of the Philippine Islands Bldg, 11th Floor, Ayala Ave, cnr Paseo de Roxas, Makati City, 1226 Metro Manila; tel. (2) 8162836; fax (2) 8916646; Ambassador MAIMUNAH Dato' Paduka Haji ELIAS.

Cambodia: Unit 7A, 7th Floor, Country Space One Bldg, Sen. Gil J. Puyat Ave, Makati City, Metro Manila; tel. (2) 8189981; fax (2) 8189983; e-mail cam.emb.ma@netasia.net; Ambassador CHOEUNG BUNTHENG.

Canada: Floors 6–8, Tower 2, RCBC Plaza, 6819 Ayala Ave, Makati City, 1200 Metro Manila; tel. (2) 8579001; fax (2) 8431082; e-mail manil@dfait-maeci.gc.ca; internet www.dfait-maeci.gc.ca/manila; Ambassador PETER SUTHERLAND.

Chile: 17th Floor, Liberty Center, 104 H. V. de la Costa St, cnr Leviste St, Salcedo Village, Makati City, 1261 Metro Manila; tel. (2) 8433461; fax (2) 8431976; e-mail echileph@mnl.sequel.net; Ambassador JORGE MONTERO.

China, People's Republic: 4896 Pasay Rd, Dasmariñas Village, Makati City, Metro Manila; tel. (2) 8443148; fax (2) 8439970; e-mail emb-chn@pacific.net.ph; internet www.china-embassy.org.ph; Ambassador WU HONGBO.

Colombia: Aurora Tower, 18th Floor, Araneta Center, Quezon City, Metro Manila; tel. (2) 9113101; fax (2) 9112846; Chargé d'affaires a.i. STELLA MÁRQUEZ DE ARANETA.

Cuba: 101 Aguirre St, cnr Trasierra St, Cacho-Gonzales Bldg Penthouse, Legaspi Village, Makati City, Metro Manila; tel. (2) 8171192; fax (2) 8164094; Ambassador RAMÓN DOMINGO ALONSO MEDINA.

Czech Republic: 30th Floor, Rufino Pacific Tower, Ayala Ave, cnr Herrera St, Makati City, Metro Manila; tel. (2) 8111155; fax (2) 8111020; e-mail manila@embassy.msv.cz; internet www.mzv.cz/manila; Ambassador STANISLAV SLAVICKY.

Egypt: 2229 Paraiso St, cnr Banyan St, Dasmariñas Village, Makati City, Metro Manila; tel. (2) 8439232; fax (2) 8439239; Ambassador SABER ABDEL KADER MANSOUR.

Finland: 21st Floor, Far East Bank and Trust Center, Sen. Gil J. Puyat Ave, Makati City, Metro Manila; tel. (2) 8915011; fax (2) 8914107; e-mail sanomat.mni@formin.fi; Ambassador RITTA RESCH.

France: Pacific Star Bldg, 16th Floor, Makati Ave, cnr Sen. Gil J. Puyat Ave, Makati City, Metro Manila; tel. (2) 8576900; fax (2) 8576948; e-mail consulat@france.com.ph; internet www.ambafrance-ph.org; Ambassador RENÉE VEYRET.

Germany: PS Bank Center, 6th Floor, 777 Paseo de Roxas, Makati City, 1226 Metro Manila; tel. (2) 8924906; fax (2) 8104703; e-mail germanembassymanila@surfshop.net.ph; internet www.germanembassy-philippines.com; Ambassador Dr AXEL WEISHAUPT.

Holy See: 2140 Taft Ave, POB 3364, 1099 Metro Manila (Apostolic Nunciature); tel. (2) 5210306; fax (2) 5211235; e-mail nuntiusp@info.com.ph; Apostolic Nuncio Most Rev. ANTONIO FRANCO (Titular Archbishop of Gallese).

India: 2190 Paraiso St, Dasmariñas Village, POB 2123, Makati City, Metro Manila; tel. (2) 8430101; fax (2) 8158151; e-mail amb@embindia.org.ph; internet www.embindia.org.ph; Ambassador NAVREKHA SHARMA.

Indonesia: 185 Salcedo St, Legaspi Village, Makati City, Metro Manila; tel. (2) 8925061; fax (2) 8925878; e-mail kbri@impactnet.com; internet www.kbrimanila.org.ph; Chargé d'affaires a.i. ALEXANDER LATURIUW.

Iran: 2224 Paraiso St, cnr Pasay Rd, Dasmariñas Village, Makati City, Metro Manila; tel. (2) 8884757; fax (2) 8884777; e-mail ambassador@iranembassy.org.ph; Ambassador JALAL KALANTARI.

Israel: Trafalgar Plaza, 23rd Floor, H. V. de la Costa St, Salcedo Village, Makati City, 1200 Metro Manila; POB 1697, Makati Central Post Office, 1256 Metro Manila; tel. (2) 8940441; fax (2) 8941027; e-mail israinfor@pacific.net.ph; Ambassador YEHOSHUA SAGI.

Italy: Zeta Bldg, 6th Floor, 191 Salcedo St, Legaspi Village, Makati City, Metro Manila; tel. (2) 8924531; fax (2) 8171436; e-mail ambitaly@iname.com; Ambassador UMBERTO COLESANTI.

Japan: 2627 Roxas Blvd, Pasay City, 1300 Metro Manila; tel. (2) 5515710; fax (2) 5515780; e-mail info@embjapan.ph; internet www.embjapan.ph; Ambassador KOJIRO TAKANO.

Korea, Republic: Pacific Star Bldg, 10th Floor, Sen. Gil J. Puyat Ave, cnr Makati Ave, Makati City, 1226 Metro Manila; tel. (2) 8116139; fax (2) 8116148; Ambassador YU MYUNG-HWAN.

Kuwait: 1230 Acacia Rd, Dasmariñas Village, Makati City, Metro Manila; tel. (2) 8876880; fax (2) 8876666; Ambassador IBRAHIM MUHANNA AL-MUHANNA.

Laos: 34 Lapu-Lapu St, Magallanes Village, Makati City, Metro Manila; tel. and fax (2) 8525759; Ambassador PHIANE PHILAKONE.

Libya: 1644 Dasmarinas St, cnr Mabolo St, Dasmariñas Village, Makati City, Metro Manila; tel. (2) 8177331; fax (2) 8177337; e-mail lpbmanila@skynet.net; Ambassador SALEM M. ADAM.

Malaysia: 107 Tordesillas St, Salcedo Village, Makati City, 1200 Metro Manila; tel. (2) 8174581; fax (2) 8163158; e-mail mwmanila@indanet.com; Chargé d'affaires a.i. MAHINDER SINGH.

Malta: 6th Floor, Cattleya Condominium, 235 Salcedo St, Legaspi Village, Makati City, Metro Manila; tel. (2) 8171095; fax (2) 8171089; e-mail syquia@intlaw.com.ph; Ambassador ENRIQUE P. SYQUIA.

Mexico: 2157 Paraiso St, Dasmariñas Village, Makati City, Metro Manila; tel. (2) 8122211; fax (2) 8929824; e-mail ebmexfil@info.com.ph; Ambassador ENRIQUE HUBBARD.

Myanmar: Xanland Center, 8th Floor, 152 Amorsolo St, Legaspi Village, Makati City, Metro Manila; tel. (2) 8172373; fax (2) 8175895; Ambassador U TIN HTUN.

Netherlands: King's Court Bldg, 9th Floor, 2129 Chino Roces Ave, POB 2448, Makati City, 1264 Metro Manila; tel. (2) 8125981; fax (2) 8154579; e-mail man@minbuza.nl; internet www.dutchembassy.ph; Ambassador ROBERT VORNIS.

New Zealand: BPI Buendia Centre, 23rd Floor, Sen. Gil J. Puyat Ave, POB 3228, MCPO, Makati City, Metro Manila; tel. (2) 8915358; fax (2) 8915353; e-mail nzmanila@nxdsl.com.ph; Ambassador ROB MOORE-JONES.

Nigeria: 2211 Paraiso St, Dasmariñas Village, Makati City, 1221 Metro Manila; POB 3174, MCPO, Makati City, 1271 Metro Manila; tel. (2) 8439866; fax (2) 8439867; e-mail embnigmanila@pacific.net.ph; Chargé d'affaires a.i. Chief OYEBOLA KUKU.

Norway: Petron Mega Plaza Bldg, 21st Floor, 358 Sen. Gil J. Puyat Ave, Makati City, 1209 Metro Manila; tel. (2) 8863245; fax (2) 8863244; e-mail emb.manila@mfa.no; internet www.norway.ph; Ambassador PÅL MOE.

Pakistan: Alexander House, 6th Floor, 132 Amorsolo St, Legaspi Village, Makati City, Metro Manila; tel. (2) 8172776; fax (2) 8400229; e-mail parepmnl@info.co.ph; Ambassador IFTIKHAR HUSSAIN KAZMI.

Palau: Splendido Gardens, 146 H. V. de la Costa St, cnr Alfaro St, Makati City, Metro Manila; tel. (2) 8130799; e-mail rop_piembassy@yahoo.com; Ambassador ANITA RECHIREI SUTA.

Panama: 10th Floor, MARC 2000 Tower, 1973 Taft Ave and San Andres St, cnr Quirino Ave, Malate, 1004 Metro Manila; tel. (2) 5212790; fax (2) 5215755; e-mail panaembassy@i-manila.com.ph; Ambassador JUAN CARLOS ESCALONA AVILA.

Papua New Guinea: 3rd Floor, Corinthian Plaza Condominium Bldg, cnr Paseo de Roxas and Gamboa St, Makati City, Metro Manila; tel. (2) 8113465; fax (2) 8113466; e-mail kundumnl@pngembmnl.com.ph; Ambassador DAMIEN DOMINIC GAMIANDU.

Peru: Unit 1604, 16th Floor, Antel Corporate Centre, 139 Valero St, Salcedo Village, Makati City, Metro Manila; tel. (2) 8138731; fax (2) 8929831; e-mail leprumanila@embassyperu.com.ph; internet www.embassyperu.com.ph; Ambassador JORGE CHAVEZ SOTO.

Portugal: 17th Floor, Units C and D, Trafalgar Plaza, 105 H. V. de la Costa St, Salcedo Village, Makati City, Metro Manila; tel. (2) 8483789; fax (2) 8483791; Ambassador Joao Caetano da Silva.

Qatar: 1601 Cypress St, Dasmariñas Village, Makati City, Metro Manila; tel. (2) 8874944; fax (2) 8876406; Ambassador Ibrahim Abdulrahman Al-Meghaiseeb.

Romania: 1216 Acacia Rd, Dasmariñas Village, Makati City, Metro Manila; tel. (2) 8439014; fax (2) 8439063; e-mail amaro@skyinet.net; Ambassador Radu Homescu.

Russia: 1245 Acacia Rd, Dasmariñas Village, Makati City, Metro Manila; tel. (2) 8930190; fax (2) 8109614; e-mail RusEmb@i-manila .com.ph; Ambassador Anatoly Nebogatov.

Saudi Arabia: Saudi Embassy Bldg, 389 Sen. Gil J. Puyat Ave Ext., Makati City, Metro Manila; tel. (2) 8909735; fax (2) 8953493; e-mail phemb@mofa.gov.sa; Ambassador Mohammad Ameen Wali.

Singapore: The Enterprise Centre, Tower I, 35th Floor, 6766 Ayala Ave, cnr Paseo de Roxas, Makati City, Metro Manila; tel. (2) 7512345; Ambassador Lim Kheng Hua.

Spain: ACT Tower, 5th Floor, 135 Sen. Gil J. Puyat Ave, Makati City, 1200 Metro Manila; tel. (2) 8183561; fax (2) 8102885; e-mail embesphh@mail.mae.es; Ambassador Ignacio Sagaz.

Sri Lanka: 2260 Avocado Ave, Dasmariñas Village, Makati City, Metro Manila; tel. (2) 8439813; fax (2) 8439813; Ambassador Ariya Bandara Rekawa.

Sweden: Equitable PCI Bank Tower II, 16th Floor, Makati Ave, Makati City, Metro Manila; POB 2322, MCPO 1263, Makati City, Metro Manila; tel. (2) 8191951; fax (2) 8153002; e-mail ambassaden .manila@foreign.ministry.se; internet www.swedemb-manila.com; Ambassador Annika Markovic.

Switzerland: Equitable Bank Tower, 24th Floor, 8751 Paseo de Roxas, Makati City, 1226 Metro Manila; tel. (2) 7579000; fax (2) 7573718; e-mail vertretung@man.rep.admin.ch; Ambassador Lise Favre.

Thailand: 107 Rada St, Legaspi Village, Makati City, 1229 Metro Manila; tel. (2) 8154220; fax (2) 8154221; e-mail thaimnl@pacific.net .ph; Ambassador Busba Bunnag.

Turkey: 2268 Paraiso St, Dasmariñas Village, Makati City, Metro Manila; tel. (2) 8439705; fax (2) 8439702; Ambassador Tanju Sumer.

United Arab Emirates: Renaissance Bldg, 2nd Floor, 215 Sakedo St, Legaspi Village, Makati City, Metro Manila; tel. (2) 8173906; fax (2) 8183577; Ambassador Mohammed Ebrahim Abdullah al-Jowaid.

United Kingdom: Locsin Bldg, 15th–17th Floors, 6752 Ayala Ave, cnr Makati Ave, Makati City, 1226 Metro Manila; tel. (2) 8167116; fax (2) 8197206; e-mail uk@info.com.ph; internet www .britishembassy.org.ph; Ambassador Paul Dimond.

USA: 1201 Roxas Blvd, Metro Manila; tel. (2) 5231001; fax (2) 5224361; e-mail manila1@pd.state.gov; internet manila.usembassy .gov; Ambassador Francis Ricciardone.

Venezuela: Unit 17A, Multinational Bancorporation Center, 6805 Ayala Ave, Makati City, Metro Manila 1226; tel. (2) 8452841; fax (2) 8452866; e-mail embavefi@compass.com.ph; Chargé d'affaires a.i. José Clavijo.

Viet Nam: 554 Pablo Ocampo St, Malate, Metro Manila; tel. (2) 5252837; fax (2) 5260472; e-mail sqvnplp@qinet.net; Ambassador Dinh Tich.

Judicial System

The February 1987 Constitution provides for the establishment of a Supreme Court comprising a Chief Justice and 14 Associate Justices; the Court may sit *en banc* or in divisions of three, five or seven members. Justices of the Supreme Court are appointed by the President from a list of a minimum of three nominees prepared by a Judicial and Bar Council. Other courts comprise the Court of Appeals, Regional Trial Courts, Metropolitan Trial Courts, Municipal Courts in Cities, Municipal Courts and Municipal Circuit Trial Courts. There is also a special court for trying cases of corruption (the Sandiganbayan). The Office of the Ombudsman (Tanodbayan) investigates complaints concerning the actions of public officials.

Supreme Court

Taft Ave, cnr Padre Faura St, Ermita, 1000 Metro Manila; tel. (2) 5268123; e-mail infos@supremecourt.gov.ph; internet www .supremecourt.gov.ph.

Chief Justice: Hilario G. Davide, Jr.

Court of Appeals

Consists of a Presiding Justice and 68 Associate Justices.

Presiding Justice: Cancio Garcia.

Islamic Shari'a courts were established in the southern Philippines in July 1985 under a presidential decree of February 1977. They are presided over by three district magistrates and six circuit judges.

Religion

In 1991 94.2% of the population were Christians: 84.1% were Roman Catholics, 6.2% belonged to the Philippine Independent Church (Aglipayan) and 3.9% were Protestants. There is an Islamic community, and an estimated 43,000 Buddhists. Animists and persons professing no religion number approximately 400,000.

CHRISTIANITY

Sangguniang Pambansa ng mga Simbahan sa Pilipinas (National Council of Churches in the Philippines): 879 Epifanio de los Santos Ave, Diliman, Quezon City, Metro Manila; tel. (2) 9288636; fax (2) 9267076; e-mail nccp-ga@philonline.com; f. 1963; 11 mem. churches, 10 assoc. mems; publishes NCCP news magazine quarterly and TUGON periodically; Gen. Sec. Sharon Rose Joy Ruiz-Duremdes.

The Roman Catholic Church

For ecclesiastical purposes, the Philippines comprises 16 archdioceses, 55 dioceses, six territorial prelatures and eight apostolic vicariates. At 31 December 2002 approximately 79.5% of the population were adherents.

Catholic Bishops' Conference of the Philippines (CBCP)
470 General Luna St, Intramuros, 1002 Metro Manila; tel. (2) 5274138; fax (2) 5274063; e-mail cbcp@info.com.ph; internet www .cbcp.com.

f. 1945; statutes approved 1952; Pres. Most Rev. Fernando R. Capalla (Archbishop of Davao).

Archbishop of Caceres: Most Rev. Leonardo Z. Legaspi, Archbishop's House, Elias Angeles St, POB 6085, 4400 Naga City; tel. (54) 4738483; fax (54) 4732800.

Archbishop of Cagayan de Oro: Most Rev. Jesus B. Tuquib, Archbishop's Residence, POB 113, 9000 Misamis Oriental, Cagayan de Oro City; tel. (88) 8571357; fax (88) 726304; e-mail orochan@cdo .weblinq.com.

Archbishop of Capiz: Most Rev. Onesimo C. Gordoncillo, Chancery Office, POB 44, 5800 Roxas City; tel. (36) 6215595; fax (36) 6211053.

Archbishop of Cebu: Cardinal Ricardo J. Vidal, Archbishop's Residence, cnr P. Gomez St and P. Burgos St, POB 52, 6000 Cebu City; tel. (32) 2530123; fax (32) 2530616; e-mail adelito@skynet.net.

Archbishop of Cotabato: Most Rev. Orlando B. Quevedo, Archbishop's Residence, Sinsuat Ave, POB 186, 9600 Cotabato City; tel. (64) 4212918; fax (64) 4211446.

Archbishop of Davao: Most Rev. Fernando R. Capalla, Archbishop's Residence, 247 Florentino Torres St, POB 80418, 8000 Davao City; tel. (82) 2271163; fax (82) 2279771; e-mail bishop-davao@skyinet.net.

Archbishop of Jaro: Most Rev. Angel N. Lagdameo, Archbishop's Residence, Jaro, 5000 Iloilo City; tel. (33) 3294442; fax (33) 3293197; e-mail abpjaro@skynet.net.

Archbishop of Lingayen-Dagupan: Most Rev. Oscar V. Cruz, Archbishop's House, 2400 Dagupan City; tel. (75) 52353576; fax (75) 5221878; e-mail oscar@rezcom.com.

Archbishop of Lipa: Most Rev. Ramon C. Argüelles, Archbishop's House, St Lorenzo Ruiz St, Lipa City, 4217 Batangas; tel. (43) 7562573; fax (43) 7562964; e-mail rcalipa@batangas.net.ph.

Archbishop of Manila: Most Rev. Gaudencio B. Rosales, Arzobispado, 121 Arzobispo St, Intramuros, POB 132, 1099 Metro Manila; tel. (2) 5277631; fax (2) 5273955; e-mail aord@mailstation.net; internet www.geocities.com/aocmanila.

Archbishop of Nueva Segovia: Most Rev. Edmundo M. Abaya, Archbishop's House, Vigan, 2700 Ilocos Sur; tel. (77) 7222018; fax (77) 7221591.

Archbishop of Ozamis: Most Rev. Jesus A. Dosado, Archbishop's House, POB 2760, 7200 Ozamis City; tel. (65) 5212820; fax (65) 5211574.

Archbishop of Palo: Most Rev. Pedro R. Dean, Archdiocesan Chancery, Bukid Tabor, Palo, 6501 Leyte; POB 173, Tacloban City,

6500 Leyte; tel. (53) 3232213; fax (53) 3235607; e-mail rcap@mozcom .com.

Archbishop of San Fernando (Pampanga): Most Rev. PACIANO B. ANICETO, Chancery House, San José, San Fernando, 2000 Pampanga; tel. (45) 9612819; fax (45) 9616772; e-mail rca@pamp.pworld.net.ph.

Archbishop of Tuguegarao: Most Rev. DIOSDADO A. TALAMAYAN, Archbishop's House, Tuguegarao, 3500 Cagayan; tel. (78) 8441663; fax (78) 8462822; e-mail rcat@cag.pworld.net.ph.

Archbishop of Zamboanga: Most Rev. CARMELO DOMINADOR F. MORELOS, Sacred Heart Center, POB 1, Justice R. T. Lim Blvd, 7000 Zamboanga City; tel. (62) 9911329; fax (62) 9932608; e-mail aofzam@jetlink.com.ph.

Other Christian Churches

Convention of Philippine Baptist Churches: POB 263, 5000 Iloilo City; tel. (33) 3290621; fax (33) 3290618; e-mail gensec@iloilo .net; f. 1935; Gen. Sec. Rev. Dr NATHANIEL M. FABULA; Pres. DONATO ENABE.

Episcopal Church in the Philippines: 275 E. Rodriguez Sr Ave, Quezon City, 1102 Metro Manila; POB 10321, Broadway Centrum, Quezon City, 1102 Metro Manila; tel. (2) 7228481; fax (2) 7211923; e-mail ecpi@info.com.ph; internet www.episcopalphilippines.net; f. 1901; six dioceses; Prime Bishop Most Rev. IGNACIO C. SOLIBA.

Iglesia Evangélica Metodista en las Islas Filipinas (Evangelical Methodist Church in the Philippines): Beulah Land, Iemelif Center, Greenfields 1, Subdivision, Marytown Circle, Novaliches, Quezon City, 1129 Metro Manila; tel. (2) 9356519; fax (2) 4185017; e-mail admin@iemelif.org; internet www.iemelif.org; f. 1909; 40,000 mems (2003); Gen. Supt Bishop NATHANAEL P. LAZARO.

Iglesia Filipina Independiente (Philippine Independent Church): 1500 Taft Ave, Ermita, 1000 Metro Manila; tel. (2) 5237242; fax (2) 5213932; e-mail ifiphil@hotmail.com; internet ifi .ph; f. 1902; 34 dioceses; 6.0m. mems; Obispo Maximo (Supreme Bishop) Most Rev. TOMAS MILLAMENA.

Iglesia ni Cristo: 1 Central Ave, New Era, Quezon City, 1107 Metro Manila; tel. (2) 9814311; fax (2) 9811111; f. 1914; 2m. mems; Exec. Minister Brother ERAÑO G. MANALO.

Lutheran Church in the Philippines: 4461 Old Santa Mesa, 1008 Metro Manila; POB 507, 1099 Metro Manila; tel. (2) 7157084; fax (2) 7142395; f. 1946; Pres. Rev. EDUARDO LADLAD.

Union Church of Manila: cnr Legaspi St and Rada St, Legaspi Village, Makati City, Metro Manila; tel. (2) 8126062; fax (2) 8172386; e-mail ucmweb@unionchurch.ph; internet www .unionchurch.ph; Senior Pastor Dr ALEXANDER B. ARONIS.

United Church of Christ in the Philippines: 877 Epifanio de los Santos Ave, West Triangle, Quezon City, Metro Manila; POB 718, MCPO, Ermita, 1099 Metro Manila; tel. (2) 9240215; fax (2) 9240207; e-mail uccpnaof@manila-online.net; f. 1948; 900,000 mems (1996); Gen. Sec. Rev. ELMER M. BOLOCON (Bishop).

Among other denominations active in the Philippines are the Iglesia Evangélica Unida de Cristo and the United Methodist Church.

ISLAM

Some 14 different ethnic groups profess the Islamic faith in the Philippines, and Muslims comprised 4.6% of the total population at the census of 1990. Mindanao and the Sulu and Tawi-Tawi archipelago, in the southern Philippines, are predominantly Muslim provinces, but there are 10 other such provinces, each with its own Imam, or Muslim religious leader. More than 500,000 Muslims live in the north of the country (mostly in, or near to, Manila).

Confederation of Muslim Organizations of the Philippines (CMOP): Metro Manila; Nat. Chair. JAMIL DIANALAN.

BAHÁ'Í FAITH

National Spiritual Assembly: 1070 A. Roxas St, cnr Bautista St, Singalong Subdiv., Malate, 1004 Metro Manila; POB 4323, 1099 Metro Manila; tel. (2) 5240404; fax (2) 5245918; e-mail nsaphil@ skyinet.net; mems resident in 129,949 localities; Chair. GIL MARVEL TABUCANON; Sec.-Gen. VIRGINIA S. TOLEDO.

The Press

The Office of the President implements government policies on information and the media. Freedom of the press and freedom of speech are guaranteed under the 1987 Constitution.

METRO MANILA

Dailies

Abante: Monica Publishing Corpn, Rooms 301–305, BF Condominium Bldg, 3rd Floor, Solana St, cnr. A. Soriano St, Intramuros, Metro Manila; tel. (2) 5273385; fax (2) 5274470; e-mail abante@ abante-tonite.com; internet www.abante.com.ph; morning; Filipino and English; Editor NICOLAS QUIJANO, Jr; circ. 417,000.

Abante Tonite: Monica Publishing Corpn, Rooms 301–305, BF Condominium Bldg, 3rd Floor, Solana St, cnr. A. Soriano St, Intramuros, Metro Manila; tel. (2) 5273385; fax (2) 5274470; e-mail tonite@abante-tonite.com; internet www.abante-tonite.com; afternoon; Filipino and English; Man. Editor NICOLAS QUIJANO, Jr; circ. 277,000.

Ang Pilipino Ngayon: 202 Railroad St, cnr 13th St, Port Area, Metro Manila; tel. (2) 401871; fax (2) 5224998; Filipino; Publr and Editor JOSE M. BUHAIN; circ. 286,452.

Balita: Liwayway Publishing Inc, 2249 China Roces Ave, Makati City, Metro Manila; tel. (2) 8193101; fax (2) 8175167; internet www .balita.org; f. 1972; morning; Filipino; Editor MARCELO S. LAGMAY; circ. 151,000.

Daily Tribune: Penthouse Suites, Plywood Industries Bldg, T. M. Kalaw St, cnr A. Mabini St, Ermita, Metro Manila; tel. (2) 5215511; fax (2) 5215522; e-mail nco@tribune.net.ph; internet www.tribune .net.ph; f. 2000; English; Publr and Editor-in-Chief NINEZ CACHO-OLIVARES.

Malaya: People's Independent Media Inc, 575 Atlanta St, Port Area, Metro Manila; tel. (2) 5277651; fax (2) 5271839; e-mail opinion@ malaya.com.ph; internet www.malaya.com.ph; f. 1983; English; Editor JOEY C. DE LOS REYES; circ. 175,000.

Manila Bulletin: Bulletin Publishing Corpn, cnr Muralla and Recoletos Sts, Intramuros, POB 769, Metro Manila; tel. (2) 5271519; fax (2) 5277534; e-mail bulletin@mb.com.ph; internet www.mb.com .ph; f. 1900; English; Publr EMILIO YAP; Editor BEN RODRIGUEZ; circ. 265,000.

Manila Standard: Leyland Bldg, 21st St, cnr Railroad St, Port Area, Metro Manila; tel. (2) 5278351; e-mail infoms@philonline.com; internet www.manilastandard.net; morning; English; Editor-in-Chief JULLIE YAP DAZA; circ. 96,000.

Manila Times: Liberty Bldg, 13th St, Port Area, Metro Manila; tel. (2) 5245664; fax (2) 5216887; e-mail newsboy1@manilatimes.net; internet www.manilatimes.net; f. 1945; morning; English; Publr, Chair. and Pres. DANTE A. ANG; Editor-in-Chief CIPRIANO S. ROXAS.

People Tonight: Philippine Journalist Inc, Railroad St, cnr 19th and 20th Sts, Port Area, Metro Manila; tel. (2) 5278421; fax (2) 5274627; f. 1978; English and Filipino; Editor FERDIE RAMOS; circ. 500,000.

People's Bagong Taliba: Philippine Journalist Inc, Railroad St, cnr 19th and 20th Sts, Port Area, Metro Manila; tel. (2) 5278121; fax (2) 5274627; Filipino; Editor MATEO VICENCIO; circ. 229,000.

People's Journal: Philippine Journalist Inc, Railroad St, cnr 19th and 20th Sts, Port Area, Metro Manila; tel. (2) 5278421; fax (2) 5274627; English and Filipino; Editor ROSAURO ACOSTA; circ. 219,000.

Philippine Daily Globe: Nova Communications Inc, 2nd Floor, Rudgen Bldg, 17 Shaw Blvd, Metro Manila; tel. (2) 6730496; Editor-in-Chief YEN MAKABENTA; circ. 40,000.

Philippine Daily Inquirer: Philippine Daily Inquirer Bldg, cnr Mascardo St and Yague St, Pasong Tamo, Makati City, 1220 Metro Manila; tel. (2) 8978808; fax (2) 8914793; e-mail feedback@inquirer .com.ph; internet www.inquirer.net; f. 1985; English; Chair. MARIXI R. PRIETO; Editor-in-Chief RIGOBERTO TIGLAO; circ. 250,000.

Philippine Herald-Tribune: V. Esguerra II Bldg, 140 Amorsolo St, Legaspi Village, Makati City, Metro Manila; tel. (2) 853711; f. 1987; Christian-orientated; Pres. AMADA VALINO.

Philippine Star: 13th and Railroad Sts, Port Area, Metro Manila; tel. (2) 5277901; fax (2) 5276851; e-mail philippinestar@hotmail .com; internet www.philstar.com; f. 1986; Editor BOBBY DE LA CRUZ; circ. 275,000.

Tempo: Bulletin Publishing Corpn, Recoletos St, cnr Muralla St, Intramuros, Metro Manila; tel. (2) 5278121; fax (2) 5277534; internet www.tempo.com.ph; f. 1982; English and Filipino; Editor BEN RODRIGUEZ; circ. 230,000.

Today: Independent Daily News, 55 Paseo de Roxas, Makati City, 1225 Metro Manila; tel. (2) 8940644; fax (2) 8131417; e-mail today@ impactnet.com; internet www.today.net.ph; f. 1993; Editor-in-Chief TEODORO L. LOCSIN, Jr; Man. Editor LOURDES MOLINA-FERNANDEZ; circ. 106,000.

United Daily News: 812 and 818 Benavides St, Binondo, Metro Manila; tel. (2) 2447171; f. 1973; Chinese; Editor-in-Chief CHUA KEE; circ. 85,000.

Selected Periodicals

Weeklies

Banawag: Liwayway Bldg, 2249 Pasong Tamo, Makati City, Metro Manila; tel. (2) 8193101; fax (2) 8175167; f. 1934; Ilocano; Editor DIONISIO S. BULONG; circ. 42,900.

Bisaya: Liwayway Bldg, 2249 Pasong Tamo, Makati City, Metro Manila; tel. (2) 8193101; fax (2) 8175167; f. 1934; Cebu-Visayan; Editor SANTIAGO PEPITO; circ. 90,000.

Liwayway: Liwayway Bldg, 2249 Pasong Tamo, Makati City, Metro Manila; tel. (2) 8193101; fax (2) 8175167; f. 1922; Filipino; Editor RODOLFO SALANDANAN; circ. 102,400.

Panorama: Manila Bulletin Publishing Corpn, POB 769, cnr Muralla and Recoletos Sts, Intramuros, Metro Manila; tel. and fax (2) 5277509; f. 1968; English; Editor RANDY V. URLANDA; circ. 239,600.

Philippine Starweek: 13th St, cnr Railroad St, Port Area, Metro Manila; tel. (2) 5277901; fax (2) 5275819; e-mail starweek@pacific.net.ph; internet www.philstar.com; English; Publr MAXIMO V. SOLIVEN; circ. 268,000.

SELECTED REGIONAL PUBLICATIONS

The Aklan Reporter: 1227 Rizal St, Kalibo, Panay, Aklan; tel. (33) 3181; f. 1971; weekly; English and Aklanon; Editor ROMAN A. DE LA CRUZ; circ. 3,500.

Baguio Midland Courier: 16 Kisad Rd, POB 50, Baguio City; English and Ilocano; Editor SINAI C. HAMADA; circ. 6,000.

Bayanihan Weekly News: Bayanihan Publishing Co, P. Guevarra Ave, Santa Cruz, Laguna; tel. (645) 1001; f. 1966; Mon.; Filipino and English; Editor ARTHUR A. VALENOVA; circ. 1,000.

Bohol Chronicle: 56 B. Inting St, Tagbilaran City, Bohol; tel. (32) 3100; internet www.boholchronicle.com; f. 1954; 2 a week; English and Cebuano; Editor and Publr ZOILO DEJARESCO; circ. 5,500.

The Bohol Times: 100 Gallares St, Tagbilaran City, Bohol; tel. (38) 4112961; fax (38) 4112656; e-mail times@bit.fapenet.org; internet www.calamay.bit.fapenet.org/btimes; Publr Dr LILIA A. BALITE; Editor-in-Chief ATTY SALVADOR D. DIPUTADO.

The Kapawa News: L. V. Moles and Jose Abad Santos Sts, Tangub, POB 365, 6100 Bacolod City; tel. and fax (34) 4441941; e-mail LM-Kapawa@eudoramail.com; f. 1966; weekly; Sat.; Hiligaynon and English; Publr and Editor HENRY G. DOBLE; circ. 2,000.

Mindanao Star: 44 Kolambagohan-Capistrano St, Cagayan de Oro City; internet www.mindanaostar.com; weekly; Editor ROMULFO SABAMAL; circ. 3,500.

Mindanao Times: UMBN Bldg, Ponciano Reyes St, Davao City, Mindanao; tel. 2273252; e-mail timesmen@mozcom.com; internet www.mindanaotimes.com.ph; daily; Publr JOSEFINA SAN PEDRO (acting); Editor-in-Chief VIC SUMALINOG; circ. 5,000.

Pagadian Times: Pagadian City, 7824 Zamboanga del Sur; tel. 586; e-mail pagtimes@mozcom.com; internet www.pagadian.mozcom.com/pagtimes; f. 1969; weekly; English; Publr PEDE G. LIU; Editor JACINTO LUMBAY; circ. 5,000.

Palihan: Diversion Rd, cnr Sanciangco St, Cabanatuan City, Luzon; f. 1966; weekly; Filipino; Editor and Publr NONOY M. JARLEGO; circ. 5,000.

Sorsogon Today: 2903 Burgos St, East District, 4700 Sorsogon; tel. and fax (56) 2111340; e-mail sortoday@yahoo.com; f. 1977; weekly; Publr and CEO MARCOS E. PARAS, Jr; circ. 2,250.

Sun Star Cebu: Sun Star Bldg, P. del Rosario St, Cebu City; tel. (32) 2546100; fax (32) 2537256; e-mail centralnewsroom@sunstar.com.ph; internet www.sunstar.com.ph/cebu; f. 1982; daily; English; Editor-in-Chief ATTY PACHICO A. SEARES; Gen. Man. ORLANDO P. CARVAJAL.

The Tribune: Maharlika Highway, 2301 Cabanatuan City, Luzon; f. 1960; weekly; English and Filipino; Editor and Publr ORLANDO M. JARLEGO; circ. 8,000.

The Valley Times: Daang Maharlika, San Felipe, Ilagan, Isabela; f. 1962; weekly; English; Editor AUREA A. DE LA CRUZ; circ. 4,500.

The Visayan Tribune: 826 Iznart St, Iloilo City; tel. (33) 75760; f. 1959; weekly; Tue.; English; Editor HERBERT L. VEGO; circ. 5,000.

The Voice of Islam: Davao City; tel. (82) 81368; f. 1973; monthly; English and Arabic; official Islamic news journal; Editor and Publr NASHIR MUHAMMAD AL'RASHID AL HAJJ.

The Weekly Negros Gazette: Broce St, San Carlos City, 6033 Negros Occidental; f. 1956; weekly; Editor NESTORIO L. LAYUMAS, Sr; circ. 5,000.

NEWS AGENCIES

Philippines News Agency: PIA Bldg, 2nd Floor, Visayas Ave, Diliman, Quezon City, Metro Manila; tel. (2) 9206551; fax (2) 9206566; e-mail philna@ops.gov.ph; internet www.pna.ops.gov.ph; f. 1973; Gen. Man. GEORGE REYES; Exec. Editor SEVERINO SAMONTE.

Foreign Bureaux

Agence France-Presse (AFP): Kings Court Bldg 2, 5th Floor, Pasong Tamo, cnr de la Rosa St, Makati City, Metro Manila; tel. (2) 8112028; fax (2) 8112664; Bureau Chief MONICA EGOY.

Agencia EFE (Spain): Unit 1006, 88 Corporate Center Bldg, 141 Sedeño St, cnr Valero St, Salcedo Village, Makati City, 1227 Metro Manila; tel. (2) 8431986; fax (2) 8431973; e-mail manila@efe.com; Bureau Chief ESTHER REBOLLO.

Associated Press (AP) (USA): S&L Bldg, 3rd Floor, 1500 Roxas Blvd, Ermita, 1000 Metro Manila; tel. (2) 5259217; fax (2) 5212430; Bureau Chief DAVID THURBER.

Deutsche Presse Agentur (dpa) (Germany): Physicians Tower Bldg, 533 United Nations Ave, Ermita 1000, Metro Manila; tel. (2) 5221919; fax (2) 5221447; Representative GIRLIE LINAO.

Inter Press Service (IPS) (Italy): Amberland Plaza, Room 510, J. Vargas Ave, Ortigas Complex, Pasig City, 1600 Metro Manila; tel. (2) 6353421; fax (2) 6353660; Correspondent JOHANNA SON.

Jiji Tsushin (Jiji Press) (Japan): Legaspi Tower, Suite 21, 3rd Floor, 2600 Roxas Blvd, Metro Manila; tel. (2) 5211472; fax (2) 5211474; Correspondent IPPEI MIYASAKA.

Kyodo News Service (Japan): Pacific Star Bldg, 4th Floor, Makati Ave, cnr Sen. Gil J. Puyat Ave, Makati City, Metro Manila; tel. (2) 8133072; fax (2) 8133914; Correspondent KIMIO OKI.

Reuters (United Kingdom): L.V. Locsin Bldg, 10th Floor, Ayala Ave, cnr Makati Ave, Makati City, 1226 Metro Manila; tel. (2) 8418900; fax (2) 8176267; Country Man. RAJU GOPALAKRISHNAN.

United Press International (UPI) (USA): Manila Pavilion Hotel, Room 526C, United Nations Ave, Ermita, 1000 Metro Manila; tel. (2) 5212051; fax (2) 5212074; Bureau Chief MICHAEL DI CICCO.

Xinhua (New China) News Agency (People's Republic of China): 705B Gotesco Twin Towers, 1129 Concepcion St, Ermita, Metro Manila; tel. (2) 5271404; fax (2) 5271410; Chief Correspondent CHEN HEGAO.

PRESS ASSOCIATION

National Press Club of the Philippines: National Press Club Bldg, Magallanes Drive, 1002 Intramuros, Metro Manila; tel. (2) 494242; f. 1952; Pres. ANTONIO ANTONIO; Vice-Pres. ROMEO DEL CASTILLO; 942 mems.

Publishers

Abiva Publishing House Inc: Abiva Bldg, 851 Gregorio Araneta Ave, Quezon City, 1113 Metro Manila; tel. (2) 7120245; fax (2) 7320308; e-mail mmrabiva@i-manila.com.ph; internet www.abiva.com.ph; f. 1937; reference and textbooks; Pres. LUIS Q. ABIVA, Jr.

Ateneo de Manila University Press: Bellarmine Bldg, Ateneo de Manila University, Katipunan Ave, Loyola Heights, Quezon City, Metro Manila; tel. (2) 4265984; fax (2) 4265909; e-mail unipress@admu.edu.ph; internet www.ateneopress.com; f. 1972; literary, textbooks, humanities, social sciences, reference books on the Philippines; Dir MARICOR E. BAYTION.

Bookman, Inc: 373 Quezon Ave, Quezon City, 1114 Metro Manila; tel. (2) 7124813; fax (2) 7124843; e-mail bookman@info.com.ph; f. 1945; textbooks, reference, educational; Pres. LINA PICACHE-ENRIQUEZ; Exec. Vice-Pres. MARIETTA PICACHE-ENRIQUEZ.

Capitol Publishing House, Inc: 13 Team Pacific Bldg, Jose C. Cruz St, cnr F. Legaspi St, Barrio Ugong, Pasig City, Metro Manila; tel. (2) 6712662; fax (2) 6712664; e-mail cacho@mozcom.com; f. 1947; Gen. Man. MANUEL L. ATIENZA.

Heritage Publishing House: 33 4th Ave, cnr Main Ave, Cubao, Quezon City, POB 3667, Metro Manila; tel. (2) 7248114; fax (2) 6471393; e-mail heritage@iconn.com.ph; internet www.iconn.com.ph/heritage; art, anthropology, history, political science; Pres. MARIO R. ALCANTARA; Man. Dir RICARDO S. SANCHEZ.

The Lawyers' Co-operative Publishing Co Inc: 1071 Del Pan St, Makati City, 1206 Metro Manila; tel. (2) 5634073; fax (2) 5642021;

e-mail lawbooks@info.com.ph; f. 1908; law, educational; Pres. ELSA K. ELMA.

Liwayway Publishing Inc: 2249 Chino Roces Ave, Makati City, Metro Manila; tel. (2) 8193101; fax (2) 8175167; magazines and newspapers; Pres. RENE G. ESPINA; Chair. DIONISIO S. BULONG.

Mutual Books Inc: 429 Shaw Blvd, Mandaluyong City, Metro Manila; tel. (2) 7257538; fax (2) 7213056; f. 1959; textbooks on accounting, management and economics, computers and mathematics; Pres. ALFREDO S. NICDAO, Jr.

Reyes Publishing Inc: Mariwasa Bldg, 4th Floor, 717 Aurora Blvd, Quezon City, 1112 Metro Manila; tel. (2) 7221827; fax (2) 7218782; e-mail reyespub@skyinet.net; f. 1964; art, history and culture; Pres. LOUIE REYES.

SIBS Publishing House Inc: 8th Floor, Globe Telecom Plaza II, Pioneer Highlands, cnr Madison St, Mandaluyong City, 1552 Metro Manila; tel. (2) 6876164; fax (2) 6871716; e-mail sibsbook@info.com.ph; internet www.sibs.com.ph; f. 1996; science, language, religion, literature and history textbooks; Pres. CARMEN MIMETTE M. SIBAL; Gen. Man. JUAN CARLOS SIBAL.

Sinag-Tala Publishers Inc: GMA Lou-Bel Plaza, 6th Floor, Chino Roces Ave, cnr Bagtikan St, San Antonio Village, Makati City, 1203 Metro Manila; tel. (2) 8971162; fax (2) 8969626; e-mail stpi@info.com.ph; internet www.sinagtala.com; f. 1972; educational textbooks; business, professional and religious books; Man. Dir LUIS A. USON.

University of the Philippines Press: Epifanio de los Santos Ave, U. P. Campus, Diliman, Quezon City, 1101 Metro Manila; tel. (2) 9252930; fax (2) 9282558; e-mail press@up.edu.ph; internet www.upd.edu.ph/~uppress; f. 1965; literature, history, political science, sociology, cultural studies, economics, anthropology, mathematics; Officer in Charge RUTH JORDANA L. PISON.

Vibal Publishing House Inc: G. Araneta Ave, cnr Maria Clara St, Talayan, Quezon City, Metro Manila; tel. (2) 7122722; fax (2) 7118852; internet www.vibalpublishing.com; f. 1955; linguistics, social sciences, mathematics, religion; CEO ESTHER A. VIBAL.

PUBLISHERS' ASSOCIATIONS

Philippine Educational Publishers' Asscn: 84 P. Florentino St, Quezon City, 1104 Metro Manila; tel. (2) 7402698; fax (2) 7115702; e-mail dbuhain@cnl.net; Pres. DOMINADOR D. BUHAIN.

Publishers' Association of the Philippines Inc: Gammon Center, Alfaro St, Salcedo Village, Makati City, Metro Manila; f. 1974; mems comprise all newspaper, magazine and book publrs in the Philippines; Pres. KERIMA P. TUVERA; Exec. Dir ROBERTO M. MENDOZA.

Broadcasting and Communications

National Telecommunications Commission (NTC): NTC Bldg, BIR Rd, East Triangle, Diliman, Quezon City, Metro Manila; tel. (2) 9244042; fax (2) 9217128; e-mail ntc@ntc.gov.ph; internet www.ntc.gov.ph; f. 1979; supervises and controls all private and public telecommunications services; Commr RONALD OLIVAR SOLIS.

TELECOMMUNICATIONS

BayanTel: Benpres Bldg, 5th Floor, Meralco Ave, cnr Exchange Rd, Pasig City, Metro Manila; tel. (2) 4493000; fax (2) 4492511; e-mail bayanserve@bayantel.com.ph; internet www.bayantel.com.ph; 359,000 fixed lines (1999); Pres. RODOLFO SALAZAR.

Bell Telecommunications Philippines (BellTel): Pacific Star Bldg, 3rd and 4th Floors, Sen. Gil J. Puyat Ave, cnr Makati Ave, Makati City, Metro Manila; tel. (2) 8400808; fax (2) 8915618; e-mail info@belltel.ph; internet www.belltel.ph; f. 1997; Pres. EDRAGDO REYES.

Capitol Wireless Inc: Dolmar Gold Tower, 6th Floor, 107 Carlos Palanca, Jr, St, Legaspi Village, Makati City, Metro Manila; tel. (2) 8159961; fax (2) 8941141; Pres. EPITACIO R. MARQUEZ.

Digital Telecommunications Philippines Inc: 110 Eulogio Rodriguez, Jr, Ave, Bagumbayan, Quezon City, 1110 Metro Manila; tel. (2) 6330000; fax (2) 6339387; provision of fixed line telecommunications services; 564,304 fixed lines (1998); Chief Exec. RICARDO J. ROMULO; Pres. JOHN GOKONGWEI.

Domestic Satellite Philippines Inc (DOMSAT): Solid House Bldg, 4th Floor, 2285 Pasong Tamo Ext., Makati City, 1231 Metro Manila; tel. (2) 8105917; fax (2) 8671677; Pres. SIEGFRED MISON.

Globe Telecom (GMCR) Inc: Globe Telecom Plaza, 57th Floor, Pioneer St, cnr Madison St, 1552 Mandaluyong City, Metro Manila; tel. (2) 7302701; fax (2) 7302586; e-mail custhelp@globetel.com.ph; internet www.globe.com.ph; 700,000 fixed and mobile telephone subscribers (1999); Pres. and CEO GERARDO C. ABLAZA, Jr.

Philippine Communications Satellite Corpn (PhilcomSat): 12th Floor, Telecoms Plaza, 316 Sen. Gil J. Puyat Ave, Makati City, Metro Manila; tel. (2) 8158406; fax (2) 8159287; Pres. MANUEL H. NIETO.

Philippine Global Communications, Inc (PhilCom): 8755 Paseo de Roxas, Makati City, Metro Manila; tel. (2) 8162851; fax (2) 8162872; e-mail aong@philcom.com; internet www.philcom.com; Pres. EVELYN SINGSON.

Philippine Long Distance Telephone Co: Ramon Cojuangco Bldg, Makati Ave, POB 2148, Makati City, Metro Manila; tel. (2) 8168883; fax (2) 8186800; e-mail media@pldt.com.ph; internet www.pldt.com.ph; f. 1928; monopoly on overseas telephone service until 1989; retains 94% of Philippine telephone traffic; 2,516,748 fixed lines (1998); Chair. ANTONIO COJUANGCO; Pres. and CEO MANUEL PANGILINAN.

Pilipino Telephone Corpn (Piltel): Bankers Center, 9th Floor, 6764 Ayala Ave, Makati City, Metro Manila; tel. (2) 8913888; fax (2) 8171121; major cellular telephone provider; 400,000 subscribers (1999); Chair. ROBERTO V. ONGPIN; Pres. and CEO NAPOLEON L. NAZARENO.

Smart Communications, Inc (SCI): Rufino Pacific Tower, 12th Floor, Ayala Ave, cnr Herrera St, Makati City, Metro Manila; tel. (2) 8110213; fax (2) 5113400; 900,000 mobile telephone subscribers (1999); Pres. ORLANDO B. VEA.

BROADCASTING

Radio

Banahaw Broadcasting Corpn: Broadcast City, Capitol Hills, Diliman, Quezon City, 3005 Metro Manila; tel. (2) 9329949; fax (2) 9318751; 14 stations; Station Man. BETTY LIVIOCO.

Bureau of Broadcast Services (BBS): Office of the Press Sec., Philippine Information Agency Bldg, 4th Floor, Visayas Ave, Quezon City, Metro Manila; tel. (2) 9242607; fax (2) 9242745; internet www.pbs.gov.ph; f. 1952 as Philippines Broadcasting Service; govt-operated; 32 radio stations; Dir RAFAEL DANTE A. CRUZ.

Cebu Broadcasting Co: FJE Bldg, 105 Esteban St, Legaspi Village, Makati City, Metro Manila; tel. (2) 8159131; fax (2) 8125592; Chair. HADRIAN ARROYO.

Far East Broadcasting Co Inc: POB 1, Valenzuela, 0560 Metro Manila; 62 Karuhatan Rd, Karuhatan, Valenzuela City, 1441 Metro Manila; tel. (2) 2921152; fax (2) 2925790; e-mail febcomphil@febc.org.ph; internet www.febc.org.ph; f. 1948; 18 stations; operates a classical music station, eight domestic stations and an overseas service in 64 languages throughout Asia; Pres. CARLOS L. PEÑA.

Filipinas Broadcasting Network: Legaspi Towers 200, Room 306, Paseo de Roxas, Makati City, Metro Manila; tel. (2) 8176133; fax (2) 8177135; Gen. Man. DIANA C. GOZUM.

GMA Network Inc: GMA Network Center, EDSA cnr Timog Ave, Diliman, Quezon City, 1103 Metro Manila; tel. (2) 9287021; fax (2) 4263925; e-mail yourgmafamily@gmanetwork.com; internet www.igma.tv; f. 1950; fmrly Republic Broadcasting System Inc; transmits nation-wide through 44 television stations; Chair., Pres. and CEO FELIPE L. GOZON; Exec. Vice-Pres. GILBERTO R. DUAVIT, Jr.

Manila Broadcasting Co: FJE Bldg, 4th Floor, 105 Esteban St, Legaspi Village, Makati City, Metro Manila; tel. (2) 8177043; fax (2) 8400763; f. 1946; 10 stations; Pres. RUPERTO NICDAO, Jr; Gen. Man. EDUARDO L. MONTILLA.

Nation Broadcasting Corpn: NBC Broadcast Center, Jacinta II Bldg, Epifanio de los Santos Ave, Guadelupe, Makati City, 1200 Metro Manila; tel. (2) 8821622; fax (2) 8821360; e-mail joey@nbc.ph; internet www.nbc.com.ph; f. 1963; 31 stations; Pres. FRANCIS LUMEN.

Newsounds Broadcasting Network Inc: Florete Bldg, Ground Floor, 2406 Nobel, cnr Edison St, Makati City, 3117 Metro Manila; tel. (2) 8430116; fax (2) 8173631; 10 stations; Gen. Man. E. BILLONES; Office Man. HERMAN BASBANO.

Pacific Broadcasting System: c/o Manila Broadcasting Co, FJE Bldg, 105 Esteban St, Legaspi Village, Makati City, Metro Manila; tel. (2) 8921660; fax (2) 8400763; Pres. RUPERTO NICDAO, Jr; Vice-Pres. RODOLFO ARCE.

PBN Broadcasting Network: Ersan Bldg, 3rd Floor, 32 Quezon Ave, Quezon City, Metro Manila; tel. (2) 7325424; fax (2) 7438162; e-mail pbn@philonline.com.ph; Pres. JORGE D. BAYONA.

Philippine Broadcasting System: FJE Bldg, 105 Esteban St, Legaspi Village, Makati City, Metro Manila; tel. (2) 8177043; fax (2) 8400763; Pres. RUPERTO NICDAO, Jr.

Philippine Federation of Catholic Broadcasters: 2307 Pedro Gil, Santa Ana, POB 3169, Metro Manila; tel. (2) 5644518; fax (2) 5637316; e-mail nomm@surfshop.net.ph; 48 radio stations and four TV channels; Pres. Fr FRANCIS LUCAS.

Radio Philippines Network, Inc: Broadcast City, Capitol Hills, Diliman, Quezon City, Metro Manila; tel. (2) 9318627; fax (2) 984322; f. 1969; seven TV stations, 14 radio stations; Pres. Edgar San Luis; Gen. Man. Felipe G. Medina.

Radio Veritas Asia: Buick St, Fairview Park, POB 2642, Quezon City, Metro Manila; tel. (2) 9390011; fax (2) 9381940; e-mail rveritas-asia@rveritas-asia.org; internet www.rveritas-asia.org; f. 1969; Catholic short-wave station, broadcasts in 17 languages; Pres. and Chair. Archbishop Gaudencio B. Rosales; Gen. Man. Fr Carlos S. Lariosa (SVD).

UM Broadcasting Network: Xanland Corporate Center, Room 7B, 152 Amorsolo St, Legaspi Village, Makati City, Metro Manila; tel. (2) 8158754; fax (2) 8173505; e-mail umbmmkt@mozcom.com; Exec. Vice-Pres. Willy Torres.

Vanguard Radio Network: J & T Bldg, Room 208, Santa Mesa, Metro Manila; tel. (2) 7161233; fax (2) 7160899; Pres. Manuel Galvez.

Television

In July 1991 there were seven originating television stations and 105 replay and relay stations. The seven originating stations were ABS-CBN (Channel 2), PTV4 (Channel 4), ABC (Channel 5), GMA (Channel 7), RPN (Channel 9), IBC (Channel 13) and SBN (Channel 21). The following are the principal operating television networks:

ABC Development Corpn: APMC Bldg, 136 Amorsolo St, cnr Gamboa St, Legaspi Village, Makati City, Metro Manila; tel. (2) 8923801; fax (2) 8128840; CEO Edward U. Tan.

ABS-CBN Broadcasting Corpn: ABS-CBN Broadcasting Center, Sgt E. Esguerra Ave, cnr Mother Ignacia Ave, Quezon City, 1103 Metro Manila; tel. (2) 9244101; fax (2) 9215888; internet www.abs-cbn.com; Chair. Eugenio Lopez III; Gen. Man. Federico M. Garcia.

AMCARA Broadcasting Network: Mother Ignacia St, cnr Sgt Esguerra Ave, Quezon City, Metro Manila; tel. (2) 4152272; fax (2) 4119646; Man. Dir Manuel Quiogue.

Banahaw Broadcasting Corpn: Broadcast City, Capitol Hills, Quezon City, 3005 Metro Manila; tel. (2) 9329949; fax (2) 9318751; Station Man. Betty Livioco.

Channel V Philippines: Sagittarius Bldg, 6th Floor, H. V. de la Costa St, Salcedo Village, Makati City, Metro Manila; tel. (2) 8173747; fax (2) 8184192; e-mail channelv@i-next.net; Pres. Joel Jimenez; Gen. Man. Mon Alcaraz.

GMA Network, Inc: GMA Network Center, Timog Ave, cnr Epifanio de los Santos Ave, Diliman, Quezon City, 1103 Metro Manila; tel. (2) 9287021; fax (2) 4263925; internet www.igma.tv; f. 1950; transmits nation-wide through 45 VHF and 2 affiliate stations and in Asia, Australia and Hawaii through Measat-2 satellite; Chair., Pres. and CEO Felipe L. Gozon; Exec. Vice-Pres. Gilberto R. Duavit, Jr.

Intercontinental Broadcasting Corpn: Broadcast City Complex, Capitol Hills, Diliman, Quezon City, Metro Manila; tel. (2) 9318781; fax (2) 9318743; 19 stations; Pres. and Chair. Boots Anson-Roa.

Maharlika Broadcasting System: Metro Manila; tel. (2) 9220880; jtly operated by the Bureau of Broadcasts and the National Media Production Center; Dir Antonio Barriero.

People's Television Network Inc (PTV4): Broadcast Complex, Visayas Ave, Quezon City, Metro Manila; tel. (2) 9206514; fax (2) 9204342; f. 1992; public television network; Chair. Lourdes I. Ilustre.

Radio Mindanao Network: State Condominium, 4th Floor, Salcedo St, Legaspi Village, Makati City, Metro Manila; tel. (2) 8191073; fax (2) 8163680; e-mail sales@rmn.ph; internet www.rmn.com.ph; f. 1952; owns and operates 50 radio and television stations; Chair. Henry Canoy; Pres. Eric S. Canoy.

Radio Philippines Network, Inc: Broadcast City, Capitol Hills, Diliman, Quezon City, Metro Manila; tel. (2) 9315080; fax (2) 9318627; 7 primary TV stations, 14 relay stations; Pres. Edgar San Luis; Gen. Man. Felipe G. Medina.

RJ TV 29: Save a Lot Bldg, Pasong Tamo Ext., Makati City, Metro Manila; tel. (2) 8942320; fax (2) 8942360; Gen. Man. Bea J. Colamonici.

Southern Broadcasting Network, Inc: Strata 200 Bldg, 22nd Floor, Ortigas Center, Emerald Ave, Pasig City, Metro Manila; tel. (2) 6365496; fax (2) 6365495; Pres. Luis B. Pacquing.

United Broadcasting Network: FEMS Tower 1, 11th Floor, 1289 Zobel Roxas, cnr South Superhighway, Malate; tel. (2) 5216138; fax (2) 5221226; Gen. Man. Joseph Hodreal.

Broadcasting Association

Kapisanan ng mga Brodkaster sa Pilipinas (KBP) (Association of Broadcasters in the Philippines): LTA Bldg, 6th Floor, 118 Perea St, Legaspi Village, Makati City, Metro Manila; tel. (2) 8151990; fax (2) 8151989; e-mail kbp@pacific.net.ph; internet www.kbp.org.ph; Chair. Cerge Remonde; Pres. Ruperto S. Niodao, Jr.

Finance

(cap. = capital; res = reserves; dep. = deposits; m. = million; brs = branches; amounts in pesos, unless otherwise stated)

BANKING

Legislation enacted in June 1993 provided for the establishment of a new monetary authority, the Bangko Sentral ng Pilipinas, to replace the Central Bank of the Philippines. The Government was thus able to restructure the Central Bank's debt (308,000m. pesos).

In May 1994 legislation was promulgated providing for the establishment in the Philippines of up to 10 new foreign bank branches over the following five years (although at least 70% of the banking system's total resources were to be owned by Philippine entities). Prior to this legislation only four foreign banks (which had been in operation when the law restricting the industry to locally owned banks was enacted in 1948) were permitted to operate. Two other foreign banks were subsequently licensed to organize locally incorporated banks with minority Filipino partners. By the end of 2002 the number of foreign banks had increased to 13; at that time some 44 principal commercial banks were operating in the Philippines.

Central Bank

Bangko Sentral ng Pilipinas (Central Bank of the Philippines): A. Mabini St, cnr Pablo Ocampo St, Malate, 1004 Metro Manila; tel. (2) 5247011; fax (2) 5231252; e-mail bspmail@bsp.gov.ph; internet www.bsp.gov.ph; f. 1993; cap. 10,000m., res 171,350m., dep. 487,608m. (Dec. 2002); Gov. Rafael B. Buenaventura; 19 brs.

Principal Commercial Banks

Allied Banking Corpn: Allied Bank Centre, 6754 Ayala Ave, cnr Legaspi St, Makati City, 1200 Metro Manila; tel. (2) 8187961; fax (2) 8160921; internet www.alliedbank.com.ph; f. 1977; cap. 495m., res 14,864m., dep. 100,834m. (Dec. 2002); Chair. Panfilo O. Domingo; Pres. Reynaldo A. Maclang; 287 brs.

Banco de Oro Universal Bank: 12 ADB Ave, Mandaluyong City, 1550 Metro Manila; tel. (2) 6366060; fax (2) 6317810; e-mail bancoro@bdo.com.ph; f. 1996; cap. 10,690m., res 77,880m., dep. 56,617m. (Dec. 2001); Chair. Teresita T. Sy; Pres. Nestor V. Tan; 125 brs.

Bank of Commerce: Bankers' Center, 6764 Ayala Ave, Makati City, 1226 Metro Manila; tel. (2) 8174906; fax (2) 8172426; e-mail bk_commerce@mp.bkcomp.bridge.com; internet www.bankcom.com.ph; f. 1983; fmrly Boston Bank of the Philippines; merged with Traders Royal Bank 2001; cap. 2,812m., dep. 8,621m. (Jan. 1999); Chair. Antonio Cojuangco; Pres. Raul B. de Mesa; 38 brs.

Bank of the Philippine Islands: BPI Bldg, Ayala Ave, cnr Paseo de Roxas, POB 1827, MCC, Makati City, 0720 Metro Manila; tel. (2) 8185541; fax (2) 8910170; e-mail macvillareal@bpi.com.ph; internet www.bpiexpressonline.com; f. 1851; merged with Far East Bank and Trust Co in April 2000; merged with DBS Bank Philippines, Inc, 2001; cap. 18,505m., res 27,103m., dep. 315,481m. (Dec. 2002); Pres. Xavier P. Loinaz; Chair. Jaime Zobel de Ayala; 340 brs.

China Banking Corpn: CBC Bldg, 8745 Paseo de Roxas, cnr Villar St, Makati City, 1226 Metro Manila; tel. (2) 8855555; fax (2) 8920220; e-mail postmaster@chinabank.com.ph; internet www.chinabank.com.ph; f. 1920; cap. 3,045m., res 2,902m., dep. 72,112m. (Dec. 2003); Chair. Gilbert U. Dee; Pres. and CEO Peter S. Dee; 141 brs.

Development Bank of the Philippines: DBP Bldg, Makati Ave, cnr Sen. Gil J. Puyat Ave, Makati City, 1200 Metro Manila; tel. (2) 8189511; fax (2) 8128089; e-mail info@devbankphil.com.ph; internet www.devbankphil.com.ph; f. 1947 as the Rehabilitation Finance Corpn; govt-owned; provides medium- and long-term loans for agricultural and industrial development; cap. 19,225m., dep. 37,404m. (March 2004); Chair. Vitaliano N. Nañagas, II; Pres. and CEO Simon R. Paterno; 77 brs.

East West Banking Corpn: 21st Floor, PBCOM Tower, 6795 Ayala Ave, cnr Herrera St, Makati City, 1226 Metro Manila; tel. (2) 8150233; fax (2) 8184155; e-mail service@eastwestbanker.com; internet www.eastwestbanker.com; f. 1994; cap. 1,341m., dep. 2,857m. (March 1997); Chair. Andrew L. Gotianum; Pres. Elrey T. Ramos; 20 brs.

Equitable PCI Bank: Equitable PCI Bank Towers, 262 Makati Ave, cnr H. V. de la Costa St, Makati City, 1200 Metro Manila; tel. (2) 8407000; fax (2) 8941893; internet www.equitablepcib.com; f. 1950; formed by a merger between Equitable Banking Corpn and Philippine Commercial International (PCI) Bank in 1999; cap. 7,270m., res 33,948m., dep. 171,360m. (Dec. 2002); Senior Exec. Vice-Pres. RENE J. BUENAVENTURA; 399 brs.

Export and Industry Bank, Inc (Exportbank): Exportbank Plaza, Chino Roces Ave, cnr Sen. Gil J. Puyat Ave, Makati City, 1200 Metro Manila; tel. (2) 8879000; fax (2) 8780000; internet www.exportbank.com.ph; merged with Urban Bank, Inc, 2002; cap. 1,638m., dep. 2,767m. (Dec. 1997); Chair. SERGIO R. ORTIZ-LUIS, Jr; Pres. BENJAMIN P. CASTILLO.

Land Bank of the Philippines: LandBank Plaza, 1598 M. de Pilar St, cnr J. Quintos St, Malate, 1004 Metro Manila; tel. (2) 5220000; fax (2) 5288556; e-mail lbp@mail.landbank.com; internet www.landbank.com; f. 1963; specialized govt bank with universal banking licence; cap. 22,100m., res 20,100m., dep. 185,200m. (Dec. 2003); Pres. and CEO GARY B. TEVES; 350 brs.

Manila Banking Corpn: Manila Bank Bldg, 6772 Ayala Ave, Makati City, 1226 Metro Manila; tel. (2) 8645000; fax (2) 8645016; internet www.manilabank.com; f. 1999; cap. 573m., res 2,433m., dep. 3,361m. (Dec. 2002); Pres. BENJAMIN S. YAMBAO; Chair. LUIS S. PUYAT.

Maybank Philippines Inc: Legaspi Towers 300, Pablo Ocampo St, cnr Roxas Blvd, Malate, 1004 Metro Manila; tel. (2) 5216169; fax (2) 5218513; e-mail mayphil@maybank.com.ph; internet www.maybank.com.my/philippines/html/frame_main.htm; f. 1961; cap. 1,671m., dep. 2,701m. (Dec. 1997); Pres. and CEO LIM HONG TAT; 44 brs.

Metropolitan Bank and Trust Co (Metrobank): Metrobank Plaza, Sen. Gil J. Puyat Ave, Makati City, 1200 Metro Manila; tel. (2) 8988000; fax (2) 8176248; e-mail metrobank@metrobank.com.ph; internet www.metrobank.com.ph; f. 1962; acquired Global Business Bank (Globalbank) 2002; cap. 32,673m., res 9,888m., dep. 377,484m. (Dec. 2003); Chair. GEORGE S. K. TY; Pres. ANTONIO S. ABACAN, Jr; 344 local brs, 6 overseas brs.

Philippine Bank of Communications: PBCOM Tower, 6795 Ayala Ave, cnr Herrera St, 1226 Makati City; tel. (2) 8307000; fax (2) 8182598; e-mail info@pbcom.com.ph; internet www.pbcom.com.ph; f. 1939; cap. 5,259m., res 1,298m., dep. 38,851m. (Dec. 2002); merged with AsianBank Corpn in early 1999; Chair. LUY KIM GUAN; Pres. ISIDRO C. ALCANTARA; 64 brs.

Philippine National Bank (PNB): PNB Financial Center, President Diosdado Macapagal Blvd, Pasay City, 1300 Metro Manila; tel. (2) 5263131; fax (2) 8331245; e-mail mainbranch@pnb.com.ph; internet www.pnb.com.ph; f. 1916; partially transferred to the private sector in 1996 and 2000, 10.93% govt-owned; cap. 22,930m., res 2,923m., dep. 135,139m. (Dec. 2002); Chair. FRANCISCO A. DIZON; Pres. and CEO LORENZO V. TAN; 324 local brs, 5 overseas brs.

Philippine Veterans Bank: PVB Bldg, 101 V. A. Rufino St, cnr de la Rosa St, Legaspi Village, Makati City, Metro Manila; tel. (2) 8943919; fax (2) 8940625; e-mail corpcomm@veteransbank.com.ph; internet www.veteransbank.com.ph; cap. 3,600m., dep. 10,400m. (June 1998); Chair. EMMANUEL DE OCAMPO; Pres. RICARDO A. BALBIDO, Jr; 38 brs.

Philtrust Bank (Philippine Trust Co): Philtrust Bank Bldg, United Nations Ave, cnr San Marcelino St, Ermita, 1045 Metro Manila; tel. (2) 5249061; fax (2) 5217309; e-mail ptc@bancnet.net; f. 1916; cap. 3,683m., res 695m., dep. 33,001m. (Dec. 2003); Pres. ANTONIO H. OZAETA; Chair. EMILIO T. YAP; 38 brs.

Prudential Bank: Prudential Bank Bldg, 6787 Ayala Ave, Makati City, 1200 Metro Manila; tel. (2) 8178981; fax (2) 8175146; e-mail feedback@prudentialbank.com; internet www.prudentialbank.com.ph; f. 1952; merged with Pilipinas Bank in May 2000; cap. 822m., res 7,791m., dep. 38,526m. (Dec. 2002); Chair. and Pres. JOSE L. SANTOS; 117 brs.

Rizal Commercial Banking Corpn: Yuchengco Tower, RCBC Plaza, 6819 Alaya Ave, Makati City, POB 2202, 1200 Metro Manila; tel. (2) 8949000; fax (2) 8949958; e-mail customer_service@rcbc.com; internet www.rcbc.com; f. 1960; cap. and res 15,512m., dep. 154,427m. (Dec. 2002); Chair. ALFONSO T. YUCHENGCO; Pres. VALENTIN A. ARANETA; 176 brs.

Security Bank Corpn: 6776 Ayala Ave, Makati City, 1200 Metro Manila; tel. (2) 8676788; fax (2) 8132069; e-mail inquiry@securitybank.com.ph; internet www.securitybank.com; f. 1951; fmrly Security Bank and Trust Co; cap. 10,066m., dep. 49,645m. (Dec. 2003); Pres. and CEO ALBERTO S. VILLAROSA; Chair. FREDERICK Y. DY; 114 brs.

Union Bank of the Philippines: SSS Makati Bldg, Ayala Ave, cnr Herrera St, Makati City, 1200 Metro Manila; tel. (2) 8920011; fax (2) 8186058; e-mail online@unionbankph.com; internet www.unionbankph.com; f. 1954; cap. 7,087m., res 6,442m., dep. 55,023m. (Dec. 2002); Chair. JUSTO A. ORTIZ; Pres. and CEO ARMAND F. BRAUN, Jr; 120 brs.

United Coconut Planters' Bank: UCPB Bldg, Makati Ave, Makati City, 0728 Metro Manila; tel. (2) 8119000; fax (2) 8119706; e-mail crc@ucpb.com.ph; internet www.ucpb.com; f. 1963; cap. 10,727m., dep. 87,425m. (Dec. 2000); Chair. DEOGRACIAS N. VISTAN; Pres. and CEO JOSE L. QUERUBIN; 178 brs.

United Overseas Bank Philippines: 17th Floor, Pacific Star Bldg, Sen. Gil J. Puyat Ave, cnr Makati Ave, Makati City, Metro Manila; tel. (2) 8788686; fax (2) 8115917; e-mail crd@uob.com.ph; internet www.uob.com.ph; f. 1999; cap. 2,731m., dep. 15,021m. (Dec. 1999); Chair. WEE CHO YAW; Pres. CHUA TENG HUI; 66 brs.

Rural Banks

Small private banks have been established with the encouragement and assistance (both financial and technical) of the Government in order to promote and expand the rural economy. Conceived mainly to stimulate the productive capacities of small farmers, merchants and industrialists in rural areas, and to combat usury, their principal objectives are to place within easy reach and access of the people credit facilities on reasonable terms and, in co-operation with other agencies of the Government, to provide advice on business and farm management and the proper use of credit for production and marketing purposes. The rural banks numbered 1,942 in 1998; their registered resources totalled 59,970m. pesos at 31 December 1998.

Thrift Banks

Thrift banks mobilize small savings and provide loans to lower income groups. The thrift banking system comprises savings and mortgage banks, stock savings and loan associations and private development banks. In 1998 there were 1,474 thrift banks; their registered resources totalled 216,440m. pesos.

Development Banks

1st e-Bank: 1st e-Bank Tower, 8737 Paseo de Roxas, Makati City, 1226 Metro Manila; tel. (2) 8158536; fax (2) 8195376; e-mail 1stecare@1stebank.com.ph; internet www.1stebank.com.ph; f. 1963 with World Bank assistance, as Private Development Corporation of the Philippines; 1992 converted into a development bank; name changed as above in May 2000; banking and lending services; financial advisory, trust and investment services; training and consultancy; insurance brokerage; cap. and res 2,079.7m., dep. 7,244.6m. (1996); Pres. and CEO CARLOS A. PEDROSA; 60 brs.

Pampanga Development Bank: MacArthur Highway, Dolores San Fernando, Pampanga, Luzon; tel. (45) 9612786; fax (45) 9633931; e-mail pdb@ag.triasia.net; originally Agribusiness Development Bank; name changed as above 1995; cap. 75.0m., res 6.6m., dep. 72.9m. (June 2003); Pres. JOSE ERIBERTO H. SUAREZ.

Foreign Banks

ANZ Banking Group Ltd (Australia and New Zealand): Ayala Triangle, Tower One, 3rd Floor, Ayala Ave, cnr Paseo de Roxas, Makati City, 1226 Metro Manila; tel. (2) 8485091; fax (2) 8485086; e-mail labrooyM2@anz.com; f. 1995; cap. 250m., dep. 2,000m. (Dec. 1999); Pres. MICHAEL LA BROOY.

Banco Santander Philippines, Inc: Tower One, 27th Floor, Ayala Triangle, Ayala Ave, cnr Paseo de Roxas, Makati City, 1200 Metro Manila; tel. (2) 7594144; fax (2) 7594190; cap. 1,351m., dep. 3,943m. (March 1997); Chair. ANA PATRICIA BOTIN; Dir and Pres. VICENTE B. CASTILLO.

Bangkok Bank Public Company Ltd (Thailand): Far East Bank Bldg, 25th Floor, Sen. Gil J. Puyat Ave, Makati City, 1200 Metro Manila; tel. (2) 8914011; fax (2) 8914037; f. 1995; cap. 266m. (Dec. 1996), dep. 647m. (March 1997); Pres. PREYAMIT HETRAKUL.

Bank of America NA (USA): BA-Lepanto Bldg, 2nd Floor, 8747 Paseo de Roxas, POB 1767, Makati City, 1257 Metro Manila; tel. (2) 8155000; fax (2) 8155895; f. 1947; cap. 210m. (Dec. 1996), dep. 5,015m. (June 1999); Sr Vice-Pres. and Country Man. JOSE L. QUERUBIN.

Bank of Tokyo-Mitsubishi Ltd (Japan): 6750 Ayala Ave, 5th Floor, Makati City, Metro Manila; tel. (2) 8921976; fax (2) 8160413; f. 1977; cap. 200m. (Dec. 1996), dep. 2,940m. (March 1997); Gen. Man. HISAO SAKASHITA.

Citibank NA (USA): Citibank Plaza, 8741 Paseo de Roxas, Makati City, 1226 Metro Manila; tel. (2) 8947700; fax (2) 8157703; f. 1948; cap. 2,536m., dep. 30,167m. (March 1997); CEO SURESH MAHARAJ; 3 brs.

Deutsche Bank AG (Germany): Ayala Triangle, Tower One, 26th Floor, Ayala Ave, cnr Paseo de Roxas, Makati City, 1226 Metro Manila; tel. (2) 8946900; fax (2) 8946901; f. 1995; cap. 625m. (Dec.

1996), dep. 2,595m. (March 1997); Chief Country Officer ENRICO CRUZ.

Hongkong and Shanghai Banking Corpn (Hong Kong): Enterprise Center, Tower I, 6766 Ayala Ave, cnr Paseo de Roxas, Makati City, 1200 Metro Manila; tel. (2) 8305300; fax (2) 8865343; internet www.asiapacific.hsbc.com; cap. HK $113,520m., res HK $42,280m., dep. HK $1,419,076m. (June 2002); CEO PAUL LAWRENCE; 5 brs.

ING Bank NV (The Netherlands): Ayala Triangle, Tower One, 21st Floor, Ayala Ave, cnr Paseo de Roxas, Makati City, 1200 Metro Manila; tel. (2) 8408888; fax (2) 8151116; f. 1995; cap. 643m., dep. 2,594m. (March 1997); Country Man. MANUEL SALAK.

International Commercial Bank of China (Taiwan): Pacific Star Bldg, Ground and 3rd Floors, Sen. Gil J. Puyat Ave, cnr Makati Ave, Makati City, 1200 Metro Manila; tel. (2) 8115807; fax (2) 8115774; cap. 188m., dep. 458m. (March 1997); Gen. Man. HERMAN C. CHEN.

Korea Exchange Bank: Citibank Tower, 33rd Floor, 8741 Paseo de Roxas, Makati City, 1229 Metro Manila; tel. (2) 8481988; fax (2) 8195377; e-mail koexbank@surfshop.net.ph; cap. 257m., dep. 271m. (March 1997); Gen. Man. TAE-HONG JIN.

Mizuho Bank Ltd (Japan): Citibank Tower, 26th Floor, Valero St, cnr Villar St, Legaspi Village, Makati City, 1229 Metro Manila; tel. (2) 8480001; fax (2) 8153770; f. 1995; cap. 202m. (Dec. 1996), dep. 858m. (March 1997); Gen. Man. TAKUJI IWASAKI.

Standard Chartered Bank (Hong Kong): 6788 Ayala Ave, Makati City, 1226 Metro Manila; tel. (2) 8867888; fax (2) 8866866; f. 1873; cap. 1,500m. (Dec. 1996), dep. 3,770m. (March 1997); CEO EIRVIN B. KNOX.

Islamic Bank

Al-Amanah Islamic Investment Bank of the Philippines: 2nd Floor, Classica Tower I, H. V. de la Costa St, Salcedo Village, Makati City, Metro Manila; tel. (2) 8164258; fax (2) 8195249; e-mail islambnk@tri-isys.com; internet www.islamicbank.com.ph; f. 1989; Chair. Dato' ZACARIA A. CANDAO; Pres. ABDUL GAFFOOR ASHROOF.

Major 'Offshore' Banks

ABN AMRO Bank, NV (Netherlands): LKG Tower, 18th Floor, 6801 Ayala Ave, Makati City, 1200 Metro Manila; tel. (2) 8842000; fax (2) 8843954; Gen. Man. CARMELO MARIA L. BAUTISTA.

American Express Bank Ltd (USA): Ayala Bldg, 11th Floor, 6750 Ayala Ave, Makati City, Metro Manila; tel. (2) 8186731; fax (2) 8172589; f. 1977; Sr Dir and Country Man. VICENTE L. CHUA.

BankBoston NA (USA): 6750 Ayala Ave, 23rd Floor, Makati City, Metro Manila; tel. (2) 8170456; fax (2) 8191251; Country Man. BENJAMIN C. SEVILLA.

Bank Dagang Nasional Indonesia: Ayala Tower and Exchange Plaza, 19th Floor, Unit B, Ayala Ave, Makati City, Metro Manila; tel. (2) 8486189; fax (2) 8486176; Man. CONSUELO N. PADILLA.

Bank of Nova Scotia (Canada): Solidbank Bldg, 9th Floor, 777 Paseo de Roxas, Makati City, 1200 Metro Manila; tel. (2) 8179751; fax (2) 8178796; f. 1977; Man. M. S. (CORITO) SEVILLA.

Bankers Trust Co (USA): Pacific Star Bldg, 12th Floor, Makati Ave, Makati City, Metro Manila; tel. (2) 8190231; fax (2) 8187349; Man. Dir JOSE ISIDRO N. CAMACHO.

BNP Paribas (France): PCIB Tower Two, Makati Ave, cnr H. V. de la Costa St, Makati City, 1227 Metro Manila; tel. (2) 8158821; fax (2) 8179237; f. 1977; Country Man. PIERRE IMHOF.

Chase Manhattan International Finance Ltd: Corinthian Plaza, 4th Floor, 121 Paseo de Roxas, Makati City, Metro Manila; tel. (2) 8113348; fax (2) 8781290; Man. HELEN S. CIFRA.

Crédit Agricole Indosuez (France): Citibank Tower, 17th Floor, 8741 Paseo de Roxas, Makati City, 1200 Metro Manila; tel. (2) 8481344; fax (2) 8481380; f. 1977; Gen. Man. MARC MEULEAU.

Crédit Lyonnais (France): Pacific Star Bldg, 14th Floor, Makati Ave, cnr Sen. Gil J. Puyat Ave, Makati City, Metro Manila; POB 1859 MCC, 3117 Makati City, Metro Manila; tel. (2) 8171616; fax (2) 8177145; f. 1981; Gen. Man. PIERRE EYMERY.

KBC Bank, NV (Belgium): Far East Bank Center, 22nd Floor, Sen. Gil J. Puyat Ave, Makati City, 1200 Metro Manila; tel. (2) 8915331; fax (2) 8915352; Gen. Man. EDWIN YAPTANGCO.

Overseas Union Bank Ltd: Corinthian Plaza, 7th Floor, Paseo de Roxas, Makati City, Metro Manila; tel. (2) 8179951; fax (2) 8113168; f. 1977; Man. TAN LYE OON.

Société Générale (France): Antel Corporate Center, 21st Floor, 139 Valero St, Salcedo Village, Makati City, Metro Manila; tel. (2) 8492000; fax (2) 8492940; f. 1980; CEO CLAUDE I. TOUITOU.

Standard Chartered Bank Ltd: 9th Floor, 6788 Ayala Ave, Makati City, Metro Manila; tel. (2) 8867888; fax (2) 8866866; Vice-Pres. and Gen. Man. IMELDA B. CAPISTRANO.

Union Bank of California (USA): ACE Bldg, 8th Floor, cnr Rada and de la Rosa Sts, Legaspi Village, Makati City, Metro Manila; tel. (2) 8923056; fax (2) 8170102; f. 1977; Branch Man. TERESITA MALABANAN.

Banking Associations

Bankers Association of the Philippines: Sagittarius Cond. Bldg, 11th Floor, H. V. de la Costa St, Salcedo Village, Makati City, Metro Manila; tel. (2) 8103858; fax (2) 8103860; Pres. CESAR E. A. VIRATA; Exec. Dir LEONILO G. CORONEL.

Chamber of Thrift Banks: Cityland 10 Condominium Tower 1, Unit 614, H. V. de la Costa St, Salcedo Village, Makati City, Metro Manila; tel. (2) 8126974; fax (2) 8127203; Pres. DIONISIO C. ONG.

Offshore Bankers' Association of the Philippines, Inc: MCPO 3088, Makati City, 1229 Metro Manila; tel. (2) 8103554; Chair. ANTONIO DE LOS ANGELES.

Rural Bankers' Association of the Philippines: RBAP Bldg, A. Soriano, Jr, Ave, cnr Arzobispo St, Intramuros, Manila; tel. (2) 5272968; fax (2) 5272980; e mail info@rbap.org; internet www.rbap.org; Pres. DANIEL B. ARCENAS.

STOCK EXCHANGES

Securities and Exchange Commission: SEC Bldg, Epifanio de los Santos Ave, Greenhills, Mandaluyong City, Metro Manila; tel. (2) 7274543; fax (2) 7254399; e-mail mis@sec.gov.ph; internet www.sec.gov.ph; f. 1936; Chair. LILIA R. BAUTISTA.

Philippine Stock Exchange: Philippine Stock Exchange Center, Exchange Rd, Ortigas Centre, Pasig City, 1605 Metro Manila; tel. (2) 6887600; fax (2) 6345113; e-mail piac@pse.org.ph; internet www.pse.com.ph; f. 1994 following the merger of the Manila and Makati Stock Exchanges; Chair. ALICIA RITA M. ARROYO; Pres. CAYETANO W. PADERANGA, Jr.

INSURANCE

At the end of 2000 a total of 156 insurance companies were authorized by the Insurance Commission to transact in the Philippines. Foreign companies were permitted to operate in the country.

Principal Domestic Companies

Ayala Life Assurance Inc: Ayala Life Bldg, 6786 Ayala Ave, Makati City, Metro Manila; tel. (2) 8885433; fax (2) 8180171; e-mail customer.service@ayalalife.com.ph; internet www.ayalalife.com.ph; Pres. ALFONSO L. SALCEDO, Jr.

BPI/MS Insurance Corpn: Ayala Life-FGU Center, 16th Floor, 6811 Ayala Ave, Makati City, 1226 Metro Manila; tel. (2) 8409000; fax (2) 8910147; e-mail insure@bpims.com; internet www.bpims.com; f. 2002 as result of merger of FGU Insurance Corpn and FEB Mitsui Marine Insurance Corpn; jt venture of Bank of the Philippine Islands and Sumitomo Insurance Co (Japan); cap. 731m. (2003); Chair. XAVIER P. LOINAZ; Pres. RYUICHI ITO.

Central Surety & Insurance Co: UniversalRe Bldg, 2nd Floor, 106 Paseo de Roxas, Legaspi Village, Makati City, 1200 Metro Manila; tel. (2) 8174931; fax (2) 8170006; f. 1945; bonds, fire, marine, casualty, motor car; Pres. FERMIN T. CASTAÑEDA.

Commonwealth Insurance Co: 10th Floor, 1st e-Bank Tower, 8737 Paseo de Roxas, Makati City, Metro Manila; tel. (2) 8187626; fax (2) 8138575; f. 1935; Pres. MARIO NOCHE.

Co-operative Insurance System of the Philippines: CISP Bldg, 80 Malakas St, Diliman, Quezon City, Metro Manila; tel. (2) 9240388; fax (2) 9240471; Chair. DOMINADOR ESTRADA; Pres. AMBROSIO M. RODRIGUEZ.

Domestic Insurance Co of the Philippines: 5th Floor, Champ Bldg, Anda Circle, Bonifacio Drive, Port Area, Manila; tel. (2) 5278181; fax (2) 5273052; e-mail gdicp@skyinet.net; f. 1946; cap. 10m.; Pres. and Chair. MAR S. LOPEZ.

Empire Insurance Co: Prudential Life Bldg, 2nd Floor, 843 Arnaiz Ave, Legaspi Village, Makati City, 1229 Metro Manila; tel. (2) 8159561; fax (2) 8152599; f. 1949; fire, bonds, marine, accident, motor car, extraneous perils; Pres. and CEO JOSE MA G. SANTOS.

Equitable Insurance Corpn: Equitable Bank Bldg, 4th Floor, 262 Juan Luna St, Binondo, POB 1103, Metro Manila; tel. (2) 2430291; fax (2) 2415768; e-mail info@equitableinsurance.com.ph; internet www.equitableinsurance.com.ph; f. 1950; fire, marine, casualty, motor car, bonds; Pres. NORA T. GO; Exec. Vice-Pres. ANTONIO C. OCAMPO.

Insular Life Assurance Co Ltd: Insular Life Corporate Center, Insular Life Drive, Filinvest Corporate City, Alabang, 1781 Muntinlupa City; tel. (2) 7711818; fax (2) 7711717; e-mail headofc@insular.com.ph; internet www.insularlife.com.ph; f. 1910; members' equity 7,251m. (Dec. 2002); Chair. and CEO VICENTE R. AYLLÓN.

Makati Insurance Co Inc: Far East Bank Center, 19th Floor, Sen. Gil J. Puyat Ave, Makati City, 1200 Metro Manila; tel. (2) 8459576; fax (2) 8915229; f. 1965; non-life; Pres. and Gen. Man. JAIME L. DARANTINAO; Chair. OCTAVIO V. ESPIRITU.

Malayan Insurance Co Inc: Yuchengco Tower, 4th Floor, 500 Quintin Paredes St, POB 3389, 1099 Metro Manila; tel. (2) 2428888; fax (2) 2412449; e-mail malayan@malayan.com; internet www.malayan.com; f. 1949; cap. 100m. (1998); insurance and bonds; Pres. YVONNE S. YUCHENGCO.

Manila Surety & Fidelity Co Inc: 66 P. Florentino, Quezon City, Metro Manila; tel. (2) 7122251; fax (2) 7124129; f. 1945; Pres. MA LOURDES V. PEÑA; Vice-Pres. EDITHA LIM.

Metropolitan Insurance Co: Ateneum Bldg, 3rd Floor, Leviste St, Salcedo Village, Makati City, Metro Manila; tel. (2) 8108151; fax (2) 8162294; f. 1933; non-life; Pres. JOSE M. PERIQUET, Jr; Exec. Vice-Pres. ROBERTO ABAD.

National Life Insurance Co of the Philippines: National Life Insurance Bldg, 6762 Ayala Ave, Makati City, Metro Manila; tel. (2) 8100251; fax (2) 8178718; f. 1933; Pres. BENJAMIN L. DE LEON; Sr Vice-Pres. DOUGLAS MCLAREN.

National Reinsurance Corpn of the Philippines: PS Bank Tower, 18th Floor, Sen. Gil J. Puyat Ave, cnr Tindalo St, Makati City, Metro Manila; tel. (2) 7595801; fax (2) 7595886; e-mail nrcp@nrcp.com.ph; internet www.nrcp.com.ph; f. 1978; Chair. WINSTON F. GARCIA; Pres. and CEO WILFRIDO C. BANTAYAN.

Paramount General Insurance Corpn: Sage House, 15th Floor, 110 Herrera St, Makati City, Metro Manila; tel. (2) 8127956; fax (2) 8133043; e-mail insure@paramount.com.ph; internet www.paramount.com.ph; f. 1950; fire, marine, casualty, motor car; Chair. PATRICK L. GO; Pres. GEORGE T. TIU.

Philippine American Life and General Insurance Co (Philamlife): Philamlife Bldg, United Nations Ave, Metro Manila; POB 2167, 0990 Metro Manila; tel. (2) 5216300; fax (2) 5217057; internet www.philamlife.com.ph; Pres. JOSE CUISIA.

Pioneer Insurance and Surety Corpn: Pioneer House Bldg, 320 Nueva St, Binondo, Metro Manila; tel. (2) 2428801; fax (2) 2421564; e-mail info@pioneer.com.ph; f. 1954; cap. 172.5m. (1997); Pres. and CEO DAVID C. COYUKIAT.

Rizal Surety and Insurance Co: Prudential Life Bldg, 3rd Floor, 843 Arnaiz Ave, Legaspi Village, Makati City, Metro Manila; tel. (2) 8403610; fax (2) 8173550; e-mail rizalsic@mkt.weblinq.com; f. 1939; fire, bond, marine, motor car, accident, extraneous perils; Chair. TOMAS I. ALCANTARA; Pres. REYNALDO DE DIOS.

Standard Insurance Co Inc: Standard Insurance Tower, 999 Pedro Gil St, cnr F. Agoncillo St, Metro Manila; tel. (2) 5223230; fax (2) 5261479; f. 1958; Chair. LOURDES T. ECHAUZ; Pres. ERNESTO ECHAUZ.

Sterling Insurance Co: Zeta II Annex Bldg, 6th Floor, 191 Salcedo St, Legaspi Village, Makati City, Metro Manila; tel. (2) 8925787; fax (2) 8183630; f. 1960; fmrly Dominion Insurance Corpn; name changed as above Nov. 2001; fire, marine, motor car, accident, engineering, bonds; Pres. RAFAEL GALLAGA.

Tico Insurance Co Inc: Trafalgar Plaza, 7th Floor, 105 H. V. de la Costa St, Salcedo Village, Makati City, 1227 Metro Manila; tel. (2) 8140143; fax (2) 8140150; f. 1937; fmrly Tabacalera Insurance Co Inc; Chair. and Pres. CARLOS CATHOLICO.

UCPB General Insurance Co Inc: 24th and 25th Floors, LKG Tower, 6801 Ayala Ave, Makati City, Metro Manila; tel. (2) 8841234; fax (2) 8841264; e-mail ucpbgen@ucpbgen.com; internet www.ucpbgen.com; f. 1989; non-life; Pres. ISABELO P. AFRICA; Chair. JERONIMO U. KILAYKO.

Universal Reinsurance Corpn: Ayala Life Bldg, 9th Floor, 6786 Ayala Ave, Makati City, Metro Manila; tel. (2) 7514977; fax (2) 8173745; f. 1949; life and non-life; Chair. JAIME AUGUSTO ZOBEL DE AYALA II; Pres. HERMINIA S. JACINTO.

Regulatory Body

Insurance Commission: 1071 United Nations Ave, Metro Manila; tel. (2) 5252015; fax (2) 5221434; e-mail oic@i-manila.com.ph; internet www.ic.gov.ph; regulates the private insurance industry by, among other things, issuing certificates of authority to insurance companies and intermediaries and monitoring their financial solvency; Commr BENJAMIN S. SANTOS.

Trade and Industry

GOVERNMENT AGENCIES

Board of Investments: 385 Sen. Gil J. Puyat Ave, Makati City, Metro Manila; tel. (2) 8976682; fax (2) 8953521; e-mail OSAC@boi.gov.ph; internet www.boi.gov.ph; Chair. CESAR V. PURISIMA; Gov. JOSE ANTONIO C. LEVISTE.

Cagayan Economic Zone Authority: Westar Bldg, 7th Floor, 611 Shaw Blvd, Pasig City, 1603 Metro Manila; tel. (2) 6365776; fax (2) 6313997; e-mail cagayanecozone@vasia.com; Administrator RODOLFO G. ALVARADO.

Clark Development Corpn: Bldg 2127, C. P. Garcia St, cnr E. Quirino St, Clark Field, Pampanga; tel. (2) 5994602; fax (2) 5992506; e-mail eya@clark.com.ph; internet www.clark.com.ph; Pres. and CEO Dr EMMANUEL Y. ANGELES.

Industrial Technology Development Institute: DOST Compound, Gen. Santos Ave, Bicutan, Taguig, 1631 Metro Manila; tel. (2) 8372071; fax (2) 8373167; e-mail epl@dost.gov.ph; internet mis.dost.gov.ph/itdi; Dir Dr ERNESTO P. LOZADA.

Maritime Industry Authority (MARINA): PPL Bldg, 1000 United Nations Ave, cnr San Marcelino St, Ermita, Metro Manila; tel. (2) 5238651; fax (2) 5242746; e-mail feedback@marina.gov.ph; internet www.marina.gov.ph; f. 1974; development of inter-island shipping, overseas shipping, shipbuilding and repair, and maritime power; Administrator OSCAR M. SEVILLA.

National Tobacco Administration: NTA Bldg, Scout Reyes St, cnr Panay Ave, Quezon City, Metro Manila; tel. (2) 3743987; fax (2) 3742505; e-mail ntamis@ph.inter.net; internet www.geocities.com/miscsdnta; f. 1987; Administrator CARLITOS S. ENCARNACION.

Philippine Coconut Authority (PCA): PCA R & D Bldg, Elliptical Rd, Diliman, Quezon City, 1104 Metro Manila; tel. (2) 9278116; fax (2) 9216173; e-mail pca_ofad@mozcom.com; internet pca.da.gov.ph; f. 1972; Administrator DANIEL M. CORONACION.

Philippine Council for Advanced Science and Technology Research and Development (PCASTRD): DOST Main Bldg, Gen. Santos Ave, Bicutan, Taguig, 1631 Metro Manila; tel. (2) 8377522; fax (2) 8373168; e-mail aal@dost.gov.ph; internet www.dostweb.dost.gov.ph/pcastrd; f. 1987; Exec. Dir Dr IDA F. DALMACIO.

Philippine Economic Zone Authority: Roxas Blvd, cnr San Luis St, Pasay City, Metro Manila; tel. (2) 5513454; fax (2) 8916380; e-mail info@peza.gov.ph; internet www.peza.gov.ph; Dir-Gen. LILIA B. DE LIMA.

Privatization and Management Office: Department of Finance, 104 Gamboa St, Legaspi Village, Makati City, 1229 Metro Manila; tel. (2) 8932383; fax (2) 8933453; e-mail pmo@eastern.com.ph; internet ecommunity.ncc.gov.ph/pmo; f. 2002 to handle the privatization of govt assets; succeeded Asset Privatization Trust; Chief Exec. RENATO V. VALDECANTOS.

Subic Bay Metropolitan Authority: SBMA Center, Bldg 229, Waterfront Rd, Subic Bay Freeport Zone, 2222 Zambales; tel. (47) 2524895; fax (47) 2523014; e-mail fcpayumo@sbma.com; internet www.sbma.com; Chair. FELICITO PAYUMO.

DEVELOPMENT ORGANIZATIONS

Bases Conversion Development Authority: 2nd Floor, Bonifacio Technology Center, 31st St, Crescent Park West, Bonifacio Global City, Taguig, 1634 Metro Manila; tel. (2) 8166666; fax (2) 8160996; e-mail bcda@bcda.gov.ph; internet www.bcda.gov.ph; f. 1992 to facilitate the conversion, privatization and development of fmr military bases; Chair. Dr FLORENCIO PADERNAL; Pres. and CEO RUFO COLAYCO.

Bureau of Land Development: DAR Bldg, Elliptical Rd, Diliman, Quezon City, Metro Manila; tel. (2) 9287031; fax (2) 9260971; Dir EUGENIO B. BERNARDO.

Capital Market Development Council: Metro Manila; Chair. CONCHITA L. MANABAT.

Co-operatives Development Authority: Benlor Bldg, 1184 Quezon Ave, Quezon City, Metro Manila; tel. (2) 3723801; fax (2) 3712077; Chair. JOSE C. MEDINA, Jr; Exec. Dir CANDELARIO L. VERZONA, Jr.

National Development Co (NDC): NDC Bldg, 8th Floor, 116 Tordesillas St, Salcedo Village, Makati City, Metro Manila; tel. (2) 8404898; fax (2) 8404862; e-mail corplan@info.com; internet www.dti.gov.ph/ndc; f. 1919; govt-owned corpn engaged in the organization, financing and management of subsidiaries and corpns incl. commercial, industrial, mining, agricultural and other enterprises assisting national economic development, incl. jt industrial ventures with other ASEAN countries; Chair. MANUEL A. ROXAS; Gen. Man. OFELIA V. BULAONG.

Philippine National Oil Co: Energy Complex, Bldg 6, 6th Floor, Merritt Rd, Fort Bonifacio, Makati City, Metro Manila; tel. (2) 5550254; fax (2) 8442983; internet www.pnoc.com.ph; f. 1973; state-owned energy development agency mandated to ensure stable and sufficient supply of oil products and to develop domestic energy resources; sales 3,482m. pesos (1995); Chair. VINCENT S. PEREZ, Jr; Pres. and CEO THELMO Y. CUNANAN.

Southern Philippines Development Authority: Basic Petroleum Bldg, 104 Carlos Palanca, Jr, St, Legaspi Village, Makati City, Metro Manila; tel. (2) 8183893; fax (2) 8188907; Chair. ROBERTO AVENTAJADO; Manila Rep. GERUDIO 'KHALIQ' MADUENO.

CHAMBERS OF COMMERCE AND INDUSTRY

American Chamber of Commerce of the Philippines: Corinthian Plaza, 2nd Floor, Paseo de Roxas, Makati City, 1229 Metro Manila; tel. (2) 8187911; fax (2) 8113081; e-mail info@amchamphilippines.com; internet www.amchamphilippines.com; Pres. TERRY J. EMRICK.

Cebu Chamber of Commerce and Industry: CCCI Center, cnr 11th and 13th Ave, North Reclamation Area, Cebu City 6000; tel. (32) 2321421; fax (32) 2321422; e-mail ccci@esprint.com; internet www.esprint.com/~ccci; f. 1921; Pres. SABINO R. DAPAT.

European Chamber of Commerce of the Philippines: PS Bank Tower, 19th Floor, Sen. Gil J. Puyat Ave, cnr Tindalo St, Makati City, 1200 Metro Manila; tel. (2) 8451324; fax (2) 8451395; e-mail info@eccp.com; internet www.eccp.com; f. 2000; 900 mems; Pres. WILLIAM BAILEY; Exec. Vice-Pres. HENRY J. SCHUMACHER.

Federation of Filipino-Chinese Chambers of Commerce and Industry Inc: Federation Center, 6th Floor, Muelle de Binondo St, POB 23, Metro Manila; tel. (2) 2419201; fax (2) 2422361; e-mail ffcccii@bell.com.ph; internet www.ffcccii.com.ph; Pres. JOHN K. C. NG; Sec.-Gen. JOAQUIN SY.

Japanese Chamber of Commerce of the Philippines: Jaycem Bldg, 6th Floor, 104 Rada St, Legaspi Village, Makati City, Metro Manila; tel. (2) 8923233; fax (2) 8150317; e-mail jccipi@jccipi.com.ph; internet www.jccipi.com.ph; Pres. MASAHARU TAMAKI.

Philippine Chamber of Coal Mines (Philcoal): Rm 1007, Princeville Condominium, S. Laurel St, cnr Shaw Blvd, 1552 Mandaluyong City; tel. (2) 5330518; fax (2) 5315513; f. 1980; Exec. Dir BERTRAND GONZALES.

Philippine Chamber of Commerce and Industry: 14th Floor, Multinational Bancorporation Centre, 6805 Ayala Ave, Makati City, Metro Manila; tel. (2) 8445713; fax (2) 8434102; e-mail pcciintr@mozcom.com; internet www.philcham.com; f. 1977; Pres. SERGIO R. ORTIZ-LUIS, Jr.

Philippine Chamber of Mines: Rm 204, Ortigas Bldg, Ortigas Ave, Pasig City, Metro Manila; tel. (2) 6354123; fax (2) 6354160; e-mail comp@vasia.com; f. 1975; Chair. GERARD H. BRIMO; Pres. ARTEMIO DISINI.

FOREIGN TRADE ORGANIZATIONS

Bureau of Export Trade Promotion: New Solid Bldg, 5th–8th Floors, 357 Sen. Gil J. Puyat Ave, Makati City, 1200 Metro Manila; tel. (2) 8990133; fax (2) 8904707; e-mail betpod@dti.gov.ph; internet tradelinephil.dti.gov.ph; Dir FERNANDO P. CALA, II.

Bureau of Import Services: Oppen Bldg, 3rd Floor, 349 Sen. Gil J. Puyat Ave, Makati City, Metro Manila; tel. (2) 8905418; fax (2) 8957466; e-mail bis@dti.gov.ph; internet www.dti.gov.ph/bis; Exec. Dir ALEXANDER B. ARCILLA.

Philippine International Trading Corpn (PITC): Philippines International Center, 46 Sen. Gil J. Puyat Ave, Makati City, 1200 Metro Manila; POB 2253 MCPO; tel. (2) 8454376; fax (2) 8454476; e-mail pitc@info.com.ph; internet www.dti.gov.ph/pitc/; f. 1973; state trading company to conduct international marketing of general merchandise, industrial and construction goods, raw materials, semi-finished and finished goods, and bulk trade of agri-based products; also provides financing, bonded warehousing, shipping, cargo and customs services; Pres. ARTHUR C. YAP.

INDUSTRIAL AND TRADE ASSOCIATIONS

Chamber of Automobile Manufacturers of the Philippines: Metro Manila; Sec.-Gen. MARIO DE GRANO.

Cotton Development Administration: Asitrust Bank Annex Bldg, 1st Floor, 1424 Quezon Ave, Quezon City, 1100 Metro Manila; tel. and fax (2) 3747427; e-mail coda_ho2003@yahoo.com; Administrator Dr EUGENIO D. ORPIA, Jr.

Fiber Industry Development Authority: Asiatrust Bank Annex Bldg, 1424 Quezon Ave, Quezon City, Metro Manila; tel. (2) 3737489; fax (2) 3737494; e-mail fida@pacific.net.ph; Administrator CECILIA GLORIA J. SORIANO.

Philippine Fisheries Development Authority: Union Square 1 Bldg, 7th Floor, 145 15th Ave, Cubao, Quezon City, Metro Manila; tel. (2) 9113829; fax (2) 9113018; f. 1976; Gen. Man. PABLO B. CASIMINA.

Semiconductor and Electronic Industries in the Philippines (SEIPI): Unit 1102, Alabang Business Tower 1, Acacia Ave, Madrigal Business Park, Ayala Alabang, Muntinlupa City, 1780 Metro Manila; tel. (2) 8078458; fax (2) 8078459; e-mail philelectronics@seipi.org; internet www.seipi.org; Exec. Dir ERNIE SANTIAGO.

EMPLOYERS' ORGANIZATIONS

Employers' Confederation of the Philippines (ECOP): ECC Bldg, 4th Floor, 355 Sen. Gil J. Puyat Ave, Makati City, Metro Manila; tel. (2) 8904845; fax (2) 8958576; e-mail ecop@webquest.com; internet www.ecop.org.ph; f. 1975; Pres. MIGUEL B. VARELA; Dir-Gen. VICENTE LEOGARDO, Jr.

Filipino Shipowners' Association: Victoria Bldg, Room 503, United Nations Ave, Ermita, 1000 Metro Manila; tel. (2) 5227318; fax (2) 5243164; e-mail filiship@info.com.ph; f. 1950; 34 mems; Chair. and Pres. CARLOS C. SALINAS; Exec. Dir AUGUSTO Y. ARREZA, Jr.

Philippine Cigar and Cigarette Manufacturers' Association: Unit 508, 1851 Dr Antonio Vasquez St, Malate, Metro Manila; tel. (2) 5249285; fax (2) 5249514; Pres. ANTONIO B. YAO.

Philippine Coconut Producers' Federation, Inc: Wardley Bldg, 2nd Floor, 1991 Taft Ave, cnr San Juan St, Pasay City, 1300 Metro Manila; tel. (2) 5230918; fax (2) 5211333; e-mail cocofed@pworld.net.ph; Pres. MARIA CLARA L. LOBREGAT.

Philippine Sugar Millers' Association Inc: 1402 Security Bank Centre, 6776 Ayala Ave, Makati City, 1226 Metro Manila; tel. (2) 8911138; fax (2) 8911144; e-mail psma@netasia-mnl.net; internet www.psma.com.ph; f. 1922; Pres. V. FRANCISCO VARUA; Exec. Dir JOSE MA T. ZABALETA.

Philippine Retailers' Association: Unit 2610, Jollibee Plaza, Emerald Ave, Ortigas Centre, Pasig City; tel. (2) 6874180; fax (2) 6360825; e-mail pra@nwave.net; internet www.philretailers.com; Pres. BIENVENIDO V. TANTOCO, III.

Textile Mills Association of the Philippines, Inc (TMAP): Ground Floor, Alexander House, 132 Amorsolo St, Legaspi Village, Makati City, 1229 Metro Manila; tel. (2) 8186601; fax (2) 8183107; e-mail tmap@pacific.net.ph; f. 1956; 11 mems; Pres. HERMENEGILDO C. ZAYCO; Chair. JAMES L. GO.

Textile Producers' Association of the Philippines, Inc: Downtown Center Bldg, Room 513, 516 Quintin Paredes St, Binondo, Metro Manila; tel. (2) 2411144; fax (2) 2411162; Pres. GO CUN UY; Exec. Sec. ROBERT L. TAN.

UTILITIES

Energy Regulatory Commission: Pacific Center Bldg, San Miguel Ave, Ortigas Center, Pasig City, 1600 Metro Manila; tel. (2) 6334556; fax (2) 6315871; e-mail info@erc.gov.ph; internet www.erc.gov.ph; Chair. RODOLFO B. ALBANO, Jr; Exec. Dir ARNIDO O. INUMERABLE.

Electricity

Davao Light and Power Co: 163 C. Bangoy, Sr, St, Davao City 8000; tel. (82) 2212191; fax (82) 2212105; e-mail davaolight@davao-online.com; internet www.davaolight.com; the country's third largest electric utility with a peak demand of 175 MW in 2000.

Manila Electric Co (Meralco): Lopez Bldg, Meralco Center, Ortigas Ave, Pasig City, 0300 Metro Manila; tel. (2) 6312222; fax (2) 6328501; e-mail wmtirona@meralco.com.ph; internet www.meralco.com.ph; f. 1903; supplies electric power to Manila and seven provinces in Luzon; largest electricity distributor, supplying 54% of total consumption in 2000; privatized in 1991, 34% govt-owned; cap. and res 54,382m., sales 85,946m. (1998); Chair. and CEO MANUEL M. LOPEZ; Pres. JESUS P. FRANCISCO.

National Power Corpn (NAPOCOR): Quezon Ave, cnr BIR Rd, Quezon City, Metro Manila; tel. (2) 9213541; fax (2) 9212468; e-mail pad@napocor.gov.ph; internet www.napocor.gov.ph; f. 1936; state-owned corpn supplying electric and hydroelectric power throughout the country; scheduled for privatization; installed capacity in 1998, 11,810 MW; sales 86,611m. pesos (Dec. 1998); 12,043 employees; Pres. ROGELIO M. MURGA; Chair. JUANITA AMATONG (Secretary of Finance).

Gas

First Gas Holdings Corpn: Benpres Bldg, 4th Floor, Exchange Rd, cnr Meralco Ave, Pasig City, Metro Manila; tel. (2) 6343428; fax (2) 6352737; internet www.firstgas.com.ph; major interests in power generation and distribution; Pres. PETER GARRUCHO.

Water

Regulatory Authority

Metropolitan Waterworks and Sewerage System: 4th Floor, Administration Bldg, MWSS Complex, 489 Katipunan Rd, Balara, Quezon City, 1105 Metro Manila; tel. (2) 9223757; fax (2) 9212887; e-mail info@mwss.gov.ph; internet www.mwss.gov.ph; govt regulator for water supply, treatment and distribution within Metro Manila; Administrator ORLANDO C. HONDRADE.

Distribution Companies

Davao City Water District: Km 5, J. P. Laurel Ave, Bajada, Davao City; tel. (82) 2219400; fax (82) 2264885; e-mail dcwd@interasia.com .ph; f. 1973; public utility responsible for the water supply of Davao City; Gen. Man. WILFRED G. YAMSON.

Manila Water: MWSS Compound, 2nd Floor, 489 Katipunan Rd, Balaran, Quezon City, Metro Manila; tel. (2) 9267999; fax (2) 9281223; e-mail info@manilawateronline.com; internet www .manilawateronline.com; f. 1998 following the privatization of Metro Manila's water services; responsible for water supply to Manila East until 2023; Pres. ANTONINO T. AQUINO.

Maynilad Water: MWSS Compound, Katipunan Rd, Balara, Quezon City, Metro Manila; tel. (2) 4353583; fax (2) 9223759; e-mail frestuar@mayniladwater.com.ph; f. 1998 following the privatization of Metro Manila's water services; responsible for water supply, sewage and sanitation services for Manila West until 2021; Pres. FIORELLO R. ESTUAR.

Metropolitan Cebu Water District: Magallanes St, cnr Lapulapu St, 6000 Cebu City; tel. (32) 2560413; fax (32) 2545391; e-mail mcwd@cvis.net.ph; internet www.cvis.net.ph/mcwd; f. 1974; public utility responsible for water supply and sewerage of Cebu City and surrounding towns and cities; Chair. RUBEN D. ALMENDRAS; Gen. Man. ARMANDO H. PAREDES.

MAJOR COMPANIES

(Amounts in pesos, unless otherwise stated)

Automobiles

Honda Cars Philippines, Inc: 105 South Main Ave, Laguna Technopark, Sta Rosa, Laguna; tel. (2) 8190440; fax (2) 8150509; internet www.hondaphil.com; f. 1990; sales 4,690m. (1994/95); 800 employees; Pres. and Gen. Man. TAKASHI HASHIGAWA.

Mitsubishi Motors Philippines Corpn: Ortigas Ave Ext., Cainta, Rizal; tel. (2) 6580911; fax (2) 6580671; e-mail a-dalida@ mitsubishi-motors.com.ph; internet www.mitsubishi-motors.com .ph; f. 1987; car assembly; sales 13,273m. (1995); 1,300 employees; Pres. MAKOTO MAEDA.

Toyota Motor Philippines Corpn: Km 15, South Superhighway, Parañaque, 1700 Metro Manila; tel. (2) 8244701; fax (2) 8244741; e-mail fbj@toyota.com.ph; internet www.toyota.com.ph; f. 1988; sales 7,648m. (1998); 1,518 employees (1998); Chair. Dr GEORGE S. K. TY; Pres. TAKESHI FUKUDA.

Cement

Alsons Cement Corpn: Alsons Bldg, 2285 Pasong Tamo Ext., Makati City, Metro Manila; tel. (2) 8175506; fax (2) 8940655; e-mail alcemir@pworld.net.ph; internet www.alsonscement.com; f. 1969; mfrs of cement and construction-related products; cap. and res 4,568m. (1998), sales 2,846m. (2000); 550 employees; Chair. and Pres. TOMAS I. ALCANTARA; Sec. A. A. PICAZO.

Bacnotan Consolidated Industries Inc: Phinma Bldg, 4th Floor, 166 Salcedo St, Legaspi Village, Makati City, Metro Manila; tel. (2) 8109526; fax (2) 8109252; e-mail phinma@bworldonline.com; internet www.phinma.com; f. 1957; holding co with subsidiaries in manufacture of cement and steel; cap. and res 5,502m., sales 7,957m. (2000); 1,220 employees; Chair. RAMON V. DEL ROSARIO, Sr; Pres. OSCAR J. HILADO.

Fortune Cement Corpn: 139 Valero St, Salcedo Village, Makati City, 1200 Metro Manila; tel. (2) 7193840; f. 1967; cap. and res 3,597m. (2000), sales 1,645m. (2001); 346 employees; Chair. RENATO SUNICO; Pres. EDGARDO R. SORIANO.

Southeast Asia Cement Holdings Inc: Chatham House Bldg, 17th Floor, 116 Valero St, cnr Herrera St, Salcedo Village, Makati City, 1229 Metro Manila; tel. (2) 8454201; fax (2) 8403559; cap. and res 4,711m., sales 2,454m. (2001); Chair. PETER J. HODDINOTT; Pres. and CEO ARLENE C. DE GUZMAN.

Union Cement Corpn: L3 Phinma Plaza, 39 Plaza Drive, Rockwell Center, Makati City, 1200 Metro Manila; tel. (2) 8981980; fax (2) 8700498; e-mail zita.diez@holcim.com; internet www.unioncement .com; f. 2000 following merger of Hi Cement Corpn, Davao Union Cement Corpn and Bacnotan Cement Corpn; acquired majority interest in Alsons Cement Corpn 2002; mfr and marketer of cement; Pres. OSCAR J. HILADO.

Coconut Products

International Copra Export Corpn: 1000 A. Mabini St, Ermita, Metro Manila; tel. (2) 5238311; fax (2) 5212098; f. 1961; wholesaler of coconut and coconut products; sales 3,771m. (1997); 350 employees; Chair. K. G. LUY; Pres. ENRIQUE LUY.

Legaspi Oil Co, Inc: UCPB Bldg, 16th Floor, Makati Ave, Makati City, 1200 Metro Manila; tel. (2) 8921961; fax (2) 8153370; f. 1929; processors of coconut oil; six subsidiaries; sales 2,892m. (1997); 150 employees; Chair. TIRSO ANTIPORDA; Pres. JEREMIAS B. BENICO.

Lu Do and Lu Ym Oleochemical Corpn: 101–103 Tupaz St, POB 18, Cebu City; tel. (32) 2531930; fax (32) 54102; f. 1948; mfrs of crude coconut oil, refined edible oil, copra meal products, corn starch, corn oil and gluten meal; net sales 260.3m. (1997); 538 employees; Pres. DOUGLAS LU YM; Exec. Vice-Pres. and Gen. Man. V. D. VELASCO; Chair. PATERNO LU YM.

Philippine Refining Co, Inc: 1351 United Nations Ave, Metro Manila; tel. (2) 504011; f. 1927; detergents, personal products, and food mfrs; processors of coconut oil; 1,550 employees; Pres. and Chair. CESAR B. BAUTISTA.

Procter and Gamble Philippine Manufacturing Corpn: Ayala Center, 14th–21st Floors, 6750 Ayala Ave, Makati City, Metro Manila; tel. (2) 8430621; fax (2) 8148551; internet www.pg.com; f. 1935; processors of coconut oil; toilet preparations and detergents; food mfrs; sales US $367.2m. (1999); 2,000 employees; Pres. JOHNIP G. CUA.

Construction

Asian Construction and Development Corpn: 2nd Floor, Union-Ajinomoto Bldg, Sen. Gil J. Puyat Ave, Makati City, Metro Manila; tel. (2) 8906337; fax (2) 8906421; e-mail asiakon@rp1.net; f. 1981; gen. engineering and construction; 1,300 employees; Chair. DEOGRACIAS G. EUFEMIO.

Cityland Development Corpn: 2nd Floor, Cityland Condominium, Tower One, POB 5000, 6815 H. V. de la Costa St, Ayala Ave, Makati City, 1226 Metro Manila; tel. (2) 8936060; fax (2) 8928656; e-mail investment@cityland.net; internet www.cityland.net; cap. and res 2,954m., sales 759m. (2000); 240 employees; Chairs VICENTE T. PATERNO, ANDREW I. LIUSON.

Construction Consultants Corpn: Zeta II Bldg, 5th Floor, Salcedo St, Makati City, 1200 Metro Manila; tel. (2) 877118; fax (2) 8185646; f. 1976; consulting, management, engineering; Pres. TEODORO GENER.

DMCI Holdings Inc: Dacon Bldg, 3rd Floor, 2281 Pasong Tamo Ext., Makati City, 1231 Metro Manila; tel. (2) 8920984; fax (2) 8167362; e-mail dmcihi@mozcom.com; internet www.dmchi.com; f. 1995; cap. and res 5,713m., sales 4,592m. (2000); 14,000 employees; Pres. ISIDRO A. CONSUNJI.

DM Consunji, Inc: DMCI Plaza, 2281 Pasong Tamo Ext., Makati City, Metro Manila; tel. (2) 8880841; fax (2) 8883053; e-mail dmci@ dmcinet.com; internet www.dmconsunji.com; construction and services; cap. and res 3,087m., sales 3,155m. (1999); 16,500 employees; Chair. DAVID M. CONSUNJI; Pres. JORGE A. CONSUNJI.

Philippine National Construction Corpn (PNCC): PNCC Bldg, Epifanio de los Santos Ave, cnr Reliance St, Mandaluyong, Metro Manila; tel. (2) 6318431; fax (2) 6315362; internet www.pncc.net; f. 1966; 80% govt-owned; construction; design engineering; steel and concrete products, heavy machinery; sales 2,733m. (1998); Chair. VICTORINO A. BASCO; Pres. and CEO ROLANDO JOSE L. MACASAET.

Electrical and Electronics

Amkor Technology Philippines, Inc: Km 22 East Service Rd, South Superhighway, Bo Cupang-Muntinlupa City, Metro Manila; tel. (2) 8507000; fax (2) 8507287; e-mail marketing@amkor.com; f. 1976; fmrly Amkor/Anam Pilipinas, Inc; mfr of semiconductors and electronic components; 4,000 employees; Pres. ANTONIO NG.

Integrated Microelectronics, Inc: North Science Ave, Special Export Processing Zone, Laguna Technopark, 4024 Binan Laguna; tel. (2) 8420542; fax (49) 5491042; e-mail sales@imiphil.com; internet www.imiphil.com; f. 1980; mfr of electronic components and related products; cap. US $21m., sales US $94m. (2003); 12,000 employees; Pres. and CEO ARTHUR R. TAN; Chair. JAIME AUGUSTO ZOBEL DE AYALA, II.

Intel Philippines MFG, Inc: MN2 Bldg, 1321 Apolinario St, Bangkal, Makati City, 1233 Metro Manila; tel. (2) 8492111; fax (2) 8492170; f. 1974; mfr of integrated circuits; 5,700 employees; Pres. JACOB A. PENA.

Ionics EMS, Inc: Jannov Bldg, 2296 Pasong Tamo Ext., Makati City, Metro Manila; tel. (2) 8167481; fax (2) 8180935; internet www

.ionics-ems.com; f. 1974; mfr of electronic equipment; sales 7,511m. (2000); 8,000 employees; Pres. LAWRENCE QUA.

Matsushita Electric Philippines Corpn (MEPCO): B. Mapandan, Ortigas Ave Ext., Taytay, Rizal, 1901 Metro Manila; tel. (2) 6352260; fax (2) 8189478; e-mail headoffice@mepco.panasonic.com.ph; internet mepco.panasonic.com.ph; f. 1967; mfr of electrical appliances; cap. and res 2,627m. (March 1998), sales 7,679m. (March 2001); 1,800 employees; Pres. YUKIHARU KUBOTA.

Panasonic Mobile Communications Corpn of the Philippines (PMCP): 102 Laguna Blvd, Laguna Technopark, 4026 Sta Rosa, Laguna; tel. (92) 8181263; fax (92) 8183303; e-mail mat_com@mozcom.com; internet www.mcp.panasonic.com.ph; f. 1987; fmrly Matsushita Communication Industrial Corpn of the Philippines; name changed as above June 2003; manufacture and sale of industrial communications products; sales 7,640m. (1998); 1,893 employees; Pres. REYNALDO S. LICO.

Solid Group Inc: Solid House, 2285 Don Chino Roces Ave, Makati City, 1231 Metro Manila; tel. (2) 8431511; fax (2) 8128273; e-mail SGI@mail.com.sen.ph; internet www.sen.com.ph; mfr of consumer electronic products carrying the Sony and Aiwa brand names: cap. and res 7,459m., sales 4,149m. (2000); 1,470 employees; Chair. SUSAN L. TAN; Dir, Pres. and CEO DAVID S. LIM.

Temic Telefunken Microelectronics Philippines, Inc: FTI Complex, Bagsakan Rd, Taguig, Makati City, Metro Manila; tel. (2) 8158635; fax (2) 8158640; f. 1974; mfr of electronics components; sales 5,907m. (1996); 3,500 employees; Gen. Man. JOSE M. FACUNDO.

Texas Instruments (Philippines), Inc: Baguio Export Processing Zone, Laokan Rd, Baguio City, Benguet; tel. (74) 8450927; fax (2) 8931960; e-mail n-veria@ti.com; internet www.ti.com; f. 1979; mfr of semiconductors; sales US $1,391.1m. (1999); 2,050 employees; Pres. NORBERTO A. VIERA.

Food and Food Products

Coca-Cola Bottlers Philippines, Inc: 19th Floor, San Miguel Properties Center, 7 St Francis St, 1550 Mandaluyong City; tel. (2) 6885888; e-mail ccbho.consumeraffairs@ccbpi.com; internet www.cocacola.com; f. 1981; beverages; sales US $708.4m. (1999); 10,240 employees; Chair. ANDRES SORIANO.

Cosmos Bottling Corpn: RFM Bldg, 5th Floor, cnr Sheridan and Pioneer Sts, Mandaluyong City, Metro Manila; tel. (2) 6318101; fax (2) 6320839; e-mail cosmosmail@cosmos.com.ph; f. 1918; manufactures, markets and distributes soft drinks; cap. and res 4,056m. (2000), sales 8,010m. (2001); 5,100 employees; Chair. JOSE MARIE A. CONCEPCION, III.

Dole Philippines, Inc: BA-Lepanto Bldg, 4th Floor, 8747 Paseo de Roxas, Makati City, Metro Manila; tel. (2) 8102601; fax (2) 8166483; f. 1963; mfr of canned food; sales 5,259.3m. (1997); 4,550 employees; Chair. PAUL CUYEGKENG.

Jollibee Foods Corpn: Jollibee Plaza, 10th Floor, Emerald and Ruby Rd, Ortigas Center, Pasig City, Metro Manila; tel. (2) 6341111; fax (2) 6358888; e-mail president@jollibee.com.ph; internet www.jollibee.com.ph; f. 1975; operation of fast food chain; cap. 5,900m., sales 20,300m. (2000); 14,243 employees; Chair., Pres. and CEO TONY TAN CAKTIONG.

Nestlé Philippines, Inc: 31 Plaza Drive, Rockwell Center, Makati City, 1200 Metro Manila; tel. (2) 8980001; fax (2) 8980089; internet www.nestle.com.ph; f. 1961; mfr of food products; sales US $856.4m. (1999); 4,500 employees; Pres. and CEO SALVADOR PIGEM.

Pure Foods Corpn: JMT Corporate Condominium, ADB Ave, Ortigas Centre, Pasig City, POB 2695, Metro Manila; tel. (2) 6341010; fax (2) 6338747; e-mail dimayuga.teodoro@purefoods.com.ph; internet www.purefoods.com.ph; f. 1956; cap. and res 3,571m., sales 12,650,375m. (2000); 3,700 employees; Chair. EDUARDO M. COJUANGCO, Jr; Pres. ENRIQUE A. GOMEZ, Jr.

RFM Corpn: RFM Bldg, Pioneer St, cnr Sheridan St, Mandaluyong City, Metro Manila; tel. (2) 6318101; fax (2) 6315039; e-mail rfmmail@rfm.com.ph; internet www.rfm.com.ph; mfrs of flour, feeds, meat, snacks, fats and oil, etc.; cap. and res 5,929m., sales US $431.3m. (1999); 12,000 employees; Chair. JOSE S. CONCEPCION, Jr; Pres. and CEO JOSE MARIE A. CONCEPCION, III.

San Miguel Corpn: 40 San Miguel Ave, Mandaluyong City, Makati City, Metro Manila; tel. (2) 6323000; fax (2) 6323099; internet www.sanmiguel.com.ph; f. 1890; breweries, food-processing, packaging; cap. and res 104,356m., sales 148,590m. (2003); 14,900 employees; Chair. EDUARDO COJUANGCO.

Swift Foods, Inc: RFM Corporate Centre, Pioneer Cnr, Sheridan, Mandaluyong City, 1603 Metro Manila; tel. (2) 6318101; fax (2) 6315064; e-mail swiftmail@swiftfoods.com.ph; internet www.rfm.com.ph; mfr of processed meat products, poultry products and commercial feeds; cap. and res 3,302m., sales 9,171m. (2000); Chair. JOSE A. CONCEPCION III.

La Tondeña Distillers, Inc: St Francis St, Mandaluyong City, 1550 Metro Manila; tel. (2) 6899100; fax (2) 7349584; e-mail csmapa@sanmiguel.com.ph; internet www.ltdi.com; f. 1987; mfr of liquors and mineral water; cap. and res 6,756m., sales 11,362m. (2002); 1,250 employees; Chair. EDUARDO M. CONJUANGCO, Jr; Pres. ARNALDO L. AFRICA.

Universal Robina Corpn: URC Bldg, 110 E. Rodriguez Ave, Bagong Ilog, Pasig City, Metro Manila; tel. (2) 6712935; fax (2) 6345276; internet www.urc.com.ph; f. 1954; mfr of snacks, chocolates, candies, biscuits, pasta and ice cream; cap. and res 14,288m. (Sept. 1998), sales 15,706m. (Sept. 2000); 7,300 employees; Chair. and CEO JAMES L. GO, Jr; Pres. JAMES L. GO.

Vitarich Corpn: 2316 Sarmiento Bldg, Pasong Tamo Ext., Makati City, Metro Manila; tel. (2) 8430236; fax (2) 8167236; internet www.vitarich.com; f. 1962; mfr of animal feeds; cap. and res 1,542m. (1997), sales 5,862m. (2000); 1,800 employees; Chair. ROGELIO M. SARMIENTO; Pres. RENATO P. SARMIENTO.

Metal Mining

Atlas Consolidated Mining and Development Corpn: Quad Alpha Centrum, 7th Floor, 125 Pioneer St, Mandaluyong City, 1554 Metro Manila; tel. (2) 6350063; fax (2) 6333759; f. 1953; mining of copper ore and recovery of by-products of gold, silver and pyrite at Cebu mines; mining of gold ore (with silver) at Masbate Gold Operations; cap. and res –4,830m., sales 66m. (2000); 10,558 employees; Chair. JOSE C. IBAZETA; Pres. ROGELIO C. SALAZAR.

Benguet Corpn: Corporate Plaza, 3rd Floor, 845 Arnaiz Ave, cnr Pasay Rd, Legaspi Village, Makati City, 1223 Metro Manila; tel. (2) 8121380; fax (2) 8136611; e-mail spp@benguetcorp.com.ph; internet www.benguetcorp.com; f. 1903; principal primary gold producer; cap. and res 1,275m., sales 406m. (1998); 480 employees; Chair., Pres. and CEO BENJAMIN PHILIP G. ROMUALDEZ.

Lepanto Consolidated Mining Co: BA-Lepanto Bldg, 21st Floor, 8747 Paseo de Roxas, Makati City, Metro Manila; tel. (2) 8159447; fax (2) 8105583; e-mail lepanto@i-next.net; f. 1936; copper, gold, silver and calcines; cap. and res 4,555m., sales 2,999m. (2000); 2,250 employees; Chair. and CEO FELIPE U. YAP; Pres. ARTEMIO F. DISINI.

Maricalum Mining Corpn: 2283 Pasong Tamo Ext., Makati City, Metro Manila; tel. (2) 864011; f. 1949; fmrly Marinduque Mining and Industrial Corpn; nickel, copper and cement production; Pres. TEODORO G. BERNARDINO.

Philex Mining Corpn: Philex Bldg, 27 Brixton St, Pasig City, 1660 Metro Manila; tel. (2) 6311381; fax (2) 6333242; e-mail philex@skyinet.net; internet www.philexmining.com.ph; mining; f. 1955; cap. and res 3,990m., sales 4,457m. (2000); 4,000 employees; Chair. and CEO GERARD H. BRIMO; Pres. LEONARDO P. JOSEF.

Petroleum

Alsons Consolidated Resources Inc: Alsons Bldg, 2nd Floor, 2286 Pasong Tamo Ext., Makati City, 1231 Metro Manila; tel. (2) 8175506; fax (2) 8940655; e-mail acrinrel@pworld.net.ph; internet www.acr-alsons.com; exploration and development of petroleum and petroleum products and gas; cap. and res 8,635m., sales 3,591m. (2000); Chair. and Pres. NICASIO I. ALCANTARA.

Caltex (Philippines) Inc: 540 Padre Faura, Ermita, Metro Manila; tel. (2) 8136001; fax (2) 8944116; internet www.caltex.com.ph; f. 1921; petroleum refining; sales US $939.5m. (1999); 878 employees; Chair. NICHOLAS C. FLORIO.

Mobil Philippines Inc: The Orient Square, 17th Floor, Emerald Ave, Ortigas, Pasig City; tel. (2) 6382333; fax (2) 8151844; sales 2,194m. (1998); Pres. C. B. RICCI.

Petron Corpn: Petron MegaPlaza, 358 Sen. Gil J. Puyat Ave, Makati City, 1200 Metro Manila; tel. (2) 8863888; fax (2) 8863064; e-mail aipestano@petron.com; internet www.petron.com; petroleum refining; 40% govt-owned, 40% owned by Saudi Arabian Oil Co; subsidiary of the state-owned Philippine National Oil Co; cap. and res 17,950m. (1999), sales 87,968m. (2000); 1,250 employees; Chair. NICASIO I. ALCANTARA.

Pilipinas Shell Petroleum Corpn: Shell House, 156 Valero St, Salcedo Village, Makati City, 1227 Metro Manila; tel. (2) 8166501; fax (2) 8166565; e-mail contact@shell.com.ph; internet www.shell.com.ph; f. 1959; petroleum refining and marketing; sales US $1,357.4m. (1999); 1,100 employees; Chair. R. WILLEMS.

Shell Gas Eastern, Inc: Shell House, 156 Valero St, Salcedo Village, Makati City, 1299 Metro Manila; tel. (2) 8166501; fax (2) 8166399; f. 1980; sales 6,803m. (1998); Pres. O. S. REYES.

Pharmaceuticals

Colgate-Palmolive Philippines: 1049 J. P. Rizal St, Makati City, 2800 Metro Manila; tel. (2) 8959444; fax (2) 8959457; f. 1947; mfr of cosmetics and toiletries; sales 4,600m. (1996); 480 employees; Pres. and Gen. Man. ROBERT GALAN.

Euro Med Laboratories Philippines, Inc: PPL Bldg, United Nations Ave, cnr San Marcelino, Metro Manila; tel. (2) 5240091; cap. and res 614,153m., sales 780,037m. (2000); Chairs Dr TOMAS P. MARAMBA, Jr, GEORGIANA S. EVIDENTE.

International Pharmaceuticals, Inc: Juan Luna Ave, Mabolo, 6000 Cebu City; tel. (32) 312685; fax (32) 310658; e-mail export@ipi-phil.com; f. 1959; mfr of pharmaceuticals, toiletries, veterinary products, soaps and detergents; 1,500 employees; Chair. and Pres. PIO CASTILLO.

Mercury Drug Corpn: 7 Mercury Ave, cnr E. Rodriguez, Jr, Bagumbayan, Quezon City, 1110 Metro Manila; tel. (2) 9115071; fax (2) 9116673; e-mail info@mercurydrug.com; internet www.mercurydrug.com; f. 1945; retailer of pharmaceuticals; sales 25,683m. (1999); 5,500 employees; Chair. MARIANO QUE.

Metro Drug Distribution, Inc: Sta Maria Industrial Estate, Manalac Ave, Bicutan Tagig, Metro Manila; tel. (2) 8372121; fax (2) 8372912; e-mail zuellig@mni.sequel.net; f. 1932; mfr of pharmaceuticals; sales 6,723m. (1995); 1,340 employees; Pres. PAUL KLEINER.

Unilever Philippines (PRC), Inc: 1351 United Nations Ave, Paco, Metro Manila; tel. (2) 5623951; fax (2) 5647259; e-mail CAS.Philippines@unilever.com; internet www.unilever.com.ph; f. 1927; mfr of pharmaceutical products, detergents, toiletries and foods; sales 8,808m. (1998); 1,100 employees; Chair. and CEO HOWARD BELTON.

United Laboratories, Inc: 66 United St, Mandaluyong City, Metro Manila; tel. (2) 6318501; fax (2) 6316774; e-mail jplamug@unilab.com.ph; internet www.unilab.com.ph; f. 1953; sales 9,475m. (1998); 2,375 employees; Pres. CARLOS EJERCITO; Chair. JOSE D. CAMPOS, Jr.

Zuellig Pharma Corpn: Zuellig Bldg, Sen. Gil J. Puyat Ave, Makati City, 1265 Metro Manila; tel. (2) 8191561; fax (2) 8431495; e-mail zpc@zuelligpharma.com.ph; internet www.zuelligpharma.com; f. 1953; sales US $413.4m. (1999); 1,700 employees; Chair. and Pres. REINER W. GLOOR.

Sugar

Central Azucarera de Tarlac: Cojuangco Bldg, 119 de la Rosa St, Makati City, Metro Manila; tel. (2) 8183911; fax (2) 8179309; f. 1927; sugar producer and mfr of sugar products; cap. and res 562m., sales 888m. (2000/01); 1,734 employees; Pres. and Chair. PEDRO COJUANGCO.

Roxas Holdings, Inc: Cacho Gonzalez Bldg, 6th Floor, 101 Aquirre St, Legaspi Village, Makati City, Metro Manila; tel. (2) 8108901; fax (2) 8179247; e-mail cadp@gbbe.com.ph; internet www.cadp.com.ph; f. 1930; fmrly Central Azucarera Don Pedro; sales 2,337m. (1997); 1,200 employees; Chair. PEDRO E. ROXAS; Vice-Chair. ANTONIO J. ROXAS.

Victorias Milling Co, Inc: 9126 Sultana St, cnr Honradez St, Makati City, 1200 Metro Manila; tel. (2) 8960381; fax (2) 8153204; f. 1919; sugar mfrs and refiners, agribusiness, property; sales 5,526m. (1994/95); 6,000 employees; Chair. B. M. VILLEGAS; Pres. M. M. MANALAC.

Textiles

Litton Textile Mills: CFC Commercial Center Bldg, Pasig Blvd, Pasig City, Metro Manila; tel. (2) 6717798; fax (2) 6339207; sales 1,634m. (1994/95); Pres. JAMES L. GO.

Ramie Textiles Inc: Boston Bank Center, 5th Floor, 6764 Ayala Ave, Makati City, Metro Manila; tel. (2) 8163301; fax (2) 8108616; f. 1956; 2,022 employees; Pres. RAMON H. DAVILA; Chair. ERNEST KAHN.

Solid Mills Inc: POB 1803, Makati City, Metro Manila; tel. (2) 8926416; fax (2) 8421631; f. 1971; 2,400 employees; sales 1,040m. (1995); Pres. PHILIP T. ANG; Chair. ANG BENG UH.

Tobacco

Compania General de Tabacos de Filipinas, SA (TABACALERA): Bldg 2, Tabacalera Compound, 900 Romualdez St, Paco, Metro Manila; tel. (2) 5223402; fax (2) 5217674; e-mail cdfmnl@i-next.net; f. 1881; import and export of tobacco; Chair. and Pres. RAFAEL MUGUIRO SARTORIUS; Country Man. SERAFIN GONZALEZ.

La Perla Industries Inc: Cheng Tsai Jun Bldg, 0165 Quirino Ave, Parañaque, Metro Manila; tel. (2) 8333211; fax (2) 8337452.

Wood and Wood Products

L. S. Sarmiento & Co, Inc: Sarmiento Bldg, 2 Pasong Tamo Ext., Makati City, Metro Manila; f. 1950; mfrs and exporters of plywood and panels; 1,500 employees; Pres. PABLO M. SARMIENTO, Jr (L. S. Sarmiento & Co Inc).

PICOP Resources Corpn: Moredel Bldg, 2nd Floor, 2280 Pasong Tamo Ext., Makati City, Metro Manila; tel. (2) 8135308; fax (2) 8410459; f. 1952; mfr of paper; cap. and res 3,352m. (1997), sales 2,417m. (2000); 2,100 employees; CEO LEONARDO T. SIGUION REYNA.

Steniel Manufacturing Corpn: Tektite Tower West, Unit 2902-B, Tektite Rd, Ortigas Center, Pasig City, Metro Manila; tel. (2) 6386286; fax (2) 6386289; e-mail info@firstpac.com.hk; internet www.steniel.com.ph; mfr of packaging products; cap. and res 1,391m. (1998), sales 1,698m. (2000); 703 employees; Chair. PERRY L. PE; Pres. GENESIS GOLDI D. GOLINGAN.

Zamboanga Wood Products Inc: GPL Bldg, Room 55, 5th Floor, 219 Sen. Gil J. Puyat Ave, Makati City, Metro Manila; tel. (2) 8159636; f. 1961; 1,090 employees; Pres. CLARITO ILLUSTRE.

Miscellaneous

Aboitiz Equity Ventures Inc: Aboitiz Corporate Center, Gov. Manuel A. Cuenco Ave, Kasambagan, 6000 Cebu City; tel. (32) 2312580; fax (32) 2314031; e-mail aev@aboitiz.com; internet www.aboitiz.com.ph; f. 1989; construction and fabrication; banking, power distribution; interests in shipping, shipyards, production of gases, operator of container terminal; cap. and res 17,036m., sales 17,951m. (2003); Chair. LUIS M. ABOITIZ, Jr; Pres. JON RAMON M. ABOITIZ.

Asian Terminals, Inc: Muelle de San Francisco St, South Harbor, Port Area, 1018 Metro Manila; tel. (2) 5278051; fax (2) 5272467; e-mail ati.sh@asianterminals.com.ph; internet www.asianterminals.com.ph; f. 1986; cap. and res 3,829m., sales 2,932m. (2000); Chair. Capt. ROGER DAVIES; Pres. RICHARD D. BARCLAY.

Atlantic, Gulf and Pacific Co, Inc: 1881 President Quirino Ave Ext., Pandacan, 1011 Metro Manila; tel. (2) 5638241; fax (2) 5632276; f. 1900; engineering, heavy industrial construction, fabrication and castings, marine repairs, offshore oil platform and marine structures, industrial manufacturing and machinery sales; sales 2,185m. (1997); Chair. LUIS I. VILLANUEVA; Pres. ROBERTO T. VILLANUEVA, Jr.

Ayala Corpn: 34th Floor, Tower One, Ayala Triangle, Ayala Ave, Makati City, Metro Manila; tel. (2) 8485643; fax (2) 8485846; e-mail acquery@ayala.com.ph; internet www.ayala.com.ph; f. 1834; conglomerate with interests in food, real estate, hotels, etc.; cap. and res 50,294m., sales 30,877m. (2002); 209 employees; Chair. JAIME ZOBEL DE AYALA.

Ayala Land Inc: Tower One, Ayala Triangle, Ayala Ave, Makati City, Metro Manila; tel. (2) 8485643; fax (2) 8486059; e-mail iru@ayalaland.com.ph; internet www.ayalaland.com.ph; real estate and hotel operations; cap. and res 35,273m., sales 14,624m. (2003); 390 employees; Chair. FERNANDO ZOBEL DE AYALA.

EEI Corpn: Topy Industries Bldg, 2nd Floor, 3 Calle Economia, Bagumbayan, Quezon City, 1110 Metro Manila; tel. (2) 6350851; fax (2) 6350861; e-mail scl@eei.com.ph; internet www.eei.com.ph; f. 1931; industrial construction; general trading; overseas construction services; cap. and res 2,151m., sales 4,532m. (2000); 11,594 employees; Pres. SAMSON C. LAZO; Chair. and CEO RIZALINO S. NAVARRO.

Guoco Holdings (Philippines) Inc: BA-Lepanto Bldg, 17th Floor, 8747 Paseo de Roxas, Makati City, Metro Manila; tel. (2) 8927912; fax (2) 8132895; principal activities of group include investment holding, real estate, manufacturing and distribution of Pepsi Cola products; cap. and res 2,607m. (1998), sales 1,091m. (June 2000); Pres. KWEK LENG HAI.

J.G. Summit Holdings Inc: 43rd Floor, Robinsons-Equitable Tower, ADB Ave, cnr Poveda St, Pasig City, 1600 Metro Manila; tel. (2) 6337641; fax (2) 6339387; internet www.jgsummit.com.ph; f. 1990; retailers of food and agro-industrial products; property, power generation, electronics manufacturing, etc; cap. and res 55,683m., sales 48,297m. (2002); 15,000 employees; Chair. and CEO JAMES L. GO; Pres. LANCE Y. GOKONGWEI.

Metro Pacific Corpn: PLDT Tower I, 2nd Floor, 6799 Ayala Ave, Makati City, Metro Manila; tel. (2) 8880888; fax (2) 8880801; e-mail metro@metropacific.com; internet www.metropacific.com; principal activities include consumer products, packaging, telecommunications; cap. and res 13,277m. (1997), sales 9,826m. (2000); Chair., Pres. and CEO MANUEL V. PANGILINAN.

Pryce Corpn: Pryce Center, 17th Floor, 1179 Chino Roces Ave, cnr Bagtikan St, Makati City, 1203 Metro Manila; tel. (2) 8994401; fax (2) 8996865; e-mail pryce@info.com.ph; internet www.philgardens.com; f. 1989; principal activities include trade in gases, and land development projects; cap. and res 813,462m., sales 704,628m. (2000); Chair. SALVADOR P. ESCAÑO.

Rustan Commercial Corpn: El Mercasol Bldg, Mandaluyong City, 1501 Metro Manila; tel. (2) 7212430; fax (2) 7212432; f. 1951; retailer; cap. and res 1,426m., sales 10,614m. (2001); 2,000 employees; Chair. BIENVENIDO TANTOCO, Jr.

TRADE UNION FEDERATIONS

In 1986 the Government established the Labor Advisory Consultation Committee (LACC) to facilitate communication between the Government and the powerful labour movement in the Philippines. The LACC granted unions direct recognition and access to the Government, which, under the Marcos regime, had been available only to the Trade Union Congress of the Philippines (KMP-TUCP). The KMP-TUCP refused to join the Committee.

In May 1994 a new trade union alliance, the Caucus for Labor Unity, was established; its members included the KMP-TUCP and three groups that had dissociated themselves from the former Kilusang Mayo Uno.

Katipunang Manggagawang Pilipino (KMP-TUCP) (Trade Union Congress of the Philippines): TUCP Training Center Bldg, TUCP-PGEA Compound, Masaya St, cnr Maharlika St, Diliman, Quezon City, 1101 Metro Manila; tel. (2) 9222185; fax (2) 9219758; e-mail tucp@easy.net.ph; internet www.tucp.org.ph; f. 1975; 1.5m. mems; Pres. Democrito T. Mendoza; Gen. Sec. Ernesto F. Herrera; 39 affiliates:

Associated Labor Unions—Visayas Mindanao Confederation of Trade Unions (ALU—VIMCONTU): ALU Bldg, Quezon Blvd, Port Area, Elliptical Rd, cnr Maharlika St, Diliman, Quezon City, 1101 Metro Manila; tel. (2) 9222185; fax (2) 9223199; f. 1954; 350,000 mems; Pres. Democrito T. Mendoza.

Associated Labor Union for Metalworkers (ALU—METAL): TUCP-PGEA Compound, Diliman, Quezon City, 1101 Metro Manila; tel. (2) 9222575; fax (2) 9223199; 29,700 mems; Pres. Cecilio T. Seno.

Associated Labor Union for Textile Workers (ALU—TEXTILE): TUCP-PGEA Compound, Elliptical Rd, Diliman, Quezon City, 1101 Metro Manila; tel. (2) 9222575; fax (2) 9223199; 41,400 mems; Pres. Ricardo I. Patalinjug.

Associated Labor Unions (ALU—TRANSPORT): 1763 Tomas Claudio St, Baclaran, Parañaque, Metro Manila; tel. (2) 8320634; fax (2) 8322392; 49,500 mems; Pres. Alexander O. Barrientos.

Associated Professional, Supervisory, Office and Technical Employees Union (APSOTEU): TUCP-PGEA Compound, Elliptical Rd, Diliman, Quezon City, 1101 Metro Manila; tel. (2) 9222575; fax (2) 9223199; Pres. Cecilio T. Seno.

Association of Independent Unions of the Philippines: Vila Bldg, Mezzanine Floor, Epifanio de los Santos Ave, Cubao, Quezon City, Metro Manila; tel. (2) 9224652; Pres. Emmanuel S. Durante.

Association of Trade Unions (ATU): Antwel Bldg, Room 1, 2nd Floor, Santa Ana, Port Area, Davao City; tel. (82) 2272394; 2,997 mems; Pres. Jorge Alegarbes.

Confederation of Labor and Allied Social Services (CLASS): Doña Santiago Bldg, TUCP Suite 404, 1344 Taft Ave, Ermita, Metro Manila; tel. (2) 5240415; fax (2) 5266011; f. 1979; 4,579 mems; Pres. Leonardo F. Agting.

Federation of Agrarian and Industrial Toiling Hands (FAITH): Kalayaan Ave, cnr Masigla St, Diliman, Quezon City, Metro Manila; tel. (2) 9225244; 220,000 mems; Pres. Raymundo Yumul.

Federation of Consumers' Co-operatives in Negros Oriental (FEDCON): Bandera Bldg, Cervantes St, Dumaguete City; tel. (32) 2048; Chair. Medardo Villalon.

Federation of Filipino Civilian Employees Association (FFCEA): 14 Murphy St, Pagasa, Olongapo City; tel. (2) 8114267; fax (2) 8114266; 21,560 mems; Pres. Roberto A. Flores.

Federation of Unions of Rizal (FUR): Perpetual Savings Bank Bldg, 3rd Floor, Quirino Ave, Parañaque, Metro Manila; tel. and fax (2) 8320110; 10,853 mems; Officer-in-Charge Eduardo Asuncion.

Lakas sa Industriya ng Kapatirang Haligi ng Alyansa (LIKHA): 32 Kabayanihan Rd Phase IIa, Karangalan Village, Pasig City, Metro Manila; tel. (2) 6463234; fax (2) 6463234; e-mail jbvlikha@yahoo.com; Pres. Jesus B. Villamor.

National Association of Free Trade Unions (NAFTU): CVC Bldg, Room 3, AD Curato St, Butuan City; tel. (8822) 3620941; 7,385 mems; Pres. Jaime Rincal.

National Congress of Unions in the Sugar Industry of the Philippines (NACUSIP): 7431-A Yakal St, Barangay San Antonio, Makati City, Metro Manila; tel. (2) 8437284; fax (2) 8437284; e-mail nacusip@compass.com.ph; 32 affiliated unions and 57,424 mems; Nat. Pres. Zoilo V. dela Cruz, Jr.

National Mines and Allied Workers' Union (NAMAWU): Unit 201, A. Dunville Condominium, 1 Castilla St, cnr Valencio St, Quezon City, Metro Manila; tel. (2) 7265070; fax (2) 4155582; 13,233 mems; Pres. Roberto A. Padilla.

Pambansang Kilusan ng Paggawa (KILUSAN): TUCP-PGEA Compound, Elliptical Rd, Diliman, Quezon City, 1101 Metro Manila; tel. (2) 9284651; 13,093 mems; Pres. Avelino V. Valerio; Sec.-Gen. Igmidio T. Ganagana.

Philippine Agricultural, Commercial and Industrial Workers' Union (PACIWU): 5 7th St, Lacson, Bacolod City; fax (2) 7097967; Pres. Zoilo V. dela Cruz, Jr.

Philippine Federation of Labor (PFL): FEMII Bldg, Suite 528, Aduana St, Intramuros, Metro Manila; tel. (2) 5271686; fax (2) 5272838; 8,869 mems; Pres. Alejandro C. Villaviza.

Philippine Federation of Teachers' Organizations (PFTO): BSP Bldg, Room 112, Concepcion St, Ermita, Metro Manila; tel. (2) 5275106; Pres. Federico D. Ricafort.

Philippine Government Employees' Association (PGEA): TUCP-PGEA Compound, Elliptical Rd, Diliman, Quezon City, Metro Manila; tel. (2) 6383541; fax (2) 6375764; e-mail eso_pgea@hotmail.com; f. 1945; 65,000 mems; Pres. Esperanza S. Ocampo.

Philippine Integrated Industries Labor Union (PIILU): Mendoza Bldg, Room 319, 3rd Floor, Pilar St, Zamboanga City; tel. (992) 2299; f. 1973; Pres. Jose J. Suan.

Philippine Labor Federation (PLF): ALU Bldg, Quezon Blvd, Port Area, Cebu City; tel. (32) 71219; fax (32) 97544; 15,462 mems; Pres. Crispin B. Gastardo.

Philippine Seafarers' Union (PSU): TUCP-PGEA Compound, Elliptical Rd, Diliman, Quezon City, 1101 Metro Manila; tel. (2) 9222575; fax (2) 9247553; e-mail psumla@info.com.ph; 10,000 mems; Pres. Democrito T. Mendoza; Gen. Sec. Ernesto F. Herrera.

Philippine Transport and General Workers' Organization (PTGWO—D): Cecilleville Bldg, 3rd Floor, Quezon Ave, Quezon City, Metro Manila; tel. (2) 4115811; fax (2) 4115812; f. 1953; 33,400 mems; Pres. Victorino F. Balais.

Port and General Workers' Federation (PGWF): Capilitan Engineering Corpn Bldg, 206 Zaragoza St, Tondo, Manila; tel. 208959; Pres. Franklin D. Butcon.

Public Sector Labor Integrative Center (PSLINK): 9723 C. Kamagong St, San Antonio Village, Makati City, Metro Manila; tel. (2) 8961573; fax (2) 9243525; 35,108 mems; Pres. Ernesto F. Herrera; Gen. Sec. Annie Geron.

United Sugar Farmers' Organization (USFO): SPCMA Annex Bldg, 3rd Floor, 1 Luzuriaga St, Bacolod City; Pres. Bernardo M. Remo.

Workers' Alliance Trade Unions (WATU): Delta Bldg, Room 300, Quezon Ave, cnr West Ave, Quezon City, Metro Manila; tel. (2) 9225093; fax (2) 975918; f. 1978; 25,000 mems; Pres. Temistocles S. Dejon, Sr.

INDEPENDENT LABOUR FEDERATIONS

The following organizations are not affiliated to the KMP-TUCP:

Associated Marine Officers and Seamen's Union of the Philippines (AMOSUP): Seaman's Centre, cnr Cabildo and Sta Potenciana Sts, Intramuros, Metro Manila; tel. (2) 495415; internet www.amosup.org/; f. 1960; 23 affiliated unions with 55,000 mems; Pres. Gregorio S. Oca.

Federation of Free Workers (FFW): FFW Bldg, 1943 Taft Ave, Malate, Metro Manila; tel. (2) 5219435; fax (2) 4006656; f. 1950; affiliated to the Brotherhood of Asian Trade Unionists and the World Confed. of Labour; 300 affiliated local unions and 400,000 mems; Pres. Ramon J. Jabar.

Lakas ng Manggagawa Labor Center: Rm 401, Femii Bldg Annex, A. Soriano St, Intramuros, Metro Manila; tel. and fax (2) 5280482; a grouping of 'independent' local unions; Chair. Oscar M. Acerson.

Manggagawa ng Komunikasyon sa Pilipinas (MKP): 22 Libertad St, Mandaluyong City, Metro Manila; tel. (2) 5313701; fax (2) 5312109; f. 1951; Pres. Pete Pinlac.

National Confederation of Labor: Suite 402, Carmen Bldg, Ronquillo St, cnr Evangelista St, Quiapo, Metro Manila; tel. and fax (2) 7334474; f. 1994 by fmr mems of Kilusang Mayo Uno; Pres. Antonio Diaz.

Philippine Social Security Labor Union (PSSLU): Carmen Bldg, Suite 309, Ronquillo St, Quiapo, Metro Manila; f. 1954; Nat. Pres. Antonio B. Diaz; Nat. Sec. Ofelia C. Alavera.

Samahang Manggagawang Pilipino (SMP) (National Alliance of Teachers and Office Workers): Fersal Condominium II, Room 33, 130 Kalayaan Ave, Quezon City, 1104 Metro Manila; tel. and fax (2) 9242299; Pres. Adelisa Raymundo.

Solidarity Trade Conference for Progress: Rizal Ave, Dipolog City; tel. and fax (65) 2124303; Pres. Nicolas E. Sabandal.

Trade Unions of the Philippines and Allied Services (TUPAS): Med-dis Bldg, Suites 203–204, Solana St, cnr Real St, Intramuros, Metro Manila; tel. (2) 493449; affiliated to the World Fed. of Trade Unions; 280 affiliated unions and 75,000 mems; Nat. Pres. DIOSCORO O. NUÑEZ; Sec.-Gen. VLADIMIR R. TUPAZ.

Transport

RAILWAYS

The railway network is confined mainly to the island of Luzon.

Light Rail Transit Authority (Metrorail): Adm. Bldg, LRTA Compound, Aurora Blvd, Pasay City, Metro Manila; tel. (2) 8320423; fax (2) 8316449; e-mail lrt.authority@lrta.gov.ph; internet www.lrta.gov.ph; managed and operated by Light Rail Transit Authority (LRTA); electrically-driven mass transit system; Line 1 (15 km, Baclaran to Monumento) began commercial operations in Dec. 1984; Line 1 South Extension (12 km, Baclaran to Bacoot) expected to start construction in 2004 and Line 2 (13.8 km, Santolan to Recto) scheduled to become fully operational by Oct. 2004; Administrator TEODORO B. CRUZ, Jr.

Philippine National Railways: PNR Management Center, Torres Bugallon St, Kalookan City, Metro Manila; tel. (2) 3654716; fax (2) 3620824; internet www.pnr.gov.ph; f. 1887; govt-owned; northern line services run from Manila to Caloocan, 6 km (although the track extends to San Fernando, La Union) and southern line services run from Manila to Legaspi, Albay, 479 km; Chair. and Gen. Man. JOSE M. SARASOLA II.

ROADS

In 2000 there were 201,994 km of roads in the Philippines, of which 30,013 km were highways and 49,992 km were secondary roads; an estimated 42,419 km of the network were paved. Bus services provided the most widely-used form of inland transport.

Department of Public Works and Highways: Bonifacio Drive, Port Area, Metro Manila; tel. (2) 5274111; fax (2) 5275635; e-mail soriquez.florante@dpwh.gov.ph; internet www.dpwh.gov.ph; responsible for the construction and maintenance of roads and bridges; Sec. FLORANTE M. SORIQUEZ.

Land Transportation Franchising and Regulatory Board: East Ave, Quezon City, Metro Manila; tel. (2) 4262505; fax (2) 4262515; internet www.ltfrb.gov.ph; f. 1987; Chair. DANTE LANTIN.

Land Transportation Office (LTO): East Ave, Quezon City, 1130 Metro Manila; tel. (2) 9219072; fax (2) 9219071; e-mail ltombox@lto.gov.ph; internet www.lto.gov.ph; f. 1987; plans, formulates and implements land transport rules and regulations, safety measures; registration of motor vehicles; issues licences; Exec. Dir BELLA G. BERMUNDO; Assistant Sec. ANNELI R. LONTOC.

SHIPPING

In 2000 there were 102 national and municipal ports, 20 baseports, 58 terminal ports and 270 private ports. The eight major ports are Manila, Cebu, Iloilo, Cagayan de Oro, Zamboanga, General Santos, Polloc and Davao.

Pangasiwaan ng Daungan ng Pilipinas (Philippine Ports Authority): Marsman Bldg, 22 Muelle de San Francisco St, South Harbour, Port Area, 1018 Metro Manila; tel. (2) 5274856; fax (2) 5274853; e-mail oscar.sevilla@ppa.gov.ph; internet www.ppa.gov.ph; f. 1977; supervises all ports within the Philippine Ports Authority port system; Gen. Man. OSCAR M. SEVILLA.

Domestic Lines

Albar Shipping and Trading Corpn: 2649 Molave St, United Parañaque 1, Parañaque, Metro Manila; tel. (2) 8232391; fax (2) 8233046; e-mail admin@albargroup.com.ph; internet www.albargroup.com.ph; f. 1974; manning agency (maritime), trading, ship husbanding; Chair. AKIRA S. KATO; Pres. JOSE ALBAR G. KATO.

Candano Shipping Lines, Inc: Victoria Bldg, 6th Floor, 429 United Nations Ave, Ermita, 2802 Metro Manila; tel. (2) 5238051; fax (2) 5211309; f. 1953; inter-island chartering and Far East, cargo shipping; Pres. and Gen. Man. JOSE CANDANO.

Carlos A. Gothong Lines, Inc: Quezon Blvd, Reclamation Area, POB 152, Cebu City; tel. (32) 211181; fax (32) 212265; Exec. Vice-Pres. BOB D. GOTHONG.

Delsan Transport Lines Inc: Magsaysay Center Bldg, 520 T. M. Kalaw St, Ermita, Metro Manila; tel. (2) 5219172; fax (2) 2889331; Pres. VICENTE A. SANDOVAL; Gen. Man. CARLOS A. BUENAFE.

Eastern Shipping Lines, Inc: ESL Bldg, 54 Anda Circle, Port Area, POB 4253, 2803 Metro Manila; tel. (2) 5277841; fax (2) 5273006; e-mail eastship@skyinet.net; f. 1957; services to Japan; Pres. ERWIN L. CHIONGBIAN; Exec. Vice-Pres. ROY L. CHIONGBIAN.

Loadstar Shipping Co Inc: Loadstar Bldg, 1294 Romualdez St, Paco, 1007 Metro Manila; tel. (2) 5238381; fax (2) 5218061; Pres. and Gen. Man. TEODORO G. BERNARDINO.

Lorenzo Shipping Corpn: Birch Tree Plaza Bldg, 6th Floor, 825 Muelle dela Industria St, Binondo, Metro Manila; tel. (2) 2457481; fax (2) 2446849; Pres. Capt. ROMEO L. MALIG.

Luzteveco (Luzon Stevedoring Corpn): Magsaysay Bldg, 520 T.M. Kalaw St, Ermita, Metro Manila; f. 1909; two brs; freight-forwarding, air cargo, world-wide shipping, broking, stevedoring, salvage, chartering and oil drilling support services; Pres. JOVINO G. LORENZO; Vice-Pres. RODOLFO B. SANTIAGO.

National Shipping Corpn of the Philippines: Knights of Rizal Bldg, Bonifacio Drive, Port Area, Metro Manila; tel. (2) 473631; fax (2) 5300169; services to Hong Kong, Taiwan, Korea, USA; Pres. TONY CHOW.

Negros Navigation Co Inc: Rufino Pacific Tower, 33rd Floor, 6784 Ayala Ave, cnr Herrera St, Makati City, Metro Manila; tel. (2) 8110115; fax (2) 8183707; Chair. DANIEL L. LACSON; Man. Dir MANUEL GARCIA.

Philippine Pacific Ocean Lines Inc: Delgado Bldg, Bonifacio Drive, Port Area, POB 184, Metro Manila; tel. (2) 478541; Vice-Pres. C. P. CARANDANG.

Philippine President Lines, Inc: PPL Bldg, 1000–1046 United Nations Ave, POB 4248, Metro Manila; tel. (2) 5249011; fax (2) 5251308; trading world-wide; Chair. EMILIO T. YAP, Jr; Pres. ENRIQUE C. YAP.

Sulpicio Lines, Inc: 1st St, Reclamation Area, POB 137, Cebu City; tel. (32) 73839; Chair. ENRIQUE S. GO; Man. Dir CARLOS S. GO.

Sweet Lines Inc: Pier 6, North Harbour, Metro Manila; tel. (2) 201791; fax (2) 205534; f. 1937; Pres. EDUARDO R. LOPINGCO; Exec. Vice-Pres. SONNY R. LOPINGCO.

Transocean Transport Corpn: Magsaysay Bldg, 8th Floor, 520 T. M. Kalaw St, Ermita, POB 21, Metro Manila; tel. (2) 506611; Pres. and Gen. Man. MIGUEL A. MAGSAYSAY; Vice-Pres. EDUARDO U. MANESE.

United Philippine Lines, Inc: UPL Bldg, Santa Clara St, Intramuros, POB 127, Metro Manila; tel. (2) 5277491; fax (2) 5271603; e-mail uplines@skyinet.net; services world-wide; Pres. FERNANDO V. LISING.

WG & A Philippines, Inc: South Harbour Center 2, cnr Railroad St, South Harbour, Metro Manila; tel. (2) 5274605; fax (2) 5360945; internet www.wgasuperferry.com; f. 1996 following the merger of William Lines, Aboitiz Shipping and Carlos A. Gothong Lines; passenger and cargo inter-island services; Pres. ENDICA ABOITIZ; Chair. W. L. CHIONGBIAN.

CIVIL AVIATION

In March 1999 there were 92 national and 103 private airports in the Philippines. In addition to the international airports in Metro Manila (the Ninoy Aquino International Airport), Cebu (the Mactan International Airport), Angeles City (the Clark International Airport), and Olongapo City (the Subic Bay International Airport), there are five alternative international airports: Laoag City, Ilocos Norte; Davao City; Zamboanga City; Gen. Santos (Tambler) City; and Puerto Princesa City, Palawan. A new international airport, in Davao City, was opened in December 2003. In March 2004 the Government announced that construction of an international airport in Iloilo City, intended to replace the city's existing airport, would begin in that month.

Air Transportation Office: MIA Rd, Pasay City, Metro Manila; tel. (2) 8799104; fax (2) 8340143; internet www.ato.gov.ph; implements govt policies for the development and operation of a safe and efficient aviation network; Dir-Gen. NILO C. JATICO.

Civil Aeronautics Board: Airport Rd, Pasay City, Metro Manila; tel. (2) 8317266; fax (2) 8336911; internet www.cab.gov.ph; exercises general supervision and regulation of, and jurisdiction and control over, air carriers, their equipment facilities and franchise; Dir MANUEL C. SAN JOSE.

Air Philippines: R1 Hangar, APC Gate 1, Andrews Ave, Nichols, Pasay City, Metro Manila; tel. (2) 8517601; fax (2) 8517922; e-mail info@airphilippines.com.ph; internet www.airphils.com; f. 1995; domestic and regional services; Chair. and Pres. WILLIAM GATCHALIAN.

Asian Spirit: G & A Bldg, 3rd Floor, 2303 Don Chino Roces Ave, Makati City, Metro Manila; tel. (2) 8403811; fax (2) 8130183; e-mail info@asianspirit.com; internet www.asianspirit.com; Man. ANTONIO BUENDIA.

Cebu Pacific Air: 30 Pioneer St, cnr Epifanio de los Santos Ave, Mandaluyong City, Metro Manila; tel. (2) 6371810; fax (2) 6379170; e-mail feedback@cebupacificair.com; internet www.cebupacificair.com; f. 1995; domestic and international services; Pres. LANCE

Gokongwei; Vice-Pres. (Corporate Planning and External Affairs) Peggy P. Vera.

Grand Air: Philippines Village Airport Hotel, 8th Floor, Pasay City, Metro Manila; tel. (2) 8312911; fax (2) 8917682; f. 1994; Pres. Rebecca Panlili.

Manila International Airport Authority (MIAA): NAIA Complex, Pasay City, Metro Manila; tel. (2) 8322938; fax (2) 8331180; e-mail gm@miaa.gov.ph; internet www.miaa.gov.ph; Gen. Man. Edgardo C. Manda.

Philippine Airlines Inc (PAL): PAL Corporate Communications Dept, PAL Center, Ground Floor, Legaspi St, Legaspi Village, Makati City, Metro Manila; tel. (2) 8316541; fax (2) 8136715; e-mail rgeccd@pal.com.ph; internet www.philippineairlines.com; f. 1941; in Jan. 1992 67% of PAL was transferred to the private sector; operates domestic, regional and international services to destinations in the Far East, Australasia, the Middle East, the USA and Canada; Chair. Lucio Tan; Pres. Avelino Zapanta.

Tourism

Tourism, although adversely affected from time to time by political unrest, remains an important sector of the economy. In 2002 arrivals totalled 1,932,677, compared with 1,796,893 in the previous year. Visitor expenditure totalled US $1,741m. in 2002 (compared with $1,723m. in 2001).

Philippine Convention and Visitors' Corpn: Legaspi Towers, 4th Floor, 300 Roxas Blvd, Metro Manila; tel. (2) 5259318; fax (2) 5216165; e-mail pcvcnet@info.com.ph; internet www.dotpcvc.gov.ph; Chair. Roberto M. Pagdanganan; Exec. Dir Daniel G. Corpuz.

Philippine Tourism Authority: Department of Tourism Bldg, T. M. Kalaw St, Ermita, 1000 Metro Manila; tel. (2) 5241032; fax (2) 5232865; e-mail info@philtourism.com; internet www.philtourism.com; Gen. Man. Robert Dean Barbers.

Defence

In August 2003 the total strength of the armed forces was estimated at 106,000: army 66,000, navy an estimated 24,000 (including 8,000 Marines and a 2000-strong Coast Guard), air force an estimated 16,000. The Citizen Armed Forces Geographical Units (CAFGU), which replaced the civil home defence force, numbered about 40,000. Military service is voluntary. In early 2002 the US Government dispatched troops to support the Philippine armed forces in a six-month counter-terrorist operation. Most of the US forces were withdrawn in July, but it was announced that some 400 were to remain in the Philippines to assist in training the military. In late 2002 the USA and the Philippines signed a five-year military agreement pledging to strengthen co-operation between the armed forces of the two countries and to facilitate the movement of heavy equipment and logistical supplies.

Defence Expenditure: Budgeted at an estimated 42,400m. pesos in 2003.

Chief of Staff of the Armed Forces: Gen. Narciso Abaya.

Chief of Staff (Army): Lt-Gen. Efren Abu.

Chief of Staff (Navy): Vice-Adm. Ernesto de Leon.

Chief of Staff (Air Force): Maj.-Gen. Jose Reyes.

Education

The 1987 Constitution commits the Government to provide free elementary and high school education; elementary education is compulsory. The organization of education is the responsibility of the Department of Education, Culture and Sports.

There are both public and private schools. The private schools are either sectarian or non-sectarian. Education in the Philippines is divided into four stages: pre-school (from the age of three); elementary school, which begins at seven years of age and lasts for six years; secondary or high school, which begins at 13 and lasts for five years (extended in 1994 from four years); and higher education, normally lasting four years. The public schools offer a general secondary curriculum and there are private schools that offer more specialized training courses. There is a common general curriculum for all students in the first and second years and more varied curricula in the third and fourth years leading to either college or technical vocational courses.

Total enrolment at pre-primary level in 1998/99 was equivalent to 30.7% (males 31.5%; females 30.0%) of children in the relevant age-group. In 1998/99 enrolment in primary education was equivalent to 113.2% (males 113.0%; females 113.3%) of children in the relevant age group, while enrolment at secondary level included 50.9% of the appropriate age-group (males 48.7%; females 53.1%). In that year total enrolment at tertiary level was equivalent to 29.5% of the relevant age-group (males 26.1%; females 32.9%). Instruction is in both English and Filipino at elementary level and English is the usual medium at the secondary and tertiary levels. In 1998/99 there were 2,349 tertiary level institutions in the country. The 2003 budget allocated 107,710m. pesos (13.0% of total national expenditure) to education.

Bibliography

GENERAL

Abueva, Jose V. (Ed.). *The Making of the Filipino Nation and Republic.* Metro Manila, University of the Philippines Press, 1999.

Adib Majul, Cesar. *The Political & Constitutional Ideas of the Philippine Revolution.* Metro Manila, University of the Philippines Press, 1999.

Muslims in the Philippines. Metro Manila, University of the Philippines Press, 1999.

Barreveld, Dirk J. *Terrorism in the Philippines: The Bloody Trail of Abu Sayyaf, Bin Laden's East Asian Connection.* San Jose, CA, Writers Club Press, 2002.

Billig, Michael S. *Barons, Brokers and Buyers: The Institutions and Cultures of Philippine Sugar.* Honolulu, HI, University of Hawaii Press, 2002.

Broad, Robin, and Cavanagh, John. *Plundering Paradise: The Struggle for the Environment in the Philippines.* Berkeley, CA, University of California Press, 1993.

Burley, T. M. *The Philippines. An Economic and Social Geography.* London, G. Bell and Sons, 1973.

Hedman, Eva-Lotta, and Sidel, John T. (Eds). *Philippine Politics and Society in the Twentieth Century: Colonial Legacies, Post-Colonial Trajectories.* London, Routledge, 2000.

McCoy, Alfred W., and de Jesus, Ed C. (Eds). *Philippine Social History.* Quezon City, Metro Manila, Ateneo de Manila University Press, 1982.

Putzel, James. *A Captive Land.* London, Catholic Institute for International Relations, 1992.

San Juan, Jr, E. *From Exile to Diaspora.* Boulder, CO, Westview Press, 1998.

Wernstedt, F. L., and Spencer, J. E. *The Philippine Island World: A Physical, Cultural and Regional Geography.* Berkeley and Los Angeles, CA, University of California, 1967.

HISTORY

Alfonso, Oscar M. *Theodore Roosevelt and the Philippines, 1897–1908.* Quezon City, University of the Philippines, 1970.

Bain, David Haward. *Sitting in Darkness: Americans in the Philippines.* Boston, MA, Houghton Mifflin, 1985.

Bankoff, Greg. *Cultures of Disaster: Society and Natural Hazards in the Philippines.* London, RoutledgeCurzon, 2002.

Bankoff, Greg, and Weekley, Kathleen. *Post-Colonial National Identity in the Philippines: Celebrating the Centennial of Independence.* London, Ashgate Publishing Company, 2002.

Bresnan, John (Ed.). *Crisis in the Philippines: The Marcos Era and Beyond.* Princeton, NJ, Princeton University Press, 1987.

The Burden of Proof. Quezon City, University of the Philippines Press, 1984.

Carlson, Keith Thor. *The Twisted Road to Freedom.* Metro Manila, University of the Philippines Press, 1996.

Connaughton, Richard. *MacArthur and Defeat in the Philippines.* New York, NY, Overlook Press, 2002.

Connaughton, R., Pimlott, J., and Anderson, D. *The Battle for Manila.* London, Bloomsbury, 1995.

Constantino, L. R. *The Snap Revolution.* Quezon City, Karrel Inc, 1986.

Constantino, R. *History of the Philippines: From the Spanish Colonization to the Second World War*. New York and London, MR Press, 1976.

Constantino, R., and Constantino, L. R. *The Philippines: The Continuing Past*. Quezon City, Foundation for Nationalist Studies, 1979.

George, T. J. S. *Revolt in Mindanao: The Rise of Islam in Philippine Politics*. Oxford University Press, 1980.

Go, Julian, and Foster, Anne. *The American Colonial State in the Philippines*. Durham, NC, Duke University Press, 2003.

Hamilton-Paterson, James. *America's Boy: The Rise and Fall of Ferdinand Marcos and Other Misadventures of US Colonialism in the Philippines*. New York, Henry Holt, 1999.

Jones, Gregg R. *Red Revolution, Inside the Philippine Guerilla Movement*. Boulder, CO, Westview Press, 1989.

Kessler, Richard. *Rebellion and Repression in the Philippines*. New Haven, CT, Yale University Press, 1990.

Lico, Gerard. *Edifice Complex: Power, Myth and Marcos State Architecture*. Honolulu, HI, University of Hawaii Press, 2003.

Linn, Brian MacAllister. *The Philippine War, 1899–1902*. Lawrence, KS, University Press of Kansas, 2002.

Magno, A., Quiros, C., and Ofreneo, R. *The February Revolution*. Quezon City, Karrel Inc, 1986.

May, Anthony. *Battle for Batangas: A Philippine Province at War*. New Haven, CT, Yale University Press, 1991.

McFerson, Hazel M. (Ed.). *Mixed Blessing: the Impact of the American Colonial Experiences on Politics and Society in the Philippines (Contributions in Comparative Colonial Studies)*. Westport, CT, Greenwood Publishing Group, 2002.

Mojares, Resil B. *The War Against the Americans: Resistance and Collaboration in Cebu, 1899–1906*. Metro Manila, University of the Philippines Press, 1999.

Muslim, M. *The Moro Armed Struggle in the Philippines*. Marawi City, Mindanao State University, 1994.

Paredus, R. R. (Ed.). *Philippine Colonial Democracy*. New Haven, CT, Yale University Press, 1988.

Reid, Robert H., and Guerrero, Eileen. *Corazon Aquino and the Brushfire Revolution*. Baton Rouge, LA, Louisiana State University Press, 1996.

Rosenberg, David A. (Ed.). *Marcos and Martial Law in the Philippines*. Ithaca, NY, and London, Cornell University Press, 1979.

Salman, Michael. *The Embarrassment of Slavery: Controversies over Bondage and Nationalism in the American Colonial Philippines*. Berkeley, CA, University of California Press, 2001.

Schirmer, D. B. *Republic or Empire: American Resistance to the Philippine War*. Mass, Schenkman Publishing Co, 1972.

Shaw, Angel V., and Francia, Luis H. (Eds). *Vestiges of War: the Philippine-American War and the Aftermath of an Imperial Dream 1899–1999*. New York, New York University Press, 2002.

Stanley, Peter W. (Ed.). *Reappraising an Empire: New Perspectives on Philippine-American History*. Cambridge, MA, Harvard University Press, 1985.

Timberman, David G. *A Changeless Land: Continuity and Change in Philippine Politics*. Singapore, Institute of Southeast Asian Studies, 1992.

Wilson, Andrew R. *Ambition and Identity: Chinese Merchant Elites in Colonial Manila, 1880–1916*. Honolulu, HI, University of Hawaii Press, 2004.

ECONOMY

Aguilar, Jr, Filomeno V. *Clash of Spirits: The History of Power and Sugar Planter Hegemony on a Visayan Island*. Metro Manila, University of the Philippines Press, 1998.

Balisacan, Arsenio M., and Hill, Hal (Eds). *Philippine Economy: Development, Policies and Challenges*. Oxford, Oxford University Press, 2002.

Bautista, Germelino M. *Natural Resources, Economic Development and the State: The Philippine Experience*. Singapore, Institute of Southeast Asian Studies, 1994.

Boyce, James K. *The Philippines: The Political Economy of Growth and Impoverishment in the Marcos Era*. Metro Manila, University of the Philippines Press, 1993.

Center for Research and Communications. *The Philippines at the Crossroads: Some Visions for the Nation*. Manila, Center for Research and Communications, 1986.

Corpuz, O. D. *An Economic History of the Philippines*. Quezon City, University of the Philippines Press, 1999.

de Dios, Emanuel (Ed.). *An Analysis of the Philippine Economic Crisis: A Workshop Report*. Quezon City, University of the Philippines, 1984.

Dolan, Ronald E. (Ed.). *The Philippines: A Country Study*. Washington, DC, Federal Reserve Division, Library of Congress, 1993.

Eaton, Kent. *Politicians and Economic Reform in New Democracies: Argentina and the Philippines in the 1990s*. Philadelphia, PA, University of Pennsylvania Press, 2002.

Eder, James F. *A Generation Later: Household Strategies and Economic Change in the Rural Philippines*. Metro Manila, University of the Philippines Press, 1999.

Estanislao, Jesus P. *The Philippine Economy: An Emerging Asian Tiger*. Singapore, Institute of Southeast Asian Studies, 1997.

Gonzalez, III, Joaquin, L. *Philippine Labour Migration: Critical Dimensions of Public Policy*. Singapore, Institute of Southeast Asian Studies, 1998.

Hodgkinson, Edith. *The Philippines to 1993: Making Up Lost Ground*. London, Economist Intelligence Unit, 1988.

Hutchcroft, Paul D. *Booty Capitalism: The Politics of Banking in the Philippines*. Ithaca, NY, Cornell University Press, 1999.

International Alert. *Breaking the Links Between Economics and Conflict in Mindanao*. London, 2003.

Jorgensen, Erika, et al. *A Strategy to Fight Poverty: The Philippines*. Washington, DC, World Bank, 1996.

Krinks, P. *The Economy of the Philippines*. London, Routledge-Curzon, 2002.

Mearl, L., et al. *Rice Economy of the Philippines*. Quezon City, University of the Philippines, 1974.

Ofreneo, Rene. *Capitalism in Philippine Agriculture*. Quezon City, Foundation for Nationalist Studies, 1980.

Philippine Center for Investigative Journalism. *Saving the Earth: The Philippine Experience*. Manila, 1992.

Power, J. H., and Sicat, G. P. *The Philippines: Industrialization and Trade Policies*. London, Oxford University Press, 1971.

Van Den Top, Gerhard. *The Social Dynamics of Deforestation in the Philippines: Actions, Options and Motivations*. Copenhagen, Nordic Institute of Asian Studies, 2002.

Villegas, Edberto M. *Studies in Philippine Economy*. Manila, Silangan Publishers, 1983.

POLITICS AND GOVERNMENT

Abinales, P. (Ed.). *The Revolution Falters: The Left in Philippine Politics after 1986*. Ithaca, NY, Cornell University Press, 1996.

Making Mindanao: Cotabato and Davao in the Formation of the Philippine Nation-State. Quezon City, Ateneo de Manila University Press, 2001.

ABS-CBN Broadcasting Corpn. *People Power 2: Lessons and Hopes*.

Alejo, Albert E. *Generating Energies in Mount Apo: Cultural Politics in a Contested Environment*. Honolulu, HI, University of Hawaii Press, 2001.

Bonner, Raymond. *Waltzing with a Dictator: The Marcoses and the Making of American Policy*. New York, Times Books, 1987.

Canoy, Reuben R. *The Counterfeit Revolution: Martial Law in the Philippines*. Manila, 1980.

Catholic Institute for International Relations. *The Philippines: Politics and Military Power*. London, 1992.

Coronel, Sheila S. (Ed.). *Pork and Other Perks*. Manila, Philippine Center for Investigative Journalism, 1998.

EDSA 2: A Nation in Revolt. Manila, AsiaPix/Anvil, 2001.

Investigating Estrada: Millions, Mansions and Mysteries. Manila, Philippine Centre for Investigative Journalism, 2001.

Delmendo, Sharon. *The Star-Entangled Banner: One Hundred Years of America in the Philippines*. Piscataway, NJ, Rutgers University Press, 2004.

Franco, Jennifer C., and Kerkvliet, B. J. T. *Elections and Democratization in the Philippines (Comparative Studies of Democratization)*. New York, NY, Garland Publishing, 2001.

Goodno, James B. *Land of Broken Promises*. London, Zed Press, 1991.

Gutierrez, Eric. *The Ties that Bind: A Guide to Business, Family and Other Interests in the House of Representatives*. Manila, Philippine Center for Investigative Journalism and Institute for Popular Democracy, 1993.

Gutierrez E., Torrente I., and Narca N. *All in the Family: A Study of Elites and Power Relations in the Philippines*. Quezon City, Institute for Popular Democracy, 1992.

Hedman, Eva-Lotta, and Sidel, John T. (Eds). *Philippine Politics and Society in the Twentieth Century: Colonial Legacies, Post-Colonial Trajectories*. London, Routledge, 2000.

Hodder, Rupert. *Between Two Worlds – Society, Politics and Business in the Philippines*. London, Curzon Press Ltd, 2002.

Kerkvliet, B. J. T. *Political Change in the Philippines: Studies of Local Politics Preceding Martial Law.* Honolulu, HI, University of Hawaii Press, 1974.

Everyday Politics in the Philippines: Class and Status Relations in a Central Village. Berkeley, CA, University of California Press, 1991.

Kudeta: The Challenge to Philippine Democracy. Philippine Center for Investigative Journalism and the Photojournalist Guild of the Philippines, 1991.

The Huk Rebellion: a Study of Peasant Revolt in the Philippines. Lanham, MD, Rowman & Littlefield, 2002.

Kerkvliet, B. J. T., and Mojares, R. (Eds). *From Marcos to Aquino: Local Perspectives on Political Transition in the Philippines.* Honolulu, HI, University of Hawaii Press, 1992.

Landé, Carl H. *Leaders, Factions and Parties: The Structure of Philippine Politics.* Yale University Press, Southeast Asia Studies, Monograph Series No. 6.

Post-Marcos Politics: A Geographical and Statistical Analysis of the 1992 Presidential Election. Singapore, Institute of Southeast Asian Studies, 1996.

May, R. J., and Nemenzo, Francisco (Eds). *The Philippines After Marcos.* Beckenham, Kent, Croom Helm, 1985.

Nadeau, Kathleen M. *Liberation Theology in the Philippines: Faith in a Revolution.* Westport, CT, Praeger Publications, 2001.

Pesigan, Guillermo, and MacDonald, Charles J. (Eds). *Old Ties and New Solidarities: Studies on Philippine Communities.* Honolulu, HI, University of Hawaii Press, 2001.

Philippine Center for Investigative Journalism. *People Power Uli! A Scrapbook About EDSA 2.* Manila, 2001.

Philippine Center for Investigative Journalism and Ateneo Center for Social Policy and Public Affairs. *1992 and Beyond: Forces and Issues in the Philippine Elections.* Manila, 1992.

Silliman, G., and Noble, L. *Organizing for Democracy: NGOs, Civil Society and the Philippine State.* Quezon City, Ateneo de Manila University Press, 1998.

Tadiar, Neferti Xina M. *Fantasy Production: Sexual Economies and Other Philippine Consequences for the New World Order.* Hong Kong, Hong Kong University Press, 2003.

Timberman, David G. *The Philippines: New Directions in Domestic Policy and Foreign Relations.* Singapore, Institute of Southeast Asian Studies, 1998.

Tordesillas, Ellen, and Hutchinson, Greg. *Hot Money, Warm Bodies: The Downfall of Philippine President Joseph Estrada.* Manila, Anvil, 2001.

Tyner, James A. *Made in the Philippines.* London, RoutledgeCurzon, 2004.

Vitug, Marites Danguilan, and Gloria, Glenda. *Under the Crescent Moon: Rebellion in Mindanao.* Manila, Ateneo Center for Social Policy and Public Affairs/Institute for Popular Democracy, 2000.

Wurfel, D. *Filipino Politics: Development and Decay.* Quezon City, Ateneo de Manila University Press, 1988.

SINGAPORE

Physical and Social Geography

HARVEY DEMAINE

The Republic of Singapore is an insular territory, with an area of 659.9 sq km (254.8 sq miles), lying to the south of the Malay peninsula, to which it is joined by a causeway, 1.2 km long, carrying a road, a railway and a water pipeline across the intervening Straits of Johor. Singapore Island, which is situated less than 1.5° north of the Equator, occupies a focal position at the turning-point on the shortest sea-route from the Indian Ocean to the South China Sea.

PHYSICAL FEATURES

The mainly granitic core of the island, which rises in a few places to summits of over 100 m, is surrounded by lower land, much of it marshy, though large areas are now intensively cultivated. Singapore City has grown up on the firmer ground adjacent to the Mt Faber ridge, the foreshore of which provides deep water anchorage in the lee of two small offshore islands, Pulau Sentosa and Pulau Brani. In recent years suburban growth has been rapid towards the north and along the eastern foreshore, and since 1961 a large expanse of mangrove swamp to the west of the dock area has been reclaimed to provide industrial estates for the Jurong Town Corporation.

The climate, like that of the Malay peninsula, is hot and humid, with no clearly defined seasons, although February is usually the sunniest month and December often the least sunny. Rainfall averages 2,367 mm annually, and the average daytime temperature is 26.6°C, falling to an average minimum of 23.7°C at night.

POPULATION AND CULTURE

According to the census of 30 June 2000, the population (including non-residents) was 4,017,733. At mid-2003 the population was officially estimated at 4,185,200, giving a population density of 6,342.2 per sq km, one of the highest in the world. Of the total population (excluding non-residents) at the 2000 census, 76.8% were Chinese, 13.9% Malay and 7.9% Indian. In 2000 the birth rate was estimated at 11.3 per 1,000 (compared with 18.4 in 1990) and the death rate at 3.8 per 1,000.

There are four official languages: Chinese (in 1990, as a first language, Mandarin was spoken by 26% of the population, Chinese dialects by 36.7%), English (used by nearly 20% in 1990), Malay and Tamil. Chinese dialects were spoken as a first language by 24% of the population in 2000.

History

C. M. TURNBULL

Revised for this edition by MICHAEL BARR

ORIGINS AND EARLY DEVELOPMENT

The island of Singapore has a record of human habitation going back possibly some 2,000 years, but its early history is obscure, and the very name of Singapura (Sanskrit for 'Lion City') is unexplained, since the lion is not native to the region. The original seaport, Temasek, may have been part of the great Sumatran maritime empire of Srivijaya, which disintegrated in the 13th century. The earliest historical chronicle, the *Malay Annals*, described 14th-century Temasek as a prosperous trading centre. At about this time it became known as Singapura. However, the island's prosperity brought it into the sphere of rival expanding empires: Thai Ayudhya and Javanese Majapahit. Attacked by both and torn further by internal strife, the city was destroyed in the final years of the 14th century. The ruler and his followers fled to found a more auspicious settlement at Melaka (Malacca), which became the centre of a renowned Empire.

Following this violent episode, Singapore remained almost deserted for more than 400 years, home of a few tribes of *orang laut* (boat people), who lived by fishing, piracy and petty trading and owed allegiance to the Malay Riau-Johor Empire, centred nearby in the Riau archipelago. Soon all that remained of the old city were a few crumbling ramparts and the neglected graves of its former rulers on the hill above the Singapore river. Riau itself was a prosperous emporium for regional commerce, but Singapore merely battened on the piratical fringe of that trade.

In the early 19th century the Straits of Melaka and the southern part of the Malay peninsula assumed a new commercial and strategic importance when Britain's East India Co sought bases to protect its China trade and to challenge the Dutch commercial monopoly in the Malay peninsula and archipelago. In 1819 Sir Stamford Raffles, an official of the East India Co, obtained permission, from the Sultan of Johor and the local chief, to establish a trading post at the mouth of the Singapore river, and five years later the two Malay chiefs ceded the island in perpetuity to the East India Co and its successors, in return for money payments and pensions.

The Straits Settlements

In 1826 the East India Co united Singapore with its two other dependencies on the west coast of the Malay peninsula: Pinang (Penang), acquired in 1786, and Melaka, ceded by the Dutch in 1824. These scattered territories remained one political entity, known as the Straits Settlements, for the next 120 years. Under the protection of the East India Co, the new port, conveniently situated and free from customs duties or restrictions, attracted traders and settlers from all over the region. They came first from the nearby ports of Riau, Melaka and Pinang, but soon others began to arrive from further afield in the Indonesian archipelago, and from Thailand, Indo-China, Burma (now Myanmar), Borneo, the Philippines, India, China and Europe.

In 1833 the East India Co lost its monopoly of the China trade, after which it had no further use for the Straits Settlements. It continued to administer them but on a severely constrained budget. The very laxness of government and freedom from taxation attracted further steady immigration and trade, but the European merchants of Singapore became increasingly dissatisfied with administrative inefficiency and the lack of representative institutions. In 1857 Singapore's merchants petitioned that the Straits Settlements be separated from India and brought under direct British rule, and in 1867 the Straits Settlements became a crown colony, with a constitution that remained basically unchanged until the Second World War. A Governor, appointed by the British Government, ruled with the assistance of an Executive and a Legislative Council, the latter comprising a majority of officials with unofficial members nominated by the Governor. Over the years the number of 'unofficials' and of Asian councillors increased, so that by the 1920s there were equal numbers of officials and 'unofficials' on the Legislative Council, with the Governor having the casting vote. By that time some 'unofficials' were nominated by the chambers of commerce but there were no elected members as such.

The transfer of the Straits Settlements to direct colonial rule was soon followed by two events which gave a new impetus to Singapore's development: the opening of the Suez Canal in 1869, and the first treaties of protection made between the British and rulers of the Malay states in 1874. With the opening of the Suez Canal, the India–Melaka Straits route, on which Singapore was situated in a dominating position, became the main highway from Europe to the Far East. The 1874 treaties with the west coast Malay states of Perak, Selangor and Sungei Ujong marked the first step in bringing all the Malay states under British protection. Singapore provided a political focal point as the base of the Governor of the Straits Settlements Colony, who was also High Commissioner of the Malay States and the chief authority for the three British-protected Borneo states of Sarawak, North Borneo and Brunei.

In the late 19th and early 20th century Singapore became the commercial and financial centre for the whole region. Growing Western interests in South-East Asia, the increasing use of steamships (which from the 1880s replaced sailing ships as the main carriers) and the development of telegraphs all put Singapore, with its fine natural sheltered harbour, at the hub of international trade in South-East Asia. It became a vital link in the chain of British ports which stretched from Gibraltar, through the Mediterranean Sea and the Indian Ocean, to the Far East.

The 60 years from the opening of the Suez Canal to the onset of the Great Depression in 1929 were a time of almost unbroken peace, steady economic expansion and population growth in Singapore. A more sophisticated administration gradually brought the population into the pale of the Government and the law courts. In the early years the different immigrant groups lived in specified districts of the town, largely supervised by their own community leaders or *kapitans*. As early as 1827 the Chinese were the biggest single community in Singapore and by the beginning of the 20th century constituted three-quarters of the population, a proportion which has remained fairly constant since that time. Most Chinese came from the troubled Guangdong and Fujian Provinces of southern China, representing a variety of dialect groups, chiefly Fukien, Cantonese, Teochew, Hakka and, in the later years of the century, Hainanese. Most were young adult men who aimed to make money abroad and return to their native China, so that, well into the 20th century, the population was transitory and shifting, with very few women, children or old people. As the years passed more women came to Singapore, and a large number of immigrants settled permanently and raised families. By the early 20th century there was a sizeable Straits-born Asian community, who were British subjects and sometimes English-educated. Some Straits-born Asians collaborated with the colonial establishment as Legislative and Municipal Councillors or Justices of the Peace, while many others worked as clerks and assistants. In the 1930s the colonial regime created a Straits civil service, Straits legal service and Straits medical service which offered the first openings to Asian British subjects in professional government work, but only at subordinate levels. All senior administrative and technical posts in Singapore were held by British European officials.

There was little interest in local politics. From the mid-1920s, a Legislative Councillor, Tan Cheng Lock, began to call for elected representation on the Legislative Council for those who had made the Straits Settlements their home, but he found little backing even among his fellow Straits Chinese. There was no indigenous nationalist movement or desire for independence and any political interest was directed more to China or India.

Immigration reached a peak in the boom year of 1927, when an all-time record of 360,000 Chinese landed in Singapore. However, the Great Depression of 1929–33 hit Singapore hard, leading to the first restrictions. In 1930 an Immigration Restriction Ordinance imposed a quota on Chinese men, but it did not affect women for some years, which encouraged the trend towards permanent family settlement.

THE SECOND WORLD WAR AND JAPANESE OCCUPATION

After the First World War, Singapore acquired a new strategic significance as a naval and military base. However, this did not deter Japan from attacking Malaya in December 1941 as part of a campaign to seize the raw materials it needed from South-East Asia. After a lightning campaign down the Malay peninsula, culminating in a week of fighting on the island itself, Singapore capitulated to the Japanese in February 1942.

Renamed Syonan, or 'Light of the South', it remained under Japanese occupation for more than three years. Japan intended to retain Singapore as a permanent colony and military base, a focal point in its proposed 'Greater East Asia Co-Prosperity Sphere', but wartime priorities and difficulties precluded the Japanese from developing this concept. They destroyed the colonial economy without replacing it by an Asian alternative, and tried to suppress the English language and colonial education without building up Japanese in its place. Japan did nothing to promote indigenous politics in Singapore, although the island for a time became the regional base for the collaborationist Indian National Army and the Indian Independence League. The occupation was a time of misery, hunger and fear, which came to an abrupt end in August 1945, when Japan surrendered after atom bombs were dropped on two of its own cities, thus sparing Singapore the ordeal of an Allied invasion.

THE POST-WAR PERIOD

In September 1945 the British set up a temporary military administration in preparation for a return to colonial government under a new structure. They aimed to unite the peninsula by creating a Malayan Union comprising all the Malay states, together with Pinang and Melaka, but separating Singapore as a crown colony. In April 1946 civil rule was restored and the Malayan Union and Singapore Crown Colony came into being. These decisions were taken because of Singapore's special position as a free port, its importance as a military base and the complications of trying to absorb it into a Malaya embarking on the road to self-government and independence. However, Singapore's separation was not intended to be permanent.

The proposals provoked protest in Singapore, leading in 1945 to the creation of its first political party, the Malayan Democratic Union (MDU), which campaigned for the island's incorporation in the Malayan Union. The party joined other opposition groups in the peninsula to form the All-Malaya Council of Joint Action, which put forward proposals calling for the inclusion of Singapore in a Malayan federation. However, these were swept aside by the intensity of peninsular Malay nationalism, reacting against the Malayan Union, the loss of state identity, and the proposed liberal granting of citizenship to immigrants. As a result, in 1948 a Federation of Malaya replaced the Malayan Union. Singapore remained separated, but this time on racial grounds, since the Malay leaders did not want to upset the ethnic balance of the federation by including a predominantly Chinese Singapore.

Constitutional Development, 1948–65

Meanwhile, Singapore was being prepared for eventual self-government, and in 1948 the first elections were held for six members of the Legislative Council. The MDU, now heavily communist-infiltrated, boycotted the election, and most of the elected seats went to the Singa-pore Progressive Party. This upper-middle-class English-educated group supported British plans for gradual constitutional reform while maintaining the economic status quo. There was little popular enthusiasm for this experiment. Voting was confined to British subjects, registration was voluntary, and the activities of the English-speaking Legislative Council were remote from the majority of the population, with their problems of poverty, unemployment, bad housing and inadequate vernacular education. The outbreak of a communist 'Emergency' in the Federation of Malaya in 1948 led to a period of severe repression of all radical politics in Singapore. While the colony was not directly involved in the revolt, the same emergency regulations were enforced. The Malayan Communist Party was proscribed in 1948 and the MDU disbanded itself.

As the Malayan 'Emergency' was brought under control, the authorities permitted a greater level of political activity in Singapore. In 1955 a new Constitution was granted in an effort to speed up constitutional reform and encourage a sense of real participation and responsibility. The new Legislative Assembly had an elected majority (25 out of 32 members), voters were registered automatically, and the leader of the largest party in the Assembly, as Chief Minister, would form a Council of Ministers responsible to the Assembly.

Two new rival left-of-centre parties were created to fight the election, both headed by lawyers: the Labour Front under David Marshall and the People's Action Party (PAP), led by Lee Kuan Yew. Unexpectedly, these two parties routed the Progressives and other conservatives, and Marshall's Labour Front formed a minority Government. However, in 1955–56 the communist-linked left wing of the PAP made a determined bid for power through the trade union movement and Chinese middle schools. For a time they even wrested control of the PAP Central Executive Committee away from Lee Kuan Yew. The Labour Front Government curbed them by using emergency regulations to imprison their leaders and by taking steps to remove genuine grievances. A 1957 Education Ordinance gave parity to the four main language streams; in the same year citizenship was offered on generous terms to nearly all residents, and a new public services commission was set up to achieve rapid localization of the civil service. In 1958 terms were agreed for full internal self-government, and at elections held in 1959 to implement this, the PAP won an outright majority (43 out of 51 seats). Lee Kuan Yew took office as Prime Minister, and the party was to remain in power into the 21st century.

The new Government committed itself to a programme of rapid industrialization and social reform, with ambitious schemes for education, housing and a far-reaching Women's Charter. It also aspired within its four-year term to achieve full independence through a merger with the Federation of Malaya, which had itself secured independence in 1957. Such a union was seen as vital to provide the military security and free access to a Malayan market, which were considered essential for Singapore's survival. However, the implementation of this policy quickly led to dissension in the party, and to challenges from the extreme left wing, which threatened to tear the party apart and plunge Singapore into chaos.

The conservative, Malay-dominated Government of the Federation of Malaya, which was originally reluctant to draw closer to Singapore because of its large Chinese population and left-wing tendencies, decided it needed to exercise direct control over its unstable neighbour. To do this, in May 1961 the Malayan Prime Minister, Tunku Abdul Rahman, proposed a closer association between the federation and Singapore, by bringing in the three Borneo territories to achieve racial equilibrium. The PAP leadership eagerly took up this idea, but the prospect of merging Singapore in a conservative, anti-communist Malayan federation so alarmed members of the party's radical left wing that in July 1961 they made a bid to topple Lee Kuan Yew's Government. Having failed by a narrow margin at a dramatic all-night session of the Assembly, the rebels then broke away to form an opposition socialist front, the Barisan Sosialis (BS).

Following an intensive, government-sponsored campaign in its favour, the Malaysia Agreement was signed in July 1963, under which Singapore was to join Malaya, Sarawak and Sabah (North Borneo) in forming the Federation of Malaysia on 31 August 1963. The implementation was deferred until mid-September to enable UN representatives to ascertain the wishes of the people of the Borneo states. Singapore took advantage of this delay to declare its own unilateral independence from colonial rule and to call an election. Having survived the 1961 crisis, the moderate element of the PAP had reorganized and strengthened the party, and profited from the success of the Malaysia merger to secure a comfortable victory.

However, the association was strained. 'Confrontation' by Indonesia, which objected to the formation of the Malaysian federation, severely damaged Singapore's trade and led to acts of violence by Indonesian saboteurs. Simultaneously, the central Government and Singapore clashed over what each perceived as undue interference in the other's internal affairs. In July and September 1964 communal riots in Singapore further strained relations with Kuala Lumpur, while the ruling Alliance Party in Malaysia objected to the PAP's contesting the 1964 general elections. Finally, Lee Kuan Yew's attempts to unite all Malaysian opposition parties brought the crisis to a head, and on 9 August 1965, the central Government forced Singapore to agree to a separation.

REPUBLIC OF SINGAPORE

In this way Singapore became independent against the declared wishes of its own leaders, although it later emerged that Prime Minister Lee Kuan Yew and Minister of Finance Goh Keng Swee had ,in fact, negotiated the terms of Singapore's supposed expulsion from the Federation, having already become disillusioned with Malaysia. Singapore then joined the UN in September 1965 and was admitted as a member of the Commonwealth the following month. The new Republic was committed to multi-racial, non-communist, democratic socialist policies, and to co-operation with Malaysia, particularly in economic and defence matters.

Politically, the transition to full independence was smooth. The machinery of government remained almost intact, with small constitutional amendments, notably the appointment of a non-executive President as Head of State. Effective power rested with the Prime Minister and his Cabinet, responsible to the single-chamber Parliament (formerly the Legislative Assembly), which was elected for a five-year term by all adult Singapore citizens. Other adjustments were more painful. Few, if any, had visualized a separate independent Singapore on the grounds that such a tiny state, with a large population and no natural resources, could neither sustain its economy nor defend itself. Initially, the British defence 'umbrella' continued to shelter Malaysia and Singapore, but in 1966 the United Kingdom decided to withdraw its bases from 'east of Suez' during the mid-1970s. Singapore introduced compulsory national service and was working to build up a credible defence force when, less than two years later, the British Government advanced the withdrawal date to 1971. The accelerated time-table threatened Singapore's economy, since the bases accounted for 20% of the country's gross national income (GNI), but the Government took advantage of the situation to call an election in 1968, when it won a mandate to pass far-reaching labour legislation curbing trade union activity, to provide the right climate for foreign investment. As hopes for access to the Malaysian market evaporated, Singapore stepped up its efforts to achieve rapid export-orientated industrialization. To attract foreign and local capital and expertise, the Government rejected doctrinaire socialism in favour of a mixed economy, largely privately owned and managed, but with a sizeable public ownership stake used by the Government to give impetus and direction to industrialization.

With independence suddenly thrust upon it, Singapore needed to create a sense of nationhood. In the early years this was done by stressing the unique qualities and differences that distinguished Singapore from its neighbours, resulting in an abrasive foreign policy. While the Republic was a founder member of the Association of South East Asian Nations (ASEAN), formed in 1967 along with Malaysia, Thailand, Indonesia and the Philippines (and joined in 1984 by Brunei, in 1995 by Viet Nam, in 1997 by Myanmar and Laos, and in 1999 by Cambodia), at first it made little practical contribution to regional co-operation. However, the British withdrawal, the sharp increase in oil prices in 1974 (petroleum refining being Singapore's largest industry), followed by the fall of South Viet Nam to the communists in 1975, induced Singapore to draw closer to its neighbours, and it subsequently led the way in supporting regional solidarity.

The Singapore Government was anxious to maintain cordial relations and to forge economic links with all countries: the West, communist states, developing countries, the Middle East, Japan, Taiwan and the People's Republic of China. Despite developing substantial economic relations with China, however, the Singapore Government waited until 1990 before establishing formal links at ambassadorial level, in order to minimize its image as an ethnic Chinese nation.

Any desire to be readmitted to the Federation of Malaysia disappeared with growing confidence in the country's viability as an independent state and the maturing sense of nationhood. After an initially troubled period, in the early 1970s the relationship between the two countries became friendlier and they maintained close co-operation in matters of mutual concern: combating subversion and illicit trading in narcotic drugs, and protecting the Straits of Melaka.

Independence was followed by eight years of international boom, when Singapore achieved an 'economic miracle' in industrialization. Official policy aimed to improve living standards without creating a welfare state: it encouraged full employment and subsidized education, housing and public health. Social reform was initially accompanied by a strict policy of family limitation and population control, together with stringent immigration curbs. In the 1960s and 1970s a vigorous education programme concentrated on promoting bilingualism in primary and secondary schooling, while developing technical skills appropriate to the Republic's economy. English became increasingly accepted as the language of development and modernization. An energetic programme of urban renewal and the construction of new townships meant that by the end of the century about 90% of households owned their homes, the highest rate of home-ownership in the world.

After the 1961 crisis, the Government pursued a consistently anti-communist policy, using the emergency regulations—a legacy of colonial times—to imprison 'hard-core' subversives. The BS refused to acknowledge Singapore's independence in 1965, and boycotted the Parliament, opting to oppose the Government by extra-constitutional means. After the communist victories in Indo-China in 1975, the extreme left wing regrouped to attempt a resurgence in Singapore and Malaysia, but the Government arrested leading cadres of the Malayan National Liberation Front, the militant satellite of the Communist Party of Malaya.

From 1968 until October 1981 the PAP held every seat in Parliament, and the party leadership maintained a remarkable cohesion, which contributed to the country's stability but tended to stifle criticism. Radio and television services were state-owned. Newspapers, all concentrated under the ownership of one public company, Singapore Press Holdings Ltd, required annual licences and inclined towards self-censorship. This concentration of power and patronage, and the acknowledgement of the considerable achievements and success of PAP policies, emphasized the lack of credible alternatives. Seven opposition parties contested the 1980 parliamentary elections, but the ruling party achieved its widest margin of victory, winning all seats for the fourth successive election and capturing nearly 78% of the votes. In two decades of remarkable economic progress, Singapore had achieved the third highest per caput income in Asia (excluding the Middle East), after Brunei and Japan. The Republic was also acquiring a sense of nationhood; the 1980 census showed that more than 78% of the population were Singapore-born. In the more relaxed climate, numbers of political prisoners were released, leaving only a small band of committed left-wingers in detention.

However, a nation that had begun to take its relatively high standard of living for granted became less tolerant of the more arbitrary aspects of PAP policies. The victory of J. B. Jeyaretnam, the Secretary-General of the opposition Workers' Party, at a parliamentary by-election in 1981 was an unsettling event for the ruling PAP. At the 1984 general election the PAP's share of the total vote declined to below 63% and the party lost a second seat, this time to Chiam See Tong, founder and Secretary-General of the Singapore Democratic Party (SDP). This reverse caused consternation among the PAP's leadership. While two opposition members in Parliament posed no serious threat to the Government (Jeyaretnam subsequently lost his seat in 1986 and was disbarred from Parliament for five years, following his conviction for perjury), and there was still no viable alternative to the PAP, the period of the party's unchallenged dominance was over. After the 1984 election Lee Kuan Yew remained as Prime Minister but delegated daily government administration to younger men, led by Goh Chok Tong, the First Deputy Prime Minister. Lee Kuan Yew's elder son, Brig.-Gen. Lee Hsien Loong, was appointed a junior minister in 1985 and Minister for Trade and Industry in the following year.

An economic recession in 1985–86 prompted the Government to review some of its policies and to adopt new strategies for economic development and social policy. In 1987 a New Population Policy was introduced, since strict population control had resulted in a decline in fertility to below replacement levels. Early marriage was encouraged, as was a return to three-child families, particularly among the more prosperous and better-educated groups. The 1990 census recorded a resident population of 2.7m., and in 1991 an official Concept Plan set a target of 4m. by 2010. The Government aimed to achieve this by raising fertility, encouraging the immigration of talented Asians, and discouraging emigration by making life in Singapore more attractive.

The Government placed even greater emphasis on the need for political stability, in order to attract investors. In 1986 press laws were extended to reduce the circulation of those foreign publications that were considered to be unduly critical of Singapore. These were invoked against several journals, and in 1990, following a dispute between the Government and three foreign publications, which had allegedly interfered in Singapore's politics, the Newspaper and Printing Presses Act was amended: all publications of which the 'contents and editorial policy were determined outside Singapore' and which dealt with politics and current events in South-East Asia would be required to obtain a ministerial licence, renewable annually and limiting the number of copies sold. Punitive damages were awarded against *The International Herald Tribune* and the author of articles that were published in 1994 implying nepotism and the Government's use of a compliant judiciary to suppress opposition by bankrupting rival opposition politicians. This increased caution in reporting on Singapore by journals that circulated locally, although Singapore continued to attract opposition from the US liberal media.

In 1987 the Government alleged that it had discovered a 'Marxist network' which had infiltrated student and church groups and the Workers' Party and arrested a group of alleged activists under the Internal Security Act, which permitted indefinite detention, without trial, of persons suspected by the authorities of subversion. Mainly young, English-speaking university-educated professionals, the detainees differed from political dissidents of former years. In 1988 the Internal Security Act was tightened to put detention beyond the review of the law courts. Despite the most vigorous campaigning for 20 years, Chiam See Tong of the SDP was again the only successful non-PAP candidate in the 1988 general election. The electorate once more responded to the Government's slogan of 'More Good Years', and economic growth resumed its upward movement. With the expansion of population and migration to new towns, constituencies were redelineated and the number of elective parliamentary seats was increased from 58 in 1967 to 83 in 1997. In 1988 MPs were allocated a role as town councillors in managing new towns, and 39 constituencies were reorganized into 13 group representation constituencies. These were contested by teams of three from the same party, including at least one candidate from a minority community. (This was changed in 1991 to a minimum of three and a maximum of four candidates, and in 1996 the maximum was raised to six, in a measure that the Government claimed would facilitate town council operations, but that also made it more difficult for opposition parties to present electoral teams.) In 1990 Parliament approved legislation enabling the Government to appoint up to six unelected MPs for a two-year term. Independent, well-educated people, successful in their own field, these Nominated Members of Parliament (NMPs) were expected to provide constructive non-partisan criticism and to improve the quality of parliamentary debate. They had the right to vote on all legislative proposals except those concerning financial and constitutional affairs.

In 1990 the Republic celebrated its 25th anniversary of independence with the slogan 'one people, one nation, one Singapore'. However, government leaders expressed concern that the popu-lation was too cosmopolitan and individualistic, and in danger of being overwhelmed by foreign cultures. Instead the authorities supported the concept of an ethnic and racial mosaic, respecting individual cultural roots. To the minority communities, however, the formulation of a new national ideology, which stressed core values of Confucian morality, family loyalty and placing society before self, smacked of Chinese cultural chauvinism. It threatened to detract from the PAP's own considerable success over the past quarter of a century in creating a modern, secular, multiracial society. The Malays, in particular, feared relegation to the status of an underclass, and responded with the formation of an Association of Muslim Professionals. In 1990 the Government ended the automatic waiver of tertiary education fees for Malays, although it awarded a S $10m. grant to help impoverished Malay students.

Over the years a 'Speak Mandarin' campaign, which had originally been adopted in 1979, had successfully diverted the Chinese-educated from using regional dialects. By the early 1990s, however, Mandarin itself was threatened by the rapidly increasing use of English, which was promoted as the essential language for modernization, particularly after 1987 (when English became the medium of instruction for all schools). From 1992, therefore, the 'Speak Mandarin' campaign concentrated on encouraging the English-educated to maintain their Mandarin.

Goh Chok Tong's Premiership

In November 1990 Lee Kuan Yew stepped down as Prime Minister in favour of Goh Chok Tong, but remained in the Cabinet as Senior Minister, kept his influential post as PAP Secretary-General, and continued to travel extensively abroad promoting Singapore's interests. Lee Hsien Loong and Ong Teng Cheong, formerly the Second Deputy Prime Minister (and also the Secretary-General of the National Trades Union Congress—NTUC), were appointed Deputy Prime Ministers of supposedly equal rank, but Lee Hsien Loong was to act in the Prime Minister's absence.

Goh offered a more open political system and, at his inauguration, invited 'fellow citizens to join me, to run the next lap together'. An official programme, entitled *The Next Lap*, which was published in 1991, promised to make Singapore 'more prosperous, gracious and interesting over the next 20 to 30 years', as befitted an affluent, well-educated society. Censorship of films, television and magazines was relaxed, the new Prime Minister embarked on an extensive programme of community visits and exhorted the public to participate and express its views. This gained him considerable support among the Malay and Indian minorities, who were traditionally wary of the PAP. The changes, however, represented a difference of style rather than of substance. Goh still stressed the virtues of hard work, meritocracy, sound education, economic growth, the need to avoid the pitfalls of a welfare state and a defence policy based on national resilience. The public was invited to offer constructive criticism, but not to advocate radical change. Restrictions on the foreign press remained in force. Although all political detainees were released from prison by June 1990, Goh refused to repeal the Internal Security Act, and it was only in November 1998 that the final residential restrictions were lifted on the movements of Chia Thye Poh, Singapore's last political prisoner, who had been arrested in 1966. Even then he was prohibited from engaging in politics.

Seeking popular endorsement for his style of government, the Prime Minister called a 'snap' election in August 1991, two years ahead of schedule, arguing that consensus and national unity precluded the need for a formal opposition. The opposition parties, in a plan conceived by Chiam See Tong, decided to contest only 40 out of the 81 seats, conceding victory to the ruling PAP by default, but inviting voters to create a strong opposition. In the event, the PAP's share of the vote dropped to 61.0% (from 63.2% in 1988), and the party won 77 seats, compared with 80 in 1988. Jeyaretnam was unable to contest the election as his disqualification still remained in force, but the Workers' Party won in one constituency while Chiam's SDP secured three seats. During the election campaign the Prime Minister warned there would be a return to a more authoritarian and paternalistic form of government if he failed to receive a popular mandate. However, Goh subsequently announced that he would continue his style of government but would implement reforms more slowly.

In November 1992 the Government announced that both Deputy Prime Ministers were suffering from cancer, although Ong Teng Cheong's case was low grade and Lee Hsien Loong gained complete remission by April 1993. In the mean time Lee Hsien Loong relinquished his office, and the illness of both men gave urgency to holding a by-election, which the Prime Minister had pledged to stage to counter Jeyaretnam's claim to have been deliberately excluded by the early timing of the 1991 general election. In an astute move, Goh held the contest in his own four-member group representation constituency in December 1992; Jeyaratnam was unable to enter the contest, since the Workers' Party did not have the required minimum of four candidates. The PAP secured 72.9% of the vote; this result, combined with the effect of Lee Hsien Loong's indisposition and Goh's unanimous election earlier in the month as Secretary-General of the PAP (replacing Lee Kuan Yew), strengthened the rather fragile personal power of the Prime Minister considerably.

Following many years of public discussion, in 1991 Parliament approved a constitutional change in the presidency. Instead of a ceremonial figurehead appointed by Parliament, the new-style President was chosen directly by the electorate, and was empowered to safeguard the large financial reserves that had been accumulated over the previous 30 years and to veto senior civil service and judicial appointments. Only those who had served as cabinet ministers, chief justice, senior civil servants or the head of a large company would be eligible as presidential candidates.

In 1993 Ong Teng Cheong was elected President for a six-year term, securing 58.7% of the votes cast. The election of Ong, who had been nominated by the NTUC and who enjoyed strong PAP support, was almost a foregone conclusion; however, the only other candidate, a former accountant-general, Chua Kim Yeow, who adopted an apolitical platform, attracted a significant 41.3% of the vote.

In January 1994 Goh reorganized the Cabinet, appointing Lee Hsien Loong to supervise both the trade and industry and the defence portfolios. By mid-1994 senior ministers were once more expressing concern about the difficulty of recruiting able people into public service and, most particularly, into the Cabinet. The consequent decision to increase ministerial salaries to a level comparable with private-sector leaders provoked some protest, especially from the SDP. The ruling party also encouraged junior MPs simultaneously to hold posts as professionals and top business executives, provided there were no conflicts of interest. The PAP subsequently sought to recruit younger parliamentary candidates of a high calibre and with the potential for ministerial office.

When the term of office of the six NMPs expired in 1994, the system was made permanent on the basis of two-year appointments, with the possibility of reappointment. A Maintenance of Parents Bill, introduced by NMP Dr Walter Woon in 1995, offered destitute parents the right to claim maintenance from their children. As a Private Member's Bill, this was open to free discussion and provoked spirited debate both in Parliament and the press. The legislation, which gained widespread support and set up a tribunal in 1996, was designed to address one of Singapore's most serious problems: the ageing of the population and the consequent concern to reduce pressure on health care and other social services by enforcing family responsibility for the older generation. The birth rate continued to decline, despite government measures to encourage marriage and child-bearing, and in 1998 the retirement age was raised from 60 to 62.

Profiting from the booming regional economy, Singapore consolidated its reputation as a leading financial centre. In 1996 the Organisation for Economic Co-operation and Development (OECD) promoted Singapore from the list of developing countries eligible for aid to the rank of 'more advanced developing country'. Goh Chok Tong and Lee Kuan Yew, however, continued to stress the Republic's vulnerability and the need for good government and a disciplined people to sustain the momentum of a dynamic economy.

Many of the benefits of economic success were passed on to the general population in the form of generous tax rebates, ongoing programmes to upgrade Housing and Development Board flats, Edusave Merit bursaries for bright students from low-income families, and contributions to citizens' Medisave accounts. An amended Women's Charter was passed, after much discussion, in 1996, which widened the provisions protecting families in case of violence and divorce. Over the years, however, rising living standards and expectations led to more open debate and dissatisfaction with the 'nanny state'. The Prime Minister and Senior Minister reacted sharply to mildly critical articles that appeared in the local press in late 1994, countering that critics should enter the political arena themselves if they wanted to set a political agenda. Leaders insisted that the elected Government must establish the limits on consultation, which should not extend to challenging those in authority as equals or to damaging respect for the Prime Minister.

Under the slogan 'Singapore 21: make it our best home', the Prime Minister contested the general election of January 1997 on three key issues: his own impressive track record; a solid programme offering political stability, economic growth, high-quality education and rising living standards; and a capable leadership being prepared for the 21st century. On the eve of the election seven opposition parties came to an agreement to avoid splitting the anti-PAP vote in most constituencies, but, despite their promising performance in the 1991 general election, the opposition was in an unstable state. The SDP was divided by personal feuds and became more confrontational under the leadership of Chee Soon Juan, a former university lecturer, who was elected Secretary-General in 1995. Chee's book, *Dare to Change*, which he had published in 1994 as a counterbalance to the PAP's *The Next Lap*, argued for greater democracy and was adopted as the SDP's policy. Articulate and vocal, the new SDP leaders received substantial coverage in the local press, but their strident anti-Government censure was frequently criticized by journalists and many of the public as unpatriotic. Meanwhile, Chiam was expelled from the party and in July 1995 formed the Singapore People's Party, which aimed to revive the original moderate policy of the early SDP. The new party sought to liberalize publishing and the internal security laws and to create a 'caring and civic society in which people could enter politics without fear'.

Unopposed in 47 of the 83 seats in the January 1997 general election, the PAP was assured of being returned to power. Nevertheless, Goh conducted an energetic campaign, emphasizing the successes of his six years in office. He insisted that the principal role of government was to improve the quality of life by providing a stable and secure environment in which citizens could better themselves. At the same time he proposed to give Singaporeans more say in municipal affairs and in determining their living environment by creating geographic communities with mayors and Community Development Councils.

The opposition parties had no specific programmes to match those of the PAP, focusing instead on negative aspects. Immediately prior to the election Senior Minister Lee Kuan Yew and his son, Deputy Prime Minister Lee Hsien Loong, had become implicated in a scandal in which they and other members of their family were revealed to be the beneficiaries of huge financial discounts, worth millions of dollars, on real estate purchases. Their defence—that the discounts were routine and in no way surreptitious—only highlighted the gulf between their lives and those of ordinary Singaporeans, making them easy political targets. Goh's declaration early in the campaign that priority in upgrading housing and community services would be given to wards that showed commitment by voting in favour of the PAP was also attacked by the opposition candidates, and attracted criticism from the US State Department. The Workers' Party, and the SDP in particular, accused the Government of using the upgrading issue as a bribe and of denying people the right to an alternative voice in Parliament. However, in a masterful political stroke, Senior Minister Lee regained the initiative by shifting the focus of the campaign and the media onto the issue of racial and religious harmony. He denounced a particularly effective Workers' Party candidate, Tang Liang Hong, as an anti-Christian Chinese chauvinist, while Tang in turn accused PAP leaders of lying. The PAP secured the constituency, with the Workers' Party winning a creditable 45% of the vote, but, much more significantly, Lee Kuan Yew had re-established his political ascendancy. Prime Minister Goh adopted a more passive style of premiership, allowing Lee and his son to dominate the policy-formation and decision-making processes.

Overall in 1997 the PAP achieved its best result since the 1984 general election, winning all but two seats, increasing its share of the contested votes to 65% (up from 61% in 1991), and regaining two opposition seats. Low Thia Kiang of the Workers' Party retained his seat with a slightly increased vote, Chiam was returned for a fourth term but as leader of the Singapore People's Party with a reduced majority, while Chee and the SDP were completely routed. Jeyaretnam was declared a Non-Constituency MP, under a constitutional amendment that provided for the highest-scoring losers to be admitted to Parliament if there were fewer than three opposition members; Non-Constituency MPs are prohibited from voting on constitutional and financial matters.

The new Cabinet incorporated no important changes in the senior posts but introduced several younger ministers. The Prime Minister appealed for unity following the election, but acrimonious court disputes ensued. The Prime Minister, Senior Minister and other PAP leaders sued Tang for defamation and he subsequently fled to Johor, declaring that he had received threats against his life. In his absence, the court ordered Tang to pay a record S $8m. in damages and costs. The Court of Appeal later substantially reduced the damages but Tang, by this time settled in Australia, refused to pay and was declared bankrupt. The PAP leadership also sued Jeyaretnam, following remarks made at an election rally concerning two police reports submitted by Tang that accused the PAP leadership of criminal conspiracy and of lying. Jeyaretnam insisted that the legal suits were an attempt to bankrupt him and thus disqualify him from Parliament. Observers from the human rights organization Amnesty International and the Geneva-based International Commission of Jurists monitored the trial, and the report subsequently made to the Commission accused the court of being compliant in its procedures. The High Court found against Jeyaretnam, but ordered the payment of only modest damages of S $20,000 (one-10th of the amount claimed by Prime Minister Goh) and 60% of the costs. The Court of Appeal, however, upheld the Prime Minister's appeal for higher damages: Jeyaretnam's damages were multiplied fivefold, to S $100,000, and full costs were awarded against him.

The number of NMPs was increased to nine in 1997, by which time Parliament comprised 93 representatives, of whom six were women and 17 were selected from the minority communities. The Prime Minister urged a change of attitude: instead of expecting every initiative to come from a few political leaders, all Singaporeans should feel responsible for solving local issues and shaping their own communities. Nine Community Development Councils were created to encompass all constituencies, including the two opposition wards; however, opposition MPs could not be chairmen of the councils or have power to disburse funds. Goh also called for adaptability to tackle new challenges. Despite impressive progress in education over the past 30 years, the Republic still suffered from a shortage of graduates. The Government planned to upgrade tertiary institutions, increase local student numbers, admit more foreign students, and urge Singaporeans working abroad to return home. To attract talented foreign settlers, the Government relaxed immigration rules and allocated affordable housing to immigrants. Under an 'Intelligent Island' programme, Singapore aimed to lead the world in connecting all households as well as businesses to a multimedia computer network. The Republic aspired to become 'the Boston of the East'. Leading tertiary institutions in the USA and Europe linked up with the National University of Singapore to run joint degree courses and exchange programmes. The existing universities (the National University of Singapore—NUS—and the Nanyang Technological University) were upgraded, and a third Singapore Management University, which was devoted to business studies, opened in 2000. In May 2002 the NUS announced that it was to establish two new campuses by 2010: one for postgraduate science and engineering, and the second to provide post-graduate medical education. A fifth polytechnic opened in 2003, bringing the total polytechnic student intake to 22,000. After his appointment as Acting Minister for Education in August 2003, Tharman Shanmugaratnam instituted a host of reforms to the education sector, some of which appeared to have weakened Singapore's commitment to Mandarin as the second language of Singapore, and even moderated the Singaporean fixation on exams and scholarships. Senior Minister Lee confirmed in June 2004 that, after 40 years of effort, he had abandoned his goal of trying to make Singapore's citizens bilingual.

In February 1998 the Government banned political parties from making videos and from promoting their views on television, with the reported aim of maintaining the propriety of political debate. At that time 22 opposition parties were registered, of which 13 were dormant and the remainder rarely met. The SDP was devastated by its general election defeat. In early 1999 Chee was convicted twice under the Public Entertainment

Act after deliberately refusing to apply for licences to make two public speeches, in which he criticized government policy; he served two brief spells in prison for refusing to pay the fines.

The Workers' Party was active immediately after the 1997 election, with Jeyaretnam meeting the public at regular weekly sessions, but he ended these when he became involved in legal proceedings. In addition to the damages and costs owed to the PAP leaders, in November 1998 10 members of a Tamil Language Committee won defamatory damages from Jeyaretnam and the Workers' Party for an article published in the party's journal, *The Hammer*. An appeal against the order was rejected, and Jeyaretnam was finally declared bankrupt in January 2001, consequently losing his non-constituency seat and becoming barred from contesting the next general election. In May 2001, at the age of 76, he relinquished the post of Secretary-General of the Workers' Party, which he had held for the past 30 years, and was succeeded by the former Assistant Secretary-General, Low Thia Khiang, MP. Meanwhile, associations began to form outside of the formal political arena. The Association of Women for Action and Research aimed to achieve full equality for women in public and private life and was influential in gaining significant amendments to the Women's Charter concerning domestic violence. The Association of Malay/Muslim Professionals sought to raise the educational, economic and social status of the Malay/Muslim community. 'Roundtable', a discussion group of young professionals, urged a review of the Societies, Internal Security, and Public Entertainments Acts.

In June 1999 President Ong Teng Cheong, whose six-year mandate was due to expire in September, announced that he did not intend to seek a second term in office, citing the difficulties he had experienced as Singapore's first elected President: notably, restrictions placed on the presidential right of veto, official obstruction of his efforts to establish the details of the Republic's financial reserves, and disappointment at not being approached to release the funds needed to help Singapore through the recession. In response, the Government insisted that net investment income lay outside of the President's jurisdiction and that the surplus accumulated during the current Parliament would be sufficient to finance measures to be implemented up to the year 2000 without the need to use reserves. In July 1999 Parliament endorsed a set of rules agreed between Ong and the Prime Minister to ensure a smooth future working relationship: it was decreed that a President could block the sale of land holdings and prohibit unreasonable disposals but could not fetter the day-to-day operations of government; furthermore, the President would be obliged to give warning before blocking any transaction, while the Government in turn undertook regularly to furnish the President with information. Public speculation about the potential impact of the appointment of a new President was quickly dispelled by Senior Minister Lee Kuan Yew, who emphasized that there could be only one centre of executive power and that the powers of the President were largely ceremonial and confined to preventing an irresponsible administration from squandering the reserves or making unsuitable appointments. S. R. Nathan, a former Singaporean ambassador to the USA whose presidential candidature was supported by the Cabinet and welcomed by the NTUC, was chosen as the new President without an election being held, since the two other potential contestants failed to meet the stringent criteria for nomination. Reviewing Singapore's position in 1999, the Prime Minister concluded that although the Republic had prospered since independence, it was still a fragile society with racial divisions. The 'Singapore 21' concept, which was the result of two years' wide-ranging consultation, was embodied in a book, entitled *Singapore 21: Together We Make The Difference*, and debated in a marathon parliamentary session in May 1999. It aspired to be a 'total vision', for the country's future in the next millennium. This went beyond mere material achievement to developing a partnership between the public and private sectors and the citizens of the Republic, with the aim of building—over one or two generations—a multi-racial nation, in which each community would have its own distinctive identity but overlap with other communities, and in which English would be the common language and equal opportunities would be enjoyed by all. The population target of 4m. envisaged in the 1991 Concept Plan was achieved in 1999, 11 years ahead of schedule, principally as the result of an influx of foreign immigrants, and the target was raised to 5m. by the middle of the 21st century.

In celebrating National Day in August 1999 Goh Chok Tong outlined a vision of Singapore as a 'world-class renaissance city', with excellence in education, arts and sport to match a world-class economy, which would attract foreign talent and encourage Singaporeans to remain in the country. President Nathan developed this theme in his inaugural address to Parliament in October and also appealed for constructive disagreement. In 1997 Goh Chok Tong had told Parliament that Singapore needed a civil society 'to harness the talents and energies of its people and to build a cohesive and vibrant nation'. In October 1999 he declared consultation and participation to be among the Government's main objectives, encouraging citizens to turn the republic into 'a nation of ideas'. In September 2000 a 'Speakers' Corner' was designated, which would be open seven days a week for any Singapore citizen to register and speak freely, provided they did not endanger law and order, respected racial and religious sensitivities, and avoided libel. Opposition politicians were wary, and an initial flurry of public interest soon petered out. The PAP leaders maintained their stance that politics was the domain of elected politicians, since they alone had a popular mandate and were accountable to the electorate. Goh Chok Tong had insisted that 'meaningful politics must mean joining a political party', and Lee Kuan Yew repeated a warning about would-be critics crossing 'out of bounds markers'. The Prime Minister insisted that the press should inform and educate but was not entitled to set the national agenda. In May 2000 a Political Donations Bill was passed, which aimed to keep Singapore politics free of foreign interference. Political parties were required to keep lists of donors, accept donations only from Singaporean and local companies, and to declare any single donation exceeding S $10,000. Associations promoting purely social causes would not be affected, but any organization would be gazetted as political if its activities related wholly or mainly to politics in Singapore or if it pressed for changes in the political or legal structure. While agreeing that foreigners should be kept out of Singapore politics, all three opposition MPs registered spirited resistance to the Bill, arguing that it would place a heavy burden on small parties to keep detailed records and would deter donors, who were often reluctant to make their support for opposition parties known.

While political stability, a sound economic infrastructure and massive foreign-exchange reserves placed Singapore in a better position than its neighbours to withstand the economic crisis that beset the region from mid-1997, nevertheless, since the economy was driven primarily by external demand, the crisis posed the most serious challenge faced by Goh's Government since it first took office. Growth decelerated markedly in 1998, tax revenue declined for the first time in a decade and key manufacturing and commercial sectors showed negative growth. The Housing and Development Board upgrading programme, which was financed out of budget surpluses, was scaled down. Both the Prime Minister and the leader of the NTUC stressed giving priority to full employment by adjusting wages in order to curb costs and to save jobs. By November 1998 Singapore was formally in recession, and the Government introduced far-reaching measures to reduce business costs, which included wage reductions and the halving of employers' Central Provident Fund (CPF) contributions from 20% to 10%. At the same time, the Government reduced corporate and property taxes, stamp duty, vehicle taxes, utility charges, Housing and Development Board rents and mortgage interest rates in order to lower costs and to ease the burden on individuals. However, it resisted pleas from the business community for more positive intervention, insisting that its priority was to continue heavily to invest in education, training programmes, defence and the economic infrastructure, in order to strengthen Singapore's future prospects of profiting from the eventual upturn in the regional economy.

By May 1999 Singapore was technically no longer in recession, and the economy experienced greater growth than expected for the rest of the year and throughout 2000. This prompted the National Wages Council to recommend wage increases, bonus payments and the restoration, at a more rapid rate then the proposed five-year period, of the 10% CPF reduction. Some 2% of the CPF decrease was restored in April 2000, and a further 4% in January 2001. In August 2000 Goh Chok Tong announced plans to revive the programme to upgrade Housing Development Board (HDB) flats.

Singapore celebrated National Day in August 2000 in an optimistic mood, but the Prime Minister warned of 'grave problems', notably falling fertility, emigration, and a widening gap between rich and poor. Despite tax incentives introduced in the mid-1980s to encourage the educated classes to have more children, the fertility rate had declined sharply to less than 1.5 and was still falling, and already one-quarter of the population were non-Singaporeans. In August 2000 further incentives were announced in the form of bonus payments and improved maternity leave for those parents producing a second and third child. To compound the fertility problem, competition from other countries to recruit talent in a global economy was encouraging many Singaporeans to emigrate, thus raising salaries among the most highly-paid echelons. This was a particular threat to the public service and, in July 2000, in order to curb rising resignations, the Government announced a radical overhaul of civil service rewards. All ministers and senior officials were given substantial salary increases and enhanced bonuses, linked partly to personal performance and partly to the national economy. Lifelong tenure was to be replaced by fixed 10-year appointments, and outstanding young officials could expect accelerated promotion. Opposition MPs objected to the large increases, urging that public service demanded some sacrifice, but the press gave cautious approval.

While the upsurge in the economy created many new jobs for professionals and technicians, unemployment continued to increase among the unskilled, who had suffered most during the 1997–98 crisis. The gap between those earning 'First World' salaries and 'Third World' wages was widening, and in the midst of affluence, the number of Singaporeans living in poverty was growing perceptibly and uncomfortably. The Government acknowledged that it was impossible to resist the economic transformation created by globalization, and that high achievers must be paid their international worth, but also that the State should help the poor and the old, while simultaneously creating jobs for the younger generation. While rejecting the Workers' Party demand for widespread welfare measures, which it believed would encourage dependency, in August 2000 the Government announced a grants programme, providing for, in particular, help for the elderly, the poor and families with young children, and including CPF and medical insurance 'top-up' sums and HDB mortgage discounts. Furthermore, there was to be continued emphasis on areas such as education, retraining and upgrading skills.

While a very high proportion of Singaporeans were educated to upper secondary and tertiary level, the Government remained concerned at the low academic standards and high drop-out rates at some private schools, notably in *madrasahs* (Islamic religious schools). A minimum level of schooling had never hitherto been enforced, for fear of offending the sensitivities of the Malay/Muslim community, and a proposal made in October 1999 to introduce six years' minimum compulsory education by 2003 and require all schools to achieve specified standards provoked heated exchanges between the Association of Muslim Professionals (AMP) and the Singapore authorities. The Association called for an end to discrimination, particularly in promotions to top military posts, and proposed a system of 'collective leadership' of 'independent-non-political' Malay leaders, as a counterweight to the 11 PAP Malay MPs. The Prime Minister insisted that there was no intention of abolishing *madrasahs*, provided they achieved an acceptable standard, and, addressing a large gathering of Muslims at the second National Convention organised by the AMP in November 2000, he objected vigorously to their parallel leadership suggestion, warning this would lead to race-based politics. In the following month the AMP abandoned its controversial proposal, a meeting with Lee Kuan Yew in March 2001 eased the tension, and Malay MPs drew up a blueprint to ensure the progress of the Malay community.

Confidence in the recovery of the economy was shaken in March 2001 when the manufacturing sector contracted for the first time in two years, and the situation worsened with the deceleration in the US economy. By July 20,000 workers faced retrenchment, and in August exports fell to a record low. The terrorist attacks on New York and Washington, DC, on 11 September 2001 destroyed immediate prospects of global recovery, and by the end of the year Singapore was suffering from the deepest recession since independence.

Appalled by the September 2001 attacks, the Singapore Government immediately pledged its support for the US effort to create a world-wide alliance to counter terrorism, urging in particular that Muslim and developing countries should be included in the coalition. The Republic undertook to co-operate on intelligence, eliminate terrorist networks and monitor financial transactions, but both the Government and the press insisted that the military front was not the most important, and that the 'war on terror' could only be won by fighting both the causes and symptoms, and, in particular, by resolving the Israeli–Palestinian conflict.

In order to gain a strong mandate and convince foreign and local investors that the PAP had the electorate's confidence to address the economic crisis and terrorist threat, the Government decided to bring forward the date of the general election (which was required to be held no later than August 2002). The election was scheduled for 3 November 2001, allowing only the minimum legal nine-day notice period, prompting protests from the opposition. After lengthy research the PAP had prepared a new electoral team, comprising prospective cabinet ministers from the senior ranks of the private sector, together with well-qualified younger candidates, including a number of women and representatives of ethnic minorities. In July 2001 three opposition parties, the Singapore People's Party, the National Solidarity Party and the Singapore Malay National Organization, had formed a Singapore Democratic Alliance (SDA), under the leadership of Chiam See Tong, but both Low Thia Khiang's Workers' Party and Chee Soon Juan's SDP refused to join the new coalition. A Workers' Party team, which planned to contest a Group Representation Constituency, was disqualified by its failure to submit the correct nomination papers, and at the elections the PAP was unopposed in 55 of the 84 elective seats and consequently returned to power with the largest number of outright victories since the 1968 election. Nevertheless, the party contested the election with customary vigour, campaigning on its good record, promising to overcome recession and to have a strong new leadership in place by the next election in five years' time. Unlike previous parliamentary contests, the campaign was not marred by racial or religious discord, but inaccurate allegations about a proposed loan to Indonesia, directed at Goh Chok Tong and Lee Kuan Yew by Chee Soon Juan, caused controversy.

Despite Goh's warning that the recession might continue for up to two years, with unemployment rising further and retrenchment being inevitable, the PAP won an overwhelming 73.3% of votes cast in the poll, the biggest margin since the 1980 election. This represented a 10% transfer of the vote since the 1997 election, and the party emerged with 82 of the elective 84 seats. Low Thia Khiang and Chiam See Tong retained their seats, but with smaller majorities. Chiam, who had represented his constituency for 17 years, was now the longest ever serving opposition member, but his majority fell to 52.4%. Low won for the third time, but his proportion of the vote had declined from 58% to 55%. As the losing candidate with the highest number of votes, Steve Chia of the SDA became a Non-Constituency MP. Although the SDP was overwhelmingly defeated, Chee Soon Juan declared that he would not relinquish the leadership of the party. However, the Prime Minister and Lee Kuan Yew refused to accept Chee's apology for the false allegations made during the general election campaign. Following protracted legal proceedings, the damages awarded against him were upheld on appeal in April 2003.

Goh announced that he would remain in office throughout the period of recession, but would resign prior to the next general election in favour of a younger team. Immediately after the election, the Prime Minister appointed his new Cabinet. The Deputy Prime Minister, Lee Hsien Loong, assumed control of the Ministry of Finance, and Tony Tan Keng Kam continued as Second Deputy Prime Minister and Minister of Defence. Five first-term MPs were appointed Ministers of State; all of them were young and from the private sector, rather than government officials or military officers as in the past, but no women were appointed to senior office.

In November 2001, implementing a concept first proposed by Goh five years earlier, Singapore was divided into five districts, each with a mayor (who would become an MP with the same rank as a Minister of State), and managed by their own Community Development Councils. The aim was to improve community spirit and to allow local government to administer social services, such as welfare, health and sports.

The Government embarked immediately upon addressing the nation's economic problems, giving greatest priority to combating unemployment, which in December 2001 reached its highest level for 15 years. 'New Singapore shares' were issued to all citizens, with a greater proportion being allocated to the poor. An Economic Review Committee was appointed, under the chairmanship of Lee Hsien Loong, with seven sub-committees chaired by prominent politicians and businessmen. The Committee's remit was to recommend long-term economic restructuring to meet the changes in the global economy, notably China's increasing influence and the transfer in emphasis to North-East Asia, which threatened to marginalize the South-East Asian markets. In April 2002 the committee recommended a major revision of taxation and employment benefits, in order to liberalize the labour market and create jobs. It proposed the reduction of direct corporate and personal taxes, including income tax, in favour of increasing indirect general sales taxes. These recommendations were implemented in the budget in May 2002, but with the stipulation that, in order to lessen the impact on the poor of transfer to indirect taxation, at least five years' financial aid would be required, such as in rebates on public housing service charges. Addressing the issue of company structure, the Economic Review Committee recommended that, in principle, the Government should remain involved only in strategic areas, such as the Port of Singapore, and withdraw gradually and methodically from enterprises that the private sector could manage. Regarding the CPF, in July 2002 the Committee proposed the imposition of a permanent 'freeze' on employers' contributions at the level of 16% for the age group of over 50 years, in order to keep jobs competitive for older workers, a measure that was accepted with reservations by the trade unions.

A parallel 'Remaking Singapore Committee' was established in February 2002, comprising younger citizens under the chairmanship of the Minister of State for National Development. Continuing the research of the Singapore 21 Committee, this body was to make a comprehensive review of political, social and cultural affairs, including the arts and censorship. In March 2002 the PAP invited its MPs to speak and vote according to their personal views, except on matters of critical national importance, such as the budget, the Constitution or security. For the first time the expertise of NMPs appointed in 2002 was extended to include the areas of community activities, the media and sports, and academia, including the President of the 'Roundtable' political discussion group, and five women. The greater relaxation in party control resulted in some unusually vigorous criticism in Parliament in July 2002 over the Government's decision to increase public transport fares, but official reaction, while conciliatory in tone, was not influenced into modifying the policy. Most of the recommendations of the Remaking Singapore Committee were finally accepted in April 2004, although there was a general sense of disappointment that those rejected were the ones that could have liberalized Singapore's political system.

The atmosphere in Singapore following the 11 September 2001 attacks on the USA remained calm. Muslim leaders condemned terrorism, and there was no harassment of Muslims by other communities. Nevertheless, the authorities feared that the peace was fragile, that Singapore itself was vulnerable to attack and that the international terrorist network al-Qa'ida, believed to be responsible for the attacks, had links with neighbouring Indonesia, Malaysia and the Philippines. A new national security secretariat was created, and in December 15 alleged terrorists were arrested, of whom 13 received two-year sentences of imprisonment under the Internal Security Act. Of these, all but one were middle-class Singapore citizens, while none of them had been educated at *madrasahs* or involved in mosque activities or Muslim community organizations, and six had served as full-time national servicemen. They were believed to be members of Jemaah Islamiah, a secret network of at least three active groups, part of a larger network in Malaysia and Indonesia, which had been established in 1995 with the aim of creating a Java-based Islamic state, incorporating Indonesia, Malaysia, Singapore, Brunei and the southern Philippines. The group was alleged to be linked to al-Qa'ida, and some Jemaah Islamiah members were trained in Afghanistan and led by Indonesian militants. The arrests foiled Jemaah Islamiah's long-planned action against targets in Singapore, including the hijacking of an aeroplane to crash into Changi Airport, attacks against US service personnel and US naval vessels, suicide bombing of US, British, Israeli and Australian diplomatic missions, and a simultaneous offensive against US embassies in Singapore, Kuala Lumpur and Jakarta.

Singapore Muslim leaders expressed strong support for the government action against terrorist suspects, and there were no signs of an adverse reaction against the Muslim community. Goh, in his Chinese New Year message in February 2002, emphasized that security was a more long-standing problem than the current economic crisis and depended on racial harmony. The Government sought even greater national cohesion, urging compliance with legislation such as that banning Malay girls in primary schools from wearing the head covering (*tudung*), which provoked protests from some parents. In January 2002 new 'grassroots' community groups, known as 'Inter-racial Confidence Circles' were established in each constituency, and Goh, together with Lee Hsien Loong and other senior ministers, organized a confidential meeting of some 1,700 community leaders to discuss the impact of the arrests of Jemaah Islamiah suspects.

During the following months further arrests were made in the Philippines and Oman of individuals suspected of involvement in the alleged conspiracy to attack Singapore, and in September 2002 there was a second wave of arrests in the Republic. A total of 21 Singaporeans were taken into custody, of whom 18 were detained for two years and three were placed under restriction orders. All 'blue-collar' workers, these were junior operatives, instructed by Singaporean and Malaysian leaders under direction from Indonesia. The arrests disrupted the Jemaah Islamiah network in Singapore, but the regional organization remained a threat. Singapore was also frustrated by the refusal of the Indonesian Government to extradite Indonesian cleric Abu Bakar Bashir, allegedly the spiritual leader of the regional Jemaah Islamiah network. In the wake of the September arrests, the Minister in charge of Muslim Affairs, Dr Yaacob Ibrahim, pleaded for citizens, Muslim and non-Muslim alike, to unite against terrorism, and in October Goh proposed a draft Code on Religious Harmony at a meeting with community leaders. In January 2003 Parliament held a three-day debate over a government 'White Paper' detailing Jemaah Islamiah and its international links, which concluded that, while Singapore could not eliminate terrorism, it could police its ideological spread, strengthen social cohesion and religious harmony, and encourage the Muslim community to curb extremism in religious education. MPs, including the three opposition party members, gave unanimous support to the Government's anti-terrorist measures, which aimed to transform Singapore into a difficult target to attack through a combination of high-profile security at vulnerable points, such as the airport, and discreet intelligence gathering.

Meanwhile, the Republic continued to suffer its longest and deepest recession since independence. Overall, the economy contracted in 2001, and hopes for a return to pre-recession levels of growth in 2002 were tempered by factors outside Singapore's control, notably concern over the state of the economies of the USA, Japan and the European Union (EU). In February 2003, for the first time in its history, the Port of Singapore Authority (PSA) made a large part of its work-force redundant, and civil service salaries were reduced further, with senior officials and ministers suffering the greatest decreases. The economy in general, and tourism in particular, suffered from the repercussions of the terrorist attack on Bali in October 2002 (see the chapter on Indonesia) and, in March 2003, from the commencement of the US-led campaign to oust the regime of Saddam Hussein in Iraq and a serious epidemic of Severe Acute Respiratory Syndrome (SARS). In mid-2003 Singapore Airlines imposed dramatic pay cuts and retrenchment. The immediate campaign in Iraq was brief, and public co-operation with the stringent measures imposed by the Government quickly contained the SARS epidemic. Although SARS was believed to have been eliminated by the end of August 2003, it re-emerged almost immediately in the form of a lone case originating in a supposedly sterile SARS laboratory. While most sectors of Singapore's economy had recovered from the repercussions of SARS before the end of the year, the threat that it, or another epidemic, could occur in Singapore again remained a new and influential factor in national planning. Although Singapore was unaffected by a regional outbreak of avian influenza in early 2004, the epidemic provided a timely reminder of this new threat. By July 2003 consumer confidence was returning and output from manufacturing rising, but Prime Minister Goh warned in his National Day message in August that Singapore could expect growth in gross domestic product (GDP) of less than 1% in 2003. However, the 2004 budget bore testament to Singapore's ongoing economic recovery. Minister of Finance Lee Hsien Loong announced tax reductions and increased investment in services, although he was still unable to promise a budgetary surplus, thus putting the budget in deficit for the third year out of the last four. He also failed to assuage ongoing popular discontent concerning increases in the cost of living brought about by rises in government charges, persistent high levels of unemployment (5.3% in 2004), and the large numbers of skilled foreign workers commonly believed to be taking jobs in the country at the expense of Singaporean citizens.

Repeatedly stressing that long-term prosperity meant the acceptance of severe measures to restructure the economy, reduce wages and promote productivity, the Government continued to negotiate bilateral Free Trade Agreements (FTAs), similar to those already concluded with Japan and New Zealand. In February 2003 Singapore signed an FTA with Australia, and in August 2003, after two years of difficult negotiations, all obstacles to the conclusion of a highly valued FTA with the USA had been overcome. The Free Trade Agreement (FTA) with Australia came into force in July 2003, and—much more significantly—the FTA with the USA was ratified by the US Congress in August. An FTA with Jordan was signed in May 2004, before the focus of negotiations moved onto the Republic of Korea. It was partly in response to the demands of these FTAs that the Government began increasing bureaucratic transparency and decreasing the extent of government ownership of the myriad of government-linked companies in the country, as well as initiating other 'liberalizing' measures, such as opening up the banking sector to greater international competition. At the same time Singapore sought to redress its overdependence on Western economies by increasing its trade with China, Taiwan and Hong Kong; in 2002, for the first time, its trade with this area exceeded that with the USA. The PSA faced competition at home from the new Malaysian port of Tanjung Pelepas, founded in 1999 on the Johor Strait. Two major customers—the Danish Maersk Line and Taiwan's Evergreen—transferred their operations to Tanjung Pelepas, but the PSA expanded its activities and by September 2002 had invested in projects in eight countries in Europe and Asia.

The ascendancy of a new, more activist leadership in the Air Line Pilots Association of Singapore (ALPA-S) in 2003 promised a more difficult round of negotiations with Singapore Airlines in 2004, but intervention by Senior Minister Lee resulted in the new leadership being subdued and one of its leading figures, Capt. Ryan Goh, being expelled from Singapore in early 2004. The revocation of Capt. Goh's Permanent Residency after having lived with his family in the country for 18 years became a topic for discussion among Singapore's citizens and the expatriate community.

After the general election of late 2001, the opposition parties took little active role in politics, supporting the Government's stance on anti-terrorism measures and acquiescing in general to its programme for the revival of the economy. In November 2002 Chiam See Tong gave up his law practice in order to concentrate fully on his parliamentary ward. He urged opposition politicians to acknowledge the failure of individual parties and the need to combine in order to produce an effective two-party system, and invited the Workers' Party and SDP to join the SDA. However, this proposal did not attract Low Thia Khiang's Workers' Party, celebrating its 45th anniversary that month, although the party continued to criticize the political system as being flawed, lacking checks and balances, and ignoring social justice in its quest for economic growth. With no MPs in Parliament, the SDP sank into further disarray when its leaders, Chee Soon Juan and Gandhi Ambalam, were arrested and fined heavily for attempting to stage a Labour Day rally without permission in the grounds of the Istana (the official residence of the President) in May 2002.

In June 2004 independent civil society was restricted even further when the Societies Act was amended in order to control more stringently the registration of societies. Since 1999 discretionary decisions concerning registration had rested with the directly elected President of Singapore. The amendment awarded this discretion to a government minister. This incremental change confirmed the pattern of restrictions on civil society that had taken place under Goh's premiership.

At the beginning of 2002 seven new ministers were awarded junior cabinet portfolios by Deputy Prime Minister Lee. Of these, three—Dr Ng Eng Hen, Khaw Boon Wan and Tharman Shanmugaratnam—were promoted to senior portfolios in August 2003, albeit initially in an 'acting' capacity. These three ministers were clearly identified as 'rising stars' in the forthcoming Government of Lee Hsien Loong. Their promotion constituted part of a broader cabinet reorganization announced in late April 2003, in which Deputy Prime Minister and former Minister of Defence Dr Tony Tan Keng Yam was appointed to the newly created position of Co-ordinating Minister of Security and Defence in the Prime Minister's Office (effective from August 2003). The acting Minister of Information, Communications and the Arts, David Lim Tik En, resigned and was replaced by Dr Lee Boon Yang, the former Minister of Manpower, who was succeeded by Dr Ng Eng Hen. Meanwhile, Khaw Boon Wan was appointed to be acting Minister of Health, replacing Lim Hng Kiang, who became a Minister in the Prime Minister's Office. The defence portfolio was allocated to the former Minister of Education, Teo Chee Hean, who was to be succeeded in an acting capacity by Tharman Shanmugaratnam.

The Accession of Lee Hsien Loong to the Premiership

In mid-2004 the succession of Lee Hsien Loong to the premiership was confirmed through a complex process whereby Goh's official endorsement of Lee as his successor was ritually confirmed by a meeting of PAP MPs and the PAP Central Executive Committee. This exercise seemed superfluous to most observers, but was intended by Goh as a step towards institutionalizing—perhaps normalizing—the selection of the Prime Minister and, to some extent, beginning the process of depersonalizing political power.

Prime Minister Lee's first Cabinet formally assumed office on 12 August 2004. The new Government was distinguished by the retention of former Prime Ministers Goh Chok Tong and Lee Kuan Yew as Senior Minister and 'Minister Mentor', respectively. Dr Tony Tan Keng Yam was retained as a Deputy Prime Minister, to be replaced by Minister of Home Affairs Wong Kan Seng in mid-2005 following the former's retirement. Prof. Shanmugam Jayakumar was also appointed a Deputy Prime Minister, becoming Singapore's first citizen of Indian origin to hold the post. Lee retained the finance portfolio, but appointed Goh Chok Tong as his replacement as Chairman of the Monetary Authority of Singapore (MAS). The ascendancy of the three 'rising stars' was confirmed, with all three being appointed full ministers. Tharman Shanmugaratnam was also promoted to the deputy chairmanship of the MAS.

The safety standards of Singapore's construction industry were placed under close scrutiny following two major incidents in April and May 2004. The collapse of the Nicoll Highway that resulted from construction work being carried out on a new railway line was particularly serious, leading to the temporary suspension of all construction work on the line. Public hearings into the episode began in mid-2004.

FOREIGN RELATIONS

Singapore's dealings with Malaysia have always been dominated by the uneasy proximity of the two countries and their mutual dependency. With this exception, the sole objective of Singapore's foreign policy has always been to develop cordial international relationships, the only significant variation being whether issues of security or profit were the primary motivation. From 2001 the increased terrorist threat gave security concerns fresh prominence. The threat gave even greater impetus to Singapore's aim to establish good relations with neighbouring countries and to encourage peace and co-operation in economic affairs and anti-terrorist operations. Regional co-operation to combat terrorism developed slowly. However, in September Malaysia arrested a key leader of a militant organization linked to Jemaah Islamiah in Singapore, and, as well as passing stringent anti-terrorism legislation, the Indonesian Government arrested Abu Bakar Bashir on charges of treason after Jemaah Islamiah was suspected of responsibility for the Bali attacks in October 2002.

During the 1990s the Republic had focused more on considerations of profit within the context of its foreign policy. It consistently supported the promotion of the proposed ASEAN Free Trade Area (AFTA), which had been agreed at the summit held in Singapore in 1992, and the fostering of regional security. The Republic welcomed the entry of Viet Nam, Myanmar, Laos and Cambodia as members of ASEAN in the 1990s. Singapore supported ASEAN's policy of 'constructive engagement' and gentle persuasion to reform Myanmar's autocratic military regime, in contrast to economic sanctions recommended by the EU. Singapore was anxious to restore the influence of ASEAN, which had failed collectively to address the 1997 regional economic crisis, while relations between its member states had suffered as a result of increased pressure, dissipating the confidence of overseas investors as a consequence. At meetings of ASEAN Singapore was foremost in securing substantial measures to increase regional trade and attract capital back to South-East Asia. At the ASEAN summit in 2000 the Republic pressed for further economic integration and persuaded ASEAN to undertake a feasibility study on proposals to link the regional economies to those of China, Japan and the Republic of Korea. The Asian economic crisis strained relations between Singapore and its immediate neighbours but, at the National Day rally in August 1998, Prime Minister Goh emphasized Singapore's dependence upon Indonesia and Malaysia. Goh also took the initiative in promoting the development of the Singapore-Johor-Riau 'Growth Triangle'. While this would benefit both Malaysia and Indonesia, it was particularly crucial to Singapore, offering the Republic's economy the opportunity to compensate for the geographical constraints of land and labour shortages.

In January 1999 Singapore signed a long-term trade agreement with Indonesia for the supply of natural gas from West Natuna to Singapore. The Singapore authorities co-operated closely with the Government of President Suharto on economic enterprises. Political relations were sometimes strained with his successors, President B. J. Habibie, who took office following the resignation of Suharto in May 1998, and Abdurrahman Wahid, who became President in 1999. However, Senior Minister Lee Kuan Yew accepted Indonesia's invitation to join an international advisory council to help accelerate the country's recovery, Singapore took the initiative in encouraging investment in Indonesia, and President Wahid and Goh launched the West Natuna gas pipeline in January 2001. Nevertheless, the Singapore Government welcomed the end of Wahid's 'wayward presidency' in July 2001, followed by the swift and peaceful transition to power of the new Indonesian President, Megawati Sukarnoputri, with her appeal for unity. In February 2002 Singapore proposed the extension of the FTA being negotiated with the USA to Indonesia, to support its manufacturing base and attract investment. Relations were strained for a time after Lee Kuan Yew publicly reprimanded the Indonesian Government for allegedly allowing terrorists to remain at large, provoking Indonesian protests of interference, but the Singapore Government welcomed the more stringent anti-terrorist laws passed by Indonesia after the Bali bombing, followed by the immediate arrest of cleric Abu Bakar Bashir in October 2002. In December 2002 Prime Minister Goh became the first foreign leader to visit Indonesia after the Bali attack; the two leaders pledged to work together to combat terrorism and to revive the region's economy. This co-operation bore fruit in February 2003, when Indonesia arrested and imprisoned Mas Selamat Kastari, an Indonesian-born Singaporean, who was believed to be the former leader of Jemaah Islamiah in Singapore. However, the terrorist bomb attack on the Marriott Hotel in central Jakarta in August 2003 negatively affected market confidence in Singapore. In 2004 Singapore began pushing for US and regional co-operation in patrolling the Melaka Straits in order to guard against the potentially disastrous effects of a hijacking or terrorist attack in this narrow stretch of water, through which passes one-third of the world's global trade and one-half of the world's oil supply. Despite US reluctance, Indonesia, Malaysia and Singapore began co-ordinating their naval patrols in July of that year.

Despite their common desire to confront terrorism, tensions continued to exist between Singapore and Malaysia. The two countries had grown closer in the mid-1990s, agreeing to settle all their differences amicably and to work together to promote investment in emerging Asian economies. In 1995 the two countries launched a bilateral defence forum, signed a Defence Co-operation Pact, and agreed on the permanent boundary of their territorial waters after 15 years of negotiations. By mid-1996 Singapore was well on the way to becoming the top investor in Malaysia, but indiscreet remarks by Lee Kuan Yew at that time provoked a ferocious response from the Malaysian press and the United Malays National Organization (UMNO), the dominant element in the Malaysian ruling coalition. After several meetings between Goh Chok Tong and the Malaysian Prime Minister, Dr Mahathir Mohamad, by early 1998 it appeared that Singapore and Malaysia had put aside their differences. The two Prime Ministers discussed ways to help restore economic confidence in the region, and agreed to co-operate in the areas of banking and finance and to encourage private investment in each other's economies. Mahathir announced that bilateral problems could be solved, given flexibility. However, friction escalated over conflicting interpretations of the Points of Agreement which had been signed between the two countries in 1990 covering the status and development of Malayan Railway land in Singapore, under which land at Tanjong Pagar, Kranji and Woodlands would be developed jointly by the two countries. In accordance with a long-standing arrangement, Singapore prepared to transfer railway customs, immigration and quarantine facilities from Tanjong Pagar to the frontier at Woodlands in August 1998; however, Malaysia refused to do likewise and ended talks with Singapore abruptly at the end of July. In September Malaysia imposed restrictions on the use of its airspace by Singaporean military aircraft, complained that Malaysian ports were losing trade to Singapore, and withdrew from a Five Power Defence Arrangement exercise, citing tensions with Singapore as well as economic difficulties. The publication of the first volume of Lee Kuan Yew's memoirs in September 1998 revived old contentions and provoked protests from the Malaysian Prime Minister and other Malay leaders in Kuala Lumpur. However, Singapore politicians continued to emphasize the need for good relations with Malaysia and the importance of maintaining racial harmony in Singapore. In November 1998 Goh visited Malaysia at the invitation of Mahathir, and at the ASEAN summit meeting in Hanoi, Viet Nam, in the following month the two Prime Ministers agreed upon an arrangement to settle all outstanding issues. Officials met several times over the next few months, but discussions were abandoned in May 1999. Contact was resumed unexpectedly by Senior Minister Lee Kuan Yew, who paid a four-day informal visit to Malaysia in August 2000 for the first time in 10 years. After receiving a very cordial welcome, including meetings with Ma-hathir and other ministers, Lee declared that, given compromise on both sides, all disputes could be settled within a few months. Lee's second volume of memoirs, published in September 2000, was warmly received in Malaysia, and the two Prime Ministers had discussions at the time of the ASEAN summit meeting in Singapore in November 2000. A joint naval exercise was held that month, and in February 2001 the Malaysian Deputy Prime Minister brought a large delegation for informal discussions, at which a compromise over the railway station was agreed in principle, and ideas were mooted for a new bridge to replace the causeway and an undersea rail tunnel. Meeting in Kuala Lumpur in August 2001, the Malaysian Prime Minister and Senior Minister Lee Kuan Yew agreed a compromise in principle on all outstanding issues, leaving the detailed arrangements to be worked out by officials.

The negotiations were, however, fraught with difficulties. The main issue of contention was the price of water supplied by Malaysia under two agreements, made in 1961 and 1962, which were guaranteed under the 1965 Separation Agreement and were due to remain in force until 2011 and 2061, respectively. Malaysia had undertaken to supply raw water at a cost of 3 sen (Malaysian cents) per 1,000 gallons and to receive treated water back from Singapore. While provision was included in the agreements to vary the price after 25 years, Malaysia had not proposed this at the relevant times in 1986 and 1987, but first raised complaints about the price early in 2001. In September of that year Singapore offered to pay 45 sen for the period between 2011 and 2061 (although not legally required to do so), and Malaysia pledged to continue the supply after 2061 at an increased price, which would be reviewed every five years. Despite a conciliatory visit by Lee Hsien Loong to Kuala Lumpur, Malaysia's national press and politicians continued what Singapore's Minister of Foreign Affairs described as 'psychological warfare'. Complaints about Singapore profiting from cheap water were followed for several months from February 2002 by vociferous objections to Singapore's reclamation projects at Tuas and Pulau Tekong in the Johor Strait. Singapore insisted that the reclamation would cause no damage and was entirely within the Republic's territorial waters. The two Governments also came into conflict over the nature of a possible replacement for the causeway.

Singapore aimed to resolve nearly all outstanding issues in one settlement, which would include water, the location of Customs, Immigration and Quarantine facilities in Singapore, the redevelopment of Malaysian railway land in Singapore, the early release of CPF money to West Malaysian contributors, the use of Malaysian airspace and the construction of a bridge to replace the causeway. However, at the first meeting of officials, held in July 2002, Malaysia argued that the water issue should be settled first; in October Prime Minister Mahathir notified Singapore that Kuala Lumpur had decided to abandon the idea of resolving all outstanding issues at the same time. Later that month officials met in Johor Baru to discuss the water issue, but no progress was made; both sides then aired their grievances in public. Although the two countries had agreed in 1998 to refer the issue of sovereignty over Pedra Branca (a rocky outcrop off the eastern Johor coast on which Singapore had constructed and maintained a lighthouse since the mid-19th century) to international arbitration, in December 2002 Mahathir accused Singapore of continuing to develop Pedra Branca. Meanwhile, Singapore criticized Malaysia for delaying the signing of the Special Agreement concerning referral of the issue, which had already been concluded, but which needed formal signature before it could be submitted to the International Court of Justice (ICJ) in the Hague, the Netherlands. In January 2003 Singapore's Minister of Foreign Affairs, Prof. Shanmugam Jayakumar, presented to Parliament the published details of all the correspondence between the two Governments, calling it 'the Republic's most significant statement on bilateral relations since independence'. Jayakumar stressed that the water dispute was not an issue of price but one of national sovereignty, since the Water Agreements were included in the Constitution, registered with the UN, and could not be changed unilaterally by either party. Kuala Lumpur objected to the release of the documents. Goh Chok Tong insisted throughout that relations with Malaysia were not beyond repair, and the Agreement referring the Pedra Branca issue to the ICJ was signed in February 2003, but the dispute continued. Meanwhile, Singapore sought to reduce its dependence on Malaysia by making the Republic self-sufficient in water after 2061, if necessary. In December 2001 approval had been given for the construction of Singapore's first desalination plant, which would produce the equivalent of 10% of the Republic's current consumption by 2005. Two new fresh water reservoirs were to be constructed and all reservoirs linked by 2009. Two plants producing recycled 'new water', which could be produced at one-half of the cost of treating sea water, began to supply industrial plants in 2003, and further plants were scheduled to become operational by 2011. Indeed, 'new water' was promoted as safe drinking water, not just for industrial use, and thousands of bottles were distributed on National Day in August 2002. Relations with Malaysia continued to be strained in 2003, but eased with the succession of Abdullah Ahmad Badawi to the Malaysian premiership. Singapore was clearly delighted to see Abdullah's UMNO-led coalition returned to power with an increased majority, at the expense of the Muslim 'fundamentalist' Parti Islam se Malaysia (PAS), at the legislative elections held in Malaysia in March 2004.

Singapore's relations with other members of ASEAN remained harmonious. Full diplomatic relations were restored between Singapore and the Philippines in January 1996, following an acrimonious dispute in the previous year, which had caused the deepest rift between ASEAN partners for more than a quarter of a century when Singapore insisted in March 1995 on hanging a Filipino woman found guilty of murder, despite pleas from President Fidel Ramos for a stay of execution. When President Joseph Estrada of the Philippines visited Singapore in October 1998 a few months after his election, the Republic welcomed his pledge to retain an open economy and pro-market policies. At the ASEAN summit meeting (held in Hanoi) in December, the two countries signed a Philippine-Singapore Action Plan to develop closer bilateral co-operation in trade.

Goh Chok Tong welcomed the efforts of the Thai Prime Minister, Thaksin Shinawatra, to revive the Thai economy, and in February 2002 proposed an extension of their existing co-operation in the form of the Singapore-Thailand Enhanced Economic Relationship (Steer). In August 2003 the Republic commended the arrest in Bangkok of Hambali, the alleged operational head of Jemaah Islamiah. At the annual ASEAN summit meeting held in Phnom-Penh, Cambodia, in November 2002 Goh Chok Tong, with the strong support of Thailand, proposed a single ASEAN market with free movement of goods and services and no tariffs.

Singapore continued to maintain friendly relations with the People's Republic of China. By 2001 earlier problems over the progress of the Sino-Singaporean Suzhou joint venture, from which Singapore had contemplated withdrawing, had been overcome, and Singapore continued to encourage investment and commercial links. Lee Kuan Yew travelled frequently to China, always receiving a warm welcome. In April 2000 Goh Chok Tong paid an official visit, reaffirming co-operation with the People's Republic, and in 2001 Singapore established institutions for closer business links, following China's admission to the World Trade Organization (WTO). In April 2002 the Chinese Vice-President, Hu Jintao (who assumed the leadership of the Chinese Communist Party in 2002 and succeeded Jiang Zemin as President in 2003), made an official visit to Singapore. He and Goh agreed to establish a high-level joint council to investigate potential areas of practical co-operation, such as an early start to negotiations for a China-ASEAN FTA, and Beijing also offered Singapore military training facilities on Hainan Island as an alternative to Taiwan. In 2003 and 2004 the Singaporean Government intensified its attention towards China, regarding it as being crucial to Singapore's prosperity as well as its 'main challenger'. Clear indications of the focus on Sino-Singaporean relations included: ministerial visits; the opening of a replica of Singapore's Raffles City in Shanghai; a new, positive focus on development of the Suzhou Industrial Estate; and plans to begin talks concerning an FTA. However, the seemingly ineluctable progress of the drive towards ever closer bilateral relations was almost derailed in July 2004 by Deputy Prime Minister Lee Hsien Loong's misjudged 'private' visit to Taiwan. The visit provoked unexpectedly strong protests from the Chinese Government, which even threatened to disrupt the scheduled beginning of FTA negotiations between the two countries.

Anxious to avoid conflict in the region, Singapore hoped that Taiwan would reach an agreement with China, with a view to reunification. Singapore had hosted the first Taiwan-China talks in 1993, but a second round of discussions, proposed for 1999, was abandoned, when the then Taiwanese President, Lee Teng-hui, declared that relations with China were 'state-to-state'. The election of President Chen Shui-bian, who represented a Taiwanese pro-independence party, prompted escalating tension between China and Taiwan in mid-2000. In October 2001 Lee Kuan Yew made a four-day visit to Taiwan (his first in six years) on his own initiative, and not at the invitation of the Taiwan Government. He was received with respect, although his views differed from those of President Chen. Lee Kuan Yew argued that Taiwan should accept the 'one China' reunification principle on the terms proposed by the Chinese Government, rather than after losing its economic prosperity to mainland China. Consequently, Singapore was critical of President Chen's proposal in August 2002 that a referendum should be held on declaring independence.

The maintenance of good relations with the USA continued to be of crucial importance to Singapore. The Republic had distanced itself from the US Government's hostile stance towards China during the Clinton Administration in 1994–2000, and during 1993–96 the Clinton Government relegated the Republic to *persona non grata* status, following the sentence of caning imposed on a young US national, who had been convicted for vandalism. However, the measures taken by the Singaporean Government to mitigate the Republic's domestic financial problems following the onset of the regional economic crisis in 1997 won praise in the US press. Goh Chok Tong was the first South-East Asian leader to meet President George W. Bush in Washington in June 2001, when he urged a permanent US interest in South-East Asia. In recognition of Singapore's strong support after the 11 September attacks, the US Government designated Singapore 'a friendly foreign country', with privileged access to technology and eligible to take part in co-operative programmes in research, development and defence. In October 2002 a new Singapore Congressional Caucus was established in the US House of Representatives, with the objective of strengthening security and economic links. The Republic warmly welcomed the Joint Declaration for Co-operation to Combat International Terrorism agreed between ASEAN and the USA and signed by the US Secretary of State, Colin Powell, in Brunei in July 2002. The agreement, reached at meetings of ASEAN ministers responsible for foreign affairs and their 13 dialogue partners, provided for sharing intelligence, blocking terrorist funds, and tightening border security, without increasing the number of US troops in the region or creating new military bases. While giving the final keynote speech at the World Economic Forum's East Asia Economic Summit, held in Kuala Lumpur in October 2002, Goh Chok Tong called for the closer economic integration of ASEAN. He urged the formation of an East Asian community, potentially similar to the EU, with a common currency and single market, in order to combat terrorism and to keep abreast of global economic change.

One of Prime Minister Goh's last foreign policy ventures before his retirement was a tour of the Middle East in 2004, whereby he hoped to progress towards opening up a new area of commercial activity for the Republic. Goh visited Egypt, Jordan and Bahrain in February of that year, before visiting Iran in July.

Economy

GAVIN PEEBLES

Revised by PETER WILSON

INTRODUCTION

Despite its small size, with a population of about 4m. people, the city-state of Singapore plays an important part in the world economy. Illustrating the nature of international fragmentation in production, amongst other aspects of globalization, Singapore presents some of the most extreme examples of economic features, has been one of the fastest growing economies in the world since the 1960s and provides evidence for different sides of the various arguments about the correct foundations for economic development. To some Singapore is an epitome of economic freedom, usually being ranked the second most economically free economy in the world after Hong Kong; able to transform itself from colonial neglect and the status of a fishing village to first world status in four decades through discipline, not democracy. To others it is an Orwellian nightmare of social and economic control by a single, all-intrusive political party that has been in power for more than 40 years, placed restrictions on many aspects of life and repressed wages in order to benefit foreign capital; its high growth rates reflect the result of massive levels of investment and not technological progress. Even institutions that rank it as economically free can say that the ruling political party controls all aspects of social and economic life.

Another distinguishing feature of this small, seemingly very vulnerable economy, is that during the Asian financial crisis of the late 1990s it suffered less than its larger neighbours. Growth stopped in 1998, but real growth rates of 6.9% and 9.7% (at revised 1995 prices) were achieved in 1999 and 2000, respectively, and unemployment did not rise significantly. In 1998 there was mild deflation, with the consumer price index (CPI) falling by 0.3%, and in the following two years the inflation rates were zero and 1.3%. However, in 2001, when the economy contracted by 1.9%, gross domestic product (GDP) at current prices declined to S $154,078m. and Singapore experienced its worst recession since independence in 1965. The economy appeared to have recovered in 2002, recording reasonable growth of 2.2%, with the main impetus coming from external demand, especially electronics exports, but momentum slowed in 2003 and there were further retrenchments of workers as the seasonally adjusted unemployment rate rose to 4.5% in March, compared with 4.2% in December 2002. External demand remained quite positive, particularly for electronics and chemicals, but domestic private investment, including in the construction sector, was weak, as was consumer spending. Whilst uncertainty over the consequences of the US-led invasion of Iraq had begun to recede by the end of May, together with concerns about oil prices, the external outlook was weakened further by the effects of the outbreak of Severe Acute Respiratory Syndrome (SARS) in March, which disrupted economic activity in many of Singapore's Asian trading partners, such as the People's Republic of China, Hong Kong and Taiwan. In Singapore itself the disease seriously affected the tourism industry and transport-related activities as foreign visitor arrivals declined sharply and domestic residents remained indoors. Following further downward revisions later in the year, the final figure for GDP growth in 2003 was a modest 1.1%. Economic growth recovered sharply in the first two quarters of 2004 compared with the corresponding periods of the previous year, reaching 7.5% and 12.5%, respectively, driven by external demand as the global recovery became more sustained. The rate of unemployment remained quite high, but stabilized at around 4.5%. Despite significant rises in oil prices in the third quarter, an expected slowdown in electronics demand and the risk of economic overexpansion in China, both official and private full-year growth forecasts were revised upwards, to between 8% and 9%. Consumer price inflation was also expected to increase, to between 1.5% and 2%, in 2004 and 2005.

Singapore has had positive net factor income flows since the mid-1980s. By 2000 gross national income (GNI) per caput was US $24,740; in terms of purchasing-power parity, this put it in the leading 10 countries in the world. Singapore has what is probably the highest gross national saving rate (equivalent to one-half of GNI), one of the lowest levels of personal consumption, a high investment rate and a very large current-account surplus. The typical recent profile of expenditures on GDP (using 2000 for illustration) is: private consumption 42%; government consumption 12%; investment 28%; and net exports of goods and services 25%. This is a very different profile from economies of the same level of per caput GNI, where private consumption is usually a larger proportion, as is government consumption, and there is less investment, slower growth and a smaller export surplus.

The dependence on exports, especially electronics, and on the USA as a major export market and investor, has always been regarded as one of the vulnerabilities of the economy. This was highlighted in 2001 when reductions in investment in the USA decreased the demand for electronic products from many Asian economies, and, as feared, Singapore sank into recession, with a contraction of 1.9% in real GDP. In an attempt to pre-empt this, in 2000 the Government had announced a 10-year plan to create a 'New Singapore', intended to form a new social contract whereby the lower paid would receive more help than usual.

Singapore both receives large inflows of direct foreign investment and is increasingly investing abroad, but has a policy of aiming for balance-of-payments surpluses and a stable trade-weighted exchange rate. Recent years have witnessed continuing balance-of-payments surpluses but at a much lower rate than previously. This contributes to the increasing official foreign reserves held by Singapore, which, on a per caput basis, are among the highest in the world. Singapore has had a low rate of inflation, and emphasis has been placed on trade. In the 1970s (using 1995 as the base year) the average annual real GDP growth rate was 9.3%, in the 1980s 7.5%, and in the 1990s 7.6%. Despite rapid growth and rising standards of living, some in the Government do not yet feel that Singapore is a developed country owing to its reliance on exports, especially of electronic products to the USA, and to its dependence on foreign capital and labour. Furthermore, they have realized that formerly successful policies of government ownership, control and protection of certain sectors are no longer viable and there has been a paradigm shift in the Government's approach to managing the economy, especially in the financial sector and in trade agreements.

DEVELOPMENT STRATEGY AND INSTITUTIONS

Interpretations of this remarkable record of growth tend to elicit comparisons between Singapore and Hong Kong. The similarities between the two city-sized economies in terms of their dimension, entrepôt role and free-trade policies, maintenance of certain British institutions and majority Chinese populations have led some commentators to assume that there must be resemblances in the nature of their economic policies. Both are sometimes hailed as examples of dynamic Asian capitalism, achieving rapid growth through free-market policies and free trade. Although both of these economies are characterized by a very high level of economic freedom, they are not at all similar in the nature of their development strategies.

Singapore is a 'planned' economy in the sense that the mobilization, co-ordination, improvement and allocation of resources are subject to strong guidance from the public sector, which also decides which sectors should be subsidized. Furthermore, a significant part of Singapore's output has been produced by public-sector bodies such as the statutory boards and hundreds of government-linked companies (GLCs). GLCs have benefited from access to cheaper bank loans, as bankers believed that they would always be supported by the Government.

The principal planning body in Singapore is the Economic Development Board (EDB), a statutory board under the Ministry of Trade and Industry (MTI), established in August 1961. The EDB's main aims became, and remain, the attraction of foreign investment into Singapore, the fostering of the development of local enterprises and, from the mid-1990s, the promotion of outflows of investment from Singapore. Singapore's industrialization drive and growth have been based on attracting world-class, foreign multinational corporations (MNCs). To address the problem of unemployment in the late 1960s, labour-intensive investments were initially encouraged. The first significant foreign investors were in the electronics industry, which has become the core of Singapore's manufacturing sector. Singapore adopted a welcoming policy towards MNCs, granting favourable tax status, allowing complete foreign ownership and imposing no restrictions on the employment of foreign workers. The EDB became a 'one stop' agency at which foreign investors could receive all necessary help and co-ordination with other agencies, enabling them to start production within a very short period of time.

The EDB's plan, Manufacturing 2000 (M2000), was to ensure that manufacturing continued to account for a minimum of 25% of GDP and more than 20% of employment. Industry 21 (I21), launched in January 1999, focused more on knowledge-intensive activities and reflected the Government's intention to build a knowledge-based economy (KBE). The strategic plan known as Manpower 21 was to co-ordinate the development of the labour force in order to meet the output target set by the EDB. The continued emphasis on manufacturing contained in I21 probably reflected the view that a large part of the labour force was not yet suitable for employment in the service sector and that a manufacturing sector that exported to a diverse range of countries was more stable than a service sector that served the regional economies only. This plan also included the aim of increasing the share of manufacturing services (such as research and development, product design and testing and marketing) from 3% to 6% of GDP by 2010. Although companies were encouraged to relocate some of their activities in neighbouring countries, the Government seemed unwilling to allow basic manufacturing activity to contract to the extent seen in Hong Kong, for example.

The EDB managed to maintain a high level of investment commitment by foreign firms despite the uncertainties caused by the Asian financial crisis of the late 1990s. In 2003 56% of investment commitments were in the electronics sector and 28% in petroleum and chemical products, a sector the Government has been developing. Historically, as much as 80% of investment in manufacturing each year came from outside Singapore. In 2003 the USA provided 32% of net investment commitments in manufacturing, the European Union (EU) 30%, Japan 18% and Singapore 17%. Total investment commitments in manufacturing reached S $7,511m. in 2003. The areas that the EDB has stressed are wafer-fabrication plants and the petrochemicals sector, alongside a desire to create knowledge-based jobs. Another initiative that the EDB has co-ordinated is the building of a fourth pillar of the economy through the development of the life sciences sector, which draws on the pharmaceuticals sector, medicine, medical equipment design and manufacture, health promotion and research in such areas as molecular biology and biotechnology. The EDB co-invests with MNCs and local firms to develop targeted sectors in Singapore and to encourage outward investment. In 1995 the EDB established the Promising Local Enterprises (PLEs) programme, which initially identified 100 local companies for support in order to increase their annual sales to S $100m. each by the year 2005. By 2000 52 PLEs had achieved revenue of more than S $100m. Tax incentives, financial assistance, innovation grants, the arrangement of strategic partnerships and the development of business contacts for the PLEs are the methods used by the EDB. In 2000 there were about 300 PLEs in both the manufacturing and service sectors that had benefited from the programme. In that year 13 PLEs were listed on the stock exchange.

Another government initiative, the Technopreneurship 21 (T21) programme, was announced in April 1999. The four aspects of this programme were: to develop a pro-enterprise environment; to develop suitable physical infrastructure; to develop a venture investment infrastructure; and to develop the education system to encourage entrepreneurship, initiative and risk taking. These policies can be seen as moves away from a development strategy based on public enterprises managed by civil servants to a more entrepreneurial one aimed at responding to the needs of the knowledge-based economy. However, much of the funding for such programmes will come from the public sector such as the Technopreneurship fund (which will also have a private-sector contribution) and the use of government funds managed by the Government of Singapore Investment Corporation Special Investments, a branch of the Government's wholly owned company, the Government of Singapore Investment Corporation Pte Ltd (GIC), which manages its reserves.

A very important aspect of Singapore's planning system has been the Government's policy towards land ownership and allocation. Various pieces of legislation, especially the Land Acquisition Act of 1966, have allowed the Government to acquire land from the private sector. Together with the transfer of land to the Government from the British forces and land reclamation, this active procurement policy increased the State's ownership of land from 44% of total land area in 1960 to about 85% by 2000. This has had a huge impact on the nature of Singapore's development. Sales of leases on state land have produced large revenues for the state budget, and the Government has been able quickly to acquire and allocate land to its chosen projects. The Urban Redevelopment Authority (URA) is the statutory board that plans and co-ordinates the development and sale of land. In June 2001 a new statutory board, the Singapore Land

Authority (SLA), was established by merging four land associated departments.

Many important sectors have been dominated by public enterprises. Electricity, water and piped gas were provided by the Public Utilities Board (PUB), a statutory board which was corporatized in October 1995. The PUB remains the water authority and is the regulator of the divested suppliers of electricity and gas. The crucial port facilities have been administered by the Port of Singapore Authority (PSA), another statutory board, corporatized in October 1997. Apart from some 50 statutory boards, the public sector consists of hundreds of government-linked companies. There are four major government-owned holding companies under the Ministry of Finance: Temasek Holdings Pte Ltd, the most important, Singapore Technologies Holdings Pte Ltd, MND Holdings Pte Ltd and Health Corporation of Singapore Pte Ltd. They own various stakes in hundreds of GLCs. Other crucial and distinctive institutions in Singapore's development include the National Wages Council (NWC), established in 1972, which brings together the Government, employers and trade unions in negotiations and advises the Government on wages policies. It makes annual recommendations on wage increases and bonus payments that, while not binding, are (when accepted by the Government) used as guidelines for negotiations over pay increases. Since the late 1980s Singapore has had a flexible wage system that pays variable components and bonuses depending on the performance of the economy (based on the growth rate of GDP) or the individual company. In the late 1990s about 16% of the typical annual salary in the private sector consisted of a variable component, and for civil servants it was at least 20%; this had risen from the overall average of 11% in 1987. The NWC was instrumental in introducing the 'high wage policy' of the late 1970s and early 1980s, which attempted to encourage companies to move to higher value-added and capital- and skill-intensive activities, and which also led to the move to cut wages in the recessionary environment of the late 1990s. In November 2001 the NWC began to advise companies under economic pressure to 'freeze' salaries. In May 2003, partly in response to the SARS epidemic, the Government announced reductions in ministerial and civil service salaries of up to 10% and accepted the NWC proposals for wage decreases for firms affected by the SARS epidemic, wage restraint for firms facing difficult industry conditions and a wage 'freeze' for most other companies affected by the general economic uncertainty. The implementation of wage reforms to structure pay around performance and market conditions, rather than employee seniority, was also to be accelerated.

Another important institution is the National Trades Union Congress (NTUC), virtually a branch of the Government, which has a government minister as its Secretary-General. At the end of 2003 there were 68 employees' trade unions, with 417,166 members, and three employers' unions, with 2,052 members. Incorporating almost all of the trade-union membership, this organization relays government policy to the unions and provides training and recreation facilities for its members. It operates the largest taxi company, a major insurance company, a chain of supermarkets (on co-operative principles) and also a radio station. This relationship ensures smooth industrial relations and an almost complete absence of industrial disputes. It is said that since 1978 there have been no strikes except for a two-day action in 1986. For five consecutive years Business Environment Risk Intelligence (Beri) ranked the Singaporean labour force as the best in the world based on four criteria: legal framework, relative productivity, worker attitude and technical skills. Singapore was judged the second best destination after Switzerland for foreign investment. Foreign investors consistently state that the nature of labour relations is the most important factor in their investment decision, ahead of tax and other incentives. In 2003 252 industrial disputes were referred for conciliation, compared with 260 in 2002. The vast majority of disputes, most of them involving wages or conditions of service, were duly resolved, with only 38 disputes referred for arbitration.

The Central Provident Fund (CPF) plays a crucial role in financing Singapore's development and in public housing policy. Established under the colonial administration in July 1955, this statutory board was intended to provide retirement benefits for civil servants, financed from contributions from both employees and employers. The system was subsequently extended to all employees, and the self-employed can opt in. Contributions are paid into three accounts, which are owned by the member, accumulate and are available for withdrawal, subject to maintaining a minimum sum, at the age of 55 or at death or on permanent disability or on leaving Singapore. Over the years, the Government has used this institution in areas of social policy by allowing members to use their funds for various approved purposes. Members have been allowed to use their funds to buy and upgrade flats built by public housing authorities, to buy private residential properties and land, for investment in approved shares and unit trusts and gold. CPF funds can also be used to finance the education of members' children at local tertiary institutions.

Furthermore, the rates of contribution to CPF funds were increased steadily, thus improving Singapore's very high saving rate. By 1994 the long-term goal of having equal contribution rates of 20% from both employer and employee had been achieved. The measures introduced by the Government in 1998 to reduce costs included an amendment of these rates, however, and the Government subsequently implied that the rates might not be restored to their previous levels. It has been estimated that by the mid-1980s about 64% of gross national saving was done by the public sector, 30% was through the CPF scheme and the remainder was voluntary saving. Discussion of the adequacy of individual private saving for retirement is a sensitive issue, and debate is not facilitated by the lack of good data on households' saving patterns and resources. Ministers have frequently stated that current savings rates are insufficient for people to finance their retirements and that the CPF scheme is inadequate and should be reformed to include private-sector participation. The returns the CPF pays on CPF funds are based on short-term interest rates and have been quite low in real terms over the last decade or so. Between 1965 and 1999, with the exception of 1993 when withdrawals of CPF funds exceeded contributions owing to the demand for SingTel shares, net contributions were positive. In 1998 they reached S $2,390m. (only 54% of the amount recorded in 1997), and in 1999 net contributions declined to only S $38m., a remarkably low figure that resulted from a 20% decrease in the amount of contributions, owing to the reduction in employers' contributions announced in 1998. Net contributions were negative in 2000 and 2001, at S $463m. and S $538m. respectively. In 2001 the amount due to members was S $92,221m. To reduce wage costs, employers' CPF rates were halved from 1 January 1999 to 10% of salary, increased by two percentage points from 1 April 2000 and by a further four percentage points from 1 January 2001. In November 2002 the Government announced that further restorations would be deferred.

In July 2002 the Deputy Prime Minister and Minister of Finance, Lee Hsien Loong, announced CPF reforms, including a reduction to the aggregate contribution rate for workers aged 50–55 years, from 36% to 32%, as part of efforts to encourage employers to retain older workers. The Government also limited CPF withdrawals for home loans to 150% of the property value, to be reduced to 120% over five years. Finally, the Government was to grant financial institutions priority over the pension board in mortgage default claims. In May 2003 further reforms were announced based on the recommendations of the Economic Review Committee (ERC). These included: a two-year deferment of the restoration of the employers' contribution rate to 20%; a reduction in the salary 'ceiling' for private-sector CPF contributions; and a further reduction in employee contributions for workers aged between 50 and 55 years, to 16% by January 2005.

Regionalization

During the early 1990s there was increased emphasis by the Government on encouraging companies to invest elsewhere in the region. The implementation of more market- and trade-orientated policies by such former centrally planned economies as China, Viet Nam, Laos and Cambodia has created opportunities for such investment. Both the private sector and GLCs, with support from such agencies as the EDB, are encouraged to take advantage of these opportunities. The EDB is keen to enter joint ventures with companies, including GLCs and MNCs, for this purpose. This sector of the Singapore economy is called the 'external wing'. The EDB also arranges training schemes for managers under its Initiatives in New Technology (Rationalization) programme. In 2002 the stock of Singapore's direct equity investment abroad amounted to S $148,252m.

Emphasis is being placed upon assistance to local firms to encourage them to venture outside Asia to such places as Mexico and Central and Eastern Europe, because of their proximity to the USA and the EU respectively. The main sectors were financial services and manufacturing, but Singapore is also developing tourist sites in the region. In 2002 approximately half of Singapore's foreign direct investment abroad was in Asia, with one-quarter of this accounted for by China. Other key regional destinations were Malaysia (17%), Hong Kong (17%) and Indonesia (12%). These investment links will boost trade, just as Japan's investment in Singapore contributed to commercial growth between those two countries. In terms of earnings from these investments, it is estimated that Singapore's net factor income from abroad constituted 11% of GDP in 2003. Some of these investments abroad have followed the Singaporean development pattern of establishing industrial parks by public-sector agencies in joint ventures and making them available for Singaporean and foreign firms. The Jurong Town Corporation (JTC) has been involved in ventures to develop a number of industrial townships in such places as China, Viet Nam, Thailand, India and Indonesia. However, there were reports that Singapore's major 7,000-ha China-Singapore Suzhou Industrial Park, not far from Shanghai, was not developing in a satisfactory manner. In 2001 the Singapore Government reduced its stake from 65% to 35%, thereby transferring ownership of the project to China. The measure followed difficulties arising from differences in business practices and expectations, as well as the existence of competition in the form of an earlier established park. Initiatives in 2003 included: 'Network Indonesia', to encourage small and medium-sized industries to invest in Indonesia; and proposals for an India-Singapore Fund to invest in opportunities created by the India-Singapore Comprehensive Economic Cooperation Agreement launched in April 2003.

LABOUR FORCE

Compared with 2000, when there was a net gain of 108,500 new workers, Singapore's labour market suffered in 2001 and 2002. According to the Ministry of Manpower, by June 2002 the labour force had reached 2.13m., of a total population of 4.2m. in that year. In 2002 women comprised 43.5% of the labour force. Manufacturing employment declined by 8,100 jobs in 2002, having fallen by 36,800 in 2001. Especially badly affected was the construction sector, where employment decreased by 35,100 and 20,500 in 2001 and 2002 respectively. Overall, employment in the service sector increased by 38,900 in 2001 and 3,700 in 2002, but there were declines in the wholesale and retail industries, financial services and business services in 2002. The seasonally adjusted unemployment rate for December 2002 fell to 4.2% and retrenchments slowed from 25,800 in the previous year to 18,900. Overall unemployment in 2002 averaged 4.4%, compared with an average of 3.3% in 2001, 3.1% in 2000, 3.5% in 1999 and 3.2% in 1998. In 2003, however, the outlook worsened significantly as growth in the global economy remained slow, owing to uncertainty about the US-led campaign to oust the regime of Saddam Hussain in Iraq and the reduction in domestic activity caused by the SARS epidemic. Total employment declined for the seventh consecutive quarter in the first three months of 2003, by 9,400 posts, more than twice the number of job losses in the previous quarter, and the seasonally adjusted unemployment rate rose to 4.5% in March. Retrenchments in 2003 totalled 15,800 and the average unemployment rate was 4.7%. With the strong recovery of the economy in early 2004 the unemployment rate began to stabilize at about 4.5% and retrenchments slowed.

The average age of Singapore's work-force rose gradually, from 36.9 years to 37.8 years, between 1997 and 2001. While the 15–29 age-group represented 41% of the work-force in 1989, by June 2002 this had declined to 25.4%. The Government responded by raising the retirement age from 60 to 62 years from 1 January 1999, with an eventual plan to raise it to 67 years.

Of the 2003 population of 4,185,200, a total of 3,437,300 were resident in Singapore as citizens or permanent residents. The 747,900 non-nationals, who had resided in the country for more than one year, thus comprised 17.9% of the total. Singapore's resident population growth rate is rather low and has been decreasing. The average annual growth rate of the resident population during 1990–2003 was 1.8%, whereas that of non-residents was 8.5%, giving a growth rate for the total population of 2.5% per year. The Government was planning for a projected population of 5.5m. by the year 2040. As in other developed countries, the Government in Singapore was very concerned about the declining birth rate, and in August 2000 announced a new incentive to encourage procreation. On the birth of a married couple's second and third child, the Government would open a Children Development Account ('baby bonus'), into which it would pay up to S $1,500 or S $3,000 annually for every second or third child, respectively, until the child reached six years of age. In addition, for every dollar placed in the account by the parents, the Government would make a matching dollar-for-dollar contribution of up to S $1,000 a year for the second and up to S $2,000 for the third child. The parents would be free to decide how the funds should be used. In August 2004 the Government announced an additional incentive payment of S $10,000 on the birth of a couple's third and fourth child. In addition, the Minister of Home Affairs was charged with specific responsibility for overseeing policies intended to boost the country's population.

Recent policy changes have allowed a higher ratio of foreigners to be employed in key sectors such as construction and ship-repairing. There have been some key appointments of foreigners in local banks and even at the Monetary Authority of Singapore (MAS). Foreign workers are employed in jobs that Singaporeans dislike, and if there is a downturn in a sector, such as construction, the foreign worker will lose his or her job. The official view is that if two workers are equally productive and it is necessary to dismiss one worker, then the foreigner should be made redundant. In 2002 and 2003 redundancies increased in the manufacturing sector, and banking and finance sectors, as mergers took place. It is possible that many of the older workers displaced from manufacturing might enter low-level service sectors, such as retailing and tourism, and displace foreign workers employed there.

STRUCTURAL CHANGE

Singapore's strategy of industrializing a basically service-based, entrepôt economy has had profound effects on its production structure. In 1965 manufacturing accounted for only 15.2% of GDP. The contributions of other sectors were: commerce 27.2%; transport and communications 11.5%; financial and business services 16.6%; utilities 2.2%; construction 6.5%; and other services 17.6%. By 1995 manufacturing had reached 24.9% of GDP, with the contributions of other sectors being: commerce 18.6%; transport and communications 11.1%; financial and business services 26.9%; utilities 1.5%; construction 6.7%; and other services 10.0%. This illustrated the increased importance of the manufacturing and financial services sectors (52% of GDP in 1995, compared with 32% in 1965) and the decline in commerce. In 2003 manufacturing accounted for 26.3% of GDP and financial and business services 24.9%.

Singapore's strategy of attracting export-orientated MNCs and high infrastructure investment and fixed capital formation has had profound implications for the structure of expenditures in the economy. In 1965 private consumption expenditure made up 79.2% of GDP, and government consumption expenditures represented 10.4%. With gross fixed capital formation at 21.1% of GDP and stock accumulation at 0.8%, net exports had to be equivalent to 12% of GDP (imports exceeded exports), showing that Singapore was borrowing from abroad in order to finance domestic expenditures, which were 112% of domestic production. Use of the CPF scheme to increase forced saving, persistent fiscal surpluses and surpluses of statutory boards have restrained consumption expenditures and have boosted national saving. By 1995 personal consumption expenditure had been reduced to 40.7% of GDP and, with government consumption expenditure at only 8.5% of GDP, Singapore's gross fixed capital formation reached 33% of GDP and net exports were equivalent to 18.1% of GDP. Net exports only became positive in the mid-1980s, and this had implications for Singapore's balance-of-payments accounts (see below). The reduction of private consumption expenditure as a proportion of GDP means that private consumption expenditures continue to grow in real terms each year but at a slower rate than that of overall output.

The distribution of income is likely to have altered in view of such structural changes. Only in 1998 did Singapore produce provisional data on the income-based estimates of GDP. Official estimates show the distribution of GDP into indirect taxes, gross operating surplus and remuneration (wages and salaries). In 1997 indirect taxes took 10% of GDP, remuneration 43% and profits 47%. This relatively low remuneration share is similar to that of Hong Kong (45.9%), but lower than that of developed economies such as the USA (58.2%) and Japan (55.3%). The profit-to-remuneration ratio in Singapore in the late 1990s (1.11 in 1997) was the highest amongst the Asian newly industrializing economies (compared with 1.05 for Hong Kong and 0.70 for Taiwan) and the developed economies (0.59 for the USA, 0.66 for Japan and 0.65 for Canada). This was interpreted as showing that the economy remained competitive and provided adequate returns for companies in Singapore. The recessionary environment of the late 1990s and the recovery had a strong impact on the structure of the economy. In 1998 33.9% of GDP was received by resident foreign companies and resident foreigners, but by 2003 this share had increased to 39.6%.

MANUFACTURING

Singapore's export-orientated policy, together with the constant pressure to move to higher levels of technology, has created a significant manufacturing sector based on a few important industries. Factories are often built for high-technology, exporting MNCs, and as such tend to be large, much bigger than those in Hong Kong. In terms of output, the average foreign company produces 12 times the amount of an average local firm. Foreign firms tend to be more profitable than local ones and produce about 76% of total manufacturing output. During 1965–84 manufacturing output increased at an average annual rate of 21%. After the global recession of 1985–86, growth resumed at a rate of 12% per annum until 1995. By 2003 manufacturing accounted for 26.3% of GDP and employed some 342,921 people. In the same year electronics comprised 31.4% of manufacturing value added, chemicals and chemical products 25.2% and transport equipment 8.2%. Fabricated metal products accounted for 4.8% of manufacturing value added, petroleum products 4.0% and printing and publishing 3.4%. By far the most important electronic products are computer disk drives, followed by communications equipment and televisions. Direct exports constituted 64% of total manufacturing output in 2003. The export performance of the manufacturing sector is strongly influenced by world electronics demand, especially from the USA.

Manufacturing was the most significant contributor to economic expansion in 2000–03. In terms of sub-sector growth, chemicals and chemical products performed best in 2003, at 10.1%, compared with 40% in 2002 and 7.6% in 2001. Electronics, which had performed well in 2000, recording growth of 25.2%, achieved expansion of only 4.1% in 2002 and 5.3% in 2003. Machinery and equipment production has also been volatile, with a growth rate of 27% in 2000 followed by a contraction of 16% in 2001, expansion of only 0.2% in 2002 and growth of just 0.8% in 2003.

Singapore is often criticized for over-reliance on electronics exports, but it must be remembered that 'electronics' is just a statistical categorization and that there has been restructuring within this sector, with less reliance on items associated with older technologies, such as personal desk-top computers, and more investments from abroad in components for new technologies such as hand phones, portable computers and so on. There are still large investment commitments for the manufacturing sector, generally in the currently favoured sectors of electronics and chemicals, and mostly from sources outside Singapore. Net investment commitments in manufacturing reached S $7,511.1m. in 2003, with 56% in electronic products and components and 27.6% in petroleum and chemical products. Investment commitments in biomedical sciences have also been increasing steadily. These inflows will continue to increase the supply-side capacity of the manufacturing sector.

MONETARY SYSTEM AND POLICY

The Singapore dollar was first issued only in June 1967, nearly two years after nationhood. Singapore had previously been part of a currency union with Malaysia and Brunei, which had been operating for 29 years. In that month the three countries issued their own currencies and the Singapore dollar was interchangeable at par with the Malaysian ringgit and the Brunei dollar. The old Malaysian currency was withdrawn from circulation in Singapore by January 1969. After 1967 the three currencies could still be used in the other two countries at par. In June 1973, when the Singapore dollar was floated, it was delinked from the Malaysian ringgit but remained exchangeable at par against the Brunei dollar, which was also delinked from the Malaysian ringgit. The Singapore and Brunei currencies (unlike the Malaysian ringgit) remain 'customary currencies', are exchangeable at par and can be used for transactions in either country.

Until October 2002 the Singapore dollar was issued by the Board of Commissioners of Currency, Singapore (BCCS), established in April 1967, and not by the MAS, established in 1971, which is Singapore's *de facto* central bank and the regulator of the banking and financial sectors. However, in October 2002 the BCCS merged with the MAS and the MAS assumed responsibility for currency issuance in Singapore. The official statement of exchange-rate policy is that the Singapore dollar is managed against an undefined 'basket' of major relevant currencies, with the aim of achieving both price stability and international competitiveness. Singapore thus has a managed floating system. Official policy since the early 1980s has been to ensure a 'strong' dollar to offset any effects of imported inflation. Trade-weighted indexes of the value of the Singapore dollar show a long-term nominal and real appreciation. Singapore has maintained persistent balance-of-payments surpluses since the early 1970s, implying excess demand for the currency at prevailing exchange rates. The MAS policy has been to intervene by selling Singapore dollars to prevent too rapid an appreciation of the local currency. This explains the steady increase in Singapore's official foreign-exchange reserves, which by June 2004 stood at S $175,204m., or US $101,863m.

INFLATION

In terms of its CPI and GDP deflator, Singapore has achieved a remarkable record of low and relatively stable inflation. Since the 1960s the average annual rate of inflation has been about 4%. Consequently, the 1999 consumer price level was only 2.9 more than its 1965 level. In comparison, the USA's 1996 price level was 4.4 more than its 1965 level. There have been significant departures from this average, however. The overall GDP deflator rose slightly, by 0.4%, in 2002 but decreased by the same amount in 2003. In 2002 the CPI declined by 0.4%, owing to the weakness of the global economy, low imported price inflation, and overcapacity in the high-technology sector. In 2003 the CPI remained subdued, rising by only 0.5%. However, inflation began to accelerate again in the first half of 2004, and it was expected to reach between 1.5% and 2.0% for the year as a whole.

The main source of inflation in the 1990s was domestic, as monetary policy (that is, exchange-rate policy) was aimed at offsetting any external inflationary pressures. Of the 2.0% increase recorded in 1997, for example, it was estimated that 1.8% was due to domestic factors. Historically, many prices have been set by public-sector suppliers, and their changes were subject to government influence. The best statistical explanation of domestic inflation is the movement in Unit Labour Costs, and these have been regularly reduced since 1998 as an anti-recessionary policy.

When aspects of asset inflation are examined, the trends in the 1990s have been very different. There were significant increases in the prices of residential properties, with prices of both private and public housing approximately quadrupling over the period 1991–96. This led some economists to warn of the possibility of Singapore suffering from the problems of a 'bubble' economy, as was the case in Japan. In May 1996 the Government introduced measures aimed at halting the price rises. These included taxing gains on properties sold within three years of purchase, higher stamp duties on such sales, extending stamp duties to other sales and limiting the extent of housing loans to 80% of a property's value. These measures made owning a house less attractive, and thus reduced the price that potential buyers would be willing to pay. By 1998 private residential property prices were about 42% lower than their 1996 peak. In 1999 they rose by 34% and in 2000 and 2001 by 2% and 1% respectively. However, prices declined by 1.5% in 2002 and again in 2003, by 0.4%.

EXTERNAL TRADE AND BALANCE OF PAYMENTS

Singapore's export-orientated development strategy has resulted in a remarkable increase in its external trade volume (goods and services), from a total of S $7,698m. in 1965 (2.6 times that year's GDP) to S $485,986m. (63 times as much in nominal terms and 3.4 times GDP) in 1997. Total exports of goods and services in 1997 were 67 times their 1965 level and imports were 59 times their 1965 level, in nominal terms. Trade decelerated in 1998 to total S $405,689m., or 2.9 times GDP. Total trade in goods declined by 7.5% in that year, but increased by 8.1% in 1999 and by 23% in 2000. In 2001, however, external trade fell by 9.4%, largely owing to a sharp downturn in global demand for electronics and the overall deceleration in the US economy. In 2002 and 2003 trade began to increase again, by 1.5% and 9.6% respectively, as a result of a recovery in the US economy, an acceleration in the global electronics industry and an improvement in demand from Asian countries, especially China. In the first half of 2004 external demand began to recover significantly, and total trade expanded at an average rate of 22%.

The composition of trade has changed. In 1965 food, crude materials and petroleum were the main export categories, contributing 51% of exports. In 1995 machinery and transport equipment comprised 66% of exports, and oil only 8.3%. Of the machinery and transport equipment exports, electronic components and parts accounted for 25%. In 1965 primary commodities comprised 55% of imports, but by 1995 machinery and transport equipment contributed 58% of imports and oil only 8.1%. Electronic components and parts accounted for 32% of machinery and transport equipment imports. Despite Singapore's establishment of a substantial manufacturing sector and the perceived decline in its entrepôt role, some economists still designate Singapore as a 're-export economy'. Strictly speaking, re-exports are those goods that are imported and then simply repackaged and exported elsewhere. In 2003 re-exports constituted 45% of total exports, compared with 61% in 1970. However, many domestic exports, such as electronics, require the import of parts. Thus, machinery and transport equipment constitute the bulk of both imports and exports. Oil has declined from constituting 15.3% of total trade in 1980 to 12% in 2003. In the latter year electronic parts and components accounted for 25% of exports and chemicals 11.8%. Of the total major sectoral exports in 2003, according to the Ministry of Trade and Industry, machinery and equipment accounted for 61.2%, followed by chemicals (11.8%), manufactured goods (3.7%) and food (1.1%). Major imports of commodities in 2003 were machinery and equipment (59.1%), manufactured goods (6.8%), chemicals (6.7%) and food (2.4%).

Singapore's exports are destined for geographically diverse countries in Asia, Europe and the Americas. This diversification has been an important factor in protecting the country to some extent from recession in one of these regions, although it did not help in 2001, when the slowdown in the USA was the main factor in Singapore's own slowdown. Singapore's main export markets in 2003 were Malaysia (15.8%), USA (13.3%), Europe (14.8%), Hong Kong (10.0%), China (7.0%) and Japan (6.7%). Trade with Malaysia is highly correlated with global sales of computer chips and consists of cross-border shippings by MNCs who have plants in both Singapore and Malaysia. Exports to the USA, Japan and Europe increased by 1.6%, 5.5% and 18.7% respectively in 2003, while those to China, the Republic of Korea and Hong Kong increased by 43.8%, 13% and 22.6% respectively.

The sources of imports are slightly different, with Japan, China and Saudi Arabia being more important as sources of products than as export markets. Relevant shares of imports for 2003 were: Malaysia 16.8%; the USA 13.9%; Japan 12.0%; Europe 15.7%; China 8.7%; Taiwan 5.1%; and Thailand 4.3%. There are no official statistics on trade between Singapore and Indonesia available from Singapore sources.

There are no taxes on international trade except for duties on vehicles, petrol, alcohol and tobacco. Singapore has signed bilateral free trade agreements (FTAs) with such countries as New Zealand, Australia, Japan and the USA. A major motive for signing these agreements was to encourage the membership of the Association of South East Asian Nations (ASEAN) to expand its trading and investment beyond the limits of the Association and the rules of the ASEAN Free Trade Area (AFTA). Pressure from the USA during negotiations for the bilateral FTA are likely to lead to further liberalization of the financial and banking sectors.

Singapore has always had a positive services balance, and the export orientation of foreign MNCs in the manufacturing sector reduced the merchandise trade deficit significantly, so that by the mid 1980s the current account became positive. The goods balance itself became positive in the mid-1990s and the current account became extremely large, reaching 24% of GNI in 1999 and 31% in 2003.

The 1970s and 1980s witnessed large surpluses on the capital and financial account because of the inflow of foreign direct investment. These inflows continue, but recently outward direct investment from Singapore has increased. The capital and financial account has been mostly negative recently, but (as it is smaller than the current-account surplus) there have been substantial overall balance-of-payments surpluses. The current-account surplus was S $33,794m. in 2002 and S $49,106m. in 2003. In 2002 the overall balance of payments recorded a surplus of S $2,286.5m, This increased to S $11,774.5m. in 2003.

EXCHANGE RATES

Owing to persistent balance-of-payments surpluses, there has been an historical tendency towards excess demand for the Singapore dollar, which has therefore appreciated against most currencies. Some of the appreciations have been very large. For example, since 1975, following the general floating of world currencies in 1973, the exchange rate against the US dollar had fallen from S $2.4895 to S $1.6755 at the end of 1997. The pound sterling decreased from S $5.0381 in 1975 to S $2.7771 at the end of 1997 and the Malaysian ringgit from S $0.9618 in 1975 (having been at par before 1973) to S $0.4307 at the end of 1997. The price of 100 Japanese yen rose from S $0.8161 in 1975 to S $1.2893 at the end of 1997.

Between the onset of the Asian financial crisis in July 1997 and mid-1998, the Singapore dollar appreciated markedly against the Thai baht (a 37% increase over 1997), Malaysian ringgit (38%), Korean won (30%) and Indonesian rupiah (410%). Given the MAS's statements that it manages the value of the Singapore dollar against a basket of currencies, it was clear that the depreciation against non-Asian countries was expected. The official position was that in 1997 the trade-weighted exchange rate remained 'relatively stable'. In 1999 the Singapore dollar continued to depreciate against the US dollar and the Japanese yen but appreciated slightly against the Deutsche mark and pound sterling. The Asian economic recovery of 1999 led to an appreciation of regional currencies against the Singapore dollar, which nevertheless remained stronger against them in relation to 1997, when they had fallen significantly. As the Malaysian ringgit is effectively pegged to the US dollar and the Hong Kong dollar is linked to the US dollar, the Singapore dollar depreciated against these two regional currencies in 1999. The marked slowdown in the economy in 2001 resulted in a depreciation of the Singapore dollar against the US dollar. In March of that year the MAS eased a number of currency controls on the Singapore dollar, as the exchange rate rose to an average of US $1 = S $1.7917. In 2002 and 2003 it remained stable, at US $1 = S $1.7906 and US $1 = S $1.7402 respectively. In the first half of 2004 the Singapore dollar began to appreciate against the US dollar, reaching US $1 = S $1.7021 by the end of June.

FISCAL POLICY

The Government is committed to a policy of ensuring significant budget surpluses, defended by the view that the country's reserves are the mainstay of Singapore and overall budget surpluses were significant in the 1990s. The Government's economic policy is pro-business, but does not necessarily favour economic freedom. The Government has consistently used tax breaks and incentives in an attempt to attract MNCs and more specialized foreign enterprises into Singapore. Its incentive schemes include: lowering the minimum tax rate to 5% under the development and expansion initiative; and providing a double tax deduction for approved research and development expenses for all service companies. The Government was confident that these measures would be effective. There is no widespread public social security or unemployment benefit scheme. Publicly funded pensions are available only to senior civil servants, members of the judiciary and parliament and military officers. The CPF scheme and private savings are supposed to provide for the retirement needs of the rest of the population, but this is privately, not publicly, funded (although the Government has occasionally 'topped up' certain accounts of citizens by small amounts in the manner of a company paying extra dividends), and government ministers admit that CPF savings are not sufficient for these needs and that more reforms to the system are necessary. The Public Assistance Scheme provides very small benefits for the aged poor, the totally disabled and chronically ill, after a means test. The unemployed are supposed to finance themselves through savings or through the support of their family. There is no capital gains tax, apart from a few exceptions, such as when house selling is considered to be trading.

An innovation to the CPF scheme announced in the 1999 budget was to allow voluntary additional contributions by employees into their CPF accounts, under a Supplementary Retirement Scheme that would be tax deductible and taxable on withdrawals. Whether this would prove to be attractive was debatable as returns paid on CPF funds had been virtually zero in real terms. Furthermore, a voluntary Special Retirement Scheme was established by 2001, enabling foreigners to make tax-deductible contributions, which would be taxed on withdrawal. In August 2000 the Government announced that it would use S $2,000m. to enhance the CPF accounts of all adult Singaporean citizens who had paid at least S $100 into their CPF accounts between 1 January 1998 and 31 December 2000. Instead of the usual method of an equal payment for all (as the 2000 budget had given), this time those earning more than S $2,000 a month would receive S $500, those earning between S $1,200 and S $2,000 would receive S $1,000 and those earning less than S $1,200 would receive S $1,500. The self-employed and those whose incomes could be ascertained from CPF records would receive the enhancement according to the type of house they lived in. The payments were to be made in two instalments, with the first being in January 2001.

In 2000 government operating revenue and tax revenue both increased by 17%. The budget announced in February 2001 took advantage of this and was welcomed as pro-local business, as it attempted to address the grievances of the small domestic production sector, which has long complained of being crowded out by government companies. The corporate tax rate was reduced from 25.5% to 24.5%, against all expectations, since a rate below 25% is considered by some countries to be unfair and to denote the country as a tax haven. The top marginal income tax rate was reduced from 28% to 26% and the property tax rate was reduced from 12% to 10%. While the corporate tax reductions applied to all companies, small ones were given generous tax exemptions, with the expectation that two-thirds of taxable companies would pay half as much tax as before. The Government announced it would allow the establishment of limited partnerships and limited liability partnerships to encourage company formation. The 2002 budget maintained an expansionary fiscal stance, but was also presented as constituting part of a longer-term strategic restructuring intended to encourage manufacturing competitiveness and entrepreneurship, attract foreign capital and talent and create high value-added jobs. The goods and services tax (GST) was raised from 3% to 4%, effective from January 2003, but this was accompanied by a range of measures to offset the rise, which included rebates on rentals and service and conservancy charges. The corporate tax rate was reduced by 2.5 percentage points, to 22%, and the personal income tax rate by four percentage points, also to 22%, in line with the Government's commitment to bring both corporate and personal tax rates down to 20% within three years. This made the corporate tax rate much lower than China's 33%, Malaysia's 28% and Taiwan's 25%, though still above Hong Kong's 16% rate. The budget also announced the introduction of 'Economic Restructuring Shares' for all adult Singaporeans, to be distributed in three tranches from January 2003.

The 2003 budget projected a deficit of S $900m. This was intended to stimulate the economy in the short term, and did not represent an abandonment of the Government's general philosophy to balance the budget and accumulate a modest surplus over the business cycle. The restoration of the employers' CPF contribution rate to 20% from 16% was deferred for two years, other reforms to the CPF were announced, including the reduction of the employee contributions for workers aged 50–55 years, and measures were introduced to lower business costs. Longer-term measures to enhance competitiveness and foster entrepreneurship were also outlined, including the appointment of a minister responsible for encouraging entrepreneurship, tax exemptions on corporate income remitted back to Singapore, and tax incentives to make it more attractive to create and hold intellectual property in Singapore. The 2004 budget was aimed at maintaining Singapore's competitiveness and reinforcing its procreation policy. Its key measures were the lowering of the corporate tax rate to 20% and the exemption from taxation of all foreign-sourced income remitted by resident individuals. Singaporeans aged 50 years and over also benefited from additional government contributions to their Medisave accounts. Overall, the Government planned for a budget deficit of S $1,350m. for fiscal year 2004, following the S $1,800m. deficit eventually recorded in 2003, but reaffirmed its commitment to balance the budget by fiscal year 2005.

FINANCIAL AND BUSINESS SERVICES

The financial part of this sector consists of banking, insurance, stockbroking, fund management and currency and futures trading and so on, and the business part of such activities as real estate, legal services, advertising, consultancies and information technology services. These two sectors' combined contribution to GDP increased from 16.6% in 1965 to 24.9% in 2003, when the sectors employed approximately 348,000 people.

The financial sector's development was aided by government policy and tax concessions. In 1968 the Government allowed the establishment within banks and merchant banks of departments known as Asian Currency Units (ACUs). These are separate legal and accounting departments of the banks in which they are located and were given special tax treatment, and this has continued over the years, contributing to their rapid growth. They operate in the Asian dollar market. At the end of 1997 their total assets/liabilities, at US $557,194m., were 1,429 times their 1970 level. In 1998 the assets of the Asian dollar market contracted by 9.6% and in 1999 by 4.6% as non-bank customers wished to borrow much less. In 2003 total assets/liabilities were US $509,145.9m., an increase of 5.5% compared with the previous year, when assets/liabilities increased by 3.7%.

Other important parts of the financial sector have been the Stock Exchange of Singapore (SES), the foreign-exchange market and the financial futures market, formerly known as the Singapore International Monetary Exchange (SIMEX). SIMEX started operations in September 1984 and established links with the Chicago Mercantile Exchange through its system of mutual offsets. It was said to be the first financial futures exchange in Asia, as it assumed the functions of the Gold Exchange of Singapore, which had earlier commenced financial futures trading. On 1 December 1999, following the demutualization of both stock and futures exchanges, SIMEX and the SES were merged to form a new company known as Singapore Exchange Limited (SGX), which presented itself as the first demutualized, integrated securities and derivatives exchange in the Asia-Pacific region. By the beginning of 2002 Singapore Exchange Securities Trading Limited (SGX-ST) had 492 listed companies and a total market capitalization of S $334,700m.

The first stages of the Asian financial crisis brought business to the financial markets in 1997, which witnessed high turnover on the foreign-exchange market and stock exchange, increased earnings and rapid growth of the financial sector. In 1998 the turnover of the stock exchange was 14% lower than in 1997. The domestic market recovered in 1999 when the value of turnover on the SGX (Equities) more than doubled, to S $151,107m. However, in 2000 turnover increased by only S $72m., and in both 2001 and 2002 it declined further. In 2003 turnover recovered to S $138,315.8, a rise of 38% compared with the previous year. In 1999 the trading volume of derivative contracts on the Singapore Exchange Derivatives Trading Limited (SGX-DT) declined by 7.2% from the record level of 1998. In June 2000 the SGX-DT launched a futures contract based on the Straits Times Index, the second product to be based on that index in Singapore. In 2001 trading volume reached an all-time high of 30,989,862 contracts, an increase of 12.5% from the previous year. Financial futures trading is dominated by the three-month Eurodollar interest rate futures, three-month Euroyen interest rate futures, Nikkei-225 stock index futures and the MSCI Taiwan stock index futures contract. The foreign-exchange market in 2001 recorded an average daily turnover of about US $95,000m., approximately 5% below the 2000 level, which itself was 19% below that of the previous year. The volume of foreign-exchange transactions in Singapore ranks it as the fourth busiest in the world. Trade is mainly of US dollars against the Japanese yen and the US dollar against the euro.

Responding to long-held beliefs that the financial sector had been over-regulated by the MAS and other authorities, stifling development and innovation, the Government established the Financial Sector Review Group in 1997, and the recommendations of its subcommittees were publicized during 1998. The Corporate Finance Committee recommended a move towards a US system of fuller disclosure by companies, with investors taking more responsibility for their own decisions. Overall, securities regulation should be removed from the SES, the Securities Industry Council and other bodies and responsibility centralized in the MAS in a 'super regulator'. It was felt that the Singapore Exchange had been taking into account too many restrictive criteria when deciding whether a listing should be allowed. Other recommendations included: consolidating securities market laws and updating them; promoting listings by foreign and smaller local companies; encouraging the use of technology; and encouraging internet trading.

In recent years the Government's policies to increase the presence of fund managers in Singapore, through favourable tax treatment and making funds available for them to manage, have begun to yield noticeable results. The Government also accepted the recommendation to issue a 10-year government bond to establish a benchmark for the yield curve. This would facilitate the corporate issue of bonds. In 2001 further issues of government bonds were made for this purpose, including a Singapore Government bond futures contract. Following the recommendations of a Committee on Banking Disclosure Standards, most of which were accepted by the Government, local banks will be required to stop the practice of keeping hidden reserves. Furthermore, the limit of S $200m. put on loans to residents by foreign banks was increased to S $300m. for each bank in any year and was subsequently further raised to S $1,000m. By the end of 2001 the Government had issued six licences for a new category of banks, known as Qualifying Full Banks (QFBs). QFBs can establish themselves in 15 locations, of which up to 10 can be branches, and share automated teller machines (ATMs) amongst themselves, but since they cannot share the ATM network of the local banks their ability to compete is still restricted.

The limit of 40% on foreign ownership of Singapore bank shares was removed in 1999, when all banks were also allowed to offer interest on current accounts and the restrictions on the size and duration of fixed deposits they could accept was removed. The view within the industry is that the pace of liberalization was rather slow and that the degree of competition had not significantly increased, while the main four foreign banks that received licences had not co-operated in setting up their own ATM network.

The Government declared its commitment to a five-year plan of consolidating the banking sector and allowing more competition. In July 1998 the Post Office Savings Bank system (a statutory board) was merged with the Development Bank of Singapore (DBS) for S $1,600m. in a significant liberalization of the banking sector. In June 2000 the Government announced that it would introduce legislation to require local banks to divest themselves of all non-core assets within three years, signalling that, as voluntary compliance with government suggestions had not been adopted, legislation would be used. These assets are mainly in the property development and manufacturing sectors. The government view is that there is only room for two domestic banks in Singapore, and in mid-2001 there was an unprecedented frenzy of hostile takeover activities, resulting in two large, private, local banks merging to become larger than the DBS.

Liberalization has been extended to the insurance industry, which had been protected by the Government. Only one company had been allowed to enter the industry since 1986, a policy that the Government now recognizes 'stifled new products and efficiency'. In 1999 and 2000 there was strong growth in the insurance sector, with most of the new business coming from outside Singapore. The domestic market continued to experience declines in rates owing to increased competition. In September 1998 Malaysia linked its currency to the US dollar and imposed capital controls. One consequence was that the holdings of some 170,000 people with shares in Malaysian companies that were traded on Singapore's Central Limit Order Book International Market (CLOB) were 'frozen'. An arrangement was reached (accepted by 94% of the owners who owned 98% of the shares), providing for the gradual release of their shares from June 2000 in weekly batches over 13 weeks. This issue and its solution elicited strong expressions of discontent from the Singapore shareholders, who were highly critical of the Stock Exchange, the Government and the Malaysian authorities.

There were 32 initial public offers in the first nine months of 2001. This represented a decline from the 47 new listings that had come to market in the same period in 2000. A total of 85 public offers took place in 2000, compared with 53 in 1999 and 24 in 1998. Six new foreign firms were listed on the Singapore Exchange in the year to August 2001, increasing the total number of locally listed foreign firms to 90. On 1 July 2002 the second stage of the Securities and Futures Act 2001 (SFA) went into effect. Under the new measures, intended to improve disclosure and streamline the listing process, companies must submit prospectuses to the MAS rather than the Singapore Exchange.

TOURISM

Although Singapore has few areas of natural beauty, it has always been an important tourist destination, mainly for those going in transit or who include a visit to Singapore as part of a trip to some combination of Indonesia, Malaysia and Thailand. In 2003 the average length of stay by a tourist was 3.2 days, having declined steadily since the early 1990s, and was much shorter than the length of visits to such places as Indonesia and the Philippines at 10 days each and Thailand at 7.4 days. Average per caput daily expenditure has also fallen since the early 1990s. Combining a modern, efficient city with diverse examples of Asian culture, Singapore has been called 'Asia for beginners'. For many years it was promoted as a 'shoppers' paradise' but, despite the continuing annual 'Great Singapore Sale' in May and June or July, this aspect has faded, mainly owing to the appreciation of the Singapore dollar. Visitors can obtain a rebate of the goods and services tax on purchases of more than S $300 in any participating retail outlet. During the economic crisis tourism, and hence the retail and hotel sectors, suffered from the strength of the Singapore dollar, the decline in incomes in neighbouring countries and the return of smoke pollution from forest fires in Indonesia. The impact on the economy continued into 1998. In 2001 the number of visitor arrivals contracted by 2.2% compared with the previous year, but arrivals rose marginally in 2002, to 7,567,039. The hotel occupancy rate fell to 74% in 2002 from 76.3% in 2001 and over 84% in 2000, but was still higher than its lowest rate of 71.3% in 1998. In 2003 arrivals contracted sharply, by 19%, to only 6,125,480, as a result of the SARS epidemic, and the hotel occupancy rate fell to 67.2%, but both began to recover in the first half of 2004.

The Tourism 21 plan, adopted by the Singapore Tourist Board (STB), featured more promotions with neighbouring Asian countries and Western Australia. One aspect of its marketing was to link tourism with Singapore's plans to become an arts hub ('Global city for the arts' being the slogan), and tourist promotions have been linked to Singapore's hosting of popular musical and theatrical events. Another important aspect of tourism development is the promotion of Singapore as the location for major international conferences, conventions and exhibitions. The successful hosting of the inaugural World Trade Organization (WTO) Ministerial Conference in December 1996 provided much publicity and experience. A new exhibition centre near Changi International Airport began operations in 1999, adding to the substantial convention facilities near the Central Business District (CBD) that were completed in the mid-1990s. Singapore has ranked as Asia's foremost convention centre since the mid-1980s and is regarded as one of the best in the world.

INFRASTRUCTURE

With such a huge volume of trade and people passing through such a small place, and with the need to provide international communications together with the building of factories, infrastructural development has been crucial to Singapore's overall progress. In the 1990s construction output accounted for an average of about 6.5% of GDP, having reached as much as 10% of GDP in 1985 as a result of a boom in public-sector housing and a decline in overall GDP in that year. Although construction output rose in 1997 and 1998, it declined continuously over the next four years. In 2003 the construction sector contributed about 5% to GDP. The Government launched Construction 21, a strategic plan to restructure Singapore's construction industry which, it has admitted, has suffered from negative productivity growth and heavy reliance on unskilled foreign workers.

There are some significant projects under way that will add considerably to supply-side capacity. To the south of Singapore extensive reclamation by the JTC continues between a group of seven islands. This will produce a 3,000 ha petrochemical complex, Jurong Island, to be completed by 2015. The area is already used for refining, petrochemical and chemical production and will become the main site for this important part of Singapore's industrial sector as envisioned in the EDB's plans. Singapore's first petroleum refinery was established by Shell in 1961 on the small island of Pulau Bukum.

Singapore obtains about one-half of its water supply from Malaysia. Whenever political tensions arise between Malaysia and Singapore, fears are expressed that Malaysia might discontinue water supplies and not renew the two supply agreements, which expire in the years 2011 and 2061, although Malaysia has stated that it will renew the contracts, whatever the circumstances. Singapore is building a water desalination plant in Tuas and has plans for two more before the year 2011. It is possible that a scheme to build a 200-km pipeline from Indonesia to deliver water to Singapore will be undertaken and is scheduled to be completed by 2005. Furthermore, treated sewage is to be processed into 'new water', which is pure enough to be used by the very high technological level wafer fabrication plants that Singapore is still attracting.

Transport

In February 1996 six new stations and 16 km of track were added to the railway system, the Mass Rapid Transit (MRT) system operated by the MRTC, which was a statutory board before September 1995, when it was reconstituted as part of the Land Transport Authority (LTA). In 2003 work was completed on the 20-km North–East line, which added 16 underground stations and connected with the existing system at two central stations, with some stations also being developed into offices and retail developments. A new 6.4-km line, linking the East–West line with Changi International Airport, was opened in 2002. Light Rapid Transit (LRT) systems were under construction to link new towns to the wider MRT network. By 2005 the MRT/LRT system was expected to cover an area 50% greater than in 2000.

Singapore's Changi International Airport is regularly voted the world's best airport. Designs for the planned third terminal were made public in 1998, and construction commenced in October 2000. The new terminal is due to open in 2006, with the two existing terminals being upgraded in the mean time. Separate facilities for new low cost airlines are also being planned. Airport development depends in part on the fortunes of Singapore Airlines (SIA), the national carrier. SIA has expanded its business interests significantly through a 49% share in the British carrier Virgin Atlantic, and by August 2004 had a fleet of 87 aircraft, with 10 on order, and a global network of 90 destinations in 40 countries. However, the decrease in traffic in 2001, owing to the economic downturn, together with the SARS epidemic in 2003 significantly affected SIA's profitability, and substantial staff retrenchments were implemented in mid-2003.

Work on a major extension of the container port at Pasir Panjang has been undertaken. Singapore was developing the port's facilities to cope with large cruise ships, as the country sought to become a regional hub for this expanding leisure industry. Singapore remains the busiest port in the world in terms of shipping tonnage and is the world's largest supplier of bunkers (fuel used by ships). However, the decision by Taiwan's Evergreen Marine Corporation in April 2002 to move its transhipment business to Tanjung Pelepas in Malaysia, following Denmark's Maersk Sealand International, which moved in 2000, has slowed down the process to privatize the PSA and could affect future port development.

In January 1998 a new toll road bridge, financed equally by Singapore and Malaysia and linking the west of Singapore to Malaysia, was opened. This was intended to reduce congestion on the existing causeway that provides road, rail and water pipeline links to Malaysia. The new link is, so far at least, surprisingly underutilized, with Malaysian commercial drivers citing the cost of the toll and the longer travelling distances as the main reasons for not using it.

Singapore's high volume of trade requires a good, uncongested road system. This has been planned since 1990 by strictly controlling the growth in the number of vehicles through a government-determined quota for vehicles sales. In order to buy a vehicle, buyers must purchase, at a twice-monthly auction, a Certificate of Entitlement (COE). In early 2000 a COE for a luxury car cost between S $43,000 and S $51,000 and for a small car about S $42,000. By March 2003 these prices had fallen to S $26,000 and S $26,999, their lowest level for many years. Despite the recovery of the economy in the first half of 2004, COE prices remained low. This non-proportionality to the price of the car has made large cars relatively cheaper in Singapore. In addition, substantial import duties contribute to high car prices.

As well as the quota system, the Government tries to influence road usage by charging drivers for access to such areas as the CBD. This has been done by requiring drivers to buy and display a licence either for daily or monthly use. In April 1998 road pricing took a technological leap with the introduction of what is said to be the world's first use of Electronic Road Pricing (ERP), although this claim has been disputed. Access to the roads currently subject to charges requires vehicles to be fitted with an in-vehicle unit (IU) on the dashboard that can be read by cameras placed at points on the restricted roads. Units were supplied and fitted free for a limited period. Drivers of cars visiting Singapore can hire the units. Charges for usage are deducted from a cash card placed in the unit. Charges for access vary according to the time of day. Overall, the Government aims for 3% growth in the vehicle population each year, adjusting the number of COEs accordingly. However, the Government has stated that if this system of road pricing keeps the roads free of congestion, it might increase the number of COEs in the future.

Communications

Efficient telecommunications are vital for a financial and trading centre such as Singapore. Plans for a more competitive telecommunications system became clearer in April 1998. After a competitive bidding process, a licence for fixed-line telecommunications, along with one for mobile phone operations, was awarded to StarHub, which is owned by Singapore Technologies Communications, Singapore Power, British Telecom and Nippon Telegraph & Telephone. Originally it was planned that StarHub would be the sole competitor for the former monopoly supplier, SingTel, from 1 April 2000 for two years. The Government compensated SingTel for earlier than agreed loss of its monopoly by granting it S $1,500m. However, in January 2000, in an unexpected move, the Government announced that it would no longer stand by its original commitment and that from 1 April it would open up the telecommunications market. The Government also stated that it would appoint advisers to suggest the extent of compensation appropriate for the two companies that would be affected and in September awarded compensation of S $859m. to SingTel and S $1,082m. to StarHub. Furthermore, it insisted that StarHub install the nation-wide infrastructure to serve the fixed-line residential customers. Despite the fear that this change of policy might discourage further foreign investment in Singapore, the Government claimed that it thought this liberalization would bring in S $3,000m. in further investment and 2,500 new jobs over three years. Within a few months of the change in policy 58 new licences had been issued, and competition had reduced phone rates and increased the variety of products. By 2003 StarHub offered a range of information communications and entertainment services, including 40 international cable channels, as well as fixed, mobile and internet platforms.

SingTel is a major Asian communications company with investments in over 20 countries, but with its two main hubs in Singapore and Australia. By 2003 it had significant strategic investments in Hong Kong, India, Indonesia, the Philippines, Taiwan and Thailand and a market capitalization of S $25,000m. Singapore's IT2000 plan includes the development of Singapore ONE (One Network for Everyone), which will link the entire population through developing a multimedia broadband infrastructure. Singapore also has a national literacy programme to encourage Singaporeans to become adept users of information communications appliances and services. Singapore ONE now has more than 250,000 users; and according to a local internet research firm, NetValue, from October 2001 to March 2002 the number of broadband users rose by 35%, to 172,000, with 71,000 households connected. Overall, there are over 35 Internet Service Providers (ISPs) and 20 International Simple Resellers (ISRs) selling overseas calls. Mobile phones are widely used in Singapore, and in April 2003 there were 3.3m. mobile subscribers, equivalent to a penetration rate of 79.2% of the population, compared with 479,000 subscribers (a 15.5% penetration rate) in April 1997. The number of internet subscribers reached over 2m. in April 2003, with a penetration rate of 48.8%, compared with a penetration rate of only 13.8% in April 1999. Research by NetValue conducted in September 2001 also indicated that one-third of those 'wired' Singaporeans now use online banking services.

The Government is promoting the growth of electronic commerce, and some international companies have located their Asian headquarters for electronic commerce in Singapore. These developments are being encouraged by the Infocommunications Development Authority of Singapore (IDA), a new statutory board which was created by the merger of the National Computer Board and the Telecommunications Authority of Singapore in December 1999. The IDA promotes greater use of information technology in Singapore, and even has a strategic planning branch that is responsible for anticipating new developments in 'hot' areas such as broadband and mobile commerce. In a bid to boost Singapore's competitiveness in international electronic commerce, the IDA has encouraged businesses to invest more heavily in the information technology sector. In 2000 total business spending in information communications and the media industry increased by 29% year-on-year to reach S $730m. The IDA itself introduced a S $30m. incentive programme in October 2000 to encourage businesses to increase their electronic commerce levels. In the same month the IDA also launched 'Wired with Wireless', a S $200m. programme to give Singapore a fully integrated wired and wireless infrastructure. In February 2003 the IDA launched its ultra-wideband (UWB) programme to bring this wideband technology to Singapore.

Training and Research

Education and training have always been an important part of Singapore's supply-side policies. The public sector dominates although there are private schools, mainly in commercial, computing and fine arts education. At the tertiary level many students can study through correspondence and internet courses with foreign universities, and the Singapore Institute of Management (SIM) offers Open University courses. At the post-secondary level the five polytechnics, four universities and the Institute of Technical Education are all statutory boards under the Ministry of Education. Since 2000 students have been able to enter the privately run Singapore Management University (SMU). In 2003 the number of students at Singapore's four universities totalled 40,095, and there were 55,376 students at the five polytechnics. In January 2003 the Ministry of Education announced that it intended to increase the university cohort participation rate from 21% to 25% by 2010 through the addition of a further 3,500 places. The National University of Singapore was to be transformed into a multi-campus university and Nanyang Technological University was to become a comprehensive university. Foreign business schools from the USA and Europe are also establishing campuses in Singapore, and the Government is exploring the possibility of opening more private universities. Recently, schools have been encouraging creative thinking, while at higher levels emphasis has been placed on research.

It has long been observed that the amount spent on research and development within Singapore has been rather low compared with other countries. In the past much of the research expenditure from which Singapore benefited was probably being made in other locations, but many MNCs and local companies have started conducting research and development in Singapore. Total research and development expenditure reached 2.1% of GDP in 2002, marginally increased from 2.0% in 2001. Most of the research and development is being carried out by the private sector, mainly in the areas of electronics, chemicals, engineering, information technology and telecommunications.

THE CURRENT SITUATION AND SINGAPORE'S PROSPECTS

The economy of Singapore is small, open and heavily dependent upon foreign capital, expertise and workers, as well as upon overseas markets. It is often argued that the Republic's manufacturing sector is too dependent upon the electronics subsector; this is a cause for concern, as the demand for electronic products is cyclical. Other sectors such as petroleum and tourism are similarly vulnerable: tourism, which is very important to both the retail and transport sectors, is susceptible to regional and international instability, as demonstrated by the impact of the terrorist attacks on the USA in September 2001. The tourism sector also suffered from the decline in visitor arrivals during the SARS epidemic in 2003 and the effects of smoke pollution in the region since the late 1990s, particularly in late 1997. This pollution has recurred on several occasions since then, but to a milder extent in Singapore than in neighbouring countries. This has prompted the Government to review the National Green Plan (launched in 1991), in which it aimed to: limit carbon dioxide emissions and phase out controlled chlorofluorocarbons (CFCs); and tighten safety procedures on the storage, handling and transport of hazardous products. The authorities have made some progress in these areas, including the ban on CFCs in 1996, the phasing out of leaded petrol in 1998 and the implementation of new anti-pollution regulations in January 2001. However, environmental degradation remained a priority, particularly with continuing problems from smog caused by forest fires. The Government has therefore presented new National Green Plan targets for 2012. Four key areas, agreed following public consultations in 2001, have been identified: 'quality living environment' (limiting air pollution, increasing the use of natural gas, reducing landfill levels through increased recycling); 'working in partnership with the community' (building greater community responsibility and gathering public feedback); reducing noise levels; and 'doing our part for the global environment' (helping to enforce international environmental commitments).

The region has shown greater political instability and violence in the last few years. Foreign demand can be rather unstable and consists of about 80% of the demand for Singapore's output. The domestic market is small, and personal consumption (at about 42% of expenditure on GDP) is much lower than in other industrialized economies. Domestic demand is also vulnerable to changes in the stock market and in property prices. The Singapore economy has been affected by the industrialization and liberalization of the economies of neighbouring countries, and by their competition for foreign direct investment, some of which has come from Singapore, but much of which has been committed by sources that might otherwise have chosen to invest in Singapore. The Government is particularly concerned about the threat posed by China as a potentially more desirable destination for foreign investments, to the detriment of ASEAN, but also recognizes the opportunities China presents as both a market and a partner in Singapore's production supply chain. In view of the above, it might be construed that the economy of Singapore is by nature very vulnerable and might therefore have been likely to suffer rather badly as a result of the Asian economic crisis that began in 1997. However, Singapore was the Asian country least affected by the crisis in terms of growth rates, inflation and unemployment; the deceleration in the economy could be attributed to the external factors that had repercussions through trade and tourism and not to any inherent problems in the domestic economy. Foreign direct investment commitments did not decline significantly during the crisis years. The supply capacity of the economy continues to expand significantly.

In 2000 the economy expanded by an impressive 9.7%, its highest rate since 1994. However, in the following year it contracted by 1.9%, thus illustrating its vulnerability, as capital investment in the USA fell in late 2000 and Singapore's exports and manufacturing output were affected very quickly. In 2002 the economy seemed to be on the path to recovery, recording a 2.2% growth in GDP, but, in the aftermath of the invasion of Iraq and the SARS crisis, growth in 2003 was only 1.1%. By the end of the second quarter of 2004 the economy appeared to be on track to record growth of about 8%–9% for the full year, although the second half of the year was widely expected to experience slower expansion than the impressive first half.

On the economic side, the Government introduced an off-budget set of policies in July 2001 similar to earlier policies aimed at reducing business costs and maintaining the profitability of firms. This package included reductions in charges by government agencies, rental rebates and grants to companies of up to S $600 a month for hiring retrenched workers. In addition, some public-sector projects were brought forward. At a value of S $2,200m., this package was about 1.4% of GDP, much less than that required at the time of the Asian financial crisis. The co-operative trade union movement, through its chain of supermarkets, decided to reduce the prices of a range of staple commodities to help the lower-paid population. The CPF and the Housing and Development Board (HDB) rescheduled mortgage payments for families that experienced financial difficulties. The Government is preparing the population for a future of slower growth, increased retrenchments and higher unemployment. Mergers in the banking sector and the opening up of the financial sector to foreign competition are likely to lead to retrenchments there. The global electronics slowdown in 2001 led many foreign MNCs to retrench workers in their foreign operations, including in Singapore. In the first three months of 2003 total employment declined for the seventh consecutive quarter, by 9,400. In the past, MNCs that closed down production operations in Singapore would often relocate them to Malaysia. The Government fears that the future destination of these jobs will be China, but the nature of globalization is illustrated by the fact that, when one US telecommunications equipment company recently closed down its manufacturing operations in Singapore, it relocated them to Ireland. The Government retains its strong commitment to helping retrenched workers learn new skills, but if the electronics sector contracts much more, they may be forced into lower-level service jobs. The Government has advised older workers not to be too selective when searching for new employment.

In the wake of the recession of 2001, Prime Minister Goh Chok Tong reiterated his aim to create a 'New Singapore' within 10 years. This 'New Singapore' would be a global city ('globapolis') with people from all over the world well connected to all parts of the globe. There would be abundant opportunities for Singaporeans and talented foreigners to work, do well and enjoy excellent recreational and artistic facilities. Singapore's economic activity would expand beyond its limited space and this restricted size would not restrain its growth but would encourage it to become one of the world's most habitable cities. Foreign advisers had long supported a move towards a more service-based economy and more entertainment and artistic venues that would attract foreign professionals.

The main policies required for this were to increase the productivity and standards of service of the domestic sector, which is behind that of the MNCs and other foreign firms in Singapore. Diversification away from electronics was required. The Prime Minister also argued that the country needed to alter its entire mindset. Singaporeans needed to be more innovative, rather than merely copying the ideas of others, and should be more willing to take risks and to expand further into the region. Towards this end he argued that Singaporeans should accept non-conformist thinking, be more tolerant of people who make mistakes in their business affairs and not be resentful of private entrepreneurs who 'make it big'. (Income distribution became more unequal during the 1990s and the Government expects the income gap to widen further.) To encourage innovation—and here the Prime Minister acknowledged the irony of using a 'top-down' approach—the Government would establish the National Innovation Council (NIC).

The Prime Minister promised that this 'New Singapore', in which competition would become increasingly fierce, would also be accompanied by a 'new social compact' with the people. The Government would continue to subsidize housing, education and healthcare. In addition, it would introduce 'New Singapore Shares' that would be distributed to all citizens and would provide a guaranteed dividend for a fixed number of years and a bonus when the economy did well. On 1 March 2002 the CPF credited a total of 12,340,000 New Singapore Shares (NSS) to the accounts of 2,080,000 NSS shareholders. Each shareholder earned a dividend ranging from 1 to 17 bonus NSS, according to the number of shares held as of mid-February.

In February 2003 the Government presented the final report of the ERC, which was established in December 2001 to provide an open assessment of economic strategy. However, many of its recommendations had already been made public throughout 2002, including advice included in the 2002/03 budget, which was presented to Parliament in May 2002. The ERC's short-term goal was to boost economic recovery and make Singapore more competitive in the region by reducing business costs. As far as the longer-term strategy was concerned, manufacturing (electronics, chemicals and biomedical sciences) and services (financial, business and medical) would together continue to provide the impetus of growth, but there would be a reduction in government-owned businesses, the development of a more creative entrepreneurial environment, and reforms to the wage system based on seniority to make it conform more to the market. There would be further liberalization in the banking and power generation sectors and the Government would have to learn how to regulate these sectors rather than being the producer itself. Furthermore, the Government might have to deal with a protracted slowdown and rising unemployment, possibly accompanied by a significant and permanent loss of manufacturing jobs, without an immediate increase in suitable service sector jobs. The Government has continually expressed the need to attract foreign talent into those sectors it hopes will expand, such as life sciences, which require skilled personnel and researchers. There is also likely to be a greater foreign presence in the financial and service sectors. The Government has envisaged a more competitive, free-market environment in which there is likely to be a greater foreign presence in areas that were reserved for local firms. Many of the recommendations of the ERC were not new but reaffirmations of existing policies.

In August 2004 Lee Hsien Loong, the son of former Prime Minister Lee Kuan Yew, succeeded Goh Chok Tong as Prime Minister. Lee Kuan Yew became 'Minister Mentor' and Goh Chok Tong Senior Minister in the new Cabinet. In his National Day speech in the same month, which was unusually long and full of personal anecdotes, Lee Hsien Loong outlined his vision of a Singapore 'full of opportunity and promise'. Many of his themes already constituted part of government policy and the new regime appeared to be very much one characterized by continuity rather than radical change. The need for Singaporeans to be more 'spontaneous' and 'unconventional' was reiterated and some restrictions on their freedom to participate in (private) debate and at 'Speakers' Corner' were to be relaxed. It was unclear, however, how significant the easing of these restrictions would be in practice. The need to increase the quality of education and reduce learning by rote was also stressed, as was the commitment of the ruling People's Action Party (PAP) to family values and procreation. Further financial help and leave from work for childcare purposes would be given and more emphasis than in the past was to be placed on the equalization of male and female medical benefits and child-raising duties. The civil service, partly in consequence, would adopt a five-day working week in order to help foster a change in attitudes towards the family and children.

With regard to the economy, Singapore would continue with restructuring, upgrading, making labour markets more flexible, and rewarding workers through the use of 'key performance indicators'. Interestingly, the new Prime Minister announced that measures would be introduced to help retrenched workers find work by making what were regarded as unpleasant jobs more attractive and better paid. In addition, the living environment for older workers, many of whom had suffered during the recent economic downturns, would be improved by, for example, increasing Medicare support.

Statistical Survey

Source (unless otherwise stated): Department of Statistics, 100 High St, 05-01 The Treasury, Singapore 179434; tel. 63327686; fax 63327689; e-mail info@singstat.gov.sg; internet www.singstat.gov.sg.

Area and Population

AREA, POPULATION AND DENSITY

Area (sq km)	659.9*
Population (census results)†	
30 June 1990	3,047,132‡
30 June 2000	
Males	2,061,800§
Females	1,955,900§
Total	4,017,733
Population (official estimates at mid-year)	
2001	4,131,200
2002	4,171,300
2003	4,185,200
Density (per sq km) at mid-2003	6,342.2

* 254.8 sq miles.

† Includes non-residents, totalling 311,264 in 1990 and 754,524 in 2000.

‡ Includes resident population temporarily residing overseas.

§ Provisional.

ETHNIC GROUPS

(at census of 30 June 2000)*

	Males	Females	Total
Chinese	1,245,782	1,259,597	2,505,379
Malays	228,174	225,459	453,633
Indians	134,544	123,247	257,791
Others	21,793	24,613	46,406
Total	1,630,293	1,632,916	3,263,209

* Figures refer to the resident population of Singapore.

BIRTHS, MARRIAGES AND DEATHS*

	Registered live births		Registered Marriages		Registered deaths	
	Number	Rate (per 1,000)	Number	Rate (per 1,000)†	Number	Rate (per 1,000)
1993	50,225	15.4	25,298	7.8	14,461	4.4
1994	49,554	14.7	24,662	7.3	14,946	4.4
1995	48,635	14.0	24,974	7.2	15,569	4.5
1996	48,577	13.2	24,111	6.6	15,590	4.2
1997	47,333	12.5	25,667	8.2	15,307	4.0
1998	43,838	11.2	23,106	7.3	15,657	4.0
1999	43,336	11.0	25,648	8.0	15,516	3.9
2000†	46,631	11.3	22,561	6.9	15,692	3.8

* Data are tabulated by year of registration, rather than by year of occurrence.

† Provisional.

Source: mainly UN, *Demographic Yearbook*.

2001: Registered marriages 22,280; Marriage rate (per 1,000) 6.7.

2002: Registered marriages 23,198; Marriage rate (per 1,000) 6.9.

Expectation of life (WHO estimates, years at birth): 79.6 (males 77.4; females 81.7) in 2002 (Source: WHO, *World Health Report*).

ECONOMICALLY ACTIVE POPULATION

('000 persons aged 15 years and over, at June of each year)

	2000*	2001	2002
Agriculture, hunting, forestry and fishing	5.1	6.3	6.1
Mining and quarrying	0.7	0.6	1.0
Manufacturing	434.9	384.0	367.6
Electricity, gas and water supply	7.1	10.4	9.0
Construction	274.0	124.9	119.1
Wholesale and retail trade; repair of motor vehicles, motorcycles and personal and household goods	286.8	303.6	304.4
Hotels and restaurants	114.5	128.3	125.3
Transport, storage and communications	196.5	228.2	218.8
Financial intermediation	96.3	108.7	107.9
Real estate, renting and business activities	226.2	243.1	237.4
Public administration and defence; compulsory social security	105.9	133.9	135.4
Education, health and social work	127.4	159.9	166.2
Other community, social and personal service activities; private households with employed persons	217.3	212.4	217.2
Extra-territorial organizations and bodies	2.1	2.4	2.3
Total employed	2,094.8	2,046.7	2,017.4
Unemployed	97.5	72.9	111.2
Total labour force	2,192.3	2,119.7	2,128.6
Males	1,324.3	1,190.4	1,202.1
Females	868.0	929.2	926.5

* Figures at census of 30 June 2000.

Source: ILO.

Health and Welfare

KEY INDICATORS

Total fertility rate (children per woman, 2002)	1.4
Under-5 mortality rate (per 1,000 live births, 2002)	4
HIV/AIDS (% of persons aged 15–49, 2003)	0.20
Physicians (per 1,000 head, 1998)	1.63
Hospital beds (per 1,000 head, 1994)	3.05
Health expenditure (2001): US $ per head (PPP)	993
Health expenditure (2001): % of GDP	3.9
Health expenditure (2001): public (% of total)	33.5
Access to water (% of persons, 2000)	100
Access to sanitation (% of persons, 2000)	100
Human Development Index (2002): ranking	25
Human Development Index (2002): value	0.902

For sources and definitions, see explanatory note on p. vi.

Agriculture

PRINCIPAL CROPS
(FAO estimates, '000 metric tons)

	2000	2001	2002
Vegetables	5.0	5.0	5.0

Source: FAO.

LIVESTOCK
(FAO estimates, '000 head, year ending September)

	2000	2001	2002
Pigs	300	300	250
Chickens	2,000	2,000	2,000
Ducks	600	600	600

Source: FAO.

LIVESTOCK PRODUCTS
('000 metric tons)

	2000	2001	2002
Pig meat*	21.3	23.2	20.1
Poultry meat*	76.0	75.4	75.5
Hen eggs	16.0	16.0	16.0

* FAO estimates.

Source: FAO.

Forestry

SAWNWOOD PRODUCTION
(FAO estimates, '000 cubic metres, incl. railway sleepers)

	1990	1991	1992
Coniferous (softwood)	5	10	5
Broadleaved (hardwood)	50	20	20
Total	55	30	25

1993–2002: Annual production as in 1992 (FAO estimates).

Source: FAO.

Fishing

(metric tons, live weight)

	2000	2001	2002
Capture	5,371	3,342	2,769
Shrimps and prawns	422	250	222
Common squids	348	186	185
Aquaculture	5,112	4,443	5,027
Indonesian snakehead	500	613	455
Milkfish	676	656	956
Green mussel	2,898	2,454	2,903
Total catch	10,483	7,785	7,796

Note: Figures exclude crocodiles, recorded by number rather than by weight. The number of estuarine crocodiles caught was: 481 in 2000; 1,041 in 2001; 1,058 in 2002.

Source: FAO.

Industry

PETROLEUM PRODUCTS
('000 metric tons)

	1998	1999	2000
Liquefied petroleum gas	1,005	972	883
Naphtha	4,149	3,710	3,933
Motor spirit (petrol)	4,840	3,959	3,973
Kerosene	1,348	859	798
Jet fuel	6,451	6,336	5,844
Gas-diesel (distillate fuel) oils	14,662	13,818	11,618
Residual fuel oil	12,136	6,918	4,497
Lubricating oils	1,133	1,326	1,347
Petroleum bitumen (asphalt)	1,352	1,370	1,246

Source: UN, *Industrial Commodity Statistics Yearbook*.

SELECTED OTHER PRODUCTS

	1988	1989	1990
Paints ('000 litres)	48,103.6	52,746.9	58,245.9
Broken granite ('000 metric tons)	6,914.0	7,007.5	6,371.7
Bricks ('000 units)	103,136	116,906	128,386
Soft drinks ('000 litres)	269,689.4	252,977.6	243,175.1
Plywood, plain and printed ('000 sq m)	31,307.0	28,871.3	26,106.9
Vegetable cooking oil (metric tons)	75,022	103,003	102,854
Animal fodder (metric tons)	110,106	115,341	104,541
Electricity (million kWh)	13,017.5	14,039.0	15,617.6
Gas (million kWh)	681.1	722.4	807.1
Cassette tape recorders ('000 sets)	15,450	14,006	18,059

Source: UN, *Industrial Commodity Statistics Yearbook*.

Plywood ('000 cu m, estimates): 280 per year in 1991–2000 (Source: FAO).

Electricity (million kWh): 29,520 in 1999; 31,665 in 2000; 33,089 in 2001 (Source: Asian Development Bank, *Key Indicators of Developing Asian and Pacific Countries*).

Finance

CURRENCY AND EXCHANGE RATES

Monetary Units
100 cents = 1 Singapore dollar (S $).

Sterling, US Dollar and Euro Equivalents (31 May 2004)
£1 sterling = S $3.1155;
US $1 = S $1.6981;
€1 = S $2.0795;
S $100 = £32.10 = US $58.89 = €48.09.

Average Exchange Rate (Singapore dollars per US $)
2001 1.7917
2002 1.7906
2003 1.7423

BUDGET
(S $ million)

Revenue*	2001	2002	2003
Tax revenue	25,109	21,484	20,736
Income tax	13,464	11,550	10,414
Corporate and personal income tax	12,602	10,926	10,028
Contributions by statutory board	862	624	386
Assets taxes	1,694	1,242	1,243
Taxes on motor vehicles	2,473	1,559	1,290
Customs and excise duties	1,840	1,697	1,802
Betting taxes	1,574	1,523	1,566
Stamp duty	835	700	648
Goods and services tax	2,013	2,098	2,724
Others	1,216	1,116	1,048
Fees and charges	4,947	3,751	3,587
Total (incl. others)	30,266	25,401	24,643

* Figures refer to operating revenue only. The data exclude investment income and capital revenue.

Expenditure	2001	2002	2003
Operating expenditure	17,846	19,244	19,236
Security and external relations	8,890	9,362	9,249
Social development	7,094	7,979	8,202
Education	4,366	4,768	4,876
Health	1,204	1,625	1,655
Community development and sports	515	526	581
Environment	413	469	452
Economic development	1,053	1,105	994
Communications and information technology*	386	—	—
Trade and industry	484	535	515
Transport*	69	379	304
Government administration	809	799	792
Development expenditure	9,999	7,877	7,953
Total	27,845	27,121	27,189

* The Ministry of Communications and Information Technology was renamed the Ministry of Transport in November 2001. Its portfolio of information technology and telecommunications was transferred to the Ministry of Information, Communications and the Arts.

INTERNATIONAL RESERVES

(US $ million at 31 December)

	2001	2002	2003
Gold and foreign exchange	74,851	81,367	94,975
IMF special drawing rights	150	177	207
Reserve position in the IMF	374	478	564
Total	75,375	82,021	95,746

Source: IMF, *International Financial Statistics*.

MONEY SUPPLY

(S $ million at 31 December)

	2001	2002	2003
Currency outside banks	11,868	12,360	12,838
Demand deposits at commercial banks	24,215	23,468	25,884
Total money	36,083	35,828	38,722

Source: IMF, *International Financial Statistics*.

COST OF LIVING

(National Consumer Price Index; base: year ending October 1998 = 100)

	2001	2002	2003
Food	101.9	101.9	102.5
Transport and communication	99.4	98.4	98.5
Clothing	97.5	97.8	98.2
Housing	101.8	99.6	99.2
Education	105.1	106.6	109.0
Health	105.7	109.1	111.3
All items (incl. others)	102.1	101.7	102.2

NATIONAL ACCOUNTS

(S $ million at current prices)

National Income and Product

	1998	1999	2000
Gross domestic product (GDP) in market prices	137,464.0	142,111.0	159,042.0
Primary incomes received from abroad	21,705.0	25,382.0	26,295.0
Less Primary incomes paid abroad	13,126.0	14,024.0	15,740.0
Gross national income (GNI)	146,043.0	153,469.0	169,597.0
Less Consumption of fixed capital	19,173.0	20,254.0	20,920.0
Net national income	126,870.0	133,215.0	148,676.0
Current transfers from abroad *Less* Current transfers paid abroad	–1,837.0	–1,972.0	–2,342.0
Net national disposable income	125,033.0	131,243.0	146,334.0

Source: UN, *National Accounts Statistics*.

Expenditure on the Gross Domestic Product

	2001	2002	2003*
Government final consumption expenditure	18,683.5	19,052.3	18,995.8
Private final consumption expenditure	67,472.8	69,212.8	68,652.3
Increase in stocks	–7,289.7	–7,260.9	–18,328.4
Gross fixed capital formation	45,586.0	40,705.0	39,573.4
Statistical discrepancy	162.5	367.4	–2,818.0
Total domestic expenditure	124,615.1	122,076.6	106,075.1
Trade in goods and services (net)	29,462.9	35,987.5	53,059.9
GDP in purchasers' values	154,078.0	158,064.1	159,135.0
GDP at constant 1995 prices	159,073.0	162,493.2	164,265.9

* Preliminary figures.

Gross Domestic Product by Economic Activity

	2001	2002	2003*
Agriculture, fishing and quarrying	177.5	165.8	163.5
Manufacturing	36,548.4	41,080.4	41,601.2
Electricity, gas and water	2,966.7	2,693.7	2,621.9
Construction	9,443.9	8,530.3	7,833.9
Wholesale and retail trade	19,078.8	19,510.7	20,686.0
Hotels and restaurants	3,628.4	3,503.0	2,998.2
Transport and communications	17,402.8	18,222.8	17,571.3
Financial services	19,075.0	18,921.2	18,357.7
Business services	22,213.7	21,641.1	21,025.5
Owner-occupied dwellings	5,610.4	5,635.7	5,647.0
Other services	18,164.0	19,279.9	19,427.3
Sub-total	154,309.6	159,184.6	157,933.5
Taxes on products	9,951.4	9,214.6	10,087.7
Less Financial intermediation services indirectly measured	10,183.0	10,335.1	8,886.2
GDP in purchasers' values	154,078.0	158,064.1	159,135.0

* Preliminary figures.

BALANCE OF PAYMENTS
(US $ million)

	2000	2001	2002
Exports of goods f.o.b.	139,861	124,443	128,374
Imports of goods f.o.b.	−127,563	−109,675	−109,825
Trade balance	12,298	14,768	18,549
Exports of services	29,098	28,855	29,701
Imports of services	−26,938	−26,886	−27,298
Balance on goods and services	14,459	16,737	20,953
Other income received	16,291	15,566	14,367
Other income paid	−16,349	−15,020	−15,512
Balance on goods, services and income	14,401	17,283	19,808
Current transfers received	127	122	122
Current transfers paid	−1,248	−1,268	−1,227
Current balance	13,280	16,137	18,704
Capital account (net)	−163	−161	−160
Direct investment abroad	−6,061	−9,548	−4,082
Direct investment from abroad	12,463	10,949	6,097
Portfolio investment assets	−11,482	−11,284	−11,374
Portfolio investment liabilities	−2,036	187	−1,272
Other investment assets	−8,452	−11,183	−289
Other investment liabilities	13,642	5,490	−4,735
Net errors and omissions	−4,386	−1,448	−1,546
Overall balance	6,806	−861	1,342

Source: IMF, *International Financial Statistics*.

External Trade

PRINCIPAL COMMODITIES
(distribution by SITC, S $ million)

Imports c.i.f.	2001	2002	2003
Mineral fuels, lubricants, etc.	26,119.3	27,199.2	30,191.4
Crude petroleum	13,589.0	13,204.0	13,951.9
Chemicals and related products	12,181.4	12,990.0	14,865.8
Basic manufactures	15,249.0	15,350.9	15,082.2
Machinery and equipment	123,932.5	122,632.4	131,707.0
Electric generators	8,826.4	8,542.1	8,662.4
Electronic components and parts	43,555.7	44,913.8	49,954.0
Aircraft and vessels	7,170.2	6,068.7	7,516.4
Miscellaneous manufactured articles	18,521.5	18,370.9	19,339.2
Total (incl. others)	207,692.1	208,311.9	222,811.1

Exports f.o.b.	2001	2002	2003
Mineral fuels, lubricants, etc.	22,472.8	23,250.7	28,050.2
Petroleum products	16,043.9	16,993.8	20,711.1
Chemicals and related products	17,631.7	20,816.5	29,571.4
Basic manufactures	8,429.6	8,776.2	9,352.5
Machinery and equipment	140,620.3	142,318.3	153,596.0
Electronic components and parts	51,127.8	54,587.2	63,418.9
Miscellaneous manufactured articles	19,040.4	19,066.1	21,362.9
Total (incl. others)	218,026.3	223,901.4	251,095.7

Source: International Enterprise Singapore.

PRINCIPAL TRADING PARTNERS
(S $ million)

Imports c.i.f.	2001	2002	2003
Australia	4,305.6	3,863.4	3,805.1
China, People's Republic	12,900.3	15,853.4	19,276.3
France	3,719.3	3,791.8	4,326.9
Germany	6,861.5	7,077.6	8,455.8
Hong Kong	4,985.0	5,075.1	5,000.0
India	2,003.1	2,075.3	2,510.2
Italy	2,350.4	2,321.2	2,775.6
Japan	28,794.1	26,079.8	26,808.3
Korea, Republic	6,842.7	7,690.5	8,637.4
Kuwait	2,431.8	3,118.0	3,682.9
Malaysia	35,974.6	37,950.9	37,527.7
Netherlands	2,047.3	2,324.8	1,898.0
Philippines	4,572.6	4,480.6	4,920.6
Saudi Arabia	7,557.7	6,828.8	6,823.4
Switzerland	3,501.7	3,322.1	3,508.0
Taiwan	8,830.6	9,530.1	11,263.0
Thailand	9,242.6	9,676.7	9,587.1
United Arab Emirates	2,440.1	2,816.1	3,285.7
United Kingdom	4,372.9	4,212.1	4,428.5
USA	34,137.0	29,515.2	31,060.2
Total (incl. others)	207,692.1	208,311.9	222,811.1

Exports f.o.b.	2001	2002	2003
Australia	5,658.9	6,029.1	8,148.1
China, People's Republic	9,545.0	12,268.1	17,638.2
France	2,850.8	2,614.3	3,508.3
Germany	7,690.9	7,257.2	7,624.8
Hong Kong	19,373.7	20,492.1	25,116.2
India	4,872.8	4,717.7	5,382.7
Japan	16,712.3	15,990.2	16,875.4
Korea, Republic	8,391.7	9,316.5	10,550.2
Malaysia	37,821.5	39,002.9	39,672.4
Netherlands	7,225.6	7,691.1	8,042.7
Philippines	5,516.1	5,438.0	5,636.1
Taiwan	11,219.5	11,086.7	12,011.8
Thailand	9,486.6	10,214.2	10,710.7
United Arab Emirates	2,055.1	2,172.2	2,527.7
United Kingdom	5,102.5	4,631.6	7,969.8
USA	33,533.6	32,935.3	33,460.1
Viet Nam	3,769.3	3,729.2	4,194.5
Total (incl. others)	218,026.3	223,901.4	251,095.7

Source: International Enterprise Singapore.

Transport

ROAD TRAFFIC
(registered vehicles)

	2001	2002	2003
Cars*	405,354	404,274	405,328
Motorcycles and scooters	130,910	131,437	134,767
Motor buses	12,624	12,707	12,653
Taxis	18,798	19,106	19,384
Goods vehicles (incl. private)	127,273	125,931	125,023
Others	13,411	13,501	13,888
Total	708,370	706,956	711,043

* Including private, company, tuition and private hire cars.

SHIPPING

Merchant Fleet
(at 31 December)

	2001	2002	2003
Number of vessels	1,729	1,768	1,761
Displacement (grt)	21,022,604	21,148,090	23,240,945

Source: Lloyd's Register-Fairplay, *World Fleet Statistics*.

International Sea-borne Shipping
(vessels of over 75 net registered tons)

	1995	1996	1997
Vessels:			
entered	104,014	117,723	130,333
cleared	104,123	117,662	130,237
Cargo ('000 freight tons):			
loaded	130,224.3	134,592	159,272
unloaded	175,259.7	179,572	188,234

Note: One freight ton equals 40 cubic feet (1.133 cubic metres) of cargo.

1998: Vessels entered 140,922; Vessels cleared 140,838.

1999: Vessels entered 141,523; Vessels cleared 141,745.

2000: Vessels entered 145,383; Vessels cleared 145,415.

Source: mainly UN, *Statistical Yearbook*.

CIVIL AVIATION

	1995	1996	1997
Passengers:			
arrived	10,919,739	11,587,394	11,915,969
departed	10,823,457	11,542,408	11,883,259
in transit	1,453,046	1,384,446	1,375,116
Mail (metric tons):			
landed	9,521	10,572	10,813
dispatched	9,543	10,380	10,978
Freight (metric tons):			
landed	577,749	622,019	696,710
dispatched	528,024	568,438	639,544

Source: Civil Aviation Authority of Singapore.

1998: Kilometres flown (million) 293; Passengers carried ('000) 13,316; Passenger-km (million) 58,174; Total ton-km (million) 10,381 (Source: UN, *Statistical Yearbook*).

1999: Kilometres flown (million) 326; Passengers carried ('000) 15,283; Passenger-km (million) 65,471; Total ton-km (million) 11,824 (Source: UN, *Statistical Yearbook*).

Tourism

FOREIGN VISITOR ARRIVALS
('000, incl. excursionists)

Country of Nationality	2001	2002	2003
Australia	550,681	538,402	392,818
China, People's Republic	497,398	670,093	568,389
Germany	166,981	157,508	121,351
Hong Kong	276,157	265,970	226,230
India	339,813	375,658	309,383
Indonesia	1,364,380	1,392,999	1,341,174
Japan	755,766	723,422	433,972
Korea, Republic	359,083	371,050	261,333
Malaysia*	578,719	548,653	439,298
Philippines	190,630	195,562	176,522
Taiwan	222,087	209,317	144,901
Thailand	260,958	263,865	235,728
United Kingdom	460,018	458,522	387,905
USA	343,805	327,647	250,612
Total (incl. others)*	7,522,163	7,567,039	6,125,480

* Figures exclude arrivals of Malaysians by land.

Tourism receipts (S $ million): 6,292.6 in 2000; 5,699.3 in 2001; 5,425.8 in 2002.

Source: Singapore Tourism Board.

Communications Media

(at 31 December)

	2000	2001	2002
Television receivers ('000 in use)	1,200	n.a.	n.a.
Telephones ('000 main lines in use)	1,946.5	1,947.5	1,927.2
Mobile cellular telephones ('000 subscribers)	2,747.4	2,991.6	3,312.6
Personal computers ('000 in use)	1,941	2,100	2,590
Internet users ('000)	1,200	1,700	2,100

Sources: International Telecommunication Union; UN, *Statistical Yearbook*.

Radio receivers ('000 in use): 2,550 in 1997.

Facsimile machines ('000 in use, estimate): 100 in 1998.

Daily newspapers: 8 (with average circulation of 1,095,000 copies) in 1996.

Non-daily newspapers: 2 in 1996.

Sources: mainly UNESCO, *Statistical Yearbook*; UN, *Statistical Yearbook*.

Education

(June 2003)

	Institutions	Students	Teachers
Primary	175	299,939	12,356
Secondary	165	206,426	11,146
Centralized Institutes	2	851	106
Junior Colleges	16	23,708	1,991
Institute of Technical Education	1	17,941	1,257*
Polytechnics	5	55,376	3,931*
National Institute of Education	1	2,953	561*
Universities	4	40,095	2,826*

* June 2000 figure.

Adult literacy rate (UNESCO estimates): 92.5% (males 96.6%; females 88.6%) in 2002 (Source: UN Development Programme, *Human Development Report*).

Directory

The Constitution

A new Constitution came into force on 3 June 1959, with the establishment of the self-governing State of Singapore. This was subsequently amended as a consequence of Singapore's affiliation to Malaysia (September 1963 to August 1965) and as a result of its adoption of republican status on 22 December 1965. The Constitution was also amended in January 1991 to provide for the election of a President by universal adult suffrage, and to extend the responsibilities of the presidency, which had previously been a largely ceremonial office. A constitutional amendment in October 1996 placed restrictions on the presidential right of veto. The main provisions of the Constitution are summarized below:

HEAD OF STATE

The Head of State is the President, elected by universal adult suffrage for a six-year term. He normally acts on the advice of the Cabinet, but is vested with certain functions and powers for the purpose of safeguarding the financial reserves of Singapore and the integrity of the Public Services.

THE CABINET

The Cabinet, headed by the Prime Minister, is appointed by the President and is responsible to Parliament.

THE LEGISLATURE

The Legislature consists of a Parliament of 84 elected members, presided over by a Speaker who may be elected from the members of Parliament themselves or appointed by Parliament although he may not be a member of Parliament. Members of Parliament are elected by universal adult suffrage for five years (subject to dissolution) in single-member and multi-member constituencies.* Additionally, up to three 'non-constituency' seats may be offered to opposition parties, in accordance with a constitutional amendment approved in 1984, while legislation approved in 1990, and amended in 1997, enables the Government to nominate up to nine additional, politically-neutral members for a term of two years; these members have restricted voting rights.

A 21-member Presidential Council, chaired by the Chief Justice, examines material of racial or religious significance, including legislation, to see whether it differentiates between racial or religious communities or contains provisions inconsistent with the fundamental liberties of Singapore citizens.

CITIZENSHIP

Under the Constitution, Singapore citizenship may be acquired either by birth, descent or registration. Persons born when Singapore was a constituent State of Malaysia could also acquire Singapore citizenship by enrolment or naturalization under the Constitution of Malaysia.

* A constitutional amendment was introduced in May 1988, whereby 39 constituencies were merged to form 13 'group representation constituencies' which would return 'teams' of three Members of Parliament. At least one member of each team was to be of minority (non-Chinese) racial origin. In January 1991 the Constitution was further amended, stipulating that the number of candidates contesting 'group representation constituencies' should be a minimum of three and a maximum of four. The maximum was increased to six by constitutional amendment in October 1996.

The Government

HEAD OF STATE

President: SELLAPAN RAMANATHAN (S. R.) NATHAN (took office 1 September 1999).

CABINET
(September 2004)

Prime Minister and Minister of Finance: Brig.-Gen. (retd) LEE HSIEN LOONG.

Senior Minister in the Prime Minister's Office: GOH CHOK TONG.

Minister Mentor: LEE KUAN YEW.

Deputy Prime Minister and Minister of Law: Prof. SHANMUGAM JAYAKUMAR.

Deputy Prime Minister and Co-ordinating Minister of Security and Defence: Dr TONY TAN KENG YAM.

Minister of Foreign Affairs: Brig.-Gen. (retd) GEORGE YONG BOON YEO.

Minister of Home Affairs: WONG KAN SENG.

Minister of Trade and Industry: LIM HNG KIANG.

Minister of Information, Communications and the Arts: Dr LEE BOON YANG.

Minister of Manpower: Dr NG ENG HEN.

Minister of Health: KHAW BOON WAN.

Minister of Education: THARMAN SHANMUGARATNAM.

Minister of Transport: YEO CHEOW TONG.

Minister in the Prime Minister's Office: LIM SWEE SAY.

Minister in the Prime Minister's Office: LIM BOON HENG.

Minister of National Development: MAH BOW TAN.

Minister of Community Development, Youth and Sports: Dr VIVIAN BALAKRISHNAN (acting).

Minister of Defence: Rear-Adm. (retd) TEO CHEE HEAN.

Minister of the Environment and Water Resources: Dr YAACOB IBRAHIM.

Minister of State for Community Development and Sports: YU-FOO YEE SHOON.

Minister of State for Finance and Transport: LIM HWEE HUA.

Second Minister of State for Finance: RAYMOND LIM SIANG KEAT (acting).

MINISTRIES

Office of the President: The Istana, Orchard Rd, Singapore 238823; internet www.istana.gov.sg.

Office of the Prime Minister: Orchard Rd, Istana Annexe, Istana, Singapore 238823; tel. 62358577; fax 67324627; e-mail goh_chok_tong@pmo.gov.sg; internet www.pmo.gov.sg.

Ministry of Community Development, Youth and Sports: 512 Thomson Rd, MCD Bldg, Singapore 298136; tel. 62589595; fax 63548324; e-mail mcds_home@mcds.gov.sg; internet www.mcds.gov.sg.

Ministry of Defence: Gombak Drive, off Upper Bukit Timah Rd, Mindef Bldg, Singapore 669645; tel. 67608844; fax 67646119; e-mail mfu@starnet.gov.sg; internet www.mindef.gov.sg.

Ministry of Education: 1 North Buona Vista Drive, MOE Bldg, Singapore 138675; tel. 68721110; fax 67755826; e-mail contact@moe.edu.sg; internet www.moe.gov.sg.

Ministry of the Environment and Water Resources: 40 Scotts Rd, Environment Bldg, 22nd Floor, Singapore 228231; tel. 67327733; fax 67319456; e-mail contact_env@env.gov.sg; internet www.env.gov.sg.

Ministry of Finance: 100 High St, 06-03 The Treasury, Singapore 179434; tel. 62259911; fax 63327435; e-mail mof_qsm@mof.gov.sg; internet www.mof.gov.sg.

Ministry of Foreign Affairs: Tanglin, Singapore 248163; tel. 63798000; fax 64747885; e-mail mfa@mfa.gov.sg; internet www.mfa.gov.sg.

Ministry of Health: 16 College Rd, College of Medicine Bldg, Singapore 169854; tel. 63259220; fax 62241677; e-mail moh_info@moh.gov.sg; internet www.moh.gov.sg.

Ministry of Home Affairs: New Phoenix Park, 28 Irrawaddy Rd, Singapore 329560; tel. 62359111; fax 62546250; e-mail mha_feedback@mha.gov.sg; internet www.mha.gov.sg.

Ministry of Information, Communications and the Arts: 140 Hill St, 02-02 MITA Bldg, Singapore 179369; tel. 62707988; fax 68379480; e-mail mita_pa@mita.gov.sg; internet www.mita.gov.sg.

Ministry of Law: 100 High St, 08-02 The Treasury, Singapore 179434; tel. 63328840; fax 63328842; e-mail mlaw_enquiry@minlaw.gov.sg; internet www.minlaw.gov.sg.

Ministry of Manpower: 18 Havelock Rd, 07-01, Singapore 059764; tel. 65341511; fax 65344840; e-mail mom_hq@mom.gov.sg; internet www.mom.gov.sg.

Ministry of National Development: 5 Maxwell Rd, 21/22-00, Tower Block, MND Complex, Singapore 069110; tel. 62221211; fax 63257254; e-mail mnd_hq@mnd.gov.sg; internet www.mnd.gov.sg.

Ministry of Trade and Industry: 100 High St, 09-01 The Treasury, Singapore 179434; tel. 62259911; fax 63327260; e-mail mti_email@mti.gov.sg; internet www.mti.gov.sg.

Ministry of Transport: 460 Alexandra Rd, 39-00 PSA Bldg, Singapore 119963; tel. 62707988; fax 63757734; e-mail mot@mot.gov.sg; internet www.mot.gov.sg.

President and Legislature

PRESIDENT

On 18 August 1999 SELLAPAN RAMANATHAN (S. R.) NATHAN was nominated as President by a state-appointed committee. S. R. Nathan was the sole candidate for the Presidency, following the rejection by the committee of the only two other potential candidates, and was officially appointed to the post on 1 September.

PARLIAMENT

Parliament House

1 Parliament Place, Singapore 178880; tel. 63326666; fax 63325526; e-mail parl@parl.gov.sg; internet www.parliament.gov.sg.

Speaker: ABDULLAH TARMUGI.

General Election, 3 November 2001

	Seats
People's Action Party	82*
Workers' Party	1
Singapore Democratic Alliance	1
Total	84

* 55 seats were unopposed.

Political Organizations

Angkatan Islam (Singapore Muslim Movement): Singapore; f. 1958; Pres. MOHAMED BIN OMAR; Sec.-Gen. IBRAHIM BIN ABDUL GHANI.

National Solidarity Party: Blk 531, Upper Cross St, Hong Lim Complex, 03-30, Singapore 050531; fax 65366388; e-mail nsp@nsp-singapore.org; internet www.nsp-singapore.org; f. 1987; joined Singapore Democratic Alliance (SDA) in July 2001; Pres. YIP YEW WENG; Sec.-Gen. STEVE CHIA.

People's Action Party (PAP): Blk 57B, PCF Bldg, 01-1402 New Upper Changi Rd, Singapore 463057; tel. 62444600; fax 62430114; e-mail paphq@pap.org.sg; internet www.pap.org.sg; f. 1954; governing party since 1959; 12-member Cen. Exec. Cttee; Chair. Dr TONY TAN KENG YAM; Sec.-Gen. GOH CHOK TONG.

Pertubuhan Kebangsaan Melayu Singapura (PKMS) (Singapore Malay National Organization): 218F Changi Rd, 4th Floor, PKM Bldg, Singapore 1441; tel. 64470468; fax 63458724; f. in 1950 as the United Malay National Organization (UMNO) of Malaysia; renamed as UMNO Singapore in 1954 and as PKMS in 1967; seeks to advance the implementation of the special rights of Malays in Singapore, as stated in the Constitution, to safeguard and promote the advancement of Islam and to encourage racial harmony and goodwill in Singapore; joined Singapore Democratic Alliance (SDA) in July 2001; Pres. Haji BORHAN ARIFFIN; Sec.-Gen. MOHD RAHIZAN YAACOB.

Singapore Democratic Alliance (SDA): Singapore; f. 2001 to contest 2001 general election; coalition of Pertubuhan Kebangsaan Melayu Singapura (PKMS), Singapore People's Party (SPP), National Solidarity Party and the Singapore Justice Party; Chair. CHIAM SEE TONG.

Singapore Democratic Party (SDP): 1357A Serangoon Rd, Singapore 328240; tel. and fax 63981675; e-mail speakup@singaporedemocrats.org; internet www.singaporedemocrats.org; f. 1980; 12-mem. Cen. Cttee; Chair. LING HOW DOONG; Sec.-Gen. CHEE SOON JUAN.

Singapore Justice Party: Singapore; f. 1972; joined Singapore Democratic Alliance (SDA) in July 2001; Pres. A. R. SUIB; Sec.-Gen. MUTHUSAMY RAMASAMY.

Singapore People's Party (SPP): Singapore; f. 1993; a breakaway faction of the SDP, espousing more moderate policies; joined Singapore Democratic Alliance (SDA) in July 2001; 12-mem. Cen. Exec. Cttee; Chair. SIN KEK TONG; Sec.-Gen. CHIAM SEE TONG.

Workers' Party: 411B Jalan Besar, Singapore 209014; tel. 62984765; fax 64544404; e-mail wp@wp.org.sg; internet www.wp.org.sg; f. 1961; active as opposition party in Singapore since 1957, merged with Barisan Sosialis (Socialist Front) in 1988; seeks to establish a democratic socialist govt with a constitution guaranteeing fundamental citizens' rights; Chair. SYLVIA LIM SWEE LIAH; Sec.-Gen. LOW THIA KHIANG.

Other parties are the Alliance Party Singapura, Barisan Sosialis (Socialist Front), the Democratic People's Party, the Democratic Progressive Party, the National Party of Singapore, the Partai Rakyat, the People's Front, the Parti Kesatuan Ra'ayat (United Democratic Party), the People's Republican Party, the Persatuan Melayu Singapura, the Singapore Chinese Party, the Singapore Indian Congress, the Singapore National Front, the United National Front, the United People's Front and the United People's Party.

Diplomatic Representation

EMBASSIES AND HIGH COMMISSIONS IN SINGAPORE

Angola: 9 Temasek Blvd, 44-03 Suntec Tower Two, Singapore 038989; tel. 63419360; fax 63419367; Ambassador ELISEA AVILA DE JESUS FIGUIEREDO.

Australia: 25 Napier Rd, Singapore 258507; tel. 68364100; fax 67337134; e-mail public-affairs-sing@dfat.gov.au; internet www.singapore.embassy.gov.au; High Commissioner GARY QUINLAN.

Austria: 600 North Bridge Rd, 24-04/05 Parkview Sq., Singapore 188778; tel. 63966350; fax 63966340; e-mail embassy.singapore@austriantrade.org; internet www.austria.org.sg; Chargé d'affaires a.i. GERHARD MESCHKE.

Bangladesh: 101 Thomson Rd, 05-04 United Sq., Singapore 307591; tel. 62550075; fax 62551824; e-mail bdoot@singnet.com.sg; internet www.bangladesh.org.sg; High Commissioner MUNSHI FAIZ AHMAD.

Belgium: 8 Shenton Way, 14-01 Temasek Tower, Singapore 068811; tel. 62207677; fax 62226976; e-mail singapore@diplobel.org; internet www.diplomatie.be/singapore; Ambassador PHILIPPE KRIDELKA.

Brazil: 101 Thomson Rd, 09-05 United Sq., Singapore 307591; tel. 62566001; fax 62566619; e-mail consular1@brazil.org.sg; internet web.singnet.com.sg/~cinbrem; Ambassador JOÃO GUALBERTO MARQUES-PORTO.

Brunei: 325 Tanglin Rd, Singapore 247955; tel. 67339055; fax 67375275; High Commissioner Pengiran Dato' Paduka Haji YUSOF BIN Pengiran KULA.

Cambodia: 152 Beach Rd, 11-05 Gateway East, Singapore 189721; tel. 62993028; fax 62993622; e-mail cambodiaembasy@pacific.net.sg; Ambassador KEM MONGKOL.

Canada: 80 Anson Rd, 14-00 and 15-01/02 IBM Towers, Singapore 079907; tel. 63253200; fax 63253297; e-mail echiesg@pacific.net.sg; internet www.dfait-maeci.gc.ca/singapore; High Commissioner (vacant).

Chile: 105 Cecil St, 25-00 The Octagon, Singapore 069534; tel. 62238577; fax 62250677; e-mail echilesg@pacific.net.sg; internet www.conchile.org.sg; Ambassador ANGEL FLISFISCH FERNÁNDEZ.

China, People's Republic: 150 Tanglin Rd, Singapore 247969; tel. 64180239; fax 64795345; e-mail chinaemb_sg@mfa.gov.cn; internet www.chinaembassy.org.sg; Ambassador ZHANG YUN.

Czech Republic: 7 Temasek Blvd, 18-02 Suntec Tower 1, Singapore 038987; tel. 63322378; fax 63322372; e-mail singapore@embassy.mzv.cz; internet www.mfa.cz/singapore; Chargé d'affaires RUDOLF HYKL.

Denmark: 101 Thomson Rd, 13-01/02 United Sq., Singapore 307591; tel. 63555010; fax 62533764; e-mail sinamb@um.dk; internet www.denmark.com.sg; Ambassador JØRGEN ØRSTRØM MØLLER.

Egypt: 75 Grange Rd, Singapore 249579; tel. 67371811; fax 67323422; Ambassador NADIA KAFAFI.

Finland: 101 Thomson Rd, 21-03 United Sq., Singapore 307591; tel. 62544042; fax 62534101; e-mail sanomat.sin@formin.fi; Ambassador RISTO R. REKOLA.

France: 101–103 Cluny Park Rd, Singapore 259595; tel. 68807800; fax 68807801; e-mail ambassadeur@france.org.sg; internet www.france.org.sg; Ambassador JEAN-PAUL REAU.

Germany: 545 Orchard Rd, 14-00 Far East Shopping Centre, Singapore 238882; tel. 67371355; fax 67372653; e-mail germany@singnet.com.sg; internet www.germanembassy-singapore.org.sg; Ambassador ANDREAS MICHAELIS.

Holy See: 55 Waterloo St 6, Singapore 187954 (Apostolic Nunciature); tel. 63372466; fax 63372466; Apostolic Nuncio Most Rev. SALVATORE PENNACCHIO (Titular Archbishop of Montemarano).

Hungary: 250 North Bridge Rd, 29-01 Raffles City Tower, Singapore 179101; tel. 68830882; fax 68830177; e-mail humemsin@singnet.com.sg; Ambassador Dr GYÖRGY NANOVFSZKY.

India: 31 Grange Rd, India House, Singapore 239702; tel. 67376777; fax 67326909; e-mail indiahc@pacific.net.sg; internet www.embassyofindia.com; High Commissioner ALOK PRASAD.

Indonesia: 7 Chatsworth Rd, Singapore 249761; tel. 67377422; fax 67375037; e-mail administrator_kbri@kbri.org.sg; internet www.kbri.org.sg; Ambassador SLAMET HIDAYAT.

Ireland: Ireland House, 541 Orchard Rd, 8th Floor, Liat Towers, Singapore 238881; tel. 62387616; fax 62387615; e-mail ireland@magix.com.sg; Ambassador HUGH SWIFT.

Israel: 58 Dalvey Rd, Singapore 259463; tel. 68349200; fax 67337008; e-mail singapor@israel.org; internet www.israelbiz.org.sg; Ambassador ITZHAK SHOHAM.

Italy: 101 Thomson Rd, 27-02 United Sq., Singapore 307591; tel. 62506022; fax 62533301; e-mail ambitaly@italyemb.org.sg; internet www.italyemb.org.sg; Ambassador GUIDO SCALICI.

Japan: 16 Nassim Rd, Singapore 258390; tel. 62358855; fax 67331039; e-mail eojs04@singnet.com.sg; internet www.sg.emb-japan.go.jp; Ambassador TAKAAKI KOJIMA.

Korea, Democratic People's Republic: 7500A Beach Rd, 09-320 The Plaza, Singapore 199591; tel. 64403498; fax 63482026; Ambassador JI JAE-SUK.

Korea, Republic: 47 Scotts Rd, 08-00 Goldbell Towers, Singapore 228233; tel. 62561188; fax 62543191; e-mail info@koreaembassy.org.sg; internet www.koreaembassy.org.sg; Ambassador RYU KWANG-SOK.

Laos: 101 Thomson Rd, 05-03A United Sq., Singapore 307591; tel. 62506044; fax 62506014; e-mail laoembsg@singnet.com.sg; Ambassador DONE SOMVORACHIT.

Malaysia: 30 Hill St, 02-01, Singapore 179360; tel. 62350111; fax 67336135; e-mail mwspore@singnet.com.sg; High Commissioner Dato' NAGALINGAM PARAMESWARAN.

Mexico: 152 Beach Rd, 06-07/08, 6th Floor, Gateway East Tower, Singapore 189721; tel. 62982678; fax 62933484; e-mail embamexsing@embamexsing.org.sg; internet www.embamexsing.org.sg; Ambassador ANDRÉS CARRAL CUEVAS.

Myanmar: 15 St Martin's Drive, Singapore 257996; tel. 67350209; fax 67356236; e-mail ambassador@mesingapore.org.sg; Ambassador U HLA THAN.

Netherlands: 541 Orchard Rd, 13-01 Liat Towers, Singapore 238881; tel. 67371155; fax 67371940; e-mail nlgovsin@singnet.com.sg; internet www.nethemb.org.sg; Ambassador H. J. VAN PESCH.

New Zealand: 391A Orchard Rd, 15-06/10 Ngee Ann City, Singapore 238873; tel. 62359966; fax 67339924; e-mail enquiries@nz-high-com.org.sg; internet www.nzembassy.com; High Commissioner RICHARD STURGE GRANT.

Nigeria: 390 Havelock Rd, 06-06 King's Centre, Singapore 169662; tel. 67321743; fax 67321742; e-mail highcommission@nigerian-singapore.org.sg; High Commissioner ALEX CHIKE ANIEGBO.

Norway: 16 Raffles Quay, 44-01 Hong Leong Bldg, Singapore 048581; tel. 62207122; fax 62202191; e-mail emb.singapore@mfa.no; internet www.norway.org.sg; Ambassador ENOK NYGAARD.

Pakistan: 1 Scotts Rd, 24-02/04 Shaw Centre, Singapore 228208; tel. 67376988; fax 67374096; e-mail parep@singnet.com.sg; internet www.parep.org.sg; High Commissioner SAJJAD ASHRAF.

Panama: 16 Raffles Quay, 41-06 Hong Leong Bldg, Singapore 048581; tel. 62218677; fax 62240892; e-mail pacosin@pacific.net.sg; Ambassador RICARDO A. ARAGON.

Peru: 390 Orchard Rd, 12-03 Palais Renaissance, Singapore 238871; tel. 67388595; fax 67388601; e-mail embperu@pacific.net.sg; Ambassador FERNANDO GUILLÉN.

Philippines: 20 Nassim Rd, Singapore 258395; tel. 67373977; fax 67339544; e-mail php@pacific.net.sg; internet www.rpsing.org; Ambassador BELEN F. ANOTA.

Poland: 435 Orchard Rd, 10-01/02 Wisma Atria, Singapore 238877; tel. 67340466; fax 67346129; e-mail polish_embassy@pacific.net.sg; internet polishembassyweb.com.sg; Chargé d'affaires a.i. FLORIAN BUKS.

Romania: 48 Jalan Harom Setangkai, Singapore 258827; tel. 64683424; fax 64683425; e-mail comofrom@starhub.net.sg; Chargé d'affaires a.i. SEVER COTU.

Russia: 51 Nassim Rd, Singapore 258439; tel. 62351834; fax 67334780; e-mail rosposol@pacific.net.sg; internet www.singapore.mid.ru; Ambassador SERGEI B. KISELEV.

Saudi Arabia: 40 Nassim Rd, Singapore 258449; tel. 67345878; fax 67385291; Ambassador Dr MOHAMAD AMIN KURDI.

South Africa: 331 North Bridge Rd, 15/01-06 Odeon Towers, Singapore 188720; tel. 63393319; fax 63396658; e-mail sahcp@singnet.com.sg; internet www.southafricahc.org.sg; High Commissioner ZANELE MAKINA.

Spain: 7 Temasek Blvd, 39-00 Suntec Tower 1, Singapore 038987; tel. 63333035; fax 63333025; e-mail singapur@mcx.es; Ambassador FRANCISCO JOSÉ RABENA BARRACHINA.

Sri Lanka: 13-07/12 Goldhill Plaza, 51 Newton Rd, Singapore 308900; tel. 62544595; fax 62507201; e-mail slhcs@lanka.com.sg; internet www.lanka.com.sg; High Commissioner AJIT JAYARATNE.

Sweden: 111 Somerset Rd, 05-01 Power Bldg, Singapore 238164; tel. 64159720; fax 64159747; e-mail swedemb@singnet.com.sg; internet www.swedenabroad.com/singapore; Ambassador TEPPO TAURIAINEN.

Switzerland: 1 Swiss Club Link, Singapore 288162; tel. 64685788; fax 64668245; e-mail vertretung@sin.rep.admin.ch; internet www.eda.admin.ch/singapore_emb/e/home.html; Ambassador DANIEL WOKER.

Thailand: 370 Orchard Rd, Singapore 238870; tel. 67372158; fax 67320778; e-mail thaisgp@singnet.com.sg; internet www.thaiembsingapore.org; Ambassador THAKUR PHANIT.

Turkey: 2 Shenton Way 10-03, SGX Centre 1, Singapore 068804; tel. 65333390; fax 65333360; e-mail turksin@singnet.com.sg; Ambassador HAYRI EROL.

Ukraine: 50 Raffles Place, 16-05 Singapore Land Tower, Singapore 048623; tel. 65356550; fax 65352116; e-mail emb_sg@mfa.gov.ua; internet www.embassy-ukraine.com; Ambassador OLEKSANDR O. HORIN.

United Arab Emirates: 96 Somerset Rd, 08-01/08 UOL Bldg, Singapore 238163; tel. 62388206; fax 62380081; e-mail emarat@singnet.com.sg.

United Kingdom: 100 Tanglin Rd, Singapore 247919; tel. 64244200; fax 64244218; e-mail commercial.singapore@fco.gov.uk; internet www.britishhighcommission.gov.uk; High Commissioner ALAN COLLINS.

USA: 27 Napier Rd, Singapore 258508; tel. 64769100; fax 64769340; e-mail singaporeusembassy@state.gov; internet singapore.usembassy.gov; Ambassador FRANKLIN L. LAVIN.

Viet Nam: 10 Leedon Park, Singapore 267887; tel. 64625938; fax 64625936; e-mail vnemb@singnet.com.sg; Ambassador DUONG VAN QUANG.

Judicial System

Supreme Court: St Andrew's Rd, Singapore 178957; tel. 63381034; fax 63379450; e-mail supcourt_qsm@supcourt.gov.sg; internet www.supcourt.gov.sg.

The judicial power of Singapore is vested in the Supreme Court and in the Subordinate Courts. The Judiciary administers the law with complete independence from the executive and legislative branches of the Government; this independence is safeguarded by the Constitution. The Supreme Court consists of the High Court and the Court of Appeal. The Chief Justice is appointed by the President if the latter, acting at his discretion, concurs with the advice of the Prime Minister. The other judges of the Supreme Court are appointed in the same way, in consultation with the Chief Justice. At 1 June 2004 there were two judges of appeal, including the Chief Justice, and 10 judges in the Supreme Court. Under a 1979 constitutional amendment, the position of judicial commissioner of the Supreme Court was created 'to facilitate the disposal of business in the Supreme Court'. A judicial commissioner has the powers and functions of a judge, and is appointed for such period as the President thinks fit. At 1 June 2004 there were two judicial commissioners in the Supreme Court.

The Subordinate Courts consist of District Courts and Magistrates' Courts. In addition, there are also specialized courts such as the Coroner's Court, Family Court, Juvenile Court, Mentions Court, Night Court, Sentencing Courts and Filter Courts. The Primary Dispute Resolution Centre and the Small Claims Tribunals are also managed by the Subordinate Courts. The Subordinate Courts have also established the Multi-Door Courthouse, which serves as a one-stop centre for the screening and channelling of any cases to the most appropriate forum for dispute resolution.

District Courts and Magistrates' Courts have original criminal and civil jurisdiction. District Courts try offences for which the maximum penalty does not exceed 10 years of imprisonment and in civil cases where the amount claimed does not exceed S $250,000. Magistrates' Courts try offences for which the maximum term of imprisonment does not exceed three years. The jurisdiction of Magistrates' Courts in civil cases is limited to claims not exceeding S $60,000. The Coroners' Court conducts inquests. The Small Claims Tribunal has jurisdiction over claims relating to a dispute arising from any contract for the sale of goods or the provision of services and any claim in tort in respect of damage caused to any property involving an amount that does not exceed S $10,000. The Juvenile Court deals with offences committed by young persons aged under 16 years.

The High Court has unlimited original jurisdiction in criminal and civil cases. In its appellate jurisdiction it hears criminal and civil appeals from the District Courts and Magistrates' Courts. The Court of Appeal hears appeals against the decisions of the High Court in both criminal and civil matters. In criminal matters, the Court of Appeal hears appeals against decisions made by the High Court in the exercise of its original criminal jurisdiction. In civil matters, the Court of Appeal hears appeals against decisions made by the High Court in the exercise of both its original and appellate jurisdiction.

With the enactment of the Judicial Committee (Repeal) Act 1994 in April of that year, the right of appeal from the Court of Appeal to the Judicial Committee of the Privy Council in the United Kingdom was abolished. The Court of Appeal is now the final appellate court in the Singapore legal system.

Attorney-General: Chan Sek Keong.

Chief Justice: Yong Pung How.

Judge of Appeal: Chao Hick Tin.

Judges of the High Court: Lai Kew Chai, M. P. H. Rubin, Kan Ting Chiu, Lai Siu Chiu, Judith Prakash, Tan Lee Meng, Choo Han Teck, Belinda Ang, Woo Bih Li, Tay Yong Kwang.

Judicial Commissioners: V. K. Rajah, Andrew Ang.

Religion

According to the 2000 census, 64.4% of ethnic Chinese, who constituted 76.8% of the population, professed either Buddhism or Daoism (including followers of Confucius, Mencius and Lao Zi) and 16.5% of Chinese adhered to Christianity. Malays, who made up 13.9% of the population, were 99.6% Muslim. Among Indians, who constituted 7.9% of the population, 55.4% were Hindus, 25.6% Muslims, 12.1% Christians and 6.3% Sikhs, Jains or adherents of other faiths. There are small communities of Zoroastrians and Jews. Freedom of worship is guaranteed by the Constitution.

BAHÁ'Í FAITH

National Spiritual Assembly: 110D Wishart Rd, Singapore 098733; tel. 62733023; fax 62732497; e-mail secretariat@bahai.org.sg; internet www.bahai.org.sg.

BUDDHISM

Buddhist Union: 28 Jalan Senyum, Singapore 418152; tel. 64435959; fax 64443280.

Singapore Buddhist Federation: 12 Ubi Ave 1, Singapore 408932; tel. 67444635; fax 67473618; e-mail buddhist@singnet.com.sg; internet www.buddhist.org.sg; f. 1948.

Singapore Buddhist Sangha Organization: 88 Bright Hill Drive, Singapore 579644.

Evangelical Fellowship of Singapore (EFOS): Singapore; f. 1980.

CHRISTIANITY

National Council of Churches: 6 Mt Sophia, Singapore 228457; tel. 63372150; fax 63360368; e-mail nccs@cyberway.com.sg; f. 1948; six mem. churches, six assoc. mems; Pres. Bishop John Tan; Gen. Sec. Rev. Canon Dr James Wong.

Singapore Council of Christian Churches (SCCC): Singapore; f. 1956.

The Anglican Communion

The Anglican diocese of Singapore (also including Indonesia, Laos, Thailand, Viet Nam and Cambodia) is part of the Province of the Anglican Church in South-East Asia.

Bishop of Singapore: The Rt Rev. Dr John Hiang Chea Chew, 4 Bishopsgate, Singapore 249970; tel. 64741661; fax 64791054; e-mail bpoffice@anglican.org.sg.

Orthodox Churches

The Orthodox Syrian Church and the Mar Thoma Syrian Church are both active in Singapore.

The Roman Catholic Church

Singapore comprises a single archdiocese, directly responsible to the Holy See. In December 2002 there were an estimated 159,132 adherents in the country, representing 3.8% of the total population.

Archbishop of Singapore: Most Rev. Nicholas Chia, Archbishop's House, 31 Victoria St, Singapore 187997; tel. 63378818; fax 63334725; e-mail nc@veritas.org.sg.

Other Christian Churches

Brethren Assemblies: Bethesda Hall (Ang Mo Kio), 601 Ang Mo Kio Ave 4, Singapore 569898; tel. 64587474; fax 64566771; e-mail bethesda@pacific.net.sg; internet www.bethesdahall.com; f. 1864; Hon. Sec. Wong Tuck Keong.

Methodist Church in Singapore: 70 Barker Rd, Singapore 309936; tel. 64784786; fax 64784794; e-mail episcopacy@methodist.org.sg; internet www.methodist.org.sg; f. 1885; 31,657 mems (July 2002); Bishop Dr Robert Solomon.

Presbyterian Church: 3 Orchard Rd, cnr Penang Rd, Singapore 238825; tel. 63376681; fax 63391979; e-mail orpcenglish@orpc.org.sg; f. 1856; services in English, Chinese (Mandarin), Indonesian and German; 2,000 mems; Chair. Rev. David Burke.

Singapore Baptist Convention: 01 Goldhill Plaza, 03-19 Podium Blk, Singapore 308899; tel. 62538004; fax 62538214; e-mail sbcnet@gala.com.sg; internet www.baptistconvention.org.sg; f. 1974; Chair. Rev. Sebastian Chua; Asst Exec. Dir Alan Phua.

Other denominations active in Singapore include the Lutheran Church and the Evangelical Lutheran Church.

HINDUISM

Hindu Advisory Board: c/o 397 Serangoon Rd, Singapore 218123; tel. 62963469; fax 62929766; e-mail heb@pacific.net.sg; Chair. Ajaib Hari Dass; Sec. E. Sanmugam.

Hindu Endowments Board: 397 Serangoon Rd, Singapore 218123; tel. 62963469; fax 62929766; e-mail heb@pacific.net.sg; internet www.gov.sg/heb; Chair. V. R. Nathan; Sec. Satish Appoo.

ISLAM

Majlis Ugama Islam Singapura (MUIS) (Islamic Religious Council of Singapore): Islamic Centre of Singapore, 273 Braddell Rd, Singapore 579702; tel. 62568188; fax 62537572; e-mail maarof@muis.gov.sg; internet www.muis.gov.sg; f. 1968; Pres. Haji Maarof Haji Salleh.

Muslim Missionary Society Singapore (JAMIYAH): 31 Lorong, 12 Geylang Rd, Singapore 399006; tel. 67431211; fax 67450610; e-mail info@jamiyah.org.sg; internet www.jamiyah.org.sg; Pres. Haji Abu Bakar Maidin; Sec.-Gen. Ismail Roziz.

SIKHISM

Sikh Ad: c/o 2 Towner Rd, 03-01, Singapore 327804; tel. and fax 62996440; Chair. Bhajan Singh.

The Press

Compulsory government scrutiny of newspaper management has been in operation since 1974. All newspaper enterprises must be public companies. In August 1986 there were more than 3,700 foreign publications circulating in Singapore. The Newspaper and Printing Presses (Amendment) Act 1986 empowers the Government to restrict the circulation of foreign periodicals that are deemed to exert influence over readers on domestic political issues. An amendment to the Newspaper and Printing Presses Act was promulgated in October 1990. Under this amendment, all publications of which the 'contents and editorial policy were determined outside Singapore' and which dealt with politics and current events in South-East Asia would be required to obtain a ministerial licence, renewable annually. The permit would limit the number of copies sold and require a deposit in case of legal proceedings involving the publication. Permits could be refused or revoked without any reason being given. In November, however, a statement was issued exempting 14 of the 17 foreign publications affected by the amendment, which came into effect in December.

DAILIES

English Language

Business Times: 1000 Toa Payoh North, Podium Level 3, Singapore 318994; tel. 63196319; fax 63198277; e-mail btnews@sph.com.sg; internet www.business-times.asia1.com.sg; f. 1976; morning; Editor ALVIN TAY; circ. 28,000 (Singapore only).

The New Paper: 1000 Toa Payoh North, Annexe Blk, Level 6, Singapore 318994; tel. 63196319; fax 63198266; e-mail tnp@asia1.com.sg; internet newpaper.asia1.com.sg; f. 1988; afternoon tabloid; Editor IVAN FERNANDEZ; circ. 120,394 in Aug. 2003 (Singapore only).

Streats: 1000 Toa Payoh North, Podium Blk, Level 1, Singapore 318994; tel. 63192421; fax 63198191; e-mail streats@sph.com.sg; f. 2000; Mon.-Fri.; morning tabloid; Editor PAUL JANSEN; circ. 280,000 in Aug. 2003 (Singapore only).

The Straits Times: 1000 Toa Payoh North, Singapore 318994; tel. 63196319; fax 67320131; e-mail stlocal@sph.com.sg; internet straitstimes.asia1.com.sg; f. 1845; morning; Editor HAN FOK KWANG; circ. 389,248 in Aug. 2003 (Singapore only).

Today: 24 Raffles Place, 28–01/06 Clifford Centre, Singapore 048621; tel. 62364889; fax 64812098; e-mail sales_enquiry@newstoday.com.sg; internet www.todayonline.com; f. 2000; free morning tabloid; Editor P. N. BALJI; circ. 250,000–300,000.

Chinese Language

Lianhe Wanbao: 1000 Toa Payoh North, Podium Blk, Level 4, Singapore 318994; tel. 63196319; fax 63198133; e-mail wanbao@sph.com.sg; f. 1983; evening; Editor KOH LIN HOE; circ. 127,021 (week), 120,879 (weekend) in Aug. 2003.

Lianhe Zaobao: 1000 Toa Payoh North, Podium Blk, Level 4, Singapore 318994; tel. 63196319; fax 63198119; e-mail zaobao@zaobao.com.sg; internet www.zaobao.com; f. 1923; Editor LIM JIM KOON; circ. 200,761.

Shin Min Daily News (S) Ltd: 1000 Toa Payoh North, Podium Blk, Level 4, Singapore 318994; tel. 63196319; fax 63198166; e-mail shinmin@sph.com.sg; f. 1967; evening; Editor TOH LAM HHAT; circ. 126,541 (week), 122,029 (weekend) in Aug. 2003.

Malay Language

Berita Harian: 1000 Toa Payoh North, Singapore 318994; tel. 63196319; fax 63198255; e-mail aadeska@sph.com.sg; internet cyberita.asia1.com.sg; f. 1957; morning; Editor MOHD GUNTOR SADALI; circ. 61,177 in Aug. 2003 (Singapore only).

Tamil Language

Tamil Murasu: 82 Genting Lane, Singapore 349567; tel. 63196319; fax 63196345; e-mail murasu4@cyberway.co.sg; internet tamilmurasu.asia1.com.sg; f. 1935; Editor CHITRA RAJARAM; circ. 7,928 (week), 15,119 (Sunday) in Aug. 2003.

WEEKLIES

English Language

The New Paper on Sunday: 1000 Toa Payoh North, Annexe Blk, Level 6, Singapore 318994; tel. 63196319; fax 63198266; e-mail tnp@asia1.com.sg; internet newpaper.asia1.com.sg; f. 1999; tabloid; Editor IVAN FERNANDEZ; circ. 152,080 in Aug. 2003.

The Sunday Times: 1000 Toa Payoh North, Singapore 318994; tel. 63195397; fax 67320131; e-mail stlocal@sph.com.sg; internet straitstimes.asia1.com.sg; f. 1931; Editor-in-Chief CHEONG YIP SENG; circ. 387,205 in Aug. 2003 (Singapore only).

Weekend Today: 24 Raffles Place, 28–01/06 Clifford Centre, Singapore 048621; tel. 62364889; fax 64812098; e-mail sales_enquiry@newstoday.com.sg; internet www.todayonline.com; f. 2002; Editor P. N. BALJI; circ. 300,000.

Malay Language

Berita Minggu: 1000 Toa Payoh North, Singapore 318994; tel. 63196319; fax 63198255; e-mail aadeska@sph.com.sg; internet cyberita.asia1.com.sg; f. 1960; Sunday; Editor MOHD GUNTOR SADALI; circ. 68,209 in Aug. 2003 (Singapore only).

SELECTED PERIODICALS

English Language

Accent: Accent Communications, 215 Intrepid Warehouse Complex, 4 Ubi Ave, Singapore 1440; tel. 67478088; fax 67472811; f. 1983; monthly; lifestyle; Senior Editor DORA TAY; circ. 65,000.

Cherie Magazine: 12 Everton Rd, Singapore 0208; tel. 62229733; fax 62843859; f. 1983; bimonthly; women's; Editor JOSEPHINE NG; circ. 20,000.

8 Days: 298 Tiong Bahru Rd, 19-01/06 Tiong Bahru Plaza, Singapore 168730; tel. 62789822; fax 62724800; e-mail feedback@8daysonline.com; internet 8days.mediacorppublishing.com; f. 1990; weekly; Editor-in-Chief TAN LEE SUN; circ. 113,258.

Her World: SPH Magazines Pte Ltd, 82 Genting Lane, Singapore 349567; tel. 63196319; fax 63196345; e-mail herworld@cyberway.com.sg; internet www.herworld.com; f. 1960; monthly; women's; Editor CAROLINE NGUI; circ. 60,883.

Her World Brides: SPH Magazines Pte Ltd, 82 Genting Lane, Singapore 349567; tel. 63196319; fax 63196345; e-mail sphmag@sph.com.sg; f. 1998; quarterly; circ. 13,193.

Home and Decor: SPH Magazines Pte Ltd, 82 Genting Lane, Singapore 349567; tel. 63196319; fax 63196345; e-mail hdecor@cyberway.com.sg; f 1981; 6 a year; home-owners; Editor SOPHIE KHO; circ. 19,814.

LIME: 65 Io Ang Mo Kio St, 01-06/08 Techpoint, Singapore 569059; tel. 64837118; fax 64837286; internet lime.mediacorppublishing.com; f. 1996; Editor PAMELA QUEK; circ. 34,947.

Mondial Collections: Singapore; f. 1990; bimonthly; arts; Chair. CHRIS CHENEY; circ. 100,000.

NSman: SAFRA National Service Association, 5200 Jalan Bukit Merah, Singapore 159468; tel. 62786011; fax 63779898; e-mail hq@safra.org.sg; internet www.safra.org.sg; f. 1972; bimonthly; publication of the SAF National Service; Gen. Man. TAN KEE BOO; circ. 130,000.

Republic of Singapore Government Gazette: SNP Corpn Ltd, 1 Kim Seng Promenade, 18-01 Great World City East Tower, Singapore 237994; tel. 68269600; fax 68203341; e-mail legalpub@snpcorp.com; internet www.myepb.com; weekly; Friday.

Reservist: 5200 Jalan Bukit Merah, Singapore 0315; tel. 62786011; fax 6273441; f. 1973; bimonthly; men's; Editor SAMUEL EE; circ. 130,000.

Singapore Medical Journal: Singapore Medical Association, Level 2, Alumni Medical Centre, 2 College Rd, Singapore 169850; tel. 62231264; fax 62247827; e-mail smj@sma.org.sg; internet www.sma.org.sg/smj; monthly; Editor Prof. W. C. G. PEH; circ. 4,500.

Times Guide to Computers: 1 New Industrial Rd, Times Centre, Singapore 536196; tel. 62848844; fax 62850161; e-mail ttd@corp.tpl.com.sg; internet www.tpl.com.sg; f. 1986; annually; computing and communications; Vice-Pres. LESLIE LIM; circ. 30,000.

Visage: Ubi Ave 1, 02-169 Blk 305, Singapore 1440; tel. 67478088; fax 67472811; f. 1984; monthly; Editor-in-Chief TENG JUAT LENG; circ. 63,000.

WEEKENDeast: 82 Genting Lane, News Centre, Singapore 349567; tel. 67401200; fax 67451022; e-mail focuspub@cyberway.com.sg; f. 1986; Editor VICTOR SOH; circ. 75,000.

Woman's Affair: 140 Paya Lebar Rd, 04-10 A-Z Bldg, Singapore 1440; tel. 67478088; fax 67479119; f. 1988; 2 a month; Editor DORA TAY; circ. 38,000.

Young Parents: SPH Magazines Pte Ltd, 82 Genting Lane, Singapore 349567; tel. 63196319; fax 63196345; e-mail yparents@cyberway.com.sg; internet www.youngparents.com.sg; f. 1986; monthly; family; Editor CRISPINA ROBERT; circ. 13,187.

Chinese Language

Characters: 1 Kallang Sector, 04-04/04-05 Kolam Ayer Industrial Park, Singapore 349276; tel. 67458733; fax 67458213; f. 1987; monthly; television and entertainment; Editor SAM NG; circ. 45,000.

The Citizen: People's Association, 9 Stadium Link, Singapore 397750; tel. 63405138; fax 63468657; monthly; English, Chinese, Tamil and Malay; Man. Editor OOI HUI MEI.

i-weekly: 298 Tiong Bahru Rd, 19-01/06 Tiong Bahru Plaza, Singapore 168730; tel. 62789822; fax 62724811; internet i-weekly.mediacorppublishing.com; f. 1981; weekly; radio and television; Editor-in-Chief LOKE TAI TAY; circ. 113,000.

Punters' Way: 4 Ubi View (off Ubi Rd 3), Pioneers and Leaders Centre, Singapore 408557; tel. 67458733; fax 67458213; e-mail pnlhldg@pcl.com.sg; f. 1977; biweekly; English and Chinese; sport; Editor T. S. PHAN; circ. 90,000.

Racing Guide: 1 New Industrial Rd, Times Centre, Singapore 1953; tel. 62848844; fax 62881186; f. 1987; 2 a week; English and Chinese; sport; Editorial Consultant BENNY ORTEGA; Chinese Editor KUEK CHIEW TEONG; circ. 20,000.

Singapore Literature: Singapore Literature Society, 122B Sims Ave, Singapore 1438; quarterly; Pres. YAP KOON CHAN; Editor LUO-MING.

Tune Monthly Magazine: Henderson Rd 06-04, Blk 203A, Henderson Industrial Park, Singapore 0315; tel. 62733000; fax 62749538; f. 1988; monthly; women's and fashion; Editor CHAN ENG; circ. 25,000.

Young Generation: SNP Media Asia Pte Ltd, 97 Ubi Avenue 4, SNP Building, Singapore 408754; tel. 67412500; fax 68463890; e-mail yg@snp.com.sg; internet www.snp.com.sg; monthly; children's; Editor EVELYN TANG; circ. 80,000.

You Weekly: SPH Magazines Pte Ltd, 82 Genting Lane, Singapore 349567; tel. 63196319; fax 63196345; e-mail sphmag@sph.com.sg; f. 2001; weekly; entertainment; circ. 80,000.

Malay Language

Manja: 10 Ang Mo Kio St 65, 01-06/08 Techpoint, Singapore 569059; tel. 64837118; fax 64812098; internet manja.mediacorppublishing.com; monthly; entertainment and lifestyle.

NEWS AGENCIES

Foreign Bureaux

Agence France-Presse (AFP): 10 Hoe Chiang Rd, 07-03 Keppel Towers, Singapore 089315; tel. 62228581; fax 62247465; e-mail afpsin@afp.com; Bureau Chief ROBERTO Z. COLOMA.

Agenzia Nazionale Stampa Associata (ANSA) (Italy): Blk 628, 8 Hougang Ave, 04-94, Singapore 530628; tel. 63855514; fax 63855279; e-mail kuttynng@pacific.net.sg; Correspondent N. G. KUTTY.

Associated Press (AP) (USA): 10 Anson Rd, 32-11 International Plaza, Singapore 079903; tel. 62201849; fax 62212753; e-mail sgutkin@ap.org; Chief of South-East Asian Services STEVEN GUTKIN.

Central News Agency Inc (CNA) (Taiwan): 78A Lorong N, Telok Kurau, Ocean Apartments, Singapore 425223; tel. 63445746; fax 63467645; e-mail cnanews@singnet.com.sg; South-East Asia Correspondent SHERMAN SHEAN-SHEN WU.

Inter Press Service (IPS) (Italy): Marine Parade Rd, 10-42 Lagoon View, Singapore 1544; tel. 64490432; Correspondent SURYA GANGADHARAN.

Jiji Tsushin (Jiji Press) (Japan): 10 Anson Rd, 24-16A International Plaza, Singapore 079903; tel. 62244212; fax 62240711; e-mail jijisp@jiji.com.sg; internet www.jiji.com; Bureau Chief AMANO KAZUTOSHI.

Kyodo News (Japan): 138 Cecil St, Cecil Court, Singapore; tel. 62233371; fax 62249208; e-mail goinrk@kyodonews.or.jp; Chief NORIKO GOI.

Reuters Singapore Pte Ltd: 18 Science Park Drive, Singapore 118229; tel. 68703080; fax 67768112; e-mail singapore.newsroom@reuters.com; Bureau Chief RICHARD HUBBARD.

Rossiiskoye Informatsionnoye Agentstvo—Vesti (RIA—Vesti) (Russia): 8 Namly Grove, Singapore 1026; tel. 64667998; fax 64690784; Correspondent MIKHAIL I. IDAMKIN.

United Press International (UPI) (USA): Singapore; tel. 63373715; fax 63389867; Man. DEAN VISSER.

Bernama (Malaysia), United News (India) and Xinhua (People's Republic of China) are also represented.

Publishers

ENGLISH LANGUAGE

Butterworths Asia: 1 Temasek Ave, 17-01 Millenia Tower, Singapore 039192; tel. 63369661; fax 63369662; e-mail sales@butterworths.com.sg; internet www.lexisnexis.com.sg/butterworths-online; f. 1932; law texts and journals; Man. Dir GRAHAM J. MARSHALL.

Caldecott Publishing Pte Ltd: 10 Ang Mo Kio St 65, 01-06/08 Techpoint, Singapore 569059; tel. 64837118; fax 64837286; f. 1990; Editorial Dir MICHAEL CHIANG; Group Editor TAN LEE SUN.

EPB Publishers Pte Ltd: Blk 162, 04-3545 Bukit Merah Central, Singapore 150162; tel. 62780881; fax 62782456; e-mail epb@sbg.com.sg; fmrly Educational Publications Bureau Pte Ltd; textbooks and supplementary materials, general, reference and magazines; English and Chinese; Gen. Man. ROGER PHUA.

FEP International Pte Ltd: 11 Arnsal Chetty Rd 03-02, Singapore 239949; tel. 67331178; fax 67375561; f. 1960; textbooks, reference, children's and dictionaries; Gen. Man. RICHARD TOH.

Flame of the Forest Publishing Pte Ltd: Blk 5, Ang Mo Kio Industrial Park 2A, 07-22/23, AMK Tech II, Singapore 567760; tel. 64848887; fax 64842208; e-mail editor@flameoftheforest.com; internet www.flameoftheforest.com; Man. Dir ALEX CHACKO.

Graham Brash Pte Ltd: 45 Kian Teck Drive, Blk 1, Level 2, Singapore 628859; tel. 62624843; fax 62621519; e-mail graham_brash@giro.com.sg; internet www.grahambrash.com.sg; f. 1947; general, academic, educational; English, Chinese and Malay; CEO CHUAN I. CAMPBELL; Dir HELENE CAMPBELL.

HarperCollins, Asia Pte Ltd: 970 Toa Payoh North, 04-24/26, Singapore 1231; tel. 62501985; fax 62501360; f. 1983; educational, trade, reference and general; Man. Dir FRANK FOLEY.

Institute of Southeast Asian Studies: 30 Heng Mui Keng Terrace, Pasir Panjang Rd, Singapore 119614; tel. 67780955; fax 67781735; e-mail publish@iseas.edu.sg; internet www.iseas.edu.sg; f. 1968; scholarly works on contemporary South-East Asia and the Asia-Pacific; Dir K. KESAVAPANY.

Intellectual Publishing Co: 113 Eunos Ave 3, 04-08 Gordon Industrial Bldg, Singapore 1440; tel. 67466025; fax 67489108; f. 1971; Man. POH BE LECK.

Marshall Cavendish International (Singapore) Pte Ltd: Times Centre, 1 New Industrial Rd, Singapore 536196; tel. 62139288; fax 62849772; e-mail tmesales@tpl.com.sg; internet www.marshallcavendish.com/academic; f. 1957; fmrly Times Media Pte Ltd; academic texts; Publr and Gen. Man. SIM KHIM HUANG; Man. Editor ANTHONY THOMAS.

Pearson Education South Asia Pte Ltd: 23/25 First Lok Yang Rd, Jurong Town, Singapore 629733; tel. 63199388; fax 63199171; e-mail asia@pearsoned.com.sg; educational; Reg. Dir LOW CHWEE LEONG.

Simon & Schuster Asia Pte Ltd: 317 Alexandra Rd, 04-01 Ikea Bldg, Singapore 159965; tel. 64764688; fax 63780370; e-mail prenhall@signet.com.sg; f. 1975; educational; Man. Dir GUNAWAN HADI.

Singapore University Press (Pte) Ltd: National University of Singapore, 31 Lower Kent Ridge Rd, Singapore 119078; tel. 67761148; fax 67740652; e-mail supbooks@nus.edu.sg; internet www.nus.edu.sg/npu; f. 1971; scholarly; Man. Dir PETER SCHOPPERT.

Stamford College Publishers: Colombo Court 05-11A, Singapore 0617; tel. 63343378; fax 63343080; f. 1970; general, educational and journals; Man. LAWRENCE THOMAS.

Times Editions Pte Ltd: Times Centre, 1 New Industrial Rd, Singapore 536196; tel. 62139288; fax 62844733; e-mail tpl@tpl.com.sg; internet www.tpl.com.sg; f. 1978; political, social and cultural books, general works on Asia; Chair. LIM KIM SAN; Pres. and CEO LAI SECK KHUI.

World Scientific Publishing Co Pte Ltd: 5 Toh Tuck Link, Singapore 596224; tel. 64665775; fax 64677667; e-mail wspc@wspc.com.sg; internet www.worldscientific.com; f. 1981; academic texts and science journals; Man. Dir DOREEN LIU.

MALAY LANGUAGE

Malaysia Press Sdn Bhd (Pustaka Melayu): Singapore; tel. 62933454; fax 62911858; f. 1962; textbooks and educational; Man. Dir ABU TALIB BIN ALLY.

Pustaka Nasional Pte Ltd: 2 Joo Chiat Rd, 05-1125 Joo Chiat Complex, Singapore 420002; tel. 67454321; fax 67452417; e-mail webmaster@pustaka.com.sg; internet www.pustaka.com.sg; f. 1965; Malay and Islamic religious books and CD-Roms; Man. Dir SYED AHMAD BIN MUHAMMED.

CHINESE LANGUAGE

Shanghai Book Co (Pte) Ltd: 231 Bain St, 02-73 Bras Basah Complex, Singapore 180231; tel. 63360144; fax 63360490; e-mail shanghaibook@pacific.net.sg; f. 1925; educational and general; Man. Dir MA JI LIN.

Shing Lee Publishers Pte Ltd: 120 Hillview Ave, 05-06/07 Kewalram Hillview, Singapore 2366; tel. 67601388; fax 67625684; e-mail shingleebook@sbg.com.sg; f. 1935; educational and general; Man. PEH CHIN HUA.

Union Book Co (Pte) Ltd: 231 Bain St 03-01, Bras Basah Complex, Singapore 0718; tel. 63380696; fax 63386306; general and reference; Gen. Man. CHOW LI-LIANG.

TAMIL LANGUAGE

EVS Enterprises: Singapore; tel. 62830002; f. 1967; children's books, religion and general; Man. E. V. SINGHAN.

Government Publishing House

SNP Corpn Ltd: 1 Kim Seng Promenade, 18-01 Great World City East Tower, Singapore 237994; tel. 68269600; fax 68203341; e-mail btsypang@snpcorp.com; internet www.snpcorp.com; f. 1973; printers and publishers; Pres. and CEO Yeo Chee Tong.

PUBLISHERS' ORGANIZATIONS

National Book Development Council of Singapore (NBDCS): 50 Geylang East Ave 1, Singapore 389777; tel. 68488290; fax 67429466; e-mail nbdcs@nbdcs.org.sg; internet www.nbdcs.org.sg; f. 1969; independent non-profit org.; promotes reading, writing and publishing and organizes the annual Asian Congress of Storytellers and Asian Children's Writers and Illustrators' Conference; offers professional training programmes through Centre for Literary Arts and Publishing; Exec. Dir Joyce Tan.

Singapore Book Publishers' Association: c/o Chopmen Publrs, 865 Mountbatten Rd, 05-28 Katong Shopping Centre, Singapore 1543; tel. 63441495; fax 63440180; Pres. N. T. S. Chopra.

Broadcasting and Communications

TELECOMMUNICATIONS

Infocomm Development Authority of Singapore: 8 Temasek Blvd, 14-00 Suntec Tower Three, Singapore 038988; tel. 62110888; fax 62112222; e-mail info@ida.gov.sg; internet www.ida.gov.sg; f. 1999 as result of merger of National Computer Board and Telecommunication Authority of Singapore; the national policy maker, regulator and promoter of telecommunications in Singapore; CEO Leong Keng Thai (acting).

Netrust Pte Ltd: 10 Collyer Quay, 09-05/06 Ocean Bldg, Singapore 049315; tel. 62121388; fax 62121366; e-mail infoline@netrust.net; internet www.netrust.net; f. 1997; the only licensed Certification Authority (CA) in Singapore, jtly formed by the National Computer Board and the Network for Electronic Transfers; the authority verifies the identity of parties doing business or communicating in cyberspace through the issuing of electronic identification certificates, in order to enable government organizations and private enterprises to conduct electronic transactions in a secure manner; CEO Foo Jong Ai.

Singapore Telecommunications Ltd (SingTel): Comcentre, 31 Exeter Rd, Singapore 239732; tel. 68383388; fax 67383769; e-mail contact@singtel.com; internet www.singtel.com; f. 1992; a postal and telecommunications service operator and a holding company for a number of subsidiaries; 68%-owned by Temasek Holdings (Private) Ltd (a government holding company), 32% transferred to the private sector; Chair. Ang Kong Hua; Pres. and CEO Lee Hsien Yang.

StarHub Pte Ltd: Head Office, 51 Cuppage Rd, 07-00 StarHub Centre, Singapore 229469; tel. 68255000; fax 67215000; e-mail customercare@starhub.com.sg; internet www.starhub.com.sg; f. 2000; telecommunications service provider; consortium includes Singapore Technologies Telemedia Pte Ltd, Singapore Power Ltd, Nippon Telegraph and Telephone Corpn (NTT) and British Telecom (BT); CEO and Pres. Terry Clontz.

Singapore Technologies Telemedia: 51 Cuppage Rd, 10-11/17, Starhub Centre, Singapore 229469; tel. 67238777; fax 67207277; e-mail contactus@stt.st.com.sg; internet www.sttelemedia.com; Pres. Lee Theng Kiat.

BROADCASTING

Regulatory Authority

Media Development Authority (MDA): MITA Bldg 04-01, 140 Hill St, Singapore 179369; tel. 68379973; fax 63368023; internet www.mda.gov.sg; f. 1994; fmrly Singapore Broadcasting Authority, name changed as above in Jan. 2003; licenses and regulates the media industry in Singapore; encourages, promotes and facilitates the development of media industries in Singapore, ensures the provision of an adequate range of media services to serve the interests of the general public, maintains fair and efficient market conduct and effective competition in the media industry, ensures the maintenance of a high standard of media services, regulates public service broadcasting; Chair. Tan Chin Nam; CEO Lim Hock Chuan.

Radio

Far East Broadcasting Associates: 30 Lorong Ampas, 07-01 Skywaves Industrial Bldg, Singapore 328783; tel. 62508577; fax 62508422; e-mail febadmin@febasgp.com; f. 1960; Chair. Goh Ewe Kheng; Exec. Dir Rev. John Chang.

Media Corporation of Singapore: Caldecott Broadcast Centre, Andrew Rd, Singapore 299939; tel. 63333888; fax 62515628; e-mail cherfern@mediacorpradio.com; internet www.mediacorpsingapore.com; f. 1994 as Singapore International Media (SIM), following the corporatization of the Singapore Broadcasting Corpn; holding co for seven operating cos—Television Corpn of Singapore (TCS), Singapore Television Twelve (STV12), Radio Corpn of Singapore (RCS), MediaCorp Studios, MediaCorp News, MediaCorp Interactive and MediaCorp Publishing; Chair. Cheng Wai Keung.

Radio Corpn of Singapore Pte Ltd (RCS): Caldecott Broadcast Centre, Radio Bldg, Andrew Rd, Singapore 299939; tel. 62518622; fax 62569533; e-mail feedback@rcs.com.sg; internet www.mediacorpradio.com; f. 1936; operates 12 domestic services—in English (five), Chinese (Mandarin) (three), Malay (two), Tamil (one) and foreign language (one)—and three international radio stations (manages Radio Singapore International (RSI)—services in English, Mandarin and Malay for three hours daily and service in Bahasa Indonesia for one hour daily); Chief Operating Officer Chua Foo Yong.

Radio Heart: Singapore; internet www.heart913.org.sg; f. 1991; first private radio station; broadcasts in English, Chinese (Mandarin), Malay and Tamil; 2 channels broadcasting a total of 280 hours.

Rediffusion (Singapore) Pte Ltd: 6 Harper Rd, 04-01/08 Leong Huat Bldg, Singapore 369674; tel. 63832633; fax 63832622; e-mail md@rediffusion.com.sg; internet www.rediffusion.com.sg; f. 1949; commercial audio wired broadcasting service and wireless digital audio broadcasting service; broadcasts two programmes in Mandarin (18 hours daily) and English (24 hours daily); Man. Dir Wong Ban Kuan.

SAFRA Radio: Defence Technology Towers, Tower B, 5 Depot Rd, 12-04, Singapore; tel. 63731924; fax 62783039; e-mail power98@pacific.net.sg; internet www.power98.com.sg/; f. 1994; broadcasts in Chinese (Mandarin) and English.

Television

CNBC Asia Business News (S) Pte Ltd: 10 Anson Rd, 06-01 International Plaza, Singapore 079903; tel. 63230488; fax 63230788; e-mail talk2us@cnbcasia.com; internet www.cnbcasia.com.sg; f. 1998; cable and satellite broadcaster of global business and financial news; US controlled; broadcasts in English (24 hours daily) and Mandarin; Pres. Paul France.

Media Corporation of Singapore: see Radio.

Singapore CableVision: Singapore; internet www.scv.com.sg; f. 1992; broadcasting and communications co, subscription television service; three channels; launched cable service with 33 channels in June 1995; offers broadband access services.

Singapore Television Twelve Pte Ltd (STV12): 12 Prince Edward Rd, 05-00 Bestway Bldg, Singapore 079212; tel. 62258133; fax 62203881; internet www.stv12.com.sg; f. 1994; terrestrial television station; 2 channels—Suria (Malay, 58 hours weekly) and Central (110.5 hours weekly); Chief Operating Officer Woon Tai Ho.

SPH MediaWorks Ltd: 82 Genting Lane, Singapore 349567; tel. 63197988; fax 67443318; e-mail mwcc@sphmediaworks.com; internet www.sphmediaworks.com; f. 2000; subsidiary of Singapore Press Holdings Ltd (SPH); two channels—Channel U (Mandarin) and Channel i (English); also owns two radio stations; Exec. Dir Wee Leong How.

Television Corpn of Singapore (TCS): Caldecott Broadcast Centre, Andrew Rd, Singapore 299939; tel. 62560401; fax 62538119; e-mail webmaster@mediacorptv.com; internet www.mediacorptv.com; f. 1994 following the corporatization of Singapore Broadcasting Corpn; four channels—TCS 5 (English), TCS 8 (Mandarin), Suria (Malay) and Central; teletext service on two channels; also owns TVMobile, Singapore's first digital television channel, and Digital TV; Chair. Cheng Wai Keung; CEO Lim Hup Seng.

Finance

(cap. = capital; res = reserves; dep. = deposits; m. = million; brs = branches; amounts in Singapore dollars)

BANKING

The Singapore monetary system is regulated by the Monetary Authority of Singapore (MAS) and the Ministry of Finance. The MAS performs all the functions of a central bank and also assumed responsibility for the issuing of currency following its merger with the Board of Commissioners of Currency in October 2002. In September 2004 there were 113 commercial banks (5 local, 108 foreign) and 48 representative offices in Singapore. Of the foreign banks, 23 had full licences, 37 had wholesale licences and 48 foreign banks had 'offshore' banking licences.

Government Financial Institution

Monetary Authority of Singapore (MAS): 10 Shenton Way, MAS Bldg, Singapore 079117; tel. 62255577; fax 62299229; e-mail webmaster@mas.gov.sg; internet www.mas.gov.sg; merged with Board of Commissioners of Currency Oct. 2002; cap. 100m., res 15,301m., dep. 99,927m. (March 2003); Chair. GOH CHOK TONG; Man. Dir KOH YONG GUAN.

Domestic Full Commercial Banks

Bank of Singapore Ltd: 18 Church St, 01-00 OCBC Centre South, Singapore 049479; tel. 63893000; fax 63893016; internet www.bankofsingapore.com; f. 1954; subsidiary of Oversea-Chinese Banking Corpn Ltd; cap. 50m., res 207m., dep. 2,984.5m. (Dec. 1997); Chair. DAVID PHILBRICK CONNER; CEO TAN NGIAP JOO.

DBS Bank (Development Bank of Singapore Ltd): 6 Shenton Way, DBS Bldg, Tower One, Singapore 068809; tel. 68788888; fax 64451267; e-mail dbs@dbs.com; internet www.dbs.com; f. 1968; 29% govt-owned; merged with Post Office Savings Bank in 1998; cap. 1,962m., res 14,288m., dep. 87,250m. (Dec. 2003); Chair. S. DHANABALAN; Vice-Chair. and CEO JACKSON TAI; 107 local brs, 9 overseas brs.

Far Eastern Bank Ltd: 156 Cecil St, 01-00 FEB Bldg, Singapore 069544; tel. 62219055; fax 62242263; internet www.uobgroup.com; f. 1959; subsidiary of United Overseas Bank Ltd; cap. 100m., res 42.5m., dep. 639.7m. (Dec. 2002); Chair. and CEO WEE CHO YAW; Pres. WEE EE CHEONG; 3 brs.

Oversea-Chinese Banking Corpn (OCBC) Ltd: 65 Chulia St, 08-00 OCBC Centre, Singapore 049513; tel. 65357222; fax 65337955; e-mail info@ocbc.com.sg; internet www.ocbc.com; f. 1932; merged with Keppel TatLee Bank Ltd in Aug. 2001; cap. 1,284.1m., res 8,774.8m., dep. 67,398.2m. (Dec. 2003); Chair. Dr CHEONG CHOONG KONG; Vice-Chair. MICHAEL WONG PAKSHONG; 63 local brs, 52 overseas brs.

United Overseas Bank Ltd: 80 Raffles Place, UOB Plaza, Singapore 048624; tel. 65339898; fax 65342334; internet www.uobgroup.com; f. 1935; merged with Overseas Union Bank Ltd in Jan. 2002 and Industrial and Commercial Bank Ltd in Aug. 2002; cap. 1,571.7m., res 7,463.6m., dep. 79,367.2m. (Dec. 2003); Chair. and CEO WEE CHO YAW; Pres. WEE EE CHEONG; 59 local brs, 21 overseas brs.

Foreign Banks

Full Commercial Banks

ABN AMRO Asia Merchant Bank (Singapore) Ltd (Netherlands): 63 Chulia St, Singapore 049514; tel. 62318888; fax 65323108; Chair. RAJAN RAY; Man. Dir GEOFFREY W. S. MCDONALD.

American Express Bank Ltd (USA): 16 Collyer Quay, Hitachi Tower, Singapore 049318; tel. 65384833; fax 65343022; Sr Country Exec. S. LACHLAN HOUGH.

Bangkok Bank Public Co Ltd (Thailand): 180 Cecil St, Bangkok Bank Bldg, Singapore 069546; tel. 62219400; fax 62255852; e-mail torpong.cha@bbl.co.th; Sr Vice-Pres. and Gen. Man. TORPHONG CHARUNGCHAREONVEJI.

Bank of America NA (USA): 9 Raffles Place, 18-00 Republic Plaza Tower 1, Singapore 048619; tel. 62393888; fax 62393188; CEO ALAN KOH; Man. Dir GOETZ EGGELHOEFER.

Bank of China (People's Republic of China): 4 Battery Rd, Bank of China Bldg, Singapore 049908; tel. 65352411; fax 65343401; Gen. Man. ZHU HUA.

Bank of East Asia Ltd (Hong Kong): 137 Market St, Bank of East Asia Bldg, Singapore 048943; tel. 62241334; fax 62251805; Gen. Man. KHOO KEE CHEOK.

Bank of India (India): 01-01, 02-01, 03-01, Hong Leong Centre, 138 Robinson Rd, Singapore 068906; tel. 62220011; fax 62254407; Chief Exec. VIJAY MEHTA; Gen. Man. B. RAMASUBRAMANIAM.

PT Bank Negara Indonesia (Persero) Tbk (Indonesia): 158 Cecil St, 01-00 to 04-00 Dapenso Bldg, Singapore 069545; tel. 62257755; fax 62254757; e-mail ptbni@starhub.net.sg; Gen. Man. MUHAMMAD YAZEED.

Bank of Tokyo-Mitsubishi Ltd (Japan): 9 Raffles Place, 01-01 Republic Plaza, Singapore 048619; tel. 65383388; fax 65388083; Gen. Man. SATOSHI SADAKATA; Dep. Gen. Man. MAKOTO NAKAGAWA.

BNP Paribas (France): 20 Collyer Quay, 18-01 Tung Centre, Singapore 049319; tel. 62103888; fax 62103671; internet www.bnpparibas.com.sg; CEO SERGE FORTI.

Citibank NA (USA): 3 Temasek Ave, 12-00 Centennial Tower, Singapore 039190; tel. 62242611; fax 6325880; internet www.citibank.com.sg; Country Corporate Officer SANJIV MISRA.

Crédit Agricole Indosuez (France): 6 Raffles Quay, 17-00 John Hancock Tower, Singapore 048580; tel. 65354988; fax 65322422; e-mail cai.singapore@sg.ca-indosuez.com; internet www.ca-indosuez.com; f. 1905; Sr Country Man. JEAN-FRANCOIS CAHET.

HL Bank (Malaysia): 20 Collyer Quay, 01-02 and 02-02 Tung Centre, Singapore 049319; tel. 65352466; fax 65339340; Country Head GAN HUI TIN.

Hongkong and Shanghai Banking Corpn Ltd (Hong Kong): 14-01 HSBC Bldg, 21 Collyer Quay, Singapore 049320; tel. 65305000; fax 62250663; e-mail HSBC@hsbc.com.sg; internet www.hsbc.com.sg; CEO (Singapore) PAUL LAWRENCE.

Indian Bank (India): 3 Raffles Place, Bharat Bldg, Singapore 048617; tel. 65343511; fax 65331651; e-mail ceibsing@mbox3.singnet.com.sg; Chief Exec. V. SRINIVASAN.

Indian Overseas Bank (India): POB 792, 64 Cecil St, IOB Bldg, Singapore 049711; tel. 62251100; fax 62244490; Chief Exec. KONIDALA PERUMAL MUNIRATHNAM.

JP Morgan Chase Bank (USA): 168 Robinson Rd, 15th Floor, 14-01 Capital Tower, Singapore 068912; tel. 68822888; fax 68821756; Sr Country Officer RAYMOND CHANG.

Maybank (Malaysia): Maybank Chambers, 2 Battery Rd 01-00, Singapore 049907; tel. 65352266; fax 65327909; internet www.maybank2u.com.sg; Sr Gen. Man. SPENCER LEE TIEN CHYE.

RHB Bank Bhd (Malaysia): 90 Cecil St 01-00, Singapore 069531; tel. 62253111; fax 62273805; Country Head LIM HUN JOO.

Southern Bank Bhd (Malaysia): 39 Robinson Rd, 01-02 Robinson Point, Singapore 068911; tel. 65321318; fax 65355366; Dir YEAP LAM YANG.

Standard Chartered Bank (UK): 6 Battery Rd, Singapore 049909; tel. 62258888; fax 64230965; internet www.standardchartered.com.sg; Group Exec. Dirs MIKE DENOMA, KAI NARGOLWALA; Chief Exec. (Singapore) EULEEN GOH.

Sumitomo Mitsui Banking Corpn (Japan): 3 Temasek Ave, 06-01 Centennial Tower, Singapore 039190; tel. 68820001; fax 68870330; Gen. Man. MASAMI TASHIRO.

UCO Bank (India): 3 Raffles Place, 01-01 Bharat Bldg, Singapore 048617; tel. 65325944; fax 65325044; e-mail general@ucobank.com.sg; Chief Exec. R. K. MUKHERJEE.

Wholesale Banks

Australia and New Zealand Banking Group Ltd (Australia): 10 Collyer Quay, 17-01/07 Ocean Bldg, Singapore 049315; tel. 65358355; fax 65396111; Regional Gen. Man. JOHN WINDERS.

Bank of Nova Scotia (Canada): 10 Collyer Quay, 15-01/09 Ocean Bldg, Singapore 049315; tel. 65358688; fax 65363325; Country Head, Vice-Pres. and Man. SEONG KOON WAH SUN.

Barclays Bank PLC (UK): 23 Church St, 13-08 Capital Sq., Singapore 049481; tel. 63953000; fax 63953139; Regional Head JAMES LOH; Country Man. QUEK SUAN KIAT.

Bayerische Hypo- und Vereinsbank AG (Germany): 30 Cecil St, 26-01 Prudential Tower, Singapore 049712; tel. 64133688; fax 65368591; Gen. Man. PETER NG POH LIAN.

Bayerische Landesbank Girozentrale (Germany): 300 Beach Rd, 37-01 The Concourse, Singapore 199555; tel. 62933822; fax 62932151; e-mail sgblb@blb.de; Gen. Man. and Sr Vice-Pres. MANFRED WOLF; Exec. Vice-Pres. HEINZ HOFFMANN.

BNP Paribas Private Bank (France): 20 Collyer Quay, 18-01 Tung Centre, Singapore 049319; tel. 62101037; fax 62103671; internet www.bnpparibas.com.sg; CEO SERGE FORTI.

Chiao Tung Bank Co Ltd (Taiwan): 80 Raffles Place, 23-20 UOB Plaza II, Singapore 048624; tel. 65366311; fax 65360680; Sr Vice-Pres. and Gen. Man. CHIANG SHENG-LI.

Commerzbank AG (Germany): 8 Shenton Way, 41-01 Temasek Tower, Singapore 068811; tel. 63110000; fax 62253943; Gen. Man. MICHAEL OLIVER.

Crédit Lyonnais (France): 3 Temasek Ave, 11-01 Centennial Tower, Singapore 039190; tel. 63336331; fax 63336332; Gen. Man. PIERRE EYMERY.

Crédit Suisse (Switzerland): 1 Raffles Link, 05-02, Singapore 039393; tel. 62126000; fax 62126200; internet www.cspb.com.sg; Br. Man. DIDIER VON DAENIKEN.

Crédit Suisse First Boston (Switzerland): 1 Raffles Link, 03/04-01 South Lobby, Singapore 039393; tel. 62122000; fax 62123100; Br. Man. ERIC M. VARVEL.

Deutsche Bank AG (Germany): 6 Shenton Way, 15-08 DBS Bldg, Tower Two, Singapore 068809; tel. 64238001; fax 62259442; Gen. Man. RONNY TAN CHONG TEE.

Dresdner Bank AG (Germany): 20 Collyer Quay, 22-00 Tung Centre, Singapore 049319; tel. 62228080; fax 62244008; Man. Dirs Andreas Ruschkowski, Raymond B. T. Koh, Piers Willis; CEO Baudouin Groonenberghs.

First Commercial Bank (Taiwan): 76 Shenton Way, 01-02 ONG Bldg, Singapore 079119; tel. 62215755; fax 62251905; e-mail fcbsin@singnet.com.sg; Gen. Man. Wu Ho-Li.

Fortis Bank SA/NV (Belgium/Netherlands): 63 Market St, 21-01, Singapore 048942; tel. 65394988; fax 65394933; Gen. Man. Gijsbert Schot.

Habib Bank Ltd (Pakistan): 3 Phillip St, 01-03 Commerce Pt, Singapore 048693; tel. 64380055; fax 64380644; e-mail gmhbl@singnet.com.sg; Regional Gen. Man. Ashraf Mahmood Wathra.

HSBC Republic Bank (Suisse) SA (Switzerland): 21 Collyer Quay, 21-01 HSBC Bldg, Singapore 049320; tel. 62248080; fax 62237146; CEO Kenneth Sit Yiu Sun.

Industrial and Commercial Bank of China (People's Republic of China): 6 Raffles Quay, 12-01 John Hancock Tower, Singapore 048580; tel. 65381066; fax 65381370; e-mail icbcsg@singnet.com.sg; Gen. Man. Xu Gang (acting).

ING Bank NV (Netherlands): 9 Raffles Place, 19-02 Republic Plaza, Singapore 048619; tel. 65353688; fax 65338329; Country Head J. Kestemont.

KBC Bank NV (Belgium): 30 Cecil St, 12-01/08 Prudential Tower, Singapore 049712; tel. 63952828; fax 65342929; Gen. Man. Thierry Mezeret.

Korea Exchange Bank (Republic of Korea): 30 Cecil St, 24-03/08 Prudential Tower, Singapore 049712; tel. 65361633; fax 65382522; e-mail kebspore@singnet.com.sg; Gen. Man. Kim Seong Jung.

Landesbank Baden-Württemberg (Germany): 25 International Business Park, 01-72 German Centre, Singapore 609916; tel. 65627722; fax 65627729; Man. Dr Wolfhart Auer van Herrenkirchen.

Mizuho Corporate Bank Ltd (Japan): 168 Robinson Rd, 13-00 Capital Tower, Singapore 068912; tel. 64230330; fax 64230012; Gen. Man. Tadao Ogoshi.

Moscow Narodny Bank Ltd (UK): 50 Robinson Rd, MNB Bldg, Singapore 068882; tel. 62209422; fax 62250140; Man. Dir Evgeny Mikhailovich Grevtsev.

National Australia Bank Ltd (Australia): 5 Temasek Blvd, 15-01 Suntec City Tower Five, Singapore 038985; tel. 63380038; fax 63380039; Gen. Man. Cheo Chai Hong.

National Bank of Kuwait SAK (Kuwait): 20 Collyer Quay, 20-00 Tung Centre, Singapore 049319; tel. 62225348; fax 62245438; Gen. Man. R. J. McKegney.

Norddeutsche Landesbank Girozentrale (Germany): 6 Shenton Way, 16-08 DBS Bldg Tower Two, Singapore 068809; tel. 63231223; fax 63230223; e-mail nordlb.singapore@nordlb.com; Gen. Man. and Regional Head Asia/Pacific Heinz Werner Frings.

Northern Trust Company (USA): 80 Raffles Place, 46th Floor, UOB Plaza 1, Singapore 048624; tel. 64376666; fax 64376609; e-mail hynl@ntrs.com; Sr Vice-Pres. Lawrence Au.

Rabobank International (Netherlands): 77 Robinson Rd, 09-00 SIA Bldg, Singapore 068896; tel. 65363363; fax 65363236; Exec. Vice-Pres. Christian H. A. M. Mol.

Royal Bank of Scotland PLC (UK): 50 Raffles Place, 08-00 Singapore Land Tower, Singapore 048623; tel. 64168600; fax 62259827; Gen. Man. Robert I. Garden.

San Paolo IMI SpA (Italy): 6 Temasek Blvd, 42/04-05 Suntec Tower Four, Singapore 038986; tel. 63338270; fax 63338252; e-mail singapore.sg@sanpaoloimi.com; Gen. Man. Lim Khai Seng.

Société Générale (France): 80 Robinson Rd, 25-00, Singapore 068898; tel. 62227122; fax 62252609; Chief Country Officer Eric Wormser.

State Street Bank and Trust Co (USA): 8 Shenton Way, 33-03 Temasek Tower, Singapore 068811; tel. 63299600; fax 62259377; Br. Man. Lee Yow Fee.

UBS AG (Switzerland): 5 Temasek Blvd, 18-00 Suntec City Tower, Singapore 038985; tel. 64318000; fax 64318188; e-mail rolf-w.gerber@wdr.com; Man. Dir and Head of Br. (Singapore) Brad Orgill.

UFJ Bank Ltd (Japan): 6 Raffles Quay, 24-01 John Hancock Tower, Singapore 048580; tel. 65384838; fax 65384636; Gen. Man. Toru Kojima.

UniCredito Italiano SpA (Italy): 80 Raffles Place, 51-01 UOB Plaza 1, Singapore 048624; tel. 62325728; fax 65344300; e-mail singaporebranch@gruppocredit.it; Gen. Man. Maurizio Brentegani.

WestLB AG (Germany): 3 Temasek Ave, 33-00 Centennial Tower, Singapore 039190; tel. 63332388; fax 63332399; Gen. Man. Teo Ee-Ngoh.

'Offshore' Banks

ABSA Bank Ltd (South Africa): 9 Temasek Blvd, 40-01 Suntec Tower Two, Singapore 038989; tel. 63331033; fax 63331066; Gen. Man. David Meadows.

Agricultural Bank of China (People's Republic of China): 80 Raffles Place, 27-20 UOB Plaza 2, Singapore 048624; tel. 65355255; fax 65387960; e-mail aboc@abchina.com.sg; Gen. Man. Wang Gang.

Arab Bank PLC (Jordan): 80 Raffles Place, 32-20 UOB Plaza 2, Singapore 048624; tel. 65330055; fax 65322150; e-mail abplc@pacific.net.com.sg; Exec. Vice-Pres. and Area Exec. Asia Pacific Kim Eun-Young.

Banca Monte dei Paschi di Siena SpA (Italy): 10 Collyer Quay, 13-01 Ocean Bldg, Singapore 0104; tel. 65352533; fax 65327996; Gen. Man. Giuseppe De Giosa.

Banca di Roma (Italy): 9 Raffles Place, 20-20 Republic Plaza II, Singapore 048619; tel. 64387509; fax 65352267; e-mail bdrsi@singnet.com.sg; Gen. Man. Mario Fattorusso.

Bank of Communications (People's Republic of China): 50 Raffles Place, 26-04 Singapore Land Tower, Singapore 048623; tel. 65320335; fax 65320339; Gen. Man. Niu Ke Rong.

PT Bank Mandiri (Persero) (Indonesia): 16 Collyer Quay, 28-00 Hitachi Tower, Singapore 049318; tel. 65320200; fax 65320206; Gen. Man. Muhadjir Sangidu.

Bank of New York (USA): 1 Temasek Ave, 02-01 Millenia Tower, Singapore 039192; tel. 64320222; fax 63374302; Sr Vice-Pres. and Man. Dir Jai Arya.

Bank of New Zealand (New Zealand): 5 Temasek Blvd, 15-01 Suntec City Tower, Singapore 038985; tel. 63322990; fax 63322991; Gen. Man. Cheo Chai Hong.

Bank of Taiwan (Taiwan): 80 Raffles Place, 28-20 UOB Plaza 2, Singapore 048624; tel. 65365536; fax 65368203; Gen. Man. Chiou Ye-Chin.

Bumiputra Commerce Bank Bhd (Malaysia): 7 Temasek Blvd, 37-01/02/03 Suntec Tower One, Singapore 038987; tel. 63375115; fax 63371335; e-mail bpsp3700@pacific.net.sg; Gen. Man. Raja Sulong Ahmad bin Raja Abdul Razak.

Canadian Imperial Bank of Commerce (Canada): 16 Collyer Quay, 04-02 Hitachi Tower, Singapore 049318; tel. 65352323; fax 65357565; Br. Man. Norman Sim Chee Beng.

Chang Hwa Commercial Bank Ltd (China): 1 Finlayson Green, 08-00, Singapore 049246; tel. 65320820; fax 65320374; Gen. Man. Yang Jih-Cheng.

China Construction Bank (People's Republic of China): 9 Raffles Place, 33-01/02 Republic Plaza, Singapore 048619; tel. 65358133; fax 65358133; internet www.ccb.com.sg; Gen. Man. Kong Yong Xin.

Chohung Bank (Republic of Korea): 50 Raffles Place, 04-02/03 Singapore Land Tower, Singapore 048623; tel. 65361144; fax 65331244; Gen. Man. Choi Heung Min.

Commonwealth Bank of Australia (Australia): 50 Raffles Place, 22-02 Singapore Land Tower, Singapore 048623; tel. 62243877; fax 62245812; Gen. Man. Robert Lewis Buchan.

Crédit Industriel et Commercial (France): 63 Market St, 15-01, Singapore 048942; tel. 65366008; fax 65367008; internet www.cic.com.sg; Gen. Man. Jean-Luc Anglada.

Crédit Lyonnais (Suisse) SA (Switzerland): 3 Temasek Ave 11-01, Centennial Tower, Singapore 039190; tel. 68320900; fax 63338590; e-mail singaporebranch@creditlyonnais.ch; Man. Dir Antoine Candiotti.

Dexia Banque Internationale à Luxembourg (Luxembourg): 9 Raffles Place, 42-01 Republic Plaza, Singapore 048619; tel. 62227622; fax 65360201; Gen. Man. Alexandre Josset.

DZ Bank AG Deutsche Zentral (Germany): 50 Raffles Place, 40-01 Singapore Land Tower, Singapore 048623; tel. 64380082; fax 62230082; Gen. Man. Klaus Gerhard Borig.

Hana Bank (Republic of Korea): 8 Cross St, 23-06 PWC Bldg, Singapore 048424; tel. 64384100; fax 64384200; Gen. Man. Choi Cheong-Il.

Hang Seng Bank Ltd (Hong Kong): 21 Collyer Quay, 14-01 HSBC Bldg, Singapore 049320; tel. 65363118; fax 65363148; e-mail sgp@hangseng.com; Country Man. Anthony Kam Ping Leung.

HSH Nordbank AG (Germany): 3 Temasek Ave, 32-03 Centennial Tower, Singapore 039190; tel. 65509000; fax 65509003; e-mail info@hsh-nordbank.com.sg; Gen. Man. and Regional Head Klaus Heiner Boritzka.

Hua Nan Commercial Bank Ltd (Taiwan): 80 Robinson Rd, 14-03, Singapore 068898; tel. 63242566; fax 63242155; Gen. Man. Oliver Hsu.

ICICI Bank Ltd (India): 9 Raffles Place, 50-01 Republic Plaza, Singapore 048619; tel. 67239288; fax 67239268; internet www.icicibank.com.sg; Chief. Exec. Suvek Nambiar.

International Commercial Bank of China (Taiwan): 6 Battery Rd, 39-03, Singapore 049909; tel. 62277667; fax 62271858; Vice-Pres. and Gen. Man. Yen-Ping Hsiang.

Korea Development Bank (Republic of Korea): 8 Shenton Way, 07-01 Temasek Tower, Singapore 068811; tel. 62248188; fax 62256540; Gen. Man. Kim Byoung Soo.

Krung Thai Bank Public Co Ltd (Thailand): 65 Chulia St, 32-05/08 OCBC Centre, Singapore 049513; tel. 65336691; fax 65330930; e-mail ktbs@pacific.net.sg; Gen. Man. Pumin Leelayoova.

Land Bank of Taiwan: UOB Plaza 1, 34-01 Raffles Place, Singapore 048624; tel. 63494555; fax 63494545; Gen. Man. Wilson W. B. Lin.

Lloyds TSB Bank PLC (UK): 3 Temasek Ave 19-01, Centennial Tower, Singapore 039190; tel. 65341191; fax 65322493; e-mail mktg@lloydstsb.com.sg; Country Head Barry Lea.

Mitsubishi Trust and Banking Corpn (Japan): 50 Raffles Place, 42-01/06 Singapore Land Tower, Singapore 048623; tel. 62259155; fax 62241857; Gen. Man. Mikio Kobayashi.

Natexis Banques Populaires (France): 50 Raffles Place, 41-01, Singapore Land Tower, Singapore 048623; tel. 62241455; fax 62248651; Gen. Man. Philippe Petitgas.

Nedcor Bank Ltd (South Africa): 30 Cecil St, 10-05 Prudential Tower, Singapore 049712; tel. 64169438; fax 64388350; e-mail nedsing@nedcor.com; Gen. Man. Brian Shegar.

Nordea Bank Finland Plc (Finland): 50 Raffles Place, 15-01 Singapore Land Tower, Singapore 048623; tel. 62258211; fax 62255469; e-mail singapore@nordea.com; Gen. Man. Thor Erling Kylstad.

Norinchukin Bank (Japan): 80 Raffles Place, 53-01 UOB Plaza 1, Singapore 048624; tel. 65351011; fax 65352883; Gen. Man. Yuji Shimauchi.

Den norske Bank ASA (Norway): 8 Shenton Way, 48-02 Temasek Tower, Singapore 068811; tel. 62206144; fax 62249743; e-mail dnb.singapore@dnb.no; Gen. Man. Pål Skoe.

Philippine National Bank (Philippines): 96 Somerset Rd, 04-01/04 UOL Bldg, Singapore 238183; tel. 67374646; fax 67374224; e-mail singapore@pnb.com.ph; Vice-Pres. and Gen. Man. Rodelo G. Franco.

Raiffeisen Zentralbank Oesterreich Aktiengesellschaft (Austria): 50 Raffles Place, 45-01 Singapore Land Tower, Singapore 048623; tel. 62259578; fax 62253973; Gen. Man. Rainer Silhavy.

Royal Bank of Canada (Canada): 20 Raffles Place, 27-03/08 Ocean Towers, Singapore 048620; tel. 65369206; fax 65322804; Gen. Man. Trevor David Wynn.

Skandinaviska Enskilda Banken AB Publ (Sweden): 50 Raffles Place, 36-01 Singapore Land Tower, Singapore 048623; tel. 62235644; fax 62253047; Gen. Man. Sven Björkman.

Siam Commercial Bank Public Company Ltd (Thailand): 16 Collyer Quay, 25-01 Hitachi Tower, Singapore 049318; tel. 65364338; fax 65364728; Vice-Pres. and Gen. Man. Nattapong Samit Ampaipisarn.

State Bank of India (India): 6 Shenton Way, 22-08 DBS Bldg Tower Two, Singapore 068809; tel. 62222033; fax 62253348; e-mail sbinsgsg@pacific.net.sg; CEO Padma Ramasubban.

Sumitomo Trust & Banking Co Ltd (Japan): 8 Shenton Way, 45-01 Temasek Tower, Singapore 068811; tel. 62249055; fax 62242873; Gen. Man. Masayuki Imanaka.

Svenska Handelsbanken AB (publ) (Sweden): 65 Chulia St, 21-01/04 OCBC Centre, Singapore 049513; tel. 65323800; fax 65344909; Gen. Man. Jan Birger Djerf.

Toronto-Dominion (South East Asia) Ltd (Canada): 15-02 Millenia Tower, 1 Temasek Ave, Singapore 039192; tel. 64346000; fax 63369500; Br. Dir Akhileshwar Lamba.

Union de Banques Arabes et Françaises (UBAF) (France): 6 Temasek Blvd, 25-04-05 Suntec Tower Four, Singapore 038986; tel. 63336188; fax 63336789; e-mail ubafsg@singnet.com.sg; Gen. Man. Benoit Gueroult.

Westpac Banking Corpn (Australia): 77 Robinson Rd, 19-00 SIA Bldg, Singapore 068896; tel. 65309898; fax 65326781; e-mail yhlee@westpac.wm.au; Gen. Man. Christopher David Rand.

Woori Bank (Republic of Korea): 5 Shenton Way, 17-03 UIC Bldg, Singapore 068808; tel. 62235855; fax 62259530; e-mail combksp@singnet.com.sg; Gen. Man. Park Dong Young.

Bankers' Association

Association of Banks in Singapore: 10 Shenton Way, 12-08 MAS Bldg, Singapore 079117; tel. 62244300; fax 62241785; e-mail banks@abs.org.sg; internet www.abs.org.sg; Dir Ong-Ang Ai Boon.

STOCK EXCHANGE

Singapore Exchange Limited (SGX): 2 Shenton Way, 19-00 SGX Centre One, Singapore 068804; tel. 62368888; fax 65356994; e-mail webmaster@sgx.com; internet www.sgx.com; f. 1999 as a result of the merger of the Stock Exchange of Singapore (SES, f. 1930) and the Singapore International Monetary Exchange Ltd (SIMEX, f. 1984); Chair. J. Y. Pillay; CEO Hsieh Fu Hua; Pres. Ang Swee Tian.

INSURANCE

The insurance industry is supervised by the Monetary Authority of Singapore (see Banking). In September 2004 there were 136 insurance companies, comprising 52 direct insurers (six life insurance, 39 general insurance, seven composite insurers), 29 professional reinsurers (three life reinsurers, 19 general reinsurers, seven composite reinsurers) and 55 captive insurers.

Domestic Companies

Life Insurance

The Asia Life Assurance Society Ltd: 2 Finlayson Green, 05-00 Asia Insurance Bldg, Singapore 049247; tel. 62243181; fax 62239120; e-mail asialife@asialife.com.sg; internet www.asialife.com.sg; f. 1948; Chair. Tan Eng Heng.

Axa Life Insurance Singapore Pte Ltd: 143 Cecil St, 03-01/10 GB Bldg, Singapore 069542; tel. 68805500; fax 68805501; e-mail comsvc@axa-life.com.sg; internet www.axa-life.com.sg; Prin. Officer Gary Harvey.

China Life Insurance Co Ltd: 105 Cecil St, 18-00 and 19-00 The Octagon, Singapore 069534; tel. 62222366; fax 62221033; Prin. Officer Shen Nan Ning.

John Hancock Life Assurance Co Ltd: 6 Raffles Quay, 21-00 John Hancock Tower, Singapore 048580; tel. 65383333; fax 65383233; e-mail customerservice@jhancock.com.sg; internet www.jhancock.com.sg; f. 1954; Pres. Malcolm J. J. Arney.

Manulife (Singapore) Pte Ltd: 491B River Valley Rd, 07-00 Valley Pt, Singapore 248373; tel. 67371221; fax 67378488; e-mail service@manulife.com; internet www.manulife.com.sg; Pres. and CEO Brian Boisvenue.

UOB Life Assurance Ltd: 156 Cecil St, 10-01 Far Eastern Bank Bldg, Singapore 069544; tel. 62278477; fax 62243012; e-mail uoblife@uobgroup.com; internet www.uoblife.com.sg; Man. Dir Raymond Kwok Chong See.

General Insurance

Allianz Insurance Company of Singapore Pte Ltd: 3 Temasek Ave, 03-01 Centennial Tower, Singapore 039190; tel. 62972529; fax 62971956; e-mail askme@allianz.com.sg; internet www.allianz.com.sg; formed by merger between Allianz Insurance (Singapore) Pte Ltd and AGF Insurance (Singapore) Pte Ltd; Man. Dir Neil Bassett Emery.

Asia Insurance Co Ltd: 2 Finlayson Green, 03-00 Asia Insurance Bldg, Singapore 049247; tel. 62243181; fax 62214355; e-mail asiains@asiainsurance.com.sg; internet www.asiainsurance.com.sg; f. 1923; cap. p.u. S $75m.; Chair. Tan Eng Heng; Man. Dir Loo Sun Mun.

Asian Securitization and Infrastructure Assurance (Pte) Ltd: 9 Temasek Blvd, 38-01 Suntec Tower 2, Singapore 038989; tel. 63342555; fax 63342777; e-mail general@asialtd.com; Dir Eleanor L. Lipsey.

Axa Insurance Singapore Pte Ltd: 143 Cecil St, 01-01 GB Bldg, Singapore 069542; tel. 63387288; fax 63382522; e-mail customer.service@axa.com.sg; internet www.axa.com.sg; CEO Bernard Marseille.

Cosmic Insurance Corpn Ltd: 410 North Bridge Rd, 04-01 Cosmic Insurance Bldg, Singapore 188726; tel. 63387633; fax 63397805; internet www.cosmic.com.sg; f. 1971; Gen. Man. Teo Lip Moh.

ECICS-COFACE Guarantee Co (Singapore) Ltd: 7 Temasek Blvd, 11-01 Suntec City Tower 1, Singapore 038987; tel. 63374779; fax 63389267; e-mail ecics@ecics.com.sg; internet www.ecics.com.sg; Chair. Kwah Thiam Hock; Asst Vice-Pres. Jonathan Tang.

First Capital Insurance Ltd: 80 Robinson Rd, 09-02/03, Singapore 068898; tel. 62222311; fax 62223547; e-mail enquiry@first-insurance.com.sg; internet www.first-insurance.com.sg; Gen. Man. RAMASWAMY ATHAPPAN.

India International Insurance Pte Ltd: 64 Cecil St, 04-00/05-00 IOB Bldg, Singapore 049711; tel. 62238122; fax 62244174; e-mail insure@iii.com.sg; internet www.iii.com.sg; f. 1987; all non-life insurance; CEO R. ATHAPPAN.

Kemper International Insurance Co (Pte) Ltd: 3 Shenton Way, 22-09 Shenton House, Singapore 068805; tel. 68369120; fax 68369121; e-mail vchia@kemper.com.sg; internet www.kemper.com.sg; Gen. Man. VIOLET CHIA.

Liberty Citystate Insurance Pte Ltd: 51 Club St, 03-00 Singapore 069428; tel. 62218611; fax 62263360; e-mail koh.peter@libertycitystate.com.sg; internet www.libertycitystate.com.sg; fmrly Citystate Insurance Pte Ltd; Man. Dir PETER KOH TIEN HOE.

Mitsui Sumitomo Insurance (Singapore) Pte Ltd: 16 Raffles Quay, 24-01 Hong Leong Bldg, Singapore 048581; tel. 62209644; fax 62256371; internet www.ms-ins.com.sg; fmrly Mitsui Marine and Fire Insurance (Asia) Private Ltd; merged with The Sumitomo Marine and Fire Insurance Co Ltd and name changed as above; Man. Dir YOSHIO IIJIMA; Dep. Gen. Man. NOBORI OMORI.

The Nanyang Insurance Co Ltd: 302 Orchard Rd, 09-01 Tong Bldg, Singapore 238862; tel. 67388211; fax 67388133; e-mail tnicl@nanyanginsurance.com.sg; internet www.nanyanginsurance.com.sg; f. 1956; Dir and Principal Officer FREDDIE YEO.

Overseas Union Insurance Ltd: 50 Collyer Quay, 02-02 Overseas Union House, Singapore 049321; tel. 62251133; fax 62246307; internet www.oub.com.sg; f. 1956; Gen. Man. PETER YAP KIM KEE; Asst Gen. Man. YEO TIAN CHU.

Royal & Sun Alliance Insurance (Singapore) Ltd: 77 Robinson Rd, 18-00 SIA Bldg, Singapore 068896; tel. 62201188; fax 64230798; e-mail rsail-sg@sg.royalsun.com; internet www.royalsunalliance.com.sg; Man. Dir and CEO EDMUND LIM.

Singapore Aviation and General Insurance Co (Pte) Ltd: 25 Airline Rd, 06-A Airline House, Singapore 819829; tel. 65423333; fax 65450221; f. 1976; Man. AMARJIT KAUR SIDHU.

Standard Steamship Owners' Protection and Indemnity Association (Asia) Ltd: 140 Cecil St, 10-02 PIL Bldg, Singapore 069540; tel. 62211060; fax 62211082; e-mail central@ctg-ap.com; internet www.standard-club.com; Prin. Officer DAVID ALWYN.

Tenet Insurance Co Ltd: 10 Collyer Quay, 04-01 Ocean Bldg, Singapore 049315; tel. 65326022; fax 65333871; Chair. ONG CHOO ENG.

The Tokio Marine & Fire Insurance Co (Singapore) Pte Ltd: 6 Shenton Way, 23-08 DBS Bldg Tower Two, Singapore 068809; tel. 62216111; fax 62240895; Man. Dir KYOZO HANAJIMA.

United Overseas Insurance Ltd: 156 Cecil St, 09-01 Far Eastern Bank Bldg, Singapore 069544; tel. 62227733; fax 62242718; internet www.uoi.com.sg; Man. Dir DAVID CHAN MUN WAI.

Yasuda Fire and Marine Insurance Co (Asia) Pte Ltd: 50 Raffles Place, 03-03 Singapore Land Tower, Singapore 048623; tel. 62235293; fax 62257947; e-mail yasudaqa@yasudaasia.com; internet www.yasudaasia.com; f. 1989; Man. Dir and Gen. Man. KOJI OTSUKA.

Zürich Insurance (Singapore) Pte Ltd: 78 Shenton Way, 06-01, Singapore 079120; tel. 62202466; fax 62255749; Prin. Officer and Man. Dir RONALD CHENG JUE SENG.

Composite Insurance

American International Assurance Co Ltd: 1 Robinson Rd, AIA Tower, Singapore 048542; tel. 62918000; fax 65385802; internet www.aia.com.sg; Prin. Officer MARK O'DELL.

Aviva Ltd: 4 Shenton Way, 01-01 SGX Centre 2, Singapore 068807; tel. 68277988; fax 68277900; internet www.aviva-singapore.com.sg; Prin. Officer KEITH PERKINS.

Great Eastern Life Assurance Co Ltd: 1 Pickering St, 13-01 Great Eastern Centre, Singapore 048659; tel. 62482000; fax 65322214; e-mail wecare@lifeisgreat.com.sg; internet www.lifeisgreat.com.sg; f. 1908; Dir and CEO TAN BENG LEE.

HSBC Insurance (Singapore) Pte Ltd: 3 Killiney Rd, 10-01/09, Winsland House 1, Singapore 239519; tel. 62256111; fax 62212188; internet www.insurance.hsbc.com.sg; Prin. Officer JASON DOMINIC SADLER.

NTUC Income Insurance Co-operative Ltd: 75 Bras Basah Rd, NTUC Income Centre, Singapore 189557; tel. 63363322; fax 63381500; e-mail inbox@income.wm.sg; internet www.income.com.sg; CEO TAN KIN LIAN; Gen. Man. ALOYSIUS TEO SENG LEE.

Overseas Assurance Corpn Ltd: 1 Pickering St, 13-01 Great Eastern Centre, Singapore 048659; tel. 62482000; fax 65322214; e-mail general@oac.com.sg; internet www.oac.com.sg; f. 1920; Chair. MICHAEL WONG PAKSHONG.

Prudential Assurance Co Singapore (Pte) Ltd: 30 Cecil St, 30-01 Prudential Tower, Singapore 049712; tel. 65358988; fax 65354043; e-mail customer.service@prudential.com.sg; internet www.prudential.com.sg; CEO TAN SUEE CHIEH.

Associations

General Insurance Association of Singapore: 103 Amoy St, Singapore 069923; tel. 62218788; fax 62272051; internet www.gia.org.sg; f. 1965; Pres. TERENCE TAN; Vice-Pres. STELLA TAN.

Life Insurance Association, Singapore: 20 Cross St, 02-07/08 China Court, China Sq. Central, Singapore 048422; tel. 64388900; fax 64386989; e-mail lia@lia.org.sg; internet www.lia.org.sg; f. 1967; Pres. RAYMOND KWOK.

Reinsurance Brokers' Association: 57A Amoy St, Singapore 069883; tel. 62245583; fax 62241091; e-mail secretariat@rbas.org.sg; internet www.rbas.org.sg; Chair. RICHARD AUSTEN.

Singapore Insurance Brokers' Association: 138 Cecil St, 15-00 Cecil Court, Singapore 069538; tel. 62227777; fax 62220022; Pres. ANTHONY LIM; Vice-Pres. DAVID LUM.

Singapore Reinsurers' Association: 85 Amoy St, Singapore 069904; tel. 63247388; fax 62248910; e-mail secretariat@sraweb.org.sg; internet www.sraweb.org.sg; f. 1979; Chair. PHUA KIA TING.

Trade and Industry

GOVERNMENT AGENCIES

Housing and Development Board: 480 Lorong 6, Toa Payoh, Singapore 310480; tel. 64901111; fax 63972070; e-mail hdbmailbox@hdb.gov.sg; internet www.hdb.gov.sg; f. 1960; public housing authority; Chair. NGIAM TONG DOW.

Singapore Land Authority (SLA): 8 Shenton Way, 26-01 Temasek Tower, Singapore 068811; tel. 63239829; fax 63239937; e-mail SLA_enquiry@sla.gov.sg; internet www.sla.gov.sg; responsible for management and development of state land resources; Chief Exec. TAN KEE YONG.

Urban Redevelopment Authority (URA): 45 Maxwell Rd, URA Centre, Singapore 069118; tel. 62216666; fax 62263549; e-mail ura_email@ura.gov.sg; internet www.ura.gov.sg; statutory board; responsible for national planning.

DEVELOPMENT ORGANIZATIONS

Agency for Science, Technology and Research (A*STAR): 10 Science Park Rd, 01/01-03 The Alpha, Singapore Science Park II, Singapore 117684; tel. 67797066; fax 67771711; e-mail astar_contact@a-star.gov.sg; internet www.a-star.gov.sg; f. 1990; fmrly National Science and Technology Board; statutory board; responsible for the development of science and technology; Chair. PHILIP YEO; Man. Dir BOON SWAN FOO.

Applied Research Corpn (ARC): independent non-profit-making research and consultancy org. aiming to facilitate and enhance the use of technology and expertise from the tertiary institutions to benefit industry, businesses and joint institutions.

Asian Infrastructure Fund (AIF): Singapore; f. 1994; promotes and directs investment into regional projects; Chair. MOEEN QURESHI.

Economic Development Board (EDB): 250 North Bridge Rd, 24-00 Raffles City Tower, Singapore 179101; tel. 63362288; fax 63396077; e-mail webmaster@edb.gov.sg; internet www.sedb.com.sg; f. 1961; statutory body for industrial planning, development and promotion of investments in manufacturing, services and local business; Chair. TEO MING KIAN; Man. Dir KO KHENG HWA.

Government of Singapore Investment Corp Pte Ltd (GSIC): 250 North Bridge Rd, 38-00 Raffles City Tower, Singapore 179101; tel. 63363366; fax 63308722; internet www.gic.com.sg; f. 1981; Chair. LEE KUAN YEW; Man. Dir LEE EK TIENG.

Infocomm Development Authority of Singapore (IDA): see under Telecommunications.

International Enterprise Singapore: 230 Victoria St, 07-00 Bugis Junction Office Tower, Singapore 188024; tel. 63376628; fax 63376898; e-mail enquiry@iesingapore.gov.sg; internet www.iesingapore.gov.sg; f. 1983 to develop and expand international trade; fmrly Trade Development Board; statutory body; Chair. STEPHEN LEE; CEO LEE YI SHYAN.

Jurong Town Corpn: 8 Jurong Town Hall Rd, Singapore 609434; tel. 65600056; fax 65655301; e-mail askjtc@jtc.gov.sg; internet www

.jtc.gov.sg; f. 1968; statutory body responsible for developing and maintaining industrial estates; Chair. LIM NEO CHIAN; CEO CHONG LIT CHEONG.

Singapore Productivity and Standards Board (PSB): PSB Bldg, 2 Bukit Merah Central, Singapore 159835; tel. 62786666; fax 62786667; e-mail queries@psb.gov.sg; internet www.psb.gov.sg; f. 1996; merger of Singapore Institute of Standards and Industrial Research and the National Productivity Board; carries out activities in areas including workforce development, training, productivity and innovation promotion, standards development, ISO certification, quality programmes and consultancy, technology application, product and process development, system and process automation services, testing services, patent information, development assistance for small and medium-sized enterprises; Chief Exec. LEE SUAN HIANG.

CHAMBERS OF COMMERCE

Singapore Business Federation (SBF): 19 Tanglin Rd, Level 7, Tanglin Shopping Centre, Singapore 247909; tel. 68276828; fax 68276807; e-mail webmaster@sbf.org.sg; internet www.sbf.org.sg; f. 2002 as result of restructuring of Singapore Federation of Chambers of Commerce and Industry; Chair. STEPHEN LEE CHING YEN; Hon. Sec.-Gen. CHENG WAI KEUNG; mems include the following:

Singapore Chinese Chamber of Commerce and Industry: 47 Hill St, 09-00 Singapore 179365; tel. 63378381; fax 63390605; e-mail corporate@sccci.org.sg; internet www.sccci.org.sg; f. 1906; promotes trade and industry and economic development of Singapore.

Singapore Confederation of Industries: The Enterprise 02-02, 1 Science Centre Rd, Singapore 609077; tel. 68263000; fax 68228828; e-mail scihq@sci.org.sg; internet www.sci.org.sg; f. 1932; Pres. ROBIN LAU CHUNG KEONG; Sec.-Gen. THAM HOCK CHEE.

Singapore Indian Chamber of Commerce and Industry: 101 Cecil St, 23-01/04 Tong Eng Bldg, Singapore 069533; tel. 62222855; fax 62231707; e-mail sicci@singnet.com.sg; internet www.sicci.com.

Singapore International Chamber of Commerce: 6 Raffles Quay, 10-01 John Hancock Tower, Singapore 048580; tel. 62241255; fax 62242785; e-mail general@sicc.com; internet www.sicc.com.sg; f. 1837.

Singapore Malay Chamber of Commerce: 72A Bussorah St, Singapore 199485; tel. 62979296; fax 63924527; e-mail smcci@singnet.com.sg; internet www.smcci.org.sg.

INDUSTRIAL AND TRADE ASSOCIATIONS

Association of Singapore Marine Industries (ASMI): 1 Maritime Sq., 09–54 Harbour Front Centre, Singapore 099253; tel. 62704730; fax 62731867; e-mail asmi@pacific.net.sg; internet www.asmi.com; f. 1968; 12 hon. mems, 70 assoc. mems, 51 ord. mems (Oct. 2003); Pres. HENG CHIANG GNEE; Exec. Dir WINNIE LOW.

Singapore Commodity Exchange (SICOM): 111 North Bridge Rd, 23-04/05 Peninsula Plaza, Singapore 179098; tel. 63385600; fax 63389116; e-mail marketing@sicom.com.sg; internet www.sicom.com.sg; f. 1968 as Rubber Association of Singapore, adopted present name 1994; to regulate, promote, develop and supervise commodity futures trading in Singapore, including the establishment and dissemination of official prices for various grades and types of rubber; provides clearing facilities; endorses certificates of origin and licences for packers, shippers and mfrs; Chair. PATRICK HAYS; Gen. Man. LIM TOH ENG.

EMPLOYERS' ORGANIZATION

Singapore National Employers' Federation: 19 Tanglin Rd, 10-01/07 Tanglin Shopping Centre, Singapore 247909; tel. 68276827; fax 68276800; e-mail webmaster@snef.org.sg; internet www.sgemployers.com; f. 1948; Pres. STEPHEN LEE CHING YEN; Vice-Pres. ALEX CHAN, BOB TAN BENG HAI, LANDIS W. HICKS.

UTILITIES

Electricity and Gas

Singapore Power Ltd: 111 Somerset Rd 16-01, Singapore Power Bldg, Singapore 238164; tel. 68238888; fax 68238188; e-mail corpcomms@spower.com.sg; internet www.spower.com.sg; incorporated in 1995 to take over the piped gas and electricity utility operations of the Public Utilities Board (see Water) which now acts as a regulatory authority for the privately owned cos; 100% owned by government holding company, Temasek Holdings Pte Ltd; subsidiaries include PowerGrid, PowerGas, Power Supply, Singapore Power International, Development Resources, Power Automation, SP E-Services, SP Systems, Singapore District Cooling and SP Telecommunications; Chair. NG KEE CHOE; Pres. and CEO KWEK SIEW JIN.

Water

Public Utilities Board: 111 Somerset Rd 15-01, Singapore 238164; tel. 62358888; fax 67313020; e-mail pub_pr@pub.gov.sg; internet www.pub.gov.sg; statutory board responsible for water supply; manages Singapore's water system to optimize use of water resources, develops additional water sources; Chair. TAN GEE PAW.

MAJOR COMPANIES

(cap. = capital; res = reserves; m. = million; amounts shown are in Singapore dollars unless otherwise stated)

Building and Building Materials

Econ International Ltd: 2 Ang Mo Kio St 64, Ang Mo Kio Industrial Park 3, Singapore 569084; tel. 64842222; fax 64842221; e-mail corp@econ.com.sg; internet www.econ.com.sg; investment holding co with subsidiaries engaged in manufacture and trading of building materials and construction; cap. and res 118m., sales 327m. (2001/02); Chair. CHEW TIONG KHENG; Exec Dir JOSEPH SIN KAM CHOI.

Hong Leong Corpn: 16 Raffles Quay, 26-00 Hong Leong Bldg, Singapore 048581; tel. 62208411; fax 62200087; e-mail inquiry@hlcorp.com.sg; internet www.hlcorp.com.sg; f. 1982; cap. and res 441m. (2000), sales 2,100m. (1995, est.); mfr and retailer of concrete, metal packaging, plastic packaging and containers; Pres. PETER H. REN; 2,400 employees.

Jurong Cement Ltd: 15 Pioneer Crescent, Jurong Town, Singapore 628551; tel. 62618816; fax 62659178; e-mail jcg@jurongcement.com; internet www.jurongcement.com; f. 1973; cap. and res 111m., sales 68m. (2000/01); mfrs of cement; Chair. TAY JOO SOON; Man. Dir TAN ENG SIM.

Lee Kim Tah Holdings Ltd: 20 Jalan Afifi, 07-01 CISCO Centre, Singapore 409179; tel. 67453318; fax 67451218; e-mail info@leekimtah.com; internet www.leekimtah.com; cap. and res 148m. (1999), sales 51m. (2001); investment holding co with subsidiaries engaged in manufacture of precast concrete building products and other building materials, building and civil engineering construction and property develolpment; Chair. LEE SOON TECK; 700 employees.

Low Keng Huat (S) Ltd: 80 Marine Parade Rd, 18-05/09 Parkway Parade, Singapore 449269; tel. 63442333; fax 63457841; e-mail lkhsl@singnet.com.sg; cap. and res 119m., sales 177m. (2000/01); property management, construction and building materials; Chair. Tan Sri Dato' LOW KENG HUAT; Man. Dir LOW KENG BOON.

Ssangyong Cement (Singapore) Ltd: 29 International Business Park, Acer Bldg, Tower B, 08-05/06, Singapore 609923; tel. 65617978; fax 65619770; e-mail ssya@cemtecasia.com.sg; internet www.cemtecasia.com.sg; cap. and res 222m., sales 79m. (2000); mfrs and retailers of cement, concrete, vermiculite and building materials; Chair. KI-HO KIM.

Tuan Sing Holdings Ltd: 30 Robinson Rd, 12-01 Robinson Towers, Singapore 048546; tel. 62237211; fax 62241085; e-mail tuansing@pacific.net.sg; internet www.tuansing.com; f. 1969; cap. and res 230m., sales 605m. (2003); investment holding co with interests in property, manufacturing, construction and trading; Chair. PATRICK YEOH KHWAI HOH; Man. Dir LOW KIAN BENG; 3,098 employees.

Wearnes International (1994) Ltd: 45 Leng Kee Rd, Singapore 159103; tel. 64716288; fax 64720009; e-mail wicorp@wearnes-intl.com.sg; internet www.wearnes-intl.com.sg; cap. and res 200m., sales 412m. (2002/03); mfrs of food equipment, retailers of building materials, suppliers of industrial, laundry, kitchen and food-processing equipment; Chair. TANG I-FANG; Man. Dir SOH YEW HOCK.

Chemicals and Petroleum

BP Singapore Pte Ltd: 396 Alexandra Rd, 18-01 BP Tower, Singapore 119954; tel. 63718888; fax 63718761; e-mail gohl@bp.com; internet www.bp.com; f. 1966; cap. and res 4,706m., sales 10,731m. (2000); retailer of petroleum products; Pres. KOH KIM WAH; 750 employees.

Caltex Singapore Pte Ltd: 30 Raffles Place, 25-00 Caltex House, Singapore 048622; tel. 65333000; fax 64391790; e-mail enquiries@caltexoil.com.sg; internet www.caltex.com.sg; f. 1989; sales 47,761m. (2000); retailing and refining of petroleum products; Pres. and CEO JACK MCKENZIE; 800 employees.

Chemical Industries (Far East) Ltd: 17 Upper Circular Rd, 05-00 Juta Bldg, Singapore 058415; tel. 65354884; fax 65344582; cap. and res 121m., sales 62m. (2001/02); mfrs and retailers of chemicals; Chair. and Man. Dir LIM SOO PENG; 120 employees.

Cosmo Oil International Pte Ltd: 6 Battery Rd, 26-06, Singapore 068809; tel. 64230709; fax 64230730; f. 1985; cap. and res 34m., sales

963m. (2000); retailer of petroleum and associated products; Man. Dir NOBUYUKI KITAMURA.

Idemitsu Lube (Singapore) Pte Ltd: 37 Pandan Rd, Singapore 609280; tel. 62655672; fax 62655610; e-mail automotive@idemitsu-ils.com.sg; internet www.idemitsu-ils.com.sg; retailer of petroleum products; Man. Dir Z. SUDA.

Itochu Petroleum Co (Singapore) Pte Ltd: 9 Raffles Place, 48-01, Singapore; tel. 65323542; fax 65323809; f. 1984; cap. and res 6m., sales 6,030m. (2000); retailer of petroleum products; Chair. JUNJI TANIUCHI.

Marubeni Singapore Petroleum: 16 Raffles Quay, 40-01 Hong Leong Bldg, Singapore 048581; tel. 62240446; fax 62215458; cap. and res 25m., sales 656m. (2000); retailer of petroleum and related products; CEO FUMIYA KOKUBU.

Mobil Oil Singapore Pte Ltd: 18 Pioneer Rd, Singapore 628498; tel. 62650000; fax 62656570; f. 1963; sales 3,286m. (1995); petroleum refining; retailing of petroleum products; Gen. Man. G. H. KOHLENBURGER; 640 employees.

Nippon Oil (Asia) Pte Ltd: 6 Battery Rd, 29-02/03, Singapore 049909; tel. 62236732; fax 62248921; f. 1980; cap. and res 30m., sales 5,245m. (2000); retailer of petroleum products; Man. Dir N. TAKAYAMA.

Nissho Iwai Petroleum Pte Ltd: 77 Robinson Rd, 33-00 SIA Bldg, Singapore 065896; tel. 64382566; fax 64382577; f. 1991; sales 3,138m. (2000/01); petroleum and petroleum products; Man. Dir SADAO EZAKI.

Shell Eastern Petroleum (Pte) Ltd, Shell Eastern Trading (Pte) Ltd, Seraya Chemicals Singapore Pte Ltd, Shell Research Eastern (Pte) Ltd, Shell Singapore Trustees Pte Ltd: Shell House, UE Sq., 83 Clemenceau Ave, Singapore 239920; tel. 63848000; fax 63848373; internet www.shell.com.sg; cap. and res 712m., sales 31,717m. (2000); Chair. LEE TZU YANG; 1,700 employees.

Singapore Petroleum Co Ltd: 1 Maritime Sq., 1010 World Trade Centre, Singapore 099253; tel. 62766006; fax 62756006; e-mail ayew@spc.singnet.sg; internet www.spc.com.sg; f. 1969; cap. and res 589m., sales 3,188m. (2003); petroleum refining, tanker transport, trading, distribution and marketing; investment in oil and gas development and production; Chair. CHOO CHIAU BENG.

SK Energy Asia Pte Ltd: 5 Shenton Way, 34-04 UIC Bldg, Singapore 068808; tel. 62201266; fax 62211225; e-mail singapore@skcorp.com; sales 3,891m. (1994/95); fmrly Yukong International Pte Ltd; petroleum products; Man. Dir C. K. LEE.

Electrical and Electronics

Acma Ltd: 17 Jurong Port Rd, Singapore 619092; tel. 62687733; fax 62663998; f. 1965; cap. and res 184.8m., sales 475.3m. (2001); mfrs, retailers and servicers of electrical and electronic equipment; Chair. QUEK SIM PIN; Man. Dir RAJ RAJEN; 200 employees.

Amcol Holdings Ltd: 9 Temasek Blvd, 42-00 Suntec Tower 2, Singapore 038989; tel. 63371722; fax 63321800; cap. and res 555m., sales 981m. (1995/96); mfrs and retailers of electronic and electrical products; Chair. SUDWIKATMONO; Man. Dir HENRY PRIBADI.

Aztech Systems Ltd: 31 Ubi Rd 1, Aztech Bldg, Singapore 408694; tel. 67417211; fax 67418678; internet www.aztech.com.sg; f. 1986; cap. and res 50.0m., sales 62.8m. (2003/04); designers, mfrs and distributors of multimedia products; Pres. and CEO MICHAEL HONG YEW MUN; 1,393 employees.

Beyonics Technology Ltd: 30 Marsiling Industrial Estate, Rd 8, Singapore 739193; tel. 63490600; fax 63490500; e-mail enquiries@sg.beyonics.com; internet www.beyonics.com; f. 1983; mfr of precision components for computer and electronics industries; investment holding; cap. and res 208.2m., sales 554.8m. (2002/03); CEO GOH CHAN PENG; 2,500 employees.

Clipsal Industries (Holdings) Ltd: 5 Fourth Chin Bee Rd, Singapore 619699; tel. 62665936; fax 62662989; e-mail clipsal@singnet.com.sg; internet www.clipsal.com.sg; f. 1991; cap. and res 228m. (2000), sales 197m. (2001); investment holding co with subsidiaries engaged in manufacture and marketing of electrical installation products; Chair. VICTOR LO CHUNG WING; Man. Dir CHAU KWOK WAI; 2,400 employees.

Conner Peripherals Pte Ltd: 151 Lorong Chuan, 06-01 New Tech Park, Singapore 556741; tel. 62845366; fax 62845060; sales 1,400m. (1994, est.); mfr of computer components.

Creative Technology Ltd: 31 International Business Park, Creative Resource, Singapore 609921; tel. 68954000; fax 68954999; internet www.creative.com; f. 1983; cap. and res US $428.8m., sales US $701.8m. (2002/03); multimedia products peripherals; Chair. and CEO SIM WONG HOO.

Elec & Eltek International Co Ltd: 8 Shenton Way, 37-02/03 Temasek Tower, Singapore 06811; tel. 62260488; fax 62202377; internet www.eleceltek.com; f. 1993; cap. and res 384m., sales 596m. (2000/01); investment holding, with subsidiaries engaged in design, mfr and distribution of circuit boards; Chair. DAVID SO CHEUNG SING; 6,638 employees.

Eltech Electronics Ltd: 28 Marsiling Lane, Singapore 739152; tel. 63729898; fax 63636556; e-mail harryng@eltech.com.sg; internet www.eltech.com.sg; cap. and res 64m., sales 167m. (2000); electronics equipment; Chair. TAN CHEOW KOON; Man. Dir TAN GEH.

General Magnetics Ltd: 625 Lorong 4, Toa Payoh, Singapore 319519; tel. 62595511; fax 62593723; e-mail sales@genmag.com.sg; internet www.genmag.com.sg; cap. and res 39.5m., sales 25.4m. (2003); mfrs and retailers of magnetic media and related products and telecommunication equipment; Chair. and Man. Dir OH LOON LIAN; 400 employees.

Giken Sakata (S) Ltd: 40 Jalan Pemimpin 04-05, Tat Ann Bldg, Singapore 577185; tel. 62599133; fax 62599822; e-mail enquiry@giken.com.sg; internet www.giken.com.sg; f. 1979; cap. and res 15m., sales 83m. (2003/04); parts for electronics equipment; Pres. EIJU YOKOTA; 4,886 employees.

GP Batteries International Ltd: 97 Pioneer Rd, Singapore 639579; tel. 65599800; fax 65599801; e-mail gpbi@gpbatteries.com.sg; internet www.gpbatteries.com.sg; f. 1964; total assets 870m., sales 827m. (2003/04); batteries and related products; Chair. and CEO ANDREW NG SUNG ON; 7,000 employees.

Hewlett-Packard Singapore (Pte) Ltd: 450 Alexandra Rd, Singapore 119960; tel. 62733888; fax 62756839; e-mail hpdirect_sgp@hp.com; internet www.hp.com.sg; f. 1970; cap. and res 10,361m., sales 13,609m. (1999/2000); mfrs of opto-electronic components, inkjet printers and peripherals, personal computers, imaging products, hand held devices and integrated circuits; Man. Dir CHIA WEE BOON; 9,000 employees.

Hitachi Home Electronics Asia (S) Pte Ltd: 438A Alexandra Rd 01-01/02, Alexandra Technopark, Singapore 119967; tel. 62419444; fax 64444572; e-mail customerservice@hitachiconsumer.com; internet www.hitachiconsumer.com.sg; f. 1972; sales 200m. (1995); mfrs of TV and radio receivers, tape-recorders, vacuum cleaners, air purifiers; Man. Dir M. MATSUNAGA; 1,000 employees.

Infineon Components Pte Ltd: 168 Kallang Way, Singapore 349253; tel. 68400888; fax 68400291; e-mail hr-asiapacific@infineon.com; internet www.infineon.com; f. 1970; cap. and res 306m., sales 3,865m. (1999/2000); mfr of electronic components; Man. Dir LOH KIN WAH; 2,000 employees.

IPC Corpn Ltd: 23 Tai Seng Drive, IPC Bldg, Singapore 535224; tel. 67442688; fax 67430691; internet www.ipc.com.sg; f. 1985; cap. and res -49m., sales 25m. (1999); mfrs and retailers of computers and electronic products; Chair. and CEO PATRICK NGIAM MIA JE.

Matsushita Electronic Components (S) Pte Ltd: 3 Bedok South Ave, Singapore 469269; tel. 64437744; fax 64453168; e-mail mesaki@singnet.com.sg; f. 1977; cap. and res 117m., sales 584m. (2000/01); mfrs of radio receivers, tape-recorders and stereophonic equipment; Man. Dir KASHIMA KELVIN; 2,200 employees.

Maxtor Peripherals (Singapore) Pte Ltd: 2 Ang Mo Kio St 63, Ang Mo Kio Ind. Park 3, Singapore 569448; tel. 64821100; fax 64815513; internet www.maxtor.com; f. 1990; cap. and res 365m., sales 4,113m. (2000); mfr of hard disk drives and other computer peripherals; Man. Dir TEH KEE HONG; 5,500 employees.

Micron Semiconductor Asia Ltd: 990 Bendemeer Rd, Singapore 339942; tel. 62903191; fax 62903690; e-mail jamesheng@micron.com; internet www.micron.com; cap. and res 357m., sales 5,351m. (1999/2000); provider of semiconductor memory solutions; Man. Dir JEN KWONG HWA; 3,000 employees.

Motorola Electronics Pte Ltd: 12 Ang Mo Kio Ind. Park 3, St 64, Singapore 569088; tel. 64812000; fax 64813081; internet www.motorola.com.sg; f. 1983; cap. and res 3,169m., sales 3,318m. (2000); mfr of mobile telephones, pagers, cable modems; Man. Dir ARTHUR CHEN WAN SHOU; 2,445 employees.

PCI Ltd: 56 Serangoon North Ave 4, Singapore 555851; tel. 64828181; fax 64820053; e-mail sales@pciltd.com.sg; internet www.pciltd.com; f. 1988; cap. and res 55m., sales 215m. (2000/01); mfr of printed circuit boards, liquid crystal displays, cordless telephones and other turnkey products; Chair. JIMMY PEH KWEE CHIM; 2,700 employees.

Pentex-Schweizer Circuits Ltd: 55 Penjuru Rd, Singapore 609140; tel. 62664311; fax 62665872; e-mail daniel_wong@pentex-schweizer.com.sg; internet www.pentex-schweizer.com.sg; cap. and res 67m., sales 133m. (2000/01); mfrs and retailers of double-sided printed circuit boards; Chair. CHRISTOPH OSKAR SCHWEIZER; 500 employees.

Philips Electronics Singapore Pte Ltd: 620A Lorong 1, Toa Payoh, Singapore 319762; tel. 63502000; fax 62533395; internet www.philips.com.sg; f. 1951; cap. and res 121m., sales 2,883m. (2000); mfrs of consumer electronic products, domestic appliances and telecommunications equipment; Chair. ROBERT MARTIJNSE; 5,570 employees.

Pioneer Electronics (S) Pte Ltd: 253 Alexandra Rd, 04-01 Comco Bldg, Singapore 159936; tel. 64721111; fax 64725333; internet www.pioneer.com.sg; f. 1988; cap. and res 191m., sales 2,203m. (2000/01); marketing and sales of electronic goods; Man. Dir K. SHIMIZU; 6,000 employees.

Samsung Asia Pte Ltd: 83 Clemenceau Ave, 08-06 UE Sq., Singapore 239920; tel. 68333402; fax 68333100; internet www.samsungelectronics.com; f. 1988; cap. and res 283m., sales 3,115m. (2001); retailers of semiconductors and telecommunications equipment; CEO KIM YIN SOO; 250 employees.

Sanyo Industries (S) Pte Ltd: 117–119 Neythal Rd, Singapore 628604; tel. 62650077; fax 62653015; internet www.sanyo-sis.com.sg; f. 1966; cap. p.u. 2m.; mfrs of electrical household appliances; Chair. K. IUE; Man. Dir NG GHIT CHEONG; 500 employees.

Seagate Technology International (Singapore) Pte Ltd: 06-01 New Tech Park, 151 Lorung Chuan, Singapore 556741; tel. 62870700; fax 64887525; f. 1987; sales 4,900m. (1994); mfr of personal computers and computer components; Vice-Pres. DON KENNEDY; 1,250 employees.

Singatronics Ltd: 506 Chai Chee Lane, Singatronics Bldg, Singapore 469026; tel. 64486211; fax 64452506; e-mail corp_hq@singatronics.com.sg; internet www.singatronics.com.sg; cap. and res 500m., sales 62m. (2001); engaged in manufacture and retailing of computer-related, telecommunication and other electronic products, property development and leasing, hotel ownership; Chair. EDDIE FOO CHIK KIN; 1,200 employees.

Solectron Technology Singapore Pte Ltd: 12 Kallang Rd, Singapore 349216; tel. 68411888; fax 63868230; cap. and res 840m., sales 4,513m. (2000); mfr of computer peripherals, communication and office equipment; Pres. CHESTER LIN CHIEN; 12,797 employees.

ST Microelectronics Asia Pacific (Pte) Ltd: 28 Ang Mo Kio Industrial Park 2, Singapore 569508; tel. 64821411; fax 64820240; e-mail charlie.loh@st.com; internet www.st.com.sg; f. 1969; cap. and res 736m., sales 4,130m. (2000); manufacturing, marketing and sales, design of integrated circuits; Man. Dir RENATO SIRTORI; 4,479 employees.

Thakral Corpn Ltd: 20 Upper Circular Rd, The Riverwalk 03-06, Singapore 058416; tel. 63368966; fax 63367225; e-mail tcloct96@thakralcorp.com.sg; internet www.thakral.com; cap. and res 167m., sales 530m. (2004); investment holding, distributor of consumer electronic and home entertainment products with operations in Hong Kong and China; Chair. KARTAR SINGH THAKRAL; Man. Dir INDERBETHAL SINGH THAKRAL; 769 employees.

Thales Electro Optics Pte Ltd: 14 Fifth Lok Yang Rd, Jurong Town, Singapore 629763; tel. 62655122; fax 62651479; e-mail customer.services@sg.thalesgroup.com; internet www.thales-eo.com.sg; cap. and res 48m., sales 218m. (1998/99); fmrly Avimo Group Ltd; designers, developers, mfrs, marketers and servicers of advanced precision optics, electro-optics instruments, laser equipment; custom plating processing of component products; Chair. BERNARD JEAN-NOEL ROCQUEMONT; Man. Dir IAN CLARK BOOTH.

Toshiba Electronics Asia (S) Pte Ltd: 460 Alexandra Rd, 21-00 PSA Bldg, Singapore 119963; tel. 62785252; fax 62715155; e-mail wee_kee_tay@tea.toshiba.ceo.jp; internet www.toshiba.co.jp; sales 2,314m. (1994/95); electronic components retailers; CEO K. HIROSE; 80 employees.

Venture Manufacturing (Singapore) Ltd: 5006 Ang Ko Mio Ave 5, 05-01/12 Techplace II, Singapore 569873; tel. 64821755; fax 64820122; e-mail info@venture.com.sg; internet www.venture-mfg.com.sg; mfrs and traders for companies in electronic and computer-related industries; cap. and res 1,341m., sales 3,170m. (2003); Chair. GOPALA ACHUTA MENON; Man. Dir WONG NGIT LIONG.

Vikay Industrial Ltd: 331 North Bridge Rd, 04-06 Odeon Towers, Singapore 188720; tel. 62700300; fax 62706989; internet www.arianecorp.com; f. 1984; mfrs of electronic products; cap. and res 13.6m., sales 4.8m. (2003); Chair. LEW SYN PAU; CEO KEA KAH KIM; 3,688 employees.

Western Digital (S) Pte Ltd: 750B Chai Chee Road Block E, Chai Chee Ind. Park, Singapore 496002; tel. 64433443; fax 64456107; sales 2,187m. (1994/95); mfr of personal computer components; Man. Dir VINCE MASTROPIETRO.

Food and Beverages

Asia Pacific Breweries (S) Pte Ltd: 21-00 Alexandra Pt, 438 Alexandra Rd, Singapore 119958; tel. 62729488; fax 62710811; e-mail apbpr@singnet.com.sg; internet www.tigerbeer.com; f. 1990; cap. and res 581.6m., sales 1,009.1m. (2000/01); brewers of beer and stout; Chair. Dr MICHAEL FAM; 650 employees.

Auric Pacific Group Ltd: 9 Temasek Blvd, 41-01 Suntec City Tower Two, Singapore 0468621; tel. 63362262; fax 63362272; f. 1988; cap. and res 254.3m., sales 125.5m. (2000); food manufacturing; Chair. JAMES T. RIADY; 2,068 employees.

Cerebos Pacific Ltd: 18 Cross St 12-01/08, China Sq. Central, Singapore 048423; tel. 62120100; fax 62262126; e-mail cerebos@singnet.sg; internet www.cerebos.com.sg; cap. and res 369m., sales 291m. (2002/03); subsidiaries engaged in manufacture and marketing of food products; Chair. TEO CHIANG LONG; Pres. and CEO EIJI KOIKE; 2,000 employees.

Cold Storage Singapore Ltd: 1 Sophia Rd, 06-38 Peace Centre, Singapore 228149; tel. 63372766; fax 63334026; internet www.coldstorage.com.sg; f. 1960; cap. and res 120m., sales 625m. (2000); retailers and distributors of food, mfrs of food and beverages; Chair. MICHAEL P. K. KOK; 3,000 employees.

Fraser and Neave Ltd: 21-00 Alexandra Pt, 438 Alexandra Rd, Singapore 119958; tel. 62729488; fax 62710811; internet www.fraserandneave.com; f. 1964; cap. and res 2,314m., sales 2,939m. (2000/01); investment holding company with subsidiaries involved in the production of soft drinks, beer, stout and dairy products, and in packaging; Pres. Tan Sri TAN CHIN TUAN; Chair. MICHAEL Y. O. FAM; 12,000 employees.

Kuok Oils and Grains Pte Ltd: 1 Kim Seng Promenade, Great World City, 05-01 Singapore 237994; tel. 65344722; fax 65325335; e-mail trading@kuokoil.com; f. 1989; cap. and res 315m., sales 4,050m. (2000); mfr of edible oils and fats; Man. Dir KWOK KIAN HAI; 130 employees.

Network Foods International Ltd: 5 Shenton Way, 29-00 UIC Bldg, Singapore 068808; tel. 67577622; fax 67570300; internet www.networkfoods.com; cap. and res 0.6m., sales 56m. (1998); mfrs and marketers of chocolate and other cocoa-based products, biscuits and confectioneries; Chair. KOK QUAN TAN; Exec. Dir WOON SEE SOON.

Prima Ltd: 201 Keppel Rd, Singapore 099419; tel. 62728811; fax 62732933; e-mail enquiry@prima.com.sg; internet www.prima.com.sg; f. 1961; cap. 38m., sales 66m. (2001); producers of wheat flour, wheat bran and pollard; fish and meat; Chair. CHENG CHIH KWONG PRIMUS; 235 employees.

QAF Ltd: 150 South Bridge Rd, 09-04 Fook Hai Bldg, Singapore 058727; tel. 65382866; fax 65386866; e-mail info@qaf.com.sg; internet www.qaf.com.sg; f. 1958; cap. and res 175m., sales 487m. (1999/2000); wholesalers and retailers of food and beverages; Chair. DIDI DAWIS; Man. Dir TAN KONG KING; 3,880 employees.

Yeo Hiap Seng Ltd: 3 Senoko Way, Singapore 758057; tel. 67522122; fax 67523122; cap. and res 377m., sales 404m. (2000); mfrs and distributors of food and beverages; Chair. ROBERT CHEE SIONG NG; Man. Dir and CEO LEONG HORN KEE.

Metals and Engineering

Amtek Engineering Ltd: 1 Kian Teck Drive, Singapore 628818; tel. 62640033; fax 62652510; e-mail sales@ammail.amtek.com.sg; internet www.amtek.com.sg; f. 1980; cap. and res 162.5m., sales 714.5m. (2000/01); mfrs and stampers of precision metal parts; precision rubber components and moulds; mechanical assemblies; computer enclosures; Chair. LEE AH BEE; Man. Dir LAI FOOK KUEN; 44,813 employees.

Carnaudmetalbox Asia Ltd: 750-D Technopark Chai Chee, 08-02/03 Chai Chee Rd, Singapore 469004; tel. 62421788; fax 62423722; internet www.sciencepark.com.sg/data/IO5.html; cap. and res 287m., sales 241m. (2000); investment holding co and mfrs of metal packaging; Chair. and CEO WILLIAM HENRY VOSS; 3,000 employees.

GB Holdings Ltd: 29 Loyang Crescent, Singapore 509015; tel. 65452929; fax 65430029; internet www.grandbanks.com; cap. and res 43m., sales 66m. (1999/2000); investment holding co with subsidiaries engaged in manufacture and retailing of diesel-powered cruisers, sailing boats; yacht repairers; Chair. ROBERT W. LIVINGSTON; 870 employees.

Hitachi Zosen Engineering Singapore Ltd: JTC Summit Bldg, 23-04, 8 Jurong Town Hall Rd, Singapore 609434; tel. 63162771; fax 63162773; f. 1970; cap. and res 170m., sales 228m. (1998); shipbuilding and repair, steel structures, industrial engineering; Chair. TAKESHI TENJIN; Man. Dir FOO MENG KEE.

Jurong Engineering Ltd: 25 Tanjong Kling Rd, Singapore 628050; tel. 62653222; fax 62684211; e-mail webmaster@jel.com.sg; internet www.jel.com.sg; cap. and res 121m., sales 179m. (2003); construction and engineering services; Chair. BOEY TAK HAP; Man. Dir and CEO OSAMU ABIKO.

Keppel Corpn Ltd: 23 Church St, 15-01 Capital Square, Singapore 049481; tel. 68857390; fax 68857391; internet www

.keppelcorporation.com; f. 1859; cap. and res 2,890m., sales 5,947m. (2003); Chair. LIM CHEE OMN; Man. Dir LOH WING SIEW; 15,947 employees.

Keppel Offshore and Marine Ltd: 50 Gul Rd, Singapore 629531; tel. 68637200; fax 62617719; internet www.keppelom.com; f. 2002 following merger of Keppel FELS Ltd and Keppel Hitachi Zosen Ltd; sales 1,460m. (2003); builders and repairers of offshore production facilities; Chair. and CEO CHOO CHIAU BENG; Man. Dir TONG CHONG HEONG; 1,970 employees.

Liang Huat Aluminium Ltd: 51 Benoi Rd, Blk 8, 08-05 Liang Huat Industrial Complex, Singapore 629908; tel. 68622228; fax 68624962; e-mail lhg_corpadm1@pacific.net.sg; internet www.lianghuatgroup .com.sg; f. 1978; cap. and res 79m. (2001), sales 32m. (2002); aluminium products; Man. Dir PETER TAN YONG KEE; 1,000 employees.

Lion Asiapac Ltd: 10 Arumugam Rd, 10-00 Lion Industrial Bldg, Singapore 409957; tel. 67459677; fax 67479493; f. 1974; cap. and res 62m., sales 89m. (2000/01); mfrs and retailers of containers, and investment holding; Chair. OTHMAN WOK; Man. Dir SAM CHONG KEEN; 286 employees.

NatSteel Ltd: 22 Tanjong Kling Rd, Jurong Town, Singapore 628048; tel. 62651233; fax 62658317; internet www.natsteel.com.sg; f. 1961; fmrly National Iron and Steel Mills; cap. and res 992m., sales 1,585m. (2001); manufacturing and marketing of iron and steel products, building materials, electronics and lime products; Chair. Dr CHAM TAO SOON; Pres. ANG KONG HUA; 18,658 employees.

Pacific Can Investment Holdings Ltd: 1 Sixth Lok Yang Rd, Singapore 628099; tel. 62656877; fax 62683280; cap. and res 9m., sales 58m. (1997/98); investment holding co with subsidiaries engaged in production of cans and components, packaging, printing; Chair. LOW HUA KIN; Man. Dir and CEO KO CHING-SHUEI.

SembCorp Industries: 30 Hill St, 05-04, Singapore 179360; tel. 67233113; fax 68223254; internet www.sembcorp.com.sg; f. 1998 following merger of Singapore Technologies Industrial Corpn and Sembawang Corpn Ltd; cap. and res 402m., sales 3,225m. (2001); investment holding with subsidiaries involved in engineering and construction, environmental engineering, logistics and utilities; Chair. PETER SEAH LIM HUAT; 9,767 employees.

Sembcorp Marine Ltd: 29 Tanjong Kling Rd, Singapore 628054; tel. 62651766; fax 62650201; internet www.sembcorpmarine.com.sg; f. 1968; cap. and res 927m., sales 1,068m. (2003); fmrly Jurong Shipyard Ltd; maritime services, shipping, civil construction, construction of oil rigs and specialized marine equipment; also active in industrial and venture sectors; Pres. TAN KWI KIN; Chair. WONG KOK SIEW; 1,800 employees.

Singapore Technologies Aerospace Ltd: 540 Airport Rd, Paya Lebar, Singapore 539938; tel. 62871111; fax 62809713; e-mail mktg .aero@stengg.com; internet www.stengg.com; cap. and res 381m. (2000), sales 1,031m. (2001); maintenance, modification and refurbishment of military and civilian aircraft, engines and components; Pres. TAY KHOK KHIANG; 4,772 employees.

Singapore Technologies Marine Ltd: 7 Benoi Rd, Singapore 629882; tel. 68612244; fax 68613028; internet www.stengg.com; f. 1968; cap. and res 93m., sales 359m. (2000); naval and commercial shipbuilding and repair, engineering equipment; Chair. TAN GUONG CHING; 1,128 employees.

Straits Trading Co Ltd (THE): 9 Battery Rd, 21-00 Straits Trading Bldg, Singapore 049910; tel. 65354722; fax 65327939; e-mail straits@stc.com.sg; f. 1887; cap. and res 5m. (2000), sales 105m. (2001); tin smelting; Pres. NORMAN IP KA CHEUNG; Chair. and CEO HOWE YOON CHONG.

Superior Multi-Packaging Ltd: 7 Benoi Sector, Singapore 629842; tel. 62683933; fax 62657151; cap. and res 66m., sales 54m. (2000); fmrly Superior Metal Printing Ltd; mfrs of metal containers and packaging; Chair. STEVEN CHAN CHENG.

United Engineers Ltd: 257 Jalan Ahmad Ibrahim, Singapore 629147; tel. 62662388; fax 62686993; cap. and res 660m., sales 397m. (2000); civil, mechanical, electrical and environmental engineering; fabrication of industrial, construction and agricultural machinery; Chair. TANG I-FANG; Man. Dir JACKSON CHEVALIER YAP KIT SIONG; 2,000 employees.

Wood and Paper Products

Hiap Moh Corpn Ltd: Block 162, Bukit Merah Central, 08-3545, Singapore 150162; tel. 62735333; fax 62789707; internet www .collection.com.hk/sta/hiapmoh/eindex.htm; cap. and res 35m., sales 50m. (2001); mfrs and traders of paper products; Chair. MICHAEL WEE SOON LOCK; Man. Dir NG KAH LIN; 267 employees.

United Pulp and Paper Co Ltd: 35 Tuas View Crescent, Singapore 637608; tel. 68610018; fax 68613318; e-mail uppgroup@singnet .com.sg; cap. and res 78m., sales 58m. (2001); investment holding co with subsidiaries engaged in manufacture and retail of paper products; Chair. HO POH CHOON; 300 employees.

Miscellaneous

Ace Dynamics Ltd: 1 Shipyard Rd, Singapore 628128; tel. 62664868; fax 62662026; e-mail adl@acedynamics.com.sg; internet www.acedynamics.com.sg; cap. and res 35.7m., sales 33.8m. (2001); investment holding co, with subsidiaries engaged in distribution of industrial welding and safety equipment and the manufacture and supply of industrial gases; Chair. and CEO STEVEN THAM WENG CHEONG; 220 employees.

Alliance Technology and Development Ltd: 139 Joo Seng Rd, 06-01, ATD Centre, Singapore 368362; tel. 67491090; fax 62825377; cap. and res 36m., sales 34m. (1998/99); investment holding co with subsidiary engaged in manufacture of contact lenses and related eye-care products, property investment and development, leisure and entertainment, hospitality and Underwater World; Chair. CHANG CHING CHUAN; Exec. Vice-Chair. and CEO CHANG WHE HAN.

AsiaMedic Ltd: 50 Orchard Rd, 08-00 Shaw House, Singapore 238868; tel. 67898888; fax 67384136; internet www.asiamedic.com .sg; f. 1987; cap. and res 11.4m., sales 12.6m. (2000); precision plastic moulding, mould manufacture and medical services; Chair. LOW CZE HONG.

Bonvests Holdings Ltd: 541 Orchard Rd, 16-00 Liat Towers, Singapore 238881; tel. 67325533; fax 67383092; internet www .bonvests.com.sg; f. 1982; cap. and res 367m., sales 229m. (2003); investment holding co with subsidiary engaged in property development and rental, hotel and fast food chain ownership and management, waste collection and cleaning services; Chair. and Man. Dir HENRY NGO; 1,992 employees.

British-American Tobacco Co (Singapore) Ltd: 15 Senoko Loop, Singapore 758168; tel. 67588555; fax 67557798; f. 1978; cap. and res 1.5m., sales 495m. (2000); investment holding co with subsidiaries engaged in manufacturing, importing and retailing tobacco products; Man. Dir PATRICK O'KEFFE; 500 employees.

Bukit Sembawang Estates Ltd: 65 Chulia St, 49-05 OCBC Centre, Singapore 049513; tel. 65361836; fax 65361858; cap. and res 168m., sales 17m. (2001/02); property development; Chair. CECIL VIVIAN RICHARD WONG.

Capitaland Commercial Ltd: 39 Robinson Rd 18-01, Robinson Point, Singapore 068911; tel. 65361188; fax 65363788; e-mail ask_us@capitalandcommercial.com; internet www.capitaland.com; f. 2000 following merger of Pidemco Land Ltd and DBS Land Ltd; cap. and res 6,078m., sales 3,830m. (2003); property investment and development; Chair. RICHARD HU TSU TAU; 2,390 employees.

Centrepoint Properties Ltd: 438 Alexandra Rd, 02-00 Alexandra Point, Singapore 119538; tel. 62732122; fax 62757732; e-mail feedback@centrepointhomes.com; internet www.centrepointhomes .com; f. 1963; cap. and res 377m., sales 543m. (2000/01); property ownership and development; Chair. Dr MICHAEL FAM.

City Developments Ltd: 36 Robinson Rd, 04-01 City House, Singapore 068877; tel. 62212266; fax 62232746; internet www.cdl.com.sg; cap. and res 3,969m., sales 2,227m. (2001); property ownership and development; Chair. KWEK LENG BENG; Man. Dir KWEK LENG JOO; 317 employees.

Comfort Group Ltd: 383 Sin Ming Drive, Singapore 575717; tel. 64576255; internet www.comfortgroup.com.sg; f. 1993; cap. and res 317m., sales 415m. (2000/01); management and investment holding co, with subsidiaries involved in the provision of transport-related services; Chair. LIM JIT POH; Man. Dir GOH CHEE WEE; 860 employees.

Datacraft Asia Ltd: 6 Shenton Way 24-11 DBS Bldg Tower Two, Singapore 068809; tel. 63237988; fax 63237933; internet www .datacraft-asia.com; cap. and res 707m., sales 997.8m. (2000/01); management and investment holding, data communication systems; Chair. DEREK PETER ALTHORP; CEO RONALD J. CATTELL; 1,800 employees.

First Capital Corpn Ltd: 20 Collyer Quay, 17-03 Tung Centre, Singapore 049319; tel. 65356455; fax 65326196; f. 1976; cap. and res 1,004m., sales 241m. (2000/01); investment holding co with principal activities consisting of investment holding and property development; Chair. SAT PAL KHATTAR; CEO QUEK CHEE HOON.

Gold Coin Services (S) Pte Ltd: 99 Bukit Timah Rd, Alfa Centre 05-03/06 Singapore 229835; tel. 63374300; fax 63376911; e-mail lps@goldcoin.com.sg; internet www.goldcoin.com.sg; f. 1983; cap. 40m., sales 347m. (1990); investment holding co with subsidiaries engaged in manufacturing animal feeds and marketing and distribution of industrial, agricultural and consumer products; Man. Dir LOUIS SCHENBERG; 203 employees.

Goldtron Ltd: 21 Serangoon North Ave 5, 06–02, Singapore 554864; tel. 67471616; fax 67413525; e-mail skong@singnet.com.sg; f. 1962; cap. and res -42m., sales 340m. (2000/01); investment holding co with subsidiaries engaged in manufacture and retail of telecommunication products; Chair. MOKHZANI BIN MAHATHIR; 1,462 employees.

GP Industries Ltd: 18-02, Millenia Tower, 1 Temasek Ave, Singapore 039192; tel. 63950850; fax 63950860; e-mail gpind@gp-industries.com; internet www.gp-industries.com; f. 1995; cap. and res 339m., sales 322m. (2002/03); mfr of car radios; Chair. VICTOR LO CHUNG WING; 4,550 employees.

Haw Par Corpn Ltd: 178 Clemenceau Ave 08-00, Haw Par Glass Tower, Singapore 239922; tel. 63379102; fax 63369232; f. 1969; cap. and res 589m., sales 126m. (2001); investment holding co with subsidiaries engaged in property, investments and manufacture of industrial products, sports products and pharmaceutical products; Chair. WEE CHO YAW; Pres. and CEO Dr HONG HAI; 715 employees.

Hotel Properties Ltd: 50 Cuscaden Rd, 08-01 HPL House, Singapore 249724; tel. 67345250; fax 62350729; internet www.hotelprop.com; cap. and res 898m., sales 512m. (2001); hotels; leisure; food retailing; Chair. PETER FU YUN SIAK; Man. Dir ONG BENG SENG; 200 employees.

Htl International Holdings Ltd: 11 Gul Circle, Singapore 629567; tel. 67475050; fax 67478497; e-mail htlsales@htl.com.sg; internet www.htlinternational.com; f. 1976; cap. and res 113.3m., sales 435.0m. (2003); furniture, fabrics; Chair. PHUA YONG PIN; Man. Dir PHUA YONG TAT.

Hwa Hong Corpn Ltd: 38/40 South Bridge Rd, Singapore 058672; tel. 65326818; fax 65326816; e-mail hwahong@pacific.net.sg; cap. and res 437m., sales 113m. (2001); investment holding co with subsidiaries engaged in general insurance and related activities, packaging of edible oil products, construction and engineering; Chair. and Man. Dir ONG CHOO ENG; 800 employees.

Jardine Cycle and Carriage Ltd: 239 Alexandra Rd, Singapore 159930; tel. 64733122; fax 64757088; internet www.jcclgroup.com; f. 1969; fmrly Cycle and Carriage Ltd; name changed as above 2004; cap. and res 1,035m., sales 4,988m. (2002); retailer of motor vehicles; property investment; Chair. ANTHONY NIGHTINGALE; Man. Dir PHILIP ENG HENG NEE; 100,000 employees.

Keppel Land Ltd: 230 Victoria St, 15-05 Bugis Junction Towers, Singapore 188024; tel. 63388111; fax 63377168; e-mail marketg@kepland.com.sg; internet www.keppelland.com.sg; cap. and res 2,240m. (2000), sales 301m. (2001); property investment, development and management; hotel and resort operations; retail management; Chair. LIM CHEE ONN; 2,010 employees.

Kim Eng Holdings Ltd: 9 Temasek Blvd, 39-00 Suntec Tower 2, Singapore 038989; tel. 63369090; fax 63396003; e-mail helpdesk@kimeng.com; internet www.kimeng.com; cap. and res 624m., sales 248m. (2003); sharebrokers, stockbrokers and investment holding co; Chair. GLORIA LEE KIM YEW; Man. Dir RONALD ANTHONY OOI THEAN YAT.

Marco Polo Developments Ltd: 501 Orchard Rd, 04-01/03 Lane Wheelock Place, Singapore 238880; tel. 67388660; fax 67359833; internet www.mpdl.com; cap. and res 857m., sales 297m. (2000/01); property development; Chair. GONZAGA W. J. LI; Man. Dir DAVID J. LAWRENCE; 600 employees.

MCL Land Ltd: 78 Shenton Way, 33-00, Singapore 079120; tel. 62218111; fax 62253383; internet www.mclland.com.sg; f. 1963; cap. and res 820m., sales 166m. (2001); investment holding, property investment and management; Chair. KEI MEI RIN; Man. Dir EE SENG LIM.

MPH Ltd: 47 Scotts Rd, 18-02 Goldbell Towers, Singapore 228233; tel. 67341515; fax 67322241; f. 1927; cap. and res 212m., sales 679m. (2000/01); fmrly Jack Chia-MPH Ltd; investment holding co; activities of subsidiaries: publication, distribution and retail of books and magazines, manufacture and distribution of confectionary products, residential home building; Chair. and Man. Dir SIMON CHEONG; 2,065 employees.

Orchard Parade Holdings Ltd: 14 Scotts Rd, 06-01 Far East Plaza, Singapore 228213; tel. 62352411; f. 1967; cap. and res 465m., sales 131m. (2001); management and investment holding, property investment and development; Chair. PHILIP NG CHEE TAT; Man. Dir CHAU TIAN CHU; 228 employees.

Overseas Union Enterprise Ltd: 333 Orchard Rd, Overseas Union Bank Bldg, Singapore 238867; tel. 67374411; fax 67372620; f. 1964; cap. and res 823m., sales 168m. (2000); advertising; hotels and leisure; property; Chair. and Man. Dir Dr LIEN YING CHOW; 1,065 employees.

Parkway Holdings Ltd: 1 Grange Rd, 11-01 Orchard Bldg, Singapore 239693; tel. 67960622; fax 67960637; e-mail enquiries_phl@parkway.com.sg; internet www.parkway.com.sg; cap. and res 431m., sales 349m. (2003); hospital ownership and management; wholesale of medicines and medical products; Chair. ANIL THADANI; Man. Dir Dr LIM CHEOK PENG.

Rothmans Industries Ltd: 15 Senoko Loop, Singapore 758168; tel. 63388998; fax 63388181; cap. and res 113m., sales 318m. (1996/97); investment holding co with subsidiaries engaged in manufacture, import and retail of cigarettes, pipe tobaccos, cigars and other tobacco-related products; Chair. ALAN YEO CHEE YEOW.

San Teh Ltd: 701 Sims Drive, 06-01 LHK Bldg, Singapore 387383; tel. 67496386; fax 67473456; f. 1979; cap. and res 300m., sales 55m. (2001); mfrs of specialized silicone rubber products; Chair. KAO SHIN PING; Man. Dir CHUANG WEN-FU; 217 employees.

Singapore Press Holdings Ltd: 1000 Toa Payoh North, Singapore 318994; tel. 63196319; fax 63198282; e-mail sphcorp@cyberway.com.sg; internet www.sph.com.sg; cap. and res 2,151m., sales 1,030m. (2000/01); publishing, printing and distribution of newspapers and magazines; Chair. LIM KIM SAN; 3,277 employees.

SM Summit Holdings Ltd: 45 Ubi Rd 1, Summit Bldg, Singapore 408696; tel. 67453288; fax 67489612; e-mail co@smsummit.com.sg; internet www.smsummit.com.sg; f. 1981; cap. and res 36.7m., sales 51.2m. (2003); investment holdings, mfrs and retailers of compact discs (CDs), digital versatile discs (DVDs) and related products; Chair. and Man. Dir LEE KERK CHONG; 470 employees.

Sumitomo Corporation (Singapore) Pte Ltd: 20 Cecil St, 23/24-01/08, Singapore 049705; tel. 65337722; fax 65339693; e-mail shinichi.watabe@sumitomocorp.co.jp; cap. and res 65m., sales 4,196m. (2000); retailing and project management; Man. Dir HISAYOSHI UNO, 151 employees.

Toepfer International Asia Pte: 100 Beach Rd, 31-01 Shaw Towers, Singapore 189702; tel. 62932366; fax 62927556; cap. and res –1,469m., sales 2,021m. (2000); sale of commodities; Chair. SOH KIM SIANG.

United Overseas Land Ltd: 101 Thomson Rd, 33-00 United Sq., Singapore 307591; tel. 62550233; fax 62529822; e-mail uol@pacific.net.sg; internet www.uol.com.sg; cap. and res 1,749m., sales 350m. (2001); property development; hotels; Chair. WEE CHO YAW; Pres. and CEO GWEE LIAN KHENG; 3,451 employees.

United Industrial Corpn Ltd: 5 Shenton Way, 02-16 UIC Bldg, Singapore 068808; tel. 62201352; fax 62240278; cap. and res 2,102m., sales 404m. (2000); investment holding co with subsidiaries engaged in manufacture of detergent products, toiletries and electronic systems; Chair. WEE CHO YAW; Pres. and CEO LIM HOCK SAN; 1,027 employees.

WBL Corpn Ltd: 65 Chulia St, 31-00 OCBC Centre, Singapore 049513; tel. 65333444; fax 65341443; e-mail info@wbl.com.sg; internet www.wbl.com.sg; cap. and res 523m., sales 1,302m. (2003); investment holding co with subsidiaries engaged in manufacture of vehicles, industrial machinery, electronic components and products, building materials and precision engineering products, property development and investment, provision of management services; Pres. TAN CHIN TUAN; Chair. TANG I-FANG.

Wing Tai Holdings Ltd: 107 Tampines Rd, Wing Tai Industrial Centre, Singapore 535129; tel. 62809111; fax 63838940; internet www.wingtaiasia.com.sg; cap. and res 1,568m., sales 150m. (2000/01); investment holding co with subsidiaries engaged in property development and hospitality, trading in fabrics, garments and architectural products and the provision of internet-related services; Chair. and Man. Dir CHENG WAI KEUNG.

TRADE UNIONS

At the end of 2003 there were 68 employees' trade unions and associations, with 417,166 members, and three employer unions, with 2,052 members.

In December 1998 the largest private-sector union, the United Workers of Electronics and Electrical Industries, had 40,433 members, while the largest public-sector union, the Amalgamated Union of Public Employees, had 20,872 members.

National Trades Union Congress (NTUC): NTUC Centre, 1 Marina Blvd 11-01, Singapore 018989; tel. 62138000; fax 63278800; internet www.ntuc.org.sg; f. 1961; 63 affiliated unions and 4 affiliated associates, 434,922 mems (May 2004); Pres. JOHN DE PAYVA; Sec.-Gen. LIM BOON HENG.

Transport

RAILWAYS

In 1993 there was 26 km of 1-m gauge railway, linked with the Malaysian railway system and owned by the Malayan Railway Pentadbiran Keretapi Tanah Melayu—KTM. The main line crosses the Johor causeway (1.2 km) and terminates near Keppel Harbour. Branch lines link it with the industrial estate at Jurong.

Construction began in 1983 on the Mass Rapid Transit (MRT) system. The first section was completed in 1987, and the remaining sections by 1990. The system extends for 83 km and consists of two lines with 48 stations (32 elevated, 15 under ground and one at ground level). The construction of a further 20-km line, the North–East Line, began in 1997; the line has 16 stations and one depot, and was completed in 2003. In 2002 the extension of the East–West line to Changi Airport was completed; the Changi Airport Line (CAL) has one underground and one elevated station. In the early 2000s construction was under way on the Circle Line, a 34-km orbital line running entirely under ground which was to link all the radial lines into the city. The line was to start at Dhoby Ghaut and terminate at Harbour Front and was to be implemented in five stages, the last of which was scheduled for completion in 2010.

Singapore's first Light Rapid Transit (LRT) system, the Bukit Panjang LRT (a 7.8-km line with 14 stations), was completed in April 1998. In January 2003 an LRT line in Sengkang also became operational, and work on a further LRT line, in Punggol, was expected to be completed in 2004.

Land Transport Authority: 460 Alexandra Rd, 28-00 PSA Bldg, Singapore 119963; tel. 63757100; fax 63757200; internet www.lta.gov.sg; Chair. FOCK SIEW WAH; Chief Exec. Maj.-Gen. HAN ENG JUAN.

ROADS

In 1999 Singapore had a total of 3,066 km of roads, of which 150 km were motorway; in that year 100% of the road network was paved. In 1990 the Government introduced a quota system to control the number of vehicles using the roads. A manual road-pricing system was introduced on one expressway in June 1996 and extended to two further expressways in May 1997. It was replaced by a system of Electronic Road Pricing (ERP) in 1998, whereby each vehicle was charged according to road use in congested areas. The ERP was first introduced on one expressway in April 1998, and was subsequently extended to include two other expressways and the Area Licensing Scheme gantry areas by early September of that year. In 2001 construction of the 12-km Kallang–Paya Lebar expressway began; an estimated 9 km of the expressway was to be underground. It was scheduled for completion in 2007.

SHIPPING

The Port of Singapore is the world's busiest in tonnage terms; Singapore handled an estimated 135,386 vessels with a total displacement of 986.4m. grt in 2003.

The Port of Singapore Authority operates six cargo terminals: Tanjong Pagar Terminal, Keppel Terminal, Brani Terminal, Pasir Panjang Terminal, Sembawang Terminal and Jurong Port.

Tanjong Pagar Terminal and Keppel Terminal have the capacity to handle 10.7m. 20-foot equivalent units (TEUs).

The third container terminal, Brani, built on an offshore island connected to the mainland by a causeway, has a capacity of 5.5m. TEUs.

Pasir Panjang Terminal, Singapore's main gateway for conventional cargo (particularly timber and rubber), has five deep-water berths and eight coastal berths and 17 lighter berths.

Sembawang Terminal has three deep-water berths and one coastal berth. This terminal is the main gateway for car carriers as well as handling steel and timber products.

Jurong Port (which handles general and dry-bulk cargo, is situated in south-western Singapore, and serves the Jurong Industrial Estate) has 20 deep-water berths and one coastal berth.

A new container terminal, at Pasir Panjang, was officially opened in 2000, following the completion of the second of four planned phases of building work. Upon completion, scheduled for 2009, the terminal was to have 49 berths and a handling capacity of 35m. TEUs.

Maritime and Port Authority of Singapore: 460 Alexandra Rd, 18-00 PSA Bldg, Singapore 119963; tel. 63751600; fax 62759247; internet www.mpa.gov.sg; f. 1996; regulatory body responsible for promotion and development of the port, overseeing all port and maritime matters in Singapore.

Port of Singapore Authority: 460 Alexandra Rd, PSA Bldg, Singapore 119963; tel. 63751600; fax 62759247; internet www.singaporepsa.com; f. 1964 as a statutory board under the Ministry of Communications; made a corporate entity in 1997 in preparation for privatization; responsible for the provision and maintenance of port facilities and services; Chair. Dr YEO NING HONG; Pres. KHOO TENG CHYE.

Major Shipping Companies

American President Lines Ltd (APL): 456 Alexandra Rd, 08-00 NOL Bldg, Singapore 119962; tel. 62789000; fax 62742113; e-mail hung-song-goh@apl.com; internet www.apl.com; container services to North and South Asia, the USA and the Middle East; Chief Operating Officer ED ALDRIDGE.

Glory Ship Management Private Ltd: 24 Raffles Place 17-01/02, Clifford Centre, Singapore 048621; tel. 65361986; fax 65361987; e-mail gene@gloryship.com.sg.

Guan Guan Shipping Pte Ltd: 2 Finlayson Green, 13-05 Asia Insurance Bldg, Singapore 049247; tel. 65343988; fax 62276776; e-mail golden@golden.com.sg; f. 1955; shipowners and agents; cargo services to East and West Malaysia, Indonesia, Pakistan, Sri Lanka, Bengal Bay ports, Persian (Arabian) Gulf ports, Hong Kong and China; Man. Dir RICHARD THIO.

IMC Shipping Co Pte Ltd: 5 Temasek Blvd, 12-01 Suntec City Tower, Singapore 038985; tel. 63362233; fax 63379715; e-mail commercial@imcshipping.com.sg; internet www.imcshipping.com; Man. Dir PETER CHEW.

Nedlloyd Lines Singapore Pte Ltd: 138 Robinson Rd, 01-00 Hong Leong Centre, Singapore 068906; tel. 62218989; fax 62255267; f. 1963; Gen. Man. F. C. SCHUCHARD.

Neptune Orient Lines Ltd: 456 Alexandra Rd, 05-00 NOL Bldg, Singapore 119962; tel. 63715300; fax 63716489; e-mail shirley_ting@ccgate.apl.com; internet www.nol.com.sg; f. 1968; liner containerized services on the Far East/Europe, Far East/North America, Straits/Australia, South Asia/Europe and South-East Asia, Far East/Mediterranean routes; tankers, bulk carriers and dry cargo vessels on charter; 36% govt-owned; Chair. LUA CHENG ENG; CEO RON WIDDOWS (acting).

New Straits Shipping Co Pte Ltd: 51 Anson Rd, 09-53 Anson Centre, Singapore 0207; tel. 62201007; fax 62240785.

Ocean Tankers (Pte) Ltd: 41 Tuas Rd, Singapore 638497; tel. 68632202; fax 68639480; e-mail admin@oceantankers.com.sg; Marine Supt V. LIM.

Osprey Maritime Ltd: 8 Cross St, 24-02/03 PWC Bldg, Singapore 048424; tel. 62129722; fax 65570450; internet www.ospreymaritime.com; CEO PETER GEORGE COSTALAS.

Pacific International Lines (Pte) Ltd: 140 Cecil St, 03-00 PIL Bldg, POB 3206, Singapore 069540; tel. 62218133; fax 62273933; e-mail sherry.chua@sgp.pilship.com; shipowners, agents and managers; liner services to South-East Asia, the Far East, India, the Red Sea, the Persian (Arabian) Gulf, West and East Africa; container services to South-East Asia; world-wide chartering, freight forwarding, container manufacturing, depot operators, container freight station operator; Exec. Chair. Y. C. CHANG; Man. Dir S. S. TEO.

Petroships Private Ltd: 460 Alexandra Rd, 25-04 PSA Bldg, Singapore 119963; tel. 62731122; fax 62732200; e-mail gen@petroships.com.sg; Man. Dir KENNETH KEE.

Singa Ship Management Private Ltd: 78 Shenton Way 16-02, Singapore 079120; tel. 62244308; fax 62235848; e-mail agency@singaship.com.sg; Chair. OLE HEGLAND; Man. Dir EILEEN LEONG.

Syabas Tankers Pte Ltd: 10 Anson Rd, 34-10 International Plaza, Singapore 0207; tel. 62259522.

Tanker Pacific Management (Singapore) Private Ltd: 1 Temasek Ave, 38-01 Millenia Tower, Singapore; tel. 63365211; fax 63375570; Man. Dir HUGH HUNG.

CIVIL AVIATION

Singapore's international airport at Changi was opened in 1981. Construction on a third terminal at Changi began in October 2000, and was scheduled to be completed in 2006, increasing the airport's total capacity to 64m. passengers a year. A second airport at Seletar operates as a base for charter and training flights.

Civil Aviation Authority of Singapore: Singapore Changi Airport, POB 1, Singapore 918141; tel. 65412222; fax 65421231; e-mail thennarasee_R@caas.gov.sg; internet www.caas.gov.sg; responsible for regulatory and advisory services, air services development, airport management and development and airspace management and organization; Chair. TJONG YIK MIN; Dir-Gen. WONG WOON LIONG.

SilkAir: Singapore Airlines Superhub, Airfreight Terminal 5, Core L, 5th Storey, 30 Airline Rd, Singapore 819830; tel. 65428111; fax 65426286; internet www.silkair.net; f. 1975; fmrly Tradewinds Private; wholly-owned subsidiary of Singapore Airlines Ltd; began scheduled services in 1989; Chair. CHEW CHOON SENG; Gen. Man. PAUL TAN.

Singapore Airlines Ltd (SIA): Airline House, 25 Airline Rd, Singapore 819829; tel. 65415880; fax 65456083; e-mail publicaffairs@singaporeair.com.sg; internet www.singaporeair.com; f. 1972; passenger services to 88 destinations in 42 countries; Chair. KOH BOON HWEE; CEO CHEW CHOON SENG.

Tiger Airways: Singapore Changi Airport, POB 82, Singapore 918143; internet www.tigerairways.com; f. 2003; 49%-owned by Singapore Airlines, 24%-owned by US co Indigo Partners, 11%-

owned by Temasek Holdings Pte Ltd; services to 10 regional destinations; Chair. WILLIAM FRANKE; CEO PATRICK GAN.

ValuAir: Singapore; f. 2003; regional destinations; Chair. LIM CHIN BENG.

Tourism

Singapore's tourist attractions include a blend of cultures and excellent shopping facilities. Tourist arrivals totalled 6,125,480 in 2003. Receipts from tourism totalled an estimated S $5,426m. in 2002.

Singapore Tourism Board: Tourism Court, 1 Orchard Spring Lane, Singapore 247729; tel. 67366622; fax 67369423; e-mail feedback@stb.com.sg; internet www.stb.com.sg; f. 1964; Chair. WEE EE-CHAO; Chief Exec. LIM NEO CHIAN.

Defence

In August 2003 the total strength of the armed forces was 72,500 (including 39,800 conscripts): army 50,000 (35,000 conscripts), navy an estimated 9,000 (1,800 conscripts), air force 13,500 (3,000 conscripts). Military service lasts 24-30 months. Army reserves numbered an estimated 312,500. Paramilitary forces of an estimated 96,300 comprised the Singapore police force, marine and Gurkha guard battalions, and a civil defence force (numbering an estimated 84,300). Singapore is a participant in the Five-Power Defence Arrangements with Malaysia, Australia, New Zealand and the United Kingdom.

Defence Expenditure: Budgeted at S $8,200m. in 2002/03.

Chief of the Defence Forces: Maj.-Gen. NG YAT CHUNG.

Chief of the Army: Maj.-Gen. DESMOND KUEK BAK CHYE.

Chief of the Air Force: Maj.-Gen. LIM KIM CHOON.

Chief of the Navy: Rear-Adm. RONNIE TAY.

Education

Education in Singapore is not compulsory. All children are entitled to at least six years' free primary education, which may be completed in eight years by less able pupils, under the New Primary Education System implemented in 1979.

In 1996 the number of children attending primary schools was equivalent to 94% of children in the relevant age-group (males 95%; females 93%) and secondary school enrolment was equivalent to 74% of children in the relevant age-group. Expenditure on education by the central Government in the 2002/03 financial year was S $4,876m. (17.9% of total expenditure).

The policy of bilingualism ensures that children are taught two languages, English and one of the other official languages, Chinese, Malay or Tamil. The option to study French, German or Japanese is offered in secondary schools to interested pupils with linguistic ability as a third language, and as a second language to pupils not of Chinese, Malay or Tamil ethnic origin.

Primary education focuses on the core subjects of English, mathematics and the mother tongue (Chinese, Malay or Tamil). After four years pupils are streamed according to their abilities into the normal bilingual course (six years' primary education), extended bilingual course (eight years) or the monolingual course (eight years). Pupils of the bilingual courses take the Primary School Leaving Examination (PSLE) on completion of their primary education and are streamed, depending on their performance in the PSLE, into the Special course (four years), the Express course (four years) or the Normal course (five years). Pupils of Special and Express courses will take the Singapore-Cambridge General Certificate of Education (GCE) 'Ordinary' Level Examination. Normal course pupils will sit the GCE 'Normal' Level Examination after four years, and those who do well may take the GCE 'O' Level after the fifth year. Based on 'O' Level results, pupils may follow a two-year or three-year pre-university course leading to the GCE 'Advanced' Level Examination. Pupils of the monolingual course may continue their education in the technical or commercial institutes under the Institute of Technical Education.

HIGHER EDUCATION

There were 10 tertiary institutions in 2003, including the National University of Singapore, the Nanyang Technological University, the Ngee Ann Polytechnic, the Singapore Polytechnic, the Nanyang Polytechnic and the Temasek Polytechnic. In June of that year the total enrolment in the universities and colleges was 98,424.

VOCATIONAL AND INDUSTRIAL TRAINING

The Institute of Technical Education provides and regulates vocational training. It conducts institutional training for school-leavers, offers part-time continuing education and training programmes and registers apprentices. It is also responsible for setting national skills standards, the conduct of public trade testing and certification of skills.

Bibliography

GENERAL

Barr, Michael D. *Lee Kuan Yew: The Beliefs Behind the Man.* Washington, DC, Georgetown University Press, 2000.

Bedlington, Stanley S. *Malaysia and Singapore: The Building of New States.* Ithaca, NY, Cornell University Press, 1978.

Buchanan, Iain. *Singapore in Southeast Asia; An Economic and Political Appraisal.* London, G. Bell, 1972.

Chan Heng Chee. *Singapore: Politics of Survival 1965–1967.* Singapore, Oxford University Press, 1971.

A Sensation of Independence: Singapore's David Marshall. Singapore, Oxford University Press, 1984.

Chee Soon Juan. *Dare to Change.* Singapore, 1994.

Singapore: My Home Too. Singapore, Melodies Press Co, 1995.

Cherian, George. *Singapore: The Air-Conditioned Nation—Essays on the Politics of Comfort and Control, 1990–2000.* Singapore, Landmark Books, 2001.

Chew, Ernest C. T., and Lee, Edwin (Eds). *A History of Singapore.* Singapore, Oxford University Press, 1991.

Chew, Melanie. *Leaders of Singapore.* Singapore, Resource Press, 1996.

Chua Beng-Huat. *Communitarian Ideology and Democracy in Singapore.* London, Routledge, 1997.

Political Legitimacy and Housing—Singapore's Stakeholder Society. London, Routledge, 1997.

Life Is Not Complete Without Shopping: Consumption Culture in Singapore. Singapore, Singapore University Press, 2003.

Clammer, John. *Singapore: Ideology, Society, Culture.* Singapore, Chapman, 1985.

da Cunha, Derek. *Debating Singapore (Reflective Essays).* Singapore, Institute of Southeast Asian Studies, 1994.

The Price of Victory: The 1997 Singapore General Election and Beyond. Singapore, Institute of Southeast Asian Studies, 1997.

Singapore in the New Millennium: Challenges Facing the City-State. Singapore, Institute of Southeast Asian Studies, 2001.

Darusman, Suryono. *Singapore and the Indonesian Revolution 1945–50: Recollections of Suryono Darusman.* Singapore, Institute of Southeast Asian Studies, 1992.

Drysdale, John. *Singapore: Struggle for Success.* Singapore, Times Books International, 1984.

Ern Ser Tan. *Does Class Matter? Social Stratification and Orientations in Singapore.* Singapore, World Scientific Publishing Co, 2004.

Farrel, H. P., and Hunter, S. *Sixty Years On: The Fall of Singapore Revisited.* Singapore, Times Academic Press, 2003.

Gomez, James. *Self-Censorship: Singapore's Shame.* Singapore, Think Center, 2000.

Hack, Karl, and Blackburn, Kevin. *Did Singapore Have to Fall?: Churchill and the Impregnable Fortress.* London, RoutledgeCurzon, 2003.

Han Fook Kwang, Fernandez, Warren, and Tan, Sumiko. *Lee Kuan Yew: The Man and his Ideas.* Singapore, 1997.

Harper, R. W. E., and Miller, Harry. *Singapore Mutiny: The Story of a Little 'Local Disturbance'.* Singapore, Oxford University Press, 1984.

Ho Khai Leong. *Shared Responsibilities, Unshared Power: The Politics of Policy-Making in Singapore.* Singapore, Times Academic Press, 2000.

Hong Liu and Sin Kiong Won. *Singapore Chinese Society in Transition: Business, Politics and Socio-Economic Change 1945–1965.* New York, NY, Peter Lang Publishing, 2004.

Kau Ah Keng, Tambayah Siok Kuan, Tan Soo Jiuan, and Jung Kwon. *Understanding Singaporeans: Values, Lifestyles, Aspirations and Consumption Behaviours.* Singapore, World Scientific Publishing Co, 2004.

Kenley, David. *New Culture in a New World: The May Fourth Movement and the Chinese Diaspora in Singapore, 1919–1932.* London, Routledge, 2003.

Khun Eng, Pearce. *State, Society and Religious Engineering: Towards a Reformist Buddhism in Singapore.* Singapore, Eastern Universities Press, 2003.

Lam Peng Er, and Tan, Kevin (Eds). *Managing Politics in Singapore: the Elected Presidency.* London and New York, Routledge, 1997.

Lee's Lieutenants. Singapore, Allen and Unwin, 1999.

Lau, Albert. *A Moment of Anguish.* Singapore, 1998.

Lee Kuan Yew. *The Battle for Merger.* Singapore, Ministry of Culture, 1961.

The Singapore Story: Memoirs of Lee Kuan Yew. Prentice Hall/Simon & Schuster Asia, Singapore, 1998.

From Third World to First: The Singapore Story: 1965–2000. Singapore, Times Media Private and Straits Times Press, 2000.

Leifer, Michael. *Singapore's Foreign Policy—Coping with Vulnerability.* London, Routledge, 2000.

Lim Pui Huen P., and Wong, Diana (Eds). *War and Memory in Malaysia and Singapore.* Singapore, Institute of Southeast Asian Studies, 1999.

Lim, R. *Got Singapore.* Singapore, Angsana Books, 2002.

Liu, Gretchen. *Singapore—A Pictorial History 1819–2000.* London, Curzon Press, 2001.

Lyons, Lenore. *State of Ambivalence: The Feminist Movement in Singapore.* Leiden, Netherlands, Brill Academic Publishers, 2004.

Mahizhnan, Arun, and Lee Tsao Yuan (Eds). *Singapore: Re-Engineered Success.* Oxford, Oxford University Press, 2002.

Makepeace, Walter, Brooke, Gilbert E., and Braddell, Roland St J. (Eds). *One Hundred Years of Singapore.* 2 vols, 1921, 1991.

Milne, R. S., and Mauzy, Diane K. *Singapore: The Legacy of Lee Kuan Yew.* Boulder, CO, Westview Press, 1990.

Singapore Politics Under the People's Action Party. London, Routledge, 2002.

Minchin, James. *No Man is an Island.* London, Unwin Hyman, 1987.

Mulliner, K. *Historical Dictionary of Singapore.* Lanham, MD, Scarecrow Press, 2002.

Mutalib, Hussin. *Parties and Politics: A Study of Opposition Parties and the PAP in Singapore.* Singapore, Eastern Universities Press, 2003.

Quah, Jon S. T., Chan Heng Chee and Seah Chee Meow (Eds). *Government and Politics of Singapore.* Singapore, Oxford University Press, 1985.

Rappa, Antonio, L. *Modernity and Consumption: Theory, Politics and the Public in Singapore and Malaysia.* Singapore, World Scientific Publishing Co, 2002.

Régnier, Philippe. *Singapore, City-State in South East Asia.* London/Honolulu, Hurst Publishers/Hawaii University Press, 1991.

Saw Swee Hock. *Asian Metropolis: Singapore in Transition.* University of Pennsylvania and Oxford University Press, 1970.

The Population of Singapore. Singapore, Institute of Southeast Asian Studies, 1999.

Selvan, T. S. *Singapore the Ultimate Island (Lee Kuan Yew's Untold Story).* Victoria, Freeway Books, 1990.

Seow, Francis. *To Catch a Tartar.* New Haven, CT, Yale University Press, 1994.

The Media Enthralled: Singapore Revisited. Boulder, CO, and London, Lynne Rienner, 1998.

Singapore Government. *The Next Lap.* Singapore, 1991.

Singh, Bilveer. *Whither PAP's Dominance? An Analysis of Singapore's 1991 General Elections.* Petaling Jaya, Pelanduk Publications, 1992.

Song Ong Siang. *One Hundred Years' History of the Chinese in Singapore.* Oxford University Press, 1923, 1967, 1984.

Suryadinata, L. *Ethnic Chinese in Singapore and Malaysia.* Singapore, Times Academic Press, 2002.

Tamney, Joseph B. *The Struggle over Singapore's Soul: Western Modernization and Asian Culture.* Berlin and New York, Walter de Gruyter, 1996.

Thio, Eunice (Ed.). *Singapore 1819–1969.* University of Singapore, Journal of South-East Asian History, March 1969.

Times Editions. *Singapore: Island. City. State.* Singapore, 1990.

Tong Chee Kiong and Lian Kwen Fee (Eds). *The Making of Singapore Sociology: Society and State.* Leiden, Netherlands, Brill Academic Publishers, 2003.

Turnbull, C. M. *A History of Singapore 1819–1988.* Oxford University Press, 1989.

A History of Malaysia, Singapore, and Brunei. Sydney, Allen and Unwin, 1989.

The Straits Settlements, 1826–67. London, Athlone Press, and Kuala Lumpur, Oxford University Press, 1972.

Dateline Singapore: 150 Years History of the Straits Times. Singapore, Singapore Press Holdings, 1995.

Vasil, Raj. *Governing Singapore: A History of National Development and Democracy.* Singapore, Institute of Southeast Asian Studies, 2000.

Warren, J. F. *Rickshaw Coolie: A People's History of Singapore.* Singapore, Oxford University Press, 1986.

Ah Ku and Karayuki-san: Prostitution in Singapore 1870–1940. Singapore, Oxford University Press, 1993.

Wilson, H. E. *Social Engineering in Singapore: Educational Policies and Social Change, 1819–1972.* Singapore, Singapore University Press, 1978.

Worthington, Ross. *Governance in Singapore.* London, Routledge-Curzon, 2002.

Wurtzburg, C. E. *Raffles of the Eastern Isles.* London, Hodder and Stoughton, 1954, 1984.

Yeo Kim Wah. *Political Development in Singapore, 1945–1955.* Singapore, Singapore University Press, 1973.

Yeoh, Brenda S. A. *Contesting Space in Colonial Singapore: Power Relations and the Urban Built Environment.* Singapore, Singapore University Press, 2003.

Yong Chin Fatt. *Tan Kah-Kee: The Making of a Legend.* Singapore, Oxford University Press, 1987.

Zahari, Said. *Dark Clouds at Dawn: A Political Memoir.* Kuala Lumpur, INSAN, 2001.

Zhang Wei-Bin. *Singapore's Modernization: Westernization and Modernizing Confucian Manifestations.* New York, Nova Science Publishers, 2002.

ECONOMY

Bhaskaran, Manu. *Re-inventing the Asian Model: The Case of the Singapore Economy.* Singapore, Times Academic Press, 2003.

Board of Commissioners of Currency. *Currency Board System: A Stop-gap Measure or a Necessity: Currency Board System Symposium '97.* Singapore, 1997.

Dent, Christopher. *The Foreign Economic Policies of Singapore, South Korea and Taiwan.* Cheltenham, Edward Elgar Publishing, 2002.

Deutsch, Antal, and Zowall, Hanna. *Compulsory Savings and Taxes in Singapore.* Singapore, Institute of Southeast Asian Studies, 1988.

Doshi, Tilak. *The Singapore Petroleum Industry.* Singapore, Institute of Southeast Asian Studies, 1989.

Freeman, Nick, Chia Siow Yue, Venkatesan, R., and Malvea, S. V. (Eds). *Growth and Development of the IT Industry in Bangalore and Singapore: A Comparative Study.* Singapore, Institute of Southeast Asian Studies, 2001.

Goh Keng Swee. *The Economics of Modernization and Other Essays.* Singapore, Asia Pacific Press, 1972.

The Practice of Economic Growth. Singapore, Federal Publications, 1977.

Wealth of East Asian Nations: Speeches and Writings by Goh Keng Swee. Singapore, Kuala Lumpur and Hong Kong, Federal Publications, 1995.

Gwartney, James, Lawson, Robert, and Block, Walter. *Economic Freedom of the World 1975–1995.* Vancouver, The Fraser Institute, 1996.

Hiroshi, Shimizu and Hitoshi, Hirakawa. *Japan and Singapore in the World Economy: Japan's Economic Advance into Singapore 1870–1965.* London, Routledge, 1999.

Huff, Gregg. *The Economic Growth of Singapore: Trade and Development in the Twentieth Century.* Cambridge, Cambridge University Press, 1994.

Krause, Lawrence B., Koh Ai Tee, and Lee Yuan. *Singapore Economy Reconsidered.* Singapore, Institute of Southeast Asian Studies, 1987.

Lee Sheng-Yi. *Public Finance and Public Investment in Singapore.* Singapore, Institute of Banking and Finance, 1978.

The Monetary and Banking Development of Singapore and Malaysia. Singapore, Singapore University Press, 3rd edn, 1990.

Lee Tsao Yuan. *Growth Triangle: The Johor-Singapore-Riau Experience.* Singapore, Institute of Southeast Asian Studies, 1991.

Lee Tsao Yuan and Low, Linda. *Local Entrepreneurship in Singapore: Private and State.* Singapore, Times Academic Press, 1990.

Lim, Chong Yah, et al. *Policy Options for the Singapore Economy.* Singapore, McGraw-Hill Book Co, 1988.

Lim Chong Yah (Ed.). *Economic Policy Management in Singapore.* Singapore, Addison-Wesley, 1996.

Lim, L., and Eng Fong, P. *Trade, Employment and Industrialization in Singapore.* Geneva, International Labour Office, 1985.

Lingle, Christopher. *Singapore: Authoritarian Capitalism: Asian Values, Free Market Illusions and Political Dependency.* Barcelona, Edicions Sirocco, 1996.

Low, Linda. *The Political Economy of a City-State: Government-Made Singapore.* Singapore, Oxford University Press, 1998.

Low, Linda, Toh Mun Heng, Soon Teck Wong, Tan Kong Yam and Hughes, Helen. *Challenge and Response: Thirty Years of the Economic Development Board.* Singapore, Times Academic Press, 1993.

Low, Linda and Aw, T. C. *Housing a Healthy, Educated and Wealthy Nation Through the CPF.* Singapore, Times Academic Press for The Institute of Policy Studies, 1997.

Luckett, Dudley G., Schulze, David L., and Wong, Raymond W. Y. *Banking, Finance and Monetary Policy in Singapore.* Singapore, McGraw-Hill, 1994.

Mirza, Hafiz. *Multinationals and the Growth of the Singapore Economy.* London, Croom Helm, 1986.

Murray, Geoffrey, and Perera, Audrey (Eds). *Singapore: The Global City State.* Richmond, Surrey, Curzon Press, 1996.

Peebles, Gavin, and Wilson, Peter. *The Singapore Economy.* Cheltenham, and Brookfield, USA, Edward Elgar Publishing, 1996.

Economic Growth and Development in Singapore: Past and Future. Cheltenham, and Brookfield, USA, Edward Elgar Publishing, 2002.

Quak Ser Khoon. *Marketing Abroad: Competitive Strategies and Market Niches for the Singapore Construction Industry.* Singapore, Institute of Southeast Asian Studies, 1991.

Rajan, Ramkishen S. (Ed.). *Sustaining Competitiveness in the New Global Economy: The Experience of Singapore.* Cheltenham, and Brookfield, USA, Edward Elgar Publishing, 2003.

Rajan, Ramkishen S., Sen, Rahul, and Siregar, Reza. *Singapore and Free Trade Agreements: Economic Relations with Japan and the United States.* Singapore, Institute of Southeast Asian Studies, 2001.

Rodan, Garry. *The Political Economy of Singapore's Industrialisation: National State and International Capital.* London, Macmillan, 1989.

Sharma, Shankar. *The Role of the Petroleum Industry in Singapore's Economy.* Singapore, Institute of Southeast Asian Studies, 1989.

Singh Sandhu, Kernial, and Wheatley, Paul (Eds). *Management of Success: The Moulding of Modern Singapore.* Singapore, Institute of Southeast Asian Studies, 1990.

Tan Chwee Huat. *Singapore Financial and Business Sourcebook.* Singapore, Singapore University Press, 2002.

Financial Services in Singapore. Singapore, Singapore University Press, 2004.

Tan Ern Ser, Yeoh, Brenda S. A., and Wang, Jennifer (Eds). *Tourism Management and Policy: Perspectives from Singapore.* Singapore, World Scientific Publishing Co, 2002.

Tan Sook Yee. *Private Ownership of Public Housing in Singapore.* Singapore, Times Academic Press, 1998.

Toh Mun Heng, and Low, Linda. *Economic Impact of the Withdrawal of the GSP on Singapore.* Singapore, Institute of Southeast Asian Studies, 1991.

Toh Mun Heng, and Tan Kong Yam (Eds). *Competitiveness of the Singapore Economy: A Strategic Perspective.* Singapore, Singapore University Press and World Scientific, 1998.

Tremewan, Peter. *The Political Economy of Social Control in Singapore.* London, St. Martin's Press, 1994.

Vasil, Raj. *Asianising Singapore: The PAP's Management of Ethnicity.* Singapore, Institute of Southeast Asian Studies, 1995.

Walter, Ingo. *High-Performance Financial Systems: Blueprint for Development.* Singapore, Institute of Southeast Asian Studies, 1993.

THAILAND

Physical and Social Geography

HARVEY DEMAINE

The Kingdom of Thailand (formerly Siam) occupies the centre of the South-East Asian mainland, bordered by Myanmar (Burma) to the west, by Laos and Cambodia to the east, and by Peninsular Malaysia to the south. Its total area is 513,115 sq km (198,115 sq miles). Most of this territory lies to the north of the Bight of Bangkok, and hence well removed from the main shipping routes across the South China Sea between Singapore and Hong Kong, though peninsular Thailand, extending south to the Malaysian border approximately at latitude 6°N, has a coastline of some 960 km facing the Gulf of Thailand, and a somewhat shorter one facing the Andaman Sea. Between these two the peninsula narrows at the isthmus of Kra to a straight-line distance of only 56 km between salt water on both sides.

PHYSICAL AND CLIMATIC ENVIRONMENT

Apart from peninsular Thailand, which (except in the far south) consists of mainly narrow coastal lowlands backed by low and well-wooded mountain ranges, the country comprises four main upland tracts—in the west, north, north-east and south-east—surrounding a large central plain drained by the principal river, the Menam Chao Phraya. Because of its position, while experiencing tropical temperatures throughout its entire area, Thailand receives relatively less rainfall than either Myanmar to the west or most parts of Indo-China to the east. In general, rainfall is highest in the south and south-east, and in the uplands of the west and, to some extent, in the higher hills in the north, but most of the rest of the country, in effect, constitutes a rain-shadow area where the total annual fall is below 1,500 mm.

The western hills are formed by a series of north–south ridges, thickly covered by tropical monsoon forest with much bamboo, and drained by the Kwei Noi and Kwei Yai rivers. Although summit levels here are only of the order of 600 m–900 m, the ridge-and-furrow pattern makes this generally inhospitable country. In the northern uplands, which represent the southernmost portion of the great Yunnan-Shan-Laos plateau, altitudes are higher than in the west, reaching an upper limit of about 1,500 m, and the upland surface is fairly well forested, although the natural cover has clearly deteriorated in many areas as a result of shifting cultivation. However, in the four parallel valleys of the Ping, Wang, Yom and Nan rivers, which flow through these uplands and subsequently converge farther south to form the Chao Phraya, there are relatively broad lowlands with a more open vegetation, now largely cleared for rice cultivation.

The north-eastern plateau, also known as the Korat plateau, is mostly of much lower altitude than the two uplands just described. On its western and southern edges it presents a continuous rim usually exceeding 300 m, and in places much higher than that, but elsewhere it consists of a relatively low and undulating surface, draining eastwards, via the Nam Si and the Nam Mun, to the Mekong, which flows along its entire northern and eastern edge.

In contrast to most of the other uplands, the Korat plateau is an area of barely adequate rain, which during the dry season presents a barren and desiccated appearance. Since the main rivers flowing across it rise within this same area of low rainfall, Korat is less favourably placed in respect of irrigation water than the central plain, which, though likewise receiving an annual rainfall of less than 1,500 mm, is well watered by the Chao Phraya system.

Because of its focal position, its fertile alluvial soils, and the well developed system of natural waterways, the central plain forms by far the most important single region within the country; and within this region, the delta, which begins about 190 km from the coast, enjoys all these advantages to a more pronounced extent.

NATURAL RESOURCES

Thailand's main natural resources lie in its agricultural potential, and in particular in the capacity of the central plain (and to a lesser extent the Korat plateau) to produce a significant surplus of rice. In addition, since the late 1950s substantial areas of upland have been opened up in these areas for the cultivation of maize, cassava (tapioca), kenaf (upland jute), beans and, more recently, cotton and pineapple. The more humid and more truly equatorial coastal plains of the southern peninsula of Thailand have similarly expanded their production of rubber. Unfortunately, this expansion has been very much at the expense of the country's timber resources, which are estimated to have contracted to less than 20% of the total area, with the once famous teak of the northern hills now in extremely short supply.

Thailand is not especially well endowed with minerals. Tin was traditionally the most important, but this was superseded as an export by gypsum in 1993. In 1995 large-scale extraction of potash began in Chaiyaphum. Various other minerals, including tungsten, lead, fluorite and lignite, are being worked, and the country's heavy dependence on energy imports has begun to lessen, following the initial exploitation of reserves of natural gas in the Gulf of Thailand. These reserves are estimated at 172,000m. cu m, and there is a prospect of a substantial increase in Thailand's total reserves, following onshore discoveries in the Nam Phong area of Khonkaen province, on the north-east plateau. Small quantities of petroleum have also been identified in the north-central plain province of Kampaengphet. In addition, massive rock-salt deposits are known to underlie the Korat plateau.

POPULATION AND ETHNIC GROUPS

According to the census of 1 April 2000, the population totalled 60,606,947, giving an average density of 118.1 per sq km. At mid-2003, according to UN estimates, the population was 62,833,000, the average density being 122.5 per sq km. Although average densities fall to between one-quarter and one-half of this in the west and north, the total area of really sparsely populated upland is small. The proportion formed by indigenous minority peoples is low. Apart from some 700,000 Muslim Malays in the far south, a smaller number of Cambodians near the eastern borders, and a total of 300,000 scattered hill peoples—Meo, Lahu, Yao, Lisu, Lawa, Lolo and Karen—mainly in the far north and west, virtually the entire indigenous population belongs to the Thai ethnic group (which also includes the Shan and Lao) and subscribes to Buddhism, predominantly of the Hinayana (Theravada) form. However, it should be added that the inhabitants of the north-east tend to be closer in speech and custom to the Lao populations on the other side of the Mekong than to those of central Thailand. Excluding the Lao groups, the largest minority in Thailand may be said to be the ethnic Chinese. However, estimates as to their proportion of the total population vary and many Chinese have been assimilated into the Thai culture. Most have become Thai citizens.

Thailand shows only a relatively limited degree of urbanization. The urban scene is totally dominated by the single great complex of Bangkok Metropolis (including Thonburi), which had a population of 6,320,174 at 1 April 2000. According to UN estimates, the population of Bangkok Metropolis had risen to 6,486,401 by mid-2003.

History

RUTH MCVEY

Revised by PATRICK JORY

PREHISTORY AND EARLY PERIOD

Recent archaeological findings show very ancient civilizations to have existed in what is now Thailand, but the earliest evidence that we have of the Thai people is as part of a population speaking related languages and inhabiting mountainous areas of what is now Yunnan Province in the People's Republic of China. These Thai-speaking groups gradually spread southward into the highland areas of present-day Laos, northern Viet Nam, north-eastern Myanmar (Burma) and northern Thailand, where many preserve their identity as 'hill tribes' which are only very partially integrated into modern nation-states. The mountainous terrain in which they lived ensured that their polities remained small and simply organized, but in some river valleys of northern Thailand the development of irrigated rice cultivation led to relatively dense population and complex states controlling water distribution systems. Southward migration of these people brought them to the edge of the great central plain of the Chao Phraya river system; there, in AD 1238, they established the first historical Thai (Siamese) Kingdom of Sukothai.

Sukothai was initially subject to the major mainland power, the Khmer (Cambodian) empire of Angkor, but the growth of its population enabled it to express an increasingly Thai character and finally to assert independence under King Ramkamheng (reigned 1283–c. 1317). He extended the Kingdom's influence against both the Khmers and the Mons to the south and west, establishing Thai power over the central plain and making it a major element in the South-East Asian state system. Ramkamheng is also credited with establishing the standard Thai writing system, which was derived, under Mon and Khmer influence, from Indian scripts, reflecting the great prestige which Indian culture and statecraft had for the ancient civilizations of South-East Asia. The religion of Sukothai was Hinayana (Theravada) Buddhism, which spread initially from Sri Lanka and became the dominant faith of the major population groups of mainland South-East Asia.

The continued southward movement of the Thais on to the central plain brought the establishment in the mid-14th century of a new centre of Siamese power, called Ayudhya, at a point on the Chao Phraya river attainable by seagoing vessels. In 1368 it conquered Angkor and established Siam as a power with interests in Cambodia. Ayudhya's great King Trailok (reigned 1448–88) pursued its extension against the Burmese to the west, Thai principalities in the north, and Malay sultanates in the southern peninsula. He greatly strengthened Siamese state organization, making the first efforts at a centralized bureaucratic structure and ranking system and codifying customary rules into law. None the less, Ayudhya's power was diffuse by modern standards, resting, as in the earliest Thai states, on the ability of a population centre (*muang*), under its lord (*chao muang*), to enforce authority over, and extract tribute from, the rural areas around it. Those who commanded larger resources in population and wealth, usually by virtue of strategic location on a river system which enabled them to control trade, established themselves as overlords; the strongest of these might make himself king. However, the monarch's control remained uncertain, and the borders of the kingdoms shifted with the waxing and waning of the central *muang*'s ability to exact loyalty from its more distant tributaries.

Ayudhya's position on the central river, near enough to the sea to become involved in the developing trade between Europe and the Far East, gave it a particular advantage in the accumulation of power. The Portuguese, then the dominant European power in South-East Asia, sent a mission to Siam soon after their conquest of Melaka (Malacca) in 1511; thereafter, Ayudhya became increasingly involved in European rivalries until, under King Narai (1657–88), it accepted a French military mission. Narai's death brought a reaction against the involvement with foreigners and inaugurated a long period of isolation. Ayudhya's reduced resources and unstable leadership enabled rising Burmese power to challenge it, and in 1767 the city was laid waste.

EMERGENCE OF A MODERN STATE

On Ayudhya's fall, a new Thai state centre was founded by the Chinese war-lord Taksin at Thonburi, near the mouth of the Chao Phraya. He was overthrown in 1782 by the house of Chakri, and the capital was moved across the river to Bangkok. The early Chakri kings were anxious to consolidate Thai power against Burmese and, increasingly, European threats. They pressed the extension of Siamese authority over Laos, western Cambodia and the northern Malay states, and sought to strengthen the central administrative structure and to substitute tax farming for the older tributary system. This concern for increased efficiency received a forceful impetus in 1855, when the Bowring Treaty, imposed by the British, deprived Thai rulers of important income from tolls and monopolies on foreign trade. Under Kings Mongkut (Rama IV, 1851–68) and Chulalongkorn (Rama V, 1868–1910), interest in Western technology and ideas grew rapidly among the royal and noble élite, and Chulalongkorn was able to implement a major reorganization of the State, which substituted a modern centralized bureaucracy and fiscal system for the old *chao muang* arrangement.

By its internal reforms and concessions to European interests, Siam was able to maintain formal independence, but it had to concede to France its claims to suzerainty over Laos and western Cambodia, and to the United Kingdom its claims over the Malay states of Kedah, Perlis, Kelantan, and Trengganu. In order to placate the Europeans, to improve its expertise, and to finance economic and administrative modernization, Siam sought foreign loans and accepted European and US advisers in key governmental posts. It attempted to prevent any one country from having predominance but, as the United Kingdom was by far the strongest power in the region, it fell effectively within the British sphere of influence.

However, at the same time that Siam suffered under unequal relationships, it prospered with the development of the international rice trade, which provided the major source of finance for the modern Thai state. The extension of rice lands led to a rapid expansion of settlement throughout the central Thai plain. Bangkok developed into a major trading centre, in which (as in colonized South-East Asia) European firms dominated large-scale international activity and immigrant Chinese took roles as shopkeepers, middlemen and labourers. Central control over Siam's remaining territories was strengthened, though not without alienating people in the north, north-east and south who were still loyal to their customary chiefs. Although unrest was rigorously suppressed in 1902, Bangkok's authority in these regions never penetrated as deeply as in the central plain.

The bureaucracy which was developed to administer the new Siam consisted, in part, of people drawn from the old nobility and partly of commoners recruited through a rapidly expanding modern educational system. Influenced by Western ideas of progress and efficiency, its members soon acquired an ethos which clashed with the principle of royal absolutism and patronage. In 1912 an attempt by young army officers to overthrow King Wachirawut (Rama VI, 1910–25) in favour of a republic had already revealed the potential for a rift between royal authority and the new bureaucratic élite. King Wachirawut responded by identifying royalty with the new ideological force of nationalism: 'Nation, Religion, King' became, and remains, the central patriotic slogan. This preserved the royal symbol but not, in the end, royal power, for tension between royal and bureaucratic authority came to a head over financial stringencies that were imposed by the collapse of the international rice market during the Great Depression. In 1932 a bloodless coup brought to an end the absolute rule of King Prajadhipok (Rama VII, 1925–35).

MILITARY-BUREAUCRATIC RULE, 1932–44

The 'Revolution' of 1932 was led by European-educated radicals of lesser noble rank, Pridi Phanomyang and Maj. (later Marshal) Phibun Songkhram, and by the highest-ranking military officer of commoner background, Col Phahon Phonphayuhasena. The leaders of this faction saw themselves as representing the Thai public interest, but they were very much divided as to what this entailed beyond substituting their own wisdom for that of royalty in managing national affairs. Pridi, influenced by French Utopian socialist ideas, presented an economic plan whose emphasis on land nationalization and abolition of private trade probably owed as much to Thai absolutist traditions as to European socialist concepts. However, his programme antagonized the conservative members of the new ruling élite, who had increased their private wealth and financial security with Siam's economic development and the decline in royal prerogatives, and who were no more willing to entrust their welfare to socialism than they were to subject it to the royal whim. They allied with still-strong monarchist forces against Pridi, who was branded a communist and went into exile.

After a period of instability, a conservative constitutionalist regime was established under Phahon. However, real power lay increasingly with the radical Phibun Songkhram, who brought back Pridi; but, at the same time, the illegality of communism was confirmed, parties were banned, and press censorship was instituted. With little possibility of organizing popular support, civilian politicians could only represent competing factions within the Bangkok élite, and power increasingly rested with the army.

Neither Pridi nor other civilian radicals had tried to mobilize mass support while they had the opportunity, for there seemed to them little possibility that people could be aroused to follow anything but the dictates of local officials and customary chiefs. Thailand's great economic development had taken place without much change in its social structure at the mass level: peasants could plant more rice than they had before with little alteration in organization or technology, and profits from the rice trade were such that the State could take its share and still allow the peasants enough to encourage them to produce a surplus. Moreover, the later Chakri kings had seen in peasant smallholders a source of social stability and so had limited landholdings and the possibility of seizing land in payment of debt; this prevented a drastic rural upheaval akin to that experienced by Burma (now Myanmar) during the Great Depression. A certain amount of rural indebtedness and absentee landholding had occurred in the area near Bangkok, largely as a result of real estate investment by members of the capital's élite, but this inspired migration to the city rather than peasant protest.

Nor did there seem to be a basis for urban political mobilization. The only sizeable town was the capital; until the 1970s it so dwarfed other urban centres in size and importance as to be the only place that counted politically. The working class was overwhelmingly Chinese; some of it was radical—indeed, a small communist party was established illegally in the early 1930s—but it was interested in Chinese affairs and had almost no involvement in Thai politics. A few journalists and writers formed a tiny intelligentsia, continuing a tradition of iconoclastic writing that had begun in the late 19th century and reached something of a height in the 1920s under Wachirawut, when the King himself took to journalism to set forth his ideas and reply to his critics. However, whatever potential they had for influencing public opinion disappeared with the introduction of the new press laws. The independent middle class was overwhelmingly Chinese, uninterested in Thai intellectual affairs and anxious to avoid the displeasure of the officials on whom depended their chance of trade and their permission to stay in the country. All power and virtually all political participation lay with an élite employed in government service, and most particularly the military part of it. Politics became, in effect, a struggle for office and patronage by 'strongmen' and their associates within the bureaucracy. Whether Thailand's outward form was constitutional or dictatorial, the basic locus of power and the means of exercising it changed little.

This 'bureaucratic polity' was inefficient as an administrative machine, since its members were more involved with internecine power struggles and the maintenance of patronage networks than they were in performing their duties. Conscious of this weakness, which they saw as evidence that Siam was insufficiently modernized, radical Thai leaders made more drastic efforts at renovation. Phibun, who became Prime Minister in 1938, took fascism as his model. He embarked on a militantly anti-Chinese and anti-Western campaign, encouraging a rise of Japanese influence, as a counter to the dominant British, and playing upon popular resentment at the Chinese minority's powerful economic position. Stressing his role as supreme leader and pioneer of a modern society, he changed the country's name from Siam to Thailand in 1939, decreed new words of greeting, and imposed modes of dress that he considered modern. He acted ruthlessly against his opponents, encouraged the beginnings of state-sponsored industrialism in order to achieve autarchy in basic military supplies, and took an expansionist course aimed at securing, with Japanese aid, the lands that had earlier been lost to the British and French.

In 1940 Thailand attacked Indo-China, then cut off from France by war, and, in a Japanese-sponsored settlement, acquired the Laotian lands west of the Mekong River and the north-western provinces of Cambodia. Japanese troops landed in peninsular Thailand on 8 December 1941, and from there spread south to Malaya. Phibun elected to become Japan's ally, declaring war on the United Kingdom and the USA in January 1942. In 1943 Japan awarded Thailand two of the Shan states which had been incorporated into British Burma and the four states that Siam had been compelled to cede to Malaya.

The fact that Thailand experienced the war as an ally of Japan rather than under its military occupation meant that it underwent far less economic, social and political upheaval than that experienced by the other South-East Asian states. None the less, the war caused considerable privation, and, as it became evident that the Japanese were going to lose, it grew increasingly unpopular. Fortunately, some prominent Thais had always opposed the Japanese alignment, among them the ambassador to the USA, Seni Pramoj. With US help, he established the Free Thai movement, which made contact with Pridi, then regent for King Ananda (Rama VIII, 1935–46, still a child and at school in Switzerland). Pridi set up an anti-Japanese resistance which enjoyed some immunity from the Thai authorities. In August 1944 Phibun was formally deposed by the National Assembly, and in September 1945 Seni Pramoj was named Prime Minister.

DEMOCRACY AND TURBULENCE, 1945–47

The initial post-war years brought considerable turmoil, particularly in the life of the capital, for times were hard and inflation rampant. The political system's principal element, the armed forces, had been removed by the failure of the Japanese alliance and by the need to maintain the good graces of the Allies. The civilian leaders were occupied with securing US support against extensive British demands for reparations (the lands that had been gained with Japanese aid were lost). Constitutional democracy was restored, and free play was given to the formation of political parties, but these groupings were little more than the personal followers of politicians in the capital. They included the conservative Democrat Party (DP), which counted among its adherents Seni Pramoj and his brother Kukrit. The parties were unable to discipline themselves, much less to agree on a common course of policy or to gain control over an increasingly corrupt and demoralized civil service.

In March 1946 Pridi became Prime Minister, but in spite of his long association with Phibun, he was still feared as a socialist by many conservatives. Phibun still had a strong following in the armed forces, which increased as he put himself forward as the spokesman for resentment at civilian corruption and mismanagement. In June the young King Ananda died in mysterious circumstances, and Phibun had a major role in assigning the blame to Pridi, who was forced to resign. In November 1947 the army seized power, in a coup led by Gen. (later Marshal) Phin Choonhavan. He acted on behalf of Phibun, who took power in his own name in April 1948.

One factor that emboldened Phibun was a shift in the international situation. Hitherto, he and the army had been the objects of Allied displeasure; but, with the 'Cold War' beginning, Burma apparently near collapse, and Indo-China embroiled in revolution, Thailand seemed a possible centre of stability. In February 1949 an attempted coup by Pridi's supporters sent that leader into exile in China. In November 1951 a bloodless coup by Phibun ended what remained of Thai democracy. The Constitution was abrogated, political parties were banned, radical leaders were imprisoned or executed, and dissent was stifled. Thailand embarked on a new international career as an anti-communist ally of the USA, receiving considerable US military and economic aid, and in 1954 becoming a founding member of the South-East Asian Treaty Organization (SEATO), which established its seat in Bangkok.

PHIBUN'S SECOND DICTATORSHIP, 1948–57

Political power in this period rested with a triumvirate composed of Phin, Phibun, and Phin's son-in-law, Gen. Phao Sriyanon, head of the police. Aside from the alliances consolidated by patronage and intermarriage among the country's leading military and bureaucratic families, there were informal (but increasingly close) ties between political power-holders and wealthy local Chinese. In the early days of his post-war rule Phibun had revived some of his economic nationalism and persecution of the Chinese minority, but in the boom that accompanied the Korean War the chance to profit from Chinese capital and economic expertise was too much to resist, and increasingly generals found themselves on the boards of directors of Chinese-run enterprises. The Chinese themselves were becoming more assimilated, by force of circumstances and by the fact that immigration from China had ceased. There was an increasing identity of interest between Thailand's political and economic leadership, but the Thai élite remained overwhelmingly in control of the bureaucracy, through which they extracted tribute and gifts from the politically weak entrepreneurs.

Increasingly, Phibun found it difficult to defend his position against the ambitions of his associates, particularly the new Bangkok army head, Gen. (later Marshal) Sarit Thanarat. He attempted to recoup his position by changing the sources of his support: in 1955, after a trip to Europe and the USA, he restored free speech, allowed the formation of political parties, and announced elections for 1957. These took place amid growing turbulence and accusations of corruption; the Government's victory resulted only in it being discredited, so blatantly was the vote rigged. Sarit, who prudently stayed clear of the campaign, used the opportunity to seize power, first installing the SEATO Secretary-General, Pote Sarasin, as Prime Minister and finally, in October 1958, assuming the leadership himself.

SARIT AND AUTHORITARIAN DEVELOPMENT, 1958–72

Sarit stamped out the brief democracy of the election campaign, and urged a restoration of traditional values, including (for the first time since the 1932 Revolution) an appeal for loyalty to the King. As it was no longer likely that royalty could challenge military rule, Sarit could restore it as a major element in a nationalist appeal for popular support. His cultural and political conservatism was accompanied, however, by a conviction that Thailand should open itself to rapid economic modernization. He removed the limits on land-holding and welcomed foreign investment. The time was propitious, for US involvement in the Viet Nam War was creating the conditions for both an economic boom and profitable foreign entanglement. Soon Thailand became a major provider of US military bases and supplies, Bangkok was a rest-and-recreation centre for US troops, and Thai troops were serving, at a price, in the Indo-China campaign.

In the countryside, US-funded road-building for security purposes rapidly increased communications and transport; peasants became deeply involved in the money economy, moved to towns to seek construction work, or sought to give their children an education that would raise them above peasant status. Both in the cities and the countryside, the old social structures were weakening. In the cities the clash of cultures, generations and interests was more obvious, but the contrast was perhaps more serious in the countryside, as hitherto lightly-ruled peasants and hill peoples in the distant provinces were confronted with officials who were determined to impose their will in the name of security and development. Rural resentment was encouraged by Chinese, Vietnamese and Pathet Lao aid to rebel groups in the north and north-east, but even in the southern peninsula unrest flared into rebellion. In 1965 a China-based Thai Patriotic Front proclaimed a 'war of national liberation', and by the end of the decade most of the provinces outside the central plain were, to some degree, insecure.

Sarit had died in 1963; he was discovered to have appropriated vast sums of public money, but the scandal was soon forgotten in a wave of nostalgia for his despotic but highly successful rule. His successors, the duumvirate Gen. Thanom Kittikachorn and Gen. Praphat Charusathien, were less dynamic leaders, though initially they prospered, not only because of the Indo-China War boom but also because of increasing Japanese investment in trade, manufacturing and agribusiness. In 1968 they sought to broaden their power base by a cautious revival of constitutional democracy. This brought more criticism than they had bargained for, and in November 1971 they again seized full power; but by now their authority was severely compromised.

The US Government's decision to withdraw from Viet Nam and to seek a *rapprochement* with China brought great uncertainty. The economic boom was ending, creating hardship in the capital, which had been swollen by rural immigrants. The great expansion of education and bureaucratic recruitment in the preceding decade had to be curtailed for lack of cash, at the cost of considerable popular resentment. In rural areas, peasants were turning from rice to better-paying cash crops, for the Government, anxious to husband its declining profit from the rice trade and to keep the domestic price low enough to satisfy the urban populace, was not offering a worthwhile price. In the more distant countryside, revolt was spreading. There was a general weariness of corrupt military rulers; within the army itself, those outside the Thanom-Praphat faction resented their long separation from profitable office, while within it annoyance grew at the slow rate of promotion and the inability of their leaders to address Thailand's problems.

THE 'DEMOCRATIC REVOLUTION', 1973–76

During 1972 student-led demonstrations began in protest at the regime. Army leaders outside the ruling faction came increasingly to feel that the military should withdraw from direct responsibility for rule, in order not to be identified with the unpopular economic measures any successor government would need to take. In October 1973, when Thanom and Praphat tried to halt the demonstrators by armed action, the army refused to support them. Additionally, King Bhumibol Adulyadej (Rama IX, 1946–) withdrew his support, thus depriving the duumvirate of any legitimacy. The military regime collapsed, and its leaders went into exile.

The upheaval of October 1973 was no more a revolution, in the sense of mass armed action, than had been the Revolution of 1932 but, like its predecessor, it marked a watershed in the development of the Thai political system. It, too, was the culmination of gradual social, economic and ideological changes which finally overwhelmed the old bases of rule. The demonstrations that brought the downfall of military rule were only the beginning of a turbulent three years of political participation. Not only students in the capital, but also those in the new provincial universities, demonstrated for social reform; organized under the National Student Centre of Thailand (NSCT), they presented increasingly radical demands to the Government and worked to organize peasants and labourers against economic exploitation and ill-treatment by officials. The peasants and workers were, in contrast to earlier democratic periods, receptive to urgings for organized action to secure redress: particularly in the central plain, peasants began to form farmers' associations and to send delegations to the capital. Workers in Bangkok's now numerous textile mills began to strike against the low pay and hard working conditions that had attracted international investment to Thai industry. Even the Buddhist monkhood, hitherto apolitical except for a generalized support for established authority, was caught up in the fervour. Younger monks, concerned at social inequality and at the growing secularism and materialism of modern Thai culture, called for the religious to protect the poor from injustice, and aided peasants in pressing their claims.

Though the democratic institutions of 1973–76 allowed the expression of such sentiments, they did very little to satisfy demands. A caretaker regime under a university rector, Sanya Thammasak, was appointed by the King following the collapse of the Thanom-Praphat regime and presided over a period of frenetic party formation, with elections held in January 1975. With great difficulty—as none of the numerous parties had a working majority—a coalition Government was formed under Kukrit Pramoj, who had broken with the DP, under his brother Seni, and now headed the Social Action Party (SAP). His regime lasted for less than a year; new elections in April 1976 brought Seni and the DP to power in a centre-right coalition, but this, too, proved unable to formulate, let alone execute, a coherent policy.

The Governments of the period were markedly more conservative than the non-parliamentary expressions of public opinion; moreover, the general electoral trend was to the right, while the demands of the petitioners, strikers and demonstrators were increasingly radical. Most workers and peasants were still politically passive and either did not participate in elections or voted as directed by village headmen or district officials. The less likely it seemed to them that they could change things by voting, the less attempt they made at challenging the establishment electorally, and the more they did so by rejecting it altogether: hence an increasingly conservative trend in voting was accompanied by an alarming rise in rural insurgency. Thais who had benefited from the recent economic development, and who had hoped for further reform, became alarmed at the gathering disorder and the possibility that economic collapse might endanger their hard-won prosperity. Increasingly, there was not only renewed conservatism in the middle classes but sympathy for a radical-right response to leftist demands. Rightist mobilization was apparent even among the Buddhist monkhood, where Kittiwutto Bhikku, hitherto noted for his efforts at founding a socially relevant monastic educational system, broke with the Buddhist precept against taking life to announce that the slaying of communists was not a sin.

The parties themselves were poorly organized and still bound largely by ties of personal loyalty and patronage. However, they were, for the first time, active in the provinces as well as in the capital, and the DP, in particular, had consolidated a national following as the representative of conservative reform. Further to the right were parties composed of leading military, bureaucratic and business figures; the most important of these was the Chart Thai (Thai Nation) party, which centred on the heirs of the old Phin-Phao-Phibun military clique. However, these parties were not merely military-bureaucratic factions, for the Thai élite was being drawn increasingly into business endeavours. A growing sector of big business—particularly the banks—was now Thai-owned and confident enough to take an open role in politics. This impelled the Thai right wing away from statist solutions; parties confronted economic dilemmas more seriously and policy differences, as well as personal loyalties, determined the élite coalitions that formed the basis of the right.

Far weaker were the parties of the left. Led by intellectuals and with little popular appeal, they were undermined by the communists, who were pursuing victory by armed action, and by students and other militants who saw extra-parliamentary pressure as the way to achieve reform. From the right they were victimized by a rising tide of counter-revolutionary violence.

As civilian governments showed themselves unable even to keep order, military officers and their civilian allies became increasingly restive. Gen. Krit Siwara, who had become Army Commander-in-Chief just before the fall of the Thanom-Praphat regime and had gained general acclaim for his refusal to fire upon the students, was unwilling to overthrow a constitutionalist system which he, as well as other moderate conservatives, saw as the best ultimate guarantee of stability. While he hesitated, military rightists gave funds and encouragement to groups advocating mass action: the Nawaphon movement, of which Kittiwutto was a prominent advocate; the strong-arm Red Gaur squads; and the rural vigilantes of the Village Scout organization. All of these took King Wachirawut's slogan of 'Nation, Religion, King' as their cry, and stressed loyalty to the monarch as the touchstone of Thai nationalism.

REPRESSION AND RESTORATION, 1976–88

Krit's death in April 1976 hastened the return to military rule, an increasingly popular solution as the urban middle classes came to see authoritarian government as preferable to paralysis and disorder. A military-dominated National Administrative Reform Council took over, banning political parties, dissolving the National Assembly, and setting up a Government which was headed by a right-wing Supreme Court judge, Thanin Kraivixien. Left-wing activists were arrested; many student leaders fled to the jungle, where they joined the communist forces. Labour unions, left-wing parties and farmers' associations were firmly suppressed, and the already marked level of violence against worker and peasant organizers grew with officially sanctioned killings. Legally-expressed dissent was silenced, but revolutionary protest grew apace, bolstered not only by new recruits but by confidence and material flowing from communism's 1975 victories in Indo-China.

The political harshness of Thanin's regime alienated a great many people who had earlier looked to a military solution, provided the communists with a good deal of talent and considerable legitimacy, and did nothing to restore the economy or Thailand's international reputation. Previous military clampdowns had brought killing and imprisonment, but these had involved only a small number of socially obscure or marginal dissidents. Now the repression affected a significant part of the population; moreover, the students who were the chief targets of the reaction were also the offspring of the ruling élite. In October 1977 the armed forces Commander-in-Chief, Gen. Kriangsak Chomanan, seized power, and restored some elements of democracy by reducing censorship, releasing detainees, and permitting the resumption of party activity under arrangements that ensured conservative control of the legislature and key government appointments.

In March 1980, having failed to find economic policies that were both realistic and acceptable to the élite, Kriangsak was replaced as Prime Minister by the Army Commander-in-Chief and Defence Minister, Gen. Prem Tinsulanonda. Prem's regime, like Kriangsak's, was based on a combination of the military and centre-right party politicians. Ultimate power lay with the armed forces, but the political parties played a major role in linking military, civilian, bureaucratic and business interests. The result was a ruling élite too powerful to dislodge but not sufficiently united to govern effectively.

None the less, in this period the evolution of foreign affairs and domestic society combined to reduce greatly the pressures on the Thai polity. Viet Nam's invasion of Cambodia had brought disunity among communist supporters, and a breach between Viet Nam and China, which greatly reduced support for the Thai communist insurgency. Refugees from Laos, Cambodia, and Viet Nam burdened Thai facilities but also provided testimony against the rigours of communist rule. Thailand found common cause with China as well as the USA in supporting the deposed Pol Pot against the Viet Nam-sponsored Heng Samrin regime. Insurgents in the north and north-east, long accustomed to material support from and refuge in neighbouring Laos, Viet Nam and China, were demoralized by their patrons' quarrel. The students who had joined them were increasingly disillusioned by the Thai Communist Party leadership's rigid insistence on the outdated Maoist doctrine of protracted rural warfare. The relative liberalism of the post-1977 regimes began to attract them, and, following the announcement of an amnesty for dissidents, many of them returned. Later Prem extended the amnesty to the rank and file of the communist insurgency; this resulted in mass defections from the rebellion in late 1982 and early 1983.

These achievements were somewhat marred by lack of success in responding to the consolidation of power by communist regimes in Indo-China. A Thai military operation to dislodge the Laotians from a disputed border territory ended ingloriously with a truce in March 1988. On the Cambodian border the Thai army showed little ability to respond to Vietnamese incursions, but the Thai Government argued strongly against any compromise on the Cambodian issue by the Association of South East Asian Nations (ASEAN), of which it had been an enthusiastic member since that group's founding in 1967.

While maintaining an unyielding stance towards its communist neighbours, the Thai Government quietly restored US military aid projects and training schemes, which had been rejected following the 1973 uprising and the defeat of the USA in Viet Nam. In late 1985 US forces began to participate in Thai military exercises on a substantial scale; in April 1986 plans were devised for the establishment of a stockpile of US weapons in Thailand, the first occasion for a US reserve of weapons in a country without US military bases.

Towards the end of the 1970s the discovery of petroleum and natural gas in the Gulf of Thailand helped revive local economic optimism. At the same time foreign investors, realizing that Thailand was not after all about to fall to communism, began to move funds into the country. In particular, Japan, seeking new sources of cheap manufacturing labour and outlets for its capital, contributed greatly to a new wave of economic development. The pattern of expansion begun in the 1960s was resumed, and the outlines of a major social transformation became more apparent. Industrial, agribusiness, and construction sectors expanded rapidly, with Thai as well as foreign business interests prominent in them. The middle and working classes grew in importance and self-confidence, pressing for such reforms as the establishment of a social security system. Secondary urban centres began to flourish, challenging the capital's monopoly on political power.

The growth of the 1980s differed from that of Sarit's time in that it lacked any clear leadership or ideology. The regimes of Kriangsak and Prem were reactive rather than dynamic, and they confined their innovative efforts to the search for a constitutional formula that would create strong yet controllable political parties. The Government attempted to remove the army from the political process while safeguarding its interests, an effort which found support in the early 1980s from the Democratic Officers Movement (DOM), whose leaders argued that this was necessary for the sake of national progress, military unity, and professionalism. Indeed, the basis of politics appeared to be changing from the old personalist ties of patron-client relations to alliances based on interest groups and political ideals. Banking, manufacturing, and agribusiness interests began to play a significant role in politics; partly because of their participation the parties seemed to acquire more substance, and young army officers began to form groups supporting particular political objectives rather than an individual patron or faction.

The effort to reform the system culminated in the constitutional changes endorsed by the general election to the House of Representatives in April 1983, which greatly reduced the power of the appointed Senate and banned the appointment of civil servants (including military officers) to positions in the Council of Ministers. Such attacks on entrenched interests did not go unchallenged, both by senior officers aspiring to national leadership and young ones who sought a more dynamic role for the army, and the State in general, in developing the country. Already in 1981 radical nationalist middle-rank officers had attempted a coup. In 1983 Gen. Arthit Kamlang-ek, Supreme Commander of the Armed Forces and Commander-in-Chief of the Army, formed an alliance with these 'Young Turks' in his attempt to consolidate military support against the constitutional reforms. This culminated in a coup attempt in September 1985, led by Col Manoon Roopkachorn and quietly supported by senior military officers. Its failure resulted in Arthit's replacement as Supreme Commander of the Armed Forces and Army Commander-in-Chief by Gen. Chavalit Yongchaiyudh, one of Prem's closest advisers and the acknowledged leader of the successful campaign against the communist insurgency.

Chavalit had once been prominent in the DOM, but soon after his elevation he began to voice opinions which seemed more in line with the ideas of the 'Young Turks'. He began to call for a state-led 'revolution' that would transform the country's political, economic, and social structure. Increasingly, he appeared as a dynamic alternative to Prem. The latter's administration had been re-endorsed by elections in July 1986, which gave the DP a strong plurality in the House of Representatives. However, the government coalition (comprising the DP, Chart Thai, the SAP and Rassadorn) was highly unstable, and the influence of political parties remained small compared with that of the armed forces. The notable growth of interest-group politics and the middle class meant increased participation but, often enough, political deadlock as well. During the 1970s and 1980s the King became a pivotal figure in political negotiations, his opinion being sought as a means of breaking the impasse resulting from conflicting interests. Prem, though excellent at the manoeuvring and compromise necessary for political survival, was not able to provide a sense of direction. As a result, what was in many ways a very successful regime was characterized by a high level of popular frustration. Out of this arose the opportunity for a revival of parliamentary government.

PARLIAMENTARY RULE, 1988–91

In July 1988 a general election for the House of Representatives was held: it was marred by vote-buying and campaign violence. The results endorsed the existing coalition Government but also revealed a severe decline in the power of the DP, which split, and the growth of the right-wing Chart Thai, which had established itself as the representative of powerful Thai business interests and had spent lavishly on the campaign. The King asked Prem to continue as Prime Minister, but he refused in favour of the leader of Chart Thai, Gen. Chatichai Choonhavan.

Chatichai was not only a retired general but also the son of Phin Choonhavan, founder of the Phin-Phao-Phibun military clique that led Thailand from 1948 to 1957. His accession was seen, however, as a major step towards democracy: for the first time since 1976, an elected leader was in charge of the Government. Chatichai was known as a politician rather than as a military man, and he was the architect of Chart Thai's business orientation. The size of the business community's financial support for Chart Thai's campaign reflected not only appreciation of the party's attitude but also a new respect for the importance of party politics and a departure from reliance on the protection of powerful individuals rather than political organizations. Chatichai's coalition Government comprised Chart Thai, the SAP, the DP, Rassadorn, the United Democratic Party and Muan Chon.

Chatichai initiated reforms that opened Thailand to a major expansion of the business sector. At the same time, he asserted a concern for developing the country's poorer regions, and secured the passage of Thailand's first social security law. In November 1988 all logging of Thailand's rain forest was banned, an act that aroused some astonishment given the very powerful interests that it affected. (In fact, however, Thailand's forested area had been almost exhausted, and the Government intended to compensate the politically well-connected logging interests with a major new field of exploitation, Myanmar—formerly Burma.)

Chatichai's innovations were particularly notable in foreign affairs. Overcoming considerable bureaucratic and military resistance, he abandoned the previously intransigent Thai stance on the Indo-China question, declaring his ambition to turn Indo-China 'from a battlefield into a market-place'. In particular, he hoped to transform the north-east of Thailand into a centre for trade and industry, linking Laos and eventually the rest of Indo-China to Thailand. In November 1988 he succeeded in settling the border dispute with Laos. The Vietnamese withdrawal from Cambodia in September 1989 presented him with an opportunity to stress, against army opposition, a willingness to improve relations with the rest of Indo-China.

In December 1988 Chavalit visited Myanmar, earning the gratitude of its internationally isolated regime. A principal purpose of his trip was to arrange for Myanma timber, hitherto smuggled via Thailand to the world market, to be exported by firms recommended by the Thai Government to Myanmar's state Timber Corporation. There ensued a rush for logging concessions by politically well-connected Thai businesses. In addition, Thai interests obtained favourable agreements for fishing rights and the exploitation of minerals, and for the supply of urgently needed goods. From having been the agrarian victim of unequal economic relationships with industrialized countries, Thailand was itself becoming the exploiter of a poorer and more backward economy.

In spite of these initiatives, Chatichai's regime began to falter in late 1989. Commerce expanded but was accompanied by a widespread feeling of unease at Thailand's precipitate capitalist development. The country was experiencing unprecedented prosperity but also growing inequality; the promises to aid the poorer regions had not been fulfilled. Public concern mounted over corruption, the high rate of inflation, traffic congestion in Bangkok, pollution and other environmental issues, drug addiction and AIDS. Once again, the demand rose for strong leadership to address the country's ills. In 1990 a series of no-confidence motions was brought against the Government in the House of Representatives, and Chatichai engaged in increasingly desperate government reshuffles. He attempted to allay military disaffection by giving Chavalit unrestricted responsibility for senior military appointments. Chavalit took advantage of this to allot key posts to the leaders of the powerful Class 5 Group, based on the 1958 graduates of Chulachomklao Military Academy.

Military Academy class membership had long been an important source of faction and patronage within the army. Previously, Prem and Kriangsak had been careful, as government leaders, to ensure that no one class dominated the military power structure. Class 5 was a particularly large and cohesive group, whose ties had been strengthened by marriages and business relationships among its principal figures. Chavalit relied for his influence over Class 5 on his generous patronage of it and especially on the presumed personal loyalty of its leader, Gen. Suchinda Kraprayoon, whom he named Deputy Commander-in-Chief of the Army in September 1989. Chavalit also acted to establish himself as an opposition figure within the parliamentary context, resigning from his official posts in 1990 and forming the New Aspiration Party (NAP) to contest the next elections. In establishing himself formally in political opposition, he had given up his official sources of military power. His place as Supreme Commander of the Armed Forces was taken by Gen. Sunthorn Kongsompong, who was regarded as favourable to Class 5. The position of Commander-in-Chief of the Army was assumed by Suchinda, whose brother-in-law, Gen. Issarapong Noonpakdi, became his deputy.

RETURN TO MILITARY RULE

The Prime Minister, alarmed at this entrenchment of Class 5, attempted to rally its opponents in the army, but only succeeded in bringing the situation to a climax. On 23 February 1991 Chatichai was seized in a bloodless military coup. A National Peace-keeping Council (NPC), headed by Sunthorn, took command of the country. Martial law was declared, the Constitution suspended and the National Assembly dissolved.

Although politically active Thais had believed that an unquestioned military intervention was no longer possible, there was little reaction to the country's 17th coup. The public was too disenchanted with the corruption and unscrupulous capitalism of Chatichai's regime to be willing to defend it. The domestic and international business interests, which had been its chief source of support, were easily persuaded by the perpetrators of the coup that they intended merely to install a more honest and competent rule of experts. The President of the Federation of Thai Industries, Anand Panyarachun, was appointed Prime Minister. Anand appointed a predominantly civilian 35-member interim Cabinet comprising respected technocrats and former ministers and including only eight members of the armed forces. It was declared that an interim National Legislative Assembly would be created to draft a new constitution and prepare for a general election. The Assembly, however, consisted overwhelmingly of senior military officers (149 of its 292 members). The other members were business executives, bureaucrats and others known for their conservatism and military connections. The new rulers dissolved trade unions in the public sector, announced a new anti-communist campaign and an offensive against crime and corruption, and declared a revival of the 'hard-line' policy on Cambodia. They did not, however, ban political parties, as had occurred following every previous military take-over.

Although parties remained legal, they were in complete disarray. An Assets Examination Committee installed by the new regime identified leaders of the major parties, in particular the former Prime Minister, Chatichai, as possessing 'unusual wealth'. Politicians seeking new allegiance were attracted to the NAP of Chavalit, who had declared himself opposed to the overthrow of parliamentary rule. To offset this, the Class 5 ally, Air Chief Marshal Kaset Rojananin (a member of the NPC), sponsored the Samakkhi Tham party as a vehicle for supporters of the new regime. In addition, Chart Thai, reportedly in exchange for the abandoning of corruption charges against its major leaders, aligned itself in support of the military regime under a new leader, Air Chief Marshal Somboon Rahong. During the next months Class 5 strengthened its hold on the government machinery, and in August 1991 Suchinda replaced Sunthorn as Supreme Commander of the Armed Forces and appointed Issarapong as Commander-in-Chief of the Army.

In December 1991 the new Constitution was proclaimed, which ensured the protection of conservative, military-bureaucratic interests. The new basic law was greeted with some popular dismay but little organized protest. Public opinion endorsed Anand's competent and honest governance, though Thais were increasingly disenchanted with the military's blatant profiting from its hold on politics.

On 22 March 1992 National Assembly elections were held under the new dispensation. They provided encouragement for neither the military nor its opponents. With a bare majority, they returned a weak coalition led by Samakkhi Tham, Chart Thai, and the SAP, and also including Prachakorn Thai and Rassadorn. None of these presented a credible candidate for Prime Minister, and finally, despite his previous assurances to the contrary, Suchinda assumed the premiership.

Rule by a non-elected military man was sanctioned by the new Constitution and was historically very familiar. Suchinda seemed to have an invincible military-political machine at his call. However, his assumption of the premiership crystallized civilian unease, and resulted in demonstrations demanding an elected Prime Minister. These culminated in a dramatic hunger strike by Maj.-Gen. (retd) Chamlong Srimuang, who had been the highly popular and effective Governor of Bangkok, and was the leader of the Palang Dharma (Righteous Force) party. On 17–20 May 1992, after unsuccessfully trying to wait out the protests, Suchinda and his chief supporters resorted to violence that led to the deaths of an estimated 100 Bangkok demonstrators. As increasingly in major Thai political crises, royal intervention was required to achieve a settlement. On 24 May (the day after the declaration of an amnesty for all 'offenders' in the recent demonstrations—including those responsible for the deaths of protesters) Suchinda submitted his resignation.

THE RE-ESTABLISHMENT OF DEMOCRATIC RULE

Following Suchinda's resignation, the five government coalition parties nominated Somboon Rahong, the leader of Chart Thai, as Prime Minister. The fact that Somboon was perceived, however, as having close links with the military leaders led to fears of new unrest, and his appointment was not confirmed. Instead, on 10 June 1992, the King again appointed Anand Panyarachun as Prime Minister. Anand announced that he would organize new elections within four months. On the same day as his appointment, the National Assembly approved constitutional amendments, reducing the powers of the non-elected Senate and stipulating that the Prime Minister must be an elected member of the Assembly. Anand formed an administration that comprised many of the apolitical figures whom he had appointed during his previous term of office. At the end of June a 'National Democratic Front' was formed by four parties that had opposed the military Government: the DP, the NAP, Palang Dharma and Ekkaparb. In July Chatichai declined the leadership of Chart Thai (offered to him after the resignation of Somboon from that post) and formed a new party, Chart Pattana (National Development).

In August 1992 Anand demoted the military leaders regarded as being responsible for the violent repression of the May demonstrations, and appointed in their place officers who had a reputation as professional soldiers of integrity. The new Commander-in-Chief of the Army, Gen. Wimol Wongwanit, promised that the army would not interfere in politics during his period of command. Anand also reduced military control of state enterprises, appointing civilians instead of military officers to head the national airline, the communications and telephone authorities, and other institutions hitherto regarded as the preserve of the armed forces.

In the general election of 13 September 1992 Thailand's oldest party, the DP, won the largest number of seats (79), while Chart Thai won 77 and Chart Pattana 60. With the other parties that belonged to the 'National Democratic Front', the DP was able to form a coalition Government, commanding 185 of the 360 seats in the House of Representatives: Chuan Leekpai, the leader of the DP, became Prime Minister. In order to increase the coalition's narrow majority, the SAP (which had won 22 seats) was also invited to join the coalition, despite its participation in the previous Government.

Major formal shifts in the way Thailand was ruled were not accompanied by clear and consistent changes in public opinion. The underlying reason for the popular rejection of both Chatichai's and Suchinda's regimes was not so much the institutional form of their rule as dismay at blatant corruption and abuse of office. Prime Minister Anand, although he served the military regime and was not himself elected, was perfectly acceptable to public opinion because he was seen as honest and competent. This public opinion was largely middle-class and located in Bangkok, characteristics which no longer reflected, as in the past, the failure of national politics to penetrate the provinces, but rather the urban middle class's growing strength. This could be seen in the importance of business opinion to the smooth acceptance of both the 1991 military coup and Suchinda's downfall. Such opinion was not yet committed to any particular set of political institutions, although it was beginning to express itself particularly through political parties, but was concerned to find competent and reasonably honest rule. At the same time, those powerful business executives who had access to political decision-makers were reluctant to forgo the advantages of the old patronage system, and they found a new source of strength in the rising power of provincial business and political leaders. These relied on relationships of patronage to make profits and to garner votes; money, not principle, was the basis for their loyalties. Since vote-buying and local 'pork-barrel' projects were the only benefits most rural people realized from government, they tended to support leaders who offered the most immediate financial return; and as 90% of the National Assembly seats went to provincial districts, this gave the politics of patronage great electoral strength.

On assuming office as Prime Minister, Chuan had declared his intention to eradicate corrupt practices, to decentralize government from Bangkok to the provinces and to pay increased attention to rural development. However, although he was regarded as an able and honest politician, he did not appear a forceful leader, and his tenure was devoted largely to holding together his unsteady coalition. There was little business-political support for his reforms, and the decentralization efforts aroused the anger of the still-powerful bureaucracy. Nevertheless, with Chuan's Government in office, the military appeared to move into the political background. Chavalit indeed served as Minister of the Interior, but this was as much due to the importance of his NAP to the ruling coalition as it was to his continuing influence in military circles. Business executives and technocrats, not military leaders, were increasingly seen as the source of dynamism and reform.

In June 1993 the position of Chuan's administration was consolidated by its decisive survival of a motion of 'no confidence' in the Cabinet introduced by Chart Pattana and Chart Thai. In early September the SAP announced that it was to merge with four opposition parties, including Chart Pattana, under the leadership of Chatichai, while remaining in the ruling coalition. This appeared to be an attempt to secure the premiership for Chatichai, as the new SAP would command more seats than any other member of the coalition. However, four days after the merger plans were announced the SAP was expelled from the Government. It was replaced by the Seritham Party, which controlled only eight seats, compared with 21 for the SAP, bringing the coalition's total representation to 193 seats.

On 31 March 1994 a joint session of the upper and lower houses of the National Assembly was held to consider government proposals to democratize the Constitution, principally by reducing the powers of the appointed Senate and broadening the method by which its members were chosen. The senators, most of whom had been appointed under the Suchinda regime, were understandably reluctant to diminish their powers; together with the opposition parties from the House of Representatives, they defeated the government plan and put forward instead proposals to strengthen the Constitution's authoritarian aspects. A crisis was avoided by the appointment of a joint committee to formulate a compromise reform programme.

In September 1994 Chamlong resumed the leadership of Palang Dharma in party elections. He subsequently persuaded the party executive committee to approve the replacement of all 11 of Palang Dharma's cabinet members. Chamlong's nominations for the cabinet posts, which included Thaksin Shinawatra, a prominent business executive, and Vichit Surapongchai, a former President of the Bangkok Bank, as Minister of Foreign Affairs and Minister of Transport and Communications respectively, provoked strong protests within the divided party, which were led by Prasong Soonsiri, who opposed the selection of non-elected candidates and objected to Chamlong's authoritarian style of leadership. The appointments were confirmed in a cabinet reorganization in late October, when Chamlong was named Deputy Prime Minister.

In December 1994 the NAP withdrew from the ruling coalition over a constitutional amendment providing for the future election of local government representatives, including village headmen. Chavalit voted against the Government in what was widely viewed as an attempt to ingratiate himself with powerful local interests who would control many votes at the next election. The NAP's defection to the opposition left the ruling coalition with a minority of seats in the House of Representatives. Chuan therefore found it necessary to invite Chatichai's Chart Pattana to join the coalition, although this severely compromised the Government's claims to represent honesty and reform.

In January 1995 Chuan finally managed to secure the approval of a joint session of the National Assembly for a series of amendments to the Constitution aimed at expanding the country's democratic base. The reforms included a lowering of the minimum voting age from 20 years to 18, a reduction in the size of the appointed military-dominated Senate to two-thirds that of the elective House of Representatives, equality for women and the prohibition of senators and members of the Government from holding monopolistic concessions with government or state bodies. This last amendment necessitated the resignation of Thaksin as Minister of Foreign Affairs, owing to his extensive business interests.

In December 1994 the Minister of Agriculture and Co-operatives, Niphon Phromphan, and his Deputy Minister, Suthep Thuaksuwan, had resigned prior to a planned motion of 'no confidence' in the Government, owing to allegations of corruption in connection with a land distribution scheme promoted by Chuan as the centre-piece of his administration's programme. Following a government investigation, in April 1995 land titles were removed from seven deed-holders, and a senior official was dismissed from the land reform department. However, opposition parties tabled a motion of 'no confidence' in the Government in May, and Chamlong announced that Palang Dharma would not support the coalition in the 'no confidence' vote. Chuan dissolved the House of Representatives on 19 May and announced that a general election would take place in early July.

The elections held on 2 July 1995 marked the triumph of provincially based money politics. Chart Thai secured the largest number of seats (92), not one of them from the more sophisticated electorate in Bangkok. The party's leader, Banharn Silapa-Archa, became Prime Minister. A self-made businessman and politician from the central Thai province of Suphanburi, he had been one of those leaders accused of being 'unusually rich' by the military junta that seized power in 1991. Banharn headed a seven-party coalition comprising Chart Thai, the NAP, Palang Dharma (led by Thaksin since late May), the SAP, Prachakorn Thai, Nam Thai (a new, overtly technocratic, business-orientated party formed in 1994 by Amnuay Viravan, a former minister and former head of Bangkok Bank) and the Muan Chon party. His Cabinet, unlike Chuan's, eschewed technocrats in favour of patronage-based politicians. Banharn presented himself as a dynamic leader, willing to ignore legalities and sensitivities in order to achieve progress. He defined his

Government's aims much as his predecessor had, for these were issues generally conceded to be urgent, even if their solutions were not agreed. However, Banharn's ability to outperform Chuan was sharply curtailed by the instability of his coalition and by his dependence on serving the interests of provincial patronage. Thus, although resolving Bangkok's infamous traffic congestion was a prime test of effective leadership, he allocated Palang Dharma's new leader, Thaksin, responsibility for the management of traffic in inner Bangkok, and gave the Prachakorn Thai leader, Samak Sundaravej, control of traffic in outer Bangkok. Such politicization of decision-making on a critical issue led to much public disillusionment, culminating in August 1995 in an unprecedented criticism of government incompetence by the King.

In September 1995 the Minister of Defence, Chavalit, took the unusual step of rejecting recommendations for promotions by the outgoing Commander-in-Chief of the Army, Wimol, and appointed his own close associate, Gen. Pramon Phalasin, to replace him. This was a set-back to Banharn, who was close to Wimol, but he managed to strengthen his position by the appointment of Gen. Viroj Saengsanit as Supreme Commander of the Armed Forces. Viroj, a Class 5 member and supporter of the 1991 coup, was considered to be unsympathetic to Chavalit. Increasingly Chavalit, who was also Deputy Prime Minister, appeared as Banharn's unspoken rival and most likely successor. His NAP party, with 57 seats, was the third largest party in the legislature and the second largest in the governing coalition. He had a strong base in the military, which he consolidated both through controlling appointments and by increasing weapons purchases and the opportunities for military participation in business. Chavalit also endeavoured to increase his support among business interests, by promoting privatization, and among the bureaucracy, by advocating strong central administration. Although he urged rule through parliamentary institutions rather than by military junta, he emphasized his view that the army had a legitimate and important role in politics.

By early 1996 it was evident that Thailand had replaced weak but relatively honest and competent rule with a weak and patently venal administration. The Banharn Government attempted to regain credibility by pursuing some of the political reform projects it had promised at its inception, and in March the Prime Minister appointed a new Senate that included only 39 military officers, a considerable step towards democratizing that institution. At the end of March the Government also announced a major programme of investment in education, which it hoped would both alleviate rural discontent and meet business demands for a better-trained work-force. It announced a plan for financial decentralization, which was intended to promote rural development. Most of the Government's efforts were blocked, however, by disagreements within the ruling coalition and by the requirements of the all-pervasive patronage system.

In May 1996 the Government faced a vote of 'no confidence' in which 10 cabinet ministers were accused by the opposition of incompetence and/or corruption. These included Banharn but not Chavalit, despite widespread criticism of the latter's role in military procurement. Evidently the opposition parties, led by the DP, were leaving a line open for future negotiations with the powerful Minister of Defence. The Government won without difficulty, as it held 233 of the 391 seats in the House of Representatives. None the less, public disillusionment eroded support for the coalition parties: this was shown in the Bangkok gubernatorial elections of 2 June, when Palang Dharma, hitherto the capital's most popular political party, lost to the independent candidate, Bhichit Rattakul. In August Palang Dharma withdrew from the Government, following a cabinet dispute over bribery in the awarding of bank licences. Speculation grew that Banharn would come to an agreement with Chatichai Choonhavan to bring Chart Pattana into the ruling coalition, a move that would both strengthen the latter's parliamentary position and support Banharn in his rivalry with the increasingly powerful Chavalit. In September Banharn was only able to secure the support of his coalition partners in a scheduled motion of 'no confidence' in his administration by undertaking to resign from the premiership to make way for another candidate. However, the inability of the coalition partners to agree on the appointment of a new Prime Minister led to the dissolution of the House of Representatives by the King on 27 September.

The general election, which was held on 17 November 1996, was characterized by extensive campaign violence and vote-buying on a massive scale. The NAP gained a plurality with 125 seats secured mostly through patronage politics in the impoverished rural north-east of the country. The DP, which secured 123 seats, appealed to the more sophisticated electorate, winning 29 of Bangkok's 37 seats, thus contributing to the reduction of Palang Dharma's representation to only one seat from 23 at the previous election.

The Chavalit Coalition

On 25 November 1996 the King formally appointed Chavalit as Prime Minister leading a coalition (comprising the NAP, Chart Pattana, the SAP, Prachakorn Thai, the Seritham Party and Muan Chon) that commanded a total of 221 seats in the House of Representatives. Initially, Chavalit's coalition appeared to offer the prospect of more stable government, after the months of incapacitating dissension between the parties of the Banharn coalition. Besides holding the Premiership, Chavalit also retained control of the powerful Ministry of Defence, a portfolio he had held in the previous Government. The key position of Minister of the Interior was awarded to Sanoh Thienthong, the former Secretary-General of Chart Thai, as a reward for Sanoh's support for Chavalit's bid for the premiership prior to the general election and his subsequent defection from the Chart Thai Party to Chavalit's NAP. Most other influential (and lucrative) ministries were awarded to NAP or Chart Pattana members. Gen. Mongkon Ampornpisit, the Supreme Commander of the Armed Forces, and Gen. Chettha Thanajaro, the Army Commander-in-Chief, were both Chavalit appointees; thus Chavalit was in a position of great control over the armed forces. This also gave Chavalit much influence in the sensitive and politicized process of annual promotions and appointments in the military.

As evidence of Chavalit's avowed intention to revive the economy, he allocated the finance portfolio and the deputy premiership with responsibility for economic affairs to Amnuay Viravan, a respected technocrat. However, Chavalit subsequently failed to give Amnuay the necessary support to overcome the obstacles imposed by other coalition members anxious to prevent the implementation of austere fiscal policies. The Government's indecisive handling of the economy contributed to a financial crisis in mid-1997, following a series of sustained attacks on the Thai baht by currency speculators. In June Amnuay resigned and was replaced by the little-known Thanong Bidaya, the President of the Thai Military Bank. At the end of July 1997 the Governor of the Central Bank, Rerngchai Marakanonda, also resigned, citing political interference. The deterioration in the state of the economy in 1997 was reflected in an increase in popular protests; in response, the Government introduced measures in June to constrain media attacks on the authorities and to limit mass protests.

At the end of August 1997 Chavalit and Thanong had an audience with the King to discuss the economic crisis. Later that month Chavalit implemented a cabinet reorganization in an attempt to regain public confidence. On the advice of the former Prime Minister, Prem, the former Minister of Finance, Virabongsa Ramangkura, an outspoken critic of the Government's handling of the economic crisis, was appointed Deputy Prime Minister with responsibility for economic affairs. Thaksin, who had resigned as leader of Palang Dharma, following the party's disastrous performance in the general election, joined the Cabinet as Deputy Prime Minister with responsibility for regional development and trade. Despite these changes the opposition threatened to introduce a motion of 'no confidence' in the Government.

For much of Chavalit's first year in office the major issue on the political agenda was the drafting of a new Constitution by the Constitutional Drafting Assembly, presided over by long-time political activist, Uthai Pimchaichon. The draft Constitution was promoted by its supporters (including the former Prime Minister, Anand Panyarachun) as providing the basis for a truly democratic system of government in Thailand. Many of the provisions sought to address the endemic problems of political corruption and vote-buying that had plagued Thai politics for decades. Some of the major amendments contained in the new draft Constitution included the reduction of the number of senators to 200; the introduction of direct elections for the Senate (in place of the system whereby senators were appointed by the Prime Minister); a clause requiring members of the National Assembly who became ministers to resign from the legislature; a requirement that the minimum level of education for a member of the National Assembly be a degree; a provision that 50,000 eligible voters could initiate investigations into corruption by members of the National Assembly; and restrictions on media censorship by the State. By September 1997 the draft Constitution had gained the support of all opposition parties and most of the print media. However, considerable opposition to the new charter came from within the Chavalit coalition and sections of the Senate. Chavalit himself reneged on his pledge to support the overwhelmingly popular draft, but then finally gave it his endorsement following the application of pressure by both business representatives and the armed forces.

Chavalit ensured his survival of a 'no confidence' motion tabled by the opposition by scheduling the censure debate prior to the vote on the draft Constitution. If defeated, Chavalit could thus dissolve the National Assembly rather than resign, delaying the vote on the draft Constitution and forcing fresh elections under the existing Constitution. In the event the Government won the 'no confidence' vote and the draft Constitution was passed on 27 September 1997. Chavalit was then prohibited by law from dissolving the National Assembly and was allowed 240 days to complete enabling legislation for the new Constitution, prior to a general election (although the King subsequently urged the accelerated completion of the necessary reforms). The new Constitution was promulgated on 11 October.

In October 1997, following the resignation of the Minister of Finance, Thanong Bidaya, Chavalit effected a reorganization of the Cabinet. However, unprecedented criticism of the Chavalit administration in the Thai media, based on the perception that the Government was out of its depth in handling the worsening economic crisis, as well as on opposition to the new draft Constitution, led in early November to Chavalit's resignation as Prime Minister. In the political manoeuvring that followed, the DP succeeded in persuading 12 members of Prachakorn Thai (later dubbed the 'Cobra faction' by the media) to defect and join the Democrats. With the support of Chart Thai and other minor parties, the DP was able to form a new eight-party coalition commanding the support of 207 members of the 393-member House of Representatives, which gave it the right to form a government.

The Government of Chuan Leekpai

On 9 November 1997 the leader of the DP, Chuan Leekpai, was appointed Prime Minister. The majority of the key portfolios in the new Cabinet were awarded to DP members. Despite some opposition from within the armed forces, Chuan himself assumed the post of Minister of Defence concurrently with the Prime Ministership, thus becoming the first non-military figure to occupy the position in recent history. The DP Secretary-General, Sanan Kajornprasart, assumed responsibility for the influential post of Minister of the Interior. The key Ministries of Finance and of Commerce were headed by the two former bankers, Tarrin Nimmanhaeminda and Supachai Panichpakdi respectively, while Surin Pitsuwan, also of the DP, became Thailand's first Muslim Minister of Foreign Affairs.

Thailand's political agenda in 1998 was dominated by the economy which, despite the change of government, showed little sign of improvement. The value of the baht continued to fall, reaching a new all-time low of 55 to the US dollar in January 1998. However, the new Government enjoyed an extended 'honeymoon' period, assisted by a supportive media, which perceived in the Chuan administration a more credible team of economic managers. Regular meetings between the IMF and Thailand's economic ministers resulted in the Government's acceptance of most of the IMF's demands for macro-economic reform in exchange for a badly-needed rescue programme of US $17,200m. As a consequence of Thailand's co-operation with the IMF, the Chuan Government received considerable international support. In March Chuan visited the USA and received strong praise from President Bill Clinton, as well as from other US political and business leaders, for the Government's handling of the economic crisis. The visit did much to repair relations with the USA (a long-standing ally of Thailand), which had become strained at the height of the financial crisis, when emphatic criticism of the USA's perceived abandonment of Thailand in the country's hour of greatest need had been voiced in the Thai press and by certain parliamentarians. The visit also afforded Chuan the opportunity to differentiate Thailand's economic predicament from the worsening situation in Indonesia, where President Suharto's intransigence towards the IMF's demands had led to social and political turmoil and was to result in his eventual resignation in May. Largely as a result of Thailand's adherence to the conditions of the IMF rescue programme, by April the currency had recovered to a rate of around 40 baht to the US dollar.

Throughout 1998 the austerity measures imposed by the Government in accordance with IMF recommendations began to take effect. Extensive redundancies of factory workers became regular occurrences. Meanwhile, government forecasts estimated that the number of unemployed would rise to 2m. by the end of 1998. Fuelled by impatience at the apparent slow progress of the Government in rectifying Thailand's economy, a perception grew in the Thai media that the administration was interested only in solving the 'problems of the rich'. Protesters from farming groups, mostly from the impoverished north-east, regularly rallied outside the National Assembly to criticize the Government's failure to address their problems. Increasing resentment towards the IMF, particularly in relation to the way in which the IMF appeared to be dictating Thailand's economic policy, was expressed in the press and intellectual circles; the IMF's policy of maintaining high interest rates, along with the pressure it was placing on the Government to privatize state enterprises, was perceived by many as an attempt to force Thailand to sell off its assets to foreign interests. A 'Thai Help Thai' campaign was launched by the Government to solicit contributions and donations from the public in an attempt to alleviate Thailand's foreign debt. Television commercials urged Thais to practise moderation in their spending habits. A 'Buy Thai' campaign encouraged Thai consumers to avoid foreign products and to purchase local ones instead, in order to improve Thailand's balance-of-payments problem. Despite these pressures, however, the Chuan Government continued to enjoy a relatively high degree of stability. In March the Government easily survived a censure debate led by Chuan's predecessor, Chavalit, which did more harm to the opposition's credibility than to that of the Government. Popular support for the Government was demonstrated at the elections to the Bangkok city council, which took place on 26 April, when the DP won 22 of the 60 seats, followed by the non-political supporters of the incumbent Governor, Bhichit Rattakul, who secured 20 seats. By August Thailand's economic situation appeared to have stabilized, although the future was to a great extent dependent on the moribund Japanese economy which, by mid-1998, was officially in recession.

New electoral legislation was passed in May 1998, enabling polls to be held six months subsequently. Despite pressure from the opposition, Chuan stated that elections would not be held until certain economic reforms had been put in place. Under the new legislation, however, it appeared likely that 12 of the members of the ruling coalition who had defected from Prachakorn Thai to help form the Government in November 1997 might lose their legal status as legislators, thus placing in jeopardy the Government's parliamentary majority. (The 12 were formally expelled from Prachakorn Thai in October 1998 but continued to form part of the coalition Government.) A new political party, Thai Rak Thai, was formed by the former Deputy Prime Minister, Thaksin Shinawatra, in July 1998.

In August 1998, following a change in the attitude of the Thai press towards Chuan from May, Queen Sirikit, in an unusual move, praised Prime Minister Chuan and urged support for the efforts of his Government to stabilize the economy. During the same period, however, the Government was afflicted by allegations of corruption. On 5 October Prime Minister Chuan reorganized his Government, bringing the Chart Pattana party into the ruling coalition. The inclusion of Chart Pattana reinforced the Government's majority in the House of Representatives, increasing the number of seats held by the coalition to 257. Five new ministers were appointed to the Cabinet, including the

Chart Pattana Secretary-General, Suwat Liptapallop, who was assigned the industry portfolio. Many of the most influential portfolios, however, were retained by the DP.

The widespread support that the Chuan Government had enjoyed in its first year began to diminish in its second year as public frustration grew over the depressed state of the economy. The Government was constantly accused in the media and in the National Assembly of 'selling out the nation' as a result of its economic policies, and a perception persisted that the Government was failing to deal decisively with a series of scandals. In December 1998 and January 1999 there was intense debate in the National Assembly and in the media concerning 11 'Economic Recovery Bills' introduced by the Government. Ostensibly designed to restore confidence in the economy, the bills were also intended partly to meet IMF recommendations for its restructuring. Among the most controversial aspects of the bills were new bankruptcy laws that made it much easier for creditors to prosecute debtors defaulting on loans. Another controversial bill allowed foreigners to hold shares in state enterprises. The Government came under strongest criticism from leading members of the Senate, most notably the upper house's President, Meechai Rachupan, who protested that the bills would lead to foreign control of the Thai economy. Senate opposition to the bills was not, however, entirely based on nationalistic motives. It was revealed that the bills would also affect the business interests of a number of influential senators. Despite such opposition, and under considerable pressure from the Government, the bills were eventually passed by the Senate in March 1999.

In December 1998 the opposition had filed a motion to remove Chuan and the Minister of Finance, Tarrin Nimmanhaeminda, from office, on the grounds that they had allegedly acted in breach of the Constitution by submitting four Letters of Intent to the IMF without first securing the approval of the House of Representatives; the motion was unsuccessful. In January 1999 a motion of censure was filed by the opposition against three DP ministers (Tarrin, the Deputy Prime Minister and Minister of the Interior, Maj.-Gen. Sanan Kajornprasart, and the Minister of Transport and Communications, Suthep Thaungsuban), who were accused by the opposition of corruption, nepotism, incompetence and dishonesty. As expected, all three ministers survived the vote on the motion; however, there was some concern that the censure debate could potentially undermine popular confidence in the Government.

The Government's standing was further damaged in December 1998 when the media uncovered evidence implicating Sanan in a land scandal in Kanchanaburi province. A more damaging issue for the personal reputation of the Prime Minister, who also held the defence portfolio, was the revelation, in March, that he had countersigned an army recommendation awarding Field Marshal Thanom Kittikachorn an honorary appointment as a Royal Guards officer. A military dictator of the 1960s and 1970s, Thanom was notorious for his role in the attempt violently to suppress the 1973 pro-democracy demonstrations in Bangkok in which some 80 (mainly student) protesters were killed. Despite considerable pressure from the media for the appointment to be withdrawn, Chuan declined to interfere on the grounds that it was an internal matter for the Armed Forces and that his signature was simply a matter of procedure; however, Thanom subsequently resigned from the position. In April 1999 Chavalit resigned as leader of the NAP; he was expected to stand for re-election to the post, however, in an attempt to outmanoeuvre his political rivals and increase his influence within the party in advance of the next general election.

In July 1999 intense infighting within the SAP concerning the allocation of cabinet positions forced the party to withdraw from the governing coalition. In the ensuing cabinet reshuffle Chart Pattana further strengthened its position. In mid-1999, meanwhile, campaigning by Thaksin Shinawatra's party, Thai Rak Thai, intensified in anticipation of an early election. Speculation that Thai Rak Thai was using financial inducements to recruit members of other parties was confirmed by the senior Democrat ministers, Chamni Sakdiset and Khunying Supatra Massdit, who opened a public debate on the issue, partly as an attempt to prevent support flowing from the Democrats to Thai Rak Thai. Despite such criticism, the media reported a groundswell of support for the new party, especially in the northern provinces (Thaksin being a native of Chiang Mai) and in November Chavalit pledged his willingness to support Thai Rak Thai in the pending general election in the event of his own NAP performing poorly at the polls. By August 1999, with public approval of the Government declining, the Democrats found themselves trying to forestall growing demands from opposition parties and numerous public figures to dissolve the House of Representatives and call fresh elections. In August 1999 a report on the state-owned Krung Thai Bank, which exposed the bank's dubious loan policies and other irregularities, was leaked to the Senate and the press. The report led to criticism of the Government for its handling of the bank, and also resulted in direct allegations of corruption being made against Tarrin, whose brother, Sirin Nimmanhaeminda, had been the President of Krung Thai Bank until his resignation in January. Although the Government was reported in October to have exonerated a number of senior Krung Thai executives, including Sirin, of allegations of inefficiency and dishonesty relating to the findings of the report, the controversy surrounding the bank continued adversely to affect the Government throughout November and early December; it constituted the main focus of a joint no-confidence motion filed by three main opposition parties against the Government in December, in which the ruling coalition was accused of mismanagement of the economy and of condoning corruption. As expected, the Government survived the censure debate, winning 229 votes against the opposition's 125. However, the debate exposed divisions within the governing coalition (particularly between the DP and Chart Thai), and also proved particularly damaging to Tarrin and to Sanan. Although polls of public opinion carried out in the early months of 2000 indicated that the DP-led Government continued to command a significant level of support, the DP suffered a reverse in the provincial elections held in February, possibly reflecting discontent amongst the rural population with the Government's refusal to subsidize crop prices.

On 4 March 2000 elections to the new 200-member Senate were held as required under the 1997 Constitution. The elections were contested by more than 1,500 candidates, and a voter participation level of 71.89% was reported. The power of the new Election Commission, which had been established under the reformist Constitution to oversee the elections, was demonstrated when it disqualified 78 of the 200 winning candidates on the grounds of either vote-buying or fraudulent conduct. The Election Commission subsequently announced that a second round of polling would be held in the 78 affected constituencies on 29 April, in which all but two of the disqualified contestants would be permitted to stand again. A total of 66 senators were duly elected, including more than one-half of those disqualified after the first round of polling. The Election Commission announced that a third round of polling would be held in June to elect senators to the 12 remaining seats, and rejected calls for the dismissal of the allegations of fraud in the interest of political continuity. (While the final results of the elections remained in dispute, none of the declared winners were able to assume their seats as they lacked the necessary quorum, resulting in a backlog of legislation and obliging the outgoing Senate to remain in office in an interim capacity.) The Election Commission also announced that a number of prominent individuals were likely to be prosecuted for their involvement in electoral misconduct. While the insistence of the Election Commission on the holding of a further round of polls led to fears that the recent constitutional reforms were perhaps overzealous and could result in the paralysis of the legislature, the disclosures of electoral fraud were welcomed by many as demonstrating the Government's commitment to the elimination of political corruption. However, it was also recognized that the disruption of the legislative process caused by the delay in the appointment of the new Senate could have implications for the timing of the pending general elections, as a number of bills scheduled to be passed before the Senate were thought likely by many observers to be critical to the electoral prospects of the ruling coalition. On 4 June the third round of polling was duly held, which resulted in the election of a further eight senators, although voter participation was officially registered at a mere 41.28%. Further allegations of electoral fraud necessitated fourth and fifth rounds of polling on 9 and 22 July respectively, both of which attracted voter turn-outs of some 31%. On 27 July the last of the 200 members of the Senate were finally endorsed, and on 1 August the new Senate was officially sworn in. A significant feature of the composition of the new Senate was the greatly reduced proportion of senators coming from military and bureaucratic backgrounds, and increased numbers from academia, the media and NGOs.

Meanwhile, in March 2000 the DP was set back by the findings of the National Anti-Corruption Commission's (NACC) investigation into Deputy Prime Minister, Minister of the Interior and Secretary-General of the DP, Sanan Kajornprasart, which, in a watershed decision, indicted Sanan on charges of falsifying his assets statement. Sanan immediately resigned from his ministerial positions and was replaced by Banyat Bantadtan; Sanan's position as Secretary-General of the DP was eventually taken by Anant Anatakul, a former permanent secretary in the Ministry of the Interior, following the former's conviction, in August, of corruption charges by the Constitutional Court, which resulted in a five-year ban from holding political position. While the NACC's findings damaged the credibility of the DP, Sanan's resignation also spared the party the difficulty of facing a new election with Sanan, whose tarnished image was increasingly seen as a liability. The Government's difficulties were, however, ameliorated by the growing disarray in the ranks of the core party of the opposition, the NAP. In March the party was hit by a scandal in which Surasak Nananukkol, the NAP's chief economic adviser and a former Deputy Minister of Finance, was arrested in the USA on charges of involvement in an oil smuggling deal with Iraq, which violated UN sanctions on Iraqi oil exports as well as US law. Furthermore, rumours abounded that in the run-up to the next election large numbers of NAP deputies, including the influential Wang Nam Yen faction controlled by Sanoh, were preparing to desert Chavalit's NAP in order to join Thai Rak Thai. In June the NAP leadership took the step of urging NAP deputies to resign from their seats in the House of Representatives in a bid to force the Government into calling an early election. The controversial strategy revealed the deep divisions within the NAP, with at least 20 NAP deputies refusing to take part in the mass resignation (although almost 100 did resign), including former NAP Secretary-General Chaturon Chaisaeng. The NAP's mass defection had little effect on the DP-led coalition, however, which now faced a greatly reduced opposition in the House of Representatives. In July a second opposition party, the SAP, also resigned from the House of Representatives.

The turmoil within the NAP was taken advantage of by Thai Rak Thai, which was emerging as the DP's likely major rival in the next election. However, Thai Rak Thai was also faced with its own difficulties. Its candidate in the Bangkok gubernatorial election, Sudarat Keyuraphan, was decisively defeated by the veteran leader of Prachakorn Thai, Samak Sundaravej. In the largest ever electoral turn-out for a gubernatorial election (58.87% according to official figures) Samak became the first candidate in the history of the election to poll over 1m. votes. The result was interpreted by many observers as a set-back to Thai Rak Thai's hopes of winning seats in Bangkok in the forthcoming legislative elections. Thai Rak Thai's image continued to be damaged by ongoing criticism in the media that it was offering financial incentives in order to recruit deputies from other parties. Several high profile deputies, including SAP leader Suwit Khunkitti and Chart Pattana's Yingphan Manasikarn, confirmed that they would run under the Thai Rak Thai banner in the next legislative elections. Concern had also been raised over Thaksin's controversial acquisition in late May of a significant stake in a television station that had been established following the pro-democracy demonstrations in 1992 as a supposedly independent media organization. In September it was announced that Thaksin was to be investigated on possible charges of corruption.

In November 2000 Prime Minister Chuan Leekpai ended months of speculation by dissolving the legislature and setting 6 January 2001 as the date for an election to the House of Representatives, the first under the 1997 Constitution. The announcement came at a time when the Government was suffering from rapidly declining public support as a result of the continued stagnation of the Thai economy, a perception in the media that the Government's policies were addressing only the problems of major businesses, and ongoing criticisms that the Government had capitulated to the demands of the IMF and the foreign investment community at the expense of national interests. However, Thai Rak Thai, which in the short time of its existence had emerged as the DP's major rival, was faced with an internal problem in the form of a major corruption scandal involving its leader, Thaksin. After a lengthy investigation, in December Thaksin was indicted by the National Counter Corruption Commission (NCCC) on charges that he and his wife had violated assets-disclosure regulations (laid out in the new Constitution) by deliberately concealing shares worth thousands of millions of baht. The NCCC alleged that Thaksin had transferred the shares to his household staff. Refusing to accept the verdict, Thaksin launched an immediate appeal to the 15-member Constitutional Court, another so-called 'independent body', like the NCCC, established under the new Constitution. If found guilty, Thaksin faced the prospect of being banned from politics for five years.

The Thai Rak Thai election campaign was characterized by its sophisticated use of the mass media, which portrayed Thai Rak Thai as the party of 'the new generation', summed up by its election slogan, 'New Ideas, New Actions'. Thaksin was represented as a successful business executive who would bring his skills from the corporate world to government, in contrast to Chuan, who was depicted as indecisive, ineffective, and mired in bureaucratic process. Thai Rak Thai also made a series of populist election promises, including: development fund grants of 1m. baht for each of Thailand's 77,000 villages in a bid to stimulate the rural economy; a public health programme that would see Thais pay just 30 baht for each hospital consultation; and a three-year debt moratorium for Thailand's indebted farmers. The DP's campaign focused on highlighting the Government's record of guiding the Thai economy out of the crisis of the late 1990s and its implementation of vital economic reforms, while at the same time accusing Thaksin of trying to buy votes at the election in the same way that he had used his enormous financial resources to recruit politicians to Thai Rak Thai.

The election result was an overwhelming victory for Thai Rak Thai. The final outcome (following a second round of elections in a number of constituencies on 29 January 2001 as a result of voting irregularities in the earlier poll) gave Thai Rak Thai 248 seats, almost double that of the party with the next highest number of seats, the DP, which secured 128.

The Government of Thaksin Shinawatra

Following Thaksin's successful negotiation of a merger between the Seritham Party and Thai Rak Thai, the party became the first in Thai history to hold an absolute majority in the House of Representatives. To further consolidate its position, Thai Rak Thai elected to form a coalition with Chart Thai and the NAP to give the Government an ample majority of 339 seats. One of the most significant outcomes of the election was the eclipse of a number of long-standing provincial political dynasties, as a result of electoral rules set out in the new Constitution that were designed to decrease the influence of smaller parties in the interests of more stable government. Thaksin impressed his dominance on the coalition Government by filling 27 out of the 36 Cabinet positions with Thai Rak Thai appointees, including many of the key ministries. Given his substantial personal financial assets, his extensive media interests, his control of the Thai Rak Thai party, and Thai Rak Thai's dominance of the coalition Government, Thaksin was thus in a more powerful political position than any previous civilian Prime Minister.

From the start, the new Government emphasized its nationalist credentials and distanced itself from the previous administration, which had been widely portrayed in the media as being too accommodating with regard to 'the West'. The new Minister of Foreign Affairs, Surakiart Sathirathai, announced a redirection of Thailand's foreign policy to focus on strengthening economic relations with Thailand's Asian neighbours, dismissing his predecessor's perceived preoccupation with broader issues of human rights and democracy, which, while winning praise from Western nations, had aggravated numerous governments within ASEAN. In a speech in February 2001 Thaksin declared that Thailand must, 'cease being a slave to the world', and abandon policies that worked against 'Thai interests'. In another controversial and widely reported address to the UN

Economic and Social Commission for Asia and the Pacific (ESCAP) in April, Thaksin appeared to indicate that Thailand would be following a more inward-looking, self-sufficient economic policy, and would give more support to 'small and medium enterprise industries'. The Prime Minister also discussed the importance of promoting greater economic integration among Asian countries. Following adverse reports in the Western media, Thaksin was later at pains to assure the international business community that Thailand would maintain an open economy and continue to welcome foreign investment.

Early in Thai Rak Thai's tenure of office, concerns were raised about apparent attempts by the Government to limit the powers of a number of the so-called 'independent bodies' established under the 1997 Constitution. Following a decision by the Election Commission (one of the most active of these bodies) to disqualify 10 Senators for alleged violations of electoral laws, the Government mounted a legal challenge to limit the Commission's extensive powers. The NCCC similarly came under pressure from the Government after its investigation into the Thaksin share concealment case. Three members of the NCCC were accused by the Government of violating constitutional regulations, which eventually resulted in the resignation of the leader of the investigation. Thaksin also expressed his intention of reviewing the powers of the Constitutional Court by making amendments to the Constitution. In May, after weeks of tension over interest rate policy, Thaksin dismissed the Governor of the central bank, Chatu Mongol Sonakul, and replaced him with Pridiyathorn Davakula. Despite the Prime Minister's insistence that the dismissal was primarily due to personality differences, concerns were raised about government interference in the independence of the Bank of Thailand. The Thai media were also targeted by the Government. Accusations were made that the Government was attempting systematically to cajole the media into presenting the Government in a favourable light. These accusations were strengthened considerably when one of Thai Rak Thai's own members of the House of Representatives, the maverick Piyanat Wacharaporn, publicly expressed concerns about government interference in the media, which prompted a sharp rebuke from the Prime Minister.

In early August 2001 months of uncertainty over Thaksin's political future finally ended when the Constitutional Court controversially acquitted him of corruption charges by an eight-to-seven ruling. The decision came after campaigns of public support for Thaksin had put pressure on the Court to return an acquittal verdict, while numerous Thai Rak Thai politicians had warned that a guilty verdict would have an adverse effect on the Thai economy, and could even lead to political violence. While the Court's decision received considerable public support, concern was expressed in certain quarters that a number of the judges might have been placed under pressure, or even bribed, into returning a verdict of acquittal. The allegations were sufficiently serious to prompt the Senate to begin an inquiry into the Court's decision.

Thaksin's increasing control over the political process was strengthened in April 2002, when the long-expected merger of the NAP, which had been a major force in Thai politics during the 1990s, with Thai Rak Thai finally took place. NAP leader Chavalit, however, retained his post as Minister of Defence and Deputy Prime Minister. Thai Rak Thai's position in the House of Representatives was further consolidated with the defection of Chart Pattana from the opposition to the Government, giving Thai Rak Thai 368 seats, out of a total of 500, with the DP, the next largest party in the House of Representatives, commanding only 129. Not only did the merger of the NAP and the defection of Chart Pattana reinforce Thai Rak Thai's dominance over the opposition, but it also gave Thaksin increased authority over factions within his own party, especially the Wang Nam Yen group of 50 legislators, controlled by veteran politician Sanoh Thienthong. The Government's increased majority enabled it to survive easily a three-day no-confidence debate in May 2002. The Prime Minister also extended his influence over the military with his appointment of Gen. Somdhat Attanand, known to have close links with the Thai Rak Thai party, to the key post of Commander-in-Chief of the Army during the annual military reshuffle, which took place in August 2002. In addition, the Government moved to strengthen its control over the bureaucracy, completing a major reform of the sector by October 2002. This included a reorganization of ministries and departments, as well as an increase in the existing number of ministries, from 14 to 20. The principal feature of the provincial administration reforms was the creation of what were dubbed 'CEO governors' through a strengthening of the powers of the provincial governor over the myriad of bureaucratic agencies at provincial and local government level. The policy was, however, criticized by the opposition and sections of the media as being an apparent reversal of the policy of decentralization favoured by the previous Democrat administration, a claim denied by the Government.

Thai Rak Thai's remarkable dominance of the House of Representatives, and Thaksin's often abrasive style of leadership, provoked accusations from the political opposition, sections of the print media, numerous academics and even members of his own party, that Thaksin was acting like a 'dictator'. In a rare public rebuke, the King singled out the Prime Minister for criticism in his birthday address on 4 December 2001, warning that Thailand was heading for 'catastrophe' and calling for increased unity amongst all politicians. The King's comments led to a sharp decline in the Prime Minister's popularity in the opinion polls and encouraged further criticism in the media. Thaksin also ran foul of the Western news media, owing partly to statements (which he had made regarding Thailand's economy soon after becoming Prime Minister) that were interpreted as being anti-Western in orientation. A report in the Hong Kong-based news weekly, the *Far Eastern Economic Review*, in March 2002, concerning the apparent tensions between the Government and the Palace, led to moves to have the two journalists involved in the publication of the article expelled from the country, a decision that was later reversed following negotiations between the Government and the *Review*. In the same month an issue of another leading current affairs weekly, *The Economist*, containing a special feature on Thailand, was banned from circulation in the country, ostensibly for portraying the Thai royal family in an unfavourable light. While both the domestic and international media protested against what they perceived to be direct attacks on the freedom of the press, Thaksin was able to gain some domestic support for the Government by representing his actions as constituting a defence of the monarchy against foreign criticism. The domestic media were also targeted by the Government. In March 2002 the Nation Multimedia Group, a publisher of newspapers in both Thai and English, as well as a broadcaster of television and radio programmes, which had been one of Thaksin's most prominent critics, was warned by the Ministry of Defence to delete political content from one radio programme after it aired an interview with an opposition deputy critical of the Government. The Nation Multimedia Group was also at the centre of another controversy when senior members were placed under investigation by the Anti-Money Laundering Office. Following an outcry in the media, the investigation was eventually abandoned. To counter what it perceived to be inaccurate media reporting, the Government initiated the establishment of a new agency intended to implement guide-lines to be followed by the state-controlled media. In October 2001 Shin Corporation, the telecommunications conglomerate which Thaksin had founded, and in which his family retained a controlling interest, increased its share in ITV, the only non-state-owned television station, from an initial 25% (which it had acquired in 2000), to 77%. There were even accusations, strongly denied by Thaksin, that the Government was using advertising accounts worth thousands of millions of baht to reward media sources that were sympathetic towards the Government. Thus, during its second year in office, the Government appeared to be complementing its dominance of the legislature with an increasingly effective control over the media.

Despite these criticisms, the Thai Rak Thai-led Government managed to maintain a relatively high level of popularity. A prominent 'social order' campaign, implemented by the Minister of the Interior, Purachai Piumsombun, one of the Prime Minister's close associates and, like Thaksin, a former police officer with a doctorate in criminology, was publicly well received. The campaign involved an uncharacteristically stringent nationwide campaign against illegal entertainment establishments. Entertainment venues violating zoning laws were closed down, opening and closing times were strictly enforced, and under-age youths were prohibited from attending such venues. Despite lobbying from the night-time entertainment industry, from members of the House of Representatives who belonged to his own party—including the powerful Sanoh Thienthong—and from sections of the military with links to the industry, the policy remained in place.

The Government's fortunes were also improved by an apparent recovery in Thailand's economic prospects. Overall, in 2002 the economy expanded at a robust rate of 5.4%, despite the depressed state of the US economy and the disruptions to the world economy that had occurred in the aftermath of the terrorist attacks on the USA on 11 September 2001. In July the Government conducted negotiations with Malaysia and Indonesia over the establishment of a rubber cartel, which would control rubber prices on the world market. Not only did the cartel promise to improve incomes for Thailand's rubber producers, it also had the potential to enhance the Government's popularity in the DP's heartland of southern Thailand, where most of the country's rubber continued to be produced.

One of the opposition's major criticisms of the Thaksin Government (and one that had been raised in the no-confidence censure motion debated in May) was that the Prime Minister was allegedly using his public office to further his private business interests. In August 2001 the House of Representatives approved legislation reducing the limit on foreign ownership of telecommunications companies from 49% to 25%. While the decision was presented in nationalist rhetoric concerning the need to protect Thailand's national interests from foreign influence—a theme that Thai Rak Thai had successfully exploited during the election campaign—the decision directly benefited Shin Corporation, the major telecommunications organization controlled by the Shinawatra family, which continued to dominate the domestic telecommunications market. Several of the accusations of a conflict of interest involved Shin Corporation's satellite services branch, Shin Satellite, or 'Sattel'. In February 2002 the Prime Minister made a sudden visit to India, where it was alleged that he negotiated a deal between Sattel and India's Department of Space. Sattel already had a substantial stake in India's satellite cable television market. In May 2002 Sattel announced a US $13m. deal with the Myanma government agency, Bagan Cybertech (headed by the son of Gen. Khin Nyunt, then First Secretary of the ruling State Peace and Development Council (SPDC), the military junta of Myanmar) to provide technology and satellite services for a telephone network in rural Myanmar. Sattel was also preparing to launch a new-generation broadband satellite system, 'iPSTAR', which would provide most of the Asian continent with high-speed telecommunications, internet and multimedia services at a fraction of the current costs. Sattel had already been awarded a contract by the Thai Government to equip 40,000 state schools and educational institutions with iPSTAR technology. According to company projections, iPSTAR would provide US $300m. in annual earnings for Sattel. Given Shin Corporation's interests in Sattel, the success or otherwise of its iPSTAR venture, and, indeed, of the company's other regional business interests, was expected to have direct implications for Thaksin's political fortunes.

Thai Rak Thai's domination of Thai politics was greatly facilitated by internal divisions within the DP, the most significant opposition party, following long-term leader Chuan Leekpai's announcement that he would stand down from the position in early 2003. Chuan's reluctance to name a clear successor split the party between supporters of Chuan's young protégé Abhisit Vejajiva, a talented performer in the House of Representatives and widely popular with the electorate, and the seasoned but uncharismatic deputy from Surat Thani, Banyat Bantadtan, who received the crucial support of the DP's powerful former Secretary-General, Sanan Kajornprasart, currently serving a five-year ban from politics on charges of falsifying his assets statement. In April 2003, in a three-way ballot for the leadership joined by outsider Arthit Urairat, Banyat emerged the victor by a narrow margin, with Abhisit assuming the post of Deputy Leader. Pradit Pattaraprasit, a close associate of Sanan, was duly appointed to the key position of Secretary-General. Despite the change in leadership, consistent rumours in the national media of party infighting and fears of a mass defection in the House of Representatives of DP members to Thai Rak Thai in the run-up to the next election, due to take place in 2005, continued to damage the party, already disadvantaged in terms of its access to the media. In May a 'no confidence' debate led by the DP against the Government, a traditional tactic used by the opposition to exploit divisions within governments in the hope of gaining defections, was one of the least effective in recent years. The debate was hampered mainly by the overwhelming number of Thai Rak Thai members of the House of Representatives, a result of the merging of other parties into Thai Rak Thai. Moreover, Thai Rak Thai was threatening to overwhelm the DP's powerbase in the south of the country, where it held every seat apart from those in the predominantly Muslim border provinces. The Government's standing in the south, home to most of Thailand's rubber producers, was greatly boosted by its successful policy of maintaining the rubber price at over 30 baht per kg. Prices for palm oil fruit and raw coffee beans, two of the south's other major exports, had also been considerably higher in recent years. On a number of visits to the south Thaksin had declared that the region would be one of the party's main targets in the next election.

Thai Rak Thai's political control was increased further by a number of other developments. In October 2002 the Government made new appointments to the Constitutional Court, the body responsible for almost ending Thaksin's political career in the previous year. The Government also attempted to increase its influence over the nominally independent Senate by attempting to bring about the removal of its President, Maj.-Gen. Manoonkrit Roobkajorn, a friend of Sanan Kajornprasart and one of the Government's most outspoken critics. Meanwhile, the Prime Minister sought to eliminate the possibility of a split in the ruling coalition by holding negotiations with Chart Pattana, the largest party in the coalition after Thai Rak Thai, over an intended merger with Thai Rak Thai scheduled to take place before the next election. Following the election of Suwat Liptapanlop, Chart Pattana's main financial supporter, to the party leadership in February, Thaksin publicly praised Suwat as a potential future prime minister.

Strong economic growth contributed to the Government's continued popularity in the opinion polls. Despite the commencement of the US-led campaign to oust the regime of Saddam Hussain in Iraq, and the regional epidemic of Severe Acute Respiratory Syndrome (SARS), which had a devastating effect on Thailand's tourism industry in the first half of 2003, the economy was on target to expand by up to 6% over the year, making it one of the strongest economies in Asia. In the first half of the year the Stock Exchange of Thailand was the best performing in the region. In July Prime Minister Thaksin announced that the Government had made the final repayment of the IMF emergency loan package negotiated during the 1997–98 economic crisis, thus liberating the Government from the economic policy directives imposed by the IMF as a condition of the loan.

Thai Rak Thai's management of disputes arising from a number of controversial major development projects also tested the party's strategy in dealing with another source of potential opposition, the so-called 'social forces', a loose network of disaffected villagers, NGOs, academics, and their supporters in the media. The movement had been brought together through the common goal of opposing state-sponsored development projects lacking in local consultation and financial transparency, involving blatant corruption and environmental destruction, and with often devastating effects on the local communities most affected by the projects. Since the democracy demonstrations in the early 1990s, the influence of this movement on national politics had grown considerably, and indeed had received a certain degree of legal protection through provisions included in the 1997 Constitution. Plans to build a power-generating station near the picturesque coastal villages of Bo Nork and Hin Krut in Prajuab Khirikhan province—a project mired both in accusations of corruption and concerns for its environmental impact, and which had become a celebrated cause in the media—were eventually postponed for three years by the Government. At the same time demands by local residents affected by the Pak Mun dam in the north-east of the country, a long-running dispute which had been paramount in highlighting the social injustices created by state development projects, were partially met by

Thaksin's decision to open the dam's sluice gates for four months of the year to allow the devastated fishing industry, the major livelihood of villagers affected by the dam, to recover. However, the Government continued with the gas pipeline project it had planned for the Chana district of Songkhla province, part of a joint development project between Thailand and Malaysia. The project had prompted huge protests from the local community, angered by the process through which the project had been conceived, with little local consultation and with no apparent concern for its environmental effects. Meanwhile, Deputy Prime Minister Chavalit Yongchaiyudh resurrected a long-delayed project to build a shipping canal across the southern Thai peninsula by awarding a US $50m. contract to a Hong Kong-based company to conduct a feasibility study.

In February 2003 the Government launched its 'war on drugs'. While the country's drugs problem had been recognized as a serious social issue since the mid-1990s, this campaign was unprecedented both in its severity and in the degree of co-ordination that it required among different government agencies. Police departments in each province were required to compile lists of known and suspected drugs-traffickers and drug addicts. The latter, with the support of their families, were required to undertake rehabilitation programmes, while suspected traffickers were given a period of grace in which they were to promise to abandon the drugs trade or face the consequences. Within days of the announcement of the campaign, 'small-time' traffickers were being killed all over the country, either murdered by more important figures in the drugs trade to silence them before police interrogation, or victims of 'extra-judicial killings' carried out by the police, ostensibly in self defence. Within a few months of the start of the campaign, more than 2,000 people had died in drugs-related killings. Alarmed by the excessive violence associated with the offensive, the UN High Commissioner for Human Rights, as well as the US Government, publicly criticized the Thai Government's handling of the campaign. In February Thailand's own human rights 'watch-dog' expressed its concern over the killings, and was immediately branded by the Government as 'unpatriotic'. However, despite the domestic and international criticism, the campaign met with considerable public support.

In May 2003 the Government announced a new policy aimed at suppressing so-called 'persons with influence', a euphemism used to describe mafia figures who wielded significant economic and political power, particularly in the provincial regions, and who often enjoyed local support. The Ministry of the Interior compiled lists of 'persons with influence' in 15 categories, including: drugs dealers; extortionists; those involved in illegal gambling and prostitution; hired killers; arms dealers; and local contractors who used intimidation to win government contracts. However, the campaign was criticized by the DP as a ploy to destroy Thai Rak Thai's political enemies, especially after the name of the brother of senior DP member Jurin Laksanavisit appeared on one of the lists.

In July 2003 the Thai public was captivated by a scandal involving the wealthy owner of a number of massage parlours in Bangkok, Chuwit Kamonvisit, and the police department. Chuwit alleged that he had been kidnapped, drugged and left on a Bangkok highway. Following the incident, he held a press conference at which he alleged that he had made regular monthly payments of 13m. baht to local police stations in order to be allowed to operate his business freely. As Chuwit made further allegations of bribery against the police, the incident escalated into a major scandal that received prominent media coverage and had repercussions throughout the entire police department. Prime Minister Thaksin, himself a former police officer, pledged that the officers involved in the scandal would be relocated or dismissed entirely from the police force, and that major reforms would be carried out to rid the force of the corruption endemic within it.

From early 2004, for the first time since coming to power in 2001, the Government's apparent invincibility was shaken as Thaksin and his Thai Rak Thai party suffered several serious set-backs. The media began to speculate that the Government's fortunes had reached their peak and that it was now on a 'downward slide'. In December 2003, in his annual birthday speech, the King once again chastised the Prime Minister, effectively admonishing him for the manner in which the 'war on drugs' had been conducted and urging the establishment of an investigation into how almost 3,000 people had died during the three-month campaign. Tensions with the Palace continued into 2004 and in March the head of one of the King's charities spoke publicly of the King's concern over 'integrated' corruption at the highest levels.

Meanwhile, the Government suffered an unexpected set-back to its plans to privatize the powerful state enterprise, the Electricity Generating Authority of Thailand (EGAT). The plan provoked strong opposition from the EGAT Labour Union (and some sections of the management), which proceeded to organize a series of mass rallies. The Union's campaign played on the public's concern that not only would the privatization lead to higher energy costs for consumers but also that it would mainly benefit government associates, as had been the case with the privatization of the government-owned petroleum corporation, the Petroleum Authority of Thailand (PTT), in 2001. On this occasion individuals close to the Government acquired substantial numbers of shares, which then proceeded to increase rapidly in value. The campaign appeared to be successful. Having earlier indicated that it was determined that the privatization of EGAT should proceed, if necessary through the use of executive decree, the Government uncharacteristically backed down and indefinitely delayed its plans. Moreover, it subsequently declared that future privatizations would have to conform to regulations preventing the acquisition of shares by government associates, an apparent acknowledgement of the public's concerns.

A second crisis occurred in the form of an outbreak of avian influenza. The virus had earlier affected China and Viet Nam, resulting in the deaths (both from the disease itself and from the resultant mass culls) of millions of chickens. The disease had also led to a number of human deaths. Despite the deaths of a significant number of chickens in Thailand in late 2003, the Government refused to acknowledge publicly that the disease had entered the country. It feared that its discovery would devastate the country's poultry industry, and, in particular, have a negative impact upon the fortunes of the major agri-business company, Charoen Pokphand Foods Public Co Ltd, which was the country's biggest producer and exporter of poultry. In January 2004, in the face of overwhelming evidence, the Government was forced to admit that the virus had indeed reached Thailand and ordered the culling of millions of birds in the affected areas. Not only did the virus have a massive impact on the country's chicken farmers—Thailand is a major exporter of chicken meat—but it also adversely affected the country's vital tourism industry, as tourists avoided travelling to countries known to be affected by the virus. As a result of the crisis the Government was forced to revise downwards its forecast for economic growth for the year.

However, in 2004 the major crisis facing the Government was the escalation of violence in the Malay-Muslim majority provinces in the far south of the country. The region had a long history of unrest and of nurturing movements for autonomy or separatism since its incorporation into the modern Thai state as a result of the Anglo-Siamese Agreement of 1909. However, in more recent years it had appeared that the region was finally being integrated peacefully into the Thai nation-state, owing to the adoption of a more religiously and culturally sensitive approach by government officials, as well as the process of democratization that had taken place in the country since the 1990s. The climate appeared to have changed, however, since the launching of the US-led 'war on terror' in 2001 and the consequent invasions of Afghanistan and Iraq. Both events provoked major protests by Muslims in the south of Thailand and calls for boycotts of US products. In 2003 Thailand began to demonstrate its increasingly close alignment with the USA. In October it accepted a US request that it provide 450 troops to contribute to the Allied occupation force in Iraq. In the same month the Government announced that it had been designated a 'major non-North Atlantic Treaty Organization (NATO) ally' by the USA, a status that would strengthen links between the Thai and US armed forces and give Thailand privileged access to US military intelligence and 'state of the art' weaponry. As the security situation in Iraq deteriorated, the wisdom of the Government's decision was questioned, especially after two Thai soldiers serving there were killed in an attack. Another factor that contributed to the growing unrest in the south was the Government's decision in 2002 to dismantle the military-led southern security council, and transfer responsibility for security matters to the police force. The decision seemed to have provoked dissent within both the armed forces and local business interests allied to them.

In January 2004 the continuing unrest in the region came to a head when it was reported that a raid had been carried out on the Rajanakharin military camp in Narathiwat province. Four soldiers were killed in the attack and over 300 weapons were seized. The Government responded by declaring martial law in all three southern provinces of Pattani, Yala and Narathiwat. However, attacks on police and government buildings and arson attacks on schools only increased. In one incident in March, 38 government posts were simultaneously attacked, only three days after a cabinet meeting had been held in the area. Killings of police officers, local government officials and villagers occurred on a daily basis. In one incident three Buddhist monks were killed by assailants wielding machetes, signalling a dramatic change in the nature of the violence, since militants had never previously targeted religious figures. Schools were a common target for the militants. In January, following an attack on two students in Narathiwat province, over 1,000 state-run schools in the region were forced to close temporarily owing to fears for the safety of the students. For their part, local Muslim groups accused the security forces of abducting and murdering villagers. The Government appeared helpless to stop the violence and was internally divided as to both its basic causes and responsibility for it. Thaksin initially blamed gangsters, local politicians and drugs traffickers, and categorically denied that it was in any way connected to international terrorism. However, public statements made by the military and the Minister of Defence appeared to contradict the Prime Minister. They attributed the violence to separatist groups, which had regrouped in the area. The spreading unrest also affected Thailand's regional relations, as Malaysia angrily denied an accusation made by the Prime Minister in February that the militants were using Malaysia as a base to launch their raids into southern Thailand.

In mid-March 2004, frustrated at the ongoing violence, Thaksin dismissed the Minister of the Interior, the Minister of Defence, the Chief of Police and the Army Commander of the Fourth (southern) Region. In the same month a prominent Muslim human rights lawyer, Somchai Neelaphaijit, who was defending three suspected members of the regional terrorist organization Jemaah Islamiah, disappeared several days after accusing the security forces of using torture on suspected Muslim militants. He was assumed to have been murdered. As events in the south seemed to be advancing out of control, Thaksin sent Deputy Prime Minister Chaturon Chaisaeng, a former student activist and pro-democracy campaigner who was regarded as a 'dove' in the Cabinet, to meet with Islamic leaders in the south. He returned having agreed a seven-point plan to end the violence, involving: the lifting of martial law; an amnesty for militants not directly involved in the killings of state officials; ending the security forces' intimidation of the *pondok* religious schools; and the gradual replacement of state officials and teachers in the region with locals. However, the plan was not received favourably by the military, which protested over the Deputy Prime Minister's accusations of its role in the torture and abduction of villagers, and the plan was eventually abandoned.

On 28 April 2004 the months of intensifying violence in the south of the country came to a climax. In pre-dawn raids militants simultaneously attacked 11 police posts in the three provinces. The security forces appeared to have been prepared for the attacks, responding by killing 107 of the attackers and capturing 17. Five security officials died in the fighting. In the bloodiest incident commandos stormed the ancient, culturally significant Kruese Mosque in Pattani province, killing all 32 militants who had retreated there. National television captured images of the military firing heavy weaponry into the mosque after the occupants refused to surrender. Aware of the damaging symbolism of the attack, the Government immediately ordered the transfer of the officer who had ordered the attack on the mosque. The suppression of the raids was the bloodiest single day of violence in recent Thai history. Subsequent media polls, however, showed overwhelming public support for the extreme measures the Government had taken against the militants. Yet the question of who had organized the attacks, the involvement of separatist organizations and the links, if any, with international groups remained unclear.

Meanwhile, another issue that began to gain increasing media attention (despite the constraints within which the media now operated) was the continuing conflict of interest between the Prime Minister's business interests and government policy. Thaksin's Advanced Information Service mobile telephone operating company, which had benefited from government legislation in 2001 limiting foreign ownership in the telecommunications sector, controlled two-thirds of the market by 2003. In February 2004 the Government approved a renegotiation of the operating conditions of ITV, the television station in which the Prime Minister's Shin Corporation had acquired a controlling stake. The licence fee the station was required to pay to the Government was reduced to a fraction of its former level. Moreover, the station, which had built its reputation on strong and independent news reporting, was permitted to increase the level of entertainment programming it aired during 'prime time' from 30% to 50%. In the months following the announcement of the amended operating conditions, ITV's share price surged by nearly 600%, almost 10 times higher than the average stock market return. Another case involved the deregulation of the low-cost air travel industry, following which the Shin Corporation was granted a licence jointly to operate a new budget airline, Air Asia. The venture also benefited from a decision by the Airports Authority of Thailand to reduce the docking fees for the new airline by 50%. Shin Corporation also gained from the Government's new policy of promoting consumer credit by launching its own credit finance company, Capital OK Co Ltd. Meanwhile, Sattel was granted tax concessions for its iPSTAR venture and other incentives worth an estimated 16,400m. baht over eight years. The Shin Corporation's total revenues for the year 2003 amounted to 90,000m. baht, a figure which was predicted to increase to 200,000m. baht by the end of 2004, and double in the year 2005. In 2003–04 Shin Corporation's annual profit increased by 84%.

Politically, Thai Rak Thai continued to expand its influence in 2004. It was by far the best performing party in the inaugural Provincial Administrative Organization elections held nationwide in March 2004. These directly-elected local government bodies had been created as part of the 1997 Constitution's mandate to promote political decentralization. When fully operational, they would be financed by a substantial 35% share of the national budget. Teams presented under the Thai Rak Thai banner won in 47 provinces, with the DP coming a distant second, winning in only 13 provinces, most of them located in the south. Further mergers of minor parties swelled the ranks of Thai Rak Thai members of the House of Representatives, in preparation for the next general election, to be held by January 2005, which Thaksin had earlier predicted Thai Rak Thai would win by 'an avalanche'. According to the Constitution, changes of allegiance between parties had to take place not less than six months prior to an election. In November 2003 Korn Dabbaransi, head of Chart Pattana, resigned from his post as party leader and joined Thai Rak Thai, taking with him 15 other Chart Pattana members. In July 2004 there was a mass defection of 23 Chart Thai members who left to join Thai Rak Thai, decimating the membership of Chart Thai, Thailand's second oldest party. In the same month, after much negotiation, Suwat Liptapanlop, the leader of Chart Pattana, and the party executive decided to disband the party founded by Chatichai Choonhavan and merge with Thai Rak Thai. The merger brought the remaining 28 members of the House of Representatives belonging to the Chart Pattana party into Thai Rak Thai. Meanwhile, former Secretary-General of the DP, Gen. Sanan Kajornprasart, sent the DP into even further disarray by quitting the party and establishing a new grouping, Mahachon. In August the victory of the DP candidate, Apirak Kosayhodin, in the Bangkok gubernatorial election was perceived to be indicative of growing popular dissatisfaction with Thai Rak Thai. The candidate favoured by Thai Rak Thai, former Secretary-General of Chart Pattana Paveena Hongsakul, finished second in the poll.

RECENT FOREIGN AFFAIRS

Thai regimes since the mid-1980s, whether military or civilian, had attempted to achieve a *modus vivendi* between established military-bureaucratic élites and the increasingly powerful interests of capitalism. Domestic and international business interests greatly circumscribed the ability of Thailand's rulers to change policy from the *laissez-faire*, export-promoting course on which the country had embarked. It was not possible, for example, for the military to restore the 'hard line' on Indo-China as Suchinda's regime had announced on seizing power in 1991. Thai business interests were already too deeply engaged in Laos and keen to penetrate Cambodia, while ASEAN and other countries were increasingly determined to reach an Indo-China settlement. Consequently, the military regime's leaders quietly acceded to Chatichai's vision of Thailand as the capitalist focus for the economically feeble socialist states that surrounded it, and they adopted policies toward Cambodia, Laos, and Myanmar that favoured the penetration of Thai interests. Under the democratic regime of Chuan Leekpai, this line was pursued with vigour. In February 1993 Thailand, Viet Nam, Laos and Cambodia signed a joint communiqué that resulted in the establishment of a Bangkok-based commission for developing the Mekong river basin. In October 1993 Thailand was accepted as a full member of the Non-aligned Movement.

Thailand's efforts to establish itself as a patron to its economically weaker neighbours were complicated in Cambodia by the involvement of powerful Thai business and military interests with the communist dissident faction, the Khmers Rouges, a lucrative source of concessions for logging and gem-mining. At the beginning of 1993 Thailand formally closed its borders to trade with the Khmers Rouges, in accordance with UN sanctions. Unofficial dealings continued, however, straining Thai relations not only with the UN but with the USA, which Thailand was otherwise eager to accommodate. In December the Government was deeply embarrassed by the seizure of a large shipment of weapons from a Thai army stockpile, which was evidently on its way to the Khmers Rouges. In January 1994 Chuan became the first Thai Prime Minister to pay an official visit to Cambodia. Thereafter, his Government made increasing efforts to disentangle Thai military and business interests from the Khmers Rouges. In December 1994 Chuan issued an order to governors and military commanders of the border provinces to cease their co-operation with Cambodian guerrillas. This was largely successful, and co-operation between the Thai and Cambodian armies on the border increased notably during 1995, although there continued to be clashes between Thai and Cambodian troops as a result of incursions into Thai territory by the latter in pursuit of Khmer Rouge guerrillas. In mid-1996 Thai interests switched to support the breakaway Khmer Rouge faction of Ieng Sary, which controlled the border area on which the gem and timber trade was centred. In the subsequent negotiations between Ieng Sary and the Cambodian Government, Thailand played a major mediating role.

Thailand refused to join in the international boycott of the Myanma military regime, proclaiming instead a policy of 'constructive engagement'. The profitability of this remained considerable, in spite of Myanmar's cancellation of many Thai logging and fishing concessions at the end of 1992. The Myanma military's increasing domination of border areas led to a new influx of refugees into Thailand in 1992–93; Thai authorities accommodated them but took care to prevent their aiding insurgents. They also moved to reduce possible clashes with the Myanma regime by joint demarcation of the border, and by supporting mediated settlements of Myanmar's ethnic rebellions. At the same time Thai diplomacy within ASEAN successfully secured a more positive approach by the regional organization to Myanmar; Thailand invited the Myanma Minister of Foreign Affairs, Ohn Gyaw, to attend an ASEAN ministerial meeting in Bangkok in August 1994.

In January and February 1995 about 10,000 members of the Kayin (formerly Karen) National Union (KNU) fled to Thailand, following an assault by Myanma government forces on the rebels' headquarters on the Moei river. Attacks on the refugee camps where the disarmed Kayins were held prompted Thailand to move the refugees to a site 10 km inside Thailand and to warn Myanmar that it would retaliate against border incursions. Bilateral relations deteriorated further in March when Myanmar closed its only land border with Thailand, in retaliation for alleged Thai support for raids in the area by the Mong Tai Army controlled by the 'opium warlord', Khun Sa. In September Chavalit visited Yangon and assured the Myanma regime that Thailand did not intend to support Khun Sa and his Shan followers.

Thailand achieved mixed results in its efforts to 'open up' its socialist neighbours. Once the countries of what Thai leaders liked to call the 'Greater Mekong Sub-region' had established connections with a range of foreign capitalist interlocutors, they tried to reduce their dependence on Thai patronage. In 1995 Cambodia gave preference to investment from Malaysia and Singapore. Myanmar became less pliable to Thai interests and likewise encouraged investment from Singapore. Relations with Viet Nam were marred by competition over fishing rights in the Gulf of Siam, leading in May 1995 to an armed clash between Thai and Vietnamese naval patrol boats. Even Laos, which normally preferred avoidance to assertion in its international relations, demanded that the headquarters of the Mekong River Commission be removed to Vientiane, and attempted to balance its Thai connections by renewed dealings with China and Viet Nam. None the less, Thailand's political and economic centrality to the mainland South-East Asian area led to increasing international acknowledgement of its regional importance.

Thailand's foreign business investments were not limited to raw material extraction from its poorer neighbours. Increasingly, major Thai banks and businesses looked abroad for profit, particularly to mainland China's booming economy. At the same time, trade with the rest of ASEAN grew rapidly, and Thailand's professed objective of making Bangkok an investment hub for mainland South-East Asia and inland southern China began to be taken seriously by foreign observers. This increasing Asian orientation had an effect on Thai-US relations, with growing tension over issues such as investment access, US import quotas, and intellectual property rights. Thailand, too, was distancing itself from its traditional patron, as its economic interests and diplomatic possibilities diversified. However, Thailand's growing attraction as a market-place and as a base for business in the ASEAN area brought increasing US investment, counterbalancing tensions aroused over other issues.

In July 1993 Thailand, Malaysia and Indonesia endorsed the concept of a 'growth triangle' that would enhance economic development in northern Sumatra, the five southernmost states of Thailand, and the north-western border states of Malaysia. For Thailand, this was an economic initiative with profound political implications: whereas once Thai authorities had been concerned to sever the cross-border connections of the largely Muslim and Malay-speaking south, they now seemed confident that economic advantage would outweigh the risk of encouraging sympathy for Muslim separatism.

Under Chavalit, Thailand's foreign policy included a number of new initiatives. The most important and controversial of these was increased defence links with the People's Republic of China. In early April 1997 Chavalit led a high-level delegation to Beijing for talks with the Chinese authorities. While agreements were reached on increasing trade relations between the two countries, media attention focused on China's pledge to strengthen defence ties.

Chavalit continued Thailand's policy of strengthening bilateral relations with its neighbours by making trips to Viet Nam, Myanmar and Cambodia. On most of these trips the agenda was dominated by security issues. There was continuing instability on Thailand's western border with Myanmar where tens of thousands of Kayin, Karenni and Mon refugees had crossed into Thailand as a result of campaigns by the Myanma military against ethnic insurgents. By mid-1997 it was estimated that as many as 100,000 refugees were housed in refugee camps inside the Thai border with Myanmar. Despite the unstable situation on Thailand's border with Myanmar, however, Thailand's support was a key factor in Myanmar's admission into ASEAN as a full member in July 1997. In Cambodia the fragile alliance between the Co-Prime Ministers, Hun Sen and Prince Ranariddh, broke down in July, with fighting erupting between forces loyal to the two rival leaders. Fierce battles between the two forces spilled over into Thailand's eastern border with Cambodia, with thousands of refugees fleeing the fighting and crossing into Thailand. Border issues preoccupied the Armed

Forces' Survey Directorate throughout 1997. Talks continued with the Laotian and Myanma Governments on disputed border demarcations, while an agreement was reached with Viet Nam regarding its maritime boundaries with Thailand.

Under the Chuan administration, there was a substantial improvement in relations with Malaysia. By April 1998 Chuan and the Malaysian Prime Minister, Mahathir Mohamad, had met an unprecedented five times in as many months to discuss regional and economic issues affecting the two countries. The most important outcome of these discussions was the agreement by Malaysia to end all support for the Muslim separatist movement in the troubled border region of southern Thailand, and to take a more assertive stance in suppressing the activity of known Muslim separatists operating on Malaysian soil. In the months following the agreement, a number of key separatist leaders were apprehended or gave themselves up to the Thai authorities. Discussions were held regarding the use of common currencies in the conduct of bilateral trade instead of relying on US dollars. In April the two Prime Ministers also witnessed an agreement signed by the two respective state-owned oil companies of Thailand and Malaysia on the development of gas reserves in the Thai-Malaysian Joint Development Area, once a disputed region between the two countries. The gas development deal was part of the Indonesia-Malaysia-Thailand Economic Triangle co-operation agreement, which gained added significance as a result of the economic slump in the region in the late 1990s.

Work on the controversial Yadana gas pipeline running between Thailand and Myanmar was approaching completion in 1998. The purpose of the pipeline was to give Thailand access to gas reserves being developed by the Myanma Government in Myanmar's Andaman Sea; however, the project attracted strong criticism from Thai environmentalists and human-rights activists, causing tension in Thailand's relations with Myanmar. Relations with Myanmar were further strained by a series of other issues, the most serious of which were the continuing incursions across the Thai–Myanma border by Myanma troops and the ongoing refugee problem resulting from continued fighting between Myanma government forces and rebel ethnic groups on the border. The Myanma ruling junta's continued restrictions on the movements and political activities of opposition leader Aung San Suu Kyi also received considerable media attention in Thailand. In early 1998, with Thailand's economic crisis resulting in increasing levels of unemployment, the large numbers of illegal migrant labourers who had been working in the country since the years of the economic boom were also becoming a major concern to the Thai Government. According to official statistics, almost 1m. illegal migrant labourers remained in Thailand, most of whom were Myanma nationals, with the remainder coming from Laos and Cambodia. In March the Government announced plans for the forced repatriation of several hundred thousand illegal Myanma migrant workers; however, the Government subsequently announced that it would reconsider its plans, as some industries were encountering difficulties in finding local replacements for the foreign workers.

In mid-1998 Thailand's new Minister of Foreign Affairs, Surin Pitsuwan, signalled a change in Thailand's approach to regional relations within ASEAN. Instead of 'constructive engagement', which effectively meant that ASEAN members were to refrain from commenting on the internal affairs of other member countries, Surin proposed a new policy of 'flexible engagement', in accordance with which ASEAN nations would have the right to discuss the domestic affairs of another member country if those affairs had an impact beyond that country's borders. The proposed change in policy had been prompted by a number of recent regional issues, including the violent events in Cambodia in July 1997, the devastating forest fires in Indonesia which had severely affected the air quality (and the tourist industries) of neighbouring countries, the continuing abuse of human rights by the military regime in Myanmar and the problem, experienced by several ASEAN nations, of large numbers of illegal foreign workers. While Surin's proposal was welcomed by US, European and Australian government officials as a sign of a new political openness in the region, it met with a generally negative response from most other ASEAN members. At the meeting of ASEAN Ministers of Foreign Affairs in Manila in July, the only regional support Thailand received for the 'flexible engagement' concept came from the Philippines, while Indonesia, Malaysia, Myanmar and Viet Nam strongly rejected any modification of ASEAN's long-standing principle of non-interference.

A major issue dominating Thailand's foreign agenda for much of 1998–99 was the impasse over the appointment of the next Director-General of the World Trade Organization (WTO). The Thai candidate, Deputy Prime Minister and Minister of Commerce, Supachai Panichpakdi, who was supported by the majority of Asian nations, was involved in a contest with former New Zealand Prime Minister, Michael Moore. Intense criticism in the Thai media regarding the apparent support of the USA for Moore turned the issue into one of national pride, and Chuan wrote to President Clinton to express his wish that the USA remain neutral with regard to the appointment. The perception that the USA was obstructing Supachai's bid for the position fuelled growing criticism of the USA by Thai politicians and the Thai media concerning the effects of the economic crisis and the perceived political influence of the IMF in Thailand. The impasse was finally resolved in July 1999 after Supachai agreed to an Australian proposal for the post to be divided into two consecutive three-year terms, with Supachai taking the second term.

Undoubtedly, the major foreign affairs issue for the Thai Government in 1999 was its high-profile participation in the UN-organized International Force for East Timor (Interfet). In the frantic diplomatic activity that followed the outbreak of militia violence in East Timor (now Timor-Leste) in the wake of the referendum on independence, held on 30 August, the Thai Government agreed to commit 1,500 troops to the Interfet force. The Thai contingent was the largest of those sent by ASEAN nations (and the second largest after the Australian contribution), and a Thai general, Maj.-Gen. Songkitti Chakrapatr, was appointed Deputy Commander of Interfet. The situation was a very sensitive one for the Thai Government, given that it risked breaking the accepted convention of non-interference in the internal affairs of an ASEAN member state. Prime Minister Chuan and the Minister of Foreign Affairs, Surin Pitsuwan, repeatedly stated that they had agreed to take part in the mission only after being invited to do so by Indonesia's President Habibie, Gen. Wiranto (then Commander-in-Chief of the Indonesian National Defence Forces) and the UN. The Thai opposition expressed concern over the cost of such an operation at a time of severe economic hardship in the country. The prominent Thai participation in Interfet contrasted with the initially reluctant responses by other South-East Asian nations, most of which sent only token forces. Within Thailand the Government was accused by the opposition and some sections of the media, especially a number of Thai Muslim news magazines, of pandering to the West and breaking ranks with ASEAN members.

In 1999 Thailand's relations with Myanmar were strained over the issue of drugs-trafficking into Thailand. By the late 1990s amphetamine addiction had come to be regarded not only as a serious social issue but also as a problem of national security. The Thai authorities publicly accused the United Wa State Army (UWSA), based in north-east Myanmar, of being behind the production and trafficking of millions of amphetamine tablets across the border into Thailand. The Wa had signed a peace agreement with the Myanma Government and were co-operating with the Myanma military in the suppression of Kayin insurgents. In August, on the advice of the National Security Council, the Thai Government closed a major border trading point at San Ton Du in Chiang Mai province, which was regarded as one of the main entry points of amphetamines into Thailand. Thai border police and military units reported intermittent clashes with suspected Wa traffickers who had crossed into Thai territory. Meanwhile, in Thailand's southern provinces bordering Malaysia, a series of violent incidents occurred in July and August, including the shooting of a policeman, and bombings and arson attacks directed against government property. The Thai authorities blamed the attacks, the first for some time, on a renewed campaign of violence carried out by Muslim separatists. Malaysia promised to co-operate in apprehending anyone suspected of involvement in the attacks who was using Malaysian territory to evade capture by the Thai authorities.

On 1 October 1999 Thai–Myanma relations were strained further when a group of heavily armed Myanma students stormed the Myanma embassy in Bangkok, taking all 89 people in the compound hostage, including 13 diplomats. The gunmen demanded that all political prisoners in Myanmar be released, that a dialogue between the military junta and other political parties be opened, and that the elected legislature be convened. The situation was further complicated as the Thai Minister of the Interior, Sanan Kajornprasart, and the Deputy Minister of Foreign Affairs, Sukhumphand Paripatra, publicly appeared to express sympathy with the pro-democracy aspirations of the students. The situation was, however, resolved within 24 hours when the group agreed to release the hostages in exchange for the Thai Government's guarantee of a safe passage by helicopter to the Thai–Myanma border. The ruling military junta in Myanmar was quick to express its displeasure at the way in which the hostage crisis had been handled by the Thai Government. In a series of retaliatory measures, it closed the border with Thailand, cancelled fishing concessions and refused to allow Thai fishing vessels into Myanma waters. Relations between the two countries deteriorated further when, in November, the Thai Government began forced repatriations of thousands of illegal Myanma migrant workers. The deterioration in relations between the two countries was eventually halted at the end of November when, following a visit by Thai Minister of Foreign Affairs, Surin Pitsuwan, Myanmar agreed to reopen both the common border and its territorial waters to Thai fishing vessels. The importance that the Thai Government now placed on cordial relations with the junta in Myanmar was evident with the handling of a second hostage crisis in January 2000: 10 armed Myanma rebels from God's Army, an ethnic Kayin faction fighting an armed insurgency against the Myanma Government, took control of a hospital in Ratchaburi Province, close to the Myanmar border, holding more than 500 patients, doctors and nurses hostage. They demanded that the shelling of Kayin positions by the Thai military should cease, that medical assistance should be given to injured Kayin fighters, and that the Thai military should cease its co-operation with the Myanma forces in suppressing Kayin military activity on the Thai–Myanma border. In contrast to its peaceful handling of the earlier siege at the Myanma embassy, the Thai Government acted decisively; a team of commandos stormed the hospital in a pre-dawn raid, killing all 10 hostage-takers (according to some hostage reports, after they had already surrendered). The brutal resolution of the incident was praised by the military Government in Myanmar, and many observers viewed the Thai Government's handling of the incident as signalling a broader campaign by the Thai authorities on dissident activity by ethnic Myanma opposition groups operating in Thailand. In another sign of a change in the formerly-relaxed Thai policy towards the tens of thousands of Myanma and Kayin refugees housed in camps in Thailand, the Government appealed to the UN to organize the rapid resettlement of the refugees to third countries.

Despite the new Thai Rak Thai Government's announcement of a redirection in foreign policy that would focus on strengthening relations with Thailand's neighbours, relations between Thailand and Myanmar reached a new low point soon after the new administration took office. In February 2001 thousands of ethnic Shan villagers were forced to cross the border into Thailand, as a result of a renewed campaign by the Myanma Government against Shan insurgents. During the fighting, Myanma forces shelled Thai villages in the border region of Chiang Rai province, as well as crossing into Thai territory on a number of occasions in pursuit of the insurgents. The Myanma Government's actions were apparently linked to public accusations by Thai government officials that the Myanma Government was aiding the flow of amphetamines produced in the ethnic Wa-controlled regions into Thailand. The Myanma regime's official daily newspaper, *The New Light of Myanmar*, published an unprecedented series of articles highly critical of Thailand, even at one point obliquely criticizing the Thai royal family, which prompted an immediate protest from the Thai Ministry of Foreign Affairs. The rapid deterioration in relations was eventually halted when Thaksin made a long-delayed official visit to Yangon in June. Thaksin's visit was followed by a further visit by the Minister of Defence, Gen. Chavalit Yongchaiyudh, known to have long-standing personal relations with a number of the Myanma junta officials. In September a junta leader, Gen. Khin Nyunt, paid a reciprocal visit to Bangkok, during which he was warmly received by the Government.

Tensions between Thailand and Myanmar persisted, however, principally owing to Thai accusations that the Myanma Government was not doing enough to curb the flow of amphetamines across the border into Thailand, and to ongoing Myanma complaints that the Thai military was assisting the ethnic insurgent group, the Shan State Army (SSA), in the Myanma military's campaign against it. In June 2002 a series of articles appeared in *The New Light of Myanmar* on the subject of Thai and Myanma history, which the Thai Government deemed to be insulting. The publication of an article critical of the late 16th-century Thai King Naresuan (who was credited with having recovered the independence of the kingdom of Ayuthaya from the Burmese) led to a major diplomatic incident. The Supreme Command of the Thai military launched an official protest against the article, which it claimed was offensive to the Thai monarchy itself, whilst radio and television stations, under the control of the Thai military, aired anti-Myanmar programmes for several days. Yangon responded by closing its border entry points with Thailand. The incident was finally closed when the Vice-Chairman of the SPDC, Gen. Maung Aye, delivered a message to the Thai Government promising to prevent the state-controlled media from publishing articles deemed to be insulting to the Thai monarchy. Thaksin's relations with the Thai military itself were also tested when he urged the army not to react to alleged border incursions by the Myanma military. Both opposition and press sources accused the Government of failing to respond more resolutely to the Myanma regime in order to protect the business interests of the Prime Minister and the Minister of Defence, Chavalit, in Myanmar.

In early July 2000 a group of some 60 Laotians and Thais, believed to have links to a rebel Lao political group, unsuccessfully attacked a customs and immigration check-point on the Thai–Laotian border. It was reported that six of the rebels had been killed during the fighting and that, after being repelled by Lao Government troops, 28 members of the group had fled into Ubon Ratchathani Province in Thailand, where they were arrested by Thai police. The Laotian authorities officially requested the extradition of the 28 'international terrorists'; however, the Thai Government insisted that the rebels would first face trial in Thailand on charges of illegally entering the country and possessing weapons of war. Representatives from the two countries subsequently met in Chong Mek, Thailand, to discuss measures to prevent similar occurrences in the future. In August around 30 Laotian soldiers moved onto the Mano I and Mano II islands in the Mekong River between the two countries, forcing numerous Thai farmers to leave. Under the Siam-Franco treaty of 1926, all islands on the Mekong River belong to Laos, and many observers believed that the occupation was an attempt on behalf of the Laotian administration to assert its territorial rights over the islands prior to the demarcation of the river boundary between the two countries.

In accordance with the Thai Rak Thai Government's declaration that it was to adopt a more regionally focused foreign policy, the Prime Minister and the Minister of Foreign Affairs paid official visits to most of Thailand's Asian neighbours, including China, Japan, Indonesia and India. In December 2001 Thaksin led his Cabinet on a week-long official trip to the USA, amidst accusations by the opposition that government policies had led to a decline in Thai–US relations. Following the visit, the opposition claimed that the US Administration had effectively rebuffed Thaksin, comparing his reception unfavourably with that accorded to the leaders of Malaysia and Indonesia during their respective visits.

Relations with Cambodia reached a new low point in February 2003 when rioters in Phnom-Penh looted and then burned down the Thai embassy and a number of Thai businesses. Cambodian police apparently did little to stop the rioters. The riots were precipitated by what appeared to have been a fabricated story in the Cambodian press that a Thai television actress, well-known in Cambodia, had made disparaging remarks about the ancient ruins at Angkor Wat, implying that they really belonged to Thailand. As the rioting continued, Thaksin threatened to send commandos into Cambodia in order to ensure the safety of Thai citizens. In the event this threat was not carried out, but Thai nationals in the country, fearful for their safety, were airlifted back to Thailand. The Thai ambassador, who had escaped from rioters by scaling the wall surrounding the embassy compound, was recalled from Cambodia, diplomatic relations downgraded and border crossings closed. The underlying cause of the riots appeared to have been the powerful and increasing role of Thai business in Cambodia's economy, which had become a theme in nationalistic speeches by Prime Minister Hun Sen in the run-up to the national elections planned for July. In the aftermath of the riots the Cambodian Government made a formal apology to the Thai Government and agreed to pay the cost of the damages, which amounted to at least 1,800m. baht.

In late May 2003 Thailand's relations with Myanmar were once again strained when a motorcade carrying Myanmar's opposition leader, Aung San Suu Kyi, and members of her party, the National League for Democracy (NLD), was attacked, allegedly by Myanma government supporters. Several people were killed or injured as a result. The incident led the SPDC to place Aung San Suu Kyi under house arrest once again, a move that provoked widespread condemnation by the international community. Surprisingly, ASEAN, which normally follows a principle of non-interference in the internal affairs of member countries, also publicly criticized the actions of Myanmar's military leadership and urged the regime to reconsider its actions. In July the US Congress voted to impose a comprehensive economic boycott on Myanmar. Later the same month the Thai Government announced that it had devised a 'road map' detailing plans for the release of Aung San Suu Kyi and a peaceful transition to democracy in Myanmar. The plan was, however, immediately rejected by the Myanma Government.

For much of the first half of 2003 Thailand's foreign policy was dominated by the US-led invasion of Iraq. Officially, the Thai Government maintained a policy of neutrality, mindful, on the one hand, of its military dependence on the USA but, on the other, of the sizeable Muslim minority resident in southern Thailand, which was overwhelmingly opposed to the war, Thai public opinion, which was also generally against the war, and the opinion of Thailand's predominantly Muslim neighbours, Malaysia and Indonesia, the governments of which were extremely critical of the USA's actions against Iraq. In June Prime Minister Thaksin flew to the USA to meet President George W. Bush for discussions on counter-terrorism operations. In the aftermath of the devastating bombing of two night-clubs in Bali, Indonesia, in October 2002, which had killed over 200 people, it had been alleged that members of Jemaah Islamiah, the organization believed to have been responsible for the bombing, might have used Thailand as a base for the planning of the operation. Thaksin's meeting with President Bush coincided with the arrests, in the southern province of Narathiwat, of three suspected members of Jemaah Islamiah, a move that provoked widespread criticism among religious leaders in the largely Muslim provinces of the south.

In September 2003 Thaksin visited Singapore and the Philippines, where he was lauded by the leaders of both countries for his apparently successful economic policies, which had been dubbed 'Thaksinomics'. In October Thailand hosted the Asia-Pacific Economic Co-operation (APEC) summit. The meeting was dominated by security issues and, particularly, the USA's proposal that trade liberalization be linked to the implementation by member governments of stricter anti-terrorism security measures. Thailand also actively pursued bilateral trade negotiations. It successfully concluded free trade agreements with China (concerning trade of fruit and vegetables), Singapore, and Australia. In March 2004 the entire Thai Cabinet travelled to Laos, where they met their Laotian counterparts to discuss the strengthening of economic relations between the two countries. In May 2004 Prime Minister Thaksin announced that he was about to buy a stake in Liverpool Football Club, in the United Kingdom. Thaksin had for some time publicly expressed his desire personally to invest in a British football team (he had earlier shown an interest in acquiring Fulham Football Club). However, from the beginning the question of how the purchase was to be financed provoked controversy. Initially, it appeared the funds would be raised by a group of affluent members of Thai Rak Thai, but subsequently it was proposed that public money be used. Thaksin even suggested establishing a public lottery with a prize of 1,000m. baht to finance the bid, a proposal that eventually had to be abandoned after numerous protests over the impact that such a lottery would have on the country's poor. While the deal had earlier been portrayed as being assured, it eventually collapsed as a result of the public opposition surrounding the financing of the deal, as well as opposition from Liverpool Football Club itself.

Economy

CHRIS DIXON

In 1997 the strong economic growth that Thailand had experienced since the mid-1980s came to an abrupt halt as a result of major currency and financial crises. In 1997 and 1998, respectively, gross domestic product (GDP) contracted by 0.4% and 9.4%, industrial output declined by 5.9% and 13.5% and unemployment rose from 2.2% in February 1997 to 5.25% at the end of 1998. This unprecedented negative growth followed speculative action against the Thai currency, the baht, and the virtual collapse of the financial and property sectors. In August 1997 the depth of the crisis forced the Government to seek assistance from the IMF, and in that month a 'rescue' programme valued at a total of some US $17,200m. was agreed. Conditions attached to this agreement included the rapid enforcement of tight monetary policy and the implementation of major liberalization and privatization programmes, particularly aimed at opening the economy to foreign activity and ownership. The implementation of this programme was accompanied by two years of recession and financial turmoil.

By the end of 1999 it was evident that the macroeconomic crisis was over; during that year GDP had expanded by 4.2%, the value of exports by 7.5%, industrial output by 7.6% and industrial capacity utilization had recovered from its 1998 record low of 52.2%, to reach 64.3%. In addition, interest rates, inflation and currency values and continued net inflows of direct investment had broadly stabilized. The economy had also been substantially opened to foreign investment and ownership (see Investment). In June Thailand departed from the IMF rescue programme, having received generally favourable comments regarding both the Government's handling of the crisis and the country's prospects for sustained economic recovery. However, in the event, this proved to be a highly optimistic view; by the end of 2000 GDP had grown by a respectable, but disappointing, 4.3%. During 2001 there was a marked slowing of growth during the first three quarters of the year, a trend that was accentuated following the impact of the 11 September 2001 terrorist attacks on the USA on the international economy. In consequence, GDP expanded by only 2.1% and export earnings contracted by 6.3%. However, it could be argued that the performance of the Thai economy during 2001 was far from disastrous given the general economic climate. To a considerable extent this can be attributed to the comparatively diverse nature of the Thai economy and to government policy (see Future Development and Prospects). These factors, together with generally improved global conditions, were critical to the subsequent high rates of GDP growth, which reached 5.4% in 2002, 6.7% in 2003 and a projected 7.0% in 2004. Despite this impressive economic expansion, it might still be premature to suggest that Thailand had returned to a situation of high and consistent growth. The recent growth had been heavily dependent on high levels of domestic consumption and state expenditure, rather than on increasing exports and expanding production (see Future Development and Prospects).

The abrupt reversal in the progress of the Thai economy during 1997 could to a degree be attributed to the weaknesses of the financial sector and to its liberalization, to poor central regulation and to ineffective macroeconomic management by the Bank of Thailand (see Finance and Banking). These problems were compounded by the failure of the Government and the central bank to take early and effective action. In 1996 there was negligible growth in export earnings, a widening balance-of-payments deficit, escalating private-sector debt, increasing short-term speculative capital movements and the 'over-heating' of the property and financial sectors. None of these was regarded as a cause for alarm, given the high rate of GDP growth and the continued influx of foreign funds. While no observers predicted the 1997 crisis and subsequent recession, from the early 1990s doubts had been raised concerning the sustainability of Thailand's high rates of growth and structural change. Particular attention focused on the inadequate infrastructure and shortages of skilled labour that were increasing the costs and difficulty of operating in Thailand at a time when the rapid re-engagement of the People's Republic of China and Viet Nam with international capital were providing cheaper alternative Asian locations for investment in labour-intensive manufacturing.

The 1997 crisis and its aftermath must be seen in the context of a remarkable period of growth and structural change lasting from 1986 to 1996. Between 1986 and 1991 Thailand became one of the fastest-growing economies in the world. During this period the value of exports grew at 30% per year and GDP at 9.2%. Subsequently, expansion decelerated slightly, but remained rapid and remarkably consistent, with GDP growth averaging 8.2% annually in 1992–96. The most striking feature of the whole period from 1986 was the expansion of manufacturing. This sector increased its share of GDP from 21% in 1985 to 30% in 1995, and its share of export earnings from 43% to 81%. Overall, from the mid-1980s Thailand experienced a period of accelerated integration into the Asian and global economies. Between 1986 and 1995 the openness ratio (the value of trade expressed as a percentage of gross national income—GNI) increased from 43.9% to 77.6%.

The rapid growth of the Thai economy attracted a great deal of attention. It was not only increasingly referred to as one of the next newly-industrializing countries (NICs), but was also presented as a new model of development—one that was more appropriate to the other less-developed Asian economies than that of the NICs. However, the rapid expansion and changing composition of GDP and export earnings had yet to be fully reflected in employment or residence patterns (see below). When set alongside the heavy dependence of post-1985 growth on the influx of foreign investment and the continued heavy reliance on imported raw materials and components, the description of 'NIC' looked remarkably premature even before the 1997 crisis.

The accelerated growth that the Thai economy had experienced during the 1986–91 period was the result of a number of interrelated factors. During 1986 the export of manufactured goods began to expand significantly. The devaluation and realignment of the baht substantially improved the competitiveness of Thai exports. Thai exporters were remarkably successful in expanding sales in established markets and penetrating new ones. This expansion initially involved the redirecting of surplus manufacturing capacity towards the export market. From 1987 there was a surge of domestic and foreign investment in export-orientated, labour-intensive manufacturing industry. Thailand has become a major destination for East Asian investment. The loss of comparative advantage in labour-intensive assembly work resulted in a relocation of these activities away from the Republic of Korea, Hong Kong and Taiwan. Thailand was seen by investors as the Asian economy of which the location, political and economic stability and low labour costs better suited their purposes.

As economic growth accelerated, the Government implemented a wide range of liberalization measures, some of which had been originally advocated during the early 1980s by the World Bank and the IMF. These included the removal of subsidies and marketing restrictions on a wide range of products, simplification of investment promotion procedures, the reduction of tariff levels and a considerable measure of financial liberalization, including the virtual elimination of foreign-exchange controls. The furtherance of such measures formed a key element in the Seventh (1992–96) and Eighth (1997–2001) National Plans. Under these Plans the Government aimed to reduce its role as the 'principal economic stimulator' by limiting public expenditure and encouraging private investment, and by becoming the 'co-ordinator, adviser and facilitator to the private sector'. To this end, a wide-ranging programme of privatization was announced. However, the programme was subject to considerable debate, opposition and delay. Overall, liberalization accompanied and furthered Thai economic growth, rather than preceding it and playing any significant role in its initiation.

Liberalization from the late 1980s has to be viewed in the context of the increased political influence of business interests that expected to gain through deregulation and the easing of access to international capital. There was also considerable pressure for liberalization from Thailand's principal trading partners and from international agencies. In addition, tariff reductions and the removal of restrictions on foreign market access and ownership became necessary conditions for membership of the Association of South East Asian Nations (ASEAN) Free Trade Area (AFTA) and the World Trade Organization (WTO). Between 1997 and 2000 liberalization was assigned a greater degree of urgency under the IMF programme. The Government of Chuan Leekpai (November 1997 to December 2000) generally followed the IMF programme closely, and was perceived as being in favour of globalization. However, almost every liberalization measure that it proposed resulted in considerable debate and resistance, sometimes accompanied by public demonstrations held with media support. As a result, many measures were either delayed or weakened, and their effectiveness was limited. Thus, reform was slow, uneven and far from consistent in its implementation. Following the election of the Thai Rak Thai-dominated Government of Thaksin Shinawatra in January 2001 (see History), economic policy statements became markedly more nationalistic. However, in the event, these statements proved to have little real substance and slow liberalization, and even slower privatization, continued at their previous inconsistent rates.

STRUCTURAL PROBLEMS

By the early 1990s the rapid growth that Thailand had experienced since 1985 had begun to expose serious long-term structural problems. The lack of infrastructure, particularly port and air freight facilities, roads, water supply and telephones, was creating serious congestion and raising production costs (see Communications and Infrastructure). Shortages, and, in some cases, the high prices of domestically produced materials and components, constrained production and accelerated the growth of imports—a problem exacerbated by the limited development of many basic industries and the continuation of tariff protection and import restrictions. Similarly, the lack of adequate technical training schemes resulted in a serious shortage of skilled labour, which began to restrict the growth of the manufacturing sector. Thailand has practically the lowest secondary and tertiary education participation rates in South-East Asia: in 2000 only 22% of the labour force had more than primary level education. In addition, the quality of secondary and tertiary education is often poor, with a curriculum that is ill suited to the requirements of a modern economy. There is a particular shortage of science and technology graduates. Overall, the limited participation in, and deficiencies of, the education system remain a major barrier to the development of more technology- and skill-intensive production, while Thailand's comparative advantage in labour-intensive manufacturing has been eroded. Since the early 1990s there have been a variety of government schemes aimed at resolving the skills shortage. The 1997 Constitution guaranteed all children free education until the age of 12, and in 1999 the National Education Act set out an ambitious three-year programme of education reform. In addition, during 2001 a Skill Development Act was formulated and a new Skill Development Fund established. Meanwhile, in 2002 the guarantee of 12 years free education was implemented, ahead of schedule. These developments have been accompanied by significant increases in funding—an 8% rise during 2003 allocated education some 24% of the state budget. However, little progress has been made in reforming the curriculum, and the funding still appeared to be inadequate for the necessary upgrading of facilities and equipment. With respect to the skills development programmes, problems of co-ordination continued to occur. During 2002 the World Bank reported that there were some 17 separate organizations involved in a range of overlapping and largely unco-ordinated educational and skills enhancement programmes. While it is significant that the Government is addressing this critical issue, certainly to a much greater extent than any of its predecessors, it has become clear that, even if the reforms are effectively implemented, there can be no rapid solution to the skills problem.

The growth in the Thai economy has been heavily concentrated in the Bangkok Metropolitan Region (BMR) and its immediate environs, reinforcing the capital's economic dominance. According to the 2000 census, 43.3% of the population lived in urban areas—a comparatively low level of urbanization considering the country's per caput GDP—but 12m. (44%) lived in the BMR. Particularly since 1986 there has been a sharp increase in the level of income disparity between the BMR and the rest of the Kingdom. In 2000 the official incidence of poverty was less than 1% in the BMR, 5% in the Central Region, 11% in the south, 12% in the north, and 28% in the north-east. Despite the establishment of various programmes aimed at promoting investment in provincial areas, there has been little success. Congestion and land shortages have begun to extend development into the fringes of the 10 provinces adjacent to the BMR. Similarly, despite the still variable infrastructural provision, there has been considerable expansion along the Eastern Seaboard. However, these developments represent the extension of the BMR rather than any dispersal of economic growth.

The 1997 Constitution set out a commitment to significant decentralization of administration, public finance and decision-making. In 1998 the ability of some 100 urban and municipal authorities to raise funds through taxation and land sales was significantly increased. Previously, only the Bangkok Metropolitan Administration (BMA) was empowered to raise significant amounts of funds. Through increased local taxation and the reallocation of central funds, it was intended that by 2006 35% of public funding would be locally controlled. However, there seemed to be considerable reluctance on the part of the Government to implement this pledge. In addition, the provisions of the 2002 bureaucratic reforms, which created five new ministries and 35 new departments, appeared to be increasing centralization and state control.

By the early 1990s the concentration of growth in the BMR had created one of the world's most polluted and congested cities. However, while regulation and enforcement remains problematic, there have been significant improvements. The fitting of catalytic converters to new cars from 1993, the banning of leaded petrol from 1996, and revised limits on exhaust emissions in 1999, have significantly improved the air quality. In 2002 the World Bank considered Bangkok to have become less polluted than either Manila or Jakarta. In addition, major transport developments (see Communications and Infrastructure) have combined to significantly reduce traffic congestion. However, it should be stressed that Bangkok remains, by any standards, a city where traffic levels and air quality leave much to be desired.

There has been serious, long-term neglect of the rural areas, and shortages of funds, vested interest and unwieldy administration have limited the impact of programmes directed at such areas as credit, water supply and land reform. In many rural areas there is an acute shortage of land, with much landlessness, illegal forest clearance and cultivation of unsuitable land (including some on watersheds, which are prone to soil erosion). However, there has been increasing official recognition of the environmental consequences of unrestrained development, as witnessed by the creation in 2002 of the new Ministry of Natural Resources and the Environment, but the controls and planning mechanisms remain weak.

The long-term growth of the Thai economy has been associated with a significant decline in the incidence of poverty. The percentage of the population living below the official poverty line declined from 23.0% in 1980 to 18.0% in 1990 and to 11.4% in 1996. However, as a result of the 1997 crisis, the overall incidence of poverty rose to 15.9% in 1999, before falling to 14.2% in 2000 and to 9.8% in 2002. The rapid decline in levels of poverty before 1997 was particularly striking, given the limited efforts directed at finding a solution to the problem. However, the success in reducing poverty has to be set against an increase in disparities of income. By 1996 the UN Development Programme (UNDP) had concluded that Thailand had one of the world's most uneven distributions of income. In addition, poverty has become heavily concentrated in rural areas. In 2000 the official incidence was 20% in rural areas, 9% in the semi-urbanized sanitary districts and 1% in urban areas. This situation was exacerbated by the 1997 crisis, which contributed to substantial remigration to rural areas in the wake of the collapse of much urban employment and the reduction of remittance flows. The returning population, as well as increasing levels of rural poverty, brought many 'urban problems' to villages, including drugs, organized crime and HIV/AIDS. The persistence of high levels of income disparity, the low income-generating capacity of agriculture, and the concentration of poverty in rural areas has precipitated large-scale protests, particularly since the 1997 crisis. These have played a significant role in increasing the concern of policy-makers with the agricultural sector (see Agriculture), but, as with education and infrastructure, there can be no quick solutions.

Even before the 1997 crisis there were signs that structural problems and the apparent inability of the Government to deal with them were casting serious doubt over the sustainability of the Kingdom's pattern of rapid economic growth and structural change. From the early 1990s there were calls from a variety of areas, including business groups, for a more pro-active, interventionist and effective Government, prepared to address training and infrastructural deficiencies and the economic and social consequences of growth. Such views have been reinforced by the 1997 crisis and its aftermath. However, the Government of Thaksin Shinawatra has been significantly more pro-active than any of its immediate predecessors and has attempted to address many of the country's ongoing structural issues.

AGRICULTURE

The agricultural sector employs some 41% of the labour force and directly supports a similar proportion of the population. Between 1980 and 2002 the sector's contribution to GDP declined from 40% to 9.4%, and its share of export earnings from 60% to 7.0%. However, Thailand is one of only two net food exporters in Asia (the other is Viet Nam). It is the world's largest exporter of rice (a position it attained in 1981), rubber and cassava, and the second largest exporter of sugar. As well as providing a continuing contribution to export earnings and supplying domestic foodstuffs and raw materials, the agricultural sector is the basis of the rapidly expanding agribusiness sector. However, Thailand is becoming a comparatively high-cost producer of agricultural goods and is losing markets to new low-cost producers. Despite this and other sectoral weaknesses, since the 1997 crisis the agricultural sector has provided an invaluable 'cushion' for a significant proportion of the population, both in terms of its contribution to export earnings and to the economy as whole.

The rapid economic growth that Thailand experienced between 1986 and 1996 posed a number of problems for the agricultural sector, particularly in the areas adjacent to the BMR. Good-quality agricultural land has been lost to housing, industry, infrastructure and leisure developments. Increasing pressure on water supplies restricted irrigation, particularly for rice, sugar cane, maize and intensive vegetable production; and increased employment opportunities resulted in rising labour costs.

Most of Thai agriculture remains unintensive in character, and until comparatively recently accommodated increases in production and population by expansion of the area cultivated. Between 1950 and 1975 the cultivated area increased at an average annual rate of 2.7%. The rate of expansion slowed to 1.2% between 1975 and 1988 and to 0.6% between 1989 and 1993. Subsequently, expansion has almost ceased, with the limited clearance for upland and tree crops, often of land that is unsuited to permanent cultivation, being offset by land passing out of cultivation owing to salination, declining fertility, erosion and the availability of alternative and more attractive sources of income. Controls over clearance have been generally ineffective, even within forestry reserves. As a result, forest cover declined rapidly, from 54.6% of the country in 1965 to an estimated 12%

in 2000. By the late 1980s the expansion of agriculture, weak controls over logging and limited replanting schemes had effectively eliminated the Thai forestry industry.

The long-term expansion has for many crops been associated with stagnant or declining yields. This is largely accounted for by the progressive clearance of less reliable and less fertile land, reinforced by low and uncertain price levels, the high cost of fertilizer, and insecure land tenure. Since the early 1980s the deceleration in the rate of expansion of the cultivated area and the emergence of land shortages in many areas has resulted in some intensification of production and increasing yields. However, yields and rates of fertilizer application for almost all the Kingdom's major crops remain among the lowest in Asia. This was particularly striking with respect to rice. Production remains remarkably extensive in character; the only significant South-East Asian rice-producing country with a lower yield than Thailand is Cambodia.

While rice continues to dominate Thai agriculture, since the early 1960s there has been significant diversification of production. The spread of such crops as maize and cassava brought much land unsuited to rice production into cultivation. Thus, the percentage of land planted with rice fell from 88% in 1950 to 68% in 1970. These developments were reinforced from the late 1970s by a second wave of diversification, which contributed to the reduction of the rice sector's share of the cultivated area to 52% in 1990 and 50% in 2000. This involved horticulture, aquaculture, tree crops—particularly rubber—and intensive livestock farming. Some of these developments involved substantial capital inputs and were associated with the emergence of a major agribusiness sector, either through direct control of production or various contractual arrangements. Particular growth areas for the export market have been frozen chickens, canned fruit (notably pineapples), frozen fruit and vegetables, fresh fruit and vegetables and cut flowers. In addition, for the domestic market, the growth of Western-style supermarkets and the rise of a comparatively wealthy middle class in the BMR have led to the development of packaging and grading of a wide range of produce. This trend has been reinforced by the increased presence of international retailing companies in the country—for example, the British supermarket chain Tesco purchased the Locus chain of shops in 1998.

In the wake of the post-1997 recession the Government has focused increased attention on the agricultural sector as a principal area for economic growth. Priority has been given to the expansion of exports, the creation of employment, and the raising of rural incomes and a series of agricultural development and promotion programmes have been implemented.

During 1998 the Cabinet approved a Master Plan for Agriculture, aimed at stimulating the production and export of rice, rubber, cassava, canned seafood and fresh and frozen shrimps. In addition, the Master Plan for Industrial Restructuring, also approved in that year, included provisions that aimed to stimulate industries that use agricultural products, notably the food, animal feed, rubber and rubber products industries. Further measures aimed at developing new markets were announced in 1999, and in 2000 the Board of Investment (BOI) implemented a new range of incentives aimed at the agricultural sector. Also during 2000 the Agricultural Sector Programme Loan scheme (originally introduced in 1998) was extended, using US $600m. of funding from the Asian Development Bank (ADB) and the Japanese Overseas Economic Co-operation Fund. This is a broad-based programme involving irrigation, land reform and extension work, all aimed at upgrading agricultural production and raising yields. Additional policies announced during 2001 included: a moratorium on farmers' debt; a renewed attempt to raise agricultural productivity; promotion of agricultural exports; and the establishment of the Village Development Fund. Under the latter, each of Thailand's 75,000 villages received $23,000 to provide credit and working capital for small-scale enterprises. However, both programmes have attracted considerable criticism on the grounds that the funds had either been misused or failed to reach their target groups. Despite this, the Government has claimed that the Village Development Fund and the debt moratorium have been of major importance in improving rural living conditions. While there has, as yet, been little evidence to support this claim, further programmes were to be implemented during 2004 to support the Government's goal of making Thailand the 'kitchen of the world'.

The Thaksin Government's initial handling of the outbreak of avian influenza that occurred in late 2003 and early 2004 has been much criticized (see Future Developments and Prospects). However, the cull of some 35m. birds in the 42 affected provinces appeared by June 2004 to have been successful. The rapid payment of compensation to farmers and the implementation of a loan scheme was suggestive of the Government's continuing commitment to the rural sector. Despite this, many farmers suffered considerable losses as a result of the epidemic and there was likely to be long-term damage to the export trade in poultry products, already suffering from competition from lower-cost producers.

Overall, while increased attention has been devoted to Thailand's rural sector, particularly under the Thaksin Government, this does have to be considered against the extent of the problems, doubts over the utility of the measures themselves, and the limited resources directed at them. The agricultural sector's share of public expenditure declined from 10.6% in 1996 to 7.6% in 1999, rising only marginally, to 8%, in 2001.

FISHING

Thailand has one of the largest marine fishing sectors in Asia, along with China, Japan and Indonesia. In 2002 there were some 50,000 registered fishing vessels, 17,000 of which were classed as deep-sea trawlers. However, in recent years over-fishing has depleted inshore fish stocks. The expansion of catches from 1.6m. metric tons in 1980 to 3.0m. tons in 1999 resulted from agreements with Oman, India and Bangladesh on the exploration of distant waters, and agreements with Indonesia and Myanmar over joint exploitation of their territorial waters. Despite the existence of these agreements, since 1992 there have been an increasing number of disputes with neighbouring countries relating to fishing by Thai vessels in their territorial waters. Since 1999 Thailand has been excluded from Myanma waters, and in May 2002 the impounding of over 1,000 Thai trawlers for illegal fishing resulted in a major dispute with Indonesia.

Since the late 1980s the most rapidly growing fishing sub-sector has been coastal aquaculture. Output increased from 151,600 metric tons in 1987 to 602,800 tons in 1999, of which 225,000 tons were giant tiger prawns. Since the mid-1990s Thailand has been the world's largest producer and exporter of prawns and shrimps, accounting for more than 30% of global production. There is increasing concern over the environmental damage caused by aquaculture, particularly where mangroves are removed to facilitate production. In addition, since 1998, in an attempt to protect rice fields from pollution, the Government has imposed a total ban on inland (freshwater) shrimp and prawn farming. While this has proved extremely difficult to enforce, many have abandoned shrimp farming in the face of the imposition of stringent testing for antibiotic content in order to comply with European Union (EU) regulations.

For the Thai market, much fish is dried, salted, fermented or turned into paste or sauce for preservation. Since the mid-1970s and, more especially, the mid-1980s the development of canning and freezing facilities has greatly expanded the export of seafood, particularly shrimp, prawn and squid. Until the late 1990s the frozen seafood sector expanded extremely rapidly, with the main export markets being Japan, the USA and the EU. However, expansion slowed in the wake of increased competition from lower-cost producers, notably China, India, Indonesia, and Viet Nam. In addition, since 1999 there have been increased tariff barriers to markets in the EU and the USA. Despite this, in 2003 some 50% of shrimp exports went to the USA. However, the trade has been further threatened by the prospect of the imposition of anti-dumping tariffs.

Thailand is the world's largest exporter of canned tuna, accounting for about 80% of world trade. The principal market, the USA, where Thai tuna accounts for 60% of sales, is jeopardized not only by increased tariffs but also by the US ban on tuna caught with drift-nets and by consumer reaction to fishing methods that are harmful to dolphins. However, Unicord, which is the largest exporter of canned tuna fish in Asia, established a firm hold on the US market in 1990 with the purchase of Bumble

Bee Seafoods, the third largest US producer. Subsequently, Unicord has shipped increasing amounts of frozen tuna fish to the USA for canning, thus avoiding the higher tariffs on the imported canned product.

The freshwater sector is highly subsistent in character, and official production figures (228,898 metric tons in 1997) are almost certainly a considerable underestimate. It remains, however, of vital importance to rural protein supply, and small-scale aquaculture is being widely promoted as a means of raising and diversifying rural incomes. Some inland commercial development has taken place with the stocking of the larger irrigation reservoirs. The Thai fisheries sector, like so much of the Kingdom's resource-based activity, continues to suffer from the long-term results of uncontrolled exploitation and underinvestment.

MINING

Despite the depressed state of the international market and declining production, tin remains Thailand's most important metallic mineral resource. In terms of mineral content, reserves are estimated at some 1m. metric tons, representing 10% of the world's total reserves. A wide range of other minerals are present in commercial quantities, notably: gold, lignite, gypsum, potash, antimony, iron, tungsten, lead, manganese, copper, zinc and a range of precious and semi-precious gem stones. Most established mineral production is small in scale, with large-scale operations confined to lead, zinc, gypsum, lignite, and, most recently, potash. Since late 2001 there has been increasing interest in the country's gold deposits, with commercial operations starting in Pichet and Loei.

The Lampang lignite field is responsible for the majority of Thai production, almost all of which is used for electricity generation. Lignite production peaked at 23.4m. metric tons in 1996. Subsequently, production has contracted, reaching only 19.5m. tons in 2002, in the wake of rising costs, concern over pollution and increased oil and gas output (see Energy). During 2001 these factors led to the cancellation of plans for two new lignite-fuelled power plants

Thailand remains an important source of a wide range of stones, including sapphires, rubies, zircon, garnet, quartz and jadeite. Since the early 1980s the expansion of production has been closely related to the development of the Thai jewellery industry, but has become increasingly incapable of supplying its needs. While, by the early 1990s, Thailand had become the world's principal exporter of jewellery, the industry had become increasingly dependent on imported stones, particularly from Sri Lanka, Myanmar and Cambodia.

The Thai tin industry comprises a small number of large dredging concerns and numerous small-scale producers, using very simple hydraulic technology. Many of these enterprises are highly marginal, reworking old alluvial deposits, and move in and out of production depending on the prevailing price levels. Since the early 1980s low world prices, increasing operating costs and the expansion of production in lower-cost producers have resulted in the almost complete collapse of the industry. Large numbers of operations have closed, and production of tin concentrates, with a metal content of some 73%, declined from a peak of 46,547 metric tons in 1980 to 14,393 tons in 1991, to 3,926 tons in 1994 and to 800 tons in 1997. Subsequently, despite the contraction of industrial production, the high cost of imports resulted in an expansion of tin output to 3,400 tons in 1999. However, this recovery proved short-lived in the context of a faltering industrial recovery and falling international prices: production declined to 2,383 tons in 2000, 1,384 tons in 2002 and approximately 1,000 tons in 2003.

MANUFACTURING

Since the early 1980s there has been rapid expansion and diversification of the Thai manufacturing sector. Between 1980 and 2002 the sector increased its shares of: export earnings, from 28.8% to 82.0%; GDP, from 21.3% to 33.9%; and the labour force, from 2.0% to 17.8%. This rapid expansion was initially led by the textile and garment sector and from the mid-1980s by electronics and electrical goods.

The electrical goods sector initially emerged during the early 1970s, led by foreign and later by Thai companies producing for the highly protected domestic market. Electronic assembly work for export began during the late 1970s, involving almost exclusively foreign capital. Expansion was rapid after 1986, with heavy investment and relocation of production by companies based particularly in Japan, the Republic of Korea, Taiwan and Singapore. In 1980 the electronics and electrical sector contributed 15.2% of the earnings from manufactured exports; this figure rose to 24.5% in 1991 and to 42.5% in 2002, reaching 34.9% of total merchandise export earnings. This rise was accompanied by significant changes in the composition of exports. In 1980 91.0% of exports were integrated circuits; however, by 2001 these accounted for only 28.1% of the total value of exports, while computers and computer parts (particularly hard disk drives) accounted for 35.7% and electrical appliances and machinery 32.2%. Nevertheless, the extent of skill and technological upgrading remains limited, and the industry is basically a labour-intensive assembly activity. Thailand cannot effectively compete with such lower-cost locations as China, particularly under prevailing conditions of oversupply in the global electronics sector. During 2002 the Taiwan-based company Hana Microelectronics, which had been operating in Thailand since the late 1980s, announced that it was cancelling a planned US $12m.–$15m. expansion of integrated circuit production in the country and transferring all of its operations to Shanghai. The company cited global market conditions and the limited progress made by Thailand over the last 10 years in upgrading basic industrial infrastructure as reasons for its decision.

Despite the relaxation of regulations and tariffs, sufficient barriers remain for Thai-owned concerns to control most of the domestic market for electrical and electronic products. The situation may well change dramatically with the full implementation of AFTA and Thailand's obligations under membership of the WTO. While many Thai companies are involved in sub-contracting, the export market remains dominated by Japanese and NIC concerns, a situation reinforced by the closure of Thai companies since 1997 and some significant foreign purchases. By 2002 there were 100 wholly Thai-owned electronics companies, 243 jointly-owned companies and 288 foreign companies operating in Thailand. Of the latter, 50% were Japanese, 16% Taiwanese, 7% Singaporean and 5% Korean.

Textiles have been a major sector since the 1960s, first for the internal market, and later for export. The expansion of the sector has been closely associated with Japanese investment through partnerships with Thai firms. From 1986 textiles were the largest source of export earnings until overtaken by the electronics sector in 1994. Textiles' share of export earnings fell from 14.6% in 1988 to 5.5% in 2002. Expansion has been impeded by the existence of a surplus in the world textile market and by the imposition of quotas and tariff barriers by the EU and the USA in particular. Considerable efforts have been made to develop new, largely 'non-quota', markets in the Middle East, Eastern Europe, Australia and Japan. These developments presented the Thai textile industry with an increasingly diversified market, which made it less vulnerable to increase in protection or reduction in quotas on the part of the EU or the USA.

The textile sector remains heavily dependent on imported raw materials and machinery. Domestic production of cotton is small and of low quality and 90% of requirements are imported. Similarly, despite the recent expansion of synthetic fibre production from 231,443 metric tons in 1990 to 768,100 tons in 2002, representing over 95% of the sector's needs, all the raw materials are imported. Despite the long-term success of textiles, increasing concern has been expressed regarding the large number of outdated plants, ineffective controls on standards, the poor development of dyeing and finishing sections and the shortage of skilled labour. In 2000 less than 25% of the 4m. workers employed in the sector were classed as 'skilled', while Thailand faced the loss of comparative advantage to such lower-cost locations as Bangladesh and China. In 1996 it was reported that the hourly labour costs in the Thai textile industry were US $1.49, compared with $0.48 in China; even after the depreciation in the value of the baht, these two countries still maintained a labour cost advantage over Thailand. During 2000 the Government began to consider the textile sector a 'problem' industry and, early in 2001, a World Bank project was implemented, aimed at encouraging investment and moves into high value-added areas of production.

The heavy dependence of the textile sector on imports is typical of much of Thai industry. However, during the 1990s the expansion of domestic steel, cement and refined petroleum products began to reduce some of the serious deficiency in basic manufacturing industry. This expansion has combined with the continuing recession to create over-capacity in a number of areas. This has become particularly acute in the oil-refining sector. During 1996 the opening of two new refineries by Shell and Caltex, and the expansion of capacity by Esso, proved sufficient for the small export trade. However, since 1997 declining demand has created serious problems, with a number of concerns becoming heavily indebted. Capacity utilization decreased from 90% in 1997 to 75% in 1999, slowly recovering to 80% by the end of 2003.

As the manufacturing sector has expanded there has been a general increase in the size of manufacturing concerns: the proportion of workers employed in plants with more than 50 employees rose from 19.1% in 1980 to 45.2% in 1996. However, the manufacturing sector remains dominated by small-scale operations. In addition, much of the expansion in textiles, garments, and other labour-intensive production has taken place through casual employment, out-work and subcontracting to large numbers of very small workshops—many neither registered nor regulated. These developments must be set against the emergence of a number of very large corporations, notably Siam City Cement, which before the 1997 crisis had a controlling interest in over 40 companies and was the largest manufacturing group in ASEAN.

Following the 1997 crisis there was some restructuring and concentration of production. Between 1997 and 2000 some 20,000 plants closed, 16% of those registered. Most of these were small or medium-sized concerns. In addition, some of the major concerns restructured their operations; the animal food manufacturer Charoen Pokphand, for example, disposed of a number of domestic and overseas subsidiaries and concentrated on its core agribusiness activities. However, there were few major corporate mergers. This reflected the continuing dominance of family ownership, low levels of capitalization, cumbersome merger, take-over and bankruptcy procedures, and the slow progress of debt restructuring (see Finance and Banking). There has been a similar inhibition of foreign mergers and take-overs, despite the relaxation of controls over foreign ownership (see Investment). Under normal trading conditions this had made mergers rare and hostile take-overs almost unheard of. In August 2002, after three years of negotiations, Siam City Cement agreed to merge its two steel concerns with the NTS Steel Group. The resultant Millennium Corporation would therefore control 25% of Thai steel-making capacity. This was by far the largest post-1997 industrial merger. However, whether this heralded the beginning of a major wave of mergers, as some observers suggested, remained highly debatable.

Following the currency depreciation of 1997, Thai manufacturing regained some of its competitiveness. However, this gave only a temporary respite, and from the latter part of 2001 there were indications that competitiveness was declining. Unless the problems of low productivity, skill shortages and infrastructural deficiencies are rapidly addressed, Thailand will be unable effectively to gain an advantage in the face of competition elsewhere in the region. Concerns have also been expressed over the impact of the progressive implementation of AFTA and obligations to the WTO. These will, in many cases, give domestic manufacturers access to cheaper components, raw materials and machinery, but will also open the national market to foreign competition, which may undermine some domestically orientated manufacturing and result in substantial business failures and loss of employment.

ENERGY

Thailand has substantial offshore natural gas resources, with 13 proven fields in the Gulf of Thailand and rather more limited onshore oil reserves in three small fields, the most successful at Kamphaengphet. Gas began to be piped ashore in 1981, and from 1990 production expanded at some 20% per year; by 2003 it had risen to 75,200m. cu ft of natural gas and 9m. metric tons of crude petroleum. The expansion of oil, gas and lignite (see Mining) production reduced Thailand's dependence on imported energy from a peak of 90% in 1992 to 60% in 2003. However, the level of imports still makes the economy vulnerable to increases in international oil prices. In consequence, there are plans to reduce imports through energy-saving campaigns and in 2002 the National Ethanol Committee was established, with the aim of promoting the mass production of ethanol for cassava and sugar. More significant in the longer term will be increased access to energy resources in neighbouring countries.

Thailand has been importing electrical power from Laos since 1978; this involves almost the entire 150,000 kW output of the Nam Ngum plant. In 2000 agreement was reached over the supply of electricity from Laos's largest hydropower plant, Nam Thuen 2, which is scheduled to commence in 2006. Further negotiations are in train to develop additional supplies from Laos and Cambodia. In addition, since 2000 the Petroleum Authority of Thailand (PTT) has imported gas from Myanmar's Yadana and Yetagun fields. Early in 2001 work started on the exploitation, with Malaysia, of the offshore gas reserves of the Joint Development Area. It was expected that the completion of gas pipelines and an associated separation plant would, by 2007, give Thailand access to gas that would increase the Kingdom's reserves by 40%. However, protests by local people and environmental groups have resulted in substantial delays; it is now unlikely that the project will be completed before 2009. Even more problematic is the exploitation of the substantial offshore gas reserves believed to lie in a 2,600 sq km area, of which the ownership is disputed with Cambodia.

In the energy sector, as with Thai infrastructure as a whole (see Communications and Infrastructure, below), the Government has increasingly looked to the private sector to finance developments. During 1996 the Electricity Generating Authority of Thailand (EGAT) reached agreement with two private companies to construct power plants. In addition, the Ratchaburi Electricity Holding Company, the largest generating plant in the country, was sold in late 2000, following the withdrawal of opposition by the labour unions. However, while both EGAT and the PTT are scheduled for privatization, progress has been slow and the subject of considerable opposition and debate. In 2001 the PTT was converted into a limited company (made possible by the 2000 Corporatization Act) and in November 30% of its shares were sold. While this raised funds, it left the Government firmly in control of policy and prices, a situation much criticized by foreign investors.

INVESTMENT

During the late 1970s and early 1980s investment was equivalent to between 25% and 28% of GDP, rising very sharply from 26% in 1986 to 40% in 1990 and then fluctuating between 38% and 41% until the 1997 crisis. The increase in the level of investment from the mid-1980s was almost entirely financed by the inflow of foreign funds, initially in the form of direct investment, but later also increasingly as loans to the private sector. Despite the rapid growth of the 1986–96 period, Thailand has not been able to generate the high levels of domestic savings and investment that characterized the economies of Singapore and the Republic of Korea, and remains heavily dependent on foreign sources of funding.

Prior to 1986 direct foreign investment in Thailand was relatively small, compared with that of other ASEAN countries. In general, Thailand was unattractive to multinational corporations and foreign investors because of poor international communications, uncertainty over the country's long-term political stability, and the absence of any significant labour cost advantages. However, between 1986 and 1991 a combination of rising costs elsewhere in the region and the appearance of economic and political stability made Thailand one of the most attractive investment locations in Asia, particularly for investors from Japan and the Asian NICs. During the period 1987–96 33.3% of net foreign investment was from Japan, 36.6% from Hong Kong, Taiwan and Singapore, 16.5% from the USA and 9.3% from the EU. Since the late 1980s there has been a general decline in the share of Japanese investors and an increase in the share of those from the NICs. There has, however, been some shift in these shares. Between 1997 and 2001 the sources of foreign investment were: Japan 26.3%, NICs 29.3%, EU 17.6% and USA 17.8%.

The foreign investment boom began to falter during the early 1990s, with inflows contracting particularly sharply during 1994. This reflected the problems associated with inadequate infrastructure, congestion, skilled labour shortages, and increased competition from Viet Nam and China. Until 1994 foreign investment had been increasingly in the manufacturing sector, which expanded its share from 20% in 1983 to 48.7% in 1993. The most significant growth was in electronics and electrical equipment, the share of which increased from 14.3% in 1983 to 44.6% in 1992. Renewed growth of foreign investment from 1995 was associated with a decline in the share of manufacturing (to 32.0% in 1996) and greater flows into the property sector, which increased its share of net inflows from 5.5% in 1993 to 33.5% in 1996. Despite the 1997 financial crisis, foreign investment was maintained at a high level until 2001, with net inflows totalling US $18,876m.—an annual average of $3,775m. This was in marked contrast with the period 1990–96, for most of which Thailand was regarded as a 'hot spot' for regional investment, with high levels of economic growth closely related to the level of net inflows, although these amounted to only $8,951m.—an annual average of $1,279m. Given the depressed and uncertain state of the Thai economy and the ongoing issues of low productivity and competitiveness, the scale of investment inflows in 1998–2001 was initially surprising. However, little of this represented 'greenfield' investment, with the majority comprising recapitalization of the corporate structure, sometimes involving debt for equity swaps and the acquisition of distressed assets. Indeed, 57.9% of inflows in 1998–2001 involved foreign acquisitions (of 10% or more of equity); this contrasted sharply with 12.2% in 1990–96.

The high level of mergers and acquisitions activity since 1998 reflected the rapid liberalization of restrictions on foreign ownership—a key condition attached to the 1997 IMF rescue package. Before the crisis foreign ownership was governed by the 1972 Alien Business Law. This required that every registered business in Thailand should have majority Thai ownership, and prohibited even minority foreign ownership in certain industries. The only exceptions to this were those US-based corporations that were permitted to have 100% foreign ownership under the 1966 Thai-US Treaty of Amity and Co-operation, and companies operating under the BOI promotional programmes (see below).

Following the conclusion of the Thai Government's agreement with the IMF in August 1997, some immediate actions were taken to liberalize foreign ownership. These included the permitting of majority foreign ownership in the distressed financial sector, and in Thai companies operating under BOI's promotional schemes. These and other *ad hoc* measures implemented during 1998 and 1999 rapidly opened up the Thai economy and were codified in the Foreign Business Act, which became effective in March 1999, replacing the 1972 legislation. While most areas became open to majority foreign ownership, foreign participation continued to be limited in some sectors on grounds of: national security; cultural considerations; environmental issues; the fact of 'Thai nationals not being ready to compete with foreigners'; and other 'special reasons'. These prohibited activities included: the media; farming; fishing; real estate; domestic transport; mining; sugar refining; rice milling; engineering; architecture services; most construction services; tourism services; low-level wholesaling and retailing; insurance; telecommunications; accountancy; law; and some brokerage services.

While Thailand has given a commitment to the WTO to allow full access to all service areas by 2006, there remained various indirect barriers to foreign purchase. These included: continuing restrictions on the direct purchase of real estate, which can sometimes prevent foreign investors from acquiring majority control; anti-trust laws, which can prevent incoming buyers from acquiring a dominant market position following a takeover; and the intricate and demanding legal requirements for mergers. Overall, the country's complex, interlocking family ownership patterns and low levels of capitalization seriously limit the possibilities for hostile take-overs.

Since 1998 there have been significant movements of foreign capital into the finance, particularly banking (see Finance and Banking), and property sectors. The share of foreign investment in the manufacturing sector rose by 34% in 1999 and by 57.3% in 2001. However, until 2000 this tended to reflect the acquisition of assets rather than 'greenfield' investment. During 2001 the mergers and acquisitions share of inflows contracted sharply, while net investment increased to US $3,866m., compared with $3,366m. in 2000. This suggested a return to significant investment in new operations and the expansion of existing ones. To some observers, the level of investment in 2001 was indicative of a major recovery in Thailand's competitive position. However, in the event, these hopes proved to be short-lived, with a sharp contraction of inflows to $1,020m. in 2002, and a limited recovery, to a provisional figure of $1,800m., in 2003.

It appears that Thailand is no longer perceived as an export platform for the more labour-intensive textile and electronics sectors. In these areas Thailand is unable to compete with Viet Nam, and, much more importantly, the People's Republic of China. However, in contrast, Thailand has come to be regarded as a significant location for motor vehicle production. During 2002 Japan's Isuzu Motors Ltd began production of trucks for export in the country, and in 2003 Toyota announced a major expansion of its Samut Prakan plant and the establishment of a research and development centre, the first of its kind in South-East Asia. In addition, Ford, Honda and Thai Yamaha announced expansion of the export capacity of their existing plants. These decisions reflected Thailand's established capacity in motor parts, some relaxation of local content rules, and the lack of a protected and promoted national car company—such as PROTON in Malaysia.

Since 1959 the BOI has co-ordinated a programme of investment promotion, under which both Thai and foreign firms were issued with 'promotional certificates', which exempted them from restrictions on repatriating profits and capital, and on foreign ownership of land, import duties and taxes on equipment. The certificates were also guarantees against nationalization or competition from state enterprises. From 1982 the promotional programme became more selective; privileges were proportional to the degree to which the firms complied with criteria on employment, the use of Thai raw materials, provincial location and the level of exports. In addition, priority was given to the promotion and financing of research into new technology and particular emphasis has been given to biotechnology, computer software, solar cells and microwave insulators. Since 1996 the Ministry of Industry has offered additional incentives to those provided by the BOI for firms producing electrical appliances and parts, precision instruments, medical equipment and vehicle parts, packaging, jewellery and processed food. During December 2001 a revised incentive programme was announced, which included tax concessions for companies establishing their regional headquarters in Thailand.

Since 1972 the BOI has also offered greater incentives to firms locating outside the BMR. In the mid-1980s the Kingdom was divided into three zones for promotional purposes, generally reflecting levels of development. Both the levels of privileges and the zones themselves have been the subject of various minor revisions and in 1998 the BOI announced the abolition of the whole system owing to its lack of success in encouraging investment outside the BMR. However, the zones were retained and were further revised in August 2000. Zone 1 comprises the BMR and has the lowest level of privileges. Zone 2 comprises the provinces of Ang Thong, Ayutthaya, Chatchoengsai, Chonburi, Kanchanaburi, Nakhon Nayok, Phuket, Ratchaburi, Rayong, Samut Songkhram, Saraburi and Suphanburi. These provinces have all received increased levels of investment and economic activity as a result of the spreading of development from the BMR. Zone 3 contains the remaining 53 provinces and has the highest level of privileges. During 2002 further modifications were implemented, particularly with respect to the promotion of industrial estates. However, there is little evidence to suggest that the BOI incentives have had any significant impact on decisions regarding location. During 2003, in an effort to stimulate manufacturing growth, any impact that the incentives might have had was seriously undermined by the extension of most of the Zone 2 and 3 privileges to Zone 1.

Despite streamlining measures and changes in regulations, the BOI has continued to be criticized for application procedures, bureaucratic delays, and the rejection of applications in order to protect established influential producers. Some critics have also seen the privileges as ineffective in attracting investment. However, the debate over the effectiveness of the promotional programmes could well become irrelevant, given Thailand's commitment to the WTO. In August 2000 the BOI revised the investment privileges, reducing the indirect subsidy element in order to conform to WTO agreements. Tax exemptions were removed for companies that do not export and the length of corporate tax 'holidays' was reduced. However, in order fully to comply with WTO requirements, many of the regulations regarding levels of exports and the use of domestically produced components and raw materials would also have to be removed. Indeed, a 2002 ruling by the WTO implied that the BOI would have to withdraw most of its remaining concessions. Subsequently, an agreement was reached that gave Thailand until 2005 to comply fully.

Until the end of 2001 the Thaksin Government appeared to take a more nationalist approach to investment than its immediate predecessors, a reflection of its election pledges. Privatization was represented as being 'privatization for Thais'; for example, the Prime Minister stated that no foreign concern would be invited to purchase shares in Thai Airways International. While such claims might have been no more than political rhetoric, they did not send reassuring messages to foreign investors. However, in December 2001 a new programme to promote foreign investment was announced; this was accompanied by commitments to invite foreign participation in the sale of state enterprises and to match foreign funds invested in the Stock Exchange of Thailand (SET). This was followed by the abandonment of plans to limit the activities of foreign retail chains and the reversal of restrictions on foreign ownership in the telecommunications sector imposed in November 2001 (see Communications and Infrastructure). Overall, there seemed to have been a general reduction in statements of economic nationalism.

FINANCE AND BANKING

To a considerable extent the 1997 crisis arose from the rapid liberalization of the financial regime from 1990 onwards without significant development of the system of central regulation. From 1991 virtually all restrictions on foreign-exchange and capital movements were removed. Lending rates were liberalized, deposit interest rates (except those for savings) were freed, and a conduit for international capital, the Bangkok International Banking Facility (BIBF), was created in March 1993. The BIBF licensed 32 foreign banks to provide 'offshore' facilities for Thailand, including 12 of the 14 foreign banks already fully licensed to operate branches within the country. This facility in particular, combined with the financial liberalization, gave Thai companies easy access to cheap international funds. The freedom of financial movements, combined with the volatile and ineffectively regulated SET, the promotion of overseas corporate bond issues and a currency that was remarkably stable from the early 1980s until 1997, encouraged speculative movements of capital. These movements became increasingly large and erratic and in 1995 the Bank of Thailand reintroduced some controls. However, these proved of limited effectiveness and, with the benefit of hindsight, it was clear that from mid-1996 an extremely unstable financial situation was emerging. The Bank of Thailand, while expressing concern over speculative movements, took no further action to regulate matters. Subsequently, much attention focused on the unsatisfactory nature of the central bank's monitoring and management procedures. During the latter part of 1996, however, speculative pressure on the baht, a reflection of its degree of linkage with the US dollar and the size of the current-account deficit, resulted in a massive intervention by the Bank of Thailand. The attempt to support the currency was doomed to failure, and in early July 1997 the baht was placed on a 'managed float'. This resulted almost immediately in a 20% depreciation. Further depreciation meant that by 12 January 1998 the baht had reached an historic low against the US dollar of 55.5, a 45% depreciation since July 1997. This was an undervaluation, and the baht then began to appreciate against the dollar, reaching 40.8 by mid-1998. The termination of the direct link between the baht and the US dollar and the exposure of the regulatory weakness of the Bank of Thailand removed two significant contributory factors to Thailand's long-term financial stability and attractiveness for investors. However, this did not result, as many expected, in a dramatic decline in foreign investment.

The 1997 financial crisis resulted in the suspension of the activities of 58 of the 91 major financial institutions (42 at the behest of the IMF), 56 of which were subsequently closed and 12 placed under state management. In addition, six banks (Siam City Bank, Bangkok Metropolitan Bank, First Bangkok City Bank, Bangkok Bank of Commerce, Union Bank of Bangkok and Laem Thong Bank) were nationalized, owing to their inability to recapitalize. The latter four were closed and the assets of the First Bangkok City Bank and the Union Bank of Bangkok were absorbed into the Krung Thai Bank (which was state-owned before the crisis). Two new state-controlled banks were established to assume control of the assets of the other closed banks and finance companies: the Bankthai and the Radanasin Bank.

Prior to the 1997 economic crisis, foreign banks had only been permitted to open three branches, one of which had to be in the provinces, and they were excluded from the purchase of majority ownership of Thai banks. The latter restriction was removed shortly after the economic crisis, permitting the sale of the Radanasin Bank to the Singapore-based United Overseas Bank (UOB) in 1999, the acquisition of the Bank of Asia by the Dutch-based ABN AMRO and the sale of the Thai Danu Bank to the Development Bank of Singapore. In addition, there were major increases in foreign participation in the Thai Farmers Bank (now the Kasikorn Bank—49.98%), the Bangkok Bank (49%), the Siam Commercial Bank (49%), and the Bank of Ayudhya (40%). While further foreign purchases had been expected, these did not materialize. Potential purchasers were discouraged by the consistently high level of non-performing loans (NPLs) in the sector, continued operating difficulties, the failure of the existing foreign banks to expand their market shares and the large losses that they had recorded. This situation was reflected in the closure during 2001 of the Thai operations of the Sukbank (of Japan), the Dresdner Bank (of Germany), and the Ibducto Bank (of Japan). During 2002 further changes to the banking system took place with the merger of the state-controlled Siam City Bank and the Bangkok Metropolitan Bank, and the privatization of the Bankthai. In addition, in April 2002 the Bank of Thailand issued a restricted licence to the Thanachart Bank; this allowed it to enable the issue of cheques and to provide overdraft facilities after five years of operation, then effectively turning it into a commercial bank. Overall, between 1998 and 2003 the structure of the Thai commercial banking sector changed radically. In 1997 there were 15 Thai-owned commercial banks in operation, one of which was state-controlled. By 2003 there were only 12 fully operational banks, four of which were foreign-owned and two under state control. Prior to the crisis 12 banks had been family-controlled; by 2002 this had been reduced to three—the Bangkok Bank, the Thai Farmers Bank and the Bank of Ayudhya—all with much reduced holdings and substantial foreign ownership. However, significant family interests have been retained in other banks, notably two of the banks under foreign control, where members of the families formerly in control remain CEOs. While the structure of the banking system has been transformed since 1997, it is important not to exaggerate the extent of foreign control. The four banks with majority foreign ownership only controlled 10.2% of branches in the country, 5.6% of deposits, 6.2% of the market and 4.7% of assets. However, foreign influence on the sector has already been very much greater than these figures would suggest. Foreign interests were believed to have played a significant part in its rapid modernization, increasing the accessibility of branches and encouraging the spreading of automatic teller machines (ATMs), online banking and the expansion and diversification of consumer credit. More significantly, the increased foreign presence reinforced the moves by the Ministry of Finance and the Bank of Thailand to introduce international forms of regulation and banking practice. This was particularly evident in the manner in which the pre-crisis debt equity levels, commonly 3:1, have been generally reduced to 1.5:1, compared with the Western normal maximum of 1:1.

In 2004 further major changes were in progress in the financial sector. In May ABN AMRO sold its 80.77% holding in the Bank of Asia to Singapore's UOB, which already held a 79%

share in UOB Radanasin Bank. It was expected that UOB would merge its two Thai operations. Indeed, this consolidation could be required under new regulations allowing foreign banks to maintain only one representative institution. Further concentration also resulted from the three-way merger of the Thai Military Bank, DBS Thai Danu Bank and the Industrial Finance Corporation of Thailand (IFCT), which was finalized in September 2004. Such mergers, which have been extremely rare in Thailand, could become a significant feature of the financial sector if the Finance Master Plan, launched in January 2003, is fully implemented. The Plan aims to transform the entire Thai financial sector, engendering competition by placing institutions on a 'level playing field', with new bank licences blurring the distinction between different types of business, and preparing for full market liberalization in 2007, in order to comply with WTO membership regulations. It is expected to result in a substantial reduction in the number of financial institutions. However, there was likely to be considerable opposition to the plan.

The crisis and subsequent process of liberalization have also significantly transformed the insurance sector. Since 1997 the number of insurance companies has expanded rapidly owing to the relaxation of restrictions on the issue of licences. Foreign investment in the sector has been stimulated by the fact that companies with up to 25% foreign ownership and an equal proportion of foreign directors were permitted to operate within it. Majority ownership and a 50% foreign directorship were permitted from 2004 in order to comply with WTO regulations. By the end of 2001 there were foreign stakes in 13 of the 25 life insurance concerns in Thailand. However, there is a long-standing exception to the controls over foreign ownership—the American International Assurance Co Ltd, which dominates the life insurance sector, commanding 50% of market share in 2001. This wholly-US-owned company has operated in Thailand since the 1960s under the terms of the 1966 Thai-US Treaty of Amity and Co-operation.

In response to the 1997–98 financial crisis a number of new institutions were set up to manage the sector's restructuring and improve regulation. The most notable of these were: the Bank of Thailand's Financial Institutions Development Fund (FIDF), which was established in order to guarantee the deposits and liabilities of the financial institutions; the Financial Restructuring Authority (FRA), which was formed to handle and sell the assets of the closed finance companies; the Asset Management Corporation (AMC), established to purchase those assets that could not be disposed of by the FRA in any other way; and the Thai Asset Management Corporation (TAMC), intended to purchase NPLs from the banking sector. Progress in the sale of assets and restructuring has been slow. This reflects the scale of the operations, the limited resources and expertise of the new financial bodies, a difficulty in finding buyers, a reluctance to raise new loans, and the cumbersome procedures in place to deal with asset sales and bankruptcy. These procedures have continued to be adhered to despite reforms and the establishment of a new Central Bankruptcy Court. Early in 2002 the World Bank reported that there were over 65,000 debt- and asset-related cases pending in the Civil Court system—a backlog that could take up to seven years to clear. Despite these issues, the Ministry of Finance estimated that the year-end level of NPLs as a percentage of financial sector loans had fallen to 17.9% in 2001, to 16.7% in 2002 and to 15.7% in 2003. This compared with a peak of 47.7% in May 1999. However, the rate of reduction of NPLs has declined, while the rate of reversion to NPL status has increased. More significantly, the World Bank has suggested that the official NPL statistics have become a poor indication of the extent of effective restructuring. The Bank concluded that, once allowances had been made for changes in the definition of NPLs adopted by the Thai system, reversions and write-offs, in mid-2002 the level of 'distressed' loans was still some 35%. The Bank of Thailand disputed this view, and in 2003 announced a plan to reduce the share of NPLs to 5% by 2006. This would involve amending the TAMC Act to allow the purchase and management of the NPLs of commercial banks.

By the end of 2002 the FIDF had accumulated debts of US $33,000m., which the Government has assumed partial responsibility for through bond issues. By the end of 2002 $21,500m.-worth of government or government-guaranteed bonds had been issued, with a further $11,300m. scheduled before 2006–07, when the FIDF was scheduled to close. While such bond issues lessened the impact on public-sector borrowing, this has expanded dramatically as a direct result of the Government's assumption of responsibility for NPLs and the debts of the banking and financial sector. According to the World Bank, public-sector debt increased from the equivalent of 37.0% of GDP in 1996 to a peak of 57.6% in 2001, before declining to 53.0% in 2002 and to 49.0% in 2003. The Ministry of Finance has asserted that some two-thirds of the increase was directly related to the debt crisis; the remainder reflected increased government expenditure (see below).

Since 1999 the Government has initiated a series of programmes in an attempt to stimulate economic recovery. Initially, US $3,000m. was directed to the support of debt-laden companies unable to secure finance from banks, many of which are still also overburdened with NPLs. A further $1,400m. programme was announced in March 2000. This included tax reductions aimed at boosting consumer demand and reducing business costs, measures to reduce energy costs and schemes aimed at generating 486,000 jobs, principally in rural areas, through small-scale infrastructure and public-works projects. The tax cuts included the reduction of value-added tax (VAT) from 10% to 7% until March 2001 (thus reversing the 1997 increase demanded by the IMF), elimination of the 1.5% VAT levied on small businesses, increased personal tax allowances and the reduction of fuel-oil tax from 17.5% to 5% (see Future Development and Prospects). During 2001 corporation tax was reduced from 30% to 25% for a five-year period. Tariffs were decreased on some 500 categories during 2000 and on some 9,000 further items during 2001. These changes reflected attempts to reduce costs, as well as moves to implement AFTA and commitments to the WTO. However, the tariff reductions increased government dependence on tax income at a time when the depressed state of the economy was exacerbating the problems of low collection rates, thus increasing the pressure on the budget deficit and public-sector borrowing, with the budget deficit as a proportion of GDP increasing from 2.2% in 2000 to 2.6% in 2001. However, the World Bank suggested that with the inclusion of all the quasi-fiscal measures and off-budget expenditure—such as the Village Development Fund—the real public-sector deficit was 3% in 2001 and would rise to some 6% in 2002. In the event, official figures suggested a contraction of the deficit, to 1.4%, during 2002 and the creation of a (provisional) 0.4% surplus in 2003. This abrupt change reflected increased revenues, particularly from VAT and corporation tax, high levels of consumption, the return of more companies to profitability, and, during 2003, expenditure below the planned level.

Prior to the 1997 economic crisis, Thailand was by no means a heavily-indebted country. Indeed, rapid growth of GDP and export earnings, combined with the emergence of government budget surpluses that enabled the repayment of some debts, reduced the debt-service ratio from 30.0% in 1986 to 10.5% in 1995. However, from 1996 a sharp contraction in export earnings and the depreciation of the currency increased the debt-service ratio to 15.5% in 1997, 22.0% in 1999 and to 20.8% by the end of 2001. There were also marked changes in the composition of debt. In 1996 38% of foreign debt was short-term, and 74% of medium- and long-term debt was owed by the private sector. By the end of 2000 short-term debt had fallen to 18.4%, and the private sector's share of medium- and long-term debt to 57.4%. These changes in the composition of debt reflected restructuring and the extent to which the Government had assumed responsibility for the debts of the financial sector. However, there has been significant repayment—some ahead of schedule—with Thailand's external debt declining from a peak of US $109,000m. (72% of GDP) in 1997, to $71,000m. (61.1%) in 2001, and $52,300m. (34.4%) by the end of 2003, with the debt-service ratio falling to 15.7%.

FOREIGN TRADE

Since the early 1980s the composition of Thailand's exports has changed dramatically. In 1980 68.4% of export earnings were derived from the agricultural sector and 28.8% from manufacturing. By 2002 the manufacturing sector's share of export earnings had grown to 85.0%, while that of agriculture had declined to 7.0%. This spectacular expansion of manufactured exports was associated with an increase in the share of import expenditure on machinery, components and raw materials (excluding fuels and lubricants), from 48.5% in 1980 to 73.7% in 2002. While between 1986 and 1995 the annual increase in export earnings averaged 18.5%, this failed to keep pace with the growth of imports. To some extent the resultant widening trade gap was offset by the growth of tourism and remittances from overseas workers. However, from the early 1990s these inflows were countered by the increasing number of Thais who travelled abroad.

The increasing trade deficit, reinforced by higher international interest rates and the level of profit repatriation from the substantial inflow of foreign business and investment that had taken place since the mid-1980s, resulted in a substantial rise in the balance-of-payments current-account deficit, from US $6,364m. in 1993 (equivalent to 5.7% of GDP) to $14,691m. in 1996 (the equivalent of 8.1% of GDP). The deficit fell sharply to $3,024m. in 1997 (the equivalent of 0.9% of GDP) in the wake of the financial crisis and a sharp decline in imports. In every subsequent year surpluses were recorded in both the balance of trade and the balance of payments. In 2003 these totalled $4,200m. and $8,400m., respectively, the latter the equivalent of 5.9% of GDP. However, these high levels of surplus were not the result of any sustained recovery of the export sector; rather, they reflected declines in imports. Between 1998 and 2003 export growth averaged a respectable 8.1%, compared with 18.6% for the period 1986–95. Only in 2000 was there a recovery of exports to their pre-1997 levels, with an increase of 27.3%. However, the continued dependence of much of the Thai export sector on imports was reflected in a similar increase in the cost of imports.

From the early 1980s onwards Thailand became increasingly dependent on the USA as a destination for exports and on Japan and the NICs as sources of imports. In 2003 Japan accounted for 14.2% of exports and 24.1% of imports, while the USA took 17.0% of exports and contributed 9.5% of imports, and the EU purchased 14.7% of exports and supplied 10.0% of imports. The contraction of many Asian-Pacific markets, combined with government and private-sector efforts in opening new markets in the Middle East and North Africa, reduced the share of Thailand's exports to the region from 44.7% in 1996 to 35.8% in 1999. However, by 2002 exports to the Asian-Pacific market had risen to 47.7%, a reflection of the recovery of some established regional markets and a significant increase in trade with China.

The Thaksin Government has made efforts to promote exports through the provision of low-cost export credits for small and medium-sized enterprises (SMEs), agreements with other producers, notably with Indonesia and Malaysia over rubber prices, and through negotiating tariff reductions in new and established markets. The latter approach has been marked since 2000 by the investigation of a variety of bilateral free trade agreements, notably with Australia, Bangladesh, Chile, China, the EU, the Republic of Korea, Mexico, New Zealand, South Africa, and the USA. How much advantage there is for Thailand in these agreements is debatable. The treaty with China, which covers only trade in fruit and vegetables, has resulted in considerable protest from Thai farmers, who are facing an influx of cheaper Chinese produce while finding it difficult to develop markets in China, owing to both costs and health restrictions. The negotiations with the USA, which began in February 2004, remained highly contentious because of US demands for full access to such restricted sectors as agriculture and telecommunications.

TOURISM

Tourism has been Thailand's largest single source of foreign exchange since 1982. In 2002 earnings from tourism were the equivalent of 6.2% of GDP. The number of tourists increased from 4.2m. in 1988 to 10.9m. in 2002. Despite this impressive long-term growth, the sector has been experiencing problems since the early 1990s. Before the 1997 crisis, rising prices had made Bangkok, Pattaya and Phuket among the most expensive locations in South-East Asia. This was compounded by the adverse publicity attendant on the drugs trade and the sex industry and the associated treatment of women and children, and by increased competition from other Asian-Pacific tourist centres. In addition, many major tour companies were beginning to avoid Bangkok because of the levels of pollution and congestion. Bangkok, none the less, remains the gateway to the southern coastal locations, despite the increase in direct international flights to provincial centres. The increased pollution of the Gulf of Thailand was also having an adverse effect on the coastal resorts. As a result, growth in the number of tourist arrivals had occurred principally in the low-spending sector, while the more expensive Bangkok hotels had become heavily dependent on business travellers.

The depreciation of the currency during 1997–98 substantially increased Thailand's attraction for tourists. In 1998 the number of arrivals increased by 7.6%, in 1999 by 10.5%, and in 2000 by 10.8%. Arrivals slowed in 2001, increasing by only 5.7%, reflecting reduced levels of international travel from, in particular, North America and Europe, in the aftermath of the terrorist attacks on the USA on 11 September 2001. To some extent this was offset by the arrival of increasing numbers of visitors from the Middle East, India and China. In 2002 the number of tourists increased by 7.3%, 60% of whom came from East Asia. As in the pre-crisis period, growth has been principally in such southern coastal resorts as Phuket and Koh Samui. Occupancy rates in Bangkok, which had a substantial surplus of rooms before the financial crisis and depends heavily on business travellers, remained depressed.

During 2003 the post-crisis growth pattern in the sector was reversed, owing to the effects of the US-led invasion of Iraq, fears of terrorist activity and the regional outbreak of Severe Acute Respiratory Syndrome (SARS). Arrivals in that year fell by 7.3%, to 10.1m. The Tourism Authority of Thailand (TAT) predicted a significant recovery during 2004, although there was little sign of this during the first half of the year, a reflection, perhaps, of the violence in the south of the country and concerns over the epidemic of avian influenza.

COMMUNICATIONS AND INFRASTRUCTURE

The rapid growth that the Thai economy experienced from 1986 put very considerable pressure on the Kingdom's infrastructure. It was apparent that even before the present period of accelerated growth there were serious deficiencies in the provision of roads, railways, port facilities, power generation, water supply and telephones. These problems are at their most acute in the BMR and its immediate environs, where the majority of the recent growth has been concentrated.

In Bangkok the lack of an effective water supply system has resulted in increased levels of exploitation of underground sources. Some 40% of the population are dependent on deep wells for their water supply. This extraction, in conjunction with the weight of modern buildings, is resulting in serious and widespread subsidence, particularly in the central and eastern areas. These movements are resulting in serious damage to buildings, bridges, roads, drainage and water supplies.

The capital's road system is inadequate in two ways. First, in terms of the area devoted to roads (12.5% of the city area, compared with 20%–25% in European cities); and second, by the low level of integration. The latter is a reflection of the largely uncontrolled manner in which the urban area has expanded. By the early 1990s the average speed of traffic was less than 7 km per hour, and the World Bank costed commuting time as the equivalent of 1.7% of GDP. While there has been some improvement, the capital remains heavily congested.

A rail mass-transit system for Bangkok was proposed as long ago as 1971. In 1993 plans were finalized and contracts awarded for three projects that would together provide 100 km of elevated track. Subsequently operations were halted, and the plans were revised to put the central parts of the system underground. Construction finally began in mid-1996 but was disrupted by further changes of plan and disputes with contractors and landowners. In December 1999 the first project, run by the Bangkok Mass Transit System and comprising a 23-km elevated railway system, began operating, three years behind schedule. However, by the end of 2002 it was still only recording an average of 300,000 daily passenger trips compared with the planned 600,000 (later reduced to 430,000). This was attributed to some continuing operational problems and, more significantly, to the limited network and the high level of fares. The 20-

km central subway project, the Blue Line, began trial operations in April 2004 and was expected to have become fully operational by August. The third line (the Hopewell project), which would have connected the other two lines to suburban areas, has been cancelled. However, during 2001 the Mass Rapid Transit Authority (MRTA) announced plans to extend the central subway project by 13 km. While serious doubts have been expressed over the financial viability of this extension, it would at last provide Bangkok with an important, if limited, mass-transit system

The port facilities at Klong Toey have long been regarded as inadequate. The opening of the new port at Laem Chabang on the Eastern Seaboard in 1992 did not provide either as high or as rapid a level of relief as expected: firstly, because of the delay in completing links with Bangkok (the road link was completed during 2000 but plans for a high-speed rail link have been suspended) and secondly, owing to inadequate provision of container and handling facilities. The latter problem has resulted, since 2002, in the development of new container and cargo facilities at Klong Toey. Plans to privatize the Port Authority of Thailand (PAT), originally scheduled for 1999, have been repeatedly delayed by concerted opposition, most recently in 2002. In July it was announced that 30% of the PAT would be sold, but this was also postponed. During 2003 a plan was announced that would separate the ownership of the facilities from port operations. The operations would be privatized, while the facilities and the majority of the labour force would remain under state control. Whether this would prove any more viable than previous proposals remained uncertain.

The increasing congestion at Don Muang airport, which handles 35.5m. passengers annually, has led to plans to produce a second facility at Nong Ngu Hao with a capacity of 40m. passengers. However, this development was seriously disrupted by the severe floods of 1995, delays in relocating residents and disputes regarding the allocation of contracts. During 1997 the project was abandoned, only to be reactivated later in the year following a change of government. Construction finally started in 2002, with a planned completion date of September 2005. While critics argued that this was becoming increasingly unfeasible, Prime Minister Thaksin repeatedly emphasized his determination that the target be met. It was expected that some decentralization of air traffic would result from the upgrading to international status of facilities at Chiang Mai, Hat Yai, Ngu Hao, Phuket and Utapoa. During 2002 flights began from Chiang Mai to Dhaka (Bangladesh) and there have been proposals to develop the northern centre as an international hub.

There have been considerable improvements in telecommunications since the early 1990s. The number of telephone lines per 100 people increased from 4.7 in 1993 to 8.4 in 1999. A major factor in the expansion of provision has been the entry of private companies since 1992. However, the Thai telephone system still compares unfavourably with that of Malaysia, which had 19.8 lines per 100 people. Similar unfavourable comparisons can be made with regard to mobile cellular telephones; there were 32 per 1,000 people in Thailand compared with 99 per 1,000 people in Malaysia. The limitations of the Thai telecommunications sector are particularly apparent with respect to internet connections, with only 2% of the population having access to facilities in 2000 and only 15% of businesses making active use of them. The high level of charges for lines remains a major disincentive, with charges in 2003 reported to be more than double those in Malaysia and the Philippines.

The telecommunications infrastructure is controlled by two state corporations, the Telephone Organization of Thailand (TOT) and the Communications Authority of Thailand (CAT). While both are scheduled for privatization, there have been repeated postponements. Considerable vested interests existed within the ranks of the Thaksin Government and its supporters who were opposed to rapid liberalization of the telecommunications sector and increased foreign participation. This was reflected during 2001 by the reduction of the permitted level of foreign ownership in the sector from 49% to 25%. This caused problems for several companies, which already had foreign partners owning more than 25% stakes. In consequence, after much debate, the decision was reversed in July 2002. Concern over vested interests has delayed the establishment of the National Telecommunications Commission (NTC), which was intended to oversee the liberalization of the sector. However, Thailand has a commitment to the WTO fully to liberalize the telecommunications sector by 2006.

Mobile cellular telephone facilities have expanded rapidly, from some 2m. in 1999 to 17m. in 2001 and 21.5m. in 2003. The sector is dominated by Prime Minister Thaksin's family-owned Shin Corporation, which controls some 62% of the market. The other major operators all have a degree of foreign ownership. Total Access Communications, with 30% of the market, is partly owned by the Norwegian company Telenor. TA Orange, with 8% of the market, is a joint Thai venture with the British company Orange, although the establishment of this operation was initially seriously hampered by the changes in foreign ownership limits in the sector.

Since the late 1980s successive governments have increasingly looked to the private sector to finance the development of infrastructure. This view was reinforced by the IMF following the 1997 financial crisis and appears to remain the official position. In addition, under prevailing economic conditions and levels of public-sector debt, the prospects of substantial government expenditure on infrastructure appear remote. More disturbing is that while there has been some success with telecommunications, the progress of privatization has been slow, and the private sector has shown little interest in providing infrastructure under the prevailing economic and regulatory conditions. However, this did not prevent the proposal of some major schemes during 2002, notably a direct road route to China (from Chiang Rai to Kunming), an additional Mekong bridge, a regional highway to link Thailand with India via Myanmar, and the Kra canal. In 2003 plans were announced for a major national road building programme, and the construction of a new city in Nakorn Nayok.

FUTURE DEVELOPMENT AND PROSPECTS

The improved performance of the economy during 2002 reflected a recovery in global trading conditions and was reinforced by the comparatively diverse nature of the Thai economy and government policy. The Thaksin Government was significantly more interventionist and pro-active in managing the economy than its immediate predecessors. In a statement made during 2002 the Government stressed its commitment to promoting economic growth and poverty elimination through expansionary policies and deficit spending. Various attempts have been made to stimulate the economy through measures such as: tax concessions; low interest rates; loose credit and direction of lending by the state banks and financial institutions; support of SMEs; low-cost public housing construction; and public expenditure in general. In January 2002 the Bank of Thailand reduced interest rates to 2.0%, in November to 1.7% and in June 2003 to a historic low of 1.25%. However, tax cuts effective from January 2003 were partly offset by increased rates of VAT and corporation tax, though the latter remains one of the lowest in the world. An ambitious programme of poverty reduction and infrastructure development was being implemented during 2004, perhaps not entirely unconnected with the legislative elections expected to take place early in 2005. In April 2004 a range of corporate and income tax incentives was announced. It was expected that the expansion of expenditure and tax concessions would be funded by increased tax yields, perhaps supplemented by some privatization, although this seemed unlikely given the Government's privatization record. While it was possible that these measures would not result in a significant budget deficit, much of the planned expenditure would, as in previous years, be off-budget.

Overall, government policies since 2000 have combined with the changes in the financial sector and the availability of surplus funds to significantly expand consumer credit. This has contributed significantly to the growth of consumption, which has been the dominant component of recent economic growth. However, growth has also been stimulated by high levels of government expenditure. While there have been signs that the contribution of consumption has fallen with the gradual recovery of private investment and exports, the World Bank perceived that the increased levels of public expenditure in 2004 were becoming the 'new driver of growth'.

Against the positive aspects of government policies there had to be set the implications of the general approach of the Thaksin

Government. Its nationalist policies and rather more forceful, if inconsistent, nationalist rhetoric seemed to have been largely abandoned by December 2001. However, this was accompanied by increasing evidence that the Thaksin Government, like many of its predecessors, tended to favour particular vested business interests rather than national and political ones. This did little to encourage foreign investors, particularly when taken in conjunction with the slowing pace of liberalization, the limited effectiveness of corporate reform, the cumbersome legal system, the large number of companies that remained burdeneded with debts and lacking sufficient operating capital and the limited progress made in solving Thailand's long-term problems of infrastructure, labour force skills and competitiveness. Indeed, these factors, together with the continuing weakness of the financial sector, lay behind the general downgrading of Thailand during 2002 in most international risk and competitiveness ratings. Furthermore, Thailand's international image was not enhanced by an increasing number of domestic complaints concerning the repression of the media, the concentration of power in the office of the Prime Minister, disputes with foreign journalists and publications, the policies towards NGOs, conflicts with the bureaucracy, the divisive nature of the policies of the Thaksin Government and heavy-handed suppression of crime. In the context of the latter, the large number of shootings by police resulted in considerable international and domestic concern and criticism. The Government's standing was further undermined by the initial mishandling of the avian influenza outbreak and the upsurge of violence in the south of the country. However, in 2004 Thaksin and his Government still appeared to retain a high level of popular support.

While the outlook for the Thai economy remains uncertain and closely linked to the fortunes of the global system, the country remains economically and politically stable with low inflation, declining unemployment and gradually expanding capacity utilization. To this must be added the positive image generated by the declining level of international debt, the generally positive governmental policy towards economic management, and the reforming of education and training and public administration. However, Thailand's ongoing weak competitive position is a cause for concern. The Kingdom appears to be no longer regarded as a favourable location for such labour-intensive activities as the manufacture of textiles, footwear, clothing, electronics and electrical goods. In these areas Thailand, like many other countries in the region, has found it increasingly difficult to compete with China, where competitiveness and attractiveness to foreign investment appear to have substantially increased following its accession to membership of the WTO in December 2001. While the Thai Government has been attempting to open new markets, notably in North Africa, and is pursuing a policy of bilateral agreements, these, however successful, will not solve the Kingdom's fundamental problems.

Statistical Survey

Source (unless otherwise stated): National Statistical Office, Thanon Larn Luang, Bangkok 10100; tel. (2) 281-8606; fax (2) 281-3815; internet www.nso.go.th.

Area and Population

AREA, POPULATION AND DENSITY

Area (sq km)	513,115*
Population (census results)†	
1 April 1990	54,548,530
1 April 2000	
Males	29,844,870
Females	30,762,077
Total	60,606,947
Population (UN estimates at mid-year)‡	
2001	61,555,000
2002	62,193,000
2003	62,833,000
Density (per sq km) at mid-2003	122.5

* 198,115 sq miles.

† Excluding adjustment for underenumeration.

‡ Source: UN, *World Population Prospects: The 2002 Revision*.

REGIONS
(2000 census)

	Area (sq km)	Population	Density (per sq km)
Bangkok	1,565	6,320,174	4,038.4
Central Region (excl. Bangkok)	102,335	14,101,530	137.8
Northern Region	169,645	11,367,826	67.0
Northeastern Region	168,855	20,759,899	122.9
Southern Region	70,715	8,057,518	113.9
Total	513,115	60,606,947	118.1

PRINCIPAL TOWNS
(population at 2000 census)

Bangkok Metropolis*	6,320,174	Pak Kret	141,788
Samut Prakan	378,694	Si Racha	141,334
Nanthaburi	291,307	Khon Kaen	141,034
Udon Thani	220,493	Nakhon Pathom	120,657
Nakhon Ratchasima	204,391	Nakhon Si Thammarat	118,764
Hat Yai	185,557	Thanya Buri	113,818
Chon Buri	182,641	Surat Thani	111,276
Chiang Mai	167,776	Rayong	106,585
Phra Padaeng	166,828	Ubon Ratchathani	106,552
Lampang	147,812	Khlong Luang	103,282

* Formerly Bangkok and Thonburi.

Mid-2003 (UN estimate, incl. suburbs): Bangkok 6,486,401 (Source: UN, *World Urbanization Prospects: The 2003 Revision*).

BIRTHS, MARRIAGES AND DEATHS*

	Registered live births		Registered marriages		Registered deaths	
	Number	Rate (per 1,000)	Number	Rate (per 1,000)	Number	Rate (per 1,000)
1993	957,832	16.5	484,569	8.4	285,731	4.9
1994	960,248	16.4	n.a.	n.a.	305,526	5.2
1995	963,678	16.2	470,751	7.9	324,842	5.5
1996	944,118	15.7	436,831	7.3	342,645	5.7
1997	897,604	14.8	396,928	6.5	303,918	5.0
1998	897,495	14.7	324,262	5.3	317,793	5.2
1999	754,685	12.3	348,803	5.7	362,607	5.9
2000	773,009	12.4	339,443	5.4	365,741	5.9

* Registration is incomplete. According to UN estimates, the average annual rates in 1990–95 were: Births 19.7 per 1,000; Deaths 6.5 per 1,000; and in 1995–2000: Births 18.2 per 1,000; Deaths 6.8 per 1,000 (Source: UN, *World Population Prospects: The 2002 Revision*).

Sources: mainly Ministry of Public Health, Ministry of the Interior, Bangkok.

Expectation of life (WHO estimates, years at birth): 69.3 (males 66.0; females 72.7) in 2002 (Source: WHO, *World Health Report*).

ECONOMICALLY ACTIVE POPULATION*
(million persons aged 15 years and over)

	2000	2001	2002
Agriculture, hunting, forestry and fishing	13.89	13.59	13.74
Mining and quarrying	0.45	0.47	0.06
Manufacturing	4.99	5.68	5.86
Electricity, gas, water and sanitary services	0.17	0.17	0.16
Construction	1.50	1.58	1.70
Trade, financing, insurance and real estate	4.89	4.49	4.75
Transport, storage and communications	0.97	1.02	1.01
Other services (incl. restaurants and hotels)	4.83	5.60	5.71
Total employed	31.29	32.17	32.99
Unemployed	1.93	1.75	1.26
Total labour force	33.22	33.92	34.25

* Excluding the armed forces.

Source: IMF, *Thailand: Statistical Appendix* (December 2003).

Health and Welfare

KEY INDICATORS

Total fertility rate (children per woman, 2002)	1.9
Under-5 mortality rate (per 1,000 live births, 2002)	28
HIV/AIDS (% of persons aged 15–49, 2003)	1.50
Physicians (per 1,000 head, 1995)	0.24
Hospital beds (per 1,000 head, 1995)	1.99
Health expenditure (2001): US $ per head (PPP)	254
Health expenditure (2001): % of GDP	3.7
Health expenditure (2001): public (% of total)	57.1
Access to water (% of persons, 2000)	80
Access to sanitation (% of persons, 2000)	96
Human Development Index (2002): ranking	76
Human Development Index (2002): value	0.768

For sources and definitions, see explanatory note on p. vi.

Agriculture

PRINCIPAL CROPS
('000 metric tons)

	2000	2001	2002
Rice (paddy)	25,844	26,514	25,611
Maize	4,462	4,466	4,211
Sorghum	148	145	145
Potatoes	91	97	97
Cassava (Manioc, Tapioca)	19,064	18,396	16,868
Dry beans	226	238	260
Soybeans (Soya beans)	312	292	289
Groundnuts (in shell)	132	129	131
Coconuts	1,400	1,396	1,418
Oil palm fruit	3,256	4,089	3,902
Kapok fruit*	138	136	139
Cabbages†	205	210	210
Tomatoes	224	246	248†
Pumpkins, squash, gourds†	210	220	220
Cucumbers and gherkins†	210	220	220
Dry onions	277	266	280
Garlic	132	126	122
Other vegetables†	1,379	1,489	1,495
Watermelons†	400	410	410
Sugar cane	49,563	60,013	62,517
Bananas†	1,750	1,750	1,800
Oranges†	325	325	340
Tangerines, mandarins, clementines, satsumas†	650	650	668
Mangoes	1,633	1,700†	1,750†
Pineapples	2,248	1,979	1,655
Papayas†	119	120	120
Other fruits†	1,119	1,147	1,130
Tobacco (leaves)	74	64	74
Natural rubber	2,378	2,424	2,404

* Unofficial figures.
† FAO estimate(s).

Source: FAO.

LIVESTOCK
('000 head, year ending September)

	2000	2001	2002
Horses	9	8	8*
Cattle	4,602	4,640	5,909
Buffaloes	1,712	1,524	1,617
Pigs	6,558	6,689	6,989
Sheep	37	43	39
Goats	144	188	178
Chickens	215,584	220,000†	228,779
Ducks	27,884	28,448	25,090
Geese	233	250*	260*

* FAO estimate.
† Unofficial figure.

Source: FAO.

LIVESTOCK PRODUCTS
('000 metric tons)

	2000	2001	2002
Beef and veal†	167	172	180
Buffalo meat†	52	58	58
Pig meat†	475	486	508
Chicken meat*	1,091	1,230	1,320
Duck meat	103*	105*	93†
Cows' milk	520	564*	580*
Hen eggs	515	504†	500†
Other poultry eggs†	293	298	304
Cattle hides (fresh)†	44	45	45
Buffalo hides (fresh)†	6	7	7

* Unofficial figure(s).
† FAO estimate(s).

Source: FAO.

Forestry

ROUNDWOOD REMOVALS
('000 cubic metres, excl. bark)

	2000	2001	2002
Sawlogs, veneer logs and logs for sleepers	158	210	300
Other industrial wood	4,576	5,053	5,500
Fuel wood*	20,553	20,396	20,250
Total	25,287	25,659	26,050

* FAO estimates.

Source: FAO.

SAWNWOOD PRODUCTION
('000 cubic metres, incl. railway sleepers)

	2000	2001	2002
Coniferous (softwood)	17	18	18*
Broadleaved (hardwood)	203	215	270
Total	220	233	288

* FAO estimate.

Source: FAO.

Fishing

('000 metric tons, live weight)

	2000	2001	2002
Capture	2,997.4	2,932.4	2,921.2
Sardinellas	164.0	192.6	193.2
Anchovies, etc.	143.1	153.6	154.3
Indian mackerels	152.9	151.2	151.6
Aquaculture	738.2	724.2	644.9*
Giant tiger prawn	305.0	276.0	160.0*
Total catch	3,735.6	3,656.6	3,566.1*

* FAO estimate.

Source: FAO.

Mining

(production in metric tons, rounded, unless otherwise indicated)

	2000	2001	2002
Tin concentrates*	2,363	2,383	1,384
Tungsten concentrates*	54	92	53
Lead ore*	24,760	800	6,500
Zinc ore*	159,093	88,664	151,575
Antimony ore*	178	40	3
Gold (kg)	—	320	4,950
Iron ore*	100	50	570,110
Gypsum ('000 metric tons)	5,830	6,191	6,326
Barite	56,180	23,559	137,469
Dolomite	625,127	871,308	933,209
Feldspar	542,991	710,543	783,773
Gemstones ('000 carats)	928	1,071	1,597
Lignite ('000 metric tons)	17,714	19,617	19,602
Fluorspar†	4,745	3,020	2,270
Manganese ore†	225	45	—
Crude petroleum (million barrels)	20.9	22.2	27.2
Natural gas (million cu m)	20,190	19,637	20,451

* Figures refer to the gross weight of ores and concentrates.
† Metallurgical grade.

Source: US Geological Survey.

Industry

SELECTED PRODUCTS

('000 metric tons, unless otherwise indicated)

	1999	2000	2001
Synthetic fibre*	696	735	727
Raw sugar	5,630	6,447	4,865
Cement	25,354	25,499	27,913
Galvanized iron sheets	299	369	434
Pulp	854	917	n.a.
Petroleum products (million litres)	41,414	41,060	n.a.
Beer (million litres)	1,042	1,165	1,238
Integrated circuits (million pieces)	5,182	7,070	n.a.
Tin metal	17	17	n.a.

* Source: Asian Development Bank, *Key Indicators of Developing Asian and Pacific Countries*.

Sources: IMF, *Thailand: Statistical Appendix* (August 2001); and Bank of Thailand, Bangkok.

Finance

CURRENCY AND EXCHANGE RATES

Monetary Units
100 satangs = 1 baht.

Sterling, Dollar and Euro Equivalents (31 May 2004)
£1 sterling = 74.25 baht;
US $1 = 40.47 baht;
€1 = 49.56 baht;
1,000 baht = £13.47 = $24.71 = €20.18.

Average Exchange Rate (baht per US $)
2001 44.432
2002 42.960
2003 41.485

Note: Figures refer to the average mid-point rate of exchange available from commercial banks. In July 1997 the Bank of Thailand began operating a managed 'float' of the baht. In addition, a two-tier market was introduced, creating separate exchange rates for purchasers of baht in domestic markets and those who buy the currency overseas.

BUDGET

(million baht, year ending 30 September)

Revenue	2000/01	2001/02	2002/03
Tax revenue	783,809.3	815,920.6	856,440.3
Direct taxes	237,600.0	259,360.0	284,400.0
Personal income tax	101,830.0	102,000.0	115,500.0
Corporate income tax	125,770.0	145,360.0	153,900.0
Indirect taxes	546,209.3	556,560.6	572,040.3
Value-added tax	226,450.0	235,790.0	235,100.0
Specific business tax	22,440.0	16,200.0	13,740.0
Tax on petroleum and petroleum products	67,000.0	67,200.0	66,800.0
Excise taxes	17,695.0	17,258.0	18,886.0
Consumption tax	99,524.0	97,376.0	114,100.0
Taxes on international trade	98,625.0	104,100.0	102,880.0
Non-tax revenue	87,330.8	87,629.4	100,109.7
State enterprises	47,510.0	40,908.5	37,000.0
Other	23,960.8	35,731.1	43,928.8
Total	871,140.0	903,550.0	956,550.0

Expenditure	2000/01	2001/02	2002/03
General governmental services	178,642.7	186,161.1	191,023.6
General public services	44,338.2	52,275.2	54,321.2
Defence	76,445.7	77,207.4	75,691.1
Public order and safety	57,858.8	56,678.5	61,011.3
Community and social services	382,398.2	425,846.8	420,442.8
Education	221,591.5	222,989.8	235,393.8
Health	65,041.2	72,769.7	78,418.7
Social security and welfare	51,562.1	70,781.0	76,244.2
Housing and community amenities	37,494.1	52,984.2	24,084.2
Economic affairs and services	205,094.5	238,763.1	207,100.8
Agriculture, forestry, fishing and hunting	73,860.4	75,497.5	73,681.5
Transportation and communication	76,851.2	64,176.7	57,765.5
Others	46,932.4	91,628.0	65,516.4
Miscellaneous and unclassified items	143,864.6	172,229.0	181,332.8
Total	910,000	1,023,000.0	999,900.0
Current expenditure	679,286.5	773,714.1	753,454.7
Capital expenditure	218,578.2	223,617.0	211,493.5
Principal repayment	12,135.3	25,668.9	34,951.8

Source: Bureau of the Budget, Office of the Prime Minister.

INTERNATIONAL RESERVES

(US $ million at 31 December)

	2001	2002	2003
Gold*	686	869	1,071
IMF special drawing rights	5	4	—
Foreign exchange	32,350	38,042	40,965
Total	33,041	38,915	42,036

* Revalued annually on the basis of the London market price.

Source: IMF, *International Financial Statistics*.

MONEY SUPPLY

('000 million baht at 31 December)

	2001	2002	2003
Currency outside banks	440.9	496.0	546.9
Demand deposits at deposit money banks	134.4	157.8	215.5
Total money (incl. others)	650.6	674.9	869.2

Source: IMF, *International Financial Statistics*.

COST OF LIVING
(Consumer Price Index; base: 1990 = 100)

	1999	2000	2001
Food (incl. beverages)	166.9	165.1	166.2
Fuel and light	152.7	164.6	177.4
Clothing (incl. footwear)	152.8	154.3	155.9
Rent	126.3	126.1	125.8
All items (incl. others)	153.2	155.6	158.1

Source: ILO, *Yearbook of Labour Statistics*.

2002 (base: 2000 = 100): All items 102.3 (Source: IMF, *International Financial Statistics*).

2003 (base: 2000 = 100): All items 104.1 (Source: IMF, *International Financial Statistics*).

NATIONAL ACCOUNTS
(million baht at current prices)

National Income and Product

	2000	2001	2002*
Compensation of employees	1,541,640	1,603,945	1,682,305
Operating surplus	2,170,713	2,257,626	2,392,664
Domestic factor incomes	3,712,353	3,861,571	4,074,969
Consumption of fixed capital	728,297	759,426	790,807
Gross domestic product (GDP) at factor cost	4,440,650	4,620,997	4,865,776
Indirect taxes, *less* subsidies	482,613	512,839	586,078
GDP in purchasers' values	4,923,263	5,133,836	5,451,854
Net factor income from abroad	–76,874	–85,069	–89,494
Gross national product	4,846,389	5,048,767	5,362,360
Less Consumption of fixed capital	728,297	759,426	790,807
National income in market prices	4,118,092	4,289,341	4,571,553

* Provisional figures.

Source: National Economic and Social Development Board, Bangkok.

Expenditure on the Gross Domestic Product

	2001	2002	2003
Government final consumption expenditure	592,700	609,800	629,200
Private final consumption expenditure	2,913,700	3,067,400	3,337,600
Increase in stocks	46,700	43,700	71,300
Gross fixed capital formation	1,178,500	1,251,600	1,427,600
Statistical discrepancy	57,300	67,400	74,000
Total domestic expenditure	4,788,900	5,039,900	5,539,700
Exports of goods and services	3,386,100	3,516,900	3,898,900
Less Imports of goods and services	3,051,600	3,123,600	3,499,600
GDP in purchasers' values	5,123,400	5,433,300	5,938,900
GDP at constant 1988 prices	3,058,700	3,224,600	3,457,700

Source: IMF, *International Financial Statistics*.

Gross Domestic Product by Economic Activity

	2001	2002*	2003*
Agriculture, hunting, forestry and fishing	468,456	510,877	579,460
Mining and quarrying	126,204	136,457	155,719
Manufacturing	1,715,280	1,848,397	2,089,434
Construction	154,193	165,796	175,008
Electricity, gas and water	166,683	175,805	191,474
Transport, storage and communications	428,576	444,292	464,072
Wholesale and retail trade	856,816	867,418	912,288
Finance, insurance, real estate and business services	315,486	338,015	372,791
Public administration and defence.	221,865	244,532	262,346
Other services†	680,277	720,265	736,470
GDP in purchasers' values	5,133,836	5,451,854	5,939,062

* Provisional figures.

† Including restaurants and hotels.

Source: Asian Development Bank, *Key Indicators of Developing Asian and Pacific Countries*.

BALANCE OF PAYMENTS
(US $ million)

	2001	2002	2003
Exports of goods f.o.b.	63,082	66,089	78,397
Imports of goods f.o.b.	–54,539	–57,008	–66,790
Trade balance	8,543	9,081	11,606
Exports of services	13,024	15,391	15,774
Imports of services	–14,610	–16,720	–18,503
Balance on goods and services	6,957	7,751	8,877
Other income received	3,833	3,356	2,988
Other income paid	–5,200	–4,696	–4,790
Balance on goods, services and income	5,591	6,411	7,075
Current transfers received	990	978	1,275
Current transfers paid	–389	–375	–385
Current balance	6,192	7,014	7,965
Direct investment abroad	–344	–106	–558
Direct investment from abroad	3,892	953	1,866
Portfolio investment assets	–360	–913	–937
Portfolio investment liabilities	–525	–694	302
Other investment assets	577	4,135	–416
Other investment liabilities	–6,897	–6,263	–8,441
Net errors and omissions	–258	1,410	736
Overall balance	2,276	5,537	518

Source: IMF, *International Financial Statistics*.

External Trade

PRINCIPAL COMMODITIES
(distribution by SITC, US $ million)

Imports c.i.f.	1999	2000	2001
Food and live animals	2,019.4	2,068.1	2,501.1
Crude materials (inedible) except fuels	2,047.3	2,556.6	2,531.3
Mineral fuels, lubricants, etc.	4,879.3	7,606.7	7,526.7
Petroleum and petroleum products	4,775.5	7,371.8	6,707.2
Crude petroleum	3,897.1	6,108.5	5,773.7
Chemicals and related products	5,448.1	6,539.4	6,419.2
Organic chemicals	1,391.2	1,777.4	1,625.6
Basic manufactures	9,816.9	10,484.2	10,117.9
Iron and steel	2,728.7	2,809.0	2,608.5
Other metal manufactures	2,509.4	2,006.0	1,937.6
Machinery and transport equipment	21,675.9	27,392.6	27,985.4
Machinery specialized for particular industries	1,219.7	1,872.0	2,025.1
Office machines and automatic data-processing equipment	2,567.7	3,821.1	3,902.2
Parts and accessories for office machines, etc.	2,138.1	3,125.3	2,789.7
Telecommunications and sound equipment	1,256.3	1,868.3	2,704.4
Other electrical machinery, apparatus, etc.	9,739.4	12,986.8	11,023.9
Switchgear, etc., and parts	1,522.9	1,835.0	1,694.3
Thermionic valves, tubes, etc.	5,928.1	8,365.8	6,791.5
Road vehicles and parts*	1,304.7	1,976.6	1,999.1
Other transport equipment	2,108.6	409.3	1,476.8
Aircraft, associated equipment and parts	1,809.4	330.5	1,371.3
Miscellaneous manufactured articles	3,011.1	3,548.4	3,457.1
Total (incl. others)	50,309.1	61,450.6	62,057.5

* Excluding tyres, engines and electrical parts.

Exports f.o.b.	1999	2000	2001
Food and live animals	9,687.7	9,662.1	9,711.9
Fish, crustaceans and molluscs	4,095.7	4,335.8	4,034.8
Fresh, chilled, frozen, salted or dried crustaceans and molluscs	1,648.3	1,865.0	1,603.9
Prepared or preserved fish, crustaceans and molluscs	2,016.0	2,057.9	2,011.9
Cereals and cereal preparations	2,165.9	1,860.3	1,865.0
Milled rice	1,906.9	1,590.3	1,548.7
Crude materials (inedible) except fuels	2,104.5	2,790.7	2,390.6
Chemicals and related products	2,942.4	4,039.4	3,723.1
Basic manufactures	6,917.5	8,157.8	7,716.2
Textile yarn, fabrics, etc.	1,833.1	1,976.5	1,902.3
Machinery and transport equipment	24,490.8	30,016.1	27,334.0
General industrial machinery, equipment and parts	1,747.5	2,066.5	2,122.6
Office machines and automatic data-processing equipment	8,211.8	8,790.4	8,003.6
Automatic data-processing machines and units	1,935.3	1,999.2	1,785.2
Parts and accessories for office machines, etc.	6,033.0	6,522.0	6,002.8
Telecommunications and sound equipment	2,997.1	4,014.4	3,510.4
Other electrical machinery, apparatus, etc.	8,153.5	10,891.5	9,267.6
Thermionic valves, tubes, etc.	4,031.1	5,876.3	4,699.9
Miscellaneous manufactured articles	9,075.9	9,954.7	9,829.2
Clothing and accessories (excl. footwear)	3,495.9	3,790.1	3,618.4
Total (incl. others)	58,423.1	68,786.7	65,113.3

Source: UN, *International Trade Statistics Yearbook*.

PRINCIPAL TRADING PARTNERS
(US $ million)

Imports c.i.f.	2001	2002	2003
Australia	1,380	1,506	1,585
China, People's Republic	3,711	4,928	6,067
Germany	2,562	2,482	2,533
Indonesia	1,364	1,559	1,774
Japan	13,881	14,902	18,267
Korea, Republic	2,121	2,527	2,919
Malaysia	3,078	3,640	4,536
Oman	1,264	1,201	n.a.
Singapore	2,854	2,904	3,269
United Arab Emirates	1,529	1,428	2,028
USA	7,198	6,197	7,190
Total (incl. others)	62,057	64,721	75,809

Exports f.o.b.	2001	2002	2003
Australia	1,358	1,640	2,168
China, People's Republic	2,863	3,553	5,710
Germany	1,568	1,533	1,801
Hong Kong	3,298	3,699	4,332
Japan	9,964	10,001	11,435
Malaysia	2,722	2,835	3,887
Netherlands	2,028	1,890	2,376
Singapore	5,287	5,554	5,876
United Kingdom	2,328	2,391	2,590
USA	13,246	13,522	13,688
Total (incl. others)	65,112	68,851	80,518

Source: Asian Development Bank, *Key Indicators of Developing Asian and Pacific Countries*.

Transport

RAILWAYS
('000)

	1991	1992	1993
Passenger journeys	86,906	87,769	87,183
Passenger-km	12,819,567	14,135,915	14,717,667
Freight (ton-km)	3,365,431	3,074,786	3,059,043
Freight carried (metric tons)	7,990	7,600	7,498

Source: State Railway of Thailand.

1994: Passenger-km ('000) 13,814,000; Freight (ton-km, '000) 3,072,000.

1995: Passenger-km ('000) 12,975,000; Freight (ton-km, '000) 3,242,000.

1996: Passenger-km ('000) 12,205,000; Freight (ton-km, '000) 3,286,000.

1997: Passenger-km ('000) 11,804,000; Freight (ton-km, '000) 3,410,000.

1998: Passenger-km ('000) 10,947,000; Freight (ton-km, '000) 2,874,000.

1999: Passenger-km ('000) 9,894,000; Freight (ton-km, '000) 2,929,000.

2000: Passenger-km ('000) 10,040,000; Freight (ton-km, '000) 3,347,000.

Source: UN, *Statistical Yearbook*.

ROAD TRAFFIC
('000 motor vehicles in use at 31 December)

	1999	2000	2001
Passenger cars	2,124	2,111	2,281
Buses and trucks	731	775	804
Vans and pick-ups	3,098	3,209	3,341
Motorcycles	13,245	13,817	15,236

Source: Ministry of Transport.

SHIPPING

Merchant Fleet
(registered at 31 December)

	2001	2002	2003
Number of vessels	568	629	671
Total displacement ('000 grt)	1,771.4	1,879.6	2,268.7

Source: Lloyd's Register-Fairplay, *World Fleet Statistics*.

International Sea-borne Freight Traffic
(Ports of Bangkok and Laem Chabang)

	2000	2001	2002
Goods loaded ('000 metric tons)	12,153	12,744	14,803
Goods unloaded ('000 metric tons)	20,055	21,912	24,755
Vessels entered	6,145	6,625	7,000

Source: Port Authority of Thailand.

CIVIL AVIATION

	1988	1989	1990
Kilometres flown	85,003,000	93,871,689	101,589,000
Passengers carried: number	6,282,438	7,394,199	8,272,516
Passengers carried: passenger-km ('000)	16,742,348	18,876,972	19,869,143
Freight carried: tons	157,964	165,502	199,399
Freight carried: ton-km ('000)	589,064	626,497	799,267
Mail carried: tons	7,636	8,893	9,287
Mail carried: ton-km ('000)	33,576	39,104	39,070

Sources: Airport Authority of Thailand and the Department of Aviation.

Kilometres flown ('000): 153,000 in 1997; 158,000 in 1998; 163,000 in 1999 (Source: UN, *Statistical Yearbook*).

Passengers carried ('000): 14,236 in 1997; 15,015 in 1998; 15,950 in 1999 (Source: UN, *Statistical Yearbook*).

Passenger-km (million): 30,827 in 1997; 34,340 in 1998; 38,345 in 1999 (Source: UN, *Statistical Yearbook*).

Total ton-km (million): 4,460 in 1997; 4,682 in 1998; 5,184 in 1999 (Source: UN, *Statistical Yearbook*).

Tourism

FOREIGN TOURIST ARRIVALS

Country of origin	2001	2002	2003
Australia	366,468	358,616	284,749
China, People's Republic	695,372	763,708	624,923
France	238,550	254,610	220,659
Germany	407,353	412,968	389,293
Hong Kong	531,300	533,798	657,458
India	206,541	253,475	230,790
Japan	1,179,202	1,233,239	1,026,287
Korea, Republic	553,441	717,361	695,034
Malaysia	1,161,490	1,297,619	1,340,193
Singapore	669,166	687,982	633,805
Sweden	224,268	222,154	210,882
Taiwan	728,953	678,511	525,916
United Kingdom	522,117	574,007	550,087
USA	494,920	519,668	469,165
Total (incl. others)	10,132,509	10,872,976	10,082,109

Receipts from tourism (US $ million): 7,146 in 2000; 6,731 in 2001; 7,902 in 2002.

Sources: Tourism Authority of Thailand; World Tourism Organization.

Communications Media

	2001	2002	2003
Telephones ('000 main lines in use)	6,049.1	6,499.8	6,600.0
Mobile cellular telephones ('000 subscribers)	7,550.0	16,117.0	n.a.
Personal computers ('000 in use)	2,003	2,461	n.a.
Internet users ('000)	3,536.0	4,800.0	6,031.3

Source: International Telecommunication Union.

Radio receivers ('000 in use): 13,959 in 1997.

Television receivers ('000 in use): 17,200 in 2000.

Facsimile machines ('000 in use): 150 in 1997.

Book production (titles, excluding pamphlets): 8,142 in 1996.

Daily newspapers: 35 (with average circulation of 2,766,000 copies) in 1994; 35 (with average circulation of 2,700,000* copies) in 1995; 30 (with average circulation of 3,808,000 copies) in 1996.

Non-daily newspapers: 280 in 1995; 320 in 1996.

* Provisional.

Sources: UNESCO, *Statistical Yearbook*; UN, *Statistical Yearbook*.

Education

(2002)

	Institutions	Teachers	Students
Office of the Permanent Secretary	—	1,720	—
Religious Affairs Department	16,006	—	1,258,201
Non-Formal Education Department	998	2,777	2,364,879
Physical Education Department	27	1,673	24,902
Fine Arts Department	16	1,146	10,416
General Education Department	2,668	125,797	2,590,392
Vocational Education Department	413	17,679	592,406
Rajamangala Institute of Technology	55	4,718	92,179
Office of the National Primary Education Commission	30,476	344,581	6,633,809
Office of the Private Education Commission	7,258	854	2,049,150
Office of Rajabhat Institutes Council	41	7,522	275,737
Mahachulalongkornrajavidyalaya University	25	—	9,292
Mahamakut Buddhist University	8	—	5,765
Mahidol Withayanuson School	1	—	877
Total (incl. others)	57,991	508,467	15,872,418

Source: Office of the Permanent Secretary, Ministry of Education.

Adult literacy rate (UNESCO estimates): 92.6% (males 94.9%; females 90.5%) in 2002 (Source: UN Development Programme, *Human Development Report*).

Directory

The Constitution

On 27 September 1997 the National Assembly approved a new Constitution, which was endorsed by the King and promulgated on 11 October. The main provisions are summarized below.

GENERAL PROVISIONS

Sovereignty resides in the people. The King as Head of State exercises power through the National Assembly, the Council of Ministers (Cabinet) and the Courts in accordance with the provisions of the Constitution. The human dignity, rights and liberty of the people shall be protected. The people, irrespective of their origin, sex or religion, shall enjoy equal protection under the Constitution. The Constitution is the supreme law of the State.

THE KING

The King is a Buddhist and upholder of religions. He holds the position of Head of the Thai Armed Forces. He selects and appoints qualified persons to be the President of the Privy Council and not more than 18 Privy Councillors to constitute the Privy Council. The Privy Council has a duty to render such advice to the King on all matters pertaining to His functions as He may consult. Whenever the King is absent from the Kingdom or unable to perform His functions, He will appoint a Regent. For the purposes of maintaining national or public safety or national economic security, or averting public calamity, the King may issue an Emergency Decree which shall have the force of an Act.

RIGHTS, LIBERTIES AND DUTIES OF THE THAI PEOPLE

All persons are equal before the law and shall enjoy equal protection under the law. Men and women shall enjoy equal rights. Unjust discrimination against a person on the grounds of origin, race, language, sex, age, physical or health condition, personal status, economic or social standing, religious belief, education or constitutionally political view, shall not be permitted. A person shall enjoy full liberty to profess a religion, a religious sect or creed, and observe religious precepts or exercise a form of worship in accordance with his or her belief, provided that it is not contrary to his or her civic duty, public order or good morals. A person shall enjoy the liberty to express his or her opinion, make speeches, write, print, publicize, and make expression by other means. A person shall enjoy an equal right to receive the fundamental education for the duration of not less than 12 years which shall be provided by the State without charge. A person shall enjoy the liberty to assemble peacefully and without arms, to unite and form an association, a union, league, co-operative, farmer group, private organization or any other group, and to unite and form a political party. A person shall enjoy an equal right to receive standard public health service, and the indigent shall have the right to receive free medical treatment from public health centres of the State. Every person shall have a duty to exercise his or her right to vote at an election. Failure to vote will result in the withdrawal of the right to vote as provided by law.

THE NATIONAL ASSEMBLY

The National Assembly consists of the House of Representatives and the Senate. The President of the House of Representatives is President of the National Assembly. The President of the Senate is Vice-President of the National Assembly. A bill approved by the National Assembly is presented by the Prime Minister to the King to be signed within 20 days from the date of receipt, and shall come into force upon its publication in the Government Gazette. If the King refuses his assent, the National Assembly must re-deliberate such bill. If the National Assembly resolves to reaffirm the bill with a two-thirds' majority, the Prime Minister shall present such bill to the King for signature once again. If the King does not sign and return the bill within 30 days, the Prime Minister shall cause the bill to be promulgated as an Act in the Government Gazette as if the King had signed it.

The House of Representatives consists of 500 members, 100 of whom are from the election on a party-list basis and 400 of whom are from the election on a constituency basis. Election is by direct suffrage and secret ballot. The list of any political party receiving votes of less than 5% of the total number of votes throughout the country shall be regarded as one for which no person listed therein is elected and such votes shall not be reckoned in the determination of the proportional number of the members of the House of Representatives. A person seeking election to the House of Representatives must be of Thai nationality by birth, be not less than 25 years of age, have graduated with not lower than a Bachelor's degree or its equivalent, except for the case of having been a member of the House of Representatives or a Senator before, and be a member of any and only one political party, for a consecutive period of not less than 90 days, up to the date of applying for candidacy in an election. The term of the House of Representatives is four years from the election day. The King has the prerogative power to dissolve the House of Representatives for a new election of members of the House. Members of the House may not renounce their party affiliation without resigning their seats.

The Senate consists of 200 members, elected by direct suffrage and secret ballot. A person seeking election to the Senate must be of Thai nationality by birth, of not less than 40 years of age, and have graduated with not lower than a Bachelor's degree or its equivalent. The term of the Senate is six years as from the election day.

When it is necessary for the interests of the State, the King may convoke an extraordinary session of the National Assembly.

The Election Commission consists of a Chairman and four other Commissioners appointed, by the King with the advice of the Senate, from persons of apparent political impartiality and integrity. Election Commissioners shall hold office for a term of seven years.

The National Human Rights Commission consists of a President and 10 other members appointed, by the King with the advice of the Senate, from persons having apparent knowledge and experiences in the protection of rights and liberties of the people. The term of office of members of the National Human Rights Commission is six years. The National Human Rights Commission has the duty to examine and report the commission or omission of acts which violate human rights or which do not comply with obligations under international treaties to which Thailand is a party.

THE COUNCIL OF MINISTERS

The King appoints the Prime Minister and not more than 35 other Ministers to constitute the Council of Ministers (Cabinet) having the duties to carry out the administration of state affairs. The Prime Minister must be appointed from members of the House of Representatives and must receive the approval of more than half the total number of members of the House. No Prime Minister or Ministers shall be members of the House of Representatives or Senators simultaneously. A Minister must be of Thai nationality by birth, of not less than 35 years of age and be a graduate with not lower than a Bachelor's degree or its equivalent.

The State shall establish the National Economic and Social Council to be charged with the duty to give advice and recommendations to the Council of Ministers on economic and social problems. A national economic and social development plan and other plans as provided by law shall obtain opinions of the National Economic and Social Council before they can be adopted and published.

LOCAL GOVERNMENT

Any locality which meets the conditions of self-government shall have the right to be formed as a local administrative organization as provided by law. A local administrative organization shall have a local assembly and a local administrative committee or local administrators. Members of a local assembly shall be elected by direct suffrage and secret ballot. A local administrative committee or local administrators shall be directly elected by the people or by the approval of a local assembly. Members of a local assembly, local administrative committee or local administrators shall hold office for a period of four years. A member of a local administrative committee or local administrator shall not be a government official holding a permanent position or receiving a salary, or an official or an employee of a state agency, state enterprise or local administration.

AMENDMENT OF THE CONSTITUTION

A motion for amendment must be proposed either by the Council of Ministers or by not less than one-fifth of the total number of members of the House of Representatives or the National Assembly as a whole. A motion for amendment must be proposed in the form of a draft Constitution Amendment and the National Assembly shall consider it in three readings. Promulgation must be approved by the votes of more than half of the total number of members of both Houses.

The Government

HEAD OF STATE

King: HM King Bhumibol Adulyadej (King Rama IX—succeeded to the throne June 1946).

PRIVY COUNCIL

Members: Gen. (retd) Prem Tinsulanonda (President), Dr Sanya Dharmasakti, M. L. Chirayu Navawongs, Dr Chaovana Nasylvanta, Thanin Kraivixien, Rear-Adm. M. L. Usni Pramoj, Air Vice-Marshal Kamthon Sindhavananda, Air Chief Marshal Siddhi Savetsila, Chulanope Snidvongs, M. R. Adulkit Kitiyakara, Gen. Pichitr Kullavanijaya, Ampol Senanarong, Chamras Kemacharu, M. L. Thawisan Ladawan, M. R. Thep Devakula, Sakda Mokkamakkul, Palakorn Suwannarat, Kasem Watanachai.

CABINET
(October 2004)

A coalition of the Thai Rak Thai and Chart Thai.

Prime Minister: Thaksin Shinawatra.

Deputy Prime Ministers: Gen. Chavalit Yongchaiyudh, Chaturon Chaisaeng, Wissanu Krea-ngam, Purachai Piemsomboon, Somsak Thepsuthin, Suwat Liptapanlop.

Minister of Defence: Gen. Samphan Boonyanant.

Minister of Labour and Social Welfare: Uraiwan Tiengthong.

Minister of Finance: Somkid Jatusripitak.

Minister of Culture: Anurak Jureemat.

Minister of Education: Adisai Photharamik.

Minister of Foreign Affairs: Surakiart Sathirathai.

Minister of the Interior: Bhokin Bhalakula.

Minister of Justice: Pongthep Thepkanjana.

Minister of Natural Resources and Environment: Suwit Khunkitti.

Minister of Energy: Prommin Lertsuridej.

Minister of Agriculture and Co-operatives: Wan Muhamad Nor Matha.

Minister of Tourism and Sports: Sonthaya Khunpluem.

Minister of Transport: Suriya Jungrungruangkit.

Minister of Commerce: Watana Muangsook.

Minister of Public Health: Sudarat Keyuraphun.

Minister of Industry: Pongsak Ruktapongpisal.

Minister of Information and Communications Technology: Surapong Suebwonglee.

Minister of Science and Technology: Korn Dabbaransi.

Minister of Social Development and Human Security: Sora-at Klinpratum.

MINISTRIES

Office of the Prime Minister: Government House, Thanon Nakhon Pathom, Bangkok 10300; tel. (2) 280-3526; fax (2) 282-8792; internet www.pmoffice.go.th.

Ministry of Agriculture and Co-operatives: Thanon Ratchadamnoen Nok, Bangkok 10200; tel. (2) 281-5955; fax (2) 282-1425; internet www.moac.go.th.

Ministry of Commerce: Thanon Sanamchai, Pranakorn, Bangkok 10200; tel. (2) 282-6171; fax (2) 280-0775; internet www.moc.go.th.

Ministry of Culture: Thanon Na Phra That, Phra Nakorn, Bangkok 10200; internet www.culture.go.th.

Ministry of Defence: Thanon Sanamchai, Bangkok 10200; tel. (2) 222-1121; fax (2) 226-3117; internet www.mod.go.th.

Ministry of Education: Wang Chankasem, Thanon Ratchadamnoen Nok, Bangkok 10300; tel. (2) 281-9809; fax (2) 281-8218; e-mail website@emisc.moe.go.th; internet www.moe.go.th.

Ministry of Energy: 17 Kasatsuk Bridge, Thanon Rama I, Rong Mueng, Patumwan, Bangkok 10330; tel. (2) 222-2593; fax (2) 222-3785; e-mail thiti@energy.go.th; internet www.energy.go.th.

Ministry of Finance: Thanon Rama VI, Samsennai, Phaya Thai, Rajatevi, Bangkok 10400; tel. (2) 273-9021; e-mail foc@vayu.mof.go.th; internet www.mof.go.th.

Ministry of Foreign Affairs: Thanon Sri Ayudhya, Bangkok 10400; tel. (2) 643-5000; fax (2) 225-6155; e-mail thaiinfo@mfa.go.th; internet www.mfa.go.th.

Ministry of Industry: 75/6 Thanon Rama VI, Ratchathewi, Bangkok 10400; tel. (2) 202-3000; fax (2) 202-3048; internet www.industry.go.th.

Ministry of Information and Communications Technology: Bangkok 10210; tel. (2) 238-5422; fax (2) 238-5423; e-mail pr@mict.go.th; internet www.mict.go.th.

Ministry of the Interior: Thanon Atsadang, Bangkok 10200; tel. (2) 222-1141; fax (2) 223-8851; internet www.moi.go.th.

Ministry of Justice: Thanon Ratchadaphisek, Chatuchak, Bangkok 10900; tel. (2) 502-8051; fax (2) 502-8059; internet www.moj.go.th.

Ministry of Labour: Thanon Mitmaitri, Dindaeng, Huay Kwang, Bangkok 10400; tel. (2) 248-5558; fax (2) 246-1520; internet www.mol.go.th.

Ministry of Natural Resources and Environment: Bangkok; e-mail monre@environnet.in.th; internet www.monre.go.th.

Ministry of Public Health: Thanon Tiwanon, Amphoe Muang, Nonthaburi 11000; tel. (2) 591-8495; fax (2) 591-8492; internet www.moph.go.th.

Ministry of Science and Technology: Thanon Phra Ram VI, Ratchathewi, Bangkok 10400; tel. (2) 246-0064; fax (2) 247-1449; internet www.moste.go.th.

Ministry of Social Development and Human Security: Bangkok 10900; tel. (2) 612-8888; e-mail society@m-society.go.th; internet www.m-society.go.th.

Ministry of Tourism and Sports: Bangkok 10330; tel. (2) 214-0120; fax (2) 216-7390; e-mail webmaster@mots.go.th; internet www.mots.go.th.

Ministry of Transport: 38 Thanon Ratchadamnoen Nok, Khet Pom Prab Sattruphai, Bangkok 10100; tel. (2) 283-3044; fax (2) 280-1714; internet www.motc.go.th.

Legislature

RATHA SAPHA
(National Assembly)

Woothi Sapha
(Senate)

A constitutional amendment was enacted in January 1995, limiting the membership of the Senate to two-thirds of that of the House of Representatives.

On 22 March 1996 the Prime Minister appointed a new Senate, comprising 260 members, only 39 of whom were active members of the armed forces. (Under the new Constitution promulgated in October 1997, the Senate was to comprise 200 directly elected members.) Elections to the new 200-member Senate were held on 4 March 2000; however, owing to the repeated incidence of electoral fraud, a second round of voting took place in April and a third round was held in a number of constituencies in June, by which time only 197 members had been elected to the new Senate, which had yet to take office pending the election of its three remaining members. A fourth round of polling was held in June/July, which resulted in the election of a further two members to the Senate. At a fifth round of voting held later that month the final member was elected to the Senate. The 200-member Senate was subsequently sworn in on 1 August 2000.

President of the Senate: SUCHON CHALEEKURE.

Sapha Poothaen Rassadorn
(House of Representatives)

Speaker of the House of Representatives and President of the National Assembly: UTHAI PHIMCHAICHON.

General Election, 6 January 2001*

Party	Seats
Thai Rak Thai	248
Democrat Party	128
Chart Thai	41
New Aspiration Party	36
Chart Pattana	29
Seritham Party	14
Social Action Party	1
Independents	3
Total	500

*The new Constitution of 1997 provided for an increase in the number of deputies in the House of Representatives from 393 to 500. Of those who gained seats in January 2001, 400 contested the election in single-member constituencies and 100 through a party-list system.

Political Organizations

Chart Thai (Thai Nation): 1 Thanon Pichai, Dusit, Bangkok 10300; tel. (2) 243-8070; fax (2) 243-8074; e-mail chartthai@chartthai.or.th; internet www.chartthai.or.th; f. 1981; right-wing; founded political reform policy; includes mems of fmr United Thai People's Party and fmr Samakkhi Tham (f. 1991); Leader BANHARN SILAPA-ARCHA.

Democrat Party (DP) (Prachatipat): 67 Thanon Setsiri, Samsen Nai, Phyathai, 10400 Bangkok; tel. (2) 270-0036; fax (2) 279-6086; e-mail admin@democrat.or.th; internet www.democrat.or.th; f. 1946; liberal; Leader BANYAT BANTADTAN; Sec.-Gen. PRADIT PATTARAPRASIT.

Ekkaparb (Solidarity): 670/104 Soi Thepnimit, Thanon Jaransanitwong, Bangpaid, Bangkok 10700; tel. (2) 424-0291; fax (2) 424-8630; f. 1989; opposition merger by the Community Action Party, the Prachachon Party, the Progressive Party and Ruam Thai; Leader CHAIYOS SASOMSAP; Sec.-Gen. NEWIN CHIDCHOB.

Mahachon: Bangkok; internet www.mahachon.or.th; f. 2004 following split in Democrat Party; Leader Gen. SANAN KAJORNPRASART.

Muan Chon (Mass Party): 630/182 Thanon Prapinklao, Bangkok 10700; tel. and fax (2) 424-0851; f. 1985; dissolved after defection of leader Capt. Chalerm Yoobamrung to New Aspiration Party (NAP); re-formed in 2002 following merger of NAP with Thai Rak Thai; Leader Gen. VORAVIT PIBOONSILP; Sec.-Gen. KAROON RAKSASUK.

Palang Dharma (PD) (Righteous Force): 445/15 Ramkhamhaeng, 39 Bangkapi, Bangkok 10310; tel. (2) 718-5626; fax (2) 718-5634; internet www.pdp.or.th; f. 1988; Leader CHAIWAT SINSUWONG; Sec.-Gen. RAVEE MASCHAMADOL.

Prachakorn Thai (Thai Citizens Party): 1213/323 Thanon Srivara, Bangkapi, Bangkok 10310; tel. (2) 559-0008; fax (2) 559-0016; f. 1981; right-wing; monarchist; Leader SAMAK SUNDARAVEJ; Sec.-Gen. YINGPAN MANSIKARN.

Social Action Party (SAP) (Kij Sangkhom): 126 Soi Ongkarak Samsen, 28 Thanon Nakhon Chaisi, Dusit, Bangkok 10300; tel. (2) 243-0100; fax (2) 243-3224; f. 1981; conservative; Leader MONTREE PONGPANIT; Sec.-Gen. SUWIT KHUNKITTI.

Thai Freedom Party: c/o House of Representatives, Bangkok; f. Aug. 2002; Leader NIT SORASIT.

Thai Party: 193 Thanon Amnauy Songkram, Dusit, Bangkok; tel. and fax (2) 669-4367; f. 1997; Leader THANABORDIN SANGSATHAPORN; Gen. Sec. Rear Adm. THAVEEP NAKSOOK.

Thai Rak Thai (Thais Love Thais): c/o House of Representatives, Bangkok; tel. (2) 668-2000; fax (2) 668-6000; e-mail public@thairakthai.or.th; internet www.thairakthai.or.th; f. 1998; merged with New Aspiration Party (NAP) and Seritham Party following 2001 general election and with Chart Pattana in Aug. 2004; Leader THAKSIN SHINAWATRA.

Thon Trakul Thai (First Thai Nation): c/o House of Representatives, Bangkok; f. Nov. 2003; Leader CHUWIT KAMONVISIT.

Groupings in armed conflict with the Government include:

National Revolutionary Front: Yala; Muslim secessionists.

Pattani United Liberation Organization (PULO): e-mail p_u_l_@hotmail.com; internet www.pulo.org; advocates secession of the five southern provinces (Satun, Narathiwat, Yala, Pattani and Songkhla); Leader HAYI DAO THANAM.

Diplomatic Representation

EMBASSIES IN THAILAND

Argentina: 16th Floor, Suite 1601, Glas Haus Bldg, 1 Soi Sukhumvit 25, Klongtoey, Bangkok 10110; tel. (2) 259-0401; fax (2) 259-0402; e-mail embtail@mozart.inet.co.th; Ambassador CARLOS FAUSTINO GARCÍA.

Australia: 37 Thanon Sathorn Tai, Bangkok 10120; tel. (2) 287-2680; fax (2) 287-2029; e-mail austembassy.bangkok@dfat.gov.au; internet www.austembassy.or.th; Ambassador MILES KUPA.

Austria: 14 Soi Nandha, off Thanon Sathorn Tai, Soi 1, Bangkok 10120; tel. (2) 303-6057; fax (2) 287-3925; e-mail bangkok-ob@bmaa.gv.at; Ambassador Dr HERBERT TRAXL.

Bangladesh: 727 Thanon Thonglor Sukhumvit, Soi 55, Bangkok 10110; tel. (2) 392-9437; fax (2) 391-8070; e-mail bdoot@samart.co.th; Ambassador SHAHED AKHTAR.

Belgium: 17th Floor, Sathorn City Tower, 175 Thanon Sathorn Tai, Tungmahamek, Sathorn, Bangkok 10120; tel. (2) 679-5454; fax (2) 679-5467; e-mail bangkok@diplobel.org; internet www.diplomatie.be/bangkok; Ambassador JAN MATTHYSEN.

Belize: 10th Floor, Pilot Pen Bldg, 331/1–3 Thanon Silom, Bangrak, Bangkok 10500; tel. (2) 636-8377; fax (2) 235-7653; e-mail belize@embelizeth.com; Ambassador DAVID A. K. GIBSON.

Bhutan: 375/1 Soi Ratchadanivej, Thanon Pracha-Uthit, Huay Kwang, Bangkok 10320; tel. (2) 274-4740; fax (2) 274-4743; e-mail bht_emb_bkk@yahoo.com; Ambassador CHENKYAB DORJI.

Brazil: 34th Floor, Lumpini Tower, 1168/101 Thanon Rama IV, Sathorn, Bangkok 10120; tel. (2) 679-8567; fax (2) 679-8569; e-mail embrasbkk@mozart.inet.co.th; internet www.brazilembassy.or.th; Ambassador MARCO ANTÔNIO DINIZ BRANDÃO.

Brunei: 132 Sukhumvit, Soi 23, Thanon Sukhumvit, Bangkok 10110; tel. (2) 204-1476; fax (2) 204-1486; Ambassador Dato' Paduka Haji MOHD YUNOS BIN Haji MOHD HUSSEIN.

Bulgaria: 64/4 Soi Charoenmitr, Sukhumvit 63, Wattana, Bangkok 10110; tel. (2) 391-6180; fax (2) 391-6182; e-mail bulgemth@asianet.co.th; Ambassador ROUMEN IVANOV SABEV.

Cambodia: 185 Thanon Ratchadamri, Lumpini, Bangkok 10330; tel. (2) 254-6630; fax (2) 253-9859; e-mail recanbot@loxinfo.co.th; Ambassador UNG SEAN.

Canada: Abdulrahim Bldg, 15th Floor, 990 Thanon Rama IV, Bangrak, Bangkok 10500; tel. (2) 636-0540; fax (2) 636-0565; e-mail bangkok@international.gc.ca; internet www.international.gc.ca/bangkok; Ambassador DENIS COMEAU.

Chile: UBC II Bldg, 591 Thanon Sukhumvit, Soi 33, Klongtoey Nua, Wattana, Bangkok 10110; tel. (2) 260-3870; fax (2) 260-4328; e-mail embajada@chile-thai.com; internet www.chile-thai.com; Ambassador LUIS ALBERTO SEPÚLVEDA.

China, People's Republic: 57 Thanon Ratchadaphisek, Bangkok 10310; tel. (2) 245-7043; internet www.chinaembassy.or.th; Ambassador ZHANG JIUHUAN.

Czech Republic: 71/6 Soi Ruamrudi 2, Thanon Ploenchit, Bangkok 10330; tel. (2) 255-3027; fax (2) 253-7637; e-mail bangkok@embassy.mzv.cz; internet www.mfa.cz/bangkok; Ambassador Dr JIŘI SITLER.

Denmark: 10 Soi Attakarn Prasit, Thanon Sathorn Tai, Bangkok 10120; tel. (2) 213-2021; fax (2) 213-1752; e-mail bkkamb@um.dk; internet www.denmark-embassy.or.th; Ambassador ULRIK HELWEG-LARSEN.

Egypt: 6 Las Colinas Bldg, 42nd Floor, Sukhumvit 21, Wattana, Bangkok 10110; tel. (2) 661-7184; fax (2) 262-0235; e-mail egyptemb@loxinfo.co.th; Ambassador TAMER ABDEL-AZIZ ABDALLA KHALIL.

Finland: Amarin Tower, 16th Floor, 500 Thanon Ploenchit, Bangkok 10330; tel. (2) 256-9306; fax (2) 256-9310; e-mail Sanomat.BAN@formin.fi; Ambassador HEIKKI TUUANEN.

France: 35 Soi Rong Phasi Kao, Thanon Charoenkrung, Bangkok 10500; tel. (2) 266-8250; fax (2) 236-7973; e-mail press@ambafrance-th.org; internet www.ambafrance-th.org/; Ambassador LAURENT AUBLIN.

Germany: 9 Thanon Sathorn Tai, Bangkok 10120; tel. (2) 287-9000; fax (2) 287-1776; e-mail rk@german-embassy.or.th; internet www.german-embassy.or.th; Ambassador ANDREAS VON STECHOW.

Greece: Thai Wah Tower II, 30th Floor, 21/159 Thanon Sathorn Tai, Bangkok 10120; tel. (2) 679-1462; fax (2) 679-1463; e-mail bagremb@ksc.th.com; Ambassador MILTIADIS HISKAKIS.

Holy See: 217/1 Thanon Sathorn Tai, POB 12–178, Bangkok 10120 (Apostolic Nunciature); tel. (2) 212-5853; fax (2) 212-0932; e-mail vatemb@mozart.inet.co.th; Apostolic Nuncio Most Rev. SALVATORE PENNACCHIO (Titular Archbishop of Montemarano).

Hungary: Oak Tower, 20th Floor, President Park Condominium, 95 Sukhumvit Soi 24, Prakhanong, Bangkok 10110; tel. (2) 661-1150; fax (2) 661-1153; e-mail huembbgk@mozart.inet.co.th; Ambassador SANDOR JOLSVAI.

India: 46 Soi Prasarnmitr, 23 Thanon Sukhumvit, Bangkok 10110; tel. (2) 258-0300; fax (2) 258-4627; e-mail indiaemb@mozart.inet.co.th; internet www.indiaemb.or.th; Ambassador L. K. PONAPPA.

Indonesia: 600–602 Thanon Phetchburi, Phyathai, Bangkok 10400; tel. (2) 252-3135; fax (2) 255-1267; e-mail kukbkk@ksc11.th.com; internet www.kbri-bangkok.com; Ambassador IBRAHIM YUSUF.

Iran: 602 Thanon Sukhumvit, between Soi 22–24, Bangkok 10110; tel. (2) 259-0611; fax (2) 259-9111; e-mail emb@mozart.inet.co.th; Ambassador RASOUL ESLAMI.

Israel: Ocean Tower II, 25th Floor, 75 Sukhumvit, Soi 19, Thanon Asoke, Bangkok 10110; tel. (2) 204-9200; fax (2) 204-9255; e-mail bangkok@israel.org; Ambassador GERSHON ZOHAR.

Italy: 399 Thanon Nang Linchee, Thungmahamek, Yannawa, Bangkok 10120; tel. (2) 285-4090; fax (2) 285-4793; e-mail ambitbkk@loxinfo.co.th; internet www.ambasciatabangkok.org; Ambassador STEFANO STARACE JANFOLLA.

Japan: 1674 Thanon Phetchburi Tadmai, Bangkok 10320; tel. (2) 252-6151; fax (2) 253-4153; internet embjp-th.org; Ambassador ATSUSHI TOKINOYA.

Kazakhstan: Suite 3E1–3E2, 3rd Floor, 139 Rimco House, Sukhumvit 63, Klongton Nua, Wattana, Bangkok 10110; tel. (2) 714-9890; fax (2) 714-7588; e-mail kzdipmis@asianet.co.th; Chargé d'affaires a.i. SAKEN SEIDUALIYEV.

Korea, Democratic People's Republic: 14 Mooban Suanlaemthong 2, Thanon Pattanakarn, Suan Luang, Bangkok 10250; tel. (2) 319-2686; fax (2) 318-6333; Ambassador O SONG CHOL.

Korea, Republic: 23 Thanon Thiam-Ruammit, Huay Kwang, Bangkok 10320; tel. (2) 247-7537; fax (2) 247-7535; e-mail korea_emb_th@yahoo.co.kr; Ambassador JEE-JOON YOON.

Kuwait: 100/44 Sathorn Nakhon Tower, 24th Floor, Thanon Sathorn Nua, Bangrak, Bangkok 10500; tel. (2) 636-6600; fax (2) 636-7363; e-mail hamood@mozart.inet.co.th; Ambassador KHALED MOHAMMAD AHMAD AL-SHAIBANI.

Laos: 520/502/1–3 Soi Sahakarnpramoon, Thanon Pracha Uthit, Wangthonglang, Bangkok 10310; tel. (2) 539-6667; fax (2) 539-3827; e-mail sabaidee@bkklaoembassy.com; internet www.bkklaoembassy.com; Ambassador HIEM PHOMMACHANH.

Malaysia: 35 Thanon Sathorn, Tungmahamek, Sathorn, Bangkok 10120; tel. (2) 679-2190; fax (2) 679-2208; e-mail mwbngkok@samart.co.th; Ambassador Dato' SYED NORULZAMAN SYED KAMARULZAMAN.

Mexico: 20/60–62 Thai Wah Tower I, 20th Floor, Thanon Sathorn Tai, Bangkok 10120; tel. (2) 285-0995; fax (2) 285-0667; e-mail mexthai@loxinfo.co.th; internet www.sre.gob.mx/tailandia; Ambassador JAVIER RAMÓN BRITO MONCADA.

Mongolia: 251 Soi Rojana, Thanon Sukhumvit 21, Klongtoey Nua, Wattana, Bangkok 10110; tel. (2) 640-8017-8; fax (2) 258-3849; e-mail mongemb@loxinfo.co.th; Ambassador LUVSANDORJ DAWAGIV.

Morocco: One Pacific Place, 19th Floor, 140 Thanon Sukhumvit, between Soi 4–6, Bangkok 10110; tel. (2) 653-2444; fax (2) 653-2449; e-mail sifambkk@mweb.co.th; Ambassador EL HASSANE ZAHID.

Myanmar: 132 Thanon Sathorn Nua, Bangkok 10500; tel. (2) 233-2237; fax (2) 236-6898; e-mail mebkk@asianet.co.th; Ambassador U MYO MYINT.

Nepal: 189 Soi 71, Thanon Sukhumvit, Bangkok 10110; tel. (2) 390-2280; fax (2) 381-2406; e-mail nepembkk@asiaaccess.net.th; internet www.royalnepaleseembassy.org; Ambassador JANAK BAHADUR SINGH.

Netherlands: 106 Thanon Witthayu, Bangkok 10330; tel. (2) 254-7701; fax (2) 254-5579; e-mail ban@minbuza.nl; internet www.netherlandsembassy.in.th; Ambassador G. J. H. C. KRAMER.

New Zealand: M Thai Tower, 14th Floor, All Season's Place, 87 Thanon Witthayu, Lumpini, Bangkok 10330; tel. (2) 254-2530; fax (2) 253-9045; e-mail nzembbkk@loxinfo.co.th; Ambassador PETER RIDER.

Nigeria: 100 Sukhumvit, Soi 38, Prakhanong, Klongtoey, Bangkok 10110; tel. (2) 391-5197; fax (2) 391-8819; e-mail nigeriabkk@hotmail.com; internet www.embnigeriabkk.com; Ambassador THOMPSON SUNDAY OLUFUNSO OLUMOKO.

Norway: UBC II Bldg, 18th Floor, 591 Thanon Sukhumvit, Soi 33, Bangkok 10110; tel. (2) 302-6415; fax (2) 262-0218; e-mail emb.bangkok@mfa.no; internet www.emb-norway.or.th/; Ambassador RAGNE BIRTE LUND.

Oman: 82 Saeng Thong Thani Tower, 32nd Floor, Thanon Sathorn Nua, Bangkok 10500; tel. (2) 639-9380; fax (2) 639-9390; Ambassador MOHAMMED YOUSUF DAWOOD SHALWANI.

Pakistan: 31 Soi Nana Nua, Thanon Sukhumvit, Bangkok 10110; tel. (2) 253-0288; fax (2) 253-0290; e-mail parepbkk@ji-net.com; Ambassador HUSSAIN BAKHSH BANGULZAI.

Panama: 14 Sarasin Bldg, 7th Floor, 14 Thanon Surasak, Bangrak, Bangkok 10500; tel. (2) 237-9008; fax (2) 237-9009; e-mail ptybkk@ksc.th.com; Ambassador RICARDO ANTONIO QUIJANO JIMÉNEZ.

Peru: Glas Haus Bldg, 16th Floor, 1 Soi Sukhumvit 25, Khet Wattana, Bangkok 10110; tel. (2) 260-6243; fax (2) 260-6244; e-mail peru@peruthai.or.th; internet www.peru.org.pe; Chargé d'affaires JOSÉ BUSTINZA.

Philippines: 760 Thanon Sukhumvit, cnr Soi 30/1, Klongtan, Klongtoey, Bangkok 10110; tel. (2) 259-0139; fax (2) 259-2809; e-mail inquiry@philembassy-bangkok.net; internet www.philembassy-bangkok.net; Ambassador ANTONIO V. RODRIGUEZ.

Poland: Sriyukon Bldg, 84 Sukhumvit, Soi 5, Bangkok 10110; tel. (2) 251-8891; fax (2) 251-8895; Ambassador BOGDAN GORALCZYK.

Portugal: 26 Bush Lane, Thanon Charoenkrung, Bangkok 10500; tel. (2) 234-2123; fax (2) 238-4275; e-mail portemb@loxinfo.co.th; Ambassador JOÃO LIMA PIMENTEL.

Romania: 20/1 Soi Rajakhru, Phaholyothin Soi 5, Thanon Phaholyothin, Phayathai, Bangkok 10400; tel. (2) 617-1551; fax (2) 617-1113; e-mail romembkk@ksc.th.com; Ambassador CRISTIAN TEODORESCU.

Russia: 78 Thanon Sap, Bangrak, Bangkok 10500; tel. (2) 234-9824; fax (2) 237-8488; e-mail rosposol@cscoms.com; internet www.thailand.mid.ru; Ambassador YEVGENY OSTROVENKO.

Saudi Arabia: Sathorn Thani Bldg, 10th Floor, 82 Thanon Sathorn Nua, Bangkok 10500; tel. (2) 639-2999; fax (2) 639-2950; Chargé d'affaires HANI ABDULLAH MOMINAH.

Singapore: Rajanakarn Bldg, 9th and 18th Floors, 183 Thanon Sathorn Tai, Bangkok 10120; tel. (2) 286-2111; fax (2) 287-2578; e-mail singemb@pacific.net.th; internet www.mfa.gov.sg/bangkok; Ambassador CHAN HENG WING.

Slovakia: Thai Wah Tower II, 22nd Floor, 21/144 Thanon Sathorn Tai, Bangkok 10120; tel. (2) 677-3445; fax (2) 677-3447; e-mail slovakemb@actions.net; Ambassador MARIÁN TOMÁŠIK.

South Africa: The Park Place, 6th Floor, 231 Soi Sarasin, Lumpini, Bangkok 10330; tel. (2) 253-8473; fax (2) 253-8477; e-mail saembbkk@loxinfo.co.th; Ambassador BUYISIWE MAUREEN PHETO.

Spain: Diethelm Towers, 7th Floor, 93/1 Thanon Witthayu, Bangkok 10330; tel. (2) 252-6112; fax (2) 255-2388; e-mail embesp@bkk3.loxinfo.co.th; internet www.embesp.or.th; Chargé d'affaires a.i. FRANCISCO DE ASIS BENITEZ SALAS.

Sri Lanka: Ocean Tower II, 13th Floor, 75/6–7 Sukhumvit, Soi 19, Bangkok 10110; tel. (2) 261-1934; fax (2) 261-1936; e-mail slemb@ksc.th.com; Ambassador J. D. A. WIJEWARDENA.

Sweden: First Pacific Place, 20th Floor, 140 Thanon Sukhumvit, Bangkok 10110; tel. (2) 263-7200; fax (2) 263-7260; e-mail ambassaden.bangkok@foreign.ministry.se; internet www.swedenabroad.com/bangkok; Ambassador JONAS HAFSTROM.

Switzerland: 35 Thanon Witthayu, Bangkok 10330; tel. (2) 253-0156; fax (2) 255-4481; e-mail vertretung@ban.rep.admin.ch; internet www.eda.admin.ch/bangkok_emb/e/home.html; Ambassador HANS-PETER ERISMANN.

Turkey: 61/1 Soi Chatsan, Thanon Suthisarn, Huay Kwang, Bangkok 10310; tel. (2) 274-7262; fax (2) 274-7261; e-mail tcturkbe@mail.cscoms.com; Ambassador MUMIN ALANAT.

Ukraine: 87 All Seasons Place, CRC Tower, 33rd Floor, Thanon Witthayu, Lumpini, Pathumwan, Bangkok 10330; tel. (2) 685-3215; fax (2) 685-3217; Ambassador IHOR V. HUMMENNYI.

United Arab Emirates: 82 Seng Thong Thani Bldg, 25th Floor, Thanon Sathorn Nua, Bangkok 10500; tel. (2) 639-9820; fax (2) 639-9818; Ambassador SALIM ISSA ALI AL-KATTAM AL-ZAABI.

United Kingdom: 1031 Thanon Witthayu, Lumpini, Pathumwan, Bangkok 10330; tel. (2) 305-8333; fax (2) 253-7121; e-mail info.bangkok@fco.gov.uk; internet www.britishembassy.gov.uk/Thailand; Ambassador DAVID WILLIAM FALL.

USA: 95 Thanon Witthayu, Bangkok 10330; tel. (2) 205-4000; fax (2) 254-2990; internet www.usa.or.th/; Ambassador RALPH (SKIP) BOYCE.

Viet Nam: 83/1 Thanon Witthayu, Lumpini, Pathumwan, Bangkok 10330; tel. (2) 251-3551; fax (2) 251-7203; e-mail vnembassy@bkk.a-net.net.th; Ambassador NGUYEN QUOC KHANH.

Judicial System

SUPREME COURT

(Sarn Dika)

Thanon Ratchadamnoen Nai, Bangkok 10200.

The final court of appeal in all civil, bankruptcy, labour, juvenile and criminal cases. Its quorum consists of three judges. However, the Court occasionally sits in plenary session to determine cases of exceptional importance or where there are reasons for reconsideration or overruling of its own precedents. The quorum, in such cases, is one-half of the total number of judges in the Supreme Court.

President (Chief Justice): ATTHANITI DISAMNART.

Vice-President: SUPRADIT HUTASINGH.

COURT OF APPEALS

(Sarn Uthorn)

Thanon Ratchadaphisek, Chatuchak, Bangkok 10900.

Appellate jurisdiction in all civil, bankruptcy, juvenile and criminal matters; appeals from all the Courts of First Instance throughout the country, except the Central Labour Court, come to this Court. Two judges form a quorum.

Chief Justice: KAIT CHATANIBAND.

Deputy Chief Justices: PORNCHAI SMATTAVET, CHATISAK THAMMASAKDI, SOMPOB CHOTIKAVANICH, SOMPHOL SATTAYA-APHITARN.

COURTS OF FIRST INSTANCE

Civil Court
(Sarn Pang)

Thanon Ratchadaphisek, Chatuchak, Bangkok 10900.

Court of first instance in civil and bankruptcy cases in Bangkok. Two judges form a quorum.

Chief Justice: PENG PENG-NITI.

Criminal Court
(Sarn Aya)

Thanon Ratchadaphisek, Kwaeng Jompol, Khet Jatujak, Bangkok 10900; tel. (2) 541-2274; fax (2) 541-2273; e-mail crim1@judiciary.go.th.

Court of first instance in criminal cases in Bangkok. Two judges form a quorum.

Chief Justice: PRADIT EKMANEE.

Central Juvenile and Family Court
(Sarn Yaowachon Lae Krobkrow Klang)

Thanon Rachini, Bangkok 10200; fax (2) 224-1546.

Original jurisdiction over juvenile delinquency and matters affecting children and young persons. Two judges and two associate judges (one of whom must be a woman) form a quorum.

Chief Justice: JIRA BOONPOJANASUNTHON.

Central Labour Court
(Sarn Rang Ngan Klang)

404 Thanon Phra Ram IV, Bangkok 10500.

Jurisdiction in labour cases throughout the country.

Chief Justice: CHAVALIT PRAY-POO.

Central Tax Court
(Sarn Phasi-Arkorn Klang)

Thanon Ratchadaphisek, Chatuchak, Bangkok 10900.

Jurisdiction over tax cases throughout the country.

Chief Justice: SANTI TAKRAL.

Provincial Courts (Sarn Changwat): Exercise unlimited original jurisdiction in all civil and criminal matters, including bankruptcy, within their own district, which is generally the province itself. Two judges form a quorum. At each of the five Provincial Courts in the south of Thailand (i.e. Pattani, Yala, Betong, Satun and Narathiwat) where the majority of the population are Muslims, there are two Dato Yutithum or Kadis (Muslim judges). A Kadi sits with two trial judges in order to administer *Shari'a* (Islamic) laws and usages in civil cases involving family and inheritance where all parties concerned are Muslims. Questions on Islamic laws and usages which are interpreted by a Kadi are final.

Thon Buri Civil Court (Sarn Pang Thon Buri): Civil jurisdiction over nine districts of metropolitan Bangkok.

Thon Buri Criminal Court (Sarn Aya Thon Buri): Criminal jurisdiction over nine districts of metropolitan Bangkok.

South Bangkok Civil Court (Sarn Pang Krungthep Tai): Civil jurisdiction over southern districts of Bangkok.

South Bangkok Criminal Court (Sarn Aya Krungthep Tai): Criminal jurisdiction over southern districts of Bangkok.

Magistrates' Courts (Sarn Kwaeng): Adjudicate in minor cases with minimum formality and expense. Judges sit singly.

Religion

Buddhism is the predominant religion, professed by more than 95% of Thailand's total population. About 4% of the population are Muslims, being ethnic Malays, mainly in the south. Most of the immigrant Chinese are Confucians. The Christians number about 305,000, of whom about 75% are Roman Catholic, mainly in Bangkok and northern Thailand. Brahmins, Hindus and Sikhs number about 85,000.

BUDDHISM

Sangha Supreme Council

The Religious Affairs Dept, Thanon Ratchadamnoen Nok, Bangkok 10300; tel. (2) 281-6080; fax (2) 281-5415.

Governing body of Thailand's 350,000 monks, novices and nuns.

Supreme Patriarch of Thailand: Somdej Phra Yansangwara.

The Buddhist Association of Thailand: 41 Thanon Phra Aditya, Bangkok 10200; tel. (2) 281-9563; f. 1934; under royal patronage; 4,183 mems; Pres. Prapasana Auychai.

CHRISTIANITY

The Roman Catholic Church

For ecclesiastical purposes, Thailand comprises two archdioceses and eight dioceses. At 31 December 2002 there were an estimated 291,720 adherents in the country, representing about 0.5% of the population.

Catholic Bishops' Conference of Thailand: 122/11 Soi Naaksuwan, Thanon Nonsi, Yannawa, Bangkok 10120; tel. (2) 681-3900; fax (2) 681-5369; e-mail cbct@ptn.ac.th; f. 1969; Pres. Cardinal Michael Michai Kitbunchu (Archbishop of Bangkok).

Catholic Association of Thailand

57 Soi Burapa, Bangrak, Bangkok 10500; tel. (2) 233-2976.

Archbishop of Bangkok: Cardinal Michael Michai Kitbunchu, Assumption Cathedral, 40 Thanon Charoenkrung, Bangrak, Bangkok 10500; tel. (2) 237-1031; fax (2) 237-1033; e-mail arcdibkk@loxinfo.co.th.

Archbishop of Tharé and Nonseng: (vacant), Archbishop's House, POB 6, Sakon Nakhon 47000; tel. (42) 711718; fax (42) 712023.

The Anglican Communion

Thailand is within the jurisdiction of the Anglican Bishop of Singapore (q.v.).

Other Christian Churches

Baptist Church Foundation (Foreign Mission Board): 90 Soi 2, Thanon Sukhumvit, Bangkok 10110; tel. (2) 252-7078; Mission Admin. Tom Williams (POB 832, Bangkok 10501).

Church of Christ in Thailand: CCT Bldg, 13th Floor, 109 Thanon Surawong, Bangkok 10500; tel. (2) 236-9400; fax (2) 238-3520; e-mail cctecume@loxinfo.co.th; f. 1934; 132,440 communicants; Moderator Rev. Dr Boonratna Boayen; Gen. Sec. Rev. Dr Sint Kimhachandra.

ISLAM

Office of the Chularajmontri: 100 Soi Prom Pak, Thanon Sukhumvit, Bangkok 10110; Sheikh Al-Islam (Chularajmontri) Haji Sawasdi Sumalayasak.

BAHÁ'Í FAITH

National Spiritual Assembly: 77/1 Soi Lang Suan 4, Thanon Ploenchit, Lumpini Patumwan, Bangkok 10330; tel. (2) 252-5355; fax (2) 254-4199; mems resident in 76 provinces.

The Press

DAILIES

Thai Language

Baan Muang: 1 Soi Pluem-Manee, Thanon Vibhavadi Rangsit, Bangkok 10900; tel. (2) 513-3101; fax (2) 513-3106; Editor Mana Praebhand; circ. 200,000.

Daily News: 1/4 Thanon Vibhavadi Rangsit, Laksi, Bangkok 10210; tel. (2) 561-1456; fax (2) 940-9875; internet www.dailynews.co.th; f. 1964; Editor Pracha Hetrakul; circ. 800,000.

Dao Siam: 60 Mansion 4, Thanon Rajdamnern, Bangkok 10200; tel. (2) 222-6001; fax (2) 222-6885; f. 1974; Editor Santi Untrakarn; circ. 120,000.

Khao Panich (Daily Trade News): 22/27 Thanon Ratchadaphisek, Bangkok 10900; tel. (2) 511-5066; Editor Samruane Sumphandharak; circ. 30,000.

Khao Sod (Fresh News): Bangkok; Editor-in-Chief Kiatichai Pongpanich; circ. 350,000.

Kom Chad Luek: 44 Moo 10, Thanon Bangna Trad, Km 4.5, Bang Na, Bangkok 10260; tel. (2) 325-5555; fax (2) 317-2071; internet www.komchadluek.com; Editor Adisal Limprungpatakit.

Krungthep Turakij Daily: Nation Multimedia Group Public Co Ltd, 44 Moo 10, Thanon Bangna-Trad, Bangna, Prakanong, Bangkok 10260; tel. (2) 317-0042; fax (2) 317-1489; e-mail ktwebmaster@bangkokbiznews.com; internet www.bangkokbiznews.com; f. 1987; Publr and Group Editor Suthichai Yoon; Editor Adisak Limprungpatanakit; circ. 75,882.

Matichon: 12 Thanon Thedsaban Naruban, Prachanivate 1, Chatuchak, Bangkok 10900; tel. (2) 589-0020; fax (2) 589-9112; e-mail matisale@matichon.co.th; internet www.matichon.co.th; f. 1977; Man. Editor Prasong Lertratanavisuth; circ. 180,000.

Naew Na (Frontline): 96 Moo 7, Thanon Vibhavadi Rangsit, Bangkok; tel. (2) 521-4647; fax (2) 552-3800; Editor Wanchai Wongmeechai; circ. 200,000.

Siam Daily: 192/8–9 Soi Vorapong, Thanon Visuthikasat, Bangkok; tel. (2) 281-7422; Editor Narong Charusophon.

Siam Post: Bangkok; internet www.siampost.co.th; Man. Dir Paisal Sricharatchanya.

Siam Rath (Siam Nation): 12 Mansion 6, Thanon Rajdamnern, Bangkok 10200; tel. (2) 622-1810; fax (2) 224-1982; e-mail siamrath@siamrath.co.th; internet www.siamrath.co.th; f. 1950; Editor Assiri Thammachot; circ. 120,000.

Sue Tuakij: Bangkok; f. 1995; evening; business news.

Thai: 423–425 Thanon Chao Khamrop, Bangkok; tel. (2) 223-3175; Editor Vichien Mana-Natheethoratham.

Thai Rath: 1 Thanon Vibhavadi Rangsit, Bangkok 10900; tel. (2) 272-1030; fax (2) 272-1324; internet www.thairath.co.th; f. 1948; Editor Pithoon Sunthorn; circ. 800,000.

English Language

Bangkok Post: Bangkok Post Bldg, 136 Soi Na Ranong, Klongtoey, Bangkok 10110; tel. (2) 240-3700; fax (2) 240-3666; e-mail editor@bangkokpost.co.th; internet www.bangkokpost.net; f. 1946; morning; Editor Veera Prateepchaikul; circ. 49,378 (Dec. 1999).

Business Day: Olympia Thai Tower, 22nd Floor, 444 Thanon Ratchadaphisek, Huay Kwang, Bangkok 10310; tel. (2) 512-3579; fax (2) 512-3565; e-mail info@bday.net; internet www.bday.net; f. 1994; business news; Man. Editor Chatchai Yenbamroong.

The Nation: 44 Moo 10, Thanon Bangna Trad, Km 4.5, Bang Na, Phra Khanong, Bangkok 10260; tel. (2) 317-0420; fax (2) 317-2071; e-mail info@nationgroup.com; internet www.nationmultimedia.com; f. 1971; morning; Publr and Group Editor Suthichai Yoon; Editor Pana Janviroj; circ. 55,000.

Thailand Times: 88 Thanon Boromrajachonnee, Taling Chan, Bangkok; tel. (2) 434-0330; Editor Wuthipong Lakkham.

Chinese Language

Kia Hua Tong Huan: 108 Thanon Suapa, Bangkok; tel. (2) 221-4182; f. 1959; Editor Suradech Kornkittichai; circ. 80,000.

New Chinese Daily News: 1022–1030 Thanon Charoenkrung, Talad-Noi, Bangkok; tel. (2) 234-0684; fax (2) 234-0684; Editor Pusadee Keetaworanart; circ. 72,000.

Sing Sian Yit Pao Daily News: 267 Thanon Charoenkrung, Bangkok 10100; tel. (2) 222-6601; fax (2) 225-4663; e-mail singpau@loxinfo.co.th; f. 1950; Man. Dir Netra Ruthaiyanont; Editor Tawee Yodpetch; circ. 70,000.

Tong Hua Daily News: 877/879 Thanon Charoenkrung, Talad-Noi, Bangkok 10100; tel. (2) 236-9172; fax (2) 238-5286; Editor CHART PAYONITHIKARN; circ. 85,000.

WEEKLIES

Thai Language

Bangkok Weekly: 533–539 Thanon Sri Ayuthaya, Bangkok 10400; tel. (2) 245-2546; fax (2) 247-3410; Editor VICHIT ROJANAPRABHA.

Mathichon Weekly Review: 12 Thanon Thedsaban Naruban, Prachanivate 1, Chatuchak, Bangkok 10900; tel. (2) 589-0020; fax (2) 589-9112; e-mail weekly@matichon.co.th; internet www.matichon.co.th/weekly; Editor RUANGCHAI SABNIRAND; circ. 150,000.

Satri Sarn: 83/35 Arkarntrithosthep 2, Thanon Prachathipatai, Bangkok 10300; tel. (2) 281-9136; f. 1948; women's magazine; Editor NILAWAN PINTONG.

Siam Rath Weekly Review: 12 Mansion 6, Thanon Rajdamnern, Bangkok 10200; Editor PRACHUAB THONGURAI.

Skul Thai: 58 Soi 36, Thanon Sukhumvit, Bangkok 10110; tel. (2) 258-5861; fax (2) 258-9130; Editor SANTI SONGSEMSAWAS.

Wattachak: 88 Thanon Boromrajachonnee, Talingchan, Bangkok 10170; tel. (2) 434-0330; fax (2) 435-0900; e-mail rattanasingha@wattachak.com; internet www.wattachak.com; Editor PRAPHAN BOONYAKIAT; circ. 80,000.

English Language

Bangkok Post Weekly Review: U-Chuliang Bldg, 3rd Floor, 968 Thanon Phra Ram Si, Bangkok 10500; tel. (2) 233-8030; fax (2) 238-5430; f. 1989; Editor ANUSSORN THAVISIN; circ. 10,782.

Business Times: Thai Bldg, 1400 Thanon Phra Ram IV, Bangkok 10110.

FORTNIGHTLIES

Thai Language

Darathai: 9-9/1 Soi Sri Ak-Sorn, Thanon Chuapleung, Tungmahamek, Sathorn, Bangkok 10120; tel. (2) 249-1576; fax (2) 249-1575; f. 1954; television and entertainment; Editor USA BUKKAVESA; circ 80,000.

Dichan: 1400 Thai Bldg, Thanon Phra Ram Si, Bangkok; tel. (2) 249-0351; fax (2) 249-9455; Editor KHUNYING TIPYAVADI PRAMOJ NA AYUDHYA.

Lalana: 44 Moo 10, Thanon Bangna-Trad, Bang Na, Bangkok 10260; tel. (2) 317-1400; fax (2) 317-1409; f. 1972; Editor NANTAWAN YOON; circ. 65,000.

Praew: 7/9–18 Thanon Arun-Amarin, Bangkoknoi, Bangkok 10700; tel. (2) 434-0286; fax (2) 433-8792; e-mail praew@amarin.co.th; internet www.amarin.com; f. 1979; women and fashion; Editorial Dir SUPAWADEE KOMARADAT; Editor NUANCHAN SUPANIMIT; circ. 150,000.

MONTHLIES

Bangkok 30: 98/5–6 Thanon Phra Arthit, Bangkok 10200; tel. (2) 282-5467; fax (2) 280-1302; f. 1986; Thai; business; Publr SONCHAI LIMTHONGKUL; Editor BOONSIRI NAMBOONSRI; circ. 65,000.

Chao Krung: 12 Mansion 6, Thanon Rajdamnern, Bangkok 10200; Thai; Editor NOPPHORN BUNYARIT.

The Dharmachaksu (Dharma-vision): Foundation of Mahamakut Rajavidyalai, 241 Thanon Phra Sumeru, Bangkok 10200; tel. (2) 629-1417; fax (2) 629-4015; e-mail books@mahamakuta.inet.co.th; internet www.mahamakuta.inet.co.th; f. 1894; Thai; Buddhism and related subjects; Editor WASIN INDASARA; circ. 5,000.

Grand Prix: 4/299 Moo 5, Soi Ladplakhao 66, Thanon Ladplakhao, Bangkhan, Bangkok 10220; tel. (2) 971-6450; fax (2) 971-6469; e-mail grandprix@grandprixgroup.com; internet www.grandprixgroup.com/gpi/maggrandprix/grandprix.asp; Editor VEERA JUNKAJIT; circ. 80,000.

The Investor: Pansak Bldg, 4th Floor, 138/1 Thanon Petchburi, Bangkok 10400; tel. (2) 282-8166; f. 1968; English language; business, industry, finance and economics; Editor TOS PATUMSEN; circ. 6,000.

Kasikorn: Dept of Agriculture, Catuchak, Bangkok 10900; tel. (2) 561-4677; fax (2) 579-5369; e-mail pannee@doa.go.th; f. 1928; Thai; agriculture and agricultural research; Editor-in-Chief PAIROJ SUWANJINDA; Editor PANNEE WICHACHOO.

Look: 1/54 Thanon Sukhumvit 30, Pra Khanong, Bangkok 10110; tel. (2) 258-1265; Editor KANOKWAN MILINDAVANIJ.

Look East: 52/38 Soi Saladaeng 2, Silom Condominium, 12th Floor, Thanon Silom, Bangkok 10500; tel. (2) 235-6185; fax (2) 236-6764; f. 1969; English; Editor ASHA SEHGAL; circ. 30,000.

Motorcycle Magazine: 4/299 Moo 5, Soi Ladplakhao 66, Thanon Ladplakhao, Bangkhan, Bangkok 10220; tel. (2) 971-6450; fax (2) 971-6469; e-mail motorcycle@grandprixgroup.com; internet www.grandprixgroup.com/gpi/magmotor/motorcycle.asp; monthly; Publr PRACHIN EAMLAMNOW; circ. 70,000.

Saen Sanuk: 50 Soi Saeng Chan, Thanon Sukhumvit 42, Bangkok 10110; tel. (2) 392-0052; fax (2) 391-1486; English; travel and tourist attractions in Thailand; Editor SOMTAWIN KONGSAWATKIAT; circ. 85,000.

Sarakadee Magazine: Bangkok; e-mail admin@sarakadee.com; internet www.sarakadee.com; Thai; events, culture and nature.

Satawa Liang: 689 Thanon Wang Burapa, Bangkok; Thai; Editor THAMRONGSAK SRICHAND.

Villa Wina Magazine: Chalerm Ketr Theatre Bldg, 3rd Floor, Bangkok; Thai; Editor BHONGSAKDI PIAMLAP.

NEWS AGENCIES

Foreign Bureaux

Agence France-Presse (AFP): 18th Floor, Alma Link Bldg, 25 Soi Chidlom, Thanon Ploenchit, Lumpini Patumwan, Bangkok 10330; tel. (2) 650-3230; fax (2) 650-3234; e-mail afpbgk@samart.co.th; Bureau Chief PHILIPPE AGRET.

Associated Press (AP) (USA): Charn Issara Tower, 14th Floor, 942/51 Thanon Phra Ram IV, POB 775, Bangkok; tel. (2) 234-5553; Bureau Chief DENIS D. GRAY.

Jiji Tsushin (Jiji Press) (Japan): Boonmitr Bldg, 8th Floor, 138 Thanon Silom, Bangkok; tel. (2) 236-8793; fax (2) 236-6800; Bureau Chief JUNICHI ISHIKAWA.

Kyodo News Service (Japan): U Chuliang Bldg, 2nd Floor, 968 Thanon Phra Ram IV, Bangkok 10500; tel. (2) 236-6822; Bureau Chief YOSHISUKE YASUO.

Reuters (UK): Maneeya Centre Bldg, 518/5 Thanon Ploenchit, 10330 Bangkok; tel. (2) 637-5500; Man. (Myanmar, Thailand and Indo-China) GRAHAM D. SPENCER.

United Press International (UPI) (USA): U Chuliang Bldg, 968 Thanon Phra Ram IV, Bangkok; tel. (2) 238-5244; Bureau Chief JOHN B. HAIL.

Viet Nam News Agency (VNA): 3/81 Soi Chokchai Ruamit, Bangkok; Bureau Chief NGUYEN VU QUANG.

Xinhua (New China) News Agency (People's Republic of China): Room 407, Capital Mansion, 1371 Thanon Phaholyothin, Saphan Kwai, Bangkok 10400; tel. and fax (2) 271-1413; Chief Correspondent YU ZUNCHENG.

PRESS ASSOCIATIONS

Confederation of Thai Journalists: 55 Mansion 8, Thanon Ratchadamnoen Klang, Bangkok 10200; tel. (2) 629-0022; fax (2) 280-0337; internet www.ctj.in.th; Pres. SMARN SUDTO; Sec.-Gen. SUWAT THONGTHANAKUL.

Press Association of Thailand: 299 Thanon Ratchasima, Dusit, Bangkok 10300; tel. (2) 241-2064; e-mail webmaster@thaipressasso.com; internet www.thaipressasso.com; f. 1941; Pres. PREECHA SAMAKKIDHAM.

There are also regional press organizations and journalists' organizations.

Publishers

Advance Media: 1400 Rama IV Shopping Centre, Klongtoey, Bangkok 10110; tel. (2) 249-0358; Man. PRASERTSAK SIVASAHONG.

Amarin Printing and Publishing Public Co Ltd: 7/9–18 Thanon Arunamarin, Bangkoknoi, Bangkok 10700; tel. (2) 434-0286; fax (2) 433-8792; e-mail info@amarin.co.th; internet www.amarin.co.th; f. 1976; general books and magazines; CEO CHUKIAT UTAKAPAN.

Bhannakij Trading: 34 Thanon Nakornsawan, Bangkok 10100; tel. (2) 282-5520; fax (2) 282-0076; Thai fiction, school textbooks; Man. SOMSAK TECHAKASHEM.

Chalermnit Publishing Co Ltd: 108 Thanon Sukhumvit, Soi 53, Bangkok 10110; tel. (2) 662-6264; fax (2) 662-6265; e-mail chalermnit@hotmail.com; internet www.chalermnit.com; f. 1937; dictionaries, history, literature, guides to Thai language, works on Thailand and South-East Asia; Man. Dir Dr PARICHART JUMSAI.

Dhamabuja: 5/1–2 Thanon Asadang, Bangkok; religious; Man. VIROCHANA SIRI-ATH.

Graphic Art Publishing: 105/19–2 Thanon Naret, Bangkok 10500; tel. (2) 233-0302; f. 1972; textbooks, science fiction, photography; CEO Mrs ANGKANA SAJJARAKTRAKUL.

Prae Pittaya Ltd: POB 914, 716–718 Wangburabha, Bangkok; tel. (2) 221-4283; general Thai books; Man. CHIT PRAEPANICH.

Praphansarn: Siam Sq., Soi 2, Thanon Phra Ram I, Bangkok; tel. (2) 251-2342; e-mail editor@praphansarn.com; internet www.praphansarn.com; Thai pocket books; Man. Dir SUPHOL TAECHATADA.

Ruamsarn (1977): 864 Wangburabha, Thanon Panurangsri, Bangkok 10200; tel. (2) 221-6483; fax (2) 222-2036; f. 1951; fiction and history; Man. PITI TAWEWATANASARN.

Ruang Silpa: 663 Thanon Samsen Nai, Bangkok; Thai pocket books; Propr DHAMNOON RUANG SILPA.

Sermvitr Barnakarn: 222 Werng Nakorn Kasem, Bangkok 10100; general Thai books; Man. PRAVIT SAMMAVONG.

Silkworm Books: 104/5 Thanon Chiang Mai-Hot, M.7, T. Suthep, Muang, Chiang Mai 50200; tel. (53) 271-889; fax (53) 275-178; e-mail silkworm@loxinfo.co.th; internet www.silkwormbooks.info; f. 1991; South-East Asian studies.

Suksapan Panit (Business Organization of Teachers' Institute): 128/1 Thanon Ratchasima, Bangkok 10300; tel. (2) 811-845; e-mail ssp@suksapan.or.th; internet www.suksapan.or.th; f. 1950; general, textbooks, children's, pocket books; Pres. PANOM KAW KAMNERD.

Suriyabarn Publishers: 14 Thanon Pramuan, Bangkok 10500; tel. (2) 234-7991; f. 1953; religion, literature, Thai culture; Man. Dir PRASIT SAETANG.

Thai Watana Panit: 599 Thanon Maitrijit, Bangkok 10100; tel. (2) 221-0111; children's, school textbooks; Man. Dir VIRA T. SUWAN.

Watana Panit Printing and Publishing Co Ltd: 31/1–2 Soi Siripat, Thanon Mahachai, Samranrat, Pranakorn, Bangkok 10120; tel. (2) 222-1016; fax (2) 225-6556; school textbooks; Man. ROENGCHAI CHONGPIPATANASOOK.

White Lotus Co Ltd: 11/2 Soi 58, Thanon Sukhumvit, POB 1141, Bangkok 10260; tel. (2) 332-4915; fax (2) 311-4575; e-mail ande@loxinfo.co.th; internet thailine.com/lotus; f. 1992; regional interests; Publisher DIETHARD ANDE.

PUBLISHERS' ASSOCIATION

Publishers' and Booksellers' Association of Thailand (PUBAT): 83/159 Moo 6, Thanon Ngam Wong Wan, Thung Song Hong, Lak Si, Bangkok 10210; tel. (2) 954-9560; fax (2) 954-9565; e-mail info@pubat.or.th; internet www.pubat.or.th; f. 1960; organizes national book fairs and provides promotional opportunities for publishers; Pres. THANACHAI SANTICHAIKUL; Gen. Sec. PLEARNPIT PRAEPANIT.

Broadcasting and Communications

TELECOMMUNICATIONS

Post and Telegraph Department: Thanon Phaholyothin, Payathai, Bangkok 10400; tel. (2) 271-0151; fax (2) 271-3511; e-mail webadmin@ptd.go.th; internet www.ptd.go.th; interim regulator of telecommunications sector pending creation of National Telecommunications Commission; Dir-Gen. SETHAPORN CUSRIPITUCK.

Advanced Info Service Public Co Ltd: 414 Thanon Phaholyothin, Shinawatra Tower I, Phayathai, Bangkok 10400; tel. (2) 299-5000; fax (2) 299-5719; e-mail callcenter@ais900.com; internet www.ais900.com; mobile telephone network operator; Chair. PAIBOON LIMPAPHAYOM.

CAT Telecom Public Co Ltd: 99 Thanon Changwattana, Laksi, Bangkok 10002; tel. (2) 573-0099; fax (2) 574-6054; e-mail pr@cattelecom.co.th; internet www.cattelecom.com; state-owned; f. 2003 following division of Communications Authority of Thailand (CAT); telephone operator; scheduled for transfer to private sector in 2004; Chair. Lt-Gen. ANUSORN TEPTADA.

Samart Corpn Public Co Ltd: Moo 4, Software Park Bldg, 38th Floor, 99 Thanon Chaengwattana, Klong Gluar, Pak-kred, Nonthaburi 11120; tel. (2) 502-6000; fax (2) 502-6043; e-mail nikhila@samartcorp.com; internet www.samartcorp.com; telecommunications installation and distribution; cap. and res 700m. baht, sales 4,692m. baht (1996); Chair. PICHAI VASANASONG; Exec. Chair. CHAROENRATH VILAILUCK.

Telecomasia Corpn Public Co Ltd: Telecom Tower, 18 Thanon Ratchadaphisek, Huay Kwang, Bangkok 10320; tel. (2) 643-1111; fax (2) 643-9669; e-mail iroffice@asianet.co.th; internet www.telecomasia.co.th; telecommunications services; Chair. DHANIN CHEARAVANONT; Pres. and CEO SUPACHAI CHEARAVANONT.

Thai Telephone and Telecommunication Public Co Ltd (TT&T): Muang Thai Phatra Complex Tower 1, 24th Floor, 252/30 Thanon Ratchadaphisek, Huay Kwang, Bangkok; tel. (2) 693-2100; fax (2) 693-2126; e-mail webeditor@ttt.co.th; internet www.ttt.co.th; f. 1992; distributors of telecommunications equipment and services; Chair. DHONGCHAI LAMSAM; Pres. and CEO PRACHUAB TANTINON.

TOT Corporation Public Co Ltd: 89/2 Moo 3, Thanon Chaengwattana, Thungsonghong, Laksi, Bangkok 10210; tel. (2) 505-1000; internet www.tot.co.th; state-owned; fmrly Telephone Organization of Thailand; telephone operator; *de facto* regulator; scheduled for transfer to private sector in 2004; Pres. SUTHAM MALILA.

Total Access Communications Public Co Ltd: Chai Bldg, 19th Floor, 333/3 Moo 4, Thanon Vibhavadi Rangsit, Ladyao Chatuchak, Bangkok 10900; tel. (2) 202-87000; fax (2) 202-8102; e-mail feedback@dtac.co.th; internet www.dtac.co.th; mobile telephone network operator; 71.6% owned by United Communication Industry Public Co Ltd; Chair. BOONCHAI BENCHARONGKUL; CEOs VICHAI BENCHARONGKUL, SIGVE BREKKE.

United Communication Industry Public Co Ltd: 499 Moo 3, Benchachinda Bldg, Thanon Vibhavadi Rangsit, Chatuchak, Bangkok 10900; tel. (2) 953-1111; fax (2) 953-0248; e-mail hr@ucom.co.th; internet www.ucom.co.th; telecommunications service provider; Chair. SRIBHUMI SUKHANETR; Pres. and Chief Exec. BOONCHAI BENCHARONGKUL.

BROADCASTING

Regulatory Authority

Radio and Television Executive Committee (RTEC): Programme, Administration and Law Section, Division of RTEC Works, Government Public Relations Dept, Thanon Ratchadamnoen Klang, Phra Nakhon Region, Bangkok 10200; constituted under the Broadcasting and TV Rule 1975, the committee consists of 17 representatives from 14 government agencies and controls the administrative, legal, technical and programming aspects of broadcasting in Thailand; regulatory functions due to be assumed by an independent body, the National Broadcasting Commission, which was in the process of being established in mid-2004.

Radio

Radio Thailand (RTH): National Broadcasting Services of Thailand, Government Public Relations Dept, Soi Aree Sampan, Thanon Phra Ram VI, Khet Phayathai, Bangkok 10400; tel. (2) 618-2323; fax (2) 618-2340; internet www.prd.go.th/mcic/radio.htm; f. 1930; govt-controlled; educational, entertainment, cultural and news programmes; operates 109 stations throughout Thailand; Dir of Radio Thailand SOMPHONG VISUTTIPAT.

Home Service: 12 stations in Bangkok and 97 affiliated stations in 50 provinces; operates three programmes; Dir CHAIVICHIT ATISAB.

External Services: f. 1928; in Thai, English, French, Vietnamese, Khmer, Japanese, Burmese, Lao, Malay, Chinese (Mandarin), German and Bahasa Indonesia; Dir AMPORN SAMOSORN.

Ministry of Education Broadcasting Service: Centre for Innovation and Technology, Ministry of Education, Bangkok; tel. (2) 246-0026; f. 1954; morning programmes for schools (Mon.–Fri.); afternoon and evening programmes for general public (daily); Dir of Centre PISAN SIWAYABRAHM.

Pituksuntirad Radio Stations: stations at Bangkok, Nakorn Rachasima, Chiangmai, Pitsanuloke and Songkla; programmes in Thai; Dir-Gen. PAITOON WAIJANYA.

Radio Saranrom: Thanon Ratchadamnoen, POB 2-131, Bangkok 10200; tel. (2) 224-4904; fax (2) 226-1825; f. 1968 as Voice of Free Asia; name changed as above in 1998; operated by the Ministry of Foreign Affairs; broadcasts in Thai; Dir of Broadcasting PAIBOON KUSKUL.

Television

Bangkok Broadcasting & TV Co Ltd (Channel 7): 998/1 Soi Sirimitr, Phaholyothin, Talad Mawchid, POB 456, Bangkok 10900; tel. (2) 278-1255; fax (2) 270-1976; e-mail prdept@ch7.com; internet www.ch7.com; commercial.

Bangkok Entertainment Co Ltd (Channel 3): 1126/1 Thanon Petchburi Tadmai, Bangkok 10400; tel. (2) 204-3333; fax (2) 204-1384; internet www.tv3.co.th; Programme Dir PRAVIT MALEENONT.

Independent Television (ITV): 19 SCB Park Plaza, East Tower 3, Thanon Rachadapisek, Bangkok; tel. (2) 937-8080; e-mail webmaster@itv.co.th; internet www.itv.co.th; f. 1996; owned by a consortium including Nation Publishing Co, Siam Commercial Bank Public Co Ltd and Daily News media group.

The Mass Communication Organization of Thailand (Channel 9): 222 Thanon Asoke Dindaeng, Bangkok 10300; tel. (2) 245-1844; fax (2) 245-1855; e-mail webmaster@mcot.or.th; internet www.mcot.or.th; f. 1952 as Thai Television Co Ltd; colour service; Dir-Gen. (vacant).

The Royal Thai Army Television HSA-TV (Channel 5): Thanon Phaholyothin, Sanam Pao, Bangkok 10400; tel. (2) 278-1470; fax (2) 279-0430; internet www.tv5.co.th; f. 1958; operates channels nationwide; Dir-Gen. Maj.-Gen. Vijit Junapart.

Television of Thailand (Channel 11): Public Relations Dept, Fortune Town Bldg, 26th Floor, Thanon Ratchadaphisek, Huay Kwang, Bangkok 10310; tel. and fax (2) 248-1601; internet www.prd.go.th/prdnew/thai/tv11; operates 16 colour stations; Dir-Gen. (vacant).

Thai Sky TV: 21 Vibhavadi Rangsit, Bangkok; tel. (2) 273-8977; pay-television; owned by Wattachak Co.

TV Pool of Thailand: c/o Royal Thai Army HSA-TV, Thanon Phaholyothin, Sanam Pao, Bangkok 10400; tel. (2) 271-0060; fax (2) 279-0430; established with the co-operation of all stations to present coverage of special events; Chair. Maj.-Gen. Soonthorn Sophonsiri.

Finance

(cap. = capital; p.u. = paid up; res = reserves; dep. = deposits; m. = million; brs = branches; amounts in baht)

BANKING

Central Bank

Bank of Thailand: 273 Thanon Samsen, Bangkhunprom, Bangkok 10200; tel. (2) 283-5353; fax (2) 280-0449; e-mail webmaster@bot.or.th; internet www.bot.or.th; f. 1942; bank of issue; cap. 20m., res 29,860m., dep. 762,846m. (Dec. 2003); Gov. Pridiyathorn Davakula; 4 brs.

Commercial Banks

Bangkok Bank Public Co Ltd: 333 Thanon Silom, Bangkok 10500; tel. (2) 231-4333; fax (2) 236-8281; internet www.bangkokbank.com; f. 1944; cap. 13,633m., res 135,400m., dep. 1,123,300m. (Dec. 2003); 49% foreign-owned; Chair. Chatri Sophonpanich; Pres. Chartsiri Sophonpanich; 628 local brs, 21 overseas brs.

Bank of Asia Public Co Ltd: 191 Thanon Sathorn Tai, Khet Sathorn, Bangkok 10120; tel. (2) 287-2211; fax (2) 287-2973; e-mail webmaster@boa.co.th; internet www.bankasia4u.com; f. 1939; cap. 50,954m., res –9,376m., dep. 147,431m. (Dec. 2003); 78.85%-owned by ABN AMRO Bank NV; Pres. and CEO Chulakorn Singhakowin; 120 brs.

Bank of Ayudhya Public Co Ltd: 1222 Thanon Rama III, Bangkok 10120; tel. (2) 296-2000; fax (2) 683-1304; e-mail webmaster@krungsri.com; internet www.krungsri.com; f. 1945; cap. 28,503m., res 36,623m., dep. 436,503m. (Dec. 2003); Chair. Krit Ratanarak; Pres. Jamlong Atikul; 384 local brs, 3 overseas brs.

Bankthai Public Co Ltd: 44 Thanon Sathorn Nua, Silom, Bangrak, Bangkok 10500; tel. (2) 638-8000; fax (2) 633-9026; e-mail bankthai_pr2@hotmail.com; internet www.bankthai.co.th; f. 1998; cap. 14,935m., res –4,249m., dep. 214,269m. (Dec. 2002); Chair. Pramon Sutivong; Pres. Phirasilp Subhapholsiri; 84 brs.

Kasikorn Bank Public Co Ltd: 1 Soi Kasikornthai, Thanon Ratburana, Bangkok 10140; tel. (2) 222-0000; fax (2) 470-1144; e-mail webmaster@kasikornbank.com; internet www.kasikornbank.com; f. 1945; fmrly Thai Farmers Bank Public Co Ltd; name changed as above April 2003; cap. 23,634m., res 62,878m., dep. 700,612m. (June 2004); 48.98% foreign-owned; Chair. Banyong Lamsam; CEO Banthoon Lamsam; Pres. Prasarn Trairatvorakul; 496 local brs, 4 overseas brs.

Krung Thai Bank Public Co Ltd (State Commercial Bank of Thailand): 35 Thanon Sukhumvit, Klongtoey, Bangkok 10110; tel. (2) 255-2222; fax (2) 255-9391; e-mail ptsirana@ktb.co.th; internet www.ktb.co.th; f. 1966; cap. 57,603m., res 4,560m., dep. 981,693m. (Dec. 2002); taken under the control of the central bank in 1998, pending transfer to private sector; merged with First Bangkok City Bank Public Co Ltd in 1999; Chair. Suphachai Phisitvanich; Pres. Viroj Nualkhair; 610 local brs, 7 overseas brs.

Siam City Bank Public Co Ltd: 1101 Thanon Petchburi Tadmai, Bangkok 10400; tel. (2) 208-5000; fax (2) 253-1240; e-mail pr@scib.co.th; internet www.scib.co.th; f. 1941; cap. 21,128m., res 7,245m., dep. 426,425m. (Dec. 2003); taken under the control of the central bank in Feb. 1998; merged with Bangkok Metropolitan Bank Public Co Ltd in April 2002; Pres. Sompol Kiatphaibool; 365 local brs, 1 overseas br.

Siam Commercial Bank Public Co Ltd: 9 Thanon Ratchadaphisek, Lardyao, Chatuchak, Bangkok 10900; tel. (2) 544-1111; fax (2) 937-7754; e-mail ccs@telecom.scb.co.th; internet www.scb.co.th; f. 1906; cap. 31,305m., res 91,337m., dep. 612,697m. (Dec. 2001); Chair. Dr Chirayu Isarangkun Na Ayuthaya; Pres. and CEO Jada Wattansiritham; 477 local brs, 4 overseas brs.

Standard Chartered Nakornthon Bank Public Co Ltd: 90 Thanon Sathorn Nua, Silom, Bangkok 10500; tel. (2) 233-2111; fax (2) 236-1968; e-mail suteel@samart.co.th; internet www.ntb.co.th; f. 1933 as Wang Lee Bank Ltd, renamed 1985; cap. 7,003m., res 255m., dep. 57,174m. (Dec. 2002); taken under the control of the central bank in July 1999, 75% share sold to Standard Chartered Bank (United Kingdom); name changed as above 1999; Chair. David George Moir; CEO Vishnu Mohan; 67 brs.

Thai Military Bank Public Co Ltd: 3000 Thanon Phahon Yothin, Bangkok 10900; tel. (2) 299-1111; fax (2) 273-7121; e-mail tmbir@tmb.co.th; internet www.tmb.co.th; f. 1957; merged with DBS Thai Danu Bank Public Co Ltd and Industrial Finance Corpn of Thailand Sept. 2004; cap. 104,080m., res –19,883m., dep. 338,646m. (Dec. 2003); Pres. and CEO Subhak Siwaraksa; 367 local brs, 3 overseas brs.

Thanachart Bank Public Co Ltd: 1st, 2nd, 14th and 15th Floors, Thonson Bldg, 900 Thanon Phoenchit, Lumpini, Patumwan, Bangkok 10330; tel. (2) 655-9000; fax (2) 655-9001; e-mail nfs_rb@nfs.co.th; internet www.thanachart.com; f. 2002; Man. Dir Suvarnapha Suvarnaprathip.

UOB Radanasin Bank Public Co Ltd: 690 Thanon Sukhumvit, Bangkok 10110; tel. (2) 260-0090; fax (2) 260-5310; e-mail customerservice@uob-radanasin.co.th; internet www.uob-radanasin.co.th; f. 1998 as the result of the merger of Laem Thong Bank Ltd with Radanasin Bank; name changed as above Nov. 1999; cap. 9,847m., res 61m., dep. 45,772m. (Dec. 2002); Man. Dir and CEO Gan Hui Beng; 37 brs.

Foreign Banks

ABN AMRO Bank NV (Netherlands): Bangkok City Tower, 3rd–4th Floors, 179/3 Thanon Sathorn Tai, Bangkok 10120; tel. (2) 679-5900; fax (2) 679-5901; e-mail wholesale.clients.internet@nl.abnamro.com; Country Man. Burno Schrieke.

Bank of America, NA (USA): Bank of America Centre, 2/2 Thanon Witthayu, Bangkok 10330; tel. (2) 251-6333; fax (2) 253-1905; f. 1949; cap. p.u. 2,000m. (May 1998); Man. Dir and Country Man. Frederick Chin.

Bank of China (People's Republic of China): Bangkok City Tower, 179 Thanon Sathorn Tai, Bangkok 10120; tel. (2) 268-1010; fax (2) 268-1020.

Bank of Nova Scotia (Canada): Ground Floor, Ploenchit Tower, 898 Thanon Ploenchit, Bangkok 10330; tel. (2) 263-0303; fax (2) 263-0150; e-mail bbangkok@scotiabank.com; Vice-Pres. and Man. K. Duangdee.

Bank of Tokyo-Mitsubishi Ltd (Japan): Harindhorn Tower, 54 Thanon Sathorn Nua, Bangrak, Bangkok 10500; tel. (2) 266-3011; fax (2) 266-3055; cap. p.u. 3,100m., dep. 17,213m. (March 1996); Dir and Gen. Man. Hiroshi Motomura.

Bharat Overseas Bank Ltd (India): 221 Thanon Rajawongse, Sumpphanthawongse, Bangkok 10100; tel. (2) 224-5412; fax (2) 224-5405; f. 1974; cap. p.u. 225m., dep. 1,254m. (March 1996); Man. K. S. Ganapathy.

BNP Paribas (France): 29th Floor, Abdulrahim Place, 990 Thanon Rama IV, Silom, Bangrak, Bangkok 10500; tel. (2) 636-1900; fax (2) 636-1935.

Citibank, NA (USA): Sangtongtanee Bldg, 82 Thanon Sathorn Nua, Bangrak, Bangkok 10500; tel. (2) 232-2000; fax (2) 639-2564; internet www.citibank.com/thailand; cap. p.u. 3,849m., dep. 9,483m. (Feb. 1996); Man. David L. Hendrix.

Crédit Agricole Indosuez (France): CA Indosuez House, 152 Thanon Witthayu, POB 303, Bangkok 10330; tel. (2) 651-4590; fax (2) 651-4586; cap. p.u. 1,400m., dep. 2,241m. (April 1996); Man. Dr Jean-François Cahet.

Deutsche Bank AG (Germany): 208 Thanon Witthayu, Bangkok 10330; tel. (2) 651-5000; fax (2) 651-5151; cap. p.u. 1,200m., dep. 3,703m. (March 1996); Man. Gerhard Heigl.

Hongkong and Shanghai Banking Corpn (HSBC) (Hong Kong): HSBC Bldg, 968 Thanon Rama IV, Silom, Bangkok 10500; tel. (2) 614-4000; fax (2) 632-4818; cap. p.u. 2,000m., dep. 8,328m. (March 1996); Man. R. J. O. Cromwell.

International Commercial Bank of China (Taiwan): 36/12 P. S. Tower Asoke, 21 Thanon Sukhumvit, Phrakhanong, Bangkok 10110; tel. (2) 259-2000; fax (2) 259-1330; e-mail accdept@icbc.co.th; cap. p.u. 500m., dep. 1,334m. (Feb. 1996); Vice-Pres. and Gen. Man. Chia-Nan Wang.

JP Morgan Chase Bank (USA): Bubhajit Bldg, 2nd Floor, 20 Thanon Sathorn Nua, Silom, Bangrak, Bangkok 10500; tel. (2) 234-5992; fax (2) 234-8386; Man. Dir RAYMOND C. C. CHANG.

Mizuho Corporate Bank Ltd (Japan): 18th Floor, Tisco Tower, 48 Thanon Sathorn Nua, Silom, Bangrak, Bangkok 10500; tel. (2) 638-0200–5; fax (2) 638-0218.

Oversea-Chinese Banking Corpn Ltd (Singapore): Charn Issara Tower, 2nd Floor, 942/80 Thanon Rama IV, Suriwongse, Bangkok 10500; tel. (2) 236-6730; fax (2) 237-7390; cap. p.u. 651m., dep. 455m. (Feb. 1996); Gen. Man. NEW JING YAN.

RHB Bank Bhd (Malaysia): Liberty Sq. Bldg, 10th Floor, 287 Thanon Silom, Bangrak, Bangkok 10500; tel. (2) 631-2010; fax (2) 631-2018; e-mail rhbbkbk@asianet.co.th; Man. LEH THIAM GUAN.

Standard Chartered Bank (UK): Sathorn Nakorn Tower, 100 Thanon Sathon Nua, Silom, Bangrak, Bangkok 10500; tel. (2) 724-4000; fax (2) 636-0536; e-mail th.standardchartered-service@th.standardchartered.com; internet www.stanchart.com/th/; cap. p.u. 1,200m., dep. 5,585m. (March 1996); CEO VISHNU MOHAN.

Sumitomo Mitsui Banking Corpn (Japan): Boon-Mitr Bldg, 138 Thanon Silom, Bangkok 10500; tel. (2) 353-8000; fax (2) 353-8100; cap. p.u. 3,200m., dep. 8,343m. (Feb. 1996); Gen. Rep. YOSHIHIRO YOSHIMURA.

Development Banks

Bank for Agriculture and Agricultural Co-operatives (BAAC): 469 Thanon Nakorn Sawan, Dusit, Bangkok 10300; tel. (2) 280-0180; fax (2) 280-0442; e-mail train@baac.or.th; internet www.baac.or.th; f. 1966 to provide credit for agriculture; cap. 22,761m., dep. 180,563m. (Dec. 2000); Exec. Chair. VARATHEP RATANAKORN; Pres. PITTAYAPOL NATTARADOL; 491 brs.

Export-Import Bank of Thailand (EXIM THAILAND): EXIM Bldg, 1193 Thanon Phaholyothin, Phayathai, Bangkok 10400; tel. (2) 271-3700; fax (2) 271-3204; e-mail info@exim.go.th; internet www.exim.go.th; f. 1993; provides financial services to Thai exporters and Thai investors investing abroad; cap. 6,500m., res 2,060m., dep. 13,424m. (Dec. 2003); Chair. PAKORN MALAKUL NA AYUDYA; Pres. SATAPORN JINICHITRA; 7 brs.

Government Housing Bank: 63 Thanon Phra Ram IX, Huay Kwang, Bangkok 10310; tel. (2) 246-0303; fax (2) 246-1789; e-mail ghb_pr@housing.ghb.co.th; internet www.ghb.co.th; f. 1953 to provide housing finance; cap. 9,483m., assets 152,973m., dep. 80,251m. (Dec. 1995); Chair. Dr SOMCHAI RICHUPAN; Pres. SIDHIJAI TANPHIPHAT; 120 brs.

Small and Medium Enterprise Development Bank of Thailand: 9th Floor, Siripinyo Bldg, 475 Thanon Sri Ayudhaya, Rajtavee, Bangkok 10400; tel. (2) 201-3700; fax (2) 201-3723–34; e-mail sme@smebank.co.th; internet www.smebank.co.th; f. 2002; fmrly Small Industrial Finance Co.

Savings Bank

Government Savings Bank: 470 Thanon Phaholyothin, Phayathai, Bangkok 10400; tel. (2) 299-8000; fax (2) 299-8490; e-mail vicheal@gsb.or.th; internet www.gsb.or.th; f. 1913; cap. 0.1m., res 8,040m., dep. 522,824m. (Dec. 2002); Chair. SOMCHAINUK ENGTRAKUL; Dir-Gen. Dr CHARNCHAI MUSIGNISARKORN; 578 brs.

Bankers' Association

Thai Bankers' Association: Lake Rachada Office Complex, 4th Floor, 195/8–10 Thanon Ratchadaphisek, Klongtoey, Bangkok 10110; tel. (2) 264-0883; fax (2) 264-0888; e-mail atmpool@tba.or.th; internet www.tba.or.th; Chair. CHARTSIRI SOPHONPANICH.

STOCK EXCHANGE

Stock Exchange of Thailand (SET): The Stock Exchange of Thailand Bldg, 62 Thanon Ratchadaphisek, Klongtoey, Bangkok 10110; tel. (2) 229-2000; fax (2) 654-5649; e-mail clientsupport@set.or.th; internet www.set.or.th; f. 1975; 27 mems; Pres. KITTIRAT NA RANONG; Chair. CHAVALIT THANACHANAN.

Securities and Exchange Commission: Diethelm Towers B, 10th/13th–16th Floors, 93/1 Thanon Witthayu, Lumpini, Patumwan, Bangkok 10330; tel. (2) 252-3223; fax (2) 256-7711; e-mail info@sec.or.th; internet www.sec.or.th; f. 1992; supervises new share issues and trading in existing shares; chaired by Minister of Finance; Sec.-Gen. THIRACHAI PHUVANAT NARANUBULA.

INSURANCE

In March 1997 the Government granted licences to 12 life assurance companies and 16 general insurance companies, bringing the total number of life insurance companies to 25 and non-life insurance companies to 83.

Selected Domestic Insurance Companies

American International Assurance Co Ltd: American International Tower, 181 Thanon Surawongse, Bangkok 10500; tel. (2) 634-8888; fax (2) 236-6452; internet www.aia.co.th; f. 1983; ordinary and group life, group and personal accident, credit life; Exec. Vice-Pres. and Gen. Man. DHADOL BUNNAG.

Ayudhya Insurance Public Co Ltd: Ploenchit Tower, 7th Floor, Thanon Ploenchit, Pathumwan, Bangkok 10330; tel. (2) 263-0335; fax (2) 263-0589; internet www.ayud.co.th; non-life; Chair. KRIT RATANARAK.

Bangkok Insurance Public Co Ltd: Bangkok Insurance Bldg, 25 Thanon Sathorn Tai, Bangkok 10120; tel. (2) 285-8888; fax (2) 677-3777; e-mail corp.comm@bki.co.th; internet www.bki.co.th; f. 1947; non-life; Chair. and Pres. CHAI SOPHONPANICH.

Bangkok Union Insurance Public Co Ltd: 175–177 Thanon Surawongse, Bangkok 10500; tel. (2) 233-6920; fax (2) 237-1856; f. 1929; non-life; Chair. MANU LIEWPAIROT.

China Insurance Co (Siam) Ltd: 36/68–69, 20th Floor, PS Tower, Thanon Asoke, Sukhumvit 21, Bangkok 10110; tel. (2) 259-3718; fax (2) 259-1402; f. 1948; non-life; Chair. JAMES C. CHENG; Man. Dir FANG RONG-CHENG.

Indara Insurance Public Co Ltd: 364/29 Thanon Si Ayutthaya, Ratchthewi, Bangkok 10400; tel. (2) 247-9261; fax (2) 247-9260; e-mail contact@indara.co.th; internet www.indara.co.th; f. 1949; non-life; Chair. PRATIP WONGNIRUND; Man. Dir SOPHON DEJTHEVAPORN.

International Assurance Co Ltd: 488/7–9 Thanon Henri Dunant, Patumwan, Bangkok 10330; tel. (2) 658-1919; fax (2) 285-8679; f. 1952; non-life, fire, marine, general; Chair. PICHAI KULAVANICH; Man. Dir SOMCHAI MAHASANTIPIYA.

Mittare Insurance Co Ltd: 295 Thanon Si Phraya, Bangrak, Bangkok 10500; tel. (2) 237-4646; fax (2) 236-1376; internet www.mittare.com; f. 1947; life, fire, marine, health, personal accident, automobile and general; fmrly Thai Prasit Insurance Co Ltd; Chair. SURACHAN CHANSRICHAWLA; Man. Dir SUKHATHEP CHANSRICHAWLA.

Ocean Life Insurance Co Ltd: 170/74–83 Ocean Tower I Bldg, Thanon Rachadapisek, Klongtoey, Bangkok 10110; tel. (2) 261-2300; fax (2) 261-3344; e-mail info@oli.co.th; internet www.oli.co.th; f. 1949; life; Chair. and Man. Dir KIRATI ASSAKUL.

Paiboon Insurance Co Ltd: Thai Life Insurance Bldg, 19th–20th Floors, 123 Thanon Ratchadapisek, Bangkok 10310; tel. (2) 246-9635; fax (2) 246-9660; f. 1927; non-life; Chair. ANUTHRA ASSAWANONDA; Pres. VANICH CHAIYAWAN.

Prudential TS Life Assurance Public Co Ltd: Sengthong Thani Tower, 30th–32nd Floors, 82 Thanon Sathorn Nua, Bangkok 10500; tel. (2) 639-9500; fax (2) 639-9699; e-mail customer.service.ptsl@ibm.net; internet www.prudential.co.th; f. 1983; Chair. Prof. BUNCHANA ATTHAKOR; Vice-Chair. DAN R. BARDIN.

Siam Commercial New York Life Insurance Public Co Ltd: 4th Floor, SCB Bldg 1, 1060 Thanon Petchburi, SCB Chidlom, Ratchathewi, Bangkok 10400; tel. (2) 655-3000; fax (2) 256-1517; e-mail don@scnyl.com; internet www.scnyl.com; f. 1976; life; Pres. and CEO C. DONALD CARDEN.

Southeast Insurance (2000) Co Ltd (Arkanay Prakan Pai Co Ltd): Southeast Insurance Bldg, 315G, 1–3 Thanon Silom, Bangrak, Bangkok 10500; tel. (2) 631-1331; f. 1946; life and non-life; Chair. CHAYUT CHIRALERSPONG; Gen. Man. NATDANAI INDRASUKHSRI.

Syn Mun Kong Insurance Public Co Ltd: 279 Thanon Srinakarin, Bangkapi, Bangkok 10240; tel. (2) 379-3140; fax (2) 731-6590; e-mail info@smk.co.th; internet www.smk.co.th; f. 1951; fire, marine, automobile and personal accident; Chair. THANAVIT DUSADEESURAPOJ; Man. Dir RUENGDEJ DUSADEESURAPOJ.

Thai Commercial Insurance Public Co Ltd: Sathorn Nakorn Tower, 25th Floor, 100/48–49 Thanon Sathorn Nua, Silom, Bangrak, Bangkok 10500; tel. (2) 236-5987; fax (2) 236-5990; f. 1940; automobile, fire, marine and casualty; Chair. SUKIT WANGLEE; Exec. Dir SUCHIN WANGLEE; Man. Dir SURAJIT WANGLEE.

Thai Health Insurance Co Ltd: 31st Floor, RS Tower, 121/89 Thanon Ratchadaphisek, Ding-Daeng, Bangkok 10400; tel. (2) 642-3100; fax (2) 642-3130; e-mail info@thaihealth.co.th; internet www.thaihealth.co.th; f. 1979; Chair. APIRAK THAIPATANAKUL; Man. Dir VARANG SRETHBHAKDI.

Thai Insurance Public Co Ltd: 34/3 Soi Lang Suan, Thanon Ploenchit, Lumpini, Pathumwan, Bangkok 10330; tel. (2) 652-2880; fax (2) 652-2870; e-mail tic@thaiins.com; internet www.thaiins.com; f. 1938; non-life; Chair. KAVI ANSVANANDA.

Thai Life Insurance Co Ltd: 123 Thanon Ratchadaphisek, Din-Daeng, Bangkok 10400; tel. (2) 247-0247; fax (2) 246-9945; e-mail corpcomm@thailife.com; internet www.thailife.com; f. 1942; life; Chair. and CEO VANICH CHAIYAWAN; Pres. APIRAK THAIPATANAGUL.

Viriyah Insurance Co Ltd: Viriya Panich Bldg, 1242 Thanon Krung Kasem, Pomprapsattrupuy, Bangkok 10100; tel. (2) 223-0851; fax (2) 224-9876; e-mail info@viriyah.co.th; internet www.viriyah.co.th; Man. Dir SUVAPORN THONGTHEW.

Wilson Insurance Co Ltd: 18th Bangkok Insurance/YWCA Bldg, 25 Thanon Sathorn Tai, Thungmahamek, Sathorn, Bangkok 10120; tel. (2) 677-3999; fax (2) 677-3978–9; e-mail wilson@wilsonins.co.th; f. 1951; fire, marine, motor car, general; Chair. CHOTE SOPHONPANICH; Pres. NOPADOL SANTIPAKORN.

Associations

General Insurance Association: 223 Soi Ruamrudee, Thanon Witthayu, Bangkok 10330; tel. (2) 256-6032–8; fax (2) 256-6039–40; e-mail general@thaigia.com; internet www.thaigia.com; Exec. Dir CHALAW FUANGAROMYA.

Thai Life Assurance Association: 36/1 Soi Sapanku, Thanon Rama IV, Thungmahamek, Sathorn, Bangkok 10120; tel. (2) 287-4596; fax (2) 679-7100; e-mail tlaa@tlaa.org; internet www.tlaa.org; Pres. SUTTI RAJITRANGSON.

Trade and Industry

GOVERNMENT AGENCIES

Board of Investment (BOI): 555 Thanon Vipavadee Rangsit, Chatuchak, Bangkok 10900; tel. (2) 537-8111; fax (2) 537-8177; e-mail head@boi.go.th; internet www.boi.go.th; f. 1958 to publicize investment potential and encourage economically and socially beneficial investments and also to provide investment information; Chair. THAKSIN SHINAWATRA; Sec.-Gen. SOMPHONG WANAPHA.

Board of Trade of Thailand: 150/2 Thanon Rajbopit, Bangkok 10200; tel. (2) 622-1860; fax (2) 225-3372; e-mail bot@tcc.or.th; internet www.thaiechamber.com/templace; f. 1955; mems: chambers of commerce, trade asscns, state enterprises and co-operative societies (large and medium-sized companies have associate membership); Chair. AIVA TAULANANDA.

Central Sugar Marketing Centre: Bangkok; f. 1981; responsible for domestic marketing and price stabilization.

Financial Sector Restructuring Authority (FSRA): 130–132 Tower 3, Thanon Witthayu, Patumwan, Bangkok 10330; tel. (2) 263-2620; fax (2) 650-9872; e-mail webmaster@fra.or.th; internet www.fra.or.th; f. 1997 to oversee the restructuring of Thailand's financial system; Chair. KAMOL JUNTIMA; Sec. Gen. MONTRI CHENVIDAYAKAM.

Forest Industry Organization: 76 Thanon Ratchadamnoen Nok, Bangkok 10200; tel. (2) 282-3243; fax (2) 282-5197; e-mail info@fio.or.th; internet www.fio.or.th; f. 1947; oversees all aspects of forestry and wood industries; Man. Col M. R. ADULDEJ CHAKRABANDHU.

Office of the Cane and Sugar Board: Ministry of Industry, Thanon Phra Ram Hok, Bangkok 10400; tel. (2) 202-3291; fax (2) 202-3293; e-mail wimonlak@narai.oie.go.th.

Rubber Estate Organization: Nabon Station, Nakhon Si Thammarat Province 80220; tel. (75) 411554; internet www.reothai.com; Man. Dir SOMPHOL ISARATHANACHAI.

DEVELOPMENT ORGANIZATIONS

Industrial Finance Corpn of Thailand (IFCT): 1770 Thanon Petchburi Tadmai, Bangkok 10320; tel. (2) 253-7111; fax (2) 253-9677; e-mail sec_ifct@ifct.th.com; internet www.ifct.co.th; f. 1959 to assist in the establishment, expansion or modernization of industrial enterprises in the private sector; organizes pooling of funds and capital market development; makes medium- and long-term loans, complementary working capital loans, investment advisory services, underwriting shares and securities and guaranteeing loans; Chair. SOMMAI PHASEE; Gen. Man. KRAITHIP KRAIRIKSH (acting).

National Economic and Social Development Board: 962 Thanon Krung Kasem, Bangkok 10100; tel. (2) 282-8454; fax (2) 628-2871; e-mail mis@nesdb.go.th; internet www.nesdb.go.th; economic and social planning agency; Sec.-Gen. CHAKRAMON PHASUKAVANICH.

Royal Developments Projects Board: Office of the Prime Minister, Bangkok; internet www.rdpb.go.th; Sec.-Gen. Dr SUMET TANTIVEJKUL.

Small Industries Finance Office (SIFO): 24 Mansion 5, Thanon Ratchadamnoen, Bangkok 10200; tel. (2) 224-1919; f. 1964 to provide finance for small-scale industries; Chair. PISAL KONGSAMRAN; Man. VICHIEN OPASWATTANA.

CHAMBER OF COMMERCE

Thai Chamber of Commerce: 150 Thanon Rajbopit, Bangkok 10200; tel. (2) 622-1880; fax (2) 225-3372; e-mail echam@thaiechamber.com; internet www.thaiechamber.com; f. 1946; 2,000 mems, 10 assoc. mems (1998); Chair Dr AJVA TAULANANDA; Pres. VICHIEN TEJAPHAIBOON.

INDUSTRIAL AND TRADE ASSOCIATIONS

Bangkok Rice Millers' Association: 14/3 Thanon Sathorn Tai, Bangkok 10120; tel. (2) 286-8298.

The Federation of Thai Industries: Queen Sirikit National Convention Center, Zone C, 4th Floor, 60 Thanon Ratchadaphisek Tadmai, Klongtoey, Bangkok 10110; tel. (2) 229-4255; fax (2) 229-4941; e-mail fa-dept@off.fti.or.th; internet www.fti.or.th; f. 1987; fmrly The Association of Thai Industries; 4,800 mems; Chair. TAWEE BUTSUNTORN.

Mining Industry Council of Thailand: Soi 222/2, Thai Chamber of Commerce University, Thanon Vibhavadi Rangsit, Dindaeng, Bangkok 10400; tel. (2) 275-7684–6; fax (2) 692-3321; e-mail miningthai@miningthai.org; internet www.miningthai.org; f. 1983; intermediary between govt organizations and private mining enterprises; Chair. YONGYOTH PETCHSUWAN; Sec.-Gen. PUNYA ADULYAPICHIT.

Rice Exporters' Association of Thailand: 37 Soi Ngamdupli, Thanon Phra Rama IV, Bangkok 10120; tel. (2) 287-2674; fax (2) 287-2678; e-mail reat@ksc.th.com; internet www.riceexporters.co.th; Pres. VICHAI SRIPRASERT; Sec.-Gen. Lt CHAREON LAOTHAMATAS.

Rice Mill Association of Thailand: 81–81/1 Trok Rongnamkheng, 24 Thanon Charoenkrung, Talat Noi, Sampanthawong, Bangkok 10100; tel. (2) 235-7863; fax (2) 234-7286; Pres. NIPHON WONGTRAGARN.

Sawmills Association: 101 Thanon Amnuaysongkhram, Dusit, Bangkok 10300; tel. (2) 243-4754; fax (2) 243-8629; e-mail smassoc@mail.cscoms.com; Pres. VITOON PONGPASIT.

Thai Diamond Manufacturers Association: 116/1 Thanon Silom, Bangkok 10500; tel. (2) 238-2718; fax (2) 266-4830; e-mail odtcbkk@loxinfo.co.th; 11-mem. board; Pres. CHIRAKITTI TANGKATHAC.

Thai Food Processors' Association: Tower 1, Ocean Bldg, 9th Floor, 170/21–22 Thanon Rajchadapisaktadmai, Klongtoey, Bangkok 10110; tel. (2) 261-2995; fax (2) 261-2996; e-mail thaifood@thaifood.org; internet www.thaifood.org.

Thai Jute Mill Association: Sivadol Bldg, 10th Floor, Rm 10, 1 Thanon Convent, Silom, Bangrak, Bangkok 10500; tel. (2) 234-1438; fax (2) 234-1439.

Thai Lac Association: 66 Soi Chalermkhetr 1, Thanon Yukul, Pomprab, Bangkok 10100; tel. (2) 233-8331.

Thai Maize and Produce Traders' Association: Sathorn Thani II Bldg, 11th Floor, 92/26–27 Thanon Sathorn Nua, Bangrak, Bangkok 10500; tel. (2) 234-4387; fax (2) 236-8413.

Thai Pharmaceutical Manufacturers Association: 188/107 Thanon Charansanitwongs, Banchanglaw, Bangkoknoi, Bangkok 10700; tel. (2) 863-5106; fax (2) 863-5108; e-mail tpma@asiaaccess.net.th; f. 1969; Pres. CHERNPORN TENGAMNUAY.

Thai Rubber Traders' Association: 57 Thanon Rongmuang 5, Pathumwan, Bangkok 10500; tel. (2) 214-3420; f. 1951; Pres. SANG UDOMJARUMANCE.

Thai Silk Association: Textile Industry Division, Soi Trimitr, Thanon Rama IV, Klongtoey, Bangkok 10110; tel. (2) 712-4328; fax (2) 391-2896; e-mail thsilkas@thaitextile.org; internet www.thaitextile.org/tsa; f. 1962; Pres. MONTIGAN LOVICHIT.

Thai Sugar Manufacturing Association: 20th Floor, Gypsum Metropolitan Tower, 539/2 Thanon Sri Ayudhya, Rajdhareé, Bangkok 10400; tel. (2) 642-5248; fax (2) 642-5251; e-mail tsma2000@cscoms.com.

Thai Sugar Producers' Association: 8th Floor, Thai Ruam Toon Bldg, 794 Thanon Krung Kasem, Pomprap, Bangkok 10100; tel. (2) 282-0990; fax (2) 281-0342.

Thai Tapioca Trade Association: Sathorn Thani II Bldg, 20th Floor, 92/58 Thanon Sathorn Nua, Silom, Bangkok 10500; tel. (2) 234-4724; fax (2) 236-6084; e-mail ttta@loxinfo.co.th; internet www.ttta-tapioca.org; Pres. SEREE DENWORALAK.

Thai Textile Manufacturing Association: 454–460 Thanon Sukhumvit, Klongton, Klongtoey, Bangkok 10110; tel. (2) 258-2023; fax (2) 260-1525; e-mail ttma@thaitextile.org; internet www.thaitextile.org/ttma.

Thai Timber Exporters' Association: Ratchada Trade Centre, 4th Floor, 410/73–76 Thanon Ratchadaphisek, Bangkok 10310; tel. (2) 278-3229; fax (2) 259-0481.

Timber Merchants' Association: 4 Thanon Yen-Arkad, Thung-Mahamek, Yannawa, Bangkok 10120; tel. (2) 249-5565.

Union Textile Merchants' Association: 562 ESPREME Bldg, 4th Floor, Soi Watgunmatuyalarm, Thanon Rajchawong, Samphanthawong, Bangkok 10100; tel. (2) 622-6711; fax (2) 622-6714; e-mail utma@thaitextile.org; internet www.thaitextile.org/utma; Chair. SURAPHOL TIYAVACHRAPONG.

UTILITIES

Electricity

Cogeneration Public Co Ltd: Grand Amarin Tower, 29th Floor, 1550 Thanon Petchburi Tadmai, Rachtavee, Bangkok 10320; tel. (2) 207-0970; fax (2) 207-0910; generation and supply of co-generation power; cap. and res 5,634m., sales 3,424m. (1996/97); Chair. PIERRE CHARLES E. SWARTENBROEK; Pres. CHANCHAI JIVACATE.

Electricity Generating Authority of Thailand (EGAT): 53 Thanon Charan Sanit Wong, Bang Kruai, Nothaburi, Bangkok 11130; tel. (2) 436-4800; fax (2) 436-4879; internet www.egat.or.th; f. 1969; scheduled for transfer to the private sector in 2004; Gov. NARONGSAK VICHETPHAN (acting); Chair. CHAIANAN SAMUDAVANIJA.

Electricity Generating Public Co Ltd (EGCO): EGCO Tower, 222 Moo 5, Thanon Vibhavadi Rangsit, Tungsonghong, Laksi, Bangkok 10210; tel. (2) 955-0955; fax (2) 955-0956; e-mail pr@egco.com; internet www.egco.com; a subsidiary of the Electricity Generating Authority of Thailand (EGAT); 59% transferred to the private sector in 1994–96; 14.9%-owned by China Light and Power Co (Hong Kong); Pres. CHALERMCHAI RATNARAK.

The Metropolitan Electricity Authority: 30 Soi Chidlom, Thanon Ploenchit, Lumpini, Pathumwan, Bangkok 10330; tel. (2) 254-9550; fax (2) 253-1424; internet www.mea.or.th; f. 1958; one of the two main power distribution agencies in Thailand; Gov. CHALIT RUENGVISECH.

The Provincial Electricity Authority: 200 Thanon Ngam Wongwan, Chatuchak, Bangkok 10900; tel. (2) 589-0100; fax (2) 589-4850; e-mail webmaster@pea.or.th; internet www.pea.or.th; f. 1960; one of the two main power distribution agencies in Thailand; Chair. PRINYA NAKCHUDTREE.

Water

Metropolitan Waterworks Authority: 18/137 Thanon Prachachuen, Don Muang, Bangkok 10210; tel. (2) 504-0123; fax (2) 503-9493; e-mail mwa1125@water.mwa.or.th; internet www.mwa.or.th; f. 1967; state-owned; provides water supply systems in Bangkok; scheduled for transfer to private sector in 2004; Gov. CHUANPIT DHAMASIRI.

Provincial Waterworks Authority: 72 Thanon Chaengwattana, Don Muang, Bangkok 10210; tel. (2) 551-1020; fax (2) 552-1547; e-mail pwait2000@yahoo.com; internet www.pwa.co.th; f. 1979; provides water supply systems except in Bangkok Metropolis; scheduled for transfer to private sector in 2005; Gov. WANCHAI GHOOPRASERT.

CO-OPERATIVES

In 1992 there were 4m. members of co-operatives, including 1.6m. in agriculture.

MAJOR COMPANIES

(cap. = capital; res = reserves; m. = million; amounts in baht, unless otherwise stated)

Advance Agro Public Co Ltd: 122 Thanon Sathorn Nua, Bangrak, Bangkok 10500; tel. (2) 267-7164; fax (2) 267-1848; e-mail pr.aa@shs.co.th; internet www.advanceagro.com; mfrs of paper; cap. and res 4,709m. (2000), sales 18,968m. (2003); Chair. NARONG SRISA-AN.

Alucon Public Co Ltd: 500 Moo 1, Soi Sirikam, 72 Thanon Sukhumvit, Tambol Samrong Nua, Amphur Muang-Samutprakarn, Bangkok 10270; tel. (2) 398-0147; fax (2) 398-3455; e-mail alucon@ksc.com.th; f. 1994; mfr of aluminium tubes and containers; cap. and res 881m., sales 1,548m. (2001); Chair. H. W. SCHNEIDER; Man. Dir TAKAAKI TAKEUCHI.

American Standard Sanitary Ware (Thailand) Public Co Ltd: 1/6 Moo 1, Thanon Phaholyothin, Km 32, Tambol Klongnung, Amphur Klongluang, Pathumthani 12120; tel. (2) 901-4455; fax (2) 901-4422; internet www.americanstandard.co.th; f. 1969; mfrs of sanitary ware; cap. and res 1,522m., sales 1,196m. (2001); Chair. CHALERMBHAND SRIVIKORN; Man. Dir NORMAN D. LIVINGSTONE; 1,300 employees.

Asia Fiber Public Co Ltd: Wall Street Tower, 27th Floor, 33/133–136 Thanon Surawong, Bangkok 10500; tel. (2) 632-7071; fax (2) 236-1982; e-mail afcny6@ksc.th.com; internet www.asiafiber.com; f. 1970; mfrs of nylon filament yarn, nylon textured yarn, filament woven fabrics; cap. and res 962m., sales 1,295m. (2002/03); Chair. PIPAT SIRIKIETSOONG; Pres. CHEN NAMCHAISIRI; 1,421 employees.

Bangchak Petroleum Public Co Ltd: 38 Thanon Srinakarin, Prawet, Bangkok 10260; tel. (2) 301-2700; fax (2) 301-2726; internet www.bangchak.co.th; f. 1984; refining and distribution of petroleum; cap. and res 2,431m., sales 48,596m. (2001); Chair. SOMMAI PHASEE; 850 employees.

Bangkok Agro-Industrial Products Public Co Ltd: C.P. Tower, 313 Thanon Silom, Bangrak, Bangkok 10500; tel. (2) 231-0231; fax (2) 238-1921; mfrs and distributors of animal feedstuffs, animal husbandry, poultry and swine breeding, fruit production; cap. and res 3,405m., sales 4,544m. (2000); Chair. JARAN CHIARAVANONT; Pres. ADIREK SRIPRATAK; 1,650 employees.

Bangkok Expressway Public Co Ltd: 238/7 Thanon Asoke-Dindaeng, Bangkapi, Huay Kwang, Bangkok 10320; tel. (2) 641-4611; fax (2) 641-4610; engaged in the construction and management of Thailand's second stage expressway project; cap. and res 13,827m., sales 5,945m. (2001); Chair. VIRABONGSA RAMANGKURA.

Bangkok Produce Merchandizing Public Co Ltd: CP Tower Bldg, 18th Floor, 313 Thanon Silom, Bangkok 10500; tel. (2) 638-2000; fax (2) 631-0989; f. 1978; production of processed chicken; cap. and res 1,970m., sales 19,148m. (2000); Chair. DHANIN CHIRAVANONT; Pres. EAM NGAMDAMRONK; 4,000 employees.

Bangkok Rubber Public Co Ltd: 611/40 Soi Watchan Nai, Bangklo, Bangkloaem, Bangkok 10120; tel. (2) 291-2577; fax (2) 291-1353; f. 1974; mfrs of sports shoes and other footwear; cap. and res 3,252m., sales 6,185m. (2001); Chair. PRAPAN SIRIRATTHAMRONG; Pres. BOONSONG TONDULYAKUL; 2,369 employees.

Bangkok Steel Industry Public Co Ltd: United Flour Mill Bldg, 5th and 7th Floors, 205 Thanon Rajawongse, Chakkawad, Sampantawongse, Bangkok 10100; tel. (2) 225-0200; fax (2) 222-7497; e-mail info@bangkoksteel.co.th; internet www.bangkoksteel.co.th; f. 1964; mfrs of steel products; cap. and res 1,600m., sales 2,234m. (2001); Chair. PRASERT TANGTRONGSAKDI; Pres. JARAY BHUMICHITRA; 1,200 employees.

Bangkok Weaving Mills Ltd: 879 Thanon Bangkok-Nonthaburi, Bangsue, Bangkok 10800; tel. (2) 586-0901; fax (2) 587-2338; e-mail bwm@thai.com; internet www.bwmthai.com; f. 1950; mfr of cotton yarn and fabrics; cap. and res 951m., sales 2,119m. (2000); Chair. NANTANA ASSAKUL; 1,900 employees.

Banpu Public Co Ltd: Grand Amarin Tower, 26th–28th Floors, 1550 Thanon Petchburi Tadmai, Bangkok 10310; tel. (2) 207-0688; fax (2) 207-0697; e-mail piyanuch_j@banpu.co.th; internet www.banpu.co.th; mineral mining and power generation; cap. and res 7,767m., sales 5,734m. (2000/01); Chair. CHIRA PANUPONG; 1,126 employees.

Berli Jucker Public Co Ltd: Berli Jucker House, 99 Soi Rubia, 42 Thanon Sukhumvit, Klongtoey, Prakanong, Bangkok 10110; tel. (2) 367-1111; fax (2) 367-1000; e-mail bjc@berlijucker.co.th; internet www.berlijucker.co.th; f. 1882; engineering; mfr and distribution of consumer, imaging and technical products and packaging; cap. and res 6,176m., sales 11,522m. (2001); Chair. PITI SITHI-AMNUAI; Pres. and CEO DAVID NICOL; 5,700 employees.

Boon Rawd Brewery Co Ltd: 999 Thanon Samsen, Dusit, Bangkok 10300; tel. (2) 669-2050; fax (2) 669-2089; e-mail nanthawan@boonrawd.co.th; internet www.boonrawd.co.th; f. 1933; mfrs of beer and soft drinks; cap. and res 7,411m., sales 10,870m. (2000); Chair. CHAMNONG BHIROMBHAKDI; 2,300 employees.

Caltex Oil (Thailand) Ltd: 14–26 Sun Towers B, 123 Thanon Vibhavadi Rangsit, Jautchak, Bangkok 10900; tel. (2) 617-6888; fax (2) 617-6828; e-mail nanthapolc@caltex.co.th; internet www.caltex.co.th; f. 1946; distributors of petroleum products; sales 20,144m. (1994); Chief Exec. STEPHEN LONG; 350 employees.

Capetronic International (Thailand) Public Co Ltd: 105 Moo 3, Thanon Bangna-Trad, Km 52 Thakham, Bangpakong, Chachoengsao 24130; tel. (38) 573-161; fax (38) 573-174; f. 1985; mfr of computer monitors and printed circuit board assembly; cap. and res 1,018m., sales 4,926m. (2001); Chair. Dr STANLEY H. S. HO; 2,200 employees.

Central Pattana Public Co Ltd: Central Plaza Lardprao, 12th–13th Floors, 1693 Thanon Phaholyothin, Lardyao, Chatuchak, Bangkok 10900; tel. (2) 937-1555; fax (2) 937-1578; e-mail cpnternt@ksc11.ksc.co.th; internet www.centralpattana.co.th; engaged in property investment, development, management and construction; cap. and res 4,123m., sales 2,950m. (2001); Chair. VANCHAI CHIRATHIVAT.

Charoen Pokphand Foods Public Co Ltd: CP Tower, 14th Floor, 313 Thanon Silom, Bangkok 10500; tel. (2) 638-2936; fax (2) 638-2942; e-mail iroffice@cp-foods.com; internet www.cp-foods.com; f. 1921; mfrs of feed for poultry, pigs and prawns; poultry and pig breeding; cap. and res 30,504m., sales 87,109m. (2003); Chair. DHANIN CHEARAVANONT; Pres. and CEO ADIREK SRIPRATAK; 3,953 employees.

Charoen Pokphand Northeastern Public Co Ltd: C.P. Tower, 14th Floor, 313 Thanon Silom, Bangkok 10500; tel. (2) 631-0671; fax

(2) 631-0863; f. 1984; animal food mfr; cap. and res 22,479m., sales 74,828m. (2001); Chair. PRASERT POONGKUMARN; Man. Dir ADIREK SRIPRATAK.

CH Karnchang Public Co Ltd: 587 Thanon Sutthisarn, Dindaeng Dindaeng, Bangkok 10400; tel. (2) 277-0460; fax (2) 275-7029; internet www.ch-karnchang.co.th; constructor and sub-contractor, engaged in general construction; cap. and res 4,110m., sales 5,982m. (2003); Chair. THAVORN TRIVISVAVET; Pres. PLEW TRIVISVAVET.

Christiani and Nielsen (Thai) PCL: 50/670 Thanon Sukhumvit, 105 (Soi La Salle) Moo 3, Bang Na, Bangkok 10260; tel. (2) 398-0158; fax (2) 398-9860; e-mail cnt@cn-thai.co.th; internet www.cn-thai.co.th; f. 1930; construction and property investment; cap. and res 312m., sales 2,287m. (2002); Chair. SANTI GRACHANGNETARA; Man. Dir DANUCH YONTARARAK; 309 employees.

Circuit Electronic Industries Public Co Ltd: 45 Moo, 12 Rojana Industrial Park, Amphoe-Uthai, Ayudhya 13210; tel. (3) 522-6280; fax (3) 522-6710; e-mail larry@cei.co.th; internet www.cei.co.th; f. 1984; mfr of integrated circuits and other electronic components; cap. and res 2,346m., sales 3,775m. (2001); Chair. SIVA NGANTHAVEE; Pres. LARRY L. WOLFF; 1,500 employees.

Delta Electronics (Thailand) Public Co Ltd: 909 Moo 4, Tambon Prakasa, Amphur Muang, Samutprakarn 10280; tel. (2) 709-2800; fax (2) 709-3790; e-mail mae@delta.co.th; internet www.delta.co.th; design and manufacture of electronic and electrical equipment; cap and res 15,348m., sales 30,557m. (2001); Chair. JAMES NG KONG MENG.

Diethelm Trading Co Ltd: 2533 Thanon Sukhumvit, Bangchak, Prakhanong, Bangkok 10250; tel. (2) 332-7140; fax (2) 332-6103; e-mail marketing@diethelmtrading.com; internet diethelmtrading.com; f. 1906; retailers of chemicals, consumer products and industrial equipment; sales 13,737m. (1994); Chair. MARK DIETHELM; 2,712 employees.

Esso (Thailand) Public Co Ltd: 3195/17–29 Thanon Phra Rama IV, Klongtoey, Bangkok 10110; tel. (2) 262-4280; fax (2) 262-4800; internet www.exxon.com; f. 1965; distributors of petroleum products; sales 89,910m. (2000); Man. Dir WILLIAM M. COLTON; 1,600 employees.

GFPT Public Co Ltd: 69/6–13 Thanon Suksawatdi, Ratburana, Bangkok 10140; tel. (2) 463-0040; fax (2) 463-5751; f. 1981; producers of frozen chicken; cap. and res 2,594m., sales 7,283m. (2001); Chair. CHAROEN SIRIMONGKOLKASEM; Man. Dir VIRACH SIRIMONGKOLKASEM; 4,856 employees.

Grammy Entertainment Public Co Ltd: 209/1 CMIC Tower-B, Thanon Asoke, Klongtoey, Bangkok 10110; tel. (2) 664-4000; fax (2) 664-0248; internet www.grammy.co.th; f. 1983; operators of media businesses, including television and radio; cap. and res 2,944m., sales 2,856m. (2001); Chair. PAIBOON DAMRONGCHAITHAM; 2,000 employees.

GSS ARRAY Technology Public Co Ltd: 94 Moo 1, Hi-Tech Industrial Estate, Thanon Banlane, A. Bang Pa-In, Ayudhya 13160; tel. (3) 535-0890; fax (3) 535-0945; f. 1985; electronics contract manufacturing; cap. and res 633m., sales 9,766m. (1998); Chair. GARY M. STICKLES, ROBERT E. ZINN; 2,700 employees.

Hemaraj Land and Development Public Co Ltd: UM Tower, 18th Floor, 9 Thanon Ramkhamhaeng, Suanluang, Bangkok 10250; tel. (2) 719-9555; fax (2) 719-9546; e-mail sales@hemaraj.com; internet www.hemaraj.com; engaged in property development; cap. and res 2,831m., sales 954m. (2001); Chair. SAWASDI HORRUNGRUANG; Man. Dir DAVID NARDONE.

Hino Motor Sales (Thailand) Ltd: 212 Moo 4, Thanon Vibhavadi Rangsit, Talad Bangkhen, Laksi, Bangkok 10210; tel. and fax (2) 900-5000; e-mail info@hinothailand.com; internet www.hinothailand.com; f. 1962; distributors of industrial motor vehicles; sales 12,170m. (1994); Pres. TARO YOSHIMURA.

Honda Automobile (Thailand) Co Ltd: 2754/1 Soi Sukhumvit 66/1, Thanon Sukhumvit, Khwang Bangna, Khet Bangna, Bangkok 10260; tel. (2) 744-7744; fax (2) 744-7755; internet www.honda.co.th; f. 1983; mfrs of motor vehicles; sales 24,895m. (1994); Chair. SAICHIRO FUJIE; Man. Dir NOBUNARI MATSUSHITA.

ICC International Public Co Ltd: 757/10 Soi Pradoo, 1 Thanon Sadhupradit, Yannawa, Bangkok 10120; tel. (2) 294-0281; fax (2) 294-3024; e-mail webmaster@icc.co.th; internet www.icc.co.th; distributors of consumer goods; cap. and res 7,780m., sales 8,546m. (2001); Chair. SOM CHATUSRIPITAK; Pres. BOONKIET CHOKWATANA; 4,319 employees.

Italian-Thai Development Public Co Ltd: 2034/132-161 Italthai Tower, Thanon Petchburi Tadmai, Bangkapi, Bangkok 10320; tel. (2) 716-1600; fax (2) 716-1488; e-mail itdmail@italian-thai.co.th; internet www.italian-thai.co.th; f. 1958; construction; cap. and res – 856m., sales 13,080m. (2001); Chair. Dr CHAIJUDH KARNASUTA; Pres. PREMCHAI KARNASUTA; 16,200 employees.

Jalaprathan Cement Public Co Ltd: 2974 Thanon Petchburi Tadmai, Bangkapi, Huay Kwang, Bangkok 10320; tel. (2) 318-7111; fax (2) 318-3469; e-mail s.sutyakorn@jccitc.com; f. 1956; mfrs of cement; cap. and res 2,099m., sales 1,520m. (2001); Chair. Gen. AYUPOON KARNASUTA; Man. Dir RAPEE SUKHAYANGA; 585 employees.

Jasmine International PCL: 6th Floor, Tower 2, Laksi Plaza, Thanon Chaengwatana, Donmueng, Bangkok 10210; tel. (2) 576-0200; fax (2) 576-0196; e-mail webmaster@jasmine.th.com; internet www.jasmine.co.th; mfrs and distributors of telecommunications equipment; cap. and res 947m., sales 4,177m. (2001); Chair. Dr ADISAI BODHARAMIK; 1,075 employees.

KGI Securities (Thailand) Public Co: 323 United Center Bldg, 9th Floor (Trading Rm), Thanon Silom, Bangkok 10500; tel. (2) 231-1111; fax (2) 231-1524; e-mail ebusiness@securities-one.com; internet www.kgieworld.co.th; finance co engaged in securities broking, dealing and investment advice; cap. and res 3,946m., sales 2,025m. (2003); 51% owned by a Taiwan-based co; Chair. ANGELO KOO JOHN YNN.

Loxley Public Co Ltd: 102 Thanon Na Ranong, Klongtoey, Bangkok 10110; tel. (2) 240-3000; fax (2) 240-3101; e-mail editor@loxley.co.th; internet www.loxley.co.th; f. 1939; information technology, infrastructure, telecommunications, consumer electronics, consumer products, chemicals, construction materials, environment, media and entertainment; cap. and res 1,198m., sales 8,428m. (2001); Chair. PIROTE LAMSAM; Man. Dir DHONGCHAI LAMSAM.

Luckytex (Thailand) Public Co Ltd: Bubhajit Bldg, 5th Floor, 20 Thanon Sathorn Nua, Silom, Bangrak, Bangkok 10500; tel. (2) 266-6600; fax (2) 238-3957; e-mail visit.h_t@toray.co.th; internet www.toray.co.th; f. 1960; mfrs of spun fabrics and yarn; cap. and res 518.4m., sales 5,285m. (2002/03); Chair. and Man. Dir KAZUYA HAYASHI; 2,424 employees.

Millennium Microtech (Thailand) Co Ltd (MMT): 17/2 Moo 18, Thanon Suwintawong, Tambon Saladang, Amphur Bangnumpriew, Cha Choeng Sao 24000; tel. (38) 845-530; fax (38) 593-229; e-mail arayale@alphatec.co.th; internet www.m-microtech.com; fmrly Alphatec Semiconductor Packaging Co Ltd; name changed as above 2003; mfr of semiconductors; Chair. JOHN LIM; Pres. APICHAT NATASILPA.

Millennium Steel Public Co Ltd: Shinawatra Tower 3, 22nd Floor, 1010 Thanon Viphavadi Rangsit, Ladyao, Chatuchak, Bangkok 10900; tel. (2) 949-2949; fax (2) 949-2889; e-mail siroroma@cementhai.co.th; f. 2002 following merger between steel-producing facilities of Siam Cement Public Co Ltd and NTS Steel Group Public Co Ltd; mfrs of steel rods and bars; Pres. SANTI CHARNKOLRAWEE.

Minebea (Thai) Ltd: Thanon Paholyothin 1, Chiangkraknoi, Ayudhya 13180; tel. (35) 361429; fax (35) 361177; internet www.minebea.co.th; f. 1984; mfrs of electrical goods and electronic components; sales 22,308m. (1994/95); Man. Dir RYUSUKE MIZUKAMI.

National Petrochemical Public Co Ltd: 30th–35th Floors, Suntowers B, 123 Thanon Vibhavadi Rangsit, Chatuchak, Bangkok 10900; tel. (2) 617-7800; fax (2) 617-7888; e-mail npc@npc.co.th; internet www.npc.co.th; f. 1984; mfr and distributor of ethylene and propylene; jetty operation and provision of storage facilities for petrochemical products; cap. and res 21,054m., sales 15,604m. (2003); Chair. MANU LAOPAIROTE; Pres. VIROJ MAVICHAK; 753 employees.

Osothsapha (Teck Heng Yoo) Co Ltd: 2100 Thanon Ram Khamhaeng, Huamark, Bangkok 10240; tel. (2) 351-1486; fax (2) 374-4028; e-mail imdnet@ksc.th.com; internet www.osotspa.com; f. 1891; mfr of energy drinks; cap. and res 3,316m., sales 9,593m. (2000); Man. Dir THANA CHAIPRASIT; 3,000 employees.

Padaeng Industry Public Co Ltd: CTI Tower, 26th–27th Floors, 191/18–25 Thanon Ratchadaphisek, Klongtoey, Bangkok 10110; tel. (2) 661-9900; fax (2) 661-9490; e-mail padaeng@inet.co.th; internet www.padaeng.co.th; f. 1981; miners and refiners of zinc ingot and zinc alloy; cap. and res 2,861m., sales 5,197m. (2001); Chair. ARSA SARASIN; CEO PINIT VONGMASA; Man. Dir ANDRE R. VAN DER HEYDEN; 759 employees.

Pakfood Public Co Ltd: 103 Soi Ruammitr, Thanon Nonsee, Yannawa, Bangkok 10120; tel. (2) 295-1991; fax (2) 295-2002; e-mail mkt@pakfood.co.th; internet www.pakfood.co.th; f. 1984; processors and exporters of frozen seafood; cap. and res 300m. (2000), sales 5,164m. (2002); Pres. YONG AREECHAROENLERT; Man. Dir WIWAT KANOKWATANAWAN; 3,847 employees.

Petroleum Authority of Thailand (PTT): 555 Thanon Vibhavadi Rangsit, Bangkok 10900; tel. (2) 537-2000; fax (2) 537-3499; internet www.ptt.or.th; f. 1978; activities relate to the development, exploitation, production and distribution of petroleum and gas; regrouped its business into two subsidiaries in February 1996: PTT Oil, comprising all domestic operations and PTT International; 100% state-owned; scheduled for transfer to the private sector; cap. and res 57,979m., sales 88,118m. (2001); Chair. MANU LEOPAIROTE; Gov. VISET CHOOPIBAN; 26,500 employees.

Phoenix Pulp and Paper Public Co Ltd: PB Tower, 16th Floor, 1000/63–67 Sukhumvit Soi 71, North Klongtoey, Wattana, Bangkok 10110; tel. (2) 391-9270; fax (2) 382-0047; e-mail vashi@phoenixpulp.com; mfrs of pulp and paper; cap. and res 5,788m., sales 4,318m. (2001); Chair. CHUMPOL NALAMLIENG; Exec. Dir VASHI T. PURSWANI; 1,300 employees.

Pranda Jewelry PLC: 333 Soi Rungsang, Thanon Bangna-Trat, Bangna, Bangkok 10260; tel. (2) 361-3311; fax (2) 399-4877; e-mail prida@pranda.co.th; producers and exporters of jewellery; cap. and res 497m., sales 2,243m. (2000); Chair. PRIDA TIASUWAN; 6,000 employees.

PTT Exploration and Production PCL (PTTEP): PTTEP Bldg, 555 Thanon Rangsit Vibhavadi, Chatuchak, Bangkok 10900; tel. (2) 537-4000; fax (2) 537-4444; e-mail webmaster@pttep.com; internet www.pttep.com; f. 1985; extraction and refining of petroleum; cap. and res 31,118m., sales 29,310m. (2001); Chair. MANU LEOPAIROTE; Pres. CHITRAPONGSE KWANGSUKSTITH.

Saha Pathanapibul Public Co Ltd: 2156 Thanon Petchburi Tadmai, Bangkapi, Huay Kwang, Bangkok 10320; tel. (2) 318-0062; fax (2) 319-1678; e-mail info@sahapat.co.th; internet www.sahapat.co.th; f. 1942; mfr and distributor of soaps, toothpastes, detergents, baby products and food products; sales 10,008m. (2001); Chair. BOONSITHI CHOKWATANA; 1,382 employees.

Saha-Union Public Co Ltd: 1828 Thanon Sukhumvit, Phrakanong, Bangkok 10250; tel. (2) 311-5111; fax (2) 332-5616; e-mail webmaster@hits.sahaunion.co.th; internet www.sahaunion.co.th; f. 1972; holding co whose subsidiaries produce textiles, plastic and rubber products, garment accessories and footwear; cap. and res 9,991m. (2000), sales 17,735m. (2001); Chair. ANAND PANYARACHUN; Pres. KAMOL KHOOSUWAN; 16,500 employees.

Sahaviriya Steel Industries Public Co Ltd: 2nd–3rd Floors, Prapawit Bldg, 28/1 Thanon Surasek, Silom Bangrak, Bangkok 10500; tel. (2) 238-3063-82; fax (2) 236-8890; internet www.ssi-steel.com; f. 1990; mfrs of hot-rolled coils; cap. and res 5,022m., sales 16,325m. (2000); Chair. MARUAY PHADOONSIDHI; Chair. WIT VIRIYAPRAPAIKIT; 719 employees.

Sanyo Universal Electric Public Co Ltd: 38 Premier Place Bldg, 3rd Floor, Thanon Srinakarin, Pravej, Bangkok 10260; tel. (2) 301-2400; fax (2) 398-4708; internet www.sanyosue.com; f. 1958; mfrs and distributors of electronic equipment and electrical appliances; cap. and res -602m. (1999), sales 4,045m. (2000); Chair. PRAMUDE BURANASIRI; 4,038 employees.

Seagate Technology (Thailand) Ltd: Moo 7, 1627 Thanon Theparak, Theparak Muang, Samutprakan 10270; tel. (2) 383-5779; fax (2) 383-5736; internet www.seagate-asia.com; f. 1983; mfrs of computer hardware; sales 32,429m. (1994); Man. Dir JIRAPANNEE SUPRATCHYA.

Serm Suk Public Co Ltd (The): Muang Thai Phatra Complex, 27th–28th Floors, 252/35–36 Thanon Ratchadaphisek, Huay Kwang, Bangkok 10310; tel. (2) 693-2255; fax (2) 693-2266; e-mail sermsuk@thai.com; f. 1952; mfrs and distributors of carbonated soft drinks; cap. and res 7,958m., sales 15,153m. (2001); Chair. PRASIT TANSETTHI; Pres. and CEO SOMCHAI BULSOOK; 8,500 employees.

Sharp Appliances (Thailand) Ltd: Moo 5, 64 T. Bang Samak, Pakong, Cachoengsao 24130; tel. (38) 538663; fax (38) 538863; f. 1987; mfrs of kitchen equipment and other electronic household appliances; sales 9,656m. (1994/95); Man. Dir YASUO SHIN.

Shell Company of Thailand Ltd, The: 10 Thanon Soonthornkosa, Klongtoey, Bangkok 10110; tel. (2) 249-0483; fax (2) 249-0489; e-mail generalenquiries@shell.co.th; internet www.shell.co.th; distributors of petroleum products; sales 44,677m. (1995, est.); Chair. TIRAPHOT VAJRABHAYA; 980 employees.

Shin Corporation Public Co Ltd: Shinawatra Tower, 414 Thanon Paholyothin, Samsennai, Payathai, Bangkok 10400; tel. (2) 299-5000; fax (2) 299-5224; e-mail investor@shincorp.com; internet www.shincorps.com; f. 1983; mfrs and distributors of mobile telephones, pagers and other telecommunications equipment; cap. and res 32,690m., sales 10,247m. (2003); Chair. BHANAPOT DAMAPONG; 1,620 employees.

Shin Satellite Public Co Ltd: 41/103 Thanon Rattanathibet, Nonthaburi 11000; tel. (2) 591-0736; fax (2) 591-0705; e-mail richardw@thaicom.net; internet www.thaicom.net; f. 1991; operation and administration of satellite projects; cap. and res 5,777m., sales 4,817m. (2001); Chair. DUMRONG KASEMSET.

Siam Cement Public Co Ltd: 1 Thanon Siam Cement, Bangsue, Bangkok 10800; tel. (2) 586-3333; fax (2) 587-2199; internet www.siamcement.com; f. 1913; parent company of Siam Cement Group, Thailand's largest industrial group, with subsidiaries manufacturing cement and construction materials, machinery, electrical goods, pulp and paper; cap. and res 66,085m., sales 122,643m. (2001); Chair. CHAOVANA NA SYLVANTA; Pres. CHUMPOL NA LAMLIENG; 25,000 employees.

Siam City Cement Public Co Ltd: 18th Floor, Mahatun Plaza Bldg, 888/180–189 Thanon Ploenchit, Bangkok 10330; tel. (2) 272-5555; fax (2) 253-2891; e-mail pr@sccc.co.th; internet www.siamcitycement.com; f. 1969; mfrs of cement and ready-mixed concrete; cap. and res 14,030m. (2000), sales 15,928m. (2001); Chair. KRIT RATANARAK; Man. Dir VINCENT BICHET; 6,200 employees.

Siam Nissan Automobile Co Ltd: 74 Moo 2 Soi, Thanon Bang-Na Trad Km 21, Yai Bang Phili, Samutprakarn 10540; tel. (2) 215-0830; fax (2) 215-9433; f. 1973; mfrs of motor vehicles; sales 20,576m. (1994); Man. Dir PHOMPONG PHORNPRAPHA.

Siam Pulp and Paper Public Co Ltd: 1 Thanon Siam Cement, Bangsue, Bangkok 10800; tel. (2) 586-3333; fax (2) 587-0738; e-mail chantimn@cementthai.co.th; internet www.cementthai.co.th; mfrs of paper pulp; cap. and res 10,060m., sales 26,994m. (2001); Chair. CHUMPOL NA LAMLIENG; Man. Dir CHAISAK SAENG-XUTO; 4,000 employees.

Siam Yamaha Co Ltd: 1 Thanon Dindang, Samseannai, Phyathai, Bangkok 10400; tel. (2) 245-7820; fax (2) 246-2779; f. 1964; mfrs of motor vehicles; sales 14,010m. (1995, est.); Pres. KASEM NARONGDEJ.

Srithai Superware Public Co Ltd: 355 Soi 36, Thanon Suksawat, Bangpakok, Rasburana, Bangkok 10140; tel. (2) 427-0088; fax (2) 871-1378; internet www.srithaisuperware.com; f. 1964; mfrs of plastic products and melamine tableware; cap. and res US $63m., sales US $73m. (2001); Chair. SUMIT LERTSUMITKUL; Pres. SANAN ANGUBOLKUL; 3,500 employees.

Surapon Foods Public Co Ltd: 247 Moo 6, Thanon Therapak, Therapak, Samutprakan, Bangkok 10270; tel. (2) 385-3038; fax (2) 385-3176; internet www.surapon.co.th; f. 1977; mfr and exporter of frozen food; sales 6,100m. (2001); Chair. and CEO SURAPON VONGVADHANAROJ; 1,300 employees.

Thai Arrow Products Ltd: 460/1-9 Soi Siam Sq. Zone 4, Rama 1 Rd, Pathumwam, Bangkok 10330; tel. (2) 254-7860; fax (2) 255-4457; f. 1963; mfrs of electrical equipment for motor vehicles; sales 10,602m. (1995); Chair. YAZUMI OISHI.

Thai Asahi Glass Public Co Ltd: 200 Moo 1, Thanon Suksawad, Phra Samut Chedi, Samut Prakan 10290; tel. (2) 815-5000; fax (2) 425-8816; e-mail tag-mkt@ksc.th.com; internet www.tag.co.th; f. 1963; mfrs of plain glass; cap. and res 838m., sales 2,173m. (2000); Pres. TADAYUKI OI; 1,000 employees.

Thai Carbon Black Public Co Ltd: Mahatun Plaza Bldg, 16th Floor, 888/162-163, Thanon Ploenchit, Lumpini, Patumwan, Bangkok 10330; tel. (2) 253-6745; fax (2) 254-9031; e-mail pawantcb@loxinfo.co.th; internet www.thaicarbon.com; chemical products; cap. and res 2,929m. (2000), sales 3,507m. (2001); Chair. KUMAR MANGALAM BIRLA.

Thai Central Chemical Public Co Ltd: Metro Bldg, 180–184 Thanon Rajawongse, Bangkok 10100; tel. (2) 225-0135; fax (2) 226-1263; e-mail tccc@thaicentral.co.th; internet www.tccthai.com; f. 1973; mfrs of chemical fertilizers; cap. and res 2,623m., sales 5,061m. (2003); Chair. KRAISRI CHATIKAVANIJ; Pres. TAKASHI IKEDA; 891 employees.

Thai Glass Industries Public Co Ltd: 15 Moo 1, Thanon Rajaburana, Bangkok 10140; tel. (2) 427-0060; fax (2) 427-6603; e-mail faxadmin@berlijucker.co.th; internet www.berlijucker.co.th; f. 1951; mfrs and distributors of glassware and glass containers; cap. and res 1,492m., sales 3,472m. (2001); Chair. Dr CHAIYUT PILUN-OWAD; 1,634 employees.

Thai Hua Rubber Public Co Ltd: 238/1 Thanon Ratchada Pisek, Huay Kwang, Bangkok 10320; tel. (2) 274-0471; fax (2) 274-0231; e-mail mkt@thaihua.com; internet www.thaihua.com; f. 1985; producer and exporter of rubber products; sales 12,260m. (2003); Chair. SANGOB PANDOKMAI; Pres. and CEO LUCKCHAI KITTIPOL.

Thai Oil Co Ltd (Thaioil): 15th Floor, Harindhorn Tower, 54 Thanon Sathorn Nua, Silom, Bangrak, Bangkok 10500; tel. (2)231-7000; fax (2) 231-7222; e-mail rhino@thaioil.co.th; internet www.thaioil.co.th; petroleum-refining; 100% state-owned; sales 37,418m. (2000); Chair. MANU LEOPAIROTE; 980 employees.

Thai Petrochemical Industry Public Co Ltd: TPI Tower, 26–56 Thanon New Chan, Tungmahamek, Bangkok 10120; tel. (2) 678-5000; fax (2) 678-5001; e-mail tpiadmin@tpigroup.co.th; internet www.tpigroup.co.th; f. 1978; mfrs of plastics; cap. and res 12,475m., sales 90,296m. (2003); declared insolvent in March 2000 and placed under rehabilitation; Chair. SUNTHORN HONGLADAROM; Chief Exec. PRACHAI LEOPHAIRATANA.

Thai Plastic and Chemical Public Co Ltd: Rajanakarn Bldg, 14th–15th Floors, 183 Thanon Sathorn Tai, Yannawa, Sathorn, Bangkok 10120; tel. (2) 676-6000; fax (2) 676-6045; e-mail tpc@thaiplastic.co.th; internet www.thaiplastic.co.th; f. 1966; mfrs of

PVC resins and compounds; cap. and res 5,725m., sales 13,695m. (2001); Chair. YOS EUARCHUKIATI; Man. Dir DHEP VONGVANICH; 730 employees.

Thai Rayon Public Co Ltd: Mahatun Plaza Bldg, 12th Floor, 888/127 Thanon Ploenchit, Lumpini, Pathumwan, Bangkok 10330; tel. (2) 627-3801; fax (2) 627-3804; e-mail rayon@thairayon.com; internet www.thairayon.com; mfrs and distributors of fibre and chemicals; cap. and res 6,989m., sales 4,300. (2003); Chair. PRACHITR YOSSUNDARA; Pres. P. M. BAJAJ; 900 employees.

Thai Union Frozen Products Public Co Ltd: 979/12 M Floor, SM Tower, Thanon Paholyothin, Samsennai-Phayathai, Bangkok 10400; tel. (2) 298-0537; fax (2) 298-0548; e-mail chansiri@mail.thaiunion.co.th; internet www.thaiuniongroup.com; f. 1988; manufacture and export of frozen and canned foods; cap. and res 7,839m., sales 35,324m. (2001); Chair. KRAISON CHANSIRI; Pres. THIRAPHONG CHANSIRI; 5,140 employees.

Thai Wah Public Co Ltd: Thai Wah Tower, 21st–22nd Floors, 21/63–66 Thanon Sathorn Tai, Bangkok 10120; tel. (2) 285-0040; fax (2) 285-0269; e-mail starch@thaiwah.com; internet www.thaiwah.com; f. 1947; mfrs of tapioca products; cap. and res 339m. (1999), sales 2,245m. (2001); Chair. HO KWON PING; Man. Dir PINAI KIATTEPPAWAN; 1,850 employees.

Thailand Fishery Coldstorage Public Co Ltd: 592 Moo 2, Thanon Thayban, Amphur Muang, Samutprakarn 10280; tel. (2) 387-1171; fax (2) 387-2227; freezers and exporters of shrimps and squids; cap. and res 271m., sales 4,082m. (1998); Chair. and Chief Exec. THAVESAKDI LAOTRAKUL; Pres. and Man. Dir YUWAREE AUKARNJANAWILAI; 885 employees.

Thonburi Automotive Assembly Plant Co Ltd: Soi 3, 72 Thanon Klang Ratchadamneon, Bowon Niwet, Phranakhon, Bangkok 10200; tel. (2) 226-0021; fax (2) 384-2246; f. 1960; mfrs of motor vehicles; sales 28,363m. (1994); Man. Dir PARKPIEN WIRIYAPANT.

Tipco Asphalt Public Co Ltd: 118/1 Thanon Ram VI, Samsen Nai, Phayathai, Bangkok 10400; tel. (2) 273-6000; fax (2) 273-6030; e-mail adm-tipc@mozart.inet.co.th; internet www.tipco.co.th; f. 1979; mfrs of construction materials; cap. and res 1,774m., sales 3,807m. (2001); Chair. PRASIT SUPSAKORN; Man. Dir SOMCHIT SERTTHIN.

Toyota Motor Thailand Co. Ltd: Moo 1, 186/1 Thanon Old Railway, Samrongtai, Phra Pradaeng, Samutprakarn 10130; tel. (2) 386-1000; fax (2) 384-1891; e-mail pr@toyota.co.th; internet www.toyota.co.th; f. 1962; mfrs and traders of motor vehicles; sales 48,980m. (1999); Pres. YOSHIAKI MURAMATSU.

TPI Polene Public Plc: 8th Floor, TPI Tower, 26/56 Thanon Chan Tadmai, Thungmahamek, Sathorn, Bangkok 10120; tel. (2) 678-5100; fax (2) 678-5000; e-mail tpiadmin@tpigroup.co.th; internet www.tpipolene.com; f. 1978; mfrs and distributors of cement and LDPE; cap. and res 22,128m., sales 4,550m. (2003); Chair. SUNTHORN HONGLADAROM; Chief Exec. PRACHAI LEOPHAIRATANA; 4,500 employees.

Tuntex (Thailand) Public Co Ltd: 54 BB Bldg, 18th Floor, Room 1812, 54 Sukhumvit 21 (Soi Asoke), Thanon Sukhumvit, Klongtoey Nua, Wattana, Bangkok 10110; tel. (2) 260-8020; fax (2) 260-8012; e-mail tuntex01@asiaaccess.th.com; internet www.tuntexthailand.com; manufacture and distribution of polyester products; cap. and res 6,702m. (1999), sales 8,433m. (2000); Chair. SHAN HUA SHEN; 2,000 employees.

Unicord Public Co Ltd: 404 Thanon Phayathai, Wangmai, Patumwan, Bangkok 10330; tel. (2) 216-0200; fax (2) 216-1468; f. 1977; mfrs of canned seafood; largest exporter of canned tuna in Asia; cap. and res 3,090m., sales 11,952m. (1995); Chair. KAMCHORN SATHIRAKUL; Pres. PORNPHUN KONUNTAKIET; 6,800 employees.

Union Mosaic Industry Public Co Ltd: Chamnan Phenjati Business Center Bldg, 29th Floor, 65 Thanon Phra Ram IX, Huay Kwang, Bangkok 10310; tel. (2) 248-7007; fax (2) 248-7005; e-mail export@umi-tiles.com; internet www.umi-tiles.com; f. 1973; mfrs of tiles; cap. and res 416m., sales 2,066m. (2003); CEO PAWEENA LAOWIWATWONG; 1,508 employees.

United Flour Mill Public Co Ltd: UFM Bldg, 9th Floor, 177–179 205 Thanon Rajawong, Sumpuntawong, Bangkok 10100; tel. (2) 226-0680; fax (2) 224-5670; f. 1961; wheat flour milling; cap. and res 315m. (2000), sales 2,033m. (2001); Chair. NIRAN SIRINAVIN; 171 employees.

Unocal Thailand Ltd: SCB Park Plaza, Tower III, 5th Floor, 19 Thanon Ratchadaphisek, Chatuchak, Bangkok 10900; tel. (2) 541-1970; fax (2) 541-1436; internet www.unocal.com; f. 1962; mining of petroleum and natural gas; sales 30,394m. (1997); Pres. BRIAN MARCOTTE.

Vanachai Group Public Co Ltd: 2/1 Thanon Pibulsongkram, Bangsue, Bangkok 10800; tel. (2) 585-4900; fax (2) 587-0516; e-mail woodtek@loxinfo.co.th; internet www.vanachai.com; manufacturer of wood-based panels; cap. and res 2,044m., sales 3,887m. (2001); Chair. SOMPON SAHAVAT.

Vinythai Public Co Ltd: Green Tower, 14th Floor, 3656/41 Thanon Rama IV, Khet Klongtoey, Bangkok 10110; tel. (2) 240-2425; fax (2) 240-1383; e-mail vinythai@solvay.com; internet www.vinythai.co.th; production and distribution of chemicals; cap. and res 7,454m., sales 6,130m. (2003); Chair. Dr CHRISTIAN DE SLOOVER; Man. Dir VINCENT DE CUYPER.

TRADE UNIONS

Under the Labour Relations Act (1975), a minimum of 10 employees are required in order to form a union; by August 1997 there were an estimated 1,028 such unions.

Confederation of Thai Labour (CTL): 25/20 Thanon Sukhumvit, Viphavill Village, Tambol Paknam, Amphur Muang, Samutprakarn, Bangkok 10270; tel. (2) 756-5346; fax (2) 323-1074; represents 44 labour unions; Pres. AMPORN BANDASAK.

Labour Congress of Thailand (LCT): 420/393–394 Thippavan Village 1, Thanon Teparak, Samrong-Nua, Muang, Samutprakarn, Bangkok 10270; tel. and fax (2) 384-6789; e-mail lct_org@hotmail.com; f. 1978; represents 224 labour unions, four labour federations and approx. 140,000 mems; Pres. PRATNENG SAENGSANK; Gen. Sec. SAMAM THOMYA.

National Congress of Private Employees of Thailand (NPET): 142/6 Thanon Phrathoonam Phrakanong, Phrakanong, Klongtoey, Bangkok 10110; tel. and fax (2) 392-9955; represents 31 labour unions; Pres. BANJONG PORNPATTANANIKOM.

National Congress of Thai Labour (NCTL): 1614/876 Samutprakarn Community Housing Project, Sukhumvit Highway Km 30, Tai Baan, Muang, Samutprakarn, Bangkok 10280; tel. (2) 389-5134; fax (2) 385-8975; represents 171 unions; Pres. PANAS THAILUAN.

National Free Labour Union Congress (NFLUC): 277 Moo 3, Thanon Ratburana, Bangkok 10140; tel. (2) 427-6506; fax (2) 428-4543; represents 51 labour unions; Pres. ANUSSAKDI BOONYAPRANAI.

National Labour Congress (NLC): 586/248–250 Moo 2, Mooban City Village, Thanon Sukhumvit, Bang Phu Mai, Mueng, Samutprakarn, Bangkok 10280; tel. and fax (2) 709-9426; represents 41 labour unions; Pres. CHIN THAPPHLI.

Thai Trade Union Congress (TTUC): 420/393–394 Thippavan Village 1, Thanon Teparak, Tambol Samrong-nua, Amphur Muang, Samutprakarn, Bangkok 10270; tel. and fax (2) 384-0438; f. 1983; represents 172 unions; Pres. PANIT CHAROENPHAO.

Thailand Council of Industrial Labour (TCIL): 99 Moo 4, Thanon Sukhaphibarn 2, Khannayao, Bungkum, Bangkok; tel. (2) 517-0022; fax (2) 517-0628; represents 23 labour unions; Pres. TAVEE DEEYING.

Transport

RAILWAYS

Thailand has a railway network of 4,041 km, connecting Bangkok with Chiang Mai, Nong Khai, Ubon Ratchathani, Nam Tok and towns on the isthmus.

State Railway of Thailand: 1 Thanon Rong Muang, Rong Muang, Pathumwan, Bangkok 10330; tel. (2) 220-4567; fax (2) 225-3801; e-mail info@railway.co.th; internet www.railway.co.th; f. 1897; 4,041 km of track in 2002; responsible for licensing a 4,044-km passenger and freight rail system, above ground; Chair. PRINYA JINDAPRASERT; Gov. CHITSANTI DHANASOBHON.

Bangkok Mass Transit System Corpn Ltd: Alma Link Bldg, 9th Floor, 25 Soi Chidlom, Thanon Ploenchit, Lumpini, Bangkok 10330; fax (2) 255-8651; responsible for the construction and management of the Skytrain, a two-line, 23.5-km elevated rail system, under the supervision of the Bangkok Metropolitan Area, the initial stage of which was opened in December 1999; Chair. KASAME CHATIKAVANIJ; Exec. Vice-Pres. DAVID RACE.

Mass Rapid Transit Authority of Thailand: 175 Thanon Rama IX, Huay Kwang, Bangkok 10320; tel. (2) 246-5733; fax (2) 246-3687; e-mail pr@mrta.co.th; internet www.mrta.co.th; a 20-km subway system was opened in Bangkok in July 2004; assigned to construct three new lines totalling 91 km in length, including: 27-km Blue Line (Hua Lamphong-Tha Phra); 24-km Orange Line (Bang Kapi—Bang Bumru); and 40-km Purple Line (Bang Yai-Rat Burana); Gov. PRAPAT CHONGSANGUAN.

ROADS

The total length of the road network was an estimated 53,436 km in 2001. A network of toll roads has been introduced in Bangkok in an attempt to alleviate the city's severe congestion problems.

Bangkok Mass Transit Authority (BMTA): 131 Thanon Thiam Ruammit, Huay Kwang, Bangkok 10310; tel. (2) 246-0973; fax (2)

247-2189; e-mail cnai.bmta@motc.go.th; internet www.bmta.co.th; controls Bangkok's urban transport system; Chair. JARUPONG REUNG-SUWAN; Dir and Sec. POKSAK SETHABUTR.

Department of Highways: Thanon Sri Ayudhaya, Ratchathevi, Bangkok 10400; tel. (2) 245-9912; e-mail doh_pr@yahoo.com; internet www.doh.mot.go.th; Dir-Gen. TERDSAK SEDTHAMANOP.

Department of Land Transport: 1032 Thanon Phaholyothin, Chatuchak, Bangkok 10900; tel. (2) 272-5671; fax (2) 272-5680; e-mail admin@dlt.motc.go.th; internet www.dlt.motc.go.th; Dir-Gen. PREECHA ORPRASIRTH.

Express Transportation Organization of Thailand (ETO): 485/1 Thanon Sri Ayudhaya, Ratchathevi, Bangkok 10400; tel. (2) 245-3231; e-mail eto_001@mot.go.th; internet www.eto.mot.go.th; f. 1947; Pres. KOVIT THANYARATTAKUL.

SHIPPING

There is an extensive network of canals, providing transport for bulk goods. The port of Bangkok is an important shipping junction for South-East Asia, and consists of 37 berths for conventional and container vessels.

Marine Department: 1278 Thanon Yotha, Talardnoi, Samphanthawong, Bangkok 10100; tel. (2) 233-1311; fax (2) 236-7148; e-mail marine@mot.go.th; internet www.md.go.th; Dir-Gen. WIT WORAKUPT.

Office of the Maritime Promotion Commission: 19 Thanon Phra Atit, Bangkok 10200; tel. (2) 281-9367; e-mail motc@motc.go.th; internet www.ompc.moto.go.th; f. 1979; Sec.-Gen. SOMSAK PIENSAKOOL.

Port Authority of Thailand: 444 Thanon Tarua, Klongtoey, Bangkok 10110; tel. (2) 269-3000; fax (2) 249-0885; e-mail patodga@loxinfo.co.th; internet www.pat.or.th; 18 berths at Bangkok Port, 12 berths at Laem Chabang Port; scheduled for transfer to private sector in 2005; Chair. WANCHAI SARATHULTHAT; Dir-Gen. MANA PATRAM.

Principal Shipping Companies

Jutha Maritime Public Co Ltd: Mano Tower, 2nd Floor, 153 Soi 39, Thanon Sukhumvit, Bangkok 10110; tel. (2) 260-0050; fax (2) 259-9825; e-mail jutha@loxinfo.co.th; services between Thailand, Malaysia, Korea, Japan and Viet Nam; Chair. Rear-Adm. CHANO PHENJATI; Man. Dir CHANET PHENJATI.

Precious Shipping Public Co Ltd: Cathay House, 7th Floor, 8/30 Thanon Sathorn Nua, Khet Bangrak, Bangkok 10500; tel. (2) 237-8700; fax (2) 633-8460; e-mail psl@preciousshipping.com; internet www.preciousshipping.com; Chair. Adm. AMNARD CHANDANAMATTHA; Man. Dir HASHIM KHALID MOINUDDIN.

Regional Container Lines Public Co Ltd: Panjathani Tower, 30th Floor, 127/35 Thanon Ratchadaphisek, Chongnonsee Yannawa, Bangkok 10120; tel. (2) 296-1088; fax (2) 296-1098; e-mail rclbkk@rclgroup.com; Chair. KUA PHEK LONG; Pres. SUMATE TANTHUWANIT.

Siam United Services Public Co Ltd: 30 Thanon Ratburana, Bangprakok, Ratburana, Bangkok 10140; tel. (2) 428-0029; fax (2) 427-6270; f. 1977; Chair. and Man. Dir MONGKHOL SIMAROJ.

Thai International Maritime Enterprises Ltd: Sarasin Bldg, 5th Floor, 14 Thanon Surasak, Bangkok 10500; tel. (2) 236-8835; services from Bangkok to Japan; Chair. and Man. Dir SUN SUNDISAMRIT.

Thai Maritime Navigation Co Ltd: Manorom Bldg, 15th Floor, 51 Thanon Phra Ram IV, Klongtoey, Bangkok 10110; tel. (2) 249-0100; fax (2) 249-0108; internet www.tmn.co.th; services from Bangkok to Japan, the USA, Europe and ASEAN countries; Chair. VORASUGDI VORAPAMORN; Vice-Chair. ORMSIN CHIVAPRUCK.

Thai Mercantile Marine Ltd: 599/1 Thanon Chua Phloeng, Klongtoey, Bangkok 10110; tel. (2) 240-2582; fax (2) 249-5656; e-mail tmmbkk@asiaaccess.net.th; f. 1967; services between Japan and Thailand; Chair. SUTHAM TANPHAIBUL; Man. Dir TANAN TANPHAIBUL.

Thai Petroleum Transports Co Ltd: 355 Thanon Sunthornkosa, POB 2172, Klongtoey, Bangkok 10110; tel. (2) 249-0255; Chair. C. CHOWKWANYUN; Man. Capt. B. HAM.

Thoresen Thai Agencies Public Co Ltd: 26/26–27 Orakarn Bldg, 8th Floor, 26–27 Soi Chidlom, Thanon Ploenchit, Kwang Lumpinee, Khet Pathumwan, Bangkok 10330; tel. (2) 254-8437; fax (2) 253-9497; e-mail tta@thoresen.com; shipowner, liner operator, shipping agent (in Thailand and Viet Nam), ship repairs, offshore and diving services; Chair. M. R. CHANDRAM S. CHANDRATAT; Man. Dir ARNE TEIGEN.

Unithai Group: 11th Floor, 25 Alma Link Bldg, Soi Chidlom, Thanon Ploenchit, Pathumwan, Bangkok 10330; tel. (2) 254-8400; fax (2) 253-3093; e-mail gerrit.d@unithai.com; internet www.unithai.com; regular containerized/break-bulk services to Europe, Africa and Far East; also bulk shipping/chartering; Chair. SIVAVONG CHANGKASIRI; CEO GERRIT J. DE NYS.

CIVIL AVIATION

Bangkok, Chiang Mai, Chiang Rai, Hat Yai, Phuket and Surat Thani airports are of international standard. U-Tapao is an alternative airport. In May 1991 plans were approved to build a new airport at Nong Ngu Hao, south-east of Bangkok, at an estimated cost of US $1,200m. Construction began in 1995, and the project was scheduled for completion in 2000. In January 1997, however, it was announced that, due to the Government's financial problems, the Nong Ngu Hao project was to be suspended; priority was, instead, to be given to the existing airport at Don Muang in Bangkok, which was to be expanded to handle 45m. passengers annually by 2007, compared with 25m. in 1997. In December 2001 construction of the passenger terminal complex of Bangkok's second airport, Suvarnabhumi International Airport, commenced. The project was scheduled for completion in 2005.

Airports Authority of Thailand: Bangkok International Airport, 171 Thanon Vibhavadi Rangsit, Don Muang, Bangkok 10210; tel. (2) 535-1111; fax (2) 535-5559; e-mail aatpr@airportthai.or.th; internet www.airportthai.or.th; f. 1998; Man. Dir Air Chief Marshal TERDSAK SUJJARUK.

Department of Civil Aviation: 71 Soi Ngarmduplee, Thanon Phra Ram IV, Tung Mahamek, Sathorn District, Bangkok 10120; tel. (2) 287-0320; fax (2) 286-1295; e-mail doa@aviation.go.th; internet www.aviation.go.th; f. 1963; Dir-Gen. CHALOR KOTCHARAT.

Air Andaman: 87 Nailert Bldg, 4th Floor, Unit 402A, Thanon Sukhumvit, Bangkok; tel. (2) 251-4905; fax (2) 655-2378; internet www.airandaman.com; f. 2000; regional, scheduled passenger and charter services; Pres. ATICHART ATHAKRAVISUNTHORN.

Bangkok Airways: 60 Queen Sirikit Nat. Convention Centre, Thanon Ratchadaphisek Tadmai, Klongtoey, Bangkok 10110; tel. (2) 229-3434; fax (2) 229-3450; e-mail reservation@bangkokair.co.th; internet www.bangkokair.com; f. 1968 as Sahakol Air, name changed as above in 1989; privately owned; scheduled and charter passenger services to regional and domestic destinations; Pres. and CEO Dr PRASERT PRASARTTONG-OSOTH.

Kampuchea Airlines (OX): 138/70 Jewellery Centre, 17th Floor, Thanon Nares, Bangrak, Bangkok 10500; tel. (2) 267-3210; fax (2) 267-3216; e-mail bondmx@yahoo.com; f. 1997; owned by Cambodian govt (51%) and Orient Thai Airlines (49%); regional passenger flights from Phnom-Penh to Hong Kong and Bangkok; CEO UDOM TANTIPRASONGCHAI.

Nok Air: 89 Thanon Vibhavadi Rangsit, Bangkok 10900; tel. (2) 513-0121; fax (2) 513-0203; e-mail public.info@thaiairways.co.th; f. 2004; 39%-owned by Thai Airways International Public Co Ltd; flights to six domestic destinations; CEO PATEE SARASIN.

One-Two-Go: 138/70 17th Floor, Jewellery Centre, Thanon Nares, Bangrak, Bangkok 10500; tel. (2) 267-2999; fax (2) 267-3217; e-mail info@orient-thai.com; internet www.onetwo-go.com; f. 2003; subsidiary of Orient Thai Airlines; low-cost domestic flights; CEO UDOM TANTIPRASONGCHAI.

Orient Thai Airlines: 138/70 17th Floor, Jewellery Centre, Thanon Nares, Bangrak, Bangkok 10500; tel. (2) 267-2999; fax (2) 267-3217; e-mail info@orient-thai.com; internet www.orient-thai.com; f. 1993 as Orient Express Air; domestic and international flights; CEO and Man. Dir UDOM TANTIPRASONGCHAI.

PB Air: 101 Thanon Samsen, Bangkok 10300; tel. (2) 669-2066; fax (2) 669-2092; internet www.pbair.com; f. 1990; scheduled domestic passenger services; Chair. PIYA BHIROM BHAKDI.

Phuket Airlines: 1168/71, 25th Floor, Lumpini Tower Bldg, Thungmahamek, Bangkok; tel. (67) 982-3840; internet www.phuketairlines.com; f. 1999; domestic passenger services linking Ranong to Bangkok, Had Yai and Phuket.

Thai AirAsia Co Ltd: Bangkok; internet www.airasia.com; f. 2004; jt venture between Shin Corpn Public Co Ltd (51%) and Malaysia's Air Asia Sdn Bhd (49%); low-cost domestic flights; CEO TASSAPON BIJLEVELD.

Thai Airways International Public Co Ltd (THAI): 89 Thanon Vibhavadi Rangsit, Bangkok 10900; tel. (2) 513-0121; fax (2) 513-0203; e-mail public.info@thaiairways.co.th; internet www.thaiair.com; f. 1960; 93% govt-owned; shares listed in July 1991, began trading in July 1992; merged with Thai Airways Co in 1988; scheduled for partial privatization in late 2000; domestic services from Bangkok to 20 cities; international services to over 50 destinations in Australasia, Europe, North America and Asia; Chair. THANONG BHIDAYA; Pres. KANOK ABHIRADEE.

Tourism

Thailand is a popular tourist destination, noted for its temples, palaces, beaches and islands. In 2003 tourist arrivals totalled 10.08m. Tourism is Thailand's largest single source of foreign exchange. Revenue from tourism, according to the World Tourism Organization, was an estimated US $7,902m. in 2002.

Tourism Authority of Thailand (TAT): Head Office: 1600 Thanon New Phetburi, Makkasan, Rachathewi, Bangkok 10400; tel. (2) 250-5500; fax (2) 253-7468; e-mail inteldiv@tat.or.th; internet www.tat.or.th; f. 1960; Chair. SONTHAYA KHUNPLERM; Gov. JUTHAMAS SIRIWAN.

Tourist Association of Northern Thailand: 51/20 Thanon Mahidol (Northern Tour), Chiang Mai 50100; tel. (53) 276-848; fax (53) 272-394; Pres. AURAWAN NIMANANDA.

Defence

In August 2003 the total strength of the armed forces was 314,200: army 190,000 (70,000 conscripts), navy 79,200 (27,000 conscripts), air force 45,000. Paramilitary forces numbered approximately 113,000, including a National Security Volunteer Corps of 50,000. Military service lasts for two years between the ages of 21 and 30 and is compulsory.

Defence Expenditure: Budgeted at an estimated 79,900m. baht for 2003/04.

Supreme Commander of the Armed Forces: Gen. CHAISITH SHINAWATRA.

Commander-in-Chief of the Army: Gen. PRAVIT WONGSUWAN.

Commander-in-Chief of the Air Force: Air Chief Marshal KHONGSAK WANTHANA.

Commander-in-Chief of the Navy: Adm. SAMPHOP AMRAPAN.

Education

Education in Thailand is free and compulsory for six years between the ages of six and 11 years. From 1996 the Government planned to extend the period of compulsory education to nine years. Furthermore, the 1997 Constitution guaranteed all children education until the age of 12 years; this was scheduled to be implemented from 2003, but in the early 2000s had fallen well behind this target. The 1999 National Education Act set out a three-year programme of educational reform. There are four types of schools: (i) Government schools established and maintained by government funds; (ii) Local schools which are usually financed by the Government; however, if they are founded by the people of the district, funds collected from the public may be used in supporting such schools; (iii) Municipal schools, a type of primary school financed and supervised by the municipality; (iv) Private schools set up and owned by private individuals under the provisions of the 1954 Private Schools Act. The National Scheme of Education provides for education on four levels: (i) Pre-School Education (nursery and kindergarten), which is not compulsory; (ii) Primary Education; (iii) Secondary Education; (iv) Higher Education. Budgetary expenditure on education in the financial year ending 30 September 2003 was estimated to be 235,393.8m. baht, or 23.5% of total planned spending.

Pre-primary education begins at three years and enrolment was equivalent to 63% of the relevant age-group in 1996. Primary education starts at the age of six and lasts for six years. In 1996 enrolment at primary level was equivalent to 87% of the relevant age group. Secondary education, which also lasts for six years, is divided into two three-year cycles. In 1996 enrolment at secondary schools was equivalent to 56% of the school-age population. In 1995 there were 20 state universities and 26 private universities and colleges in Thailand, offering both undergraduate and graduate courses in all fields. Other higher education establishments include the various Military and Police Academies providing a standard of training equivalent to that of civil establishments, and teacher-training establishments. Enrolment in higher education was equivalent to 22% of the relevant age-group in 1995. The rates of secondary and tertiary education participation in Thailand are among the lowest in South-East Asia.

Bibliography

GENERAL

Askew, Marc. *Bangkok: Place, Practice and Representation.* London, Routledge, 2002.

Brummelhuis, Hanten, and Kemp, J. (Eds). *Strategies and Structures in Thai Society.* Amsterdam, University of Amsterdam Antropologisch-Sociologisch Centrum, 1984.

Connors, Michael Kelly. *Democracy and National Identity in Thailand.* London, RoutledgeCurzon, 2002.

Dearden, Phillip (Ed.). *Environmental Protection and Rural Development in Thailand.* Bangkok, White Lotus, 2002.

Delang, Claudio. *Living at the Edge of Thai Society: The Karen in the Highlands of Northern Thailand.* London, RoutledgeCurzon, 2003.

Fordham, Graham. *A New Look at Thai Aids: Perspectives from the Margin.* New York, NY, Berghahn Books, 2004.

Goscha, Christopher E. *Thailand and the Southeast Asian Networks of the Vietnamese Revolution, 1885–1954.* London, RoutledgeCurzon, 1998.

Ishii, Yoneo. *Sangha, State and Society: Thai Buddhism in History.* Honolulu, HI, University of Hawaii Press, 1986.

Ishii, Yoneo (Ed.). *Thailand, A Rice-Growing Society.* Honolulu, HI, University of Hawaii Press, 1978.

Jeffrey, Leslie Ann. *Sex and Borders: Gender, National Identity, and Prostitution Policy in Thailand.* Vancouver, BC, University of British Columbia Press, 2002.

Kislenko. *Culture and Customs of Thailand.* Westport, CT, Greenwood Press, 2004.

Klausner, William J. *Thai Culture in Transition: Collected Writings.* Bangkok, Siam Society, 1998.

McKinnon, J., and Vienne, B. (Eds). *Hill Tribes Today: Problems in Change.* Bangkok and Paris, White Lotus-ORSTOM, 1991.

McKinnon, John, and Wanat Bhruksari. *Highlanders of Thailand.* Kuala Lumpur, Oxford University Press, 1983.

Piprell, Collin, and Boyd, Ashley J. *Thailand's Coral Reefs: Nature Under Threat.* Bangkok, White Lotus, 1996.

Rajadhon, Phya Anuman. *Popular Buddhism in Siam and other Essays on Thai Studies.* Bangkok, Thai Inter-Religions Commission for Development, 1985.

Reynolds, Craig I. (Ed.). *National Identity and its Defenders: Thailand, 1939–1989.* Victoria, Monash University, 1993.

Sanitsuda, Ekachai. *Behind the Smile: Voices of Thailand.* Bangkok, Thai Development Support Committee and Post Publishing, 1990.

Seidenfaden, E. *The Thai Peoples.* Bangkok, The Siam Society, 1963.

Skinner, G. W. *Chinese Society in Thailand.* Ithaca, NY, Cornell University Press, 1957.

Skinner, William G., and Kirsch, Thomas A. (Eds). *Change and Persistence in Thai Society.* Ithaca, NY, Cornell University Press, 1975.

Takaya, Yoshikazu. *Agricultural Development of a Tropical Delta: A Study of the Chao Phraya Delta.* Honolulu, HI, University of Hawaii Press, 1987.

Tambiah, S. J. *World Conqueror and World Renouncer: A Study of Buddhism and Polity in Thailand Against a Historical Background.* Cambridge University Press, 1975.

Tanabe, Shigeharu, and Keyes, Charles F. (Eds). *Cultural Crisis and Social Memory: Modernity and Identity in Thailand and Laos.* London, RoutledgeCurzon, 2002.

Tapp, Nicholas. *Sovereignty and Rebellion: The White Hmong of Northern Thailand.* Singapore, Oxford University Press, 1990.

Wijeyewardene, Gehan, and Chapman, E. C. (Eds). *Patterns and Illusions: Thai History and Thought.* Canberra, Australian National University and Singapore, Institute of Southeast Asian Studies, 1993.

HISTORY

Brailey, Nigel (Ed.). *Two Views of Siam on the Eve of the Chakri Reformation.* Arran, Kiscadale Publications, 1990.

Bunnag, Tej. *The Provincial Administration of Siam 1892–1915.* Oxford University Press, 1977.

Callahan, William A. *Imagining Democracy—Reading 'The Events of May' in Thailand.* Singapore, Institute of Southeast Asian Studies, 1999.

Chakrabongse, Prince Chula. *Lords of Life: the Paternal Monarchy of Bangkok, 1782–1932.* New York, Taplinger and London, Alvin Redman, 1960.

Chaloemtiarana, Thak (Ed.). *Thai Politics 1932–1957.* Bangkok, Social Sciences Association of Thailand, 1978.

Chaloemtiarana, Thak. *Thailand: The Politics of Despotic Paternalism.* Bangkok, Social Sciences Association of Thailand, 1979.

Fineman, Daniel Mark. *A Special Relationship—The United States and Military Government in Thailand, 1947–1958.* Honolulu, HI, University of Hawaii Press, 1997.

Haseman, J. B. *The Thai Resistance Movement During World War II.* Seattle, WA, University of Washington Press, 2002.

Hong Lysa. *Thailand in the Nineteenth Century: Evolution of the Economy and Society.* Singapore, Institute of Southeast Asian Studies, 1984.

Jackson, Peter A. *Buddhadasa: Theravada Buddhism and Modernist Reform in Thailand.* Seattle, WA, University of Washington Press, 2003.

LeBlanc, Marcel (Ed.). *History of Siam in 1688.* Seattle, WA, University of Washington Press, 2004.

Mead, Kullada Kesboonchoo. *The Rise and Decline of Thai Absolutism.* London, RoutledgeCurzon, 2004.

Peleggi, Maurizio. *Lords of Things: The Fashioning of the Siamese Monarchy's Modern Image.* Honolulu, HI, University of Hawaii Press, 2002.

Stowe, Judith A. *Siam Becomes Thailand: A Story of Intrigue.* London, Hurst, 1991.

Suksamran, Somboon. *Political Buddhism in Southeast Asia: The Role of the Sangha in the Modernisation of Thailand.* London, Hurst, 1977.

Terweil, B. J. *A History of Modern Thailand 1767–1942.* St Lucia, University of Queensland Press, 1983.

Wood, W. A. R. *A History of Siam.* Bangkok, Chalermnit, 1959.

Wright, Joseph J. *The Balancing Act: A History of Modern Thailand.* Oakland, Pacific Rim Press, 1991.

Wyatt, D. K. *Thailand: A Short History.* New Haven, CT, Yale University Press, 1984.

Young, Ernest. *The Kingdom of the Yellow Robe.* Oxford University Press, 1982.

ECONOMICS AND POLITICS

Arghiros, Daniel. *Democracy, Development and Decentralization in Provincial Thailand.* London, Curzon Press, 2001.

Barbier, Edward B., and Sathirathai, Suthawan (Eds). *Shrimp Farming and Mangrove Loss in Thailand.* Cheltenham, Edward Elgar Publishers, 2004.

Bello, Walden, Cunningham, Shea, and Li, Kheng Poh. *A Siamese Tragedy—Development and Disintegration in Modern Thailand.* London, Zed Books, 1999.

Brown, A. *Labour, Politics and the State in Industrialising Thailand.* London, RoutledgeCurzon, 2003.

Brown, I. *The Elite and the Economy in Siam c. 1890–1920.* Singapore, Oxford University Press, 1989.

The Creation of the Modern Ministry of Finance in Siam, 1885–1910. Basingstoke, Macmillan, 1992.

Dhiravegin, Likhit. *Demi Democracy.* Singapore, Times Academic Press, 1993.

Dixon, Chris. *The Thai Economy: Uneven Development and Internationalisation.* London, Routledge, 1999.

Girling, J. L. S. *Thailand: Society and Politics.* Ithaca, NY, and London, Cornell University Press, 1981.

Glassman, Jim. *Thailand at the Margins: Internationalization of the State and the Transformation of Labour.* Oxford, Oxford University Press, 2004.

Hewison, K. *The Development of Capital and the Role of the State in Thailand.* New Haven, CT, Yale University Press, 1988.

Power and Politics in Thailand. Manila, Journal of Contemporary Asia Publishers, 1990.

Political Change in Thailand: Democracy and Participation. London and New York, Routledge, 1997.

Hirst, P. *Development Dilemmas in Rural Thailand.* Singapore, Oxford University Press, 1990.

Ingram, J. C. *Economic Change in Thailand 1850–1970.* 2nd Edn, Stanford, CA, Stanford University Press, 1971.

Kerdphol, Gen. Saiyud. *The Struggle for Thailand: Counter-Insurgency 1965–1985.* Bangkok, South Research Centre, 1986.

Keyes, Charles F. *Thailand: Buddhist Kingdom as Modern Nation-State.* Boulder, CO, Westview Press, 1987.

Kulick, Elliott, and Wilson, Dick. *Thailand's Turn: Profile of a New Dragon.* New York, St Martin's Press, 1993.

Laothamatus, Anek. *Business Associations and the New Political Economy of Thailand.* Singapore, Institute of Southeast Asian Studies, 1993.

McCargo, Duncan. *Politics and the Press in Thailand: Media Machinations.* London, Routledge, 2000.

McCargo, Duncan (Ed.). *Reforming Thai Politics.* Singapore, Institute of Southeast Asian Studies, 2002.

McVey, Ruth T. (Ed.). *Money and Power in Provincial Thailand.* Singapore, Institute of Southeast Asian Studies, 2000.

Morell, D., and Chai-Anan Samudvanij. *Thailand: Reform, Reaction and Revolution.* Cambridge, MA, Oelgeschlager, Gunn and Hain, 1981.

Muscat, Robert J. *Thailand and the United States: Development, Security, and Foreign Aid.* New York, Columbia University Press, 1990.

na Pombhejra, Vichitvong. *Readings in Thailand's Political Economy.* Bangkok Printing Enterprise, 1978.

Neher, C. D. *Modern Thai Politics: From Village to Nation.* Cambridge, MA, Schenkman, 1979.

Ockey, James. *Making Democracy: Leadership, Class, Gender, and Political Participation in Thailand.* Honolulu, HI, University of Hawaii Press, 2004.

Parnwell, M. (Ed.). *Uneven Development in Thailand.* Aldershot, Avebury, 1995.

Phongpaichit, Pasuk, and Baker, Chris. *Thailand: Economy and Politics.* Kuala Lumpur, Oxford University Press, 1996.

Thailand's Boom and Bust. Chiang Mai, Silkworm Books, 1998.

Thailand's Crisis. Chiang Mai, Silkworm Books, 2000.

Thaksin: The Business of Politics in Thailand. Copenhagen, NIAS Press, 2004.

Phongpaichit, Pasuk, and Piriyarangsan, Sungsidh. *Corruption and Democracy in Thailand.* Chiang Mai, Silkworm Books, 1998.

Phongpaichit, Pasuk, Piriyarangsan, Sungsidh, and Treerat, Nualnoi. *Guns, Girls, Gambling, Ganja—Thailand's Illegal Economy and Public Policy.* Chiang Mai, Silkworm Books, 1999.

Pian, Kobkua Suwannathat. *Kings, Country and Constitutions: Thailand's Political Development 1932–2000.* London, RoutledgeCurzon, 2002.

Rigg, Jonathan (Ed.). *Counting the Costs: Economic Growth and Environmental Change in Thailand.* Singapore, Institute of Southeast Asian Studies, 1996.

Samudvanija, Chai-Anan. *The Thai Young Turks.* Singapore, Institute of Southeast Asian Studies, 1982.

Suehiro, Akira. *Capital Accumulation in Thailand 1955–1985.* Tokyo, Centre for East Asian Cultural Studies, Toyo Bunko, 1989.

Suksamran, Somboon. *Buddhism and Politics in Thailand.* Singapore, Institute of Southeast Asian Studies, 1982.

Unger, Daniel. *Building Social Capital: Fibers, Finance, and Infrastructure.* Cambridge University Press, Cambridge, 1998.

Unphakorn, P., et al. *Finance, Trade and Economic Development in Thailand.* Bangkok, 1973.

Warr, Peter. *Thailand Beyond the Crisis.* London, Routledge, 2002.

Wedel, Yuengrat. *The Thai Radicals and the Communist Party.* Singapore, Maruzen Asia, 1983.

Wilson, Ara. *The Intimate Economies of Bangkok: Tomboys, Tycoons and Avon Ladies in the Global City.* Berkeley, CA, University of California Press, 2004.

Xuto, Somsakdi (Ed.). *Government and Politics of Thailand.* Hong Kong, Oxford University Press, 1987.

TIMOR-LESTE

(EAST TIMOR)

Physical and Social Geography

The Democratic Republic of Timor-Leste (known as East Timor until its accession to independence on 20 May 2002), which is styled Timor Loro Sa'e (Timor of the rising sun) in the principal indigenous language, Tetum, occupies the eastern half of the island of Timor, which lies off the north coast of Western Australia and extends between 8° 15′ and 10° 30′ S and 123° 20′ and 127° 10′ E. The western half of the island is Indonesian territory. In addition to the eastern half of Timor island, the territory also includes an enclave around Oecusse (Oekussi) Ambeno on the north-west coast of the island, and the islands of Ataúro (Pulo Cambing) and Jaco (Pulo Jako). Timor-Leste occupies an area of 14,609 sq km (5,641 sq miles).

Timor's climate is dominated by intense monsoon rain, succeeded by a pronounced dry season. The north coast of the island has a brief rainy season from December to February; the south coast a double rainy season from December to June, with a respite in March. The mountainous spine of the island experiences heavy rains that feed torrential floods. Once every three to four years, however, the climatic phenomenon known as El Niño is likely to subject the island to serious drought.

Very irregular, rugged hills and mountains form the core of the island, which is split by a longitudinal series of depressions and by small, discontinuous plateaux. There are many extinct volcanoes. Some good soils have been formed from the older volcanic rock, but the dominant soil consists of soft clay, which does not support heavy vegetation.

The indigenous peoples are of mixed origin. The aboriginal population is composed mainly of Melanesians, who probably resulted from the fusion of a basic Papuan stock with immigrant Asian elements. Evidence exists, also, of an Australoid strain. These peoples were displaced from the more favoured areas by subsequent arrivals from Indonesia, while communities of Chinese and other Asians gained control of much of the commerce conducted on the island. In a UN registration process completed in June 2001, the population of East Timor totalled 737,811, about 90% of whom lived in rural areas. More than 250,000 people were displaced by the conflict of 1999. In mid-2001 more than 113,000 remained in refugee camps in neighbouring West Timor. In 2004 many had yet to return to Timor-Leste.

History

ROBERT CRIBB

Only fragments are known of the early history of Timor. The island's name means 'east' and for at least a millennium Timor appears to have remained on the eastern fringe of the Indonesian commercial world, a source of sandalwood for trade to India and China and of slaves for markets in the archipelago. Indian and Islamic cultural influences on Timor were meagre, however, and there is no evidence of literacy or of large-scale state formation before 1600. Rather, it would appear that the island was divided among a fluctuating number of small polities headed by powerful chiefs, later known as *liurai*. Early European accounts report that these polities were grouped into two federations, generally referred to as the Wehale (Belu) and the Sonbai. While the nature of these federations is not clear, they certainly did not constitute co-ordinated political units. Timor's sandalwood attracted Portuguese interest in the mid-16th century, but the Portuguese preferred to establish their bases in the relative security of neighbouring Solor and Larantuka, rather than on the Timor coast itself. During the next century, however, Dominican missionaries converted many *liurai* to Catholicism, and the coastal regions of the island came increasingly under the domination of the so-called Topasses, or 'Black Portuguese'. The Topasses were descendants of Portuguese and other Western and Asian soldiers who married local women; they were rough adventurers who soon established a sphere of influence in western Timor. In the 1640s the Portuguese authorities attempted to assert control over the island by constructing a fort at Kupang at the western end of the island. This fort was no sooner constructed than it was seized in 1653 by the Dutch East Indies Company, which named it Castle Concordia. As Topass power grew under the rival de Hornay and da Costa families, however, the Dutch were unable to expand their power beyond the environs of Kupang. Portuguese influence in the region increased with the arrival of Catholic refugees from Makassar in Sulawesi, which the Dutch had conquered in 1660, but Timor itself remained firmly under the control of the Topasses. Successive Topass leaders received formal appointment as governors of the island from the Governor-General in Goa, and occasionally paid tax or tribute to the Portuguese crown from proceeds of trade in sandalwood, slaves, beeswax, gold and horses. In 1702 the Portuguese shifted their headquarters from Larantuka to Lifau, on the northern coast of west Timor, and appointed outsiders as governors. These governors led a miserable existence, besieged by the Topasses and harassed by the Dominicans, until 1769 when Governor António de Menezes moved his office to what was then the small settlement of Dili, further east along the coast. The Dutch, meanwhile, defeated a Topass attack on Kupang in 1749 and began slowly to expand their hegemony over the western part of the island, although they were never able to subdue the Topass strongholds around Lifau and in the inland region of Noimuti.

For about a century, neither colonial power did any more than continue to exploit the traditional trade of the island. However, from the mid-19th century, the development of modern colonialism and the fear of losing their colonies to newer, more dynamic colonial powers led both the Portuguese and the Dutch to begin to develop the island. The Portuguese began a programme of road-building, which enabled them to exercise closer control both of the *liurai* and of the general population. Coffee plantations were established, and a poll tax was imposed on all Timorese to encourage the growing of cash crops. Eventually the authority of the *liurai* was formally abolished, although they remained powerful local figures. In 1859 the Portuguese and the Dutch agreed to consolidate their territorial holdings, the Portuguese giving up Larantuka and outposts on other islands in exchange for the establishment of demarcated borders on Timor itself. Under a further treaty, in 1902 the two powers exchanged several small territories in the interior for the sake of a neater border, although the area around Lifau remained as a Portuguese enclave on the north coast, with the name Oecusse (Oekussi). Portugal's interference with the powers of the *liurai* led to a revolt in 1910–12 under Dom Boaventura, but the Timorese were defeated with the assistance of troops sent from Mozambique.

Portugal's cultural influence on East Timor was relatively limited. Although there was an official policy of encouragement of the adoption of Portuguese culture and Catholicism, the colonial authorities were often suspicious of the Catholic Church, and missionaries were banned from the colony for 40 years from 1834. As for general education, the colonial budget was minimal, and few funds were available for schooling. In consequence, even in 1950 adult literacy in East Timor was estimated at less than 5%, while less than 0.5% of the indigenous population of East Timor was classified as *civilizado*, that is, speaking Portuguese and having an income sufficient to maintain a 'civilized' life style. The remainder of the colonial élite was European, Chinese and mestizo. Portugal declared its neutrality at the start of the Second World War, and apprehension quickly grew in Australia and the Netherlands Indies that the Portuguese authorities in East Timor might accept a Japanese presence in the territory, much as the Vichy French authorities had allowed the Japanese access to French Indo-China. To forestall this possibility, Australia landed troops in Dili in mid-December 1941, despite Portuguese objections. These troops were not numerous enough to resist Japanese landings in mid-February 1942, but they retreated into the interior and undertook highly effective guerrilla warfare until they were evacuated in January 1943. The Japanese occupation appears to have been a very difficult time for the Timorese themselves, partly because of the guerrilla war and Allied and Japanese bombing, and partly because the impoverished territory was cut off from supplies of cloth and other consumer goods.

PRESSURES FOR DECOLONIZATION

By the end of the Second World War in 1945, Portugal's continued tenure of East Timor was by no means assured. Indonesian nationalists considered, and then rejected, the possibility of claiming the territory as a part of the new Indonesian Republic. More seriously, Australia proposed taking over the territory, perhaps with a UN mandate, to ensure that the island could serve as a base for a more effective forward line of defence in the event of another attack from the north. It appears that Portugal was able to stave off the Australian threat only by negotiating its intention to return to East Timor against NATO's interest in having access to the Azores, a Portuguese possession strategically located in the North Atlantic. During the 1950s, however, Portugal came under more general pressure to decolonize East Timor, pressure that even delayed the country's entry into the UN in 1956. In 1951 Portugal declared East Timor, along with its other colonies, to be an overseas province, the subjects of which had the same (limited) political rights as metropolitan Portuguese. The UN, however, continued to consider the former colonies as non-self-governing territories. Whereas Portugal's African territories seemed destined for eventual independence, its three small Asian colonies, including East Timor, all appeared likely to be absorbed by their larger neighbours. India occupied Goa in 1961, and in 1968 China forced Portugal to acknowledge Macao as Chinese territory under Portuguese administration. Indonesian leaders appear to have assumed that East Timor would eventually be absorbed into Indonesia, but until 1968 Indonesia took no action to claim the territory, the Indonesian Government being occupied with the pursuit and consolidation of the archipelago's claim to formerly Dutch West New Guinea. However, Indonesian intelligence forces may have sponsored a brief rebellion in the territory in 1959 and a Government-in-exile shortly afterwards.

In April 1974, however, the future of East Timor was abruptly placed on the international political agenda following a coup by the armed forces in Lisbon. The new Portuguese Government lifted political restrictions and foreshadowed major political changes, which appeared to include the possibility of independence, with regard to the country's colonies. Within a month, two new Timorese political parties had emerged: the União Democrática Timorense (UDT, Timorese Democratic Union), which was led by plantation owners and senior officials from the Portuguese administration, advocated democratization and eventual independence from Portugal, while the Associação Social Democrática Timorense (Timorese Social Democratic Association), which drew its membership from amongst younger professionals and intellectuals, argued for a more rapid transition to independence and for more extensive social reforms. A third party, the Associação Popular Democrática Timorense (Apodeti, Timorese Popular Democratic Association), which appears to have been sponsored from the outset by Indonesian intelligence organizations, proposed integration with Indonesia. The UDT was initially the most popular of the parties, but during 1974 it gradually lost support to the Associação Social Democrática Timorense, which adopted an increasingly radical profile. In September 1974 the latter renamed itself Frente Revolucionária do Timor Leste Independente (Fretilin, Revolutionary Front for an Independent East Timor) and claimed to be the sole representative of the East Timorese people.

The assumption that East Timor's natural destiny was absorption by Indonesia remained dominant throughout the international community, however: Portugal was preoccupied with internal political difficulties and with the decolonization of its African colonies, and other international powers had no natural interest in the territory, while the Australian Government believed that East Timor was too small and too poor for independence. In June 1974 the Indonesian Minister of Foreign Affairs formally stated that Indonesia respected East Timor's right to independence and had no intention of taking over the territory. The growing popularity and leftward political shift of Fretilin, however, added political weight to the arguments of Indonesian military and intelligence groups already in favour of annexation. Furthermore, less than a decade earlier, the Indonesian military had violently suppressed the Indonesian Communist Party, and in 1974 communist forces in Indo-China were clearly gaining influence. Under these circumstances, the Indonesian military feared that Fretilin would seek to establish a communist regime in East Timor and that such a regime would provide a base for 'subversion' in Indonesia itself. It appears that in about July 1974, therefore, sections of the Indonesian military intelligence initiated what became known as 'Operation Komodo', a broad-based strategy intended to secure the integration of East Timor into Indonesia. This strategy included the provision of funding and logistical assistance for supporters of integration, the promotion of the perception that East Timor was incapable of managing its own independence and the initiation of thinly concealed preparations for armed intervention in the territory.

In January 1975 local Portuguese officials, concerned by Indonesia's shift away from the acceptance of East Timorese independence, persuaded Fretilin and the UDT to form a coalition as the basis for a national transitional government to oversee the territory's passage to independence. From March local elections were held in several areas, most of which were won by Fretilin supporters; the colonial authorities tentatively scheduled the territory's accession to independence for the end of 1976. These arrangements had not been ratified by the Portuguese metropolitan government, however, which called a conference in Macao in May 1975 to discuss the decolonization of East Timor. Fretilin declined to attend the conference, apparently because its leaders regarded decolonization as a process that should be led from within East Timor itself, rather than by Lisbon, and because the movement objected to the inclusion of Apodeti in the negotiations. At about this time, Indonesian intelligence apparently warned UDT leaders that Indonesia would invade East Timor in order to prevent 'Communist' Fretilin from coming to power. The UDT responded first by pulling out of the coalition with Fretilin in May and then by staging a coup in Dili on 11 August with the assistance of the police force. The Portuguese Governor, Col Mário Lemos Pires, was under official instructions not to intervene and subsequently withdrew to the offshore island of Ataúro. Fretilin sympathizers in the local army units, however, launched a counter-attack, retaking Dili by 27 August and driving the remaining UDT forces across the border into Indonesian-controlled West Timor by the last week of September. From among the somewhat ramshackle collection of East Timorese parties and individuals displaced by Fretilin, Indonesian intelligence then assembled a coalition to demand integration. Meanwhile, Indonesian special forces, disguised as anti-Fretilin guerrillas, began to move into the territory, capturing the border town of Batugade on 8 October. However, difficult terrain and determined resistance by Fretilin forces, who had seized some NATO weaponry from the Portuguese, meant that Indonesian progress was slow. Only in late

November, after the fall of the town of Atabae, did Fretilin conclude that Indonesian conquest was likely. On 28 November, in an attempt to galvanize domestic and international support, Fretilin declared East Timor's independence as the Democratic Republic of East Timor, with Francisco Xavier do Amaral as President. Indonesia responded with a naval and airborne attack on Dili on 7 December. The attack took place one day after the departure from Jakarta of the US President, Gerald Ford, who had been on an official visit to Indonesia. The US President and the Secretary of State, Henry Kissinger, were subsequently acknowledged to have given approval to the Indonesian invasion. The operation was officially claimed to have been the work of East Timorese opposed to Fretilin, assisted by Indonesian 'volunteers'; however, the invasion was in fact carried out by regular marines and troops from Indonesia's élite strategic reserve (KOSTRAD). In the period during and after the capture of Dili, Indonesian troops killed several hundred East Timorese civilians suspected of offering resistance or supporting Fretilin, and this pattern was repeated on a smaller scale as the Indonesian troops fanned out across the territory to take other centres. On 17 December Indonesia sponsored a 'provisional Government' of East Timor led by Apodeti and a number of UDT leaders, with Arnaldo dos Reis Araújo as acting Governor. In May 1976 a 'People's Assembly' of 37 specially selected delegates formally petitioned Indonesia for integration, and on 17 July President Suharto of Indonesia formally declared the territory as the country's 27th province.

INDONESIAN RULE AND EAST TIMORESE RESISTANCE

Many observers expected that Indonesia would quickly achieve full control of East Timor and that the East Timorese would soon adjust to the new administration. In the event, however, Fretilin offered effective military resistance to the Indonesian armed forces in the countryside and retained a broad base of support amongst the East Timorese. Fretilin's initial military success was due to its access to modern weapons from the former Portuguese forces, to the fact that some of its troops had gained previous battle experience in Portugal's African colonies, and to the suitability of the East Timorese terrain for guerrilla warfare. Fretilin's popular support was based principally on the extensive political work carried out by the movement in rural areas since September 1974. In a society deprived of education, Fretilin activists had carried out extensive rural literacy programmes. They had also promoted the development of agricultural and trading co-operatives, which challenged the unpopular economic power of Chinese shopkeepers and wholesalers. At the same time, the activists refrained from suggesting major changes in traditional society, and so were able to win the support of many influential *liurai*. Reports on the three months of Fretilin administration in East Timor prior to the Indonesian attack on Dili suggest that Fretilin officials were generally efficient and humane. The contrasting brutality of Indonesian troops during the invasion, moreover, further alienated many East Timorese from the Indonesian cause. However, continuing warfare in the territory, together with Indonesia's resettling of villagers into strategic hamlets between 1977 and 1979, led to a famine in which perhaps 100,000 people died (of an original population of about 650,000). By the end of the resettlement campaign, Indonesia had succeeded in destroying the founding leadership of Fretilin and believed that the territory was under control.

Although the USA and Australia had made it clear that they would not intervene against Indonesia's occupation of East Timor, the UN Security Council passed a resolution on 23 December 1975 urging Indonesia's withdrawal from the territory and East Timorese self-determination. Indonesia refused to co-operate with a visit by the UN Secretary-General's special representative, Winspeare Guicciardi, and the Security Council passed a further resolution in April 1976 demanding that Indonesia withdraw from East Timor. There was, however, no significant international interest in pursuing the issue to the extent of the imposition of sanctions or other hostile measures. From 1976 to 1982 the UN General Assembly passed annual resolutions affirming the right of the East Timorese to self-determination and independence. Australia, however, gave *de jure* recognition to Indonesia's annexation of the territory in 1979.

Indonesia's policy in the territory was to suppress the independence movement while hoping that the slow acculturation of young East Timorese to Indonesian rule would gradually erode the resistance base. Substantial development aid was also allocated to the province, producing a dramatic improvement in communications infrastructure and education. Bahasa Indonesia came to be widely spoken, and by 1993 more than 1,000 East Timorese were studying at Indonesian universities. Many East Timorese sought employment elsewhere in Indonesia, particularly in Bali. In December 1988 Indonesia opened East Timor to foreign tourists, and in April 1990 it disbanded the special military command in the territory. In April 1991 the Indonesian Government announced that only 200 Fretilin guerrillas remained and that the Indonesian security forces would not pursue them because they represented no danger. Throughout this period, however, Indonesian oppression of the East Timorese population continued: both the European Community (EC, now European Union—EU) and the international human rights organization, Amnesty International, found compelling evidence of widespread killing and systematic torture. Such Indonesian brutality created deep resentment among the East Timorese people, including those of the younger generation who remembered nothing but Indonesian rule. External interests, moreover, came to dominate the province's economy. A military-controlled company obtained an effective monopoly on the coffee crop—the province's main export commodity—while other Indonesian interests dominated the construction and service industries. An indefinite number of Indonesians migrated to the province, and these migrants tended to dominate the administration and to control the lower reaches of the East Timorese economy, so that only a relatively small proportion of the benefits of the province's economic growth reached indigenous East Timorese.

In the mid-1980s resistance re-emerged, led by the Fretilin commander, José Alexandre 'Xanana' Gusmão. This resistance both encouraged, and was encouraged by, renewed international support for the East Timorese cause. Portugal in particular reasserted its claim—which was supported by the UN—that it was legally the administering power in East Timor, and increasingly used its position in the EU to press the East Timorese case. The former Portuguese colonies in Africa were also sympathetic to East Timor's cause, and the issue remained a persistent source of tension in Indonesia's international relations. In 1990 the Indonesian Minister of Foreign Affairs, Ali Alatas, began discussions with Portugal, through the office of the UN Secretary-General, in the hope of reaching a solution that would allow Indonesia to gain international recognition as the legitimate governing power in East Timor, possibly by finding some special constitutional status for the territory within Indonesia. These negotiations, however, were overtaken by political events within East Timor and also within Indonesia itself.

On 12 November 1991 Indonesian security forces fired on a demonstration at the funeral in Dili of a Fretilin sympathizer, killing between 100 and 180 people. A further 100 witnesses were said to have been summarily executed shortly afterwards. However, foreign news crews had been present, and film of the massacre was smuggled out of the country and widely broadcast. Although the Indonesian armed forces initially claimed that only 19 had died and that the troops involved had been 'provoked' by Fretilin supporters, intense international pressure led President Suharto to establish a separate inquiry, which found that 50 had died and 90 had 'disappeared' in the incident. This rare public criticism of the army led to the court martial and conviction of 10 military personnel and the dismissal of two senior army officers. Widespread criticism followed, however, concerning the disparity between the sentences given to protesters and to members of the armed forces.

In November 1992 the resistance suffered a major set-back when Xanana Gusmão was captured near Dili. In May 1993 he was found guilty of rebellion, conspiracy, attempting to establish a separate state and illegal possession of arms, and was condemned to life imprisonment. Following a plea for clemency, however, the sentence was commuted to 20 years by Suharto in August. In June 1997 the senior guerrilla leader, David Alex, was apprehended by Indonesian forces and died in a military hospital soon after his capture. The circumstances of his death were highly controversial, with the resistance claiming that Alex had been tortured or poisoned. In March 1998 Konis Santana, the military commander and acting leader of Fretilin, died as the result of an accident; he was subsequently replaced as acting leader by Taur Matan Ruak.

Despite military successes, a senior officer of the Indonesian armed forces acknowledged in early 1994 that Indonesia had failed to win the support of the East Timorese, and suggested that it would take another two generations until Indonesian rule could be accepted. Even this gloomy prognosis was made to seem optimistic by the growing religious dimension of the conflict. The predominantly Catholic East Timorese were deeply offended by incidents such as the mistreatment of nuns, the desecration of a church, and general anti-Catholic remarks by Muslim Indonesian officials, while Indonesians, reluctant to see genuine nationalism behind the Timorese resistance, became increasingly inclined to blame the territory's recalcitrance on Catholic separatism. These issues were complicated by the growing numbers of Muslim Indonesian residents of the province, who sometimes felt targeted by East Timorese demonstrators.

The UN sent a special investigator, Bacre Waly Ndiaye, to report on conditions in East Timor in July 1994; his conclusion that a climate of fear and suspicion dominated the territory subsequently encouraged the UN Secretary-General, Boutros Boutros-Ghali, to organize contacts and then talks between Timorese groups for and against integration with Indonesia, in the hope that these discussions might lead to a consensus on the best future for the territory. In mid-December 1998 the UN special envoy, Jamsheed Marker, visited East Timor and held talks with Xanana Gusmão and the acting Bishop of Dili, Carlos Ximenes Belo.

Within the EU, Portugal was the most vociferous in condemning Indonesia and lobbying for a UN-supervised referendum in East Timor; in July 1992 Portugal blocked an economic co-operation treaty between the Association of South East Asian Nations (ASEAN) and the EC on these grounds. Portugal also began proceedings against Australia in the International Court of Justice (ICJ), seeking a ruling against the so-called Timor Gap Treaty, concluded between Australia and Indonesia in 1991. The Treaty provided a legal framework for petroleum and gas exploration in the maritime zone between Australia and East Timor, which had not been covered by earlier Indonesian-Australian treaties. Portugal claimed that the agreement infringed both Portuguese sovereignty and the East Timorese right to self-determination. (Only Australia was named because Indonesia does not come under the court's jurisdiction.) In a judgment brought in June 1995, however, the Court ruled that it could not exercise jurisdiction because the central issue was the legality of actions by Indonesia, which had refused to present a case. Formal contacts between the Indonesian and Portuguese Governments, especially over the status of Portuguese culture in East Timor, took place in 1995 and 1996 but were for the most part inconclusive. In January 1996 Portugal began direct satellite television broadcasts to East Timor.

In June 1995, March 1996 and October 1997, All-Inclusive Intra-East Timorese Dialogues (AIETD) were held in Austria under UN auspices. Participants pressed for better protection of human rights in the territory, and in October 1997 adopted the name Loro Sai (Sa'e) for East Timor. On repeated occasions after late 1995, groups of young East Timorese entered foreign embassies in Jakarta to request political asylum; most were allowed to leave Indonesia. International awareness of East Timor was heightened in October 1996 when Bishop Carlos Ximenes Belo and resistance leader José Ramos Horta were jointly awarded the Nobel Peace Prize. The award especially enhanced Ramos Horta's campaign to seek international support for East Timorese self-determination, but there was little diplomatic movement until July 1997 when, with the apparent approval of President Suharto, President Nelson Mandela of South Africa met the imprisoned Xanana Gusmão for informal discussions. Despite high international hopes and an apparently amicable meeting between Mandela and Suharto on the issue, however, the Mandela initiative ended without result.

The accession of Bucharuddin Jusuf (B. J.) Habibie to the presidency of Indonesia following the downfall of President Suharto in May 1998 immediately raised expectations that East Timor would be dealt with as part of a more general plan to address problems left by the outgoing regime. Demonstrations against Indonesian rule continued, and the new President publicly suggested both that the territory might be given a new 'special' status within Indonesia and that troops might be withdrawn. Xanana Gusmão, however, was not among the numerous political prisoners released by the new regime immediately after the fall of Suharto, and there was initially no indication that Indonesia was prepared to contemplate independence for East Timor. In July speculation that the territory's status might be altered in the near future led thousands of non-East Timorese to flee to neighbouring East Nusa Tenggara province. At the end of the same month, the withdrawal from East Timor of hundreds of Indonesian troops began. In early August it was announced that Indonesia and Portugal had agreed to hold discussions on the possibility of 'wide-ranging' autonomy for the province.

On 27 January 1999, however, Indonesia surprised observers by announcing that, if East Timor rejected the autonomy programme that was being negotiated, it would consider allowing the province to become independent. Although the Indonesian Government was initially determined that the decision on the future of East Timor should not be reached on the basis of a referendum, it soon agreed to a UN-supervised poll in which all East Timorese would vote on whether to accept the autonomy proposals offered or to opt for independence, and signed an agreement with Portugal to this effect on 5 May, with the poll scheduled to be held on 8 August. Few observers had previously imagined that such a concession would be made, particularly since the Indonesian military was believed to view the relinquishing of East Timor as having dangerous implications for overall national security and as an insult to army prestige. However, the Indonesian Armed Forces commander, Gen. Wiranto, was reportedly receptive to arguments that the continued garrisoning of East Timor would weaken the army's capacity to maintain order elsewhere in the archipelago; some Muslim leaders were also reported to favour the removal of East Timor's predominantly Catholic population from the Indonesian body politic. Although all Indonesians resident in East Timor were to be permitted to vote in the poll, non-Timorese citizens constituted only a small minority of the population of 830,000 and since Timorese living in exile were also to be allowed to participate, a victory for the supporters of independence appeared likely. However, a virulent campaign of violence and intimidation waged by a number of anti-independence militia groups based within the territory threatened proceedings and appeared to cast some doubt on the certainty of a vote in favour of independence. The militias were accused of carrying out summary killings, kidnappings, looting, harassment and the forced recruitment of young East Timorese in order to sabotage the poll. Furthermore, it emerged that the Indonesian military itself was supporting, encouraging and training a number of the militias.

The violence continued to escalate throughout April and May 1999. In April anti-independence militia members shot and hacked to death 57 people in a churchyard in the town of Liquiça; further massacres were reported to have occurred in other areas of the territory, including Dili. Also in April Xanana Gusmão (who in February had been moved from Cipinang prison in Jakarta to serve out the remainder of his 20-year sentence under effective house arrest in the capital) abruptly reversed his previous position on the conflict in response to increasing violence by anti-independence militias and urged guerrilla fighters in Falintil (the military wing of Fretilin) to resume their struggle. The escalating violence in the territory, together with logistical difficulties, led the UN to postpone the referendum to 21 August and then to 30 August. On 18 June the rival pro-independence and integrationist factions signed a peace accord urging a cease-fire and disarmament in advance of the scheduled referendum.

Despite the continuing intimidation and violence, the referendum proceeded on 30 August 1999. About 98.5% of the electorate participated in the poll, which resulted in an overwhelming rejection, by 78.5% of voters, of autonomy proposals and an endorsement of independence for East Timor. The announcement of the result, however, precipitated a rapid descent into anarchy. As pro-Jakarta militias embarked upon a campaign of murder and destruction, which many observers believed to be premeditated, hundreds of civilians were killed, thousands were forced to flee their homes, and many buildings were destroyed in arson attacks. Anti-independence activists stormed the residence of the Nobel laureate, Bishop Carlos Ximenes Belo, evicting at gunpoint some 6,000 refugees who had sought shelter in the compound, and then burned down the home of the bishop. Bishop Ximenes Belo was evacuated to Australia, while Xanana Gusmão took refuge in the British embassy in Jakarta, following his release from house arrest. Thousands of terrified civilians besieged the UN compound in Dili, the premises of other international agencies, churches and police stations in a desperate search for protection from the indiscriminate attacks of the militias.

On 7 September 1999 martial law was declared in the territory, and a curfew was imposed. The violence continued unabated, however, and in mid-September, with international concern rising, the Indonesian Government yielded to pressure and agreed to permit the deployment of a multinational peace-keeping force. Following a visit to Jakarta by the UN High Commissioner for Human Rights, Mary Robinson, the Indonesian President also agreed to the holding of an international inquiry into whether the country's army was responsible for the perpetration of the atrocities. As the massacre of innocent civilians continued, thousands of refugees were airlifted to safety in northern Australia, along with remaining employees of the UN (many local staff members of the UN Mission in East Timor (UNAMET) having been among the victims of the violence). Meanwhile, aid agencies warned that as many as 300,000 East Timorese people faced starvation if humanitarian assistance were not urgently provided: in East Timor itself the number of displaced persons was estimated at 200,000, many of whom were in hiding in the mountains, while a further 100,000 were believed to have been driven into neighbouring West Timor, where their fate was unknown.

The first contingent of several thousand peace-keeping troops, forming the International Force for East Timor (Interfet), landed in the territory on 20 September 1999. Led by Australia, which committed 4,500 troops, the force gradually restored order. Other substantial contributions to the operation were made by the Philippines and Thailand, each of which provided 1,000 soldiers. A week later the Indonesian armed forces formally relinquished responsibility for security to the multinational force. At the end of October, after 24 years as an occupying force, the last Indonesian soldiers left East Timor. Amid scenes of jubilation in Dili, Xanana Gusmão, who had recently returned to his homeland, was able personally to witness the Indonesian commanders' final departure.

UNTAET ADMINISTRATION

On 19 October 1999 the result of the referendum was ratified by the Indonesian legislature, thus permitting East Timor's accession to independence to proceed. Shortly thereafter, on 25 October, the UN Security Council established the UN Transitional Administration in East Timor (UNTAET) as an integrated peace-keeping operation fully responsible for the administration of East Timor during its transition to independence. UNTAET, with an initial mandate extending until 31 January 2001, was to exercise all judicial and executive authority in East Timor, to undertake the establishment and training of a new police force, and to assume responsibility for the co-ordination and provision of humanitarian assistance and emergency rehabilitation; Interfet was scheduled to be replaced as soon as possible by UNTAET's military component. Meanwhile, the UN also began a large-scale emergency humanitarian relief effort. The UN administration in East Timor faced two major areas of difficulty in the months following the referendum, however. First, in an already poverty-stricken territory, the restoration of basic services after the major destruction of infrastructure during the violence of September 1999 proved to be extremely difficult. Because of the large-scale destruction of houses, many people displaced in the violence were unable to return to their homes. The Indonesian professionals who had provided medical, agricultural, educational and technical services had all fled by the time the UN took control of East Timor, and foreign aid workers were not numerous enough adequately to replace them. In addition, although development aid worth US $523m. had been promised to East Timor by various sources in December 1999, by March 2000 only $22m. had actually been provided. While construction work supplied some employment opportunities, there was little activity in other areas of the economy, and some sources estimated the unemployment rate in the territory to be as high as 80%. Because the Indonesian rupiah (which was itself extremely unstable in the late 1990s) remained the legal currency in the territory, East Timor's economy continued to be adversely affected by fluctuations in the Indonesian economy. Crime rates also increased, but could not be dealt with effectively because of the lack of any proper judicial system and prisons. In May 2000 only 700 of a total 1,610 international police officers promised for the civilian police force (CivPol) established by the UN had been deployed in the territory.

The second major area of difficulty faced by the UN administration concerned the various conflicts that emerged in East Timorese society following the referendum, as a consequence of the previous 25 years of Indonesian occupation and the violence of 1999. In particular, there were conflicts over the ownership of land that had changed hands as a result of political pressures, and tensions surrounding the issue of whether skilled East Timorese who had co-operated with the Indonesian authorities should be placed in positions of political responsibility in the newly independent East Timor. Although Xanana Gusmão urged the national reconciliation of all East Timorese, tensions were often acute at local level, especially when alleged East Timorese members of pro-Indonesia militia groups returned home. There was considerable discussion over whether the reconciliation of the East Timorese people would be best served by placing militia leaders on trial, by announcing a general amnesty for those involved in the violence, or by installing some form of 'truth and reconciliation commission' as in South Africa.

Following his popularly acclaimed return to Dili in October, Xanana Gusmão met the UNTAET Transitional Administrator, Sérgio Vieira de Mello, in November, and reportedly communicated the concerns of local East Timorese organizations that they were being marginalized by UNTAET officials. In late November he visited Jakarta in order to establish relations with the Indonesian Government, and in early December he visited Australia, where he met with representatives of the Australian Government to discuss the Timor Gap Treaty. A memorandum outlining an agreement temporarily to preserve the provisions of the original Timor Gap Treaty was signed in February 2000 by the Australian Government and East Timor's UN administrators to enable the exploitation of resources in the area to continue; Indonesia had ceased to be party to the original Treaty when it relinquished control of East Timor in October 1999. In May 2000 Australia announced that it was willing to consider the further renegotiation of the Treaty, possibly with a view to conceding rights to a larger share of the area's resources to East Timor.

On 1 December 1999 José Ramos Horta returned to East Timor after 24 years of exile. Ramos Horta, who commanded much popular support, urged the East Timorese people to show forgiveness towards their former oppressors and called for reconciliation between Indonesia and East Timor. On 2 December UNTAET established a 15-member National Consultative Council (NCC), comprising representatives of UNTAET itself, the National Council of Timorese Resistance (CNRT, the nationalist 'umbrella' organization including Fretilin), the Catholic church and groups that had formerly supported integration with Indonesia. The NCC's mandate was both to monitor the UNTAET administration and to advise on preparations for full independence. The NCC decided that Portuguese, rather than English, Bahasa Indonesia or the indigenous lingua franca, Tetum, would become the national language. However, other key issues such as how the East Timorese armed forces should be constructed, and the nature of the future electoral system, remained undecided. On 23 February 2000 the transfer of command of military operations in East Timor from Interfet to the UNTAET peace-keeping force was completed. The UNTAET force included contingents supplied by Australia, Bangladesh, Brazil, Canada, Fiji, Ireland, Jordan, Kenya, the Republic of Korea, Malaysia, New Zealand, Pakistan, the Philippines and Portugal.

In late January 2000 a panel appointed by the Indonesian Government to investigate human rights abuses in East Timor delivered its report to the Indonesian Attorney-General. The panel reportedly named 24 individuals whom it recommended should be prosecuted for their alleged involvement in violations of human rights in the territory. One of those named was the former Minister of Defence and Security and Commander-in-Chief of the Indonesian armed forces, Gen. Wiranto, who had since been appointed Co-ordinating Minister for Politics and Security in the Indonesian Government; also named were a number of senior military officers, as well as leaders of the pro-Jakarta militias responsible for the extreme violence perpetrated during the period following the referendum. However, pro-independence leaders in East Timor strongly criticized the report as inadequate. Also in late January the International Commission of Inquiry in East Timor recommended that the UN establish an independent international body to investigate allegations of human rights violations in East Timor, and an international tribunal to deal with the cases of those accused by the investigators. In February the recently appointed President of Indonesia, Abdurrahman Wahid, visited East Timor and publicly apologized for the atrocities committed by the Indonesian armed forces during the Republic's occupation of the territory. In the same month Wahid reaffirmed the commitment of the Indonesian Government to the prosecution of any individuals implicated in the violation of human rights in East Timor. In the same month Wahid suspended Gen. Wiranto from the Indonesian Government. The UN Secretary-General, Kofi Annan, made an official visit to East Timor in mid-February, during which he pledged that investigations into violations of human rights in the territory would be carried out. The President of Portugal, Dr Jorge Sampaio, also made a three-day official visit to East Timor in that month, during which he expressed the solidarity of the people of Portugal with the East Timorese.

A number of mass graves containing the bodies of suspected victims of the violence perpetrated by the anti-independence militias both before and after the holding of the referendum in August 1999 were discovered in East Timor (including two in the Oecusse enclave) in late 1999 and early 2000. In early December 1999 Sonia Picado Sotela, the Chair of the International Commission of Inquiry in East Timor, confirmed that the team of UN investigators had discovered evidence of 'systematic killing'. The UN announced in March 2000 that at least 627 East Timorese had been killed in the violence that followed the referendum, although this estimate was likely to increase as investigations continued. By May 2000 approximately 161,000 East Timorese refugees had returned to the territory under official auspices, leaving about 150,000 in 185 camps in Indonesian-controlled West Timor. Although there were fears that these camps were dominated by pro-Indonesia militias who might use them as a base for attacks on East Timor, UNTAET was reluctant to press for the refugees' immediate return because of a lack of facilities to house and feed them. In September 2000 the office of the UN High Commissioner for Refugees (UNHCR) temporarily suspended its relief operations in West Timor following the brutal murder of three of the organization's representatives in the region by pro-Indonesia militia members. During early 2001 the repatriation of refugees resumed, in order to enable people to register to vote in the Constituent Assembly elections scheduled for August. In July the Indonesian authorities formally asked all remaining refugee families to choose between repatriation and resettlement elsewhere in Indonesia. Only 1.1%, representing 1,250 people, chose to return to East Timor. Some observers attributed this figure to intimidation in the camps and to reports of the harassment of returnees in East Timor, but others pointed out that most of those who wished to return had probably already done so.

On 5 April 2000 UNTAET signed an agreement with the Indonesian Government relating to Indonesia's co-operation in efforts to resolve judicial and human rights issues. The agreement allowed for the extradition of Indonesians to East Timor for trial on charges relating to the violence of 1999. Relations between East Timor and Indonesia remained tense, however. In July a 17-member team from the Indonesian Attorney-General's Office visited East Timor to investigate a limited number of cases of human rights violations relating to the violence that occurred in East Timor in 1999. In August 2000, however, an amendment to Indonesia's Constitution appeared to contradict the assurances previously made to the international community by President Wahid concerning Indonesia's ability to conduct its own independent investigation into atrocities committed in East Timor by members of the country's armed forces. The amendment excluded military personnel from retroactive prosecution and (despite the suggestion of senior Indonesian legislators that the amendment would probably not apply to crimes such as genocide, war crimes and terrorism) was perceived by many international observers as a serious threat to the possibility of the prosecution of members of the Indonesian military believed responsible for recent human rights violations in East Timor. After the murder of the three UNHCR aid workers in West Timor, Indonesia agreed in October 2000 to disarm militias still active in its territory, but only a few hundred antiquated weapons were surrendered. In the same month, moreover, Indonesia formally refused to extradite the militia leader, Eurico Guterres, accused of playing a leading role in the Liquiça massacre, promising that he would be tried under Indonesian law. After a three-month trial in early 2001 Guterres was finally convicted of weapons offences and sentenced to six months' imprisonment. In view of the time already spent in custody, however, he was released after serving only 23 days.

In June 2000 an agreement was reached between UNTAET and East Timorese leaders on the formation of a new transitional coalition Government, in which the two sides were to share political responsibility. The Cabinet of the new transitional Government initially included four East Timorese cabinet ministers: João Carrascalão, President of the UDT and a Vice-President of the CNRT, was allocated responsibility for infrastructure; Mari Alkatiri, Secretary-General of Fretilin, was appointed Minister for Economic Affairs; Father Filomeno Jacob was appointed to oversee social affairs; and Ana Pessôa was placed in charge of internal administration. The new Cabinet also included four international representatives. Mariano Lopes da Cruz, an East Timorese national, was appointed as Inspector-General. It was reported that Xanana Gusmão, whilst having no formal position in the new Government, was to be consulted informally by Sérgio Vieira de Mello—who was to retain ultimate control over the approval of any draft legislation proposed to the Cabinet—with respect to all political decisions. On 14 July UNTAET approved the establishment of a 'National Council'—which consisted of 33 East Timorese representatives from the political, religious and private sectors—to advise the new Cabinet. The current NCC was to be dissolved at the first session of the new Council. On 20 October, however, the composition of an expanded, 36-member National Council was announced; the new East Timor National Council was to replace the 15-member NCC, which held its last meeting on that day. On 23 October Xanana Gusmão was elected leader of the new National Council. Earlier in the same month José Ramos Horta was appointed to the Cabinet of the transitional Government as Minister of Foreign Affairs, increasing the number of ministers in the Cabinet to nine.

Both Ramos Horta and Xanana Gusmão stressed their commitment to good relations with Indonesia, downplaying the need to prosecute Indonesian military commanders for their role in the violence and refusing to support the independence movements in the Indonesian provinces of Aceh and Irian Jaya (now Papua). In March 2000 Indonesia agreed to allow a land corridor to link the enclave of Oecusse to the rest of East Timor. In April 2001 East Timor and Indonesia reached an agreement on maintaining border security, and in April 2002 the two sides began a formal border demarcation survey. As interim Minister for Economic Affairs, Mari Alkatiri renegotiated the Timor Gap Treaty with Australia. In July 2001 the two sides reached the Timor Sea Arrangement, under which production from the so-called Joint Petroleum Development Area in the Timor Sea (formal ownership of which was still undecided) was to be divided 90:10 between East Timor and Australia. Although this arrangement appeared to favour East Timor, and was presented as a future economic mainstay for the new country, some observers argued that most of the indirect benefits of the Arrangement would flow to Australia.

On 20 August 2000, meanwhile, Xanana Gusmão retired as the Military Commander of Falintil to concentrate on his political role in the process of guiding East Timor towards full independence in advance of the general election, subsequently scheduled for August 2001. Xanana Gusmão relinquished control of the guerrilla army to his deputy, Taur Matan Ruak. Falintil was formally disbanded on 1 February 2001. Some 650 of its members were recruited into the newly formed East Timor Defence Force (Falintil-ETDF) and given immediate training as regular troops. The first of these trainees graduated in June, and Taur Matan Ruak was sworn in as Brigadier-General and commander of Falintil-ETDF. The remainder of Falintil was demobilized.

Although there was a widespread feeling that only Xanana Gusmão had public support for the future post of President of East Timor, he himself repeatedly said that he would not be a candidate, and on 28 March 2001 he resigned as chair of the National Council. In the vote for Gusmão's successor on 9 April, José Ramos Horta (reportedly supported by UNTAET) was defeated by Manuel Carrascalão, who criticized UNTAET for what he described as 'colonialist' practices. In August, after all the newly registered political parties declared Xanana Gusmão to be their preferred candidate and also in response to the encouragement of the international community, he announced that he would in fact contest the presidential election, scheduled for 2002.

On 4 July 2001 the 16 parties that had registered for the elections agreed on a National Unity Pact, which included a promise of mutual respect during the campaign and to honour the election results. Approximately 380,000 East Timorese were eligible to vote for the 88 members of a Constituent Assembly (13 of them chosen as district representatives, the rest by proportional representation). A transition period began on 15 July: the National Council was dissolved, members of the Transitional Cabinet who intended to play a political role resigned from their posts, and formal campaigning began. The Constituent Assembly was to draft a constitution for East Timor and was expected to complete its work before the expiry of the UN mandate on 31 January 2002.

At the election for the Constituent Assembly, conducted on 30 August 2001, Fretilin garnered 57% of the votes cast and secured a total of 55 seats (including 12 of the 13 district seats). Of the remaining 33 seats, the Partido Democrático (PD) won seven, the Partido Social Democrata (PSD) six and the Associação Social-Democrata Timorense (ASDT) also six. The composition of the new Cabinet, headed by Mari Alkatiri, was announced in September. Alkatiri, a member of East Timor's small Muslim community, who had lived in exile from 1975 to 1999, retained the portfolio of economic affairs, while José Ramos Horta continued as Minister of Foreign Affairs. Fretilin was allocated a total of nine cabinet posts. Two positions were occupied by PD members, with the remaining nine posts being assigned to various independents and experts.

The Constituent Assembly was initially expected to prepare East Timor's Constitution within 90 days, but the final document was not approved until 22 March 2002. The Constitution provided for parliamentary government with a five-year term and with a largely symbolic, but popularly elected, President. A Standing Committee of Parliament was designated to act on behalf of the legislature when it was not in session. The assembly revived the name Democratic Republic of East Timor, used by the short-lived independent Fretilin Government in late 1975, and declared 28 November 1975 as the date of independence. Although the Constitution provided for an elected parliament of 52–65 members, the 88-member Constituent Assembly declared itself the first National Parliament of the new republic. The Constitution designated Portuguese and Tetum as official languages and permitted East Timorese to hold dual citizenship. It also provided for the separation of church and state but specifically refrained from outlawing discrimination on the basis of sexual orientation, although many other forms of discrimination were banned.

In January 2002 a Commission for Reception, Truth and Reconciliation (Comissão de Acolhimento, Verdade e Reconciliação—CAVR) was established with a two-year mandate to address the difficult problem of reconciling those responsible for violence in the period from April 1974 to October 1999 with their victims and their victims' families. The aims of the commission were both to describe, acknowledge and record past human rights abuses along the lines of the post-apartheid South African Truth and Reconciliation Commission, and to devise procedures that would facilitate reconciliation at village level. The CAVR was not empowered to provide amnesties, and serious cases were to be dealt with in the courts. In Jakarta, meanwhile, in March 2002 the former governor of East Timor, Abílio Soares, was placed on trial, charged with knowingly permitting the mass violence of August–September 1999. Altogether, 18 high-ranking Indonesian officials were indicted, including three generals. The trial of Maj.-Gen. Tono Suratman, military commander of the province at the time of the violence, began in July 2002. When convicted in mid-August, Abílio Soares received a prison sentence of only three years, provoking international condemnation. Furthermore, the acquittal of the former police chief of East Timor, Timbo Salaen, along with several other officers, prompted human rights groups to demand the UN's intervention in the process.

Xanana Gusmão's overwhelming popularity was confirmed on 14 April 2002, when he was elected President of East Timor with nearly 83% of the votes cast. The only other candidate was Francisco Xavier do Amaral, who had served as President briefly in 1975 and who had declared his candidacy only for the sake of providing an alternative.

East Timor finally achieved independence on 20 May 2002 in a ceremony attended by the UN Secretary-General, Kofi Annan, and the Indonesian President, Megawati Sukarnoputri. From this date the nation officially became known as the Democratic Republic of Timor-Leste. The celebrations were marred, however, by the unauthorized arrival of six Indonesian warships in Dili harbour just before the ceremony, ostensibly to guard Megawati. The action was widely seen as an attempt by sections of the Indonesian military to create an incident that would force Megawati to abandon her visit. In the event, however, the warships withdrew peacefully. President Gusmão swore in a new 24-member Cabinet headed by Prime Minister Mari Alkatiri, and the UNTAET administration formally came to an end.

INDEPENDENCE

Independent Timor-Leste faced several political problems. The ownership of property emerged as a major issue, as confiscation and forced sales in the Indonesian period had often led to several people having rival but legitimate claims to the same property. The issue of law and order was a problem, both because high unemployment encouraged robbery and because demobilized freedom fighters who were not included in the defence force formed a significant social group that readily accused the Government of betraying the ideals of the independence struggle. Tension was also apparent between the older Portuguese-speaking generation of leaders and a younger generation educated under Indonesian rule, for whom English and Bahasa Indonesia were the preferred languages for international communication. Two days of riots occurred in Dili in early December 2002. Supermarkets, shops, hotels and the house of the Prime Minister, Mari Alkatiri, were burnt, and at least two people were killed when police opened fire on protesting students, before UN troops restored order.

The UN presence remained in Timor-Leste on a reduced scale in the form of the UN Mission of Support in East Timor (UNMISET), headed by a former Indian diplomat, Kamalesh Sharma. UNMISET's mandate, due to last two years only, was to maintain continuity in policing, to pay particular attention to gender and HIV policies and to supervise a continuing military presence of (initially) about 5,000 international troops. Following the Dili riots of December 2002, the UN agreed to keep two peace-keeping battalions near the country's border with Indonesia and to maintain an international police unit for an additional year, although the Timorese authorities began to assume responsibility from the UN for the administration of border controls in August 2003.

Timor-Leste's first political opinion poll, conducted in November 2003, showed that support for the Fretilin Government of Mari Alkatiri had begun to decline. Reports of corruption and ineffectiveness in the management of health, justice and education provided a basis for resentment, which was compounded by the perception that the Government was dominated by a Portuguese-speaking élite that had spent the occupation years outside East Timor. Another poll suggested that 39% of the population felt that they were worse off than under Indonesian rule. Although Alkatiri's insistence on national ownership of Timor-Leste's resources was popular, many of his policies were believed to be discouraging the foreign investment needed to create jobs. Many observers noted a sharp personal tension between President Gusmão and Prime Minister Alkatiri. Gusmão fulfilled his election promise to act as a 'watch-dog' by accusing parliamentarians of 'irresponsibility' after two sessions of the legislature failed to achieve a quorum and by criticizing the Government for failing to uphold the editorial independence of the state-run radio and television stations.

The most serious issue facing the new Government, however, was growing insecurity. Low wages and high levels of unemployment in Dili and other major towns encouraged a culture of street gangs and rising levels of urban crime. Resentful former members of Falintil who had not been integrated into Falintil-ETDF formed the core of many of these gangs. Moreover, a perception that Gusmão loyalists had been given preferential access to the defence force engendered political dissent within the gangs. The new Timor-Leste army, intended to consist of 1,500 regular soldiers and 1,500 reservists, was deployed in border regions to combat infiltration by former militia from West Timor. Complaints were soon heard of arrests without warrant and beatings in custody carried out by the army.

In December 2003 the Popular Council for the Defence of the Democratic Republic of East Timor (Conselho Popular pela Defesa da República Democrática de Timor Leste—CPD-RDTL), composed of dissident former guerrillas, launched a low-key rebellion against the Government in border regions. Rejecting the use of Portuguese and the role of former exiles in government, the organization announced that it would take power after the UN withdrawal. Sections of the CPD-RDTL, however, were brutally crushed in police operations, which aroused renewed fears about potential indiscipline in the police force once UN supervision was removed. Police used tear gas and rubber bullets to disperse a rally of veterans in Dili in July 2004.

Increasingly, the new Timor-Leste police force, numbering 2,800, came to be regarded as a political rival of the defence force. Unlike the army, the core of the police force had been recruited largely from the occupation-era police, who were considered to have essential expertise in managing law and order. Although in September 2003 the UN transferred to the police responsibility for law and order, except in Dili, many observers feared on the one hand that the new force was unprepared to carry out its responsibilities after the UN withdrawal scheduled for May 2004, and on the other hand that it was too strong and independent of civilian supervision. During 2003 several East Timorese leaders expressed the hope that some form of UN security presence would remain after the scheduled end of the UN mandate in May 2004. The UN Security Council duly extended UNMISET's mandate for a further year, but reduced its presence to 604, while full responsibility for policing and external security in the country was transferred to the Government of Timor-Leste.

President Gusmão's authority, by contrast, remained solid, and he showed a fine sense of judgement in keeping his offices, widely known as the Palace of the Ashes, in the burnt-out shell of a former motor vehicle registry and in reminding Timorese not to expect a swift transition to prosperity. None the less, tension continued over the issue of reconciliation, with President Gusmão favouring a general amnesty, while some in Fretilin sought a more punitive approach. In the absence of an amnesty, a UN-sponsored Special Panel for Serious Crimes (SPSC), created in Dili in 2001, continued to indict people for crimes carried out before and just after the independence vote. Some Fretilin leaders were angered in December 2003, when the CAVR held four days of hearings into the events of 1974–76, when a civil war between East Timorese had provided the pretext for the Indonesian invasion and atrocities had been carried out on both sides. The SPSC had charged 367 people by the end of 2003, of whom it had convicted 40. Of those charged, however, 280 remained free in Indonesia, including at least 32 Indonesian commanders and the former military chief, Gen. Wiranto. Indonesia's own courts, established in response to international pressure, tried only 18 military officers and civilian officials in 2002–03, of whom only six were convicted. In May 2004 the Timor-Leste Parliament narrowly adopted a proposal to discuss a general amnesty for all crimes committed up to 31 March, including the so-called 'serious crimes' perpetrated by Indonesian troops and anti-independence militias in 1999. However, similar bills had failed to gain approval in 2001 and 2003.

In July 2003 a ruling by Supreme Court judge Claudio Ximenes that the legal system adopted at independence was invalid because it was not derived from previous Portuguese law cast doubt over the whole legal system. Parliament responded in September by passing a law establishing Indonesian, not Portuguese, law as the country's applicable subsidiary legal system, on the grounds that neither courts nor police in Timor-Leste were familiar with Portuguese law. None the less, a critical shortage of judges meant that cases often did not reach the courts until two years after the arrest of suspects.

INTERNATIONAL RELATIONS

President Gusmão visited Jakarta in early July 2002, when he and President Megawati announced a plan to promote economic co-operation between the two countries. The two sides also agreed to establish a joint commission to consider Indonesian claims for compensation for assets lost in the territory when it became independent, although Timor-Leste intimated that it might refer to Indonesian exploitation and destruction of property in resisting those claims. Indonesia's claims referred both to infrastructure, such as roads, which it constructed during its rule of the territory, and to the private property of Indonesian citizens abandoned in 1999. The new Government's priority was to ensure good relations with Indonesia, so that issues of trade, border demarcation, militia remnants in West Timor and access to the Oecusse enclave could be resolved easily. Accordingly, Timor-Leste repeatedly assured Indonesia that it did not support the separatist movements in Aceh, Papua and elsewhere. Timorese leaders were also mild in their public criticism of the acquittal of many Indonesian soldiers accused of atrocities in the former East Timor, and of what were seen as light sentences passed on those who were convicted. None the less, there was evidence that Indonesian army elements continued to arm and train militias for incursions across the border from West Timor. At the same time, therefore, the Government of Timor-Leste was keen to anchor the country more firmly in the broader region. It signalled that it would seek membership of ASEAN, although Singapore was reportedly opposed to the entry of another economically weak country into the association. It also expressed interest in joining the Pacific Islands Forum. Timor-Leste sought good relations with the People's Republic of China, which had previously recognized it in 1975 and which was the first country to establish formal diplomatic relations in 2002. Timor-Leste became a member of the UN in September 2002, having previously joined the World Bank, the IMF and the Asian Development Bank (ADB).

In September 2003 Timor-Leste began formal negotiations with Indonesia to demarcate their shared land and maritime borders, including the terms of East Timorese overland access to the Oecusse enclave and provision for border passes to enable local residents to visit cross-border markets. Indonesia continued to claim compensation for property abandoned in East Timor in 1999, estimated to be worth €212m., while the Timor-Leste Government argued that accepting such a claim would make Indonesia liable to far greater claims over its actions during the occupation. Indonesia offered to provide military training to Falintil-EDTF, but there was widespread public hostility to this proposal. In December 2003 an Indonesian warship shelled the disputed island of Fatu Sinai (Pulau Batek), off the Oecusse enclave, and in August 2004 Timor-Leste acknowledged Indonesian sovereignty over the island. Smuggling across the border remained a major irritant. In May 2004 the SPSC in Dili finally issued an arrest warrant for Gen. Wiranto, then a candidate for the Indonesian presidency. Both President Gusmão and the Prosecutor-General, Longuinhos Monteiro, described the warrant as a mistake, and stated that the Timor-Leste Government would not enforce it. In July the two countries signed an agreement demarcating 90% of the land boundary.

Relations with Australia were made difficult by disagreement over the maritime boundaries in the Timor Sea. Australia's maritime border with Indonesia in the Timor Sea, settled in 1972, followed the continental shelf and gave a large part of the

Sea to Australia. There had been no such agreement with Portugal, and during the Indonesian occupation of East Timor the unregulated space between the territory and Australia was known as the Timor Gap. Although Australia would have preferred to draw a line following the continental shelf, international law had shifted in such cases to preferring borders drawn along a median line between the two coasts. Pending resolution of this complex issue, Australia and Indonesia had reached an agreement for regulating access to gas resources in the Gap. The agreement created a zone of co-operation (ZOC), the income from which was to be shared by the two countries. This agreement became invalid with East Timor's departure from Indonesia, and the Timorese authorities signalled that they intended to claim the full zone up to the median line, including areas that had been allocated to Australia under the Timor Gap Treaty. The lucrative Greater Sunrise field, which was being exploited by Australia, was within this claimed area. In response, in March 2002 Australia formally withdrew the issue from the jurisdiction of the ICJ and the international tribunal established under the 1982 UN Convention on the Law of the Sea. On 20 May, the day of East Timor's independence, the two countries signed an interim Timor Sea Treaty, which largely preserved the previous arrangement but gave a larger share of revenue to East Timor. In July, however, the new Timor-Leste Parliament approved a law establishing an Exclusive Economic Zone extending for 200 nautical miles in all directions, subject to future negotiations with Indonesia and Australia. Timor-Leste based its claim on international law and on its need for income from resources for its own economic development, and sought a rapid permanent settlement of the border issue. Australia and Timor-Leste later signed further agreements for revenue-sharing in the former ZOC (now called the Joint Petroleum Development Area, JPDA), but some observers accused Australia of both delaying the formal boundary negotiations and accelerating exploitation of areas it would almost certainly lose to Timor-Leste under international law. Reserves in the disputed area were estimated to be worth approximately US $8,000m. Despite these tensions, Australia remained a major aid donor to Timor-Leste and, in view of the declining security situation within Timor-Leste, Australian aid was increasingly directed towards strengthening the police force. In August 2003 Australia and Timor-Leste signed a memorandum of understanding to collaborate in combating terrorism.

Meanwhile, the future of an estimated 28,000 East Timorese still in refugee camps in West Timor remained unresolved. President Gusmão urged them to return to their homeland and Indonesia announced plans to close the remaining camps in August 2002, but many refugees were reluctant to move, in some cases fearing retribution for their alleged roles in the 1999 violence. In January 2003 the UN announced plans to fund the settlement of the refugees on the nearby island of Sumba. Timor-Leste agreed to sponsor a series of so-called reconciliation meetings between August and November, at which refugees could gain confidence in the prospect of life in independent Timor-Leste, but several meetings were cancelled, apparently in order to exclude former pro-Indonesia militia groups from the process. In October, moreover, 26 Timor-Leste citizens sought political asylum in Indonesia, claiming that they had been persecuted for their earlier sympathies with Indonesia; they were deported to Timor-Leste in December. There were also inconclusive discussions over the status of hundreds of Indonesians who had stayed in Timor-Leste after independence, intending to become citizens, but who had been excluded under the country's descent-based citizenship criteria. Tension with Australia also arose over the Australian Government's decision to return to Timor-Leste 1,600 East Timorese who had taken refuge in Australia in 1999, even though Gusmão had stated that Timor-Leste was not yet economically ready to accept them. It was reported that hundreds of young East Timorese had obtained Portuguese citizenship, on the basis of which they were seeking work elsewhere within the EU.

Economy

BROAD ECONOMIC TRENDS

The Portuguese Colonial Era

Until the early 1970s Timor-Leste, then still known by its pre-independence name of East Timor, was a remote outpost of the Portuguese colonial empire. During the 400 years or so of Portuguese colonial rule, which began in the latter half of the 16th century, the territory was largely neglected, with little progress being recorded in terms of economic or social development. Even after a change in its official status to that of an 'overseas province' of Portugal after the Second World War, the territory fared little better, reflecting in part the relative economic backwardness and resource constraints of Portugal itself. The situation was exacerbated by the political upheavals of the early 1970s, both in East Timor itself and in the metropolitan centre, which led eventually to Indonesian military intervention in 1975 and the formal integration of East Timor by Indonesia as the country's 27th province in 1976.

Indonesian Administration

The period of Indonesian rule, which ended amid chaos, brutal violence and widespread destruction in September 1999, was one of undeniable economic progress in East Timor, albeit in the context of severe political and social repression. The resentments engendered among the East Timorese population as a result of the oppressive Indonesian rule were reinforced by a number of economic grievances. These included the large-scale immigration into the province from other parts of the Indonesian archipelago of traders and professionals who were perceived to have gained a disproportionate advantage from the improvement in East Timor's economic circumstances and to have deprived the local population of its fair share of the benefits of this development. This perception was heightened by the tight control that the Indonesian military and business interests close to former President Suharto of Indonesia soon established over the territory's principal resource, namely, its production of high-quality arabica coffee.

Indonesian rule over East Timor became increasingly untenable in the 1990s in the face of growing pressure, from both within East Timor and abroad, for Indonesia to permit the territory to exercise its right to self-determination. This pressure increased as Indonesia descended into political, social and economic instability in the aftermath of the financial crisis that afflicted much of South-East Asia during 1997–98. Bacharuddin Jusuf (B. J.) Habibie, appointed President of Indonesia in May 1998, was eventually forced to permit a referendum among the East Timorese population as to whether they would prefer their territory to become independent or to remain a province of Indonesia. This decision had profound implications for the economy of East Timor. Soon after the announcement of the forthcoming referendum, the Indonesian military and other conservative factions within the Indonesian political élite mobilized armed militia gangs in East Timor to promote the pro-Indonesia cause. When, despite the mounting coercion and intimidation exercised by these groups, more than 78% of the population of East Timor voted in a referendum held on 30 August 1999 to sever the territory's ties with Indonesia, the response of these groups was uncompromising. In the killing and destruction that followed the announcement of the result of the referendum, they razed much of East Timor to the ground, completely devastating the territory's already weak economic infrastructure. At the same time, many of the non-indigenous professionals, tradesmen and administrators fled the territory, further undermining its capacity for recovery.

UN Transitional Administration

The havoc wreaked by the 'scorched earth' policy pursued by the pro-Indonesia militias in East Timor and their patrons outside the territory following the referendum of 1999 prompted a forceful international reaction. The UN Security Council subsequently agreed to the urgent dispatch of an Australian-led international peace-keeping force under the name of the International Force for East Timor (Interfet). By the time Interfet entered East Timor in September 1999 and began to restore law and order, the territory, which was already one of the poorest in Asia in both economic and social terms, lay in ruins. According to the UN, at least 627 people were killed and more than 75% of the population displaced in the violence that followed the referendum. In addition, some 70% of the territory's physical infrastructure, which had been targeted with particular vehemence, had been destroyed, mostly by arson. The fixed assets demolished as a result included buildings, installations and equipment.

Inevitably, this rampage prompted a sharp decline in all of East Timor's principal economic and social indicators, as exemplified by a World Bank estimate that gross domestic product (GDP) per head had decreased by almost 50% in the weeks following the referendum. A subsequent report prepared by the UN and the World Bank pointed out that with a GDP per head of US $424 in 1998, poverty rates had already been more than twice the average of those prevailing in Indonesia. East Timor thus had no effective economic reserves with which to counter the destruction of assets and livelihoods that followed the announcement of the referendum results in 1999.

Once Interfet had succeeded in restoring a degree of peace and law and order in East Timor, the responsibility for preparing the territory for independence fell to the UN Transitional Administration in East Timor (UNTAET), which was established in October 1999 by the UN Security Council. This task included rehabilitating the economic and institutional infrastructure, and laying the groundwork for the territory's sustainable development in the future.

The work of UNTAET was facilitated greatly by the compassion generated within the international donor community and by the extensive exposure given by the international media to the events surrounding the referendum on independence. Within two months of the landing of Interfet troops in Dili, the capital, a high-level meeting was called between the East Timorese national leadership and donor representatives in Tokyo, Japan, in mid-December 1999, at which donors pledged a total of US $523m. for an ambitious relief and reconstruction programme for East Timor. This sum comprised $157m. in support of an initial humanitarian relief programme and $366m. in support of a longer-term programme to promote governance, administrative capacity-building and economic and social reconstruction.

The implementation of the humanitarian component of the programme had been largely completed by mid-2000, with the most pressing humanitarian needs having been met through relief operations undertaken by a variety of UN agencies and non-governmental organizations (NGOs). By the end of April 2000 these measures had resulted in the distribution of more than 35,000 metric tons of rice, 9,700 shelter kits and 50,000 sheets of roofing material for use as emergency shelter. The World Health Organization (WHO) and the UN Population Fund had by this time also issued sufficient medical supplies and reproductive health kits to cover the needs of the local population.

With regard to the longer-term reconstruction component of the programme, the initial focus of operations following the Tokyo meeting was the establishment of a transitional administration capable of supporting the implementation of the programme. UNTAET was granted wide-ranging legislative and executive powers in support of its mandate to develop basic national institutions and to recruit, train and empower a corps of East Timorese civil servants to manage these institutions. To this end, UNTAET co-operated closely with the East Timorese leadership through the National Council of Timorese Resistance (more commonly known by its Portuguese acronym, CNRT) and the National Consultative Council (NCC). In its attempts to develop a suitable civil service infrastructure, however, UNTAET was faced with a disappointingly small pool of qualified personnel from which to draw its recruits. This was due partly to the fact that much of the territory's provincial civil service during the era of Indonesian rule had been staffed by non-East Timorese, and partly to the fact that many of the East Timorese who had worked for the Indonesian civil service had either fled the territory following the violence in 1999 or were distrusted by the new administration. To overcome this difficulty and provide a facility for the development of East Timor's future administrators, UNTAET opened a Civil Service Academy in May 2000.

The mandate of UNTAET ended with East Timor's transition to independence on 20 May 2002, upon which it became known as Timor-Leste, or Timor Loro Sa'e in the inigenous language, Tetum. Notwithstanding the many challenges it faced, the UN administration was widely acknowledged to have achieved its objectives with a relatively high degree of success. Apart from its efforts in the restoration of governance structures, UNTAET had succeeded in rehabilitating much of the territory's devastated physical and social infrastructure by the end of its mandate, and set the scene for a broader economic recovery. This success was underlined by an IMF report issued in June 2002, which noted that economic conditions in East Timor had improved steadily since mid-2000, with real GDP growth rates of 15% and 18% having been recorded in 2000 and 2001, respectively, as against a contraction of 35% in 1999. As a result, overall output had been restored to near pre-crisis levels by mid-2002.

Following the completion of UNTAET's mandate in May 2002, the UN established a successor mission, the UN Mission of Support in East Timor (UNMISET), which was intended to deliver a number of specific services to sustain the viability of the country's state institutions over a two-year period, ending in May 2004, which was subsequently extended for a further year. Under this framework, the UN was authorized to recruit some 300 international advisers to maintain the country's political, economic and social stability, and to promote its development. These experts provided on-the-job skills transfer, an in-country training programme and external support to various policy- and decision-making bodies, including, in the economic sphere, the Banking and Payments Authority and the Ministry of Planning and Finance.

Post-independence Development Strategy

By the latter half of 2001 the progress in post-conflict rehabilitation and the approaching deadline for the transfer of authority to an independent government prompted UNTAET and the newly installed Second Transitional Government to turn their attention to issues of longer-term economic development. A Planning Commission was therefore established to oversee the preparation of the National Development Plan shortly after the swearing in of the new Council of Ministers on 20 September 2001. The plan was drafted on the basis of an exhaustive country-wide consultative process in order to engage the people of East Timor as fully as possible. To heighten the sense of public ownership of this plan, a popular version entitled *East Timor to 2020: Our Nation, Our Future* was issued on 11 May 2002 and distributed to every household in East Timor with a view to promoting a common national understanding of the respective roles and responsibilities of the Government and the community in achieving economic growth and development.

The principal objectives of the National Development Plan are to reduce poverty and promote economic growth. It urges a combination of: capacity-building measures (especially in the early implementation phases); co-ordinated efforts by various developmental participants (government, civil society, community organizations and NGOs) to reduce poverty; and a number of sector-specific strategies to overcome the impediments to poverty reduction and economic development. To achieve these goals, the plan proposes a phased approach: in the short term, priority will continue to be given to the building of an appropriate legislative framework and institutional capacities, as well as the further development of infrastructure, education and health; over the longer term, the foundations laid by these efforts will be built upon to achieve sustainable development funded by the anticipated increase in oil and gas revenues.

The National Development Plan came into operation following Timor-Leste's accession to independence on 20 May 2002, and is intended to be implemented on a rolling basis consistent with the country's annual budgets. For this purpose, the formulation of annual action plans (AAPs) and quarterly reporting matrices (QRMs) was initiated with the commencement of the

2002/03 fiscal year on 1 June 2002. In addition, efforts were made to determine the prioritization and sequencing of the activities foreseen in the plan, and to prepare a 'road map' for its implementation over the following four years. At the same time the Government also began to formulate Timor-Leste's national Millennium Development Goals, based on the objectives established by the world's heads of state at the Millennium Summit held at the UN headquarters in New York, USA, in September 2000.

ECONOMIC STRUCTURE AND TRENDS

Upon achieving independence in 2002, Timor-Leste became the poorest country in Asia, with an estimated annual income of US $431 per head. While this figure matched, or even slightly exceeded, the corresponding figure of $424 recorded in 1998, some $400 of it was estimated to have been contributed by foreign donors. The sectoral distribution of GDP has traditionally been very heavily biased towards agriculture, which in 1998–99 accounted for more than 40% of the total, although this figure fell to approximately 25% in the following years as a result of the destruction of the country's productive base in 1999. Manufacturing has consistently accounted for less than 3% of GDP, although the overall contribution of industry (including mining, construction and utilities) has fluctuated between 15% and 20% in recent years.

As indicated above, the economy contracted sharply in 1999 as a result of the political violence of that year, but recovered between 2000 and 2001 in response to the financial and technical support received from the international community through UNTAET. Since then, however, the gradual downsizing of the UN presence in Timor-Leste has led to a corresponding deceleration in the economy, with the construction, services and financial sectors being particularly badly affected. The negative impact of these developments was exacerbated by a downturn in agricultural production owing to unfavourable weather conditions, as a result of which GDP was estimated to have declined by 1.1% in 2002 (although precise figures cannot be determined due to the absence of a sufficiently well-developed statistical system). This economic downturn aggravated further the already difficult employment situation, especially in urban areas and amongst younger sections of the population, and contributed to public riots in Dili in December 2002. It persisted into 2003, as donor activities continued to slow down, with some estimates suggesting a contraction of up to 3% in real GDP in that year, but a stabilization was anticipated for 2004, with projections published by the IMF indicating renewed growth of some 1% in that year, supported primarily by an increase in agricultural output.

NATURAL AND HUMAN RESOURCES

Timor-Leste has a total area of 14,609 sq km, including the coastal enclave of Oecusse (Oekussi) in the western half of the island of Timor, which is surrounded on all its land frontiers by Indonesian territory. The country has a rugged, mountainous terrain, relatively poor soils and a comparatively dry climate, which restricts its agricultural potential. It also has very limited onshore mineral deposits, although there are indications of significant crude petroleum reserves in the waters off its southern coasts.

The population of Timor-Leste is relatively small, albeit not precisely quantifiable. Estimates prepared during the Portuguese era, which were based on numbers submitted to the colonial authorities by *liurai* (village headmen), but which are acknowledged to be unreliable, indicated a total population of 624,564 in 1973. The subsequent civil war and Indonesian military intervention took a heavy toll on the local population, both directly, through acts of war, and indirectly, through the hunger and disease that ensued. The first formal population census under Indonesian rule, conducted in 1980, thus showed a total population of 555,350. By the time of the next census in 1990 this number had risen to 747,750, and an inter-censal survey carried out in 1995 suggested a further increase in the population to 843,100. This latter overall figure included a substantial, but unknown, number of non-indigenous people, however, many of whom left East Timor in the unrest preceding and following the referendum on independence held in August 1999, together with some 213,000 East Timorese who had fled the violence to Indonesian-governed West Timor. In a UN registration process completed in June 2001, the population of East Timor totalled 737,811.

The most widely accepted estimates suggested that in mid-2002 the country had an overall population of some 750,000, with some 50,000 East Timorese still living as refugees in West Timor. The number of these refugees returning to East Timor accelerated particularly rapidly from the beginning of 2002 onwards, as the political and economic situation in the country stabilized and the Indonesian Government ceased food distributions to the refugee camps. In an attempt to resolve the situation, the Government of Timor-Leste agreed to offer repatriation assistance to the remaining refugees until the end of August 2002.

AGRICULTURE

Agriculture forms the mainstay of Timor-Leste's economy, accounting for the employment of more than 70% of its labour force. The country has historically produced a variety of food and commercial crops, and also has a history of animal husbandry and fisheries. Like the rest of the economy, however, the agricultural sector suffered serious dislocations as a result of the upheavals that followed the vote for independence from Indonesia in 1999. A major reconstruction effort was therefore introduced by UNTAET, with financial support from the World Bank, for the agricultural sector. This effort comprised both a rapid-delivery component to facilitate the recovery of agricultural production levels, and a number of longer-term measures to ensure the sustained growth of the sector. To enable the achievement of its more immediate goal, the programme provided for: the restoration of productive assets to farmers; the community-based repair of rural feeder roads and small irrigation systems; the provision of agricultural credits; the establishment of local radio services to provide information to farmers; and the development of five pilot agricultural service centres managed by private entrepreneurs or NGOs to provide support services to farmers.

According to a World Bank report, by mid-2003 some 20,650 rural families had received support from this programme, with irrigation facilities covering an area of 7,742 ha and 109 km of rural access roads having been rehabilitated by this time. In addition, three agricultural service centres had been established to boost rice and coffee production and marketing, as well as to assist in the sale and processing of candlenut into tung oil. With regard to the longer-term objectives of the programme, major irrigation rehabilitation works and a substantial initiative to support coffee production and marketing were initiated. Measures were also implemented to initiate watershed management and reafforestation measures, and to help local fishermen to restore their fishing capacity. These measures had resulted in the recovery of both agricultural and fisheries production to pre-1999 levels by the time the country gained independence in mid-2002.

Given the importance of the agricultural sector to the national economy, the post-independence Government was expected to continue to give high priority to its development. Thus, the National Development Plan stated that, 'in the medium term the agricultural sector, more than most, provides opportunities for economic growth, exports, employment and improvements in social welfare throughout East Timor', and projected that, 'by 2020 East Timor will have sustainable, competitive and prosperous agricultural, fisheries and forestry industries that support improved living standards for the nation's people'. To achieve this goal, the Plan provided for the Government, through the newly established Ministry of Agriculture, Forestry and Fisheries, to 'efficiently deliver services to agricultural, fishing and forestry communities in East Timor...that support improved productivity, income earning potential and exports and that, therefore, support improved social welfare in the rural areas of the nation'.

Food Crops

The principal food crops grown in Timor-Leste comprise maize (which is the chief staple of the local population) and rice, although small quantities of such other staples as cassava, sweet potatoes and soybeans are also produced in the country.

Data published by the Indonesian Central Statistical Agency suggested that East Timor produced 36,848 metric tons of rice in 1998. This marked a significant increase over the corresponding figure of 8,005 tons in 1976, by which time the political unrest of the preceding years had resulted in food crop production falling to approximately one-quarter of the level achieved in 1970 (which itself represented the peak level achieved in food crop production during the colonial period). The increase in production between 1976 and 1998 was attributable mainly to the extensive rehabilitation and expansion of irrigation systems. Meanwhile, production data for non-rice staples in 1998 indicated an output of 58,857 tons of maize, 32,092 tons of cassava, 11,549 tons of sweet potatoes, and 672 tons of soybeans.

The production of food crops was severely disrupted by the political upheavals and unrest of 1999, as a result of which UNTAET estimated that output of both rice and maize had declined to some 70% of pre-conflict levels in 2000. This decrease was attributed to a combination of factors, including a reduction in the planted area and dislocation of rural labour owing to the violence linked to the independence ballot, the destruction of irrigation schemes, and the loss of services from government and private-sector suppliers. To overcome these difficulties UNTAET accorded high priority to supporting a recovery in food crop production and introduced a number of programmes to bring more land under cultivation and to increase yields. These included restitution of farming household assets, rehabilitation of rural feeder roads and irrigation systems, and fertilizer demonstrations.

By early 2002 these programmes had enabled UNTAET to meet its ambitious target of increasing food production to 80%–120% of 1997 levels. Although food production has remained steady in the years since independence, the prospects for future growth are constrained in the short term by the low productivity and high production costs of the food crop subsector. These difficulties were acknowledged in a statement made by the Minister of Agriculture, Forestry and Fisheries, Estanislau Aleixo da Silva, at the 25th anniversary session of the Governing Council of the International Fund for Agricultural Development (IFAD) in February 2003, although he also reiterated his Government's commitment to the promotion of food production as a means of assuring food security.

Cash Crops

The main commercial crop grown in Timor-Leste is coffee, although small plantations of coconut, cloves and cocoa have also been established. In addition to the low-value robusta coffees grown throughout the Indonesian archipelago, Timor-Leste also produces substantial quantities of the high-value arabica variety in its higher elevations. The country's coffee plantations were originally established during the Portuguese colonial era, when they accounted for some 90% of its export earnings. Production declined sharply in the transitional period of the mid-1970s as a result of the departure of Portuguese agricultural experts and the virtual abandonment of the coffee plantations. This led to a major replanting effort being initiated by the Indonesian authorities, with more than 1m. arabica and robusta seedlings being planted in the first four years of Indonesian rule. By 1994 coffee production had reached almost 8,800 metric tons, although it declined to less than 5,000 tons in 1997–98 as a result of the economic crisis affecting Indonesia and the effects of the drought caused by the weather phenomenon known as El Niño. In 1999 coffee production increased to around 10,000 tons, and much of the crop survived the destruction that followed the referendum.

One important feature of the East Timorese coffee industry during the era of Indonesian rule was the controversial marketing system employed by the Indonesian authorities: soon after the territory was integrated into Indonesia, a monopoly was established over the coffee market by PT Denok, a company associated with the Indonesian military, through the seizure of Portuguese-era coffee plantations and the forced sale of coffee by smallholders to PT Denok at artificially low prices. As a result of pressure from the US Government, this monopoly began to disintegrate in the mid-1990s, at which time a new system was put in place linking the Indonesian system of rural co-operatives to the US National Co-operative Business Association (NCBA) and providing local smallholders with significantly higher prices for their output. As a result of this arrangement, which remained in place even after the extensive political changes of 1999, the NCBA accounted for some 25% of the sale of East Timorese coffee. Another major purchaser of East Timorese coffee was the US-based Starbucks Corporation, cited in the media as one of East Timor's best clients.

One serious difficulty faced by East Timorese coffee producers as a result of the violence of 1999 was that all of the 'wet processing' facilities used for the processing of high-grade arabica coffee were badly damaged and inoperable, meaning that farmers were forced to 'dry-process' their coffee, which then fetched a much lower price. Although the coffee sector consequently did not escape entirely the widespread destruction of 1999, it remains in relatively good condition and was seen by both UNTAET and the national leadership of Timor-Leste as one of the prime sources of potential future export earnings and fiscal revenues.

The coffee industry has also benefited from the fact that local trees have never been treated with pesticides or chemicals, thus allowing the coffee to gain organic certification and increasing its value on the environmentally-conscious Western markets. By the end of 2000 all of Timor-Leste's 19 coffee-producing subdistricts had been declared organic zones. Some estimates suggested that the value of the country's coffee exports could rise to US $50m. per year in the near future, equivalent to almost one-half of its estimated GDP (of $119m.) in 1999.

Apart from coffee, other important cash crops produced in Timor-Leste include cloves, coconuts and cocoa. These were developed largely during the era of Indonesian rule in an effort to diversify the territory's economic base away from its heavy dependence on coffee, which suffered sharp price fluctuations in the 1980s and early 1990s. Another cash crop developed under Indonesian rule was sugar cane, which is currently cultivated mainly by smallholders over an area of some 10,000 ha. Small quantities of candlenut, coconut, Manila hemp and vanilla are also grown. The National Development Plan provided for additional emphasis to be given to the cultivation of such cash crops as a source of employment and income creation in rural areas and of export earnings for the whole nation.

Animal Husbandry and Fisheries

Animal husbandry has traditionally played a significant role in the agricultural economy of Timor-Leste, but was particularly badly affected by the political unrest of the mid-1970s. Concerted efforts made during the period of Indonesian rule to rebuild herds had a considerable impact, however, with official statistics indicating that the total number of cattle in the territory more than doubled in the four years between 1978 and 1982. By the late 1980s the number of cattle in East Timor was officially estimated at approximately 63,600, with the number rising further to some 72,200 by the mid-1990s. The animal husbandry sector also suffered serious dislocations as a result of the violence that followed the independence ballot. UNTAET began to rehabilitate the sector through the restitution of livestock to poor farmers and the introduction of a wide-ranging livestock vaccination campaign in 2000. By mid-2003 this campaign had resulted in the vaccination of more than 100,000 cattle and buffalo and nearly 250,000 pigs, increasing the vaccination rates for these animals to 78% and 97%, respectively. In addition, more than 3,000 cattle and 71,000 chickens had been distributed to farmers and the poor.

The waters around Timor-Leste have rich fisheries resources, which were largely neglected during Portuguese rule. Efforts to develop a viable commercial fishing industry during the period of Indonesian administration had a significant impact, however, and by the mid-1990s the territory produced more than 2m. metric tons of saltwater fish and 400 tons of freshwater fish per year. In 2000 the further development of the fisheries industry was widely perceived to represent one of the most promising sources of potential additional revenue. Draft strategic guidelines for the rehabilitation and development of the aquaculture and seafood industries were thus formulated in that year under the UNTAET administration, and subsistence fishing had recovered by 2002 with the assistance of a number of programmes administered by NGOs and the UN. The prospects for a commercialization of Timor-Leste's fisheries were also being examined through a definition of the country's maritime boundaries with Indonesia and Australia, increased research into its offshore fish resources, training of industry personnel and the development of appropriate marketing networks. As a result of these measures, the National Development Plan proposed a 50% overall increase in production and a 25% increase in industry employment within five years. A major step in this direction was taken in mid-2003 with the completion of an extensive project for the rehabilitation of the Hera fisheries port about 16 km to the east of Dili. Over five years, it is estimated the port will yield an extra 1,360 tons of catch, providing the population of Timor-Leste with more than one-half of the recommended daily consumption of fish per head.

Forestry

Although Timor-Leste was officially reported to have a forested area of 170,484 ha in 1994, the forestry sector suffered from unsustainable land management and the clearing of land for agricultural use and fuel wood in the 1980s and 1990s, which has resulted in serious deforestation and associated soil degradation and erosion. Its principal products are such local timbers as cendana, eucalyptus, redwood and lontar palm. Sandalwood, which once covered many of the country's mountains and hills, is viewed as another potentially important forestry crop, but is in need of extensive replanting. By 2000 this replanting had already been initiated, and an effort is being made to undertake a comprehensive inventory of the country's forestry resources and an analysis of their potential for development. In order to ensure the sustainable exploitation of these resources, the National Development Plan proposed the formulation of a Forestry Master Management Plan, and a number of specific action plans based on that master plan.

MINING

Onshore Resources

Timor-Leste's main onshore mineral resource is high-grade, high-quality marble. Exploitation of this resource had already begun in the 1980s and 1990s, when marble was employed in the construction of many hotels, office buildings and shopping malls in the Indonesian capital, Jakarta. In addition, the country is also believed to have some reserves of silver and manganese.

Offshore Oil and Gas

The most notable mineral resources of Timor-Leste are petroleum and natural gas, which are located mainly in the Timor Sea to the south of the country in the so-called 'Timor Gap' area between Timor-Leste and Australia. In order to be able to proceed with the development of these resources, and despite the uncertainty surrounding the international status of these waters, the Governments of Australia and Indonesia signed a controversial petroleum production revenue-sharing agreement known as the Timor Gap Treaty in 1989. Under Australian pressure, this agreement implicitly demarcated the maritime border between the then Indonesian province of East Timor and Australia not along the median line between the two countries' coastlines, as provided for by the UN Convention on the Law of the Sea, but at the southern edge of the Timor Trough, an undersea trench lying considerably closer to Timor than to Australia, thereby giving Australia control over the larger share of the offshore petroleum and gas resources in that area. Operationally, the agreement provided for the establishment of a 'zone of co-operation' in the waters between East Timor and Australia, which was divided into three areas, and decreed that each country had the right to 90% of the petroleum revenues derived from the areas nearest to its own shores, with the other country having the right to the remaining 10%; in addition, both countries were to share equally the revenues derived from the central area known as 'zone of co-operation A' (ZOCA).

The signing of this agreement resulted in exploration and production concessions being awarded to a number of oil companies. Several of the fields covered by these concessions—the Elang, Kakatua and Kakatua North fields in the south-western corner of ZOCA—have already entered into production, with the first petroleum being extracted in July 1998. However, the royalties generated by these fields for the contracting states are still relatively small, at some US $2m.– $3m. Two much larger fields, Bayu-Undan, lying 18 km to the south of the existing fields, and Greater Sunrise, lying partially within the north-eastern corner of ZOCA, remain to be developed. Bayu-Undan has estimated reserves of about 400m. barrels of natural gas condensate and other liquid hydrocarbons, and has the potential to generate tens of millions of US dollars in revenue, while Greater Sunrise has proved and probably recoverable reserves of some 8,400,000m. cu ft of natural gas and 320m. barrels of condensate.

Following the withdrawal of Indonesia from East Timor, the Indonesian Government acknowledged in October 1999 that it no longer retained any jurisdiction over the Timor Gap area. Negotiations were subsequently initiated involving UNTAET, Australia and the East Timorese national leadership on the amendment of the Timor Gap Treaty. Following extensive negotiations, provisional arrangements for the sharing of the hydrocarbon resources in the maritime border area between Australia and East Timor were concluded in July 2001, when a new Timor Sea Treaty was initialled by the Australian Government, UNTAET and a representative of the interim Government of East Timor. This agreement covers an area of 75,000 sq km between Timor-Leste and Australia, which is now known as the Joint Petroleum Development Area (JPDA). Although the agreement ostensibly assigns 90% of the revenue from the area to Timor-Leste and only 10% to Australia, it is widely acknowledged to be far less favourable to Timor-Leste than these figures suggest, owing to a number of complex issues relating to the measurement of petroleum production, the definition of by-products, the taxation of the petroleum production and shipping activities, and the proposed siting of the Timor Sea natural gas processing industry in Darwin rather than in Timor-Leste.

As this new agreement was negotiated by UNTAET on behalf of Timor-Leste, it needed to be ratified by the country's post-independence Government. After tough negotiations with the Australian Government in the run-up to independence, the Second Transitional Government of East Timor, established after the elections of August 2001, agreed to accept the Timor Sea Treaty, which was duly ratified shortly after the country gained sovereign independence in May 2002. In ratifying the Treaty, however, the Government made it clear that it had done so without prejudice to its maritime claims, and in July 2002 both the National Parliament and the President of the newly independent country stressed their determination to seek a delineation of the maritime boundary between Timor-Leste and Australia midway between the shorelines of the two countries. This would place the known Elang, Kakatua, Kakatua North, Bayu-Undan, Laminaria-Corallina and Greater Sunrise petroleum and natural gas fields firmly within the territorial waters of Timor-Leste, as they are located much closer to its shores than to those of Australia. In the event, however, the Government of Timor-Leste was forced to accept the original Treaty in March 2003, following prolonged and acrimonious negotiations with the Australian Government, and made little headway in subsequent negotiating rounds in November 2003 and April 2004. In particular, Australia has refused to accept the concept of a midway boundary, having earlier withdrawn from the International Convention on the Law of the Sea and from the jurisdiction of the International Court of Justice (ICJ) on maritime boundary issues. As a result, Timor-Leste stood to lose control over four-fifths of the Greater Sunrise field to Australia, which, according to some estimates, could result in a loss of some US $26,500m. in revenues over the duration of the field's productive life. Not surprisingly, the uncompromising stance of the Australian Government has generated great bitterness at all levels in Timor-Leste.

MANUFACTURING

No manufacturing base of any significance was established in East Timor during the periods of either Portuguese or Indonesian rule, and any enterprises that had been established were destroyed in the aftermath of the independence plebiscite. Thus manufacturing accounted for only 2.8% of GDP in 2002, with the bulk of the existing output coming from the weaving of traditional cloth and the production of furniture in relatively small enterprises. A study by the UN noted further that the immediate potential for manufacturing was limited, given the country's shortage of skilled labour, relatively high local living costs and wages, and poor transport links. The report therefore concluded that the nation's best options appeared to be in foreign investment in textiles and footwear.

INFRASTRUCTURE

Transport and Communications

The transport and communications network established during the period of Portuguese rule was very limited, and was intended mainly to enable the colonial administration to achieve its tax-collection objectives and to facilitate the transport of the territory's coffee crop to external markets. The upgrading and expansion of these facilities became a major focus of government policy during the subsequent period of Indonesian administration, and considerable progress was achieved in this regard. Particularly significant results were recorded in the development of land transport services. Whereas only one paved road, 20 km in length and located in the capital, Dili, was built in East Timor during the era of Portuguese rule, by the mid-1990s the Indonesian Government had built more than 3,800 km of roads, including 428 km of paved highways, as well as 18 bridges. This achievement was accompanied by a sharp increase in the number of motor vehicles registered, as well as by the establishment of a number of bus routes linking East Timor's towns and villages. Progress was also made by the Indonesian Government in the development of East Timor's air transport infrastructure. In 1981 the construction of the Komoro airport in Dili was completed to complement the airport built at Baucau in the east of the territory. During Indonesian rule these airports were served regularly by the two state-owned Indonesian airlines, PT Garuda Indonesia and PT Merpati Nusantara Airlines. East Timor was also connected to the rest of the Indonesian archipelago by regular scheduled passenger and cargo ship services.

East Timor's telecommunications system was also improved substantially under the Indonesian administration. Following the launching of the Palapa series of Indonesian telecommunications satellites from the mid-1970s, East Timor became linked through telephone and television circuits to all other Indonesian provinces and to the world beyond. To increase the efficiency of the telecommunications system, a new automatic telephone switching system was installed in Dili.

The transport and communications infrastructure was very badly damaged by the events of 1999. The road system fell into particularly serious disrepair, as the result of a combination of the impact of the post-referendum violence, the deceleration in repair and maintenance activities caused by the Indonesian economic crisis of 1997–98 and the subsequent wear imposed on the network by the heavy military vehicles of Interfet and the UNTAET peace-keeping force. Similarly, East Timor's ports and airports also suffered serious damage as a result of inadequate maintenance, the destruction of equipment and excessively heavy use in 1999–2000. High priority was therefore given to the restoration of the transport and communications infrastructure.

UNTAET launched a wide-ranging emergency rehabilitation programme soon after it assumed administrative control, and subsequently initiated a longer-term programme for the sustainable development of the sector. By the end of 2000 both the airport and the seaport of Dili had become operational, and significant progress had been made in road repairs and maintenance, especially with regard to the major arterial roads needed to ensure access to all of Timor-Leste's important population centres. With these objectives achieved, UNTAET established comprehensive business plans for the aviation, ports and roads sectors by mid-2001, and also began the reconstruction of public buildings. Increasingly, moreover, it moved its rehabilitation efforts beyond the capital, Dili. Despite the progress made by UNTAET, it was widely acknowledged that much remained to be done, and the National Development Plan drafted by the new Government placed high priority on the restoration and/or establishment of required physical capabilities and public services. Special emphasis is being given in the current phase of infrastructure development to an improvement of the road network, port facilities, power generation and distribution, and water supply and sanitation.

SOCIAL DEVELOPMENT

Health

The healthcare situation in East Timor prior to the violence that followed the referendum in 1999 was poor, with key indicators of the population's overall state of health, such as life expectancy and infant mortality rates, showing that the territory lagged far behind Indonesia. At the time there were eight small district hospitals in the territory and one central referral hospital in Dili. The sector was managed by the Provincial Health Department, and the total health sector work-force numbered around 5,000 (including staff employed by the East Timor Provincial Health Department, staff directly employed by the Ministry of Health in Jakarta, and non-technical and administrative staff). In addition to the government-administered system, a loose network of 31 healthcare centres also existed, managed by various religious groups; these clinics provided basic curative care and some preventative services. There were also several private clinics and some Indonesian military health facilities.

Extensive destruction of health infrastructure by pro-Indonesia militias in 1999 caused the total breakdown of the healthcare system. The situation was exacerbated by the complete loss of all healthcare equipment and medication and by the departure of senior health staff from the central, district and subdistrict levels of the healthcare system as a result of the violence, with a reported 130 of the territory's 160 doctors having left East Timor. Following Indonesia's withdrawal from East Timor in September 1999, immediate humanitarian assistance for health was provided by the International Committee of the Red Cross and various NGOs, supported by several UN agencies and, initially, some Interfet medical staff. Significant efforts were subsequently made to repair the damage inflicted during the period of post-referendum violence, with NGOs playing a particularly important role in this process. By June 2000 some 80 new health facilities had been established, supplemented by a number of mobile clinics. In addition, an operating Interim Health Authority was established in March 2000, while a training programme was initiated for district health officers and health laboratory services were also developed.

To speed up the recovery of the health services, a wide-ranging Health Sector Rehabilitation and Development Programme was launched in June 2000 by UNTAET, with financial support from the World Bank and a number of bilateral donors. This programme was expected to cost some US $38m. over a three-year period. It aimed to restore access to basic health services in the transitional period between the provision of humanitarian relief and the development of the health system, to begin the development of an effective health policy and health system, and to build local administrative capacities to implement and manage the health programme.

By mid-2002 almost 90 community health centres had been established to provide basic services throughout the country, and a proposal had been submitted to the Cabinet for the rehabilitation/reconstruction of Dili National Hospital and five regional hospitals. This was accompanied by the establishment of a new medical warehouse to supply medicines and other medical consumables to all districts of East Timor in February 2002. As the construction of community health centres continued into 2002–03, moreover, the World Bank reported that, by mid-2003, 85% of the country's children under five years of age had been vaccinated, and that 90% of the population lived within two hours' walking distance of a hospital or clinic. This was followed in mid-2004 by an announcement by the UN Children's Fund (UNICEF) that immunization coverage had increased by more than one-half, to 60%, and Vitamin A supplements were by that time being provided to 99% of children under five years of age. The UNICEF report also noted that mobile registration teams, which had already registered more than 17,000 children, were expanding across the country. The National Development Plan continued to allocate high priority to the health sector, with a special focus on primary health care, preventive medicine and the prevention of health hazards, and on reproductive health. This focus on primary health care appeared entirely justified in view of data released by UNICEF in the above-mentioned report, which suggested that Timor-Leste had the highest infant and maternal mortality rates in the region (88 per 1,000 live births and 830 per 100,000 live births, respectively) and a fertility rate of 7.4 children per household. Given the acute shortage of qualified doctors in Timor-Leste, the health system appeared certain to remain dependent upon external assistance for some time to come.

Education

Timor-Leste's education system had been developed to a certain extent under the Indonesian administration. Prior to the violence of 1999, there were 788 primary schools (including 140 private schools, most of which were administered by the Roman Catholic Church), 114 junior secondary schools, 37 senior secondary schools, and 17 vocational and technical schools. At the tertiary level, more than 1,500 East Timorese students were reported to have obtained university scholarships, mainly to the Indonesian universities of Malang, Jakarta and Denpasar. The University of East Timor was established in 1986. The rate of illiteracy in East Timor remained high during the period of Indonesian rule, however, at more than 50% in 1999, and the education provided by the system was often of poor quality, with technical education being particularly weak. Many teachers (the majority of whom, with the exception of teachers in the primary education sector, were non-East Timorese) were inadequately trained and poorly paid, and basic teaching equipment was also in short supply.

The education system was devastated by the extensive post-referendum violence. Approximately 95% of schools and other educational institutions were destroyed, and a large percentage (perhaps 70%–80%) of senior administrative staff and secondary school teachers was lost from the education system. Their departure also reduced the literacy rate in the territory to a mere 43% in 1999. The rehabilitation of the education system (to include the repair of classrooms, the supply of basic teaching and learning resources, and the establishment of a programme of teacher training) was therefore a priority concern of UNTAET. Soon after assuming control, the transitional administration developed a comprehensive three-stage School System Revitalization Programme (SSRP) with World Bank support to build upon the voluntarist measures already taken by local communities to revive the school system. The first stage of this programme, comprising a US $13.9m. Emergency School Readiness Project (ESRP), was launched in June 2000. It resulted in the rehabilitation of 2,780 classrooms by mid-2003, together with the necessary furniture, books and other instructional materials, the building of new schools to replace those too badly damaged to be repaired, and measures to improve the quality of instruction. Subsequent stages of the SSRP would provide for further improvements to the school system in order to enhance the quality of education and attract the increased participation of parents, teachers and the broader community in the school improvement process.

At the tertiary level, the university was reopened in January 2001, with provisions being made for the enrolment of 4,500 students in degree programmes and 3,000 students in bridging courses. In addition, scholarships were provided to some 500 students in 2000 to attend courses at foreign universities. These activities were supplemented by various training programmes for civil servants, administrators, teachers and other professionals. With Portuguese having been restored, along with the local Tetum, as one of the country's two official languages, these programmes also included substantial Portuguese-language training components, funded primarily by Portugal.

Following independence, UNMISET helped the newly established Ministry of Education, Culture, Youth and Sport to formulate an educational policy framework and a legislative foundation for the education system, including the drafting of appropriate educational laws and regulations. UNMISET also assisted in updating educational plans, preparing project proposals and establishing monitoring systems.

FINANCE

Foreign Aid and Trade

The devastation caused by the post-referendum violence left East Timor heavily dependent on external financial assistance for its economic and social reconstruction. The international community responded vigorously to these needs; a first high-level meeting between the East Timorese national leadership and donor representatives was held in mid-December 1999 in Tokyo. This meeting was followed by seven others—in Lisbon, Portugal, in June 2000, Brussels, Belgium, in December 2000, Dili in March 2001, Canberra, Australia, in June 2001, Oslo, Norway, in December 2001 and Dili in May 2002 and June 2003—as well as a 'virtual' meeting in December 2002. While underlining the donors' commitment to a co-ordinated effort to support the recovery and sustainable development of Timor-Leste, these meetings were intended to monitor the performance of the technical co-operation activities financed by the foreign assistance, and to make appropriate adjustments where necessary.

At the first meeting in December 1999, the donors pledged US $523.2m. for a major three-year relief and reconstruction programme for East Timor (see above). These pledges were subsequently supplemented by a variety of other bilateral and multilateral commitments. Although the disbursement of these funds was initially somewhat slow as a result of the inevitable delays in formulating specific projects for their expenditure, the rate of disbursement began to accelerate significantly during the latter part of UNTAET's administration, when the availability of these foreign funds played a pre-eminent role in driving East Timor's economic recovery. The degree of the country's resulting dependence on external financial support was highlighted by the fact that of the estimated GDP per head, $431 in 2002, some $400 had been contributed by foreign donors. Moreover, most of these foreign funds were made available in the form of grants rather than loans, in order to avoid burdening the newly independent country, one of the poorest in the world, with significant levels of foreign debt.

To maximize the developmental impact of the externally provided resources, the donor community agreed to co-ordinate their disbursement largely through two trust funds, the Consolidated Fund for East Timor (CFET), administered by UNTAET, and the Trust Fund for East Timor (TFET), administered by the World Bank in partnership with the Asian Development Bank (ADB). The former was intended principally to fund the administrative costs of government and to strengthen the capacity of the Timorese administration-in-waiting, while the latter was intended to finance economic reconstruction and development. Both were scheduled to be phased out by 2003 or 2004, upon completion of the post-crisis rehabilitation and recovery tasks for which they were established.

By mid-2003 12 donor countries and the European Union (EU) had pledged more than US $168m. to these two funds, of which $142m. had been placed in the TFET. The principal activities undertaken under the TFET included: the community empowerment and local governance project; the emergency infrastructure rehabilitation project; the small enterprise project; the health sector rehabilitation and development programme; the emergency school rehabilitation project and a follow-on project to improve the quality of education; the agriculture rehabilitation project; the water supply and sanitation project; and the microfinance project.

Together, these activities have helped to restore the country's physical infrastructure and social welfare system, as well as helping to promote entrepreneurship and support a recovery in the livelihoods of the country's citizens. Given Timor-Leste's limited capacity for generating domestic revenues in the short term, however, the country was expected to remain dependent on external financial support for at least a few more years until its hydrocarbon resources began to generate sufficient revenues.

In recognition of Timor-Leste's short-term funding constraints, the May 2002 donors' meeting in Dili agreed to provide further financial assistance in support of the National Development Plan. This resulted in the establishment of a Transitional Support Programme (TSP) by the World Bank in July 2002, which provided a grant of US $5m. and served as an anchor for a larger multi-donor support package, with a value of $31m. for 2003. The TSP had four components, covering: the establishment of a framework for poverty reduction planning; the creation of an institutional and legal framework for open democratic governance, and an enabling environment for the private sector; measures to ensure the orientation of public spending towards the poor and a strengthening of expenditure management controls; and an improvement of cost recovery in the power sector.

Timor-Leste's international trade performance since 1999 has reflected its developmental trends. The reconstruction efforts after 1999 prompted a sharp rise in imports of goods and services, which increased rapidly, reaching US $261m. by 2001. The slowdown in economic activity resulted in a significant decline in imports, to only $239m., in 2002, and a further contraction, to $174m., in 2003. Some 97% of these imports were funded by official transfers, with export earnings amounting to a mere $4m.–$6m. per year in 2000–02, before rising modestly, to $7m., in 2003. Almost one-half of these export earnings were accounted for by coffee, the price of which continued to be depressed during this period.

Fiscal Policy

As a relatively poor agricultural economy, dependent on coffee as its only significant export commodity, Timor-Leste has historically been unable to cover its financial requirements. During the period of Indonesian rule it was dependent on external transfers for approximately 85% of its recurrent and capital expenditure. This dependence on external funding was greatly increased following the Indonesian withdrawal as a result of the collapse in domestic revenue-generating capacity and the sharp rise in capital spending required to rebuild the devastated infrastructure. UNTAET therefore placed considerable emphasis on establishing a framework for sustainable public finances and developing the associated financial management systems and human resources. By mid-2000 a Central Fiscal Authority had already been established as the precursor to a Ministry of Finance, which was given the responsibility of developing local capacities in the area of tax administration.

The first consolidated budget for East Timor, covering the 2000/01 fiscal year (from 1 July 2000), was presented at the second donors' meeting in Lisbon in June 2000. As subsequently revised in November 2000, this budget proposed a total expenditure of US $60.8m. and domestic revenues of $25.1m. The resulting cash deficit of $35.7m. was expected to be fully covered by grants pledged by external donors. The introduction of this budget was accompanied by the adoption of a series of measures relating to tax and user charges intended to generate the required domestic revenues, the achievement of which was also supported by a significant increase in oil revenues from the Timor Gap area. At the same time, an institutional framework was created for the responsible fiscal management of the economy. The task of the execution of the budget, including expenditure control and monitoring, was assigned to a new Treasury agency within the Central Fiscal Authority, which was established to develop the foundations for sound and transparent fiscal management and subsequently evolved into a full Ministry of Finance. In addition, the 2000/01 budget provided for the establishment of an Economic Development Agency, to be responsible for business registration and the creation of an appropriate regulatory framework.

The budgets for 2001/02 and 2002/03, endorsed by the donors at the meetings held in Canberra in June 2001 and in Dili in May 2002, respectively, followed a broadly similar pattern to that of 2000/01 in that they projected significant deficits expected to be financed in large part from external sources. However, despite some subsequent revisions arising from a slower than expected growth of hydrocarbon revenues, the medium-term fiscal framework developed for 2003/04 to 2006/07 suggests a gradual reversal of this pattern. According to this plan, the large fiscal deficits of recent years were to be steadily reduced, and then give way to surpluses, once significant oil revenues begin to accrue from the development of the Bayu-Undan and, later, the Greater Sunrise natural gas and condensate fields. Preliminary estimates suggest that, even on the basis of the current revenue-sharing agreements with Australia, the Timor Sea reserves may yield up to US $3,000m. in petroleum revenue for Timor-Leste within 20 years. The prospect of such a hydrocarbon 'bonanza' in the foreseeable future has given rise to fears that it might destabilize the economy of Timor-Leste. Responding to these fears, the Government has indicated that it plans to save a significant proportion of its petroleum revenues, and to establish an 'oil fund' for this purpose, which will provide an insurance against unforeseen problems in the short term and a sustainable flow of income once the country's hydrocarbon resources have been depleted in the longer term.

Money and Banking

The devastation wreaked in the aftermath of the independence referendum also took its toll on the financial services sector in East Timor, which had suffered an almost total collapse as a result of the withdrawal of the Indonesian banks after September 1999. Although some progress has been made in reviving the banking system, it remains weak. In early 2001 UNTAET established a Central Payments Office, subsequently renamed the Banking and Payments Authority, as a precursor to a central bank charged with the responsibilities of currency management, the making of payments and receipts on behalf of the East Timorese administration, and bank licensing and supervision. In addition to the Banking and Payments Authority, which acts as a *de facto* central bank, there were only two financial institutions in operation in Timor-Leste in mid-2003—the Australia and New Zealand Banking Group and the Banco Nacional Ultramarino of Portugal—although this number was increased later that year, when the Indonesian PT Bank Mandiri commenced operations. The prospects for a rapid expansion of the banking sector remain uncertain, even though several other large foreign banks have expressed an interest in opening branches in Timor-Leste. In addition, there has been some local interest in the establishment of micro-financing institutions, most of which are being established under the auspices of a project funded under TFET and managed by the ADB. Both of the two existing institutions were established in 2000 and provide mainly foreign-exchange services, although one began to provide loans to the private sector in mid-2000 under a small enterprise development project funded by the World Bank. The available data suggested that although current-account deposits had already risen by the end of 2000 to a level approximately equivalent to those held in September 1999 by the Indonesian banks then operating in East Timor, very little commercial bank credit had been extended, with the exception of loans issued under the TFET-funded small enterprise development project, owing to lack of adequate collateral. According to data published by the Government in mid-2003, the volume of bank deposits had increased to US $57m. by February of that year, although domestic lending amounted to barely 10% of deposits, with most of the loans being used for housing. The majority of bank deposits continued to be invested abroad, owing to the lack of domestic collateral-based lending opportunities and of any enforcement mechanism.

The lack of a single universally accepted currency also posed a problem for an extended period of time. Although UNTAET chose the US dollar as East Timor's interim currency in January 2000, three other currencies—the Indonesian rupiah, the Portuguese escudo and the Australian dollar—remained in common use. The Indonesian rupiah retained its influence as the currency used by the population at large, although its use was affected by the demonetization of several high-value banknotes by the Indonesian central bank in August 2000, which resulted in the emergence of a steeply discounted secondary market for such notes in East Timor. The Australian dollar, meanwhile, continued to circulate mainly in Dili, where a relatively large expatriate Australian community had established itself. Despite its status as the official currency, the use of the US dollar by the local population was initially hampered by the slow execution of budgetary expenditures, the scarcity of low-denomination notes and coins, and the unfamiliarity of the currency. After mid-2001, however, the increased disbursement of TFET funds in particular resulted in an accelerated adoption of the currency, and by mid-2002 it was reported that the US dollar had begun to circulate widely throughout the country. In mid-2003, however, the Government reported that it planned to issue local coins of small denomination in order to 'consolidate the currency system' and 'augment the limited supply of US coins and eventually to supplant them'. In this context the Government also noted that its inability to make relative price and wage adjustments through monetary and exchange-rate policies as a result of the dollarization of the economy had made it difficult for Timor-Leste to compete with its neighbours under the prevailing wage and price regimes.

ECONOMIC PROSPECTS

Timor-Leste's relatively strong economic performance in 2000 and 2001, supported by both the externally funded economic rehabilitation measures and the presence of a large and relatively wealthy expatriate community, was unlikely to be sustained in the short term, as the international donor community phased out its activities following the country's accession to independence in 2002. Most observers consequently anticipated a deceleration in the rate of economic growth for at least the next three to four years until oil revenues from the Timor Sea became available, with estimates published by the ADB suggesting a contraction of GDP by 1.1% in 2002 and by a further 3.0% in 2003. In its submission to the donors' meeting in June 2003, the Government of Timor-Leste projected a decline of up to 5% in the 2002/03 fiscal year, and of a further 2% in 2003/04 before the onset of a slight recovery in 2004/05, with growth of 1% forecast. Recently released data for 2003 indicate that a contraction of up to 3% in real GDP did in fact take place in that year.

In the mean time the country was expected to remain heavily dependent on international support to finance its still substantial developmental needs in almost all social and economic fields. The weakening of the resource constraints resulting from the availability of substantial petroleum revenues by around 2006 will permit some acceleration of the development process, although this will continue to be restrained by insufficient technical and institutional capacities. Over time, however, these revenues will enable Timor-Leste to overcome many of its current weaknesses and emerge as a prosperous hydrocarbon-based economy.

Statistical Survey

Sources (unless otherwise stated): Indonesian Central Bureau of Statistics, Jalan Dr Sutomo 8, Jakarta 10710, Indonesia; tel. (21) 363360; fax (21) 3857046; internet www.bps.go.id; IMF, *East Timor: Establishing the Foundations of Sound Macroeconomic Management*.

AREA AND POPULATION

Area: 14,609 sq km (5,641 sq miles).

Population: 555,350 at census of 31 October 1980; 747,750 (males 386,939, females 360,811) at census of 31 October 1990; *2001* (UN registration): 737,811.

Density (2001): 50.5 per sq km.

Principal Towns (population in 2000): Dili (capital) 48,200; Dare 17,100; Baucau 14,200; Maliana 12,300; Ermera 12,000. Source: Stefan Helders, *World Gazetteer* (internet www.world-gazetteer.com). *Mid-2003* (UN estimate, incl. suburbs): Dili 48,731 (Source: UN, *World Urbanization Prospects: The 2003 Revision*).

Births and Deaths (UN estimates, annual averages): Birth rate per 1,000: 43.3 in 1985–90; 36.9 in 1990–95; 28.1 in 1995–2000. Death rate per 1,000: 20.4 in 1985–90; 17.5 in 1990–95; 14.6 in 1995–2000. Source: WHO, *World Health Report*.

Expectation of Life (WHO estimates, years at birth): 57.5 (males 54.8; females 60.5) in 2002. Source: WHO, *World Health Report*.

Economically Active Population (survey, persons aged 10 years and over, August 1993): Total employed 336,490; Unemployed 5,397; *Total labour force* 341,887. *Mid-2002* (estimates, '000): Agriculture, etc. 318; Total labour force 391 (Source: FAO).

HEALTH AND WELFARE

Key Indicators

Total Fertility Rate (children per woman, 2002): 3.9.

Under-5 Mortality Rate (per 1,000 live births, 2002): 126.

Health Expenditure (2001): % of GDP: 9.8.

Health Expenditure (2001): public (% of total): 9.0.

Human Development Index (2002): ranking: 158.

Human Development Index (2002): value: 0.436.

For sources and definitions, see explanatory note on p. vi.

AGRICULTURE, ETC.

Principal Crops ('000 metric tons, 2002): Rice (paddy) 54; Maize 94; Cassava (Manioc) 56*; Sweet potatoes 13*; Groundnuts (in shell) 4*; Cabbages 1*; Other vegetables 1*; Bananas 1*; Other fruit 1*. Source: FAO.

* FAO estimate.

Livestock (FAO estimates, '000 head, 2002): Cattle 165; Sheep 22; Goats 50; Pigs 345; Horses 45; Buffaloes 70; Chickens 2,100. Source: FAO.

Livestock Products (FAO estimates, '000 metric tons, 2002): Pig meat 10.0; Beef and buffalo meat 1.6; Hen eggs 5.6. Source: FAO.

Fishing (metric tons): Total catch 362 in 2000; 356 in 2001; 350 (FAO estimate) in 2002. Source: FAO.

FINANCE

Currency and Exchange Rate: United States currency is used: 100 cents = 1 US dollar ($). *Sterling and Euro Equivalents* (31 May 2004): £1 sterling = US $1.8347; €1 = US $1.2246; US $100 = £54.50 = €81.66.

Budget (estimates, US $ million, year ending 30 June 2003): *Revenue*: Domestic revenue 17.4 (Direct taxes 4.8; Indirect taxes 10.7; Non-tax revenue 1.9); Timor Sea revenue 26.7 (Tax revenues 24.6; Royalties and interest 2.1); Total 44.1 (excl. Grants 33.0). *Expenditure:* Recurrent expenditure 58.2 (Salaries and wages 22.5; Goods and services 35.7); Capital expenditure 12.4; Total 70.6. Source: mostly IMF, *Timor-Leste: Selected Issues and Statistical Appendix* (July 2003).

Money Supply (estimate, US $ million, 31 December 2002): Demand deposits 40.0; Savings deposits 10.2; Time deposits 5.3; Total broad money 55.5. Source: IMF, *Timor-Leste: Selected Issues and Statistical Appendix* (July 2003).

Cost of Living (Consumer Price Index; base: year ending April 2000 = 100): 109.6 in 2000; 113.5 in 2001; 113.1 in 2002. Source: IMF, *Timor-Leste: Selected Issues and Statistical Appendix* (July 2003).

Gross National Product (US $ million at current prices): 327.7 in 2000; 397.1 in 2001; 394.2 in 2002. Source: IMF, *Timor-Leste: Selected Issues and Statistical Appendix* (July 2003).

Gross Domestic Product by Economic Activity (US $ million at current prices, 2002): Agriculture, forestry and fishing 102.6; Mining and quarrying 3.5; Manufacturing 10.4; Electricity, gas and water 3.1; Construction 57.7; Trade, restaurants and hotels 34.0; Transport and communications 30.7; Finance, rents and business services 26.1; Public administration and defence 107.3; Private services 2.6; *GDP in purchasers' values* 378.0. Source: IMF, *Timor-Leste: Selected Issues and Statistical Appendix* (July 2003).

Balance of Payments (estimates, US $ million, 2002): Exports of goods 6; Imports of goods −186; *Trade balance* −180; Services and other income (net) −50; *Balance on goods, services and income* −230; Current transfers 267; *Current balance* 37; Capital transfers 83; Other capital flows (net) −78; Net errors and omissions −18; *Overall balance* 23. Source: IMF, *Timor-Leste: Selected Issues and Statistical Appendix* (July 2003).

EXTERNAL TRADE

Principal Commodities (US $ million, 1998): *Imports*: Rice 13; Other foodstuffs 22; Petroleum products 14; Construction materials 20; Total (incl. others) 135. *Exports*: Products of agriculture, forestry and fishing 51 (Food crops 8, Other crops 28, Livestock 12); Products of manufacturing 3; Total (incl. others) 55. *2003*: Total imports 222.0; Total exports 142.7 (Source: Asian Development Bank, *Key Indicators of Developing Asian and Pacific Countries)*.

Principal Trading Partner (US $ million, 1998): *Imports*: Indonesia 135. *Exports*: Indonesia 53.

Oil/gas receipts (US $ million, estimates): 6.5 in 2000; 11.9 in 2001; 16.1 in 2002. Source: IMF, *Timor-Leste: Selected Issues and Statistical Appendix* (July 2003).

TOURISM

Tourist Arrivals ('000 foreign visitors, excl. Indonesians, at hotels): 0.8 in 1996; 1.0 in 1997; 0.3 in 1998. Figures exclude arrivals at non-classified hotels: 204 in 1996; 245 in 1997; 41 in 1998.

EDUCATION

Adult Literacy Rate (UNESCO estimate): 58.6% in 2002. Source: UN Development Programme, *Human Development Report*.

Directory

The Constitution

The Constitution of the Democratic Republic of East Timor was promulgated by the Constituent Assembly on 22 March 2002 and became effective on 20 May 2002, when the nation formalized its independence. (From this date the country elected to be known by its official name, the Democratic Republic of Timor-Leste.) The main provisions of the Constitution are summarized below:

FUNDAMENTAL PRINCIPLES

The Democratic Republic of East Timor is a democratic, sovereign, independent and unitary state. Its territory comprises the historically defined eastern part of Timor island, the enclave of Oecussi, the island of Ataúro and the islet of Jaco. Oecussi Ambeno and Ataúro shall receive special administrative and economic treatment.

The fundamental objectives of the State include the following: to safeguard national sovereignty; to guarantee fundamental rights and freedoms; to defend political democracy; to promote the building of a society based on social justice; and to guarantee the effective equality of opportunities between women and men.

Sovereignty is vested in the people. The people shall exercise the political power through universal, free, equal, direct, secret and periodic suffrage and through other forms stated in the Constitution.

In matters of international relations, the Democratic Republic of East Timor shall establish relations of friendship and co-operation with all other peoples. It shall maintain privileged ties with countries whose official language is Portuguese.

The State shall recognize and respect the different religious denominations, which are free in their organization. Tetum and Portuguese shall be the official languages.

FUNDAMENTAL RIGHTS, DUTIES, LIBERTIES AND GUARANTEES

All citizens are equal before the law and no one shall be discriminated against on grounds of colour, race, marital status, gender, ethnic origin, language, social or economic status, political or ideological convictions, religion, education and physical or mental condition. Women and men shall have the same rights and duties in family, political, economic, social and cultural life. Rights, freedoms and safeguards are upheld by the State and include the following: the right to life; to personal freedom, security and integrity; to habeas corpus; to the inviolability of the home and of correspondence; to freedom of expression and conscience; to freedom of movement, assembly and association; and to participate in political life. Freedom of the press is guaranteed.

Rights and duties of citizens include the following: the right and the duty to work; the right to vote (at over 17 years of age); the right to petition; the right and duty to contribute towards the defence of sovereignty; the freedom to form trade unions and the right to strike; consumer rights; the right to private property; the duty to pay taxes; the right to health and medical care; the right to education and culture.

ORGANIZATION OF THE POLITICAL POWER

Political power lies with the people. The organs of sovereignty shall be the President of the Republic, the National Parliament, the Government and the Courts. They shall observe the principle of separation and interdependence of powers. There shall be free, direct, secret, personal and regular universal suffrage. No one shall hold political office for life.

PRESIDENT OF THE REPUBLIC

The President of the Republic is the Head of State and the Supreme Commander of the Defence Force. The President symbolizes and guarantees national independence and unity and the effective functioning of democratic institutions. The President of the Republic shall be elected by universal, free, direct, secret and personal suffrage. The candidate who receives more than half of the valid votes shall be elected President. Candidates shall be original citizens of the Democratic Republic of East Timor, at least 35 years of age, in possession of his/her full faculties and have been proposed by a minimum of 5,000 voters. The President shall hold office for five years. The President may not be re-elected for a third consecutive term of office.

The duties of the President include the following: to preside over the Supreme Council of Defence and Security and the Council of State; to set dates for elections; to convene extraordinary sessions of the National Parliament; to dissolve the National Parliament; to promulgate laws; to exercise the functions of the Supreme Commander of the Defence Force; to veto laws; to appoint and dismiss the Prime Minister and other Government members; to apply to the Supreme Court of Justice; to submit relevant issues of national interest to a referendum; to declare a State of Emergency following the authorization of the National Parliament; to appoint and dismiss diplomatic representatives; to accredit foreign diplomatic representatives; to declare war and make peace with the prior approval of the National Parliament.

COUNCIL OF STATE

The Council of State is the political advisory body of the President of the Republic. It is presided over by the President of the Republic and comprises former Presidents of the Republic who were not removed from office, the Speaker of the National Parliament, the Prime Minister, five citizens elected by the National Parliament and five citizens nominated by the President of the Republic.

NATIONAL PARLIAMENT

The National Parliament represents all Timorese citizens, and shall have a minimum of 52 and a maximum of 65 members, elected by universal, free, direct, equal, secret and personal suffrage for a term of five years. The duties of the National Parliament include the following: to enact legislation; to confer legislative authority on the Government; to approve plans and the Budget and monitor their execution; to ratify international treaties and conventions; to approve revisions of the Constitution; to propose to the President of the Republic that issues of national interest be submitted to a referendum. The legislative term shall comprise five legislative sessions, and each legislative session shall have the duration of one year.

GOVERNMENT

The Government is the supreme organ of public administration and is responsible for the formulation and execution of general policy. It shall comprise the Prime Minister, the Ministers and the Secretaries of State, and may include one or more Deputy Prime Ministers and Deputy Ministers. The Council of Ministers shall comprise the Prime Minister, the Deputy Prime Ministers, if any, and the Ministers. It shall be convened and presided over by the Prime Minister. The Prime Minister shall be appointed by the President of the Republic. Other members of the Government shall be appointed by the President at the proposal of the Prime Minister. The Government shall be responsible to the President and the National Parliament. The Government's programme shall be submitted to the National Parliament for consideration within 30 days of the appointment of the Government.

JUDICIARY

The Courts are independent organs of sovereignty with competence to administer justice. There shall be the Supreme Court of Justice and other courts of law, the High Administrative, Tax and Audit Court, other administrative courts of first instance and military courts. There may also be maritime courts and courts of arbitration.

It is the duty of the Public Prosecutors to represent the State. The Office of the Prosecutor-General shall be the highest authority in public prosecution and shall be presided over by the Prosecutor-General, who is appointed and dismissed by the President of the Republic. The Prosecutor-General shall serve for a term of six years.

ECONOMIC AND FINANCIAL ORGANIZATION

The economic organization of East Timor shall be based on the co-existence of the public, private, co-operative and social sectors of ownership, and on the combination of community forms with free initiative and business management. The State shall promote national investment. The State Budget shall be prepared by the Government and approved by the National Parliament. Its execution shall be monitored by the High Administrative, Tax and Audit Court and by the National Parliament.

NATIONAL DEFENCE AND SECURITY

The East Timor defence force—Falintil-ETDF—is composed exclusively of national citizens and shall be responsible for the provision of military defence to the Democratic Republic of East Timor. There shall be a single system of organization for the whole national territory. Falintil-ETDF shall act as a guarantor of national independence, territorial integrity and the freedom and security of the population against any external threat or aggression. The police shall guarantee the internal security of the citizens.

The Superior Council for Defence and Security is the consultative organ of the President of the Republic on matters relating to defence and security. It shall be presided over by the President of the Republic and shall include a higher number of civilian than military entities.

GUARANTEE AND REVISION OF THE CONSTITUTION

Declaration of unconstitutionality may be requested by: the President of the Republic; the Speaker of the National Parliament; the Prosecutor-General; the Prime Minister; one-fifth of the Members of the National Parliament; the Ombudsman.

Changes to the Constitution shall be approved by a majority of two-thirds of Members of Parliament and the President shall not refuse to promulgate a revision statute.

FINAL AND TRANSITIONAL PROVISIONS

Confirmation, accession and ratification of bilateral and multilateral conventions, treaties, agreements or alliances that took place before the Constitution entered into force shall be decided by the respective bodies concerned; the Democratic Republic of East Timor shall not be bound by any treaty, agreement or alliance not thus ratified. Any acts or contracts concerning natural resources entered into prior to the entry into force of the Constitution and not subsequently confirmed by the competent bodies shall not be recognized.

Indonesian and English shall be working languages, together with the official languages, for as long as is deemed necessary.

Acts committed between 25 April 1974 and 31 December 1999 that can be considered to be crimes of humanity, of genocide or of war shall be liable to criminal proceedings within the national or international courts.

The Government

Until 20 May 2002 all legislative and executive authority in East Timor was exercised by the UN Transitional Administration in East Timor (UNTAET). Under its guidance the Constituent Assembly promulgated a new Constitution on 22 March 2002 and the people of East Timor elected their first President on 14 April. Having satisfied both these preconditions East Timor was formally granted independence on 20 May, at which time the Constituent Assembly transformed itself into East Timor's first Parliament and the Government was formally inaugurated by the President. The Government was largely composed of the same Cabinet members who had constituted the pre-independence Council of Ministers.

Upon independence UNTAET was succeeded by the UN Mission of Support in East Timor (UNMISET), which was to remain in the country for an initial period of one year. UNMISET's mandate comprised the following elements: to provide support to core administrative structures critical to the political stability and viability of the new country; to assist in interim law enforcement and public security; to aid in the development of a new law enforcement agency—the East Timor Police Force (ETPF); and to contribute to the maintenance of internal and external security. Downsizing of the mission was to take place as rapidly as possible and it was intended that, over a period of two years, all operational responsibilities would be fully devolved to the East Timorese authorities. In May 2003 the UN Security Council extended the mandate of UNMISET for one further year and, in May 2004, for an additional year.

HEAD OF STATE

President: JOSÉ ALEXANDRE (XANANA) GUSMÃO (took office 20 May 2002).

CABINET
(September 2004)

Prime Minister and Minister for Development and the Environment: MARI BIN AMUDE ALKATIRI.

Deputy Prime Minister: ANA MARIA PESSÔA PEREIRA DA SILVA PINTO.

Senior Minister for Foreign Affairs and Co-operation: JOSÉ RAMOS HORTA.

Minister for Justice: DOMINGOS SARMENTO.

Minister for Planning and Finance: MADALENA BRITES BOAVIDA.

Minister for Internal Administration: ROGÉRIO TIAGO LOBATO.

Minister for Health: RUI MARIA DE ARAÚJO.

Minister for Transport, Communications and Public Works: OVÍDIO DE JESUS AMARAL.

Minister for Education, Culture, Youth and Sport: ARMINDO MAIA.

Minister for Agriculture, Forestry and Fisheries: ESTANISLAU ALEIXO DA SILVA.

Secretary of State for Defence: ROQUE RODRIGUES.

Secretary of State for Labour and Solidarity: ARSÊNIO PAIXÃO BANO.

Secretary of State for Commerce and Industry: ARLINDO RANGEL.

Secretary of State for Tourism, Environment and Investment and Acting Secretary of State for Mineral Resources and Power Policy: JOSÉ TEXEIRA.

Secretary of State for the Council of Ministers: GREGORIO JOSÉ DA CONCEIÇÃO FERREIRA DE SOUSA.

Secretary of State for Parliamentary Affairs: ANTONINHO BIANCO.

Secretary of State for Public Works: JOÃO BAPTISTA FERNANDES ALVES.

Secretary of State for Education, Culture, Youth and Sports: VIRGILIO SMITH.

Secretary of State for Water and Sanitation: EGÍDIO DE JESÚS.

There are, in addition, 10 Vice-Ministers.

MINISTRIES

Office of the President: Palácio das Cinzas, Kaikoli, Dili; tel. 3339011; e-mail presidente-tl@easttimor.minihub.org.

Office of the Prime Minister: Government Palace, Rua Av. Presidente Nicolau Lobato, Dili; tel. 7230087; e-mail mail@primeministerandcabinet.gov.tp; internet www.pm.gov.tp.

Ministry of Agriculture, Forestry and Fisheries: Dili; e-mail agriculture@gov.east-timor.org.

Ministry of Development and the Environment: Dili.

Ministry of Education, Culture, Youth and Sport: Dili; e-mail education@gov.east-timor.org.

Ministry of Foreign Affairs and Co-operation: Edif. GPA 1, Ground Floor, Rua Av. Presidente Nicolau Lobato, POB 6, Dili; tel. 3339600; fax 3339025; e-mail mnecdratl@yahoo.com; internet www.mfac.gov.tp.

Ministry of Health: Palácio das Reparticões, Edif. 3, Unit 4, POB 374, Dili; tel. 3322467; fax 3325189; e-mail health@gov.east-timor.org.

Ministry of the Interior: Dili.

Ministry of Justice: Av. Jacinto Candido, Dili; e-mail moj@moj.gov-rdtl.org; internet www.moj.gov-rdtl.org.

Ministry of Planning and Finance: Dili; e-mail finance@gov.east-timor.org.

Ministry of State Administration: Dili.

Ministry of Transport, Communications and Public Works: Dili; e-mail transport@gov.east-timor.org.

President and Legislature

PRESIDENT

Presidential Election, 14 April 2002

Candidate	Votes	%
José Alexandre Gusmão	301,634	82.69
Francisco Xavier do Amaral	63,146	17.31
Total	364,780	100.00

NATIONAL PARLIAMENT

A single-chamber Constituent Assembly was elected by popular vote on 30 August 2001. Its 88 members included 75 deputies elected under a national system of proportional representation and one representative from each of Timor-Leste's 13 districts, elected under a 'first-past-the-post' system. Upon independence on 20 May 2002 the Constituent Assembly became the National Parliament; it held its inaugural session on the same day.

Speaker: FRANCISCO GUTERRES.

General Election, 30 August 2001

	Seats
Frente Revolucionária do Timor Leste Independente (Fretilin)	55
Partido Democrático (PD)	7
Associação Social-Democrata Timorense (ASDT)	6
Partido Social Democrata (PSD)	6
Klibur Oan Timor Asuwain (KOTA)	2
Partido Democrata Cristão (PDC)	2
Partido Nacionalista Timorense (PNT)	2
Partido do Povo de Timor (PPT)	2
União Democrática Timorense (UDT)	2
Partido Democrata Cristão de Timor (UDC/PDC)	1
Partido Liberal (PL)	1
Partido Socialista de Timor (PST)	1
Independent	1
Total	88

Political Organizations

Associação Popular Democrática de Timor Pro Referendo (Apodeti Pro Referendo) (Pro Referendum Popular Democratic Association of Timor): c/o Frederico Almeida Santos Costa, CNRT Office, Balide, Dili; tel. 3324994; f. 1974 as Apodeti; adopted present name in August 2000; fmrly supported autonomous integration with Indonesia; Pres. FREDERICO ALMEIDA SANTOS COSTA.

Associação Social-Democrata Timorense (ASDT) (Timor Social Democratic Association): Av. Direitos Humanos Lecidere, Dili; tel. 3983331; f. 2001; Pres. FRANCISCO XAVIER DO AMARAL.

Barisan Rakyat Timor Timur (BRTT) (East Timor People's Front): fmrly supported autonomous integration with Indonesia; Pres. FRANCISCO LOPES DA CRUZ.

Conselho Popular pela Defesa da República Democrática de Timor Leste (CPD-RDTL) (Popular Council for the Defence of the Democratic Republic of East Timor): opp. the Church, Balide, Dili; tel. 3481462; f. 1999; promotes adoption of 1975 Constitution of Democratic Republic of East Timor; Spokesperson CRISTIANO DA COSTA.

Frente Revolucionária do Timor Leste Independente (Fretilin) (Revolutionary Front for an Independent East Timor): Rua dos Mártires da Pátria, Dili; tel. 3321409; internet www.geocities.com/SoHo/Study/4141/; f. 1974 to seek full independence for East Timor; entered into alliance with the UDT in 1986; Pres. FRANCISCO GUTERRES; Sec. for International Relations JOSÉ RAMOS HORTA.

Klibur Oan Timor Asuwain (KOTA) (Association of Timorese Heroes): Rua dos Mártires da Pátria, Fatuhada, Dili; tel. 3324661; e-mail clementinoamaral@hotmail.com; f. 1974 as pro-integration party; currently supports independence with Timorese traditions; Pres. CLEMENTINO DOS REIS AMARAL (acting).

Movement for the Reconciliation and Unity of the People of East Timor (MRUPT): f. 1997; Chair. MANUEL VIEGAS CARRASCALÃO.

Partai Liberal (PL) (Liberal Party): Talbessi Sentral, Dili; tel. 3786448; Pres. ARMANDO JOSÉ DOURADO DA SILVA.

Partido Democrata Cristão (PDC) (Christian Democrat Party): Former Escola Cartilha, Rua Quintal Kiik, Bairo Economico, Dili; tel. 3324683; e-mail arlindom@octa4.net.au; f. 2000; Pres. ANTÓNIO XIMENES.

Partido Democrático (PD) (Democratic Party): 1 Rua Democracia, Pantai Kelapa, Dili; tel. 3608421; e-mail flazama@hotmail.com; Pres. FERNANDO DE ARAÚJO.

Partido Democratik Maubere (PDM) (Maubere Democratic Party): Blk B II, 16 Surikmas Lama Kraik, Fatumeta, Dili; tel. 3184508; e-mail pdm_party@hotmail.com; f. 2000; Pres. PAOLO PINTO.

Partido Nacionalista Timorense (PNT) (Nationalist Party of Timor): Dili; tel. 3323518; Pres. Dr ABÍLIO ARAÚJO (acting).

Partido do Povo de Timor (PPT) (Timorese People's Party): Dili; tel. 3568325; f. 2000; pro-integration; supported candidacy of Xanana Gusmão for presidency of East Timor; Pres. Dr JACOB XAVIER.

Partido Republika National Timor Leste (PARENTIL) (National Republic Party of East Timor): Perumnar Bairopite Bob Madey Ran, Fahan Jalam, Ailobu Laran RTK; tel. 3361393; Pres. FLAVIANO PEREIRA LOPEZ.

Partido Social Democrata Timor Lorosae (PSD) (Social Democrat Party of East Timor): Apartado 312, Correios de Dili, Dili; tel. 3357027; e-mail psdtimor@hotmail.com; f. 2000; Pres. MÁRIO VIEGAS CARRASCALÃO.

Partido Socialista de Timor (PST) (Socialist Party of Timor): Rua Colegio das Madras, Balide, Dili; tel. 3560246; e-mail kaynaga@hotmail.com; Marxist-Leninist Fretilin splinter group; Pres. AVELINO DA SILVA.

Partido Trabalhista Timorense (PTT) (Timor Labour Party): 2B Rua Travessa de Befonte, 2 Bairro Formosa, Dili; tel. 3322807; f. 1974; Pres. PAULO FREITAS DA SILVA.

União Democrática Timorense (UDT) (Timorese Democratic Union): Palapagoa Rua da India, Dili; tel. 3881453; e-mail uchimov2001@yahoo.com; internet www.unitel.net/udttimor; f. 1974; allied itself with Fretilin in 1986; Pres. JOÃO CARRASCALÃO; Sec.-Gen. DOMINGOS OLIVEIRA.

União Democrata-Cristão de Timor (UDC/PDC) (Christian Democratic Union of Timor): 62 Rua Americo Thomaz, Mandarin, Dili; tel. 3325042; f. 1998; Pres. VINCENTE DA SILVA GUTERRES.

Diplomatic Representation

EMBASSIES IN TIMOR-LESTE

Australia: Av. dos Mártires da Pátria, Dili; tel. 3322111; fax 3323615; Ambassador MARGARET TWOMEY.

Brazil: Rua Governador Serpa Rosa, POB 157, Farol, Dili; tel. 3324203; fax 3324620; e-mail esctimor@office.net.au; Ambassador Dr KYWAL DE OLIVEIRA.

China, People's Republic: Av. Governador Serpa Rosa, Farol, Dili; tel. 3325163; fax 3325166; e-mail chinaemb_tp@mfa.gov.cn; Ambassador SHAO GUANFU.

Indonesia: Farol, Palapaco, POB 207, Dili; tel. 3317107; fax 3312332; e-mail kukridil@hotmail.com; Ambassador AHMED BEY SOFWAN.

Japan: Av. do Portugal, Pantai Kelapa Dili, POB 175, Dili; tel. 3323131; fax 3323130; e-mail japrepet@yahoo.co.jp; Ambassador HIDEAKI ASAHI.

Korea, Republic: Av. de Portugal, Motael, Dili; tel. 3321635; fax 3323636; e-mail koreadili@mofat.go.kr; Ambassador RYU JIN-KYU.

Malaysia: Rua Almirante Américo Thomás, Mandarin, Dili; tel. 3311141; fax 3321805; e-mail mwdili@timortelecom.tp; Ambassador ABDULLAH FAIZ BIN MOHD ZAIN.

Portugal: Av. Presidente Nicolau Lobato, Dili; tel. 3312155; fax 3312526; e-mail embaixada.portugal@embpor.tp; Ambassador RUI QUARTIM DOS SANTOS.

Thailand: Suite 355–357, Central Maritime Hotel, Av. dos Direitos Humanos, Dili; tel. 3311605; fax 3311607; e-mail diliemb@mfa.go.th; Ambassador KULKUMUT SINGHARA NA AYUDHAYA.

United Kingdom: Pantai Kelapa, Av. do Portugal, Dili; tel. 3312652; fax 3312652; e-mail dili.fco@gtnet.gov.uk; Ambassador TINA REDSHAW.

USA: Av. Dr Sérgio Vieira de Mello, Farol, Dili; tel. 3324684; fax 3313206; Ambassador GROVER JOSEPH REES.

Judicial System

Until independence was granted on 20 May 2002 all legislative and executive authority with respect to the administration of the judiciary in East Timor was vested in UNTAET. During the transitional period of administration a two-tier court structure was established, consisting of District Courts and a Court of Appeal. The Constitution, promulgated in March 2002, specified that Timor-Leste should have three categories of courts: the Supreme Court of Justice and other law courts; the High Administrative, Tax and Audit Court and other administrative courts of first instance; and military courts. The judiciary would be regulated by the Superior Council of the Judiciary, the function of which would be to oversee the judicial sector and, in particular, to control the appointment, promotion, discipline and dismissal of judges. The effectiveness of the newly established judicial system was severely impaired by Timor-Leste's lack of human and material resources. In July 2002 there were only 22 judges in Timor-Leste, none of whom possessed more than two years of legal experience. In July 2003 the Court of Appeal was reconstituted. However, in the same month a ruling by its President that Timor-Leste's law should be based on that of Portugal and not

Indonesia (as was currently the case) threatened to have serious consequences for the legal system.

Court of Appeal: Dili; Pres. CLAUDIO XIMENES.

Office of the Prosecutor-General: Dili; Prosecutor-General LONGUINHOS MONTEIRO; Dep. Prosecutor-General AMANDIO BENEVIDES.

Religion

In 2002 it was estimated that about 93.1% of the total population were Roman Catholic.

CHRISTIANITY

The Roman Catholic Church

Timor-Leste comprises the dioceses of Dili and Baucau, directly responsible to the Holy See. In December 2002 the country had an estimated 767,209 Roman Catholics.

Bishop of Baucau: Most Rev. BASILIO DO NASCIMENTO, Largo da Catedral, Baucau 88810; tel. 4121209; fax 4121380.

Bishop of Dili: Most Rev. ALBERTO RICARDO DA SILVA, Av. dos Direitos Humanos, Bidau Lecidere, CP 4, Dili 88010; tel. 3321177.

Protestant Church

Igreja Protestante iha Timor Lorosa'e: Jl. Raya Comoro, POB 1186, Dili 88110; tel. and fax 3323128; f. 1988 as Gereja Kristen Timor Timur (GKTT); adopted present name 2000; Moderator Rev. FRANCISCO DE VASCONCELOS; 30,000 mems.

The Press

The office and printing plant of the largest newspaper in East Timor under Indonesian rule, *Suara Timor Timur* (Voice of East Timor, circulation of 8,000), were destroyed prior to the referendum on independence in August 1999. The Constitution promulgated in March 2002 guarantees freedom of the press in Timor-Leste.

Lalenok (Mirror): Rua Gov. Celestino da Silva, Farol, Dili; tel. 3321607; e-mail lalenok@hotmail.com; f. 2000; publ. by Kamelin Media Group; Tetum; 3 a week; Dir-Gen. and Chief Editor VIRGÍLIO DA SILVA GUTERRES; Editor JOSÉ MARIA POMPELA; circ. 300.

Lian Maubere: Dili; f. 1999; weekly.

Suara Timor Lorosae: Dili; e-mail stl_redaksi@yahoo.com; internet www.suaratimorlorosae.com; daily.

The Official Gazette of East Timor: Dili; f. 1999 by UNTAET; forum for publication of all govt regulations and directives, acts of organs or institutions of East Timor and other acts of public interest requiring general notification; published in English, Portuguese and Tetum, with translations in Bahasa Indonesia available on request.

Tais Timor: Dili; f. 2000; fmrly published by the Office of Communication and Public Information (OCPI) of the UN; Tetum, English, Portuguese and Bahasa Indonesia; every fortnight; distributed free of charge; circ. 75,000.

Talit@kum: Dili; f. 2000; news magazine; publ. by Kdadlak Media Group; Bahasa Indonesia; monthly; Chief Editor HUGO FERNANDEZ.

Timor Post: Rua D. Aleixo Corte Real No. 6, Dili; f. 2000; managed by editors and staff of the former *Suara Timor Timur*; Bahasa Indonesia, Tetum, Portuguese and English; daily; Man. Editor OTELIO OTE; Chief Editor HUGO DE COSTA; circ. 600.

PRESS ASSOCIATION

Timor Lorosae Journalists Association (TLJA): Jalan Kaikoli, Dili; tel. 3324047; fax 3327505; e-mail timorjournal@eudoramail.com; f. 1999; Co-ordinator OTELIO OTE; Pres. VIRGÍLIO DA SILVA GUTERRES.

Broadcasting and Communications

TELECOMMUNICATIONS

Prior to the civil conflict in 1999, East Timor's telephone lines totalled about 12,000. Telecommunications transmission towers were reported to have suffered significant damage in the civil conflict. Indonesia withdrew its telecommunication services in September. A cellular telephone network service was subsequently provided by an Australian company, Telstra. In July 2002 the Government granted a consortium led by Portugal Telecom a 15-year concession permitting it to establish and operate Timor-Leste's telecommunications systems. Under the terms of the concession the consortium agreed to provide every district in Timor-Leste with telecommunications services at the most inexpensive tariffs viable within 15 months. At the expiry of the concession in 2017 the telecommunications system was to be transferred to government control. In February 2003 Timor-Leste Telecom assumed control of the country's cellular telephone network from Telstra.

Timor-Leste Telecom (TT): Sala No. 7, Hotel Timor, Av. dos Mártires da Pátria, Dili; fax 3322245; e-mail info@timortelecom.tp; internet www.timortelecom.tp; f. 2003; jt venture mainly operated by Portugal Telecom; provides telecommunications services in Timor-Leste.

BROADCASTING

Following independence UNMISET transferred control of public television and radio in Timor-Leste to the new Government. In 2003 a Public Broadcasting Service was established, controlled by an independent board of directors.

Radio

There are 18 radio stations operating in Timor-Leste, including community radio stations for each of the country's 13 districts. In May 2002 Radio UNTAET, launched by the UN, became Radio Timor-Leste, which broadcasts in four languages to an estimated 90% of Timor-Leste's population. The Roman Catholic Church also operates a radio station, Radio Kamanak, while a third populist station, Voz Esperança, broadcasts in Dili. A fourth radio station, Radio Falintil FM, also operates in Dili and, in October 2003, the Christian station Voice FM was established. In 2000 the US radio station, Voice of America, began broadcasting to Timor-Leste seven days a week in English, Portuguese and Bahasa Indonesia.

Television

TV Timor-Leste (TVTL): Dili; f. 2000 as Televisaun Timor Lorosa'e by UNTAET; adopted present name in May 2002; broadcasts in Tetum and Portuguese; Dir-Gen. VIRGÍLIO DA SILVA GUTERRES.

Finance

In January 2000 a Central Fiscal Authority was established by the UNTAET Transitional Administrator. This later became the country's Ministry of Planning and Finance. On 24 January the National Consultative Council formally adopted the US dollar as Timor-Leste's transitional currency; the process of dollarization had been largely completed by early 2003. In November 2003 a new local currency was introduced, in the form of low-denomination coins (*centavos*), which were intended to facilitate transactions by the Timorese population and were fully exchangeable with the dollar. The new coins could not, however, be used outside Timor-Leste.

BANKING

Timor-Leste's banking system collapsed as a result of the violent unrest that afflicted the territory in 1999. In late 1999 Portugal's main overseas bank, Banco Nacional Ultramarino, opened a branch in Dili. In January 2001 the Australia and New Zealand Banking Group also opened a branch in the capital. During 2003 PT Bank Mandiri of Indonesia became the third foreign bank operating in the country. In February 2001 the East Timor Central Payments Office was officially opened. This was succeeded in November of that year by the Banking and Payments Authority, which was intended to function as a precursor to a central bank.

Banking and Payments Authority (BPA): Av. Bispo Medeiros, POB 59, Dili; tel. 3313718; fax 3313716; e-mail bancocentraltimor@yahoo.com; inaugurated Nov. 2001; regulates and supervises Timor-Leste's financial system, formulates and implements payments system policies, provides banking services to Timor-Leste's administration and foreign official institutions, manages fiscal reserves; fmrly Central Payments Office; Gen. Man. LUÍS QUINTANEIRO; Dep. Gen. Mans ABRAAO VASCONSELOS, SAM ROBINSON.

Foreign Banks

Australia and New Zealand Banking Group Ltd (ANZ) (Australia): POB 264, Unit 2, 17–19 Rua José Maria Marques, Dili; tel. 3324800; fax 3324822; e-mail anzeasttimor@anz.com; internet www.anz.com/TimorLeste; retail and commercial banking services; Gen. Man. KIRK MCNAMARA; Chief Operating Officer RICHARD HARE.

Banco Nacional Ultramarino (Portugal): 11–12 Rua Presidente Nicolau Lobato, Dili; tel. 3323385; fax 3323678; e-mail timor@bnu.com.mo; internet www.bnu.pt; Dir-Gen. Dr CORREIA PINTO.

PT Bank Mandiri (Persero) (Indonesia): Dili.

Trade and Industry

UTILITIES

In the early 2000s Timor-Leste's total generating capacity amounted to some 40 MW. As a result of the civil conflict in 1999, some 13–23 power stations were reported to require repairs ranging from moderate maintenance to almost complete rehabilitation. The rehabilitation of the power sector was ongoing in 2004, with funding largely provided by foreign donors. By July 2003 31 generators had been restored, supplying electricity to Dili, as well as to 12 districts and 33 subdistricts.

Electricidade de Timor-Leste (EDTL): Dili; govt dept; responsible for power generation, distribution and financial management of power sector in Timor-Leste; transferred to external management in 2002.

Transport

ROADS

The road network in Timor-Leste is poorly designed and has suffered from long-term neglect. In December 1999 the World Bank reported that some 57% of the country's 1,414 km of paved roads were in poor or damaged condition. Many gravel roads are rough and potholed and are inaccessible to most vehicles. Some repair and maintenance work on the road network was being carried out in 2004, using funding supplied by external donors.

SHIPPING

Timor-Leste's maritime infrastructure includes ports at Dili, Carabela and Com, smaller wharves at Oecusse (Oekussi) and Liquiça (Likisia), and slip-landing structures in Oecusse, Batugade and Suai. In November 2001, following its reconstruction, the management of the port at Dili was transferred to the Government. The port was expected to be a significant source of revenue.

CIVIL AVIATION

Timor-Leste has two international airports and eight grass runways. At mid-2000 operators of international flights to East Timor included Qantas Airways and Air North of Australia. In June 2001 Dili Express Pte was the first Timor-based company to begin international flights, with a service to Singapore.

Defence

In October 1999, following the intervention of an international peace-keeping force, Interfet, the last of the occupying Indonesian armed forces departed from East Timor. Interfet was subsequently replaced by the armed forces of the UN Transitional Administration for East Timor (UNTAET). At independence on 20 May 2002 UNTAET was succeeded by the UN Mission of Support in East Timor (UNMISET). At 31 July 2004 the UNMISET peace-keeping force comprised: 422 military personnel; 43 military observers; and 139 civilian police. It was supported by 260 international and 585 local civilians. The maximum authorized strength was 5,000 troops, including 120 military observers and 1,250 civilian police. Provision was also made for 455 international civilian staff, 100 experts for a Civilian Support Group, 977 locally recruited staff and 241 UN Volunteers. The peace-keeping force was gradually to be replaced, through a two-year process of phased withdrawal, by an East Timor Defence Force (Falintil-ETDF), comprising a light infantry force of 1,500 regulars and 1,500 reservists. The first battalion of 600 members was recruited in February 2001 from the ranks of the former guerrilla army, Falintil, and the first ETDF troops were deployed into the peace-keeping structure in early August. A second deployment followed in September 2001. On 23 July 2002 the first battalion of Falintil-ETDF assumed responsibility from UNMISET's peace-keeping force for the district of Lautem—the first such transfer since the UN's arrival in the country in 1999. In August 2003, according to Western estimates, Falintil-ETDF comprised 650 army personnel, including a naval element of 36. In May 2004 UNMISET transferred all responsibility for policing and external security in the country to the Government of Timor-Leste.

Commander-in-Chief: Brig.-Gen. TAUR MATAN RUAK.

Force Commander: Lt-Gen. Datuk KHAIRUDDIN MAT YUSOF (Malaysia).

Chief Military Observer: Brig.-Gen. PEDRO ROCHA PENA MADEIRA (Portugal).

Chief of Civilian Police: Commr SANDRA PEISLEY (Australia).

Education

Prior to the civil conflict in 1999 there were 167,181 primary students at the 788 schools, 32,197 junior secondary students at 114 schools and 14,626 senior secondary students at 37 schools. Primary school enrolment included 70% of those in the relevant age-group and secondary enrolment about 39%. The 17 vocational and technical schools had 4,347 students. Some 4,000 students were engaged in further education at the university and the polytechnic in Dili, while an additional several thousand East Timorese were following courses at Indonesian universities. It was estimated that, as a result of the civil conflict in 1999, 75%–80% of primary and secondary schools were either partially or completely destroyed. Illiteracy rates were reported to be above 50% in 1999, with higher rates recorded for females than for males. In 2001 an estimated 236,000 students attended 900 schools in the country, with some 5,000 students engaged in further education at the National University of East Timor (now the National University of Timor-Leste). By July 2001 UNTAET estimated that the number of usable classrooms within East Timor (following a massive rehabilitation programme that restored 2,000 classrooms to a basic operational level) was only 14% below the 1999 pre-violence level. In June 2002 the Emergency Schools Readiness Project, funded by the Trust Fund for East Timor (TFET), was completed, having renovated 535 schools and 2,780 classrooms since August 2000. It was succeeded by the Fundamental School Quality Project, under which 65 primary schools were to be constructed during 2002/03. The budget of the Central Fund for East Timor (CFET) allocated an estimated US $17.1m. to the education sector in 2002/03, 24.2% of total expenditure.

Bibliography

See also Indonesia

Aarons, Mark, and Domm, Robert. *East Timor: A Western Made Tragedy*. Sydney, Left Book Club, 1992.

Aditjondro, George J. *In the Shadow of Mount Ramelau: The Impact of the Occupation of East Timor*. Indonesian Documentation and Information Centre, 1994.

Alkatiri, Mari. *Statement on the National Development Plan for East Timor*. Donors' Meeting on East Timor, Dili, 14–15 May 2002.

Andersen, Tim. *Aidwatch Background Paper: Main Features of the Timor Sea Agreements*. Aidwatch, Sydney, May 2002.

Allied Geographical Section. *Area Study of Portuguese Timor*. London, 1943.

Asian Development Bank. *Eighth Progress Report on Timor-Leste*. Manilla, 2004.

Auburn, F. M., Ong, David, and Forbes, Vivian L. *Dispute Resolution and the Timor Gap Treaty*. Nedlands, WA, Indian Ocean Centre for Peace Studies, University of Western Australia, 1994.

Ball, Desmond, and McDonald, Hamish. *Death in Balibo, Lies in Canberra*. St Leonards, NSW, Allen & Unwin, 2000.

Bijlmer, Hendricus Johannes Tobias. *Outlines of the Anthropology of the Timor-Archipelago*. Weltevreden, Kolff, 1929.

Breen, Bob. *Mission Accomplished: East Timor*. St Leonards, NSW, Allen & Unwin, 2001.

Bruce, Robert H. *US Response to the East Timor Massacre: Historical Grounds for Scepticism about a Suggested Remedy*. Nedlands, WA, Indian Ocean Centre for Peace Studies, University of Western Australia, 1992.

Budiardjo, Carmel, and Liem Soei Liong. *The War against East Timor*. London, Zed Press, and Leichhardt, NSW, Pluto Press, 1984.

Campaign for Independent East Timor. *East Timor on the Road to Independence: A Background Report*. Sydney, 1974.

Cardoso, Luís. *The Crossing: A Story of East Timor*. London, Granta Books, 2000.

Carey, Peter, and Bentley, G. Carter (Eds). *East Timor at the Crossroads: The Forging of a Nation.* Honolulu, HI, University of Hawaii Press, 1995.

Catholic Commission for Justice, Development and Peace. *The Church and East Timor: A Collection of Documents by National and International Catholic Church Agencies.* Melbourne, 1993.

Catholic Institute for International Relations. *International Law and the Question of East Timor.* London, 1995.

Cristalis, Irena. *Bitter Dawn: East Timor—A People's Story.* London, Zed Press, 2002.

Cultural Survival. *East Timor, Five Years after the Indonesia Invasion: Testimony Presented at the Decolonization Committee of the United Nations' General Assembly, October 1980.* Cambridge, MA, 1981.

Da Costa, Helder, and Soesastro, Hadi. 'Building East Timor's Economy' in *Comparing Experiences with State Building in Asia and Europe: The Cases of East Timor, Bosnia and Kosovo.* Council for Asia Europe Co-operation (CAEC), internet www.caec-asiaeurope.org/conference/publications/costasoesastro.pdf.

Da Silva, E. A. *East Timor Agriculture: Strategic Issues and Policy Directions, Statement made at the 25th Anniversary Session of the International Fund for Agricultural Development.* Rome, 2003.

Departemen Luar Negeri, Indonesia. *Decolonization in East Timor.* Jakarta, Department of Information, 1977.

Departemen Penerangan, Indonesia. *Government Statements on the East-Timor Question.* Jakarta, 1975.

Department of Foreign Affairs and Trade, Australia. *Briefing Papers: International Court of Justice, Portugal v. Australia (concerning East Timor).* Canberra, 1995.

Department of Information, Indonesia. *The Province of East Timor: Development in Progress.* Jakarta.

Downie, Sue, and Kingsbury, Damien (Eds). *The Independence Ballot in East Timor: Report of the Australian Volunteer Observer Group.* Clayton, Vic, Monash Asia Institute, 2001.

Dunn, James. *East Timor: The Balibo Incident in Perspective.* Broadway, NSW, The Australian Centre for Independent Journalism, 1995.

Timor: A People Betrayed. Sydney, ABC Books, 1996.

East Timor: A Rough Passage to Independence. Double Bay, NSW, Longueville Books, 2003.

Fitzpatrick, Daniel. *Land Claims in East Timor.* Canberra, Asia Pacific Press, 2002.

Fox, James J., and Babo Soares, Dionísio (Eds). *East Timor: Out of the Ashes, the Destruction and Reconstruction of an Emerging State.* Bathurst, NSW, Crawford House, 1999.

Glover, Ian. *Archaeology in Eastern Timor, 1966–67.* Canberra, Department of Prehistory, Research School of Pacific Studies, Australian National University, 1986.

Greenlees, Don, and Garran, Robert. *Deliverance: The Inside Story of East Timor's Fight for Freedom.* Crow's Nest, NSW, Allen & Unwin, 2002.

Gunn, Geoffrey C. *Wartime Portuguese Timor: The Azores Connection.* Clayton, Vic, Monash University, Centre of Southeast Asian Studies, 1988.

Timor Loro Sae: 500 Years. Macao, Livros do Oriente, 1999.

A Critical View of Western Journalism and Scholarship on East Timor. Sydney, Journal of Contemporary Asia Publishers, 1994.

East Timor and the United Nations: The Case for Intervention. Lawrenceville, NJ, Red Sea Press, 1997.

Gusmão, Xanana. *To Resist Is To Win: The Autobiography of Xanana Gusmão.* Richmond, Vic, Aurora Books, 2000.

Hainsworth, Paul, and McCloskey, Stephen (Eds). *The East Timor Question: The Struggle for Independence from Indonesia.* New York, NY, I. B. Tauris, 2000.

Head, Mike. 'New Timor Gap Treaty Secures Australian Control of Oil and Gas Projects', World Socialist Web Site, 11 July 2001, www.wsws.org/articles/2001/jul2001/timo-j11.shtml.

Hill, Hal, and Saldanha, João (Eds). *East Timor and Economic Development.* Singapore, Institute of Southeast Asian Studies, 2001.

East Timor: Development Challenges for the World's Newest Nation. Singapore, Institute of Southeast Asian Studies; and Canberra, Asia Pacific Press, Australian National University, 2001.

Hill, Helen. *Stirrings of Nationalism in East Timor: Fretilin 1974–1978: the Origins, Ideologies and Strategies of a Nationalist Movement.* Otford, NSW, Otford Press, 2002.

Hoadley, J. Stephen. *The Future of Portuguese Timor: Dilemmas and Opportunities.* Singapore, Institute of Southeast Asian Studies, 1975.

Inbaraj, Sonny. *East Timor: Blood and Tears in Asean.* Chiang Mai, Silkworm Books, 1995.

International Commission of Jurists. *Tragedy in East Timor: Report on the Trials in Dili and Jakarta.* Geneva, 1992.

Report of the Trial of Xanana Gusmão in Dili, East Timor. Geneva, 1993.

International Monetary Fund. *Staff Statement.* Donors' Meeting on East Timor, Dili, 14–15 May 2002.

Jardine, Matthew. *East Timor: Genocide in Paradise.* Tucson, AZ, Odonian Press, 1995.

Joint Assessment Mission. *East Timor—Building a Nation: A Framework for Reconstruction and Development.* November 1999.

Agriculture Background Paper.

Health and Education Background Paper.

Macro-Economics Background Paper.

Jolliffe, Jill. *East Timor: Nationalism and Colonialism.* St Lucia, Qld, University of Queensland Press, 1978.

Cover-up: the Inside Story of the Balibo Five. Carlton North, Vic, Scribe, 2001.

Jonsson, Gabriel (Ed.). *East Timor: Nation-building in the 21st Century.* Stockholm, Center for Pacific Asia Studies, Stockholm University, 2003.

Kim, Insu, and Schwarz, Stephen. *Birth of a Nation: East Timor Gains Independence, Faces Challenges of Economic Management, Poverty Alleviation.* IMF Survey, Volume 31, No. 11, 10 June 2002.

Kingsbury, Damien (Ed.). *Guns and Ballot Boxes: East Timor's Vote for Independence.* Melbourne, Vic, Monash Asia Institute, 2000.

Kohen, Arnold, and Taylor, John. *An Act of Genocide: Indonesia's Invasion of East Timor.* London, Tapol, 1979.

Kohen, Arnold S. *From the Place of the Dead: The Epic Struggles of Bishop Belo of East Timor.* New York, St Martin's Press, 1999.

Krieger, Heike (Ed.). *East Timor and the International Community: Basic Documents.* Cambridge, Cambridge University Press, 1997.

Lennox, Rowena. *Fighting Spirit of East Timor: The Life of Martinho da Costa Lopes.* Sydney, Pluto Press, 2000.

Marker, Jamsheed. *East Timor: A Memoir for the Negotiations for Independence.* Jefferson, NC, McFarland and Co, 2003.

Martin, Ian. *Self-determination in East Timor: The United Nations, the Ballot, and International Intervention.* Boulder, CO, Lynne Rienner Publishers, 2001.

Martinkus, John. *A Dirty Little War.* Milsons Point, NSW, Random House, 2001.

McDonald, Hamish, et al. *Masters of Terror: Indonesia's Military and Violence in East Timor in 1999.* Canberra, Australian National University, Strategic and Defence Studies Centre, 2002.

Metzner, Joachim K. *Man and Environment in Eastern Timor: A Geoecological Analysis of the Baucau-Viqueque Area as a Possible Basis for Regional Planning.* Canberra, Development Studies Centre, Australian National University, 1997.

Mubyarto, et al. *East Timor: The Impact of Integration: An Indonesian Socio-anthropological Study.* Northcote, Vic, Indonesia Resources and Information Program (IRIP), 1991.

Nicol, Bill. *Timor: The Stillborn Nation.* Melbourne, Vic, Widescope International, 1978.

Orentlicher, Diane. *Human Rights in Indonesia and East Timor.* New York, Human Rights Watch, 1988.

Ormeling, Ferdinand Jan. *The Timor Problem: A Geographical Interpretation of an Underdeveloped Island.* Groningen, Wolters, 1955.

Pinto, Constâncio, and Jardine, Matthew. *East Timor's Unfinished Struggle: Inside the Timorese Resistance.* Boston, MA, South End Press, 1997.

Ramos Horta, Arsénio. *The Eyewitness: Bitter Moments in East Timor Jungles.* Singapore, Usaha Quality Printers.

Ramos Horta, José. *Funu: Unfinished Saga of East Timor.* Trenton, NJ, Red Sea Press, 1987.

Retbøll, Torben (Ed.). *East Timor, Indonesia, and the Western Democracies: A Collection of Documents.* Copenhagen, International Work Group for Indigenous Affairs (IWGIA), 1980.

Roff, Sue Rabbitt. *East Timor: A Bibliography, 1970–1993.* Canberra, Peace Research Centre, 1994.

Timor's Anschluss: Indonesian and Australian Policy in East Timor, 1974–76. Leviston, NY, Edwin Mellen Press, 1992.

Rohland, Klaus. *Opening Remarks.* Donors' Meeting on East Timor, Dili, 14–15 May 2002.

Rothwell, Donald R., and Tsamenyi, Martin (Eds). *The Maritime Dimensions of Independent East Timor.* NSW, Centre for Maritime Policy, University of Wollongong, 2000.

Rowland, Ian. *Timor: including the Islands of Roti and Ndao*. Oxford and Santa Barbara, CA, Clio Press, 1992.

Schlicher, Monika. *Portugal in Ost-Timor: Eine Kritische Untersuchung zur Portugiesischen Kolonialgeschichte in Ost-Timor*. Hamburg, Abera, 1996.

Schulte Nordholt, H. G. *The Political System of the Atoni of Timor*. The Hague, M. Nijhoff, 1971.

Sherlock, Kevin P. *A Bibliography of Timor: Including East (formerly Portuguese) Timor, West (formerly Dutch) Timor, and the Island of Roti*. Canberra, Australian National University, 1980.

East Timor: Liurais and Chefes de Suco; Indigenous Authorities in 1952. Darwin, NT, 1983.

Sherman, Tom. *Second Report on the Deaths of Australian-based Journalists in East Timor in 1975*. Canberra, Department of Foreign Affairs and Trade, 1999.

Singh, Bilveer. *East Timor, Indonesia, and the World: Myths and Realities*. Kuala Lumpur, ADPR Consult (M), 1996.

Smith, Michael G., and Dee, Moreen. *Peacekeeping in East Timor: The Path to Independence*. Boulder, CO, Lynne Rienner Publishers, 2003.

Sousa Saldanha, João Mariano de. *The Political Economy of East Timor Development*. Jakarta, Pustaka Sinar Harapan, 1994.

Sousa Saldanha, João Mariano de (Ed.). *An Anthology: Essays on the Political Economy of East Timor*. Casuarina, NT, Centre for South-east Asian Studies, Northern Territory University, 1995.

Stepan, Sasha. *Credibility Gap: Australia and the Timor Gap Treaty*. Canberra, Australian Council for Overseas Aid, 1990.

Subroto, Hendro. *Eyewitness to Integration of East Timor*. Jakarta, Pustaka Sinar Harapan, 1997.

Suter, Keith. *East Timor and West Irian*. London, Minority Rights Group, 1982.

Sword, Kirsty, and Walsh, Pat (Eds). *'Opening up': Travellers' Impressions of East Timor 1989–1991*. Fitzroy, Vic, Australia East Timor Association, 1991.

Tanter, Richard, Selden, Mark, and Shalom, Stephen R. (Eds). *Bitter Flowers, Sweet Flowers: East Timor, Indonesia, and the World Community*. Lanham, MD, Rowman & Littlefield Publishers, 2001.

Tapol, The Indonesian Human Rights Campaign. *East Timor: United Nations Resolutions, 1975–1982*. Thornton Heath, Surrey, Tapol, 1991.

Taylor, John G. *East Timor: The Price of Freedom*. New York, Zed Books, 1999.

Teklkamp, Gerard J. *De Ekonomische Struktuur van Portugees-Timor in de Twintigste Eeuw: Een Voorlopige Schets*. Amsterdam, Centrale Bibliotheek, Koninklijk Instituut voor de Tropen, 1975.

Turner, Michele. *Telling East Timor: Personal Testimonies 1942–1992*. Kensington, NSW, New South Wales University Press, 1992.

United Nations Security Council. *Report of the Secretary-General on the United Nations Transitional Administration in East Timor*. United Nations, New York, 17 April 2002.

United Nations Transitional Administration in East Timor (UNTAET), World Bank and International Monetary Fund. *Background Paper for Donors' Meeting on East Timor, Lisbon, Portugal, June 22–23, 2000*. Washington, DC, World Bank, 2000.

Valdivieso, Luís M., et al. *East Timor: Establishing the Foundations of Sound Macroeconomic Management*. Washington, DC, IMF, August 2000.

Valdivieso, Luís M., and Lopez-Mejia, Alejandro. *East Timor: Macroeconomic Management on the Road to Independence*. Finance and Development, Volume 38, No. 1, March 2001.

Valdivieso, Luís M. *Staff Statement*. Donors' Meeting for East Timor, Canberra, 14–15 June 2001.

Vieira de Mello, Sérgio. *Statement*. Donors' Meeting on East Timor, Dili, 14–15 May 2002.

Way, Wendy (Ed). *Australia and the Indonesian Incorporation of Portuguese Timor. 1974–1976*. Carlton, Vic, Melbourne University Press, 2000.

Wiarda, Siqueira. *The Portuguese in Southeast Asia: Malacca, Moluccas, East Timor*. Hamburg, Abera Verlag, 1997.

Winters, Rebecca. *Buibere: Voice of East Timorese Women*. Nightcliff, Darwin, NT, East Timor International Support Centre, 1999.

Winters, Rebecca, and Kelly, Brian. *Children of the Resistance: the Current Situation in East Timor as seen through the Eyes of two Australian Tourists*. Darwin, NT, Australians for a Free East Timor, 1996.

World Bank, *Background Paper for Donors' Meeting on East Timor*. Donors' Meeting on East Timor, Dili, 14–15 May 2002.

Website: www.timorseaoffice.gov.tp/enindex.htm.

VIET NAM

Physical and Social Geography

HARVEY DEMAINE

The Socialist Republic of Viet Nam covers a total area of 329,247 sq km (127,123 sq miles) and lies along the western shore of the South China Sea, bordered by the People's Republic of China to the north, by Laos to the west and by Cambodia to the south-west. The capital is Hanoi.

PHYSICAL FEATURES AND CLIMATE

The fundamental geographical outlines of the country are determined by the deltas and immediate hinterlands of the Mekong and Songkoi (Red River) which are linked by the mountain backbone and adjacent coastal lowlands of Annam.

Of the two rivers which are thus of major significance in the geography of Viet Nam, the Songkoi, rising, like the Mekong, in south-western China, is much the shorter, and its delta, together with that of a series of lesser rivers, forms a total area of some 14,500 sq km, which is less than one-half that of the great Mekong delta in the south. The north of Viet Nam also includes a much more extensive area of rugged upland, mainly in the north and west, which represents a southward continuation of the Yunnan and adjacent plateaux of south-western China, and forms an inhospitable and sparsely populated divide, some 900–1,500 m above sea-level, between North Viet Nam and northern Laos.

Both the Songkoi and its main right-bank tributary, the Songbo (Black River), flow in parallel north-west/south-east gorges through this upland before their confluence some 100 km above the apex of the delta, while a third main river, the Songma, also follows a parallel course still further to the south, beyond which rises the similarly north-west/south-east inclining Annamite chain or Cordillera. This, in relief if not in structure, constitutes a further prolongation of the massive upland system already described, and extends without a break to within about 150 km of the Mekong delta.

With an average breadth of 150 km, and an extremely rugged and heavily forested surface, at many points exceeding 1,500 m in altitude, the Annamite chain, which lies mainly in southern Viet Nam, provides an effective divide between the Annam coast and the middle Mekong valley of southern Laos and eastern Cambodia. Moreover, from the Porte d'Annam (latitude 18°N) southwards, the chain not only reaches to within a few miles of the coast, but also sends off a series of spurs, which terminate in rocky headlands overlooking the sea. Thus, along the 1,000-km stretch of coast between latitudes 18° and 11°N, the continuity of the coastal plain is repeatedly interrupted and it dwindles to an average width of less than 16 km and often to less than half that figure; but thereafter it broadens out to merge with the vast deltaic plain of the Mekong and its associated natural waterways, the whole forming an almost dead flat surface covering some 37,800 sq km.

In forming the western hinterland of the South China Sea virtually from the tropic of Cancer to within 9°N of the Equator, Viet Nam might be assumed to be wholly within the zone of the tropical monsoon climate. The greater part of the country does merit such a designation, and the city of Hué, practically at the mid-point of the coastal zone, has a mean monthly temperature range from 20°C in January to 30°C in August, and a total rainfall of 2,600 mm, of which 1,650 mm falls between September and November. However, the Songkoi delta in the north is not strictly tropical in the climatological sense, as, owing to its exposure to cold northern air during the season of the north-east monsoon, it experiences a recognizable cool season from December to March, and in both January and February the mean monthly temperatures in Hanoi are only 17°C. This fact is of great importance since the cooler weather gives greater effectiveness to the 130–150 mm of rain that fall during these months, and so makes it possible to raise a 'winter' as well as a summer crop of rice in this part of the country.

NATURAL RESOURCES

In terms of agricultural potential, the two river deltas are of overriding importance. The uplands, until recently, have offered far less opportunity for supporting population, not only because of the extremely restricted prospects for rice cultivation but also because of their intensely malarial character. Owing to increasing pressure on land resources in the Tongking delta in particular, great efforts are being made to develop the uplands for cash cropping, although there are suggestions that this may be at the expense of environmental stability. In respect of mineral wealth, on the other hand, the uplands, particularly in the north, contain a wide variety of lesser metallic ores, and also some useful apatite (a source of phosphates), though economically the most important mineral is in the anthracite field of Quang-Yen immediately to the north-east of the Songkoi delta. Reserves here are put at some 20m. metric tons, and total coal reserves are estimated at 3,500m. metric tons. Petroleum has been discovered off shore in the north and south of the country, and the Bach Ho oilfield, 160 km to the east of Ho Chi Minh City, came into operation in 1986. There are also plans to exploit reserves of natural gas.

POPULATION AND ETHNIC GROUPS

The population of Viet Nam totalled 76,323,173 at the census of 1 April 1999, according to provisional figures, (compared with the April 1989 census total of 64,411,713). By mid-2003 the population had increased further to 80,670,000, according to official estimates, and there was an average population density of 245.0 per sq km. This average, however, is highly misleading, firstly because the great majority of the population (and practically all of the Vietnamese proper) live within the lowlands, and secondly because these regions comprise rather more than one-third of the total area of southern Viet Nam, but barely one-sixth of that of northern Viet Nam. Thus, whereas most of the vast Mekong delta supports rural densities varying from 40 to 200 per sq km, the comparable figures for the Songkoi delta are between four and five times as great. The South, however, shows the higher degree of urbanization. In mid-2003 the largest town in Viet Nam was Ho Chi Minh City, in the South, with a population of 4,850,717, while the largest towns in the North, the capital Hanoi and its port, Haiphong, had populations of 3,977,202 and 1,754,537, respectively. The distribution of population in Viet Nam changed considerably following a vast resettlement programme: 3.5m. people were moved after 1975 and 2.1m. after 1981, mostly to Viet Nam's New Economic Zones or to the Central Highlands.

Vietnamese, who are ethnically related to the southern Chinese, form 80% of the population. There are also significant minority groups, notably the Tai in the North (numbering some 2m.), some 750,000 Hmong and related groups and a number of smaller groups of people in the Central Highlands, usually known as *montagnards* and numbering up to 1m. There are perhaps 600,000 Cambodians along Viet Nam's south-western border and a now uncertain number of Chinese, still primarily concentrated in the southern city of Cholon, but much reduced in numbers by migration.

History

RALPH SMITH

Revised for this edition by ZACHARY ABUZA

EARLY HISTORY

Recent archaeological work suggests that Viet Nam had a distinctive bronze age civilization, perhaps by 1000 BC. The present northern Viet Nam became incorporated into the Chinese Han empire in 112 BC and remained effectively a province of China for the ensuing millennium. During these centuries there occurred a gradual interaction of Vietnamese and Chinese ideas and institutions, and the Chinese language was introduced, although it did not supplant Vietnamese. The Chinese religions of Daoism, Confucianism and Mahayana Buddhism also became widespread.

With the dissolution of the Tang empire in the early part of the 10th century, Viet Nam came under the rule of an independent southern dynasty. However, this regime's army was defeated by the Vietnamese in 931 and 938, and it is from the latter year that the Vietnamese usually date their independence. During the next four centuries Viet Nam (or Dai Viet) gradually developed into a strong and fairly centralized state, with a capital in Hanoi and institutions modelled on those of China. Under the dynasties of the Ly (1009–1225) and Tran (1225–1400), Dai Viet was strong enough to resist further Chinese attempts at reconquest: by the Song in 1075–77, and by the Yuan (Mongols) in the years 1282–88. A new Chinese attack succeeded in 1407, but the Chinese were once more driven out by Le Loi, founder of the Le dynasty (1428–1789).

Meanwhile, the region now comprising central Viet Nam belonged to the kingdom of Champa, with a culture strongly influenced by Hindu-Buddhist culture, while the Mekong delta belonged to Angkor (cf. Cambodia). Dai Viet annexed areas of Cham territory in 1070, 1306 and 1470, reducing Champa to a small principality which was finally extinguished in 1695.

CONFLICTS, DIVISION AND UNIFICATION

The 15th century was the 'golden age' of Confucian culture in Viet Nam and a period of general stability. In the early 16th century, however, the kingdom began to break up under the stress of conflict between rival clans, one of which established the Mac dynasty (1527–92). After intermittent civil war in the 16th century, the next 100 years saw a more lasting division of Viet Nam between two powerful clans: the Trinh in the north and the Nguyen in central Viet Nam. It was the Nguyen who led the way in a further southward expansion, beginning about 1658, which established their control over the greater part of what is now southern Viet Nam. Also in this period, Christian missions were established in both parts of Viet Nam.

In the 1770s, however, the Tay Son rebellion cast the entire country into a state of civil war. Although the Tay Son established a new dynasty and defeated a Chinese invasion in 1789, they could not bring new stability. The last Nguyen survivor recaptured Saigon (now Ho Chi Minh City) and from there went on to conquer the centre and north by 1802. He proclaimed himself Emperor at Hué, with the title Gia Long, and for the first time in its history the present area of Viet Nam was brought under the control of a single ruler. For a brief period Minh Mang (1820–41) also extended his control to embrace much of Cambodia. During the reign of Tu Duc (1847–83), however, the Vietnamese were confronted by the challenge of French colonialism.

In 1858–59 a Franco-Spanish fleet attacked Tourane (Da Nang), and then Saigon. French annexation of three provinces of Cochin-China (the southernmost part of Viet Nam) was formally recognized by Tu Duc in 1862. In 1867 the French annexed another three provinces. A new Franco-Vietnamese treaty, signed in 1874, recognized French possession of all Cochin-China and permitted the French to trade in Tongking (Tonkin). In 1883 the French imposed a treaty of 'protection' on the Vietnamese empire. Two years later, after a border war with China, the French secured Chinese recognition of their protectorate.

FRENCH RULE AND VIETNAMESE OPPOSITION

By 1885 the whole of Viet Nam was under French colonial rule: Cochin-China, in the south, as a directly administered colony; Tongking and Annam (the north and centre) as protectorates. In 1887 they were united with Cambodia to form the Union Indochinoise, to which Laos was added in 1893. By 1901 the administration of the Union had become an effective central authority with financial control over the whole of Indo-China.

The French administration promoted the mining of coal and minerals in Tongking, the cultivation of rubber and other plantation crops in the hill areas, the export of rice from Cochin-China, and the construction of several railways. They also created several new towns, notably Saigon and Hanoi, which for a time were alternate capitals of the Union and which attracted the growth of a small urban, French-educated élite among the Vietnamese population. In the countryside, on the other hand, the peasantry enjoyed few benefits from French rule. Their taxes were increased several times from about 1897, but they were given little help in increasing productivity. During the 1930s, with the economic slump, followed by a new inflation, the economic condition of the peasantry declined seriously and many of the poor and landless villagers became permanently indebted to their richer neighbours. The political developments of the 1930s and 1940s must be seen against this background of social conflict. At the same time the development of French education tended to undermine traditional values and to accentuate economic inequalities.

Opposition to French rule was essentially 'traditional' before about 1900: a long sequence of local risings led by scholars or military men who had had some status in the old society, but who had no effective means of defeating the new regime. During the first decade of the 20th century a new opposition movement began, strongly influenced by the model of Japanese modernization and by Chinese revolutionary aspirations. However, an anti-taxation revolt and a nationalist education movement in 1907–08 were both suppressed by the French. The two most prominent opposition leaders of this period, Phan Chau Trinh and Phan Boi Chau, spent long periods in exile.

The growth of constitutionalist opposition to Western colonial rule throughout Asia, from about 1916, was reflected in Viet Nam by the Constitutionalist Party, founded in Cochin-China in 1917. The party participated in elections for the Colonial Council, but was not allowed to operate outside Cochin-China: the French did not permit a nation-wide constitutional opposition movement comparable to the Congress Party in India. Consequently, from about 1925 Vietnamese nationalism began to produce revolutionary organizations, some of which became communist. In 1930 it also found expression in a number of open revolts, quickly suppressed. A peasant movement that emerged in mid-1930 in parts of Cochin-China, and in Nghe An, Ha Tinh and Annam, reached more serious proportions. It was organized largely by the Communist Party of Indo-China (CPIC), which was formally established in 1930. The movement, generally known as the 'Nghe Tinh Soviets', continued until 1931, when it was suppressed with the use of French air power. As a result many communist leaders (including several future members of the North Vietnamese Politburo) spent the years 1930–36 in prison.

In 1936 the Popular Front Government in France allowed an amnesty of political prisoners and a measure of press freedom and open political activity. Among those who took advantage of it to organize meetings and to write articles were Pham Van Dong, Vo Nguyen Giap and Dang Xuan Khu (Truong Chinh).

Meanwhile, the CPIC held its first National Congress at Macau in 1935, and a clandestine leadership was created by Le Hong Phong, who had trained in Moscow. During the same period Trotskyist and Stalinist movements emerged. Following the outbreak of war in Europe in 1939, the various left-wing groups were again banned, but in the North, at least, the communists managed to maintain a secret network of cells throughout the war period, under the leadership of Truong Chinh.

A different type of political movement, which developed in the South during the 1920s and 1930s, was associated with a number of religious sects, including the Caodaists, formally established in 1926, and the Hoa Hao Buddhists. These groups tended to draw political inspiration and support from Japan. Also during the 1930s there was a revival of more orthodox Mahayana Buddhism in all regions of Viet Nam, although it did not play an active role in politics until the 1960s. In addition, Roman Catholicism made considerable progress under the French, so that by 1954 there were probably over 2m. Vietnamese adherents.

REVOLUTION, WAR AND PARTITION

In July 1941 the Japanese advanced into southern Indo-China. Administrative control, however, was left in French hands, with a French admiral acting as Governor and accepting the authority of Vichy France. Among the Vietnamese, pro-Japanese groups gained support, notably the Cochin-Chinese sects and the new Dai Viet (Great Viet Nam) Party. The Japanese finally deposed the French administration and disarmed its forces in March 1945, in order to prevent its playing any role that would assist the allies in the closing stages of the war. At Japan's behest, the Emperor Bao Dai revoked the treaties of 1884–85 which had made Annam and Tongking into protectorates, and proclaimed their independence. It was not until August, however, that he was permitted to revoke the treaties concerning Cochin-China, signed in 1862 and 1874. In April 1945 Bao Dai appointed a new Government, led by the pro-Japanese Tran Trong Kim, which made some effort to institute reforms. However, in wartime conditions, with most communications destroyed by US bombing, this regime was unable to control the country or to relieve the famine which had afflicted northern Viet Nam since late 1944.

Meanwhile, in southern China, a number of Vietnamese political groups were active, of which the communists were by far the most effective. Their leader in China was Ho Chi Minh, who as Nguyen Ai Quoc had played a significant part in founding the CPIC and who now returned from Moscow as its Comintern-appointed leader. In June 1941 the Chinese-based leaders and the leaders of the party still active in Tongking decided to found the Viet Nam Doc Lap Dong Minh Hoi (Revolutionary League for the Independence of Viet Nam), usually known as the Viet Minh. During the next few years the communists survived a series of French punitive campaigns against them.

In early 1945, after a period of imprisonment at the hands of the Chinese Nationalists, Ho Chi Minh made contact with the US forces operating behind Japanese lines, and obtained their assistance in expanding his base areas inside Tongking. When the Japanese removed the French administration, the Viet Minh soon emerged as the most effective political force within the country. After the Japanese surrender, the Viet Minh quickly gained control of many provinces throughout the country. On 24 August Bao Dai abdicated in their favour, and on 2 September Ho Chi Minh, as President of the new provisional Government, read the declaration of independence which marked the foundation of the Democratic Republic of Viet Nam. In the next month, the Viet Minh leaders carried out the first stages of what amounted to a political and social revolution. They were stronger in the North, where they had been active throughout the war. In the South they had only a loosely arranged front organization, in which their allies and rivals included the formerly pro-Japanese sects and various other small groups. Their provisional committee took over in Saigon but was not very firmly established in power.

The Allied powers had agreed at the Potsdam Conference in August 1945 on the temporary occupation of Viet Nam by British and Chinese Nationalist forces, respectively south and north of the 16th parallel. In the South, Franco-British co-operation enabled the French to recover control in Saigon and to land reinforcements there in October. By the end of the year the British zone was virtually under French control, apart from a number of rural areas where guerrillas held out; the British themselves left in early 1946. In the North, the French were obliged to negotiate first with the Chinese, who agreed to withdraw in March 1946, and then with Ho Chi Minh's Government in Hanoi. Elections for a National Assembly were held in the North, and some parts of the South, in January 1946. The Viet Minh dominated the polls, although they reserved a number of seats for Nationalist members, and they had the upper hand in the coalition Government formed in February. The withdrawal of the Chinese troops in the next few months left the Nationalists with very little power.

Negotiations between the French and the Viet Minh permitted the return of a French army to Tongking in early 1946, but a semi-formal *modus vivendi* broke down, when Cochin-China was granted independence. In May 1946 Ho Chi Minh led a delegation to France for further discussions, which collapsed by the end of the year. An attempted Viet Minh rising in Hanoi on 19 December was defeated by the French, and the Viet Minh withdrew to the countryside to plan a more protracted guerrilla war. The French began a series of negotiations with Vietnamese anti-communists (including Bao Dai) which led to the establishment of the Associated State of Viet Nam in 1949. If the French could recover firm control over the country, it was to be united under the new regime, with France retaining control of defence and other important policy areas. The new state had a succession of Prime Ministers between then and 1954, and was internationally recognized by the Western powers in early 1950. At about the same time, China and the USSR recognized the Democratic Republic of Viet Nam.

In 1950–51 communist victories in neighbouring China resulted in closer links with Viet Nam, while the USA responded by offering aid to the French side. The CPIC re-emerged in 1951, renamed as the Viet Nam Workers' (Lao Dong) Party, and held its second National Congress that year. The conflict assumed a 'cold war' dimension. Despite defeats in early 1951, the Viet Minh armies gained in strength, with the steady supply of Chinese material: the military conflict culminated in the siege of Dien Bien Phu, a fortress-encampment near the Laos border, which fell to the Viet Minh in May 1954.

In May 1954 an international conference on Indo-China began in Geneva. On 21 July a cease-fire agreement was signed by representatives of the French and Viet Minh high commands, and an international declaration by 14 governments set forth conditions for an eventual political settlement, emphasizing that politically Viet Nam remained one country. In effect, however, two zones were created, to be administered by the two existing Vietnamese Governments: that of the Democratic Republic in Hanoi and that of the State of Viet Nam in Saigon, with latitude 17°N as the boundary between them. A period of 300 days was allowed for regrouping, during which almost 130,000 people moved north, while as many as 900,000 people (mainly Roman Catholics) moved to the South.

SOUTH VIET NAM AND THE WAR

The State of Viet Nam, originally within the French Union, concluded an independence agreement with France in June 1954. After the cease-fire agreements at Geneva in July, French forces withdrew, leaving the state's jurisdiction limited to the zone south of 17°N. Complete sovereignty was transferred by France in December. For a little over 20 years after the Geneva agreements, the area south of the 17th parallel remained a separate state with an anti-communist administration. A French community continued to look after its own interests there, notably the large rubber estates north of Saigon. However, economically, culturally and, above all, militarily, South Viet Nam moved out of the French ambit and into the US sphere of influence, developing much closer relations with other US-aligned states: the Philippines, Thailand and the Republic of Korea. It very soon became dependent for its survival on US aid.

From 1954 to 1963 Ngo Dinh Diem and his brother, Ngo Dinh Nhu, who headed the secret security forces, dominated the politics of South Viet Nam. Coming from a family of Roman Catholic mandarins (a combination characteristic of the colonial period), Diem had been a minister at Hué briefly in 1933 but had since become strongly anti-French, as well as being deeply anti-communist. He became Prime Minister at the behest of the USA in June 1954. His position was initially very insecure; by October 1955, however, he was strong enough to hold a referendum, the result of which enabled him to depose Bao Dai and to proclaim himself President of the Republic of Viet Nam. He also repudiated the Geneva declaration and rejected plans to hold elections which it had envisaged for July 1956. During the next few years, he attempted to destroy the Viet Minh (or, as he termed it, Viet Cong) network in the South. Many communists and their sympathizers who had remained in the South in 1954 were imprisoned. In May 1959 the communist leadership in Hanoi authorized a political, and limited military, campaign in the South. In December 1960 the communists created the National Front for the Liberation of South Viet Nam (NLF), to unite opposition to Diem.

In 1961–62, determined to prevent any communist advance in South-East Asia, President Kennedy provided US troops (numbering 8,000 by the end of 1962) to act as advisers to the South Vietnamese army. US confidence in Diem, whose increasingly brutal tactics included the systematic repression of Buddhist monks, was beginning to decline, and in November 1963 he was overthrown by a military coup with tacit US approval. Diem himself was killed during the coup. However, a change of government in Saigon did not provide protection against a growing rural guerrilla movement. During the latter half of 1963 the communists adopted a more offensive strategy and by the end of the year the situation of the Saigon Government was precarious. French proposals to neutralize Viet Nam were unacceptable to the USA; and, soon after the death of President Kennedy in November, his successor, Lyndon Johnson, committed the USA to a policy of defending South Viet Nam at all costs.

It did not prove easy to create a stable and legitimate regime to succeed Diem. Following a second military coup in January 1964 Gen. Nguyen Khanh assumed power. In February 1965 he was ousted by a group of younger officers led by the air force commander, Air Gen. Nguyen Cao Ky. A nominally civilian government under Phan Huy Quat was followed, in June 1965, by a new military regime, with Lt-Gen. Nguyen Van Thieu as Head of State and Nguyen Cao Ky as Prime Minister.

By that time the USA had decided to reinforce the Saigon Government by sending its own combat troops to fight in Viet Nam. An incident involving US ships in the Gulf of Tongking in August 1964 enabled President Johnson to obtain virtually unrestricted authority from the US Congress, and he used the power granted to him at that time to deal with the Vietnamese situation without any formal declaration of war. The number of US forces in Viet Nam increased from 23,000 at the beginning of 1965 to more than 500,000 by March 1968; in addition, contingents were sent from the Republic of Korea, Australia, the Philippines and Thailand. As a result, the conflict escalated into a war of major proportions, with the communists obliged to send regular North Vietnamese troops to the South, and to rely increasingly on aid from China and the USSR. The USA commenced aerial bombardment of the North in March 1965.

A measure of political stability gradually returned to Saigon, when Nguyen Cao Ky defeated a Buddhist revolt in early 1966. A new Constitution for South Viet Nam came into effect in 1967. As a result, Gen. Nguyen Van Thieu was elected President in September 1967, with Gen. Ky as his Vice-President; elections were also held for the upper and lower houses of a legislature based on the US model. Thieu remained President from October 1967 until April 1975.

Despite their successes in the countryside, communist leaders aimed to cause heavy damage in the urban areas. In January–February 1968 the communists launched an offensive to coincide with the lunar New Year (known as *Tet*). It was on a larger scale than any previous operation, and included attacks on Saigon, Hué and many other towns. There was also heavy fighting just south of the 17th parallel. The *Tet* offensive, although a military defeat, was a political victory that forced the USA to reconsider its policy. Faced with a major financial crisis, and increasing opposition to the fighting, President Johnson decided against a further expansion of the war. Discussions between US and North Vietnamese representatives began in Paris in May. In October Washington and Hanoi agreed to extend the negotiations and the USA thereupon ended its bombing raids against the North. Heavy bombing continued, however, to be an essential part of US strategy in the South, and the talks led to no further decrease in the fighting.

In January 1969, after the transfer of power in the USA from President Johnson to Richard Nixon, the informal talks in Paris were transformed into a formal conference between representatives of the USA, North Viet Nam, South Viet Nam and the NLF. In June 1969 the NLF was supplemented by the creation of a new Provisional Revolutionary Government of South Viet Nam. During 1969–70 the conflict continued: although the USA began to withdraw its own troops, fears that the North Vietnamese might take advantage of the withdrawal led to a sequence of moves by the USA which tended, in fact, to intensify the war. The most spectacular was the invasion of Cambodia (q.v.) in April 1970, following US-supported moves to overthrow the Government of Prince Sihanouk in Phnom-Penh. Equally important was the 'Lam Son' operation of February 1971 in Laos, which damaged North Vietnamese supply lines even though it ended in South Vietnamese retreat. By this time the communist war effort in the South was heavily dependent on the presence of North Vietnamese regular troops.

In March 1972, with US forces reduced to about 95,000, the North Vietnamese launched a new offensive, which led to some of the most intense fighting of the war. The US Government reacted by renewing its bombing of the North. By September it was clear that the situation had reached an impasse, while in the USA itself there was mounting pressure to bring the war to a speedy end. Secret meetings in Paris between President Nixon's foreign policy adviser, Dr Henry Kissinger, and Le Duc Tho of the Vietnamese Politburo, which had taken place several times since 1969, began to show progress, following Dr Kissinger's visit to Moscow in September 1972. In December, however, discussions collapsed, and, in order to pressurize the Vietnamese Politburo to resume negotiations, US planes conducted the heaviest bombing raids of the war against North Viet Nam. Only after that was the cease-fire agreement finally signed in Paris in January 1973.

The Paris Agreement provided for the complete withdrawal of all US troops from Viet Nam, together with the return of US prisoners of war, by the end of March 1973. That part of the agreement was largely fulfilled. However, the remaining terms of the agreement, including provisions for political freedom in the South and the creation of a National Council of Reconciliation and Concord were virtually ignored during the next two years. For the USA the war was over. Since 1961 they had suffered 45,941 combat deaths and over 10,000 deaths from other causes in Viet Nam, as well as 150,000 casualties. In the same period, it was estimated, nearly 2m. Vietnamese on both sides had been killed in the war.

For the Vietnamese the war was not yet over: the Paris Agreement provided for a cease-fire, without any requirement that North Vietnamese forces be withdrawn from the South. Nor was there any provision for action by the USA to enforce the agreement, or to end the fighting in neighbouring Cambodia, and in July 1973 the US Congress made any further US military action in Indo-China illegal. The international commission set up to supervise the cease-fire was not able to prevent frequent outbreaks of fighting between the two sides, while the North Vietnamese were now free to undertake a final offensive. Following the fall of the entire province of Phuoc Long in January 1975, the pace was accelerated. By the end of March the communists controlled Hué and Da Nang and were advancing southwards along the coast. In April they threatened Saigon. Thieu resigned, to be succeeded for a few days by his Vice-President and then by Gen. Duong Van Minh. On 30 April the last members of the US embassy and other personnel were evacuated and the communists entered Saigon, which they renamed Ho Chi Minh City.

NORTH VIET NAM

During the same 20-year period following the Geneva agreements, Viet Nam north of the 17th parallel underwent a political and social revolution under a communist-led regime. A land reform programme, carried out in 1953–56, proved controversial, with an estimated 15,000 people killed during its implementation. Its object was to eliminate as a class the landlords and rich peasants who—often through money-lending as much as through actual ownership of land—had dominated the economic life of the villages. Land was confiscated and redistributed, and in many cases the former owners were publicly disgraced, imprisoned or killed. This social revolution in the countryside was followed in 1959–60 by a movement for the widespread introduction of co-operatives. During the 10 years from 1956 a series of national economic plans made some progress towards developing agricultural and industrial production, using a limited amount of material and technical aid from the major communist countries. However, these programmes were impeded by the separation from the southern half of the country.

There was an inevitable conflict of priorities between the desire to implement a socialist revolution in the North, and the policy of 'completing the revolution' in the South and bringing about national reunification. It was probably not finally settled until 1960, when the Hanoi leaders decided on a combination of support for the NLF in the South, and the simultaneous implementation of a Five-Year Plan (1961–65) in the North. Meanwhile, important changes in leadership took place in 1956–57, with the removal of Truong Chinh as party Secretary-General and the emergence (by October 1957) of Le Duan as a major figure in Hanoi; previously he had led the party in the South.

At the beginning of 1960 a new Constitution came into effect, and in May elections were held for a new National Assembly to take the place of that elected in 1946. The third National Congress of the Lao Dong Party, in September 1960, established the broad framework of policy within which the country was to develop, despite the ravages of war, over the next 15 years. The policy adopted in 1960 assumed a substantial measure of economic and military aid from the USSR. However, North Viet Nam, which also aimed to continue good relations with China, faced a growing division between the two major communist powers after 1959. The escalation of the war in the South in 1965 made it imperative to secure more Soviet aid, but Hanoi sought to maintain good relations with both Moscow and Beijing for the next 10 years.

The North was involved in the war as the 'rear area' of the NLF, following a decision in 1960 that the South Vietnamese should, in principle, achieve their own revolution, with as much support from the North as they needed. By 1973 there were 150,000 regular North Vietnamese troops in the South, but their status there continued to be denied. The war seriously disrupted life in the North during the years of US bombing, from March 1965 until November 1968, and again in 1972. During 1965–66, to defend the economy against this aerial bombardment, the Hanoi Government devised a system of decentralized activity which may well have had long-term significance for economic development. However, by 1969, when Ho Chi Minh died, it became concerned about the possibility that in a more liberal system, capitalist agriculture might again become established. During 1969–70 Truong Chinh led a movement to expand the co-operative system and to limit the encroachments of individualism. Nevertheless, during the early 1970s it became clear that Le Duan was the dominant leader in the party; he placed more emphasis on economic modernization than on ideological purity. Even after the Paris Agreement of January 1973, there continued to be a strong emphasis on combining defence and production.

THE SOCIALIST REPUBLIC OF VIET NAM

The period from the fall of Saigon in April 1975 to mid-1976 was one of transition. At first it appeared that the northern and southern parts of the country might continue to have a separate existence for several years, although it was clear that the ultimate authority in both halves was a single united party, with its Political Bureau and Secretariat in Hanoi and a southern bureau in Ho Chi Minh City. In May 1975 revolutionary committees were created at all levels in the South. The South was recognized as having its own distinctive problems, arising from the fact that its 'national democratic' revolution had still to be completed, whereas the North was already in the stage of 'socialist' revolution. A first priority in the South was to bring the economy under control, and, following the closure of private banks in August, the traders who had dominated the Saigon economy under Nguyen Van Thieu were accused of hoarding commodities and their stores were confiscated; their savings (in piastres) were rendered valueless by the introduction of a new currency which could be regulated more easily. Although it was still permissible for small capitalist enterprises to continue operating, the state took steps to control, if not to own, capitalist enterprises left behind by foreign investment. Foreign banks and enterprises were no longer allowed. Former members of the South Vietnamese army and civil service were obliged to undergo 're-education' courses, sometimes in camps where study was combined with labour. More than 300,000 Saigonese officials were sentenced to long-term re-education.

In November 1975 a reunification conference was held in Ho Chi Minh City, presided over by Truong Chinh (representing the North) and Pham Hung (representing the South), which formally decided that reunification should take place through elections to be held throughout the country during 1976. Accordingly, a single National Assembly was elected in April, and when it met on 2 July it declared the inauguration of the Socialist Republic of Viet Nam, with its capital in Hanoi. The Assembly also established a committee to draft a new Constitution. The new Government included a few members of the former Provisional Revolutionary Government of South Viet Nam, but it was dominated for the most part by the leaders of the former Democratic Republic and, in effect, of the Political Bureau of the Viet Nam Workers' Party, which was restyled the Communist Party of Viet Nam (Dang Cong San Viet Nam) at the fourth Party Congress in December. The Congress decided that the immediate economic priorities were the development of agriculture and light industry, with heavy industry (the ultimate goal) temporarily assuming secondary importance.

In 1976 Viet Nam joined the IMF and the World Bank, and in 1977 it sought Western investment in Vietnamese industrial projects. During the same year there were moves towards the 'normalization' of relations between Viet Nam and the USA, which abandoned its opposition to Vietnamese membership of the UN, to which Viet Nam was subsequently admitted in September 1977. However, the US Congress had voted in May to forbid aid to Viet Nam, while the Vietnamese, until mid-1978, made the establishment of full diplomatic relations conditional on aid. A trend towards increased dependence on the USSR began in 1977, and in June 1978 Viet Nam became a full member of the (now-defunct) Soviet trading bloc, the Council for Mutual Economic Assistance (CMEA). During the same period, a new phase of economic transformation began, and led eventually to the abolition of private trading and street markets in the South and the unification of the currencies (March–April 1978). These changes were accompanied by a programme to convert agriculture to a system of co-operatives and to transform private industry in the South, and to step up the 'redeployment' of labour from cities to the New Economic Zones (NEZs).

The economic decisions of mid-1977 were part of a much more fundamental shift in Viet Nam's orientation towards the USSR and away from the People's Republic of China. When Pham Van Dong visited Beijing in June of that year, the Chinese Government presented a memorandum suggesting that Viet Nam had not honoured agreements made in the 1950s concerning Sino-Vietnamese borders (both on land and in the Gulf of Bac Bo), and the problem of citizenship of Chinese residents living in Viet Nam. A third issue was China's insistence on maintaining close relations with the Pol Pot regime in Cambodia, which by this time was positively hostile towards Viet Nam. The Vietnamese wanted to draw Phnom-Penh into close ties with Viet Nam and Laos, in an 'Indo-Chinese Federation', whereas China wished to maintain direct relations with Cambodia and Laos independently of Hanoi, thereby thwarting Soviet ambitions in the region.

From September 1977 there was a growing military confrontation between Viet Nam and Cambodia (or Kampuchea, as it was known between 1976 and 1989). It became clear in late 1978 that the Chinese were unable or unwilling to escalate their own military involvement, and it was impossible for the Cambodians to hold out indefinitely. In December the Vietnamese invaded, ostensibly in support of the Kampuchean National United Front for National Salvation, and gained control of Phnom-Penh and other major centres during the early part of

1979, enabling a pro-Vietnamese Government under Heng Samrin to take power (although Pol Pot forces continued to resist in more remote areas). In February 1979 this Government signed a treaty of friendship with Viet Nam, comparable to that already signed with Laos, and was willing to accept the continuing presence of Vietnamese forces which by mid-1984 were estimated at 160,000 to 180,000.

By that time relations with the People's Republic of China had deteriorated drastically. The socialist measures that were introduced in early 1978 had especially affected the Chinese community in southern Viet Nam, which dominated the commercial sector. This fact, combined with fears among Chinese in northern Viet Nam that sooner or later there might be a Sino-Vietnamese war, led to a major exodus of Chinese residents. Between April and July 1978 (when China closed its border), 160,000 of them had fled by land into China; others began to leave by sea, hoping to find refuge elsewhere in South-East Asia. Until this point China had continued to give economic aid to Viet Nam, but between May and July 1978 they abandoned all projects, leaving the Vietnamese more dependent than ever on the USSR and Eastern Europe. In November 1978 Le Duan and Pham Van Dong signed a formal treaty of friendship and co-operation with the USSR. In due course it became evident that the Soviet navy had been granted facilities at Cam Ranh Bay, once a major US base. Over the next few years, the USSR was able to convert it into a naval base for its own use. By 1984 there were said to be several thousand military and civilian advisers from the USSR in Viet Nam.

China was dissatisfied with Viet Nam's invasion of Cambodia, with its orientation towards Moscow, and with its treatment of Chinese residents; by early 1979 there were also serious frontier clashes. In February China sent a punitive force into Viet Nam, and fierce fighting lasted for a month; the Chinese withdrew in March. Negotiations held in April and May produced no formal agreement but it was decided in May to start the exchange of prisoners of war.

Meanwhile, Vietnamese policy towards the Chinese residents (also known as 'Hoa' people) had become so severe that vast numbers of them were taking to ships in the South China Sea, hoping to reach land or to be picked up by larger vessels. It was said that the Vietnamese authorities had now ceased to object to their leaving, but charged large sums in gold or Western currencies for exit papers. More Chinese also crossed into the People's Republic of China at this time. By mid-1979 it was estimated that there were 200,000 such refugees from Viet Nam in China, and perhaps another 200,000 had reached other countries of South-East Asia, Hong Kong, Taiwan or Australia. Many thousands more were thought to have drowned at sea. As a result of a UN conference in July, Viet Nam agreed to allow an Orderly Departure Programme, sponsored by the UN High Commissioner for Refugees (UNHCR): by June 1992 354,500 people had left the country in this way. Illegal departures continued, however.

Important political changes occurred during 1980–81, starting with a government reshuffle in January 1980. In December, after four years of debate, a new Constitution was finally adopted, under which a newly elected National Assembly met in July 1981 and appointed both a State Council (which was to constitute a collective presidency) and a new Council of Ministers. Truong Chinh became President of the former, and therefore Head of State, while Pham Van Dong remained Chairman of the Council of Ministers (Prime Minister) with slightly reduced powers. Le Duan, still dominant in the Communist Party, was not included in either of the principal state organs. The party's fifth Congress, twice postponed, was eventually held at the end of March 1982. It became clear that the party was deeply divided and that some elements in the leadership were being blamed for the country's severe economic failures. When the Congress finally met, a number of prominent figures were removed from the new Politburo (including Gen. Vo Nguyen Giap) and others from the Central Committee. A number of younger men were promoted, but the generation of the 1950s had still not finally relinquished control.

In the early 1980s tension between Viet Nam and Thailand led to greater involvement of the member states of the Association of South East Asian Nations (ASEAN) in the Cambodian question. ASEAN countries promoted an alliance between Pol Pot's Khmer Rouge forces and the non-communist groups, led by Prince Sihanouk and Son Sann, with the result that those three elements formed the coalition Government-in-exile of Democratic Kampuchea in June 1982. The Democratic Kampuchean resistance forces continued to oppose the presence of Vietnamese troops, and were able to establish base camps along the border between Cambodia and Thailand, with the support of the ASEAN states, China and the USA. In early 1985 the Vietnamese captured and destroyed the bases of all three resistance groups, but were unable to defeat the more mobile guerrilla forces. In 1984 Viet Nam announced that it anticipated the total withdrawal of its troops from Cambodia within five to 10 years, and in August 1985 the deadline was established as 1990. However, there appeared to be little reason to believe that either the Government in Hanoi or that in Phnom-Penh genuinely desired a compromise with the forces of Democratic Kampuchea. Meanwhile, the Chinese continued to exert pressure on Viet Nam's northern frontier, forcing Hanoi to keep a substantial part of its armed forces in that area.

REFORM INITIATIVES

By 1985 relations between Viet Nam and the USSR were so close that it was impossible for the changes that followed the emergence of Mikhail Gorbachev as Soviet leader not to have some impact in Hanoi. In mid-1985 the USSR promised to increase its aid to Viet Nam, notably in the fields of petroleum exploration and refining, petrochemicals and metallurgy, while emphasizing the need to apply in Viet Nam the same kind of economic restructuring as was taking place elsewhere in the socialist world: a reduction of the role of centralized planning and state subsidies and the introduction of 'socialist accounting'. The Eighth Plenum of the Central Committee, convened in Hanoi in June, decided to implement such reforms; but a poorly managed change of currency in September led to rampant inflation, for which the reformers were blamed. The policy of reform was thus slowed throughout 1986.

Significant changes in the Vietnamese leadership occurred during 1986–87. In July 1986 Le Duan died, and was succeeded as General Secretary of the Communist Party by Truong Chinh, who also retained the position of President (the first occasion on which the top party and state posts had been combined in Viet Nam since the death of Ho Chi Minh). At the sixth Party Congress in December 1986, it was announced that the three most senior members of the Politburo, Truong Chinh himself, Pham Van Dong and Le Duc Tho, would all retire, to become advisers to the Central Committee. Nguyen Van Linh was elected the new General Secretary of the party. Linh had played an important role in the South from 1945 to 1975 and had then become party leader in Ho Chi Minh City from 1975 to 1986. He had been responsible for a number of economic experiments in the South, but in 1982 had been removed from the Politburo, only to resume his place there in June 1985, when the need for reform was again accepted. In February 1987 the Minister of Defence, Van Tien Dung, was forced to retire and was succeeded by another veteran of the southern struggle, Le Duc Anh. At that time, several other appointments were made as part of a major restructuring of government ministries in the economic field. At that time a new National Assembly, elected in April, appointed Vo Chi Cong as President of the State Council (and Head of State) and Pham Hung as Chairman of the Council of Ministers. They too were southern veterans, as was Vo Van Kiet, the reform-orientated head of the State Planning Commission and senior Deputy Chairman of the Council of Ministers. Thus, in the course of one year, the incumbents of all the highest positions had changed, and there was a significant shift of power in favour of southerners, who tended to be more supportive of the economic reform programme.

Viet Nam also had to adjust to the new Soviet policy towards Asia, first indicated in a comprehensive way by Gorbachev in July 1986 in a major speech delivered in Vladivostok (USSR). He advocated not only an improvement in Sino-Soviet relations, which by then was already in progress, but also an eventual restoration of relations between China and Viet Nam, which in turn required a concerted effort to resolve the conflict in Cambodia. Incidents on the Sino-Vietnamese border decreased in number in the second half of the year. More fighting occurred, however, in January 1987, when a major battle on the Yunnan sector of the border was alleged by the Vietnamese to have resulted in the deaths of more than 1,500 Chinese troops. The Chinese leadership now regarded Cambodia as the principal obstacle to a normalization of Sino-Soviet relations. Beijing insisted that the USSR should at least put pressure on Viet Nam to withdraw its troops, instead of continuing to provide the support that enabled the Vietnamese occupation to continue. The Cambodian situation also remained an obstacle to any attempt by Viet Nam itself to open up its own economy to foreign investment. The USA, Japan, the ASEAN countries and the European Community (EC, now the European Union—EU) had all made it clear that the lifting of their embargo on economic relations with Viet Nam, imposed in 1979–80, was conditional upon the withdrawal of Vietnamese troops from Cambodia and upon the conclusion of a comprehensive political settlement. In March 1986 an eight-point peace proposal, submitted by the coalition Government of Democratic Kampuchea, was summarily rejected by the Vietnamese, although it was later regarded as the first step towards negotiations. It called for a supervised withdrawal of Vietnamese troops, followed by the establishment of a quadripartite coalition government including the Phnom-Penh administration and the three resistance groups. In January 1987 the Phnom-Penh regime announced its willingness to negotiate with the resistance groups, but this was immediately rejected by the Khmers Rouge. Viet Nam's urgent need of Western aid, and increasing pressure from the USSR, prompted an announcement in May 1988 that Viet Nam would repatriate 50,000 (of an estimated total of 100,000) troops from Cambodia by the end of 1989. The Vietnamese military high command left Cambodia in June, placing the remainder of the Vietnamese forces under Cambodian control. In July Viet Nam was represented at an informal meeting in Jakarta with the four Cambodian factions and the six members of ASEAN. In January 1989 a commitment to repatriate all Vietnamese troops by September was declared, provided that a political settlement had been achieved in Cambodia.

Meanwhile, the Vietnamese leadership appeared committed to the advancement of reform, although the continuing influence of 'hardliners' remained an obstacle to progress. Following the death of the Chairman of the Council of Ministers, Pham Hung, a meeting of the National Assembly in June took the unprecedented step of nominating Vo Van Kiet, an advocate of political reform, to oppose the Central Committee's more conservative candidate, Do Muoi (ranked third in the Politburo), in the election for the chairmanship of the Council of Ministers. Do Muoi was elected by a majority to the post, but Vo Van Kiet received unexpectedly strong support (36% of the votes). The 'retired' Le Duc Tho continued to exert considerable influence until his death in October 1990. Widespread dissatisfaction with the slow progress towards 'renovation' (*doi moi*) of the economy, often due to the refusal of officials to implement new policies, led to the removal of hundreds of cadres from government posts. In March 1989 a reshuffle of senior economic ministers took place, adjusting the balance further towards reform and strengthening the position of the party leader, Nguyen Van Linh. In the late 1980s he implemented a number of significant economic reforms.

While economic reforms were undertaken, there were few concurrent measures to liberalize the political system. Municipal elections to provincial and district councils took place in November 1989. Under new legislation, adopted during a meeting of the National Assembly in June, candidates who were not members of the Communist Party were allowed to participate for the first time. (Large numbers of non-Communist Party candidates won posts in November 1994 at the subsequent local government elections.) In December 1989 the National Assembly approved legislation that imposed new restrictions on the press. Open dissension towards the policy of the Communist Party also became a criminal offence. In late 1989 progress towards political reform under *doi moi* was adversely affected by government concern over the demise of socialism in Eastern Europe. At a meeting of the Central Committee in March 1990, tension over the issue of political pluralism resulted in the dismissal of a member of the Politburo, Tran Xuan Bach, who had openly advocated political reform. In early April the Council of State announced an extensive ministerial reorganization.

Following the death in October 1990 of Le Duc Tho, division within the Government on issues of political and economic policy increased. In early 1991, after the publication of articles criticizing government policy and the socialist system, the Communist Party increased surveillance of dissidents and ordered the media to publish retaliatory articles condemning party critics. The seventh party Congress took place in June, following a series of delays which were attributed to increasing dissension within the party. The Congress approved two draft reports presented by the Central Committee, one of which comprised a 10-year programme of economic liberalization, while the other reaffirmed the party's commitment to a socialist political system. Seven members of the Politburo, including Nguyen Van Linh, were retired from their posts. Do Muoi was elected General Secretary of the Communist Party. There seemed to be a consensus within the party leadership, however, that a measure of political reform had become vital, both for the success of economic reform and in order to obviate any potential challenge to the party's monopoly of political power. There was also a notable attempt at the seventh Congress to emphasize the distinctiveness and national character of Vietnamese communism. While Viet Nam's leaders robustly and repeatedly rejected any idea of political pluralism following the collapse of the USSR in December 1991, they remained aware that the success of the country's economic reforms required a move away from archaic Leninist political structures.

The ninth session of the National Assembly (eighth legislature), which took place between late July and early August 1991, elected the reformist, Vo Van Kiet, to replace Do Muoi as Chairman of the Council of Ministers. The National Assembly considered extensive amendments to the Constitution, whereby the power of the premier and individual ministers would be greatly increased. The postponement of decisions about these constitutional amendments, at a Plenum of the party Central Committee in December, indicated that conservatives remained unhappy with the proposed democratic changes. The amendments were finally ratified by the National Assembly in April 1992. Like the previous (1980) Constitution, it emphasized the central role of the Communist Party; however, it stipulated that the party must be subject to the law. While affirming adherence to a state-regulated socialist economic system, the new Constitution guaranteed protection for foreign investment in Viet Nam, and permitted foreign travel and overseas investment for Vietnamese. Land was to remain the property of the State, although the right to long-term leases, which could be inherited or sold, was granted. The National Assembly was to be reduced in number, but was to have greater power. The Council of State was to be replaced by a single President as Head of State: he or she would be responsible for appointing (subject to approval of the National Assembly) a Prime Minister and senior members of the judiciary, and would command the armed forces (which were now obliged under the Constitution to defend the socialist regime). The new Constitution was to enter into effect after the forthcoming general election.

Under the provisions of the new Constitution, elections to the Ninth National Assembly took place on 19 July 1992. The delays in enacting the legislative amendments, however, confirmed the fears of many in the party leadership that political reform would inevitably undermine the Communist Party's monopoly of power. The elections and the manner in which they were conducted emphasized the continuing influence of party conservatives. While there was a choice of candidates in all constituencies (a total of 601 candidates contested 395 seats), about 90% were members of the ruling Communist Party. Even this figure was misleading. In the past, candidates had to be sponsored by the Viet Nam Fatherland Front, an official body encompassing mass organizations, such as the trade unions. Under the new Constitution, however, private individuals were permitted to stand for election without the endorsement of the Fatherland Front. In practice, only two independent candidates contested the elections, both of whom failed to win a seat, while 41 others who had applied subsequently withdrew their candidature or were disqualified. Only 105 deputies, however, were re-elected from the Eighth National Assembly.

At the first session of the new National Assembly (ninth legislature) in September 1992 the conservative Gen. Le Duc Anh, a former Minister of National Defence and a high-ranking member of the Politburo, was elected to the new post of President, and Nguyen Thi Binh, a member of the National Assembly, was elected as Vice-President (the first woman to hold such a senior position). Vo Van Kiet was re-elected as Prime Minister. At the beginning of October the National Assembly approved the membership of the new National Defence and Security Council and the new Cabinet. The anticipated streamlining of bureaucracy did not take place; indeed, several posts were upgraded to ministerial rank.

During 1993 and subsequently there were indications that the National Assembly was beginning to play a greater role in the political life of Viet Nam. In July, after lengthy debate, the Assembly adopted legislation on agriculture that stopped short of recognizing private ownership of land, but liberalized landholding rights for the country's 55m. peasants. Under the law, farmers were to be permitted to purchase, sell, transfer and rent as well as inherit the right to use land, even though it was to remain the property of the State. The new legislation marked a further step in the phasing-out of collective agriculture. Nevertheless, intense disputes over access to land and land usage continued. In February 1995 the Prime Minister amended the 1993 land law by decree in order to curtail land speculation. This was followed in March by the declaration that rice land could not be converted to other purposes, except under special circumstances. Both actions led to critical comments by deputies in the National Assembly.

Despite the country's economic liberalization, however, there was no toleration of political dissent. In August 1993 14 people were convicted of having conspired to overthrow the Government, and sentenced to terms of imprisonment, and in November several Buddhist monks were imprisoned for having, it was alleged, incited anti-Government demonstrations. In late 1994 two senior Buddhist dissidents, Thich Huyen Quang and Thich Quang Do, were taken into custody. In August of the same year political dissidents associated with the Tan Dai Viet Party (Movement for National Unification and Democracy Building) and officials of the opposition Unified Buddhist Church of Viet Nam were put on trial separately. Arrests and imprisonments for dissent continued throughout 1995 and 1996.

In January 1994, at a mid-term conference of the Communist Party, Do Muoi praised the country's recent economic achievements and its expanding relations with foreign countries, but emphasized the continuing prevalence of poverty, the increase in corruption and crime, and the inadequacy of the educational system; he also strongly criticized the party for disunity and poor organization. Immediately before the conference four new members were appointed to the Politburo, including the Minister of Foreign Affairs, Nguyen Manh Cam, and during the conference 20 new members (all under the age of 55) were elected to the party's Central Committee. This reflected a recruitment drive during 1994, when some 60,000 new members joined the Communist Party (an increase of 20%, compared with the previous year).

In June 1994 the National Assembly approved new legislation, including a labour law which guaranteed the right of workers to strike (providing the 'social life of the community' was not adversely affected). Strikes subsequently occurred in some southern provinces and, in August, the first incident of industrial action in Hanoi was reported. At a Plenum of the Central Committee of the party, held in late July–early August, Do Muoi reasserted the need to strengthen Viet Nam's 'open-door' policy in order to attract wider foreign investment; attention was also drawn to the increasing income differential between the urban and rural economies.

In October 1994 the National Assembly held a 12-day session to consider its legislative programme for 1995; priority was assigned to further administrative and legislative reforms, including the draft civil code. This priority was reaffirmed at a Plenum of the Central Committee of the Communist Party, held in January 1995. In March–April the National Assembly finally approved legislation on state enterprises, determining various categories of enterprise and the degree of government control over ownership. At its end-of-year session in November, the legislature finally adopted the civil code whose 12 drafts had been under consideration since 1980. The Assembly also approved a major restructuring of ministerial portfolios: eight separate ministries (or state commissions) were merged into three new ministries (of agriculture and rural development, of industry, and of planning and investment).

The end-of-year Plenum (November 1995) of the Communist Party adopted the confidential draft documents to be presented to the eighth national Congress scheduled for the following year. Included was a controversial amendment to the party statutes, which abolished the Secretariat and replaced it with a more powerful Standing Board of the Politburo. This drew attention to the party leadership issue, and provoked heated disagreement among conservatives and reformers. In February 1996 the conservatives launched a campaign against 'social evils', including pornography, prostitution, gambling and drug abuse.

Conservative party leaders also made a concerted effort to bring about the retirement of key reformers and to capture the party's most senior leadership post. Signs of internal party division soon emerged in public. Prime Minister Vo Van Kiet was attacked by conservatives for a confidential memorandum he had written in August 1995, shortly after the normalization of relations with the USA, in which he downgraded the importance of ideology, argued for the curtailment of party interference in day-to-day state affairs, and recommended thoroughgoing reform of state-owned enterprises. The conservatives countered in October in a document widely circulated within party ranks. They argued that the normalization of relations with the USA would result in greater interference in Viet Nam's internal affairs. Viet Nam, they asserted, was the target of a 'hostile campaign of peaceful evolution' aimed at overthrowing Communist Party rule. In April the party Central Committee's 10th Plenum released to the public draft policy documents to be presented to the eighth Congress and, in an unexpected decision, announced the expulsion from the party of leading conservative Nguyen Ha Phan for 'having committed serious mistakes in his past activities'. Phan was in charge of the party's economic commission and also of a special personnel commission charged with examining the suitability of delegates to the eighth Congress. Shortly after Phan's demise, his ideological and political mentor, senior Politburo member Dao Duy Tung, was reportedly placed under house arrest. The party leadership then moved to reselect delegates to the forthcoming national Congress.

At the Eighth Congress (June–July 1996) party delegates acted to restore internal party unity. The three senior leaders, Secretary-General Do Muoi, State President Le Duc Anh and Prime Minister Vo Van Kiet were retained in office, despite strong rumours that they would step down. Their retention underscored the party's desire to maintain political stability and policy continuity. The new Politburo was expanded to 19 members, the largest in the party's history. Delegates also endorsed, with several notable modifications, the compromise policy documents drafted in late 1995 by the reformist and conservative camps. The new documents made clear that Viet Nam would not embark on any destabilizing political reforms. Political pluralism and multi-party democracy were condemned in fulsome terms. Priority was given to the state sector, which was to continue to dominate the economy, but private-sector activities were also to be allowed. Viet Nam was to continue its 'open-door' policy. Priority was to be given to improving the legal and legislative base to support continued foreign direct investment. A key theme emerging from the eighth Congress was the stress on economic modernization and industrialization in the years to come. In an unexpected move, the delegates voted to delete the powers accorded to the Politburo Standing Board from the draft party statutes. The statutes were then approved.

In 1995 the party launched a major recruitment drive to mark the 65th anniversary of its founding. Figures released a year later indicated that although the party had increased its membership by 85,900 new members (the highest intake since 1986) to a total of 2.13m., recruitment had not kept pace with population growth. More women, ethnic minorities and better-educated individuals were recruited, but figures for the under-30 age group showed a decline. Party membership in the over-60 category increased slightly. The percentage of urban party members increased, while that of peasant farmers fell from one-third to just over one-quarter of total membership. Official figures released in 1996 and 1997 indicated that membership of the party increased in those years by 90,000 and 103,000 respectively. Of the members that joined in 1997—the highest number to join the party in any single year for a decade—24.3% were women, 12.8% were from ethnic minority groups and 19% were college graduates. In August 1998 it was reported that more than 47,600 new members had been admitted to the party in the first six months of that year.

In 1996 two high-profile corruption cases received extensive media attention. The first involved Tamexco, a trading firm with connections to the Ho Chi Minh City party apparatus. Several leading officers of the Viet Nam Commercial Bank were also implicated. In January 1997 the Ho Chi Minh City Court sentenced four of the 15 defendants to death by firing squad; this decision was upheld in March by the Appeals Court, and in January 1998 three of the four who had been sentenced to death were executed. The second corruption case involved directors of the Minh Phung Export Garment Company and Epco, a trading firm: the head of the former engaged in real estate speculation using collateral provided by the latter for use in securing a loan from the Viet Nam Commercial Bank. The fraud was revealed when Minh Phung's director could not repay the bank loan and it was discovered that the collateral was missing. In early 1998, in an illustration of the Government's commitment to countering corruption, the Communist Party announced that in 1997 it had disciplined and expelled some 18,000 members, and sentenced 469 to terms of imprisonment. Further members of the party were reported to have been disciplined over corruption-related offences in July 1998. Also in 1998 it was reported that allegations of corruption made against one of the members of the Politburo Standing Board, Pham The Duyet, were under investigation by Communist Party leaders. Political analysts alternatively speculated that the emergence of the allegations suggested a power struggle within the Politburo or served as evidence of a rift between the current leadership of the Communist Party and the party 'old guard'.

In November 1996 the 10th session of the Ninth National Assembly brought about a major cabinet reorganization (three ministerial-level posts were abolished) and reshuffle that reflected personnel changes on the new party Central Committee selected by the eighth national Congress. In all, 12 positions were vacated and eight new ministers appointed. The National Assembly also voted to remove Le Thanh Dao from his post as head of the Supreme People's Organ of Control because of corruption in his business dealings. It also decided to increase the number of provinces from 53 to 61 (since several were becoming too powerful), and raised the status of Da Nang (directly responsible to the central government).

The party convened the second plenary session of the Central Committee in December 1996. In late January 1997 the Politburo issued an anti-corruption resolution. The next meeting of the Central Committee was held against a background of widespread rural unrest in central Viet Nam, especially in the province of Thai Binh. Owing to repression of both the foreign and domestic media, it was only in September 1997 that detailed reports concerning the unrest appeared in the party and army newspapers. These sources indicated that, in May, an estimated 3,000 farmers from 128 villages had marched on a district office to protest against the imposition of numerous taxes and fees as well as corruption by local officials. In June 1997 rural protests escalated in intensity and spread to the provincial capital. According to official reports issued later, these incidents were serious in nature, and 'extremist actions', such as physical assault on local officials and their property, were alleged to have occurred. Special police units were dispatched to restore law and order; party leaders in Hanoi also responded by sending a member of the Politburo to make two inspection visits. In November, as unrest continued, the Government pledged to provide financial aid to Thai Binh province. In July 1998 more than 30 people were sentenced to terms of imprisonment for their involvement in the unrest. In order to combat continued instability, the Politburo authorized local police and army units to establish temporary detainment camps, where individuals could be held for up to two years without trial.

The Third Plenum of the Central Committee, held in June 1997, considered these events and passed a major resolution on 'developing the people's right to be masters of the nation'; in other words, in order to regain peasant support the party decided to implement policies that would give farmers greater influence at the basic level. At the same time, more scope was to be given to representatives from 'people's organizations' at local level in the political process. These initiatives were designed to provide checks and balances on local authorities. The Third Plenum also made the decision to expand the party's Central Control Commission by including two new members, to enhance its effectiveness in dealing with abuse by party and state officials. More importantly, the Third Plenum adopted a second major resolution, which set out the guide-lines for developing a personnel strategy for cadres employed by the State. Finally, the Third Plenum also decided on the procedures to be followed in the forthcoming elections for the 10th legislature (1997–2002) of the National Assembly. On the eve of the Plenum it was announced that President Le Duc Anh, Prime Minister Vo Van Kiet and nine other cabinet-level officials, as well as Secretary-General Do Muoi, would not be standing for re-election. According to foreign press reports, the Plenum endorsed Phan Van Khai, Deputy Prime Minister, to replace Vo Van Kiet; it also endorsed the retention of Nong Duc Manh as Chairman of the National Assembly Standing Committee. The Plenum could not, however, reach consensus on Le Duc Anh's replacement as President.

Viet Nam's National Assembly elections proved to be a watershed. Under a revised electoral law, the National Assembly was expanded from 395 to 450 seats organized in multi-member constituencies, which were contested by a record 663 candidates. The party-controlled pre-selection process was modified to enable more independent (or self-nominated) candidates to stand than previously. On 20 July 1997 the elections resulted in the selection of a younger and better-educated legislature. The number of women and ethnic minority deputies rose markedly, while the number of non-party deputies nearly doubled from 8% to 15% of the total. Significantly, 11 independent candidates qualified to stand, of which three were elected.

The convening of the first session of the 10th legislature was preceded by the holding of the party Central Committee's Fourth Plenum in September. This meeting once again considered nominees for the office of President, as well as other leadership posts. There were two main contenders for the presidency, namely Nguyen Manh Cam, the Minister of Foreign Affairs, and Doan Khue, the Minister of National Defence. Once again, however, deadlock ensued, but a compromise was eventually reached: Deputy Prime Minister Tran Duc Luong, a technocrat, was selected as a candidate. The National Assembly convened in September and duly endorsed the party's nominees. Tran Duc Luong was elected President after Doan Khue stepped aside. President Luong then nominated Nguyen Thi Binh as Vice-President, Nong Duc Manh as Chairman of the Assembly's Standing Committee and Phan Van Khai as Prime Minister; all were elected.

Prime Minister Khai then successfully proposed that the number of Deputy Prime Ministers be expanded from three to five and that each be given responsibility for an enlarged portfolio. Nguyen Tan Dung, at 48 the youngest member of the Politburo, was elected as First Deputy Prime Minister with responsibility for internal and economic affairs; Nguyen Manh Cam, who retained his previous post as Minister for Foreign Affairs, was given responsibility for foreign political and economic relations; Nguyen Cong Tan, the former Minister for Agriculture and Rural Development, was made responsible for rural development and water resources; Phan Gia Khiem, the former Minister for Science, Technology and the Environment, was charged with overseeing culture, education and environmental matters; and Ngo Xuan Loc, the former Minister of Construction, was placed in charge of industry, construction and transport. Prime Minister Khai also proposed a Cabinet consisting of 31 persons (29 men and two women). Included in this number were seven new ministers. Unexpectedly, however, the National Assembly rejected Phan Van Khai's nominee for Governor of the State Bank of Viet Nam, Cao Sy Kiem. Kiem's rejection was due to his implication in several scandals and debt-management problems in the banking sector during the previous year. In May 1998 Deputy Prime Minister Nguyen Tan Dung was appointed Governor of the State Bank of Viet Nam.

At the enlarged Fourth Plenum of the Central Committee, which met in December 1997, Do Muoi resigned and was replaced as party General Secretary by Le Kha Phieu. (Phieu had served as a political officer in the army before being appointed to the party Secretariat in 1992 and subsequently being placed in charge of the party's Internal Political Protection Commission.) Following the resignation from the Politburo of Do Muoi, Le Duc Anh and Vo Van Kiet, four new members—Phan Dien, Nguyen Minh Triet, Nguyen Phu Trong and Gen. Pham Thanh Ngan—were appointed. These appointments strengthened the conservative influence within the party. Immediately after his election, Phieu reshuffled the Politburo Standing Board and effected a major reorganization of the military hierarchy. Tran Duc Luong, Phan Van Khai, Nong Duc Manh and Pham The Duyet were appointed to the Politburo Standing Board.

In addition to the leadership question, the party Central Committee's Fourth Plenum also considered economic issues. The Committee resolved to accord priority to the improvement of Viet Nam's economic efficiency and international competitiveness and to mobilize domestic capital; particular emphasis was to be placed on the labour-intensive, export-orientated processing industries. It was also decided to continue with the equitization of state-owned enterprises although at a cautious pace. The Government was ordered to reduce spending, balance the budget and practise thrift. The party Central Committee held its fifth plenary session in July 1998. Two principal issues were discussed: namely, the challenge posed to Vietnamese identity and culture by the forces of globalization and the impact on the country's economy of the Asian financial crisis.

In August 1998 the release of more than 5,000 prisoners under a general amnesty was announced by the Government. Among those released were prominent dissidents Doan Viet Hoat and Nguyen Doan Que; the release of both men was, however, reported to be conditional upon their immediate exile from Viet Nam. Following his release, Hoat sought exile in the USA while Que, who refused to emigrate, was placed under house arrest. In early September 1999 it was announced that the Government was to free a further 1,712 prisoners and to reduce the sentences of more than 4,000 others under an amnesty to mark the 54th anniversary of the country's declaration of independence. Meanwhile, in October 1997 the Special Rapporteur on religious intolerance of the UN Commission on Human Rights visited Viet Nam to investigate assertions made by the Vietnamese authorities that greater freedom of religious expression was being tolerated in the country. At the end of his visit, he stated that his investigation had been obstructed by government officials and that he had been prevented from meeting key religious dissidents.

Viet Nam's deteriorating economic circumstances were addressed by the Communist Party Central Committee's Sixth Plenum (first session) in October 1998. Although the final communiqué noted that Viet Nam's 'economic growth rate has slowed, and output and trade have become less effective', it declared, none the less, that Viet Nam had achieved considerable success in maintaining political stability and in the attainment of a 6% expansion of the country's gross domestic product (GDP) in 1998. The Plenum accorded priority to agriculture and rural development, and noted that Viet Nam continued to possess plentiful untapped domestic resources that could be used to stimulate development. It indicated that an effort would also be made to revitalize rural co-operatives in an attempt to reassert central control following a period of prolonged decentralization. During the Sixth Plenum Viet Nam's leaders once again demonstrated their unwillingness to adopt the comprehensive reform programme being urged upon them by international financial institutions and donor countries. Their central concern instead was to maintain internal stability and party unity, while preserving Vietnamese culture and values in the face of the intrusive forces of globalization. These concerns were heightened by a marked rise in citizens' complaints about corruption and continued rural unrest during 1998. In December the international donor community pledged US $2,700m. in aid to Viet Nam; an additional 'package' of $500m. was made conditional upon Viet Nam's adoption of accelerated reforms.

After two postponements, the Communist Party Central Committee held the second session of the Sixth Plenum from 25 January to 2 February 1999. The session was held against a background of several important developments. First, the Communist Party announced that an exceptionally high number of new members—106,000—had been recruited in 1998, a level of intake that exceeded that achieved in all years since 1986, when *doi moi* was adopted; more than half of the newly recruited members were under 30 years of age. Second, the party was faced with a new uprising of internal party dissent, spearheaded by very senior, retired party officials such as Gen. Tran Do, the former head of the party's Ideology and Culture Commission. As a consequence of Tran Do's criticisms, he was expelled from the party in January 1999, provoking renewed protests. Third, the session was held amidst speculation, precipitated by the death of Politburo member and former Minister of Defence, Doan Khue, that forecast major changes in the party and state leadership. However, no consensus could be reached as to who would replace him on the ruling body. During its second session, the Sixth Plenum considered the best ways and means by which the efficiency and administrative performance of the Communist Party in the implementation of Viet Nam's reform programme might be improved, and focused in particular upon the need to counter degradation in the party's ranks caused by corruption, excessive bureaucracy, 'individualism', and internal disunity. The Plenum resolved to launch a three-year 'criticism and self-criticism' campaign, designed to rid the party of its degenerate members and to restore unity. Party officials were charged with the task of establishing guidelines on the question of the extent to which a party member and his or her family members should be allowed to participate in private economic activities; a committee was also set up to formulate a plan to reduce party and state bureaucracies. The discussion of any reorganization of the party leadership was, however, postponed by the Plenum. In June the official Vietnamese media published the text of a Politburo decision prohibiting party members from speaking out publicly against party decisions or inciting others to do so; party members also continued to be banned from distributing documents at variance with official party policy. The Central Committee's Seventh Plenum met in Hanoi in August. It focused on issues relating to the organization and apparatus of the political system, wages and social allowances funded by the state budget, and preparations for the forthcoming ninth national party congress, which was scheduled for the first quarter of 2001.

Two large-scale corruption trials were held in Viet Nam in 1999. The trial of 74 people accused of smuggling (more than one-half of whom were reported to be former government officials) opened in late March; two of the defendants were subsequently sentenced to death, and the remainder to terms of imprisonment, in April. In May 77 businessmen, bankers and government officials appeared in court charged with the fraudulent procurement of state loans; subsequently, in early August, six of the defendants were sentenced to death and six others to life imprisonment, while the remainder received various other terms of imprisonment.

The political situation in Viet Nam throughout 1999–2000 was affected by the continued debate between reformist and conservative party members over the pace and extent of the economic reform programme. The Vietnamese economy remained weak in 1998–2000, failing to provide sufficient employment for the rapidly increasing labour force. Foreign investment fell dramatically in the late 1990s, as a result of the severe negative impact of the Asian economic crisis. Although Viet Nam received substantial international assistance in June 2000, the World Bank and IMF again withheld lending until the Government adopted a 'more comprehensive approach to reform'. However, the leadership continued to resist any attempts radically to reform the economy, preferring not to jeopardize political stability.

Fierce debates over changes in leadership and policies caused the seventh party Plenum, held in August 1999, to be delayed on five occasions. One of the most controversial issues was a proposal to rationalize the Government by eliminating 15% of posts. However, the reform programme suffered several reverses. In September 1999 the authorities took a number of suppressive measures against dissent. Three Buddhist monks were arrested, some 24 people were sentenced to terms of imprisonment for allegedly attempting to overthrow the Government, while a leading political dissident, Gen. Tran Do, was denied permission to publish a newspaper. Then, at a meeting in October, the Politburo rejected the bilateral trade agreement with the USA that had been reached in principle in July (see below). Continued corruption scandals assumed particular significance in the later part of the year. In October the Customs

Chief, Phan Van Dinh, was dismissed, and 22 department officials were charged with corruption in December. In November the Ministry of Finance conducted a government audit and discovered evidence of considerable misuse of public funds. In December Phan Van Khai's principal aide, Nguyen Thai Nguyen, was detained, while the Deputy Prime Minister, Tran Xuan Gia, was dismissed for his involvement in a development project in Hanoi, which had been associated with malpractice. The head of the illicit United Buddhist Church, Thich Huyen Quang, was detained in that month, after meeting with a US diplomat. In December 1999 the National Assembly adopted a new penal code; significantly, the number of capital crimes was reduced from 44 to 29.

At the Eighth Plenum in January 2000 there was an extensive government reorganization: Vu Khoan replaced Truong Dinh Tuyen as Minister of Trade, while the hitherto Deputy Foreign Minister, Nguyen Dy Nien, replaced Nguyen Manh Cam as Minister of Foreign Affairs (although Manh Cam remained in his post of Deputy Prime Minister). There were also several changes in the party leadership. A further increase in membership was announced; the party remained divided, however. Frustrated at conservative opposition to the reform programme, Khai tendered his resignation in March 2000, although it was rejected. The Ninth Plenum, held in April 2000, began to make preparations for changes in leadership and the drafting of the political report for the Ninth Party Congress, provisionally scheduled for March 2001.

Nation-wide celebrations were held for the 25th anniversary of the fall of Saigon and the unification of the country. For the first time since 1975, public recognition of the NLF's contribution was made, and a nation-wide amnesty for 12,000 prisoners was granted. In May 2000 the authorities of Ho Chi Minh City, having become frustrated with bureaucratic delays in Hanoi, which, they believed, were acting as a disincentive to foreign investors, announced that they would unilaterally approve foreign investment if the approval process in Hanoi continued for more than two weeks. The National Assembly had also become more critical of the Government following the economic crisis, censuring several cabinet ministers and questioning many government policies in May.

At the 10th Plenum in June 2000 the future of state ownership was debated, although none of the anticipated leadership reorganizations were discussed. Proposed changes to the party's Constitution, which would allow the party General Secretary to serve concurrently as the head of state, were raised, but were subsequently rejected.

Massive peasant protests erupted in the politically sensitive Central Highlands region in February 2001, prompting great concern on the part of the Government, which rapidly dispatched troops to quell the unrest. The demonstrators belonged to ethnic minority groups, and many of them had been allied with, and served as mercenaries for, the USA during the war. Although the Government claimed that it had defeated a local guerrilla group, known as the United Front for the Liberation of Oppressed Races, in the early 1990s, it contended that this movement, and indirectly the US Government, had organized the demonstrations. However, the unrest had, in fact, erupted for domestic reasons; Hanoi had forced migration to the sparsely populated border regions since 1975, disturbing the ethnic balance there, while a further 3.6m. Vietnamese had migrated to the area in later years to establish coffee plantations. The Government was prepared for more unrest, since the Son La hydroelectric project was expected to force the relocation of some 100,000 additional people to the Central Highlands region.

The continuing campaign against corruption dominated political life throughout the year. In September 2000 the Government dispatched missions to 18 of the country's 61 provinces to investigate complaints of local-level corruption. There were several dismissals of prominent officials throughout the year. The head of the Communist Party's Committee for Ethnic Minorities and Mountainous Areas, who had held the office since 1982, was reprimanded for corruption and mismanagement in the agency, which had disbursed little of the allocated funds of some US $100m. to ethnic minority groups. Two provincial party leaders were replaced in early 2001. Some 160,000 party members were investigated, and the Government announced that 43% of the examined party organs had been involved in corrupt practices. (Viet Nam was rated as the most corrupt country in Asia.)

The most significant political event in 2001 was the Ninth Party Congress, at which the General Secretary of the Communist Party, Le Kha Phieu, was replaced. Phieu had been criticized for his mismanagement of the economy, nepotism and misuse of the security services for clandestine surveillance of political rivals. In an attempt to establish a more youthful leadership, an age limit of 65 years for election to the post was established. Throughout the last quarter of 2000 and the first quarter of 2001 there was intense political debate, as Le Kha Phieu struggled to retain his position. After the rural unrest in the Central Highlands, however, Phieu was able to convince the Politburo that the country needed political stability in times of crisis. (Although he was over the 65-year age limit, an exception for principal party officials was made.) At the party Plenum in March the Minister of Defence and the Chief of the General Staff department, both Central Committee members, received the party's highest reprimand for 'management shortcomings'. It was later revealed that the two, who were close allies of Phieu, had been disciplined for allowing the military intelligence service to use clandestine surveillance against Phieu's rivals. At the 12th Plenum in early April the Politburo voted to re-elect Phieu, but the Central Committee, in an unprecedented action, subsequently rejected the Politburo's endorsement of him. (The Central Committee, which had become dominated by provincial leaders, had been critical of the Government's management of the economy.) The Central Committee overwhelmingly elected the hitherto Chairman of the National Assembly and a prominent advocate of reforms, Nong Duc Manh, as the new General Secretary. The Ninth Congress elected a Central Committee (which had been reduced from 170 members to 150) and Politburo (reduced from 19 members to 15). The military lost several of its seats on the Politburo, although it maintained its level in the Central Committee. Both the incumbent party President, Tran Duc Luong, and Prime Minister, Phan Van Khai, were re-elected to their posts. In June a member of the Politburo, Nguyen Van An, was elected as the new Chairman of the National Assembly.

The Ninth Congress took place amidst continued economic recession. The Government ordered suppressive measures against dissidents prior to the Congress, which intensified in strength throughout the second half of 2001 and into 2002. Ha Si Phu, a dissident Catholic priest, Father Nguyen Van Ly, and the leader of the clandestine Unified Buddhist Church, Thich Quang Do, were all placed under house arrest, while Le Quang Liem, the 82-year-old leader of the Hoa Hao Buddhist sect, was also detained. In October 2001 Father Ly was sentenced to 10 years' imprisonment; he was subsequently adopted by the human rights organization Amnesty International as a 'Prisoner of Conscience'. The Government continued to view organized religion as a threat to its monopoly of power, and, in particular, made efforts to suppress the activities of religious leaders, who avoided party control. Evangelical Protestantism, the fastest-growing religion in the country, was granted official status and legalized. Unrest continued in the Central Highlands region, following the trial, conducted in secret, of 14 members of ethnic minorities who had protested against the Government in February 2001, and received sentences of between six and 12 years in prison, after being denied legal council. The dissident community suffered a further reverse, when, in August 2002, the most prominent critic of the Government, Gen. Tran Do, died. Although a representative of the Communist Party Central Committee gave a eulogy at the funeral of the war hero (who was himself a former member of the Central Committee), this included the statement that Do 'had made mistakes at the end of his life'. Unusually, a protest erupted during the funeral, following Tran Do's son's speech, in which he rejected the official party eulogy and verdict.

Despite the Government's uncompromising stance on dissent, there were several signs that gradual political reform was under way. At the Fourth Plenum in November 2001, the Central Committee protested at the December 2000 border treaty with the People's Republic of China, which had been negotiated by the ousted General Secretary, Le Kha Phieu, and there were calls for the National Assembly to reject the treaty. At the Fifth Plenum, in March 2002, the Communist Party enshrined the role of the private sector in the economy and allowed party members themselves to engage in private entrepreneurial activities. At the same Plenum, General Secretary Nong Duc Manh urged the decentralization of political power and the empowerment of 'grassroots' political organizations. At the final session of the 10th National Assembly in March, the delegates prepared for the parliamentary elections, scheduled for 19 May. For the first time private businessmen were permitted to contest the elections. There was also greater transparency in the election process—the Viet Nam News Agency published details of all candidates. Although 85% of the candidates were party members, they were better educated and more professional. At the March 2002 session the National Assembly adopted legislation requiring that at least 25% of the deputies become full-time members, thereby acknowledging the need for more professional and skilled lawmakers. The legislation also increased educational requirements for deputies.

At the elections to the 11th National Assembly on 19 May 2002, the 498 seats (increased from the previous 450-member Assembly) were contested by 759 candidates, only 13 of whom were independent. Of the candidates who were not members of the Communist Party, 51 secured seats, while independent candidates won two, the remainder being taken by party members. The first session of the 11th National Assembly, which took place from July to August 2002, approved a significant reorganization in the leadership. Although President Tran Duc Luong, Prime Minister Phan Van Khai and the Chairman of the National Assembly, Nguyen Van An, were all re-elected to their positions, Vu Khoan was promoted from the Ministry of Trade to one of the deputy premierships, while the hitherto Minister of Public Security, Le Minh Huong, was replaced by the party's top disciplinarian and Politburo member, Le Hong Anh. The 'hardline' Politburo member, Truong Quang Duoc, also became the Vice-Chairman of the National Assembly, reinforcing the party's control over the increasingly independent and assertive legislature.

Corruption continued to provoke extreme controversy within the Communist Party. In July 2002 two members of the party's élite Central Committee were expelled, owing to their connections to the notorious criminal leader, Truong Van Cam, known as Nam Cam. In June 2003 a Vietnamese court sentenced Nam Cam to death following a corruption trial that was unprecedented both in the freedom given to the media to report on it, and in the scale of the corruption itself. The trial, which had some 163 defendants, led to the imprisonment of 16 government officials. Perhaps its most notable aspect was the spirited defence the defendants' legal teams were allowed to present, and in particular their ability to challenge evidence. Government officials announced that there would be a second phase of the investigation. Deputy Prime Minister Nguyen Minh Triet, who had assumed a leading role in effecting the arrest of Nam Cam and those officials who had abetted him, gained in prominence.

In 2003–04 promoting political stability in the Central Highlands remained a priority for the Vietnamese leadership. Party General Secretary Nong Duc Manh made several inspection trips to the region in 2003, urging its leaders to accelerate reform and to encourage foreign direct investment in the face of an ongoing drought that had affected 40,000 ha of coffee plantations. Prime Minister Phan Van Khai also travelled to the region in April 2003, concerned that poverty levels in the area were not being reduced at a sufficiently rapid rate. In March land pressure and resentment intensified further when the resettlement of the first of 51,000 residents of Son La province to make way for a hydroelectric dam began in the Central Highlands. Mass demonstrations attended by 10,000–30,000 minority hill tribe Christians in the Central Highlands city of Buon Me Thuot were brutally suppressed in mid-April 2004. Other crackdowns on dissent occurred in Dak Lak, Gia Lai and Dak Nong provinces. Although the number of people wounded and killed remained unknown, as the Vietnamese Government closed the region to foreign journalists, one Western human rights organization reported that 10 people had been killed by the security forces. Rural issues remained a priority in other parts of the country as well; in October 2002 there was a riot over failed land compensation payments, in which 11 people were injured. The National Assembly continued to play an important role in legal codification. At the third session of the 11th National Assembly in May 2003 the legislature prepared eight new laws, covering: border demarcation; statistics and accounting; value-added tax (VAT); corporate incomes; the Viet Nam State Bank; and a special consumption tax. The National Assembly also passed a new law intended to give lower-level people's committees more authority and autonomy. It also drafted a new law concerning its own supervisory role that would give it unprecedented power over the entire state system. The legislation, which was adopted on 20 May by an overwhelming margin (81.5%), was intended to make state agencies and ministries more accountable and to ensure that the Constitution and legal code were enforced uniformly across Viet Nam. Legislators felt that the law was necessary as, in practice, the National Assembly's supervisory role was ineffective. The new law not only increased the power of the standing committees of the National Assembly, but also increased its ability to dismiss government officials, including senior officials. In another event that augured well for political reform prospects, seven ministers, including Deputy Prime Minister Nguyen Tan Dung, were questioned by National Assembly deputies and forced to address petitions from voters on live television during a three-day hearing. Many delegates expressed concern over the slow pace of reform.

The Vietnamese Government's inconsistent attitude towards the dissident community reflected both its commitment to maintaining its one-party rule and countenancing no opposition, and pressure from the USA. The US Congress drafted two bills in 2003 that focused on Viet Nam's human rights situation—the Viet Nam Freedom of Information Bill and the Viet Nam Human Rights Bill. The legislation was intended to block all non-humanitarian aid to Viet Nam. Although neither bill was likely to be passed, both angered the Vietnamese leadership. The report on religious freedom in Viet Nam issued by the congressionally-mandated US Commission on International Religious Freedom (USCIRF) in May 2003 was rejected as wrongful and slanderous and as being blatant intervention in the country's internal affairs. The USCIRF described religious freedoms in the country as worsening and listed Viet Nam as being a country of particular concern. Following the publication of the report, President Tran Duc Luong conferred the nation's highest honour, the First Class Independence Order, on Thich Tam Tich of the official Viet Nam Buddhist Church. Soon afterwards, in June 2003, the leading dissident monk of the outlawed Unified Buddhist Church, Thich Quang Do, was released, two months before the end of his sentence. Likewise, Father Ly's prison sentence, imposed in September 2001, was reduced by one-third. In a shift of policy, the Government became far more aggressive towards the country's secular dissidents. In November 2002 an individual was imprisoned for four years for questioning the Sino-Vietnamese border agreement (of which many National Assembly delegates also disapproved); in the same month another dissident was imprisoned for 12 years on charges of espionage. In March 2003 dissident Nguyen Dan Que was arrested for the alleged possession and distribution of documents. In June Pham Hong Son, a pharmaceutical company employee, was sentenced to a 13-year prison term for disseminating anti-Government propaganda and seeking to advocate a multi-party system of government, although, in an unusual move, the Supreme Court subsequently reduced his sentence to five years' imprisonment and three years under house arrest. The harshest measures were taken against the leaders of the February 2001 uprising in the Central Highlands. In October 2002 a Montagnard, handed over to Vietnamese officials by the Cambodian Government, was sentenced to nine years in prison. Cambodian officials had repatriated some 400 people since the uprising. In May 2003 an additional 15 provocateurs were sentenced to prison, with the leaders receiving terms of between seven and 10 years.

The dissident situation slightly worsened from 2003, in addition to the crackdowns in the Central Highlands. The leaders of the banned Unified Buddhist Church, Thich Huyen Quang and his deputy Thich Quang Do both remained in detention, increasingly ill and frail, the latter having been again placed under house arrest in October 2003. Nguyen Vu Binh, a former journalist, was convicted of espionage and sentenced to seven years' imprisonment in December for using the internet to criticize the Government and gathering anti-Government information and documents for 'reactionary organizations' in exile to help them oppose the Government. Binh may have been intentionally targeted as he submitted testimony on the human rights situation in Viet Nam to the US Congress in 2002. Although statistics on the death penalty were not made public in Viet Nam, where 27 offences were punishable by death, according to Amnesty International, more than 100 people were sentenced to death and a further 60 executed in 2003, with both figures representing an increase of at least 100% over the previous year.

In July 2003 Nong Duc Manh gave a speech to the Central Committee in which he stressed the importance of strengthening national defence, including the assessment of both domestic and external threats to the country. The most immediate concern for the leadership was the sudden outbreak of Severe Acute Respiratory Syndrome (SARS) in the region in early 2003. The Government displayed decisive leadership during the crisis and garnered high praise for the Vietnamese medical establishment's effectiveness in handling the outbreak of the disease and efficiently organizing the quarantine of infected individuals. The unprecedented display of leadership, decisiveness and transparency during the 45-day outbreak was instrumental in enabling Viet Nam to become the first country to succeed in containing the disease. The Government quickly established the National SARS Prevention Committee, which had considerable authority and resources at its disposal. An allocation of 30,000m. dông was made available to combat the disease, and an additional US $520,000 was received in aid from Japan. Although there were 68 SARS cases in Viet Nam, only six deaths occurred and it did not spread to the general population beyond the French Hospital where the first case occurred. However, the SARS outbreak had a severe impact on the economy; the number of tourist arrivals fell sharply in 2003. The Government also became more transparent with regard to the rapid spread of HIV/AIDS in Viet Nam. There were 63,000 carriers of HIV in the country, and the rate of new carriers was increasing by an annual rate of 1.5%. The National Assembly warned that only drastic measures could contain the epidemic. In May the Government announced a five-year HIV education programme, the stated goal of which was to reduce the growth rate to 1%.

The Central Committee of the Communist Party held its major mid-term conference at its Ninth Plenum in January 2004, half-way through its five-year term, which was due to expire in the first quarter of 2006. The Central Committee identified six key tasks for 2004–05, including achieving economic growth of 8% per year, privatizing state-owned enterprises, decreasing rural poverty, and maintaining political and economic stability. The Ninth Plenum also focused on the continued unrest in the Central Highlands and warned that 'international reactionary forces' would invoke the need for democracy and human rights as a pretext to interfere in Viet Nam's internal affairs. The Ninth Plenum also disciplined four senior party officials, including the Minister of Agriculture and Rural Development and three provincial officials, for fraud, financial irregularities, and for failing to control the situation in the Central Highlands.

RECENT FOREIGN RELATIONS

In the 1980s severe unemployment and poverty remained urgent problems. Refugees—'economic migrants' in addition to victims of political oppression—continued to leave Viet Nam for South-East Asia or for Hong Kong. The increasing reluctance of Western countries to provide resettlement opportunities led to a meeting in Malaysia in 1989, at which representatives of 30 nations drafted a programme of action. Members of ASEAN subsequently ceased to accept refugees for automatic resettlement and instituted a screening procedure to distinguish genuine refugees from economic migrants. A similar procedure had been in effect since June 1988 in Hong Kong, and an agreement had been signed in November, whereby Viet Nam agreed to accept voluntary repatriation of refugees from Hong Kong, funded by UNHCR and the United Kingdom, with UN supervision to protect the returning refugees from punitive measures by the Vietnamese Government. Between March 1989 and March 1995 more than 73,000 Vietnamese were voluntarily repatriated. In June 1989, at a UN-sponsored conference, Western countries committed themselves to the resettlement of the 55,000 'boat people' who had arrived in reception countries before March 1989 (June 1988 in the case of Hong Kong). The USA and Viet Nam agreed to expand the Orderly Repatriation Programme. Despite pressure from the United Kingdom and Hong Kong (where Vietnamese refugees totalled 43,000 in June 1989), the conference rejected proposals of forced repatriation to Viet Nam. In December 1989 the United Kingdom initiated a programme of mandatory repatriation, which, however, was halted as a result of continued opposition from the USA and Viet Nam's refusal to accept deportations. In July 1990 the USA accepted in principle a UN plan, which involved the 'involuntary' repatriation of Vietnamese classed as economic migrants who did not actively oppose deportation. In September the United Kingdom, Viet Nam and UNHCR reached an agreement, whereby the Vietnamese Government would no longer refuse refugees who were repatriated 'involuntarily'.

In October 1991 the British and Vietnamese Governments concluded an agreement on the deportation of about 300 Vietnamese refugees who had been voluntarily repatriated but had returned to Hong Kong. Later in that month the agreement was extended to include all Vietnamese who were not considered to be political refugees. The policy appeared to have deterred all but a few Vietnamese from fleeing to Hong Kong during 1992. In that year more than 12,000 Vietnamese returned from Hong Kong under the voluntary repatriation programme. At the end of 1993 there remained about 31,000 Vietnamese in Hong Kong camps, and a further 20,000 in Indonesia, Malaysia and other parts of South-East Asia. In February 1994 representatives of 31 nations attended a conference in Geneva on the issue of Indo-Chinese refugees. It was agreed that political conditions in Viet Nam were now sufficiently favourable to permit the repatriation of the remaining 51,000 boat people by the end of 1995. In February 1995 Viet Nam and the Philippines reached an agreement on the orderly return of boat people who had not qualified as refugees under international law. A further international conference on the issue of Vietnamese refugees was held in Geneva in March 1995. It was decided to end the UN programme for boat people and to support the repatriation of those remaining in camps, either voluntarily or through the Orderly Repatriation Programme. However, the 1995 deadline was not met. Refugees in Hong Kong delayed the process of repatriation through a series of demonstrations in detention camps as well as court appeals. Approximately 6,000 boat people volunteered to return to Viet Nam under a US State Department programme that offered hope of resettlement in the USA only after an immigration interview had been conducted in Viet Nam.

In 1996 decisive steps were taken by the international community to resolve the long-standing problem of Vietnamese refugees in South-East Asia. In mid-January a UN-sponsored meeting in Bangkok agreed that all camps outside Hong Kong should be cleared of refugees by June 1996. UNHCR terminated its financial assistance on 30 June. Throughout the year Indonesia, Malaysia, the Philippines, Singapore and Thailand took steps to empty their camps. Vietnamese boat people were offered the choice of returning voluntarily or being forcibly repatriated. This new policy provoked rioting and violent resistance by camp inmates in the Philippines, Thailand and Malaysia. While Malaysia succeeded in emptying its camps of those who refused voluntary return, the Philippines abandoned forcible repatriation after violent protests. China brought pressure to bear on Hong Kong to empty its camps before 30 June 1997 when the United Kingdom was to relinquish control of the territory. Funding for the Hong Kong refugee camps was terminated at the end of 1996; the EU, however, agreed to provide a further US $20m. to assist in resettlement, in addition to a $75m. ongoing programme to assist with the reintegration of returnees. In March 1997 6,000 refugees remained in Hong Kong, a number that was reduced to 1,800 by 1 July 1997. Viet Nam continued to co-operate with the Government in Hong Kong in meeting its obligations under the Orderly Departure Programme following the return of Hong Kong to Chinese sovereignty in 1997. The number of boat people declined to fewer than 2,000 in 1999. In April 1999 it was reported that a total of 13,495 refugees had been repatriated from Hong Kong to Viet Nam since October 1991 under the Orderly Repatriation Programme. In July 1999 Hong Kong announced that the remaining Vietnamese refugees would be given permanent residency, while a group of 500 recent arrivals would be returned home as illegal immigrants. Also in July the EU announced that it was terminating its support for the Returnee Assistance Programme, which had provided credit and job-training schemes for returning boat people. In December 1998 it was reported that UNHCR had announced the closure of its office in Ho Chi Minh City; the agency was, however, to maintain a small office in Hanoi. Viet Nam and the USA were co-operating in a programme under which boat people in South-East Asian camps who agreed to return to Viet Nam would be interviewed for resettlement in the USA. Approximately 2,000 persons per month were interviewed during 1998. In February 2000 the remaining 1,400 refugees in Hong Kong were granted permanent residency.

The necessity of securing greater foreign investment, through the removal of the international embargo on economic relations with Viet Nam, placed the Vietnamese Government under increasing pressure to fulfil its commitment to withdraw all Vietnamese troops from Cambodia in 1989. In September Viet Nam unilaterally withdrew its 26,000 remaining forces from Cambodia, but without international verification. In July 1990 the USA finally acknowledged that all Vietnamese troops had departed. The Khmer Rouge, however, repeatedly charged that Vietnamese troops remained in the country. The Cambodian conflict was eventually settled by the intervention of the five permanent members of the UN Security Council in support of regional diplomatic efforts. Intense diplomatic efforts throughout 1990 and 1991 led to the convening of the Paris International Conference on Cambodia in October 1991. This conference adopted a comprehensive political agreement on Cambodia (see the chapter on Cambodia).

Following the elections in Cambodia in May 1993, the status of Vietnamese residents became a political issue between Cambodia and Viet Nam. The Vietnamese Government was concerned by the Cambodian National Assembly's adoption in August 1994 of a new immigration law, which, it claimed, contravened international agreements on human rights and might lead to discrimination against the Vietnamese minority in Cambodia. In the following month such concerns appeared substantiated by two incidents involving the persecution and killing of Vietnamese villagers in Cambodia by Khmer Rouge guerrillas. In December Prince Norodom Ranariddh, the First Prime Minister of Cambodia, visited Viet Nam. The two countries agreed to resolve their outstanding problems on the basis of previous joint communiqués. Relations improved further in December 1995 when King Norodom Sihanouk paid an official visit to Viet Nam, his first in 20 years. In the following year, however, the question of relations with Viet Nam became a controversial issue in internal Cambodian politics. In January 1996 Norodom Ranariddh alleged that Viet Nam had encroached into Svay Rieng province and relocated border markers. Vietnamese local authorities subsequently claimed that Cambodian troops had opened fire on unarmed civilians without provocation. Two months later Ranariddh threatened military action to curb what he termed Viet Nam's 'border invasion'. This prompted a visit by Prime Minister Vo Van Kiet in mid-April to defuse the situation. Both parties agreed to follow a set of procedures under which negotiation of border incidents would be dealt with first by local and then by provincial officials before being raised at national level. High-level talks on the border issue were held in Ho Chi Minh City in May 1996, but in July Ranariddh once again charged that Viet Nam was encroaching on Cambodian territory. Viet Nam refrained from involving itself in internal Cambodian affairs in the aftermath of the July 1997 action by Second Prime Minister Hun Sen, which ousted First Prime Minister Ranariddh from power. Viet Nam initially supported Cambodia's entry into ASEAN, but soon accepted the ASEAN consensus to delay Cambodia's membership until 'free and fair' elections were held. Representatives of Viet Nam were included in a delegation sent by ASEAN to observe elections held in Cambodia in July 1998. In June 1999 the General Secretary of the Central Committee of the Vietnamese Communist Party, Lt-Gen. Le Kha Phieu, paid an official visit to Cambodia at the invitation of King Norodom Sihanouk; subsequently, in July, the Chairman of the Central Committee of the Cambodian People's Party, Chea Sim, made an official friendship visit to Viet Nam. Although there were several high-level exchanges of visits between 1999 and 2001, no agreement was reached on either the border demarcation or on the ethnic Vietnamese in Cambodia. Relations between the two countries became strained in February 2001, when, following the Vietnamese Government's suppression of unrest in the Central Highlands region, large numbers of protesters fled to Cambodia seeking sanctuary. Viet Nam demanded their return, but under US and UN pressure, the Cambodian authorities allowed 24 refugees to emigrate to the USA, following a review of their cases. At mid-2001 many protesters remained in Cambodia, and UNHCR had failed to receive guarantees from the Vietnamese Government that they would not be prosecuted for involvement in the unrest if they were returned.

Senior officials from Viet Nam's Ministry of Public Security and Cambodia's Ministry of the Interior met in Phnom-Penh in December 2003 to discuss ways to increase co-operation on border security. Also in that month King Norodom Sihanouk received a delegation of the Communist Party of Viet Nam, led by Nguyen Van Son, a member of the party's Central Committee and Director of its Commission for External Relations, yet the two sides remained far apart on the issue of unrest in the Central Highlands.

Meanwhile, in the aftermath of the collapse of the USSR and the ending of the 'Cold War', along with the settlement in Cambodia in 1991, the 'cornerstones' of Vietnamese foreign policy—alliances with the USSR, Eastern Europe, Laos and Cambodia—were broken. Viet Nam readjusted its foreign policy so decisively that ideological factors now appeared to play little or no role in the Government's considerations. Viet Nam has opened relations with former enemies and, after a short period of readjustment, placed relations with old allies on a new footing.

In 1989–90 the dramatic political changes in Eastern Europe prompted renewed efforts by Viet Nam to restore political and economic links with the People's Republic of China. The October 1991 comprehensive political agreement on Cambodia, which *inter alia* called on all external parties to cease military support for the Cambodian protagonists, removed the last remaining major obstacle to the restoration of normal relations. In early November the resumption of political relations between the two countries was formally announced during a Sino-Vietnamese 'summit' meeting in Beijing. Territorial disputes between the two countries, however, remained unresolved. In May 1992 Viet Nam made a formal protest against China's unilateral decision to grant a US company petroleum exploration rights in an area in the Gulf of Tongking, which was claimed as territorial waters by Viet Nam. In June discussions took place in Yogyakarta, Indonesia, concerning the sovereignty of the Spratly Islands which have been the subject of a protracted dispute involving Viet Nam, China, the Philippines, Malaysia, Taiwan and Brunei, which claim all or part of the archipelago (believed to be rich in mineral resources). In July Viet Nam protested when Chinese troops landed on another islet in the Spratlys. (Viet Nam maintained a military presence on 21 of the islands, China on seven.)

In September 1992 China agreed to discuss land border disputes and the disagreement over the Gulf of Tongking, but ruled out negotiations on the Spratlys. The discussions, which took place in October, failed to achieve significant progress. However, in November–December the Chinese Premier, Li Peng, paid an official visit to Viet Nam (the first by a Chinese head of government since 1971); new accords on economic, scientific and cultural co-operation were concluded, and the two Governments agreed to accelerate the negotiations on disputed territory. The first round of border discussions was duly held in August 1993, and in October the two countries concluded an agreement to avoid the use of force in resolving territorial disputes. However, armed confrontation between Chinese and Vietnamese ships in the disputed waters was only narrowly avoided in April 1994, and in July two Chinese warships blocked access to Vietnamese oil-prospecting facilities on one of the Spratly Islands. In September Vice-Prime Minister Phan Van Khai visited Beijing to discuss the matter, and agreement was reached to continue seeking a negotiated settlement. In the following month China renewed diplomatic protests at Viet Nam's decision to invite foreign oil companies to develop areas of the Gulf of Tongking. These protests came on the eve of a visit to Hanoi in November by the Chinese President and Communist Party General Secretary, Jiang Zemin. This represented the highest-level visit by a Chinese official since the resumption of normal bilateral relations in late 1991. The joint communiqué issued at the end of the visit called for an early settlement of territorial disputes by peaceful means. Various economic agreements were also signed, including the establishment of a joint trade commission. In the following month, Viet Nam and China held the third round of talks on border and territorial issues, which concentrated on demarcating the Gulf of Tongking. Problems in Sino-Vietnamese relations were exacerbated in early 1995, when it was discovered that the Chinese had occupied Mischief Reef, located in an area of the Spratlys claimed by the Philippines. This was the first such incident involving China in a direct dispute with an ASEAN member, and it served as a welcome opportunity for Viet Nam to demonstrate its solidarity with ASEAN prior to its membership of the Association (see below).

Sino-Vietnamese relations have alternated between the exchange of high-level delegations (and progress on economic matters) and contention over contested territory. In November 1995 Do Muoi led a senior party delegation to China to accelerate the pace of economic relations. Although Viet Nam and China had signed some 20 economic and commercial agreements since the normalization of relations in 1991, the only tangible result of this visit was the decision to reopen rail links between Kunming in Yunnan Province with Hanoi and Haiphong in February 1996. Despite the fact that cross-border trade was valued officially at US $378m. in 1996, in 1998 freight trade along the reopened rail line from Kunming to Haiphong had yet to develop. In November 1995 the first round of talks was held on maritime issues between expert groups of the two countries. In June 1996 Premier Li Peng visited Viet Nam to attend the eighth party Congress. Relations became strained once again, however, when Viet Nam announced the award of a petroleum exploration and production rights contract to Conoco, a US company, in an offshore block claimed by China. The seventh meeting of the joint Sino-Vietnamese working group on the delineation of the Gulf of Tongking took place in Hanoi in August 1996. At the end of the year, however, tensions flared again when an armed Chinese naval vessel came to the assistance of a boat in the Gulf of Tongking, which Vietnamese authorities had stopped on suspicion of smuggling; two Vietnamese patrol boats were seized by the Chinese.

In August 1997 Viet Nam and China held their fifth round of talks on the land border issue and agreed to reach a settlement by 2000. In January 1998 Viet Nam accused China of a 'serious violation' of border treaties when Chinese workers reclaimed land along a border river causing a diversion of water into Vietnamese territory; however, Vietnamese officials denied that border tensions increased in 1998. A further round of talks between the two countries on the land border issue was held in September 1998.

In mid-October 1998 it was announced that the Vietnamese Prime Minister, Phan Van Khai, would pay his first official visit to China later the same month. On the eve of Khai's departure, however, the official Chinese news agency, Xinhua, issued a critical review of the state of the Vietnamese economy. During Khai's visit to Beijing, the Vietnamese Prime Minister and his Chinese counterpart, Zhu Rongji, discussed the implementation of economic reforms in their respective countries, concluding jointly that the maintenance of 'socialist stability' was the most important consideration for both Viet Nam and China, thereby underscoring the main concern of the Communist Party that too rapid an introduction of economic reforms could lead to political instability. The Chinese Vice-President, Hu Jintao, visited Viet Nam in December 1998 to attend the China-ASEAN leadership meeting. While in Hanoi, Hu held talks with his Vietnamese counterparts, emphasizing the need for the resolution of border issues and for closer bilateral co-operation between Viet Nam and China; an agreement on economic and technical co-operation was also signed, in which China pledged to provide 10m. yuan in non-refundable aid to Viet Nam. During the visit of the Vietnamese Communist Party General Secretary, Le Kha Phieu, to Beijing in February 1999, a further agreement on economic and technical co-operation, valued at 20m. yuan, was signed. In discussions held between Phieu and Chinese President Jiang Zemin, both parties reiterated their determination to conclude a border agreement by the end of the year and to agree on the demarcation of the Gulf of Tongking by 2000. In 1999–2000 some 80 high-level visits were made by both sides. A significant land border agreement came into effect in January 2000. At an ASEAN summit meeting in September, chaired by Viet Nam, member countries and China reached consensus on some major principles of a regional code of conduct governing activities in the South China Sea.

Relations continued to progress in 2000, and in that year bilateral trade between the two states reached US $2,000m. Following success in efforts to resolve the bilateral territorial dispute, President Tran Duc Luong travelled to China in December 2000 to sign an 'agreement on the demarcation of the territorial waters, exclusive economic zone, and continental shelf in the Gulf of Tongking'. A second agreement on fishing co-operation in the Gulf of Tongking was signed at the same time. The agreement did not cover the Spratly Islands, which continued to be problematic to bilateral relations. In February Viet Nam announced that it was to establish a civilian administration for the Spratly Islands. The Chinese Government responded angrily, while Viet Nam claimed that China was acting more aggressively in the South China Sea. Relations between the two nations continued to progress, however. Although high-level exchanges continued throughout 2001 and 2002, there was considerable mistrust of China and dissatisfaction over the border agreement. At the Central Committee's Fourth Plenum in November 2001, members urged the National Assembly to reject the treaty. In order to placate the Chinese Government, General Secretary Nong Duc Manh travelled to Beijing in that month and reaffirmed the land border treaty and the demarcation of the Gulf of Tongking. A general framework agreement on economic and technical co-operation was also signed during the visit, although no further issues were discussed. In February 2002 the Chinese President, Jiang Zemin, travelled to Viet Nam (his first visit since 1994) and offered some $12.1m. in credit and grants to Viet Nam. Party Leader Nong Duc Manh reciprocated the visit in April 2003 and met with Hu Jintao and Jiang Zemin, Chairman of the Central Military Commission. Although China expressed displeasure that the Vietnamese National Assembly had not ratified the border agreement, economic ties were becoming stronger. Bilateral trade increased from $2,100m. in 2000 to $3,500m. in 2002, with a goal of $5,000m. by 2005. China continued its aid programme to upgrade the Thai Nguyen Iron and Steel and the Bac Giang Nitrogen Fertilizer Plants. It also waived $49m. of Vietnamese debt.

In the early 1990s Viet Nam sought to consolidate relations with the successor states of the USSR, as well as with the newly democratic countries of Eastern Europe. In June 1994 Prime Minister Vo Van Kiet visited Ukraine, Kazakhstan and Russia at the head of a large Vietnamese delegation. The visit focused on economic relations, and Viet Nam's outstanding debt, as well as the issue of Vietnamese workers residing in republics of the former USSR. Defence ties were also discussed, including continued Russian access to Cam Ranh Bay (finally agreed in mid-1995), the supply of spare parts for Viet Nam's military, and co-operation in a new security system for Asia. While in Moscow, Kiet renegotiated the 1978 Soviet-Vietnamese treaty of friendship and co-operation. The new document merely stipulated bilateral consultations in the event of a crisis. In February 1996 the commander of the Russian Pacific Fleet visited Viet Nam and announced Russia's intention 'to maintain and develop' Cam Ranh Bay after the expiry of the access agreement in 2004. In November 1997 the Russian Prime Minister, Viktor Chernomyrdin, paid an official visit to Viet Nam, where he witnessed the signing of four bilateral co-operation agreements on tourism, trade and science, oil and gas exploitation, and veterinary sciences. He also discussed with his counterparts the sale of military aircraft and the question of Viet Nam's outstanding debt. President Tran Duc Luong paid a reciprocal visit to Russia and to Belarus in August 1998. While in Moscow, President Luong signed agreements on oil, gas and energy, weapons sales and support services, narcotics control, and legal relations, but was unable to resolve the issue of Hanoi's outstanding debt to Moscow.

Relations with Russia improved significantly in 2001. A visit by Phan Van Khai to Moscow in September 2000 contributed greatly to resolving the issue of Viet Nam's debt to Russia. In February 2001 the Russian President, Vladimir Putin, visited Viet Nam, prompting further rapid exchanges of bilateral visits. Putin, who was welcomed warmly in the country, resolved the long-standing bilateral dispute by drastically reducing the amount of Viet Nam's debt to the former USSR from US $11,000m. to about $1,600m., and extended the terms of repayment to 23 years. Viet Nam was henceforth to pay Russia $100m. per year. Both countries pledged to renew their historical friendship and signed a 'Strategic Partnership Declaration'. Putin indicated his desire to extend the lease on the naval and signals intelligence facility in Cam Ranh Bay, but the Vietnamese Government demanded a much higher rent. The Minister of Foreign Affairs, Nguyen Dy Nien, visited Moscow in June 2001, with an offer of $2,000m. for a 10-year lease, which the Russians rejected. In mid-2001 Viet Nam indicated that it would not renew the lease and would develop the naval base into an export-processing zone. Russia formally transferred the facilities of the naval base in Cam Ranh Bay to Viet Nam in July 2002, prior to the expiry of the lease. Nevertheless, bilateral relations remained strong, owing to increased economic co-operation and military sales. Bilateral trade reached $570m. in 2001, and the heads of state of both countries pledged to increase the total to $1,000m. in 2002. The Russian Government provided $100m. in credit for two hydroelectric installations, while Russian companies invested $1,500m. in 38 projects, making Russia Viet Nam's ninth largest foreign investor. Its most important investment, a joint-venture petroleum refinery at Dung Quat, valued at $1,300m., has fallen greatly behind schedule. The 20-year-old joint venture, Vietsovpetro, accounted for 80% of Viet Nam's crude petroleum output, and the Dung Quat refinery was expected to provide 50% of Viet Nam's demand for refined petroleum products. Russian military sales to Viet Nam have advanced significantly. In 2002 Viet Nam purchased 50 surface-to-air defence missiles, valued at $64m., while it was reported that the Russian Government envisaged the establishment of a regional maintenance centre for its aircraft in Viet Nam. Viet Nam's Deputy Minister of Foreign Affairs, Nguyen Van Bang, travelled to Russia in December 2003, followed, in May 2004, by Nguyen Dy Nien, who signed an agreement with the Russian Government to expand co-operation to new areas, particularly high-technology industry. Later in May President Tran Duc Luong made an official visit to Moscow, at the invitation of President Putin, his second since 1998.

Viet Nam's relations with ASEAN have been transformed from confrontation to accommodation to membership. Agreement was reached on the repatriation of refugees from ASEAN countries (see above). In October 1991, after the Cambodian peace agreement had been signed, Vo Van Kiet visited Indonesia, Singapore and Thailand, where he signed a number of economic agreements. In addition, Singapore announced the resumption of trade relations with Viet Nam. In January 1992 Vo Van Kiet visited Malaysia, and in late February he visited the remaining ASEAN members, Brunei and the Philippines.

Trade and investment from various ASEAN states has grown considerably. Between 1990 and 1997 bilateral trade between Viet Nam and ASEAN increased at an annual average rate of 26.9%. ASEAN investment increased 10-fold between 1991 and 1994, thus comprising 16% of total direct foreign investment. By 1996 Singapore, Malaysia and Thailand were involved in 274 projects with a paid-up capital of $4,300m. Singapore was the largest ASEAN investor, with 153 projects capitalized at nearly $2,600m., and the third largest overall foreign investor. In August 1996 the ASEAN Secretary-General told Vietnamese leaders that they would be required to draw up a master plan to eliminate import quotas and other non-tariff barriers for approval by ASEAN economic ministers. However, in the following month ASEAN's plans to adopt an agreement outlining a mechanism for the settlement of trade disputes were stymied when Vietnamese officials refused to sign, claiming that they had no authority to do so. Viet Nam has overlapping territorial claims with several ASEAN states. Viet Nam and Indonesia have held 17 rounds of talks regarding disputed territory near the Natuna Islands, but with no result. In September 1992 Viet Nam reached agreement with Malaysia and Thailand on co-operation in the use of waters in the Gulf of Thailand. As a result of continuing friction between Thai and Vietnamese fishing boats, in April 1996 the two countries agreed to establish a joint commission to resolve the matter. In early September Nong Duc Manh, Chairman of the National Assembly's Standing Committee, visited Bangkok to discuss this issue. However, in mid-October another incident took place between Thai fishing trawlers and a Vietnamese naval vessel, in which two Thais were killed. This incident prompted Thailand and Viet Nam to intensify their efforts to reach a *modus vivendi* for the maritime area in which their territorial claims conflicted. Finally, in October 1997 Viet Nam and Thailand reached an agreement to delimit their maritime boundary (continental shelf and exclusive economic zone) in the Gulf of Thailand. This agreement was expected to lead to joint naval patrols, fishery surveys and petroleum and gas exploration. In 1996 Viet Nam and Malaysia reached an agreement jointly to develop an area in the Gulf of Thailand where they had overlapping claims not involving a third country. In the same year Viet Nam and the Philippines began discussions on the joint development of their overlapping territorial claims in the South China Sea. In March 1998 President Tran Duc Luong visited Malaysia and Singapore, where he sought assistance in the drafting of Viet Nam's individual plan of action that was a prerequisite for the country's potential admission to the Asia-Pacific Economic Co-operation forum (APEC).

In July 1992 Viet Nam signed the 1976 ASEAN Treaty on Amity and Co-operation, and was granted observer status at its annual meeting of foreign ministers. In July 1994 Viet Nam became a founding member of the ASEAN Regional Forum (ARF), a security discussion group. On 28 July 1995 Viet Nam became ASEAN's seventh member and announced its intention to join the ASEAN Free Trade Area (AFTA). ASEAN members granted Viet Nam a three-year grace period, until 2003, to lower its tariff barriers. As an ASEAN member, Viet Nam attended the first 'summit' meeting of all 10 heads of government in South-East Asia held in Bangkok in December 1995. At the 31st annual ASEAN ministerial meeting in July 1998, Viet Nam opposed a Thai initiative to modify ASEAN's long-standing policy of non-interference in the internal affairs of member countries in favour of a policy of 'flexible engagement'. Viet Nam hosted the sixth ASEAN summit in Hanoi in December of that year and joined other regional states in endorsing the South-East Asia Nuclear-Weapon Free Zone Treaty. In November 1996 ASEAN members successfully supported Viet Nam's application for membership of APEC; formal admission took place in November 1998. Relations with ASEAN remained strained, owing to continued pressure from Thailand and the Philippines for greater intervention in the internal affairs of other member states. From July 2000 to July 2001 Viet Nam held the rotating chairmanship of ASEAN, but was criticized for its continued support of Myanmar and refusal to challenge ASEAN's non-interference principle. Relations with Thailand appeared stable, however, and, following the discovery of a bomb at the Vietnamese embassy in Bangkok, the Thai Government committed itself to preventing groups from using Thailand as a base of operations against the Vietnamese Government. At the ASEAN summit in July in Hanoi, the organization urged wealthier member nations to assist the development of the poorer ones. Viet Nam exhibited concern that its development was again behind that of the other states in the region. Earlier in March the Deputy Prime Minister, Nguyen Tan Dung, travelled to Singapore to promote foreign investment and to demonstrate a pledge to the Singaporean business community of Viet Nam's commitment to reform. The newly elected Indonesian President, Megawati Sukarnoputri, visited Viet Nam in August 2001. In May 2002 President Tran Duc Luong visited Myanmar, where he met with the leader of the State Peace and Development Council (SPDC), Than Shwe. ASEAN states continued to be economically important for Viet Nam, which received some US $330m. in total ASEAN investment in 2001.

The common threat of the SARS epidemic and other important economic initiatives focused Viet Nam's diplomatic attention on its fellow ASEAN states in 2003. In March Singapore's Prime Minister Goh Chok Tong visited the country. In the same month Than Shwe also travelled to Viet Nam, reflecting Myanmar's increasing political closeness. The Commander-in-Chief of the Thai armed forces visited Viet Nam in May 2003, but economic ties continued to drive the relationship. Both bilateral trade and Thai investment in Viet Nam increased, and by 2003 Thailand had become one of the top 10 foreign investors in Viet Nam, with 112 projects worth US $1,160m. The Senior Minister for Foreign Affairs and Co-operation of Timor-Leste (known as East Timor until it acceded to independence in May 2002), José Ramos Horta, paid his first state visit in March. Although Viet Nam did not endorse the new nation's bid to join ASEAN, it did promise to support ASEAN co-operation with Timor-Leste.

Relations with ASEAN remained paramount to Viet Nam in 2003–04, and Prime Minister Phan Van Khai attended the ninth ASEAN summit in Bali, Indonesia, in October 2003. In December Thailand and Viet Nam held a joint cabinet meeting and signed eight separate agreements, including a framework agreement on economic co-operation; the two Governments held a further series of bilateral meetings in February 2004. Newly elected Malaysian Prime Minister Abdullah Ahmad Badawi travelled to Viet Nam in January 2004 to meet Phan Van Khai and announce an ambitious plan for enhanced bilateral co-operation. The Singaporean Minister of Defence, Rear-Adm. Teo Chee Hean, led an official friendship visit to Viet Nam in December 2003. Viet Nam and Singapore signed a comprehensive co-operation framework document in March 2004, during a four-day visit by Phan Van Khai to Singapore, aimed at expanding their already close political and economic ties. Singapore is the largest investor in Viet Nam, with more than US $7,000m. in 283 projects, including an industrial park and training centre. A Singaporean-Vietnamese business forum was held in conjunction with the Prime Minister's visit. A delegation of the National Assembly of Viet Nam, led by Chairman Nguyen Van An, visited the Philippines and then Singapore in April 2004. In that month the Minister of National Defence, Lt-Gen. General Pham Van Tra, led a friendship visit to Brunei and Singapore.

Viet Nam's relations with other Asian countries also improved markedly after 1991. Diplomatic relations with the Republic of Korea (already an important trading partner and investor in Viet Nam) were restored in December 1992; they had remained severed since the fall of Saigon in 1975. Defence attachés were exchanged in 1996, and in November of that year the President of the Republic of Korea, Kim Young-Sam, visited Hanoi. The South Korean Prime Minister, Lee Han-Dong, also visited in April 2002. Prime Minister Phan Van Khai and his South Korean counterpart signed three agreements in September 2003 on encouraging investment, the extradition of criminals and co-operation on legal assistance. In December the South Korean Minister of Foreign Affairs and Trade, Yoon Young-Kwan, travelled to Viet Nam for talks with President Tran Duc Luong, Prime Minister Phan Van Khai and his counterpart, Nguyen Dy Nien, demonstrating the continuing importance of economic relations between the two countries. South Korea is Viet Nam's fourth largest foreign investor, with total capital investment worth US $3,920m., and Viet Nam's fifth largest trading partner, with two-way trade turnover reaching $2,750m. in 2002. However, Viet Nam has been careful not to antagonize its former ally, the Democratic People's Republic of Korea, and in mid-2002 President Tran Duc Luong travelled to its capital to meet with Kim Jong Il. A North Korean trade mission travelled to Hanoi and held a round of talks in September 2003 and in mid-2004 Viet Nam donated 1,000 metric tons of rice to North Korea.

In September 1993 full diplomatic relations were re-established between Viet Nam and Japan. In the mid-1990s Japan became Viet Nam's largest aid donor and principal trading partner. Between 1991 and 1998 Japan pledged US $3,200m. of development assistance, most of which was in the form of 'soft' loans related to infrastructure projects. In late March 1999 the Vietnamese Prime Minister, Phan Van Khai, made an official visit to Japan; each country extended 'most favoured nation' status to the other. In April, in response to a written request from Prime Minister Khai, Japan informed South-East Asian Ministers of Finance that it would include Viet Nam in the $30,000m. Miyazawa Plan; Viet Nam was allocated $160m. in concessional loans under this plan. The Japan Bank for International Cooperation provided Viet Nam with a $54m. loan for the Bai Chay Bridge near Haiphong, the destination of much Japanese foreign investment. In 2001 Japan provided $540m. in aid for five additional infrastructure projects. By this time Viet Nam had received more than $7,000m. in aid from Japan since 1992, 40% of all bilateral aid committed to Viet Nam. Japan was Viet Nam's third largest investor, with more than $4,000m. allocated to 335 projects. Bilateral trade in 2001 totalled more than $5,000m. Viet Nam negotiated a new investment deal with Japan in 2003, and a bilateral agreement on the liberalization and protection of investment was signed in November of that year. Viet Nam lifted the visa requirement for Japanese tourists, of whom there were 209,730 in 2003. Prime Minister Phan Van Khai held talks with his Japanese counterpart, Junichiro Koizumi, in December to exchange views on measures to increase economic investment and trade co-operation between the two countries. Ties have expanded beyond economic relations, and Senior Lt-Gen. Phung Quang Thanh, the Deputy Minister of National Defence and Chief of General Staff of the Army, led a high-level military delegation to the People's Republic of China and Japan in October 2003.

In July 1998 the Prime Minister of Laos, Gen. Sisavat Keobounphan, made an official visit to Viet Nam, and in June 1999 the President of Viet Nam, Tran Duc Luong, visited Laos. In mid-2000, following ethnic and political violence in Laos, the Lao army commander visited Viet Nam to appeal for military assistance. The Vietnamese Government agreed to increase aid to Laos, but publicly refused to send troops to assist in the suppression of dissident activity (despite reports of the transport of Vietnamese forces to Laos). Viet Nam and Laos continued their close relations and exchanged many high-level visits throughout 2000–01, culminating in sending reciprocal delegations to their respective party congresses. The Ministers of Defence of the two countries signed a protocol for co-operation and mutual assistance in February 2001. The Vietnamese Government remained extremely concerned over the spread of ethnic minority unrest from Laos into Viet Nam and intensified military aid. The new party General Secretary, Nong Duc Manh, made a prominent visit to Laos in July 2001, and Laos continued to be Viet Nam's most important ally. In May 2002 the Lao Head of State, Khamtay Siphandone, made a visit to Hanoi to celebrate 40 years of diplomatic relations and the 25th anniversary of the Treaty of Friendship and Co-operation, demonstrating that the 'special relationship' between the two states remained very strong, despite the fact that bilateral trade was in 2000 valued at only US $177m. Viet Nam was at that time the 14th largest foreign investor in Laos and an agreement to reduce the taxation on Vietnamese investments was reached in late 2001. Viet Nam remained the most important source of education and technical training for Lao officials. In 2003 Minister of Public Security Le Hong Anh travelled to Laos and signed new agreements on security co-operation, including several agreements on bilateral military co-operation. In December 2003 Lao fourth Deputy Prime Minister Bousone Boupavanh travelled to Hanoi to meet First Deputy Prime Minister Nguyen Tan Dung as well as the General Secretary of the Communist Party, Nong Duc Manh. In that month a 37-km road section of Highway 18B, built by Viet Nam, was officially opened to traffic in Laos's southern Attopeu province. The road is the first phase of a $48m. project on construction of the 113-km Highway 18B, funded with a preferential credit provided by the Vietnamese Government. Viet Nam's Minister of Agricuture and Rural Development, Le Huy Ngo, and his Lao counterpart, Siane Saphanthong, signed several joint development agreements in April 2004, in which Viet Nam pledged more than $100,000 to support food production in Laos.

In January 1999 the Vietnamese Deputy Prime Minister and Minister of Foreign Affairs, Nguyen Manh Cam, paid an official visit to India to attend the ninth session of the Intergovernmental Committee for Economic, Cultural, Scientific and Technical Co-operation. On the eve of his visit it was announced that Viet Nam and India had reached agreement on co-operation over nuclear energy. Close military relations between the two countries were consolidated in 2000, with the visit to Viet Nam of the Indian Minister of Defence. Agreement was reached on joint naval exercises, while Viet Nam pledged to assist the Indian army in guerrilla warfare training. The Indian Prime Minister, Atal Bihari Vajpayee, visited Viet Nam in March 2001, although plans for an agreement on naval construction co-operation were not realized. General Secretary Nong Duc Manh travelled to

India in April 2003, the highest level visit by a Vietnamese leader for a number of years. The Indian naval ship *Magar* arrived at Sai Gon Port, Ho Chi Minh City, in December 2003. In early 2004 the Indian Chairman of the Joint Chiefs of Staff Committee, Adm. Madhvendra Singh, arrived in Hanoi for official consultations.

Relations between Viet Nam and the USA remained in part an exception to the general improvement in Hanoi's relations with the outside world during 1991–94, owing to the lingering problem of the more than 2,600 US servicemen listed as 'missing in action' (MIA) during the Viet Nam War. In 1986 and 1987 agreements were reached for Vietnamese co-operation in the search for missing servicemen, in return for a promise of humanitarian aid from the USA. The Vietnamese Government also agreed in principle to permit the emigration to the USA of former re-education camp detainees and their families, subject to assurances that they would not engage in 'anti-Vietnamese' activities. Between November 1988 and January 1989 the remains of 86 soldiers were returned to the US Government. In October 1990 the Vietnamese Minister of Foreign Affairs, Nguyen Co Thach, met leading US officials in Washington, DC (the first visit by a Vietnamese delegate in 15 years) to discuss the UN peace plan for Cambodia, and the possible release of information concerning the MIA. The meeting represented a significant move towards the establishment of diplomatic relations between the two countries. In early 1991 a US representative was stationed in Hanoi to supervise inquiries into MIA, the first official US presence in Viet Nam since 1975. In April the USA proposed a four-stage programme for the establishment of normal diplomatic relations with Viet Nam, conditional on Vietnamese co-operation in reaching a diplomatic settlement in Cambodia, and in accounting for the remaining MIA. In December 1991 the USA ended restrictions on US citizens travelling to Viet Nam. In March 1992 relations improved further when the US Assistant Secretary of State, Richard Solomon, visited Hanoi, the most senior US official to do so since 1986. In exchange for enhanced co-operation on the MIA issue, Solomon announced that the USA would grant Viet Nam US $3m. in humanitarian aid. He reiterated, however, that there would be no deviation from the four-stage programme for normalization. Despite the public urging of allies such as Thailand, Australia and the United Kingdom, in September 1992 the USA extended the trade embargo against Viet Nam for a further year.

Expectations that the coming to power in the USA of a Democratic administration under President Bill Clinton (who took office in January 1993) would lead to the early normalization of relations between Viet Nam and the USA were not realized. Nevertheless, in July the USA did withdraw its long-standing opposition to loans to Viet Nam from the IMF, the World Bank and the Asian Development Bank (ADB). As a result, in October (having paid its arrears to the IMF with the help of donor countries) Viet Nam was granted US $223m. in credits by the IMF, and large loans from the ADB (for irrigation) and the World Bank (for education and road repairs) followed. In December a conference of international aid organizations and governments agreed to provide $1,860m. in aid to Viet Nam over the ensuing year, and in the same month creditor nations agreed to reschedule part of Viet Nam's external debt. The resumption of large-scale external aid was expected to give impetus to the improvement of the country's infrastructure, regarded as essential for its economic development. However, despite increased Vietnamese endeavours to assist in the search for MIA and despite the stationing of US diplomats in Hanoi from August 1993, President Clinton announced in September a renewal of the embargo against Viet Nam (although US businesses were to be permitted greater access to the country).

Viet Nam was not kept isolated from the international community during the period when the US embargo was in effect. Australia was the first Western country to restore direct development assistance, a decision that was made in October 1991 in anticipation of a comprehensive political settlement in Cambodia. Australia has consistently maintained its position among foreign investors in Viet Nam since then. In May 1993 Vo Van Kiet visited Australia, and reportedly agreed to receive a delegation of parliamentarians. In the following year the Australian Prime Minister paid a return visit. The question of a parliamentary consultative delegation 'on human rights' proved to be a contentious matter, and it was not until 1995 that such a mission was permitted to make a brief visit. Later in the year Australia's Governor-General visited Hanoi, while Do Muoi visited Australia (and also New Zealand). Bilateral relations with Viet Nam were briefly strained when the new Australian Government, elected in March 1996, sought to withdraw the financial support for the My Thuan bridge in the Mekong Delta, which the previous Government had pledged. This policy was reversed during a visit to Hanoi by the new Australian Minister for Foreign Affairs. In 1997 Australia quietly opened a regional security dialogue with defence officials in Hanoi; the first formal dialogue took place in May 1998.

In 1998 bilateral trade between Australia and Viet Nam increased by 39%, reaching a record US $1,000m. The Vietnamese Prime Minister, Phan Van Khai, paid an official visit to Australia in March 1999; at the same time, Australia announced it was to increase its official development aid to Viet Nam to $236m. for the 1998–2001 period; a separate agreement on co-operation in the areas of education and training was also signed. The completion of the My Thuan bridge in mid-2000 was a significant point in bilateral relations between Australia and Viet Nam. Australia signed a Defence Co-operation Agreement with Viet Nam in 2000, and from 1999 commenced training Vietnamese security forces, with the aim of improving co-operation in law enforcement and establishing a narcotics intelligence database. Australia also funded a 10-year $15m. rural development programme in Quang Ngai province to alleviate rural poverty. Australia pledged $47.5m. in official development aid in 2003. In a bid to improve bilateral ties with Australia, President Tran Duc Luong commuted the death sentence of the convicted drugs-trafficker Le My Linh, an Australian national. The Australian Minister for Foreign Affairs, Alexander Downer, visited Hanoi in late 2003, where premier Phan Van Khai expressed his gratitude for Australian assistance in providing education, health care and safe water. A Vietnamese high-ranking military delegation, led by Senior Lt-Gen. Phung Quang Thanh, the Deputy Minister of National Defence and Chief of General Staff of the Army, paid an official visit to Australia in September 2003, at the invitation of the Chief of the Australian Defence Force, Gen. Peter Cosgrove, the first official visit of Viet Nam's army commander to Australia since bilateral diplomatic ties were established in 1973.

In February 1993 President Mitterrand of France became the first Western Head of State to visit Viet Nam since its reunification. In November 1997 President Jacques Chirac of France made an official visit to Viet Nam, which was hosting a summit meeting of French-speaking countries. In 1995 Viet Nam and the EU signed a co-operation agreement which included a provision for the protection of human rights. Germany's relations with Viet Nam have been strained by the issue of Vietnamese illegal residents in Germany who refuse to return home. Some Vietnamese gangs were implicated in the smuggling of cigarettes, while others attracted publicity as a result of violent gang warfare; Viet Nam has been reluctant to accept them back. In January 1995 Germany pledged DM 100m. in development assistance (suspended since 1990) and an equivalent amount in export credits in return for which Viet Nam agreed to accept, by the year 2000, 40,000 of an estimated 95,000 Vietnamese living in Germany. The remaining 55,000 would be given official residence permits. In January 1999 Germany agreed to waive the debt of US $21.6m. owed to the country by Viet Nam as a contribution towards the financing of environmental protection measures in Viet Nam. In June Germany pledged continued support for a programme to assist Vietnamese nationals returning to Viet Nam from Germany. In February 1999 the Danish Minister of Foreign Affairs, Niels Helveg Petersen, visited Viet Nam and announced Denmark's contribution of $111m. towards the funding of landmine clearance; a 'soft' loan agreement of $40m. for 1999–2000 was also signed. In September Prime Minister Phan Van Khai paid official visits to Sweden, Finland, Norway and Denmark, where he signed a number of aid agreements. The party General Secretary visited France and Italy in May 2000 (one of his first overseas visits to non-communist states). In April of that year Canada suspended diplomatic relations with Viet Nam, after the Vietnamese authorities executed a Canadian citizen for drugs-trafficking (despite evidence, provided by the Canadian Government, indicating the victim's innocence). The Vietnamese Minister of Foreign Affairs, Nguyen Dy Nien, travelled to Canada in September 2003 to celebrate the 30th anniversary of the establishment of diplomatic relations between the two countries. EU officials were angered at Viet Nam's support for Myanmar during its chairmanship of ASEAN. The EU has steadily increased its economic co-operation with Viet Nam, agreeing in 2002 to provide $150m. in financial assistance over the following five years for poverty reduction and sustainable development in the countryside. By 2002 the EU had provided over $2,400m. in aid to Viet Nam (75% in grants, 25% in loans), making it the third largest donor, after Japan and the World Bank. In that year Germany alone provided DM 63m. in official development assistance, increased from an average of DM 6.1m. in the early 1990s.

General Secretary Nong Duc Manh visited France in May 2003, and in the same month German Chancellor Gerhard Schroeder spent two days in Viet Nam, during which he extended DM 16.4m. in development assistance. Bilateral trade between Viet Nam and Germany reached US $1,300m. in 2002. In November 2003 Viet Nam and France signed two credit agreements worth some $23m. for projects on air equipment and environmental protection and pledged further to open markets to foster economic development. France was Viet Nam's second largest market after Germany among the EU countries, and French investors had pledged more than $2,000m. for 130 projects. In August 2003, in Belgium, Deputy Prime Minister Vu Khoan held a meeting with EU trade commissioner Pascal Lamy, who affirmed the EU's policy to enhance co-operation with Viet Nam and its support for Vietnamese accession to the World Trade Organization (WTO). Vu Khoan and Lamy agreed on measures to implement the February 2003 agreement on increasing quotas for Viet Nam's garment exports to the EU market. The British Secretary of State for Foreign and Commonwealth Affairs, Jack Straw, and the British Minister of State for Trade visited Viet Nam in September 2003 and March 2004, respectively, the latter to promote bilateral ties and discuss British investment in Viet Nam. Most significantly, the United Kingdom became Viet Nam's largest non-refundable aid donor, with the Department for International Development's 2004–06 programme for the country, under which non-refundable aid to Viet Nam was to increase from $35m. in 2003 to $100m. in 2005.

In February 1994 President Bill Clinton announced the lifting of the US economic embargo. He expressed the hope that this would encourage the Vietnamese to give further assistance in the investigations of MIAs (the number of whom was now believed to be about 1,600). In May Viet Nam and the USA agreed to open liaison offices in their respective capitals; the USA opened its office in Hanoi in February 1995. The restoration of full diplomatic relations was finally announced by President Clinton on 11 July, and in the following month the new US embassy was inaugurated in Hanoi. However, owing to a delay in the approval of President Clinton's nominee for the ambassadorial position, Douglas Peterson, the posting of the US ambassador was delayed until April 1997.

By 1995 the USA had become the sixth largest foreign investor in Viet Nam, and maintained that ranking in 1996, with US $1,300m. invested in 64 projects. In April 1997 the USA signed an agreement with Viet Nam settling a $150m. debt issue outstanding since the end of the Viet Nam War. The USA stated, however, that Viet Nam would be required to undertake a series of measures, such as protecting intellectual property and agreeing to a trade accord, before it would be granted Most Favoured Nation (MFN) trading status (now Normal Trade Relations—NTR). Although an agreement on intellectual property rights was signed, negotiations on a trade agreement made little progress in the face of US demands that Viet Nam open its markets, remove various trade barriers, abandon investment licensing and grant 'national treatment' to US banks, law firms and advertising agencies. In March 1998, however, Viet Nam's co-operation with the USA on a resettlement programme for Vietnamese 'boat people' in camps in South-East Asia (see above) resulted in President Clinton's decision to waive the Jackson-Vanik Amendment to the 1974 Trade Act; this amendment prohibits the USA from trading with or providing investment funds to countries that do not permit free emigration. As a result of the waiver, the Overseas Private Investment Corporation (OPIC) and the Export-Import Bank were permitted to commence operations in Viet Nam. (OPIC provides special insurance cover for US firms doing business in Viet Nam, while the Export-Import Bank provides loans to US companies to assist their exports to Viet Nam.) During 1998 Viet Nam and the USA intensified negotiations on trade and aviation agreements.

In February 1999 President Clinton affirmed that Viet Nam was 'fully co-operating in good faith' in all areas of bilateral relations in which the USA itself sought progress (particularly in the area of joint searches for the remains of MIAs). Viet Nam also intensified its co-operation with the USA on immigration issues in 1999: following some initial difficulties, the Vietnamese Government gave permission for refugees in overseas camps to return to Viet Nam for processing, in preparation for emigration to the USA under the Resettlement of Vietnamese Returnees Programme, and also expedited the processing of former re-education camp detainees. Improved Vietnamese co-operation in these areas led to President Clinton's agreement to the second waiver of the Jackson-Vanik Amendment, as a consequence of which US export promotion and investment support programmes were permitted to recommence. In 1999 Viet Nam and the USA made substantial progress in their negotiations of a bilateral trade agreement. The most contentious issues during the final round of negotiations were those of non-tariff barriers and tariff reduction schedules. Once enacted, the Agreement would allow Vietnamese exports access to the US market on an NTR basis. The World Bank estimated that Vietnamese exports to the USA would double in the first year following the enactment of the Agreement, from US $470m. to nearly $1,000m. In July the USA and Viet Nam announced that they had finally reached agreement in principle upon the terms of the bilateral trade agreement. It was expected that, once all the outstanding technical details had been completed, the Agreement would finally be signed in September and then submitted to the US Congress for approval. However, a visit to Viet Nam by the US Secretary of State, Madeleine Albright, in September was reported not to have proceeded well, and hopes that the trade agreement would be signed at the APEC summit held in the same month in Auckland, New Zealand, were dispelled. In October the Politburo rejected the trade accord. The Vietnamese Government attempted to renegotiate certain provisions of the agreement, although the USA steadfastly refused, offering only to clarify certain aspects. The USA made efforts to continue discussions on the issue, with a series of high-level visits in the first half of 2000.

The US Secretary of Defense, William Cohen, travelled to Viet Nam in April 2000. Although the visit represented progress in improving bilateral military links, the Vietnamese Government appeared reluctant to support Cohen's suggestion that Viet Nam and the USA were potential allies against 'growing hegemony' in the region and to allow port visits of US naval vessels. The USA praised Viet Nam for its co-operation regarding the search for MIAs, although Hanoi made little progress in appealing to the US Government for aid in the search. A resolution by the US Congress condemning the communist regime in April also prompted hostility. In mid-2000 the Vietnamese authorities, which had sentenced 86 drugs-traffickers to death in 1999, pledged to co-operate with the USA in measures to prevent the passage of illicit drugs through the country. Nevertheless, bilateral relations suffered certain reverses. In May Hanoi charged a dissident, Ha Si Phu, with treason, prompting five other dissidents to submit a joint appeal to the National Assembly. His case was supported by several US legislators, infuriating the Vietnamese Government, which rejected their intervention as interference in its internal affairs. The nomination of the dissident Buddhist monk, Thich Quang Do, for the Nobel Peace Prize also antagonized the Politburo, which increasingly urged continued vigilance against foreign conspiracies to undermine the regime through 'peaceful evolution'. Viet Nam attempted to gain credibility for its claim of being committed to the improvement of human rights with its election in May to the UN Economic and Social Council's Human Rights and Social Development Committee; however, allegations of continued violations of human and religious rights persisted.

In July 2000 the newly appointed Minister of Trade, Vu Khoan, travelled to Washington, DC, and, as expected, the bilateral trade agreement was finally signed. Viet Nam was accorded NTR status, effectively lowering tariff rates on exports from 40% to 3%. The US Government ceded to Vietnamese demands that foreign companies would only be allowed to become minority partners in the telecommunications sector, but in return accepted a three-year period of introduction for other components of the accord, as opposed to the four years that had been previously negotiated. The agreement laid the foundations for Vietnamese membership of the WTO, which Viet Nam hoped to join by 2003 (although its planned accession was subsequently postponed until 2005). Both the US Congress and the Vietnamese National Assembly delayed ratification of the bilateral trade agreement for domestic political reasons. The US Congress finally ratified the agreement in August 2001, while the National Assembly endorsed it in November.

President Bill Clinton visited Viet Nam in November 2000, the first visit by a US President to a unified Viet Nam. Clinton had an austere official reception and there was little ceremony surrounding the visit, although he was welcomed by the public. Military co-operation suffered a reverse in April, when a helicopter carrying a mission of US and Vietnamese military personnel to search for MIA remains crashed in the Quang Binh province, killing all passengers. The crash prompted renewed controversy over the efficacy of continuing the searches. As of mid-2001 65 joint missions had been carried out and the remains of more than 600 US military personnel had been repatriated.

Viet Nam continued to react with hostility to what it considered unwarranted interference in its internal affairs. In February 2001 the USCIRF conducted hearings on Viet Nam and, concluding that the Vietnamese Government practised systematic religious repression, recommended that the World Bank suspend financing. At the same time the US Government granted asylum to 24 members of ethnic minority groups, who had fled to Cambodia after leading the protests in the Central Highlands. In September the US Congress adopted the Viet Nam Human Rights Act, thereby angering the Vietnamese Government. Meanwhile, the effects of Agent Orange, a chemical defoliant that had been sprayed over the country in large quantities by the US armed forces during the Viet Nam War, remained a contentious political issue. In June 1999 it was reported that a nation-wide census was to be held in order to assess the health of Vietnamese who had been exposed to the defoliant; the chemical was believed to be responsible for widespread health problems, including genetic damage to unborn children. In 2000 Viet Nam authorized the US Government to dispatch a team of researchers to study the impact of Agent Orange; in July the US Ford Foundation for the first time granted aid to Vietnamese victims of Agent Orange. In March 2002 US and Vietnamese government scientists held their first meeting to discuss the effects of the defoliant. The US scientists denied that the chemical had caused severe long-term health problems, and the US Government refused to accept any financial liability. The USA and Viet Nam undertook to co-operate on further research.

Viet Nam responded cautiously to the US Government's announcement of a 'war on terror', following the attacks against New York and Washington, DC, on 11 September 2001. While condemning the terrorist acts against the USA and supporting UN and ASEAN resolutions, Viet Nam was highly critical of the US use of military force in Afghanistan. Nevertheless, Viet Nam granted US military aircraft overflight rights to Vietnamese territory. The alleged conspiracy by terrorists linked to the al-Qa'ida (Base) organization (the network of fundamentalist Islamist militants believed to be responsible for the September acts) to attack US naval vessels in Singapore prompted the US Commander-in-Chief of Pacific Forces, Adm. Dennis Blair, to make the issue of port access to Cam Ranh Bay a priority. The Vietnamese Government had rejected initial requests from both the USA and China for access to the former Russian base, but had not refused to permit port visits.

There were a number of set-backs in relations with the USA in 2003. Although the USA supported Viet Nam's admission to the WTO, in April the two countries signed an agreement that would limit Vietnamese textile exports to the USA to an annual growth rate of 7%. Viet Nam was very unhappy with the agreement as exports had been increasing rapidly, from US $48m. in 2001 to $900m. in 2002, and it had already exported goods worth $500m. in the first quarter of 2003. The Vietnamese leadership had anticipated that half of its $3,200m. of textile exports would go to the USA. Viet Nam was also angered by a number of measures taken by the US Congress which it perceived to constitute blatant interference in its internal affairs. These included the Viet Nam Freedom of Information Bill and the Viet Nam Human Rights Bill, which sought to reduce non-humanitarian aid. The Vietnamese Government openly condemned the US-led campaign to oust the regime of Saddam Hussain in Iraq in early 2003 and expressed sympathy for the Iraqi people, while urging the UN to take a lead with regard to events in the country.

Despite ongoing differences over Iraq, ties between the USA and Viet Nam were strengthened from 2003. The US Assistant Secretary of State for Population, Refugees and Migration, Arthur E. Dewey, arrived in Hanoi in August 2003 and was allowed to visit Viet Nam's Central Highlands region, which had been the scene of mass demonstrations in February 2001. Also in August 2003 a 19-member delegation of the US House of Representatives Special Committee on Intelligence, led by Porter Goss, its Chairman, met with members of the Vietnamese National Assembly. Minister of Foreign Affairs Nguyen Dy Nien travelled to the USA in September with the stated aim of building 'a framework for long-term and stable relations'. Nien held talks with his US counterpart, Colin Powell, and met with congressional leaders. His visit was shortly followed by a five-day visit by a delegation led by the Minister of Planning and Investment, Vo Hong Phuc, which made efforts to attract more US investment. Two years after the bilateral trade agreement was ratified, Vietnamese exports to the USA had increased two-fold, while US goods imported into Viet Nam had increased by 30%. Most significantly, US Secretary of Defense Donald Rumsfeld hosted his Vietnamese counterpart, Pham Van Tra, in November 2003, the first visit by a Vietnamese military chief since the end of the Viet Nam War in 1975. While no agreements were signed on military co-operation, the Vietnamese urged the USA to do more to help the 2m. Vietnamese suffering from exposure to Agent Orange. The USA requested, although it did not receive, rights to make port visits to Cam Ranh Bay. Viet Nam, sensitive to reactions from China, resisted this, although the first port visit by a US naval ship, USS *Vandegrift*, took place in Ho Chi Minh City in November 2003. In February 2004 Adm. Thomas Fargo, head of the US Pacific Command, made a four-day visit to Viet Nam to improve military co-operation, and visited the naval facility at Cam Ranh Bay. Fargo held talks with the Vietnamese Deputy Minister of National Defence and Chief of General Staff of the Army, Senior Lt-Gen. Phung Quang Thanh. In December 2003 a high-ranking Vietnamese delegation, led by Deputy Prime Minister Vu Khoan, travelled to Washington, DC, for talks with US government officials and members of Congress, including Secretary of State Colin Powell, Assistant to the President for National Security Affairs Condoleezza Rice, Trade Representative Robert Zoellick, Secretary of Agriculture Ann Veneman, and Acting Secretary of Commerce Samuel Bodman. The most significant achievement of the visit was the conclusion of an aviation treaty that would allow direct passenger and cargo flights between the two countries for the first time since the Viet Nam War. Trade between the two countries had reached around US $5,000m. annually by this time.

The former Prime Minister and Vice-President of South Viet Nam Nguyen Cao Ky made a symbolic return to Viet Nam to celebrate *Tet* in February 2004. He was the highest-ranking member of the former South Vietnamese regime to return to Viet Nam.

Economy

ADAM McCARTY

Revised for this edition by COLIN STELEY

INTRODUCTION

Viet Nam was a country ruled by colonists, divided, or at war for most of the 20th century. Unification of the country in 1976 was followed by the invasion of Cambodia in 1978, and a subsequent brief but violent war with the People's Republic of China. This legacy had profound consequences for economic development in general and for attempts to impose central planning in particular. Central planning implementation occurred in North Viet Nam in the 1950s, and less extensively in South Viet Nam after 1976. However, this was both brief and incomplete.

Viet Nam's economic transition, which is typically identified as having been initiated in 1986, was a relatively smooth period of structural adjustment and stabilization—at least for the first decade. The state sector was never large in Viet Nam, where 80% of the population lived in rural areas and were mostly engaged in the cultivation of rice. Furthermore, by 1986 the planned part of the economy was thoroughly undermined. The collapse of trading relations with the Council for Mutual Economic Assistance (CMEA—which Viet Nam joined as a full member in 1978) forced a restructuring of the modest state-owned enterprise sector, which disposed of almost one-quarter of its labour force during 1988–92. Institutional reforms also generated growth in the urban household services economy, which offset the increase in the numbers of unemployed. The shift to markets was relatively easy without the burden of a large military-industrial complex, and goods hitherto exported to the CMEA found new Western markets without difficulty.

During the 1990s the economy of Viet Nam underwent a transformation. Structural changes occurred during 1987–91, but changes in institutions were slower to occur, and the legacy of administrative control continued. Furthermore, important services, such as accounting, banking and the legal system, remained highly rudimentary. In some regards, therefore, 'transition' involves a generational process.

Before Reunification

In 1954 Viet Nam was divided into two zones at the 17th parallel. In the North, under the rule of the Viet Nam Workers' Party (now the Communist Party of Viet Nam), a central economic plan following the Stalinist model was adopted, and was implemented much as it had been in China, commencing with a campaign to collectivize agricultural production. The purging of rich peasants and landlords in the late 1950s suppressed opposition to collectivization and extended Communist Party economic control to the village level. The absence of a campaign of enforced collectivization in the South after reunification, similar to that which had taken place in the North, was a major factor in the failure to collectivize agriculture in the Mekong delta area. The inefficiency of collectivization had also become evident. Yields of paddy rice per ha in the North had fallen from 2,290 kg in 1959 to 1,840 kg in 1960, accounting for most of the decline in total rice production, from 5.19m. metric tons in 1959 to 4.18m. tons in 1960. Average annual rice production during 1960–69 remained low (4.2m. tons). Collectivization had placed the entire Northern economy under pressure.

Until 1976 the economy of South Viet Nam was dominated by the inflow of US military and civilian funding. Aid funds provided infrastructure. Industrial development was pursued through import substitution. Farmers benefited by selling their agricultural surpluses. The non-traded goods sector, particularly services, grew rapidly as consumer demand increased. The effects of war, import substitution and the growth of services resulted in rapid urbanization. The market economy was vigorous but overdeveloped. When US forces withdrew, therefore, the consumer demand that had supported the economy collapsed.

After Reunification

After reunification in 1976 Viet Nam was a poor country attempting to recover from a prolonged period of war. Only 21% of the population lived in urban areas. The North could barely feed its population of 25m., owing to the stagnation of agricultural production that had resulted from two decades of war and central planning. Between 1958 and 1975 the Northern population had grown by 64%, while the production of staples had increased by only 11%. There was little industrial and almost no infrastructural and service development. Markets were clandestine and hence ill developed. Although market transactions persisted to a considerable degree in the South, reunification resulted in an immediate decline in living standards.

The national economic strategy for recovery was to integrate the two economies by imposing Northern central planning upon the South. A five-year Plan, which aimed to 'achieve basic socialist transition in the South', was initiated in 1976. All financial activity was brought under the control of the central bank, and a campaign to collectivize the villages of the Mekong delta commenced. The 1976–80 Plan failed as a result of the inherent inefficiencies of planning and the unwillingness to take measures that would collectivize and extract an economic surplus from Southern agriculture. Amid resistance to collectivization in the South, Viet Nam's output of staple foods declined by 9%, and the paddy rice yield per ha by 18% in 1976–78. Farmers refused to harvest crops, or refused to sell them at low official prices, slaughtered their livestock, and abandoned land. It was estimated that over 1m. metric tons of rice was fed to pigs each year. By the end of the decade the Mekong delta remained largely uncollectivized. Most of the production collectives and co-operatives that did exist had limited functions. The planned industrial growth from enhanced North-South economic integration also failed to materialize. Weak infrastructure, together with barriers to internal trade (tolls and travel permits), increased transport and transaction costs. Output figures recorded a decline: in 1980 total quantities produced of coal, cement, steel, fertilizer, cloth, and paper were all below 1976 levels. State enterprises were reportedly operating at only 30%–50% of their productive capacity.

Failure of the 1976–80 Plan was already apparent in early 1978, but the military demands of Viet Nam's invasion of Cambodia, and the subsequent decline in foreign aid and the war with China, precipitated a crisis. Reductions in aid, particularly Chinese assistance, decreased supplies of cheap inputs for state enterprises. Viet Nam was still attempting to maintain an army of more than 1m. persons until the late 1980s. Unplanned economic activity proliferated. State enterprises had to work outside the plan in order to provide minimal living wages for employees, and workers found additional jobs to supplement those incomes. Central state management weakened when cheap inputs disappeared. The central authorities did not have an economic surplus to support the state sector, which itself did not produce a surplus.

Liberalization after 1978 was a spontaneous process, which placed great pressure on authorities to sanction economic realities and to permit further compromises to central planning. The first official steps towards liberalization were taken at the Communist Party's Sixth Plenum in September 1979, which adopted plans to stimulate agricultural production. Procurement quotas for paddy rice, and the agricultural tax, were fixed for five years, and state procurement prices were increased. Above-quota production could be sold on the free market. Markets were officially sanctioned, so that extensive legal market activity began to co-exist with central planning (unlike most other centrally-planned economies).

During the period of the five-year Plan, instead of imposing an already diminished Northern development model on the South,

Southern practices were extended throughout Viet Nam. The economy that emerged in the 1980s was as much a product of 'southernizing' the North, as it was of 'northernizing' the South. After the economic crisis of 1978–79 the State periodically tried to re-establish the dominance of central planning, but was too weak to do so. Owing to the inherent inefficiencies of planning, economic resources were under-utilized, and hence even modest liberalization was able to generate strong supply responses. Periods of growth after liberalization measures were viewed by the authorities as opportunities to reimpose central control, which only precipitated further crises. The liberalization initiated in 1979 continued into the early 1980s. Co-operatives were allocated production quotas above which output could continue to be sold on markets or at 'negotiated' state prices that were close to market prices and included some bartering. Central-planning controls over state enterprises were also officially relaxed. In January 1981 a 'Three-Plan System', which legalized the already extensive market activities of state enterprises, was introduced.

During the early 1980s a confusing policy of pragmatic liberalization and attempts to recentralize economic authority were implemented. Without a domestic economic surplus, the State could expand only through aid or by reducing the real incomes paid to workers. Both strategies were pursued. In response, state enterprises expanded 'unplanned' activities to supplement official salaries, and employees increasingly found additional work. It was reported that the supplementary incomes earned by state workers increased after 1979, reaching almost 60% of total income in 1986, but declined slowly thereafter.

In 1976 45% of total budgetary revenue was derived from external sources. This share declined as relations with China and industrial countries worsened, but rose again to 41% in 1980, after the USSR increased its support. The trade deficit with the CMEA countries increased from 380m. roubles in 1980 to 721m. roubles in 1985 (when Viet Nam's CMEA exports covered only about one-third of imports). Foreign assistance was never large in relation to the whole economy, but because it was entirely directed to the state sector, its supporting role was vital. Without foreign assistance, the State could not have maintained its position during the 1980s. By 1985 the state sector was again under strain. Agricultural production had stagnated, real incomes had declined, and 'unplanned' economic activity continued to increase. The recentralizing policies pursued after 1982 again undermined the economy. Increases in the money supply induced inflation in the free market, which state prices had to follow. This resulted in a so-called 'leading effect' of the parallel, or open, market that made the transition process in Viet Nam differ from that of the USSR or other socialist countries in Eastern Europe.

In September 1985, in an unsuccessful attempt to solve the problem of high free-market prices, the authorities increased state prices and introduced a currency reform. The price rises were accompanied by increases in loans to state enterprises and in wages. The monetary reform introduced a new dông. Since personal savings had been predominantly in gold and US dollars, the main impact of this reform was to eliminate the cash holdings of state enterprises. Without commodity convertibility, the bank deposits of state enterprises had little direct influence on the price level, but a significant increase in budgetary deficits and a sharp rise in inflation were inevitable. In 1984 budgetary revenue had covered 81% of expenditure; in 1985 receipts covered only 55% of expenditure. In 1986 the general price index rose by 487%.

Official planners were unable to disregard the markets. The State maintained constant prices until the breakdown of planning in 1978–79. In the late 1970s market inflation was running at about 30% annually, eroding the real value of state procurement prices. Official price increases above the prevailing market inflation rate were consequently required in 1981–82 to raise the real state price being paid for agricultural produce. Thereafter, state prices followed free market trends closely. The failure to maintain strong central planning made it possible to undertake drastic price reform in 1989. Repressed inflation associated with a monetary 'overhang' had been eliminated by free market exchanges (commodity convertibility) and annual inflation of more than 300% during the preceding three years. Free market prices had come to dominate economic activity, with the notable exception of key industrial sector inputs, where low prices were maintained through the subsidized trading and aid relationship with CMEA countries. Thus, the issue of prices was already significant when the Vietnamese authorities began dismantling the old planning system.

The 'reforms' of 1985 were the last attempt to maintain the system of central planning. The failure of the 1985 policies strengthened the position of reformers who advocated a more market-orientated economy. The pro-market forces included a rising commercial interest group within the state sector and Southern liberals within the Communist Party. The death of the General Secretary of the Communist Party, Le Duan, in July 1986, created the opportunity for fundamental change within the existing political structure. The Sixth Party Congress of December 1986 was an ideological turning-point, which initiated the transition to a market economy in Viet Nam.

The Transition to a Market Economy: The late 1980s and 1990s

In fact, the Sixth Party Congress in December 1986 introduced a period of policy disarray. Liberalization continued to be selective, and the planned economy remained protected. When interprovincial trade barriers were abolished in 1987, rice prices declined. Central Committee policy decrees issued during 1987–88 relaxed control over foreign investment, land, foreign trade, banking, state industrial management, the private and household sectors, and agriculture. Reforms accelerated in 1988, when the co-operative method of agricultural production was abandoned in favour of household production, and small private enterprises were officially encouraged. However, the fundamental contradictions of maintaining a two-price system remained. The budget deficit grew to 7.1% of gross domestic product (GDP) in 1988, and the trade deficit increased as the State tried to compensate for the rapid decline in aid. Most of the deficit was financed by central bank borrowings. Inflation remained high, at an annual rate of about 300%. In early 1989 any residual planning still in place had to be abandoned.

In March 1989 the Vietnamese Government introduced several stabilization measures. Although direct subsidies and price controls on electricity, coal and some other products remained, most were removed. Government expenditure for the first half of the year was tightened, declining from 14.1% of GDP in 1988 to 12.3% in 1989. Bank interest rates were raised to high positive levels. They were accompanied by credit restrictions on lending to state enterprises. The foreign-exchange rate was brought close to the black market equilibrium rate. Gold trading was liberalized, while a relaxation of trade regulations facilitated a rapid increase in rice exports. The effect was impressive. Inflation, which increased at a monthly rate of 13.6% during 1988, declined to around 2.5% in 1998, and this level was maintained into mid-1990.

The 1989 reforms addressed the difficulties of money flow, rather than any perceived money stock problem (the 'monetary overhang'), which had been the focus of the 1978 and 1985 monetary reforms. Money supply (M1) rose from 9.2% to 15.2% of GDP. Budget and credit policies had been tightened as indicated, and strong monetary growth was absorbed by a one-off 'monetization' of the economy and 'sterilization' of increased household deposits in the banking system. Credit limits restricted lending to state enterprises, while the underdeveloped legal and banking systems further constrained non-state lending options.

After years of near hyperinflation, the 'monetization' of the economy and of household saving portfolios in early 1989 prompted a revival of confidence in the domestic currency. In a significant change in savings behaviour, households shifted into monetary assets. The real purchasing power of gold and US dollars declined by about 25% in the three months to May 1989, as households shifted into dông and bank deposits. During the 10 months from March 1989 household deposits in the commercial banks increased from 0.8% to 3.8% of GDP. Universal poverty and the distortions of the period up to 1986 resulted in the high supply responses to later reforms, particularly to the ending of active discrimination against private enterprise, but also in a continuing distrust of the banking system. Non-dông forms of saving were still preferred. In 1993 a national survey reported that only 6% of the Vietnamese population used banks.

The 'monetization' of the economy in 1989 helped to limit the inflationary impact of radical price reforms and monetary expansion. The estimated GDP growth of 8% in 1989 resulted from supply responses in the agricultural sector and also in services, as well as from reaction to changes introduced in 1988, which more than offset the 2.8% decline in industrial production. The fiscal expansion of late 1989 continued into 1990. Inflation returned, and real interest rates again became negative in August of that year. Despite increases in budget revenues, notably from petroleum exports in 1990, inflation stood at 68%. The fundamental causes (money-financed budget deficits and state enterprise subsidies) continued. The Government reacted by reducing current and capital expenditures, with the consequence that the budget deficit decreased from 6% of GDP in 1990 to 2.5% of GDP in 1991.

At a meeting of the Central Committee of the Communist Party in December 1991, macroeconomic stability and control of inflation were declared the priority policy objectives to 1995. An anti-inflationary campaign followed, characterized by large-scale selling of dollar and gold reserves to draw dông from circulation. This, and the delayed response to tighter fiscal policies in 1991, reduced the annual inflation rate to 18% for 1992. Indirect subsidies to state enterprises were also lowered significantly in 1992. The momentum was maintained, with the effect that inflation declined to 5% in 1993, although it rose to 14% in 1994. Relative stability of the exchange rate, and in gold buying and selling, was also maintained from 1991. The interest rate (the monthly rate for three-month deposits) was lowered progressively by the central bank, from 2.0% to 1.4%.

Control of inflation continued to improve. Diversified financing of the deficit, including experimental bond floats, reduced the growth of money supply. The growth of M2 (M1 and sight deposits) was estimated at 28% for 1994, compared with 19% in 1993 and 34% in 1992. With non-inflationary financing of the budget deficit from 1992 to 1994, the annual inflation rate did not return to pre-1992 levels. In 1993 current budget revenues increased by 13%, owing, in part, to the rise in prices of petroleum, postal services and electricity; although these areas continued to be lightly taxed by international standards. Steady increases in revenue from external trade taxes from 1991 were also beneficial.

Macroeconomic stability in the mid-1990s became more threatened by trade and exchange rate policies than by fiscal management. In 1995 the trade deficit grew to 11.5% of GDP. This rose to about 19% of GDP in 1996, with imports exceeding exports by US $3,200m. These deficits were only partly covered by aid and foreign investment inflows. A regulated contraction of imports was the policy response that achieved a reduction in the trade deficit in the late 1990s, at the cost of lowering imports.

Growth and Development after the Asian Crisis

The Asian financial crisis, which had severe effects on the economies of most of the countries in the region in late 1997, had relatively little immediate impact on those countries without open capital accounts and foreign-currency rationing. Thus, despite the Asian financial crisis and apparent hesitation in Viet Nam to adopt further structural change, the economy continued to grow quite rapidly. Major economic indicators showed a general recovery from 1998, with real GDP growing by around 5.5% in 2000. Total export earnings grew by 25% in that year (owing, in part, to an increase in the international prices for some agricultural products and crude petroleum), and non-oil exports increased by a robust 16% in 1999 and 2000. Sustained high rates of development would, however, require a return of foreign investment, an increased rate of gross savings as a percentage of GDP, and development of a private corporate sector in Viet Nam.

In 2001, despite external weaknesses, real GDP growth slowed only slightly, to an estimated 5%, supported by robust domestic demand. Inflation was still very low (0.8%), with non-food prices nearly unchanged. Export growth decreased sharply, to 4% from 25% in 2000, owing to weak manufactured exports and low non-oil commodity prices. The external current account was, however, in surplus, at the equivalent of 2.25% of GDP, owing to an even sharper slowdown in imports. With a strengthening of foreign direct investment (FDI) inflows (US $1,000m. in 2002, compared with $800m. in 2001), gross reserves rose to the equivalent of around nine weeks of prospective imports.

With the outbreak of Severe Acute Respiratory Syndrome (SARS) in early 2003, Viet Nam's tourism industry, which had been performing strongly owing to the country's new-found image as a safe haven, was adversely affected. Tourist arrivals declined sharply and hotel occupancy rates decreased from nearly full capacity to only 40% by early April. By the end of April WHO declared Viet Nam to be free of SARS. Tourist numbers recovered in the first half of 2004, with over 1.39m. foreign arrivals, an increase of 29.6% in comparison with the corresponding period of the previous year, according to the Viet Nam National Administration of Tourism.

In the period following the 1997 Asian financial crisis, and throughout the regional SARS epidemic, real GDP growth in Viet Nam never fell below 4% as the country experienced a relatively stable period of economic growth compared with other countries in the region. This was in part a result of its protectionist economic policies. Even as it began to liberalize its markets, Viet Nam continued to enjoy relative economic success, achieving a real GDP growth rate of 7% for the first half of 2004, even though this fell short of the government target of 8%. This was partly due to a second bout of avian influenza in the country in mid 2004.

In the first half of 2004 there was considerable growth in the industrial sector (which grew by 10% compared with the first six months of 2003), fuelled by expansion in the mining industry, while growth in the processing, gas and power industries declined compared with the corresponding period of the previous year. In particular, the construction sector grew by just 7.3%, compared with growth of 10.6% in the first half of 2003. The services sector grew by 7% in the first half of 2004, compared with growth of 6.4% in the first six months of 2003.

Since the Enterprise Law came into effect in 2000, the domestic private sector has witnessed a significant expansion; by 2004 some 136,000 private businesses had been established, with a total investment of approximately US $12,000m. This has led to an increase in the private share of investment, from 20% in 2000 to 26.5% in 2003, and has significantly increased the share of private investment in GDP; capital accumulation by the formal private sector amounted to 9% of GDP in 2002. In 2003 more than 30,000 private enterprises in Viet Nam applied for business registration, each with an average registered capital of around 2,100m. dông, exceeding the 2002 average of 1,500m. dông.

Crude petroleum, garments and footwear continued to make the greatest contribution to exports in 2004. Having registered 31% growth in 2003, compared with the previous year, in the first half of 2004 their contribution had increased by 19.8% in comparison with the first six months of 2003.

FDI also displayed an upward trend, with the prospect of political stability and the potential of exports to the USA under the bilateral trade agreement concluded in 2001. In addition, the National Assembly has issued several pieces of taxation legislation, creating equal conditions for both local and foreign businesses. The new legislation, especially the Law on Corporate Income Tax, decreased tax incentives for new projects at industrial parks, in order that the Ministry of Planning and Investment would have to submit to the Government for approval a list of projects that should attract investment incentives.

POLITICAL ECONOMY

The strategy of the Ninth Party Congress (held in April 2001) to lay the foundations for the country to proceed along a path to industrialization was implemented through a number of 'long term plans and strategies', addressing a range of issues, including several which targeted economic and social development. The plans attracted significant criticism for being too broad in their scope, failing to prioritize their goals and lacking linkages between components of the process, from inputs through to outputs. The 10-year strategy for socio-economic development included the Government's Comprehensive Poverty Reduction and Growth Strategy (CPRGS), the fifth and sixth five-year socio-economic development plans, the 'Vietnamization' of the Millennium Development Goals (eight socio-economic goals adopted by 189 of the UN member states at the organization's 2000 Millennium Summit), and the expansion of foreign economic relations and the improvement of their performance, amongst others.

In June 2002, following a series of consultations at the national, provincial, and village level, the Vietnamese Government launched the CPRGS, taking a new, pro-active and more comprehensive approach to poverty reduction and growth, in line with the Government's development strategy. The CPRGS outlined three broad objectives: (i) the achievement of high levels of economic growth through a transition to a market economy—this laid out the Government's agenda for structural reform and concrete plans for their implementation; (ii) the achievement of an equitable, socially inclusive and sustainable pattern of growth—this was reflected in detailed plans for implementing sectoral economic and social policies; (iii) the adoption of a modern public administration, legal and governance system—this aimed to facilitate the design and implementation of the policies and programmes necessary for attaining the first two goals. These objectives, combined with the Millennium Development Goals, formed Viet Nam's Development Goals, which aimed to prioritize both the set of core development goals that would establish an improved accountability framework for the Government, and the CPRGS, which drew on national plans while being compatible with international commitments.

The five-year plan listed a number of actual objectives, for the most part derived from outcomes of the existing 10-year plan. Resources were mobilized in order to create an immediate transformation in the economic and production structure of all industries and sectors. This would lead towards the enhancement of the linkage efficiency between production and markets for consumption, meeting the demands for food, accommodation, transportation and education of the people while maintaining the principle of independence and sovereignty in socio-economic development.

By 2005 total GDP was targeted to have reached twice its 1995 level. The annual rate of GDP growth was projected to be 7.5% during the 2001–05 plan. This objective included: an increase of 4% in the annual GDP of the agriculture, hunting, forestry and fishing sector, increasing its contribution to total GDP to 20–21%; annual growth of an additional 10.4% in the industry and construction sector, increasing its contribution to 38–39% of total GDP; and an increase of 6.8% in the services sector, in order that it contribute 41–42% of total GDP by 2005.

Viet Nam's Government has made clear its determination to join the World Trade Organization (WTO) by 2005. Whether this ambition could be realized or not depended upon the outcome of the ninth round of negotiations, scheduled to take place at the end of 2004. The country had to pass through 10 such rounds before it could be admitted. Viet Nam needed to revise at least 30 laws and ordinances before it would be eligible for admission, and much depended upon the preparations made by the sensitive agricultural sector and its capacity to adapt.

In 2004 the overall fiscal deficit was projected to be equivalent to 4.6% of GDP, a decline of 0.2% compared with the previous year, thus maintaining the Government's moderate expansionary stance and the growth of domestic demand. Even though this was not a dramatic change, over the previous five years there had been a significant, and somewhat worrying, increase in government spending. In 1999 the overall deficit was equivalent to 0.1% of GDP. The Government was under pressure to increase investment across the board, and particularly under foreign pressure to reform state-owned enterprises (SOEs). In 2004 Viet Nam's revenue to GDP ratio was around 23%, an exceptionally high ratio for the East and South-East Asian region. As deriving a large proportion of its revenue from SOEs conflicted with Viet Nam's gradual move to a market economy, there was a need for the Government to devise alternative methods to increase its revenue base in order to expand expenditure while maintaining its overall deficit at under the government target of 5% of GDP. While admission to the WTO and other trade policies would have mixed effects on revenue, the Government has standardized the corporation tax rate for both domestic and foreign companies at 28%, included goods previously exempt from value-added tax (VAT), and placed a consumption tax on automobiles.

By September 2004 the rate of inflation over the previous 12 months had almost doubled, from 3.2% to 6%, exceeding the Government's stated target of 5%. This increase was mainly attributable to the effect on food prices of the outbreak of avian influenza in the country in early 2004. As sales of most poultry were barred by the Government, the price of all other foods rose. As food constituted 48% of the consumer price index (CPI), it was inevitable that the increase would occur. However, further contributing factors to inflation were: an increase in government wages; an increase in the prices of pharmaceutical products and building materials; the depreciation of the dông owing to the increasing price of imported goods; and the general economic drive, which increased domestic demand by 9.4%.

Several key policy measures were undertaken during the first half of 2004. A number of taxation reforms were implemented, including: the harmonization of corporate rates at 25%; the reduction of VAT to either 5% or 10%; and the reduction of the highest rate of personal income tax to 40%. In line with its goals of implementing the CPRGS to the provincial level and preparing the next five-year Plan, the Ministry of Planning and Investment circulated a note specifying the requirements for socio-economic planning. The Government broadened the scope of the CPRGS by incorporating additional aims regarding the development of large-scale infrastructure. In order to intensify the implementation of the CPRGS, efforts were being made to integrate it with local and provincial planning. In addition, efforts were made to develop the domestic primary bond market, in order to address the problem of raising medium- to long-term finance for development—for example, the issuance of bonds worth US $160m. for education and $578m. for infrastructure projects (the secondary market is yet to be developed). The business environment should further flourish, as a new Land Law was passed in 2004 intended to promote FDI. In addition, new legislation concerning competition and bankruptcy was enacted in 2004, and work on harmonizing enterprise and investment laws was under way. In order to regulate SOEs and state-owned commercial banks (SOCBs), a new state-owned debt trading and management company came into operation in February 2004. Further equalization of SOEs and a detailed study intended systematically to estimate the number of non-performing loans of SOCBs were also announced.

Meanwhile, in order further to improve governance in Viet Nam, in December 2003 the Government signed the UN Convention against Corruption and participated in the Asian Development Bank-Organisation for Economic Co-operation and Development (ADB-OECD) Anti-Corruption Initiative Conference in Kuala Lumpur, Malaysia. It also drafted a decree concerning the accountability of heads of government agencies and offices; strengthened the supervision of public affairs research implementation; announced the national introduction of the 'one-stop shop' model for simplified administrative procedures and service delivery; expanded the block grant scheme for public expenditure management; and trained civil servants, having instituted computerization in state administration.

EMPLOYMENT

During the early years of economic transition, from 1986 to 1991, substantial structural change took place in Viet Nam's labour force. Total state sector employment decreased by about 250,000 persons each year during 1989–91. About 500,000 soldiers were demobilized, and returning worker migrants added to the labour force. Furthermore, the number of persons of working age increased by 3.7% per year (39% of the Vietnamese population was under 15 years of age in 1990). During this 'first phase', the expanding household sector absorbed the growing supply of workers. Unemployment and underemployment, while severe, were never at socially dangerous levels. Unemployment during this period was estimated at between 1.9m. to 6m. people. During the 'second phase', beginning around 1992–93, change in the structure of employment was less dramatic. In 2000 the threat of large-scale unemployment and underemployment remained a serious problem for Viet Nam, both in terms of political stability and efficient resource use. The sustained growth of the private household sector absorbed some of this labour force growth, but the ability of the economy to generate jobs with higher value-added was essential.

The 'open door' policy has resulted in the creation of a foreign-invested sector in Viet Nam. Total employment in firms with foreign investment increased from 146,000 in 1996 to 218,350 in 2000. In 1996 foreign-invested firms accounted for about one-quarter of industrial output, while employing approximately 5% of the industrial labour force. However, employment in the foreign-invested sector remained modest, accounting for less than 1% of the labour force in 1999. This was partly because of much higher levels of productivity in the foreign-invested firms, but also because much of the inflow of foreign investment was channelled into heavy and capital-intensive industries, with highly protected domestic markets but little capacity for job creation. Moreover, foreign-invested firms are required to hire labour through officially sanctioned employment agencies, which adds an estimated 8% to the cost of hiring. In May 2002 the National Assembly approved a new labour law giving foreign-invested enterprises more flexibility in hiring labour. However, generally policies, either directly or indirectly, remained biased towards the development of capital-intensive industries, a fact that seriously hindered the development of a process conducive to job creation.

During 1996–2000 total employment increased by 2.3m. people, of which the state sector absorbed nearly 700,000, or 33%, while the non-state sector recruited about 1.2m., accounting for more than 50% of total job creation over that period. By 2000 the private sector accounted for about 40% of total industrial jobs, and in the subsequent two years this proportion increased to about 51%. This expansion could be attributed to the aftermath of the Asian financial crisis, as a large proportion of new FDI was transferred from countries in the Asia-Pacific region, targeting low-technology and high-employment sectors such as footwear and garments. Both NSE (non-state enterprise) and foreign-invested components of the private sector showed remarkable increases, as employment growth rates for each sector reached over 60%. This was primarily due to the new Enterprise Law and the type of FDI. From 2002 onwards employment growth rates for NSEs and foreign-invested companies began to diverge. However, there is no statistical evidence of this, as a number of tariffs and quotas have limited the production levels of local export industries, government policies have prevented full foreign ownership of resources and limited banking laws, foreign-owned enterprises have fulfilled their market potential and, therefore, growth rates should have slowed down. This was likely to be only a temporary phenomenon, as markets would open up again when Viet Nam joined the WTO and trade liberalization was in progress, taking advantage of the country's comparative advantage in terms of labour. Meanwhile, NSEs probably enjoyed a period of growth as opportunities for local small businesses were enhanced by taxation reforms, new business laws, and banking reforms.

In 2004 migration policy had two significant impacts. The first was the exodus of rural populations to Viet Nam's two main industrial centres: Ho Chi Minh City and Haiphong. Both recorded increases in unemployment rates in 2003–04, while the urban labour force increased by 1% compared with the rural labour force for the entire country. The other impact was on cross-national migratory labour. As Viet Nam was renowned for its cheap labour force, a number of Asian countries were hiring Vietnamese labour, a total of 40,000 workers in the first 10 months of 2004. The Government has felt that it was losing too much human capital and implemented laws making it difficult to export labour; out of 152 existing labour export businesses, 20 were expected to shut down as a result.

INDUSTRIAL ECONOMY

Only a limited range of industrial activity was allowed to develop during the colonial period, to avoid competition with French imports. Apart from mining, the only significant industries in Indo-China before 1940 were those of cement at Haiphong and textiles at Nam Dinh and Hanoi. After 1954 the Vietnamese aimed to develop both existing and new industries. In the South there was some development of light industry in the late 1950s and early 1960s, mainly related to the production or assembly of consumer goods, and a cement works was established at Ha Tien, near the Cambodian border.

'Socialist construction' entailed the creation of a heavy industrial base for the new economy. However, US bombing destroyed the development of industry before 1965. After 1968 a gradual decline in the emphasis on heavy industry was perceived. One of the principal aims of the second National Plan was to expand light and consumer goods industries, in an effort to relieve the shortages of basic commodities, which were causing further discontent among the long-suffering population. In practice, however, these policies made little progress until the advent of economic reforms in the 1980s. Although emphasis continued to be placed on heavy industry, the expansion of light industry, which began in the early 1980s, was furthered in the 1990s. During 1993–96 industrial growth averaged a substantial 13.6% per year, slowing to 12.1% in 1998. Following recovery from the 1997 Asian financial crisis, industrial output expanded faster, achieving the highest growth rate in three years (1998–2000), of 15.7%. In the first quarter of 2001 industry expanded by 14.4% compared with the first three months of 2000. Between 1991 and 2000, the average annual industrial growth rate was 11.2%.

In the early 1990s electronics and automobile production were still at an early stage of development. Both were import-substitution industries protected by tariffs and non-tariff barriers, and had fulfilled the demand of the domestic market by 2000. In 1993 foreign companies assembling electronics goods in Viet Nam included Daewoo, Hitachi and Philips. Beginning in 1995, the sector expanded considerably, with output of assembled television sets almost trebling in the period 1995–98, to reach 364,000 pieces in the latter year. The 30%–40% annual growth rate in the electronics industry of the early 1990s had declined to about 10% by 2000. There has been no strategy for electronics production since the 1990s, resulting in foreign-invested producers, most of whom focus on assembly rather than production, holding nearly 90% of the domestic electronics sector's investment capital and export revenue. According to the Association of South East Asian Nations (ASEAN) Free Trade Agreement (AFTA), by 2005 imports of electronic products made by ASEAN countries would not be taxed. This has resulted in many electronics joint ventures in Viet Nam reducing their production in the country in preparation for withdrawal.

In 1998 212,000 motorcycles were assembled in Viet Nam. This increased to 509,000 in 1999, and to an estimated 1.6m. in 2000. In 2001 the Government decided to reduce the number of motorcycles on the road, which totalled 11m. in that year, owing to increases in traffic accidents and pollution levels, and thereby implemented a law limiting ownership to one motorcycle per person. In 2003 motorcycle production declined to 1.2m., and 350,000 remained in stock. This trend continued in the first six months of 2004, with 651,000 motorcycles being produced, while 400,000 remained unsold. In 2004 there were a total of 52 companies producing motorcycles in Viet Nam, seven of them with foreign investment, and a total capacity of 3m. units per year.

By 1997 14 foreign joint ventures in the automobile industry had been licensed. Investment totalled US $940m., with assembly capacity estimated at 144,660 cars annually, and with 45 different models. Of the 14 joint ventures, nine companies (Mekong Corporation, Vina Star Motor Corporation, Viet Nam Motor Corporation, Mercedes-Benz Viet Nam, Ford Viet Nam, Toyota Viet Nam, Isuzu Viet Nam, Suzuki Corporation and the Daihatsu Motor Company of Japan) were in production. Only 6,882 cars were assembled in 1999 (compared with 6,404 in 1998), which was less than 10% of projected domestic demand for the period 2000–05. In 2000 and early 2001 there was a surge in second-hand car imports (about 10,000 units over the first half of 2001). Only 11 vehicle assembly joint ventures remained in operation in 2000, with an estimated total annual output of 12,500 units; most joint ventures were therefore running at as little as 10% capacity. The automobile industry has become the fastest-growing industrial sub-sector, with sales increasing by 41% in 2001, 36% in 2002, and 66% in 2003, with total sales reaching 35,000 units. Even though Viet Nam is one of Asia's smallest automobile markets, there was a shortage of supply owing to an 80% increase in taxes planned for 2005. The Vietnam Automobile Manufacturers' Association believed that, because of this tax, in 2005 there would be a 30%–40% drop in sales.

In the first eight months of 2004 growth of 15.4% was recorded in Viet Nam's industrial sector. Of this, 11.8% was generated by SOEs, while NSEs registered a record 21.6% growth. Coal, processed seafood, cotton and crude petroleum were Viet Nam's best-performing industries, while power generation, rolled steel production, and diesel engines all performed poorly compared with previous years. The industries that contributed most to the sector, owing to their high prices, included fertilizer, construction, steel, cement, and other industry-orientated materials. The Government stated that, in order to maintain growth into 2005, industries would have to reduce their production costs, raise their production capacity, explore non-quota markets, and stabilize steel and fertilizer prices.

MINING AND ENERGY

Mining

Interest in mineral resources was one of the factors that attracted the French to Tonkin in the late 1880s. Their exploitation was an important feature of the colonial economy, and was further developed by the Japanese during their brief control of the region during 1940–45. The Vietnamese themselves continued this development from the 1950s with the assistance of Soviet geologists. Viet Nam has commercially viable reserves of coal, iron ore, bauxite, chromite, copper, tin, titanium, zinc, gold, apatite and gemstones, although many of these remained under-exploited. Coal reserves are located in the north-eastern province of Quang Ninh and in Bac Thai province in the Thai Nguyen Basin. Proven reserves of coal were estimated to be approximately 3,500m. metric tons, although the actual figure for reserves of all types of coal (bituminous coal, anthracite, lignite and peat) was possibly as high as 30,000m. tons. Iron ore deposits of around 500m. tons have been located in a mine at Thac Khe. Proven bauxite reserves were estimated to be 4,000m. tons, mainly in the southern province of Lam Dong. Deposits of zinc and lead have been located in the North, although the size of the reserves remained unknown in 2000. Son La province is currently active in mining slate for both domestic use and export.

The most important mining activity arises from the coal deposits. Output expanded considerably in the early 1970s, but stagnated for some years after 1977, following the widespread withdrawal of Chinese labourers after the deterioration of Sino-Vietnamese relations. Production reached 6.9m. metric tons in 1988, but averaged only 4.4m. tons per year in 1989–91. Production subsequently recovered, with annual coal output averaging 6.8m. tons in 1993–96, increasing to 10.7m. tons in 1998, and slightly decreasing to 9.6m. tons in 1999. The Viet Nam Coal Corporation, Vinacoal (which assumed sole responsibility for production and distribution in 1994), adopted the target of increasing Viet Nam's annual coal production to 12m. tons in 2000. However, external market conditions permitted the production of only 10.8m. tons. In 1999 mines under the control of Vinacoal were being closed. Exports of coal declined for three years in succession from 1977, to only 600,000 tons in 1980, and in 1987 the level of exports was estimated at only 233,000 tons. By 1990, however, coal exports had recovered to 788,500 tons. Further rapid increases were subsequently achieved, and in 1996 coal exports were 3.6m. tons, valued at US $115m., before a decline to 3.2m. tons ($102m.) in 1998 and 3m. tons in 2000. Since the 1980s the Government has sought to promote foreign investment in the mining sector. However, this has remained at a relatively low level despite the 1996 mining law.

In 2004 a fourth credit loan from Finland's Nordic Investment Bank, of US $15m., for new and modern equipment to extract and refine coal was expected to lead to a further increase in coal production. In the first nine months of the year total coal production increased by 37% compared with the corresponding period of the previous year, reaching 15.74m. metric tons. This led to a 35% increase in local consumption, while coal exports increased by a dramatic 65%, to 6.73m. tons. However, as a result of the increase in coal production, the Government has issued environmental warnings concerning the illegal export of coal across the Sino–Vietnamese border, where it attracts a price 10 times higher per ton. Aside from the lack of government revenue resulting from this activity, the Government claimed that it would lead to unsustainable exploitation of Viet Nam's coal reserves and have a concomitant effect upon the environment.

Petroleum and Natural Gas

In the mid-1990s Viet Nam's production level of crude petroleum was approximately equivalent to that of a minor producer such as Romania, but still well below that of neighbouring Malaysia and Indonesia. (In 2000 Viet Nam's petroleum reserves were estimated to be about one-third of those of Indonesia.) During 1995–98 petroleum production increased by 22% annually, reaching 12.5m. metric tons in 1998, but production increased by only 7% between 1999 and 2000, reaching 16.3m. tons. As a result of increases in the international price, the value of Viet Nam's petroleum exports rose from US $1,200m. in 1998 to $2,100m. in 1999. In some respects, the industry had yet to fulfil initial expectations. Between 1988 and 1999 the Government allocated 35 production-sharing contracts to foreign oil companies, investing a total of approximately $3,100m. However, in terms of commercially viable finds, the results have been disappointing. In 1995 less than 25% of the country's petroleum blocks had been allocated, and it was believed that some of the most promising areas had yet to be explored. Nearly all the major international oil companies were represented in Viet Nam, including those from the USA, following the removal of the US trade embargo in early 1994.

Viet Nam also aimed to exploit its natural gas resources, with proven reserves of 300,000m. cu m and prospective reserves of 1,700,000m. cu m. Since production at the offshore gas field of Tien Hai reached its maximum in 1987, Viet Nam produced only associated gas from Bach Ho (nearly all of which was burnt off). However, in 1995 a gas pipeline from Bach Ho to Ba Ria (near the oil centre of Vung Tau) was completed, thereby allowing the utilization of gas for power generation, and gas production increased by 154% during 1995–98. A three-stage programme for the industry has been drafted, with 14,000m. cu m of gas expected to be produced in the final phase (2005–10).

In 2004 the Viet Nam National Petroleum Corporation (Petrovietnam) was accelerating the construction of four major projects: Viet Nam's largest oil refinery; a gas-electricity-fertilizer complex; an oil refining-petrochemical complex; and the longest oil pipeline to be built in the country. These projects were intended, as in the coal industry, to exploit the available natural resources Viet Nam has to offer, untapped owing to the previous lack of available investment, technical knowledge and heavy machinery. In 2003 production of petroleum and natural gas increased by 43%, to 13m. metric tons, earning more than US $3,520m. This represented the attainment of 75.8% of the Government's production target for the year, which stood at 17.5m. tons of petroleum and 5,700m. cu m of natural gas. At its peak, the price of crude petroleum on the global market reached a 20-year high of $42 per barrel. Although the Government's decision to raise petrol prices in June 2004 was certain to have an adverse effect on transport services in the future, the prime objective was to allow the local market to adjust to international prices as Viet Nam became more competitive on the world stage.

Power Generation

By 1975 the total power generation of both zones of the country was only 3,000m. KWh. Production in 1980 was estimated at 3,720m. KWh. Although the situation has improved since the mid-1980s, there were frequent power shortages, particularly in the South, during the 1990s. With energy demand rising at a rate of about 20% per year, electricity output has failed to keep pace. To meet demand additional output was largely contributed by gas turbine generators at the Ba Ria-Vung Tau power plant, and the Da Nhim and Tri An hydroelectric power stations. Further important developments were the construction of a new generator at the hydroelectric power station at Hoa Binh, the completion of the 1,500-km North-South power line and two hydroelectric power stations in the centre and the South; YALY and Thacmo. Total electricity output rose from 19,253m. kWh in 1997 to 21,847m. kWh in 1998. It was projected that generating capacity would reach 33,000m. kWh by 2002. Some 19% of the electricity generated by power stations was lost in transmission in 1997. However, electricity production increased by 13% in 2000 and by 15% in 2001. During 1998–99 the Government also announced its intention to reduce the country's dependence on hydroelectric power, anticipating greater reliance on gas and coal to generate electricity.

The Government aimed to supply electricity to 80% of rural households by the end of 2005 (compared with 63% in 1995), and in 2004 it supplied 85% of households. Another target achieved ahead of schedule was the accomplishment of producing 50,000m. KWh of electricity per year, 40,000 KWh being used for commercial purposes to drive the country's 7% annual GDP growth. Even though the Government passed a law in 2004 allowing a free market to establish competitive local prices for electricity, the 32 independent companies were subsidiaries of the Viet Nam Electricity Corpn (ENV) and were only participating in a pilot programme mimicking the private electricity market in order to realize true electricity costs.

TRANSPORT AND COMMUNICATIONS

In 2000 a significant factor that continued to constrain development was the inadequacy of the country's infrastructure. During the French colonial period, Viet Nam was provided with a railway system, the main lines of which linked Hanoi to Ho Chi Minh City (then Saigon) and Haiphong to Yunnan Province in the People's Republic of China. After 1954 the Chinese assisted in the construction of a third major route between Hanoi and the Chinese province of Guangxi via Lang Son. The lines to China, however, were closed towards the end of 1978, and further stretches were damaged, as a result of the war with China during February–March 1979. Only in February 1996 did improved Sino-Vietnamese relations lead to the reopening of the cross-border rail link. In 1995 the Government estimated that investment of more than US $3,670m. was required for the modernization and expansion of the railway system. Moreover, of 1,813 railway bridges, 33 are shared with road transport and many of them have been in use for more than 100 years. Currently, the national railway system has six single-track routes, totalling 3,260 km, equivalent to 0.04 km per head of population. This is only about one-third of the density of other low-income countries. Construction of the Yen Vien–Cai Lan railway route was scheduled to start in September 2004, with an investment of over 39,000m. dông, raised from government bonds. The 129.2 km route will start from Yen Vien station and run through Pha Lai and Ha Long stations to Cai Lan port station. It will have five new stations, a locomotive and a carriage maintenance service. Once completed in 2008, the railway route will allow passenger trains to run at 120 km per hour and cargo trains at 80 km per hour.

Road transport became increasingly important in the 1960s, even though roads and bridges suffered heavy damage during the war. In the South, the USA constructed a new highway between Ho Chi Minh City and Bien Hoa, as well as a large number of lesser roads. However, the road transport sector continued to be perceived as a 'weak link' in the economy. This could explain why over the last three decades a huge proportion of state investment has gone into road construction and maintenance. A large proportion of the US $13,040m. pledged by foreign aid donors during 1993–98 was allocated to improving the road network. This assistance was most apparent in the upgraded North–South nation-wide Road No. 1 and the newly-constructed Hanoi–Haiphong highway. Most road bridges in Viet Nam had a maximum weight capacity of just 10 metric tons, and of the country's 105,500 km of highways only 25% were paved. A six-lane highway, linking Hanoi with Noi Bai airport, was opened in 1994. From 1998 to 2002 85% of total state investment went into roads. In 2002 alone 98,800m. dông ($634m.) was spent on upgrading and constructing highways, which was three-and-a-half times more than the amount allocated in 1998 for the same purpose. However, the condition and quality of roads and highways have deteriorated both in cities and the countryside, with the State being able to meet less than 30% of demands for maintenance. In 2004 maintenance alone cost $150m., and this was expected to increase to $195m. in 2005 owing to the construction of new roads, including the recently completed Ho Chi Minh Highway. In 2004 Viet Nam had a total of 218,527 km of roads, of which 17,300 km were highways and 17,425 km were provincial asphalt roads.

The principal port facilities before 1975 were at Haiphong, Da Nang and Ho Chi Minh City. By 1978 projects had been initiated to construct a new port at Cua Lo, as well as to modernize Haiphong in order to increase its capacity to 2.7m. tons per year. Upgrading work was ongoing in the mid-1990s and, pending its completion, severe congestion in Haiphong was expected to continue. In the southern-central provinces, it was decided to modernize the ports of Quy Nhon and Nha Trang, to enable them to cope with timber exports. In the region of the Mekong delta the development of a new container facility at Ho Chi Minh City and the upgrading of several of the river ports to take ocean-going vessels were planned. Haiphong and Can Tho ports required dredging to reduce chronic silting, while new container terminals were also planned at Haiphong, and Tan Thuan. Vung Tau port was due for extensive upgrading. Although large amounts of foreign financial aid and investment have been pledged, their disbursement and implementation were delayed by problems associated with bureaucracy and the low level of domestic capital mobilization. Aid disbursements reportedly totalled US $5,300m. during 1993–98 (equivalent to 39% of commitments). Viet Nam has a national plan for the development of its seaport system until 2010, which focuses on three fields: national fleet development, port development (concentrating on deep-sea development), and developing the shipbuilding industry. Until 2004 all three fields had seen little forward development. A five-year programme for port development, started in 1999, achieved little success. Viet Nam boasts of having 100 seaports along its coast, but because of a weak port system and a poor turnaround of goods, big ships are unable to dock and unload even small quantities of merchandise. The national commercial fleet is highly uncompetitive for two reasons: it is old and weak, with ships an average of 20 years of age; and as government policy favours awarding government contracts to local companies and local importation laws differ from international standards, Vietnamese fleets and companies are finding it very difficult to establish contracts with foreign ports and companies. An additional port in Cai Lan was to be developed to deal with these issues.

As a result of rapid economic and political developments in recent years, by 1997 more than 17 foreign airlines served Viet Nam. In the early 1990s a major programme of airport development was under way, with Noi Bai (Hanoi), Tan Son Nhat (Ho Chi Minh City) and Da Nang airports serving as the principal hubs. Tan Son Nhat is scheduled to accommodate 10m. passengers and 1m. tons of freight annually by 2010. An upgrade to the international terminal at Tan Son Nhat was completed in June 2001. The number of passengers carried by Vietnamese airlines increased from 500,000 in 1990 to more than 2.7m. in 1996 and to over 2.8m. in 2000, while the volume of freight carried increased from 4,000 tons in 1990 to 50,100 tons in 1997. Viet Nam Airlines had one of the most modern fleets in the world in 1998, with 22 aircraft less than one year old, and expanded its operation to cover 20 domestic and 20 international routes. In 2003 Viet Nam Airlines acquired four modern long-range Boeing 777s (two purchased and two on lease) to modernize its fleet and open more direct routes to Europe and America. It reported revenue of US $488.3m. in the first half of 2004, an increase of 14% compared with the first six months of 2003, contributing just over $9m. to the state budget. The company flew 22,000 flights, serving 2.3m. passengers, and carried 45,143 tons of cargo and postal parcels in the first six months of 2004, an increase of 42% and 21.5%, respectively, over the corresponding period of 2003. Domestic passengers accounted for 1.3m.; the remaining 1m. were international clients. Viet Nam Airlines' market share in the international aviation market increased by 40.5% in the first half of 2004 compared with the first six months of 2003.

With the implementation of reforms, Viet Nam's telecommunications network also received attention. Viet Nam ranks among the top 30 countries in the developing world, with over 2m. telephone subscribers. In 2002–04 Viet Nam integrated internet and mobile cellular telephone services into its infrastructure. In 1991 only two telephones per 1,000 people were recorded; by 2004 the ratio had increased to 150 land-line users per 1,000 people, between 180 and 200 mobile cellular telephones per 1000 people, and 500–600 internet users per 1,000 people. The telecommunications industry has been permitted to form a more competitive structure, and has attracted smaller companies into the market by allowing them to set their own, more competitive, prices. This would force the two largest companies, government SOEs, which controlled 99% of the industry, to lower prices. A 25% increase in subscribers in August 2004 compared with the previous month was due to a massive industry-wide reduction in tariffs. There are currently five mobile cellular telephone service providers in Viet Nam, worth US $180m. per year, and growing at an annual average of 28%.

The significantly expanding services sector contributed US $15,000m. to Viet Nam's GDP in the first eight months of 2004. Ho Chi Minh City contributed 59% as Viet Nam's leading commercial hub, with Hanoi accounting for 33% while all other major urban areas contributed a little over 1%. The Vietnamese Government realizes the potential of the services sector, investing $160m. in it over the course of 2004 in an attempt to improve the quality of the sector. The competition between SOEs and private firms was inherently unfair, owing to the potential for SOEs to undercut costs and place pressure on the private sector within the sector.

AGRICULTURE

Agricultural Reform

The productive economy in Viet Nam remained reliant on the agriculture, forestry, and fisheries sector, which employed 60.7% of the labour force and contributed 21.8% of GDP in 2003. Although the contribution of agriculture to GDP has steadily declined from 1990 onwards, the sector enjoyed a 3.2% growth rate in the first eight months of 2004. Thus, the performance of the agricultural sector has largely been determined by the direction of political and policy reforms since 1976. The contribution of the agricultural sector to a country's GDP typically declines in relation to the process of development, although agricultural output rises in absolute terms. However, this was not the case in Viet Nam, under its system of central planning. In the late 1980s the agricultural share of GDP rose as a result of official price increases rather than output gains, and despite falling productivity indicators. It was not until the 1990s that a clear trend of increasing aggregate output and productivity, and a declining share of GDP, became evident.

Total food production increased by 50% during 1988–97 (and by 30% on a per caput basis), while livestock numbers and industrial crop production increased even faster. This extraordinary performance enabled Viet Nam to become a major rice exporter, and led to a steady decline in the incidence of poverty. Liberalization of markets and a strengthening of property rights have generated strong supply responses in the agriculture sector during transition. Long-term investments in annual and perennial crops by households have had a dramatic impact on living standards. The benefits, however, have not been evenly distributed. More farmers have lost their land, while those who were growing only enough crops to survive have not been incorporated in the reforms. Improving the situation of the poorest households was viewed as a more complex task than that involving those who were capable of producing a surplus for markets.

In 2004 an ongoing battle against the USA concerning the dumping of shrimps on the US market was finally moving in favour of Viet Nam. US shrimp—or prawn—producers have claimed that low-priced shrimp from Viet Nam and other Asian and Latin American countries were being dumped at below cost on the US market. Producers were faced with a 98% import duty, as the market had become the target of US protectionists. Shrimp is Viet Nam's largest seafood export, and the USA was its largest shrimp export market in 2002, accepting US $467m., or 48%, of total Vietnamese shrimp exports.

Government agricultural policy in 2004 focused on Viet Nam's ability to improve the production chain of industrial processes for agricultural products made for international markets. This was expected to improve the 3.1% growth in sectoral GDP experienced in 2003 to the 4.6% growth projected for 2004. Another agricultural shift that had taken place was from paddy farming to aquaculture—producing shrimp for world markets at premium prices. With the improving economic conditions, meat and poultry production has increased to meet rising local demand, while production of rice, Asia's staple food, was expected to increase marginally due to increasing export demands. However, there remained a domestic supply shortage of basic goods such as milk and urea, amongst others.

Land

In 2000 it appeared that land scarcity was continuing to increase in most regions of the country; 30% of the land area is categorized as unused or barren. Most was once forested and now lies fallow, and much is badly degraded. An average farming family of five cultivated about 0.5 ha of land, which is the lowest level in the world, with farmers in the Mekong and Central Highlands averaging 1 ha. The People's Committees had less scope to reallocate land, or to find unused land for newly-formed families. Families with three or more children (often the poorer families) could no longer expect to receive more land on this account. A new class of landless families was thus being created, who worked for wages on other people's land, who might find off-farm activities, or who joined the increasing flow of people seeking work in urban areas. In 1996 at least 2m. rural residents were officially estimated to have migrated to cities in search of employment, increasing pressure on land prices.

Enacted in 1993, the Land Law has been revised twice, in 1998 and 2001. In principle it recognizes five rights: the right to transfer, exchange, lease, inherit and mortgage land ownership. However, the definition of land ownership, management and usage is somewhat blurred, causing large numbers of civil disputes, and deterring local and foreign investment and the development of the land market in Viet Nam. Recent developments, which have seen the economy move further towards a market-orientated one, and the demands of urbanization make it imperative that the Land Law again be revised. Areas of concern that have been addressed in the latest round of legislation include: a uniform land administration regime, ensuring clearer administrative procedures; and equal opportunities for both foreign and domestic land users, ending the current dual-pricing system involving government and market-determined prices. The decree to implement the law was expected to be completed at the end of 2004. Preliminary responses have been mixed. Land planning and assignment continued to be significant areas of concern, which have not been addressed owing to conflicts with the Vietnamese Constitution, which treats private land as being communally owned. This mainly affected rural areas, while in urban areas or those proximate to them, where foreign investor interests lay, it was difficult to know how land purchasing would be affected, as foreign investors still could not exercise their land-use rights to obtain mortgages from offshore lenders. Either way, the price of land would have to be reduced dramatically if foreign companies were to establish themselves. Around urban areas, especially Hanoi, the lack of variety of investment options has intensified the demand for land, dramatically increasing prices even though supply outweighed demand by a ratio of approximately four to one.

BANKING AND FINANCE

In contrast to the negligible role that the financial sector played in centrally-planned economies, financial intermediation is a crucial factor in the functioning of markets. Reform of the financial sector is consequently a test of the transition to a market economy. The Vietnamese financial sector remained dominated by a few large banks. In the late 1990s the banking sector contributed only 2% of GDP, reflecting its undeveloped nature.

In July 2004, on the 10th anniversary of the foundation of the Viet Nam Bank Association (VNBA), branches of foreign banks and joint-venture banks were invited to join the association, being barred, however, from having voting rights for the board of directors. The leading state-owned banks were attempting to create more favourable conditions in which to deal with the increasing pressures on the Government's financial deficit by issuing bank stocks. The Government intended that, after 2010, commercial banks should operate according to international standards governing capital, management and information technology in order to meet the needs of the economy.

Interest Rate Policy

Prior to 1989 government policy kept interest rates strongly negative with obligatory enterprise and household deposits. Early 1989 rates were set at levels that were strongly positive in real terms, and it was announced that future changes would take explicit account of inflation. The correlation between interest rates and inflation during 1989 was very weak; it was not until after 1991 that interest rate movements became discernibly related to inflation trends. Even so, a deposit of a three-month rolling term would have accrued a real interest return of 132% over the 30 months between July 1992 and December 1994. In 2000 the central bank, the State Bank of Viet Nam (SBV), introduced a more flexible interest rate policy, allowing more freedom for commercial banks to determine the price of their loans. In the first half of the year the SBV announced an interest rate band with a base rate of 0.7% per month. From August 2000 credit institutions were allowed to offer any rate up to the new ceiling rate; consequently, banks could price their loans differently to prime and non-prime customers, subject to the ceiling. Lending in foreign currency was anchored to the Singapore inter-bank market. However, as the global economy is linked to that of the USA, as interest rates were rising in 2003–04, Viet Nam increased interest rates in order to remain competitive against the USA. On 1 June 2001 the SBV removed the interest rate ceiling on both US dollar loans and dông loans. Deregulating the limit on these rates was a key step towards decentralizing the financial system. It was announced in August that locally based companies could borrow foreign exchange from overseas lenders at rates negotiated by the two sides. In November the SBV reduced the reserve requirement for foreign deposits from 15% to 10%, with a view to helping commercial banks equipped to supply the high demand for dollars. The SBV also permitted the state-owned Bank for Investment and Development of Vietnam (Vietinde Bank) and the Vietnam Export-Import Commercial Joint-Stock Bank (Eximbank) to provide foreign-exchange options. The maximum value of transactions was US $500,000, and the permitted transaction period was between seven days and three months. Following that, the SBV allowed local commercial banks to provide interest rate swaps that allowed fixed-rate borrowing and imposed a limited period of three years on such arrangements. The granting of these permissions, together with the SBV's launch of new bank notes and increasing use of automated teller machines (ATMs), which numbered 500 nation-wide in 2004, illustrated the growing sophistication of the Vietnamese banking sector. In addition, the use of credit and debit cards, introduced by commercial banks, witnessed a dramatic increase, with only 20,000 cards nation-wide in 2002 increasing to 600,000 by September 2004.

Government Banks

A move towards financial sector reform was made in 1988 when the Agriculture Bank of Viet Nam (VBA—and now VBARD—Bank for Agriculture and Rural Development) and the Industrial and Commercial Bank of Viet Nam were created as commercial banks from departments of the SBV. They subsequently dominated commercial deposits and lending. Two existing small specialized banks, the Bank for Foreign Trade of Viet Nam (Vietcombank) and Vietinde Bank, constituted the formal financial sector. Establishing the two-tier banking system in 1988 had little effect until monetization in 1989, when the World Bank considered that deposit growth caused a 'transformation of the degree of penetration of the banking system in the economy'. Deposits increased to 3.8% of GDP, but by international standards this was still low. In 1991 the ADB considered the financial sector to be a major constraint on economic growth. Although the first round of banking reforms of 1988–91 were conducted and the structure of the banking system has diversified and fiscal control has improved, the current banking system remained considerably behind that of the region. Total bank credits do not exceed 22% of GDP and bank deposits do not exceed 8% of GDP. Four SOCBs have continued to dominate total lending and policy lending.

In the late 1990s a detailed restructuring plan for the four large SOCBs was created, with the development of a phased recapitalization fund from the Government. The plan began with financial audits of the large SOCBs by international auditors. Just 70% of recapitalization capital came from the Government, with the remainder coming from bond issues. In 2000 a Debt Resolution Committee subsequently identified the extent of existing bad debts, and facilitated the payment of some outstanding debts, in some cases through liquidation proceedings. The Government also authorized the cancellation of some bank debts.

Meanwhile, the restructuring of joint-stock banks (JSBs) was being conducted through their closure, merging or rehabilitation. The restructuring of JSBs has gained momentum after a long delayed start. By February 2002 13 JSBs had been closed or merged, reducing the number of JSBs from 52 to 38. Several JSBs have been rehabilitated using stakeholders to provide additional capital.

There are a number of further improvements needed in the financial sector. Lending procedures have recently been simplified by: broadening the scope of financial leasing and creating a more attractive environment for domestic and foreign leasing firms; allowing domestic banks to set up internal systems for clearing payment transactions without the involvement of the SBV but with its permission; allowing JSBs and foreign banks to receive collateral in the form of land, land-use rights or land certificates and savings; and providing the legal framework for cross-border payments, recognizing that international practices can be used to govern cross-border transactions if Vietnamese law does not require otherwise.

The main concern for the chief executives of the SOCBs was international integration, which would force them to compete with stronger foreign banks. With Viet Nam becoming a signatory to a number of international commitments, such as AFTA and the bilateral trade agreement (FTA) with the USA, and as a result of negotiations with the WTO, the country's banking sector in general, and the SOCB sector in particular, have realized that there is no alternative but to restructure. The four large SOCBs have advanced by restructuring their human resources, developing services, expanding networks, and, in particular, applying international standards to their management systems.

Apart from introducing a wide range of new banking services, such as retail banking, personal loans or credit and debit cards, all four banks recently joined the real estate market by providing or contributing loans and capital for both local and foreign companies to build housing and even luxury resorts. They also established some affiliates, such as securities companies and finance leasing companies, or partially entered the insurance market. Vietcombank has become well known for its currency trading activities, while VBARD, the largest SOCB in terms of its branch network, is strong in providing loans to farmers in remote and mountainous areas. The two remaining banks, Vietinde Bank and Vietincombank, have acquired a reputation for lending to industrial and investment projects. All have thus found measures to improve their financial status. However, it appeared that this would not be enough. It has become evident that profit levels were insufficient at most SOCBs. According to their 2003 annual reports, the four large SOCBs have recorded annual average after-tax profits of about 400,000m. dông (US $25m.). Although such figures appeared in their annual reports, the true figures remained a matter of speculation.

Stock Exchange and Regulatory Framework

Efficient financial intermediation has been severely constrained by the lack of appropriate monitoring capacity at the SBV. In 1991 it was reported that no department at the Bank was responsible for the collection and processing of reports of joint-venture banks and that it did not even have complete lists of joint-stock and housing banks in Viet Nam. By 2000 comprehensive supervision of all financial sector organizations was the responsibility of the SBV, although it lacked powers of direct intervention. Implementation of the Bank's limited authority continued to be constrained by skill shortages and poor administrative procedures. Its supervisory capacity was also dependent on the perceived credibility of its role as an autonomous agent of monetary policy. In 2000 this role was compromised by the necessity to act as a 'frequent lender' to the state commercial banks that were unable to make profits from financial intermediation. In 2000 regulations were also issued providing for banks to calculate provisions against their non-performing loans on a quarterly basis. A legal framework for creditor rights and assets resolution was to be established, which would secure transparency and accountability and enable all banks to become more competitive.

In 2002 regulations on financial reporting systems for all credit institutions were issued to set out the bases and principles for preparing financial statements. The SBV also issued regulations concerning the manner in which bank accounts with the SBV and other banks in Viet Nam were to be opened and used. The regulations applied to credit institutions, state treasuries, foreign banks operating overseas and monetary and international organizations that opened accounts with the SBV, as well as to organizations and individuals that opened accounts with banks operating in Viet Nam.

In October 2004 the Ministry of Finance issued Decision 73, which regulates the organization and operation of investment funds in securities and fund-management companies. The regulations provided legal guide-lines regarding the implementation of Decree 144 concerning securities and the stock market. The regulations concerned two types of funds, public funds as well as those with members, with the former required to issue certificates to the public. The other funds must have at least 49 members contributing money to them. All funds must be at least 5,000m. dông. The regulations were expected to give impetus to the development of the national stock market, which was still small.

Viet Nam has one public fund, the Vietnam Securities Investment Fund (VF1), run by the VietFund Management Company, a joint venture between the Saigon Thuong Tin Commercial Joint Stock Bank (Sacombank) and the British company Dragon Capital. Although a licence was granted in March 2004 for the establishment of VF1, enabling it to issue 300,000m. dông in the form of certificates to the public, the fund was only permitted to list on the stock exchange in September. By the end of August bonds represented 60.4% of the fund's investment, while listed shares accounted for almost 19% and unlisted shares nearly 16%. In the future investors would be given another option, in the form of shares in the Saigon Fuel Company (SFC).

TRADE

Buoyant trade growth has been a remarkable feature of Viet Nam's transition. The value of total trade increased steadily from 1988, with only a slight levelling-out in 1990 when trade with the CMEA collapsed. Using the exchange rates of the Bretton Woods institutions, the total volume of CMEA trade was not large prior to 1987 and subsequently decreased. Relations with the CMEA permitted Viet Nam to run a large trade deficit for only a short period to 1988. After 1988 trade with the convertible area was always larger. The collapse of the CMEA in 1990–91, therefore, had no significant adverse effects on the economy as a whole. The impact was mainly in the loss of subsidies to state industrial enterprises.

Viet Nam's total trade as a percentage of GDP increased from 32% to 88% in the eight years to 1999. Exports rose erratically to US $14,400m. (46% of GDP) in 2000, while import growth was steady and, in some years, excessive. In 2001 the value of exports reached $15,100m., an annual growth rate of 4.5%, which represented a decrease from 2000, owing to weak external demand and low commodity prices. Of the growth in total export earnings in 2000, 60% was attributable to exports to Asian markets outside ASEAN, mainly Japan and China ($2,622m. and $1,534m. respectively). The switch to hard currency exporting was achieved largely because Viet Nam's main exports were primary products that found an international market without difficulty. In 2003 and 2004 Viet Nam's exports increased steadily. Export revenues from the country's four main commodities have exceeded the $2,000m. threshold, most significantly crude petroleum and clothing exports, which raised $3,820m. and $3,690m., respectively, in 2003. The principal destinations for Viet Nam's exports in 2003 included: the USA, Japan, China, Australia and Singapore, each accounting for more than $1,000m. of the country's revenues from exports.

According to the Ministry of Trade, Viet Nam's commodities have so far been exported to more than 200 countries and territories nation-wide, although its exports in recent years have been concentrated on the USA and other markets in the North American and European regions, while ignoring key markets in Japan, China, Hong Kong and Taiwan. Trade with Japan, instead of achieving an annual growth rate of 17%–18% by 2010, as targeted, had declined to 14.4% in 2003, compared with 15%–16% in 2000. The growth rate of processed exports with high added value was also far behind the Government's target. Under the strategy, policy makers forecast that mineral and agricultural export growth would reduce sharply, from 24.4% and 23.4%, respectively, in 2000, to 9.3% and 21.6%, respectively, in 2005, although in 2004 growth was 19.9% and 22.1%, respectively. In the first nine months of 2004 Viet Nam experienced an export boom, with exports gaining an estimated US $19,800m. in value, an increase of 27.2% compared with the corresponding period of 2003, and more than twice the National Assembly's target of an 11–12% increase for the whole year.

In 2001 imports grew more slowly than exports, expanding at a rate of 2.3% in 2001, mainly driven by imports of construction materials such as steel and glass, cars and motorcycles. In the first nine months of 2004 imports grew by 21.3%, costing an estimated $22,470m. Amongst Viet Nam's most expensive imports in that period were five aircraft, costing a total of $488m., and 8.315m. metric tons of fuel, worth $2,560m., an increase of 12.2% compared with the previous year.

In 2001 the trade deficit stood at approximately US $1,000m., equivalent to 3% of GDP. The trade deficit was partly financed by aid, which constituted $1,100m. in loans and $400m. in grants in that year. Generally, it was thought likely that the trade deficit would widen again if the international price of petroleum declined and the exchange rate remained relatively fixed at US $1 = 14,200 dông. However, the dông had lost 3.5% of its value against the US dollar in 2001. In the first nine months of 2004 an increase in the global prices of key commodities drove the trade deficit higher, to $3,389m.; the deficit was predicted to reach $4,000–$4,500 by the end of the year, as imports for the full year were expected to total approximately $30,000m. Experts predicted, however, that such a high trade deficit could be controlled, because in relation to exports the proportion was only 20%.

Trade Policies

Viet Nam's trade and investment policies can still be characterized as 'export-led protectionism', whereby import substitution is encouraged with trade protection, and export industries are promoted by providing subsidies to counterbalance the high relative costs of intermediary products. This should not be interpreted simply as a model similar to that of the Republic of Korea, but rather as protectionism based on central planning, combined with a crude interpretation of export-led development.

The legacy of central planning is, most importantly, a strong belief in self-sufficiency through import substitution. Another important aspect is the continuing belief in the role of the Government to manage and to make frequent adjustments to the whole economy, and a consequent disregard for the effects of price distortions and uncertainty created by this approach in a market economy.

A tariff code was introduced in 1993 and had over 20 different rates ranging from zero to 200%, making it too complex to be effectively administered. The new rates were generally higher than those set in 1991, but for many goods they were still in an acceptable range, according to the World Bank. The average (unweighted) rate was 12%. Almost 80% of tariff lines were set at 20% or lower. In 1994 the requirement for import permits for all but 15 products (refined petroleum, fertilizer, cement, and some consumer items) was eliminated. An import duty exemption scheme was improved in 1994, with the result that a company was not required to pay duties on those imports used to produce exports. A process of imposing international standards of free trade on Viet Nam began in 1996, following the country's admission to AFTA. This transition was expected to take many years to be effective, particularly in view of the allowances made for poor developing countries like Viet Nam. Quantitative import restrictions on eight out of 19 groups of products were removed in 2000. In addition, a plan for tariff reductions to be implemented under AFTA during 2001–06 was approved, whereby most tariff lines would be reduced to 20% by early 2003 and to 5% by early 2006.

Customs duties comprised 25% of government revenue by the late 1990s. This, however, was only one of the objectives assigned to this policy instrument. Consequently, the tariff schedule was complex and constantly changing. The number of tariff rates decreased from 36 in 1995 to 31 in 1996, then increased to 35 in 1997, but was subsequently reduced to 26 rates, ranging from 0% to 60%, in 1998. The schedule in February 1998 identified 3,163 separate items, and within one item there may be many relevant minimum import prices applied. Despite the ASEAN agreement to remove all surcharges by 1996, Viet Nam maintained surcharges (export taxes) on a small number of important products. Customs surcharges were subsequently applied for various periods of time to imported petroleum, some types of iron and steel, and fertilizer. Surcharges were also applied on exports of coffee, unprocessed cashew nuts and rubber in the late 1990s, purportedly as part of a price stabilization fund process. The surcharge rates have been changed frequently. Between mid-1994 and 2000 the rates on petroleum products were changed 14 times. Rates on petroleum products differed between North and South Viet Nam until 1996. Rates between 2% and 10% were applied to eight different types of iron and steel products.

In 1998 foreign-exchange management controls became a more prominent instrument of economic policy in Viet Nam, and these were used to meet macroeconomic and specific trade and industry policy objectives. The increased emphasis on financial controls preceded the removal of import licensing, and was most probably a response to the Asian financial crisis and dwindling capital inflows. That is, an emerging shortage of foreign capital promoted a strengthening of foreign-exchange controls that, as usual, were then also employed to pursue other development strategy objectives. In 1999 foreign-exchange controls were slightly relaxed, by reducing the foreign-exchange surrender requirement on foreign-exchange earnings from 80% to 50% or 40%. Vietnamese trade policy reforms of the late 1990s might be interpreted as generally protectionist, although in some instances they were merely codifying existing policies. The most obvious trend of the trade reform process has really been one of contradictory liberalization: both protectionist and liberalizing measures were being introduced every year. Publication of the tariff reduction schedule and the abolition of import licences in 1998 appeared to indicate one course, while foreign-exchange controls, combined with the gradual increase in the use of quotas, bans and surcharges, pointed in the other direction.

Viet Nam has established diplomatic relations with 165 countries and has signed bilateral trade agreements with 72 of them. The country joined AFTA and became a member of the Asia-Europe Meeting (ASEM) Forum in 1996. In November 2001 Viet Nam signed a bilateral trade agreement with the USA. All these moves towards international integration were preparatory steps towards the country's membership of the WTO. From 1 July 2003 the country applied tariff reductions to 755 imported commodities, reducing them to between 5% and 20%. The cuts made no significant impact on the domestic economy, particularly on prices, as some had initially forecast.

Viet Nam also benefited from its tariff reduction 'road map'. Some countries granted preferential tax rates from zero to 5% on almost all Vietnamese exports and Singapore completely removed tariffs on Vietnamese goods. The ASEAN meeting of ministers of finance was successful, with the outcome that it was decided to issue a regional bond in future and establish an ASEAN economic community by 2020. The country's achievements in general, and in the process of integration into the world economy in particular, have been highly valued by international partners. As a result, the international donor community pledged to grant Viet Nam $2,800m. in 2004, the highest figure to date. Viet Nam has signed 85 commercial agreements with foreign countries in a bid to expand export markets and boost production. The country has attracted more than US $40,000m. in FDI and nearly $20,000m. in official development assistance (ODA).

FOREIGN INVESTMENT

The promotion of foreign investment was an early and relatively straightforward stage in Viet Nam's reform process. Viet Nam enacted its first foreign investment law in December 1987, with amendments in June 1990 and in December 1992. Although the law was intended to reassure foreign investors with regard to their rights, and although it provided the usual incentives in the form of tax concessions and other privileges, it was ineffective until a realistic exchange rate was established in 1989. Even then, the weak legal framework, the continuing US trade and investment embargo, and the lengthy process of approving foreign investment projects prevented a significant rise in commitments until the mid-1990s. As in China, foreign investment began informally in labour-intensive export production outside the approval procedure and statistical structure.

Remarkably, in 1999 the Paris Club, an informal grouping of creditor nations, rescheduled Viet Nam's debt to the Russian Federation, lowering the country's debt burden and debt ratio. Other positive developments have included progress on the economic reform process, the achievement of relative political stability, and, in 2001, the signing of a bilateral trade agreement with the USA. In 2001 a major international credit ratings agency upgraded Viet Nam from 'negative' to 'stable' in its rankings. Investors were encouraged by: a reduction in the number of items foreign-invested enterprises were expected to export from 24 to 14, including tiles, ceramics, footwear, electric fans, plastic products, and common paints; the granting of permission to foreign-invested enterprises to engage in exports of coffee, material, certain wood products, and certain textiles and garments; the allowing of mortgaging of land by foreign bank branches in Viet Nam; the granting of permission for the automatic registration of export-oriented foreign investment; the making of provision for government guarantee for infrastructure projects; the amendment of the Law on Petroleum, in order to improve the investment and regulatory environment for foreign investment in the oil and gas sector; the permitting of overseas Vietnamese to hold land-use rights, decentralizing control and monitoring land-use rights to enhance the functioning of the real estate market; the permitting of joint ventures and foreign banks to take collateral in the form of land-use rights and land certificates; and the provision of detailed guidelines and comprehensive listing of all necessary documentation allowing foreign-invested enterprises to mortgage land-use rights. New rules allowing overseas Vietnamese to buy homes also raised the volume of inward remittances.

Serious obstacles to foreign investment remained, however, in the form of provisions that allowed minority partners, often state enterprises or government entities, powers of veto in a joint venture. Foreign investment projects required the approval of several investment agencies. Although the central Government processed all large projects directly, the approval process remained lengthy, arbitrary and tedious by regional standards. The absence of legislation governing the implementation of contractual agreements continued. In this situation, the lack of an adequate dispute resolution mechanism was a serious problem. Local authorities had considerable power and their approval was as necessary as that of central agencies. Disputes over problems such as site clearance for hotels caused prolonged delays. Property rights remained unclear. These weaknesses increased the evident risks of investing in a country that continued to subscribe to 'market socialism'. In June 2000, the Law on Foreign Investments was amended to include covered land issues and transfer rights, but was not particularly broad or encouraging.

In 2000 an initial focus on petroleum and gas projects was replaced by one on manufacturing and infrastructure projects. Seven projects in construction, with a total investment capital of US $9,100m., were licensed in the first seven months of 2001. Investments in hotels and tourism remained strong throughout the 1990s, with the total capital in this sector increasing to $3,506m. by July 2001. More than two-thirds of investment came from Asian countries in the 1990s. Taiwan was the largest investor, followed by Hong Kong and the Republic of Korea. Total Japanese commitments included 110 projects, with total capital pledges of $1,700 million by July 1995. This represented 10% of total commitments, which was less than that from Western Europe (14%), but greater than that from the USA (6%). By July 2001 Japanese commitments totalled $3,995m. in 311 projects, with more than 67% implemented, and the USA ranked 13th, investing $986m., with more than 45% of projects implemented. The bilateral trade agreement with the USA increased FDI in labour-intensive industries (services, light industry, food, and construction). By mid-2001 Singapore was the largest foreign investor in Viet Nam, with a total invested capital of $6,809m., followed by Taiwan and Japan.

In 2004 six foreign-invested companies were to be part of the Ministry of Planning and Investment's first pilot equitization programme. The pilot equitization programme would give foreign investors the right to invest directly in Viet Nam. It would also play an important role in enabling policy makers to perfect the legal systems related to equitized foreign-invested enterprises. Until the end of the year the Ministry would receive equitization applications from foreign-invested enterprises to submit to the Government. Under a government decision taken in December 2003, only foreign-invested enterprises with capital of between US $1m. and $70m. would be allowed to equitize, which observers have claimed hampers the process. Another hurdle for the equitization process was the inherent risk of going public. Many foreign-invested enterprises had not applied for equitization because they were waiting for a new law to safeguard their capital and assets. This was just part of the overwhelming bureaucracy which has been an impediment to the potential Viet Nam has to attract FDI. However, regionally China is the dominant player, attracting most FDI as it has just become a member of the WTO, providing assurance for investors of adherence to certain rules and regulations. This accounted for the haste with which Viet Nam was driving towards WTO membership in 2005.

There was huge potential for further FDI in Viet Nam. By September 2004 the Ministry of Planning and Investment had licensed 518 new FDI projects, with a capitalization of over $1,600m. It has also permitted existing projects to expand their investment capital by $1,400m. Forty-six projects, worth $2,200m., were awaiting licences from the Ministry. Seventeen years after the implementation of the Foreign Investment Law, Viet Nam had, by the end of September 2004, seen 5,995 FDI projects licensed, of which more than 4,910 projects are still operational and are valued at over $44,700m., including 118 run by overseas Vietnamese, with a total registered capital of $362.6m.

SOCIAL WELFARE

Following near famine conditions in 1988, and with one-quarter of Vietnamese people in a situation of food poverty in the early 1990s, the number of Vietnamese in poverty has since declined by 50% according to the World Bank. Alleviating poverty requires access to markets and facilities, and education. The Central Highlands and the Mekong delta have suffered the most from weak facilities. There were few all-weather roads penetrating rural areas, which are now the target of most foreign aid. Integrated efforts were required to link rural people to markets (using roads and bridges), to improve access to facilities and resources (by improving land rights) and to strengthen personal capacities and information flows (through the improvement of education).

With such progress in poverty alleviation, it would appear, however, that rapid economic growth in Viet Nam has not been accompanied by a markedly worsening income distribution. According to Viet Nam Living Standards Survey data, greater inequality in urban areas, particularly in northern Viet Nam, resulted in a marginal national increase. Vietnamese survey data demonstrated erratic and marginal changes in income distribution throughout Viet Nam during 1994–97, with an apparent reversal of the inequality trend in 1997. Therefore, both the absolute level of inequality and the trend did not appear to be matters for concern, and by international standards Viet Nam remained a relatively equitable society.

In 2004 the poverty rate declined to 8.3%, from 10.5% in the previous year, and the child malnutrition rate of 26% had fallen from 28.4% in 2003. According to reports by IMF country staff, poverty has declined in all regions of Viet Nam. A report titled 'Poverty and inequality in Vietnam: Spatial patterns and geographic determinants', published in 2003, concluded that the regions with the highest poverty rates—the North East, North West, and Central Highlands—are so sparsely populated that the number of poor people living in them is relatively small. In contrast, the densely populated cities and the deltas account for a greater absolute number of poor despite their low poverty rates. Inequality has risen but is still modest by international standards. A 'Comprehensive Poverty Reduction and Growth Strategy' was implemented in November 2003: it is a document that elaborates upon all the general objectives, institutional arrangements, policies and solutions of the 10-Year Strategy and two Five-Year Plans and translates them into detailed specific action plans.

PROSPECTS FOR THE 21ST CENTURY

Viet Nam's transition is undoubtedly one of success. Incomes have risen, poverty has declined, and farmers have benefited. This success, however, is largely due to the failure to implement extensive central planning after 1976. Furthermore, as a poor developing country, with largely unprocessed agricultural exports, and a labour force willing and able to work at wages slightly above subsistence level, Viet Nam differs significantly from most transitional economies. Structural transformation of Viet Nam's 'residual element' of central planning was therefore achieved during the first decade of transition without dramatic adverse effects, with the exception of a short-term decrease in industrial production. It was reported, however, that social services and support systems also experienced a decline, followed by gradual recovery.

Viet Nam's reform process, it has been suggested, decelerated in the late 1990s. Nevertheless, transition is a long process, and overall progress in that decade was impressive. Structural reform in Viet Nam between 1988 and about 1992 was considerable: macroeconomic stabilization, state enterprise reform (the sector shedding 22% of its work-force during 1989–91), a reorientation of international trade, and an increase in private rural and urban household economic activity owing largely to fundamental institutional reforms. Reforms in the latter part of the 1990s inevitably appeared comparatively less impressive, but the country now seems to be heading from an agriculture-based economy towards an industrial one. With the support of oil to help ensure a balanced development, Viet Nam appeared positioned to enjoy a strong economic performance over the next decades.

Statistical Survey

Sources (unless otherwise stated): General Statistical Office of the Socialist Republic of Viet Nam, 2 Hoang Van Thu, Hanoi; tel. (4) 8464385; fax (4) 264345; Communist Party of Viet Nam.

Area and Population

AREA, POPULATION AND DENSITY

Area (sq km)	329,247*
Population (census results)	
1 April 1989	64,411,713
1 April 1999	
Males	37,469,117
Females	38,854,056
Total	76,323,173
Population (official estimates at mid-year)	
2001	78,686,000
2002	79,715,000
2003	80,670,000
Density (per sq km) at mid-2003	245.0

* 127,123 sq miles.

ADMINISTRATIVE DIVISIONS
(2001)

Provinces and cities	Area (sq km)	Population ('000)	Density (per sq km)
North East	65,327	9,037	138
Ha Giang	7,884	626	79
Tuyen Quang	5,868	693	118
Cao Bang	6,691	502	75
Lang Son	8,305	715	86
Lao Cai	8,057	617	77
Yen Bai	6,883	700	102
Bac Kan	4,857	283	58
Thai Nguyen	3,540	1,062	300
Phu Tho	3,520	1,288	366
Bac Giang	3,822	1,522	398
Quang Ninh	5,900	1,030	175
North West	35,637	2,313	65
Lai Chau	16,919	616	36
Son La	14,055	922	66
Hoa Binh	4,663	774	166
Red River Delta	14,799	17,243	1,165
Hanoi	921	2,842	3,086
Hai Phong	1,523	1,711	1,123
Vinh Phuc	2,192	1,116	509
Hung Yen	1,638	1,091	666
Hai Duong	851	1,671	1,964
Ha Tay	1,648	2,432	1,476
Bac Ninh	923	958	1,038
Thai Binh	1,371	1,815	1,324
Nam Dinh	1,384	1,916	1,384
Ha Nam	1,544	800	518
Ninh Binh	804	892	1,109
North Central Coast	51,504	10,188	198
Thanh Hoa	11,106	3,510	316
Nghe An	16,490	2,914	177
Ha Tinh	6,056	1,285	212
Quang Binh	8,052	813	101
Quang Tri	4,746	589	124
Thua Thien-Hué	5,054	1,079	213
South Central Coast	33,066	6,694	202
Quang Nam	10,407	1,403	135
Da Nang	1,255	715	570
Quang Ngai	5,135	1,206	235
Binh Dinh	6,026	1,492	248
Phu Yen	5,045	811	161
Khanh Hoa	5,198	1,066	205
Central Highlands	54,475	4,330	79
Gia Lai	15,496	1,048	68
Kon Tum	9,615	331	34
Dak Lak	19,599	1,901	97
Lam Dong	9,765	1,050	108
South East	34,733	12,362	356
Ho Chi Minh City	2,095	5,378	2,567
Ninh Thuan	3,360	532	158
Binh Duong	2,718	768	283
Binh Phuoc	6,856	708	103
Tay Ninh	4,028	990	246
Binh Duong	2,696	768	285
Dong Nai	5,895	2,067	351
Binh Thuan	7,828	1,080	138
Ba Ria-Vung Tau	1,975	839	425
Mekong River Delta	39,706	16,519	416
Long An	4,492	1,348	300
Dong Thap	3,238	1,593	492
An Giang	3,406	2,099	616
Tien Giang	2,367	1,636	691
Ben Tre	2,316	1,308	565
Vinh Long	1,475	1,023	694
Tra Vinh	2,215	989	447
Can Tho	2,986	1,852	620
Soc Trang	3,223	1,213	376
Kien Giang	6,269	1,543	246
Bac Lieu	2,524	757	300
Ca Mau	5,195	1,158	223
Total	329,247	78,686	239

PRINCIPAL TOWNS
(estimated population, excl. suburbs, at mid-1992)

Ho Chi Minh City (formerly Saigon)	3,015,743*	Nam Dinh	171,699
Hanoi (capital)	1,073,760	Qui Nhon	163,385
Haiphong	783,133	Vung Tau	145,145
Da Nang	382,674	Rach Gia	141,132
Buon Ma Thuot	282,095	Long Xuyen	132,681
Nha Trang	221,331	Thai Nguyen	127,643
Hué	219,149	Hong Gai	127,484
Can Tho	215,587	Vinh	112,455
Cam Pha	209,086		

* Including Cholon.

Source: UN, *Demographic Yearbook*.

Mid-2003 (UN estimates, incl. suburbs): Ho Chi Minh City 4,850,717; Hanoi 3,977,202; Haiphong 1,754,537 (Source: UN, *World Urbanization Prospects: The 2003 Revision*).

BIRTHS AND DEATHS
(UN estimates, annual averages)

	1985–90	1990–95	1995–2000
Birth rate (per 1,000)	32.5	28.3	21.5
Death rate (per 1,000)	9.3	8.1	7.0

Source: UN, *World Population Prospects: The 2002 Revision*.

Expectation of life (WHO estimates, years at birth): 69.6 (males 67.1; females 72.2) in 2002 (Source: WHO, *World Health Report*).

EMPLOYMENT
('000 persons)

	1995	1996	1997
Agriculture	24,537	24,543	24,814
Forestry	228	232	
Industry*	3,379	3,442	3,656
Construction	996	975	977
Trade and catering	2,290	2,677	2,672
Transport	512	799	856
Communications	56	57	
Science	38	39	41
Education	973	994	999
Culture, arts and sport	94	96	96
Public health	358	359	296
Other activities	1,128	1,579	2,587
Total employed	34,590	35,792	36,994

* Comprising manufacturing, mining and quarrying, electricity, gas and water.

Source: IMF, *Vietnam: Statistical Appendix* (July 1999).

1998 ('000): Total employment 35,233, of which 24,504 in agriculture, fisheries, and forestry; 3,279 in industry; 878 in construction; 2,372 in trade; 859 in transport and communications; and 1,250 in education, health, science and arts (Source: IMF, *Vietnam: Statistical Appendix* (December 2003)).

1999 ('000): Total employment 35,976, of which 24,792 in agriculture, fisheries, and forestry; 3,392 in industry; 908 in construction; 2,538 in trade; 894 in transport and communications; and 1,275 in education, health, science and arts (Source: IMF, *Vietnam: Statistical Appendix* (December 2003)).

2000 ('000): Total employment 36,702, of which 25,045 in agriculture, fisheries, and forestry; 3,507 in industry; 939 in construction; 2,714 in trade; 929 in transport and communications; and 1,299 in education, health, science and arts (Source: IMF, *Vietnam: Statistical Appendix* (December 2003)).

2001 (estimates, '000): Total employment 37,676, of which 25,305 in agriculture, fisheries, and forestry; 3,644 in industry; 1,068 in construction; 2,904 in trade; 1,026 in transport and communications; and 1,357 in education, health, science and arts (Source: IMF, *Vietnam: Statistical Appendix* (December 2003)).

2002 (estimates, '000): Total employment 38,715, of which 25,573 in agriculture, fisheries, and forestry; 3,787 in industry; 1,216 in construction; 3,106 in trade; 1,133 in transport and communications; and 1,418 in education, health, science and arts (Source: IMF, *Vietnam: Statistical Appendix* (December 2003)).

Health and Welfare

KEY INDICATORS

Total fertility rate (children per woman, 2002)	2.3
Under-5 mortality rate (per 1,000 live births, 2002)	39
HIV/AIDS (% of persons aged 15–49, 2003)	0.40
Physicians (per 1,000 head, 1998)	0.48
Hospital beds (per 1,000 head, 1997)	1.67
Health expenditure (2001): US $ per head (PPP)	134
Health expenditure (2001): % of GDP	5.1
Health expenditure (2001): public (% of total)	28.5
Access to water (% of persons, 2000)	56
Access to sanitation (% of persons, 2000)	73
Human Development Index (2002): ranking	112
Human Development Index (2002): value	0.691

For sources and definitions, see explanatory note on p. vi.

Agriculture

PRINCIPAL CROPS
('000 metric tons)

	2000	2001	2002
Rice (paddy)	32,530	32,108	34,447
Maize	2,006	2,162	2,511
Potatoes	316	316	377
Sweet potatoes	1,611	1,654	1,704
Cassava (Manioc)	1,986	3,509	4,438
Dry beans	145	161	144
Other pulses*	100	95	95
Cashew nuts†	270	293	516
Soybeans (Soya beans)	149	174	201
Groundnuts (in shell)	355	363	397
Coconuts	885	892	915
Dry onions	210*	223	223*
Other vegetables*	6,064	6,774	6,752
Bananas	1,125	1,126	1,044
Oranges	427	428	442
Watermelons	200	245	245*
Mangoes	177	179	209
Pineapples	291	285	348
Other fruit*	2,322	2,431	2,640
Sugar cane	15,044	14,657	16,824
Coffee (green)	803	841	689
Tea (made)	70	76	90
Tobacco (leaves)	27	32	34
Natural rubber	291	313	331

* FAO estimate(s).
† Unofficial figures.

Source: FAO.

LIVESTOCK
('000 head, year ending September)

	2000	2001	2002
Horses	127	113	111
Cattle	4,128	3,900	4,063
Buffaloes	2,897	2,808	2,814
Pigs	20,194	21,800	23,170
Goats	544	572	622
Chickens*	137,300	152,670	163,100
Ducks*	58,800	65,430	69,900

* Unofficial figures.

Source: FAO.

LIVESTOCK PRODUCTS
('000 metric tons)

	2000	2001	2002
Beef and veal	92.3	97.8	102.5
Buffalo meat*	92.5	96.8	98.9
Pig meat	1,409.0	1,515.3	1,653.6
Chicken meat	295.7	308.0	338.4
Duck meat*	69.6	77.4	81.6
Cows' milk	54.5	64.7	78.5
Buffaloes' milk*	30.0	30.0	31.0
Poultry eggs	185.4	200.5	226.5*
Cattle hides (fresh)*	15.7	15.7	15.7
Buffalo hides (fresh)*	17.2	18.0	18.4

* FAO estimate(s).

Source: FAO.

Forestry

ROUNDWOOD REMOVALS
(FAO estimates, '000 cubic metres, excl. bark)

	2000	2001	2002
Sawlogs, etc.:			
Coniferous	184	184	184
Broadleaved	2,387	2,387	2,387
Other industrial wood	350	350	350
Fuel wood (all broadleaved)	26,686	26,615	26,547
Total	29,607	29,536	29,468

Source: FAO.

SAWNWOOD PRODUCTION
('000 cubic metres, incl. railway sleepers)

	2000	2001	2002
Total (all broadleaved)	2,950	2,950*	2,950*

* FAO estimate.

Source: FAO.

Fishing

('000 metric tons, live weight)

	2000	2001	2002
Capture*	1,450.6	1,490.3	1,508.0
Freshwater fishes	169.0*	132.3	148.2
Marine fishes	927.2	1,033.6	1,025.8
Prawns and shrimps*	81.7	90.0	90.0
Cephalopods*	180.0	130.0	130.0
Aquaculture*	510.6	518.5	518.5
Freshwater fishes*	381.2	390.0	390.0
Total catch*	1,961.1	2,008.8	2,026.5

* FAO estimate(s).

Source: FAO.

Mining

('000 metric tons)

	2000	2001	2002*
Coal	11,609	12,962	15,900
Chromium ore	76	80	80
Phosphate rock	785	750	770
Salt (unrefined)	590	575	600
Crude petroleum ('000 barrels)	115,373	120,464	136,700

* Estimates.

Source: US Geological Survey.

Industry

SELECTED PRODUCTS
('000 metric tons, unless otherwise indicated)

	1999	2000	2001
Raw sugar	947	1,209	1,067
Beer (million litres)	690	779	871
Cigarettes (million packets)	2,147	2,836	3,075
Fabrics (million metres)	322	356	410
Chemical fertilizers	1,143	1,210	1,065
Insecticides	22	20	20
Bricks (million)	7,831	9,087	9,811
Cement	10,489	13,298	16,073
Crude steel	1,375	1,583	1,914
Diesel motors (pieces)	15,347	15,623	18,721
Television receivers ('000)	903	1,013	1,126
Electric engines (pieces)	38,091	45,855	53,442
Bicycle tyres ('000)	18,326	20,675	21,658
Bicycle tubes ('000)	21,544	21,917	22,997
Rice milling equipment (pieces)	12,136	12,484	18,298
Transformers (pieces)	10,264	13,535	15,664
Electric energy (million kWh)	23,599	26,682	30,673

Finance

CURRENCY AND EXCHANGE RATES

Monetary Units

100 xu = 1 new dông.

Sterling, Dollar and Euro Equivalents (31 May 2004)

£1 sterling = 28,887.4 dông;
US $1 = 15,745.0 dông;
€1 = 19,281.3 dông;
100,000 new dông = £3.46 = $6.35 = €5.19.

Average Exchange Rate (new dông per US $)

2001	14,725.2
2002	15,279.5
2003	15,509.6

Note: The new dông, equivalent to 10 former dông, was introduced in September 1985.

BUDGET
('000 million dông)

Revenue	2000	2001	2002
Tax revenue	72,787	80,305	87,515
Corporate income tax	22,240	25,032	26,060
Individual income tax	1,831	1,930	1,975
Value-added tax (VAT)	17,072	18,916	23,015
Special consumption tax	5,250	6,304	6,890
Natural resources tax	7,487	7,558	6,780
Taxes on trade	13,437	16,200	20,000
Non-tax revenue	15,096	14,695	15,015
Fees and charges	4,905	5,250	5,985
Net profit after tax	6,710	6,700	5,903
Other	2,782	3,600	2,410
Capital revenues	838	600	670
Grants	2,028	1,900	2,000
Resources transfer revenue	n.a.	3,400	1,700
Total	90,749	100,900	106,900

Expenditure	2000	2001	2002
Current expenditure	70,127	77,020	78,784
General administrative services	8,089	7,970	7,210
Economic services	5,796	7,880	6,988
Social services	30,694	35,793	38,460
Education	9,910	12,222	13,581
Training	2,767	3,488	4,034
Health	3,453	4,120	4,460
Pensions and social relief	10,739	12,190	12,260
Other services (incl. defence)	22,034	21,094	19,882
Interest on public debt	3,514	4,283	6,244
Expenditure on investment and development	29,624	36,390	39,000
Capital expenditure	26,211	32,486	35,050
Others	3,413	3,904	3,950
Others	n.a.	n.a.	2,700
Resource transfer	3,400	1,700	n.a.
Total	103,151	115,110	120,484

Source: Ministry of Finance, Hanoi.

INTERNATIONAL RESERVES
(US $ million at 31 December)

	2001	2002	2003
Gold*	90.6	110.8	135.0
IMF special drawing rights	14.6	—	2.2
Foreign exchange	3,660.0	4,121.0	6,222.0
Total	3,765.2	4,231.8	6,359.2

* National valuation.

Source: IMF, *International Financial Statistics*.

MONEY SUPPLY
('000 million đồng at 31 December)

	2001	2002	2003
Currency outside banks	66,320	74,263	90,584
Demand deposits at banks	46,089	51,066	66,441
Total money	112,408	125,329	157,025

Source: IMF, *International Financial Statistics*.

COST OF LIVING
(Consumer Price Index; base: 1990 = 100)

	1999	2000	2001
Food	467.4	449.5	443.2
All items	425.8	419.0	417.7

Source: ILO.

All items (base: 2000 = 100): 99.6 in 2001; 103.4 in 2002; 106.6 in 2003 (Source: IMF, *International Financial Statistics*).

NATIONAL ACCOUNTS
('000 million đồng at current prices)

Expenditure on the Gross Domestic Product

	2001	2002	2003
Government final consumption expenditure	30,463	33,390	41,770
Private final consumption expenditure	312,144	348,747	392,951
Increase in stocks	9,732	11,155	12,826
Gross fixed capital formation	140,301	166,828	199,654
Total domestic expenditure	492,640	560,120	647,201
Exports of goods and services	262,846	304,262	365,394
Less Imports of goods and services	273,828	331,946	411,119
Sub-total	481,658	532,436	601,476
Statistical discrepancy	–360	3,326	4,110
GDP in purchasers' values	481,298	535,762	605,586
GDP at constant 1994 prices	292,535	313,247	335,989

Gross Domestic Product by Economic Activity

	2001	2002	2003
Agriculture, hunting, forestry and fishing	111,858	123,383	132,193
Mining and quarrying	44,345	46,153	57,070
Manufacturing	95,211	110,285	125,984
Electricity, gas and water	16,028	18,201	23,241
Construction	27,931	31,558	35,638
Trade	67,788	75,617	83,397
Transport, storage and communications	19,431	21,095	22,589
Finance	8,762	9,763	10,881
Public administration	40,894	44,940	53,762
Other community, social and personal services	49,049	54,767	60,831
Total	481,298	535,762	605,586

Source: Asian Development Bank, *Key Indicators of Developing Asian and Pacific Countries*.

BALANCE OF PAYMENTS
(US $ million)

	2000	2001	2002
Exports of goods f.o.b.	14,448	15,027	16,706
Imports of goods f.o.b.	–14,073	–14,546	–17,760
Trade balance	375	481	–1,054
Exports of services	2,702	2,810	2,948
Imports of services	–3,252	–3,382	–3,698
Balance on goods and services	–175	–91	–1,804
Other income received	331	318	167
Other income paid	–782	–795	–888
Balance on goods, services and income	–626	–568	–2,525
Current transfers received	1,732	1,250	1,921
Current balance	1,106	682	–604
Direct investment from abroad	1,298	1,300	1,400
Other investment assets	–2,089	–1,197	624
Other investment liabilities	475	268	66
Net errors and omissions	–680	–847	–1,038
Overall balance	110	206	448

Source: IMF, *International Financial Statistics*.

External Trade

SELECTED COMMODITIES
(distribution by SITC, US $ million)

Imports c.i.f.	2000	2001	2002
Food and live animals	627	834	939
Beverages and tobacco	103	108	149
Crude materials (inedible) except fuels	591	690	816
Mineral fuels, etc.	2,121	1,970	2,166
Chemicals and related products	2,402	2,490	2,933
Basic manufactures	3,402	3,729	5,415
Machinery and transport equipment	4,711	4,865	5,758
Miscellaneous manufactured goods	1,586	1,447	1,427
Total (incl. others)	15,637	16,218	19,745

Exports f.o.b.	2000	2001	2002
Food and live animals	3,780	4,052	4,118
Mineral fuels, etc.	3,825	3,469	3,568
Machinery and transport equipment	1,276	1,399	1,337
Miscellaneous manufactured goods	4,052	4,408	5,691
Total (incl. others)	14,483	15,029	16,706

2003 (US $ million): Total imports c.i.f. 25,227; Total exports f.o.b. 20,176.

Source: Asian Development Bank, *Key Indicators of Developing Asian and Pacific Countries*.

PRINCIPAL TRADING PARTNERS
(US $ million)

Imports c.i.f.	2001	2002	2003
China, People's Republic	1,606.2	2,364.9	3,496.4
Germany	396.7	556.6	676.4
Hong Kong	537.6	845.4	1,076.6
Indonesia	289.1	432.2	482.7
Japan	2,183.1	2,348.7	2,688.6
Korea, Republic	1,886.8	2,464.2	2,810.3
Malaysia	464.4	730.8	833.5
Singapore	2,478.3	2,289.7	2,653.2
Thailand	792.3	1,042.0	1,394.7
USA	411.3	638.3	1,456.8
Total (incl. others)	16,216.2	20,014.2	24,863.8

Exports f.o.b.	2001	2002	2003
Australia	1,041.8	1,171.8	1,545.3
China, People's Republic	1,417.4	1,013.2	1,323.5
France	468.4	503.1	580.6
Germany	721.8	1,001.0	1,202.4
Japan	2,509.8	2,299.5	2,896.9
Korea, Republic	406.1	427.6	487.6
Netherlands	364.5	379.8	461.4
Singapore	1,043.7	852.7	930.9
United Kingdom	511.6	666.7	902.9
USA	1,065.7	2,349.8	4,463.2
Total (incl. others)	15,013.8	15,713.9	20,678.8

Source: Asian Development Bank, *Key Indicators of Developing Asian and Pacific Countries*.

Transport

RAILWAYS
(traffic)

	1999	2000	2001
Passengers carried (million)	9.3	9.8	10.6
Passenger-km (million)	2,722.0	3,199.3	3,426.1
Freight carried ('000 metric tons)	5,146.0	6,258.2	6,456.7
Freight ton-km (million)	1,445.5	1,955.0	2,054.4

Source: *Vietnam Statistics Yearbook 2002*.

ROAD TRAFFIC

	1999	2000	2001
Passengers carried (million)	588.4	621.3	655.4
Passenger-km (million)	22,053.3	23,192.4	24,237.7
Freight carried (million metric tons)	132.1	141.1	151.5
Freight ton-km (million)	7,159.8	7,888.5	8,095.4

Source: *Vietnam Statistics Yearbook 2002*.

INLAND WATERWAYS

	1999	2000	2001
Passengers carried (million)	125.7	126.5	133.9
Passenger-km (million)	2,109.7	2,136.9	2,484.1
Freight carried (million metric tons)	40.0	43.0	48.5
Freight ton-km (million)	3,967.8	4,267.6	4,672.4

Source: *Vietnam Statistics Yearbook 2002*.

SHIPPING

Merchant Fleet
(registered at 31 December)

	2001	2002	2003
Number of vessels	700	721	735
Total displacement ('000 grt)	1,073.6	1,130.5	1,250.8

Source: Lloyd's Register-Fairplay, *World Fleet Statistics*.

International Sea-Borne Shipping
(estimated freight traffic, '000 metric tons)

	1988	1989	1990
Goods loaded	305	310	303
Goods unloaded	1,486	1,510	1,510

Source: UN, *Monthly Bulletin of Statistics*.

CIVIL AVIATION
(traffic on scheduled services)

	1999	2000	2001
Passengers carried ('000)	2,699	2,806	3,853
Passenger-km (million)	4,042	4,383	6,111
Freight carried (million metric tons)	40.0	43.0	48.5
Total ton-km (million)	3,967.8	4,267.6	4,672.4

Source: *Vietnam Statistics Yearbook 2002*.

Tourism

TOURIST ARRIVALS BY NATIONALITY

Country	2001	2002	2003
Australia	84,085	96,624	93,292
Cambodia	76,620	69,538	84,256
China, People's Republic	672,846	724,385	693,423
France	99,700	111,546	86,791
Japan	204,860	279,769	209,730
Korea, Republic	75,167	105,060	130,076
Laos	40,696	37,237	75,396
Malaysia	26,265	46,086	48,662
Taiwan	200,061	211,072	207,866
United Kingdom	64,673	69,682	63,348
USA	230,470	259,967	218,928
Total (incl. others)	2,330,050	2,627,988	2,428,735

Source: Vietnam National Administration of Tourism.

Tourism receipts (US $ million): 88 in 1997; 86 in 1998 (Source: World Bank).

Communications Media

	2000	2001	2002
Television receivers ('000 in use)	14,750	n.a.	n.a.
Telephones ('000 main lines in use)	2,542.7	3,049.9	3,929.1
Mobile cellular telephones ('000 subscribers)	788.6	1,251.2	1,902.4
Personal computers ('000 in use)	700	700	800
Internet users ('000)	200	1,010	1,500
Newspapers and magazines ('000 copies sold)	580,000	635,044	n.a.
Book production:			
Titles	9,487	11,455	n.a.
Copies (million)	177.6	166.5	n.a.

Sources: International Telecommunication Union; *Vietnam Statistics Yearbook 2002*.

Facsimile machines ('000 in use): 31 in 1999.

Radio receivers ('000 in use): 8,200 in 1997.

Book production: 8,186 titles (169,800,000 copies) in 1995.

Daily newspapers: 10 (with estimated circulation of 300,000 copies) in 1996.

2003: Mobile cellular telephones ('000 subscribers) 2,742.0; Telephones ('000 main lines in use) 4,402.0; Internet users ('000) 3,500.

Sources: mainly UN, *Statistical Yearbook*; and UNESCO, *Statistical Yearbook*.

Education

(2001/02, '000)

	Teachers	Students
Pre-primary	103.8	2,171.8
Primary	359.9	9,315.3
Secondary	334.2	8,560.3
Higher	31.4	873.0

Source: *Vietnam Statistics Yearbook 2002*.

Adult literacy rate (UNESCO estimates): 92.7% (males 94.5%; females 90.9%) in 2001 (Source: UN Development Programme, *Human Development Report*).

Directory

The Constitution

On 15 April 1992 the National Assembly adopted a new Constitution, a revised version of that adopted in December 1980 (which in turn replaced the 1959 Constitution of the Democratic Republic of Viet Nam). The National Assembly approved amendments to 24 articles of the Constitution on 12 December 2001. The main provisions of the Constitution (which originally entered into force after elections in July 1992) are summarized as follows:

POLITICAL SYSTEM

All state power belongs to the people. The Communist Party of Viet Nam is a leading force of the state and society. All party organizations operate within the framework of the Constitution and the law. The people exercise power through the National Assembly and the People's Councils.

ECONOMIC SYSTEM

The State develops a multi-sectoral economy, in accordance with a market mechanism based on state management and socialist orientations. All lands are under state management. The State allots land to organizations and individuals for use on a stabilized and long-term basis: they may transfer the right to the use of land allotted to them. Individuals may establish businesses with no restrictions on size or means of production, and the State shall encourage foreign investment. Legal property of individuals and organizations, and business enterprises with foreign invested capital, shall not be subjected to nationalization.

THE NATIONAL ASSEMBLY

The National Assembly is the people's highest representative agency, and the highest organ of state power, exercising its supreme right of supervision over all operations of the State. It elects the President and Vice-President, the Prime Minister and senior judicial officers, and ratifies the Prime Minister's proposals for appointing members of the Government. It decides the country's socio-economic development plans, national financial and monetary policies, and foreign policy. The term of each legislature is five years. The National Assembly Standing Committee supervises the enforcement of laws and the activities of the Government. Amendments to the Constitution may only be made by a majority vote of at least two-thirds of the Assembly's members.

THE PRESIDENT OF THE STATE

The President, as Head of State, represents Viet Nam in domestic and foreign affairs. The President is elected by the National Assembly from among its deputies, and is responsible to the National Assembly. The President's term of office is the same as that of the National Assembly. He or she is Commander-in-Chief of the people's armed forces, and chairs the National Defence and Security Council. The President asks the National Assembly to appoint or dismiss the Vice-President, the Prime Minister, the Chief Justice of the Supreme People's Court and the Chief Procurator of the Supreme People's Organ of Control. According to resolutions of the National Assembly or of its Standing Committee, the President appoints or dismisses members of the Government, and declares war or a state of emergency.

THE GOVERNMENT

The Government comprises the Prime Minister, the Vice-Prime Ministers, ministers and other members. Apart from the Prime Minister, ministers do not have to be members of the National Assembly. The Prime Minister is responsible to the National Assembly, and the term of office of any Government is the same as that of the National Assembly, which ratifies the appointment or dismissal of members of the Government.

LOCAL GOVERNMENT

The country is divided into provinces and municipalities, which are subordinate to the central Government; municipalities are divided into districts, precincts and cities, and districts are divided into villages and townships. People's Councils are elected by the local people.

JUDICIAL SYSTEM

The judicial system comprises the Supreme People's Court, local People's Courts, military tribunals and other courts. The term of office of the presiding judge of the Supreme People's Court corresponds to the term of the National Assembly, and he or she is responsible to the National Assembly. The Supreme People's Organ of Control ensures the observance of the law and exercises the right of public prosecution. Its Chief Procurator is responsible to the National Assembly. There are local People's Organs of Control and Military Organs of Control.

The Government

HEAD OF STATE

President: TRAN DUC LUONG (elected by the 10th National Assembly on 24 September 1997; re-elected by the 11th National Assembly on 24 July 2002).

Vice-President: TRUONG MY HOA.

CABINET
(September 2004)

Prime Minister: PHAN VAN KHAI.

First Deputy Prime Minister: NGUYEN TAN DUNG.

Deputy Prime Ministers: VU KHOAN, PHAM GIA KHIEM.

Minister of National Defence: Lt-Gen. PHAM VAN TRA.

Minister of Public Security: LE HONG ANH.

Minister of Foreign Affairs: NGUYEN DY NIEN.

Minister of Justice: UONG CHU LUU.

Minister of Finance: NGUYEN SINH HUNG.

Minister of Labour, War Invalids and Social Welfare: NGUYEN THI HANG.

Minister of Education and Training: Prof. NGUYEN MINH HIEN.

Minister of Public Health: TRAN THI TRUNG CHIEN.

Minister of Culture and Information: PHAM QUANG NGHI.

Minister of Construction: NGUYEN HONG QUAN.

Minister of Transport and Communications: DAO DINH BINH.

Minister of Agriculture and Rural Development: CAO DUC PHAT (acting).

Minister of Fisheries: Dr TA QUANG NGOC.

Minister of Industry: HOANG TRUNG HAI.

Minister of Trade: TRUONG DINH TUYEN.

Minister of Planning and Investment: VO HONG PHUC.

Minister of Internal Affairs: DO QUANG TRUNG.

Minister of Science and Technology: HOANG VAN PHONG.

Minister of Natural Resources and the Environment: MAI AI TRUC.

Minister of Posts and Telecommunications: DO TRUNG TA.

State Inspector-General: QUACH LE THANH.

Minister, Chairman of the Ethnic Minorities and Mountain Region Commission: KSOR PHUOC.

Minister, Head of the Government Office: DOAN MANH GIAO.

Minister, Chairman of the Committee for Physical Training and Sports: NGUYEN DANH THAI.

Minister, Chairman of the Committee for Population, Family and Children: LE THI THU.

MINISTRIES AND COMMISSIONS

Ministry of Agriculture and Rural Development: 2 Ngoc Ha, Ba Dinh District, Hanoi; tel. (4) 8468161; fax (4) 8454319; e-mail bnn@fpt.vn; internet www.mard.gov.vn.

Ministry of Construction: 37 Le Dai Hanh, Hai Ba Trung District, Hanoi; tel. (4) 9760271; fax (4) 9762153; e-mail bxd-vp@hn.vnn.vn.

Ministry of Culture and Information: 51–53 Ngo Quyen, Hoan Kiem District, Hanoi; tel. (4) 8255349; fax (4) 8267101; internet www.cinet.vnn.vn.

Ministry of Education and Training: 49 Dai Co Viet, Hai Ba Trung District, Hanoi; tel. (4) 8694795; fax (4) 8694085; e-mail intlaff@iupui.edu; internet www.edu.net.vn.

Ministry of Finance: 8 Phan Huy Chu, Hoan Kiem District, Hanoi; tel. (4) 9341540; fax (4) 8262266; e-mail ird-mof@hn.vnn.vn; internet www.mof.gov.vn.

Ministry of Fisheries: 10 Nguyen Cong Hoan, Hanoi; tel. (4) 7716269; fax (4) 7716702; e-mail mofi@mofi.gov.vn; internet www.mofi.gov.vn.

Ministry of Foreign Affairs: 1 Ton That Dam, Ba Dinh District, Hanoi; tel. (4) 1992000; fax (4) 8445905; e-mail bc.mfa@mofa.gov.vn; internet www.mofa.gov.vn.

Ministry of Industry: 54 Hai Ba Trung, Hoan Kiem District, Hanoi; tel. (4) 8258311; fax (4) 8265303; internet www.moi.gov.vn.

Ministry of Internal Affairs: 37A Nguyen Binh Khiem, Hanoi; tel. (4) 9780870; fax (4) 9781005; e-mail ngoc-hien@hn.vnn.vn.

Ministry of Justice: 58–60 Tran Phu, Hanoi; tel. (4) 7336213; fax (4) 8431431.

Ministry of Labour, War Invalids and Social Welfare: 12 Ngo Quyen, Hoan Kiem District, Hanoi; tel. (4) 8246137; fax (4) 8248036; e-mail vvnd@fpt.vn.

Ministry of National Defence: 1A Hoang Dieu, Ba Dinh District, Hanoi; tel. (4) 8468101; fax (4) 8265540.

Ministry of Natural Resources and the Environment: 83 Nguyen Chi Thanh, Dong Da District, Hanoi; tel. (4) 8357910; fax (4) 8352191.

Ministry of Planning and Investment: 2 Hoang Van Thu, Ba Dinh District, Hanoi; tel. (4) 8453027; fax (4) 8234453; internet www.mpi.gov.vn.

Ministry of Posts and Telecommunications: 18 Nguyen Du, Hanoi; tel. (4) 9435602; fax (4) 8226590; e-mail nmhong@mpt.gov.vn; internet www.mpt.gov.vn.

Ministry of Public Health: 138A Giang Vo, Ba Dinh District, Hanoi; tel. (4) 8464051; fax (4) 8460701; internet www.moh.gov.vn.

Ministry of Public Security: 44 Yet Kieu, Hanoi; tel. (4) 8226602; fax (4) 8260774.

Ministry of Science and Technology: 39 Tran Hung Dao, Hoan Kiem District, Hanoi; tel. (4) 9439731; fax (4) 8252733; internet www.moste.gov.vn.

Ministry of Trade: 21 Ngo Quyen, Hanoi; tel. (4) 8262538; fax (4) 8264696; e-mail webmaster@mot.gov.vn; internet www.mot.gov.vn.

Ministry of Transport and Communications: 80 Tran Hung Dao, Hoan Kiem District, Hanoi; tel. (4) 8254012; fax (4) 8267291; e-mail tic_mot@hn.vnn.vn; internet www.mt.gov.vn.

National Committee for Population, Family and Children: 12 Ngo Tat To, Hanoi; tel. (4) 8260020; fax (4) 8258993; e-mail htqt@vietnam.org.vn.

State Inspectorate: 28 Tang Bat Ho, Hanoi; tel. (4) 8254497.

NATIONAL DEFENCE AND SECURITY COUNCIL

President: TRAN DUC LUONG.

Vice-President: PHAN VAN KHAI.

Members: NGUYEN VAN AN, LE HONG ANH, Lt-Gen. PHAM VAN TRA, NGUYEN DY NIEN.

Legislature

QUOC HOI
(National Assembly)

Elections to the 11th National Assembly were held on 19 May 2002. The new Assembly comprised 498 members (compared with 450 in the previous Assembly), elected from among 759 candidates; 51 non-party candidates were elected.

Standing Committee

Chairman: NGUYEN VAN AN.

Vice-Chairmen: NGUYEN VAN YEU, NGUYEN PHUC THANH, TRUONG QUANG DUOC.

Political Organizations

Dang Cong San Viet Nam (Communist Party of Viet Nam): 1 Hoang Van Thu, Hanoi; e-mail cpv@hn.vnn.vn; internet www.cpv.org.vn; f. 1976; ruling party; fmrly the Viet Nam Workers' Party, f. 1951 as the successor to the Communist Party of Indo-China, f. 1930; c. 2.2m. mems (1996); Gen. Sec. of Cen. Cttee NONG DUC MANH.

POLITBURO

Members: NONG DUC MANH, TRAN DUC LUONG, PHAN VAN KHAI, NGUYEN MINH TRIET, NGUYEN TAN DUNG, LE MINH HUONG, NGUYEN PHU TRONG, PHAN DIEN, LE HONG ANH, TRUONG TAN SANG, Lt-Gen. PHAM VAN TRA, NGUYEN VAN AN, TRUONG QUANG DUOC, TRAN DINH HOAN, NGUYEN KHOA DIEM.

SECRETARIAT

Members: NONG DUC MANH, LE HONG ANH, NGUYEN VAN AN, TRAN DINH HOAN, NGUYEN KHOA DIEM, LE VAN DUNG, TONG THI PHONG, TRUONG VINH TRONG, VU KHOAN, NGUYEN VAN CHI.

Ho Chi Minh Communist Youth Union: 60 Ba Trieu, Hanoi; tel. (4) 9435709; fax (4) 9348439; e-mail cydeco@hn.vnn.vn; f. 1931; 4m. mems; First Sec. HOANG BINH QUAN.

People's Action Party (PAP): Hanoi; Leader LE VAN TINH.

Viet Nam Fatherland Front: 46 Trang Thi, Hanoi; f. 1930; replaced the Lien Viet (Viet Nam National League), the successor to Viet Nam Doc Lap Dong Minh Hoi (Revolutionary League for the Independence of Viet Nam) or Viet Minh; in 1977 the original org. merged with the National Front for the Liberation of South Viet Nam and the Alliance of National, Democratic and Peace Forces in

South Viet Nam to form a single front; 200-member Cen. Cttee; Pres. Presidium of Cen. Cttee PHAM THE DUYET; Gen. Sec. TRAN VAN DANG.

Vietnam Women's Union: 39 Hang Chuoi, Hanoi; tel. (4) 9717225; fax (4) 9713143; e-mail VWUnion@netnam.org.vn; f. 1930; 11.4m. mems; Pres. HA THI KHIET.

Diplomatic Representation

EMBASSIES IN VIET NAM

Algeria: 13 Phan Chu Trinh, Hanoi; tel. (4) 8253865; fax (4) 8260830; e-mail aldjazairvn@hn.vnn.vn; Ambassador TEWFIK ABADA.

Argentina: Daeha Business Centre, 360 Kim Ma, Ba Dinh District, Hanoi; tel. (4) 8315262; fax (4) 8315577; e-mail embarg@hn.vnn.vn; internet www.embargentina.org.vn; Ambassador TOMÁS FERRARI.

Australia: 8 Dao Tan, Ba Dinh District, Hanoi; tel. (4) 8317755; fax (4) 8317711; e-mail ausemb@fpt.vn; internet www.ausinvn.com; Ambassador JOE THWAITES.

Austria: Prime Centre, 8th Floor, 53 Quang Trung, Hai Ba Trung District, Hanoi; tel. (4) 9433050; fax (4) 9433055; e-mail hanoi-ob@bmaa.gv.at; Ambassador Dr JOSEF MÜLLNER.

Bangladesh: 7th Floor, Daeha Business Centre, 360 Kim Ma, Ba Dinh District, Hanoi; tel. (4) 7716625; fax (4) 7716628; Ambassador ASHRAF UD DOUTA.

Belarus: 52 Tay Ho, Tay Ho District, Hanoi; tel. (4) 8290494; fax (4) 7197125; e-mail vietnam@belembassy.org; Ambassador ALYAKSANDR KUTSALAY.

Belgium: 9th Floor, Somerset Grand Hanoi, 49 Hai Ba Trung, Hanoi; tel. (4) 9346179; fax (4) 9346183; e-mail hanoi@diplobel.org; Ambassador PHILIPPE JOTTARD.

Brazil: 14 Thuy Khue, T72 Hanoi; tel. (4) 8430817; fax (4) 8432542; e-mail ambviet@netnam.org.vn; Ambassador CHRISTIANO WHITAKER.

Brunei: 27 Quang Trung, Hoan Kiem District, Hanoi; tel. (4) 9435249; fax (4) 9435201; e-mail bruemviet@hn.vnn.vn; Ambassador Dato' Paduka Haji ALI Haji HASSAN.

Bulgaria: Van Phuc Quarter, Nui Truc, Hanoi; tel. (4) 8452908; fax (4) 8460856; e-mail bulemb@hn.vnn.vn; Ambassador GUEORGUI MIHOV.

Cambodia: 71 Tran Hung Dao, Hanoi; tel. (4) 9424788; fax (4) 9423225; e-mail arch@fpt.vn; Ambassador VAR SIM SAMRETH.

Canada: 31 Hung Vuong, Hanoi; tel. (4) 7345000; fax (4) 7345049; e-mail hanoi@dfait-maeci.gc.ca; internet www.dfait-maeci.gc.ca/vietnam; Ambassador RICHARD LECOQ.

China, People's Republic: 46 Hoang Dieu, Hanoi; tel. (4) 8453736; fax (4) 8232826; e-mail eossc@hn.vnn.vn; Ambassador QI JIANGUO.

Cuba: 65A Ly Thuong Kiet, Hanoi; tel. (4) 9424775; fax (4) 9422426; e-mail embacuba@netnam.org.vn; Ambassador FREDESMAN TURRÁ GONZÁLEZ.

Czech Republic: 13 Chu Van An, Hanoi; tel. (4) 8454131; fax (4) 8233996; e-mail hanoi@embassy.mzv.cz; internet www.mfa.cz/hanoi; Ambassador LUBOS NOVÝ.

Denmark: 19 Dien Bien Phu, Hanoi; tel. (4) 8231888; fax (4) 8231999; e-mail hanamb@um.dk; internet www.dk-vn.dk; Ambassador PETER LYSHOLT HANSEN.

Egypt: 63 To Ngoc Van, Quang An, Tay Ho District, Hanoi; tel. (4) 8294999; fax (4) 8294997; e-mail arabegypt@ftp.vn; Ambassador ABDALLAH OMAR ALARNOSY.

Finland: 6th Floor, Central Bldg, 31 Hai Ba Trung, Hanoi; tel. (4) 8266788; fax (4) 8266766; e-mail sanomat.han@formin.fi; Ambassador KARI ALANKO.

France: 57 Tran Hung Dao, Hanoi; tel. (4) 9437719; fax (4) 9437236; e-mail ambafrance@hn.vnn.vn; internet www.ambafrance-vn.org; Ambassador JEAN-FRANÇOIS BLAREL.

Germany: 29 Tran Phu, Hanoi; tel. (4) 8453836; fax (4) 8453838; e-mail germanemb.hanoi@fpt.vn; internet www.germanembhanoi.org.vn; Ambassador CHRISTIAN-LUDWIG WEBER-LORTSCH.

Hungary: Daeha Business Centre, 12th Floor, 360 Kim Ma, Ba Dinh District, Hanoi; tel. (4) 7715714; fax (4) 7715716; e-mail hungemb@hn.vnn.vn; Ambassador Dr DÉNES SZÁSZ.

India: 58–60 Tran Hung Dao, Hanoi; tel. (4) 8244990; fax (4) 8244998; e-mail india@netnam.org.vn; Ambassador NEELAKANTAN RAVI.

Indonesia: 50 Ngo Quyen, Hanoi; tel. (4) 8253353; fax (4) 8259274; e-mail komhan@hn.vnn.vn; internet www.indonesia-hanoi.org.vn; Ambassador ARTAULI RATNA MENARA PANGGABEAN TOBING.

Iran: 54 Tran Phu, Ba Dinh District, Hanoi; tel. (4) 8232068; fax (4) 8232120; e-mail iriemb@fpt.vn; Ambassador HOSSEIN MOLLA-ABDOLLAHI.

Iraq: 66 Tran Hung Dao, Hanoi; tel. (4) 9424141; fax (4) 9424055; e-mail iraqyia@hn.vnn.vn; Ambassador SALAH AL-MUKHTAR (designate).

Israel: 68 Nguyen Thai Hoc, Hanoi; tel. (4) 8433141; fax (4) 8435760; e-mail infor@emisrael-vn.org; Ambassador AVRAHAM NIR.

Italy: 9 Le Phung Hieu, Hoan Kiem District, Hanoi; tel. (4) 8256256; fax (4) 8267602; e-mail embitaly@embitaly.org.vn; internet www.embitalyvietnam.org; Ambassador ALFREDO MATACOTTA CORDELLA.

Japan: 27 Lieu Giai, Ba Dinh District, Hanoi; tel. (4) 8463000; fax (4) 8463043; Ambassador NARIO HATTORI.

Korea, Democratic People's Republic: 25 Cao Ba Quat, Hanoi; tel. (4) 8453008; fax (4) 8231221; Ambassador PAK UNG-SOP.

Korea, Republic: 4th Floor, Daeha Business Centre, 360 Kim Ma, Ba Dinh District, Hanoi; tel. (4) 8315111; fax (4) 8315117; e-mail korembviet@mofat.go.kr; Ambassador YOO TAE-HYUN.

Laos: 22 Tran Binh Trong, Hanoi; tel. (4) 9424576; fax (4) 8228414; Ambassador VILAYVANH PHOMKEHE.

Libya: A3 Van Phuc Residential Quarter, Hanoi; tel. (4) 8453379; fax (4) 8454977; e-mail libpbha@yahoo.com; Secretary SALEM ALI SALEM DANNAH.

Malaysia: 16th Floor, Fortuna Tower, 6B Lang Ha, Hanoi; tel. (4) 8313400; fax (4) 8313402; e-mail mwhanoi@hn.vnn.vn; Ambassador Dato' AHMAD ANUAR ABDUL HAMID.

Mexico: 14 Thuy Khue, T-11, Hanoi; tel. (4) 8470948; fax (4) 8470949; e-mail embvietnam@sre.gob.mx; Ambassador FEDERICO URRUCHUA DURAND.

Mongolia: 39 Tran Phu, Hanoi; tel. (4) 8453009; fax (4) 8454954; e-mail mongembhanoi@hn.vnn.vn; Ambassador GANBOLD BAASANJAV.

Myanmar: A3 Van Phuc Diplomatic Quarter, Hanoi; tel. (4) 8453369; fax (4) 8452404; e-mail myan.emb@fpt.vn; Ambassador U KHIN AUNG.

Netherlands: Daeha Office Tower, 6th Floor, 360 Kim Ma, Ba Dinh District, Hanoi; tel. (4) 8315650; fax (4) 8315655; e-mail han@minbuza.nl; internet www.netherlands-embassy.org.vn; Ambassador GERBEN S. DE JONG.

New Zealand: Level 5, 63 Ly Thai To, Hanoi; tel. (4) 8241481; fax (4) 8241480; e-mail nzembhan@fpt.vn; Ambassador MICHAEL CHILTON.

Norway: 7th Floor, Suite 701–702, Metropole Centre, 56 Ly Thai To, Hanoi; tel. (4) 8262111; fax (4) 8260222; Ambassador PER G. STAVNUM.

Pakistan: 8th Floor, Daeha Business Centre, 360 Kim Ma, Ba Dinh District, Hanoi; tel. (4) 7716420; fax (4) 7716418; e-mail parep-hanoi@hn.vnn.vn; Ambassador MUHAMMAD YOUSAF ALI.

Philippines: 27B Tran Hung Dao, Hanoi; tel. (4) 9437873; fax (4) 9435760; e-mail phvn@hn.vnn.vn; Ambassador VICTORIANO M. LECAROS.

Poland: 3 Chua Mot Cot, Hanoi; tel. (4) 8452027; fax (4) 8236914; Ambassador WIESLAW SCHOLZ.

Romania: 5 Le Hong Phong, Hanoi; tel. (4) 8452014; fax (4) 8430922; e-mail romambhan@fpt.vn; Ambassador CONSTANTIN LUPEANU.

Russia: 191 La Thanh, Hanoi; tel. (4) 8336991; fax (4) 8336995; e-mail moscow.vietnam@hn.vnn.vn; Ambassador ANDREI ALEXEEVICH TATARINOV.

Singapore: 41–43 Tran Phu, Hanoi; tel. (4) 8233965; fax (4) 7337627; e-mail singemb@hn.vnn.vn; internet www.mfa.gov.sg/hanoi; Ambassador TAN SENG CHYE.

Slovakia: 6 Le Hong Phong, Hanoi; tel. (4) 8454334; fax (4) 8454145; e-mail zuskemb@hn.vnn.vn; Ambassador ANTON HAJDUK.

South Africa: 4th Floor, 59A Ly Thai To, Hanoi; tel. (4) 9340888; fax (4) 9360680; e-mail saembassy@pressclub.netnam.vn; Chargé d'affaires a.i. ELIZABETH ERASMUS.

Spain: 15th Floor, Daeha Business Centre, 360 Kim Ma, Ba Dinh District, Hanoi; tel. (4) 7715207; fax (4) 7715206; e-mail embespvn@mail.mae.es; Ambassador GONZALO ORTIZ Y DÍEZ-TORTOSA.

Sri Lanka: 55B Tran Phu, Ba Dinh District, Hanoi; tel. (4) 7341894; fax (4) 7341897; e-mail slembvn@fpt.vn; Ambassador MUSTHAFA M. JAFFEER.

Sweden: 2 Nui Truc, Van Phuc, Khu Ba Dinh, Hanoi; tel. (4) 8454824; fax (4) 8232195; e-mail ambassaden.hanoi@sida.se; internet www.hanoi.embassy.ud.se; Ambassador MARIE SJÖLANDER.

Switzerland: 44B Ly Thuong Kiet, 15th Floor, Hanoi; tel. (4) 9346589; fax (4) 9346591; e-mail vertretung@han.rep.admin.ch; Ambassador THOMAS FELLER.

Thailand: 63–65 Hoang Dieu, Hanoi; tel. (4) 8235092; fax (4) 8235088; e-mail thaiemhn@netnam.org.vn; Ambassador KRIT KRAICHITTI.

Turkey: 4th Floor, North Star Bldg, 4 Da Truong, Hanoi; tel. (4) 8222460; fax (4) 8222458; e-mail turkeyhn@fpt.vn; Ambassador KAYNA INAL.

Ukraine: 49 Nguyen Dy, Hanoi; tel. (4) 9432764; fax (4) 9432766; e-mail ukraine@hn.vnn.vn; internet webua.net/viet; Ambassador PAVLO SULTANSKY.

United Kingdom: Central Bldg, 31 Hai Ba Trung, Hanoi; tel. (4) 9360500; fax (4) 9360561; e-mail behanoi@fpt.vn; internet www.uk-vietnam.org; Ambassador ROBERT GORDON.

USA: 7 Lang Ha, Ba Dinh District, Hanoi; tel. (4) 7721500; fax (4) 7721510; e-mail irchano@pd.state.gov; internet usembassy.state.gov/vietnam; Ambassador RAYMOND BURGHARDT.

Judicial System

Supreme People's Court

48 Ly Thuong Kiet, Hanoi.

The Supreme People's Court in Hanoi is the highest court and exercises civil and criminal jurisdiction over all lower courts. The Supreme Court may also conduct trials of the first instance in certain cases. There are People's Courts in each province and city which exercise jurisdiction in the first and second instance. Military courts hear cases involving members of the People's Army and cases involving national security. In 1993 legislation was adopted on the establishment of economic courts to consider business disputes. The observance of the law by ministries, government offices and all citizens is the concern of the People's Organs of Control, under a Supreme People's Organ of Control. The Chief Justice of the Supreme People's Court and the Chief Procurator of the Supreme People's Organ of Control are elected by the National Assembly, on the recommendation of the President.

Chief Justice of the Supreme People's Court: NGUYEN VAN HIEN.

Chief Procurator of the Supreme People's Organ of Control: HA MANH TRI.

Religion

Traditional Vietnamese religion included elements of Indian and all three Chinese religions: Mahayana Buddhism, Daoism and Confucianism. Its most widespread feature was the cult of ancestors, practised in individual households and clan temples. In addition, there were (and remain) a wide variety of Buddhist sects, the sects belonging to the 'new' religions of Caodaism and Hoa Hao, and the Protestant and Roman Catholic Churches. The Government has guaranteed complete freedom of religious belief.

BUDDHISM

In the North a Buddhist organization, grouping Buddhists loyal to the Democratic Republic of Viet Nam, was formed in 1954. In the South the United Buddhist Church was formed in 1964, incorporating several disparate groups, including the 'militant' An-Quang group (mainly natives of central Viet Nam), the group of Thich Tam Chau (mainly northern emigrés in Saigon) and the southern Buddhists of the Xa Loi temple. In 1982 most of the Buddhist sects were amalgamated into the state-approved Viet Nam Buddhist Church (which comes under the authority of the Viet Nam Fatherland Front; number of adherents estimated at 7% of total population in 1991). The Unified Buddhist Church of Viet Nam is an anti-Government organization.

Viet Nam Buddhist Church: Pres. Exec. Council Most Ven. THICH TRI TINH; Gen. Sec. THICH MING CHAU.

Unified Buddhist Church of Viet Nam: Patriarch THICH HUYEN QUANG.

CAODAISM

Formally inaugurated in 1926, this is a syncretic religion based on spiritualist seances with a predominantly ethical content, but sometimes with political overtones. A number of different sects exist, of which the most politically involved (1940–75) was that of Tay Ninh. Another sect, the Tien Thien, was represented in the National Liberation Front from its inception. There are an estimated 2m. adherents, resident mainly in the South.

Leader: Cardinal THAI HUU THANH.

CHRISTIANITY

In 1991 the number of Christian adherents represented an estimated 7% of the total population.

The Roman Catholic Church

The Roman Catholic Church has been active in Viet Nam since the 17th century, and since 1933 has been led mainly by Vietnamese priests. Many Roman Catholics moved from North to South Viet Nam in 1954–55, but some remained in the North. The total number of adherents was estimated at 5,550,186 in December 2002, representing 6.7% of the population. For ecclesiastical purposes, Viet Nam comprises three archdioceses and 22 dioceses.

Committee for Solidarity of Patriotic Vietnamese Catholics: 59 Trang Thi, Hanoi; Pres. Rev. VUONG DINH AI.

Bishops' Conference

Conférence Episcopal du Viet Nam, 40 Pho Nha Chung, Hanoi; tel. (4) 8254424; fax (4) 8285920; e-mail ttgmhn@hn.vnn.vn.

f. 1980; Pres. Most Rev. PAUL NGUYEN VAN HOA (Bishop of Nha Trang).

Archbishop of Hanoi: Cardinal PAUL JOSEPH PHAM DINH TUNG, Archevêché, 40 Pho Nha Chung, Hanoi; tel. (4) 8254424; fax (4) 8285920; e-mail ttgmhn@hn.vnn.vn.

Archbishop of Ho Chi Minh City: Cardinal JEAN-BAPTISTE PHAM MINH MÂN, Archevêché, 180 Nguyen Dinh Chieu, Ho Chi Minh City 3; tel. (8) 9303828; fax (8) 9300598.

Archbishop of Hué: Most Rev. ETIENNE NGUYEN NHU THE, Archevêché, 37 Phan Dinh Phung, Hué; tel. (54) 823100; fax (54) 833656; e-mail tgmhue@dng.vnn.vn.

The Protestant Church

Introduced in 1920 with 500 adherents; the total number is estimated at 180,000.

HOA HAO

A new manifestation of an older religion called Buu Son Ky Huong, the Hoa Hao sect was founded by Nguyen Phu So in 1939, and at one time claimed 1.5m. adherents in southern Viet Nam.

ISLAM

The number of Muslims was estimated at 50,000 in 1993.

The Press

The Ministry of Culture and Information supervises the activities of newspapers, news agencies and periodicals.

DAILIES

Hanoi

Le Courrier du Viet Nam: 33 Le Thanh Tong, Hanoi; tel. (4) 9334587; fax (4) 8258368; e-mail courrier@vnagency.com.vn; internet lecourrier.vnagency.com.vn; French; publ. by the Viet Nam News Agency; Editor-in-Chief NGUYEN TUONG.

Hanoi Moi (New Hanoi): 44 Le Thai To, Hoan Kiem District, Hanoi; tel. (4) 8253067; fax (4) 8248054; e-mail hanoimoi@hanoimoi.com.vn; internet www.hanoimoi.com.vn; f. 1976; organ of Hanoi Cttee of the Communist Party of Viet Nam; Editor HO XUAN SON; circ. 35,000.

Lao Dong (Labour): 15/167 Tay Son, Hanoi; tel. (4) 5330305; fax (4) 5332525; e-mail laodong@fpt.vn; internet www.laodong.com.vn; f. 1929; organ of the Viet Nam General Confederation of Labour; Editor-in-Chief VUONG VAN VIET; circ. 80,000.

Nhan Dan (The People): 71 Hang Trong, Hoan Kiem District, Hanoi; tel. (4) 8254231; fax (4) 8255593; e-mail toasoan@nhandan.org.vn; internet www.nhandan.org.vn; f. 1946; official organ of the Communist Party of Viet Nam; Editor-in-Chief DINH THE HUYNH; circ. 180,000.

Quan Doi Nhan Dan (People's Army): 7 Phan Dinh Phung, Hanoi; tel. (4) 8254118; f. 1950; organ of the armed forces; Editor NGUYEN PHONG HAI; circ. 60,000.

Tin Tuc (News): 5 Ly Thuong Kiet, Hanoi; tel. (4) 8252931; internet www.tintucvietnam.com; f. 1982; publ. by the Viet Nam News Agency; afternoon; Vietnamese.

Viet Nam News: 11 Tran Hung Dao, Hanoi; tel. (4) 8222884; fax 9424908; e-mail vnnews@vnagency.com.vn; internet www.vietnamnews.vnanet.vn; f. 1991; English; publ. by the Viet Nam News Agency; Editor-in-Chief TRAN MAI HUONG; circ. 25,000 (Mon.-Sat.), 20,000 Sun. edn (Viet Nam News Sunday); circ. 60,000.

Ho Chi Minh City

Sai Gon Giai Phong (Saigon Liberation): 432 Nguyen Thi Minh Khai, Ho Chi Minh City; tel. (8) 8395942; fax (8) 8334183; e-mail sggp@hcm.vnn.vn; internet www.sggp.org.vn; f. 1975; organ of Ho Chi Minh City Cttee of the Communist Party of Viet Nam; Editor-in-Chief Vu Tuat Viet; circ. 100,000.

Saigon Times: 35 Nam Ky Khoi Nghia, District 1, Ho Chi Minh City; tel. (8) 8295936; fax (8) 8294294; e-mail sgt@hcm.vnn.vn; internet www.saigontimes.com.vn; f. 1991; Vietnamese and English; business issues; Editor-in-Chief Vo Nhu Lanh.

PERIODICALS

Dai Doan Ket (Great Unity): 66 Ba Trieu, Hanoi; tel. (4) 8262420; f. 1977; weekly; organ of the Viet Nam Fatherland Front; Editor Nguyen Quang Canh.

Dau Tu: 175 Nguyen Thai Hoc, Hanoi; tel. (4) 8450537; fax (4) 8457937; e-mail vir@hn.vnn.vn; internet www.vir.com.vn; 3 a week business newspaper publ. in Vietnamese; Editor-in-Chief Nguyen Phu Ky; circ. 40,000.

Dau Tu Chung Khoan: 175 Nguyen Thai Hoc, Hanoi; tel. (4) 8450537; fax (4) 8457937; e-mail vir@hn.vnn.vn; internet www.vir.com.vn; weekly; stock market news publ. in Vietnamese; Editor-in-Chief Nguyen Phu Ky; circ. 40,000.

Giao Duc Thoi Dai (People's Teacher): 29B Ngo Quyen, Hoan Kiem District, Hanoi; tel. (4) 8241781; fax (4) 9345611; e-mail gdtd@fpt.vn; internet www.gdtd.com.vn; f. 1959; weekly; organ of the Ministry of Education and Training; Editor Nguyen Ngoc Chu.

Giao Thong-Van Tai (Communications and Transport): 1 Nha Tho, Hanoi; tel. (4) 8255387; f. 1962; weekly; Thursday; organ of the Ministry of Transport and Communications; Editor Ngo Duc Nguyen; circ. 10,000.

Hoa Hoc Tro (Pupils' Flowers): 5 Hoa Ma, Hanoi; tel. (4) 8211065; internet hhtonline.net; weekly; Editor Nguyen Phong Doanh; circ. 150,000.

Khoa Hoc Ky Thuat Kinh Te The Gioi (World Science, Technology and Economy): 5 Ly Thuong Kiet, Hanoi; tel. (4) 8252931; f. 1982; weekly.

Khoa Hoc va Doi Song (Science and Life): 70 Tran Hung Dao, Hanoi; tel. (4) 8253427; f. 1959; weekly; Editor-in-Chief Tran Cu; circ. 30,000.

Nghe Thuat Dien Anh (Cinematography): 65 Tran Hung Dao, Hanoi; tel. (4) 8262473; f. 1984; fortnightly; Editor Dang Nhat Minh.

Nguoi Cong Giao Viet Nam (Vietnamese Catholic): 59 Trang Thi, Hanoi; tel. (4) 8256242; internet www.vietcatholic.org; f. 1984; weekly; organ of the Cttee for Solidarity of Patriotic Vietnamese Catholics; Editor-in-Chief So Chi.

Nguoi Dai Bieu Nhan Dan (People's Deputy): 35 Ngo Quyen, Hanoi; tel. (4) 08046231; fax (4) 08046659; e-mail ndbnd@hn.vnn.vn; f. 1988; bi-weekly; disseminates resolutions of the National Assembly and People's Council; Editor-in-Chief Ho Anh Tai (acting); circ. 2m.

Nguoi Hanoi (The Hanoian): 19 Hang Buom, Hanoi; tel. (4) 8255662; f. 1984; Editor Vu Quan Phuong.

Nha Bao Va Cong Luan (The Journalist and Public Opinion): 59 Ly Thai To, Hanoi; tel. (4) 8253609; fax (4) 8250797; f. 1985; monthly review; organ of the Viet Nam Journalists' Asscn; Editor-in-Chief Phan Duoc Toan; circ. 40,000.

Nong Nghiep Viet Nam (Viet Nam Agriculture): 14 Ngo Quyen, Hanoi; tel. (4) 8256492; fax (4) 8252923; f. 1987; weekly; Editor Le Nam Son.

Outlook: 11 Tran Hung Dao, Hanoi; tel. (4) 8222884; fax (4) 9424908; e-mail vnnews@vnagency.com.vn; internet www.vietnamnews.vnanet.vn; f. 2002; monthly news magazine; Editor-in-Chief Tran Mai Huong; circ. 6,000.

Phu Nu (Woman): Vietnam Women's Union, International Relations Dept, 39 Hang Chuoi, Hanoi; e-mail VWunion@netnam.org.vn; f. 1997; fortnightly; women's magazine; circ. 100,000.

Phu Nu Thu Do (Capital Women): 72 Quan Su, Hanoi; tel. (4) 8247228; fax (4) 8223989; f. 1987; weekly; magazine of the Hanoi Women's Union; Editor-in-Chief Mai Thuc.

Phu Nu Viet Nam (Vietnamese Women): 39 Hang Chuoi, Hanoi; tel. (4) 8253500; weekly; magazine of the Vietnamese Women's Union; Editor-in-Chief Phuong Minh.

Suc Khoe Va Doi Song (Health and Life): 138 A Giang Vo, Hanoi; tel. (4) 8443144; weekly; published by the Ministry of Public Health; Editor Le Thau; circ. 20,000.

Tap Chi Cong San (Communist Review): 1 Nguyen Thuong Hien, Hanoi; tel. (4) 9422061; fax (4) 8222846; e-mail bbttccs@hn.vnn.vn; internet www.tapchicongsan.org.vn; f. 1955 as *Hoc Tap*; fortnightly; political and theoretical organ of the Communist Party of Viet Nam; Editor-in-Chief Ha Dang; circ. 50,000.

Tap Chi Nghiên Cuu Van Hoc (Literature Research Magazine): 20 Ly Thai To, Hanoi; tel. (4) 8252895; e-mail tcvapmail@vnn.vn; monthly; published by the Institute of Literature; Editor-in-Chief Phan Trong Thuong.

Tap Chi San Khau (Theatre Magazine): 51 Tran Hung Dao, Hanoi; tel. (4) 9434423; fax (4) 9434293; e-mail trongkhoi@hn.vnn.vn; f. 1973; monthly; Editor Ngo Thao.

Tap Chi Tac Pham Van Hoc: 65 Nguyen Du, Hanoi; tel. (4) 8252442; f. 1987; monthly; organ of the Viet Nam Writers' Asscn; Editor-in-Chief Nguyen Dinh Thi; circ. 15,000.

Tap Chi Tu Tuong Van Hoa (Ideology and Culture Review): Hanoi; f. 1990; organ of the Central Committee Department of Ideology and Culture; Editor Pham Huy Van.

The Thao Van Hoa (Sports and Culture): 5 Ly Thuong Kiet, Hanoi; tel. (4) 8267043; fax (4) 8264901; f. 1982; weekly; Editor-in-Chief Nguyen Huu Vinh; circ. 100,000.

The Thao Viet Nam (Viet Nam Sports): 5 Trinh Hoai Duc, Hanoi; f. 1968; weekly; Editor Nguyen Hung.

Thieu Nhi Dan Toc (The Ethnic Young): 5 Hoa Ma, Hanoi; tel. (4) 9317133; bi-monthly; Editor Pham Thanh Long; circ. 60,000.

Thieu Nien Tien Phong (Young Pioneers): 5 Hoa Ma, Hanoi; tel. (4) 9317133; three a week; Editor Pham Thanh Long; circ. 210,000.

Thoi Bao Kinh Te Viet Nam: 175 Nguyen Thai Hoc, Hanoi; tel. (4) 8452411; fax (4) 8432755; f. 1993; 2 a week; Editor-in-Chief Pavef Daonguyencat; circ. 37,000.

Thoi Trang Tre (New Fashion): 12 Ho Xuan Huong, Hanoi; tel. (4) 8254032; fax (4) 8226002; f. 1993; monthly; Editor Vu Quang Vinh; circ. 80,000.

Thuong Mai (Commerce): 100 Lo Duc, Hanoi; tel. (4) 8263150; f. 1990; weekly; organ of the Ministry of Trade; Editor Tran Nam Vinh.

Tien Phong (Vanguard): 15 Ho Xuan Huong, Hanoi; tel. (4) 8264031; fax (4) 8225032; f. 1953; four a week; organ of the Ho Chi Minh Communist Youth Union and of the Forum of Vietnamese Youth; Editor Duong Xuan Nam; circ. 165,000.

Van Hoa (Culture and Arts): 26 Dien Bien Phu, Hanoi; tel. (4) 8257781; f. 1957; fortnightly; Editor Phi Van Tuong.

Van Nghe (Arts and Letters): 17 Tran Quoc Toan, Hanoi; tel. (4) 8264430; f. 1949; weekly; organ of the Vietnamese Writers' Union; Editor Huu Thinh; circ. 40,000.

Van Nghe Quan Doi (Army Literature and Arts): 4 Ly Nam De, Hanoi; tel. (4) 8254370; f. 1957; monthly; Editor Nguyen Tri Huan; circ. 50,000.

Viet Nam Business Forum: 9 Dao Duy Anh, Dong Da District, Hanoi; tel. and fax (4) 5743063; e-mail vbfhn@hn.vnn.vn; internet www.vibforum.vcci.com.vn; weekly magazine in English; publ. by the Viet Nam Chamber of Commerce and Industry; Editor-in-Chief Doan Duy Khuong.

Viet Nam Courier: 5 Ly Thuong Kiet, Hanoi; tel. (4) 8261847; fax (4) 8242317; weekly; English; publ. by the Viet Nam News Agency; Editor-in-Chief Nguyen Duc Giap.

Viet Nam Cultural Window: 46 Tran Hung Dao, Hanoi; tel. (4) 8253841; fax (4) 8269578; e-mail vncw@hn.vnn.vn; f. 1998; every two months; English; Dir Than Doan Lam.

Viet Nam Economic Times: 175 Nguyen Thai Hoc, Hanoi; tel. (4) 8452411; fax (8) 8356716; e-mail vet@hn.vnn.vn; internet www.vneconomy.com.vn; f. 1994; twice weekly; in Vietnamese (with bi-monthly circ. 38,900; edn in English); Editor-in-Chief Prof. Dao Nguyen Cat.

Viet Nam Investment Review (VIR): 175 Nguyen Thai Hoc, Hanoi; tel. (4) 8450537; fax (4) 8457937; e-mail vir@hn.vnn.vn; internet www.vir.com.vn; f. 1990; weekly; business newspaper publ. in English; Editor-in-Chief Nguyen Phu Ky; circ 40,000.

Vietnam Pictorial: 11 Tran Hung Dao, Hanoi; tel. (4) 9332296; fax (4) 9332291; e-mail vnpictorial@vnagency.com.vn; internet www.vietnampictorial.vnanet.vn; f. 1954; monthly online, in Vietnamese, English, French, Chinese, Spanish and Russian; fmrly Viet Nam Review; Editor-in-Chief Nguyen Vinh Quang; circ. 138,000.

Viet Nam Renovation: Hanoi; f. 1994; quarterly magazine on reform of the agricultural sector; in Vietnamese, Chinese and English.

Viet Nam Social Sciences: 27 Tran Xuan Soan, Hanoi; tel. (4) 9784578; fax (4) 9783869; e-mail 21.6.tapchikhxh@fpt.vn; f. 1984; every 2 months; publ. in English and Vietnamese; organ of Viet Nam Social Academy; Editor-in-Chief Dr Le Dinh Cuc.

Vietnamese Studies: 46 Tran Hung Dao, Hanoi; tel. (4) 8253841; fax (4) 8269578; e-mail thegioi@hn.vnn.vn; f. 1964; quarterly; English and French edns; Dir TRAN DOAN LAM.

NEWS AGENCIES

Viet Nam News Agency (VNA): 5 Ly Thuong Kiet, Hanoi; tel. (4) 8255443; fax (4) 8252984; e-mail btk@vnagency.com.vn; internet www.vnagency.com.vn; mem. of Organization of Asian and Pacific News Agencies; Dir-Gen. LE QUOC TRUNG.

Foreign Bureaux

Agence France-Presse (AFP): 76 Ngo Quyen, BP 40, Hanoi; tel. (4) 8252045; fax (4) 8266032; e-mail afphanoi@fpt.vn; Bureau Chief PHILLIPPE PERDRIAU.

Informatsionnoye Telegrafnoye Agentstvo Rossii— Telegrafnoye Agentstvo Suverennykh Stran (ITAR—TASS) (Russia): Trung Tam Da Nganh, Thanh Xuan Bac, Dong Da District, Hanoi; tel. and fax (4) 8541381; e-mail tassvn@mail.ru; Bureau Chief IOURI A. DENISSOVITCH.

Kyodo News Service (Japan): Room 304, 8 Tran Hung Dao, Hanoi; tel. (4) 8259622; fax (4) 8255848; Bureau Chief KAZUHISA MIYAKE.

Polska Agencja Prasowa (PAP) (Poland): B5 Van Phuc Residential Quarter, Hanoi; tel. (4) 8252601; Chief TOMASZ TRZCINSKI.

Prensa Latina (Cuba): 66 Ngo Thi Nham, Hanoi; tel. (4) 9434366; fax (4) 9434866; e-mail plvietnam@hn.vnn.vn; Correspondent FELIX ALBISU CRUZ.

Reuters (UK): Room 402, 8 Tran Hung Dao, Hanoi; tel. (4) 8259623; fax (4) 8268606; e-mail hanoi.newsroom@reuters.com.

Rossiiskoye Informatsionnoye Agentstvo—Novosti (RIA—Novosti) (Russia): 55A Tran Phu, Hanoi; tel. (4) 8431607; fax (4) 8230001; e-mail riahanoi@hn.vnn.vn; internet www.rian.ru; f. 1955; Bureau Chief ANDREI P. SHAMSHIN.

Xinhua (New China) News Agency (People's Republic of China): 6 Khuc Hao, Hanoi; tel. (4) 8232521; fax (4) 8452913; Chief Correspondent ZHANG JIAXIANG.

PRESS ASSOCIATION

Viet Nam Journalists' Association: 59 Ly Thai To, Hanoi; tel. (4) 8269747; fax (4) 8250797; e-mail vja@hn.vnn.vn; f. 1950; asscn of editors, reporters and photographers working in the press, radio, television and news agencies; 11,000 mems (2000); Pres. HONG VINH; Vice-Pres. DINH PHONG.

Publishers

All publishing enterprises are controlled by the Ministry of Culture and Information.

Am Nhac Dia Hat (Music) Publishing House: 61 Ly Thai To, Hoan Kiem District, Hanoi; tel. (4) 8256208; f. 1986; produces cassettes, videocassettes, books and printed music; Dir PHAM DUC LOC.

Cong An Nhan Dan (People's Public Security) Publishing House: 167 Mai Hac De, Hai Ba Trung District, Hanoi; tel. (4) 8260910; f. 1981; managed by the Ministry of the Interior; cultural and artistic information, public order and security; Dir PHAM VAN THAM.

Giao Thong Van Tai (Communications and Transport) Publishing House: 80B Tran Hung Dao, Hanoi; tel. (4) 8255620; f. 1983; managed by the Ministry of Transport and Communications; Dir TO KHANH THO.

Khoa Hoc Va Ky Thuat (Science and Technology) Publishing House: 70 Tran Hung Dao, Hanoi; tel. (4) 9424786; fax (4) 8220658; e-mail todanghai@hn.vnn.vn; internet www.nxbkhkt.com.vn; f. 1960; scientific and technical works, guide books, dictionaries, popular and management books; Dir Prof. Dr TO DANG HAI.

Khoa Hoc Xa Hoi (Social Sciences) Publishing House: 61 Phan Chu Trinh, Hanoi; tel. (4) 8255428; f. 1967; managed by the Institute of Social Science; Dir Dr NGUYEN DUC ZIEU.

Kim Dong Publishing House: 62 Ba Trieu, Hanoi; tel. (4) 9434730; fax (4) 8229085; e-mail kimdong@hn.vnn.vn; internet www.nxbkimdong.com.vn; f. 1957; children's; managed by the Ho Chi Minh Communist Youth Union; Dir PHAM QUANG VINH; Editor-in-Chief LE THI DAT.

Lao Dong (Labour) Publishing House: 54 Giang Vo, Hanoi; tel. (4) 8515380; f. 1945; translations and political works; managed by the Viet Nam General Confederation of Labour; Dir LE THANH TONG.

My Thuat (Fine Arts) Publishing House: 44B Hamlong, Hanoi; tel. (4) 8253036; f. 1987; managed by the Plastic Arts Workers' Association; Dir TRUONG HANH.

Nha Xuat Ban Giao Duc (Education) Publishing House: 81 Tran Hung Dao, Hanoi; tel. (4) 8220554; fax (4) 8262010; f. 1957; managed by the Ministry of Education and Training; Dir PHAM VAN AN; Editor-in-Chief Prof. NGUYEN NHU Y.

Nha Xuat Ban Hoi Nha Van (Writers' Association) Publishing House: 65 Nguyen Du, Hoan Kiem District, Hanoi; tel. and fax (4) 8222135; f. 1957; managed by the Vietnamese Writers' Association; Editor-in-Chief and Dir (acting) NGO VAN PHU.

Nong Nghiep (Agriculture) Publishing House: DH 14, Phoung Mai Ward, Dong Da District, Hanoi; tel. (4) 8523887; f. 1976; managed by the Ministry of Agriculture and Rural Development; Dir DUONG QUANG DIEU.

Phu Nu (Women) Publishing House: 16 Alexandre De Rhodes, Hanoi; tel. (4) 8294459; f. 1957; managed by the Vietnamese Women's Union; Dir TRAN THU HUONG.

Quan Doi Nhan Dan (People's Army) Publishing House: 25 Ly Nam De, Hanoi; tel. (4) 8255766; managed by the Ministry of National Defence; Dir DOAN CHUONG.

San Khau (Theatre) Publishing House: 51 Tran Hung Dao, Hanoi; tel. (4) 8264423; f. 1986; managed by the Stage Artists' Association.

Su That (Truth) Publishing House: 24 Quang Trung, Hanoi; tel. (4) 8252008; fax (4) 8251881; f. 1945; managed by the Communist Party of Viet Nam; Marxist-Leninist classics, politics and philosophy; Dir TRAN NHAM.

Thanh Nien (Youth) Publishing House: 270 Nguyen Dinh Chieu, District 3, Hanoi; tel. and fax (4) 8222612; f. 1954; managed by the Ho Chi Minh Communist Youth Union; Dir BUI VAN NGOI.

The Duc The Thao (Physical Education and Sports) Publishing House: 7 Trinh Hoai Duc, Hanoi; tel. (4) 8256155; f. 1974; managed by the Ministry of Culture and Information; Dir NGUYEN HIEU.

The Gioi Publishers: 46 Tran Hung Dao, Hanoi; tel. (4) 8253841; fax (4) 8269578; e-mail thegioi@hn.vnn.vn; f. 1957; foreign language publications; managed by the Ministry of Culture and Information; Dir MAI LY QUANG.

Thong Ke (Statistics) Publishing House: 96 Thuy Khe, Hanoi; tel. (4) 8257814; f. 1980; managed by the Gen. Statistical Office; Dir NGUYEN DAO.

Van Hoa (Culture) Publishing House: 43 Lo Duc, Hanoi; tel. (4) 8253517; f. 1971; managed by the Ministry of Culture and Information; Dir QUANG HUY.

Van Hoc (Literature) Publishing House: 49 Tran Hung Dao, Hanoi; tel. (4) 8252100; f. 1948; managed by the Ministry of Culture and Information; Dir LU HUY NGUYEN.

Xay Dung (Building) Publishing House: 37 Le Dai Hanh, Hanoi; tel. (4) 8268271; fax (4) 8215369; f. 1976; managed by the Ministry of Construction; Dir NGUYEN LUONG BICH.

Y Hoc (Medicine) Publishing House: 4 Le Thanh Ton, Phan Chu Trinh, Hoan Kiem District, Hanoi; tel. (4) 8255281; e-mail xuatbanyhoc@netnam.vn; managed by the Ministry of Public Health; Dir HOANG TRONG QUANG.

Broadcasting and Communications

TELECOMMUNICATIONS

Directorate General for Posts and Telecommunications (DGPT): Department of Science-Technology and International Cooperation, 18 Nguyen Du, Hanoi; tel. (4) 8226580; fax (4) 8226590; industry regulator; Sec.-Gen. Dr MAI LIEM TRUC.

Board of the Technical and Economic Programme on Information Technology: 39 Tran Hung Dao, Hanoi; e-mail nyenet@itnet.gov.vn; Gen. Dir Dr DO VAN LOC.

Army Electronics and Communications Corporation: military-owned communications company; awarded a licence in 1998 to provide national telephone services, including a fixed network and mobile and paging systems.

Saigon Post and Telecommunications Service Corpn: 45 Le Duan, District 1, Ho Chi Minh City; tel. (8) 4040608; fax (8) 4040609; e-mail saigonpostel@saigonpostel.com.vn; internet www.saigonpostel.com.vn; f. 1995; partially owned by the State; nationwide post and telecommunications services; Chair. TRAN THANH LONG; Gen. Dir TRINH DINH KHUONG.

Viet Nam Posts and Telecommunications Corporation (VNPT): 18 Nguyen Du, Hai Ba Trung District, Hanoi; tel. (4) 8265104; fax (4) 8255851; e-mail infocen2@hn.vnn.vn; internet www.vnpt.com.vn; f. 1995; state-owned communications company; Chair. VU VAN LUAN; Pres. and CEO PHAM LONG TRAN.

RADIO

In 1998 there were 390 FM stations, 567 district radio stations and 6,505 commune radio stations in Viet Nam. The Ministry of Culture and Information is responsible for the management of radio services.

Voice of Viet Nam (VOV): 58 Quan Su, Hanoi; tel. (4) 8255669; fax (4) 8261122; e-mail qhqt.vov@hn.vnn.vn; internet www.vov.org.vn; f. 1945; four domestic channels in Vietnamese; two foreign service channels in English, Japanese, French, Khmer, Laotian, Spanish, Thai, Cantonese, Mandarin, Indonesian, Vietnamese and Russian; Dir-Gen. VU VAN HIEN.

TELEVISION

At the end of 1994 there were 53 provincial television stations and 232 relay stations in Viet Nam. The Ministry of Culture and Information is responsible for the management of television services.

Viet Nam Television (VTV): 43 Nguyen Chi Thanh, Hanoi; tel. (4) 8318123; fax (4) 8318124; e-mail webmaster@vtv.org.vn; internet www.vtv.org.vn; television was introduced in South Viet Nam in 1966 and in North Viet Nam in 1970; broadcasts from Hanoi (via satellite) to the whole country and Asia region; Vietnamese, French, English; Dir-Gen. VU VAN HIEN.

Finance

(cap. = capital; res = reserves; dep. = deposits; m. = million; brs = branches)

BANKING

In early 1990 the Government established four independent commercial banks, several joint-venture banks, and introduced legislation to permit the operation of foreign banks in Viet Nam. At the end of 1998 the Vietnamese banking system comprised six state-owned banks, four joint-venture banks, 24 foreign bank branches, 51 joint-stock commercial banks, 977 People's Credit Funds, two joint-stock finance companies, three corporation-subordinated finance companies and eight leasing companies. At the beginning of 2002 there were 27 foreign bank branches and 42 representative offices of foreign financial institutions in Viet Nam.

Central Bank

State Bank of Viet Nam: 47–49 Ly Thai To, Hanoi; tel. (4) 9342524; fax (4) 8268765; e-mail icd.boc@fpt.vn; f. 1951; central bank of issue; provides a national network of banking services and supervises the operation of the state banking system; Gov. LE DUC THUY; 61 brs and sub-brs.

State Banks

Bank for Agriculture and Rural Development: 2 Lang Ha, Ba Dinh District, Hanoi; tel. (4) 8313710; fax (4) 8313717; e-mail qhqt@fpt.vn; internet www.vbard.com; f. 1988; cap. 3,844,915m. dông, res –4,215,110m. dông, dep. 82,607,320m. dông (Dec. 2002); Chair. NGUYEN QUOC TOAN; Gen. Man. VAN SO LE; 1,291 brs.

Bank for Foreign Trade of Viet Nam (Vietcombank): 198 Tran Quang Khai, Hanoi; tel. (4) 8265503; fax (4) 8269067; e-mail webmaster@vietcombank.com.vn; internet www.vietcombank.com.vn; f. 1963; authorized to deal in foreign currencies and all other international banking business; cap. 1,080,611m. dông, res 859,138m. dông, dep. 71,891,861m. dông (Dec. 2001); Chair. LE DAC CU; Dir-Gen. VU VIET NGOAN; 23 brs.

Bank for Investment and Development of Vietnam (Vietinde Bank): 194 Tran Quang Khai, Hanoi; tel. (4) 8266966; fax (4) 8266959; e-mail bidv@hn.vnn.vn; internet www.bidv.com.vn; Chair. PHUNG THI VAN ANH; Gen. Dir TRINH NGOC HO.

Industrial and Commercial Bank of Viet Nam (VIETINCOMBANK): 108 Tran Hung Dao, Hanoi; tel. (4) 9421066; fax (4) 9421143; e-mail webmaster@icb.com.vn; internet www.icb.com.vn; f. 1987; state-owned; authorized to receive personal savings, extend loans, issue stocks and invest in export-orientated cos and jt ventures with foreigners; cap. 2,100,000m. dông, res 898,174m. dông, dep. 59,283,956m. dông (Dec. 2002); Chair. NGUYEN VAN BINH; Gen. Dir PHAM HUY HONG; 116 brs.

Saigon Bank for Industry and Trade: 2C Pho Duc Chinh, District 1, Ho Chi Minh City; tel. (8) 9143183; fax (8) 9143193; e-mail saigonbank@hcm.vnn.vn; internet www.saigonbank.com.vn; cap. 250,000m. dông, res 36,826m. dông, dep. 1,702,469m. dông (Dec. 2003); specializes in trade and industry activities; Dir DUONG XUAN MINH; 14 brs and sub-brs.

Vietnam Export-Import Commercial Joint-Stock Bank (Vietnam Eximbank): 7 Le Thi Hong Gam, District 1, Ho Chi Minh City; tel. (8) 8210055; fax (8) 8216913; e-mail icbr.eximbank@hcm.vnn.vn; internet www.eximbank.com.vn; f. 1989; authorized to undertake banking transactions for the production and processing of export products and export-import operations; cap. 300,000m. dông, res 26,798m. dông, dep. 3,927,635m. dông (Dec. 2002); Chair. NGUYEN THANH LONG; Dir-Gen. NGUYEN GIA DINH; 3 brs.

Viet Nam Technological and Commercial Joint-Stock Bank (Techcombank): 15 Dao Duy Tu, Hoan Kiem District, Hanoi; tel. (4) 8243941; fax (4) 8250545; internet www.techcombank.com.vn; CEO NGUYEN DUC VINH.

Joint-Stock and Other Banks

CBD (Codo Rural Share Commercial Bank): Co Do, Thoi Dong, O Mon, Can Tho Province; tel. (71) 61642; Dir TRAN NGOC HA.

Chohung Vina Bank: 3-5 Ho Tung Mau, District 1, Ho Chi Minh City; tel. (8) 8291581; fax (8) 8291583; e-mail hcmc.fvb@hcm.vnn.vn; f. 1993; jt venture between the Bank for Foreign Trade of Viet Nam and Korea First Bank; cap. US $20.0m., res US $1.2m., dep. US $71.9m. (Dec. 2001); Chair. VIET NGOAN VU; Dir-Gen. SHIN-SEONG KANG.

DS Bank (Dongthap Commercial Joint-Stock Bank): 48 Rad 30/4, Cao Lanh Town, Dong Thap Province; tel. (67) 51441; fax (67) 51878; Dir HOANG VAN TU.

Ficombank (Denhat Joint-Stock Commercial Bank): 67A Le Quang Sung, District 6, Ho Chi Minh City; tel. (8) 5857089; fax (8) 8557093; Dir VAN DUONG.

Indovina Bank Ltd: 39 Ham Nghi, District 1, Ho Chi Minh City; tel. (8) 8224995; fax (8) 8230131; e-mail ivbhcm@hcm.vnn.vn; internet www.indovinabank.com.vn; f. 1990; jt venture of the Cathay United Bank (Taiwan) and the Industrial and Commercial Bank of Viet Nam; also has brs in Hanoi, Haiphong, Binh Duong and Can Tho; cap. US $20m. (2001), res US $2.7m., dep. US $148m. (2003); Chair. PHAM HUY HUNG; Gen. Dir JAN YEI FONG.

Maritime Bank: 5A Nguyen Tri Phuong, Haiphong; tel. (3) 1823076; fax (3) 1823063; e-mail msb@msb.com.vn; internet www.msb.com.vn; f. 1991; cap. 109,310m. dông (2003); Pres. TRAN BA VINH; 9 brs.

Phuong Nam Bank (Phuong Nam Commercial Joint-Stock Bank): 258 Minh Phung, District 11, Ho Chi Minh City; tel. (8) 9606050; fax (8) 9606047; e-mail phuongnambank@hcm.vnn.vn; internet www.phuongnambank.com.vn; f. 1993; Dir HOANG VAN TOAN.

Quedo Joint-Stock Bank: 1-3-5 Can Giuoc, District 8, Ho Chi Minh City; tel. (8) 8562418; fax (8) 8553596; Dir BAO LAN.

VID Public Bank: Ground Floor, Hanoi Tungshing Square, 2 Ngo Quyen, Hanoi; tel. (4) 8268307; fax (4) 8268228; e-mail vpb.han@hn.vnn.vn; f. 1992; jt venture between the Bank for Investment and Development (Viet Nam) and the Public Bank Berhad (Malaysia); commercial bank; cap. US $20m. (2001); Chair. TRAN ANH TUAN (acting); Gen. Dir TAY HONG HENG; 5 brs.

VinaSiam Bank: 2 Pho Duc Chinh, District 1, Ho Chi Minh City; tel. (8) 8210557; fax (8) 8210585; e-mail vsb@hcm.vnn.vn; internet www.vinasiambank.com; f. 1995; jt venture between Bank for Agriculture and Rural Development, Siam Commercial Bank (Thailand) and Charoen Pokphand Group (Thailand); cap. US $15m., res US $2.8m., dep. US $18.1m. (2002); Chair. LE VAN SO; Gen. Man. VIROJ THANAPITAK.

VP Bank (Viet Nam Commercial Joint-Stock Bank for Private Enterprises): 18B Le Thanh Tong, Hanoi; tel. (4) 8245246; fax (4) 8260182; Chair. LAM HOANG LOC.

Foreign Banks

Australia and New Zealand Banking Group Ltd (Australia): 14 Le Thai To, Hanoi; tel. (4) 8258190; fax (4) 8258188; e-mail ahmada@anz.com; Gen. Man. ADIL AHMAD; also has br. in Ho Chi Minh City.

Bangkok Bank Public Co Ltd (Thailand): Harbour View Tower, 35 Nguyen Hue, District 1, Ho Chi Minh City; tel. (8) 8214396; fax (8) 8213772; e-mail bblhcm@hcm.vnn.vn; Man. WITTAYA SUPATANAKUL.

Bank of Tokyo-Mitsubishi Ltd (Japan): The Landmark, 8th Floor, 5B Ton Duc Thang, District 1, Ho Chi Minh City; tel. (8) 8231560; fax (8) 8231559.

Crédit Agricole Indosuez (France): Somerset Chancellor Court, 4th Floor, 21–23 Nguyen Thi Minh Khai, District 1, Ho Chi Minh City; tel. (8) 8295048; fax (8) 8296065; e-mail cai.vietnam@hcm.vnn.vn; Gen. Man. OLIVIER PRECHAC.

Crédit Lyonnais SA (France): Han Nam Officetel, 4th Floor, 65 Nguyen Du, District 1, Ho Chi Minh City; tel. (8) 8299226; fax (8)

8296465; e-mail contact_vnhcmc@creditlyonnais.fr; f. 1993; Man. Ngoc Vu Han; also has br. in Hanoi.

Natexis Banques Populaires (France): Rm 16-02, Prime Ctr, 53 Quang Trung, Hanoi; tel. (4) 9433667; fax (4) 9433665; e-mail natexis-vn@hcm.vnn.vn; Rep. Mme Ut.

Shinhan Bank (Republic of Korea): Yoco Bldg, 7th Floor, 41 Nguyen Thi Minh Khai, District 1, Ho Chi Minh City; tel. (8) 8230012; fax (8) 8230009; Gen. Man. Hae-Soo Kim.

STOCK EXCHANGE

Securities Trading Centre: 45–47 Chuong Duong, Ho Chi Minh City; f. July 2000 by the State Securities Commission (see Development Organization).

INSURANCE

In January 1994 it was announced that foreign insurance companies were to be permitted to operate in Viet Nam and in 1995 it was announced that Baoviet's monopoly of the insurance industry was to be ended. A number of new insurance companies were subsequently established. By September 2003 there were 17 insurance companies operating in the country, of which three were state-owned, three were joint-stock, four were wholly foreign-owned and seven were joint ventures.

Allianz General Insurance Vietnam: Unit 4, 8th Floor, The Metropolitan, 235 Dong Khoi, District 1, Ho Chi Minh City; tel. (8) 8245050; fax (8) 8245054; e-mail jnr@allianzagf.com; f. 1999; fmrly Allianz–AGF Insurance Co Ltd; non-life.

Aon Inchibrok Insurance Services Co Ltd: Vietcombank Tower, 198 Tran Quang Khai, Hanoi; tel. (4) 8244828; fax (4) 8243983; f. 1994; fmrly Inchibrok Insurance; jt venture between Baoviet and Aon Corpn (USA).

Bao Long (Nha Rong Joint-Stock Insurance Co): 185 Dien Bien Phu, District 1, Ho Chi Minh City; tel. (8) 8239219; fax (8) 8239223; internet www.nharong.com; cap. 22,000m. dông (1994); Dir Tran Van Binh.

Bao Minh CMG Life Insurance Co Ltd: Level 3, Saigon Riverside Office Centre, 2a–4a Ton Duc Thang, District 1, Ho Chi Minh City; tel. (8) 8291919; fax (8) 8293131; e-mail baominhcmg@baominhcmg.com.vn; internet www.baominhcmg.com.vn; f. 1999; jt venture between Bao Minh Insurance Co and CMG Colonial Mutual Life Assurance Society (Australia); Gen. Dir Rod Carkeet.

Bao Minh Insurance Co (Ho Chi Minh City Insurance Co): 26 Ton That Dam, District 1, Ho Chi Minh City; tel. (8) 8294180; fax (8) 8294185; internet www.baominhvn.com; f. 1995; non-life.

Baoviet (Viet Nam Insurance Co): 35 Hai Ba Trung, Hoan Kiem District, Hanoi; tel. (4) 8262632; fax (4) 8257188; e-mail service@baoviet.com.vn; internet www.baoviet.com.vn; f. 1965; property and casualty, personal accident, liability and life insurance; total assets 8,817,000m. dông (2004); CEO Trinh Thanh Hoan; Chair. Prof. Truong Moc Lam.

Manulife (Vietnam) Ltd: 12th Floor, Diamond Plaza, 34 Le Duan, District 1, Ho Chi Minh City; tel. (8) 8257730; fax (8) 8257718; e-mail manulifevn_info@manulife.com; internet www.manulife.com.vn; f. 1999; fmrly Chinfon-Manulife Life Insurance Co Ltd; first wholly foreign-owned life insurance co to operate in Viet Nam; Gen. Dir David Matthews.

Petro Viet Nam Insurance Co (PV Insurance): 154 Hguyen Thai Hoc, Ba Dinh District, Hanoi; tel. (4) 7335588; fax (4) 7336284; e-mail info@pv-insurance.com; internet www.pv-insurance.com; f. 1996; non-life insurance; Man. Dir Le Van Hung.

PJICO Insurance (Petrolimex Joint-Stock Insurance Co): 22 Lang Ha, Dong Da District, Hanoi; tel. (4) 7760865; fax (4) 7760868; e-mail pjico@petrolimex.com.vn; internet www.pjico.com.vn; f. 1995; non-life; Gen. Exec. Dir Tran Nghia Vinh.

Viet Nam International Assurance (VIA): 8th Floor, The Landmark, 5b Ton Duc Thang, District 1, Ho Chi Minh City; tel. and fax (8) 8221340; e-mail hcm@via.com.vn; internet www.via.com.vn; f. 1996; jt-venture co, 51% owned by Baoviet, 49% owned by Tokyo Marine and Fire Insurance Co (Japan); non-life insurance and reinsurance for foreign cos.

Trade and Industry

GOVERNMENT AGENCIES

State Financial and Monetary Council (SFMC): f. 1998; established to supervise, review and resolve matters relating to national financial and monetary policy.

Vinacontrol (The Viet Nam Superintendence and Inspection Co): 54 Tran Nhan Tong, Hanoi; tel. (4) 9433840; fax (4) 9433844; e-mail vinacontrolvn@hn.vnn.vn; internet www.vinacontrol.com.vn; f. 1957; brs in all main Vietnamese ports; controls quality and volume of exports and imports and transit of goods, and conducts inspections of deliveries and production processes; price verification, marine survey, damage survey, claim settling and adjustment; Gen. Dir Le Viet Su.

DEVELOPMENT ORGANIZATION

State Securities Commission: Hanoi; internet www.ssc.gov.vn; f. 1997; responsible for developing the capital markets, incl. the establishment of a stock exchange; 14 mems; Chair. Nguyen Duc Quang.

CHAMBER OF COMMERCE

VCCI (Viet Nam Chamber of Commerce and Industry): 9 Dao Duy Anh, Hanoi; tel. (4) 5742022; fax (4) 5742030; e-mail vcci@hn.vnn.vn; internet www.vcci.com.vn; f. 1963; offices in Ho Chi Minh City, Da Nang, Haiphong, Can Tho, Vung Tau, Nha Trang, Nghe An, Thanh Hoa and Vinh; promotes business and investment between foreign and Vietnamese cos; protects interests of businesses; organizes training activities, exhbns and fairs in Viet Nam and abroad; provides information about and consultancy in Viet Nam's trade and industry; represents foreign applicants for patents and trade mark registration; issues certificates of origin and other documentation; helps domestic and foreign businesses to settle disputes by negotiation or arbitration; in 1993 foreign businesses operating in Viet Nam and Vietnamese businesses operating abroad were permitted to become assoc. mems; Pres. and Chair. Dr Vu Tien Loc; Sec.-Gen. Pham Gia Tuc; associated organizations: Viet Nam International Arbitration Centre, Viet Nam General Average Adjustment Committee, Advisory Board.

Viet Nam International Arbitration Centre: 9 Dao Duy Anh, Hanoi; tel. (4) 5742021; fax (4) 5743001; e-mail viac-vcci@hn.vnn.vn; adjudicates in disputes concerning both domestic and international economic relations.

INDUSTRIAL AND TRADE ORGANIZATIONS

Agrex Saigon (Agricultural Products and Foodstuffs Export Co): 58 Vo Van Tan, District 3, Ho Chi Minh City; tel. (8) 9306606; fax (8) 9303451; e-mail agr-ckc@hcm.vnn.vn; internet www.agrexsaigon.com; f. 1976; exports agricultural produce, coffee, frozen foods and aquatic products; Gen. Dir Duong Ky Hung.

Agrimex (Viet Nam National Agricultural Products Corpn): 173 Hai Ba Trung, District 3, Ho Chi Minh City; tel. (8) 8241049; fax (8) 8291349; e-mail agrimex@hcm.fpt.vn; f. 1956; imports and exports agricultural products; Gen. Dir Nguyen Bach Tuyet.

Airimex (General Civil Aviation Import-Export and Forwarding Co): Le Van Kim, Gia Lam District, Hanoi; tel. (4) 8271939; fax (4) 8271925; e-mail airimex@fpt.vn; f. 1989; imports and exports aircraft, spare parts and accessories for aircraft and air communications; Gen. Dir Pham Doa Hong.

An Giang Afiex Co (An Giang Agriculture and Foods Import and Export Co): 34–36 Hai Ba Trung, Long Xuyen Town, An Giang Province; tel. (76) 841021; fax (76) 843199; e-mail xnknstpagg@hcm.vnn.net; f. 1992; mfr and sale of agricultural products, also beverages; Dir Pham Van Bay.

Artexport–Hanoi (Viet Nam Handicrafts and Art Articles Export-Import Corpn): 31–33 Ngo Quyen, Hanoi; tel. (4) 8252760; fax (4) 8259275; e-mail artexport.hn@fpt.vn; internet www.artexport.com.vn; f. 1964; deals in craft products and art articles; Gen. Dir Do Van Khoi.

Barotex (Viet Nam National Bamboo and Rattan Export-Import Co): 15 Ben Chuong Duong, District 1, Ho Chi Minh City; tel. (8) 8295544; fax (8) 8295352; e-mail barotexsg@saigonnet.vn; f. 1971; specializes in art and handicrafts made from natural materials, sports shoes, ceramic and lacquer wares, gifts and other housewares, fibres, agricultural and forest products; Gen. Dir Ta Quoc Toan.

Bimson Cement Co: Lam Son Hamlet, Bim Con Town, Thanh Hoa Province; tel. (52) 824242; fax (37) 824046; mfr of cement; Dir Le Van Chung.

Binh Tay Import-Export Co (BITEX): 78–82 Hau Giang, District 6, Ho Chi Minh City; tel. (8) 8559669; fax (8) 8557846; e-mail bitex@hcm.vnn.vn; trade in miscellaneous goods; Dir Nguyen Van Thien.

B12 Petroleum Co: Cai Lan, Bay Chay Sub-District, Ha Long City, Quang Ninh Province; tel. (33) 846360; fax (33) 846349; distribution of petroleum products; Dir Vu Ngoc Hai.

Centrimex (Viet Nam National General Import-Export Corpn): 48 Tran Phu, Nha Trang City, Khanh Hoa; tel. (58) 821239; fax (58) 821914; e-mail centrimex@dng.vnn.vn; internet www.centrimexhn

.com.vn; f. 1986; exports and imports goods for five provinces in the south-central region of Viet Nam; Gen. Dir LE VAN NGOC.

Coalimex (Viet Nam National Coal Export-Import and Material Supply Corpn): 47 Quang Trung, Hanoi; tel. (4) 9423166; fax (4) 9422350; e-mail coalimex@fpt.vn; internet www.coalimex.com.vn; f. 1982; exports coal, imports mining machinery and equipment; Gen. Dir NINH XUAN SON.

Cocenex (Central Production Import-Export Corpn): 80 Hang Gai, Hanoi; tel. (4) 8254535; fax (4) 8294306; f. 1988; Gen. Dir BUI THI THU HUONG.

Coffee Supply, Processing and Materials Co: 38B Nguyen Bieu, Nha Trang City, Khanh Hoa Province; tel. (58) 21176; coffee producer.

Cokyvina (Post and Telecommunication Equipment Import-Export Service Corpn): 178 Trieu Viet Vuong, Hanoi; tel. (4) 9782362; fax (4) 9782368; e-mail cokyvina@hn.vnn.vn; f. 1987; imports and exports telecom equipment, provides technical advice on related subjects, undertakes authorized imports, jt ventures, jt co-ordination and co-operation on investment with foreign and domestic economic organizations; Dir NGUYEN KIM KY.

Constrexim (Viet Nam Construction Investment and Export-Import Holdings Corpn): 39 Nguyen Dinh Chieu, Hanoi; tel. (4) 9744836; fax (4) 9742701; e-mail constrexim@fpt.com; internet www.constrexim.com.vn; f. 1982; exports and imports building materials, equipment and machinery; undertakes construction projects in Viet Nam and abroad, and production of building materials with foreign partners; also involved in investment promotion and projects management, real estate development, and human resources development and training; Gen. Dir NGUYEN QUOC HIEP.

Culturimex (State Enterprise for the Export and Import of Works of Art and other Cultural Commodities): 22B Hai Ba Trung, Hanoi; tel. (4) 8252226; fax (4) 8259224; e-mail namson@fpt.vn; f. 1988; exports cultural items and imports materials for the cultural industry; Gen. Dir NGUYEN LAI.

Dau Tieng Rubber Corpn: Dau Tieng Townlet, Dau Tieng District, Binh Duong Province; tel. (650) 561847; fax (650) 561488; e-mail dtrubber@hcm.vnn.vn; f. 1981; planting, processing and export of natural rubber; Man. Dir LE VAN KHOA.

Epco Ltd (Export Import and Tourism Co Ltd): 1 Nyuyen Thuong Hien, District 3, Ho Chi Minh City; tel. (8) 8324392; fax (8) 8324744; f. 1986; processes seafood; tourism and hotel business; Gen. Dir NGUYEN LOC RI.

Foocosa (Food Co Ho Chi Min City): 57 Nguyen Thi Minh Khai, District 1, Ho Chi Minh City; tel. (8) 9309070; fax (8) 9304552; e-mail foocosa@hcm.vnn.vn; internet www.foocosa.com.vn; mfr and distributor of food products (rice, instant noodles, porridge, sauces, biscuits); Dir NGO VAN TAN.

Forexco (Forest Products Export Co): Dien Ngoc Village, Dien Ban District, Quangnam Province; tel. (510) 843595; fax (510) 843619; e-mail forexcoqnam@dng.vnn.vn; internet www.forexcoqnam.com; f. 1986; mfr and exporter of furniture and other wood products; Gen. Dir NGUYEN THANG CANH.

Garmex Saigon (Saigon Garment Manufacturing Import-Export Co): 213 An Duong Vuong, District 5, Ho Chi Minh City; tel. (8) 8557166; fax (8) 8557299; e-mail gmsg@hcm.fpt.vn; internet www.garmexsaigon.com; f. 1993; garment production and export; Gen. Dir LE QUANG HUNG.

Genecofov (General Co of Foods and Services): 64 Ba Huyen Thanh Quan, District 3, Ho Chi Minh City; tel. (8) 9325366; fax (8) 9325428; e-mail gecofov@hcm.fpt.vn; f. 1956; import and export of food products, handicrafts and ceramics, garage services, vehicle trading; under the Ministry of Commerce; Dir TO VAN PHAT.

Generalexim (Viet Nam National General Export-Import Corpn): 46 Ngo Quyen, Hoan Kiem, Hanoi; tel. (4) 8264009; fax (4) 8259894; e-mail gexim@generalexim.com.vn; f. 1981; export and import on behalf of production and trading organizations, also garment processing for export and manufacture of toys; Gen. Dir HOANG TUAN KHAI.

Generalimex (Viet Nam National General Import-Export Corpn): 66 Pho Duc Chinh, Ho Chi Minh City; tel. (8) 8292990; fax (8) 8292968; e-mail generalimex@hcm.fpt.vn; exports of agricultural products and spices, imports of machinery, vehicles, chemicals and fertilizers; Gen. Dir NGUYEN VAN HOANG.

Geruco (Viet Nam General Rubber Corpn): 236 Nam Ky Khoi Nghia, District 3, Ho Chi Minh City; tel. (8) 9325235; fax (8) 9327341; e-mail grc-imexdept@hcm.vnn.vn; internet www.vngeruco.com; merged with Rubexim (rubber export-import corpn) in 1991; manages and controls the Vietnamese rubber industry, including the planting, processing and trading of natural rubber and rubber wood products; also imports chemicals, machinery and spare parts for the industry; Dir-Gen. LE QUANG THUNG.

Haprosimex (Hanoi General Production and Import-Export Company): 22 Hang Luoc, Hanoi; tel. (4) 8266601; fax (4) 8264014; e-mail hapro@fpt.vn; internet www.hapro.com.vn; specializes in handicrafts, textiles, clothing and agricultural and forestry products; Gen. Dir NGUYEN CU TAM.

Hatien Cement Co No 1: Km 8 Highway Hanoi, Thu Duc, Ho Chi Minh City; tel. (8) 8966608; fax (8) 8967635; mfr of cement; Dir NGUYEN NGOC ANH.

Hatien Cement Co No 2: Kien Luong Town, Ha Tien, Kien Giang Province; tel. (77) 53004; fax (77) 53005; mfr of cement; Dir NGUYEN MANH.

Haugiang Petrolimex (Haugiang Petrol and Oil Co): 21 Cach Mang Thang 8, Can Tho City, Can Tho Province; tel. (71) 21657; fax (71) 12746; distributor of fuel; Dir TRINH MANG THANG.

Hoang Thach Cement Co: Minh Tan Hamlet, Kim Mon, Hai Hung Province; tel. (32) 821092; fax (32) 821098; sale of construction materials; Dir NGUYEN DUC HOAN.

Machinoimport (Viet Nam National Machinery Export-Import Corpn): 8 Trang Thi, Hoan Kiem, Hanoi; tel. (4) 8253703; fax (4) 8254050; e-mail machino@hn.vnn.vn; internet www.machinoimport.com.vn; f. 1956; imports and exports machinery, spare parts and tools; consultancy, investment, jt-venture, and manufacturing services; comprises 12 cos; Chair. NGUYEN TRAN DAT; Gen. Dir HO QUANG HIEU.

Marine Supply (Marine Technical Materials Import-Export and Supplies): 276A Da Nang, Ngo Quyen, Haiphong; tel. (31) 847308; fax (31) 845159; f. 1985; imports and exports technical materials for marine transportation industry; Dir PHAN TRANG CHAN.

Mecanimex (Viet Nam National Mechanical Products Export-Import Co): 37 Trang Thi, Hoan Kiem District, Hanoi; tel. (4) 8257459; fax (4) 9349904; e-mail mecahn@fpt.vn; exports and imports mechanical products and hand tools; Gen. Dir TRAN BAO GIOC.

Minexport (Viet Nam National Minerals Export-Import Corpn): 35 Hai Ba Trung, Hanoi; tel. (4) 8253674; fax (4) 8253326; e-mail minexport@fpt.vn; internet www.minexport.com; f. 1956; exports minerals and metals, quarry products, chemical products; imports metals, chemical products, industrial materials, fuels and oils, fertilizers; Gen. Dir VO TRONG CUONG.

Nafobird (Viet Nam Forest and Native Birds, Animals and Ornamental Plants Export-Import Enterprises): 64 Truong Dinh, District 3, Ho Chi Minh City; tel. (8) 8290211; fax (8) 8293735; f. 1987; exports native birds, animals and plants, and imports materials for forestry; Dir VO HA AN.

Naforimex (Hanoi Forest Products Export-Import and Production Corpn): 19 Ba Trieu, Hoan Kiem District, Hanoi; tel. (4) 8261255; fax (4) 8259264; e-mail naforimexhanoi@fpt.vn; f. 1960; imports chemicals, machinery and spare parts for the forestry industry and water supply network; exports oils, forest products, gum benzoin and resin; CEO NGUYEN BA HUNG.

Packexport (Viet Nam National Packaging Technology and Import-Export Co): 31 Hang Thung, Hanoi; tel. (4) 8262792; fax (4) 8269227; e-mail packexport-vn@vnn.vn; f. 1976; manufactures packaging for domestic and export demand, and imports materials for the packaging industry; Gen. Dir TRINH LE KIEU.

Petec-Trading and Investment Corpn: 70 Ba Huyen Thanh Quan, District 3, Ho Chi Minh City; tel. (8) 8299299; fax (8) 8299686; f. 1981; imports equipment and technology for oil drilling, exploration and oil production, exports crude petroleum, rice, coffee and agricultural products; invests in silk, coffee, financial and transport sectors; Gen. Dir TRAN HUU LAC.

Petrol and Oil Co (Zone 1): Duc Gliang Town, Gia Lam, Hanoi; tel. (4) 8271400; fax (4) 8272432; sales of oil and gas; Dir PHAN VAN DU.

Petrolimex (Viet Nam National Petroleum Corpn): 1 Kham Thien, Dong Da District, Hanoi; tel. (4) 8512603; fax (4) 8519203; e-mail xttm@petrolimex.com.vn; internet www.petrolimex.com.vn; f. 1956; import, export and distribution of petroleum products and liquefied petroleum gas; Chair. NGUYEN MANH TIEN.

Petrolimex Saigon Petroleum Co (Zone 2): 15 Le Duan, District 1, Ho Chi Minh City; tel. (8) 8292081; fax (8) 8222082; sales of petroleum products; Dir TRAN VAN THANG.

Petrovietnam (Viet Nam National Petroleum Corpn): 22 Ngo Quyen, Hoan Kiem District, Hanoi; tel. (4) 8252526; fax (4) 8265942; e-mail webmaster@hn.pv.com.vn; internet www.petrovietnam.com.vn; f. 1975; exploration and production of petroleum and gas; Chair. HO SI THOANG; Pres. and CEO NGO THUONG SAN.

Saigon Brewery Co: 187 Nguyen Chi Thanh, District 5, Ho Chi Minh City; tel. (8) 8396342; fax (8) 8296856; producer of beer; Dir HOANG CHI QUY.

Seaco (Sundries Electric Appliances Co): 64 Pho Duc Chinh, District 1, Ho Chi Minh City; tel. (8) 8210961; fax (8) 8210974; deals in miscellaneous electrical goods; Dir MAI MINH CUONG.

Seaprodex Hanoi (Hanoi Sea Products Export-Import Co): 20 Lang Ha, Hanoi; tel. (4) 8344437; fax (4) 8354125; e-mail seaprodexhn@hn.vnn.vn; Dir DANG DINH BAO.

Seaprodex Saigon (Ho Chi Minh City Sea Products Import-Export Corpn): 87 Ham Nghi, District 1, Ho Chi Minh City; tel. (8) 8293669; fax (8) 8224951; e-mail seaprodex.saigon@bdvn.vnd.net; internet www.seaprodexsg.com; f. 1978; exports frozen and processed sea products; imports machinery and materials for fishing and processing; Gen. Dir PHUNG QUOC MAN.

SJC (Saigon Jewellery Co): 115 Nguyen Cong Tru, District 1, Ho Chi Minh City; tel. (8) 9144056; fax (8) 9144057; e-mail sjc@hcm.vnn.vn; internet www.sjcvn.com; f. 1998; manufacturing, processing and trading of gold, gemstones, silver and jewellery; Man. Dir NGUYEN THANH LONG.

Technimex (Viet Nam Technology Export-Import Corpn): 70 Tran Hung Dao, Hanoi; tel. (4) 8256751; fax (4) 8220377; e-mail hanoi.technimex@bdvn.vnmail.vnd.net; f. 1982; exports and imports machines, equipment, instruments, etc.; Dir NGUYEN HUY BINH.

Technoimport (Viet Nam National Complete Equipment and Technics Import-Export Corpn): 16–18 Trang Thi, Hanoi; tel. (4) 8254974; fax (4) 8254059; e-mail technohn@netnam.vn; internet www.technoimport.com.vn; f. 1959; imports and exports equipment, machinery, transport equipment, spare parts, materials and various consumer commodities; exports products by co-investment and jt-venture enterprises; provides consulting services for trade and investment, transport and forwarding services; acts as import-export brokering and trading agents; Gen. Dir VU CHU HIEN.

Terraprodex (Corpn for Processing and Export-Import of Rare Earth and Other Specialities): 35 Dien Bien Phu, Hanoi; tel. (4) 8232010; fax (4) 8256446; f. 1989; processing and export of rare earth products and other minerals; Dir TRAN DUC HIEP.

Tocontap Saigon (Viet Nam National Sundries Import-Export Co): 35 Le Quy Don, District 3, Ho Chi Minh City; tel. (8) 9321062; fax (8) 9325963; e-mail tocontapsaigon@hcm.vnn.vn; internet www.tocontapsaigon.com; f. 1956; imports and exports apparel, agricultural products, art and handicrafts and sundries; Gen. Dir LE THI THANH HUONG.

Vegetexco (Viet Nam National Vegetables and Fruit Export-Import Corpn): 2 Trung Tu, Dong Da District, Hanoi; tel. (4) 5740779; fax (4) 8523926; e-mail vegetexcovn@fpt.vn; internet www.vegetexcovn.tripod.com; f. 1971; exports fresh and processed vegetables and fruit, spices and flowers, and other agricultural products; imports vegetable seeds and processing materials; Gen. Dir LE VAN ANH.

Vietgoldgem (Viet Nam National Gold, Silver and Gemstone Corpn): 23B Quang Trung, Hoan Kiem District, Hanoi; tel. (4) 8260453; fax (4) 8260405; sales of precious metals and gems; Dir NGUYEN DUY HUNG (acting).

Vietrans (Viet Nam National Foreign Trade Forwarding and Warehousing Corpn): 13 Ly Nam De, Hoan Kiem District, Hanoi; tel. (4) 8457801; fax (4) 8455829; e-mail vietrans@hn.vnn.vn; internet www.vietrans.com.vn; f. 1970; agent for forwarding and transport of exports and imports, diplomatic cargoes and other goods, warehousing, shipping and insurance; Gen. Dir THAI DUY LONG.

Vietranscimex (Viet Nam Transportation and Communication Import-Export Co): 22 Nguyen Van Troi, Phu Nhuan District, Ho Chi Minh City; tel. (8) 8442993; fax (8) 8445240; exports and imports specialized equipment and materials for transportation and communication; Gen. Dir PHAM QUANG VINH.

Viettronimex (Viet Nam Electronics Import-Export Corpn): 74–76 Nguyen Hue, District 1, Ho Chi Minh City; tel. (8) 8298201; fax (8) 8294873; e-mail vtr@hcm.vnn.vn; f. 1981; imports and exports electronic goods; Dir NGUYEN HUU THINH.

Vigecam (Viet Nam General Corpn of Agricultural Materials): 16 Ngo Tat To, Dong Da District, Hanoi; tel. (4) 8231972; fax (4) 7474647; exports and imports agricultural products; Gen. Dir TRAN VAN KANH.

Viglacera (Viet Nam Glass and Ceramics Corpn): 628 Hoang Hoa Tham, Ba Dinh District, Hanoi; tel. (4) 8326982; fax (4) 7613292; e-mail vgc@hn.vnn.vn; internet www.viglacera.com.vn; f. 1974; mfr of building materials; Gen. Dir DINH QUANG HUY.

Vimedimex II (Viet Nam National Medical Products Export-Import Corpn): 246 Cong Quynh, District 1, Ho Chi Minh City; tel. (8) 8398441; fax (8) 8325953; e-mail vietpharma@hcm.vnn.vn; internet www.vietpharm.com.vn; f. 1984; exports and imports medicinal and pharmaceutical materials and products, medical instruments; Gen. Dir NGUYEN TIEN HUNG.

Vinacafe (Viet Nam National Coffee Import-Export Corpn): 5 Ong Ich Khiem, Ba Dinh District, Hanoi; tel. (4) 8235449; fax (4) 8456422; e-mail vinacafe@hn.vnn.vn; internet www.vinacafe.com.vn; f. 1995; exports coffee, and imports equipment and chemicals for coffee production; Chair. (vacant).

Vinachem (Viet Nam National Chemical Corpn): 1A Trang Tien, Hoan Kiem District, Hanoi; tel. (4) 8240551; fax (4) 8252995; e-mail vinachem@hn.vnn.vn; internet www.vinachem.com.vn; production, import and export of chemicals and fertilizers.

Vinachimex (Viet Nam National Chemicals Import-Export Corpn): 4 Pham Ngu Lao, Hanoi; tel. (4) 8256377; fax (4) 8257727; f. 1969; exports and imports chemical products, minerals, rubber, fertilizers, machinery and spare parts; Dir NGUYEN VAN SON.

Vinafilm (Viet Nam Film Import, Export and Film Service Corpn): 73 Nguyen Trai, Dong Da District, Hanoi; tel. (4) 8244566; f. 1987; export and import of films and video tapes; film distribution; organization of film shows and participation of Vietnamese films in international film festivals; Gen. Man. NGO MANH LAN.

Vinafimex (Viet Nam National Agricultural Produce and Foodstuffs Import and Export Corpn): 58 Ly Thai To, Hanoi; tel. (4) 8255768; fax (4) 8255476; e-mail fime@hn.vnn.vn; f. 1984; exports cashews, peanuts, coffee, rubber and other agricultural products, and garments; imports malt, fertilizer, insecticide, seeds, machinery and equipment, etc.; Pres. NGUYEN TOAN THANG; Gen. Dir NGUYEN VAN THANG.

Vinafood Hanoi (Hanoi Food Import-Export Co): 6 Ngo Quyen, Hoan Kiem District, Hanoi; tel. (4) 8256771; fax (4) 8258528; f. 1988; exports rice, maize, tapioca; imports fertilizers, insecticides, wheat and wheat flour; Dir NGUYEN DUC HY.

Vinalivesco (Vietnam National Livestock Corporation): 519 Minh Khai, Hanoi; tel. (4) 8626763; fax (4) 8623645; f. 1996; imports and exports animal and poultry products, animal feeds and other agro-products, and foodstuffs; Gen. Dir NGUYEN VAN KHAC.

Vinamilk (Viet Nam Milk Co): 36–38 Ngo Duc Ke, District 1, Ho Chi Minh City; tel. (8) 8298054; fax (8) 8238117; internet www.hcm.fpt.vn/adv/vinamilk/vn; producer of dairy products; Dir MAI TRIUE LIEN.

Vinapimex (Viet Nam Paper Corpn): 25A Ly Thuong Kiet, Hanoi; tel. (4) 8260143; fax (4) 8260381; f. 1995; production and marketing of paper; Gen. Dir NGUYEN NGOC MINH.

Vinaplast (Viet Nam Plastics Corpn): 92–94 Ly Tu Trong, District 1, Ho Chi Minh City; tel. (8) 8298232; fax (8) 8291841; e-mail vnplast@hcm.vnn.vn; f. 1976; import and export of products for plastic processing industry; production and trade of plastic products; Gen. Dir NGUYEN MINH TAM.

Vinasteel (Viet Nam National Steel Corpn): D2 Ton That Tung, Dong Da District, Hanoi; tel. (4) 8525537; fax (4) 8262657; distributor of metal products; Dir NGO HUY PHAN.

Vinataba (Viet Nam National Tobacco Corpn): 25A Ly Thuong Kiet, Hoan Kiem District, Hanoi; tel. (4) 8265778; fax (4) 8265777; internet www.vinataba.com.vn; mfr of tobacco products; Dir NGUYEN NAM HAI.

Vinatea (Viet Nam National Tea Development Investment and Export-Import Co): 46 Tang Bat Ho, Hanoi; tel. (4) 8212005; fax (4) 8212663; e-mail thitruong@vinatea.com.vn; exports tea, imports tea-processing materials; Gen. Dir (vacant).

Vinatex (Viet Nam National Textile and Garment Corpn): 25 Ba Trieu, Hoan Kiem District, Hanoi; tel. (4) 8257700; fax (4) 8262269; e-mail vinatexhn@vinatex.com.vn; internet www.vinatex.com; f. 1995; imports raw material, textile and sewing machinery, spare parts, accessories, dyestuffs; exports textiles, ready-made garments, carpets, jute, silk; Gen. Dir VU DUC THINH.

VNCC (Viet Nam National Cement Corpn—Vinacement): 228 Le Duan, Dong Da District, Hanoi; tel. (4) 8510593; fax (4) 8512778; f. 1980; manufactures and exports cement and clinker; Gen. Dir NGUYEN VAN HANH.

Vocarimex (National Co for Vegetable Oils, Aromas, and Cosmetics of Viet Nam): 58 Nguyen Binh Khiem, District 1, Ho Chi Minh City; tel. (8) 8294513; fax (8) 8290586; e-mail vocar@hcm.vnn.vn; f. 1976; producing and trading vegetable oils, oil-based products and special industry machinery; packaging; operating port facilities; Gen. Dir DUONG THI NGOC TRINH.

Xunhasaba (Viet Nam State Corpn for Export and Import of Books, Periodicals and other Cultural Commodities): 32 Hai Ba Trung, Hanoi; tel. (4) 8262989; fax (4) 8252860; e-mail xunhasaba@hn.vnn.vn; internet www.xunhasaba.com.vn; f. 1957; exports and imports books, periodicals, postage stamps, greetings cards, calendars and paintings; Dir HA TRIEU KIEN.

UTILITIES

Electricity

Power Co No 1 (PC1): 20 Tran Nguyen Han, Hoan Kiem, Hanoi; tel. (4) 8255074; fax (4) 8244033; e-mail anhdn@pc1.com.vn; manages the generation, transmission and distribution of electrical power in northern Viet Nam; Dir DO VAN LOC.

Power Co No 2 (PC2): 72 Hai Ba Trung, District 1, Ho Chi Minh City; tel. (8) 8297151; fax (8) 8299680; manages the distribution of electrical power in southern Viet Nam; Dir NGUYEN VAN GIAU.

Power Co No 3 (PC3): 315 Trung Nu Vuong, Hai Chau District, Da Nang; tel. (511) 621028; fax (511) 625071; f. 1975; manages the generation, transmission and distribution of electrical power in central Viet Nam; Gen. Dir TA CANH.

Viet Nam Electricity Corpn (EVN): produces electrical power; Gen. Dir DAO VAN HUNG.

Water

Hanoi Water Business Co: 44 Yen Phu, Hanoi; tel. (4) 8292478; fax (4) 8294069; f. 1954; responsible for the supply of water to Hanoi and its five urban and two suburban districts; Dir Gen. BUI VAN MAT.

Ho Chi Minh City Water Supply Co: 1 Cong Truong Quoc Te, District 3, Ho Chi Minh City; tel. (8) 8291974; fax (8) 8241644; e-mail hcmcwater@hcm.vnn.vn; f. 1966; manages the water supply system of Ho Chi Minh City; Dir VO DUNG.

CO-OPERATIVES

Viet Nam Co-operatives Alliance: 77 Nguyen Thai Hoc, Ba Dinh, Hanoi; tel. (4) 7330779; fax (4) 8431883; e-mail vca@hn.vnn.vn; internet www.vietnamcoop.org; f. 1993; fmrly Viet Nam Co-operatives Council; Pres. Dr NGUYEN TIEN QUAN.

TRADE UNIONS

Tong Lien doan Lao dong Viet Nam (Viet Nam General Confederation of Labour): 82 Tran Hung Dao, POB 627, Hanoi; tel. and fax (4) 8253781; e-mail doingoaitld@hnn.vnn.vn; f. 1929; merged in 1976 with the South Viet Nam Trade Union Fed. for Liberation; 4,000,000 mems; Pres. CU THI HAU; Vice-Pres NGUYEN AN LUONG, DANG NGOC CHIEN, DO DUC NGO, NGUYEN DINH THANG.

Cong Doan Nong Nghiep Cong Nghiep Thu Pham Viet Nam (Viet Nam Agriculture and Food Industry Trade Union): Hanoi; f. 1987; 550,000 mems.

National Union of Building Workers: 12 Cua Dong, Hoan Kiem, Hanoi; tel. (4) 8253781; fax (4) 8281407; f. 1957; Pres. NGUYEN VIET HAI.

Vietnam National Union of Industrial Workers: 54 Hai Ba Trung, Hanoi; tel. (4) 9344426; fax (4) 8245306; f. 1997; Pres. VU TIEN SAU.

Vietnam National Union of Post and Telecoms Workers: 30 Hang Chuoi, Hai Ba Trung, Hanoi; tel. (4) 9713514; fax (4) 9720236; f. 1947; Chair. HOANG DUY CAN.

Transport

RAILWAYS

Duong Sat Viet Nam (DSVN) (Viet Nam Railways): 118 Le Duan, Hanoi; tel. (4) 8220537; fax (4) 9424998; e-mail vr.hn.irstd@fpt.vn; internet www.vr.com.vn; 2,600 km of main lines (1996); lines in operation are: Hanoi–Ho Chi Minh City (1,726 km), Hanoi–Haiphong (102 km), Hanoi–Dong Dang (167 km), Hanoi–Lao Cai (296 km), Hanoi–Thai Nguyen (75 km), Thai Nguyen–Kep–Bai Chay (106 km); Gen. Dir Dr NGUYEN HUU BANG.

ROADS

In 1999 there were an estimated 93,300 km of roads, of which 25.1% were paved. In 1995 the Government announced plans to upgrade 430 km of the national highway which runs from Hanoi in the north to Ho Chi Minh City in the south. In 1997 the Government announced plans to build a new 1,880-km north–south highway in the east of the country. In 2001 the Government also announced plans to upgrade the section of the national highway that runs between the northern provinces of Hoa Binh and Son La.

SHIPPING

The principal port facilities are at Haiphong, Da Nang and Ho Chi Minh City. In 2002 the Vietnamese merchant fleet (721 vessels) had a combined displacement totalling 1,130,500 grt. In 1994 there were 150 shipping companies with a combined capacity of 800,000 dwt.

Cong Ty Van Tai Duong Bien Viet Nam (VOSCO) (Viet Nam Ocean Shipping Co): 215 Tran Quoc Toan, Ngo Quyen District, Haiphong; tel. (31) 846951; fax (31) 845107; controlled by the Viet Nam Gen. Dept of Marine Transport; Dir TRAN VAN LAM; Dir-Gen. TRAN XUAN NHON.

Dai Ly Hang Hai Viet Nam (VOSA) (VOSA group of companies): Unit 1003, 10th Floor, Harbour View Tower, 35 Nguyen Hue, District 1, Ho Chi Minh City; tel. (8) 9140422; fax (8) 8214919; e-mail vosagroup@hcm.vnn.vn; internet www.vosagroup.com; f. 1957; fmrly the Viet Nam Ocean Shipping Agency; controlled by the Viet Nam National Shipping Lines (VINALINES); in charge of merchant shipping; arranges ship repairs, salvage, passenger services, air and sea freight forwarding services; main br. offices in Haiphong and Ho Chi Minh City; brs in Hanoi, Da Nang, Ben Thuy, Qui Nhon, Quang Ninh, Nha Trang, Vung Tau and Can Tho; Dir-Gen. HA CAM KHAI (acting).

Transport and Chartering Corpn (Vietfracht): 74 Nguyen Du, Hanoi; tel. (4) 9422355; fax (4) 9423679; e-mail vfhan@hn.vnn.vn; internet www.vietfracht.com.vn; f. 1963; ship broking, chartering, ship management, shipping agency, international freight forwarding; consultancy services, import-export services; Gen. Dir DO XUAN QUYNH.

Viet Nam Sea Transport and Chartering Co (Vitranschart): 428–432 Nguyen Tat Thanh, District 4, Ho Chi Minh City; tel. (8) 9404027; fax (8) 9404711; e-mail vitrans@fmail.vnn.vn; Dir VO PHUNG LONG.

CIVIL AVIATION

Viet Nam's principal airports are Tan Son Nhat International Airport (Ho Chi Minh City) and Thu Do (Capital) International Airport at Noi Bai (Hanoi). They cater for both overseas and domestic traffic. Airports at Da Nang, Hué, Nha Trang, Da Lat, Can Tho, Haiphong, Muong Thanh (near Dien Bien Phu), Buon Ma Thuot and Lien Khuong handle domestic traffic. In 1997 it was announced that numerous abandoned wartime airstrips were to be repaired and returned to service under a programme scheduled for completion in 2010, by which time the Government also aims to expand the annual capacity of Tan Son Nhat to 30m. passengers and 1m. tons of freight. In January 2004 plans were announced for the construction of a new international airport at Long Thang, in Dong Nai province. The airport was to have four runways and four terminals, the first of which were scheduled to become operational in 2011. Five further airports to handle domestic traffic were scheduled to open in 2004, at Ca Mau, Can Tho, Con Son, Chu Lai and Cam Ranh, bringing the total number of civil airports in Viet Nam to 21 by the end of that year.

Viet Nam Airlines: Gialem Airport, Hanoi; tel. (4) 8732732; fax (4) 8272291; internet www.vietnamair.com.vn; fmrly the Gen. Civil Aviation Admin. of Viet Nam, then Hang Khong Viet Nam; wholly state-owned; operates domestic passenger services from Hanoi and from Ho Chi Minh City to the principal Vietnamese cities, and international services to 18 countries; Pres. and CEO NGUYEN XUAN HIEN; Chair. NGUYEN SY HUNG.

Pacific Airlines: 112 Hong Ha, Tan Binh District, Ho Chi Minh City; tel. (8) 8450090; fax (8) 8450085; internet www.pacificairlines.com.vn; f. 1991; operates charter cargo flights, also scheduled passenger and cargo services to Taiwan and Hong Kong; Man. Dir DUONG CAO THAI NGUYEN; Chair. LUONG THE PHUC.

Tourism

In the 1990s the Vietnamese Government encouraged tourism as a source of much-needed foreign exchange, with the objective of attracting some 4m. tourists annually by the year 2005. About 2.4m. foreign tourists visited Viet Nam in 2003. In 1998 revenue from tourism totalled US $86m.

Viet Nam National Administration of Tourism (VNAT): 80 Quan Su, Hanoi; tel. (4) 9421061; fax (4) 9424115; e-mail icdvnat@vietnamtourism.com; internet www.vietnamtourism.com; f. 1960; Gen. Dir NGO HUY PHUONG.

Hanoi Tourism Service Company (HANOI TOSERCO): 8 To Hien Thanh, Hanoi; tel. (4) 9780004; fax (4) 8226055; e-mail hanoitoserco@hn.vnn.vn; internet www.tosercohanoi.com; f. 1998; manages the development of tourism, hotels and restaurants in the capital and other services including staff training; Dir TRAN TIEN HUNG.

Saigon Tourist: 23 Le Loi, District 1, Ho Chi Minh City; tel. (8) 8225887; fax (8) 8291026; e-mail saigontourist@sgtourist.com.vn; internet www.saigon-tourist.com; f. 1975; tour operator; Gen. Dir DO VAN HOANG.

Unitour (Union of Haiphong Tourist Co): 40 Tran Quang Khai, Haiphong; tel. (31) 8247295.

Viet Nam Tourism in Haiphong: 60A Dien Bien Phu, Haiphong; tel. and fax (31) 823651; f. 1960; hotel services, travel arrangements, visa service; Dir NGUYEN HUU BONG.

Defence

In August 2003 the total strength of the armed forces was an estimated 484,000: army 412,000; navy 42,000; air force 30,000. Men are subject to a two-year minimum term of compulsory military service between 18 and 35 years of age. Paramilitary forces number 4m.–5m. and include the urban People's Self Defence Force and the rural People's Militia. Border defence troops number an estimated 40,000. The defence budget for 2003 was estimated at US $2,300m.

Commander-in-Chief of the Armed Forces: TRAN DUC LUONG.

Chief of General Staff (Army): PHUNG QUANG THANH.

Commander of the Navy: Adm. DO XUAN CONG.

Education

Primary education, which is compulsory, begins at six years of age and lasts for five years. Secondary education, beginning at the age of 11, lasts for up to seven years, comprising a first cycle of four years and a second cycle of three years. In 1999/2000 some 96.3% of children in the relevant age-group were enrolled in primary schools, while total secondary enrolment included 61.4% of children in the appropriate age-group (boys 64.4%; girls 58.3%). In 2001/02 an estimated total of 9,315,300 pupils attended primary schools, of which there were 22,199 in 1999/2000. In 2001/02 there were 8,560,300 students in secondary education; a further 177,600 students were enrolled in technical secondary education in 1998/99. In 1999/2000 total enrolment at tertiary level was equivalent to 9.7% of students in the relevant age-group (males 11.2%; females 8.1%). In 1998/99 there were 123 universities and colleges of higher education. In 2001/02 these had a total enrolment of 873,000 students and 31,400 teachers. In 1989 Viet Nam's first private college since 1954, Thang Long College, was opened in Hanoi to cater for university students. Of total planned budgetary expenditure by the central Government in 2002, 13,581,000m. dông (11.3% of total expenditure) was allocated to education.

Bibliography

See also Cambodia and Laos

Alpert, William T. (Ed.). *The Vietnamese Economy and its Transformation to an Open Market System*. New York, NY, M. E. Sharpe, 2004.

Amer, Ramses. *The Ethnic Chinese in Vietnam and Sino-Vietnamese Relations*. Petaling Jaya, Forum, 1992.

Anderson, Kym. *Vietnam's Transforming Economy and WTO Accession*. Singapore, Institute of Southeast Asian Studies, 1999.

Ang Cheng Guan. *The Vietnam War from the Other Side*. London, RoutledgeCurzon, 2002.

Ending the Vietnam War: The Vietnamese Communists' Perspective. London, RoutledgeCurzon, 2003.

Bao Ninh. *The Sorrow of War*. London, Secker and Warburg, 1994.

Beresford, Melanie. *Vietnam: Politics, Economics and Society*. London, Pinter, 1988.

National Unification and Economic Development in Vietnam. Basingstoke, Macmillan, 1989.

Beresford, Melanie, and Dang Phong. *Economic Transition in Vietnam—Trade and Aid in the Demise of a Centrally Planned Economy*. Cheltenham, Edward Elgar, 2000.

Beresford, Melanie, and Angie Ngoc Tran (Eds). *Reaching for the Dream: Challenges of Sustainable Development in Vietnam*. Copenhagen, NIAS Press, 2003.

Bertram, Trent, and McCarty, Adam. *Enterprise Reform Project: Labour Market and Employment Issues*. Hanoi, Ministry of Planning and Investment/Asian Development Bank, 1997.

Bilton, Michael, and Sim, Kevin. *Four Hours in My Lai*. London, Viking, 1992.

Binh Tran Nam, and Chi Do Pham (Eds). *The Vietnamese Economy*. London, RoutledgeCurzon, 2002.

Boothroyd, Peter, and Pham Xuan Nam (Eds). *Socio-Economic Renovation in Vietnam—The Origin, Evolution and Impact of Doi Moi*. Singapore, Institute of Southeast Asian Studies, 2000.

Braestrup, P. (Ed.). *Vietnam as History: Ten Years after the Paris Peace Accords*. Washington, DC, University Press of America, for the Wilson Center, 1984.

Brazier, Chris. *Vietnam: The Price of Peace*. Oxford, Oxfam, 1992.

Brownmiller, Susan. *Seeing Vietnam: Encounters of the Road and Heart*. New York, HarperCollins, 1994.

Bui Tin. *Following Ho Chi Minh: Memoirs of a North Vietnamese Colonel*. New South Wales, Crawford House, 1996.

Butler, D. *The Fall of Saigon*. New York City, Simon and Schuster, 1986.

Buttinger, J. *Vietnam, a Political History*. London, André Deutsch, 1969.

Vietnam: a Dragon Embattled. London, Pall Mall Press, 1967.

Cable, James. *The Geneva Conference of 1954 on Indochina*. London, Macmillan, 1986.

Chanda, Nayan. *Brother Enemy—The War After the War: A History of Indochina since the Fall of Saigon*. San Diego, New York and London, Harcourt Brace Jovanovich, 1986.

Chen, K. *Vietnam and China, 1938–54*. Princeton, NJ, Princeton University Press, 1969.

Chesneaux, J., Boudarel, G., Hemery, D., et al. *Tradition et Révolution au Vietnam*. Paris, Editions Anthropos, 1971.

Dahm, Henrich. *French and Japanese Economic Relations with Vietnam since 1975*. Richmond, Curzon Press, 1999.

Dalloz, J. *La Guerre d'Indochine 1945–1954*. Paris, Le Seuil, 1987.

Decaro, Peter A. *Rhetoric of Revolt: Ho Chi Minh's Discourse for Revolution*. Westport, CT, Praeger Publishers, 2002.

Devillers, P. *Histoire du Viet Nam de 1940 à 1952*. Paris, Le Seuil, 1952.

Devillers, P., and Lacouture, J. *End of a War, Indochina 1954*. London, Pall Mall Press, 1969.

Dien, K. *Population and Ethno-Demography in Vietnam*. Washington, DC, University of Washington Press, 2003.

Dollar, David, Glewwe, Paul, and Agrawal, Nisha (Eds). *Economic Growth, Poverty, and Household Welfare in Vietnam*. Washington, DC, World Bank, 2004.

Duiker, William J. *The Rise of Nationalism in Vietnam, 1900–1941*. Ithaca, NY, Cornell University Press, 1976.

Historical Dictionary of Vietnam. Metuchen, NJ, Scarecrow Press, 1989.

Victory in Vietnam: the Official History of the People's Army of Vietnam. Lawrence, KS, University Press of Kansas, 2002.

Eisen, Arlene. *Women and Revolution in Vietnam*. London, Zed Books, 1985.

Engelman, Larry. *Tears before the Rain: An Oral History of the Fall of South Vietnam*. New York, Oxford University Press, 1991.

Evans, Grant, and Rowley, Kelvin. *Red Brotherhood at War: Indochina Since the Fall of Saigon*. London, Verso Editions, 1985 (revised edn 1990).

Fall, B. *The Two Viet Nams, a Political and Military Analysis*. 2nd Edn, London, Pall Mall Press, 1967.

Street Without Joy: Insurgency in Indochina, 1946–63. London, Pall Mall Press, 1963.

Hell in a very Small Place: the Siege of Dien Bien Phu. London, Pall Mall Press, 1967.

Fforde, Adam. *The Limits of National Liberation: Economic Management and the Re-Unification of the Democratic Republic of Vietnam*. London, Croom Helm, 1984.

The Agrarian Question in North Viet Nam 1974–1979: A Study of Cooperator Resistance to State Policy. New York, NY, M. E. Sharpe, 1990.

Doi Moi: Ten Years after the 1986 Party Congress. Political and Social Change Monograph 24. Canberra, Australian National University, 1997.

Fforde, Adam, and de Vylder, Stefan. *From Plan to Market: The Economic Transition in Viet Nam.* Boulder, CO, Westview Press, 1996.

Forbes, Dean K., Hull, Terrence H., Marr, David G. and Brogan, Brian (Eds). *Doi Moi: Vietnam's Renovation—Policy and Performance. Political and Social Change Monograph 14.* Canberra, Australian National University, 1991.

Franklin, H. Bruce. *M.I.A. or Mythmaking in America.* New York, Lawrence Hill Books, 1992.

Gainsborough, Martin. *Changing Political Economy of Vietnam: The Case of Ho Chi Minh City.* London, RoutledgeCurzon, 2002.

Gelb, L. H., and Betts, R. K. *The Irony of Vietnam.* Washington, DC, Brookings Institution, 1979.

Gibbons, William C. *The US Government and the Vietnam War* (2 vols). Princeton University Press, 1986.

Gough, K. *Political Economy in Vietnam.* Berkeley, CA, Folklore Institute, 1990.

Grant, Zalin. *Over the Beach.* New York, W. W. Norton, 1987.

Griffiths, Philip Jones. *Agent Orange: Collateral Damage in Vietnam.* London, Trolley, 2003.

Hammer, E. J. *The Struggle for Indochina.* 2nd Edn, Stanford, CA, Stanford University Press, 1966.

Hannah, Norman B. *The Key to Failure: Laos and the Vietnam War.* New York, Madison Books, 1989.

Hardy, Andrew. *Red Hills: Migrants and the State in the Highlands of Vietnam.* Honolulu, HI, University of Hawaii Press, 2002.

Haughton, Dominique, Haughton, Jonathan, Bales, Sarah, Truong Thi Kim Chuyen, and Nguyen Nguyet Nga (Eds). *Health and Wealth in Vietnam (An Analysis of Household Living Standards).* Singapore, Institute of Southeast Asian Studies, 1999.

Herring, G. C. *America's Longest War: The United States and Vietnam 1950–1975.* New York, Wiley, 1979.

Hiebert, Murray. *Vietnam Notebook.* Hong Kong, Review Publishing Co Ltd, 1993 (revised edn 1998).

Chasing the Tigers—A Portrait of the New Vietnam. London, Kodansha Europe, 1997.

Hodgkin, T. *Vietnam, the Revolutionary Path.* London, Macmillan, 1981.

Hue-Tam Ho Tai. *Millenarianism and Peasant Politics in Vietnam.* Cambridge, MA, Harvard University Press, 1983.

Hurst, Steven. *The Carter Administration and Vietnam.* London, Macmillan, 1996.

Huynh Kim Khanh. *Vietnamese Communism 1925–1945.* Ithaca, NY, Cornell University Press, 1982.

Hy Van Luong (Ed.). *Postwar Vietnam: Dynamics of a Transforming Society.* Singapore, Institute of Southeast Asian Studies, 2003.

International Monetary Fund. *Vietnam Economic Development.* Hanoi, International Monetary Fund, 1998.

Isaacs, A. R. *Without Honour: Defeat in Vietnam and Cambodia.* Baltimore, Johns Hopkins University Press, 1983.

Kahin, G. McT. *Intervention. How America became involved in Vietnam.* New York, Alfred Knopf, 1986.

Karnow, Stanley. *Vietnam, A History.* Harmondsworth, Penguin Books, 1984.

Kerkvliet, Benedict, Heng, Russell, and David Koh Wee Hock (Eds). *Getting Organized in Vietnam: Moving in and Around the Socialist State.* Singapore, Institute of Southeast Asian Studies, 2003.

Kerkvliet, Benedict, and Porter, Doug (Eds). *Vietnam's Rural Transformation.* Boulder, CO, Westview Press, 1995.

Kerkvliet, Benedict, and Marr, David G. *Beyond Hanoi: Local Government in Vietnam.* Singapore, Institute of Southeast Asian Studies, 2004.

Kimura, Tetsusaburo. *The Vietnamese Economy 1975–86.* Tokyo, Institute of Developing Economies, 1989.

Kleinen, John. *Facing The Future, Reviving The Past. A Study of Social Change in a Northern Vietnamese Village.* Singapore, Institute of Southeast Asian Studies, 1999.

Kolko, Gabriel. *Vietnam: Anatomy of a Peace.* London, Routledge, 1997.

Lacouture, J. *Ho Chi Minh.* London, Penguin Books, 1968.

Lancaster, D. *The Emancipation of French Indochina.* London, Oxford University Press, 1961.

Le Manh Hung. *The Impact of World War II on the Economy of Vietnam, 1939–45.* Singapore, Times Academic Press. 2004.

Le Thanh Khoi. *Le Viet Nam, Histoire et Civilisation.* Paris, Editions de Minuit, 1955.

Leung, Suiwah (Ed.). *Vietnam Assessment: Creating a Sound Investment Climate.* Singapore, Institute of Southeast Asian Studies, 1996.

Lewy, G. *America in Vietnam.* New York, Oxford University Press, 1978.

Ljunggren, Börje (Ed.). *The Challenge of Reform in Indochina.* Cambridge, MA, Harvard Institute for International Development, 1993.

Lockhart, Greg. *Nation in Arms: The Origins of the People's Army of Vietnam.* Sydney, Allen and Unwin, 1989.

Logevall, Frederik. *Choosing War: the Lost Chance for Peace and the Escalation of War in Vietnam.* Berkeley, CA, University of California Press, 2001.

Luibrand, Annette. *Transition in Vietnam: Impact of the Rural Reform Process on an Ethnic Minority.* Frankfurt, Peter Lang, 2002.

Luong, Hy Van. *Revolution in the Village: Tradition and Transformation in North Vietnam, 1925–1988.* Honolulu, University of Hawaii Press, 1992.

(Ed.). *Postwar Vietnam: Dynamics of a Transforming Society.* Singapore, Institute of Southeast Asian Studies, 2003.

McAleavy, H. *Black Flags in Vietnam.* London, Allen and Unwin, 1968.

McAlister, J. T. *Viet Nam, the Origins of Revolution.* London, Allen Lane, 1970.

McCargo, Duncan. *Rethinking Vietnam.* London, RoutledgeCurzon, 2004.

McCoy, A. W. *The Politics of Heroin in South-East Asia.* New York, Harper and Row, 1972.

McHale, Shawn Frederick. *Print and Power: Buddhism, Confucianism and Communism in the Making of Modern Vietnam.* Honolulu, HI, University of Hawaii Press, 2003.

Macdonald, Peter. *Giap: The Victor in Vietnam.* London, 4th Estate, 1994.

McKelvey, Robert S. *A Gift of Barbed Wire: America's Allies Abandoned in South Vietnam.* Seattle, WA, University of Washington Press, 2002.

McNamara, Robert S. *In Retrospect: The Tragedy and Lessons of Vietnam.* New York, NY, Times Books, 1995.

Mai, Pham Hoang. *Foreign Direct Investment and Development in Vietnam: Policy Implications.* Singapore, Institute of Southeast Asian Studies, 2003.

Malarney, Shaun. *Culture, Ritual and Revolution in Vietnam.* London, RoutledgeCurzon, 2002.

Mangold, Tom, and Penycate, John. *The Tunnels of Cu Chi.* Sevenoaks, Kent, Hodder and Stoughton, 1985.

Maraniss, David. *They Marched into Sunlight: War and Peace, Vietnam and America October 1967.* Pymble, NSW, Simon and Schuster Australia, 2003.

Marr, D. G. *Vietnamese Anticolonialism, 1885–1925.* Berkeley, CA, University of California, 1971.

Vietnamese Tradition on Trial, 1920–45. Berkeley, CA, University of California, 1981.

Vietnam 1945: The Quest for Power. Berkeley, CA, University of California, 1996.

Marr, David G., and White, Christine (Eds). *Postwar Vietnam: Dilemmas in Socialist Development.* Southeast Asia Program, Ithaca, Cornell University, 1988.

Mio, Tadashi (Ed.). *Indochina in Transition: Confrontation or Co-prosperity.* Japan Institute of International Affairs, Tokyo, 1989.

Moise, Edwin. *Land Reform in China and North Vietnam: Consolidating the Revolution at the Village Level.* Chapel Hill, NC, University of North Carolina Press, 1983.

Tonkin Gulf and the Escalation of the Vietnam War. Chapel Hill, NC, University of North Carolina Press, 1996.

Morley, James W. and Masashi, Nishihara (Eds). *Vietnam Joins the World.* New York, NY, M. E. Sharpe, 1997.

Morris, Stephen J. *Why Vietnam Invaded Cambodia (Political Culture and the Causes of War).* Cambridge, Cambridge University Press, 1998.

Neale, Jonathan. *The American War.* London, Bookmarks Publications, 2001.

Ng Chee Yuen, Freeman, Nick J., and Hiep Huynh, Frank. *State-Owned Enterprise Reform in Vietnam: Lessons from Asia.* Singapore, Institute of Southeast Asian Studies, 1996.

Nghia M. Vo. *The Bamboo Gulag: Political Imprisonment in Communist Vietnam.* Jefferson, NC, McFarland and Co, 2003.

Ngo Vinh Long. *Before the Revolution: the Vietnamese Peasants under the French.* Cambridge, MA, MIT Press, 1973.

Nguyen Long, with Kendall, Harry H. *After Saigon Fell: Daily Life under the Vietnamese Communists.* Los Angeles, University of California Press, 1981.

Nguyen Long Thanh Nam. *Hoa Hao Buddhism in the Course of Vietnam's History.* Hauppauge, NY, Nova Science Publishers, 2004.

Nguyen, Nicholas. *Vietnam: The Second Revolution.* Brighton, In Print Publishing, 1996.

Nguyen Tien Hung. *Economic Development of Socialist Vietnam, 1955–1980.* New York, Praeger Publishers, 1977.

Nguyen Tien Hung and Schecter, Jerrold L. *The Palace File.* New York, Harper & Row, 1987.

Nguyen Van Canh. *Vietnam under Communism, 1975–1982.* Stanford, CA, Hoover Institution Press, 1983.

Nixon, Richard. *No More Vietnams.* New York, NY, Arbor House, 1985.

O'Rourke, Dara. *Community-Driven Regulation: Balancing Development and the Environment in Vietnam.* Cambridge, MA, MIT Press, 2004.

Osborne, M. E. *The French Presence in Cochin-China and Cambodia: Rule and Response (1859–1905).* Ithaca, NY, Cornell University Press, 1969.

Palmer, General Bruce. *The 25-Year War.* New York, Harper and Row Publishers, 1985.

Palmujoki, Eero. *Vietnam and the World: Marxist-Leninist Doctrine and the Changes in International Relations, 1975–93.* London, Macmillan, 1997.

Patti, A. L. A. *Why Vietnam? Prelude to America's Albatross.* University of California, 1981.

Pelley, Patricia M. *Postcolonial Vietnam.* Durham, NC, Duke University Press, 2002.

The Pentagon Papers: the Defense Department History of US Decision-making on Vietnam.'Senator Gravel Edition', 4 vols. Boston, MA, Beacon Press, 1971.

Pham Hoang Mai. *Foreign Direct Investment and Development in Vietnam: Policy Implications.* Singapore, Institute of Southeast Asian Studies, 2003.

Pike, D. *Viet Cong, the Organization and Techniques of the National Liberation Front of South Vietnam.* Cambridge, MA, MIT Press, 1966.

PAVN: People's Army of Viet Nam. Oxford, Brassey's Defence Publishers, 1986.

Vietnam and the Soviet Union. Boulder, CO, and London, Westview Press, 1987.

Popkin, Samuel L. *The Rational Peasant: The Political Economy of Rural Society in Viet Nam.* Los Angeles, CA, University of California Press, 1979.

Porter, G. *Vietnam: The Politics of Bureaucratic Socialism.* Ithaca, NY, Cornell University Press, 1993.

Prados, John. *The Blood Road—The Ho Chi Minh Trail and the Vietnam War.* New York, NY, John Wiley & Sons, 1999.

Quinn-Judge, Sophie. *Ho Chi Minh: the Missing Years.* Berkeley, CA, University of California Press, 2002.

Ronnas, Per, and Ramamurthy, Bhargavi (Eds). *Entrepreneurship in Vietnam: Transformation and Dynamics.* Singapore, Institute of Southeast Asian Studies, 2001.

Ronnas, Per, and Sjöberg, O. *Socio-Economic Development in Vietnam; The Agenda for the 1990s.* Stockholm, Swedish International Development Authority, 1991.

Ronnas, Per and Sjöberg, O. (Eds). *Doi Moi—Economic Reforms and Development Policies in Vietnam.* Stockholm, Swedish International Development Authority, 1990.

Salemink, Oscar. *The Ethnography of Vietnam's Central Highlanders.* London, RoutledgeCurzon, 2002.

Salisbury, Harrison. *Vietnam Reconsidered.* New York, Harper and Row Publishers, 1985.

Sansom, R. L. *The Economics of Insurgency in the Mekong Delta of Vietnam.* Cambridge, MA, MIT Press, 1970.

Schoenl, William (Ed.). *New Perspectives on the Vietnam War: Our Allies' Views.* Lanham, MD, University Press of America, 2002.

Shaplen, R. *The Lost Revolution in Vietnam, 1945–65.* London, 1965.

Smith, R. *Viet Nam and the West.* London, Heinemann Educational, 1968.

Smith, R. B. *An International History of the Vietnam War: Vol. 1—Revolution versus Containment, 1955–1961; Vol. 2—The Struggle for South-east Asia, 1961–65; Vol. 3—Making of a United War, 1965–66.* London, Macmillan, 1983, 1985, 1990.

Snepp, F. *Decent Interval, an Insider's Account of Saigon's Indecent End.* New York, Random House, 1978.

Sorley, Lewis. *A Better War: the Unexamined Victories and Final Tragedy of America's Last Years in Vietnam.* New York, NY, Harcourt Brace & Co., 1999.

Spector, Ronald H. *After Tet: The Bloodiest Year in Vietnam.* New York, Vintage Books, 1994.

Stern, Lewis M. *Renovating the Vietnamese Communist Party* Singapore, Institute of Southeast Asian Studies, 1993.

Tana, Li. *Peasants on the Move: Rural–Urban Migration in the Hanoi Region.* Singapore, Institute of Southeast Asian Studies, 1996.

Taylor, Philip. *Goddess on the Rise: Pilgrimage and Popular Religion in Vietnam.* Honolulu, HI, University of Hawaii Press, 2004.

Templer, Robert. *Shadows and Wind: A View of Modern Vietnam.* London, Little Brown, 1998.

Thakur, R., and Thayer, C. A. *Soviet Relations with India and Vietnam.* London, Macmillan, 1992.

Than, Mya, and Tan, Joseph L. H. (Eds). *Vietnam's Dilemmas and Options: The Challenge of Economic Transition in the 1990s.* Singapore, Institute of Southeast Asian Studies, 1993.

Thayer, Carlyle A. *War by Other Means: National Liberation and Revolution in Vietnam.* Sydney, Allen and Unwin, 1989.

The Vietnam People's Army Under Doi Moi. Pacific Strategic Paper No. 7, Singapore, Institute of Southeast Asian Studies, 1994.

Thayer, Carlyle A., and Amer, Ramses (Eds). *Vietnamese Foreign Policy in Transition.* Singapore, Institute of Southeast Asian Studies, 1999.

Thayer, Carlyle A., and Marr, David G. (Eds). *Vietnam and the Rule of Law. Political and Social Change Monograph No. 19.* Canberra, Australia National University, 1993.

Thierry d'Argenlieu, Admiral. *Chronique d'Indochine, 1945–1947.* Paris, Albin Michel, 1986.

Thomas, Mandy, and Drummond, Lisa (Eds). *Consuming Urban Culture in Contemporary Vietnam.* London, RoutledgeCurzon, 2003.

Thrift, Nigel, and Forbes, Dean. *The Price of War: Urbanisation in Vietnam 1954–85.* Sydney, Allen and Unwin, 1986.

Tonnesson, Stein. *The Vietnamese Revolution of 1945: Roosevelt, Ho Chi Minh and de Gaulle in a World at War.* Oslo/London, International Peace Research Institute/Sage Publications, 1992.

Tran, Khanh. *The Ethnic Chinese and Economic Development in Vietnam.* Singapore, Institute of Southeast Asian Studies, 1993.

Tran Thi Que. *Vietnam's Agriculture: The Challenges and Achievements.* Singapore, Institute of Southeast Asian Studies, 1998.

Tran Tri Vu. *Lost Years. My 1,632 Days in Vietnamese Re-education Camps.* Berkeley, CA, Institute of East Asian Studies, University of California, 1989.

Truong Buu Lam. *Patterns of Vietnamese Response to Foreign Intervention 1858–1900.* New Haven, CT, Yale University Press, 1967.

Truong Chinh. *Primer for Revolt.* B. B. Fall (Ed.), New York, Praeger, 1963.

Truong Nhu Tong. *Journal of a Viet-Cong.* London, Cape, 1986.

Turley, William S. (Ed.). *Vietnamese Communism in Comparative Perspective.* Boulder, CO, Westview Press, 1980.

Turley, W. S. *The Second Indo-China War: A Short Political and Military History.* Boulder, CO, Westview Press, 1986.

Turley, W. S., and Selden, M. (Eds). *Reinventing Vietnamese Socialism.* Boulder, CO, Westview Press, 1993.

Turner, Robert, F. *Vietnamese Communism: Its Origins and Development.* Stanford, Hoover Institution Press, 1975.

Van Dyke, J. M. *North Vietnam's Strategy for Survival.* Palo Alto, CA, 1972.

Vickerman, A. *The Fate of the Peasantry: A Premature 'Transition to Socialism' in the Democratic Republic of Vietnam.* New Haven, CT, Yale University Press, 1987.

Vo Dai Luoc (Ed.). *Investment and Trade Policies and the Development of Major Industries in Vietnam.* Nha Xuat Ban Khoa hoc Xa hoi, Hanoi, 1998.

Vo Nguyen Giap. *People's War, People's Army.* New York, Praeger, 1962.

Vo Nhan Tri. *Croissance Economique de la République Démocratique du Viet Nam.* Hanoi, 1967.

Vietnam's Economic Policy Since 1975. Singapore, Institute of Southeast Asian Studies, 1990.

Vu Tuan Anh. *Development in Vietnam: Policy Reforms and Economic Growth.* Singapore, Institute of Southeast Asian Studies, 1994.

Weller, Keith. *The Birth of Vietnam.* Berkeley, CA, University of California Press, 1985.

Werner, Jayne and Huynh, Luu Doan (Eds). *The Vietnam War—Vietnamese and American Perspectives.* New York, NY, M. E. Sharpe, 1993.

Wiegersma, N. *Vietnam: Peasant Land, Peasant Revolution: Patriarchy and Collectivity in the Rural Economy.* New York, St. Martin's Press, 1988.

Wiest, Andy. *The Vietnam War 1956–1975.* London, Routledge-Curzon, 2003.

Willbanks, James H. *Abandoning Vietnam: How America Left and South Vietnam Lost its War.* Lawrence, KS, University Press of Kansas, 2004.

Willenson, K. *The Bad War: An Oral History of the Vietnam War.* New York, New American Library, 1987.

Williams, Michael C. *Vietnam at the Crossroads.* London, Royal Institute of International Affairs/Pinter Publishers, 1992.

Windrow, Martin. *The Last Valley: Dien Bien Phu and the French Defeat in Vietnam.* London, Wiedenfeld and Nicholson, 2004.

Woodside, A. B. *Vietnam and the Chinese Model: a Comparative Study of Vietnamese and Chinese Government in the first half of the Nineteenth Century.* Cambridge, MA, Harvard University Press, 1971.

Community and Revolution in Modern Vietnam. Boston, MA, Houghton and Mifflin, 1976.

Zasloff, J., and Brown, McA. (Eds). *Communism in Indochina: New Perspectives.* Lexington, MA, Heath and Co, 1975.

PART THREE

Regional Information

REGIONAL ORGANIZATIONS

THE UNITED NATIONS

Address: United Nations, New York, NY 10017, USA.

Telephone: (212) 963-1234; **fax:** (212) 963-4879; **internet:** www.un.org.

The United Nations (UN) was founded on 24 October 1945. The organization, which has 191 member states, aims to maintain international peace and security and to develop international co-operation in addressing economic, social, cultural and humanitarian problems. The principal organs of the UN are the General Assembly, the Security Council, the Economic and Social Council (ECOSOC), the International Court of Justice and the Secretariat. The General Assembly, which meets for three months each year, comprises representatives of all UN member states. The Security Council investigates disputes between member countries, and may recommend ways and means of peaceful settlement: it comprises five permanent members (the People's Republic of China, France, Russia, the United Kingdom and the USA) and 10 other members elected by the General Assembly for a two-year period. The Economic and Social Council comprises representatives of 54 member states, elected by the General Assembly for a three-year period: it promotes co-operation on economic, social, cultural and humanitarian matters, acting as a central policy-making body and co-ordinating the activities of the UN's specialized agencies. The International Court of Justice comprises 15 judges of different nationalities, elected for nine-year terms by the General Assembly and the Security Council: it adjudicates in legal disputes between UN member states.

Secretary-General: Kofi Annan (Ghana) (1997–2006).

MEMBER STATES IN THE FAR EAST AND AUSTRALASIA

(with assessments for percentage contributions to UN budget for 2004–06, and year of admission)

Australia	1.592	1945
Brunei	0.034	1984
Cambodia	0.002	1955
China, People's Republic*	2.053	1945
Fiji	0.004	1970
Indonesia	0.142	1950
Japan	19.468	1956
Kiribati	0.001	1999
Korea, Democratic People's Republic	0.010	1991
Korea, Republic	1.796	1991
Laos	0.001	1955
Malaysia	0.203	1957
Marshall Islands	0.001	1991
Micronesia, Federated States	0.001	1991
Mongolia	0.001	1961
Myanmar	0.010	1948
Nauru	0.001	1999
New Zealand	0.221	1945
Palau	0.001	1994
Papua New Guinea	0.003	1975
Philippines	0.095	1945
Samoa	0.001	1976
Singapore	0.388	1965
Solomon Islands	0.001	1978
Thailand	0.209	1946
Timor-Leste	0.001	2002
Tonga	0.001	1999
Tuvalu	0.001	2000
Vanuatu	0.001	1981
Viet Nam	0.021	1977

* From 1945 to 1971 the Chinese seat was occupied by the Republic of China (confined to Taiwan since 1949).

SOVEREIGN COUNTRY NOT IN THE UNITED NATIONS

Republic of China (Taiwan)

Diplomatic Representation

PERMANENT MISSIONS TO THE UNITED NATIONS

(October 2004)

Australia: 150 East 42nd St, 33rd Floor, New York, NY 10017; tel. (212) 351-6600; fax (212) 351-6610; e-mail australia@un.int; internet www.australiaun.org; Permanent Representative John Dauth.

Brunei: 771 First Ave, New York, NY 10017; tel. (212) 697-3465; fax (212) 697-9889; e-mail info@bruneimission-ny.org; Permanent Representative Abdul Ghafor Shofry.

Cambodia: 866 United Nations Plaza, Suite 420, New York, NY 10017; tel. (212) 223-0676; fax (212) 223-0425; e-mail cambodia@un.int; internet www.un.int/cambodia; Permanent Representative Ouch Borith.

China, People's Republic: 350 East 35th St, New York, NY 10016; tel. (212) 655-6100; fax (212) 634-7626; e-mail chinamission_un@fmprc.gov.cn; internet www.china-un.org; Permanent Representative Wang Guangya.

Fiji: 630 Third Ave, 7th Floor, New York, NY 10017; tel. (212) 687-4130; fax (212) 687-3963; e-mail fiji@un.int; Permanent Representative Isikia Rabici Savua.

Indonesia: 325 East 38th St, New York, NY 10016; tel. (212) 972-8333; fax (212) 972-9780; e-mail ptri@indonesiamission-ny.org; internet www.indonesiamission-ny.org; Permanent Representative Rezlan Ishar Jenie.

Japan: 866 United Nations Plaza, 2nd Floor, New York, NY 10017; tel. (212) 223-4300; fax (212) 751-1966; e-mail mission@un-japan.org; internet www.un.int/japan; Permanent Representative Koichi Haraguchi.

Korea, Democratic People's Republic: 820 Second Ave, 13th Floor, New York, NY 10017; tel. (212) 972-3105; fax (212) 972-3154; e-mail dprk@un.int; Permanent Representative Pak Gil Yon.

Korea, Republic: 335 East 45th St, New York, NY 10017; tel. (212) 439-4000; fax (212) 986-1083; e-mail korea@un.int; internet www.un.int/korea; Permanent Representative Kim Sam-Hoon.

Laos: 317 East 51st St, New York, NY 10022; tel. (212) 832-2734; fax (212) 750-0039; e-mail lao@un.int; internet www.un.int/lao; Permanent Representative Alounkèo Kittikhoun.

Malaysia: 313 East 43rd St, New York, NY 10017; tel. (212) 986-6310; fax (212) 490-8576; e-mail malaysia@un.int; internet www.un.int/malaysia; Permanent Representative Rastam Mohamad Isa.

Marshall Islands: 800 Second Ave, 18th Floor, New York, NY 10017; tel. (212) 983-3040; fax (212) 983-3202; e-mail marshallislands@un.int; Permanent Representative Alfred Capelle.

Micronesia, Federated States: 820 Second Ave, Suite 17A, New York, NY 10017; tel. (212) 697-8370; fax (212) 697-8295; e-mail fsmun@fsmgov.org; internet www.fsmgov.org/fsmun; Permanent Representative Masao Nakayama.

Mongolia: 6 East 77th St, New York, NY 10021; tel. (212) 737-3874; fax (212) 861-9460; e-mail mongolia@un.int; internet www.un.int/mongolia; Permanent Representative Baatar Choisuren.

Myanmar: 10 East 77th St, New York, NY 10021; tel. (212) 535-1310; fax (212) 737-2421; e-mail myanmar@un.int; Permanent Representative Kyaw Tint Swe.

Nauru: 800 Second Ave, Suite 400D, New York, NY 10017; tel. (212) 937-0074; fax (212) 937-0079; Permanent Representative Vinci Neil Clodumar.

New Zealand: 1 United Nations Plaza, 25th Floor, New York, NY 10017; tel. (212) 826-1960; fax (212) 758-0827; e-mail nz@un.int; internet www.nzmissionny.org; Permanent Representative Don MacKay.

Palau: New York; e-mail palau@un.int; Permanent Representative STUART BECK.

Papua New Guinea: 201 East 42nd St, Suite 405, New York, NY 10017; tel. (212) 557-5001; fax (212) 557-5009; e-mail png@un.int; Permanent Representative ROBERT GUBA AISI.

Philippines: 556 Fifth Ave, 5th Floor, New York, NY 10036; tel. (212) 764-1300; fax (212) 840-8602; e-mail misunphil@idt.net; internet www.un.int/phillipines; Chargé d'affaires a.i. LAURO L. BAJA.

Samoa: 800 Second Ave, Suite 400J, New York, NY 10017; tel. (212) 599-6196; fax (212) 599-0797; e-mail samoa@un.int; Permanent Representative ALI'IOAIGA FETURI ELISAIA.

Singapore: 231 East 51st St, New York, NY 10022; tel. (212) 826-0840; fax (212) 826-2964; e-mail sgpun@prodigy.net; internet www.mfa.gov.sg/newyork; Permanent Representative KISHORE MAHBUBANI.

Solomon Islands: 800 Second Ave, Suite 8008, New York, NY 10017; tel. (212) 599-6193; fax (212) 661-8925; e-mail solomonislands@un.int; Permanent Representative COLIN D. BECK.

Thailand: 351 East 52nd St, New York, NY 10022; tel. (212) 754-2230; fax (212) 754-2535; e-mail thailand@un.int; Permanent Representative LAXANACHANTORN LAOHAPHAN.

Timor-Leste: 866 Second Ave, 9th floor, New York, NY 10017; tel. (212) 759-3675; fax (212) 759-4196; e-mail timor-leste@un.int; internet www.un.int/timor-leste; Permanent Representative JOSÉ LUÍS GUTERRES.

Tonga: 800 Second Ave, Suite 400J, New York, NY 10017; tel. (212) 599-6190; fax (212) 808-4975; e-mail tongaunmission@aol.com; Permanent Representative S. TU'A TAUMOEPEAU-TUPOU.

Tuvalu: 800 Second Ave, Suite 400B, New York, N.Y. 10017; tel. (212) 490-0534; fax (212) 808-4975; Permanent Representative ENELE SOSENE SOPOAGA.

Vanuatu: 866 United Nations Plaza, 3rd Floor, New York, NY 10017; tel. (212) 425-9600; fax (212) 425-9653; e-mail vanuatu@un.int; Chargé d'affaires a.i. SELWYN ARUTANGAI.

Viet Nam: 866 United Nations Plaza, Suite 435, New York, NY 10017; tel. (212) 644-0594; fax (212) 644-5732; e-mail vietnamun@aol.com; internet www.un.int/vietnam; Permanent Representative LE LUONG MINH.

OBSERVERS

Asian-African Legal Consultative Organization: 404 East 66th St, Apt 12C, New York, NY 10021; tel. (212) 734-7608; e-mail aalco@un.int; Permanent Representative K. BHAGWAT-SINGH (India).

Commonwealth Secretariat: 800 Second Ave, 4th Floor, New York, NY 10017; tel. (212) 599-6190, fax (212) 808-4975; e-mail comsec@thecommonwealth.org.

International Committee of the Red Cross: 801 Second Ave, 18th Floor, New York, NY 10017; tel. (212) 599-6021; fax (212) 599-6009; e-mail nyc@icrc.org; Head of Delegation GEORGES PACLISANU.

Organization of the Islamic Conference: 130 East 40th St, 5th Floor, New York, NY 10016; tel. (212) 883-0140; fax (212) 883-0143; e-mail oic@un.int; internet www.un.int/oic; Permanent Representative MOKHTAR LAMANI.

World Conservation Union—IUCN: 406 West 66th St, New York, NY 10021; tel. and fax (212) 734-7608.

The African, Caribbean and Pacific Group of States, the Economic Co-operation Organization and the Pacific Islands Forum are among several intergovernmental organizations that have a standing invitation to participate as observers but do not maintain permanent offices at the UN.

United Nations Information Centres/Services

Australia: POB 4045, 46–48 York St, 5th Floor, Sydney, NSW 2001; tel. (2) 9262-5111; fax (2) 9262-5886; e-mail unic@un.org.au; internet www.un.org.au; also covers Fiji, Kiribati, Nauru, New Zealand, Samoa, Tonga, Tuvalu and Vanuatu.

Indonesia: Gedung Surya, 14th Floor, 9 Jalan M. H. Thamrin Kavling, Jakarta 10350; tel. (21) 3983-1011; fax (21) 3983-1010; e-mail unicjak@cbn.net.id.

Japan: UNU Bldg, 8th Floor, 53–70 Jingumae S-chome, Shibuya-ku, Tokyo 150 0001; tel. (3) 5467-4451; fax (3) 5467-4455; e-mail unic@untokyo.jp; internet www.unic.or.jp; also covers Palau.

Myanmar: POB 230, 6 Natmauk Rd, Yangon; tel. (1) 292619; fax (1) 292911; e-mail unic.myanmar@undp.org.

Philippines: NEDA Bldg, Ground Floor, 106 Amorsolo St, Legaspi Village, Makati City, Manila; tel. (2) 8920611; fax (2) 8163011; e-mail infocentre@unicmanila.org; also covers Papua New Guinea and Solomon Islands.

Thailand: ESCAP, United Nations Bldg, Rajdamnern Ave, Bangkok 10200; tel. (2) 288-1234; fax (2) 288-1000; e-mail unisbkk.unescap@un.org; internet www.unescap.org/unic; also covers Cambodia, Hong Kong, Laos, Malaysia, Singapore and Viet Nam.

Economic and Social Commission for Asia and the Pacific—ESCAP

Address: United Nations Bldg, Rajadamnern Nok Ave, Bangkok 10200, Thailand.

Telephone: (2) 288-1234; **fax:** (2) 288-1000; **e-mail:** unisbkk.unescap@un.org; **internet:** www.unescap.org.

The Commission was founded in 1947 to encourage the economic and social development of Asia and the Far East; it was originally known as the Economic Commission for Asia and the Far East (ECAFE). The title ESCAP, which replaced ECAFE, was adopted after a reorganization in 1974. From 2002 ESCAP's administrative structures and programme activities underwent a process of intensive restructuring.

MEMBERS

Afghanistan	Korea, Democratic People's Republic	Philippines
Armenia	Korea, Republic	Russia
Australia	Kyrgyzstan	Samoa
Azerbaijan	Laos	Singapore
Bangladesh	Malaysia	Solomon Islands
Bhutan	The Maldives	Sri Lanka
Brunei	Marshall Islands	Tajikistan
Cambodia	Micronesia, Federated States	Thailand
China, People's Republic	Mongolia	Timor-Leste
Fiji	Myanmar	Tonga
France	Nauru	Turkey
Georgia	Nepal	Turkmenistan
India	Netherlands	Tuvalu
Indonesia	New Zealand	United Kingdom
Iran	Pakistan	USA
Japan	Palau	Uzbekistan
Kazakhstan	Papua New Guinea	Vanuatu
Kiribati		Viet Nam

ASSOCIATE MEMBERS

American Samoa	Hong Kong	Northern Mariana Islands
Cook Islands	Macao	
French Polynesia	New Caledonia	
Guam	Niue	

Organization

(October 2004)

COMMISSION

The Commission meets annually at ministerial level to examine the region's problems, to review progress, to establish priorities and to decide upon the recommendations of the Executive Secretary or the subsidiary bodies of the Commission.

Ministerial and intergovernmental conferences on specific issues may be held on an *ad hoc* basis with the approval of the Commission, although no more than one ministerial conference and five intergovernmental conferences may be held during one year.

COMMITTEES AND SPECIAL BODIES

The following advise the Commission and help to oversee the work of the Secretariat:

Committee on Poverty Reduction: has sub-committees on Poverty Reduction Practices; and Statistics.

Committee on Managing Globalization: has sub-committees on International Trade and Investment; Transport Infrastructure and Facilitation and Tourism; Environment and Sustainable Development; and Information, Communications and Space Technology.

Committee on Emerging Social Issues: has sub-committees on Socially Vulnerable Groups; and Health and Development.

Special Body on Least-Developed and Land-locked Developing Countries: meets every two years.

Special Body on Pacific Island Developing Countries: meets every two years.

In addition, an Advisory Committee of permanent representatives and other representatives designated by members of the Commission functions as an advisory body.

SECRETARIAT

The Secretariat operates under the guidance of the Commission and its subsidiary bodies. It consists of the Office of the Executive Secretary and two servicing divisions, covering administration and programme management, in addition to the following substantive divisions: Emerging Social Issues; Environment and Sustainable Development; Information, Communication and Space Technology; Poverty and Development; Statistics; Trade and Investment; and Transport and Tourism.

The Secretariat also includes a Poverty Centre, a Least Developed Countries Co-ordination Unit, and the UN information service/Bangkok.

Executive Secretary: KIM HAK-SU (Republic of Korea).

SUB-REGIONAL OFFICE

ESCAP Pacific Operations Centre—EPOC: Private Mail Bag 9004, Port Vila, Vanuatu; tel. 23458; fax 23921; e-mail escap@vanuatu.com.vu; f. 1984, to provide effective advisory and technical assistance at a sub-regional level and to identify the needs of island countries; under the Commission-wide reform process implemented in the early 2000s responsibility for EPOC's work programme was to be placed under ESCAP'S Poverty and Development sub-programme; it was also agreed that the Centre would be relocated to Suva, Fiji; Dir NIKENIKE VUROBARAVU.

Activities

ESCAP acts as a UN regional centre, providing the only intergovernmental forum for the whole of Asia and the Pacific, and executing a wide range of development programmes through technical assistance, advisory services to governments, research, training and information.

In 1992 ESCAP began to reorganize its programme activities and conference structures in order to reflect and serve the region's evolving development needs. The approach that was adopted focused on regional economic co-operation, poverty alleviation through economic growth and social development, and environmental and sustainable development. In May 2002, having considered the recommendations of an intergovernmental review meeting held in March, ESCAP determined to implement a further restructuring of its conference structures and thematic priorities. Three main thematic programmes were identified: poverty reduction (comprising sub-programmes on poverty and development; and statistics), managing globalization (with sub-programmes on trade and investment; environment; and space technology); and emerging social issues (with sub-programmes on health and development, gender and development, and population and social integration).

Emerging social issues: The sub-programme has three sections: Health and development, Gender and development, and Population and social integration. The sub-programme's main objective is to assess and respond to regional trends and challenges in social policy and human resources development, with particular emphasis on the planning and delivery of social services and training programmes for disadvantaged groups, including the poor, youths, women, the disabled, and the elderly. Aims to strengthen the capacity of public and non-government institutions to address the problems of such marginalized social groups and to foster partnerships between governments, the private sector, community organizations and all other involved bodies. Undertakes work through the health and development section on controlling the spread of HIV/AIDS, drug abuse and combating sexual exploitation. Implements global and regional mandates, such as the Programme of Action of the World Summit for Social Development and the Jakarta Plan of Action on Human Resources Development. The Biwako Millennium Framework for Action towards an Inclusive, Barrier-free and Rights-based Society for Persons with Disabilities in Asia and the Pacific was adopted by ESCAP as a regional guide-line underpinning the Asian and Pacific Decade of Disabled Persons (2003–12). In 1998 ESCAP initiated a programme of assistance in establishing a regional network of Social Development Management Information Systems (SOMIS). ESCAP collaborated with other agencies towards the adoption, in November 2001, of a Regional Platform on Sustainable Development for Asia and the Pacific. The Commission undertook regional preparations for the World Summit on Sustainable Development, which was held in Johannesburg, South Africa, in August–September 2002. Also prepares specific publications relating to population. Implements global and regional mandates, such as the Programme of Action of the International Conference on Population and Development. The Secretariat co-ordinates the Asia-Pacific Population Information Network (POPIN). The fifth Asia and Pacific Population Conference, sponsored by ESCAP, was held in Bangkok, Thailand, in December 2002.

Environment and sustainable development: Concerned with strengthening national capabilities to achieve environmentally-sound and sustainable development by integrating economic concerns, such as the sustainable management of natural resources, into economic planning and policies. The sub-programme is responsible for implementation of the Regional Action Programme for Environmentally Sound and Sustainable Development for the period 2001–05. Other activities have included the promotion of integrated water resources development and management, including water quality and conservation and a reduction in water-related natural disasters; strengthening the formulation of policies in the sustainable development of land and mineral resources; the consideration of energy resource options, such as rural energy supply, energy conservation and the planning of power networks; and promotion of the use of space technology applications for environmental management, natural disaster monitoring and sustainable development.

Information, communications and space technology: The work of the sub-programme is undertaken by the following sections: ICT Policy, ICT Applications, and Space Technology Applications. Aims to strengthen capacity for access to and the application of ICT and space technology. The sub-programme supports the development of cross-sectoral policies and strategies, and also supports regional co-operation aimed at sharing knowledge between advanced and developing economies and in areas such as cyber-crime and information security.

Poverty and development: The work of the sub-programme is undertaken by the following sections: Development Policy, Socio-economic Anaylsis and Poverty Reduction. The sub-programme aims to increase the understanding of the economic and social development situation in the region, with particular attention given to the attainment of the UN Millennium Development Goals, sustainable economic growth, poverty alleviation, the integration of environmental concerns into macroeconomic decisions and policy-making processes, and enhancing the position of the region's disadvantaged economies. The sub-programme is responsible for the provision of technical assistance, and the production of relevant documents and publications.

Statistics: Provides training and advice in priority areas, including national accounts statistics, poverty indicators, gender statistics, population censuses and surveys, and the strengthening and management of statistical systems. Supports co-ordination throughout the region of the development, implementation and revision of selected international statistical standards. Disseminates comparable socio-economic statistics, with increased use of the electronic media, promotes the use of modern information and communications technology (ICT) in the public sector and trains senior-level officials in the effective management of ICT.

Trade and investment: Provides technical assistance and advisory services. Aims to enhance institutional capacity-building; gives special emphasis to the needs of least-developed, land-locked and island developing countries, and to economies in transition in accelerating their industrial and technological advancement, promoting their exports, and furthering their integration into the region's economy; supports the development of electronic commerce and other information technologies in the region; and promotes the intra-regional and inter-subregional exchange of trade, investment and technology through the strengthening of institutional support services such as regional information networks.

Transport and tourism: Aims to develop inter- and intra-regional transport links to enhance trade and tourism, mainly through implementation of an Asian Land Transport Infrastructure Development (ALTID) programme. ALTID projects include the development of the Trans-Asian Railway and of the Asian Highway road network. Other activities are aimed at improving the planning process in developing infrastructure facilities and services, in accordance with the Regional Action Programme (Phase II, 2002–06) of the New Delhi Action Plan on Infrastructure Development in Asia and the Pacific, which was adopted at a ministerial conference held in October 1996, and at enhancing private sector involvement in national infrastructure development through financing, management, operations and risk-sharing. A Ministerial Conference on Infrastructure Development was organized by ESCAP in November 2001. The meeting concluded a memorandum of understanding, initially signed by ESCAP, Kazakhstan, the Republic of Korea, Mongolia and Russia, to facilitate the transport of container goods along the Trans-Asian Railway. The sub-programme aims to reduce the adverse environmental impact of the provision of infrastructure facilities and to promote more equitable and easier access to social amenities. Tourism concerns include the development of human resources, improved policy planning for tourism development, greater investment in the industry, and minimizing the environmental impact of tourism. In November 2003 ESCAP approved a new initiative, the Asia-Pacific Network for Transport and Logistics Education and Research (ANTLER), to comprise education, training and research centres throughout the region.

Throughout all the sub-programmes, ESCAP aims to focus particular attention on the needs and concerns of least-developed, land-locked and island developing nations, and economies in transition in the region.

CO-OPERATION WITH THE ASIAN DEVELOPMENT BANK

In July 1993 a memorandum of understanding was signed by ESCAP and the Asian Development Bank, outlining priority areas of co-operation between the two organizations. These were: regional and sub-regional co-operation; issues concerning the least-developed, land-locked and island developing member countries; poverty alleviation; women in development; population; human resource development; the environment and natural resource management; statistics and data bases; economic analysis; transport and communications; and industrial restructuring and privatization. The two organizations were to co-operate in organizing workshops, seminars and conferences, in implementing joint projects, and in exchanging information and data on a regular basis. A new memorandum of understanding between the two organizations was signed in May 2004 with an emphasis on achieving poverty reduction throughout the region.

ASSOCIATED BODIES

Asian and Pacific Centre for Transfer of Technology: APCTT Bldg, POB 4575, Qutab Institutional Area, New Delhi 110 016, India; tel. (11) 26966509; fax (11) 26856274; e-mail infocentre@apctt.org; internet www.apctt.org; f. 1977 to assist countries of the ESCAP region by strengthening their capacity to develop, transfer and adopt technologies relevant to the region, and to identify and to promote regional technology development and transfer; Dir SE-JUN MOON; publs *Asia Pacific Tech Monitor*, *VATIS Updates on Biotechnology*, *Food Processing*, *Ozone Layer Protection*, *Non-Conventional Energy*, and *Waste Technology* (each every 2 months), *International Technology and Business Opportunities: Catalogue* (quarterly).

Asian and Pacific Centre for Agricultural Engineering and Machinery—APCAEM: China International Science & Technology Convention Centre, 12 Yumin Rd, Madian, Deshengmenwai, Chaoyang District, Beijing 100029, People's Republic of China; e-mail juokslahti@un.org; internet www.apcaem.org; f. 1977 as Regional Network for Agricultural Engineering and Machinery, elevated to regional centre in 2002; aims to reduce poverty by enhancing technical co-operation throughout the region, and promotes agricultural engineering and machinery and the region's agro-based biotechnologies. Mems: Bangladesh, People's Republic of China, India, Indonesia, Iran, Nepal, Pakistan, Philippines, Republic of Korea, Sri Lanka, Thailand, Viet Nam; Dir TAPIO JUOKSLAHTI.

ESCAP/WMO Typhoon Committee: c/o UNDP, POB 7285, ADC, Pasay City, Metro Manila, Philippines; tel. (632) 3733443; fax (2) 3733419; e-mail tcs@philonline.com; f. 1968; an intergovernmental body sponsored by ESCAP and the World Meteorological Organization for mitigation of typhoon damage. It aims at establishing efficient typhoon and flood warning systems through improved meteorological and telecommunication facilities. Other activities include promotion of disaster preparedness, training of personnel and co-ordination of research. The committee's programme is supported from national resources and also by UNDP and other international and bilateral assistance. Mems: Cambodia, People's Republic of China, Hong Kong, Japan, Democratic People's Republic of Korea, Republic of Korea, Laos, Macao, Malaysia, Philippines, Singapore, Thailand, USA, Viet Nam; Co-ordinator of Secretariat Dr ROMAN L. KINTANAR.

Statistical Institute for Asia and the Pacific: JETRO-IDE Building, 2-2 Wakaba 3-chome, Mihama-ku, Chiba-shi, Chiba 2618787, Japan; tel. (43) 2999782; fax (43) 2999780; e-mail staff@unsiap.or.jp; internet www.unsiap.or.jp; f. 1970; trains government statisticians; prepares teaching materials, provides facilities for special studies and research of a statistical nature, assists in the development of training on official statistics in national and sub-regional centres; Dir TOMAS P. AFRICA (Philippines).

UNESCAP Centre for Alleviation of Poverty through Secondary Crops' Development in Asia and the Pacific: Jalan Merdeka 145, Bogor 16111, Indonesia; tel. (251) 343277; fax (251) 336290; e-mail capsa@uncapsa.org; internet www.uncapsa.org; f. 1981 as CGPRT Centre, current name adopted April 2004; initiates and promotes socio-economic and policy research, training, dissemination of information and advisory services to enhance the living conditions of rural poor populations reliant on secondary crop agriculture; Dir (vacant); publs *Palawija News* (quarterly), working paper series, monograph series and statistical profiles.

WMO/ESCAP Panel on Tropical Cyclones: Technical Support Unit (TSU), c/o Headquarters, Pakistan Meteorological Dept, POB 1214, H-8/2, Islamabad, Pakistan; tel. (51) 9257314; fax (51) 4432588; e-mail tsupmd@hotmail.com; internet www.tsuptc-wmo.org; f. 1972 to mitigate damage caused by tropical cyclones in the Bay of Bengal and the Arabian Sea. Mems: Bangladesh, India, the Maldives, Myanmar, Oman, Pakistan, Sri Lanka, Thailand; TSU Co-ordinator Dr QAMAR-UZ-ZAMAN CHAUDHRY.

Finance

For the two-year period 2004–05 ESCAP's proposed programme budget, an appropriation from the UN budget, was US $64.7m. The regular budget is supplemented annually by funds from various sources for technical assistance.

Publications

Annual Report.
Agro-chemicals News in Brief (quarterly).
Asia-Pacific Development Journal (2 a year).
Asia-Pacific in Figures (annually).
Asia-Pacific Population Journal (quarterly).
Asia-Pacific Remote Sensing and GIS Journal (2 a year).
Atlas of Mineral Resources of the ESCAP Region.
Bulletin on Asia-Pacific Perspectives (annually).
Confluence (water resources newsletter, 2 a year).
Economic and Social Survey of Asia and the Pacific (annually).
Environmental News Briefing (every 2 months).
ESCAP Energy News (2 a year).
ESCAP Human Resources Development Newsletter (2 a year).
ESCAP Population Data Sheet (annually).
ESCAP Tourism Newsletter (2 a year).
Fertilizer Trade Information Monthly Bulletin.
Foreign Trade Statistics of Asia and the Pacific (every 2 years).
Government Computerization Newsletter (irregular).
Industry and Technology Development News for Asia and the Pacific (annually).
Poverty Alleviation Initiatives (quarterly).
Regional Network for Agricultural Machinery Newsletter (3 a year).
Small Industry Bulletin for Asia and the Pacific (annually).
Social Development Newsletter (2 a year).
Space Technology Applications Newsletter (quarterly).
Statistical Indicators for Asia and the Pacific (quarterly).
Statistical Newsletter (quarterly).
Statistical Yearbook for Asia and the Pacific.
Trade and Investment Information Bulletin (monthly).
Transport and Communications Bulletin for Asia and the Pacific (annually).
Water Resources Journal (quarterly).
Bibliographies; country and trade profiles; commodity prices; statistics.

United Nations Children's Fund—UNICEF

Address: 3 United Nations Plaza, New York, NY 10017, USA.
Telephone: (212) 326-7000; **fax:** (212) 887-7465; **e-mail:** info@unicef.org; **internet:** www.unicef.org.

UNICEF was established in 1946 by the UN General Assembly as the UN International Children's Emergency Fund, to meet the emergency needs of children in post-war Europe. In 1950 its mandate was expanded to respond to the needs of children in developing countries. In 1953 the General Assembly decided that UNICEF should continue its work, as a permanent arm of the UN system, with an emphasis on programmes giving long-term benefits to children everywhere, particularly those in developing countries. In 1965 UNICEF was awarded the Nobel Peace Prize.

Organization

(October 2004)

EXECUTIVE BOARD

The Executive Board, as the governing body of UNICEF, comprises 36 member governments from all regions, elected in rotation for a three-year term by ECOSOC. The Board establishes policy, reviews programmes and approves expenditure. It reports to the General Assembly through ECOSOC.

SECRETARIAT

The Executive Director of UNICEF is appointed by the UN Secretary-General in consultation with the Executive Board. The administration of UNICEF and the appointment and direction of staff are the responsibility of the Executive Director, under policy directives laid down by the Executive Board, and under a broad authority delegated to the Executive Director by the Secretary-General. In December 2003 there were more than 7,000 UNICEF staff positions, of which about 85% were in the field.

Executive Director: CAROL BELLAMY (USA).

REGIONAL OFFICE

UNICEF has a network of eight regional and 126 field offices serving 157 countries and territories. Its offices in Tokyo, Japan, and Brussels, Belgium, support fund-raising activities; UNICEF's supply division is administered from the office in Copenhagen, Denmark. A research centre concerned with child development is based in Florence, Italy.

East Asia and the Pacific: POB 2-154, Bangkok 10200, Thailand; tel. (2) 2805931; fax (2) 2803563; e-mail eapro@unicef.org.

NATIONAL COMMITTEES

UNICEF is supported by 37 National Committees, mostly in industrialized countries, whose volunteer members, numbering more than 100,000, raise money through various activities, including the sale of greetings cards. The Committees also undertake advocacy and awareness campaigns on a number of issues and provide an important link with the general public.

Activities

UNICEF is dedicated to the well-being of children, adolescents and women and works for the realization and protection of their rights within the frameworks of the Convention on the Rights of the Child, which was adopted by the UN General Assembly in 1989 and by 2004 was almost universally ratified, and of the Convention on the Elimination of All Forms of Discrimination Against Women, adopted by the UN General Assembly in 1979. Promoting the full implementation of the Conventions, UNICEF aims to ensure that children world-wide are given the best possible start in life and attain a good level of basic education, and that adolescents are given every opportunity to develop their capabilities and participate successfully in society. The Fund also continues to provide relief and rehabilitation assistance in emergencies. Through its extensive field network in some 157 developing countries and territories, UNICEF undertakes, in co-ordination with governments, local communities and other aid organizations, programmes in health, nutrition, education, water and sanitation, the environment, gender issues and development, and other fields of importance to children. Emphasis is placed on low-cost, community-based programmes. UNICEF programmes are increasingly focused on supporting children and women during critical periods of their life, when intervention can make a lasting difference, i.e. early childhood, the primary school years, adolescence and the reproductive years. Priorities include early years development, immunization strategies, girls' education, combating the spread and impact of HIV/AIDS, and strengthening the protection of children against violence, exploitation and abuse. These priorities are guided by the relevant UN Millennium Development Goals adopted by world leaders in 2000, and by the 'A World Fit for Children' declaration and plan of action endorsed by the UN General Assembly special session on Children in 2002 (see below).

UNICEF was instrumental in organizing the World Summit for Children, held in September 1990 and attended by representatives from more than 150 countries, including 71 heads of state or government. The Summit produced a Plan of Action which recognized the rights of the young to 'first call' on their countries' resources and formulated objectives for the year 2000, including: (i) a reduction of the 1990 mortality rates for infants and children under five years by one-third, or to 50–70 per 1,000 live births, whichever is lower; (ii) a reduction of the 1990 maternal mortality rate by one-half; (iii) a reduction by one-half of the 1990 rate of severe malnutrition among children under the age of five; (iv) universal access to safe drinking water and to sanitary means of excreta disposal; and (v) universal access to basic education and completion of primary education by at least 80% of children. UNICEF supported the efforts of governments to achieve progress towards these objectives. The Fund served as the substantive secretariat for, and played a leading role in helping governments and other partners prepare for, the UN General Assembly special session on Children, which was held in May 2002 to assess the outcome of the 1990 summit and to determine a set of actions and objectives for the next 10 years. At the session the General Assembly adopted a declaration entitled 'A World Fit for Children', reaffirming its commitment to the agenda of the 1990 summit, and outlining a plan of action that resolved to achieve as yet unmet World Summit goals by 2010 and to work towards the attainment by 2015 of 21 new goals and targets supporting UN Millennium Development Goals in the areas of education, health and the protection of children. The latter included: a reduction of mortality rates for infants and children under five by two-thirds; a reduction of maternal mortality rates by three-quarters; a reduction by one-third in the rate for severe malnutrition among children under the age of five; and enrolment in primary education by 90% of children. UNICEF's medium-term strategic plan for 2002–05 incorporated the following aims: 80% immunization coverage for children in 80% of countries globally; a reduction by one-half in the number of deaths caused by measles; and the certified global eradication of poliomyelitis. A database of economic and social indicators is available at www.unicef.org/information/databases.

In 2000 UNICEF launched a new initiative, the Global Movement for Children—comprising governments, private- and public-sector bodies, and individuals—which aimed to rally world-wide support to improve the lives of all children and adolescents. In April 2001 a 'Say Yes for Children' campaign was adopted by the Global Movement, identifying 10 critical actions required to further its objectives. These were: eliminating all forms of discrimination and exclusion; putting children first; ensuring a caring environment for every child; fighting HIV/AIDS; eradicating violence against and abuse and exploitation of children; listening to children's views; universal education; protecting children from war; safeguarding the earth for children; and combating poverty. UNICEF was to co-ordinate the campaign.

UNICEF, in co-operation with other UN agencies, promotes universal access to and completion of basic and good quality education. The Fund, with UNESCO, UNDP, UNFPA and the World Bank, co-sponsored the World Conference on Education for All, held in Thailand in March 1990, and undertook efforts to achieve the objectives formulated by the conference, which included the elimination of disparities in education between boys and girls. UNICEF participated in and fully supports the objectives and framework for action adopted by the follow-up World Education Forum in Dakar, Senegal, in April 2000. The Fund supports education projects in sub-Saharan Africa, South Asia and countries in the Middle East and North Africa, and implements a Girls' Education Programme in more than 100 countries, which aims to increase the enrolment of girls in primary schools. More than 120m. children world-wide, of whom nearly 53% are girls, remain deprived of basic education. In December 2002 UNICEF initiated the '25 by 2005' initiative, which aimed to eliminate gender disparities in education in 25, mainly African and Asian, countries by 2005. In 2002–03 a UNICEF-led initiative facilitated the return to school of more than 1m. Afghan girls.

Through UNICEF's efforts the needs and interests of children were incorporated into Agenda 21, which was adopted as a plan of action for sustainable development at the UN Conference on Environment and Development, held in June 1992. In mid-1997, at the UN General Assembly's Special Session on Sustainable Development, UNICEF highlighted the need to improve safe water supply, sanitation and hygiene, and thereby reduce the risk of diarrhoea and other water-borne diseases, as fundamental to fulfilment of child rights. The Fund supports initiatives to provide the benefits of safe water, sanitation and hygiene education to communities in some 90 developing countries; and participates in the inter-agency initiative Focusing Resources on Effective School Health (FRESH). UNICEF also works with UNEP to promote environment issues of common concern and with the World Wide Fund for Nature to support the conservation of local ecosystems.

UNICEF aims to break the cycle of poverty by advocating for the provision of increased development aid to developing countries, and aims to help poor countries obtain debt relief and to ensure access to basic social services. To this end it supports NetAid, an internet-based strategy to promote sustainable development and combat extreme poverty. UNICEF is the leading agency in promoting the 20/20 initiative, which was endorsed at the World Summit for Social Development, held in Copenhagen, Denmark, in March 1995. The initiative encourages the governments of developing and donor countries to allocate at least 20% of their domestic budgets and official development aid respectively, to healthcare, primary education and low-cost safe water and sanitation.

UNICEF estimates that the births of some 50m. children annually are not officially registered, and promotes universal registration in order to prevent the abuse of children without proof of age and nationality, for example through trafficking, forced labour, early marriage and military recruitment. It estimates that some 246m. children are involved in exploitative labour and that annually around 1.2m. children are trafficked. The Fund, which vigorously opposes the exploitation of children as a violation of their basic human rights, works with ILO and other partners to promote an end to exploitative and hazardous child labour, and supports special projects to provide education, counselling and care for the estimated 250m. children between the ages of five and 14 years working in developing countries. UNICEF played a major role at the World Congress against Commercial Sexual Exploitation of Children, held in Stockholm, Sweden, in 1996, which adopted a Declaration and Agenda for Action to end the sexual exploitation of children. UNICEF also actively participated in the International Conference on Child Labour held in Oslo, Norway, in November 1997. The Conference adopted an Agenda for Action to eliminate the worst forms of child labour, including slavery-like practices, forced labour, commercial sexual exploitation and the use of children in drugs-trafficking and other hazardous forms of work. UNICEF supports the 1999 ILO Worst Forms of Child Labour Convention, which aims at the prohibition and elimination of the worst forms of child labour. In 1999 UNICEF launched a global initiative, Education as a Preventive Strategy Against Child Labour, with the aim of providing education to children forced to miss school because of work. The Fund helped to draft and promotes full ratification and implementation of an Optional Protocol to the Convention of the Rights of the Child concerning the sale of children, child prostitution and pornography, which was adopted in May 2000 and entered into force in January 2002. UNICEF co-sponsored and actively participated in the Second Congress Against Commercial Sexual Exploitation of Children held in Yokohama, Japan, in December 2001.

Child health is UNICEF's largest programme sector, accounting for some 40% of programme expenditure in 2000. UNICEF estimates that around 11m. children under five years of age die each year, mainly in developing countries, and the majority from largely preventable causes. UNICEF has worked with WHO and other partners to increase global immunization coverage against the following six diseases: measles, poliomyelitis, tuberculosis, diphtheria, whooping cough and tetanus. In 2002 UNICEF, in partnership with WHO, governments and other partners, helped to immunize more than 500m. children under five years of age in 53 countries against polio. In 1999 UNICEF, WHO, the World Bank and a number of public- and private-sector partners launched the Global Alliance for Vaccines and Immunization (GAVI), which aimed to protect children of all nationalities and socio-economic groups against vaccine-preventable diseases. GAVI's strategy included improving access to sustainable immunization services, expanding the use of existing vaccines, accelerating the development and introduction of new vaccines and technologies and promoting immunization coverage as a focus of international development efforts. Between 2001–04 GAVI helped to protect around 35m. children in developing countries against hepatitis B. UNICEF and WHO also work in conjunction on the Integrated Management of Childhood Illness programme to control diarrhoeal dehydration, a major cause of death among children under five years of age in the developing world. UNICEF-assisted programmes for the control of diarrhoeal diseases promote the low-cost manufacture and distribution of prepackaged salts or home-made solutions. The use of 'oral rehydration therapy' has risen significantly in recent years, and is believed to prevent more than 1m. child deaths annually. During 1990–2000 diarrhoea-related deaths were reduced by one-half. UNICEF also promotes the need to improve sanitation and access to safe water supplies in developing nations in order to reduce the risk of diarrhoea and other water-borne diseases (see 20/20 initiative, above). To control acute respiratory infections, another leading cause of death in children under five in developing countries, UNICEF works with WHO in training health workers to diagnose and treat the associated diseases. As a result, the level of child deaths from pneumonia and other respiratory infections has been reduced by one-half since 1990. Around 1m. children die from malaria every year, mainly in sub-Saharan Africa. In October 1998 UNICEF, together with WHO, UNDP and the World Bank, inaugurated a new global campaign, Roll Back Malaria, to fight the disease. UNICEF supports control programmes in more than 30 countries. In 2002 UNICEF provided more than 4m.

insecticide-treated nets as protection against malarial mosquitoes. The Fund also provides vitamin A supplements in support of its immunization programmes.

According to UNICEF estimates, around 27% of children under five years of age are underweight, while each year malnutrition contributes to about one-half of the child deaths in that age group and leaves millions of others with physical and mental disabilities. More than 2,000m. people world-wide (mainly women and children in developing countries) are estimated to be deficient in one or more essential vitamins and minerals, such as vitamin A, iodine and iron. UNICEF supports national efforts to reduce malnutrition, for example, fortifying staple foods with micronutrients, widening women's access to education, improving household food security and basic health services, and promoting sound child-care and feeding practices. Since 1991 more than 15,000 hospitals in at least 136 countries have been designated 'baby-friendly', having implemented a set of UNICEF and WHO recommendations entitled '10 steps to successful breast-feeding'. In 1996 UNICEF expressed its concern at the impact of international economic embargoes on child health, citing as an example the extensive levels of child malnutrition recorded in Iraq. UNICEF remains actively concerned at the levels of child malnutrition and accompanying diseases in Iraq and in the Democratic People's Republic of Korea, which has also suffered severe food shortages.

UNICEF estimates that more than 500,000 women die every year during pregnancy or childbirth, largely because of inadequate maternal healthcare. For every maternal death, approximately 30 further women suffer permanent injuries or chronic disabilities as a result of complications during pregnancy or childbirth. With its partners in the Safe Motherhood Initiative—UNFPA, WHO, the World Bank, the International Planned Parenthood Federation, the Population Council, and Family Care International—UNICEF promotes measures to reduce maternal mortality and morbidity, including improving access to quality reproductive health services, educating communities about safe motherhood and the rights of women, training midwives, and expanding access to family planning services.

UNICEF is concerned at the danger posed by HIV/AIDS to the realization of children's rights. It is estimated that one-half of all new HIV infections occur in young people. At the end of 2003 2.5m. children under 15 were living with HIV/AIDS world-wide. Some 700,000 children under 15 were newly infected during that year, while 500,000 died as a result of AIDS. It was estimated that about one-half of all new HIV infections during 2003 occurred in young people, aged 15–24. It is believed that more than 14m. children world-wide have lost one or both parents to AIDS since the start of the epidemic. UNICEF's priorities in this area include prevention of infection among young people, reduction in mother-to-child transmission, care and protection of orphans and other vulnerable children, care and support for children, young people and parents living with HIV/AIDS. In 2002 the Fund supported HIV/AIDS peer education programmes for young people in 71 countries. UNICEF works closely in this field with governments and co-operates with other UN agencies in the Joint UN Programme on HIV/AIDS (UNAIDS), which became operational on 1 January 1996. In July 2002 UNICEF, UNAIDS and WHO jointly produced a study entitled *Young People and HIV/AIDS: Opportunity in Crisis*, examining young people's sexual behaviour patterns and knowledge of HIV/AIDS.

At December 2002 it was estimated that some 2.1m. young people (aged from 15–24) were living with HIV/AIDS in Asia and the Pacific.

UNICEF provides emergency relief assistance, supports education, health, mine-awareness and psychosocial activities and helps to demobilize and rehabilitate child soldiers in countries and territories affected by violence and social disintegration. It assists children orphaned or separated from their parents and made homeless through armed conflict. In recent years several such emergency operations have been undertaken, including in Afghanistan, Angola, Burundi, Democratic Republic of the Congo, Iran, Iraq, Kosovo, Liberia, Sierra Leone and Sudan. In 2004 UNICEF was providing emergency assistance to children in 55 countries. In 1999 UNICEF adopted a Peace and Security Agenda to help guide international efforts in this field. Emergency education assistance includes the provision of 'Edukits' in refugee camps and the reconstruction of school buildings. In the area of health the Fund co-operates with WHO to arrange 'days of tranquility' in order to facilitate the immunization of children in conflict zones. Psychosocial assistance activities include special programmes to assist traumatized children and help unaccompanied children to be reunited with parents or extended families.

An estimated 300,000 children are involved in armed conflicts as soldiers, porters and forced labourers. UNICEF encourages ratification of the Optional Protocol to the Convention on the Rights of the Child on the involvement of children in armed conflict, which was adopted by the General Assembly in May 2000 and entered into force in February 2002, and bans the compulsory recruitment of combatants below 18 years. The Fund also urges states to make unequivocal statements endorsing 18 as the minimum age of voluntary recruitment to the armed forces. UNICEF was an active participant in the so-called 'Ottawa' process (supported by the Canadian Government) to negotiate an international ban on anti-personnel land-mines which, it was estimated, killed and maimed between 8,000 and 10,000 children every year. The Convention on the Prohibition of the Use, Stockpiling, Production and Transfer of Anti-Personnel Mines and on their Destruction was adopted in December 1997 and entered into force in March 1999. By September 2004 the Convention had been ratified by 143 countries. UNICEF is committed to campaigning for its universal ratification and full implementation, and also supports mine-awareness campaigns.

Finance

UNICEF is funded by voluntary contributions from governments and non-governmental and private-sector sources. UNICEF's income is divided into contributions for 'regular resources' (used for country programmes of co-operation approved by the Executive Board, programme support, and management and administration costs) and contributions for 'other resources' (for special purposes, including expanding the outreach of country programmes of co-operation and ensuring capacity to deliver critical assistance to women and children, for example during humanitarian crises). Total income in 2003 amounted to US $1,688m., compared with $1,454m. in 2002, of which 67% was from governments and intergovernmental organizations.

Total expenditure in 2003 amounted to US $1,480m., compared with $1,273m. in 2002.

Publications

Facts and Figures (in English, French and Spanish).

The State of the World's Children (annually, in Arabic, English, French, Russian and Spanish and about 30 other national languages).

UNICEF Annual Report (in English, French and Spanish).

UNICEF at a Glance (in English, French and Spanish).

Reports and studies; series on children and women; nutrition; education; children's rights; children in wars and disasters; working children; water, sanitation and the environment; analyses of the situation of children and women in individual developing countries.

United Nations Development Programme—UNDP

Address: One United Nations Plaza, New York, NY 10017, USA.

Telephone: (212) 906-5295; **fax:** (212) 906-5364; **e-mail:** hq@undp.org; **internet:** www.undp.org.

The Programme was established in 1965 by the UN General Assembly. Its central mission is to help countries to eradicate poverty and achieve a sustainable level of human development, an approach to economic growth that encompasses individual well-being and choice, equitable distribution of the benefits of development, and conservation of the environment. UNDP advocates for a more inclusive global economy.

Organization

(October 2004)

UNDP is responsible to the UN General Assembly, to which it reports through ECOSOC.

EXECUTIVE BOARD

The Executive Board is responsible for providing intergovernmental support to, and supervision of, the activities of UNDP and the UN Population Fund (UNFPA). It comprises 36 members: eight from

Africa, seven from Asia, four from eastern Europe, five from Latin America and the Caribbean and 12 from western Europe and other countries.

SECRETARIAT

In recent years UNDP has implemented a process aimed at restructuring and improving the efficiency of its administration. Offices and divisions at the Secretariat include: an Operations Support Group; Offices of the United Nations Development Group, the Human Development Report, Audit and Performance Review, and Communications; and Bureaux for Crisis Prevention and Recovery, Resources and Strategic Partnerships, Development Policy, and Management. Five regional bureaux, all headed by an assistant administrator, cover: Africa; Asia and the Pacific; the Arab states; Latin America and the Caribbean; and Europe and the Commonwealth of Independent States. There is also a Division for Global and Interregional Programmes.

Administrator: MARK MALLOCH BROWN (United Kingdom).

Associate Administrator: Dr ZÉPHIRIN DIABRÉ (Burkina Faso).

Assistant Administrator and Director, Regional Bureau for Asia and the Pacific: HAFIZ PASHA.

COUNTRY OFFICES

In almost every country receiving UNDP assistance there is an office, headed by the UNDP Resident Representative, who usually also serves as UN Resident Co-ordinator, responsible for the co-ordination of all UN technical assistance and operational development activities, advising the Government on formulating the country programme, ensuring that field activities are undertaken, and acting as the leader of the UN team of experts working in the country. The offices function as the primary presence of the UN in most developing countries.

OFFICES OF UNDP REPRESENTATIVES IN THE FAR EAST AND AUSTRALASIA

Cambodia: 53 rue Pasteur, Boeung Keng Kang, POB 877, Phnom-Penh; tel. (23) 216167; fax (23) 216257; e-mail registry.kh@undp.org; internet www.un.org.kh/undp.

China, People's Republic: 2 Liangmahe Nan Lu, 100600 Beijing; tel. (10) 65323731; fax (10) 65322567; e-mail registry.cn@undp.org; internet www.unchina.org/undp.

Fiji: Private Mail Bag, Tower Level 6, Reserve Bank Bldg, Pratt St, Suva; tel. 331-2500; fax 330-1718; e-mail fo.fiji@undp.org; internet www.undp.org.fj; also covers French Polynesia, Kiribati, Marshall Islands, Federated States of Micronesia, Nauru, New Caledonia, Palau, Solomon Islands, Tonga, Tuvalu, Vanuatu and Wallis and Futuna.

Indonesia: POB 2338, 14 Jalan M. H. Thamrin, Jakarta 10240; tel. (21) 314-1308; fax (21) 314-4251; e-mail media.id@undp.org; internet www.un.or.id/undp.

Japan: UNU Bldg, 8th Floor, 5-53-70, Jingumae, Shibuya-ku, Tokyo 150-0001; tel. (3) 5467-4751; fax (3) 5467 4753; e-mail fo.jpn@undp.org; internet www.undp.or.jp/indexe.htm.

Korea, Republic: 730 Hannam 2-Dong, Yongsan-ku, Seoul; tel. (2) 790-9562; fax (2) 749-1417; e-mail registry.kr@undp.org; internet www.undp.or.kr.

Laos: POB 345, Phonekheng Rd 01004, Vientiane; tel. (21) 213390; fax (21) 212029; e-mail info.lao@undp.org; internet www.undplao.org.

Malaysia: Wisma UN, Block C, Kompleks Pejabat Damansara, Jalan Dungun, Damansara Heights, 50490 Kuala Lumpur; tel. (3) 2095-9122; fax (3) 2095-2870; e-mail registry.my@undp.org.my; internet www.undp.org.my; also covers Brunei and Singapore.

Mongolia: POB 49/1009, Ulan Bator; tel. (1) 327585; fax (1) 326221; e-mail registry.mn@undp.org; internet www.un-mongolia.mn/~undp.

Myanmar: POB 650, No. 6 Natmauk Rd, Tamwe Township, Yangon 11181; tel. (1) 542911; fax (1) 292739; e-mail registry.mm@undp.org; internet www.mm.undp.org.

Papua New Guinea: POB 1041, ADF House, 3rd Level, Musgrave St, Port Moresby; tel. 3212877; fax 3211224; e-mail fo.png@undp.org; internet www.undp.org.pg.

Philippines: 7F NEDA Sa Makati Bldg, 106 Amorsolo St, Legaspi Village, 1229 Makati City; tel. (2) 8920611; fax (2) 8164061; e-mail registry.ph@undp.org; internet www.undp.org.ph.

Samoa: Lauofo Meti's Bldg, Four Corners, Matautu-uta, Private Mail Bag, Apia; tel. 23670; fax 23555; e-mail registry.ws@undp.org; internet www.undp.org.ws; also covers Cook Islands, Niue and Tokelau.

Thailand: GPO Box 618, UN Bldg, 12th Floor, Rajdamnern Nok Ave, Bangkok 10200; tel. (2) 2881234; fax (2) 2800556; e-mail registry.th@undp.org; internet www.undp.or.th; also covers Hong Kong.

Timor-Leste: Caicoli St 14, Dili; tel. (390) 312481; fax (390) 312408; e-mail registry.tp@undp.org; internet www.undp.east-timor.org.

Viet Nam: 25–29 Phan Bôi Chau, Hanoi; tel. (4) 9421495; fax (4) 9422267; e-mail registry@undp.org.vn; e-mail registry@undp.org.vn; internet www.undp.org.vn.

Activities

As the world's largest source of grant-funded technical assistance for developing countries, UNDP provides advisory and support services to governments and UN teams. Assistance is mostly non-monetary, comprising the provision of experts' services, consultancies, equipment and training for local workers, including fellowships for advanced study abroad. UNDP supports programme countries in attracting aid and utilizing it efficiently. The Programme is committed to allocating some 88% of its regular resources to low-income developing countries. Developing countries themselves contribute significantly to the total project costs in terms of personnel, facilities, equipment and supplies.

Since the mid-1990s UNDP has strengthened its focus on results, streamlining its management practices and promoting clearly defined objectives for the advancement of sustainable human development. Under 'UNDP 2001', an extensive internal process of reform initiated during the late 1990s, UNDP placed increased emphasis on its activities in the field and on performance and accountability, focusing on the following priority areas: democratic governance; poverty reduction; crisis prevention and recovery; energy and environment; promotion of information and communications technology; and combating HIV/AIDS. In 2001 UNDP established six Thematic Trust Funds, covering each of these areas, to enable increased support of thematic programme activities. Gender equality and the provision of country-level and co-ordination services are also important focus areas. In accordance with the more results-oriented approach developed under the 'UNDP 2001' process the Programme introduced a new Multi-Year Funding Framework (MYFF), of which the first phase covered the period 2000–03 and the second phase 2004–07. The MYFF outlines the country-driven goals around which funding is to be mobilized, integrating programme objectives, resources, budget and outcomes. It provides the basis for the Administrator's Business Plans for the same duration and enables policy coherence in the implementation of programmes at country, regional and global levels. A Results-Oriented Annual Report (ROAR) was produced for the first time in 2000 from data compiled by country offices and regional programmes. It was hoped that UNDP's greater focus on performance would generate increased voluntary contributions from donors, thereby strengthening the Programme's core resource base. In September 2000 the first ever Ministerial Meeting of ministers of development co-operation and foreign affairs and other senior officials from donor and programme countries, convened in New York, USA, endorsed UNDP's shift to a results-based orientation.

From the mid-1990s UNDP also determined to assume a more active and integrative role within the UN system-wide development framework. UNDP Resident Representatives—usually also serving as UN Resident Co-ordinators, with responsibility for managing inter-agency co-operation on sustainable human development initiatives at country level—were to play a focal role in implementing this approach. In order to promote its co-ordinating function UNDP allocated increased resources to training and skill-sharing programmes. In 1997 the UNDP Administrator was appointed to chair the UN Development Group (UNDG), which was established as part of a series of structural reform measures initiated by the UN Secretary-General, with the aim of strengthening collaboration between all UN funds, programmes and bodies concerned with development. The UNDG promotes coherent policy at country level through the system of UN Resident Co-ordinators (see above), the Common Country Assessment mechanism (CCA, a country-based process for evaluating national development situations), and the UN Development Assistance Framework (UNDAF, the foundation for planning and co-ordinating development operations at country level, based on the CCA). Within the framework of the Administrator's Business Plans for 2000–03 a new Bureau for Resources and Strategic Partnerships was established to build and strengthen working partnerships with other UN bodies, donor and programme countries, international financial institutions and development banks, civil society organizations and the private sector. The Bureau was also to serve UNDP's regional bureaux and country offices through the exchange of information and promotion of partnership strategies.

UNDP has a catalyst and co-ordinating function as the focus of UN system-wide efforts to achieve the so-called Millennium Development Goals (MDGs), pledged by governments attending a summit meeting of the UN General Assembly in September 2000. The objectives included a reduction by 50% in the number of people with an income of less than US $1 a day and those suffering from hunger and lack of safe drinking water by 2015. Other commitments made concerned equal access to education for girls and boys, the provision of universal primary education, the reduction of maternal mortality by 75%, and the reversal of the spread of HIV/AIDS and other diseases. UNDP plays a leading role in efforts to integrate the MDGs into all aspects of the UN activities at country level. The Programme supports the formulation of MDG Reports for all developing countries.

UNDP aims to help governments to reassess their development priorities and to design initiatives for sustainable human development. UNDP country offices support the formulation of national human development reports (NHDRs), which aim to facilitate activities such as policy-making, the allocation of resources and monitoring progress towards poverty eradication and sustainable development. In addition, the preparation of Advisory Notes and Country Co-operation Frameworks by UNDP officials helps to highlight country-specific aspects of poverty eradiction and national strategic priorities. In January 1998 the Executive Board adopted eight guiding principles relating to sustainable human development that were to be implemented by all country offices, in order to ensure a focus to UNDP activities. A network of nine Sub-regional Resource Facilities (SURFs) has been established to strengthen and co-ordinate UNDP's technical assistance services. Since 1990 UNDP has published an annual *Human Development Report*, incorporating a Human Development Index, which ranks countries in terms of human development, using three key indicators: life expectancy, adult literacy and basic income required for a decent standard of living. In 1997 a Human Poverty Index and a Gender-related Development Index, which assesses gender equality on the basis of life expectancy, education and income, were introduced into the Report for the first time.

UNDP's activities to facilitate poverty eradication include support for capacity-building programmes and initiatives to generate sustainable livelihoods, for example by improving access to credit, land and technologies, and the promotion of strategies to improve education and health provision for the poorest elements of populations (with a focus on women and girls). In 1996 UNDP launched the Poverty Strategies Initiative (PSI) to strengthen national capacities to assess and monitor the extent of poverty and to combat the problem. All PSI projects were to involve representatives of governments, the private sector, social organizations and research institutions in policy debate and formulation. In 1997 a UNDP scheme to support private-sector and community-based initiatives to generate employment opportunities, MicroStart, became operational. With the World Bank, UNDP helps governments of developing countries applying for international debt relief to draft Poverty Reduction Stategy Papers.

In May 2001 UNDP signed a memorandum of understanding with the Asian Development Bank to provide for strengthened collaboration, in particular in joint activities relating to poverty measurement and assessment.

Approximately one-quarter of all UNDP programme resources support national efforts to ensure efficient and accountable governance and to build effective relations between the state, the private sector and civil society, which are essential to achieving sustainable development. UNDP undertakes assessment missions to help ensure free and fair elections and works to promote human rights, a transparent and competent public sector, a competent judicial system and decentralized government and decision-making. Within the context of the UN System-wide Special Initiative on Africa, UNDP supports the Africa Governance Forum which convenes annually to consider aspects of governance and development. In July 1997 UNDP organized an International Conference on Governance for Sustainable Growth and Equity, which was held in New York, USA. At the World Conference on Governance held in Manila, the Philippines, in May–June 1999, UNDP sponsored a series of meetings held on the subject of Building Capacities for Governance. In April of that year UNDP and the Office of the High Commissioner for Human Rights launched a joint programme to strengthen capacity-building in order to promote the integration of human rights issues into activities concerned with sustainable human development.

In September 1999 the Regional Bureau for Asia and the Pacific inaugurated a Regional Governance Programme, with the aim of supporting the implementation of reforms required at national level for the advancement of good governance. During 2001 UNDP undertook a large-scale civic education campaign to ensure a peaceful general election in East Timor (now Timor-Leste), which was conducted in late August.

UNDP plays a role in developing the agenda for international cooperation on environmental and energy issues, focusing on the relationship between energy policies, environmental protection, poverty and development. UNDP supports the development of national programmes that emphasize the sustainable management of natural resources, for example through its Sustainable Energy Initiative, which promotes more efficient use of energy resources and the introduction of renewable alternatives to conventional fuels. UNDP is also concerned with forest management, the aquatic environment and sustainable agriculture and food security. Within UNDP's framework of urban development activities the Local Initiative Facility for Urban Environment (LIFE) undertakes small-scale environmental projects in low-income communities, in collaboration with local authorities and community-based groups. Other initiatives include the Urban Management Programme and the Public–Private Partnerships Programme for the Urban Environment, which aimed to generate funds, promote research and support new technologies to enhance sustainable environments in urban areas. In 1996 UNDP initiated a process of collaboration between city authorities world-wide to promote implementation of the commitments made at the 1995 Copenhagen summit for social development (see below) and to help to combat aspects of poverty and other urban problems, such as poor housing, transport, the management of waste disposal, water supply and sanitation. The first Forum of the so-called World Alliance of Cities Against Poverty was convened in October 1998, in Lyon, France. The second Forum took place in April 2000 in Geneva, Switzerland, the third Forum in April 2002 in Huy, Belgium, and the fourth was held in March–April 2004 in Rome, Italy.

In 2003 UNDP launched an initiative to support developing countries to obtain sovereign credit ratings and thereby mobilize resources from private capital markets.

UNDP collaborates with other UN agencies in countries in crisis and with special circumstances to promote relief and development efforts, in order to secure the foundations for sustainable human development and thereby increase national capabilities to prevent or pre-empt future crises. In particular, UNDP is concerned to achieve reconciliation, reintegration and reconstruction in affected countries, as well as to support emergency interventions and management and delivery of programme aid. In 1995 the Executive Board decided that 5% of total UNDP regular resources be allocated to countries in 'special development situations', i.e. urgently requiring major, integrated external support. Special development initiatives include the demobilization of former combatants, rehabilitation of communities for the sustainable reintegration of returning populations, the restoration and strengthening of democratic institutions, and clearance of anti-personnel landmines. UNDP has established a mine action unit within its Bureau for Crisis Prevention and Recovery in order to strengthen national de-mining capabilities. UNDP is seeking to incorporate conflict prevention into its development strategies. The Conflict-related Development Analysis (CDA), building upon conflict assessment activities undertaken during 2001–02 in countries including Guatemala, Guinea-Bissau, Nepal, Nigeria and Tajikistan, is being developed as a tool for country offices to use in formulating and analyzing programmes in conflict zones. UNDP is the focal point within the UN system for strengthening national capacities for natural disaster reduction (prevention, preparedness and mitigation relating to natural, environmental and technological hazards). UNDP's Disaster Management Programme oversees the system-wide Disaster Management Training Programme. In February 2004 UNDP introduced a Disaster Risk Index that enabled vulnerability and risk to be measured and compared between countries and demonstrated the correspondence between human development and death rates following natural disasters.

In March 1998, in response to the financial crisis that emerged in South-East Asia in late 1997, UNDP initiated a programme of Special Assistance to Countries in Economic Crisis in Asia, which aimed to assist governments with planning policies to address the social consequences of the economic downturn. UNDP undertook rehabilitation activities in East Timor (now Timor-Leste) prior to its independence in May 2002. In 2004 UNDP was supporting mine action efforts in Cambodia and Laos.

UNDP supports the Tumen River Area Development Programme, under which the People's Republic of China, the Democratic People's Republic of Korea, the Republic of Korea, Mongolia and Russia collaborate to study the feasibility of joint development activity. In December 1995 the Governments agreed to pursue more formal arrangements for promoting economic and technical co-operation in the Tumen River Economic Development Area. A Secretariat, to administer an intergovernmental Consultation Commission and a Co-ordination Committee, was subsequently established in Beijing, People's Republic of China. In April 1995 the Governments of Cambodia, Laos, Thailand and Viet Nam signed an agreement for sustainable development in the Mekong River Basin, following two years of negotiations, during which UNDP provided technical support and encouragement. UNDP provided financing for the preparatory assistance phase of the implementation of the agreement, and supports the institutional capacity development of the Mekong River Commission as an independent intergovernmental organization to supervise the development, conservation and management of the river basin and its resources.

UNDP is a co-sponsor, jointly with WHO, the World Bank, UNICEF, UNESCO, UNODC, ILO, UNFPA, WFP and UNHCR, of the Joint UN Programme on HIV/AIDS (UNAIDS), which became operational on 1 January 1996. UNAIDS co-ordinates UNDP's HIV and Development Programme. UNDP regards the HIV/AIDS pandemic as a major challenge to development, and advocates for making HIV/AIDS a focus of national planning; supports decentralized action against HIV/AIDS at community level; helps to strengthen national capacities at all levels to combat the disease; and aims to link support for prevention activities, education and treatment with broader development planning and responses. UNDP places a particular focus on combating the spread of HIV/AIDS through the promotion of women's rights.

Within the UN system UNDP also has responsibility for co-ordinating activities following global UN conferences. In March 1995 government representatives attending the World Summit for Social Development, which was held in Copenhagen, Denmark, approved initiatives to promote the eradication of poverty, to increase and reallocate official development assistance to basic social programmes and to promote equal access to education. The Programme of Action adopted at the meeting advocated that UNDP support the implementation of social development programmes, co-ordinate these efforts through its field offices and organize efforts on the part of the UN system to stimulate capacity-building at local, national and regional levels. The PSI (see above) was introduced following the summit. A special session of the General Assembly to review the implementation of the summit's objectives was convened in June 2000. Following the UN Fourth World Conference on Women, held in Beijing, People's Republic of China, in September 1995, UNDP led inter-agency efforts to ensure the full participation of women in all economic, political and professional activities, and assisted with further situation analysis and training activities. (UNDP also created a Gender in Development Office to ensure that women participate more fully in UNDP-sponsored activities.) In June 2000 a special session of the General Assembly (Beijing + 5) was convened to review the conference. UNDP played an important role, at both national and international levels, in preparing for the second UN Conference on Human Settlements (Habitat II), which was held in İstanbul, Turkey, in June 1996 the (see UN Human Settlements Programme). At the conference UNDP announced the establishment of a new facility, which was designed to promote private-sector investment in urban infrastructure. A special session of the UN General Assembly, entitled Istanbul + 5, was held in June 2001 to report on the implementation of the recommendations of the Habitat II conference.

UNDP aims to ensure that, rather than creating an ever-widening 'digital divide', ongoing rapid advancements in information technology are harnessed by poorer countries to accelerate progress in achieving sustainable human development. UNDP advises governments on technology policy, promotes digital entrepreneurship in programme countries and works with private-sector partners to provide reliable and affordable communications networks. The Bureau for Development Policy operates the Information and Communication Technologies for Development Programme, which aims to promote sustainable human development through increased utilization of information and communications technologies globally. The Programme aims to establish technology access centres in developing countries. A Sustainable Development Networking Programme focuses on expanding internet connectivity in poorer countries through building national capacities and supporting local internet sites. UNDP has used mobile internet units to train people even in isolated rural areas. In 1999 UNDP, in collaboration with an international communications company, Cisco Systems, and other partners, launched NetAid, an internet-based forum (accessible at www.netaid.org) for mobilizing and co-ordinating fundraising and other activities aimed at alleviating poverty and promoting sustainable human development in the developing world. With Cisco Systems and other partners, UNDP has worked to establish academies of information technology to support training and capacity-building in developing countries. By September 2003 88 academies had been established. UNDP and the World Bank jointly host the secretariat of the Digital Opportunity Task Force, a partnership between industrialized and developing countries, business and non-governmental organizations that was established in 2000. UNDP is a partner in the Global Digital Technology Initiative, launched in 2002 to strengthen the role of information and communications technologies in achieving the development goals of developing countries. In January 2004 UNDP and Microsoft Corporation announced an agreement to develop jointly information and communication technology (ICT) projects aimed at assisting developing countries to achieve the MDGs.

In 1996 UNDP implemented its first corporate communications and advocacy strategy, which aimed to generate public awareness of the activities of the UN system, to promote debate on development issues and to mobilize resources by increasing public and donor appreciation of UNDP. UNDP sponsors the International Day for the Eradication of Poverty, held annually on 17 October.

Finance

UNDP and its various funds and programmes are financed by the voluntary contributions of members of the United Nations and the Programme's participating agencies, as well as through cost-sharing by recipient governments and third-party donors. In 2004–05 total voluntary contributions were projected at US $3,500m., of which a projected $1,700m. constituted regular (core) resources and $1,807m. third-party co-financing and thematic trust fund income. Cost-sharing by programme country governments was projected at $2,100m., bringing total resources (both donor and local) to a projected $5,600m.

Publications

Annual Report of the Administrator.

Choices (quarterly).

Global Public Goods: International Co-operation in the 21st Century.

Human Development Report (annually, also available on CD-ROM).

Poverty Report (annually).

Results-Oriented Annual Report.

Associated Funds and Programmes

UNDP is the central funding, planning and co-ordinating body for technical co-operation within the UN system. A number of associated funds and programmes, financed separately by means of voluntary contributions, provide specific services through the UNDP network. UNDP manages a trust fund to promote economic and technical co-operation among developing countries.

CAPACITY 2015

UNDP initiated Capacity 2015 at the World Summit for Sustainable Development, which was held in August–September 2002. Capacity 2015 aims to support developing countries in expanding their capabilities to meet the Millennium Development Goals pledged by governments at a summit meeting of the UN General Assembly in September 2000.

GLOBAL ENVIRONMENT FACILITY—GEF

The GEF, which is managed jointly by UNDP, the World Bank and UNEP, began operations in 1991 and was restructured in 1994. Its aim is to support projects concerning climate change, the conservation of biological diversity, the protection of international waters, reducing the depletion of the ozone layer in the atmosphere, and (since October 2002) arresting land degradation and addressing the issue of persistent organic pollutants. The GEF acts as the financial mechanism for the Convention on Biological Diversity and the UN Framework Convention on Climate Change. UNDP is responsible for capacity-building, targeted research, pre-investment activities and technical assistance. UNDP also administers the Small Grants Programme of the GEF, which supports community-based activities by local non-governmental organizations, and the Country Dialogue Workshop Programme, which promotes dialogue on national priorities with regard to the GEF. Some 32 donor countries pledged US $2,920m. for the third periodic replenishment of GEF funds (GEF-3), covering the period 2002–06. During 1991–2003 the GEF allocated $4,500m. in grants and raised $14,474m. in co-financing from other sources in support of more than 1,400 projects.

The GEF administers a regional project on the prevention and management of marine pollution in East Asian seas and a South Pacific Biodiversity Programme.

Chair. and CEO: Dr Leonard Good (Canada).

MONTREAL PROTOCOL

Through its Montreal Protocol Unit UNDP collaborates with public and private partners in developing countries to assist them in eliminating the use of ozone-depleting substances (ODS), in accordance with the Montreal Protocol to the Vienna Convention for the Protection of the Ozone Layer, through the design, monitoring and evaluation of ODS phase-out projects and programmes. In particular, UNDP provides technical assistance and training, national capacity-building and demonstration projects and technology transfer investment projects. By mid-2003, through the Executive Committee of the Montreal Protocol, UNDP had implemented projects and activities resulting in the elimination of 33,529 metric tons of ODS.

UNDP DRYLANDS DEVELOPMENT CENTRE—DDC

The Centre, based in Nairobi, Kenya, was established in February 2002, superseding the former UN Office to Combat Desertification and Drought (UNSO). (UNSO had been established following the conclusion, in October 1994, of the UN Convention to Combat Desertification in Those Countries Experiencing Serious Drought and/or Desertification, Particularly in Africa; in turn, UNSO had replaced the former UN Sudano-Sahelian Office.) The DDC was to focus on the following areas: ensuring that national development planning takes account of the needs of dryland communities, particularly in poverty reduction strategies; helping countries to cope with the effects of climate variability, especially drought, and to prepare for future climate change; and addressing local issues affecting the utilization of resources.

Director: PHILIP DOBIE (United Kingdom).

UNITED NATIONS CAPITAL DEVELOPMENT FUND—UNCDF

The Fund was established in 1966 and became fully operational in 1974. It invests in poor communities in least-developed countries through local governance projects and microfinance operations, with the aim of increasing such communities' access to essential local infrastructure and services and thereby improving their productive capacities and self-reliance. UNDCF encourages participation by local people and local governments in the planning, implementation and monitoring of projects. The Fund aims to promote the interests of women in community projects and to enhance their earning capacities. In 1998 the Fund nominated 15 less-developed countries in which to concentrate subsequent programmes. A Special Unit for Microfinance (SUM), established in 1997 as a joint UNDP/UNCDF operation, was fully integrated into UNCDF in 1999. UNDCF/SUM helps to develop financial services for poor communities and supports UNDP's MicroStart initiative. UNCDF's annual programming budget amounts to some US $40m.

Officer-in-Charge: HENRIETTE KEIJZERS.

UNITED NATIONS DEVELOPMENT FUND FOR WOMEN—UNIFEM

UNIFEM is the UN's lead agency in addressing the issues relating to women in development and promoting the rights of women worldwide. The Fund provides direct financial and technical support to enable low-income women in developing countries to increase earnings, gain access to labour-saving technologies and otherwise improve the quality of their lives. It also funds activities that include women in decision-making related to mainstream development projects. In 2001 UNIFEM's Trust Fund in Support of Actions to Eliminate Violence Against Women (established in 1996) provided grants to 21 national and regional programmes. During 1996–2003 the Trust Fund awarded grants totalling US $7.4m. in support of 157 initiatives in more than 80 countries. UNIFEM has supported the preparation of national reports in 30 countries and used the priorities identified in these reports and in other regional initiatives to formulate a Women's Development Agenda for the 21st century. Through these efforts, UNIFEM played an active role in the preparation for the UN Fourth World Conference on Women, which was held in Beijing, People's Republic of China, in September 1995. UNIFEM participated at a special session of the General Assembly convened in June 2000 to review the conference, entitled Women 2000: Gender Equality, Development and Peace for the 21st Century (Beijing + 5). In March 2001 UNIFEM, in collaboration with International Alert, launched a Millennium Peace Prize for Women. UNIFEM maintains that the empowerment of women is a key to combating the HIV/AIDS pandemic, in view of the fact that women and adolescent girls are often culturally, biologically and economically more vulnerable to infection and more likely to bear responsibility for caring for the sick. In March 2002 UNIFEM launched a three-year programme aimed at making the gender and human rights dimensions of the pandemic central to policy-making in ten countries. A new online resource (www.genderandaids.org) on the gender dimensions of HIV/AIDS was launched in February 2003. UNIFEM was a co-founder of WomenWatch (accessible online at www.un.org/womenwatch), a UN system-wide resource for the advancement of gender equality. Programme expenditure in 2003 totalled $27.0m.

Headquarters

304 East 45th St, 15th Floor, New York, NY 10017, USA; tel. (212) 906-6400; fax (212) 906-6705; e-mail unifem@undp.org; internet www.unifem.org.

Director: NOELEEN HEYZER (Singapore).

UNITED NATIONS VOLUNTEERS—UNV

The United Nations Volunteers is an important source of middle-level skills for the UN development system supplied at modest cost, particularly in the least-developed countries. Volunteers expand the scope of UNDP project activities by supplementing the work of international and host-country experts and by extending the influence of projects to local community levels. UNV also supports technical co-operation within and among the developing countries by encouraging volunteers from the countries themselves and by forming regional exchange teams comprising such volunteers. UNV is involved in areas such as peace-building, elections, human rights, humanitarian relief and community-based environmental programmes, in addition to development activities.

The UN International Short-term Advisory (UNISTAR) Programme, which is the private-sector development arm of UNV, has increasingly focused its attention on countries in the process of economic transition. Since 1994 UNV has administered UNDP's Transfer of Knowledge Through Expatriate Nationals (TOKTEN) programme, which was initiated in 1977 to enable specialists and professionals from developing countries to contribute to development efforts in their countries of origin through short-term technical assignments.

At the end of August 2004 4,517 UNVs were serving in 132 countries. At that time the total number of people who had served under the initiative amounted to more than 30,000 in some 140 countries.

Headquarters

POB 260111, 53153 Bonn, Germany; tel. (228) 8152000; fax (228) 8152001; e-mail information@unvolunteers.org; internet www.unv.org.

Executive Co-ordinator: AD DE RAAD (Netherlands).

United Nations Environment Programme—UNEP

Address: POB 30552, Nairobi, Kenya.

Telephone: (20) 621234; **fax:** (20) 624489; **e-mail:** cpiinfo@unep.org; **internet:** www.unep.org.

The United Nations Environment Programme was established in 1972 by the UN General Assembly, following recommendations of the 1972 UN Conference on the Human Environment, in Stockholm, Sweden, to encourage international co-operation in matters relating to the human environment.

Organization

(October 2004)

GOVERNING COUNCIL

The main functions of the Governing Council, which meets every two years, are to promote international co-operation in the field of the environment and to provide general policy guidance for the direction and co-ordination of environmental programmes within the UN system. It comprises representatives of 58 states, elected by the UN General Assembly, for four-year terms, on a regional basis. The Council is assisted in its work by a Committee of Permanent Representatives.

HIGH-LEVEL COMMITTEE OF MINISTERS AND OFFICIALS IN CHARGE OF THE ENVIRONMENT

The Committee was established by the Governing Council in 1997, with a mandate to consider the international environmental agenda and to make recommendations to the Council on reform and policy issues. In addition, the Committee, comprising 36 elected members, was to provide guidance and advice to the Executive Director, to enhance UNEP's collaboration and co-operation with other multilateral bodies and to help to mobilize financial resources for UNEP.

SECRETARIAT

Offices and divisions at UNEP headquarters include the Office of the Executive Director; the Secretariat for Governing Bodies: Offices for Evaluation and Oversight, Programme Co-ordination and Management, and Resource Mobilization; and divisions of communications and public information, early warning and assessment, policy development and law, policy implementation, technology and industry and economics, regional co-operation and representation, environmental conventions, and Global Environment Facility co-ordination.

Executive Director: Dr KLAUS TÖPFER (Germany).

REGIONAL OFFICES

Asia and the Pacific: UN Bldg, 10th Floor, Rajdamnern Ave, Bangkok 10200, Thailand; tel. (2) 288-1870; fax (2) 280-3829; e-mail uneproap@un.org; internet www.roap.unep.org.

UNEP China Office: 2 Liangmahe Nanlu, Beijing 100600, People's Republic of China; tel. (10) 6532-3731 (ext. 227); e-mail xiak@un.org.

OTHER OFFICES

Convention on International Trade in Endangered Species of Wild Fauna and Flora—CITES: 15 chemin des Anémones, 1219 Châtelaine, Geneva, Switzerland; tel. (22) 9178139; fax (22) 7973417; e-mail cites@unep.ch; internet www.cites.org; Sec.-Gen. WILLEM WOUTER WIJNSTEKERS (Netherlands).

Global Programme of Action for the Protection of the Marine Environment from Land-based Activities: POB 16227, 2500 BE The Hague, Netherlands; tel. (70) 3114460; fax (70) 3456648; e-mail gpa@unep.nl; internet www.gpa.unep.org; Co-ordinator Dr VEERLE VANDEWEERD.

Regional Co-ordinating Unit for East Asian Seas: UN Bldg, 10th Floor, Rajdamnern Ave, Bangkok 10200, Thailand; tel. (2) 288-1234; fax (2) 267-0808; e-mail kirkman.unescap@un.org; internet www.unep.org/water/regseas/easian/htm; Co-ordinator HUGH KIRKMAN.

Secretariat of the Basel Convention: CP 356, 13–15 chemin des Anémones, 1219 Châtelaine, Geneva, Switzerland; tel. (22) 9178218; fax (22) 7973454; e-mail sbc@unep.ch; internet www.basel.int; Exec. Sec. SACHIKO KUWABARA-YAMAMOTO.

Secretariat of the Convention on Biological Diversity: World Trade Centre, 393 St Jacques St West, Suite 300, Montréal, QC, Canada H2Y 1N9; tel. (514) 288-2220; fax (514) 288-6588; e-mail secretariat@biodiv.org; internet www.biodiv.org; Exec. Sec. HAMDALLAH ZEDAN.

Secretariat of the Multilateral Fund for the Implementation of the Montreal Protocol: 1800 McGill College Ave, 27th Floor, Montréal, QC, Canada H3A 3J6; tel. (514) 282-1122; fax (514) 282-0068; e-mail secretariat@unmfs.org; internet www.multilateralfund.org; Chief MARIA NOLAN (United Kingdom).

Secretariat of the UN Framework Convention on Climate Change: Haus Carstanjen, Martin-Luther-King-Str. 8, 53175 Bonn, Germany; tel. (228) 815-1000; fax (228) 815-1999; e-mail secretariat@unfccc.de; internet www.unfccc.de; Exec. Sec. JOKE WALLER-HUNTER (Netherlands).

UNEP/CMS (Convention on the Conservation of Migratory Species of Wild Animals) Secretariat: Martin-Luther-King-Str. 8, 53175 Bonn, Germany; tel. (228) 8152402; fax (228) 8152449; e-mail secretariat@cms.int; internet www.cms.int; Exec. Sec. ROBERT HEPWORTH.

UNEP Chemicals: International Environment House, 11–13 chemin des Anémones, 1219 Châtelaine, Geneva, Switzerland; tel. (22) 9178192; fax (22) 7973460; e-mail chemicals@unep.ch; internet www.chem.unep.ch; Dir JAMES B. WILLIS.

UNEP Division of Technology, Industry and Economics: Tour Mirabeau, 39–43, Quai André Citroën, 75739 Paris Cédex 15, France; tel. 1-44-37-14-41; fax 1-44-37-14-74; e-mail unep.tie@unep.fr; internet www.uneptie.org/; Dir MONIQUE BARBUT (France).

UNEP International Environmental Technology Centre—IETC: 2–110 Ryokuchi koen, Tsurumi-ku, Osaka 538-0036, Japan; tel. (6) 6915-4581; fax (6) 6915-0304; e-mail ietc@unep.or.jp; internet www.unep.or.jp; Dir STEVE HALLS.

UNEP Ozone Secretariat: POB 30552, Nairobi, Kenya; tel. (20) 623850; fax (20) 623913; e-mail ozoneinfo@unep.org; internet www.unep.org/ozone/; Exec. Sec. MARCO GONZALEZ (Costa Rica).

UNEP Secretariat for the UN Scientific Committee on the Effects of Atomic Radiation: Vienna International Centre, Wagramerstrasse 5, POB 500, 1400 Vienna, Austria; tel. (1) 26060-4330; fax (1) 26060-5902; e-mail norman.gentner@unvienna.org; internet www.unscear.org; Sec. Dr NORMAN GENTNER.

Activities

UNEP serves as a focal point for environmental action within the UN system. It aims to maintain a constant watch on the changing state of the environment; to analyse the trends; to assess the problems using a wide range of data and techniques; and to promote projects leading to environmentally sound development. It plays a catalytic and co-ordinating role within and beyond the UN system. Many UNEP projects are implemented in co-operation with other UN agencies, particularly UNDP, the World Bank group, FAO, UNESCO and WHO. About 45 intergovernmental organizations outside the UN system and 60 international non-governmental organizations have official observer status on UNEP's Governing Council, and, through the Environment Liaison Centre in Nairobi, UNEP is linked to more than 6,000 non-governmental bodies concerned with the environment. UNEP also sponsors international conferences, programmes, plans and agreements regarding all aspects of the environment.

In February 1997 the Governing Council, at its 19th session, adopted a ministerial declaration (the Nairobi Declaration) on UNEP's future role and mandate, which recognized the organization as the principal UN body working in the field of the environment and as the leading global environmental authority, setting and overseeing the international environmental agenda. In June a special session of the UN General Assembly, referred to as 'Rio + 5', was convened to review the state of the environment and progress achieved in implementing the objectives of the UN Conference on Environment and Development (UNCED), held in Rio de Janeiro, Brazil, in June 1992. The meeting adopted a Programme for Further Implementation of Agenda 21 (a programme of activities to promote sustainable development, adopted by UNCED) in order to intensify efforts in areas such as energy, freshwater resources and technology transfer. The meeting confirmed UNEP's essential role in advancing the Programme and as a global authority promoting a coherent legal and political approach to the environmental challenges of sustainable development. An extensive process of restructuring and realignment of functions was subsequently initiated by UNEP, and a new organizational structure reflecting the decisions of the Nairobi Declaration was implemented during 1999. UNEP played a leading role in preparing for the World Summit on Sustainable Development (WSSD), held in August–September 2002 in Johannesburg, South Africa, to assess strategies for strengthening the implementation of Agenda 21. Governments participating in the conference adopted the Johannesburg Declaration and WSSD Plan of Implementation, in which they strongly reaffirmed commitment to the principles underlying Agenda 21 and also pledged support to all internationally-agreed development goals, including the UN Millennium Development Goals adopted by governments attending a summit meeting of the UN General Assembly in September 2000. Participating governments made concrete commitments to attaining several specific objectives in the areas of water, energy, health, agriculture and fisheries, and biodiversity. These included a reduction by one-half in the proportion of people world-wide lacking access to clean water or good sanitation by 2015, the restocking of depleted fisheries by 2015, a reduction in the ongoing loss in biodiversity by 2010, and the production and utilization of chemicals without causing harm to human beings and the environment by 2020. Participants determined to increase usage of renewable energy sources and to develop by 2005 integrated water resources management and water efficiency plans. A large number of partnerships between governments, private sector interests and civil society groups were announced at the conference.

In May 2000 UNEP sponsored the first annual Global Ministerial Environment Forum (GMEF), held in Malmö, Sweden, and attended by environment ministers and other government delegates from more than 130 countries. Participants reviewed policy issues in the field of the environment and addressed issues such as the impact on the environment of population growth, the depletion of earth's natural resources, climate change and the need for fresh water supplies. The Forum issued the Malmö Declaration, which identified the effective implementation of international agreements on environmental matters at national level as the most pressing challenge for policy-makers. The Declaration emphasized the importance of mobilizing domestic and international resources and urged increased co-operation from civil society and the private sector in achieving sustainable development. The second GMEF, held in Nairobi in February 2001, addressed means of strengthening international environmental governance, establishing an Open-Ended Intergovernmental Group of Ministers or Their Representatives (IGM) to prepare a report on possible reforms. GMEF-3, held in Cartagena, Colombia, in February 2002, considered UNEP's participation in the forthcoming WSSD, with a focus on environmental guidance issues.

ENVIRONMENTAL ASSESSMENT AND EARLY WARNING

The Nairobi Declaration resolved that the strengthening of UNEP's information, monitoring and assessment capabilities was a crucial element of the organization's restructuring, in order to help establish priorities for international, national and regional action, and to ensure the efficient and accurate dissemination of emerging environmental trends and emergencies.

In 1995 UNEP launched the Global Environment Outlook (GEO) process of environmental assessment. UNEP is assisted in its analysis of the state of the global environment by an extensive network of collaborating centres. The first *Global Environment Outlook, GEO-I,* was published in January 1997, the second, *GEO 2000,* in September 1999, and *GEO-3* in May 2002. From 2003 reports on the process were to be issued annually. The following regional and national *GEO* reports have been produced: *Africa Environment Outlook* (2002), *Brazil Environment Outlook* (2002), *Latin America and the Caribbean—Environment Outlook 2000*, *Caucasus Environment Outlook* (2002), *North America's Environment* (2002), *Pacific Islands Environment Outlook* (1999), and *Western Indian Ocean Environment Outlook* (1999). UNEP is leading a major Global International Waters Assessment (GIWA) to consider all aspects of the world's water-related issues, in particular problems of shared transboundary waters, and of future sustainable management of water resources. UNEP is also a sponsoring agency of the Joint Group of Experts on the Scientific Aspects of Marine Environmental Pollution and contributes to the preparation of reports on the state of the marine environment and on the impact of land-based activities on that environment. In November 1995 UNEP published a Global Biodiversity Assessment, which was the first comprehensive study of biological resources throughout the world. The UNEP—World Conservation Monitoring Centre (UNEP—WCMC), established in June 2000, provides biodiversity-related assessment. UNEP is a partner in the International Coral Reef Action Network—ICRAN, which was established in 2000 to manage and protect coral reefs world-wide. In June 2001 UNEP launched the Millennium Ecosystems Assessment, which was expected to be completed in 2004. Other major assessments under way in 2002 included GIWA (see above); the Assessment of Impact and Adaptation to Climate Change; the Solar and Wind Energy Resource Assessment; the Regionally-Based Assessment of Persistent Toxic Substances; the Land Degradation Assessment in Drylands; and the Global Methodology for Mapping Human Impacts on the Biosphere (GLOBIO) project.

UNEP's environmental information network includes the Global Resource Information Database (GRID), which converts collected data into information usable by decision-makers. The UNEP-INFOTERRA programme facilitates the exchange of environmental information through an extensive network of national 'focal points'. By September 2004 177 countries were participating in the network. Through UNEP-INFOTERRA UNEP promotes public access to environmental information, as well as participation in environmental concerns. UNEP aims to establish in every developing region an Environment and Natural Resource Information Network (ENRIN) in order to make available technical advice and manage environmental information and data for improved decision-making and action-planning in countries most in need of assistance. UNEP aims to integrate its information resources in order to improve access to information and to promote its international exchange. This has been pursued through UNEPnet, an internet-based interactive environmental information- and data-sharing facility, and Mercure, a telecommunications service using satellite technology to link a network of 16 earth stations throughout the world.

UNEP's information, monitoring and assessment structures also serve to enhance early-warning capabilities and to provide accurate information during an environmental emergency.

POLICY DEVELOPMENT AND LAW

UNEP aims to promote the development of policy tools and guide-lines in order to achieve the sustainable management of the world environment. At a national level it assists governments to develop and implement appropriate environmental instruments and aims to co-ordinate policy initiatives. Training workshops in various aspects of environmental law and its applications are conducted. UNEP supports the development of new legal, economic and other policy instruments to improve the effectiveness of existing environmental agreements.

UNEP was instrumental in the drafting of a Convention on Biological Diversity (CBD) to preserve the immense variety of plant and animal species, in particular those threatened with extinction. The Convention entered into force at the end of 1993; by October 2004 187 countries and the European Community were parties to the CBD. The CBD's Cartagena Protocol on Biosafety (so called as it had been addressed at an extraordinary session of parties to the CBD convened in Cartagena, Colombia, in February 1999) was adopted at a meeting of parties to the CBD held in Montréal, Canada, in January 2000, and entered into force in September 2003; by October 2004 the Protocol had been ratified by 108 countries and the European Community. The Protocol regulates the transboundary movement and use of living modified organisms resulting from biotechnology in order to reduce any potential adverse effects on biodiversity and human health. It establishes an Advanced Informed Agreement procedure to govern the import of such organisms. In January 2002 UNEP launched a major project aimed at supporting developing countries with assessing the potential health and environmental risks and benefits of genetically-modified (GM) crops, in preparation for the Protocol's entry into force. In February the parties to the CBD and other partners convened a conference, in Montréal, to address ways in which the traditional knowledge and practices of local communities could be preserved and used to conserve highly-threatened species and ecosystems. The sixth conference of parties to the CBD, held in April 2002, adopted detailed voluntary guide-lines concerning access to genetic resources and sharing the benefits attained from such resources with the countries and local communities where they originate; a global work programme on forests; and a set of guiding principles for combating alien invasive species. UNEP supports co-operation for biodiversity assessment and management in selected developing regions and for the development of strategies for the conservation and sustainable exploitation of individual threatened species (e.g. the Global Tiger Action Plan). It also provides assistance for the preparation of individual country studies and strategies to strengthen national biodiversity management and research. UNEP administers the Convention on International Trade in Endangered Species of Wild Flora and Fauna (CITES), which entered into force in 1975.

A new CITES Tiger Enforcement Task Force met for the first time in New Delhi, India, in April 2001; all trade in tigers and tiger parts is prohibited under CITES. In May UNEP launched the Great Apes Survival Project (GRASP), which supports, in 23 countries in South-East Asia and Africa, the conservation of orang-utans, gorillas, chimpanzees and bonobos.

In October 1994 87 countries, meeting under UN auspices, signed a Convention to Combat Desertification (see UNDP Drylands Development Centre), which aimed to provide a legal framework to counter the degradation of drylands. An estimated 75% of all drylands have suffered some land degradation, affecting approximately 1,000m. people in 110 countries. UNEP continues to support the implementation of the Convention, as part of its efforts to protect land resources. UNEP also aims to improve the assessment of dryland degradation and desertification in co-operation with governments and other international bodies, as well as identifying the causes of degradation and measures to overcome these.

UNEP is the lead UN agency for promoting environmentally sustainable water management. It regards the unsustainable use of water as the most urgent environmental and sustainable development issue, and estimates that two-thirds of the world's population will suffer chronic water shortages by 2025, owing to rising demand for drinking water as a result of growing populations, decreasing quality of water because of pollution, and increasing requirements of industries and agriculture. In 2000 UNEP adopted a new water policy and strategy, comprising assessment, management and co-ordination components. The Global International Waters Assessment (see above) is the primary framework for the assessment component. The management component includes the Global Programme of Action (GPA) for the Protection of the Marine Environment from Land-based Activities (adopted in November 1995), and UNEP's freshwater programme and regional seas programme. The GPA for the Protection of the Marine Environment for Land-based Activities focuses on the effects of activities such as pollution on freshwater resources, marine biodiversity and the coastal ecosystems of small-island developing states. UNEP aims to develop a similar global instrument to ensure the integrated management of freshwater resources. It promotes international co-operation in the management of river basins and coastal areas and for the development of tools and guide-lines to achieve the sustainable management of freshwater and coastal resources. UNEP provides scientific, technical and administrative support to facilitate the implementation and co-ordination of 14 regional seas conventions and 13 regional plans of action, and is developing a strategy to strengthen collaboration in their implementation. The new water policy and strategy emphasizes the need for improved co-ordination of existing activities. UNEP aims to play an enhanced role within relevant co-ordination mechanisms, such as the UN open-ended informal consultation process on oceans and the law of the sea.

In 1996 UNEP, in collaboration with FAO, began to work towards promoting and formulating a legally binding international convention on prior informed consent (PIC) for hazardous chemicals and pesticides in international trade, extending a voluntary PIC procedure of information exchange undertaken by more than 100 governments since 1991. The Convention was adopted at a conference held in Rotterdam, Netherlands, in September 1998, and entered into force in February 2004. It aims to reduce risks to human health and the environment by restricting the production, export and use of hazardous substances and enhancing information exchange procedures.

In conjunction with UN-Habitat, UNDP, the World Bank and other organizations and institutions, UNEP promotes environmental concerns in urban planning and management through the Sustainable Cities Programme, as well as regional workshops concerned with urban pollution and the impact of transportation systems. In 1994 UNEP inaugurated an International Environmental

Technology Centre (IETC), with offices in Osaka and Shiga, Japan, in order to strengthen the capabilities of developing countries and countries with economies in transition to promote environmentally-sound management of cities and freshwater reservoirs through technology co-operation and partnerships.

UNEP has played a key role in global efforts to combat risks to the ozone layer, resultant climatic changes and atmospheric pollution. UNEP worked in collaboration with the World Meteorological Organization to formulate the UN Framework Convention on Climate Change (UNFCCC), with the aim of reducing the emission of gases that have a warming effect on the atmosphere, and has remained an active participant in the ongoing process to review and enforce the implementation of the Convention and of its Kyoto Protocol. UNEP was the lead agency in formulating the 1987 Montreal Protocol to the Vienna Convention for the Protection of the Ozone Layer (1985), which provided for a 50% reduction in the production of chlorofluorocarbons (CFCs) by 2000. An amendment to the Protocol was adopted in 1990, which required complete cessation of the production of CFCs by 2000 in industrialized countries and by 2010 in developing countries; these deadlines were advanced to 1996 and 2006, respectively, in November 1992. In 1997 the ninth Conference of the Parties (COP) to the Vienna Convention adopted a further amendment which aimed to introduce a licensing system for all controlled substances. The eleventh COP, meeting in Beijing, People's Republic of China, in November–December 1999, adopted the Beijing Amendment, which imposed tighter controls on the import and export of hydrochlorofluorocarbons, and on the production and consumption of bromochloromethane (Halon-1011, an industrial solvent and fire extinguisher). The Beijing Amendment entered into force in December 2001. A Multilateral Fund for the Implementation of the Montreal Protocol was established in June 1990 to promote the use of suitable technologies and the transfer of technologies to developing countries. UNEP, UNDP, the World Bank and UNIDO are the sponsors of the Fund, which by 30 June 2004 had approved financing for more than 4,300 projects in 134 developing countries at a cost of US $1,631m. Commitments of $474m. were made to the fifth replenishment of the Fund, covering the three-year period 2003–05.

POLICY IMPLEMENTATION

UNEP's Division of Environmental Policy Implementation incorporates two main functions: technical co-operation and response to environmental emergencies.

With the UN Office for the Co-ordination of Humanitarian Assistance (OCHA), UNEP has established a joint Environment Unit to mobilize and co-ordinate international assistance and expertise for countries facing environmental emergencies and natural disasters. In mid-1999 UNEP and UN-Habitat jointly established a Balkan Task Force (subsequently renamed UNEP Balkans Unit) to assess the environmental impact of NATO's aerial offensive against the Federal Republic of Yugoslavia (now Serbia and Montenegro). In November 2000 the Unit led a field assessment to evaluate reports of environmental contamination by debris from NATO ammunition containing depleted uranium. A final report, issued by UNEP in March 2001, concluded that there was no evidence of widespread contamination of the ground surface by depleted uranium and that the radiological and toxicological risk to the local population was negligible. It stated, however, that considerable scientific uncertainties remained, for example as to the safety of groundwater and the longer-term behaviour of depleted uranium in the environment, and recommended precautionary action. In December 2001 UNEP established a new Post-conflict Assessment Unit, which replaced, and extended the scope of, the Balkans Unit. In 2004 the Post-conflict Assessment Unit was undertaking activities in Afghanistan as well as the Balkans, and was compiling desk assessments of the state of the environment in Iraq and the Palestinian territories.

UNEP, together with UNDP and the World Bank, is an implementing agency of the Global Environment Facility (GEF), which was established in 1991 as a mechanism for international co-operation in projects concerned with biological diversity, climate change, international waters and depletion of the ozone layer. UNEP services the Scientific and Technical Advisory Panel, which provides expert advice on GEF programmes and operational strategies.

TECHNOLOGY, INDUSTRY AND ECONOMICS

The use of inappropriate industrial technologies and the widespread adoption of unsustainable production and consumption patterns have been identified as being inefficient in the use of renewable resources and wasteful, in particular in the use of energy and water. UNEP aims to encourage governments and the private sector to develop and adopt policies and practices that are cleaner and safer, make efficient use of natural resources, incorporate environmental costs, ensure the environmentally sound management of chemicals, and reduce pollution and risks to human health and the environment. In collaboration with other organizations and agencies UNEP works to define and formulate international guide-lines and agreements to address these issues. UNEP also promotes the transfer of appropriate technologies and organizes conferences and training workshops to provide sustainable production practices. Relevant information is disseminated through the International Cleaner Production Information Clearing House. UNEP, together with UNIDO, has established 27 National Cleaner Production Centres to promote a preventive approach to industrial pollution control. In October 1998 UNEP adopted an International Declaration on Cleaner Production, with a commitment to implement cleaner and more sustainable production methods and to monitor results; the Declaration had 443 signatories at April 2004, including representatives of 52 national governments. In 1997 UNEP and the Coalition for Environmentally Responsible Economies initiated the Global Reporting Initiative, which, with participation by corporations, business associations and other organizations and stakeholders, develops guidelines for voluntary reporting by companies on their economic, environmental and social performance. In April 2002 UNEP launched the 'Life-Cycle Initiative', which aims to assist governments, businesses and other consumers with adopting environmentally-sound policies and practice, in view the upward trend in global consumption patterns.

UNEP provides institutional servicing to the Basel Convention on the Control of Transboundary Movements of Hazardous Wastes and their Disposal, which was adopted in 1989 with the aim of preventing the disposal of wastes from industrialized countries in countries that have no processing facilities. In March 1994 the second meeting of parties to the Convention determined to ban the exportation of hazardous wastes between industrialized and developing countries. The third meeting of parties to the Convention, held in 1995, proposed that the ban should be incorporated into the Convention as an amendment. The resulting so-called Ban Amendment (prohibiting exports of hazardous wastes for final disposal and recycling from states and/or parties also belonging to OECD and, or, the European Union, and from Liechtenstein, to any other state party to the Convention) required ratification by three-quarters of the 62 signatory states present at the time of adoption before it could enter into effect; by October 2004 the Ban Amendment had been ratified by 50 parties. In 1998 the technical working group of the Convention agreed a new procedure for clarifying the classification and characterization of specific hazardous wastes. The fifth full meeting of parties to the Convention, held in December 1999, adopted the Basel Declaration outlining an agenda for the period 2000–10, with a particular focus on minimizing the production of hazardous wastes. At October 2004 the number of parties to the Convention totalled 162. In December 1999 132 states adopted a Protocol to the Convention to address issues relating to liability and compensation for damages from waste exports. The governments also agreed to establish a multilateral fund to finance immediate clean-up operations following any environmental accident.

The UNEP Chemicals office was established to promote the sound management of hazardous substances, central to which has been the International Register of Potentially Toxic Chemicals (IRPTC). UNEP aims to facilitate access to data on chemicals and hazardous wastes, in order to assess and control health and environmental risks, by using the IRPTC as a clearing house facility of relevant information and by publishing information and technical reports on the impact of the use of chemicals.

In 2003 Pollutant Release and Transfer Registers (PRTRs), for collecting and disseminating data on toxic emissions, were in effect in Japan and the Republic of Korea. Work was progressing towards the introduction of a PRTR in Taiwan.

UNEP's OzonAction Programme works to promote information exchange, training and technological awareness. Its objective is to strengthen the capacity of governments and industry in developing countries to undertake measures towards the cost-effective phasing-out of ozone-depleting substances. UNEP also encourages the development of alternative and renewable sources of energy. To achieve this, UNEP is supporting the establishment of a network of centres to research and exchange information of environmentally-sound energy technology resources.

REGIONAL CO-OPERATION AND REPRESENTATION

UNEP maintains six regional offices. These work to initiate and promote UNEP objectives and to ensure that all programme formulation and delivery meets the specific needs of countries and regions. They also provide a focal point for building national, sub-regional and regional partnership and enhancing local participation in UNEP initiatives. Following UNEP's reorganization a co-ordination office was established at headquarters to promote regional policy integration, to co-ordinate programme planning, and to provide necessary services to the regional offices.

UNEP provides administrative support to several regional conventions, for example the Lusaka Agreement on Co-operative Enforcement Operations Directed at Illegal Trade in Wild Flora and Fauna, which entered into force in December 1996 having been concluded under UNEP auspices in order to strengthen the implementation of the CBD and CITES in Eastern and Central Africa. UNEP also organizes conferences, workshops and seminars at national and regional levels, and may extend advisory services or technical assistance to individual governments.

CONVENTIONS

UNEP aims to develop and promote international environmental legislation in order to pursue an integrated response to global environmental issues, to enhance collaboration among existing convention secetariats, and to co-ordinate support to implement the work programmes of international instruments.

UNEP has been an active participant in the formulation of several major conventions (see above). The Division of Environmental Conventions is mandated to assist the Division of Policy Development and Law in the formulation of new agreements or protocols to existing conventions. Following the successful adoption of the Rotterdam Convention in September 1998, UNEP played a leading role in formulating a multilateral agreement to reduce and ultimately eliminate the manufacture and use of Persistent Organic Pollutants (POPs), which are considered to be a major global environmental hazard. The agreement on POPs, concluded in December 2000 at a conference sponsored by UNEP in Johannesburg, South Africa, was adopted by 127 countries in May 2001; it entered into force in May 2004, three months after its ratification by the requisite 50 states in February of that year.

UNEP has been designated to provide secretariat functions to a number of global and regional environmental conventions (see above for list of offices).

COMMUNICATIONS AND PUBLIC INFORMATION

UNEP's public education campaigns and outreach programmes promote community involvement in environmental issues. Further communication of environmental concerns is undertaken through the media, an information centre service and special promotional events, including World Environment Day, photography competitions, and the awarding of the Sasakawa Prize (to recognize distinguished service to the environment by individuals and groups) and of the Global 500 Award for Environmental Achievement. In 1996 UNEP initiated a Global Environment Citizenship Programme to promote acknowledgment of the environmental responsibilities of all sectors of society.

Finance

UNEP derives its finances from the regular budget of the United Nations and from voluntary contributions to the Environment Fund. A budget of US $119.9m. was authorized for the two-year period 2002–03, of which $100m. was for programme activities, $14.9m. for management and administration, and $5m. for fund programme reserves.

Publications

Annual Report.
APELL Newsletter (2 a year).
Cleaner Production Newsletter (2 a year).
Climate Change Bulletin (quarterly).
Connect (UNESCO-UNEP newsletter on environmental degradation, quarterly).
Earth Views (quarterly).
Environment Forum (quarterly).
Environmental Law Bulletin (2 a year).
Financial Services Initiative (2 a year).
GEF News (quarterly).
Global Environment Outlook (every 2–3 years).
Global Water Review.
GPA Newsletter.
IETC Insight (3 a year).
Industry and Environment Review (quarterly).
Leave it to Us (children's magazine, 2 a year).
Managing Hazardous Waste (2 a year).
Our Planet (quarterly).
OzonAction Newsletter (quarterly).
Tierramerica (weekly).
Tourism Focus (2 a year).
UNEP Chemicals Newsletter (2 a year).
UNEP Update (monthly).
World Atlas of Biodiversity.
World Atlas of Coral Reefs.
World Atlas of Desertification.
Studies, reports, legal texts, technical guide-lines, etc.

United Nations High Commissioner for Refugees—UNHCR

Address: CP 2500, 1211 Geneva 2 dépôt, Switzerland.
Telephone: (22) 7398111; **fax:** (22) 7397312; **e-mail:** unhcr@unhcr.ch; **internet:** www.unhcr.ch.

The Office of the High Commissioner was established in 1951 to provide international protection for refugees and to seek durable solutions to their problems.

Organization

(October 2004)

HIGH COMMISSIONER

The High Commissioner is elected by the United Nations General Assembly on the nomination of the Secretary-General, and is responsible to the General Assembly and to the UN Economic and Social Council (ECOSOC).

High Commissioner: Ruud Lubbers (Netherlands).
Deputy High Commissioner: Wendy Chamberlain (USA).

EXECUTIVE COMMITTEE

The Executive Committee of the High Commissioner's Programme (ExCom), established by ECOSOC, gives the High Commissioner policy directives in respect of material assistance programmes and advice in the field of international protection. In addition, it oversees UNHCR's general policies and use of funds. ExCom, which comprises representatives of 57 states, both members and non-members of the UN, meets once a year.

ADMINISTRATION

Headquarters include the Executive Office, comprising the offices of the High Commissioner, the Deputy High Commissioner and the Assistant High Commissioner. There are separate offices for the Inspector General, the Special Envoy in the former Yugoslavia, and the Director of the UNHCR liaison office in New York. The other principal administrative units are the Division of Communication and Information, the Department of International Protection, the Division of Resource Management, and the Department of Operations, which is responsible for the five regional bureaux covering Africa; Asia and the Pacific; Europe; the Americas and the Caribbean; and Central Asia, South-West Asia, North Africa and the Middle East. At July 2003 there were 251 UNHCR field offices in 115 countries. At that time UNHCR employed 6,235 people, including short-term staff, of whom 5,325 (or 85%) were working in the field.

OFFICES IN THE FAR EAST AND AUSTRALASIA

Regional Office for Australia, New Zealand, Papua New Guinea and the South Pacific: 9 Terrigal Crescent, O'Malley, ACT 2606, Australia; tel. (2) 6290-1355; fax (2) 6290-1315; e-mail aul@unhcr.ch.

People's Republic of China: Tayuan Diplomatic Office, Bldg 14, Liangmahe Nanlu, Beijing 100600, People's Republic of China; e-mail chibe@unhcr.ch; also covers Mongolia.

Indonesia: POB 4505, Jakarta, Indonesia; e-mail insja@unhcr.ch.

Japan: 4–14, Akasaka 8-chome, Minato-ku, Tokyo 107, Japan; tel. (3) 3475-1615; fax (3) 3475-1647; e-mail jpnto@unhcr.ch; also covers Republic of Korea.

Activities

The competence of the High Commissioner extends to any person who, owing to well-founded fear of being persecuted for reasons of race, religion, nationality or political opinion, is outside the country of his or her nationality and is unable or, owing to such fear or for reasons other than personal convenience, remains unwilling to accept the protection of that country; or who, not having a nationality and being outside the country of his or her former habitual residence, is unable or, owing to such fear or for reasons other than personal convenience, is unwilling to return to it. This competence may be extended, by resolutions of the UN General Assembly and decisions of ExCom, to cover certain other 'persons of concern', in addition to refugees meeting these criteria. Refugees who are assisted by other UN agencies, or who have the same rights or obligations as nationals of their country of residence, are outside the mandate of UNHCR.

In recent years there has been a significant shift in UNHCR's focus of activities. Increasingly UNHCR has been called upon to support people who have been displaced within their own country (i.e. with similar needs to those of refugees but who have not crossed an international border) or those threatened with displacement as a result of armed conflict. In addition, greater support has been given to refugees who have returned to their country of origin, to assist their reintegration, and UNHCR is working to enable local communities to support the returnees, frequently through the implementation of Quick Impact Projects (QIPs).

UNHCR has been increasingly concerned with the problem of statelessness and promotes new accessions to the 1954 Convention Relating to the Status of Stateless Persons and the 1964 Convention on the Reduction of Statelessness. It is estimated that as many as 9m. people world-wide may have no legal nationality.

At December 2003 the refugee population world-wide provisionally totalled 9.7m. UNHCR was also concerned with 1.1m. recently returned refugees, 4.2m. internally displaced persons (IDPs), 995,000 asylum seekers, 233,000 returned IDPs and 912,000 others.

World Refugee Day, sponsored by UNHCR, is held annually on 20 June.

INTERNATIONAL PROTECTION

As laid down in the Statute of the Office, UNHCR's primary function is to extend international protection to refugees and its second function is to seek durable solutions to their problems. In the exercise of its mandate UNHCR seeks to ensure that refugees and asylum-seekers are protected against *refoulement* (forcible return), that they receive asylum, and that they are treated according to internationally recognized standards. UNHCR pursues these objectives by a variety of means that include promoting the conclusion and ratification by states of international conventions for the protection of refugees. UNHCR promotes the adoption of liberal practices of asylum by states, so that refugees and asylum-seekers are granted admission, at least on a temporary basis.

The most comprehensive instrument concerning refugees that has been elaborated at the international level is the 1951 United Nations Convention relating to the Status of Refugees. This Convention, the scope of which was extended by a Protocol adopted in 1967, defines the rights and duties of refugees and contains provisions dealing with a variety of matters which affect the day-to-day lives of refugees. The application of the Convention and its Protocol is supervised by UNHCR. Important provisions for the treatment of refugees are also contained in a number of instruments adopted at the regional level. These include the 1969 Convention Governing the Specific Aspects of Refugee Problems adopted by OAU (now AU) member states in 1969, the European Agreement on the Abolition of Visas for Refugees, and the 1969 American Convention on Human Rights.

UNHCR has actively encouraged states to accede to the 1951 United Nations Refugee Convention and the 1967 Protocol: 145 states had acceded to either or both of these basic refugee instruments by October 2004. An increasing number of states have also adopted domestic legislation and/or administrative measures to implement the international instruments, particularly in the field of procedures for the determination of refugee status. UNHCR has sought to address the specific needs of refugee women and children, and has also attempted to deal with the problem of military attacks on refugee camps, by adopting and encouraging the acceptance of a set of principles to ensure the safety of refugees. In recent years it has formulated a strategy designed to address the fundamental causes of refugee flows. In 2001, in response to widespread concern about perceived high numbers of asylum-seekers and large-scale international economic migration and human trafficking, UNHCR initiated a series of Global Consultations on International Protection with the signatories to the 1951 Convention and 1967 Protocol, and other interested parties, with a view to strengthening both the application and scope of international refugee legislation. A consultation of 156 Governments, convened in Geneva, in December, reaffirmed commitment to the central role played by the Convention and Protocol. The final consultation, held in May 2002, focused on durable solutions and the protection of refugee women and children. Subsequently, based on the findings of the Global Consultations process, UNHCR developed an Agenda on Protection with six main objectives: strengthening the implementation of the 1951 Convention and 1967 Protocol; the protection of refugees within broader migration movements; more equitable sharing of burdens and responsibilities and building of capacities to receive and protect refugees; addressing more effectively security-related concerns; increasing efforts to find durable solutions; and meeting the protection needs of refugee women and children. The Agenda was endorsed by the Executive Council in October 2002. In September of that year the High Commissioner for Refugees launched the *Convention Plus* initiative, which aims to address contemporary global asylum issues by developing, on the basis of the Agenda on Protection, international agreements and measures to supplement the 1951 Convention and 1967 Protocol.

ASSISTANCE ACTIVITIES

The first phase of an assistance operation uses UNHCR's capacity of emergency response. This enables UNHCR to address the immediate needs of refugees at short notice, for example, by employing specially trained emergency teams and maintaining stockpiles of basic equipment, medical aid and materials. A significant proportion of UNHCR expenditure is allocated to the next phase of an operation, providing 'care and maintenance' in stable refugee circumstances. This assistance can take various forms, including the provision of food, shelter, medical care and essential supplies. Also covered in many instances are basic services, including education and counselling.

As far as possible, assistance is geared towards the identification and implementation of durable solutions to refugee problems—this being the second statutory responsibility of UNHCR. Such solutions generally take one of three forms: voluntary repatriation, local integration or resettlement in another country. Where voluntary repatriation, increasingly the preferred solution, is feasible, the Office assists refugees to overcome obstacles preventing their return to their country of origin. This may be done through negotiations with governments involved, or by providing funds either for the physical movement of refugees or for the rehabilitation of returnees once back in their own country.

When voluntary repatriation is not an option, efforts are made to assist refugees to integrate locally and to become self-supporting in their countries of asylum. This may be done either by granting loans to refugees, or by assisting them, through vocational training or in other ways, to learn a skill and to establish themselves in gainful occupations. One major form of assistance to help refugees re-establish themselves outside camps is the provision of housing. In cases where resettlement through emigration is the only viable solution to a refugee problem, UNHCR negotiates with governments in an endeavour to obtain suitable resettlement opportunities, to encourage liberalization of admission criteria and to draw up special immigration schemes. During 2003 an estimated 26,000 refugees were resettled under UNHCR auspices.

In the early 1990s UNHCR aimed to consolidate efforts to integrate certain priorities into its programme planning and implementation, as a standard discipline in all phases of assistance. The considerations include awareness of specific problems confronting refugee women, the needs of refugee children, the environmental impact of refugee programmes and long-term development objectives. In an effort to improve the effectiveness of its programmes, UNHCR has initiated a process of delegating authority, as well as responsibility for operational budgets, to its regional and field representatives, increasing flexibility and accountability. An Evaluation and Policy Analysis Unit reviews systematically UNHCR's operational effectiveness.

In June 2004 UNHCR became the tenth co-sponsor of UNAIDS.

EAST ASIA AND THE PACIFIC

In June 1989 an international conference was convened by UNHCR in Geneva to discuss the ongoing problem of refugees and displaced persons in and from the Indo-Chinese peninsula. The participants adopted the Comprehensive Plan of Action for Indo-Chinese Refugees (CPA), which provided for the 'screening' of all Vietnamese arrivals in the region to determine their refugee status, the resettlement of 'genuine' refugees and the repatriation (described as voluntary 'in the first instance') of those deemed to be economic migrants. A steering committee of the international conference met regularly to supervise the plan. In March 1996 UNHCR confirmed that it was to terminate funding for the refugee camps (except those in Hong Kong) at the end of June to coincide with the formal conclusion of the CPA; however, it pledged to support transitional arrangements regarding the completion of the repatriation process and maintenance of the remaining Vietnamese 'non-refugees' during the post-CPA phase-out period, as well as to continue its support for the reintegration and monitoring of returning nationals in Viet Nam and Laos. The prospect of forcible repatriation provoked rioting and violent protests in many camps throughout the region. By mid-1996 more than 88,000 Vietnamese and 22,000 Laotians had returned to their countries of origin under the framework of the CPA, with Malaysia and Singapore having completed the repatriation process. In late July the Philippines Government agreed to permit the remaining camp residents to settle permanently in that country. In September the remaining Vietnamese refugees detained on the island of Galang, in Indonesia, were repatriated, and in February 1997 the last camp for Vietnamese refugees in Thailand was formally closed. In mid-June of that year the main Vietnamese detention camp in Hong Kong was closed. However, the scheduled repatriation of all remaining Vietnamese before the transfer of sovereignty of the territory to the People's Republic of China (PRC) at the end of June was not achieved. In early 1998 the Hong Kong authorities formally terminated the policy of granting a port of first asylum to Vietnamese 'boat people'. In February 2000 UNHCR, which had proposed the integration of the remaining Vietnamese as a final durable solution to the situation, welcomed a decision by the Hong Kong authorities to offer permanent residency status to the occupants of the last remaining Vietnamese detention camp (totalling 973 refugees and 435 'non-refugees'). By the end of May, when the camp was closed, more than 200 Vietnamese had failed to apply for residency. At 31 December 2003 UNHCR was providing assistance to an estimated further 297,219 Vietnamese refugees in mainland PRC. In 1995, in accordance with an agreement concluded with the Chinese Government, UNHCR initiated a programme to redirect its local assistance to promote long-term self-sufficiency in the poorest settlements, including support for revolving-fund rural credit schemes. UNHCR favours the local integration of the majority of the Vietnamese refugee population in the PRC as a durable solution to the situation.

The conclusion of a political settlement of the conflict in Cambodia in October 1991 made possible the eventual repatriation of some 370,000 Cambodian refugees and displaced persons by April 1993. Meanwhile, however, thousands of ethnic Vietnamese (of whom there were estimated to be 200,000 in Cambodia) fled to Viet Nam, as a result of violence perpetrated against them by Cambodian armed groups. In March 1994 25,000 supporters of the Khmers Rouges in Cambodia fled across the border into Thailand, following advances by government forces. The refugees were immediately repatriated by the Thai armed forces into Khmer Rouge territory, which was inaccessible to aid agencies. In July 1997 armed conflict between opposing political forces in northern Cambodia resulted in large-scale population movement. A voluntary repatriation programme was initiated in October, and in late March 1999 UNHCR announced that the last Cambodian refugees had left camps in Thailand, the majority having been repatriated to north-western Cambodia. A UNHCR programme was initiated to monitor the welfare of returnees and assist in their reintegration; this was terminated at the end of 2000. At 31 December 2003 there were still some 15,360 Cambodian refugees in Viet Nam. In January 2002 UNHCR signed an agreement with the Governments of Viet Nam and Cambodia for the safe repatriation of an estimated 1,000 Montagnards who had fled from the Central Highland provinces of Viet Nam during 2001. UNHCR was to be permitted unlimited access to the region to assist and monitor the return process. In March 2002, however, UNCHR withdrew from the agreement owing to alleged intimidation of refugees and UN staff.

From April 1991 increasing numbers of Rohingya Muslims in Myanmar fled into Bangladesh to escape brutality and killings perpetrated by the Myanma armed forces. UNHCR launched an international appeal for financial aid for the refugees, at the request of Bangladesh, and collaborated with other UN agencies in providing humanitarian assistance. UNHCR refused to participate in a programme of repatriation of the Myanma refugees agreed by Myanmar and Bangladesh, on the grounds that no safe environment existed for them to return to. In May 1993 a memorandum of understanding between UNHCR and Bangladesh was signed, whereby UNHCR would be able to monitor the repatriation process and ensure that people were returning of their own free will. In November a memorandum of understanding, signed with the Myanma Government, secured UNHCR access to the returnees. The first refugees returned to Myanmar with UNHCR assistance at the end of April 1994. They and all subsequent returnees were provided with a small amount of cash, housing grants and two months' food rations, and were supported by several small-scale reintegration projects. Attempts by UNHCR to find a local solution for those unwilling to return to Myanmar have been met with resistance by the Bangladeshi Government. By the end of December 2003 an estimated 19,743 Myanma refugees still remained in camps in Bangladesh and were receiving basic care from UNHCR.

In the early 1990s members of ethnic minorities in Myanmar attempted to flee attacks by government troops into Thailand; however, the Thai Government refused to recognize them as refugees or to offer them humanitarian assistance. In December 1997 Thailand and Myanmar agreed to commence 'screening' the refugees to determine those who had fled persecution and those who were economic migrants. By the end of 2003 there were an estimated 118,762 people in camps along the Myanma-Thai border, the majority of whom were Karen (Kayin) refugees.

In April 1999, following the announcement by the Indonesian Government, in January, that it would consider a form of autonomy or independence for East Timor, some 26,000 Indonesian settlers left their homes as a result of clashes between opposing groups and uncertainty regarding the future of the territory. The popular referendum on the issue, conducted at the end of August, and the resulting victory for the independence movement, provoked a violent reaction by pro-integration militia. UNHCR, along with other international personnel, was forced to evacuate the territory in early September. At that time there were reports of forced mass deportations of East Timorese to West Timor, while a large number of others fled their homes into remote mountainous areas of East Timor. In mid-September UNHCR staff visited West Timor to review the state of refugee camps, allegedly under the control of militia, and to persuade the authorities to permit access for humanitarian personnel. It was estimated that 250,000–260,000 East Timorese had fled to West Timor, of whom some 230,000 were registered in 28 camps at the end of September. At that time there were also an estimated 190,000–300,000 people displaced within East Timor, although the International Committee of the Red Cross estimated that a total of 800,000 people, or some 94% of the population, had been displaced, or deported, during the crisis. The arrival of multinational troops, from 20 September, helped to stabilize the region and enable the safe receipt and distribution of food supplies, prompting several thousands to return from hiding. Most homes, however, along with almost all other buildings in the capital, Dili, had been destroyed. In early October UNHCR, together with the International Organization for Migration, initiated a repatriation programme for the refugees in West Timor. However, despite an undertaking by the Indonesian Government in mid-October that it would ensure the safety of all refugees and international personnel, persistent intimidation by anti-independence militia impeded the registration and repatriation processes. UNHCR initially aimed to complete the repatriation programme by mid-2001, prior to the staging of elections to a Constituent Assembly by the UN Transitional Administration in East Timor (UNTAET, which assumed full authority over the territory in February 2000). However, in September 2000 UNHCR suspended its activities in West Timor, following the murder by militiamen there of three of its personnel. A UN Security Council resolution, adopted soon afterwards, deplored this incident and strongly urged the Indonesian authorities to disable the militia and to guarantee the future security of all refugees and humanitarian personnel. In mid-September UNTAET and the Indonesian Government signed a Memorandum of Understanding on co-operation in resolving the refugee crisis. However, despite a subsequent operation by the Indonesian security forces to disarm the militia, intimidation of East Timorese refugees reportedly persisted, and UNHCR did not redeploy personnel to West Timor. The Office did, however, liaise with other humanitarian organizations to facilitate continuing voluntary repatriations, which have been encouraged by the Indonesian authorities. UNHCR's operation in East Timor aimed to promote the safe voluntary repatriation of refugees, monitor returnees, support their reintegration through the implementation of QIPs, pursue efforts towards sustainable development and the rehabilitation of communities, and to promote reconciliation and respect for human rights. In mid-May East Timor (now Timor-Leste) achieved independence. At that time almost 205,000 East Timorese refugees were reported to have returned since October 1999, while more than 50,000 were believed to remain in West Timorese camps. UNHCR and the newly-elected administration were co-operating to encourage further repatriation, as well as to assist with Timor-Leste's accession to the international instruments of protection and with the development of new national refugee protection legislation. On 31 December 2002 UNHCR terminated the refugee status of people who fled East Timor in 1999.

CO-OPERATION WITH OTHER ORGANIZATIONS

UNHCR works closely with other UN agencies, intergovernmental organizations and non-governmental organizations (NGOs) to increase the scope and effectiveness of its operations. Within the UN system UNHCR co-operates, principally, with the World Food Programme in the distribution of food aid, UNICEF and the World Health Organization in the provision of family welfare and child immunization programmes, OCHA in the delivery of emergency humanitarian relief, UNDP in development-related activities and the preparation of guide-lines for the continuum of emergency assistance to development programmes, and the Office of the UN High Commissioner for Human Rights. UNHCR also has close working relationships with the International Committee of the Red Cross and the International Organization for Migration. In 2003 UNHCR worked with 514 NGOs as 'implementing partners', enabling UNHCR to broaden the use of its resources while maintaining a co-ordinating role in the provision of assistance.

TRAINING

UNHCR organizes training programmes and workshops to enhance the capabilities of field workers and non-UNHCR staff, in the following areas: the identification and registration of refugees; people-orientated planning; resettlement procedures and policies; emergency response and management; security awareness; stress management; and the dissemination of information through the electronic media.

Finance

The United Nations' regular budget finances a proportion of UNHCR's administrative expenditure. The majority of UNHCR's programme expenditure (about 98%) is funded by voluntary contributions, mainly from governments. The Private Sector and Public Affairs Service aims to increase funding from non-governmental donor sources, for example by developing partnerships with foundations and corporations. Following approval of the Unified Annual Programme Budget any subsequently-identified requirements are managed in the form of Supplementary Programmes, financed by separate appeals. The total Unified Annual Programme Budget for 2004 was projected at US $954.9m.

Publications

Refugees (quarterly, in English, French, German, Italian, Japanese and Spanish).

Refugee Resettlement: An International Handbook to Guide Reception and Integration.

Refugee Survey Quarterly.

Sexual and Gender-based Violence Against Refugees, Returnees and Displaced Persons: Guide-lines for Prevention and Response.

The State of the World's Refugees (every 2 years).

UNHCR Handbook for Emergencies.

Press releases, reports.

Statistics

PERSONS OF CONCERN TO UNHCR IN THE FAR EAST AND AUSTRALASIA

('000 persons, at 31 December 2003*)

Country	Refugees	Asylum-seekers	Returned refugees	Others of concern†
Australia	56.0	3.0	—	—
China, People's Republic	299.4	0.0	—	—
Indonesia	0.2	0.0	—	16.7
Malaysia	0.4	9.2	—	69.3
Thailand	119.1	2.7	—	0.0
Viet Nam	15.4	—	—	—

* Figures are provided mostly by governments, based on their own records and methods of estimations. Countries with fewer than 10,000 persons of concern to UNHCR are not listed.

† Mainly internally displaced persons (IDPs) or recently-returned IDPs.

United Nations Peace-keeping

Address: Department of Peace-keeping Operations, Room S-3727-B, United Nations, New York, NY 10017, USA.

Telephone: (212) 963-8077; **fax:** (212) 963-9222; **internet:** www.un.org/Depts/dpko/.

United Nations peace-keeping operations have been conceived as instruments of conflict control. The UN has used these operations in various conflicts, with the consent of the parties involved, to maintain international peace and security, without prejudice to the positions or claims of parties, in order to facilitate the search for political settlements through peaceful means such as mediation and the good offices of the Secretary-General. Each operation is established with a specific mandate, which requires periodic review by the Security Council. United Nations peace-keeping operations fall into two categories: peace-keeping forces and observer missions.

Peace-keeping forces are composed of contingents of military and civilian personnel, made available by member states. These forces assist in preventing the recurrence of fighting, restoring and maintaining peace, and promoting a return to normal conditions. To this end, peace-keeping forces are authorized as necessary to undertake negotiations, persuasion, observation and fact-finding. They conduct patrols and interpose physically between the opposing parties. Peace-keeping forces are permitted to use their weapons only in self-defence.

Military observer missions are composed of officers (usually unarmed), who are made available, on the Secretary-General's request, by member states. A mission's function is to observe and report to the Secretary-General (who, in turn, informs the UN Security Council) on the maintenance of a cease-fire, to investigate violations and to do what it can to improve the situation.

The UN's peace-keeping forces and observer missions are financed in most cases by assessed contributions from member states of the organization. In recent years a significant expansion in the UN's peace-keeping activities has been accompanied by a perpetual financial crisis within the organization, as a result of the increased financial burden and some member states' delaying payment. At 30 April 2004 outstanding assessed contributions to the peace-keeping budget amounted to some US $1,270m.

UNITED NATIONS MISSION OF SUPPORT IN EAST TIMOR—UNMISET

Address: Headquarters: Dili, Timor-Leste.

Special Representative of the UN Secretary-General and Head of Mission: Sukehiro Hasegawa (Japan).

Force Commander: Lt-Gen. Khairuddin Mat Yusof (Malaysia).

UNMISET was established on 20 May 2002, initially with a mandate for one year, when Timor-Leste (previously known as East Timor) attained its independence. It succeeded the UN Transitional Administration in East Timor (UNTAET), which had been established in October 1999 to govern the territory on an interim basis following the outcome of a popular referendum in favour of independence from Indonesia. UNMISET was to provide assistance to core administrative structures, to extend interim law enforcement and public security services, to assist the development of a national police force, and to contribute to the maintenance of external and internal security. The civilian component of UNMISET was to include a Serious Crimes Unit and a Human Rights Unit, as well as focal points for gender and HIV/AIDS. The Mission was mandated to devolve all operational responsibilities to the Timor-Leste authorities within two years. In March 2003 the UN Secretary-General reported that, owing to an escalation in insecurity (including riots and attacks by armed groups) in Timor-Leste from November 2002, and the current early stage in the development of the national police force, original plans swiftly to downsize the mission should be revised. He proposed adjustments to the configuration and military strategy of UNMISET, aimed at facilitating the restoration of stability. In December 2003 the Timor-Leste National Police Force assumed responsibility for routine policing throughout the country. In February 2004 the UN Secretary-General recommended that UNMISET's mandate should be extended for one year after its scheduled expiry in May in order to contribute further assistance to the judicial system, core administrative structures and the development of the new national police force, and also to provide a continuing security presence. In May responsibility for policing and external security was transferred from UNMISET to the Timor-Leste Government. In that same month the Security Council extended the mission, with a modified mandate, for a six-month period, with a view to extending it for a final period terminating in May 2005. In May 2004 the UN Security Council determined to reduce the mission's authorized strength to 58 civilian advisors, 157 civilian police advisors, 42 military liaison officers, 310 troops and a 125-person International Response Unit.

At 30 September 2004 UNMISET comprised 429 troops, 147 civilian police officers and 42 military observers, supported by 814 international and local civilian personnel. The General Assembly budget appropriation to the Special Account for the mission amounted to US $85.3m. for the period 1 July 2004–30 June 2005.

World Food Programme—WFP

Address: Via Cesare Giulio Viola 68, Parco dei Medici, 00148 Rome, Italy.

Telephone: (06) 6513-1; **fax:** (06) 6513-2840; **e-mail:** wfpinfo@wfp.org; **internet:** www.wfp.org.

WFP, the principal food aid organization of the United Nations, became operational in 1963. It aims to alleviate acute hunger by providing emergency relief following natural or man-made humanitarian disasters, and supplies food aid to people in developing countries to eradicate chronic undernourishment, to support social development and to promote self-reliant communities.

Organization

(October 2004)

EXECUTIVE BOARD

The governing body of WFP is the Executive Board, comprising 36 members, 18 of whom are elected by the UN Economic and Social Council (ECOSOC) and 18 by the Council of the Food and Agriculture Organization (FAO). The Board meets four times each year at WFP headquarters.

SECRETARIAT

WFP's Executive Director is appointed jointly by the UN Secretary-General and the Director-General of FAO and is responsible for the management and administration of the Programme. At December 2003 there were 8,770 permanent staff members. WFP administers some 87 country offices, in order to provide operational, financial and management support at a more local level, and has established seven regional bureaux, located in Bangkok, Thailand (for Asia), Cairo, Egypt (for the Middle East, Central Asia and the Mediterranean), Rome, Italy (for Eastern Europe), Managua, Nicaragua (for Latin America and the Caribbean), Yaoundé, Cameroon (for Central Africa), Kampala, Uganda (for Eastern and Southern Africa), and Dakar, Senegal (for West Africa).

Executive Director: James T. Morris (USA).

Activities

WFP is the only multilateral organization with a mandate to use food aid as a resource. It is the second largest source of assistance in the UN, after the World Bank group, in terms of actual transfers of resources, and the largest source of grant aid in the UN system. WFP handles more than one-third of the world's food aid. WFP is also the largest contributor to South–South trade within the UN system, through the purchase of food and services from developing countries. WFP's mission is to provide food aid to save lives in refugee and other emergency situations, to improve the nutrition and quality of life of vulnerable groups and to help to develop assets and promote the self-reliance of poor families and communities. WFP aims to focus its efforts on the world's poorest countries and to provide at least 90% of its total assistance to those designated as 'low-income food-deficit'. At the World Food Summit, held in November 1996, WFP endorsed the commitment to reduce by 50% the number of undernourished people, no later than 2015. During 2003 WFP food assistance benefited some 104.2m. people world-wide (compared with 72m. in 2002), of whom 16.2m. received aid through development projects, 61.2m. through emergency operations, and 26.8m. through Protracted Relief and Recovery Operations (see below). Total food deliveries in 2003 amounted to 4.6m. metric tons, compared with 3.7m. metric tons in 2002.

WFP aims to address the causes of chronic malnourishment, which it identifies as poverty and lack of opportunity. It emphasizes the role played by women in combating hunger, and endeavours to address the specific nutritional needs of women, to increase their access to food and development resources, and to promote girls' education. It also focuses resources on supporting the food security of households and communities affected by HIV/AIDS and on promoting food security as a means of mitigating extreme poverty and vulnerability and thereby combating the spread and impact of HIV/AIDS. In February 2003 WFP and the Joint UN Programme on HIV/AIDS (UNAIDS) concluded an agreement to address jointly the relationship between HIV/AIDS, regional food shortages and chronic hunger, with a particular focus on Africa, South-East Asia and the Caribbean. In October of that year WFP became a co-sponsor of UNAIDS. WFP urges the development of new food aid strategies as a means of redressing global inequalities and thereby combating the threat of conflict and international terrorism.

WFP food donations must meet internationally-agreed standards applicable to trade in food products. In May 2003 WFP's Executive Board approved a new policy on donations of genetically-modified (GM) foods and other foods derived from biotechnology, determining that the Programme would continue to accept donations of GM/biotech food and that, when distributing it, relevant national standards would be respected.

In the early 1990s there was a substantial shift in the balance between emergency relief ('food-for-life') and development assistance ('food-for-growth') provided by WFP, owing to the growing needs of victims of drought and other natural disasters, refugees and displaced persons. By 1994 two-thirds of all food aid was for relief assistance and one-third for development, representing a direct reversal of the allocations five years previously. In addition, there was a noticeable increase in aid given to those in need as a result of civil war, compared with commitments for victims of natural disasters. Accordingly, WFP has developed a range of mechanisms to enhance its preparedness for emergency situations and to improve its capacity for responding effectively to situations as they arise. A new programme of emergency response training was inaugurated in 2000, while security concerns for personnel was incorporated as a new element into all general planning and training activities. Through its Vulnerability Analysis and Mapping (VAM) project, WFP aims to identify potentially vulnerable groups by providing information on food security and the capacity of different groups for coping with shortages, and to enhance emergency contingency-planning and long-term assistance objectives. In 2003 VAM field units were operational in more than 50 countries. WFP also co-operates with other UN agencies including FAO (collaborating on 77 projects in 41 countries in 2003), IFAD (collaborating on 21 projects in that year), UNHCR and UNICEF. The key elements of WFP's emergency response capacity are its strategic stores of food and logistics equipment, stand-by arrangements to enable the rapid deployment of personnel, communications and other essential equipment, and the Augmented Logistics Intervention Team for Emergencies (ALITE), which undertakes capacity assessments and contingency-planning. During 2000 WFP led efforts, undertaken with other UN humanitarian agencies, for the design and application of local UN Joint Logistics Centre facilities, which aimed to co-ordinate resources in an emergency situation. In 2001 a new UN Humanitarian Response Depot was opened in Brindisi, Italy, under the direction of WFP experts, for the storage of essential rapid response equipment. In that year the Programme published a set of guidelines on contingency planning.

Through its development activities, WFP aims to alleviate poverty in developing countries by promoting self-reliant families and communities. Food is supplied, for example, as an incentive in development self-help schemes and as part-wages in labour-intensive projects of many kinds. In all its projects WFP aims to assist the most vulnerable groups and to ensure that beneficiaries have an adequate and balanced diet. Activities supported by the Programme include the settlement and resettlement of groups and communities; land reclamation and improvement; irrigation; the development of forestry and dairy farming; road construction; training of hospital staff; community development; and human resources development such as feeding expectant or nursing mothers and schoolchildren, and support for education, training and health programmes. No individual country is permitted to receive more than 10% of the Programme's available development resources. During 2001 WFP initiated a new Global School Feeding Campaign to strengthen international co-operation to expand educational opportunities for poor children and to improve the quality of the teaching environment. In December 2003 WFP launched a *19-Cents-a-day* campaign to encourage donors to support its school feeding activities (19 cents being the estimated cost of one school lunch). During that year school feeding projects benefited 15.2m. children in 69 countries.

Following a comprehensive evaluation of its activities, WFP is increasingly focused on linking its relief and development activities to provide a continuum between short-term relief and longer-term rehabilitation and development. In order to achieve this objective, WFP aims to integrate elements that strengthen disaster mitigation into development projects, including soil conservation, reafforestation, irrigation infrastructure, and transport construction and rehabilitation; and to promote capacity-building elements within relief operations, e.g. training, income-generating activities and environmental protection measures. In 1999 WFP adopted a new Food Aid and Development policy, which aims to use food assistance both to cover immediate requirements and to create conditions conducive to enhancing the long-term food security of vulnerable populations. During that year WFP began implementing Protracted Relief and Recovery Operations (PRROs), where the emphasis is on fostering stability, rehabilitation and long-term development for victims of natural disasters, displaced persons and refugees. PRROs are introduced no later than 18 months after the initial emergency operation and last no more than three years. When undertaken in collaboration with UNHCR and other international agencies, WFP has responsibility for mobilizing basic food commodities and for related transport, handling and storage costs. The 14 PRROs undertaken in 2003 involved the provision of 1.68m. metric tons of food, at a cost of some US $946.5m..

In 2003 WFP operational expenditure in Asia amounted to US $399.2m. (12% of total operational expenditure in that year), including $189.4m. for emergency relief operations, $115.5m. for PRROs, and $68.4m. for agricultural, rural and human resource development projects. In recent years WFP has been actively concerned with the food situation in the Democratic People's Republic of Korea (DPRK), which has required substantial levels of emergency food supplies, owing to natural disasters and consistently poor harvests. By mid-1999 an estimated 1.5m.–3.5m. people had died of starvation in the DPRK since 1995. WFP's eighth emergency operation in the DPRK, covering the period January–December 2003, aimed to provide food aid to 6.4m. people, at a cost of $201m. In February of that year WFP and other UN agencies urged continuing international support for humanitarian activities in the DPRK in order to maintain progress achieved since 1999 in reducing the rate of malnutrition. In February 2004 WFP appealed for 485,000 metric tons of commodities, amounting to $171m., to meet the food needs of some 6.5m. people in the DPRK. During 2003 WFP undertook its first programme in Asia to support people with HIV/AIDS, focusing on a community-based project for 4,000 households in Cambodia. A pilot programme to distribute food to some 400 families affected by HIV/AIDS in Myanmar was initiated in March 2004.

Finance

The Programme is funded by voluntary contributions from donor countries, intergovernmental bodies such as the European Commission, and the private sector. Contributions are made in the form of commodities, finance and services (particularly shipping). Commitments to the International Emergency Food Reserve (IEFR), from which WFP provides the majority of its food supplies, and to the Immediate Response Account of the IEFR (IRA), are also made on a voluntary basis by donors. WFP's operational expenditures in 2003 amounted to US $3,275.3m. Contributions by donors in that year totalled $2,600.0m, of which $1,389.1m. was for the IEFR.

Publications

Annual Report.
Food and Nutrition Handbook.
School Feeding Handbook.

Food and Agriculture Organization of the United Nations—FAO

Address: Viale delle Terme di Caracalla, 00100 Rome, Italy.

Telephone: (06) 5705-1; **fax:** (06) 5705-3152; **e-mail:** fao.hq@fao.org; **internet:** www.fao.org.

FAO, the first specialized agency of the UN to be founded after the Second World War, aims to alleviate malnutrition and hunger, and serves as a co-ordinating agency for development programmes in the whole range of food and agriculture, including forestry and fisheries. It helps developing countries to promote educational and training facilities and the creation of appropriate institutions.

Organization

(October 2004)

CONFERENCE

The governing body is the FAO Conference of member nations. It meets every two years, formulates policy, determines the Organization's programme and budget on a biennial basis, and elects new members. It also elects the Director-General of the Secretariat and the Independent Chairman of the Council. Every other year, FAO also holds conferences in each of its five regions (Africa, Asia and the Pacific, Europe, Latin America and the Caribbean, and the Near East).

COUNCIL

The FAO Council is composed of representatives of 49 member nations, elected by the Conference for staggered three-year terms. It is the interim governing body of FAO between sessions of the Conference. The most important standing Committees of the Council are: the Finance and Programme Committees, the Committee on Commodity Problems, the Committee on Fisheries, the Committee on Agriculture and the Committee on Forestry.

SECRETARIAT

The number of FAO staff at mid-2004 was some 3,450, of whom 1,450 were professional staff and 2,000 general service staff. About one-half of the Organization's staff were based at headquarters. Work is supervised by the following Departments: Administration and Finance; General Affairs and Information; Economic and Social Policy; Agriculture; Forestry; Fisheries; Sustainable Development; and Technical Co-operation.

Director-General: Jacques Diouf (Senegal).

REGIONAL AND SUB-REGIONAL OFFICES

Regional Office for Asia and the Pacific: Maliwan Mansion, Phra Atit Rd, Bangkok 10200, Thailand; tel. (662) 697-4000; fax (662) 697-4445; e-mail fao-rap@fao.org; internet www.fao.or.th; Regional Rep. He Changchui.

Sub-regional Office for the Pacific Islands: Private Mail Bag, Apia, Samoa; tel. 22127; fax 22126; e-mail fao-sapa@fao.org; Sub-regional Rep. V. A. Fuavao.

JOINT DIVISION AND LIAISON OFFICES

Joint FAO/IAEA Division of Nuclear Techniques in Food and Agriculture: Wagramerstrasse 5, 1400 Vienna, Austria; tel. (1) 2600-0; fax (1) 2600-7; e-mail official.mail@iaea.org; Dir James D. Dargie.

Japan: 6F Yokohama International Organizations Centre, Pacifico-Yokohama, 1-1-1, Minato Mirai, Nishi-ku, Yokohama 220-0012; tel. (45) 222-1101; fax (45) 222-1103.

United Nations: Suite DC1-1125, 1 United Nations Plaza, New York, NY 10017, USA; tel. (212) 963-6036; fax (212) 963-5425; e-mail fao-lony@field.fao.org; Dir Howard W. Hjort.

Activities

FAO aims to raise levels of nutrition and standards of living by improving the production and distribution of food and other commodities derived from farms, fisheries and forests. FAO's ultimate objective is the achievement of world food security, 'Food for All'. The organization provides technical information, advice and assistance by disseminating information; acting as a neutral forum for discussion of food and agricultural issues; advising governments on policy and planning; and developing capacity directly in the field.

In November 1996 FAO hosted the World Food Summit, which was held in Rome and was attended by heads of state and senior government representatives of 186 countries. Participants approved the Rome Declaration on World Food Security and the World Food Summit Plan of Action, with the aim of halving the number of people afflicted by undernutrition, at that time estimated to total 828m. world-wide, by no later than 2015. A review conference to assess progress in achieving the goals of the summit, entitled World Food Summit: Five Years Later, held in June 2002, reaffirmed commitment to this objective, which is also incorporated into the UN Millennium Development Goal of eradicating extreme poverty and hunger. During that month FAO announced the formulation of a global 'Anti-Hunger Programme', which aimed to promote investment in the agricultural sector and rural development, with a particular focus on small farmers, and to enhance food access for those most in need, for example through the provision of school meals, schemes to feed pregnant and nursing mothers and food-for-work programmes. In late 2003 FAO reported that an estimated 842m. people world-wide were undernourished; of these 798m. resided in developing countries.

In November 1999 the FAO Conference approved a long-term Strategic Framework for the period 2000–15, which emphasized national and international co-operation in pursuing the goals of the 1996 World Food Summit. The Framework promoted interdisciplinarity and partnership, and defined three main global objectives: constant access by all people to sufficient nutritionally adequate and safe food to ensure that levels of undernourishment were reduced by 50% by 2015 (see above); the continued contribution of sustainable agriculture and rural development to economic and social progress and well-being; and the conservation, improvement and sustainable use of natural resources. It identified five corporate strategies (each supported by several strategic objectives), covering the following areas: reducing food insecurity and rural poverty; ensuring enabling policy and regulatory frameworks for food, agriculture, fisheries and forestry; creating sustainable increases in the supply and availability of agricultural, fisheries and forestry products; conserving and enhancing sustainable use of the natural resource base; and generating knowledge. In November 2001 the FAO Conference adopted a medium-term plan covering 2002–07, based on the Strategic Framework.

FAO organizes an annual series of fund-raising events, 'TeleFood', some of which are broadcast on television and the internet, in order to raise public awareness of the problems of hunger and malnutrition. Since its inception in 1997 public donations to TeleFood have exceeded US $12m., financing nearly 1,600 'grass-roots' projects in more than 120 countries. The projects have provided tools, seeds and other essential supplies directly to small-scale farmers, and have been especially aimed at helping women.

In 1999 FAO signed a memorandum of understanding with UNAIDS on strengthening co-operation. In December 2001 FAO, IFAD and WFP determined to strengthen inter-agency collaboration in developing strategies to combat the threat posed by the HIV/AIDS epidemic to food security, nutrition and rural livelihoods. During that month experts from those organizations and UNAIDS held a technical consultation on means of mitigating the impact of HIV/AIDS on agriculture and rural communities in affected areas.

In September 2004 FAO published *Recommendations for the Prevention, Control and Eradication of Highly Pathogenic Avian Influenza (HPAI) in Asia*. In the same month, following new outbreaks of the disease in the People's Republic of China, Cambodia, Viet Nam, Malaysia and Thailand, FAO and the World Health Organization declared the avian influenza epidemic to be a 'crisis of global importance'. FAO was working closely with the World Organisation for Animal Health to study and help to contain the disease.

The Technical Co-operation Department has responsibility for FAO's operational activities, including policy development assistance to member countries; investment support; and the management of activities associated with the development and implementation of country, sub-regional and regional programmes. The Department manages the technical co-operation programme (TCP, which funds 13% of FAO's field programme expenditures), and mobilizes resources.

AGRICULTURE

FAO's most important area of activity is crop production, accounting annually for about one-quarter of total field programme expenditure. FAO assists developing countries in increasing agricultural production, by means of a number of methods, including improved seeds and fertilizer use, soil conservation and reforestation, better water resource management techniques, upgrading storage facilities, and improvements in processing and marketing. FAO places special emphasis on the cultivation of under-exploited traditional food crops, such as cassava, sweet potato and plantains.

In 1985 the FAO Conference approved an International Code of Conduct on the Distribution and Use of Pesticides, and in 1989 the Conference adopted an additional clause concerning 'Prior Informed Consent' (PIC), whereby international shipments of newly banned or restricted pesticides should not proceed without the agreement of importing countries. Under the clause, FAO aims to inform governments about the hazards of toxic chemicals and to urge them to take proper measures to curb trade in highly toxic agrochemicals while keeping the pesticides industry informed of control actions. In 1996 FAO, in collaboration with UNEP, publicized a new initiative which aimed to increase awareness of, and to promote international action on, obsolete and hazardous stocks of pesticides remaining throughout the world (estimated in 2001 to total some 500,000 metric tons). In September 1998 a new legally-binding treaty on trade in hazardous chemicals and pesticides was adopted at an international conference held in Rotterdam, Netherlands. The so-called Rotterdam Convention required that hazardous chemicals and pesticides banned or severely restricted in at least two countries should not be exported unless explicitly agreed by the importing country. It also identified certain pesticide formulations as too dangerous to be used by farmers in developing countries, and incorporated an obligation that countries halt national production of those hazardous compounds. The treaty entered into force in February 2004. FAO was co-operating with UNEP to provide an interim secretariat for the Convention. In July 1999 a conference on the Rotterdam Convention, held in Rome, established an Interim Chemical Review Committee with responsibility for recommending the inclusion of chemicals or pesticide formulations in the PIC procedure. As part of its continued efforts to reduce the environmental risks posed by over-reliance on pesticides, FAO has extended to other regions its Integrated Pest Management (IPM) programme in Asia and the Pacific on the use of safer and more effective methods of pest control, such as biological control methods and natural predators (including spiders and wasps), to avert pests. In February 2001 FAO warned that some 30% of pesticides sold in developing countries did not meet internationally accepted quality standards. A revised International Code of Conduct on the Distribution and Use of Pesticides, adopted in November 2002, aimed to reduce the inappropriate distribution and use of pesticides and other toxic compounds, particularly in developing countries.

In the Far East and Australasia the production of 'miracle rice', a new hybrid, has been encouraged, in order to help satisfy the rapidly-growing demand for the region's most important food crop. In 1999 FAO and the International Rice Research Instituteagreed to strengthen their collaboration to promote the development and use of hybrid rice, which studies conducted in the People's Republic of China had shown to yield up to 20% more than conventional varieties. In May an FAO-organized intergovernmental meeting, convened in Bangkok, recommended the establishment of a Seed Network for Asia and the Pacific to strengthen local seed production and distribution, including the harmonization of seed regulations, support for technical developments, and improvements to seed supply and marketing.

FAO's Joint Division with the International Atomic Energy Agency (IAEA) tests controlled-release formulas of pesticides and herbicides that gradually free their substances and can limit the amount of agrochemicals needed to protect crops. The Joint FAO/IAEA Division is engaged in exploring biotechnologies and in developing non-toxic fertilizers (especially those that are locally available) and improved strains of food crops (especially from indigenous varieties). In the area of animal production and health, the Joint Division has developed progesterone-measuring and disease diagnostic kits, of which thousands have been delivered to developing countries. FAO's plant nutrition activities aim to promote nutrient management, such as the Integrated Plant Nutritions Systems (IPNS), which are based on the recycling of nutrients through crop production and the efficient use of mineral fertilizers.

The conservation and sustainable use of plant and animal genetic resources are promoted by FAO's Global System for Plant Genetic Resources, which includes five databases, and the Global Strategy on the Management of Farm Animal Genetic Resources. An FAO programme supports the establishment of gene banks, designed to maintain the world's biological diversity by preserving animal and plant species threatened with extinction. FAO, jointly with UNEP, has published a document listing the current state of global livestock genetic diversity. In June 1996 representatives of more than 150 governments convened in Leipzig, Germany, at a meeting organized by FAO (and hosted by the German Government) to consider the use and conservation of plant genetic resources as an essential means of enhancing food security. The meeting adopted a Global Plan of

Action, which included measures to strengthen the development of plant varieties and to promote the use and availability of local varieties and locally-adapted crops to farmers, in particular following a natural disaster, war or civil conflict. In November 2001 the FAO Conference adopted the International Treaty on Plant Genetic Resources for Food and Agriculture, which was to provide a framework to ensure access to plant genetic resources and to related knowledge, technologies and funding. The Treaty entered into force on 29 June 2004, having received the required number of ratifications (40) by signatory states.

An Emergency Prevention System for Transboundary Animal and Plant Pests and Diseases (EMPRES) was established in 1994 to strengthen FAO's activities in the prevention, early warning of, control and, where possible, eradication of pests and highly contagious livestock diseases (which the system categorizes as epidemic diseases of strategic importance, such as rinderpest or foot-and-mouth; diseases requiring tactical attention at international or regional level, e.g. Rift Valley fever; and emerging diseases, e.g. bovine spongiform encephalopathy—BSE). EMPRES has a desert locust component, and has published guide-lines on all aspects of desert locust monitoring. FAO has assumed responsibility for technical leadership and co-ordination of the Global Rinderpest Eradication Programme (GREP), which has the objective of eliminating the disease by 2010. Following technical consultations in late 1998, an Intensified GREP was launched. In November 1997 FAO initiated a Programme Against African Trypanosomiasis, which aimed to counter the disease affecting cattle in almost one-third of Africa. EMPRES promotes Good Emergency Management Practices (GEMP) in animal health. The system is guided by the annual meeting of the EMPRES Expert Consultation.

FAO's organic agriculture programme provides technical assistance and policy advice on the production, certification and trade of organic produce. In July 2001 the FAO/WHO Codex Alimentarius Commission adopted guide-lines on organic livestock production, covering organic breeding methods, the elimination of growth hormones and certain chemicals in veterinary medicines, and the use of good quality organic feed with no meat or bone meal content.

ENVIRONMENT

At the UN Conference on Environment and Development (UNCED), held in Rio de Janeiro, Brazil, in June 1992, FAO participated in several working parties and supported the adoption of Agenda 21, a programme of activities to promote sustainable development. FAO is responsible for the chapters of Agenda 21 concerning water resources, forests, fragile mountain ecosystems and sustainable agriculture and rural development. FAO was designated by the UN General Assembly as the lead agency for co-ordinating the International Year of Mountains (2002), which aimed to raise awareness of mountain ecosystems and to promote the conservation and sustainable development of mountainous regions.

FISHERIES

FAO's Fisheries Department consists of a multi-disciplinary body of experts who are involved in every aspect of fisheries development from coastal surveys, conservation management and use of aquatic genetic resources, improvement of production, processing and storage, to the compilation and analysis of statistics, development of computer databases, improvement of fishing gear, institution-building and training. In November 1993 the FAO Conference adopted an agreement to improve the monitoring and control of fishing vessels operating on the high seas that are registered under 'flags of convenience', in order to ensure their compliance with internationally accepted marine conservation and management measures. In March 1995 a ministerial meeting of fisheries adopted the Rome Consensus on World Fisheries, which identified a need for immediate action to eliminate overfishing and to rebuild and enhance depleting fish stocks. In November the FAO Conference adopted a Code of Conduct for Responsible Fishing, which incorporated many global fisheries and aquaculture issues (including fisheries resource conservation and development, fish catches, seafood and fish processing, commercialization, trade and research) to promote the sustainable development of the sector. In February 1999 the FAO Committee on Fisheries adopted new international measures, within the framework of the Code of Conduct, in order to reduce over-exploitation of the world's fish resources, as well as plans of action for the conservation and management of sharks and the reduction in the incidental catch of seabirds in longline fisheries. The voluntary measures were endorsed at a ministerial meeting, held in March and attended by representatives of some 126 countries, which issued a declaration to promote the implementation of the Code of Conduct and to achieve sustainable management of fisheries and aquaculture. In March 2001 FAO adopted an international plan of action to address the continuing problem of so-called illegal, unreported and unregulated fishing (IUU). In that year FAO estimated that about one-half of major marine fish stocks were fully exploited, one-quarter under-exploited, at least 15% over-exploited, and 10% depleted or recovering from depletion. IUU was estimated to account for up to 30% of total catches in certain fisheries. In October FAO and the Icelandic Government jointly organized the Reykjavik Conference on Responsible Fisheries in the Marine Ecosystem, which adopted a declaration on pursuing responsible and sustainable fishing activities in the context of ecosystem-based fisheries management (EBFM). EBFM involves determining the boundaries of individual marine ecosystems, and maintaining or rebuilding the habitats and biodiversity of each of these so that all species will be supported at levels of maximum production. FAO promotes aquaculture (which contributes almost one-third of annual global fish landings) as a valuable source of animal protein and income-generating activity for rural communities. In February 2000 FAO and the Network of Aquaculture Centres in Asia and the Pacific (NACA) jointly convened a Conference on Aquaculture in the Third Millennium, which was held in Bangkok, Thailand, and attended by participants representing more than 200 governmental and non-governmental organizations. The Conference debated global trends in aquaculture and future policy measures to ensure the sustainable development of the sector. It adopted the Bangkok Declaration and Strategy for Aquaculture Beyond 2000.

FORESTRY

FAO focuses on the contribution of forestry to food security, on effective and responsible forest management and on maintaining a balance between the economic, ecological and social benefits of forest resources. The Organization has helped to develop national forestry programmes and to promote the sustainable development of all types of forest. FAO administers the global Forests, Trees and People Programme, which promotes the sustainable management of tree and forest resources, based on local knowledge and management practices, in order to improve the livelihoods of rural people in developing countries. FAO's Strategic Plan for Forestry was approved in March 1999; its main objectives were to maintain the environmental diversity of forests, to realize the economic potential of forests and trees within a sustainable framework, and to expand access to information on forestry.

In Asia and the Pacific FAO's Forests, Trees and People Programme is implemented in collaboration with organizations and institutions in the People's Republic of China, Indonesia, Philippines, Thailand and Viet Nam.

NUTRITION

The International Conference on Nutrition, sponsored by FAO and WHO, took place in Rome in December 1992. It approved a World Declaration on Nutrition and a Plan of Action, aimed at promoting efforts to combat malnutrition as a development priority. Since the conference, more than 100 countries have formulated national plans of action for nutrition, many of which were based on existing development plans such as comprehensive food security initiatives, national poverty alleviation programmes and action plans to attain the targets set by the World Summit for Children in September 1990. In October 1996 FAO, WHO and other partners jointly organized the first World Congress on Calcium and Vitamin D in Human Life, held in Rome. In January 2001 a joint team of FAO and WHO experts issued a report concerning the allergenicity of foods derived from biotechnology (i.e. genetically modified—GM—foods). In July the Codex Alimentarius Commission agreed the first global principles for assessing the safety of GM foods, and approved a series of maximum levels of environmental contaminants in food. FAO and WHO jointly convened a Global Forum of Food Safety Regulators in Marrakesh, Morocco, in January 2002. In April the two organizations announce a joint review of their food standards operations, including the activities of the Codex Alimentarius Commission.

PROCESSING AND MARKETING

An estimated 20% of all food harvested is lost before it can be consumed, and in some developing countries the proportion is much higher. FAO helps reduce immediate post-harvest losses, with the introduction of improved processing methods and storage systems. It also advises on the distribution and marketing of agricultural produce and on the selection and preparation of foods for optimum nutrition. Many of these activities form part of wider rural development projects. Many developing countries rely on agricultural products as their main source of foreign earnings, but the terms under which they are traded are usually more favourable to the industrialized countries. FAO continues to favour the elimination of export subsidies and related discriminatory practices, such as protectionist measures that hamper international trade in agricultural commodities. FAO has organized regional workshops and national projects in order to help member states to implement World Trade Organization regulations, in particular with regard to agricultural policy, intellectual property rights, sanitary and phytosanitary measures, technical barriers to trade and the international standards of the Codex Alimentarius. FAO evaluates new market trends and helps to develop improved plant and animal quarantine procedures. In

November 1997 the FAO Conference adopted new guide-lines on surveillance and on export certification systems in order to harmonize plant quarantine standards. FAO participates in PhAction, a forum of 12 agencies that was established in 1999 to promote post-harvest research and the development of effective post-harvest services and infrastructure.

FOOD SECURITY

FAO's policy on food security aims to encourage the production of adequate food supplies, to maximize stability in the flow of supplies, and to ensure access on the part of those who need them. In 1994 FAO initiated the Special Programme for Food Security (SPFS), designed to assist low-income countries with a food deficit to increase food production and productivity as rapidly as possible, primarily through the widespread adoption by farmers of improved production technologies, with emphasis on areas of high potential. FAO was actively involved in the formulation of the Plan of Action on food security that was adopted at the World Food Summit in November 1996, and was to be responsible for monitoring and promoting its implementation. In March 1999 FAO signed agreements with IFAD and WFP that aimed to increase co-operation within the framework of the SPFS. A budget of US $10.5m. was allocated to the SPFS for the two-year period 2004–05. In 2004 the SPFS was operational in 100 countries, of which 42 were in Africa. About 70 of these countries were categorized as 'low-income food-deficit'. The Programme promotes South-South co-operation to improve food security and the exchange of knowledge and experience. By September 2003 28 bilateral co-operation agreements were in force, for example, between Egypt and Cameroon and Viet Nam and Benin.

FAO's Global Information and Early Warning System (GIEWS), which become operational in 1975, maintains a database on and monitors the crop and food outlook at global, regional, national and sub-national levels in order to detect emerging food supply difficulties and disasters and to ensure rapid intervention in countries experiencing food supply shortages. It publishes regular reports on the weather conditions and crop prospects in sub-Saharan Africa and in the Sahel region, issues special alerts which describe the situation in countries or sub-regions experiencing food difficulties, and recommends an appropriate international response. FAO's annual publication *State of Food Insecurity in the World* is based on data compiled by the Organization's Food Insecurity and Vulnerability Information and Mapping Systems programme.

In October 2003 GIEWS issued a special report on the food production situation in the Democratic People's Republic of Korea (DPRK). A report on the situation in Timor-Leste was published in June.

FAO INVESTMENT CENTRE

The Investment Centre was established in 1964 to help countries to prepare viable investment projects that will attract external financing. The Centre focuses its evaluation of projects on two fundamental concerns: the promotion of sustainable activities for land management, forestry development and environmental protection, and the alleviation of rural poverty. In 2002–03 157 projects were approved, representing a total investment of more than US $5,000m.

EMERGENCY RELIEF

FAO works to rehabilitate agricultural production following natural and man-made disasters by providing emergency seed, tools, and technical and other assistance. Jointly with the United Nations, FAO is responsible for WFP, which provides emergency food supplies and food aid in support of development projects. FAO's Division for Emergency Operations and Rehabilitation was responsible for preparing the emergency agricultural relief component of the 2004 UN inter-agency appeals for 23 countries and regions.

New projects approved by the Division for Emergency Operations and Rehabilitation in 2004 included emergency assistance for prevention, investigation, control and surveillance of avian influenza in Cambodia, the People's Republic of China (PRC), Indonesia and Laos; the emergency supply of basic agricultural inputs to flood-affected farmers in the PRC; support for a double-cropping programme in the Democratic People's Republic of Korea (DPRK); livelihood support for resettling IDPs and vulnerable populations in northern Maluku, Indonesia, and agriculture-based livelihood recovery support in resettlement areas of West Timor; and emergency supply of agricultural inputs for typhoon-affected communities in Northern Mindanao, Philippines, and emergency assistance for the rehabilitation of sustainable agriculture and fisheries in landslide- and flood-affected areas of that country. Under the UN inter-agency appeal for the DPRK in 2004 FAO appealed for US $3.5m. in support of double-cropping and horticultural crops production, and assistance for the co-ordination of emergency, rehabilitation and recovery interventions aimed at establishing agricultural and food security.

INFORMATION

FAO collects, analyses, interprets and disseminates information through various media, including an extensive internet site. It issues regular statistical reports, commodity studies, and technical manuals in local languages (see list of publications below). Other materials produced by the FAO include information booklets, reference papers, reports of meetings, training manuals and audio-visuals.

FAO's internet-based interactive World Agricultural Information Centre (WAICENT) offers access to agricultural publications, technical documentation, codes of conduct, data, statistics and multimedia resources. FAO compiles and co-ordinates an extensive range of international databases on agriculture, fisheries, forestry, food and statistics, the most important of these being AGRIS (the International Information System for the Agricultural Sciences and Technology) and CARIS (the Current Agricultural Research Information System). Statistical databases include the GLOBEFISH databank and electronic library, FISHDAB (the Fisheries Statistical Database), FORIS (Forest Resources Information System), and GIS (the Geographic Information System). In addition, FAOSTAT provides access to updated figures in 10 agriculture-related topics. The AGORA (Access to Global Online Research in Agriculture) initiative, launched in November 2003 by FAO and other partners, aims to provide free or low-cost access to more than 400 scientific journals in agriculture, nutrition and related fields for researchers from developing countries.

In June 2000 FAO organized a high-level Consultation on Agricultural Information Management (COAIM), which aimed to increase access to and use of agricultural information by policy-makers and others. The second COAIM was held in September 2002; a third meeting, scheduled to be held in June 2004, was postponed.

World Food Day, commemorating the foundation of FAO, is held annually on 16 October.

FAO Councils and Commissions

(Based at the Rome headquarters unless otherwise indicated)

Asia and Pacific Commission on Agricultural Statistics: c/o FAO Regional Office, Maliwan Mansion, Phra Atit Rd, Bangkok 10200, Thailand; f. 1962 to review the state of food and agricultural statistics in the region and to advise member countries on the development and standardization of agricultural statistics; 25 member states.

Asia and Pacific Plant Protection Commission: c/o FAO Regional Office, Maliwan Mansion, Phra Atit Rd, Bangkok 10200, Thailand; f. 1956 (new title 1983) to strengthen international co-operation in plant protection to prevent the introduction and spread of destructive plant diseases and pests; 25 member states.

Asia-Pacific Fishery Commission: c/o FAO Regional Office, Maliwan Mansion, Phra Atit Rd, Bangkok 10200, Thailand; f. 1948 to develop fisheries, encourage and co-ordinate research, disseminate information, recommend projects to governments, propose standards in technique and management measures; 20 member states.

Asia-Pacific Forestry Commission: internet www.apfcweb.org; f. 1949 to advise on the formulation of forest policy, and review and co-ordinate its implementation throughout the region to exchange information and advise on technical problems; 29 member states.

FAO/WHO Codex Alimentarius Commission: internet www.codexalimentarius.net; f. 1962 to make proposals for the co-ordination of all international food standards work and to publish a code of international food standards; established Intergovernmental Task Force on Foods Derived from Biotechnology in 1999; Trust Fund to support participation by least-developed countries was inaugurated in February 2003; 165 member states.

International Rice Commission: internet www.fao.org/ag/AGP/AGPC/doc/field/commrice/welcome.htm; f. 1949 to promote national and international action on production, conservation, distribution and consumption of rice, except matters relating to international trade; 60 member states.

Regional Animal Production and Health Commission for Asia and the Pacific: c/o FAO Regional Office, Maliwan Mansion, Phra Atit Rd, Bangkok 10200, Thailand; internet www.aphca.org; f. 1973 to promote livestock development in general, and national and international research and action with respect to animal health and husbandry problems in the region; 15 member states.

Finance

FAO's Regular Programme, which is financed by contributions from member governments, covers the cost of FAO's Secretariat, its Technical Co-operation Programme (TCP) and part of the cost of several special action programmes. The proposed budget for the two years 2004–05 totalled US $749m. Much of FAO's technical assistance programme is funded from extra-budgetary sources, predominantly by trust funds that come mainly from donor countries and international financing institutions. The single largest contributor is the United Nations Development Programme (UNDP).

Publications

Animal Health Yearbook.
Commodity Review and Outlook (annually).
Environment and Energy Bulletin.
Ethical Issues in Food and Agriculture.
Fertilizer Yearbook.
Food Crops and Shortages (6 a year).
Food Outlook (5 a year).
Food Safety and Quality Update (monthly; electronic bulletin).
Forest Resources Assessment.
Plant Protection Bulletin (quarterly).
Production Yearbook.
Quarterly Bulletin of Statistics.
The State of Food and Agriculture (annually).
The State of Food Insecurity in the World (annually).
The State of World Fisheries and Aquaculture (every two years).
The State of the World's Forests (every 2 years).
Trade Yearbook.
Unasylva (quarterly).
Yearbook of Fishery Statistics.
Yearbook of Forest Products.
World Animal Review (quarterly).
World Watch List for Domestic Animal Diversity.
Commodity reviews; studies, manuals.

International Bank for Reconstruction and Development—IBRD (World Bank)

Address: 1818 H St, NW, Washington, DC 20433, USA.
Telephone: (202) 473-1000; **fax:** (202) 477-6391; **e-mail:** pic@worldbank.org; **internet:** www.worldbank.org.

The IBRD was established in December 1945. Initially it was concerned with post-war reconstruction in Europe; since then its aim has been to assist the economic development of member nations by making loans where private capital is not available on reasonable terms to finance productive investments. Loans are made either directly to governments, or to private enterprises with the guarantee of their governments. The World Bank, as it is commonly known, comprises the IBRD and the International Development Association (IDA). The affiliated group of institutions, comprising the IBRD, the IDA, the International Finance Corporation (IFC), the Multilateral Investment Guarantee Agency (MIGA) and the International Centre for Settlement of Investment Disputes (ICSID, see below), is now referred to as the World Bank Group.

Organization

(October 2004)

Officers and staff of the IBRD serve concurrently as officers and staff in the IDA. The World Bank has offices in New York, Brussels, Paris (for Europe), Frankfurt, London, Geneva and Tokyo, as well as in more than 100 countries of operation. Country Directors are located in some 30 country offices.

BOARD OF GOVERNORS

The Board of Governors consists of one Governor appointed by each member nation. Typically, a Governor is the country's finance minister, central bank governor, or a minister or an official of comparable rank. The Board normally meets once a year.

EXECUTIVE DIRECTORS

The general operations of the Bank are conducted by a Board of 24 Executive Directors. Five Directors are appointed by the five members having the largest number of shares of capital stock, and the rest are elected by the Governors representing the other members. The President of the Bank is Chairman of the Board.

PRINCIPAL OFFICERS

The principal officers of the Bank are the President of the Bank, four Managing Directors, three Senior Vice-Presidents and 24 Vice-Presidents.

President and Chairman of Executive Directors: James D. Wolfensohn (USA).

Vice-President, East Asia and the Pacific Regional Office: Jemal-ud-din Kassum (Tanzania).

Activities

FINANCIAL OPERATIONS

IBRD capital is derived from members' subscriptions to capital shares, the calculation of which is based on their quotas in the International Monetary Fund. At 30 June 2003 the total subscribed capital of the IBRD was US $189,567m., of which the paid-in portion was $11,478m. (6.1%); the remainder is subject to call if required. Most of the IBRD's lendable funds come from its borrowing, on commercial terms, in world capital markets, and also from its retained earnings and the flow of repayments on its loans. IBRD loans carry a variable interest rate, rather than a rate fixed at the time of borrowing.

IBRD loans usually have a 'grace period' of five years and are repayable over 15 years or fewer. Loans are made to governments, or must be guaranteed by the government concerned, and are normally made for projects likely to offer a commercially viable rate of return. In 1980 the World Bank introduced structural adjustment lending, which (instead of financing specific projects) supports programmes and changes necessary to modify the structure of an economy so that it can restore or maintain its growth and viability in its balance of payments over the medium term.

The IBRD and IDA together made 240 new lending and investment commitments totalling US $18,513.2m. during the year ending 30 June 2003, compared with 225 (amounting to $19,519.4m.) in the previous year. During 2002/03 the IBRD alone approved commitments totalling $11,230.7m. (compared with $11,451.8m. in the previous year). Disbursements by the IBRD in the year ending 30 June 2003 amounted to $11,921m.

IBRD operations are supported by medium- and long-term borrowings in international capital markets. During the year ending 30 June 2003 the IBRD's net income amounted to US $5,344m.

The World Bank's primary objectives are the achievement of sustainable economic growth and the reduction of poverty in developing countries. In the context of stimulating economic growth the Bank promotes both private-sector development and human resource development and has attempted to respond to the growing demands by developing countries for assistance in these areas. In

March 1997 the Board of Executive Directors endorsed a 'Strategic Compact' to increase the effectiveness of the Bank in achieving its central objective of poverty reduction. The reforms included greater decentralization of decision-making, and investment in front-line operations, enhancing the administration of loans, and improving access to information and co-ordination of Bank activities through a knowledge management system comprising four thematic networks: the Human Development Network; the Environmentally and Socially Sustainable Development Network; the Finance, Private Sector and Infrastructure Development Network; and the Poverty Reduction and Economic Management Network. In 2000/01 the Bank adopted a new Strategic Framework which emphasized two essential approaches for Bank support: strengthening the investment climate and prospects for sustainable development in a country, and supporting investment in the poor. In September 2001 the Bank announced that it was to join the UN as a full partner in implementing the so-called Millennium Development Goals (MDGs), and was to make them central to its development agenda. The objectives, which were approved by governments attending a special session of the UN General Assembly in September 2000, represented a new international consensus to achieve determined poverty reduction targets. These included reducing by 50% the number of people with an income of less than US $1 a day and those suffering from hunger and lack of safe drinking water by 2015, achieving education for all, reducing maternal mortality, and combating HIV/AIDS, malaria and other major diseases. The Bank was closely involved in preparations for the International Conference on Financing for Development, which was held in Monterrey, Mexico, in March 2002. The meeting adopted the Monterrey Consensus, which outlined measures to support national development efforts and to achieve the MDGs. During 2002/03 the Bank, with the IMF, undertook to develop a monitoring framework to review progress in the MDG agenda.

The Bank's efforts to reduce poverty include the compilation of country-specific assessments and the formulation of country assistance strategies (CASs) to review and guide the Bank's country programmes. Since August 1998 the Bank has published CASs, with the approval of the government concerned. In 1998/99 the Bank's Executive Directors endorsed a Comprehensive Development Framework (CDF) to effect a new approach to development assistance based on partnerships and country responsibility, with an emphasis on the interdependence of the social, structural, human, governmental, economic and environmental elements of development. The Framework, which aimed to enhance the overall effectiveness of development assistance, was formulated after a series of consultative meetings organized by the Bank and attended by representatives of governments, donor agencies, financial institutions, non-governmental organizations, the private sector and academics.

In December 1999 the Bank introduced a new approach to implement the principles of the CDF, as part of its strategy to enhance the debt relief scheme for heavily indebted poor countries (see below). Applicant countries were requested to formulate a national strategy to reduce poverty, to be presented in the form of a Poverty Reduction Strategy Paper (PRSP). In cases where there might be some delay in issuing a full PRSP, it was permissible for a country to submit a less detailed 'interim' PRSP (I-PRSP) in order to secure the preliminary qualification for debt relief. During 2002/03 the Bank considered 15 PRSPs and seven progress reports. In 2000/01 the Bank introduced a new Poverty Reduction Support Credit to help low-income countries to implement the policy and institutional reforms outlined in their PRSP. The first credits were approved for Uganda and Viet Nam in May and June respectively. In January 2002 a PRSP public review conference, attended by more than 200 representatives of donor agencies, civil society groups, and developing country organizations was held as part of an ongoing review of the scheme by the Bank and the IMF. During 2002/03 the Bank undertook initiatives to support the development of national poverty reduction strategies and to strengthen its own related assistance activities.

In September 1996 the World Bank/IMF Development Committee endorsed a joint initiative to assist heavily indebted poor countries (HIPCs) to reduce their debt burden to a sustainable level, in order to make more resources available for poverty reduction and economic growth. A new Trust Fund was established by the World Bank in November to finance the initiative. The Fund, consisting of an initial allocation of US $500m. from the IBRD surplus and other contributions from multilateral creditors, was to be administered by IDA. In early 1999 the World Bank and IMF initiated a comprehensive review of the HIPC initiative. In June the G-7 and Russia, meeting in Cologne, Germany, agreed to increase contributions to the HIPC Trust Fund and to cancel substantial amounts of outstanding debt, and proposed more flexible terms for eligibility. In September the Bank and IMF reached an agreement on an enhanced HIPC scheme, with further revenue to be generated through the revaluation of a percentage of IMF gold reserves. It was agreed that, in order to qualify for debt relief and additional concessional lending, countries were to formulate a PRSP, and should demonstrate prudent financial management in the implementation of the strategy for at least one year. Those countries still deemed to have an unsustainable level of debt at the pivotal 'decision point' of the process were to qualify for assistance. In the majority of cases a sustainable level of debt was targeted at 150% of the net present value (NPV) of the debt in relation to total annual exports (compared with 200%–250% under the original HIPC scheme). Other countries with a lower debt-to-export ratio were to be eligible for assistance under the initiative, providing that their export earnings were at least 30% of GDP (lowered from 40%) and government revenue at least 15% of GDP (reduced from 20%).

There are three HIPCs in the Far East and Australasia region: Laos, Myanmar and Viet Nam.

During 2000/01 the World Bank strengthened its efforts to counter the problem of HIV and AIDS in developing countries. In November 2001 the Bank appointed its first Global HIV/AIDS Adviser. In September 2000 a new Multi-Country HIV/AIDS Programme for Africa (MAP) was launched, in collaboration with UNAIDS and other major donor agencies and non-governmental organizations. Some US $500m. was allocated to the initiative and was used to support efforts in seven countries. In February 2002 the Bank approved an additional $500m. for a second phase of MAP, which was envisaged to assist HIV/AIDS schemes in a further 12 countries, as well as regional activities. A MAP initiative for the Caribbean, with a budget of $155m., was launched in 2001.

In addition to providing financial services, the Bank also undertakes analytical and advisory services, and supports learning and capacity-building, in particular through the World Bank Institute (see below), the Staff Exchange Programme and knowledge-sharing initiatives. The Bank has supported efforts, such as the Global Development Gateway, to disseminate information on development issues and programmes, and, since 1988, has organized the Annual Bank Conference on Development Economics (ABCDE) to provide a forum for the exchange and discussion of development-related ideas and research. In September 1995 the Bank initiated the Information for Development Programme (InfoDev) with the aim of fostering partnerships between governments, multilateral institutions and private-sector experts in order to promote reform and investment in developing countries through improved access to information technology.

TECHNICAL ASSISTANCE

The provision of technical assistance to member countries has become a major component of World Bank activities. The economic and sector work (ESW) undertaken by the Bank is the vehicle for considerable technical assistance and often forms the basis of CASs and other strategic or advisory reports. In addition, project loans and credits may include funds earmarked specifically for feasibility studies, resource surveys, management or planning advice, and training. The Economic Development Institute has become one of the most important of the Bank's activities in technical assistance. It provides training in national economic management and project analysis for government officials at the middle and upper levels of responsibility. It also runs overseas courses aiming to build up local training capability, and administers a graduate scholarship programme.

The Bank serves as an executing agency for projects financed by the UN Development Programme. It also administers projects financed by various trust funds.

Technical assistance (usually reimbursable) is also extended to countries that do not need Bank financial support, e.g. for training and transfer of technology. The Bank encourages the use of local consultants to assist with projects and stimulate institutional capability.

The Project Preparation Facility (PPF) was established in 1975 to provide cash advances to prepare projects that may be financed by the Bank. In December 1994 the PPF's commitment authority was increased from US $220m. to $250m. In 1992 the Bank established an Institutional Development Fund (IDF), which became operational on 1 July; the purpose of the Fund was to provide rapid, small-scale financial assistance, to a maximum value of $500,000, for capacity-building proposals.

ECONOMIC RESEARCH AND STUDIES

In the 1990s the World Bank's research, conducted by its own research staff, was increasingly concerned with providing information to reinforce the Bank's expanding advisory role to developing countries and to improve policy in the Bank's borrowing countries. The principal areas of current research focus on issues such as maintaining sustainable growth while protecting the environment and the poorest sectors of society, encouraging the development of the private sector, and reducing and decentralizing government activities.

The Bank chairs the Consultative Group on International Agricultural Research (CGIAR), which was founded in 1971 to raise financial support for international agricultural research work for improving crops and animal production in developing countries; it supports 16 research centres.

CO-OPERATION WITH OTHER ORGANIZATIONS

The World Bank co-operates with other international partners with the aim of improving the impact of development efforts. It collaborates with the IMF in implementing the HIPC scheme and the two agencies work closely to achieve a common approach to development initiatives. The Bank has established strong working relationships with many other UN bodies, in particular through a mutual commitment to poverty reduction objectives. In May 2000 the Bank signed a joint statement of co-operation with the OECD. The Bank holds regular consultations with other multilateral development banks and with the European Union with respect to development issues. The Bank-NGO Committee provides an annual forum for discussion with non-governmental organizations (NGOs). Strengthening co-operation with external partners was a fundamental element of the Comprehensive Development Framework, which was adopted in 1998/99 (see above). In 2001/02 a Partnership Approval and Tracking System was implemented to provide information on the Bank's regional and global partnerships.

In June 1995 the World Bank joined other international donors (including regional development banks, other UN bodies, Canada, France, the Netherlands and the USA) in establishing a Consultative Group to Assist the Poorest (CGAP), which was to channel funds to the most needy through grass-roots agencies. An initial credit of approximately US $200m. was committed by the donors. The Bank manages the CGAP Secretariat, which is responsible for the administration of external funding and for the evaluation and approval of project financing. The CGAP provides technical assistance, training and strategic advice to microfinance institutions and other relevant bodies. As an implementing agency of the Global Environment Facility (GEF) the Bank assists countries to prepare and supervise GEF projects relating to biological diversity, climate change and other environmental protection measures. It is an example of a partnership in action which addresses a global agenda, complementing Bank country assistance activities.

In 1997 a Partnerships Group was established to strengthen the Bank's work with development institutions, representatives of civil society and the private sector. The Group established a new Development Grant Facility, which became operational in October, to support partnership initiatives and to co-ordinate all of the Bank's grant-making activities. Also in 1997 the Bank, in partnership with the IMF, UNCTAD, UNDP, the World Trade Organization (WTO) and International Trade Commission, established an Integrated Framework for Trade-related Assistance to Least Developed Countries, at the request of the WTO, to assist those countries to integrate into the global trading system and improve basic trading capabilities.

In March 1998 the Bank helped to organize the first Asia Development Forum, which convened some 300 representatives of government, the private sector and academia to discuss the region's prospects for economic recovery. A second Forum was held in June 2000, in Singapore, and the third, organized by the bank with the Asian Development Bank, the ADB Institute, and ESCAP to enhance further development capacity, was convened in June 2001, in Bangkok, Thailand. A fourth Forum was held in Seoul, Republic of Korea, in November 2002, on the theme of trade and poverty.

The Bank is a lead organization in providing reconstruction assistance following natural disasters or conflicts, usually in collaboration with other UN agencies or international organizations, and through special trust funds. The Bank is a trustee, with the Asian Development Bank, of the Trust Fund for East Timor, which was established in December 1999, with donations of some US $147m., to channel support for reconstruction projects and preparations for independence and the post-independence period. The Bank hosts, every six months, a Timor-Leste and Development Partners Meeting. In response to the extreme financial difficulties that confronted several Asian economies in 1997/98 the Bank established a Special Financial Operations Unit to help to alleviate the consequences of the crisis in all affected countries. The Bank pledged some $16,000m., in addition to its regular lending programme, to lessen the social consequences of the crisis by protecting social services and strengthening social security and other public funding for the poor and disadvantaged, mainly in Thailand and Indonesia. The Bank also helped to formulate programmes to strengthen legal and institutional frameworks and to restructure the financial services sector and corporate governance. In June 1998 a Japan Social Development Fund, with resources amounting to some $93m., was established by the Japanese Government to fund social programmes to assist the poorest communities affected by the financial crisis and other grassroots capacity-building activities. In April the Bank agreed to administer a new $40m. Asian Financial Crisis Response Trust Fund, which had been established at the second Asia-Europe Meeting to provide support in 70 activities in seven countries, mainly relating to corporate and banking sector reforms, at a total cost of some $43.8m. A second phase of the so-called ASEM Trust Fund became operational in March 2001, focusing on social welfare and safety nets, and financing and corporate restructuring. Contributions to the Fund were expected to total more than $38m.

The Bank conducts co-financing and aid co-ordination projects with official aid agencies, export credit institutions, and commercial banks. During the year ending 30 June 2003 a total of 103 IBRD and IDA projects involved co-financers' contributions amounting to US $3,000m.

EVALUATION

The Operations Evaluation Department is an independent unit within the World Bank. It conducts Country Assistance Evaluations to assess the development effectiveness of a Bank country programme, and studies and publishes the results of projects after a loan has been fully disbursed, so as to identify problems and possible improvements in future activities. In addition, the department reviews the Bank's global programmes and produces the *Annual Review of Development Effectiveness*. In 1996 a Quality Assurance Group was established to monitor the effectiveness of the Bank's operations and performance.

In September 1993 the Bank established an independent Inspection Panel, consistent with the Bank's objective of improving project implementation and accountability. The Panel, which became operational in September 1994, was to conduct independent investigations and report on complaints from local people concerning the design, appraisal and implementation of development projects supported by the Bank. By early 2004 the Panel had received 27 formal requests for inspection and had recommended investigations in 13 of those cases.

IBRD INSTITUTIONS

World Bank Institute (WBI): founded in March 1999 by merger of the Bank's Learning and Leadership Centre, previously responsible for internal staff training, and the Economic Development Institute (EDI), which had been established in 1955 to train government officials concerned with development programmes and policies. The new Institute aimed to emphasize the Bank's priority areas through the provision of training courses and seminars relating to poverty, crisis response, good governance and anti-corruption strategies. During 2002/03 WBI activities reached some 58,000 participants. The Institute has continued to support a Global Knowledge Partnership, which was established in 1997 to promote alliances between governments, companies, other agencies and organizations committed to applying information and communication technologies for development purposes. Under the EDI a World Links for Development programme was also initiated to connect schools in developing countries with partner establishments in industrialized nations via the internet. In 1999 the WBI expanded its programmes through distance learning, a Global Development Network, and use of new technologies. A new initiative, Global Development Learning Network (GDLN), aimed to expand access to information and learning opportunities through the internet, videoconferences and organized exchanges. In 2002/03 there were 61 GDLN centres. At mid-2004 formal partnership arrangements were in place between some 120 learning centres and public, private and non-governmental organizations; Vice-Pres. Frannie Léautier (Tanzania/France).

International Centre for Settlement of Investment Disputes (ICSID): founded in 1966 under the Convention of the Settlement of Investment Disputes between States and Nationals of Other States. The Convention was designed to encourage the growth of private foreign investment for economic development, by creating the possibility, always subject to the consent of both parties, for a Contracting State and a foreign investor who is a national of another Contracting State to settle any legal dispute that might arise out of such an investment by conciliation and/or arbitration before an impartial, international forum. The governing body of the Centre is its Administrative Council, composed of one representative of each Contracting State, all of whom have equal voting power. The President of the World Bank is (*ex officio*) the non-voting Chairman of the Administrative Council. At November 2003 140 countries had signed and ratified the Convention to become ICSID Contracting States. By mid-2004 the Centre had concluded 83 cases, while 78 were pending; Sec.-Gen. Roberto Dañino (Peru).

Publications

Abstracts of Current Studies: The World Bank Research Program (annually).

Annual Report on Operations Evaluation.

Annual Report on Portfolio Performance.

Annual Review of Development Effectiveness.

EDI Annual Report.

Global Commodity Markets (quarterly).

Global Development Finance (annually, also on CD-Rom and online).

Global Economic Prospects (annually).

ICSID Annual Report.

ICSID Review—Foreign Investment Law Journal (2 a year).

Joint BIS-IMF-OECD-World Bank Statistics on External Debt (quarterly, also available on the internet at www.worldbank.org/data/jointdebt.html).

New Products and Outreach (EDI, annually).

News from ICSID (2 a year).

Poverty Reduction and the World Bank (annually).

Poverty Reduction Strategies Newsletter (quarterly).

Research News (quarterly).

Staff Working Papers.

Transition (every 2 months).

World Bank Annual Report.

World Bank Atlas (annually).

World Bank Economic Review (3 a year).

The World Bank and the Environment (annually).

World Bank Research Observer.

World Development Indicators (annually, also on CD-Rom and online).

World Development Report (annually, also on CD-Rom).

Statistics

IBRD LOANS APPROVED IN THE FAR EAST AND AUSTRALASIA, 1 JULY 2002–30 JUNE 2003
(US $ million)

Country	Purpose	Amount
China, People's Republic	Second Anhui highway specific investment loan	250.0
	Shanghai urban environment	200.0
	Second Tianjin urban development and environment specific investment loan	150.0
	Power sector in Jiangsu Province	145.0
	Hubei Xiaogan Xiangfan highway	250.0
	Third Xinjiang highway project	150.0
Indonesia	Water resources and irrigation sector management*	25.0
	Third Kecamatan development loan*	204.3
	Java Bali power sector restructuring and strengthening loan	141.0
	Health workforce and services*	31.1
	Private provision of infrastructure technical assistance loan	17.1
Philippines	KALAHI-CIDSS basic infrastructure specific investment loan	100.0
	Second agrarian reform communities development specific investment loan	50.0
	Autonomous Region of Muslim Mindanao social fund	33.6

* Joint IBRD/IDA funded project.

Source: *World Bank Annual Report 2003*.

International Development Association—IDA

Address: 1818 H Street, NW, Washington, DC 20433, USA.

Telephone: (202) 473-1000; **fax:** (202) 477-6391; **internet:** www.worldbank.org/ida.

The International Development Association began operations in November 1960. Affiliated to the IBRD, IDA advances capital to the poorer developing member countries on more flexible terms than those offered by the IBRD.

Organization

(October 2004)

Officers and staff of the IBRD serve concurrently as officers and staff of IDA.

President and Chairman of Executive Directors: James D. Wolfensohn (*ex officio*).

Activities

IDA assistance is aimed at the poorer developing countries (i.e. those with an annual GNP per capita of less than US $865 in 2002 dollars were to qualify for assistance in 2003/04) and support their poverty reduction strategies. Under IDA lending conditions, credits can be extended to countries whose balance of payments could not sustain the burden of repayment required for IBRD loans. Terms are more favourable than those provided by the IBRD; credits are for a period of 35 or 40 years, with a 'grace period' of 10 years, and carry no interest charges. At mid-2003 81 countries were eligible for IDA assistance, including several small-island economies with a GNP per head greater than $865, but which would otherwise have little or no access to Bank funds, and 15 so-called 'blend borrowers' which are entitled to borrow from both the IDA and IBRD. IDA administers a Trust Fund, which was established in November 1996 as part of a World Bank/IMF initiative to assist heavily indebted poor countries (HIPCs).

IDA's total development resources, consisting of members' subscriptions and supplementary resources (additional subscriptions and contributions), are replenished periodically by contributions from the more affluent member countries. Discussions on the 13th replenishment of IDA funds commenced in February 2001, and for the first time involved representatives of borrowing countries, civil society and other public groups. A final commitment, providing for some US $23,000m. in resources for the period 1 July 2002–30 June 2005, was concluded in early July 2002 by some 38 donor countries. The IDA-13 lending framework was to emphasize the following objectives: promoting sound policies for growth and poverty reduction; ensuring effective assistance and measurable results; improving co-ordination, transparency, and consultation; and providing for substantial replenishment of resources. The replenishment programme also provided for greater use of grants to address the problems of the poorest recipient countries, for example those most vulnerable to debt, those in post-conflict situations, as well as reconstruction projects after a natural disaster and HIV/AIDS programmes.

During the year ending 30 June 2003 IDA credits totalling US $7,282.5m. were approved, compared with $8,067.6m. in the previous year. Of the total new lending in 2002/03 some $1,232m. (or 17%) was in the form of grants for the poorest or most vulnerable countries.

Publication

Annual Report.

Statistics

IDA CREDITS APPROVED IN THE FAR EAST AND AUSTRALASIA, 1 JULY 2002–30 JUNE 2003
(US $ million)

Country	Purpose	Amount
Cambodia	Provincial and periurban water and sanitation credit/grant	16.9/3.0
	Rural investment and local governance	22.0
	Health sector support investment and maintenance credit/grant	17.2/9.8
Indonesia	Water resources and irrigation sector management*	45.0
	Third Kecamatan development credit*	45.5
	Health workforce and services	74.5
Laos	Sustainable forestry for rural development	9.9
	Second land titling credit	14.8
Mongolia	Economic capacity-building technical assistance credit	7.5
Samoa	Telecommunications and post reform technical assistance credit	4.5
Viet Nam	Second poverty reduction support credit	100.0
	Public financial management reform credit	54.3
	Primary education for disadvantaged children	138.8

In addition, three IDA grants, totalling US $6.5m., were extended to Timor-Leste under the Bank-administered Trust Fund, and a partial risk guarantee, amounting to $75m., was provided in support of a power facility project in Viet Nam.

* Joint IBRD/IDA funded project.

Source: *World Bank Annual Report 2003*.

International Finance Corporation—IFC

Address: 2121 Pennsylvania Ave, NW, Washington, DC 20433, USA.

Telephone: (202) 473-3800; **fax:** (202) 974-4384; **e-mail:** information@ifc.org; **internet:** www.ifc.org.

IFC was founded in 1956 as a member of the World Bank Group to stimulate economic growth in developing countries by financing private-sector investments, mobilizing capital in international financial markets, and providing technical assistance and advice to governments and businesses.

Organization

(October 2004)

IFC is a separate legal entity in the World Bank Group. Executive Directors of the World Bank also serve as Directors of IFC. The President of the World Bank is *ex officio* Chairman of the IFC Board of Directors, which has appointed him President of IFC. Subject to his overall supervision, the day-to-day operations of IFC are conducted by its staff under the direction of the Executive Vice-President.

PRINCIPAL OFFICERS

President: JAMES D. WOLFENSOHN (USA).

Executive Vice-President: PETER L. WOICKE (Germany).

Director, East Asia and the Pacific Department: JAVED HAMID.

OFFICES IN THE FAR EAST AND AUSTRALASIA

Cambodia: 113 Norodom Blvd, Sangkat Chaktomuk, Phnom Penh; tel. (23) 210922; fax (23) 215157; Country Man. DEEPAK KHANNA.

China, People's Republic: 9/F Tower B, Fuhua Mansion, 8 Chaoyangmen Beidajie, Dongcheng District, Beijing 100027; tel. (10) 65544191; fax (10) 65544192; Country Man. KARIN FINKELSTON.

Indonesia: Stock Exchange Bldg, Tower II, 13th Floor, Jalan Jenderal Sudirman Kav. 52-53, 12190 Jakarta; tel. (21) 5299-3001; fax (21) 5299-3002; Country Man. GERMAN VEGARRA.

Japan: Fukoku Seimei Bldg, 10th Floor, 2-2-2, Uchisaiwai-cho, Chiyoda-ku, Tokyo 100-0011; tel. (3) 3597-6657; fax (3) 3597-6698; Dir MOTOHARU FUJIKURA.

Korea, Republic: Youngpoong Bldg, 11th Floor, Chongro-ku, Seoul 110-110; tel. (2) 399-0905; fax (2) 399-0915; Resident Rep. DEEPAK KHANNA.

Philippines: Tower One, 11th Floor, Ayala Triangle, Ayala Ave, Makati 1200, Manila; tel. (2) 8487333; fax (2) 8487339; Country Man. VIPUL BHAGHAT.

Thailand: Diethelm Tower A, 17th Floor, 93/1 Wireless Rd, Bangkok 10330; tel. (2) 650-9253; fax (2) 650-9259; Country Man. MICHAEL HIGGINS.

Viet Nam: 63 Ly Thai To, 7th Floor, Hoan Kiem, Hanoi; tel. (4) 9342282; fax (4) 9342289; Country Man. DEEPAK KHANNA.

Activities

IFC aims to promote economic development in developing member countries by assisting the growth of private enterprise and effective capital markets. It finances private sector projects, through loans, the purchase of equity, quasi-equity products, and risk management services, and assists governments to create conditions that stimulate the flow of domestic and foreign private savings and investment. IFC may provide finance for a project that is partly state-owned, provided that there is participation by the private sector and that the project is operated on a commercial basis. IFC also mobilizes additional resources from other financial institutions, in particular through syndicated loans, thus providing access to international capital markets. IFC provides a range of advisory services to help to improve the investment climate in developing countries and offers technical assistance to private enterprises and governments.

To be eligible for financing, projects must be profitable for investors, as well as financially and economically viable, must benefit the economy of the country concerned, and must comply with IFC's environmental and social guide-lines. IFC aims to promote best corporate governance and management methods and sustainable business practices, and encourages partnerships between governments, non-governmental organizations and community groups. In 2001/02 IFC developed a Sustainability Framework to help to assess the longer-term economic, environmental and social impact of projects. The first Sustainability Review was published in mid-2002. In 2002/03 IFC assisted 10 international banks to draft a voluntary set of guide-lines (the Equator Principles), based on IFC's environmental, social and safeguard monitoring policies, to be applied to their global project finance activities. By January 2004 a further 10 banks had signed up to the Equator Principles.

IFC's authorized capital is US $2,450m. At 30 June 2003 paid-in capital was $2,360m. The World Bank was originally the principal source of borrowed funds, but IFC also borrows from private capital markets. IFC's net income amounted to $487m. in 2002/03, compared with $215m. in the previous year.

In the year ending 30 June 2003 project financing approved by IFC amounted to US $5,449m. for 186 projects (compared with $5,835m. for 223 projects in the previous year). Of the total approved, $3,991m. was for IFC's own account, while $1,458m. was in the form of loan syndications and underwriting of securities issues and investment funds by more than 100 participant banks and institutional investors. Generally, the IFC limits its financing to less than

25% of the total cost of a project, but may take up to a 35% stake in a venture (although never as a majority shareholder). Disbursements for IFC's account amounted to $2,959m. in 2002/03 (compared with $1,498m. in the previous year).

The largest proportion of investment commitments in 2002/03 was allocated to Latin America and the Caribbean (43%). Europe and Central Asia received 28%, East Asia and the Pacific 12%, South Asia 8%, Middle East and North Africa 6%, and sub-Saharan Africa 3%. In 2002/03 one-half of total financing committed (50%) was for financial services. Other financing included transportation, warehousing and utilities (11%), oil, gas, mining and chemicals (8%) and food and beverages (6%).

Since the economic crisis in east Asia in the late 1990s IFC has focused its activities in East Asia and the Pacific on restructuring and strengthening corporate and financial sectors in market economies and promoting the development of market institutions in transition economies. IFC's other priority areas in the region have been to finance and advise small and medium-sized enterprises, to support private participation in infrastructure and social services, to promote use of information technology, communications and software; and to encourage corporate governance. During the financial year 2002/03 IFC approved total financing of US $583m. for 31 projects in seven countries in East Asia and the Pacific. IFC operations in the region in in that year included support for the banking sector in Viet Nam and the Republic of Korea, and equity investment in a commercial bank in western People's Republic of China; financing of capital investment in a water company in the Philippines; development of oil palm production in Indonesia; and a regional facility to support the recovery of emerging markets.

Since 1990 IFC has undertaken risk-management services, in order to assist institutions to avoid financial risks that arise from changes in interest rates, in exchange rates or in commodity prices. In 2002/03 IFC approved four risk-management projects for companies and banks, bringing the total number of projects approved since 1990 to 114 in 40 countries.

IFC's Private Sector Advisory Services (PSAS), jointly managed with the World Bank, advises governments and private enterprises on policy, transaction implementation and foreign direct investment. The Foreign Investment Advisory Service (FIAS), also jointly operated and financed with the World Bank, provides advice on promoting foreign investment and strengthening the country's investment framework at the request of governments. During 2002/03 FIAS completed 49 advisory projects. At the end of that year the service had assisted more than 125 countries since it commenced operations in 1986. Under the Technical Assistance Trust Funds Program (TATF), established in 1988, IFC manages resources contributed by various governments and agencies to provide finance for feasibility studies, project identification studies and other types of technical assistance relating to project preparation. By mid-2003 contributions to the TATF programme totalled US $178m. and more than 1,250 technical assistance projects had been approved.

IFC has helped to establish several regional facilities which aim to assist small-scale entrepreneurs to develop business proposals and generate funding for their projects. The South Pacific Project Facility, based in Sydney, Australia, was established in 1991, mainly to assist local businesses in IFC Pacific Island member countries. A separate office in Port Moresby, Papua New Guinea, was opened in 1997. The Mekong Project Development Facility was inaugurated in 1995, and became operational in 1997, specifically to support the development of small and medium-sized enterprises (SMEs) in Cambodia, Laos and Viet Nam. In June 2000 IFC approved the establishment of a China Project Development Facility to support the development of SMEs within the People's Republic of China. The Facility's headquarters in Chengdu, Sichuan Province, became operational in May 2002. In October IFC established a SouthAsia Enterprise Development Facility to extend technical assistance to SMEs in Bangladesh, Bhutan, Nepal and neighbouring states in northeastern India. An Indonesia Enterprise Development Facility, based in Bali, Indonesia, was established in 2002/03.

CHINA PROJECT DEVELOPMENT FACILITY (CPDF)

The CPDF was approved in 2000 to support the development of small and medium-sized enterprises in the interior of the People's Republic of China, focusing initially on Sichuan province. The headquarters became operational in May 2002.

Headquarters: Room 2716, CCB Sichuan Bldg, No. 88, Ti Du St, Chengdu, Sichuan Provoince, People's Republic of China 610016; tel. (28) 8676-6622; fax (28) 8676-7362; internet www.cpdf.org; Gen. Man. ERIC SIEW HEW SAM.

INDONESIA ENTERPRISE DEVELOPMENT FACILITY (IEDF)

The IEDF was established in 2002/03, with financing by IFC, Australia, Japan, the Netherlands and Switzerland. The Facility aimed to support the development of small and medium-sized enterprises in eastern Indonesia and contribute to sustainable economic growth in that region.

Headquarters: Bali, Indonesia; tel. (361) 265350; fax (361) 265352; Gen. Man. CHRIS RICHARDS.

MEKONG PROJECT DEVELOPMENT FACILITY (MPDF)

The MPDF was established in November 1996 to support the development of small and medium-sized enterprises in Cambodia, Laos and Viet Nam, mainly through technical assistance, training and the mobilization of capital. The three core programmes of the Facility are investment evaluation and promotion; institution building; and support for the Mekong Financing Line, which was to provide additional credit for enterprises in the region. IFC contributes US $4m. for the administration of the Facility and provided $5m. for the establishment of the Mekong Financing Line.

Headquarters: 21-23 N. T. Minh Khai, Ho Chi Minh City, Viet Nam; tel. (8) 8235266; fax (8) 8235271; internet www.mpdf.org; Gen. Man. MARIO FISCHEL.

SOUTH PACIFIC PROJECT FACILITY (SPPF)

The South Pacific Project Facility (SPPF) was established in 1991 and assists entrepreneurs in IFC's South Pacific island member countries (Fiji, Kiribati, Marshall Islands, the Federated States of Micronesia, Palau, Papua New Guinea, Samoa, Solomon Islands, Tonga and Vanuatu) to establish new businesses or to expand or diversify existing ones. While SPPF does not fund projects directly, it contributes finance towards market, technical and feasibility studies, and helps to raise resources from other sources. SPPF also helps in locating appropriate technology and provides technical assistance during the implementation of projects. An SPPF mission was opened in Port Moresby, Papua New Guinea, in 1997.

Headquarters: Level 18, CML Bldg, 14 Martin Pl., Sydney, NSW 2000, GPO Box 1612, Sydney, NSW 2000, Australia; tel. (2) 9223-7773; fax (2) 9223-2553; e-mail sppf-info@ifc.org; Gen. Man. DENISE ALDOUS.

Publications

Annual Report.

Emerging Stock Markets Factbook (annually).

Impact (quarterly).

Lessons of Experience (series).

Results on the Ground (series).

Review of Small Businesses (annually).

Discussion papers and technical documents.

Multilateral Investment Guarantee Agency—MIGA

Address: 1818 H Street, NW, Washington, DC 20433, USA.

Telephone: (202) 473-6163; **fax:** (202) 522-2630; **internet:** www.miga.org.

MIGA was founded in 1988 as an affiliate of the World Bank. Its mandate is to encourage the flow of foreign direct investment to, and among, developing member countries, through the provision of political risk insurance and investment marketing services to foreign investors and host governments, respectively.

Organization

(October 2004)

MIGA is legally and financially separate from the World Bank. It is supervised by a Council of Governors (comprising one Governor and one Alternate of each member country) and an elected Board of Directors (of no less than 12 members).

President: JAMES D. WOLFENSOHN (USA).

Executive Vice-President: YUKIKO OMURA (Japan).

Activities

The convention establishing MIGA took effect in April 1988. Authorized capital was US $1,082m. In April 1998 the Board of Directors approved an increase in MIGA's capital base. A grant of $150m. was transferred from the IBRD as part of the package, while the capital increase (totalling $700m. callable capital and $150m. paid-in capital) was approved by MIGA's Council of Governors in April 1999. A three-year subscription period then commenced, covering the period April 1999–March 2002 (later extended to March 2003). At 30 June 2003 97 countries had subscribed $655.1m. (or 88%) of the new capital increase. At that time total subscriptions to the capital stock amounted to $1,771.7m., of which $338.9m. was paid-in.

MIGA guarantees eligible investments against losses resulting from non-commercial risks, under four main categories:

(i) transfer risk resulting from host government restrictions on currency conversion and transfer;

(ii) risk of loss resulting from legislative or administrative actions of the host government;

(iii) repudiation by the host government of contracts with investors in cases in which the investor has no access to a competent forum;

(iv) the risk of armed conflict and civil unrest.

Before guaranteeing any investment, MIGA must ensure that it is commercially viable, contributes to the development process and is not harmful to the environment. During the fiscal year 1998/99 MIGA and IFC appointed the first Compliance Advisor and Ombudsman to consider the concerns of local communities directly affected by MIGA or IFC sponsored projects. In February 1999 the Board of Directors approved an increase in the amount of political risk insurance available for each project, from US $75m. to $200m.

During the year ending 30 June 2003 MIGA issued 59 investment insurance contracts for 37 projects with a value of US $1,372m., compared with 58 contracts valued at $1,357m. in the previous financial year. The amount of direct investment associated with the contracts in 2002/03 totalled approximately $3,900m. (compared with $4,700m. in 2001/02). Since 1988 the total investment facilitated amounted to some $49,700m. in 85 countries, through 656 contracts.

MIGA works with local insurers, government agencies and other organizations to promote insurance in a country, to ensure a level of consistency among insurers and to support capacity-building within the insurance industry. By mid-2003 MIGA had signed memoranda of understanding with 33 partners.

MIGA also offers technical assistance and investment marketing services to help to promote foreign investment in developing countries and in transitional economies, and to disseminate information on investment opportunities. In October 1995 MIGA established a new network on investment opportunities, which connected investment promotion agencies (IPAs) throughout the world on an electronic information network. The so-called IPA*net* aimed to encourage further investments among developing countries, to provide access to comprehensive information on investment laws and conditions and to strengthen links between governmental, business and financial associations and investors. A new version of IPA*net* was launched in 1997 (and can be accessed at www.ipanet.net). In June 1998 MIGA initiated a new internet-based facility, 'PrivatizationLink', to provide information on investment opportunities resulting from the privatization of industries in developing economies. In October 2000 a specialized facility within the service was established to facilitate investment in Russia (russia.privatizationlink.com). During 2000/01 an office was established in Paris, France, to promote and co-ordinate European investment in developing countries, in particular in Africa and Eastern Europe. In March 2002 MIGA opened a regional office, based in Johannesburg, South Africa. In September a new regional office was inaugurated in Singapore, in order to facilitate foreign investment in Asia.

In April 2002 MIGA launched a new service, 'FDIXchange', to provide potential investors, advisors and financial institutions with up-to-date market analysis and information on foreign direct investment opportunities in emerging economies (accessible at www.fdixchange.com). An FDIXchange Investor Information Development Programme was launched in January 2003. In January 2004 a new FDI Promotion Centre became available on the internet (www.fdipromotion.com) to facilitate information exchange and knowledge-sharing among investment promotion professionals, in particular in developing countries.

Publications

Annual Report.

Investment Promotion Quarterly (electronic news update).

MIGA News (quarterly).

International Fund for Agricultural Development—IFAD

Address: Via del Serafico 107, 00142 Rome, Italy.

Telephone: (06) 54591; **fax:** (06) 5043463; **e-mail:** ifad@ifad.org; **internet:** www.ifad.org.

IFAD was established in 1977, following a decision by the 1974 UN World Food Conference, with a mandate to combat hunger and eradicate poverty on a sustainable basis in the low-income, food-deficit regions of the world. Funding operations began in January 1978.

Organization

(October 2004)

GOVERNING COUNCIL

Each member state is represented in the Governing Council (the Fund's highest authority) by a Governor and an Alternate. Sessions are held annually with special sessions as required. The Governing Council elects the President of the Fund (who also chairs the Executive Board) by a two-thirds majority for a four-year term. The President is eligible for re-election.

EXECUTIVE BOARD

Consists of 18 members and 18 alternates, elected by the Governing Council, who serve for three years. The Executive Board is responsible for the conduct and general operation of IFAD and approves loans and grants for projects; it holds three regular sessions each year. An independent Office of Evaluation reports directly to the Board.

The governance structure of the Fund is based on the classification of members. Membership of the Executive Board is distributed as follows: eight List A countries (i.e. industrialized donor countries), four List B (petroleum-exporting developing donor countries), and six List C (recipient developing countries), divided equally among the three Sub-List C categories (i.e. for Africa, Europe, Asia and the Pacific, and Latin America and the Caribbean).

President and Chairman of Executive Board: LENNART BÅGE (Sweden).

Vice-President: CYRIL ENWEZE (Nigeria).

Activities

IFAD provides financing primarily for projects designed to improve food production systems in developing member states and to strengthen related policies, services and institutions. In allocating resources IFAD is guided by: the need to increase food production in the poorest food-deficit countries; the potential for increasing food production in other developing countries; and the importance of improving the nutrition, health and education of the poorest people in developing countries, i.e. small-scale farmers, artisanal fishermen, nomadic pastoralists, indigenous populations, rural women, and the rural landless. All projects emphasize the participation of beneficiaries in development initiatives, both at the local and national level. Issues relating to gender and household food security are incorporated into all aspects of its activities. IFAD is committed to achieving the so-called Millennium Development Goals, pledged by governments attending a special session of the UN General Assembly in September 2000, and, in particular, the objective to reduce by 50% the proportion of people living in extreme poverty by 2015. In 2001 the Fund introduced new measures to improve monitoring and impact evaluation, in particular to assess its contribution to achieving the Millennium Goals. IFAD's Strategic Framework for 2002–06 reiterates its commitment to enabling the rural poor to overcome their poverty. Accordingly, the Fund's efforts were to focus on the following objectives: strengthening the capacity of the rural poor and their organizations; improving equitable access to productive natural resources and technology; and increasing access to financial services and markets. Within this Framework the Fund has also formulated regional strategies for rural poverty reduction, based on a series of regional poverty assessments. In 2003 a new Policy Division was established under the External Affairs Department. The new Division was to co-ordinate policy work at the corporate level and aimed to launch a Policy Forum in 2004, comprising IFAD senior management and staff.

IFAD is a leading repository in the world of knowledge, resources and expertise in the field of rural hunger and poverty alleviation. In 2001 it renewed its commitment to becoming a global knowledge institution for rural poverty-related issues. Through its technical assistance grants, IFAD aims to promote research and capacity-building in the agricultural sector, as well as the development of technologies to increase production and alleviate rural poverty. In recent years IFAD has been increasingly involved in promoting the use of communication technology to facilitate the exchange of information and experience among rural communities, specialized institutions and organizations, and IFAD-sponsored projects. Within the strategic context of knowledge management, IFAD has supported initiatives to support regional electronic networks, such as ENRAP (see below) in Asia and the Pacific and FIDAMERICA in Latin America and the Caribbean, as well as to develop other lines of communication between organizations, local agents and the rural poor.

IFAD is empowered to make both grants and loans. Grants are limited to 7.5% of the resources committed in any one financial year. Loans are available on highly concessionary, intermediate and ordinary terms. Highly concessionary loans carry no interest but have an annual service charge of 0.75% and a repayment period of 40 years, including a 10-year grace period. Intermediate term loans are subject to a variable interest charge, equivalent to 50% of the interest rate charged on World Bank loans, and are repaid over 20 years. Ordinary loans carry a variable interest charge equal to that charged by the World Bank, and are repaid over 15–18 years. In 2003 highly concessionary loans represented some 76% of total lending in that year. In order to increase the impact of its lending resources on food production, the Fund seeks as much as possible to attract other external donors and beneficiary governments as co-financiers of its projects. In 2003 external cofinancing accounted for some 17.5% of all project funding, while domestic contributions, i.e. from recipient governments and other local sources, accounted for almost 25.8%.

IFAD's development projects usually include a number of components, such as infrastructure (e.g. improvement of water supplies, small-scale irrigation and road construction); input supply (e.g. improved seeds, fertilizers and pesticides); institutional support (e.g. research, training and extension services); and producer incentives (e.g. pricing and marketing improvements). IFAD also attempts to enable the landless to acquire income-generating assets: by increasing the provision of credit for the rural poor, it seeks to free them from dependence on the capital market and to generate productive activities.

In addition to its regular efforts to identify projects and programmes, IFAD organizes special programming missions to certain selected countries to undertake a comprehensive review of the constraints affecting the rural poor, and to help countries to design strategies for the removal of these constraints. In general, projects based on the recommendations of these missions tend to focus on institutional improvements at the national and local level to direct inputs and services to small farmers and the landless rural poor. Monitoring and evaluation missions are also sent to check the progress of projects and to assess the impact of poverty reduction efforts.

The Fund supports projects that are concerned with environmental conservation, in an effort to alleviate poverty that results from the deterioration of natural resources. In addition, it extends environmental assessment grants to review the environmental consequences of projects under preparation. In October 1997 IFAD was appointed to administer the Global Mechanism of the Convention to Combat Desertification in those Countries Experiencing Drought and Desertification, particularly in Africa, which entered into force in December 1996. The Mechanism was envisaged as a means of mobilizing and channelling resources for implementation of the Convention. A series of collaborative institutional arrangements were to be concluded between IFAD, UNDP and the World Bank in order to facilitate the effective functioning of the Mechanism. In May 2001 the Global Environmental Facility approved IFAD as an executing agency.

During 2003 IFAD approved lending for four operations in Asia and the Pacific, involving loans amounting to US $93.5m. (or 23.2% of the total committed in that year); one of the projects was to be implemented in Cambodia and the remainder in South Asian countries.

In mid-1998 IFAD inaugurated the Electronic Networking for Rural Asia/Pacific Projects (ENRAP), initially as a pilot project in eight countries in the region to bring the benefits of the internet to rural development projects. A Poverty Alleviation Training Programme for Asia and the Pacific (PATAP) was initiated in 1995. In 1998 IFAD undertook a series of case studies to assess the impact of the Asian economic crisis on the rural poor, and to formulate appropriate responses. At the end of that year IFAD was involved in a project to rehabilitate crops and livestock following extensive flooding in the Democratic People's Republic of Korea.

In February 1998 IFAD inaugurated a new Trust Fund to complement the multilateral debt initiative for Heavily Indebted Poor Countries (HIPCs). The Fund was intended to assist IFAD's poorest members deemed to be eligible under the initiative to channel resources from debt repayments to communities in need. In February 2000 the Governing Council approved full participation by IFAD in the enhanced HIPC debt initiative agreed by the World Bank and IMF in September 1999.

During 1998 the Executive Board endorsed a policy framework for the Fund's provision of assistance in post-conflict situations, with the aim of achieving a continuum from emergency relief to a secure basis from which to pursue sustainable development. In July 2001 IFAD and UNAIDS signed a memorandum of understanding on developing a co-operation agreement. A meeting of technical experts from IFAD, FAO, WFP and UNAIDS, held in December, addressed means of mitigating the impact of HIV/AIDS on food security and rural livelihoods in affected regions.

During the late 1990s IFAD established several partnerships within the agribusiness sector, with a view to improving performance at project level, broadening access to capital markets, and encouraging the advancement of new technologies. Since 1996 it has chaired the Support Group of the Global Forum on Agricultural Research (GFAR), which facilitates dialogue between research centres and institutions, farmers' organizations, non-governmental bodies, the private sector and donors. In October 2001 IFAD became a co-sponsor of the Consultative Group on International Agricultural Research (CGIAR).

Finance

In accordance with the Articles of Agreement establishing IFAD, the Governing Council periodically undertakes a review of the adequacy of resources available to the Fund and may request members to make additional contributions. The sixth replenishment of IFAD funds, covering the period 2003–04 and amounting to US $560m., was approved in February 2003 and became effective in December. The provisional budget for administrative expenses for 2004 amounted to $57m.

Publications

Annual Report.
IFAD Update (2 a year).
Rural Poverty Report 2001.
Staff Working Papers (series).

Statistics

PROJECT IN EAST ASIA AND THE PACIFIC APPROVED IN 2003

Country	Purpose	Loan amount (SDR m.*)
Cambodia . . .	Rural poverty reduction project in Prey Veaeng and Svay Rieng	10.9

* The value of the SDR—Special Drawing Right—at 31 December 2003 was US $1.485970.

International Monetary Fund—IMF

Address: 700 19th St, NW, Washington, DC 20431, USA.

Telephone: (202) 623-7300; **fax:** (202) 623-6278; **e-mail:** publicaffairs@imf.org; **internet:** www.imf.org.

The IMF was established at the same time as the World Bank in December 1945, to promote international monetary co-operation, to facilitate the expansion and balanced growth of international trade and to promote stability in foreign exchange.

Organization

(October 2004)

Managing Director: RODRIGO DE RATO Y FIGAREDO (Spain).

First Deputy Managing Director: ANNE KRUEGER (USA).

Deputy Managing Directors: TAKATOSHI KATO (Japan), AGUSTÍN CARSTENS (Mexico).

Director, Asia and Pacific Department: DAVID BURTON (United Kingdom).

BOARD OF GOVERNORS

The highest authority of the Fund is exercised by the Board of Governors, on which each member country is represented by a Governor and an Alternate Governor. The Board normally meets annually. The voting power of each country is related to its quota in the Fund. An International Monetary and Financial Committee (IMFC, formerly the Interim Committee) advises and reports to the Board on matters relating to the management and adaptation of the international monetary and financial system, sudden disturbances that might threaten the system and proposals to amend the Articles of Agreement.

BOARD OF EXECUTIVE DIRECTORS

The 24-member Board of Executive Directors is responsible for the day-to-day operations of the Fund. The USA, the United Kingdom, Germany, France and Japan each appoint one Executive Director. There is also one Executive Director from the People's Republic of China, Russia and Saudi Arabia, while the remainder are elected by groups of the remaining countries.

REGIONAL OFFICE

Regional Office for Asia and the Pacific: 21F Fukoku Seimei Bldg, 2-2-2, Uchisaiwai-cho, Chiyodu-ku, Tokyo 100, Japan; tel. (3) 3597-6700; fax (3) 3597-6705; e-mail oap@imf.org; f. 1997; Dir HIROYUKI HINO (Japan).

Activities

The purposes of the IMF, as defined in the Articles of Agreement, are:

(i) To promote international monetary co-operation through a permanent institution which provides the machinery for consultation and collaboration on monetary problems;

(ii) To facilitate the expansion and balanced growth of international trade, and to contribute thereby to the promotion and maintenance of high levels of employment and real income and to the development of members' productive resources;

(iii) To promote exchange stability, to maintain orderly exchange arrangements among members, and to avoid competitive exchange depreciation;

(iv) To assist in the establishment of a multilateral system of payments in respect of current transactions between members and in the elimination of foreign exchange restrictions which hamper the growth of trade;

(v) To give confidence to members by making the general resources of the Fund temporarily available to them, under adequate safeguards, thus providing them with the opportunity to correct maladjustments in their balance of payments, without resorting to measures destructive of national or international prosperity;

(vi) In accordance with the above, to shorten the duration of and lessen the degree of disequilibrium in the international balances of payments of members.

In joining the Fund, each country agrees to co-operate with the above objectives. In accordance with its objective of facilitating the expansion of international trade, the IMF encourages its members to accept the obligations of Article VIII, Sections two, three and four, of the Articles of Agreement. Members that accept Article VIII undertake to refrain from imposing restrictions on the making of payments and transfers for current international transactions and from engaging in discriminatory currency arrangements or multiple currency practices without IMF approval. By the end of 2003 157 members had accepted Article VIII status.

The financial crises of the late 1990s, notably in several Asian countries, Brazil and Russia, contributed to widespread discussions concerning the strengthening of the international monetary system. In April 1998 the Executive Board identified the following fundamental aspects of the debate: reinforcing international and domestic financial systems; strengthening IMF surveillance; promoting greater availability and transparency of information regarding member countries' economic data and policies; emphasizing the central role of the IMF in crisis management; and establishing effective procedures to involve the private sector in forestalling or resolving financial crises. During 1999/2000 the Fund implemented several measures in connection with its ongoing efforts to appraise and reinforce the global financial architecture, including, in March 2000, the adoption by the Executive Board of a strengthened framework to safeguard the use of IMF resources. During 2000 the Fund established the IMF Center, in Washington, DC, which aimed to promote awareness and understanding of its activities. In September the Fund's new Managing Director announced his intention to focus and streamline the principles of conditionality (which links Fund financing with the implementation of specific economic policies by the recipient countries) as part of the wider reform of the international financial system. A comprehensive review was undertaken, during which the issue was considered by public forums and representatives of civil society. New guide-lines on conditionality, which ,*inter alia*, aimed to promote national ownership of policy reforms and to introduce specific criteria for the implementation of conditions given different states' circumstances, were approved by the Executive Board in September 2002. In 2000/01 the Fund established an International Capital Markets Department to improve its understanding of financial markets and a separate Consultative Group on capital markets to serve as a forum for regular dialogue between the Fund and representatives of the private sector.

In early 2002 a position of Director for Special Operations was created to enhance the Fund's ability to respond to critical situations affecting member countries. In February the newly-appointed Director immediately assumed leadership of the staff team working with the authorities in Argentina to help that country to overcome its extreme economic and social difficulties. In September the IMFC approved further detailed consideration of a sovereign debt restructuring mechanism (SDRM), which aimed to establish a procedure to enable countries with an unsustainable level of debt to renegotiate loans more effectively. In January 2003 the IMF hosted a conference for representatives from the financial sector and civil society and other public officials and academics to discuss aspects of the SDRM. In April, after further discussion of the issue by the Board of Directors, the IMFC stated that the SDRM would not be implemented, although other means of orderly resolution of financial crises were to remain under consideration. In their meeting the Directors determined that the Fund promote more actively the use of Collective Action Clauses in international bond contracts, as a voluntary measure to facilitate debt restructuring should the need arise.

SURVEILLANCE

Under its Articles of Agreement, the Fund is mandated to oversee the effective functioning of the international monetary system. Accordingly, the Fund aims to exercise firm surveillance over the exchange rate policies of member states and to assess whether a country's economic situation and policies are consistent with the objectives of sustainable development and domestic and external stability. The Fund's main tools of surveillance are regular, bilateral consultations with member countries conducted in accordance with Article IV of the Articles of Agreement, which cover fiscal and monetary policies, balance of payments and external debt developments, as well as policies that affect the economic performance of a country, such as the labour market, social and environmental issues and good governance, and aspects of the country's capital accounts, and finance and banking sectors. In April 1997, in an effort to improve the value of surveillance by means of increased transparency, the Executive Board agreed to the voluntary issue of Press Information Notices (PINs) (on the internet and in *IMF Economic Reviews),* following each member's Article IV consultation with the Board, to those member countries wishing to make public the Fund's views. Other background papers providing information on and analysis of economic developments in individual countries continued to be made available. In addition, World Economic Outlook discussions are held, normally twice a year, by the Executive Board to assess policy implications from a multilateral perspective and to monitor global developments.

The rapid decline in the value of the Mexican peso in late 1994 and the financial crisis in Asia, which became apparent in mid-1997, focused attention on the importance of IMF surveillance of the economies and financial policies of member states and prompted the Fund to enhance the effectiveness of its surveillance and to encourage the full and timely provision of data by member countries in order to maintain fiscal transparency. In April 1996 the IMF established the Special Data Dissemination Standard (SDDS), which was intended to improve access to reliable economic statistical information for member countries that have, or are seeking, access to international capital markets. In March 1999 the IMF undertook to strengthen the Standard by the introduction of a new reserves data template. By July 2004 57 countries had subscribed to the Standard. In December 1997 the Executive Board approved a new General Data Dissemination System (GDDS), to encourage all member countries to improve the production and dissemination of core economic data. The operational phase of the GDDS commenced in May 2000. By July 2004 78 countries had participated in the GDDS. The Fund maintains a Dissemination Standards Bulletin Board (accessible at dsbb.imf.org), which aims to ensure that information on SDDS subscribing countries is widely available.

In April 1998 the then Interim Committee adopted a voluntary Code of Good Practices on Fiscal Transparency: Declaration of Principles, which aimed to increase the quality and promptness of official reports on economic indicators, and in September 1999 it adopted a Code of Good Practices on Transparency in Monetary and Financial Policies: Declaration of Principles. The IMF and World Bank jointly established a Financial Sector Assessment Programme (FSAP) in May 1999, initially as a pilot project, which aimed to promote greater global financial security through the preparation of confidential detailed evaluations of the financial sectors of individual countries. It remained under regular review by the Boards of Governors of the Fund and World Bank. As part of the FSAP, Fund staff may conclude a Financial System Stability Assessment (FSSA), addressing issues relating to macroeconomic stability and the strength of a country's financial system. A separate component of the FSAP are Reports on the Observance of Standards and Codes (ROSCs), which are compiled after an assessment of a country's implementation and observance of internationally recognized financial standards. By August 2004 FSAP reports had been completed for 49 countries or regions, including for Hong Kong, Japan, Republic of Korea, Pakistan and Singapore.

In March 2000 the IMF Executive Board adopted a strengthened framework to safeguard the use of IMF resources. All member countries making use of Fund resources were to be required to publish annual central bank statements audited in accordance with internationally accepted standards. It was also agreed that any instance of intentional misreporting of information by a member country should be publicized. In the following month the Executive Board approved the establishment of an Independent Evaluation Office (IEO) to conduct objective evaluations of IMF policy and operations. The Office commenced activities in July 2001. During 2002/03 it conducted assessments of the prolonged used of IMF resources, fiscal adjustment in IMF-supported programmes and an evaluation of the IMF's role in the capital account crises in Brazil, Indonesia and the Republic of Korea.

In April 2001 the Executive Board agreed on measures to enhance international efforts to counter money-laundering, in particular through the Fund's ongoing financial supervision activities and its programme of assessment of offshore financial centres. In November the IMFC, in response to the terrorist attacks against targets in the USA, which had occurred in September, resolved, *inter alia*, to strengthen the Fund's focus on surveillance, and, in particular, to extend measures to counter money-laundering to include the funds of terrorist organizations. It determined to accelerate efforts to assess offshore centres and to provide technical support to enable poorer countries to meet international financial standards. In July 2002 the Executive Board endorsed Fund participation in a pilot programme of assessments with respect to efforts to counter money-laundering and the financing of terrorism. The programme, undertaken with the World Bank, the Financial Action Task Force and other regional supervisory bodies, commenced in October. By October 2003 some 41 jurisdictions had been assessed; subsequently 12 further assessments had been completed by mid-2004.

QUOTAS

Membership and Quotas in the Far East and Australasia
(million SDR*)

Country	October 2004
Australia	3,236.4
Brunei	215.2
Cambodia	87.5
China, People's Republic	6,369.2
Fiji	70.3
Indonesia	2,079.3
Japan	13,312.8
Kiribati	5.6
Korea, Republic	1,633.6
Laos	52.9
Malaysia	1,486.6
Marshall Islands	3.5
Micronesia, Federated States	5.1
Mongolia	51.1
Myanmar	258.4
New Zealand	894.6
Palau	3.1
Papua New Guinea	131.6
Philippines	879.9
Samoa	11.6
Singapore	862.5
Solomon Islands	10.4
Timor-Leste	8.2
Thailand	1,081.9
Tonga	6.9
Vanuatu	17.0
Viet Nam	329.1

*The Special Drawing Right (SDR) was introduced in 1970 as a substitute for gold in international payments, and was intended eventually to become the principal reserve asset in the international monetary system. Its value (which was US $1.48461 at 20 October 2004 and averaged $1.39883 in 2003) is based on the currencies of the five largest exporting countries. Each member is assigned a quota related to its national income, monetary reserves, trade balance and other economic indicators; the quota approximately determines a member's voting power and the amount of foreign exchange it may purchase from the Fund. A member's subscription is equal to its quota. In January 1998 the Board of Governors adopted a resolution in support of an increase, under the Eleventh General Review, of some 45% in total quotas, subject to approval by member states constituting 85% of total quotas (as at December 1997). Sufficient consent had been granted by January 1999 to enable the overall increase in quotas to enter into effect. All members were then granted until 30 July to consent to the higher quotas. The Twelfth General Review was concluded at the end of January 2003 without an increase in quotas. At August 2004 total quotas in the Fund amounted to SDR 212,794.0m.

RESOURCES

Members' subscriptions form the basic resource of the IMF. They are supplemented by borrowing. Under the General Arrangements to Borrow (GAB), established in 1962, the 'Group of Ten' industrialized nations (G-10—Belgium, Canada, France, Germany, Italy, Japan, the Netherlands, Sweden, the United Kingdom and the USA) and Switzerland (which became a member of the IMF in May 1992 but which had been a full participant in the GAB from April 1984) undertake to lend the Fund as much as SDR 17,000m. in their own currencies, to assist in fulfilling the balance-of-payments requirements of any member of the group, or in response to requests to the Fund from countries with balance-of-payments problems that could threaten the stability of the international monetary system. In 1983 the Fund entered into an agreement with Saudi Arabia, in association with the GAB, making available SDR 1,500m., and other borrowing arrangements were completed in 1984 with the Bank for International Settlements, the Saudi Arabian Monetary Agency, Belgium and Japan, making available a further SDR 6,000m. In 1986 another borrowing arrangement with Japan made available SDR 3,000m. In May 1996 GAB participants concluded an agreement in principle to expand the resources available for borrowing to SDR 34,000m., by securing the support of 25 countries with the financial capacity to support the international monetary system. The so-called New Arrangements to Borrow (NAB) was approved by the Executive Board in January 1997. It was to enter into force, for an initial five-year period, as soon as the five largest potential creditors participating in NAB had approved the initiative and the total credit arrangement of participants endorsing the scheme had reached at least SDR 28,900m. While the GAB credit arrangement was to remain in effect, the NAB was expected to be the first facility to be activated in the event of the Fund's requiring supplementary resources. In July 1998 the GAB was activated for the first time in more than 20 years in order to provide funds of up to US $6,300m. in support of an IMF emergency assistance package for Russia (the first time the GAB had been used for a non-participant). The NAB became effective in November, and was used for the first time as part of an extensive programme of support for Brazil, which was adopted by the IMF in early December. (In March 1999, however, the activation was cancelled.) In November 2002 NAB participants agreed to renew the arrangement for a further five-year period from November 2003.

DRAWING ARRANGEMENTS

Exchange transactions within the Fund take the form of members' purchases (i.e. drawings) from the Fund of the currencies of other members for the equivalent amounts of their own currencies. Fund resources are available to eligible members on an essentially short-term and revolving basis to provide members with temporary assistance to contribute to the solution of their payments problems. Before making a purchase, a member must show that its balance of payments or reserve position makes the purchase necessary. Apart from this requirement, reserve tranche purchases (i.e. purchases that do not bring the Fund's holdings of the member's currency to a level above its quota) are permitted unconditionally.

With further purchases, however, the Fund's policy of 'conditionality' means that a member requesting assistance must agree to adjust its economic policies, as stipulated by the IMF. All requests other than for use of the reserve tranche are examined by the Executive Board to determine whether the proposed use would be consistent with the Fund's policies, and a member must discuss its proposed adjustment programme (including fiscal, monetary, exchange and trade policies) with IMF staff. Purchases outside the reserve tranche are made in four credit tranches, each equivalent to 25% of the member's quota; a member must reverse the transaction by repurchasing its own currency (with SDRs or currencies specified by the Fund) within a specified time. A credit tranche purchase is usually made under a 'Stand-by Arrangement' with the Fund, or under the Extended Fund Facility. A Stand-by Arrangement is normally of one or two years' duration, and the amount is made available in instalments, subject to the member's observance of 'performance criteria'; repurchases must be made within three-and-a-quarter to five years. An Extended Arrangement is normally of three years' duration, and the member must submit detailed economic programmes and progress reports for each year; repurchases must be made within four-and-a-half to 10 years. A member whose payments imbalance is large in relation to its quota may make use of temporary facilities established by the Fund using borrowed resources, namely the 'enlarged access policy' established in 1981, which helps to finance Stand-by and Extended Arrangements for such a member, up to a limit of between 90% and 110% of the member's quota annually. Repurchases are made within three-and-a-half to seven years. In October 1994 the Executive Board approved a temporary increase in members' access to IMF resources, on the basis of a recommendation by the then Interim Committee. The annual access limit under IMF regular tranche drawings, Stand-by Arrangements and Extended Fund Facility credits was increased from 68% to 100% of a member's quota, with the cumulative access limit remaining at 300% of quota. The arrangements were extended, on a temporary basis, in November 1997.

In addition, special-purpose arrangements have been introduced, all of which are subject to the member's co-operation with the Fund to find an appropriate solution to its difficulties. The Compensatory Financing Facility (CCF) provides compensation to members whose export earnings are reduced as a result of circumstances beyond their control, or which are affected by excess costs of cereal imports. In December 1997 the Executive Board established a new Supplemental Reserve Facility (SRF) to provide short-term assistance to members experiencing exceptional balance-of-payments difficulties resulting from a sudden loss of market confidence. The SRF was activated immediately to provide SDR 9,950m. to the Republic of Korea, as part of a Stand-by Arrangement amounting to SDR 15,550m. (at that time the largest amount ever committed by the Fund). In July 1998 SDR 4,000m. was made available to Russia under the SRF and, in December, some SDR 9,100m. was extended to Brazil under the SRF as part of a new Stand-by Arrangement. In January 2001 some SDR 2,100m. in SRF resources were approved for Argentina as part of an SDR 5,187m. Stand-by Arrangement augmentation. (In January 2002 the Executive Board approved an extension of one year for Argentina's SRF repayments.) The SDR 22,821m. Stand-by credit approved for Brazil in September 2002 included some SDR 7,600m. committed under the SRF. In April 1999 an additional facility, the Contingent Credit Lines (CCL), was established to provide short-term financing on similar terms to the SRF in order to prevent more stable economies being affected by adverse international financial developments and to maintain investor confidence. No funds were ever committed under the CCL, however, and in November 2003 the Executive Board resolved to allow the facility to terminate, as scheduled, at the end of that month. The Board requested further consideration of other precautionary arrangements to limit the risk of financial crises.

In October 1995 the Interim Committee of the Board of Governors endorsed recent decisions of the Executive Board to strengthen IMF financial support to members requiring exceptional assistance. An Emergency Financing Mechanism was established to enable the IMF to respond swiftly to potential or actual financial crises, while additional funds were made available for short-term currency stabilization. (The Mechanism was activated for the first time in July 1997, in response to a request by the Philippines Government to reinforce the country's international reserves, and was subsequently used during that year to assist Thailand, Indonesia and the Republic of Korea, and, in July 1998, Russia.) Emergency assistance was also to be available to countries in a post-conflict situation, in addition to existing arrangements for countries having been affected by natural disasters, to facilitate the rehabilitation of their economies and to improve their eligibility for further IMF concessionary arrangements.

In November 1999 the Fund's existing facility to provide balance-of-payments assistance on concessionary terms to low-income member countries, the Enhanced Structural Adjustment Facility, was reformulated as the Poverty Reduction and Growth Facility (PRGF), with greater emphasis on poverty reduction and sustainable development as key elements of growth-orientated economic strategies. Assistance under the PRGF (for which 77 countries were deemed eligible) was to be carefully matched to specific national requirements. Prior to drawing on the facility each recipient country was, in collaboration with representatives of civil society, non-governmental organizations and bilateral and multilateral institutions, to develop a national poverty reduction strategy, which was to be presented in a Poverty Reduction Strategy Paper (PRSP). PRGF loans carry an interest rate of 0.5% per year and are repayable over 10 years, with a five-and-a-half-year grace period; each eligible country is normally permitted to borrow up to 140% of its quota (in exceptional circumstances the maximum access can be raised to 185%). A PGRF Trust replaced the former ESAF Trust.

The PRGF supports, through long-maturity loans and grants, IMF participation in a joint initiative, with the World Bank, to provide exceptional assistance to heavily indebted poor countries (HIPCs), in order to help them to achieve a sustainable level of debt management. The initiative was formally approved at the September 1996 meeting of the Interim Committee, having received the support of the 'Paris Club' of official creditors, which agreed to increase the relief on official debt from 67% to 80%. Resources for the HIPC initiative are channelled through the PRGF Trust. In early 1999 the IMF and World Bank initiated a comprehensive review of the HIPC scheme, in order to consider modifications of the initiative and to strengthen the link between debt relief and poverty reduction. A consensus emerged among the financial institutions and leading industralized nations to enhance the scheme, in order to make it available to more countries, and to accelerate the process of providing debt relief. In September the IMF Board of Governors expressed its commitment to undertaking an off-market transaction of a percentage of the Fund's gold reserves (i.e. a sale, at market prices, to central banks of member countries with repayment obligations to the Fund, which were then to be made in gold), as part of the funding arrangements of the enhanced HIPC scheme; this was undertaken during the period December 1999–April 2000. Under the enhanced initiative it was agreed that countries seeking debt relief should first formulate, and successfully implement for at least one year, a national poverty reduction strategy (see above). By September 2003 a total of US $31,428m. in NPV terms had been committed, of which the Fund's share was $2,517m.

During 2002/03 the IMF approved funding commitments for new arrangements amounting to SDR 30,571m., compared with SDR 41,287m. in the previous year. Of the total amount, SDR 28,597m. was committed under 10 new Stand-by Arrangements and the augmentation of one already in place (for Uruguay). An arrangement amounting to SDR 22,821m., approved in September 2002 in support of the Brazilian Government's efforts to secure economic and financial stability, was the largest ever stand-by credit agreed by the Fund. Ten new PRGF arrangements were approved in 2002/03, and an existing commitment was augmented, amounting to SDR 1,180m. During 2002/03 members' purchases from the general resources account amounted to SDR 21,784m., compared with SDR 29,194m. in the previous year, with the main users of IMF resources being Brazil (SDR 15,316m.) and Turkey (SDR 2,246m.). Outstanding IMF credit at 30 April 2003 totalled SDR 72,879m., compared with SDR 58,699m. as at the previous year.

TECHNICAL ASSISTANCE

Technical assistance is provided by special missions or resident representatives who advise members on every aspect of economic management, while more specialized assistance is provided by the IMF's various departments. In 2000/01 the IMFC determined that technical assistance should be central to IMF's work in crisis prevention and management, in capacity-building for low-income countries, and in restoring macroeconomic stability in countries following a financial crisis. Technical assistance activities subsequently underwent a process of review and reorganization to align them more closely with IMF policy priorities and other initiatives, for example the Financial Stability Assessment Programme. In 2002/03 assistance in institution-building after a period of conflict was extended to Afghanistan, Iraq and Timor-Leste. The majority of technical assistance is provided by the Departments of Monetary and Exchange Affairs, of Fiscal Affairs and of Statistics, and by the IMF Institute. The Institute, founded in 1964, trains officials from member countries in financial analysis and policy, balance-of-payments methodology and public finance; it also gives assistance to national and regional training centres.

In May 1998 an IMF—Singapore Regional Training Institute (an affiliate of the IMF Institute) was inaugurated, in collaboration with the Singaporean Government, in order to provide training for officials from the Asia-Pacific region. The IMF is a co-sponsor, with UNDP and the Japan administered account, of the Joint Vienna Institute, which was opened in the Austrian capital in October 1992 and which trains officials from former centrally-planned economies in various aspects of economic management and public administration. In January 1999 the IMF, in co-operation with the African Development Bank and the World Bank, announced the establishment of a Joint Africa Institute, in Abidjan, Côte d'Ivoire, which was to offer training to officials from African countries from the second half of the year. The IMF Institute also co-operates with other established regional training centres and institutes in order to refine its delivery of technical assistance and training services. During 2000/01 the Institute established a new training programme with government officials in the People's Republic of China and agreed to establish a regional training centre for Latin America in Brazil. A Caribbean Regional Technical Assistance Centre (CARTAC), located in Barbados, began operations in November 2001. In October 2002 an East African Regional Technical Assistance Centre (East AFRITAC), based in Dar es Salaam, Tanzania, was inaugurated as the first of a planned series of sub-regional centres in Africa with the objective of developing local capacity in economic and financial management. A second AFRITAC was opened in Mali in May 2003. In September 2002 the IMF signed a memorandum of understanding with the African Capacity Building Foundation to strengthen collaboration, in particular within the context of a new IMF Africa Capacity-Building Initiative.

Publications

Annual Report.

Balance of Payments Statistics Yearbook.

Direction of Trade Statistics (quarterly and annually).

Emerging Markets Financing (quarterly).

Finance and Development (quarterly).

Financial Statements of the IMF (quarterly).

Global Financial Stability Report (2 a year).

Government Finance Statistics Yearbook.

IMF Commodity Prices (monthly).

IMF Research Bulletin (quarterly).

IMF Survey (2 a month).

International Financial Statistics (monthly and annually, also on CD-ROM).

Joint BIS-IMF-OECD-World Bank Statistics on External Debt (quarterly).

Quarterly Report on the Assessments of Standards and Codes.

Staff Papers (3 a year).

World Economic Outlook (2 a year).

Other country reports, economic and financial surveys, occasional papers, pamphlets, books.

United Nations Educational, Scientific and Cultural Organization—UNESCO

Address: 7 place de Fontenoy, 75352 Paris 07 SP, France.

Telephone: 1-45-68-10-00; **fax:** 1-45-67-16-90; **e-mail:** scg@unesco.org; **internet:** www.unesco.org.

UNESCO was established in 1946 'for the purpose of advancing, through the educational, scientific and cultural relations of the peoples of the world, the objectives of international peace and the common welfare of mankind'.

Organization

(October 2004)

GENERAL CONFERENCE

The supreme governing body of the Organization, the Conference meets in ordinary session once in two years and is composed of representatives of the member states.

EXECUTIVE BOARD

The Board, comprising 58 members, prepares the programme to be submitted to the Conference and supervises its execution; it meets twice or sometimes three times a year.

SECRETARIAT

Director-General: Koïchiro Matsuura (Japan).

CO-OPERATING BODIES

In accordance with UNESCO's constitution, national Commissions have been set up in most member states. These help to integrate work within the member states and the work of UNESCO.

REGIONAL OFFICES

Regional Office for Science and Technology for South-East Asia: UN Building (2nd Floor), Jalan M. H. Thamrin 14, Tromol Pos 1273/JKT, Jakarta 10002, Indonesia; tel. (21) 3141308; fax (21) 3150382; e-mail uhjak@unesco.or.id; internet www.unesco.or.id; Dir Prof. Stephen Hill.

UNESCO Apia Office: POB 5766, Matautu, Apia, Samoa; tel. 24276; fax 22253; e-mail apia@unesco.org; f. 1984; represents Australia, New Zealand and the Pacific Island states; Dir Edna Tait; publ. *UNESCO Pacific News* (2 or 3 a year).

UNESCO Asia and Pacific Regional Bureau for Communication and Information: B-5/29, Safdarjung Enclave, New Delhi 110057, India; tel. (11) 671-3000; fax (11) 671-3001; e-mail newdelhi@unesco.org; internet unescodelhi.nic.in; f. 1948 as UNESCO New Delhi Office; aims to widen access to information and knowledge resources, with a special focus on reaching the poor or illiterate people and other marginalized groups; Dir Prof. Moshen Tawfik; publs *Annual Report*, *Library Accessions List* (monthly), *Quarterly Newsletter*.

UNESCO Asia and Pacific Regional Bureau for Education: 920 Sukhumvit Rd, POB 967, Bangkok 10110, Thailand; tel. (2) 391-0577; fax (2) 391-0866; e-mail bangkok@unescobkk.org; internet www.unescobkk.org; Dir Sheldon Schaeffer.

Activities

In November 2001 the General Conference approved a medium-term strategy to guide UNESCO during the period 2002–07. The Conference adopted a new unifying theme for the organization: 'UNESCO contributing to peace and human development in an era of globalization through education, the sciences, culture and communication'. UNESCO's central mission as defined under the strategy was to contribute to peace and human development in the globalized world through its four programme domains (Education, Natural and Social and Human Sciences, Culture, and Communication and Information), incorporating the following three principal dimensions: developing universal principles and norms to meet emerging challenges and protect the 'common public good'; promoting pluralism and diversity; and promoting empowerment and participation in the emerging knowledge society through equitable access, capacity-building and knowledge-sharing. Programme activities were to be focused particularly on supporting disadvantaged and excluded groups or geographic regions. The organization aimed to decentralize its operations in order to ensure more country-driven programming. UNESCO's overall work programme for 2002–03 comprised the following major programmes: education; natural sciences; social and human sciences; culture; and communication and information. Basic education; fresh water resources and ecosystems; the ethics of science and technology; diversity, intercultural pluralism and dialogue; and universal access to information, especially in the public domain, were designated as the priority themes. The work programme incorporated two transdisciplinary projects—eradication of poverty, especially extreme poverty; and the contribution of information and communication technologies to the development of education, science and culture and the construction of a knowledge society. UNESCO aims to promote a culture of peace. The UN General Assembly designated UNESCO as the lead agency for co-ordinating the International Decade for a Culture of Peace and Non-Violence for the Children of the World (2001–10), with a focus on education, and the UN Literacy Decade (2003–12). In 2004 UNESCO was the lead agency in promoting the International Year to Commemorate the Struggle Against Slavery and its Abolition. In the implementation of all its activities UNESCO aims to contribute to achieving the UN Millennium Development Goal (MDG) of halving levels of extreme poverty by 2015, as well as other MDGs concerned with education and sustainable development (see below).

EDUCATION

Since its establishment UNESCO has devoted itself to promoting education in accordance with principles based on democracy and respect for human rights. The Associated Schools Project (ASPnet—comprising some 7,500 institutions in 174 countries in 2004) has, since 1953, promoted the principles of peace, human rights, democracy and international co-operation through education.

In March 1990 UNESCO, with other UN agencies, sponsored the World Conference on Education for All. 'Education for All' was subsequently adopted as a guiding principle of UNESCO's contribution to development. In April 2000 several UN agencies, including UNESCO and UNICEF, and other partners sponsored the World Education Forum, held in Dakar, Senegal, to assess international progress in achieving the goal of 'Education for All' and to adopt a strategy for further action (the 'Dakar Framework'), with the aim of ensuring universal basic education by 2015. The Forum launched the Global Initiative for Education for All. The Dakar Framework emphasized the role of improved access to education in the reduction of poverty and in diminishing inequalities within and between societies. UNESCO was appointed as the lead agency in the implementation of the Framework. UNESCO's role in pursuing the goals of the Dakar Forum was to focus on co-ordination, advocacy, mobilization of resources, and information-sharing at international, regional and national levels. It was to oversee national policy reforms, with a particular focus on the integration of 'Education for All' objectives into national education plans, which were to be produced by all member countries by 2002. UNESCO's work programme on Education for 2002–03 aimed to promote an effective follow-up to the Forum and comprised the following two main components: Basic education for all: Meeting the commitments of the Dakar World Education Forum; and Building knowledge societies through quality education and a renewal of education systems. 'Basic Education for All', signifying the promotion of access to learning opportunities throughout the lives of all individuals, including the most disadvantaged, was designated as the principal theme of the programme and was deemed to require urgent action. The second part of the strategy was to improve the quality of educational provision and renew and diversify education systems, with a view to ensuring that educational needs at all levels were met. This component included updating curricular programmes in secondary education, strengthening science and technology activities and ensuring equal access to education for girls and women. (UNESCO supports the UN Girls' Education Initiative, established following the Dakar Forum.) The work programme focused on the importance of knowledge, information and communication in the increasingly globalized world, and the significance of education as a means of empowerment for the poor and of enhancing basic quality of life.

UNESCO advocates 'Literacy for All' as a key component of 'Education for All', regarding literacy as essential to basic education and to social and human development. In December 2001 the UN General Assembly appointed UNESCO to be the co-ordinating agency of the UN Literacy Decade (2003–12), which aimed to formulate an international plan of action to raise literacy standards throughout the world.

In December 1993 the heads of government of nine highly-populated developing countries (Bangladesh, Brazil, the People's Republic of China, Egypt, India, Indonesia, Mexico, Nigeria and Pakistan), meeting in New Delhi, India, agreed to co operate, with the objective of achieving comprehensive primary education for all children and of expanding further learning opportunities for children and adults. By September 1999 all of the so-called 'E-9' (or Education-9) countries had officially signed the 'Delhi Declaration' issued by the meeting. UNESCO is working towards the UN MDGs of eliminating gender disparity in primary and secondary education by 2005 and attaining universal primary education in all countries by 2015.

Within the UN system, UNESCO is responsible for providing technical assistance and educational services in the context of emergency situations. This includes providing education to refugees and displaced persons, as well as assistance for the rehabilitation of national education systems.

UNESCO is concerned with improving the quality, relevance and efficiency of higher education. It assists member states in reforming their national systems, organizes high-level conferences for Ministers of Education and other decision-makers, and disseminates research papers. A World Conference on Higher Education was convened in October 1998 in Paris, France. The Conference adopted a World Declaration on Higher Education for the 21st Century, incorporating proposals to reform higher education, with emphasis on access to education, and educating for individual development and active participation in society. The Conference also approved a framework for Priority Action for Change and Development of Higher Education, which comprised guide-lines for governments and institutions to meet the objectives of greater accessibility, as well as improved standards and relevancy of higher education.

UNESCO supports the Asia-Pacific Network for International Education and Values Education to secure this objective. Two co-operative programmes, the Asia-Pacific Programme of Education for All (APPEAL) and the Asia-Pacific Programme of Educational Innovation for Development (APEID), are responsible for undertaking UNESCO's educational strategies in the region. In 1991 an Asia-Pacific Co-operative Programme for Reading Promotion and Book Development was established, in order to support literacy initiatives, the development of primary reading materials, rural reading centres and mobile libraries, as well as support to national publishing industries. In May 1997 an Asia Regional Literacy Forum, convened in Manila, Philippines, urged greater focus on literacy issues in the region. A second Forum was held in New Delhi, in February 1998, organized jointly by UNESCO, the National Literacy Mission of India and the International Literacy Institute (itself a co-operative venture of UNESCO and the University of Pennsylvania, USA), and attended by government representatives and officials from 22 Asian countries. A third forum was held in Beijing, People's Republic of China, in October 1999. In March–April 1998 UNESCO hosted a conference entitled 'Education in the 21st Century in the Asia and Pacific Region', held in Melbourne, Australia, at which it was agreed to establish working parties on a series of concerns, including access to technology and the role of higher education in preparing for the future.

The April 2000 World Education Forum recognized the global HIV/AIDS pandemic to be a significant challenge to the attainment of 'Education for All'. UNESCO, as a co-sponsor of UNAIDS, takes an active role in promoting formal and non-formal preventive health education.

NATURAL SCIENCES

In November 1999 the General Conference endorsed a Declaration on Science and the Use of Scientific Knowledge and an agenda for action, which had been adopted at the World Conference on Science, held in June–July 1999, in Budapest, Hungary. UNESCO was to co-ordinate the follow-up to the conference and, in conjunction with the International Council for Science, to promote initiatives in international scientific partnership. The following were identified as priority areas of UNESCO's work programme on Natural Sciences for 2002–03: Science and Technology: Capacity-building and Management; and Sciences, Environment and Sustainable Development. Water Security in the 21st Century was designated as the principal theme, involving addressing threats to water resources and their associated ecosystems. UNESCO was the lead UN agency involved in the preparation of the first *World Water Development Report*, issued in March 2003. In that year the UNESCO Institute for Water Education was inaugurated in Delft, the Netherlands. UNESCO was a joint co-ordinator of the International Year of Freshwater (2003), which aimed to raise global awareness of the importance of improving the protection and management of freshwater resources. The Science and Technology component of the programme focused on the follow-up of the World Conference on Science, involving the elaboration of national policies on science and technology; strengthening science education; improving university teaching and enhancing national research capacities; and reinforcing international co-operation in mathematics, physics, chemistry, biology, biotechnology and the engineering sciences. UNESCO aims to contribute to bridging the divide between community-held traditional knowledge and scientific knowledge. UNESCO supports the UN MDG concerning the implementation, by 2005, of national strategies for sustainable development with a view to achieving by 2015 the reversal of current trends in the loss of environmental resources.

UNESCO aims to improve the level of university teaching of the basic sciences through training courses, establishing national and regional networks and centres of excellence, and fostering co-operative research. In carrying out its mission, UNESCO relies on partnerships with non-governmental organizations and the world scientific communities. With the International Council of Scientific Unions and the Third World Academy of Sciences, UNESCO operates a short-term fellowship programme in the basic sciences and an exchange programme of visiting lecturers. In September 1996 UNESCO initiated a 10-year World Solar Programme, which aimed to promote the application of solar energy and to increase research, development and public awareness of all forms of ecologically-sustainable energy use.

UNESCO has over the years established various forms of intergovernmental co-operation concerned with the environmental sciences and research on natural resources, in order to support the recommendations of the June 1992 UN Conference on Environment and Development and, in particular, the implementation of 'Agenda 21' to promote sustainable development. The International Geological Correlation Programme, undertaken jointly with the International Union of Geological Sciences, aims to improve and facilitate global research of geological processes. In the context of the International Decade for Natural Disaster Reduction (declared in 1990), UNESCO conducted scientific studies of natural hazards and means of mitigating their effects and organized several disaster-related workshops. The International Hydrological Programme considers scientific aspects of water resources assessment and management; and the Intergovernmental Oceanographic Commission focuses on issues relating to oceans, shorelines and marine resources, in particular the role of the ocean in climate and global systems. The IOC has been actively involved in the establishment of a Global Coral Reef Monitoring Network and is developing a Global Ocean Observing System. An initiative on Environment and Development in Coastal Regions and in Small Islands is concerned with ensuring environmentally-sound and sustainable development by strengthening management of the following key areas: freshwater resources; the mitigation of coastline instability; biological diversity; and coastal ecosystem productivity. UNESCO hosts the secretariat of the World Water Assessment Programme on freshwater resources.

UNESCO's Man and the Biosphere Programme supports a world-wide network of biosphere reserves (comprising 440 sites in 97 countries in June 2004), which aim to promote environmental conservation and research, education and training in biodiversity and problems of land use (including the fertility of tropical soils and the cultivation of sacred sites). In October 2002 UNESCO announced that the 138 biospheres in mountainous areas would play a leading role in a new Global Change Monitoring Programme aimed at assessing the impact of global climate changes. Following the signing of the Convention to Combat Desertification in October 1994, UNESCO initiated an International Programme for Arid Land Crops, based on a network of existing institutions, to assist implementation of the Convention.

SOCIAL AND HUMAN SCIENCES

UNESCO is mandated to contribute to the world-wide development of the social and human sciences and philosophy, which it regards as of great importance in policy-making and maintaining ethical vigilance. The structure of UNESCO's Social and Human Sciences programme takes into account both an ethical and standard-setting dimension, and research, policy-making, action in the field and future-oriented activities. UNESCO's work programme for 2002–03 on Social and Human Sciences comprised three main components: The ethics of science and technology; Promotion of human rights, peace and democratic principles; and Improvement of policies relating to social transformations and promotion of anticipation and prospective studies. The priority Ethics of Science and Technology element aimed to reinforce UNESCO's role as an intellectual forum for ethical reflection on challenges related to the advance of science and technology; oversee the follow-up of the Universal Declaration on the Human Genome and Human Rights (see below); promote education in science and technology; ensure UNESCO's role in promoting good practices through encouraging the inclusion of ethical guiding principles in policy formulation and reinforcing international networks; and to promote international co-operation in human sciences and philosophy. The Social and Human Sciences programme had the main intellectual and conceptual responsibility for the transdisciplinary theme 'eradication of poverty, especially extreme poverty'.

UNESCO aims to promote and protect human rights and acts as an interdisciplinary, multicultural and pluralistic forum for reflection on issues relating to the ethical dimension of scientific advances, for example in biogenetics, new technology, and medicine. In May 1997 the International Bioethics Committee, a group of 36 specialists who meet under UNESCO auspices, approved a draft version of a Universal Declaration on the Human Genome and Human Rights, in an attempt to provide ethical guide-lines for developments in human genetics. The Declaration, which identified some 100,000 hereditary genes as 'common heritage', was adopted by the UNESCO General Conference in November and committed states to promoting the dissemination of relevant scientific knowledge and co-operating in genome research. In October 2003 the General Conference adopted an International Declaration on Human Genetic Data, establishing standards for scientists working in that field. UNESCO hosts the secretariat of the 18-member World Commission on the Ethics of Scientific Knowledge and Technology (COMEST), which aims to serve as a forum for the exchange of information and ideas and to promote dialogue between scientific communities, decision-makers and the public. COMEST met for the first time in April 1999 in Oslo, Norway. Its second meeting, which took place in December 2001 in Berlin, Germany, focused on the ethics of energy, fresh water and outer space. The third meeting of COMEST, held in Rio de Janeiro, Brazil, in December 2003, inaugurated a new regional-focused approach. An extraordinary meeting of COMEST was held in May 2004.

In 1994 UNESCO initiated an international social science research programme, the Management of Social Transformations (MOST), to promote capacity-building in social planning at all levels of decision-making. UNESCO sponsors several research fellowships in the social sciences. In other activities UNESCO promotes the rehabilitation of underprivileged urban areas, the research of socio-cultural factors affecting demographic change, and the study of family issues.

UNESCO aims to assist the building and consolidation of peaceful and democratic societies. An international network of institutions and centres involved in research on conflict resolution is being established to support the promotion of peace. Other training, workshop and research activities have been undertaken in countries that have suffered conflict. An International Youth Clearing House and Information Service (INFOYOUTH) aims to increase and consolidate the information available on the situation of young people in society, and to heighten awareness of their needs, aspirations and potential among public and private decision-makers. UNESCO also focuses on the educational and cultural dimensions of physical education and sport and their capacity to preserve and improve health. Fundamental to UNESCO's mission is the rejection of all forms of discrimination. It disseminates scientific information aimed at combating racial prejudice, works to improve the status of women and their access to education, and promotes equality between men and women.

CULTURE

In undertaking efforts to preserve the world's cultural and natural heritage UNESCO has attempted to emphasize the link between culture and development. In November 2001 the General Conference adopted the UNESCO Universal Declaration on Cultural Diversity, which affirmed the importance of intercultural dialogue in establishing a climate of peace. The work programme on Culture for 2002–03 included the following interrelated components: Reinforcing Normative Action in the Field of Culture; Protecting Cultural Diversity and Promoting Cultural Pluralism and Intercultural Dialogue; and Strengthening Links between Culture and Development. The focus was to be on all aspects of cultural heritage, and on the encouragement of cultural diversity and dialogue between cultures and civilizations. UNESCO was designated as the lead agency for co-ordinating the UN Year for Cultural Heritage, celebrated in 2002. In January 2002 UNESCO inaugurated a new six-year initiative, Global Alliance on Cultural Diversity, to promote partnerships between governments, non-governmental bodies and the private sector with a view to supporting cultural diversity through the strengthening of cultural industries and the prevention of cultural piracy. In 2004 UNESCO was formulating a draft International Convention on the Protection of the Diversity of Cultural Contents and Artistic Expressions.

UNESCO's World Heritage Programme, inaugurated in 1978, aims to protect historic sites and natural landmarks of outstanding universal significance, in accordance with the 1972 UNESCO Convention Concerning the Protection of the World Cultural and Natural Heritage, by providing financial aid for restoration, technical assistance, training and management planning. At October 2004 the 'World Heritage List' comprised 788 sites in 134 countries, of which 611 had cultural significance, 154 were natural landmarks, and 23 were of 'mixed' importance. UNESCO also maintains a list of 'World Heritage in Danger'. UNESCO participated in the successful campaign to safeguard the Buddhist temple at Borobudur, Indonesia (1973–83). The organization is assisting in the preservation of numerous historical and natural sites in the Far East and Australasia region, including the Great Barrier Reef (Australia), the Great Wall of China and the Imperial tombs of the Ming and Qing dynasties (People's Republic of China), the Himeji Castle (Japan), Sukhothai (Thailand) and Luang Prabang (Laos).

The formulation of a Declaration against the Intentional Destruction of Cultural Heritage was authorized by the General Conference in November 2001. In addition, the November General Conference adopted the Convention on the Protection of the Underwater Cultural Heritage, covering the protection from commercial exploitation of shipwrecks, submerged historical sites, etc., situated in the territorial waters of signatory states. UNESCO also administers the 1954 Hague Convention on the Protection of Cultural Property in the Event of Armed Conflict and the 1970 Convention on the Means of Prohibiting and Preventing the Illicit Import, Export and Transfer of Ownership of Cultural Property. In 1992 a World Heritage Centre was established to enable rapid mobilization of international technical assistance for the preservation of cultural sites. Through the World Heritage Information Network (WHIN), a world-wide network of more than 800 information providers, UNESCO promotes global awareness and information exchange.

UNESCO supports efforts for the collection and safeguarding of humanity's non-material 'intangible' heritage, including oral traditions, music, dance and medicine. In May 2001 UNESCO awarded the title of 'Masterpieces of the Oral and Intangible Heritage of Humanity' to 19 cultural spaces (i.e. physical or temporal spaces hosting recurrent cultural events) and popular forms of expression deemed to be of outstanding value. UNESCO produces an *Atlas of the World's Languages in Danger of Disappearing*. The most recent edition, issued in February 2002, reported that of some 6,000 languages spoken world-wide, about one-half were endangered. In October 2003 the UNESCO General Conference adopted a Convention for the Safeguarding of Intangible Cultural Heritage, which provided for the establishment of an intergovernmental committee and for participating states to formulate national inventories of intangible heritage.

UNESCO encourages the translation and publication of literary works, publishes albums of art, and produces records, audiovisual programmes and travelling art exhibitions. It supports the development of book publishing and distribution, including the free flow of books and educational material across borders, and the training of editors and managers in publishing. UNESCO is active in preparing and encouraging the enforcement of international legislation on copyright.

In December 1992 UNESCO established the World Commission on Culture and Development, to strengthen links between culture and development and to prepare a report on the issue. The first World Conference on Culture and Development was held in June 1999, in Havana, Cuba. Within the context of the UN's World Decade for Cultural Development (1988–97) UNESCO launched the Silk Roads Project, as a multi-disciplinary study of the interactions among cultures and civilizations along the routes linking Asia and Europe, and established an International Fund for the Promotion of Culture, awarding two annual prizes for music and the promotion of arts. In April 1999 UNESCO celebrated the completion of a major international project, the *General History of Africa*.

COMMUNICATION AND INFORMATION

In 2001 UNESCO introduced a major programme, 'Information for All', as the principal policy-guiding framework for the Communication and Information sector. The organization works towards establishing an open, non-exclusive knowledge society based on information-sharing and incorporating the socio-cultural and ethical dimensions of sustainable development. It promotes the free flow of, and universal access to, information, knowledge, data and best practices, through the development of communications infrastructures, the elimination of impediments to freedom of expression, and the promotion of the right to information; through encouraging international co-operation in maintaining libraries and archives; and through efforts to harness informatics for development purposes and strengthen member states' capacities in this field. Activities include assistance with the development of legislation and training programmes in countries where independent and pluralistic media are emerging; assistance in the monitoring of media independence, pluralism and diversity; promotion of exchange programmes and study tours; and improving access and opportunities for women in the media. UNESCO recognizes that the so-called global 'digital divide', in addition to other developmental differences between countries, generates exclusion and marginalization, and that increased participation in the democratic process can be attained through strengthening national communication and information capacities. UNESCO promotes the upholding of human rights in the use of cyberspace. The organization participated in the first phase of the World Summit on the Information Society, held in Geneva, Switzerland, in December 2003. The work programme on Communication and Information for 2002–03 comprised the following components: Promoting equitable access to information and knowledge, especially in the public domain, and Promoting freedom of expression and strengthening communication capacities. UNESCO's Memory of the World project aims to preserve in digital form, and thereby to promote wide access to, the world's documentary heritage.

In regions affected by conflict UNESCO supports efforts to establish and maintain an independent media service. This strategy is largely implemented through an International Programme for the Development of Communication (IPDC). In Cambodia, Haiti and Mozambique UNESCO participated in the restructuring of the media in the context of national reconciliation and in Bosnia and Herzegovina it assisted in the development of independent media. In December 1998 the Israeli-Palestinian Media Forum was established, to foster professional co-operation between Israeli and Palestinian journalists. IPDC provides support to communication and media development projects in the developing world, including the establishment of news agencies and newspapers and training editorial and technical staff. Since its establishment in 1982 IPDC, has financed some 1,000 projects in more than 130 countries.

In March 1997 the first International Congress on Ethical, Legal and Societal Aspects of Digital Information ('INFOethics') was held in Monte Carlo, Monaco. At the second INFOethics Congress, held in October 1998, experts discussed issues concerning privacy, confidentiality and security in the electronic transfer of information. In November 2000 a third INFOethics conference was held, on the theme of the 'Right to universal access to information in the 21st century'. UNESCO maintains an Observatory on the Information Society, which provides up-to-date information on the development of new information and communications technologies, analyses major trends, and aims to raise awareness of related ethical, legal and societal issues. A UNESCO Institute for Information Technologies in Education was established in Moscow, Russia in 1998. In 2001 the UNESCO Institute for Statistics was established in Montréal, Canada.

Finance

UNESCO's activities are funded through a regular budget provided by contributions from member states and extrabudgetary funds from other sources, particularly UNDP, the World Bank, regional banks and other bilateral Funds-in-Trust arrangements. UNESCO co-operates with many other UN agencies and international non-governmental organizations.

UNESCO's Regular Programme budget for the two years 2002–03 was US $544.4m., the same as for the previous biennium. Extra-budgetary funds for 2002–03 were estimated at $320m.

Publications

(mostly in English, French and Spanish editions; Arabic, Chinese and Russian versions are also available in many cases)

Atlas of the World's Languages in Danger of Disappearing.

Copyright Bulletin (quarterly).

Encyclopedia of Life Support Systems (internet-based).

International Review of Education (quarterly).

International Social Science Journal (quarterly).

Museum International (quarterly).

Nature and Resources (quarterly).

Prospects (quarterly review on education).

UNESCO Courier (monthly, in 27 languages).

UNESCO Sources (monthly).

UNESCO Statistical Yearbook.

World Communication Report.

World Educational Report (every 2 years).

World Heritage Review (quarterly).

World Information Report.

World Science Report (every 2 years).

Books, databases, video and radio documentaries, statistics, scientific maps and atlases.

World Health Organization—WHO

Address: Ave Appia 20, 1211 Geneva 27, Switzerland.
Telephone: (22) 7912111; **fax:** (22) 7913111; **e-mail:** info@who.int; **internet:** www.who.int.

WHO, established in 1948, is the lead agency within the UN system concerned with the protection and improvement of public health.

Organization

(October 2004)

WORLD HEALTH ASSEMBLY

The Assembly meets in Geneva, once a year; it is responsible for policy making and the biennial programme and budget; appoints the Director-General, admits new members and reviews budget contributions.

EXECUTIVE BOARD

The Board is composed of 32 health experts designated by, but not representing, their governments; they serve for three years, and the World Health Assembly elects 10–12 member states each year to the Board. It meets at least twice a year to review the Director-General's programme, which it forwards to the Assembly with any recommendations that seem necessary. It advises on questions referred to it by the Assembly and is responsible for putting into effect the decisions and policies of the Assembly. It is also empowered to take emergency measures in case of epidemics or disasters.

Chairman: D. Á. Gunnarson (Iceland).

SECRETARIAT

Director-General: Dr Jong-Wook Lee (Republic of Korea).

Assistant Directors-General: Denis Aitken (United Kingdom) (Director of the Office of the Director-General), Liu Peilong (People's Republic of China) (Adviser to the Director-General), Anarfi Asamoa-Baah (Ghana) (Communicable Diseases), Kazem Behbehani (Kuwait) (External Relations and Governing Bodies), Jack C. Chow (USA) (HIV/AIDS, TB and Malaria), Timothy G. Evans (Canada) (Evidence and Information for Policy), Catherine le Galès-Camus (France) (Non-Communicable Diseases and Mental Health), Kerstin Leitner (Germany) (Sustainable Development & Healthy Environments), Vladimir Lepakhin (Russia) (Health Technology and Pharmaceuticals), Anders Nordström (Sweden) (General Management), Joy Phumaphi (Botswana) (Family & Community Health).

REGIONAL OFFICES

Each of WHO's six geographical regions has its own organization consisting of a regional committee representing the member states and associate members in the region concerned, and a regional office staffed by experts in various fields of health.

South-East Asia: World Health House, Indraprastha Estate, Mahatma Gandhi Rd, New Delhi 110002, India; tel. (11) 3370804; fax (11) 3379507; e-mail pandeyh@whosea.org; internet www.whosea.org; Dir Dr Uton Muchtar Rafei (Indonesia).

Western Pacific: POB 2932, Manila 1000, Philippines; tel. (2) 5288001; fax (2) 5211036; e-mail postmaster@who.org.ph; internet www.wpro.who.int; Dir Dr Shigeru Omi (Japan).

Activities

WHO's objective is stated in the constitution as 'the attainment by all peoples of the highest possible level of health'. 'Health' is defined as 'a state of complete physical, mental and social well-being and not merely the absence of disease and infirmity'. In November 2001 WHO issued the International Classification of Functioning, Disability and Health (ICF) to act as an international standard and guide-lines for determining health and disability.

WHO acts as the central authority directing international health work, and establishes relations with professional groups and government health authorities on that basis.

It provides, on request from member states, technical and policy assistance in support of programmes to promote health, prevent and control health problems, control or eradicate disease, train health workers best suited to local needs and strengthen national health systems. Aid is provided in emergencies and natural disasters.

A global programme of collaborative research and exchange of scientific information is carried out in co-operation with about 1,200 national institutions. Particular stress is laid on the widespread communicable diseases of the tropics, and the countries directly concerned are assisted in developing their research capabilities.

It keeps diseases and other health problems under constant surveillance, promotes the exchange of prompt and accurate information and of notification of outbreaks of diseases, and administers the International Health Regulations. It sets standards for the quality control of drugs, vaccines and other substances affecting health. It formulates health regulations for international travel.

It collects and disseminates health data and carries out statistical analyses and comparative studies in such diseases as cancer, heart disease and mental illness.

It receives reports on drugs observed to have shown adverse reactions in any country, and transmits the information to other member states.

It promotes improved environmental conditions, including housing, sanitation and working conditions. All available information on effects on human health of the pollutants in the environment is critically reviewed and published.

Co-operation among scientists and professional groups is encouraged. The organization negotiates and sustains national and global partnerships. It may propose international conventions and agreements, and develops and promotes international norms and standards. The organization promotes the development and testing of new technologies, tools and guide-lines. It assists in developing an informed public opinion on matters of health.

HEALTH FOR ALL

WHO's first global strategy for pursuing 'Health for all' was adopted in May 1981 by the 34th World Health Assembly. The objective of 'Health for all' was identified as the attainment by all citizens of the world of a level of health that would permit them to lead a socially and economically productive life, requiring fair distribution of available resources, universal access to essential health care, and the promotion of preventive health care. In May 1998 the 51st World Health Assembly renewed the initiative, adopting a global strategy in support of 'Health for all in the 21st century', to be effected through regional and national health policies. The new approach was to build on the primary health care approach of the initial strategy, but was to strengthen the emphasis on quality of life, equity in health and access to health services. The following have been identified as minimum requirements of 'Health for All':

Safe water in the home or within 15 minutes' walking distance, and adequate sanitary facilities in the home or immediate vicinity;

Immunization against diphtheria, pertussis (whooping cough), tetanus, poliomyelitis, measles and tuberculosis;

Local health care, including availability of essential drugs, within one hour's travel;

Trained personnel to attend childbirth, and to care for pregnant mothers and children up to at least one year old.

WHO's technical programmes are divided into the following groups, or 'clusters': Communicable Diseases; Non-communicable Diseases and Mental Health; Family and Community Health; Sustainable Development and Healthy Environments; Health Technology and Pharmaceuticals; and Evidence and Information for Policy. In 2004–05 the following areas of work were designated as organization-wide priorities: malaria; TB; cancer, cardiovascular diseases and diabetes; tobacco; mental health; making pregnancy safer and children's health; HIV/AIDS; health and environment; food safety; health systems, including essential medicines; and blood safety. In 2000 WHO adopted a new corporate strategy, entailing a stronger focus on performance and programme delivery through standardized plans of action, and increased consistency and efficiency throughout the organization.

The Tenth General Programme of Work, for the period 2002–05, defined a policy framework for pursuing the principal objectives of building healthy populations and combating ill health. The Programme took into account: increasing understanding of the social, economic, political and cultural factors involved in achieving better health and the role played by better health in poverty reduction; the increasing complexity of health systems; the importance of safeguarding health as a component of humanitarian action; and the need for greater co-ordination among development organizations. It incorporated four interrelated strategic directions: lessening excess mortality, morbidity and disability, especially in poor and marginalized populations; promoting healthy lifestyles and reducing risk factors to human health arising from environmental, economic, social and behavioural causes; developing equitable and financially fair health systems; and establishing an enabling policy and an institutional environment for the health sector and promoting an effective health dimension to social, economic, environmental and development policy.

COMMUNICABLE DISEASES

WHO identifies infectious and parasitic communicable diseases as a major obstacle to social and economic progress, particularly in developing countries, where, in addition to disabilities and loss of productivity and household earnings, they cause nearly one-half of all deaths. Emerging and re-emerging diseases, those likely to cause epidemics, increasing incidence of zoonoses (diseases passed from animals to humans either directly or by insects) attributable to environmental changes, outbreaks of unknown etiology, and the undermining of some drug therapies by the spread of antimicrobial resistance are main areas of concern. In recent years WHO has noted the global spread of communicable diseases through international travel, voluntary human migration and involuntary population displacement.

WHO's Communicable Diseases group works to reduce the impact of infectious diseases world-wide through surveillance and response; prevention, control and eradication strategies; and research and product development. Combating malaria and tuberculosis (TB) are organization-wide priorities and, as such, are supported not only by their own areas of work but also by activities undertaken in other areas. The group seeks to identify new technologies and tools, and to foster national development through strengthening health services and the better use of existing tools. It aims to strengthen global monitoring of important communicable disease problems. The group advocates a functional approach to disease control. It aims to create consensus and consolidate partnerships around targeted diseases and collaborates with other groups at all stages to provide an integrated response. In April 2000 WHO and several partner institutions in epidemic surveillance established a Global Outbreak Alert and Response Network. Through the Network WHO aims to maintain constant vigilance regarding outbreaks of disease and to link world-wide expertise to provide an immediate response capability. From March 2003 WHO, through the Network, was co-ordinating the international investigation into the global spread of Severe Acute Respiratory Syndrome (SARS), a previously unknown atypical pneumonia. From the end of that year WHO was monitoring the spread through several Asian countries of zoonotic Avian Influenza.

A Ministerial Conference on Malaria, organized by WHO, was held in October 1992, attended by representatives from 102 member countries. The Conference adopted a plan of action for the 1990s for the control of the disease, which kills an estimated 1m. people every year and affects a further 300m.–500m. Some 90% of all cases are in sub-Saharan Africa. WHO assists countries where malaria is endemic to prepare national plans of action for malaria control in accordance with its Global Malaria Control Strategy, which emphasizes strengthening local capabilities, for example through training, for effective health control. In July 1998 WHO declared the control of malaria a priority concern, and in October the organization formally launched the 'Roll Back Malaria' programme, in conjunction with UNICEF, the World Bank and UNDP, which aimed to halve the prevalence of malaria by 2010. Emphasis was to be placed on strengthening local health systems and on the promotion of inexpensive preventive measures, including the use of bednets treated with insecticides. The global Roll Back Malaria partnership, linking governments, development agencies, and other parties, aims to mobilize resources and support for controlling the disease. WHO, with several private- and public-sector partners, supports the development of more effective anti-malaria drugs and vaccines through the 'Medicines for Malaria' venture.

In 1995 WHO established a Global Tuberculosis Programme to address the challenges of the TB epidemic, which had been declared a global emergency by the Organization in 1993. According to WHO estimates, one-third of the world's population carries the TB bacillus, and 2m.–3m. people die from the disease each year. WHO provides technical support to all member countries, with special attention given to those with high TB prevalence, to establish effective national tuberculosis control programmes. WHO's strategy for TB control includes the use of DOTS (direct observation treatment, short-course), standardized treatment guide-lines, and result accountability through routine evaluation of treatment outcomes. Simultaneously, WHO is encouraging research with the aim of further disseminating DOTS, adapting DOTS for wider use, developing new tools for prevention, diagnosis and treatment, and containing new threats such as the HIV/TB co-epidemic. In March 1999 WHO announced the launch of a new initiative, 'Stop TB', in partnership with the World Bank, the US Government and a coalition of non-governmental organizations, which aimed to promote DOTS to ensure its use in 85% of detected cases by 2005 (compared with around one-quarter in 1999). The global target for case detection by 2005 was 70%. However, inadequate control of DOTS in some areas, leading to partial and inconsistent treatments, has resulted in the development of drug-resistant and, often, incurable strains of the disease. The incidence of so-called multidrug-resistant TB (MDR-TB) strains, that are unresponsive to the two main anti-TB drugs, has risen in recent years. During 2001 WHO was developing and testing DOTS-Plus, a strategy for controlling the spread of MDR-TB in areas of high prevalence. In 2001 WHO estimated that more than 8m. new cases of TB were occurring world-wide each year, of which the largest concentration was in south-east Asia. It envisaged a substantial increase in new cases by 2005, mainly owing to the severity of the HIV/TB co-epidemic. TB is the principal cause of death for people infected with the HIV virus and an estimated one-third of people living with HIV/AIDS globally are co-infected with TB. A Global Fund to Fight AIDS, TB and Malaria was established, with WHO participation, in 2001. In March 2001 the Global TB Drug Facility was launched under the 'Stop TB' initiative; this aimed to increase access to high-quality anti-TB drugs for sufferers in developing countries. In October the 'Stop TB' partnership announced a Global Plan to Stop TB, which envisaged the expansion of access to DOTS; the advancement of MDR-TB prevention measures; the development of anti-TB drugs entailing a shorter treatment period; and the implementation of new strategies for treating people with HIV and TB.

One of WHO's major achievements was the eradication of smallpox. Following a massive international campaign of vaccination and surveillance (begun in 1958 and intensified in 1967), the last case was detected in 1977 and the eradication of the disease was declared in 1980. In May 1996 the World Health Assembly resolved that, pending a final endorsement, all remaining stocks of the smallpox virus were to be destroyed on 30 June 1999, although 500,000 doses of smallpox vaccine were to remain, along with a supply of the smallpox vaccine seed virus, in order to ensure that a further supply of the vaccine could be made available if required. In May 1999, however, the Assembly authorized a temporary retention of stocks of the virus until 2002. In late 2001, in response to fears that illegally-held virus stocks could be used in acts of biological terrorism (see below), WHO reassembled a team of technical experts on smallpox. In January 2002 the Executive Board determined that stocks of the virus should continue to be retained, to enable research into more effective treatments and vaccines.

In 1988 the World Health Assembly declared its commitment to the eradication of poliomyelitis by the end of 2000 and launched the Global Polio Eradication Initiative. In August 1996 WHO, UNICEF and Rotary International, together with other national and international partners, initiated a campaign to 'Kick Polio out of Africa', with the aim of immunizing more than 100m. children in 46 countries against the disease over a three-year period. In 2000 WHO adopted a strategic plan for the eradication of polio covering the period 2001–05, which envisaged the effective use of National Immunization Days (NIDs) to secure global interruption of polio transmission by the end of 2002, with a view to achieving certification of the global eradication of polio by the end of 2005. (In conflict zones so-called 'days of tranquility' have been negotiated to facilitate the implementation of NIDs.) Meanwhile, routine immunization services were to be strengthened. A post-certification immunization policy for polio was to be formulated. By the end of 2001 the number of confirmed polio cases world-wide had declined to 483 in 10 countries, from 35,000 in 125 countries in 1988 (the actual number of cases in 1988 was estimated at around 350,000). In 2001 575m. children in 94 countries world-wide were immunized through the use of NIDS. In that year Vitamin A was also administered during NIDS in some 60 countries in order to combat nutritional deficiencies in children. By December 2002, however, the number of confirmed cases of polio stood at 1,924, 1,599 of which were in India. Six other countries were still known to be or suspected of being polio endemic at that time: Afghanistan, Egypt, Niger, Nigeria, Pakistan and Somalia. WHO has declared the following regions 'polio-free': the Americas (1994); Western Pacific (2000); and Europe (2002).

WHO is committed to the elimination of leprosy (the reduction of the prevalence of leprosy to less than one case per 10,000 population). The use of a highly effective combination of three drugs (known as multi-drug therapy—MDT) resulted in a reduction in the number of leprosy cases world-wide from 10m.–12m. in 1988 to 597,000 in 2000. The number of countries having more than one case of leprosy per 10,000 had declined to 15 by 2000, compared with 122 in 1985. In 2000 the world-wide leprosy prevalence rate stood at 1.4 cases per 10,000 people, although the rate in the 11 most endemic countries was 4.5 cases per 10,000. India has more than one-half of all active leprosy cases. The Global Alliance for the Elimination of Leprosy, launched in November 1999 by WHO, in collaboration with governments of affected countries and several private partners, including a major pharmaceutical company, aims to bring about the eradication of the disease by the end of 2005, through the continued use of MDT treatment. In July 1998 the Director-General of WHO and representatives of more than 20 countries, meeting in Yamoussoukro, Côte d'Ivoire, signed a declaration on the control of another mycobacterial disease, Buruli ulcer.

The Special Programme for Research and Training in Tropical Diseases, established in 1975 and sponsored jointly by WHO, UNDP and the World Bank, as well as by contributions from donor countries, involves a world-wide network of some 5,000 scientists working on the development and application of vaccines, new drugs, diagnostic kits and preventive measures, and an applied field research on practical community issues affecting the target diseases.

The objective of providing immunization for all children by 1990 was adopted by the World Health Assembly in 1977. Six diseases (measles, whooping cough, tetanus, poliomyelitis, tuberculosis and diphtheria) became the target of the Expanded Programme on Immunization (EPI), in which WHO, UNICEF and many other organizations collaborated. As a result of massive international and national efforts, the global immunization coverage increased from 20% in the early 1980s to the targeted rate of 80% by the end of 1990. This coverage signified that more than 100m. children in the developing world under the age of one had been successfully vaccinated against the targeted diseases, the lives of about 3m. children had been saved every year, and 500,000 annual cases of paralysis as a result of polio had been prevented. In 1992 the Assembly resolved to reach a new target of 90% immunization coverage with the six EPI vaccines; to introduce hepatitis B as a seventh vaccine (with the aim of an 80% reduction in the incidence of the disease in children by 2001); and to introduce the yellow fever vaccine in areas where it occurs endemically.

In June 2000 WHO released a report entitled 'Overcoming Antimicrobial Resistance', in which it warned that the misuse of antibiotics could render some common infectious illnesses unresponsive to treatment. At that time WHO issued guide-lines which aimed to mitigate the risks associated with the use of antimicrobials in livestock reared for human consumption.

NON-COMMUNICABLE DISEASES AND MENTAL HEALTH

The Non-communicable Diseases and Mental Health group comprises departments for the surveillance, prevention and management of uninfectious diseases, such as those arising from an unhealthy diet, and departments for health promotion, disability, injury prevention and rehabilitation, mental health and substance abuse. Surveillance, prevention and management of non-communicable diseases, tobacco, and mental health are organization-wide priorities.

Tobacco use, unhealthy diet and physical inactivity are regarded as common, preventable risk factors for the four most prominent non-communicable diseases: cardiovascular diseases, cancer, chronic respiratory disease and diabetes. WHO aims to monitor the global epidemiological situation of non-communicable diseases, to co-ordinate multinational research activities concerned with prevention and care, and to analyse determining factors such as gender and poverty. In mid-1998 the organization adopted a resolution on measures to be taken to combat non-communicable diseases; their prevalence was anticipated to increase, particularly in developing countries, owing to rising life expectancy and changes in lifestyles. For example, between 1995 and 2025 the number of adults affected by diabetes was projected to increase from 135m. to 300m. In 2001 chronic diseases reportedly accounted for about 59% of the estimated 56.5m. total deaths globally and for 46% of the global burden of disease. In February 1999 WHO initiated a new programme, 'Vision 2020: the Right to Sight', which aimed to eliminate avoidable blindness (estimated to be as much as 80% of all cases) by 2020. Blindness was otherwise predicted to increase by as much as twofold, owing to the increased longevity of the global population. In co-operation with the International Association for the Study of Obesity (IASO), WHO has studied obesity-related issues. The International Task Force on Obesity, affiliated to the IASO, aims to encourage the development of new policies for managing obesity. WHO and FAO jointly commissioned an expert report on the relationship of diet, nutrition and physical activity to chronic diseases, which was published in March 2003.

In co-operation with the IASO WHO has studied obesity-related issues affecting countries in Asia and the Pacific. In October 2001 the first Asia-Oceania Conference on Obesity was convened in Japan under the auspices of IASO; a second conference was held in Kuala Lumpur, Malaysia, in September 2003, and a third conference was scheduled to be held in Taipei, Taiwan, in February 2005.

WHO's programmes for diabetes mellitus, chronic rheumatic diseases and asthma assist with the development of national initiatives, based upon goals and targets for the improvement of early detection, care and reduction of long-term complications. WHO's cardiovascular diseases programme aims to prevent and control the major cardiovascular diseases, which are responsible for more than 14m. deaths each year. It is estimated that one-third of these deaths could have been prevented with existing scientific knowledge. The programme on cancer control is concerned with the prevention of cancer, improving its detection and cure and ensuring care of all cancer patients in need. In 1998 a five-year programme to improve cancer care in developing countries was established, sponsored by private enterprises.

The WHO Human Genetics Programme manages genetic approaches for the prevention and control of common hereditary diseases and of those with a genetic predisposition representing a major health importance. The Programme also concentrates on the further development of genetic approaches suitable for incorporation into health care systems, as well as developing a network of international collaborating programmes.

WHO works to assess the impact of injuries, violence and sensory impairments on health, and formulates guide-lines and protocols for the prevention and management of mental problems. The health promotion division promotes decentralized and community-based health programmes and is concerned with developing new approaches to population ageing and encouraging healthy life-styles and self-care. It also seeks to relieve the negative impact of social changes such as urbanization, migration and changes in family structure upon health. WHO advocates a multi-sectoral approach—involving public health, legal and educational systems—to the prevention of injuries, which represent 16% of the global burden of disease. It aims to support governments in developing suitable strategies to prevent and mitigate the consequences of violence, unintentional injury and disability. Several health promotion projects have been undertaken, in collaboration between WHO regional and country offices and other relevant organizations, including: the Global School Health Initiative, to bridge the sectors of health and education and to promote the health of school-age children; the Global Strategy for Occupational Health, to promote the health of the working population and the control of occupational health risks; Community-based Rehabilitation, aimed at providing a more enabling environment for people with disabilities; and a communication strategy to provide training and support for health communications personnel and initiatives. In 2000 WHO, UNESCO, the World Bank and UNICEF adopted the joint Focusing Resources for Effective School Health (FRESH Start) approach to promoting life skills among adolescents.

In July 1997 the fourth International Conference on Health Promotion (ICHP) was held in Jakarta, Indonesia, where a declaration on 'Health Promotion into the 21st Century' was agreed. The fifth ICHP was convened in June 2000, in Mexico City, Mexico.

Mental health problems, which include unipolar and bipolar affective disorders, psychosis, epilepsy, dementia, Parkinson's disease, multiple sclerosis, drug and alcohol dependency, and neuropsychiatric disorders such as post-traumatic stress disorder, obsessive compulsive disorder and panic disorder, have been identified by WHO as significant global health problems. Although, overall, physical health has improved, mental, behavioural and social health problems are increasing, owing to extended life expectancy and improved child mortality rates, and factors such as war and poverty. WHO aims to address mental problems by increasing awareness of mental health issues and promoting improved mental health services and primary care.

The Substance Abuse department is concerned with problems of alcohol, drugs and other substance abuse. Within its Programme on Substance Abuse (PSA), which was established in 1990 in response to the global increase in substance abuse, WHO provides technical support to assist countries in formulating policies with regard to the prevention and reduction of the health and social effects of psychoactive substance abuse. PSA's sphere of activity includes epidemiological surveillance and risk assessment, advocacy and the dissemination of information, strengthening national and regional prevention and health promotion techniques and strategies, the development of cost-effective treatment and rehabilitation approaches, and also encompasses regulatory activities as required under the international drugs-control treaties in force.

The Tobacco or Health Programme aims to reduce the use of tobacco, by educating tobacco-users and preventing young people from adopting the habit. In 1996 WHO published its first report on the tobacco situation world-wide. According to WHO, about one-third of the world's population aged over 15 years smoke tobacco, which causes approximately 3.5m. deaths each year (through lung cancer, heart disease, chronic bronchitis and other effects). In 1998 the 'Tobacco Free Initiative', a major global anti-smoking campaign, was established. In May 1999 the World Health Assembly endorsed the formulation of a Framework Convention on Tobacco Control (FCTC) to help to combat the increase in tobacco use (although a number of tobacco growers expressed concerns about the effect of the convention on their livelihoods). The draft Framework Convention was finalized in March 2003 and was adopted by the World Health Assembly in May. The greatest increase in tobacco use is forecast to occur in developing countries.

FAMILY AND COMMUNITY HEALTH

WHO's Family and Community Health group addresses the following areas of work: child and adolescent health, research and programme development in reproductive health, making pregnancy safer, women's health, and HIV/AIDS. Making pregnancy safer and HIV/AIDS are organization-wide priorities. The group's aim is to improve access to sustainable health care for all by strengthening health systems and fostering individual, family and community development. Activities include newborn care; child health, including promoting and protecting the health and development of the child through such approaches as promotion of breast-feeding and use of the mother-baby package, as well as care of the sick child, including diarrhoeal and acute respiratory disease control, and support to women and children in difficult circumstances; the promotion of safe motherhood and maternal health; adolescent health, including the promotion and development of young people and the prevention of specific health problems; women, health and development, including addressing issues of gender, sexual violence, and harmful traditional practices; and human reproduction, including research related to contraceptive technologies and effective methods. In addition, WHO aims to provide technical leadership and co-ordination on reproductive health and to support countries in their efforts to ensure that people: experience healthy sexual development and maturation; have the capacity for healthy, equitable and responsible relationships; can achieve their reproductive intentions safely and healthily; avoid illnesses, diseases and injury related to sexuality and reproduction; and receive appropriate counselling, care and rehabilitation for diseases and conditions related to sexuality and reproduction.

In September 1997 WHO, in collaboration with UNICEF, formally launched a programme advocating the Integrated Management of Childhood Illness (IMCI), following successful regional trials in more than 20 developing countries during 1996–97. IMCI recognizes that pneumonia, diarrhoea, measles, malaria and malnutrition cause some 70% of the approximately 11m. childhood deaths each year, and recommends screening sick children for all five conditions, to obtain a more accurate diagnosis than may be achieved from the results of a single assessment. WHO's Division of Diarrhoeal and Acute Respiratory Disease Control encourages national programmes aimed at reducing childhood deaths as a result of diarrhoea, particularly through the use of oral rehydration therapy and preventive measures. The Division is also seeking to reduce deaths from pneumonia in infants through the use of a simple case-management strategy involving the recognition of danger signs and treatment with an appropriate antibiotic.

The HIV/AIDS epidemic represents a major threat to human well-being and socio-economic progress. Some 95% of those known to be infected with HIV/AIDS live in developing countries, and AIDS-related illnesses are the leading cause of death in sub-Saharan Africa. At December 2003 an estimated 40m. people world-wide were living with HIV/AIDS (including some 2.5m. children under 15 years); 5m. were newly infected during that year. WHO's Global Programme on AIDS, initiated in 1987, was concluded in December 1995. A Joint UN Programme on HIV/AIDS (UNAIDS) became operational on 1 January 1996, sponsored by WHO and other UN agencies. The UNAIDS secretariat is based at WHO headquarters. WHO established an Office of HIV/AIDS and Sexually-Transmitted Diseases in order to ensure the continuity of its global response to the problem, which included support for national control and education plans, improving the safety of blood supplies and improving the care and support of AIDS patients. In addition, the Office was to liaise with UNAIDS and to make available WHO's research and technical expertise. HIV/AIDS are an organization-wide priority. Sufferers of HIV/AIDS in developing countries have often failed to receive advanced antiretroviral (ARV) treatments that are widely available in industrialized countries, owing to their high cost. In May 2000 the World Health Assembly adopted a resolution urging WHO member states to improve access to the prevention and treatment of HIV-related illnesses and to increase the availability and affordability of drugs. A WHO-UNAIDS HIV Vaccine Initiative was launched in that year. In July a meeting of the Group of Seven industrialized nations and Russia (G-8), convened in Genoa, Italy, announced the formation of a new Global Fund to Fight AIDS, TB and Malaria (as previously proposed by the UN Secretary-General and recommended by the World Health Assembly). The Fund, a partnership between governments, UN bodies (including WHO) and other agencies, and private-sector interests, aimed in 2004 to disburse US $623m. in grants to prevention and treatment programmes in around 50 countries. In June 2001 governments participating in a special session of the UN General Assembly on HIV/AIDS adopted a Declaration of Commitment on HIV/AIDS. WHO, with UNAIDS, UNICEF, UNFPA, the World Bank, and major pharmaceutical companies, participates in the 'Accelerating Access' initiative, which aims to expand access to care, support and ARVs for people with HIV/AIDS. In March 2002, under its 'Access to Quality HIV/AIDS Drugs and Diagnostics' programme, WHO published a comprehensive list of HIV-related medicines deemed to meet standards recommended by the Organization. In April WHO issued the first treatment guide-lines for HIV/AIDS cases in poor communities, and endorsed the inclusion of HIV/AIDS drugs in its *Model List of Essential Drugs* (see below) in order to encourage their wider availability. The secretariat of the International HIV Treatment Access Coalition, founded in December of that year by governments, non-governmental organizations, donors and others to facilitate access to ARVs for people in low and middle income countries, is based at WHO headquarters. WHO, jointly with UNAIDS and the Global Fund to Fight AIDS, TB and Malaria, supports the so-called 'three-by-five' target of providing 3m. people in developing countries with ARVs by the end of 2005. WHO supports governments in developing effective health-sector responses to the HIV/AIDS epidemic through enhancing the planning and managerial capabilities, implementation capacity, and resources of health systems. In February 2003 WHO and FAO jointly published a manual on nutritional care for people living with HIV/AIDS.

At December 2003 an estimated 7.4m. people in Asia and the Pacific were reported to have HIV/AIDS, of whom an estimated 1m. were newly infected during the year. There are serious nation-wide HIV/AIDS epidemics in Cambodia (which has a national prevalence rate of about 3%), Myanmar and Thailand (with a national prevalence rate of about 2%), while the People's Republic of China has serious localized epidemics in certain provinces and autonomous regions, including Guangxi Zhuang, Guangdong, Henan, Sichuan, Yunnan and Xinjiang.

In March 1996 WHO's Centre for Health Development opened at Kobe, Japan. The Centre researches health developments and other determinants to strengthen policy decision-making within the health sector.

Joint UN Programme on HIV/AIDS (UNAIDS): 20 ave Appia, 1211 Geneva 27, Switzerland; tel. (22) 7913666; fax (22) 7914187; e-mail unaids@unaids.org; internet www.unaids.org; established in 1996 to lead, strengthen and support an expanded response to the global HIV/AIDS pandemic; activities focus on prevention, care and support, reducing vulnerability to infection, and alleviating the socioeconomic and human effects of HIV/AIDS; launched the Global Coalition on Women and AIDS in Feb. 2004; co-sponsors: WHO, UNICEF, UNDP, UNFPA, UNODC, ILO, UNESCO, the World Bank, WFP, UNHCR; Exec. Dir Peter Piot (Belgium).

SUSTAINABLE DEVELOPMENT AND HEALTHY ENVIRONMENTS

The Sustainable Development and Healthy Environments group focuses on the following areas of work: health in sustainable development; nutrition; health and environment; food safety; and emergency preparedness and response. Food safety is an organization-wide priority.

WHO promotes recognition of good health status as one of the most important assets of the poor. The Sustainable Development and Healthy Environment group seeks to monitor the advantages and disadvantages for health, nutrition, environment and development arising from the process of globalization (i.e. increased global flows of capital, goods and services, people, and knowledge); to integrate the issue of health into poverty reduction programmes; and to promote human rights and equality. Adequate and safe food and nutrition is a priority programme area. WHO collaborates with FAO, the World Food Programme, UNICEF and other UN agencies in pursuing its objectives relating to nutrition and food safety. An estimated 780m. people world-wide cannot meet basic needs for energy and protein, more than 2,000m. people lack essential vitamins and minerals, and 170m. children are estimated to be malnourished. In December 1992 WHO and FAO hosted an international conference on nutrition, at which a World Declaration and Plan of Action on Nutrition was adopted to make the fight against malnutrition a development priority. Following the conference, WHO promoted the elaboration and implementation of national plans of action on nutrition. WHO aims to support the enhancement of member states' capabilities in dealing with their nutrition situations, and addressing scientific issues related to preventing, managing and monitoring protein-energy malnutrition; micronutrient malnutrition, including iodine deficiency disorders, vitamin A deficiency, and nutritional anaemia; and diet-related conditions and non-communicable diseases such as obesity (increasingly affecting children, adolescents and adults, mainly in industrialized countries), cancer and heart disease. In 1990 the World Health Assembly resolved to eliminate iodine deficiency (believed to cause mental retardation); a strategy of universal salt iodization was launched in 1993. In collaboration with other international agencies, WHO is implementing a comprehensive strategy for promoting appropriate infant, young child and maternal nutrition, and for dealing effectively with nutritional emergencies in large populations. Areas of emphasis include promoting health-care practices that enhance successful breast-feeding; appropriate complementary feeding; refining the use and interpretation of body measurements for assessing nutritional status; relevant information, education and training; and action to give effect to the International Code of Marketing of Breast-milk Substitutes. The food safety programme aims to protect human health against risks associated with biological and chemical contaminants and additives in food. With FAO, WHO establishes food standards (through the work of the Codex Alimentarius Commission and its subsidiary committees) and evaluates food additives, pesticide residues and other contaminants and their implications for health. The programme provides expert advice on such issues as food-borne pathogens (e.g. listeria), production methods (e.g. aquaculture) and food biotechnology (e.g. genetic modification). In July 2001 the Codex Alimentarius Commission adopted the first global principles for assessing the safety of genetically-modified (GM) foods. In March 2002 an intergovernmental task force established by the Commission finalized 'principles for the risk analysis of foods derived from biotechnology', which were to provide a framework for assessing the safety of GM foods and plants. In the following month WHO and FAO announced a joint review of their food standards operations. In February 2003 the FAO/WHO Project and Fund for Enhanced Participation in Codex was launched to support the participation of poorer countries in the Commission's activities.

WHO's programme area on environment and health undertakes a wide range of initiatives to tackle the increasing threats to health and well-being from a changing environment, especially in relation to air pollution, water quality, sanitation, protection against radiation, management of hazardous waste, chemical safety and housing hygiene. Some 1,100m. people world-wide have no access to clean drinking water, while a further 2,400m. people are denied suitable sanitation systems. WHO helped launch the Water Supply and Sanitation Council in 1990 and regularly updates its *Guidelines for Drinking Water Quality*. In rural areas, the emphasis continues to be on the provision and maintenance of safe and sufficient water supplies and adequate sanitation, the health aspects of rural housing, vector control in water resource management, and the safe use of agrochemicals. In urban areas, assistance is provided to identify local environmental health priorities and to improve municipal governments' ability to deal with environmental conditions and health problems in an integrated manner; promotion of the 'Healthy City' approach is a major component of the Programme. Other Programme activities include environmental health information development and management, human resources development, environmental health planning methods, research and work on problems relating to global environment change, such as UV-radiation. A report considering the implications of climate change on human health, prepared jointly by WHO, WMO and UNEP, was published in July 1996. The WHO Global Strategy for Health and Environment, developed in response to the WHO Commission on Health and Environment which reported to the UN Conference on Environment and Development in June 1992, provides the framework for programme activities. In December 2001 WHO published a report on the relationship between macroeconomics and health.

WHO's work in the promotion of chemical safety is undertaken in collaboration with ILO and UNEP through the International Programme on Chemical Safety (IPCS), the Central Unit for which is located in WHO. The Programme provides internationally-evaluated scientific information on chemicals, promotes the use of such information in national programmes, assists member states in establishment of their own chemical safety measures and programmes, and helps them strengthen their capabilities in chemical emergency preparedness and response and in chemical risk reduction. In 1995 an Inter-organization Programme for the Social Management of Chemicals was established by UNEP, ILO, FAO, WHO, UNIDO and OECD, in order to strengthen international co-operation in the field of chemical safety. In 1998 WHO led an international assessment of the health risk from bendocine disruptors (chemicals which disrupt hormonal activities).

Following the major terrorist attacks perpetrated against targets in the USA in September 2001, WHO focused renewed attention on the potential deliberate use of infectious diseases, such as anthrax and smallpox, or of chemical agents, in acts of biological or chemical terrorism. In September 2001 WHO issued draft guide-lines entitled 'Health Aspects of Biological and Chemical Weapons'.

Within the UN system, WHO's Department of Emergency and Humanitarian Action co-ordinates the international response to emergencies and natural disasters in the health field, in close co-operation with other agencies and within the framework set out by the UN's Office for the Co-ordination of Humanitarian Affairs. In this context, WHO provides expert advice on epidemiological surveillance, control of communicable diseases, public health information and health emergency training. Its emergency preparedness activities include co-ordination, policy-making and planning, awareness-building, technical advice, training, publication of standards and guide-lines, and research. Its emergency relief activities include organizational support, the provision of emergency drugs and supplies and conducting technical emergency assessment missions. The Division's objective is to strengthen the national capacity of member states to reduce the adverse health consequences of disasters. In responding to emergency situations, WHO always tries to develop projects and activities that will assist the national authorities concerned in rebuilding or strengthening their own capacity to handle the impact of such situations In May 2001 WHO participated with governments and other international agencies in a joint exercise to evaluate national and international procedures for responding to a nuclear emergency.

Under the UN's Consolidated Appeals Process for 2002, launched in November 2001, WHO requested funding of US $7.2m. for ensuring availability of essential drugs, supporting immunization activities, provision of basic health services, training, and control of communicable diseases in the Democratic Republic of Korea, and $1.7m. for strengthening health services, disease surveillance and mental health care and psychosocial support in Indonesia.

HEALTH TECHNOLOGY AND PHARMACEUTICALS

WHO's Health Technology and Pharmaceuticals group, made up of the departments of essential drugs and other medicines, vaccines and other biologicals, and blood safety and clinical technology, covers the following areas of work: essential medicines—access, quality and rational use; immunization and vaccine development; and world-wide co-operation on blood safety and clinical technology. Blood safety and clinical technology are an organization-wide priority.

In January 1999 the Executive Board adopted a resolution on WHO's Revised Drug Strategy which placed emphasis on the inequalities of access to pharmaceuticals, and also covered specific aspects of drugs policy, quality assurance, drug promotion, drug donation, independent drug information and rational drug use. Plans of action involving co-operation with member states and other international organizations were to be developed to monitor and analyse the pharmaceutical and public health implications of international agreements, including trade agreements. In April 2001 experts from WHO and the World Trade Organization participated in a workshop to address ways of lowering the cost of medicines in less developed countries. In the following month the World Health Assembly adopted a resolution urging member states to promote equitable access to essential drugs, noting that this was denied to about one-third of the world's population. WHO participates with other partners in the 'Accelerating Access' initiative, which aims to expand access to antiretroviral drugs for people with HIV/AIDS (see above).

WHO reports that 2m. children die each year of diseases for which common vaccines exist. In September 1991 the Children's Vaccine Initiative (CVI) was launched, jointly sponsored by the Rockefeller Foundation, UNDP, UNICEF, the World Bank and WHO, to facilitate the development and provision of children's vaccines. The CVI has as its ultimate goal the development of a single oral immunization shortly after birth that will protect against all major childhood diseases. An International Vaccine Institute was established in Seoul, Republic of Korea, as part of the CVI, to provide scientific and technical services for the production of vaccines for developing countries. In September 1996 WHO, jointly with UNICEF, published a comprehensive survey, entitled *State of the World's Vaccines and Immunization*. In 1999 WHO, UNICEF, the World Bank and a number of public- and private-sector partners formed the Global Alliance for Vaccines and Immunization (GAVI), which aimed to expand the provision of existing vaccines and to accelerate the development and introduction of new vaccines and technologies, with the ultimate goal of protecting children of all nations and from all socio-economic backgrounds against vaccine-preventable diseases.

WHO supports states in ensuring access to safe blood, blood products, transfusions, injections, and health-care technologies.

EVIDENCE AND INFORMATION FOR HEALTH POLICY

The Evidence and Information for Health Policy group addresses the following areas of work: evidence for health policy; health information management and dissemination; and research policy and promotion and organization of health systems. Through the generation and dissemination of evidence the Evidence and Information for Health Policy group aims to assist policy-makers assess health needs, choose intervention strategies, design policy and monitor performance, and thereby improve the performance of national health systems. The group also supports international and national dialogue on health policy.

WHO co-ordinates the Health InterNetwork Access to Research Initiative (HINARI), which was launched in July 2001 to enable relevant authorities in developing countries to access more than 2,000 biomedical journals through the internet at no or greatly reduced cost, in order to improve the world-wide circulation of scientific information; some 28 medical publishers participate in the initiative.

Finance

WHO's regular budget is provided by assessment of member states and associate members. An additional fund for specific projects is provided by voluntary contributions from members and other sources, including UNDP and UNFPA.

A regular budget of US $901.5m. was proposed for 2004–05, of which some 11.1%, or $96.2m., was provisionally allocated to South-East Asia and 8.5% ($74.1m.) to the Western Pacific.

Publications

Action against Infection (newsletter).

Bulletin of WHO (monthly).

Environmental Health Criteria.

International Digest of Health Legislation (quarterly).

International Classification of Functioning, Disability and Health—ICF.

International Statistical Classification of Diseases and Related Health Problems (Tenth Revision, 1992–1994, versions in 37 languages).

Model List of Essential Drugs (biennially).

Weekly Epidemiological Record.

WHO Drug Information (quarterly).

WHO Model Formulary.

World Health Report (annually).

World Health Statistics Annual.

Technical report series; catalogues of specific scientific, technical and medical fields available.

Other UN Organizations Active in the Region

OFFICE FOR THE CO-ORDINATION OF HUMANITARIAN AFFAIRS—OCHA

Address: United Nations Plaza, New York, NY 10017, USA.

Telephone: (212) 963-1234; **fax:** (212) 963-1312; **e-mail:** ochany@un.org; **internet:** ochaonline.un.org.

The Office was established in January 1998 as part of the UN Secretariat, with a mandate to co-ordinate international humanitarian assistance and to provide policy and other advice on humanitarian issues. It administers the Humanitarian Early Warning System, as well as Integrated Regional Information Networks (IRIN) to monitor the situation in different countries and a Disaster Response System. A complementary service, Reliefweb, which was launched in 1996, monitors crises and publishes information on the internet.

Under-Secretary-General for Humanitarian Affairs and Emergency Relief Co-ordinator: JAN EGELAND (Norway).

UNITED NATIONS OFFICE ON DRUGS AND CRIME—UNODC

Address: Vienna International Centre, POB 500, 1400 Vienna, Austria.

Telephone: (1) 26060-0; **fax:** (1) 26060-5866; **e-mail:** unodc@unodc.org; **internet:** www.unodc.org.

The Office was established in November 1997 (as the UN Office of Drug Control and Crime Prevention) to strengthen the UN's integrated approach to issues relating to drug control, crime prevention and international terrorism. It comprises two principal components: the United Nations Drug Programme and the Crime Programme.

Executive Director: ANTONIO MARIA COSTA (Italy).

OFFICE OF THE UNITED NATIONS HIGH COMMISSIONER FOR HUMAN RIGHTS—OHCHR

Address: Palais Wilson, 52 rue de Paquis, 1201 Geneva, Switzerland.

Telephone: (22) 9179290; **fax:** (22) 9179022; **e-mail:** infodesk@ohchr.org; **internet:** www.ohchr.org.

The Office is a body of the UN Secretariat and is the focal point for UN human-rights activities. Since September 1997 it has incorporated the Centre for Human Rights. The High Commissioner is the UN official with principal responsibility for UN human rights activities.

High Commissioner: LOUISE ARBOUR (Canada).

UNITED NATIONS HUMAN SETTLEMENTS PROGRAMME—UN-Habitat

Address: POB 30030, Nairobi, Kenya.

Telephone: (20) 621234; **fax:** (20) 624266; **e-mail:** infohabitat@unhabitat.org; **internet:** www.unhabitat.org.

UN-Habitat was established, as the United Nations Centre for Human Settlements, in October 1978 to service the intergovernmental Commission on Human Settlements. It became a full UN programme on 1 January 2002, serving as the focus for human settlements activities in the UN system.

Executive Director: ANNA KAJUMULO TIBAIJUKA (Tanzania).

UNITED NATIONS CONFERENCE ON TRADE AND DEVELOPMENT—UNCTAD

Address: Palais des Nations, 1211 Geneva 10, Switzerland.

Telephone: (22) 9171234; **fax:** (22) 9070043; **e-mail:** info@unctad.org; **internet:** www.unctad.org.

UNCTAD was established in 1964. It is the principal organ of the UN General Assembly concerned with trade and development, and is the focal point within the UN system for integrated activities relating to trade, finance, technology, investment and sustainable development. It aims to maximize the trade and development opportunities of developing countries, in particular least-developed countries, and to assist them to adapt to the increasing globalization and liberalization of the world economy. UNCTAD undertakes consensus-building activities, research and policy analysis and technical co-operation.

Secretary-General: RUBENS RICÚPERO (Brazil).

UNITED NATIONS POPULATION FUND—UNFPA

Address: 220 East 42nd St, New York, NY 10017, USA.

Telephone: (212) 297-5020; **fax:** (212) 297-4911; **internet:** www.unfpa.org.

Created in 1967 as the Trust Fund for Population Activities, the UN Fund for Population Activities (UNFPA) was established as a Fund of the UN General Assembly in 1972 and was made a subsidiary organ of the UN General Assembly in 1979, with the UNDP Governing Council (now the Executive Board) designated as its governing body. In 1987 UNFPA's name was changed to the United Nations Population Fund (retaining the same acronym).

Executive Director: THORAYA A. OBAID (Saudi Arabia).

UN Specialized Agencies

INTERNATIONAL ATOMIC ENERGY AGENCY—IAEA

Address: POB 100, Wagramerstrasse 5, 1400 Vienna, Austria.

Telephone: (1) 26000; **fax:** (1) 26007; **e-mail:** official.mail@iaea.org; **internet:** www.iaea.org.

The Agency was founded in 1957 as an autonomous intergovernmental organization, although it is administratively part of the UN system and reports annually to the UN General Assembly. Its main objectives are to enlarge the contribution of atomic energy to peace, health and prosperity throughout the world, and to ensure that materials and services provided by the Agency are not used to further any military purpose.

Director-General: Dr MOHAMMAD EL-BARADEI (Egypt).

INTERNATIONAL CIVIL AVIATION ORGANIZATION—ICAO

Address: 999 University St, Montréal, QC H3C 5H7, Canada.

Telephone: (514) 954-8219; **fax:** (514) 954-6077; **e-mail:** icaohq@icao.org; **internet:** www.icao.int.

ICAO was founded in 1947, on the basis of the Convention on International Civil Aviation, signed in Chicago, in 1944, to develop the techniques of international air navigation and to help in the planning and improvement of international air transport.

Secretary-General: TAÏEB CHÉRIF (Algeria).

Regional Office for Asia and the Pacific: POB 11, Samyaek Ladprao, Bangkok 10900, Thailand; tel. (2) 537-8189; fax (2) 537-

8199; e-mail icao_apac@bangkok.icao.int; internet www.icao.int/apac; Dir LALIT B. SHAH.

INTERNATIONAL LABOUR ORGANIZATION—ILO

Address: 4 route des Morillons, 1211 Geneva 22, Switzerland.

Telephone: (22) 7996111; **fax:** (22) 7988685; **e-mail:** ilo@ilo.org; **internet:** www.ilo.org.

ILO was founded in 1919 to work for social justice as a basis for lasting peace. It carries out this mandate by promoting decent living standards, satisfactory conditions of work and pay and adequate employment opportunities. Methods of action include the creation of international labour standards; the provision of technical co-operation services; and training, education, research and publishing activities to advance ILO objectives.

Director-General: JUAN O. SOMAVÍA (Chile).

Regional Office for Asia and the Pacific: POB 1759, Bangkok 2, Thailand; tel. (2) 288-1710; fax (2) 288-3062; e-mail bangkok@ilo.org.

INTERNATIONAL MARITIME ORGANIZATION—IMO

Address: 4 Albert Embankment, London, SE1 7SR, United Kingdom.

Telephone: (20) 7735-7611; **fax:** (20) 7587-3210; **e-mail:** info@imo.org; **internet:** www.imo.org.

The Inter-Governmental Maritime Consultative Organization (IMCO) began operations in 1959, as a specialized agency of the UN to facilitate co-operation among governments on technical matters affecting international shipping. Its main aims are to improve the safety of international shipping, and to prevent pollution caused by ships. IMCO became IMO in 1982.

Secretary-General: EFTHIMIOS MITROPOULOS (Greece).

INTERNATIONAL TELECOMMUNICATION UNION—ITU

Address: Place des Nations, 1211 Geneva 20, Switzerland.

Telephone: (22) 7305111; **fax:** (22) 7337256; **e-mail:** itumail@itu.int; **internet:** www.itu.int.

Founded in 1865, ITU became a specialized agency of the UN in 1947. It acts to encourage world co-operation for the improvement and use of telecommunications, to promote technical development, to harmonize national policies in the field, and to promote the extension of telecommunications throughout the world.

Secretary-General: YOSHIO UTSUMI (Japan).

UNITED NATIONS INDUSTRIAL DEVELOPMENT ORGANIZATION—UNIDO

Address: Vienna International Centre, POB 300, 1400 Vienna, Austria.

Telephone: (1) 260260; **fax:** (1) 2692669; **e-mail:** unido@unido.org; **internet:** www.unido.org.

UNIDO began operations in 1967 and became a specialized agency in 1985. Its objectives are to promote sustainable and socially equitable industrial development in developing countries and in countries with economies in transition. It aims to assist such countries to integrate fully into global economic system by mobilizing knowledge, skills, information and technology to promote productive employment, competitive economies and sound environment.

Director-General: CARLOS ALFREDO MAGARIÑOS (Argentina).

UNIVERSAL POSTAL UNION—UPU

Address: Weltpoststr., 3000 Berne 15, Switzerland.

Telephone: (31) 3503111; **fax:** (31) 3503110; **e-mail:** info@upu.int; **internet:** www.upu.int.

The General Postal Union was founded by the Treaty of Berne (1874), beginning operations in July 1875. Three years later its name was changed to the Universal Postal Union. In 1948 UPU became a specialized agency of the UN. It aims to develop and unify the international postal service, to study problems and to provide training.

Director-General: EDOUARD DAYAN (France) (from 1 Jan. 2005).

WORLD INTELLECTUAL PROPERTY ORGANIZATION—WIPO

Address: 34 chemin des Colombettes, 1211 Geneva 20, Switzerland.

Telephone: (22) 3389111; **fax:** (22) 7335428; **e-mail:** wipo.mail@wipo.int; **internet:** www.wipo.int.

WIPO was established in 1970. It became a specialized agency of the UN in 1974 concerned with the protection of intellectual property (e.g. industrial and technical patents and literary copyrights) throughout the world. WIPO formulates and administers treaties embodying international norms and standards of intellectual property, establishes model laws, and facilitates applications for the protection of inventions, trademarks etc. WIPO provides legal and technical assistance to developing countries and countries with economies in transition and advises countries on obligations under the World Trade Organization's agreement on Trade-Related Aspects of Intellectual Property Rights (TRIPS).

Director-General: Dr KAMIL IDRIS (Sudan).

WORLD METEOROLOGICAL ORGANIZATION—WMO

Address: 7 bis, ave de la Paix, 1211 Geneva 2, Switzerland.

Telephone: (22) 7308111; **fax:** (22) 7308181; **e-mail:** ipa@wmo.int; **internet:** www.wmo.int.

WMO was established in 1950 and was recognized as a Specialized Agency of the UN in 1951, aiming to improve the exchange of information in the fields of meteorology, climatology, operational hydrology and related fields, as well as their applications. WMO jointly implements, with UNEP, the UN Framework Convention on Climate Change.

Secretary-General: MICHEL JARRAUD (France).

WORLD TOURISM ORGANIZATION

Address: Capitán Haya 42, 28020 Madrid, Spain.

Telephone: (91) 5678100; **fax:** (91) 5713733; **e-mail:** omt@world-tourism.org; **internet:** www.world-tourism.org.

The World Tourism Organization was established in 1975 and was recognized as a Specialized Agency of the UN in December 2003. It works to promote and develop sustainable tourism, in particular in support of socio-economic growth in developing countries.

Secretary-General: FRANCESCO FRANGIALLI (France).

ASIA-PACIFIC ECONOMIC CO-OPERATION—APEC

Address: 35 Heng Mui Keng Terrace, Singapore 119616.

Telephone: 67756012; **fax:** 67756013; **e-mail:** info@apec.org; **internet:** www.apec.org.

Asia-Pacific Economic Co-operation (APEC) was initiated in November 1989, in Canberra, Australia, as an informal consultative forum. Its aim is to promote multilateral economic co-operation on issues of trade and investment.

MEMBERS

Australia	Japan	Philippines
Brunei	Korea, Republic	Russia
Canada	Malaysia	Singapore
Chile	Mexico	Taiwan*
China, People's Republic	New Zealand	Thailand
Hong Kong	Papua New Guinea	USA
Indonesia	Peru	Viet Nam

*Admitted as Chinese Taipei.

Note: APEC has three official observers: the Association of South East Asian Nations Secretariat; the Pacific Economic Co-operation Council; and the Pacific Islands Forum Secretariat (formerly the South Pacific Forum Secretariat). Observers may participate in APEC meetings and have full access to all related documents and information.

Organization

(October 2004)

ECONOMIC LEADERS' MEETINGS

The first meeting of APEC heads of government was convened in November 1993, in Seattle, Washington, USA. Subsequently, each annual meeting of APEC ministers of foreign affairs and of economic affairs has been followed by an informal gathering of the leaders of the APEC economies, at which the policy objectives of the grouping are discussed and defined. The 11th Economic Leaders' Meeting was held in October 2003, in Bangkok, Thailand, and the 12th Meeting was scheduled to be held in November 2004 in Santiago, Chile.

MINISTERIAL MEETINGS

APEC ministers of foreign affairs and ministers of economic affairs meet annually. These meetings are hosted by the APEC Chair, which rotates each year, although it was agreed, in 1989, that alternate Ministerial Meetings were to be convened in an ASEAN member country. A Senior Officials' Meeting (SOM) convenes regularly between Ministerial Meetings to co-ordinate and administer the budgets and work programmes of APEC's committees and working groups. Other meetings of ministers are held on a regular basis to enhance co-operation in specific areas.

SECRETARIAT

In 1992 the Ministerial Meeting, held in Bangkok, Thailand, agreed to establish a permanent secretariat to support APEC activities, and approved an annual budget of US $2m. The Secretariat became operational in February 1993. The Executive Director is appointed from the member economy chairing the group and serves a one-year term. A Deputy Executive Director is appointed by the member economy designated to chair APEC in the following year.

Executive Director: Mario Artaza (Chile).

Deputy Executive Director: Choi Seok Young (Republic of Korea).

COMMITTEES AND GROUPS

Budget and Management Committee—BMC: f. 1993 as Budget and Administrative Committee, present name adopted 1998; advises APEC senior officials on budgetary, administrative and managerial issues. The Committee reviews the operational budgets of APEC committees and groups, evaluates their effectiveness and conducts assessments of group projects.

Committee on Trade and Investment—CTI: f. 1993 on the basis of a Declaration signed by ministers meeting in Seattle, Washington, USA, in order to facilitate the expansion of trade and the development of a liberalized environment for investment among member countries; undertakes initiatives to improve the flow of goods, services and technology in the region. An Investment Experts' Group was established in 1994, initially to develop non-binding investment principles. In May 1997 an APEC Tariff Database was inaugurated, with sponsorship from the private sector. A new Market Access Group was established in 1998 to administer CTI activities concerned with non-tariff measures. In 2001 the CTI finalized a set of non-binding Principles on Trade Facilitation. The development of the nine principles was intended to help eliminate procedural and administrative impediments to trade and to increase trading opportunities. A Trade Facilitation Action Plan was approved in 2002. In 2003 a Transparency By 2005 strategy was adopted to incorporate transparency standards.

Economic Committee—EC: f. 1994 following an agreement, in November, to transform the existing *ad hoc* group on economic trends and issues into a formal committee; aims to enhance APEC's capacity to analyse economic trends and to research and report on issues affecting economic and technical co-operation in the region. In addition, the Committee is considering the environmental and development implications of expanding population and economic growth.

SOM Committee on Economic and Technical Co-operation—ESC: f. 1998 to assist the SOM with the co-ordination of APEC's economic and technical co-operation programme (ECOTECH); monitors and evaluates project implementation and also identifies initiatives designed to strengthen economic and technical co-operation in infrastructure. The ESC co-ordinated the development of APEC's Human-Capacity Building Strategy in 2001.

In addition, the following Working Groups promote and co-ordinate practical co-operation between member countries in different activities: Agricultural technical co-operation; Energy; Fisheries; Human resources development; Industrial science and technology; Marine resource conservation; Small and medium enterprises (SMEs); Telecommunications and Information; Tourism; Trade promotion; and Transportation. (See below for more detailed information.)

ADVISORY COUNCIL

APEC Business Advisory Council (ABAC): Philamlife Tower, 43rd Floor, 8767 Paseo de Roxas, Makati City 1226, Philippines; tel. (2) 8454564; fax (2) 8454832; e-mail abacsec@pfgc.ph; internet www.abaconline.org; an agreement to establish ABAC, comprising up to three senior representatives of the private sector from each APEC member economy, was concluded at the Ministerial Meeting held in November 1995. ABAC was mandated to advise member states on the implementation of APEC's Action Agenda and on other business matters, and to provide business-related information to APEC fora. ABAC meets three or four times each year and holds a dialogue with APEC economic leaders prior to their annual informal meeting. ABAC's first meeting, convened in June 1996 in Manila, the Philippines, resolved to accelerate the liberalization of regional trade. In 1998 ABAC focused on measures to alleviate the effects of the financial crisis in Asia, in particular, by enhancing confidence in the private sector, as well as efforts to support SMEs, to develop electronic commerce in the region and to advise on APEC Individual Action Plans (IAPs, see below). In 2000 ABAC addressed the relevance of APEC to the challenges of globalization and, in its annual report to APEC leaders, issued several recommendations, including support for a new round of multilateral trade negotiations; enhancement of the IAP process, with increased implementation of electronic IAPs (e-IAPS); implementation of a proposed food system for member states; the establishment of an Institute of Directors Network; and the adoption of a regulatory framework conducive to the development of e-commerce. In 2001 ABAC concentrated on the challenges posed by globalization; the impact on APEC members of the global economic slowdown; and capacity-building in financial systems. ABAC's declared theme for 2002 was Sharing Development to Reinforce Global Security. At a meeting in February the Council considered the relationship between global security and its work to facilitate trade and investment flows. In April ABAC organized a Business on Alert Workshop, which addressed security issues in the business sector. In 2003 ABAC called on APEC to reaffirm its commitment to achieving the Bogor Declaration objectives (see below) and advancing the World Trade Organization's Doha Development Agenda (see below); Exec. Dir Rene Muga.

Activities

APEC was initiated in 1989 as a forum for informal discussion between the then six ASEAN members and their six dialogue partners in the Pacific, and, in particular, to promote trade liberalization in the Uruguay Round of negotiations, which were being conducted under the General Agreement on Tariffs and Trade

(GATT). The Seoul Declaration, adopted by ministers meeting in the Republic of Korea in November 1991, defined the objectives of APEC.

ASEAN countries were initially reluctant to support any more formal structure of the forum, or to admit new members, owing to concerns that it would undermine ASEAN's standing as a regional grouping and be dominated by powerful non-ASEAN economies. In August 1991 it was agreed to extend membership to the People's Republic of China, Hong Kong and Taiwan (subject to conditions imposed by the People's Republic of China, including that a Taiwanese official of no higher than vice-ministerial level should attend the annual meeting of ministers of foreign affairs). Mexico and Papua New Guinea acceded to the organization in November 1993, and Chile joined in November 1994. The summit meeting held in November 1997 agreed that Peru, Russia and Viet Nam should be admitted to APEC at the 1998 meeting, but imposed a 10-year moratorium on further expansion of the grouping.

In September 1992 APEC ministers agreed to establish a permanent secretariat. In addition, the meeting created an 11-member non-governmental Eminent Persons Group (EPG), which was to assess trade patterns within the region and propose measures to promote co-operation. At the Ministerial Meeting in Seattle, Washington, USA, in November 1993, members agreed on a framework for expanding trade and investment among member countries, and to establish a permanent committee (the CTI, see above) to pursue these objectives.

In August 1994 the EPG proposed the following timetable for the liberalization of all trade across the Asia-Pacific region: negotiations for the elimination of trade barriers were to commence in 2000 and be completed within 10 years in developed countries, 15 years in newly-industrialized economies and by 2020 in developing countries. Trade concessions could then be extended on a reciprocal basis to non-members in order to encourage world-wide trade liberalization, rather than isolate APEC as a unique trading bloc. In November 1994 the meeting of APEC heads of government adopted the Bogor Declaration of Common Resolve, which endorsed the EPG's timetable for free and open trade and investment in the region by the year 2020. Other issues incorporated into the Declaration included the implementation of GATT commitments in full and strengthening the multilateral trading system through the forthcoming establishment of the World Trade Organization (WTO), intensifying development co-operation in the Asia-Pacific region and expanding and accelerating trade and investment programmes.

During 1995 meetings of APEC officials and other efforts to substantiate the trade liberalization agreement revealed certain differences among members regarding the timetable and means of implementing the measures, which were to be agreed upon at the 1995 Economic Leaders' Meeting. The principal concern, expressed notably by the USA, focused on whether tariff reductions were to be achieved by individual trade liberalization plans or based on some reciprocal or common approach. In August the EPG issued a report, to be considered at the November Leaders' Meeting, which advocated acceleration of tariff reductions and other trade liberalization measures agreed under GATT; the establishment of a dispute mediation service to reduce and settle regional trade conflicts; and a review of new trade groupings within the APEC region. Further proposals for the implementation of the Bogor Declaration objectives were presented, in September, by the Pacific Business Forum, comprising APEC business representatives. The recommendations included harmonization of product quality, the establishment of one-stop investment agencies in each APEC country, training and technology transfers and the implementation of visa-free business travel by 1999. In November 1995 the Ministerial Meeting decided to dismantle the EPG, and to establish an APEC Business Advisory Council (ABAC), consisting of private-sector representatives.

In November 1995 APEC heads of government, meeting in Osaka, Japan, adopted an Action Agenda as a framework to achieve the commitments of the Bogor Declaration. Part One of the Agenda identified action areas for the liberalization of trade and investment and the facilitation of business, for example, customs procedures, rules of origin and non-tariff barriers. It incorporated agreements that the process was to be comprehensive, consistent with WTO commitments, comparable among all APEC economies and non-discriminatory. Each member economy was to ensure the transparency of its laws, regulations and procedures affecting the flow of goods, services and capital among APEC economies and to refrain from implementing any trade protection measures. A second part of the Agenda was to provide a framework for further economic and technical co-operation between APEC members in areas such as energy, transport, infrastructure, SMEs and agricultural technology. In order to resolve a disagreement concerning the inclusion of agricultural products in the trade liberalization process, a provision for flexibility was incorporated into the Agenda, taking into account diverse circumstances and different levels of development in APEC member economies. Liberalization measures were to be implemented from January 1997 (i.e. three years earlier than previously agreed) and were to be subject to annual reviews. A Trade and Investment Liberalization and Facilitation Special Account was established to finance projects in support of the implementation of the Osaka Action Agenda. In May 1996 APEC senior officials met in Cebu, the Philippines, to review Individual Action Plans—IAPs, annual reports submitted by each member state on progress in the implementation of trade liberalization measures—and to achieve some coherent approach to tariff liberalization prior to the Leaders' Meeting in November.

In November 1996 the Economic Leaders' Meeting, held in Subic Bay, the Philippines, approved the Manila Action Plan for APEC (MAPA), which had been formulated at the preceding Ministerial Meeting, held in Manila. MAPA incorporated the IAPs and other collective measures aimed at achieving the trade liberalization and co-operation objectives of the Bogor Declaration, as well as the joint activities specified in the second part of the Osaka Agenda. Heads of government also endorsed a US proposal to eliminate tariffs and other barriers to trade in information technology products by 2000 and determined to support efforts to conclude an agreement to this effect at the forthcoming WTO conference; however, they insisted on the provision of an element of flexibility in achieving trade liberalization in this sector.

The 1997 Economic Leaders' Meeting, held in Vancouver, Canada, in November, was dominated by concern at the financial instability that had affected several Asian economies during 1997. The final declaration of the summit meeting endorsed a framework of measures that had been agreed by APEC deputy ministers of finance and central bank governors at an emergency meeting convened in the previous week in Manila, the Philippines (the so-called Manila Framework for Enhanced Asian Regional Co-operation to Promote Financial Stability). The meeting, attended by representatives of the IMF, the World Bank and the Asian Development Bank, committed all member economies receiving IMF assistance to undertake specified economic and financial reforms, and supported the establishment of a separate Asian funding facility to supplement international financial assistance (although this was later rejected by the IMF). APEC ministers of finance and governors of central banks were urged to accelerate efforts for the development of the region's financial and capital markets and to liberalize capital flows in the region. Measures were to include strengthening financial market supervision and clearing and settlement infrastructure, the reform of pension systems, and promoting co-operation among export credit agencies and financing institutions. The principal item on the Vancouver summit agenda was an initiative to enhance trade liberalization, which, the grouping insisted, should not be undermined by the financial instability in Asia. The following 15 economic sectors were identified for 'early voluntary sectoral liberalization' ('EVSL'): environmental goods and services; fish and fish products; forest products; medical equipment and instruments; toys; energy; chemicals; gems and jewellery; telecommunications; oilseeds and oilseed products; food; natural and synthetic rubber; fertilizers; automobiles; and civil aircraft. The implementation of EVSL was to encompass market opening, trade facilitation, and economic and technical co-operation activities. The heads of government subsequently requested the authorities in each member state to formulate details of tariff reductions in these sectors by mid-1998, with a view to implementing the measures in 1999. (In June 1998, however, ministers of trade, meeting in Malaysia, failed to conclude an agreement on early tariff reductions, in part owing to Japan's reluctance to liberalize trade in fish and forest products.) In Vancouver APEC Economic Leaders also declared their support for an agreement to liberalize financial services (which was successfully negotiated under the auspices of the WTO in December 1997) and for the objective of reducing the emission of 'greenhouse gases', which was under consideration at a global conference, held in Kyoto, Japan, in December (resulting in the adoption of the Kyoto Protocol to the UN Framework Convention on Climate Change).

In May 1998 APEC finance ministers met in Canada to consider the ongoing financial and economic crisis in Asia and to review progress in implementing efforts to alleviate the difficulties experienced by several member economies. The ministers agreed to pursue activities in the following three priority areas: capital market development, capital account liberalization and strengthening financial systems (including corporate governance). The region's economic difficulties remained the principal topic of discussion at the Economic Leaders' Meeting held in Kuala Lumpur, Malaysia, in November. A final declaration reiterated their commitment to co-operation in pursuit of sustainable economic recovery and growth, in particular through the restructuring of financial and corporate sectors, promoting and facilitating private-sector capital flows, and efforts to strengthen the global financial system. The meeting endorsed a proposal of ABAC to establish a partnership for equitable growth, with the aim of enhancing business involvement in APEC's programme of economic and technical co-operation. Other initiatives approved included an Agenda of APEC Science and Technology Industry Co-operation into the 21st Century (for which the People's Republic of China announced it was to establish a special fund), and an Action Programme on Skills and Development in APEC. Japan's persisting opposition to a reduction of tariffs in the fish and forestry sectors again prevented the conclusion of tariff negotiations under the EVSL scheme, and it was therefore agreed that responsibility for managing the tariff reduction element of the initiative should be transferred to the WTO. The meeting was divided by political differences regarding human rights, and in particular, the treatment by the Malaysian authorities of the imprisoned former Deputy Prime Minister, Anwar Ibrahim. A declaration of support for the democratic reform movement in Malaysia by the US representative dominated discussions at the start of the summit meeting and provoked a formal complaint from the Malaysian Government.

In September 1999 political dialogue regarding the civil conflict in East Timor (now Timor-Leste) dominated the start of the annual meetings of the grouping, held in Auckland, New Zealand, although the issue remained separate from the official agenda. Ministers of foreign affairs, convened in emergency session, declared their support for the establishment of a multinational force, under UN auspices, to restore peace in the territory and determined to provide humanitarian and technical assistance to facilitate the process of reconstruction and rehabilitation. The Economic Leaders' Meeting considered measures to sustain the economic recovery in Asia and endorsed the APEC Principles to Enhance Competition and Regulatory Reform (for example, transparency, accountability, non-discrimination) as a framework to strengthen APEC markets and to enable further integration and implementation of the IAPs. The meeting endorsed a report prepared during the year by an *ad hoc* task force concerning an ABAC proposal for the development of an APEC food system. Also under discussion was the forthcoming round of multilateral trade negotiations, to be initiated by the WTO. The heads of government proposed the objective of completing a single package of trade agreements within three years and endorsed the abolition of export subsidies for agricultural products. The meeting determined to support the efforts of the People's Republic of China, Russia, Taiwan and Viet Nam to accede to WTO membership.

The Economic Leaders' Meeting for 2000, held in Brunei in November, urged that an agenda for the now-stalled round of multilateral trade negotiations should be formulated without further delay. The meeting endorsed a plan of action to promote the utilization of advances in information and communications technologies in member economies, for the benefit of all citizens. It adopted the aim of tripling the number of people in the region with access to the internet by 2005, and determined to co-operate with business and education sector interests to attract investment and expertise in the pursuit of this goal. A proposal that the Democratic People's Republic of Korea be permitted to participate in APEC working groups was approved at the meeting.

The 2001 Economic Leaders' Meeting, held in October, in Shanghai, People's Republic of China, condemned the terrorist attacks against targets in the USA of the previous month and resolved to take action to combat the threat of international terrorism. The heads of government declared terrorism to be a direct challenge to APEC's vision of free, open and prosperous economies, and concluded that the threat made the continuing move to free trade, with its aim of bolstering economies, increasing prosperity and encouraging integration, even more of a priority. The meeting emphasized the importance of sharing the benefits of globalization. The leaders also expressed their determination to address the effects on APEC countries of the prevailing global economic downturn, advocating timely policy actions in the coming year to rebuild confidence and boost growth. Human capacity-building was a central theme. The meeting adopted the Shanghai Accord, which identified development goals for APEC during its second decade and clarified measures for achieving the Bogor goals within the agreed timetable. The meeting also outlined the e-APEC Strategy developed by the e-APEC Task Force established after the Brunei Economic Leaders' meeting. Considering issues of entrepreneurship, structural and regulatory reform, competition, intellectual property rights and information security, the strategy aimed to facilitate technological development in the region. Finally, the meeting adopted a strategy document relating to infectious diseases in the Asia Pacific region, which aimed to promote a co-ordinated response to combating HIV/AIDS and other infectious diseases.

In September 2002 a meeting of APEC ministers of finance was held in Los Cabos, Mexico. Ministers discussed the importance of efforts to combat money-laundering and the financing of terrorism. The meeting also focused on ways to strengthen global and regional economic growth, to advance fiscal and financial reforms and to improve the allocation of domestic savings for economic development. The theme of the 2002 Economic Leaders' Meeting, held in the following month in Los Cabos, Mexico, was 'Expanding the Benefits of Co-operation for Economic Growth and Development—Implementing the Vision'. The meeting issued a statement on the implementation of APEC standards of transparency in trade and investment liberalization and facilitation. Leaders also issued a statement on fighting terrorism and promoting growth. In February 2003 a Counter Terrorism Task Force was established to co-ordinate implementation of the leaders' statement. In the same month the first conference to promote the Secure Trade in the APEC Region (STAR) initiative was convened in Bangkok, Thailand, and attended by representatives of all APEC member economies as well as senior officers of private sector companies and relevant international organizations. A second STAR conference was held in March 2004, in Chile.

In September 2003 APEC finance ministers, meeting in Phuket, Thailand, emphasized the importance of the following in achieving their goal of sustainable growth in the APEC region: good governance within the private and public sectors, stable and efficient financial markets, greater economic integration and openness among member economies, and heightened vigilance towards money-laundering the financing of terrorism. The 2003 Economic Leaders' Meeting, convened in mid-October, in Bangkok, Thailand, considered means of advancing the WTO's stalled Doha round of trade negotiations, emphasizing the central importance of its development dimension, and noted progress made hitherto in facilitating intra-APEC trade. The meeting also addressed regional security issues, reiterating the Community's commitment to ensuring the resilience of APEC economies against the threat of terrorism. The Leaders adopted the Bangkok Declaration on Partnership for the Future which identified the following areas as priority concerns for the group: the promotion of trade and investment liberalization; enhancing human security; and helping people and societies to benefit from globalization. The Bangkok meeting also issued a statement on health security, which expressed APEC's determination to strengthen infrastructure for the detection and prevention of infectious diseases, as well as the surveillance of other threats to public health, and to ensure a co-ordinated response to public health emergencies, with particular concern to the outbreak, earlier in the year, of Severe Acute Respiratory Syndrome (SARS). Health security activities were to be implemented by a newly-established *ad hoc* Task Force, which subsequently convened for the first time in Taiwan (Chinese Taipei), in April 2004.

WORKING GROUPS

APEC's structure of working groups aims to promote practical and technical co-operation in specific areas, and to help implement individual and collective action plans in response to the directives of the Economic Leaders and meetings of relevant ministers. APEC recognizes sustainable development as a key issue cross-cutting all forum activities. In 1997 APEC leaders declared their commitment to the integration of women into the mainstream of APEC activities. A Gender Focal Point Network (GFPN) was established in May 2003 as a mechanism for the implementation of the Framework for the Integration of Women in APEC.

Agricultural Technical Co-operation: Formally established as an APEC Expert's Group in 1996, and incorporated into the system of working groups in 2000. The group aims to enhance the role of agriculture in the economic growth of the region and to promote co-operation in the following areas: conservation and utilization of plant and animal genetic resources; research, development and extension of agricultural biotechnology; processing, marketing, distribution and consumption of agricultural products; plant and animal quarantine and pest management; development of an agricultural finance system; sustainable agriculture; and agricultural technology transfer and training. The group has primary responsibility for undertaking recommendations connected with the implementation of the proposed APEC food system. The group has conducted projects on human resource development in post-harvest technology and on capacity-building, safety assessment and communication in biotechnology.

Energy: Responsible for the development of the energy component of the 1995 Action Agenda. APEC ministers responsible for energy convened for the first time in 1996 to discuss major energy challenges confronting the region and to provide guidance for the working group. The group's main objectives were determined as: the enhancement of regional energy security and improvement of the fuel supply market for the power sector; the development and implementation of programmes of work promoting the adoption of environmentally sound energy technologies and promoting private-sector investment in regional power infrastructure; the development of energy efficiency guide-lines; and the standardization of testing facilities and results. In March 1999 the group resolved to establish a business network to improve relations and communications with the private sector. The first meeting of the network took place in April. In October 1998 APEC energy ministers, meeting in Okinawa, Japan, emphasized the role of the energy sector in stimulating economic activity and stressed the need to develop infrastructure, improve energy efficiency and accelerate the development of natural gas reserves. In May 2000 ministers meeting in San Diego, California, USA, launched the APEC 21st Century Renewable Energy Initiative, which aims to encourage co-operation in and advance the utilization of renewable energy technologies, envisaging the establishment of a Private Sector Renewable Energy Forum. In July 2002 APEC energy ministers, convened in Mexico City, proposed several initiatives, including the cross-border interconnection of natural gas pipelines and improvements in energy security. Meeting in Portland, Oregon, USA, in June 2003 ministers agreed on a framework to implement APEC's Energy Security Initiative. The first meeting of ministers responsible for mining was convened in Santiago, Chile, in June 2004.

Fisheries: Aims to maximize the economic benefits and sustainability of fisheries resources for all APEC members. Recent concerns include food safety, the quality of fish products and resource management. In 1996 the group initiated a four-year study on trade and investment liberalization in the sector, in the areas of tariffs, non-tariff barriers and investment measures and subsidies. In 1997 the group organized two technical workshops on seafood inspection systems, and conducted a workshop addressing destructive fishing techniques. The first APEC Aquaculture Forum, which considered the sustainable development of aquaculture in the region and the development of new markets for APEC fish products, was held in Taipei, Taiwan, in June 1998. In May 1999 new guide-lines were adopted to encourage the participation of the private sector in the activities of the working group. The group's first business forum was convened in July 2000. In April 2002 the first ocean-related ministerial meeting, held in Seoul, Republic of Korea, adopted the Seoul Oceans Declaration, detailing recommendations on marine environmental protection and integrated coastal management.

Human Resources Development: Established in 1990, comprising three networks: the Capacity Building Network, with a focus on human capacity-building, including management and technical skills development and corporate governance; the Education Network, promoting effective learning systems and supporting the role of education in advancing individual, social and economic development; and the Labour and Social Protection Network, concerned with promoting social integration through the strengthening of labour markets, the development of labour market information and policy, and improvements in working conditions and social safety net frameworks. The working group undertakes activities through these networks to implement ministerial and leaders' directives, as well as the 'Medium Term Strategic Priorities', which were formulated in January 1997. A voluntary network of APEC study centres links higher education and research institutions in member economies. In January 1998 the working group established a Task Force on the Human Resource and Social Impacts of the Financial Crisis. Private-sector participation in the working group has been strengthened by the establishment of a network of APEC senior executives responsible for human resources management. Recent initiatives have included a cyber-education co-operation project, a workshop on advanced risk management, training on the prevention and resolution of employment and labour disputes, and an educators' exchange programme on the use of information technology in education.

Industrial Science and Technology: Aims to contribute to sustainable development in the region, improve the availability of information, enhance human resources development in the sector, improve the business climate, promote policy dialogue and review and facilitate networks and partnerships. Accordingly, the group has helped to establish an APEC Virtual Centre for Environmental Technology Exchange in Japan; a Science and Technology Industrial Parks Network; an International Molecular Biology Network for the APEC Region; an APEC Centre for Technology Foresight, based in Thailand; and the APEC Science and Technology Web, an online database. During 1997 and 1998 the group formulated an APEC Action Framework on Emerging Infectious Diseases and developed an Emerging Infections Network (EINet), based at the University of Washington, Seattle, USA.

Marine Resource Conservation: Promotes initiatives within APEC to protect the marine environment and its resources. In 1996 a five-year project was initiated for the management of red tide and harmful algal blooms in the APEC region. An APEC Action Plan for Sustainability of the Marine Environment was adopted by ministers responsible for the environment, meeting in June 1997. The Plan aimed to promote regional co-operation, an integrated approach to coastal management, the prevention, reduction and control of marine pollution, and sustainable development. Efforts were also being undertaken to establish an Ocean Research Network of centres of excellence in the Pacific. Strategies to encourage private-sector participation in promoting the sustainable management of marine resources were endorsed by the group in June 2000. Four main themes were identified for the Action Plan in the 21st century: balancing coastal development and resource protection; ensuring sustainable fisheries and aquaculture; understanding and observing the oceans and seas; and promoting economic and technical co-operation in oceans management. The Seoul Oceans Declaration, endorsed in April 2002 (see above), established the direction of the group's future activities. A new strategic framework for the group was adopted in May 2004, at a meeting held in Puerto Varas, Chile.

Small and Medium Enterprises: The group was established in 1995, as the Ad Hoc Policy Level Group on Small and Medium Enterprises, with a temporary mandate to oversee all APEC activities relating to SMEs. It supported the establishment of an APEC Centre for Technical Exchange and Training for Small and Medium Enterprises, which was inaugurated at Los Baños, near Manila, the Philippines, in September 1996. A five-year action plan for SMEs was endorsed in 1998. The group was redesignated as a working group, with permanent status, in 2000. In 2000 and 2001 the group considered issues relating to globalization, innovation, human resource development, information technology and e-commerce, financing, and the forming of strategic alliances with other SMEs and larger firms. In August 2002 the group's action plan was revised to include an evaluation framework to assist APEC and member economies in identifying and analysing policy issues. In the same month a sub-group specialising in micro-enterprises was established. The first APEC Incubator Forum was held in July-August 2003, in Taiwan (Chinese Taipei), to promote new businesses and support their early development. During 2003 the Group undertook efforts to develop a special e-APEC Strategy for SMEs.

Telecommunications and Information: Incorporates four steering groups concerned with different aspects of the development and liberalization of the sector—Liberalization; Business facilitation; Development co-operation; and Human resource development. Activities are guided by directives of ministers responsible for telecommunications, who first met in 1995, in the Republic of Korea, and adopted a Seoul Declaration on Asia Pacific Information Infrastructure (APII). The second ministerial meeting, held in Gold Coast, Australia, in September 1996, adopted more detailed proposals for liberalization of the sector in member economies. In June 1998 ministers, meeting in Singapore, agreed to remove technical barriers to trade in telecommunications equipment (although Chile and New Zealand declined to sign up to the arrangement). At their fourth meeting, convened in May 2000 in Cancún, Mexico, telecommunications ministers approved a programme of action that included measures to bridge the 'digital divide' between developed and developing member economies, and adopted the APEC Principles on International Charging Arrangements for Internet Services and the APEC Principles of Interconnection. The fifth ministerial meeting, held in May 2002, issued a Statement on the Security of Information and Communications Infrastructures; a compendium of IT security standards has been disseminated in support of the Statement.

Tourism: Established in 1991, with the aim of promoting the long-term sustainability of the tourism industry, in both environmental and social terms. The group administers a Tourism Information Network and an APEC International Centre for Sustainable Tourism, which in mid-2003 comprised 11 member countries. In 1998 the group initiated a project to assess the impact of the Asian financial crisis on regional tourism and to identify strategies to counter any negative effects. The first meeting of APEC ministers of tourism, held in the Republic of Korea in July 2000, adopted the Seoul Declaration on the APEC Tourism Charter. The group's work plan is based on four policy goals inherent in the Seoul Declaration, namely, the removal of impediments to tourism business and investment; increased mobility of visitors and increased demand for tourism goods and services; sustainable management of tourism; and enhanced recognition of tourism as a vehicle for economic and social development. At a meeting of the working group in April 2001, APEC and the Pacific Asia Travel Association (PATA) adopted a Code for Sustainable Tourism. The Code is designed for adoption and implementation by a variety of tourism companies and government agencies. It urges members to conserve the natural environment, ecosystems and biodiversity; respect local traditions and cultures; conserve energy; reduce pollution and waste; and ensure that regular environmental audits are carried out. In November 2001 the working group considered the impact on tourism of the September terrorist attacks on the USA. The group noted that while the short-term negative effects could be substantial, a return to long-term growth could be expected in the following year. The group advocated work to improve the quality and timeliness of tourism data, to enable accurate assessment of the situation.

Trade Promotion: Aims to promote trade, as a key aspect of regional economic co-operation, through activities to enhance trade financing, skills and training, information and networking (for example, through the establishment of APECNet, providing information to the business community via the internet, accessible at www.apecnet.org.sg), and co-operation with the private sector and public agencies, including trade promotion organizations. Organizes an APEC International Trade Fair.

Transportation: Undertakes initiatives to enhance the efficiency and safety of the regional transportation system, in order to facilitate the development of trade. The working group focuses on three main areas: improving the competitiveness of the transportation industry; promoting a safe and environmentally-sound regional transportation system; and human resources development, including training, research and education. The group has published surveys, directories and manuals on all types of transportation systems, and has compiled an inventory on regional co-operation on oil spills preparedness and response arrangements. A Road Transportation Harmonization Project aims to provide the basis for common standards in the automotive industry in the Asia-Pacific region. The group has established an internet database on ports and the internet-based Virtual Centre for Transportation Research, Development and Education. It plans to develop a regional action plan on the implementation of Global Navigation Satellite Systems, in consultation with the relevant international bodies. In 2003 the group was involved in discussions on means of enhancing security throughout international and domestic supply chains in the region while facilitating cross-border movement of legitimate commerce. The group was also responsible for overseeing various projects in respect of counter-terrorism. A Special Task Force was established to assist member economies to implement a new International Ship and Port Facility Security Code, sponsored by the International Maritime Organization, which entered into force on 1 July 2004. In April 2004 an Aviation Safety Experts' Group met for the first time since 2000.

Publications

ABAC Report to APEC Leaders (annually).

APEC Business Travel Handbook.

APEC Economic Outlook (annually).

APEC Economies Beyond the Asian Crisis.

APEC Energy Handbook (annually).

APEC Energy Statistics (annually).

Foreign Direct Investment and APEC Economic Integration (irregular).

Guide to the Investment Regimes of the APEC Member Economies.

Key APEC Documents (annually).

Towards Knowledge-based Economies in APEC.

Trade and Investment Liberalization in APEC.

Working group reports, regional directories, other irregular surveys.

ASIAN DEVELOPMENT BANK—ADB

Address: 6 ADB Ave, Mandaluyong City, 0401 Metro Manila, Philippines; POB 789, 0980 Manila, Philippines.

Telephone: (2) 6324444; **fax:** (2) 6362444; **e-mail:** information@adb.org; **internet:** www.adb.org.

The ADB commenced operations in December 1966. The Bank's principal functions are to provide loans and equity investments for the economic and social advancement of its developing member countries, to give technical assistance for the preparation and implementation of development projects and programmes and advisory services, to promote investment of public and private capital for development purposes, and to respond to requests from developing member countries for assistance in the co-ordination of their development policies and plans.

MEMBERS

There are 45 member countries and territories within the ESCAP region and 18 others (see list of subscriptions below).

Organization

(October 2004)

BOARD OF GOVERNORS

All powers of the Bank are vested in the Board, which may delegate its powers to the Board of Directors except in such matters as admission of new members, changes in the Bank's authorized capital stock, election of Directors and President, and amendment of the Charter. One Governor and one Alternate Governor are appointed by each member country. The Board meets at least once a year. The 37th Bank Annual Meeting was convened in Juju Island, Republic of Korea, in May 2004.

BOARD OF DIRECTORS

The Board of Directors is responsible for general direction of operations and exercises all powers delegated by the Board of Governors, which elects it. Of the 12 Directors, eight represent constituency groups of member countries within the ESCAP region (with about 65% of the voting power) and four represent the rest of the member countries. Each Director serves for two years and may be re-elected.

Three specialized committees (the Audit Committee, the Budget Review Committee and the Inspection Committee), each comprising six members, assist the Board of Directors in exercising its authority with regard to supervising the Bank's financial statements, approving the administrative budget, and reviewing and approving policy documents and assistance operations.

The President of the Bank, though not a Director, is Chairman of the Board.

Chairman of Board of Directors and President: TADAO CHINO (Japan) (until 31 January 2005).

Vice-Presidents: JOSEPH B. EICHENBERGER (USA), KHEMPHENG PHOLSENA (Laos), LIQUIN JIN (People's Republic of China), GEERT VAN DER LINDEN (Netherlands).

ADMINISTRATION

The Bank had 2,311 staff at 31 December 2003.

On 1 January 2002 the Bank implemented a new organizational structure. The reorganization aimed to strengthen the Bank's country and sub-regional focus, as well as its capacity for poverty reduction and implementing its long-term strategic framework. Five regional departments cover East and Central Asia, the Mekong, the Pacific, South Asia, and South East Asia. Other departments and offices include Private Sector Operations, Central Operations Services, Regional and Sustainable Development, Strategy and Policy, Cofinancing Operations, and Economics and Research, as well as other administrative units.

There are Bank Resident Missions in Afghanistan, Azerbaijan, Bangladesh, Cambodia, the People's Republic of China, India, Indonesia, Kazakhstan, Kyrgyzstan, Laos, Mongolia, Nepal, Pakistan, Papua New Guinea, Sri Lanka, Tajikistan, Uzbekistan and Viet Nam, all of which report to the head of the regional department. In addition, the Bank has established a country office in the Philippines, Extended Missions in Gujarat, India and in Thailand, a Special Office in Timor-Leste, a Liaison Office in Turkmenistan, and a South Pacific Regional Mission, based in Vanuatu. Representative Offices are located in Tokyo, Japan, Frankfurt am Main, Germany (for Europe), and Washington, DC, USA (for North America).

Secretary: BINDU N. LOHANI.

General Counsel: ARTHUR M. MITCHELL.

INSTITUTE

ADB Institute—ADBI: Kasumigaseki Bldg, 8th Floor, 2–5 Kasumigaseki 3-chome, Chiyoda-ku, Tokyo 100-6008, Japan; tel. (3) 3593-5500; fax (3) 3593-5571; e-mail info@adbi.org; internet www.adbi.org; f. 1997 as a subsidiary body of the ADB to research and analyse long-term development issues and to disseminate development practices through training and other capacity-building activities; Dean Dr PETER MCCAWLEY (Australia).

FINANCIAL STRUCTURE

The Bank's ordinary capital resources (which are used for loans to the more advanced developing member countries) are held and used entirely separately from its Special Funds resources (see below). A fourth General Capital Increase (GCI IV), amounting to US $26,318m. (or some 100%), was authorized in May 1994. At the final deadline for subscription to GCI IV, on 30 September 1996, 55 member countries had subscribed shares amounting to $24,675.4m.

At 31 December 2003 the position of subscriptions to the capital stock was as follows: authorized US $51,996.6m.; 'callable' subscribed $48,339.5m.

The Bank also borrows funds from the world capital markets. Total borrowings during 2003 amounted to US $4,141m. (compared with $6,145m. in 2002). At 31 December 2003 total outstanding borrowings amounted to $26,359m.

In July 1986 the Bank abolished the system of fixed lending rates, under which ordinary operations loans had carried interest rates fixed at the time of loan commitment for the entire life of the loan. Under the new system the lending rate is adjusted every six months, to take into account changing conditions in international financial markets.

SPECIAL FUNDS

The Asian Development Fund (ADF) was established in 1974 in order to provide a systematic mechanism for mobilizing and administering resources for the Bank to lend on concessionary terms to the least-developed member countries. In 1998 the Bank revised the terms of ADF. Since 1 January 1999 all new project loans are repayable within 32 years, including an eight-year grace period, while quick-disbursing programme loans have a 24-year maturity, also including an eight-year grace period. The previous annual service charge was redesignated as an interest charge, including a portion to cover administrative expenses. The new interest charges on all loans are 1%–1.5% per annum. During 2003 47 ADF loans were approved, amounting to US $1,379m.

Successive replenishments of the Fund's resources amounted to US $809m. for the period 1976–78, $2,150m. for 1979–82, $3,214m. for 1983–86, $3,600m. for 1987–90, $4,200m. for 1992–95, and $6,300m. for 1997–2000. In September 2000 25 donor countries pledged $2,910m. towards the ADF's seventh replenishment (ADF VIII), which totalled $5,650m. to provide resources for the period 2001–04; repayments of earlier ADF loans were to provide the remaining $2,740m. ADF VIII became effective in June 2001. Negotiations on a further replenishment (ADF IX) were initiated in October 2003; a donors' meeting was convened in Tokyo, in December.

The Bank provides technical assistance grants from its Technical Assistance Special Fund (TASF). By the end of 2003, the Fund's total resources amounted to US $1,005.5m., of which $901.4m. had been utilized or committed. The Japan Special Fund (JSF) was established in 1988 to provide finance for technical assistance by means of grants, in both the public and private sectors. The JSF aims to help developing member countries restructure their economies, enhance the opportunities for attracting new investment, and recycle funds. The Japanese Government had committed a total of 99,200m. yen (equivalent to some $852.7m.) to the JSF by the end of 2003. During 2003 the Bank approved 66 technical assistance projects for the JSF, amounting to $36.9m. An Asian Currency Crisis Support Facility (ACCSF) was operational for the three-year period March 1999–March 2002, as an independent component of the JSF to provide additional technical assistance, interest payment assistance and guarantees to countries most affected by financial instability, i.e. Indonesia, Republic of Korea, Malaysia, Philippines and Thailand. At the end of 2002 the Japanese Government, as the sole financier of the fund, had contributed 27,500m. yen (some $241.0m.) to the ACCSF. The Japanese Government funds the Japan Scholarship Program, under which 1,621 scholarships had been awarded to recipients from 34 member countries between 1988 and 2003, and the ADB Institute Special Fund, which was established to finance the ADB Institute's initial operations. By 31 December 2003 cumulative commitments to the Special Fund amounted to 10,100m. yen (or $83.1m.). In May 2000 the Japan Fund for Poverty Reduction was established, with an initial contribution of 10,000m. yen (approximately $92.6m.) from the Japanese Government, to support ADB-financed poverty reduction and social development activities. By the end of 2003 cumulative commitments to the Fund totalled $302.4m. and 38 projects, amounting to $116m., had been approved for implementation. A Japan Fund for Information and Communication Technology (ICT) was established in July 2001, for a three-year period, to promote the advancement and use of ICT in developing member countries. The Fund was established with an initial contribution of 1,273m. yen (or $10.7m.) from the Japanese Government.

The majority of grant funds in support of the Bank's technical assistance activities are provided by bilateral donors under channel financing arrangements (CFAs), the first of which was negotiated in 1980. CFAs may also be processed as a thematic financing tool, for example concerned with renewable energy, water or poverty reduction, enabling more than one donor to contribute. In 2002 a new multi-donor thematic CFA with a focus on poverty reduction was introduced. In 2003 two new co-operative funds were established under CFA terms, relating to gender and development and to poverty and the environment.

Activities

Loans by the ADB are usually aimed at specific projects. In responding to requests from member governments for loans, the Bank's staff assesses the financial and economic viability of projects and the way in which they fit into the economic framework and priorities of development of the country concerned. In 1987 the Bank adopted a policy of lending in support of programmes of sectoral adjustment, not limited to specific projects; such loans were not to exceed 15% of total Bank public sector lending. In 1999 the Board of Directors increased the ceiling on programme lending to 20% of the annual total. In 1985 the Bank decided to expand its assistance to the private sector, hitherto comprising loans to development finance institutions, under government guarantee, for lending to small and medium-sized enterprises; a programme was formulated for direct financial assistance, in the form of equity and loans without government guarantee, to private enterprises. In 1992 a Social Dimensions Unit was established as part of the central administrative structure of the Bank, which contributed to the Bank's increasing awareness of the importance of social aspects of development as essential components of sustainable economic growth. During the early 1990s the Bank also aimed to expand its role as project financier by providing assistance for policy formulation and review and promoting regional co-operation, while placing greater emphasis on individual country requirements. In accordance with its medium-term strategy for 1995–98 the Bank resolved to promote sound development management, by integrating into its operations and projects the promotion of governance issues, such as capacity-building, legal frameworks, and openness of information. During that period the Bank also introduced a commitment to assess development projects for their impact on the local population and to avoid all involuntary resettlement where possible and established a formal procedure for grievances, under which the Board may authorize an inspection of a project, by an independent panel of experts, at the request of the affected community or group. In 1998 the Bank approved a new anticorruption strategy.

The currency instability and ensuing financial crises affecting many Asian economies in the second half of 1997 and in 1998 prompted the Bank to reflect on its role in the region. The Bank resolved to strengthen its activities as a broad-based development institution, rather than solely as a project financier, through lending policies, dialogue, co-financing and technical assistance. A Task Force on Financial Sector Reform was established to review the causes and effects of the regional financial crisis. The Task Force identified the Bank's initial priorities as being to accelerate banking and capital market reforms in member countries, to promote market efficiency in the financial, trade and industrial sectors, to promote good governance and sound corporate management, and to alleviate the social impact of structural adjustments. In mid-1999 the Bank approved a technical assistance grant to establish an internet-based Asian Recovery Information Centre, within a new Regional Monitoring Unit, which aimed to facilitate access to information regarding the economic and social impact of the Asian financial crisis, analyses of economic needs of countries, reform programmes and monitoring of the economic recovery process. In November the Board of Directors approved a new overall strategy objective of poverty reduction, which was to be the principal consideration for all future Bank lending, project financing and technical assistance. The strategy incorporated key aims of supporting sustainable, grass-roots based economic growth, social development and good governance. The Board also approved a health sector policy, to concentrate resources on basic primary healthcare, and initiated reviews of the Bank's private sector strategy and the efficiency of resident missions. During 2000 the Bank began to refocus its country strategies, projects and lending targets to complement the poverty reduction strategy. In addition, it initiated a process of wide-ranging discussions to formulate a long-term strategic framework for the next 15 years, based on the target of reducing by 50% the incidence of extreme poverty by 2015, one of the so-called Millennium Development Goals identified by the UN General Assembly. The framework, establishing the operational priorities and principles for reducing poverty, was approved in March 2001. At the same time a medium-term strategy, for the period 2001–05, was approved, which aimed to enhance the development impact of the Bank's assistance and to define the operational priorities within the context of the strategic agenda.

On 1 June 2004 the Bank approved a new policy to provide rehabilitation and reconstruction assistance following disasters or other emergencies. The policy also aimed to assist developing member countries with prevention, preparation and mitigation of the impact of future disasters.

In 2003 the Bank approved 85 loans for 66 projects amounting to US $6,104.8m. (compared with $5,658m. for 70 projects in 2002). Loans from ordinary capital resources in 2003 totalled $4,725.7m., while loans from the ADF amounted to $1,379.2m. Private-sector operations approved amounted to $562.7m., including loans of $187.0m., equity investments of $35.7m. and guarantees of $240.0m. The largest proportion of assistance, amounting to some 42% of total lending, was allocated to transport and communications projects. The largest borrowers were India (25% of total lending), the People's Republic of China (24%), and Pakistan (14%). Disbursements of loans during 2003 amounted to $3,816.2m., bringing cumulative disbursements to $70,186.3m.

In 2003 grants approved for technical assistance (e.g. project preparation, consultant services and training) amounted to US $176.5m. for 315 projects, with $86.7m. deriving from the TASF, $36.9m. from the JSF, and $52.9m. from other bilateral and multilateral sources. The Bank's Operations Evaluation Office prepares reports on completed projects, in order to assess achievements and problems. In April 2000 the Bank announced that, from 2001, some new loans would be denominated in local currencies, in order to ease the repayment burden on recipient economies.

The Bank co-operates with other international organizations active in the region, particularly the World Bank group, the IMF, UNDP and APEC, and participates in meetings of aid donors for developing member countries. In May 2001 the Bank and UNDP signed a memorandum of understanding (MOU) on strategic partnership, in order to strengthen co-operation in the reduction of poverty, for example the preparation of common country assessments and a common database on poverty and other social indicators. Also in 2001 the Bank signed an MOU with the World Bank on administrative arrangements for co-operation, providing a framework for closer co-operation and more efficient use of resources. In early 2002 the Bank worked with the World Bank and UNDP to assess the preliminary needs of the interim administration in Afghanistan, in preparation for an International Conference on Reconstruction Assistance to Afghanistan, held in late January, in Tokyo. The Bank pledged to work with its member governments to provide highly concessional grants and loans of some US $500m. over two-and-a-half years, with a particular focus on road reconstruction, basic education, and agricultural irrigation rehabilitation. A new policy concerning co-operation with non-governmental organizations (NGOs) was approved by the Bank in 1998.

Finance

Internal administrative expenses totalled US $252.6m. in 2003. The budget for 2004 amounted to $279.5m.

Publications

ADB Business Opportunities (monthly).
ADB Institute Newsletter.
ADB Review (6 a year).
Annual Report.
Asian Development Outlook (annually).
Asian Development Review (2 a year).
Basic Statistics (annually).
Key Indicators of Developing Asian and Pacific Countries (annually).
Law and Policy Reform Bulletin (annually).
Loan Disbursement Handbook.
Studies and technical assistance reports, information brochures, guide-lines, sample bidding documents, staff papers.

Statistics

SUBSCRIPTIONS AND VOTING POWER
(31 December 2003)

Country	Subscribed capital (% of total)	Voting power (% of total)
Regional:		
Afghanistan	0.034	0.345
Australia	5.851	4.998
Azerbaijan	0.450	0.677
Bangladesh	1.032	1.143
Bhutan	0.006	0.322
Cambodia	0.050	0.357
China, People's Republic	6.516	5.530
Cook Islands	0.003	0.320
Fiji	0.069	0.372
Hong Kong	0.551	0.758
India	6.402	5.439
Indonesia	5.507	4.723
Japan	15.781	12.942
Kazakhstan	0.816	0.970
Kiribati	0.004	0.321
Korea, Republic	5.094	4.393
Kyrgyzstan	0.302	0.559
Laos	0.014	0.329
Malaysia	2.754	2.520
The Maldives	0.004	0.321
Marshall Islands	0.003	0.320
Micronesia, Federated States	0.004	0.321
Mongolia	0.015	0.330
Country— continued		
Myanmar	0.551	0.758
Nauru	0.004	0.321
Nepal	0.149	0.436
New Zealand	1.558	1.560
Pakistan	2.203	2.080
Palau	0.003	0.320
Papua New Guinea	0.095	0.393
Philippines	2.409	2.245
Samoa	0.003	0.320
Singapore	0.344	0.593
Solomon Islands	0.007	0.323
Sri Lanka	0.586	0.787
Taiwan	1.101	1.199
Tajikistan	0.290	0.549
Thailand	1.377	1.419
Timor-Leste	0.010	0.325
Tonga	0.004	0.321
Turkmenistan	0.256	0.522
Tuvalu	0.001	0.319
Uzbekistan	0.681	0.862
Vanuatu	0.007	0.323
Viet Nam	0.345	0.594
Sub-total	63.241	64.878
Non-regional:		
Austria	0.344	0.593
Belgium	0.344	0.593
Canada	5.289	4.549
Denmark	0.344	0.593
Finland	0.344	0.593
France	2.354	2.200
Germany	4.374	3.871
Italy	1.828	1.780
Luxembourg	0.344	0.593
Netherlands	1.037	1.147
Norway	0.344	0.593
Portugal	0.344	0.593
Spain	0.344	0.593
Sweden	0.344	0.593
Switzerland	0.590	0.790
Turkey	0.344	0.593
United Kingdom	2.065	1.970
USA	15.781	12.942
Sub-total	36.759	35.122
Total	100.000	100.000

LENDING ACTIVITIES BY SECTOR

	Loan Approvals (US $ million)		
	2003		1968–2003
Sector	Amount	%	%
Agriculture and natural resources	391.90	6.42	17.06
Energy	756.70	12.40	20.33
Finance	483.00	7.91	14.17
Industry and non-fuel minerals	0.00	—	3.16
Social infrastructure	1,130.51	18.52	16.24
Transport and communications	2,577.70	42.22	21.88
Multi-sector and others	765.00	12.54	7.17
Total	6,104.81	100.00	100.00

LENDING ACTIVITIES BY COUNTRY
(US $ million)

	Loans approved in 2003		
Country	Ordinary Capital	ADF	Total
Afghanistan	—	150.00	150.00
Azerbaijan	—	22.00	22.00
Bangladesh	286.00	246.00	532.00
Bhutan	—	9.40	9.40
Cambodia	—	98.26	98.26
China, People's Republic	1,488.00	—	1,488.00
Fiji	47.00	—	47.00
India	1,532.00	—	1,532.00
Indonesia	187.40	74.20	261.60
Kazakhstan	34.60	—	34.60
Kyrgyzstan	—	15.50	15.50

Country—*continued*	Loans approved in 2003 Ordinary Capital	ADF	Total
Laos	—	46.00	46.00
The Maldives	—	6.00	6.00
Mongolia	—	29.50	29.50
Nepal	—	94.00	94.00
Pakistan	692.70	178.00	870.70
Philippines	183.76	—	183.76
Samoa	—	8.00	8.00
Sri Lanka	90.00	185.29	275.29
Tajikistan	—	38.00	38.00
Thailand	40.00	—	40.00
Uzbekistan	99.20	—	99.20
Viet Nam	—	179.00	179.00
Regional	45.00	—	45.00
Total	4,725.66	1,379.15	6,104.81

Source: *ADB Annual Report 2003*.

ASSOCIATION OF SOUTH EAST ASIAN NATIONS—ASEAN

Address: 70A Jalan Sisingamangaraja, POB 2072, Jakarta 12110, Indonesia.

Telephone: (21) 7262991; **fax:** (21) 7398234; **e-mail:** public@aseansec.org; **internet:** www.aseansec.org.

ASEAN was established in August 1967 in Bangkok, Thailand, to accelerate economic progress and to increase the stability of the South-East Asian region.

MEMBERS

Brunei	Malaysia	Singapore
Cambodia	Myanmar	Thailand
Indonesia	Philippines	Viet Nam
Laos		

Organization

(October 2004)

SUMMIT MEETING

The highest authority of ASEAN, bringing together the heads of government of member countries. The first meeting was held in Bali, Indonesia, in February 1976. The 30th anniversary of the founding of ASEAN was commemorated at an informal gathering of heads of government in Kuala Lumpur, Malaysia, in December 1997. The ninth summit meeting was held in Bali, Indonesia, in October 2003, and the 10th meeting was scheduled to be held in Vientiane, Laos, in November 2004.

MINISTERIAL MEETINGS

The ASEAN Ministerial Meeting (AMM), comprising ministers of foreign affairs of member states, meets annually, in each member country in turn, to formulate policy guide-lines and to co-ordinate ASEAN activities. These meetings are followed by 'post-ministerial conferences' (PMCs), where ASEAN ministers of foreign affairs meet with their counterparts from countries that are 'dialogue partners' as well as with ministers from other countries. Ministers of economic affairs also meet once a year, to direct ASEAN economic co-operation. Joint Ministerial Meetings, consisting of ministers of foreign affairs and of economic affairs are convened prior to a summit meeting, and may be held at the request of either group of ministers. Other ministers meet regularly to promote co-operation in different sectors.

STANDING COMMITTEE

The Standing Committee normally meets every two months. It consists of the minister of foreign affairs of the host country and ambassadors of the other members accredited to the host country.

SECRETARIATS

A permanent secretariat was established in Jakarta, Indonesia, in 1976 to form a central co-ordinating body. The Secretariat comprises four bureaux relating to: Programme Co-ordination and External Relations; Trade, Industry and Services; Investment, Finance and Surveillance; and Economic and Functional Co-operation. The Secretary-General holds office for a five-year term, and is assisted by two Deputy Secretaries-General. In each member country day-to-day work is co-ordinated by an ASEAN National Secretariat.

Secretary-General: ONG KENG YONG (Singapore).

Deputy Secretaries-General: Dr WILFRIDO V. VILLACORTA (Philippines), Pengiran MASHOR Pengiran AHMAD (Brunei).

COMMITTEES AND SENIOR OFFICIALS' MEETINGS

Ministerial meetings are serviced by 29 committees of senior officials, supported by 122 technical working groups. There is a network of subsidiary technical bodies comprising sub-committees, expert groups, *ad hoc* working groups and working parties.

To support the conduct of relations with other countries and international organizations, ASEAN committees (composed of heads of diplomatic missions) have been established in 15 foreign capitals: those of Australia, Belgium, Canada, the People's Republic of China, France, Germany, India, Japan, the Republic of Korea, New Zealand, Pakistan, Russia, Switzerland, the United Kingdom and the USA. There is also an ASEAN committee in New York (USA).

Activities

ASEAN was established in 1967 with the signing of the ASEAN Declaration, otherwise known as the Bangkok Declaration, by the ministers of foreign affairs of Indonesia, Malaysia, the Philippines, Singapore and Thailand. In February 1976 the first ASEAN summit meeting adopted the Treaty of Amity and Co-operation and the Declaration of ASEAN Concord. Brunei joined the organization in January 1984, shortly after attaining independence. Viet Nam was admitted as the seventh member of ASEAN in July 1995. Laos and Myanmar joined in July 1997 and Cambodia was formally admitted in April 1999, fulfilling the organization's ambition to incorporate all 10 countries in the sub-region. In October 2003 ASEAN leaders adopted a declaration known as 'Bali Concord II', which committed signatory states to the creation of an ASEAN Economic Community, an ASEAN Security Community and an ASEAN Sociocultural Community.

TRADE AND ECONOMIC CO-OPERATION

A Basic Agreement on the Establishment of ASEAN Preferential Trade Arrangements was concluded in 1977, but by mid-1987 the system covered only about 5% of trade between member states, since individual countries were permitted to exclude any 'sensitive' products from preferential import tariffs. In December 1987 the meeting of ASEAN heads of government resolved to reduce such exclusions to a maximum of 10% of the number of items traded and to a maximum of 50% of the value of trade, over the next five years (seven years for Indonesia and the Philippines).

In January 1992 heads of government, meeting in Singapore, signed an agreement to create an ASEAN Free Trade Area (AFTA) by 2008. In accordance with the agreement, a common effective preferential tariff (CEPT) scheme came into effect in January 1993. The CEPT covered all manufactured products, including capital goods, and processed agricultural products (which together accounted for two-thirds of intra-ASEAN trade), but was to exclude unprocessed agricultural products. Tariffs were to be reduced to a maximum of 20% within a period of five to eight years and to 0%–5% during the subsequent seven to 10 years. Fifteen categories were designated for accelerated tariff reduction, including vegetable oils, rubber products, textiles, cement and pharmaceuticals. Member states were, however, still to be permitted exclusion for certain 'sensitive' products. In October 1993 ASEAN trade ministers agreed to modify the CEPT, with only Malaysia and Singapore having adhered to the original tariff reduction schedule. The new AFTA programme, under which all member countries except Brunei were scheduled to begin tariff reductions from 1 January 1994, substantially enlarged the number of products to be included in the tariff-reduction process (i.e. on the so-called 'inclusion list') and reduced the list of products eligible for protection. In September 1994 ASEAN ministers of economic affairs agreed to accelerate the implementation of AFTA, advancing the deadline for its entry into operation from 2008 to 1 January 2003. Tariffs were to be reduced to 0%–5% within seven to 10 years, or within five to eight years for products designated for accelerated tariff cuts. In July 1995 Viet Nam was admitted as a member of ASEAN and was granted until 2006 to implement the AFTA trade agreements. In December 1995 heads of government, at a meeting convened in Bangkok, Thailand, agreed to extend liberalization to certain service industries, including banking, telecommunications and tourism. In July 1997 Laos and Myanmar became members of ASEAN and were granted a 10-year period, from 1 January 1998, to comply with the AFTA schedule.

In December 1998, meeting in Hanoi, Viet Nam, heads of government approved a Statement on Bold Measures, detailing ASEAN's strategies to deal with the economic crisis that had prevailed in the region since late 1997. These included incentives to attract investors, for example a three-year exemption on corporate taxation, accelerated implementation of the ASEAN Investment Area (AIA, see below), and advancing the AFTA deadline, for the original six members, to 2002, with some 85% of products to be covered by the arrangements by 2000, and 90% by 2001. It was envisaged that the original six members and the new members would achieve the elimination of all tariffs by 2015 and 2018, respectively. The Hanoi Plan of Action, which was also adopted at the meeting as a framework for the development of the organization over the period 1999–2004, incorporated a series of measures aimed at strengthening macroeconomic and financial co-operation and enhancing economic integration. In April 1999 Cambodia, on being admitted as a full member of ASEAN, signed an agreement to implement the tariff reduction programme over a 10-year period, commencing 1 January 2000. Cambodia also signed a declaration endorsing the commitments of the 1998 Statement on Bold Measures. In May 2000 Malaysia was granted a special exemption to postpone implementing tariff reductions on motor vehicles for two years from 1 January 2003. In November 2000 a protocol was approved permitting further temporary exclusion of products from the CEPT scheme for countries experiencing economic difficulties. On 1 January 2002 AFTA was formally realized among the original six signatories (Brunei, Indonesia, Malaysia, the Philippines, Singapore and Thailand), which had achieved the objective of reducing to less than 5% trade restrictions on 96.24% of products on the inclusion list. Some 98.36% of tariff lines for the core six AFTA members were on the inclusion list at that time. Tariffs on trade in products on the inclusion list for these countries averaged less than 2.9% in 2002.

To complement AFTA in facilitating intra-ASEAN trade, member countries are committed to the removal of non-tariff barriers (such as quotas), the harmonization of standards and conformance measures, and the simplification and harmonization of customs procedures. In June 1996 the Working Group on Customs Procedures completed a draft legal framework for regional co-operation, designed to simplify and harmonize customs procedures, legislation and product classification. The agreement was signed in March 1997 at the inaugural meeting of ASEAN finance ministers. (Laos and Myanmar signed the customs agreement in July and Cambodia assented to it in April 1999.) In 2001 ASEAN finalized its system of harmonized tariff nomenclature. Implementation began in 2002, with training on the new system to be given to public- and private-sector officials over the course of the year.

At the seventh summit meeting, held in Brunei, in November 2001 heads of government noted the challenges posed by the severe global economic slowdown, at a time when ASEAN countries were beginning to emerge from the 1997–98 crisis. Members discussed moving beyond the group's existing free-trade and investment commitments by deepening market liberalization. Specifically, it was proposed that negotiations on the liberalization of intra-ASEAN trade in services be accelerated. The third round of negotiations on liberalizing trade in services began at the end of 2001; it was scheduled to be completed by 2004. Members also agreed to start negotiations on mutual recognition arrangements for professional services. The summit meeting stated that tariff preferences would be extended to ASEAN's newer members from January 2002, under the ASEAN Integration System of Preferences (AISP), thus allowing Cambodia, Laos, Myanmar and Viet Nam tariff-free access to the more developed ASEAN markets earlier than the previously agreed target date of 2010.

In November 2000 heads of government endorsed an Initiative for ASEAN Integration (IAI), which aimed to reduce economic disparities within the region through effective co-operation, in particular, in training and other educational opportunities. In July 2002 the AMM endorsed an IAI Work Plan, which focused on the following priority areas: human resources development; infrastructure; information and communications technology (ICT); and regional economic integration. The Plan was to be implemented over the six-year period 2002–08. An IAI Development Co-operation Forum was held in August.

At the eighth summit meeting, held in November 2002, in Phnom-Penh, Cambodia, ASEAN heads of state considered the Phnom-Penh Agenda, comprising the following four elements: collaboration with an ongoing initiative to promote economic co-operation in the Greater Mekong subregion, with a view to accelerating the pace of ASEAN integration; promoting ASEAN as a single tourist destination; solidarity for peace and security, especially in combating terrorism; and progress in sustainable natural resources management, including the ratification of the Kyoto Protocol to the UN Framework Convention on Climate Change.

In 1991 ASEAN ministers discussed a proposal of the Malaysian Government for the formation of an economic grouping, to be composed of ASEAN members, the People's Republic of China, Hong Kong, Japan, the Republic of Korea and Taiwan. In July 1993 ASEAN ministers of foreign affairs agreed a compromise, whereby the grouping was to be a caucus within APEC, although it was to be co-ordinated by ASEAN's meeting of economy ministers. In July 1994 ministers of foreign affairs of nine prospective members of the group held their first informal collective meeting; however, no progress was made towards forming the proposed East Asia Economic Caucus. There was renewed speculation on the formation of an East Asian grouping following the onset of the Asian financial crisis of the late 1990s. At an informal meeting of leaders of ASEAN countries, China, Japan and the Republic of Korea, held in November 1999, all parties (designating themselves 'ASEAN + 3') issued a Joint Statement on East Asian Co-operation, in which they agreed to strengthen regional unity, and addressed the long-term possibility of establishing an East Asian common market and currency. Meeting in May 2000, in Chiang Mai, Thailand, ASEAN + 3 ministers of economic affairs proposed the establishment of an enhanced currency-swap mechanism, enabling countries to draw on liquidity support to defend their economies during balance-of-payments difficulties or speculative currency attacks and to prevent future financial crises. In July ASEAN + 3 ministers of foreign affairs convened an inaugural formal summit in Bangkok, Thailand, and in October ASEAN + 3 economic affairs ministers agreed to hold their hitherto twice-yearly informal meetings on an institutionalized basis. In November an informal meeting of ASEAN + 3 leaders approved further co-operation in various sectors and initiated a feasibility study into a proposal to establish a regional free trade area. In May 2001 ASEAN + 3 ministers of economic affairs endorsed a series of projects for co-operation in ICT, environment, small and medium-sized enterprises, Mekong Basin development, and harmonization of standards. In the same month the so-called Chiang Mai Initiative on currency-swap arrangements was formally approved by ASEAN + 3 finance ministers. A meeting of the ASEAN + 3 leaders was held alongside the seventh ASEAN summit in November 2001. Malaysia offered to host an ASEAN + 3 secretariat; however, the establishment of a formal secretariat for the grouping remained under discussion in 2004. In October 2001 ASEAN + 3 agriculture and forestry ministers met for the first time, and discussed issues of poverty alleviation, food security, agricultural research and human resource development. The first meeting of ASEAN + 3 tourism ministers was held in January 2002. In July ASEAN + 3 ministers of foreign affairs declared their support for other regional initiatives, namely an Asia Co-operation Dialogue, which was initiated by the Thai Government in June, and an Initiative for Development in East Asia (IDEA), which had been announced by the Japanese Government in January. An IDEA ministerial meeting was convened in Tokyo, in August. ASEAN + 3 ministers of labour convened in May 2003. In the following month, at an extraordinary session, ASEAN + 3 health ministers declared the region free of Severe Acute Respiratory Syndrome (SARS). In August a meeting of ASEAN + 3 finance ministers agreed to establish a Finance Co-operation Fund, to be administered by the ASEAN Secretariat; the Fund was to support ongoing economic reviews relating to projects such as the Chiang Mai Initiative. In September the sixth consultation between ASEAN + 3 ministers of economic affairs was held, at which several new projects were endorsed, including two on e-commerce.

Bali Concord II, adopted in October 2003 (see above), affirmed commitment to existing ASEAN economic co-operation frameworks, including the Hanoi Plan of Action (and any subsequently agreed regional plans of action) and the IAI, and outlined plans for the creation, by 2020, of an integrated ASEAN Economic Community (AEC), entailing: the harmonization of customs procedures and technical regulations by the end of 2004; the removal of non-tariff trade barriers and the establishment of a network of free-trade zones by 2005; and the progressive withdrawal of capital controls and strengthening of intellectual property rights. An ASEAN legal unit was to be established to strengthen and enhance existing dispute settlement systems. The free movement of professional and skilled workers would be facilitated by standardizing professional requirements and simplifying visa procedures, with the adoption of a single ASEAN visa requirement envisaged by 2005. In 2004 ASEAN economic and trade ministers worked closely to produce a 'road map' for the integration of 11 sectors identified as priority areas in the AEC plan of action: agriculture, air travel, automotive, electronics, fisheries, healthcare, information technology, rubber, textiles, tourism and wood. In July the ASEAN Ministerial Meeting reviewed progress in preparing the Vientiane Action Programme, the proposed successor to the Hanoi Plan of Action.

INDUSTRY

The ASEAN-Chambers of Commerce and Industry (CCI) aims to enhance ASEAN economic and industrial co-operation and the participation in these activities of the private sector. In March 1996 a permanent ASEAN-CCI secretariat became operational at the ASEAN Secretariat. The first AIA Council-Business Sector Forum was convened in September 2001, with the aim of developing alliances between the public and private sectors. The seventh ASEAN summit in November resolved to encourage the private sector to convene a regular ASEAN Business Summit. It was also agreed to set up an ASEAN Business Advisory Council.

The ASEAN Industrial Co-operation (AICO) scheme, initiated in 1996, encourages companies in the ASEAN region to undertake joint manufacturing activities. Products derived from an AICO arrangement benefit immediately from a preferential tariff rate of 0%–5%. The AICO scheme superseded the ASEAN industrial joint venture scheme, established in 1983. The attractiveness of the scheme is expected slowly to diminish as ASEAN moves towards the full implementation of the CEPT scheme. ASEAN has initiated studies of new methods of industrial co-operation within the grouping, with the aim of achieving further integration.

The ASEAN Consultative Committee on Standards and Quality (ACCSQ) aims to promote the understanding and implementation of quality concepts, considered to be important in strengthening the economic development of a member state and in helping to eliminate trade barriers. ACCSQ comprises three working groups: standards and information; conformance and assessment; and testing and calibration. In September 1994 an *ad hoc* Working Group on Intellectual Property Co-operation was established, with a mandate to formulate a framework agreement on intellectual property co-operation and to strengthen ASEAN activities in intellectual property protection. ASEAN aims to establish, by 2004, a regional electronic database, to strengthen the administration of intellectual property. ASEAN is also developing a Regulatory Trademark Filing System, as a first step towards the creation of an ASEAN Trademark System.

In 1988 the ASEAN Fund was established, with capital of US $150m., to provide finance for portfolio investments in ASEAN countries, in particular for small and medium-sized enterprises (SMEs). The Hanoi Plan of Action, which was adopted by ASEAN heads of state in December 1998, incorporated a series of initiatives to enhance the development of SMEs, including training and technical assistance, co-operation activities and greater access to information.

FINANCE, BANKING AND INVESTMENT

In 1987 heads of government agreed to accelerate regional financial co-operation, to support intra-ASEAN trade and investment. They adopted measures to increase the role of ASEAN currencies in regional trade, to assist negotiations on the avoidance of double taxation, and to improve the efficiency of tax and customs administrators. An ASEAN Reinsurance Corporation was established in 1988, with initial authorized capital of US $10m. In December 1995 the summit meeting proposed the establishment of an ASEAN Investment Area (AIA). Other measures to attract greater financial resource flows in the region, including an ASEAN Plan of Action for the Promotion of Foreign Direct Investment and Intra-ASEAN Investment, were implemented during 1996.

In February 1997 ASEAN central bank governors agreed to strengthen efforts to combat currency speculation through the established network of foreign-exchange repurchase agreements. However, from mid-1997 several Asian currencies were undermined by speculative activities. Subsequent unsuccessful attempts to support the foreign-exchange rates contributed to a collapse in the value of financial markets in some countries and to a reversal of the region's economic growth, at least in the short term, while governments undertook macroeconomic structural reforms. In early December ASEAN ministers of finance, meeting in Malaysia, agreed to liberalize markets for financial services and to strengthen surveillance of member country economies, to help prevent further deterioration of the regional economy. The ministers also endorsed a proposal for the establishment of an Asian funding facility to provide emergency assistance in support of international credit and structural reform programmes. At the informal summit meeting held later in December, ASEAN leaders issued a joint statement in which they expressed the need for mutual support to counter the region's financial crisis and urged greater international assistance to help overcome the situation and address the underlying problems. The heads of government also resolved to accelerate the implementation of the AIA.

In July 1998 the ASEAN Ministerial Meeting endorsed the decisions of finance ministers, taken in February, to promote greater use of regional currencies for trade payments and to establish an economic surveillance mechanism. In October ministers of economic affairs, meeting in Manila, the Philippines, signed a framework agreement on the AIA, which was expected to provide for equal treatment of domestic and other ASEAN direct investment proposals within the grouping by 2010, and of all foreign investors by 2020. The meeting also confirmed that the proposed ASEAN Surveillance Process (ASP), to monitor the economic stability and financial systems of member states, would be implemented with immediate effect, and would require the voluntary submission of economic information by all members to a monitoring committee, to be based in Jakarta, Indonesia. The ASP and the framework agreement on the AIA were incorporated into the Hanoi Plan of Action, adopted by heads of state in December 1998. The December summit meeting also resolved to accelerate reforms, particularly in the banking and financial sectors, in order to strengthen the region's economies, and to promote the liberalization of the financial services sector.

In March 1999 ASEAN ministers of trade and industry, meeting in Phuket, Thailand, as the AIA Council, agreed to open their manufacturing, agriculture, fisheries, forestry and mining industries to foreign investment. Investment restrictions affecting those industries were to be eliminated by 2003 in most cases, although Laos and Viet Nam were granted until 2010. In addition, ministers adopted a number of measures to encourage investment in the region, including access to three-year corporate income-tax exemptions, and tax allowances of 30% for investors. The AIA agreement formally entered into force in June 1999, having been ratified by all member countries. Under the agreement, member countries submitted individual action plans for 2000–04, noting specific action to be taken in the areas of investment promotion, facilitation and liberalization. In September 2001 ministers agreed to accelerate the full realization of the AIA for non-ASEAN investors in manufacturing, agriculture, forestry, fishing and mining sectors. The date for full implementation was advanced to 2010 for the original six ASEAN members and to 2015 for the newer members.

In November 2001 ASEAN heads of state considered the difficulties facing member countries as a result of the global economic and political uncertainties following the terrorist attacks on the USA in September. The summit meeting noted the recent decline in foreign direct investment and the erosion of the region's competitiveness. Short-term priorities were stated to be the stimulation of economies to lessen the impact of reduced external demand, and the adoption of appropriate fiscal and monetary policies, together with a renewed commitment to structural reform.

In April 2002 ASEAN economy ministers signed an agreement to facilitate intra-regional trade in electrical and electronic equipment by providing for the mutual recognition of standards (for example, testing and certification). The agreement was also intended to lower the costs of trade in those goods, thereby helping to maintain competitiveness.

SECURITY

In 1971 ASEAN members endorsed a declaration envisaging the establishment of a Zone of Peace, Freedom and Neutrality (ZOPFAN) in the South-East Asian region. This objective was incorporated in the Declaration of ASEAN Concord, which was adopted at the first summit meeting of the organization, held in Bali, Indonesia, in February 1976. (The Declaration also issued guide-lines for co-operation in economic development and the promotion of social justice and welfare.) Also in February 1976 a Treaty of Amity and Co-operation was signed by heads of state, establishing principles of mutual respect for the independence and sovereignty of all nations, non-interference in the internal affairs of one another and settlement of disputes by peaceful means. The Treaty was amended in December 1987 by a protocol providing for the accession of Papua New Guinea and other non-member countries in the region; it was reinforced by a second protocol, signed in July 1998. In July 2004 Pakistan and Indonesia acceded to the Treaty.

In December 1995 ASEAN heads of government, meeting in Bangkok, Thailand, signed a treaty establishing a South-East Asia Nuclear-Weapon Free Zone (SEANWFZ). The treaty was also signed by Cambodia, Myanmar and Laos. It was extended to cover the offshore economic exclusion zones of each country. On ratification by all parties, the Treaty was to prohibit the manufacture or storage of nuclear weapons within the region. Individual signatories were to decide whether to allow port visits or transportation of nuclear weapons by foreign powers through territorial waters. The Treaty entered into force on 27 March 1997. ASEAN senior officials were mandated to oversee implementation of the Treaty, pending the establishment of a permanent monitoring committee. In July 1999 the People's Republic of China and India agreed to observe the terms of the SEANWFZ.

In January 1992 ASEAN leaders agreed that there should be greater co-operation on security matters within the grouping, and that ASEAN's post-ministerial conferences (PMCs) should be used as a forum for discussion of questions relating to security with dialogue partners and other countries. In July 1992 the ASEAN Ministerial Meeting issued a statement calling for a peaceful resolution of the dispute concerning the strategically significant Spratly Islands in the South China Sea, which are claimed, wholly or partly, by the People's Republic of China, Viet Nam, Taiwan, Brunei, Malaysia and the Philippines. (In February China had introduced legislation that defined the Spratly Islands as belonging to its territorial waters.) Viet Nam's accession to ASEAN in July 1995, bringing all the Spratly Islands claimants except China and Taiwan into the grouping, was expected to strengthen ASEAN's position of negotiating a multilateral settlement on the islands. In 1999 ASEAN established a special committee to formulate a code of conduct for the South China Sea to be observed by all claimants to the Spratly Islands. A draft code of conduct was approved in November 1999. China, insisting that it would adopt the proposed code only as a set of guide-lines and not as a legally-binding document, resolved not to strengthen its presence on the islands. In 2000 and 2001 it participated in discussions with ASEAN officials concerning the document. In November 2002 ASEAN and China's foreign ministers adopted a Declaration on the Conduct of Parties in the South China Sea, agreeing to promote a peaceful environment and durable solutions for the area, to resolve territorial disputes by peaceful means, to refrain from undertaking activities that would aggravate existing tensions (such as settling unpopulated islands and reefs), and to initiate a regular dialogue of defence officials.

In July 1997 ASEAN ministers of foreign affairs reiterated their commitment to the principle of non-interference in the internal affairs of other countries. However, the group's efforts in Cambodia (see below) marked a significant shift in diplomatic policy towards one of 'constructive intervention', which had been proposed by Malaysia's Deputy Prime Minister in recognition of the increasing interdependence of the region. At the Ministerial Meeting in July 1998 Thailand's Minister of Foreign Affairs, supported by his Philippine counterpart, proposed that the grouping formally adopt a policy of 'flexible engagement'. The proposal, based partly on concerns that the continued restrictions imposed by the Myanma authorities on dissident political activists was damaging ASEAN relations with its dialogue partners, was to provide for the discussion of the affairs of other member states when they have an impact on neighbouring countries. While rejecting the proposal, other ASEAN ministers agreed to pursue a more limited version, referred to as 'enhanced interaction', and to maintain open dialogue within the grouping. In September 1999 the unrest prompted by the popular referendum on the future of East Timor (now Timor-Leste) and the resulting humanitarian crisis highlighted the unwillingness of some ASEAN member states to intervene in other member countries and undermined the political unity of the grouping. A compromise agreement, enabling countries to act on an individual basis rather than as representatives of ASEAN, was formulated prior to an emergency meeting of ministers of foreign affairs, held during the APEC meetings in Auckland, New Zealand. Malaysia, the Philippines, Singapore and Thailand declared their support for the establishment of a multinational force to restore peace in East Timor and committed troops to participate in the Australian-led operation. At their informal summit in November 1999 heads of state approved the establishment of an ASEAN Troika, which was to be constituted as an *ad hoc* body comprising the foreign ministers of the Association's current, previous and future chairmanship with a view to providing a rapid response mechanism in the event of a regional crisis.

On 12 September 2001 ASEAN issued a ministerial statement on international terrorism, condemning the attacks of the previous day in the USA and urging greater international co-operation to counter terrorism. The seventh summit meeting in November issued a Declaration on a Joint Action to Combat Terrorism. This condemned the September attacks, stated that terrorism was a direct challenge to ASEAN's aims, and affirmed the grouping's commitment to strong measures to counter terrorism. The summit encouraged member countries to sign (or ratify) the International Convention for the Suppression of Financing of Terrorism, to strengthen national mechanisms against terrorism, and to work to deepen co-operation, particularly in the area of intelligence exchange; international conventions to combat terrorism would be studied to see if they could be integrated into the ASEAN structure, while the possibility of developing a regional anti-terrorism convention was discussed. The summit noted the need to strengthen security co-operation to restore investor confidence. In its Declaration and other notes, the summit explicitly rejected any attempt to link terrorism with religion or race, and expressed concern for the suffering of innocent Afghanis during the US military action against the Taliban authorities in Afghanistan. The summit's final Declaration was worded so as to avoid any mention of the US action, to which Muslim ASEAN states such as Malaysia and Indonesia were strongly opposed. Several ASEAN countries offered to assist in peace-keeping and reconstruction in Afghanistan, following the removal of the Taliban and establishment of an interim authority. In November 2002 the eighth summit meeting adopted a Declaration on Terrorism, reiterating and strengthening the measures announced in the previous year. In January 2003 ASEAN police chiefs proposed the establishment of a network of national anti-terrorism task forces, to work in close co-operation with one another.

In June 1999 the first ministerial meeting to consider issues relating to transnational crime was convened. Regular meetings of senior officials and ministers were subsequently held. The third ministerial meeting, in October 2001, considered initiatives to combat transnational crime, which was defined as including terrorism, trafficking in drugs, arms and people, money-laundering, cyber-crime, piracy and economic crime. In May 2002 ministers responsible for transnational crime issues convened a Special Ministerial Meeting on Terrorism, held in Kuala Lumpur, Malaysia. The meeting approved a work programme to implement a plan of action to combat transnational crime, including information exchange, the development of legal arrangements for extradition, prosecution and seizure, the enhancement of co-operation in law enforcement, and the development of regional security training programmes. In a separate initiative Indonesia, Malaysia and the Philippines signed an agreement on information exchange and the establishment of communication procedures. Cambodia acceded to the agreement in July.

ASEAN Regional Forum—ARF: In July 1993 the meeting of ASEAN ministers of foreign affairs sanctioned the establishment of a forum to discuss and promote co-operation on security issues within the region, and, in particular, to ensure the involvement of the People's Republic of China in regional dialogue. The ARF was informally initiated during that year's PMC, comprising the ASEAN countries, its dialogue partners (at that time Australia, Canada, the EC, Japan, the Republic of Korea, New Zealand and the USA), and the People's Republic of China, Laos, Papua New Guinea, Russia and Viet Nam. The first formal meeting of the ARF was conducted in July 1994, following the Ministerial Meeting held in Bangkok, Thailand, and it was agreed that the ARF would be convened on an annual basis. The 1995 meeting, held in Brunei, in August, attempted to define a framework for the future of the Forum. It was perceived as evolving through three stages: the promotion of confidence-building (including disaster relief and peace-keeping activities); the development of preventive diplomacy; and the elaboration of approaches to conflict. The 19 ministers of foreign affairs attending the meeting (Cambodia participated for the first time) recognized that the ARF was still in the initial stage of implementing confidence-building measures. The ministers, having conceded to a request by China not to discuss explicitly the Spratly Islands, expressed concern at overlapping sovereignty claims in the region. In a further statement, the ministers urged an 'immediate end' to the testing of nuclear weapons, then being undertaken by the French Government in the South Pacific region. The third ARF, convened in July 1996, which was attended for the first time by India and Myanmar, agreed a set of criteria and guiding principles for the future expansion of the grouping. In particular, it was decided that the ARF would only admit as participants countries that had a direct influence on the peace and security of the East Asia and Pacific region. The ARF held in July 1997 reviewed progress made in developing the first two 'tracks' of the ARF process, through the structure of inter-sessional working groups and meetings. The Forum's consideration of security issues in the region was dominated by concern at the political situation in Cambodia; support was expressed for ASEAN mediation to restore stability within that country. Myanmar and Laos attended the ARF for the first time. Mongolia was admitted into the ARF at its meeting in July 1998. India rejected a proposal that Pakistan attend the meeting to discuss issues relating to both countries' testing of nuclear weapons. The meeting ultimately condemned the testing of nuclear weapons in the region, but declined to criticize specifically India and Pakistan. In July 1999 the ARF warned the Democratic People's Republic of Korea (DPRK) not to conduct any further testing of missiles over the Pacific. At the seventh meeting of the ARF, convened in Bangkok, Thailand, in July 2000, the DPRK was admitted to the Forum. The meeting considered the positive effects and challenges of globalization, including the possibilities for greater economic interdependence and for a growth in transnational crime. The eighth ARF meeting in July 2001 in Hanoi, Viet Nam, pursued these themes, and also discussed the widening development gap between nations. The meeting agreed to enhance the role of the ARF Chairman, enabling him to issue statements on behalf of ARF participants and to organize events during the year. In March and April 2002 ARF workshops were held on financial measures against terrorism and on the prevention of terrorism, respectively. The ninth ARF meeting, held in Bandar Seri Begawan, Brunei, in July, assessed regional and international security developments, and issued a statement of individual and collective intent to prevent any financing of terrorism. The statement included commitments by participants to freeze the assets of suspected individuals or groups, to implement international financial standards and to enhance co-operation and the exchange of information. In October the Chairman, on behalf of all ARF participants, condemned the terrorist bomb attacks committed against tourist targets in Bali, Indonesia. Pakistan joined the ARF process as its 24th participant in July 2004.

Since 2000 the ARF has published the *Annual Security Outlook*, to which participating countries submit assessments of the security prospects in the region.

EXTERNAL RELATIONS

ASEAN's external relations have been pursued through a dialogue system, initially with the objective of promoting co-operation in economic areas with key trading partners. The system has been expanded in recent years to encompass regional security concerns and co-operation in other areas, such as the environment. The ARF (see above) emerged from the dialogue system, and more recently the formalized discussions of ASEAN with Japan, China and the Republic of Korea (ASEAN + 3) has evolved as a separate process with its own strategic agenda.

European Union: In March 1980 a co-operation agreement was signed between ASEAN and the European Community (EC, as the EU was known prior to its restructuring on 1 November 1993), which provided for the strengthening of existing trade links and increased co-operation in the scientific and agricultural spheres. A Joint Co-operation Committee met in November (and annually thereafter). An ASEAN-EC Business Council was launched in December 1983, and three European Business Information Councils have since been established, in Malaysia, the Philippines and Thailand, to promote private-sector co-operation. The first meeting of ministers of economic affairs from ASEAN and EC member countries took place in October 1985. In December 1990 the Community adopted new guidelines on development co-operation, with an increase in assistance to Asia, and a change in the type of aid given to ASEAN members, emphasizing training, science and technology and venture capital, rather than assistance for rural development. In October 1992 the EC and ASEAN agreed to promote further trade between the regions, as well as bilateral investment, and made a joint declaration in support of human rights. An EU-ASEAN Junior Managers Exchange Programme was initiated in November 1996, as part of efforts to promote co-operation and understanding between the industrial and business sectors in both regions. An ASEAN-EU Business Network was established in Brussels in 2001, to develop political and commercial contacts between the two sides.

In May 1995 ASEAN and EU senior officials endorsed an initiative to strengthen relations between the two economic regions within the framework of an Asia-Europe Meeting of heads of government (ASEM). The first ASEM was convened in Bangkok, Thailand, in March 1996, at which leaders approved a new Asia-Europe Partnership for Greater Growth. The second ASEM summit meeting, held in April 1998, focused heavily on economic concerns. In February 1997 ministers of foreign affairs of countries participating in ASEM met in Singapore. Despite ongoing differences regarding human rights issues, in particular concerning ASEAN's granting of full membership status to Myanmar and the situation in East Timor (which precluded the conclusion of a new co-operation agreement), the Ministerial Meeting issued a final joint declaration, committing both sides to strengthening co-operation and dialogue on economic, international and bilateral trade, security and social issues. A protocol to the 1980 co-operation agreement was signed, enabling the participation of Viet Nam in the dialogue process. In November 1997 a session of the Joint Co-operation Committee was postponed and later cancelled, owing to a dispute concerning objections by the EU to the participation of Myanmar. A compromise agreement, allowing Myanma officials to attend meetings as 'silent' observers, was concluded in November 1998. However, a meeting of the Joint Co-operation Committee, scheduled to take place in January 1999, was again cancelled, owing to controversy over perceived discrimination by the EU against Myanmar's status. The meeting was finally convened in Bangkok, Thailand, in late May. The third ASEM summit meeting was convened in Seoul, Korea in October 2000. In December an ASEAN-EU Ministerial Meeting was held in Vientiane, Laos. Both sides agreed to pursue dialogue and co-operation and issued a joint declaration that accorded support for the efforts of the UN Secretary-General's special envoy towards restoring political dialogue in Myanmar. Myanmar agreed to permit an EU delegation to visit the country, and opposition leaders, in early 2001. In September 2001 the Joint Co-operation Committee met for the first time since 1999 and resolved to strengthen policy dialogue, in particular in areas fostering regional integration. Four new EU delegations were to be established—in Cambodia, Laos, Myanmar and Singapore. At the 14th ASEAN-EU Ministerial Meeting, held in Brussels, Belgium, in January 2003, delegates adopted an ASEAN-EU Joint Declaration on Co-operation to Combat Terrorism. In February 2003 the EU awarded ASEAN €4.5m. under the ASEAN-EU Programme on Regional Integration Support (APRIS) to enhance progress towards establishing AFTA. In April the EU proposed the creation of a regional framework, the Trans-Regional EU-ASEAN Trade Initiative (TREATI), to address mutual trade facilitation, investment and regulatory issues. It was suggested that the framework might eventually result in a preferential trade agreement. In January 2004 a joint statement was issued announcing a 'road map' for implementing the TREATI and an EU-ASEAN work plan for that year.

People's Republic of China: Efforts to develop consultative relations between ASEAN and China were initiated in 1993. Joint Committees on economic and trade co-operation and on scientific and technological co-operation were subsequently established. The first formal consultations between senior officials of the two sides were held in April 1995. In July 1996, in spite of ASEAN's continued concern at China's territorial claims to the Spratly Islands in the South China Sea, China was admitted to the PMC as a full dialogue partner. In February 1997 a Joint Co-operation Committee was established to co-ordinate the China-ASEAN dialogue and all aspects of relations between the two sides. Relations were further strengthened by the decision to form a joint business council to promote bilateral trade and investment. China participated in the informal summit meeting held in December, at the end of which both sides issued a joint statement affirming their commitment to resolving regional disputes through peaceful means. A second meeting of the Joint Co-operation Committee was held in March 1999. China was a participant in the first official ASEAN + 3 meeting of foreign ministers, which was convened in July 2000. An ASEAN-China Experts Group was established in November, to consider future economic co-operation and free-trade opportunities. The Group held its first meeting in April 2001 and proposed a framework agreement on economic co-operation and the establishment of an ASEAN-China free-trade area within 10 years (with differential treatment and flexibility for newer ASEAN members). Both proposals were endorsed at the seventh ASEAN summit meeting in November 2001. China also agreed to grant preferential tariff treatment for some goods from Cambodia, Laos and Myanmar. In November 2002 an agreement on economic co-operation was concluded by the ASEAN member states and China. The Framework Agreement on Comprehensive Economic Co-operation between ASEAN and China entered into force in July 2003,and envisaged the establishment of an ASEAN-China Free Trade Area (ACFTA) by 2010. The Agreement provided for strengthened co-operation in key areas including agriculture, information and telecommunications, and human resources development. It was also agreed to implement the consensus of the Special ASEAN-China Leaders' Meeting on SARS, held the previous April, and set up an ASEAN + 1 special fund for health co-operation. In October China acceded to the Treaty on Amity and Co-operation and signed a joint declaration with ASEAN on Strategic Partnership for Peace and Prosperity on strengthening co-operation in politics, economy, social affairs, security and regional and international issues. It was also agreed to continue consultations on China's accession to the SEANWFZ and to expedite the implementation of the Joint Statement on Co-operation in the Field of Non-Traditional Security Issues and the Declaration on the Conduct of Parties in the South China Sea.

Japan: The ASEAN-Japan Forum was established in 1977 to discuss matters of mutual concern in trade, investment, technology transfer and development assistance. The first meeting between the two sides at ministerial level was held in October 1992. At this meeting, and subsequently, ASEAN requested Japan to increase its investment in member countries and to make Japanese markets more accessible to ASEAN products, in order to reduce the trade deficit with Japan. Since 1993 ASEAN-Japanese development and cultural co-operation has expanded under schemes including the Inter-ASEAN Technical Exchange Programme, the Japan-ASEAN Co-operation Promotion Programme and the ASEAN-Japan Friendship Programme. In December 1997 Japan, attending the informal summit meeting in Malaysia, agreed to improve market access for ASEAN products and to provide training opportunities for more than 20,000 young people in order to help develop local economies. In December 1998 ASEAN heads of government welcomed a Japanese initiative, announced in October, to allocate US $30,000m. to promote economic recovery in the region. At the same time the Japanese Prime Minister announced a further package of $5,000m. to be made available as concessionary loans for infrastructure projects. In mid-2000 a new Japan-ASEAN General Exchange Fund (JAGEF) was established to promote and facilitate the transfer of technology, investment and personnel. In November 1999 Japan, along with the People's Republic of China and the Republic of Korea, attending an informal summit meeting of ASEAN, agreed to strengthen economic and political co-operation with the ASEAN countries, to enhance political and security dialogue, and to implement joint infrastructure and social projects. Japan participated in the first official ASEAN + 3 meeting of foreign ministers, which was convened in July 2000. An ASEAN-Japan Experts Group, similar to that for China, was to be established to consider how economic relations between the two sides can be strengthened. In recent years Japan has provided ICT support to ASEAN countries, and has offered assistance in environmental and health matters and for educational training and human resource development (particularly in engineering). In October 2003 ASEAN and Japan signed a Framework for Comprehensive Partnership. In mid-December Japan acceded to the Treaty of Amity and Co-operation in Southeast Asia and concluded a joint action plan with ASEAN with provisions on reinforcing economic integration within ASEAN and enhancing competitiveness, and on addressing terrorism, piracy and other transnational issues. A joint declaration was also issued on starting discussions in 2005 on the possibility of establishing an ASEAN-Japan FTA by 2012.

Australia and New Zealand: In 1999 ASEAN and Australia undertook to establish the ASEAN-Australia Development Co-operation Programme (AADCP), to replace an economic co-operation programme which had begun in 1974. In August 2002 the two sides signed a formal memorandum of understanding on the AADCP. It was to comprise three core elements, with assistance amounting to $A45m.: a Program Stream, to address medium-term issues of economic integration and competitiveness; a Regional Partnerships Scheme for smaller collaborative activities; and the establishment of a Regional Economic Policy Support Facility within the ASEAN Secretariat. All components were to be fully operational by 2003. Co-operation relations with New Zealand are based on the Inter-Institutional Linkages Programme and the Trade and Investment Promotion Programme, which mainly provide assistance in forestry development, dairy technology, veterinary management and legal aid training. An ASEAN-New Zealand Joint Management Committee was established in November 1993, to oversee the implementation of co-operation projects. New Zealand's English Language Training for Officials Programme is among the most important of these projects. In September 2001 ASEAN ministers of economic affairs signed a Framework for Closer Economic Partnership (CEP) with their counterparts from Australia and New Zealand (the Closer Economic Relations—CER—countries), and agreed to establish a Business Council to involve the business communities of all countries in the CEP. The CEP was perceived as a first step towards the creation of a free-trade area between ASEAN and CER countries. The establishment of such an area would strengthen the grouping's bargaining position regionally and multilaterally, and bring benefits such as increased foreign direct investment and the possible relocation of industry.

Other countries: The USA gives assistance for the development of small and medium-sized businesses and other projects, and supports a Center for Technology Exchange. In 1990 ASEAN and the USA established an ASEAN-US Joint Working Group, whose purpose is to review ASEAN's economic relations with the USA and to identify measures by which economic links could be strengthened. In recent years, dialogue has increasingly focused on political and security issues. In early August 2002 ASEAN ministers of foreign affairs met with their US counterpart, and signed a Joint Declaration for Co-operation to Combat International Terrorism. At the same time, the USA announced the ASEAN Co-operation Plan, which was to include activities in the fields of ICT, agricultural biotechnology, health, disaster response and training for the ASEAN Secretariat. ASEAN-Canadian co-operation projects include fisheries technology, the telecommunications industry, use of solar energy, and a forest seed centre. A Working Group on the Revitalization of ASEAN-Canada relations met in February 1999. At a meeting in Bangkok, Thailand, in July 2000, the two sides agreed to explore less formal avenues for project implementation.

In July 1991 the Republic of Korea was accepted as a 'dialogue partner' in ASEAN, and in December a joint ASEAN-Korea Chamber of Commerce was established. In 1995 co-operation projects on human resources development, science and technology, agricultural development and trade and investment policies were implemented. The Republic of Korea participated in ASEAN's informal summit meetings in December 1997 and November 1999 (see above), and took part in the first official ASEAN + 3 meeting of foreign ministers, convened in July 2000. The Republic's assistance in the field of ICT has become particularly valuable in recent years. In March 2001, in a sign of developing co-operation, ASEAN and the Republic of Korea exchanged views on political and security issues in the region for the first time. The ASEAN-Korea Work Programme for 2001–03 covers, among other areas, the environment, transport, science and technology and cultural sectors.

In July 1993 both India and Pakistan were accepted as sectoral partners, providing for their participation in ASEAN meetings in sectors such as trade, transport and communications and tourism. An ASEAN-India Business Council was established, and met for the first time, in New Delhi, in February 1995. In December 1995 the ASEAN summit meeting agreed to enhance India's status to that of a full dialogue partner; India was formally admitted to the PMC in July 1996. At a meeting of the ASEAN-India Working Group in March 2001 the two sides agreed to pursue co-operation in new areas, such as health and pharmaceuticals, social security and rural development. The fourth meeting of the ASEAN-India Joint Co-operation Committee in January 2002 agreed to strengthen co-operation in these areas and others, including technology. The first ASEAN-India consultation between ministers of economic affairs, which took place in September, resulted in the adoption as a long-term objective, of the ASEAN-India Regional Trade and Investment Area. The first ASEAN-India summit at the level of heads of state was held in Phnom Penh, Cambodia, in November. An ASEAN-Pakistan Joint Business Council met for the first time in February 2000. In early 2001 both sides agreed to co-operate in projects relating to new and renewable energy resources, ICT, agricultural research and transport and communications. In October 2003 India acceded to the Treaty on Amity and Co-operation and signed a joint Framework Agreement on Comprehensive Economic Co-operation, which was to enter into effect in July 2004. The objectives of the Agreement included: strengthening and enhancing economic, trade and investment co-operation; liberalizing and promoting trade in goods and services; and facilitating economic integration within ASEAN. Various initiatives were discussed in the fields of agriculture, biotechnology, and human resources development; and both sides adopted the Joint Declaration for Co-operation to Combat International Terrorism. It was also agreed that negotiations would begin on establishing an ASEAN-India Regional Trade and Investment Area (RTIA), including a free-trade area, by 2016.

In March 2000 the first ASEAN-Russia business forum opened in Kuala Lumpur, Malaysia. In 2003 an economic co-operation agreement between Russia and ASEAN was under consideration.

Indo-China: In July 1992 Viet Nam and Laos signed ASEAN's Treaty on Amity and Co-operation and subsequently participated in ASEAN meetings and committees as observers. Viet Nam was admitted as a full member of ASEAN in July 1995. In that month Myanmar signed the Treaty on Amity and Co-operation. In July 1996 ASEAN granted Myanmar observer status and admitted it to the ARF, despite the expression of strong reservations by (among others) the Governments of Australia, Canada and the USA owing to the human rights situation in Myanmar. In November ASEAN heads of government, attending an informal summit meeting in Jakarta, Indonesia, agreed to admit Myanmar as a full member of the grouping at the same time as Cambodia and Laos. While Cambodia's membership was postponed, Laos and Myanmar were admitted to ASEAN in July 1997.

Cambodia was accorded observer status in July 1995. In May 1997 ASEAN ministers of foreign affairs confirmed that Cambodia, together with Laos and Myanmar, was to be admitted to the grouping in July of that year. In mid-July, however, Cambodia's membership was postponed owing to the deposition of Prince Ranariddh, and the resulting civil unrest. Later in that month Cambodia's *de facto* leader, Second Prime Minister Hun Sen, agreed to ASEAN's pursuit of a mediation role in restoring stability in the country and in preparing for democratic elections. In early August the ministers of foreign affairs of Indonesia, the Philippines and Thailand, representing ASEAN, met Hun Sen to confirm these objectives. A team of ASEAN observers joined an international monitoring mission to supervise the election held in Cambodia in July 1998. International approval of the conduct of the election, and consequently of Hun Sen's victory, prompted ASEAN to agree to reconsider Cambodia's admission into the Association. In December, following the establishment of a coalition administration in Cambodia, the country was welcomed, by the Vietnamese Government, as the 10th member of ASEAN, despite an earlier meeting of ministers of foreign affairs failing to reach a consensus decision. Its formal admission took place on 30 April 1999.

In June 1996 ministers of ASEAN countries, and of the People's Republic of China, Cambodia, Laos and Myanmar adopted a framework for ASEAN-Mekong Basin Development Co-operation. The initiative aimed to strengthen the region's cohesiveness, with greater co-operation on issues such as drugs-trafficking, labour migration and terrorism, and to facilitate the process of future expansion of ASEAN. Groups of experts and senior officials were to be convened to consider funding issues and proposals to link the two regions, including a gas pipeline network, rail links and the establishment of a common time zone. In December 1996 the working group on rail links appointed a team of consultants to conduct a feasibility study of the proposals. The completed study was presented at the second ministerial conference on ASEAN-Mekong Basin Development Co-operation, convened in Hanoi, Viet Nam, in July 2000. At the November 2001 summit China pledged US $5m. to assist with navigation along the upper stretches of the Mekong

River, while other means by which China could increase its investment in the Mekong Basin area were considered. At the meeting the Republic of Korea was invited to become a core member of the grouping. Other growth regions sponsored by ASEAN include the Brunei, Indonesia, Malaysia, Philippines, East ASEAN Growth Area (BIMP-EAGA), the Indonesia, Malaysia, Singapore Growth Triangle (IMS-GT), and the West-East Corridor within the Mekong Basin Development initiative.

AGRICULTURE, FISHERIES AND FORESTRY

In October 1983 a ministerial agreement on fisheries co-operation was concluded, providing for the joint management of fish resources, the sharing of technology, and co-operation in marketing. In July 1994 a Conference on Fisheries Management and Development Strategies in the ASEAN region resolved to enhance fish production through the introduction of new technologies, aquaculture development, improvements of product quality and greater involvement by the private sector.

Co-operation in forestry is focused on joint projects, funded by ASEAN's dialogue partners, which include a Forest Tree Seed Centre, an Institute of Forest Management and the ASEAN Timber Technology Centre. In April 1995 representatives of the ASEAN Secretariat and private-sector groups met to co-ordinate the implementation of a scheme to promote the export of ASEAN agricultural and forestry products. In recent years ASEAN has urged member countries to take action to prevent illegal logging in order to prevent the further degradation of forest resources.

ASEAN holds an emergency rice reserve, amounting to 87,000 metric tons, as part of its efforts to ensure food security in the region. There is an established ASEAN programme of training and study exchanges for farm workers, agricultural experts and members of agricultural co-operatives. During 1998 ASEAN was particularly concerned with the impact of the region's economic crisis on the agricultural sector, and the possible effects of climatic change. In September ministers of agriculture and forestry, meeting in Hanoi, Viet Nam, endorsed a Strategic Plan of Action on ASEAN Co-operation in Food, Agriculture and Forestry for 1999–2004. The Plan focused on programmes and activities aimed at enhancing food security, the international competitiveness of ASEAN food, agriculture and forestry products, promoting the sustainable use and conservation of natural resources, encouraging greater involvement by the private sector in the food and agricultural industry, and strengthening joint approaches on international and regional issues. An ASEAN Task Force has been formed to harmonize regulations on agricultural products derived from biotechnology. In December 1998 heads of state resolved to establish an ASEAN Food Security Information Service to enhance the capacity of member states to forecast and manage food supplies. In 1999 agriculture ministers endorsed guide-lines on assessing risk from genetically modified organisms (GMOs) in agriculture, to ensure a common approach. In 2001 work was undertaken to increase public and professional awareness of GMO issues, through workshops and studies.

MINERALS AND ENERGY

The ASEAN Centre for Energy (ACE), based in Jakarta, Indonesia, provides an energy information network, promotes the establishment of interconnecting energy structures among ASEAN member countries, supports the development of renewable energy resources and encourages co-operation in energy efficiency and conservation. An ASEAN energy business forum is held annually and attended by representatives of the energy industry in the private and public sectors. Efforts to establish an ASEAN electricity grid were initiated in 1990. An ASEAN Interconnection Masterplan Study Working Group was established in April 2000 to formulate a study on the power grid. In November 1999 a Trans-ASEAN Gas Pipeline Task Force was established. In July 2002 ASEAN ministers of energy signed a Memorandum of Understanding to implement the pipeline project, involving seven interconnections. The meeting also approved initial plans for the implementation of the regional power grid initiative. ASEAN has forged partnerships with the EU and Japan in the field of energy, under the ASEAN Plan of Action for Energy Co-operation, running from 1999–2004.

A Framework of Co-operation in Minerals was adopted by an ASEAN working group of experts in August 1993. The group has also developed a programme of action for ASEAN co-operation in the development and utilization of industrial minerals, to promote the exploration and development of mineral resources, the transfer of mining technology and expertise, and the participation of the private sector in industrial mineral production. The programme of action is implemented by an ASEAN Regional Development Centre for Mineral Resources, which also conducts workshops and training programmes relating to the sector. In September 2002 ASEAN + 3 ministers of energy agreed joint framework for energy co-operation.

TRANSPORT AND COMMUNICATIONS

ASEAN aims to promote greater co-operation in the transport and communications sector, and in particular, to develop multi-modal transport; to harmonize road transport laws and regulations; to improve air space management; to develop ASEAN legislation for the carriage of dangerous goods and waste by land and sea; and to achieve interoperability and interconnectivity in telecommunications. The summit meeting of December 1998 agreed to work to develop a trans-ASEAN transportation network by 2000, comprising principal routes for the movement of goods and people. (The deadline for the full implementation of the agreement was subsequently moved to the end of 2000.) In September 1999 ASEAN ministers of transport and communications resolved to establish working groups to strengthen co-operation within the sector and adopted a programme of action for development of the sector in 1999–2004. By September 2001, under the action programme, a harmonized road route numbering system had been completed, a road safety implementation work plan agreed, and two pilot courses, on port management and traffic engineering and safety, had been adopted. A Framework Agreement on Facilitation of Goods in Transit entered into force in October 2000. In September 2002 ASEAN transport ministers signed Protocol 9 on Dangerous Goods, one of the implementing protocols under the framework agreement, which provided for the simplification of procedures for the transportation of dangerous goods within region using internationally accepted rules and guide-lines.

ASEAN is seeking to develop a Competitive Air Services Policy, possibly as a first step towards the creation of an ASEAN Open Skies Policy. In September 2000 a meeting of transport ministers agreed to embark on a study to formulate maritime shipping policy, to cover, *inter alia*, issues of transshipment, the competitiveness of ports, liberalization, and the integration of maritime shipping into the overall transport system. In October 2001 ministers approved the third package of commitments for the air and transport sectors under the ASEAN framework agreement on services (according to which member countries were to liberalize the selling and marketing of air and maritime transport services). The summit meeting held in November reaffirmed the large-scale Singapore–Kunming rail link as a priority transport project. Emphasis was also put on smaller-scale (and cheaper) projects in 2001. In September 2002 ASEAN senior transport officials signed the Memorandum of Understanding on Air Freight Services, which represented the first stage in full liberalization of air freight services in the region.

In October 1999 an e-ASEAN initiative was launched, aiming to promote and co-ordinate e-commerce and internet utilization. In November 2000 the informal meeting of ASEAN heads of government approved an e-ASEAN Framework Agreement to further the aims of the initiative. The Agreement incorporated commitments to develop and strengthen ASEAN's information infrastructure, in order to provide for universal and affordable access to communications services. Tariff reduction on ICT products was to be accelerated, with the aim of eliminating all tariffs in the sector by 2010. In July 2001 ministers of foreign affairs discussed measures for the economic liberalization of ICT products and for developing ICT capabilities in poorer member countries. In the same month the first meeting of ASEAN ministers responsible for telecommunications was held, in Kuala Lumpur, Malaysia, during which a Ministerial Understanding on ASEAN co-operation in telecommunications and ICT was signed. In September ASEAN ministers of economic affairs approved a list of ICT products eligible for the elimination of duties under the e-ASEAN Framework Agreement. This was to take place in three annual tranches, commencing in 2003 for the six original members of ASEAN and in 2008 for the newer member countries. During 2001 ASEAN continued to develop a reference framework for e-commerce legislation; it aimed to have e-commerce legislation in place in all member states by 2003. The second ASEAN telecommunications ministerial meeting was held in Manila, the Philippines, in August 2002. The meeting issued a declaration incorporating commitments to exploit ASEAN's competitive position in the field of ICT and to fulfil the obligations of the e-ASEAN Framework Agreement. The ministers also resolved to enhance co-operation with China, Japan and Korea with regard to ICT.

SCIENCE AND TECHNOLOGY

ASEAN's Committee on Science and Technology (COST) supports co-operation in food science and technology, meteorology and geophysics, microelectronics and ICT, biotechnology, non-conventional energy research, materials science and technology, space technology applications, science and technology infrastructure and resources development, and marine science. There is an ASEAN Science Fund, used to finance policy studies in science and technology and to support information exchange and dissemination.

The Hanoi Plan of Action, adopted in December 1998, envisaged a series of measures aimed at promoting development in the fields of science and technology, including the establishment of networks of science and technology centres of excellence and academic institutions, the creation of a technology scan mechanism, the promotion of public- and private-sector co-operation in scientific and technological (particularly ICT) activities, and an increase in research on strategic technologies. In September 2001 the ASEAN Ministerial Meeting on Science and Technology, convened for its first meeting since 1998, approved a new framework for implementation of ASEAN's Plan of Action on Science and Technology during the period 2001–04. The Plan aimed to help less developed member countries become competitive in the sector and integrate into regional co-operation activities. In September 2003 ASEAN and the People's Republic of China inauguarated a Network of East Asian Think-tanks to promote scientific and technological exchange.

ENVIRONMENT

A ministerial meeting on the environment in April 1994 approved long-term objectives on environmental quality and standards for the ASEAN region, aiming to enhance joint action in addressing environmental concerns. At the same time, ministers adopted standards for air quality and river water to be achieved by all ASEAN member countries by 2010. In June 1995 ministers agreed to co-operate to counter the problems of transboundary pollution.

In December 1997 ASEAN heads of state endorsed a Regional Haze Action Plan to address the environmental problems resulting from forest fires, which had afflicted several countries in the region throughout that year. A Haze Technical Task Force undertook to implement the plan in 1998, with assistance from the UN Environment Programme. In March ministers of the environment requested international financial assistance to help mitigate the dangers of forest fires in Indonesia, which had suffered an estimated US $1,000m. in damage in 1997. Sub-regional fire-fighting arrangement working groups for Sumatra and Borneo were established in April 1998 and in May the Task Force organized a regional workshop to strengthen ASEAN capacity to prevent and alleviate the haze caused by the extensive fires. A pilot project of aerial surveillance of the areas in the region most at risk of forest fires was initiated in July. In December heads of government resolved to establish an ASEAN Regional Research and Training Centre for Land and Forest Fire Management by 2004. In March 2002 members of the workings groups on sub-regional fire-fighting arrangements for Sumatra and Borneo agreed to intensify early warning efforts and surveillance activities in order to reduce the risks of forest fires. In June ASEAN ministers of the environment signed an Agreement on Transboundary Haze Pollution, which was intended to provide a legal basis for the Regional Haze Action Plan. The Agreement, which entered into force in November 2003, required member countries to co-operate in the prevention and mitigation of haze pollution, for example, by responding to requests for information by other states and facilitating the transit of personnel and equipment in case of disaster. The Agreement also provided for the establishment of an ASEAN Co-ordination Centre for Transboundary Haze Pollution Control. An ASEAN Specialized Meteorological Centre (ASMC) based in Singapore, plays a primary role in long-range climatological forecasting, early detection and monitoring of fires and haze.

In April 2000 ministers adopted a Strategic Plan of Action on the Environment for 1999–2004. Activities under the Plan focus on issues of coastal and marine erosion, nature conservation and biodiversity, the implementation of multilateral environmental agreements, and forest fires and haze. Other ASEAN environmental objectives include the implementation of a water conservation programme and the formation and adoption of an ASEAN protocol on access to genetic resources. An ASEAN Regional Centre for Biodiversity Conservation (ARCBC) was established in February 1999. It held several workshops in 2000 and 2001, on issues including genetically modified organisms, access to genetic resources and data sharing. In May 2001 environment ministers launched the ASEAN Environment Education Action Plan (AEEAP), with the aim of making citizens 'environmentally literate', and willing and able to participate in sustainable regional development. In November 2003 ASEAN + 3 ministers of environment agreed to prioritize environmental activities in ten key areas: environmentally sustainable cities; global environmental issues; land and forest fires and transboundary haze pollution; coastal and marine environment; sustainable forest management; freshwater resources; public awareness and environmental education; promotion of green technologies and cleaner production; sustainable development monitoring and reporting.

SOCIAL DEVELOPMENT

ASEAN concerns in social development include youth development, the role of women, health and nutrition, education, labour affairs and disaster management. In December 1993 ASEAN ministers responsible for social affairs adopted a Plan of Action for Children, which provided a framework for regional co-operation for the survival, protection and development of children in member countries. ASEAN supports efforts to combat drug abuse and illegal drugs-trafficking. It aims to promote education and drug-awareness campaigns throughout the region, and administers a project to strengthen the training of personnel involved in combating drug abuse. In October 1994 a meeting of ASEAN Senior Officials on Drug Matters approved a three-year plan of action on drug abuse, providing a framework for co-operation in four priority areas: preventive drug education; treatment and rehabilitation; law enforcement; and research. In July 1998 ASEAN ministers of foreign affairs signed a Joint Declaration for a Drug-Free ASEAN, which envisaged greater co-operation among member states, in particular in information exchange, educational resources and legal procedures, in order to eliminate the illicit production, processing and trafficking of narcotic substances by 2020. (This deadline was subsequently advanced to 2015.)

The seventh ASEAN summit meeting, held in November 2001, declared work on combating HIV and AIDS to be a priority. The second phase of a work programme to combat AIDS and provide help for sufferers was endorsed at the meeting. Heads of government expressed their readiness to commit the necessary resources for prevention and care, and to attempt to obtain access to cheaper drugs. More than 1.5m. people in South-East Asia were known to be infected with HIV. An ASEAN task force on AIDS has been operational since March 1993.

In December 1998 ASEAN leaders approved a series of measures aimed at mitigating the social impact of the financial and economic crises that had affected many countries in the region. Plans of Action were formulated on issues of rural development and poverty eradication, while Social Safety Nets, which aimed to protect the most vulnerable members of society, were approved. The summit meeting emphasized the need to promote job generation as a key element of strategies for economic recovery and growth. The fourth meeting of ministers of social welfare in August 2001 noted the need for a holistic approach to social problems, integrating social and economic development. The summit meeting in November considered the widening development gap between ASEAN members and concluded that bridging this gap was a priority, particularly with respect to developing human resources and infrastructure and providing access to ICT.

In January 1992 the ASEAN summit meeting resolved to establish an ASEAN University Network (AUN) to hasten the development of a regional identity. A draft AUN Charter and Agreement were adopted in 1995. The Network aims to strengthen co-operation within the grouping, develop academic and professional human resources and transmit information and knowledge. The 17 universities linked in the network carry out collaborative studies and research programmes. At the seventh ASEAN summit in November 2001 heads of government agreed to establish the first ASEAN University, in Malaysia.

TOURISM

National tourist organizations from ASEAN countries meet regularly to assist in co-ordinating the region's tourist industry, and a Tourism Forum is held annually to promote the sector. The first formal meeting of ASEAN ministers of tourism was held in January 1998, in Cebu, the Philippines. The meeting adopted a Plan of Action on ASEAN Co-operation in Tourism, which aimed to promote intra-ASEAN travel, greater investment in the sector, joint marketing of the region as a single tourist destination and environmentally sustainable tourism. In January 1999 the second meeting of ASEAN ministers of tourism agreed to appoint country co-ordinators to implement various initiatives, including the designation of 2002 as 'Visit ASEAN Millennium Year'; research to promote the region as a tourist destination in the 21st century, and to develop a cruise-ship industry; and the establishment of a network of ASEAN Tourism Training Centres to develop new skills and technologies in the tourism industry by 2001. The third meeting of tourism ministers, held in Bangkok, Thailand, in January 2000, agreed to reformulate the Visit ASEAN Millennium Year initiative as a long-term Visit ASEAN programme. This was formally launched in January 2001 at the fourth ministerial meeting. The first phase of the programme, implemented in 2001, promoted brand awareness through an intense marketing effort; the second phase, initiated at the fifth meeting of tourism ministers, held in Yogyakarta, Indonesia, in January 2002, was to direct campaigns towards end-consumers. Ministers urged member states to abolish all fiscal and non-fiscal travel barriers, to encourage tourism, including intra-ASEAN travel. A seminar on sustainable tourism was held in Malaysia in October 2001. In November 2002 the eighth summit of heads of state adopted a framework agreement on ASEAN co-operation in tourism, aimed at facilitating domestic and intra-regional travel. ASEAN national tourism organizations signed an implementation plan for the agreement in May 2003, when they also announced a Declaration on Tourism Safety and Security. In December 2003 an online heritage network was established with a view to boosting tourism by promoting the region's cultural diversity.

CULTURE AND INFORMATION

Regular workshops and festivals are held in visual and performing arts, youth music, radio, television and films, and print and interpersonal media. In addition, ASEAN administers a News Exchange and provides support for the training of editors, journalists and information officers. In 2000 ASEAN adopted new cultural strategies, with the aims of raising awareness of the grouping's objectives and achievements, both regionally and internationally. The strategies included: producing ASEAN cultural and historical educational materials; promoting cultural exchanges (especially for young people); and achieving greater exposure of ASEAN cultural activities and issues in the mass media.

In July 1997 ASEAN ministers of foreign affairs endorsed the establishment of an ASEAN Foundation to promote awareness of the organization and greater participation in its activities; this was inaugurated in July 1998 and is based at the ASEAN secretariat building (www.aseanfoundation.org).

Publications

Annual Report.

Annual Security Report.

ASEAN Investment Report (annually).

ASEAN State of the Environment Report (1st report: 1997; 2nd report: 2000).

Business ASEAN (quarterly).

Public Information Series, Briefing Papers, Documents Series, educational materials

THE COMMONWEALTH

Address: Commonwealth Secretariat, Marlborough House, Pall Mall, London, SW1Y 5HX, United Kingdom.

Telephone: (20) 7747-6500; **fax:** (20) 7930-0827; **e-mail:** info@commonwealth.int; **internet:** www.thecommonwealth.org.

The Commonwealth is a voluntary association of 53 independent states (at October 2004), comprising more than one-quarter of the world's population. It includes the United Kingdom and most of its former dependencies, and former dependencies of Australia and New Zealand (themselves Commonwealth countries). All Commonwealth countries accept Queen Elizabeth II as the symbol of the free association of the independent member nations and as such the Head of the Commonwealth.

MEMBERS IN THE FAR EAST AND AUSTRALASIA

Australia	Tonga	Tuvalu
Kiribati	Fiji*	Samoa
Nauru	Malaysia	Vanuatu
Papua New Guinea	New Zealand	
	Solomon Islands	

* In October 1987 Fiji's membership was declared to have lapsed (following the proclamation of a republic there). It was readmitted in October 1997, but was suspended from participation in meetings of the Commonwealth in June 2000. Fiji was formally readmitted to Commonwealth meetings in December 2001 following the staging of free and fair legislative elections in August–September.

DEPENDENCIES

Australia	New Zealand	United Kingdom
Christmas Island	Cook Islands	Pitcairn Islands
Cocos (Keeling) Islands	Niue	
Coral Sea Islands Territory	Tokelau	
Norfolk Island		

Organization

(October 2004)

The Commonwealth is not a federation: there is no central government nor are there any rigid contractual obligations such as bind members of the United Nations.

The Commonwealth has no written constitution but its members subscribe to the ideals of the Declaration of Commonwealth Principles unanimously approved by a meeting of heads of government in Singapore in 1971. Members also approved the 1977 statement on apartheid in sport (the Gleneagles Agreement); the 1979 Lusaka Declaration on Racism and Racial Prejudice; the 1981 Melbourne Declaration on relations between developed and developing countries; the 1983 New Delhi Statement on Economic Action; the 1983 Goa Declaration on International Security; the 1985 Nassau Declaration on World Order; the Commonwealth Accord on Southern Africa (1985); the 1987 Vancouver Declaration on World Trade; the Okanagan Statement and Programme of Action on Southern Africa (1987); the Langkawi Declaration on the Environment (1989); the Kuala Lumpur Statement on Southern Africa (1989); the Harare Commonwealth Declaration (1991); the Ottawa Declaration on Women and Structural Adjustment (1991); the Limassol Statement on the Uruguay Round of multilateral trade negotiations (1993); the Millbrook Commonwealth Action Programme on the Harare Declaration (1995); the Edinburgh Commonwealth Economic Declaration (1997); the Fancourt Commonwealth Declaration on Globalization and People-centred Development (1999); the Coolum Declaration on the Commonwealth in the 21st Century: Continuity and Renewal (2002); and the Aso Rock Commonwealth Declaration and Statement on Multilateral Trade (2003).

MEETINGS OF HEADS OF GOVERNMENT

Commonwealth Heads of Government Meetings (CHOGMs) are private and informal and operate not by voting but by consensus. The emphasis is on consultation and exchange of views for co-operation. A communiqué is issued at the end of every meeting. Meetings are normally held every two years in different capitals in the Commonwealth. The last meeting was held in Abuja, Nigeria, in December 2003. The next meeting was scheduled to be held in Malta in 2005.

OTHER CONSULTATIONS

Meetings at ministerial and official level are also held regularly. Since 1959 finance ministers have met in a Commonwealth country in the week prior to the annual meetings of the IMF and the World Bank. Meetings on education, legal, women's and youth affairs are held at ministerial level every three years. Ministers of health hold annual meetings, with major meetings every three years, and ministers of agriculture meet every two years. Ministers of finance, trade, labour and employment, industry, science, and the environment also hold periodic meetings. The first meeting of Commonwealth ministers of tourism was convened in March 2004.

Senior officials—cabinet secretaries, permanent secretaries to heads of government and others—meet regularly in the year between meetings of heads of government to provide continuity and to exchange views on various developments.

COMMONWEALTH SECRETARIAT

The Secretariat, established by Commonwealth heads of government in 1965, operates as an international organization at the service of all Commonwealth countries. It organizes consultations between governments and runs programmes of co-operation. Meetings of heads of government, ministers and senior officials decide these programmes and provide overall direction. A Board of Governors, on which all eligible member governments are represented, meets annually to review the Secretariat's work and approve its budget. The Board is supported by an Executive Committee which convenes four times a year to monitor implementation of the Secretariat's work programme. The Secretariat is headed by a secretary-general, elected by heads of government.

In 2002 the Secretariat was restructured in accordance with a two-year Strategic Plan (2002–04) which aimed to strengthen the effectiveness of the organization to meet the priorities determined by the meeting of Heads of Government held in Coolum, Australia, in March 2002. Under the reorganization the number of deputy secretaries-general was reduced from three to two. Certain work divisions were amalgamated, while new units or sections, concerned with youth affairs, human rights and good offices, were created to strengthen further activities in those fields. Accordingly, the new divisional structure was as follows: Legal and constitutional affairs; Political affairs; Corporate services; Communications and public affairs; Strategic planning and evaluation; Economic affairs; Governance and institutional development; Social transformation programmes; and Special advisory services. In addition there were units responsible for human rights, youth affairs, and project management and referrals, and an Office of the Secretary-General.

The new programme strategies of the Secretariat for 2002–04 were: Good offices, democracy and human rights; Good governance and the rule of law; Gender and youth; Achieving Millennium Development Goals; Vulnerability of small states; and Secretariat governance, management and communications.

Secretary-General: Rt Hon. DONALD (DON) C. MCKINNON (New Zealand).

Deputy Secretaries-General: FLORENCE MUGASHA (Uganda), WINSTON A. COX (Barbados).

Activities

INTERNATIONAL AFFAIRS

In October 1991 heads of government, meeting in Harare, Zimbabwe, issued the Harare Commonwealth Declaration, in which they reaffirmed their commitment to the Commonwealth Principles declared in 1971, and stressed the need to promote sustainable development and the alleviation of poverty. The Declaration placed emphasis on the promotion of democracy and respect for human rights and resolved to strengthen the Commonwealth's capacity to assist countries in entrenching democratic practices. In November 1995 Commonwealth heads of government, convened in New Zealand, formulated and adopted the Millbrook Commonwealth Action Programme on the Harare Declaration, to promote adherence by member countries to the fundamental principles of democracy and human rights (as proclaimed in the 1991 Declaration). The Programme incorporated a framework of measures to be pursued in support of democratic processes and institutions, and actions to be taken in response to violations of the Harare Declaration principles, in particular the unlawful removal of a democratically-elected government. A Commonwealth Ministerial Action Group on the Harare Declaration (CMAG) was to be established to implement this process and to assist the member country involved to comply with the Harare principles. On the basis of this Programme, the leaders suspended Nigeria from the Commonwealth with immediate effect, following the execution by that country's military Government of nine environmental and human rights protesters and a series of other violations of human rights. The meeting determined to expel Nigeria from the Commonwealth if no 'demonstrable progress' had been made towards the establishment of a democratic authority by the time of the next summit meeting. In addition, the Programme formulated measures to promote sustainable development in member countries, which was considered to be an important element in sustaining democracy, and to facilitate consensus-building within the international community.

In December 1995 CMAG convened for its inaugural meeting in London. The Group, comprising the ministers of foreign affairs of Canada, Ghana, Jamaica, Malaysia, New Zealand, South Africa, the United Kingdom and Zimbabwe (with membership to be reconstituted periodically), commenced by considering efforts to restore democratic government in the three Commonwealth countries then under military regimes, i.e. The Gambia, Nigeria and Sierra Leone. At the second meeting of the Group, in April 1996, ministers commended the conduct of presidential and parliamentary elections in Sierra Leone and the announcement by The Gambia's military leaders to proceed with a transition to civilian rule. In June a three-member CMAG delegation visited The Gambia to reaffirm Commonwealth support of the transition process in that country and to identify possible areas of further Commonwealth assistance. In August the Gambian authorities issued a decree removing the ban on political activities and parties, although shortly afterwards they prohibited certain parties and candidates involved in political life prior to the military take-over from contesting the elections. CMAG recommended that in such circumstances no Commonwealth observers should be sent to either the presidential or parliamentary elections, which were held in September 1996 and January 1997 respectively. Following the restoration of a civilian Government in early 1997, CMAG requested the Commonwealth Secretary-General to extend technical assistance to The Gambia in order to consolidate the democratic transition process. In April 1996 it was noted that the human rights situation in Nigeria had continued to deteriorate. CMAG, having pursued unsuccessful efforts to initiate dialogue with the Nigerian authorities, outlined a series of punitive and restrictive measures (including visa restrictions on members of the administration, a cessation of sporting contacts and an embargo on the export of armaments) that it would recommend for collective Commonwealth action in order to exert further pressure for reform in Nigeria. Following a meeting of a high-level delegation of the Nigerian Government and CMAG in June, the Group agreed to postpone the implementation of the sanctions, pending progress on the dialogue. (Canada, however, determined, unilaterally, to impose the measures with immediate effect; the United Kingdom did so in accordance with a decision of the European Union to implement limited sanctions against Nigeria.) A proposed CMAG mission to Nigeria was postponed in August, owing to restrictions imposed by the military authorities on access to political detainees and other civilian activists in that country. In September the Group agreed to proceed with the visit and to delay further a decision on the implementation of sanction measures. CMAG, without the participation of the representative of the Canadian Government, undertook its ministerial mission in November. In July 1997 the Group reiterated the Commonwealth Secretary-General's condemnation of a military coup in Sierra Leone in May, and decided that the country's participation in meetings of the Commonwealth should be suspended pending the restoration of a democratic government.

In October 1997 Commonwealth heads of government, meeting in Edinburgh, United Kingdom, endorsed CMAG's recommendation that the imposition of sanctions against Nigeria be held in abeyance pending the scheduled completion of a transition programme towards democracy by October 1998. It was also agreed that CMAG be formally constituted as a permanent organ to investigate abuses of human rights throughout the Commonwealth. Jamaica and South Africa were to be replaced as members of CMAG by Barbados and Botswana, respectively.

In March 1998 CMAG commended the efforts of the Economic Community of West African States (ECOWAS) in restoring the democratically-elected Government of President Ahmed Tejan Kabbah in Sierra Leone, and agreed to remove all restrictions on Sierra Leone's participation in Commonwealth activities. Later in that month, a representative mission of CMAG visited Sierra Leone to express its support for Kabbah's administration and to consider the country's needs in its process of reconstruction. At the CMAG meeting held in October members agreed that Sierra Leone should no longer be considered under the Group's mandate; however, they urged the Secretary-General to continue to assist that country in the process of national reconciliation and to facilitate negotiations with opposition forces to ensure a lasting cease-fire. A Special Envoy of the Secretary-General was appointed to co-operate with the UN, ECOWAS and the OAU (now African Union—AU) in monitoring the implementation of the Sierra Leone peace process, and the Commonwealth has supported the rebuilding of the Sierra Leone police force. In September 2001 CMAG recommended that Sierra Leone be removed from its remit, but that the Secretary-General should continue to monitor developments there.

In April 1998 the Nigerian military leader, Gen. Sani Abacha, confirmed his intention to conduct a presidential election in August, but indicated that, following an agreement with other political organizations, he was to be the sole candidate. In June, however, Abacha died suddenly. His successor, Gen. Abdulsalam Abubakar, immediately released several prominent political prisoners, and in early July agreed to meet with the Secretaries-General of the UN and the Commonwealth to discuss the release of the imprisoned opposition leader, Chief Moshood Abiola. Abubakar also confirmed his intention to abide by the programme for transition to civilian rule by October. In mid-July, however, shortly before he was to have been liberated, Abiola died. The Commonwealth Secretary-General subsequently endorsed a new transition programme, which provided for the election of a civilian leader in May 1999. In October 1998 CMAG, convened for its 10th formal meeting, acknowledged Abubakar's efforts towards restoring a democratic government and recommended that member states begin to remove sanctions against Nigeria and that it resume participation in certain Commonwealth activities. The Commonwealth Secretary-General subsequently announced a programme of technical assistance to support Nigeria in the planning and conduct of democratic elections. Staff teams from the Commonwealth Secretariat observed local government, and state and governorship elections, held in December and in January 1999, respectively. A Commonwealth Observer Group was also dispatched to Nigeria to monitor preparations and conduct of legislative and presidential elections, held in February. While the Group reported several irregularities in the conduct of the polling, it confirmed that, in general, the conditions had existed for free and fair elections and that the elections were a legitimate basis for the transition of power to a democratic, civilian government. In April CMAG voted to readmit Nigeria to full membership on 29 May, upon the installation of the new civilian administration.

In 1999 the Commonwealth Secretary-General appointed a Special Envoy to broker an agreement in order to end a civil dispute in Honiara, Solomon Islands. An accord was signed in late June, and it was envisaged that the Commonwealth would monitor its implementation. In October a Commonwealth Multinational Police Peace Monitoring Group was stationed in Solomon Islands; this was renamed the Commonwealth Multinational Police Assistance Group in February 2000. Following further internal unrest, however, the Group was disbanded. In June CMAG determined to send a new mission to Solomon Islands in order to facilitate negotiations between the opposing parties, to convey the Commonwealth's concern and to offer assistance. The Commonwealth welcomed the peace accord concluded in Solomon Islands in October, and extended its support to the International Peace Monitoring Team which was established to oversee implementation of the peace accords. CMAG welcomed the conduct of parliamentary elections held in Solomon Islands in December 2001. CMAG removed Solomon Islands from its agenda in December 2003 but was to continue to receive reports from the Secretary General on future developments.

In June 1999 an agreement was concluded between opposing political groups in Zanzibar, having been facilitated by the good offices of the Secretary-General; however, this was only partially implemented. In mid-October a special meeting of CMAG was convened to consider the overthrow of the democratically-elected Government in Pakistan in a military coup. The meeting condemned the action as a violation of Commonwealth principles and urged the new authorities to declare a timetable for the return to democratic rule. CMAG also resolved to send a four-member delegation, comprising the ministers of foreign affairs of Barbados, Canada, Ghana and Malaysia, to discuss this future course of action with the military regime. Pakistan was suspended from participation in meetings of the Commonwealth with immediate effect. The suspension, pending the restoration of a democratic government, was endorsed by heads of government, meeting in November, who requested that CMAG keep the situation in Pakistan under review. At the meeting, held in Durban, South Africa, CMAG was reconstituted to comprise the ministers of foreign affairs of Australia, Bangladesh, Barbados, Botswana, Canada, Malaysia, Nigeria and the United Kingdom. It was agreed that no country would serve for more than two consecutive two-year terms. CMAG was requested to remain actively involved in the post-conflict development and rehabilitation of Sierra Leone and the process of consolidating peace. In addition, it was urged to monitor persistent violations of the Harare Declaration principles in all countries. Heads of government also agreed to establish a new ministerial group on Guyana and to reconvene a ministerial committee on Belize, in order to facilitate dialogue in ongoing territorial disputes with neighbouring countries. The meeting established a 10-member Commonwealth High Level Review Group to evaluate the role and activities of the Commonwealth. In 2000 the Group initiated a programme of consultations to proceed with its mandate and established a working group of experts to consider the Commonwealth's role in supporting information technology capabilities in member countries.

In June 2000, following the overthrow in May of the Fijian Government by a group of armed civilians, and the subsequent illegal detention of members of the elected administration, CMAG suspended Fiji's participation in meetings of the Commonwealth pending the restoration of democratic rule. In September, upon the request of CMAG, the Secretary-General appointed a Special Envoy to support efforts towards political dialogue and a return to democratic rule in Fiji. The Special Envoy undertook his first visit in December. In December 2001, following the staging of democratic legislative elections in August–September, Fiji was readmitted to Commonwealth meetings on the recommendation of CMAG. Fiji was removed from CMAG's agenda in May 2004, although the Group determined to continue to note developments there, as judgments were still pending in the Fiji Supreme Court on unresolved matters concerning the democratic process.

In March 2001 CMAG resolved to send a ministerial mission to Zimbabwe, in order to relay to the government the Commonwealth's concerns at the ongoing violence and abuses of human rights in that country, as well as to discuss the conduct of parliamentary elections and extend technical assistance. The mission was rejected by the Zimbabwe Government, which queried the basis for CMAG's intervention in the affairs of an elected administration. In September, under the auspices of a group of Commonwealth foreign ministers partly derived from CMAG, the Zimbabwe Government signed the Abuja Agreement, which provided for the cessation of illegal occupations of white-owned farms and the resumption of the rule of law, in return for financial assistance to support the ongoing process of land reform in that country. In January 2002 CMAG expressed strong concern at the continuing violence and political intimidation in Zimbabwe. The summit of Commonwealth heads of government convened in early March (see below) also expressed concern at the situation in Zimbabwe, and, having decided on the principle that CMAG should be permitted to engage with any member Government deemed to be in breach of the organization's core values, mandated a Commonwealth Chairperson's Committee on Zimbabwe to determine appropriate action should an impending presidential election (scheduled to be held during that month) be found not to have been conducted freely and fairly. Following the publication by a Commonwealth observer team of an unfavourable report on the conduct of the election, the Committee decided to suspend Zimbabwe from meetings of the Commonwealth for one year. In March 2003 the Committee concluded that the suspension should remain in force pending consideration by the next summit of heads of government.

In March 2002, meeting in Coolum, near Brisbane, Australia, Commonwealth heads of government adopted the Coolum Declaration on the Commonwealth in the 21st Century: Continuity and Renewal, which reiterated commitment to the organization's principles and values. Leaders at the meeting condemned all forms of terrorism; welcomed the Millennium Development Goals declared by the UN; called on the Secretary-General to constitute a high-level expert group on implementing the objectives of the Fancourt Declaration; pledged continued support for small states; and urged renewed efforts to combat the spread of HIV/AIDS. The meeting adopted a report on the future of the Commonwealth drafted by the High Level Review Group. The document recommended strengthening the Commonwealth's role in conflict prevention and resolution and support of democratic practices; enhancing the good offices role of the Secretary-General; better promoting member states' economic and development needs; strengthening the organization's role in facilitating member states' access to international assistance; and promoting increased access to modern information and communications technologies. The meeting expanded CMAG's mandate to enable the Group to consider action against serious violations of the Commonwealth's core values perpetrated by elected administrations (such as that in Zimbabwe, see above) as well as by military regimes. At the summit CMAG was reconstituted to comprise the ministers of foreign affairs of Australia, the Bahamas, Bangladesh, Botswana, India, Malta, Nigeria and Samoa.

A Commonwealth team of observers dispatched to monitor legislative and provincial elections that were held in Pakistan, in October 2002, found them to have been well-organized and conducted in a largely transparent manner. The team made several recommendations on institutional and procedural issues. CMAG subsequently expressed concern over the promulgation of new legislation in Pakistan following the imposition earlier in the year of a number of extra-constitutional measures. CMAG determined that Pakistan should continue to be suspended from meetings of the Commonwealth, pending a review of the role and functioning of its democratic institutions. In May 2003 CMAG welcomed Pakistan's progress in establishing democratic institutions and agreed to review that country's suspension from the Commonwealth at its next meeting. In November 2002 a Commonwealth Expert Group on Papua New Guinea, established in the previous month to review the electoral process in that country (in view of unsatisfactory legislative elections that were conducted there in July), made several recommendations aimed at enhancing the future management of the electoral process.

In December 2003 the meeting of Heads of Government, held in Abuja, Nigeria, resolved to maintain the suspension of Pakistan and Zimbabwe from participation in Commonwealth meetings. President Mugabe of Zimbabwe responded by announcing his country's immediate withdrawal from the Commonwealth and alleging a pro-Western bias within the grouping. Support for Zimbabwe's position was declared by a number of members, including South Africa, Mozambique, Namibia and Zambia. A Commonwealth committee, consisting of six heads of government, was established to monitor the situation in Zimbabwe and only when the committee believed sufficient progress had been made towards consolidating democracy and promoting development within Zimbabwe would the Commonwealth be consulted on readmitting the country.

In concluding the 2003 meeting heads of government issued the Aso Rock Commonwealth Declaration which emphasized their commitment to strengthening development and democracy, and incorporated clear objectives in support of these goals. Priority areas identified included efforts to eradicate poverty and attain the Millennium Development Goals, strengthening democratic institutions, empowering women, promoting the involvement of civil society, combating corruption and recovering assets (for which a working group was to be established), facilitating finance for development, efforts to address the spread of HIV/AIDS and other diseases, combating the illicit trafficking in human beings, and promoting education. The leaders also adopted a separate statement on multilateral trade, in particular in support of the stalled Doha round of World Trade Organization negotiations.

Political Affairs Division: assists consultation among member governments on international and Commonwealth matters of common interest. In association with host governments, it organizes the meetings of heads of government and senior officials. The Division services committees and special groups set up by heads of government dealing with political matters. The Secretariat has observer status at the United Nations, and the Division manages a joint office in New York to enable small states, which would otherwise be unable to afford facilities there, to maintain a presence at the United Nations. The Division monitors political developments in the Commonwealth and international progress in such matters as disarmament and the Law of the Sea. It also undertakes research on matters of common interest to member governments, and reports back to them. The Division is involved in diplomatic training and consular co-operation.

In 1990 Commonwealth heads of government mandated the Division to support the promotion of democracy by monitoring the preparations for and conduct of parliamentary, presidential or other elections in member countries at the request of national governments. In the two years from mid-2001 Commonwealth observer groups attended elections in nine countries. Experts were dispatched to monitor voter registration in Malawi in January 2004 and to observe a general election in Antigua and Barbuda in March.

Under the reorganization of the Secretariat in 2002 a Good Offices Section was established within the Division to strengthen and support the activities of the Secretary-General in addressing political conflict in member states and in assisting countries to adhere to the principles of the Harare Declaration. The Secretary-General's good offices may be directed to preventing or resolving conflict and assisting other international efforts to promote political stability.

Human Rights Unit: undertakes activities in support of the Commonwealth's commitment to the promotion and protection of fundamental human rights. It develops programmes, publishes human rights materials, co-operates with other organizations working in the field of human rights, in particular within the UN system, advises the Secretary-General, and organizes seminars and meetings of experts. The Unit aims to integrate human rights standards within all divisions of the Secretariat.

LAW

Legal and Constitutional Affairs Division: promotes and facilitates co-operation and the exchange of information among member governments on legal matters and assists in combating financial and organized crime, in particular transborder criminal activities. It administers, jointly with the Commonwealth of Learning, a distance training programme for legislative draftsmen and assists governments to reform national laws to meet the obligations of international conventions. The Division organizes the triennial meeting of ministers, Attorneys General and senior ministry officials concerned with the legal systems in Commonwealth countries. It has also initiated four Commonwealth schemes for co-operation on extradition, the protection of material cultural heritage, mutual assistance in criminal matters and the transfer of convicted offenders within the Commonwealth. It liaises with the Commonwealth Magistrates' and Judges' Association, the Commonwealth Legal Education Association, the Commonwealth Lawyers' Association (with which it helps to prepare the triennial Commonwealth Law Conference for the practising profession), the Commonwealth Association of Legislative Counsel, and with other international non-governmental organizations. The Division provides in-house legal advice for the Secretariat. The *Commonwealth Law Bulletin*, published twice a year, reports on legal developments in and beyond the Commonwealth. The Division promotes the exchange of information regarding national and international efforts to combat serious commercial crime through a quarterly publication, *Commonwealth Legal Assistance News* and the *Crimewatch* bulletin.

The Heads of Government meeting, held in Coolum, Australia, in March 2002 endorsed a Plan of Action for combating international terrorism. A Commonwealth Committee on Terrorism, convened at ministerial level, was subsequently established to oversee its implementation.

A new expert group on good governance and the elimination of corruption in economic management convened for its first meeting in May 1998. In November 1999 Commonwealth heads of government endorsed a Framework for Principles for Promoting Good Governance and Combating Corruption, which had been drafted by the group. The conference of heads of government that met in Coolum in March 2002, endorsed a Commonwealth Local Government Good Practice Scheme, to be managed by the Commonwealth Local Government Forum (established in 1995).

ECONOMIC CO-OPERATION

In October 1997 Commonwealth heads of government, meeting in Edinburgh, the United Kingdom, signed an Economic Declaration that focused on issues relating to global trade, investment and development and committed all member countries to free-market economic principles. The Declaration also incorporated a provision for the establishment of a Trade and Investment Access Facility within the Secretariat in order to assist developing member states in the process of international trade liberalization and promote intra-Commonwealth trade.

In May 1998 the Commonwealth Secretary-General appealed to the Group of Eight industrialized nations to accelerate and expand the initiative to ease the debt burden of the most heavily indebted poor countries—HIPCs) (see World Bank and IMF). In October Commonwealth finance ministers, convened in Ottawa, Canada, reiterated their appeal to international financial institutions to accelerate the HIPC initiative. The meeting also issued a Commonwealth Statement on the global economic crisis and endorsed several proposals to help to counter the difficulties experienced by several countries. These measures included a mechanism to enable countries to suspend payments on all short-term financial obligations at a time of emergency without defaulting, assistance to governments to attract private capital and to manage capital market volatility, and the development of international codes of conduct regarding financial and monetary policies and corporate governance. In March 1999 the Commonwealth Secretariat hosted a joint IMF-World Bank conference to review the HIPC scheme and initiate a process of reform. In November Commonwealth heads of government, meeting in South Africa, declared their support for measures undertaken by the World Bank and IMF to enhance the HIPC initiative. At the end of an informal retreat the leaders adopted the Fancourt Commonwealth Declaration on Globalization and People-Centred Development, which emphasized the need for a more equitable spread of wealth generated by the process of globalization, and expressed a renewed commitment to the elimination of all forms of discrimination, the promotion of people-centred development and capacity-building, and efforts to ensure developing countries benefit from future multilateral trade liberalization measures. In June 2002 the Commonwealth Secretary-General urged more generous funding of the HIPC initiative. Meetings of ministers of finance from Commonwealth African member countries participating in the HIPC initiative have been convened in Lilongwe, Malawi in February 2002, in London, the United Kingdom, in September, and in Dar es Salaam, Tanzania, in March 2003. The Secretariat aims to assist HIPCs and other small economies through its Debt Recording and Management Sytem, which was first used in 1985 and updated in 2002.

In February 1998 the Commonwealth Secretariat hosted the first Inter-Governmental Organizations Meeting to promote co-operation between small island states and the formulation of a unified policy approach to international fora. A second meeting was convened in March 2001, where discussions focused on the forthcoming WTO ministerial meeting and OECD's harmful tax competition initiative. In September 2000 Commonwealth finance ministers, meeting in Malta, reviewed the OECD initiative and agreed that the measures, affecting many member countries with offshore financial centres, should not be imposed on governments. The ministers mandated the involvement of the Commonwealth Secretariat in efforts to resolve the dispute; a joint working group was subsequently established by the Secretariat with the OECD. In April 2002 a meeting on international co-operation in the financial services sector, attended by representatives of international and regional organizations, donors and senior officials from Commonwealth countries, was held under Commonwealth auspices in Saint Lucia.

The first meeting of governors of central banks from Commonwealth countries was held in June 2001 in London.

Economic Affairs Division: organizes and services the annual meetings of Commonwealth ministers of finance and the ministerial group on small states and assists in servicing the biennial meetings of heads of government and periodic meetings of environment ministers. It engages in research and analysis on economic issues of interest to member governments and organizes seminars and conferences of government officials and experts. The Division undertook a major programme of technical assistance to enable developing Commonwealth countries to participate in the Uruguay Round of multilateral trade negotiations and has assisted the African, Caribbean and Pacific (ACP) group of countries in their trade negotiations with the European Union. It continues to help developing countries to strengthen their links with international capital markets and foreign investors. The Division also services groups of experts on economic affairs that have been commissioned by governments to report on, among other things, protectionism; obstacles to the North-South negotiating process; reform of the international financial and trading system; the debt crisis; management of technological change; the impact of change on the development process; environmental issues; women and structural adjustment; and youth unemployment. A separate section within the Division addresses the specific needs of small states and provides technical assistance. The work of the section covers a range of issues including trade, vulnerability, environment, politics and economics. A Secretariat Task Force services a Commonwealth Ministerial Group of Small States which was established in 1993 to provide strategic direction in addressing the concerns of small states and to mobilise support for action and assistance within the international community. The Economic Affairs Division also co-ordinates the Secretariat's environmental work and manages the Iwokrama International Centre for Rainforest Conservation and Development.

The Division played a catalytic role in the establishment of a Commonwealth Equity Fund, initiated in September 1990, to allow developing member countries to improve their access to private institutional investment, and promoted a Caribbean Investment Fund. The Division supported the establishment of a Commonwealth Private Investment Initiative (CPII) to mobilize capital, on a regional basis, for investment in newly-privatized companies and in small and medium-sized businesses in the private sector. The first regional fund under the CPII was launched in July 1996. The Commonwealth Africa Investment Fund (Comafin), was to be managed by the United Kingdom's official development institution, the

Commonwealth Development Corporation, to assist businesses in 19 countries in sub-Saharan Africa, with initial resources of US $63.5m. In August 1997 a fund for the Pacific Islands was launched, with an initial capital of $15.0m. A $200m. South Asia Regional Fund was established at the Heads of Government Meeting in October. In October 1998 a fund for the Caribbean states was inaugurated, at a meeting of Commonwealth finance ministers. The 2001 summit of Commonwealth heads of government authorized the establishment of a new fund for Africa (Comafin II): this was inaugurated in March 2002, and attracted initial capital in excess of $200m.

SOCIAL WELFARE

Social Transformation Programmes Division: consists of three sections concerned with education, gender and health.

The **Education Section** arranges specialist seminars, workshops and co-operative projects, and commissions studies in areas identified by ministers of education, whose three-yearly meetings it also services. Its present areas of emphasis include improving the quality of and access to basic education; strengthening the culture of science, technology and mathematics education in formal and non-formal areas of education; improving the quality of management in institutions of higher learning and basic education; improving the performance of teachers; strengthening examination assessment systems; and promoting the movement of students between Commonwealth countries. The Section also promotes multi-sectoral strategies to be incorporated in the development of human resources. Emphasis is placed on ensuring a gender balance, the appropriate use of technology, promoting good governance, addressing the problems of scale particular to smaller member countries, and encouraging collaboration between governments, the private sector and other non-governmental organizations.

The **Gender Affairs Section** is responsible for the implementation of the 1995 Commonwealth Plan of Action on Gender and Development, which was endorsed by the Heads of Government in order to achieve gender equality in the Commonwealth. The main objective of the Plan is to ensure that gender is incorporated into all policies, programmes, structures and procedures of member states and of the Commonwealth Secretariat. A further gender equality plan, 'Advancing the Commonwealth Agenda in the New Millennium', covers the period 2000–05. The Section also addresses specific concerns such as the integration of gender issues into national budgetary processes, increasing the participation of women in politics and conflict prevention and resolution (with the objective of raising the level of female participation to 30%), and the promotion of human rights, including the elimination of violence against women and girls.

The **Health Section** organizes ministerial, technical and expert group meetings and workshops, to promote co-operation on health matters, and the exchange of health information and expertise. The Section commissions relevant studies and provides professional and technical advice to member countries and to the Secretariat. It also supports the work of regional health organizations and promotes health for all people in Commonwealth countries.

Youth Affairs: A Youth Affairs Unit, reporting directly to a Deputy Secretary-General, was established within the Secretariat in 2002.

The Unit administers the **Commonwealth Youth Programme (CYP)**, which was initiated in 1973 to promote the involvement of young people in the economic and social development of their countries. The CYP, funded through separate voluntary contributions from governments, was awarded a budget of £2.3m. for 2002–03. The Programme's activities are centred on four key programmes: Youth enterprise and development; Youth networks and governance; Youth participation; and Youth work, education and training. Regional centres are located in Zambia (for Africa), India (for Asia), Guyana (for the Caribbean), and Australia (for the South Pacific). The Programme administers a Youth Study Fellowship scheme, a Youth Project Fund, a Youth Exchange Programme (in the Caribbean), and a Youth Service Awards Scheme. It also holds conferences and seminars, carries out research and disseminates information. In 1995 a Commonwealth Youth Credit Initiative was launched, in order to provide funds, training and advice to young entrepreneurs. In May 1998 a Commonwealth ministerial meeting, held in Kuala Lumpur, Malaysia, approved a new Plan of Action on Youth Empowerment to the Year 2005.

In March 2002 Commonwealth heads of government approved the Youth for the Future initiative to encourage and use the skills of young people throughout the Commonwealth. It was to comprise four main components: Youth enterprise development; Youth volunteers; Youth mentors; and Youth leadership awards.

TECHNICAL ASSISTANCE

Commonwealth Fund for Technical Co-operation—CFTC: f. 1971 to facilitate the exchange of skills between member countries and to promote economic and social development; it is administered by the Commonwealth Secretariat and financed by voluntary subscriptions from member governments. The CFTC responds to requests from member governments for technical assistance, such as the provision of experts for short- or medium-term projects, advice on economic or legal matters, in particular in the areas of natural resources management and public-sector reform, and training programmes. The CFTC also administers the Langkawi awards for the study of environmental issues, which is funded by the Canadian Government; the CFTC budget for 2003–04 amounted to £23m.

CFTC activities are mainly implemented by the following divisions:

Governance and Institutional Development Division: strengthens good governance in member countries, through advice, training and other expertise in order to build capacity in national public institutions. The Division administers the Commonwealth Service Abroad Programme (CSAP), which is funded by the CFTC. The Programme extends short-term technical assistance through highly qualified volunteers. The main objectives of the scheme are to provide expertise, training and exposure to new technologies and practices, to promote technology transfers and sharing of experiences and knowledge, and to support community workshops and other grassroots activities.

Special Advisory Services Division: advises on economic and legal issues, such as debt and financial management, natural resource development, multilateral trade issues, export marketing, trade facilitation, competitiveness and the development of enterprises.

Finance

The Secretariat's budget for 2002–03 was £11.4m. Member governments meet the cost of the Secretariat through subscriptions on a scale related to income and population.

Publications

Commonwealth Currents (quarterly).

International Development Policies (quarterly).

Report of the Commonwealth Secretary-General (every 2 years).

Numerous reports, studies and papers (catalogue available).

Commonwealth Organizations

(In the United Kingdom, unless otherwise stated)

PRINCIPAL BODIES

Commonwealth Business Council: 18 Pall Mall, London, SW1Y 5LU; tel. (20) 7024-8200; fax (20) 7930-3944; e-mail info@cbcglobelink.org; internet www.cbcglobelink.org, f. 1997 by the Commonwealth Heads of Government Meeting to promote co-operation between governments and the private sector in support of trade, investment and development; the Council aims to identify and promote investment opportunities, in particular in Commonwealth developing countries, to support countries and local businesses to work within the context of globalization, to promote capacity-building and the exchange of skills and knowledge (in particular through its Information Communication Technologies for Development programme), and to encourage co-operation among Commonwealth members; promotes good governance; supports the process of multilateral trade negotiations and other liberalization of trade and services; represents the private sector at government level; Dir-Gen. and CEO Dr MOHAN KAUL.

Commonwealth Foundation: Marlborough House, Pall Mall, London, SW1Y 5HY; tel. (20) 7930-3783; fax (20) 7839-8157; e-mail geninfo@commonwealth.int; internet www.commonwealthfoundation.com; f. 1966; intergovernmental body promoting people-to-people interaction, and collaboration within the non-governmental sector of the Commonwealth; supports non-governmental organizations, professional associations and Commonwealth arts and culture; awards an annual Commonwealth Writers' Prize; funds are provided by Commonwealth govts; Chair. GRAÇA MACHEL (Mozambique); Dir COLIN BALL (United Kingdom); publ. *Commonwealth People* (quarterly).

Commonwealth of Learning—COL: 1055 West Hastings St, Suite 1200, Vancouver, BC V6E 2E9, Canada; tel. (604) 775-8200;

fax (604) 775-8210; e-mail info@col.org; internet www.col.org; f. 1987 by Commonwealth Heads of Government to promote the devt and sharing of distance education and open learning resources, including materials, expertise and technologies, throughout the Commonwealth and in other countries; implements and assists with national and regional educational programmes; acts as consultant to international agencies and national governments; conducts seminars and studies on specific educational needs; COL is financed by Commonwealth governments on a voluntary basis; in 1999 heads of government endorsed an annual core budget for COL of US $9m; Pres. and CEO Sir JOHN DANIEL (Canada/UK); publs *Connections*, *EdTech News*.

The following represents a selection of other Commonwealth organizations:

AGRICULTURE AND FORESTRY

Commonwealth Forestry Association: POB 142, Bicester, OX1 6ZJ; tel. (1865) 820935; fax (1865) 324805; e-mail cfa@cfa-international.org; internet www.cfa-international.org; f. 1921; produces, collects and circulates information relating to world forestry and promotes good management, use and conservation of forests and forest lands throughout the world; mems: 1,000; Chair. Prof. J. BURLEY; publs *International Forestry Review* (quarterly), *Commonwealth Forestry News* (quarterly), *Commonwealth Forestry Handbook* (irregular).

Standing Committee on Commonwealth Forestry: Forestry Commission, 231 Corstorphine Rd, Edinburgh, EH12 7AT; tel. (131) 314-6137; fax (131) 316-4344; e-mail libby.jones@forestry.gsi.gov.uk; f. 1923 to provide continuity between Confs, and to provide a forum for discussion on any forestry matters of common interest to mem. govts which may be brought to the Cttee's notice by any mem. country or organization; 54 mems; 2005 Conference: Sri Lanka; Sec. LIBBY JONES; publ. *Newsletter* (quarterly).

COMMONWEALTH STUDIES

Institute of Commonwealth Studies: 28 Russell Sq., London, WC1B 5DS; tel. (20) 7862-8844; fax (20) 7862-8820; e-mail ics@sas.ac.uk; internet www.sas.ac.uk/commonwealthstudies; f. 1949 to promote advanced study of the Commonwealth; provides a library and meeting place for postgraduate students and academic staff engaged in research in this field; offers postgraduate teaching; Dir Prof. TIMOTHY SHAW; publs *Annual Report*, *Collected Seminar Papers*, *Newsletter*, *Theses in Progress in Commonwealth Studies*.

COMMUNICATIONS

Commonwealth Telecommunications Organization: Clareville House, 26–27 Oxendon St, London, SW1Y 4EL; tel. (20) 7930-5511; fax (20) 7930-4248; e-mail info@cto.int; internet www.cto.int; f. 1967 as an international development partnership between Commonwealth and non-Commonwealth governments, business and civil society organizations; aims to help to bridge the digital divide and to achieve social and economic development by delivering to developing countries knowledge-sharing programmes in the use of information and communication technologies (ICT) in the specific areas of telecommunications, IT, broadcasting and the internet; CEO Dr EKWOW SPIO-GARBRAH; publs *CTO Update* (quarterly), *Annual Report*, *Research Reports*.

EDUCATION AND CULTURE

Association of Commonwealth Universities—ACU: John Foster House, 36 Gordon Sq., London, WC1H 0PF; tel. (20) 7380-6700; fax (20) 7387-2655; e-mail info@acu.ac.uk; internet www.acu.ac.uk; f. 1913; promotes international co-operation and understanding; provides assistance with staff and student mobility and development programmes; researches and disseminates information about universities and relevant policy issues; organizes major meetings of Commonwealth universities and their representatives; acts as a liaison office and information centre; administers scholarship and fellowship schemes; operates a policy research unit; mems: 500 universities in 36 Commonwealth countries or regions; Sec.-Gen. Dr JOHN ROWETT; publs include *Yearly Review* (annually), *Commonwealth Universities Yearbook* (annually), *ACU Bulletin* (quarterly), *Report of the Council of the ACU* (annually), *Who's Who of Executive Heads: Vice-Chancellors, Presidents, Principals and Rectors*, *International Awards*, student information papers (study abroad series).

Commonwealth Association for Education in Journalism and Communication—CAEJC: c/o Faculty of Law, University of Western Ontario, London, ON N6A 3K7, Canada; tel. (519) 6613348; fax (519) 6613790; e-mail caejc@julian.uwo.ca; f. 1985; aims to foster high standards of journalism and communication education and research in Commonwealth countries and to promote co-operation among institutions and professions; c. 700 mems in 32 Commonwealth countries; Pres. Prof. SYED ARABI IDID (Malaysia); Sec. Prof. ROBERT MARTIN (Canada); publ. *CAEJAC Journal* (annually).

Commonwealth Association of Science, Technology and Mathematics Educators—CASTME: 7 Lion Yard, Tremadoc Rd, London, SW4 7NQ; tel. (20) 7819–3932; fax (20) 7720–5403; e-mail ann.powell@lect.org.uk; internet www.castme.org; f. 1974; special emphasis is given to the social significance of education in these subjects; organizes an Awards Scheme to promote effective teaching and learning in these subjects, and biennial regional seminars; Hon. Sec. Dr LYN HAINES; publ. *CASTME Journal* (quarterly).

Commonwealth Council for Educational Administration and Management: International Educational Leadership and Management Centre, Lincoln School of Management, Lincoln University Campus, Brayford Pool, Lincoln LNG 7TS; tel. (1522) 886071; fax (1522) 886023; f. 1970; aims to foster quality in professional development and links among educational administrators; holds nat. and regional confs, as well as visits and seminars; mems: 24 affiliated groups representing 3,000 persons; Pres. Prof. ANGELA THODAY; publ. *International Studies in Educational Administration* (2 a year).

Commonwealth Institute: 230 Kensington High St, London, W8 6NQ; tel. (20) 7603-4535; fax (20) 7603-4525; e-mail info@commonwealth.org.uk; internet www.commonwealth.org.uk; f. 1893 as the Imperial Institute; restructured as an independent pan-Commonwealth agency Jan. 2000; governed by a Bd of Trustees elected by the Bd of Governors; Commonwealth High Commissioners to the United Kingdom act as *ex-officio* Governors; the Inst. houses a Commonwealth Resource and Literature Library and a Conference and Events Centre; supplies educational resource materials and training throughout the United Kingdom; provides internet services to the Commonwealth; operates as an arts and conference centre, running a Commonwealth-based cultural programme; a five-year strategic plan, entitled 'Commonwealth 21', was inaugurated in 1998; in 2004 the Institute, in collaboration with Cambridge University, established a Centre of Commonwealth Education; Chair. JUDITH HANRATTY; Chief Exec. DAVID FRENCH; publ. *Annual Review*.

League for the Exchange of Commonwealth Teachers: 7 Lion Yard, Tremadoc Rd, London, SW4 7NQ; tel. (20) 7498-1101; fax (20) 7720-5403; e-mail info@lect.org.uk; internet www.lect.org.uk; f. 1901; promotes educational exchanges between teachers in Australia, the Bahamas, Barbados, Bermuda, Canada, Guyana, India, Jamaica, Kenya, Malawi, New Zealand, Pakistan, South Africa and Trinidad and Tobago; Dir ANNA TOMLINSON; publs *Annual Report*, *Exchange Teacher* (annually).

HEALTH

Commonwealth Medical Association: BMA House, Tavistock Sq., London, WC1H 9JP; tel. (20) 7272-8492; fax (20) 7272-1663; e-mail office@commat.org; f. 1962 for the exchange of information; provision of tech. co-operation and advice; formulation and maintenance of a code of ethics; liaison with WHO and other UN agencies on health issues; meetings of its Council are held every three years; mems: medical asscns in Commonwealth countries; Pres. Dr P. KRISHAN; Sec. (vacant).

Commonwealth Pharmaceutical Association: 1 Lambeth High St, London, SE1 7JN; tel. (20) 7572-2364; fax (20) 7572-2508; e-mail admin@commonwealthpharmacy.org; internet www.commonwealthpharmacy.org; f. 1970 to promote the interests of pharmaceutical sciences and the profession of pharmacy in the Commonwealth to maintain high professional standards, encourage links between members and the creation of nat. asscns; and to facilitate the dissemination of information; holds confs (every four years) and regional meetings; mems: pharmaceutical asscns from over 40 Commonwealth countries; Pres. GRACE ALLEN YOUNG; publ. *Quarterly Newsletter*.

Commonwealth Society for the Deaf: 34 Buckingham Palace Rd, London, SW1W 0RE; tel. (20) 7233-5700; fax (20) 7233-5800; e-mail sound.seekers@btinternet.com; internet www.sound-seekers.org.uk; undertakes initiatives to establish audiology services in developing Commonwealth countries, including mobile clinics to provide outreach services; aims to educate local communities in aural hygiene and the prevention of ear infection and deafness; provides audiological equipment and organizes the training of audiological maintenance technicians; conducts research into the causes and prevention of deafness; Chief Exec. GARY WILLIAMS; publ. *Annual Report*.

Sight Savers International: Grosvenor Hall, Bolnore Rd, Haywards Heath, West Sussex, RH16 4BX; tel. (1444) 446600; fax (1444) 446688; e-mail generalinformation@sightsavers.org.uk; internet www.sightsavers.org; f. 1950 to prevent blindness and restore sight in developing countries, and to provide education and community-based training for incurably blind people; operates in collaboration with local partners, with high priority given to training local staff; Chair. Sir JOHN COLES; Dir RICHARD PORTER; publ. *Sight Savers News*.

INFORMATION AND THE MEDIA

Commonwealth Broadcasting Association: 17 Fleet St, London, EC4Y 1AA; tel. (20) 7583-5550; fax (20) 7583-5549; e-mail cba@cba.org.uk; internet www.cba.org.uk; f. 1945; gen. confs are held every two years (2004: Fiji); mems: 97 in 57 countries; Pres. GEORGE VALARINO; Sec.-Gen. ELIZABETH SMITH; publs *Commonwealth Broadcaster* (quarterly), *Commonwealth Broadcaster Directory* (annually).

Commonwealth Institute: see under Education and Culture.

Commonwealth Journalists Association: internet www.commonwealthjournalists.org; f. 1978 to promote co-operation between journalists in Commonwealth countries, organize training facilities and confs, and foster understanding among Commonwealth peoples; Pres. RAY EKPU; Exec. Dir LAWRIE BREEN; publ. *Newsletter* (3 a year).

Commonwealth Press Union (Association of Commonwealth Newspapers, News Agencies and Periodicals): 17 Fleet St, London, EC4Y 1AA; tel. (20) 7583-7733; fax (20) 7583-6868; e-mail lindsay@cpu.org.uk; internet www.cpu.org.uk; f. 1950; promotes the welfare of the Commonwealth press; provides training for journalists and organizes biennial confs; mems: c. 700 newspapers, news agencies, periodicals in 42 Commonwealth countries; Exec. Dir LINDSAY ROSS; publ. *Annual Report*.

LAW

Commonwealth Lawyers' Association: c/o Institute of Commonwealth Studies, 28 Russell Sq., London, WC1B 5DS; tel. (20) 7862-8824; fax (20) 7862-8816; e-mail cla@sas.ac.uk; internet www.commonwealthlawyers.com; f. 1983 (fmrly the Commonwealth Legal Bureau); seeks to maintain and promote the rule of law throughout the Commonwealth, by ensuring that the people of the Commonwealth are served by an independent and efficient legal profession; upholds professional standards and promotes the availability of legal services; assists in organizing the triennial Commonwealth law confs; Exec. Sec. CLAIRE MARTIN; publs *The Commonwealth Lawyer*, *Clarion*.

Commonwealth Legal Advisory Service: c/o British Institute of International and Comparative Law, Charles Clore House, 17 Russell Sq., London, WC1B 5DR; tel. (20) 7636-5802; fax (20) 7323-2016; e-mail bicl@dial.pipex.com; f. 1962; financed by the British Institute and by contributions from Commonwealth govts; provides research facilities for Commonwealth govts and law reform commissions; Chair. Rt Hon. Lord BROWNE-WILKINSON; publ. *New Memoranda* series.

Commonwealth Legal Education Association: c/o Legal and Constitutional Affairs Division, Commonwealth Secretariat, Marlborough House, Pall Mall, London, SW1Y 5HX; tel. (20) 7747-6415; fax (20) 7747-6406; e-mail clea@commonwealth.int; internet www.clea.org.uk; f. 1971 to promote contacts and exchanges and to provide information regarding legal education; Gen. Sec. JOHN HATCHARD; publs *Commonwealth Legal Education Association Newsletter* (3 a year), *Directory of Commonwealth Law Schools* (every 2 years).

Commonwealth Magistrates' and Judges' Association: Uganda House, 58/59 Trafalgar Sq., London, WC2N 5DX; tel. (20) 7976-1007; fax (20) 7976-2395; e-mail info@cmja.org; internet www.cmja.org; f. 1970 to advance the administration of the law by promoting the independence of the judiciary, to further education in law and crime prevention and to disseminate information; confs and study tours; corporate membership for asscns of the judiciary or courts of limited jurisdiction; assoc. membership for individuals; Pres. Hon. Chief Justice RICHARD BANDA; Exec. Vice-Pres. MICHAEL A. LAMBERT; publs *Commonwealth Judicial Journal* (2 a year), *CMJA News*.

PARLIAMENTARY AFFAIRS

Commonwealth Parliamentary Association: Westminster House, Suite 700, 7 Millbank, London, SW1P 3JA; tel. (20) 7799-1460; fax (20) 7222-6073; e-mail hq.sec@cpahq.org; internet www.cpahq.org; f. 1911 to promote understanding and co-operation between Commonwealth parliamentarians; organization: Exec. Cttee of 32 MPs responsible to annual Gen. Assembly; 170 brs in national, state, provincial and territorial parliaments and legislatures throughout the Commonwealth; holds annual Commonwealth Parliamentary Confs and seminars; also regional confs and seminars; Sec.-Gen. Hon. DENIS MARSHALL; publ. *The Parliamentarian* (quarterly).

PROFESSIONAL AND INDUSTRIAL RELATIONS

Commonwealth Association of Architects: 66 Portland Pl., London, W1N 4AD; tel. (20) 7490-3024; fax (20) 7253-2592; e-mail info@comarchitect.org; internet comarchitect.org; f. 1964; an asscn of 38 socs of architects in various Commonwealth countries; objectives: to facilitate the reciprocal recognition of professional qualifications; to provide a clearing house for information on architectural practice; and to encourage collaboration. Plenary confs every three years; regional confs are also held; Exec. Dir TONY GODWIN; publs *Handbook, Objectives and Procedures: CAA Schools Visiting Boards, Architectural Education in the Commonwealth* (annotated bibliography of research), *CAA Newsnet* (2 a year), a survey and list of schools of architecture.

Commonwealth Association for Public Administration and Management—CAPAM: 1075 Bay St, Suite 402, Toronto, ON M5S 2B1, Canada; tel. (416) 920-3337; fax (416) 920-6574; e-mail capam@capam.ca; internet www.capam.comnet.mt/; f. 1994; aims to promote sound management of the public sector in Commonwealth countries and to assist those countries undergoing political or financial reforms; an international awards programme to reward innovation within the public sector was introduced in 1997, and is awarded every 2 years; more than 1,200 individual mems and 80 institutional memberships in some 80 countries; Pres. Hon. JOCELYNE BOURGON (Canada); Exec. Dir ART STEVENSON (Canada).

Commonwealth Trade Union Council: Congress House, 23–28 Great Russell St, London, WC1B 3LS; tel. (20) 7467-1301; fax (20) 7436-0301; e-mail info@commonwealthtuc.org; internet www.commonwealthtuc.org; f. 1979; links trade union national centres (representing more than 30m. trade union mems) throughout the Commonwealth; promotes the application of democratic principles and core labour standards, works closely with other international trade union orgs; Dir ANNIE WATSON; publ. *Annual Report*.

SCIENCE AND TECHNOLOGY

Commonwealth Engineers' Council: c/o Institution of Civil Engineers, One Great George St, London, SW1P 3AA; tel. (20) 7222-7722; fax (20) 7222-7500; e-mail international@ice.org.uk; f. 1946; the Council meets every two years to provide an opportunity for engineering institutions of Commonwealth countries to exchange views on collaboration; there is a standing cttee on engineering education and training; organizes seminars on related topics; Sec.-Gen. J. A. WHITWELL.

Commonwealth Geological Surveys Forum: c/o Commonwealth Science Council, CSC Earth Sciences Programme, Marlborough House, Pall Mall, London, SW1Y 5HX; tel. (20) 7839-3411; fax (20) 7839-6174; e-mail comsci@gn.apc.org; f. 1948 to promote collaboration in geological, geochemical, geophysical and remote sensing techniques and the exchange of information; Geological Programme Officer Dr SIYAN MALOMO.

SPORT

Commonwealth Games Federation: 4th Floor, 26 Upper Brooke Street, London, W1K 7QE; tel. (20) 7491-8801; fax (20) 7409-7803; e-mail info@thecgf.com; internet www.thecgf.com; the Games were first held in 1930 and are now held every four years; participation is limited to competitors representing the mem. countries of the Commonwealth; held in Manchester, United Kingdom, in 2002; mems: 72 affiliated bodies; Pres. HRH The Earl of WESSEX; CEO MICHAEL HOOPER.

YOUTH

Commonwealth Youth Exchange Council: 7 Lion Yard, Tremadoc Rd, London, SW4 7NQ; tel. (20) 7498-6151; fax (20) 7622-4365; e-mail mail@cyec.org.uk; internet www.cyec.org.uk; f. 1970; promotes contact between groups of young people of the United Kingdom and other Commonwealth countries by means of educational exchange visits, provides information for organizers and allocates grants; provides host governments with technical assistance for delivery of the Commonwealth Youth Forum, held every two years (2005: Malta); 222 mem. orgs; Chief Exec. V. S. G. CRAGGS; publs *Contact* (handbook), *Exchange* (newsletter), *Final Communiqués* (of the Commonwealth Youth Forums), *Safety and Welfare* (guide-lines for Commonwealth Youth Exchange groups).

Duke of Edinburgh's Award International Association: Award House, 7-11 St Matthew St, London, SW1P 2JT; tel. (20) 7222-4242; fax (20) 7222-4141; e-mail sect@intaward.org; internet www.intaward.org; f. 1956; offers a programme of leisure activities for young people, comprising Service, Expeditions, Physical Recreation, and Skills; operates in more than 60 countries (not confined to the Commonwealth); International Sec.-Gen. DAVID MANSON; publs *Award World* (2 a year), *Annual Report*, handbooks and guides.

MISCELLANEOUS

Commonwealth Countries League: 96 High Street, Hampton Wick, Kingston-upon-Thames KT1 4DQ; tel. (20) 8943-3001; fax (20) 8458-0763; e-mail info@ccl-int.org.uk; internet www.ccl-int.org.uk; f. 1925 to secure equal opportunities and status between men and women in the Commonwealth, to act as a link between Commonwealth women's orgs, and to promote and finance secondary education of disadvantaged girls of high ability in their own countries, through the CCL Educational Fund; holds meetings with speakers and an annual conf., organizes the annual Commonwealth Fair for fund-raising; individual mems and affiliated socs in the Commonwealth; Exec. Chair. LEOLYNN JONES; publs *CCL Newsletter* (3 a year), *Annual Report*.

Commonwealth War Graves Commission: 2 Marlow Rd, Maidenhead, Berks, SL6 7DX; tel. (1628) 634221; fax (1628) 771208; internet www.cwgc.org; casualty and cemetery enquiries e-mail casualty.enq@cwgc.org; f. 1917 (as Imperial War Graves Commission); responsible for the commemoration in perpetuity of the 1.7m. members of the Commonwealth Forces who died during the wars of 1914–18 and 1939–45; provides for the marking and maintenance of war graves and memorials at some 23,000 locations in 150 countries; mems: Australia, Canada, India, New Zealand, South Africa, United Kingdom; Pres. HRH The Duke of KENT; Dir-Gen. RICHARD KELLAWAY.

Joint Commonwealth Societies' Council: c/o Royal Commonwealth Society, 18 Northumberland Ave, London, WC2N 5BJ; tel. (20) 7930-6733; fax (20) 7930-9705; e-mail jcsc@rcsint.org; internet www.commonwealthday.com; f. 1947; provides a forum for the exchange of information regarding activities of mem. orgs which promote understanding among countries of the Commonwealth; co-ordinates the distribution of the Commonwealth Day message by Queen Elizabeth, organizes the observance of the Commonwealth Day and produces educational materials relating to the occasion; mems: 13 unofficial Commonwealth orgs and four official bodies; Chair. Sir PETER MARSHALL; Sec. Sir DAVID THORNE.

Royal Commonwealth Ex-Services League: 48 Pall Mall, London, SW1Y 5JG; tel. (20) 7973-7263; fax (20) 7973-7308; internet www.commonwealthveterans.org.uk; links the ex-service orgs in the Commonwealth, assists ex-servicemen of the Crown and their dependants who are resident abroad; holds triennial confs; 57 mem. orgs in 48 countries; Grand Pres. HRH The Duke of EDINBURGH; publ. *Annual Report*.

Royal Commonwealth Society: 18 Northumberland Ave, London, WC2N 5BJ; tel. (20) 7930-6733; fax (20) 7930-9705; e-mail info@rcsint.org; internet www.rcsint.org; f. 1868 to promote international understanding of the Commonwealth and its people; organizes meetings and seminars on topical issues, and cultural and social events; library housed by Cambridge University Library; more than 10,000 mems; Chair. Baroness PRASHAR; Dir STUART MOLE; publs *Annual Report*, *Newsletter* (3 a year), conference reports.

Royal Over-Seas League: Over-Seas House, Park Place, St James's St, London, SW1A 1LR; tel. (20) 7408-0214; fax (20) 7499-6738; e-mail info@rosl.org.uk; internet www.rosl.org.uk; f. 1910 to promote friendship and understanding in the Commonwealth; clubhouses in London and Edinburgh; membership is open to all British subjects and Commonwealth citizens; Chair. Sir COLIN IMRAY; Dir-Gen. ROBERT F. NEWELL; publ. *Overseas* (quarterly).

Victoria League for Commonwealth Friendship: 55 Leinster Sq., London, W2 4PU; tel. (20) 7243-2633; fax (20) 7229-2994; f. 1901; aims to further personal friendship among Commonwealth peoples and to provide hospitality for visitors; maintains Student House, providing accommodation for students from Commonwealth countries; has brs elsewhere in the UK and abroad; Chair. JOHN KELLY; Gen. Sec. JOHN M. W. ALLAN; publ. *Annual Report*.

EUROPEAN UNION*

MEMBERS

Austria	Greece	Netherlands
Belgium	Hungary	Poland
Cyprus	Ireland	Portugal
Czech Republic	Italy	Slovakia
Denmark	Latvia	Slovenia
Estonia	Lithuania	Spain
Finland	Luxembourg	Sweden
France	Malta	United Kingdom
Germany		

The European Union (EU)'s relations with the developed countries of the Far East and Australasia have been dominated by problems of competition in trade. The traditional agricultural exports to the United Kingdom by Australia and New Zealand clash with EU preferences, although the EU has made some concessions on meat and butter. Regular consultations are held with Australia at ministerial level on co-operation in the areas of trade, industry, science and technology, energy, development assistance and the environment. In March 1994 an EU-Australia wine agreement entered into effect, entailing improved access for Australian wines to EU markets and the phasing out of the use of European geographical indicators by Australian producers. Negotiations to conclude a new partnership and co-operation agreement between the EU and Australia were suspended in late 1996, owing to a dispute regarding the inclusion of a human-rights clause. A joint declaration, committing both sides to greater political, cultural and economic co-operation, was signed in June 1997. During 1996 extensive negotiations between the New Zealand Government and EU officials failed to conclude an agreement on import duties for a substantial sector of its butter industry and related products. In March 1997 New Zealand took the dispute to the World Trade Organization (WTO), which later ruled against the EU's policy of subjecting the disputed sector to a different tariff regime, and in June 1999 the European Commission agreed to comply with the ruling. A joint declaration detailing areas of co-operation and establishing a consultative framework to facilitate the development of these was signed in May 1999. Mutual recognition agreements were also signed with Australia and New Zealand in 1999, with the aim of facilitating bilateral trade in industrial products. In 2001 a National Europe Centre, based at the Australian National University in Canberra, was established jointly by the EU and the University to consolidate Australia-EU relations.

In 1987 an EC-Japan Centre for Industrial Co-operation was established in Tokyo. (An office in Brussels was opened in June 1996.) However, the EU's substantial trade deficit with Japan was a continuing source of friction between the two sides. In 1990 a joint committee was established, with the aim of reducing the deficit. In July 1991 the heads of government of Japan and the Community signed a joint declaration on closer co-operation, to be undertaken not only in economic matters, but also with regard to political questions and security; annual meetings of heads of government were to take place. Under an agreement concluded in the same month exports of Japanese cars to the EC/EU were limited until the end of 1999, with all restrictions abolished by the EU from 1 January 2000. The agreement exempted cars produced in Europe by Japanese companies. During 1993 Japan undertook to reduce significantly its volume of exports to the EC. In early 1994 the EU initiated a scheme to promote European exports to Japan by providing advice and financial support to companies undertaking trade missions to Japan, and organizing training programmes. During 1995 the EU's ongoing policy of enhancing political dialogue with Japan and of holding regular low-profile negotiations between EU and Japanese government and business representatives on trade and economic co-operation was considered to be an important factor contributing to its steadily improving trade balance with Japan. Nevertheless, that country's restrictive trading practices continued to be a source of concern. In October 1996 the WTO upheld a long-standing complaint brought by the EU that Japanese taxes on alcoholic spirits discriminated against certain European products. Other areas of dispute were the treatment of foreign shipping companies in Japanese ports, trade restrictions on access to the Japanese photographic film and paper markets and policies for trade in semiconductors. In April 1997 the EU reintroduced a reference price for semiconductors in order to protect the European electronic industry from alleged underpriced imports from Japan and the Republic of Korea. In January 1998 an EU–Japan summit meeting was held, followed by a meeting at ministerial level in October. Subsequent summits aimed to strengthen political dialogue and deepen negotiations, and the 10th summit, held in December 2001, adopted an action plan aimed at fortifying co-operation over the following 10 years, in the areas of peace and security, economic and trade partnership, global and other challenges, and cultural dialogue. At the 12th summit, held in May 2003, progress in the implementation of the action plan was reviewed and a statement on Japan-EU initiatives on investment was issued. The 13th summit, convened in June 2004, again assessed progress in the implementation of the action plan, and

*The European Union (EU) was formally established on 1 November 1993 under the Treaty on European Union; prior to that date it was known as the European Community (EC).

adopted a joint declaration on disarmament and non-proliferation, a co-operation framework for the promotion of Japan–EU two-way investment, and a statement on co-operation in ICT matters. The meeting also issued documentation on a Japan-EU joint initiative for the enforcement of intellectual property rights in Asia.

Textiles exports by Asian countries have caused concern in the EU, owing to the depressed state of its own textiles industry. During 1982 bilateral negotiations were held under the Multi-Fibre Arrangement (MFA) with Asian producers, notably Hong Kong, the Republic of Korea and Macao. Agreements were eventually reached involving reductions in clothing quotas, 'anti-surge' clauses to prevent flooding of European markets, and measures to be imposed in the event of fraud. In 1986 new bilateral negotiations were held under the MFA and agreements were reached with the principal Asian textile exporters; in most cases a slight increase in quotas was permitted by the EC. In 1993 the MFA, which was due to expire on 31 December, was extended by one year. In accordance with the WTO's transitional Agreement on Textiles and Clothing (ATC), which replaced the MFA on 1 January 1995, the EU was gradually to liberalize its textiles trade regime and the quotas determined under the MFA were to be phased out in four stages over a 10-year period. The ATC was to expire at the end of December 2004.

A trade agreement with the People's Republic of China (PRC), signed in 1978, was replaced in May 1985 by a new trade and economic co-operation agreement. In June 1989, following the violent repression of the Chinese pro democracy movement, EC heads of government agreed to impose sanctions on the PRC, including the cancellation of official contacts and the suspension of loans by member states. In October 1990 the Community resolved that relations with the PRC should be 'progressively normalized'. Regular meetings continued to be held between Chinese and European representatives, focusing on issues relating to human rights, economic reforms in the PRC and bilateral trade relations. The EU supported the PRC's involvement in the international community and, in particular, its application for membership of the WTO. In November 1994 a PRC-Europe International Business School was inaugurated in Shanghai. In 1995 the EU sought to focus on the PRC as a key area for export promotion, and five events were held in the PRC to assist European companies to gain access to new market opportunities. A joint declaration was signed with the PRC in October 1996 to strengthen co-operation in the energy sector. In October 1997 senior representatives of the EU and the Chinese authorities signed a memorandum on future co-operation. The first meeting of the two sides at the level of heads of government was convened in April 1998. In November the President of the Commission made an official visit to the PRC and urged that country to remove trade restrictions imposed on European products. In the same month the EU and Hong Kong signed a co-operation agreement to combat drugs-trafficking and copyright piracy. The agreement was the first international accord to be signed by the territory since its reversion to Chinese sovereignty in July 1997. An agreement on science and technology was signed by the EU and the PRC in January 1999. A second meeting of EU-PRC heads of state was convened in December. In May 2000 the PRC and the EU concluded a bilateral trade agreement, removing a major barrier to WTO accession by the PRC (which was eventually achieved in December 2001). The EU announced that it would continue disbursing aid to the PRC after accession, to help smooth the country's transition to a market economy. A third EU-PRC summit meeting was held in Beijing in October 2000. At the fourth summit, in September 2001, the two sides agreed to strengthen and widen political dialogue and to continue their dialogue on human rights. Negotiations on an EU-PRC maritime transport agreement were also formally opened in September. EU–PRC trade amounted to €115,270m. in 2002. In March of that year the European Commission approved a strategy document setting out a framework for co-operation between the EU and the PRC over the period 2002–06, and in September the fifth EU-PRC summit discussed trade relations and future co-operation on illegal immigrants and tourism. At the sixth EU-PRC summit, held in Beijing in October 2003, two agreements were signed instigating a new dialogue on industrial policy and confirming the PRC's participation in the EU/European Space Agency 'Galileo' civil navigation and positioning project; in addition, a memorandum of understanding was initialled, paving the way for Chinese tourist groups to travel to the EU more easily. It was envisaged that the EU would commit up to €250m. to co-operation projects in the PRC during 2002–06.

In November 1992 the EC and the Republic of Korea signed an arrangement on scientific and technical co-operation; the EC also ended a suspension of tariff preferences for imports from the Republic of Korea, following that country's decision to end discrimination against the Community in the field of intellectual property. Further removal of trade barriers by the Republic of Korea prompted the EU, in October 1994, to announce that a new bilateral trade agreement was to be negotiated. During negotiations, initiated in May 1995, the EU expressed concern at restrictions on European car exports to the Republic of Korea, and advocated deregulation of that country's financial sector. A customs co-operation agreement was initialled with the Republic of Korea in July 1996. In October the two sides signed a framework trade and co-operation agreement, aiming to increase co-operation in trade and industry, scientific research, technology and environmental protection. In October 1997 the parties signed an agreement regarding a reciprocal opening of markets for telecommunications equipment, following a protracted dispute, which had led the EU to lodge a complaint with the WTO. In September 1997, however, the Commission submitted a further complaint to the WTO, accusing the Republic of Korea of tax discrimination against European alcoholic spirits exporters. In May 2000 the EU noted that the extremely low prices offered by shipyards in the Republic of Korea were causing severe pressures in the global market. A dialogue was opened with the Republic of Korea on this issue; however, in September 2002 it was announced that the negotiations had failed and that the EU was to register the case with the WTO. The framework trade and co-operation agreement entered into force in April 2001.

In September 1997 the European Atomic Energy Community (Euratom—an integral part of the EU) formally acceded to the executive board of the Korean Peninsula Energy Development Organization (KEDO), which aims to enhance nuclear safety and reduce the threat of nuclear proliferation from the energy programme of the Democratic People's Republic of Korea (DPRK). Political dialogue between the EU and the DPRK was initiated in 1998. In March 2001 the EU determined to strengthen its efforts to establish peace and stability in the Korean peninsula. A high-level EU delegation visited the DPRK in May. Later in that month, the EU announced that it would establish diplomatic relations with the DPRK, with the aim of facilitating inter-Korean reconciliation and alleviating food insecurity in the DPRK. In late 2002, after the DPRK had admitted pursuing a clandestine nuclear weapons programme, in contravention of its international non-proliferation commitments, the EU Council of Ministers stated that failure to resolve the nuclear issue would jeopardize the future development of EU-DPRK relations.

In June 1992 the EC signed trade and co-operation agreements with Mongolia and Macao, with respect for democracy and human rights forming the basis for the envisaged co-operation. The sixth EU-Mongolia co-operation council met in Brussels in November 2002. A co-operation accord was formally signed with Viet Nam in July 1995, under which the EU agreed to increase quotas for Vietnamese textile products by 15%, support the country's efforts to join the WTO and provide aid for environmental and public management projects. The agreement, which came into effect on 1 June 1996, incorporated a commitment by Viet Nam to guarantee human rights and allowed for the gradual repatriation of some 40,000 Vietnamese refugees from Germany, who lost their legal status (granted by the Government of East Germany) after reunification. In October 1997 the EU agreed to increase Viet Nam's textile quotas by some 30%, in exchange for improved market access for EU exports. The agreement was ratified by both sides in September 1998. A third textile and clothing agreement was signed in October 2000, allowing increased access for Viet Nam to EU markets. Non-preferential co-operation agreements (without financial protocols) were initialled with Laos and Cambodia in November 1996 and signed in April 1997. The agreement with Laos entered into force on 1 December; however, the agreement with Cambodia was postponed, owing to adverse political developments in that country. In 1998 the EU provided financial assistance to support preparations for a general election in Cambodia, and dispatched a 200-member team to monitor the poll, which was conducted in July. The co-operation agreement with Cambodia finally came into force in November 1999. The first meeting of the EU-Laos Joint Committee was held in June 1998 and the first meeting of the EU-Cambodia Joint Committee in May 2000, when an agreement on textiles was also signed. In June 2003 the EU and Cambodia concluded a bilateral market access agreement in respect of Cambodia's application for WTO membership.

Concern at the human-rights situation in Myanmar and the lack of progress towards establishing a democratic government in that country has been a source of friction between the EU and Asian governments. In October 1996 the EU adopted a Common Position on Myanmar, reaffirming sanctions previously imposed against that country concerning arms exports (1990), co-operation in defence matters (1991) and all but humanitarian bilateral aid provision. New strict limits on entry visas for Myanma officials were also imposed, in view of Myanmar's recent refusal to permit a Commission delegation to investigate allegations of forced labour in certain government industries. In March 1997 ministers of foreign affairs agreed to revoke Myanmar's special trade privileges under the Generalized System of Preferences. In November a meeting of EU and ASEAN senior officials was postponed, owing to Myanmar's insistence (then as a full member of the ASEAN grouping) that it should attend with full observer status. After several further delays, the meeting was finally convened, with Myanmar as a 'silent' observer, in late May 1999. The Common Position has subsequently been strengthened, including, in April 2000, the freezing of funds held abroad by those affected by the visa ban. During 2003 the EU issued statements of concern at restrictions on freedom of expression and political detentions in Myanmar, in particular at the increased harassment of the leader of the National League for Democracy, Daw Aung San Suu Kyi. The EU Common Position on Myanmar was most recently renewed in April 2004. A Myanma delegation was permitted to attend the fifth ASEM summit meeting held in October 2004 (see below).

In 2000 the EU contributed €19m. to a Trust Fund established by the World Bank to finance reconstruction activities in East Timor (now Timor-Leste). In December the EU hosted the third multilateral conference of donors to East Timor. In July 2001 the EU agreed to fund an observer mission to monitor the elections (held in the following month) to a new Constituent Assembly in the territory. An EU observer team monitored the presidential election held in April 2002 prior to Timor-Leste's accession to independence in the following month. During 2002 the EU provided €650,000 in development aid for Timor-Leste.

Relations between the EU and ASEAN are based on the Co-operation Agreement of 1980. Under this agreement, joint committee meetings are held approximately every 18 months. The 14th joint committee meeting took place in Brussels in September 2001. There are also regular ministerial meetings and post-ministerial conferences. In addition, the EU is a member of the ASEAN Regional Forum (ARF), a security grouping designed to promote peace and stability, established in 1994. In December 1994 the European Council endorsed a new strategy for Asia, which recognized the region's increasing economic and political importance and pledged to strengthen bilateral and regional dialogue. The strategy aimed to enhance the development of trade and investment, promote peace and security, and assist the less-developed countries in Asia. In May 1995 ASEAN and EU senior officials endorsed an initiative to convene an Asia-Europe Meeting of heads of government (ASEM). The first meeting, attended by heads of state and government of EU member states, ASEAN countries, Japan, the PRC and the Republic of Korea was held in March 1996 in Bangkok, Thailand. The meeting concluded a new Asia-Europe Partnership for Greater Growth, which aimed to strengthen links between the two economic regions in order to contribute to long-term prosperity and regional and global stability. In addition to measures to promote an expansion in trade and investment, the agreement incorporated provisions for co-operation in human resources development, technology and science and for educational and cultural exchanges. A controversial linkage of human rights criteria to trade, demanded by some non-governmental organizations prior to the meeting and discussed previously by European Governments, was avoided: the final statement of the meeting agreed on the promotion of 'fundamental rights', but incorporated a commitment for non-intervention in internal affairs. Among the initiatives agreed in order to pursue the objectives of the Partnership were the establishment of an Asia-Europe Business Forum, to promote private-sector trade and investment, an Asia-Europe Foundation, based in Singapore, for academic and cultural exchanges, and a Centre for Environmental Technology, based in Thailand. The meeting also proposed the initiation of a transport project linking the trans-Asian and trans-European railway networks. The first meeting of ministers of foreign affairs of countries participating in ASEM was held in Singapore, in February 1997. The meeting reviewed co-operation initiatives between the two regions and agreed to consolidate the process of economic and political dialogue. An agreement was signed to provide for the establishment of the Asia-Europe Foundation. The second ASEM summit, convened in April 1998, was dominated by concerns regarding the economic and financial situation in Asia, and both sides' declared intention to prevent a return to protectionist trading policies. A special statement, issued at the end of the meeting, identified the need for economic reform in individual countries and urged a reinforcement of international financial institutions. The meeting established an ASEM Asian Financial Crisis Response Trust Fund, under the auspices of the World Bank, to alleviate the social impact of the crisis. Other initiatives adopted by ASEM were an Asia–Europe Co-operation Framework, to guide, focus and co-ordinate political, economic and financial co-operation, a Trade Facilitation Action Plan, and an Investment Promotion Action Plan, incorporating a new Investment Experts Group. The meeting resolved to promote efforts to strengthen relations in all areas, and to set up a series of working bodies; however, it was decided not to establish a permanent secretariat for the ASEM arrangement. In September 2000 an ASEM senior officials' meeting agreed to permit the DPRK to participate in future ASEM co-operation initiatives. ASEM heads of government convened for the third time in Seoul, Republic of Korea, in October. ASEM III welcomed the ongoing *rapprochement* between the two Korean nations, declared a commitment to the promotion of human rights, and endorsed several initiatives related to globalization and information technology, including an Initiative to Address the Digital Divide and the creation of a Trans-Eurasia Information Network. Initiatives aimed at combating money-laundering, corruption and transnational crime were also adopted. The meeting established a new Asia-Europe Co-operation Framework (AECF), identifying ASEM's principles and priorities for the next 10 years. The fourth ASEM summit meeting, held in September 2002, in Copenhagen, Denmark, adopted a declaration and a co-operation programme on combating international terrorism, and determined to convene an ASEM seminar on the issue. Leaders participating in the summit also agreed to establish an *ad hoc* consultative mechanism that would enable senior officials of member countries to hold extraordinary meetings to address relevant international events. The summit endorsed a conference on cultures and civilizations, adopted a declaration on achieving peaceful reconciliation in the Korean peninsula, declared its commitment to the new round of WTO multilateral trade negotiations agreed at Doha, Qatar in November 2001, and authorized the creation of an ASEM Task Force on Closer Economic Partnership to address interregional trade, investment and finance.

The fifth ASEM summit, held in early October 2004 in Hanoi, Vietnam, was attended for the first time by Cambodia, Laos and (with representation below the level of head of state) Myanmar, as well as by the 10 new member states admitted to the EU on 1 May 2004. The meeting adopted the Hanoi Declaration on Closer ASEM Economic Partnership, establishing a series of measures for strengthening ASEM economic ties, and an ASEM Declaration on Dialogue among Cultures and Civilizations.

In September 2001 the EU adopted a new Communication on relations with Asia for the coming decade. Representing an updating of the 1994 strategy, this focused on strengthened partnership, particularly in the areas of politics, security, trade and investment. It aimed to reduce poverty and to promote democracy, good governance and the rule of law throughout the region. Partnerships and alliances on global issues were to be forged. A fundamental aim was to strengthen the EU's presence in Asia, promoting mutual awareness and knowledge on both sides. To this end, the EU established a scholarship scheme in the PRC and increased the number of EU delegation offices in the region.

During 2003 total assistance granted to Asia through the European Community Humanitarian Office (ECHO) totalled €117m., including €17m. for humanitarian assistance in the DPRK and some €6m. for assisting Myanma refugees living in the border regions of Thailand. A €4m. action plan for South-East Asia was adopted under the EU's Natural Disaster Prevention and Preparedness Programme (Dipecho) in 2000. The plan aimed to improve local capabilities and set up early-warning systems. In December 2002 the Commission approved total funding of some €6.6m. within the framework of Dipecho to help, among others, vulnerable communities in South-East Asia prepare for natural disasters.

European Union-ACP Partnership

From 1976 to February 2000 the principal means of co-operation between the Community and developing countries were the Lomé Conventions, concluded by the EU and African, Caribbean and Pacific (ACP) countries. The latter include 15 Pacific states: the Cook Islands, Fiji, Kiribati, Marshall Islands, Federated States of Micronesia, Nauru, Niue, Palau, Papua New Guinea, Samoa, Solomon Islands, Timor-Leste, Tonga, Tuvalu and Vanuatu, as well as 48 African and 16 Caribbean nations.

The first Lomé Convention (Lomé I), which was concluded at Lomé, Togo, in February 1975 and came into force on 1 April 1976, replaced the Yaoundé Conventions and the Arusha Agreement and was designed to provide a new framework of co-operation, taking into account the varying needs of developing ACP countries. Under Lomé I, the Community committed ECU 3,052.4m. for aid and investment, through the European Development Fund (EDF) and the European Investment Bank (EIB). Provision was made for over 99% of ACP (mainly agricultural) exports to enter the EC market duty free, while certain products competing directly with Community agriculture, such as sugar, were given preferential treatment but not free access. The Stabex (Stabilization of Export Earnings) scheme was designed to help developing countries to withstand fluctuations in the price of their agricultural products, by paying compensation for reduced export earnings.

The second Lomé Convention (January 1981–28 February 1985) envisaged Community expenditure of ECU 5,530m.; it extended some of the provisions of Lomé I, and introduced new fields of co-operation, including a new scheme, Sysmin, to safeguard exports of mineral products. Lomé III provided a total of ECU 8,500m. in assistance to the ACP states over the five years from March 1985, representing little or no increase, in real terms, over the amount provided by Lomé II.

The fourth Lomé Convention entered partially into force (trade provisions) on 1 March 1990, and fully into force on 1 September 1991. It was to cover the 10-year period 1990–99 (but was subsequently extended until February 2000). The financial protocol for 1990–95 made commitments of ECU 12,000m., of which

ECU 10,800m. was from the EDF (including ECU 1,500m. for Stabex and ECU 480m. for Sysmin) and ECU 1,200m. from the EIB. Under Lomé IV the obligation of most of the ACP states to contribute to the replenishment of Stabex resources, including the repayment of transfers made under the first three Conventions, was removed. In addition, special loans made to ACP member countries were to be cancelled, except in the case of profit-orientated businesses. Other innovations included the provision of assistance for structural adjustment programmes, measures to avoid increasing recipient countries' indebtedness (e.g. by providing Stabex and Sysmin assistance in the form of grants, rather than loans), and increased support for the private sector, environmental protection and control of population growth.

In September 1993 the EC announced plans to revise and strengthen its relations with the ACP countries under the Lomé Convention. A mid-term review of the Convention was initiated in May 1994. The EU reiterated its intention to maintain the Convention as an aid instrument. It emphasized, however, that stricter conditions relating to the awarding of aid would be imposed, based on standards of human rights, human resource development and environmental protection. In February 1995 a joint EU-ACP ministerial council, which was scheduled to conclude the negotiations, was adjourned, owing to significant disagreement among EU member states concerning reimbursement of the EDF for the period 1995–2000. In June EU heads of government reached an agreement, which was subsequently endorsed by an EU-ACP ministerial group. The accord was to provide ECU 14,625m. for the second phase of Lomé IV, with ECU 12,967m. allocated from the EDF and ECU 1,658m. in loans from the EIB. Agreement was also reached on revision of the country-of-origin rules for manufactured goods; expansion of the preferential system of trade for ACP products; a new protocol on the sustainable management of forest resources; and a joint declaration on support for the banana industry. The revised Convention was signed in November, in Mauritius. The new agreement included a reference to the observance of human rights, respect for democracy and the rule of law as essential elements of the preferential trading arrangement accorded under the Convention. Financial resources were to be made available to support institutional and administrative reforms to strengthen these principles in contracting states. In addition, the Convention provided for reforms in the administration of aid, including assistance to support structural economic adjustments. The revised Convention formally entered into force on 1 June 1998, having been ratified by all of the then 15 EU member states.

In March 1997 the Commission proposed the provision of debt relief assistance worth ECU 25m. per year for the period 1997–2000 to the 11 heavily indebted poor countries (as identified by the World Bank and IMF) that formed part of the ACP group. The funding was intended to support international efforts to reduce outstanding debt and to encourage economic prospects. In May 2001 the EU announced that it would cancel all outstanding debts arising from its trade accords with former colonies of member states. It was believed that the loans made under the Lomé Conventions could be as high as US $200m. In November 1997 the first summit meeting of heads of state of ACP countries was held, in Libreville, Gabon. The principal issues under consideration at the meeting were the strategic challenges confronting the ACP group of countries and, in particular, relations with the EU beyond 2000, when Lomé IV was scheduled to expire. The summit mandated ACP ministers of finance and of trade and industry to organize a series of regular meetings in order to strengthen co-ordination within the grouping

Formal negotiations on the conclusion of a successor agreement to the Lomé Convention were initiated in September 1998 and concluded in February 2000; the new partnership accord was signed by ACP and EU Heads of State and Government in June, in Cotonou, Benin. The so-called Cotonou Agreement covered the period 2000–20 and was subject to revision every five years. (The Agreement entered into force on 1 April 2003 following ratification by the then 15 EU member states and more than the requisite two-thirds of the ACP countries; however, many of its provisions had been applicable for a transitional period since August 2000.) It comprised the following main elements: increased political co-operation; the enhanced participation of civil society in ACP-EC partnership affairs; a strong focus on the reduction of poverty (addressing the economic and technical marginalization of developing nations was a primary concern); a reform of the existing structures for financial co-operation; and a new framework for economic and trade co-operation. Under the provisions of the new accord, the EU was to negotiate free-trade arrangements (replacing the previous non-reciprocal trade preferences) with the most developed ACP countries (excluding South Africa, which has a special arrangement, see below) during 2000–08; these would be structured around a system of regional free-trade zones, and would be designed to ensure full compatibility with WTO provisions. Once in force, the agreements would be subject to revision every five years. An assessment to be conducted in 2004 would identify those mid-ranking ACP nations also capable of entering into such free-trade deals. Meanwhile, the least-developed ACP nations were to benefit from an EU initiative to allow free access for most of their products by 2005. The preferential agreements currently in force would be retained initially (phase I), in view of a waiver granted by the WTO; thereafter ACP–EU trade was to be gradually liberalized over a period of 12–15 years (phase II). It was envisaged that Stabex and Sysmin would be eliminated gradually. In February 2001 the EU agreed to phase out trade barriers on imports of everything but military weapons from the world's 48 least-developed countries, 39 of which were in the ACP group. Duties on sugar, rice, bananas and some other products were to remain until 2009. The review process was initiated by the ACP-EU Council of Ministers, meeting in Gaborone, Botswana, in May 2004.

A financial protocol was attached to the Cotonou Agreement which indicated the funds available to the ACP through the EDF. The first financial protocol, covering the initial five-year period from March 2000, provided a total budget of €13,500m., of which €1,300m. was allocated to regional co-operation and €2,200m. was for the new investment facility for the development of the private sector. In addition, uncommitted balances from previous EDFs amounted to a further €2,500m. The first meeting of the ACP-EU Joint Parliamentary Assembly following the signing of the Cotonou Agreement was held in Brussels in October 2000. Resolutions were adopted on subjects including the banana dispute and AIDS. The meeting also called for increased funding for decentralized co-operation to be made available in the EU budget. In total, the EU provided €2,543m. in financing for ACP countries in 2002. Humanitarian aid to the ACP was projected at €165m. in 2004.

At the fourth plenary session of the joint ACP-EU Joint Parliamentary Assembly, held in Cape Town, South Africa, in March 2002, the ninth EDF was discussed. The plan was to be organized along sub-regional lines, with countries grouped into West Africa, Central Africa, the Caribbean, the Pacific region, Eastern and Southern Africa (partly via the Southern Africa Development Community). One major programme set up on behalf of the ACP countries and financed by the EDF was the new ProInvest programme, which was launched in 2002, with funding of €110m. over a seven-year period. The first general stage of negotiations for Economic Partnership Agreements (EPAs), involving discussions with all ACP countries regarding common procedures, began in September 2002; the Cotonou Agreement provided for a system of EPAs to replace all non-reciprocal trade preferences with the most developed ACP countries. The regional phase of EPA negotiations to establish a new framework for trade and investment commenced in October 2003, including discussions with the Pacific region from September 2004. In May 2003 Cuba, which had been admitted to the ACP in December 2000, was granted observer status at the ACP-EU Council of Ministers. Cuba withdrew its application to join the Cotonou Agreement in July 2003.

ACP-EU INSTITUTIONS

Council of Ministers: one minister from each signatory state; one co-chairman from each of the two groups; meets annually.

Committee of Ambassadors: one ambassador from each signatory state; chairmanship alternates between the two groups; meets at least every six months.

Joint Assembly: EU and ACP are equally represented; attended by parliamentary delegates from each of the 79 ACP countries and an equal number of members of the European Parliament; one co-chairman from each of the two groups; meets twice a year.

Secretariat of the ACP-EU Council of Ministers: 175 rue de la Loi, 1048 Brussels, Belgium; tel. (2) 285-61-11; fax (2) 285-74-58.

Centre for the Development of Enterprise—CDE: 52 ave Hermann Debroux, 1160 Brussels, Belgium; tel. (2) 679-18-11; fax (2) 675-19-03; e-mail info@cde.int; internet www.cde.int; f. 1977 (as the Centre for the Development of Industry: present name adopted 2001) to encourage investment in the ACP states by providing contacts and advice, holding promotion meetings and helping to finance feasibility studies; manages the Pro€Invest programme; Dir FERNANDO MATOS ROSA.

Technical Centre for Agricultural and Rural Co-operation: Postbus 380, 6700 AJ Wageningen, Netherlands; tel. (317) 467100; fax (317) 460067; e-mail cta@cta.int; internet www.agricta.org; f. 1983 to provide ACP states with better access to information, research, training and innovations in agricultural development and extension; Dir CARL B. GREENIDGE.

ACP INSTITUTIONS

ACP Council of Ministers.

ACP Committee of Ambassadors.

ACP Secretariat: ACP House, 451 ave Georges Henri, 1200 Brussels, Belgium; tel. (2) 743-06-00; fax (2) 735-55-73; e-mail info@acp.org; internet www.acpsec.org; Sec.-Gen. JEAN-ROBERT GOULONGANA (Gabon).

ORGANIZATION OF THE ISLAMIC CONFERENCE—OIC

Address: Kilo 6, Mecca Rd, POB 178, Jeddah 21411, Saudi Arabia.

Telephone: (2) 690-0001; **fax:** (2) 275-1953; **e-mail:** oiccabinet@arab.net.sa; **internet:** www.oic-oci.org.

The Organization was formally established in May 1971, when its Secretariat became operational, following a summit meeting of Muslim heads of state at Rabat, Morocco, in September 1969, and the Islamic Foreign Ministers' Conference in Jeddah in March 1970, and in Karachi, Pakistan, in December 1970.

MEMBERS

Afghanistan	Indonesia	Qatar
Albania	Iran	Saudi Arabia
Algeria	Iraq	Senegal
Azerbaijan	Jordan	Sierra Leone
Bahrain	Kazakhstan	Somalia
Bangladesh	Kuwait	Sudan
Benin	Kyrgyzstan	Suriname
Brunei	Lebanon	Syria
Burkina Faso	Libya	Tajikistan
Cameroon	Malaysia	Togo
Chad	The Maldives	Tunisia
Comoros	Mali	Turkey
Côte d'Ivoire	Mauritania	Turkmenistan
Djibouti	Morocco	Uganda
Egypt	Mozambique	United Arab Emirates
Gabon	Niger	
The Gambia	Nigeria	Uzbekistan
Guinea	Oman	Yemen
Guinea-Bissau	Pakistan	
Guyana	Palestine	

Note: Observer status has been granted to Bosnia and Herzegovina, the Central African Republic, Thailand, the Muslim community of the 'Turkish Republic of Northern Cyprus', the Moro National Liberation Front (MNLF) of the southern Philippines, the United Nations, the African Union, the Non-Aligned Movement, the League of Arab States, the Economic Co-operation Organization, the Union of the Arab Maghreb and the Co-operation Council for the Arab States of the Gulf.

Organization

(October 2004)

SUMMIT CONFERENCES

The supreme body of the Organization is the Conference of Heads of State, which met in 1969 at Rabat, Morocco, in 1974 at Lahore, Pakistan, and in January 1981 at Mecca, Saudi Arabia, when it was decided that summit conferences would be held every three years in future. An extraordinary summit conference was convened in Doha, Qatar, in March 2003, to consider the situation in Iraq. The 10th Conference was held in Putrajaya, Malaysia, in October 2003.

CONFERENCE OF MINISTERS OF FOREIGN AFFAIRS

Conferences take place annually, to consider the means for implementing the general policy of the Organization, although they may also be convened for extraordinary sessions.

SECRETARIAT

The executive organ of the Organization, headed by a Secretary-General (who is elected by the Conference of Ministers of Foreign Affairs for a four-year term, renewable only once) and four Assistant Secretaries-General (similarly appointed).

Secretary-General: Dr ABDELOUAHED BELKEZIZ (Morocco).

At the summit conference in January 1981 it was decided that an International Islamic Court of Justice should be established to adjudicate in disputes between Muslim countries. Experts met in January 1983 to draw up a constitution for the court; however, by 2004 it was not yet in operation.

STANDING COMMITTEES

Al-Quds Committee: f. 1975 to implement the resolutions of the Islamic Conference on the status of Jerusalem (Al-Quds); it meets at the level of foreign ministers; maintains the Al-Quds Fund; Chair. King MUHAMMAD VI OF MOROCCO.

Standing Committee for Economic and Commercial Co-operation—COMCEC: f. 1981; Chair. AHMET NECDET SEZER (Pres. of Turkey).

Standing Committee for Information and Cultural Affairs—COMIAC: f. 1981; Chair. ABDOULAYE WADE (Pres. of Senegal).

Standing Committee for Scientific and Technological Co-operation—COMSTECH: f. 1981; Chair. Gen. PERVEZ MUSHARRAF (Pres. of Pakistan).

Other committees comprise the Islamic Peace Committee, the Permanent Finance Committee, the Committee of Islamic Solidarity with the Peoples of the Sahel, the Eight-Member Committee on the Situation of Muslims in the Philippines, the Six-Member Committee on Palestine, and the *ad hoc* Committee on Afghanistan. In addition, there is an Islamic Commission for Economic, Cultural and Social Affairs and OIC contact groups on Bosnia and Herzegovina, Kosovo, Jammu and Kashmir, and Sierra Leone.

Activities

The Organization's aims, as proclaimed in the Charter that was adopted in 1972, are:

(i) To promote Islamic solidarity among member states;

(ii) To consolidate co-operation among member states in the economic, social, cultural, scientific and other vital fields, and to arrange consultations among member states belonging to international organizations;

(iii) To endeavour to eliminate racial segregation and discrimination and to eradicate colonialism in all its forms;

(iv) To take necessary measures to support international peace and security founded on justice;

(v) To co-ordinate all efforts for the safeguard of the Holy Places and support of the struggle of the people of Palestine, and help them to regain their rights and liberate their land;

(vi) To strengthen the struggle of all Muslim people with a view to safeguarding their dignity, independence and national rights; and

(vii) To create a suitable atmosphere for the promotion of co-operation and understanding among member states and other countries.

The first summit conference of Islamic leaders (representing 24 states) took place in 1969 following the burning of the Al Aqsa Mosque in Jerusalem. At this conference it was decided that Islamic governments should 'consult together with a view to promoting close co-operation and mutual assistance in the economic, scientific, cultural and spiritual fields, inspired by the immortal teachings of Islam'. Thereafter the foreign ministers of the countries concerned met annually, and adopted the Charter of the Organization of the Islamic Conference in 1972.

At the second Islamic summit conference (Lahore, Pakistan, 1974), the Islamic Solidarity Fund was established, together with a committee of representatives which later evolved into the Islamic Commission for Economic, Cultural and Social Affairs. Subsequently, numerous other subsidiary bodies have been set up (see below).

ECONOMIC CO-OPERATION

A general agreement for economic, technical and commercial co-operation came into force in 1981, providing for the establishment of joint investment projects and trade co-ordination. This was followed by an agreement on promotion, protection and guarantee of investments among member states. A plan of action to strengthen economic co-operation was adopted at the third Islamic summit conference in 1981, aiming to promote collective self-reliance and the development of joint ventures in all sectors. In 1994 the 1981 plan of action was revised; the reformulated plan placed greater emphasis on private-sector participation in its implementation. Although several meetings of experts were subsequently held to discuss some of the 10 priority focus areas of the plan, little progress was achieved in implementing it during the 1990s.

The fifth summit conference, held in 1987, approved proposals for joint development of modern technology, and for improving scientific and technical skills in the less developed Islamic countries. The first international Islamic trade fair was held in Jeddah, Saudi Arabia, in March 2001.

In 1991 22 OIC member states signed a framework agreement concerning the introduction of a system of trade preferences among member states. It was envisaged that, if implemented, this would represent the first step towards the eventual establishment of an

Islamic common market. In May 2001 the OIC Secretary-General urged increased progress in the ratification of the framework agreement. An OIC group of experts was considering the implications of the proposed creation of such a common market.

CULTURAL CO-OPERATION

The Organization supports education in Muslim communities throughout the world, and was instrumental in the establishment of Islamic universities in Niger and Uganda. It organizes seminars on various aspects of Islam, and encourages dialogue with the other monotheistic religions. Support is given to publications on Islam both in Muslim and Western countries. The OIC organizes meetings at ministerial level to consider aspects of information policy and new technologies.

HUMANITARIAN ASSISTANCE

Assistance is given to Muslim communities affected by wars and natural disasters, in co-operation with UN organizations, particularly UNHCR. The countries of the Sahel region (Burkina Faso, Cape Verde, Chad, The Gambia, Guinea, Guinea-Bissau, Mali, Mauritania, Niger and Senegal) receive particular attention as victims of drought. In April 1999 the OIC resolved to send humanitarian aid to assist the displaced ethnic Albanian population of Kosovo and Metohija, in southern Serbia. Several member states have provided humanitarian assistance to the Muslim population affected by the conflict in Chechnya. During 2001 the OIC was providing emergency assistance to Afghanistan, and in October established an Afghan People Assistance Fund. The OIC also administers a Trust Fund for the urgent return of refugees and the displaced to Bosnia and Herzegovina. A resolution on the status of refugees in the Muslim world that was adopted by the 10th OIC summit meeting, held in October 2003, urged all member states to accede to the 1951 UN Convention on the Status of Refugees.

POLITICAL CO-OPERATION

Since its inception the OIC has called for vacation of Arab territories by Israel, recognition of the rights of Palestinians and of the Palestine Liberation Organization (PLO) as their sole legitimate representative, and the restoration of Jerusalem to Arab rule. The 1981 summit conference called for a *jihad* (holy war—though not necessarily in a military sense) 'for the liberation of Jerusalem and the occupied territories'; this was to include an Islamic economic boycott of Israel. In 1982 Islamic ministers of foreign affairs decided to establish Islamic offices for boycotting Israel and for military co-operation with the PLO. The 1984 summit conference agreed to reinstate Egypt (suspended following the peace treaty signed with Israel in 1979) as a member of the OIC, although the resolution was opposed by seven states.

In August 1990 a majority of ministers of foreign affairs condemned Iraq's recent invasion of Kuwait, and demanded the withdrawal of Iraqi forces. In August 1991 the Conference of Ministers of Foreign Affairs obstructed Iraq's attempt to propose a resolution demanding the repeal of economic sanctions against the country. The sixth summit conference, held in Senegal in December, reflected the divisions in the Arab world that resulted from Iraq's invasion of Kuwait and the ensuing war. Twelve heads of state did not attend, reportedly to register protest at the presence of Jordan and the PLO at the conference, both of which had given support to Iraq. Disagreement also arose between the PLO and the majority of other OIC members when a proposal was adopted to cease the OIC's support for the PLO's *jihad* in the Arab territories occupied by Israel, in an attempt to further the Middle East peace negotiations.

In August 1992 the UN General Assembly approved a non-binding resolution, introduced by the OIC, that requested the UN Security Council to take increased action, including the use of force, in order to defend the non-Serbian population of Bosnia and Herzegovina (some 43% of Bosnians being Muslims) from Serbian aggression, and to restore its 'territorial integrity'. The OIC Conference of Ministers of Foreign Affairs, which was held in December, demanded anew that the UN Security Council take all necessary measures against Serbia and Montenegro, including military intervention, in order to protect the Bosnian Muslims.

A report by an OIC fact-finding mission, which in February 1993 visited Azad Kashmir while investigating allegations of repression of the largely Muslim population of the Indian state of Jammu and Kashmir by the Indian armed forces, was presented to the 1993 Conference. The meeting urged member states to take the necessary measures to persuade India to cease the 'massive human rights violations' in Jammu and Kashmir and to allow the Indian Kashmiris to 'exercise their inalienable right to self-determination'. In September 1994 ministers of foreign affairs, meeting in Islamabad, Pakistan, agreed to establish a contact group on Jammu and Kashmir, which was to provide a mechanism for promoting international awareness of the situation in that region and for seeking a peaceful solution to the dispute. In December OIC heads of state approved a resolution condemning reported human rights abuses by Indian security forces in Kashmir.

In July 1994 the OIC Secretary-General visited Afghanistan and proposed the establishment of a preparatory mechanism to promote national reconciliation in that country. In mid-1995 Saudi Arabia, acting as a representative of the OIC, pursued a peace initiative for Afghanistan and issued an invitation for leaders of the different factions to hold negotiations in Jeddah.

A special ministerial meeting on Bosnia and Herzegovina was held in July 1993, at which seven OIC countries committed themselves to making available up to 17,000 troops to serve in the UN Protection Force in the former Yugoslavia (UNPROFOR). The meeting also decided to dispatch immediately a ministerial mission to persuade influential governments to support the OIC's demands for the removal of the arms embargo on Bosnian Muslims and the convening of a restructured international conference to bring about a political solution to the conflict. In December 1994 OIC heads of state, convened in Morocco, proclaimed that the UN arms embargo on Bosnia and Herzegovina could not be applied to the Muslim authorities of that Republic. The Conference also resolved to review economic relations between OIC member states and any country that supported Serbian activities. An aid fund was established, to which member states were requested to contribute between US $500,000 and $5m., in order to provide further humanitarian and economic assistance to Bosnian Muslims. In relation to wider concerns the conference adopted a Code of Conduct for Combating International Terrorism, in an attempt to control Muslim extremist groups. The code commits states to ensuring that militant groups do not use their territory for planning or executing terrorist activity against other states, in addition to states refraining from direct support or participation in acts of terrorism. In a further resolution the OIC supported the decision by Iraq to recognize Kuwait, but advocated that Iraq comply with all UN Security Council decisions.

In July 1995 the OIC contact group on Bosnia and Herzegovina (at that time comprising Egypt, Iran, Malaysia, Morocco, Pakistan, Saudi Arabia, Senegal and Turkey), meeting in Geneva, declared the UN arms embargo against Bosnia and Herzegovina to be 'invalid'. Several Governments subsequently announced their willingness officially to supply weapons and other military assistance to the Bosnian Muslim forces. In September a meeting of all OIC ministers of defence and foreign affairs endorsed the establishment of an 'assistance mobilization group' which was to supply military, economic, legal and other assistance to Bosnia and Herzegovina. In a joint declaration the ministers also demanded the return of all territory seized by Bosnian Serb forces, the continued NATO bombing of Serb military targets, and that the city of Sarajevo be preserved under a Muslim-led Bosnian Government. In November the OIC Secretary-General endorsed the peace accord for the former Yugoslavia, which was concluded, in Dayton, USA, by leaders of all the conflicting factions, and reaffirmed the commitment of Islamic states to participate in efforts to implement the accord. In the following month the OIC Conference of Ministers of Foreign Affairs, convened in Conakry, Guinea, requested the full support of the international community to reconstruct Bosnia and Herzegovina through humanitarian aid as well as economic and technical co-operation. Ministers declared that Palestine and the establishment of fully-autonomous Palestinian control of Jerusalem were issues of central importance for the Muslim world. The Conference urged the removal of all aspects of occupation and the cessation of the construction of Israeli settlements in the occupied territories. In addition, the final statement of the meeting condemned Armenian aggression against Azerbaijan, registered concern at the persisting civil conflict in Afghanistan, demanded the elimination of all weapons of mass destruction and pledged support for Libya (affected by the US trade embargo). Ministers determined that an intergovernmental group of experts should be established in 1996 to address the situation of minority Muslim communities residing in non-OIC states.

In December 1996 OIC ministers of foreign affairs, meeting in Jakarta, Indonesia, urged the international community to apply pressure on Israel in order to ensure its implementation of the terms of the Middle East peace process. The ministers reaffirmed the importance of ensuring that the provisions of the Dayton Peace Agreement for the former Yugoslavia were fully implemented, called for a peaceful settlement of the Kashmir issue, demanded that Iraq fulfil its obligations for the establishment of security, peace and stability in the region and proposed that an international conference on peace and national reconciliation in Somalia be convened. The ministers elected a new Secretary-General who confirmed that the organization would continue to develop its role as an international mediator. In March 1997, at an extraordinary summit held in Pakistan, OIC heads of state and of government reiterated the organization's objective of increasing international pressure on Israel to ensure the full implementation of the terms of the Middle East peace process. An 'Islamabad Declaration' was also adopted, which pledged to increase co-operation between members of the OIC. In June the OIC condemned the decision by the US House of

Representatives to recognize Jerusalem as the Israeli capital. The Secretary-General of the OIC issued a statement rejecting the US decision as counter to the role of the USA as sponsor of the Middle East peace plan.

In early 1998 the OIC appealed for an end to the threat of US-led military action against Iraq arising from a dispute regarding access granted to international weapons inspectors. The crisis was averted by an agreement concluded between the Iraqi authorities and the UN Secretary-General in February. In March OIC ministers of foreign affairs, meeting in Doha, Qatar, requested an end to the international sanctions against Iraq. Additionally, the ministers urged all states to end the process of restoring normal trading and diplomatic relations with Israel pending that country's withdrawal from the occupied territories and acceptance of an independent Palestinian state. In April the OIC, jointly with the UN, sponsored new peace negotiations between the main disputing factions in Afghanistan, which were conducted in Islamabad, Pakistan. In early May, however, the talks collapsed and were postponed indefinitely. In September the Secretaries-General of the OIC and UN agreed to establish a joint mission to counter the deteriorating security situation along the Afghan–Iranian border, following the large-scale deployment of Taliban troops in the region and consequent military manoeuvres by the Iranian authorities. They also reiterated the need to proceed with negotiations to conclude a peaceful settlement in Afghanistan. In December the OIC appealed for a diplomatic solution to the tensions arising from Iraq's withdrawal of co-operation with UN weapons inspectors, and criticized subsequent military air-strikes, led by the USA, as having been conducted without renewed UN authority. An OIC Convention on Combating International Terrorism was adopted in 1998. An OIC committee of experts responsible for formulating a plan of action for safeguarding the rights of Muslim communities and minorities met for the first time in 1998.

In early April 1999 ministers of foreign affairs of the countries comprising OIC's contact group met to consider the crisis in Kosovo. The meeting condemned Serbian atrocities being committed against the local Albanian population and urged the provision of international assistance for the thousands of people displaced by the conflict. The group resolved to establish a committee to co-ordinate relief aid provided by member states. The ministers also expressed their willingness to help to formulate a peaceful settlement and to participate in any subsequent implementation force. In June an OIC Parliamentary Union was inaugurated; its founding conference was convened in Tehran, Iran.

In early March 2000 the OIC mediated contacts between the parties to the conflict in Afghanistan, with a view to reviving peace negotiations. Talks, held under OIC auspices, ensued in May. In November OIC heads of state attended the ninth summit conference, held in Doha, Qatar. In view of the significant deterioration in relations between Israel and the Palestinian (National) Authority during late 2000, the summit issued a Declaration pledging solidarity with the Palestinian cause and accusing the Israeli authorities of implementing large-scale systematic violations of human rights against Palestinians. The summit also issued the Doha Declaration, which reaffirmed commitment to the OIC Charter and undertook to modernize the organization's organs and mechanisms. Both the elected Government of Afghanistan and the Taliban sent delegations to the Doha conference. The summit determined that Afghanistan's official participation in the OIC, suspended in 1996, should not yet be reinstated. In early 2001 a high-level delegation from the OIC visited Afghanistan in an attempt to prevent further destruction of ancient statues by Taliban supporters.

In May 2001 the OIC convened an emergency meeting, following an escalation of Israeli–Palestinian violence. The meeting resolved to halt all diplomatic and political contacts with the Israeli government, while restrictions remained in force against Palestinian-controlled territories. In June the OIC condemned attacks and ongoing discrimination against the Muslim Community in Myanmar. In the same month the OIC Secretary-General undertook a tour of six African countries—Burkina Faso, The Gambia, Guinea, Mali, Niger and Senegal—to promote co-operation and to consider further OIC support for those states. In August the Secretary-General condemned Israel's seizure of several Palestinian institutions in East Jerusalem and aerial attacks against Palestinian settlements. The OIC initiated high-level diplomatic efforts to convene a meeting of the UN Security Council in order to discuss the situation.

In September 2001 the OIC Secretary-General strongly condemned major terrorist attacks perpetrated against targets in the USA. Soon afterwards the US authorities rejected a proposal by the Taliban regime that an OIC observer mission be deployed to monitor the activities of the Saudi Arabian-born exiled militant Islamist fundamentalist leader Osama bin Laden, who was accused by the US Government of having co-ordinated the attacks from alleged terrorist bases in the Taliban-administered area of Afghanistan. An extraordinary meeting of OIC ministers of foreign affairs, convened in early October, in Doha, Qatar, to consider the implications of the terrorist atrocities, condemned the attacks and declared its support for combating all manifestations of terrorism within the framework of a proposed collective initiative co-ordinated under the auspices of the UN. The meeting, which did not pronounce directly on the recently-initiated US-led military retaliation against targets in Afghanistan, urged that no Arab or Muslim state should be targeted under the pretext of eliminating terrorism. It determined to establish a fund to assist Afghan civilians. In February 2002 the Secretary-General expressed concern at statements of the US administration describing Iran and Iraq (as well as the Democratic People's Republic of Korea) as belonging to an 'axis of evil' involved in international terrorism and the development of weapons of mass destruction. In early April OIC foreign ministers convened an extraordinary session on terrorism, in Kuala Lumpur, Malaysia. The meeting issued the 'Kuala Lumpur Declaration', which reiterated member states' collective resolve to combat terrorism, recalling the organization's 1994 code of conduct and 1998 convention to this effect; condemned attempts to associate terrorist activities with Islamists or any other particular creed, civilization or nationality, and rejected attempts to associate Islamic states or the Palestinian struggle with terrorism; rejected the implementation of international action against any Muslim state on the pretext of combating terrorism; urged the organization of a global conference on international terrorism; and urged an examination of the root causes of international terrorism. In addition, the meeting strongly condemned Israel's ongoing military intervention in areas controlled by the Palestinian (National) Authority. The meeting adopted a plan of action on addressing the issues raised in the declaration. Its implementation was to be co-ordinated by a 13-member committee on international terrorism. Member states were encouraged to sign and ratify the Convention on Combating International Terrorism in order to accelerate its implementation. In June ministers of foreign affairs, meeting in Khartoum, Sudan, issued a declaration reiterating the OIC call for an international conference to be convened, under UN auspices, in order clearly to define terrorism and to agree on the international procedures and mechanisms for combating terrorism through the UN. The conference also repeated demands for the international community to exert pressure on Israel to withdraw from all Palestinian-controlled territories and for the establishment of an independent Palestinian state. It endorsed the peace plan for the region that had been adopted by the summit meeting of the League of Arab States in March.

In June 2002 the OIC Secretary-General expressed his concern at the escalation of tensions between Pakistan and India regarding Kashmir. He urged both sides to withdraw their troops and to refrain from the use of force. In the following month the OIC pledged its support for Morocco in a territorial dispute with Spain over the small island of Perejil, but called for a negotiated settlement to resolve the issue.

An extraordinary summit conference of Islamic leaders convened in Doha, Qatar, in early March 2003 to consider the ongoing Iraq crisis welcomed the Saddam Hussain regime's acceptance of UN Security Council Resolution 1441 and consequent co-operation with UN weapons inspectors, and emphatically rejected any military strike against Iraq or threat to the security of any other Islamic state. The conference also urged progress towards the elimination of all weapons of mass destruction in the Middle East, including those held by Israel. In May the 30th session of the Conference of Ministers of Foreign Affairs, entitled 'Unity and Dignity', issued the Tehran Declaration, in which it resolved to combat terrorism and to contribute to preserving peace and security in Islamic countries. The Declaration also pledged its full support for the Palestinian cause and rejected the labelling as 'terrorist' of those Muslim states deemed to be resisting foreign aggression and occupation. The 10th OIC summit meeting, held in October in Putrajaya, Malaysia, issued the Putrajaya Declaration, in which Islamic leaders resolved to enhance Islamic states' role and influence in international affairs. The leaders adopted a plan of action that entailed: reviewing and strengthening OIC positions on international issues; enhancing dialogue among Muslim thinkers and policy-makers through relevant OIC insitutions; promoting constructive dialogue with other cultures and civilizations; completing an ongoing review of the structure and efficacy of the OIC Secretariat; establishing a working group to address means of enhancing the role of Islamic education; promoting among member states the development of science and technology, discussion of ecological issues, and the role of information communication technology in development; improving mechanisms to assist member states in post-conflict situations; and advancing trade and investment through data-sharing and encouraging access to markets for products from poorer member states.

Finance

The OIC's activities are financed by mandatory contributions from member states. The budget for 2002/03 totalled US $11.4m.

Subsidiary Organs

Islamic Centre for the Development of Trade: Complexe Commercial des Habous, ave des FAR, BP 13545, Casablanca, Morocco; tel. (2) 314974; fax (2) 310110; e-mail icdt@icdt.org; internet www.icdt.org; f. 1983 to encourage regular commercial contacts, harmonize policies and promote investments among OIC mems; Dir-Gen. ALLAL RACHDI; publs *Tijaris: International and Inter-Islamic Trade Magazine* (bi-monthly), *Inter-Islamic Trade Report* (annually).

Islamic Jurisprudence (Fiqh) Academy: POB 13917, Jeddah, Saudi Arabia; tel. (2) 667-1664; fax (2) 667-0873; internet www.fiqhacademy.org.sa; f. 1982; Sec.-Gen. SHEIKH MOHAMED HABIB IBN AL-KHODHA.

Islamic Solidarity Fund: c/o OIC Secretariat, POB 178, Jeddah 21411, Saudi Arabia; tel. (2) 680-0800; fax (2) 687-3568; f. 1974 to meet the needs of Islamic communities by providing emergency aid and the finance to build mosques, Islamic centres, hospitals, schools and universities; Chair. Sheikh NASIR ABDULLAH BIN HAMDAN; Exec. Dir ABDULLAH HERSI.

Islamic University in Uganda: POB 2555, Mbale, Uganda; Kampala Liaison Office: POB 7689, Kampala; tel. (45) 33502; fax (45) 34452; e-mail iuiu@info.com.co.ug; tel. (41) 236874; fax (41) 254576; f. 1988 to meet the educational needs of Muslim populations in English-speaking African countries; mainly financed by OIC; Principal Officer Prof. MAHDI ADAMU.

Islamic University of Niger: BP 11507, Niamey, Niger; tel. 723903; fax 733796; f. 1984; provides courses of study in *Shari'a* (Islamic law) and Arabic language and literature; also offers courses in pedagogy and teacher training; receives grants from Islamic Solidarity Fund and contributions from OIC member states; Rector Prof. ABDELALI OUDHRIRI.

Islamic University of Technology—IUT: GPO Box 3003, Board Bazar, Gazipur 1704, Dhaka, Bangladesh; tel. (2) 980-0960; fax (2) 980-0970; e-mail vc@iut-dhaka.edu; internet www.iutoic-dhaka.edu; f. 1981 as the Islamic Centre for Technical and Vocational Training and Resources, named changed to Islamic Institute of Technology in 1994, current name adopted in June 2001; aims to develop human resources in OIC mem. states, with special reference to engineering, technology, tech. and vocational education and research; 224 staff and 1,000 students; library of 23,000 vols; Vice-Chancellor Prof. Dr FAZLI ILAHI; publs *News Bulletin* (annually), annual calendar and announcement for admission, reports, human resources development series.

Research Centre for Islamic History, Art and Culture—IRCICA: POB 24, Beşiktaş 80692, İstanbul, Turkey; tel. (212) 2591742; fax (212) 2584365; e-mail ircica@superonline.com; internet www.ircica.org; f. 1980; library of 50,000 vols; Dir-Gen. Prof. Dr EKMELEDDİN IHSANOĞLU; publs *Newsletter* (3 a year), monographical studies.

Statistical, Economic and Social Research and Training Centre for the Islamic Countries: Attar Sok 4, GOP 06700, Ankara, Turkey; tel. (312) 4686172; fax (312) 4673458; e-mail oicankara@sesrtcic.org; internet www.sesrtcic.org; f. 1978; Dir-Gen. ERDINÇ ERDÜN; publs *Journal of Economic Co-operation among Islamic Countries* (quarterly), *InfoReport* (quarterly), *Statistical Yearbook* (annually).

Specialized Institutions

International Islamic News Agency—IINA: King Khalid Palace, Madinah Rd, POB 5054, Jeddah 21422, Saudi Arabia; tel. (2) 665-8561; fax (2) 665-9358; e-mail iina@cyberia.net.sa; internet www.islamicnews.org; f. 1972; distributes news and reports daily on events in the Islamic world, in Arabic, English and French; Dir-Gen. ABDULWAHAB KASHIF.

Islamic Development Bank: POB 5925, Jeddah 21432, Saudi Arabia; tel. (2) 636-1400; fax (1) 636-6871; e-mail idbarchives@isdb.org.sa; internet www.isdb.org; f. 1975; promotes the economic and social development of OIC member countries and Muslim communities in non-member countries; provides assistance in the form of loans and grants for technical aid, in accordance with the principles of the Islamic *Shari'a* (sacred law); Pres. and Chair. Dr AHMED MOHAMED ALI; Sec.-Gen. Dr ABDERRAHIM OMRANA.

Islamic Educational, Scientific and Cultural Organization—ISESCO: BP 755, Rabat 10104, Morocco; tel. (7) 772433; fax (7) 772058; e-mail cid@isesco.org.ma; internet www.isesco.org.ma; f. 1982; Dir-Gen. Dr ABDULAZIZ BIN OTHMAN AT-TWAIJRI; publs *ISESCO Newsletter* (quarterly), *Islam Today* (2 a year), *ISESCO Triennial*.

Islamic States Broadcasting Organization—ISBO: POB 6351, Jeddah 21442, Saudi Arabia; tel. (2) 672-1121; fax (2) 672-2600; e-mail isbo@isbo.org; internet www.isbo.org; f. 1975; Sec.-Gen. HUSSEIN AL-ASKARY.

Affiliated Institutions

International Association of Islamic Banks—IAIB: King Abdulaziz St, Queen's Bldg, 23rd Floor, Al-Balad Dist, POB 9707, Jeddah 21423, Saudi Arabia; tel. (2) 651-6900; fax (2) 651-6552; f. 1977 to link financial institutions operating on Islamic banking principles; activities include training and research; mems: 192 banks and other financial institutions in 34 countries; Sec.-Gen. SAMIR A. SHAIKH.

Islamic Chamber of Commerce and Industry: POB 3831, Clifton, Karachi 75600, Pakistan; tel. (21) 5874756; fax (21) 5870765; e-mail icci@icci-oic.org; internet icci-oic.org; f. 1979 to promote trade and industry among member states; comprises nat. chambers or feds of chambers of commerce and industry; Sec.-Gen. AQEEL AHMAD AL-JASSEM.

Islamic Committee for the International Crescent: POB 17434, Benghazi, Libya; tel. (61) 95823; fax (61) 95829; f. 1979 to attempt to alleviate the suffering caused by natural disasters and war; Sec.-Gen. Dr AHMAD ABDALLAH CHERIF.

Islamic Solidarity Sports Federation: POB 5844, Riyadh 11442, Saudi Arabia; tel. and fax (1) 482-2145; f. 1981; Sec.-Gen. Dr MOHAMMAD SALEH GAZDAR.

Organization of Islamic Capitals and Cities—OICC: POB 13621, Jeddah 21414, Saudi Arabia; tel. (2) 698-1953; fax (2) 698-1053; e-mail secrtriat@oicc.org; internet www.oicc.org; f. 1980 to promote and develop co-operation among OICC mems, to preserve their character and heritage, to implement planning guide-lines for the growth of Islamic cities and to upgrade standards of public services and utilities in those cities; Sec.-Gen. OMAR ABDULLAH KADI.

Organization of the Islamic Shipowners' Association: POB 14900, Jeddah 21434, Saudi Arabia; tel. (2) 663-7882; fax (2) 660-4920; e-mail oisa@sbm.net.sa; f. 1981 to promote co-operation among maritime cos in Islamic countries; In 1998 mems approved the establishment of a new commercial venture, the Bakkah Shipping Company, to enhance sea transport in the region; Sec.-Gen. Dr ABDULLATIF A. SULTAN.

World Federation of Arab-Islamic Schools: POB 3446, Jeddah, Saudi Arabia; tel. (2) 670-0019; fax (2) 671-0823; f. 1976; supports Arab-Islamic schools world-wide and encourages co-operation between the institutions; promotes the dissemination of the Arabic language and Islamic culture; supports the training of personnel.

PACIFIC COMMUNITY

Address: BP D5, 98848 Nouméa Cédex, New Caledonia.

Telephone: 26-20-00; **fax:** 26-38-18; **e-mail:** spc@spc.int; **internet:** www.spc.org.nc.

In February 1947 the Governments of Australia, France, the Netherlands, New Zealand, the United Kingdom, and the USA signed the Canberra Agreement establishing the South Pacific Commission, which came into effect in July 1948. (The Netherlands withdrew from the Commission in 1962, when it ceased to administer the former colony of Dutch New Guinea, now Papua, formerly known as Irian Jaya, part of Indonesia.) In October 1997 the 37th South Pacific Conference, convened in Canberra, Australia, agreed to rename the organization the Pacific Community, with effect from 6 February 1998. The Secretariat of the Pacific Community (SPC) services the Community, and provides research, technical advice, training and assistance in economic, social and cultural development to 22 countries and territories of the Pacific region. It serves a population of about 6.8m., scattered over some 30m. sq km, more than 98% of which is sea.

MEMBERS

American Samoa
Australia
Cook Islands
Fiji
France
French Polynesia
Guam
Kiribati
Marshall Islands
Federated States of Micronesia
Nauru
New Caledonia
New Zealand
Niue
Northern Mariana Islands
Palau
Papua New Guinea
Pitcairn Islands
Samoa
Solomon Islands
Tokelau
Tonga
Tuvalu
United Kingdom
USA
Vanuatu
Wallis and Futuna Islands

Organization

(October 2004)

CONFERENCE OF THE PACIFIC COMMUNITY

The Conference is the governing body of the Community (replacing the former South Pacific Conference) and is composed of representatives of all member countries and territories. The main responsibilities of the Conference, which meets annually, are to appoint the Director-General, to determine major national or regional policy issues in the areas of competence of the organization and to note changes to the Financial and Staff Regulations approved by the Committee of Representatives of Governments and Administrations (CRGA).

COMMITTEE OF REPRESENTATIVES OF GOVERNMENTS AND ADMINISTRATIONS (CRGA)

This Committee comprises representatives of all member states and territories, having equal voting rights. It meets annually to consider the work programme evaluation conducted by the Secretariat and to discuss any changes proposed by the Secretariat in the context of regional priorities; to consider and approve any policy issues for the organization presented by the Secretariat or by member countries and territories; to consider applicants and make recommendations for the post of Director-General; to approve the administrative and work programme budgets; to approve amendments to the Financial and Staff Regulations; and to conduct annual performance evaluations of the Director-General.

SECRETARIAT

The Secretariat of the Pacific Community—SPC—is headed by a Director-General, a Senior Deputy Director-General and a Deputy Director-General, based in Suva, Fiji. Three administrative Divisions cover Land Resources, Marine Resources and Social Resources. The Secretariat also provides information services, including library facilities, publications, translation and computer services. The organization has about 250 staff members.

Director-General: Lourdes Pangelinan (Guam).

Senior Deputy Director-General: Dr Jimmie Rodgers (Solomon Islands).

Deputy Director-General: Yves Corbel (France).

Regional Office: Private Mail Bag, Suva, Fiji; tel. 3370733; fax 3370021; e-mail spcsuva@spc.org.fj.

Activities

The SPC provides, on request of its member countries, technical assistance, advisory services, information and clearing-house services aimed at developing the technical, professional, scientific, research, planning and management capabilities of the regional population. The SPC also conducts regional conferences and technical meetings, as well as training courses, workshops and seminars at the regional or country level. It provides small grants-in-aid and awards to meet specific requests and needs of members. In November 1996 the Conference agreed to establish a specific Small Islands States fund to provide technical services, training and other relevant activities. The organization's three programme divisions are: land resources, marine resources and social resources. The Pacific Community oversees the maritime programme and telecommunications policy activities of the Pacific Islands Forum Secretariat.

In 1998 the SPC adopted a Corporate Plan for 1999–2003, the main objectives of which included developing national capabilities in 'value-adding' technology; enhancing the integration of cross-sectoral issues (such as economics, gender, culture and community education) into national planning and policy-making processes; and developing a co-ordinated human resources programme as a focal point for providing information, advice and support to the regional population. The 1999 Conference, held in Tahiti in December, adopted the 'Déclaration de Tahiti Nui', a mandate that detailed the operational policies and mechanisms of the Pacific Community, taking into account operational changes not covered by the founding Canberra Agreement. The Déclaration was regarded as a 'living document' that would be periodically revised to record subsequent modifications of operational policy. The SPC has signed memoranda of understanding with WHO, the Forum Fisheries Agency, the South Pacific Regional Environment Programme (SPREP), and several other partners. The organization participates in meetings of the Council of Regional Organizations in the Pacific (CROP, see under Pacific Islands Forum Secretariat). Representatives of the SPC, SPREP and the South Pacific Applied Geoscience Commission hold periodic 'troika' meetings to develop regional technical co-operation and harmonization of work programmes.

LAND RESOURCES

The land resources division comprises two major programmes: agriculture (incorporating advice and specific activities in crop improvement and plant protection; animal health and production services; and agricultural resource economics and information); and forestry (providing training, technical assistance and information in forestry management and agroforestry). Objectives of the agriculture programme, based in Suva, Fiji, include the promotion of land and agricultural management practices that are both economically and environmentally sustainable; strengthening national capabilities to reduce losses owing to crop pests (insects, pathogens and weeds) and animal diseases already present, and to prevent the introduction of new pests and diseases; and facilitating trade through improved quarantine procedures. A Regional Animal Health Service was established in 1991. A Pacific Regional Agricultural Programme (PRAP), funded by the European Union (EU), was introduced by eight member states in 1990. The SPC assumed responsibility for administering PRAP in 1998. Funding through PRAP and the Australian Government enabled the establishment of the Regional Germplasm Centre in March 1999, which was to address the issue of conservation of plant genetic resources in the region. In 2003 an EU-funded Development of Sustainable Agriculture in the Pacific (DSAP) project was initiated to assist 10 member countries to implement sustainable agriculture measures and to improve food production and security. A further six Pacific countries joined the programme in 2004. The Regional Forestry Programme, based at the Pacific Islands Forum Secretariat, was formally established in January 2000. The programme's strategic plan for 2001–04 stated the following as among its key objectives: strengthening the capacity of member states to formulate and implement sound forestry polices and practices; promoting the application and adoption of multiple land-use systems; promoting awareness and participation on the part of local communities in the management, use and protection of forests; and providing a focal point for advocacy, collaboration, information dissemination and resource mobilization. The programme also incorporates the Forests and Trees Project, funded by the Australian Government, and the Pacific–German Regional Forestry Project.

In September 2004 the SPC hosted the first joint conference of Pacific Ministers of Agriculture and of Forestry to reflect activities to integrate the two activities under a single management of the Land Resources Division. The meeting considered a new strategic plan for the division for 2005–08, which focused on the following themes: Food security and health; Sustainable agriculture and forestry; and Biosecurity and trade facilitation. The meeting was followed by the first combined regional meeting of the Heads of Agriculture and Forestry Services.

MARINE RESOURCES

The SPC aims to support and co-ordinate the sustainable development and management of inshore fisheries resources in the region, to undertake scientific research in order to provide member governments with relevant information for the sustainable development and management of tuna and billfish resources in and adjacent to the South Pacific region, and to provide data and analytical services to national fisheries departments. The main components of the Community's fisheries activities are the Coastal Fisheries Programme (CFP), the Oceanic Fisheries Programme (OFP), and the Regional Maritime Programme (RMP). The CFP is divided into the following sections: community fisheries (research and assessment of and development support for people occupied in subsistence and artisanal fisheries); fisheries training; sustainable fisheries development; reef fisheries assessment and management; fisheries information; and post-harvest development (offering advice and training in order to improve handling practices, storage, seafood product development, quality control and marketing). The OFP consists of the following three sections: statistics and monitoring; tuna ecology and biology; and stock assessment and modelling. The statistics and monitoring section maintains a database of industrial tuna fisheries in the region. The OFP contributed research and statistical information for the formulation of the Convention for the Conservation and Management of Highly Migratory Fish Stocks in the Western and Central Pacific, which aimed to establish a regime for the sustainable management of tuna reserves. The Convention was opened for signature in Honolulu in September 2000, and entered into force in June 2004. In March 2002 the SPC and European Commission launched a Pacific Regional Oceanic and Coastal Fisheries Project (PROCFISH). The oceanic component of the project aimed to assist the OFP with advancing knowledge of tuna fisheries ecosystems, while the coastal element was to produce the first comparative regional baseline assessment of reef fisheries. The RMP advises member governments in the fields of policy, law and training. In early 2002 the RMP launched the model Pacific Islands Maritime Legislation and Regulations as a framework for the development of national maritime legislation. The SPC administers the Pacific Island Aquaculture Network, a forum for promoting regional aquaculture development.

The SPC hosts the Pacific Office of the WorldFish Center (the International Centre for Living Aquatic Resources Management—ICLARM); the SPC and the WorldFish Center have jointly implemented a number of projects.

SOCIAL RESOURCES

The Social Resources Division comprises the Public Health Programme and the Socio-economic Programme (including sub-programmes and sections on statistics; population and demography; rural energy development; youth issues; culture; women's and gender equality; community education training; and media training).

The Public Health Programme aims to implement health promotion programmes; to assist regional authorities to strengthen health information systems and to promote the use of new technology for health information development and disease control (for example, through the Public Health Surveillance and Disease Control Programme); to promote efficient health services management; and to help all Pacific Islanders to attain a level of health and quality of life that will enable them to contribute to the development of their communities. The Public Health Services also work in the areas of non-communicable diseases and nutrition (with particular focus on the high levels of diabetes and heart disease in parts of the region); environmental health, through the improvement of water and sanitation facilities; and reducing the incidence of HIV/AIDS and other sexually-transmitted infections (STIs), tuberculosis, and vector-borne diseases such as malaria and dengue fever. The SPC operates a project (mainly funded by Australia and New Zealand), to prevent AIDS and STIs among young people through peer education and awareness. In August 2004 a new grants scheme was launched to fund the development and implementation of national HIV/AIDS and STI strategic plans. A Lifestyle Health Section aims to assist member countries to improve and sustain health, in particular in nutrition.

The Statistics Programme assists governments and administrations in the region to provide effective and efficient national statistical services through the provision of training activities, a statistical information service and other advisory services. A Regional Meeting of Heads of Statistics facilitates the integration and co-ordination of statistical services throughout the region, while the Pacific Regional Information System (PRISM), initiated by the National Statistics Office of the Pacific Islands and developed with British funding, provides statistical information about member countries and territories. Activities of a new Pacific Regional Poverty Programme were to be integrated into PRISM in 2004.

The Population and Demography Programme provides technical support in population, demographic and development issues to member governments, other SPC programmes, and organizations active in the region. The Programme aims to assist governments effectively to analyse data and utilize it into the formulation of national development policies and programmes. The Programme organizes national workshops in population and development planning, provides short-term professional attachments, undertakes demographic research and analysis, and disseminates information.

The Pacific Youth Bureau (PYB) co-ordinates the implementation of the Pacific Youth Strategy 2005, which aims to develop opportunities for young people to play an active role in society. The PYB provides non-formal education and support for youth, community workers and young adults in community development subjects and provides grants to help young people find employment. It also advises and assists the Pacific Youth Council in promoting a regional youth identity. The Pacific Women's Bureau (PWB) aims to promote the social, economic and cultural advancement of women in the region by assisting governments and regional organizations to include women in the development planning process. The PWB also provides technical and advisory services, advocacy and management support training to groups concerned with women in development and gender and development, and supports the production and exchange of information regarding women.

The Cultural Affairs Programme aims to preserve and promote the cultural heritage of the Pacific Islands. The Programme assists with the training of librarians, archivists and researchers and promotes instruction in local languages, history and art at schools in member states and territories. The SPC acts as the secretariat of the Council of Pacific Arts, which organizes the Festival of Pacific Arts on a four-yearly basis. The ninth Festival was held in July 2004, in Palau.

The SPC regional office in Suva, Fiji, administers a Community Education Training Centre (CETC), which conducts a seven-month training course for up to 40 women community workers annually, with the objective of training women in methods of community development so that they can help others to achieve better living conditions for island families and communities. The Regional Media Centre provides training, technical assistance and production materials in all areas of the media for member countries and territories, community work programmes, donor projects and regional non-governmental organizations. The Centre comprises a radio broadcast unit, a graphic design and publication unit and a TV and video unit.

In 2000 the SPC's Information Technology and Communication Unit launched ComET, a satellite communications project aimed at linking more closely the organization's headquarters in New Caledonia and regional office in Fiji. The Information and Communications Programme is developing the use of modern communication technology as an invaluable resource for problem-solving, regional networking, and uniting the Community's scattered, often physically isolated, island member states. In conjunction with the Secretariat of the Pacific Islands Forum the SPC convened the first regional meeting of Information and Communication Technology workers, researchers and policy-makers in August 2001. The meeting addressed strategies for the advancement of new information technologies in member countries.

Finance

The organization's core budget, funded by assessed contributions from member states, finances executive and administrative expenditures, the Information and Communications Programme, and several professional and support positions that contribute to the work of the three programme divisions (i.e. Land Resources, Marine Resources and Social Resources). The non-core budget, funded mainly by aid donors and in part by Community member states, mostly on a contractual basis, finances the SPC's technical services. Administrative expenditure for 2004 amounted to CFP 3,211m.

Publications

Annual Report.

Fisheries Newsletter (quarterly).
Pacific Aids Alert Bulletin (quarterly).
Pacific Island Nutrition (quarterly).
Regional Tuna Bulletin (quarterly).
Report of the Conference of the Pacific Community.
Women's Newsletter (quarterly).

Technical publications, statistical bulletins, advisory leaflets and reports.

PACIFIC ISLANDS FORUM

Address: c/o Pacific Islands Forum Secretariat, Private Mail Bag, Suva, Fiji.

Telephone: 3312600; **fax:** 3305573; **e-mail:** info@forumsec.org.fj; **internet:** www.forumsec.org.fj.

MEMBERS

Australia	Niue
Cook Islands	Palau
Fiji	Papua New Guinea
Kiribati	Samoa
Marshall Islands	Solomon Islands
Federated States of Micronesia	Tonga
Nauru	Tuvalu
New Zealand	Vanuatu

Note: New Caledonia was admitted as an observer at the Forum in 1999. Timor-Leste was granted 'special observer' status in 2002 and French Polynesia was admitted as an observer in August 2004.

The Pacific Islands Forum (which changed its name from South Pacific Forum in October 2000) was founded as the gathering of Heads of Government of the independent and self-governing states of the South Pacific. Its first meeting was held on 5 August 1971, in Wellington, New Zealand. It provides an opportunity for informal discussions to be held on a wide range of common issues and problems and meets annually or when issues require urgent attention. The Forum has no written constitution or international agreement governing its activities nor any formal rules relating to its purpose, membership or conduct of meeting. Decisions are always reached by consensus, it never having been found necessary or desirable to vote formally on issues. In October 1994 the Forum was granted observer status by the General Assembly of the United Nations.

Since 1989 each Forum has been followed by 'dialogues' with representatives of other countries with a long-term interest in and commitment to the region. In October 1995 the Forum Governments suspended France's 'dialogue' status, following that country's resumption of the testing of nuclear weapons in French Polynesia. France was reinstated as a 'dialogue partner' in September 1996. In 2004 'dialogue partners' comprised Canada, the People's Republic of China, France, India, Indonesia, Japan, the Republic of Korea, Malaysia, Philippines, the United Kingdom, the USA, and the European Union. Thailand was to participate in the dialogue meetings from 2005.

The South Pacific Nuclear-Free Zone Treaty (Treaty of Rarotonga), prohibiting the acquisition, stationing or testing of nuclear weapons in the region, came into effect in December 1986, following ratification by eight states. The USSR signed the protocols to the treaty (whereby states possessing nuclear weapons agree not to use or threaten to use nuclear explosive devices against any non-nuclear party to the Treaty) in December 1987 and ratified them in April 1988; the People's Republic of China did likewise in December 1987 and October 1988 respectively. In July 1993 the Forum petitioned the USA, the United Kingdom and France, asking them to reconsider their past refusal to sign the Treaty in the light of the end of the 'Cold War'. In July 1995, following the decision of the French Government to resume testing of nuclear weapons in French Polynesia, members of the Forum resolved to increase diplomatic pressure on the three Governments to sign the Treaty. In October the United Kingdom, the USA and France announced their intention to accede to the Treaty, by mid-1996. Following France's decision, announced in January 1996, to end the programme four months earlier than scheduled, representatives of the Governments of the three countries signed the Treaty in March.

In 1990 five of the Forum's smallest island member states formed an economic sub-group to address their specific concerns, in particular economic disadvantages resulting from a poor resource base, absence of a skilled work-force and lack of involvement in world markets. In September 1997 the 28th Forum, convened in Rarotonga, the Cook Islands, endorsed the inclusion of the Marshall Islands as the sixth member of the Smaller Island States (SIS) sub-group. Representatives of the grouping, which also includes Kiribati, the Cook Islands, Nauru, Niue and Tuvalu, meet regularly. In February 1998 senior Forum officials, for the first time, met with representatives of the Caribbean Community and the Indian Ocean Commission, as well as other major international organizations, to discuss means to enhance consideration and promotion of the interests of small island states. Small island member states have been particularly concerned about the phenomenon of global warming and its potentially damaging effects on the region (see below).

At the 24th Forum, held in Yaren, Nauru, in August 1993 it was agreed that effective links needed to be established with the broader Asia-Pacific region, with participation in Asia-Pacific Economic Co-operation (APEC), where the Forum has observer status, to be utilized to the full. The Forum urged an increase in intra-regional trade and asked for improved opportunities for Pacific island countries exporting to Australia and New Zealand. New Caledonia's right to self-determination was supported. Environmental protection measures and the rapid growth in population in the region, which was posing a threat to economic and social development, were also discussed by the Forum delegates.

The 25th Forum was convened in Brisbane, Australia, in August 1994 under the theme of 'Managing Our Resources'. In response to the loss of natural resources as well as of income-earning potential resulting from unlawful logging of timber by foreign companies, Forum members agreed to impose stricter controls on the exploitation of forestry resources and to begin negotiations to standardize monitoring of the region's resources. The Forum also agreed to strengthen its promotion of sustainable exploitation of fishing stocks, reviewed preparations of a convention to control the movement and management of radioactive waste within the South Pacific and discussed the rationalization of national airlines, on a regional or sub-regional basis, to reduce operational losses.

The 26th Forum, held in Madang, Papua New Guinea, in September 1995, was dominated by extreme hostility on the part of Forum Governments to the resumption of testing of nuclear weapons by France in the South Pacific region. The decision to recommence testing, announced by the French Government in June, had been instantly criticized by Forum Governments. The 26th Forum reiterated their demand that France stop any further testing, and also condemned the People's Republic of China for conducting nuclear tests in the region. The meeting endorsed a draft Code of Conduct on the management and monitoring of indigenous forest resources in selected South Pacific countries, which had been initiated at the 25th Forum; however, while the six countries concerned committed themselves to implementing the Code through national legislation, its signing was deferred, owing to an initial unwillingness on the part of Papua New Guinea and Solomon Islands. The Forum did adopt a treaty to ban the import into the region of all radioactive and other hazardous wastes, and to control the transboundary movement and management of these wastes (the so-called Waigani Convention). The Forum agreed to reactivate the ministerial committee on New Caledonia, comprising Fiji, Nauru and Solomon Islands, which was to monitor political developments in that territory prior to its referendum on independence, scheduled to be held in 1998. In addition, the Forum resolved to implement and pursue means of promoting economic co-operation and long-term development in the region. In December 1995 Forum finance ministers, meeting in Port Moresby, Papua New Guinea, discussed the issues involved in the concept of 'Securing Development Beyond 2000' and initiated an assessment project to further trade liberalization efforts in the region.

The 27th Forum, held in Majuro, the Marshall Islands, in September 1996, supported the efforts of the French Government to improve relations with countries in the South Pacific and agreed to readmit France to the post-Forum dialogue. The Forum meeting recognized the importance of responding to the liberalization of the global trading system by reviewing the region's economic tariff policies, and of assisting members in attracting investment for the development of the private sector. The Forum advocated that a meeting of economy ministers of member countries be held each year. The Forum was also concerned with environmental issues: in particular, it urged the ratification and implementation of the Waigani Convention by all member states, the promotion of regional efforts to conserve marine resources and to protect the coastal environment, and the formulation of an international, legally-binding agreement to reduce emissions by industrialized countries of so-called 'greenhouse gases'. Such gases contribute to the warming of the earth's atmosphere (the 'greenhouse effect') and to related increases in global sea-levels, and have therefore been regarded as a threat to low-lying islands in the region. The Forum requested the ministerial committee on New Caledonia (established by the 1990 Forum to monitor, in co-operation with the French authorities, political developments in the territory) to pursue contacts with all parties there and to continue to monitor preparations for the 1998 referendum.

In July 1997 the inaugural meeting of Forum economy ministers was convened in Cairns, Australia. It formulated an Action Plan to encourage the flow of foreign investment into the region by committing members to economic reforms, good governance and the implementation of multilateral trade and tariff policies. The meeting also commissioned a formal study of the establishment of a free-trade agreement between Forum island states. The 28th Forum, held in Rarotonga, the Cook Islands, in September, considered the economic challenges confronting the region. However, it was marked by a failure to conclude a common policy position on mandatory targets for reductions in 'greenhouse gas' emissions, owing to an ongoing dispute between Australia and other Forum Governments.

The 29th Forum, held in Pohnpei, Federated States of Micronesia, in August 1998, considered the need to pursue economic reforms and to stimulate the private sector and foreign investment in order to ensure economic growth. Leaders reiterated their support for efforts to implement the economic Action Plan and to develop a framework for a free-trade agreement, and endorsed specific recommendations of the second Forum Economic Ministers Meeting, which was held in Fiji, in July, including the promotion of competitive telecommunications markets, the development of information infrastructures and support for a new economic vulnerability index at the UN to help determine least developed country status. The Forum was also concerned with environmental issues, notably the shipment of radioactive wastes, the impact of a multinational venture to launch satellites from the Pacific, the need for ongoing radiological monitoring of the Mururoa and Fangataufa atolls, and the development of a South Pacific Whale Sanctuary. The Forum adopted a Statement on Climate Change, which urged all countries to ratify and implement the gas emission reductions agreed upon by UN member states in December 1997 (the so-called Kyoto Protocol of the UN Framework Convention on Climate Change), and emphasized the Forum's commitment to further measures for verifying and enforcing emission limitation.

In October 1999 the 30th Forum, held in Koror, Palau, endorsed in principle the establishment of a regional free-trade area (FTA), which had been approved at a special ministerial meeting held in June. The FTA was to be implemented from 2002 over a period of eight years for developing member countries and 10 years for smaller island states and least developed countries. The Forum requested officials from member countries to negotiate the details of a draft agreement on the FTA (the so-called Pacific Island Countries Trade Agreement—PICTA), including possible extensions of the arrangements to Australia and New Zealand. The heads of government adopted a Forum Vision for the Pacific Information Economy, which recognized the importance of information technology infrastructure for the region's economic and social development and the possibilities for enhanced co-operation in investment, job creation, education, training and cultural exchange. Forum Governments also expressed concern at the shipment of radioactive waste through the Pacific and determined to pursue negotiations with France, Japan and the United Kingdom regarding liability and compensation arrangements; confirmed their continued support for the multinational force and UN operations in East Timor (now Timor-Leste); and urged more countries to adopt and implement the Kyoto Protocol to limit the emission of 'greenhouse gases'. In addition, the Forum agreed to rename the grouping (hitherto known as the South Pacific Forum) the Pacific Islands Forum, to reflect the expansion of its membership since its establishment; the new designation took effect at the 31st Forum, which was convened in Tarawa, Kiribati, at the end of October 2000.

At the 31st Forum the heads of government discussed the escalation in regional insecurity that had occurred since the previous Forum. Concern was expressed over the unrest that prevailed during mid-2000 in Fiji and Solomon Islands, and also over ongoing political violence in the Indonesian province of Irian Jaya (now Papua). The Forum adopted the Biketawa Declaration, which outlined a mechanism for responding to any future such crises in the region while urging members to undertake efforts to address the fundamental causes of instability. The detrimental economic impact of the disturbances in Fiji and Solomon Islands was noted. The Forum endorsed a proposal to establish a Regional Financial Information Sharing Facility and national financial intelligence units, and welcomed the conclusion in June by the European Union and ACP states (which include several Forum members) of the Cotonou Agreement. The Forum also reiterated support for the prompt implementation of the Kyoto Protocol.

In August 2001, convened at the 32nd Forum, in Nauru, nine regional leaders adopted PICTA, providing for the establishment of the FTA (as envisaged at the 30th Forum). A related Pacific Agreement on Closer Economic Relations (PACER), envisaging the phased establishment of a regional single market including the signatories to PICTA and Australia and New Zealand, was also adopted. The Forum expressed concern at the refusal of the USA (responsible for about one-quarter of world-wide 'greenhouse gas' emissions) to ratify the Kyoto Protocol. In response to an ongoing initiative by the Organisation of Economic Co-operation and Development (OECD) to eliminate the operation of 'harmful tax systems', the Forum reaffirmed the sovereign right of nations to establish individual tax regimes and urged the development of a new co-operative framework to address issues relating to financial transparency. (OECD had identified the Cook Islands, the Marshall Islands, Nauru and Niue as so-called 'tax havens' lacking financial transparency and had demanded that they impose stricter legislation to address the incidence of international money-laundering on their territories.) Forum leaders also reiterated protests against the shipment of radioactive materials through the region.

The 33rd Forum, held in Suva, Fiji, in August 2002, adopted the Nasonini Declaration on Regional Security, which recognized the need for immediate and sustained regional action to combat international terrorism and transnational crime, in view of the perceived increased threat to global and regional security following the major terrorist attacks perpetrated against targets in the USA in September 2001. Regional leaders also approved a Pacific Island Regional Ocean Policy, which aimed to ensure the future sustainable use of the ocean and its resources by Pacific Island communities and external partners. In addition the Forum urged OECD to adopt a more flexible approach in the implementation of its ongoing harmful tax initiative; welcomed the third assessment report on climate change issued in 2001 by the WMO/UNEP Intergovernmental Panel on Climate Change and reiterated demands for a world-wide reduction in greenhouse gas emissions; invited member island states to declare their coastal waters as whale sanctuaries; and urged the development of a Pacific Regional Plan of Action against HIV/AIDS.

Regional leaders convened at the 34th Forum, held in August 2003, in Auckland, New Zealand, commended the swift response by member countries and territories in deploying a Regional Assistance Mission in Solomon Islands, which had been approved by Forum ministers of foreign affairs at a meeting held in Sydney, Australia, in June, in accordance with the Biketawa Declaration. The Forum also welcomed the entry into force of the PICTA and PACER trade agreements and agreed, in principle, that the USA and France should become parties to both agreements. Regional leaders adopted a set of Forum Principles of Good Leadership, establishing key requirements for good governance, including respect for law and the system of government, and respect for cultural values, customs and traditions, and for freedom of religion. The leaders determined that a review of the activities of the Forum and its Secretariat should be undertaken, and established an Eminent Persons Group to initiate that process.

In April 2004 a Special Leaders' Retreat was convened in Auckland, New Zealand, in order to consider the future activities and direction of the Forum. The meeting concluded that the following four areas were key priorities for the grouping: economic growth; sustainable development; security; and good governance. The heads of government mandated the development of a Pacific Plan to strengthen co-operation throughout Forum countries and to achieve their objectives, which included the sustainable management of resources, the observation of democratic values and good governance, economic sustainability, peaceful relations with neighbours, and improved communications throughout the region. They also recognized the specific needs of Small Island States. The 35th Forum was held in August, in Apia, Samoa. Leaders commended the work already being undertaken on the Pacific Plan by a newly-established Task Force, and anticipated a substantial document to be presented to the 36th Forum. Leaders adopted new Principles on Regional Transport Services, based on the results of a study requested by the 34th Forum, 'to improve the efficiency, effectiveness and sustainability of air and shipping services'. The 35th Forum also reiterated the importance of fisheries to the region's economy and population, and resolved to pursue an increase in sustainable returns through greater participation of resource-owning countries in the fishing industry. Forum leaders also approved an HIV/AIDS Regional Strategy and requested its implementation with immediate effect, resolved to support a request from the Government of Nauru for economic and technical assistance in accordance with the Biketawa Declaration (2000), and determined to review the region's response to national disasters and other emergencies, with particular concern given to the recovery efforts of Niue following a devastating cyclone which struck the territory in January.

Pacific Islands Forum Secretariat

Address: Private Mail Bag, Suva, Fiji.

Telephone: 3312600; **fax:** 3301102; **e-mail:** info@forumsec.org.fj; **internet:** www.forumsec.org.fj.

The South Pacific Bureau for Economic Co-operation (SPEC) was established by an agreement signed on 17 April 1973, at the third meeting of the South Pacific Forum (now Pacific Islands Forum) in Apia, Western Samoa (now Samoa). SPEC was renamed the South Pacific Forum Secretariat in 1988; this, in turn, was redesignated as the Pacific Islands Forum Secretariat in October 2000.

Organization

(October 2004)

FORUM OFFICIALS COMMITTEE

The Forum Officials Committee is the Secretariat's executive board. It comprises representatives and senior officials from all member countries. It meets twice a year, immediately before the meetings of the Pacific Islands Forum and at the end of the year, to discuss in detail the Secretariat's work programme and annual budget.

SECRETARIAT

The Secretariat undertakes the day-to-day activities of the Forum. It is headed by a Secretary-General, with a staff of some 70 people drawn from the member countries. The Secretariat comprises the following four Divisions: Corporate Services; Development and Economic Policy; Trade and Investment; and Political, International and Legal Affairs.

Secretary-General: GREGORY (GREG) URWIN (Australia).

Deputy Secretary-General: IOSEFA MAIAVA (Samoa).

Activities

The Secretariat's aim is to enhance the economic and social well-being of the people of the South Pacific, in support of the efforts of national governments.

The Secretariat's trade and investment services extend advice and technical assistance to member countries in policy, development, export marketing, and information dissemination. Trade policy activities are mainly concerned with improving private sector policies, for example investment promotion, assisting integration into the world economy (including the provision of information and technical assistance to member states on WTO-related matters and supporting Pacific Island ACP states with preparations for negotiations on trade partnership with the EU under the Cotonou Agreement), and the development of businesses. The Secretariat aims to assist both island governments and private sector companies to enhance their capacity in the development and exploitation of export markets, product identification and product development. A regional trade and investment database is being developed. The Trade and Investment Division of the Secretariat co-ordinates the activities of the regional trade offices located in Australia, New Zealand and Japan (see below). A representative trade office in Beijing, People's Republic of China, opened in January 2002. A Forum office was opened in Geneva, Switzerland, in 2004 to represent member countries at the WTO.

In 1981 the South Pacific Regional Trade and Economic Co-operation Agreement (SPARTECA) came into force. SPARTECA aimed to redress the trade deficit of the Pacific Island countries with Australia and New Zealand. It is a non-reciprocal trade agreement under which Australia and New Zealand offer duty-free and unrestricted access or concessional access for specified products originating from the developing island member countries of the Forum. In 1985 Australia agreed to further liberalization of trade by abolishing (from the beginning of 1987) duties and quotas on all Pacific products except steel, cars, sugar, footwear and garments. In August 1994 New Zealand expanded its import criteria under the agreement by reducing the rule of origin requirement for garment products from 50% to 45% of local content. In response to requests from Fiji, Australia agreed to widen its interpretation of the agreement by accepting as being of local content manufactured products that consist of goods and components of 50% Australian content. A new Fiji/Australia Trade and Economic Relations Agreement (AFTERA) was concluded in March 1999 to complement SPARTECA and compensate for certain trade benefits that were in the process of being withdrawn.

Two major regional trade accords signed by Forum heads of state in August 2001 entered into force in April 2003 and October 2002, respectively: the Pacific Island Countries Trade Agreement (PICTA), providing for the establishment of a Pacific Island free-trade area (FTA); and the related Pacific Agreement on Closer Economic Relations (PACER), incorporating trade and economic co-operation measures and envisaging an eventual single regional market comprising the PICTA FTA and Australia and New Zealand. It was envisaged that negotiations on free-trade agreements between Pacific Island states and Australia and New Zealand, with a view to establishing the larger regional single market envisaged by PACER, would commence within eight years of PICTA's entry into force. SPARTECA (see above) would remain operative pending the establishment of the larger single market, into which it would be subsumed. Under the provisions of PACER, Australia and New Zealand were to provide technical and financial assistance to PICTA signatory states in pursuing the objectives of PACER. In September 2004 Forum trade officials adopted a Regional Trade Facilitation Programme, within the framework of PACER, which included measures concerned with customs procedures, quarantine, standards and other activities to harmonize and facilitate trade between Pacific Island states and Australia and New Zealand, as well as with other international trading partners.

In April 2001 the Secretariat convened a meeting of seven member island states—the Cook Islands, the Marshall Islands, Nauru, Niue, Samoa, Tonga and Vanuatu—as well as representatives from Australia and New Zealand, to address the regional implications of the OECD's Harmful Tax Competition Initiative. The meeting requested the OECD to engage in conciliatory negotiations with the listed Pacific Island states. The August Forum reiterated this stance, proclaiming the sovereign right of nations to establish individual tax regimes, and supporting the development of a new co-operative framework to address financial transparency concerns. In December the Secretariat hosted a workshop for officials from nine member states concerned with combating financial crime. The workshop was attended and sponsored by several partner organizations and bodies, including the IMF. In 2004 the Secretariat undertook work to prepare a Pacific Plan that had been mandated a the Special Leaders' Retreat held in April.

The Political, International and Legal Affairs Division of the Secretariat organizes and services the meetings of the Forum, disseminates its views, administers the Forum's observer office at the United Nations, and aims to strengthen relations with other regional and international organizations, in particular APEC and ASEAN. The Division's other main concern is to promote regional co-operation in law enforcement and legal affairs, and it provides technical support for the drafting of legal documents and for law enforcement capacity-building. In 1997 the Secretariat undertook an assessment to survey the need for specialist training in dealing with money laundering in member countries. In recent years the Forum Secretariat has been concerned with assessing the legislative reforms and other commitments needed to ensure implementation of the 1992 Honiara Declaration on Law Enforcement Co-operation. The Division assists member countries to ratify and implement the 1988 UN Convention against Illicit Trafficking in Narcotic Drugs and Psychotropic Substances. In December 1998 the Secretariat initiated a five-year programme to strengthen regional law enforcement capabilities, in particular to counter cross-border crimes such as money-laundering and drugs-trafficking. All member states, apart from Australia and New Zealand, were to participate in the initiative. In December 2001 the first ever Forum Election Observer Group was dispatched to monitor legislative elections held in Solomon Islands. In July 2004 a joint mission, with officials from the Commonwealth Secretariat, observed elections to the National Assembly in Vanuatu. A similar joint observer mission was to monitor elections in Nauru, in October. At the end of 2001 a conference of Forum immigration ministers expressed concern at rising levels of human-trafficking and illegal immigration in the region, and recommended that member states become parties to the 2000 UN Convention Against Transnational Organized Crime. In February 2003 the Forum hosted a seminar of Pacific Island law-makers under a regional initiative to draft legislation on combating money laundering and the financing of terrorism. A Pacific Transnational Crime Co-ordination Centre was established in Suva, Fiji, in 2004, to enhance and gather law enforcement intelligence generated by national transnational crime units. Under the proposed Pacific Plan, to be formulated in 2004–05, the Forum Secretariat requested the establishment of a Pacific Islands Regional Security Technical Co-operation Unit to support legislative efforts in, *inter alia*, transnational organized crime, counter-terrorism and financial intelligence.

The Secretariat helps to co-ordinate environmental policy. With support from the Australian Government, it administers a network of stations to monitor sea-levels and climate change throughout the Pacific region. In recent years the Secretariat has played an active role in supporting regional participation at meetings of the Conference of the Parties to the UN Framework Convention on Climate Change.

The Development and Economic Policy Division of the Secretariat aims to co-ordinate and promote co-operation in development activities and programmes throughout the region. The Division administers a Short Term Advisory Service, which provides consultancy services to help member countries meet economic development priorities, and a Fellowship Scheme to provide practical training in a range of technical and income-generating activities. A Small Island Development Fund aims to assist the economic development of this sub-group of member countries (i.e. the Cook Islands, Kiribati, the Marshall Islands, Nauru, Niue and Tuvalu) through project financing. A separate fellowship has also been established to provide training to the Kanak population of New Caledonia, to assist in their social, economic and political development. The Division aims to assist regional organizations to identify development priorities and to provide advice to national governments on economic analysis, planning and structural reforms. The Secretariat chairs the Council of Regional Organizations in the Pacific (CROP), an *ad hoc* committee comprising the heads of eight regional organizations, which aims to discuss and co-ordinate the policies and work programmes of the various agencies in order to avoid duplication of or omissions in their services to member countries.

The Secretariat services the Pacific Group Council of ACP states receiving assistance from the EU, and in 1993 a joint unit was established within the Secretariat headquarters to assist Pacific ACP countries and regional organizations in submitting projects to the EU for funding. In October 2002 the Secretariat signed, on behalf of the Pacific Island ACP states, a framework agreement providing for EU-Pacific co-operation over a period of five years, and making a €29m. contribution in support of trade and economic integration (in particular the implementation of PICTA), the development of human resources (with particular emphasis on basic education), and the fisheries sector.

The Forum established the Pacific Forum Line and the Association of South Pacific Airlines (see below), as part of its efforts to promote co-operation in regional transport. On 1 January 1997 the work of the Forum Maritime Programme, which included assistance for regional maritime training and for the development of regional maritime administrations and legislation, was transferred to the regional office of the South Pacific Commission (renamed the Pacific Community from February 1998) at Suva. At the same time responsibility for the Secretariat's civil aviation activities was transferred to individual countries, to be managed at a bilateral level. Telecommunications policy activities were also transferred to the then South Pacific Commission at the start of 1997. In May 1998 ministers responsible for aviation in member states approved a new regional civil aviation policy, which envisaged liberalization of air services, common safety and security standards and provisions for shared revenue. During 2003–04 the Forum Secretariat worked with member countries to ensure their compliance with a new International Shipping and Port Facility Code. In accordance with the Principles on Regional Transport Services, which were adopted by Forum Leaders in August 2004, the Secretariat was to support efforts to enhance air and shipping services, as well as develop a regional digital strategy.

Finance

The Governments of Australia and New Zealand each contribute some one-third of the annual budget and the remaining amount is shared by the other member Governments. Extra-budgetary funding is contributed mainly by Australia, New Zealand, Japan, the EU and France. The Forum's 2004 budget amounted to FJ $15.4m.

Publications

Annual Report.

Forum News (quarterly).

Forum Trends.

Forum Secretariat Directory of Aid Agencies.

South Pacific Trade Directory.

SPARTECA (guide for Pacific island exporters).

Reports of meetings; profiles of Forum member countries.

Associated and Affiliated Organizations

Association of South Pacific Airlines—ASPA: POB 9817, Nadi Airport, Nadi, Fiji; tel. 6723526; fax 6720196; f. 1979 at a meeting of airlines in the South Pacific, convened to promote co-operation among the member airlines for the development of regular, safe and economical commercial aviation within, to and from the South Pacific; mems: 16 regional airlines, two associates; Chair. SEMISI TAUMOEPEAU; Sec.-Gen. GEORGE E. FAKTAUFON.

Forum Fisheries Agency—FFA: POB 629, Honiara, Solomon Islands; tel. (677) 21124; fax (677) 23995; e-mail info@ffa.int; internet www.ffa.int; f. 1979 to promote co-operation in fisheries among coastal states in the region; collects and disseminates information and advice on the living marine resources of the region, including the management, exploitation and development of these resources; provides assistance in the areas of law (treaty negotiations, drafting legislation, and co-ordinating surveillance and enforcement), fisheries development, research, economics, computers, and information management; the FFA was closely involved in the legal process relating to the establishment of a Western and Central Pacific Fisheries Commission, scheduled to be inaugurated in December 2004; a Vessel Monitoring System, to provide automated data collection and analysis of fishing vessel activities throughout the region, was inaugurated by the FFA in 1998; on behalf of its 16 member countries, the FFA administers a multilateral fisheries treaty, under which vessels from the USA operate in the region, in exchange for an annual payment; Dir VICTORIO UHERBELAU; publs *FFA News Digest* (every two months), *FFA Reports*, *MCS Newsletter* (quarterly), *Tuna Market Newsletter* (monthly).

Pacific Forum Line: POB 796, Auckland, New Zealand; tel. (9) 356-2333; fax (9) 356-2330; e-mail info@pflnz.co.nz; internet www.pflnz.co.nz; f. 1977 as a joint venture by South Pacific countries, to provide shipping services to meet the special requirements of the region; operates three container vessels; conducts shipping agency services in Australia, Fiji, New Zealand and Samoa, and stevedoring in Samoa; Chair. T. TUFUI; CEO W. J. MACLENNAN.

Pacific Islands Centre—PIC: Sotobori Sky Bldg, 5th Floor, 2-11 Ichigayahonmura-cho, Shinjuku-ku, Tokyo 162-0845, Japan; tel. (3) 3268-8419; fax (3) 3268-6311; e-mail info@pic.or.jp; internet www.pic.or.jp; f. 1996 to promote and to facilitate trade, investment and tourism among Forum members and Japan; Dir AKIRA OUCHI.

Pacific Islands Trade and Investment Commission (Australia Office): Level 11, 171 Clarence St, Sydney, NSW 20010, Australia; tel. (2) 9290-2133; fax (2) 9299-2151; e-mail info@pitic.org.au; internet www.sptc.gov.au; f. 1979; assists Pacific Island Governments and business communities to identify market opportunities in Australia and promotes investment in the Pacific Island countries; Senior Trade Commr AIVU TAUVASA (Papua New Guinea).

Pacific Islands Trade and Investment Commission (New Zealand Office): Flight Centre, 48 Emily Pl., Auckland, New Zealand; tel. (9) 3020465; fax (9) 3776642; e-mail info@pitic.org.nz; internet www.pitic.org.nz; Senior Trade Commr PARMESH CHAND.

OTHER REGIONAL ORGANIZATIONS

Agriculture, Food, Forestry and Fisheries

(For organizations concerned with agricultural commodities, see Commodities)

Asian Vegetable Research and Development Center—AVRDC: POB 42, Shanhua, Tainan 741, Taiwan; tel. (6) 5837801; fax (6) 5830009; e-mail avrdcbox@netra.avrdc.org.tw; internet www.avrdc.org.tw; f. 1971; aims to enhance the nutritional well-being and raise the incomes of the poor in rural and urban areas of developing countries, through improved varieties and methods of vegetable production, marketing and distribution; runs an experimental farm, laboratories, gene-bank, greenhouses, quarantine house, insectarium, library and weather station; provides training for research and production specialists in tropical vegetables; exchanges and disseminates vegetable germplasm through regional centres in the developing world; serves as a clearing-house for vegetable research information; and undertakes scientific publishing; mems: Australia, France, Germany, Japan, Republic of Korea, Philippines, Taiwan, Thailand, USA; Dir-Gen. Dr Thomas A. Lumpkin; publs *Annual Report*, *Technical Bulletin*, *Proceedings*, *Centerpoint* (quarterly).

CAB International—CABI: Wallingford, Oxon, OX10 8DE, United Kingdom; tel. (1491) 832111; fax (1491) 833508; e-mail cabi@cabi.org; internet www.cabi.org; f. 1929 as the Imperial Agricultural Bureaux (later Commonwealth Agricultural Bureaux), current name adopted in 1985; aims to improve human welfare world-wide through the generation, dissemination and application of scientific knowledge in support of sustainable development; places particular emphasis on sustainable agriculture, forestry, human health and the management of natural resources, with priority given to the needs of developing countries; compiles and publishes extensive information (in a variety of print and electronic forms) on aspects of agriculture, forestry, veterinary medicine, the environment and natural resources, Third World rural development and others; maintains regional centres in the People's Republic of China, India, Kenya, Malaysia, Pakistan, Switzerland, Trinidad and Tobago, and the United Kingdom; mems: 40 countries; Dir-Gen. Dr Denis Blight.

CABI Bioscience: Bakeham Lane, Egham, Surrey, TW20 9TY, United Kingdom; tel. (1491) 829080; fax (1491) 829100; e-mail bioscience.egham@cabi.org; internet www.cabi-bioscience.org; f. 1998 by integration of the following four CABI scientific institutions: International Institute of Biological Control; International Institute of Entomology; International Institute of Parasitology; International Mycological Institute; undertakes research, consultancy, training, capacity-building and institutional development measures in sustainable pest management, biosystematics and molecular biology, ecological applications and environmental and industrial microbiology; maintains centres in Kenya, Malaysia, Pakistan, Switzerland, Trinidad and Tobago, and the United Kingdom; Dir David Dent.

Commission for the Conservation of Southern Bluefin Tuna: Unit 1, J.A.A. House, 19 Napier Close, Deakin, Canberra, ACT 2600, Australia; tel. (2) 6282-8396; fax (2) 6282-8407; e-mail bmacdonald@ccsbt.org; internet www.ccsbt.org; f. 1994 when the Convention for the Conservation of Southern Bluefin Tuna (signed in May 1993) entered into force; aims to promote sustainable management and conservation of the southern bluefin tuna; holds an annual meeting and annual scientific meeting; collates relevant research, scientific information and data; encourages non-member countries and bodies to co-operate in the conservation and optimum utilization of Southern Bluefin Tuna through accession to the Convention or adherence to the Commission's management arrangements; mems: Australia, Japan, Republic of Korea, New Zealand; Exec. Sec. Brian Macdonald.

Indian Ocean Tuna Commission—IOTC: POB 1011, Victoria, Mahé, Seychelles; tel. 225494; fax 224364; e-mail iotcsecr@seychelles.net; internet www.seychelles/net/iotc; f. 1993 by FAO, as the successor to the Indo-Pacific Tuna Development and Management Programme (f. 1982); technical activities include sampling, tagging, methodology, information on tuna stocks, conversion factors, biological parameters, other research; 2001 budget US $1.1m; mems: Australia, People's Republic of China, European Union, Eritrea, France, India, Japan, Republic of Korea, Madagascar, Mauritius, Malaysia, Oman, Pakistan, Seychelles, Sudan, Sri Lanka, Thailand, United Kingdom; Sec. David Ardill (Mauritius).

Inter-American Tropical Tuna Commission—IATTC: 8604 La Jolla Shores Drive, La Jolla, CA 92037-1508, USA; tel. (858) 546-7100; fax (858) 546-7133; e-mail rallen@iattc.org; internet www.iattc.org; f. 1950; administers two programmes, the Tuna-Billfish Programme and the Tuna-Dolphin Programme. The principal responsibilities of the Tuna-Billfish Programme are to study the biology of the tunas and related species of the eastern Pacific Ocean to determine the effects of fishing and natural factors on their abundance, to recommend appropriate conservation measures in order to maintain stocks at levels which will afford maximum sustainable catches, and to collect information on compliance with Commission resolutions. The functions of the Tuna-Dolphin Programme are to monitor the abundance of dolphins and their mortality incidental to purse-seine fishing in the eastern Pacific Ocean, to study the causes of mortality of dolphins during fishing operations and promote the use of techniques and equipment that minimize these mortalities, to study the effects of different fishing methods on the various fish and other animals of the pelagic ecosystem, and to provide a secretariat for the International Dolphin Conservation Programme; mems: Costa Rica, Ecuador, El Salvador, France, Guatemala, Japan, Mexico, Nicaragua, Panama, Peru, Spain, USA, Vanuatu, Venezuela; Dir Robin Allen; publs *Bulletin* (irregular), *Annual Report*, *Fishery Status Report*, *Stock Assessment Report* (annually), *Special Report* (irregular).

International Crops Research Institute for the Semi-Arid Tropics—ICRISAT: Patancheru 502 324, Andhra Pradesh, India; tel. (40) 23296161; fax (40) 23241239; e-mail icrisat@cgiar.org; internet www.icrisat.org; f. 1972 to promote the genetic improvement of crops and for research on the management of resources in the world's semi-arid tropics, with the aim of reducing poverty and protecting the environment; research covers all physical and socio-economic aspects of improving farming systems on unirrigated land; Dir Dr William D. Dar (Philippines); publs *ICRISAT Report* (annually), *SAT News* (2 a year), *International Chickpea and Pigeonpea Newsletter*, *International Arachis Newsletter*, *International Sorghum and Millet Newsletter* (annually), information and research bulletins.

International Rice Research Institute—IRRI: Los Baños, Laguna, DAPO Box 7777, Metro Manila, Philippines; tel. (2) 8450563; fax (2) 8911292; e-mail irri@cgiar.org; internet www.irri.org; f. 1960; conducts research on rice, with the aim of developing technologies of environmental, social and economic benefit; works to enhance national rice research systems and offers training; operates Riceworld, a museum and learning centre about rice; maintains a library of technical rice literature; organizes international conferences and workshops; Dir-Gen. Dr Ronald P. Cantrell; publs *Rice Literature Update*, *Hotline*, *Facts about IRRI*, *News about Rice and People*, *International Rice Research Notes*.

International Whaling Commission—IWC: The Red House, 135 Station Rd, Impington, Cambridge, CB4 9NP, United Kingdom; tel. (1223) 233971; fax (1223) 232876; e-mail secretariat@iwcoffice.com; internet www.iwcoffice.org; f. 1946 under the International Convention for the Regulation of Whaling, for the conservation of world whale stocks; reviews the regulations covering whaling operations; encourages research; collects, analyses and disseminates statistical and other information on whaling. A ban on commercial whaling was passed by the Commission in July 1982, to take effect three years subsequently (in some cases, a phased reduction of commercial operations was not completed until 1988). A revised whale-management procedure was adopted in 1992, to be implemented after the development of a complete whale management scheme; mems: governments of 49 countries; Chair. Prof. Bo Fernholm (Sweden); Sec. Dr Nicola Grandy; publ. *Annual Report*.

Network of Aquaculture Centres in Asia and the Pacific—NACA: POB 1040, Kasetsart Post Office, Bangkok 10903, Thailand; tel. (2) 561-1728; fax (2) 561-1727; e-mail publications@enaca.org; internet www.enaca.org; f. 1990; promotes the development of aquaculture in the Asia and Pacific region through development planning, interdisciplinary research, regional training and information; mems: Australia, Bangladesh, Cambodia, People's Republic of China, Hong Kong, India, Democratic People's Republic of Korea, Malaysia, Myanmar, Nepal, Pakistan, Philippines, Sri Lanka, Thailand and Viet Nam; Dir-Gen. PEDRO B. BUENOS; publs *NACA Newsletter* (quarterly), *Aquaculture Asia* (quarterly).

North Pacific Anadromous Fish Commission: 889 W. Pender St, Suite 502, Vancouver, BC V6C 3B2, Canada; tel. (604) 775-5550; fax (604) 775-5577; e-mail secretariat@npafc.org; f. 1993; mems: Canada, Japan, Republic of Korea, Russia, USA; Exec. Dir VLADIMIR FEDORENKO; publs *Annual Report*, *Newsletter* (2 a year), *Statistical Yearbook*, *Scientific Bulletin*, *Technical Report*.

Western and Central Pacific Fisheries Commission: c/o Ministerial Task Force on IUU Fishing, OECD, 2 rue André Pascal, Paris Cédex 16, France; f. 2004 under the Convention for the Conservation and Management of Highly Migratory Fish Stocks in the Western and Central Pacific, which entered into force in June of that year, six months after the deposit of the 13th ratification; inaugural session scheduled to be convened in December, in Pohnpei, Federated States of Micronesia; Head of Interim Secretariat MICHAEL LODGE.

WorldFish Center (International Centre for Living Aquatic Resources Management—ICLARM): Jalan Batu Maung, Batu Maung, 11960 Bayan Lepas, Penang, Malaysia; POB 500, 10670 Penang; tel. (4) 626-1606; fax (4) 626-5530; e-mail worldfishcenter@cgiar.org; internet www.worldfishcenter.org; f. 1973; became a mem. of the Consultative Group on International Agricultural Research—CGIAR in 1992; aims to contribute to food security and poverty eradication in developing countries through the sustainable development and use of living aquatic resources; carries out research and promotes partnerships; Dir-Gen. MERYL J. WILLIAMS; publ. *NAGA* (quarterly newsletter).

Arts and Culture

Organization of World Heritage Cities: 15 rue Saint-Nicolas, Québec, QC G1K 1MB, Canada; tel. (418) 692-0000; fax (418) 692-5558; e-mail secretariat@ovpm.org; internet www.ovpm.org; f. 1993 to assist cities inscribed on the UNESCO World Heritage List to implement the Convention concerning the Protection of the World Cultural and Natural Heritage (1972); promotes co-operation between city authorities, in particular in the management and sustainable development of historic sites; holds a General Assembly, comprising the mayors of member cities, at least every two years; mems: 208 cities world-wide; Sec.-Gen. DENIS RICARD.

Royal Asiatic Society of Great Britain and Ireland: 60 Queen's Gardens, London, W2 3AF, United Kingdom; tel. (20) 7724-4742; fax (20) 7706-4008; e-mail info@royalasiaticsociety.org; internet www.royalasiaticsociety.org; f. 1823 for the study of history and cultures of the East; mems: c. 700, branch societies in Asia; Pres. Prof. A. J. STOCKWELL; Sec. ADRIAN P. THOMAS; publ. *Journal* (3 a year).

Commodities

Asian and Pacific Coconut Community—APCC: 3rd Floor, Lina Bldg, Jalan H. R. Rasuna Said Kav. B7, Kuningan, Jakarta 10002, Indonesia; POB 1343, Jakarta 10013; tel. (21) 5221712; fax (21) 5221714; e-mail apcc@indo.net.id; internet www.apccsec.sg; f. 1969 to promote and co-ordinate all activities of the coconut industry, to achieve higher production and better processing, marketing and research; organizes annual Coconut Technical Meeting (COCOTECH); mems: Fiji, India, Indonesia, Kiribati, Malaysia, Marshall Islands, Federated States of Micronesia, Papua New Guinea, Philippines, Samoa, Solomon Islands, Sri Lanka, Thailand, Vanuatu, Viet Nam; assoc. mem.: Palau; Chair. IOTEBA REDFERN; Exec. Dir Dr P. RATHINAM; publs *Cocomunity* (fortnightly), *CORD* (2 a year), *Statistical Yearbook*, *Cocoinfo International* (2 a year).

Association of Natural Rubber Producing Countries—ANRPC: Bangunan Getah Asli, 148 Jalan Ampang, 7th Floor, 50450 Kuala Lumpur, Malaysia; tel. (3) 2611900; fax (3) 2613014; e-mail anrpc@capo.jaring.my; f. 1970 to co-ordinate the production and marketing of natural rubber, to promote technical co-operation amongst members and to bring about fair and stable prices for natural rubber; holds seminars, meetings and training courses on technical and statistical subjects; a joint regional marketing system has been agreed in principle; mems: India, Indonesia, Malaysia, Papua New Guinea, Singapore, Sri Lanka, Thailand, Viet Nam; Sec.-Gen. G. W. S. K. DE SILVA; publs *ANRPC Statistical Bulletin* (quarterly), *ANRPC Newsletter*.

International Coffee Organization—ICO: 22 Berners St, London, W1T 4DD, United Kingdom; tel. (20) 7580-8591; fax (20) 7580-6129; e-mail info@ico.org; internet www.ico.org; f. 1963 under the International Coffee Agreement, 1962, which was renegotiated in 1968, 1976, 1983, 1994 (extended in 1999) and 2001; aims to improve international co-operation and provide a forum for intergovernmental consultations on coffee matters; to facilitate international trade in coffee by the collection, analysis and dissemination of statistics; to act as a centre for the collection, exchange and publication of coffee information; to promote studies in the field of coffee; and to encourage an increase in coffee consumption; mems: 40 exporting and 15 importing countries; Chair. of Council JACQUES THINSY (Belgium); Exec. Dir NÉSTOR OSORIO (Colombia).

International Cotton Advisory Committee—ICAC: 1629 K St, NW, Suite 702, Washington, DC 20006-1636, USA; tel. (202) 463-6660; fax (202) 463-6950; e-mail secretariat@icac.org; internet www.icac.org; f. 1939 to observe developments in world cotton; to collect and disseminate statistics; to suggest measures for the furtherance of international collaboration in maintaining and developing a sound world cotton economy; and to provide a forum for international discussions on cotton prices; mems: 43 countries; Exec. Dir Dr TERRY TOWNSEND (USA); publs *Cotton This Month*, *Cotton: Review of the World Situation*, *Cotton: World Statistics*, *The ICAC Recorder*.

International Grains Council—IGC: 1 Canada Sq., Canary Wharf, London, E14 5AE, United Kingdom; tel. (20) 7513-1122; fax (20) 7513-0630; e-mail igc@igc.org.uk; internet www.igc.org.uk; f. 1949 as International Wheat Council, present name adopted in 1995; responsible for the administration of the International Grains Agreement, 1995, comprising the Grain Trade Convention (GTC) and the Food Aid Convention (FAC, under which donors pledge specified minimum annual amounts of food aid for developing countries in the form of grain and other eligible products); aims to further international co-operation in all aspects of trade in grains, to promote international trade in grains, and to achieve a free flow of this trade, particularly in developing member countries; seeks to contribute to the stability of the international grain market; acts as a forum for consultations between members; provides comprehensive information on the international grain market; mems: 25 countries and the EU; Exec. Dir G. DENIS; publs *World Grain Statistics* (annually), *Wheat and Coarse Grain Shipments* (annually), *Report for the Fiscal Year* (annually), *Grain Market Report* (monthly), *IGC Grain Market Indicators* (weekly).

International Jute Study Group—IJSG: 145 Monipuriparu, Near Farmgate, Tejgaon, Dhaka 1215, Bangladesh; tel. (2) 9125581; fax (2) 9125248; e-mail ijoinf@bdmail.net; f. 2002 as successor to International Jute Organization (f. 1984 in accordance with an agreement made by 48 producing and consuming countries in 1982, under the auspices of UNCTAD); aims to improve the jute economy and the quality of jute and jute products through research and development projects and market promotion; Sec.-Gen T. NANADAKUMAR.

International Pepper Community—IPC: 4th Floor, Lina Bldg, Jalan H. R. Rasuna Said, Kav. B7, Kuningan, Jakarta 12920, Indonesia; tel. (21) 5224902; fax (21) 5224905; e-mail ipc@indo.net.id; internet www.ipcnet.org; f. 1972 for promoting, co-ordinating and harmonizing all activities relating to the pepper economy; mems: Brazil, India, Indonesia, Malaysia, Federated States of Micronesia, Papua New Guinea, Sri Lanka, Thailand; Exec. Dir Dr K. P. G. MENON; publs *Pepper Statistical Yearbook*, *International Pepper News Bulletin* (quarterly), *Directory of Pepper Exporters*, *Directory of Pepper Importers*, *Weekly Prices Bulletin*, *Pepper Market Review*.

International Rubber Study Group: Heron House, 109–115 Wembley Hill Rd, Wembley, HA9 8DA, United Kingdom; tel. (20) 8900-5400; fax (20) 8903-2848; e-mail irsg@rubberstudy.com; internet www.rubberstudy.com; f. 1944 to provide a forum for the discussion of problems affecting synthetic and natural rubber and to provide statistical and other general information on rubber; mems: 17 governments; Sec.-Gen. Dr A. F. S. BUDIMAN (Indonesia); publs *Rubber Statistical Bulletin* (every 2 months), *Rubber Industry Report* (every 2 months), *Proceedings of International Rubber Forums* (annually), *World Rubber Statistics Handbook*, *Key Rubber Indicators*, *Rubber Statistics Yearbook* (annually), *Rubber Economics Yearbook* (annually), *Outlook for Elastomers* (annually).

International Silk Association: 34 rue de la Charité, 69002 Lyon, France; tel. 4-78-42-10-79; fax 4-78-37-56-72; e-mail isa-silk.ais-sole@wanadoo.fr; f. 1949 to promote closer collaboration between all branches of the silk industry and trade, develop the consumption of silk, and foster scientific research; collects and disseminates information and statistics relating to the trade and industry; organizes biennial congresses; mems: employers' and technical organizations in 40 countries; Gen. Sec. X. LAVERGNE; publs *ISA Newsletter* (monthly), congress reports, standards, trade rules, etc.

International Spice Group: c/o International Trade Centre (UNCTAD/WTO), 54–56 rue de Montbrillant, 1202 Geneva, Switzerland; tel. (22) 7300101; fax (22) 7300254; e-mail itcreg@intracen.org; f. 1983 to provide a forum for producers and consumers of spices; works to increase the consumption of spices; mems: 33 producer countries, 15 importing countries; Chair. HERNAL HAMILTON (Jamaica).

International Sugar Organization: 1 Canada Sq., Canary Wharf, London, E14 5AA, United Kingdom; tel. (20) 7513-1144; fax (20) 7513-1146; e-mail exdir@isosugar.org; internet www.isosugar.org;

administers the International Sugar Agreement (1992), with the objectives of stimulating co-operation, facilitating trade and encouraging demand; aims to improve conditions in the sugar market through debate, analysis and studies; serves as a forum for discussion; holds annual seminars and workshops; sponsors projects from developing countries; mems: 71 countries producing some 83% of total world sugar; Exec. Dir Dr PETER BARON; publs *Sugar Year Book*, *Monthly Statistical Bulletin*, *Market Report and Press Summary*, *Quarterly Market Outlook*, seminar proceedings.

International Tea Committee Ltd—ITC: Sir John Lyon House, 5 High Timber St, London, EC4V 3NH, United Kingdom; tel. (20) 7248-4672; fax (20) 7329-6955; e-mail info@intteacomm.co.uk; internet www.intteacomm.co.uk; f. 1933 to administer the International Tea Agreement; now serves as a statistical and information centre; in 1979 membership was extended to include consuming countries; producer mems: national tea boards or asscns in eight countries; consumer mems: United Kingdom Tea Asscn, Tea Asscn of the USA Inc., Comité européen du thé and the Tea Council of Canada; assoc. mems: Netherlands and UK ministries of agriculture, Cameroon Development Corpn; Chair. M. J. BUNSTON; publs *Annual Bulletin of Statistics*, *Monthly Statistical Summary*.

International Tea Promotion Association—ITPA: c/o Tea Board of Kenya, POB 20064, 00200 Nairobi, Kenya; tel. (20) 572421; fax (20) 562120; e-mail teaboardk@kenyaweb.com; internet www.teaboard.or.ke; f. 1979; mems: eight countries; Chair. GEORGE M. KIMANI; publ. *International Tea Journal* (2 a year).

International Tobacco Growers' Association—ITGA: Apdo 5, 6001-081 Castelo Branco, Portugal; tel. (272) 325901; fax (272) 325906; e-mail itga@mail.telepac.pt; internet www.tobaccoleaf.org; f. 1984 to provide a forum for the exchange of views and information of interest to tobacco producers; mems: 22 countries producing over 80% of the world's internationally traded tobacco; Pres. ALBERT KAMULAGA (Malawi); Exec. Dir ANTONIO ABRUNHOSA (Portugal); publs *Tobacco Courier* (quarterly), *Tobacco Briefing*.

International Tropical Timber Organization—ITTO: International Organizations Center, 5th Floor, Pacifico-Yokohama, 1-1-1, Minato-Mirai, Nishi-ku, Yokohama 220-0012, Japan; tel. (45) 223-1110; fax (45) 223-1111; e-mail itto@itto.or.jp; internet www.itto.or.jp; f. 1985 under the International Tropical Timber Agreement (1983); a new treaty, ITTA 1994, came into force in 1997; provides a forum for consultation and co-operation between countries that produce and consume tropical timber, and is dedicated to the sustainable development and conservation of tropical forests; facilitates progress towards 'Objective 2000', which aims to move as rapidly as possible towards achieving exports of tropical timber and timber products from sustainably managed resources; encourages, through policy and project work, forest management, conservation and restoration, the further processing of tropical timber in producing countries, and the gathering and analysis of market intelligence and economic information; mems: 33 producing and 26 consuming countries and the EU; Exec. Dir Dr MANOEL SOBRAL FILHO; publs *Annual Review and Assessment of the World Timber Situation*, *Tropical Timber Market Information Service* (every 2 weeks), *Tropical Forest Update* (quarterly).

Organization of the Petroleum Exporting Countries—OPEC: 1020 Vienna, Obere Donaustrasse 93, Austria; tel. (1) 211-12-279; fax (1) 214-98-27; e-mail info@opec.org; internet www.opec.org; f. 1960 to unify and co-ordinate members' petroleum policies and to safeguard their interests generally; holds regular conferences of member countries to set reference prices and production levels; conducts research in energy studies, economics and finance; provides data services and news services covering petroleum and energy issues; mems: Algeria, Indonesia, Iran, Iraq, Kuwait, Libya, Nigeria, Qatar, Saudi Arabia, United Arab Emirates, Venezuela; Sec.-Gen. a.i. Dr PURNOMO YUSGIANTORO (Indonesia); publs *Annual Report*, *Annual Statistical Bulletin*, *OPEC Bulletin* (monthly), *OPEC Review* (quarterly), *Monthly Oil Market Report*.

OPEC Fund for International Development: Postfach 995, 1011 Vienna, Austria; tel. (1) 515-64-0; fax (1) 513-92-38; e-mail info@opecfund.org; internet www.opecfund.org; f. 1976 by mem. countries of OPEC, to provide financial co-operation and assistance for developing countries; in 2003 commitments amounted to US $315.8m.; Dir-Gen. SULEIMAN J. AL-HERBISH (Saudi Arabia); publs *Annual Report*, *OPEC Fund Newsletter* (3 a year).

Development and Economic Co-operation

Afro-Asian Rural Development Organization—AARDO: No. 2, State Guest Houses Complex, Chanakyapuri, New Delhi 110 021, India; tel. (11) 4100475; fax (11) 4672045; e-mail aardohq@nde.vsnl.net.in; internet www.aardo.org; f. 1962 to act as a catalyst for the co-operative restructuring of rural life in Africa and Asia and to explore opportunities for the co-ordination of efforts to promote rural welfare and to eradicate hunger, thirst, disease, illiteracy and poverty; carries out collaborative research on development issues; organizes training; encourages the exchange of information; holds international conferences and seminars; awards 100 individual training fellowships at nine institutes in Egypt, India, Japan, the Republic of Korea, Malaysia and Taiwan; mems: 13 African countries, 14 Asian countries, one African associate; Sec.-Gen. ABDALLA YAHIA ADAM; publs *Afro-Asian Journal of Rural Development*, *Annual Report*, *AARDO Newsletter* (2 a year).

Asian and Pacific Development Centre: Pesiaran Duta, POB 12224, 50770 Kuala Lumpur, Malaysia; tel. (3) 6511088; fax (3) 6510316; e-mail info@apdc.po.my; internet www4.jaring.my/apdc; f. 1980; undertakes research and training, acts as clearing-house for information on development and offers consultancy services, in co-operation with national institutions; current programme includes assistance regarding the implementation of national development strategies; the Centre aims to promote economic co-operation among developing countries of the region for their mutual benefit; mems: 19 countries and two associate members; Dir Dr M. NOOR HARUN; publs *Annual Report*, *Newsletter* (2 a year), *Asia-Pacific Development Monitor* (quarterly), studies, reports, monographs.

Asia-Pacific Mountain Network—APMN: c/o International Centre for Integrated Mountain Development, GPO Box 3226, Kathmandu, Nepal; tel. (1) 525313; fax (1) 524509; e-mail baden@zhk.l-card.msk; internet www.apmn.mtnforum.org; f. 1995; forum for the production and dissemination of information on sustainable mountain development, reducing the risk of mountain disasters, economic development, the elimination of poverty, and cultural heritage; mems: about 1,000, including mems from Russia and Central Asia.

Association of Development Financing Institutions in Asia and the Pacific—ADFIAP: Skyland Plaza, 2nd Floor, Sen. Gil J. Puyat Ave, Makati City, Metro Manila, 1200 Philippines; tel. (2) 8161672; fax (2) 8176498; e-mail inquire@adfiap.org; internet www.adfiap.org; f. 1976 to promote the interests and economic development of the respective countries of its member institutions, through development financing; mems: 65 institutions in 30 countries; Chair. ISOA KALOUMAIRA (Fiji); Sec.-Gen. ORLANDO P. PEÑA (Philippines); publs *Asian Banking Digest*, *Journal of Development Finance* (2 a year), *ADFIAP Newsletter*, *ADFIAP Accompli*, *DevTrade Finance*.

Colombo Plan: Bank of Ceylon Merchant Tower, 28 St Michael's Rd, Colombo 03, Sri Lanka; tel. (1) 564448; fax (1) 564531; e-mail cplan@slt.lk; internet www.colombo-plan.org; f. 1950 by seven Commonwealth countries, to encourage economic and social development in Asia and the Pacific; the Plan comprises the Programme for Public Administration, to provide training for officials in the context of a market-orientated economy; the Programme for Private Sector Development, which organizes training programmes to stimulate the economic benefits of development of the private sector; a Drug Advisory Programme, to encourage regional co-operation in efforts to control drug-related problems, in particular through human resources development; a programme to establish a South-South Technical Co-operation Data Bank, to collate, analyse and publish information in order to facilitate South-South co-operation; and a Staff College for Technician Education (see below). All programmes are voluntarily funded; developing countries are encouraged to become donors and to participate in economic and technical co-operation activities among developing mems; mems: 24 countries; Sec.-Gen. U. SARAT CHANDRAN (India); publs *Annual Report*, *Colombo Plan Focus* (quarterly), Consultative Committee proceedings (every 2 years).

Colombo Plan Staff College for Technician Education: POB 7500, Domestic Airport Post Office, NAIA, Pasay City 1300, Philippines; tel. (2) 631-0991; fax (2) 631-0996; e-mail cpsc@skyinet.net; internet www.cpsc.org.ph; f. 1973 with the support of member Governments of the Colombo Plan; aims to enhance the development of technician education systems in developing mem. countries; Dir MAN-GON PARK; publ. *CPSC Quarterly*.

Council of Regional Organizations in the Pacific—CROP: c/o Private Mail Bag, Suva, Fiji; tel. 3312600; fax 3301102; f. as South Pacific Organizations' Co-ordinating Committee; renamed 1999; aims to co-ordinate work programmes in the region and improve the efficiency of aid resources; holds annual meetings; chairmanship alternates between the participating organizations; first meeting of subcommittee on information technologies convened in 1998; mems: Pacific Islands Development Programme, Secretariat of the Pacific Community, South Pacific Applied Geoscience Commission, Pacific Islands Forum Secretariat, South Pacific Regional Environment Programme, Forum Fisheries Agency, the Tourism Council of the South Pacific and the University of the South Pacific; Chair. NOEL LEVI.

Developing Eight—D-8: Muşir Fuad Paşa Yalisi, Eski Tersane, Emirgan, Cad. 90, 80860 İstanbul, Turkey; tel. (212) 2775513; fax (212) 2775519; internet www.mfa.gov.tr/d-8; inaugurated at a meeting of heads of state in June 1997; aims to foster economic co-operation between member states and to strengthen the role of developing countries in the global economy; project areas include trade and industry, agriculture, human resources, telecommunications, rural development, finance (including banking and privatization), energy, environment, and health; fourth Summit meeting, convened in Tehran, Iran, in Feb. 2004, addressed the need for improved implementation by mem. states of specific projects and the facilitation of intra-D-8 trade; mems: Bangladesh, Egypt, Indonesia, Iran, Malaysia, Nigeria, Pakistan, Turkey; Exec. Dir AYHAN KAMEL.

Foundation for the Peoples of the South Pacific, International—FSPI: POB 951, Port Vila, Vanuatu; tel. 22915; fax 24510; e-mail fspi@fsp.org.vu; internet www.oneworld.org/fsp; f. 1965; provides training and technical assistance for self-help community development groups and co-operatives; implements long-term programmes in sustainable forestry and agriculture, the environment, education, nutrition, women in development, child survival, and fisheries; mems: non-governmental affiliates operating in Australia, Fiji, Kiribati, Papua New Guinea, Samoa, Solomon Islands, Tonga, Tuvalu, United Kingdom, USA, Vanuatu; Regional Man. KATHY FRY; publs *Annual Report*, *News* (quarterly), technical reports (e.g. on intermediate technology, nutrition, teaching aids).

International Centre for Integrated Mountain Development—ICIMOD: 4/80 Jawalakhel, Lalitpur, GPO Box 3226, Kathmandu, Nepal; tel. (1) 525313; fax (1) 524509; e-mail dits@icimod.org.np; internet www.icimod.org.sg; f. 1983; an autonomous organization sponsored by regional member countries and by the governments of Nepal, Germany, Switzerland, Austria, Netherlands and Denmark, to help promote an economically and environmentally sound ecosystem and to improve the living standards of the population in the Hindu Kush-Himalaya; aims to serve as a focal point for multi-disciplinary documentation, training and applied research, and as a consultative centre in scientific and practical matters pertaining to mountain development; participating countries: Afghanistan, Bangladesh, Bhutan, People's Republic of China, India, Myanmar, Nepal, Pakistan; Dir-Gen. J. GABRIEL CAMPBELL.

Mekong River Commission—MRC: 364 M. V. Preah Monivong, Sangkat Phsar Doerm Thkouv, Khan Chamkar Mon, POB 1112, Phnom Penh, Cambodia; tel. (23) 720979; fax (23) 720972; e-mail mrcs@mrcmekong.org; internet www.mrcmekong.org; f. 1995 as successor to the Committee for Co-ordination of Investigations of the Lower Mekong Basin (f. 1957); aims to promote and co-ordinate the sustainable development and use of the resources of the Mekong River Basin for navigational and non-navigational purposes, in order to assist the social and economic development of member states and preserve the ecological balance of the basin. Provides scientific information and policy advice; supports the implementation of strategic programmes and activities; organizes an annual donor consultative group meeting; maintains regular dialogue with Myanmar and the People's Republic of China; mems: Cambodia, Laos, Thailand, Viet Nam; Chief Exec. JOERN KRISTENSEN; publ. *Mekong News* (quarterly).

Pacific Basin Economic Council—PBEC: 900 Fort St, Suite 1080, Honolulu, HI 96813, USA; tel. (808) 521-9044; fax (808) 521-8530; e-mail info@pbec.org; internet www.pbec.org; f. 1967; an assen of business representatives aiming to promote business opportunities in the region, in order to enhance overall economic development; advises governments and serves as a liaison between business leaders and government officials; encourages business relationships and co-operation among members; holds business symposia; mems: 20 economies (Australia, Canada, Chile, People's Republic of China, Colombia, Ecuador, Hong Kong, Indonesia, Japan, Republic of Korea, Malaysia, Mexico, New Zealand, Peru, Philippines, Russia, Singapore, Taiwan, Thailand, USA); Chair. SUCK-RAI CHO; Pres. DALTON TANONAKA; publs *Pacific Journal* (quarterly), *Executive Summary* (annual conference report).

Pacific Economic Co-operation Council—PECC: 4 Nassim Rd, Singapore 258372; tel. 67379823; fax 67379824; e-mail info@pecc.org; internet www.pecc.org; f. 1980; an independent, policy-orientated organization of senior research, government and business representatives from 25 economies in the Asia-Pacific region; aims to foster economic development in the region by providing a forum for discussion and co-operation in a wide range of economic areas; holds a General Meeting every 2 years; mems: Australia, Brunei, Canada, Chile, the People's Republic of China, Colombia, Ecuador, Hong Kong, Indonesia, Japan, the Republic of Korea, Malaysia, Mexico, Mongolia (assoc. mem.), New Zealand, Peru, Philippines, Russia, Singapore, Taiwan, Thailand, USA, Viet Nam and the Pacific Islands Forum; French Pacific Territories (assoc. mem.); Dir-Gen. EDUARDO PEDROSA (acting); publs *Issues PECC* (quarterly), *Pacific Economic Outlook* (annually), *Pacific Food Outlook* (annually).

Pacific Islands Development Program: 1601 East-West Rd, Honolulu, HI 96848, USA; tel. (808) 944-7724; fax (808) 944-7670; e-mail halapuas@ewc.hawaii.edu; internet pidp.ewc.hawaii.edu; f. 1980; promotes regional development by means of education, research and training; mems: 22 Pacific islands; Dir Dr SITIVENI HALAPUA.

Pacific Trade and Development Conference: c/o Australia-Japan Research Centre, John Crawford Bldg, Australian National University, Canberra, ACT 0200, Australia; tel. (2) 6125-3780; fax (2) 6125-0767; e-mail paftad.sec@anu.edu.au; internet ajrcnet.anu.edu.au/paftad; f. 1968; holds annual conference for discussion of regional trade policy issues by senior economists and experts; Exec. Officer ANDREW DEANE.

US-Pacific Island Nations Joint Commercial Commission: c/o Pacific Islands Development Program, 1601 East-West Rd, Honolulu, HI 96848, USA; tel. (808) 944-7721; fax (808) 944-7721; e-mail kroekers@ewc.hawaii.edu; internet pidp.ewc.hawaii.edu/jcc/; f. 1993 to promote mutually beneficial commercial and economic relations between the independent Pacific island nations and the USA; mems: Pacific island members of the Pacific Islands Forum and the USA; publ. *JCC Trade Links* (quarterly).

Economics and Finance

Asian Clearing Union—ACU: 207/1 Pasdaran Ave, POB 15875/7177, Tehran, Iran; tel. (21) 2842076; fax (21) 2847677; e-mail acusecret@cbi.ir; internet www.asianclearingunion.org; f. 1974 to provide clearing arrangements, whereby members settle payments for intra-regional transactions among the participating central banks, on a multilateral basis, in order to economize on the use of foreign exchange and promote the use of domestic currencies in trade transactions among developing countries; part of ESCAP's Asian trade expansion programme; the Central Bank of Iran is the Union's agent; in September 1995 the ACU unit of account was changed from SDR to US dollars, with effect from 1 January 1996; mems: central banks of Bangladesh, Bhutan, India, Iran, Myanmar, Nepal, Pakistan, Sri Lanka; Sec.-Gen. BAHEREH MIRZAEI-TEHRANI; publs *Annual Report*, *Newsletter* (monthly).

Asian Reinsurance Corporation: 17th Floor, Tower B, Chamnan Phenjati Business Center, 65 Rama 9 Rd, Huaykwang, Bangkok 10320, Thailand; tel. (2) 245-2169; fax (2) 248-1377; e-mail asianre@asianrecorp.com; internet www.asianrecorp.com; f. 1979 by ESCAP with UNCTAD, to operate as a professional reinsurer, giving priority in retrocessions to national insurance and reinsurance markets of member countries, and as a development organization providing technical assistance to countries in the Asia-Pacific region; cap. (auth.) US $15m., (p.u.) $8m.; mems: Afghanistan, Bangladesh, Bhutan, People's Republic of China, India, Iran, Republic of Korea, Philippines, Sri Lanka, Thailand; Gen. Man. A. S. MALABANAN.

Association of Asian Confederations of Credit Unions—AACCU: 36/2 Moo 3, Soi Malee Suanson Ramkanheang Rd, Bangkapi, Bangkok 10240, Thailand; tel. (2) 374-3170; fax (2) 374-5321; e-mail accuran@ksc.th.com; internet www.aaccu.net; links and promotes credit unions in Asia, provides research facilities and training programmes; mems: in 20 Asian countries; Pres. CHARLES YIP WAI KWONG (Hong Kong); CEO RANJITH HETTIARACHICHI (Thailand); publs *ACCU News* (every 2 months), *Annual Report and Directory*.

Financial Action Task Force on Money Laundering (FATF) (Groupe d'action financière sur le blanchiment de capitaux—GAFI): 2 rue André-Pascal, 75775 Paris Cédex 16, France; tel. 1-45-24-82-00; fax 1-45-24-85-00; e-mail contact@fatf.gafi.org; internet www1.oecd.org/fatf; f. 1989, on the recommendation of the Group of Seven industrialized nations (G-7), to develop and promote policies to combat money laundering and the financing of terrorism; formulated a set of recommendations for member countries to implement; established regional task forces in the Caribbean, Asia-Pacific, Europe, Africa and South America; mems: 31 countries, the European Commission, and the Co-operation Council for the Arab States of the Gulf; Pres. CLAES NORGREN (Sweden); Exec. Sec. PATRICK MOULETTE; publ. *Annual Report*.

Insurance Institute for Asia and the Pacific: Zapote Rd, Alabang, Metro Manila, Philippines; tel. (2) 8420691; fax (2) 8420692; f. 1974 to provide insurance management training and conduct research in subjects connected with the insurance industry; publ. *IIAP Journal* (quarterly).

Education

Asian and South Pacific Bureau of Adult Education—ASPBAE: c/o MAAPL, Eucharistic Congress Bldg No. 3, 9th Floor, 5 Convent St, Colaba, Mumbai 400 039, India; tel. (22) 22021391; fax (22) 22832217; e-mail aspbae@vsnl.com; internet www.aspbae.org; f. 1964 to assist non-formal education and adult literacy; organizes training courses and seminars; provides material and advice relating to adult education; mems in 36 countries and territories; Sec.-Gen. MARIA-LOURDES ALMAZAN-KHAN; publ. *ASPBAE News* (3 a year).

Asian Confederation of Teachers: c/o FIT, 55 Abhinav Apt, Mahturas Rd Extn, Kandivli, Mumbai 400 067, India; tel. (22) 8085437; fax (22) 6240578; e-mail vsir@hotmail.com; f. 1990; mems in 10 countries and territories; Pres. MUHAMMAD MUSTAPHA; Sec.-Gen. VINAYAK SIRDESAI.

Asian Institute of Technology—AIT: POB 4, Klong Luang, Pathumthani 12120, Thailand; tel. (2) 516-0110; fax (2) 516-2126; e-mail jlarmand@ait.ac.th; internet www.ait.ac.th; f. 1959; Master's, Doctor's and Diploma programmes are offered in four schools: Advanced Technologies, Civil Engineering, Environment, Resources and Development, and Management; specialized training is provided by the Center for Library and Information Resources (CLAIR), the Continuing Education Center, the Center for Language and Educational Technology, the Regional Computer Center, the AIT Center in Viet Nam (based in Hanoi) and the Swiss-AIT-Viet Nam Management Development Program (in Ho Chi Minh City); other research and outpost centres are the Asian Center for Engineering Computations and Software, the Asian Center for Research on Remote Sensing, the Regional Environmental Management Center, the Asian Center for Soil Improvement and Geosynthetics and the Urban Environmental Outreach Center; there are four specialized information centres (on ferro-cement, geotechnical engineering, renewable energy resources, environmental sanitation) under CLAIR; the Management of Technology Information Center conducts short-term courses in the management of technology and international business; Pres. Prof. JEAN-LOUIS ARMAND; publs *AIT Annual Report*, *Annual Report on Research and Activities*, *AIT Review* (3 a year), *Prospectus*, other specialized publs.

Association of South-East Asian Institutions of Higher Learning—ASAIHL: Secretariat, Ratasastra Bldg 2, Chulalongkorn University, Henri Dunant Rd, Bangkok 10330, Thailand; tel. (2) 251-6966; fax (2) 253-7909; e-mail oninnat@chula.ac.th; internet www.seameo.org/asaihl; f. 1956 to promote the economic, cultural and social welfare of the people of South-East Asia by means of educational co-operation and research programmes; and to cultivate a sense of regional identity and interdependence; collects and disseminates information, organizes discussions; mems: 160 university institutions in 14 countries; Pres. Prof. Tan Sri Dr SYED JALALUDIN SYED SALIM; Sec.-Gen. Dr NINNAT OLANVORAVUTH; publs *Newsletter*, *Handbook* (every 3 years).

International Union for Oriental and Asian Studies: c/o Közraktar u. 12a 11/2, 1093 Budapest, Hungary; f. 1951 by the 22nd International Congress of Orientalists under the auspices of UNESCO, to promote contacts between orientalists throughout the world, and to organize congresses, research and publications; mems: in 24 countries; Sec.-Gen. Prof. GEORG HAZAI.

Southeast Asian Ministers of Education Organization—SEAMEO: M. L. Pin Malakul Bldg, 920 Sukhumvit Rd, Bangkok 10110, Thailand; tel. (2) 391-0144; fax (2) 381-2587; e-mail secretariat@seameo.org; internet www.seameo.org; f. 1965 to promote co-operation among the Southeast Asian nations through projects in education, science and culture; SEAMEO has 15 regional centres including: BIOTROP for tropical biology, in Bogor, Indonesia; INNOTECH for educational innovation and technology; an Open-Learning Centre in Indonesia; RECSAM for education in science and mathematics, in Penang, Malaysia; RELC for languages, in Singapore; RIHED for higher education development in Bangkok, Thailand; SEARCA for graduate study and research in agriculture, in Los Baños, Philippines; SPAFA for archaeology and fine arts in Bangkok, Thailand; TROPMED for tropical medicine and public health with regional centres in Indonesia, Malaysia, Philippines and Thailand and a central office in Bangkok; VOCTECH for vocational and technical education; the SEAMEO Training Centre in Ho Chi Minh City, Viet Nam; and the SEAMEO Centre for History and Tradition (CHAT) in Yangon, Myanmar; mems: Brunei, Cambodia, Indonesia, Laos, Malaysia, Philippines, Singapore, Thailand, Viet Nam; assoc. mems: Australia, Canada, France, Germany, Netherlands, New Zealand; Pres. Dr EDILBERTO DE JESUS (Philippines); Dir Dr ARIEF S. SADIMAN (Indonesia) (until 31 Dec. 2004); publs *Annual Report*, *Journal of Southeast Asian Education*.

University of the South Pacific: Suva, Fiji; tel. 3313900; fax 3301305; internet www.usp.ac.fj; f. 1968; comprises three campuses (in Fiji, Samoa and Vanuatu), five schools (agriculture, science, humanities, law and social sciences), 11 extension centres in different countries and seven institutes; mems: Cook Islands, Fiji, Kiribati, Marshall Islands, Nauru, Niue, Samoa, Solomon Islands, Tokelau, Tonga, Tuvalu, Vanuatu; Vice-Chancellor SAVENACA SIWATIBAU; publs *USP Annual Report*, *USP Beat* (every 2 weeks), *USP Calendar* (annually).

World Scout Bureau/Asia Pacific Region: POB 4050, MCPO 1280, Makati City, Philippines; tel. (2) 8180984; fax (2) 8190093; e-mail wsb@apr.scout.org; internet www.apr.scout.or.jp; f. 1956 to further the Scout Movement in the Asia-Pacific region by promoting the spirit of brotherhood, unity of purpose, co-operation and mutual assistance amongst Scout organizations within the region; conducts training courses, seminars, workshops, youth forums, jamborees and conferences; mems: 19m. Scouts in 23 countries; Regional Dir ABDULLAH RASHEED; publs *Asia-Pacific Scouting Newsletter* (monthly), *Regional Director's Report*.

Environmental Conservation

IUCN—The World Conservation Union: 28 rue Mauverney, 1196 Gland, Switzerland; tel. (22) 9990000; fax (22) 9990002; e-mail mail@hq.iucn.org; internet www.iucn.org; f. 1948, as the International Union for Conservation of Nature and Natural Resources; supports partnerships and practical field activities to promote the conservation of natural resources, to secure the conservation of biological diversity as an essential foundation for the future; to ensure wise use of the earth's natural resources in an equitable and sustainable way; and to guide the development of human communities towards ways of life in enduring harmony with other components of the biosphere, developing programmes to protect and sustain the most important and threatened species and eco-systems and assisting governments to devise and carry out national conservation strategies; maintains a conservation library and documentation centre and units for monitoring traffic in wildlife; mems: more than 1,000 states, government agencies, non-governmental organizations and affiliates in some 140 countries; Pres. YOLANDA KAKABADSE NAVARRO (Ecuador); Dir-Gen. ACHIM STEINER; publs *World Conservation Strategy*, *Caring for the Earth*, *Red List of Threatened Plants*, *Red List of Threatened Species*, *United Nations List of National Parks and Protected Areas*, *World Conservation* (quarterly), *IUCN Today*.

South Pacific Regional Environment Programme—SPREP: POB 240, Apia, Samoa; tel. 21929; fax 20231; e-mail sprep@sprep.org.ws; internet www.sprep.org.ws; f. 1978 by the South Pacific Commission (where it was based; now Pacific Community), the South Pacific (now Pacific Islands) Forum, ESCAP and UNEP; formally established as an independent institution in June 1993; aims to promote regional co-operation in environmental matters, to assist members to protect and improve their shared environment, and to help members work towards sustainable development; mems: 22 Pacific islands, Australia, France, New Zealand, USA; Dir ASTERIO TAKESY (Federated States of Micronesia); publs *SPREP Newsletter* (quarterly), *CASOLink* (quarterly), *La lettre de l'environnement* (quarterly), *South Pacific Sea Level and Climate Change Newsletter* (quarterly).

WWF International: ave du Mont-Blanc, 1196 Gland, Switzerland; tel. (22) 3649111; fax (22) 3645358; e-mail info@wwfint.org; internet www.panda.org; f. 1961 (as World Wildlife Fund), name changed to World Wildlife Fund for Nature 1986, current nomenclature adopted 2001; aims to stop the degradation of the natural environment, conserve bio-diversity, ensure the sustainable use of renewable resources, and promote the reduction of both pollution and wasteful consumption; addresses six priority issues: forests, fresh water programmes, endangered seas, species, climate change, and toxins; has identified, and focuses its activities in, 200 'ecoregions' (the 'Global 200'), believed to contain the best part of the world's remaining biological diversity; actively supports and operates conservation programmes in 90 countries; mems: 28 national organizations, four associates, c. 5m. individual mems world-wide; Pres. Chief EMEKA ANYAOKU (Nigeria); Dir-Gen. CLAUDE MARTIN; publs *Annual Report*, *Living Planet Report*.

Government and Politics

Afro-Asian Peoples' Solidarity Organization—AAPSO: 89 Abdel Aziz Al-Saoud St, POB 11559-61 Manial El-Roda, Cairo, Egypt; tel. (2) 3636081; fax (2) 3637361; e-mail aapso@idsc.net.eg; f. 1958; acts among and for the peoples of Africa and Asia in their struggle for genuine independence, sovereignty, socio-economic development, peace and disarmament; mems: national committees and affiliated organizations in 66 countries and territories, assoc. mems in 15 European countries; Sec.-Gen. NOURI ABDEL RAZZAK HUSSEIN (Iraq); publs *Solidarity Bulletin* (monthly), *Socio-Economic Development* (3 a year), *Human Rights Newsletter* (6 a year).

Alliance of Small Island States—AOSIS: c/o 800 Second Ave, Suite 400D, New York, NY 10017, USA; tel. (212) 599-6196; fax (212) 599-0797; e-mail samoa@un.int; internet www.sidsne.org/aosis; f. 1990 as an *ad hoc* intergovernmental grouping to focus on the special problems of small islands and low-lying coastal developing states; mems: 43 island nations; Chair. TUILOMA NERONI SLADE (Samoa); publ. *Small Islands, Big Issues*.

ANZUS: c/o Dept of Foreign Affairs and Trade, Locked Bag 40, Queen Victoria Terrace, Canberra, ACT 2600, Australia; tel. (2) 6261-9111; fax (2) 6273-3577; internet www.dfat.gov.au; the ANZUS Security Treaty was signed in 1951 by Australia, New Zealand and the USA, and ratified in 1952 to co-ordinate partners' efforts for collective defence for the preservation of peace and security in the Pacific area, through the exchange of technical information and strategic intelligence, and a programme of exercises, exchanges and visits. In 1984 New Zealand refused to allow visits by US naval vessels that were either nuclear-propelled or potentially nuclear-armed, and this led to the cancellation of joint ANZUS military exercises: in 1986 the USA formally announced the suspension of its security commitment to New Zealand under ANZUS. Instead of the annual ANZUS Council meetings, bilateral talks were subsequently held every year between Australia and the USA. ANZUS continued to govern security relations between Australia and the USA, and between Australia and New Zealand; security relations between New Zealand and the USA were the only aspect of the treaty to be suspended. Senior-level contacts between New Zealand and the USA resumed in 1994. The Australian Govt invoked the Anzus Security Treaty for the first time following the international terrorist attacks against targets in the USA that were perpetrated in September 2001.

Comunidade dos Países de Língua Portuguesa—CPLP (Community of Portuguese-Speaking Countries): rua S. Caetano 32, 1200-829 Lisbon, Portugal; tel. (1) 392-8560; fax (1) 392-8588; e-mail comunicacao@cplp.org; internet www.cplp.org; f. 1996; aims to produce close political, economic, diplomatic and cultural links between Portuguese-speaking countries and to strengthen the influence of the Lusophone commonwealth within the international community; mems: Angola, Brazil, Cape Verde, Guinea-Bissau, Mozambique, Portugal, São Tomé e Príncipe, Timor-Leste; Exec. Sec. Luis Fonseca (Cape Verde).

Eastern Regional Organization for Public Administration—EROPA: One Burgundy Plaza, Suite 12M, 307 Katipunan Ave, Loyola Heights, Quezon City 1105, Metro Manila, Philippines; tel. (2) 4338175; fax (2) 4349223; e-mail eropa@eropa.org.ph; internet www.eropa.org.ph; f. 1960 to promote regional co-operation in improving knowledge, systems and practices of governmental administration, to help accelerate economic and social development; organizes regional conferences, seminars, special studies, surveys and training programmes. There are three regional centres: Training Centre (New Delhi), Local Government Centre (Tokyo), Development Management Centre (Seoul); mems: 12 countries, 111 organizations/groups, 483 individuals; Chair. Karina C. David; Sec.-Gen. Patricia A. Sto Tomas (Philippines); publs *EROPA Bulletin* (quarterly), *Asian Review of Public Administration* (2 a year).

Melanesian Spearhead Group: c/o Ministry of Foreign Affairs, PMB 051, Port Vila, Vanuatu; tel. 22913; fax 23142; f. 1986 to promote political and cultural co-operation among the Melanesian peoples; supports independence process in New Caledonia; first Melanesian arts festival held in 1991; in July 1993 a free-trade agreement was signed, awarding each member country most-favoured nation status for all trade; heads of state or of government meet every two years; regular meetings of Group trade and economic officials are also held; in July 1999 members resolved to establish a permanent secretariat for the Group, based in Vanuatu; mems: Fiji, Papua New Guinea, Solomon Islands, Vanuatu; Front de Libération Nationale Kanake Socialiste (New Caledonia); Chair. Roch Wamytan.

Non-aligned Movement—NAM: c/o Permanent Representative of Malaysia to the UN, 313 East 43rd St, New York, NY 10016, USA (no permanent secretariat); tel. (212) 986-6310; fax (212) 490-8576; e-mail malaysia@un.int; internet www.namkl.org.my; f. 1961 by a meeting of 25 Heads of State, with the aim of linking countries that had refused to adhere to the main East/West military and political blocs; co-ordination bureau established in 1973; works for the establishment of a new international economic order, and especially for better terms for countries producing raw materials; maintains special funds for agricultural development, improvement of food production and the financing of buffer stocks; South Commission promotes co-operation between developing countries; seeks changes in the United Nations to give developing countries greater decision-making power; holds summit conference every three years; 13th conference (February 2003): Kuala Lumpur, Malaysia; mems: 116 countries.

Shanghai Co-operation Organization—SCO: f. 2001, replacing the Shanghai Five (f. 1996 to address border disputes); comprises People's Republic of China, Kazakhstan, Kyrgyzstan, Russia, Tajikistan and Uzbekistan; aims to achieve security through mutual co-operation: promotes economic co-operation and measures to eliminate terrorism and drugs-trafficking; agreement on combating terrorism signed June 2001; f. a Convention on the Fight against Terrorism, Separatism and Extremism signed June 2002; a SCO anti-terrorism centre was to be established in Bishkek, Kyrgyzstan; holds annual summit meeting (June 2004: Tashkent, Uzbekistan).

Industrial and Professional Relations

Brotherhood of Asian Trade Unionists—BATU: FFW Bldg, 1943 Taft Ave, Malate, Manila, Philippines; tel. (2) 5240709; fax (2) 5218335; e-mail batunorm@iconn.com.ph; f. 1963 as the regional body in Asia of the World Confederation of Labour, to develop mutual co-operation among Asian trade unionists through exchanges of information, conferences, and educational activities; mems: 5m. in 15 countries; Pres. Juan C. Tan (Pres. Emeritus, Federation of Free Workers, the Philippines); publs *BATU Monitor* (2 a month), *BATU Research and Education Journal*.

International Confederation of Free Trade Unions-Asian and Pacific Regional Organization—ICFTU-APRO: 73 Bras Basoh Rd, 4th Floor, Singapore 189556; tel. 2226294; fax 2217380; e-mail gs@icftu-apro.org; internet www.icftu-apro.org; f. 1951; sub-regional office in New Delhi, India; mems: 33m. in 39 organizations in 29 countries; Gen. Sec. Noriyuki Suzuki; publs *Asian and Pacific Labour* (monthly), *ICFTU-APRO Labour Flash* (2 a week).

South Pacific and Oceanic Council of Trade Unions—SPOCTU: 3rd Floor, TLC Bldg, 16 Peel St, Brisbane 4101, Australia; tel. (7) 3846-1806; fax (7) 3846-4968; e-mail spoctu@ozemail.com.au; f. 1989 as sub-regional body of ICFTU-APRO; Chair. Jennie George; Exec. Officer Rod Ellis; publ. *Pacific Unionist* (newsletter).

Law

Asian-African Legal Consultative Organization—AALCO: E-66, Vasant Marg, Vasant Vihar, New Delhi 110057, India; tel. (11) 26152251; fax (11) 26152041; e-mail mail@aalco.org; internet www.aalco.org; f. 1956 to consider legal problems referred to it by member countries and to serve as a forum for Afro-Asian co-operation in international law, including international trade law, and economic relations; provides background material for conferences, prepares standard/model contract forms suited to the needs of the region; promotes arbitration as a means of settling international commercial disputes; trains officers of member states; has permanent UN observer status; mems: 46 countries; Pres. Choi Young-Jin (Republic of Korea); Sec.-Gen. Dr Wafik Zaher Kamil (Egypt).

Law Association for Asia and the Pacific—LAWASIA: LAWASIA Secretariat, GPO Box 980, Brisbane, Qld 4001, Australia; tel. (7) 3222-5888; fax (7) 3222-5850; e-mail lawasia@lawasia.asn.au; internet www.lawasia.asn.au; f. 1966; provides an international, professional network for lawyers to update, reform and develop law within the region; comprises six Sections and 21 Standing Committees in Business Law and General Practice areas, which organize speciality conferences; also holds a biennial conference (2003: Tokyo, Japan); mems: national orgs in 23 countries; 2,500 mems in 55 countries; publs *Directory* (annually), *Journal*(annually), *LAWASIA Update* (quarterly).

Medicine and Health

Asia Pacific Academy of Ophthalmology—APAO: c/o Prof. Arthur S. M. Lim, Eye Clinic Singapura, 6A Napier Rd, 02-38 Gleneagles Annexe Block, Gleneagles Hospital, Singapore 258500; tel. 64666666; fax 67333360; f. 1956; holds Congress every two years; Pres. Ian Constable (Australia); Sec.-Gen. Prof. Arthur S. M. Lim (Singapore).

Asia Pacific Dental Federation—APDF: c/o Dr J. Annan, 16 The Terrace, Wellington, New Zealand; tel. 4472-5516; fax 4472-5448; e-mail jannan@apdf.info; internet www.apdf.info; f. 1955 to establish closer relationships among dental asscns in Asian and Pacific countries and to encourage research on dental health in the region; holds congress every year; mems: 26 national dental asscns; Pres. Dr J. C. Tumaneng; Sec.-Gen. Dr Jeff Annan.

International Association for the Study of Obesity—IASO: 231 North Gower St, London, NW1 2NS, United Kingdom; e-mail inquiries@iaso.org; internet www.iaso.org; f. 1986; supports research into the prevention and management of obesity throughout the world and disseminates inormation regarding disease and accompanying health and social issues; incorporates the International Obesity Task Force; international congress every four years (2002: Sao Paulo, Brazil); Exec. Dir Kate Baillie.

Pan-Pacific Surgical Association: POB 61479 Honolulu, HI 96839, Hawaii, USA; tel. (808) 941-1010 ; fax (808) 536-4141; e-mail ppsa.info@panpacificsurgical.org; internet www.panpacificsurgical.org; f. 1929 to bring together surgeons to exchange scientific knowledge relating to surgery and medicine, and to promote the improvement and standardization of hospitals and their services and facilities; congresses are held every two years; mems: 2,716 regular, associate and senior mems from 44 countries; Pres. JOHN WONG; Chair. THOMAS KOSASA.

World Self-Medication Industry—WSMI: Centre International de Bureaux, 13 chemin du Levant, 01210 Ferney-Voltaire, France; tel. 4-50-28-47-28; fax 4-50-28-40-24; e-mail dwebber@wsmi.org; internet www.wsmi.org; Dir-Gen. Dr DAVID E. WEBBER.

Posts and Telecommunications

Asia-Pacific Telecommunity: No. 12/49, Soi 5, Chaengwattana Rd, Thungsonghong, Bangkok 10210, Thailand; tel. (2) 573-0044; fax (2) 573-7479; e-mail aptmail@aptsec.org; internet www.aptsec.org; f. 1979 to cover all matters relating to telecommunications in the region; mems: Afghanistan, Australia, Bangladesh, Brunei, the People's Republic of China, India, Indonesia, Iran, Japan, the Republic of Korea, Laos, Malaysia, Maldives, Myanmar, Nauru, Nepal, Pakistan, the Philippines, Singapore, Sri Lanka, Thailand, Viet Nam; assoc. mems: Cook Islands, Hong Kong; two affiliated mems each in Indonesia, Japan and Thailand, three in the Republic of Korea, four in Hong Kong, one in Maldives and six in the Philippines; Exec. Sec. JONG-SOON LEE.

Asian-Pacific Postal Union: Post Office Bldg, 1000 Manila, Philippines; tel. (2) 470760; fax (2) 407448; f. 1962 to extend, facilitate and improve the postal relations between the member countries and to promote co-operation in the field of postal services; mems: 23 countries; Dir JORGE SARMIENTO; publs *Annual Report*, *Exchange Program of Postal Officials*, *APPU Newsletter*.

Pacific Telecommunications Council—PTC: 2454 S. Beretania St, 302 Honolulu, HI 96826, USA; tel. (808) 941-3789; fax (808) 944-4874; e-mail info@ptc.org; internet www.ptc.org; f. 1980 to promote the development, understanding and beneficial use of telecommunications and information systems/services throughout the Pacific region; provides forum for users and providers of communications services; sponsors annual conference and seminars; mems: 650 (corporate, government, academic and individual); Pres. BRUCE DRAKE (Canada).

Press, Radio and Television

Asia-Pacific Broadcasting Union—ABU: POB 1164, 59700 Kuala Lumpur, Malaysia; tel. (3) 22823592; fax (3) 22825292; e-mail sg@abu.org.my; internet www.abu.org.my; f. 1964 to foster and co-ordinate the development of broadcasting in the Asia-Pacific area, to develop means of establishing closer collaboration and co-operation among broadcasting orgs, and to serve the professional needs of broadcasters in Asia and the Pacific; holds annual General Assembly; mems: 102 in 50 countries and territories; Pres. KATSUJI EBISAWA (Japan); Sec.-Gen. HUGH LEONARD; publs *ABU News* (every 2 months), *ABU Technical Review* (every 2 months).

Confederation of ASEAN Journalists: Gedung Dewan Pers, 4th Floor, 34 Jalan Kebon Sirih, Jakarta 10110, Indonesia; tel. (21) 3453131; fax (21) 3453175; e-mail aseancaj@mega.net.id; f. 1975; holds General Assembly every two years, Press Convention, workshops; mems: journalists' asscns in Brunei, Indonesia, Laos, Malaysia, Philippines, Singapore, Thailand and Viet Nam; observers: journalists' asscns in Cambodia and Myanmar; Perm. Sec. ABDUL RAZAK; publs *The ASEAN Journalist* (quarterly), *CAJ Yearbook*.

Organization of Asia-Pacific News Agencies—OANA: c/o Xinhua News Agency, 57 Xuanwumen Xidajie, Beijing 100803, People's Republic of China; tel. (10) 3074762; fax (10) 3072707; internet www.oananews.com; f. 1961 to promote co-operation in professional matters and mutual exchange of news, features, etc. among the news agencies of Asia and the Pacific via the Asia-Pacific News Network (ANN); mems: Anadolu Ajansi (Turkey), Antara (Indonesia), APP (Pakistan), Bakhtar (Afghanistan), BERNAMA (Malaysia), BSS (Bangladesh), ENA (Bangladesh), Hindustan Samachar (India), IRNA (Iran), ITAR-TASS (Russia), Kaz-TAG (Kazakhstan), KABAR (Kyrgyzstan), KCNA (Korea, Democratic People's Republic), KPL (Laos), Kyodo (Japan), Lankapuvath (Sri Lanka), Montsame (Mongolia), PNA (Philippines), PPI (Pakistan), PTI (India), RSS (Nepal), Samachar Bharati (India), TNA (Thailand), UNB (Bangladesh), UNI (India), Viet Nam News Agency, Xinhua (People's Republic of China), Yonhap (Republic of Korea); Pres. GUO CHAOREN; Sec.-Gen. MIKHAIL GUSMANN.

Press Foundation of Asia: POB 1843, S & L Bldg, 3rd Floor, 1500 Roxas Blvd, Manila, Philippines; tel. (2) 5233223; fax (2) 5224365; e-mail pfa@pressasia.org; internet www.pressasia.org; f. 1967; an independent, non-profit-making organization governed by its newspaper members; acts as a professional forum for about 200 newspapers in Asia; aims to reduce cost of newspapers to potential readers, to improve editorial and management techniques through research and training programmes and to encourage the growth of the Asian press; operates *Depthnews* feature service; mems: 200 newspapers; Exec. Chair. MAZLAN NORDIN (Malaysia); Chief Exec. MOCHTAR LUBIS (Indonesia); publs *Pressasia* (quarterly), *Asian Women* (quarterly).

Religion

Christian Conference of Asia—CCA: 96, 2nd District, Pak Tin Village, Mei Tin Rd, Shatin, NT, Hong Kong; tel. 26911068; fax 26924378; e-mail cca@cca.org.hkt; internet www.cca.org.hk; f. 1957 (present name adopted 1973) to promote co-operation and joint study in matters of common concern among the Churches of the region and to encourage interaction with other regional Conferences and the World Council of Churches; mems: more than 100 churches and councils of churches from 18 Asian countries; Gen. Sec. Dr AHN JAE WOONG; publ. *CCA News* (quarterly).

Muslim World League—MWL (Rabitat al-Alam al-Islami): POB 537, Makkah, Saudi Arabia; tel. (2) 5600919; fax (2) 5601319; e-mail info@muslimworldleague.org; internet www.muslimworldleague.org; f. 1962; aims to advance Islamic unity and solidarity, and to promote world peace and respect for human rights; provides financial assistance for education, medical care and relief work; has 30 offices throughout the world; Sec.-Gen. Dr ABDULLAH BIN ABDULMOSHIN AL-TURKI; publ *Al-Aalam al Islami* (weekly, Arabic), *Dawat al-Haq* (monthly).

Pacific Conference of Churches: POB 208, 4 Thurston St, Suva, Fiji; tel. 3311277; fax 3303205; e-mail pacific@is.com.fj; f. 1961; organizes assembly every five years, as well as regular workshops, meetings and training seminars throughout the region; mems: 36 churches and councils; Moderator Pastor REUBEN MAGEKON; Gen. Sec. Rev. VALAMOTU PALU.

World Fellowship of Buddhists: 616 Benjasiri Pk, Soi Medhinivet off Soi Sukhumvit 24, Bangkok 10110, Thailand; tel. (2) 661-1284; fax (2) 661-0555; e-mail wfb-hq@asianet.co.th; internet www.wfb-hq.org; f. 1950 to promote strict observance and practice of the teachings of the Buddha; holds General Conference every 2 years; 146 regional centres in 37 countries; Pres. PHAN WANNAMETHEE; Hon. Sec.-Gen. PHALLOP THAIARRY; publs *WFB Journal* (6 a year), *WFB Review* (quarterly), *WFB Newsletter* (monthly), documents, booklets.

World Hindu Federation: c/o Dr Jogendra Jha, Pashupati Kshetra, Kathmandu, Nepal; tel. (1) 470182; fax (1) 470131; e-mail hem@karki.com.np; f. 1981 to promote and preserve Hindu philosophy and culture and to protect the rights of Hindus, particularly the right to worship; executive board meets annually; mems: in 45 countries and territories; Sec.-Gen. Dr JOGENDRA JHA (Nepal); publ. *Vishwa Hindu* (monthly).

Science

Co-ordinating Committee for Coastal and Offshore Geoscience Programmes in East and Southeast Asia—CCOP: OMO Bldg, 2nd Floor, 110/2 Sathorn Nua Rd, Bangrak, Bangkok 10500, Thailand; tel. (2) 234-3578; fax (2) 237-1221; e-mail ccopts@ccop.or.th; internet www.ccop.or.th; f. 1966 as a regional intergovernmental organization, to promote and co-ordinate geoscientific programmes concerning the exploration of mineral and hydrocarbon resources and environmentally sound coastal zone management in the offshore and coastal areas of member nations; works in partnership with developed nations which have provided geologists and geophysicists as technical advisers; receives aid from co-operating countries, and other sources; mems: Cambodia, People's Republic of China, Indonesia, Japan, Republic of Korea, Malaysia, Papua New Guinea, Philippines, Singapore, Thailand, Viet Nam; 14 co-operating countries; Dir SAHNG-YUP KIM; publs *CCOP Newsletter* (quarterly), *Technical Bulletin*, *Technical Publication*, *CCOP Map Series*, *Proceedings of Annual Session*, digital dataset/CD-ROM series, other technical reports.

Federation of Asian Scientific Academies and Societies—FASAS: c/o Malaysian Scientific Association (MSA), Room 1, 2nd Floor, Bangunan Sultan, Salahuddin Abdul Aziz Shah, 16 Jalan Utara, POB 48, 46700 Petaling Jaya, Malaysia; tel. (3) 7954-1644; fax (3) 7957-8930; e-mail malsci@tm.net.my; f. 1984 to stimulate regional co-operation and promote national and regional self-reliance in science and technology, by organizing meetings, training and research programmes and encouraging the exchange of scientists and of scientific information; mems: national scientific academies and societies from Afghanistan, Australia, Bangladesh, People's Republic of China, India, Republic of Korea, Malaysia, Nepal, New

Zealand, Pakistan, Philippines, Singapore, Sri Lanka, Thailand; Pres. Prof. TING-KUEH SOON (Philippines); Sec. Prof. INDIRA NATH (India).

International Council for Science—ICSU: 51 blvd de Montmorency, 75016 Paris, France; tel. 1-45-25-03-29; fax 1-42-88-94-31; e-mail secretariat@icsu.org; internet www.icsu.org; f. 1919 as International Research Council; present name adopted 1931; new statutes adopted 1996; to co-ordinate international co-operation in theoretical and applied sciences and to promote national scientific research through the intermediary of affiliated national organizations; General Assembly of representatives of national and scientific members meets every three years to formulate policy. The following committees have been established: Cttee on Science for Food Security, Scientific Cttee on Antarctic Research, Scientific Cttee on Oceanic Research, Cttee on Space Research, Scientific Cttee on Water Research, Scientific Cttee on Solar-Terrestrial Physics, Cttee on Science and Technology in Developing Countries, Cttee on Data for Science and Technology, Programme on Capacity Building in Science, Scientific Cttee on Problems of the Environment, Steering Cttee on Genetics and Biotechnology and Scientific Cttee on International Geosphere-Biosphere Programme. The following services and Inter-Union Committees and Commissions have been established: Federation of Astronomical and Geophysical Data Analysis Services, Inter-Union Commission on Frequency Allocations for Radio Astronomy and Space Science, Inter-Union Commission on Radio Meteorology, Inter-Union Commission on Spectroscopy, Inter-Union Commission on Lithosphere; national mems: academies or research councils in 98 countries; scientific mems and assocs: 26 international unions and 28 scientific associates; Pres. W. ARBER; Sec.-Gen. H. A. MOONEY; publs *ICSU Yearbook*, *Science International* (quarterly), *Annual Report*.

Pacific Science Association: 1525 Bernice St, POB 17801, Honolulu, HI 96817; tel. (808) 848-4139; fax (808) 847-8252; e-mail psa@bishop.bishop.hawaii.org; internet www.pacificscience.org; f. 1920 to promote co-operation in the study of scientific problems relating to the Pacific region, more particularly those affecting the prosperity and well-being of Pacific peoples; sponsors Pacific Science Congresses and Inter-Congresses; mems: institutional representatives from 35 areas, scientific societies, individual scientists. Tenth Inter-Congress: Guam, 2001; 20th Congress: Bangkok, Thailand, 2003; Pres. Dr R. GERARD WARD (Japan); Exec. Sec. Dr LUCIUS G. ELDREDGE; publ. *Information Bulletin* (2 a year).

South Pacific Applied Geoscience Commission—SOPAC: Private Mail Bag, GPO, Suva, Fiji; tel. 3381377; fax 3370040; e-mail director@sopac.org; internet www.sopac.org; f. 1972; activities, which aim to assist member countries to assess their natural resources and to achieve self-sufficiency in the geosciences, include the following: assessment of coastal, onshore and offshore mineral potential; promotion of wave energy and geothermal energy; coastal protection assessment, monitoring and management advice; seabed mapping; water resources and sanitation; workshops and seminars; organizes workshops, provides technical services and degree and certificate studies; mems: Australia, Cook Islands, Fiji, French Polynesia (assoc.), Guam, Kiribati, Marshall Islands, Federated States of Micronesia, New Caledonia (assoc.), New Zealand, Niue, Papua New Guinea, Samoa, Solomon Islands, Tonga, Tuvalu, Vanuatu; Dir ALFRED SIMPSON (Fiji); publs *Annual Report*, *SOPAC News*, various technical reports and bulletins.

Social Sciences

Eastern Regional Organisation for Planning and Housing: POB 10867, 50726 Kuala Lumpur, Malaysia; tel. (3) 718-7068; fax (3) 718-3931; f. 1958 to promote and co-ordinate the study and practice of housing and regional town and country planning; maintains offices in Japan, India and Indonesia; mems: 57 organizations and 213 individuals in 28 countries; Sec.-Gen. JOHN KOH SENG SIEW; publs *EAROPH News and Notes* (monthly), *Town and Country Planning* (bibliography).

International Peace Academy—IPA: 777 United Nations Plaza, New York, NY 10017, USA; tel. (212) 687-4300; fax (212) 983-8246; e-mail ipa@ipacademy.org; internet www.ipacademy.org; f. 1970 to promote the prevention and settlement of armed conflicts between and within states through policy research and development; educates government officials in the procedures needed for conflict resolution, peace-keeping, mediation and negotiation, through international training seminars and publications; off-the-record meetings are also conducted to gain complete understanding of a specific conflict; Chair. RITA E. HAUSER; Pres. DAVID M. MALONE; publ. *Annual Report*.

Social Welfare and Human Rights

International Federation of Red Cross and Red Crescent Societies: 17 Chemin des Crêts, Petit-Saconnex, CP 372, 1211 Geneva 19, Switzerland; tel. (22) 7304222; fax (22) 7330395; e-mail secretariat@ifrc.org; internet www.ifrc.org; f. 1919 to prevent and alleviate human suffering and to promote humanitarian activities by national Red Cross and Red Crescent societies; conducts relief operations for refugees and victims of disasters, co-ordinates relief supplies and assists in disaster prevention; Pres. JUAN MANUEL SUÁREZ DEL TORO RIVERO (Spain); Sec.-Gen. MARKKU NISKALA (Finland); publs *Annual Report*, *Red Cross Red Crescent* (quarterly), *Weekly News*, *World Disasters Report*, *Emergency Appeal*.

International Organization for Migration—IOM: 17 route des Morillons, CP 71, 1211 Geneva 19, Switzerland; tel. (22) 7179111; fax (22) 7986150; e-mail info@iom.int; internet www.iom.int; f. 1951 as Intergovernmental Cttee for Migration; name changed in 1989; a non-political and humanitarian organization, activities include the handling of orderly, planned migration to meet the needs of emigration and immigration countries and the processing and movement of refugees, displaced persons etc. in need of international migration services; mems: 105 countries; 68 international governmental and non-governmental organizations; an additional 29 countries and 68 international governmental and non-governmental organizations hold observer status; Dir-Gen. BRUNSON MCKINLEY (USA); publs include *International Migration* (quarterly) and *IOM News* (quarterly, in English, French and Spanish).

Médecins sans frontières—MSF: 39 rue de la Tourelle, 1040 Brussels, Belgium; tel. (2) 280-18-81; fax (2) 280-01-73; internet www.msf.org; f. 1971; independent medical humanitarian org. composed of physicians and other members of the medical profession; aims to provide medical assistance to victims of war and natural disasters; operates longer-term programmes of nutrition, immunization, sanitation, public health, and rehabilitation of hospitals and dispensaries; awarded the Nobel peace prize in Oct. 1999; mems: national sections in 18 countries in Europe, Asia and North America; Pres. Dr ROWAN GILLIES; Sec.-Gen. MARINE BUISONNIERE; publ. *Activity Report* (annually).

Pan-Pacific and South East Asia Women's Association—PPSEAWA: POB 119, Nuku'alofa, Tonga; tel. 24003; fax 41404; e-mail nanasi@kalianet.to; internet www.ppseawa.org; f. 1928 to foster better understanding and friendship among women in the region, and to promote co-operation for the study and improvement of social conditions; holds international conference every three years; mems: 19 national member organizations; Pres. HRH Princess NANASIPAU'U TUKU'AHO; publ. *PPSEAWA Bulletin* (2 a year).

Sport and Recreations

International Federation of Association Football (Fédération internationale de football association—FIFA): FIFA House, Hitzigweg 11, POB 85, 8030 Zürich, Switzerland; tel. (1) 3849595; fax (1) 3849696; e-mail media@fifa.org; internet www.fifa.com; f. 1904 to promote the game of association football and foster friendly relations among players and national asscns; to control football and uphold the laws of the game as laid down by the International Football Association Board to prevent discrimination of any kind between players; and to provide arbitration in disputes between national asscns; organizes World Cup competition every four years; mems: 204 national asscns, six continental confederations; Pres. JOSEPH S. BLATTER (Switzerland); Gen. Sec. URS LINSI; publs *FIFA News* (monthly), *FIFA Magazine* (every 2 months) (both in English, French, German and Spanish), *FIFA Directory* (annually), *Laws of the Game* (annually), *Competitions' Regulations* and *Technical Reports* (before and after FIFA competitions).

Technology

World Association of Industrial and Technological Research Organizations—WAITRO: c/o SIRIM Berhad, 1 Persiaran Dato' Menteri, Section 2, POB 7035, 40911 Shah Alam, Malaysia; tel. 5544-6635; fax 5544-6735; e-mail info@waitro.sirim.my; internet www.waitro.org; f. 1970 by the UN Industrial Development Organization to organize co-operation in industrial and technological research; provides financial assistance for training and joint activities; arranges international seminars; facilitates the exchange of information; mems: 200 research institutes in 80 countries; Pres. BJORN LUNDBERG (Sweden); Contact MOSES MENGU; publs *WAITRO News* (quarterly), *WAITRO News* (quarterly).

Tourism

Pacific Asia Travel Association—PATA: Unit B1, 28th Floor, Siam Tower, 989 Rama 1 Rd, Pratumwan, Bangkok 10330, Thailand; tel. (2) 658-2000; fax (2) 658-2010; e-mail patabkk@pata.org; internet www.pata.org; f. 1951; aims to enhance the growth, value and quality of Pacific Asia travel and tourism for the benefit of PATA members; holds annual conference and travel fair; divisional offices in Germany, Australia (Sydney), USA; mems: more than 1,200 governments, carriers, tour operators, travel agents and hotels; publs *PATA Compass* (every 2 months), *Statistical Report* (quarterly), *Forecasts Book*, research reports, directories, newsletters.

South Pacific Tourism Organization: POB 13119, Suva, Fiji; tel. 3304177; fax 3301995; e-mail info@spto.org; internet www.tcsp.com; fmrly the Tourism Council of the South Pacific; aims to foster regional co-operation in the development, marketing and promotion of tourism in the island nations of the South Pacific; receives EU funding and undertakes sustainable activities; mems: 13 countries in the South Pacific; Chief Exec. LISIATE AKOLO; publ. *Weekly Newsletter*.

Trade and Industry

Asian Productivity Organization: Hirakawacho Daiichi Seimei Bldg 2F, 1-2-10 Hirakawa-cho, Chiyoda-ku, Tokyo 102–0093, Japan; tel. (3) 5226-3920; fax (3) 3226-3950; e-mail apo@apo-tokyo.org; internet www.apo-tokyo.org; f. 1961 as non-political, non-profit making, non-discriminatory regional intergovernmental organization with the aim of contributing to the socio-economic development of Asia and the Pacific through productivity promotion; activities cover industry, agriculture and service sectors, with the primary focus on human resources development; five key areas are incorporated into its activities: knowledge management, green productivity, strengthening small and medium enterprises, integrated community development and development of national productivity organizations; serves its members as a think tank, catalyst, regional adviser, institution builder and clearing house; mems: 20 countries; Sec.-Gen. SHIGEO TAKENAKA; publs *APO News* (monthly), *Annual Report*, *APO Asia-Pacific Productivity Data and Analysis* (annually), other books and monographs.

Cairns Group: (no permanent secretariat); internet www.cairnsgroup.org; f. 1986 by major agricultural exporting countries; aims to bring about reforms in international agricultural trade, including reductions in export subsidies, in barriers to access and in internal support measures; represents members' interests in WTO negotiations; mems: Argentina, Australia, Bolivia, Brazil, Canada, Chile, Colombia, Costa Rica, Fiji, Guatemala, Indonesia, Malaysia, New Zealand, Paraguay, Philippines, South Africa, Thailand, Uruguay; Chair. MARK VAILE (Australia).

Confederation of Asia-Pacific Chambers of Commerce and Industry—CACCI: 9th Floor, 3 Sungshou Rd, Taipei 110, Taiwan; tel. (2) 27255663; fax (2) 27255665; e-mail cacci@ttn.net; internet www.cacci.org.tw; f. 1966; holds biennial conferences to examine regional co-operation; liaises with governments to promote laws conducive to regional co-operation; serves as a centre for compiling and disseminating trade and business information; encourages contacts between businesses; conducts training and research; mems: national chambers of commerce and industry in 22 countries in the region, also affiliate and special mems; Dir-Gen. Dr WEBSTER KIANG; publs *CACCI Profile* (monthly), *CACCI Journal of Commerce and Industry* (2 a year).

International Co-operative Alliance—ICA: Regional Office for Asia and Pacific: E-4, Defence Colony, Ring Rd, New Delhi 110 024, India; tel. (11) 4694989; fax (11) 4694964; e-mail icaroap@vsnl.com; internet www.icarop.org.sg; f. 1960; promotes economic relations and encourages technical assistance among the national co-operative movements; represents the ICA in other regional forums; holds courses, seminars and conferences, and maintains the Co-operative Information Resource Centre (holding more than 20,000 volumes); administers the ICA Domus Trust (f. 1988) which, *inter alia*, supports the propagation of co-operative principles, publication of material for the study and teaching of co-operation, education and training activities and the promotion of collaboration between co-operatives and the state; mem. orgs: 67 in 28 countries of the region; Regional Dir SHIL KWAN LEE; publs *Annual Report*, *Asia and Pacific Co-op News*, *Co-op Dialogue*, *Review of International Co-operation*.

Pacific Power Association: Private Mail Bag, Suva, Fiji; tel. 3306022; fax 3302038; e-mail ppa@ppa.org.fj; internet www.ppa.org.fj; f. 1992 to facilitate co-operation and development of regional power utilities; represents the Pacific Islands' power sector at international meetings; mems: 25 power utilities throughout the region; 45 allied mems.; Exec. Dir TONY NEIL; publ. *Pacific Power* (quarterly).

South-East Asia Iron and Steel Institute: 2E Block 2, 5th Floor, Worldwide Business Park, Jalan Tinju 13/50, 40675 Shah Alam, Selangor Darul Ehson, Malaysia; tel. (3) 55191102; fax (3) 55191159; e-mail seaisi@seaisi.org; internet www.seaisi.org; f. 1971 to further the development of the iron and steel industry in the region, encourage regional co-operation, provide advisory services and a forum for the exchange of knowledge, establish training programmes, promote standardization, collate statistics and issue publications; mems: more than 800 in 26 countries; Sec.-Gen. IAIN BARTHOLOMEW; publs *SEAISI Quarterly Journal*, *SEAISI Directory* (annually), *Iron and Steel Statistics* (annually, for each member country), *Newsletter* (monthly), country reports.

Transport

Association of Asia Pacific Airlines: 9th Floor, Kompleks Antarabangsa, Jalan Sultan Ismail, 50250 Kuala Lumpur, Malaysia; tel. (3) 2145-5600; fax (3) 2145-2500; e-mail info@aapa.org.my; internet www.aapairlines.org; f. 1966 as Orient Airlines Assn; present name adopted in 1997; represents the interests of Asia Pacific airlines; encourages the exchange of information and increased co-operation between airlines; seeks to develop safe, efficient, profitable and environmentally friendly air transport; mems: 17 scheduled international airlines; Dir-Gen. RICHARD T. STIRLAND; publs *Annual Report*, *Annual Statistical Report*, *Monthly International Statistics*, *Orient Aviation* (10 a year).

Youth and Students

Asia Students Association: 353 Shangai St, 14/F, Kowloon, Hong Kong; tel. 23880515; fax 27825535; e-mail asasec@netvigator.com; f. 1969; aims to promote students' solidarity in struggling for democracy, self-determination, peace, justice and liberation; conducts campaigns, training of activists, and workshops on human rights and other issues of importance; there are Student Commissions for Peace, Education and Human Rights; mems: 40 national or regional student unions in 25 countries and territories; Secretariat LINA CABAERO (Philippines), STEVEN GAN (Malaysia), CHOW WING-HANG (Hong Kong); publs *Movement News* (monthly), *ASA News* (quarterly).

WFUNA Youth: c/o Palais des Nations, 16 ave Jean-Tremblay, 1211 Geneva 10, Switzerland; tel. (22) 7985850; fax (22) 7334838; internet www.wfuna-youth.org; f. 1948 by the World Federation of United Nations Associations (WFUNA) as the International Youth and Student Movement for the United Nations (ISMUN), independent since 1949; an international non-governmental organization of students and young people dedicated especially to supporting the principles embodied in the United Nations Charter and Universal Declaration of Human Rights; encourages constructive action in building economic, social and cultural equality and in working for national independence, social justice and human rights on a world-wide scale; maintains regional offices in Austria, France, Ghana, Panama and the USA; mems: asscns in 53 countries world-wide; Pres. ALYSON KELLY.

MAJOR COMMODITIES OF ASIA AND THE PACIFIC

Note: For each of the commodities in this section, there are generally two statistical tables: one relating to recent levels of production, and one indicating recent trends in prices. Each production table shows estimates of output for the world and for the countries covered by this volume. In addition, the table lists the main producing countries of the Far East and Australasia and, for comparison, the leading producers from outside the region. In most cases, the table referring to prices provides indexes of export prices, calculated in US dollars. The index for each commodity is based on specific price quotations for representative grades of that commodity in countries that are major traders (excluding countries of Eastern Europe and the former USSR).

Aluminium and Bauxite

Aluminium is the second most abundant metallic element in the earth's crust after silicon, comprising about 8% of the total. However, it is much less widely used than steel, despite having about the same strength and only half the weight. Aluminium has important applications as a metal because of its lightness, ease of fabrication and other desirable properties. Other products of alumina (aluminium oxide) are materials in refractories, abrasives, glass manufacture, other ceramic products, catalysts and absorbers. Alumina hydrates are used for the production of aluminium chemicals, fire retardant in carpet backing, and industrial fillers in plastics and related products.

The major markets for aluminium are in transportation, building and construction, electrical machinery and equipment, consumer durables and the packaging industry, which in the late 1990s accounted for about 20% of all aluminium use. Although the production of aluminium is energy-intensive, its light weight results in a net saving, particularly in the transportation industry. About one-quarter of aluminium output is consumed in the manufacture of transport equipment, particularly road motor vehicles and components, where the metal is increasingly being used as a substitute for steel. In the early 1990s steel substitution accounted for about 16% of world aluminium consumption, and it has been forecast that aluminium demand by the motor vehicle industry alone could more than double, to reach 5.7m. metric tons in 2010, from around 2.4m. tons in 1990. Aluminium is valued by the aerospace industry for its weight-saving characteristics and for its low cost relative to alternative materials. Aluminium-lithium alloys command considerable potential for use in this sector, although the traditional dominance of aluminium in the aerospace industry was under challenge during the 1990s from 'composites' such as carbonepoxy, a fusion of carbon fibres and hardened resins, whose lightness and durability can exceed that of many aluminium alloys.

Until recently world markets for finished and semi-finished aluminium products were dominated by six Western producers—Alcan (Canada), Alcoa, Reynolds, Kaiser (all USA), Pechiney (France) and algroup (Switzerland). Proposals for a merger between Alcan, algroup and Pechiney, and between Alcoa and Reynolds, were announced in August 1999. However, the proposed terms of the Pechiney-Alcan-algroup merger encountered opposition from the European Commission, on the grounds that the combined grouping could restrict market competition and adversely affect the interests of consumers. The tripartite merger plan was abandoned in April 2000, although Alcan and algroup were permitted to merge in October. In 2003, having agreed to meet conditions imposed by the European Commission and the US Department of Justice in respect of safeguarding free competition, Alcan was permitted to purchase Pechiney. In the USA Alcoa Inc. and Reynolds Metals Co merged in mid-2000. In 2002, after its purchase of Germany's VAW, Norway's Norsk Hydro became the world's third largest integrated aluminium concern. Prior to the mergers detailed above the level of dominance of the six major Western producers had been reduced by a significant geographical shift in the location of alumina and aluminium production to areas where cheap power is available, such as Australia, Brazil, Norway, Canada and Venezuela. The Gulf states of Bahrain and Dubai, with the advantage of low energy costs, also produce primary aluminium. Since the mid-1990s Russia has also become a significant force in the world aluminium market (see below), and in 2000 the country's principal producers, together with a number of plants located in the Commonwealth of Independent States, merged to form the Russian Aluminium Co, whose facilities are now the source of some 70% of Russian and more than 10% of global primary aluminium output. Sual is Russia's other major producer.

Bauxite is the principal aluminium ore, but nepheline syenite, kaolin, shale, anorthosite and alunite are all potential alternative sources of alumina, although not currently economic to process. Of all bauxite mined, approximately 85% is converted to alumina (Al_2O_3) for the production of aluminium metal. The developing countries, in which at least 70% of known bauxite reserves are located, supply some 50% of the ore required. The industry is structured in three stages: bauxite mining, alumina refining and smelting. While the high degree of 'vertical integration' (i.e. the control of successive stages of production) in the industry means that a significant proportion of trade in bauxite and alumina is in the form of intra-company transfers, and the increasing tendency to site alumina refineries near to bauxite deposits has resulted in a shrinking bauxite trade, there is a growing free market in alumina, serving the needs of the increasing number of independent (i.e. non-integrated) smelters.

The alumina is separated from the ore by the Bayer process. After mining, bauxite is fed to process directly if mine-run material is adequate (as in Jamaica), or else it is crushed and beneficiated. Where the ore 'as mined' presents handling problems, or weight reduction is desirable, it may be dried prior to shipment.

At the alumina plant the ore is slurried with spent-liquor directly, if the soft Caribbean type is used, or, in the case of other types, it is ball-milled to reduce it to a size which will facilitate the extraction of the alumina. The bauxite slurry is then digested with caustic soda to extract the alumina from the ore while leaving the impurities as an insoluble residue. The digest conditions depend on the aluminium minerals in the ore and the impurities. The liquor, with the dissolved alumina, is then separated from the insoluble impurities by combinations of sedimentation, decantation and filtration and the residue washed to minimize the soda losses. The clarified liquor is concentrated and the alumina precipitated by seeding with hydrate. The precipitated alumina is then filtered, washed and calcined to produce alumina. The ratio of bauxite to alumina is approximately 1.95:1.

The smelting of the aluminium is generally by electrolysis in molten cryolite. Because of the high consumption of electricity by this process, alumina is usually smelted in areas where low-cost electricity is available. However, most of the electricity now used in primary smelting in the Western world is generated by hydroelectricity—a renewable energy source.

The recycling of aluminium is economically (as well as environmentally) desirable, as the process uses only 5% of the electricity required to produce a similar quantity of primary aluminium. Aluminium that has been recycled from scrap currently accounts for almost 30% of the total annual world output of primary aluminium. With the added impetus of environmental concerns, considerable growth occurred world-wide in the recycling of used beverage cans (UBC) during the 1990s. In the middle of that decade, according to aluminium industry estimates, the recycling rate of UBC amounted to at least 55% world-wide.

In 2003, according to the International Aluminium Institute (IAI), world output of primary aluminium totalled an estimated 21.9m. metric tons, of which East and South Asian producers (including the People's Republic of China, Japan, the Democratic People's Republic of Korea, the Republic of Korea and Indonesia), together with Australia and New Zealand, accounted for about 4.7m. tons. The USA normally accounts for more than one-quarter of total aluminium consumption (excluding communist and former communist countries), and was for long the world's principal producing country. In 2001, however, US output of primary aluminium was surpassed by that of Russia and China. In 2002 Canadian production was, in addition to that of Russia and China, estimated to have overtaken US output.

Aluminium consumption has been advancing significantly in East Asia and India in recent years. In the early 2000s China was high among the world's major consumers of aluminium. Its requirements had been forecast to rise by about 60% in 1995–2000, to 2.5m. metric tons annually by 2000. In the event, domestic consumption totalled 3.4m. tons in that year. China's domestic aluminium production capacity has undergone significant expansion in recent years, and was expected to rise to more than 7m. tons by the end of 2003, compared with 4.2m. tons at the end of 2001. In 2006, provided expansion plans are approved by the government, production capacity will rise to about 10m. tons. In 2003 output of primary aluminium was forecast at 5.4m. tons. In 2002, according to data cited by the USGS, China became a net exporter of aluminium and aluminium alloys, with sales thereof exceeding imports of equivalent metals by more than 200,000 tons. In 2001, owing to a surplus of supplies, the domestic price of aluminium was reported to have fallen by 12% compared with 2000. In 2003, however, producers reportedly succeeded in maintaining the domestic price at the level of the previous year. As Chinese exports have been forecast to rise, they are likely in future to affect international price levels for aluminium. In 2001 Aluminium Corpn of China (Chalco), a subsidiary of China Aluminium Industry Corpn (Chinalco), was listed on the Hong Kong and New York stock markets. In the same year a 'strategic alliance' was announced between Chalco and Alcoa Inc. of the USA, which would take the form of a joint venture to expand the refining and smelting capacity of the Pingguo Aluminium Co. In 2003 the establishment of the Aluminium Industry Group of China was reported to have received government approval. The Group was to comprise some 67 member organizations, including aluminium producers and trading companies, centred around Chalco and its subsidiaries. The new Aluminium Industry Group would aim, *inter alia*, to stabilize the domestic price of aluminium and to facilitate Chinese competition on the international market for the metal.

Japan was the world's leading importer of unwrought aluminium in the late 1990s. The country produces only small amounts of primary aluminium, and is thus almost entirely reliant on imports to meet domestic requirements metal. In 2002, according to the USGS, these imports (comprising both ingots and alloys) totalled some 2.6m. metric tons, at a cost of about US $3,590m. According to industry estimates, in 2002 some 46% of primary aluminium imported by Japan was supplied under long-term contracts by smelter projects overseas in which Japanese companies were substantial investors. Russia, however, in whose aluminium sector Japanese companies had no equity interest, was the principal supplier of aluminium and aluminium alloy (26.4%) in 2002, followed by Australia (22.4%). Domestic demand for primary aluminium was reported to have declined slightly, by about 0.2%, to some 2m. tons, in 2002 compared with 2001. Japanese exports of primary aluminium reportedly amounted to 12,447 tons in 2002, with a value of US $26.1m. China was the principal market for Japanese aluminium ingots in 2002, absorbing 37.4% of the total exported, while Indonesia was the destination for about 30% of Japanese exports of aluminium alloys.

The Republic of Korea does not produce primary aluminium, but has facilities for the manufacture of rolled aluminium products—as does Japan. In 1999 Alcan and a local wire-manufacturing company formed a Korean-based joint venture, Alcan Taihan Aluminum Co, to market such products throughout the Asia-Pacific region. In 2000 Alcan became the leading Asian manufacturer of aluminium rolled products through the acquisition of Aluminium of Korea Ltd (Koralu). China and Russia were the principal suppliers of aluminium ingots to Korea in 2002.

Australia is by far the world's largest producer and exporter of bauxite and alumina. In 2003 the country accounted for about 38% of the world's estimated bauxite production and, in 2002, for some 32% of estimated global output of alumina. Australia's bauxite mines produced an estimated 55m. metric tons of the ore in 2003, compared with about 54m. tons in 2002. The country's alumina refineries include the huge Worsley facility in Western Australia, reported to be probably the lowest-cost refinery in the world. Expansion of the Worsley plant, increasing its annual capacity from 1.7m. tons to 3.1m. tons (so making it the world's largest alumina refinery), was completed in 2000. At that time Alcoa owned 56% of the Worsley refinery, but the company was required to dispose of this interest, among several assets, as a precondition, imposed by competition authorities, for its merger with Reynolds (see above). Australia's output of alumina increased slightly from about 16.3m. tons in 2001 to an estimated 16.4m. tons in 2002. The country has extensive bauxite deposits in Western Australia, the Northern Territory and Queensland. Owing to the easy availability of bauxite and alumina (Australia's bauxite reserves, estimated at 4,400m. tons, account for some 20% of the world's identified reserves), and of low-cost power (vast coal resources), the country provides a desirable location for aluminium smelters. In 2002 the combined output of primary aluminium of Australia's six smelters was estimated at about 1.8m. tons. Output of primary aluminium was estimated to have risen slightly, to about 1.9m. tons, in 2003. In 2002 Alcoa of Australia, the world's largest alumina-producing company, owned and operated one smelter at Point Henry, Victoria; and managed another, at Portland, Victoria, in which it held a 45% share. In addition, the company owned one, and held a substantial interest in another, alumina refinery in Western Australia, where, near Perth, it also owned a bauxite mine. In 2001/02 the aluminium industry was Australia's sixth largest export sector, generating revenue of $A4,414m., representing 3.6% of total earnings from merchandise exports.

New Zealand has a single aluminium smelter, commissioned in 1971 at Tiwai Point, near Invercargill, South Island. The plant is operated by Comalco (New Zealand) Ltd, a subsidiary of Rio Tinto plc of the United Kingdom. A major expansion of the smelter was completed in 1996, increasing annual capacity to 313,000 metric tons of primary aluminium. By the end of 2001 yearly capacity had been increased further, to 320,000 tons. In that year more than 90% of the smelter's production was exported, mainly to Japan, the Republic of Korea and other Asian destinations.

Indonesia has bauxite mines in western Kalimantan (Borneo) and in the Riau archipelago, east of Sumatra. In the main producing area of Bintan Island (part of the Riau group), where bauxite had been mined since 1935, however, the sole bauxite mine (controlled by PT Aneka Tambang) was to close in 2003 owing to depletion of reserves. Studies pertaining to the development of an alumina project in Tayan, western Kalimantan, were reported to have been completed in 2002, but arrangements to finance the project remained under discussion. The project will include the construction of a bauxite mine and a chemical alumina plant. The new mine is expected eventually to produce 5m. metric tons of bauxite annually, while the alumina plant will have an annual capacity of some 300,000 tons. Indonesia has one major aluminium smelter, capable of producing 225,000 tons per year, although in 2001 and 2002 it was forced to operate at less than its full capacity owing to inadequate power supplies. The plant forms part of the Asahan Development Project, along the Asahan river on the island of Sumatra. The establishment of the smelter, which began operating in 1982, was intended as the first stage of a fully integrated national aluminium industry. Nippon Asahan Aluminium Co Ltd of Japan owns 59% of the venture, PT Indonesia Asahan Aluminium Co (PT Inalum), and takes a proportionate share of the plant's output. National production of primary aluminium declined to an estimated 160,000 tons in 2002, compared with 180,000 tons in 2001.

In 2002 Viet Nam National Minerals Corpn (VIMICO) submitted for government approval a feasibility study for the construction of a bauxite/aluminium complex in Lam Dong Province in the country's central highlands where bauxite reserves have been estimated at more than 4,000m. metric tons. China Nonferrous Metals Corpn was reported to be considering participation in joint Sino-Vietnamese projects to develop a bauxite mine and to construct an alumina refinery in Dak Lak Province, also in the central highlands.

Production of Bauxite
(crude ore, '000 metric tons)

	2001	2002*
World total (excl. USA)	137,000*	144,000
Far East	11,101*	13,323
Oceania	53,285	54,024
Leading regional producers		
Australia	53,285	54,024
China, People's Repub.	9,800*	12,000
Indonesia	1,237	1,283
Malaysia	64	40
Other leading producers		
Brazil	13,178	13,900
Greece	2,052	2,492
Guinea†	15,700*	15,700
Guyana†	1,985	2,000
India	7,864	9,274
Jamaica†	12,370	13,119
Kazakhstan	3,685	4,377
Russia	4,000*	3,800
Suriname	4,512	4,500
Venezuela	4,526	5,000

* Estimated production.
† Dried equivalent of crude ore.
Source: US Geological Survey.

Although world demand for aluminium advanced by an average of 3% annually from the late 1980s until 1994, industrial recession began, in 1990, to create conditions of oversupply. Despite the implementation of capacity reductions at an annual rate of 10% by the major Western producers, stock levels began to accumulate. The supply problem was exacerbated by a rapid rise, beginning in 1991, of exports by the USSR and its successor states, which had begun to accumulate substantial stocks of aluminium as a consequence of the collapse of the Soviet arms industry. The requirements of these countries for foreign exchange to restructure their economies led to a rapid acceleration in low-cost exports of high-grade aluminium to Western markets. These sales caused considerable dislocation of the market and involved the major Western producers in heavy financial losses. Producing members of the European Community (EC, now the European Union—EU) were particularly severely affected, and in August 1993 the EC imposed quota arrangements, under which aluminium imports from the former USSR were to be cut by 50% for an initial three-month period, while efforts were made to negotiate an agreement that would reduce the flow of low-price imports and achieve a reduction in aluminium stocks (by then estimated to total 4.5m. metric tons world-wide).

These negotiations, which involved the EC, the USA, Canada, Norway, Australia and Russia (but in which the minor producers, Brazil, the Gulf states, Venezuela and Ukraine were not invited to take part), began in October 1993. Initially, the negotiations made little progress, and in November the market price of high-grade aluminium ingots fell to an 11-year 'low'. Following further meetings in January 1994, however, a memorandum of understanding (MOU) was finalized on a plan whereby Russia was to 'restructure' its aluminium industry and reduce its output by 500,000 metric tons annually. By March the major Western aluminium producers had agreed to reduce annual production by about 1.2m. tons over a maximum period of two years. Additionally, Russia was to receive US $2,000m. in loan guarantees. The MOU provided for participants to monitor world aluminium supplies and prices on a regular basis. In March the EU quota was terminated. By July the world price had recovered by about 50% on the November 1993 level.

The successful operation of the MOU, combined with a strong recovery in world aluminium demand, led to the progressive reduction of stock levels and to a concurrent recovery in market prices during 1994 and 1995. Consumption in the Western industrialized countries and Japan rose by an estimated 10.3% to 17.3m. metric tons, representing the highest rate of annual growth since 1983. This recovery was attributed mainly to a revival in demand from the motor vehicle sector in EU countries and the USA, and to an intensified programme of public works construction in Japan. Increased levels of demand were also reported in China and in the less industrialized countries of the South and East Asia region. Demand in the industrialized countries advanced by 11% in 1994, and by 2.2% in 1995.

Levels of world aluminium stocks were progressively reduced during 1994, and by late 1995 it was expected that the continuing fall in stock levels could enable Western smelters to resume full capacity operation during 1996. In May 1996 stock levels were reported to have fallen to their lowest level since March 1993, and exports of aluminium from Russia, totalling 2m. metric tons annually, were viewed as essential to the maintenance of Western supplies. Meanwhile, progress continued to be made in arrangements under the MOU for the modernization of Russian smelters and their eventual integration into the world aluminium industry. International demand for aluminium rose by approximately 0.2% in 1995 and by 0.8% in 1996. In 1997 aluminium consumption by industrialized countries advanced by an estimated 5.4%. This growth in demand was satisfied by increased primary aluminium production, combined with sustained reductions in world levels of primary aluminium stocks. Demand in 1998, however, was adversely affected by the economic crisis in East Asia, and consumption of aluminium in established market economy countries (EMEC) rose by only 0.1%: the lowest growth in aluminium demand since 1982. However, consumption in the EMEC area increased by an estimated 3.9% in 1999, with demand for aluminium rising strongly in the USA and in much of Asia. Compared with 1998, growth in consumption was, however, reduced in Europe and Latin America. World-wide, the fastest-growing sector of aluminium demand in 1999 was the transport industry (the largest market for the metal), whose consumption rose by about 9%. In 2000, according to USGS data, production of primary aluminium grew by 3.4%, to 24.4m. tons. At the end of that year, according to the IAI, total world inventories (comprising unwrought aluminium, unprocessed scrap, metal in process and finished, semi-fabricated metal) had declined slightly, compared with the previous year. Stocks of primary aluminium held by the London Metal Exchange (LME), meanwhile, had fallen heavily. Demand from the USA and Asia was characterized by the USGS as weak during 2000, especially during the second half of the year, while European demand remained firm. In 2001 IAI data indicated a slight fall in total world inventories of aluminium, while, conversely, those of primary aluminium held by the LME rose substantially. Production of primary aluminium in 2001 was slightly lower than in 2000. In 2002, however, production of primary aluminium was estimated to have risen by about 6.6%. In that year, according to analysis by Alcan, consumption of primary aluminium in the West rose by more than 3.5% to almost 20m. tons. IAI data indicated a decline in total aluminium stocks of about 2.8% in 2002. In 2003, when prices fell to their lowest level ever in real (i.e. constant US dollar) terms, it was evident that reductions in costs had enabled larger, integrated producers to safeguard the viability of their enterprises. (Older and, generally, smaller producers, meanwhile, had been forced into closure.) As evidence of this, analysts cited plans to create substantial new primary metal capacity world-wide up to 2010. According to IAI data, world inventories of aluminium rose by 1.3% in 2003.

Export Price Index for Aluminium
(base: 1980 = 100)

	Average	Highest month(s)	Lowest month(s)
1990	93		
1995	104		
2000	88	95 (Jan., Feb.)	83 (April, May)
2001	82	92 (Jan.)	73 (Oct.)
2002	77	80 (March)	74 (Aug., Sept.)

In November 1993 the price of high-grade aluminium (minimum purity 99.7%) on the London Metal Exchange (LME) was quoted at US $1,023.5 (£691) per metric ton, its lowest level for about eight years. In July 1994 the London price of aluminium advanced to US $1,529.5 (£981) per ton, despite a steady accumulation in LME stocks of aluminium, which rose to a series of record levels, increasing from 1.9m. tons at mid-1993 to 2.7m. tons in June 1994. In November 1994, when these holdings had declined to less than 1.9m. tons, the metal was traded at US $1,987.5 (£1,269) per ton.

In January 1995 the LME price of aluminium rose to US $2,149.5 (£1,346) per metric ton, its highest level since 1990. The aluminium price was reduced to US $1,715.5 (£1,085) per ton in May 1995, but recovered to US $1,945 (£1,219) in July. It retreated to US $1,609.5 (£1,021) per ton in October. Meanwhile, on 1 May the LME's stocks of aluminium were below 1m. tons for the first time since January 1992. In October 1995 these holdings stood at 523,175 tons, their lowest level for more than four years and only 19.7% of the June 1994 peak. Thereafter, stock levels moved generally higher, reaching 970,275 tons in October 1996. During that month the London price of aluminium fell to US $1,287 (£823) per ton, but later in the year it exceeded US $1,500.

In March 1997 the London price of aluminium reached US $1,665.5 (£1,030) per metric ton. This was the highest aluminium price recorded in the first half of the year, despite a steady decline in LME stocks of the metal. After falling to less than US $1,550 per ton, the London price of aluminium rose in August to US $1,787.5 (£1,126). In that month the LME's aluminium holdings were reduced to 620,475 tons, but in October they reached 744,250 tons. Stocks subsequently decreased, but at the end of December the metal's price was US $1,503.5 (£914) per ton, close to its lowest for the year. The average price of aluminium on the LME in 1997 was 72.5 US cents per lb, compared with 68.3 US cents per lb in 1996 and 81.9 US cents per lb in 1995.

The decline in aluminium stocks continued during the early months of 1998, but this had no major impact on prices, owing partly to forecasts of long-term oversupply. In early May the LME's holdings stood at 511,225 metric tons, but later that month the price of aluminium fell to less than US $1,350 per ton. In June LME stocks increased to more than 550,000 tons, and in early July the price of the metal was reduced to US $1,263.5 (£768) per ton. In early September the LME's holdings decreased to about 452,000 tons (their lowest level since July 1991), and the aluminium price recovered to US $1,409.5 (£840) per ton. However, the market remained depressed by a reduction in demand from some consuming countries in Asia, affected by the economic downturn, and in December 1998 the London price of aluminium declined to US $1,222 (£725) per ton. For the year as a whole, the average price was 61.6 US cents per lb.

During the first quarter of 1999 the aluminium market continued to be oversupplied, and in March the London price of the metal fell to US $1,139 (£708) per metric ton: its lowest level, in terms of US currency, since early 1994. Later that month the LME's stocks of aluminium rose to 821,650 tons. By the end of July 1999 these holdings had been reduced to 736,950 tons, and the price of aluminium had meanwhile recovered to US $1,433.5 (£902) per ton. Stock levels later rose, but, following the announcement in August of proposed cost-cutting mergers between major producers (see above), the aluminium price continued to increase, reaching US $1,626.5 (£1,009) per ton at the end of the year. The steady rise was also attributable to a sharp increase in the price of alumina in the second half of the year. Following an explosion at (and subsequent closure of) a US alumina refinery in July, the price of the material advanced from US $160 per ton in that month to US $400 per ton (its highest level for 10 years) in December. However, the average LME price of aluminium for the year (61.7 US cents per lb) was almost unchanged from the 1998 level.

With alumina remaining in short supply, prices of aluminium continued to rise during the opening weeks of 2000, and in late January the London quotation (the official closing offer price at the end of morning trading on the LME) reached US $1,745 per metric ton: its highest level for more than two years. However, the LME's stocks of the metal also increased, reaching 868,625 tons in February. The London price of aluminium declined to US $1,397 per ton in April, but recovered to US $1,599 in July. Throughout this period there was a steady decrease in LME holdings, which were reduced to less than 700,000 tons in April, under 600,000 tons in May and below 500,000 tons in July. At the end of July aluminium stocks were 461,975 tons: only 53% of the level reached in February. In August stocks fell below 400,000 tons, and in December to less than 300,000 tons at one point. During August–December the average monthly London quotation ranged between US $1,600.8 (£1,116) per ton, recorded in September, and US $1,464.9 (£1,070) per ton, recorded in December.

At the end of January 2001 the London quotation rose to US $1,737 (£1,176) per metric ton, thus approaching the highest level recorded in 2000. By the end of January stocks had recovered to 394,075 tons. By the end of February LME holdings stood at 483,200 tons, and the London price had declined to US $1,553 (£1,069) per ton. In both March and April, on a month-on-month basis, stocks fell, while prices were generally weaker, the London quotation reaching US $1,540 (£1,073) on 27 April. Although the London quotation rose as high as US $1,586 (£1,112) on 4 May, prices fell precipitously in June–November, reaching a low of US $1,243 (£866) on 7 November, but recovered to $1,430 at the end of the latter month. The sustained decline was attributed to slow economic growth worldwide, which had caused the market to be over-supplied, aggravated by the events in New York, USA, on 11 September. For the whole of 2001 the LME average monthly 'spot' price for high-grade aluminium was 65.5 US cents per pound.

On 28 March 2002 LME stocks rose to 1,029,400 metric tons. The London quotation weakened in both April and May, falling to US $1,318 (£903) on 23 May. In early June the price recovered to US $1,398 (£998), but had fallen to US $1,364.5 per ton by the end of the month, when stocks of aluminium held by the LME totalled more than 1.2m. tons. By the end of July the price of aluminium traded on the LME had fallen to US $1,310 per ton, by which time stocks held by the Exchange had risen to 1,291,000 tons. On 14 August the London quotation fell to US $1,279 per ton, but had recovered to US $1,293.5 per ton by the end of the month. On 13 August, meanwhile, stocks rose to 1,300,125 tons, and were to remain above 1,290,000 tons until the first day of October, when they fell to 1,288,200 tons. On 11 September the London quotation recovered to US $1,340.5 per ton, but subsequently weakened, ending the month at US $1,280.5 per ton. At the end of the first week of October the price of aluminium fell to US $1,275.5 per ton, but it had risen to US $1,337.5 per ton by 31 October. From early November the London quotation began to recover somewhat, reaching US $1,370.5 per ton on 4 November and ending the month at US $1,378 per ton. This upward movement continued into December: on 13 December the London quotation was just short of US $1,400 per ton. Stocks declined further in the final month of 2002, falling to 1,238,000 tons on 12 December and ending the year at 1,241,350 tons. The price of aluminium, meanwhile, declined to US $1,344.5 per ton on 31 December.

On 22 January 2003 the London quotation for primary aluminium closed at more than US $1,400 per metric ton for the first time since 22 March 2002, and it had risen further, to US $1,247 per ton, by the end of the month. By the end of January 2003 stocks held by the LME had fallen to 1,199,550 tons and these continued to fall thereafter until about the middle of February. With the exception of 19 February, the London quotation closed at more than US $1,400 per ton on each day of that month. The price weakened during March, however, at the same time as stocks were rising. On 31 March the London quotation closed at US $1,350 per ton, while stocks were recorded at 1,252,775 tons. The London price continued to fall until around mid-April, subsequently strengthening to finish the month at US $1,356.5 per ton. From around mid-May this stronger trend became more pronounced, with the London quotation closing at more than US $1,400 per ton on each day of the month after 12 May. By the end of May stocks of the metal held by the LME had declined to 1,130,625 tons. On 16 June stocks fell to 1,115,150 tons. Prices remained stable throughout June, the London quotation standing at US $1,389 per ton at the end of that month.

On 14 July 2003 the London quotation for primary aluminium closed at US $1,463 per metric ton. Prices were thereafter generally weaker until towards the end of the month. On 28 July the closing price for the metal was US $1,484.5 per ton. On 1 August the London quotation closed at US $1,505 per ton, the first time a closing price in excess of US $1,500 per ton had been recorded since May 2001. On 29 August, the last trading day of that month, the London price closed at US $1,432 per ton, having fallen as low as US $1,427.5 in the interim. On 11 September the price declined to US $1,378 per ton, but it had recovered by the end of the month to US $1,407.5 per ton. Generally, from October until the end of 2003, the London price strengthened, reaching US $1,552 per ton on 5 December and ending the year at US $1,592.5 per ton. During the second half of 2003 stocks of aluminium held by the LME rose steadily. At the end of July they totalled 1,304,450 tons; by the end of December they had increased to 1,423,275 tons.

On 2 January 2004 the London quotation for primary aluminium closed at US $1,600 per metric ton, the highest price recorded since February 2001. On 30 January a closing price of US $1,636.5 per ton was recorded. The London quotation strengthened further in February, closing at US $1,754 per ton on 18 February, but thereafter declined somewhat to end the month at US $1,702 per ton. The London price remained above US $1,625 per ton throughout March, ending the month at US $1,688.5 per ton. Sharp increases occurred from early April, and on 16 April a closing price greater than US $1,800 per ton (US $1,802) was recorded for the first time in the 2000s. By 10 May, however, the quotation had fallen to US $1,575 per ton. On 2 June a closing price of US $1,703.5 was recorded, rising to US $1,721 per ton on 21 June. Stocks of aluminium held by the LME rose as high as 1,453,125 tons in January. From February until the end of June, however, they declined steadily. At the end of February they totalled 1,393,675 tons, but had fallen to 940,200 tons by the end of June.

In early July 2004 the London quotation for primary aluminium rose steadily, reaching US $1,759 per metric ton on 9 July. Later in the month the price weakened, however, and was quoted at US $1,687 per ton on 30 July. The quotation rose above US $1,700 per ton again in mid-August, but by the end of the month had declined slightly, to US $1,688.5 per ton. Stocks declined further throughout July and August, falling to 753,850 tons at the end of the latter month.

The International Aluminium Institute (IAI), based in London, is a global forum of producers of aluminium dedicated to the development and wider use of the metal. In 2004 the IAI had 26 member companies, representing every part of the world, including Russia and the People's Republic of China, and responsible for about 80% of global primary aluminium production and a significant proportion of the world's secondary output.

Cassava (Manioc, Tapioca, Yuca) (*Manihot esculenta*)

Cassava is a perennial woody shrub, up to 5 m in height, which is cultivated mainly for its enlarged, starch-rich roots, although the young shoots and leaves of the plant are also edible. The plant can be harvested at any time from seven months to three years after planting. A native of central and South America, cassava is now one of the most important food plants in all parts of the tropics (except at the highest altitudes), having a wide range of adaptation for rainfall (500–8,000 mm per year). Cassava is also well adapted to low-fertility soils, and grows where other crops will not. It is produced mainly on marginal agricultural land, with virtually no input of fertilizers, fungicides or insecticides.

The varieties of the plant fall into two broad groups, bitter and sweet cassava, formerly classed as two separate species, *M. utilissima* and *M. dulcis* or *aipi*. The roots of the sweet variety are usually boiled and then eaten. The roots of the bitter variety are either soaked, pounded and fermented to make a paste, or dried, as in the case of 'gaplek' in Indonesia, or given an additional roasting to produce 'gari'. They can also be made into flour and starch, or dried and pelletized as animal feed. In Indonesia a high-fructose syrup is being produced from cassava starch in order to reduce sugar imports.

The cassava plant contains two toxic substances, linamarin and lotaustralin, in its edible roots and leaves which release the poison cyanide, or hydrocyanic acid, when plant tissues are damaged. Sweet varieties of cassava produce as little as 20 mg of acid per kg of fresh roots, whereas bitter varieties may produce more than 1,000 mg per kg. Although traditional methods of food preparation are effective in reducing cyanogenic content to harmless levels, if roots of bitter varieties are underprocessed and the diet lacks protein and iodine (as occurs during famines and wars), cyanide poisoning can cause fatalities. Despite the disadvantages of the two toxins, some farmers prefer to cultivate the bitter varieties, possibly because the cyanide helps to protect the plant from potential pests, and possibly because the texture of certain food products made from bitter varieties is preferred to that of sweet cassavas.

Cassava is the most productive source of carbohydrates and produces more calories per unit of land than any single cereal crop. Cassava is a staple source of carbohydrates and forms an essential part of the diet throughout tropical areas. Although the nutrient content of the roots consists almost entirely of starch, the leaves are high in vitamins, minerals and protein. A plot of cassava may be left unattended in the ground for two years after maturity without deterioration of the roots and the plant is resistant to prolonged drought, so the crop is valued as a famine reserve. The roots are highly perishable after harvest and, if not consumed immediately, must be processed into flour, starch, pellets, etc.

While the area under cassava has expanded considerably in recent years, there is increasing concern that the rapid expansion of cassava root planting may threaten the fertility of the soil and subsequently other crops. Under cropping systems where no fertilizer is used, cassava is the last crop in the succession because of its particular adaptability to infertile soils and its high nutrient use-efficiency in yield terms (although there is now evidence to suggest that cassava yields increase with the use of fertilizer). With the exception of potassium, cassava produces more dry matter with fewer nutrients than most other food crops. Soil fertility is not threatened by cassava itself, but rather by the cultivation systems which employ it without fertilizer use.

Production of Cassava
('000 metric tons)

	2002	2003*
World total	186,391*	189,100
Far East	45,902*	49,601
Oceania	189*	177
Leading regional producers		
China, People's Repub.†	3,901	3,901
Indonesia	16,913	18,474
Philippines	1,626	1,400‡
Thailand	16,868	18,430
Viet Nam	4,438	5,229
Other leading producers		
Angola	5,620	5,69
Brazil	23,066	22,236
Congo, Dem. Repub.	14,929	14,929†
Ghana	9,731	10,000†
India†	7,000	7,100
Mozambique	5,925	6,150‡
Nigeria	34,476	33,379
Paraguay	4,430	3,900†
Tanzania	6,888	6,888†
Uganda	5,373	5,400†

* Provisional figure(s).
† FAO estimate(s).
‡ Unofficial figure.

Some interest has been shown in the utilization of cassava as an industrial raw material as well as a food crop. Cassava has the potential to become a basic energy source for ethyl alcohol (ethanol), a substitute for petroleum. 'Alcogas' (a blend of cassava alcohol and petrol) can be mixed with petrol to provide motor fuel, while the high-protein residue from its production can be used for animal feed. The possibility of utilizing cassava leaves and stems (which represent about 50% of the plant and are normally discarded as cattle-feed concentrates) has also received scientific attention.

Thailand is the world's leading exporter of aggregate dry cassava products (tapioca), accounting, for instance, for about 80% of shipments of dried cassava world-wide and about 88% of shipments of cassava starch in 2002. Used in the preparation of animal feed, in combination with a protein source, cassava provides a cheap substitute for cereals such as barley and maize. In 2002 the main importers of cassava products (as forecast by FAO on the basis of provisional data) were the People's Republic of China and the member states of the European Union (EU). As, owing to its Common Agricultural Policy, cereal prices had risen within the European Community (EC, now the EU), the search for cheaper alternatives and the fact that cassava was not subject to the variable levy led to an increase in imports of feed-grade cassava (mostly tapioca pellets from Thailand). Following protests from cereal producers (notably in France), the EC decided in 1979 to eliminate subsidies on compound animal feed, with the aim of reducing the use of cassava-based products. Thailand subsequently agreed to adopt voluntary export quotas for its tapioca products to EC countries. However, during the mid-1990s Thailand's cassava exports to the EU declined, largely as a result of adverse weather conditions and lower yields. Since 1994 the Government of Thailand has been offering cassava growers incentives to plant alternative crops. In 2002, according to FAO, international trade in cassava products (dry weight) was forecast to decline by 19%, compared with the previous year, to slightly less than 6m. metric tons. In spite of a marginal increase in trade in cassava in the form of flour and starch, trade in cassava chips and pellets was forecast to decline by about one-third. Much of this decline was concentrated in the EU member states, whose imports of cassava feed products were estimated to have fallen by more than 40% as a consequence of the low price of EU-produced feed grains. In 2002 as (for the first time) in 2001, developing countries in the Far East represented the major market for cassava. Thai exports of cassava products were forecast to fall to only 5.7m. tons (dry weight of chips or pellets) in 2002, compared with an estimated 7.1m. tons in 2001. In mid-2003 FAO forecast growth in international trade in cassava products in that year, in view of a likely increase in exportable supplies from Thailand. Exports of pellets and chips from Thailand in January–mid-April 2003 were reported to have risen by 3% compared with the corresponding period of 2002. The pattern of trade in the first four months of 2003 appeared to indicate that Far Eastern destinations would continue to displace the EU as the world's major market for cassava products. From January until the first week of May the European Commission had reportedly issued import certificates for only some 500,000 tons of cassava pellets, about 400,000 tons less than in the corresponding period of 2002. The decline was attributed, once again, to competitive feed grain prices within the EU.

Within South-East Asia demand increased during the 1990s for dried starch derived from cassava for use in the production of textiles and paper, and by manufacturers of processed foods. Viet Nam, with a total area under cassava of 283,000 ha, has obtained private Taiwanese and Japanese investment in the construction of a substantial facility for cassava-processing at Dong Nai.

Export Price Index for Cassava
(base: 1980 = 100)

	Average	Highest month(s)	Lowest month(s)
1990	78		
1995	85		
2000	36	*	*
2001	36	*	*
2002	36	*	*

* The monthly index remained constant at 36 throughout 2000, 2001 and 2002.

During 1993 the average monthly import price of hard cassava pellets at the port of Rotterdam, Netherlands, declined from US $160 per metric ton in January to US $120 per ton in November. The annual average was US $137 per ton, compared with US $183 in 1992. Prices continued downward in later years, and fell very sharply in 1997. In 1999 the average import price of hard cassava pellets at Rotterdam was US $102 per ton. A further sharp decline was experienced in 2000, in which year the average price was US $84 per ton, and in 2001 the average price fell to only US $82 per ton. In 2002 the average price recovered to US $90 per ton. During the first four months of 2003 an average price of US $94 per ton was recorded. In January–August 1999 the average international price of cassava starches and flours (f.o.b. Bangkok) was US $176 per ton (the lowest price since 1988), compared with an average price of US $305 per ton in the corresponding period of 1998. This price declined further throughout 2000, to an annual average of US $158 per ton, although a recovery in January–March 2001 restored prices to the previous year's average in that period. For the whole of 2001 an average of US $173 per ton was recorded, and this rose further, to US $184 per ton, in 2002. In 2003 the average international price of cassava starches and flours was US $182 per ton.

Coal

Coal is a mineral of organic origin, formed from the remains of vegetation over millions of years. There are several grades: anthracite, the hardest coal with the highest proportion of carbon, which burns smoke-free; bituminous and sub-bituminous coal, used for industrial power: some is made into coke when the volatile matter is driven off by heating; and lignite or brown coal, the lowest grade and nearest to the peat stage. Anthracite and bituminous coal are classed as 'hard' coal. Coal gas is made from brown coal, but is not widely used for energy except in the republics of the former USSR.

Geographically, coal is one of the most evenly distributed of the fossil fuels. Of estimated world proven reserves of 984,453m. metric tons at the end of 2003 (comprising 519,062m. tons of anthracite and bituminous coal and 465,391m. tons of sub-bituminous coal and lignite), about 21% were located in the Far East and Australasia, some 9% in South Asia (India and Pakistan), 26% in North America, 23% in the republics of the former USSR and 13% in Europe.

During the period 1978–2002, according to the World Coal Institute, annual world production of hard coal (excluding sub-bituminous coal and lignite) increased from 2,619m. metric tons to 3,837m. tons: a rise of more than 46%. High levels of output and demand were maintained during the 1990s, owing to the increasing use of coal world-wide as the primary fuel for electricity generation. In 2001 coal accounted for about 39% of this demand, which is expanding at an annual rate of about 3%. Environmentalists, however, have increasingly criticized the large-scale use of fossil fuels as a prime causative factor in 'acid rain' pollution and the warming of the global atmosphere by accretions of carbon gases. In 2003 coal was estimated to account for about 45% of primary energy consumption in the Asia-Pacific region, compared with some 43% in 2002. Among the countries with the highest levels of utilization of coal for electricity generation are Australia (an estimated 77% in 2002) and the People's Republic of China (76% in 2001). In 2003 the Asia-Pacific region was the principal world consumer of coal, accounting for an estimated 50.7% of global demand.

As the greater part of coal output is consumed within the producing country, only a relatively small proportion of coal production enters world trade. In 2002 an estimated 436m. metric tons of steam coal (for power generation) was traded internationally, together with about 187m. tons of coking coal (for use in metallurgical industries). About 90% of total world consumption of coal is both mined and used in the same producing countries, mainly the USA and China. In 2003 China was estimated to have accounted for more than one-third of world coal production and for about 31% of world consumption.

China, with identified reserves of 114,500m. metric tons (including sub-bituminous coal and lignite) at the end of 2003, is the world's largest producer of hard coal. About 75% of these deposits are located in the north and north-west regions. Only about 7% of China's proven reserves of anthracite and bituminous coal can be surface-mined, however. In 1992 the Government began to implement a comprehensive restructuring of the coal industry, in which uneconomic and dangerous mines were closed, the industry's labour force was reduced from 3m. to 2.5m. by 1995, and controls on the selling prices of coal were abolished. The reorganization continued in the later 1990s, with many small, unprofitable and unsafe coal-mines, including some operated by township enterprises, being forced to close. As a result, China's annual production of coal (including brown coal) was reduced from 1,374m. tons in 1996 to 1,044m. tons in 1999. During the latter year about 31,000 small mines (some of them operating illegally) were closed, leaving fewer than 38,000 functioning at the end of 1999. The programme continued in 2000, with a further 18,900 coal-mines scheduled for closure. In May it was reported that the Chinese Government aimed to limit coal production to 870m. tons in 2000, with output restrictions imposed on large state-owned collieries as well as on small mines. In the event, production reportedly totalled about 999m. tons. In 2001, however, output rose to about 1,099m. tons, and in 2002 record production, of some 1,450m. tons, was recorded. Output was reported to have been raised in response to greater demand from the cement, metallurgy and power sectors in 2002. Mine safety remained a major problem: more than 6,000 Chinese miners were killed in accidents in 2002, of which by far the majority occurred in county-level mines. It was reported to be common for mines that had been closed on safety grounds to resume production, illegally, after 'closure' owing to the substantial profits that they were able to realize. It is hoped that under the government's ongoing restructuring programme mine safety will be improved. In 2002 the large, state-owned mines that have resulted from restructuring were reported to have accounted for 51% of total output, compared with only 39% in 1996. In 2003 Chinese coal production was estimated to have risen to a new record level of 1,667m. tons. Traditionally China has exported only a very small proportion of its coal. The greater efficiency and improved competitiveness of the Chinese coal industry that has been achieved in recent years, however, has at the same time led to a substantial increase in foreign sales, in particular to Japan, the Republic of Korea and the Republic of China, whose markets had hitherto generally been dominated by Australia. In 2002, according to the World Coal Institute, China's exports totalled 85.8m. tons (72m. tons of steam coal and 13.8m. tons of coking coal), compared with about 90m. tons in 2001. As recently as 1998, however, China's total annual exports of coal amounted to only about 32m. tons. According to the Energy Information Administration (EIA) of the US Department of Environment, China also emerged as a significant importer of coal in 2002, purchasing in total some 14m. tons. Imported coal was reportedly available to southern coastal electricity producers at cheaper prices than domestic supplies.

Production of Coal*
(million metric tons)

	1999	2000*
Leading regional producers		
Australia	338.9	347.2
China, People's Repub.	1,449.6	1,667.0
Indonesia	103.4	114.6
Viet Nam	15.4	19.0
Other leading producers		
Germany	208.2	205.0
India	359.3	367.3
Kazakhstan	73.7	84.7
Poland	161.9	162.8
Russia	255.4	274.8
South Africa	220.2	238.8
Ukraine	82.9	80.3
USA	992.3	970.0

* Commercial solid fuels only, comprising bituminous coal and anthracite (hard coal), and lignite and brown (sub-bituminous) coal.

Source: BP, *Statistical Review of World Energy 2004*.

Australia is also an important coal producer in the region, having overtaken the USA in 1986 as the world's leading coal exporter, and accounting in 2002 for about 21% of world trade in steam coal and about 57% of world trade in coking coal. Although Australia possesses less than 10% of the world's coal reserves, it has become a major exporter because of comparatively low domestic demand for coal (which supplies about 43% of domestic primary energy), and the mineral's location in unpopulated areas, inexpensive to mine. More than one-half of Australia's coal output is exported, about 40% of this to Japan, mainly for use in its steel industry. Coal is Australia's principal export commodity, and coal exports were valued at $A11,946m. (representing 10.1% of the country's total export earnings) in the year to June 2003. Until recently Australian coal-exporting companies had been forced to accept lower prices in their negotiations with Japanese power utilities (for sales of steam coal) and steel-producers (for coking coal). For the year to March 2000 contract prices for steam coal were reduced by 13%, and those for coking coal by 18%. For the Japanese fiscal year (April–March) 2000/2001 contract prices for steam coal were reduced by a further 4%, to US $28.75 per metric ton (f.o.b.); and contract prices for coking coal by a further 5%, to US $39.75 per ton (f.o.b.). For the Japanese fiscal year 2001/02, however, this downward trend was checked, with contract prices for steam coal rising by 20%, and those for high-quality coking coal by 7.5%. For the Japanese fiscal year 2003/04, according to the EIA, the reference price for steam coal was US $24.27 per ton (f.o.b. port of exit, nominal dollars) and that for coking coal US $41.91 per ton (f.o.b. port of exit, nominal dollars). Negotiations under way in late 2003–early 2004 indicated that the contract price for steam coal would be up to 50% higher and that for coking coal 20%–50% higher in the Japanese fiscal year 2004/05. A substantial increase in the value of the Australian dollar relative to its US counterpart, however, would attenuate the benefits of anticipated higher prices to Australian producers. To increase the Australian industry's competitiveness, mining companies have restructured their activities, reducing employment at Australian mines by almost 30% in the 24 months to June 1999, and raising productivity by about 40%. Australia is involved with Japan in a joint project to establish two coal liquefaction plants. A promising future was foreseen for this process, with petroleum prices rising (as they were when the project was envisaged) and conventional reserves rapidly being depleted.

In recent years Indonesia has emerged as a significant regional producer and exporter of coal, of which its total reserves were estimated at 5,370m. metric tons at the end of 2003. These reserves are in part formed by a variety with an exceptionally low content of sulphur and ash, the causative factors in air pollution resulting from coal combustion. Output of this coal, which has attracted interest in the USA, Japan, Spain and the Scandinavian countries, totalled 2m. tons in 1991 and was forecast to rise to 8m. tons annually by the late 1990s. Total Indonesian coal production was forecast to reach a level of 120m. tons annually by 2002, as East Asian consumers reduced their reliance on exporters outside the region. Actual output in 2002, however, was only about 103m. tons. In 2003 production increased to some 115m. tons. The government now aims to raise annual output to 120m. tons in 2006. In 2002 Indonesia was the world's fourth largest exporter of steam coal (after Australia, China and South Africa), exporting some 66m. tons in that year.

The world's leading coal importer is Japan, which is seeking to reduce its dependence on petroleum for energy. In 2003, however, coal accounted for only about 22% of Japan's total primary energy supply. The exploitation of Japan's proven coal reserves, estimated to total only 773m. metric tons at 31 December 2003, has proved too expensive for domestic supplies to be able to compete with much cheaper imported coal. The country's coal industry has been in decline since the mid-1980s, with uneconomic mines closing, despite government provisions of subsidies and 'guide-lines' seeking to persuade the steel industry to purchase domestic coal at prices far above those on the international market.

Export Price Index for Coal
(base: 1980 = 100)

	Average	Highest month(s)	Lowest month(s)
1990	108		
1995	110		
2000	74	82 (Jan.)	69 (Oct., Nov.)
2001	83	94 (July)	74 (March)
2002	82	89 (Dec.)	72 (Feb., March)

At the end of 2003 Viet Nam's exploitable reserves of high-quality anthracite and bituminous coal totalled 150m. metric tons. In the late 1990s the country emerged as a leading world supplier of anthracite to the Japanese steel industry and to Western European countries, where its low ash, nitrogen, phosphorous and sulphur content facilitate compliance with measures being implemented to protect the environment. In 2002, according to the USGS, Viet Nam's exports of anthracite totalled 5.9m. tons, generating US $149m. in revenue. In the previous year Viet Nam's exports of anthracite amounted to some 4.3m.

tons, representing about 36% of all anthracite traded world-wide in that year. Japan and the EU member states remained the most important markets for this coal in 2001, but it was also shipped to ASEAN destinations and to North American markets, including the USA. Vietnamese anthracite is reportedly confronted by increased competition from cheaper Chinese supplies in Japanese and European markets. The mining, distribution and export of coal is undertaken mainly by state-controlled Viet Nam National Coal Corpn (Vinacoal).

Australia's exports of coal (including briquettes) had an average value of US $37.35 per metric ton in 1994, rising to US $40.08 per ton in 1995. The average price per ton was US $43.14 in 1996, but it declined to US $41.42 in 1997 and to US $37.04 in 1998. During the first three months of 1999 Australia's coal exports earned an average of only US $34.68 per ton, and prices subsequently fell further. Most Australian coal shipments are delivered to overseas buyers under long-term contracts at negotiated prices, but in 1999 international coal prices on the 'spot' market (for prompt delivery) were at their lowest levels for more than 20 years. In September hard coking coal was being traded in Asia at about US $32 per ton. International 'spot' prices remained depressed throughout most of 2000, but, from the beginning of the final quarter of that year, began to rise dramatically in response to increased demand for US thermal coal from US electricity generators. By mid-2001 'spot' prices of US coal were reported to have doubled—in some cases trebled—since late 2000. By mid-2001, as a result of the weakness of the Australian dollar, coal prices expressed in terms of the Australian dollar were almost 80% higher than at the beginning of 2000. In 2003, according to the EIA, a further substantial increase in the export price of coal was recorded as a result of greater demand world-wide and of a reduction in supplies from China towards the end of the year. The EIA, quoting data reported by McCloskey Coal Information Services, indicated a 'spot' price of US $33.57 per ton (f.o.b.) in December 2003 for steam coal shipped from Newcastle, Australia, compared with US $22.52 per ton (f.o.b.) in December 2002. In late July 2004, according to industry sources, the 'spot' price for steam coal at Newcastle rose to US $60.8 per ton, 35% higher than the Japanese contract price negotiated for 2004/05.

Cocoa (*Theobroma cacao*)

This tree, up to 14 m tall, originated in the tropical forests of Central and South America. The first known cocoa plantations were in southern Mexico around AD 600, although the crop may have been cultivated for some centuries earlier. Cocoa first came to Europe in the 16th century, after Spanish explorers had found the beans being used in Mexico as a form of primitive currency as well as the basis of a beverage. The Spanish and Portuguese introduced cocoa into Africa—on the islands of Fernando Póo (now Bioko), in Equatorial Guinea, and São Tomé and Príncipe—at the beginning of the 19th century. At the end of the century the tree was established on the African mainland, first in Ghana and then in other west African countries. In the Asia-Pacific region, Criollo cocoa was introduced to Indonesia in 1560, but its cultivation remained confined to the island of Java. In the late 19th century, after the failure of the coffee crop on Java in 1880 had given impetus to the adoption of cocoa by farmers, the exposure of Criollo stock to Forastero cocoa led to the development of a robust Trinitario variant producing the so-called Java A bean. Elsewhere, Trinitario cocoa (originally from Trinidad) was first planted in Sri Lanka in 1834 and was subsequently taken from there to Singapore, Fiji and Samoa.

Cocoa is now widely grown in the tropics, usually at altitudes of less than 300 m above sea-level, where it needs a fairly high rainfall and good soil. The cocoa tree has a much shallower tap root than, for example, the coffee bush, making cocoa more vulnerable to dry weather. Cocoa trees can take up to four years from planting before producing sufficient fruit for harvesting. They may live to 80 years or more, although the fully productive period is usually about 20 years. The tree is highly vulnerable to pests and diseases, and it is also very sensitive to climatic changes. Its fruit is a large pod, about 15–25 cm in length, which at maturity is yellow in some varieties and red in others. The ripe pods are cut from the tree, where they grow directly out of the trunk and branches. When opened, cocoa pods disclose a mass of seeds (beans) surrounded by white mucilage. After harvesting, the beans and mucilage are scooped out and fermented. Fermentation lasts several days, allowing the flavour to develop. The mature fermented beans, dull red in colour, are then dried, ready to be bagged as raw cocoa which may be further processed or exported.

Cultivated cocoa trees may be broadly divided into three groups. Most cocoas belong to the Amazonian Forastero group, which now accounts for more than 80% of world cocoa production. It includes the Amelonado variety, suitable for chocolate manufacturing. Criollo cocoa is not widely grown and is used only for luxury confectionery. The third group is Trinitario, which comprises about 15% of world output and is cultivated mainly in Central America and the northern regions of South America.

Cocoa processing takes place mainly in importing countries. The processes include shelling, roasting and grinding the beans. Almost half of each bean after shelling consists of a fat called cocoa butter. In the manufacture of cocoa powder for use as a beverage, this fat is largely removed. Cocoa is a mildly stimulating drink, because of its caffeine content, and, unlike coffee and tea, is highly nutritious.

The most important use of cocoa is in the manufacture of chocolate, of which it is the main ingredient. About 90% of all cocoa produced is used in chocolate-making, for which extra cocoa butter is added, as well as other substances such as sugar—and milk in the case of milk chocolate. Proposals that were initially announced in December 1993 (and subsequently amended in November 1997) by the consumer countries of the European Union (EU), permitting chocolate-manufacturers in member states to add as much as 5% vegetable fats to cocoa solids and cocoa fats in the manufacture of chocolate products, have been perceived by producers as potentially damaging to the world cocoa trade. In 1998 it was estimated that the implementation of this plan could reduce world demand for cocoa beans by 130,000–200,000 metric tons annually. In July 1999, despite protests from Belgium, which, with France, Germany, Greece, Italy, Luxembourg, the Netherlands and Spain, prohibits the manufacture or import of chocolate containing non-cocoa-butter vegetable fats, the European Commission cleared the way to the abolition of this restriction throughout the EU countries. The implementation of the new regulations took effect in May 2000.

Production of Cocoa Beans
('000 metric tons)

	2002	2003*
World total	3,107	3,257
Far East	514	480†
Oceania	47	47
Leading regional producers		
Indonesia‡	450	426
Malaysia	48	48†
Papua New Guinea‡	42	42
Other leading producers		
Brazil	175	171
Cameroon	125‡	125†
Colombia	48	47†
Côte d'Ivoire	1,225‡	1,225†
Dominican Repub.	50	50†
Ecuador	88	89
Ghana	341	475‡
Mexico	46	48
Nigeria	340	380†

* Provisional figure(s).
† FAO estimate(s).
‡ Unofficial figure(s).

In 2003, according to FAO, the Far East and Oceania provided an estimated 16.2% of the total world cocoa crop of 3,257,065 metric tons and was thus second largest producing region after (West) Africa. In that year Indonesia was the world's third largest producer of cocoa beans, after Côte d'Ivoire and Ghana. After coffee and sugar, cocoa is the most important agricultural export commodity in international trade. Recorded world exports (excluding re-exports) of cocoa beans totalled an estimated 2,446,624 tons in 2002, of which countries in the Far East and Oceania accounted for 433,964 tons (18%). Within the region, Indonesia is by far the most important exporter, accounting for about 84% of its total foreign sales in 2002. Indonesia thus ranked as the world's second largest exporter of cocoa beans in 2002, after Côte d'Ivoire, whose exports totalled an estimated 1,004,283 tons in that year, and ahead of Ghana (310,738 tons), Nigeria (180,723 tons), Cameroon (129,210 tons) and Ecuador (55,598 tons).

The principal importers of cocoa are developed countries with market economies, which generally account for more than 80% and sometimes for as much as 90% of cocoa imports from developing countries. Recorded world imports of cocoa beans in 2002 totalled an estimated 2,246,102 metric tons. The principal importing countries in that year were the Netherlands (with 495,238 tons, representing 22% of the total), the USA (323,257 tons) and Germany (205,174 tons). The Dutch cocoa-grinding industry is long established, having developed out of the country's dominance of world trade in cocoa in the 17th century.

In Indonesia, as noted above, the cultivation of cocoa began in the second half of the 16th century, but for many years remained confined to Java and to North Sumatra where, according to the International Cocoa Organization (ICCO), cocoa plantations covered some 6,500 ha prior to the Second World War. Output at that time amounted to about 2,000 metric tons annually, but increased gradually as cultivation commenced on Sulawesi and Kalimantan. In 1975 a national programme to expand smallholder cultivation of cocoa was initiated involving the extensive distribution of seedlings of the Upper Amazon Interclonal Hybrid cocoa variety. By 1980 cocoa was being cultivated on some 37,000 ha in Indonesia, and by 1988 the area under the crop had risen to more than 135,000 ha. Output rose accordingly, from some 10,000 tons in 1980 to more than 110,000 tons in 1989, thus having increased by an annual average rate of more than 30% in 1980–89. In 1997 Indonesia overtook Ghana as the world's second largest producer of cocoa beans after Côte d'Ivoire, but it has not retained that rank in every subsequent year. Since 1999 Indonesia's annual output of cocoa has usually been substantially in excess of 400,000 tons. It is likely that it would have risen still further had it not been for infestation by the cocoa pod borer, *Conopomorpha cramerella*, a pest that in the late 1990s reportedly beset up to one-fifth of all Indonesian cocoa. By means of cultural control (regular, thorough harvesting with machetes), biological control (through the black (cocoa) ant *Dolichoderus thoracicus* and the parasitoid wasp *Trichogrammatoidea bactrae fumata*), and behavioural control (trapping the adult moths into which the cocoa pod borer beetle develops using sex pheromones) infestations have been contained, but the cocoa bod borer nevertheless remains the single greatest constraint on the expansion of Indonesian cocoa production. In 2003, with output of an estimated 426,000 tons, Indonesia was the world's third largest producer of cocoa after Côte d'Ivoire and Ghana. The cultivation of cocoa in Indonesia is dominated by smallholders, who accounted for more than 75% of the total area under cocoa in the late 1990s. Plantations, both government-owned and private, accounted for the remainder.

In the Far East, after Indonesia, Malaysia is the most important producer of cocoa, output there totalling an estimated 48,000 metric tons in 2003. The area under cocoa was estimated at about 120,000 ha in the early 2000s, when, according to the Cocoa Producers' Alliance (COPAL), exports of cocoa accounted for some 2% of Malaysia's total foreign exchange earnings. As in Indonesia, production of cocoa grew considerably in the 1980s, in response to the government's policy of encouraging it. From some 35,000 tons in 1980, output increased to a peak of 243,000 tons in 1989, thus having increased by an annual average rate of 24% in 1980–89. In the 1990s, however, as a consequence of the depressed international price of cocoa and the superior profitability of alternative crops such as oil palm, output declined precipitously. Malaysian cocoa production has continued to decline since 2000, to the point at which, in 2003, it was only some 10,000 tons greater than before the expansion programme of the 1980s was pursued. Malaysian exports of cocoa have fallen as output has declined. In 1988, at some 190,000 tons, they were more than three times greater than Indonesia's foreign sales of cocoa. In 2002, however, Malaysia's exports of cocoa had dwindled to only an estimated 21,000 tons. Local annual cocoa processing capacity is estimated at 100,000 tons, much of which now has to be supplied by imports. In the early 2000s, according to COPAL, the government's second National Master Plan and its third National Agricultural Policy both included among their objectives the revival of cocoa production.

Elsewhere in the region, Papua New Guinea accounts for the bulk of Oceania's output of cocoa, with estimated production of 42,000 metric tons in 2003. The Philippines is a modest producer of cocoa, with annual output of approximately 6,000–8,000 tons, as are some of the small island territories of the Pacific, notably Solomon Islands and Vanuatu.

World prices for cocoa are highly sensitive to changes in supply and demand, making its market position volatile. Negotiations to secure international agreement on stabilizing the cocoa industry began in 1956. Full-scale cocoa conferences, under United Nations auspices, were held in 1963, 1966 and 1967, but all proved abortive. A major difficulty was the failure to agree on a fixed minimum price. In 1972 the fourth UN Cocoa Conference took place in Geneva and resulted in the first International Cocoa Agreement (ICCA), adopted by 52 countries, although the USA, the world's principal cocoa importer at that time, did not participate. The ICCA took formal effect in October 1973. It operated for three quota years and provided for an export quota system for producing countries, a fixed price range for cocoa beans and a buffer stock to support the agreed prices. In accordance with the ICCA, the ICCO, based in London, was established in 1973. In March 2004 the membership of the 2001 ICCA (see below) comprised 30 countries (12 exporting members, 18 importing members), representing about 80% of world cocoa production and some 60% of world cocoa consumption. The European Union is also an intergovernmental party to the 2001 Agreement. However, neither the USA, a leading importer of cocoa, nor Indonesia, a major producer and exporter, are members. The governing body of the ICCO is the International Cocoa Council (ICC), established to supervise implementation of the ICCA. It is planned to relocate the ICCO to Abidjan, Côte d'Ivoire.

A second ICCA operated during 1979–81. It was followed by an extended agreement, which was in force in 1981–87. A fourth ICCA took effect in 1987. (For detailed information on these agreements, see *Africa South of the Sahara 1991*.) During the period of these ICCAs, the effective operation of cocoa price stabilization mechanisms was frequently impeded by a number of factors, principally by crop and stock surpluses, which continued to overshadow the cocoa market in the early 1990s. In addition, the achievement of ICCA objectives was affected by the divergent views of producers and consumers, led by Côte d'Ivoire, on one side, and by the USA, on the other, as to appropriate minimum price levels. Disagreements also developed over the allocation of members' export quotas and the conduct of price support measures by means of the buffer stock (which ceased to operate during 1983–88), and subsequently over the disposal of unspent buffer stock funds. The effectiveness of financial operations under the fourth ICCA was severely curtailed by the accumulation of arrears of individual members' levy payments, notably by Côte d'Ivoire and Brazil. The fourth ICCA was extended for a two-year period from October 1990, although the suspension of the economic clauses relating to price support operations rendered the agreement ineffective in terms of exerting any influence over cocoa market prices.

Preliminary discussions on a fifth ICCA, again held under UN auspices, ended without agreement in May 1992, when consumer members, while agreeing to extend the fourth ICCA for a further year (until October 1993), refused to accept producers' proposals for the creation of an export quota system as a means of stabilizing prices, on the grounds that such arrangements would not impose sufficient limits on total production to restore equilibrium between demand and supply. Additionally, no agreement was reached on the disposition of cocoa buffer stocks, then totalling 240,000 metric tons. In March 1993 ICCO delegates abandoned efforts to formulate arrangements whereby prices would be stabilized by means of a stock-withholding scheme. At a further negotiating conference in July, however, terms were finally agreed for a new ICCA, to take effect from October, subject to its ratification by at least five exporting countries (accounting for at least 80% of total world exports) and by importing countries (representing at least 60% of total imports). Unlike previous commodity agreements sponsored by the UN, the fifth ICCA aimed to achieve stable prices by regulating supplies and promoting consumption, rather than through the operation of buffer stocks and export quotas.

The fifth ICCA, operating until September 1998, entered into effect in February 1994. Under the new agreement, buffer stocks totalling 233,000 metric tons that had accrued from the previous ICCA were to be released on the market at the rate of 51,000 tons annually over a maximum period of four-and-a-half years, beginning in the 1993/94 crop season. At a meeting of the ICCO, held in October 1994, it was agreed that, following the completion of the stocks reduction programme, the extent of stocks held should be limited to the equivalent of three months' consumption. ICCO members also assented to a voluntary reduction in output of 75,000 tons annually, beginning in 1993/94 and terminating in 1998/99. Further measures to achieve a closer balance of production and consumption, under which the level of cocoa stocks would be maintained at 34% of world grindings during the 1996/97 crop year, were introduced by the ICCO in September 1996. The ICCA was subsequently extended until September 2001. In April 2000 the ICCO agreed to implement measures to remedy low levels of world prices (see below), which were to centre on the elimination of sub-grade cocoa in world trade: these cocoas were viewed by the ICCO as partly responsible for the downward trend in prices. In mid-July Côte d'Ivoire, Ghana, Nigeria and Cameroon disclosed that they had agreed to destroy a minimum of 250,000 tons of cocoa at the beginning of the 2000/2001 crop season, with a view to assisting prices to recover and to 'improving the quality of cocoa' entering world markets.

A sixth ICCA was negotiated, under the auspices of the UN, in February 2001. Like its predecessor, the sixth ICCA aims to achieve stable prices through the regulation of supplies and the promotion of consumption. The Agreement took effect on 1 October 2003. In December, in accordance with its provisions, the ICC was to establish a Consultative Board on the World Cocoa Economy, a private-sector board with a mandate to 'contribute to the development of a sustainable cocoa economy; identify threats to supply and demand and propose action to meet the challenges; facilitate the exchange of information on production, consumption and stocks; and advise on other cocoa-related matters within the scope of the Agreement'.

Export Price Index for Cocoa
(base: 1980 = 100)

	Average	Highest month(s)	Lowest month(s)
1990	49		
1995	55		
2000	35	38 (Jan.)	32 (Nov.)
2001	40	48 (Dec.)	36 (Feb., June, July)
2002	66	32 (Oct.)	49 (Jan.)

As the table above indicates, international prices for cocoa have generally been very low in recent years. In 1992 the average of the ICCO's daily prices (based on selected quotations from the London and New York markets) was US $1,099.5 per metric ton (49.9 US cents per lb), its lowest level since 1972. The annual average price per ton subsequently rose steadily, reaching US $1,456 in 1996 and US $1,619 in 1997. The average rose in 1998 to US $1,676 per ton, its highest level since 1987, but slumped in 1999 to US $1,140 (a fall of 32.0%). In 1996 the monthly average ranged from US $1,339 per ton (in March) to US $1,538 (June). In 1997 it varied from US $1,373 per ton (February) to US $1,770 (September). The average increased in May 1998 to US $1,794 per ton (its highest monthly level since February 1988), but fell in December to US $1,515. In 1999 the highest monthly average was US $1,455 per ton in January, and the lowest was US $919 in December. The comparable figure for February 2000 was only US $859 per ton: the lowest monthly average since March 1973.

On the London Commodity Exchange (LCE) the price of cocoa for short-term delivery increased from £637 (US $983) per metric ton in May 1993 to £1,003.5 in November, but it later retreated. In July 1994, following forecasts that the global production deficit would rise, the price reached £1,093.5 (US $1,694) per ton.

In late February 1995 the London cocoa quotation for March delivery stood at £1,056.5 per metric ton, but in March the price was reduced to £938 (US $1,498). The downward trend continued, and in late July the LCE cocoa price was £827.5 (US $1,321) per ton. Prices under short-term contracts remained below £1,000 per ton until the end of December, when the 'spot' quotation (for immediate delivery) stood at £847.5 (US $1,319) per ton.

During the first quarter of 1996 London cocoa prices continued to be depressed, but in April the short-term quotation rose to more than £1,000 per metric ton. In May the LCE 'spot' price reached £1,104.5 (US $1,672) per ton. Cocoa prices had increased in spite of the ICCO's forecast that supply would exceed demand in 1995/96, following four consecutive years of deficits. In July 1996, however, the 'spot' quotation in London declined from £1,049 (US $1,630) per ton to £924 (US $1,438). In December the 'spot' price was reduced to £848.5 (US $1,419) per ton.

In January and February 1997 short-term quotations for cocoa were at similarly low levels, but in March the 'spot' price on the London market rose from £894.5 (US $1,449) per metric ton to £1,012.5 (US $1,621) in less than two weeks. On 1 July the London 'spot' price stood at £1,143 (US $1,895) per ton: its highest level, in terms of sterling, for more than nine years. Three weeks later, however, the price declined to £963 (US $1,615) per ton. By late August international cocoa prices had recovered strongly, in response to fears that crops would suffer storm damage, and in early September the LCE 'spot' quotation reached £1,133.5 (US $1,798) per ton, while prices for longer-term contracts were at their highest for almost a decade. Thereafter, the trend in cocoa prices was generally downward. In the first half of December, however, the London price advanced from £987 (US $1,663) per ton to £1,117 (US $1,824).

In February 1998 the LCE price of cocoa for short-term delivery was reduced to less than £1,000 per metric ton. Following political unrest in Indonesia and forecasts of an increased global supply deficit, the cocoa market rallied in the first half of May, with the London 'spot' price rising from £1,072.5 (US $1,787) per ton to £1,140 (US $1,857). Meanwhile, cocoa under long-term contracts was being traded at more than £1,200 per ton. However, in late June the London price of cocoa for July delivery declined to £1,002.5 per ton. During July the 'spot' price reached £1,070 (US $1,752) per ton, before easing to £1,026.5 (US $1,684). London cocoa prices remained above £1,000 per ton until late September, when the 'spot' quotation fell to £970 (US $1,651). Later in the year there was a steady downward trend, and in late December the price of cocoa was about £860 (US $1,440) per ton.

During the early weeks of 1999 the London cocoa market remained relatively stable, but in March the 'spot' price declined to £803 (US $1,307) per metric ton. The slump later intensified, following forecasts of plentiful crops and a weakening in consumption trends, and by late May the London price of cocoa had fallen to only £602.5 (US $962) per ton. Prices subsequently rallied, and in June, with the EU failing to resolve an impasse over common rules on chocolate products (see above), the quotation for July delivery reached £819 per ton. In July, after the EU agreed to allow chocolate manufacturers to include vegetable fats, the 'spot' price of cocoa eased to £694 (US $1,089) per ton, although it later recovered to £754 (US $1,194). A further decline in cocoa prices ensued, and in September the 'spot' quotation fell to £601.5 (US $975) per ton. After a slight recovery, the downward trend continued. In November the London cocoa price for short-term delivery was reduced to £527.5 per ton. In December the 'spot' quotation reached £570.5 (US $926) per ton, but later in the month the price retreated to £530.5 (US $854): its lowest level, in terms of sterling, since 1992.

Despite the coup in Côte d'Ivoire in December 1999, the cocoa market weakened further during the opening weeks of 2000, and in late February the London price for short-term delivery stood at only £509 per metric ton. Meanwhile, the equivalent New York price of cocoa fell in February to only US $734 per ton: its lowest level for more than 25 years. In March the London 'spot' quotation advanced to £598.5 (US $940) per ton, before easing to £549 (US $874). Comparable prices in May ranged from £575.5 (US $880) to £606.5 (US $911) per ton, and those in July were between £582 (US $881) and £599 (US $907). In August the LCE price of cocoa for short-term delivery fell to £564 per ton. In

September, however, the London 'spot' quotation rallied, ranging between £586 (US $855) and £593 (US $860) in that month. In December a further downward movement occurred, the 'spot' quotation in that month ranging between £556 (US $803) and £534 (US $774). Early in the same month the New York second position 'futures' price declined to US $707 per ton, its lowest level for 27 years.

In January 2001 the London price of cocoa for short-term delivery rose to £833 per metric ton, this upward movement being attributed to a steeper decline in deliveries from Côte d'Ivoire than had earlier been forecast. This recovery was sustained in February, when speculative buying and estimates of potential shortages in supply boosted the short-term price to £945. In March, however, as a result of improved forecasts of production by Côte d'Ivoire, the London 'spot' quotation declined, ranging between £799 (US $1,153) and £917 (US $1,346) per ton. Comparable prices in May ranged from £745 (US $1,068) to £799 (US $1,136). In early May fund and speculative buying, together with renewed pessimism about the level of production in Côte d'Ivoire, boosted the price of the London July 'futures' contract to a three-week high of £815 per ton at one point. In June, however, the London price of cocoa for short-term delivery declined steadily, falling to £676 per ton late in the month. During the first two weeks of July the London 'spot' market quotation ranged between £728 (US $1,024) per ton and £674 (US $947) per ton. In August reports of a poor conclusion to the main crop in Côte d'Ivoire caused the price of the second 'futures' position on the London market to rise above £750 per ton, but this recovery was cancelled out in September by substantial selling of new crop West African cocoa. Late in the month forecasts of reduced production by Côte d'Ivoire in 2001/02 caused the London December 'futures' contract to rise as high as £758 per ton at one point, the rally attributed to short-covering by commodity funds. In early October the price of the London second position 'futures' contract rose to £838 per ton, the highest price since mid-May. This strengthening of the second position contract was, once again, attributed to fund buying. London 'futures' prices increased rapidly from mid-November, in response to renewed pessimism with regard to production by Côte d'Ivoire, and forecasts of a deficit in the season's cocoa crop of some 200,000 tons. In the third week of November London 'futures' prices reached their highest levels for three years. The closing price of the London March contract on 16 November was £952 per ton, representing an increase of some £200 per ton over the preceding 14 days.

The upward trend in prices continued during 2002, amid further reports of poor weather in West Africa and political unrest in the principal cocoa-growing areas of Côte d'Ivoire. The maintenance by one trading company of an unusually 'long' position in cocoa, thereby causing scarcity elsewhere in the market, exacerbated this trend throughout the first half of 2002. In March the price of the London second position contract reached £1,261 per metric ton, the highest price for a second position contract since September 1987. By the end of July 2002 the London 'futures' price had risen to £1,317. At the end of 2002 the price of the London March contract was more than £1,600 per ton. Prices subsequently fell back, however, as a result of confidence that any disruptions to supply arising from the situation in Côte d'Ivoire would be manageable.

In late January 2003, according to the ICCO, the price of cocoa was propelled upwards as a consequence of intensified civil conflict in Côte d'Ivoire. By mid-February the price of cocoa had risen to £1,494 per metric ton in London, and to US $2,357 per ton in New York. The subsequent perception that the civil conflict in Côte d'Ivoire was abating caused terminal prices to weaken in both London and New York towards the end of February, however. In May 2003, according to the ICCO, cocoa terminal prices fell persistently: by the end of the month London futures prices had declined below £1,000 per ton, while New York futures prices were less than US $1,500 per ton. From mid-June prices began to recover in response to forecasts of lower cocoa production in West Africa in 2003/04, in combination with indications that consumption in Europe was rising. In early August, however, cocoa futures prices fell to their lowest levels for the year so far, falling to £943 per ton in London and to US $1,418 per ton in New York. In the second half of August, however, concern arose over the possible impact of drought on West African crops in 2003/04 and over the effect of dry weather and pests on the crop in Sulawesi, the main producing area in Indonesia. By the end of August cocoa futures prices had risen to £1,177 per ton in London and to US $1,742 per ton in New York. During the first three weeks of September markets were subjected to correction, however, as, among other things, commodity funds liquidated their speculative positions. Falls in the price of cocoa were succeeded, in the final week of September, by further upward movement that was sparked by a deterioration in the political situation in Côte d'Ivoire.

COPAL, with headquarters in Lagos, Nigeria, had 10 members in 2003, including Malaysia. COPAL was formed in 1962 with the aim of preventing excessive price fluctuations by regulating the supply of cocoa. In 2002 members of COPAL accounted for about 76% of world cocoa production. COPAL has acted in concert with successive ICCAs.

The principal centres for cocoa-trading in the industrialized countries are the London Cocoa Terminal Market, in the United Kingdom, and the New York Coffee, Sugar and Cocoa Exchange, in the USA.

Coconut (*Cocos nucifera*)

The coconut palm is a tropical tree, up to 25 m tall, with a slender trunk surmounted by a feathery crown of leaves. The geographical origins of the tree are thought to be in the Asia-Pacific region. Its presence in most coastal areas and on many islands in the tropics is largely due to man, who introduced it to West Africa and the Americas. The tree's fruits first appear after about six years, though the palm may not reach full bearing until it is about 20 years old. It may continue fruiting for a further 60 years. (Hybrid varieties have advanced the time of initial fruiting from the sixth to the fourth year, and the onset of full bearing from the 20th to the 10th year.) The fruits, green at first but turning yellow as they ripen, are often left to fall naturally, but, as many are then over-ripe, harvesting by hand is widely practised.

Coconut, the most important of all cultivated palms, is frequently a smallholder crop, found mainly in small plots around houses and in gardens, although in the Philippines the average coconut farm covers 4–7 ha in area. The plant's fruit, fronds and wood provide many thousands of families with a cash income as well as basic necessities such as food, drink, fuel and shelter. The palms grow with little or no attention where conditions are favourable. More than 80 varieties are known, divided broadly into tall palms, produced by cross-pollination, and dwarf palms, which are self-pollinating. The sap of the coconut palm itself can be evaporated to produce sugar or fermented to make an alcoholic drink known as 'toddy'. This may be distilled to produce a spirit called 'arrack'.

Production of Coconuts
('000 metric tons)

	2002	2003*
World total	53,313	52,940
Far East	43,028	42,605
Oceania	1,715	1,770
Leading regional producers		
Indonesia‡	16,086	15,630
Malaysia†	738	740
Papua New Guinea	513‡	570†
Philippines	13,683	13,700†
Thailand	1,418	1,420†
Vanuatu	207‡	207†
Viet Nam	915	920
Other leading producers		
Brazil	2,892	2,834
India†	9,500	9,500
Mexico†	959	959
Mozambique†	265	265
Sri Lanka	1,818	1,850†
Tanzania†	370	370

* Provisional figures.
† FAO estimate(s).
‡ Unofficial figure(s).

Coconut oil is a rich source of medium-chain triglycerides (MCT), whose applications in medical nutrition include infant milk formulas and foods for persons unable to digest and assimilate fats. Its other food applications include use as a flavouring and also as an ingredient to prolong the shelf life of certain food products.

All parts of the fruit have their uses. Beneath the outer skin is a thick layer of fibrous husk. The fibres can be combed out to produce coir (from the Malay word *kayar*, which means 'cord'), a material used for making ropes, coconut matting, brushes, mattresses and upholstery (see below). Inside the husk is the nut—what people in temperate areas think of as a 'coconut' since the whole fruit is not usually imported. The nut has a hard shell, inside which is a thin white fleshy layer of edible 'meat'. The nut's hollow interior is partially filled with a liquid called 'coconut water' which is gradually absorbed as the fruit ripens. This 'water' is a refreshing and nutritious drink when taken from a young nut (7–8 months), while that from more mature nuts can be prepared as a soft drink and is also used in the production of yeast, alcohol, wine and vinegar. The so-called 'coconut milk' is the white, creamy extract obtained after pressing freshly grated coconut 'meat'. Coconut flour, a by-product of 'coconut milk', is a useful nutritional source of dietary fibre. The shells are mainly utilized as fuel, but small quantities are used to make containers, ornaments, ladles and buttons, and pulverized shells can be used as filler in plastics moulding, plywood and mosquito coil repellants. Raw coconut shell is a more efficient fuel after it has been carbonized into charcoal and, on further processing, can be converted into the still more efficient activated carbon, which finds a market in highly industrialized countries concerned with pollution control.

After harvesting, the fruits are split open, the husk removed and the nuts usually broken open. The 'meat' is sometimes eaten directly or used to prepare desiccated coconut, widely used in the bakery and confectionery trades. However, by far the most important economic product of the plant is obtained by drying the 'meat' into copra, either in the sun or in a kiln which may be heated by burning the coconut shells. The dried copra is the source of coconut oil, used mainly in the manufacture of soap, detergent and cosmetics, and also as a cooking oil and in margarine production. As technology advances, more uses for coconut oil are being developed. Recent experiments have shown that it can be converted into diesel fuel, and a programme of conversion might alleviate the financial burden on countries that are heavily dependent on petroleum imports. The residue left after the extraction of oil from copra is a valuable oilcake for feeding livestock, particularly dairy cattle.

Production of Copra
('000 metric tons)

	2002	2003*
World total	5,008	5,361
Far East	3,597	3,950
Oceania	183	187
Leading regional producers		
Indonesia	1,216	1,272†
Malaysia	58†	64‡
Papua New Guinea†	81	85
Philippines	2,010	2,300†
Thailand‡	66	66
Vanuatu	23	23‡
Viet Nam	234†	234‡
Other leading producers		
India†	700	725
Mexico	203	185
Mozambique‡	45	45
Sri Lanka	59	58‡

* Provisional figures.
† Unofficial figure(s).
‡ FAO estimate(s).

Good copra has an oil content of about 64%. Most extraction is done in the coconut-growing countries, although there is a substantial trade in copra to countries that extract the oil themselves. Although the largest producer of copra, the Philippines was overtaken as the largest exporter in 1992 by Papua New Guinea, whose copra sales represented 28% of the world export trade in 2001. The eclipse of the Philippines as a copra exporter was attributable to increased levels of crushings of its copra into crude coconut oil for export. In 2002 Papua New Guinea was overtaken in turn as the largest exporter of copra by Indonesia, whose sales in that year accounted for about one-third of the world export trade. Japan was formerly the main importer of copra, but it has been overtaken by Germany, which in the mid-1990s accounted, on average, for about 20%–25% of all imports. That share has declined in more recent years, however. In 2001, albeit exceptionally, Japanese imports again exceeded those of Germany. Germany regained primacy in 2002, but accounted for only 11% of copra imports world-wide in that year. Germany's partners within the European Union (EU) also constitute a major market for copra, coconut oil, desiccated coconut and copra meal.

Coconut oil has encountered competition from the development of more productive annual oilseed crops, such as soybeans and rapeseed in the northern hemisphere, and from oil palm in the tropics. Up to 7 metric tons of oil per ha can be produced from oil palm, compared with a maximum of 3.25 tons from coconuts. The necessity for the production of copra prior to the extraction of oil has further eroded the competitiveness of coconuts in the world market for vegetable oils. In 2002 the Philippines accounted for about 53% of world exports of coconut oil. The USA is the main importer, absorbing, on average, slightly less than one-quarter of total world imports in recent years. In 2003, according to provisional figures from FAO, world production of coconut oil was 3,378,818 tons, including 2,488,754 tons from Far East Asia and 93,281 tons from Oceania. The major producing countries in that year were the Philippines (1,486,308 tons) and Indonesia (750,000 tons).

The Philippines is the world's largest exporter of coconut products, and about one-third of its population is directly or indirectly dependent on the coconut sector for a livelihood. In 1979 the Philippine Government agreed to the operation of a monopoly of the coconut industry by a group of private producers who controlled 80% of the country's coconut-milling capacity, and accounted for more than 50% of coconut exports. The monopoly was dismantled in 1985, and the export of coconut products was opened to private enterprises. At that time, 10 coconut trading co-operatives were established, with a common majority shareholder, the government-controlled United Coconut Planters Bank. This has, however, given rise to criticism that the abolition of the former monopoly was more apparent than real, and coconut farmers have been reluctant to conduct business with the new trading firms. Government financial support for coconut-replanting ceased in 1986. In 1999 more than 3m. ha in the Philippines was planted with coconuts.

International promotion of biodegradable products during the late 1990s generated a revival of demand for coir products, of which India is the largest producer and exporter, principally to markets in the EU and the USA, which in 2002 accounted for 86% of world imports of mats, matting and rugs, etc., made from coir. Indonesia and the Philippines, each of whose coconut output exceeds that of India, both lack the facilities to carry out coir production.

The Asian and Pacific Coconut Community, with headquarters in Jakarta, Indonesia, was established in 1969. Its 15 members account for more than 90% of world production of coconuts.

The import price of Philippine copra in Europe has varied widely in recent years. In 1997 the price at European ports averaged US $433.75 per metric ton, but this declined to US $411.03 in 1998. In 1999 the price averaged US $462.27 per ton, but this declined steeply in 2000, to only US $308.92 per ton. In 2001 the average price fell sharply again, to only US $195.55 per ton. During the first quarter of 2002 the average import price of Philippine copra at European ports recovered to US $228.29 per ton. The average price rose further, to US $245.23 per ton, in April, and again, to US $263.26 per ton, in May.

The import price of Philippine coconut oil has also fluctuated considerably. The quotation (c.i.f. Rotterdam) rose from US $720 per metric ton in September 1996 to US $810 in December. It declined to US $550 per ton in August 1997, but recovered to US $670 in September. The price of coconut oil was reduced to US $510 per ton in January 1998, but rose to US $755 in May. It subsequently eased, but advanced to US $777.5 per ton by the end of the year. In May 1999, with coconut oil supplies very scarce, the price increased to US $945 per ton, its highest level since December 1984. The surge in prices was short-lived, and in July 1999 the Rotterdam quotation fell to US $625 per ton. The price of coconut oil recovered in October to US $780 per ton and ended the year at US $700. The market for coconut oil was depressed in the first half of 2000, with plentiful supplies of the commodity (and of palm oil) available. In July coconut oil was traded in Europe at only US $380 per ton. The average import price of coconut oil of Philippine origin (c.i.f. Rotterdam) was US $450.3 per ton in 2000, and fell further in 2001, to only US $318.1 per ton. In 2002, however, a substantial recovery occurred and an average price of US $421 per ton was recorded. In December 2003 the import price of coconut oil rose as high as US $583 per ton, and for the whole of 2003 an average price of US $467.3 per ton was recorded. In April 2004 the Rotterdam quotation rose to US $736 per ton. According to FAO, prices in the oilcrop complex (in which copra is included) were influenced in the October 2003–September 2004 marketing year by very tight supplies of soybeans and by slower growth in output of palm oil.

Export Price Index for Copra
(base: 1980 = 100)

	Average	Highest month(s)	Lowest month(s)
1990	51		
1995	97		
2000	68	93 (Jan.)	47 (Oct.)
2001	45	52 (Aug.)	40 (March)
2002	59	67 (Dec.)	49 (Jan.)

Export Price Index for Coconut Oil
(base: 1980 = 100)

	Average	Highest month(s)	Lowest month(s)
1990	48		
1995	96		
2000	64	94 (Jan.)	47 (Dec.)
2001	46	52 (Aug.)	41 (Feb., March)
2002	60	69 (Dec.)	52 (Jan., March)

Coffee (*Coffea*)

This is an evergreen shrub or small tree, generally 5 m–10 m in height, indigenous to Asia and tropical Africa. Wild trees grow to 10 m, but cultivated shrubs are usually pruned to a maximum of 3 m. The dried seeds (beans) are roasted, ground and brewed in hot water to provide the most popular of the world's non-alcoholic beverages. Coffee is drunk in every country in the world, and its consumers comprise an estimated one-third of the world's population. Although it has little nutrient value, coffee acts as a mild stimulant, owing to the presence of caffeine, an alkaloid also present in tea and cocoa.

There are about 40 species of *Coffea*, most of which grow wild in the eastern hemisphere. The species of economic importance are *C. arabica* (native to Ethiopia), which, in the early 2000s, accounted for about 60%–65% of world production, and *C. canephora* (the source of robusta coffee), which accounted for almost all of the remainder. Arabica coffee is more aromatic but robusta, as the name implies, is a stronger plant. Coffee grows in the tropical belt, between 20°N and 20°S, and from sea-level to as high as 2,000 m above. The optimum growing conditions are found at 1,250 m–1,500 m above sea-level, with an average temperature of around 17°C and an average annual rainfall of 1,000 mm–1,750 mm. Trees begin bearing fruit three to five years after planting, depending upon the variety, and give their maximum yield (up to 5 kg of fruit per year) from the sixth to the 15th year. Few shrubs remain profitable beyond 30 years.

Arabica coffee trees are grown mostly in the American tropics and supply the largest quantity and the best quality of coffee beans. In Africa and Asia arabica coffee is vulnerable in lowland areas to a serious leaf disease and consequently cultivation has been concentrated on highland areas. Some highland arabicas, such as those grown in Kenya, have a high reputation for quality.

The robusta coffee tree, grown mainly in east and west Africa, and in the Far East, has larger leaves than arabica, but the beans are generally smaller and of lower quality and, consequently, fetch a lower price. However, robusta coffee has a higher yield than arabica as the trees are more resistant to disease. Robusta is also more suitable for the production of soluble ('instant') coffee. About 60% of African coffee is of the robusta variety. Soluble coffee accounts for more than one-fifth of world coffee consumption.

Each coffee berry, green at first but red when ripe, usually contains two beans (white in arabica, light brown in robusta) which are the commercial product of the plant. To produce the best quality arabica beans—known in the trade as 'mild' coffee—the berries are opened by a pulping machine and the beans fermented briefly in water before being dried and hulled into green coffee. Much of the crop is exported in green form. Robusta beans are generally prepared by dry-hulling. Roasting and grinding are usually undertaken in the importing countries, for economic reasons and because roasted beans rapidly lose their freshness when exposed to air.

Apart from beans, coffee produces a few minor by-products. When the coffee beans have been removed from the fruit, what remains is a wet mass of pulp and, at a later stage, the dry material of the 'hull' or fibrous sleeve that protects the beans. Coffee pulp is used as cattle feed, the fermented pulp makes a good fertilizer and coffee bean oil is an ingredient in soaps, paints and polishes.

Production of Green Coffee Beans
('000 bags, each of 60 kg, coffee years, ICO members only)

	2002/03	2003/04
World total	120,925	101,324
Far East	19,818	18,692
Oceania	1,108	1,200
Leading regional producers		
Indonesia	6,785	6,012
Papua New Guinea	1,108	1,200
Philippines	721	433*
Thailand	757	997
Viet Nam	11,555	11,250
Other leading producers		
Brazil	48,480	28,820
Cameroon	801	1,150*
Colombia	11,889	11,750*
Costa Rica	1,936	2,120
Côte d'Ivoire	2,483	2,325*
Ecuador	732	804
El Salvador	1,442	1,252*
Ethiopia	3,693	4,333
Guatemala	4,070	3,000*
India	4,565	4,508
Honduras	2,497	2,913*
Mexico	4,000	4,550*
Nicaragua	1,199	1,150*
Peru	2,900	2,525
Uganda	2,910	3,100*

* Estimate.

Source: International Coffee Organization.

More than one-half of the world's coffee is produced on small-holdings of less than 5 ha. In many producing countries, and especially in Africa, coffee is almost entirely an export crop, with little domestic consumption. Green coffee accounts for some 96% of all the coffee that is exported, with soluble and roasted coffee comprising the balance. Tariffs on green/raw coffee are usually low or non-existent, but those applied to soluble coffee may be as high as 30%. The USA is the largest single importer, although its volume of coffee purchases was overtaken in 1975 by the combined imports of the (then) nine countries of the European Community (EC, now the European Union—EU).

After petroleum, coffee is the major raw material in world trade, and the single most valuable agricultural export of the tropics. Latin America (with 60.4% of estimated world output, excluding minor producers, in 2003/04) is the leading coffee-growing region. Africa, which formerly ranked second, was overtaken in 1992/93 by Asian producers. In 2003/04 Asian countries accounted for 24.2% of the estimated world coffee crop, compared with 15.4% for African countries.

In the Far East and Australasia Viet Nam has emerged, in a very short period of time, as a major producer and exporter of coffee. In 1980 coffee was grown on only 20,000 ha, but by 1998 the area cultivated had been increased to 300,000 ha. (Over the same period the yield obtained per hectare rose from 0.6 to 1.6 metric tons.) The increase in the area cultivated has taken place in the context of a programme of economic reforms that, with regard to the agricultural sector, has emphasized the expansion of cash crop production. Viet Nam planned to increase the area sown to coffee by a further 100,000 ha by 2005 and was proceeding, in the early 2000s, with a project to bring 40,000 ha in the north of the country into the production of arabica coffee. However, it is as an exporter of robusta that Viet Nam has become most significant to world markets. In 1998/99 the country was the largest exporter of robusta in the world, with shipments (mainly to the USA and the EU) of some 400,000 tons, equivalent to 7.7% of the total quantity of coffee exported world-wide. In that year Viet Nam was the third largest exporter of all coffees, after Brazil and Colombia. Exports increased in 1999/2000 to about 460,000 tons, equivalent to some 9% of all coffee exported world-wide, and, again, in 2000/2001, when, at almost 700,000 tons, they represented 13% of global coffee exports. In 2001 exports of coffee contributed 2.6% of Viet Nam's estimated total export earnings. In the 1999/2000 crop year, with production of 11.6m. bags, Viet Nam overtook Indonesia as the world's largest producer of robusta, and Colombia as the world's second largest producer of coffee overall, and maintained those ranks in the two subsequent crop years, although total production declined by 11.5%, to 13.1m. bags, in 2001/02. In 2002/03 Viet Nam's production of coffee was reduced by 11.5%, to 11.6m. bags. In that year Viet Nam's exports totalled 11.8m. bags, or some 700,000 tons, and the country thus remained the second largest exporter of all coffees after Brazil. In 2003/04 Vietnamese output of coffee was estimated to have declined by about 2.6% to some 11.3m. bags. The country's exports in that crop year were estimated at about 11.6m. bags, or almost 700,000 tons, equivalent to about 13.7% of all coffee exported world-wide and second in quantity only to those of Brazil. Viet Nam has been accused of contributing, by rapidly increasing its output, to the glut of supplies available for export that has plunged coffee production into crisis world-wide (see below). In mid-2002, as part of global attempts to resolve that crisis, the Vietnamese Government was reportedly pursuing an initiative to reduce the area sown to coffee, with the aim of lowering output to about 10m. bags by around 2005.

Until the 1999/2000 crop year, when it was overtaken by Viet Nam, Indonesia was the largest producer of coffee in the Far East and Australasia. Production, mainly of robusta varieties, had by the late 1990s risen approximately threefold since the late 1960s, although it continued to be based on smallholdings rather than large estates. Until 1999/2000 Indonesia was the world's largest producer of robusta. In that crop year output of all coffees (Indonesia also produces high quality arabicas) amounted to about 345,000 metric tons, a decline of about 29% compared with 1998/99. In 2000/01 production recovered to about 420,000 tons, but fell to about 410,000 tons in 2001/02. Output declined slightly, to about 407,000 tons, in 2002/03, and was estimated to have fallen more substantially, to about 361,000 tons, in 2003/04. In 1998/99 Indonesian exports of coffee totalled about 336,000 tons. In 1999/2000 they declined to some 306,000 tons, but rose to 321,000 tons in 2000/01. In the 2001/02 crop year Indonesia's exports of coffee declined further, to some 315,000 tons, and fell substantially, to only some 257,000 tons in 2002/03. In 2003/04 Indonesia's exports of coffee were estimated at approximately the same level as in the previous crop year. The importance of coffee in Indonesia's total foreign trade has declined as that of petroleum and manufactured goods has risen. In 2001 exports of coffee contributed less than 1% of the country's total export earnings.

Papua New Guinea ranks as the fourth largest producer of coffee in the Far East and Australasia. In the 2003/04 crop year production was estimated at about 72,000 metric tons, compared with some 66,000 tons in 2002/03. In the late 1990s arabicas accounted for about 97% of all coffee cultivated. Production, initially based on large estates, but subsequently dominated by smallholdings, began to expand in the early 1950s. In the mid-1990s Germany was the principal export market for Papua New Guinea's coffee, followed by Australia, the United Kingdom and the USA. Papua New Guinea exported some 78,000 tons of coffee in 1998/99, and about the same quantity again in 1999/2000. In 2000/01 exports declined to about 63,000 tons, and, at 65,000 tons, remained at approximately the same level in the 2001/02 crop year. In the 2002/03 crop year Papua New Guinea's exports of coffee again totalled some 63,000. The country's foreign sales of coffee were estimated to have risen to about 69,000 tons in 2003/04. In 2003 exports of coffee accounted for 3.8% of Papua New Guinea's total export earnings. In the 2000/01 and 2001/02 crop years the Philippines produced about 46,000 tons of coffee. (Production has declined somewhat since the late 1990s. In 1997/98, for instance, output totalled 54,000 tons.) In 2002/03 output declined to about 43,000 tons, and was estimated to have fallen very substantially, by almost 40%, to about 26,000 tons in the 2003/04 crop year. Robusta accounts for about 90% of all coffee produced in the Philippines. Output was adversely affected by drought in the 1998/99 crop year. The Philippines had already, in the previous crop year, become a net importer of coffee, and imports rose substantially in 1998/99.

Effective international attempts to stabilize coffee prices began in 1954, when a number of producing countries made a short-term agreement to fix export quotas. After three such agreements, a five-year International Coffee Agreement (ICA), covering both producers and consumers, and introducing a quota system, was signed in 1962. This led to the establishment, in 1963, of the International Coffee Organization (ICO), with its headquarters in London. In September 2004 the International Coffee Council, the highest authority of the ICO, comprised 72 members (i.e. participants in the 2001 ICA—43 exporting countries and 29 importing countries). In that month it was announced that the USA, which had withdrawn from the ICO in 1993, had begun the formal process of acceding to the 2001 ICA. Successive ICAs took effect in 1968, 1976, 1983, 1994 and 2001 (see below), but the system of export quotas to stabilize prices was abandoned in July 1989. During each ICA up to and including that implemented in 1994, contention arose over the allocation of members' export quotas, the operation of price support mechanisms, and, most importantly, illicit sales by some members of surplus stocks to non-members of the ICO (notably to the USSR and to countries in Eastern Europe and the Middle East). These 'leaks' of low-price coffee, often at less than one-half of the official ICA rate, also found their way to consumer members of the ICO through free ports, depressing the general market price and making it more difficult for exporters to fulfil their quotas.

The issue of coffee export quotas became further complicated in the 1980s, as consumer tastes in the main importing market, the USA, and, to a lesser extent, in the EC moved away from the robustas exported by Brazil and the main African producers in favour of the milder arabica coffees grown in Central America. Disagreements over a new system of quota allocations, taking account of coffee by variety, had the effect of undermining efforts in 1989 to preserve the economic provisions of the ICA, pending the negotiation of a new agreement. The ensuing deadlock between consumers and producers, as well as among the producers themselves, led in July to the collapse of the quota system and the suspension of the economic provisions of the ICA. The administrative clauses of the agreement, however, continued to operate and were subsequently extended until October 1993, pending an eventual settlement of the quota issue and the entry into force of a successor ICA.

With the abandonment of the ICA quotas, coffee prices fell sharply in world markets, and were further depressed by a substantial accumulation of coffee stocks held by consumers. The response by some Latin American producers was to seek to revive prices by imposing temporary suspensions of exports; this strategy, however, merely increased losses of coffee revenue. By early 1992 there had been general agreement among the ICO exporting members that the export quota mechanism should be revived. However, disagreements persisted over the allocation of quotas, and in April 1993 it was announced that efforts to achieve a new ICA with economic provisions had collapsed. In the following month Brazil and Colombia, the two largest coffee producers at that time, were joined by some Central American producers in a scheme to limit their coffee production and exports in the 1993/94 coffee year. Although world consumption of coffee exceeded the level of shipments, prices were severely depressed by surpluses of coffee stocks totalling 62m. bags (each of 60 kg), with an additional 21m. bags held in reserve by consumer countries. Prices, in real terms, stood at historic 'lows'.

In September 1993 the Latin American producers announced the formation of an Association of Coffee Producing Countries (ACPC) to implement an export-withholding, or coffee retention, plan. The Inter-African Coffee Organization (IACO, see below), whose membership includes Côte d'Ivoire, Kenya and Uganda, agreed to join the Latin American producers in a new plan to withhold 20% of output whenever market prices fell below an agreed limit. With the participation of Asian producers, a 28-member ACPC was formally established. (Angola and Zaire, now the Democratic Republic of the Congo, were subsequently admitted to membership.) With headquarters in London, its signatory member countries numbered 28 in 2001, 14 of which were ratified. Production by the 14 ratified members in 1999/2000 accounted for 61.4% of coffee output world-wide.

The ACPC coffee retention plan came into operation in October 1993 and gradually generated improved prices; by April 1994 market quotations for all grades and origins of coffee had achieved their highest levels since 1989. In June and July 1994 coffee prices escalated sharply, following reports that as much as 50% of the 1995/96 Brazilian crop had been damaged by frosts. In July 1994 both Brazil and Colombia announced a temporary suspension of coffee exports. The onset of drought following the Brazilian frosts further affected prospects for its 1994/95 harvest, and ensured the maintenance of a firm tone in world coffee prices during the remainder of 1994.

The intervention of speculative activity in the coffee 'futures' market during early 1995 led to a series of price falls, despite expectations that coffee consumption in 1995/96, at a forecast 93.4m. bags, would exceed production by about 1m. bags. In an attempt to restore prices, the ACPC announced in March 1995 that it was to modify the price ranges of the export withholding scheme. In May the Brazilian authorities, holding coffee stocks of about 14.7m. bags, introduced new arrangements under which these stocks would be released for export only when the 20-day moving average of the ICO arabica coffee indicator rose to about US $1.90 per lb. Prices, however, continued to decline, and in July Brazil joined Colombia, Costa Rica, El Salvador and Honduras in imposing a reduction of 16% in coffee exports for a one-year period. Later in the same month the ACPC collectively agreed to limit coffee shipments to 60.4m. bags from July 1995 to June 1996. This withholding measure provided for a decrease of about 6m. bags in international coffee exports during this period. In July 1997 the ACPC announced that the export withholding programme was to be replaced by arrangements for the restriction of exports of green coffee. Total exports for 1997/98 were to be confined to 52.75m. bags. Following the withdrawal, in September 1998, of Ecuador from the export restriction scheme (and subsequently from the ACPC) and the accession of India to membership in September 1999, there were 14 ratified member countries participating in the withholding arrangements. The continuing decline in world coffee prices (see below) prompted the ACPC to announce in February 2000 that it was considering the implementation of a further scheme involving the withholding of export supplies. In the following month the members indicated their intention to withdraw supplies of low-grade beans (representing about 10% of annual world exports), and on 19 May they announced arrangements under which 20% of world exports would be withheld until the ICO 15-day composite price reached 95 US cents per lb (at that time the composite price stood at 69 US cents per lb). Retained stocks would only be released when the same indicator price reached 105 US cents per lb. Five non-member countries, Guatemala, Honduras, Mexico, Nicaragua and Viet Nam, also signed a so-called London Agreement, pledging to support the retention plan. Implementation of the plan, which had a duration of up to two years, was initiated by Brazil in June, with Colombia following in September. In December 2000 the ACPC identified a delay in the full implementation of the retention plan as one of the factors that had caused the average ICO composite indicator price in November to fall to its lowest level since April 1993, and the ICO robusta indicator price to its lowest level since August 1969. In May 2001 the ACPC reported that exchange prices continued to trade at historical 'lows'. Their failure to recover, despite the implementation of the retention plan, was partly attributed to the hedging of a proportion of the 7m. bags of green coffee retained by that time. On the physical market, meanwhile, crop problems and the implementation of the retention plan had significantly increased differentials for good quality coffees, in particular those of Central America. In April 2001 the ICO daily composite indicator price averaged 47.13 US cents per lb (compared with an average of 64.24 US cents per lb for the whole of 2000, the lowest annual average since 1973), the lowest monthly average since September 1992. In October 2001 the ACPC announced that it would dissolve itself in January 2002. The Association's relevance had been increasingly compromised by the failure of some of its members to comply with the retention plan in operation at that time, and by some members' inability to pay operating contributions to the group owing to the depressed state of the world market for coffee. Upon its dissolution, the group's members announced that they would consider establishing a successor organization if prices returned to a level permitting this.

In June 1993 the members of the ICO agreed to a further extension of the ICA, to September 1994. However, the influence of the ICO, from which the USA withdrew in October 1993, was increasingly perceived as having been eclipsed by the ACPC. In 1994 the ICO agreed provisions for a new ICA, again with primarily consultative and administrative functions, to operate for a five-year period, until September 1999. In November of that year it was agreed to extend this limited ICA until September 2001. A successor ICA took effect, provisionally, in October 2001. By late May 2002 the new ICA had been endorsed by 37 members of the International Coffee Council (25 exporting members and 12 importing members). Among the principal objectives of the ICA of 2001 were the promotion of international co-operation with regard to coffee, and the provision of a forum for consultations, both intergovernmental and with the private sector, with the aim of achieving a reasonable balance between world supply and demand in order to guarantee adequate supplies of coffee at fair prices for consumers, and markets for coffee at remunerative prices for producers.

In February 1995 five African Producers (Burundi, Kenya, Rwanda, Tanzania and Uganda) agreed to participate in coffee price guarantee contract arrangements sponsored by the Eastern and Southern Africa Trade and Development Bank under the auspices of the Common Market for Eastern and Southern Africa (COMESA). This plan seeks to promote producer price guarantees in place of stock retention schemes. The contract guarantee arrangements would indemnify producers against reductions below an agreed contract price.

Export Price Index for Coffee
(base: 1980 = 100)

	Average	Highest month(s)	Lowest month(s)
1990	46		
1995	83		
2000	44	55 (Jan.)	33 (Dec.)
2001	30	34 (Feb.)	25 (Oct.)
2002	28	32 (Nov.)	25 (Aug.)

International prices for coffee beans in the early 1990s were generally at very low levels, even in nominal terms (i.e. without taking inflation into account). On the London Commodity Exchange (LCE) the price of raw robusta coffee for short-term delivery fell in May 1992 to US $652.5 (£365) per metric ton, its lowest level, in terms of dollars, for more than 22 years. By December the London coffee price had recovered to US $1,057.5 per ton (for delivery in January 1993). The LCE quotation eased to US $837 (£542) per ton in January 1993, and remained within this range until August, when a sharp increase began. The coffee price advanced in September to US $1,371 (£885) per ton, its highest level for the year. In April 1994 a further surge in prices began, and in May coffee was traded in London at more than US $2,000 per ton for the first time since 1989. In late June 1994 there were reports from Brazil that frost had damaged the potential coffee harvest for future seasons, and the LCE quotation exceeded US $3,000 per ton. In July, after further reports of frost damage to Brazilian coffee plantations, the London price reached US $3,975 (£2,538) per ton. Market conditions then eased, but in September, as a drought persisted in Brazil, the LCE price of coffee increased to US $4,262.5 (£2,708) per ton: its highest level since January 1986. In December 1994, following forecasts of a rise in coffee production and a fall in consumption, the London quotation for January 1995 delivery stood at US $2,481.5 per ton.

The coffee market later revived, and in March 1995 the LCE price reached US $3,340 (£2,112) per metric ton. However, in early July coffee traded in London at US $2,400 (£1,501) per ton, although later in the month, after producing countries had announced plans to limit exports, the price rose to US $2,932.5 (£1,837). During September the LCE 'spot' quotation (for immediate delivery) was reduced from US $2,749 (£1,770) per ton to US $2,227.5 (£1,441), but in November it advanced from US $2,370 (£1,501) to US $2,739.5 (£1,786). Coffee for short-term delivery was traded in December at less than US $2,000 per ton, while longer-term quotations were considerably lower.

In early January 1996 the 'spot' price of coffee in London stood at US $1,798 (£1,159) per metric ton, but later in the month it reached US $2,050 (£1,360). The corresponding quotation rose to US $2,146.5 (£1,401) per ton in March, but declined to US $1,844.5 (£1,220) in May. The 'spot' contract in July opened at US $1,730.5 (£1,112) per ton, but within four weeks the price fell to US $1,487 (£956), with the easing of concern about a threat of frost damage to Brazilian coffee plantations. In November the 'spot' quotation rose to US $1,571 (£934) per ton, but it slumped to US $1,375.5 (£819) within a week. By the end of the year the London price of coffee (for delivery in January 1997) had been reduced to US $1,259 per ton.

In early January 1997 the 'spot' price for robusta coffee stood at only US $1,237 (£734) per metric ton, but later in the month it reached US $1,597.5 (£981). The advance in the coffee market continued in February, but in March the price per ton was reduced from US $1,780 (£1,109) to US $1,547.5 (£960) within two weeks. In May coffee prices rose spectacularly, in response to concerns about the scarcity of supplies and fears of frost in Brazil. The London 'spot' quotation increased from US $1,595 (£986) per ton to US $2,502.5 (£1,526) by the end of the month. Meanwhile, on the New York market the price of arabica coffee for short-term delivery exceeded US $3 per lb for the first time since 1977. However, the rally was short-lived, and in July 1997 the London price for robusta coffee declined to US $1,490 (£889) per ton. In the first half of November the coffee price rose from US $1,445 (£862) per ton to US $1,658 (£972). During December the price for January 1998 delivery reached US $1,841 per ton, but a week later it decreased to US $1,657.

The coffee market rallied in January 1998, with the London 'spot' quotation rising from US $1,746.5 (£1,066) per metric ton to US $1,841 (£1,124). Coffee prices for the corresponding contract in March ranged from US $1,609 (£977) per ton to US $1,787 (£1,065). Following reports of declines in the volume of coffee exports by producing countries (owing to inadequate rainfall), the upward trend in prices continued in April, with the price of robusta for short-term delivery reaching US $1,992 per ton. In the first half of May there was another surge in prices (partly as a result of political unrest in Indonesia, the main coffee-producing country in Asia at that time), with the London quotation rising from US $1,881.5 (£1,129) per ton to US $2,202.5 (£1,351). Later in the month, however, the price was reduced to US $1,882.5 (£1,155) per ton. Coffee prices subsequently fell further, and in late July the London 'spot' contract stood at only US $1,505.5 (£909) per ton, before recovering to US $1,580 (£963). The quotation per ton for September delivery reached US $1,699.5 at the beginning of August, having risen by US $162 in a week. In September the 'spot' price advanced from US $1,640 (£974) per ton to US $1,765 (£1,036) a week later. In late October a further sharp rise in coffee prices began, following storm damage in Central America, and in November the London 'spot' quotation for robusta increased from US $1,872.5 (£1,123) per ton to US $2,142.5 (£1,278). Meanwhile, trading in other contracts continued at less than US $1,800 per ton until December, when the London price of coffee (for delivery in January 1999) rose to US $1,977.

Coffee prices retreated in January 1999, with the London 'spot' quotation falling from US $1,872.5 (£1,131) per metric ton to US $1,639 (£995). During March the price was reduced from US $1,795.5 (£1,111) per ton to US $1,692.5 (£1,030), but it recovered to US $1,795 (£1,112) within a week. As before, the market for longer-term deliveries was considerably more subdued, with coffee trading mainly within a range of US $1,490–$1,590 per ton. Thereafter, a generally downward trend was evident, and in May the 'spot' price declined to US $1,376.5 (£850) per ton, although it reached US $1,536.5 (£962) by the end of the month. The advance was short-lived, with prices for most coffee contracts standing at less than US $1,400 per ton in late June. The 'spot' price in July fell to only US $1,255 (£805) per ton. In August the London quotation for September 'futures', which had been only US $1,282.5 per ton in July, rose to US $1,407. However, the 'spot' price in September retreated from US $1,323 (£825) per ton to US $1,212.5 (£754). In October the price for short-term delivery was reduced to less than US $1,200 per ton. The 'spot' quotation in November advanced from US $1,212 (£736) per ton to US $1,399.5 (£866). Prices strengthened further in December, with the London quotation for short-term delivery reaching US $1,557 per ton. Meanwhile, the market for longer-term contracts was more stable, with prices remaining below US $1,400 per ton. In that month the Brazilian Government's forecast for the country's coffee output in the year beginning April 2000 was higher than some earlier predictions, despite fears that the crop would have been damaged by the unusually dry weather there since September 1999. For 1999 as a whole average prices of robusta coffee declined by 18.3% from the previous year's level, while arabica prices fell by 23.2%.

In January 2000 the 'spot' price of coffee in London rose strongly towards the end of the month, increasing from US $1,401.5 (£848) per metric ton to US $1,727.5 (£1,067) within a week. However, prices of coffee 'futures' continued to be much lower: at the end of January the quotation for March delivery was US $1,073.5 per ton. In February prices of robusta coffee 'futures' were below US $1,000 per ton for the first time for nearly seven years. In March the 'spot' quotation eased from US $993 (£628) per ton to US $944 (£593). Prices continued to weaken in April, with the quotation for short-term delivery falling to less than US $900 per ton. The 'spot' price in May declined to US $891.5 (£602) per ton, but it then recovered to US $941 (£639). Another downward movement ensued, and by early July the London 'spot' quotation stood at only US $807 (£532) per ton. Later that month prices briefly recovered, owing to concerns about the possible danger of frost damage to coffee crops in Brazil. The 'spot' quotation rose to US $886.5 (£585) per ton, while prices of coffee 'futures' advanced to more than US $1,000. However, the fear of frost was allayed, and on the next trading day the 'spot' price of coffee in London slumped to US $795 (£525) per ton: its lowest level, in terms of US currency, since September 1992.

The weakness in the market was partly attributed to the abundance of supplies, particularly from Viet Nam, which has substantially increased its production and export of coffee in recent years (see above). By mid-2000 Viet Nam had overtaken Indonesia to become the world's leading supplier of robusta coffee and was rivalling Colombia as the second largest coffee-producing country. Viet Nam and Mexico were the most significant producers outside the ACPC, but their representatives supported the Association's plan for a coffee retention scheme to limit exports and thus attempt to raise international prices. The plan was also endorsed by the Organisation africaine et malgache du café (OAMCAF), a Paris-based grouping of nine African coffee-producing countries.

In the first week of September 2000 the 'spot' market quotation rallied to US $829 (£577) per metric ton, remaining at this level until 21 September, when another downward movement occurred. Towards the end of the month the 'spot' quotation declined to US $776 (£530) per ton.

At the beginning of November 2000 the London 'spot' quotation stood at only US $709 (£490) per metric ton, and it was to decline steadily throughout the month, reaching US $612 (£432) on 30 November. High consumer stocks and uncertainty about the size of the Brazilian crop were cited as factors responsible for the substantial decline in November, when the average ICO robusta indicator price fell to its lowest level since August 1969.

In early January 2001 the 'spot' quotation on the London market rallied, rising as high as US $677 (£451) per metric ton. This recovery, which was attributed to concern about the lack of availability of new-crop Central American coffees and reports that producers in some countries were refusing to sell coffee for such low prices, was sustained, broadly, until March, when the downward trend resumed. On 23 March the London 'spot' quotation declined to only US $570 (£399) per ton. On 17 April the London price of robusta coffee 'futures' for July delivery declined to a life-of-contract 'low' of US $560 per ton, the lowest second-month contract price ever recorded.

By May 2001 the collapse in the price of coffee had been described as the deepest crisis in a global commodity market since the 1930s, with prices at their lowest level ever in real terms. The crisis was regarded, fundamentally, as the result of an ongoing increase in world production at twice the rate of growth in consumption, this over-supply having led to an overwhelming accumulation of stocks. During May the London 'spot' quotation fell from US $584 (£407) per metric ton to US $539 (£378) per ton.

In June 2001 producers in Colombia, Mexico and Central America were reported to have agreed to destroy more than 1m. bags of low-grade coffee in a further attempt to boost prices. The ACPC hoped that this voluntary initiative would eventually be adopted by all of its members. By this time the ACPC's retention plan was widely regarded as having failed, with only Brazil, Colombia, Costa Rica and Viet Nam having fully implemented it.

In early July 2001 the price of the robusta coffee contract for September delivery fell below US $540 per metric ton, marking a record 30-year 'low'. At about the same time the ICO recorded its lowest composite price ever, at 43.80 US cents per lb. Despite a recovery beginning in October, the average composite price recorded by the ICO for 2001 was 45.60 US cents per lb, 29% lower than the average composite price (64.25 US cents per lb) recorded in 2000. In 2001 coffee prices were at their lowest level since 1973 in nominal terms, and at a record low level in real terms. The decline in the price of robusta coffees was especially marked in 2001, the ICO recording an average composite price of only 27.54 US cents per lb, compared with 41.41 US cents per lb in 2000, and 67.53 US cents per lb in 1999. In 1996–98 the ICO average composite price for robusta varieties had averaged 81.11 US cents per lb. In 2002 the ICO recorded an average composite price for all coffees of 47.74 US cents per lb, 4.7% higher than the average composite price recorded in 2001. The average composite price for robustas recovered to 30.02 US cents per lb, the monthly price rising consistently and substantially during the final four months of the year. In 2003 the average composite price rose by 8.7%, compared with the previous year, to 51.91 US cents per lb. During the first seven months of 2004 the ICO composite price averaged 60.13 US cents per lb. In January an average of 58.69 US cents per lb was recorded, but in June the composite price rose to 64.28 US cents per lb. In July the average composite price was 58.46 US cents per lb. Despite these improvements, coffee production remained economically unviable for growers in many producer countries in mid-2004. In an assessment of the market in the 2001/02 crop year, the ICO noted that the ongoing crisis had arisen as a result of an imbalance between supply and demand for coffee. While total production in that year was forecast to reach 113m. bags (about 6.8m. tons), world consumption stood at only 106m. bags (about 6.4m. tons). World stocks of coffee, moreover, stood at about 40m bags, some 2.4m. tons. The ICO attributed the glut to the rapid expansion in Vietnamese output and to new plantations in Brazil. Within the framework of the ICA that entered into force in October 2001, the ICO has proposed that the following steps should be taken to address the crisis affecting coffee markets, without resort to market regulation: quality improvement; with effect from October 2002, the ICO has introduced a new global Coffee Quality-Improvement Programme that sets minimum grading standards and maximum moisture content for coffee exports. One aim of the programme is to reduce the current surplus in supplies by eliminating sub-standard coffees from the market; diversification, in the form of action to reduce the dependence of some farmers on coffee by encouraging additional or alternative activities and greater coffee product segmentation; production monitoring, with the ICO to act as an information centre for member countries' production programmes; promotion, especially in new markets such as Russia and China, to boost consumption of coffee; the elimination of tariff and other barriers to trade in all forms of coffee.

The IACO was formed in 1960, with its headquarters at Abidjan in Côte d'Ivoire. In 2003 the IACO represented 25 producer countries, all of which, except Benin and Liberia, were also members of the ICO. The aim of the IACO is to study common problems and to encourage the harmonization of production.

Cotton (*Gossypium*)

This is the name given to the hairs that grow on the epidermis of the seed of the plant genus *Gossypium*. The initial development of the cotton fibres takes place within a closed pod, called a boll, which, after a period of growth of about 50–75 days (depending upon climatic conditions), opens to reveal the familiar white tufts of cotton hair. After the seed cotton has been picked, the cotton fibre, or lint, has to be separated from the seeds by means of a mechanical process, known as ginning. Depending upon the variety and growing conditions, it takes about three metric tons of seed cotton to produce one ton of raw cotton fibre. After ginning, a fuzz of very short cotton hairs remains on the seed. These are called linters, and may be removed and used in the manufacture of paper, cellulose-based chemicals, explosives, etc.

About one-half of the cotton produced in the world is used in the manufacture of clothing, about one-third is used for household textiles, and the remainder for numerous industrial products (tarpaulins, rubber reinforcement, abrasive backings, filters, high-quality papers, etc.).

The official cotton 'season' (for trade purposes) runs from 1 August to 31 July of the following year, and quantities are measured in both metric tons and bales; for statistical purposes, one bale of cotton is 226.8 kg (500 lb) gross or 217.7 kg (480 lb) net.

The price of a particular type of cotton depends upon its availability relative to demand and upon characteristics related to yarn quality and suitability for processing. These include fibre length, fineness, cleanliness, strength and colour. The most important of these is length. Generally speaking, the length of the fibre determines the quality of the yarn produced from it, with the longer fibres being preferred for the finer, stronger and more expensive yarns.

Cotton is the world's leading textile fibre. However, with the increased use of synthetics, cotton's share in the world's total consumption of fibre declined from 48% in 1988 to only 39% in 1998. About one-third of the decline in its market share is attributable to increases in the real cost of cotton relative to prices of competing fibres, and about two-thirds of the loss in market share is attributable to other factors. Expanded use of chemical fibre filament yarn (yarn that is not spun but is extruded in a continuous string) in domestic textiles, such as carpeting, accounts for much of the rest of the loss in market share for cotton. The break-up of the Council for Mutual Economic Assistance (the communist countries' trading bloc) in 1990, and of the USSR in 1991, led to substantial reductions in cotton consumption in those countries and also contributed to cotton's declining share of the world market. Officially enforced limits on the use of cotton in the People's Republic of China (which accounts for about 25%–30% of cotton consumption world-wide) have also had an impact on the international market.

The area devoted to cotton cultivation totalled 31m.–36m. ha between the 1950s and the early 1990s, accounting for about 4% of world cropped area. During the mid-1980s, however, world cotton consumption failed to keep pace with the rate of growth in production, and the resultant surpluses led to a fall in prices, which had serious consequences for those countries that rely on cotton sales for a major portion of their export earnings. In the mid-1990s, despite improvements in world price levels, cotton cultivation came under pressure from food crop needs, and world-wide the harvested areas under cotton declined from 35.9m. ha in 1995/96 to 33.2m. ha in 1998/99. In 1999/2000 the harvested areas under cotton declined further, to 32.4m. ha, and again, in 2000/01, to 32.3m. ha. According to the US Department of Agriculture (USDA), the area under cotton world-wide rose to about 34m. ha in 2001/02, but declined to only 30.6m. ha in 2002/03. Data released by USDA in August 2004 estimated the area under cotton world-wide at 32.4m. ha in 2003/04, and forecast that it would increase to 35.1m. ha in 2004/05.

Cotton is a major crop in several countries in the Asian and Pacific regions, and exports of yarn, fabric and garments are also a source of income to many countries, especially in South-East

Asia. The leading producers of cotton lint in the Asia-Pacific region are the People's Republic of China, and Australia. The production of cotton in most of the countries in the region has expanded gradually, in parallel with world production, but in China and Australia the rate of growth has been dramatic: in the case of China production rose from around 2m. metric tons annually in the mid-1970s to a yearly output averaging almost 4.5m. tons in the period 1985–96. From a large net importer of cotton in the China was transformed in the late 1970s and early 1980s into a significant net exporter. However, with production stable and consumption rising, the country became once again, after the late 1980s, a regular net importer of cotton. In 1994/95–1996/97 China accounted for about one-quarter of all imports world-wide, more, by far, than any other country. In August 1998, however, the Government announced that guaranteed prices for domestic cotton were to be abolished by late 1999. This measure encouraged the textile industry to reduce its purchases of imported cotton in favour of less costly domestic supplies. Cotton imports into China (excluding Hong Kong) declined from 399,000 tons in 1997/98 to only 78,000 tons in 1998/99, 25,000 in 1999/2000, and to 52,000 tons in 2000/01. In 2001/02 they rose to 98,000 tons. In the following crop year a very substantial increase, to 681,000 tons, occurred. USDA estimated that in 2003/04 Chinese imports had increased to as much as 1.9m. tons, but forecast that they would decline to some 1.4m. tons in 2004/05. Chinese cotton consumption has risen very sharply in recent years as a consequence of the rapid expansion of the country's textile industry, in particular its export-orientated sectors. USDA forecast in September 2004 that in the 2004/05 marketing year Chinese cotton consumption would account for one-third of total consumption world-wide. The country's exports of cotton, which had been at negligible levels during the three years 1995/96 to 1997/98, reached 148,000 tons in 1998/99, and in 1999/2000 totalled 368,000 tons, despite a decline in cotton production (to 3.8m. tons) in 1999.

According to FAO, Australia's annual cotton crop expanded from 99,000 metric tons in 1980/81 to some 608,000 tons in 1996/97. Production increased to about 666,000 tons in 1997/98, to about 724,000 tons in 1998/99 and to about 740,000 tons in 1999/2000. In 2000/01 output rose again, to more than 800,000 tons, but declined slightly, to 745,000 tons in 2001/02. In 2002/03, according to FAO data, Australia's output of cotton fell very substantially, by some 48%, to only 386,000 tons. In 2003/04 production was estimated to have once again fallen by a considerable margin, to 287,000 tons. Australian consumption of cotton was only about 40,000 tons per year in the late 1990s, so almost all of the country's output is exported. Exports of cotton contributed 1.3% of Australia's total export earnings in the year to June 2002.

Production of Cotton Lint
('000 metric tons, excluding linters)

	2002	2003*
World total	18,268	19,529
Far East	4,620	5,021
Oceania	386	287
Leading regional producers		
Australia	386	287
China, People's Repub.	4,916	5,200
Other leading producers		
Argentina	62	65†
Brazil†	713	726
Egypt	285†	250‡
Greece†	355	333
India	1,583	2,100†
Pakistan	1,736	1,690†
Syria†	245	283
Turkey	850	946†
USA	3,747	3,968
Uzbekistan	1,008	946

* Provisional figures.
† Unofficial figure(s).
‡ FAO estimate.

The leading exporters of cotton in 2003/04, according to USDA estimates, were the USA (with sales of some 3m. metric tons), Uzbekistan (659,000 tons), Australia (468,000 tons), Mali (256,000 tons) and Greece (255,000 tons). Other important exporters in that year were Burkina Faso (207,000 tons) and Brazil (201,000 tons). Prominent importers of cotton in 2003/04, in the light of USDA estimated data, were the People's Republic of China (purchasing some 1.9m. tons), Turkey (479,000 tons), Indonesia (468,000 tons), Pakistan (403,000 tons), Mexico (397,000 tons), Thailand (370,000 tons) and Bangladesh (335,000 tons). In the late 1980s and early 1990s Russia was the world's major importer of cotton, with an annual intake of more than 1m. tons, but imports have since declined to, generally, about 350,000 tons annually. China (see above) was the foremost cotton-importing country in the mid-1990s, thereby accumulating large stocks of the fibre, but it subsequently reduced these holdings.

Although co-operation in cotton affairs has a long history, there have been no international agreements governing the cotton trade. Proposals in recent years to link producers and consumers in price stabilization arrangements have been opposed by the USA (the world's largest cotton exporter), and by Japan and the European Union. The International Cotton Advisory Committee (ICAC), an inter-governmental body established in 1939 with its headquarters in Washington, DC, publishes statistical and economic information and provides a forum for consultation and discussion among its 43 members.

Export Price Index for Cotton Lint
(base: 1980 = 100)

	Average	Highest month(s)	Lowest month(s)
1990	93		
1995	103		
2000	62	66 (Dec.)	53 (Jan.)
2001	51	65 (Jan.)	41 (Oct.)
2002	47	55 (Dec.)	42 (May)

The British city of Liverpool is the historic centre of cotton-trading activity, and international cotton prices are still collected by organizations located in Liverpool. However, almost no US cotton has been imported through the port of Liverpool in recent years. Consumption in the textile industry in the United Kingdom has fallen to approximately 14,000 metric tons per year, most of which comes from Africa, Greece, Spain and Central Asia. The price for Memphis cotton, from the USA, quoted in international markets is c.i.f. North European ports, of which Bremen, in Germany, is the most important.

The average price for Memphis Territory cotton in North Europe rose to US $2,848 per metric ton in June 1995, owing to an increase in imports by the People's Republic of China, combined with declines in production in India, Turkey and Pakistan. Prices have trended down since, averaging US $2,175 per ton during all of 1994/95, US $2,088 in 1995/96 and US $1,826 in 1996/97. The average Memphis quote c.i.f. North Europe during December 1997 fell to US $1,756 per ton. The market was depressed further during the next two years. In 1999, according to the World Bank, the price of Memphis cotton (US origin, c.i.f. North Europe) averaged US $1,228 per ton. In 2000, however, the average price recovered to US $1,463 per ton in response to increased demand and stagnant production. In January–March 2001, the price of Memphis cotton averaged US $1,467 per ton, but it declined to only US $1,175 per ton in the subsequent quarter. In January–August 2001 the average price was US $1,271 per ton, with an average of US $1,126 per ton in the final month of that period.

The principal Liverpool index of cotton import prices in North Europe is based on an average of the cheapest five quotations from a selection of styles of medium-staple fibre. In 1999 the index recorded an average offering price of 53.1 US cents per lb: its lowest annual level since 1986. On a monthly basis, the average price in December 1999 was only 44.4 US cents per lb. The decline in prices was attributed to the plentiful availability of cotton, with high levels of production resulting in large stocks (more than 9m. metric tons world-wide in recent years). In 2000 the index recorded an average offering price of 59.0 US cents per lb. On a monthly basis, the average price in December 2000 was 65.8 US cents per lb. Slow growth in production and stronger demand for cotton were the main reasons cited for the recovery in prices. In 2001, however, the decline in prices resumed and in that year the index recorded an average offering price of only 48

US cents per lb. This declined further in 2002, to only 46 US cents per lb. Provisional data indicated that the Liverpool index had recorded an average offering price of 63 US cents per lb in 2003, cotton prices having risen sharply late in that year as a consequence of lower Chinese output following natural disasters. In the first quarter of 2004 the average offering price was estimated to have risen as high as 74 US cents per lb, but declined to 68 US cents per lb in the second quarter of the year. Weaker prices after March 2004 were attributed to a decline in Chinese domestic cotton prices that had occurred as a result of lower-than-expected domestic consumption in 2003/04, and in anticipation of substantially increased output in the 2004/05 crop year. In July 2004 an estimated average offering price of 57 US cents per lb was recorded. The offering price declined further in August, to only 53 US cents per lb.

Groundnut (Peanut, Monkey Nut, Earth Nut) (*Arachis hypogaea*)

This is not a true nut, although the underground pod, which contains the kernels, forms a more or less dry shell at maturity. The plant is a low-growing annual herb introduced from South America.

Each groundnut pod contains between two and four kernels, enclosed in a reddish skin. The kernels are very nutritious because of their high content of both protein (about 30%) and oil (40%–50%). In tropical countries the crop is grown partly for domestic consumption and partly for export. Whole nuts of selected large dessert types, with the skin removed, are eaten raw or roasted. Peanut butter is made by removing the skin and germ and grinding the roasted nuts. The most important commercial use of groundnuts is the extraction of oil. Groundnut oil is used as a cooking and salad oil, as an ingredient in margarine and, in the case of lower-quality oil, in soap manufacture. The oil is faced with strong competition from soybean, cottonseed and sunflower oils—all produced in the USA. In 2003 groundnut oil was the fifth most important of soft edible oils in terms of production. In 2002 its position in terms of exports was ninth, accounting for less than 1% of total world exports of food oils.

An oilcake, used in animal feeding, is manufactured from the groundnut residue left after oil extraction. However, trade in this groundnut meal is limited by health laws in some countries, as groundnuts can be contaminated by a mould which generates toxic and carcinogenic metabolites, the most common and most dangerous of which is aflatoxin B_1. The European Union (EU) bans imports of oilcake and meal for use as animal feed which contain more than 0.03 mg of aflatoxin per kg. The meal can be treated with ammonia, which both eliminates the aflatoxin and enriches the cake. Groundnut shells, which are usually incinerated or simply discarded as waste, can be converted into a low-cost organic fertilizer, which has been produced since the early 1970s.

Production of Groundnuts
(in shell, '000 metric tons)

	2002	2003*
World total	33,303	35,658
Far East	17,471	16,163
Oceania	26	41
Leading regional producers		
China, People's Repub.	14,895	13,447
Indonesia	1,267	1,377
Myanmar	723	730†
Thailand	112	132‡
Viet Nam	400	400
Other leading producers		
Argentina	517	316
Chad†	450	450
Congo, Dem. Repub.	355	355‡
India	4,363	7,500†
Nigeria	2,699	2,700‡
Senegal	501	900‡
Sudan	1,267	1,200‡
USA	1,506	1,880

* Provisional figures.
† Unofficial figure(s).
‡ FAO estimate.

In recent years some 90% of the world's groundnut output has come from developing countries. In 2003, according to preliminary figures from FAO, world output of groundnut oil was 5,776,863 metric tons, including 2,316,090 tons from Far East Asia and 2,103 tons from Oceania. The major producing countries in that year were China (2,064,500 tons) and India (1,685,000 tons).

Export Price Index for Groundnuts
(base: 1980 = 100)

	Average	Highest month(s)	Lowest month(s)
1990	102		
1995	92		
2000	58	65 (Dec.)	55 (June)
2001	59	65 (April, May)	48 (Dec.)
2002	49	64 (Dec.)	42 (May)

Export Price Index for Groundnut Oil
(base: 1980 = 100)

	Average	Highest month(s)	Lowest month(s)
1990	113		
1995	116		
2000	84	93 (Jan.)	77 (Sept.)
2001	80	82 (March–June)	78 (Aug.–Oct.)
2002	81	99 (Dec.)	75 (June–Aug.)

In recent years, as the tables above make clear, prices for groundnut oil have generally been more volatile than those for groundnuts. The average import price of groundnut oil at the port of Rotterdam, Netherlands, declined from US $909.4 per metric ton in 1998 to US $787.7 in 1999. It declined further, to US $713.7 per ton, in 2000, and to US $680.3 per ton in 2001. In 2002 the average price rose slightly, to US $687.1 per ton. From late 2002 the average import price of groundnut oil at Rotterdam rose precipitously, averaging US $718 per ton, US $771 per ton and US $845 per ton, respectively, in the final three months of the year. In March 2003 an average price of US $1,195 per ton was recorded, and in July this rose further, to US $1,397 per ton. For the whole of 2003 an average price of US $1,243.2 per ton was recorded. One factor behind this sharp increase was anticipation of a substantial reduction in groundnut production in 2002/03 as a consequence of weather-related damage to crops in Argentina, India, Senegal and the USA. Additionally, from the final quarter of 2003 the price of groundnut oil has been supported by higher demand for, and lower supply of, soybeans world-wide. During the first six months of 2004 the average import price of groundnut oil at Rotterdam ranged between US $1,150 per ton (June) and US $1,237 per ton (April).

The market for edible groundnuts is particularly sensitive to the level of production in the USA, which provides about one-half of world import requirements. In each of the years 1987–91 the People's Republic of China was the leading exporter of green (unroasted) groundnuts, but in 1992 the USA was the main exporting country. In 1991 China exported 427,693 metric tons of groundnuts, valued at US $360.3m. Exports of this commodity in 1992 declined to 303,606 tons, earning only US $190.3m. The average price per ton fell from US $842 in 1991 to US $627 in 1992. China regained its leading position in 1993, although its exports of groundnuts increased only slightly, to 315,905 tons. In 1994 China's groundnut exports totalled 480,935 tons, valued at US $315.0m. (US $655 per ton). In 1997 the USA again overtook China as the world's leading exporter of groundnuts, Chinese exports in that year declining to 171,474 tons, compared with 351,068 tons in 1996. In each year since then China has been the leading exporter, often by a substantial margin. In 2002, according to FAO, Chinese exports of groundnuts totalled 520,689 tons.

The average annual price of edible groundnuts (any origin, US runner, 40%–50% shelled basis, c.i.f. Rotterdam) was US $909 per metric ton in 1995. The average price rose to US $962 per ton in 1996, and in both 1997 and 1998 an average price of US $990 per ton was recorded. In 1999, however, the annual average price declined by more than 15%, to US $836 per ton, recovering marginally, to US $838 per ton, in 2000. In 2001 the import price declined slightly, to US $835 per ton, and it fell again, more substantially, to US $751 per ton in 2002. From late 2002 the import price at Rotterdam rose steeply, averaging US $943 per ton in November. In March 2003 an average price of US $1,000 per ton was recorded and the monthly average remained at, or very close to, that level until July, when it fell slightly, to US $948 per ton. For the whole of 2003 an average price of US $975.7 per ton was recorded. In the first half of 2004 the import price of groundnuts at Rotterdam ranged between US $970 per ton (April, May and June) and US $1,000 per ton (January, February and March).

Jute

Jute fibres are obtained from *Corchorus capsularis* and *C. olitorius*. Jute-like fibres include a number of jute substitutes, the main ones being kenaf or mesta and roselle (*Hibiscus* spp.) and Congo jute or paka (*Urena lobata*). The genus *Corchorus* includes about 40 species distributed throughout the tropics, with the largest number of species being found in Africa.

Jute flourishes in the hot damp regions of Asia. Commercial fibre varies from yellow to brown in colour and consists of tow (bunches of strands), which is pressed into bales of 181.4 kg (400 lb) after it has been retted (softened).

Jute has a number of uses. The relatively cheap, hard-wearing fibre is used to manufacture sacking for the storage of grains, cocoa, coffee and other food crops, which currently accounts for about 75% of consumption. The remainder of the crop is taken by manufacturers of carpet backing and other quality users. The finest jute standards are spun into carpet yarn and woven into curtains and wall coverings; in the area of quality goods jute is less under threat from synthetics. Jute is mixed with wool, after treatment with petroleum-derived softening agents (see below), and processed into cheap clothing fabrics or blankets in developing Asian countries. In its traditional market, however, jute has, despite its biodegradability, encountered strong competition from lighter synthetic materials. Its use as packaging is under challenge from wood and plywood, and has been diminishing as a consequence of the trend towards bulk handling containerization. The main jute-producing countries, however, have continued to stress the environmental desirability of jute use, and anxieties about the health aspects of petroleum use in the retting process have been overcome by the introduction of castor oil as the softening agent.

The People's Republic of China is the most important regional producer of jute and jute-like fibres, with estimated output of 155,000 metric tons in 2002/03 and 165,000 tons in 2003/04. China has been a significant importer of raw jute in recent years: in 1998 it purchased more than 100,000 tons. However, Chinese imports fell precipitously in 1999, when they totalled only about 9,000 tons, and in the two subsequent years they totalled, respectively, only some 6,000 and 7,000 tons. In 2002 Chinese imports of jute were estimated to have risen to about 41,000 tons. China's exports of jute products have averaged about 7,000 tons annually in recent years. Myanmar, with estimated output of about 42,000 tons in 2003/04, Thailand, whose output was estimated at some 36,000 tons in that year, and Viet Nam, with estimated output of about 21,000 tons, are the other significant producers (but not significant exporters) of jute in the Far East. Thai imports of raw jute rose substantially in 2000, when they totalled more than 42,000 tons, compared with only about 11,000 tons in 1999, and a mere 200 tons in 1998. In 2001 Thai imports of raw jute declined slightly, to 38,000 tons, and they fell again in 2002, to some 15,000 tons. Thai imports were estimated at about 37,000 tons in 2003.

The longer-term outlook for the jute industry in Far East Asia, as well as other producing areas, depends, in large part, on the achievement of higher crop yields. Prospects exist for an expansion in world demand for jute in the growing market for 'environmentally friendly' jute packaging and products, and for high-quality blended jute fabrics for wall-hangings and furnishings.

The average export price of raw jute at Bangladeshi ports increased from US $271 per metric ton in 1993 to US $454 per ton in 1996. However, the annual average declined to US $302 per ton in 1997 and to US $259 in 1998. It recovered to US $279 per ton in 1999, and remained at that level, on average, in 2000. In 2001 the average price of BWD-grade jute (f.o.b. Mongla) rose sharply, to US $331 per ton. In 2002, however, the average price slumped to its lowest level, US $185 per ton, for some 20–25 years, owing to high levels of stocks from the 2001/02 crop year and estimated record production in India in 2002/03. The average export price of Bangladeshi BWD-grade jute recovered to US $242 per ton in 2003. In January–March 2004 the export price of BWD-grade jute remained constant at US $245 per ton, rising to US $290 per ton in April and remaining at that level in May–August.

Export Price Index for Jute
(base: 1980 = 100)

	Average	Highest month(s)	Lowest month(s)
1990	113		
1995	105		
2000	72	84 (Jan.)	63 (Sept.)
2001	94	108 (July)	84 (Jan.)
2002	82	99 (Jan.–March)	69 (Oct.)

FAO and the United Nations Conference on Trade and Development (UNCTAD) provide international forums for the discussion of jute developments. Jute is one of the crops eligible for assistance from the Common Fund for Commodities (CFC, established by an UNCTAD agreement in 1989), which aims to stabilize commodity prices. An international jute (and allied fibres) agreement (IJA) was negotiated in 1982 by 48 producing and consuming countries, but lapsed in 1984, when the requisite number of consumer countries failed to ratify the agreement. A subsequent IJA was negotiated, under the auspices of UNCTAD, in 1989. The International Jute Organization (IJO), which administered the agreement, was based in Dhaka, Bangladesh. It also conducted research and development projects, and promoted the competitiveness of jute in relation to synthetic substitutes. In 1998 the IJO's membership comprised five exporting countries (Bangladesh, the People's Republic of China, India, Nepal and Thailand) and 20 importing countries, together with the European Union (EU). In January 1999, however, India announced its withdrawal from the IJO, which had declined to appoint India's nominee for the post of executive director. Thailand, citing its domestic economic problems, withdrew from the IJO in March. India rejoined the organization in December, but Thailand remained outside. Meanwhile, some EU members believed that the IJO had failed to promote trade in jute, and consequently favoured the organization's disbandment.

Following preparatory meetings at IJO headquarters, the UN Conference on Jute and Jute Products was convened in March 2000, under UNCTAD auspices, in Geneva, Switzerland. However, the EU and its member countries (which together accounted for about 30% of the IJO's financing) declined to attend this meeting or a resumed session in Dhaka in early April. A draft text for an 'International Instrument of Co-operation', to succeed the IJA, was supported by jute-producing countries, but, without the consent of major consumers, no agreement could be concluded. At a subsequent meeting of the IJO's governing council, the EU rejected adoption of the successor 'instrument'. As a result, the IJA expired in April, at the end of its term, and the IJO entered a liquidation process. However, a working group, including representatives of producing countries, was formed to consider future international co-operation on jute affairs. It was hoped that these countries would continue to obtain support for jute projects from the CFC. In March 2001, under UNCTAD auspices, the UN Conference on Jute and Jute Products established the International Jute Study Group (IJSG) as the successor entity to the IJO. *Inter alia*, the objectives of the IJSG are to provide an effective framework for international co-operation among its members with regard to all relevant aspects of the world jute economy; and to promote the expansion of international trade in jute and jute products. Membership of the IJSG, which was to have its headquarters in Dhaka, was to be open to all states (and to the EU) with an interest in the production and consumption of, and international trade in, jute and jute products. The terms of reference of the IJSG formally entered into force in April 2002 once a number of states together accounting for 60% of trade (imports and exports combined) in jute and jute products had given notice of their formal acceptance of the IJSG's terms of reference. Another forum for jute co-operation is FAO's Intergovernmental Group on Jute, Kenaf and Allied Fibres, which normally convenes every two years. The Group convened most recently in July 2003, in a joint meeting with the Intergovernmental Group on Hard Fibres. It is next due to meet in December 2004.

Production of Jute, Kenaf and Allied Fibres
('000 metric tons)

	2002	2003*
World total	3,273	3,232
Far East	281	436
Leading regional producers		
China, People's Repub.	159‡	180†
Myanmar	47	50†
Thailand	46	62†
Viet Nam	20	13
Other leading producers		
Bangladesh	801	801†
India	2,051	1,976
Nepal	17	18†
Russia†	47	48
Uzbekistan†	20	20

* Provisional figures.
† FAO estimate(s).
‡ Unofficial figure.

Maize (Indian Corn, Mealies) (*Zea mays*)

Maize is one of the world's three principal cereal crops, with wheat and rice. Originally from America, maize has been dispersed to many parts of the world. The main varieties are dent maize (which has large, soft, flat grains) and flint maize (which has round, hard grains). Dent maize is the predominant type world-wide. Maize may be white or yellow (there is little nutritional difference), but the former is often preferred for human consumption. Maize is an annual crop, planted from seed, which matures within three to five months. It requires a warm climate and ample water supplies during the growing season. Genetically modified varieties of maize, with improved resistance against pests, are now being developed, particularly in the USA, and also in Argentina and the People's Republic of China. Ecological and consumer concerns are, however, placing in question the further commercialization of genetically modified varieties.

Maize is an important foodstuff in parts of Far East Asia, such as Indonesia, the Philippines and southern China, where the climate precludes the extensive cultivation of wheat. It tends, however, to be replaced by wheat in diets, as disposable incomes rise; maize consumption per caput has been declining in most of the developing countries. In some countries the grain is ground into a meal, mixed with water, and boiled to produce a gruel or porridge. In other areas it is made into (unleavened) corn bread or breakfast cereals. Maize is also the source of an oil, which is used in cooking.

The high starch content of maize makes it highly suitable as a compound feed ingredient, especially for pigs and poultry. Animal feed is the main use of maize in the USA, Europe and Japan, and large quantities are now also used for feed in developing countries in Far East Asia, Latin America and, to some extent, in North Africa. Maize also has a variety of industrial uses, including the preparation of ethyl alcohol (ethanol), which may be added to petrol to produce a blended motor fuel. In addition, maize is a source of dextrose and fructose, which can be used as artificial sweeteners, many times sweeter than sugar. The amounts used for these purposes depend, critically, on prices relative to petroleum, sugar and other potential raw materials. Maize cobs, previously discarded as a waste product, may be used as feedstock to produce various chemicals (e.g. acetic acid and formic acid).

In recent years world maize output has averaged about 600m. metric tons annually; it reached a record 615m. tons in 1998. The USA is by far the largest producer, with harvests of about 244m. tons per year. It is, however, subject to occasional major crop reverses in years of drought, as in 1995 (when output fell to only about 188m. tons). Far East Asia has accounted, on average, for between one-fifth and one-quarter of world maize production in recent years. The major producer in the region, and the world's second largest, is the People's Republic of China. Assisted by the expanded use of hybrid varieties, China's annual maize harvest increased by more than 50% in the 1980s, enabling the country to become, for a few seasons, a significant exporter of maize, until the rapid growth in domestic feed requirements drew level with supplies in the mid-1990s. New efforts to stimulate production, including higher support prices, resulted in a record crop of 128m. tons in 1996, and the accumulation of substantial maize stocks, which enabled supplies to be maintained without recourse to imports, following a much smaller harvest in 1997. Output recovered to a new record level of 133m. tons in 1998, but it declined to 128m. tons in 1999 and to only 106m. tons in 2000. In 2001 Chinese production recovered to about 114m. tons, and in 2002 totalled estimated 121m. tons. Output was estimated at 114m. tons in 2003.

Production of Maize
('000 metric tons)

	2002	2003*
World total	604,162	638,043
Far East	145,034	139,986
Oceania	620	486
Leading regional producers		
China, People's Repub.	121,499	114,175
Indonesia	9,654	10,910
Korea, Dem. People's Repub.	1,651	1,725
Philippines	4,319	4,478
Thailand	4,230	4,500†
Viet Nam	2,511	2,934
Other leading producers		
Argentina	15,000	15,040
Brazil	35,933	47,809
Canada	8,999	9,587
France	16,440	11,898
Hungary	6,121	4,534
India	11,167	14,800†
Italy	10,824	8,978
Mexico	19,299	19,652
Romania	8,400	9,577
Serbia and Montenegro	5,597	3,826
South Africa	10,049	9,714
USA	228,805	256,905

* Provisional figures.
† Unofficial figure.

In Indonesia maize is a secondary crop, usually planted in the dry season after the main rice crop. The high cost of seed and other inputs deters farmers from planting hybrid varieties, although the Government is encouraging the expansion of maize output to meet the country's rapidly growing requirements for poultry feed. Production has averaged about 10m. metric tons annually in recent years. Imports of maize for feed were increasing until the Asian economic crisis of 1997 severely reduced meat consumption and feed requirements. Despite a small crop, there was an exportable surplus in 1997/98. About one-half of the Philippine crop (which has averaged more than 4m. tons annually in recent years) consists of white flint maize, used for human consumption. Feed use (for pigs and poultry) is increasing rapidly, but rising import costs, following the depreciation of the Philippine peso, inhibited growth in demand in 1997/98. Maize was formerly an important export crop in Thailand, but the areas under maize declined sharply in the late 1980s because of the greater profitability of other crops, such as sugar cane. By the mid-1990s Thailand had to resort to imports in most years to satisfy its domestic feed requirements. In 1997–2002 production was steady at about 4.3m. tons annually, compared with about 3.8m. tons per year, on average, in 1990–95. In 2003 Thai production totalled an estimated 4.5m. tons.

World trade in maize totalled about 85m. metric tons in 2002, compared with some 79m. tons in 2001, and some 82m. tons in 2000. In the mid-1990s growth in trade was curtailed by adverse economic conditions in eastern Asia, which severely impacted on the region's meat consumption and, consequently, on its demand for animal feed. Recovery in the Asian economies, together with growing demand from Latin America and North Africa, subsequently restored that demand.

The USA is invariably the largest maize-exporting country, with an average market share of about 60% in recent years. US exports reached 60m. metric tons in 1995/96, but have subsequently been lower (approximately 40m.–50m. tons annually), owing to increased competition from Argentina, the People's Republic of China and a number of eastern and central European countries. US exports totalled an estimated 48m. tons in 2002. Argentina, usually the second largest supplier, has benefited considerably from liberalized marketing systems and improved port facilities. Argentinian exports of maize reached a record level of 12m. tons in 1998, but declined to only 8m. tons in 1999. Exports recovered to 10.8m. tons in 2000, and totalled about 10.9m. tons in 2001 and 9.5m. tons in 2002. As a result of steeply rising output, China became a leading exporter of maize in the early 1990s, with sales attaining 11m. tons in 1993. One of its principal markets was Japan, whose coastal feed mills can receive shipments direct from China at low transport costs. The growth in China's own feed grain needs, together with smaller harvests, subsequently reduced the surplus, although exports continued from southern ports as a result of the high cost of moving grain overland to China's feed deficit regions. Exports by China were only 4.3m. tons in 1999, but they recovered to 10.5m. tons in 2000 before declining again, to about 6m. tons, in 2001. In 2002 China's foreign sales of maize were estimated to have risen to a new record level of 11.7m. tons.

Far East Asia has long been a major market for maize, and its importance has increased significantly since the 1980s with the establishment of livestock (and particularly poultry) industries in many countries of the region. Total imports of maize in Far East Asia rose to a record level in the mid-1990s, but fell back later in that decade as a result of the region's economic difficulties. Japan is usually the world's leading maize-importing country. Its purchases accelerated sharply in the 1970s, but advanced more slowly in the 1980s. Since the mid-1990s Japanese purchases have stabilized at, approximately, 16m.–16.5m. tons annually. Feed users in the Republic of Korea are willing to substitute other grains, particularly feed wheat, for maize when prices are attractive, and maize imports are therefore variable, although averaging about 8.3m. tons annually in recent years. Taiwan imported more than 6m. tons in 1995/96, but the volume subsequently fell, following an outbreak of swine disease that reduced pig numbers by 40%. Nevertheless, the country continues regularly to import about 5m. tons of maize annually.

Few countries other than the USA and the People's Republic of China usually hold more than minimal carry-over stocks of maize. The USA accumulated massive stocks in the mid-1980s, the peak carry-over being 124m. metric tons at the end of August 1987. Government support programmes were successful in discouraging surplus production, but several poor harvests also contributed to the depletion of these stocks, which were reduced to only 11m. tons at the close of the 1995/96 marketing year. Domestic requirements in the USA have continued to increase, although three successive plentiful crops in 1996, 1997 and 1998, together with increased competition from other exporting countries, resulted in an appreciable build-up in US stocks in the second half of the 1990s. Carry-over stocks at the end of the 1998/99 marketing year reached 45m. tons, declined to 44m. tons in 1999/2000 and rose again, to 45m. tons, in 2000/01. They fell substantially, to 31m. tons, in 2002/03, and declined again, to 27m. tons, at the end of the 2003/04 marketing year, their lowest level since 1975/76, as a consequence of strong growth in maize consumption, that outstripped production, and the heavy depletion of Chinese stocks.

Export Price Index for Maize
(base: 1980 = 100)

	Average	Highest month(s)	Lowest month(s)
1990	92		
1995	80		
2000	54	59 (Jan.–April)	47 (Sept.)
2001	59	63 (Jan.)	56 (June)
2002	67	80 (Sept.)	56 (April)

Export prices of maize are mainly influenced by the level of supplies in the USA, and the intensity of competition between the exporting countries. Record quotations were achieved in April 1996, when the average price of US No. 2 Yellow Corn (f.o.b. Gulf Ports) reached US $210 per metric ton. However, the quotation declined rapidly over the following months, falling as low as US $117 per ton in November 1996. For the whole of 1996 the quotation averaged US $165 per ton. The average annual quotation declined to US $117 per ton in 1997, to US $102 per ton in 1998, to US $92 per ton in 1999, and to US $88 per ton in 2000. In 2001 an average price of US $90 per ton was recorded, and the average rose in 2002 to US $99 per ton, and to US $105 per ton in 2003. During the first seven months of 2004 the average monthly price of US No. 2 Yellow Corn (f.o.b. Gulf ports) ranged between US $107.6 per ton, recorded in July, and US $134.7 per ton, recorded in April.

Nickel

Nickel is a silvery, malleable metal which occurs in various types of ores, the most important being sulphide ores and laterite ores. Sulphide ores are generally mined by underground methods, while laterites are usually mined by open-cast methods. Nickel in sulphide ores is often found in conjunction with copper, cobalt and platinum group metals (PGM); in laterites it is often found with iron, chromium and cobalt. The sulphide ore is subjected to crushing and grinding, with final treatment by the flotation method. The resulting concentrate is roasted and smelted in furnaces to remove the bulk of the sulphur, leaving a matte of nickel and copper. The nickel and copper can then be separated from each other by a variety of processes and sent to an electrolytic refinery. The primary product of processing laterites is ferronickel; a few ferronickel operations also produce limited amounts of matte by the deliberate addition of pyrite or sulphur to the ore at the drying stage. Nickel production is an energy-intensive operation, and the cost of energy to the industry has risen steeply in recent years. During the late 1990s, however, several Australian laterite nickel producers were conducting pilot projects in the processing of ores by high-pressure acid leaching, which, if shown to be successful, could substantially reduce production costs. The world's land-based reserves of nickel ore in 2003 were estimated by the US Geological Survey (USGS) at 62m. metric tons on an elemental basis, of which about 24.8m. tons were in sulphide deposits and 37.2m. tons in laterite ores. The distinction is significant, as sulphide ores are cheaper to process. According to USGS estimates, the largest proportion of nickel reserves are in Australia (35.5% of the world total), followed by Russia (10.6%), Cuba (9%), Canada (8.4%), Brazil (7.3%), New Caledonia (7.1%), South Africa (6%) and Indonesia (5.2%).

Nickel is used in a wide range of alloys where it imparts corrosion resistance and high-temperature strength. Its most important use is in steel production, and the majority of high-tensile steels contain nickel. On a world-wide basis, the stainless steel industry accounts for more than 60% of nickel consumption. Nickel continues to be a significant constituent of much military equipment. Alloys containing a high proportion of nickel are used in power and chemical plants. The metal also has high-temperature applications, such as gas turbines, and rocket engines for the aerospace industry.

Extensive deposits of ferromanganese nodules—also containing nickel, copper and cobalt—are known to exist on the floor of the world's oceans. A conservative estimate assesses world nodule reserves at 290m. metric tons of nickel, 240m. tons of cobalt and 6,000m. tons of manganese. However, quite apart from the question of rights of recovery tortuously negotiated at the UN Conference on the Law of the Sea, experts believe that the extraction of metals from the nodules is unlikely to be commercially practicable for many years.

Nickel prices in Western countries have traditionally been governed by the four companies that, historically, were the largest producers—International Nickel (Inco), Falconbridge (both of Canada), Western Mining (of Australia) and the French-controlled Société Le Nickel (SLN). In 2002 Norilsk Nickel of Russia was the world's single largest producer, and its output, combined with that of the four aforementioned companies and that of BHP Billiton plc of the United Kingdom, accounted for about 66% of world production. Inco (the Western world's largest producer) has traditionally supplied almost one-third of the nickel requirements of the major industrial countries.

The major nickel-producing countries in the Asia-Pacific region are Australia, New Caledonia and Indonesia. New Caledonia possesses the world's largest identified deposits of nickel-bearing laterite. Substantial deposits of nickel exist in Western Australia, where the Leinster mine has reserves of 333m. metric tons of underground ore, grading 1.8% nickel, and about 2.9m. tons of open-pit ore, grading 1.5% nickel. There are deposits at Mt Keith (at least 91m. tons of reserves, grading about 0.5% nickel) and also at Yakabindie (more than 289m. tons, with an average grading of about 0.6% nickel). The development of reserves in the Murrin Murrin district, estimated at 119m. tons (grading 1.4% nickel), commenced in 1997. The refinery associated with the Murrin Murrin open-cast mine, which is 60% owned by Anaconda Nickel Ltd of Australia, had a planned initial capacity to produce 45,000 tons of nickel, and 3,000 tons of cobalt, per year. Nickel briquet output commenced in 1999, but, owing to a subsequent need to re-engineer parts of the plant, production capacity was not expected to be reached until the first half of 2002. Murrin Murrin is the largest of three projects in Western Australia that have been pioneering the potentially low-cost process of producing nickel and cobalt from laterite deposits by high-pressure acid leaching (see above). In the second phase of Anaconda's Murrin Murrin project it is planned to expand annual production capacity to 106,000 tons for nickel and 7,600 tons for cobalt. The value of Australia's exports of nickel ores and concentrates and of intermediate products of nickel was $A1,410m. in the year ending June 2002, representing 1.2% of the total value of merchandise exports.

In the longer term Australia's relatively high-cost production could be faced with a considerable challenge in world nickel markets from the commercial exploitation of vast, and as yet undeveloped, deposits near Voisey's Bay, in north-eastern Canada. This prospect has been stated to contain as much as 150m. metric tons of nickel ore, with the estimated low-cost productive capacity to provide one-third of all new nickel mine output within 10 years of its entry into production. After protracted negotiations, it was announced in 2002 that Inco—which acquired the rights to develop the Voisey's Bay nickel deposits in 1996—had concluded an agreement with the government of Newfoundland and Labrador to develop (via a subsidiary, Voisey's Bay Nickel Co Ltd) the prospect, at a cost of US $1,900m.

Production of Nickel Ore
(nickel content, metric tons)

	2001	2002
World total*	1,340,000	1,340,000
Far East*	180,899	203,062
Oceania	314,554	310,650
Leading regional producers		
Australia	197,000	211,000
China, People's Repub.*	51,500	54,500
Indonesia	102,000	122,000*
New Caledonia	117,554	99,650
Philippines	27,359	26,532
Other leading producers		
Botswana	18,585	20,005
Brazil	47,097	45,029
Canada	194,058	178,338
Colombia	52,962	58,196
Cuba	72,620	73,000*
Dominican Repub.	39,120	38,859
Russia*	325,000	310,000
South Africa	36,443	38,546

* Estimated production. World and Far East totals and other estimated data are rounded.

Source: US Geological Survey.

In 2003, according to provisional data, New Caledonia tied with Indonesia as the world's fourth largest producer of nickel in terms of mine output. The territory's dominant production company is SLN, 60% owned by the French government-controlled mining conglomerate Eramet, 30% by the three provinces of New Caledonia, and 10% by Nishin Steel of Japan. In 2001 SLN accounted for 47% of New Caledonia's total mine output of nickel. The company exploits deposits of nickel-bearing ores at four mining centres, all on La Grande Terre, New Caledonia's main island. Mining at a fifth SLN site, also on La Grande Terre, has been contracted to an independent company. The output of ore from mines operated by SLN and its sub-contractors totalled 3,580,000 metric tons (wet weight) in 2001, compared with 3,710,000 tons in 2000. Ore from these mines is sent for processing in the company's smelter at Doniambo, near Nouméa, in the south of the island. Eramet is the world's leading company producing ferronickel, and SLN's Doniambo smelter is the largest ferronickel plant in the world. In 2001 output from Doniambo was a record 58,973 tons of nickel, comprising 45,912 tons of nickel in ferronickel and 13,061 tons of nickel in matte. In 2001 SLN initiated a programme that, at an estimated cost of US $180m., will increase the annual refining capacity of the Doniambo smelter by 25%, to 75,000 tons (ferronickel and matte), before 2006.

Products from the Doniambo plant are destined for export. In 2002 the nickel industry provided some 10% of New Caledonia's gross domestic product and an estimated 80% of all foreign exchange earnings. During the late 1990s more than one-half of the nickel ore from the territory's mines was exported without further processing, and there has been considerable local support for proposals to construct additional smelting facilities. However, New Caledonia's heavy dependence on the nickel industry has resulted in periods of economic recession, and, since the mid-1980s, mining operations have been intermittently overshadowed by political unrest. During the 1980s and 1990s a major issue in discussions concerning the political future of New Caledonia was the extent of participation by the indigenous Melanesian people (Kanaks) in the territory's development, including its major industry. The Société Minière du Sud-Pacifique (SMSP), controlled by Kanak interests, is the most important of the six independent nickel-mining companies in New Caledonia. During negotiations over increased autonomy for the territory, local representatives sought an expansion of indigenous involvement in the nickel sector. In 1996 the SMSP and Falconbridge formed a partnership aimed at constructing a ferronickel plant in the Kanak-dominated north of the main island. In February 1998, in response to the demands of Kanak political leaders to make such a project viable, the French Government, Eramet, the SMSP and others signed the Bercy accords, whereby Eramet was to relinquish control of its promising site at Koniambo, in the north, in exchange for the Poum mine, operated by the SMSP, in the south. As the Koniambo prospect was estimated to contain about double the volume of nickel at the Poum site, the French Government agreed to pay compensation of some 1,000m. French francs to Eramet for the reduction in its reserves. In April the SMSP formed a joint venture with Falconbridge to develop the Koniambo nickel deposits. The agreement on exchanging nickel resources helped to facilitate the conclusion of the Nouméa Accord, signed in May and approved by referendum in November, which provided for a gradual transfer of powers from metropolitan France to local institutions. In compliance with the Nouméa Accord, an agreement was reached in February 1999 to enable the transfer of 30% of SLN's share capital to a newly created company representing local interests, the Société Territoriale Calédonienne de Participation Industrielle (STCPI), to be owned by the development companies of the three New Caledonian provinces. In July 2000, following two years of negotiations, New Caledonia's political leaders signed an agreement on the formation of the STCPI. The transfer of shares, reducing Eramet's interest in SLN from 90% to 60%, took place in September.

Meanwhile, in September 1998 Falconbridge began test-drilling at the Koniambo site, and by the end of 2000 the company had identified an estimated 151m.metric tons of ore, grading 2.58% nickel. The eventual aim was to construct a smelter producing 54,000 tons of nickel (contained in ferronickel) per year, and it was hoped that a feasibility study for the project could be completed by the end of 2002. Under the Bercy accords, however, control of the Koniambo deposits will revert to SLN unless construction of the smelter has begun by the end of 2005.

In 2002 Inco began the construction of a mining and processing complex at Goro, in the southern part of La Grande Terre, to exploit nickel-cobalt deposits totalling more than 200m. metric tons of ore, averaging 1.6% nickel, using new pressure acid leaching and solvent extraction technologies. Operations at the new plant, which will have an initial annual capacity of 54,000 tons (nickel in oxide), were initially scheduled to begin towards the end of 2004. In late 2002, however, Inco announced that it was slowing work at Goro pending the completion of a comprehensive review of the project that the company had initiated in response to rising estimates of its capital costs.

In Indonesia the main nickel-producing companies are PT International Nickel Indonesia (PT Inco), an Inco subsidiary, and the government-controlled PT Aneka Tambang (PT Antam). The largest of PT Antam's nickel mines is on Gebe Island, east of Halmahera, in the Maluku (Moluccas) archipelago. This mine has an estimated 27m. metric tons of reserves, with an average nickel content of 2.2%. Its output is divided, roughly equally, between high-grade saprolitic ore and lower-grade limonitic ore. Further large deposits of potentially exploitable ore containing nickel and cobalt have been located on Gag Island (about 40 km south-east of Gebe), in the province of West Papua (formerly Irian Jaya). In 1996 PT Antam formed a joint venture with BHP Minerals Ltd of Australia (now BHP Billiton) to evaluate prospects for developing these reserves. BHP, which has a 75% interest in the project, has estimated that Gag Island has total reserves (including inferred resources) of about 240m. tons of laterite ore, grading 1.35% nickel. In 2000 BHP announced provisional agreement with Falconbridge for the division, on an equal basis, of BHP's interest in the project and the consequent sharing of development costs. Falconbridge withdrew from the project in late 2001, however, although PT Antam has retained its 25% interest. PT Antam is collaborating in a separate joint venture to develop nickel and cobalt deposits at several locations on Halmahera, north of the island's Weda Bay. Its partner is a Canadian prospecting company, Weda Bay Minerals Inc, which has a 90% interest in the project. In May 1999 Weda Bay Minerals reported that tests in the areas of exploration had shown indicated and inferred reserves of 117m. tons of ore, grading 1.36% nickel. These reserves were subsequently reassessed at 202m. tons of ore, grading 1.37% nickel. In 2003 PT Inco and PT Antam planned to initiate the exploration component of a joint project to develop nickel reserves in east Pomalaa, Kolaka regency, south-east Sulawesi province. It was anticipated that production from the project, which included the construction of a new smelter, would begin in 2005.

The reserves of nickel of the People's Republic of China were estimated at 1.1m. metric tons in 2003. Mined output of nickel in 2002 was estimated at 54,500 tons. Elsewhere in the Far East, the reserves of nickel of the Philippines were estimated at 940,000 tons in 2003. Philippine mined output of nickel totalled an estimated 26,532 tons in 2002.

Japan and the USA are the world's largest nickel consumers, and Japan is, additionally, a major producer of refined nickel, all from imported materials. The republics of the former USSR and the European Union (EU), especially Germany, are also important consumers. Consumption in China, the Republic of Korea and the Republic of China (Taiwan) advanced strongly in the 1990s, however, and it was forecast in 1996 that East Asia could account for almost one-quarter of world nickel consumption by 2005.

In April 1979 dealing in nickel began on the London Metal Exchange (LME), despite opposition from the major producers, who feared that speculative free-market trading would increase volatility in prices. These fears have proved to be justified, and since the late 1970s nickel-producers have been operating in a highly cyclical trading environment, with wide and unpredictable fluctuations on the free market. However, the amount of nickel traded in this way is still relatively small. Producers typically obtain about 20 US cents per lb more for nickel that they supply than for metal that is traded on the LME. This arrangement, however, has been affected since 1991 by significant sales of nickel cathodes by the leading Russian mining company, Norilsk Nickel, which is the world's largest nickel-producer (accounting for about 19% of total refinery production in 2001). In the 1990s Norilsk provided about 80% of Russia's nickel output. In November 1998 the company approved a 10-year development programme, intended to modernize its mining, smelting and refining operations, conducted under licence, in northern Russia. It was hoped to raise some US $3,000m.–$5,000m. in foreign investment in order to finance the programme. In 2003 Norilsk approved a modernization programme, to be pursued up to 2015 at an estimated cost of US $3,600m.–$5,300m., under which, *inter alia*, the company's annual nickel production capacity would be increased by 10%, to 240,000 metric tons.

Export Price Index for Nickel
(base: 1980 = 100)

	Average	Highest month(s)	Lowest month(s)
1990	191		
1995	156		
2000	153	196 (March)	117 (Nov.)
2001	95	115 (Jan.)	78 (Oct.)
2002	114	127 (Nov., Dec.)	96 (Jan., Feb.)

On the LME the price of nickel in January 1995 stood at US $10,160 (equivalent to £6,376 sterling) per metric ton, its highest level since 1990. The surge in prices occurred despite a rise in LME stocks of nickel to a record 151,254 tons in November 1994. The London quotation was reduced to US $6,947.5 (£4,393) per ton in May 1995. It recovered to US $9,200 (£5,962) per ton in August, but eased to US $7,630 (£4,851) in October. The nickel price advanced to US $8,885 (£5,618) per ton in November and ended the year at US $7,930 (£5,106). Meanwhile, LME stocks of the metal steadily declined, reaching 44,556 tons in December.

The nickel market remained volatile in the first few weeks of 1996, while the LME's holdings continued to diminish. In early January the London price of nickel fell to US $7,455 (£4,803) per metric ton, but at the beginning of February it reached US $8,625 (£5,699). Thereafter, prices moved generally lower, and in August the London quotation was US $6,837.5 (£4,426) per ton. In the mean time LME stocks of nickel were reduced to 31,998 tons in July. The price of nickel recovered in September to US $7,625 (£4,897) per ton, but declined in December to US $6,310 (£3,732).

The LME's holdings of the metal rose to 49,776 metric tons in January 1997, but decreased to 45,174 tons in February. The London price of nickel advanced to US $8,220 (£5,082) per ton in March, but fell to US $5,862.5 (£3,480) at the end of the year. Meanwhile, LME stocks steadily expanded, reaching 67,056 tons in November. For 1997 as a whole, the average nickel price was US $7,053 (about £4,300) per ton.

Nickel prices continued to decline during the early months of 1998, and in May the London quotation fell to less than US $5,000 per metric ton. In September, despite a decrease in LME stocks of nickel to about 58,300 tons, the price of the metal was reduced to US $4,000 (£2,375) per ton. Stock levels later rose, and in October the LME nickel price declined to US $3,755 (£2,204) per ton. In November, following a strike by nickel miners in New Caledonia, the price recovered to US $4,230 (£2,549) per ton. However, in December, with LME stocks standing at about 63,500 tons, the London price of nickel fell to only US $3,745 (£2,218) per ton: its lowest level, in term of US currency, since 1987. The slump in the nickel market was attributed partly to a fall in demand for stainless steel, particularly from Asian countries experiencing financial problems.

The LME's stocks of nickel increased to 66,222 metric tons in February 1999, but declined to less than 50,000 tons in August. Meanwhile, there was a sustained rally in the nickel market, encouraged by a recovery in demand, and in early September the London price reached US $7,350 (£4,577) per ton. A week later, the LME's holdings of the metal were reduced to 48,900 tons. After rising somewhat, levels of nickel stocks resumed their downward movement in October and fell to 46,158 tons in November, as demand for stainless steel continued to strengthen. Meanwhile, prices for the metal moved higher, and the LME quotation at the end of the year was US $8,470 (£5,553) per ton: a rise of more than 100% in 12 months. The strong advance in nickel prices during 1999 was influenced by the closure of some high-cost production facilities and by technical problems at innovatory laterite nickel developments in Australia. Meanwhile, there was a strike at a major production plant in Canada.

These trends in the nickel market continued during the early weeks of 2000. In March, during a seven-week strike at SLN's operations in New Caledonia, the LME price of nickel (the official closing offer price at the end of morning trading on the LME) reached US $10,660 per metric ton: its highest level, in terms of US currency, for more than nine years. While prices of the metal rose, stocks of nickel steadily declined, with the LME's holdings reduced to less than 40,000 tons in February and to below 30,000 tons in April. The London price of nickel eased to US $9,360 per ton in April, but advanced to US $10,600 in May. The nickel price subsequently retreated, although demand remained strong, and in late July the LME quotation was US $7,500 per ton. This downward price movement occurred despite a continuing decline in LME stocks of nickel, which fell to less than 20,000 tons in June and to only 15,684 tons in July. During the remainder of 2000 LME holdings of nickel continued to decline steadily, falling to only 9,678 tons in December. During August–December the average monthly London quotation ranged between US $8,634 (£6,017) per ton, recorded in September, and US $7,310 (£5,000) per ton in December.

In early 2001 LME holdings of nickel continued to decline, falling to 8,304 metric tons in March. From April, however, they rose steadily, reaching 12,156 tons in May, 15,984 tons in June, 16,674 tons in July and 17,688 tons in August. In September LME stocks fell as low as 16,272 tons, ending the month at 16,716 tons, but thereafter, for the remainder of the year, they rose consistently, reaching 18,180 tons in October and 19,002 tons in November, ending the year at 19,188 tons. In 2001 the average monthly London quotation for nickel ranged between US $7,057 (£4,950) per ton, recorded in May, and US $4,822 (£3,321) per ton in September. LME holdings of nickel increased substantially in January 2002, reaching 24,564 tons at one point. They declined in February, to 21,174 tons, and, again, in March, to 17,004 tons. During April LME nickel holdings fell as low as 16,674 tons early in the month, but subsequently began to accumulate, rising as high as 20,778 tons. Stocks rose again, to 27,690 tons, in May, and to 29,130 tons in June before declining to 27,984 tons before the end of that month. In July LME holdings declined to 23,670 tons. During the first seven months of 2002 the average monthly London quotation for nickel ranged between US $7,139 per ton (£4,595) in July, and US $6,026 per ton (£4,237) in February.

Oil Palm (*Elaeis guineensis*)

This tree, which is native to West Africa, is widely cultivated, mainly on plantations, in the Far East. The entire fruit is of use commercially; palm oil is made from its pulp, and palm kernel oil from the seed. Palm oil is a versatile product and, because of its very low acid content (4%–5%), is almost all used in food. It is used in margarine and other edible fats; as a 'shortener' for pastry and biscuits; as an ingredient in ice cream and in chocolate; and in the manufacture of soaps and detergents. Palm kernel oil, which is similar to coconut oil, is also used for making soaps and fats. The sap from the stems of the tree is used to make palm wine, an intoxicating beverage.

Palm oil can be produced virtually through the year once the palms have reached oil-bearing age, which takes about five years. The palms continue to bear oil for 30 years or more and the yield far exceeds that of any other oil plant, with 1 ha of oil palms producing as much oil as 6 ha of groundnuts or 10–12 ha of soybeans. However, it is an intensive crop, needing considerable investment and skilled labour.

During the 1980s palm oil accounted for more than 15% of world production of vegetable oils (second only to soybean oil), owing mainly to a substantial expansion in Malaysian output. Assisted by high levels of demand from Pakistan and the People's Republic of China, palm oil considerably increased its share of world markets for vegetable oils in the early 1990s. In the late 1990s palm oil exports continued substantially to exceed those of soybean oil in international trade: in 2002 more than twice the quantity of palm oil than of soybean oil was exported worldwide. The increase in output of palm oil has posed a particular challenge to the soybean industry in the USA, which has since the mid-1970s greatly reduced its imports of palm oil. In 1988, in response to health reports that both palm and coconut oils tended to raise levels of cholesterol (a substance believed to promote arteriosclerosis in the body), several leading US food-processors announced that they were to discontinue the use of these oils. The scientific validity of these reports, however, has been vigorously challenged by palm oil producers.

In 2003, according to provisional figures, world output of palm products totalled 7.5m. metric tons of kernels and a record 28m. tons of palm oil. The Far East was the main producing area, providing 5.9m. tons of palm kernels and 24.6m. tons of palm oil.

The leading producer and exporter of palm oil is Malaysia, which, in recent years, has accounted on average for about 60% of world trade in this commodity. More than 90% of the country's production is exported, and in 2001 earnings from palm oil exports represented 2.7% of Malaysia's total export earnings. The most important markets for Malaysian palm oil are India, the European Union, the USA, Japan and Pakistan. The country accounted for about 48% of estimated world production of palm oil in 2003, compared with 46.3% in 2002. Output in 2003 amounted to 13.4m. metric tons, compared with 11.9m. tons in 2002.

Because of continued growing world demand, the Malaysian Government has actively encouraged the industry; in 1977 a Palm Oil Registration and Licensing Authority (PORLA) was established to supervise all aspects of production—research, replanting and marketing. In 1980 a palm oil exchange, offering contracts for trading in fixed amounts at specified dates, was established in Kuala Lumpur—a logical development from Malaysia's dominant position in the world palm oil trade. Since the early 1990s, however, growth in Malaysia's palm oil output has been inhibited by severe labour shortages on the oil palm plantations, and government consent has been given to the limited recruitment of foreign workers. As oil palm production costs are low, and sales contracts are denominated in US dollars, Malaysian producers benefited from an upward trend in prices during the Asian economic crisis of the late 1990s, attributable in part to restrictions on Indonesian supplies entering world trade (see below). In 1999 a number of growers' groups merged to form the Malaysian Palm Oil Association. In May 2000 a new statutory body, the Malaysian Palm Oil Board (absorbing PORLA), became operational.

In Indonesia, the second largest producer and exporter, the area under oil palm has expanded rapidly in recent years. In 1990 approximately 715,000 ha were harvested and a further 400,000 ha had been planted. In the early 1990s about 100,000–150,000 ha per year were being brought under oil palm cultivation, and production of palm oil, which had been forecast to reach 4.9m. metric tons annually by 2000, reached this target in 1996. Output rose to 5.4m. tons in 1997, to 5.9m. tons in 1998, to 6.6m. tons in 1999 and to 7.3m. tons in 2000. In 2003 Indonesia produced an estimated 10.2m. tons of palm oil, compared with 9.4m. tons in 2002, and 8.1m. tons in 2001. In 2002 Indonesia derived 3.7% of its total export revenue from sales of palm oil. The Asian economic crisis, which caused a sharp depreciation in the value of the Indonesian rupiah and led to steep rises in the government-subsidized domestic price of cooking oils, prompted the Indonesian Government to impose a three-month ban on oil palm exports from December 1997. This embargo was renewed in March 1998 and replaced in May by a 40% tax on oil palm exports. The tax was increased to 60% in July, but was reduced to 40% in early 1999. During the 1990s oil palm cultivation was encouraged in Myanmar and Papua New Guinea. At an estimated 325,000 tons in 2003, Papua New Guinea's output had more than doubled since 1990.

Production of Palm Kernels
('000 metric tons)

	2002	2003*
World total	7,059	7,503
Far East	5,520	5,948
Oceania	90	91
Leading regional producers		
China†	56	56
Indonesia	2,053	2,187
Malaysia	3,269	3,550
Papua New Guinea‡	82	83
Thailand‡	126	138
Other leading producers		
Brazil§	120	121†
Colombia	116	125‡
Congo, Dem. Repub.†	81	81
Nigeria†	608	610

* Provisional figures.
† FAO estimate(s).
‡ Unofficial figure(s).
§ Babassu kernels.

Production of Palm Oil
('000 metric tons)

	2002	2003*
World total	25,721	28,078
Far East	22,126	24,452
Oceania	350	359
Leading regional producers		
China†	220	220
Indonesia‡	9,350	10,200
Malaysia	11,909	13,354‡
Papua New Guinea‡	316	3250
Thailand‡	590	620
Other leading producers		
Colombia	528	527‡
Congo, Dem. Repub.‡	170	175
Côte d'Ivoire	276‡	276†
Ecuador	241	244
Nigeria	908	910†

* Provisional figures.
† FAO estimate(s).
‡ Unofficial figure(s).

Internationally, palm oil is faced with sustained competition from the other major edible lauric oils—soybean, rapeseed and sunflower oils—and these markets are subject to a complex and changing interaction of production, stocks and trade. In the longer term, prospects for palm oil exporters (particularly the higher-cost producers in sub-Saharan Africa) do not appear favourable. Technological advances in oil palm cultivation, particularly in the introduction of laboratory-produced higher-yielding varieties (HYVs), may also militate against the smaller-scale producer, as, for economic and technical reasons, many HYVs can be produced only on large estates, exposing small-holder cultivators to increasingly intense price pressure.

During September 1996 the import price of Malaysian palm oil in the Netherlands (c.i.f. Rotterdam) declined from US $572.5 per metric ton to US $502.5 per ton. The price advanced to US $585 per ton in February 1997, but fell to US $467.5 in July. Subsequently the market revived, and in late December palm oil traded at US $565 per ton. The upward movement continued in the early months of 1998, and in May the import price of palm oil reached US $800 per ton. In June the price eased to US $592.5 per ton. The Rotterdam price of palm oil stayed within this range for the remainder of 1998, ending the year at US $605 per ton. For the year as a whole, the European import price (c.i.f. north-west Europe) averaged US $671.08 per ton, compared with US $545.83 in 1997. However, from January 1999 there was strong downward pressure on palm oil prices, as demand fell while supplies remained plentiful. By early August the Rotterdam import price had been reduced to only US $300 per ton. The sharp fall in prices was partly due to a Chinese decision to import oilseeds rather than the oils derived from them, in an attempt to protect China's domestic processing industry. For 1999 as a whole the European import price of palm oil averaged US $436.00 per ton: 35% below the previous year's level. In 2000 the average European import price declined by almost 29%, to US $310.3 per ton. A further decline, of 7.9%, in the average import price, to US $285.7 per ton, occurred in 2001. In 2002, however, the average European import price rose by some 37%, to US $390.3 per ton. Prices continued to rise in 2003, when an average price of US $443.3 per ton (13.6% higher than the average price in 2003) was recorded. In the first half of 2004 an average monthly import price of US $512 per ton was recorded.

Export Price Index for Palm Oil
(base: 1980 = 100)

	Average	Highest month(s)	Lowest month(s)
1990	49		
1995	107		
2000	53	64 (April)	44 (Oct., Nov.)
2001	49	62 (Aug.)	40 (May)
2002	67	79 (Aug.)	56 (May)

Petroleum

Crude oils, from which petroleum fuel is derived, consist essentially of a wide range of hydrocarbon molecules which are separated by distillation in the refining process. Refined oil is treated in different ways to make the different varieties of fuel. More than four-fifths of total world oil supplies are used as fuel for the production of energy in the form of power or heating.

Petroleum, together with its associated mineral fuel, natural gas, is extracted both from onshore and offshore wells in many areas of the world. The dominant producing region is the Middle East, whose proven reserves in December 2003 accounted for 63.2% of known world deposits of crude petroleum and natural gas liquids. The Middle East accounted for 29.6% of world output in 2003. The combined proven reserves of Australia, Brunei, the People's Republic of China, Indonesia, Malaysia, Papua New Guinea, Thailand and Viet Nam amounted to 5,500m. metric tons of proven reserves (3.5% of the world total) at the end of 2003, in which year those countries accounted for some 8.9% of world production.

From storage tanks at the oilfield wellhead, crude petroleum is conveyed, frequently by pumping for long distances through large pipelines, to coastal depots where it is either treated in a refinery or delivered into bulk storage tanks for subsequent shipment for refining overseas. In addition to pipeline transportation of crude petroleum and refined products, natural (petroleum) gas is, in some areas, also transported through networks of pipelines. Crude petroleum varies considerably in colour and viscosity, and these variations are a determinant both of price and of end-use after refining.

In the refining process, crude petroleum is heated until vaporized. The vapours are then separately condensed, according to their molecular properties, passed through airless steel tubes and pumped into the lower section of a high, cylindrical tower, as a hot mixture of vapours and liquid. The heavy unvaporized liquid flows out at the base of the tower as a 'residue' from which is obtained heavy fuel and bitumen. The vapours passing upwards then undergo a series of condensation processes that produce 'distillates', which form the basis of the various petroleum products.

The most important of these products is fuel oil, composed of heavy distillates and residues, which is used to produce heating and power for industrial purposes. Products in the kerosene group have a wide number of applications, ranging from heating fuels to the powering of aviation gas turbine engines. Gasoline (petrol) products fuel internal combustion engines (used mainly in road motor vehicles), and naphtha, a gasoline distillate, is a commercial solvent that can also be processed as a feedstock. Propane and butane, the main liquefied petroleum gases, have a wide range of industrial applications and are also used for domestic heating and cooking.

Petroleum is the leading raw material in international trade. World-wide demand for this commodity totalled 78.1m. barrels per day (b/d) in 2003. The world's 'published proven' reserves of petroleum and natural gas liquids at 31 December 2003 were estimated to total 156,700m. metric tons, equivalent to about 1,147,700m. barrels (1 metric ton is equivalent to approximately 7.3 barrels, each of 42 US gallons or 34.97 imperial gallons, i.e. 159 litres).

Indonesia, China and Malaysia are the major petroleum exporters in the Far East and Australasia, and Indonesia is the world's leading exporter of liquefied natural gas (LNG). Japan, the world's second largest consumer of petroleum, is the dominant importer in the region. China, the Republic of Korea and Taiwan are also substantial consumers of petroleum products.

Within the region generally, there has been a rising level of exploration activity by international petroleum companies in recent years. Joint ventures have increasingly been sought by countries committed to policies of state control and central economic planning. These countries include not only China and Viet Nam, but also Myanmar and Cambodia.

Private-sector exploration operations have long been a central feature of the petroleum industry in Australia, Indonesia and Malaysia. Thailand's oil reserves have been boosted by small new discoveries off shore, in the Gulf of Thailand, in recent years and were estimated at 100m. metric tons in December 2003. The country's consumption fell in 1997–2001, in part owing to government efforts to reduce oil imports by increasing taxes on petroleum products. In both 2002 and 2003, however, consumption rose substantially in spite of the government's efforts to restrain it through fiscal measures. In 2003 Viet Nam produced, on average, about 372,000 b/d of petroleum at six oilfields. Almost all of the country's output is exported to destinations including Japan, the Republic of Korea and the USA, as Viet Nam has yet to develop a domestic refining industry. (The construction of a refinery in Quang Ngai province was expected to be completed in 2004, although it may not become fully operational until 2007.) The country's recoverable reserves of petroleum were estimated at 300m. tons in December 2003. Since 1998 Viet Nam has granted exploration concessions to more than 30 foreign companies, although some have subsequently terminated their operations, owing to disappointing results and difficulties related to regulation. In 2000 a consortium comprising Conoco, KNOC and SK Corpn of the Republic of Korea and France's Geopetrol announced a substantial discovery in the Cuu Long Basin, where commercial production was expected to reach 100,000 b/d by mid-2004. Viet Nam has asserted exploration claims over a large area of the South China Sea, but these are disputed by the People's Republic of China, the Philippines, Indonesia, Brunei, Malaysia and Taiwan. In 1992 the Chinese Government awarded a contract to a US company to explore for petroleum in a sector of this area, where substantial deposits of hydrocarbons are believed to exist. China has also pursued a dispute with Japan over the sovereignty of the Diaoyu (Senkaku) group of eight uninhabited islands in the East China Sea, which are believed to lie in an area of offshore production potential. The Republic of Korea and the Democratic People's Republic of Korea are also involved in disputes with China over the demarcation of exploration rights in the Yellow Sea continental shelf.

Production of Crude Petroleum
('000 metric tons, including natural gas liquids)

	2002	2003
World total	3,561,700	3,697,000
Far East and Oceania*	333,500	329,700
Leading regional producers		
Australia	31,600	26,600
Brunei	10,200	10,500
China, People's Repub.	166,900	169,300
Indonesia	63,000	57,500
Malaysia	36,700	38,800
Viet Nam	17,300	18,000
Other leading producers		
Canada	134,000	141,900
India	36,800	36,700
Iran	168,800	190,100
Iraq	99,700	65,900
Kuwait	91,800	110,200
Mexico	178,400	188,800
Nigeria	98,600	107,200
Norway	157,300	153,000
Russia	379,600	421,400
Saudi Arabia	417,300	474,800
United Arab Emirates	100,400	117,800
United Kingdom	115,900	105,600
USA	346,900	341,100
Venezuela	165,400	153,400

* Figures are the sums of output by the leading regional producers and Thailand only.

Source: BP, *Statistical Review of World Energy 2004*.

In 2002 the People's Republic of China overtook Japan as the world's second largest consumer of petroleum after the USA. China retained that rank in 2003, when consumption totalled about 275m. metric tons. Some analysts have forecast that Chinese consumption may rise to more than 500m. tons annually by 2020. The country has been a net importer of petroleum since 1993 and domestic production is oriented, primarily, towards meeting domestic consumption. The Daqing oilfield in Heilongjiang Province accounts for about 30% of domestic production, but output there is declining owing to natural depletion. China has sought foreign participation to assist in the extension of production at its second largest oilfield at Liaohe in the north east of the country, and in late 2000 the Government removed some of the regulatory obstacles to the collaboration of foreign oil companies with their Chinese counterparts. Chinese production capacity is currently, by and large, located onshore, and the development of offshore reserves is a focus of government energy strategy. Recent exploration activity has reportedly been concentrated on areas in the Bohai Sea, to the east of Tianjin, and the Pearl river mouth. In mid-2004 a substantial new discovery of oil in the area of the existing Shengli field earlier in the year was reportedly being assessed. Realizing that its dependence on imports is likely to increase, the state-owned China National Petroleum Corpn (CNPC) has purchased interests in various foreign exploration and production projects, notably a 60% share in Aktobemunaigaz of Kazakhstan. In 2004 China and Kazakhstan agreed to collaborate in the construction of a pipeline to supply western Chinese refineries with Kazakh crude oil. China has also held talks with Russia regarding the feasibility of constructing a pipeline link to allow China to import Russian crude oil. China's recoverable petroleum reserves were estimated to total 3,200m. tons in December 2003. In early 1998 the Governments of China and Taiwan approved an agreement whereby their state-owned petroleum corporations would conduct joint offshore exploration activities in the South China Sea.

China's petroleum industry is undergoing considerable change as the country fulfils obligations and attempts to meet challenges arising from membership of the World Trade Organization (WTO). In the late 1990s a degree of vertical integration was introduced through the transfer of petroleum and other energy assets to the CNPC and the China Petrochemical Corporation (SINOPEC). CNPC (via a subsidiary), SINOPEC and the China National Offshore Oil Corpn (CNOOC) all raised capital in the early 2000s by selling minority stakes for purchase via the New York and Hong Kong stock exchanges. Several of the major multinational oil companies, which will need Chinese partners before they can begin operating in Chinese markets, invested substantially in these share offerings. In 2003 a State Energy Administration was established to regulate China's oil and other energy industries.

In 1989 the Governments of Australia and Indonesia established a 'zone of co-operation' in an area of the Timor Sea known as the 'Timor Gap', to allow petroleum exploration and development to proceed, on a profit-sharing basis, in a disputed maritime area covering 62,000 sq km. This agreement was unsuccessfully challenged in the International Court of Justice by Portugal, acting on behalf of the people of its former colony of East Timor. Indonesia, whose reserves are becoming depleted and which is consequently expected to become a net importer of petroleum during 2005–10, is actively promoting exploration for new reserves in remote areas of its eastern archipelago. In 1999 it was estimated that total reserves in the 'zone of co-operation' exceeded 100m. barrels of petroleum. Following the vote for independence in East Timor, the UN, as the body responsible for a transitional administration there, signed a treaty with Australia in February 2000 to extend the terms of the 1989 agreement, although it appeared unlikely that any significant production would commence before 2003. It is hoped that production from a number of major new petroleum projects, due to begin or having already begun in 2003–06, will, in the short term at least, arrest the decline in Indonesian output. The West Seno field, off shore of East Kalimantan, was reported to be producing 40,000 b/d in mid-2004 and was expected to produce 60,000 b/d on the completion of a second phase of development in 2005. Capacity at the Belanek project in West Natuna (in the South China Sea) was to have been raised substantially by 2004. The Banyu Urip oilfield in Java, with anticipated production capacity of 100,000 b/d, is expected to come on stream in 2006. The Cepu field in Java, with reserves estimated at more than 660m. barrels, may also become operational by 2006.

At the end of 2003 Indonesia's state oil company, Pertamina, lost its monopoly on domestic production, and in mid-2004 its monopoly on the distribution of petroleum products was also due to be terminated. This liberalization will increase the scope for participation in the Indonesian petroleum industries of foreign companies, which already play a substantial role in parts of the sector. The government has reportedly stated its intention to expose the petroleum industry to full competition by 2005. Pertamina itself became a limited liability company in 2003 and is due to be fully privatized by 2006.

In 1991 the Government of Cambodia invited foreign tenders for 26 onshore and offshore blocks, and further exploration concessions were granted in the following year. In 1994 foreign capital was sought for the exploration of three additional offshore blocks and for 19 onshore blocks along the Ton Le Sap lake area and along the Mekong river, and 10 additional blocks were made available in 1996. To date, however, only four wells have been drilled, of which three have been found to contain some petroleum or natural gas. In 1990 Myanmar signed onshore exploration contracts with 10 foreign petroleum companies, although exploration activity had ceased by mid-1994, following disappointing results. Progress has been made, however, in the development of the Yadana offshore field of liquefied petroleum gas, whose recoverable reserves have been estimated at 160,000m. cu m. Initial exports of natural gas from the Yadana field to Thailand commenced in 1998. The offshore Yetagun field, with estimated recoverable reserves of 34,000m. cu m, began production in May 2000.

Crude Petroleum Distillation Capacity, 2002
('000 metric tons per year)

Australia	42,400
Brunei	450
China, People's Repub.	226,400
Indonesia	49,650
Japan	239,300
Korea, Dem. People's Repub.	3,550
Korea, Repub.	128,000
Malaysia	25,750
Myanmar	1,600
New Zealand	5,300
Philippines	21,000
Singapore	62,950
Taiwan	46,000
Thailand	34,100
Total (Far East and Australasia)	886,450

Source: US Energy Information Administration.

Petroleum consumption in the Asia-Pacific region (including Australia, China, the (Chinese) Special Administrative Region of Hong Kong, Indonesia, Japan, Malaysia, Philippines, Singapore, the Republic of Korea, Taiwan and Thailand) increased by an annual average of 3.3% during the period 1993–2003. Singapore has become the world's third largest refining centre (after Rotterdam and Houston), with a capacity of more than 1m. b/d. The Republic of Korea, Indonesia and Thailand have been expanding their refinery capacity to enable them to export refined products. Japan has been increasing its primary refining capacity, in an effort to reduce its reliance on imports, and in 2001 Viet Nam commenced the construction of a refinery with a capacity of some 140,000 b/d in Quang Ngai province in order to lessen import dependence from 2005. About 2m. b/d of new refinery capacity came on stream during 1996–98 from the expansion of existing facilities and the opening of new refinery plants in the Philippines, Thailand, the Republic of Korea, Malaysia, Taiwan and the People's Republic of China.

In 1993 the People's Republic of China became, for the first time in 30 years, a net importer of crude petroleum and petroleum products. China's petroleum import requirements have subsequently increased, and have been forecast to reach 100m. metric tons annually by 2010 (compared with an estimated 33m. tons in 1997), unless domestic production can be substantially expanded. In the late 1990s China's refineries were operating at less than 80% of full capacity, and more than 100 small refineries were closed down. The development of refining capacity is currently focused on the improvement of existing facilities rather than the construction of new ones. In particular, China needs to expand its capacity for refining heavy Middle Eastern crude petroleums, which will be imported in greater quantities in future. In the mid-1990s the Far East and Australasia region relied upon imports from the Middle East for about 71% of its crude petroleum requirements. The region's total imports of petroleum were expected to exceed 9m. b/d in the late 1990s, and these requirements were forecast to rise to 19m. b/d by 2010, of which the Middle East was projected to account for about 90%. Regional demand for LNG is also advancing rapidly. It has been forecast that increasing requirements for electricity use, and delays in the development of nuclear and coal-fired power plants, could raise levels of Far East demand to 90m. tons of LNG annually by 2010, compared with consumption of 41.9m. tons in 1991.

International petroleum prices are strongly influenced by the Organization of the Petroleum Exporting Countries (OPEC), founded in 1960 to co-ordinate the production and marketing policies of those countries whose main source of export earnings is petroleum. In 2004 OPEC had 11 members, of which Indonesia was the sole participant in the Far East and Australasia.

Export Price Index for Crude Petroleum
(base: 1980 = 100)

	Average	Highest month(s)	Lowest month(s)
1990	68		
1995	54		
2000	86	96 (Sept.)	73 (April, Dec.)
2001	72	82 (May)	55 (Nov., Dec.)
2002	76	88 (Dec.)	58 (Jan.)

The average price of crude petroleum was about US $20 per barrel in 1992, declined to only about US $14 in late 1993, and recovered to more than US $18 in mid-1994. Thereafter, international petroleum prices remained relatively stable until the early months of 1995. The price per barrel reached about US $20 again in April and May 1995, but eased to US $17 later in the year. In April 1996 the London price of the standard grade of North Sea petroleum for short-term delivery rose to more than US $23 per barrel, following reports that stocks of petroleum in Western industrialized countries were at their lowest levels for 19 years. After another fall, the price of North Sea petroleum rose in October to more than US $25 per barrel, its highest level for more than five years. The price per barrel was generally in the range of US $22–$24 for the remainder of the year.

Petroleum traded at US $24–$25 per barrel in January 1997, but the short-term quotation for the standard North Sea grade fell to less than US $17 in June. The price increased to more than US $21 per barrel in October, in response to increased tension in the Persian (Arabian) Gulf region. However, the threat of an immediate conflict in the region subsided, and petroleum prices eased. The market was also weakened by an OPEC decision, in November, to raise the upper limit on members' production quotas, and by the severe financial and economic problems affecting many countries in eastern Asia. By the end of the year the price for North Sea petroleum had again been reduced to less than US $17 per barrel.

Petroleum prices declined further in January 1998, with the standard North Sea grade trading at less than US $15 per barrel. Later in the month the price recovered to about US $16.5 per barrel, but in March some grades of petroleum were trading at less than US $13. Later that month three of the leading exporting countries—Saudi Arabia, Venezuela and Mexico (not an OPEC member)—agreed to reduce petroleum production, in an attempt to revive prices. In response, the price of North Sea petroleum advanced to about US $15.5 per barrel. Following endorsement of the three countries' initiative by OPEC, however, there was widespread doubt that the proposals would be sufficient to have a sustained impact on prices, in view of the existence of large stocks of petroleum. Under the plan, OPEC members and five other countries (including Mexico and Norway) agreed to reduce their petroleum output between 1 April and the end of the year. The proposed reductions totalled about 1.5m. b/d (2% of world production), with Saudi Arabia making the greatest contribution (300,000 b/d). In early June, having failed to make a significant impact on international prices, the three countries that had agreed in March to restrict their production of petroleum announced further reduction in output, effective from 1 July. However, petroleum prices continued to be depressed, and in mid-June some grades sold for less than US $12 per barrel. A new agreement between OPEC and other producers, concluded later that month, envisaged further reductions in output, totalling more than 1m. b/d, but this attempt to stimulate a rise in petroleum prices had little effect. The price on world markets was generally in the range of US $12–$14 per barrel over the period July–October. Subsequently there was further downward pressure on petroleum prices, and in December the London quotation for the standard North Sea grade was below US $10 per barrel for the first time since the introduction of the contract in 1986. For 1998 as a whole, the average price of North Sea petroleum was US $13.37

per barrel, more than 30% less than in 1997 and the lowest annual level since 1976. In real terms (i.e. taking inflation into account), international prices for crude petroleum in late 1998 were at their lowest level since the 1920s.

During the early months of 1999 there was a steady recovery in the petroleum market. In March five leading producers (including Mexico) announced plans to reduce further their combined output by about 2m. b/d. Later that month an OPEC meeting agreed reductions in members' quotas totalling 1.7m. b/d (including 585,000 b/d for Saudi Arabia), to operate for 12 months from 1 April. The new quotas represented a 7% decrease from the previous levels (applicable from July 1998). At the same time, four non-members agreed voluntary cuts in production, bringing total proposed reductions in output to about 2.1m. b/d. In May the London price of North Sea petroleum rose to about US $17 per barrel. After easing somewhat, the price advanced again, reaching more than US $20 per barrel in August. The upward trend continued, and in late September the price of North Sea petroleum (for November delivery) was just above US $24 per barrel. After easing somewhat, prices rose again in November, when North Sea petroleum (for delivery in January 2000) was traded at more than US $25 per barrel. The surge in prices followed indications that, in contrast to 1998, the previously agreed limits on output were, to a large extent, being implemented by producers and thus having an effect on stock levels. Surveys found that the rate of compliance among the 10 OPEC countries participating in the scheme to restrict production was 87% in June and July 1999, although it fell to 83% in October.

International prices for crude petroleum rose steadily during the opening weeks of 2000, with OPEC restrictions continuing to operate and stocks declining in industrial countries. In early March the London price for North Sea petroleum exceeded US $31.5 per barrel, but later in the month nine OPEC members agreed to restore production quotas to pre-March 1999 levels from 1 April 2000, representing a combined increase of about 1.7m. b/d. The London petroleum price fell in April to less than US $22 per barrel, but the rise in OPEC production was insufficient to increase significantly the stocks held by major consuming countries. In June the price of North Sea petroleum rose to about US $31.5 per barrel again, but later that month OPEC ministers agreed to a further rise in quotas (totalling about 700,000 b/d) from 1 July. By the end of July the North Sea petroleum price was below US $27 per barrel, but in mid-August it rose to more than US $32 for the first time since 1990 (when prices had surged in response to the Iraqi invasion of Kuwait). In New York in the same month, meanwhile, the September contract for light crude traded at a new record level of US $33 per barrel at one point. The surge in oil prices in August was attributed to continued fears regarding supply levels in coming months, especially in view of data showing US inventories to be at their lowest level for 24 years, and of indications by both Saudi Arabia and Venezuela that OPEC would not act to raise production before September. Deliberate attempts to raise the price of the expiring London September contract were an additional factor.

In early September 2000 the London price of North Sea petroleum for October delivery climbed to a new 10-year high of US $34.55, reflecting the view that any production increase that OPEC might decide to implement would be insufficient to prevent tight supplies later in the year. In New York, meanwhile, the price of light crude for October delivery rose beyond the US $35 per barrel mark. This latest bout of price volatility reflected the imminence of an OPEC meeting at which Saudi Arabia was expected to seek an agreement to raise the Organization's production by at least 700,000 b/d, in order to stabilize the market. In the event, OPEC decided to increase production by 800,000 b/d, with effect from 1 October, causing prices in both London and New York to ease. This relaxation was short-lived, however. Just over a week after OPEC's decision was announced the price of the New York October contract for light crude closed at US $36.88 per barrel, in response to concerns over tension in the Persian (Arabian) Gulf area between Iraq and Kuwait. The same contract had at one point risen above US $37 per barrel, its highest level for 10 years. These latest increases prompted OPEC representatives to deliver assurances that production would be raised further in order to curb price levels regarded as economically damaging in the USA and other consumer countries. Towards the end of September the London price of North Sea petroleum for November delivery fell below US $30 per barrel for the first time in a month, in response to the decision of the USA to release petroleum from its strategic reserve in order to depress prices.

In the first week of October 2000, however, the price of the November contract for both North Sea petroleum traded in London and New York light crude had stabilized at around US $30 per barrel, anxiety over political tension in the Middle East preventing the more marked decline that had been anticipated. This factor exerted stronger upward pressure during the following week, when the London price of North Sea petroleum for November delivery rose above US $35 per barrel for the first time since 1990. In early November 2000 crude oil continued to trade at more than US $30 per barrel in both London and New York, despite the announcement by OPEC of a further increase in production, this time of 500,000 b/d, and a lessening of political tension in the Middle East. Prices were volatile throughout November, with those of both London and New York contracts for January delivery remaining in excess of US $30 per barrel at the end of the month.

During the first week of December 2000 the price of the January 'futures' contract in both London and New York declined substantially, mainly in response to an unresolved dispute between the UN and Iraq over the pricing of Iraqi oil. On 8 December the closing London price of North Sea petroleum for January delivery was US $26.56 per barrel, while the equivalent New York price for light crude was US $28.44. Trading during the second week of December was characterized by further declines, the London price of North Sea petroleum for January delivery falling below US $25 per barrel at one point. Analysts noted that prices had fallen by some 20% since mid-November, and OPEC representatives indicated that the Organization might decide to cut production in January 2001 if prices fell below its preferred trading range of US $22–$28 per barrel. At US $23.51 per barrel on 14 December, the price of the OPEC 'basket' of crudes was at its lowest level since May. During the third week of December the price of the OPEC 'basket' of crudes declined further, to US $21.64 per barrel. Overall, during December, the price of crudes traded in both London and New York declined by some US $10 per barrel, and remained subject to pressure at the end of the month.

Continued expectations that OPEC would decide to reduce production later in the month caused 'futures' prices to strengthen in the first week of January 2001. The London price of North Sea petroleum for February delivery closed at US $25.18 on 5 January, while the corresponding price for light crude traded in New York was US $27.95. Prices remained firm in the second week of January, again in anticipation of a decision by OPEC to reduce production. Paradoxically, prices fell immediately after OPEC's decision to reduce production by 1.5m. b/d was announced on 17 January. However, it was widely recognized that the reduction had been factored into markets by that time.

Oil prices rose significantly at the beginning of February 2001, although the gains were attributed mainly to speculative purchases rather than to any fundamental changes in market conditions. On 2 February the London price of North Sea petroleum for March delivery closed at US $29.19 per barrel, while the corresponding price for light crude traded in New York was US $31.19. On 8 February prices rose to their highest levels for two months, the London price for North Sea petroleum for March delivery exceeding US $30 per barrel at one point. The upward movement came in response to a forecast, issued by the US Energy Information Administration (EIA), that the 'spot' price for West Texas Intermediate (WTI—the US 'marker' crude) would average close to US $30 per barrel throughout 2001. During the remainder of the month prices in both London and New York remained largely without direction.

During the early part of March 2001 oil prices in both London and New York drifted downwards while it remained unclear whether OPEC would decide to implement a further cut in production, and what the effect of such a reduction might be. On 9 March the London price of North Sea petroleum for April delivery was US $26.33 per barrel, while the corresponding price for light crude traded in New York was US $28.01 per barrel. By late March prices in both London and New York had declined further, the London price of North Sea petroleum for May delivery closing near to US $24.82 per barrel on 30 March, while the corresponding price for New York light crude was US $26.35.

During the first week of April 2001 crude oil prices on both sides of the Atlantic strengthened in response to fears of a gasoline shortage in the USA later in the year. On 6 April the London price of North Sea petroleum for May delivery was US $25.17 per barrel, while New York light crude for delivery in May closed at US $27.06. Strong demand for crude oil by US gasoline refiners was the strongest influence on markets throughout April as gasoline 'futures' rose markedly. Towards the end of April the price of the New York contract for gasoline for May delivery rose to the equivalent of US $1.115 per gallon, higher than the previous record price recorded in August 1990. Fears of a shortage of gasoline supplies in the USA remained the key influence on oil markets in early May 2001, with 'futures' prices rising in response to successive record prices for gasoline 'futures'. On 4 May 2001 the London price of North Sea petroleum for June delivery was US $28.19 per barrel, while that of the corresponding contract for light crude traded in New York closed at US $28.36. Prices were prevented from rising further during the week ending 4 May by the report of an unexpected increase in US inventories of crude oil. A further check came in the following week, when the International Energy Agency (IEA) reduced its forecast of world growth in demand for crude oil by 300,000 b/d to 1.02m. b/d. On 18 May, however, demand for crude oil by gasoline refiners raised the price of New York crude for June delivery to its highest level for three months, while the price of the July contract rose above US $30 per barrel. On 5 June, at an extraordinary conference held in Vienna, Austria, OPEC voted to defer a possible adjustment of its production level for one month, noting that stocks of both crude oil and products were at a satisfactory level, the market in balance and that the year-to-date average of the OPEC reference 'basket' of crudes had been US $24.8 per barrel (i.e. within the trading range of US $22–$28 per barrel targeted by the Organization). OPEC nevertheless decided to hold a further extraordinary conference in early July in order to take account of future developments. At that meeting OPEC oil ministers once again opted to maintain production at the prevailing level, emphasizing that they would continue to monitor the market and take further measures, if deemed necessary, to maintain prices within the Organization's preferred trading range. The conference appealed to other oil exporters to continue to collaborate with OPEC in order to minimize price volatility and safeguard stability. Towards the end of the month, as prices declined steadily towards (and briefly below) US $23 per barrel, the Secretary-General of OPEC, Dr Ali Rodríguez Araque, indicated that he was consulting OPEC ministers regarding the possibility of holding a further extraordinary conference early in August. Two days later, on 25 July, OPEC agreed to reduce production by a further 1m. b/d, to 23.2m. b/d, with effect from 1 September; the Organization reiterated that it was retaining the option to convene an extraordinary meeting if the market warranted it (this latest reduction, which had been agreed without a full meeting of OPEC, had been ratified by oil ministers by telephone). The Organization again expressed confidence that its action in reducing output would be matched by non-OPEC producing/exporting countries, and recognized in particular Mexico's support for its efforts. While there was general agreement that the production cut would reduce inventories, the consensus remained that demand would also decline in view of the prevailing world economic outlook. The extent of this decline remained the subject of much speculation, with a statement by Rodríguez in mid-July affirming that growth in demand would average 850,000 b/d in 2001, seeking to counter a forecast by the IEA of demand growth averaging 450,000 b/d. A decline of 1.1% in US inventories of crude oil, reported by the American Petroleum Institute (API) in late July 2001, apparently indicated that US demand was resisting, for the time being, a deceleration in economic growth, and was cited as the main reason for a recovery in the price of Brent blend North Sea petroleum, to US $24.97 per barrel, on 1 August. A further decline of the same order, reported on 7 August, raised the price of Brent to US $27.94 per barrel, and that of the OPEC 'basket' of crudes to US $24.99 per barrel. Throughout most of the remainder of August declining US inventories of both crude and refined products appeared to suggest a strength of demand that belied pessimistic assessments of US (and global) economic prospects in the near term, combining with anticipation of the reduction of OPEC production by 1m. b/d from 1 September, to support the price of the Brent reference blend and the OPEC 'basket' of crudes.

Immediately following the suicide attacks carried out against US targets in New York and Washington, DC, on 11 September 2001, as the price of Brent blend North Sea petroleum rose above US $30 per barrel, OPEC's Secretary-General was swift to emphasize the Organization's commitment to 'strengthening market stability and ensuring that sufficient supplies are available to satisfy market needs', by utilizing its spare capacity, if necessary. However, there was virtual unanimity among commentators, following the attacks, that their effect would be to worsen the prospects of the global economy, if not plunge it into recession, causing a considerable decline in demand for oil. By mid-October 2001 the price of the OPEC 'basket' of crudes had remained below US $22 per barrel, the minimum price the Organization's market management strategy was designed to sustain, since late September, and it was clear that, at the risk of adding to recessionary pressures, OPEC would have to implement a further cut in production if it was to bring the price back into its preferred trading range. In late October Venezuela, Iran, Saudi Arabia, the UAE and non-OPEC Oman all declared themselves in favour of a further cut in production. However, diplomacy undertaken by President Hugo Chávez Frías of Venezuela and Saudi Arabia's Minister of Petroleum and Mineral Resources, Ali ibn Ibrahim an-Nuaimi, had apparently made no progress in achieving its objective of persuading key non-OPEC producers Mexico, Norway and Russia to support the Organization's management strategy.

In the first week of November 2001 the price of Brent blend North Sea petroleum fell to fractionally above US $19 per barrel, while that of OPEC's 'basket' of crudes declined to US $17.56 per barrel. In Vienna, on 14 November, an extraordinary meeting of the OPEC conference observed that 'as a result of the global economic slowdown and the aftermath of the tragic events of 11 September 2001, in order to achieve a balance in the oil market, it will be necessary to reduce the supply from all oil producers by a further 2m. b/d, bringing the total reduction in oil supply to 5.5m. b/d from the levels of January 2001, including the 3.5m. b/d reduction already effected by OPEC this year. In this connection, and reiterating its call on other oil exporters to co-operate so as to minimize price volatility and ensure market stability, the Conference decided to reduce an additional volume of 1.5m. b/d, effective 1 January 2002, subject to a firm commitment from non-OPEC oil producers to cut their production by a volume of 500,000 b/d simultaneously'. The meeting acknowledged the positive response of Mexico and Oman to OPEC's efforts to balance the market. However, it was widely recognized that the success of a collaboration of the kind envisaged depended on the co-operation of Russia. Prior to the extraordinary conference held in November, Russia had indicated that it would be willing to reduce its production, estimated at more than 7m. b/d, by no more than 30,000 b/d, far less than would be necessary for OPEC's strategy to be effective. It was not until early December that Russia's Prime Minister announced the country's commitment to reducing its exports of crude by up to 150,000 b/d, and it was uncertain, in any case, whether a reduction of that magnitude could be enforced, owing to the Russian Government's lack of control over Russia's oil industry.

On 18 December 2001 the price of the OPEC 'basket' of crudes fell to US $16.62 per barrel. Thereafter, in December, however, the price recovered, in response to commitments by major non-OPEC producers to collaborate with the Organization by reducing either output or exports. On 28 December, at a consultative meeting of the OPEC conference, convened in Cairo, Egypt, OPEC confirmed its decision to implement a reduction of 1.5m. b/d in its overall production from 1 January 2002, having received assurances that Angola, Mexico, Norway, Oman and Russia would reduce their output, or, in the case of Russia, exports, of crude petroleum by a total of 462,500 b/d. BP's *Statistical Review of World Energy 2002* noted that, for the whole of 2001, the price of Brent blend North Sea petroleum had averaged US $24.77 per barrel. The average price of the reference blend was substantially less than US $20 per barrel in October–December 2001, however. In 2001, for the first time since 1993, consumption of oil world-wide declined, albeit marginally.

In January 2002 the decline in the price of the OPEC 'basket' of crudes was halted for the first time since the suicide attacks in September 2001. According to the Organization's own data, the 'basket' price rose by 4.6% in January 2002, compared with December 2001, but was 24% lower when considered on a year-on-year basis. The price of Brent blend North Sea petroleum, meanwhile, averaged US $19.48 per barrel in January. Reviewing the state of the market in January 2002, OPEC noted that it was still too early to assess the effectiveness of the reduction in output and exports that had begun on 1 January. The recovery in the average price of the OPEC 'basket' had been uneven throughout the month. Low demand, as indicated by data published by the EIA and the API, recording rises in US crude inventories, and either rises or lower-than-expected declines in inventories of distillate products, combined, in the second week of January with uncertainty regarding Russia's commitment to reducing its exports by 150,000 b/d (as it had pledged to do) to exert pressure on prices. During the third week of January, however, prices were supported, according to OPEC, by the strength of product prices, and by the USA's decision to add 22m. barrels of crude to its Strategic Petroleum Reserve. Among other factors, apparently good adherence by OPEC members to the revised quotas announced in December 2001 helped to sustain the upward trend in the final week of January 2002.

In February 2002 the average price of the OPEC 'basket' of crudes rose for the second consecutive month, recording an increase of 3.1% compared with January. As OPEC noted in its review of crude price movements in February, however, the average price of the 'basket' was 25.7% lower when considered on a year-on-year basis. The price of the OPEC 'basket' rose steadily during the first two weeks of February, supported, among other factors, by reports of an explosion at oil-gathering facilities in Kuwait that was initially expected to remove some 600,000 b/d from the market; and by increased political tension between the USA and Iraq. In the third week of February the OPEC 'basket' price moved up and down, but, overall, was weaker compared with the previous week. Continued doubts over Russia's commitment to reducing its exports was one factor that exerted downward pressure on prices. In the final week of the month prices strengthened considerably as a result of the interplay between reportedly higher product inventories, renewed political tension between the West and Iraq, and a dispute between the Venezuelan Government and employees of Petróleos de Venezuela.

In early March 2002 a meeting took place between a delegation of senior OPEC representatives and Russian government and energy officials. OPEC's objective at the meeting was to persuade Russia to continue to limit its exports of crude to 150,000 b/d during the second quarter of 2002. OPEC regarded the continued limitation of Russia exports as imperative if market stability was to be maintained at a time of seasonally weak markets for oil. However, Russia's initial commitment to reduce its exports applied only to the first quarter of 2002, with any continuation of the restriction subject to a review of market conditions. Norway had, by this time, already agreed to continue to restrict its production of crude to 150,000 b/d during April–June 2002. Moreover, it was clear in March that the market stability measures undertaken since the beginning of the year had been successful. The price of the OPEC 'basket' of crudes rose by 20% in March, compared with the previous month, although it was 5% lower when considered on a year-on-year basis. As OPEC noted, in its review of markets for crude in March, prices rose consistently throughout the month. Evidence of economic recovery in the USA was a positive factor in early March—data published by the API indicated declining US inventories of gasoline and distillate products—as was the apparent likelihood of Russia agreeing to carry over into the second quarter the restriction applied to its exports of crude. In the second week of March an intensification of the conflict between Israel and the Palestinians, and OPEC's announcement that it would maintain production at the prevailing level until the end of June at least, were additional factors that supported prices. In the third week of March prices rose above US $25 per barrel, owing, according to OPEC, to technical factors that were subsequently cancelled out by profit-taking. The upward movement continued towards the end of March, when positive inventory data combined with optimism about the sustainability of economic recovery in the USA to support prices.

From late March 2002 the escalating conflict between Israel and the Palestinians was cited as a key factor supporting crude prices. On 1 April, for the first time since 11 September 2001, the price of the OPEC 'basket' of crudes rose above US $25 per barrel. On 8 April Iraq added to the increased political tension in the Middle East by suspending its exports of crude for a period of 30 days in response to attacks by Israeli armed forces against Palestinian targets in the West Bank. Although Iraq's action was of little real consequence for crude markets, since other producers could easily compensate for the loss of its exports if necessary, it came at a time when a number of countries, including Iraq itself, Iran and Libya, had expressed their support for an embargo to be placed on the supply of oil by OPEC, in support of the Palestinian struggle against Israeli occupation. It was generally acknowledged in April that prices were inflated by a so-called war premium of some US $4–$6 per barrel, without which they would decline towards the lower end of OPEC's preferred trading range. Political upheaval in Venezuela also lent a degree of volatility to prices in mid-April, when the brief removal of President Chávez from power caused them to decline sharply. Following Chávez's reinstatement as President, the key market influence for the remainder of the month was the perception, supported by a reported decline in US inventories of crude, that demand for oil was growing in response to improved economic conditions in the USA.

A combination of apparently stronger US demand and tighter supplies was regarded as the most significant determinant of the direction of markets for crude in early May 2002. On 7 May the API announced that US inventories of crude had declined by some 4.5m. barrels in the week to 3 May, and on the day of the announcement the price of Brent blend North Sea petroleum reached US $27.14 per barrel, an increase of more than US $1 per barrel compared with the previous week. Among other factors contributing to the tightening of supplies at this time was OPEC's apparent decision not to compensate for the 30-day suspension of Iraqi exports. In the second week of May prices declined somewhat in response to data published by the US Department of Energy that indicated an increase in US inventories of crude. Towards the end of the month prices weakened again, in response to the publication of data that appeared to cast doubt on the strength of the US economic recovery.

Prices remained under pressure at the beginning of June 2002, owing to the publication of data indicating a further, unexpected increase in US inventories of crude and of distillate products in late May. Iraqi exports had also risen substantially in late May and early June. Most commentators appeared to agree with OPEC, representatives of whose members referred to a balanced market for crude in statements released early in the month, and indicated that OPEC would not alter its production quotas at its forthcoming extraordinary ministerial conference. In its assessment of the oil market in June, the IEA noted that geopolitical factors (i.e. violence in the Middle East) were now perceived as less of a risk to supplies of crude and predicted that prices would continue to weaken. US inventories of crude and gasoline were reported to have declined slightly in the first week of June. Crude stocks fell again during the second week of June, but this decline was balanced by substantial increases in inventories of gasoline and distillate products, indicating the ongoing weakness of economic recovery in the USA. At an extraordinary ministerial conference, held in Vienna on 26 June, OPEC, as expected, agreed to maintain production at the prevailing level until the end of September 2002. The Organization noted that its 'reduction measures during 2001 and 2002, supported by similar measures from some non-OPEC producers over the first half of the year, had restored relative market balance'. At the same time, OPEC observed that 'the relative strength in current market prices is partially a reflection of the prevailing political situation rather than solely the consequence of market fundamentals', and undertook to continue carefully to monitor market conditions and to take further action, if necessary, to maintain market stability.

The price of OPEC's 'basket' of crudes eased in the final week of July and the first week of August 2002, but rose to about US $26 per barrel in the second week of August owing to increased political tension in the Middle East. The threat of US military action against Iraq continued to support prices during the remainder of August and the first two weeks of September. Prior to the OPEC ministerial conference held in Osaka, Japan, on 19 September, Iraq's expressed willingness to allow the return of UN weapons inspectors caused prices to weaken, but they were subsequently supported by the Organization's decision to maintain production at the prevailing level until 12 December 2002, when the conference would meet again to review the market. The average price of the OPEC 'basket' of crudes in September, at US $27.38 per barrel, was the third highest recorded for that month since 1984. At one point during the final week of the month the closing price of the 'basket', at US $28.11 per barrel, exceeded the upper limit of the Organization's targeted trading range for the first time since November 2000. The average 'spot' quotation for Brent increased by US $1.6 to US $28.28 per barrel. WTI for immediate delivery, meanwhile, traded at an average price of US $29.52 per barrel, compared with US $28.41 per barrel in August. In its review of the markets for crude petroleum in September, OPEC noted that the price of the 'basket' had averaged US $23.48 per barrel during the first nine months of the year: US $1.23 per barrel lower than the average price recorded during the corresponding period of 2001. In the final week of September 2002 the front-month (October) WTI 'futures' contract traded on the New York Mercantile Exchange (NYMEX) rose to its highest level for 19 months, closing above US $30 per barrel at one point, having displayed considerable volatility during the month in accordance with the perceived likelihood of US military action against Iraq. In October 2002 the average 'spot' quotation for the crudes comprising OPEC's reference 'basket' declined slightly, by US $0.06, to US $27.32 per barrel, for the first time since July. The average weekly price of the 'basket' reached its highest level, US $28.24 per barrel, in the first week of the month, but declined quite steeply, by US $1 per barrel and US $1.45 per barrel, in, respectively, the third and final weeks. The average 'spot' quotation for Brent declined by US $0.59 to US $27.69 per barrel, while that of WTI fell by US $0.52 to US $29.0 per barrel. Accordingly, during the first 10 months of 2002 the cumulative average price of the OPEC reference 'basket' was US $23.91 per barrel, US $0.02 below the average price recorded during the corresponding period of 2001. In its monthly review, OPEC again identified political developments pertaining to the Middle East, in particular a statement by US President George W. Bush early in the month which appeared to lessen the likelihood of US military action against Iraq, as a key influence—at the expense of fundamental factors, such as a sharp and greater-than-anticipated increase in both OPEC and non-OPEC supplies—on markets for crude petroleum. The price of 'futures' contracts for crude petroleum responded similarly to the perceived reduction in political tension, in combination with a reported increase in OPEC production and rising US stocks. On NYMEX the front-month WTI 'futures' contract declined from US $30.83 per barrel (1 October) to only US $27.22 per barrel (31 October).

In November 2002 the average price of the OPEC reference 'basket' declined steeply, by more than US $3, to US $24.29 per barrel. In the same month, nevertheless, the year-to-date average price of the 'basket' rose above the corresponding price for 2001 for the first time, reaching US $23.94 per barrel. The price of the 'basket' was at its weakest in the second week of the month. During the second half of November quotations recovered to the extent that the average price of the 'basket' re-entered OPEC's targeted price range, by a narrow margin, in the final week of the month. In its monthly assessment of markets for crude oil, OPEC attributed the steep decline in prices in the first half of November to the dissipation of the so-called 'political/war premium' after the UN Security Council's approval, and Iraq's subsequent unconditional acceptance of, Resolution 1441. Their recovery, in the second half of the month, was attributed to the perception that OPEC would take action to curb over-production, and to colder weather conditions in Northern America and Northern Asia. The average 'spot' quotation for Brent fell by US $3.7 to US $23.99 per barrel in November, while that of WTI fell by US $2.69 to US $26.31 per barrel. The front-month NYMEX WTI 'futures' contract, meanwhile, fell to a 'low' of US $25.16 per barrel on 13 November, subsequently recovering by some US $2 per barrel before the end of the month.

During December 2002 the average price of the OPEC reference 'basket' of crudes rose by more than US $4 to US $28.39 per barrel, its highest level for two years. Average 'spot' quotations rose consistently during the month, in particular during the third and final weeks. By the end of the month the average price exceeded US $30 per barrel, one of the highest levels ever recorded in December. The average 'spot' quotation for Brent rose by US $4.84 to US $28.83 per barrel, while that of WTI increased by US $3.35 to US $29.66 per barrel. In its monthly review of markets for crude petroleum, in addition to the continued threat of military conflict in Iraq, OPEC identified declining crude oil inventories (especially in the USA) and a sharp fall in Venezuelan production and exports as a consequence of strike action as the principal market influences. Before an extraordinary meeting of the OPEC conference took place in Vienna on 12 December, there was speculation that the Organization would seek to reassert its credibility by increasing its formal quotas (or overall 'ceiling') while, at the same time, making clear its intent to bring actual production into line with the new (raised) production level in order to restore discipline. In its assessment of market conditions prior to the extraordinary conference, the IEA concluded that an increase of some 1.5m. b/d in OPEC production that had occurred over the previous three months had probably been necessary in order to avert astronomical prices and to prevent a perilous depletion of industrial stocks at a time of geopolitical tension as winter approached. In the event, as some commentators had predicted, the decision was taken at the conference to raise the production 'ceiling' for member states (excluding Iraq) from 21.7m. b/d to 23m. b/d, with effect from 1 January 2003, and to take steps to ensure that *de facto* production was reduced to within the new 'ceiling'. In its *Statistical Review of World Energy 2003*, BP noted that, as a result of production restraint and various unforeseen disruptions, OPEC's output had fallen by some 1.8m. b/d, or 6.4%, in the course of 2002. Oil demand in 2002 had been exceptionally weak for the third consecutive year, with consumption growing by only 290,000 b/d.

At US $30.34 per barrel, the average price of the OPEC reference 'basket' in January 2003 was the highest recorded for that month since 1983. Declines in the average 'spot' quotation during the first two weeks of the month were offset by an increase in the third week and a further increase, followed by a correction, in the final week. By the end of January the average price of the 'basket' had been above US $28 per barrel—the upper limit of OPEC's targeted trading range—for more than 33 consecutive days. The average 'spot' quotation for Brent increased by US $2.48 to US $31.31 per barrel, while that of WTI rose by US $3.42 to US $33.08 per barrel. The continued rise in the price of crude petroleum was attributed by OPEC to the combination of preparations for military action against Iraq in the Middle East, ongoing strike action by Venezuelan oil workers and a consequent steep decline in US inventories of crude petroleum, and cold weather conditions in the northern hemisphere. In response to these key market characteristics, at an extraordinary meeting of the OPEC conference convened in Vienna on 12 January, the Organization agreed to raise its production 'ceiling' by 1.5m. b/d to 24.5m. b/d, with effect from 1 February 2003. Production under the new 'ceiling' was to be distributed as follows (b/d, former production level in parentheses): Algeria 782,000 (735,000); Indonesia 1,720,000 (1,192,000); Iran 3,597,000 (3,377,000); Kuwait 1,966,000 (1,845,000); Libya 1,312,000 (1,232,000); Nigeria 2,018,000 (1,894,000); Qatar 635,000 (596,000); Saudi Arabia 7,963,000 (7,476,000); UAE 2,138,000 (2,007,000); Venezuela 2,819,000 (2,647,000).

During February 2003 the average price of the OPEC reference 'basket' rose by a further US $1.20 per barrel to US $31.45: the third highest average price recorded in February since 1982 and US $12.65 per barrel higher than in February 2002. 'Spot' quotations rose on a weekly basis throughout the month. The average 'spot' quotation for Brent increased by US $1.24 to US $32.54 per barrel, while that for WTI rose by US $2.55 to US $35.63 per barrel. As in the previous month, prices were boosted by the continued likelihood of war in Iraq and by very low inventories of crude petroleum and products in the USA. In its overview of market conditions in February, the IEA noted that, 'The issues of high oil prices, stocks and spare capacity have assumed a greater urgency in advance of a potential military invasion of Iraq'. Despite an increase in production of some 2m. b/d in February (of which OPEC had contributed 1.5m. b/d), producers had been unable to restrain prices and their options for further action were now limited by the consequent significant reduction in surplus production capacity.

Markets for crude petroleum were subject to a correction in March 2003. As the US-led military operation against the regime of Saddam Hussain in Iraq commenced, the so-called 'war premium', which had been a key characteristic of markets for many months, evaporated. During March the average price of the OPEC reference 'basket' fell to US $29.78 per barrel, US $1.76 per barrel lower than in February. Even so, this was the highest average price recorded in March for 20 years, and the cumulative average price for the first quarter of 2003 was, at more than US $30 per barrel, the highest ever recorded. The average 'spot' quotation for Brent fell by US $1.56 to US $30.98 per barrel, while that of WTI declined by US $1.75 to US $33.88 per barrel. The correction to prices occurred in spite of ongoing or recent disruptions to supplies from Iraq, Venezuela and Nigeria and apparently reflected consumers' confidence that measures taken by producers (such as the strategic locating of crude in major consuming areas) to offset these disruptions would be effective. At a meeting of the OPEC ministerial conference held in Vienna on 11 March it was agreed to maintain the Organization's production at its existing level, which was deemed adequate, in view of the restoration of Venezuelan production to normal levels, to meet demand.

The price of crude petroleum declined even more sharply in April 2003, the average price of the OPEC reference 'basket' falling by almost 15%, compared with the previous month, to US $25.34 per barrel. However, as OPEC noted in its monthly market review, despite the steep, consecutive monthly declines in March and April, the average price remained solidly within the Organization's targeted trading range and, indeed, the cumulative average price for the first four months of 2003 exceeded that of the corresponding period of 2002 by some US $2.79 per barrel, almost 37%. In April the greatest decline occurred in the final week of the month, when the 'basket' lost some 7% of its value, the average price having moved both up and down during the preceding three weeks. The average 'spot' quotation of Brent declined by US $5.91 to US $25.07 per barrel, while that of WTI fell by US $5.48 to US $28.40 per barrel. OPEC noted that the US-led military campaign in Iraq remained the key influence on markets for crude, with prices weakening as the likelihood of protracted hostilities diminished. Other factors that exerted downward pressure on markets for crude during the second half of April were the gradual return of Nigerian light-sweet crude to the market and the collapse of European refiners' margins. At a consultative meeting of the OPEC conference held in Vienna on 24 April, it was decided to reduce the Organization's actual production by 2m. b/d and to set a new 'ceiling' for output at 25.4m. b/d, effective from 1 June 2003. Quotas within the new 'ceiling' were as follows (b/d): Algeria 811,000; Indonesia 1,317,000; Iran 3,729,000; Kuwait 2,038,000; Libya 1,360,000; Nigeria 2,092,000; Qatar 658,000; Saudi Arabia 8,256,000; UAE 2,217,000; Venezuela 2,923,000.

In May 2003 the average price of OPEC's reference 'basket' of crude oils rose by US $0.26 to US $25.60 per barrel. In its monthly review of markets for crude, the Organization noted that the cumulative average for 2003, at US $28.37 per barrel, was almost 30% higher than the average for the corresponding period of 2002. The value of the 'basket' increased consistently throughout the month, in particular in the second week when it rose by 6.4%. The average 'spot' quotation of Brent increased by US $0.72 to US $25.79 per barrel, while that of WTI declined by US $0.17 to US $28.23 per barrel. OPEC noted the declining influence of events in Iraq and the re-establishment of fundamental factors as the key market drivers. The most important of these was the low level of US stocks of crude, reformulated gasoline and distillates.

On 11 June 2003, at an extraordinary meeting of the OPEC conference convened in Doha, Qatar, it was agreed to maintain production at the prevailing level of 25.4m. b/d, with strict compliance. The conference noted that, while markets had been stable since OPEC had reduced its actual production to 25.4m. b/d and remained well supplied, prices had recently displayed an upward trend as a consequence of the slower-than-anticipated recovery in Iraqi output and unusually low inventory levels. The average price of the OPEC reference 'basket' did, in fact, rise substantially in June: by 4.5%, to US $26.74 per barrel. At the end of June the cumulative average price of the 'basket' stood at US $28.11 per barrel, 27% higher than the average recorded for the first half of 2002. The greatest increase in the value of OPEC crudes occurred in the second week of the month, when the price of the 'basket' rose by 2.5%. This, together with smaller increases in the first and final weeks of June, was sufficient to compensate for a 5% decline in the price in the third week of the month. The average price of Brent rose by US $1.65 to US $27.44 per barrel, while that of WTI increased by US $2.48 to US $30.71 per barrel. Prices were supported in June by a further decline in US stocks of crude petroleum, especially in the early part of the month. OPEC's decision, on 11 June, to maintain production at the prevailing level had been anticipated to a large extent, and its influence on prices was regarded as relatively insignificant. Unanticipated delays in the recovery of Iraqi production were another factor that supported prices in June.

In July 2003 the average price of the OPEC reference 'basket' was US $27.43 per barrel, 2.5% higher than the average price recorded in June. During the first seven months of 2003, accordingly, the cumulative average price of the 'basket' was US $27.99 per barrel, some 24% higher than the average price recorded in the corresponding period of 2002 and just below the upper limit of OPEC's targeted trading range. In August the price of the 'basket' increased further, averaging US $28.63 per barrel. The cumulative average price for the first eight months of 2003, at US $28.07 per barrel, was, accordingly, 22% higher than that recorded in the corresponding period of 2002.

At a meeting of the OPEC conference held in Vienna on 24 September 2003, and attended by an Iraqi delegation headed by Iraq's newly appointed Minister of Oil, Ibrahim Bahr al-Ulum, the decision was taken to reduce the Organization's production ceiling to 24.5m. b/d, with effect from 1 November. This decision was made in light of OPEC's assessment of markets for crude as well supplied, and its observation that 'only normal, seasonal growth in demand [was] expected for the fourth quarter. . .'. As a result of continued increases in non-OPEC output and an ongoing recovery in Iraqi supplies, stocks were reported to be rapidly approaching normal seasonal levels. Furthermore, the supply/demand balance in the final quarter of 2003 and first quarter of 2004 indicated a 'contra-seasonal stock build-up' which, it was feared, could destabilize markets. OPEC's decision aimed to avert that threat, and the Organization appealed to non-OPEC producers to support it by likewise restraining increases in output.

In September 2003 the average price of the OPEC reference 'basket' of crude oils declined sharply, by US $2.31, to US $26.32 per barrel. In the same month the average 'spot' quotation of Brent fell by US $2.46, to US $27.32 per barrel, while that of WTI declined by as much as US $3.05, to US $27.32 per barrel. The decline in the average price of the OPEC 'basket' was most marked in the first two weeks of the month, when it averaged, respectively, US $27.61 per barrel and US $26.42 per barrel. By 25 September the cumulative decline for the month amounted to more than 12%, but in the final week of September an increase of US $1.45 per barrel was recorded. In its review of market developments in September, OPEC noted that the price of crude petroleum had hitherto drawn support from firm speculative US gasoline prices. From the beginning of the month, however, these had declined to a surprising extent, thus removing that support. At mid-September prices were regarded as 'steady', the major fundamental influence being concern over the level of US and EU heating oil stocks. The Organization naturally defended its decision to reduce the production ceiling to 24.5m. b/d—which had taken speculators by surprise—as a 'reasoned response to market fundamentals and as a proactive effort . . . to accommodate the return of Iraqi production'.

In October 2003 the average price of the OPEC 'basket' increased by US $2.22 per barrel, to US $28.54. Taking this increase into account, the cumulative average price of the OPEC 'basket' in 2003 stood at US $27.91 and was thus approaching the upper limit—US $28 per barrel—of OPEC's targeted trading range. The average 'spot' quotation of Brent, meanwhile, rose by US $2.53 per barrel, to US $29.85, and that of WTI by US $1.88, to US $30.43 per barrel. Analysts attributed the surge in prices in October to the continued psychological effect of the Organization's unexpected decision in September to reduce production. Another factor in the first three weeks of October was the perception that US stocks of heating oil and distillates, though rising, would be inadequate to meet demand during the long, severe winter that was forecast. In the final week, however, prices fell in response to more realistic formulations of the supply/demand equation likely to pertain in coming months.

The average price of the OPEC 'basket' declined marginally, by US $0.09, to US $28.45 per barrel, in November 2003. The average 'spot' quotation for Brent declined by US $1.17 per barrel, to US $28.68, while that of WTI rose by US $0.51, to US $30.94 per barrel. Quotations, which had been relatively weak in the second half of October, strengthened during most of November, before weakening again at the end of the month. The most influential fundamental factor remained continued concern regarding the level of US stocks of heating oil as the winter approached. In mid-November speculation was identified as the factor behind surges in the prices of WTI and Brent. In its review of market developments in November, OPEC noted that the speculative rally was fuelled by 'fears of inadequate crude oil and product inventories in the USA and Europe, preliminary figures showing OPEC-10 [i.e. all OPEC members except Iraq] was implementing the September 24 Agreement calling for production cuts, and the dramatic increase in speculators' long positions at the NYMEX, which indicates that the market expects prices to rise in the future'. The speculative rally ended at the close of the month with profit-taking by market participants and reduction of exposure in advance of a forthcoming extraordinary meeting of OPEC.

On 4 December 2003, at the extraordinary conference held in Vienna, OPEC decided to maintain production at its current level until further notice. During the month the average price of the OPEC 'basket' of crude oils rose by US $0.99 per barrel, to US $29.44, its highest level since March. At the same time, the average quotation of Brent rose by US $1.14, to US $29.82 per barrel, and that of WTI by US $1.21, to US $32.15. In its review of market developments in December, OPEC noted that the cumulative average 'spot' quotation of its reference 'basket' in 2003 was, at US $28.10, the highest nominal annual average since 1984. During December 2003 prices were initially supported by very strong Asian, in particular Chinese, demand for petroleum products, by declines in US commercial inventories of crude and by indications that economic recovery was well established in the USA. As the month progressed these factors were reinforced by cold weather conditions. BP noted, in its *Statistical Review of World Energy 2004*, that oil prices in 2003 had been at their highest level in nominal terms (i.e. without taking inflation into account) for 20 years. Consumption world-wide had also risen strongly, by 2.1%. Despite interruptions to the output of Iraq and Venezuela, OPEC production had increased substantially in 2003, by some 1.9m. b/d.

In January 2004 the average price of the OPEC reference 'basket' of crudes rose by US $0.89 per barrel, to US $30.33. The average 'spot' quotation of Brent rose by US $1.51, to US $31.33, in that month, while the average quotation of WTI increased by as much as US $2.18, to US $34.33. The surge in the price of WTI was attributed to a combination of very low US inventories of crude oil, which in mid-January reportedly fell below the minimum operational level that had been established in 1998, and very cold weather in eastern areas of the USA early in the month. In spite of the substantial discount in the price of North Sea and West African crudes relative to WTI, deliveries of these crudes to US markets were restricted throughout most of January 2004 by very high freight rates.

On 10 February 2004, at an extraordinary conference held in Algiers, Algeria, OPEC decided to reduce its production 'ceiling' from 24.5m. b/d to 23.5m. b/d, with effect from 1 April 2004. This decision was taken in response to projections of a 'significant supply surplus in the seasonally low demand second quarter [of 2004]', in order to avert downward pressure on prices. Production under the new 'ceiling' was to be distributed as follows (b/d, former production level in parentheses): Algeria, 750,000 (782,000); Indonesia, 1,218,000 (1,270,000); Iran, 3,450,000 (3,597,000); Kuwait, 1,886,000 (1,966,000); Libya, 1,258,000 (1,312,000); Nigeria, 1,936,000 (2,018,000); Qatar, 609,000 (635,000); Saudi Arabia, 7,638,000 (7,963,000); UAE, 2,051,000 (2,138,000); Venezuela, 2,704,000 (2,819,000). During February the average price of the OPEC 'basket' of crudes declined by US $0.77, to US $29.56 per barrel. The average 'spot' quotation of Brent declined by US $0.68 per barrel, to US $30.65, in February, while that of WTI rose by US $0.29, to US $34.62 per barrel. In its review of market developments in February, OPEC indicated the increasing importance as an influence on markets for crude of the US market for gasoline, which faced a potential supply shortage owing to a combination of steady and rising demand, the inability of Asia-Pacific refiners to supply it, and low domestic (US) inventories.

In March 2004 the average price of the OPEC reference 'basket' increased by US $2.49, to US $32.05 per barrel. It was the first time that an average price in excess of US $32 per barrel had been recorded since October 1990. The average 'spot' quotation of Brent rose by US $3.05, to US $33.70 per barrel in March 2004, while that of WTI increased by US $1.97, to US $36.59 per barrel. According to OPEC, the US gasoline market remained the key influence on markets for crude, while strong global demand for crude petroleum and economic growth were other important factors. At a conference held in Vienna on 31 March OPEC confirmed that it would adjust its production 'ceiling' to 23.5m. b/d from the beginning of April, in accordance with the decision announced in February. In the view of the Organization, prevailing high prices for crude petroleum did not reflect supply/demand fundamental factors, but rather were 'predominantly a consequence of long positions of market speculators in the futures markets coupled with a tightening in the US gasoline market in some regions, and exacerbated by uncertainties arising from prevailing geopolitical concerns . . .'.

The average price of the OPEC reference 'basket' in April 2004, at US $32.35 per barrel, was the second highest ever recorded—the highest had been US $34.32, registered in October 1990. The average 'spot' quotation of Brent fell by US $0.47 in April, to US $33.23 per barrel, while that of WTI rose by US $0.21, to US $36.80. OPEC indicated that the cumulative average price of its reference 'basket' up to 30 April 2004 was US $31.13 per barrel, compared with US $29.02 per barrel for the corresponding period of 2003. The continued strength of markets for crude was attributed to, among other things, low US inventories of gasoline, in particular of reformulated gasoline (RFG), stocks of which were reportedly some 32% lower at 30 April 2004 than at 30 April 2003, and some 41% lower than the five-year average, according to the EIA. Continued strong US demand for gasoline was likely to face further pressure, as it rose during the summer months, from new specifications, introduced from January 2004, banning the use of methyl-tertiary-butane ether for the production of RFG in the states of California, Connecticut and New York. OPEC also identified very strong demand on the part of the People's Republic of China for gasoline and its consequent reduced availability for export as an additional factor supporting gasoline markets. In addition to these fundamental factors, greater concern over unrest in petroleum-producing countries had reportedly led to increased speculative activity on crude markets, pushing prices further upward.

In May 2004 the average price of the OPEC 'basket' of crudes, at US $36.27, was the highest ever recorded. At the same time the average 'spot' quotation of Brent rose by US $4.48 per barrel to US $37.71, while that of WTI increased by US $3.31 per barrel to US $40.11. At the end of May the cumulative average price of the OPEC reference 'basket' had reached US $32.11 per barrel, an increase in excess of 13% compared with the corresponding period of 2003. OPEC indicated that prices had continued to be propelled upwards by 'tight gasoline markets, especially in the USA, where new and more stringent specifications have created operational bottlenecks'. Gasoline consumption was reported to be have been some 4.5% higher in 2004 than in 2003, and it was noted that the increase in demand had occurred before the onset of the US 'driving season'. OPEC also acknowledged the influence on speculative activity of fears of a disruption to supplies. Increasingly, in view of 'understated world oil demand for the present year, which has been revised up as much as 1m. b/d according to many market analysts,' markets had begun to question whether supplies would be sufficient to meet seasonal demand in the final part of the year, which was perceived as likely to approach total world production capacity. Nevertheless, OPEC concluded that 'the market is well supplied with crude and the current high oil prices are rooted in exuberant speculations by the futures market on perceptions of possible supply disruptions'.

In early June 2004, following attacks by militant Islamists, who were suspected of having links with al-Qa'ida, on foreign workers in the Saudi Arabian city of al-Khobar, an important centre for the Saudi oil industry, the price of crude petroleum traded in the USA rose to the record level of US $42.45 per barrel, while that of Brent approached US $40 per barrel. On 3 June, at an extraordinary conference held in Beirut, Lebanon, OPEC decided to raise its production 'ceiling' to 25.5m. b/d with effect from 1 July, and to 26m. b/d with effect from 1 August. The Organization noted that prices had continued to escalate, in spite of its efforts to ensure that markets were well supplied, as a result of continued growth in demand in the USA and the People's Republic of China, geopolitical tensions, problems in respect of refining and distribution in some consuming regions and more stringent product specifications. Production under the new 'ceiling' was to be distributed as follows from 1 July (b/d): Algeria, 814,000; Indonesia, 1,322,000; Iran, 3,744,000; Kuwait, 2,046,000; Libya, 1,365,000; Nigeria, 2,101,000; Qatar, 661,000; Saudi Arabia, 8,288,000; UAE, 2,225,000; Venezuela, 2,934,000. From 1 August the distribution of OPEC production was to be: Algeria, 830,000; Indonesia, 1,347,000; Iran, 3,817,000; Kuwait, 2,087,000; Libya, 1,392,000; Nigeria, 2,142,000; Qatar, 674,000; Saudi Arabia, 8,450,000; UAE, 2,269,000; Venezuela, 2,992,000.

In the immediate aftermath of OPEC's announcement of its new production 'ceilings' the price of both US-traded crudes and of Brent declined. However, some analysts expressed doubts over whether OPEC's action was sufficient to exercise a sustained calming effect on markets, noting that most of its members were already producing at close to capacity, in some cases in breach of prevailing quotas. The new 'ceilings' would thus, in the view of those observers, simply legitimize over-production. Representatives of some OPEC member states, meanwhile, conceded that the new production limits would not necessarily bring prices back within the Organization's preferred trading range, but would counter any perception of shortages.

In July 2004 the average price of the OPEC reference 'basket' rose to a new record level of US $36.29 per barrel—in June it had declined slightly in comparison with May, to US $34.61 per barrel, none the less the highest average price recorded in the month of June for 22 years. In July the average price of Brent increased by US $3.21 per barrel to US $38.33, while that of WTI rose by US $2.51 per barrel to US $40.69. The cumulative average price of the OPEC reference basket for January–July 2004 rose accordingly to US $33.45 per barrel, some 18% higher than the average price recorded in the corresponding period of 2003. In its assessment of markets for crude in July, OPEC attempted to clarify the market's perception of tightness in supplies, indicating that while there had been a scarcity of light, sweet crudes, 'it is also true that sour crudes are inundating the market . . .'. OPEC also noted that the global refining system was operating at close to full capacity and concluded its assessment by indicating that, apparently, 'the market has entered a new reality, one where tightness in upstream spare capacity due to lack of capacity expansion and surprisingly robust oil demand growth promises to set the scene for a new market dynamic'.

In August 2004 the price of light crude petroleum traded in New York rose as high as US $49.40 per barrel, while that of Brent, the UK reference blend, traded at US $45.15 per barrel at one point. The immediate cause of the latest price increases was identified as the escalation of unrest in Iraq, where it was feared that supplies would be disrupted by attacks by insurgents that had reportedly targeted that country's oil facilities. A longer-term factor was the continued acceleration of US and Chinese demand. It was reported at the time of the price increases in August that Chinese imports of crude petroleum had increased by 40% in the first seven months of 2004 compared with the corresponding period of 2003.

Rice (*Oryza sativa*)

Rice is the staple food in most of the countries of monsoon Asia, and about 90% of the total world area under rice lies within the region. Rice is the main food crop because it is well suited to Asian climatic conditions, producing high yields of a nutritious grain where other cereal crops will not readily grow. Wet rice cultivation is typically associated with the alluvial lowlands of monsoon Asia, but rice will tolerate a wide range of geographic, climatic and ecological conditions and is even grown under upland cultivation.

There are two cultivated species of rice, *Oryza sativa* and *O. glaberrima*. *O. sativa*, which is native to tropical Asia, is widely grown in tropical and semi-tropical areas, while the cultivation of *O. glaberrima* is limited to the high rainfall zone of West Africa. In Asia and Africa unmilled rice is referred to as 'paddy', but 'rough' rice is the common appellation in the West. After removal of the outer husk, it is called 'brown' rice. After the grain is milled to remove the bran layers, it is described as 'milled' rice. As rice loses 30%–40% of its weight in the milling process, most rice is traded in the milled form, to minimize shipping expenses.

Rice is an annual grass belonging to the same family as (and having many similar characteristics to) small grains such as wheat, oats, rye and barley. It is principally the semi-aquatic nature of rice that distinguishes it from other grain species, and this is an important factor in determining its place of origin, its dominant role in monsoon Asia and its extension to other environments. Rice varieties may broadly be classified into two main groups: *indica* and *japonica*. (There is also an intermediate or *java* type, cultivated in parts of Indonesia.) However, many rice varieties currently being grown are improved crosses of *indica* and *japonica* rices. The *indica* group, prevalent in South and South-East Asia, and covering a high proportion of the total rice area of Asia, has been associated with low yields and primitive production techniques. The *japonica* types (which predominate in East Asia), while not inherently more productive than the *indica* types, are more responsive to natural and artificial fertilizers and give higher average yields.

Production of Paddy Rice
('000 metric tons)

	2002	2003*
World total	569,527	589,126
Far East	341,644	336,048
Oceania	1,310	413
Leading regional producers		
Cambodia	3,823	4,300
China, People's Repub.	176,342	166,417†
Indonesia	51,490	52,079
Japan	11,111	9,740
Korea, Dem. People's Repub.	2,186	2,284
Korea, Repub.	6,687	6,068†
Malaysia	2,091	2,145
Myanmar†	22,780	24,640
Philippines	13,271	14,031
Taiwan	1,803	1,617†
Thailand	26,057	27,000†
Viet Nam	34,447	34,519
Other leading producers		
Bangladesh†	37,851	38,060
Brazil	10,457	10,199
Egypt	5,600†	5,800‡
India	107,600	132,013†
Nepal†	4,133	4,155
Nigeria	3,192	4,952
Pakistan	6,718	6,751†
Sri Lanka	2,859	3,071
USA	9,569	9,034

* Provisional figures.
† Unofficial figure(s).
‡ FAO estimate.

Underlying the low average rice yields in South-East Asia (including Laos, Cambodia, Viet Nam and Thailand), compared with those in East Asia (the People's Republic of China, the Koreas, Japan and Taiwan), are the lack of modern varieties which are adapted to local conditions, and the shortage of associated technology, including fertilizers and pesticides, and of adequate and timely supplies of water. Conventional rice varieties respond to increased fertilizer usage by producing more leaf and stalk instead of grain, causing the plant to lodge (fall over), decreasing net yields.

During the 1960s the International Rice Research Institute (IRRI), based in the Philippines, developed a series of stiff-stemmed, semi-dwarf varieties, bearing upright leaves, that respond positively to high rates of fertilizer application and other improved cultural practices. These improved varieties may yield as much as 10 metric tons of paddy rice per ha, while old varieties may yield less than 1 ton per ha. Most of the newly developed varieties are resistant to insect pests, diseases and some soil problems, although much work remains to be done in the selection of resistant varieties. Agronomists at the IRRI, and in national programmes, are continually developing varieties that will tolerate drought, flood, deep water and suboptimum temperatures, and high-yielding varieties (HYVs) which are designed for areas without costly irrigation or where water is scarce. Farmers cultivating these varieties may expect a reduced risk of crop failures, and are more likely to invest in other production inputs. In the late 1990s the IRRI was developing new HYVs that, it believed, could increase harvest yields by 20%–30% by the early 2000s. Hybrid rice technology developed by the IRRI has clearly demonstrated yields that are 1–1.5 tons per ha higher than modern inbred varieties commonly cultivated in such countries as the Philippines and Viet Nam. In 2003, according to the IRRI, the area under rice hybrids increased to 600,000 ha in Viet Nam and to 100,000 in the Philippines. Indonesia was reported to have also begun to commercialize this technology on, initially, some 5,000 ha. It was forecast by the IRRI in 1997 that annual world rice production would have to advance by 70% by 2025 in order to keep pace with current rates of population growth.

China developed its own high-yielding semi-dwarf varieties in the late 1950s, and these were widely disseminated in the country by the mid-1960s, prior to the release of the first IRRI varieties. Of recent interest has been the development in Hunan Province of a true hybrid rice which increases yields by as much as 20%. In 1999 it was reported that a new HYV, which could increase yields by 30%, had been developed in China. In 2000, according to the IRRI, hybrid varieties were planted on more than 16m. ha throughout China, a little more than one-half of the total area planted to rough rice in that year. A major limitation of the hybrids (apart from the high cost of seed production) is their rather long maturation period, which limits their suitability for intensive cropping patterns. The area under rough rice in China declined by about 1.8m. ha in 1991–95 (owing to factory construction, road construction and urban development, which necessitated the use of land previously used for rice cultivation), and in 1995–2000 a further 780,000 ha were taken out of rice production. However, volume output has been assisted by incentives for growers, such as the reduction of required quotas and permission for families and individuals to conduct business with production units. Although China was exporting more than 1m. metric tons of rice annually in the late 1980s, shortcomings in the country's internal transport structure necessitated imports, mainly from Thailand, of more than 200,000 tons per year. Chinese production has declined considerably since 1998, when serious floods reduced output by more than 2m. tons. More significant than the occasional adverse growing conditions that affect Chinese production, however, has been a process of agricultural reform which, according to FAO, has caused the contraction of, in particular, early and late Chinese rice crops. China's output of rice was officially forecast to increase to some 180.7m. tons in 2004, an increase of 12% compared with the previous year. According to FAO, the anticipated increase in output was due to rising market prices and government incentives to producers.

The world's leading exporter of rice in recent years has been Thailand. The average export price of Thai milled white rice ('Thai 100% B second grade', f.o.b. Bangkok) increased from US $336 per metric ton in 1995 to US $352.1 per ton in 1996, but declined to US $317 in 1997. The price averaged US $316 per ton in 1998, but subsequently slumped. to US $251.7 in 1999, US $206.7 in 2000, and US $177.4 in 2001. The decrease in prices was partly attributable to the plentiful supply of rice, with abundant stocks available, particularly in China and India. While world output of rice advanced in 1999/2000, the volume of trade declined, as many major importing countries, assisted by favourable weather, increased production. In 2002 the average export price of Thai milled white rice recovered to US $196.9 per ton. In 2003 an average price of US $200.9 per ton was recorded. In January–July 2004, on a monthly basis, the average price ranged between US $200.5 per ton (January) and US $252.5 per ton (March).

Export Price Index for Rice
(base: 1980 = 100)

	Average	Highest month(s)	Lowest month(s)
1990	74		
1995	84		
2000	74	82 (Jan., Feb.)	67 (Dec.)
2001	62	66 (Jan., Feb.)	59 (Nov., Dec.)
2002	58	61 (Jan.)	57 (Aug.–Dec.)

Thailand became the world's leading rice exporter in 1981, and rice has remained the country's principal agricultural export commodity, accounting for 2.4% of its export income in 2001. Owing to the depressed state of the world rice market since the early 1980s and the achievement of self-sufficiency by a number of Asian countries, Thailand increased its exports of rice largely at the expense of the USA. However, the decision by the USA to include rice in its Food Security Act, which came into operation in 1986 (making US-grown rice eligible for subsidized export credits), significantly affected the level of sales from Asian rice producers by undercutting the price of their exports. During the late 1980s, however, Thailand substantially expanded sales of its high quality rice to the European Community (now the European Union, EU), Russia, South Africa, Saudi Arabia and Iran. Thailand's dominance in the world rice market, in which it accounted for about 27% (7.3m. metric tons) of all rice exported in 2002, compared with 29% (7.7m. tons) in 2001 has, according to FAO, come under increasing pressure in recent years from India and Viet Nam.

Viet Nam's harvests have been greatly stimulated by the Government's recent willingness to grant long-term land tenure to farmers. In 1989 Viet Nam became the world's third largest rice exporter, after Thailand and the USA, and in that year sales by Viet Nam of low-grade rice at prices substantially below those of Thai exporters prompted the Thai Government to introduce a programme of domestic rice subsidies. Vietnamese rice exports, which totalled 1.6m. metric tons in 1990, advanced to almost 2m. tons in 1992. In 1995 Viet Nam's exports of about 2m. tons positioned the country as the world's fourth largest exporter of rice, after the USA, India and Thailand. In 1996 exports of rice provided 16.7% of Viet Nam's foreign revenue, and in 1997 the country became, after Thailand, the world's second largest exporter of rice. In early 1998, however, the Vietnamese Government imposed temporary controls on rice export volumes in order to maintain the security of domestic supplies. Nevertheless, exports for the full year were, at 3.7m. tons, 5.7% higher than those of the previous year, and, in 1999, rose to a record level of 4.5m. tons, contributing 8.9% of Viet Nam's total export earnings. In 2000, however, Viet Nam's exports of rice declined by more than 20%, to only 3.5m. tons. According to FAO, Viet Nam's rice exports totalled 3.7m. tons in 2001. In that year they contributed 3.9% of the country's total export earnings. Viet Nam's exports of rice were estimated to have fallen to 3.2m. tons in 2002. Exports of rice have, in recent years, constituted a key component of the economy of Myanmar, accounting for 10.9% of export revenue in 1995/96. In 1998/99, however, this proportion fell to only 2.4%. The total area under paddy rice in 1995/96 was estimated at 6.2m. ha, accounting for 47.5% of all land under cultivation. Production advanced strongly during the 1990s, largely as a result of the increased use of HYVs. Export growth, however, has been inhibited by the relatively low quality of Myanmar's rice in relation to that of competitors such as Thailand, and to the enforced curtailment of exports by the Government in order to avert domestic shortages. Myanmar's exports of rice totalled only 28,000 tons in 1997, but rose to more than 120,000 tons in 1998. In 1999 exports declined substantially, to only some 54,000 tons, but rose in 2000 to more than 250,000 tons, and again, very substantially, to 939,000 tons, in 2001. In 2002 FAO estimated Myanmar's exports of rice at 900,000 tons.

Indonesia, the region's third largest rice producer, sustained a fairly sharp fall in output, to about 49m. metric tons, during 1997/98, following a severe drought brought about by the climatic phenomenon known as 'El Niño'. Production remained at about 49m. tons in 1998/99, but has since risen slightly, totalling an estimated 52m. tons in 2003. The country's rice imports, principally from Thailand and Viet Nam, rose from about 350,000 tons in 1997 to about 2.9m. tons in 1998. They increased further, approaching 5m. tons, in 1999, but declined in 2000 to only 1.4m. tons. In 2001 Indonesia's rice imports fell to some 640,000 tons, but were estimated at close to 2m. tons in 2002. Indonesia, which maintained self-sufficiency in rice during 1984–93, experienced a rapid advance in domestic rice requirements during the 1990s, and became the world's principal rice importer. The country is expected to remain a net importer of rice for the foreseeable future. In recent years Australia has emerged as a significant exporter of rice, especially to Japan, although its trade is on a much smaller scale than the major exporters of the Far East. In 2002 Australia's exports of rice totalled an estimated 331,000 tons, according to FAO, compared with about 615,000 tons in 2001 and some 620,000 tons in 2000.

World exports in rice (usually in a milled or semi-milled form), which averaged about 20m. metric tons annually in the mid-1990s, fell from 25.2m. tons in 1999 to only 23.6m. tons in 2000. In 2001, according to FAO, total world exports of rice recovered to 26.8m. tons, and they were estimated to have advanced further, to 27.5m. tons, in 2002. In 2003 FAO estimated world imports of milled rice at 28.1m. tons, but forecast that these would decline to 26.5m. tons in 2004. The bulk of rice production is consumed in the producing countries, and international trade generally accounts for less than 5% of world output. Governments are the principal traders, and five countries dominate the export market in most years: Thailand, Viet Nam, the USA, China and India. Middle Eastern and African countries have traditionally been the most important markets for exports of rice from the Far East. Japan, a substantial producer of rice, has customarily enforced a strict policy of self-sufficiency, and domestic production is heavily subsidized by the government, which formerly imposed a total prohibition on commercial imports of rice. The prohibition was strongly challenged by leading exporters (notably the USA) and was temporarily relaxed in September 1993, following a substantial shortfall in the domestic rice harvest. As part of the Uruguay Round of negotiations under the General Agreement on Tariffs and Trade, Japan and the Republic of Korea accepted provisions (with effect from 1 January 1995) allowing foreign access to their rice markets. Rice imports by Japan accounted for only about 4% of domestic requirements in 1995 and, according to USDA, for about 7.5% in 2000/2001. Imported rice supplied 1% of consumption requirements by the Republic of Korea in 1995, accounted for less than 3% in 2000/2001, according to USDA, and was not expected to meet more than 4% of domestic needs by 2004.

Rubber (*Hevea brasiliensis*)

Rubber cultivation is suited to both estate and smallholder methods of farming but productivity on rubber estates is greater as it is the estates that have pioneered both the development of cultivation techniques and the improvement of the clones, or selected high-yield strains. These may either be planted as seedlings or propagated by grafting on to seedlings of ordinary trees (root stock) and planted out subsequently.

Of total estimated world production of 7,437,129 metric tons of natural rubber in 2003, Asian countries accounted for more than 90%. The principal producers are Thailand, Indonesia, Malaysia, India, the People's Republic of China, Viet Nam and Sri Lanka. In terms of export revenue, rubber is of particular significance to Indonesia, Thailand, Malaysia and Sri Lanka.

Natural rubber can be produced from plants other than *Hevea*. One of these is guayule, a desert shrub found in Mexico and the south-western USA. Commercial rubber production from guayule has been undertaken at various times, notably during periods of restricted supplies of *Hevea*. The US Department of Agriculture (USDA) has conducted research into the development and commercialization of guayule as a domestic source of natural rubber.

In 2003, according to the International Rubber Study Group (IRSG), natural rubber accounted for some 41% of world consumption of new rubber, the remainder being met by petroleum-based synthetic rubber. Almost 75% of total rubber output, both natural and synthetic, is used in motor vehicles, particularly tyres, which themselves account for about 70% of synthetic rubber demand. Epoxidized natural rubber (ENR), which is natural rubber treated with peracids (which are made by treating acetic acid with hydrogen peroxide and formic acid) to enhance its ability to resist stress, was developed in the United Kingdom and Malaysia as an alternative to synthetic rubber for use in the manufacture of car tyres.

The IRSG assessed that in 2003 world consumption of natural and synthetic rubber increased by 5.4% to 19.3m. metric tons. Output of natural rubber in that year increased by 8.8%, compared with 2002, to 8m. tons, according to the same source, while that of synthetic rubber rose by 4.9% to 11.5m. tons. In early 2004 the IRSG forecast that world rubber consumption would increase by 3.8% in 2004, compared with the previous year, and by 4.9% in 2005 (compared with 2004). At the same time, the IRSG forecast that production of natural rubber would increase by 7.5% in both 2004 and 2005.

In 1975 the Association of Natural Rubber Producing Countries (ANRPC), whose members (Malaysia, Thailand, Indonesia, Singapore, Sri Lanka, India, Viet Nam and Papua New Guinea) account for more than 80% of world rubber output (according to FAO), initiated plans to establish a buffer stock of natural rubber and to 'rationalize' supplies by keeping surplus stocks off the market. The operation of the buffer stock and the supply rationalization scheme was to be entrusted to an International Natural Rubber Council (INRC), which was established in 1978. Action to implement this plan was deferred, however, in the hope that discussions under the auspices of the UN Conference on Trade and Development (UNCTAD) would result in a more broadly-based price stabilization agreement involving both producers and consumers. In 1979 the UNCTAD conference of 55 countries reached accord on the terms of an International Natural Rubber Agreement (INRA), which became fully operational in 1982, administered by an International Natural Rubber Organization (INRO), with headquarters in Kuala Lumpur, Malaysia.

The first INRA (which, with two extensions, remained in force until January 1989, when a successor agreement, operating on similar principles, took effect) provided for an adjustable price range (quoted in combined Malaysian/Singaporean dollars and cents), maintained by means of a buffer stock. In January 1993 negotiations began, under the auspices of the INRO, to seek the formulation of a new agreement, to take effect at the expiry of the current INRA in December 1993. By mid-1993, however, wide areas of disagreement over the operation of pricing provisions of a third INRA remained unresolved. In November, after producers agreed to a 5% reduction in reference prices for buffer stock operations, the INRC decided to resume negotiations for a third INRA and to extend the current agreement by one year. These negotiations, which continued during 1994, resulted in the adoption, in February 1995, of a third INRA, to take effect from late December, on expiry of the current agreement (although full ratification was not expected to be completed until early 1996). Subsequent delays in ratification arrangements, however, caused INRA provisions technically to lapse, although the INRO secretariat continued to exercise administrative functions. At a meeting held in March 1996 the producing and consuming countries agreed to extend the date for ratification until 31 July. Although Indonesia, Thailand, Sri Lanka and Malaysia, jointly accounting for more than 90% of world rubber output at that time, had assented to the third INRA, continuing delays in ratification by a number of consumer countries, notably the USA (which represented about 29% of world rubber consumption), prompted a further extension, to 28 December, of the finalization of the agreement. The third INRA, which entered into force in March 1997, again provided for a guaranteed reference price, maintained by means of a buffer stock, the reference price being subject to review every 12 months. The duration of the agreement was four years, with the option of two extensions of one year each. The effective operation of successive INRAs was attributable, in large part, to each agreement's acceptance by virtually all producing and consuming countries. However, only about 40% of rubber consumed is natural rubber, and INRA provisions did not cover trading in synthetic rubber.

Production of Natural Rubber
('000 metric tons, dry weight)

	2002	2003*
World total	7,021	7,437
Far East	5,657	6,646
Oceania	4	4
Leading regional producers		
China, People's Repub.	527	550†
Indonesia	1,630	1,792
Malaysia	589	589†
Philippines‡	87	88
Thailand	2,456	2,615‡
Viet Nam	298	314
Other leading producers		
Brazil	96‡	96†
Côte d'Ivoire	123‡	123†
India	650	694‡
Liberia	108‡	108†
Nigeria	112	112†
Sri Lanka	91	92

* Provisional figures.
† FAO estimate.
‡ Unofficial figure(s).

In June 1992 the membership of ANRPC announced that it was to seek the creation of a single, centralized open market, probably to be based in Singapore, as a means of counteracting the large volume of private transactions between smallholder groups and consumers (estimated to represent more than 70% of the world's natural rubber trade), which the ANRPC blamed for depressed price levels. During 1993 Indonesia, Malaysia and Thailand expressed dissatisfaction with the operational record of the INRO, which, they stated, had failed to ensure adequate financial returns on rubber sales.

The persistence of depressed market conditions for natural rubber (see below) intensified the discontent of these producers with the operation of the third INRA, particularly in relation to the price support and buffer stock arrangements as operated by the INRO. Criticism was led by Malaysia and Thailand, which, during July and August 1998, indicated that they were actively considering withdrawal from the INRO, on the grounds that the guaranteed intervention price provided by the INRA was too low, and that buffer stock increases were insufficient to counteract falling prices. It was reported that Thailand and Malaysia, together with Indonesia, were contemplating withdrawal from the INRO, with the intention of forming a producers' organization that would implement voluntary supply restriction arrangements, independent of the INRO, under which production of natural rubber would be limited to 20% below world demand. The possibility was also discussed of forming a new regional rubber exchange, and of a co-ordinated marketing system to reduce competion among producers. In October 1998 Malaysia gave the requisite 12 months' notice of withdrawal from the INRO. Although Indonesia subsequently affirmed its intention to remain in the organization, Thailand announced in March 1999 that it would terminate its membership with effect from March 2000. It was widely believed that the INRO would be unable to remain in existence without the participation of Thailand, which provided about 40% of the organization's total financial contributions from producer members.

The Government of Thailand, meanwhile, indicated that it was formulating a plan to restructure its national rubber sector, with a view to doubling export revenue from this source during the period 2000–04. It was understood that this plan included a price intervention mechanism, which was to be financed in part from the retrieval of the rubber stocks held by the INRO, whose value in early 1999 was estimated at US $65m.

In April 1999 the INRO met in Kuala Lumpur, Malaysia, but was unable to persuade Malaysia and Thailand to reconsider their decision to withdraw, and in August Sri Lanka announced that it was also to terminate its membership, leaving only Indonesia, Nigeria and Côte d'Ivoire as producer members: at that time the INRO had also grouped 16 importing members, including the European Union (EU), the USA and Japan. Following a further meeting in September, the INRO announced that it was to disband with effect from mid-October, and scheduled a further meeting for December to determine the means and timing of the disposal of its rubber stockpile, estimated at 138,000 metric tons. The timetable for the disposal was strongly disputed, with Thailand and Malaysia advocating a swift sale, while the other INRO members favoured the gradual disposal of stocks to prevent price disruption; it was subsequently announced that all INRO-owned rubber would be sold by the end of June 2001. Thailand and Malaysia, meanwhile, announced that they were to co-ordinate their operations in purchasing rubber direct from growers. It was also intended to harmonize the two countries' stock disposals, and to maintain a minimum market price. Efforts to include Indonesia in this agreement were, however, unsuccessful.

In August 2002 Indonesia, Malaysia and Thailand signed an agreement establishing the International Tripartite Rubber Corpn (ITRC), with the objective of regulating the international price of rubber. This was to be achieved through ITRC purchases of rubber on world markets, and, in the event of prices falling below a certain level, from south-east Asian rubber farmers. In December 2001 the three founding countries had agreed to reduce their exports of rubber by 10% and undertaken to implement further reductions over the next two years; and to immediately reduce their production of rubber by 4%. In March 2003 Viet Nam joined the ITRC, which was renamed the International Rubber Corpn (IRC). Almost immediately after the establishment of the ITRC the price of rubber began to rise (see below), and it appeared uncertain, in early 2003, to what extent the members of the new organization would be prepared to abide by their undertaking in respect of production and exports. Both Thailand and Malaysia were reported at that time to be considering increasing their production.

Thailand, since 1993 the world's leading producer and exporter of natural rubber, carried out the replanting of more than 50% of its rubber-producing areas during the period 1961–93. During 1971–2000 Thailand's annual production of rubber increased at an average annual rate of 6.7%. The emphasis on high-yielding stock has resulted in a substantial improvement in the general quality of latex output, which has increased Thailand's competitiveness in the production of motor vehicle tyres and has led to the rapid expansion of Thailand's rubber goods industry. More than 80% of Thailand's rubber output is consigned for export, principally to Germany, the People's Republic of China, the USA, Singapore and the Republic of Korea. In 2002, according to FAO, Thailand's exports of natural rubber totalled some 730,000 tons, compared with some 684,000 tons in 2001. In 2001 2.1% of Thailand's total export earnings were derived from exports of natural rubber. During the late 1990s the Government was implementing pilot schemes to extend the cultivation of rubber into the eastern and north-eastern regions of the country. As in Malaysia (see below), most rubber production in Thailand is carried out by smallholder farmers.

Since 1995 Malaysia's output of rubber has ranged, approximately, between 550,000–1m. metric tons annually, representing 7%–17% of world production. Owing to the low international price of rubber, and a consequent acceleration in the conversion of rubber land to other crops, output declined from 1.1m. tons in 1996 to an estimated 590,000 tons in 2003. Estate production has been falling since the 1960s, owing to a shift in emphasis from rubber to oil palm and cocoa. Smallholders have become the principal producers of rubber, as the large estates have expanded the cultivation of these alternative crops, which guarantee a substantially higher return than that of rubber. As a result of government encouragement (in the form of tax incentives and higher replanting grants), average smallholder yields have risen to 700–800 kg per ha, and smallholders now account for about 70% of Malaysian rubber production (about 82% of the area planted with rubber is the property of smallholders). In 1999 the total area under rubber cultivation was estimated to be 1.5m. ha, of which 85% was located in Peninsular Malaysia. Malaysia's exports have generally accounted for about 10%–20% of the world rubber trade in recent years, according to FAO. In the mid-1990s the largest market for Malaysian rubber was the EU (accounting for 27% of total rubber exports), followed by the Republic of Korea (14%) and the USA (13%). In 2001 exports of natural rubber latex, natural rubber and gums accounted for less than 1% of Malaysia's total export earnings.

Indonesia, once the world's main source of natural rubber, has a larger area planted to rubber (3.03m. ha) than Malaysia, but only about 15% of its rubber area was planted to high-yielding varieties, resulting in average yields about 40% of those of Malaysia. More than 90% of Indonesian rubber production has traditionally been exported. In 2003, with production estimated at 1.7m. metric tons by FAO, Indonesia was the world's second largest producer of rubber, after Thailand.

Export Price Index for Natural Rubber
(base: 1980 = 100)

	Average	Highest month(s)	Lowest month(s)
1990	60		
1995	112		
2000	48	52 (Feb.)	45 (Jan., Dec.)
2001	41	44 (Jan., June)	36 (Dec.)
2002	54	63 (Sept.)	41 (Jan.)

Ribbed smoked sheets (RSS) are the principal source of reference for rubber prices in commodity markets. In April 1993 the import price of RSS in the London rubber market stood at £560 per metric ton (for delivery in June), but by September it had risen to £620 per ton (for October delivery). After easing somewhat, the London rubber price advanced strongly in 1994, rising in July to £980 per ton (for September delivery). It eased in September to £855 per ton (for October), but increased in October to £952.5 (for November). The price of rubber retreated in early November to £857.5 per ton (for December delivery), but later that month a strong upward movement in prices, resulting from growth in demand, began. In December the London quotation reached £1,000 per ton for the first time. The surge in prices continued in the early weeks of 1995, and in February rubber traded at a record £1,230 per ton (for March delivery). This peak was matched in March (for April delivery). Thereafter, the rubber price moved generally downward, falling to less than £1,000 per ton in June and to only £837.5 (for September delivery) in early August. The London quotation reached £1,000 per ton again in October, and rose in December to £1,150 (for delivery in January 1996). In April 1996 the London price of rubber was below £1,000 per ton, and in December it declined to £792.5 (for delivery in January 1997). During the first half of 1997 the downward trend continued, and by mid-year the rubber price had fallen to £657.5 per ton (for August delivery). The rubber market remained depressed, and in October the London quotation was reduced to £545 per ton (for December delivery). The price remained at this level for the remainder of the year. For 1997 as a whole, despite the rise in demand (see above), prices for all grades of rubber on the London market declined by an average of 47%.

In January 1998 the London price of RSS (for delivery in February) fell to only £427.5 per metric ton. The slump in the rubber market was partly a consequence of the economic crisis affecting eastern Asia, including many rubber-producing countries. In February the rubber price recovered to £550 per ton (for March delivery), but in June it was reduced to £450 (for delivery in July or August). The price was maintained within this range for the remainder of 1998, ending the year at £460 per ton (for delivery in January 1999).

During the early months of 1999 there was further downward movement in rubber prices, and in March the London quotation (for May delivery) declined to £412.5 per metric ton. The market subsequently recovered, but in early August, with the INRO on the verge of collapse (see above), the price of RSS (for delivery in September) fell to only £397.5 per ton, its lowest level since early 1976. Later in August 1999 rubber prices in Tokyo, Japan (the world's largest market for trading rubber 'futures' contracts), were at their lowest for 30 years. The London price of rubber advanced in November to £530 per ton (for December delivery), after unfavourable weather had reduced output. On a 'spot' price basis, the London quotation averaged £451.3 per ton for the whole of 1999, rising to £511.0 per ton, on average, in 2000. In 2001 the London 'spot' quotation for RSS averaged £476.3 per ton. During the first six months of 2002 the average London 'spot' price for RSS ranged between £460.0 per ton in January and £852.5 per ton in June. The very substantial increase in the price of physical rubber in June (in May the price had averaged only £523.5 per ton) occurred, according to the IRSG, as a result of unfavourable weather conditions, defaults by rubber shippers on their forward contracts, and the forced return of buyers to 'spot' markets.

The International Rubber Study Group (IRSG) is an intergovernmental body which was established in 1944 to provide a forum for the discussion of problems affecting the production and consumption of, and trade in, both natural and synthetic rubber. In 2004 the contributing members of the IRSG were 17 in number. In 2003 IRSG members collectively accounted for about 52% of all natural and synthetic rubber consumed worldwide, and for 74% of global natural rubber and 65% of global synthetic rubber production. The Group's secretariat, based in London, regularly publishes current statistical information on rubber production, consumption and trade.

Sugar

Sugar is a sweet crystalline substance which may be derived from the juices of various plants. Chemically, the basis of sugar is sucrose, one of a group of soluble carbohydrates which are important sources of energy in the human diet. It can be obtained from trees, including the maple and certain palms, but virtually all manufactured sugar is derived from two plants, sugar beet (*Beta vulgaris*) and sugar cane, a giant perennial grass of the genus *Saccharum*.

Production of Sugar Cane
('000 metric tons)

	2002	2003*
World total	1,338,169	1,333,253
Far East	245,164	234,283
Oceania	35,849	39,757
Leading regional producers		
Australia	32,260	36,012†
China, People's Repub.	92,203	92,370
Indonesia†	25,530	25,600
Philippines	27,203	25,835†
Thailand	74,258	64,408†
Other leading producers		
Argentina‡	19,250	19,250
Brazil	363,721	386,232
Colombia‡	35,800	36,600
Cuba	34,700	22,902
India	297,208	289,630
Mexico	45,635	45,127
Pakistan	48,042	52,056
South Africa	22,928	20,601‡
USA	32,253	31,301

* Provisional figures.
† Unofficial figure(s).
‡ FAO estimate(s).

Sugar cane, found in tropical areas, grows to a height of up to 5 m. The plant is native to Polynesia, but its distribution is now widespread. It is not necessary to plant cane every season as, if the root of the plant is left in the ground, it will grow again in the following year. This practice, known as 'ratooning', may be continued for as long as three years, after which yields begin to decline. Cane is ready for cutting 12–24 months after planting, depending on local conditions. Much of the world's sugar cane is still cut by hand, but rising costs are hastening the change-over to mechanical harvesting. The cane is cut as close as possible to the ground, and the top leaves, which may be used as cattle fodder, are removed.

After cutting, the cane is loaded by hand or by machine into trucks or trailers and towed directly to a factory for processing. Sugar cane rapidly deteriorates after it has been cut and should be processed as soon as possible. At the factory the cane passes first through shredding knives or crushing rollers, which break up the hard rind and expose the inner fibre, and then to squeezing rollers, where the crushed cane is subjected to high pressure and sprayed with water. The resulting juice is heated and lime is added for clarification and the removal of impurities. The clean juice is then concentrated in evaporators. This thickened juice is next boiled in steam-heated vacuum pans until a mixture or 'massecuite' of sugar crystals and 'mother syrup' is produced. The massecuite is then spun in centrifugal machines to separate the sugar crystals (raw cane sugar) from the residual syrup (cane molasses).

Production of Sugar Beets
('000 metric tons)

	2002	2003*
World total	256,946	233,487
Far East	16,918	10,351
Leading regional producers		
China, People's Repub.	12,820	6,190
Japan	4,098	4,161
Other leading producers		
France	33,450	29,238
Germany	26,794	26,400†
Poland	13,434	10,900
Turkey	16,523	13,090
Ukraine	14,453	13,340
USA	25,145	27,764

* Provisional figures.
† FAO estimate.

The production of beet sugar follows the same process, except that the juice is extracted by osmotic diffusion. Its manufacture produces white sugar crystals which do not require further refining. In most producing countries, it is consumed domestically, although the European Union (EU), which accounts for about 13% of total world sugar production, is a net exporter of white refined sugar. Beet sugar usually accounts for more than one-third of world production. Production data for sugar cane and sugar beet generally cover all crops harvested, except crops grown explicitly for feed. The third table covers the production of raw sugar by the centrifugal process. In the late 1990s global output of non-centrifugal sugar (i.e. produced from sugar cane which has not undergone centrifugation) was about 14m. metric tons per year. The main producer of non-centrifugal sugar, almost all of which is used for local consumption, is India, with an output of 9.9m. tons in 1998.

Production of Centrifugal Sugar
(raw value, '000 metric tons)

	2002	2003*
World total	145,505	146,091
Far East	22,811	23,224
Oceania	5,355	5,756
Leading regional producers		
Australia	4,987	5,371
China†	11,754	11,112
Indonesia	1,869	1,910†
Philippines†	1,965	2,237
Thailand	5,947	6,631†
Other leading producers		
Brazil†	23,810	24,780
Colombia	2,523	2,574†
Cuba	3,603	2,205
France	5,139	4,282†
Germany	4,395	4,158†
India	20,475	22,140
Mexico†	5,180	5,240
Pakistan†	3,507	4,004
Poland	2,250	1,899
South Africa	2,626	2,626‡
Turkey†	2,110	1,875
Ukraine	1,554	1,576†
USA	7,602	8,118

* Provisional figures.
† Unofficial figure(s).
‡ FAO estimate.

Most of the world's output of raw cane sugar is sent to refineries outside the country of origin, unless the sugar is for local consumption. India, Thailand, Brazil and Cuba are among the few cane-producers that export part of their output as refined sugar. The refining process further purifies the sugar crystals and eventually results in finished products of various grades, such as granulated, icing or castor sugar. The ratio of refined to raw sugar is usually about 0.9:1.

As well as providing sugar, quantities of cane are grown in some countries for seed, feed, fresh consumption, the manufacture of alcohol and other uses. Molasses may be used as cattle feed or fermented to produce alcoholic beverages for human consumption, such as rum, a distilled spirit manufactured in Caribbean countries. Sugar cane juice may be used to produce ethyl alcohol (ethanol). This chemical can be mixed with petroleum derivatives to produce fuel for motor vehicles. The steep rise in the price of petroleum after 1973 made the large-scale conversion of sugar cane into alcohol economically attractive (particularly to developing nations), especially as sugar, unlike petroleum, is a renewable source of energy. Several countries developed alcohol production by this means in order to reduce petroleum imports and to support cane-growers. The blended fuel used in cars is known as 'gasohol', 'alcogas' or 'green petrol'. The pioneer in this field was Brazil, which operates the largest 'gasohol' production programme in the world. During the late 1990s the use of ethanol by Brazilian motorists increased rapidly, as, despite the additional costs involved in adapting motor vehicles to ethanol consumption, the price of this fuel, which can be mixed with petrol or used in undiluted form, had fallen (owing to high levels of stocks) to less than one-half of the price of petrol. In 2003 India was reported to have initiated a project to produce ethanol from molasses and sugar cane juice for fuel purposes. In the Far East and Australasia unsuccessful attempts to establish 'gasohol' production were made in the Philippines and Papua New Guinea, while plans to initiate a 'gasohol' project in Thailand have been indefinitely postponed.

After the milling of sugar, the cane has dry fibrous remnants known as bagasse, which is usually burned as fuel in sugar mills but can be pulped and used for making fibreboard, particle board and most grades of paper. As the costs of imported wood pulp have risen, cane-growing regions have turned increasingly to the manufacture of paper from bagasse. In view of rising energy costs, some countries are encouraging the use of bagasse as fuel for electricity production in order to reduce expenditure of foreign exchange on imports of petroleum. Another by-product, cachaza (which was formerly discarded), is currently being utilized as an animal feed.

In recent years sugar has encountered increased competition from other sweeteners, including maize-based products, such as isoglucose (a form of high-fructose corn syrup or HFCS), and chemical additives, such as saccharine, aspartame (APM) and xylitol. Consumption of HFCS in the USA was equivalent to about 42% of the country's sugar consumption in the late 1980s, while in Japan and the Republic of Korea HFCS accounted for 19% and 25%, respectively, of domestic sweetener use. APM was the most widely used high-intensity artificial sweetener in the early 1990s, although its market dominance was under challenge from sucralose, which is about 600 times as sweet as sugar (compared with 200–300 times for other intense sweeteners) and is more resistant to chemical deterioration than APM. In the late 1980s research was being conducted in the USA to formulate means of synthesizing thaumatin, a substance derived from the fruit of a west African plant, *Thaumatoccus daniellii*, which is several thousand times as sweet as sugar. If, as has been widely predicted, thaumatin can be commercially produced, it could obtain a substantial share of the markets for both sugar and artificial sweeteners. In 1998 the USA approved the domestic marketing of sucralose, the only artificial sweetener made from sugar. Sucralose was stated to avoid many of the taste problems associated with other artificial sweeteners.

The major sugar producers in East and South-East Asia are the People's Republic of China, Thailand, Pakistan, Indonesia and the Philippines. Of these, only China and Indonesia were net importers rather than net exporters in the late 1990s. The main exporters in the Far East and Australasia are Australia, the Philippines and Thailand; the region's main importers are Indonesia and Japan. Indonesia, following unsuccessful efforts to achieve self-sufficiency in sugar, has become one of the main buyers of Indian sugar. Indonesia's total imports were estimated at about 1m. metric tons (raw value) in 2002, having risen as high as 2.3m. tons in 1999. Thailand, after seeking a reduction in sugar production to encourage a change-over to other crops, has promoted a controlled expansion of sugar production. However, sugar exports (both raw and refined) accounted for less than 1% of Thailand's export revenue in 2001. Thailand's exports of sugar (raw value) totalled an estimated 4.2m. tons in 2002, compared with 3.3m. tons in 2001. In the case of Australia, sugar is the most important export crop after wheat. The export trade accounts for about 80% of the country's annual raw sugar production. Sugar (including sugar preparations and honey) accounted for 1.5% of Australian export earnings in 1997/98. In 1998/99 and 1999/2000, however, the contribution of sugar to total export revenues was only 0.2%. The sugar industry in the Philippines was, for many years, an important source of foreign-exchange earnings. Since 1983, however, the country's output of raw sugar has reflected depressed price levels and curbs on production. In the mid-1990s the Government of Viet Nam was seeking foreign investment in the country's sugar sector, with the aim of reducing domestic dependence on sugar imports.

The first International Sugar Agreement (ISA) was negotiated in 1958, and its economic provisions operated until 1961. A second ISA did not come into operation until 1969. It included quota arrangements and associated provisions for regulating the price of sugar traded on the open market, and established the International Sugar Organization (ISO) to administer the agreement. However, the USA and the six original members of the European Community (EC, now the EU) did not participate in the ISA, and, following its expiry in 1974, it was replaced by a purely administrative interim agreement, which remained operational until the finalization of a third ISA, which took effect in 1978. The new agreement's implementation was supervised by an International Sugar Council (ISC), which was empowered to establish price ranges for sugar-trading and to operate a system of quotas and special sugar stocks. Owing to the reluctance of the USA and EC countries (which were not a party to the agreement) to accept export controls, the ISO ultimately lost most of its power to regulate the market, and since 1984 the activities of the organization have been restricted to compiling statistics and providing a forum for discussion between producers and consumers. Subsequent ISAs, without effective regulatory powers, have been in operation since 1985. (For detailed information on the successive agreements, see *The Far East and Australasia 1991.*)

In tandem with world output of cane and beet sugars, stock levels are an important factor in determining the prices at which sugar is traded internationally. These stocks, which were at relatively low levels in the late 1980s, increased significantly in the 1990s, although not, according to USDA data, in each successive trading year (September–August). In the early 1990s rises in stocks were due partly to the disruptive effects of the Gulf War on demand in the Middle East (normally a major sugar-consuming area), and were also a result of considerably increased production in Mexico and the Far East. Another factor was the increased area under sugar cane and beet in the EU and in Australia. According to data released by USDA, world stocks of sugar declined from about 39m. metric tons in 2000/2001 to some 36m. tons in 2001/02. In 2002/03, when world sugar production totalled about 149m. tons and consumption some 138m. tons, stocks of sugar held world-wide rose to about 41m. tons. These increases were succeeded, in 2003/04, when world sugar production was estimated at about 142m. tons and consumption at about 140m. tons, by a decline in stocks held to an estimated 36m. tons. USDA forecast, on the basis of estimated world production of 142.4m. tons and estimated world consumption of some 141.95m. tons, that world sugar stocks would decline again, to about 31m. tons, in 2004/05.

Most of the world's sugar output is traded at fixed prices under long-term agreements. On the free market, however, sugar prices often fluctuate with extreme volatility.

In February 1992 the import price of raw cane sugar on the London Market was only US $193.0 (equivalent to £107.9 sterling) per metric ton, its lowest level, in terms of US currency, for more than four years. The London price of sugar advanced to US $324.9 (£211.9) per ton in May 1993, following predictions of a world sugar deficit in 1992/93. Concurrently, sugar prices in New York were at their highest levels for three years. The London sugar price declined to US $237.2 (£157.9) per ton in August 1993, but increased to US $370.6 (£237.4) in December 1994.

Export Price Index for Sugar
(base: 1980 = 100)

	Average	Highest month(s)	Lowest month(s)
1990	45		
1995	48		
2000	28	34 (Oct.)	20 (March)
2001	30	34 (Jan., May)	24 (Oct.)
2002	25	29 (Dec.)	21 (June)

The rise in sugar prices continued in January 1995, when the London quotation for raw sugar reached US $378.1 (£243.2) per metric ton, its highest level, in dollar terms, since early 1990. Meanwhile, the London price of sugar was US $426.5 (£274.4) per ton, the highest quotation for four-and-a-half years. The surge in the sugar market was partly a response to forecasts of a supply deficit in the 1994/95 season. Prices later eased, and in April 1995 raw sugar was traded in London at US $326.0 (£202.1) per ton. The price recovered to US $373.1 (£235.6) per ton in June, but retreated to US $281.5 (£179.4) in September.

During the early months of 1996 the London price of raw sugar reached about US $330 per metric ton. In May the price was reduced to US $262.9 (£175.3) per ton, but in July it rose to US $321.4 (£206.3), despite continued forecasts of a significant sugar surplus in 1995/96. The sugar price fell to US $256.3 (£156.4) per ton in December.

In January 1997 the price of raw sugar declined to US $250.8 (£154.0) per metric ton, but in August it reached US $288.5 (£179.1). In terms of US currency, the London sugar price advanced later in the year, standing at US $298.5 (£178.3) per ton in December.

International sugar prices moved generally downward in the first half of 1998. In June, following forecasts that world output of raw sugar in 1997/98 would be significantly higher than in the previous crop year (leading to a further rise in surplus stocks), the London price was reduced to US $185.0 (£113.1) per metric ton. In July the quotation for raw sugar recovered to US $222.6 (£135.0) per ton, but in September, following forecasts of a higher surplus in 1998/99, the price fell to only US $168.8 (£100.2). The sugar price stayed within this range for the remainder of 1998, ending the year at US $196.9 (£118.3) per ton.

In January 1999 the London price of raw sugar reached US $217.2 (£132.4) per metric ton, but from February the free market was affected by strong downward pressure, as sugar supplies remained plentiful. In April the London sugar price was reduced to only US $127.3 (£79.0) per ton, its lowest level, in terms of US currency, since late 1986. Meanwhile, the short-term quotation in the main US sugar market declined to about 4.6 US cents per lb, also a 12-year 'low'. The slump in international sugar prices was partly a consequence of the severe economic problems affecting Russia (normally the world's leading importer of sugar) and countries in eastern Asia. The decline in demand from these areas coincided with high output, as a result of generally favourable growing conditions, in sugar-producing countries. The London price recovered to US $166.2 (£105.1) per ton in June 1999, but declined to US $131.4 (£84.4) in July. A renewed advance took the price to US $184.3 (£111.4) per ton in October. At the end of the year the London sugar quotation stood at US $162.4 (£100.8) per ton.

During the early weeks of 2000 the international sugar market was weak, and at the end of February the London price of raw sugar was reduced to US $126.5 (£80.1) per metric ton, its lowest level, in terms of US currency, for more than 13 years. However, there was subsequently a strong recovery in sugar prices, in response to reports of reductions in output and planting, while demand rose. At the end of July the London quotation was US $264.0 (£176.3) per ton. Meanwhile, the New York price of sugar rose during that month to its highest level for more than two years. In early August the price of white sugar traded in London rose to its highest level, US $279.0 (£185.3) per ton, for more than two-and-a-half years. The average price for raw sugar on the main New York market was in excess of 11 US cents per lb in August, the first time this level had been achieved since February 1998. These recoveries were not sustained, however, and by 22 September the London quotation for raw sugar had fallen to US $226.5 (£155.3) per ton. The London quotation for raw sugar remained within the established range until 10 October, when it rose to US $266.5 (£183.9) per ton. On 17 October the London quotation stood at US $281.5 (£195.5) per ton, but thereafter the price fell back to and remained within the previously established range, ending the year at US $249.4 (£166.9) per ton. A persistent characteristic of the market from October 2000 was the erosion of the premium at which white sugars normally trade above raw supplies. In January 2001 the ISO forecast that the premium would remain depressed for several months to come, owing to the abundance of supplies of white sugar and the availability of substantial quantities for export from India. At the end of the first quarter of 2001 the London quotation for raw sugar stood at US $216.7 per ton, declining further, to US $205.3 (£143.9) per ton on 3 April. The average New York price for March 2001 was 9.6 US cents per lb. At the end of May the London price had advanced somewhat, to US $221.8 per ton. However, prices subsequently declined sharply, particularly in October, when the New York price averaged 7.2 US cents per lb, compared with 8.6 US cents per lb in September. Reports of damage caused by 'Hurricane Michelle' to the crops of several Caribbean countries, most notably Cuba, supported a brief recovery in November, before the downward trend continued into 2002. The New York price again declined below 6 US cents per lb, before recovering above that level in August.

The Group of Latin American and Caribbean Sugar Exporting Countries (GEPLACEA) complements the activities of the ISO (whose 71 members, on the basis of data for 2002, account for 83% of world sugar production, 65% of world sugar consumption, 92% of world sugar exports and 36% of world sugar imports) as a forum for co-operation and research. The USA withdrew from the ISO in 1992, following a disagreement over the formulation of members' financial contributions. The USA had previously provided about 9% of the ISO's annual budget.

Tea (*Camellia sinensis*)

Tea is a beverage made by infusing in boiling water the dried young leaves and unopened leaf-buds of the tea plant, an evergreen shrub or small tree. Black and green tea are the most common finished products. The former accounts for the bulk of the world's supply and is associated with machine manufacture and plantation cultivation, which guarantees an adequate supply of leaf to the factory. The latter, produced mainly in the People's Republic of China and Japan, is grown mostly on smallholdings, and much of it is consumed locally. There are two main varieties of tea, the China and the Assam, although hybrids may be obtained, such as Darjeeling. In this survey, wherever possible, data on production and trade relate to made tea, i.e. dry, manufactured tea. Where figures have been reported in terms of green (unmanufactured) leaf, appropriate allowances have been made to convert the reported amounts to the approximate equivalent weight of made tea.

Total recorded tea exports by producing countries achieved successive records in each of the years 1983–90. World exports (excluding transactions between former Soviet republics) declined in 1991 and 1992, but rose by 13.7% in 1993. However, the total fell by 10.3% in 1994. Export volume increased again in 1995 and 1996, rising further, to 1,203,785 tons, in 1997. World tea exports reached a new record of 1,304,896 metric tons in 1998, but eased to 1,260,190 tons in 1999. In 2000 export volume attained a new record level of 1,328,395 tons, a total exceeded in 2001 when world export volume amounted to 1,388,640 tons. Exports of tea world-wide attained a new record level, of 1,422,952 tons, in 2002, but were estimated to have fallen by 3.1%, to 1,378,739 tons, in 2003. Global production of tea reached an unprecedented level in 1998, with record crops in all of the major producing countries (India, China, Kenya and Sri Lanka). In 1999, however, world output declined (from 2,991,001 tons in 1998) to 2,908,450 tons, although China and Sri Lanka again reported record crops. In 2000 production increased to 2,908,525 tons, spurred, once again, by record crops in China and Sri Lanka. In 2001, for the first time, world production of tea exceeded 3m. tons. Record crops in China and Kenya contributed to an increase in global output of 4.6%, to 3,041,018 tons. In 2002 China and Sri Lanka again achieved record production, but world output of tea rose only marginally, to 3,056,026 tons. In 2003 production of tea world-wide increased by 1.3%, to an estimated 3,096,707 tons. China was estimated to have achieved record production in 2003, for an eighth successive year. Exports from China (whose sales include a large proportion of green tea) have exceeded those of India in every year since 1996.

Production of Made Tea
('000 metric tons)

	2002	2003
World total*	3,056.0	3,096.7
Far East*	1,112.3	1,129.6
Oceania*	7.6	7.9
Leading regional producers		
China, People's Repub.[1]	745.4	770.0*
Indonesia[2]	172.8	168.0*
Japan[3]	84.2	87.0*
Taiwan[4]	20.3	21.0*
Viet Nam*[5]	84.0	78.0
Other leading producers		
Argentina*[6]	58.0	56.0
Bangladesh*	52.9	56.8
India*[7]	826.2	857.1
Iran*	53.0	50.0
Kenya	287.0	293.7
Malawi	39.2	41.7
Sri Lanka	310.6	303.3
Turkey*	142.0	127.0

* Provisional.
1 Mainly green tea (546,124 in 2002).
2 Including green tea (about 40,000 tons in 2002).
3 All green tea.
4 Crude tea.
5 Including green tea (about 33,000 tons in 2002).
6 Twelve months beginning 1 May of year stated.
7 Including a small quantity of green tea (about 8,500 tons in 2002).

Source: International Tea Committee, *Supplement to Annual Bulletin of Statistics 2003*.

For many years the United Kingdom was the largest single importer of tea. However, the country's annual consumption of tea per caput, which amounted to 4.55 kg in 1958, has declined in recent years, averaging 2.46 kg in 1994–96 and 1995–97, before recovering marginally, to 2.51 kg, in 1996–98. In 1997–99, however, consumption fell back to 2.44 kg; again, to 2.33 kg, in 1998–2000; and again, to 2.27 kg, in 1999–2001. In 2000–02 annual consumption of tea per caput in the United Kingdom averaged 2.26 kg. A similar trend has been observed in other developed countries, but to a notably lesser extent in the Republic of Ireland, the world's largest per caput consumer of tea, where annual consumption per person advanced from 3.17 kg in 1994–96 to 3.23 kg in 1995–97, before evidencing a slight decline, to 2.95 kg, in 1996–98. Irish annual consumption per person declined again, to 2.78 kg, in 1997–99, and to 2.69 kg in 1998–2000, before recovering to 2.71 kg in 1999–2001. In 2000–02 average annual consumption of tea per caput in the Republic of Ireland rose again, to 2.76 kg. From the late 1980s consumption and imports expanded significantly in the developing countries (notably Middle Eastern countries) and, particularly, in the USSR, which in 1989 overtook the United Kingdom as the world's principal tea importer. However, internal factors, following the break-up of the USSR in 1991, caused a sharp decline in tea imports by its successor republics; as a result, the United Kingdom regained its position as the leading tea importer in 1992. In 1993 the former Soviet republics (whose own tea production had fallen sharply) once more displaced the United Kingdom as the major importer, but in 1994 the United Kingdom was again the principal importing country. Since 1995, however, imports by the former USSR (excluding the Baltic states) have exceeded those of the United Kingdom. In 2003 the former USSR imported an estimated 204,000 metric tons of tea, accounting for 15.9% of the world market, followed by the United Kingdom, with 125,279 tons (9.8%), Pakistan (118,309 tons, or 9.2%) and the USA (94,143 tons, or 7.3%). Other major importers of tea in 2003 were Afghanistan, Egypt, Iran and Morocco.

Export Price Index for Tea
(base: 1980 = 100)

	Average	Highest month(s)	Lowest month(s)
1990	92		
1995	67		
2000	84	88 (Jan.)	79 (June)
2001	70	83 (Jan.)	65 (June)
2002	68	73 (Sept., Oct.)	63 (May)

Much of the tea traded internationally is sold by auction, principally in the exporting countries. Until declining volumes brought about their termination in June 1998 (Kenya having withdrawn in 1997, and a number of other exporters, including Sri Lanka and Malawi, having established their own auctions), the weekly London auctions had formed the centre of the international tea trade. At the London auctions, five categories of tea were offered for sale: 'low medium' (based on a medium Malawi tea), 'medium' (based on a medium Assam and Kenyan tea), 'good medium' (representing an above-average East African tea), 'good' (referring to teas of above-average standard) and (from April 1994) 'best available'. The average price of all grades of tea sold at London auctions stood at £1,238 sterling per metric ton in 1993, but fell to £1,192 per ton in 1994 and to £1,036 per ton in 1995. The annual average price in 1995 was the lowest since 1988. The average rose to £1,135 per ton in 1996 and to £1,351 in 1997. For the first half of 1998, before the auctions ceased, the average price realized was £1,458 per ton. During 1993 the monthly average reached £1,564 per ton in January, but it declined to £1,100 in June and then recovered to £1,256 in November. The average was £1,102 per ton in January 1994, but this rose to £1,284 in June and September. Quotations then moved generally lower, and in July 1995 the average London tea price was only £907.5 per ton, its lowest monthly level since August 1988. At one auction in September 1995 the price of 'medium' tea was reduced to only £780 per ton, its lowest level, even in nominal terms (i.e. without taking inflation into account), since 1976. The monthly average increased to £1,170.5 per ton in December 1995, but fell to £1,031 in July 1996. It moved generally higher in later months, reaching £1,528 per ton in August 1997, following reports that drought had reduced Kenya's tea crop. After easing somewhat, the average London price of tea rose in January 1998 to £1,905 per ton (its highest monthly level since March 1985), in response to news of serious flooding in eastern Africa. At one of that month's auctions the price of 'medium' tea reached £2,000 per ton, its highest level since 1985. However, the average London tea price retreated in May 1998 to £1,014 per ton. At the end of June, with the prospect of a record Kenyan crop, the quotation for 'medium' tea at the final London auction was £980 per ton. Based on country of origin, the highest-priced tea at London auctions during 1989–94 was that from Rwanda, which realized an average of £1,613 per ton in the latter year. The quantity of tea sold at these auctions declined from 43,658 tons in 1990 to 11,208 tons in 1997.

The National Native Produce and Animal By-Products Import and Export Corpn of the People's Republic of China is the sole exporter of tea produced in that country. Of the other exporting countries covered by this volume, only Indonesia has established a noteworthy tea auction. Even so, the quantity of tea traded annually at the Jakarta auction is very small in comparison with that traded at its main African (Mombasa, Kenya) and South Asian (Colombo, Sri Lanka) counterparts. It does, however, frequently surpass the quantity traded annually at the smallest of India's auctions, at Coimbatore. Total annual sales at the Jakarta auction amounted to 37,432 metric tons in 2002, compared with 41,802 tons in 2001 and 33,219 tons in 2000. The average price per ton of tea sold at the Jakarta auction in 2002 was US $1,011, compared with US $967 in 2001 and US $1,195 in 2000.

An International Tea Agreement (ITA), signed in 1933 by the governments of India, Ceylon (now Sri Lanka) and the Netherlands East Indies (now Indonesia), established the International Tea Committee (ITC), based in London, as an administrative body. Although ITA operations ceased after 1955, the ITC has continued to function as a statistical and information centre. In 2004 there were seven producer/exporter members (the tea boards or associations of Kenya, Malawi, India, Indonesia, Bangladesh and Sri Lanka, and the tea sub-chamber of the Chinese Chamber of Commerce), four consumer members, eight associate members and six corporate members.

In 1969 the FAO Consultative Committee on Tea was formed and an exporters' group, meeting under this committees auspices, set export quotas in an attempt to stabilize tea prices. These quotas have applied since 1970 but are generally regarded as too liberal to have more than a minimal effect on prices. The perishability of tea complicates the effective operation of a buffer stock. India, while opposing the revival of a formal ITA to regulate supplies and prices, has advocated greater co-operation between producers to regulate the market. The International Tea Promotion Association (ITPA), founded in 1979 and based in Nairobi, Kenya, comprises eight countries (excluding, however, India and Sri Lanka), accounting for almost one-third of world exports of black tea.

Tin

The world's known tin reserves, estimated by the US Geological Survey to total 6.1m. metric tons in 2003, are located mainly in the equatorial zones of Asia, in central South America and in Australia. Cassiterite is the only economically important tin-bearing mineral, and it is generally associated with tungsten, silver and tantalum minerals. There is a clear association of cassiterite with igneous rocks of granitic composition, and 'primary' cassiterite deposits occur as disseminations, or in veins and fissures in or around granites. If the primary deposits are eroded, as by rivers, cassiterite may be concentrated and deposited in 'secondary', sedimentary deposits. These secondary deposits form the bulk of the world's tin reserves. The ore is treated, generally by gravity method or flotation, to produce concentrates prior to smelting.

Tin owes its special place in industry to its unique combination of properties: low melting point, the ability to form alloys with most other metals, resistance to corrosion, non-toxicity and good appearance. Its main uses are in tinplate (about 40% of world tin consumption), in alloys (tin-lead solder, bronze, brass, pewter, bearing and type metal), and in chemical compounds (in paints, plastics, medicines, coatings and as fungicides and insecticides). In the late 1990s a number of possible new applications for tin were under study: these included its use in fire-retardant chemicals, and as an environmentally preferable substitute for cadmium in zinc alloy anti-corrosion coatings on steel. The possible development of a lead-free tin solder was also receiving consideration.

Production of Tin Concentrates
(tin content, metric tons)

	2001	2002*
World total (excl. USA)	308,000*	249,000
Far East	197,624*	144,659
Oceania	9,602	6,268
Regional producers		
Australia	9,602	6,268
China, People's Repub.*	95,000	80,000
Indonesia	90,000*	54,000
Laos	400*	400
Malaysia	4,972	4,215
Myanmar	230*	240
Thailand	2,522	1,104
Viet Nam*	4,500	4,700
Other leading producers		
Bolivia	12,298	15,242
Brazil	12,500	13,000
Peru	69,696	65,400
Portugal	1,200*	1,000
Russia*	4,500	2,900

* Estimated production.

Source: US Geological Survey.

Most South-East Asian tin production comes from gravel-pump mines and dredges, working alluvial deposits, although there are underground and open-cast mines working hard-rock deposits. In 2002 virtually all of Australia's tin output came from the low-cost underground Renison Bell mine that Murchison United operates in western Tasmania, and the alluvial

Ardlethan mine in New South Wales, where Marlborough Resources commenced production in August 2001. In February 2000 Murchison United and the Tasmanian State Government agreed to fund jointly a six-year programme of exploration and development to secure the long-term future of the Renison operation. Murchison gave notice of the possible sale of the Renison Bell mine in 2002. Australia's concentrate is sent to Malaysia and Thailand for smelting. Concentrates from Myanmar and Laos are mainly exported to Malaysia, Europe and the republics of the former USSR. Indonesia, Malaysia and Thailand now smelt virtually the whole of their mine production to produce primary tin metal. In 2002 the Indonesian government imposed a ban on the export of tin and tin concentrates as part of an attempt to prevent smuggling of the metal into Singapore, where a higher price was obtainable for tin ore. Malaysia Smelting Corpn reportedly completed the purchase in 2002 of Indonesia's PT Koba Tin from Iluka Resources Ltd (Australia). In 2004 PT Timah of Indonesia, the world's largest tin producer, planned to construct a smelter on Kundur Island, Riau province, in order to process local ore. An independent tin smelter is in operation in Singapore.

Production of Tin Metal
(metric tons, incl. secondary)

	2001	2002*
World total	295,000*	278,000
Far East	212,742*	199,689
Oceania	1,470*	911
Regional producers		
Australia	1,470*	911
China, People's Repub.	105,000*	93,000
Indonesia	53,470	53,000
Japan	668	659
Malaysia	30,417	30,000
Thailand	21,357	21,500
Viet Nam	1,800*	1,500
Other leading producers		
Belgium	8,000*	6,000
Bolivia	11,300	10,976
Brazil	12,600*	12,800
Peru	38,182	35,828
Russia	3,600	3,650
USA	6,696	6,413

* Estimated production.

Source: US Geological Survey.

According to the US Geological Survey, the Far East and Australasia accounted for an estimated 72% of world output of tin concentrates in 2002. However, since the 1970s, tin has sharply declined in importance as an export commodity in Malaysia and Thailand, which had traditionally been among the region's main producers. In Thailand, where the metal provided 12.9% of the country's export earnings in 1967, its share was only 0.6% in 1988, and tin remained at the margin of total exports in the early 2000s. In 1989 exports of tin metal (including tin alloys) provided 1.7% of Malaysia's total export earnings. The comparable proportion was 1.2% in 1990 and only 0.2% in 1997. By contrast, sales of tin accounted for 23% of Malaysia's total exports, by value, in 1965. The depressed level of world tin prices since late 1985 (see below) has accelerated the industry's decline, although Malaysia possesses substantial resources of tin and remains the leading world centre for tin smelting. In 1993, following three years of financial losses on its tin-mining operations, Malaysia Mining Corpn, which accounted for the bulk of that country's tin production, terminated these activities. Malaysia's tin output subsequently declined sharply, from 10,384 metric tons in 1993 to 5,065 tons in 1997. In 1998, however, the Malaysian Government took action to encourage the reactivation of dormant mines. In that year Malaysia recorded an increase in tin production for the first time since 1990. Output increased further, to 7,340 tons, in 1999, but declined, to 6,307 tons, in 2000. In 2001 production fell, substantially, to 4,972 tons, and it was estimated to have declined again, to only 4,215 tons. The People's Republic of China, now by far the world's largest tin (primary metal) producer, announced in late 1997 that it was to initiate a three-year mining project in Qinghai Province to develop additional tin ore production capacity of 450,000 tons annually. China's tin ore output rose by about 4% in 1998, to 70,100 tons, and again in 1999, by 14.3%, to 80,100 tons. In 2000 production increased for a third consecutive year, this time by 24.1%, reaching 99,400 tons. In 2001, however, output was estimated to have fallen by more than 20%, to 79,000 tons, owing, in part, to the longer-than-expected closure for maintenance of the Yunnan Tin Industry Group Co operation that was in turn the result of unfavourable market conditions. Output declined again, by about 16%, to an estimated 80,000 tons, in 2002.

Over the period 1956–85, much of the world's tin production and trade was covered by successive international agreements, administered by the International Tin Council (ITC), based in London. The aim of each successive International Tin Agreement (ITA), of which there were six, was to stabilize prices within an agreed range by using a buffer stock to regulate the supply of tin. (For detailed information on the last of these agreements, see *The Far East and Australasia 1991*.) The buffer stock was financed by producing countries, with voluntary contributions by some consuming countries. 'Floor' and 'ceiling' prices were fixed, and market operations conducted by a buffer stock manager who intervened, as necessary, to maintain prices within these agreed limits. For added protection, the ITA provided for the imposition of export controls if the 'floor' price was being threatened. The ITA was effectively terminated in October 1985, when the ITC's buffer stock manager informed the London Metal Exchange (LME) that he no longer had the funds with which to support the tin market. The factors underlying the collapse of the ITA included its limited membership (Bolivia and the USA, leading producing and consuming countries, were not signatories) and the accumulation of tin stocks which resulted from the widespread circumvention of producers' quota limits. The LME responded by suspending trading in tin, leaving the ITC owing more than £500m. to some 36 banks, tin smelters and metals traders. The crisis was eventually resolved in March 1990, when a financial settlement of £182.5m. was agreed between the ITC and its creditors. The ITC was itself dissolved in July.

These events lent new significance to the activities of the Association of Tin Producing Countries (ATPC), founded in 1983 by Malaysia, Indonesia and Thailand and later joined by Bolivia, Nigeria, Australia and Zaire (now the Democratic Republic of the Congo). Immediately prior to the withdrawal of Australia and Thailand at the end of 1996 (see below), members of the ATPC, which is based in Kuala Lumpur, Malaysia, accounted for about two-thirds of world mine production. The ATPC, which was intended to operate as a complement to the ITC and not in competition with it, introduced export quotas for tin for the year from 1 March 1987. Brazil and China agreed to co-operate with the ATPC in implementing these supply restrictions, which were renegotiated to cover succeeding years, with the aim of raising prices and reducing the level of surplus stocks. The ATPC membership also took stringent measures to control smuggling. Brazil and China (jointly accounting for more than one-third of world tin production) both initially held observer status at the ATPC. (China became a full member in 1994, but Brazil has remained as an observer—together with Peru and Viet Nam.) China and Brazil agreed to participate in the export quota arrangements, for which the ATPC had no formal powers of enforcement.

The ATPC members' combined export quota was fixed at 95,849 metric tons for 1991, and was reduced to 87,091 tons for 1992. However, the substantial level of world tin stocks, combined with depressed demand, led to mine closures and reductions in output, with the result that members' exports in 1991 were below quota entitlements. The progressive depletion of stock levels led to a forecast by the ATPC, in May 1992, that export quotas would be removed in 1994 if these disposals continued at their current rate. The ATPC had previously set a target level for stocks of 20,000 tons, representing about six weeks of world tin consumption. Projections that world demand for tin would remain at about 160,000 tons annually, together with continued optimism about the rate of stock disposals, led the ATPC to increase its members' 1993 export quota to 89,700 tons. The persistence, however, of high levels of annual tin exports by China (estimated to have totalled 30,000 tons in 1993 and 1994, compared with its ATPC quota of 20,000 tons), together with sales of surplus defence stocks of tin by the US Government, necessitated a reduction of the quota to 78,000 tons for 1994. In late 1993 prices had fallen to a 20-year 'low' and world tin stocks were estimated at 38,000–40,000 tons, owing partly to the non-observance of quota limits by Brazil and China, as well as to increased production by non-ATPC members. World tin stocks resumed their rise in early 1994, reaching 48,000 tons in June. However, the effects of reduced output, from both ATPC and non-ATPC producing countries, helped to reduce stock levels to 41,000 tons at the end of December. In 1995 exports by ATPC members exceeded the agreed voluntary quotas by 10%, and in May 1996, when world tin stocks were estimated to have been reduced to 20,000 tons, the ATPC suspended its quota arrangements. Shortly before the annual meeting of the ATPC was convened in September, Australia and Thailand announced their withdrawal from the organization, on the grounds that its activities had ceased to be effective in maintaining price levels favourable to tin-producers. Although China and Indonesia indicated that they would continue to support the ATPC, together with Bolivia, Malaysia and Nigeria (Zaire had ceased to be an active producer of tin), the termination of its quota arrangements in 1996, together with the continuing recovery in the tin market, indicated that its future role would be that of a forum for tin-producers and consumers. Malaysia, Australia and Indonesia left the ATPC in 1997, and Brazil became a full member in 1998. In June 1999, when the organization's headquarters were moved from Kuala Lumpur to Rio de Janeiro, Brazil, the membership comprised Brazil, Bolivia, China, the Democratic Republic of the Congo and Nigeria. It was hoped that Peru would join the ATPC following its relocation to South America.

The success, after 1985, of the ATPC in restoring orderly conditions in tin trading (partly by the voluntary quotas and partly by working towards the reduction of tin stockpiles) unofficially established it as the effective successor to the ITC as the international co-ordinating body for tin interests. The International Tin Study Group (ITSG), comprising 36 producing and consuming countries, was established by the ATPC in 1989 to assume the informational functions of the ITC. In 1991 the secretariat of the United Nations Conference on Trade and Development (UNCTAD) assumed responsibility for the publication of statistical information on the international tin market.

Export Price Index for Tin
(base: 1980 = 100)

	Average	Highest month(s)	Lowest month(s)
1990	42		
1995	37		
2000	29	32 (Jan.)	27 (Oct.–Dec.)
2001	23	27 (Jan., Feb.)	19 (Sept., Oct.)
2002	22	23 (June, July, Oct., Dec.)	19 (Feb.)

Although transactions in tin contracts were resumed on the LME in 1989, the Kuala Lumpur Commodity Exchange (KLCE), in Malaysia, had become established as the main centre for international trading in the metal. In September 1993 the price of tin (ex-works) on the KLCE stood at only 10.78 Malaysian ringgits (RM) per kg, equivalent to £2,756 (or US $4,233) per metric ton. Measured in terms of US currency, international prices for tin were at their lowest level for 20 years, without taking inflation into account. In October the Malaysian tin price recovered to RM 12.71 per kg (£3,376 or US $4,990 per ton), but in November it declined to RM 11.60 per kg (£3,056 or US $4,545 per ton).

The rise in tin prices continued in the early weeks of 1994, and in February the Malaysian quotation reached RM 15.15 per kg (£3,684 or US $5,449 per metric ton). In early March the metal's price eased to RM 14.00 per kg (£3,449 or US $5,146 per ton), but later in the month it rose to RM 15.01 per kg (£3,715 or US $5,515 per ton). The KLCE price was reduced in August to RM 12.99 per kg (£3,285 or US $5,092 per ton), but in early November it stood at RM 16.06 per kg (then £3,896 or US $6,265 per ton).

The Malaysian tin price advanced strongly in January 1995, reaching RM 16.38 per kg (£4,018 or US $6,407 per metric ton). However, in early March the price declined to RM 13.33 per kg (about £3,300 or US $5,200 per ton). The tin market later revived again, and in August the KLCE price of the metal reached RM 17.60 per kg (£4,569 or US $7,060 per ton). International tin prices were then at their highest level for nearly six years. The recovery in the market was attributed partly to a continuing decline in tin stocks, resulting from a reduction in exports from China and in disposals from the USA's strategic reserves of the metal. However, the KLCE quotation was reduced to RM 15.36 per kg (£3,851 or US $6,090 per ton) in October.

In the early months of 1996 there was a generally downward movement in international tin prices, and in March the Malaysian quotation fell to RM 15.28 per kg (£3,947 or US $6,015 per metric ton). In April the KLCE price of tin recovered to RM 16.24 per kg (£4,294 or US $6,496 per ton). A further decline ensued, and in mid-December tin was traded at RM 14.31 per kg (£3,388 or US $5,668 per ton).

During the first six months of 1997 the price of tin on the KLCE declined by nearly 8%. The tin market moved further downward in July, with the Malaysian price falling to RM 13.52 per kg (£3,185 or US $5,398 per metric ton). Later that month, after the value of the ringgit had depreciated, the equivalent of the KLCE tin price (RM 13.76 per kg) was only £3,107 (US $5,199) per ton. In late August the Malaysian price stood at RM 14.70 per kg, but this was equivalent to only £3,134 (US $5,061) per ton. In terms of local currency, the Malaysian tin price increased considerably in subsequent weeks, but most of the upward movement was the consequence of a decline in the ringgit's value in relation to other currencies. By the end of September the KLCE price of tin had risen to RM 18.13 per kg. This price was reportedly maintained over the first eight trading days of October, during which time the ringgit recovered some of its value. At the end of the period the Malaysian tin price was equivalent to £3,656 or US $5,927 per ton. The KLCE price of tin continued to move upward in subequent weeks, but the Malaysian currency's value declined again. At the end of the year the tin price stood at RM 20.61 per kg, equivalent to £3,218 or US $5,295 per ton. During the second half of 1997 the tin price, expressed in ringgits, increased by 50.5%, but the equivalent in US dollars was 2.4% lower.

The KLCE price of tin reportedly held steady at RM 20.61 per kg for the first nine trading days of January 1998. Meanwhile, the ringgit's depreciation continued. At the currency's lowest point, the tin price was equivalent to only £2,722 or US $4,399 per metric ton. Later in January the tin price advanced to RM 23.43 per kg. It remained at this level for several days, as the value of the ringgit recovered, so that in early February the price was equvalent to £3,557 or US $5,843 per ton. Later in February the Malaysian tin price eased to RM 19.27 per kg (£3,122 or US $5,120 per ton), but in June it rose to RM 24.50 per kg. This quotation was maintained over three trading days, during which the value of the ringgit rose appreciably. As a result, the equivalent of the KLCE price increased from less than US $6,000 per ton to US $6,234 (£3,757). Tin prices later eased, and at the end of September the metal was traded in Malaysia at RM 20.02 per kg, equivalent to £3,100 (US $5,268) per ton. Earlier that month a fixed exchange rate for the Malaysian currency (US $1 = RM 3.8; the rate was subsequently altered slightly on a number of occasions, and stood at US $1 = RM 3.775 in June 2002) was introduced. In early November the KLCE was merged with the Malaysia Monetary Exchange to form the Commodity and Monetary Exchange of Malaysia (COMMEX Malaysia), also based in Kuala Lumpur. During the same week the Malaysian tin price rose to RM 20.65 per kg (£3,269 or US $5,434 per ton), but in December it declined to RM 19.10 per kg (£2,990 or US $5,026 per ton).

In January 1999, with tin stocks in plentiful supply, the COMMEX price of the metal was reduced to only RM 18.77 per kg, equivalent to £2,990 (US $4,939) per metric ton. However, tin prices rose steadily thereafter, and in May the Malaysian quotation reached RM 21.42 per kg (£3,500 or US $5,637 per ton). After easing somewhat, the price advanced in November to RM 22.10 per kg (£3,594 or US $5,816 per ton). It ended the year at RM 21.41 per kg (£3,496 or US $5,634 per ton).

In January 2000 the COMMEX price of tin rose to RM 22.84 per kg, equivalent to £3,664 (US $6,011) per metric ton. By early June the price was reduced to RM 19.94 per kg (£3,488 or US $5,247 per ton), although later in the month it recovered to RM 20.71 per kg (£3,582 or US $5,450 per ton). The price generally remained around RM 20.00 per kg for the remainder of that year, reaching a maximum of RM 20.78 per kg (£3,845 or US $5,433 per ton) in September and a minimum of RM 19.42 per kg (£3,483 or US $5,077 per ton) in October, ending the year at RM 19.66 per kg (£3,545 or US $5,208 per ton).

Trading in January 2001 began with the price somewhat lower than at the end of December, the average price for the first week of trading being RM 19.29 per kg (equivalent to £3,349 or US $5,110 per metric ton). A further decline was experienced in March, when the price fell below RM 19 per kg. Prices rallied briefly in early May, before declining below RM 18 per kg in mid-June. A sharp decline began in July, the COMMEX falling as low as RM 13.34 per kg (£2,363 or US $3,534 per ton) in September. A strong recovery in November (the price exceeding RM 16.00 per kg at one point) faltered in December, the price being RM 14.95 (£2,691 or US $3,960 per ton) on the year's last trading day.

The COMMEX price was generally more stable in the first six months of 2002 than it had been in the previous year, experiencing a gradual upward trend. By the end of June the price of tin had reached RM 16.62 (£2,863 or US $4,403 per metric ton).

The International Tin Research Institute (ITRI), founded in 1932 and based in London, England, promotes scientific research and technical development in the production and use of tin.

Tobacco (*Nicotiana tabacum*)

Tobacco originated in South America and was used in rituals and ceremonials or as a medicine; it was smoked and chewed for centuries before its introduction into Europe, the Middle East, Africa and the Indian sub-Continent in the 16th century. The generic name *Nicotiana* denotes the presence of the alkaloid nicotine in its leaves. The most important species in commercial tobacco cultivation is *N. tabacum*. Another species, *N. rustica*, is widely grown, but on a smaller scale, to yield cured leaf for snuff or simple cigarettes and cigars.

In 1999, according to FAO, about 4.3m. ha were being farmed world-wide for tobacco. Commercially grown tobacco (from *N. tabacum*) can be divided into four major types—flue-cured, air-cured (including burley, cigar, light and dark), fire-cured and sun-cured (including oriental)—depending on the procedures used to dry or 'cure' the leaves. Each system imparts specific chemical and smoking characteristics to the cured leaf, although these may also be affected by other factors, such as the type of soil on which the crop is grown, the type and quantity of fertilizer applied to the crop, the cultivar used, the spacing of the crop in the field and the number of leaves left at topping (the removal of the terminal growing point). Each type is used, separately or in combination, in specific products (e.g. flue-cured in Virginia cigarettes). All types are grown in Asia.

As in other major producing areas, local research organizations in Asia have developed new cultivars with specific, desirable chemical characteristics, disease-resistance properties and improved yields. Almost all tobacco production in the Far East is from smallholdings; there is no cultivation of the crop on estates, as is common with tea. Emphasis has been placed on improving yields by the selection of cultivars, by the increased use of fertilizers and by the elimination or reduction of crop loss (through use of crop chemicals) and on reducing requirements for hand labour through the mechanization of land preparation and the use of crop chemicals. Harvesting continues to be entirely a manual operation, as the size of farmers' holdings and the cost of harvesting devices (now commonly used in the USA and Canada) preclude such development in Asia. The flue-curing process requires energy in the form of oil, gas, coal or wood. To ensure that supplies of wood are continuously renewed, the tobacco industry in several countries encourages the planting of trees.

The principal type of tobacco cultivated by Asian farmers is flue-cured. Of the countries producing this tobacco in the Far East, the most important are the People's Republic of China, Japan and the Koreas. China is the world's principal producer, as well as the largest consumer, of tobacco, and, while most of its output is still retained for local use, exports expanded by an annual average rate of about 14% in 1990–2002, totalling 169,203 metric tons in the latter year, when China ranked as the world's third largest exporter of unmanufactured tobacco. Production in China in recent years has benefited from significant changes in the organization of, and in the methods of cultivation that are practised in, tobacco-growing areas. The tobacco sector is also a significant domestic employer: according to the International Tobacco Growers' Association (ITGA, see below), about 16m. people were engaged in tobacco growing in China in the late 1990s.

Production of Tobacco Leaves
(farm sales weight, '000 metric tons)

	2002	2003*
World total	6,229	6,195
Far East	3,018	2,856
Oceania	9	9
Leading regional producers		
China, People's Repub.	2,454	2,308‡
Indonesia	134	135‡
Japan	58	60‡
Korea, Dem. People's Repub.†	64	64
Korea, Repub.	48	48†
Myanmar	49	49†
Philippines	50	51‡
Thailand	74	65†
Viet Nam	33	33
Other leading producers		
Bangladesh	38	40†
Brazil	670	649
Greece	127	121‡
India	385	595‡
Italy	122	106‡
Malawi	69	70†
Pakistan	95	95
Turkey	153	152
USA	399	377
Zimbabwe	178	178†

* Provisional figures.
† FAO estimate(s).
‡ Unofficial figure(s).

The only significant regional producers of burley tobacco are Indonesia, Japan (for domestic consumption) and the Koreas (for domestic consumption and for export), although Thailand is increasing its output of this type. Production of dark air-cured tobacco is limited mainly to China and Indonesia. Asia is not a significant producer of fire-cured and oriental tobaccos.

Japan is unable to produce sufficient flue-cured tobacco for domestic consumption, owing mainly to pressure on the availability of land, and thus has to rely on substantial and increasing imports. Korean flue-cured is now accepted on international markets. In both Thailand and the Koreas domestic requirements are likely to maintain pressure on production. The level of exports from these countries has stabilized (sales by the Democratic People's Republic of Korea have fallen very substantially since 1998), and this position is unlikely to be reversed despite the increased supplies of flue-cured tobacco now available on international markets from Brazil (the world's principal tobacco exporter), Malawi and Zimbabwe, although the price competitiveness of Korean flue-cured does not enhance its prospects as an exporter.

Of the other tobacco-producing countries in the Far East, Indonesia is a significant producer and exporter of cigar tobaccos, yet still requires imported flue-cured tobacco to sustain its domestic market, although increasing quantities of domestic flue-cured are being grown on Sulawesi, Bali and Lombok. Domestic tobacco production is increasing in Malaysia, under the aegis of the National Tobacco Board. The Philippines produces both flue-cured and air-cured tobacco acceptable to world markets.

Growing conditions in New Zealand and Australia are wholly dissimilar to those in other countries in the Asia-Pacific region, and neither country is a significant producer in world terms. However, both countries contain large tobacco holdings, mechanization is considerable and problems related to the cultivation of the crop, such as the occurrence of blue mould in Australia, are unique. Domestic production, mainly flue-cured, is encouraged through government legislation stipulating a minimum quantity of such tobacco to be incorporated into tobacco products that are marketed in Australia and New Zealand.

Export Price Index for Tobacco
(base: 1980 = 100)

	Average	Highest month(s)	Lowest month(s)
1990	125		
1995	132		
2000	141		
2001	141	*	*
2002	141	*	*

* The monthly index remained constant at 141 in 2000–02.

More than one-quarter of world tobacco production is traded internationally. Until 1993, when it was overtaken by Brazil, the USA was the world's principal tobacco-exporting country. According to the US Department of Agriculture (USDA), the value of US exports of unmanufactured tobacco (burley and flue-cured) was some US $1,050m. in 2002.

The International Tobacco Growers' Association (ITGA), with headquarters in Castelo Branco, Portugal, was formed in 1984 by growers' groups in Argentina, Brazil, Canada, Malawi, the USA and Zimbabwe. In 2004 its members numbered 25 countries, including China, Indonesia, Malaysia and the Philippines. ITGA members account for more than 80% of the world's internationally traded tobacco. The Association provides a forum for the exchange of information among tobacco producers, conducts research and publishes studies on tobacco issues.

Wheat (*Triticum*)

The most common species of wheat, *T. vulgare*, includes hard, semi-hard and soft varieties which have different milling characteristics but which, in general, are suitable for bread-making. Another species, *T. durum*, is grown mainly in semi-arid areas, including regions bordering the Mediterranean Sea. This wheat is very hard and is suitable for the manufacture of semolina, the basic ingredient of pasta and couscous. A third species, spelt (*T. spelta*), is also included in production figures for wheat. It is grown in very small quantities in parts of Europe and is used as animal feed.

Although a most adaptable crop, wheat does not thrive in hot and humid climates, and it requires timely applications of water (either through rainfall or irrigation). Wheat is an important crop in most countries in Asia north of the Tropic of Cancer, wherever the terrain is favourable and sufficient water is available. The most concentrated producing areas in the Asia-Pacific region are to be found in Pakistan, northern India and eastern parts of the People's Republic of China.

World wheat production declined during the 1990s, albeit only marginally, at an average rate of less than 0.1% a year. The fall was largely due to the sharp decline in agricultural output in the former USSR, excluding which the trend in world production growth has been upward. Wheat production is highly variable from year to year. Part of the variation is attributable to weather conditions, particularly rainfall, in the main producing areas, but national policies on support for producers have also been a major influence. In the 1990s several major wheat-producing countries, including leading exporters, pursued policies of market deregulation and began to remove the links between producers' support and the financial returns from particular commodities. This encouraged their farmers more readily to switch between crops according to expected relative market returns. After 1996, for example, when wheat was in short supply on world markets, output was stimulated in many growing areas, and a record 613m. metric tons was harvested in 1997. Unfavourable weather, particularly in Russia and China, reduced output to about 593m. tons in 1998, but, owing to low growth in consumption, exporting countries' stocks remained high, and farmers' returns fell. This discouraged plantings for the 1999 season, when production totalled only about 588m. tons. Production declined further in 2000, to some 586m. tons, but increased, to about 591m. tons, in 2001. In 2002 wheat production world-wide fell sharply, to only about 573m. tons, partly as a result of substantial declines in Australian and North American output. A further sharp decline occurred in 2003, when estimated output totalled only about 556m. tons. FAO forecast production of wheat world-wide of some 595m. tons in 2004. A sharp recovery in European output, together with a slight increase in Asian production, was expected to compensate for declines in production in North America and Oceania in 2004. In China, even though plantings had declined, it was anticipated that the achievement of record yields would raise output of wheat by 5%, to some 91m. tons. FAO forecast that inventories of wheat held world-wide for crop years ending in 2005 would total 140m. tons, a decline of 10% from the Organization's initial revised estimates.

Wheat production in Far East Asia has increased at a considerably faster rate than output in the world as a whole. In the mid-1970s the region's harvest averaged about 47m. metric tons per year, or approximately 12% of the world total, but by the late 1990s it averaged 116m. tons (approximately one-fifth of the total). This growth was attributable, in large part, to an accelerated rate of expansion of wheat output in China, brought about by extensive irrigation, better supplies of fertilizers and, perhaps most important of all, the introduction of cash incentives to farmers. Local abundance gave rise to storage problems in some areas in the mid-1980s, and, as a result, some producers lost interest in wheat, and, instead, cultivated cash crops. However, the subsequent introduction of stricter marketing regulations, higher prices payable to wheat producers and more abundant fertilizer supplies led to a resumption of growth. By the mid-1990s China, with annual output averaging about 105m. tons, was by far the world's leading wheat-producing country.

Australia is one of the world's major wheat-exporting countries, and in the year to June 2003 sales of wheat earned the country $A3,036m., representing 2.6% of total merchandise export earnings. Production varies considerably from year to year, however. The incidence of drought in the main producing areas is a major determinant (the 1994 wheat crop was reduced by 45% as the result of drought conditions), but plantings also reflect world market conditions, as Australian wheat-farming receives no government subsidies. Production in 2000 and 2001, at, respectively, 22.2m. metric tons and an estimated 23.8m. tons, was somewhat higher than the average for recent years. In 2002, however, output fell drastically, to only 10m. tons, as a consequence of drought in all of Australia's main wheat-growing regions. In 2003 production was estimated to have recovered to almost 25m. tons. In 2004 output was forecast to decline somewhat, to about 22m. tons, as a result of inadequate rainfall in some producing areas. The only other country in the Pacific area with significant wheat production is New Zealand, which is usually self-sufficient in wheat.

World consumption, which has, in the long term, been increasing at a similar rate to production, varies much less from year to year. Food use is increasing at about 1.5% per year. Most of the increase is in Far East Asia (including the South Asian sub-region). Of approximately 415m. metric tons of wheat used for direct human food world-wide, Far East Asia accounts for about one-half, and China for one-half of that. Wheat food use has been expanding at the expense of rice: its growth is associated with rising consumer incomes and an increasing number of fast-food outlets. The trend was sustained in 1997/98, despite the much higher costs of wheat imports in a number of countries such as Indonesia: one reason is that rice became even more expensive, while retail prices of wheat were, in some cases, stabilized by government subsidies. Substantial amounts of wheat are used for feed in Europe and, when prices are favourable, in North America. Substantial quantities were also used for feed in the 1980s in the former Soviet Union, but this volume has decreased sharply, in response to the diminution in livestock numbers. In Far East Asia some wheat is used for feed in Japan, while the Republic of Korea imports wheat for feed when prices are low in comparison with those of coarse grains such as sorghum and maize (corn).

Wheat is the principal cereal in international trade. Amounts exported in recent years have ranged between 100m. and 120m. metric tons annually, including wheat flour, durum wheat and semolina, feed wheat, and wheat of bread-making quality (some of which, however, was used to make other food products, such as noodles). FAO forecast that world trade in wheat would total

98.5m. tons in 2004/05. The main exporters are the USA (whose share declined from about 30% of the total in the first half of the 1990s to about 20% in 2002), the European Union (EU), Canada, Australia and Argentina. In 2001/02 and 2002/03 the EU was not only a major exporter of wheat, but also the world's biggest importer of the cereal. EU imports were forecast to decline substantially in 2003/04, however, owing to the introduction of a system of import quotas designed to curb purchases of cheap Ukrainian and Russian wheat; and again, in 2004/05, owing to increased domestic output. It was anticipated that the EU would be confronted by strong competition in disposing of its exportable surplus of wheat in 2004/05, owing to the prevailing low international price of wheat in combination with the strength of the euro relative to the US dollar. Developed countries were formerly the principal consumers, but the role of developing countries as importers has been steadily increasing and they now regularly account for approximately two-thirds of world imports.

Far East Asia has long been one of the world's principal wheat-importing regions. Imports rose considerably during the 1980s, but reduced purchases by China caused a temporary setback in the early 1990s. Japan has been a major importer since the 1960s. Its imports, of about 6m. metric tons annually, represent an important element of stability in the world wheat trade. Despite the scale of its own production, China is usually a major wheat importer. However, following purchases of 12.6m. tons in 1995/96 and 9.2m. tons in 1996/97, its imports have subsequently averaged only about 2m. tons annually, owing to large harvests, abundant stocks and a slower rate of growth in wheat consumption. Chinese wheat imports rose to about 4m. tons in 2003/04, however, according to FAO, and were forecast to increase to some 7m. tons in 2004/05 in spite of higher domestic production.

The Republic of Korea also buys varying amounts of wheat. Its basic food needs account for more than 2m. metric tons annually, but it also purchases additional amounts, of up to 4m. tons, for animal feed if the price is competitive with maize and other feed grains. Several countries in the region that grow no wheat, such as Indonesia, Malaysia and the Philippines, are now major importers of wheat.

Production of Wheat
('000 metric tons, including spelt)

	2002	2003*
World total	573,513	556,349
Far East	91,504	67,326
Oceania	10,361	25,221
Leading regional producers		
Australia	10,059	24,900
China, People's Repub.	90,290	86,100
Other leading producers		
Argentina	12,300	14,530
Canada	16,198	23,552
France	38,934	30,582
Germany	20,818	19,296
Iran	12,450	12,900†
India	72,786	65,129
Pakistan	18,227	18,210
Russia	50,609	34,062
Turkey	19,500	19,000
Ukraine	20,556	19,000
United Kingdom	15,973	14,288
USA	44,063	63,590

* Provisional figures.
† Unofficial figure.

A lengthy period of low international wheat prices, and heavily-subsidized competition among the major exporting countries, was interrupted in 1995–96 after exporting countries' stocks had fallen to their lowest levels for 20 seasons. Prices rose to record levels, but the resultant stimulus to output led to a rapid accumulation of stocks, and by 1998 prices were again very low as competition intensified between exporting countries. Because direct export subsidies were limited under international trade agreements, much of this competition took the form of offers of credit to importing countries.

The export price (f.o.b. Gulf ports) of US No. 2 Hard Winter, one of the most widely traded wheat varieties, stood at around US $140 per metric ton in mid-1994. At that time, countries benefiting from export subsidies (e.g. much of North Africa) could expect to buy wheat for at least US $40 per ton less. The peak, reached in April 1996, was almost US $300 per ton, but by mid-1997 the price had fallen by as much as 50%, to around US $150 per ton. Subdued import demand, and the disposal of surpluses by a number of countries in Central and Eastern Europe, as well as by Ukraine, further depressed prices in late 1997. Markets remained subdued in 1998 and early 1999, as a consequence of high levels of export stocks and a lack of new markets. In June 1999 US Hard Winter varieties were traded at the US Gulf at US $110 per ton. At the close of 1999 the price fell to around US $105 per ton, although a measure of recovery took place in the early months of 2000, in response to uncertainty regarding crop prospects in the USA, together with some revival in import demand. In 1999 overall the average export price (f.o.b. Gulf ports) of US No. 2 Hard Winter was US $114 per ton, while in 2000 the price averaged US $119 per ton, rising to US $130 per ton in 2001. An average price of US $151 per ton was recorded in 2002. In 2003 the average price declined marginally, to US $150 per ton. During the first seven months of 2004 the average monthly export price of US No. 2 Hard Winter wheat ranged between US $154 per ton (July) and US $173 per ton (April).

Export Price Index for Wheat
(base: 1980 = 100)

	Average	Highest month(s)	Lowest month(s)
1990	82		
1995	94		
2000	67	74 (Dec.)	64 (April, Aug.)
2001	72	74 (Jan., May)	71 (March, April, June, Aug.)
2002	82	103 (Oct.)	70 (March–May)

Since 1949 nearly all world trade in wheat has been conducted under the auspices of successive international agreements, administered by the International Wheat Council (IWC) in London. The early agreements involved regulatory price controls and supply and purchase obligations, but such provisions became inoperable in more competitive market conditions, and were abandoned in 1972. The IWC subsequently concentrated on providing detailed market assessments to its members and encouraging them to confer on matters of mutual concern. A new Grains Trade Convention, which entered into force in July 1995, gave the renamed International Grains Council (IGC) a wider mandate to consider all coarse grains as well as wheat. This facilitates the provision of information to member governments, and enhances their opportunities to hold consultations. In addition, links between governments and industry are strengthened at an annual series of grain conferences sponsored by the IGC. In mid-2002 the IGC had eight individual exporting members, together with the member states of the EU (which are counted as a single member), and 20 importing members. Membership in Far East Asia and Oceania comprises Australia, Japan and the Republic of Korea.

A new Food Aid Convention (FAC), replacing an earlier agreement introduced in 1995, was brought into force on 1 July 1999 as one of the constituent instruments of the International Grains Agreement, 1995 (the other being the Grains Trade Convention, 1995). Members of the Food Aid Committee, which administers the FAC, had resolved to renegotiate the existing Convention in accordance with the recommendations relating to Least-developed and Net Food-Importing Developing Countries adopted by members of the World Trade Organization at their Singapore Conference in December 1996, and in the light of the Declaration on World Food Security and the Plan of Action adopted by the Rome World Food Summit in the same year. They also took into account that, in recent years, there had been significant changes in the food aid policies of several donor countries.

The main objective of the new Convention is 'to contribute to world food security and to improve the ability of the international community to respond to emergency food situations and other food needs of developing countries'. FAC members will make quality food aid available to developing countries with the greatest needs on a consistent basis, regardless of fluctuations in world food prices and supplies. Particular importance is attached to ensuring that food aid is directed to the alleviation of poverty and hunger among the most vulnerable groups.

As a framework of international co-operation between food aid donors, the new FAC aims at achieving greater efficiency in all aspects of food aid operations. FAC members will place greater emphasis on the monitoring and evaluation of the impact and effectiveness of their food aid operations. They are also committed to support the efforts of recipient countries to develop and implement their own food security strategies.

Cereals will continue to represent the bulk of aid supplied. However, the list of products which may be supplied has been broadened beyond cereals and pulses to include edible oil, skimmed milk powder, sugar, seeds and products which are a component of the traditional diet of vulnerable groups in developing countries or of supplementary feeding programmes (e.g. micro-nutrients).

Members' total minimum annual commitments are approximately the same as under the previous FAC, amounting to over 5m. tons of commodities in wheat equivalent. The cost of transporting and delivering food aid beyond the f.o.b. stage will, to the extent possible, be borne by the donors.

ACKNOWLEDGEMENTS

We gratefully acknowledge the assistance of the following organizations in the preparation of this section: British Petroleum, the Foreign Agricultural Service of the US Department of Agriculture, the International Aluminium Institute, the International Cocoa Organization, the International Coffee Organization, the International Monetary Fund, the International Rice Research Institute, the International Rubber Study Group, the International Tea Committee, the International Tobacco Growers' Association, the Food and Agriculture Organization of the UN (FAO), the US Geological Survey, the US Department of Energy and the World Coal Institute.

Source for Agricultural Production Tables (unless otherwise indicated): Food and Agricultural Organization of the UN (FAO).

Source for Export Price Indices: UN, *Monthly Bulletin of Statistics*.

CALENDARS, TIME RECKONING, AND WEIGHTS AND MEASURES

The Islamic Calendar

The Islamic era dates from 16 July 622, which was the beginning of the Arab year in which the *Hijra* ('flight' or migration) of the prophet Muhammad (the founder of Islam), from Mecca to Medina (in modern Saudi Arabia), took place. The Islamic or *Hijri* Calendar is lunar, each year having 354 or 355 days, the extra day being intercalated 11 times every 30 years. Accordingly, the beginning of the *Hijri* year occurs earlier in the Gregorian Calendar by a few days each year. Dates are reckoned in terms of the *anno Hegirae* (AH) or year of the Hegira (*Hijra*). The Islamic year 1425 AH began on 22 February 2004.

The year is divided into the following months:

1. Muharram	30 days	7. Rajab	30 days
2. Safar	29 days	8. Shaaban	29 days
3. Rabia I	30 days	9. Ramadan	30 days
4. Rabia II	29 days	10. Shawwal	29 days
5. Jumada I	30 days	11. Dhu'l-Qa'da	30 days
6. Jumada II	29 days	12. Dhu'l-Hijja	29 or 30 days

The *Hijri* Calendar is used for religious purposes throughout the Islamic world, and is the official calendar in Saudi Arabia. In most Arab countries it is used in conjunction with the Gregorian Calendar for official purposes, but in Indonesia, Malaysia and Pakistan the Gregorian Calendar has replaced it.

PRINCIPAL ISLAMIC FESTIVALS

New Year: 1st Muharram. The first 10 days of the year are regarded as holy, especially the 10th.

Ashoura: 10th Muharram. Celebrates the first meeting of Adam and Eve after leaving Paradise, also the ending of the Flood and the death of Husain, grandson of the prophet Muhammad. The feast is celebrated with fairs and processions.

Mouloud (Birth of Muhammad): 12th Rabia I.

Leilat al-Meiraj (Ascension of Muhammad): 27th Rajab.

Ramadan (Month of Fasting).

Id al-Fitr or Id al-Saghir or Küçük Bayram (The Small Feast): Three days beginning 1st Shawwal. This celebration follows the constraint of the Ramadan fast.

Id al-Adha or Id al-Kabir or Büyük Bayram (The Great Feast, Feast of the Sacrifice): Four days beginning on 10th Dhu'l-Hijja. The principal Islamic festival, commemorating Abraham's sacrifice and coinciding with the pilgrimage of Muslims to Mecca. Celebrated by the sacrifice of a sheep, by feasting and by donations to the poor.

Islamic Year	1424	1425	1426
New Year	4 March 2003	22 Feb. 2004	10 Feb. 2005
Ashoura	13 March 2003	2 March 2004	19 Feb. 2005
Mouloud	14 May 2003	2 May 2004	21 April 2005
Leilat al-Meiraj	24 Sept. 2003	12 Sept. 2004	2 Sept. 2005
Ramadan begins	27 Oct. 2003	15 Oct. 2004	5 Oct. 2005
Id al-Fitr	26 Nov. 2003	14 Nov. 2004	4 Nov. 2005
Id al-Adha	1 Feb. 2004	21 Jan. 2005	11 Jan. 2006

Note: Local determinations may vary by one day from those given here.

The Chinese Calendar

China has both lunar and solar systems of dividing the year. The lunar calendar contains 12 months of 29 or 30 days, and in each period of 19 years seven intercalary months are inserted at appropriate intervals. In order not to disturb the 12-month cycle, each of these extra months bears the same title as that which preceded it. The intercalary months may not be introduced after the first, 11th or 12th month of any year.

The solar year which is used by the agrarian community of China begins regularly on 5 February of the Gregorian calendar, and is divided into 24 sections of 14, 15 or 16 days. This calendar is not upset by the discrepant cycle of the moon, and is therefore suitable for the regulation of agriculture.

Until the revolution of 1911, years were named according to a 60-year cycle, made up of 10 stems (*Ban*) and 12 branches (*Ji*). Each year of the cycle has a composite name composed of a different combination of stem and branch. Similar 60-year cycles of year-names have been in use in Thailand and Japan at various times and are still used in Hong Kong, Malaysia and Singapore.

Since 1911 years have been dated from the revolution as Years of the Republic; AD 2005 is the 94th year of the republican era. The Republic of China has been confined to Taiwan since 1949. In the People's Republic of China the Gregorian system is used.

Japan has used the Gregorian system since 1873, but a National Calendar has also been introduced, derived from the traditional date of accession of the first emperor, Zinmu, in 660 BC. The year AD 2005 corresponds to 2664 of this era.

Buddhist Calendars

The Buddhist era (BE) is attributed to the death of Gautama Buddha, historically dated at about 483 BC. The era in use is dated, in fact, from 544 BC, making the year 2005 of the Christian era equal to 2548 of the Buddhist era.

In South-East Asia there is widespread use of a lunar year of 354 days, with months of alternately 29 or 30 days, and with extra (intercalary) months approximately every third year. Under this system, New Year may fall in either April or March. In Myanmar (formerly Burma), New Year is regularly on 13 April. The Burmese era, the *Khaccapancha*, is ascribed to the ruler Popa Sawrahan, and begins in AD 638. The year 2005 of the Gregorian Calendar is equivalent to 1367 BE.

Thailand used the Burmese calendar until 1889, when a new civil era was introduced, commemorating the centenary of the first king of Bangkok. Since 1909 a calendar based on the year 543 BC (traditionally the year of Gautama Buddha's attainment of nirvana) has been in official use. The months have been adapted to correspond with those of the Gregorian Calendar, but New Year is on 1 April every year. In this calendar, called the *Pra Putta Sakarat*, AD 2005 is equivalent to 2547.

FESTIVALS

The principal festivals in the Buddhist calendars are the New Year and the spring and autumn equinox, and local festivals connected with important pagodas.

Standard Time

The following table gives the standard time adopted in the various countries and territories covered in this book, in relation to Greenwich Mean Time (GMT).

+6½	+7	+8	+9	+9½
Cocos (Keeling Is) Myanmar (fmrly Burma)	Cambodia Christmas Island Indonesia (Sumatra, Java, Madura, West and Central Kalimantan) Laos Mongolia (western) Thailand Viet Nam	Australia (Western Australia) Brunei China, People's Republic Hong Kong Indonesia (Bali, East and South Kalimantan, West Timor, Celebes/Sulawesi) Macao Malaysia Mongolia (central) Philippines Singapore Taiwan	Timor-Leste (fmrly East Timor) Indonesia (Maluku/Moluccas, Papua) Japan Korea Mongolia (eastern) Palau	Australia (Northern Territory, South Australia)

+10	+11	+11½	+12	+12¾
Australia (the ACT, NSW, Queensland, Tasmania, Victoria) Guam Northern Mariana Is Federated States of Micronesia (western) Papua New Guinea (excl. North Solomons Province)	Federated States of Micronesia (eastern) New Caledonia Solomon Is Vanuatu	Norfolk Island	Fiji Kiribati (excl. Millennium (fmrly Caroline) Island Marshall Is Nauru New Zealand Tuvalu Wallis and Futuna Is Wake Island	New Zealand (Chatham Is)

+13	–11	–10	–8
Kiribati (Millennium (fmrly Caroline) Island) Tonga	American Samoa Midway Island Niue Samoa	Cook Is French Polynesia (except Gambier Is and Marquesas Is) Tokelau	Pitcairn Is

Weights and Measures

(The following tables indicate the values of the principal weights and units of measurement which are in common use as alternatives to the metric and imperial systems)

WEIGHT

Unit	Country	Metric equivalent	Imperial equivalent
Acheintaya	Myanmar	163.33 kg	360.1 lb
Arroba	Philippines	11.5 kg	25.35 lb
Baht, Bat, Kyat or Tical	Myanmar, Thailand (Old Chinese system)	7.56 g	0.266 oz
	Siamese system	14.11 g	0.529 oz
Beittha or Viss	Myanmar	1.63 kg	3.601 lb
Candareen or Fen	China (Old system)	0.378 g	0.0133 oz
	China (New system)	0.283 g	0.010 oz
	Hong Kong	0.378 g	0.0133 oz
Candy	Myanmar	81 metric tons	80 tons
Catty, Gin, Jin, Kan, Kati, Katti, Kin, Kon or Zhang	China (Old system)	0.603 kg	1.333 lb
	China (New system)	0.5 kg	1.102 lb
	Hong Kong	0.603 kg	1.333 lb
	Indonesia	0.617 kg	1.362 lb
	Japan	0.6 kg	1.323 lb
	Malaysia and Singapore	0.603 kg	1.333 lb
	Thailand (Old Chinese system)	0.603 kg	1.333 lb
	Siamese	0.603 kg	1.333 lb
Dan	China (Old system)	60.48 kg	133.3 lb
	China (New system)	50 kg	110.23 lb
Fan	China (Old system)	0.378 g	0.0133 oz
	China (New system)	0.311 g	0.011 oz
	Hong Kong	0.378 g	0.0133 oz
Hyaku-mé	Japan	374.85 g	13.226 oz
Koyan	Malaysia and Singapore	2,419 kg	5,333.3 lb
Kwan or Kan	Japan	3.749 kg	8.267 lb
Liang or Tael	China (Old system)	37.8 g	1.333 oz
	China (New system)	31.18 g	1.102 oz
	Hong Kong	37.8 g	1.333 oz
Mace	China (Old system)	3.78 g	0.133 oz
	China (New system)	3.11 g	0.110 oz
	Hong Kong	3.78 g	0.133 oz
Me	Japan	3.78 g	0.133 oz
Momme	Japan	3.78 g	0.133 oz
Neal	Cambodia	0.6 kg	1.323 lb
Ngamus	Myanmar	8.2 g	0.288 oz
Nijo	Japan	15.02 g	0.529 oz
Picul, Pikul, Picol, Taam or Tam	China (Old system)	60.48 kg	133.33 lb
	China (New system)	50 kg	110.23 lb
	Hong Kong	60.48 kg	133.33 lb
	Indo-China (Old Chinese system)	60.48 kg	133.33 lb
	Indo-China (Siamese system)	60 kg	132.277 lb
	Indonesia	61.76 kg	136.16 lb
	Japan	60 kg	132.276 lb
	Malaysia and Singapore	60.48 kg	133.33 lb
	Sabah and Sarawak	61.53 kg	135.64 lb
	Philippines	63.25 kg	139.44 lb
	Thailand (Old Chinese system)	60.48 kg	133.33 lb
	Thailand (Siamese system)	60.55 kg	135.5 lb
Pyi	Myanmar	2.13 kg	4.69 lb
Quintal	Indonesia	100 kg	220.462 lb
	Philippines	46 kg	101.4 lb
Tahil	Malaysia and Singapore	37.8 g	1.33 oz
Viss	Myanmar	1.63 kg	3.601 lb

LENGTH

Unit	Country	Metric equivalent	Imperial equivalent
Ch am am	Cambodia	24.9 cm	9.84 in
Ch'ek or Foot	Hong Kong—by statute	37.16 cm	14.625 in
	Hong Kong—in practice	35.82 cm	14.14 in
Cheung	Hong Kong	3.698 m	4.063 yd
Chi	China (Old system)	91 cm	35.814 in
	China (New system)	33.27 cm	13.123 in
Cho	Japan (length)	109.12 cm	119.302 yd
Chung	South Korea		
Cubit	Myanmar	45.72 cm	18 in
Cun	China (Old system)	3.58 cm	1.41 in
	China (New system)	3.33 cm	1.312 in
	Hong Kong	3.58 cm	1.41 in
El, Ell or Ella	Indonesia	68.8 cm	27.08 in
	Malaysia and Singapore	0.914 m	1 yd
	Sabah	0.914 m	1 yd
Fen	China	0.33 cm	0.13 in
Garwode	Myanmar	20.44 km	12.727 miles
Hat	Cambodia	50 cm	19.68 in
Hiro	Japan	1.516 m	1.657 yd
Jo	Japan	3.032 in	3.314 yd
Kawtha	Myanmar	5.116 km	3.182 miles
Ken	Japan	1.82 m	1.988 yd
	South Korea		
Keup	Thailand	23.04 cm	9.07 in
Lan	Myanmar	1.83 m	2 yd
Li or Lei	China (Old system)	645–681 m	706–745 yd
	China (New system)	500 m	546.8 yd
	Hong Kong	645–681 m	706–745 yd
	South Korea	3.926 km	2.44 miles
Niew	Thailand	2.09 cm	0.820 in
Oke thapa	Myanmar	64 m	70 yd
Paal	Java	1,506 m	1,647 yd
	Sumatra	1,851.7 m	2,025 yd
Palgate	Myanmar	2.54 cm	1 in
Pulgada	Philippines	2.31 cm	0.914 in
Ri	Japan—length	3.926 km	2.44 miles
	Japan—marine measure	1.85 km	1 nautical mile
Sawk	Myanmar	50.29 cm	19.8 in
Sen	Thailand	40 m	43.74 yd
Shaku	Japan	30.3 cm	11.93 in
Sun	Japan	3.02 cm	1.193 in
Taing	Myanmar	3.911 km	2.43 miles
Tar	Myanmar	3.2 m	3.5 yd
Taung	Myanmar	45.72 cm	18 in
Tjengkal	Indonesia	3.66 m	4 yd
Wah	Thailand	2 m	2.19 yd
Yote	Thailand	16 km	9.942 miles
Yuzamar	Myanmar	75.83 km	47.121 miles
Zhang	China	3.34 m	3.45 yd

CAPACITY

Unit	Country	Metric equivalent	Imperial equivalent
Bag	Myanmar	122.75 litres (l)	27 galls
Cavan	Philippines	75 l	16.5 galls
Chupa	Philippines	0.37 l	0.0825 gall.
Chupak	Malaysia and Singapore	1.14 l	0.25 gall.
Ganta	Philippines	3 l	0.660 gall.
Gantang	Malaysia and Singapore	4.55 l	1 gall.
	Indonesia (used mainly for rice)	8.58 l	1.887 galls
Go	Japan	0.17 l	1.27 gills
Gwe	Myanmar	20.45 l	4.5 galls
Koku	Japan	180.38 l	39.682 galls
Kwe	Myanmar	20.45 l	4.5 galls
Kwien	Thailand	2,000 l	439.95 galls
Pau	Singapore	0.28 l	0.5 pint
Pyi	Myanmar	2.56 l	0.56 galls
Sale	Myanmar	0.64 l	0.14 gall.
Sat	Thailand	20 l	4.40 galls
Sayut	Myanmar	5.12 l	1.12 galls
Seik	Myanmar	10.24 l	2.250 galls
Shaku	Japan	0.017 l	0.127 gill
Sho	Japan	1.8 l	0.397 gall.
Suk	South Korea	175.8 l	38.682 galls
Tanan	Thailand	1 l	0.22 gall.
Tin, Tunn, Tin-han or basket	Myanmar (Thamardi)	40.91 l	9 galls
Tinaja	Philippines	48 l	10.56 galls
To	Japan	17.76 l	3.97 galls

AREA

Unit	Country	Metric equivalent	Imperial Equivalent
Bahoe or Bouw	Indonesia	0.709 ha	1.7536 acres
Bu	Japan	3.3 sq m	3.95 sq yds
Cho	Japan (square measure)	1 ha	2.47 acres
Chungbo or Jongbo	Korea	1 ha	2.47 acres
Jemba	Malaysia and Singapore	13.38 sq m	16 sq yds
Mu	China	0.065 ha	0.165 acre
Ngan	Thailand	400 sq m	478.4 sq yds
Paal	Indonesia	227.08 ha	561.16 acres
Qing	China (New system)	6.66 ha	16.47 acres
Rai	Thailand	0.16 ha	0.395 acre
Se	Japan	99.17 sq m	118.61 sq yds
Tan	Japan	0.1 ha	0.247 acre
Tsubo	Japan	3.3 sq m	3.95 sq yds

Metric to Imperial Conversions

Metric units	Imperial units	To convert metric into imperial units multiply by:	To convert imperial into metric units multiply by:
Weight:			
Gram	Ounce (Avoirdupois)	0.035274	28.3495
Kilogram (kg)	Pound (lb)	2.204623	0.453592
Metric ton ('000 kg)	Short ton (2,000 lb)	1.102311	0.907185
	Long ton (2,240 lb)	0.984207	1.016047
	(The short ton is in general use in the USA, while the long ton is normally used in the UK and the Commonwealth.)		
Length:			
Centimetre (cm)	Inch	0.3937008	2.54
Metre (m)	Yard (= 3 feet)	1.09361	0.9144
Kilometre (km)	Mile	0.62137	1.609344
Volume:			
Cubic metre (cu m)	Cubic foot	35.315	0.028317
	Cubic yard	1.30795	0.764555
Capacity:			
Litre (l)	Gallon (= 8 pints)	0.219969	4.54609
	Gallon (US)	0.264172	3.78541
Area:			
Square metre (sq m)	Square yard	1.19599	0.836127
Hectare (ha)	Acre	2.47105	0.404686
Square kilometre (sq km)	Square mile	0.386102	2.589988

Systems of Measurement

WEIGHT

China	Picul = 100 Catty
	Catty = 16 Liang
	Liang = 10 Mace
	Mace = 10 Fan or Candareen
Japan	Picul = 16 Kwan (Kan)
	Kwan = 16 Hyaku-mé
	Hyaku-mé = 20 Nijo
	Nijo = 5 Me or Mommé
Malaysia	Koyan = 40 Picul
Myanmar	Candy = 500 Acheintaya
	Acheintaya = 10 Beittha or Viss
	Beittha or Viss = 200 Ngamus
Philippines	Quintal = 4 Arroba
Thailand	Picul = 100 Catty or Kon
	Catty = 40 Baht or Kyat

LENGTH

Cambodia	Hat = 2 Chamam
China	Li = 1,500 Chi
	Chi = 10 Cun
	Cun = 10 Fen
Hong Kong	Cheung = 10 Chek
	Chek = Chi (see China)
Japan	Jo = 2 Hiro
	Hiro = 5 Shaku
	Shaku = 10 Sun
Myanmar	Garwoke = 4 Kawtha
	Oke thapa = 20 Tar
	Tar = 7 Cubit or Taung
	Lan = 4 Cubit or Taung
	Cubit = 18 Palgate
Thailand	Sen = 20 Wah
	Wah = 8 Keup

CAPACITY

Japan	To = 10 Sho
	Sho = 10 Go
	Go = 10 Shaku
Myanmar	Bag = 3 Tin
	Tin = 2 Kwe
	Kwe = 2 Seik
	Seik = 2 Sayut
	Sayut = 2 Pyi
Philippines	Cavan = 25 Ganta
	Tinaja = 15 Ganta
	Ganta = 8 Chupa
Thailand	Kwien = 100 Sat
	Sat = 20 Tanan

AREA

China	Qing = 100 Mu
Indonesia	Paal 320 Bahoe or Bouw
Japan	Cho = 10 Bu or Tan
	Bu = 2 Se
	Se = 30 Tsubo
Thailand	Ngan = 2½ Rai

RESEARCH INSTITUTES

ASSOCIATIONS AND INSTITUTIONS STUDYING THE FAR EAST AND AUSTRALASIA

(See also Regional Organizations in Part Three)

AUSTRALIA

The Asia-Australia Institute, The University of New South Wales: Sydney, NSW 2052; tel. (2) 9385-9111; fax (2) 9385-9220; e-mail aai@unsw.edu.au; internet www.aai.unsw.edu.au; f. 1990; Chair. Prof. STEPHEN FITZGERALD; Dir SUNG LEE (acting).

Asia Research Centre on Social, Political and Economic Change: Murdoch University, Murdoch, WA 6150; tel. (8) 9360-2263; fax (8) 9310-4944; e-mail t.dent@murdoch.edu.au; internet wwwarc.murdoch.edu.au; f. 1991; provides analysis of contemporary East and South-East Asia; Dir Prof. GARRY RODAN.

Asian Economics Centre: Dept of Economics, University of Melbourne, Vic 3010; tel. (3) 8344-3880; fax (3) 8344-6899; e-mail s.jayasuriya@unimelb.edu.au; internet www.ecom.unimelb.edu.au/dept/AEC; fmrly Asian Business Centre; Dir Assoc. Prof. SISIRA JAYASURIYA; Dep. Dir Assoc. Prof. LISA CAMERON.

Asian Studies Program, La Trobe University: Vic 3086; tel. (3) 9479-1315; fax (3) 9479-1880; e-mail asianstudies@latrobe.edu.au; internet www.latrobe.edu.au/www/asianstudies/; f. 1994; Assoc. Dean JOHN FITZGERALD.

Australia-Japan Research Centre: Australian National University, Canberra, ACT 0200; tel. (2) 6125-3730; fax (2) 6125-0767; e-mail ajrc@anu.edu.au; internet apseg.anu.edu.au; f. 1980; Exec. Dir Prof. GORDON DE BROUWER; publs include *Pacific Economic Papers* (monthly), *Asia Pacific Economics and Politics*, *APEC Economies Newsletter* (monthly).

Australian Institute of International Affairs: 32 Thesiger Court, Deakin, ACT 2600; tel. (2) 6282-2133; fax (2) 6285-2334; e-mail ceo@aiia.asn.au; internet www.aiia.asn.au; f. 1933; 1,400 mems; brs in all States; Pres. NEAL BLEWETT; Exec. Dir CHARLES STUART; publs *The Australian Journal of International Affairs* (3 a year), *Australia in World Affairs* (every 5 years).

Centre for Asian Studies, University of Adelaide: University of Adelaide, Ligertwood Bldg, Adelaide, SA 5005; tel. (8) 8303-5815; fax (8) 8303-4388; e-mail asian.studies@adelaide.edu.au; internet www.adelaide.edu.au/humss/asian; specializes in China and Japan.

Centre for International Economic Studies: University of Adelaide, Adelaide, SA 5005; tel. (8) 8303-5672; fax (8) 8223-1460; e-mail cies@adelaide.edu.au; internet www.adelaide.edu.au/cies; f. 1989; conducts theoretical and policy-orientated research with particular reference to the Asia-Pacific region; publs include monographs; Exec. Dir Prof. KYM ANDERSON.

Centre for Japanese Economic Studies: Dept of Economics, Macquarie University, Sydney, NSW 2109; tel. (2) 9850-7444; fax (2) 9850-8586; e-mail craig.freedman@efs.mq.edu.au; internet www/econ.mq.edu.au/CJES; f. 1991; Dir Dr CRAIG FREEDMAN; publs include working papers.

Centre for Southeast Asian Studies: Northern Territory University, Darwin 0909; tel. (8) 8946-6862; fax (8) 8946-6977; e-mail ian.walters@ntu.edu.au; f. 1986; Dir Dr IAN WALTERS; publs include newsletter, papers and monographs.

Centre for the Study of Australia-Asia Relations (CSAAR): Griffith University, Nathan, Qld 4111; tel. (7) 3875-7916; fax (7) 3875-7956; e-mail D.McNamara@ais.gu.edu.au; internet www.gu.edu.au/guris/csaar/cshome.htm; f. 1978; conducts research relating to Asian migration to Australia, security in the Asia-Pacific region, Asian perceptions of Australia and Australian-Asian economic relations; Dir Assoc. Prof. RUSSELL TROOD.

Contemporary China Centre: Research School of Pacific and Asian Studies, Australian National University, Canberra, ACT 0200; tel. (2) 6125-4150; fax (2) 6257-3642; e-mail ccc@coombs.anu.edu.au; internet www.rspas.anu.edu.au/ccc/home.htm; f. 1970; analyses post-1949 China; Dir Prof. JONATHAN UNGER; publs include *The China Journal*, *Contemporary China Papers*.

Faculty of Asian Studies, Australian National University: Canberra, ACT 0200; tel. (2) 6125-0006; fax (2) 6125-0745; e-mail executive.officer.asian.studies@anu.edu.au; internet www.anu.edu.au/asianstudies; Dean Prof. ANTHONY C. MILNER.

Monash Asia Institute: Monash University, Clayton, Vic 3800; tel. (3) 9905-2124; fax (3) 9905-5370; e-mail marika.vicziany@adm.monash.edu.au; internet www.arts.monash.edu.au/mai; f. 1988; incorporates China Research Centre, National Centre for South Asian Studies, Japanese Studies Centre, Malaysian Studies Centre, Asia Pacific Health and Nutrition Centre and Centre of Southeast Asian Studies; Dir Prof. MARIKA VICZIANY.

National Thai Studies Centre: Faculty of Asian Studies, Australian National University, Canberra, ACT 0200; tel. (2) 6125-4661; fax (2) 6125-0745; e-mail ntsc@anu.edu.au; internet www.anu.edu.au/thaionline; f. 1991; Dir CAVAN HOGUE; publs include Thai language teaching materials, *Thailand Information Papers*.

Research Institute for Asia and the Pacific (RIAP): RIAP Level 2, 353 Abercrombie St, University of Sydney, NSW 2006; tel. (2) 9351-8547; fax (2) 9351-8562; e-mail enquiries@riap.usyd.edu.au; internet www.riap.usyd.edu.au; research on human resource development, information technology, media, the environment, public administration, regional security and managing systems that underpin institutional capacity; Dir Dr STEPHANIE FAHEY.

Research School of Pacific and Asian Studies: Australian National University, Canberra, ACT 0200; tel. (2) 6125-2183; fax (2) 6249-1893; e-mail director.rspas@anu.edu.au; internet rspas.anu.edu.au; publs include *The China Journal*, *Bulletin of Indonesian Economic Studies*, *Canberra Papers on Strategy and Defence*, *Pacific Linguistics*, *East Asian History*, *Contemporary China Papers*, *Journal of Pacific History*, *Working Papers in Trade and Development Studies* and monograph series; Dir Prof. JAMES J. FOX.

School of Asian Languages and Studies, University of Tasmania: Private Bag 91, Hobart, Tasmania 7001; tel. (3) 6226-2342; fax (3) 6226-7813; e-mail H.Umeoka@utas.edu.au; internet www.utas.edu.au/docs/asian_languages/MG/html; f. 1990; fmrly Asia Centre; offers courses in Asian Studies, Chinese, Indonesian and Japanese; Heads of School Prof. BARBARA HAILEY, Dr MARSHALL CLARK.

School of European, Asian and Middle Eastern Languages and Studies, University of Sydney: Brennan Bldg, University of Sydney, NSW 2006; tel. (2) 9351-3382; fax (2) 9351-2319; e-mail SEAMELS@arts.usyd.edu.au; internet www.arts.usyd.edu.au; f. 1991; includes the depts of Asian Studies, Chinese Studies, Japanese and Korean Studies, Indian Subcontinental Studies, and Southeast Asian Studies; Head Prof. ANTONY STEPHENS.

AUSTRIA

Afro-Asiatisches Institut in Wien (Afro-Asian Institute in Vienna): 1090 Vienna, Türkenstrasse 3; tel. (1) 3105145; fax (1) 3105145-312; e-mail office@aai-wien.at; internet www.aai-wien.at; f. 1959; religious, cultural, economic and scientific exchanges between Austria and African and Asian countries; assistance to students from Africa and Asia; public relations, lectures, seminars; Rector Dr mag. KONSTANTIN SPIEGELFELD.

Institut für Ostasienwissenschaften der Universität Wien: 1090 Vienna, Spitalgasse 2-4; tel. (1) 4277-43801; fax (1) 4277-9438; e-mail ostasien@univie.ac.at; internet www.univie.ac.at/ostasien/; f. 1965; Japanese, Chinese and Korean studies; Dir Dr SEPP LINHART; publ. *Institutsbericht* (annual).

BELGIUM

European Institute for Asian Studies (EIAS): 35 rue des Deux Eglises, 1040 Brussels; tel. (2) 230-8122; fax (2) 230-5402; e-mail eias@eias.org; internet www.eias.org; Chair. Prof. Dr LUDO CUYVEES; Sec.-Gen. DICK GUPWELL; publs *The EurAsia Bulletin* (monthly), *EIAS Briefing Papers*.

Institut Orientaliste: Faculté de Philosophie et Lettres, Université Catholique de Louvain, Collège Erasme, 1348 Louvain-la-Neuve; tel. (10) 474958; fax (10) 479169; e-mail ori@glor.ucl.ac.be; internet www.fltr.ucl.ac.be/FLTR/GLOR/ORI; f. 1936; Pres. Prof. RENÉ LEBRUN; publs *Le Muséon* (2 a year), *Bibliothèque du Muséon, Publications de l'Institut Orientaliste de Louvain (PIOL), Corpus Scriptorum Christianorum Orientalium (CSCO), Bulletin de Liaison GLOR* (4 a year).

Institut Royal des Relations Internationales: 69 rue de Namur, 1000 Brussels; tel. (2) 223-4114; fax (2) 223-4116; e-mail info@irri-kiib.be; internet www.irri-kiib.be; f. 1947; research in international relations, economics, law and politics; specialized library containing 700 vols and 200 periodicals; archives; holds lectures and conferences; Pres. Viscount E. DAVIGNON; Dir-Gen. CLAUDE MISSON; publs *Studia Diplomatica* (every 2 months).

BRAZIL

Centro de Estudos Afro-Asiáticos (CEAA): Praça Pio X 7, 9°, 20040-020, Rio de Janeiro, RJ; tel. (21) 516-7157; fax (21) 518-2798; e-mail beluce@candidomendes.br; f. 1973; Deputy Dir BELUCE BELLUCCI; publ. *Estudos Afro-Asiáticos*.

CAMBODIA

Cambodia Development Resource Institute (CDRI): POB 622, 56 Street 315, Tuol Kork, Phnom-Penh; tel. (23) 368053; fax (23) 366094; e-mail cdri@camnet.com.kh; internet www.cdri.org.kh; f. 1990; research into socio-economic issues of importance to Cambodia and the region.

Economic Institute of Cambodia: POB 614, 6 Street 288, Beung Keng Kang I, Phnom-Penh; Dir SOK HACH.

CANADA

Asian Pacific Research and Resource Center: Department of Law, Carleton University, Ottawa, ON K1S 5B6; tel. (613) 520-3690; fax (613) 520-4467; e-mail paul-davidson@carleton.ca; f. 1989; conducts research in commerce and investment in the Asia-Pacific region; Dir Prof. PAUL J. DAVIDSON; publs monographs, working papers.

Canadian Council for International Co-operation: 1 Nicholas St, Suite 300, Ottawa, ON K1N 7B7; tel. (613) 241-7007; fax (613) 241-5302; e-mail info@ccic.ca; internet www.ccic.ca; f. 1968; information and training centre for international development and forum for voluntary agencies; 100 mems; Pres. and CEO GERRY BARR; publs include newsletter and directory of NGOs working overseas.

Canadian Institute of International Affairs: 205 Richmond St West, Suite 302, Toronto, ON M5V 1V3; tel. (416) 977-9000; fax (416) 977-7521; e-mail mailbox@ciia.org; internet www.ciia.org; f. 1928; research in international relations; library containing 8,000 vols; Chair. ROY MACLAREN; Pres. and CEO DOUGLAS GOOLD; publs *Behind the Headlines* (quarterly), *International Journal* (quarterly), *Annual Report*.

Centre for Developing-Area Studies: McGill University, 3715 rue Peel, Montréal, QC H3A 1X1; tel. (514) 398-3507; fax (514) 398-8432; e-mail pub.cdas@mcgill.ca; internet www.mcgill.ca/cdas; Dir Dr ROSALIND BOYD; publs include *Labour, Capital and Society* (English and French—2 a year), *Discussion Paper Series*.

Department of Asian Studies, University of British Columbia: Vancouver, BC V6T 1Z2; tel. (604) 822-0019; fax (604) 822-8937; e-mail astudies@interchange.ubc.ca; internet www.asia.ubc.ca; f. 1961; instruction and research in East, South and South-East Asia; Head Dr PETER NOSCO.

Institute of Asian Research: University of British Columbia, C. K. Choi Bldg, 1855 West Mall, Vancouver, BC V6T 1Z2; tel. (604) 822-4688; fax (604) 822-5207; e-mail iar@interchange.ubc.ca; internet www.iar.ubc.ca; f. 1978; Dir PITMAN B. POTTER; publ. *Asia Pacific Report* (2 a year).

International Development Research Centre: POB 8500, Ottawa K1G 3H9; tel. (613) 236-6163; fax (613) 563-2476; e-mail info@idrc.ca; internet www.idrc.ca; f. 1970 as a public corpn to support scientific research aimed at helping communities in the developing world find solutions to social, environmental and economic problems; has regional offices in Singapore, India, Uruguay, Egypt, Kenya and Senegal; Pres. MAUREEN O'NEIL; publs *Reports* (online magazine) and books.

Lester Pearson International: 1321 Edward St, Halifax, NS B3H 3H5; tel. (902) 494-2038; fax (902) 494-1216; e-mail lpi@dal.ca; internet www.dal.ca/lpi; f. 1985; admin. inst. responsible for organization and supervision of international activities of Dalhousie Univ.; publs include *LPI Update* (fortnightly) and occasional paper series.

PEOPLE'S REPUBLIC OF CHINA

China Centre for International Studies (CCIS): 22 Xianmen Dajie, POB 1744, Beijing 100017; tel. (10) 63097083; fax (10) 63095802; f. 1982; conducts research on international relations and problems; organizes academic exchanges; Dir-Gen. LI LUYE.

China-Europe International Business School (CEIBS): 699 Hong Feng Lu, Shanghai 201206; tel. (21) 28905890; fax (21) 28905678; e-mail info@mail.ceibs.edu.cn; internet www.ceibs.edu; Exec. Pres. Dr ALBERT BENNETT; Dean Prof. WILLIAM A. FISCHER.

China Institute of Contemporary International Relations (CICIR): 2A Wanshousi, Haidian, Beijing 100081; tel. (10) 8418640; fax (10) 8418641; f. 1980; research on international development and peace issues; Pres. SHEN QURONG; publ. *Contemporary International Relations* (monthly).

China Institute for International Strategic Studies: Beijing; Dir-Gen. XU XIN.

Chinese Academy of Social Sciences: 5 Jianguomen Nei Da Jie, Beijing 100732; tel. (10) 65137744; fax (10) 65138154; f. 1977; Pres. HU SHENG; Sec.-Gen. GUO YONGCAI; comprises more than 40 research institutes and centres and one graduate school.

Chinese People's Institute of Foreign Affairs: 71 Nanchizi Jie, Beijing 100006; tel. (10) 65131824; e-mail cpifa@public.bta.net.cn; Pres. MEI ZHAORONG; Sec.-Gen. XU SHAOHAI.

Institute of Asia-Pacific Studies, Chinese Academy of Social Sciences: 3 Zhang Zizhong Lu, Beijing 100007; tel. (10) 64063042; fax (10) 64063041; e-mail aprccass@public3.bta.net.cn; f. 1988; theoretical and policy-orientated research in economic, political, social and cultural studies; Dir ZHANG YUNLING.

Institute of Asian-Pacific Studies, Shanghai Academy of Social Sciences: 7/622 Huai Hai Zhong Lu, Shanghai 200020; tel. (21) 63271170; fax (21) 63270004; e-mail jmzhou@fudan.ihep.ac.cn; internet www.jcie.or.jp/thinknet/research_instit/china/Inst_AP_Shang.html; f. 1990; Dir Dr ZHOU JIANMING.

Institute for Southeast Asian Studies, Jinan University: Shipai, Guangzhou 8510632; tel. (20) 8516511; fax (20) 8516941; Dir CHEN QIAOZHI.

Institute of World Economy: Fudan University, 220 Handan Lu, Shanghai 200433; tel. (21) 65492222; fax (21) 65491875; Dirs Prof. WU YIKANG, MA XOXIANG.

International Business Research Institute: University of International Business and Economics, Hui Xin Dong Jie, He Ping Jie N., Beijing 100029; tel. (10) 64225522; fax (10) 64212022; Dir Prof. TENG DEXIANG.

International Trade Research Institute: 17 Hubei Lu, Qingdao, Shangdong; tel. and fax (531) 6270179; f. 1980; conducts research on developments in international commerce and economics; Dir Chen Chingchang; publs *International Economic and Trade Information*, *Shandong Foreign Trade*.

Research Institute for International Economic Co-operation (RIIEC): 28 An Wai Dong Hou Xiang, Beijing; tel. (10) 64212106; fax (10) 64212175; f. 1980; Dir Yang Jinbo; publ. *International Economic Co-operation*.

Research Institute of International Politics: Beijing University, 1 Loudouqiao, Haidian, Beijing 100871; tel. (10) 62554002; fax (10) 62564095; Dir Liang Shoude.

Shanghai Institute of International Issues (SIIS): 1 Alley, 845 Ju Lu, Shanghai 200040; tel. (21) 62471148; fax (21) 62472272; f. 1960; Dir Chen Peiyao; publs *Survey of International Affairs* (annually), *Guoji Zhanwang* (2 a month), papers, monographs.

Shanghai University of Finance and Economics, Asian Economic Research Institute: 777 Guoding Lu, Shanghai 200433; tel. and fax (21) 65361955; f. 1993; Chair. Yi Juanqiu.

Taiwan Economy Research Institute: Nankai University, Balitai, Tianjin 200071; tel. (22) 63358825; fax (22) 63344853; Dir Prof. Bao Juemin.

CROATIA

Institute for International Relations: 10000 Zagreb, POB 303, ul. Ljudevita Farkaša Vukotinovića 2/II; tel. (1) 4826522; fax (1) 4828361; e-mail ured@irmo.hr; f. 1963; affiliated to the University of Zagreb and the Ministry of Science and Technology; Dir Prof. Dr Mladen Staničić; publs include *Croatian International Relations Review* (quarterly), *Culturelink* (3 a year).

CZECH REPUBLIC

Czech Society for Eastern Studies: c/o Oriental Institute, Academy of Sciences, Pod vodárenskou věží 4, 182 08 Prague 8; tel. (2) 66052483; fax (2) 86581897; e-mail aror@orient.cas.cz; f. 1958; 71 mems; Pres. Lübica Obuchová.

Oriental Institute: Academy of Sciences of the Czech Republic, Pod vodárenskou věží 4, 182 08 Prague 8; tel. (2) 66052492; fax (2) 7987260; e-mail orient@orient.cas.cz; internet www.orient.cas.cz; f. 1922; research on languages, social and economic aspects, etc. of Asia and Africa; Chinese library of 66,177 vols, general library of more than 190,000 vols; Dir Dr Josef Kolmas (acting); publs *Archiv orientální* (quarterly), *Nový Orient* (monthly).

DENMARK

Center for Udviklingsforskning (Centre for Development Research): Gammel Kongevej 5, 1610 Copenhagen V; tel. 33-85-46-00; fax 33-25-81-10; e-mail cdr@cdr.dk; internet www.cdr.dk; f. 1969 to promote and undertake research in the economic, social and political problems of developing countries; library of 50,000 vols; Dir Poul Engberg-Pedersen; publs include *Researching Development* (in English, quarterly newsletter), *CDR Library Papers* (irregular), *Den ny Verden* (quarterly), *CDR Research Reports* (in English, irregular), *CDR Working Papers* (in English, irregular).

Nordic Institute of Asian Studies: Leifsgade 33, 2300 Copenhagen S; tel. 32-54-88-44; fax 32-96-25-30; e-mail sec@nias.ku.dk; internet www.eurasia.nias.ku.dk; f. 1967; known until 1988 as the Scandinavian Institute of Asian Studies; non-profit org. funded through Nordic Council of Ministers; research and documentation centre for the study of modern Asian societies and cultures from the perspective of the humanities and social sciences, with a multi-disciplinary profile; library of 24,000 vols and 750 periodicals; Chair. of Bd Dr Clemens Stubbe Østergaard; Dir Prof. Dr Per Ronnås; publs include *NIASnytt/Nordic Newsletter of Asian Studies* (quarterly), monographs, anthologies and research reports.

FIJI

The Fiji Society: POB 1205, Suva; f. 1936; concerned with subjects of historic and scientific interest to Fiji and other Islands of the Pacific; Pres. Ivan Williams; publ. *Transactions* (irregular).

Pacific Institute of Management and Development: University of the South Pacific, POB 1168, Suva; tel. 3212079; fax 3303229; e-mail isas@usp.ac.fj; internet www.usp.ac.fj/ISAS.

FINLAND

World Institute for Development Economics Research of the United Nations University (UNU/WIDER): Katajanokanlaituri 6B, 00160 Helsinki; tel. (9) 6159911; fax (9) 61599333; e-mail wider@wider.unu.edu; internet www.wider.unu.edu; f. 1984; research and training centre of the UNU (Japan); publs include *WIDER Discussion Papers*, *Policy Briefs*, *Research Papers*, *Annual Lectures*, *WIDER Angle*(Newsletter).

FRANCE

Centre des Hautes Etudes sur l'Afrique et l'Asie Modernes: 13 rue du Four, 75006 Paris; tel. 1-43-26-96-90; fax 1-40-51-03-58; internet www.ccfr.bnf.fr; f. 1936; library of 16,000 vols; Dir J. P. Doumenge; publs *La Lettre du Cheaam* (quarterly), *Notes africaines, asiatiques et caraïbes* (irregular).

Institut National des Langues et Civilisations Orientales: 2 rue de Lille, 75343 Paris Cedex 07; tel. 1-49-26-42-74; fax 1-49-26-42-99; internet www.inalco.fr; courses in 88 languages; research; information centre; organizes international exchanges; specializes in international relations; Pres. Gilles Delouche; publs research serials, textbooks, translations, etc.

Musée National des Arts Asiatiques Guimet: 6 place d'Iéna, 75116 Paris; tel. 1-56-52-53-00; fax 1-56-52-53-54; internet www.museeguimet.fr; f. 1889; library of 100,000 vols; art, archaeology, religions, literature and music of India, Central Asia, Tibet, Pakistan, Viet Nam, China, Korea, Japan, Cambodia, Thailand, Laos, Myanmar (fmrly Burma) and Indonesia; Dir Jean-François Jarrige; Librarian F. Macouin; publs *Annales du Musée Guimet*, *Arts Asiatiques*.

Société Asiatique: 3 rue Mazarine, 75006 Paris; tel. and fax 1-44-41-43-14; internet www.aibl.fr/fr/asie/present.html; f. 1829; library of 100,000 vols; 750 mems; Pres. Daniel Gimaret; publs *Journal Asiatique* (2 a year), *Nouveaux Cahiers*.

Unité Formation et Recherche (UFR) Langues et Civilisations de l'Asie Orientale: Université de Paris VII, 2 place Jussieu, 75005 Paris; tel. 1-44-27-57-81; fax 1-44-27-78-98; f. 1971; Dir Cécile Sakai.

FRENCH POLYNESIA

Te Fare Tauhiti Nui—Maison de la culture: 646 Blvd Pomaré, BP 1709, Papeete; tel. 544544; fax 428569; e-mail tauhiti@mail.pf; f. 1980; promotes culture locally and abroad; sponsors many public and private cultural events; library of 9,500 vols, children's library of 7,000 vols; Dir Georges Estall.

GERMANY

AAI-Abteilung für Sprache und Kultur Japans, Universität Hamburg: 20146 Hamburg, Edmund-Siemers-Allee 1, Ostflügel; tel. (40) 42838-4884; fax (40) 42838-6200; e-mail japanologie@uni-hamburg.de; internet www.uni-hamburg.de/fachbereiche-einrichtungen/japanologie/index.html; f. 1914; research into Japanese studies; 43,500 vols; Dir Prof. Dr R. Schneider; publs include *Oriens Extremus*, *Nachrichten*, *Kagami* and monograph series *MOAG*.

China-Institut: Johann Wolfgang Goethe-Universität, 60054 Frankfurt am Main, Senckenberganlage 31; e-mail B.Sude@em.uni-frankfurt.de; Dir Prof. Dr Tsung-Tung Chang.

Deutsche Gesellschaft für Asienkunde eV (German Association for Asian Studies): 20148 Hamburg, Rothenbaumchaussee 32; tel. (40) 445891; fax (40) 4107945; e-mail post@asienkunde.de; internet www.asienkunde.de; f. 1967; promotion and co-ordination of contemporary Asian research; 670 mems; Pres. Dr Theo Sommer; Sec. Christine Berg; publ. *ASIEN. Deutsche Zeitschrift für Politik, Wirtschaft und Kultur* (quarterly).

Deutsche Gesellschaft für Auswärtige Politik eV (German Society for Foreign Affairs): 53113 Bonn, Adenauerallee 131,

Postfach 1425, 53004 Bonn; tel. (228) 2675-0; fax (228) 2675-173; e-mail DGAP@compuserve.com; internet www.dgap.org; f. 1955; promotes research on problems of international politics; library of 50,000 vols; 1,600 mems; Pres. Dr WERNER LAMBY; Exec. Vice-Pres. Dr IMMO STABREIT; Dir Research Inst. Prof. Dr KARL KAISER; publs *Internationale Politik* (monthly), *Die Internationale Politik* (annually).

Deutsche Morgenländische Gesellschaft eV (DMG) (German Oriental Society): Orientalisches Seminar, Islamwissenschaft/Turkologie, 79085 Freiburg/Brsg., Werthmannplatz 3; tel. (761) 2033159; fax (761) 2033152; e-mail Jens.Peter.Laut@orient.uni-freiburg.de; internet www.dmg-web.de; f. 1845; sponsors research and holds meetings and lectures in the field of Oriental studies; 650 mems; Sec. Prof. Dr JENS PETER LAUT; publs include *Zeitschrift, Abhandlungen für die Kunde des Morgenlandes*.

Forschungsinstitut für Wirtschaftliche Entwicklungen im Pazifikraum eV (Asia-Pacific Economic Research Institute): University of Duisburg-Essen, Campus Duisburg, 47048 Duisburg, Lotharstrasse 65; tel. (203) 3792357; fax (203) 3791786; e-mail iwbheiduk@uni-duisburg.de; internet www.uni-duisburg.de/FB5/VWL/IWB/fip.htm; f. 1985; research on economic developments in the Pacific region; Dirs Prof. GÜNTER HEIDUK, Prof. WERNER PASCHA, Prof. CARSTEN HERRMANN-PILLATH.

Institut für Asienkunde (Asian Affairs): 20148 Hamburg, Rothenbaumchaussee 32; tel. (40) 4288740; fax (40) 4107945; e-mail ifa@ifa.duei.de; internet www.duei.de/ifa; f. 1956; research and documentation into all aspects of contemporary South, South-East and East Asia; Pres. Dr Dr W. RÖHL; Dir Prof. Dr MONIKA SCHÄDLER.

Museum für Islamische Kunst (Islamic Art and Antiquities): Staatliche Museen zu Berlin Preussischer Kulturbesitz, 10178 Berlin, Bodestrasse 1-3; tel. (30) 20905400; fax (30) 20905402; e-mail isl@smb.spk-berlin.de; internet www.smb.spk-berlin.de/isl; f. 1904; Dir Prof. Dr CLAUS-PETER HAASE.

Museum für Ostasiatische Kunst (Far Eastern Art): Staatliche Museen zu Berlin-Preussischer Kulturbesitz, 14195 Berlin, Takustrasse 40; tel. (30) 8301381; fax (30) 8301501; e-mail oak@smb.spk-berlin.de; internet www.smb.spk-berlin.de; f. 1906; Dir Prof. Dr W. VEIT.

HONG KONG

Asia-Pacific Financial Markets Research Centre: City University of Hong Kong, 83 Tat Chee Ave, Kowloon; tel. 27887940; fax 27888806; internet www.cityu.edu.hk/ref/index.htm; f. 1983; research into financial markets in the Asia-Pacific region; Dir Prof. RICHARD HO; publ. *Technical Reports* (monthly).

The Asia-Pacific Institute of Business: The Chinese University of Hong Kong, Li Dak Sum Bldg, 2nd Floor, Shatin, New Territories; tel. 26097428; fax 26035136; e-mail apib@cuhk.edu.hk; f. 1990; provides research, consultancy and educational courses on business and economic issues concerning Hong Kong and the Asia-Pacific region; Exec. Dir Prof. LESLIE YOUNG; publ. *The Hong Kong Securities Industry*.

Centre of Asian Studies: The University of Hong Kong, Pokfulam Rd; tel. 28592460; fax 25593185; e-mail casgen@hku.hk; internet www.hku.hk/cas; f. 1967; traditional and contemporary China, Hong Kong, East, North-East and South-East Asia; Dir Prof. S. L. WONG; publs occasional papers and monographs, bibliographies and research guides.

Hong Kong Institute of Asia-Pacific Studies: Chinese University of Hong Kong, Esther Lee Bldg, Shatin, New Territories; tel. 26098777; fax 26035215; e-mail hkiaps@cuhk.edu.hk; internet www.cuhk.edu.hk/hkiaps/homepage.htm; f. 1990; Dir Prof. YUE-MAN YEUNG; publs occasional papers, research monographs, seminar series.

Institute of Chinese Studies: The Chinese University of Hong Kong, Shatin, New Territories; tel. 26097394; fax 26035149; e-mail ics@cuhk.edu.hk; internet www.cuhk.edu.hk/ics; f. 1967; comprises Art Museum, Research Center for Translation, Center for Chinese Archaeology and Art, Chinese Language Research Center, Research Center for Contemporary Chinese Culture and Chinese Ancient Text (CHANT) Database Project; Dir Prof. JENNY SO FONG SUK; publs include *ICS Journal* (annually), *Renditions* (2 a year), *The Chinese Language Newsletter* (quarterly) and *Twenty-First Century* (every 2 months).

HUNGARY

Magyar Tudományos Akadémia Világgazdasági Kutató Intézete (Institute for World Economics of the Hungarian Academy of Sciences): 1014 Budapest, Országház u. 30; tel. (1) 224-6760; fax (1) 224-6761; e-mail vki@vki.hu; internet www.vki.hu; f. 1967; library of 103,000 vols; Dir ANDRAS INOTAI; publs *Working Papers* (c. 15 a year, in English), *Kihívások* (10–15 a year, in Hungarian), *Muhelytanulmányok* (10–15 a year, in Hungarian).

INDIA

Centre for Development Studies: Prasanth Nagar Rd, Ulloor, Trivandrum 695 011; tel. (471) 2448881; fax (471) 447137; e-mail somannair@cds.ac.in; internet www.cds.edu; f. 1971; promotes interdisciplinary research and academic instruction in disciplines relevant to development issues; library of 114,500 vols; Dir K. P. KANNAN; Librarian-in-charge JOSEPH KURIEN.

Centre for South, Central South-East Asian and South-West Pacific Studies: Jawaharlal Nehru University, New Mehrauli Rd, New Delhi 110 067; tel. (11) 26167676; Chair. Prof. BALADAS GHOSHAL.

Indira Gandhi Institute of Development Research (IGIDR): Gen. A. K. Vaidya Marg, Goregaon (E), Mumbai 400 065; tel. (22) 28400919; fax (22) 28402752; internet www.igidr.ac.in/; f. 1986; Dir Dr R. RADHAKRISHNA.

Indian Society for Afro-Asian Studies: 297 Saraswati Kunj, Indraprastha Extension, Mother Dairy Rd, New Delhi 110 092; tel. (11) 22248246; fax (11) 22425698; e-mail isaas@giasdl01; f. 1980; research and promotion of co-operation among African and Asian countries; Pres. LALIT BHASIN; Gen. Sec. Dr DARAMPAL; publs monographs, papers.

Institute of Economic Growth: University Enclave, Delhi 110 007; tel. (11) 27667260; fax (11) 27667410; e-mail system@ieg.ernet.in; internet ieg.nic.in; f. 1958; research into the problems of social and economic development of South and South-East Asia; environmental economics, health economics, agriculture, rural development, macroeconomic analysis; library and documentation services; Dir Prof. B. B. BHATTACHARYA; publs include *Contributions to Indian Sociology: New Series* (3 a year).

Namgyal Institute of Tibetology: Gangtok, Sikkim; e-mail nitsikkim@yahoo.co.in; internet www.tibetology.com; f. 1958; research centre for study of Mahayana (Northern Buddhism) and Himalayan cultures; library of Tibetan literature (canonical of all sects and secular) in MSS and xylographs; museum of icons and art objects; Pres. V. RAMA RAO (The Governor of Sikkim); Dir TASHI DENSAPA; publs include *Bulletin of Tibetology* (2 a year) and publications in Tibetan, Sanskrit and English.

Nava Nalanda Mahavihara: PO Nalanda, Bihar 803 111; tel. (6112) 281820; e-mail nnmdirector@sify.com; f. 1951; under Dept of Culture, Ministry of Tourism and Culture; postgraduate studies and research in Pali, Buddhist studies, ancient Indian and Asian history and philosophy, Tibetan, Sanskrit, Chinese and Japanese; library of 45,000 vols; Dir Dr RAVINDRA PANTH; publs include *Nava Nalanda Mahavihara Research* (annually).

RIS (Research and Information System) for the Non-Aligned and Other Developing Countries: Zone IV-B, 4th Floor, India Habitat Centre, Lodhi Rd, New Delhi 110 003; tel. (11) 24682176; fax (11) 24682173; e-mail dgoffice@ris.org.in; internet www.ris.org.in; trade and development issues; Dir-Gen. Dr NAGESH KUMAR; publs include *South Asia Development and Cooperation Reports* (every 2 years), *World Trade and Development Reports* (every 2 years), *South Asia Economic Journal* (2 a year), *Asian Biotechnology and Development Review* (3 a year), *New Asia Monitor* (quarterly).

INDONESIA

Centre for Social and Cultural Studies: Indonesian Institute of Sciences, Gedung Widya Graha Lt. 11, Jalan Jenderal

Gatot Subroto, POB 4492, Jakarta 12044, tel. (21) 511542; fax (21) 5701232.

Centre for Strategic and International Studies (CSIS): Jalan Tanah Abang III/23-27, Jakarta 10160; tel. (21) 3865532; fax (21) 3847517; e-mail csis@pacific.net.id; f. 1971; undertakes policy-orientated studies in international and domestic matters in collaboration with the industrial, commercial, political, legal and journalistic communities of Indonesia; interdisciplinary research projects; Chair. HADI SOESASTRO; publs include *The Indonesian Quarterly*, *Analisis CSIS* (4 a year).

Indonesian Institute of World Affairs: c/o Universitas Indonesia, Salemba Raya 4, Jakarta; tel. (21) 882955; Chair. Prof. SUPOMO; Sec. SUDJATMOKO.

Institute for Regional Economic Research: Universitas Andalas, Kampus Limau Manis, Padang 26163, West Sumatra; tel. (751) 71389; fax (751) 71085; Dir SYAHRUDDIN.

Perpustakaan Nasional RI (National Library of Indonesia): Jalan Salemba Raya, No 28A, Jakarta Pusat 10430; tel. (21) 3101411; fax (21) 3103554; internet www.pnri.go.id; f. 1989; 844,480 vols, 25,000 maps, 10,857 microfilms and microfiches, 10,000 MSS; Dir MASTINI HARDJOPRAKOSO; publs include *National Bibliography*.

Pusat Bahasa (National Centre for Language): Jalan Daksinapati Barat IV, POB 2625, Rawamangun, Jakarta 13220; tel. and fax (21) 4706678; f. 1975; attached to the Ministry of National Education; language policy and planning, research on language teaching and literature; library of 85,000 vols; Dir Dr DENDY SUGONO; Librarian AGNES SANTI; publs include *Informasi Pustaka Kebahasaan* (quarterly), *Bahasa dan Sastra* (quarterly).

IRAN

Asia Institute: University of Shiraz, Shiraz; tel. (71) 32111; Dir Dr Y. M. NAVABI; publs *Bulletin*, monographs.

Institute for Political and International Studies: Shaheed Bahonar Ave, Shaheed Aghaii St, POB 19395-1793, Tajrish, Tehran; tel. (21) 2571010; fax (21) 270964; f. 1983; research and information on international relations, economics, law and Islamic studies; library of 22,000 vols; publs include *Iranian Journal of International Affairs* (quarterly).

ISRAEL

Harry S. Truman Research Institute for the Advancement of Peace: The Hebrew University of Jerusalem, Mt Scopus, Jerusalem 91905; tel. (2) 5882300; fax (2) 5828076; e-mail mstruman@mscc.huji.ac.il; internet www.truman.huji.ac.il; f. 1965; conducts and sponsors social science and historical research, organizes conferences and publs works on many regions, including Asia; Dir Prof. EYAL BEN ARI.

Institute of Asian and African Studies: The Hebrew University of Jerusalem, Mt Scopus, Jerusalem 91905; tel. (2) 5883659; fax (2) 5883658; e-mail AsiaAfrica@h2.hum.cc.huji.ac.il; internet asiafrica.huji.ac.il; f. 1926 as Institute of Oriental Studies; provides degree and postgraduate courses, covering history, social sciences and languages, in Chinese, Japanese, Korean, Tibetan and South Asian studies; Dir Prof. MEIR BAR-ASHER; Sec. YEHUDIT MAGEN.

International Institute-Histadrut (incorporating Afro-Asian Institute): Bet Berl 44905, Kfar Saba; tel. (9) 7612323; fax (9) 7456962; e-mail frohlich@peoples.org.il; f. 1958 to train leadership for trade unions, co-operatives, community orgs, women's and youth groups, and for social and economic aspects of development, etc. in regions throughout the world, incl. Asia and Pacific; library of 35,000 vols; Chair. SAMI BEN YAISH; Dir and Prin. Dr YEHUDAH PAZ.

ITALY

The Bologna Center of the Paul H. Nitze School of Advanced International Studies of the Johns Hopkins University: Via Belmeloro 11, 40126 Bologna; tel. (051) 2917811; fax (051) 228505; e-mail Registrar@jhubc.it; internet www.jhubc.it; f. 1955; graduate studies in international affairs and economics; Dir MARISA R. LINO; publs *Bologna Center Catalogue*, *EuroSAIS* (alumni newsletter), occasional papers series.

Institute of Economic and Social Studies for East Asia: Università Commerciale Luigi Bocconi, Via U. Gobbi 5, 20136 Milan; tel. (02) 58363317; fax (02) 58363309; e-mail Isesao@uni-bocconi.it; internet www.isesao.uni-bocconi.it; Dir Prof. C. FILIPPINI.

Istituto Affari Internazionali (IAI): Via Angelo Brunetti 9, 00186 Rome; tel. (06) 3224360; fax (06) 3224363; f. 1965; Dir GIANNI BONVICINI; publs include *The International Spectator* (English, quarterly).

Istituto Italiano per l'Africa e l'Oriente (ISIAO): Via V. Aldrovandi 16, 00197 Rome; tel. (06) 328551; fax (06) 3224358; e-mail info@isiao.it; internet www.isiao.it; f. 1995; Pres. Prof. GHERARDO GNOLI; publs include *Africa* (quarterly), *Cina* (annually), *East and West* (quarterly), *Il Giappone* (annually).

Istituto per gli Studi di Politica Internazionale (ISPI): Palazzo Clerici, Via Clerici 5, 20121 Milan; tel. (02) 8633131; fax (02) 8692055; e-mail ispizi@tim.it; internet www.ispinet.it; f. 1933 for the promotion of the study and knowledge of all problems concerning international relations; seminars at postgraduate level; library of 100,000 vols; Pres. BORIS BIANCHERI; publs *ISPI Relazioni Internazionali* (quarterly), papers (20 a year).

Istituto Universitario Orientale (Oriental University Institute): Dept of Asian Studies, Piazza San Domenico Maggiore 12, 80134 Naples; tel. (081) 5517855; fax (081) 5517852; f. 1732; library of 200,000 vols, 2,000 periodical titles; Dir Prof. UGO MARAZZI; publs include four series (*Maior, Minor, Serie Tre* and *Baluchistan Monograph Series*) and *Annali*.

JAPAN

Ajia Daigaku (Institute for Asian Studies): Asia University, 5-24-10 Sakai, Musashino-shi, Tokyo 180; tel. (422) 54-3111; fax (422) 36-1083; e-mail koryu@asia-u.ac.jp; internet www.asia-u.ac.jp/english/main/asian.htm; f. 1973; Dir S. SAITO; publs *Journal* (annually), *Bulletin of Institute for Asian Studies* (quarterly).

Ajia Seikei Gakkai (Japan Association for Asian Studies): c/o Prof. Amako, School of International Politics, Economics and Business, Aoyama Gakuin University, 4-4-25 Shibuya, Shibuya-ku, Tokyo 150; f. 1953; 1,040 mems; Pres. TOSHIO WATANABE; publ. *Aziya Kenkyu* (quarterly).

The Center for Southeast Asian Studies: 46 Shimoadachicho, Yoshida, Sakyo-ku, Kyoto 606-8501; tel. (75) 753-7300; fax (75) 753-7350; attached to Kyoto Univ.; Dir Prof. NARIFUMI MAEDA TACHIMOTO; publs include *Southeast Asian Studies* (quarterly), monographs (English and Japanese, irregular).

Centre for Asian and Pacific Studies: Seikei University, 3-3-1, Kichijoji-Kitamachi, Musashino-shi, Tokyo 180-8633; tel. (422) 37-3549; fax (422) 37-3866; e-mail caps@jim.seikei.ac.jp; internet www.seikei.ac.jp/university/caps/; f. 1981; conducts and finances research on modernization, industrialization and structural changes in the Asia-Pacific region; publs include *Review of Asian and Pacific Studies*.

The Centre for East Asian Cultural Studies for UNESCO, Toyo Bunko (Oriental Library): Honkomagome 2-28-21 Bunkyo-ku, Tokyo 113; tel. (3) 3942-0124; fax (3) 3942-0120; internet www.toyo-bunko.or.jp/ceacs/; f. 1961; Dir YONEO ISHII; publs include *Asian Research Trends*, *Bibliotheca Codicum Asiaticorum*, *East Asian Cultural Studies*, *Recent Archaeological Discoveries in Asia*, *Historic Cities of Asia* series, translations of historical documents, monographs, bibliographies, directories.

The Economic Research Institute for Northeast Asia (ERINA): Nihonseimei Masayakoji Bldg, 6-1178-1, Kamiokawamae-dori, Niigata 951-8068; tel. (25) 222-3141; fax (25) 222-9505; internet www.erina.or.jp; f. 1993.

Nihon Boeki Shinkokiko Ajia Keizai Kenkyusho (Institute of Developing Economies, Japan External Trade Organization—JETRO): 3-2-2 Wakaba, Mihama-ku, Chiba-shi, Chiba 261-8545; tel. (43) 299-9500; fax (43) 299-9724; e-mail info@ide.go.jp; internet www.ide.go.jp; Ajia Keizai Kenkyusho est. 1958, merged with JETRO 1998; research on economic and related subjects in Asia and other developing areas; aims to promote economic co-operation and to improve trade relations between

Japan and the developing countries; 247 mems; Pres. Masahisa Fujita; library of 550,000 vols; publs include *Ajia Keizai* (Japanese, monthly), *Ajiken Warudo Torendo* (World Trends, Japanese, monthly), *The Developing Economies* (English, quarterly), occasional papers series (English, irregular).

Nihon-Indogaku-Bukkyogakkai (Japanese Association of Indian and Buddhist Studies): c/o Dept of Indian Philosophy and Buddhist Studies, Graduate School of Humanities and Sociology, University of Tokyo, 7-3-1 Hongo, Bunkyo-ku, Tokyo 113; tel. (3) 3813-5903; f. 1951; 2,500 mems; Pres. Sengaku Mayeda; publ. *Indogaku Bukkyogaku Kenkyu* (Journal of Indian and Buddhist Studies).

Nihon Keizai Kenkyu Senta (Japan Center for Economic Research): Nikkei Kayabacho Bldg, 2-6-1 Nihonbashi Kayabacho, Chuo-ku, Tokyo 103-0025; tel. (3) 3639-2801; fax (3) 3639-2839; internet www.jcer.or.jp; f. 1963; 370 corporate and 300 individual mems; library of 37,000 vols and 2,000 titles; Chair. Yutaka Kosai; Pres. Naohiri Yashiro; publs *Nihon Keizai Kenkyu Senta Kaiho* (semi-monthly), *Nihon Keizai Kenkyu* (annually), *Economic Forecast Series*, *International Conference Series*.

Nihon Kokusai Mondai Kenkyusho (The Japan Institute of International Affairs): Tokyo; tel. (3) 3503-7261; fax (3) 3595-1755; internet www.jiia.or.jp; f. 1959; Chair. Yoshizane Iwasa; Pres. Nobuo Matsunaga; publs include *Kokusai Mondai* (International Affairs, monthly), *Japan Review of International Affairs* (4 a year), newsletters, books and monographs.

Nomura Research Institute (NRI): Shin Otemachi Bldg, 2-2-1 Otemachi, Chiyoda-ku, Tokyo 100-0004; tel. (3) 5255-1852; fax (3) 5255-9315; internet www.nri.co.jap; f. 1965; research into Japanese and international economic and security issues; Pres. Shozo Hashimoto; publ. *NRI Policy Research*.

Research Institute for the Study of Languages and Cultures of Asia and Africa (ILCAA): Tokyo University of Foreign Studies, Asahicho 3-11-1, Fuchu, Tokyo 183-8534; tel. (42) 330-5600; fax (42) 330-5610; e-mail director@aa.tufs.ac.jp; internet www.aa.tufs.ac.jp; f. 1964; library of 65,445 vols; Dir Koji Miyazaki; publs *Journal of Asian and African Studies* (2 a year), *Newsletter* (3 a year).

Tōhō Gakkai (Institute of Eastern Culture): 2-4-1 Nishi-Kanda, Chiyoda-ku, Tokyo 101-0065; tel. (3) 3262-7221; fax (3) 3262-7227; e-mail iec@tohogakkai.com; internet www.tohogakkai.com; f. 1947; 1,550 mems; Chair. Prof. Yoshio Togawa; Sec.-Gen. Hiroshi Yanase; publs include *Acta Asiatica*, *Tōhōgaku*, *Transactions of the International Conference of Eastern Studies*.

Tokyo Daigaku Toyo Bunka Kenkyujo (Institute of Oriental Culture, the University of Tokyo): 7-3-1 Hongo, Bunkyo-ku, Tokyo 113-0033; tel. (3) 5841-5830; fax (3) 5841-5898; internet ioc.u-tokyo.ac.jp; f. 1941; Dir Prof. Akihiko Tanaka; publ. *Tōyō bunka kenkyūjo kiyō* (2 a year), *Tōyō bunka* (annually).

Toyo Bunko (Modern China Research Committee): Honkomagome 2-28-21 Bunkyo-ku, Tokyo 113-0021; tel. (3) 3942-0121; f. 1962; collection on modern China; Asian studies library of 870,000 vols; publ. *Kindai Chugoku Kenkyu Iho* (annually).

United Nations Centre for Regional Development: Nagono 1-47-1, Nakamura-ku, Nagoya 450-0001; tel. (52) 561-9377; fax (52) 561-9375; e-mail rep@uncrd.or.jp; internet www.uncrd.or.jp/; f. 1971; undertakes training, research, consultation and information exchange on regional development issues affecting developing countries; Dir Yo Kimura; publs *Regional Development Dialogue* (2 a year), *Regional Development Studies* (annually).

REPUBLIC OF KOREA

Asiatic Research Center: Korea University, Anam-dong 5 ga-1, Sungbuk-gu, Seoul; tel. (2) 926-1926; fax (2) 924-9132; e-mail arcku@korea.ac.kr; internet www.asiacenter.or.kr; f. 1957; Dir Prof. Hong Sung-Chick; publ. *Journal of Asiatic Studies*.

Ilmin International Relations Institute: Korea University, Anam-dong, Seoul; Pres. Han Sung Joo.

Institute of Economic Research, Korea University: 1, 5-ga, Anam-dong, Sungbuk-gu, Seoul 136-701; tel. (2) 3290-2200; fax (2) 926-3601; e-mail eghwang@kuccnx.korea.ac.kr; f. 1957; Dir Prof. Eui-Gak Hwang.

Institute of Economic Research, Seoul National University: Sinlim-dong, Kwanak-gu, Seoul 151-742; tel. (2) 877-1629; fax (2) 926-3601; f. 1961.

Institute for Far Eastern Studies: Kyungnam University, 28-42 Samchung-dong, Chongno-gu, Seoul 110-230; tel. (2) 3700-0725; fax (2) 3700-0707; e-mail ifes@kyungnam.ac.kr; internet www.ifes.kyungnam.ac.kr; f. 1972; publs monographs, 2 journals (on international and regional studies, 2 a year).

Institute of Korean Studies: Yonsei University, 134 Shinchon-dong, Sudaemun-gu, Seoul 120-749; tel. (2) 361-3502; fax (2) 365-0937; f. 1948; Dir Nam Ki Shim; publ. *Dong Bang Hak Chi* (quarterly).

Dankook University Institute of Oriental Studies (DIOS): 8 Hannam-dong, Yongsan-gu, Seoul 140-714; tel. (2) 709-2234; fax (2) 798-3010; internet www.dankook.ac.kr/~oriental; f. 1970; library of 40,000 vols; Dir Prof. Kim Sang-Bai; publs include journal and newsletter.

Korea Development Institute: POB 113, Cheongryang, Seoul 130-012; tel. (2) 958-4114; fax (2) 961-5092; e-mail kdi-guide@kdiux.kdi.re.kr; internet www.kdiux.kdi.re.kr; f. 1971; conducts research in order to help maintain high economic growth and price stability; library of 120,000 vols; Pres. Lee Jin-Soon; *KDI Economic Outlook* (quarterly), *KDI Journal and Economic Policy* (quarterly).

Korea Institute for Economics and Technology (KIET): POB 205, Cheongryang, Seoul; tel. (2) 962-6211; fax (2) 963-8540; f. 1982.

Research Institute of Asian Economics: Haeyoung Bldg, Rm 304, 148 Anguk-dong, Chongno-gu, Seoul; tel. (2) 738-1875; fax (2) 720-2367; Pres. Shin Tae-Whan.

Research Institute of Oriental Culture: Sung Kyun Kwan University, 53, 3-ga, Myung Ryun-dong, Chongno-gu, Seoul 110-745; Dir Woo-Sung Lee.

Samsung Economic Research Institute (SERI): 7th–8th Floors, Hanil Group Bldg, 191 Hangangro, 2-ga, Yongsan-gu, Seoul; tel. (2) 3780-8000; e-mail homemst@seri21.org; internet seriecon.seri-samsung.org; f. 1986; Pres. Choi Woo Sock.

MACAO

Centre of Macao Studies: University of Macao, University Hill, Taipa, POB 3001, Macao.

Macao Development Strategy Research Centre: Rua de Xangai 175, Edif. da Associação Comercial Chinesa, 19/F; tel. 780124; fax 780565; e-mail cpedm@macau.ctm.net; internet www.cpedm.com; f. 1997.

Macau Ricci Institute: Av. Cons. Ferreira d'Almeida, 95-E, Macao; tel. 532536; fax 568274; e-mail instituto@riccimac.org; internet www.riccimac.org; f. 1999; Dir Luís Sequeira; publs *Macau Ricci Institute Studies*, *Chinese Cross Currents*(quarterly).

MALAYSIA

Asian and Pacific Development Centre: Persiaran Duta, POB 12224, 50770 Kuala Lumpur; tel. (3) 6511088; fax (3) 6510316; e-mail info@apdc.po.my; internet www.apdc.com.my/apdc; f. 1980; Dir Dr Mohd Noor Haji Harun; publs annual report, newsletter, and monographs.

Dewan Bahasa dan Pustaka (Institute of Language and Literature, Malaysia): POB 10803, 50926 Kuala Lumpur; fax (3) 21444460; internet www.dbp.gov.my; f. 1956 to develop and enrich the Malay language; to develop literary talent particularly in Malay; to print, publish or promote publication in Malay and other languages to standardize spelling and pronunciation and devise appropriate technical terms in Malay; Chair. Dato' Hj. Abdul Rahim bin Abu Bakar; Dir-Gen. Dato' Hj. A. Aziz Deraman; publs include magazines, textbooks, higher learning books and general books.

Pusat Dotumentasi Melayu DBP (Malay Documentation Centre): tel. (3) 2481011 Ext. 201; fax (3) 2442081; 120,000 vols, 3,000 periodicals, audio-visual materials, special collection on Malay language and literature; Chief Librarian Mrs ROHANI RUSTAM; publs *Mutiara Pustaka* (annually), subject bibliography (occasional).

Institute for Development Studies (Sabah): Block C, Suite 7 CFO1, 7th Floor, Karamunsing Complex, 88300 Kota Kinabalu, Sabah; tel. (88) 246166; fax (88) 234707; e-mail info@ids.org.my; internet www.ids.org.my; publs include *Borneo Review* (2 a year).

Institute of Malaysian International Studies (IKMAS): Universiti Kebangsaan Malaysia, Selangor, 43600 Bangi; tel. (3) 8293205; fax (3) 8261022; e-mail ikmas@pkrisc.cc.ukm.my; f. 1995; Dir Dr OSMAN RANI HASSAN.

Institute of Strategic and International Studies (ISIS): 1, Pesiaran Sultan Salahuddin, POB 12424, 50778 Kuala Lumpur; tel. (3) 2939366; fax (3) 2913210; e-mail jawhar@isis.po.my; internet www.isis.org.my/isis; f. 1983; undertakes studies in strategic and policy issues directly relevant to national interests and public welfare; Chair. and CEO Tan Sri Dr NOORDIN SOPIEE; Dir-Gen. Dato' MOHAMED JAWHAR HASSAN.

Malaysian Branch of the Royal Asiatic Society: 130M Jalan Thamby Abdullah, off Jalan Tun Sambanthan (Brickfields), 50470 Kuala Lumpur; tel. (3) 22748345; fax (3) 22743458; e-mail mbras@tm.net.my; f. 1877; 1,086 mems; Pres. Datuk ABDULLAH BIN ALI; Hon. Sec. Datuk Haji BURHANUDDIN BIN AHMAD TAJUDIN; publs include *Journal* (2 a year), monographs, reprints.

Malaysian Institute of Economic Research (MIER): 9th Floor, Menara Dayabumi, Jalan Sultan Hishamuddin, 50050 Kuala Lumpur; tel. (3) 22730214; fax (3) 22730197; e-mail ariff@mier.po.my; internet www.mier.org.my; f. 1985; conducts policy- and business-orientated research; Exec. Dir MOHAMMED ARIFF.

South East Asian Central Bank—Research and Training Centre: Lorong University A, 59100 Kuala Lumpur; tel. (3) 7568622; fax (3) 7574616; f. 1972; research into financial and economic affairs; publs include *Economic Survey of the SEACEN Countries*.

MEXICO

Centro de Estudios de Asia y Africa (Centre for Asian and African Studies): El Colegio de México, Camino al Ajusco 20, Pedregal Sta Teresa, Apdo 10740, México, DF; tel. (5) 449-3000; fax (5) 645-0464; e-mail webmaster@colmex.mex; internet www.colmex.mx/centros/ceaa/index.htm; Dir Prof. BENJAMÍN PRECIADO; publ. *Estudios de Asia y Africa* (quarterly).

MONGOLIA

Centre of Strategic Studies: POB 870, Ulan Bator; tel. 353034; f. 1990; research into national and international security and strategy, with emphasis on North-East Asia; Dir S. DZORIG; publs include *Security and Development Issues* (2 a year).

Institute of Oriental and International Studies: Academy of Sciences, Ulan Bator; fax 322613; Dir Dr TS. BATBAYAR; publs *East-West* (in Mongolian with summaries and contents in English; quarterly), *Mongolian Journal of International Affairs* (in English, annually).

MYANMAR

(formerly Burma)

Burma Research Society: Universities' Central Library, University PO, Yangon; f. 1910 to promote cultural and scientific studies and research relating to Myanmar and neighbouring countries; 1,040 mems; Pres. U HTIN GYI; Hon. Sec. Dr SHEIN; publ. *Journal* (2 a year).

Institute of Economics: Pyay Rd, University Estate, Kamayat Township, 11041 Yangon; tel. (1) 32433; f. 1964; library of 70,000 vols; Rector Dr THAN NYUN.

NEPAL

Centre for Nepal and Asian Studies: Tribhuvan University, POB 3757, Kirtipur, Kathmandu; tel. (1) 4331740; fax (1) 4331184; e-mail cnastu@htp.com/np; f. 1972; conducts political and social research both in Nepalese and Asian contexts; Exec. Dir Prof. Dr TIRTHA PRASAD MISHRA; publs include *Contributions to Nepalese Studies* (2 a year), monographs, bibliographies, occasional papers.

United Nations Regional Centre for Peace and Disarmament in Asia: United Nations Development Programme, UN Bldg, Pulchowk, POB 107, Kathmandu; tel. (1) 4524366; fax (1) 4523991; f. 1988; Dir J. BERKE.

NETHERLANDS

International Institute for Asian Studies (IIAS): POB 9515, 2300 RA Leiden; tel. (71) 5272227; fax (71) 5274162; e-mail iias@let.leidenuniv.nl; internet www.iias.nl; based in Leiden and Amsterdam; post-doctoral research centre; Dir Prof. W. A. L. STOKHOF; publ. *IIAS Newsletter*.

Koninklijk Instituut voor Taal-, Land- en Volkenkunde (Royal Institute of Linguistics and Anthropology): Reuvensplaats 2, POB 9515, 2300 RA Leiden; tel. (71) 5272295; fax (71) 5272638; e-mail kitlv@kitlv.nl; internet www.kitlv.nl; f. 1851 to collect and catalogue books and other documents, to carry out and support research, and to publish books and journals on South-East Asia (especially Indonesia), the Pacific area and the Caribbean region (in particular Suriname and the Netherlands Antilles); 2,044 mems; library of 280,000 titles; Pres. Prof. P. J. M. NAS; Dir Prof. G. J. OOSTINDIE; publs include *Bijdragen: New West Indian Guide* (quarterly), *Verhandelingen*, *Bibliotheca Indonesica*, *Working Papers Series*, *Excerpta Indonesica* (online), bibliographies, translations.

NEW CALEDONIA

Société des Etudes Mélanésiennes: Musée Neo-Calédonien, BP 2393, Nouméa; tel. 27-23-42; fax 28-41-43; f. 1938; anthropology; publ. *Etudes Mélanésiennes* (annually).

NEW ZEALAND

Institute of Polynesian Languages and Literatures: POB 6965, Wellesley St, Auckland 1; tel. (9) 372-5094; e-mail rongosf@internet.co.nz; f. 1995; charitable trust aiming to promote study and use of Polynesian languages and literatures through publication of research, etc.; Dir Dr STEVEN ROGER FISCHER; publ. *Rongorongo Studies: A Forum for Polynesian Philology* (2 a year).

Macmillan Brown Centre for Pacific Studies: University of Canterbury, PB 4800, Christchurch; tel. (3) 364-2957; fax (3) 364-2002; e-mail mbcps@pacs.canterbury.ac.nz; internet www.pacs.canterbury.ac.nz/; Chair. Dr JOHN HENDERSON; Dir Prof. KAREN NEVO.

New Zealand Asia Institute: University of Auckland, PB 92019, Auckland; tel. (9) 373-7599; fax (9) 308-2312; e-mail nzai@auckland.ac.nz; internet www.auckland.ac.nz/nzai; f. 1995; Dir Dr JAMES KEMBER.

New Zealand Geographical Society: c/o School of Geography and Environmental Science, University of Auckland, PB 92019; tel. (9) 373-7599, ext. 88464; fax (9) 373-7434; e-mail nzgs@auckland.ac.nz; internet www.nzgs.co.nz; f. 1944; 6 brs; Pres. Prof. PETER HOLLAND; Sec. Prof. RICHARD LE HERON; publs include *New Zealand Geographer* (2 a year), *New Zealand Journal of Geography* (2 a year), *NZGS Newsletter* (2 a year), *Conference Proceedings* (every 2 years).

New Zealand Institute of Economic Research: POB 3479, Wellington; tel. (4) 4721-880; fax (4) 4721-211; e-mail econ@nzier.org.nz; internet www.nzier.org.nz; f. 1958; independent non-profit making organization; research into NZ economic development; quarterly forecast of the national economy and annual medium-term industry outlook; quarterly survey of business opinion; economic investigations on contract basis; Dir Dr BRENT LAYTON; Chair. MICHAEL WALLS; Sec. A. G. FROGGATT; publs *Quarterly Predictions*, *Quarterly Survey of Business Opinion*, *New Zealand Industry Outlook*, Research Monographs, discussion papers.

New Zealand Institute of International Affairs: c/o Victoria University of Wellington, POB 600, Wellington 2; tel. (4) 463-5356; fax (4) 463-6568; e-mail nziia@vuw.ac.nz; internet www.vuw.ac.nz/nziia; f. 1934 to examine international questions, particularly in relation to Asia and the South Pacific; Pres. Sir

KENNETH KEITH; Dir BRIAN LYNCH; publ. *New Zealand International Review* (6 a year).

Polynesian Society: c/o Maori Studies Dept, University of Auckland, PB 92019, Auckland 1; e-mail jps@auckland.ac.nz; f. 1892 to promote the study of the anthropology, ethnology, philology, history and antiquities of the New Zealand Maori and other Pacific Island peoples; library; 1,103 mems; Pres. Prof. Sir HUGH KAWHARU; publs *Memoirs*, *Journal* (quarterly), *Maori Monographs*, *Maori Texts*.

PAKISTAN

Islamic Research Institute: POB 1035, Islamabad 44000; tel. (51) 850751; fax (51) 853360; f. 1960 to conduct and co-ordinate research in Islamic studies; organizes seminars, conferences etc.; library of 75,000 vols, 1,300 microfilms and microfiches, 258 MSS; 170 cassettes; Dir-Gen. Dr ZAFAR ISHAQ ANSARI; publs include *Al-Dirasat al-Islamiyyah* (Arabic, quarterly), *Islamic Studies* (English, quarterly), *Fikr-o-Nazar* (Urdu, quarterly), monographs.

Pakistan Institute of International Affairs: Aiwan-e-Sadar Rd, POB 1447, Karachi 74200; tel. (21) 5682891; fax (21) 5686069; f. 1947 to promote interest and research in international affairs; library of over 28,000 vols; over 600 mems; Chair. FATEHYAB ALI KHAN; Sec. Dr S. ADIL HUSAIN; publs *Pakistan Horizon* (quarterly), books and monographs.

PAPUA NEW GUINEA

Institute of National Affairs: POB 1530, Port Moresby; tel. 3211044; fax 3217223; e-mail inapng@daltron.com.pg; f. 1978; Head MICHAEL MANNING; publs discussion and working papers.

National Research Institute of Papua New Guinea: POB 5854, Boroko; tel. 3260300; fax 3260213; e-mail nri@global.net.pg; internet www.nri.org.pg; f. 1989 by amalgamation of Institute of Applied Social and Economic Research, Institute of Papua New Guinea Studies and Education Research Unit; promotion of research into social, legal, military, political, economic, educational, environmental and cultural issues in PNG; library of 13,000 vols; Dir Dr THOMAS WEBSTER; publs *Monographs*, *Discussion Papers*, *Bibliography*, *Post Courier Index* (annually), *Trends in Development* (2 a year), special and educational reports, etc.

THE PHILIPPINES

Asian Center: University of the Philippines, Diliman, Quezon City, Metro Manila 1101; tel. (2) 9261841; fax (2) 9261821; e-mail asiancenter@ac.upd.edu.ph; f. 1955 as Institute of Asian Studies; Dean Dr ARMANDO MALAY, Jr; publs include *Asian Studies* (annually), monographs, occasional papers and books.

Asia-Pacific Peace Research Association: 41 Rajah Matanda, Project 4, Quezon City, Metro Manila 1109; tel. (2) 9139255; fax (2) 9136435; e-mail appra@csi.com.ph; Sec.-Gen. MARY SOLEDAD PERPIÑAN.

Centre for Research and Communication: Unit 1103, Pacific Centre Bldg, San Miguel Ave, Ortigas Centre, Pasig City 1600; tel. (2) 6311284; fax (2) 6336741; e-mail crcfi@mnl.sequel.net; Pres. ENRIQUE P. ESTEBAN; Exec. Vice-Pres. Dr EMILIO T. ANTONIO, Jr.

Cultural Center of the Philippines: CCP Complex, Roxas Blvd, Pasay City, Metro Manila; tel. (2) 8321125; fax (2) 8340471; e-mail ccp@culturalcenter.gov.ph; internet www.culturalcenter.gov.ph; f. 1966 to preserve, develop and promote Philippine arts and culture; Pres. NESTOR O. JARDIN.

Institute of Philippine Culture: Ateneo de Manila University, Loyola Heights, Quezon City, POB 154, Metro Manila 1099; tel. (2) 4266067; fax (2) 4265660; e-mail add.ipc@pusit.admu.edu.ph; f. 1960; conducts research into rural and urban poverty, agrarian reform, irrigation, community health, coastal resources, forestry and women's affairs; assists development agencies; trains agency personnel and local communities in the use of research methodologies; library of 3,166 vols, 4,070 reprints and vertical file, databank; Dir Dr GERMELINO M. BAUTISTA; publs include *IPC Papers*, *IPC Monograph Series*, *IPC Final Reports*, *IPC Reprints* (all irregular).

Research Institute for Mindanao Culture: Xavier University, Ateneo de Cagayan, POB 24, Cagayan de Oro City 9000; tel. and fax (8822) 723228; e-mail lburton@xu.edu.ph; f. 1957 to study and assist the development of north Mindanao, the Philippines in general and their peoples; Dir Dr ERLINDA M. BURTON; publs *RIMCU Updates* (quarterly), periodicals.

Research Institute for Politics and Economics: Polytechnic University of the Philippines, Anonas St, Santa Mesa, Manila; tel. (2) 616775; fax (2) 7161143; Dir Prof. DANILO CUETO.

POLAND

Institute of Developing Countries: University of Warsaw, ul. Karowa 20, 00-324 Warsaw; tel. (22) 55-23-237; fax (22) 55-23-227; e-mail ikr@uw.edu.pl; internet www.ikr.uw.edu.pl; f. 1962; undergraduate and postgraduate studies; interdisciplinary research on developing countries; Dir Prof. URSZULA ŻUŁAWSKA.

Institute of Oriental Studies: 00-927 Warsaw, ul. Krakowskie Przedmieście 26/28; tel. (22) 55-20-349; e-mail dyrekcja@orient.uw.edu.pl; internet www.orient.uw.edu.pl/~pto; f. 1922; Pres. Prof. JOLANTA SIERAKOWSKA-DYNDO; Sec. MARIA KOZŁOWSKA; publ. *Przeglad Wschodni*.

Komitet Nauk Orientalistycznych PAN (Committee for Oriental Studies of the Polish Academy of Sciences): 00-927 Warsaw, ul. Krakowskie Przedmieście 26/28; tel. (22) 849-49-02; f. 1952; Asian and African studies, particularly social sciences; Pres. Prof. Dr MIECZYSŁAW J. KÜNSTLER; publs *Rocznik Orientalistyczny* (2 a year), series *Prace orientalistyczne* (irregular).

PORTUGAL

Museu Etnográfico da Sociedade de Geografia de Lisboa (Ethnographical Museum): Rua Portas de Santo Antão 100, 1100 Lisbon; tel. (1) 3425068; fax (1) 3464553; e-mail soc.geografia.lisboa@clix.pt; f. 1875; native arts, arms, clothing, musical instruments, statues of navigators and historians, relics of voyages of discovery, scientific instruments; Dir Prof. JOÃO PEREIRA NETO; Curator Dr MANUEL CANTINHO.

RUSSIA

Centre for Japanese Studies: Russian Academy of Sciences, 117218 Moscow, pr. Nakhimovsky 32; tel. (095) 124-08-35; fax (095) 310-70-56; e-mail ifes@cemi.rssi.ru; f. 1922; part of the Institute of Far Eastern Studies; Chair. Dr VICTOR N. PAVLYATENKO.

Institute of Far Eastern Studies: Russian Academy of Sciences, 117218 Moscow, pr. Nakhimovsky 32; tel. (095) 124-08-35; fax (095) 310-70-56; e-mail ifes@cemi.rssi.ru; f. 1966; Dir Prof. MICHAEL L. TITARENKO; publs include *Far Eastern Affairs* (6 a year).

Institute of World Economy and International Relations: 117859 Moscow, ul. Profsoyuznaya 23; tel. (095) 120-43-32; fax (095) 310-70-27; e-mail ineir@sovam.com; f. 1956; Dir VLADEN A. MARTYNOV; publs include *Otnosheniya* (monthly).

Moscow State Institute of International Relations: 117454 Moscow, Vernadskogo pr. 76; tel. (095) 434-91-58; fax (095) 434-90-66; f. 1944; library of 718,000 vols; Rector ANATOLII V. TURKUNOV; publ. *Moscow Journal of International Law*.

SINGAPORE

Asian Mass Communication Research and Information Centre (AMIC): Singapore; tel. 6251506; fax 62534535; e-mail amicline@singnet.com.sg; internet sunsite.nus.edu.sg/amic/; f. 1971; library of 13,000 books, periodicals and conference papers; Sec.-Gen. VIJAY MENON; publs *ACMB* (6 a year), *Asian Journal of Communication* (2 a year), *Media Asia* (quarterly).

The China Society: 61 Club St, Singapore 069436; f. 1948 to promote the study of Chinese culture; 106 mems; Pres. SIMON K. C. HU; publ. *Journal of the China Society*.

East Asian Institute: AS5, Level 4, 10 Kent Ridge Drive, Singapore 119260; tel. 67791037; fax 67793409; e-mail eaiwgw@nus.edu.sg; internet www.nus.edu.sg/NUSinfo/EAI; Dir Prof WANG GUNGWU.

Institute of Policy Studies: 1 Hon Sui Sen Drive, Hon Sui Sen Memorial Library Bldg, Kent Ridge Drive, Singapore 117588; tel. 67792633; fax 67770700; Dir Dr LEE TSAO YUAN.

Institute of Southeast Asian Studies: 30 Heng Mui Keng Terrace, Pasir Panjang, Singapore 119614; tel. 67780955; fax 67781735; e-mail admin@iseas.edu.sg; internet www.iseas.edu.sg; f. 1968 for the promotion of research into the problems of economic development, stability and security and political and social change in South-East Asia; library of 300,000 vols; Dir K. KESAVAPANY; publs *Southeast Asian Affairs* (annually), *Regional Outlook* (annually), *Contemporary Southeast Asia* (3 a year), *SOJOURN, Journal of Social Issues in Southeast Asia* (2 a year), *ASEAN Economic Bulletin* (3 a year).

Singapore Institute of International Affairs: 6 Nassim Rd, Singapore 258373; tel. 67349600; fax 67336217; e-mail siia@pacific.net.sg; f. 1961; organizes seminars, lectures and conferences; 120 mems; Chair. SIMON TAY; Dir Dr LEE LAI TO.

SOLOMON ISLANDS

Solomon Islands National Museum and Cultural Centre: POB 313, Honiara; tel. 23351; associated with the Cultural Asscn of the Solomon Islands; research into all aspects of Solomons culture; Dir LAWRENCE FOANAOTA; publs *Journal*, *Custom Stories*.

SPAIN

Centro de Estudios de Asia Oriental (Center for East Asian Studies): Universidad Autónoma, Campus de Cantoblanco, 28049 Madrid; tel. (91) 3974695; fax (91) 3975278; e-mail ceao@uam.es; internet www.uam.es/ceao; f. 1992; promotes research relating to the economy, society and culture of East Asian countries, with particular emphasis on China and Japan; Dir TACIANA FISAC; publs include *Boletín* (annually).

Chinese Studies Centre: Universidad Autónoma de Barcelona, 08193 Barcelona; tel. (93) 5811374; fax (93) 5811037; f. 1989; Dir Dr SEÁN GOLDEN.

SWEDEN

Centre for East and Southeast Asian Studies, Lund University: Box 792, 220 07, Lund; tel. (46) 222-30-40; fax (46) 222-30-41; e-mail info@ace.lu.se; internet www.ace.lu.se; f. 1997; Dir Prof. ROGER GREATREX; publ. *Working Papers in Contemporary Asian Studies*.

Centrum för stillahavsasienstudier (Center for Pacific Asia Studies): 106 91 Stockholm; tel. (8) 162897; fax (8) 168810; e-mail cpas@orient.su.se; internet www.cpas.su.se; f. 1984; attached to Stockholm University; Dir Dr MASAKO IKEGAMI.

European Institute of Japanese Studies (EIJS): Stockholm School of Economics, Box 6501, 113 83 Stockholm; tel. (8) 736-93-64; fax (8) 31-30-17; e-mail asia@hhs.se; f. 1992; Dir Prof. MAGNUS BLOMSTRÖM; publs papers, monographs, reports.

SWITZERLAND

Institut universitaire d'Etudes du Développement: 20 rue Rothschild, CP 136, 1211 Geneva 21; tel. (22) 906-5940; fax (22) 906-5947; internet www.iued.unige.ch; f. 1961; a centre of higher education, training and research into development problems, incl. those of Asia; conducts courses, seminars and practical work; Dir JEAN-LUC MAURER; publs *Cahiers de l'IUED*, *Annuaire suisse de politique de développement*, *Itinéraires*.

Modern Asia Research Centre (MARC): Graduate Institute of International Studies, 63 rue de Lausanne, CP 36, 1211 Geneva 21; tel. (22) 9085820; fax (22) 7383996; e-mail regnier@hei.unige.ch; internet www.hei.unige.ch/marc; f. 1971; Dir Dr PHILIPPE RÉGNIER; publs books, research reports, occasional papers, articles and newsletters.

Schweizerische Asiengesellschaft (Swiss Asia Society): c/o Ostasiatisches Seminar, Universität Zürich, Zürichbergstrasse 4, 8032 Zürich; tel. (1) 634-3181; fax (1) 634-4921; e-mail office@oas.unizh.ch; internet www.sagw.ch/dt/Mitglieder/outer.aspzid=40; f. 1939; 185 mems; Pres. Prof. Dr R. H. GASSMAN; publs *Asiatische Studien / Etudes Asiatiques* (quarterly), *Schweizer Asiatische Studien / Etudes Asiatiques Suisses* (series).

Schweizerisches Institut für Auslandforschung (Swiss Institute of International Studies): Seilergraben 49, 8001 Zürich; tel. (1) 632-6362; fax (1) 632-1947; Dir Prof. DIETER RULOFF; publ. *Sozialwissenschaftliche Studien* (annually).

TAIWAN

Academia Historica: 406 Sec. 2, Pei Yi Rd, Hsintien, Taipei; tel. (2) 22171535; fax (2) 22171640; contains national archives, library, documents; engaged in preparing history of China since 1894; Pres. CHU SHAO-HWA.

Asia and World Institute (AWI): 10th Floor, 102 Kuang Fu South Rd, Taipei; f. 1976; research into international relations, and North American, Asian and Pacific, and European affairs; Dir Dr PHILLIP M. CHEN; publs *AWI Digest*, *AWI Lecture and Essay Series*, *AWI Monograph Series*.

Chia Hsin Foundation: 96 Chung Shan Rd, North, Sec. 2, Taipei 10449; tel. (2) 2523-1461; fax (2) 2523-1204; f. 1963 for the promotion of culture in Taiwan; operates nationally in the fields of the arts, social studies, science and medicine, law and education, through research projects, courses, conferences, etc.; Chair. Dr YEN CHEN-HSING; Sec. TSENG WU-HSIONG.

Chung-hua Institution for Economic Research: 75 Chang Hsing St, Taipei 106; tel. (2) 27356006; fax (2) 27356035; e-mail mai@rs930.cier.edu.tw; internet www.cier.edu.tw; Pres. MAI CHAO-CHENG.

Graduate Institute of International Business: National Taiwan University, 1 Roosevelt Rd, Sec. 4, Taipei; tel. (2) 23638399; e-mail yljaw@mba.ntu.edu.tw; Dir YI-LONG JAW.

Institute of Economics, Academia Sinica: Nankang, Taipei 11529; tel. (2) 27822791; fax (2) 27853946; e-mail service@ieas.econ.sinica.edu.tw; internet www.sinica.edu.tw/econ; f. 1970; Pres. Dr LEE YUAN-TSEH.

Institute of International Relations: 64 Wan Shou Rd, Wenshan District, Taipei 116; tel. (2) 29394921; fax (2) 29378607; e-mail dschen@cc.nccu.edu.tw; internet www.iir.nccu.edu.tw/; f. 1953; research on international relations and mainland Chinese affairs; Dir SZU-YIN HO; publs *Mainland China Studies* (monthly), *Issues and Studies* (every 2 months).

Taiwan Institute of Economic Research (TIER): 7th Floor, 16-8 Tehwei St, Taipei; tel. (2) 5865000; fax (2) 5946317; internet www.tier.org.tw; f. 1976; Dir RU WONG-I.

THAILAND

Asian Institute of Technology (AIT): POB 4, Klong Luang, Pathumthani 12120; tel. (2) 516-0110; fax (2) 516-2126; e-mail jlarmand@ait.ac.th; internet www.ait.ac.th; f. 1959; Pres. Prof. JEAN-LOUIS ARMAND; publs *AIT Annual Report*, *Annual Report on Research and Activities*, *AIT Review* (3 a year), *Prospectus*, other specialized publs.

Institute of Asian Studies (IAS): 7th Floor, Prajadhipok-Rambhai Barni Bldg, Chulalongkorn University, Thanon Phyathai, Bangkok 10330; tel. (2) 251-5199; fax (2) 255-1124; e-mail ias@chula.ac.th; f. 1967; promotion of interdisciplinary Asian studies; Dir Dr SUPANG CHANTAVANICH; publs monographs, journals.

Institute of East Asian Studies: Thammasat University, Rangsit Campus, Pathum Thani 12121; tel. (2) 564-5000; fax (2) 564-4777; e-mail ieas@tu.ac.th; internet www.asia.tu.ac.th; f. 1984; Dir Assoc. Prof. YUPHA KLANGSUWAN.

Siam Society: 131 Soi 21 (Asoke), Thanon Sukhumvit, Bangkok 10110; tel. (2) 661-6470; fax (2) 258-3491; e-mail info@siam-society.org; internet www.siam-society.org; f. 1904 to promote interest and research in art, science and cultural affairs of Thailand and neighbouring countries; library of 30,000 vols; Pres. BILAIBHAN SAMPATISIRI; Hon. Sec. Dr WORAPHAT ARTHAYUKTI; publs include journals and *The Natural History Bulletin*.

Thailand Development Research Institute (TDRI): 565 Ramkhamhaeng 39 (Thepleela 1), Bangkapi, Bangkok 10310; tel. (2) 718-5460; fax (2) 718-5461; e-mail publications@tdri.or.th; internet www.info.tdri.or.th; f. 1984; independent policy research on Thailand's economic and social development issues; Pres. Dr CHALONGPHOB SUSSANGKARN.

TIMOR-LESTE

Timor Institute of Development Studies (TIDS): Rua Maucocomate, Becora, POB 181, Dili; tel. and fax 3323889; e-mail admin@tids-et.org; internet www.tids-et.org; f. 1997; fmrly East Timor Study Group (ETSG); promotes research and teachings in democracy and social change, economics and management, and agriculture and applied technology; Exec. Dir Dr João M. Saldanha; publ. *Observer* (every 2 months).

UNITED KINGDOM

Asia Pacific and Africa Collections (British Library): 96 Euston Rd, London, NW1 2DB; tel. (20) 7412-7873; fax (20) 7412-7641; e-mail oioc-enquiries@bl.uk; internet www.bl.uk/collections/asiapacificafrica.html; f. 1801; c. 1,200,000 European and Oriental printed books, 175,000 vols and files of archival records (1600–1948), 70,000 Oriental MSS, European MSS, 14,000 vols and boxes, 15,500 British paintings and drawings relating principally to India and the East, 11,000 Oriental drawings and miniatures, 2,900 prints and over 250,000 photographs; Head Graham W. Shaw; publs *Guide* and catalogues of the collections.

Asia Research Centre: London School of Economics and Political Science, Houghton St, London, WC2A 2AE; tel. (20) 7955-7388; fax (20) 7955-7591; e-mail c.s.lee2@lse.ac.uk; internet www.lse.ac.uk/Depts/asia; f. 1995; conducts social science research on Asia; Dir Dr Christopher R. Hughes.

British Association for Chinese Studies: BACS Secretariat, University of Essex, Colchester, CO5 3SQ; tel. (1206) 872543; fax (1206) 873408; e-mail adnin@bacsuk.org.uk; internet www.bacsuk.org.uk; f. 1976; provides forum for academics and others in the field; holds annual conference; Pres. Dr Harriet Evans; Adm. Sec. Lynn Baird; publs *Newsletter* (2 a year), *Annual Bulletin*.

Catholic Institute for International Relations (CIIR): Unit 3, Canonbury Yard, 190a New North Rd, London, N1 7BJ; tel. (20) 7354-0883; fax (20) 7359-0017; e-mail ciir@ciir.org; internet www.ciir.org; f. 1940; information and research on developing countries; recruits professionals for development projects overseas; Exec. Dir Christine Allen.

Centre for Asia-Pacific Studies, Nottingham Trent University: Faculty of Humanities, Nottingham Trent University, Clifton Lane, Nottingham, NG11 8NS; tel. (115) 848-3175; fax (115) 848-6385; e-mail neil.renwick@ntu.ac.uk; internet human.ntu.ac.uk/caps; Dirs Neil Renwick, Dr Roy Smith.

Contemporary China Institute: School of Oriental and African Studies, University of London, Thornhaugh St, Russell Sq., London, WC1H 0XG; tel. (20) 7898-4736; e-mail phil.deans@soas.ac.uk; f. 1968; Dir Dr Phil Deans; publs *Research Notes and Studies* (short monographs) and CCI/Oxford University Press series.

Department of East Asian Studies, University of Leeds: Leeds, LS2 9JT; tel. (113) 343-3460; fax (113) 343-6741; internet www.leeds.ac.uk/east-asian/home; Head of Dept Dr Mark Williams.

Faculty of Oriental Studies: University of Oxford, Pusey Lane, Oxford, OX1 2LE; tel. (1865) 278200; fax (1865) 278190; e-mail orient@orinst.ox.ac.uk; internet www.orinst.ox.ac.uk; f. 1960; Sec. Charlotte Vinnicombe; comprises Oriental Institute and Institute for Chinese Studies.

Far Eastern Section, Asian Department, Victoria and Albert Museum: South Kensington, London, SW7 2RL; tel. (20) 7942-2244; fax (20) 7942-2252; e-mail fareast@vam.ac.uk; internet www.vam.ac.uk; f. 1970; perm. displays and exhbns of art from China, Japan and Korea; lectures and research; Chief Curator Rose Kerr.

Institute of Commonwealth Studies: 28 Russell Sq., London, WC1B 5DS; tel. (20) 7862-8844; fax (20) 7862-8820; e-mail ics@sas.ac.uk; internet www.sas.ac.uk/commonwealthstudies/; f. 1949 to promote advanced study of the Commonwealth; provides a library and meeting place for postgraduate student and academic staff engaged in research on this field; offers postgraduate teaching; Dir Prof. Tim Shaw; publs *Annual Report*, *Collected Seminar Papers*, *Newsletter*, *Theses in Progress in Commonwealth Studies*.

Institute of Development Studies at The University of Sussex: Brighton, Sussex, BN1 9RE; tel. (1273) 606261; fax (1273) 621202; e-mail ids@ids.ac.uk; internet www.ids.ac.uk/ids; f. 1966; teaching and research on international development; Dir Keith Bezanson; publs include *IDS Bulletin* (quarterly), *IDS Discussion Papers*, *Annual Report*, *Research Reports*.

Nissan Institute of Japanese Studies, University of Oxford: 27 Winchester Rd, Oxford, OX2 6NA; tel. (1865) 274570; fax (1865) 274574; e-mail jane.baker@nissan.ox.ac.uk; internet www.nissan.ox.ac.uk; f. 1981; devoted to the study of modern Japan; Dir Dr A. Waswo; publs *Nissan Institute/Routledge Japanese Studies* series, *Nissan Occasional Papers* series.

Oriental Ceramic Society: 30b Torrington Sq., London, WC1E 7JL; tel. (20) 7636-7985; fax (20) 7580-6749; e-mail ocs-london@beeb.net; f. 1921 to increase knowledge and appreciation of Eastern ceramic and other arts; Pres. Carol Michaelson; Sec. Jean Martin.

Overseas Development Institute: Costain House, 111 Westminster Bridge Rd, London, SE1 7JD; tel. (20) 7922-0300; fax (20) 7922-0399; e-mail odi@odi.org.uk; internet www.odi.org.uk; f. 1960 as a research centre and forum for the discussion of development issues; publishes its research findings in books and working papers; library of over 15,000 vols; Chair. Baroness Jay; Dir Simon Maxwell; publs include *Development Policy Review* (quarterly), *Disasters: The Journal of Disaster Studies and Management*.

Percival David Foundation of Chinese Art: 53 Gordon Sq., London, WC1H 0PD; tel. (20) 7387-3909; fax (20) 7383-5163; e-mail sp17@soas.ac.uk; internet www.pdfmuseum.org.uk; f. 1952; major collection of Chinese ceramics dating 10th-18th century, some of which were fmrly in the possession of the Chinese Imperial family; colloquies on art and archaeology in Asia; subscription reference library of over 5,000 vols; publs monographs, catalogues, *Guide* to the Collection, colloquy reports, postcards and slides.

Royal Asiatic Society of Great Britain and Ireland: 60 Queen's Gardens, London, W2 3AF; tel. (20) 7724-4741; e-mail info@royalasiaticsociety.org; internet www.royalasiaticsociety.org; f. 1823 for the study of the history, sociology, institutions, customs, languages and art of Asia; c. 700 mems; c. 700 subscribing libraries; brs in various Eastern cities; library of 50,000 vols and 1,500 MSS; Curator Alison Ohta; Librarian Kathy Lazenbatt; publs *Journal* and monographs.

Royal Commonwealth Society: 18 Northumberland Ave, London, WC2N 5BJ; tel. (20) 7930-6733; fax (20) 7930-9705; e-mail info@rcsint.org; internet www.rcsint.org; f. 1868 to promote international understanding of the Commonwealth and its people; organizes meetings and seminars on topical issues, and cultural and social events; library housed by Cambridge University Library; Chair. Sir Michael McWilliam; Dir Stuart Mole; publs *Annual Report*, *Newsletter* (3 a year), conference reports.

Royal Institute of International Affairs: Chatham House, 10 St James's Sq., London, SW1Y 4LE; tel. (20) 7957-5700; fax (20) 7957-5710; e-mail contact@riia.org; internet www.riia.org; f. 1920 to facilitate the scientific study of international questions; c. 3,000 mems; Chair. Dr DeAnne Julius; Dir Prof. Victor Bulmer-Thomas; publs include *International Affairs* (five times a year), *The World Today* (monthly), *Chatham House Papers*, *Annual Report*, etc.

Royal Society for Asian Affairs: 2 Belgrave Sq., London, SW1X 8PJ; tel. (20) 7235-5122; e-mail info@rsaa.org.uk; internet www.rsaa.org.uk; f. 1901; 1,200 mems; library of c. 6,500 vols; Pres. Lord Denman; Chair. Sir Harold Walker; Sec. N. J. M. Cameron; publ. *Journal* (3 a year).

St Antony's College Asian Studies Centre: Oxford, OX2 6JF; tel. and fax (1865) 274559; e-mail asian@sant.ox.ac.uk; internet www.sant.ox.ac.uk/areastudies/asian.shtml; f. 1954; devoted to the comparative study of modern Asia; Dir Dr Mark Rebick.

School of African and Asian Studies, University of Sussex: Falmer, Brighton, Sussex, BN1 9SJ; tel. (1273) 678422;

fax (1273) 678640; e-mail international@sussex.ac.uk; internet www.sussex.ac.uk/units/publications/sabroad_2002/adba.shtml; Dean Dr MICHAEL JOHNSON.

School of Development Studies, University of East Anglia: Norwich, NR4 7TJ; tel. (1603) 592807; fax (1603) 451999; e-mail dev.general@uea.ac.uk; internet www.uea.ac.uk/dev; f. 1970; teaching, research and advisory work; Dean Prof. MICHAEL STOCKING.

School of East Asian Studies (SEAS), University of Sheffield: Floor 5, Arts Tower, Western Bank, Sheffield, S10 2TN; tel. (114) 222-8400; fax (114) 222-8432; e-mail SEAS@sheffield.ac.uk; internet www.seas.ac.uk; Japanese, Korean and Chinese studies; Chair. Prof. TIMOTHY WRIGHT.

School of Oriental and African Studies (SOAS): University of London, Thornhaugh St, Russell Sq., London, WC1H 0XG; tel. (20) 7637-2388; fax (20) 7436-3844; e-mail study@soas.ac.uk; internet www.soas.ac.uk; f. 1916; includes Centre of Chinese Studies, Centre of South Asian Studies, Centre of South East Asian Studies, Centre of Korean Studies, Japan Research Centre, Contemporary China Institute, Centre for the Study of Japanese Religion, Centre for the Study of the Literature of Asia and Africa, Centre of East Asian Law; Dir Prof. COLIN BUNDY; publs *The Bulletin*, *Calendar*, *China Quarterly*, *Annual Report*.

School of Oriental and African Studies Library: Thornhaugh St, Russell Sq., London, WC1H 0XG; tel. (20) 7898-4163; fax (20) 7898-4159; e-mail libenquiry@soas.ac.uk; internet www.soas.ac.uk/library/index.cfm; f. 1916; c. 1,000,000 vols and pamphlets; 4,700 current periodicals, 54,000 maps, 6,500 microforms, 2,800 MSS and private papers collections, extensive missionary archives; all covering Asian and African languages, literatures, philosophy, religions, history, law, cultural anthropology, art and archaeology, social sciences, geography and music; Head of Library Services ANNE POULSON.

UNITED STATES OF AMERICA

American Oriental Society: Near East Division, Hatcher Graduate Library, University of Michigan, Ann Arbor, MI 48109-1205; tel. (734) 647-4760; fax (734) 763-6743; e-mail jrodgers@umich.edu; internet www.umich/edu/~aos; f. 1842; 1,400 mems; library of 23,000 vols; Pres. (vacant); Sec.-Treas. JONATHAN RODGERS; publs *Journal of the American Oriental Society* (quarterly), *AOS Monograph Series* (irregular).

The Asia Foundation: POB 193223, San Francisco, CA 94119; tel. (415) 982-4640; fax (415) 392-8863; e-mail info@asiafound.org; internet www.asiafoundation.org; br. in Washington, DC, and 13 offices throughout Asia; private, non-profit NGO; f. 1951; library of 3,370 vols; collaborates with partners from public and private sectors to support leadership and institutional development, exchanges, dialogue, technical assistance, research and policy engagement related to governance and law, economic reform and development, women's political participation and regional relations; Pres. WILLIAM P. FULLER; publs *Annual Report*.

Asia-Pacific Center for Security Studies: 2058 Maluhia Rd, Honolulu, HI 96815; tel. (808) 971-8900; fax (808) 971-8999; e-mail pao@apcss.org; internet www.apcss.org; f. 1995; Pres. H. C. STACKPOLE.

Asia/Pacific Research Center: Stanford University, Encina Hall, Rm E301, Stanford CA 94305; tel. (650) 723-9741; fax (650) 723-6530; e-mail Asia-Pacific-Research-Center@stanford.edu; internet aparc.stanford.edu; Dir Prof. ANDREW G. WALDER.

Asia Society: 725 Park Ave, New York, NY 10021; tel. (212) 288-6400; fax (212) 517-8315; e-mail webmaster@asiasoc.org; internet www.asiasociety.org; f. 1956; carries out educational and cultural programmes to increase awareness of the arts, history and contemporary affairs of Asia; regional brs in Washington, DC, Houston, Los Angeles, Hong Kong and Melbourne; 8,000 mems; Pres. NICHOLAS PLATT; publs *Archives of Asian Art*, exhibition catalogues, contemporary affairs publications.

Asian Art Museum of San Francisco/Chong-Moon Lee Center for Asian Art and Culture: 200 Larkin St, San Francisco, CA 94102; tel. (415) 581-3500; fax (415) 581-4700; e-mail pr@asianart.org; internet www.asianart.org; f. 1966; museum and centre of research and publication on outstanding collections representing the countries and cultures of Asia; library of 35,000 vols; Dir Dr EMILY SANO.

Asian Cultural Council: 437 Madison Ave, New York, NY 10022; tel. (212) 812-4300; fax (212) 812-4299; e-mail acc@accny.org; internet www.asianculturalcouncil.org; f. 1980; supports cultural exchanges in the visual and performing arts between the USA and Asia; publicly supported; Dir RALPH SAMUELSON.

Association for Asian Studies (AAS): 1021 E. Huron St, Ann Arbor, MI 48104; tel. (734) 665-2490; fax (734) 665-3801; internet www.aasianst.org; f. 1941; Pres. JAMES L. WATSON; publs include *Asian Studies Newsletter*, *Journal of Asian Studies* (quarterly), *Education About Asia* (3 a year) and monographs.

The Brookings Institution: 1775 Massachusetts Ave, NW, Washington, DC 20036-2188; tel. (202) 797-6000; fax (202) 797-6004; e-mail brookinfo@brook.edu; internet www.brook.edu; f. 1916; research, education and publishing in economics, govt, and foreign policy; maintains Social Science Computation Center; education division, Center for Public Policy Education, organizes conferences and seminars; 50 resident scholars; library of c. 85,000 vols; Pres. STROBE TALBOTT; publs *Annual Report*, *The Brookings Review* (quarterly), *Brookings Papers on Economic Activity* (3 a year), *Brookings Papers on Education Policy* (annually), *Brookings/Wharton Papers on Financial Policy* (annually).

Center for Asia-Pacific Policy: RAND, 1700 Main St, Santa Monica, CA 90407-2138; tel. (310) 393-0411; fax (310) 451-6960; e-mail Nina.Hachigian@rand.org; internet www.rand.org/nsrd/capp/; Dir NINA HACHIGIAN.

Center for Chinese Studies, University of Michigan: Suite 3668 SSWB, 1080 South University, Ann Arbor, MI 48109-1106; tel. (734) 764-6308; fax (734) 764-5540; e-mail chinese.studies@umich.edu; internet www.umich.edu/~iinet/ccs; f. 1961; library of more than 610,000 vols, reels of microfilm and sheets of microfiches; Dir JAMES Z. LEE; publs include *Michigan Monographs in Chinese Studies*, *Twentieth Century China Journal*.

Center for East Asian Studies, Stanford University: Bldg 50, Main Quad, Stanford University, Stanford, CA 94305-2034; tel. (650) 723-3362; fax (650) 725-3350; e-mail connie.chin@forsythe.stanford.edu; internet www.stanford.edu/dept/CEAS; f. 1973; Dir Prof. JEAN OI.

Center for Japanese Studies, University of Michigan: 3603 International Institute, 1080 S. University, Ann Arbor, MI 48109; tel. (734) 764-6307; fax (734) 936-2948; e-mail umcjs@umich.edu; internet www.umich.edu/~iinet/cjs; f. 1947; library of 224,000 vols, reels of microfilm, microfiche 2,000 periodical titles (Asia Library Japanese collection); Dir JOHN LIE; publs *Michigan Classics in Japanese Studies*, *Michigan Papers in Japanese Studies*, *Michigan Monograph Series in Japanese Studies*, etc.

Center for Pacific Islands Studies: School of Hawaiian, Asian and Pacific Studies, University of Hawaii at Mānoa, Moore Hall 215, 1890 East-West Rd, Honolulu, HI 96822; tel. (808) 956-7700; fax (808) 956-7053; e-mail cpis@hawaii.edu; internet www.hawaii.edu/cpis; Dir DAVID HANLON.

Center for the Pacific Rim: University of San Francisco, 2130 Fulton St, CA 94117-1080; tel. (415) 422-6357; fax (415) 422-5933; e-mail pacrim@usfca.edu; internet www.pacificrim.usfca.edu; f. 1988; Exec. Dir Dr BARBARA K. BUNDY.

Centers for South and Southeast Asian Studies, University of Michigan: 1080 S University, Ste 3640, Ann Arbor, MI 48109-1106; tel. (734) 764-0352; fax (734) 936-0996; e-mail csseas@umich.edu; internet www.umich.edu/~iinet/csseas; Dir (South Asia) SUMATHI RAWASWAMY; Dir (Southeast Asia) JUDITH BECKER; Dir (Southeast Asia Business Program, Business School) LINDA LIM (tel. (734) 763-5796); publs *Michigan Papers on South and Southeast Asia*, *Michigan Studies of South and Southeast Asia*, *Michigan Studies in Buddhist Literature* (occasional papers), *Journal of Asian Business* (quarterly).

Center of International Studies: Woodrow Wilson School of Public and International Affairs, Princeton University, Bendheim Hall, Princeton, NJ 08544-1022; tel. (609) 258-4851; fax (609) 258-3988; e-mail wwswww@princeton.edu; internet www

.wws.princeton.edu/~cis; f. 1951; international relations and national development research, including analysis of international security and the political economy; Dir AARON L. FRIEDBERG; publs include *World Politics* (quarterly), monographs.

Center on Japanese Economy and Business: Columbia University, Business School, 321 Uris Hall, MC 5968, 3022 Broadway, New York, NY 10027-6902; tel. (212) 854-3976; fax (212) 678-6958; e-mail htp1@columbia.edu; internet www.columbia.edu/cu/business/japan; Dir HUGH PATRICK.

Columbia University East Asian Institute: Mail Code 3333, Columbia University, 420 West 118th St, New York, NY 10027; tel. (212) 854-2592; fax (212) 749-1497; e-mail eaiinfo@columbia.edu; internet www.sipa.columbia.edu/eai; f. 1949; Dir XIAOBO LU.

Cornell University East Asia Program (EAP): 140 Uris Hall, Ithaca, NY 14853-7601; tel. (607) 255-6222; fax (607) 255-1388; e-mail cueap@cornell.edu; internet www.einaudi.cornell.edu/eastasia; f. 1950 for the development of instruction and research on East Asia; library of 460,000 vols; 180 graduate students; Dir Prof. JOHN B. WHITMAN; publs *Cornell East Asia Series (CEAS)*.

Cornell University Southeast Asia Language and Area Center: 180 Uris Hall, Ithaca, NY 14853; tel. (607) 255-2378; fax (607) 254-5000; e-mail seap@cornell.edu; internet www.einaudi.cornell.edu/SoutheastAsia; f. 1950 for the development of instruction and research on South-East Asia; library of 212,060 vols, 19,950 periodicals and 841 newspapers; Dir Prof. THAK CHALOEMTIARANA.

Council on Foreign Relations, Inc: 58 East 68th St, New York, NY 10021; tel. (212) 434-9400; fax (212) 861-1789; e-mail communications@cfr.org; internet www.cfr.org; f. 1921; 3,600 mems; Pres. RICHARD N. HAASS; publs include *Foreign Affairs* (every 2 months).

Council on Regional Studies: 218 Palmer Hall, Princeton University, Princeton, NJ 08544; tel. (609) 258-4720; f. 1961; Chair. Prof. EZRA N. SULEIMAN.

East-West Center—Center for Cultural and Technical Interchange between East and West, Inc: 1601 East-West Rd, Honolulu, HI 96848; tel. (808) 944-7111; fax (808) 944-7970; e-mail ewcinfo@EastWestCenter.org; internet www.eastwestcenter.org; f. 1960 by Congress to promote better relations and understanding among the nations and peoples of Asia, the Pacific and the USA through co-operative study, training and research; conducts multidisciplinary programmes on environmental policy, population, resources, development, journalism and Pacific Islands development; provides awards and grants to scholars, journalists, graduate students, and managers to participate in the Center's studies; Pres. Dr CHARLES MORRISON.

Freer Gallery of Art: 12th St and Jefferson Drive, SW, Washington, DC 20560; tel. (202) 357-4880; fax (202) 663-9105; internet www.si.edu/asia; f. 1906, opened 1923; conducts research on the major collections of Asian and late 19th and early 20th century American art, gift of the late Charles L. Freer; art collection of 28,000 objects; library of 55,000 vols, 75,000 slides; Dir Dr MILO BEACH; publs include *Artibus Asiae*, *Ars Orientalis*.

Henry M. Jackson School of International Studies: University of Washington, Seattle, WA 98195; tel. (206) 543-4370; fax (206) 685–0668; e-mail jsis@u.washington.edu; internet www.jsis.artsci.washington.edu; Dir ANAND YANG.

The Jamestown Foundation: 4516 43rd Street NW, Washington, DC 20016; tel. (202) 483-8888; internet www.jamestown.org; f. 1984; Pres. GLEN E. HOWARD; publs include *Terrorism Monitor*.

John King Fairbank Center for East Asian Research (Harvard University): 1737 Cambridge St, Cambridge, MA 02138; tel. (617) 495-4046; fax (617) 495-9976; internet www.fas.harvard.edu/~fairbank; Dir Prof. ELIZABETH J. PENY; publs incl. *Papers on Chinese History*, *Harvard Studies on Taiwan*, etc.

Library of International Relations: 565 W Adams, Chicago, IL 60661; tel. (312) 906-5600; fax (312) 906-5685; f. 1932; financed by voluntary contributions; stimulates interest and research in international problems; conducts seminars and offers special services to businesses and academic institutions; library of 560,000 items.

The Mongolia Society: 322 Goodbody Hall, Indiana University, 1011 E 3rd St, Bloomington, IN 47405-7005; tel. (812) 855-4078; fax (812) 855-7500; e-mail monsoc@indiana.edu; internet www.indiana.edu/~mongsoc; f. 1961 to promote and further the study of Mongolia, its history, language and culture; Pres. Dr HENRY G. SCHWARZ; publs include *Mongolian Studies: Journal of the Mongolia Society*, *Mongolia Survey*, *Newsletter*, *Bulletin*, *Special Papers* and *Occasional Papers* series; dictionaries.

Museum of Fine Arts, Boston: 465 Huntington Ave, Boston, MA 02115; tel. (617) 267-9300; fax (617) 236-0362; f. 1870; has dept of Asiatic art with major collection of Chinese, Japanese, Indian and South-East Asian sculpture, painting, prints and decorative art works dating from Neolithic period to modern times; library of 65,000 books, periodicals and pamphlets; Dir Dr MALCOLM A. ROGERS; Curator WU TUNG.

National Bureau of Asian Research (NBR): 4518 University Way NE, Suite 300, Seattle, Washington 98105; tel. (206) 632-7370; fax (206) 632-7487; e-mail nbr@nbr.org; internet www.nbr.org; conducts research into policy-related issues in Asia; Pres. RICHARD J. EUINGS; publs include NBR book series, *NBR Analysis*, *AccessAsia Review*.

Princeton University, East Asian Studies Department: Princeton, NJ 08544; tel. (609) 452-5905; e-mail collcutt@princeton.edu; internet www.princeton.edu/~eastasia; Chair. MARTIN COLLCUTT.

Arthur M. Sackler Gallery: 1050 Independence Ave, SW, Washington, DC 20560; tel. (202) 633-0488; fax (202) 357-4911; internet www.asia.si.edu; f. 1982, opened 1987; exhbns, research and education in Asian art; art collection of 2,500 objects, including inaugural gift from the late Arthur M. Sackler; library of 85,000 vols, 90,000 slides; Dir Dr JULIAN RABY.

St John's University Institute of Asian Studies: 8000 Utopia Parkway, Jamaica, NY 11439; tel. (718) 990-6582; fax (718) 990-1881; e-mail linj@stjohns.edu; Dir Dr JOHN LIN.

School of Hawaiian, Asian and Pacific Studies: University of Hawaii, Moore Hall 309, 1890 East-West Rd, Honolulu, HI 96822; tel. (808) 956-8324; fax (808) 956-6345; e-mail edgara@hawaii.edu; internet www.hawaii.edu/shaps; administers Centers for Chinese, Hawaiian, Japanese, Korean, Pacific Islands, Philippine, South Asian, South-East Asian and Buddhist Studies; Interim Dean EDGAR A. PORTER.

School of International and Public Affairs: Columbia University, 420 West 118th St, New York, NY 10027; tel. (212) 854-4604; Dean LISA ANDERSON.

Seton Hall University Department of Asian Studies: South Orange, NJ 07079; tel. (973) 761-9464; fax (973) 740-1843; library of 40,000 vols; Chair. WINSTON YANG.

The Sigur Center for Asian Studies: George Washington University, 1957 E St, NW, Suite 503, Washington, DC 20052; tel. (202) 994-5886; fax (202) 994-6096; e-mail gsigur@gwu.edu; internet www.gwu.edu/~sigur.

University of Arizona Department of East Asian Studies: Tucson, AZ 85721; tel. (520) 621-7505; e-mail vancet@u.arizona.edu; internet www.arizona.edu/~EAS; Head Dr TIMOTHY VANCE.

University of California at Berkeley, Department of East Asian Languages and Cultures: 104 Durant Hall, Berkeley, CA 94720-2230; tel. (510) 642-3480; fax (510) 642-6031; e-mail ealang@berkeley.edu; internet www.ealc.berkeley.edu; Head of Dept Prof. MACK HORTON.

University of California Department of South and Southeast Asia Studies: 7233 Dwinelle Hall, Berkeley, CA 94720-2540; tel. (510) 642-4564; fax (510) 643-2959; e-mail casmaoff@socrates.berkeley.edu; internet ls.berkeley.edu/dept/sseas.

University of Illinois Center for East Asian and Pacific Studies: 910 S. Fifth St, Champaign, IL 61801; tel. (217) 333-7273; fax (217) 244-5729; f. 1965; Dir GEORGE T. YU.

University of Kansas Center for East Asian Studies: Bailey Hall, 1440 Jayhawk Blvd, Rm 202, Lawrence, KS 66045-7574;

tel. (785) 864-3849; fax (785) 864-5034; e-mail ceas@ku.edu; internet www.ceas.ku.edu; Dir Dr ELAINE GERBERT; publs research and reference series.

University of Pittsburgh Department of East Asian Languages and Literatures: 702 Old Engineering Hall, Pittsburgh, PA 15260; tel. (412) 624-5568; fax (412) 624-3458; e-mail hnara@pitt.edu; internet www.pitt.edu/~deall; Chair. HIROSHI NARA.

University of Southern California East Asian Studies Center: VKC 263, Los Angeles, CA 90089-0046; tel. (213) 740-2991; fax (213) 740-8409; e-mail easc@usc.edu; internet www.usc.edu/dept/LAS/EASC; f. 1962; Dir OTTO SCHNEPP; publs occasional papers.

University of Southern California-University of California (Los Angeles) Joint East Asian Language and Area Studies National Resource Center: VKC 263, Los Angeles, CA 90089-0046; tel. (213) 740-2991; fax (213) 740-8409; e-mail easc@usc.edu; internet www.usc.edu/dept/LAS/EASC; f. 1976; Dir Dr GORDON BERGER; publ. *Newsletter*.

Yale University Southeast Asia Studies: Yale University, POB 208206, New Haven, CT 06520-8206; tel. (203) 432-3431; fax (203) 432-3432; e-mail seas@yale.edu; internet www.yale.edu/seas/contactSEAS.htm; library of 260,000 vols; Chair. J. JOSEPH ERRINGTON; Librarian RICHARD RICHIE; publs *The Vietnam Forum*, *Lac Viet* series, *Southeast Asia Monograph* series.

VIET NAM

Hoi Phat Hoc Nam Viet (Association for Buddhist Studies): Xa-Loi Pagoda, 89 Ba Huyen Thanh Quan, Ho Chi Minh City; f. 1950 to further the study and practice of Buddhism; 30,000 mems; library of 5,000 vols; Pres. Dr CAO VAN TRI; publs *Tu Quang*, *Le Ngoc Diep*.

Institute of Economics: Commission for Social Sciences, 27 Tran Xuan Soan St, Hanoi; tel. (4) 254774; fax (4) 259071; f. 1960; library of 8,000 vols; publs *Economic Studies Review* (2 a year), *Nghien Cuu Kinh Te* (2 a year).

SELECT BIBLIOGRAPHY—BOOKS

See also bibliographies at end of relevant chapters in Part Two.

Abbott, Jason P. *Developmentalism and Dependency in Southeast Asia: The Case of the Automotive Industry*. London, RoutledgeCurzon, 2002.

Abegglen, James C. *Sea Change: Pacific Asia as the New World Industrial Center*. New York, NY, The Free Press, 1994.

Abuza, Zachary. *Militant Islam in Southeast Asia: Crucible of Terror*. Boulder, CO, Lynne Rienner Publishers, 2003.

Acharya, Amitav. *Constructing a Security Community in Southeast Asia—ASEAN and the Problem of Regional Order*. London, Routledge, 2000.

The Quest for Identity: International Relations of South-East Asia. Oxford, Oxford University Press, 2001.

Agrawal, Pradeep, et al. *Economic Restructuring in East Asia and India: Perspectives on Policy Reform*. Singapore, Institute of Southeast Asian Studies, 1995.

Agrawal, Pradeep, Gokran, Subir V., Mishira, Veena, Parikh, Kirit S., and Sen, Kunal. *Policy Regimes and Industrial Competitiveness—A Comparative Study of East Asia and India*. Singapore, Institute of Southeast Asian Studies, 2000.

Ahuja, Vinod, Bidani, Benud, Ferrera, Francisco, and Walton, Michael. *Everyone's Miracle: Revisiting Poverty and Inequality in East Asia*. Washington, DC, World Bank, 1997.

Alagappa, Muthiah (Ed.). *Political Legitimacy in Southeast Asia—The Quest for Moral Authority*. Cambridge, Cambridge University Press, 1998.

Coercion and Governance: The Declining Political Role of the Military in Asia. Stanford, CA, Stanford University Press, 2001.

Alatas, Ali. *A Voice for Peace*. Singapore, Institute of Southeast Asian Studies, 2001.

Aldrich, Richard J., Rawnsley, Gary D., and Rawnsley, Ming-Yeh T. (Eds). *The Clandestine Cold War in Asia, 1945–65*. Ilford, Frank Cass, 2000.

Anderson, Benedict. *The Spectre of Comparisons: Nationalism, Southeast Asia and the World*. London, Verso, 1999.

Andersson, Martin, and Gunnarsson, Christer. *Development and Structural Change in Asia-Pacific: Globalising Miracles or the End of a Model?* London, RoutledgeCurzon, 2003.

Andrews, John. *The Asian Challenge: Looking Beyond 2000*. Hong Kong, Longman, 1992.

Andrews, Tim, Baldwin, Bryan J., and Chompusri, Nartnalin. *The Changing Face of Multinationals in South East Asia*. London, Routledge, 2002.

Ang Swee Hoon, et al. *Surviving the New Millennium: Lessons from the Asian Crisis*. Singapore, McGraw Hill, 1999.

Antlöv, Hans, and Ngo Tak-Wing (Eds). *The Cultural Construction of Politics in Asia*. Richmond, Curzon Press, 1999.

Antons, Christoph (Ed.). *Law and Development in East and South-East Asia*. Richmond, Curzon Press, 2001.

Aoki, Masahiko, Kim Hyung-Ki and Okuno-Fujiwara, Masahiro. *The Role of the Government in East Asian Economic Development: Comparative Institutional Analysis*. Oxford, Oxford University Press, 1997.

Ariff, Mohamed (Ed.). *APEC & Development Co-operation*. Singapore, Institute of Southeast Asian Studies, 1998.

Ariff, Mohamed, and Khalid, Ahmed M. *Liberalization, Growth and Transitional Economics in Asia*. Cheltenham, Edward Elgar Publishing, 1999.

Ariff, Mohamed, and Tan Loong-Hoe (Eds). *The Uruguay Round: ASEAN Trade Policy Options*. Singapore, Institute of Southeast Asian Studies, 1988.

Arndt, Heinz W., and Hill, Hal (Eds). *Southeast Asia's Economic Crisis—Origins, Lessons and the Way Forward*. Singapore, Institute of Southeast Asian Studies, 1999.

Arrighi, Giovanni, Takeshi, Hamashita, and Selden, Mark (Eds). *The Resurgence of East Asia*. London, RoutledgeCurzon, 2003.

ASEAN Secretariat. *ASEAN Economic Co-operation: Transition and Transformation*. Singapore, Institute of Southeast Asian Studies, 1997.

Ashton, David, Green, Francis, James, Donna, and Sung, Johnny. *Education and Training for Development in East Asia—The Political Economy of Skill Formation in Newly Industrialised Economies*. London, Routledge, 1999.

Asian Development Bank. *Asian Development Outlook*. Manila, ADB, annually.

A Continent in Change: Thirty Years of the Asian Development Bank. Manila, ADB, 1997.

East ASEAN Growth Area: Brunei Darussalam, Indonesia, Malaysia, Philippines (Vols I to VII). Manila, ADB, 1997.

Emerging Asia—Changes and Challenges. Manila, ADB, 1997.

The Future of Asia in the World Economy. Manila, ADB, 1998.

The Global Trading System and Developing Asia. Manila, ADB, 1997.

Growth Triangles in Asia—A New Approach to Regional Economic Cooperation. Manila, ADB, 1998.

Key Indicators of Developing Asian and Pacific Countries. Manila, ADB, annually.

Small Countries, Big Lessons: Governance and the Rise of East Asia. Manila, ADB, 1996.

Athukorala, Prema-Chandra. *Trade Policy Issues in Asian Development*. London, Routledge, 1998.

Athukorala, Prema-Chandra, Manning, Chris, and Wickramasekara, Piyasiri. *Growth, Employment and Migration in Southeast Asia—Structural Change in the Greater Mekong Countries*. Cheltenham, Edward Elgar Publishing, 2000.

Atkins, William. *The Politics of Southeast Asia's New Media*. Richmond, Curzon Press, 2001.

Backman, Michael. *Asian Eclipse: Exposing the Dark Side of Business in Asia (Revised Edn)*. Singapore, John Wiley & Sons, 2001.

Bahamonde, Ramón. *International Policy Institutions Around the Pacific Rim: A Directory of Resources in East Asia, Australasia, and the Americas*. Boulder, CO, Lynne Rienner, 1998.

Bale, Chris. *Faces in the Crowd: A Journey in Hope*. Hong Kong, Chinese University Press, 1999.

Bales, Kevin. *Disposable People: New Slavery in the Global Economy*. Berkeley, CA, University of California Press, 2000.

Barlow, Colin (Ed.). *Institutions and Economic Change in Southeast Asia—The Context of Development from the 1960s to the 1990s*. Cheltenham, Edward Elgar Publishing, 1999.

Barr, Michael D. *Cultural Politics and Asian Values—The Tepid War*. London, RoutledgeCurzon, 2002.

Bauer, Joanne R., and Bell, Daniel A. (Eds). *The East Asian Challenge for Human Rights*. Cambridge, Cambridge University Press, 1999.

Beeson, Mark. *Reconfiguring East Asia—Regional Institutions and Organizations After the Crisis*. London, RoutledgeCurzon, 2002.

Berger, Mark, and Borer, Douglas (Eds). *The Rise of East Asia—Critical Visions of the Pacific Century*. London, Routledge, 1997.

Berry, James, and McGreal, Stanley (Eds). *Cities in the Pacific Rim*. London, Routledge, 1999.

Bert, Wayne. *The United States, China and Southeast Asian Security: A Changing of the Guard?* New York, NY, Palgrave Macmillan, 2003.

Beyrer, Chris. *War in the Blood: Sex, Politics and AIDS in Southeast Asia*. London, Zed Books, 1998.

Biers, Dan (Ed.). *Crash of '97—How the Financial Crisis is Reshaping Asia*. Hong Kong, Review Publishing, 1998.

Bishop, Ryan, Phillips, John, and Yeo Wei Wei. *Postcolonial Urbanism: The Southeast Asia Supplement*. New York, NY, Routledge, 2003.

Boomgaard, Peter, and Brown, Ian (Eds). *Weathering the Storm: The Economies of Southeast Asia in the 1930s Depression*. Singapore, Institute of Southeast Asian Studies, 2000.

Booth, Ann, and Ash, Robert (Eds). *The Economies of Asia 1945–1998*. London, Routledge, 1999.

Booth, Martin. *The Dragon Syndicates: The Global Phenomenon of the Triads*. New York, NY, Doubleday, 1999.

Borrus, Michael, Ernst, Dieter, and Haggard, Stephan (Eds). *International Production Networks in Asia—Rivalry or Riches*. London, Routledge, 2000.

Bowie, Anthony, and Unger, Daniel. *The Politics of Open Economies—Indonesia, Malaysia, the Philippines, and Thailand*. Cambridge, Cambridge University Press, 1997.

Bracken, Paul. *Fire in the East: The Rise of Asian Military Power and the Second Nuclear Age*. London, HarperCollins, 1999.

Bresnan, John. *From Dominoes to Dynamoes: The Transformation of Southeast Asia*. New York, NY, Council on Foreign Relations, 1995.

Bridges, Brian. *Europe and the Challenge of the Asia Pacific—Change, Continuity and Crisis*. Cheltenham, Edward Elgar Publishing, 1999.

Brown, David. *The State and Ethnic Politics in Southeast Asia*. London, Routledge, 1996.

Brown, Michael E., and Ganguli, Sumit (Eds). *Government Policies and Ethnic Relations in Asia and the Pacific*. Cambridge, MA, Massachusetts Institute of Technology, 1998.

Burnett, Alan. *The Western Pacific: The Challenge of Sustainable Growth*. Sydney, Allen and Unwin, 1994.

Calder, Kent E. *Asia's Deadly Triangle: How Arms, Energy and Growth Threaten to Destabilize Asia-Pacific*. London, Nicholas Brealey Publishing, 1996.

Camilleri, Joseph A. *States, Markets and Civil Society in Asia Pacific*. The Political Economy of the Asia-Pacific Region, Vol. I, Cheltenham, Edward Elgar Publishing, 2000.

Regionalism in the New Asia-Pacific Order. The Political Economy of the Asia-Pacific Region, Vol. II, Cheltenham, Edward Elgar Publishing, 2003.

Campos, Ed, and Root, Hilton L. *The Key to the Asian Miracle: Making Shared Growth Credible*. Washington, DC, Brookings Institution, 1996.

Capie, David, and Evans, Paul. *The Asia-Pacific Security Lexicon*. Singapore, Institute of Southeast Asian Studies, 2002.

Carpenter, William M., and Wiencek, David G. (Eds). *Asian Security Handbook—An Assessment of Political-Security Issues in the Asia-Pacific Region*. Armonk, NY, M. E. Sharpe, 1996.

Carr, Brian, and Mahalingam, Indira (Eds). *Companion Encyclopedia of Asian Philosophy*. London, Routledge, 1996.

Case, William. *Politics in Southeast Asia—Democracy or Less*. London, RoutledgeCurzon, 2002.

Comparing Politics in Southeast Asia. London, Sage Publications, 2003.

Cauquelin, Josiane, Lim, Paul, and Mayer-Konig, Birgit (Eds). *Asian Values—Encounter with Diversity*. Richmond, Curzon Press, 1998.

Chan Heng Chee (Ed.). *The New Asia-Pacific Order*. Singapore, Institute of Southeast Asian Studies, 1997.

Chang Ching-Cheng, Mendelsohn, Robert, and Shaw, Daigee. *Global Warming and the Asian Pacific*. Cheltenham, Edward Elgar Publishing, 2003.

Chang Ho-Joon. *The Political Economy of Industrial Policy*. Basingstoke, Macmillan Press, 1996.

Chee Soon Juan. *To Be Free: Stories from Asia's Struggle Against Oppression*. Monash Asia Institute, Vic, 1999.

Chen, Min. *Asian Management Systems: Chinese, Japanese and Korean Styles of Business*. London and New York, NY, Routledge, 1995.

Chern, Wen S., Carter, Colin A., and Shei Shun-yi (Eds). *Food Security in Asia: Economics and Policies*. Cheltenham, Edward Elgar Publishing, 2000.

Cheung, Anthony, and Scott, Ian. *Governance and Public Sector Reform in Asia—Paradigm Shift or Business as Usual?* London, RoutledgeCurzon, 2002.

Chia Lin Sien (Ed.). *Southeast Asia Transformed: A Geography of Change*. Singapore, Institute of Southeast Asian Studies, 2003.

Chia Lin Sien, Goh, Mark, and Tongzon, Jose. *Southeast Asian Regional Port Development: A Comparative Analysis*. Singapore, Institute of Southeast Asian Studies, 2003.

Chia Siow Yue (Ed.). *APEC: Challenges and Opportunities*. Singapore, Institute of Southeast Asian Studies, 1995.

Chia Siow Yue and Lim, Jamus Jerome (Eds). *Information Technology in Asia: New Development in Paradigms*. Singapore, Institute of Southeast Asian Studies, 2002.

Chia Siow Yue and Pacini, Marcello (Eds). *ASEAN in the New Asia*. Singapore, Institute of Southeast Asian Studies, 1998.

Chia Siow Yue and Tan, Joseph L. H. (Eds). *ASEAN in the WTO: Challenges and Responses*. Singapore, Institute of Southeast Asian Studies, 1996.

ASEAN in the New Millennium. Singapore, Institute of Southeast Asian Studies, 1999.

Chibusawal, Masahide. *Pacific Asia in the 1990s*. London, Routledge, 1991.

Chowdhury, Anis, and Islam, Iyanatul. *The Newly Industrializing Economies of East Asia*. London, Routledge, 1993.

Asia-Pacific Economies. London, Routledge, 1997.

Chowdhury, Anis, and Islam, Iyanatul (Eds). *Beyond the Asian Crisis—Pathways to Sustainable Growth*. Cheltenham, Edward Elgar Publishing, 2001.

Christie, Clive J. *A Modern History of Southeast Asia: Decolonization, Nationalism and Separatism*. Singapore, Institute of Southeast Asian Studies, 1996.

Christie, Kenneth, and Roy, Denny. *The Politics of Human Rights in East Asia*. London, Pluto Press, 2001.

Chu Yun-Peng and Hill, Hal (Eds). *The Social Impact of the Asian Financial Crisis*. Cheltenham, Edward Elgar Publishing, 2001.

Chu Yun-Peng and Wu Rong-I (Eds). *Business, Markets and Government in the Asia-Pacific—Competition Policy, Convergence and Pluralism*. London, Routledge, 1998.

Chua Beng-Huat. *Consumption in Asia—Lifestyle and Identities*. London, Routledge, 2000.

Clark, Cal, and Roy, K. C. *Comparing Development Patterns in Asia*. Boulder, CO, Lynne Rienner, 1997.

Clifford, Mark, and Engardio, Pete. *Meltdown—Asia's Boom, Bust and Beyond*. Paramus, NJ, Prentice Hall Press, 2000.

Cobbold, Richard. *The World Reshaped: Fifty Years after the War in Asia*. Basingstoke, Macmillan Press, 1996.

Coedès, G. *The Indianized States of Southeast Asia*. Honolulu, HI, University of Hawaii Press, 1970.

Collignon, Stefan, Pisani-Ferry, Jean, and Park Yung Chul (Eds). *Exchange Rate Policies in Emerging Asian Countries*. London, Routledge, 1999.

Collins, Alan. *The Security Dilemmas of Southeast Asia*. Singapore, Institute of Southeast Asian Studies, 2001.

Security and Southeast Asia: Domestic, Regional and Global Issues. Boulder, CO, Lynne Rienner Publishers, 2003.

Connors, Michael, Dosch, Jörn, and Davison, Remy. *The New Global Politics of the Asia Pacific*. London, RoutledgeCurzon, 2003.

Cooney, Sean, Lindsey, Tim, and Zhu Ying (Eds). *Law and Labour Market Regulation in South East Asia*. London, Routledge, 2002.

Corden, Max. *The Asian Crisis—Is There A Way Out?* Singapore, Institute of Southeast Asian Studies, 1999.

Cotterell, Arthur. *East Asia: From Chinese Predominance to the Rise of the Pacific Rim.* London, John Murray, 1993.

Council for Asia-Europe Cooperation (CAEC). *Asia-Europe Cooperation: Beyond the Financial Crisis.* Singapore, Institute of Southeast Asian Studies, 2000.

Crozier, Brian. *South-East Asia in Turmoil.* Baltimore, MD, Penguin Books, 1966.

Cumings, Bruce. *Parallax Visions: Making Sense of American-East Asian Relations at the End of the Century.* Durham, NC, Duke University Press, 1999.

Da Cunha, Derek (Ed.). *Southeast Asian Perspectives on Security.* Singapore, Institute of Southeast Asian Studies, 2000.

Dadush, Uri, Uzan, Mark, and Dasgupta, Dipak. *Private Capital Flows in the Age of Globalization: The Aftermath of the Asian Crisis.* Cheltenham, Edward Elgar Publishing, 2000.

Dahles, Heidi, and van den Muijzenberg, Otto (Eds). *Capital and Knowledge in Asia: Changing Power Relations.* London, RoutledgeCurzon, 2003.

Das, Dilip K. *The Asia-Pacific Economy.* Basingstoke, Macmillan Press, 1996.

Davidson, P. J. *ASEAN: The Evolving Legal Framework for Economic Co-Operation.* Singapore, Times Academic Press, 2003.

De Brouwer, Gordon, and Pupphavesa, Wisarn (Eds). *Asia Pacific Financial Deregulation.* London, Routledge, 1999.

De Brouwer, Gordon, and Yunjong Wang. *Financial Governance in East Asia.* London, RoutledgeCurzon, 2003.

De Jonge, Huub, and Kaptein, Nico. *Transcending Borders: Arabs, Politics, Trade and Islam in Southeast Asia.* Singapore, Institute of Southeast Asian Studies, 2004.

Delhaise, Philippe F. *Asia in Crisis: The Implosion of the Banking and Finance Systems.* Singapore, John Wiley & Sons, 1999.

Dent, Christopher M. *The European Union and East Asia—An Economic Relationship.* London, Routledge, 1999.

The Foreign Economic Policies of Singapore, South Korea and Taiwan. Cheltenham, Edward Elgar Publishing, 2002.

Dent, Christopher M., and Huang, David W. F. (Eds). *Northeast Asian Regionalism—Lessons from the European Experience.* London, RoutledgeCurzon, 2002.

Devan, Janadas (Ed.). *Southeast Asia: Challenges of the 21st Century.* Singapore, Institute of Southeast Asian Studies, 1994.

DeWitt, David B. *Globalization, Development and Security in Southeast Asia: The International Political Economy of New Regionalisms.* Burlington, VT, Ashgate Publishing Company, 2003.

Dickinson, David G., Ford, Jim L., Fry, Maxwell L., Mullineux, Andrew W., and Sen, Somnath. *Finance, Governance and Economic Performance in Pacific and South East Asia.* Cheltenham, Edward Elgar Publishing, 2000.

Dickinson, David, et al. (Eds). *Financial Sector Reform and Economic Development in Pacific and South East Asia.* Cheltenham, Edward Elgar Publishing, 2000.

Dixon, Chris. *Southeast Asia in the World Economy; A Regional Geography.* Cambridge, Cambridge University Press, 1993.

Dixon, Chris, and Drakakis-Smith, David. *Economic and Social Development in Pacific Asia.* London, Routledge, 1994.

Dobbs-Higginson, Michael S. *Asia Pacific: A View on Its Role in the New World Order.* Hong Kong, Longman, 1993.

Asia Pacific: Its Role in the New World Disorder. London, William Heinemann, 1994.

Dobson, Wendy, and Chia Siow Yue (Eds). *Multinationals and East Asian Integration.* Singapore, Institute of Southeast Asian Studies, 1997.

Drysdale, Peter (Ed.). *Reform and Recovery in East Asia.* London, Routledge, 2000.

The New Economy in East Asia and the Pacific. London, Routledge, 2003.

Drysdale, Peter, and Dong Dong Zhang (Eds). *Japan and China: Rivalry or Cooperation in East Asia?* Canberra, Asia Pacific Press, 2000.

Drysdale, Peter, Zhang Yunling and Ligang Song (Eds). *APEC and Liberalisation of the Chinese Economy.* Canberra, Asia Pacific Press, 2000.

Dunung, Sanjyot P. *Doing Business in Asia: The Complete Guide.* Singapore, Simon and Schuster, 1995.

Dupont, Alan. *East Asia Imperilled: Transnational Challenges to Security.* Cambridge, Cambridge University Press, 2001.

Dutta, M. *Economic Regionalization in the Asia-Pacific—Challenges to Economic Cooperation.* Cheltenham, Edward Elgar Publishing, 1999.

Dwyer, Denis. *South-East Asian Development: Geographical Perspectives.* Harlow, Longman Scientific and Technical, 1990.

East Asia Analytical Unit. *Overseas Chinese Business Networks in Asia.* Parkes, ACT, Australian Department of Foreign Affairs and Trade, 1995.

Eccleston, Bernard, Dawson, Michael, and McNamara, Deborah (Eds). *The Asia-Pacific Profile.* London, Routledge, 1998.

Edmonds, Christopher M. (Ed.). *Reducing Poverty in Asia: Emerging Issues in Growth, Targeting and Measurement.* Cheltenham, Edward Elgar Publishing, 2003.

Eichengreen, Barry. *Toward a New International Financial Architecture: A Practical Post-Asia Agenda.* Washington, DC, Institute for International Economics, 1999.

Ellings, Richard J., and Simon, Sheldon W. (Eds). *Southeast Asian Security in the New Millennium.* Armonk, NY, M. E. Sharpe, 1996.

Elmhurst, Becky, and Saptari, Rainer (Eds). *Labour in Southeast Asia—Local Processes in a Globalizing World.* Richmond, Curzon Press, 2001.

Emmers, Ralf. *Cooperative Security and the Balance of Power in ASEAN and the ARF.* London, RoutledgeCurzon, 2003.

Engelbert, Thomas, and Kubitscheck, Hans Dieter (Eds). *Ethnic Minorities and Politics in Southeast Asia.* New York, NY, Peter Lang, 2004.

Engholm, Christopher. *Doing Business in Asia's Booming 'China Triangle'.* New Jersey, NJ, Prentice Hall, 1995.

Ernst, Dieter, Ganiatsos, Tom, and Mytelka, Lynn (Eds). *Technological Capabilities and Export Success in Asia.* London, Routledge, 1998.

Evans, Grant, Hutton, Christopher, and Kuah Khun Eng (Eds). *Where China Meets Southeast Asia: Social and Cultural Change in the Border Regions.* Singapore, Institute of Southeast Asian Studies, 2000.

Evers, Hans-Dieter, and Korff, Rüdiger. *Southeast Asian Urbanism: The Meaning and Power of Social Space.* Singapore, Institute of Southeast Asian Studies, 2001.

Fahn, James David. *A Land on Fire: The Environmental Consequences of the South-East Asian Boom.* Boulder, CO, Westview Press, 2003.

Fallows, James. *Looking at the Sun: The Rise of the New East Asian Economic and Political System.* New York, NY, Vintage Books, 1994.

Fane, George. *Capital Mobility, Exchange Rates and Economic Crises.* Cheltenham, Edward Elgar Publishing, 2000.

Farrington, John, and Lewis, David (Eds). *Nongovernmental Organizations and the State in Asia: Rethinking Roles in Sustainable Agricultural Development.* London and New York, NY, Routledge, 1993.

Feinberg, Richard E. (Ed.). *APEC as an Institution: Multilateral Governance in the Asia-Pacific.* Singapore, Institute of Southeast Asian Studies, 2003.

Feinberg, Richard E., and Zhao Ye (Eds). *Assessing APEC's Progress: Trade, Ecotech, and Institutions.* Singapore, Institute of Southeast Asian Studies, 2001.

Findlay, Christopher, Chia Lin Sien and Singh, Karmjit (Eds). *Asia Pacific Air Transport: Challenges and Policy Reforms.* Singapore, Institute of Southeast Asian Studies, 1997.

Fisher, Charles A. *South-East Asia: A Social, Economic and Political Geography*. London, Methuen, 1964.

Flynn, Dennis O., Frost, Lionel, and Latham, A. J. H. (Eds). *Pacific Centuries—Pacific and Pacific Rim Economic History since the 16th Century*. London, Routledge, 1998.

Francks, Penelope, Boestel, Joanna, and Kim Choo Hyop. *Agriculture and Economic Development in East Asia—From Growth to Protectionism in Japan, Korea and Taiwan*. London, Routledge, 1999.

Freeman, Nick (Ed.). *Financing Southeast Asia's Economic Development*. Singapore, Institute of Southeast Asian Studies, 2003.

Freeman, Nick, and Bartels, Frank (Eds). *The Future of Foreign Investment in Southeast Asia*. London, RoutledgeCurzon, 2003.

Friedman, Edward. *An Asian Way? The Politics of Democratization: Generalizing East Asian Experiences*. Boulder, CO, Westview Press, 1995.

Fry, M. J. *Foreign Direct Investment in Southeast Asia: Differential Impacts*. Singapore, Institute of Southeast Asian Studies, 1993.

Fu Tsu-Tan, Huang, Cliff J., and Lovell, C. A. Knox (Eds). *Productivity and Economic Performance in the Asia-Pacific Region*. Cheltenham, Edward Elgar Publishing, 2002.

Fukuchi, Takao, and Kagami, Mitsuhiro (Eds). *Perspectives on the Pacific Basin Economy: A Comparison of Asia and Latin America*. Tokyo, Institute of Developing Economies and the Asian Club Foundation, 1990.

Funston, John (Ed.). *Government & Politics in Southeast Asia*. Singapore, Institute of Southeast Asian Studies, 2002.

Gamble, Andrew, and Payne, Anthony. *Regionalism and World Order*. Basingstoke, Macmillan Press, 1996.

Ganesan, N. *Bilateral Tensions in Post-Cold War ASEAN*. Singapore, Institute of Southeast Asian Studies, 1999.

Ganguly, Rajat, and Macduff, Ian. *Ethnic Conflict and Secessionism in South and Southeast Asia: Causes, Dynamics, Solutions*. London, Sage Publications, 2003.

Garnaut, Ross. *Asian Market Economies: Challenges of a Changing International Environment*. Singapore, Institute of Southeast Asian Studies, 1996.

Open Regionalism and Trade Liberalization: An Asia-Pacific Contribution to the World Trade System. Singapore, Institute of Southeast Asian Studies, 1996.

Garnaut, Ross (Ed.). *Sustaining Export-Oriented Development—Ideas from East Asia*. Cambridge, Cambridge University Press, 1995.

Garran, Robert. *Tigers Tamed: The End of the Asian Miracle*. Honolulu, HI, University of Hawaii Press, 1999.

Gaul, Karen K., and Hiltz, Jackie (Eds). *Landscapes and Communities on the Pacific Rim*. Armonk, NY, M. E. Sharpe, 2000.

Gibney, Frank. *The Pacific Century: America and Asia in a Changing World*. New York, NY, Charles Scribner's Sons, 1993.

Gilson, Julie. *Asia Meets Europe*. Cheltenham, Edward Elgar Publishing, 2002.

Glover, Ian, and Bellwood, Peter. *South East Asia: An Archaeological History*. London, RoutledgeCurzon, 2003.

Godemont, François. *The New Asian Renaissance—From Colonialism to the Post-Cold War*. London, Routledge, 1997.

The Downsizing of Asia. London, Routledge, 1998.

Goldman, Merle, and Gordon, Andrew (Eds). *Historical Perspectives on Contemporary East Asia*. Cambridge, MA, Harvard University Press, 2000.

Goldstein, Morris. *The Asian Financial Crisis: Causes, Cures and Systemic Implications*. Washington, DC, International Monetary Fund, 1998.

Gomez, Edmund. *Political Business in East Asia*. London, Routledge, 2001.

Gonzalez III, Joaquin L., Lauder, Kathleen, and Melles, Brenda. *Opting for Partnership: Governance Innovations in Southeast Asia*. Singapore, Institute of Southeast Asian Studies, 1999.

Goodman, David S. G. (Ed.). *Towards Recovery in Pacific Asia*. London, Routledge, 1999.

Goodman, Roger, White, Gordon, and Kwon Huck-Ju. *The East Asian Welfare Model—Welfare, Orientalism and the State*. London, Routledge, 1998.

Gosling, David. *Religion and Ecology in India and Southeast Asia*. London, Routledge, 2001.

Gough, Leo. *Asia Meltdown: The End of the Miracle*. Oxford, Capstone Publishing, 1998.

Gourevitch, Peter A. (Ed.). *The Pacific Region: Challenges to Policy and Theory*. Newbury Park, CA, Sage Publications, 1989.

Gray, Sidney J., Purcell, William R., and McGaughey, Sara L. (Eds). *Asia-Pacific Issues in International Business*. Cheltenham, Edward Elgar Publishing, 2001.

Greenough, Paul R., and Lowenhaupt Tsing, Anna (Eds). *Nature in the Global South: Environmental Projects in South and Southeast Asia*. Durham, NC, Duke University Press, 2003.

Haacke, Jürgen. *ASEAN's Diplomatic and Security Culture—Origins, Developments and Prospects*. London, RoutledgeCurzon, 2002.

Haggard, Stephan. *The Political Economy of the Asian Financial Crisis*. Washington, DC, Institute for International Economics, 2000.

Halib, Mohamed, and Huxley, Tim (Eds). *An Introduction to Southeast Asian Studies*. Singapore, Institute of Southeast Asian Studies, 1997.

Hall, D. G. E. *A History of South-East Asia*. Basingstoke, Macmillan Press, 1981.

Hamada, Koichi, Matsushita, Mitsuo, and Komura, Chikara (Eds). *Dreams and Dilemmas: Economic Friction and Dispute Resolution in the Asia-Pacific*. Singapore, Institute of Southeast Asian Studies, 2000.

Han Sung-Joo (Ed.). *Changing Values in Asia: Their Impact on Governance and Development*. Singapore, Institute of Southeast Asian Studies, 1999.

Harris, Paul G. *Global Warming and East Asia: The Domestic and International Politics of Climate Change*. London, RoutledgeCurzon, 2003.

Harris, Stuart, and Cotton, James (Eds). *The End of the Cold War in Northeast Asia*. Boulder, CO, Lynne Rienner, 1991.

Harrison, Brian. *South-East Asia: A Short History*. London, Macmillan Press, and New York, NY, St Martin's Press, 1955.

Harvie, Charles, and Lee Boon-Chye. *Globalisation and SMEs in East Asia*. Cheltenham, Edward Elgar Publishing, 2002.

The Role of SMEs in National Economies in East Asia. Cheltenham, Edward Elgar Publishing, 2002.

Hay, M. Cameron. *Remembering to Live (Southeast Asia: Politics, Meaning and Memory)*. Ann Arbor, MI, University of Michigan Press, 2003.

Hefner, Robert W., and Horvatich, Patricia (Eds). *Islam in an Era of Nation-States—Politics and Religious Renewal in Muslim South-East Asia*. Honolulu, HI, University of Hawaii Press, 1997.

Hellmann, Donald C., and Pyle, Kenneth B. (Eds). *From APEC to Xanadu—Creating a Viable Community in the Post-Cold War Pacific*. Armonk, NY, M. E. Sharpe, 1997.

Henders, Susan J. (Ed.). *Democratization and Identity: Regimes and Ethnicity in East and Southeast Asia (Global Encounters)*. Lanham, MD, Lexington Books, 2003.

Henderson, Callum. *Asia Falling? Making Sense of the Asian Currency Crisis and Its Aftermath*. Singapore, McGraw-Hill Book Co, 1998.

Heng, Russell Hiang Khng (Ed.). *Media Fortunes, Changing Times: ASEAN States in Transition*. Singapore, Institute of Southeast Asian Studies, 2002.

Heng, Russell Hiang Khng, and Hew, Dennis (Eds). *Regional Outlook: Southeast Asia 2004–2005*. Singapore, Institute of Southeast Asian Studies, 2004.

Hersh, Jacques. *The USA and the Rise of East Asia since 1945*. New York, NY, St Martin's Press, 1995.

Heryanto, Ariel, and Mandal, Sumit K. (Eds). *Challenging Authoritarianism in Southeast Asia: Comparing Indonesia and Malaysia*. London, RoutledgeCurzon, 2003.

Hian Teck Hoon. *Trade, Jobs and Wages*. Cheltenham, Edward Elgar Publishing, 2000.

Higgott, Richard, Leaver, Richard, and Ravenhill, John (Eds). *Pacific Economic Relations in the 1990s: Conflict or Cooperation?* Boulder, CO, Lynne Rienner, 1993.

Hill, Hal (Ed.). *The Economic Development of Southeast Asia*. Cheltenham, Edward Elgar Publishing, 2002.

Hilsdon, Anne-Marie, et al. (Eds). *Human Rights and Gender Politics—Asia-Pacific Perspectives*. London, Routledge, 2000.

Hirono, Ryokichi (Ed.). *Asian Development Experience Vol. 3: Regional Co-operation in Asia*. Singapore, Institute of Southeast Asian Studies, 2003.

Hirsch, Philip, and Warren, Carol (Eds). *The Politics of Environment in Southeast Asia*. London, Routledge, 1998.

Hiscock, Geoff. *Asia's Wealth Club*. London, Nicholas Brealey Publishing, 1997.

Ho, Alfred Kuo-liang. *The Far East in World Trade: Developments and Growth since 1945*. New York, NY, F. A. Praeger, 1967.

Holloway, Nigel. *Japan in Asia: The Economic Impact on the Region*. Hong Kong, Review Publishing, 1991.

Hooker, M. Barry (Ed.). *Law and the Chinese in Southeast Asia*. Singapore, Institute of Southeast Asian Studies, 2002.

Horowitz, Shale, and Heo, Uk (Eds). *The Political Economy of International Financial Crisis—Interest Groups, Ideologies and Institutions*. Singapore, Institute of Southeast Asian Studies, 2001.

Horton, Susan. *Women and Industrialization in Asia*. London, Routledge, 1995.

Hsiung, James C. (Ed.). *Asia Pacific in the New World Politics*. Boulder, CO, Lynne Rienner, 1993.

Hu Teh-Wei and Hsieh Chee-Ruey (Eds). *The Economics of Health Care in Asia-Pacific Countries*. Cheltenham, Edward Elgar Publishing, 2002.

Huang Yasheng. *FDI in China—An Asian Perspective*. Singapore, Institute of Southeast Asian Studies, 1998.

Hughes, Helen (Ed.). *Achieving Industrialization in East Asia*. Cambridge, Cambridge University Press, 1990.

Hughes, Helen. *East Asia: Is there an East Asian Model?* Canberra, Australian National University, Research School of Pacific Studies, 1993.

Hutanuwatr, Pracha, and Manivannan, Ramu (Eds). *The Asian Future. Dialogues for Change Vols 1 & 2*. London, Zed Books, 2004.

Hutchison, Jane, and Brown, Andrew (Eds). *Organising Labour in Globalising Asia*. London, Routledge, 2001.

Ichimura, Shinichi. *Political Economy of Japanese and Asian Development*. Singapore, Institute of Southeast Asian Studies, 1998.

Ikeo, Aiko. (Ed.). *Economic Development in Twentieth-Century East Asia—The International Context*. London, Routledge, 1996.

Institute of Southeast Asian Studies. *Regional Outlook—Southeast Asia 2001–2002*. Singapore.

Asia-Europe Cooperation: Beyond the Financial Crisis. Singapore, 2000.

International Library of Sociology. *The Sociology of East Asia*. London, Routledge, 1998.

International Monetary Fund. *The IMF's Response to the Asian Crisis*. Washington, DC, IMF, 1999.

Jackson, Karl D. (Ed.). *Asian Contagion—The Causes and Consequences of a Financial Crisis*. Singapore, Institute of Southeast Asian Studies, 1999.

James, David L. *The Executive Guide to Asia-Pacific Communications*. St Leonards, NSW, Allen and Unwin, 1995.

James, William E., Naya, Seiji, and Meier, Gerald M. *Asian Development: Economic Success and Policy Lessons*. San Francisco, CA, International Center for Economic Growth, 1987.

Japan Environmental Council, Teranishi, Shunichi, and Awagi, Takehisa. *The State of the Environment in Asia 1999/2000*. London, Routledge, 2000.

Jayasuriya, Kanishka (Ed.). *Law, Capitalism and Power in Asia—The Rule of Law and Legal Institutions*. London, Routledge, 1998.

Jeshurun, Chandran (Ed.). *China, India, Japan and the Security of Southeast Asia*. Singapore, Institute of Southeast Asian Studies, 1993.

Jomo, K. S. *Southeast Asia's Misunderstood Miracle: Industrial Policy and Economic Development in Thailand, Malaysia and Indonesia*. Boulder, CO, Westview Press, 1997.

Jomo, K. S. (Ed.). *Tigers in Trouble: Financial Governance, Liberalisation and Crises in East Asia*. London, Zed Books, 1999.

Manufacturing Competitiveness in Asia. London, RoutledgeCurzon, 2003.

Southeast Asian Paper Tigers? London, RoutledgeCurzon, 2003.

Jomo, K. S., and Folk, Brian C. *Ethnic Business: Chinese Capitalism in Southeast Asia*. London, Routledge, 2003.

Jones, Eric, Frost, Lionel, and White, Colin. *Coming Full Circle: An Economic History of the Pacific Rim*. Boulder, CO, Westview Press, 1994.

Kahn, Joel S. (Ed.). *Southeast Asian Identities: Culture and the Politics of Representation in Indonesia, Malaysia, Singapore and Thailand*. Singapore, Institute of Southeast Asian Studies, 1998.

Kapur, Ashok. *Regional Security Structures in Asia*. London, RoutledgeCurzon, 2002.

Kawagoe, Toshihiko, and Sekiguchi, Sueo (Eds). *East Asian Economies: Transformation and Challenges*. Singapore, Institute of Southeast Asian Studies, 1997.

Keenan, Faith (Ed.). *The Aftershock—How an Economic Earthquake is Rattling Southeast Asian Politics*. Hong Kong, Review Publishing, 1998.

Kelly, David, and Reid, Anthony (Eds). *Asian Freedoms—The Idea of Freedom in East and Southeast Asia*. Cambridge, Cambridge University Press, 1998.

Kemenade, Willem van. *China, Hong Kong, Taiwan, Inc.* New York, NY, Alfred A. Knopf, 1997.

Kershaw, Roger. *Monarchy in South East Asia—The Faces of Tradition in Transition*. London, Routledge, 2000.

Khan, Ilyas. *Underdogs in Overdrive: 10 Insanely Great Ideas For the Asian Techno-preneur*. Singapore, John Wiley & Sons, 2001.

Kim Young Jeh (Ed.). *The New Pacific Community in the 1990s*. Armonk, NY, M. E. Sharpe, 1996.

King, Victor T. *Environmental Challenges in South-East Asia*. Richmond, Curzon Press, 1998.

The Modern Anthropology of Southeast Asia: An Introduction. London, RoutledgeCurzon, 2002.

Kinnvall, Catarina, and Jonsson, Kristina. *Globalization and Democratization in Asia—The Construction of Identity*. London, RoutledgeCurzon, 2002.

Kohama, Hirhisa (Ed.). *Asian Development Experience Vol. 1: External Factors for Asian Development*. Singapore, Institute of Southeast Asian Studies, 2003.

Korhonen, Pekka. *Japan and Asia-Pacific Integration: Pacific Romances 1968–1996*. London, Routledge, 1998.

Kratoska, Paul H. (Ed.). *South East Asia*. London, Routledge, 2001.

Kreinin, Mordechai E., and Plummer, Michael G. (Eds). *Economic Integration and Asia—The Dynamics of Regionalism in Europe, North America and the Asia-Pacific*. Cheltenham, Edward Elgar Publishing, 2001.

Krugman, Paul. *The Return of Depression Economics*. New York, NY, W. W. Norton & Co, 1999.

Kwak Tae-Hwan and Olsen, Edward A. (Eds). *The Major Powers in Northeast Asia: Seeking Peace and Security*. Boulder, CO, Lynne Rienner, 1996.

Kwan, C. H. *Economic Interdependence in the Asia-Pacific Region—Towards a Yen Bloc*. London, Routledge, 1994.

Kwan, C. H., Vandenbrink, D., and Chia Siow Yue (Eds). *Coping with Capital Flows in East Asia*. Singapore, Institute of Southeast Asian Studies, 1998.

Lall, Sanjaya. *Learning from the Asian Tigers: Studies in Technology and Industrial Policy*. Basingstoke, Macmillan Press, 1995.

Laothamatas, Anek (Ed.). *Democratization in Southeast and East Asia*. Singapore, Institute of Southeast Asian Studies, 1997.

Lasserre, Philippe, and Schütte, Helmut. *Strategies for Asia Pacific*. Basingstoke, Macmillan Press, 1995.

Latham, A. J. H., and Kawakatsu, Heita (Eds). *Asia-Pacific Dynamism 1550–2000*. London, Routledge, 2000.

Leaman, Oliver (Ed.). *Encyclopedia of Asian Philosophy*. London, Routledge, 2000.

Lee, Chung H. *Financial Liberalization and the Economic Crisis in Asia*. London, RoutledgeCurzon, 2002.

Lee, Kyung Tae. *Globalization and the Asia Pacific Economy*. London, Routledge, 2002.

Lee Sang-Gon and Ruffini, Pierre-Bruno (Eds). *The Global Integration of Europe and East Asia—Studies of International Trade and Investment*. Cheltenham, Edward Elgar Publishing, 1999.

Lee, Yok-shiu F., and So, Alvin Y. (Eds). *Asia's Environmental Movements—Comparative Perspectives*. Armonk, NY, M. E. Sharpe, 1999.

Leifer, Michael. *Dictionary of the Modern Politics of South-east Asia*. London, Routledge, 2000.

Asian Nationalism. London, Routledge, 2000.

Li Kui Wai. *Capitalist Development and Economism in East Asia—The Rise of Hong Kong, Singapore, Taiwan and South Korea*. London, Routledge, 2002.

Lim, Robyn. *The Geopolitics of East Asia*. London, Routledge-Curzon, 2003.

Lingle, Christopher. *Rise and Decline of the Asian Century: False Starts on the Path to the Global Millennium*. Aldershot, Ashgate Publishing, 3rd edn, 2000.

Lintner, Bertil. *Blood Brothers: The Criminal Underworld of Asia*. Basingstoke, Palgrave Macmillan, 2003.

Lipton, Michael, and Osmani, Siddiqur. *The Quality of Life in Emerging Asia*. Brighton, University of Sussex, 1996.

Liu Fu-kuo and Régnier, Philippe (Eds). *Regionalism in East Asia*. London, RoutledgeCurzon, 2002.

Lovell, David W. (Ed.). *Asia-Pacific Security: Policy Challenges*. Singapore, Institute of Southeast Asian Studies, 2003.

Low, Linda. *ASEAN Economic Co-operation and Challenges*. Singapore, Institute of Southeast Asian Studies, 2004.

Lukauskas, Arvid J., and Rivera-Batiz, Francisco L. *The Political Economy of the East Asian Crisis and its Aftermath—Tigers in Distress*. Cheltenham, Edward Elgar Publishing, 2001.

MacIntyre, Andrew (Ed.). *Business and Government in Industrialising Asia*. St Leonards, NSW, Allen and Unwin, 2nd edn, 2001.

Mack, Andrew, and Ravenhill, John. *Talking Shop? Pacific Co-operation: Building Economic and Security Regimes in the Asia-Pacific Region*. St Leonards, NSW, Allen and Unwin, 1994.

Mackerras, Colin (Ed.). *East and Southeast Asia: A Multi-disciplinary Survey*. Boulder, CO, Lynne Rienner, 1995.

Maddison, Angus, Rao, Prasada D. S., and Shepherd, William (Eds). *The Asian Economies in the Twentieth Century*. Cheltenham, Edward Elgar Publishing, 2002.

Mahathir Mohamad. *A New Deal for Asia*. Kuala Lumpur, Pelanduk Publications, 1999.

Mahbubani, Kishore. *Can Asians Think?* Singapore, Times Books International, 1998.

Maidment, Richard, and Mackerras, Colin (Eds). *Culture and Society in the Asia-Pacific*. London, Routledge, 1998.

Maidment, Richard, Goldblatt, David, and Mitchell, Jeremy (Eds). *Governance in the Asia-Pacific*. London, Routledge, 1998.

Mallet, Victor. *The Trouble with Tigers: The Rise and Fall of Southeast Asia*. London, HarperCollins, 1999.

Manning, Robert A. *The Asian Energy Factor: Myths and Dilemmas of Energy, Security and the Pacific Future*. Hampshire, Palgrave, 2001.

Margolis, Eric. *War at the Top of the World—The Struggle for Afghanistan, Kashmir and Tibet*. London, Routledge, 2001.

Masina, Pietro P. (Ed.). *Rethinking Development in East Asia—From Illusory Miracle to Economic Crisis*. Richmond, Curzon Press, 1999.

Masuyama, Seiichi, and Vandenbrink, Donna (Eds). *Towards a Knowledge-Based Economy: East Asia's Changing Industrial Geography*. Singapore, Institute of Southeast Asian Studies, 2003.

Masuyama, Seiichi, Vandenbrink, Donna, and Chia Siow Yue (Eds). *Industrial Policies in East Asia*. Singapore, Institute of Southeast Asian Studies, 1997.

East Asia's Financial Systems—Evolution and Crisis. Singapore, Institute of Southeast Asian Studies, 1999.

Restoring East Asia's Dynamism. Singapore, Institute of Southeast Asian Studies, 2000.

Industrial Restructuring in East Asia. Singapore, Institute of Southeast Asian Studies, 2001.

Maull, Hanns, Segal, Gerald, and Wanandi, Jusuf (Eds). *Europe and the Asia-Pacific*. London, Routledge, 1998.

McCargo, Duncan. *Media and Politics in Pacific Asia*. London, Routledge, 2002.

McDougall, Derek. *The International Politics of the New Asia Pacific*. Singapore, Institute of Southeast Asian Studies, 1997.

McGrew, Anthony, and Brook, Christopher (Eds). *Asia-Pacific in the New World Order*. London, Routledge, 1998.

McLeod, Ross H., and Garnaut, Ross (Eds). *East Asia in Crisis—From Being a Miracle to Needing One?* London, Routledge, 1998.

McNally, Christopher A., and Morrison, Charles E. (Eds). *Asia Pacific Security Outlook: 2002*. Washington, DC, Brockings Institution, 2002.

McTurnan Kahin, George. *Southeast Asia: A Testament*. London, RoutledgeCurzon, 2002.

Meyer, Karl, and Brysac, Shareen. *Tournament of Shadows: The Great Game and the Race for Empire in Asia*. London, Little, Brown, 2001.

Miller, Sally M., Latham, A. J. H., and Flynn, Dennis O. (Eds). *Studies in the Economic History of the Pacific Rim*. London, Routledge, 1997.

Milner, Anthony. *Region, Security and the Return of History*. Singapore, Institute of Southeast Asian Studies, 2003.

Mirza, Hafiz, and Wee Kee Hwee (Eds). *Transnational Corporate Strategies in the ASEAN Region*. Cheltenham, Edward Elgar Publishing, 2000.

Montes, Manuel F. *The Currency Crisis in Southeast Asia (Updated Edition)*. Singapore, Institute of Southeast Asian Studies, 2000.

Montes, Manuel F., and Popov, Vladimir V. *The Asian Crisis Turns Global*. Singapore, Institute of Southeast Asian Studies, 1999.

Morley, James W. (Ed.). *Driven by Growth—Political Change in the Asia-Pacific Region*. Armonk, NY, M. E. Sharpe, 1998.

Murphey, Rhoads. *East Asia: A New History*. London, Longman, 1997.

Naisbitt, John. *Megatrends Asia*. London, Nicholas Brealey Publishing, 1995.

Nakamura, Mitsuo, Siddique, Sharon, and Bajunid, Omar Farouk (Eds). *Islam and Civil Society in Southeast Asia*. Singapore, Institute of Southeast Asian Studies, 2001.

Narine, Shaun. *Explaining ASEAN: Regionalism in Southeast Asia*. Boulder, CO, Lynne Rienner Publishers, 2002.

Neary, Ian. *Human Rights in Japan, South Korea and Taiwan*. London, RoutledgeCurzon, 2002.

Noland, Marcus. *Pacific Basin Developing Countries: Prospects for the Future*. Washington, DC, Institute for International Economics, 1990.

Nomura Research Institute and Institute of Southeast Asian Studies. *The New Wave of Foreign Direct Investment in Asia*. Singapore, Institute of Southeast Asian Studies, 1995.

Ohmae, Kenichi. *The End of the Nation State: The Rise of Regional Economies*. New York, NY, The Free Press, 1995.

Olds, Kris, et al. (Eds). *Globalisation and the Asia-Pacific—Contested Territories*. London, Routledge, 1999.

Osborne, Milton. *Southeast Asia: An Illustrated Introductory History*. St Leonards, NSW, Allen and Unwin, 7th Edn, 1998.

The Mekong—Turbulent Past, Uncertain Future. Boston, MA, Atlantic Monthly Press, 2000.

Pacific Business Forum. *The Osaka Action Plan: Roadmap to Realizing the APEC Vision*. Singapore, Asia-Pacific Economic Co-operation Secretariat, 1995.

Pape, Wolfgang. *East Asia 2000 and Beyond*. Richmond, Curzon Press, 1998.

Parnwell, Michael, and Bryant, Raymond. *Environmental Change in South-East Asia—People, Politics and Sustainable Development*. London, Routledge, 1996.

Pempel, T. J. (Ed.). *The Politics of the Asian Economic Crisis*. Ithaca, NY, Cornell University Press, 1999.

Peou, Sorpong. *The ASEAN Regional Forum and Post-Cold War IR Theories—A Case for Constructive Realism?* Singapore, Institute of Southeast Asian Studies, 1999.

Petri, Peter A. (Ed.). *Regional Co-operation and Asian Recovery*. Singapore, Institute of Southeast Asian Studies, 2000.

Phillips, David R. (Ed.). *Ageing in the Asia-Pacific Region—Issues, Policies and Future Trends*. London, Routledge, 2000.

Phillips, David R., and Chan, Alfred C. M. (Eds). *Ageing and Long-term Care: National Policies in the Asia-Pacific*. Singapore, Institute of Southeast Asian Studies, 2003.

Pinches, Michael (Ed.). *Culture and Privilege in Capitalist Asia*. London, Routledge, 1999.

Piper, Nicola, and Uhlin, Anders. *Transnational Activism in Asia*. London, Routledge, 2003.

Pomfret, Richard. *Asian Economies in Transition: Reforming Centrally Planned Economies*. Cheltenham, Edward Elgar Publishing, 1996.

Preston, P. W., and Gilson, Julie (Eds). *The European Union and East Asia—Inter-Regional Linkages in a Changing Global System*. Cheltenham, Edward Elgar Publishing, 2001.

Raina, Vinod, Chowdhury, Amit, and Chowdhury, Samit. *The Dispossessed: Victims of Development in Asia*. Hong Kong, Arena Press, 1999.

Ravi, Srilata, Rutten, Mario, and Goh, Beng-Lan (Eds). *Asia in Europe, Europe in Asia*. Singapore, Institute of Southeast Asian Studies, 2004.

Reid, Anthony J. S. *Southeast Asia in the Age of Commerce 1450–1680*, Volume II, *Expansion and Crisis*. London, Yale University Press, 1995.

The Last Stand of Asian Autonomies: Responses to Modernity in the Diverse States of Southeast Asia and Korea, 1750–1900. Basingstoke, Macmillan Press, 1997.

Charting the Shape of Early Modern Southeast Asia. Cheltenham, Edward Elgar Publishing, 2000.

Ries, Phillippe. *The Asian Storm—Asia's Economic Crisis Examined*. Tokyo, Tuttle Publishing, 2001.

Rigg, Jonathan. *Southeast Asia—The Human Landscape of Modernization and Development*. London, Routledge, 2002.

Robin, Jeffrey (Ed.). *Asia: The Winning of Independence: the Philippines, India, Indonesia, Vietnam, Malaya*. London, Macmillan Press, 1981.

Robison, Richard, and Goodman, David S. G. (Eds). *The New Rich in Asia—Mobile Phones, McDonald's and Middle-Class Revolution*. London, Routledge, 1996.

Robison, Richard, et al. (Eds). *Politics and Markets in the Wake of the Asian Crisis*. London, Routledge, 1999.

Robles, Alfredo C. *The Political Economy of Interregional Relations: ASEAN and the EU (The International Political Economy of New Regionalisms S.)*Burlington, VT, Ashgate, 2004.

Rock, Michael T. *Pollution Control in East Asia: Lessons from Newly Industrializing Economies*. Singapore, Institute of Southeast Asian Studies, 2002.

Rodan, Garry. *Transparency and Authoritarian Rule in Southeast Asia: Singapore and Malaysia*. London, RoutledgeCurzon, 2004.

Rodan, Garry (Ed.). *Political Oppositions in Industrialising Asia*. London, Routledge, 1996.

Rodan, Garry, Hewison, Kevin, and Robison, Richard (Eds). *The Political Economy of South-east Asia*. Oxford, Oxford University Press, 2001.

Rodrigo, G. Chris. *Technology, Economic Growth and Crises in East Asia*. Cheltenham, Edward Elgar Publishing, 2000.

Rohwer, Jim. *Asia Rising*. New York, NY, Simon and Schuster, 1996.

Ross, Robert S. (Ed.). *East Asia in Transition—Toward a New Regional Order*. Armonk, NY, M. E. Sharpe, 1995.

Rowen, Henry S. (Ed.). *Behind East Asian Growth—The Political and Social Foundations of Prosperity*. London, Routledge, 1997.

Rugman, Alan M., and Boyd, Gavin (Eds). *Deepening Integration in the Pacific Economies—Corporate Alliances, Contestable Markets and Free Trade*. Cheltenham, Edward Elgar Publishing, 1999.

Rüland, Jürgen (Ed.). *The Dynamics of Metropolitan Management in Southeast Asia*. Singapore, Institute of Southeast Asian Studies, 1997.

Rüland, Jürgen, Manske, Eva, and Draguhn, Werner (Eds). *Asia-Pacific Economic Cooperation (APEC)—The First Decade*. London, RoutledgeCurzon, 2002.

Rutton, Mario. *Rural Capitalists in Asia—A Comparative Analysis on India, Indonesia and Malaysia*. Richmond, Curzon Press, 2001.

Sardesai, D. R. *South-East Asia*. Basingstoke, Macmillan Press, 1997.

Southeast Asia: Past and Present. Boulder, CO, Westview Press, 2003.

Sen, Amartya. *Beyond the Crisis: Development Strategies in Asia*. Singapore, Institute of Southeast Asian Studies, 1999.

Sen, Krishna, and Stivens, Maila (Eds). *Gender and Power in Affluent Asia*. London, Routledge, 1998.

Sen, Rahul. *Free Trade Agreements in Southeast Asia*. Singapore, Institute of Southeast Asian Studies, 2004.

Sharma, Kishor (Ed.). *Trade Policy, Growth and Poverty in Asian Developing Countries*. London, Routledge, 2003.

Shastri, Amita, and Wilson, A. Jeyaratnam. *The Post-colonial State of South Asia: Political and Constitutional Problems*. Richmond, Curzon Press, 2000.

Shibata, Hirofumi, and Ihori, Toshihiro (Eds). *The Welfare State, Public Investment and Growth*. Singapore, Institute of Southeast Asian Studies, 1998.

Shigetomi, Shinichi (Ed.). *The State and the NGOs: Perspective from Asia*. Singapore, Institute of Southeast Asian Studies, 2002.

Shimomura, Yasutami (Ed.). *Asian Devlopment Experience Vol. 2: The Role of Governance in Asia*. Singapore, Institute of Southeast Asian Studies, 2003.

Sien Chia Lin (Ed.). *Southeast Asia Transformed: A Geography of Change*. Singapore, Institute of Southeast Asian Studies, 2001.

Simon, Sheldon W. (Ed.). *East Asian Security in the Post-Cold War Era*. Armonk, NY, M. E. Sharpe, 1993.

Singh, Daljit, and Chin Kin Wah (Eds). *Southeast Asian Affairs 2004*. Singapore, Institute of Southeast Asian Studies, 2004.

Singh, Daljit, and Freeman, Nick J. (Eds). *Regional Outlook: Southeast Asia 2000-2001*. Singapore, Institute of Southeast Asian Studies, 2001.

Singh, Daljit, and Tin Maung Maung Than (Eds). *Regional Outlook—Southeast Asia 1999–2000*. Singapore, Institute of Southeast Asian Studies, 1999.

Slater, Jim, and Strange, Roger (Eds). *Business Relationships with East Asia—The European Experience*. London, Routledge, 1997.

Smith, David W., and Dixon, Chris (Eds). *Economic and Social Development in Pacific Asia*. London, Routledge, 1993.

Smith, Heather (Ed.). *The Economic Development of Northeast Asia*. Cheltenham, Edward Elgar Publishing, 2002.

Solidum, Estrella D. *Politics of ASEAN: An Introduction to Southeast Asian Regionalism*. Singapore, Times Academic Press, 2003.

Souchou, Yao (Ed.). *House of Glass: Culture, Modernity, and the State in Southeast Asia*. Singapore, Institute of Southeast Asian Studies, 2000.

Starrs, Roy (Ed.). *Asian Nationalism in an Age of Globalization*. Richmond, Curzon Press, 2001.

Steinberg, David Joel (Ed.). *In Search of Southeast Asia—A Modern History (Revised Edn)*. Honolulu, HI, University of Hawaii Press, 1987.

Sudo, Sueo. *The International Relations of Japan and Southeast Asia—Forging a New Regionalism*. London, Routledge, 2001.

Sugiyama, Shinya, and Grove, Linda. *Commercial Networks in Modern Asia*. Richmond, Curzon Press, 2000.

Suryadinata, Leo (Ed.). *Nationalism and Globalization: East and West*. Singapore, Institute of Southeast Asian Studies, 2000.

Ethnic Relations and Nation-Building in Southeast Asia. Singapore, Institute of Southeast Asian Studies, 2004.

Sutter, Robert, G. *East Asia and the Pacific: Challenges for US Policy*. Boulder, CO, Westview Press, 1995.

Tan, Joseph L. H. (Ed.). *AFTA in the Changing International Economy*. Singapore, Institute of Southeast Asian Studies, 1996.

Human Capital Formation as an Engine of Growth: The East Asian Experience. Singapore, Institute of Southeast Asian Studies, 1999.

Tan, Joseph L. H., and Luo Zhao Hong (Eds). *ASEAN-China Economic Relations—In the Context of Pacific Economic Development and Co-operation*. Singapore, Institute of Southeast Asian Studies, 1992.

Tarling, Nicholas. *Nations and States in South East Asia*. Cambridge, Cambridge University Press, 1998.

Imperialism in South East Asia. London, Routledge, 2001.

South-East Asia: A Modern History. Oxford, Oxford University Press, 2001.

Tarling, Nicholas (Ed.). *The Cambridge History of Southeast Asia* (2 vols). Cambridge, Cambridge University Press, 1993.

Tay, Simon S. C., et al. (Eds). *Reinventing ASEAN*. Singapore, Institute of Southeast Asian Studies, 2002.

Taylor, R. H. (Ed.). *The Politics of Elections in Southeast Asia*. Cambridge, Cambridge University Press, 1997.

Terry, Edith. *How Asia Got Rich—Japan and the Asia Miracle*. Armonk, NY, M. E. Sharpe, 2000.

Than, Mya (Ed.). *ASEAN Beyond the Regional Crisis: Challenges and Initiatives*. Singapore, Institute for Southeast Asian Studies, 2001.

Than, Mya, and Gates, Carolyn L. (Eds). *ASEAN Enlargement: Impacts and Implications*. Singapore, Institute of Southeast Asian Studies, 2000.

Thompson, Grahame. *Economic Dynamism in the Asia-Pacific*. London, Routledge, 1998.

Tien Hung-mao and Cheng Tun-jen (Eds). *The Security Environment in the Asia-Pacific*. Armonk, NY, M. E. Sharpe, 2000.

Timmer, Marcel. *The Dynamics of Asian Manufacturing—A Comparative Perspective in the Late Twentieth Century*. Cheltenham, Edward Elgar Publishing, 2000.

Tinker, Irene, and Summerfield, Gale (Eds). *Women's Rights to House and Land: China, Laos, Vietnam*. Boulder, CO, Lynne Rienner, 1999.

Tipton, Frank B. *The Rise of Asia*. Basingstoke, Macmillan, 1998.

Toh Thian Ser (Ed.). *Megacities, Labour and Communications*. Singapore, Institute of Southeast Asian Studies, 1998.

Tongzon, Jose L. *The Economies of Southeast Asia—Before and After the Crisis*. Cheltenham, Edward Elgar Publishing, 2002.

Tow, William T. *Asia-Pacific Strategic Relations: Seeking Convergent Security*. Cambridge, Cambridge University Press, 2002.

Tran Van Hoa. *Economic Developments and Prospects in the ASEAN*. Basingstoke, Macmillan Press, 1997.

Tran Van Hoa (Ed.). *The Asia Recovery—Issues and Aspects of Development, Growth, Trade and Investment*. Cheltenham, Edward Elgar Publishing, 2001.

Competition Policy and Global Competitiveness in Major Asian Economies. Cheltenham, Edward Elgar Publishing, 2003.

Trocki, Carl A. (Ed.). *Gangsters, Democracy and the State in Southeast Asia*. Ithaca, NY, Southeast Asia Program Publications, 1998.

Opium, Empire and the Global Political Economy—A Study of the Asian Opium Trade, 1750–1950. London, Routledge, 1999.

United Nations. *Women in Asia and the Pacific 1985/93*. New York, NY, 1995.

Asian and Pacific Developing Economies and the First WTO Ministerial Conference: Issues of Concern. New York, NY, 1996.

Assessing the Potential and Direction of Agricultural Trade within the ESCAP Region. New York, NY, 1996.

Trade Prospects for the Year 2000 and Beyond for the Asian and Pacific Region. New York, NY, 1996.

BISTEC-EC (Bangladesh, India, Sri Lanka, Thailand—Economic Cooperation) Development Programme. New York, NY, 1998.

Van Ness, Peter (Ed.). *Debating Human Rights—Critical Essays from the United States and Asia*. London, Routledge, 1998.

Vatikiotis, Michael. *Political Change in Southeast Asia: Trimming the Banyan Tree*. London, Routledge, 1996.

Debatable Land: Stories from Southeast Asia. Greenfield, MA, Talisman, 2001.

Verma, Anil, Lansbury, Russell, and Kochan, Thomas (Eds). *Employment Relations in Asian Economies*. London, Routledge, 1995.

Vines, Stephen. *The Years of Living Dangerously: Asia—from Financial Crisis to the New Millennium*. London, Orion Business Book, 2000.

Vogel, Ezra F. *The Four Little Dragons: The Spread of Industrialization in East Asia*. Cambridge, MA, Harvard University Press, 1991.

Wade, R. *Governing the Market: Economic Theory and the Role of Government in East Asian Industrialization*. Princeton, NJ, Princeton University Press, 1990.

Wang Gungwu. *The Chinese Overseas: From Earthbound China to the Quest for Autonomy*. Cambridge, MA, Harvard University Press, 2000.

Warner, Malcolm. *Culture and Management in Asia*. London, RoutledgeCurzon, 2003.

Watanabe, Koji (Ed.). *Engaging Russia in Asia Pacific*. Singapore, Institute of Southeast Asian Studies, 2000.

Weber, Maria. *Reforming Economic Systems in Asia—A Comparative Analysis of China, Japan, South Korea, Malaysia and Thailand*. Cheltenham, Edward Elgar Publishing, 2001.

Wee, C. J. W. L. *Local Cultures and the 'New Asia': The State, Culture, and Capitalism in Southeast Asia*. Singapore, Institute of Southeast Asian Studies, 2002.

Weightman, Barbara A. *Dragons and Tigers: A Geography of South, East and Southeast Asia*. Singapore, John Wiley & Sons, 2003.

Weiss, Julian (Ed.). *Tigers' Roar—Asia's Recovery and its Impact*. Armonk, NY, M. E. Sharpe, 2001.

West, Philip, Levine, Steven I., and Hiltz, Jackie. *America's Wars in Asia: A Cultural Approach to History and Memory*. Armonk, NY, M. E. Sharpe, 1998.

Wilkinson, Barry. *Labour and Industry in the Asia-Pacific: Lessons from the Newly Industrialized Countries*. Berlin and New York, NY, Walter de Gruyter, 1995.

Williams, Louise. *Wives, Mistresses and Matriarchs: Asian Women Today*. St Leonards, NSW, Allen & Unwin, 1999.

Wolters, O. W. *History, Culture, and Region in Southeast Asian Perspectives (Revised Edn)*. Ithaca, NY, Southeast Asia Program Publications, 1999.

Woodiwiss, Anthony. *Globalization, Human Rights and Labour Law in Pacific Asia*. Cambridge, Cambridge University Press, 1998.

World Bank. *East Asia: The Road to Recovery*. Washington, DC, World Bank, 1998.

World Bank Policy Research Report. *The East Asian Miracle: Economic Growth and Public Policy*. New York, NY, Oxford University Press, 1993.

Woronoff, Jon. *Asia's Miracle Economies*. New York, NY, and London, M. E. Sharpe, 1992.

Wu Yanrui. *The Macroeconomics of East Asian Growth*. Cheltenham, Edward Elgar Publishing, 2002.

Wurfel, David, and Burton, Bruce (Eds). *Southeast Asia in the New World Order: The Political Economy of a Dynamic Region*. Basingstoke, Macmillan Press, 1996.

Yahuda, Michael. *The International Politics of Asia-Pacific Since 1945*. London, Routledge, 2000.

Yamazawa, Ippei. *Economic Integration in the Asia Pacific Region*. London, Routledge, 2000.

(Ed.). *Asia Pacific Economic Cooperation (APEC)—Challenges and Tasks for the Twenty-first Century*. London, Routledge, 2000.

Yencken, David, Fien, John, and Sykes, Helen (Eds). *Environment, Education and Society in the Asia-Pacific—Local Traditions and Global Discourses*. London, Routledge, 2000.

Yeoh, Brenda S. A., Teo, Peggy, and Huang, Shirlena (Eds). *Gender Politics in the Asia-Pacific Region*. London, Routledge, 2002.

Yeung, Henry Wai-Chung. *Transnational Corporations and Business Networks—Hong Kong Firms in the ASEAN Region*. London, Routledge, 1998.

Yeung, May T., Perdikis, Nicholas, and Kerr, William A. *Regional Trading Blocs in the Global Economy—The EU and ASEAN*. Cheltenham, Edward Elgar Publishing, 1999.

Yeung Yue-man (Ed.). *Pacific Asia in the 21st Century*. Hong Kong, The Chinese University Press, 1994.

Yong Tan Tai and Kudaisya, Gyanesh. *The Aftermath of Partition in South Asia*. London, Routledge, 2000.

Yu, George T. *Asia's New World Order*. Basingstoke, Macmillan Press, 1997.

Zhang Yumei. *Pacific Asia*. London, Routledge, 2002.

Zhao Shuisheng. *Dynamics of Power Competition in East Asia*. Basingstoke, Macmillan Press, 1997.

SELECT BIBLIOGRAPHY—PERIODICALS

Acta Asiatica. Bulletin of the Institute of Eastern Culture (Tōhō Gakkai), 2-4-1 Nishi-Kanda, Chiyoda-ku, Tokyo 101-0065 Japan; tel. (3) 3262-7221; fax (3) 3262-7227; e-mail iec@tohogakkai.com; internet www.tohogakkai.com; f. 1961; in English; 2 a year.

Archaeology in Oceania. The University, Sydney, NSW 2006, Australia; tel. (2) 9351-2666; fax (2) 9351-7488; e-mail d.koller@oceania.usyd.edu.au; f. 1966; archaeology and physical anthropology; 3 a year; Editor J. PETER WHITE.

Archiv orientální. Journal of African and Asian Studies of the Oriental Institute of the Czech Academy of Sciences, Pod vodárenskou věží 4, 182 08 Prague 8, Czech Republic; tel. (2) 6605-2483; fax (2) 8658-1897; e-mail aror@orient.cas.cz; f. 1929; book reviews and notes; contributions in English or French and German; quarterly; Editor Dr STANISLAVA VAVROUŠKOVÁ.

Artibus Asiae. Museum Rietberg, Gablerstr. 15, 8002 Zürich, Switzerland; tel. (1) 2063131; fax (1) 2063132; e-mail artibus.asiae@rietb.stzh.ch; in co-operation with the Arthur M. Sackler Gallery, Smithsonian Institution, Washington, DC, USA; f. 1925; Asian art and archaeology; illustrated; 2 a year; Editor-in-Chief Dr FRANÇOIS LOUIS.

Arts of Asia. 1309 Kowloon Centre, 29–39 Ashley Rd, Kowloon, Hong Kong; tel. 23762228; fax 23763713; e-mail info@artsofasianet.com; internet www.artsofasianet.com; f. 1971; 6 a year; Publr and Editor TUYET NGUYET.

Arts Asiatiques. Musée National des Arts Asiatiques Guimet, 6 Place d'Iéna, 75116 Paris, France; tel. 1-56-52-53-00; fax 1-56-52-53-54; internet www.museeguimet.fr; f. 1924; annually.

ASEAN Briefing. GPO Box 10874, Hong Kong; fax (USA) (708) 570-7421; politics and economics; monthly.

ASEAN Economic Bulletin—A Journal of Asian and Pacific Economic Affairs. Institute of Southeast Asian Studies, 30 Heng Mui Keng Terrace, Pasir Panjang, Singapore 119614; tel. 67780955; fax 67756259; e-mail publish@iseas.edu.sg; internet bookshop.iseas.edu.sg; f. 1968; economic issues in the Asia-Pacific region; Man. Editor TRIENA ONG.

The Asia Letter. GPO Box 10874, Hong Kong; fax (USA) (708) 570-7421; f. 1964; politics and economics; weekly; also *China Letter, Indonesia Letter, Japan Letter, Philippine Letter*, etc.

Asia and Pacific Review. World of Information, CEB Ltd, 2 Market St, Saffron Walden, Essex CB10 1HZ, United Kingdom; tel. (1799) 521150; fax (1799) 524805; e-mail info@worldinformation.com; internet www.worldinformation.com.

Asia Pacific Consensus Forecasts. Consensus Economics Inc, 53 Upper Brook St, London, W1K 2LT, United Kingdom; tel. (20) 7491-3211; fax (20) 7409-2331; internet www.consensuseconomics.com; monthly; Editor SUYIN KAN.

Asia-Pacific Economic Review. EMBA Inc, POB 1363, Canberra, ACT 2601, Australia; tel. (2) 6258-1330; fax (2) 6258-3285; e-mail colin.hargreaves@anu.edu.au; f. 1995; 3 a year; Editors COLIN HARGREAVES, PETER C. B. PHILLIPS.

Asia Pacific Exchange News. Fitzroy House, 13–17 Epworth St, London EC2A 4DL, United Kingdom; tel. (20) 7825-8000; fax (20) 7251-2725; f. 1996; newsletter for global investors; monthly; Editor DAVID H. WEBSTER.

Asia-Pacific Review. 4 Park Sq., Milton Park, Abingdon, Oxon, OX14 4RN, United Kingdom; tel. (1235) 828600; fax (1235) 829000; internet www.tandf.co.uk/journals; f. 1994; 2 a year; Editor SUNNA TROTT.

Asia Pacific Viewpoint. Institute of Geography, Victoria University of Wellington, Box 600, Wellington, New Zealand; tel. (4) 472-1000, ext. 5029; fax (4) 495-5127; e-mail apv@vuw.ac.nz; internet www.blackwellpublishers.co.uk/journals/apv; f. 1996 (to replace *Pacific Viewpoint*); 3 a year; Managing Editor Dr WARWICK E. MURRAY.

Asiamoney and Finance. Euromoney Publications, Nestor House, Playhouse Yard, London, EC4V 5EX, United Kingdom; tel. (20) 7779-8992; fax (20) 7779-8873; e-mail asiamoney@dial.pipex.com; monthly.

Asian Affairs. Journal of the Royal Society for Asian Affairs, 2 Belgrave Sq., London, SW1X 8PJ, United Kingdom; tel. (20) 7235-5122; fax (20) 7259-6771; e-mail info@rsaa.org.uk; internet www.rsaa.org.uk; covers economic, cultural and political matters relating to the Near and Middle East, Central Asia, South and South-East Asia, and the Far East; 3 a year; Editor MICHAEL SHERINGHAM.

Asian Almanac. POB 2737, Singapore 9047; tel. 64816047; fax 64816092; f. 1963; weekly; abstracts of Asian affairs; Editor VEDAGIRI T. SAMBANDAN.

Asian Ethnicity. 4 Park Sq., Milton Park, Abingdon, Oxon, OX14 4RN, United Kingdom; tel. (1235) 828600; fax (1235) 829000; internet www.tandf.co.uk/journals; 3 a year; Editor-in-Chief Prof. COLIN MACKERRAS.

Asian News Digest. A-126, Niti Bagh, New Delhi 110049, India; tel. (11) 6565140; fax (11) 6862857; f. 1955 as Asian Recorder, renamed 2000; record of Asian events; weekly; Editor A. K. B. MENON.

Asian-Pacific Economic Literature. Asia Pacific School of Economics and Government, Australian National University, Canberra, ACT 0200, Australia; tel. (2) 6125-8258; fax (2) 6125-8448; e-mail apel@anu.edu.au; abstracting and survey journal; 2 a year; Editor Prof. RON DUNCAN.

Asian Perspective. Institute for Far Eastern Studies, Kyungnam University, 28-42 Samchung-dong, Chongno-ku, Seoul 110-230, Republic of Korea; tel. (2) 3700-0725; fax (2) 3700-0707; e-mail ifes@kyungnam.ac.kr; internet www.ifes.kyungnam.ac.kr; f. 1977; regional and international affairs; 4 a year; Editor-in-Chief MELVIN GURTOV.

Asian Profile. POB 1211, Metrotown Regional Post Office, 1800 Kingsway, Burnaby, BC V5H 4J8, Canada; tel. (604) 8211321; fax (604) 2760813; e-mail asianprofile@asianresearchservice.com; internet www.asianresearchservice.com; f. 1973; multidisciplinary study of Asian affairs; 6 a year; Editor NELSON LEUNG.

Asian Research Trends: A Humanities and Social Science Review. Centre for East Asian Cultural Studies for UNESCO, Toyo Bunko, Honkomagome 2-28-21, Bunkyo-ku, Tokyo 113, Japan; tel. (3) 3942-0124; fax (3) 3942-0120; internet www.toyo.bunko.or.jp/ceacs/; f. 1991; annually; Editor Prof. YONEO ISHII.

Asian Studies. Asian Center, University of the Philippines, Diliman, Quezon City, Metro Manila 1101, Philippines; tel. (2) 9261841; fax (2) 9261821; e-mail asiancenter@ac.upd.edu.ph; annually.

Asian Studies Review. Asian Studies Centre, The University of Queensland, Qld 4072, Australia; tel. (7) 3365-6762; fax (7) 3365-6811; e-mail asiareview@mailbox.uq.edu.au; internet www.coombs.anu.edu.au/SpecialProj/ASAA/as-review.html; journal of the Asian Studies Association of Australia; 4 a year; Editor Prof. KAM LOUIE.

Asian Survey. Institute of East Asian Studies, University of California, Rm 408, 6701 San Pablo, Oakland, CA 94608, USA; tel. (510) 642-0978; fax (510) 643-9930; e-mail asiasrvy@uclink.berkeley.edu; f. 1961; every 2 months; Editor LOWELL DITTMER.

Asian Thought and Society: An International Review. Dept of Political Science, State University of New York, Oneonta, NY 13820, USA; tel. (607) 431-3553; fax (607) 431-2107; f. 1976; analysis of social structures and changes in Pacific and South Asian countries with special reference to social and intellectual development in historical perspective; 3 a year; Editor-in-Chief I. J. H. TS'AO.

The Asian Wall Street Journal. 25/F Central Plaza, 18 Harbour Rd, GPOB 9825, Hong Kong; tel. 25737121; fax 28345291; internet www.wsj.com; f. 1976; daily; Editor REGINALD CHUA.

Asiatische Studien/Etudes Asiatiques. Publ. by Verlag Peter Lang, Moosstr. 1, Postfach 350, CH-2542 Pieterlen, Switzerland; tel. (32) 376-1717; fax (32) 376-1727; e-mail info@peterlang.com; internet www.peterlang.com; journal of the Swiss Asia Society; Editor R. GASSMANN, Ostasiatisches Seminar der Universität Zürich, Zürichbergstrasse 4, 8032 Zürich, Switzerland; tel. (1) 634-3181; fax (1) 634-4921; e-mail office@oas.unizh.ch; 4 a year.

ASIEN. Deutsche Gesellschaft für Asienkunde eV, 20148 Hamburg, Rothenbaumchaussee 32, Germany; tel. (40) 445891; fax (40) 4107945; e-mail post@asienkunde.de; internet www.asienkunde.de; quarterly; Editor CHRISTINE BERG.

Australian Geographer. Geographical Society of New South Wales Inc, POB 602, Gladesville, NSW 2111, Australia; tel. (2) 9817-3647; fax (2) 9817-4592; e-mail geog@idx.com.au; internet www.es.my.edu.au/societies/GSNSW/gsnsw01.htm; f. 1928; 3 a year; Editors Assoc. Prof. JAMES FORREST, Dr GEOFF HUMPHREYS, Dr PAULINE MCGUIRK.

Australian Journal of International Affairs. Journal of the Australian Institute of International Affairs, 32 Thesiger Court, Deakin, ACT 2600, Australia; f. 1947; 4 a year; Editor Prof. WILLIAM TOW (e-mail w.tow@griffith.edu.au).

Australian Journal of Politics and History. Blackwell Publishing Asia, 550 Swanston St, Carlton, Vic 3053, Australia; e-mail iward@mailbox.uq.edu.au; f. 1955; Australia and modern Europe; 3 a year; Editors IAN WARD, ANDREW BONNELL.

BCA China Analyst. 1002 Sherbrooke St West, Suite 1600, Montréal, Qué H3A 3L6, Canada; tel. (514) 499-9706; fax (514) 499-9709; e-mail circ@bcapub.com; trends in economic conditions, assessment of investment opportunities and risks in North and South-East Asia; monthly; Chair. and Editor-in-Chief J. ANTHONY BOECKH.

Beijing Review. 24 Baiwanzhuang Lu, Beijing 100037, People's Republic of China; tel. (10) 68996252; fax (10) 68326628; e-mail contact@bjreview.com.cn; internet www.bjreview.com.cn; f. 1958; current affairs; in English, French, Spanish, German and Japanese; weekly; Editor-in-Chief LII HAIBO.

Beiträge zur Japanologie. Institut für Ostasienwissenschaften der Universität Wien, Japanologie, AAKH-Campus, Hof 2, Spitalgasse 2–4, 1090 Vienna, Austria; tel. (1) 4277-43801; fax (1) 4277-9438; e-mail japanologie.ostasien@univie.ac.at; internet www.univie.ac.at/Japanologie/bzj-list.htm; irregular; Editor S. LINHART.

Boletín del Centro de Estudios de Asia Oriental. Universidad Autónoma de Madrid, 28049 Madrid, Spain; tel. (91) 397-4112; fax (91) 397-4123; e-mail asociación.orientalistas@uam.es; annually.

Bulletin de l'Ecole Française d'Extrême-Orient (BEFEO). Ecole Française d'Extrême-Orient, 22 ave du Président Wilson, 75116 Paris, France; tel. 1 53-70-18-37; fax 1-53-70-87-39; internet www.efeo.fr/; f. 1901; annually.

Bulletin of Indonesian Economic Studies. Economics Division, Research School of Pacific and Asian Studies, Australian National University, Canberra, ACT 0200, Australia; tel. (2) 6125-2370; fax (2) 6125-3700; e-mail editor.bies@anu.edu.au; internet www.tandf.co.uk/journals/titles/00074918.asp; 3 a year; Editor ROSS H. MCLEOD.

Bulletin of the School of Oriental and African Studies. School of Oriental and African Studies, University of London, Thornhaugh St, Russell Sq., London, WC1H OXG, United Kingdom; tel. (20) 7637-2388; fax (20) 7436-3844; e-mail Bulletin@soas.ac.uk; internet www.soas.ac.uk; publ. for the School of Oriental and African Studies, University of London; f. 1917; 3 a year.

Business News IndoChina. POB 9794, Hong Kong; tel. 28800307; fax 28561184; provides independent information about business in Viet Nam, Cambodia, Laos and Myanmar; Publr JOHN COPE.

The China Journal. Contemporary China Centre, RSPAS, Australian National University, Canberra, ACT 0200, Australia; tel. (2) 6125-4150; fax (2) 6257-3642; e-mail ccc@coombs.anu.edu.au; internet www.anu.edu.au/RSPAS/CCC/journal.htm; f. 1979; 2 a year; Editors ANITA CHAN, JONATHAN UNGER.

The China Quarterly. School of Oriental and African Studies, University of London, Thornhaugh St, Russell Sq., London, WC1H 0XG, United Kingdom; tel. (20) 7898-4063; fax (20) 7898-4849; e-mail chinaq@soas.ac.uk; internet www.soas.ac.uk/cq; f. 1960; all aspects of 20th century China, including Taiwan and the overseas Chinese; quarterly; Editor JULIA STRAUSS.

China Report. Sage Publications India Pvt Ltd, B-42 Panchsheel Enclave, POB 4109, New Delhi 110 017, India; tel. and fax (11) 26491290; e-mail journalsubs@indiasage.com; internet www.indiasage.com; publ. for the Centre for the Study of Developing Societies, Delhi, India; f. 1964; topical notes and research articles, book reviews, translations and documentation; quarterly; Editor SINHA BHATTACHARJEA.

China Watch. Orbis Publications LLC, 1924 47th St NW, Washington, DC 20007, USA; tel. (202) 298-7936; fax (202) 298-7938; e-mail orbis@orbischina.com; internet www.orbischina.com; politics, business and economy; monthly; Editor DEREK SCISSORS.

Chinese Cross Currents. Macau Ricci Institute, Av. Cons. Ferreira d'Almeida, 95-E, Macao; tel. 532536; fax 568274; e-mail currents@riccimac.org; internet www.riccimac.org; f. 2004; quarterly; in Chinese and English; Editor YVES CAMUS.

The Chinese Economy. M. E. Sharpe Inc, 80 Business Park Drive, Armonk, NY 10504, USA; tel. (914) 273-1800; fax (914) 273-2106; e-mail custserve@mesharpe.com; internet www.mesharpe.com; f. 1967; translations from Chinese economics journals and official party publications; 6 a year; Editor HUNG GAY FUNG.

Chinese Education and Society. M. E. Sharpe Inc, 80 Business Park Drive, Armonk, NY 10504, USA; tel. (914) 273-1800; fax (914) 273-2106; e-mail custserv@mesharpe.com; internet www.mesharpe.com; f. 1968; translations from Chinese sources; surveys and information about the Chinese educational system; every 2 months; Editors STANLEY ROSEN, GERARD A. POSTIGLIONE.

Chinese Law and Government. M. E. Sharpe Inc, 80 Business Park Drive, Armonk, NY 10504, USA; tel. (914) 273-1800; fax (914) 273-2106; e-mail custserv@mesharpe.com; internet www.mesharpe.com; f. 1968; translations of Chinese scholarly works and political policy documents; every 2 months; Editor JAMES TONG.

Chinese Sociology and Anthropology. M. E. Sharpe Inc, 80 Business Park Drive, Armonk, NY 10504, USA; tel. (914) 273-1800; fax (914) 273-2106; e-mail custserv@mesharpe.com; internet www.mesharpe.com; f. 1968; translations of contemporary studies of social issues; quarterly; Editors GREGORY GULDIN, ZHOU DAMING.

Chinese Studies in History. M. E. Sharpe Inc, 80 Business Park Drive, Armonk, NY 10504, USA; tel. (914) 273-1800; fax (914) 273-2106; e-mail custserv@mesharpe.com; internet www.mesharpe.com; f. 1967; translations of articles; quarterly; Editor LI YU-NING.

CHINOPERL Papers. The Conference on Chinese Oral and Performing Literature, c/o Prof. Joseph Lam, 402 Burton Memorial Tower, University of Michigan, Ann Arbor, MI 48019-1270, USA; tel. (734) 647-9471; fax (734) 647-1897; e-mail jsclam@umich.edu; f. 1969; Chinese oral and performing literature; annually.

Cina. Istituto Italiano per l'Africa e l'Oriente, Via Ulisse Aldrovandi 16, 00197 Rome, Italy; fax (06) 4873138; f. 1956; art, science and thought in contemporary China; annually; Editor Prof. LIONELLO LANCIOTTI.

Contemporary Chinese Thought. M. E. Sharpe Inc, 80 Business Park Drive, Armonk, NY 10504, USA; tel. (914) 273-1800; fax (914) 273-2106; e-mail custserv@mesharpe.com; internet www.mesharpe.com; f. 1969; translations of articles from Chinese sources; quarterly; Editor CARINE DEFOORT.

The Contemporary Pacific. Centre for Pacific Islands Studies, University of Hawaii at Manoa, 1890 East West Rd, 215 Moore Hall, Honolulu, HI 96822, USA.

Contemporary Southeast Asia—A Journal of International and Strategic Issues. Institute of Southeast Asian Studies, 30 Heng Mui Keng Terrace, Pasir Panjang, Singapore 119614; tel. 67780955; fax 67756259; e-mail pubsunit@iseas.edu.sg; internet bookshop.iseas.edu.sg; 3 a year.

The Developing Economies. Nihon Boeki Shinkokiko Ajia Keizai Kenkyusho (Institute of Developing Economies, Japan External Trade Organization), 3-2-2 Wakaba, Mihama-ku, Chiba-shi, Chiba 261-8545, Japan; tel. (43) 299-9500; fax (43) 299-9726; e-mail journal@ide.go.jp; internet www.ide.go.jp; f. 1962; in English; quarterly.

Development Bulletin. Development Studies Network, Research School of Social Sciences, Australian National University, Canberra, ACT 0200, Australia; tel. (2) 6125-2466; fax (2) 6125-9785; e-mail devnetwork@anu.edu.au; internet devnet.anu.edu.au; development issues in the Pacific and South-East Asia; quarterly.

Development and Socio-Economic Progress. Afro-Asian People's Solidarity Organization, 89 Abdel Aziz Al-Saoud St, POB 11559, El Malek El Saleh, Cairo, Egypt; tel. (2) 3636081; fax (2) 3637361; e-mail aapso@idsc.net.eg; 3 a year; Exec. Editor E. A. Vidyaskera; Editor-in-Chief Nouri Abdel Razzak Hussein.

Dong Bang Hak Chi (Journal of Korean Studies). Institute of Korean Studies, Yonsei University, Sudaemun-ku, Seoul 120-749, Republic of Korea; tel. (2) 361-3502; fax (2) 365-0937; f. 1948; in Korean; quarterly; Dir Nam Ki Shim.

East and West. Istituto Italiano per l'Africa e l'Oriente, Via Ulisse Aldrovandi 16, 00197 Rome, Italy; fax (6) 3225348; internet www.isiao.it; f. 1950; in English; quarterly; Editor Gherardo Gnoli.

East Asian Review. Institute for East Asian Studies, 508-143 Jungnung 2-dong, Sungbuk-gu, Seoul 136-851, Republic of Korea; tel. (2) 917-4976; fax (2) 919-5360; e-mail eastasia@ieas.or.kr; internet www.ieas.or.kr; f. 1978; quarterly; Editor Kim Hyeong-ki.

Eastern Economist. United Commercial Bank Bldg, Parliament St, New Delhi 110001, India; f. 1943; economic and financial weekly; Editor V. Balasubramanian.

Economic Record. School of Economics and Finance, Curtin University of Technology, GPOB 1987, Perth, WA 6845, Australia; tel. (8) 9266-2035; fax (8) 9266-3026; e-mail blochh@cbs.curtin.edu.au; f. 1925; quarterly; journal of Economic Soc. of Australia; Editor Prof. H. Bloch; circ. 3,500.

Estudios de Asia y Africa. Centre for Asian and African Studies, El Colegio de México, Camino al Ajusco 20, Pedregal de Sta Teresa, México DF 01000, Mexico; tel. (5) 5449-3000, ext. 3116; fax (5) 5645-0464; e-mail bprecia@colmex.mx; quarterly; Editor Benjamín Preciado Solís.

Far Eastern Economic Review. GPO Box 160, Hong Kong; tel. 25737121; fax 25031530; e-mail review@feer.com; internet www.feer.com; f. 1946; weekly; circ. 100,060; Editor David Plott.

Harvard Journal of Asiatic Studies. Harvard-Yenching Institute, 2 Divinity Ave, Cambridge, MA 02138, USA; tel. (617) 495-2758; fax (617) 495-7798; f. 1936; 2 a year; Editor Joanna Handlin Smith.

History of Development Studies. Asia Pacific Press, Australian National University, Canberra, ACT 0200, Australia; tel. (2) 6125-0178; fax (2) 6257-2886; e-mail books@asiapacificpress.com; internet www.asiapacificpress.com; irregular; Editor Maree Tait.

Hitotsubashi Journal of Arts and Sciences. Japan Publication Trading Co Ltd, POB 5030, Tokyo International, Tokyo 100-3191, Japan; fax (3) 3292-0410; e-mail serials@jptco.co.jp; f. 1960; annually; Editors Katsumi Hashinawa, Hajime Machida.

Hitotsubashi Journal of Economics. Japan Publication Trading Co Ltd, POB 5030, Tokyo International, Tokyo 100-3191, Japan; fax (3) 3292-0410; e-mail serials@jptco.co.jp; f. 1960; 2 a year; Editors N. Abe, J. Ishikawa, T. Iwaisako, M. Sato.

Hong Kong Law Journal. 1517 Two Pacific Place, 88 Queensway, Hong Kong; tel. 25260318; fax 25224721; f. 1971; 3 a year; Editor-in-Chief Denis Chang.

Indonesia and the Malay World. c/o School of Oriental and African Studies, University of London, Thornhaugh St, Russell Sq., London, WC1H 0XG, United Kingdom; tel. (20) 7637-2388; fax (20) 7436-3844; internet www.soas.ac.uk; 3 a year; Chair. Doris Johnson.

The Indonesian Quarterly. Centre for Strategic and International Studies, Jalan Tanah Abang III/23-27, Jakarta 10160, Indonesia; tel. (21) 386-5532; fax (21) 384-7517; e-mail bandoro@csis.or.id; internet www.csis.or.id; f. 1972; Editor Bantarto Bandoro.

Inter-Asia Cultural Studies. 4 Park Sq., Milton Park, Abingdon, Oxon, OX14 4RN, United Kingdom; tel. (1235) 828600; fax (1235) 829000; internet www.tandf.co.uk; 3 a year; Exec. Editors Chen Kuan-Hsing, Chua Beng Huat.

Internationales Asienforum. Arnold-Bergstraesser-Institut, Windausstr. 16, 79110 Freiburg, Germany; tel. (761) 888780; fax (761) 8887878; e-mail abifr@abi.uni-freiburg.de; internet www.arnold-bergstraesser.de; political and socio-economic developments, Asian studies; 4 a year.

Japan Forum. 4 Park Sq., Milton Park, Abingdon, Oxon, OX14 4RN, United Kingdom; tel. (1235) 828600; fax (1235) 829000; e-mail info.asian@routledge.co.uk; internet www.tandf.co.uk; f. 1989; 2 a year; Editor Mark Williams.

Japan Quarterly. Asahi Shimbun Publishing Co, 5-3-2 Tsukiji, Chuo-ku, Tokyo 104-8011, Japan; tel. (3) 5541-8699; fax (3) 5541-8700; e-mail jpnqtrly@mx.asahi-np.co.jp; f. 1954; in English; quarterly; Editor-in-Chief Takenobu Etsuo.

Japanese Annual of International Law. The International Law Association of Japan, Kenkyushitsu, Faculty of Law, University of Tokyo, 3-1 Hongo 7-chome, Bunkyo-ku, Tokyo 113, Japan; f. 1957; Editors-in-Chief Prof. Soji Yamamoto, Akira Kotera.

The Japanese Economy. M. E. Sharpe Inc, 80 Business Park Drive, Armonk, NY 10504, USA; tel. (914) 273-1800; fax (914) 273-2106; e-mail custserv@mesharpe.com; internet www.mesharpe.com; f. 1972; 4 a year; translations from Japanese sources; Editor Walter Hatch.

Japanese Studies. 4 Park Sq., Milton Park, Abingdon, Oxon, OX14 4RN, United Kingdom; tel. (1235) 828600; fax (1235) 829000; e-mail j.snodgrass@uws.edu.au; internet www.tandf.co.uk/journals/titles/10371397.asp; 3 a year; Editor Judith Snodgrass.

Jernal Undang-Undang / Journal of Malaysian and Comparative Law. c/o Faculty of Law, University of Malaya, 50603 Kuala Lumpur, Malaysia; tel. (3) 79676511; fax (3) 79573239; e-mail jmcl@um.edu.my; internet www.um.edu.my/law/jmcl.htm; f. 1974; in English and Malay; annually; Gen. Editor Assoc. Prof. Dr Gan Ching Chuan.

Journal of the American Oriental Society. American Oriental Society, Hatcher Graduate Library, University of Michigan, Ann Arbor, MI 48109-1205, USA; tel. (734) 647-4760; f. 1842; quarterly; Editor E. Gerow, Reed College, Portland, OR 97202, USA.

Journal of the Asia Pacific Economy. 4 Park Sq., Milton Park, Abingdon, Oxon, OX14 4RN, United Kingdom; tel. (1235) 828600; fax (1235) 829000; e-mail enquiry@tandf.co.uk; internet www.tandf.co.uk/journals; 3 a year; Editors A. Chowdhury, C. Kirkpatrick, I. Islam, S. Rashid.

Journal of Asian History. Harrassowitz Verlag, 65174 Wiesbaden, Germany; tel. (611) 530901; fax (611) 530999; e-mail verlag@harrassowitz.de; internet www.harrassowitz.de/verlag; f. 1967; in English, French, German, Russian; 2 a year; Editor Denis Sinor, Goodbody Hall, Indiana University, Bloomington, IN 47405, USA; tel. (812) 855-0959; fax (812) 323-1944; e-mail sinord@indiana.edu.

Journal of Asian Studies. Association for Asian Studies, 1021 E Huron St, Ann Arbor, MI 48104, USA; f. 1941; in English; quarterly; selected articles on history, arts, social sciences, philosophy and contemporary issues; extensive book reviews.

Journal of Asiatic Studies. Asiatic Research Center, Korea University, Anam-dong 5-ga 1, Sungbuk-gu, Seoul, Republic of Korea; tel. (2) 926-1926; fax (2) 924-9123; e-mail syongcho@

kuccnx.korea.ac.kr; internet www.asiacenter.or.kr; f. 1958; 2 a year; Editor CHOI SANG-YONG.

Journal Asiatique. La Société Asiatique, 3 rue Mazarine, 75006 Paris, France; tel. and fax 1-44-41-43-14; f. 1822; covers all phases of Oriental research; 2 a year; Editor ANNA SCHERRER-SCHAUB.

Journal of Australian Studies. University of Queensland Press, POB 6042, St Lucia, Qld 4067, Australia; tel. (7) 3365-2452; fax (7) 3365-1988; e-mail rosiec@uqp.uq.edu.au; internet www.uqp.uq.edu.au; 4 a year.

Journal of Chinese Economic and Business Studies. 4 Park Sq., Milton Park, Abingdon, Oxon, OX14 4RN, United Kingdom; tel. (1235) 828600; fax (1235) 829000; e-mail enquiry@tandf.co.uk; internet www.tandf.co.uk/journals; f. 2003; 3 a year; official journal of the Chinese Economic Association; Man. Ed. XIAMING LIU.

Journal of Chinese Philosophy. Blackwell Publishing Co, POB 11071, Honolulu, HI 96828, USA; f. 1974; quarterly; Editor CHENG CHUNG-YING; Dept of Philosophy, University of Hawaii, Honolulu, HI 96822, USA; tel. (808) 956-6081.

Journal of Contemporary Asia. POB 592, Manila 1099, Philippines; f. 1970; social, political and economic affairs; in English; quarterly; Editors PETER LIMQUECO, BRUCE MCFARLANE.

Journal of Contemporary China. 4 Park Sq., Milton Park, Abingdon, Oxon, OX14 4RN, United Kingdom; tel. (1235) 828600; fax (1235) 829000; e-mail journals.orders@tandf.co.uk; internet www.tandf.co.uk/journals; 4 a year; Editor SUISHENG ZHAO.

Journal of the Economic and Social History of the Orient. Brill Academic Publishing, POB 9000, 2300 PA Leiden, Netherlands; tel. (71) 5353500; fax (71) 5317532; internet www.brill.nl; f. 1957; in English, French and German; 4 a year; Editor-in-Chief HARRIET T. ZURNDORFER, Sinologisch Instituut, Leiden University, POB 9515, 2300 PA Leiden; tel. (71) 5272522; fax (71) 5272615.

Journal of the Hong Kong Branch of the Royal Asiatic Society. GPO Box 3864, Hong Kong; tel. and fax 28137500; e-mail membership@royalasiaticsociety.org.hk; internet www.royalasiaticsociety.org.hk; f. 1960; Hong Kong and South China studies, especially pre-1946 local history, ethnography etc.; annually; Hon. Editor Dr P. HALLIDAY; Pres. Dr P. H. HASE.

Journal of International and Area Studies. Institute of International Affairs, Graduate School of International Studies, Seoul National University, 56-1 Shinrim-dong San, Gwanak-ku, Seoul 151-742, Republic of Korea; tel. (2) 880-8975; fax (2) 874-7368; internet gnis.snu.ac.kr/publication/d-jias/d-jiasindex.htm; international affairs, including Korean studies; 2 a year; Editor Prof. GEUN LEE.

Journal of Japanese Studies. Society for Japanese Studies, University of Washington, Box 353650, Seattle, WA 98195-3650, USA; tel. (206) 543-9302; fax (206) 685-0668; e-mail jjs@u.washington.edu; internet depts.washington.edu/jjs; f. 1974; 2 a year; Co-Editors MARIE ANCHORDOGUY, WHITTIER TREAT.

Journal of the Malaysian Branch of the Royal Asiatic Society. 130M Jalan Thamby Abdullah, off Jalan Tun Sambanthan, Brickfields, 50470 Kuala Lumpur, Malaysia; tel. (3) 22748345; fax (3) 22743458; e-mail mbras@tm.net.my; f. 1877; 2 a year; Hon. Editor Dr CHEAH BOON KHENG.

Journal of the Oriental Society of Australia. Dept of Japanese and Korean Studies, A18, School of Languages and Cultures, University of Sydney, Sydney, NSW 2006, Australia; tel. (2) 9351-2869; fax (2) 9351-2319; e-mail slc@arts.usyd.edu.au; internet www.arts.usyd.edu.au/departs/japanese; f. 1961; annually; Editor LEITH MORTON.

Journal of Oriental Studies. Dept of Chinese, University of Hong Kong, Pokfulam Rd, Hong Kong; tel. 28597923; fax 28581334; e-mail joshkusu@hkucc.hku.hk; internet www.hku.hk/chinese/jos; 2 a year; Chief Editor Prof. C. Y. SIN.

Journal of Pacific History. Australian National University, Canberra, ACT 0200, Australia; tel. (2) 6125-3145; fax (2) 6125-5525; e-mail enquiry@tandf.co.uk; internet www.tandf.co.uk; f. 1966; 3 a year; Exec. Editor JENNIFER TERRELL.

Journal of the Polynesian Society. c/o Dept of Maori Studies, University of Auckland, PB 92019, Auckland, New Zealand; tel. (9) 373-7999 ext. 7463; fax (9) 373-7409; e-mail jps@auckland.ac.nz; internet www.arts.auckland.ac.nz/ant/jps/polsoc.html; f. 1892; study of the peoples of the Pacific area; quarterly; Editor JUDITH HUNTSMAN; circ. 1,100.

Journal of the Royal Asiatic Society of Great Britain and Ireland. 60 Queen's Gardens, London, W2 3AF, United Kingdom; tel. (20) 7724-4742; e-mail info@royalasiaticsociety.org; internet www.royalasiaticsociety.org; f. 1834; covers all aspects of oriental research; 3 a year; Editor Dr SARAH ANSARI.

Journal of the Siam Society. 131 Soi 21 (Asoke), Thanon Sukhumvit, Bangkok 10110, Thailand; tel. (2) 661-6470; fax (2) 258-3491; e-mail info@siam-society.org; internet www.siam-society.org; f. 1904; most articles in English; also in Thai, French and German; at least 2 a year; Hon. Editor Dr DHIRAVAT NA POMBEJRA.

Journal of Southeast Asian Studies. Dept of History, National University of Singapore, 11 Arts Link, Singapore 117570; tel. 67723839; fax 67742528; e-mail hisjseas@nus.edu.sg; f. 1970; 2 a year; Editor PAUL H. KRATOSKA.

Kansai University Review of Business. Kansai University, Suita, Osaka, Japan; annually; Editor TOSHIAKI KAMEI.

Kansai University Review of Economics. Kansai University, Suita, Osaka, Japan; annually; Editor YASUO MURATA.

Keio Economic Studies. Keio Economic Society, 2-15-45 Mita, Minato-ku, Tokyo 108-8345, Japan; tel. (3) 3453-4511; fax (3) 5427-1578; f. 1963; 2 a year; Editor Prof. MIKI SEKO.

Korea Observer. The Institute of Korean Studies, CPO Box 3410, Seoul 100-634, Republic of Korea; tel. (2) 569-5574; fax (2) 564-1190; f. 1968; quarterly; Editor EUN HO LEE.

Kyoto University Economic Review. Faculty of Economics, Kyoto University, Sakyo-ku, Kyoto, Japan; f. 1926; 2 a year.

Lawasia Journal. Law Association for Asia & the Pacific, GPOB 980, Brisbane, Qld 4001, Australia; tel. (7) 3222-5888; fax (7) 3222-5850; e-mail lawasia@lawasia.asn.au; internet www.lawasia.asn.au; f. 1966; legal issues related to Asia and the Pacific; annually; Editors Prof. E. ANGHTERSON, Prof. JESSE WU.

Melanesian Law Journal. Faculty of Law, University of Papua New Guinea, POB 317, University 134, NCD, Papua New Guinea; tel. 3267516; fax 3267187; e-mail Law.Publications@upng.ac.pg; f. 1971; annually; Editor G. LINGE.

Modern Asian Studies. Cambridge University Press, The Edinburgh Bldg, Shaftesbury Rd, Cambridge, CB2 2RU, United Kingdom; tel. (1223) 312393; e-mail journals-subscriptions@cambridge.org; internet www.journals.cambridge.org; f. 1967; quarterly; Editor Dr GORDON JOHNSON.

Modern China. Sage Publications Inc, 2455 Teller Rd, Newbury Park, CA 91320, USA; tel. (805) 499-0721; fax (805) 499-0871; e-mail huang@history.ucla.edu; f. 1975; history and social sciences; quarterly; Editor PHILIP C. C. HUANG.

Mongolian Studies. The Mongolia Society, 322 Goodbody Hall, Indiana University, 1011 E 3rd St, Bloomington, IN 47405-7005, USA; tel. (812) 855-4078; fax (812) 855-7500; e-mail monsoc@indiana.edu; internet www.indiana.edu/~mongsoc; f. 1974; Mongolia and Inner Asia of all periods; annually; Man. Editor CHRISTOPHER ATWOOD.

Monumenta Nipponica. Sophia University, 7-1 Kioi-cho, Chiyoda-ku, Tokyo 102-8554, Japan; tel. (3) 3238-3544; fax (3) 3238-3835; e-mail kw-nakai@sophia.ac.jp; internet monumenta.cc.sophia.ac.jp; f. 1938; studies in Japanese culture; quarterly; Editor KATE WILDMAN NAKAI.

Monumenta Serica. Arnold-Janssen-Strasse 20, 53757 St Augustin, Germany; tel. (2241) 237431; fax (2241) 206770; e-mail monumenta.serica@t-online.de; internet www.monumenta-serica.de; f. 1935; journal of oriental studies; annually; Editor ROMAN MALEK.

Le Muséon. Université Catholique de Louvain, place Blaise-Pascal 1, 1348 Louvain-la-Neuve, Belgium; tel. (10) 473793; fax (10) 472001; e-mail coulie@ori.ucl.ac.be; internet www.fltr.ucl.ac.be/FLTR/GLOR/ORI; f. 1881; oriental studies; 2 double vols a year; Editor Prof. B. COULIE.

Oceania. 110 Darlington Rd, Sydney University, Sydney, NSW 2006, Australia; tel. (2) 9351-2666; fax (2) 9351-7488; e-mail oceania@arts.usyd.edu.au; f. 1930; anthropology; quarterly; Editor Dr NEIL MACLEAN.

Oriens Extremus. Zeitschrift für Sprache, Kunst und Kultur des Länder des Fernen Ostens, Harrassowitz Verlag, 65174 Wiesbaden, Germany; tel. (611) 530901; fax (611) 530999; e-mail verlag@harrassowitz.de; f. 1954; in German and English; annually; Editors R. SCHNEIDER, H. STUMPFELDT, B. J. TERWIEL.

Oriental Art Magazine Ltd. 90 Park Ave, Suite 1700, New York, NY 10016, USA; 260 Orchard Rd, 10-10 The Heeren, Singapore 238855; tel. (65) 7379931; fax (65) 7373190; e-mail orientalart@orientalartmag.com; f. 1949; quarterly; Editor AILEEN LAU TAN.

Pacific Affairs. Ste. 164-1855 West Mall, Vancouver, BC, V6T 1Z2, Canada; tel. (604) 822-4534; fax (604) 822-9452; e-mail enquiry@pacificaffairs.ubc.ca; internet www.pacificaffairs.ubc.ca; f. 1927; current political, economic, social and diplomatic issues of the Asia-Pacific region; research articles, book reviews; quarterly; Editor Dr TIMOTHY CLARK; Man. Editor JACQUELINE GARNETT.

Pacific Economic Bulletin. Asia Pacific Press, Australian National University, Canberra, ACT 0200, Australia; tel. (2) 6249-4705; fax (2) 6257-2886; e-mail peb@anu.edu.au; internet peb.anu.edu.au; 3 a year; Editors RON DUNCAN, MAREE TAIT.

Pacific Historical Review. Pacific Coast Branch, American Historical Assn, 487 Cramer Hall, Portland State University, Portland, OR 97207-0751, USA; tel. (503) 725-8230; fax (503) 725-8235; e-mail phr@pdx.edu; internet www.ucpress.edu/journals/phr; f. 1932; Editors DAVID A. JOHNSON, CARL ABBOTT, SUSAN WLADAVER-MORGAN.

Pacific Report. POB 25, Red Hill, ACT 2603, Australia; tel. (2) 6295-6363; fax (2) 6260-8492; e-mail pacrep@ozemail.com.au; f. 1988; 24 a year; Editor KENNETH RANDALL.

The Pacific Review. Centre for the Study of Globalisation and Regionalisation, University of Warwick, Coventry, CV4 7AL, United Kingdom; tel. (24) 7657-2533; fax (24) 7657-2548; e-mail R.Higgott@warwick.ac.uk; internet www.csgr.org; f. 1988; quarterly; Editor Prof. RICHARD HIGGOTT.

Pacific Studies. Institute of Polynesian Studies, Brigham Young University, Box 1979, Hawaii, USA.

Perspectives on Global Development and Technology. Brill Academic Publishers, Leiden, Netherlands; tel. (71) 5353500; fax (71) 5317532; e-mail cs@brill.nl; internet www.brill.nl; f. 2002; 4 a year; Editor R. PATTERSON.

Philippine Studies. Ateneo de Manila University Press, POB 154, Manila 1099, Philippines; tel. (2) 4261238; fax (2) 4265909; e-mail ccastro@ateneo.edu.ph; f. 1953; quarterly; Editor-in-Chief FILOMENO V. AGUILAR Jr.

Przegląd Orientalistyczny. Polskie Towarzystwo Orientalistyczne, Redakcja, 00-927 Warsaw, Krakowskie Przedmieście 26/28, Instytut Orientalistyczny UW, Poland; tel. (22) 55-20-353; quarterly; Editor Dr hab. DANUTA STASIK.

Regional Outlook: Southeast Asia. Institute of Southeast Asian Studies, 30 Heng Mui Keng Terrace, Pasir Panjang, Singapore 119614; tel. 68702447; fax 67756259; e-mail pubsunit@iseas.edu.sg; internet bookshop.iseas.edu.sg; annually; Editors RUSSELL HENG HIANG KHNG, DENIS HEW.

Review of Indonesian and Malaysian Affairs (RIMA). Dept of Southeast Asian Studies, SEAMELS, The University of Sydney, NSW 2006, Australia; tel. (2) 9351-2681; fax (2) 9351-2319; e-mail rima@asia.usyd.edu.au; internet www.arts.usyd.edu.au/arts/departs/asia/rima; f. 1967; biannually; Editorial Collective LINDA CONNOR, ADRIAN VICKERS, KEITH FOUCHER.

Rocznik Orientalistyczny. Warsaw University Oriental Institute, 00927 Warsaw 64, ul. Krakowskie Przedmieście 26/28, Poland; e-mail mmdziekan@poczta.onet.pl; f. 1915; 2 a year; Editor-in-Chief MAREK M. DZIEKAN.

Singapore Journal of Tropical Geography. Dept of Geography, National University of Singapore, 1 Arts Link, Singapore 117570; tel. 68743853; fax 67773091; e-mail geolimkl@nus.edu.sg; internet www.blackwellpublishing.com/journals/sjtg; f. 1953; 3 a year; Editor SHIRLENA HUANG.

Sojourn: Journal of Social Issues in Southeast Asia. Institute of Southeast Asian Studies, 30 Heng Mui Keng Terrace, Pasir Panjang, Singapore 119614; tel. 68702447; fax 67756259; e-mail pubsunit@iseas.edu.sg; internet bookshop.iseas.edu.sg; 2 a year.

Southeast Asian Affairs. Institute of Southeast Asian Studies, 30 Heng Mui Keng Terrace, Pasir Panjang, Singapore 119614; tel. 67780955; fax 67756259; e-mail pubsunit@iseas.edu.sg; internet bookshop.iseas.edu.sg; f. 1968; annually; Man. Editor TRIENA ONG.

Southeast Asian Journal of Social Science. Times Academic Press, Times Media Pte Ltd, Times Centre, 1 New Industrial Rd, Singapore 536196; tel. 62848844; fax 62889254; e-mail fps@corp.tpl.com.sg; internet www.timesone.com.sg/te; f. 1973; 2 a year; Editor CHAN KWOK BUN; Dept of Sociology, National University of Singapore, Kent Ridge Crescent, Singapore 119260; tel. 67723822; fax 67779579.

South East Asia Research. School of Oriental and African Studies, University of London, Thornhaugh St, Russell Sq., London, WC1H 0XG, United Kingdom; tel. (20) 7323-6146; fax (20) 7436-6046; f. 1993; 2 a year.

Terrorism Monitor. The Jamestown Foundation, 4516 43rd Street NW, Washington, DC 20016; tel. (202) 483-8888; e-mail pubs@jamestown.org; internet www.jamestown.org; f. 2003; fortnightly; Man. Editor JULIE SIRRS.

Third World Quarterly. Dept of Geography, Royal Holloway, University of London, Egham, Surrey, TW20 0EX, United Kingdom; fax (20) 8947-1243; internet www.tandf.co.uk/journals; 8 a year; Editor SHAHID QADIR.

The Tibet Journal. Library of Tibetan Works and Archives, Gangchen Kyishong, Dharamshala 176 215, India; tel. (1892) 22467; fax (1892) 23723; e-mail tibjournal@gov.tibet.net; f. 1975; quarterly; Editor SONAM TSERING.

Toho Gakuho Journal of Oriental Studies. Institute for Research in Humanities, Kyoto University, Ushinomiyacho, Yoshida, Sakyo-ku, Kyoto 606-8501, Japan; annually.

Tokyo Business Today. Toyo Keizai Inc, 1-2-1 Nihonbashi Hongoku-cho, Chuo-ku, Tokyo 103, Japan; tel. (3) 3246-5655; f. 1934; Japan's business and finance; monthly; Editor HIROSHI FUKUNAGA.

Tonan Ajia Kenkyu (Southeast Asian Studies). The Center for Southeast Asian Studies, Kyoto University, 46 Shimoadachi-cho, Yoshida, Sakyo-ku, Kyoto 606-8501, Japan; tel. (75) 753-7344; fax (75) 753-7356; e-mail editorial@cseas.kyoto-u.ac.jp; internet www.cseas.kyoto-u.ac.jp; quarterly; Man. Editor MARIKO YONEGAWA.

T'oung Pao (International Journal of Chinese Studies). Brill Academic Publishing, POB 9000, 2300 PA Leiden, Netherlands; tel. (71) 5353500; fax (71) 5317532; internet www.brill.nl; f. 1890; Chinese studies; also available online; 2 double issues a year; Editors BAREND J. TER HAAR, PIERRE-ETIENNE WILL.

Toyogaku Bunken Ruimoku (Annual Bibliography of Oriental Studies). Documentation and Information Center for Chinese Studies, Institute for Research in Humanities, Kyoto University, Higashioguracho, Kitashirawaka, Sakyo-ku, Kyoto 606-8265, Japan; e-mail ruimoku@kanji.zinbun.kyoto-u.ac.jp; internet wwww.kanji.sinbun.kyoto-u.ac.jp/db/CHINA3/; annual bibliography; in Japanese, Chinese and European languages.

Transactions of the Korea Branch of the Royal Asiatic Society. CPOB 255, Seoul, Republic of Korea; tel. (2) 763-9483; fax (2) 766-3796; e-mail info@raskorea.org; f. 1900; annually; Gen. Man. SUE J. BAE.

Viet Nam Investment Review. 175 Nguyen Thai Hoc, Hanoi, Viet Nam; tel. (4) 8450537; fax (4) 8457937; e-mail vir@hn.vnn.vn; internet www.vir.com.vn; in English language; weekly; Editor-in-Chief NGUYEN TRI DUNG.

Wiener Zeitschrift für die Kunde Südasiens (WZKS) (Vienna Journal of South Asian Studies). Institute for South Asian, Tibetan and Buddhist Studies—South Asian Studies, A-1090 Vienna, University Campus, Spitalgasse 2, Hof 4/2.1, Austria; tel. (1) 4277-43511; fax (1) 4277-9435; e-mail istb@univie.ac.at; internet www.univie.ac.at/istb; f. 1957; published by the Austrian Academy of Sciences Press; annually; Editors G. OBERHAMMER, K. PREISENDANZ, CH. H. WERBA.

Working Papers in Trade and Development. Division of Economics, Research School of Pacific and Asian Studies, Australian National University, Canberra, ACT 0200, Australia; tel. (2) 6125-2188; fax (2) 6125-3700; e-mail seminars.economics@anu.edu.au; irregular.

Yazhou Zhoukan. 15F, Block A, Ming Pao Industrial Centre, 18 Ka Yip St, Chai Wan, Hong Kong; tel. 25155358; e-mail yzzk@mail.mingpao.com; internet www.yzzk.com; f. 1987; global Chinese news weekly; Editor-in-Chief YAU LOP POON.

INDEX OF REGIONAL ORGANIZATIONS

(Main reference only)

Index of Territories